Textbook of Biochemistry, Biotechnology, Allied and Molecular Medicine

Fourth Edition

Edited by

G.P. TALWAR

Docteur es Sciences
F.A.M.S., F.A.Sc., F.N.A.
F.R.C.O.G. (London a.e), F.W.A.A.S.
Member La Academia Nacional de Medicina Mexico

Director, Talwar Research Foundation, New Delhi
Formerly Director, National Institute of Immunology,
Professor and Head, Department of Biochemistry
All India Institute of Medical Sciences, New Delhi

SEYED E. HASNAIN

Ph.D., D.Sc. (h.c.), D.Med.Sc. (h.c.), F.N.A., F.T.W.A.S, M.L.

Professor
Kusuma School of Biological Sciences
Indian Institute of Technology Delhi
Former Vice Chancellor
University of Hyderabad

SHIV KUMAR SARIN

M.D., D.M., D.Sc. (Hon.), F.N.A., F.N.A.Sc.

Senior Professor and Head of Hepatology
Director, Institute of Liver and Biliary Sciences
New Delhi

PHI Learning Private Limited

Delhi-110092
2016

₹ 1995.00

TEXTBOOK OF BIOCHEMISTRY, BIOTECHNOLOGY, ALLIED AND MOLECULAR MEDICINE
Fourth Edition
Edited by G.P. Talwar, Seyed E. Hasnain, and Shiv Kumar Sarin

ISBN-978-81-203-5125-7

The export rights of this book are vested solely with the publisher.

Eighth Printing (Fourth Edition)   ...   ...   **January, 2016**

Published by Asoke K. Ghosh, PHI Learning Private Limited, Rimjhim House, 111, Patparganj Industrial Estate, Delhi-110092 and Printed by Rajkamal Electric Press, Plot No. 2, Phase IV, HSIDC, Kundli-131028, Sonepat, Haryana.

Contents

Section V
VITAMINS AND MINERALS

Section VI
GASTROINTESTINAL TRACT AND HEPATOBILIARY SYSTEM

Section VII
PROBIOTICS

Section VIII
HUMAN GENETICS, BIOCHEMICAL BASIS OF INHERITANCE, EXPRESSION OF GENETIC INFORMATION, AND GENETIC ENGINEERING

Section IX
CELL REPLICATION AND CANCER

Section X
STEM CELLS AND REGENERATIVE THERAPY

Section XI
HORMONES

Section XV
EMERGING INFECTIONS AND VACCINES UNDER DEVELOPMENT

Section XVI
VACCINES

Section XVII
COMPLEMENTARY, ALLIED AND INTEGRATED MEDICINE

Foreword
(To the first edition)

Sir Frederick Gowland Hopkins once called the cell the theatre of life. I am reminded of this statement as I go through the *Textbook of Biochemistry and Human Biology* written by Professor Talwar and his colleagues. It begins appropriately with a section on Chemical Architecture of the Living System and then goes on to consider a diversity of fields with membranes and receptors, endorphins and encephalins and prostaglandins at one end of the spectrum to such topics as biochemistry of common intestinal parasites, protein-calorie malnutrition and principles of chemotherapy at the other end.

And yet this is not a disjointed ensemble of unrelated topics. It has a central theme and a direction. Biochemistry is treated not as an end in itself but as a part of human biology. Basic knowledge has been linked, wherever appropriate, to the applicability of that knowledge for understanding physiological processes in the human body. Inter-relatedness and regulations of biological processes are emphasized all through. When the biochemistry and physiology of haemopoiesis are being discussed, the problem of anaemia is also presented. Biology of reproduction is linked to the control of fertility in human populations. The elements of immunology are not considered in isolation but related to the immunological conquest of communicable diseases. Hormones are linked to behaviour. Genetic engineering is discussed with a vision of the future possibilities in biology and medicine. Indeed, this is a work of trans-disciplinary synthesis where excellence and relevance are sought to be combined.

To whom is this book addressed? I would say to all students of human biology. Undergraduates, medical and non-medical, will find that it provides up-to-date and assimilable information. Postgraduates can see gaps and uncertainties in our present knowledge. Basic scientists and clinicians alike will find here the grammar of biomedical science, the core of basic knowledge required for the intelligent and enlightened teaching and practice of medical sciences.

This work is written by Indian authors, themselves actively engaged in teaching and research, and the whole work is tinged with the background of health and disease in this part of the world. The medical student will see the practical value of basic knowledge. Professor Talwar and his colleagues must be congratulated for producing what I consider a truly monumental and unique work, thereby rendering a most valuable service to biomedical sciences in the developing world.

V. RAMALINGASWAMI
M.D., D.Sc. (Oxon), F.R.C.P. (London)
F.R.C. Path. (London)
Director-General,
Indian Council of Medical Research;
Chairman,
Editorial Cell for Core Books in Medicine,
National Book Trust,
India

Preface

The very title of the Fourth Edition evokes the stupendous changes that the new edition brings to the book, which had carved its place as a worthy textbook and an informative reference not only for the students of medical sciences but also for the scholars of biosciences and biotechnology in many universities in India and abroad.

Life sciences sector has emerged as an interdisciplinary amalgam. The traditional biochemistry restricted to chemical structure and metabolism is no longer adequate for understanding body functions and dysfunctions. Also, our knowledge of several organs and organ systems has expanded. Accordingly, the section devoted to this concept has been enlarged to include knowledge of more organs, each written by experts in the field weaving in the normal organ function with ailments and dysfunction. In the new edition, the book includes new chapters on "Kidney and Lung Functions and Dysfunctions", and "Heart Normal Functions and Dysfunctions". New chapter on "Eye and Vision and Molecular Basis of Cataract and Glaucoma" with the applications of stem cells in restoring vision brings home current developments in the field. Gastrointestinal Tract and Hepatobiliary System have been covered in four chapters, two chapters on normal physiology and dysfunction, a whole chapter on "Fatty Liver and its Implications" and another on "Hepatitis Viruses". This edition would therefore be of immense use to the medical students not only in their preclinical phase but also throughout medical education including their postgraduate years.

Another significant feature of the present edition is the inclusion of elements of allied disciplines. For instance, chapters on "Teeth and Gums", which are the prerequisites for dental medicine, add value to the book. These are as much of concern as other parts of the body. New Sections on "Stem Cells" and "Probiotics" have been introduced. The former have already found clinical applications in regenerative therapy, whereas the latter are increasingly recognized as friendly living microorganisms not only for performing useful functions in GI tract and healthy vagina but also for their emerging role in infection, immune response and metabolic disorders.

The Section on "Human Genetics and Biochemical Basis of Inheritance" has been appropriately expanded to reflect latest knowledge of various domains. Besides, DNA Fingerprinting for identity establishment, epigenetics, micro-RNA and siRNA, their role in gene expression are included as full chapters. A chapter on "Chromatin Modification and its Association with Human Diseases" exemplifies our understanding of disease regulation. Genetic engineering and expression systems for producing recombinant proteins, synthetic biology and their applications in more nutritive and higher food production as well as in enabling biological synthesis of drugs for malaria such as Artemesinin and Biofuels have also been covered

in this section. Also, included are chapter on Bioterrorism; Metabolomics and Reproductive Cloning.

Ability to reproduce is an essential definition of living entities, thus an entire Section carrying nine chapters on male and female reproductive systems, fertilization, implantation, pregnancy, and lactation are duly covered, as also a chapter on "Assisted Reproductive Technology" which enables infertile couples to have a child. Equally important are the currently available methods, as well as those under development for control of fertility, which are covered in a full chapter.

Immunology has been given due attention by upgrading knowledge in this fast growing area. A major expansion is on vaccines. Immunization of children with DPT reduced their mortality drastically. No wonder, these are being made tetra or pentavalent vaccines. Also, Polio vaccine has been discussed that enabled the elimination of this disease from India, though it is present in some neighbouring countries. Chapters on Vaccines for cholera, malaria, influenza and tuberculosis are contributed by experts working in these areas. The immunotherapeutic vaccine against multi-bacillary leprosy approved by both the Drugs Controller General of India and USFDA may likely find additional applications in the treatment of tuberculosis and some cancers. In addition, vaccines under development against emerging infections, Dengue, Chikungunya and Japanese encephalitis are included.

Given that modern Medicine, despite its stupendous advances cannot presently provide cure for all ills, patients frequently resort to alternate systems of treatment. Hence, chapters on Ayurveda, Homeopathy, Unani and Herbal Medicines, and Yoga have been introduced to enlighten practioners.

The Fourth Edition has two co-editors, who have contributed many chapters to this compendium. In addition, experts from various specialized fields have made contributions to cover diverse subjects. We feel that the present Edition of the book will be of great interest to both students and the teaching fraternity.

Acknowledgements

The Editors would like to acknowledge the assistance of their research scholars and co-workers Nishu Atrey, Priyanka Singh, Kanika Kapur, Dr. Sonam Grover, Paras Gupta, and Kuldeep Dalal in accomplishing various tasks for the completion of this multi-author book.

G.P. TALWAR
SEYED E. HASNAIN
SHIV KUMAR SARIN

List of Contributors

Abhinav, K.V.
Molecular Biophysics Unit
Indian Institute of Science
Bangalore

Agarwal, Akansha
Department of Nephrology
AIIMS, New Delhi

Agarwal, Sanjay Kumar
Professor
Department of Nephrology
AIIMS, New Delhi

Agarwal, Amita
Department of Clinical Immunology
Sanjay Gandhi Postgraduate Institute of
Medical Sciences (SGRIMS)
Lucknow

Aggarwal, Bharat B.
Professor
Department of Experimental Therapeutics
University of Texas
M.D. Anderson Cancer Center
Houston, Texas, USA

Aggarwal, Rakesh
Professor
Department of Gastroenterology, Sanjay Gandhi
Postgraduate Institute of Medical Sciences
Lucknow

Aggarwal, Sameer
AIIMS, New Delhi

Ahmed, Rafi
Director
Emory Vaccine Center
Emory University, Atlanta, USA

Alam, Seema
Professor
Department of Pediatric Hepatology
Institute of Liver and Biliary Sciences, New Delhi

Ali, Sher
Professor, Interdisciplinary Centre, Jamia Millia Islamia
New Delhi

Anand, Sandhya
Stem Cell Biology Department
National Institute for Research in Reproductive Health
Parel, Mumbai

Arora, Rajesh
Scientist
Defence Research and Development Organization
DRDO Bhawan
New Delhi

Awasthi, Divya
Associate Consultant
Department of Gynecological Laproscopy and Hysteroscopy
Max Super Specialty Hospital, Mohali

Balasubramanian, A.S.
Professor
Christian Medical College
Vellore

Balasubramanyam, D.
Director of Research
L.V. Prasad Eye Institute, Hyderabad

Banerjea, Akhil C.
Staff Scientist
Laboratory of Virology
National Institute of Immunology, New Delhi

Banerjea, Animesh
Staff Scientist
National Institute of Immunology, New Delhi

Banerjee, Rinti
Professor
Indian Institute of Technology Bombay
Mumbai

Banerjee, Sharmistha
Associate Professor
Department of Biochemistry
School of Life Sciences
University of Hyderabad
Hyderabad

Bansal, Manish
Consultant Cardiologist
Medanta: The Medicity
Gurgaon, Haryana

Basu, Rajatava
University of Alabama
Birmingham, UK

Batish, Virender K.
Emeritus Scientist
National Dairy Research Institute
Karnal, Haryana

Benjamin, Jaya
Assistant Professor
Department of Clinical Nutrition
Institute of Liver and Biliary Sciences
New Delhi

Bharati, Kaushik
Senior Program Officer
Translational Health Science and Technology Institute
(THSTI) Faridabad, Haryana

Bhargava, Balram
Professor
Department of Cardiology
Cardiothoracic Sciences Centre
AIIMS, New Delhi

Bhartiya, Deepa
Scientist & Head
National Institute for Research in Reproductive Health
Mumbai

Bhaskar, Sangeeta
Scientist
Systems Immunology Laboratory
National Institute of Immunology
New Delhi

Bhat, Sumrita
University of Calgary
Calgary, Canada

Bhatla, Neerja
Additional Professor
Department of Obstetrics & Gynecology
AIIMS, New Delhi

Bhatnagar, R.S.
University of California
San Francisco, USA

Bhattacharya, Indrashis
National Institute of Immunology (NII)
New Delhi

Bijlani, Ramesh
Former Professor
Department of Physiology
AIIMS, New Delhi

Bindu, M. Kutty
Professor, Department of Neurophysiology, NIMHANS
Bangalore

Biswas, S.
Assistant Professor
Department of Research, Institute of Liver and Biliary Sciences
New Delhi

Blasko, G.
Head, Department of Pharmaceutical Management &
Organization, Debrecen University, Postgraduate Medical
School and Sanofi-Chinoin Ltd.
Budapest, Hungary

Blasko, Susan Turi
Head
Division of Pathological Pregnancy
Department of Obstetrics & Gynecology
Army Hospital, Hungary

Chatterjee, Snehajyoti
Jawaharlal Centre for Advanced Scientific Research
(JNCASR), Bangalore

(Late) Chaudhury, Ranjit Roy
Chairman
Apollo Hospitals Educational & Research Foundation
(AHERF)
New Delhi

Chiittawar, Sachin
Assistant Professor
Gandhi Medical College
Bhopal

Chitnis, Chetan
Head, Malaria Parasite Biology & Vaccines Unit Institute
Pasteur
Paris, France

(Late) Chuttani, P.N.
Postgraduate Institute of Medical Education & Research
(PGI), Chandigarh

Das, Bhabatosh
Department of Gastroenterology and Human Nutrition
AIIMS, New Delhi

Das, H.K.
Honorary Scientist
Division of Biochemistry
Indian Agricultural Research Institute, New Delhi

Das, Nibhriti
Former Professor
Department of Biochemistry
AIIMS, New Delhi

Das, Shantanabha
Senior Research Fellow
Department of Infectious Disease and Immunology
Indian Institute of Chemical Biology
Kolkata

Datta, Asis
Professor of Eminence
National Institute of Plant Genome Research
New Delhi

Deepak, K.K.
Professor
Department of Physiology
AIIMS, New Delhi

Deobagkar, Deepti
Director
Bioinformatics Centre
University of Pune, Pune

Dhar, Ruby
Post-doctoral Scholar
University of Chicago
Chicago, USA

(Late) Ganguli, N.C.
National Dairy Research Institute (NDRI)
Karnal

Ghosh, Amit
INSA Senior Scientist
National Institute of Cholera and Enteric Diseases
Kolkata

Ghosh, Debabrata
Department of Physiology
AIIMS, New Delhi

Ghosh, Sumit
Senior Scientist
Central Institute of Medicinal and Aromatic Plants
Lucknow

Goel, B.K.
Senior Consultant and Head, Department of Biochemistry
Batra Hospital, New Delhi

Govindaraj, Vijayakumar
Department of Biochemistry
Indian Institute of Science,
Bangalore

Grover, Sunita
Principal Scientist
Molecular Biology Unit Dairy Microbiology Divisio
National Dairy Research Institute, Karnal
Haryana

Gupta, S.K.
Scientist
Reproductive Cell Biology Laboratory
National Institute of Immunology
New Delhi

Gupta, Somesh
Professor
Department of Dermatology and Venereology
AIIMS, New Delhi

Gupta, Subash C.
Department of Experimental Therapeutics
University of Texas
M.D. Anderson Cancer Center
Houston, Texas, USA

Gupta, Yashdeep
Department of Medicine
Government Medical College and Hospital
Chandigarh

Hajela, Neerja
Head
Yakult Danone India Pvt. Ltd.
New Delhi

Hasnain, Seyed E.
Professor
Kusuma School of Biological Sciences
Indian Institute of Technology (IIT) Delhi

Jailkhani, B.L.
Senior Consultant
North-Eastern Region–Biotechnology Program Management
Centre (NER-BPMC), New Delhi

Jain, S.K.
Head & Dean
Faculty of Science
Jamia Hamdard
New Delhi

Jameel, Shahid
Chief Executive Officer
The Wellcome Trust/DBT India Alliance
Hyderabad

John, T. Jacob
Retired Professor
Department of Virology
Christian Medical College
Vellore

Joshi, Y.K.
Senior Professor
Department of Hepatology and Clinical Nutrition
Institute of Liver and Biliary Sciences, New Delhi

Kapoor, Richa
National Institute of Immunology
New Delhi

Kar, Rajarshi
Assistant Professor
Department of Biochemistry
University College of Medical Sciences, Delhi

Katare, Deepshika P.
Professor & Assistant Director
Amity Institute of Biotechnology
Noida

Kaur, Gurvinder
Scientist II
Department of Histocompatibility and Immunogenetics
AIIMS, New Delhi

Khan, Asim Ali
Jamia Hamdard, New Delhi

Khanna, Navin
Group Leader
Mammalian Biology Division
International Centre for Genetic Engineering and
Biotechnology (ICGEB), New Delhi

Kharbanda, O.P.
Professor
Centre for Dental Education and Research
AIIMS, New Delhi

Kotru, Mrinalini
Assistant Professor
University College of Medical Sciences
University of Delhi, New Delhi

Kumar, Ajay
Department of Biochemistry
G.S.V.M., Kanpur

Kumar, Anupam
Assistant Professor
Department of Hepatology
Institute of Liver and Biliary Sciences
New Delhi

Kumar, Ashish
National Institute of Immunology (NII)
New Delhi

Kumar, Ashok
Professor
Department of Biological Sciences and Bio-engineering
Indian Institute of Technology Kanpur

Kumar, Lalit
Professor
Department of Medical Oncology
AIIMS, New Delhi

Kumar, Manoj
Additional Professor
Department of Hepatology
Institute of Liver and Biliary Sciences
New Delhi

Kumar, Ramesh
Assistant Professor
Department of Hepatology
Institute of Liver and Biliary Sciences
New Delhi

Kundu, Tapas Kumar
Professor
Jawaharlal Centre for Advanced Scientific Research
(JNCASR), Bangalore

Luthra, Kalpana
Associate Professor
Department of Biochemistry
AIIMS, New Delhi

Majumdaar, Partha P.
Director
National Institute of Biomedical Genomics
Kalyani, West Bengal

Majumdar, Subeer
Staff Scientist
National Institute of Immunology (NII)
New Delhi

Makharia, Govind K.
Associate Professor
Department of Gastroenterology and Human Nutrition
AIIMS, New Delhi

Malhotra, O.P.
Former Professor
Department of Biochemistry
Banaras Hindu University, Varanasi

Malhotra, Rajesh
Department of Orthopedics
AIIMS, New Delhi

Mathur, Rajendra Prasad
Senior Consultant
Department of Nephrology and Renal Transplant Services
Institute of Liver and Biliary Sciences
New Delhi

Mehra, Narinder Kumar
Former Professor & Head
Department of Histocompatibility & Immunogenetics
AIIMS, New Delhi

Mishra, Seema
Assistant Professor
Department of Biochemistry
School of Life Sciences
University of Hyderabad
Hyderabad

Modak, Rahul
Assistant Professor
KIIT School of Biotechnology
Bhubaneswar, Odisha

Modesto, Karen M.
Associate Research Scientist and Scholar
Yale University
New Haven, Connecticut

Mohareer, Krishnaveni
SERB Fast Track Scientist
Department of Biochemistry
School of Life Sciences
University of Hyderabad
Hyderabad

Moudgil, K.D.
Department of Microbiology & Immunology
University of Maryland
School of Medicine, Baltimore, USA

Mukhopadhyay, Asok
Staff Scientist
National Institute of Immunology
New Delhi

Mustafa, Abu Salim
Professor
Department of Microbiology
Faculty of Medicine
Kuwait University, Kuwait

Nagvenkar, Punam
Stem Cell Biology Department
National Institute for Research in Reproductive Health
Parel, Mumbai

Nair, G.B.
Director
Translational Health Science and Technology Institute
Faridabad

Nandi, Dipankar
Professor
Department of Biochemistry
Indian Institute of Science
Bangalore

Narula, Jagat
Philip J. and Harriet L. Goodhart Chair in Cardiology
Chairman, Department of Cardiology, St. Luke's &
Roosevelt Hospitals at Mount Sinai
Associate Dean, Arnhold Institute for Global Health at
Mount Sinai
Professor of Medicine & Radiology, Icahn School of
Medicine at Mount Sinai
Director, Cardiovascular Imaging, Mount Sinai Health
System
New York, NY 10029
Zena and Michael A. Wiener Cardiovascular Institute
New York, USA

Natarajan, K.
Professor, School of Life Sciences, JNU, New Delhi

Naz, Rajesh
Professor
Department of Obstetrics & Gynecology
West Virginia University, Morgantown, USA

Negi, D.S.
Center for DNA Fingerprinting and Diagnostics
Hyderabad

Parameswaran, Venkat
Professor, Diabetes and Endocrine Services
Royal Hobart Hospital, Australia

Pathak, Deepali
Post-Doctoral Fellow
National Institute of Immunology
New Delhi

Patwardhan, Bhushan
Professor
Interdisciplinary School of Health Sciences
University of Pune
Pune

Philips, Cyriac Abby
Senior Fellow
Department of Hepatology
Institute of Liver and Biliary Sciences,
New Delhi

Prakash, Gaurav
Department of Medical Oncology
AIIMS, New Delhi

Prakash, Prachee
Synthetic Genomics Inc
La Jolla, California, USA

Prasad, Sahdeo
Department of Experimental Therapeutics
University of Texas
M.D. Anderson Cancer Center
Houston, Texas, USA

Punjani, Sonal
Department of Orthodontics
AIIMS, New Delhi

Ragesh, R.
Department of Medicine
AIIMS, New Delhi

Raghupati, Raj
Professor
Faculty of Medicine
Kuwait University, Kuwait

Rahman, Seyed Asad
Scientist, EMBL-EBI
Hinxton, Cambridge, UK

Rai, Shweta
Department of Obstetrics & Gynecology
AIIMS, New Delhi

Rakshit, Srabanti
Department of Biochemistry
Indian Institute of Science
Bangalore

Rao, A.J.
Professor
Indian Institute of Science, Bangalore

Rao, B.S. Shankarnarayan
Assistant Professor
Department of Neurophysiology
National Institute of Mental Health & Neuroscience
(NIMHANS), Bangalore

Rao, PVL
Director, DRDO-BU-CLS
Coimbatore, India

Ravindra, P.N.
Department of Neurophysiology
National Institute of Mental Health & Neuroscience
(NIMHANS), Bangalore

Roy, Subimal
Emeritus
Department of Pathology
Sir Ganga Ram Hospital, New Delhi

Roy, Syamlal
Professor
Department of Infectious Diseases & Immunology
Indian Institute of Chemical Biology
Kolkata

Rusia, Usha
Professor
Department of Pathology,
University College of Medical Sciences, Delhi

Saini, Neeru
Scientist, CSIR-Institute of Genomics and Integrative
Biology, New Delhi

Saraya, Anoop
Professor
Department of Gastroenterology
AIIMS, New Delhi

Sarin, Shiv Kumar
Senior Professor, Hepatology, and
Director, Institute of Liver and Biliary Sciences
New Delhi

Saxena, Renu
Professor & Head
Department of Hematology
AIIMS, New Delhi

Selvamurthy, W.
President (Honorary) Amity Science
Technology and Innovation Foundation and Chair Professor
for Life Sciences, Amity University,
Noida

Sen Gupta, Sourav
Translational Health Science and Technology Institute
Faridabad

Sen, Sudip
Assistant Professor
Department of Biochemistry
AIIMS, New Delhi

Senapati, Parijat
Jawaharlal Centre for Advanced Scientific Research
(JNCASR), Bangalore

Sengupta, Jayshree
Former Professor
Department of Physiology
AIIMS, New Delhi

Sengupta, Partho P.
Associate Professor of Medicine
Director of Cardiac Ultrasound Research and Interventional
Echocardiography
The Zena and Michael A. Weiner Cardiovascular Institute
Icahn School of Medicine at Mount Sinai
New York

Shaha, Chandrima
Director
National Institute of Immunology (NII)
New Delhi

Sharan, Jitendra
Centre for Dental Education and Research
AIIMS, New Delhi

Sharma, Alpana
Additional Professor
Department of Biochemistry
AIIMS, New Delhi

Sharma, Animesh
Department of Medicine
AIIMS, New Delhi

Sharma, Hanish
Department of Gastroenterology and Human Nutrition
AIIMS, New Delhi

Sharma, K.N.
Former Professor, Department of Physiology
University College of Medical Sciences (UCMS)
New Delhi

Sharma, N.
Postgraduate Institute of Medical Education & Research
(PGI), Chandigarh

Sharma, N.C.
Ex-Vice President, Biotechnology Division of Cadila
Pharmaceuticals, Ahmedabad

Sharma, Sangeeta
Professor, Department of Psycho Neuro Pharmacology
Institute of Human Behaviour and Allied Sciences
New Delhi

Sharma, Shail K.
Former Professor
Department of Biochemistry, AIIMS, New Delhi

Sharma, Shvetank
Assistant Professor
Department of Research
Institute of Liver and Biliary Sciences
New Delhi

Sharma, Surendra K.
Head
Department of Medicine
AIIMS, New Delhi

Shembekar, Nachiket
National Institute of Immunology,
New Delhi

Singh, A.K.
Principal Scientist
Dairy Technology Division
National Dairy Research Institute
Karnal, Haryana

Singh, Archana
Assistant Professor
Department of Biochemistry
AIIMS, New Delhi

Singh, Lal
Homoeopathy Physician & Consultant
Amar HomoeoHealers
New Delhi

Singh, Namrata
Department of Gastroenterology
AIIMS, New Delhi

Singh, Neeta
Former Professor, Department of Biochemistry
AIIMS, New Delhi

Singh, T.P.
Professor
Department of Biophysics
AIIMS, New Delhi

Singla, Parul
Consultant, Department of Biochemistry
Sir Ganga Ram Hospital
New Delhi

Sinha, Aditi
Assistant Professor
Department of Paediatrics
AIIMS, New Delhi

Sinha, Uma
Department of Intensive Care Medicine, AMRI Hospital
Kolkata

Sinharoy, Indranil
Department of Intensive Care Medicine, AMRI Hospital
Kolkata

Sood, S.K.
Professor and Senior Consultant
Department of Hematology and Quality Assurance
Sir Ganga Ram Hospital
New Delhi

Sriraman, Kalpana
Research Officer
The Foundation for Medical Research
Mumbai

Srivastava, A.K.
Director
National Dairy Research Institute (NDRI)
Karnal, Haryana

Srivastava, L.M.
Professor, Senior Consultant & Head
Department of Biochemistry
Sir Ganga Ram Hospital
New Delhi

Srivastava, Tapasya
Department of Genetics, University of Delhi
South Campus, New Delhi

Sudarshan, Deepthi
Jawaharlal Centre for Advanced Scientific Research
(JNCASR)
Bangalore

Sung, Bokyung
Department of Experimental Therapeutics
University of Texas
M.D. Anderson Cancer Center
Houston, Texas,
USA

Swaminathan, S.
International Centre for Genetic Engineering and
Biotechnology
New Delhi

Talha, S.M.
University of Turku
Turku, Finland

Talwar, G.P.
Director Research
Talwar Research Foundation,
New Delhi

Tandon, Nikhil
Professor
Department of Endocrinology and Metabolism
AIIMS, New Delhi

Tyagi, Rakesh K.
Professor
Special Center for Molecular Medicine
JNU, New Delhi

Venkatesha, S.H.
Research Fellow
Department of Microbiology & Immunology
University of Maryland
School of Medicine, Baltimore,
USA

Verma, Sunil
Assistant Professor
Department of Cardiology
AIIMS, New Delhi

Vijayan, M.
Emeritus Professor
Molecular Biophysics Unit
Indian Institute of Science,
Bangalore

Vrati, Sudhanshu
Dean
Translational Health Science and Technology Institute
Faridabad

Wadhwa, Neerja
National Institute of Immunology (NII)
New Delhi

Figure 4.13

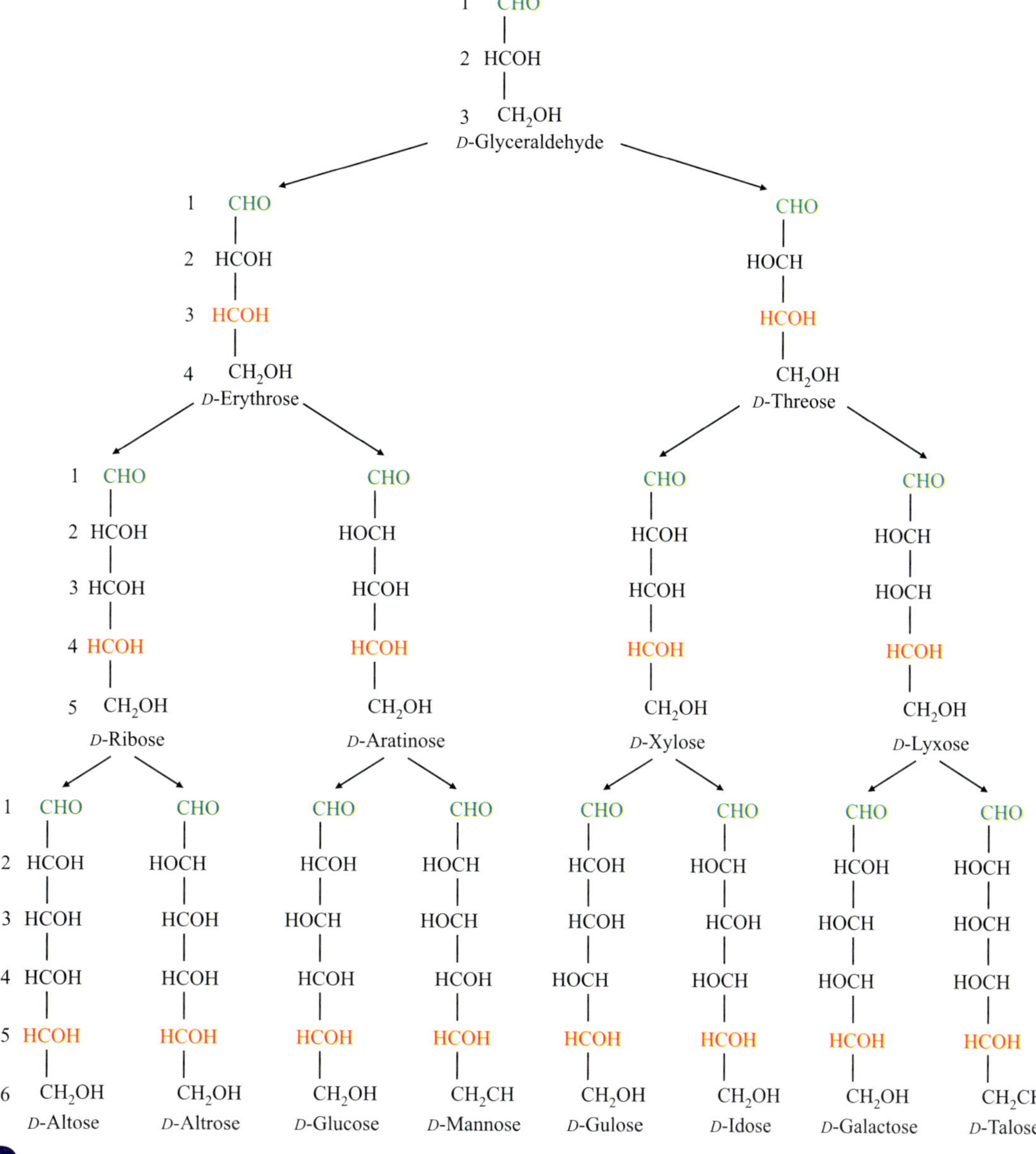

Figure 6.6

1 CH₂OH
2 C=O
3 CH₂OH
Dihydroxyacetone
1 CH₂OH
2 C=O
3 HCOH
CH₂OH
D-Erythrulose
1 CH₂OH
2 C=O
3 HCOH
4 HCOH
5 CH₂OH
D-Ribulose
CH₂OH
C=O
HOCH
HCOH
CH₂OH
D-Xylulose
1 CH₂OH
2 C=O
3 HCOH
4 HCOH
5 HCOH
6 CH₂OH
D-Psicose
CH₂OH
C=O
HOCH
HCOH
HCOH
CH₂OH
D-Fructose
CH₂OH
C=O
HCOH
HOCH
HCOH
CH₂OH
D-Sorbose
CH₂OH
C=O
HOCH
HOCH
HCOH
CH₂OH
D-Tagatose
1 CHO
2 HO—C*—H
3 H—C*—OH
4 H—C*—H
5 HO—C*—H
6 CH₂OH
L-Glucose
Left oriented —OH
1 CHO
2 H—C*—OH
3 HO—C*—H
4 H—C*—OH
5 H—C*—OH
6 CH₂OH
D-Glucose
Right oriented —OH

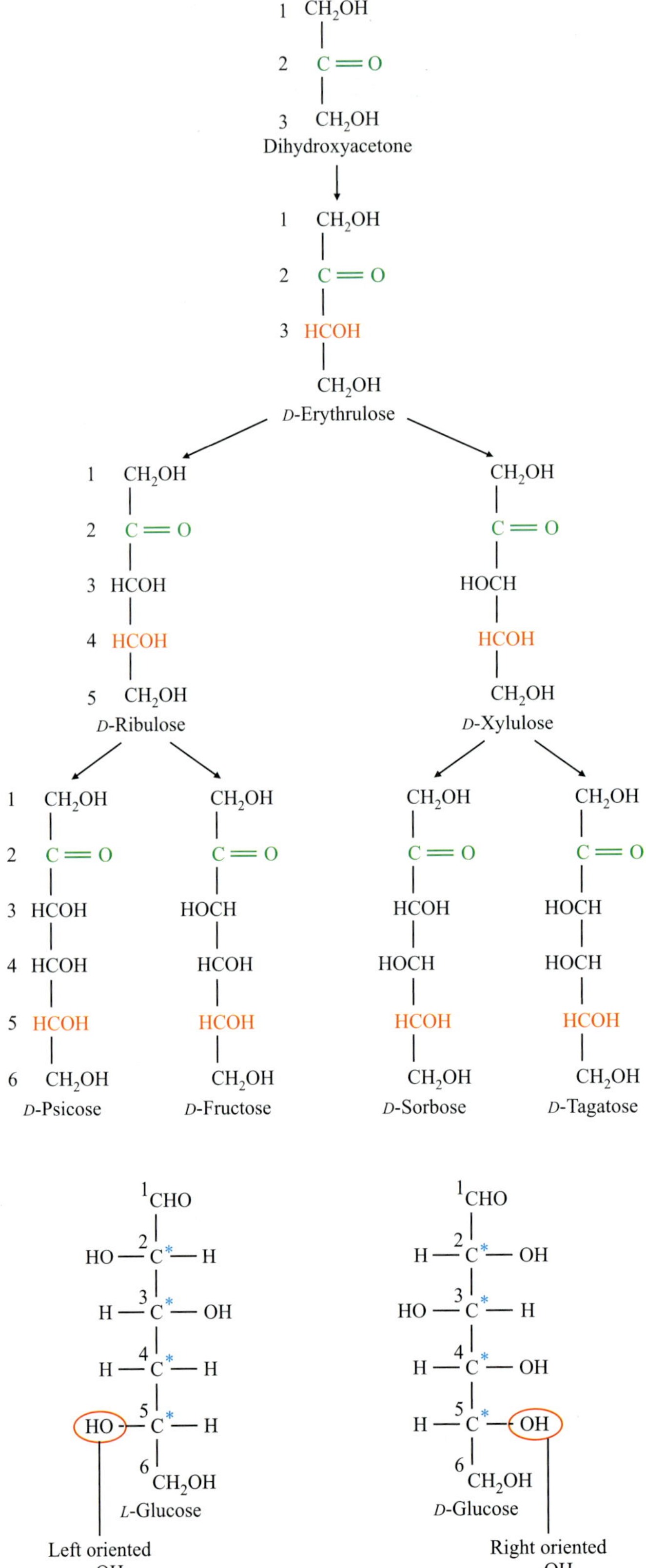

Figure 6.7

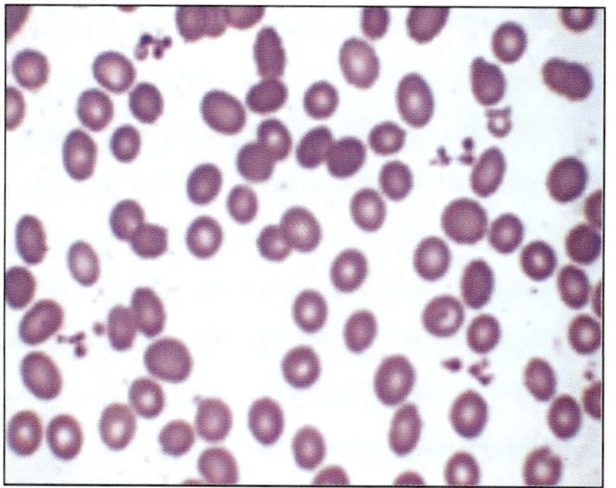

Figure 32.1

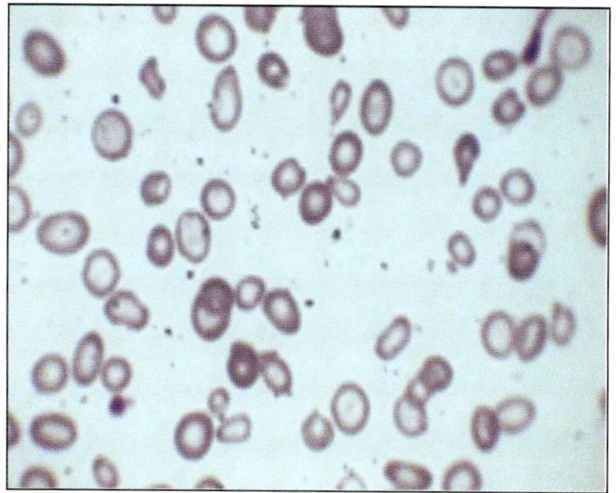

Figure 32.2

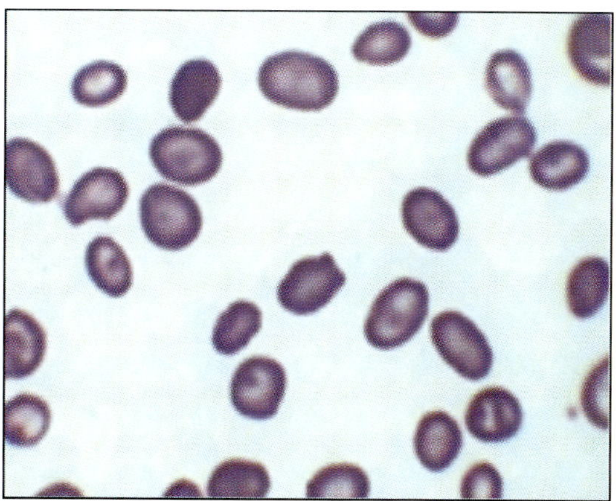

Figure 32.3

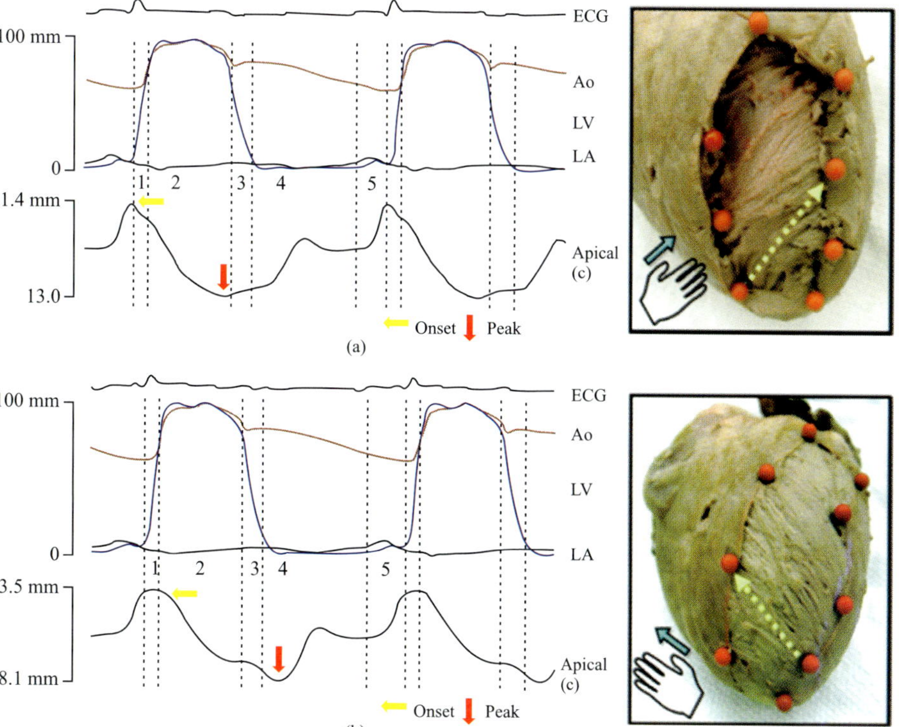

Figure 33.2

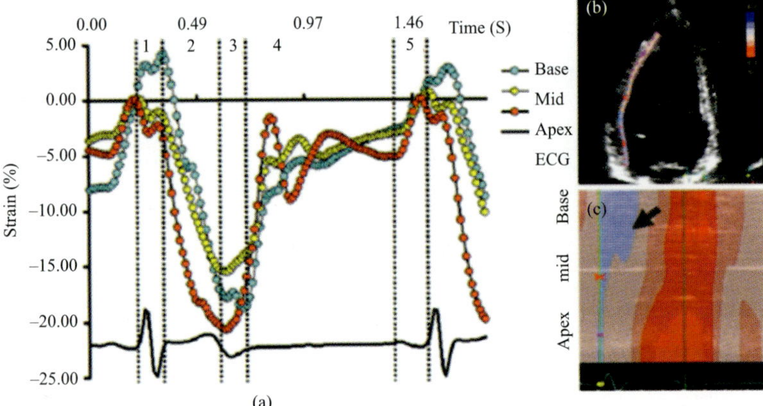

Figure 33.4

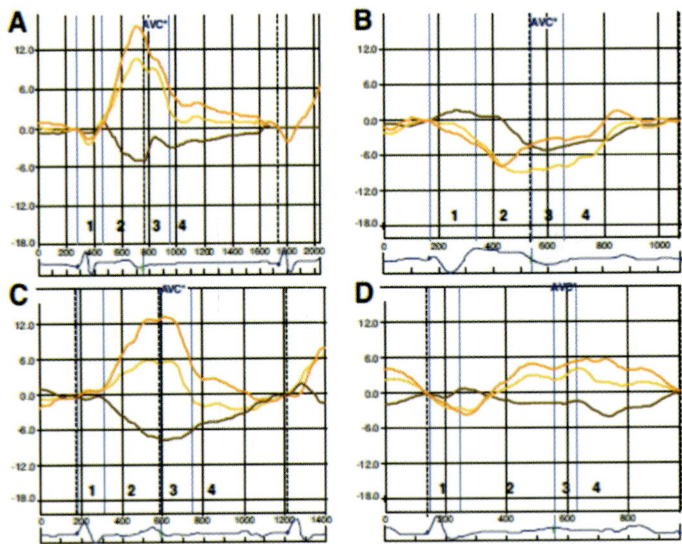

Figure 33.5

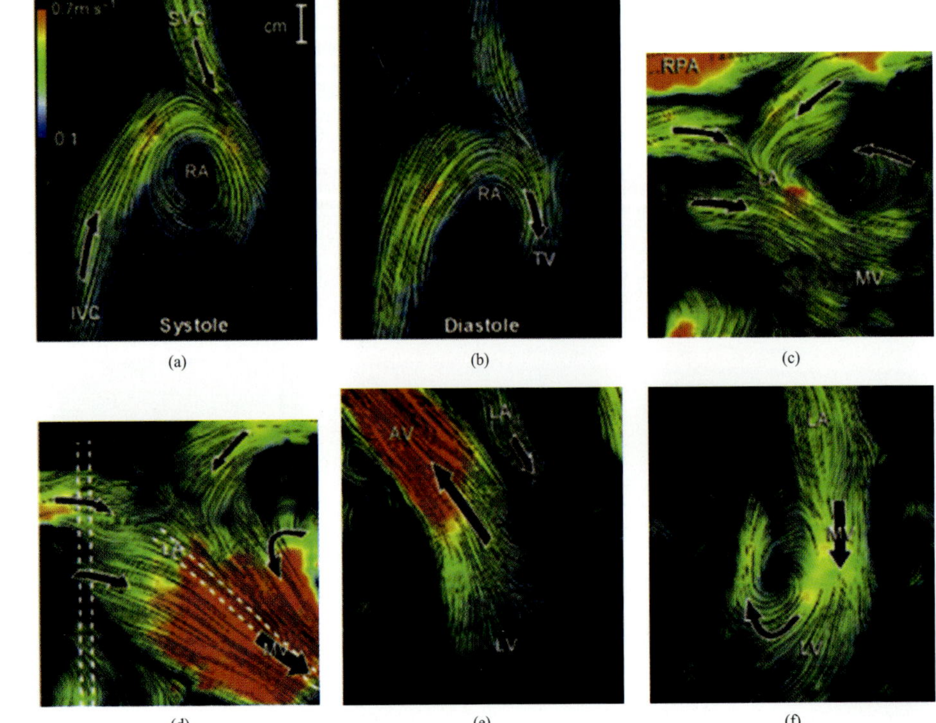

Figure 33.6

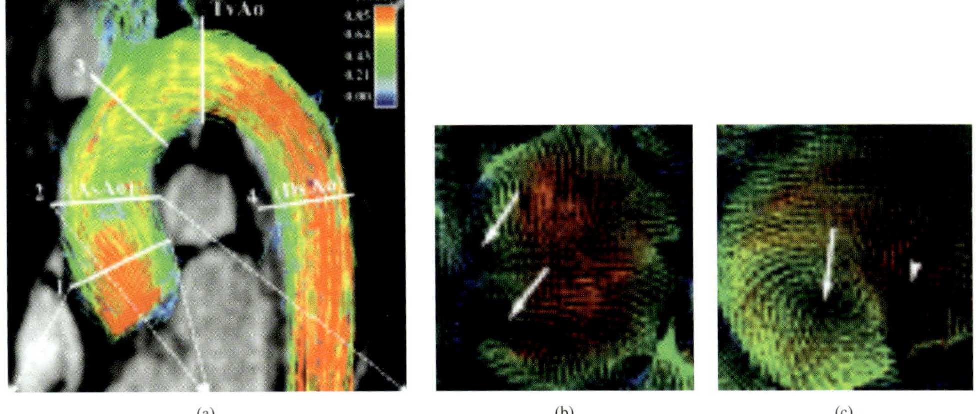

(a)   (b)   (c)

Figure 33.7

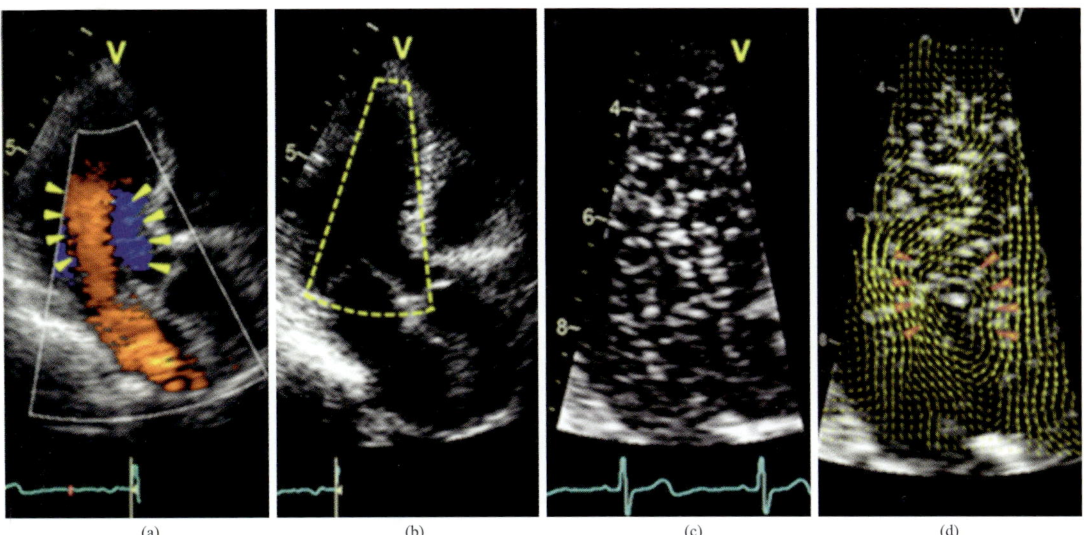

(a)   (b)   (c)   (d)

Figure 33.8

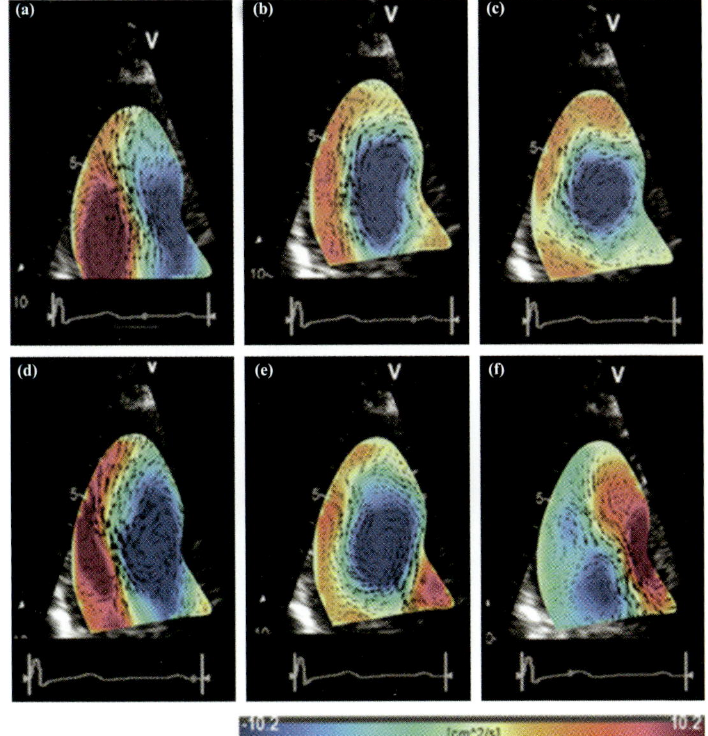

Figure 33.9

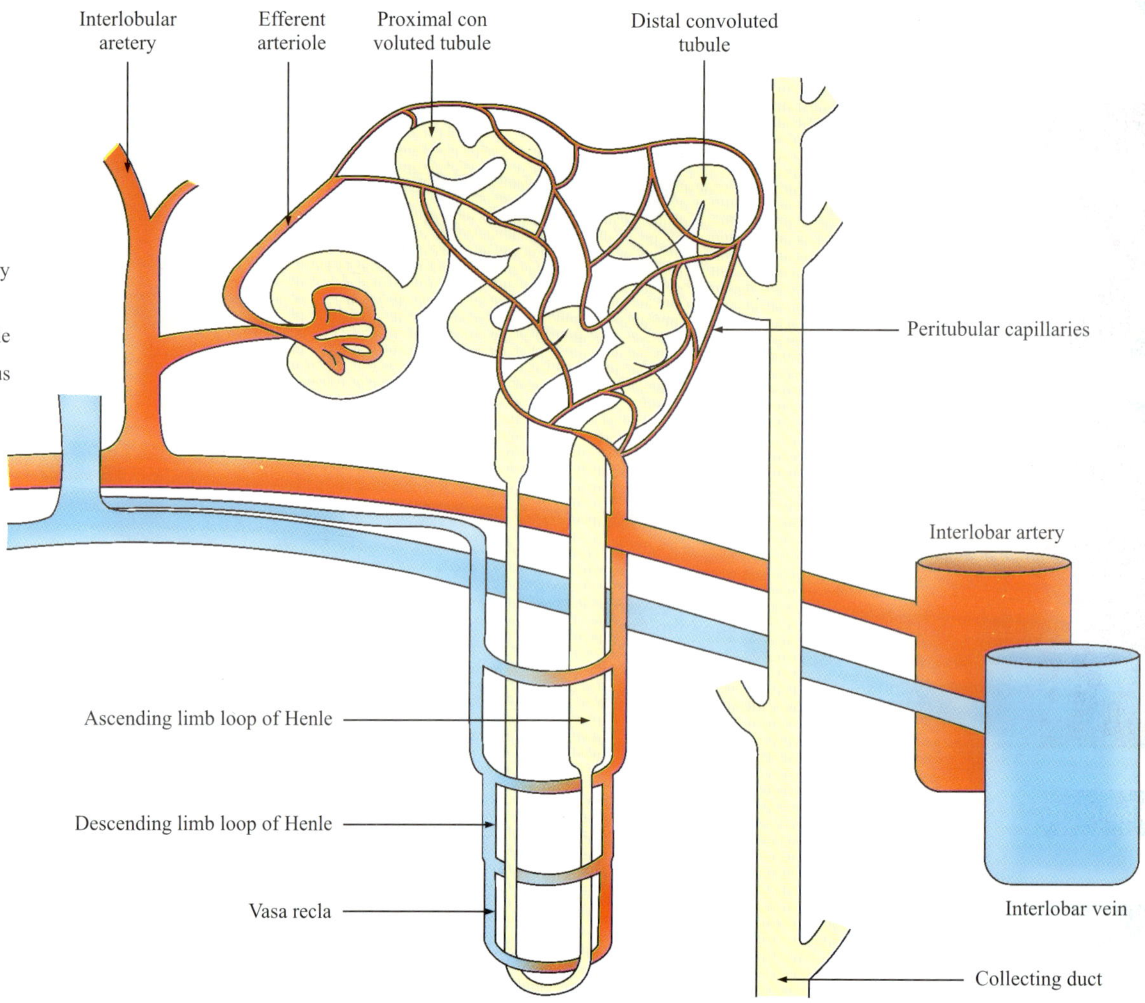

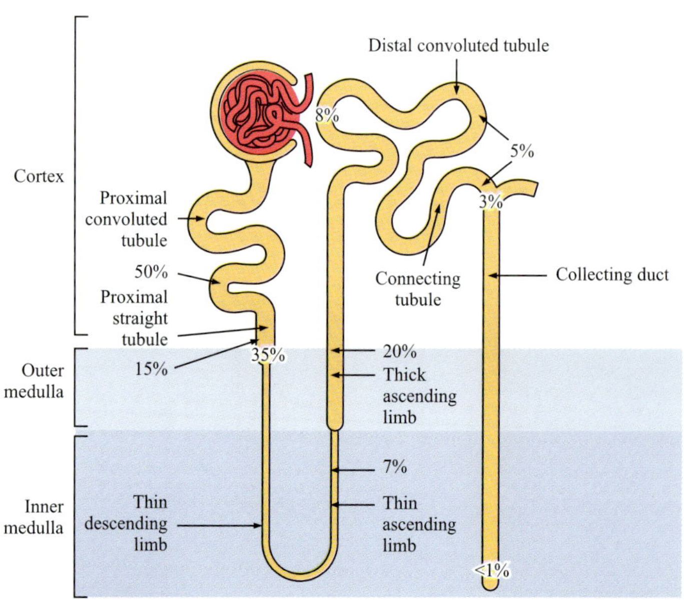

Figure 34.3

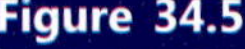

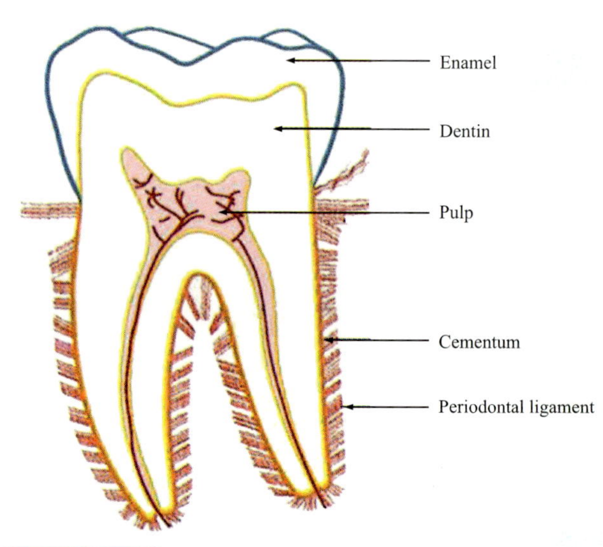

Figure 34.5

Figure 36.1

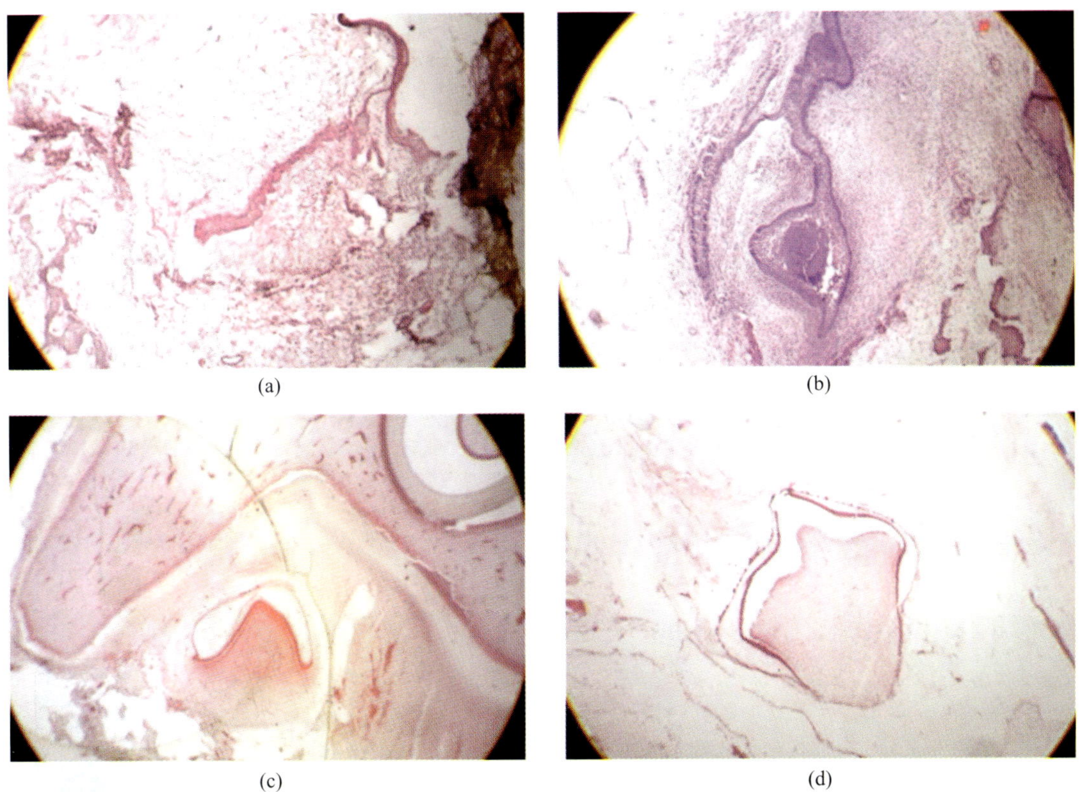

(a)

(b)

(c)

(d)

Figure 36.2

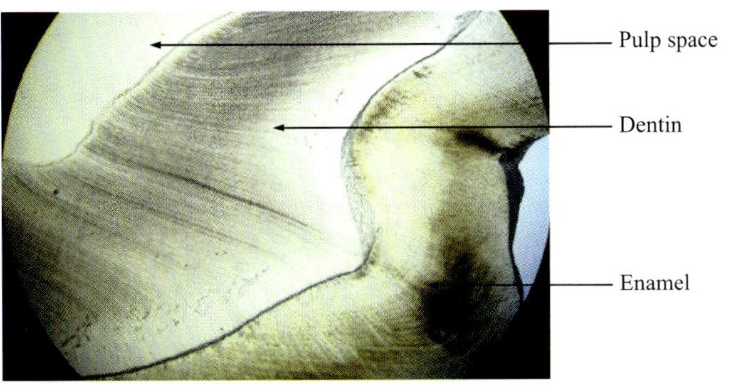

Figure 36.3

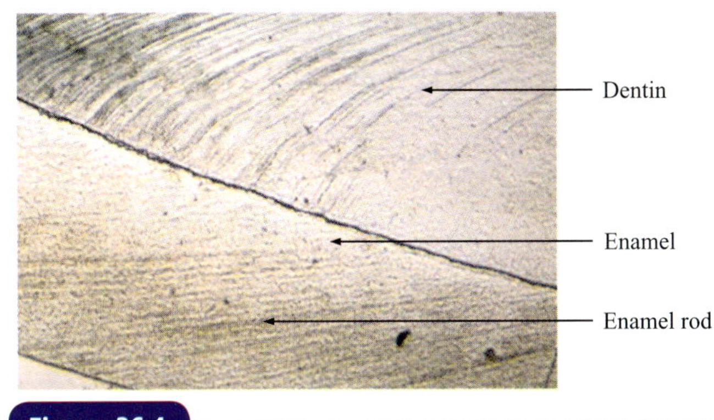

Figure 36.4

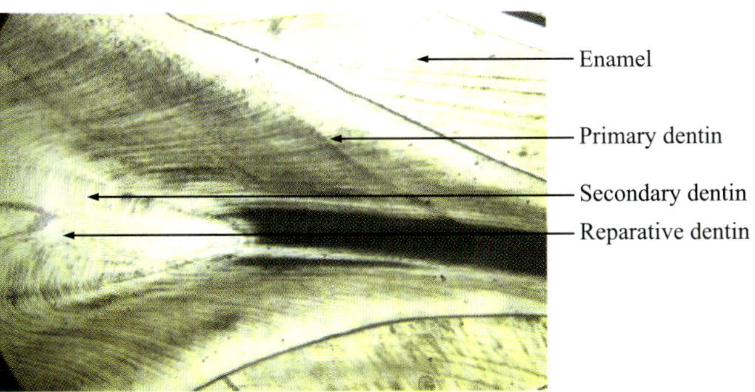

Figure 36.5

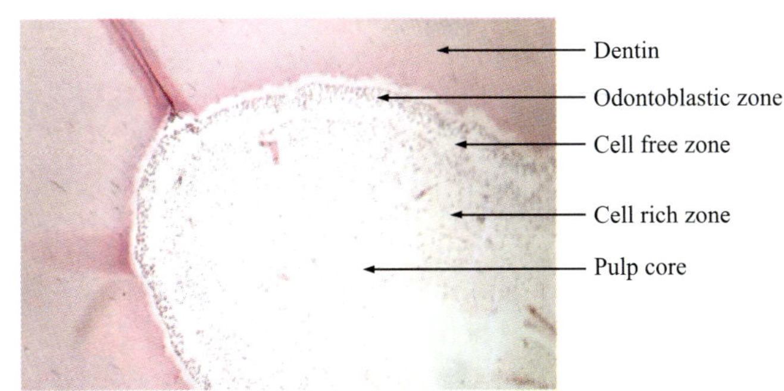

Figure 36.6

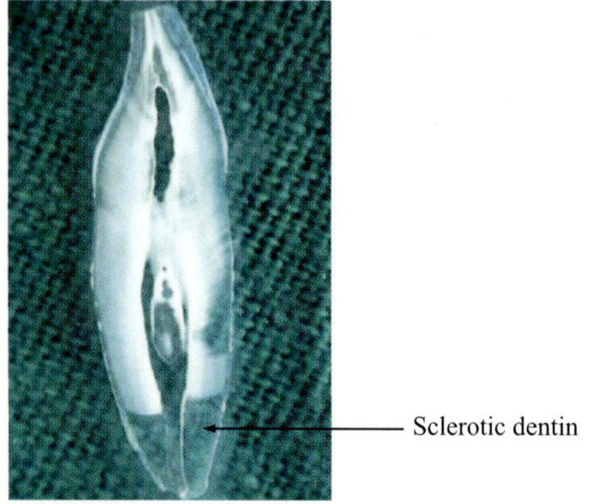

Figure 36.8

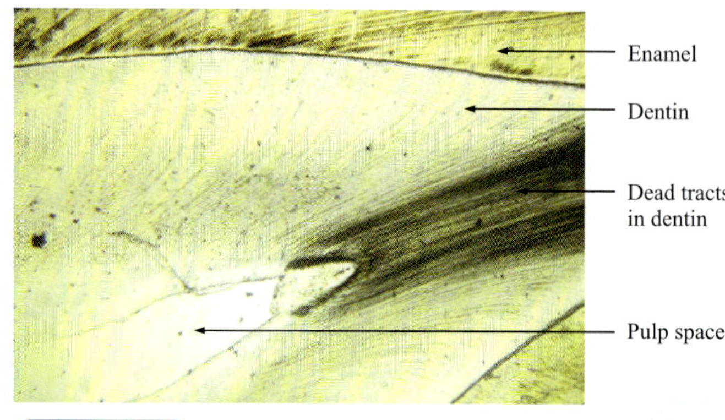

Figure 36.9

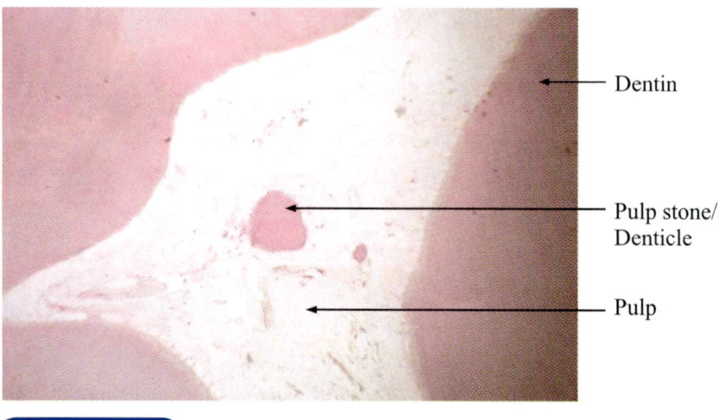

Figure 36.10

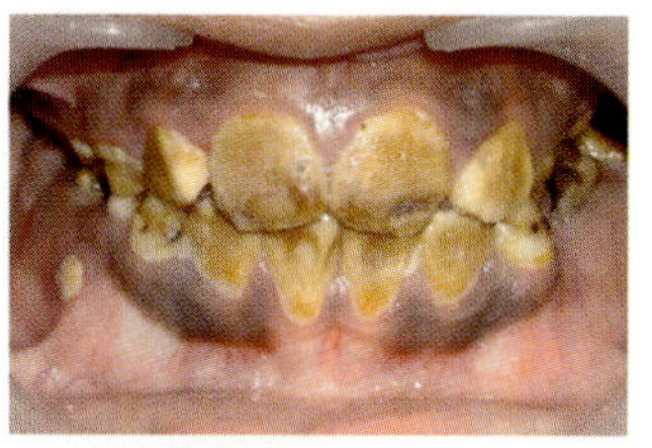

Figure 36.11

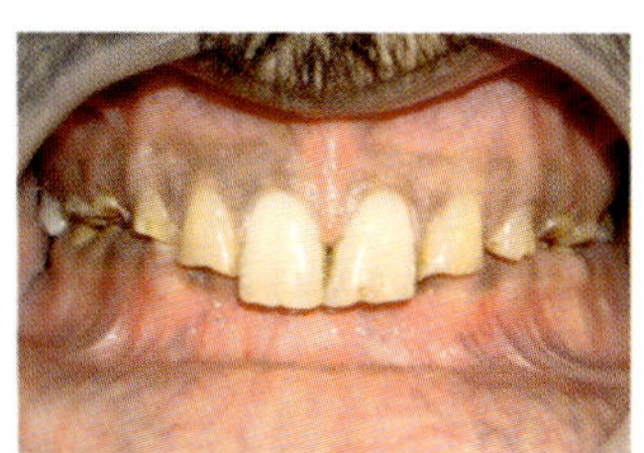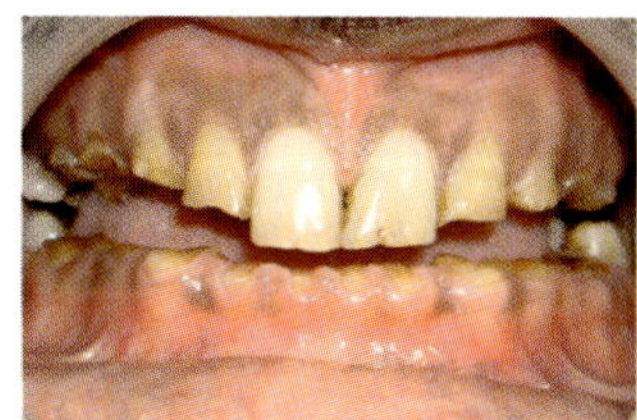

Figure 36.12

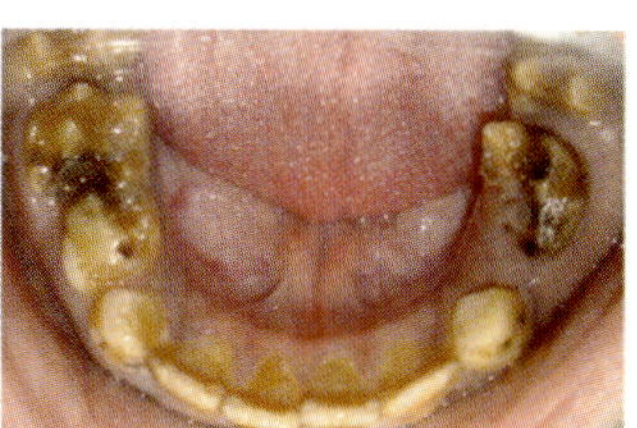

Figure 36.13

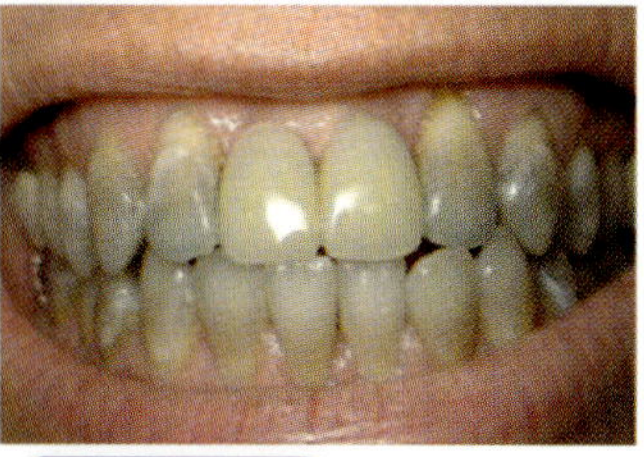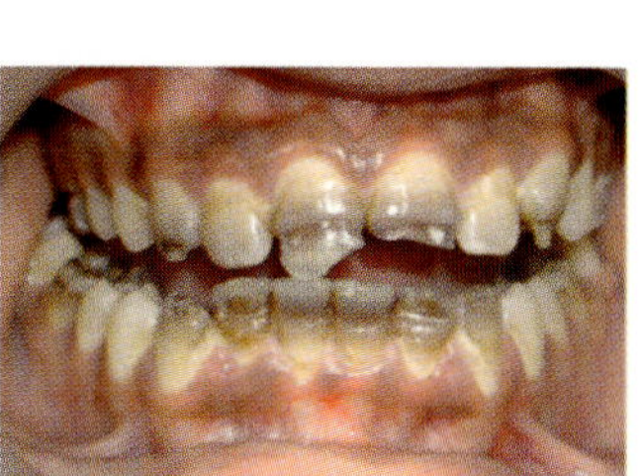

Figure 36.14

Figure 37.1

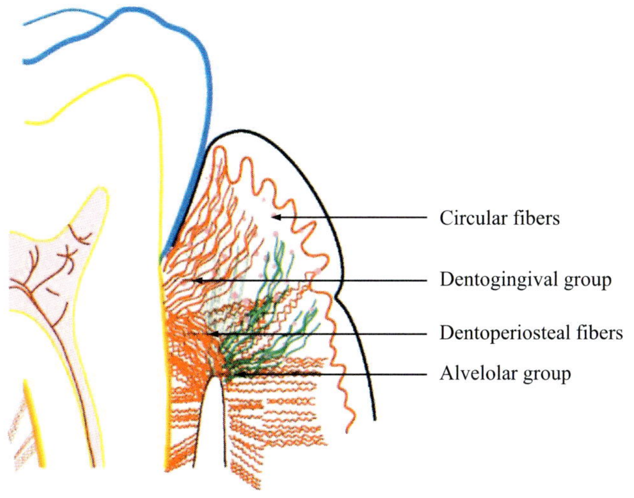

Figure 37.2

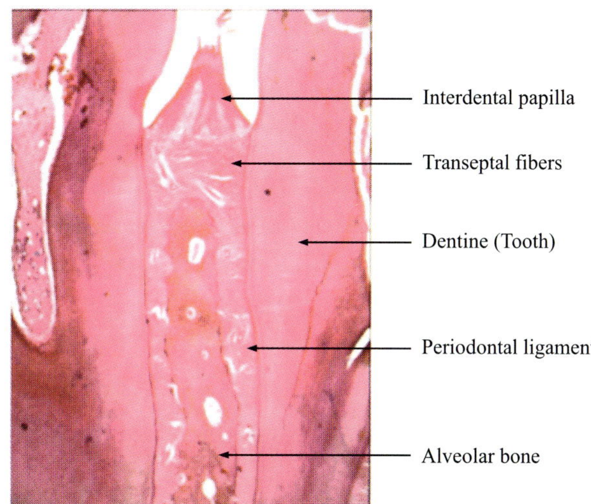

Figure 37.3

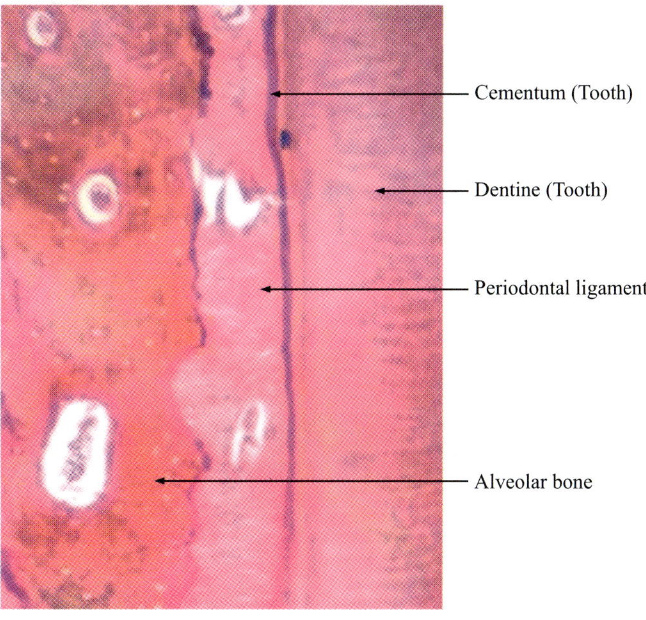

Figure 37.4

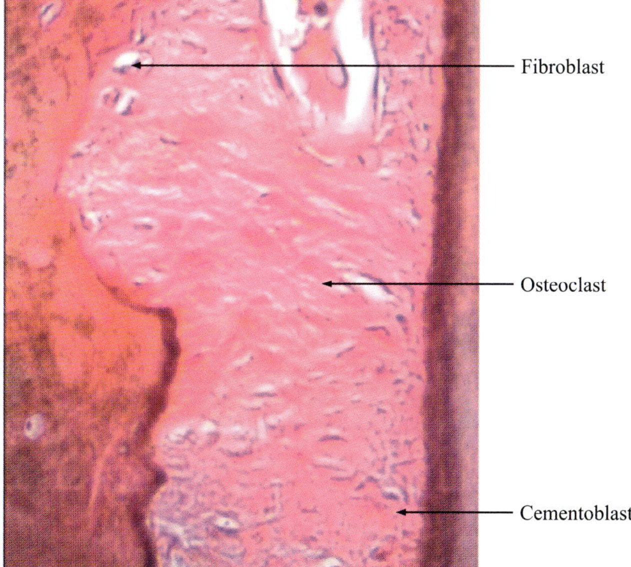

Figure 37.5

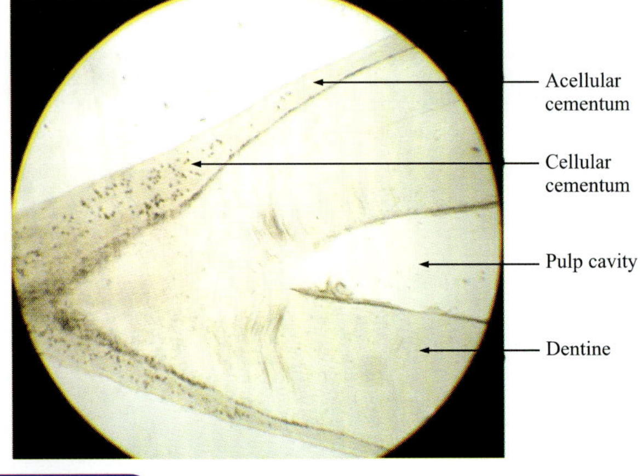

Figure 37.8

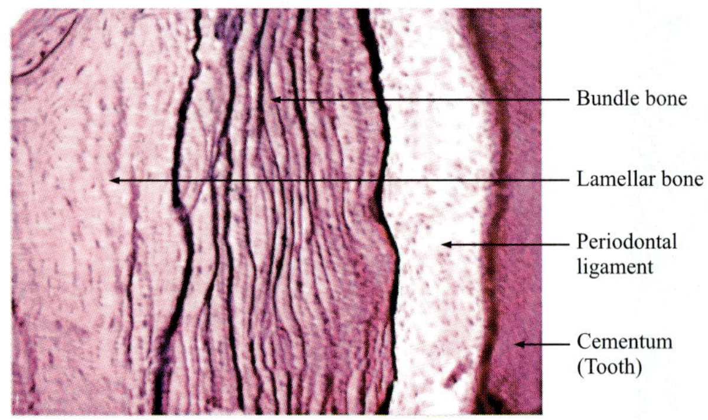

Figure 37.10

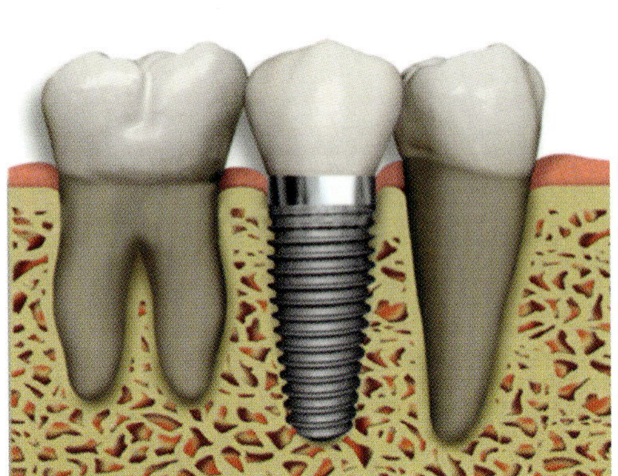

Figure 37.13

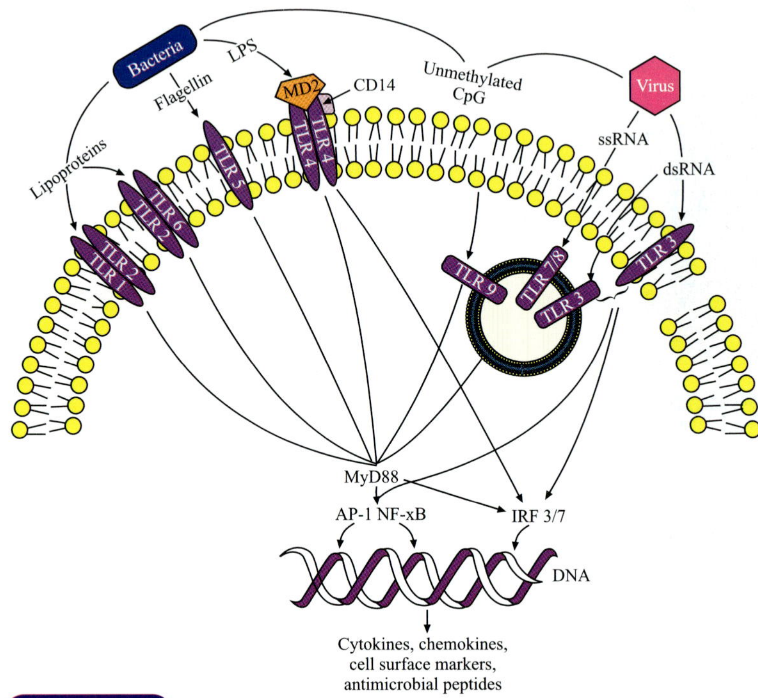

Figure 38.4

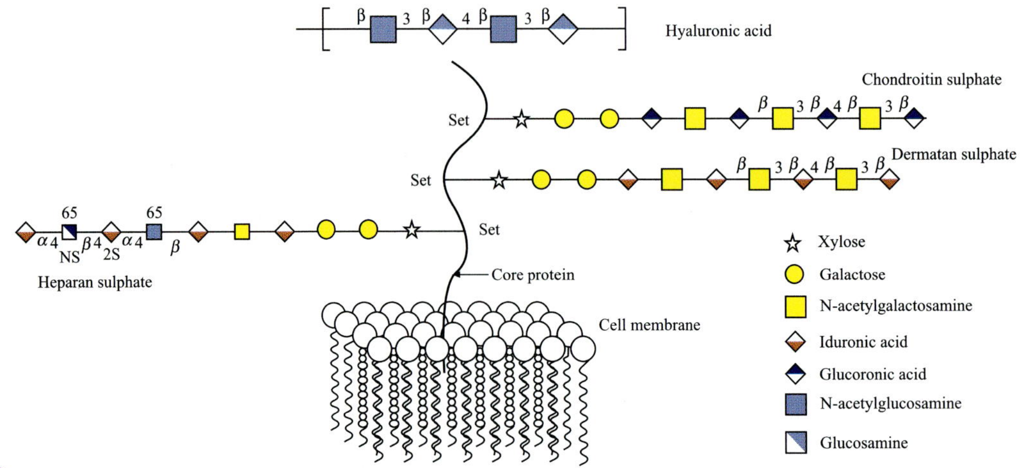

Figure 38.8

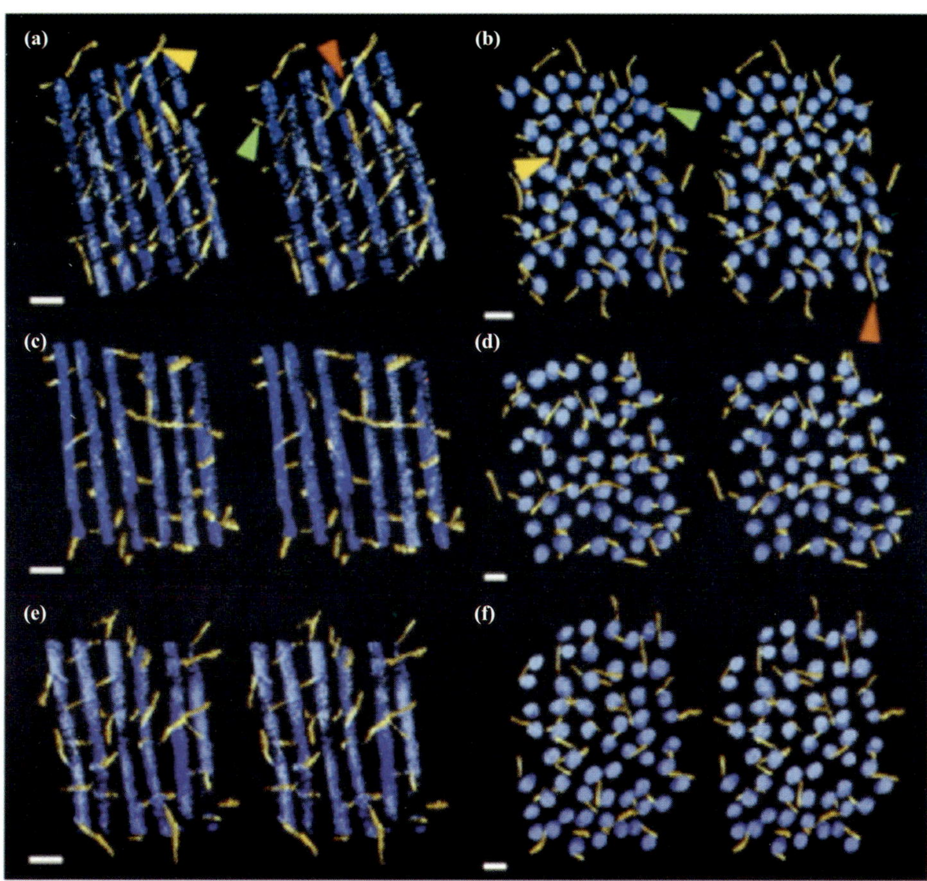

Figure 38.9

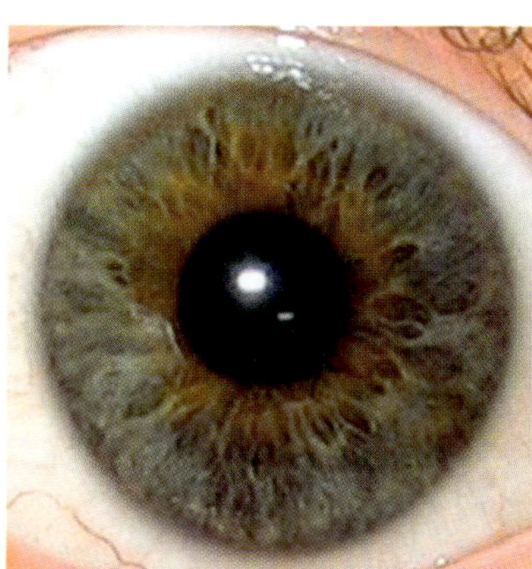

Figure 38.11

How IRIS scanners record identities

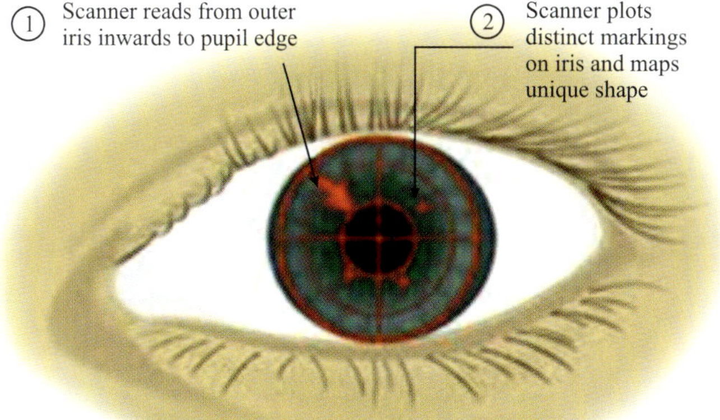

Figure 38.13

Figure 38.17

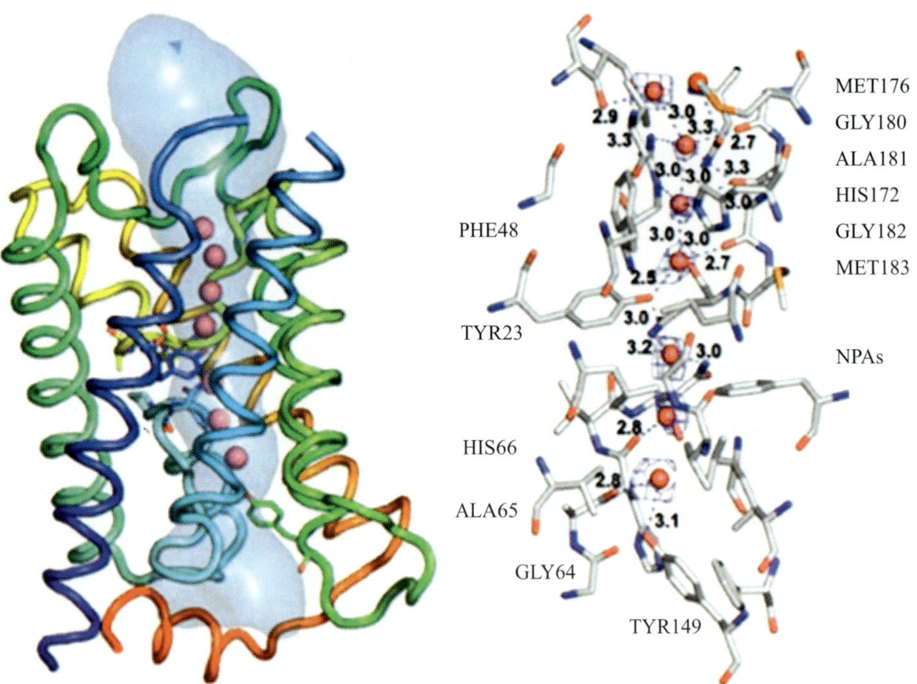

Figure 38.15

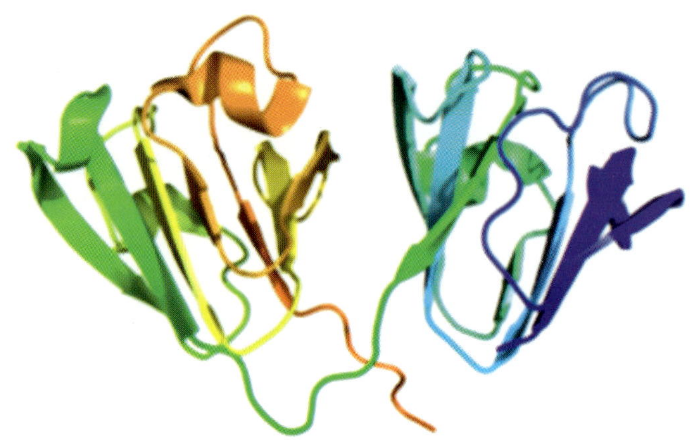

Figure 38.18

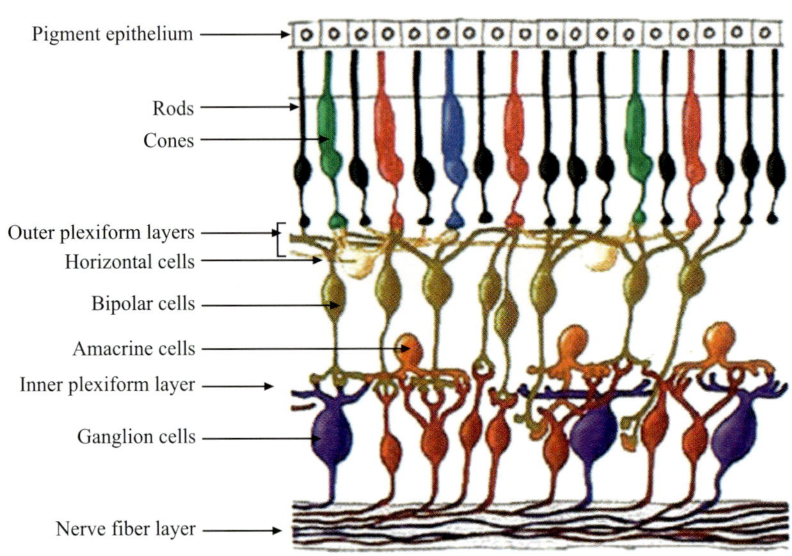

Figure 38.21

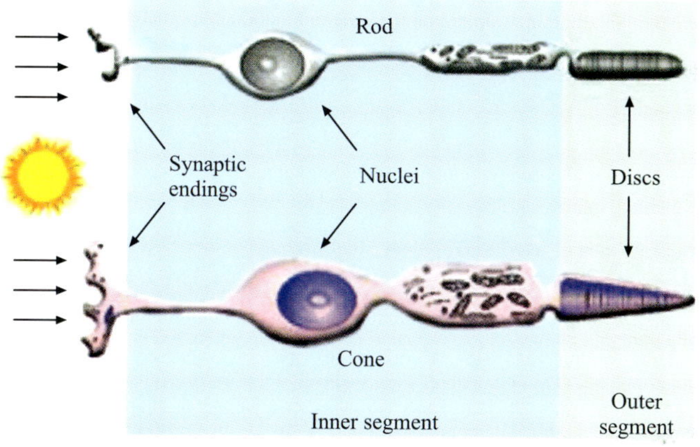

Figure 38.22

Figure 38.32

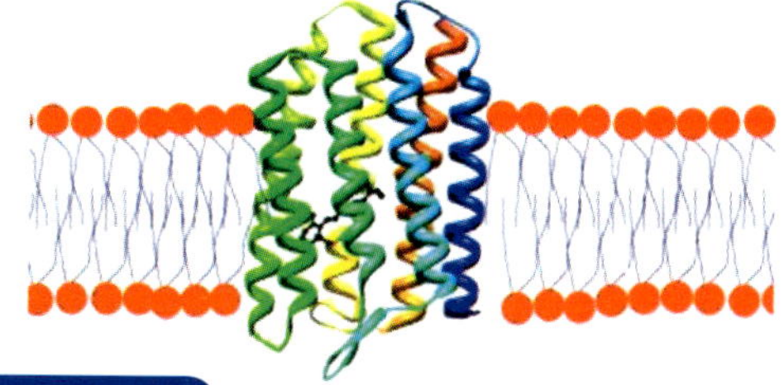

Figure 38.24(a)

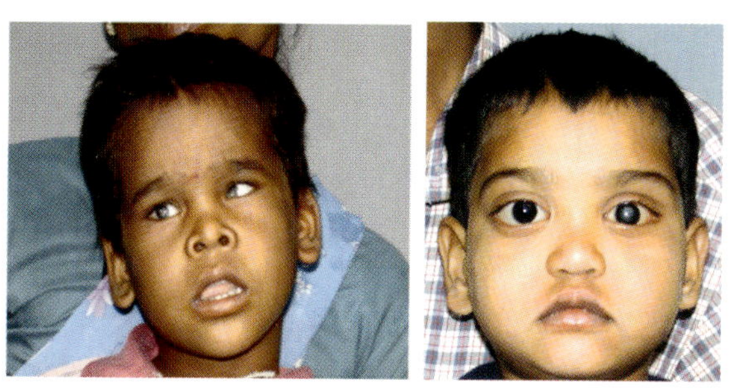

Figure 38.28

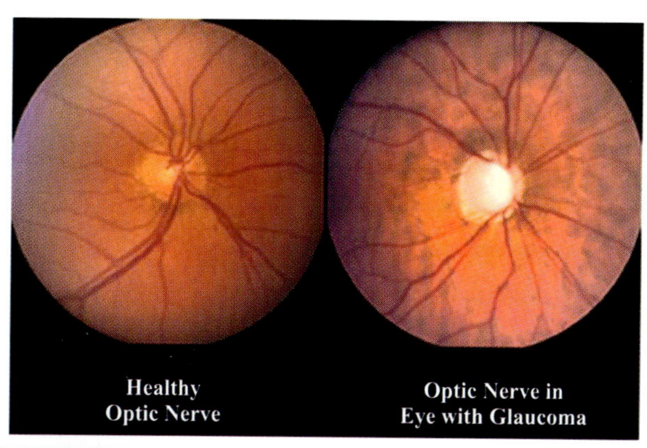

Figure 38.34

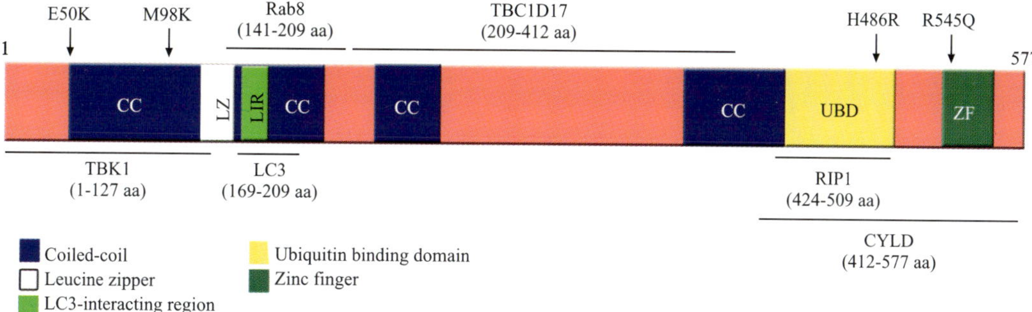

Figure 38.37

Figure 38.38

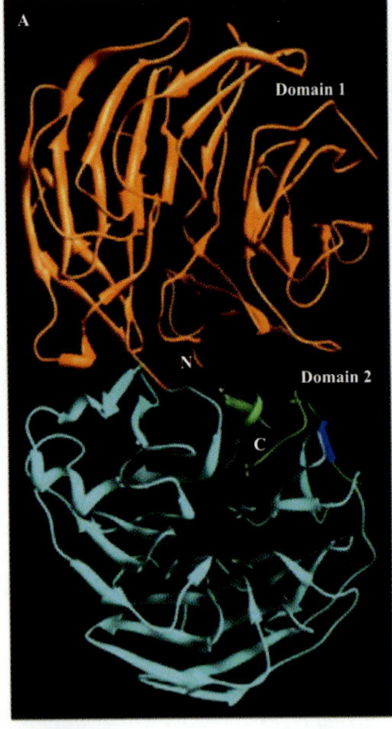

Figure 38.40

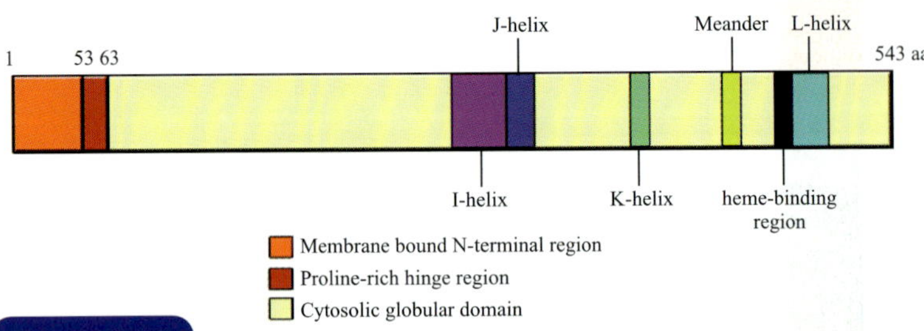

Figure 38.41

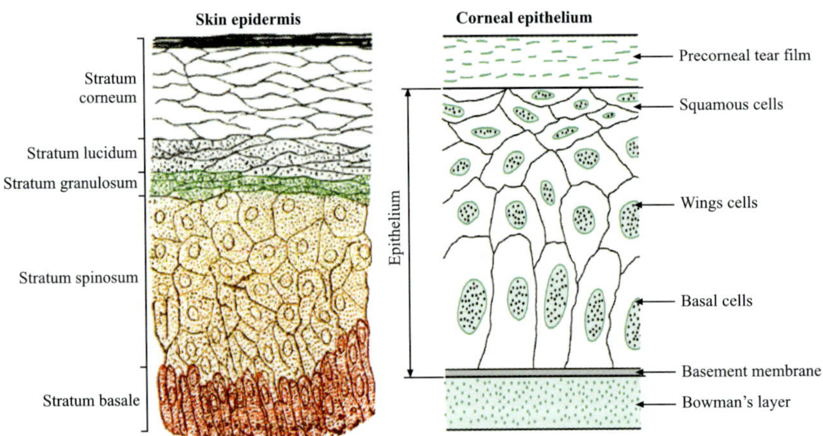

Figure 38.46

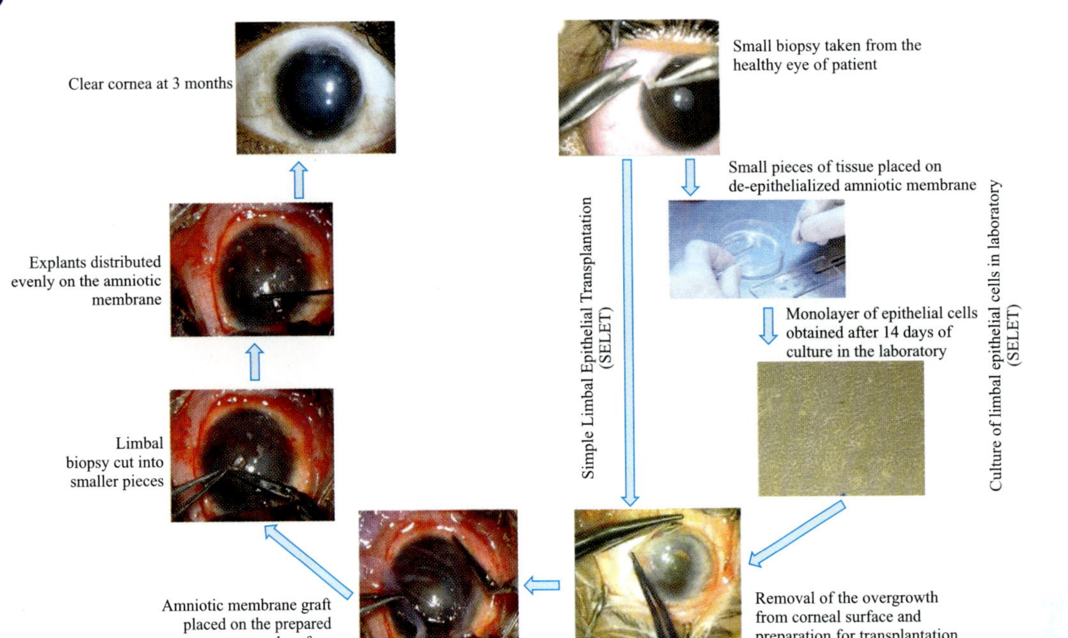

Figure 38.48

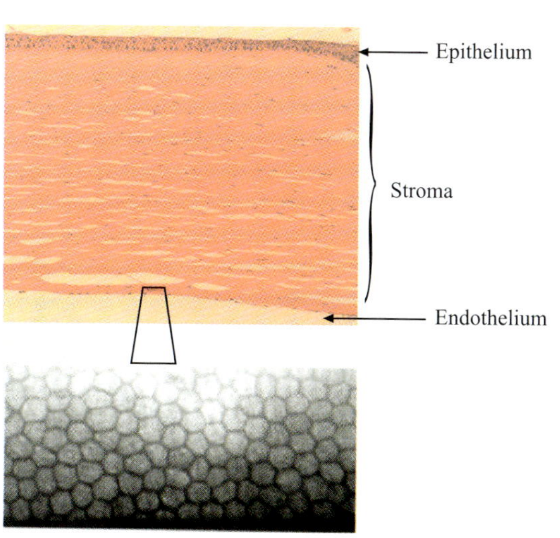

Figure 38.49

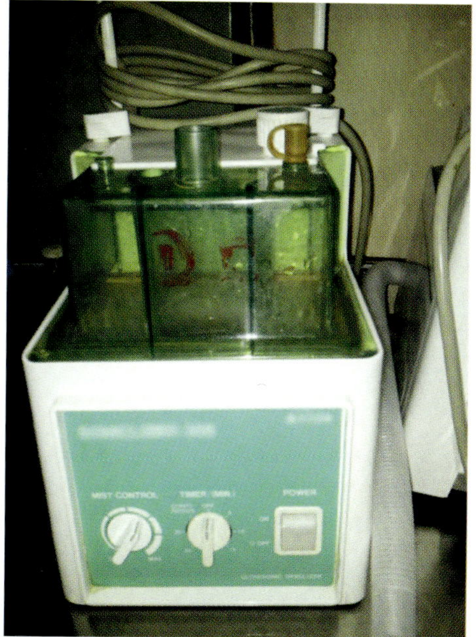

Figure 39.7

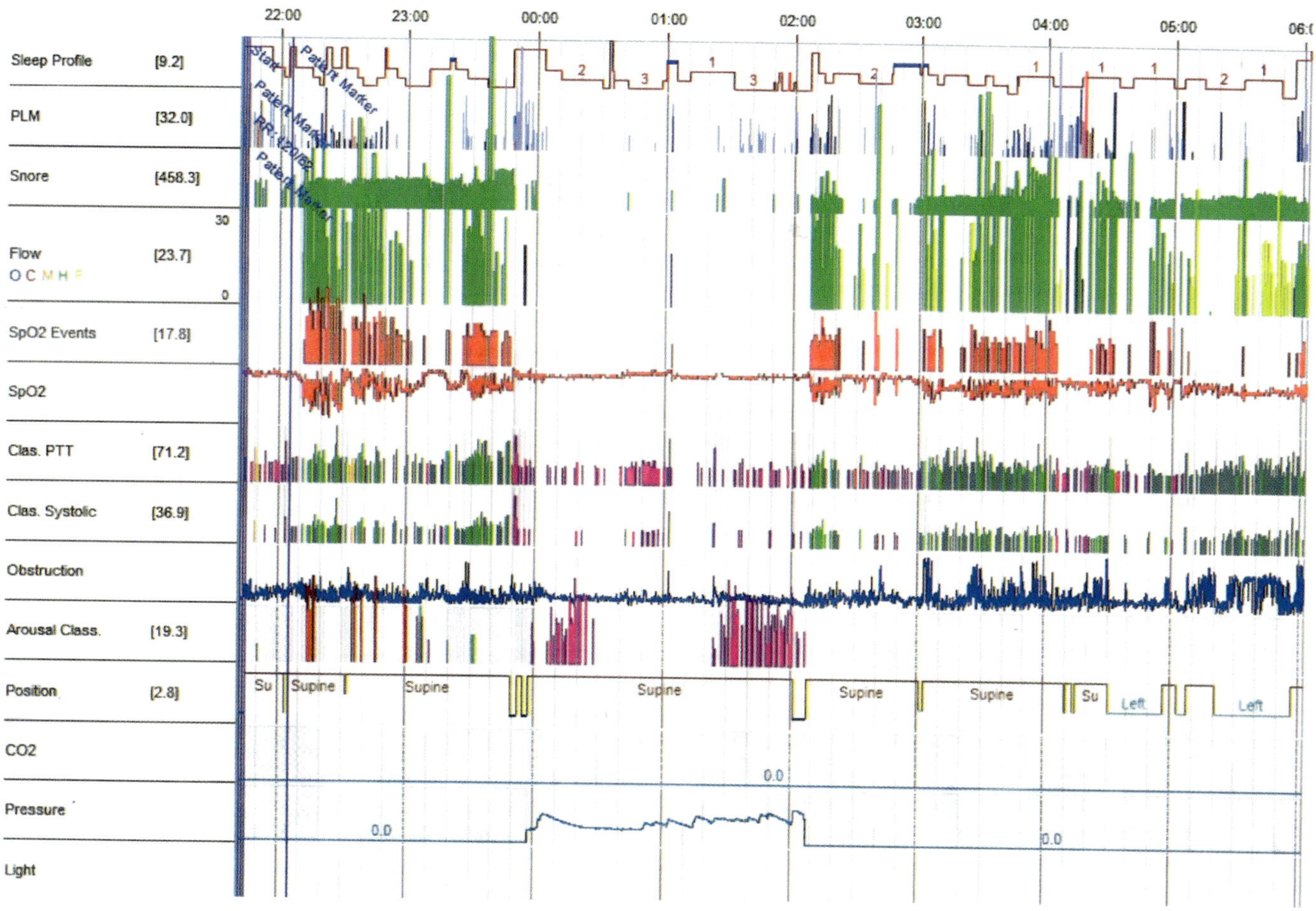

Figure 39.8

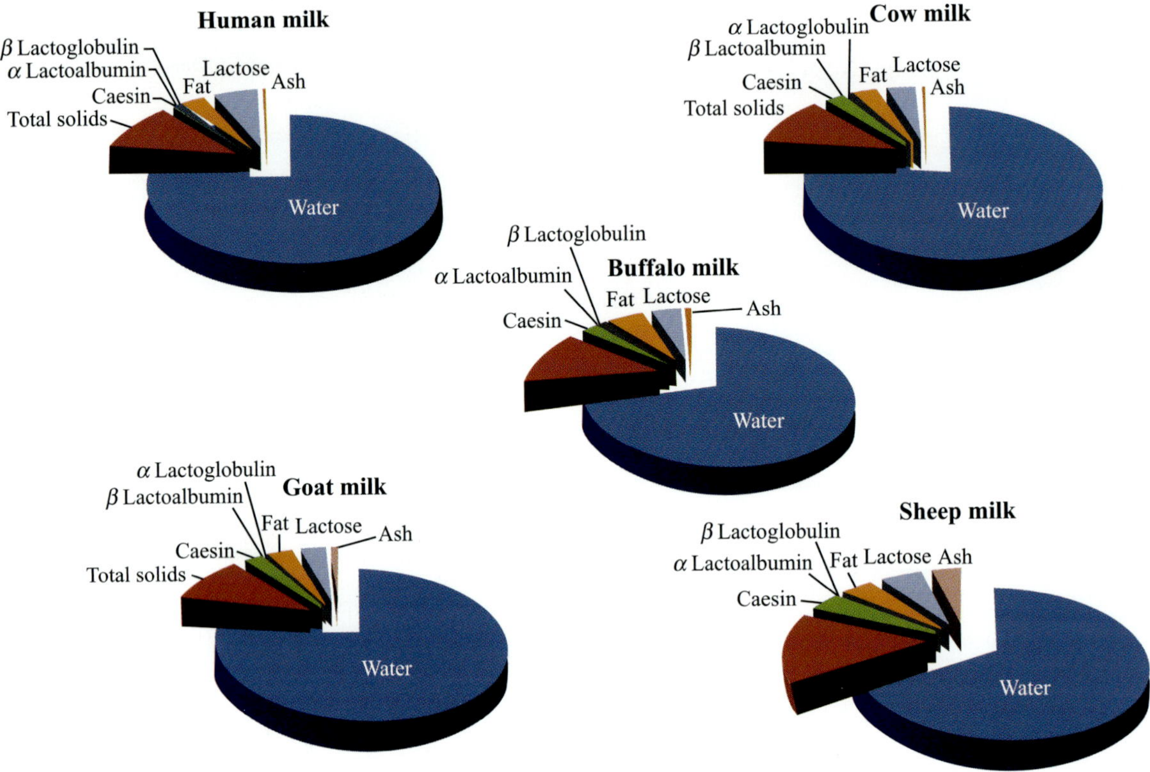

Figure 42.1

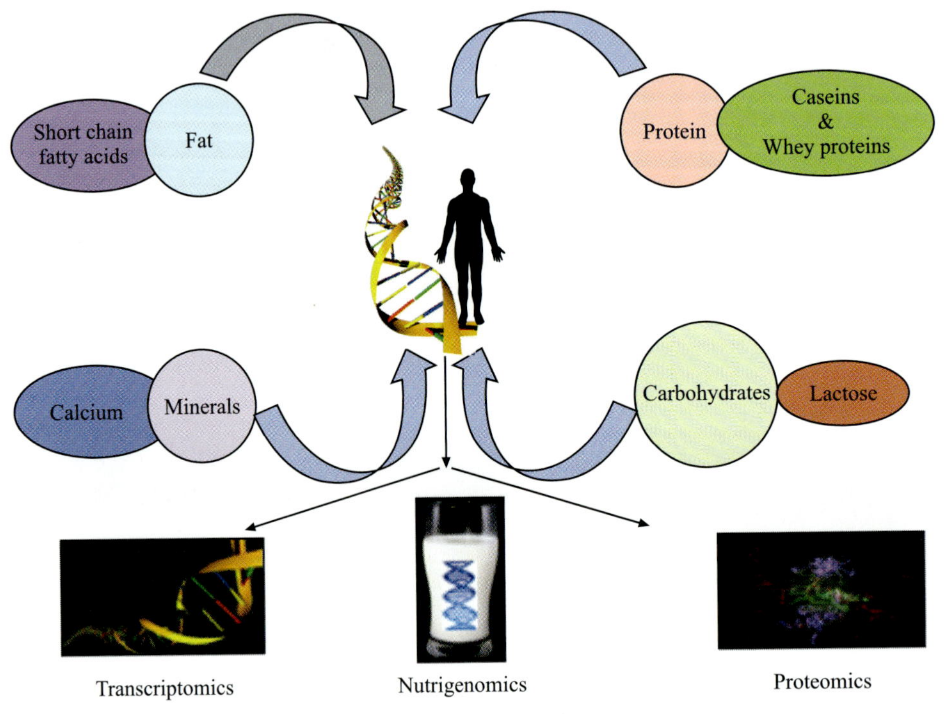

Figure 42.2

SECTION I

CHEMICAL ARCHITECTURE
OF THE LIVING SYSTEMS

The human body is made up of a myriad of molecules. What are these? What are their properties and characteristics? How are they aligned to each other to make possible the living state? This section gives the basic information on some of these questions.

Matter in appropriate physico-chemical state of organization acquires the traits of living substances. In this section, we intend to explore and discover some of these features.

زندگی کیا ہے عناصر میں ظہورِ ترتیب
موت کیا ہے اُنہیں اجزا کا پریشاں ہونا

ज़िन्दगी क्या है अनासिर में ज़हूरे तरतीब।
मौत क्या है इन्हीं अजज़ा का परेशां होना।।

What is life? An organized state of some elements.
What is death? A state of disorder of these components!

1

Elements, Molecules and Cells

G.P. Talwar

CONTENTS

I. ELEMENTS

Of about 118 elements known, the living systems are composed of about 25. Most of them are relatively light elements with atomic numbers below 30 (Zn). Among these, four elements, namely, *carbon, hydrogen, nitrogen* and *oxygen* are preeminent. They are universally present and account for over 95% of the total mass of many living organisms. Other important constituents are calcium (Ca), phosphorus (P), potassium (K), sulphur (S), sodium (Na), magnesium (Mg), chlorine (Cl) and iron (Fe). Besides these, half a dozen to a dozen elements are present in trace amounts in various organisms. A summary of the elements commonly found in the living organisms together with some comments on their role in the body is given in Table 1.1. It is interesting to note that a myriad of living forms with a wide diversity of traits is created from molecules made up of a few elements.

The major elements that constitute the back-bone of living organisms are not necessarily the commonest encountered on the crust of the earth. For example, silicon is abundant in the earth's crust, but carbon is the element that has proved to be biologically fit for synthesis of the organic compounds. Similarly, hydrogen and nitrogen have been selected out and are present in living organisms in higher amounts than in the inanimate matter around us. Oxygen is abundant in the earth's crust (49.4% by weight) and is also the richest element in the human body (65% by weight).

These *four* elements (C, H, N and O) have some common characteristics. They are light, and can form fairly strong covalent bonds by sharing electrons with each other or with other elements. Hydrogen needs one electron, oxygen two, nitrogen three and carbon four to complete their outer electron shells. These four elements are amongst the lightest capable of forming covalent bonds. As the strength of a covalent bond is inversely proportional to the atomic weights of the bonded atoms, these elements seem to have been selected by the living organisms as biologically fit, capable of forming the strongest covalent bonds.

Carbon: It is unique among elements for the number and variety of compounds that it can form. Over a quarter million have been identified and described. It has four valencies ascribing it sufficient potential to react and form compounds. With the formation of four covalent bonds by electron pair sharing, the octet of outer electron shells is completed, rendering a state of extreme stability. The octet cannot be increased, thus saturated carbon atom cannot coordinate either as a donor or acceptor. The carbon compounds are therefore inert and stable except in presence of catalysts.

Silicon cannot replace carbon. The reason is that the silicon compounds do not have the stability of those of carbon. It is in particular not possible to form stable compounds with long chains of silicon atoms. Silicon is not limited to a covalence of four. This value can rise to six, hence silicon can engage in coordinate interactions. Moreover, the energy in formation of Si—Si bond is of the order of 42 kcal which is nearly half as much as the energy in the formation of C—C, C—H, C—Cl bonds (~82 kcal). It is thus twice as easy to break Si—Si bonds than C—C bonds. Carbon can interact with other elements to give rise to long polymeric macromolecules with sufficient stability to serve as structural backbones of living molecules. Since carbon atoms readily form covalent bonds with oxygen, hydrogen, nitrogen and sulphur, a large number of different types of functional groups can be introduced into the structure of organic molecules.

Oxygen: It is the most abundant constituent of living systems, which was not present in free form in the primordial atmosphere of the earth. It is believed to have been formed around 1500–2500 million years ago as a result of the origin of photosynthetic cells, the blue-green algae. The first oxygen utilizing organisms arose later, after the increase in the oxygen content of the atmosphere. A level of 10% oxygen was attained only about 400 million years back and the present level of about 20% around 2 million years ago, when the first *Homo sapiens* appeared. It can be visualized that plants have a vital contribution to make, in generation of oxygen from combined form and in maintenance of the oxygen content of the atmosphere, without which aerobic life would be impossible. In higher organisms such as man, organs like the brain are highly dependent on a continuous supply of oxygen. Its deprivation for a time longer than a few minutes can lead to irreversible damage and death. This dependence has grown with evolution and establishment of oxidative metabolism in organisms, as a more efficient way of obtaining energy from the nutrients. For example, if glucose is utilized by anaerobic pathway in cells, one molecule of this nutrient can furnish two high energy bonds. On the other hand, if it is utilized in circumstances in which oxygen is available, it yields 38 high energy bonds. The yield is therefore 19 times higher, a factor that has played a selective role in evolution of oxidative pathways for obtaining energy in higher cells.

Hydrogen: While oxygen is required for oxidative metabolism, the energy is obtained in higher cells by the oxidation of hydrogen and not carbon. Energy yielding substrates are dehydrogenated and the H atoms removed are oxidized through a series of steps culminating in the formation of H_2O. Energy generated on account of the oxidative reactions is trapped in form of high energy phosphate bonds of ATP.

Nitrogen: Proteins, one of the most essential constituents of all living systems, are built up of nitrogen containing compounds. So are nucleic acids, porphyrins, etc. There is a constant requirement of nitrogen compounds for growth and wear and tear of cells. Nitrogen is taken by man in the form of dietary proteins of animal or vegetable origin. Raw material of animal proteins is derived from the vegetable proteins. The soil fixes nitrogen and the plants convert it into amino acids and proteins. These are taken in, utilized, converted and excreted in form of urea and ammonia. The latter return to soil and the cycle continues.

The function of other elements is described in Table 1.1. Many of the elements listed in category III and IV are present in trace amounts. These minor components, however, play a major role. Many of them have been found to act as cofactors and activators of enzymes.

II. WATER

Water is the miraculous compound without which life is not possible. Man can live without food for days, but not without water. It is a major constituent of the body. Water accounts for about 70 to 80% of weight in most soft tissues (muscles, brain, liver, etc.). Blood plasma has 93% H_2O, but erythrocytes only 60%. Hair, horn and the solid portion of the bone have relatively low water content. The average water content of the body as a whole is about 61% with a range of 55 to 67%.

The water content of tissues decreases slightly with age: young tissues have more water than the old. The available water reserves in obese (fatty) people are lower than those in lean people. Consequently, the former are a greater risk in surgical operations than the latter.

Structure

Although water is usually denoted by the formula H_2O, in reality it exists as a poly H_2O lattice. This is formed by hydrogen bonds between neighbouring molecules, the proton being shared by the two electronegative oxygens. Figure 1.1 shows the disposition of atoms in water. H—O—H angle is nearly 105°. The powerful attraction of the O nucleus tends

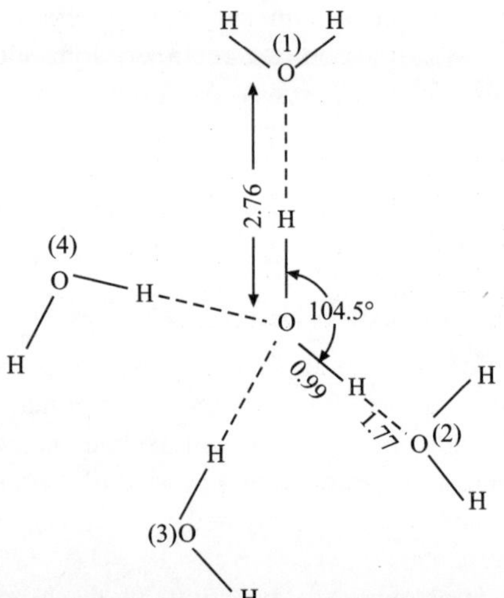

Figure 1.1 Orientation of hydrogen and oxygen of water molecules in ice. Figure also shows the hydrogen bonds (dotted lines) formed between neighbouring water molecules. In ice, four water molecules get associated in a manner that the oxygen (1), (2), (3) and (4) lie at the corners of a regular tetrahedron. Molecules (1) and (2) as well as the central molecule lie in the plane of the paper. Molecule (3) lies above this plane and molecule (4) below it.

TABLE 1.1 Chemical Elements in the Living Systems

Category	Element	Atomic no.	% by weight in the human body	Comments
Category I Most abundant Account for over 95% of the total constituents	O	8	65	Present in all living forms; indispensable.
	C	6	18	The essential backbone of constituents in all living matter.
	H	1	10	Universal, indispensable; its oxidation in biological systems provides energy.
	N	7	3	Universal and indispensable.
Category II Commonly present	Ca	20	2.01	Mineral component of the bones; also important in excitation of muscles and nerve cells.
	P	15	1.16	Besides structural role in bones, it is an important component of cell membranes and biopolymers—DNA, RNA; important in intermediary metabolism. Energy is conserved and transferred in biological systems in the form of phosphate bonds.
	K	19	0.20	Retained by cells in higher concentration than in the extracellular compartment; important in development of resting potentials across the membrane.
	S	16	0.196	Important role in the folding and conformation of proteins; present in coenzyme A, glutathione and other biologically active compounds.
	Na	11	0.109	The predominant ionic constituent of blood plasma; important role in generation of action potentials in nerve cells; in transport across membranes, in acid and base balance of body.
	Mg	12	0.036	Activator and cofactor for many enzymes; role in energy transfer reactions mediated by ATP.
	Fe	26		Preeminent role in transport of oxygen as part of haemoglobin; constituent of cellular oxidative systems and enzymes cytochromes, catalase, peroxidases, etc.
Category III Essential trace elements present in almost all living systems	Cl	17		Important anion of extracellular fluids.
	Mn	25		Important cofactor for many enzymes.
	Co	27		Constituent of vitamin B_{12}.
	Cu	29		Cofactor of some proteins and enzymes; oxygen transport in marine species.
	Zn	30		Constituent of important enzymes such as carbonic anhydrase, carboxypeptidase, alcohol dehydrogenase, etc.
Category IV Not universal in all living forms but present in some species	I	53		Important constituent of thyroid hormones in higher organisms.
	F	9		An important minor constituent of vertebrate teeth, harmful in higher amounts.
	Mo	42		Cofactor for some enzymes.
	B	5		An essential constituent of plants.
	^{90}Sr	38		A radioactive isotope, product of nuclear fission present in stratosphere, concentration on earth increasing with radioactive fall-outs; passed from soil to vegetation to cows to milk to humans, competes with Ca for concentration in bones.

to draw electrons away from the protons, leaving the region around it with a net positive charge. Each proton is bound to an oxygen nucleus by a strong covalent bond and in general also to another oxygen nucleus by a much weaker hydrogen bond. Each water molecule tends to H-bond with three to four neighbouring molecules. The proton is, however, a light and mobile structure, far more than any other atomic nucleus. It hops from one nucleus to its neighbour for intervals of a fraction of a second creating the OH^- and H_3O^+ from two molecules of H_2O (Figure 1.2). H_3O^+ would donate a proton and OH^- would accept it. One kg (55.5 moles) of H_2O at 25°C contains only 10^{-7} moles of H_3O^+ ions.

Figure 1.3 shows the *lattice of H_2O molecules* in liquid water. Much of the structure depicted in Figure 1.1 is retained

with only 15% of the total H bonds breaking. The half-life of H bonds in liquid H_2O is very short, e.g., of the order of 10^{-10} to 10^{-11} seconds. The bonds are forming and breaking. Liquid water is thus a fast flickering cluster of H_2O molecules. The high heat of vaporization of water is an evidence of the presence of fairly substantial intramolecular cohesive forces among the water molecules.

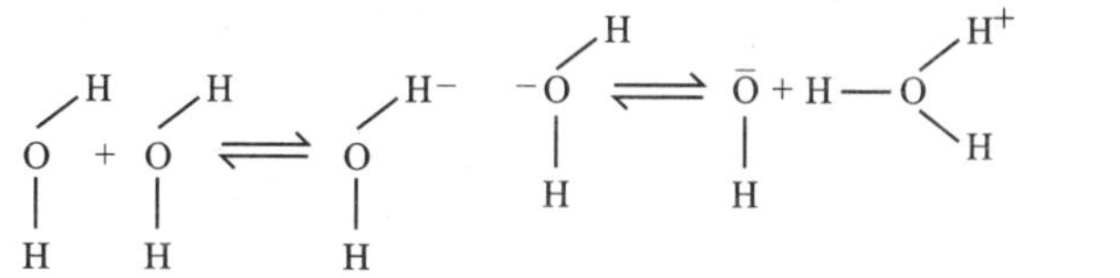

Figure 1.2 Formation of OH^- and H_3O^+ from two molecules of H_2O.

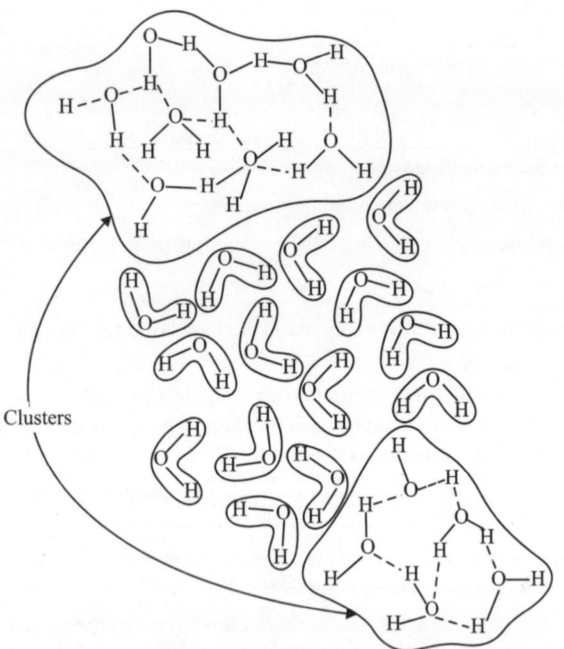

Figure 1.3   Hydrogen bonded clusters and unbonded molecules in liquid water according to flickering-cluster theory of Rank and Wen.

Properties

Water has unique physical and chemical properties, eminently suited for life processes. Compared to other solvents, water has a high melting and boiling point and surface tension. It has a high heat of vaporization, with the result that the vaporization of H_2O from skin causes cooling. Because of a high specific heat (i.e., the amount of heat necessary to raise the temperature of a unit quantity of H_2O by $1°C$), it provides a buffer capacity for maintaining the constancy of temperature in the body.

Water is an excellent solvent. Because of its dipolar nature, many substances are dispersed and salts are readily dissolved. It can engage in hydrogen bonding with non-ionic substances such as the hydroxyl groups of sugars, rendering them soluble.

The interaction of H_2O molecules with vital biopolymers such as proteins and nucleic acids is important for many of their biological properties. Hydrophilic and hydrophobic portions of the macromolecule contribute not only to the overall conformation of these molecules but also influence their interactions with other enzymes and intermolecular aggregates.

The distribution of water in various compartments of the body, as well as the changes that take place in various disorders are discussed in Chapter 91.

A number of diseases that we suffer from in India (cholera, typhoid, viral hepatitis, enteric disorders) are contracted from drinking of contaminated water. Provision of proper drinking water should have a far-reaching impact on public health and in prevention of a number of communicable diseases.

III.   CARBON DIOXIDE

In the preceding subsection, we discussed the important role of H_2O. The oxygen compound of carbon has an equally pertinent role in living systems. CO_2 is a gas at the ordinary temperature and pressure. It is unique amongst gases to distribute itself almost equally in concentration per unit volume between air and water. There is thus a perpetual exchange of CO_2 between atmosphere and waters of the world.

CO_2 has capacity to undergo reversible hydration to form the weak acid H_2CO_3. This reaction is of utmost importance in the transport of CO_2 from tissues to the expired air. It is also a vital reaction in the maintenance of acid and base balance in the body. This reaction also supplies the protons for generation of HCl in the stomach.

An average man produces about 1 kg of CO_2 as a result of metabolism in the body every day. Much of it is excreted via the expired air, thanks to the volatility of CO_2. It would, however, be wrong to believe that CO_2 is only an excretory product. It is also utilized as a building block for synthesis of a number of body constituents such as fatty acids (see Chapter 13), purines and pyrimidine nucleotides, etc.

CO_2 is a valuable source of carbon for plants, a process that releases also oxygen for enrichment of the atmosphere.

IV.   ABIOTIC GENESIS OF BIOMOLECULES

Not long back there was a belief that the living systems have a 'Vital' force that distinguishes them qualitatively from the non-living matter. This belief has since been largely shattered by the systematic evidence that has accumulated, showing an inherent relationship between simple compounds and the living entities. What is more, it has been possible to bring about the transition of simple compounds into biomolecules experimentally. The following questions can be raised and what follows is an attempt to answer them.

1. Could simple inorganic compounds and gases present in the primeval atmosphere give rise to organic molecules under the conditions prevalent at that time?
2. How did simple organic molecules get converted into biopolymers of the type present in the living systems?
3. How was the genesis of complex structures achieved? How did they acquire the properties of catalysis and replication?
4. What are the characteristics of life? Where lies the barrier between the inert and the living matter? Has life been created in the test tube?

Conversion of Inorganic to Organic Molecules

In early 1950s, many laboratories succeeded in getting a wide range of organic molecules, e.g., amino acids, porphyrins, pyrimidines, purines, fatty acids, hydrocarbons, carbohydrates, etc., by subjecting simple gases such as NH_3, CH_4, H_2, CO_2 and H_2O to high temperatures, ultraviolet radiations and/or electrical sparks. Temperatures used ranged from 80–150°C. Intermittent sparks and/or UV irradiations were produced in a manner simulating the primeval atmosphere. The products obtained were a mixture of substances like glycine, alanine, sarcosine, aminobutyric acid, aspartic acid, glutamic acid, formic acid, lactic acid, hydroxybutyric acid, urea, adenine, porphyrins, etc.

HCN is believed to be the intermediary in formation of nitrogenous compounds and of important condensing agents.

Formation of Biopolymers

Proteins, nucleic acids and polysaccharides are formed from the monomeric building blocks by condensation reactions with the removal of water. There are two general ways in which this could have been brought about in the prebiotic conditions. One of the ways would be to carry out this reaction under anhydrous conditions, e.g., at temperatures above the boiling point of water. The second way would be to carry out the reaction in presence of condensing agents that can effectively cope with the elements of water removed by the combination of the monomers. Removal of H_2O would be important to prevent the reversibility of the reaction and to ensure good yield of the polymers. Both types of reactions appear to have been utilized in the evolutionary cycle. Temperatures above 100°C can well be visualized to prevail in the primeval era and in volcanic eruptions. Thermal formation of polymers of amino acids has been demonstrated by Fox and co-workers by heating mixtures of amino acids at temperatures above 100°C. Carbodiimide and other derivatives (cyanogen) formed from simple gases and HCN in prebiotic conditions act as excellent condensing agents. Polyphosphates (polymetaphosphates) and their esters formed in local high temperature zones can also bring about the formation of polypeptides from amino acids.

Catalytic Capacity

When a mixture of amino acids is subjected to temperature above 100°C, a mixture of proteinoids is obtained. The molecular weight of these proteinoids can be more than 10,000 daltons. These have been fractionated and some of them have

$$CH_4 + NH_3 \longrightarrow HCN + 3H_2$$

$$H_2N\!-\!C\!\equiv\!N$$
(Cyanamide)

$$NH_2$$

$$N\!\equiv\!C\!-\!NH\!-\!C\!-\!NH$$
(Dicyandiamide)

Amino acids → Proteins
Purines → Nucleic acids
Porphyrins → Chlorophyll animal protoporphyrin
Condensing agents (Carbodiimide)

been found to possess catalytic properties. Fox and coworkers obtained polypeptides that had esterase activity. Dhar at C.D.R.I., Lucknow, has prepared random polypeptides that have lysozyme like activity. Enzymatic functions thus emerge with the polymerization of amino acids into proteins. With time, the more efficient and useful configurations were selected out and replicated.

Replication

An important property of living entities is the capacity to reproduce itself. At the molecular level, it can be conceived to be a consequence of the *"template" function of nucleic acids*. Polynucleotides emerged with the condensation in the first instance of bases with ribose and polyphosphoric acids to yield nucleotides, followed by polymerization to polynucleotides. An important trait of the nucleic acids is the observance of complementarity rule. Adenine (A) is recognized by uracil (U), guanine (G) by cytosine (C), and vice versa. In the presence of poly U, the polymerization of poly A by chemical condensation methods is selectively accelerated and the product is a double homopolymer poly AU. There is a ready hydrogen bonding between complementary bases. (A codes for U, U for A; G codes for C and C for G). A polynucleotide structure can therefore be replicated exactly by a two stage process, first by making of a complementary base sequence, which in turn by the same principles, would give rise to a base sequence identical to the parent template molecule.

Initial template:

...A—U—G—C—C—G—U—A

Complementary copy:

...U—A—C—G—G—C—A—U

Product using complementary copy as the template which is identical to the initial copy:

...A—U—G—C—C—G—U—A

Nucleic acids have, therefore, the property of replication. Although the presence of a polynucleotide in the reaction mixture influences the type of bases that preferentially get condensed, this process became more efficient and ordained with the evolution of catalytic enzymes (proteins) that recognized the template and helped in the synthesis of complementary sequences. Nucleoprotein complexes have, therefore, the inherent capacity for replication.

Origin of Life

Among the attributes ascribed to living entities are the following:

(i) Capacity for reproduction of its own kind;
(ii) Metabolic capacities for deriving energy from the environment and utilization for endogenous purposes;
(iii) Sensitivity to environmental agents.

The last of these is discernible in proteins. The conformation of a protein molecule changes by changes in pH, temperature, ionic strength, and nature of small molecules in the medium. With the changes in conformation, the biological activity of the molecule also changes.

The metabolic activities of a living cell reside in the set of enzymes that it possesses. It is further well established from the experiments of Buchner and Liebig, that enzymes are capable of catalysing the reactions in cell-free isolated state and are not dependent on the total intact living cell for their activity. The catalytic function is a molecular property of the enzyme or protein.

Molecules of the nucleic acids type have the capacity to serve as "informational" molecules. They can act as templates and dictate the synthesis of their own type. Thus, *the three major characteristics of living systems can be fulfilled by the physico-chemical attributes of biomolecules of the type discussed above.* The only requirement is to juxtapose them in an appropriate order to generate a functional living unit.

There are two major theories of the origin of life. The first of these was proposed several decades ago by the Russian academician Oparin. According to him, the "primitive cells" originated with the formation of boundary around proteinic molecules with catalytic properties. Coacervate droplets of a highly hydrated polymer from the primitive broth, could have been formed spontaneously by a phase separation between a microdroplet having higher polymer concentration than the surrounding aqueous environment with low concentration of the polymer. The coacervate droplets have a tendency to concentrate the polymer. They would thus grow in size with the accretion of the material from the environment. Beyond a certain size, they would have the tendency to split into smaller units. The progeny in turn would act as microcoacervate droplets and continue the process of concentration and growth. Fox and coworkers have prepared "Microspheres" from proteinic concentrates that demonstrate the growth in size by concentration of the polymer. Formation of "buds" akin to those seen in yeast have been noticed in these structures.

If a microsphere or coacervate droplet containing a proteinic enzyme such as phosphorylase is prepared, it can make a starch-like material within its boundary from the glucose–1–phosphate present in the medium. When the "microspheres" are made of two proteinic enzymes such as phosphorylase and amylase, the droplets would not only utilize glucose–1–phosphate to make starch, but will excrete to the medium, maltose, a product of the amylase action on the synthesized polysaccharide.

These experiments, therefore, demonstrate the feasibility of the creation of microstructure in an environment with metabolic functions. They are, however, devoid of template function and multiply by a random splitting process.

The Gene Hypothesis

An alternate theory was proposed by the geneticist Muller, who conceived one or more genes as the nuclei around which living forms originated. The chemical nature of the genetic material is well established. It is DNA in most of the cases but RNA in some viruses. Evidence for the formation of nucleic acids from constituents of the primitive atmosphere under abiotic conditions has been provided by the work of Ceylonese-born scientist Ponnamperuma and collaborators.

The gene hypothesis would be in consonance with some of the fundamental properties of living systems, namely that of orderly replication and faithful transmission of the hereditary traits to the daughter cells.

Viruses that have the property of giving rise to their own kind, can be considered as the simplest entities on the *border of the living and inert matter*. The simplest among viruses are particles composed of just nucleic acids and proteins. An appropriate alignment of these two vital biomolecules constitutes a unit capable of exact multiplication. It is true that the simple viruses are devoid of a complete set of enzymes, hence they do not have independent metabolism. They require, thus, the host cell machinery, which they are, nonetheless, capable of utilizing for their own propagation.

An analysis of the microorganism kingdom reveals a steady continuum in terms of size and degree of complexity between the simplest of the viruses, more complex viruses, rickettsia, bacteria and animal cells.

A few years ago, Kornberg and co-workers succeeded in synthesizing in the test tube a bacteriophage (bacterial virus) particle, that was infective and capable of replication in the host bacteria. If viruses be accepted to possess the minimal traits of living entities, it can be said that "life" at this stage of development can be created *in vitro*.

V. STRUCTURE OF AN ANIMAL CELL

A cell is the unit of a living system. It is a cosmos in itself. Conceptually it is a purposeful aggregate of different types of molecules ensuring the survival and growth of the whole structure. Nutrients are taken up from the surrounding environment by mechanisms discussed elsewhere. They are broken down step by step and the energy liberated by the breakdown of the covalent bonds is trapped in the form of high energy phosphate bonds (ATP) that are utilizable for synthesis of various constituents needed by the cell. These processes have been discussed in Chapter 10 of the book. ATP is also required for various other functions of the cell, such as the maintenance of resting membrane potentials and thereby the trait of excitability in muscle and nerve cell, mechanical action, e.g., contraction and relaxation of muscles (Chapter 21), ameboid movement of phagocytic cells, etc.

Figures 1.4(a) and (b) show the typical ultrastructure of a mammalian cell. Cells are bounded by a *membrane* called the plasma membrane. Plasma membrane is not merely an inert envelope that holds the constituents of the cell but it is an active interface between the internal and the external environment. It regulates the nature and amount of metabolites and ions, that are taken up by cells. A number of enzymes important in regulation of cellular activity, e.g., adenyl cyclase, and Na-K-Mg-ATPase are membrane linked (see Chapter 2). Membranes bear cell specific "Markers" or chemical determinants that enable the recognition of the cell by its own type and by others. If liver and kidney cells are grown in a mixed culture, they form distinct colonies, each recognizing its own type and forming an organized pattern of their own type. One of the abnormalities in malignant cells is the loss of self-recognition. In culture they grow haphazardly. B lymphocytes, precursors of the antibody forming cells, have immunoglobulin determinants on their surface that enable their recognition by the antigens. Cells responding to pituitary tropic hormones have "receptors" or chemical determinants on the surface that bind specifically the hormone, an important event in triggering of the target cell by the hormone for a specific task [see Chapter in Section XI (11)].

Membranes are built up of lipids, phospholipids, proteins and glycoproteins. Their structure and organization is discussed in Chapter 2. They usually carry a net negative charge at the physiological pH. Malignant cells have a higher charge due principally to more of sialic acid residues.

The nucleus contains the genes (DNA) organized in the chromatin structure. The chromatin has DNA besides in basic proteins, the arginine-rich and the lysine-rich histones. Non-histone acidic proteins are also present in the chromatin which are believed to exercise a regulatory role in the expression of the genetic information. The nucleolus is the region where ribosomal RNA is synthesized and processed. Extranucleolar chromatin codes for the messenger RNA. Nucleus has a double membrane surrounding it, which is interspersed at points permitting the passage of components synthesized in the nucleus. Nuclear-cytoplasmic communication in both directions is a regulated process and has influence on gene expression.

Mitochondria are rounded or cigar-like organelles with cristae. They contain a battery of oxidative and phosphorylating enzymes. They are the power house of the cell where nutrients are oxidized and ATP is generated. Mitochondria contain a small amount of DNA. A part of the mitochondrial components originate from the expression of mitochondrial DNA, the others being products formed from the genetic information in DNA of the nucleus of the cell. Mitochondrial protein synthesis is sensitive to antibiotics like chloramphenicol. They resemble bacteria in some of their traits. The number of mitochondria increases in a cell according to needs.

Endoplasmic reticulum is a tubular structure studded with ribonucleoprotein granules called the ribosomes. Ribosomes are the sites of protein synthesis. Membrane-linked ribosomes are engaged in synthesis of proteins destined for excretion by the cell. They pass via the tubular structure to the Golgi apparatus from where they are secreted out. Proteins get loaded with carbohydrate reticulum contains a number of detoxicating enzymes whose content increases with the administration of drugs.

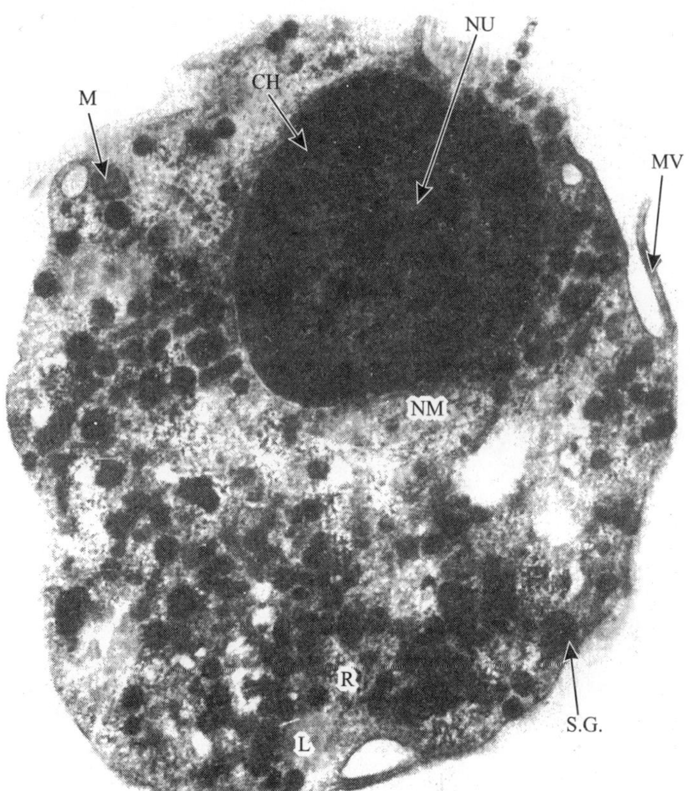

Figure 1.4(a) Electron micrograph of isolated human placental cell (× 18,000). The cytoplasm is loaded with secretory granules (SG) and ribosomes (R). A few mitochondria (M) and lipid droplets (L) are also seen. Cytoplasmic projections, the microvilli (MV) are present towards the nuclear end of the cell. The nuclear membrane (NM) which encloses the chromatin (CH) and nucleolus (NU) separates nucleus from rest of the cytoplasm.

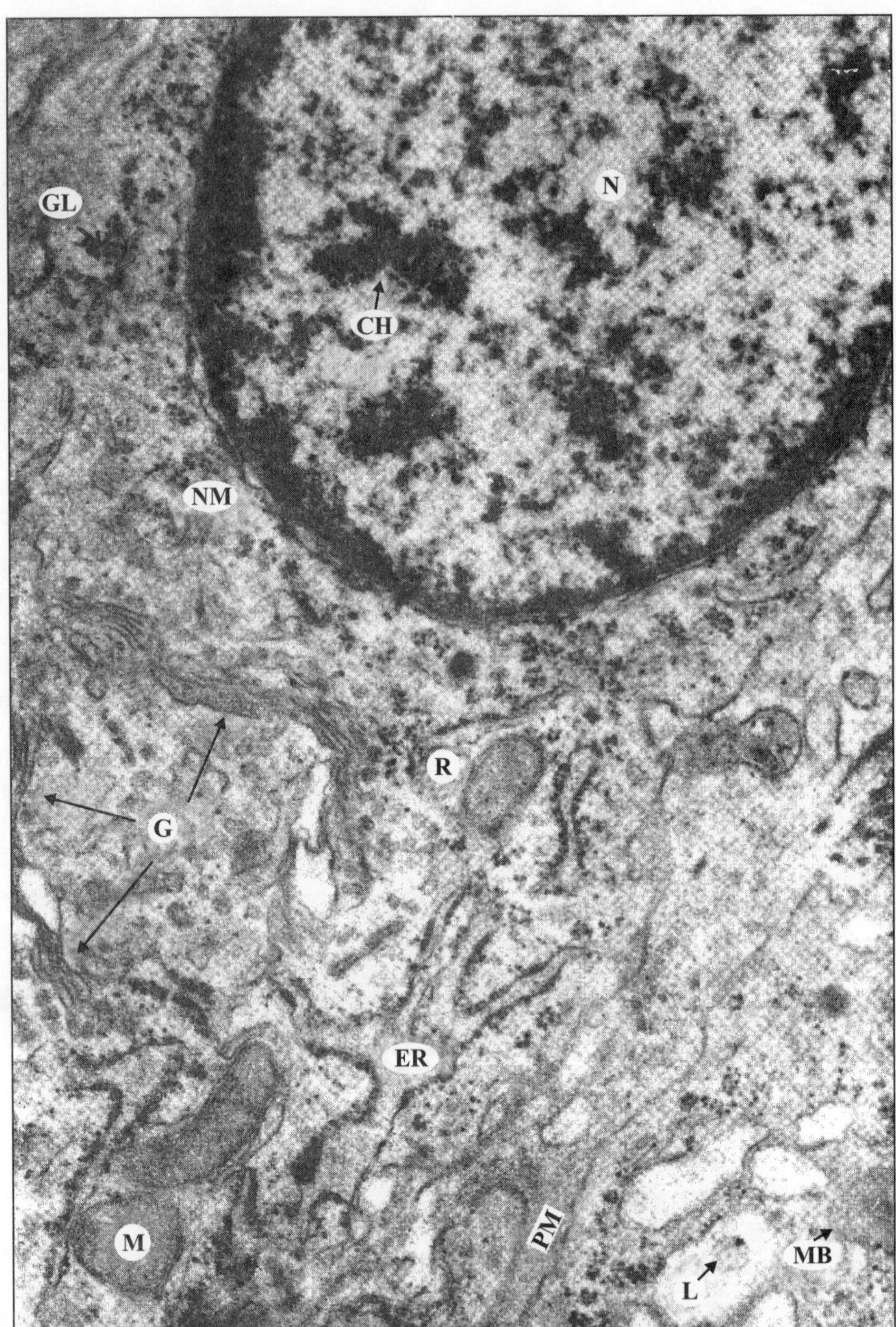

Figure 1.4(b)   Electron micrograph of monkey hepatocytes showing various cell organelles × 25,000. N, Nucleus; CH, Chromatin: NM, Nuclear membrane; G, Golgi complex: M, Mitochondria; PM, Plasma membrane; MB, Microbody; ER, Endoplasmic reticulum; R, Ribosomes; GL, Glycogen; L, Lysosome.

(Micrograph, Dr P.D. Gupta)

Lysosomes are particles slightly lighter than mitochondria in their sedimentation properties. They contain hydrolytic enzymes and constitute a scavenging mechanism for digestion of the ingested material. They are particularly numerous in macrophages.

Some cells such as the eosinophil granulocytes contain small particles rich in peroxidase, secretory cells (nerve cells, hormone producing cells) have usually a complement of vesicles containing the neurotransmitters or hormones meant for export.

Cells may also contain in the cytoplasm storage products such as glycogen or fats. The cell sap is rich in proteins and enzymes and has a pool of metabolites and building blocks.

SUGGESTIONS FOR FURTHER READING

Alberts B., Bray D., Lewis J., Raff M., Roberts K. and Watson J. (1994), *Molecular Biology of the Cell,* 3rd ed., New York, Garland.

Eisenberg D. and Kauzmann W. (1969), Advanced Treatment of the Physical Chemistry of Water, *The Structure and Properties of Water,* Oxford University Press, Fair Lawn, NJ.

Fox S.W. and Dose K. (1977), *Molecular Evolution and the Origin of Life,* New York, Dekker.

Frieden E. (1972), The Chemical Elements of Life, *Sci Am.,* **227,** 52–64, July 1972.

Likens G.E., Wright R.F., Galloway J.N. and Butler T.J. (1979), Acid Rain, *Sci. Am.,* **241,** 43–51, October 1979.

Ponnamperuma C. (Ed.) (1972), *Exobiology,* North Holland, Amsterdam.

Ponnamperuma C. (Ed.) (1981), *Comets and the Origin of Life,* D. Reidel Publ. Co.

Przybylski A.T., Stratton W.P., Syrem R.M., and Fox S.W. (1982), Membrane, Action and Oscillatory Potentials in Simulated Proto Cells, *Naturwissenschajten,* **69,** 561–563.

Roger Lewin (1982), Molecules Come to Darwin's Aid, *Science,* **216,** 1091–1092.

Szent Gyorgyi A. (1972), *The Living State,* New York, Academic Press.

2

Membrane Structure and Function

Neeta Singh

CONTENTS

I. INTRODUCTION

The ability of a cell to discriminate in its chemical exchanges with the environment is fundamental to life, and it is the plasma membrane that makes this selectivity possible. Normal cellular function depends on normal membrane. Membrane is a structure that defines and controls the composition of the cell that it encloses. Sixty percent of lean body mass is water. The membranes form boundaries/closed compartments around the cell, separating the intracellular fluid, which constitutes two-thirds of the total body water, from the extracellular fluid, which constitutes one-third of the total body water. The ionic composition of intracellular fluid and extracellular fluid differs greatly from one another. The intracellular fluid is rich in potassium and magnesium, with phosphate being the major anion, whereas the extracellular fluid is rich in sodium and calcium, with chloride the major anion. Membranes have asymmetric lipid bilayer with distinct inner and outer surfaces/leaflets. The plasma membrane shows selective permeability to various substrates and ions. The cell also contains intracellular membranes that serve distinct functions in the formation of various intracellular organelles, e.g., some of the intracellular organelles such as nucleus, mitochondria, lysosomes, endoplasmic reticulum and Golgi bodies are also surrounded by intracellular membranes or are composed of membranes. The contents of plasma membrane vary according to the nature of the membrane. Different cellular membranes have different compositions, with different functions. Membranes are essentially made up of lipids, carbohydrates and proteins and exist in a fluid state. The carbohydrates of the membranes are attached to either lipids, forming glycolipids, or to proteins, forming glycoproteins. The oligosaccharides play an important role in cell–cell recognition. The lipid and protein composition of the membrane varies from one cell type to another, as well as within the intracellular compartments that are defined by intracellular membranes. The protein concentration can vary from 20% to as much as 70% in a particular membrane. Membranes undergo turnover of lipids and proteins at different rates. There are 20 different types of membranes in a mammalian cell, performing different functions. Lipids and some proteins exhibit lateral diffusion in the plane of their membranes. However, the transverse movement of lipids across the membrane, that is, flip-flop movement, is very slow and is rare, and it does not occur in the case of membrane proteins. The plasma membrane helps in transport of various gases, solvents and solutes by simple diffusion, facilitated diffusion, i.e., passive transport or active transport. The functions of the plasma membrane include separating one cell from another by forming closed compartments, having selective permeability which is provided by channels and pumps for ions and substrates and exchanging materials with extracellular environment by exocytosis, endocytosis and gap junctions. They help in cell–cell interaction in transmembrane signaling. They localize enzymes, e.g., the ectoenzymes $5'$ nucleotidase and alkaline phosphatase are present on the outer part of the plasma membrane, they function as integral elements in excitation response coupling and they provide sites for energy transduction, e.g., photosynthesis and oxidative phosphorylation. Various signals are transmitted across membranes. Specific biochemical signals such as neurotransmitters, hormones and immunoglobulins bind to specific receptors (integral proteins) exposed to the outside of the cell membrane and transmit information through these membranes to the cytoplasm of the cell. This process is known as transmembrane signaling which involves a number of signaling molecules, e.g., cyclic AMP, calcium, phosphatidyl inositol and diacylglycerol. Changes in the membrane structure, e.g., caused by ischemia, can affect the water balance and ion fluxes and therefore every process within the cell.

II. COMPOSITION OF CELL MEMBRANE

Membrane Lipids

The basic structure of all membranes is a lipid bilayer made up of two leaflets, inner and outer leaflets, which in turn are made up of phospholipids, in which the polar hydrophilic head faces the external or the internal side that is exposed to the aqueous environment. The hydrophobic nonpolar tails are oriented towards each other at the centre of the membrane. The lipid components of the membranes are of three major classes: glycerophospholipids, glycosphingolipids (carbohydrate containing lipids) and cholesterol. Sphingolipids and glycerophospholipids constitute the largest percentage of the lipids. The fatty acid constituents in these lipids generally contain an even number of chained fatty acids, having chain length of 16 or 18 carbon atoms. The fatty acids can be saturated or unsaturated, with one or more cis double bonds. Saturated fatty acids have straight tails, whereas unsaturated fatty acids have kinked tails. More kinked tails make membranes less tightly packed and more fluid. The phospholipids can be phosphatidyl ethanolamine, phosphatidyl choline, phosphatidyl serine or phosphatidyl inositol. Another major class of phospholipids present in the membranes is sphingomyelin, which has sphingosine backbone instead of glycerol. Sphingomyelin is more prominently present in myelin sheets. Cholesterol, a sterol, is generally present in the plasma membrane of mammalian cells, but it is also found to some extent in the nuclear and mitochondrial membranes, as well as in the Golgi complex. Cholesterol intercalates within the phospholipids of the membrane and its hydroxyl group faces the outside, whereas the rest of the molecule is within the leaflet. The glycosphingolipids are made up of sugar and a backbone of ceramide. They include cerebrosides and gangliosides. The lipids in the membrane contain both hydrophobic and hydrophilic regions, and hence they are called amphipathic. Thus, the polar head of the phospholipids and hydroxyl group of cholesterol face the aqueous environment and so also the sugar groups of glycosphingolipids. Detergents are amphipathic molecules. Some of them are used to solubilize membrane proteins. The hydrophobic end of the detergent binds to the hydrophobic region of the proteins, displacing their bound lipids. The polar head of the detergent is free and brings the proteins into solution as detergent–protein complexes (Figure 2.1).

The lipid bilayer of the membrane exists as a sheet in which the hydrophobic regions of the phospholipids are protected from the aqueous environment, whereas the hydrophilic regions are in the water. Thus, the lipid bilayer of the plasma membrane is impermeable to water–soluble molecules but permeable to lipid–soluble molecules. Gases such as oxygen and carbon dioxide and small molecules such as water can diffuse through the plasma membrane. Different ions have different permeability coefficients which correlate with their solubilities in nonpolar solvents. Water has a high permeability coefficient because of its small size and relative lack of charge. Hydrophobic substances can readily traverse the plasma membrane to enter the cells. The hydrophilic molecules are transported across the plasma membrane with the help of membrane proteins, which form channels for small molecules and ions and serve as transporters for large molecules. Membrane phospholipids act as a solvent for the membrane proteins. The hydrocarbon tails of lipids result in steric limitations to their packing such that they will form disc–like micelles. The structure of these disc–like micelles results from the interaction of the hydrophobic tails of the lipids and the exposure of the polar head to the aqueous environment. This orientation results in a lipid bilayer formation and these bilayers are essentially two–dimensional fluids. The lipid components of the bilayer can diffuse laterally, and this lateral diffusion occurs rapidly. The lipids in the bilayer can also undergo transverse diffusion, also called a

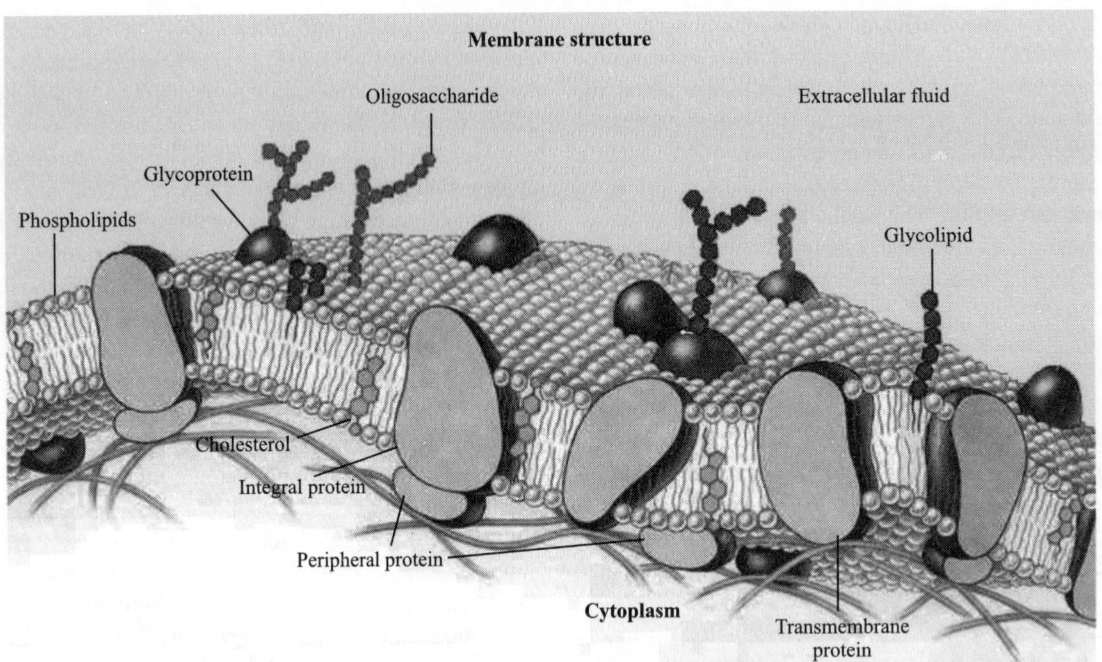

Figure 2.1 Structure of the membrane and its composition.

flip-flop movement, where the lipid diffuses from one surface to the other. Since the flip-flop movement requires the polar head group to pass the hydrophobic hydrocarbon core of the lipid bilayer, the process is very rare. Some enzymes facilitate the flip-flop process and are called flippases and floppases.

Lipid rafts: The cell membrane also contains certain specialized structures such as lipid rafts and caveolae. They are subdomains in lipid bilayers rich in cholesterol, sphingolipids, glycosphingolipids, saturated phospholipids and some proteins. They also contain some GPI–linked proteins. The lipid rafts are present on certain areas of the outer leaflet of the cell membrane. They are slightly thicker than the rest of the plasma membrane, which is generally 7 nm in thickness. Caveolae can be derived from lipid rafts and they contain the protein caveolin-1, which may be involved in their formation. This is a dynamic protein and each monomer of this protein is anchored to the inner leaflet of the plasma membrane by three palmityl molecules. They appear as flask–shaped projections of the cell membrane facing the cytosol. The proteins present in caveolae include various components of the signal transduction pathways, e.g., G proteins, insulin receptor, folate receptor and endothelial nitric oxide synthase. The lipid rafts are also involved in endocytosis, in signal transduction mechanisms, IgE signaling, insulin receptor, EGF receptor and T and B cell antigen receptor signaling. Caveolae are flask–shaped indentations on the region of lipid rafts that are involved in membrane transport and signal transduction. Transport of IgA from the luminal side occurs by caveolae–mediated transcytosis. The fusion and budding of viral particles are mediated by caveolae.

Membrane Proteins

Biological membranes contain proteins, glycoproteins (carbohydrates attached to proteins) and lipoproteins (lipids attached to proteins). They are amphipathic and form an integral part of the membrane by having hydrophilic regions protruding at the inside and outside faces of the membrane but have a hydrophobic region traversing the hydrophobic core of the lipid bilayer. This hydrophobic region contains essentially hydrophobic amino acids which form an alpha helical structure. Some proteins are anchored to only one leaflet of the membrane bilayer by covalent linkage to palmitate or myristate fatty acid of the lipids. Different membranes can have different proteins. The number of different proteins varies from 10 to 100 in the plasma membrane. The proteins are the major functional molecules of the membranes and comprise of enzymes, channels, pumps, structural components, antigens, receptors, etc. Membrane proteins differ in the degree to which they span lipid bilayers. The membrane proteins can be classified into two categories **peripheral proteins** and **integral membrane proteins,** based on the strength of their association with the membrane (Figure 2.2).

Peripheral proteins: The peripheral proteins, also called extrinsic proteins, are only loosely/weakly associated with the membrane. They do not enter the lipid bilayer and are generally found on the cytosolic face of the plasma membrane or on the luminal surface of the subcellular organelle membranes, i.e., they can be attached to the outer or inner surface/leaflet. They are either bound to the phospholipid polar head or to integral membrane proteins by non-covalent interactions. Since they

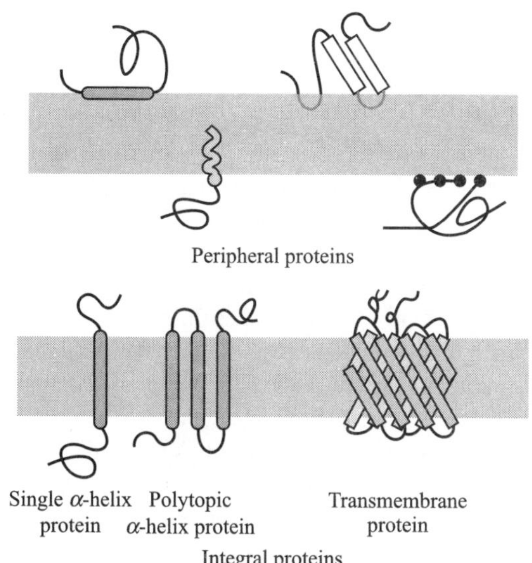

Figure 2.2 The figure shows different kinds of membrane proteins and their structures.

lack hydrophobic interaction with the membrane, they can be detached from the membrane much more easily using polar solvents. GPI anchored proteins are often attached to the outer surface of plasma membrane at micro domains called lipid rafts. They include cell identity markers. Dissociation of peripheral proteins can be achieved through treatment with a solution of high pH or high salt concentration. Other examples of peripheral proteins are the regulatory protein subunits of ion channels and transmembrane receptors, certain kinases and G proteins.

Integral proteins: They are also called intrinsic proteins and are tightly bound to the membrane through hydrophobic interactions and are inserted into or penetrate the lipid bilayer. They show a strong interaction with the membrane as their amino acids having hydrobhobic side chains interact with the fatty acyl groups of the phospholipid bilayer and are thus permanently bonded to the membrane. These proteins can be dissociated from the membranes with the help of detergents, non-polar solvents, or denaturing agents. There are two basic categories of integral proteins, i.e., **transmembrane proteins** or **monotypic proteins.** They are generally globular and amphipathic in nature. The lipid portion of a lipoprotein anchors the protein to the membrane either through interaction with the lipid bilayer directly or through interactions with integral membrane proteins. The lipids are isoprenoids such as farnesyl and geranyl residues, fatty acids such as palmitic acid, myristic acid and glycosyl phosphatidly ionositol (GPI). Transmembrane proteins include ion channels, transporters, receptors for hormones, growth factors, neurotransmitters, membrane enzymes, linkers, cellular junctions and tissue specific antigens. Monotopic proteins are permanently bound to the membrane but only from one side. Many of these proteins are enzymes and include cyclooxygenase and carnitine O-palmitoyltransferase. Anti-inflammatory drugs, such as aspirin and ibuprofen, relieve symptoms of inflammation and pain by inhibiting this enzyme. The latter is a mitochondria

transferase enzyme that participates in the metabolism of palmitoylcarnitine into palmitoyl-CoA.

Structure of membrane proteins: Membrane proteins may cross the membrane only once or several times, weaving in and out. Membrane proteins can be made of alpha-helices, or beta–pleated sheets or a combination of both. Membrane- spanning alpha-helices are the most common structural motifs. Examples of alpha-helical proteins include bacteriorhodopsin having seven alpha-helices, voltage-gated ion channels, such as potassium and chloride channels present in the outer membranes of Gram-negative bacteria, cell wall of Gram-positive bacteria, outer membrane of mitochondria and chloroplasts. Transmembrane proteins are further categorized into Type I and Type II. The difference being that the N-terminal is on the exterior of the membrane in Type I, whereas the C-terminal is present on the exterior in case of Type II.

Membrane proteins are also made up of beta pleated sheets which curl up to form channel proteins, e.g., Porin which is a cross cellular membrane protein acting as a pore through which molecules can diffuse. The outside surface is hydrophobic, while the inside of the channel is hydrophillic and is filled with water. Many membrane proteins have quaternary structures with repeating subunits stacked over each other, e.g., cytochrome b6f. The quarternary structure imparts stability and genetic efficiency as all the subunits of a quaternary protein are coded by one gene. The oligomerization helps to increase the functional output as the membrane proteins are closely packed in the lipid bilayer without coming into contact with other proteins.

Functions of membrane proteins: The functions of membrane proteins include transport of substances across membranes, enzymatic activity, e.g., smooth endoplasmic reticulum, signal transduction, e.g., cell communication, intracellular joining, e.g., intercellular junctions in animal cells, cell–cell recognition, e.g., attachment to the cytoskeleton and extracellular matrix. The membrane proteins participate in various biological processes, such as cell signal transduction pathways, play a role in controlling a wide array of gradients, such as chemical, electrical, and mechanical gradients, and are responsible for cell structure during key cellular events such as cell division. Owing to their many functions in the membrane, they are in high concentration on the surface of the membrane. They also act as channels that move specific molecules in and out of the membrane. Membrane proteins can be associated with the membrane of a cell or an organelle. Membrane proteins can be attached to both the outside and the inside of the cell membrane in a variety of ways. One involves irreversible covalent modification. Both Ras (a GTPase) and Src (protein tyrosine kinase) are known to be modified in this manner and both are involved in signal transduction pathways, but upon covalent attachment of a lipid group, they become attached to the inner face of the cytoplasmic membrane. Transport proteins are selective in what they allow to cross membranes just as enzymes are selective in what substrates they act upon. Similar to the proteins in the cytoplasm, proteolytic processing of membrane proteins is important for the cellular functions in both the cytoplasm and in the lipid bilayer. The intramembrane proteases such as Site-2 intramembrane metalloprotease and serine intramembrane protease have their active sites entrenched in the membrane and cleave a membrane protein by recognizing specific sequence of residues as they do not have an aqueous environment.

III. MEMBRANE ASYMMETRY

Membranes are asymmetric structures and tend to have a different composition on one side than the other, with proteins having different orientations in the membrane, thus making the outer leaflet different from the inner leaflet. Membrane asymmetry allows cells to automatically differ their intracellular environment from that existing extracellularly. This inside–outside asymmetry in the membrane is also due to the external location of the carbohydrates attached to the membrane proteins. Some proteins are located exclusively on the outside and some on the inside of the membrane. There is also regional heterogeneity in the membranes, e.g., that occurring at gap junctions, tight junctions, synapses and lipid rafts. There is also transverse, i.e., inside–outside, asymmetry of the phospholipid, e.g., phosphatidyl choline and sphingomyelin are located mainly on the outer leaflet whereas phosphatidyl serine and phosphatidyl ethanolamine are located on the inner leaflet. Translocases (flippases) are the enzymes that transfer certain phospholipids from the inner leaflet to the outer leaflet. The carbohydrates, attached to whether lipids or proteins, are generally exclusively found on the outer surface of the membrane, further providing asymmetry to the plasma membrane. Most of the membrane proteins are integral proteins which are asymmetrically distributed across the membrane bilayer. Peripheral proteins are bound to the hydrophilic regions of specific integral proteins and the polar heads of phospholipids and do not interact with the hydrophobic cores of phospholipids in the bilayer. For example, ankyrin is a peripheral protein present in the RBC membrane and a cytoskeleton protein spectrin is bound to it, which plays an important role in the maintenance of the biconcave shape of the RBC.

Fluid Mosaic Model of Membrane (Fluidity of Membrane)

Biological membranes are flexible. This flexibility is attained by their fluidity. The fluid mosaic structure of membranes was proposed by Singer and Nicholson in 1972 and is now widely accepted. The model proposes that certain specific integral proteins are randomly distributed in the plasma membrane, they span the membrane and have a globular structure. The fluid mosaic model allows lateral movements called the lateral diffusion, and sometimes the transverse diffusion or flip-flop movement can occur. The phase changes, and thus the fluidity of the membrane largely depends on the composition of the lipids of the membrane. The hydrophobic fatty acids chains present in the phospholipid are aligned to form a stiff/rigid structure. As the temperature increases, these hydrophobic chains undergo a transition from the ordered state to a disordered state taking on a fluid–like structure. The temperature at which the membrane structure undergoes this transition is known as the transition temperature (T_m). The longer and more saturated the fatty acids, the stronger is their interaction and this causes higher T_m, i.e., higher temperatures are needed to increase the fluidity of the membrane. However, the unsaturated fatty acids, which exist in the *cis* form, tend to increase the fluidity of the membrane by decreasing the compactness of the fatty acid chain packing. The phospholipids present in the cell membrane contain at least one unsaturated fatty acid molecule which is generally attached at the second carbon atom of the phospholipid. To maintain fluidity at lower temperatures, organisms use phospholipids containing increasing degrees of unsaturation in their fatty acids. Cholesterol is an amphipathic lipid found particularly in animal cell membranes. Cholesterol content of the plasma membrane can also modify the membrane fluidity. If the cholesterol concentration is high, the membrane becomes less fluid on the outer side but more fluid in the hydrophobic core. The effect of cholesterol on membrane fluidity is also affected by temperature. At temperatures below the T_m, cholesterol interferes with the interaction of the hydrocarbon chains of the fatty acids and hence increases fluidity and therefore the permeability, but at temperatures above the T_m, it limits fluidity as it is more rigid than the hydrocarbon chains of the fatty acids and cannot move in the membrane and thus decreases fluidity. The fluidity of the membrane has also an impact on its functions. As the membrane fluidity increases, the permeability of the membrane to water and small hydrophilic molecules also increases. The decrease in membrane fluidity may affect the activities of receptors and ion channels, e.g., LCAT deficiency is seen in Alzheimer's disease and hypertension. Increase in reactive oxygen species increases cytosolic calcium and lipid peroxidation, which adversely affect membrane fluidity.

Lateral diffusion is the movement of phospholipids laterally, which is very rapid, unless there is restriction by special interaction. A molecule of phospholipid can move several micrometers in one 1s. The lateral movement of the integral proteins also increases with an increase in the membrane fluidity. For example, changes in membrane fluidity affect the function of the insulin receptor, and so it can bind more insulin. At normal body temperature, i.e., at 37°C, the plasma membrane is present in a fluid state. Flip-flop or transverse diffusion occurs when transition of a molecule from one membrane surface to the other occurs. It is a very slow process compared to the lateral diffusion. It happens only once in several hours. The enzyme flippase catalyzes the transfer of amino phospholipid across the membrane, whereas the enzyme floppase catalyzes the outward movement, which is ATP dependent. This is observed in the role of ABC proteins mediating the efflux of cholesterol and extrusion of drugs from cells. The multidrug resistance protein, i.e., MDR associated P-glycoprotein, is a floppase. MDR-1 pumps out a variety of drugs and is implicated in cancer drug resistance.

IV. SPECIAL MEMBRANE STRUCTURE— CELLULAR JUNCTIONS

The other structures present on the cell surface include tight junctions, desmosomes, gap junctions and adheren junctions. They help in communicating between adjacent cells, transfer of small molecules between cells and also in anchorage of the cells.

Tight junctions are found on the plasma membrane, e.g., below the apical surface of epithelial cells, and are composed of specialized proteins, such as occluding and adhesion molecules. They seal neighbouring cells together in the epithelial sheet to prevent the diffusion of macromolecules from one cell to the other. They permit calcium and other small molecules to pass from one cell to another through narrow hydrophilic pores. They help in communication between two cells. Absence of tight junctions results in loss of contact inhibition, as is seen in cancer cells. Tight junctions also seal subepithelial spaces of organs from the lumen.

Gap junctions are structures that allow direct flow of small water-soluble ions and molecules (up to 1.2 KDa) from one cell to the other, allowing whole organs to be continuous from within. They are made up of a family of proteins called connexins that form a hexagonal structure consisting of 12 such proteins. Six connexins form a hemichannel and join to another hemichannel structure of the neighbouring cell to make a connexon channel. One gap junction contains several connexons. Different connexins are found in different tissues. One major function of gap junctions is to ensure a supply of nutrients to cells of an organ that are not in direct contact with the blood supply. Mutations in genes encoding connexons have been found in cardiovascular abnormalities and to be associated with a certain type of deafness.

Adheren junctions or anchoring junctions join an actin bundle in one cell to a similar bundle in the neighbouring cell. Desmosomes join the intracellular filaments in one cell with that of its neighbours, hemidesmosome joins the intracellular filaments in one cell to the basal lamina, and integrin holds basal lamina and triggers intracellular signaling. The structure of the cell is maintained by the cytoskeleton present underneath the plasma membrane. Cytoskeleton is responsible for imparting shape to the cells, for its mortality and for chromosome movement during cell division. It is made up of a network of microtubules and microfilaments which contain the protein spectrin and ankyrin. Tubules consist of polymers of tubulin.

V. ACTIVITIES OF BIOLOGICAL MEMBRANES

Although biological membranes contain various types of proteins and lipids, their distribution between the two different surfaces of the bilayer is asymmetric. This asymmetric distribution results in the generation of highly specialized subdomains within the membrane. In addition, there are highly specialized membrane structures such as endoplasmic reticulum, Golgi apparatus and vesicles. Membrane–bound proteins, e.g., growth factor receptor are processed as they transit through the endoplasmic reticulum to the Golgi apparatus and finally to the plasma membrane. As these proteins transit to the surface of the cell, they undergo a series of processing events/post translational modifications that includes glycosylation. The vesicles that pinch off the Golgi apparatus are called coated vesicles. The membranes of coated vesicles are surrounded by specialized scaffolding proteins that interact with extracellular environment. Clathrin–coated vesicles contain the protein clathrin and are involved in transmembrane protein function; GPI–linked protein and secreted protein transit to the plasma membrane. Clathrin–coated vesicles are also involved in the process of endocytosis, such as the one which occurs when the LDL receptor binds plasma LDL for uptake by the liver. Some cells have membrane compositions that are unique to one surface of the cell versus the other. For example, epithelial cells have a membrane surface that interacts with the luminal cavity of the organ, e.g., the apical surface and another surface that interacts with the surrounding cells, which is known as the basolateral surface. These two surfaces do not intermix and contain different compositions of lipids and proteins. Most eukaryotic cells are in contact with their neighbouring cells and these interactions are the basis of the formation of organs. Cells are in metabolic contact with one another and this is brought about by specialized tubular structures in the intercellular channels called gap junctions.

Membrane Channels

Because of the predominant lipid nature of biological membranes, many types of molecules are unable to diffuse through them especially charged ions and hydrophilic compounds. This is overcome by the presence of specialized channels and transporters. A channel/pore facilitates the translocation of molecules or ions across the membrane. It facilitates diffusion in both directions depending on the concentration gradient. It can also regulate the passage of molecules by opening and closing of the passageway.

There are three types of membrane channels. The α-type has homo- or hetero-oligomeric structures consisting of many dissimilar proteins. Molecules can move through channels down their concentration gradients without the input of energy. Alpha channels can show specificity with respect to the kind of molecule being transferred across the membrane. There can also be variation from one tissue to another in the kind of channel used for transport of the same molecule. For example, there are 15 different voltage-regulated channels in humans for the transfer of potassium. The transport of molecules through α-type channels can occur by different mechanisms, e.g., by voltage gated channels, by phosphorylation of the channel protein, by changes in intracellular Ca^{2+} by G-proteins and by organic modulators.

Ion channels: The transport of various electrolytes through these channels is very fast. The permeability of a channel depends on the size, the extent of hydration and the extent of charge density on the ion. Cation channels/pores have an average diameter of 5–8 nm. Most cells have a variety of specific channels for sodium, potassium, calcium and chloride. The sodium channel consists of four sub-units and each sub-unit consists of six alpha helical transmembrane domains. The amino and carboxy terminals are located in the cytoplasm. The membranes in the nerve cells contain ion channels that are responsible for the generation of action potentials. Ion channels are important for nerve impulse propagation, synaptic transmission and secretion of biologically active substances from the cells. The activity of some of the channels is controlled by neurotransmitters. The channels remain closed and open in response to the stimulus, thereby allowing rapid flux of ions down their concentration gradient. Ion channels can be gated and these gates are responsible for controlling their opening and closing. There are **ligand-gated channels**, in which a specific molecule binds to a receptor and opens the channel. **Voltage-gated channels** open or close in response to changes in the membrane potential. **Mechanically-gated channels** respond to mechanical stress, such as pressure or touch. Membrane channels are usually conserved across species. Mutations in genes encoding them can cause specific diseases. Their activities are affected by various drugs.

Ligand-gated channels: They open on binding of an effector molecule, i.e., the binding of a ligand to its receptor, resulting in opening or closing of the channel. The ligand can be an extracellular signaling molecule or an intracellular molecule/messenger. Acetylcholine receptor is an example of the ligand-gated channel. It is present in postsynaptic membrane. It is made up of five subunits containing acetylcholine binding sites and the ion channel. Acetylcholine released from the presynaptic region binds with the receptors on the postsynaptic region, which triggers the opening of the channel and influx of sodium. This generates an action potential in the postsynaptic nerve. The channel opens only for a very short duration of time (milliseconds) because the acetylcholine gets rapidly degraded by the enzyme acetylcholine esterase.

Voltage-gated channels: They open on membrane depolarization, i.e., the channels open or close in response to changes in the membrane potential. The membrane potential change, i.e., the voltage difference, opens the channels for a very short duration (25 ms). Voltage–gated channels pass from closed state to open state, to inactivated state, to depolarization. The channel cannot reopen until it is reprimed by depolarization of the membrane. Examples of voltage-gated channels are sodium channels and potassium channels which are involved in the conduction of nerve impulses in the nerve cells.

Calcium channels: They are present in the sarcoplasmic reticulum membrane. Upon appropriate stimulation, the calcium channels are opened, leading to an increased level of calcium in the cytosol of muscle cells. Hence, calcium channel blockers are used in the management of hypertension.

Channelopathies: These are a group of disorders resulting from abnormalities in the membrane proteins forming pores or channels. The examples are cystic fibrosis (defect in chloride channel), Liddle's syndrome (defect in sodium channel) and periodic paralysis (defect in potassium channel).

Ionophores: They are membrane shuttles for specific ions. They increase the permeability of membrane to ions by acting as channel formers. Ionophores contain hydrophilic centers to which specific ions can bind and on the periphery they are surrounded by hydrophobic regions. Examples are valigomycin which is a mobile ion carrier and gramicidin which is a channel former. These antibiotics are produced by certain microorganisms. Valigomycin allows potassium to enter the mitochondria and dissipates the proton gradient, and hence it acts as an uncoupler of an electron transport chain. Some microbial toxins, e.g., diphtheria toxin, can produce pores in cellular membranes, thus allowing macromolecules to traverse the membrane.

Aquaporins: Aquaporin proteins (AP) are basically a family of α-channels responsible for the transport of water across membranes by simple diffusion. They are present in some membranes, e.g., in RBCs, collecting tubules of the kidney. These aquaporins permit only the passage of water but not that of other ions. Ten aquaporin proteins have been identified in humans (AP1 to AP9). Another related family of proteins known as aqua glyceroporins help in transport of water as well as other small molecules. AP9 is an aqua glyceroporin found in humans. The aquaporins are homo-tetramers and each monomer consists of six transmembrane α-helical domains forming a distinct water pore. Aquaporins are abundantly expressed in the kidneys, with AP1, AP7, and AP8 being present in the proximal tubule and AP2, AP3, AP4, AP6, and AP8 being present in the collecting ducts. Several diseases are reported to be associated with loss of renal aquaporins, e.g., nephrogenic diabetes insipidus (NDI), acquired hypokalemia, and hypercalcemia where there is reduced expression of AP2.

Porins: Their transmembrane domain consists of β-strands forming a β-barrel structure. The mitochondrial porins are present on the outer mitochondrial membrane, are voltage-gated anion channels which help maintain mitochondrial homeostasis and are also involved in mitochondrial mediated apoptosis.

The third class of membrane channels are the pore-forming toxins first identified in bacteria. However, very few of these proteins are expressed in mammalian cells, e.g., the antibiotic proteins, defensins which are pore-forming channels present in epithelial and hematopoietic cells. They play a role in host defense against microbes.

Membrane Transporters

Since the membrane lipid bilayers are selectively permeable, a number of mechanisms exist by which substances are transported across them. There are three basic types of movements across membranes: simple diffusion, passive transport and active transport. Both passive and active transport mechanisms involve transporter proteins and show specificity for ions, sugars and amino acids. Transporters are distinguished from channels because they catalyze/mediate the movement of ions and molecules by physically binding to and moving the substance across the membrane. Thus act as enzymes. Their activity can be measured by the same kinetic parameters as applied to the study of enzyme kinetics and can also be affected by both competitive and noncompetitive inhibitors. Transporters exhibit specificity for the molecule being transported. They are also known as carriers, permeases, translocators and translocases. Transporters, based on their stoichiometry, are classified as **uniporters, symporters** and **antiporters.** The action of transporters is divided into two classes: passive-mediated transport (also called facilitated diffusion) and active transport. Facilitated diffusion involves the transport of specific molecules from an area of high concentration to one of low concentration, which results in an equilibration of the concentration gradient. Glucose transporters are a good example of passive-mediated (facilitated diffusion) transporters and so also the K^+ channels. In contrast, active transporters transport specific molecules from an area of low concentration to that of high concentration. Because this process is thermodynamically unfavourable, the process must be coupled to an exergonic process, i.e., hydrolysis of ATP. There are many different classes of transporters that couple the hydrolysis of ATP to the transport of specific molecules. In general, these transporters are referred to as ATPases. These ATPases are so named because they are autophosphorylated by ATP during the transport process. There are four different types of ATPases that function in eukaryotes. E-type ATPases are cell surface transporters that hydrolyze a range of nucleoside di- and triphosphates that includes extracellular ATP. These transporters derive their nomenclature from the fact that they are involved in extracellular transport. The activity of the E-type ATPases is dependent on Ca^{2+} or Mg^{2+}. There are at least three classes of E-type ATPases. F-type ATPases function in the translocation of H^+ in the mitochondria during the process of oxidative phosphorlyation. F-type transporters contain rotary motors. The nomenclature of F-type ATPases derives from phosphorylation factor. They are also referred to as H^+-transporting ATPases or F0F1-ATPase. The F0 subunit is the rotary core of the ATPase that is connected to the F1 catalytic core. P-type ATPases are mostly found in the plasma membrane and are involved in the transport of H^+, K^+, Na^+, Ca^{2+}, Cd^{2+}, Cu^{2+}, Mg^{2+}, Co^{2+}, Ag^{2+}, and Zn^{2+}. These transporters represent one of the largest families found in both prokaryotes and eukaryotes. The P-type ATPases are grouped into five classes designated as P_1–P_5, with several classes further divided into subclasses designated as A, B, C, etc. The P-type ATPases contain a core cytoplasmic domain structure that includes a phosphorylation domain (P domain), a nucleotide binding domain (N domain) and an actuator domain (A domain). The P-type ATPases also possess 10 transmembrane helices termed M1–M10, where helices M1–M6 comprise the core of the membrane transport domain. The P-type ATPases are also referred to as E1-E2 ATPases. V-type ATPases are located in acidic vesicles and lysosomes and have homology to the F-type ATPases and also contain rotary motors. The V nomenclature is derived from the fact that these transporters are located in vacuoles. The V-type ATPases are involved in the processes of neurotransmitter release, protein trafficking, receptor-mediated endocytosis and active transport of metabolites. A-type ATPases are archaeal bacterial transporters that function as the F-type class of ATPases.

Na^+, K^+-ATPases: One of the most thoroughly studied classes of ATPases are the Na^+, K^+-ATPases found in plasma membranes. These transporters, sometimes called sodium pump/Na^+, K^+-pumps, are involved in the transport of Na^+ out of the cell and of K^+ into the cells. The extrusion of Na^+ allows cells to control their water content. This pump has a relatively low affinity for potassium intracellularly, but high affinity for sodium ions. On binding, sodium ions stimulate ATP hydrolysis. Phosphorylation by ATP of three sodium binding sites on the cytoplasmic side of the cell induces a conformational change in this integral protein pump, leading to transfer of these three sodium ions from inside of the cell to the outside. The pump is no longer shaped in a manner that will allow attachment of intracellular sodium ions. Thus, the pump conformation change on ATP hydrolysis also results in a change in pump affinity for sodium ions. Sodium ions are consequently free to diffuse away from the pump and since they are now presented extracellularly, they diffuse in extracellular environment. This conformation change in the pump also results in an increased affinity of the pump for potassium ions and two molecules of extracellular potassium bind to the pump, resulting in dephosphorylation and transfer of the potassium ions across the membrane to the inside of the cell. Thus, the bound potassium ions are carried across the membrane and presented intracellularly. Relaxation of the conformation again changes the pump affinity, resulting in lowered potassium affinity and raised sodium affinity. Thus, three sodium ions are transported out of the cell and two potassium ions enter the cell. This results in net loss of positive charge from the cytoplasm. The amount of charge lost from the cytoplasm increases as more sodium and potassium ions are pumped. This creates an electrochemical gradient, i.e., an imbalance in the charge between inside and outside of the cell, making the inside more negative. Electrochemical gradients can be harnessed to do work. The charge difference between outside and inside of a cell is known as membrane potential. This membrane potential serves as the "electro" portion of the electrochemical gradient. Membrane potentials serve cells essentially as batteries, i.e., as stored energy. This is the basis for the electrochemical excitability of nerve cells. In fact, it is this transporter action that is the major requirement for ATP production from glucose oxidation in the central nervous system. The Na$^+$, K$^+$-ATPases belong to the P$_2$ class. These ATPases are composed of two subunits (α and β). The α-subunit binds ATP and both Na$^+$ and K$^+$ ions and contains the phosphorylation sites typical of the P-type ATPases. P-type ATPases are also subject to additional phosphorylation events via other kinases. The smaller β-subunit is absolutely necessary for activity of the complex and is responsible for facilitating the plasma membrane localization and activation of the α-subunit. Several isoforms of both α- and β-subunits are identified that exhibit different kinetic parameters and tissue distribution. In addition to the autophosphorylation site in the P domain of Na$^+$, K$^+$-ATPases, these pumps are subject to additional regulatory phosphorylation events catalyzed by protein kinase A and protein kinase C. The stimulus for these enzyme-mediated phosphorylation of Na$^+$, K$^+$-ATPases is activation of associated receptors. Adrenergic, cholinergic and dopaminergic receptor agonists result in PKA-mediated phosphorylation of the pumps. Activation of the prostaglandin E receptors has also been shown to lead to PKA-mediated phosphorylation of neuronal Na$^+$, K$^+$-ATPases. The cardiac drug digitalis inhibits Na$^+$, K$^+$-ATPase by binding to the extracellular side. Digoxin and ouabain, the cardiotonic drugs, bind to the alpha sub-units of this pump and act as competitive inhibitors to potassium ion binding to the pump. This inhibition of the pump leads to an increase in sodium levels inside the cell and extrusion of calcium from the myocardial cell. This enhances the contractibility of the cardiac muscle and thus improves heart function. Binding of these various compounds to the pump results in activation of various kinases such as Src and PI3K, resulting in modulation of cell adhesion and growth. The proton pump is simply ATP-driven active transport in which the substance pumped across the membrane is a hydrogen ion. In the epithelial cells of the stomach, gastric acid is produced by hydrogen potassium ATPase, which is an electrogenic pump.

Calcium pump: This is found in the sarcoplasmic reticulum of skeleton muscle and regulates the calcium concentration and muscle contraction. This pump is dependent on ATP. In resting muscle, the calcium concentration around the muscle fiber is low. On stimulation by a nerve impulse, there is sudden release of large amounts of calcium, which trigger muscle contraction. The calcium pump helps in removing cytosolic calcium and maintaining low concentration of cytosolic calcium, so that the muscle can receive the next signal. For each ATP hydrolyzed, two calcium ions are transported.

ABC family of transporters: They comprise the ATP-binding cassette transporter (ABC) superfamily. All members of this family contain a conserved ATP-binding domain and use the energy of ATP hydrolysis to transport various molecules across all cell membranes. There are 48 known members of the ABC transporter superfamily and they are divided into seven subfamilies based on phylogenetic analyses. These seven subfamilies are designated ABCA through ABCG. The most

typical efflux pump in the cell membrane is P-glycoprotein. High levels of both P-glycoprotein and breast cancer resistant pump (BCRP) are present in both normal and cancer stem cells. Overexpression of ATP-binding cassette transporter is shown to be responsible for multidrug resistance (MDR) via ATP-dependent drug efflux pumps.

The solute carrier (SLC) family of transporters includes over 300 proteins functionally grouped into 47 families. The SLC family of transporters includes facilitative transporters, primary and secondary active transporters, ion channels and aquaporins.

Clinical significance of transporter defects: Defects in the expression and function of membrane transporters lead to the manifestation of numerous clinical disorders. ABCA1 is involved in the transport of cholesterol out of cells when high density lipoproteins (HDLs) are bound to their cell surface receptor SR-B1. One important consequence of the activity of ABCA1 in macrophages is that the efflux of cholesterol results in a suppression of inflammatory responses triggered by macrophages that have become foam cells due to cholesterol uptake. Defects in ABCA1 result in Tangier disease, which is characterized by two clinical hallmarks: enlarged lipid-laden tonsils and low serum HDL. ABCB4 is a member of the P-glycoprotein family of multidrug resistance transporters. Defects in the ABCB4 gene are associated with six liver diseases: progressive familial intrahepatic cholestasis type 3 (PFIC3), adult biliary cirrhosis, transient neonatal cholestasis, drug-induced cholestasis, intrahepatic cholestasis of pregnancy and low phospholipid-associated cholelithiasis syndrome. ATP7A and ATP7B are copper transporting ATPases that are related to SLC31A1. Defects in ATP7A results in Menkes disease and defects in ATP7B are associated with Wilson disease. Deficiency in SLC6A19 leads to Hartnup disease which results from impaired transport of neutral amino acids across epithelial cells in renal proximal tubules and intestinal mucosa. Symptoms include transient manifestations of pellagra-like light-sensitive rash, cerebellar ataxia and psychosis. SLC30A8 is involved in efflux of zinc, and polymorphisms in the gene encoding this transporter are associated with an increased risk of the development of Type-2 diabetes. The functional defect in this protein results in impaired pancreatic β-cell function, leading to defects in insulin secretion.

Membrane Transport

The transport systems are described based on the number of molecules moved and their direction, or according to the movement, whether it is towards equilibrium or away from equilibrium. Molecules can be transferred across the membrane by passive transport down electrochemical gradients. This can be done by simple diffusion or facilitated diffusion. Passive transport is driven by transmembrane gradient of the substrate, whereas active transport always occurs against an electric or chemical gradient, and hence it requires energy, usually in the form of ATP. Substances such as gases, e.g., oxygen and carbon dioxide, can pass by simple diffusion at a rate dependent on their concentration gradients. Lipophilic molecules also diffuse across membranes at a rate that is directly proportional to the solubility of the compound in non-polar solvents. Although water can diffuse across biological membranes, the physiological need for rapid equilibrium across plasma membrane is met by aquaporins. Hydrophilic molecules move across the cell membrane by facilitated transport or active transport. Uniport allows movement of a single molecule in either direction, i.e., inside and out of the cell, depending on the concentration present. Symport allows two molecules to move in the same direction and antiport allows two molecules to move in opposite directions. The factors which affect the net diffusion of a substance include (i) its concentration gradient across the membrane; (ii) its electrical potential, solutes move towards the solution that has an opposite charge; (iii) the permeability coefficient; (iv) the hydrostatic pressure gradient, increased pressure will increase the movement across the membrane; (v) temperature, increase in temperature will increase the frequency of collisions between the solutes and hence increase transport across the membrane Figure 2.3(a) and 2.3(b).

Passive Transport

Uniporters are the proteins that move molecules in passive transport. They can be channel proteins or carrier proteins. Channel proteins open in response to a stimulus and let molecules flow through freely. Carrier proteins bind to a molecule, making it hydrophobic enough to cross the membrane.

Simple diffusion: Solutes and gases can enter the cells passively by simple diffusion and this does not require energy (ATP) or a carrier protein. It is a slow process and depends on the solubility of the solutes in lipids. Certain hydrophobic molecules can also readily diffuse across the cell membrane. Simple diffusion involves the passive flow of a solute from a higher concentration to a lower concentration due to random thermal movements in blood capillaries, which are one-cell thick, small molecules, similar to water which moves by passive diffusion. Blood pressure aids matters by pushing water and dissolved solutes out of the capillaries through tiny pores between capillary cells, a process called filtration from higher pressure to lower pressure.

Facilitated diffusion: This is also the passive transport of a solute from a higher concentration to a lower concentration but it is mediated with the help of a specific protein called transporter. Molecules that are insoluble in the lipid bilayer and some solutes diffuse down electrochemical gradients across membranes more rapidly than expected from their size, charge or partition coefficient. This is because they are assisted in crossing the cell membrane through special transport proteins. Transport/carrier proteins have specific binding sites for solute, the carrier is saturable and thus it the has maximum rate of transport which is similar to the V_{max} for enzymes, there is a

binding constant for solute and structurally similar molecules competitively block transport. Facilitated diffusion can occur bidirectionally, the process is reversible, it does not require energy, and the rate of transport is more rapid than simple diffusion. The carrier molecule can exist in two conformational states, i.e., the ping state and the pong state. In the pong state, the active sites of the carrier protein are exposed to the outside and the solute binds to them. On binding, a conformational change takes place. In the ping state, the active sites face the inside of the cell where the concentration of the solute is less and this causes release of the solute molecule and then the protein reverts to the pong state. A ping-pong mechanism explains facilitated diffusion. By this mechanism, the inward flow is facilitated but outward flow is inhibited. The rate at which the solute enters a cell by facilitated diffusion depends on (i) concentration gradient across membrane, (ii) amount of carrier protein available, (iii) rapidity of solute–carrier interaction and (iv) rapidity of conformational change. Examples of facilitated diffusion are amino acids and ions. Hormones can regulate facilitated diffusion by changing the number of the carrier proteins available. For example, insulin increases glucose transport in adipose tissue and muscles by recruiting transporters. Insulin also increases amino acid transport in liver and other tissues. Growth hormone increases amino acid transport in all cells. There are at least five different carrier proteins for amino acids in animal cells, each being specific for a group of closely related amino acids and operates via sodium-symport system. The other types of passive transport, which do not require proteins because the molecules diffuse directly through the cell membrane, are osmosis and filtration.

Osmosis: Besides solutes, even solvents can show concentration gradients. Water tends to flow from the side of a selectively permeable membrane (permeable to water but not to the solute) that has less solute (higher water concentration) to the side of the membrane that has more solute (lower water concentration), i.e., water will tend to flow down its concentration gradient from regions of high water concentrations to regions of low water concentration. Thus, the number of solute particles dissolved tends to be more important in determining which direction water will flow, rather than the chemical nature of the solute. This movement of water across selectively permeable membranes down the water concentration gradient, i.e., from a region of low solute concentration to a region of high solute concentration is called osmosis. The rate of passage of a water molecule through a semipermeable membrane is proportional to the number of water molecule collisions as well as the hole size. Osmolarity is a solute's concentration measured in terms of its impact on osmosis and is considered a colligative property of solute particles, one that varies as a function of the solute particle number rather than its size or shape.

Tonicity: When a membrane separates two solutions, one side with a higher solute concentration than the other, the side with the higher solute concentration is said to be hypertonic

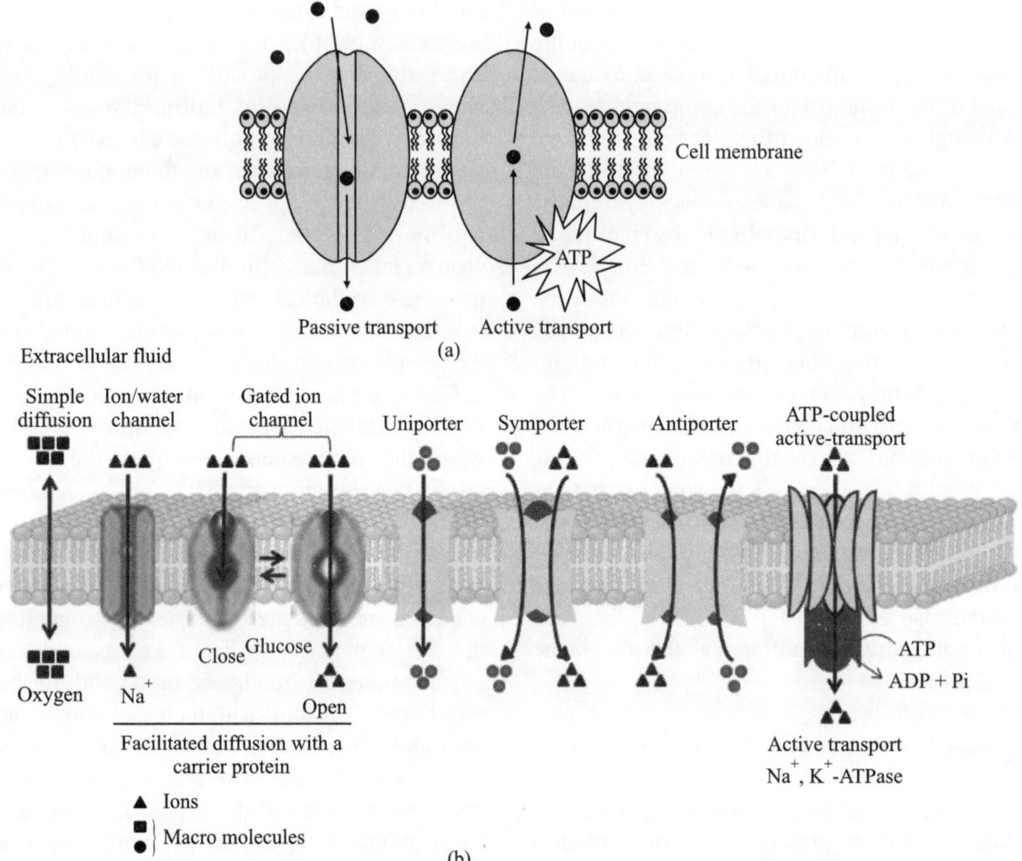

Figure 2.3   (a) Different modes of membrane transport (b) Transfer of molecules by simple diffusion, facilitated diffusion and active transport.

and the side with the lower solute concentration is said to be hypotonic. If both sides have the same solute concentration, they are said to be isotonic. However, the underlying mechanism of osmosis is the flow of water from a region of high water concentration to a region of low water concentration. The prefix "hyper" and "hypo" refer to the relative solute concentrations. The hypertonic solution has too much solute, i.e., more than the reference solution, and the hypotonic solution has too little solute, i.e., less than the reference solution. Normal animal cells are bathed in an isotonic solution. Placement of an animal cell in a hypertonic solution causes the cell to shrink, i.e., the water is lost from the cell by osmosis, whereas placement in a hypotonic solution causes it to take up water and then burst or lyse. When a plant or bacterial cell is placed in a hypertonic environment, it will show shrinkage and this shrinkage is called plasmolysis, which inhibits growth, and this principle is used for preservation of foodstuffs in highly osmotic solutions, such as salt and sugar solutions, as they impede most microbial growth.

Active Transport

Active transport is the transport of a solute across a membrane against a concentration gradient, i.e., from a lower concentration to a higher concentration, and thus it requires energy which can

be chemical energy in the form of ATP or there can be specific transporters such as pumps. It is usually related to accumulating molecules that the cell needs, such as ions, glucose and amino acids. Up to 40% of a cell's total energy may be used for active transport. It is an endergonic process and is unidirectional. The energy required can be in the form of ATP or in the form of electrochemical gradients and can also be obtained from light or electron movement. The movement of substances is in a direction that is away from equilibrium. Cells involved in active transport have a large number of mitochondria near plasma membrane at which active transport is taking place. In the active transport, a molecule or an ion combines with a carrier and this alters the shape of the carrier molecule. There are two types of active transport, primary and secondary. Both involve going against a concentration gradient using ATP, but they differ in how the ATP is used by the protein. If the process uses chemical energy such as ATP, it is called **primary active transport**. If it involves the use of an electrochemical gradient, it is called **secondary active transport**. Four major classes of ATP–driven active transporters are reported. The Na^+, K^+-ATPase, Ca^{2+}-ATPase of muscle, mitochondrial ATP synthase, CFTR protein which is a chloride channel transporter and is responsible for cystic fibrosis.

The neuronal cell membrane maintains asymmetry between the inside and the outside by electrical potential due to the presence voltage-gated channels. On stimulation by a specific synaptic membrane receptor, channels present in the membrane are opened to allow the influx of sodium and calcium with or without the efflux of potassium, so that the voltage difference rapidly collapses and that segment of the membrane is depolarized. But, by the action of the ion pumps present in the membrane, the gradient is quickly restored. When large areas of the membrane are depolarized, the electrochemical disturbance propagates in a wave-like manner, thereby generating a nerve pulse. Myelin in sheets around the nerve fibers acts as electrical insulators and speeds up the propagation of the wave signal by allowing ions to flow in and out of the membrane only where the membrane is free from the insulation. The myelin in membrane has a high content of lipids compared to the proteins and thus acts as an excellent insulator. Diseases such as multiple sclerosis are characterized by demyelination and impaired nerve conduction. Thus, transmission of nerve impulses also involves ions and pumps.

Primary active transport: ATP is expended to move a molecule up its concentration gradient. An example of this is the sodium–potassium pump, which pumps both ions against their concentration gradients in order to create a membrane voltage potential. Most of the enzymes that perform this kind of transport are transmembrane ATPases. A primary ATPase present in all forms of life is the sodium–potassium pump, which helps maintain the cell potential. Other sources of energy for primary active transport include redox energy, e.g., electron transport chain that uses the reduction energy of NADH to move photons across the inner mitochondrial membrane against their concentration gradient, and photon/ light energy, e.g., the proteins involved in photosynthesis use light energy. Examples of ATP using primary active transport are P-type ATPase (sodium potassium pump, calcium pump, proton pump), F-ATPase (mitochondrial ATP synthase), V-ATPase (vacuolar ATPase) and ABC (ATP binding cassette) transporter (MDR, CFTR).

Secondary active transport: ATP is not directly coupled to the molecule of interest in secondary active transport. Instead, another molecule is moved up its concentration gradient, which generates an electrochemical gradient. This electrochemical potential difference created by pumping ions out of the cell is used to transport the molecule of interest down the electrochemical gradient, e.g., sodium calcium exchangers. Although this process also consumes ATP to generate that gradient, the energy is not directly used to move the molecule across the membrane, and hence it is known as secondary active transport. The two main types of proteins involved in secondary active transport are antiporters and symporters.

Co-transport: If the transfer of one molecule depends on simultaneous or sequential transfer of another molecule, it is called co-transport. In this the movement of the substance against a concentration gradient is coupled with movement of a second substance down the concentration gradient. Hence, co-transporters use the gradient of one substrate created by active transport to drive the movement of the other substrate. The sodium gradient produced by Na^+–K^+-ATPase is used to drive the transport of a number of important metabolites. The co-transport system can be a symport or an antiport. Much of the active transport is not necessarily powered by ATP. Instead, it is powered by membrane potentials, i.e., electrochemical gradients.

Antiport: In this, the molecules move in opposite directions. Antiporter transports one substance in one direction and at the same time co-transports another substance in the opposite direction. One substance flows from a higher concentration to a lower concentration which yields entropic energy to drive the transport of the other substance from a lower concentration to a higher concentration. An example is the sodium–calcium exchanger, which removes calcium ions from the cell while allowing sodium ions back into the cell. The sodium is pumped out by the sodium–potassium pump, which generates the concentration gradient required for this to work. Many cells contain a calcium ATPase which can function at lower intracellular concentrations of calcium and sets the normal/ resting concentration of calcium. However, ATPase exports calcium ions much more slowly compared to sodium–calcium exchanger. This exchanger comes into play when the calcium concentration rises steeply and thus enables rapid recovery. This shows that a single type of ion can be transported by several enzymes which may not be active all the time but may exist to meet the specific intermittent needs of the cell. Another example of antiport is chloride–bicarbonate exchanger in the RBC.

Symport: In this the transporter carries two molecules in the same direction across the membrane. This usually works by allowing a molecule to move down its electrochemical gradient. The other molecule piggybacks off that movement and goes against its concentration gradient. Thus, one of the two substances is transported in the direction of its concentration gradient utilizing the energy derived from the transport of the second substance (mostly sodium, potassium and hydrogen), down its concentration gradient. Examples are the sodium–sugar transporters and sodium–amino acid transporters. Active transport often takes place in the inner lining of the small intestine. Glucose can be transported by several mechanisms. Different glucose transporters are involved and vary in different tissues. In skeleton muscle and adipocytes, glucose enters the cells by a specific transport system that is enhanced by insulin, whereas glucose transport in the small intestine is carried out by sodium–glucose symporter, which is located at the apical surface. Glucose symport SGLT1 co-transports one glucose molecule into the cell for every two sodium ions it imports into the cell. This symporter is also located in the heart, brain, testes and trachea, as well as in the proximal tubule in each nephron in the kidneys. This plays an important role in glucose rehydration therapy, and defects in SGLT1 prevent reabsorption of glucose causing familial renal glucosuria. Thus, sodium enters into the cell down its electrochemical gradient and drags glucose with it. Hence, the greater the sodium gradient, the more the glucose enters the cell. If sodium levels are low in the extracellular fluid, the transport of glucose can stop. This sodium–glucose symporter is dependent on gradients generated by sodium potassium ATPase, which maintains a low intracellular concentration of sodium. A similar mechanism is involved in the transport of amino acids across the apical lumen in polarised cells found in the kidney and intestine. The movement of glucose involves a glucose uniporter named GLUT2. In cases of diarrhea associated with cholera, there is a massive loss of fluids due to watery stools in a short duration of time, resulting in severe dehydration. Therefore, the oral rehydration therapy given to such patients primarily consists of glucose and sodium chloride, as developed by WHO. The transport of glucose and sodium across the intestinal epithelium via osmosis forces the movement of water from the lumen of the intestine into the intestinal cells, thereby resulting in rehydration. If glucose or sodium chloride is given by itself, this would not be effective. Phlorhizin is an inhibitor of sodium-dependent co-transport of glucose in the proximal convoluted tubules of the kidney and it produces renal damage, resulting in renal glycosuria. Transport of amino acids is also an example of symport. In Hartnup's disease, there is a defect in the symport mechanism for transfer of amino acids in the intestine and renal tubules. In cystinuria, the reabsorption of cystine in the kidneys is abnormal.

Macromolecules are transported across the cell membrane by endocytosis and exocytosis. Both endocytosis and exocytosis involve the vesicle formation.

Endocytosis: In this the cells take up macromolecules such as proteins, polysaccharides and polynucleotides. The cell membrane folds around the material present outside the cell and the particle is ingested. Endocytosis involves invagination and ingestion of parts of the plasma membrane. Endocytotic vesicles are formed when segments of the plasma membrane invaginate, enclosing a substrate/solute along with a small volume of extracellular fluid. The vesicle then pinches off as the fusion of plasma membrane occurs. Most endocytotic vesicles fuse with primary lysosomes to form secondary lysosomes which contain hydrolytic enzymes and are the suicide sacs of the cells. The macromolecular contents are digested to give rise to smaller molecules, such as amino acids, simple sugars and nucleotides. Endocytosis requires energy which is generally provided by ATP. Cytoplasmic contractile microfilaments take part in this kind of movement. DNA transfection also depends on endocytosis where calcium phosphate is used because calcium stimulates endocytosis and precipitates DNA, which makes it a better molecule to be endocytosed. The endocytosis can be **phagocytosis** (particle engulfment), **pinocytosis** (fluid uptake into the cell) or **receptor-mediated endocytosis**. Endocytosis provides mechanism for regulating membrane components, e.g., hormone receptors.

Phagocytosis: This occurs in specialized cells such as macrophages and granulocytes. This involves the engulfment of large particles such as bacteria, viruses, cells and cell debris by macrophages and granulocytes. They extend pseudopodia to surround the particle to form phagosomes. These, in turn, fuse with lysosomes to form phagolysosomes containing the particle which is then digested. Macrophages are very active in phagocytosis and may ingest 25% of their volume per hour and, thus, may internalize 3% of their plasma membrane every minute or the entire membrane every 30 min. Phagocytosis also leads to a respiratory burst.

Pinocytosis: This is a property of all cells and leads to fluid uptake. Absorptive pinocytosis is a receptor-mediated selective process responsible for the uptake of macromolecules for which there are a definite number of binding sites present on the cell membrane. These receptors permit the selective concentration of ligands from the medium and increase the rate at which specific molecules can enter the cell. The vesicles formed during this process are derived from invagination/pits that are coated on the cytoplasmic side with a filamentous protein and are called clathrin-coated pits. In humans, pinocytosis occurs in the small intestine when the cells engulf the fat droplets. The infants, intestinal lining ingests breast milk by pinocytosis, allowing mother's protective antibodies to enter baby's bloodstream.

Receptor-mediated endocytosis: Many hormones are taken up by the cell through receptor-mediated mechanism. The phospholipid P1P2 plays an important role in vesicle assembly. The protein dynamin, which both binds and has GTPase activity, is necessary for the pinching off/internalization of clathrin-coated vesicles from the cell surface. For example, LDL binds to LDL receptor and the complex is internalized with the help of clathrin-coated pits, which contain the LDL receptor, leading to the formation of LDL and its receptor containing endocytic vesicles. These vesicles fuse with lysosomes of the cell, releasing the receptor which is then recycled back to the surface of the plasma membrane, whereas the LDL is degraded in the lysosomes. The clathrin-coated pits constitute 2% of surface of some cells. They also help in absorption of cholesterol. Disorders of the LDL and its internalization have been observed. Similarly, extracellular glycoproteins carry specific carbohydrate recognition signals that are bound by membrane receptor molecules which play a role similar to that of a LDL receptor. The dark side is that many viruses, such as influenza virus, polio virus, hepatitis virus and HIV, bind to their specific receptors on the cell membrane and are taken up by caveolae-mediated process. Thus, these viruses mediate their damage by receptor-mediated endocytosis and cause several diseases, such as hepatitis, polio and AIDS.

Exocytosis: Most cells release macromolecules to the exterior of the cell by exocytosis. The hormones can act as signals for exocytosis. They bind to a cell surface receptor and thereby induce a local and transient change in calcium concentration, triggering exocytosis. In exocytosis, under appropriate stimulus, the secretory vesicles or vacuoles fuse with the plasma membrane. The movement is performed by the cytoplasmic contractile microtubules. The outer membrane of vesicle fuses with outer plasma membrane, while the cytoplasmic side of the vesicle fuses with the cytoplasmic side of plasma membrane. Thus, the contents of vesicles are externalized. The molecules released by exocytosis have three fates. They can attach to the cell surface and become peripheral proteins, e.g., antigens, they can become the part of the extracellular matrix, e.g., collagen and GAG, or they can enter extracellular fluid and signal other cells, e.g., insulin, and PTH. Catecholamines are packed in granules and proceed within cells to be released upon appropriate stimulation. Release of trypsinogen by the pancreatic cells, release of insulin by beta cells of the islets of Langerhans and release of acetylcholine from presynaptic cholinergic nerves are all examples of exocytosis. Exocytosis is also involved in membrane remodelling when components synthesized in Golgi bodies are carried in vesicles to plasma membrane.

Membrane Protein Associated Disorders

Since membranes are located on the cell surface as well as surround many organelles, and are involved in many processes, mutations in their protein constituents result in many disorders. Membrane proteins constitute ion channels, transporters, enzymes, receptors and structural components. Many of these are glycosylated, and mutations affecting this process may alter their functions. Some diseases associated with membrane abnormalities are familial hypercholesterolemia (mutations in gene encoding the LDL receptor), cystic fibrosis (mutations in gene encoding CFTR protein, a chloride transporter), Wilson's disease (mutations in the gene encoding a copper-dependent ATPase), congenital long QT syndrome (mutations in gene encoding ion channels in the heart), hereditary spherocytosis (mutations in the gene encoding the structural protein spectrin in RBC membrane), metastasis of cancer cells (abnormalities in the oligosaccharide chains of membrane glycoproteins and glycolipids) and I-cell disease (mutations in the gene encoding GlcNAc phosphotransferase, resulting in absence of the mannose-6-phosphate signal for lysosomal localization of certain hydrolases and thus affects the lysosomal function), Leber's hereditary optic neuropathy (LHON) (mutations in gene encoding mitochondrial membrane proteins involved in oxidative phosphorylation), leading to neurologic and other problems. Other membrane protein associated disorders include formation of autoantibodies to the acetylcholine receptor in skeletal muscle resulting in Myasthenia gravis. Various ion channels in the membranes are affected in ischemia. Snake Venom contains phospholipase which affects lysophospholipids and thus the membrane function. The membrane disorders can be due to mutations affecting receptors, transporters, ion channels, enzymes, structural protein, altered or defective glycosylation of glycoproteins, etc. Cystic fibrosis is a recessive genetic disorder prevalent in Europe and North America. It is characterized by chronic bacterial infections of the airways and sinuses, maldigestion of fat due to pancreatic exocrine insufficiency, infertility in males due to abnormal development of the vas deferens and elevated levels of chloride in sweat. This is due to mutations in a gene encoding an ion channel protein called cystic fibrosis transmembrane regulator protein (CFTR), which is a cyclic AMP-regulated chloride transporter.

Membrane and its Environment

Membrane plays an important role in the human body. Changes in the environment of membrane proteins affects their structure and function. Study of the influenza virus, especially the M2 protein has provided some insights. M2 is a homotetramer having three functional domains, i.e., the N-terminal, the TM helix and the C-terminal. Drugs targeted to block the TM helix domain prevented proton conductance and disabled the virus. Experiments done to study the structure of M2 protein in three different environments, showed that the protein was stable in a lipid bilayer environment. On adding the drug amantadine, the protein showed a fourfold symmetry structure indicating stability of the protein structure in the presence of amantadine. On comparing the crystal structures of this protein at different pH, it showed a range of conformational states. NMR studies showed that the amino acids of this protein interact to minimize electrostatic potentials and that water allowed hydrogen bond exchange. Thus studying the environmental changes on the protein structure can help develop drugs which can inhibit the influenza virus. However the major obstacles in this direction are that there are different kinds of membranes in our body, each having a different structure and function which is dependent on their membrane proteins and their respective amino acid composition. Moreover the integral membrane proteins are present in a heterogeneous environment which is difficult to mimic with the existing methodologies. It is very difficult to obtain membrane mimetic environments that support the native structure, stability and functions of a membrane protein, and the stability between proteins and lipids. Studies using Membrane Protein Complexes have been a step forward in this direction as these complexes allow the membrane proteins to retain their functions. Chaperones play a major role in the formation of these complexes as they act as physical assembly factors by interacting with the proteins and prevent unproductive interactions from occurring. Membrane protein complexes undergo dynamic changes as a mechanism for regulating damaged subunits within the complexes.

SUMMARY

The outer cell membrane results in the formation of a typical cell and is also referred to as plasma membrane. It is a dynamic structure, undergoes turnover and its composition changes throughout the life of a cell. Membranes are made up of an asymmetric lipid bilayer with distinct inner and outer surfaces. They are essentially made up of lipids, proteins and a small amount of carbohydrates. The lipids include phospholipids and cholesterol. The phospholipids contain both hydrophobic and hydrophilic regions and hence are called amphipathic. The hydrophilic heads face the external or internal side and the hydrophobic tails are oriented at the centre of the membrane. The membrane proteins include peripheral proteins, integral proteins and transmembrane proteins. Microdomains on the membrane rich in glycosphingolipids and cholesterol are called lipid rafts and are involved in transmembrane signaling events. The membrane exists in a fluid mosaic state. The nature of the fatty acids present in phospholipids defermines membrane fluidity. Unsaturated fatty acids increase fluidity, whereas saturated fatty acids decrease it. Cholesterol also alters fluidity. Lipids and some proteins exhibit lateral diffusion in the plane of their membranes. Transverse movement of lipids across the membrane, known as flip-flop movement, it also occurs, but is rare. The plasma membrane helps in transport of various gases, solvents and solutes by passive transport or active transport. Passive transport does not require the input of any energy and includes simple diffusion and facilitated diffusion. The former does not require any carrier proteins, whereas the latter does. In the passive transport, the movement of the molecules is down their concentration gradient. Active transport drives substances against their concentration gradient using energy and requires carrier proteins. The selective permeability to the membrane is provided by carriers, such as channels and pumps, for ions and various substrates. The ion channels can be ligand-gated or voltage-gated channels. Aquaporins are water channels. Transport systems are classified as (i) uniport when a single solute is carried across the membrane, (ii) symport when two solutes are carried in the same direction and (iii) antiport when two solutes or ions move in opposite directions across the membrane. Membranes help in exchanging macromolecules by endocytosis and exocytosis. Endocytosis can be further divided into pinocytosis (uptake of fluids), phagocytosis (uptake of bacteria by macrophages) and receptor-mediated endocytosis (uptake of LDL by LDL receptor). Membrane functions include transport of various molecules, cell–cell interaction, cell–cell recognition, attachment to the cytoskeleton and extracellular matrix, endocytosis, exocytosis, enzyme localization and signal transduction. Alterations of membrane components or mutations in the genes encoding the membrane proteins can lead to a number of diseases, such as cystic fibrosis, Hartnup disease and I-cell disease.

SUGGESTIONS FOR FURTHER READING

Allen T.W. and Separovic F. (2012), Membrane Protein Structure and Function, *Biochim. Biophy. Acta (BBA)—Biomembranes*, **1818** (2), 125. http://dx.doi.org/10.1016/j.bbamem.2011.12.015.

Elofsson A. and von Heijne G. (2007), Membrane Protein Structure: Prediction versus Reality, *Annu. Rev. Biochem.*, **76**, 125–140.

Hiller S., Abramson J., Mannella C., Wagner G., Zeth K. (2010), The 3D Structures of VDAC Represent a Native Conformation, *Trends Biochem. Sci.*, **35**(9), 514–251.

Pfeffer S.R. (2007), Unsolved Mysteries in Membrane Traffic. *Annu. Rev. Biochem.*, **76**, 629–645.

Smith A.W. (2012), Lipid–protein Interactions in Biological Membranes: A Dynamic Perspective, *Biochim. Biophys. Acta*, **1818**(2), 172–177.

Wright E.M., Hirayama B.A. and Loo D.F. (2007), Active Sugar Transport in Health and Disease, *Inte. Med.*, **261**(1), 32-43.

3

Proteins
Chemistry–Structure–Conformation

T.P. Singh

<table>
<tr><td colspan="2">CONTENTS</td></tr>
<tr><td>I.</td><td>Introduction</td></tr>
<tr><td>II.</td><td>Building Blocks</td></tr>
<tr><td>III.</td><td>Forces in Protein Structures</td></tr>
<tr><td>IV.</td><td>Peptide Bond and Formation of Proteins</td></tr>
<tr><td>V.</td><td>Primary Structure of Protein</td></tr>
<tr><td>VI.</td><td>Secondary Structure of Protein</td></tr>
<tr><td>VII.</td><td>Tertiary Structure of Protein</td></tr>
<tr><td>VIII.</td><td>Quaternary Structure of Protein</td></tr>
<tr><td>IX.</td><td>Methods of Protein Purification</td></tr>
<tr><td></td><td>Summary</td></tr>
</table>

I. INTRODUCTION

Proteins are the most widely distributed biological macromolecules which perform numerous functions in the living organisms. Some of these functions include biological catalysis by proteins known as enzymes, immune action by antibodies, storage and transport of oxygen and other metal ions by haemoglobin and transferring and signaling processes by hormones. The diverse biological functions of these complex protein molecules are governed by their three-dimensional structures and the precise shapes, sizes and chemical features of their binding sites. Therefore, it is necessary to understand the chemical nature and structure of protein molecules. So far, only 50,000 protein structures as determined using the technique of X-ray crystallography are available although millions of proteins are synthesized in nature. To understand proteins in a complete and concise manner, we ought to know the structures of all of them. Though the task may be challenging, it is certainly achievable. For studying the chemical nature and determining their detailed three-dimensional structures, proteins must be first obtained in pure form. Because many proteins are present in the complex biological milieu, their isolation from the source requires special techniques of separation. In addition, special methods of purification are also needed to achieve the required homogeneity.

Protein molecules are linear polymers of 20 naturally occurring amino acids. In order to attain stability, the linear protein chains fold into globular structures by maximizing attractive intramolecular interactions and at the same time by minimizing repulsive forces due to short contacts. A stable and functionally active folded protein possesses a unique structure with a well-defined binding pocket for interacting with other molecules, specifically to perform accurate biological functions. This chapter provides a preliminary information about possible patterns of folding and a brief description of the methods of protein purification. Because of the well-known chemical nature of the protein main chain, it is possible to suggest its possible modes of folding.

II. BUILDING BLOCKS

The different levels of the protein structure consist of primary, secondary, tertiary and quaternary structures. All proteins possess primary structure, secondary structure and tertiary structure, but not all proteins have the quaternary structure, which is the result of evolution of proteins over time. The structure of a protein is linked with its function and disrupting the structure leads to loss of its function. Hence, the protein structure has to be studied in detail in order to understand the function of the molecules.

There are 20 naturally occurring amino acids which are found in proteins. The amino acids are the building blocks of proteins. As the name suggests, an amino acid contains both an amine and a carboxylic functional group. Structurally, each amino acid consists of a central C alpha (Cα) atom which is tetrahedrally attached to an amino group, a carboxylic group, an H-atom and a side chain (R-group) (Figure 3.1).

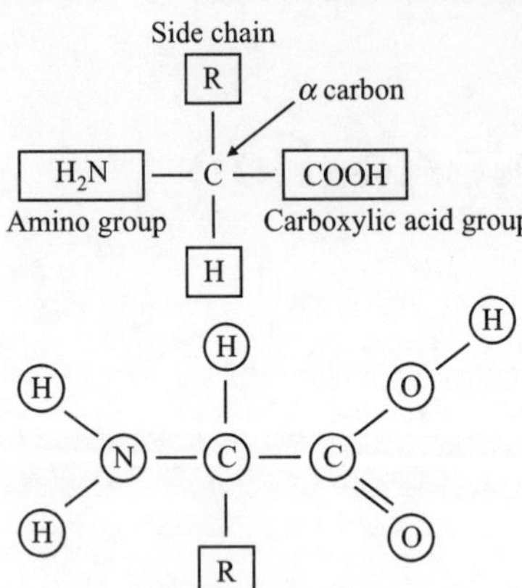

Figure 3.1   The amino acid.

The side chain or the R-group is present in all amino acids except glycine. As a result, all amino acids except glycine display optical activity. The mean mass of the standard amino acids is roughly 110 Da. In nature, there are 20 proteinogenic amino acids. These amino acids are used to form proteins and are called amino acid residues (Figure 3.2). The amino acids can be denoted by their full names, three-letter codes or one letter code (Table 3.1).

TABLE 3.1   **Three Ways of Representing Amino Acids**

S.No.	Amino acid	Three letter code	One letter code
1.	Alanine	Ala	A
2.	Arginine	Arg	R
3.	Asparagine	Asn	N
4.	Aspartic acid	Asp	D
5.	Cysteine	Cys	C
6.	Glutamine	Glin	Q
7.	Glutamine acid	Glu	E
8.	Glycine	Gly	G
9.	Histidine	His	H
10.	Isoleucine	Ie	I
11.	Leucine	Leu	L
12.	Lysine	Lys	K
13.	Methionine	Met	M
14.	Phenylalanine	Phe	F
15.	Proline	Pro	P
16.	Serine	Ser	S
17.	Threonine	Thr	T
18.	Tryptophan	Trp	W
19.	Tyrosine	Tyr	Y
20.	Valine	Val	V

Though these 20 amino acids can be classified in many ways, there are two types of classifications which are relevant for protein structures:

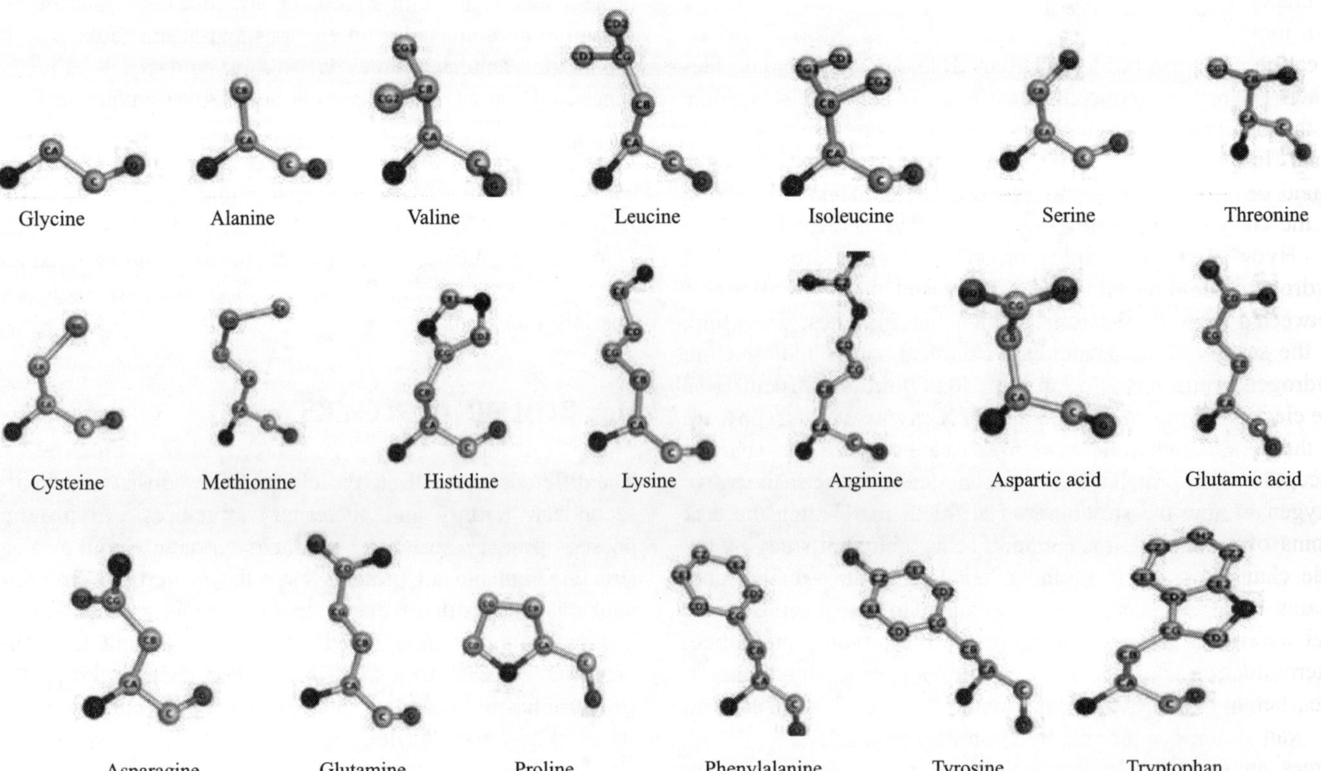

Figure 3.2   The structures of amino acid residues in proteins.

1. **Amino acid classification according to polarity of the side chain.** The 20 amino acids can be divided into two classes according to the polarity of the side chain attached to the central Cα atom. Polar amino acids can be further subclassified into acidic polar amino acids if the side chain contains a carboxylic acid and basic polar amino acids if the side chain contains an amino group (Table 3.2).
2. **Amino acid classification according to the structure of side chain.** The 20 amino acids can be divided into seven classes according to the structure of the side chain attached to the central Cα atom. These seven classes are amino acids with aliphatic side chains; aromatic side chains; sulfur side chains; side chains containing hydroxyl group; basic side chains; acidic side chains and side chain containing imino acid (Table 3.3).

III. FORCES IN PROTEIN STRUCTURES

The forces that influence the protein structure are of two types: covalent and non-covalent bonds. The covalent bonds come into existence upon sharing of electrons between two atoms and are the most potent chemical bonds contributing to protein structure. They can be found in three situations in a protein: (i) between atoms of the same amino acid, (ii) as peptide bonds that connect two amino acid residues and (iii) as disulfide bonds between cysteine residues. The attractive non-covalent bondscan be classified into three classes: electrostatic interactions, consisting of both ionic and hydrogen bonds, hydrophobic interactions and van der Waals forces.

An ionic bond is formed when two oppositely charged ions, a cation and an anion, get attracted electrostatically to each other. The strength of an ionic bond depends on the difference in the electronegativity between the two atoms involved in the bond. In a protein molecule, most of the charged groups are found on the surface of the protein, though some are found in the core of the protein.

Hydrogen bonds are so named because they need a hydrogen atom to come into existence. They are formed between one electronegative atom and a hydrogen atom bonded to the second electronegative atom. Hydrogen bonds need a hydrogen bond donor, which is the hydrogen atom attached to the electronegative atom, and a hydrogen bond acceptor, which is the second electronegative atom. Both the donor and the acceptor need to be electronegative atoms, such as nitrogen, oxygen or fluorine. The hydrogen bonds can occur between atoms of two amino acid side chains, atoms of amino acid side chains and water molecules, atoms of amino acid side chains and protein backbone atoms, or atoms of backbone and water molecules. Both ionic and hydrogen bonds can be intermolecular (i.e., between two molecules) or intramolecular (i.e., between residues of the same molecule).

van der Waals forces, also known as London dispersion forces, are weak and transitory forces which can occur between any two atoms. These forces occur due to the presence of

TABLE 3.2 The Classification of Amino Acids Based on Polarity

Classification	Amino acids
Aliphase side chain	Glycline
	Alamino
	Valine
	Leucino
	Isoleucine
Aromaffic side chain	Tyrosine
	Phenylalamine
	Tryptophan
Sulfur containing side chain	Cysteine
	Methionine
Hydrodyl or amino group side chains	Serine
	Threomine
	Asparagine
	Glutamine
Basic side chain	Arginine
	Lysine
	Histidine
Acidic side chain	Asparfic acid
	Glutamine acid
Imino side chain	Proline

TABLE 3.3 The Clubbing of Amino Acids Based on Gross Features

Classification	Amino acids
Nonpolar	Glycline
	Alanine
	Valine
	Leucino
	Isoleucine
	Proline
	Methionine
	Phenylalanine
	Tryptophan
Polar	Serine
	Threomine
	Asparagine
	Glutamine
	Cysteine
	Tyrosine
Acidic (Polar)	Aspartic acid
	Glutamic acid
Basic (Polar)	Lysine
	Arginine
	Histidine

a fluctuating electron cloud around each atom, which leads to the generation of a temporary electric dipole. When the electric dipoles between two atoms are complementary, and at an appropriate distance, attractive van der Waals forces come into existence. However, if the two atoms are too close, their electron clouds repel each other and the attractive forces do not exist any more. The appropriate distance between two atoms leading to van der Waals attractions may differ in the case of each atom due to varying atomic radii.

Hydrophobic interactions occur between two areas of hydrophobic amino acid residues which tend to exclude the water molecules and converge together to form the hydrophobic core. This hydrophobic core is well buried in the interior of the protein and is inaccessible to the surrounding water molecules.

IV. PEPTIDE BOND AND FORMATION OF PROTEINS

A peptide bond is a chemical bond that connects amino acids to each other. A peptide bond essentially results from a dehydration synthesis reaction. It is formed between two amino acids when the carboxyl group of one amino acid reacts with the amino group of the other, releasing a molecule of water (H_2O). Peptide bond is the resulting CO—NH bond and the resulting molecule is an amide (Figure 3.3).

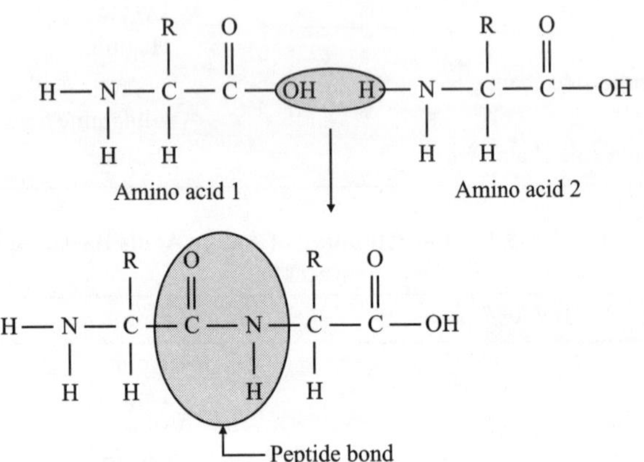

Figure 3.3 The reaction between amino acids for peptide synthesis.

The four-atom functional group —C(═O) NH— is called an amide group or a peptide group. Polypeptides and proteins are chains of amino acids held together by peptide bonds. The peptide bond is a partial double bond, and hence the C—N bond length of the peptide is 10% shorter than that found in usual C—N amine bonds.

The partial double bond nature renders the amide group planar and rigid, making all the atoms that are involved in the peptide bond lie in a flat plane. Peptides are defined as chains of about 50 or less amino acids, while polypeptides are longer peptides containing 50–100 amino acids. A polypeptide chain that contains more than 100 amino acids is called protein.

V. PRIMARY STRUCTURE OF PROTEIN

The primary structure of the protein (Figure 3.4) is its most basic and simplistic structural representation. The primary structure of a protein is its linear sequence of amino acids starting from the amino terminal (N) end and ending at the carboxy terminal (C) end. The amino acids are linked to each other by peptide bonds.

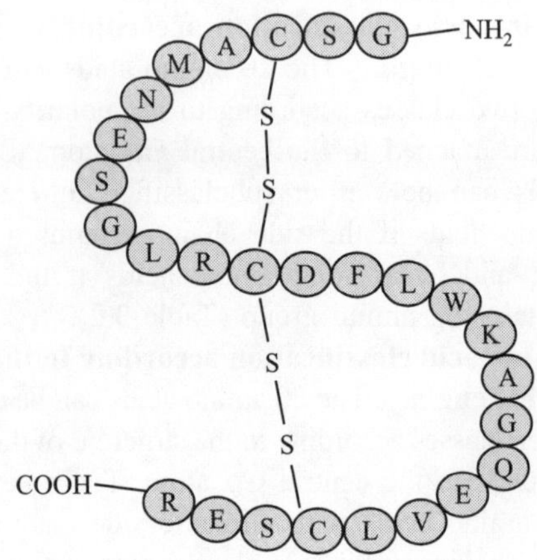

Figure 3.4 The primary structure of the protein.

Apart from the amino acid sequence, the primary structure also describes the location of any disulfide (—S—S—) bridges. A disulfide bond, also called an SS bond or a disulfide bridge, is a covalent bond that results from the coupling of two thiol groups (—S—H—). Disulfide bonds have a major function in stabilizing the folding of proteins. On a physical level, they bring together two non-bonded segments of the protein towards each other.

The disulfide bond also contributes to the formation of hydrophobic pockets in the protein since hydrophobic residues may get attracted towards its hydrophobic core, excluding the water molecules in the vicinity. Hence, this leads to the stabilization of secondary structures since water molecules, due to their property of attacking hydrogen bonds, break up secondary structure. The primary structure of the protein leads to the formation of protein backbone.

The protein backbone is formed by joining amino acids to each other linked by a peptide bond. The peptide backbone consists of repeating units of NH_2, CH, C═O; NH_2, CH, C═O; and so on.

There are three dihedral or torsion angles in the peptide backbone (Figure 3.4) that finally define the secondary structure of the protein: φ(C′—N—C$^\alpha$—C′), which controls the C′–C′ distance; ψ(N—C$^\alpha$—C′—N), which controls the N–N distance; and ω(C$^\alpha$—C′—N—C$^\alpha$), which controls the C$^\alpha$—C$^\alpha$ distance. Among the three angles, ω is very rigid due to partial double bond nature of the peptide bond and it has usually a value of 180°. Thus, the values of only φ and ψ decide the secondary structure of a protein.

VI. SECONDARY STRUCTURE OF PROTEIN

The secondary structure is the second level of the protein structural organization and is essentially defined by two parameters: the torsional angles, ϕ and ψ of the backbone atoms of the amino acid residues and the hydrogen bonds between main chain atoms. The most commonly found secondary structures in proteins are helices and beta (β)-sheets.

Helices

Helices are repetitive secondary structures, mainly because the values of their backbone torsional angles, ϕ and ψ tend to repeat. The three parameters which describe the helix about its axis are n (the number of residues per turn), h (the rise per residue) and p (the rise per turn). In nature, three types of helices are found: 3_{10}-helix, alpha (α)-helix and (π)-helix (Figure 3.5). The three helices differ from each other in their ϕ, ψ angles and in various other physical and structural chacteristics (Table 3.4).

Owing to the differences in their structural natures the appearance of these helices varies too. The 3_{10}-helix is the narrowest one, because of which van der Waals interactions across the helical axis may cause repulsion, while the π-helix is the broadest helix. The large radius of the π-helix results in the polypeptide backbone being no longer in van der Waals contact across the helical axis. The intermediate radius of the α-helix allows favourable van der Waals interactions across the helical axis.

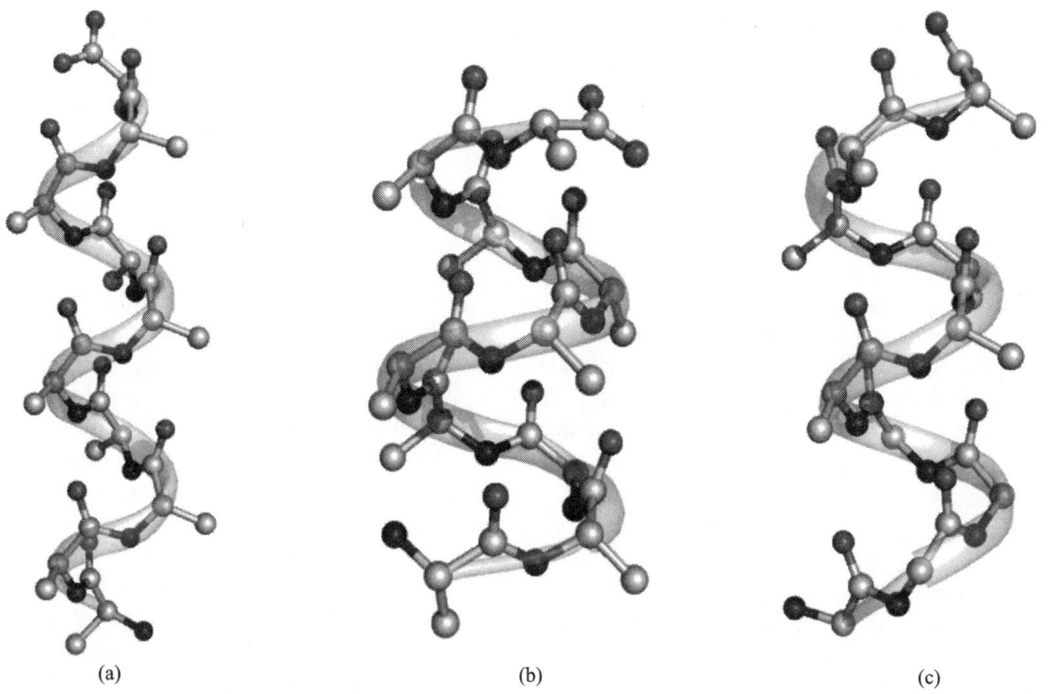

(a)   (b)   (c)

Figure 3.5 The three types of helices found in proteins: (a) 3_{10}-helix, (b) π-helix and (c) α-helix.

TABLE 3.4 The Characteristics of Secondary Structures in Proteins

Helix parameter	3_{10}-*helix*	α-*helix* (3.6_{13})	π-*helix* (4.4_{16})
Bonding between backbone N—H and C—O groups	Every backbone N—H group donates a hydrogen bond to the backbone C=O group of the amino acid three residues earlier	Every backbone N—H group donates a hydrogen bond to the backbone C=O group of the amino acid four residues earlier	Every backbone N—H group donates a hydrogen bond to the backbone C=O group of the amino acid five residues earlier
ϕ, ψ	−74, −4	−57.8, −47	−57.1, −69.1
N	3.0	3.6	4.4
Atoms in H-bonded loop	10	13	16
Position of ϕ, ψ in Ramachandran plot	ϕ, ψ lie at the edge of an allowed, minimum energy region of the Ramachandran map	ϕ, ψ lie at the centre of Ramachandran plot	ϕ, ψ lie at the very edge of an allowed, minimum energy region of the Ramachandran map
Radius	Radius 1.9 Å	Radius 2.3 Å	Radius 2.8 Å
Frequency	Infrequent (3.4%) α-helix sometimes begins or ends with 3_{10}-helix	Abundant (32–38%)	Rare As the 3_{10}-helix, one turn of π-helix is sometimes found at the ends of regular alpha helices

The same has an impact on the side chains of the residues of these helices. In the narrow 3_{10}-helix, the side chains are in identical azimuthal positions, leading to potential steric interference, while the side chains in α-helix are well staggered, which leads to minimization of steric interference among the side chains. As the diameter of the α-helix is similar to the width of the major groove in DNA, α-helices are found in DNA binding motifs, for instance, helix-turn-helix motifs, leucine zipper motifs and zinc finger motifs.

Owing to its nature and propensity, the 3_{10}-helix [Figure 3.5(a)] is relatively infrequent in nature and is mainly found at the beginning or end of α-helix. Similarly, the π-helix [Figure 3.5(b)] is the rarest form of helices found in nature and is sometimes found at the ends of regular alpha- helices. On the other hand, the α-helix [Figure 3.5(c)] is the most abundant form due to its structural and functional stability.

β-sheets

The basic unit of β-sheet is a single β-strand. A β-strand is formed when a single continuous stretch of amino acids, involved in hydrogen bonds, adopts an extended conformation. An assembly or ordered number of such strands that are hydrogen bonded to each other is called a β-sheet. The alignment of the β-strands in a β-sheet is in such a way that $C\alpha$ atoms of the residues in the adjacent strands face each other, and hence their side chains point in the same direction. The extensive hydrogen bonding network between strands happens, and so amino groups in the backbone of the first strand make hydrogen bonds with the carboxyl groups in the backbone of the next strand. Since the β-strand is extended in nature, the ϕ and ψ angles are always $-120°$ and $120°$, respectively. There are two types of β-sheets in nature—parallel [Figure 3.6(a)] and antiparallel [Figure 3.6(b)].

(i) Parallel β-sheets. β-sheets are called parallel when all the N-termini of the strands are oriented in the same direction. The values of dihedral angles φ and ψ in parallel β-sheets are about and $-120°$ and $115°$ respectively. Since the inter-strand hydrogen bonding pattern is non-planar in parallel β-sheets arrangement, this arrangement is less stable as compared to antiparallel β-sheets.

(ii) Antiparallel β-sheets. β-sheets are called antiparallel when the N-terminus of one strand is adjacent to the C-terminus of the next. The peptide backbone dihedral angles φ and ψ are about $-140°$ and $135°$, respectively. Since the inter-strand hydrogen bonding pattern is planar in antiparallel β-sheets arrangement, this arrangement is more stable, and hence it is more favourable.

VII. TERTIARY STRUCTURE OF PROTEIN

The tertiary structure is the third level of protein structure. It is formed when the primary structure of the protein makes inter- and intra-hydrogen bonding to form the secondary and supersecondary structures which fold into the final shape (Figure 3.7).

This level of protein structure depends on the number and the kind of attractive interactions between R-groups or side chains of the amino acid residues in the protein.

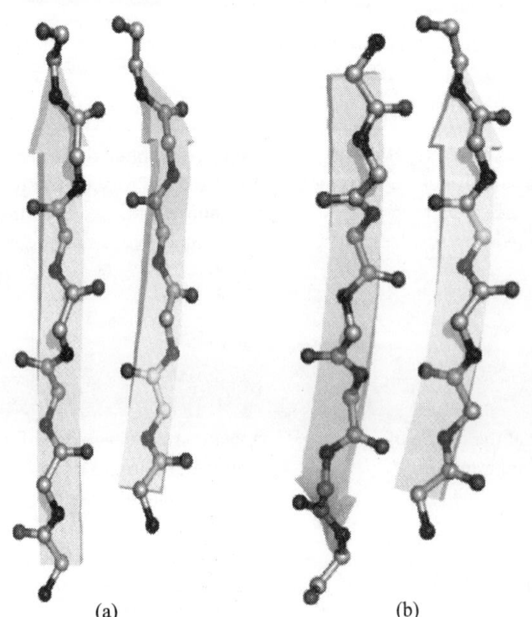

Figure 3.6 The two types of β-sheets found in proteins: (a) parallel and (b) antiparallel.

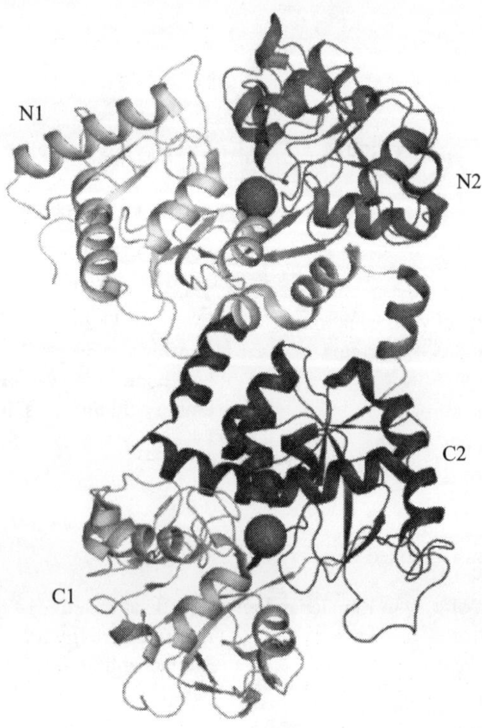

Figure 3.7 Tertiary structure of a monomeric antimicrobial, iron-binding protein, lactoferrin, as determined by X-ray crystallography, having four domains N1 (yellow), N2 (pink), C1 (green) and C2 (blue).

These attractive interactions between the side chains result in the formation of folds in the overall protein chain. The four types of attractive interactions found between the side chains are ionic bonds between charged R-groups, hydrogen bonds between polar R-groups, disulfide bonds between two cysteine R-group and hydrophobic interactions between hydrophobic R-groups.

The significance of the tertiary structure in a protein is very high, since the overall function of the protein is dependent on its tertiary structure. Disrupting the tertiary structure, which is known as denaturation of proteins, results in the loss of the function for most proteins.

The tertiary structure of a protein is also described more effectively as domains. Domains are localized and independent units found in a protein and are formed when several motifs come together and pack in a stable manner. Just as amino acids are a unit of proteins, domains can be considered to be the unit of a tertiary structure. Each domain consists of a centralized hydrophobic core formed by hydrophobic residues which cluster towards each other, excluding out water, with the polar residues towards the outside of the domain, interacting with water molecules.

A domain usually consists of an average of about 50–300 amino acid residues, though this number varies over families of proteins. There are four domains found within the protein tertiary structure based on the kind of secondary structures they contain: all α-domains, all β-domains, $\alpha + \beta$ domains and α/β domains. Each domain may have its own independent function, for instance, a domain may be involved in ligand binding, may contain a catalytic site in the case of enzymes, may traverse a cell membrane or may be involved in protein–protein interactions.

VIII. QUATERNARY STRUCTURE OF PROTEIN

The quaternary structure, the fourth level, is the final arrangement of a protein structure. It can be formed once all the lower levels, the primary, secondary, supersecondary and tertiary structures, of a protein are formed. However, not all proteins contain the quaternary structure. This is because the quaternary structure of a protein is formed only if the protein has more than one subunit. The quaternary structure of a protein describes the way independently folded protein subunits are arranged in a multi-subunit association.

The quaternary structure is similar to tertiary structure in the way that it is stabilized by the same non-covalent interactions and disulfide bonds. However, quaternary structure differs from tertiary structure because of the presence of more than one subunit in the former. A protein quaternary structure may contain identical or non-identical subunits.

An example of a protein containing identical subunits is peptidoglycan recognition protein [Figure 3.8(a)]. In the case of proteins containing different subunits, each subunit has a different function, making the protein a multifunctional

protein. Since one individual protein chain is called a monomer, anything more than one chain is called multimer. If it has two subunits, it is called a dimer, if it has three, then it is called a trimer and so on. Unlike a tertiary structure which is stabilized by different kinds of interactions, the quaternary structure of a protein is mainly stabilized by only the hydrophobic interactions generated by isolated hydrophobic patches on the surface of the protein subunits. However, once the hydrophobic interactions make the quaternary structure, hydrogen bonding may also be involved in the stability of the overall structure.

The quaternary structure is an evolved form of the protein structure and offers many advantages over the lower forms of structure. The major benefits of a quaternary structure of a protein are improved activity due to an increase in the number and the surface of active sites, cooperativity generated by inter-subunit contact and generation of allosteric sites within a protein.

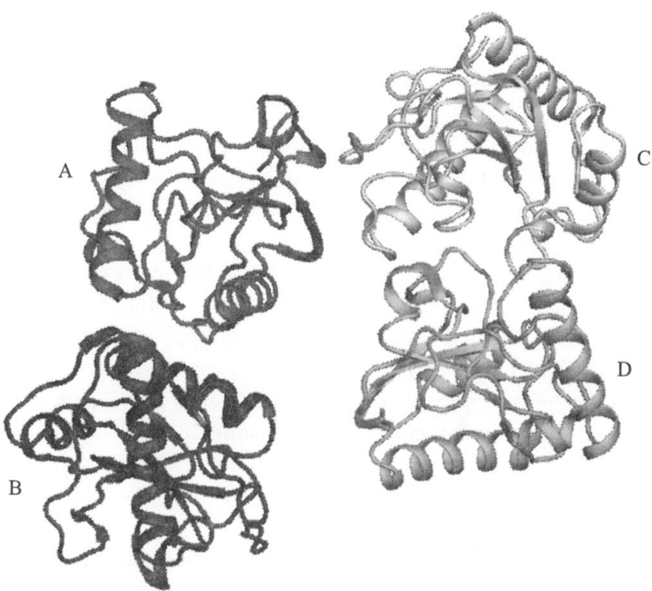

Figure 3.8 Quaternary structure of an antimicrobial protein peptidoglycan recognition protein as determined by X-ray crystallography showing its four subunits A, B, C and D.

IX. METHODS OF PROTEIN PURIFICATION

Protein purification is the primary step to determine the structure of the protein and characterize its function and interactions. It is a technique that involves isolation of a single type of protein from indigenous samples or a crude mixture. Proteins can be separated from one another according to its size, charge, solubility, binding ability and biological activity. The most widely used and effective method for protein purification is using different chromatographic techniques. The basic principle of chromatography is the fluid phase (mobile phase) containing protein molecules. The solution containing a mixture of proteins is passed through a packed column containing various solid materials (stationary phase). Based on the physicochemical properties of the various proteins, they can be separated by either the time required to pass through the column or different conditions required to elute the protein from the column. The isolated proteins can be detected by measuring their absorbance at 280 nm.

The three most commonly used chromatographic techniques are gel filtration, ion exchange and affinity chromatography.

Gel-Filtration Chromatography

Gel filtration chromatography separates protein molecules based on their size. The stationary phase consists of porous beads made up of an insoluble but highly hydrated natural polymers such as dextran and agarose or synthetic polymers such as polyacrylamide that span a relatively narrow size range. Gels of different pore sizes can be formed from these polymers by cross-linking to form a three-dimensional network.

In general, proteins of a relatively smaller size can enter and leave the gel beads repeatedly, spending more time inside the beads, compared to the larger protein molecules which will not enter the gel pores because a smaller volume is accessible to them. Thus, large proteins will elute rapidly from the column in the void volume unlike the smaller proteins. Hence, proteins will elute from the column in the order of decreasing molecular size.

The elution time of a particular protein is linearly dependent on the logarithm of its molecular mass, within the range of mass that can be fractionated by the specific stationary phase. The advantage of gel filtration is high retention of the biomolecular activity of the protein molecules due to its gentle non-adsorptive interaction with the sample.

Ion Exchange Chromatography

Ion exchange chromatography technique separates protein molecules based on their net charge. Separation depends upon the reversible adsorption of the charged protein molecules to oppositely charged moieties of the immobilized phase. An ion exchanger consists of an insoluble matrix to which charged groups are covalently bound. Depending on the functional ionic groups of the stationary phase surface, this type of chromatography is further subdivided into cation exchange chromatography and anion exchange chromatography. In cation exchange chromatography, positively charged groups of the protein molecules are attracted to negatively charged groups of the solid matrix. On the other hand, in anion exchange chromatography, negatively charged molecules interact with the positively charged solid support. The most common functional groups in anion exchangers are diethylaminoethyl (DEAE), quaternary aminoethyl (QAE), etc. and those of cation exchangers are carboxymethyl (CM) and sulfopropyl (SP). The binding affinities of protein molecules depend on different degrees of interaction with the ion exchanger due to differences in their net charges, charge densities and distribution

of charge on their surfaces, as well as pH and ionic strength of the solvent. The bound protein molecules are eluted by increasing the salt concentration or changing the pH of the solvent. The salt ions compete with the protein for binding sites on the resin. The weakly charged proteins are eluted at a comparatively lower salt concentration than the highly charged proteins which require higher salt strength for elution.

Affinity Chromatography

Affinity chromatography separates proteins based on their high affinities for specific chemical groups of a ligand coupled to a chromatography matrix. This chromatography is extensively used for separation of proteins based on interactions between enzyme and substrate, antigen and antibody, receptor and ligand, etc. The ligand or its derivative to which the desired protein binds is covalently attached to the stationary phase. Hydrophilic and neutral matrices are preferred for affinity chromatography to hinder nonspecific interaction of proteins with the gel matrix itself. Biological interactions between ligand and protein can be owing to electrostatic or hydrophobic interactions, van der Waals forces and hydrogen bonding. The functional group of the ligand used for coupling should not be essential for its binding to the protein molecules so that its specific binding affinity is retained.

The protein can be eluted specifically using a competitive ligand with a solution containing a high concentration of the ligand, which can compete with the bound ligand for binding sites on the protein or non-specifically by changing the pH, ionic strength or polarity which decreases the binding affinity of the protein and the ligand.

SUMMARY

Proteins are biological macromolecules which are responsible for a variety of biological processes in the living organisms. They are linear long chain polymers and are made up of 20 standard amino acids. The protein chain is also known as a polypeptide chain. Amino acids, the building blocks of proteins have different chemical and structural properties. Different proteins contain different sequences of amino acids. The knowledge of the sequence of amino acids in a polypeptide chain is known as the primary structure of a protein. A polypeptide chain without the side chains is known as the backbone of a protein. Based on the natures of atoms and bonds in the backbone, the protein backbone can adopt stable helical and extended structures which are known as secondary structures. In order to acquire the stability, a protein chain folds into a compact three-dimensional shape which is known as its tertiary structure. In order to perform specific functions, proteins acquire different tertiary structures. Proteins can function as monomers, dimers, trimers, etc. The stable structures which contain more than one polypeptide chain are known as quaternary structures.

SUGGESTIONS FOR FURTHER READING

Bezkorovainy A. (1970), *Basic Protein Chemistry*, C.C. Thomas, Springfield.

Blundell T.L. and Johnson L.N. (1976), *Protein Crystallography*, Academic Press, London.

Dickerson R.E. and Geis I. (1969), *The Structure and Action of Proteins*, Harper and Row, New York.

Eisenhaber F., Persson B. and Argos P. (1995), Protein Structure Prediction: Recognition of Primary, Secondary and Tertiary Structural Features from Amino Acid Sequence, *Crit. Rev. Biochem. Mol. Biol.*, 30, 1–94.

Murzin A.Z. (1996), Structural Classification of Proteins: New Superfamilies, *Curr. Opin. Struct. Biol.*, 6, 386–394.

Neurath H. (Ed), *The Proteins: Composition, Structure and Function*, Academic Press, New York, 1963–6, Volumes I-IV.

Sherwood D. and Cooper J. (2011), *Crystals, X-rays and Proteins*, Oxford University Press.

4

X-ray Crystallography and Protein Structure

M. Vijayan and K.V. Abhinav

I. INTRODUCTION

Properties of materials and functions of molecules depend upon their structures. To take a simple example, diamond and graphite have entirely different properties because they have different three-dimensional structures (arrangement of atoms in them), although both are made up of carbon (Figure 4.1). To take another example from a different domain, the function of hemoglobin, which picks up molecular oxygen from lungs and releases it in tissues, and brings back carbon dioxide, is dependent on the molecular structure of the protein. Therefore, it is important to know the structure of materials, whether they are inorganic, organic or biological. For about a century, the method of choice for determining the structure of matter at the atomic level has been X-ray crystallography.

X-rays were discovered by Roentgen in 1895, for which he was awarded the Nobel Prize in 1901. We now know them as highly penetrating electromagnetic radiation. Also, we know about the medical applications of X-rays such as in radiology and CT scan. Their utility for determining the structure of matter was first realized when von Laue and his colleagues in Germany demonstrated in 1912 that crystals diffract X-rays. However, X-ray crystallography, as the discipline involving the study of structure of matter using diffraction of X-rays by crystals is called, was nurtured in the early stages by William Bragg and Lawrence Bragg, the father and son team, in England. Laue and Braggs were subsequently awarded the Nobel Prize for their efforts in the field. Lawrence Bragg was 25 years old when he received the Nobel Prize in 1915 and he has remained the youngest person to have received the Prize. The first structure to be determined using X-ray crystallography, by Lawrence Bragg, was that of sodium chloride. Over the past century, hundreds of thousands of crystal structures have been determined. A number of Nobel Prizes have been awarded to many others during the past few decades for work involving X-ray crystallography.

II. CRYSTALS: INORGANIC, ORGANIC AND BIOLOGICAL

(a)   (b)

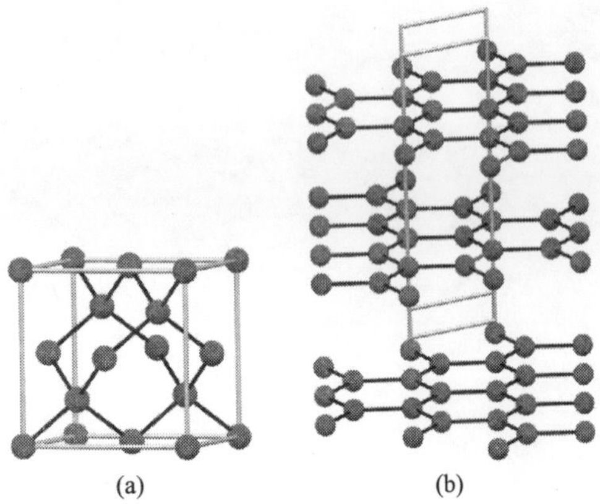

Figure 4.1 The structures of diamond (a) and graphite (b). [The spheres represent carbon atoms.]

Matter can normally exist as gas, liquid or solid. In the gaseous state, the average distance between particles, which could be an atom, a molecule or a collection of molecules, is much higher than the dimensions of the particle itself, and therefore the interaction energy between particles is low. In a liquid, obtained, say, by lowering the temperature of a gas, the particles are in contact with one another, but the interaction energy is not high enough to prevent their movement. In the solid state, the interaction energy is high enough to prevent motion except for thermal vibrations. The particles then have fixed positions in space. In such a situation, the total interaction energy tends to be a minimum when the particles are arranged in a three-dimensional periodic fashion. Other things being equal, a system prefers an arrangement in which the energy (strictly speaking, free energy) is a minimum. A region of space where particles are arranged in a three-dimensional periodic fashion is called a crystal (Figure 4.2). Thus, crystallinity is a property of matter in the solid state. Some materials, such as common salt or sugar, form single crystals. Some others like a piece of metal is a conglomeration of small crystals. The translational periodicity of a crystal can naturally be specified by three independent vectors.

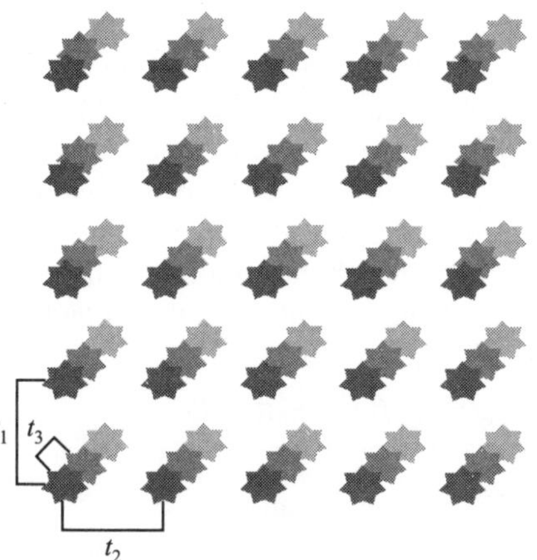

Figure 4.2   The periodicity of a crystal. [The star represents the motif which repeats (in principle) infinitely in space.]

The particles are tightly packed in normal inorganic and organic crystals. In fact, the arrangement of ions or molecules in them is substantially determined by considerations of close packing. The situation is different in the crystals of biological macromolecules like proteins (Figure 4.3). About 50% of protein crystals are typically occupied by the aqueous solution from which the protein is crystallized. The protein molecules and water in them are in equilibrium. In ambient conditions, protein crystals collapse if water in it is allowed to evaporate. In fact, the level of crowding in a protein crystal is similar to that in a typical cell, except that a crystal is usually made up of a single species of molecules or molecular assemblies, while a cell contains a large number of different types of molecules.

III.   METHODOLOGY

Diffraction of X-rays by crystals can perhaps be easily understood in comparison with that of visible light by a

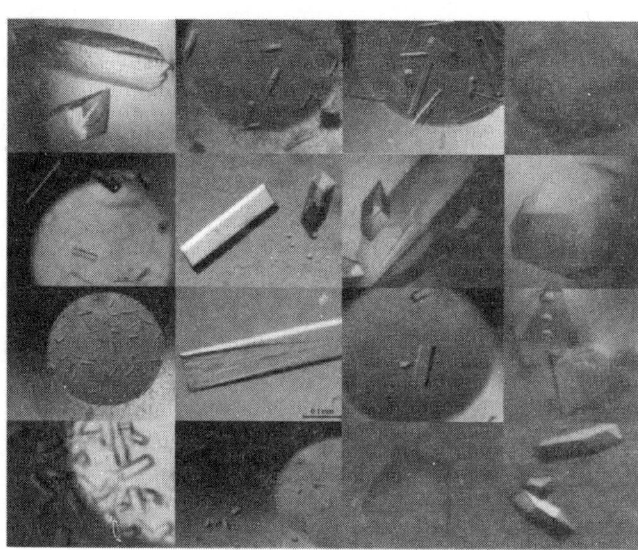

Figure 4.3   A collage of protein crystals.

transmission optical grating. In simple terms, an optical grating is a rectangular glass plate of a few centimeters in size on which equally spaced grooves or rulings are made, typically with a few hundred grooves per millimeter [Figure 4.4(a)]. The distance between adjacent grooves or the periodicity of the grating may be designated as t. We can take a simple case where monochromatic light, of wavelength λ, is incident on the grating in a direction perpendicular to the plate. Each groove scatters the incident radiation (scattering is the fundamental process. The result of the scattering is called reflection, refraction, interference, diffraction, etc. depending on the circumstance). Then, the path difference between radiation scattered by adjacent grooves in a direction θ is $t \cos\theta$ [Figure 4.4(b)]. A diffraction maximum results when n is an integer in

$$t \cos\theta = n\lambda \qquad \ldots(1)$$

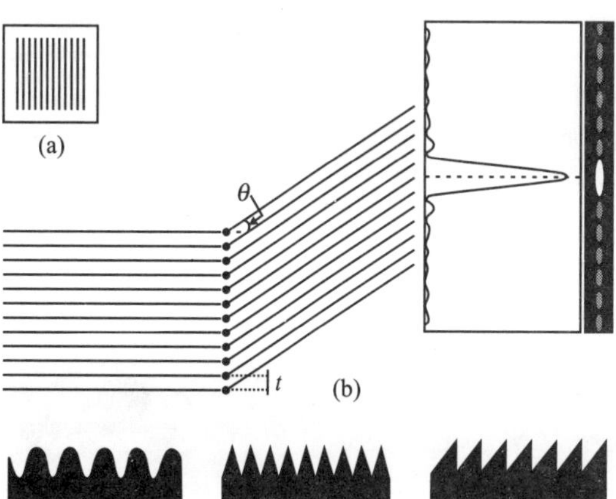

Figure 4.4   (a) Schematic representation of an optical grating. Each line represents a groove (b) Diffraction by a grating. Each dot represents the cross section of a groove. The view is perpendicular to that shown in (a) (c) Typical shapes of grooves.

Thus, for a given wavelength of the incident radiation, the angular positions at which diffraction maxima occur are governed by the periodicity of the grooves in the grating. Diffraction maxima are referred to as first-order maximum, second-order maximum, etc when n = 1, 2, etc., respectively. The distribution of intensity among diffraction maxima is governed by the "structure" of the groove [Figure 4.4(c)]. Thus, to sum up, the angular positions at which diffraction maxima occur are determined by the periodicity of the grating, while the intensities are determined by the structure of the repeating motif (groove). Conversely, the periodicity of the grating can be deduced from the experimentally observed angular positions of the diffraction maxima, while the structure of the motif can be determined by making use of the intensities (strictly speaking, amplitudes) of the maxima.

X-ray diffraction by crystals is analogous to diffraction of light by a grating, except for a couple of specific differences. A grating has one-dimensional periodicity, whereas crystals have three-dimensional periodicity. Therefore, in the latter case, the following equations similar to (1) have to be simultaneously satisfied for diffraction maxima to occur:

$$\left.\begin{array}{l} t_1 \cos\theta = h\lambda \\ t_2 \cos\theta = k\lambda \\ t_3 \cos\theta = l\lambda \end{array}\right\} \qquad ...(2)$$

where t_1, t_2 and t_3 are the three periodicities of the crystal and h, k and l are integers. The Laue equations given in (2) are represented in terms of Bragg reflections and also re-interpreted in reciprocal space. However, the three equations state the basic condition for diffraction maxima to occur. When recorded, the diffraction maxima produced by an optical grating appear as lines, whereas those produced by a crystal appear as spots (Figure 4.5). Typically, the X-ray diffraction pattern from an organic crystal consists of several hundreds or thousands of maxima or reflections, while that from a protein crystal is made up of thousands or tens of thousands of reflections. The wavelength of light is a few thousand angstrom units (Å) and the periodicity of an optical grating is normally around 10,000 Å or more. The wavelength of X-rays used for diffraction studies is usually around 1 Å, while periodicities in crystals range from a few angstroms to a few hundred

angstroms. Furthermore, the repeating motif in a grating is a groove, whereas that in a crystal, it is one or a collection of atoms, ions or molecules. Again, in a manner analogous to the case of a grating, the angular positions at which diffraction maxima occur are determined by the periodicity of the crystal, whereas the structure of the motif (roughly speaking, the molecular structure in the case of biomolecules) can be deduced from the intensities of diffraction maxima.

The way the structure of a molecule is elucidated can be explained in terms of an optical analogy. In an optical microscope, the light scattered by the object in the form of wavelets is collected by the objective lens to form the image of the object with a suitable magnification (Figure 4.6). The lens combines the wavelets, taking into account their amplitudes and phase angles. There are no X-ray lenses. Therefore, the wavelengths resulting from diffraction are combined using a Fourier summation of the form

$$\rho(X, Y, Z) = \frac{1}{V}\Sigma\Sigma\Sigma \mathbf{F}_{hkl}\, e^{2\pi i(hX + kY + lZ)} \qquad ...(3)$$

where $\rho(X, Y, Z)$ is the electron density at position X, Y, Z in the unit cell (the parallelepiped enclosing the repeating motif), V is the volume of the unit cell and $\mathbf{F}_{hkl}$ is the amplitude, with an appropriate phase angle, of the diffracted wavelet specified by induces h, k and l defined in equation (2). Peaks in the electron density map represent the positions of atoms (Figure 4.7). If the positions of atoms which constitute the structure are known, $\mathbf{F}_{hkl}$ can be readily calculated using the formula

$$\mathbf{F}_{hkl} = \sum_{j=1}^{N} f_j \exp\{2\pi i(hx_j + ky_j + lz_j)\} \qquad ...(4)$$

where f_j is the scattering amplitude of X-rays from atom j and x_j, y_j and z_j are the coordinates of the same atom. $\mathbf{F}_{hkl}$, called the structure factor, can also be written as

$$\mathbf{F}_{hkl} = \mathbf{F}_{hkl}\, e^{i\alpha hkl} \qquad ...(5)$$

where F_{hkl} is the magnitude of the amplitude of the wavelet specified by h, k, l and αhkl is its phase angle. F_{hkl} is the square root of the intensity of the diffraction maximum specified by h, k, l.

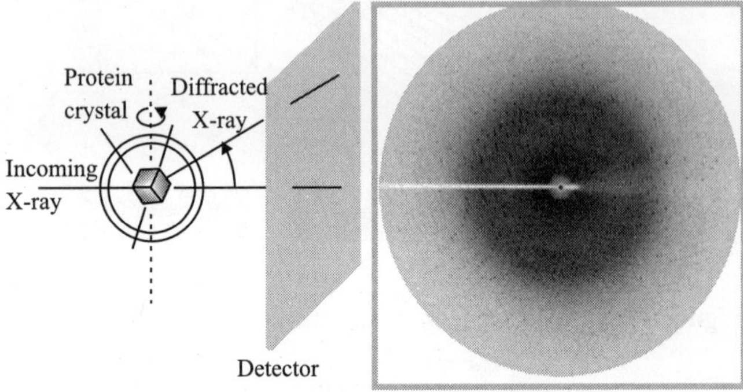

Figure 4.5   Diffraction from a crystal along with a diffraction image.

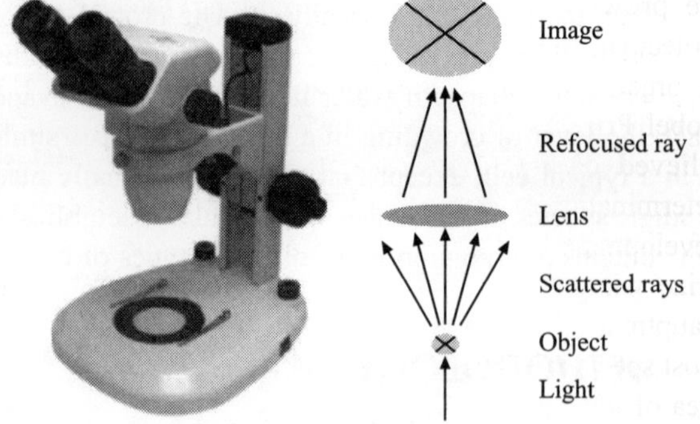

Figure 4.6   Optical microscope and a schematic representation of image formation.

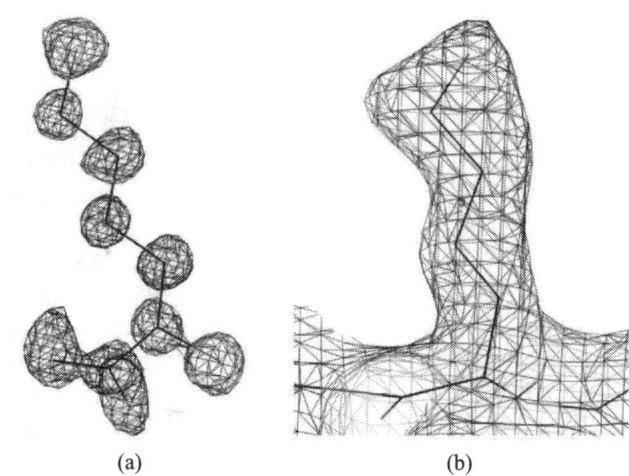

(a)   (b)

Figure 4.7 (a) Electron density at 0.8 Å resolution for lysine in the small molecular structure of L-lysine D-tartrate. (b) Density for a lysyl residue at 1.9 Å in *Mycobacterium* tuberculosis peptidyl-tRNA hydrolase.

The intensities can be measured, and therefore $\mathbf{F}_{hkl}$ can be readily obtained. However, the phase factor is lost in measurement. Therefore, the magnitude of the amplitudes of diffracted wavelets can be experimentally determined. However, to compute the electron density using (3), we need to know the phase angles as well. This is the well-known phase problem in crystallography. All the methods of structure determination in crystallography are methods for deducing phase angles. The methods for phase determination in biological macromolecular crystallography include the multiple isomorphous method, the molecular replacement method and anomalous dispersion methods. Once the phase angles are known, the electron density, and hence the structure, can be readily calculated using (3).

IV. BRIEF HISTORY

The emphasis was on the structure determination of inorganic compounds during the early days of crystallography. The emphasis shifted to organic compounds to an extent in the 1920s and substantially from the 1930s. The structure determination of cholesterol, penicillin and vitamin B_{12}, particularly vitamin B_{12}, in the 1950s and the 1960s, by Dorothy Hodgkin, defined the prowess of X-ray crystallography in determining the molecular structure and indeed often the molecular formula of organic compounds. Dorothy Hodgkin was awarded the Nobel Prize for her efforts in 1964. It is often said that she relieved organic chemists from the drudgery of structure determination of molecules. The process was completed by the development of the so-called direct methods, another Nobel Prize winning effort, by Jerome and Isabella Karle, Herbert Hauptman, David Sayre, Michael Woolfson and others. The most spectacular applications of crystallography now are in the area of molecular biology. Macromolecular crystallography is central to structural molecular biology and indeed to modern biology itself.

The beginning of macromolecular crystallography can be traced to the preliminary X-ray studies on pepsin crystals in 1934 by J.D. Bernal and his then student Dorothy Crowfoot (subsequently Hodgkin) at Cambridge. That was a time when even the nature of proteins was not completely known. It was an act of courage and foresight on the part of pioneers like Bernal to have embarked on crystallographic studies on proteins. After obtaining her doctorate, Dorothy Crowfoot returned to Oxford and the first independent problem she took up in 1935 was the preliminary studies on insulin crystals. In the meantime, an Austrian refugee, Max Perutz, joined J.D. Bernal and published preliminary diffraction studies on hemoglobin crystals in 1937. A few more preliminary macromolecular crystallography studies were carried out during that period, but it took a long time for definitive structure determinations to emerge.

In the meantime, diffraction studies of fibrous proteins were being pursued by W.T. Astbury and others. Results of such studies and molecular modeling, independently or together, began to yield major results. Models of the α-helix and the β-sheet were proposed by Linus Pauling and his colleagues in early 1950s. The celebrated double-helical structure of DNA was proposed by Francis Crick and James Watson in 1953. This was closely followed by the elucidation of the triple helical coiled coil structure of collagen in 1955 by G.N. Ramachandran and Gopinath Kartha (Figure 4.8). With that, the basic structure of the three classes of fibrous proteins was established.

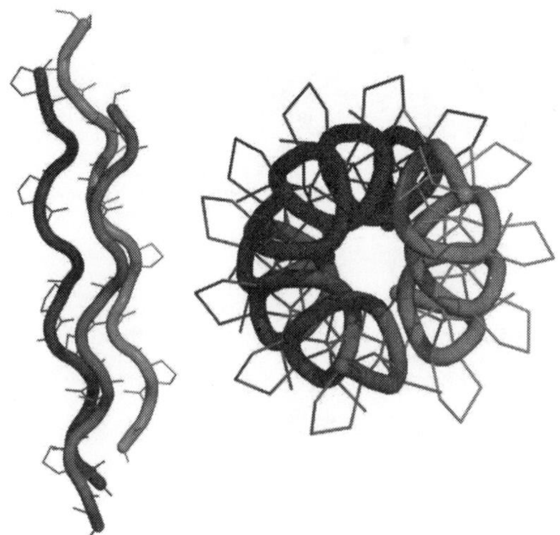

Figure 4.8 Perpendicular views of the structure of collagen.

V. MACROMOLECULAR CRYSTALLOGRAPHY IN FULL BLOOM

It was only by 1960 that the first structures of globular proteins, those of myoglobin and hemoglobin, became available (Figure 4.9). Their structures fetched the Nobel Prize to John Kendrew and Max Perutz. The third protein and the first enzyme to be X-ray analyzed, by David Phillips and his colleagues at the Royal Institution, London, was lysozyme (Figure 4.10).

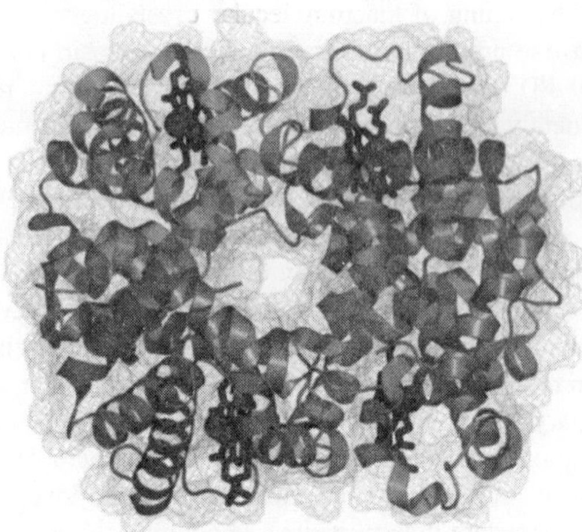

Figure 4.9 Cartoon representation of the structure of hemoglobin. The α- and (β-chains are given in different colors. The locations of the heme and the oxygen molecules are also indicated (Protein Data Bank (PDB) ID: 1GZY).

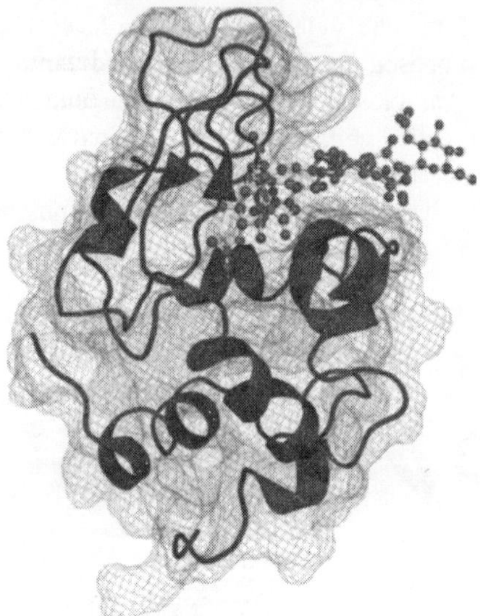

Figure 4.10 Structure of lysozyme along with bound hexa–NAG (PDB ID: 132L)

The next protein structure to emerge was that of ribonuclease A through the efforts of Gopinath Kartha and his colleagues at Buffalo, USA. The s tructures of many proteolytic enzymes subsequently became available in the 1960s. The much waited structure of insulin became available from Dorothy Hodgkin's laboratory in Oxford by the end of the decade (Figure 4.11).

Efforts to derive general principles governing protein structures also began in the 1960s. The Ramachandran map, proposed from the Madras University in 1963, is an outstanding result that emanated from these efforts. The mutual orientation of two adjacent planar peptide units is determined by two torsion angles, (φ and ψ about the N–C^α and C^α–C bonds involving the α-carbon which connects the two peptide units [Figure 4.12(a)]. Thus, the main chain conformation of a polypeptide chain can be completely specified by the φ, ψ angles at each α-carbon. On the basis of available information on the minimum distances to which different pairs of non-bonded atoms (C...C, C...N, N...O, etc.) can approach each other, Ramachandran and his colleagues devised a map indicating the limits of the values of φ, ψ. This map is widely used to validate protein structures and describe them in the simplest terms [Figure 4.12(b)]. Many other matrices have been devised to describe protein structure, but none of them has superseded the Ramachandran map in its simplicity and effectiveness.

From the 1970s, macromolecular crystallography has registered steady progress across the world. The progress has been explosive during the past couple of decades, propelled by several technological and scientific advances. Synchrotron facilities and position-sensitive detectors made intensity data collection fast and efficient. Computer power grew by leaps and bounds. There has been remarkable progress in software development as well. Macromolecular crystallography was in step with the developments and sometimes outpaced molecular biology and biochemistry. A major problem in crystallography was the requirement of large samples. This was obviated by recombinant technology. Molecular biology techniques enabled structural biologists to modify residues or regions of proteins at will and examine the structural and functional consequences

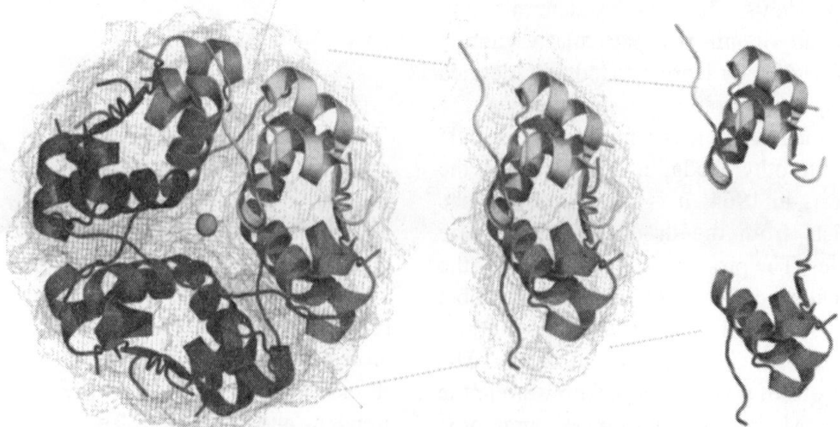

Figure 4.11 Structures of an insulin hexamer, dimer and monomer (PDB ID: 1MSO). Insulin is stored as zinc bound hexamers in the pancreas. The predominant species that circulate in the blood is a dimer. Insulin binds to its receptor as a monomer.

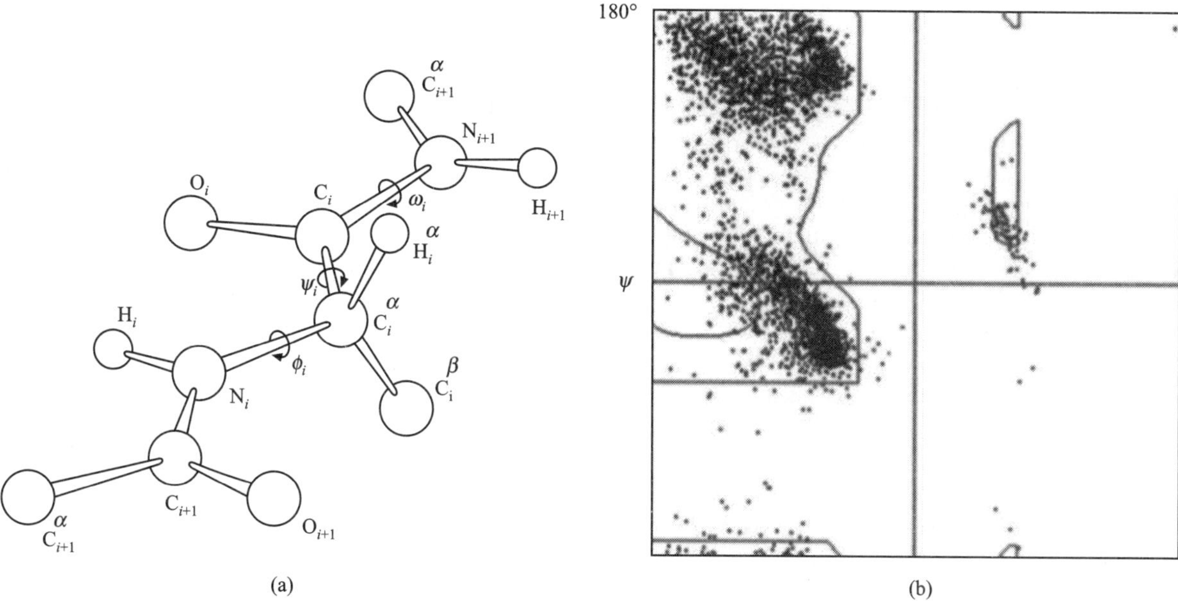

Figure 4.12 (a) Two planar peptide units hinged at the *i*th α-carbon atom. Reproduced with permission from *Resonance* pp 48-56, October 2001 (b) Ramachandran map generated from 21 selected high resolution structures. Open circles correspond to glycine residues. Reproduced from *Biog. Mem. Fell. R.Soc* **51**, 367-377 (2005).

of such modifications. Availability of genome sequences gave rise to major structural genomics or proteomics programmes involving the X-ray analysis of cassettes of large numbers of proteins. X-ray crystallography has also been used to understand large macromolecular assemblies, the largest so far being the ribosome (Figure 4.13). Ada Yonath, Tom Steitz and Venki Ramakrishnan were awarded the Nobel Prize for the structural solution of ribosome.

The structure solution of a protein molecule involves the determination of the positional coordinates x, y, z and a displacement parameter B, which indicates both the amplitude of vibration and the positional disorder of each atom in it. Upon solution of the structure and before the publication of the results, these coordinates are expected to be deposited in Protein Data Bank (PDB). From a handful in the 1970s, the number of accessions in the PDB now amounts to several tens of thousands. As of 2 October 2012, 78,760 sets of coordinates have been deposited in the PDB, of which 69,845 have resulted exclusively from X-ray crystallographic studies. There are 8411 sets produced using NMR and 504 resulting from other approaches. The 78,760 sets include the coordinates of mutants, other minor variants, ligand complexes, etc. of the same protein. If two proteins are treated as essentially the same when they have a sequence identity of 90% or more between them, the number of sets drastically reduces to 34,138. A reasonable estimate of the number of independent proteins with known three-dimensional structure can be obtained if the cut-off value is reduced to 30%. At this cut-off value, the number reduces to 19,485. This number represents only a small fraction of the total number of independent proteins likely to be available in nature. The fact that a large proportion of open reading frames remains to be annotated bears testimony

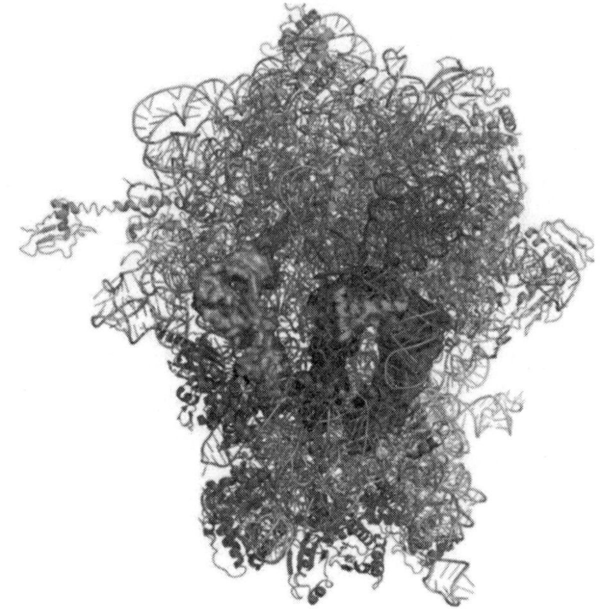

Figure 4.13 Structure of prokaryotic ribosome. (PDB IDs: 2WDK, 2WDL). RNAs in 30S and 50S subunits are shown in slate and orange, respectively. The protein components in the subunits are in blue and green, respectively. 5S RNA of the large subunit in is purple. Aminoacyl tRNAS are in surface representation. Those at the A site, the P-site and the E-site are at the right, in the middle and at the left, respectively. (*see Plate 1 for colour figure*)

to this statement. Furthermore, much still needs to be done even on most proteins of known three-dimensional structure for elucidating the structural basis of their function and biological role. Thus, macromolecular crystallography is a vibrant field full of promise. At the same time, the results that have already been obtained have had a great impact on biology. There is hardly any aspect of modern molecular biology which has not been illuminated by the results of X-ray crystallography. The impact of crystallography on medicine has been substantial. For instance, the leads for almost all the known drugs for AIDS and influenza have emanated from crystallographic studies. Structure-based design has now come to stay as an important component of drug development. The impact of crystallography is felt in many other areas of medicine, such as vaccine design and therapeutic proteins. Thus, X-ray crystallography is central to modern biology and medicine and it is likely to remain so for a long time to come.

SUMMARY

Properties of materials and functions of molecules depend upon their structures. The method of choice for determining structure at the atomic level is X-ray crystallography. The historical development of the subject and the basic principles of the methodology have been dealt with. The most spectacular applications of crystallography in recent times have been on biological macromolecules, particularly proteins and their assemblies. The progress of macromolecular crystallography from its early days is traced and the current status of work in the area is outlined. Macromolecular crystallography today impacts almost all aspects of modern biology and is an essential component of biological research at the molecular level.

SUGGESTIONS FOR FURTHER READING

Books

Blundell T.L., Johnson L.N. (1976), *Protein Crystallography*, Academic Press.

Branden C. and Tooze J. (1999), *Introduction to Protein Structure*, 2nd edition, Garland Science.

Liljas A. (2009), *Textbook of Structural Biology, World Scientific Publishing Company.*

Rhodes G. (2006), *Crystallography Made Crystal Clear: A Guide for Users of Macromolecular Models*, 3rd edition, Academic Press.

Rupp B. (2009), *Modern Biomolecular Crystallography: Principles, Practice and Application to Sturural Biology, Garland Science.*

Stout G.H., Jensen L.H. (1989), *X-ray Structure Determination: A Practical Guide*, 2nd edition, Macmillan.

Selected Publications

Adams M.J., Blundell T.L., Dodson E.J., Dodson G.G., Vijayan M., Baker E.N., Harding M.M., Hodgkin, DC, Rimmer B. and Sheat S. (1969), Structure of Rhombohedral 2 Zinc Insulin Crystals, *Nature*, 224, 491–495.

Arora A., Chandra N.R., Das A., Gopal B., Mande S.C., Prakash B., Ramachandran R., Sankaranarayanan R., Sekar K., Suguna K., Tyagi A.K., Vijayan M. (2011), Structural Biology of *Mycobacterium tuberculosis* proteins: The Indian Efforts, *Tuberculosis*, 91, 456–468.

Banerjee R., Mande S.C., Ganesh V., Das K., Dhanaraj V., Mahanta S.K., Suguna K., Surolia A., and Vijayan M. (1994), Crystal Structure of Peanut Lectin, a Protein with an Unusual Quaternary Structure, *Proc. Natl. Acad. Sci.*, USA 91, 227–231.

Berman H.M., Battistuz T., Bhat T.N., Bluhm W., Bourne P.E., Burkhardt K., Feng Z., Gilliland G.L., Iype L., Jain S., Fagan P., Marvin J., Padilla D., Ravichandran V., Schneider B., Thanki N., Weissig H., Westbrook J.D. and Zardecki C. (2002), The Protein Data Bank, *Acta Crystallogar D*, 58, 899–907.

Colman P.M. (2009), New antivirals and drug resistance, *Annu. Rev. Biochem.*, 78, 95–118.

Kendrew J.C., Dickerson R.E., Strandberg B.E., Hart R.G., Davies D.R., Phillips D.C. and Shore V.C. (1967), Structure of Myoglobin: A Three-Dimensional Fourier Synthesis at 2 Å Resolution, *Nature* 185, 422–427.

Perutz M.F., Wilkinson A.J., Paoli M., Dodson D.D. (1998), The Stereochemical Mechanism of the Cooperative Effects in Hemoglobin Revisited, *Annu. Rev. Biophys. Biomol. Struct.* 27, 1–34.

Phillips D.C. (1967), The Hen Egg-white Lysozyme Molecule, *Proc. Natl. Acad. Sci.*, USA 57, 483495.

Ramachandran G.N. and Kartha G. (1955), Structure of Collagen, *Nature*, 176, 593–595.

Ramakrishnan V. (2002), Ribosome Structure and the Mechanism of Translation, *Cell*, 108, 557–572.

5

Collagen

R.S. Bhatnagar

CONTENTS

I. INTRODUCTION

The collagens are a closely related group of proteins which together comprise nearly one-third of the total protein mass of the human body. The collagens are present in every tissue and organ in the body and are a major determinant of tissue architecture and tensile strength. In addition, they are involved in numerous regulatory interactions with cells governing cell proliferation, differentiation, adhesion and migration. The many diverse functions of collagen derive directly from their unusual polymer-like structure, a unique conformation and the ability to form highly ordered aggregates under physiological conditions. Complexes of collagen with other extracellular macromolecules impart characteristic properties to tissues as diverse as bone, skin, cartilage and blood vessels. Because of their involvement in the ontogenesis, structure, and function of all tissues in the body, any impairment in their synthesis, degradation and physicochemical characteristics can serve as a focus of pathologic change.

II. GENETIC POLYMORPHISM AND DISTRIBUTION OF COLLAGENS

The collagen present in skeletal structures, skin and tendons is the preponderant species in the body and for a long time, it was thought to be a homogeneous protein. The triple helix of "collagen" was shown to be made up of two identical chains, termed $\alpha 1$, and a distinct third chain, $\alpha 2$. Improved

isolation and analytical techniques revealed the presence of other collagen chains which differed significantly from $\alpha 1$ and $\alpha 2$. The originally described collagen was then assigned the term "Type 1 Collagen" and subsequently discovered collagens assigned Roman numerals to identify them. Some of these collagens are tissue-specific, and others also modulated as a function of the developmental stage of the tissue and pathologic states. The different collagen chains and the tissues in which they are located are summarized in Table 5.1. Collagens Types I, II, III and V are associated with the matrix and are described as interstitial collagens. Type IV collagen is present exclusively in basement membranes.

Genetically distinct collagens differ in their amino acid composition (Table 5.2) as well as their interaction properties. These differences are important because they determine the function of collagen as well as its state in tissues. Thus, Type I collagen usually forms fibrillar arrangements with fibers of high tensile strength. This type of collagen has relatively little carbohydrate associated with the α-chains. Type II collagen forms fibrous aggregates which are sparse in comparison to Type I collagen. This type of collagen has a large amount of carbohydrate covalently bound to the α-chains. Type III collagen unlike Types I and II, has significant amounts of cysteine in the α-chains. This type of collagen is very hard to isolate because it forms strong complexes with other connective tissue components. Type III collagen can form fibrils, but less readily than Type I. Interestingly, Type III collagen has been found only in association with Type I collagen and only in tissues which support an endothelium or epithelium.

TABLE 5.1 Collagen Types and Tissue Distribution

Collagen type	Chain composition	Tissues source
I	$\alpha1(I)_2\,\alpha2$	Bone, cartilage, muscle, skin, tendons, intervertebral disc, annulus fibros, synovium, lung, kidney (stroma), intestine, placenta, nerve, lymph nodes. Dentin, periodontal ligament, papilla and pulp, gingiva.
II	$\alpha1(II)_3$	Cartilage, lung (tracheal and bronchial cartilage), intervertebral disc, nucleus pulposus, bone (growth plate), eye (cartilaginous sclera, vitreous body).
III	$\alpha1(III)_3$	Lung parenchyma, muscle, heart (muscle and valve), skin, umbilical cord, placenta, amnion and chorion, spleen, liver, kidney, lymph nodes, nerve, fibrous sclera, eye (iris, nictitating membrane) tooth (Papilla and pulp), periodontal ligament, gingiva.
IV	$\alpha1(IV)_3$	Basement membranes—alveolar, capillaries, glomerular, skin, liver, placenta, lens capsule, tumors, Descemet's membrane.
V	$\alpha1(V)_2\,\alpha2(V)$ $\alpha1(V)\,\alpha2(V)\,\alpha3(V)$	Basement membranes (as above cartilage, bone, tendon, synovia, muscle, aorta, nerve skin, cornea, lung parenchyma, placenta, uterus, gingiva.
$\alpha1,\,\alpha2,\,\alpha3$	Not known	Cartilage, annulus fibrosus, nucleus fibrosus.
7–S	Not known	Lens capsule, tumor, placenta.

TABLE 5.2 Amino Acid Composition of Some Human Collagen Chains (Residues per thousand amino acid residues

	$\alpha1(I)$	$\alpha2$	$\alpha1(II)$	$\alpha1(III)$	$\alpha1(IV)$
AMINO ACID					
3-hydroxyproline	—	—	—	—	9
4-hydroxyproline	107	103	108	125	97
Aspartic Acid	45	44	42	42	54
Threonine	17	18	20	13	25
Serine	31	30	28	39	36
Glutamic Acid	78	74	91	71	85
Proline	116	111	105	107	62
Glycine	335	340	330	350	326
Alanine	108	96	99	96	41
Valine	21	28	20	14	29
Cysteine	0	0	0	2	8
Methionine	7	5	10	8	13
Isoleucine	10	16	11	13	24
Leucine	22	30	26	22	55
Tyrosine	1	3	1	3	13
Phenylalanine	10	9	13	8	27
Hydroxylysine	10	10	22	5	39
Lysine	27	21	17	6	16
Histidine	4	10	2	30	10
Arginine	48	47	51	46	31

The composition of human lung $\alpha1(I)$ and $\alpha2$ chains summarized from Seyer, et al. *J. Clin. Invest.* **57**, 1498 (1976). $\alpha1(II)$ from human nucleus pulposus, from Osebold and Pedrini, *Biochem. Biophys. Acta,* **434**, 390 (1976), $\alpha1(III)$ from Chung and Miller, *Science,* **183**, 1200 (1974) and human kidney $\alpha1(IV)$ from Trelstad and Lawley, *Biochem. Biophys. Res. Comm.,* **76**, 326 (1977).

Type IV collagen is the collagen of basement membranes. It is very closely associated with non-collagenous proteins and proteoglycans and can be isolated and solubilized only with great difficulty. Amino acid analyses show that Type IV collagen is strikingly different from other collagen types. It contains greater amounts of hydroxyproline, hydroxylysine and cysteine than the other collagens. Type IV collagen contains 3-hydroxyproline in addition to 4-hydroxyproline, which appears in all collagens (Figure 5.1). Type IV collagens from basement membranes of different organs may have different composition. Type V collagen is a minor component of tissues in which it has been found, and it has not been well characterized. There are many similarities between Types IV and V collagens including a low alanine content and a high hydroxylysine content. The hydroxylysines in both Types IV and V collagens are highly glycosylated. In addition to Types I–V collagens, numerous minor collagenous proteins have been reported. They have not been well characterized and their function remains undefined.

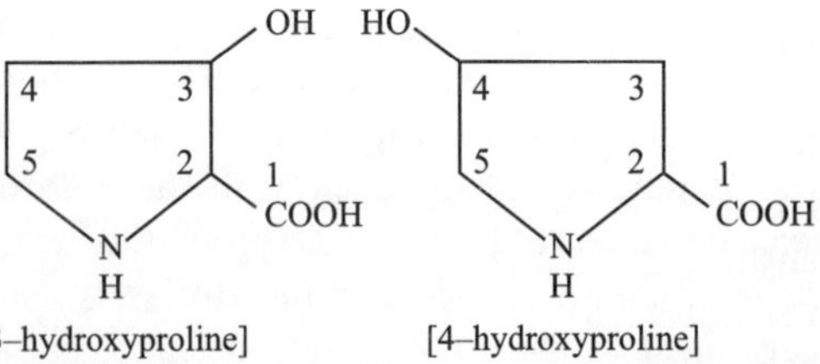

Figure 5.1 Structural formulae of 3- and 4-hydroxyproline.

The characteristic mechanical properties of a given tissue can be correlated with its collagen chain composition. The chain composition varies with age and with disease. Type III collagen appears in larger amounts in fetal and developing tissues, but in adult tissues, Type I collagen is the predominant species. The chain composition is often altered in diseased tissues. In injured or regenerating adult tissues such as skin, lungs, and in connective tissue in fibrotic liver, Type III collagen synthesis is enhanced, consistent with other observations which suggest that regenerating tissues go through stages similar to fetal development. In many diseased conditions an unusual collagen species characterized by triple-helices which are made up of three $\alpha 1$ (I) chains, is observed. The biological role of this collagen $\alpha 1$ $(I)_3$, if any, is not known. It may be speculated that such collagen may be more susceptible to proteolysis and therefore undergo turnover at higher rates; or it may have aberrant physicochemical properties which may interfere with normal function and association with other structural macromolecules. Although normal cartilage is characterized by Type II collagen, $\alpha 1$ $(II)_3$, which is relatively resistant to the tissue's endogenous collagenase activity, arthritic cartilage appears to synthesize Type I collagen $\alpha 1(I)_2 \alpha 2$, which is degraded by collagenase. This offers one possible explanation for cartilage destruction in arthritic disease.

III. MORPHOLOGICAL CHARACTERISTICS

In the body, Types I, II and III collagens always occur in the form of fibers of varying dimension, Type I forming the largest fibers. Large fibers presumably of Type I collagen appear to the naked eye as glistening pearly white strands. Under the high power of the microscope or at low level magnification in the electron microscope they appear as beaded thread-like structures. When stained with appropriate electrondense materials, the fibers exhibit a highly regular, transversely banded pattern with a periodicity or repeat distance, D, of 68 nm (Figures 5.2 and 5.3). This axial repeat is nearly a constant not only within the same organism but also across all phyla, as seen in the electron micrographs of collagen fibers derived from the spine of sea urchin, an invertebrate, rat tail tendons, or from human liver (Figure 5.2). Collagen fibers are essentially cylindrical in cross-section, with diameters ranging from 0.4 to 400 nm (4 Å to 4000 Å). Fibers in various tissues are present in bundles, or interwoven in different patterns appropriate to the properties of the tissue and consistent with its mechanical function. Skin derives its elasticity and resilience from a felt-like network of Type I collagen fibers. In contrast, the rigidity and the ability to bear high impact and mechanical stress in bone arises from the presence of collagen in complexes with proteoglycans and hydroxyapatite crystals. Orthogonal ply-like arrangements in the wall of blood vessels and parallel fibers in sheets with the fiber axes in adjacent sheets, at right angles, help in maintaining a firm control on shape and provide elasticity at the same time.

The fibers are highly ordered aggregates of the individual molecules of collagen which are rigid, rod-like structures, each approximately 300 nm long and 1.5 nm in diameter. The 68 nm axial repeat (D) is generated by the staggering of the molecules along their long axis. The length of the molecule is equal to 4.4 D. A two-dimensional arrangement in which adjacent molecules are staggered by the repeat distance D, was suggested by Petruska and Hodge (Figure 5.3). This type of arrangement led to a structure with gaps of some 30 nm between ends of the molecules and an overlap region. This explains the one light and one dark band seen in low power electron micrographs. The two-dimensional re-presentation

explains the banded appearance of the fibers observed in electron micrographs, but does not fully account for the three-dimensional packing of collagen molecules into fibers. It has been suggested that four to six collagen molecules may associate into staggered, covalently stabilized assemblies called limiting microfibrils (Figure 5.4). The limiting microfibrils aggregate to give rise to the fibrils. The mode of packing of the limiting microfibrils into fibers has not been fully elucidated. Lateral and end-to-end aggregates of the limiting microfibrils may give rise to thick fibrils or ribbon-like structures which may further aggregate to give rise to fibers, in a packing arrangement similar to multistranded ropes.

The fiber structure is stabilized by the formation of covalent crosslinks between adjacent collagen molecules. The tensile strength of the fibers is dependent on the formation of these cross-links. Mature collagen fibers are often found to be covered with a protective coat of proteoglycan molecules and the interactions between collagen fibers and other extracellular macromolecules contribute to the stabilization of fiber structure.

IV. STRUCTURE OF THE COLLAGEN MOLECULE

The collagen molecule is an elongated rod like structure, consisting of three similar polypeptide chains, called the α-chains, packed in a coiled-coil arrangement (Figure 5.5). The unique conformational properties of this structure give it unusual rigidity. Each α-chains consists of approximately 1050 residues, nearly one-fourth of which are imino residues. Because of the predominant stereochemical influences of the iminopeptide bond (Figure 5.6) each α-chain has a left-handed helix related to poly-proline. This helix is characterized by three amino acid residues per turn, and the residues are separated by 0.29 nm; one complete turn of the helix has a height of 0.93 nm. These dimensions indicate that the peptide back bone is more extended when compared to the α-helical conformation found in other proteins. (The α-helix has 3.6 residues per turn, a mean residue distance of 0.15 nm and the height of a turn is 0.54 nm). The three chains in the collagen molecule are coiled around each other in a right-handed triple helix. The three chains are parallel to each other and they are equidistant from the axis of the triple helical molecule. The pitch of the triple helix is approximately 10.4 nm.

The conformational characteristics of collagen are directly related to its unusual primary structure. Except for very small regions called telopeptides at the N-terminal and the C-terminal ends, each chain contains glycine in every third position. When the three chains are coiled together, all the glycine residues line up along the major axis. This in effect locks the three chains together and contributes to the stability of the triple helix.

Another conformation stabilizing influence in collagen is derived from the stereochemical features of the imino peptide bonds (Figure 5.6). Peptide links involving proline and hydroxyproline have severe limitations on rotation because of the rigid and bulky pyrrolidine ring astride the peptide bond,

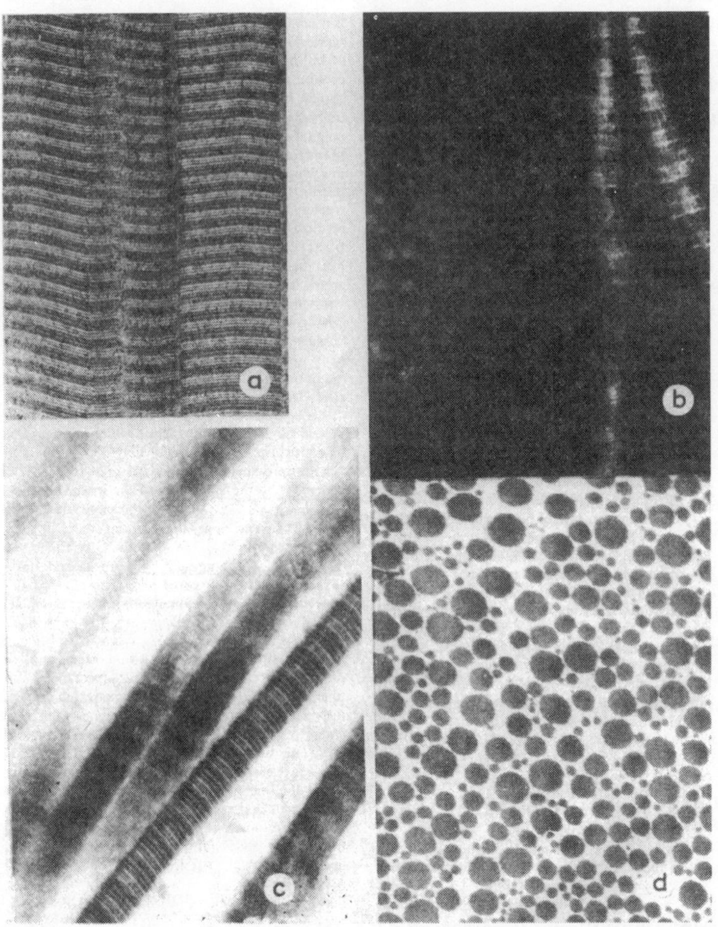

Figure 5.2   Collagen fibres as seen in the electron microscope. (a) Rat tail collagen stained with uranyl acetate and lead citrate (× 63,500). (b) Human liver collagen. Stained with phosphotungstate (× 67,200). (c) Collagen from the base of spine of sea urchin (Stronglocentratus purpurators) (× 95,550). (d) Cross-section through a rat tail tendon, showing the circular profile of the fibres (× 27,500). Electron micrographs including Figure 5.3., courtesy of Ms. Lynne Calonico, University of California, San Francisco.

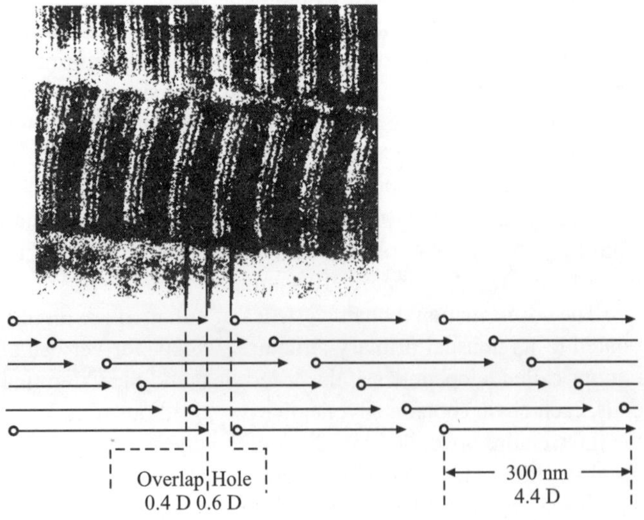

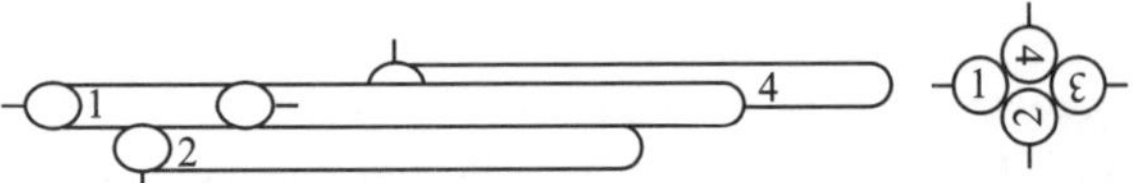

Figure 5.3   Two-dimensional representation of packing of collagen molecules into fibres. This explains the 68 nm periodic banding pattern, however, it does not explain three-dimensional fiber structure.

Figure 5.4   Schematic representation of three- dimensional packing of individual collagen molecules into a limiting microfibril. Four or more molecules are assembled into the limiting microfibril such that starting with molecule 2, each successive molecule is axially rotated by 90° and staggered by 1 D (68 nm) with respect to the preceding molecule. This allows each molecule to retain identical additional interaction sites for interacting with molecules in adjacent limiting microfibrils. Axial and radial addition of limiting microfibrils gives rise to fibrils and fibers.

and the stereochemical interactions of the pyrrolidine ring with the neighboring residue on its N-terminal side. These restrictions characterize and determine the conformation of the collagen chain. The interactions between imino residues in different chains in the triple helix also add stability to the triple helical conformation.

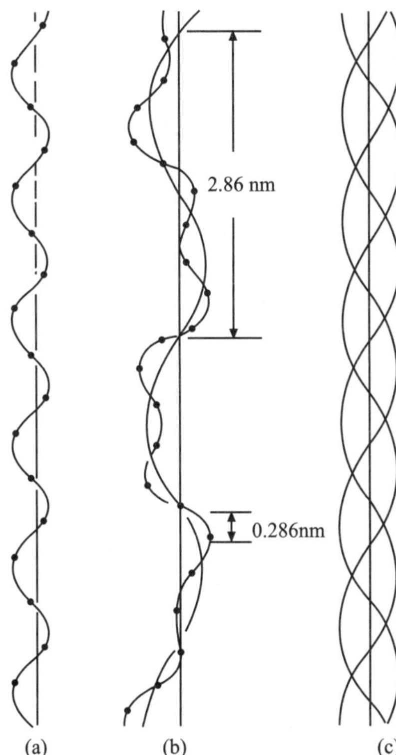

Figure 5.5 Arrangement of three α-chains in a collagen molecule. (a) Single α-chain, showing a left handed polyproline II helix, with a pitch of 0.9 nm about its axis. The points on the helix represent amino acid residues, every third one being glycine. (b) Packing of an α-chain in the collagen molecule. The axis in the individual chains are wound around the triple helix axis in a helix with a pitch of 2.86 nm. There are ten residues per turn, that is, the inter-residue distance along the triple helical axis is 0.286 nm. (c) The axes of three α-chains wound around the major axis, the axis of the triple helix.

The triple helix is also stabilized by an extensive network of hydrogen bonds, and contains two hydrogen bonds for every three residues. Hydrogen bonding occurs between $-C=O$ and $-NH$ groups of adjacent chains. An additional set of hydrogen bonds is contributed by hydroxyproline. The OH group of hydroxyproline participates in inter-chain hydrogen bonding through water molecules which in effect form bridges between the chains. The hydrogen bonds contributed by hydroxyproline play an important role in the stability of collagen since the triple helix is destroyed at temperatures below the physiological 37°, if the synthesis of hydroxyproline is inhibited.

Other important factors in stabilizing the triple helix include electrostatic or ionic interactions between oppositely charged residues on adjacent chains, van der Waal attraction, hydrophobic inter-actions and covalent cross-linking.

V. PRIMARY STRUCTURE OF COLLAGEN

The unique properties of collagen are related to its unusual amino acid composition and sequence. The amino acid composition of collagen extracted from various human tissues is summarized in Table 5.2. Collagen contains more glycine than any other vertebrate protein, one-third of all residues being glycine. As discussed below, glycine appears in every third position in the sequence. A composite sequence of rat and calf skin x-1 chains is presented in Figure 5.7. Approximately 95% of the collagen chain, with the exception of the small atypical regions at the N- and C-terminal ends, appears to be a polymer of glycine-led tripeptides $(Gly-X-Y)_n$. Collagen also contains more imino residues than other proteins, proline and hydroxyproline accounting for nearly one-fourth of the total residues. The occurrence of hydroxyproline and hydroxylysine is almost unique in collagen. Hydroxyproline and hydroxylysine are always followed on their C-terminal by glycine, or if one reads the sequence as a glycine-led tripeptide, these residues always appear in the position Y: -Gly-X-Hyp-Gly-X-Hyl-Gly-X-Y, etc. Alanine accounts for nearly one-tenth of all residues in collagen. Collagen contains very few aromatic residues and does not contain tryptophan. Collagen is also poor in sulfur containing amino acids, only 6–8 methionine residues being present per chain. Cysteine is present only in Type III and Type IV collagen chains.

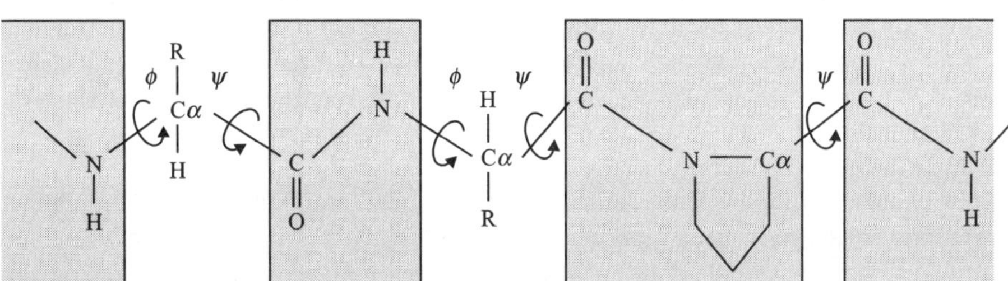

Figure 5.6 Restricted rotation around imino peptide bonds. The peptide bond atoms lie in a plane (boxed area) and the entire unit acts as a rigid structure. The only rotational freedom in a polypeptide chain is at N–Cα and Cα–C ° bonds. The dihedral angle at N–Cα, referred to as ϕ, is restricted to a value $\simeq -60°$ in case of proline by the imino ring across the peptide bond. Because of the interactions of the entire planar peptide unit with neighboring residues, Cα–C° bond angle, known as ψ also has a limited range and in case of collagen $\psi \simeq +150°$. This conformation described by $\phi = -60$, $\psi = 150°$ prevails throughout the collagen helix.

```
PGLU-LlU-SER-Tyr-gly-tyr-ASP-GLU-LYS-SER-THR-GLY-ILE-SeR-VAL-PRO-                                              N1-10

1-1011 (Helical Region)

GLY-PRo-met-GlY-PRO-ser-oly-pro-ArG-GLy-leu-hyp-gly-PRo-HYp-GLY-ALA-HYp-gLy-pRO-GLn-gly-pHe-GLn-GLy-pro-hyp-                27

gLY-gLU-HYP-CLY-glu-hYP-gLY-ALA-SER-gLY-PRo-mET-GLY-PRO-ARg-gLY-PRO-HYP-GLY-PRO-HYP-GLY-LYS-ASn-GLY-AsP-AsP-                54

gLY-GlU-ALA-gLY-LyS-PrO-GLY-aRG-HYP-GLY-GLN-ARG-GLY-PRO-HYP-GLY-PRO-gLN-GLY-ALA-ARG-GLY-LEu-hYP-GLY-thr-aLA-                81

GLY-LEU-HYP-GLY-met-hYL-GlY-HiS-ArG-GLY-PHE-SER-gLY-LEU-ASP-gLY-ALA-LYS-GLY-ASn-ALA-GLY-PrO-ALA-GLY-PRo-lyS-               100

GLY-GLU-HYP-gLY-SER-HYP-gLY-gLu-ASn-GLY-ALA-HYP-GLY-GLN-mET-gLY-PRO-ARG-GLY-LEU-HYP-GLY-GLU-ARG-GLY-ARG-HYP-              135

GLY-PRO-HYP-GLY-SER-ALA-GLY-aLA-ARg-GLY-ASP-AsP-GLY-ALA-VaL-GLY-ALA-ALA-GLY-PRO-HYP-GLY-PRO-THR-GLY-pRO-THR-              162

GLY-PRO-HYP-gLY-PHE-HYP-GLY-ALA-ALA-GLY-ALA-LYS-GLY-GLU-AlA-GLY-PRO-GLN-GLY-ALA-ARG-GLY-SER-GLU-GLY-PRO-GLn-              189

GLY-VAL-ARG-GLY-GLU-HYP-GLY-PRO-HYP-GLY-PRO-ALA-gLY-ALA-ALA-GLY-PRO-ALa-GLY-ASN-HYP-GlY-ALA-ASP-GLY-GLn-hYP-              216

GLY-ALA-LYS-GLY-aLA-ASN-GLY-ALa-HYP-GLY-ILE-ALA-GLY-AlA-HyP-GLy-PHE-HYP-GLY-aLA-aRG-GLY-PRO-SER-GLY-prD-GLn-               243

GLY-PRO-SEr-GLY-ALA-HYP-GLY-PRO-LYS-GLY-ASN-SER-GLY-GlU-HYP-GLY-ALA-HYP-GLY-ASN-LyS-GLY-ASp-THR-GLY-aLA-lyS-              270

GLY-GLU-HYP-GLY-PRO-ALa-GLY-VAL-GLn-GLY-PRO-HYP-GLy-PRO-aLA-GLY-GLU-GLu-GLY-LYS-aRG-GLY-ALA-ARG-GLY-GLU-hYP-              297

gLY-PRO-SER-GLY-lEU-HYP-GLY-PRO-HYP-GLY-GLU-ARG-GLY-GLY-hYP-GLY-SER-ARG-GLY-PHE-HYP-GLY-ALa-ASP-GLY-VaL-ALA-               324

GLy-PRO-LYS-GLY-PRO-ALA-GLY-GLU-ARG-GLY-SER-HYP-GLY-PRO-ala-GLY-PrO-LYS-GLY-SER-HYP-GLY-GLU-aLA-GLY-arG-Hyp-               351

GLY-gLU-ALA-GLY-lEU-HYP-GLY-ALA-LYS-GLY-LEU-THR-GLY-SER-HYP-GLy-sER-HYP-GLy-PRo-ASP-GLY-LYS-THR-GLY-PRo-hYP-              378

GLY-PRO-ALA-GLY-GLX-ASX-GLy-ARG-HYP-GLY-PrO-aLA-GLY-PRO-hYp-GLY-ALA-ARG-GLY-GLU-ALA-GLY-VAL-mET-GLY-PHE-HyP-              405

gLY-PRO-LYS-GLY-AlA-ALA-GLY-GLU-HYP-GLY-LYS-ALA-GLY-GLU-ARG-GLY-vAL-hyP-gLY-PRO-HYp-GLY-ALA-VAL-GLY-PRo-aLA-               432

GLY-LYS-AsP-GLY-GLU-ALA-GLY-ALA-GLN-GLY-Pro-HYP-GLY-PRO-ALa-GLY-PRO-ALA-GLY-GLU-ARG-GLY-GLU-GLN-GLY-PRO-aLA-              459

GLy-SER-HYP-GLy-PHE-GLn-GLY-LEU-HYP-GLY-PRO-alA-GLY-PRO-HYP-GLY-GLU-ALA-GLY-LYS-HyP-GLY-GLU-Gln-GLY-VAL-hYP-              486

GLY-ASP-LEU-GLY-ALA-HYP-GLY-prO-SER-GlY-ALa-ARG-GLY-GLU-ARg-GLY-PHE-HYP-GLy-GLu-ARG-Gly-VAL-GLU-gLy-PrO-hyP-              513

GLY-PRO-aLA-GLY-PRO-ARG-GLy-ALa-aSN-GLY-aLA-HYP-GLY-ASn-ASP-GLY-ALA-lYS-GLY-ASP-ALA-gLY-ALa-HYP-GLY-ala-hyP-              540

GLY-SeR-GLN-gLY-ALA-HYP-GLY-LEU-GLN-GLY-mEt-HYP-GLY-GLU-ARG-GLY-aLA-ALA-gLY-LEU-glY-pro- LYS-GLY-aSp-arG-                  567

GLY-ASP-aLA-gLY-PRO-LYS-GLY-ALA-ASP-gLY-ALA-PRO-GLY-LYS-ASP-GLY-VAL-ARG-GLY-LEU-Thr-GLY-PRO-ILE-GLY-PRo-hYP-              594

GLY-PRO-ALA-GLY-ALA-HYP-GLY-AsP-LYS-gLY-GLU-ALA-GLY-PRO-SER-GLY-PRO-ALA-GLY-ThR-arG-gLY-ALA-HYP-GLY-aSP-arG-              621

gLY-GLU-HYP-GLY-PRO-HYP-gLY-PRO-ALA-GLy-PHE-ALA-GLY-PRO-HYP-GLY-ALA-ASP-GLY-GLn-HYP-GLY-ALA-LYS-GLY-GLU-hyP-              648

GLY-ASP-ALA-GLY-ALA-LYS-GLY-ASP-ALA-GLY-PRO-HyP-GLY-PRO-ALA-GLY-PRO-ALA-GLY-PRO-HYP-GLY-PRO-iLE-GLY-ASN-VAL-              675

GLY-ALA-HYp-gLY-PRO-HYL-GLY-ALA-ARG-GLY-SER-ALA-GLy-PRO-HYP-GLY-ALA-THR-GLY-PHE-HYP-GLY-ALA-ALA-GLY-ApG-VAL-              702

GLY-GLU-THr-GLY-PRO-ALA-gLY-ARG-HYP-gLY-gLU-VAL-GLY-PRO-HYP-GLY-PRO-HYP-GLY-PrO-ALA-GLY-GLU-LYS-GLY-ALA-hYP-              756

GLY-ALA-ASP-gLY-PrO-ALA-GLY-ALA-HYP-GLY-THR-PrO-GLY-PrO-GLn-GLY-ILE-ALA-GLY-GLN-ARG-GLY-VAL-VAL-GLY-LEU-hYP-              783

GLY-GLN-ARG-GlY-GLU-ARG-GLY-PHE-HYP-GLY-LEU-HYP-GLY-PRO-SER-GLY-GLU-HYP-GLY-LYS-GLn-GLY-PRO-SER-GLY-ALA-SER-              810

GLY-GLU-ARG-GLY-PRO-HYP-GLY-PRO-mET-GLY-PRO-HYP-GLY-LEU-ALA-GLY-PRO-HYP-GLY-GLU-SER-GLY-ARG-GLU-GLY-ALA-hYP-              837

GLY-ALA-gLU-GLY-SER-HYP-GLY-ARG-ASP-GLY-SER-HYP-GLY-ALA-LYS-GLY-ASP-ARG-GLY-GLU-THR-GLY-PRO-ALA-GLY-ALa-HYP-              864

GLY-PRO-HYP-GLY-ALA-HYP-GLY-ALA-HYP-GLY-PRO-VAL-GLY-PRO-ALA-GLY-LYS-SER-GLY-ASP-ARG-GLY-GLU-THR-GLY-PRO-aLA-              891

GLY-PRO-IlE-gLY-PRO-VAL-GLY-PRO-ALA-GLY-ALA-ARG-GLY-PRO-ALA-GLY-PRO-GLN-GLY-PRO-ARG-GLY-ASX-HYL-GLY-gLX-tHR               918

Gly-GLX-GLX-GLY-ASX-ARG-GLY-ILE-HYL-GLY-hiS-ARG-GLY-PHE-SER-GLY-LeU-GLN-GLY-PRO-HYP-GLY-PRO-HYP-GLY-SER-HYP-              945

gly-gLU-GLn-gLY-PRO-SER-GLY-aLA-SER-GLY-PRO-ALA-gLY-PRO-ARG-GLY-PRO-HYP-GLY-SER-ALA-GLY-SER-HYP-GLY-LYS-ASp-              972
```

Figure 5.7 Sequence of α-1 (I) chain, showing the recurring, polymer-like features of collagen. This sequence is a composite of calf skin collagen (residues 1–551) and rat skin collagen (residues 552–819). There are great homologies in the sequence of each chain, between different species and only minor differences exist between the triple helical regions of collagens of different species. There are major inter-species differences in the sequence of the "non-helical", telepeptide regions at the N- and C-terminals.

In addition to amino acids, collagen also contains carbohydrates. Although exotic sugars have been reported to be covalently linked to collagen, the only carbohydrates whose presence has been confirmed in all collagens, are galactose and glucose, present as disaccharide appendages bound to hydroxylysine OH groups in O-glycosidic linkages as: Glucosyl-Galactosyl-O-Hydroxylysine. The amount of covalently bound carbohydrates in collagen varies from tissue to tissue and differs between different chain types. The carbohydrate appendages have been ascribed a role in regulating fibrillogenesis. They may regulate interactions between collagen molecules and between collagen and other structural macromolecules.

VI. BIOSYNTHESIS OF COLLAGEN

Synthesis of collagen is a major aspect of developing tissues and it is intimately involved in tissue differentiation, growth and remodeling. Young tissues exhibit high rates of collagen synthesis. As the tissues mature in adults, synthesis continues as part of normal tissue turnover. The highest rates of collagen turnover are observed in weight-bearing bones, lungs and in periodontal tissues. Rates of collagen synthesis are regulated in reproductive tissues by hormonal influences. Collagen synthesis is elevated under conditions requiring remodeling and replacement of tissues, for instance, during the post-partum involution of the uterus and during tissue repair such as wound and fracture healing. Collagen is also synthesized at elevated rates in pathological conditions such as fibrosis in lungs in the liver.

Cell Which Synthesize Collagen

Collagen is synthesized by cells of mesodermal origin, collectively referred to as fibroblasts. In the highly differentiated state, these cells acquire characteristics especially suited to the chemistry of the tissue. Interesting examples of this are seen in chondroblasts in cartilage, odontoblasts in the teeth and osteoblasts in the bones. Collagen synthesizing cells are characterized by an extensive rough endoplasmic reticulum (RER) and well developed Golgi apparatus. Like other secretory proteins, collagen is synthesized on the RER, on membrane-bound polyribosomes, and the completed molecules are released into the cisternae from which they pass into the Golgi and secreted into the extracellular space. Because of its unusual physiochemical properties and chemical composition the synthesis of collagen involves many steps not observed in the biosynthesis of other proteins. For instance, collagen is not soluble under physiological conditions, and aggregation after formation occurs very soon after the molecules are synthesized. Collagen must be kept in a soluble state until it is ready for fibrillogenesis. Another important consideration is the presence of hydroxyproline, hydroxylysine and hydroxylysine glycosides in the collagen molecule. There are no genetic code for these unusual residues which are present only in collagen. They must be synthesized by specific mechanisms. The assembly of the triple helix also presents a unique problem. The synthesis of collagen occurs in several discrete steps which contribute to the unusual characteristics of collagen (Figure 5.8).

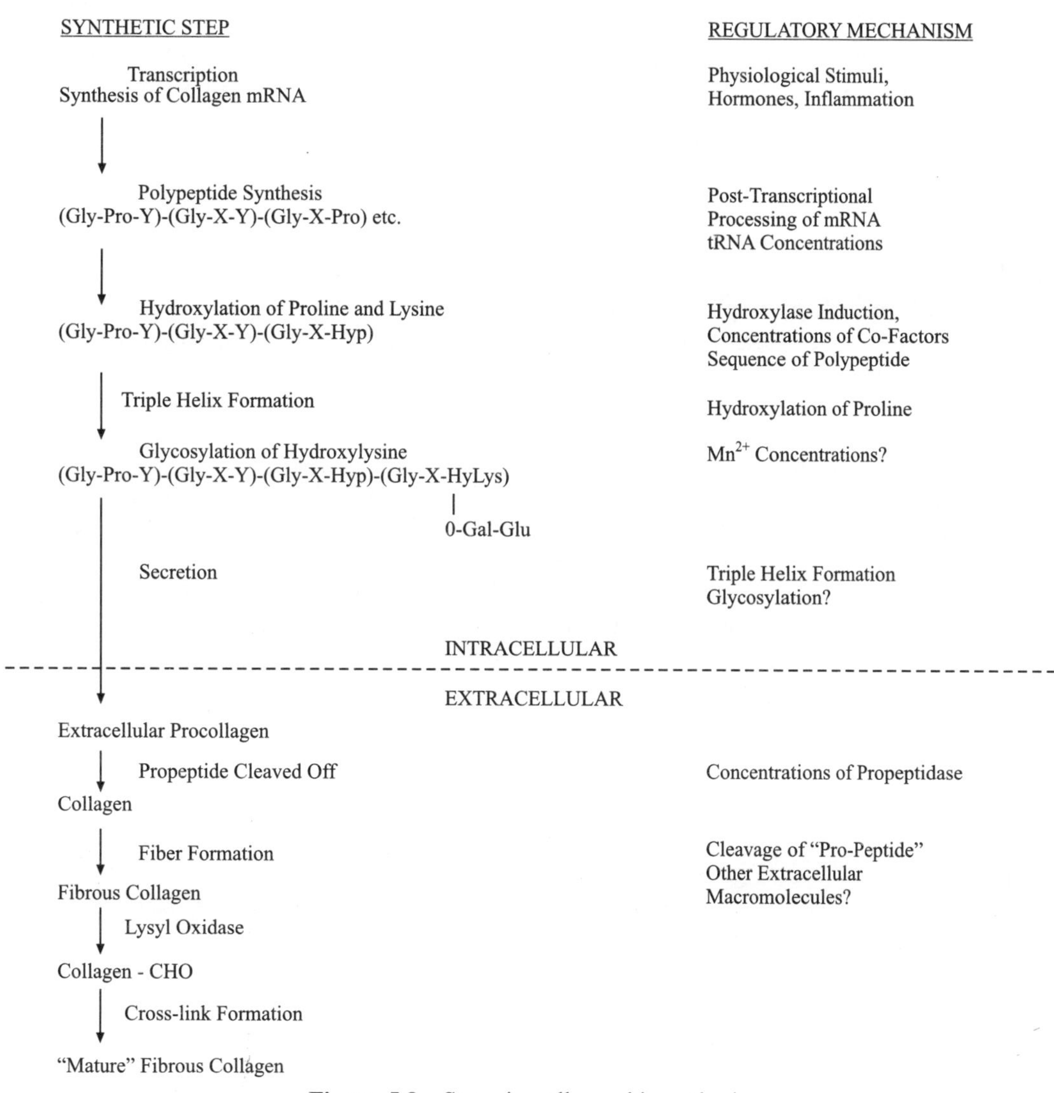

Figure 5.8 Steps in collagen biosynthesis.

Synthesis of Procollagen Chains

Since collagens are always extracellular and they are often transported to sites distant from the cell of origin, before they are incorporated into assemblies such as fibers and complexes with other macro-molecules, it is necessary for the newly synthesized collagen to exist in a soluble "transport" form. This is achieved by the synthesis of collagens in the form of precursor molecules called procollagens, which are considerably larger than the final, functional form of the protein. The newly synthesized $\alpha 1$ chain of Type I procollagen is nearly 1.5 times larger than the corresponding collagen chain. There are approximately 150 amino acid residues in the N-terminal extension, and nearly 300 residues in the carboxyl terminal extension of the pro chain. The N-terminal extension peptide contains a globular region near the N-terminus, followed by a short, triple helical region, a non-triple helical segment, and then the characteristic triple helical region. The C-terminal extension is globular, without any regions with collagen like conformation. Both the N-terminal and C-terminal propeptides contain cysteine. In the N-terminal extension, the cysteine residues are coupled only in intrachain disulfide bridges, whereas the C-terminal cysteines are involved both in inter- and intra-chain disulfide links. The C-terminal propeptide also contains sugars not found elsewhere in collagen, two residues of N-Acetylglucosamine and ten residues of mannose. A small amount of sugar is also associated with the N-terminal extension.

Several functions have been ascribed to the pro-extensions of nascent collagen. A very important part of their function has to do with keeping collagen in a dissolved form; the propeptides appear to block premature assembly into fibrils. The propeptides may also direct orderly fibrillogenesis. They may allow the alignment of the three chains in register to facilitate triple helix formation, and increase the efficiency of helix formation by serving as sites of nucleation. Cleaved propeptides appear to regulate the synthesis of procollagen by cells, by feedback inhibition.

The N-terminal proextension peptides are cleaved soon after procollagen is secreted into the extracellular space by a protease which appears to have specificity for collagens. The N-terminal peptides are removed at about the time procollagen leaves the cell, whereas the C-terminal peptides are cut off at a later time.

The three chains of a collagen molecule are synthesized on different polyribosomes in the same cell, and the synthesis of each chain is directed by a distinct gene, in contrast to proteins such as insulin which are synthesized as a very large product of a single gene and then split off into separate subunit chains. The genes for collagen are monocistronic.

Post-transnational Modification of Procollagen

There is no genetic code for hydroxyproline or for hydroxylysine, and these residues are synthesized by the post-translational hydroxylation of proline or lysine already incorporated into specific sequences:

-x-Pro-Gly-   -x-Hyp-Gly-
-x-Lsy-Gly-   -x-Hyl-Gly-

4-Hyp or Hyl are formed by the hydroxylation of proline or lysine residues which have a glycine residue on their C-terminal side. The synthesis of 3-Hyp occurs only if the proline residue is followed by another proline residue:

-Pro-Pro-Gly   -3Hyp-4Hyp-Gly-

The enzymes which hydroxylate proline and lysine are called prolyl hydroxylase and lysyl hydroxylase respectively. Because the mechanism of these enzymes involves the insertion of both atoms from an O_2 molecule into substrates, they are classified as oxygenases. The two reactions are very similar; both require O_2 which supplies the oxygen atom in the hydroxyl group and α-keto glutaric acid which undergoes oxidative decarboxylation to yield succinate in stoichiometric amounts to the hydroxy-proline and hydroxylysine formed (Figure 5.9). The reaction also requires ferrous ions and ascorbic acid as cofactors. The Fe^{2+} serves as the prosthetic group of the enzyme and it is involved in the reductive fixation of O_2. Ascorbic acid participates in the generation of superoxide free radicals which reduce the iron prosthetic group back to Fe^{2+} state after it has been oxidized to Fe^{3+} form after each hydroxylation event. The intracellular hydroxylation step can be experimentally inhibited by denying O_2 to the tissue or by removing the enzyme bound iron with a chelating agent. Hydroxylation is initiated while the nascent chains are growing on the polyribosomes.

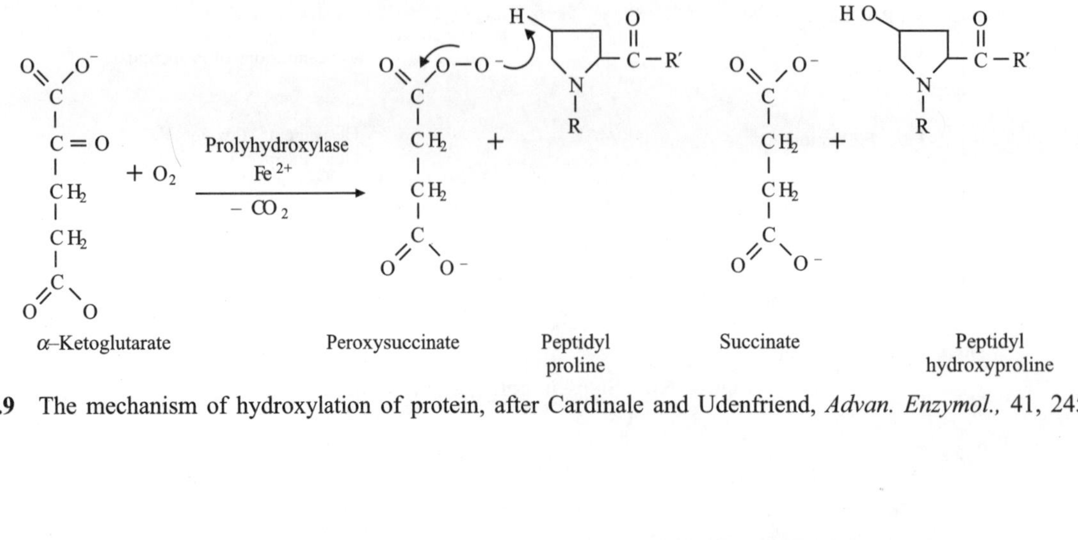

α–Ketoglutarate   Prolyhydroxylase, Fe 2+, + O2, − CO2   Peroxysuccinate   Peptidyl proline   Succinate   Peptidyl hydroxyproline

Figure 5.9 The mechanism of hydroxylation of protein, after Cardinale and Udenfriend, *Advan. Enzymol.*, 41, 245 (1974).

Hydroxylation of proline is essential for the formation of the triple helix under physiological conditions. If the hydroxylation reaction is inhibited, the denaturation temperature of collagen is lowered below physiological temperature. Unhydroxylated collagen is therefore susceptible to non-specific proteolysis.

After the hydroxylation of proline and lysine, which presumably occurs in the cytoplasm, the collagen molecules are transported to the plasma membrane where the glycosylation of hydroxylysine occurs. In the first step, a molecule of galactose is bound to the hydroxyl group in an o-glycosidic linkage. In the second step, glucose is bound to the galactose. Glycosylation of collagen may also be initiated on the polyribosomes. Since all secretory proteins contain covalently bound carbohydrate, it has been suggested that the glycosylation steps are essential for the secretion of collagen. The covalently bound carbohydrate in the triple helical region, as well as in the pro-extensions, may facilitate the transfer of the polar protein molecule through the highly non-polar environment of the plasma membrane. Unhydroxylated collagen is secreted poorly, and accumulates inside the cell. This led to the discovery that only triple helical collagen can be secreted since a unhydroxylated collagen does not generate a stable triple helix at physiological temperature.

The processing of procollagen to collagen requires the cleavage of the propeptide extensions. At least two specific enzymes have been described, one which acts on the N-terminal extension and another which cleaves the C-terminal extension. Both procollagen aminoprotease and procollagen carboxyprotease require a divalent cation, such as Ca^{2+} for their activity. Fiber formation does not occur unless the proextensions are cleaved.

VII.   CROSS-LINKING OF COLLAGEN

Covalent cross-links play a major role in the structure and function of collagen. A set of intra-molecular cross-links joins the three α-chains together in the collagen molecule, and the molecules themselves are stabilized in the fibers by an extensive network of inter-molecular cross-links. Any decrease in the number of these cross-links reduces the tensile strength of collagen and contributes to tissue fragility and major functional impairment.

Cross-linking in most proteins is a consequence of the formation of disulfide bridges between specific cysteine residues. This type of cross-links do not occur in Type I collagen which is the most abundant species in adult tissues, and which contributes to tissues' mechanical strength. Mature Type I collagen does not contain cysteine, nor does Type II collagen. Disulfide bridges may play a role in the aggregated states of Type III and Type IV collagens. These collagens have not been assigned mechanical function and the role of disulfide bridges in their structure is not clear.

The most important cross-links in collagen are derived from specific lysine and hydroxylysine residues. The synthesis of these cross-links is described in Figure 5.10. In the first step

in cross-link formation, a lysine or hydroxylysine residue is oxidatively deaminated by the copper-requiring enzyme lysyl oxidase. The resulting aldehyde residues, α-amino adipic acid-δ-semialdehyde (allysine) or α-amino adipic, -δ-hydroxy, -δ-semialdehyde (δ-hydroxyallysine) are very reactive and can react spontaneously with adjacent reactive groups to form cross-links.

Intra-molecular cross-links are formed by the aldol condensation of allysine residues in the N-terminal telopeptide regions of adjacent α-chains, resulting in the formation of an unsaturated aldehyde.

Inter-molecular cross-link formation involves condensation between allysine or hydroxyallysine residues in N- or C-terminal telopeptides of one molecule with internal lysine or hydroxylysine residues on an adjacent molecule. This condensation results in the formation of labile Schiff base or aldimine type cross-links which may be subsequently reduced. Because of the possible combinations, four different Schiff base or aldimine cross-links can be produced from the various condensations of lysine, hydroxylysine, allysine and hydroxyallysine. A complex cross-link, syndesine is derived by Schiff's base formation between hydroxylysine and hydroxyallysine, followed by an Amadori rearrangement which gives a stable ketonic structure.

The number of cross-links is small in the young, rapidly growing animal. This facilitates the rapid trunover of collagen necessary for the growth and remodelling process. With increasing age, collagen becomes more and more cross-linked. Increasing cross-linking markedly alters the properties of collagen. The most remarkable changes are seen in the extractability of collagen which is decreased, and the tensile strength which is increased. Cross-linking of collagen is inhibited in lathyrism, a disorder observed in populations in certain areas where the seeds of the sweet pea, Lathyrus odoratus, are used as part of the diet. The active principle in these seeds is a bifunctional compound, β-amino-propionitrile. This compound inhibits the lysyl oxidase reaction. The consequence of this is a large decrease in the tensile strength of collagen, resulting in fragility of bones, skin and other connective tissue structures and many developmental disorders. Cross-linking is also decreased in copper deficiency. Administration of certain drugs or compounds which bind to the carbonyl groups of allysine or hydroxyallysine, also prevents cross-link formation resulting in severe connective tissue disorders.

VIII.   DEGRADATION OF COLLAGEN

Collagen undergoes extensive degradation during normal physiological processes involving growth, development and remodeling of various tissues and organs. Rates of collagen breakdown are enhanced during many pathological conditions. Table 5.3 lists some of these processes. The breakdown of collagen is reflected in thinning and frayed appearance of collagen fibers and it culminates in the loss of periodicity and the eventual disappearance of the fibers. Degradation of collagen releases hydroxyproline and hydroxylysine which appear in body fluids, serum and urine. Since these hydroxyamino acids are present in significant amounts only in collagen, are synthesized only during collagen synthesis, and are not reincorporated into protein, their concentrations in serum and urine reflect the state of connective tissue. Increased urinary and serum hydroxyproline levels have been associated with the physiological and aberrant health states enumerated in Table 5.3, and are often used as a diagnostic index for disease.

The triple helix of native collagen is not susceptible to lysis by proteolytic enzymes. The unique structural features and physicochemical properties of collagen provide a safeguard against monomeric collagen (individual molecules) is structural material in the body. The triple helix of monomeric collagen (individual molecules) is denatured at 39°C, polymeric collagen (cross-linked molecules) in fibers at 50–60°C. These temperatures are unattainable under physiological conditions, and therefore complicated procedures must be employed to lyse collagen. There are extensively regulated biochemical mechanisms in the body which are invoked when and where collagenolysis is required.

Microbial Collagenase

Collagen degradation occurs in infected tissues and specific collagenolytic enzymes may be released by infecting microorganisms. Collagenases are produced by certain bacteria, e.g., Clostridium hystolyticum, Pseudomonas aeruginosa, Bacteroides melano-genicus and Mycobacterium tuberculosis; and fungi, e.g., Streptomyces madurae, Trichophyton schoenleinii and Aspergillus oryzae. The extracellular collagenase produced by microorganisms can cleave any form of collagen into small peptides. The best studied of these collagenases is the one from C. hystolyticum. It requires Ca^{2+} for activity and is inhibited by chelating agents, cysteine, and serum albumin. Microorganism collagenases cleave collagen at the bond indicated below:

-Gly-X-Y-Gly-X-Y-Gly-X-Y-Gly

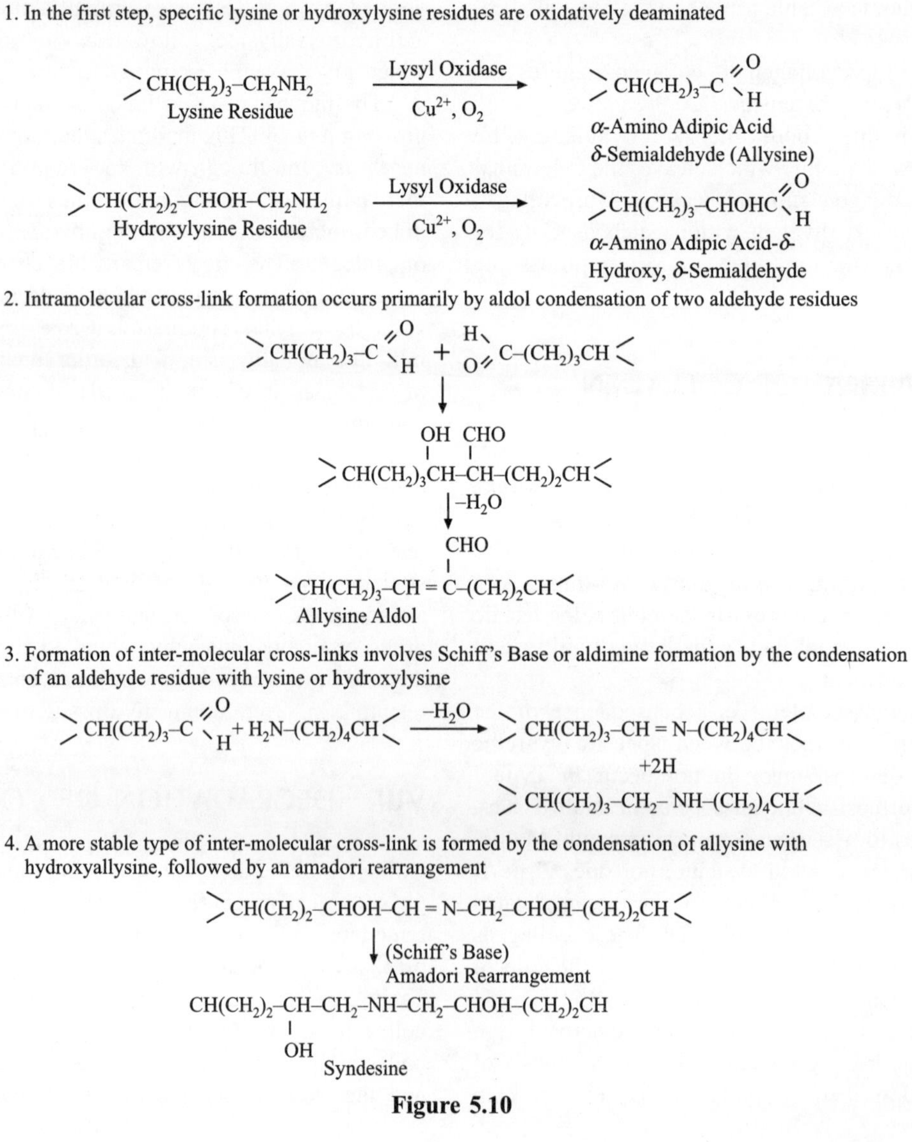

1. In the first step, specific lysine or hydroxylysine residues are oxidatively deaminated

$$>CH(CH_2)_3-CH_2NH_2 \xrightarrow[Cu^{2+},\ O_2]{Lysyl\ Oxidase} >CH(CH_2)_3-C\overset{O}{\underset{H}{\diagdown}}$$

Lysine Residue → α-Amino Adipic Acid δ-Semialdehyde (Allysine)

$$>CH(CH_2)_2-CHOH-CH_2NH_2 \xrightarrow[Cu^{2+},\ O_2]{Lysyl\ Oxidase} >CH(CH_2)_3-CHOHC\overset{O}{\underset{H}{\diagdown}}$$

Hydroxylysine Residue → α-Amino Adipic Acid-δ-Hydroxy, δ-Semialdehyde

2. Intramolecular cross-link formation occurs primarily by aldol condensation of two aldehyde residues

$$>CH(CH_2)_3-C\overset{O}{\underset{H}{\diagdown}} + \overset{H}{\underset{O}{\diagup}}C-(CH_2)_3CH<$$

$$\downarrow$$

$$>CH(CH_2)_3\underset{|}{CH}-\underset{|}{CH}-(CH_2)_2CH< \qquad \overset{OH\ \ CHO}{}$$

$$\downarrow -H_2O$$

$$>CH(CH_2)_3-CH=\underset{|}{C}-(CH_2)_2CH< \qquad \overset{CHO}{}$$

Allysine Aldol

3. Formation of inter-molecular cross-links involves Schiff's Base or aldimine formation by the condensation of an aldehyde residue with lysine or hydroxylysine

$$>CH(CH_2)_3-C\overset{O}{\underset{H}{\diagdown}}+H_2N-(CH_2)_4CH< \xrightarrow{-H_2O} >CH(CH_2)_3-CH=N-(CH_2)_4CH<$$

$$+2H$$

$$>CH(CH_2)_3-CH_2-NH-(CH_2)_4CH<$$

4. A more stable type of inter-molecular cross-link is formed by the condensation of allysine with hydroxyallysine, followed by an amadori rearrangement

$$>CH(CH_2)_2-CHOH-CH=N-CH_2-CHOH-(CH_2)_2CH<$$

$$\downarrow (Schiff's\ Base)$$

Amadori Rearrangement

$$CH(CH_2)_2-CH-CH_2-NH-CH_2-CHOH-(CH_2)_2CH$$
$$|$$
$$OH$$

Syndesine

Figure 5.10

TABLE 5.3   Increased Collagenolysis

I. Normal Physiological Resorption of Collagen

   A. Normal metabolic turnover
   B. Embryogenesis, growth and aging
      1. Metamorphic changes in tadpole tail, skin, gills
      2. Embryo morphogenesis, tissue remodeling, growth processes, organ remodeling
      3. Aging, involution of tissues and organs
   C. Bone and growth remodeling
      1. Bone and cartilage resorption
      2. Resorption in tissue culture
      3. Tooth eruption, periodontal fibers
      4. Tooth resorption, shedding
   D. Collagen resorption related to reproduction
      1. Postpartum involution of uterus and cervix
      2. Gravid uterus remodeling, decidual reaction
      3. Nidation of ovum
      4. Endometrial change in estrus cycle
      5. Ovariectomy-induced involution of the uterus
      6. Relaxation of pubic symphysis and cervix
      7. Ovarian follicle growth and rupture
      8. Reichert's membrane resorption
      9. Mammary gland involution

II. Pathological Disorders Involving Collagen Resorption

   A. Bone disorders
      1. Osteoporosis due to age or immobilization
      2. Paget's disease
      3. Fanconi's syndrome
      4. Rickets
      5. Osteomalacia
      6. Hyperthyroidism
      7. Hypothyroidism
      8. Metastatic bone tumors
      9. Disappearing bone disease
     10. Periodonatal disease
     11. Tooth resorption, "pink tooth"
   B. Connective tissue and skin disorders
      1. Osteoarthritis
      2. Rheumatoid arthritis
      3. Sjogren's syndrome
      4. Chondrolytic perichondritis
      5. Scleroderma
      6. Actinic elastosis, basal cell carcinoma
      7. Keloid regression
      8. Skin atrophies
   C. Scurvy
   D. Miscellaneous
      1. Tumor invasion and regression
      2. Lathyrism
      3. Marfan's syndrome
      4. Ulcerations
      5. Cushing's disease

III. Experimentally-induced resporpion of collagen

   A. Normal healing processes
      1. Skin and wound repair
      2. Burns
      3. Fracture callus remodeling
      4. Regeneration in lower animals
   B. Resorption of implanted collagen
      1. Catgut
      2. Collagen fibers, sponges, tendons
      3. Skin, tendon, and other tissue grafts
   C. Introduction of foreign materials into the organisms
      1. Carrageenin granuloma
      2. Polyvinyl sponge implant
      3. Silicosis
      4. Miscellaneous substances
   D. Chemical substances which cause skin collagen loss
      1. Croton oil, alkali and other lesions
      2. Distal dermal loss of collagen
   E. Experimental cirrhosis
   F. Miscellaneous
      1. Limb atrophy
      2. Starvation, protein deficiency
      3. Thyroid gland resorption

The significance of bacterial collagenases in tissue destruction is questionable since the extent of degradation observed in tissues often cannot be correlated with the small amount of enzyme produced by microorganisms. It is more likely that endotoxins and other bacterial products may invoke the production of collagenase by the host tissue. Bacterial products have been shown to induce host collagenase.

Endogenous Collagenolytic Mechanisms in the Vertebrate Body

Collagen degradation is an essential aspect of development, growth and remodeling of tissues and organs and is a part of numerous normal physiological processes such as the eruption of teeth, post-partum involution of the uterus and reproductive mechanisms as well as pathological processes such as periodontal disease. Several different mechanisms appear to be involved in the breakdown of collagen in the body.

Lysosomal Cathepsins

One component of the body's collagenolytic machinery are the lysosomal cathepsins. Cathepsins may be involved in the degradation of fibrous collagen, intracellularly after phagocytosis and possibly extracellularly as well, under extreme conditions. Lysosomal cathepsins such as Cathepsin B1, from liver lysosomes, operate at subphysiological pH (3.0–5.0). Cathepsin B1 is activated by EDTA and cysteine in contrast to the bacterial collagenases. This enzyme attacks collagen molecules at the N-terminal non-helical region and then cleaves at multiple sites in the helical region. The lysis of collagen may be facilitated by its denaturation at the low pH of the reaction. Some examples of lysosomal cathepsin catalyzed breakdown of collagen are seen in the involuting uterus, healing wounds, arthritis, tooth eruption and enternal resorption of dentin. The action of anti-inflammatory drugs such as aspirin and corticosteroids in arthritis and other pathological states, may be related to the stabilization of the lysosomal membranes by these agents, thus preventing the release of collagenase.

"True" Collagenases

Possibly the most important mechanism of collagen degradation involves a highly specific group of enzymes called collagenases which do not carry out the breakdown of collagen to completion, but only initiate and facilitate the process. A collagenase is defined as an enzyme which cleaves the collagen molecule in the helical portion at physiological pH and temperature. The collagen molecule must be in its native triplehelical conformation. Collagenases are present in numerous normal physiological and pathological systems, e.g. skin, bone and cartilage, corneal epithelium and stroma, granulocytes, PMNs, rheumatoid nodules, skin wounds, rheumatoid synovium, synovial fluids and the uterus. Collagenase activity is prominent in periodontal tissues where it is elaborated by the gingiva, epithelial cells, connective tissue cells (fibroblasts), PMN's and alveolar bone. Collagenase is also associated with tooth movement in orthodontia, in the germs of erupting teeth and the roots of resorbing teeth.

Collagenase always seems to occur in an inactive form in tissues, therefore, it was not discovered until about ten years ago when it was found that fragments of resorbing tail taken from metamorphosing tadpoles secreted collagenase like activity in *in vitro* cultures. This led to the discovery of the enzyme in numerous mammalian tissues. Apparently tissues in culture synthesize and secrete collagenase which is not inactivated because circulating inhibitory proteins are not present in the system.

Unlike microbial collagenases, the animal collagenase makes one single split across all three chains in the triple helical collagen molecule at a point 1/4th of the length of the triple helix from the C-terminal end. The two resulting pieces have denaturation temperatures lower than the body temperature and are rapidly denatured. The denatured fragments are then lysed by non-specific proteases (Figure 5.11). The fragments may also be ingested by phagocytes, macrophages, etc., and lysed intracellularly. This type of slow, controlled degradation of collagen occurs during remodeling processes. Collagenase or its proenzyme may be released into the extracellular space where it is activated and initiates the process of collagenolysis. In situations such as the very rapid resorption of collagen in the uterus after parturition, an additional means of degradation, the intralysosomal hydrolysis of large segments of insoluble collagen by Cathepsin B1 contributes to the breakdown of collagen.

Degradation of Polymeric Collagen

The mechanisms described above degrade monomeric collagen quite efficiently, but are not effective against native collagen fibers as they occur in the tissues. Collagen fibers are often coated with protein polysaccharide complexes. Cartilage proteoglycans have been shown to inhibit collagenase. The complex between collagen and protein polysaccharides must therefore, be disassociated first.

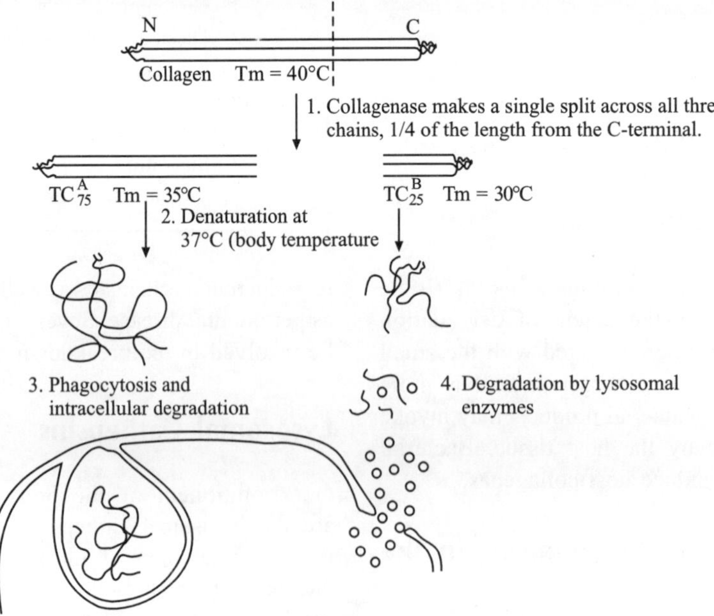

Figure 5.11 Degradation of collagen in the body. Native collagen is not lysed by proteases which attack other proteins. Collagen degradation in animal tissue is initiated by collagenase, an enzyme which cleaves the triple helical molecule into two fragments, one 3/4th and the other 1/4th of the original length. The two fragments are denatured at body temperature and are lysed by other mechanisms.

This may require the concerted action of proteases to remove the coreprotein of protein polysaccharide, followed by specific hydrolases for the polysaccharide. This would leave the collagen fibers untouched. However, further proteolytic action may cleave the cross-linking regions of some of the collagen molecules and release them in a form susceptible to collagenase action (Figure 5.11).

Regulation of Collagen Degradation

Collagen is protected by several levels of controls on collagenase action. These controls range from the synthesis and activation of collagenase to the state of collagen itself.

Regulation of collagenase synthesis is subject to regulatory controls similar to those operative for other enzymes. Collagenase is induced in affected tissues by a variety of stimuli and stresses. In bones, parathyroid hormone and hyperoxia may induce increased synthesis of the enzyme. Corticosteroids including hydrocortisone increase collagenase activity in cells in culture and in the periodontium. Other hormones, e.g. prednisone, progesterone, cortisone and estradiol reduce collagenase synthesis. Bacterial endotoxins cause increased collagenase synthesis in macrophages. Collagenase synthesis in fibroblasts is increased by chemical mediators of inflammation, prostaglandins and lymphokines.

Collagenase is synthesized, stored intra-cellularly or released into the extracellular millieu in an inactive or latent form, procollagenase. Procollagenase is a zymogen which when treated with proteases such as trypsin, results in the cleavage of a large peptide fragment and activation of the enzyme. Most cells produce a protease which activates procollagenase, and this may serve to localize the action of collagenolytic activity. A complicated mechanism for the activation of collagenase has been suggested in case of bone resorption where a specific procollagenase activating protease was shown itself to be synthesized as an inactive proenzyme, and needed proteolytic activation before it could act on procollagenase.

Collagenase is inhibited by cysteine and serum α-globulins. Strong complexes which are inactive are formed between collagenase and α_2-macro-globulin (α_2M) or α_1-antitrypsin. Certain protein rich salivary glycoproteins may also play a role in neutralizing collagenase around periodontal tissues.

Finally, as indicated above, the state and nature of the substrate also affects the action of collagenase. Thus, complex formation with proteoglycans protects collagen from collagenase. Increased cross-linking in collagen also increases the resistance to collagenase action. Surprisingly, synovial collagenase does not attack Type II collagen which is the principal component of cartilage.

An interesting aspect of collagenolysis is related to the temperature coefficient of the reaction. A 10-fold increase in the rate of lysis occurs on increasing the reaction temperature from 30° to 36°C. Normal knee joints have a temperature between 30.5°–33°C, whereas inflamed joints have a temperature of up to 38°C. It is possible that one of the harmful effects of inflammation is the increased rate of collagen degradation resulting from increased temperature.

IX. INTERACTION OF COLLAGEN WITH CELLS

Collagen plays an essential role in many regulatory processes, and these functions involve interaction between cells and collagen.

Involvement of Collagen in Cell Differentiation

Collagen is intimately involved in cell and organotypic differentiation and in developing tissues it forms the morpho-genetic template to which cells align. Collagen mediates the intercellular flow of information during epithelial-mesenchymal inter-action, regulates cell movement, cell anchoring and proliferation of cells. Disturbances in collagen synthesis, or in its physico-chemical properties are a major cause of developmental disorders and teratogenesis.

Collagen-Platelet Interaction

Collagen plays a very important role in hemostasis. When a blood vessel is injured, collagen fibres are exposed. Platelets bind to exposed collagen and are activated. An early consequence of this binding is that platelets adhering to collagen become sticky and aggregate, forming a plug blocking the flow of blood through the injury in the blood vessel. Activation of the platelets by collagen leads to the cascade of reactions in blood clotting.

Interaction of Collagen with Cells of the Immune System

Autologous collagen is apparently able to induce an immune response in some humans and in experimental animals. This interaction between collagen and immune cells results in autoimmune disorders of connective tissue, such as rheumatoid arthritis. Recent studies have shown that T cells from mice immunized with collagen, and from patients with rheumatoid arthritis interact with collagen *in vitro* and imitate physiologically significant reactions. The conformation of collagen apparently plays a role in this process.

Interaction of Collagen with Fibronectin

Fibronectin is a glycoprotein present on the surface of many cells, particularly fibroblasts, and is also present in the serum. The interaction of collagen with cells is mediated through fibronectin. Collagen and fibronectin interact, forming a complex. The cell binds to this complex. There appears to be a high degree of specificitiy in the binding of collagen to fibronectin, Type III collagen being the most efficient. The site of binding is localized to the same region of the collagen molecule which interacts with vertebrate collagenase.

6

Chemistry of Carbohydrates and Lipids

Neeru Saini

CONTENTS

Carbohydrates and lipids are present in all plants and animal tissues and are essential for the maintenance of living organisms. Nearly all the energy needed by the human body is provided by the oxidation of carbohydrates and lipids. Carbohydrates are usually used for short-term storage, whereas lipids are used for long-term storage. Both these are bound to proteins and have important structural and regulatory functions. Carbohydrates are hydrophilic (water-liking) molecules, easy to transport around the body and are more rapidly digested, and so their energy is useful if the body requires energy fast. On the other hand, lipids are insoluble in water and soluble in organic solvents, which makes them more difficult to transport and digest.

I. CARBOHYDRATES

Carbohydrates are the most abundant organic components in most fruits, vegetables, legumes and cereal grains, and they provide texture and flavour in many processed foods. Carbohydrates are defined as polyhydroxy aldehydes or ketones. Carbohydrates that have aldehyde function are called aldoses and those with a ketone function are called ketoses (Table 6.1). Carbohydrates are given names which end with the suffix "ose" in order to differentiate them from non-carbohydrates compounds. Some examples are glucose, fructose, sucrose, amylose, cellulose, starch, etc. Carbohydrates consist of carbon, hydrogen and oxygen, usually with a hydrogen : oxygen atom ratio of 2:1 (as in water), and have an empirical formula $C_m(H_2O)_n$, where m could be different from n. Some carbohydrates contain other elements such as nitrogen and sulphur. The most common carbohydrate is glucose ($C_6H_{12}O_6$).

The carbohydrates are divided into four chemical groupings: monosaccharides, disaccharides, oligosaccharides and polysaccharides.

II. MONOSACCHARIDES

The simplest form of carbohydrates is the monosaccharide. 'Mono' means 'one' and 'saccharide' means 'sugar'. The **monosaccharides** are white, crystalline solids that contain a single aldehyde or ketone functional group. They are subdivided into two classes—*aldoses* and *ketoses*—on the basis of whether they are aldehydes or ketones. They are also classified as triose, tetrose, pentose, hexose or heptose on the basis of whether they contain three, four, five, six or seven carbon atoms (Figure 6.1). Glyceraldehyde is an aldotriose and dihydroxy-acetone is a ketotriose. Similarly, ribose is an aldohexose and fructose is a ketohexose. The smallest molecules generally termed carbohydrates are glyceraldehyde and dihydroxy-acetone.

In 1891, **Emil Fischer** made the arbitrary assignments of D and L to the enantiomers of glyceraldehyde.

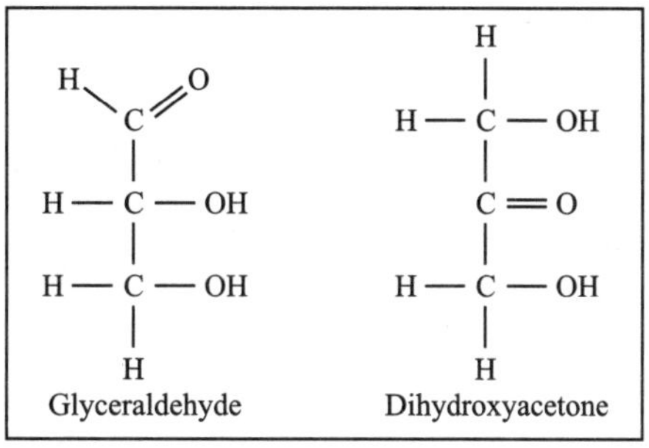

Trioses

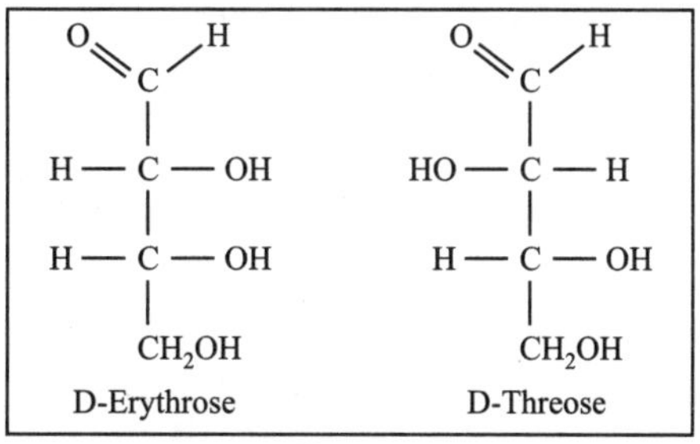

Tetroses

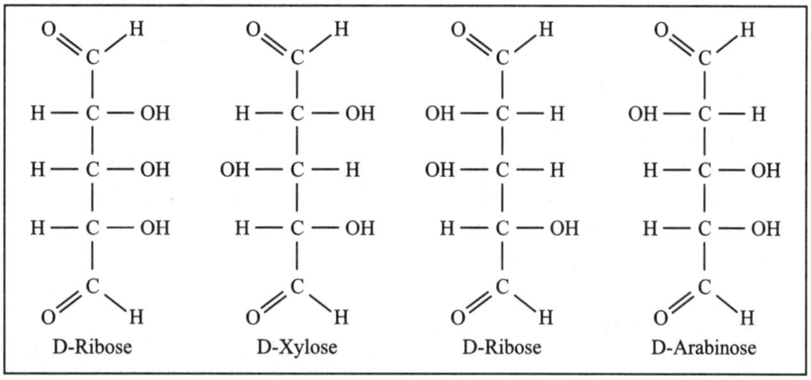

Pentoses

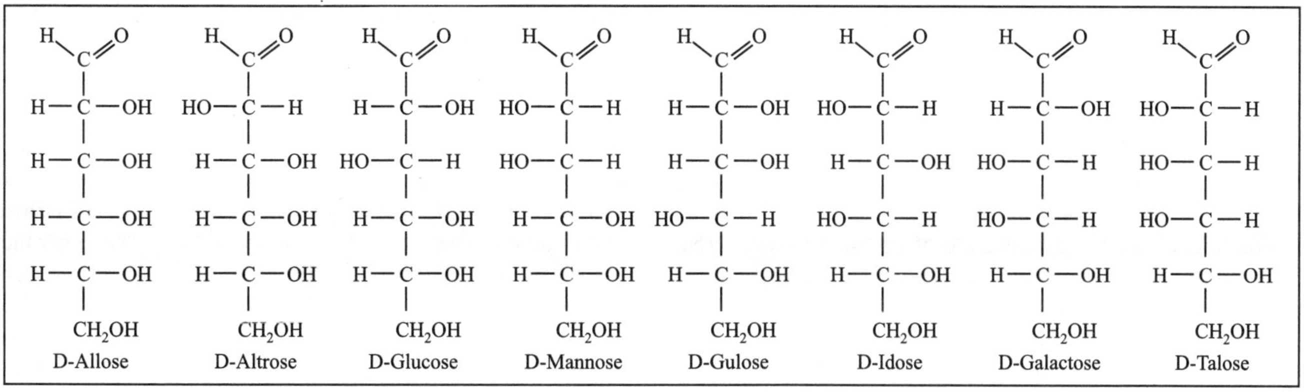

Hexoses

Figure 6.1 Structures of trioses, tetroses, pentoses and hexoses.

TABLE 6.1 **Classification of Carbohydrates by Length of Carbon Chain**

Number of carbons	Aldoses		Ketoses
Three	Trioses $C_3H_6O_3$	$D-$ and $L-$glycerose or glyceraldehyde	Dihydroxyacetone
Four	Tetroses $C_4H_8O_4$	$D-$ and $L-$erythrose $D-$ and $L-$threose	$D-$ and $L-$xyloke-tose
Five	Pentoses $C_5H_{10}O_5$	$D-$ and $L-$arabinose $D-$ and $L-$xylose $D-$ and $L-$ribose $D-$ and $L-$lyxose	$D-$ and $L-$lyxoke-tose
Six	Hexoses $C_6H_{12}O_6$	$D-$ and $L-$glucose $D-$ and $L-$mannose $D-$ and $L-$galactose $D-$ and $L-$ gulose $D-$ and $L-$ idose $D-$ and $L-$talose $D-$ and $L-$altrose $D-$ and $L-$allose	$D-$ and $L-$fructose $D-$ and $L-$sorbose

Sugars of the same series (aldohexoses) have very similar structures and they differ only in their configuration. *Configurations of a sugar represent the actual distribution in space of the atoms of a molecule of the sugar. The molecular formula* $C_6H_{12}O_6$ represents 16 different aldosugars which have the same basic carbon skeleton but have different configurations.

Let us take an example of glyceraldehydes. Glyceraldehyde has one asymmetrical carbon atom (indicated by *), and because of that two different forms of the molecule are possible (Figure 6.2). These two forms are indicated by $D-$ and L-glyceraldehydes, meaning right and left glyceraldehyde. The letters D and L refer to the asymmetrical carbon atom which is located the farthest away from the aldehyde or keto group (here one after the lowest atom).

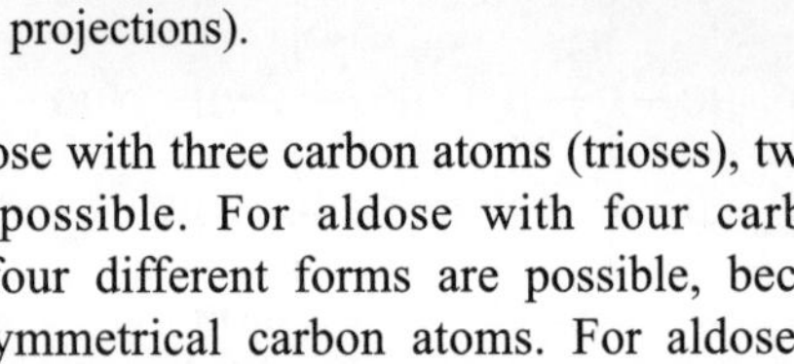

Figure 6.2 Configuration of $D-$ and L-Glyceraldehyde (Fischer projections).

For aldose with three carbon atoms (trioses), two different forms are possible. For aldose with four carbon atoms (tetroses), four different forms are possible, because there are two asymmetrical carbon atoms. For aldose with five C-atoms (pentoses) eight different and that with six C-atoms (hexoses) sixteen different forms are possible, respectively. The different aldoses until six carbon atoms are represented. The aldehyde group is represented in green. These sugars

have the D-configuration and are indicated in red. For each D-aldose, there is also an L-form.

Just as the case of glyceraldehydes, every sugar has two absolute configurations (Figure 6.3). The absolute configuration of any sugar is expressed by comparison to D-glyceraldehyde, which is taken as the standard. Thus, in all D-sugars, the terminal group has the same absolute configuration as the corresponding terminal group of D-glyceraldehydes. In a similar way, all L-sugars are related to L-glyceraldehydes. Almost all the naturally occurring sugars belong to the D-series, whereas most of the L-sugars do not exist in nature and are obtained only synthetically.

The absolute configuration of a molecule is three dimensional and can be easily represented by building a three-dimensional model. For the purpose of writing, a two-dimensional representation known as the *Fischer projection* is usually used. The $D-$ and L-configurations of glyceraldehydes shown in Figure 6.2 are the Fischer projections.

If we assume that the asymmetric carbon atom of glyceraldehydes occupies the centre of a regular tetrahedron [Figure 6.4(a)], then the four groups H, OH, CHO and CH_2OH will occupy the four corners of the tetrahedron. Let us imagine that the asymmetric carbon atom is on the plane of the paper [Figure 6.4(b) and (c)] and that we are looking at it straight from above. The horizontal lines of the projection [Figure 6.4(d)] now represent the bonds (H—OH) coming towards us out of the plane of the paper and the vertical lines represent bonds (—CHO and CH_2OH) going away from us, behind the plane of the paper.

The same principles are applied to each asymmetric carbon atom when more complex sugars such as D-glucose (Figure 6.5) are represented by the Fischer projection.

Figure 6.6 shows the stereochemical relations of aldoses and Figure 6.7 shows the stereochemical relations of ketoses.

Enantiomers, Diastereoisomers, and Epimers

Owing to the fact that carbohydrates contain multiple stereocenters, many isomers are possible, including enantiomers, diastereoisomers and epimers. Two carbohydrates are said to be **enantiomers** if they are nonsuperimposable mirror images of one another. An example of an enantiomer is the $D-$ and L-isomers of glucose, as shown in Figure 6.7.

The second type of isomer seen in carbohydrates are **diastereoisomers**. Carbohydrates are classified as diastereomers if their chiral carbons are connected to exactly the same substrates but at differing configurations ('R' or 'S'). Unlike an enantiomer, diastereomers are not object and mirror image. Two carbohydrates that are diastereoisomers are D-glucose and D-altrose (Figure 6.8).

Another type of isomer that carbohydrates can take on are **epimers**. Epimers are two diastereomers that differ only at one stereocenter. As shown in Figure 6.9, D-glucose and D-mannose are an example of an epimer with respect to carbon atom 2 and D-glucose and D-galactose are an example of an epimer with respect to carbon atom 4.

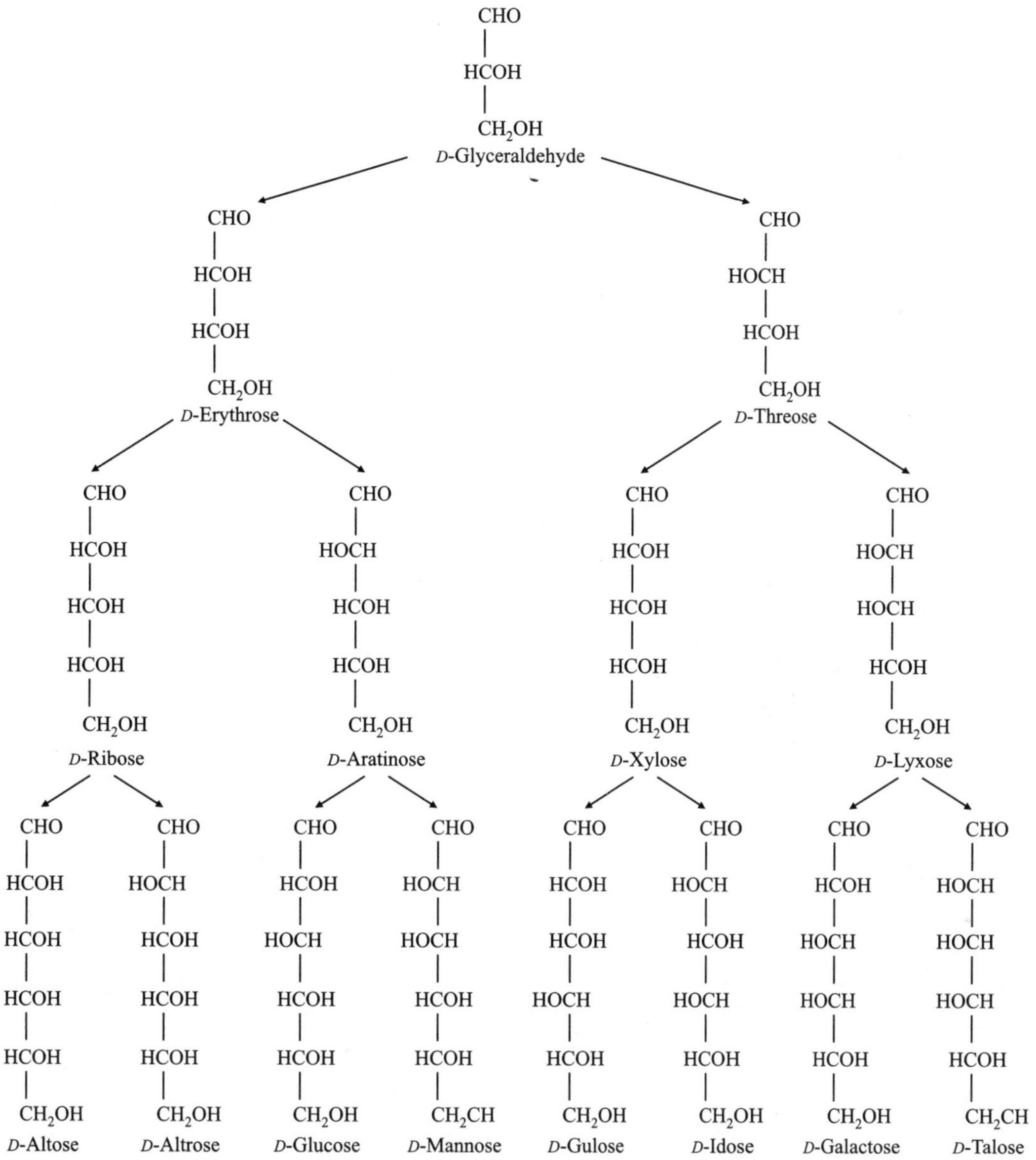

D-Glyceraldehyde
CHO
|
HCOH
|
CH₂OH

D-Erythrose
CHO
|
HCOH
|
HCOH
|
CH₂OH

D-Threose
CHO
|
HOCH
|
HCOH
|
CH₂OH

D-Ribose: CHO — HCOH — HCOH — HCOH — CH₂OH
D-Aratinose: CHO — HOCH — HCOH — HCOH — CH₂OH
D-Xylose: CHO — HCOH — HCOH — HCOH — CH₂OH
D-Lyxose: CHO — HOCH — HOCH — HCOH — CH₂OH

D-Altose: CHO — HCOH — HCOH — HCOH — HCOH — CH₂OH
D-Altrose: CHO — HOCH — HCOH — HCOH — HCOH — CH₂OH
D-Glucose: CHO — HCOH — HOCH — HCOH — HCOH — CH₂OH
D-Mannose: CHO — HOCH — HOCH — HCOH — HCOH — CH₂CH
D-Gulose: CHO — HCOH — HCOH — HOCH — HCOH — CH₂OH
D-Idose: CHO — HOCH — HCOH — HOCH — HCOH — CH₂OH
D-Galactose: CHO — HCOH — HOCH — HOCH — HCOH — CH₂OH
D-Talose: CHO — HOCH — HOCH — HOCH — HCOH — CH₂CH

Figure 6.3 The absolute configuration of any sugar is expressed by comparison to D-glyceraldehyde which is taken as the standard.

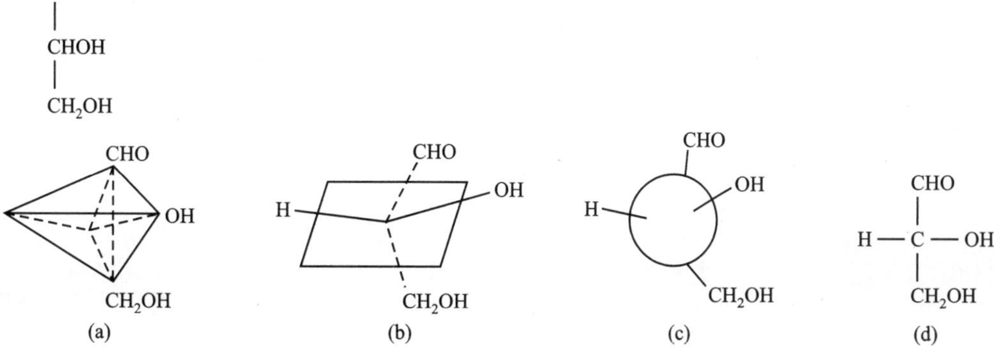

Figure 6.4 The absolute configuration of *D*-glyceraldehyde. (a) Tetrahedral arrangement of groups. (b, c) Geometric projection of groups. (d) Fischer projection.

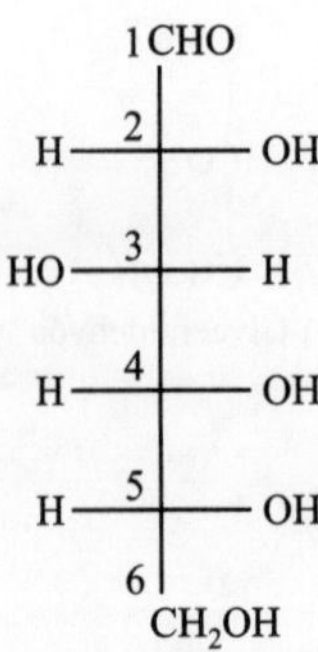

Figure 6.5 Fischer projection of glucose.

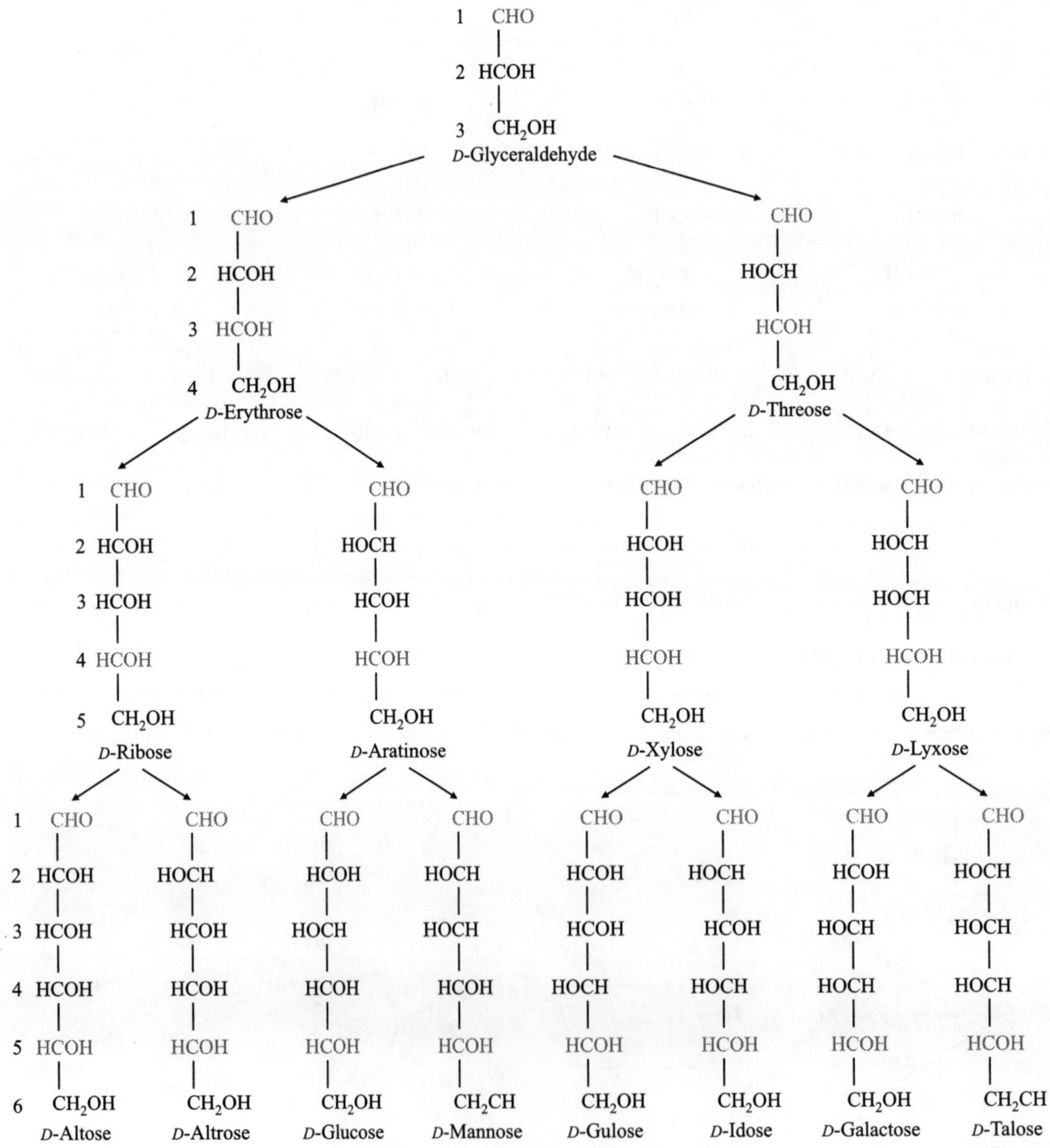

Figure 6.6 Stereochemical relations of D-aldoses containing three, four, five and six carbon atoms. These sugars are D-aldoses because they contain an aldehyde group (shown in green) and have the configuration of D-glyceraldehyde at their farthest asymmetric centre (shown in red). (*see Plate 1 for colour figure*)

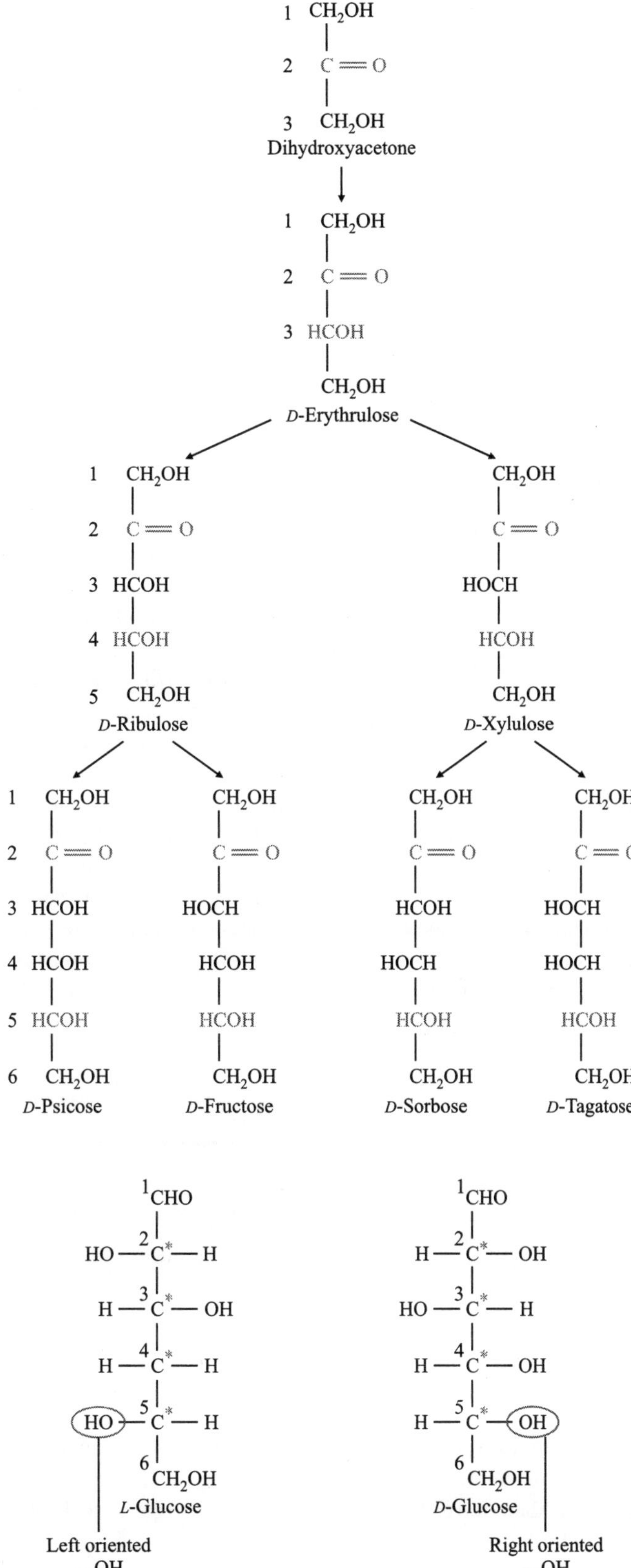

Figure 6.7　Stereochemical relations of *D*-ketoses containing three, four, five and six carbon atoms. These sugars are *D*-ketoses because they contain a keto group (shown in green) and have the configuration of *D*-glyceraldehyde at their farthest asymmetric centre (shown in red). (*see Plate 2 for colour figure*)

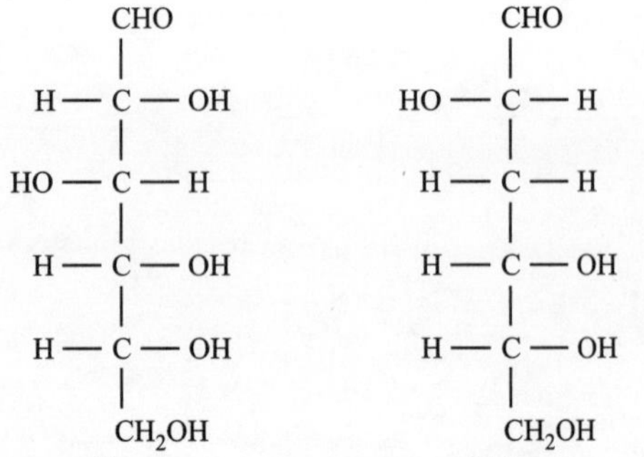

D-Glucose *D*-Altrose

Figure 6.8 Diastereoisomers.

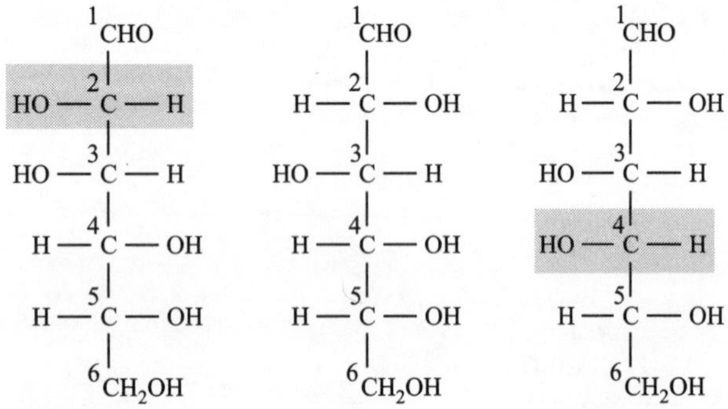

D-Mannose (epimer at C-2) *D*-Glucose *D*-Galactose (epimer at C-4)

Figure 6.9 Epimers of glucose.

Cyclic Structures

So far we have represented monosaccharides as linear molecules, but many of them also adopt cyclic structures. This conversion occurs because of the ability of aldehydes and ketones to react with alcohols (Figure 6.10).

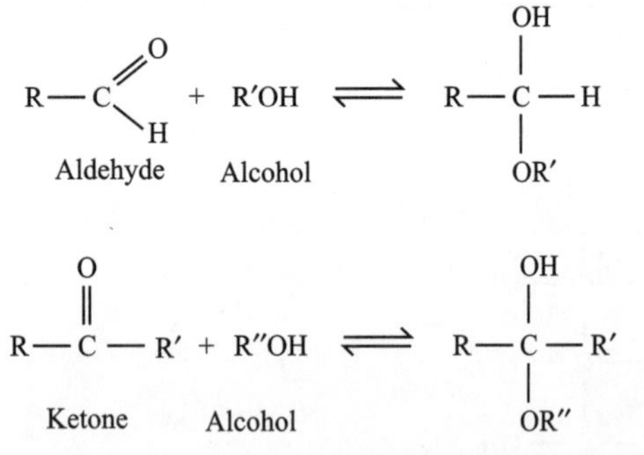

Figure 6.10 Reaction of aldehydes and ketones with alcohol.

In some cases, OH and carbonyl groups on the same molecule are able to react with one another in an intramolecular reaction. Thus, monosaccharides larger than tetroses exist mainly as cyclic compounds (Figure 6.11). You might wonder why the aldehyde reacts with the OH group on the fifth carbon atom rather than the OH group on the second carbon atom next to it. This is because cyclic alkanes containing five or six carbon atoms in the ring are the most stable. The same is true for monosaccharides that form cyclic structures: rings consisting of five or six carbon atoms are the most stable.

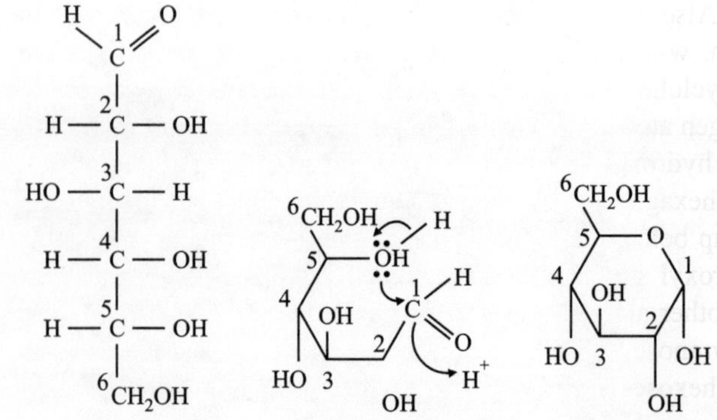

(a) Fischer projection (b) Three-dimensional representation (c) Cyclic monosaccharide

Figure 6.11 Cyclization of *D*-Glucose.

When a straight-chain monosaccharide forms a cyclic structure, the carbonyl oxygen atom may be pushed up or down, giving rise to two stereoisomers. The structure shown with the OH group on the first carbon atom projected downward represents the ***alpha (α) form***. The structure on the right side, with the OH group on the first carbon atom pointed upward, is the ***beta (β) form***. These two stereoisomers of a cyclic monosaccharide are known as ***anomers***; they differ in structure around the anomeric carbon—that is, the carbon atom that was the carbonyl carbon atom in the straight-chain form. The fact that glucose can exist in two distinct forms accounts for the formation of two different methylglucosides, methyl *α–D*-glucoside and methyl *β–D*-glucoside (as shown in Figure 6.12) The two isomers of methylglucoside, unlike the isomer of glucose, do no undergo interconversion in aqueous solution. Thus, the configuration of the sugar is fixed by glucoside formation.

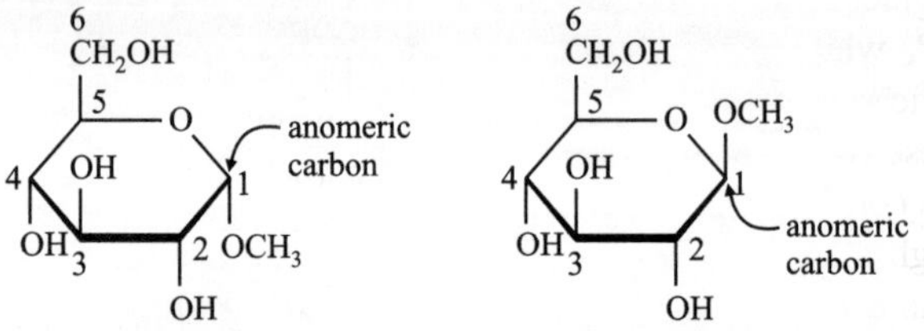

Methyl *α-D*-glucopyranoside Methyl *β-D*-glucopyranoside

Figure 6.12 Anomers of methylglucosides.

In the cyclic structure of glucose, the oxygen bridge formed between first and fifth carbon atoms establishes a six-membered ring which resembles that of pyran. The two isomers of glucose are shown in Figure 6.13. They are called *α-D*-glucopyranose and *β-D*-glucopyranose.

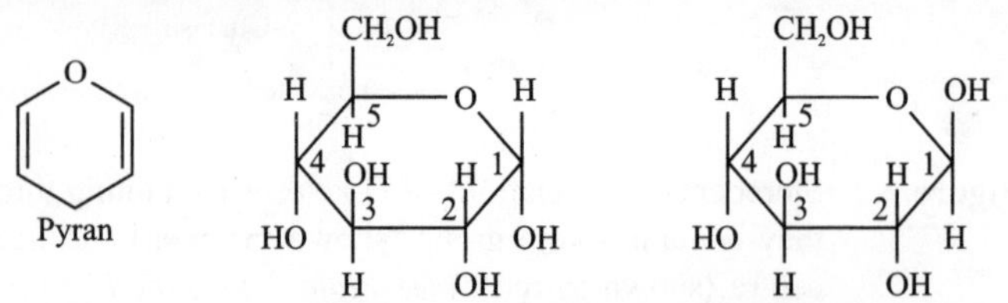

α-D-(+)-Glucopyranose *β-D*-(+)-Glucopyranose

Figure 6.13 Anomers of glucose in cyclic forms.

Also, their structures are usually written in the hexagonal form, with the ring skeleton resembling the carbon skeleton of cyclohexane. The six corners of the hexagon represent the oxygen atom and carbon atoms 1–5 which take part in the ring. The hydroxyl group on C–1 has different arrangements in the two hexagonal structures: α–D-glucopyranose has the hydroxyl group below the plane of the ring and β-D-glucopyranose has hydroxyl group above the plane. As discussed for glucose, the other aldohexoses can also exist in cyclic α-pyranose and β-pyranose. In contrast to glucose and other aldohexoses, the ketohexoses and aldopentoses predominantly exist in cyclic forms with five-membered rings. The five-membered rings of these structures resemble that of furan. Hence, such a five-membered cyclic structure is referred to as the furanose structure.

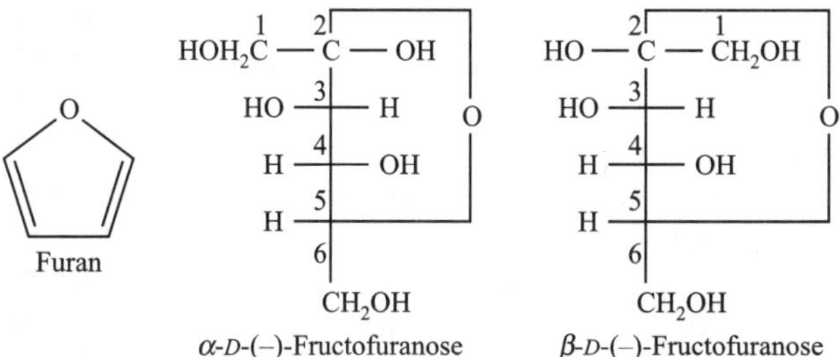

Figure 6.14   Anomers of fructofuranose.

Mutarotation

The two stereoisomeric forms of glucose, α-D-glucose and β-D-glucose, exist in separate crystalline forms and thus have different melting points and specific rotations. For example, α-D-glucose has a m.p. of 419 K with a specific rotation of +112°, while β-D-glucose has a m.p. of 424 K and has a specific rotation of +19 (Figure 6.15). However, when either of these two forms is dissolved in water and allowed to stand, it gets converted into an equilibrium mixture of α- and β-forms through a small amount of the open chain form.

As a result of this equilibrium, the specific rotation of a freshly prepared solution of α-D-glucose gradually decreases from +112° to +52.7° and that of β-D-glucose gradually increases from +19° to +52.7°.

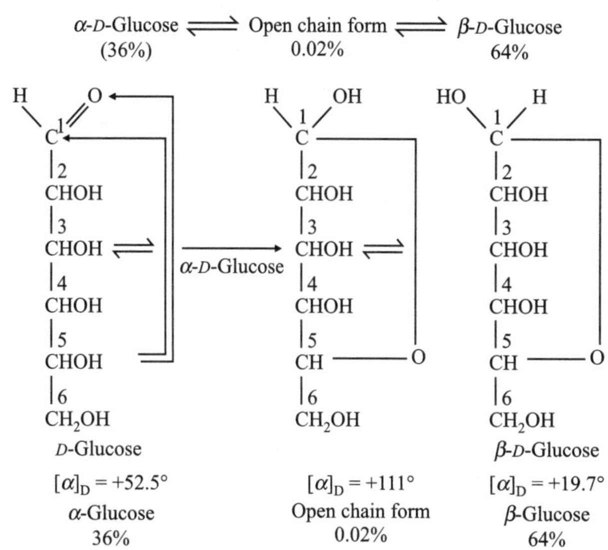

Figure 6.15   Mutarotation of glucose.

$$\alpha\text{-}D\text{-Glucose} \rightleftharpoons \text{Equilibrium mixture} \rightleftharpoons \beta\text{-}D\text{-Glucose}$$
$$[\alpha]_D = +112° \qquad [\alpha]_D = +52.7° \qquad [\alpha]_D = +19°$$

where $[a]_D$ = specific rotation.

This change in specific rotation of an optically active compound in solution with time to an equilibrium value is called mutarotation. During mutarotation, the ring opens and then recloses either in the inverted position or in the original position, giving a mixture of α- and β-forms. All reducing carbohydrates, i.e., monosaccharides and disaccharides (maltose, lactose, etc.), undergo mutarotation in aqueous solution.

Glycosidic Bond

A glycosidic bond is a type of covalent bond that joins a carbohydrate (sugar) molecule to another group, which may or may not be another carbohydrate.

An aldehyde reacts with 1 equiv of an alcohol to form hemiacetal. Similarly, a ketone reacts with an alcohol to form hemiketal and ketal (Figure 6.16).

When an aldose forms a cyclic structure by reaction between aldehyde and hydroxyl groups, a hemiacetal is formed. A distinctive feature of this type of hemiacetal is

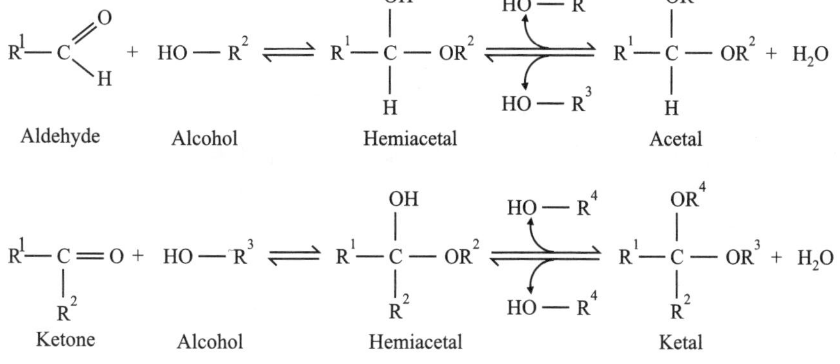

Figure 6.16   Formation of hemiacetal and hemiketal.

that about the –OR′ and –R substitutions of the hemiacetal carbon stem from the same molecule as an examination of the cyclic structure of glucose would reveal. On the other hand, hemiketal is the product when a ketone forms the cyclic structure (e.g., fructose structure). The hemiacetal carbon of an aldose and the hemiketal carbon of a ketose are both referred to as anomeric carbon atom. The residual hydroxyl group (on the anomeric carbon) of a hemiacetal can react with an alcohol to yield an acetal. This is what occurs when methanol reacts with glucose. The product, a mixed acetal where one alcoholic contribution comes from outside while the other is intermolecular, is called a *glucoside*. Compounds of this type formed by the reaction between an alcohol and the hydroxyl group of the anomeric carbon of a sugar are known as glycosides. The bond established by such a reaction is known as the glycosidic bond. Both aldoses and ketoses form glycoside. Thus, glucose forms methylglucopyranoside and fructose forms methyl fructofuranoside.

A. Derivatives of Monosaccharides

Monosaccharides undergo incorporation into oligo- and polysaccharides. Individual monosaccharides can also undergo a variety of transformations. One important modification of monosaccharides is the formation of deoxy sugars. The most biologically significant of the deoxy sugars is *β-D-2-deoxyribose* (see Figure 16.17), in which the C-2 hydroxyl group of ribose is replaced with hydrogen. This deoxy sugar is the sugar component of DNA.

In the amino sugars, an amino group replaces one or more of the hydroxyl groups. The most common of these sugars are *D*-glucosamine and *D*-galactosamine, both of which have an amino group in place of the hydroxyl group on the second carbon.

Often, these amino groups are acetylated to give *N*-acetyl sugars, such as *N*-acetylglucosamine and *N*-acetylgalactosamine. These sugars are important components of larger poly-

saccharides. Other important amino sugars are muramic acid and *N*-acetylneuraminic acid, which are components of the oligosaccharides of glycoproteins and glycolipids, and of bacterial cell walls. Muramic acid and *N*-acetylneuraminic acid are glucosamines, which are linked at C-3 or C-1 to three-carbon acids. Muramic acid is formed via an ether linkage between the C-3 of glucosamine and the hydroxyl group of lactic acid. *N*-Acetyl-*D*-neuraminic acid results from the formation of a C–C bond between the C-1 of *N*-acetyl-*D*-mannosamine and the C-3 of phosphoenolpyruvate. This and other derivatives of neuraminic acid are collectively called sialic acids and are widely found in bacteria and animals.

Finally, monosaccharides often form **ester** linkages with phosphate and sulfate ions. In fact, it is rare to find free monosaccharide in cells. Glucose is the only unmodified monosaccharide that exists in substantial quantities in living things, where it exists primarily extracellularly.

Sugar acids: Sugar acids are derived from sugars by the oxidation of aldehyde group or primary alcoholic group. Gluconic acid is derived from glucose by the oxidation of aldehyde group at carbon atom 1 and glucuronic acid is derived from glucose by oxidation of the primary alcoholic group at carbon atom 6.

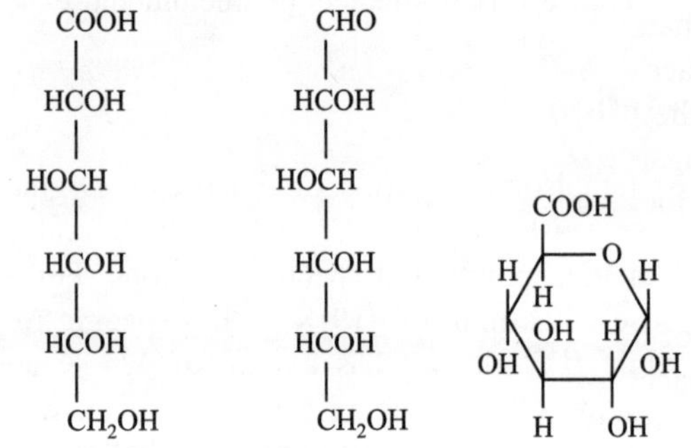

Gluconic acid (hexonic acid)   Glucuronic acid (hexonic acid)   Glucuronic acid

2-Deoxy-*D*-ribose   *β-L*-rhamnose   *D*-Glucosamine   *N*-Acetyl-*D*-glucosamine

β-L-Fucose   *D*-Galactosamine

Deoxy sugars   Deoxy amino sugars

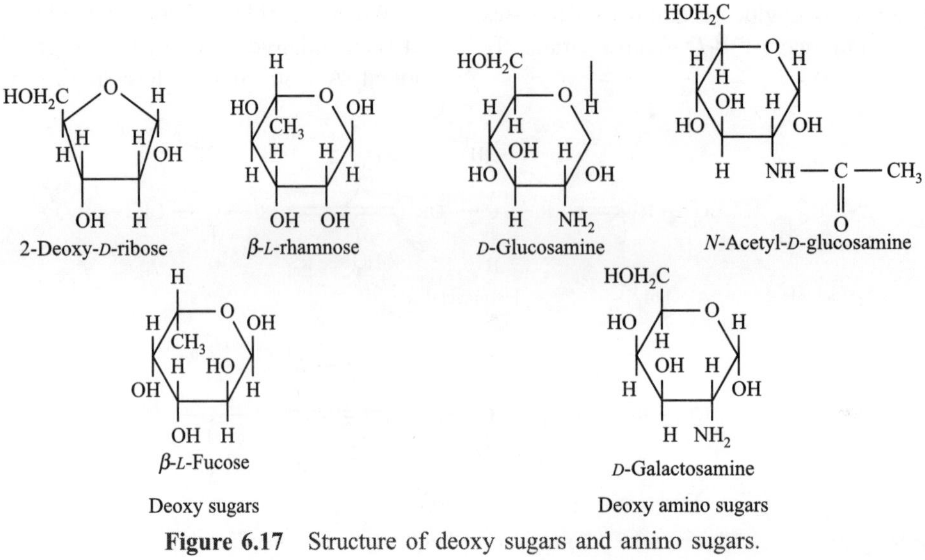

Figure 6.17 Structure of deoxy sugars and amino sugars.

Sugar alcohols or polyols: Most sugar alcohols are reduction products of aldoses and ketoses. Sorbitol and mannitol are the examples. *myo*-Inositol is a natural cyclic polyol and is of considerable interest in biochemistry.

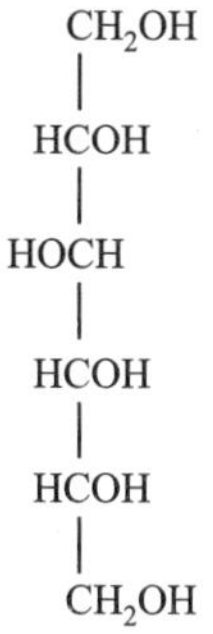

CH_2OH
HCOH
HOCH
HCOH
HCOH
CH_2OH

Sorbitol

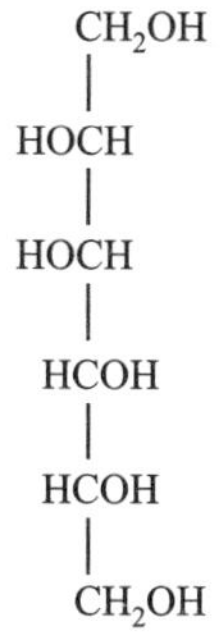

CH_2OH
HOCH
HOCH
HCOH
HCOH
CH_2OH

Mannitol

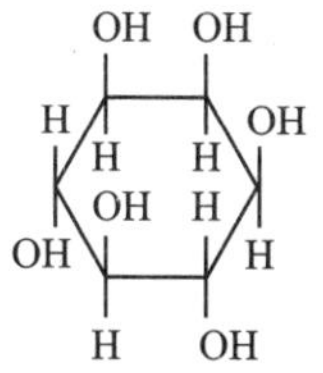

OH OH
H
OH
H H
OH H
OH
H
H OH

myo-Inositol

B. Oligosaccharides

A monosaccharide can react, through one of its hydroxyl groups, with the hydroxyl on the anomeric carbon of another monosaccharide and form a glycoside, which is a disaccharide. An oligosaccharide (oligo = few) is formed when a few monosaccharides are linked together through glycosidic bonds. The term 'oligosaccharide' is used to refer to disaccharides, trisaccharides and the higher homologues containing upto 10 monosaccharide residues. Upon hydrolysis, the oligosaccharides yield the constituent monosaccharides. Those which contain free sugar (hemiacetal) groups exist in α- and β-forms, just as the monosaccharides do.

Some disaccharides are commonly found in nature, whereas the higher homologues are very rare in nature. Some of the common disaccharides are with their constituent monosaccharides are given in the following table. Those which possess free sugar groups are reducing sugars such as the monosaccharides. The reducing property is exhibited by all sugars which have the free sugar (hemiacetal or hemiketal) group. In some disaccharides, the glycosidic bond is between the anomeric carbon atoms of the two monosaccharides. Such disaccharides do not have free sugar groups and are non-reducing sugars (e.g., sucrose).

Disaccharides ($C_{12}H_{22}O_{11}$)	*Constituent monosaccharides*
I. Reducing sugars	
lactose (milk sugar)	Glucose, galactose
Maltose	Glucose, glucose
Cellobiose	Glucose, glucose
Melibiose	Glucose, galactose
II. Non-reducing sugars	
Sucrose	Glucose, fructose
Trehalose	Glucose, glucose

Sucrose: Sucrose occurs in varying amounts in the juices of plants, such as sugarcane, sugar beets, pineapple, sugar maple and many others. It is the most abundantly distributed of the sugars. The common sugar of commerce and the kitchen (cane sugar or beet sugar) is, in fact, sucrose. Hydrolysis of sucrose by dilute acid, or by the enzyme invertase or sucrase, produces one molecule each of the constituent sugars glucose and fructose. Sucrose is not a reducing sugar and in general does not give the reactions characteristic of the sugar group (hemiacetal group or hemiketal group). The fact that sucrose has no free sugar group indicates that the linkage of glucose to fructose involves 1-hydroxyl (anomeric carbon,*) of glucose and 2-hydroxyl (anomeric carbon,*) of fructose.

Sucrose exhibits a positive (dextro-) optical rotation in a polarimeter and when it is hydrolysed, the optical rotation changes to negative (levo-). This is because *D*-fructose, one of the products of hydrolysis, is more levorotatory than the other, *D*-glucose, which is dextrorotatory.

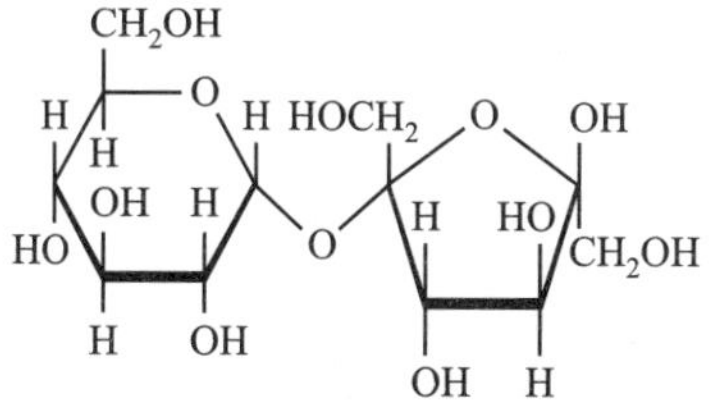

CH_2OH
O
H
H
OH H
HO
H OH
H $HOCH_2$
O
O
OH
H HO
CH_2OH
OH H

Sucrose

The hydrolysis of sucrose is known as inversion of sucrose because of the inversion of sign of rotation during the reaction, and equimolar reaction of glucose, and the fructose formed is known as invert sugar.

Lactose: This disaccharide occurs in milk to the extent of about 5 per cent and is commercially obtained from milk. Upon hydrolysis, lactose yields one molecule of galactose and one molecule of glucose.

The glycosidic bond in lactose is between 1-hydroxyl (anomeric carbon) of galactose and 4-hydroxyl of glucose (β-1,4). Hence, it is also named as 4-0-β-*D*-galactopyranosyl-*D*-glycopyranose. The hemiacetal part of glucose is free. Hence, lactose is a reducing sugar and gives reactions characteristic of free sugar groups.

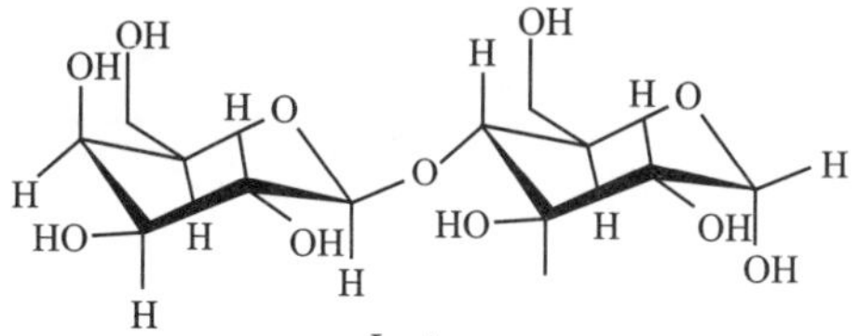

OH
OH
H O
H
HO
H
OH
H
H
O
OH
H
H O
H
HO
H
OH
OH

Lactose

Maltose: Maltose is composed of two glucose units and formed when starch is hydrolysed by the enzyme amylase or diastase. It is the product of action of salivary amylase and pancreatic amylase upon starch in the process of digestion. Commercial malt sugar contains maltose which upon acid hydrolysis yields two molecules of *D*-glucose. Maltase, an enzyme which catalyses the hydrolysis of α-glycosidic linkages only, and not β-glycosidic linkages, hydrolyses maltose to give *D*-glucose. The glycosidic bond in maltose is in the α-configuration between 1-hydroxyl and 4-hydroxyl of the two glucose residues (α-1,4). Hence, it is also named as 4-0-α-*D*-

glucopyranose. The hemiacetal part of one of the glucose residues is free. Hence, maltose is a reducing sugar and gives reactions characteristic of free sugar group. Maltose, with its α-1,4-glycosidic bond, is one of the several glycosylglucose disaccharides known.

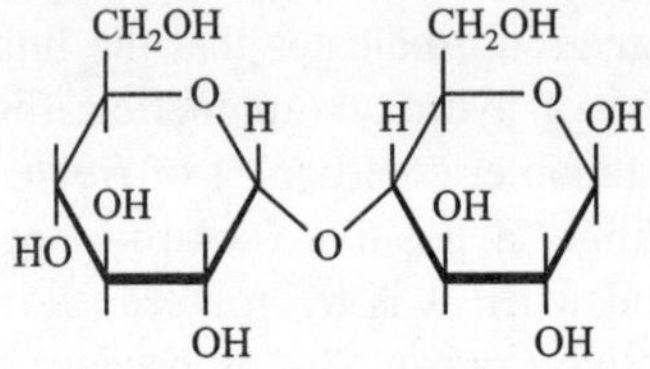

Maltose

Cellobiose is another form with β-1,4-glycosidic bond. In gentiobiose, the link is β-1,6, whereas in isomaltose, it is α-1,6.

C. Polysaccharides

Carbohydrates composed of 10 or more monosaccharide units are generally classified as polysaccharides. They are polymers in which the monosaccharides are joined together by glycosidic linkages. The bulk of all carbohydrates of nature exists as polysaccharides, which on hydrolysis yield monosaccharides or products related to monosaccharides, most frequently *D*-glucose. Others include *D*-mannose, *D*- and *L*–galactose, *L*-arabinose, *D*-glucoronic acid, *D*- and *L*-galactose, *L*-arabinose, *D*-glucoronic acid, *D*-glucosamine, etc., various polysaccharide differ not only in monosaccharide composition but also in structural features and molecular size. Some are linear polymers and others are branched. Some typical polysaccharides are starches, glycogens, celluloses, agar, gum Arabic, mucopolysaccharides, chitin, etc. Polysaccharides can be classified as homopolysaccharides and heteropolysaccharides.

Homopolysaccharides

A homosaccharide is made up of a single kind of monosaccharide. Starches, glycogens, and celluloses, which are made up of only glucose residues, are example of homopolysaccharides.

Starches: The starches occur widespread as reserve carbohydrate in tubers such as potatoes, in many fruits, grains and seeds. The starches usually occur as grains which may be spherical, oval or irregular in shape. In the grains, the starch is arranged in concentric layers. When starch grains are treated with boiling water, the substance present at the centre passes into the solution, but the greater part of the grain is not soluble. This insoluble portion swells as it absorbs water and the whole mass becomes starch paste. Both the soluble portion and the insoluble portion are heterogeneous mixtures. The soluble fraction is referred to as amylose and insoluble fraction as amylopectin. Most starches contain 80–90 per cent amylopectin and 10–20 per cent amylose. Amylose and amylopectin can be separated by taking advantage of the difference in solubility in water. Both amylose and amylopectin are polymers of glucose and upon hydrolysis with acid, they give *D*-glucose as the product.

Structure of amylose: Amylose is an unbranched long chain polymer in which the glucose residues are linked through α-1,4-glycosidic linkages. The amylose structures may be regarded as a repeated maltose structure with a free sugar group (acetal group) at one end. This is also known as reducing end (*n*th residue), whereas the opposite end (first end) is referred to as the non-reducing end. Any particular preparation of amylose usually consists of a mixture of populations of molecules which differ widely in chain length (number of glucose residues per chain).

Structure of amylopectin: Amylopectin is also made of chains of glucose residues, but the chains are highly branched.

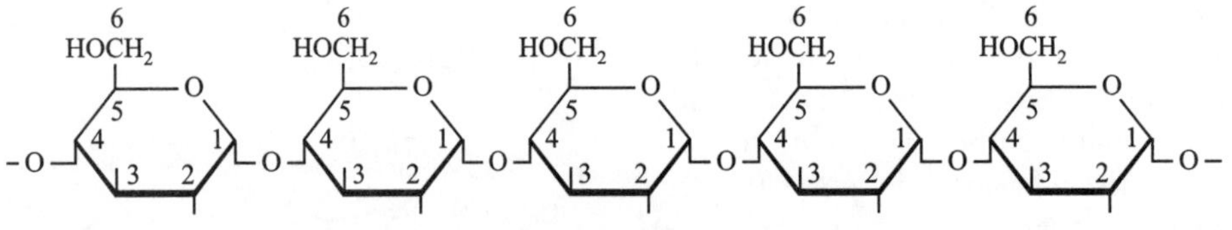

Amylose

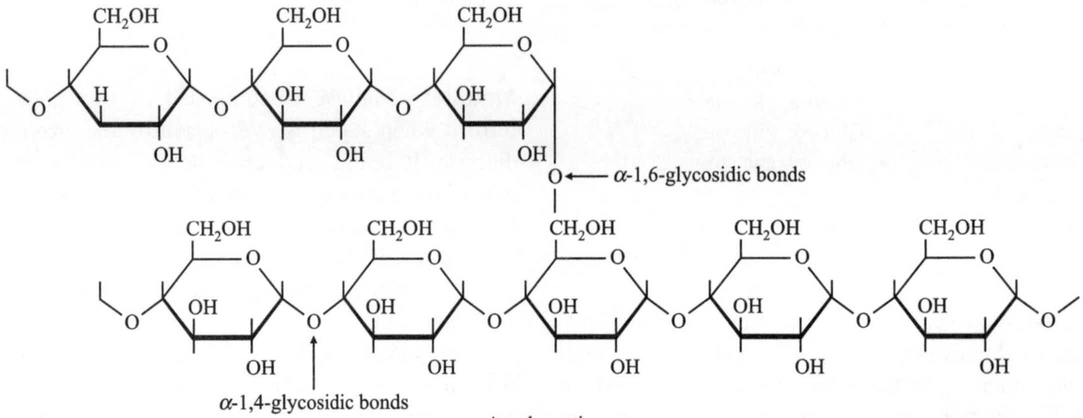

α-1,6-glycosidic bonds

α-1,4-glycosidic bonds

Amylopectin

The residue situated at the branching point is substituted not only on carbon 4 but also on carbon 6.

Action of amylases upon starch: Amylases are enzymes of plant and animal origin and they hydrolyse starch. There are two kinds of amylases, α-amylases and β-amylases. α-Amylases act on amylose and amylopectin in a random fashion. Initially, more central linkages are cleaved and smaller polysaccharides are formed. As the reaction proceeds, these are further hydrolysed to maltose and glucose.

β-Amylase hydrolyzes amylose and amylopectin in an orderly fashion. It acts from one end of the polymer and cleaves off two glucose residues at a time as a maltose unit. As the reaction proceeds, amylose is completely hydrolysed to maltose. However, with the amylopectin, the reaction stops as a branch point is approached. The terminal parts of the branches are digested away as maltose units in an orderly fashion and the central core is left behind as the enzyme is blocked at 1,6–glycosidic linkages or branch points. The

polysaccharide fragment that remains after such incomplete hydrolysis is called a *dextrin*.

Both amylose and amylopectin give characteristic colour reactions with iodine. Amylose produces a blue-black colour (dextrins formed from amylopectin by β-amylase are highly branched and give red colour with iodine. These are called *erythrodextrins*. Dextrins of a relatively small molecular size that do not give a colour with iodine are called *achrodextrins*). The colour produced by reaction with iodine is used as an indication of the degree of branching of starch.

Starch is readily hydrolyzed by dilute mineral acid with ultimate formation of glucose in quantitative yield. The course of hydrolysis may be followed by the gradual change in colour produced by iodine: blue-black to purple to red to colourless.

Celluloses: Cellulose is a polysaccharide consisting of a linear chain of several hundred to over ten thousand β-$(1 \rightarrow 4)$-linked D-glucose units. Cellulose is an important structural

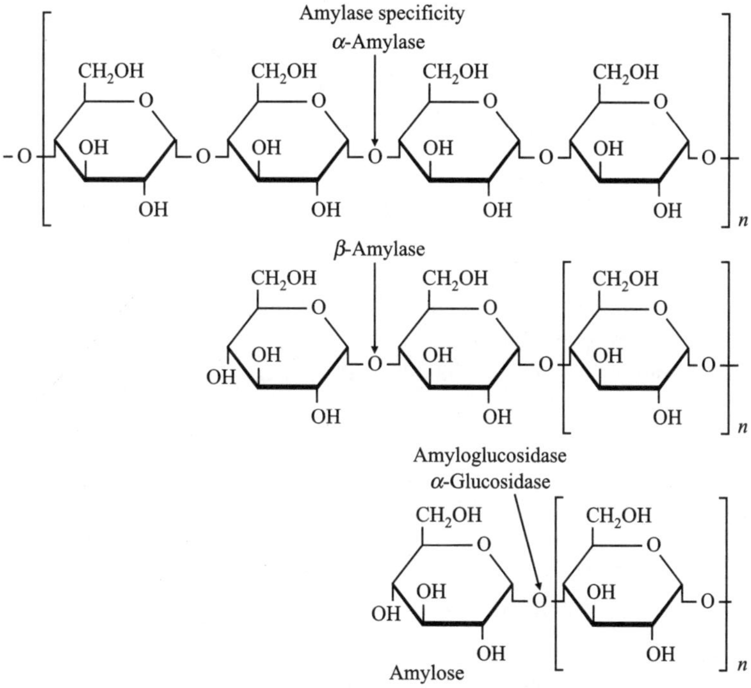

Amylase specificity

Amyloglucosidase
α-Glucosidase

Amylose
Polymer of α-(1-4)-D-glycopyranosyl units

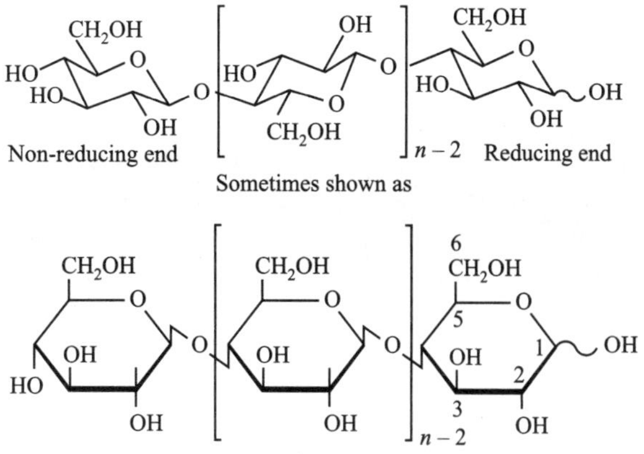

Sometimes shown as

Cellulose

component of the primary cell wall of green plants, many forms of algae and oomycetes. Some species of bacteria secrete it to form biofilms. Cellulose is the most abundant organic polymer on Earth. The purest form of cellulose is usually obtained from cotton. Cellulose is a polymer made up of glucose residues. Upon hydrolysis, cellulose yields *D*-glucose as the product. However, on partial hydrolysis with acid, cellulose is obtained as one of the products. Cellulose molecules are not branched and consist essentially of long chains with glucose residues linked in repeating sequence of cellobiose structures. The celluloses obtained from different sources differ in molecular size, but they are all made of glucose.

Glycogen: Glycogen is a polysaccharide that is the principal storage form of glucose (Glc) in animal and human cells. Glycogen is found in the form of granules in the cytosol in many cell types. Hepatocytes (liver cells) have the highest concentration of glycogen–upto 8% of the fresh weight in well fed state or 100-120 g in an adult. In the muscles, glycogen is found in a much lower concentration (1% of the muscle mass), but the total amount exceeds that in liver. Small amounts of glycogen are found in the kidneys, and even smaller amounts are found in certain glial cells in the brain and white blood cells. Glycogen plays an important role in the glucose cycle.

Glycogen is a branched biopolymer consisting of linear chains of glucose residues, with further chains branching off every 10 glucoses or so. Glucoses are linked together linearly by $\alpha(1 \rightarrow 4)$ glycosidic bonds from one glucose to the next. Branches are linked to the chains they are branching off from by $\alpha(1 \rightarrow 6)$ glycosidic bonds between the first glucose of the new branch and a glucose on the stem chain. Structure of glycogen is similar to that of amylopectin. The average molecular weights of glycogen preparations vary from 270,000 top 100,000,000.

Glycogen

Chitin: Chitin, a $\beta(1,4)$-linked homopolymer of *N*-acetyl-glucosamine, is a simple polysaccharide that is present in the cell walls of all fungi. The only chemical difference from cellulose is the replacement of a hydroxyl group at C-2 with an acetylated amino group. Chitin forms extended fibres similar to those of cellulose band as cellulose is indigestible by vertebrate animals. Chitin is the principal component of the hard exoskeletons of nearly a million species of anthropds, e.g., insects, lobsters and crabs, and is probably the second most abundant polysaccharide, next to cellulose, in nature.

Chitin

Heteropolysaccharides

Heteropolysaccharides contain two or more different kinds of monosaccharides. Usually they provide extracellular support for organisms of all kingdoms: the bacteria cell envelope or the matrix that holds individual cells together in animal tissues and provides protection, shape and support to cells, tissues and organs.

Heteropolysaccharides provide extracellular support to very different organisms, from bacteria to humans, together with fibrous proteins, such as collagen, elastin, fibronectin, laminin and others. Heteropolysaccharides are the most important components of the extracellular matrix. Hyaluronic acid, chondroitin sulfates and dermatan sulfates are important heteropolysaccharides in the extracellular matrix. These heteropolysaccharides usually are formed by the repetition of a disaccharide unit of an amino sugar and an acid sugar. Hyaluronic acid consists of *N*-acetyl-*D*-glucosamine and *D*-glucuronic acid in equimolar proportions.

D. Reactions of Carbohydrates

Carbohydrates are characterised by the reactions of the component functional groups alcohols, aldehydes or ketones and acetals or ketals. Glucose exhibits the characteristic reactions of hydroxyl and aldehyde groups, whereas fructose exhibits those of the hydroxyl and keto groups. The reactions of oligo- and polysaccharides depend upon their constituent monosaccharides. A clear understanding of the reactions of carbohydrates in general can be gained by studying the reactions of monosaccharide.

Reactions of Monosaccharides

Osazone

Phenylhydrazine ($C_6H_5NHNH_2$) reacts with carbons #1 and #2 of reducing sugars (monosaccharides) to form derivatives called osazones. The formation of these distinctive crystalline derivatives is useful for comparing the structures of sugars. The reaction involves the carbonyl carbon (aldehyde or keto group) and the next adjacent carbon. The reaction is carried out by adding a mixture of phenylhydrazine hydrochloride and sodium acetate to the sugar solution and heating in a water bath. The corresponding phenylhydrazone is first formed, which

$$\underset{\text{Aldose + Phenylhydrazine}}{\overset{\displaystyle \begin{array}{c} HC=O \\ | \\ HC-OH \\ | \\ R \end{array}}{} \;+\; H_2N-\overset{\displaystyle H}{\underset{\displaystyle}{N}}-C_6H_5} \longrightarrow \underset{\text{Phenylhydrazone}}{\overset{\displaystyle \begin{array}{c} HC=N-\overset{H}{N}-C_6H_5 \\ | \\ HC-OH \\ | \\ R \end{array}}{}}$$

$$\overset{\displaystyle \begin{array}{c} HC=N-\overset{H}{N}-C_6H_5 \\ | \\ H-C-OH \\ | \\ R \end{array}}{} + H_2N-\overset{H}{N}-C_6H_5 \longrightarrow \overset{\displaystyle \begin{array}{c} HC=N-\overset{H}{N}-C_6H_5 \\ | \\ C-O \\ | \\ R \end{array}}{} + C_6H_5-NH_2-NH_3$$

$$\overset{\displaystyle \begin{array}{c} HC=N-\overset{H}{N}-C_6H_5 \\ | \\ H-C=O \\ | \\ R \end{array}}{} + H_2N-\overset{H}{N}-C_6H_5 \longrightarrow \underset{\text{Osazone}}{\overset{\displaystyle \begin{array}{c} HC=N-\overset{H}{N}-C_6H_5 \\ | \\ C-N-\overset{H}{N}-C_6H_5 \\ | \\ R \end{array}}{}}$$

The reaction with ketose is very similar:

$$\underset{\text{Ketose}}{\overset{\displaystyle \begin{array}{c} H_2COH \\ | \\ C=O \\ | \\ R \end{array}}{}} \xrightarrow{\;C_6H_5-NH-NH_2\;} \underset{\text{Phenylhydrazone}}{\overset{\displaystyle \begin{array}{c} H_2COH \\ | \\ O=N-NH-C_6H_5 \\ | \\ R \end{array}}{}}$$

$$\overset{\displaystyle \begin{array}{c} H_2C-OH \\ | \\ C=N-NH-C_6H_5 \\ | \\ R \end{array}}{} \xrightarrow{\;C_6H_5-NH-NH_2\;} \overset{\displaystyle \begin{array}{c} HC=O \\ | \\ C=N-NH-C_6H_5 \\ | \\ R \end{array}}{} \xrightarrow{\;C_6H_5-NH-NH_2\;}$$

$$\underset{\text{Osazone}}{\overset{\displaystyle \begin{array}{c} HC=N-NH-C_6H_5 \\ | \\ C=N-NH-C_6H_5 \\ | \\ R \end{array}}{}}$$

is then converted to the corresponding osazone by further reaction with phenylhydrazine. Osazone is the final product. The osazones have characteristic crystal structures, melting points and precipitation times and are valuable in the identification of sugars. The reaction is given not only by monosaccharides but also by other carbohydrates which have the free sugar group (hemiacetal or reducing end group). Thus, lactose, which has a free sugar group, forms osazone whereas no osazone is formed by sucrose which does not have the free group.

Methylation

The hydroxyl groups of carbohydrates can be methylated by methylating reagents such as dimethyl sulphate or methyl iodide. The number of methyl groups taken by a molecule is a measure of the number of hydroxyl groups possessed by the carbohydrate.

Acetylation

The hydroxyl groups of sugars and other carbohydrates undergo acetylation upon treatment with reagents such as acetyl chloride or acetic anhydride. The number of acetyl groups taken up per molecule is a measure of the number of hydroxyl groups possessed by the carbohydrates. The glucose forms pentaacetate.

Reduction to form sugar alcohols

Both aldoses and ketoses may be reduced to the corresponding polyhydroxy alcohols by reducing agents such as sodium amalgam, nascent hydrogen produced by electrolysis or hydrogen at high pressure in the presence of a catalyst. On reduction, glucose yields sorbitol, mannose yields mannitol and fructose yields mannitol and sorbitol.

Oxidation

Many oxidizing agents oxidize aldoses. Sodium hypoiodite, NaOI (or I_2 + NaOH), oxidizes aldoses to give the sodium salt of aldonic acids by the conversion of aldehyde group to carboxyl group. Thus, glucose is oxidized to gluconic acid and ketoses are not oxidized by NaOH, and hence this is a method for distinguising between aldoses and ketoses. When an aldose is oxidized in such a way that the aldehyde group is protected while the primary alcoholic group undergoes oxidation, the product formed is the corresponding alduronic acid. Thus, glucuronic acid is obtained from glucose.

Nitric acid, under proper conditions, oxidizes both aldehyde and primary alcoholic groups of an aldose to give the corresponding dicarboxylic acid. Thus, gives glucaric acid. Monosaccharides and other carbohydrates which possess free sugar group are susceptible to oxidation by cupric ion, silver ion and ferricyanide ion in an alkaline solution. Alkaline copper sulfate is thus reduced to form red cuprous oxide and ammoniacal silver nitrate is reduced to metallic silver by reducing sugars.

Action of alkalies

When glucose is treated with dilute alkali at low temperatures, it results in the formation of a mixture of glucose, mannose and fructose. The same mixture is formed when any of these sugars is allowed to stand in dilute alkali. Thus in dilute akali, interconversion of related sugars takes place. The property is the characteristic of all carbohydrates possessing free sugar groups.

Action of acids

Polysaccharides are hydrolysed into their constituent monosaccharides by boiling with dilute (0.5–1.0 N) mineral acids, such as hydrochloric or sulfuric acids. Monosaccharides are stable to these hot dilute acids. When the concentration of the acid is increased to several normalities, not only the polysaccharides are hydrolyzed but also the monosaccharides undergo dehydration. From pentoses, the product obtained is furfural.

With hexoses, 5-hydroxymethylfurfural is first formed, which on further heating is transformed to levulinic acid. The aldehydic products of these reactions readily polymerize to give brown tars. They also condense with various phenols, carbazole and anthrone to give characteristically coloured products. Many of the colour tests for sugars depend upon such condensations.

III. TESTS FOR CARBOHYDRATES

One of the main biochemical tests that distinguishes carbohydrates from proteins is Molisch's test. All carbohydrate classes give this test positive.

Molisch test

Procedure: To 2 ml of sugar solution, 2 drops of fresh 10% α-naphthol in alcohol is added and mixed. Then 2 ml of concentrated sulphuric acid is added slowly to form a layer in the mixture. A red violet ring is formed between the two layers.

Theory: This is a very sensitive general reaction for carbohydrates. On adding concentrated sulphuric acid to the aqueous solution of carbohydrate containing alcoholic solution of α-naphthol, a deep violet colour appears at the junction of the two liquids. Concentrated sulphuric acid hydrolyses lycosidic bonds of carbohydrate to give monosaccharides which are then dehydrated to an aldehyde known as furfural that undergoes reaction with α–naphthol to give an unstable condensation product of deep violet colour.

Anthrone test

Procedure: To 2 ml of anthrone reagent (0.2% in conc. H_2SO_4), 0.2 ml of sugar solution is added. A blue-green colour is given by glucose and polysaccharides containing glucose. Different colours are obtained with different sugars. This test is generally used for hexoses and hexose containing polysaccharides.

Theory: The anthrone test is a colorimetric test that causes blue-green colour to appear when sugar is present in a sample. Carbohydrates are dehydrated with concentrated H_2SO_4 to form "furfural", which condenses with anthrone to form a green coloured complex that can be measured colorimetrically at 620 nm or measured by using a red filter. Anthrone reacts with dextrins, monosaccharides, disaccharides, polysaccharides, starch, gums and glycosides. Nevertheless, the anthrone method is widely used for the determination of starch and soluble sugars in plant materials.

Bial's test

Procedure: 5 ml of Bial's reagent (solution of orcinol in 30% HCl containing a little ferric chloride) is heated to boiling and a few drops of the sugar solution are added. Pentoses and aldohexuronic acids (which are converted to pentoses) give a green colour.

Theory: Pentoses can be distinguished from hexoses by this test. A pentose, if present, will be dehydrated to form furfural, which then reacts with the orcinol to generate a coloured substance. The solution will turn bluish and a precipitate may form. This test is named after Manfred Bial, German physician.

Phloroglucinol–hydrochloric acid test

Procedure: Equal volumes of sugar solution and a solution of phloroglucinol in hydrochloric acid are mixed and boiled. Pentoses give a cherry red colour. Galactose also gives this test. Hexuronic acids may be distinguished from pentoses by this test.

Theory: Lignin, which is a natural component of wood, is commonly detected by this test. The cinnamaldehyde end groups of lignin appear to react with phloroglucinol–HCl to give a red-violet colour. Although the reaction is not very sensitive, because of the ease of staining, this procedure is still often used as one of the tests for the presence of lignin in plant cell wall.

Naphtho-resorcinol test

Procedure: To 5 ml of the sugar solution in 1N HCl, 1 ml of 1% alcoholic solution of naphtho-resorcinol is added and heated for 5 min in a boiling water bath. The mixture is cooled and extracted with ether. A violet-red colour in the ether extract indicates the presence of hexuronic acids.

Theory: Hexuronic acids may be distinguished from pentoses by this test.

Seliwanoff's test (Resorcinol hydrochloric acid test)

Procedure: Seliwanoff's reagent is prepared by adding 3.5 ml of 0.5% resorcinol to 12 ml of conc. HCl and diluting to 35 ml with water. Add freshly prepared 5 ml of Seliwanoff's reagent to 1 ml of sugar solution in a test tube. Heat the test tube in boiling water for 2–10 min. Ketohexoses give red colour. Ketopentose gives blue-green colour. Aldoses do not give colour within 2 min.

Theory: It is a test to distinguish ketose from aldose. Ketoses dehydrate very rapidly under acidic conditions to give furfural, which reacts with resorcinol (1,3-dihydroxybenzene) to give a coloured product. Ketohexoses give red colour and ketopentoses give blue-green colour. Aldoses take a longer time to produce colour because under the same conditions, aldoses form furfural slowly, probably because β-elimination is required before dehydration to furfural. Therefore, prolonged heating should be avoided.

Benzidine test

Procedure: Add 2 drops of sugar solution to 1 ml of benzidine solution, boil and cool quickly. Pentoses give a violet colour.

Barfoed's Test

Procedure: Take 1 ml of sugar solution in a test tube and add 5 ml Barfoed's reagent (copper acetate and acetic acid). Heat the content of the test tube in a water bath to boiling for 5 min. Sheer orange-red precipitate is formed in 5–7 min by reducing monosaccharides (aldoses and ketoses) and in 7–12 min by reducing disaccharides.

Theory: Barfoed's reagent, cupric acetate in acetic acid, is slightly acidic and is balanced so that it can be reduced only by monosaccharides and not by less powerful reducing sugars. Disaccharides may also react with this reagent, but the reaction is much slower when compared to monosaccharides. Non-reducing disaccharides such as sucrose may give slight precipitates due to hydrolysis and formation of reducing monosaccharide.

Iodine test

Procedure: Make a suspension of starch (0.5 g) in 5 ml water and pour it in 50 m boiling water to get an aqueous colloidal solution. A drop of the solution of a polysaccharide acidified with HCL is added to a solution of iodine in KI. The appearance of blue colour indicates the presence of starch.

Theory: Starch gives blue colour with iodine solution due to the formation of a complex known as starch–iodide complex. Starch is present in wheat, rice, maize, potatoes, etc. However, glycogen and erythrodextrin give a red colour. Starches, glycogens and dextrins can be distinguished from other carbohydrates by this test.

Osazone Formation

Procedure: Acidify 2 ml of sugar solution with 5 drops of glacial acetic acid. Add 1 ml of phenylhydrazine solution and 1 ml of sodium acetate solution and mix well. Place the test tube in a boiling water bath for 30 min and cool. The osazone crystals formed are isolated and their crystalline shapes and melting points are determined. Osazones of monosaccharides (glucose and fructose) are formed on hot after about 15 min and appear to have the same crystal shape needles under the microscope. Osazones of reducing disaccharides (maltose and lactose) are formed after a longer time (up to 30 min) and crystal appears slowly after cooling and can be distinguished under the microscope because of following features:

- Lactose gives crystals in the form of a tuft of needles.
- Maltose gives crystals in the form of broad needles.

Theory: Compounds containing the $—CO—CHOH$ group form crystalline osazone compounds when heated with phenylhydrazine. Osazone crystals have a characteristic shape under the light microscope and help in the identification of the sugar type. The reaction is stepwise: first, phenylhydrazine reacts with carbonyl group of the sugar to form phenylhydrazone, which then reacts with two further molecules of phenylhydrazine to form the osazone.

IV. LIPIDS

A lipid is that fraction of animal and vegetable tissues which can be extracted with organic solvents such as ether, chloroform, benzene and petroleum ether. This is an operational definition and not based on structure. Lipids are a heterogeneous group of structurally related and unrelated compounds. A wide variety of lipids occur in animal and plant tissues and they are purified from these sources. Many lipids are also obtained synthetically. The prominent lipids are classified on the basis of their chemical composition in the following way:

A. Fatty acids
B. Lipids containing glycerol

1. Neutral fats
2. Phospholipids

C. Lipids not containing glycerol

1. Sphingolipids
2. Aliphatic alcohols and waxes
3. Terpenes
4. Steroids

D. Lipids in complex with other classes of compounds

1. Cerebrosides
2. Gangliosides

A. The Fatty Acids

To understand the structure and properties of many classes of lipids, first it is necessary to study the structure and properties of fatty acids. Fatty acids are not only found in ester or amide linkages in many classes of lipids but are also found in the free form to a lesser degree. The naturally occurring fatty acids are monocarboxylic acids with unbranched hydrocarbon chains. Most fatty acids have an even number of carbon atoms, although some contain an odd number of carbon atoms. Some are saturated and others are unsaturated. The unsaturated fatty acids may contain one or more double bonds.

Saturated fatty acids: The properties of saturated fatty acids depend partly on the polar hydroxyl group and partly on the non-polar hydrocarbon chain. Thus, acetic and propionic acids are miscible with water. Butyric and caproic acids have limited solubility. Higher members of the series are insoluble in the water although they are readily soluble in non-polar organic solvents. Boiling points and generally points of fatty acid rise with increasing chain length. Saturated fatty acids of less than 10 carbon atoms are liquids at room temperature, whereas, the higher members are solids.

Fatty acids are present in both animal and vegetable fats. Animal fats are generally rich in saturated fatty acids and vegetables are rich in unsaturated fatty acids. Palmitic acid (C_{16}) is generally the most abundant saturated fatty acid in animal fats, with stearic acid (C_{18}) second in amount (Table 6.2). Fatty acids of 10 carbon atoms and less are rarely present in animal fats.

TABLE 6.2   Some Saturated Fatty Acids

Common Name	Systematic Name	Structural Formula	Lipid Numbers
Propionic acid	Propanoic acid	CH_3CH_2COOH	C3:0
Butyric acid	Butanoic acid	$CH_3(CH_2)_2COOH$	C4:0
Valeric acid	Pentanoic acid	$CH_3(CH_2)_3COOH$	C5:0
Caproic acid	Hexanoic acid	$CH_3(CH_2)_4COOH$	C6:0
Enanthic acid	Heptanoic acid	$CH_3(CH_2)_5COOH$	C7:0
Caprylic acid	Octanoic acid	$CH_3(CH_2)_6COOH$	C8:0
Pelargonic acid	Nonanoic acid	$CH_3(CH_2)_7COOH$	C9:0
Capric acid	Decanoic acid	$CH_3(CH_2)_8COOH$	C10:0
Undecylic acid	Undecanoic acid	$CH_3(CH_2)_9COOH$	C11:0
Lauric acid	Dodecanoic acid	$CH_3(CH_2)_{10}COOH$	C12:0
Tridecylic acid	Tridecanoic acid	$CH_3(CH_2)_{11}COOH$	C13:0
Myristic acid	Tetradecanoic acid	$CH_3(CH_2)_{12}COOH$	C14:0
Pentadecylic acid	Pentadecanoic acid	$CH_3(CH_2)_{13}COOH$	C15:0
Palmitic acid	Hexadecanoic acid	$CH_3(CH_2)_{14}COOH$	C16:0
Margaric acid	Heptadecanoic acid	$CH_3(CH_2)_{15}COOH$	C17:0
Stearic acid	Octadecanoic acid	$CH_3(CH_2)_{16}COOH$	C18:0

Unsaturated fatty acids: This group of acids is characterized by having one or more double bonds or ethylenic groups. They are classified as monoenoic, dienoic, trienoic and polyenoic acids on the basis of the number of double bonds they possess. Those which possess several double bonds are also known as polyunsaturated fatty acids. The position of the unsaturation is usually indicated by the prefixing number in the systematic name, the carboxyl group being C-1. Thus, 9-Octadecenoic acid refers to a 18–carbon acid with a double bond between carbon atoms 9 and 10.

TABLE 6.3 Some Unsaturated Fatty Acids

Common Name	Chemical Structure
Myristoleic acid	$CH_3(CH_2)_3CH=CH(CH_2)_7COOH$
Palmitoleic acid	$CH_3(CH_2)_5CH=CH(CH_2)_7COOH$
Sapienic acid	$CH_3(CH_2)_8CH=CH(CH_2)_4COOH$
Oleic acid	$CH_3(CH_2)_7CH=CH(CH_2)_7COOH$
Linoleic acid	$CH_3(CH_2)_4CH=CHCH_2CH=CH(CH_2)_7COOH$
α-Linolenic acid	$CH_3CH_2CH=CHCH_2CH=CHCH_2CH=CH(CH_2)_7COOH$
Arachidonic acid	$CH_3(CH_2)_4CH=CHCH_2CH=CHCH_2C=CHCH_2CH=CH(CH_2)_3COOH$[NIST]

Unsaturation alters certain properties of a fatty acid. In general, the melting point is greatly lowered and solubility in non-polar solvents is enhanced. The common unsaturated fatty acids found in nature are liquids at room temperature.

Geometric isomerization of unsaturated fatty acids: The presence of double bond in unsaturated fatty acids gives rise to the possibility of *cis-trans* isomerization. Thus, 9-Octadecenoic acid exists in *cis-trans* forms. This *cis* form is oleic acid of nature and the *trans* form is known as elaidic acid. These acids have different melting points and other physical constants. The double bonds are of the *cis* configuration in most known naturally occurring unsaturated fatty acids.

cis Double bond: oleic acid

trans Double bond: elaidic acid

Other fatty acids: Branched-chain fatty acids, hydroxy fatty acids and fatty acids containing cyclic structure (e.g., fatty acids with cyclopentenyl ring at the end of the chain) are all found to exist rarely in nature.

Reaction of fatty acids: The reactions exhibited by fatty acids depend on the functional groups they possess. Thus, an unsaturated fatty acid exhibits reactions characteristic of carboxyl group and double bond.

1. Fatty acids dissociate in solution and are neutralized by alkali to form salts.
2. They react with alcohols and amines.
3. Unsaturated fatty acids (and their esters) can be hydrogenated in the presence of a catalyst such as Pt, Pd, Ni, Cu to give corresponding saturated fatty acids.

$$-CH=CH- + H_2 \longrightarrow -CH_2-CH_2$$

4. Unsaturated fatty acids and their esters form addition compounds with halogens such as as Br_2, I_2 and IBr

$$-CH=CH- + IBr \longrightarrow CHI-CHBr-$$

5. Saturated fatty acids are relatively resistant to oxidation, but unsaturated fatty acids are oxidised more easily. Unsaturated fatty acids are oxidised by O_2 when they are exposed to air, which occurs slowly and spontaneously. The reaction involves oxidation at the double bond, yielding a variety of products.

Soaps

Soaps are salts of long chain fatty acids. The soaps of commerce are sodium and potassium salts of fatty acids such as palmitic, stearic and oleic acids. They are soluble in water and are good emulsifying and cleansing agents. The soaps of most other metals are insoluble. For example, calcium and magnesium salts are insoluble. When commercial soap is dissolved in hard water containing calcium and magnesium salts, the corresponding calcium and magnesium soaps are precipitated, resulting in the wastage of soap. The soaps of commerce are usually prepared by the hydrolysis (saponification) of fats with NaOH. Hence, the fatty acid composition of the soap depends upon the fatty acid composition of the fat used.

The cleansing action of soap is due to the presence of both water-soluble and fat-soluble groups in the same molecule. Soap can act as detergent as it has the capacity to disrupt greasy and oily materials into stable minute droplets or micelles to give an emulsion. The hydrocarbon part of the soap interacts with the greasy or oily material, while the carboxyl part interacts with the water. Thus, in a stable micelle, the polar carboxyl groups of the soap are on the periphery, while the non-polar hydrocarbon chains are in the interior with the fat soluble material. Alcohols, alkyl sulphates and alkyl sulphonates containing long hydrocarbon chains are also used as detergents as they have both polar and long non-polar groups in their molecules.

$R-CH_2-OH$	Alcohol
$R-CH_2-O-SO_3Na$	Alkyl sulfate
$R-CH_2-SO_3Na$	Alkyl sulfonate

B. Lipids Containing Glycerol

Neutral fats: The neutral fats comprise the most abundant groups of lipids in nature, being especially abundant in nuts, seeds and the fat deposits of animals. These are esters of fatty acids with glycerol. One, two, or all three of the hydroxyl groups of glycerol may be esterified with fatty acids to give mono-, di- and triglycerides. Quantitatively, the triglycerides represent the most significant group of neutral fats, although mono- and diglycerides are also present in nature. The glycerides are of two types: simple glycerides and mixed glycerides. Simple glycerides are these in which glycerol hydroxyls are esterified by identical fatty acids. The mixed glycerides contain two or three different fatty acids. In order to simplify the naming of these compounds, the three carbons of glycerol are designated as α, β, α' or 1, 2, 3, respectively.

$$3 \quad \alpha \quad CH_2OH$$
$$|$$
$$2 \quad \beta \quad CHOH$$
$$|$$
$$1 \quad \alpha' \quad CH_2OH$$

Glycerol

The generic formula of a triglyceride is shown as

$$CH_2 - OOC - R$$
$$|$$
$$CH - OOC - R'$$
$$|$$
$$CH_2 - OOC - R''$$

where RCO_2H, $R'CO_2H$, and $R''CO_2H$ represent molecules of either the same or different fatty acids.

Properties of fats

Fat is a waxy substance found in multiple foods, including nuts, seeds, animal meats and dairy products. Fats are practically insoluble in water and, with the exception of castor oil, are insoluble in cold alcohol and only sparingly soluble in hot alcohol. They are soluble in ether, carbon disulfide, chloroform, carbon tetrachloride, petroleum benzine, and benzene. Fats have no distinct melting points or solidifying points because they are such complex mixtures of glycerides each of which has a different melting point. Saturation and increasing chain length of fatty acids tend to increase the melting point. Thus, vegetables fats (oils) which are richer in unsaturated fatty acids are solids at room temperature. Pure glycerides have no colour, odour or taste. The presence of these properties in a fat is the result of foreign substances dissolved in the glycerides. Since glycerides are esters, they give the reactions of esters. Those which contain double bonds in the fatty acid chains show the characteristic properties of unsaturation. The presence of hydroxylated fatty acids causes them to give reactions of the hydroxyl group. The following are some of the important reactions given by fats:

I. *Hydrolysis:* Glycerides can be hydrolyzed to glycerol and fatty acids. The hydrolysis is catalyzed by both acids and bases.
The hydrolysis of fats effected by alkali (NaOH or KOH) is commonly known as saponification. When fats are saponified with alkali, soaps and

glycerol, both of which are soluble in water, are formed as products.

II. *Hydrogenation:* Fats which contain unsaturated fatty acids can be hydrogenated in the presence of nickel as a catalyst. The unsaturated fats are usually liquids and upon hydrogenation, they become solids. Commercial solid cooking fats, such as vanaspati, are obtained by hydrogenation of vegetable oils.

III. *Halogenation:* Halogenation is a chemical reaction that involves the reaction of a compound, usually an organic compound, with a halogen. Halogens such as Br_2, I_2 and IBr can be added to the double bonds of unsaturated fats. Two atoms of halogen add to each double bond, producing a saturated halogenated glyceride.

IV. *Acetylation:* Fats containing hydroxylated fatty acids react with acetic anhydride and other acylating agents to form the corresponding esters.

V. *Oxidation:* Unsaturated fats can be oxidized at the double bond by a variety of oxidizing agents. When they are exposed to air, oxygen slowly oxidizes them at their double bonds, leading to different products. Certain natural oils, such as linseed oil, containing large proportions of highly unsaturated fats, when exposed to air and light, absorb oxygen and form products which polymerize to produce insoluble hard films. Such oils are used in the production of paints and varnishes. The chemical processes occurring in the hardening or drying of oils are very complex.

VI. *Rancidity of fats:* The unpleasant odour and taste developed by most natural fats upon exposure to air and aging is referred to rancidity. Rancidity may be due to the oxidation of unsaturated fats by oxygen which leads to the formation of products having unpleasant odour and taste. Rancidity may also be caused by hydrolysis of glycerides into free fatty acids and glycerol or monoglycerides and diglycerides.

Characterization of fats

The characterization of fats is generally based on properties such as specific gravity, melting point, refractive index and viscocity and on values of saponification number, iodine number, acetyl number, acid number and Reichert-Meissl number.

Saponification number is defined as the number of milligrams of KOH required to saponify 1 g of fat. The approximate mean molecular weight of the fatty acids in a triacylglycerol can be obtained from the saponification number using the formula

$$M_r = \frac{3 \times 56 \times 1000}{\text{Saponification number}}$$

Iodine number is the number of grams of iodine absorbed by 100 g of fat. In actual practice, Br_2, ICl and IBr are preferentially used as they are more active and give more correct results. However, the results are always calculated in terms of idoine and are expressed as iodine numbers.

$$H_2C - O - CO - R \qquad HO \;|\; H \qquad\qquad H_2C - OH$$
$$|$$
$$HC - O - CO - R \; + \; HO \;|\; H \;\longrightarrow\; 3RCOOH \; + \; HC - OH$$
$$| \qquad\qquad\qquad\qquad\qquad \text{Fatty acid} \qquad |$$
$$H_2C - O - CO - R \qquad HO \;|\; H \qquad \text{or acids} \qquad H_2C - OH$$

Glyceride   Glycerol

Acetyl number is the number of milligrams of KOH required to netralize the free fatty acids present in 1 g of fat.

Reichert-Meissl number of volatile fatty acid number is the number of millilitres of 0.1N alkali required to neutralize the volatile acids removed by steam distillation from 5 g of fat which is saponified and acidified.

Phospholipids: Phospholipids are essential molecules that are found in cellular membranes. There is a small but very important difference in the structure of phospholipids compared to triglycerides. Instead of having three fatty acids attached to the glycerol molecule, one is replaced by a phosphate group made up of phosphorous, oxygen and hydrogen. The phosphate group is hydrophilic, or it is attracted to water, in contrast to the rest of the molecule. The fact that one part of the molecule attracts water while the rest repels it affects the role of phospholipids in the cell membrane. They are found in every cellular organism studies and form major lipid components of membranes. Egg yolk, certain seeds and brain tissue are all rich in phospholipids. Tissues differ widely in their content of phospholipids. Brain is particularly rich in this class of lipid. All the different phospholipids are derived from α-glycerol phosphate.

The phospholipids can be grouped into the following types:

1. Phosphatidylcholine PC (lecithins)
2. Phosphatidylethanolamines (cephalins)
3. Phosphatidylserines 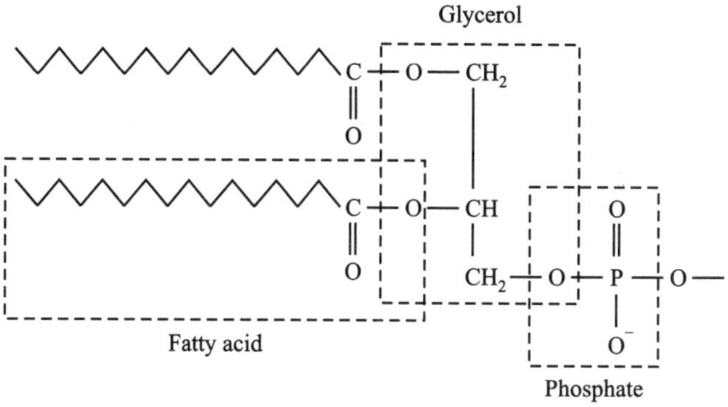
4. Other phosphatides
5. Plasmalogens

The phospholipids of the first four classes are also known as phosphatides as they are all derived from L-α-phosphatidic acid. In the phosphatides, the phosphoric acid group of the phosphatidic acid is also bound in ester linkage to other compounds, such as choline, ethanolamine, serine, inositol, phosphoinositol and glycerol.

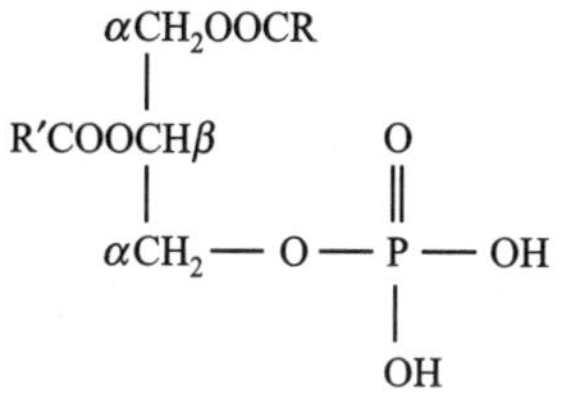

L-α-Phosphatidic acid

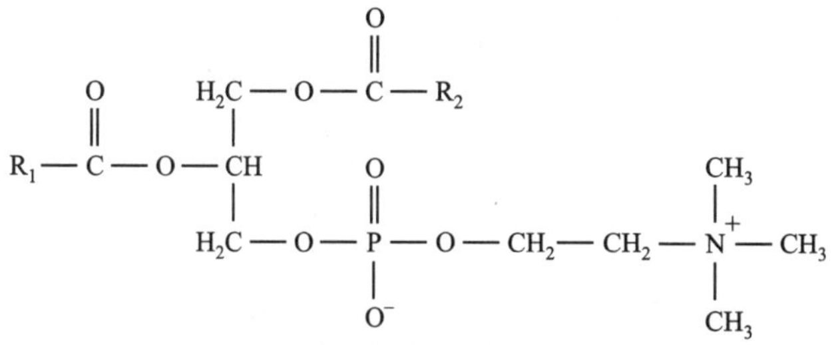

Phosphatidylcholine

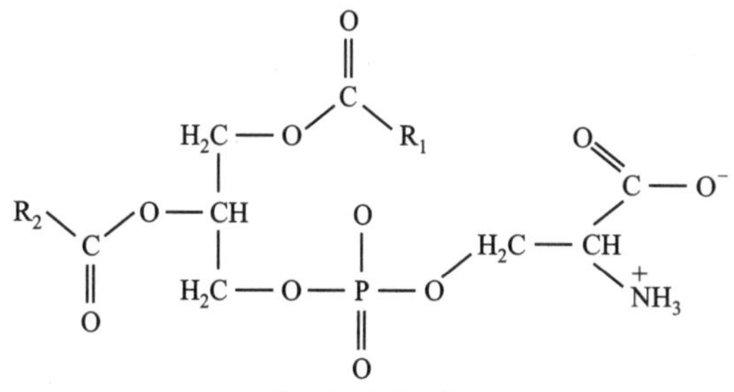

Phosphatidylserine

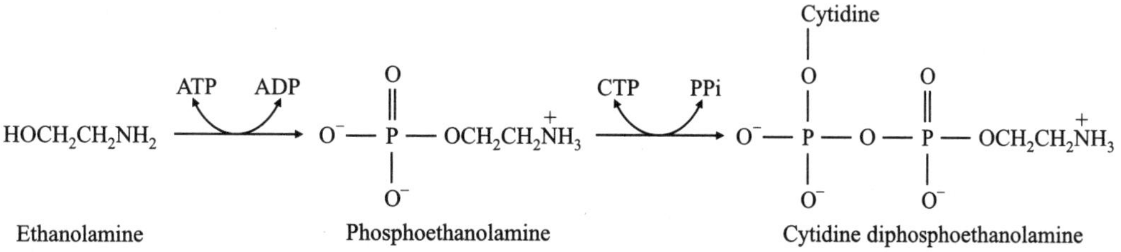

Ethanolamine — Phosphoethanolamine — Cytidine diphosphoethanolamine

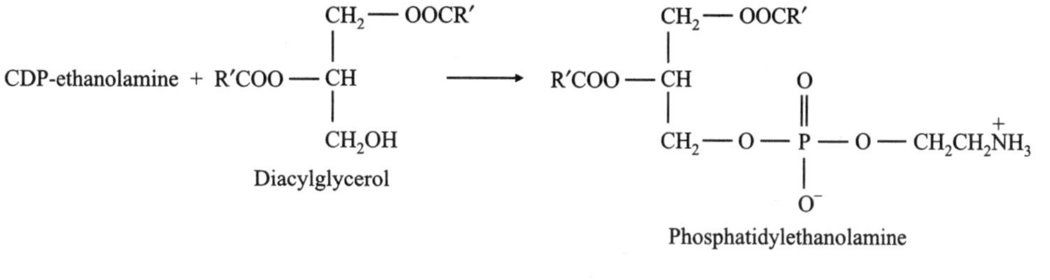

Diacylglycerol

Phosphatidylethanolamine

Phosphatidylcholine (lecithins): They all contain choline and different members differ in their fatty acid components. Most people normally ingest 3–6 g of lecithin a day through eggs, soy and meats. Vegetables, fruits and grains contain very little lecithin. PC is the most abundant phospholipid component in all cells. PC levels in brain cell membranes decline with age.

$$
\begin{array}{l}
\quad\quad\quad\quad CH_2-O-\overset{\overset{\displaystyle O}{\|}}{C}-R_1 \\
R_2-\overset{\overset{\displaystyle O}{\|}}{C}-O-CH \\
\quad\quad\quad\quad CH_2-O-\underset{\underset{\displaystyle O^-}{|}}{\overset{\overset{\displaystyle O}{\|}}{P}}-O-CH_2CH_2\overset{+}{N}(CH_3)_3
\end{array}
$$

Phosphatidylcholine

Phosphatidylethanolamine (PE) (cephalins) and phosphatidylserine (PS): It is the most abundant phospholipid in animals and plant lipids and it is frequently the main lipid component of microbial membranes. It can amount to 20% of liver phospholipids and as much as 45% of those of brain; higher proportions are found in mitochondria than in other organelles. As such, it is obviously a key building block of membrane bilayers. Phosphatidylethanolamine and phosphatidylserine resemble lecithins but contain ethanolamine and serine, respectively.

$$
\begin{array}{l}
\quad\quad\quad\quad CH_2-O-\overset{\overset{\displaystyle O}{\|}}{C}-R_1 \\
R_2-\overset{\overset{\displaystyle O}{\|}}{C}-O-CH \\
\quad\quad\quad\quad CH_2-O-\underset{\underset{\displaystyle O^-}{|}}{\overset{\overset{\displaystyle O}{\|}}{P}}-O-CH_2CH_2\overset{+}{N}H_3
\end{array}
$$

Phosphatidylethanolamine

$$
\begin{array}{l}
\quad\quad\quad\quad CH_2-O-\overset{\overset{\displaystyle O}{\|}}{C}-R_1 \\
R_2-\overset{\overset{\displaystyle O}{\|}}{C}-O-CH \\
\quad\quad\quad\quad CH_2-O-\underset{\underset{\displaystyle O^-}{|}}{\overset{\overset{\displaystyle O}{\|}}{P}}-O-CH_2\underset{\underset{\displaystyle COO^-}{|}}{C}HNH_3^+
\end{array}
$$

Phosphatidylserine

Other phosphatides: Other phosphatides which occur in nature are derived from phosphatidic acid by the attachment of compounds such as threonine, methyl ethanolamine, hydroxyproline, inositol, phosphoinositol, glycerol and polyglycerol to the phosphoric acid through ester linkage. Inositides (inositol phosphatide or phosphatidyl inositol), phosphoinositides, glycerol phosphatides (phosphatidyl glycerols) and polyglycerol phosphatides are all examples.

$$
\begin{array}{l}
\quad\quad\quad\quad CH_2-O-\overset{\overset{\displaystyle O}{\|}}{C}-R_1 \\
R_2-\overset{\overset{\displaystyle O}{\|}}{C}-O-CH \\
\quad\quad\quad\quad CH_2-O-\underset{\underset{\displaystyle O^-}{|}}{\overset{\overset{\displaystyle O}{\|}}{P}}-O\text{—(inositol ring)}
\end{array}
$$

Phosphatidylinositol (PI)

Plasmalogens: Plasmalogens make up an appreciable part of the total phospholipids in brain and in muscle. Upon hydrolysis, they yield 1 mol each of long chain aliphatic aldehyde, fatty acid, glycerol phosphate and a nitrogenous base. The aldehyde is linked to the glycerol through vinyl ether group (—O—CH=CH—).

Platelet-activating factor (PAF): It is released from white blood cells called basophils and stimulates platelet aggregation and the release of serotonin from platelets. PAF functions as a mediator of hypersensitivity, acute inflammatory reactions and anaphylactic shock. PAF is synthesized in response to the formation of antigen–IgE complexes on the surfaces of basophils, neutrophils, eosinophils, macrophages and monocytes. The synthesis and release of PAF from cells leads to platelet aggregation and the release of serotonin from platelets. PAF also produces responses in liver, heart, smooth muscle, and uterine and lung tissues.

$$
\begin{array}{l}
H_2C-O-CH_2-(CH_2)_{16}-CH_3 \\
HC-O-\underset{\underset{\displaystyle O}{\|}}{C}-CH_3 \\
H_2C-O\text{——}\underset{\underset{\displaystyle O}{|}}{\overset{\overset{\displaystyle O}{\|}}{P}}-O-CH_2CH_2\overset{+}{N}(CH_3)_3
\end{array}
$$

C. Lipids not Containing Glycerol

Sphingolipids: These are lipids derived from sphingosine, which is a long chain C-18 aliphatic nitrogenous base. Although the C-18 sphingosine is most abundant in sphingolipids, other homologues of sphinogosines such as C-16, C-17, C-19 and C-20 are also found among the naturally occurring sphingolipids.

$$
\begin{array}{l}
\quad\quad OH \quad\quad\quad\quad OH \\
\quad\quad | \quad\quad H \quad\quad | \\
H_2C-C-CH \\
\quad\quad | \quad\quad\quad\quad | \\
\quad H_3N^+ \quad\quad CH \\
\quad\quad\quad\quad\quad\quad \| \\
\quad\quad\quad\quad\quad\quad HC \\
\quad\quad\quad\quad\quad\quad | \\
\quad\quad\quad\quad\quad (CH_2)_{12} \\
\quad\quad\quad\quad\quad\quad | \\
\quad\quad\quad\quad\quad CH_3
\end{array}
$$

Sphingosine

The following are some classes of Sphingolipids:

I. *Ceramides:* Among the most simple sphingolipids are the ceramides (sphingosine and a fatty acid). These are *n*-acyl fatty acid derivatives of sphingosine and widely distributed in small amounts in plant and animal tissues. The other sphingolipids are derivatives of ceramides.

$$CH_3(CH_2)_{12}CH = CH - \overset{\overset{\displaystyle OH}{|}}{C} - H$$
$$R - \underset{\underset{\displaystyle O}{\|}}{C} - HN - \overset{}{\underset{\underset{\displaystyle CH_2OH}{|}}{C}} - H$$

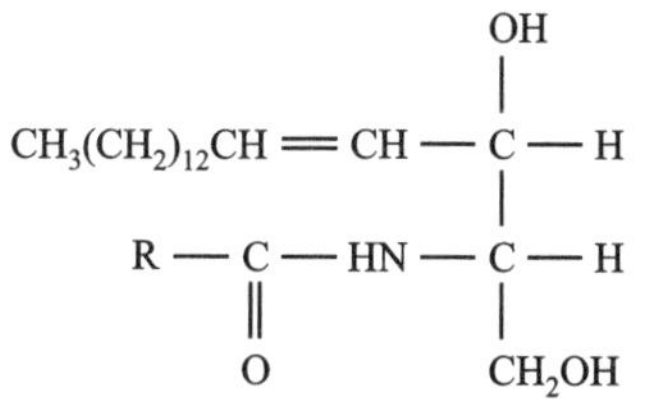

Ceramide

II. *Sphingomyelin:* Sphingomyelins are present in large amounts in brain and nerve tissues and in smaller quantities in other tissues and blood. They may be regarded as phosphorylcholine derivatives of ceramides.

III. *Cerebrosides:* Cerebrosides are sphingolipids made by attaching a sugar to a ceramide. Cerebrosides are ceramide monosaccharides occurring most abundantly in the myelin sheath of nerves. On hydrolysis, they give one molecule each of sphingosine, a fatty acid and a hexose. The hexose is most frequently *D*-galactose, and less commonly *D*-glucose. The hexoses are linked to the terminal hydroxyl group of sphingosine (of ceramide) through glycosidic bond. Galactose containing cerebroside is referred to as galactolipid or galactocerebroside. These are neutral glycolipids.

Similarly, a glucose containing cerebroside is referred to as glucolipid or glucocerebroside.

IV. *Gangliosides:* The term *ganglioside* was first applied by the German scientist Ernst Klenk in 1942 to lipids newly isolated from ganglion cells of the brain. A **ganglioside** is a molecule composed of a glycosphingolipid (ceramide and oligosaccharide) with one or more sialic acids (e.g. *n*-acetylneuraminic acid, NANA) linked on the sugar chain. NeuNAc, an acetylated derivative of the carbohydrate sialic acid, makes the head groups of gangliosides anionic at pH 7, which distinguishes them from globosides.

Aliphatic alcohols and waxes: Significant quantities of long chain aliphatic alcohols exist in nature along with their lipids. Cetyl alcohol, $CH_3(CH_2)_{14}CH_2OH$, is an example. Such esters of alcohols other than glycerol are known as waxes.

Terpenes: Numerous terpenes occur in plants and animals. They are usually long chain compounds derived from the smaller isoprene unit. Carotenes, vitamin A and essential oils such as pinene, geraniol, camphor and menthol are the examples of terpenes.

Steroids: A large number of naturally occurring compounds belong to the class of compounds known as steroids. Steroids as other lipids are usually insoluble in water, but are soluble in organic solvents and fats. Structurally, they are similar and all of them have the perhydrocyclopentanophenanthrene ring system as the nucleus. This ring system consists of three fused six-membered rings (A, B, C) and one five-membered ring (D).

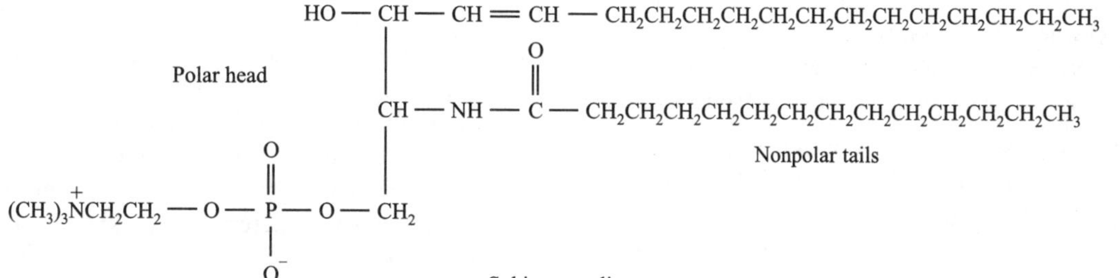

Sphingomyelin

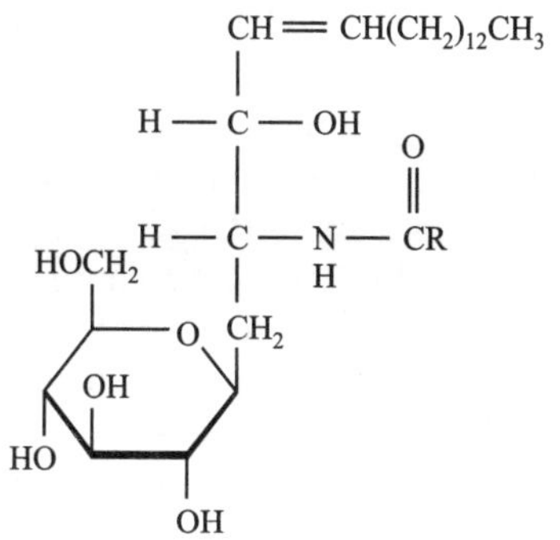

Glucocerebroside

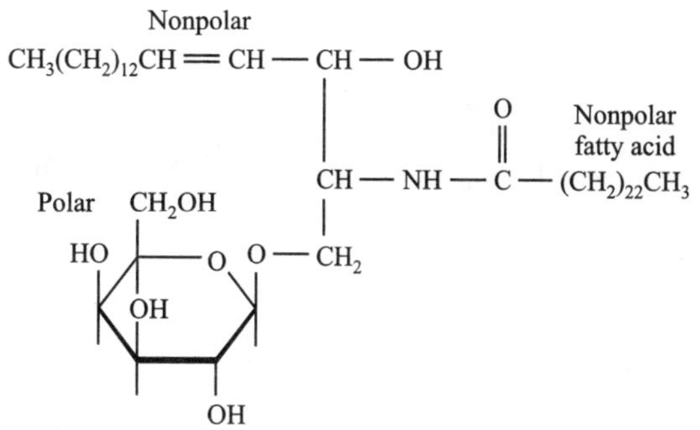

Galactocerebroside

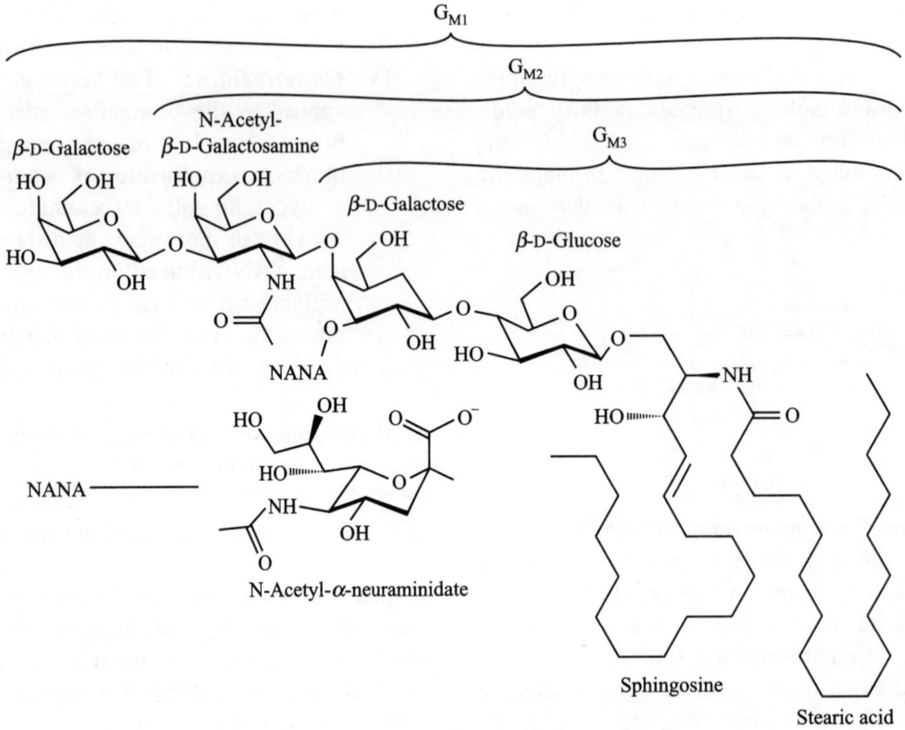

G_{M1}

G_{M2}

G_{M3}

β-D-Galactose

N-Acetyl-β-D-Galactosamine

β-D-Galactose

β-D-Glucose

NANA

NANA

N-Acetyl-α-neuraminidate

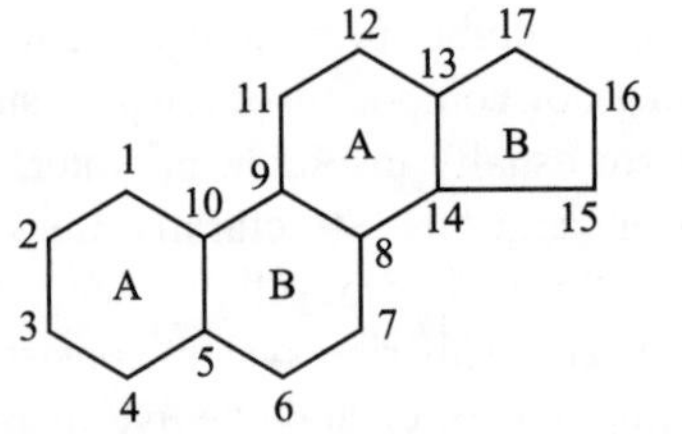

Sphingosine

Stearic acid

Perhydrocyclopentanophenanthrene ring system

Some of the natural steroids that possess the perhydrocyclopentanophenanthrene ring system as the basic nucleus are the sterols (of which cholesterol is a member), the bile acids, the male and female sex hormones, namely, androgens and estrogens, the adrenocortical hormones, the cardiac glycosides and some alkaloids.

Cholesterol: It is present in all of our cell walls, providing watertight integrity and structural support, and is especially essential to electrically conductive nerve and brain cells. It is found in all animal tissues but not in plant tissues. About 140 g of cholesterol may be present in an adult human, but its concentration in tissues varies widely. It is present in gallstone, brain tissues and spinal cord in large amounts.

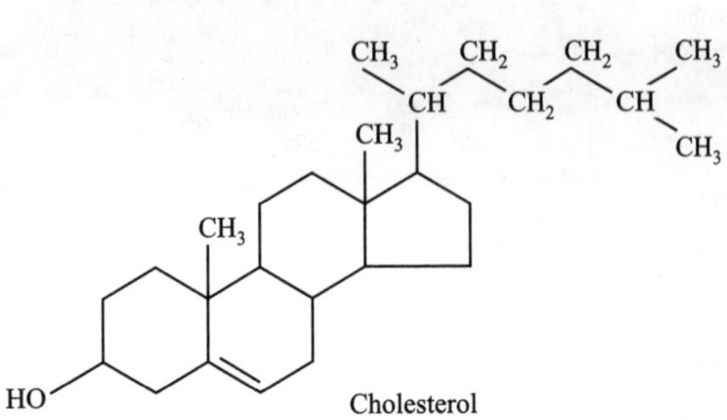

Cholesterol

Cholesterol has several asymmetric carbon atoms, and about 512 stereoisomers are possible. Only few of these are found in nature.

Cholesterol crystallizes as white shining plates. It melts at 149°–150°. It has no taste and no odour. Upon exposure to air and light, it is slowly oxidized to a mixture of products. It is insoluble in water, alkali and acids. It is readily soluble in organic solvents, such as ether, benzene, chloroform and petroleum ether.

D. Lipids in combination of other classes of compounds.

In nature, lipids are found in combination with other classes of compounds, such as carbohydrates, proteins and peptides. E.g., Lipopolysaccharides (LPS) and lipoproteins.

Lipopolysaccharides also known as lipoglycans, are large molecules consisting of a lipid and a polysaccharide joined by a covalent bond; they are found in the outer membrane of Gram-negative bacteria, act as endotoxins and elicit strong immune responses in animals.

Lipoproteins: A **lipoprotein** is a biochemical assembly that contains both proteins and lipids. The lipids or their derivatives may be covalently or non-covalently bound to the proteins. Many enzymes, transporters, structural proteins, antigens, adhesins and toxins are lipoproteins. Examples include the high-density lipoproteins (HDL) and low-density lipoproteins (LDL), which enable fats to be carried in the bloodstream, and the transmembrane proteins of the mitochondrion and the chloroplast. Bacterial lipoprotein function of lipoprotein particles is to transport water-insoluble lipids (fats) and cholesterol around the body in the blood.

7

Chemistry and Biological Significance of Important Conjugated Constituents of Cells

A.S. Balasubramanian

CONTENTS

I. AMINO SUGARS AND SIALIC ACIDS

Glucose and homopolymers of glucose, have tended to overshadow the significance of other sugars because of their wide occurrence and biological importance. In recent years, however, other heteropolymers, lipids and proteins containing sugars have also been attracting considerable attention. These have been localised in the intracellular and extracellular components or on the cell surface. Their involvement in the transport mechanism across membranes, as antigenic components or as other important structural components, indicate their significance in biological systems.

A. Amino Sugars

The most commonly occurring amino sugars are the hexosamines, glucosamine, galactosamine and mannosamine (Figure 7.1), in which the $-NH_2$ group is attached to the second

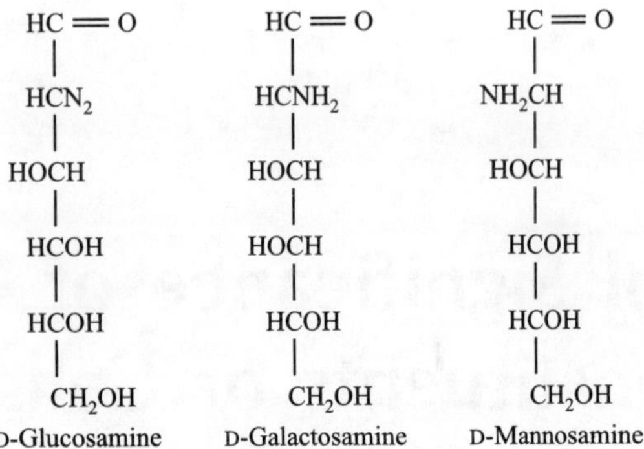

Figure 7.1 Structure of amino sugars.

carbon atom of the parent hexoses. In biological systems these hexosamines occur usually as their N-acetyl derivatives (except in the case of heparin). Amino sugars are components of glycoproteins, mucopolysaccharides (glycosaminoglycans) and gangliosides.

B. Sialic Acids

Sialic acids were discovered in bovine submaxillary mucin by Blix in 1936 and in nervous tissues by Klenk in 1941. The sialic acids (or acyl neuraminic acids) are straight chained nine-carbon compounds in which pyruvic acid occupies the first three carbon atoms and a derivative of mannosamine (N-acetyl, N-glycolyl, etc.) occupies the rest of the 6 carbon atoms (Figure 7.2). Though the parent compound neuraminic acid as such has not been detected in biological systems a number of its derivatives (Figure 7.2), collectively called sialic acids, are known to exist. The most commonly occurring neuraminic acid derivative is N-acetyl neuraminic acid.

Sialic acids are components of glycoproteins and gangliosides. The role of sialic acid in imparting special function to cell surfaces will be discussed under glycoprotein and gangliosides.

C. Metabolic Pathways involving Amino Sugars and Sialic Acids

The key step for the biosynthesis of all of the nitrogenous sugars, including the sialic acids, is the transfer of the amide nitrogen of glutamine to fructose-6-P giving glucosamine-6-P and glutamate. The glucosamine-6-P then serves as a precursor for the synthesis of other glucosamine derivatives, which in turn are converted to mannosamine derivatives, and the latter to derivatives of the sialic acids (Figure 7.3).

II. GLYCOPROTEINS

A. Chemistry and Functions

Glycoproteins are defined as proteins containing covalently bound carbohydrates. Many biologically important proteins such as enzymes, hormones and antibodies are glycoproteins in nature.

All glycoproteins have the same basic structure, viz., a polypeptide chain or chains to which are linked one or more carbohydrate groups. The carbohydrate composition of some well-known glycoproteins is shown in Table 7.1. In glycoproteins, two types of linkages by which the carbohydrate unit is attached to the polypeptide chain are well known. These are (a) the N-acetylglucosamine linkage involving the amide nitrogen of asparagine. Examples—Human chorionic gonadotropin and transferrin; (b) the O-glycosidic linkage involving the hydroxyl group of serine and threonine. Examples—Bovine submaxillary mucin and ovine submaxillary mucin. In both these examples N-acetylgalactosamine of the carbohydrate portion is involved in the linkage. Galactose and xylose are also known to be O-glycosidically linked. The two types of linkages described above are shown in Figure 7.4. The serum type of glycoproteins generally contain a few large branched oligosaccharide units

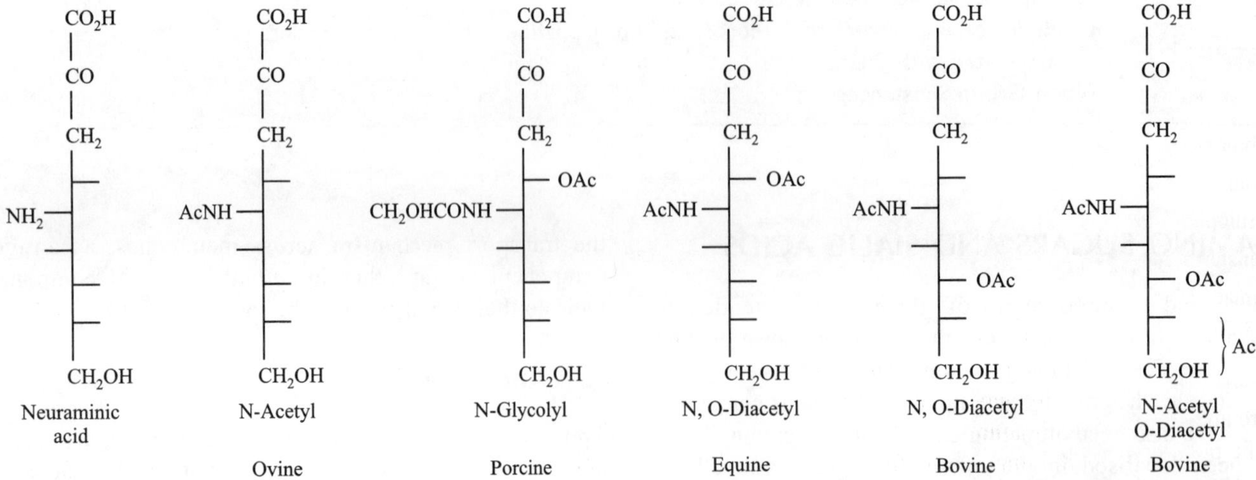

Figure 7.2 Structures of the sialic acids. Ac = Acetyl.

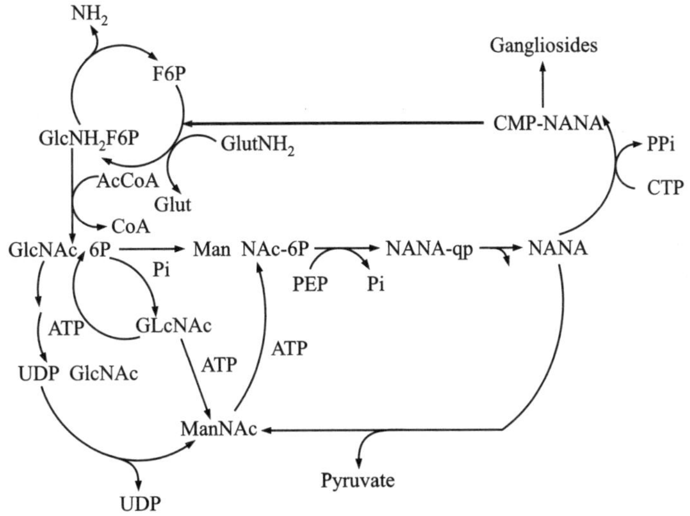

Figure 7.3 Metabolism of amino sugars and N-acetyl neuraminic acid. —Indicates feed back control. F6P = Fructose – 6 – phosphate; $GlutNH_2$ = Glutamine; Glut = Glutamic acid; $GlcNH_2$ = Glucosamine; GlcNAc = N-acetyl glucosamine; ManNAc = N-acetyl mannosamine; NANA = N-acetyl neuraminic acid.

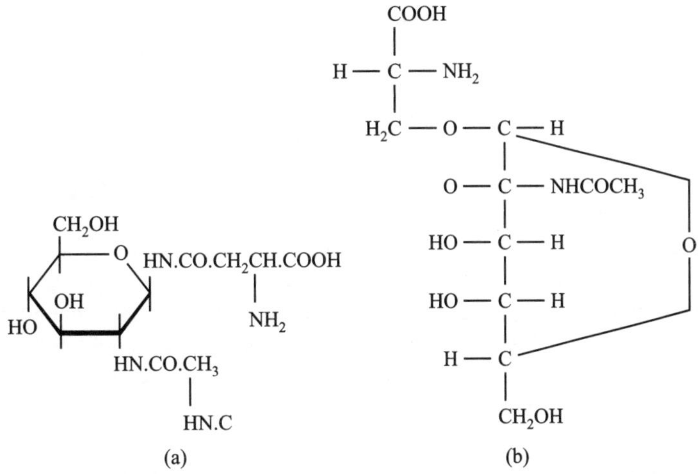

Figure 7.4 Carbohydrate-protein linkage in glycoproteins.

TABLE 7.1 Carbohydrate Composition of Some Glycoproteins

Glycoproteins	Moles of sugar/mole of glycoprotein						
	Sialic acid	*Fucose*	*Glucose*	*Galactose*	*Mannose*	*Glucosamine*	*Galactosamine*
Submaxillary glycoprotein (ovine)	800						800
Fetuin	12			12	9	18	
Ceruloplasmin	9	2		14	22	18	
Transferrin	4			4	8	8	
Human chorionic gonadotropin	8	1		9	9	11	3

linked to the protein core by the type (a) linkage. In the mucins, there are as many as 800 short chained oligosaccharides linked to the polypeptide through the type (b) linkage (Figure 7.5).

The commonly occurring carbohydrate residues in glycoprotein include N-Acetyl-glucosamine, N-acetylgalactosamine mannose and galactose. Sialic acid when present in glycoprotein occupies a terminal position where it is linked usually to galactose by a $2 \rightarrow 6$ or $2 \rightarrow 3$ glycosidic linkage. Fucose, a methyl pentose, when present also occupies the terminal position.

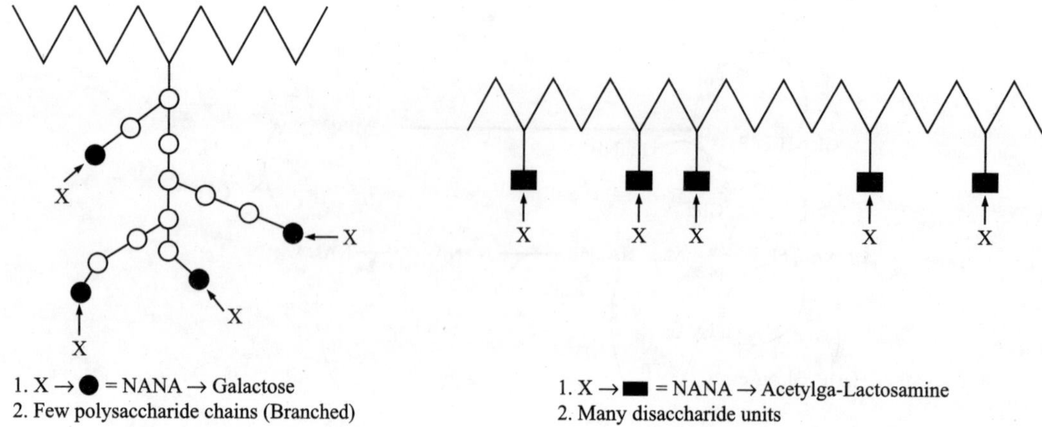

Figure 7.5   Schematic structures of serum type of glycoproteins and mucins.

B.   Biosynthesis of Glycoproteins

The intracellular site of synthesis of the carbohydrate portion of glycoprotein and its attachment to the polypeptide chain has received much attention from various workers. Whereas the synthesis of the protein takes place in the ribosomes, the addition of the carbohydrate seems to take place at a different site. Since the carbohydrates are attached to the protein in an extra-ribosomal system, these complex, carbohydrate containing polymers are formed in a totally different manner as compared to the polymers such as proteins and nucleic acids.

In the proteins and nucleic acids, relatively simple non-specific enzyme systems build each polymer in the class, and genetic information (or specificity) is determined by the genetic message, or template. By contrast, the oligosaccharide chains of glycoproteins are not synthesized by a template mechanism. Instead, the expression of the appropriate genes for the synthesis of each type of molecule, apparently results in the formation of a complex of enzymes, i.e., in each case a specific complex of glycosyl transferases. The monosaccharide units are transferred from their nucleotide derivatives to appropriate acceptor glycoproteins and each of these transfer reactions is catalyzed by a different glycosyl transferase.

The donors for the sugars are their activated derivatives (nucleotides linked sugars) such as UDP or GDP derivatives or CMP-sialic acid shown in Figure 7.6. Since the addition of the monosaccharides to the polypeptide is taking place in absence of template mechanisms, the sequential addition of the monosaccharides depend on the specificity of each of the

glycosyl transferases for a particular acceptor molecule. Mn^{2+} is essential for the optimal activity of all enzymes.

Roseman and his group have concluded from their studies that glycosyltransferases catalysing the synthesis of the terminal trisaccharide of the glycoprotein orosomucoid (viz. sialyl-galactosyl-N-acyl glucosaminyl), are all located in the same subcellular organelle and that this organelle is the Golgi apparatus (J.B.C. 245 (1970), p. 1090–1100).

C.   N-Asparagine linked Glycoproteins and their Biosynthesis

The asparagine linked oligosaccharides of eukaryotic glycoproteins have heterogenous structures that fall into two main categories, 'high mannose' and 'complex'. Both classes have inner core of three mannose and two N-acetylglucosamine residues ($Man_3GlcNAc_2$) at their reducing terminal (See box enclosed by dotted line in Figure 7.7). They differ in that the high mannose glycoproteins contain additional alpha-linked mannose residues (usually 2 to 6) while the complex oligosaccharides carry other external sugars (Figure 7.7). Both types of oligosaccharides have a common biosynthetic origin.

The biosynthesis of asparagine linked oligosaccharides involves dolichol pyrophosphate linked intermediates. An "en bloc" transfer of a preformed oligosaccharide, pyrophosphoryldolichol intermediate, to an asparagine residue on the nascent polypeptide takes place. The strongest evidence that lipid (dolichol) linked oligosaccharides serve as the source of asparagine linked glycoproteins have come from studies

MONOSACCHARIDE	"ACTIVATED" MONOSACCHARIDE	
GALACTOSE	UDP–GALACTOSE	
MANNOSE	GDP–MANNOSE	
ACETYLGLUCOSAMINE	UDP–ACETYLGLUCOSAMINE	GLYCOPROTEIN
ACETYLGALACTOSAMINE	UDP–ACETYLGALACTOSAMINE	
FUCOSE	GDP–FUCOSE	
SIALIC ACID	CMP–SIALIC ACID	

Figure 7.6   Monosaccharides and their activated derivatives.

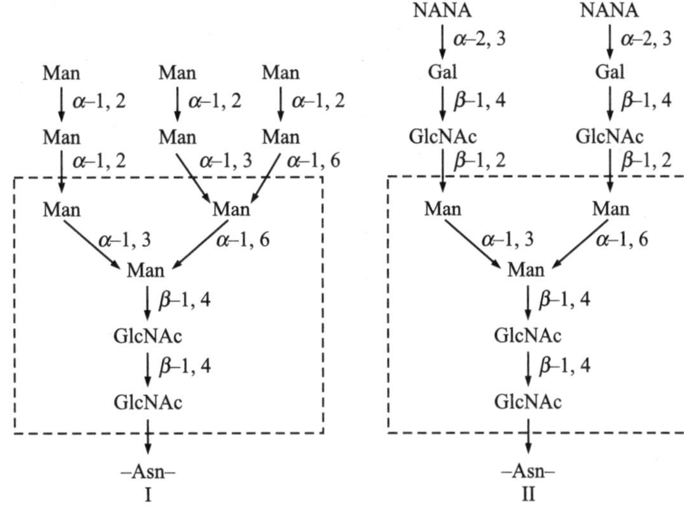

Figure 7.7 Structure of high mannose (I) and complex (II) oligosaccharide chains.

with inhibitors and mutant cell lines. Drugs that inhibit lipid linked oligosaccharide assembly have invariably been found to block glycosylation of asparagine residues as well. Both high mannose and complex oligosaccharides are absent in proteins from inhibited cells. Similarly, in mutant cells that synthesize truncated lipid linked oligosaccharides, newly synthesized proteins acquire the abnormal oligosaccharides.

Synthesis of the oligosaccharide chain starts with the transfer of two N-acetylglucosamine residues from the sugar nucleotide to dolichol phosphate (Figure 7.8). Both the first and second N-acetylglucosamine residues are donated by UDP-GlcNAc. The beta-linked mannose is then transferred via GDP-mannose giving the structure of DoI-P-P-GlcNAc-

GlcNAc-Man. Mannosylphosphoryldolichol serves as the mannosyl donor for the addition of the outer linked alpha-mannose residues. The last step in the synthesis of lipid linked oligosaccharides is the addition of three glucose units via dolichol-P-glucose giving the structure shown in Figure 7.8. The addition of three glucose residues is necessary for transfer to the protein. The glucosylated oligosaccharide has a much higher rate of transfer than the unglucosylated one.

The transfer of the glucosylated oligosaccharide to the protein is followed by the immediate removal of the three glucose residues by glucosidases. At this stage there are two possibilities. It can either remain as a high mannose oligosaccharide, or be processed into a complex oligosaccharide.

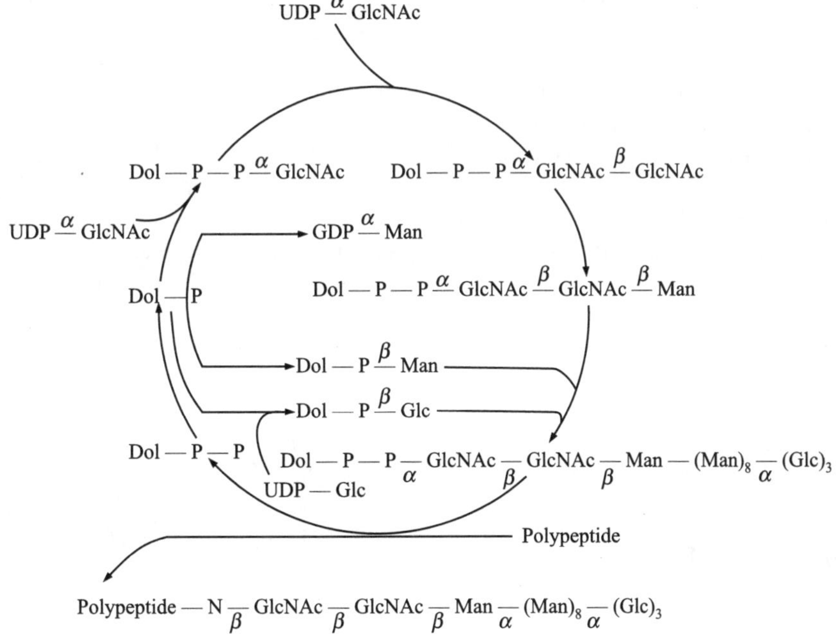

Figure 7.8 Schematic pathway for the assembly oligosaccharide chain on the carrier lipid dolicholphosphate and the *en bloc* transfer of the precursor oligosaccharide to a specific asparagine residue of an acceptor protein, in the lumen of endoplasmic reticulum.

In the processing of the complex oligosaccharide, the steps involving the removal of the four alpha-1, 2 linked outer mannose residues by mannosidases and the attachment of N-acetylglucosamine residues from UDP-GlcNAc take place. Various sugars such as galactose and NANA may now be added via sugar nucleotides giving rise to the diverse spectrum of biantennary complex oligosaccharides. Many complex oligosaccharides contain more than two external branches. The presence of triantennary and tetraantennary carbohydrate chains have been detected on proteins.

D. Physiological Significance of Glycoproteins

Most of the mammalian extracellular proteins contain covalently bound carbohydrates. Eylar and Soodak have proposed that the secretion of a protein necessitates the presence of carbohydrates in it. For example, two types of the enzyme invertase are known in yeast cell, one internal cytoplasmic invertase devoid of any carbohydrate, and an external invertase which is cell bound and outside the cytoplasmic membrane, and which contains mannose and glucosamine.

Another important function of glycoproteins arises from the presence of sialic acid in many of them. Sialic acid of glycoproteins in cell surfaces plays a vital role in physiological processes. For example, the work of Ashwell and his coworkers shows that the sialic acid of liver cell membrane is essential for the removal of asialoproteins (glycoprotein devoid of sialic acid) from the circulation. Formerly it was shown that glycoprotein hormones like human chorionic gonadotropin, follicle stimulating hormone, and leutinizing hormone, lose their biological activity when the sialic acid is removed from them. This loss of biological activity happens, actually, because they are removed from circulation by the liver as shown by Asiwell, et al.

Cuatrecasas has observed that neuraminidase treated fat cells are unable to respond to insulin. Influenza virus has been shown to bind to cell surface through the sialic acid of the cell membrane.

Based on these and other known functions, we can sum up the role of sialic acid in glycoproteins as follows:

(a) regulating the migration of transfused rat lymphocytes.
(b) influencing cell agglutination and intracellular adhesion.
(c) determining the transport of both K^+ and protein across the membrane.
(d) inhibiting sperm penetration in rat ova.
(e) determining the biological half-life of sialoproteins.

III. MUCOPOLYSACCHARIDES (GLYCOSAMINOGLYCANS)

The term, mucopolysaccharide, was first introduced by Meyer to describe hexosamine containing polysaccharides of animal origin. More recently, Jeanloz has proposed a new set of systematic names for the various mucopolysaccharides. The new and old names are given in Table 7.2. The new nomenclature will be adopted throughout the text hereafter.

TABLE 7.2   New and Old Names of Mucopolysaccharides

New	Old
Glycosaminoglycan	Mucopolysaccharide
Hyaluronic acid	Hyaluronic acid
Chondroitin 4-sulphate	Chondroitin sulphate A
Chondroitin 6-sulphate	Chondroitin sulphate C
Dermatan sulphate	Chondroitin sulphate B
Heparan sulphate	Heparitin sulphate
Keratan sulphate	Kerato sulphate
Heparin	Heparin
Proteoglycan or Protein-polysaccharide	Chondromucoprotein

A.   Chemistry

Glycosaminoglycans are the structural components of various types of connective tissue. As shown in Table 7.2, the glycosaminoglycans consist of hyaluronic acid, chondroitin-4-sulphate, chondroitin-6-sulphate, dermatan sulphate, keratan sulphate, heparan sulphate and heparin. They are comprised of linear chains of repeating disaccharide units made up of a hexosamine and a hexuronic acid (an exception is keratan sulphate in which the disaccharide unit contains a galactose residue in the place of hexuronic acid). The disaccharide repeating units of various glycosaminoglycans are shown in Figure 7.9. The monosaccharide units and the mode of linkage in the glycosaminoglycans is shown in Table 7.3. Some of the notable features are that, (a) hyaluronic acid is devoid of sulphate, (b) dermatan sulphate contains L-iduronic acid instead of D-glucuronic acid, (c) Keratan sulphate (whose structure is not yet well established) contains galactose in the place of uronic acid, (d) heparan sulphate and heparin have alpha-1 $\rightarrow$ 4 linkages, (e) heparin contains N-sulphated glucosamine, whereas, heparan sulphate contains both N-sulphated and N-acetylated glucosamine.

Recent work indicates that dermatan sulphate contains glucuronic acid also to a small extent and that heparin and heparan sulphate may contain iduronic acid to a minor extent. The distribution of sulphate in the heteropolysaccharide is not uniform. Near the linkage region the sulphate content is less than that at the nonreducing end. The distribution of the various glycosaminoglycans in some tissues is shown in Table 7.4. In addition, their presence has been demonstrated in several other tissues as well. All the glycosaminoglycans except hyaluronic acid are covalently linked to protein in the various tissues. The linkage of glycosaminoglycans to protein is through a trisaccharide residue: galactose – galactose – xylose (Figure 7.10). The xylose is attached to a serine or threonine hydroxyl group of the protein moiety.

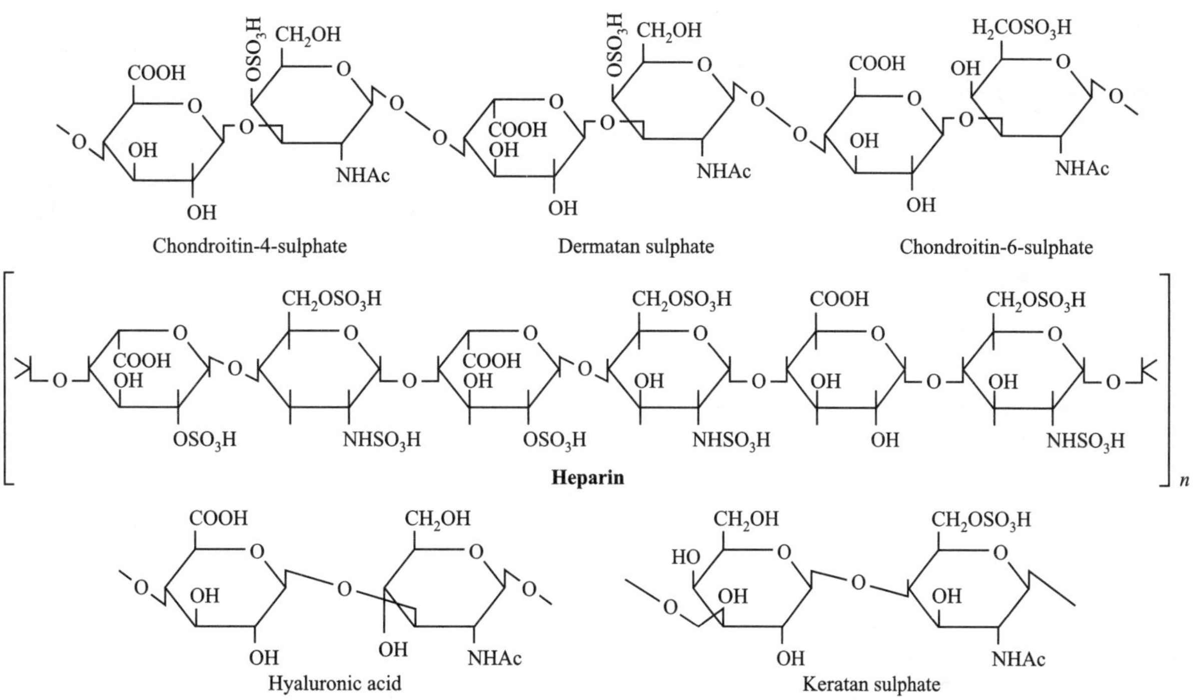

Figure 7.9 Structure of disaccharide units of glycosaminoglycans.

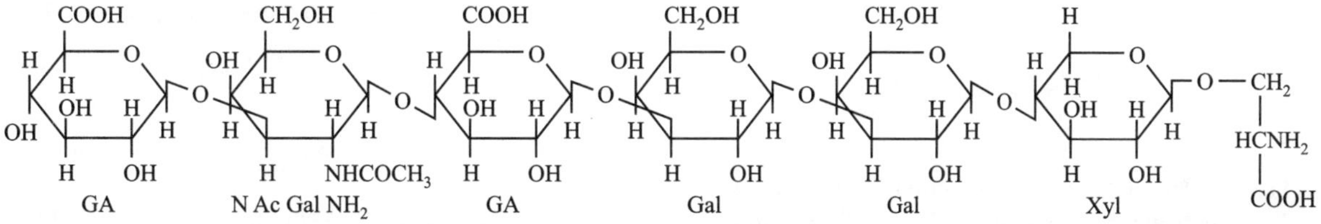

Figure 7.10 Protein-glycosaminoglycan linkage. GA = Glucuronic acid; NAcGalNH$_2$ = N-acetyl gatactosamine; Gal = Galactose; Xyl = Xylose.

TABLE 7.3 Monosaccharide Units and Mode of Linkage in Glycosaminoglycans

Nature of sugar and linkage	Hyaluronic acid	Chondroitin-4-sulphate	Chondroitin-6-sulphate	Dermatan sulphate	Heparan sulphate and Heparin
Aminosugar	N-Acetyl-D-glucosamine	N-Acetyl-D-galactosamine-4-sulphate	N-Acetyl-D-galactosa-mine-6-sulphate	N-Acetyl-D-Galactosa-mine-4-sulphate	N-sulphated as well as N-acetylated D-glucosamine
	and	and	and	and	and
Uronic acid	D-glucuronic acid	D-glucuronic acid	D-glucuronic acid	L-iduronic acid	D-glucuronic acid
Uronidic bond	b1------3	b1------3	b1------3	a1------3	a1------4
Hexosaminidic bond	b1------4	b1------4	b1------4	a1------4	a1------4

TABLE 7.4 Tissue Distribution of Glycosaminoglycans

Glycosaminoglycans	Tissue
Hyaluronic acid	Rheumatic nodules, synovial fluid, vitreous humour, umbilical cord
Chondroitin-4-sulphate	Cartilage, aorta, bone cornea, ligamentum nuchae
Dermatan sulphate	Aorta, heart valve, skin, tendon, ligamentum nuchae
Chondroitin-6-sulphate	Cartilage, nucleus pulposus, tendon
Keratan sulphate	Cornea, bone, nucleus pulposus
Heparan sulphate	Liver, lung
Heparin	Liver, lung and mast cells

B.   Biosynthesis

The major aspects of biosynthesis of glycosaminoglycans may be enumerated under the following heads:

1. Synthesis of the intermediates for glycosaminoglycan biosynthesis.
2. Synthesis of the protein-polysaccharide linkage region.
3. Chain elongation by sequential addition of the hexosamines and uronic acid residues.
4. Sulfation of the glycosaminoglycans.

The intermediates for the synthesis of a glycosaminoglycans like chondroitin sulphate are UDP-N-acetyl galactosamine and UDP-glucuronic acid, the biosynthesis of which is presented in Figures 7.11a and 7.11b.

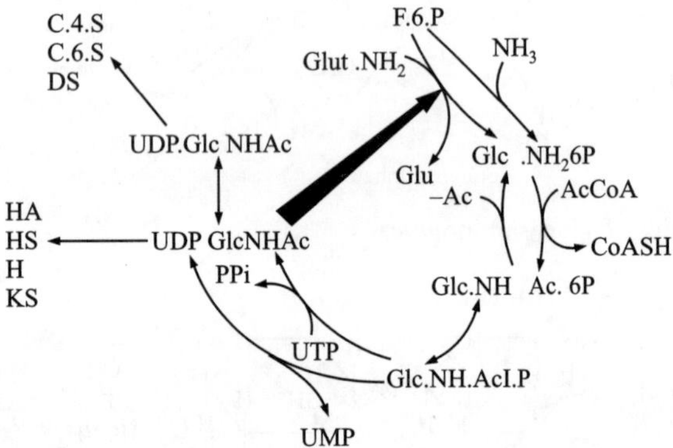

Figure 7.11a   Formation of UDP-N-acetyl hexo-samines. Indicates feed back control, F–6–P = Fructose–6–Phosphate; GlutNH$_2$ = Glutamine; Glu = Glutamic acid; GlcNH$_2$ = Glucosamine; GlcNHAc = N-acetyl glucosamine; GalNHAc = N-acetyl galactosamine; HA = Hyaluronic acid; HS = Heparan sulphate; H = Heparin; KS = Kerato-sulphate; C–4–S = Chondroitin–4–sulphate; C–6–S = Chondroitin–6–sulphate; DS = Dermatan sulphate.

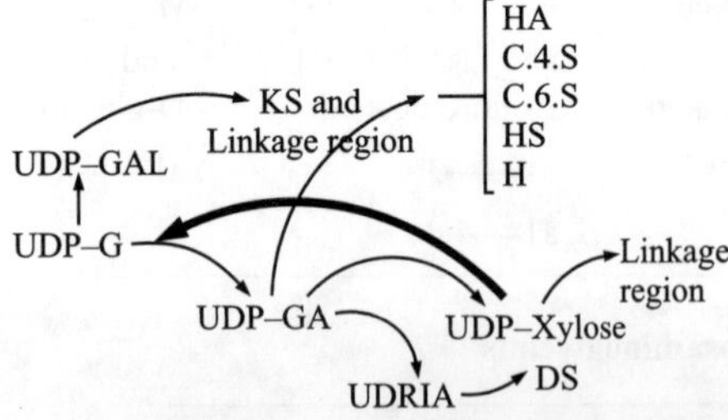

Figure 7.11b   Formation of UDP-glucuronic acid. Indicates feed back control. UDPG = Uridine diphosphoglucose; GA = Glucuronic acid; Gal = Galactose; IA = Iduronic acid; For other abbreviations see Figure 9a.

The synthesis of the linkage region involves the addition of xylose from UDP-xylose to a serine residue of the appropriate acceptor protein. Subsequently the galactose moieties are added on from UDP-galactose to complete the linkage of xylose-galactose-galactose. Uronic acid and hexosamine are then added on sequentially to the chain from their respective UDP-derivatives.

The sulphation of glycosaminoglycans occurs in close proximity to polymerization. The active donor of sulphate is 3′-phosphoadenosine-phosphosulphate (PAPS) (Figure 7.12) the formation of which from ATP and inorganic sulphate is shown in Figure 7.13.

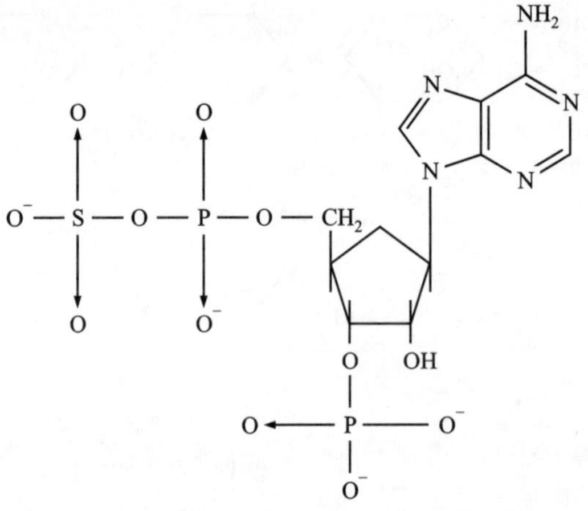

3′-Phosphoadenosine-5′-Phosphosulphate (PAPS)

Figure 7.12   Structure of 3′-phosphoadenosine-5′-phosphosulphate.

$$ATP + SO_4^{2-} \xrightleftharpoons{\text{ATP–Sulfurylase}} APS + PP_1$$

$$APS + ATP \xrightarrow{\text{APS–Kinase}} PAPS + ADP$$

$$\overline{2ATP + SO_4^{2-} \xrightarrow{\hspace{2cm}} PAPS + ADP + PP_1}$$

Figure 7.13   Enzymatic synthesis of PAPS, APS = Adenosine 5′-phosphosulphate.

The reactions involved in the synthesis of glycosaminoglycans, as just discussed, have been demonstrated in embryonic cartilage tissues and are depicted in Figure 7.14.

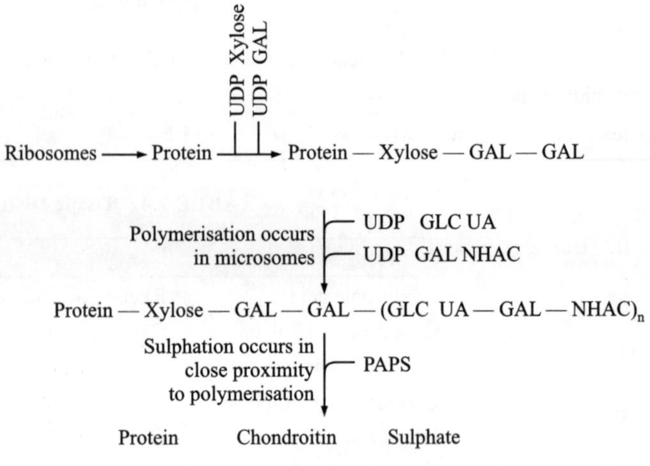

Ribosomes ⟶ Protein ⟶ Protein — Xylose — GAL — GAL   (UDP Xylose, UDP GAL)

Polymerisation occurs in microsomes — UDP GLC UA, UDP GAL NHAC

Protein — Xylose — GAL — GAL — (GLC UA — GAL — NHAC)$_n$

Sulphation occurs in close proximity to polymerisation — PAPS

Protein   Chondroitin   Sulphate

Figure 7.14   Biosynthesis of chondroitin sulphate, GlcUA = Glucuronic acid; GalNHAc = N-acetyl galactosamine.

In the synthesis of heparin, where N-sulphated hexosamine is a component, it has been demonstrated that initially a polymer containing N-acetylglucosamine is formed, and subsequently the acetyl groups of N-acetyl glucosamine in the polymer are replaced by sulphate groups from PAPS.

C. Degradation

Glycosaminoglycans have been shown to be degraded by the enzymes from various sources. Hyaluronidases and chondroitinases from different sources show differing substrate specificity; the testicular hyaluronidase which has been extensively used in the degradation of glycosaminoglycans can split hyaluronic acid and chondroitin sulphates to simpler oligo and tetrasaccharides. From liver, a lysosomal hyaluronidase has been isolated. An enzyme heparinase derived from *Flavobacterium heparinum* acts on dermatan sulphate, heparan sulphate, and heparin, converting them to simpler products. Further degradation of the products formed by hyaluronidase action are catalysed by glucuronidases, hexosaminidases and sulphatases (Figure 7.15).

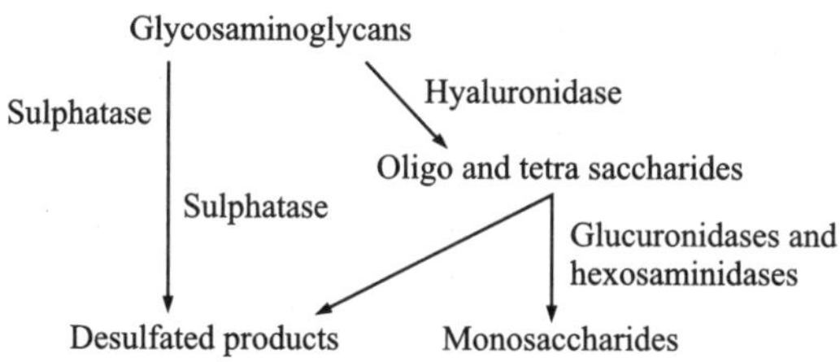

Figure 7.15 Degradation of glycosaminoglycans.

D. Physiological Functions

Glycosaminoglycans linked to proteins, impart the structural rigidity to many connective tissue structures. It has been demonstrated that upon injection of the proteolytic enzyme papain to a rabbit, its erect ears dropped within a few hours, because of the digestion of the protein portion of the glycosaminoglycan chains present in the ear cartilage. The strongly anionic character of the various glycosaminoglycans has been suggested to be responsible for their ion binding or water binding characteristics. The abnormal accumulation of different glycosaminoglycans in some genetic disorders also points towards the significance of these compounds in relation to the mental and physical development of human beings.

E. Mucopolysaccharidoses

There are several types of genetic disorders associated with the abnormal accumulation in tissues (mainly liver, kidney and brain), and excessive excretion in urine, of particular types of glycosaminoglycans. These disorders are termed mucopolysaccharidoses. They are tabulated in Table 7.5. A number of findings related to defects in the enzymatic degradation of glycosaminoglycans, have been suggested to be the case for these mucopolysaccharidoses.

IV. GLYCOLIPIDS

Glycolipids are lipids containing bound carbohydrates. The most important glycolipids are cerebrosides, sulfatides and gangliosides. Blix and Klenk isolated from the human brain several glycolipids which were later identified as galacto-cerebrosides, sulphatides and gangliosides. The structures of these are shown in Figures 7.16a, 7.16b and 7.16c. The major fatty acids in gangliosides are palmitic and stearic acids, whereas in the galactocerebrosides and sulfatides long chain fatty acids like lignoceric acid and hydroxylignoceric acid are the predominant fatty acids. There are four major gangliosides known to be present in the normal human brain. They are the monosialoganglioside, two isomers of disialoganglioside and trisialoganglioside. The linkage between sialic acid (NANA) and galactose in these compounds is 2–3 and in the gangliosides GD1b and GT1 the linkage between the two molecules of NANA is a 2–8 linkage.

TABLE 7.5 Genetic Mucopolysaccharidoses with Mucopolysacchariduria

	Types of disease	Clinical findings	Genetic	Glycosaminoglycans accumulated and excreted
I.	Hurler syndrome	Early clouding of cornea, Dwarfism, Bony deformities, Progressive mental deterioration	Autosomal recessive	Dermatan sulphate and Heparan sulphate
II.	Hunter syndrome	No clouding of cornea, milder symptoms as compared to Hurler syndrome	X-linked recessive	Dermatan sulphate and Heparan sulphate
III.	Sanfilippo syndrome	Severe mental retardation	Autosomal recessive	Heparan sulphate
IV.	Morquio syndrome	Severe bone changes of distinctive types, cloudy cornea, aortic regurgitation, intellect normal or mildly impaired	Autosomal recessive	Keratan sulphate
V.	Scheie syndrome	Stiff joints, coarse facies, cloudy cornea, aortic regurgitation, intellect normal or mildly impaired	Autosomal recessive	Dermatan sulphate
VI.	Maroteaux-Lamy syndrome	Severe osseous and corneal change, normal intellect	Autosomal recessive	Dermatan sulphate

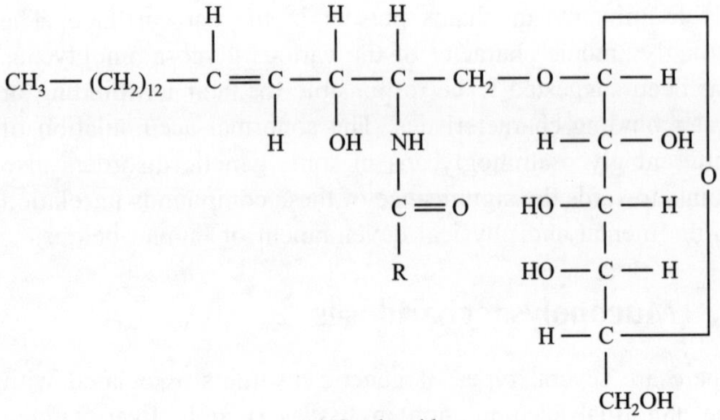

(a) Cerebroside

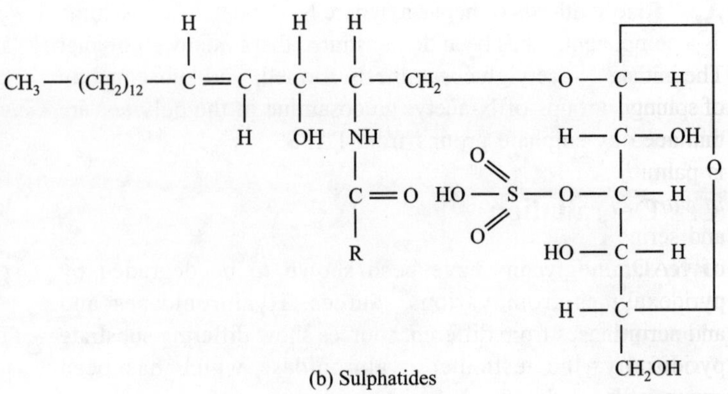

(b) Sulphatides

Chemical structure	Generic name	Svennerholm nomenclature
NANA $(2\rightarrow3)$ Gal $(\beta,1\rightarrow4)$ Glu $(\beta,1\rightarrow1)$ Cer	Monosialoganglioside	GM_3
NANA $(2\rightarrow8)$ NANA $(2\rightarrow3)$ Gal $(\beta1\rightarrow4)$ Glu $(\beta,1\rightarrow1)$ Cer	Disialosyl – Lactosyl Ceramide	GD_3
GalNAc $(\beta,1\rightarrow4)$ Gal $(\beta,1\rightarrow4)$ Glu $(\beta,1\rightarrow1)$ Cer $\binom{3}{\uparrow\ 2}$ NANA	Monosialosyl – N – Triglycosyl Ceramide	GM_2
GalNAc $(\beta,1\rightarrow4)$ Gal $(\beta,1\rightarrow4)$ Glu $(\beta,1\rightarrow1)$ Cer $\binom{3}{\uparrow\ 2}$ NANA $(8\rightarrow2)$ NANA	Disialosyl – N – Triglycosyl Ceramide	GD_2
Gal $(\beta,1\rightarrow3)$ GalNAc $(\beta,1\rightarrow4)$ Gal $(\beta,1\rightarrow4)$ Glu $(\beta,1\rightarrow1)$ Cer $\binom{3}{\uparrow\ 2}$ $\binom{3}{\uparrow\ 2}$ NANA NANA	Disialosyl – N – Tetraglycosyl Ceramide	GD1a
Gal $(\beta,1\rightarrow3)$ GalNAc $(\beta,1\rightarrow4)$ Gal $(\beta,1\rightarrow4)$ Glu $(\beta,1\rightarrow1)$ Cer $\binom{3}{\uparrow\ 2}$ NANA $(8\rightarrow2)$ NANA	Diasialosyl – N – Tetraglycosyl Ceramide	GD1b
Gal $(\beta,1\rightarrow3)$ GalNAc $(\beta,1\rightarrow4)$ Gal $(\beta,1\rightarrow4)$ Glu $(\beta,1\rightarrow1)$ Cer $\binom{3}{\uparrow\ 2}$ $\binom{3}{\uparrow\ 2}$ NANA NANA $(8\rightarrow2)$ NANA	Trisialosyl – N – Tetraglycosyl Ceramide	GT1

(c) Gangliosides of Mammalian Brain

Figure 7.16   Structure of cerebroside, sulphatide and gangliosides, R = Fatty acid residue; Cer = Ceramide. Glu = Glucose; Gal = Galactose; GalNAc = N-acetyl galactosamine; NANA = N-acetyl neuraminic acid.

A. Biosynthesis of Sphingosine

The initial step in the biosynthesis of glycolipids is the formation of sphingosine. Earlier work from various laboratories has shown that one of the precursors of sphingosine is serine and the other is palmitic acid. Recently, Braun and Snell have demonstrated *in vitro* synthesis of dihydrosphingosine from palmityl CoA and serine by a particulate fraction from yeast in the presence of NADPH. The enzyme shows absolute requirement for pyridoxal phosphate. According to their findings palmityl CoA and serine undergo enzymatic condensation in the presence of pyridoxal phosphate, with the displacement of the carboxyl group of serine and the CoA moiety of palmityl CoA, to yield a 3-keto intermediate which is reduced by NADPH to form dihydrosphingosine. For the synthesis of sphingosine, unsaturated fatty acyl CoA is the precursor. Thus, desaturation reactions involved in the sphingosine biosynthesis appear to occur at the fatty acyl CoA and not at the dihydrosphingosine level. The biosynthetic pathway of sphingosine is shown in Figure 7.17. The existence of this biosynthetic pathway in mammalian brain tissue has also been recently established.

B. Biosynthesis of Cerebroside

The biosynthesis of galactocerebroside occurs by the following steps:

Formation of ceramide by acylation of sphingosine and transfer of galactose from UDP-galactose to ceramide to form cerebroside (Figure 7.18).

Sulphation of cerebroside to form cerebroside-3-sulphate: It has been shown that there is an enzyme in the brain which transfers sulphate from active sulphate (PAPS) to galacto-cerebroside to form cerebroside-3-sulphate. This indicates that there is a direct conversion of cerebroside to cerebroside-3-sulphate. (Reaction No. 2, in Figure 7.18).

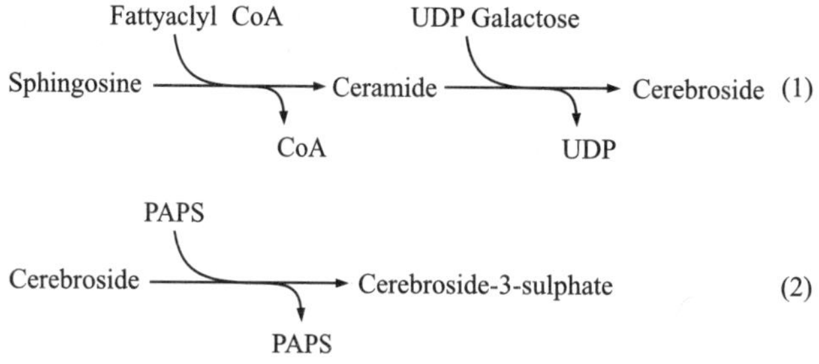

Figure 7.18 Formation of ceramide, cerebroside and sulphate.

C. Biosynthesis of Gangliosides

The biosynthesis of gangliosides involves a glycosyltransferase-mediated stepwise addition of monosaccharides from their UDP derivatives, and NANA from CMP-NANA, to ceramide. The evidence available indicates the pathway of biosynthesis of gangliosides (Figure 7.19). It has been proposed, that the glycosyltransferase systems catalysing the transfer of glycosyl units to the acceptor, determine the sequential specificity of these complex glycolipids. Each transferase is specific for its acceptor and the product of each step is the preferred substrate

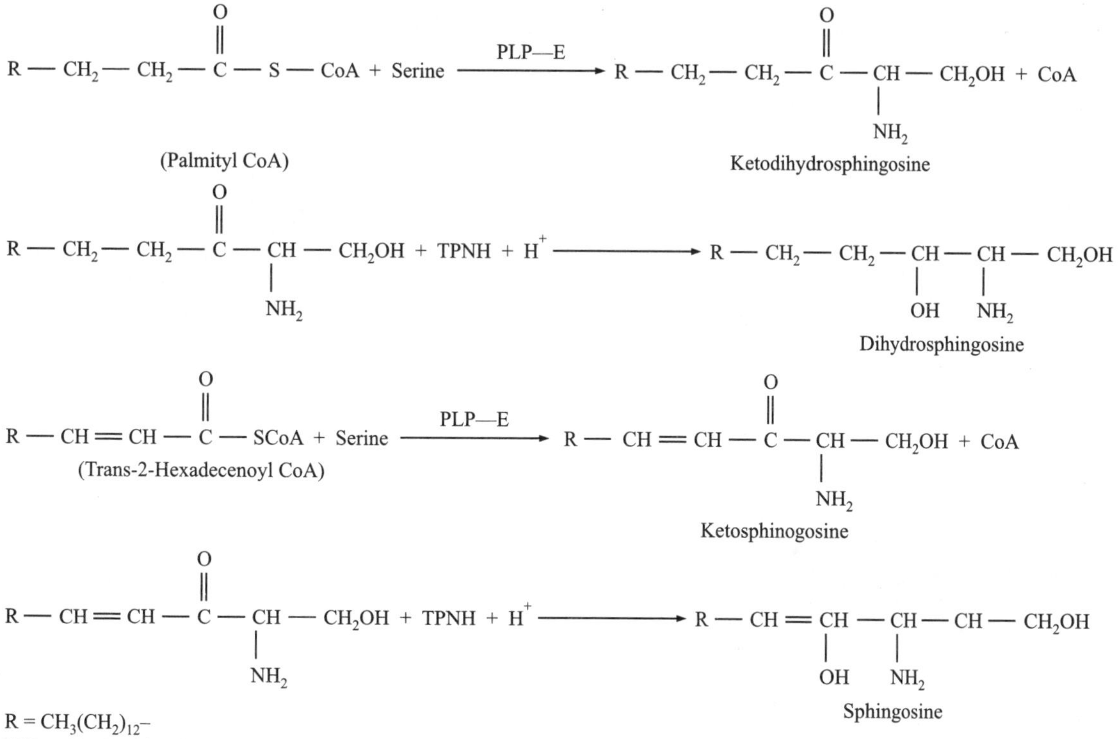

R = CH_3(CH_2)_{12}—
PLP — E = Pyridoxal phosphate dependent enzyme

Figure 7.17 Biosynthesis of sphingosine and dihydrosphingosine.

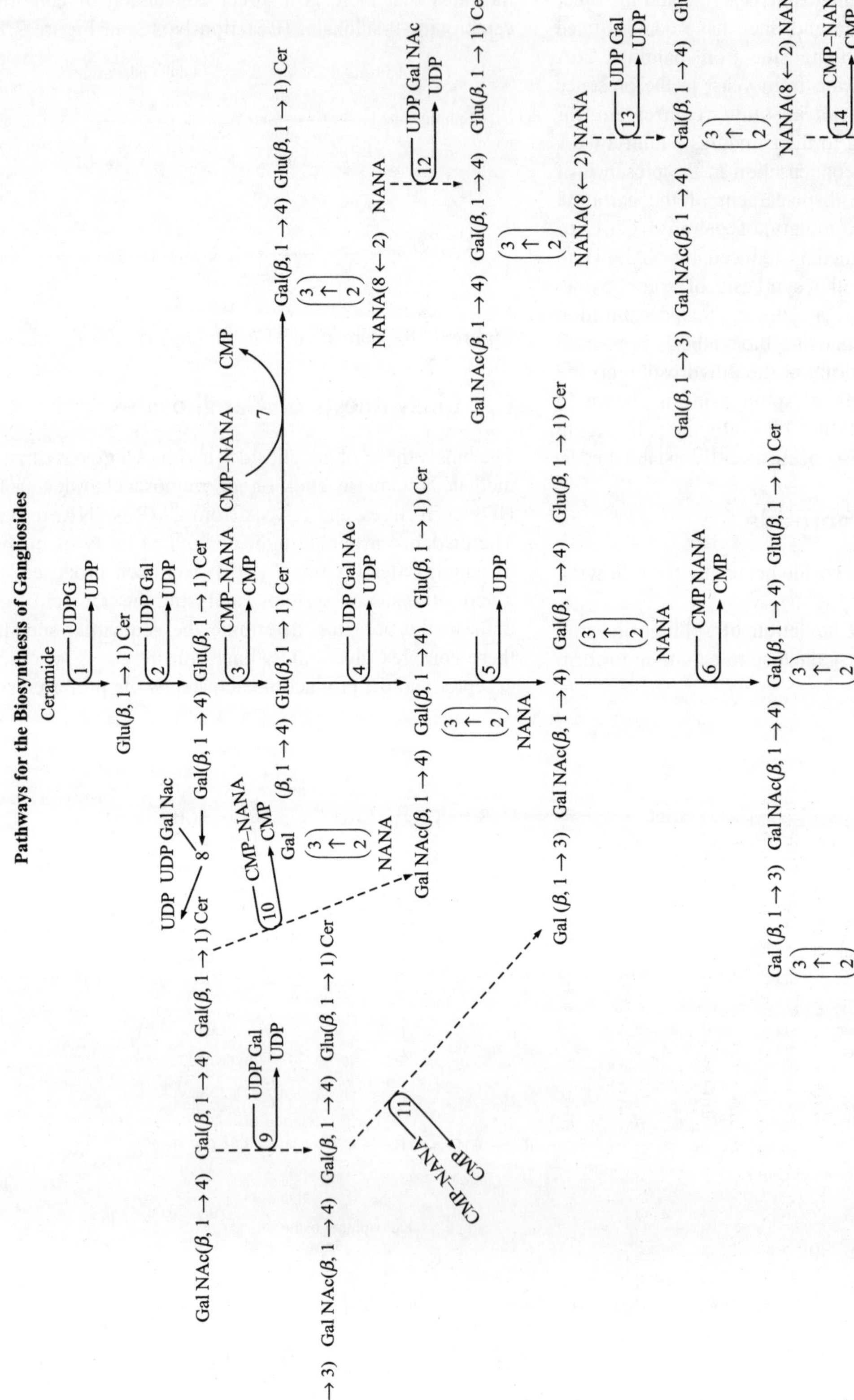

Figure 7.19 Biosynthesis of gangliosides, Cer = Ceramide; Glu = Glucose; Gal = Galactose; GalNAc = N-acetyl galactosamine; NANA = N-acetyl neuraminic acid.

for the next reaction. This proposal is similar to what has been suggested for the synthesis of the carbohydrate portion of glycoproteins.

D. Degradation of Glycolipids

The first step of cerebroside-3-sulphate degradation is catalysed by an enzyme—arylphatase A—which is localised in lysosomes to form cerebroside and inorganic sulphate. The next step of the degradation is by the enzyme galactocerebrosidase. This enzyme catalyses the cleavage of galactocerebroside to ceramide and galactose. In the case of ganglioside degradation, stepwise removal of NANA and the sugars from galactosidases and hexosaminidases, results in the final product glucocerebroside. Glucocerebroside is converted to ceramide by the action of β-glucosidase. The ceramide is further degraded to sphingosine and fatty acid by an enzyme known as ceramidase. Sphingosine thus formed by the action of ceramidase undergoes phosphorylation to give sphingosine-1-phosphate by an enzyme—sphingosine kinase. The sphingosine-1-phosphate is cleaved further to palmitaldehyde and phosphoryl ethanolamine by an enzyme—sphingosine-1-phosphate aldolase—which is a pyridoxal phosphate dependent enzyme. It is interesting to note that the end product of this sphingosine degradation could be effectively utilised in the biosynthesis of various phospholipids such as phosphatidyl ethanolamine, whereas, the palmitaldehyde may go to form plasmalogens. This degradation of cerebroside-3-sulphate and sphingosine is summarized in Figure 7.20.

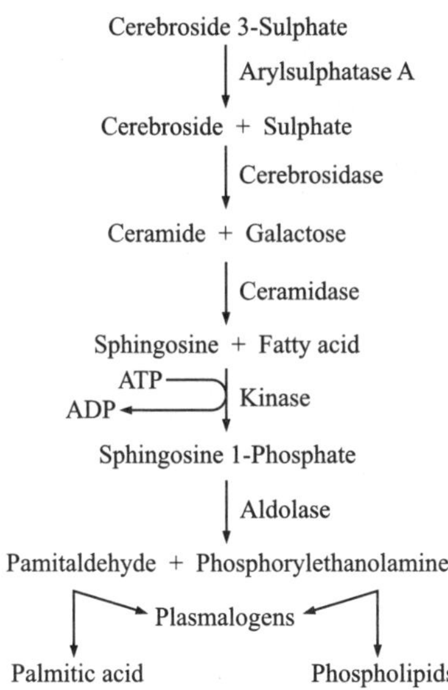

Figure 7.20 Biodegradation of cerebroside-3-sulphate.

E. Genetic Disorders of Ganglioside Metabolism

There are a number of genetic disorders associated with ganglioside metabolism. One of the earliest of such diseases investigated by Klenk himself is the Tay-Sachs disease in which there is an abnormal accumulation of the Tay-Sachs ganglioside in the brain and other tissues. Other disorders collectively known as gangliosidoses, are the GM_1 gangliosidosis and the GM_2 gangliosidosis.

The genetic disorder in Tay-Sachs ganglio-sidosis has been identified to be due to a deficiency of the enzyme β-N-acetylglucosaminidase as shown by O'Brien and his coworkers (Table 7.6). Two isoenzymes of hexosaminidase, namely, hexosaminidases A and B exist, and one of these, viz. hexosaminidase A was found to be specifically deficient in Tay-Sachs disease.

TABLE 7.6   Enzymatic Defects in Ganglioside Metabolism

Disease	*Enzyme deficiency*
Generalized gangliosidosis	GM1 ganglioside galactosidase
Tay-Sachs disease	Hexosaminidase A (GM$_2$ ganglioside N-acetyl galactosaminidase)
Variant form of Tay-Sachs disease	Total hexosaminidase activity
Gaucher's disease	Glucocerebroside glucosidase

In GM_1 gangliosidosis, there is an abnormal increase in GM_1 ganglioside as well as a keratan sulphate like mucopolysaccharide, in the tissues of patients. A deficiency of the enzyme β-galactosidase has been postulated to be the genetic defect in this disease (Table 7.6).

F. Physiological Role of Gangliosides

In the central nervous system the gangliosides are present in significant quantities in the grey matter. Their subcellular localization in the synaptosomal fraction of brain has given rise to speculations regarding the role of gangliosides in the synaptic transmission of nerve impulses. McIlwain demonstrated that the negatively charged sialic acid moiety of gangliosides is essential for the excitability of electrically stimulated cerebral cortical slices. Addition of basic proteins like protamine or histone were found to inhibit the excitability because they neutralised the negative charge of the gangliosides, and this inhibition could be reversed by addition of excess gangliosides to the medium containing the cerebral slices. It has been noted that stimulation of the central nervous system electrically or by drugs results in changes in brain gangliosides. The abnormal increase in gangliosides in certain genetic disorders also indicates the significance of these glycolipids in the metabolism of nervous tissues.

G. Biological Importance of Cerebrosides and Sulphatides, and Genetic Disorders Associated with Them

Cerebrosides and sulphatides are important constituents of myelin, a membranous structure found covering the nerve axon of the central and peripheral nervous system. Variation in the concentration of cerebrosides and sulphatides is found in the leukodystrophies and demyelinating diseases in which the myelin is damaged. Table 7.7 shows two such genetic diseases in which there is an abnormal accumulation of one of these lipids, and the corresponding enzyme deficiencies which cause these diseases.

TABLE 7.7 **Genetic Diseases Associated with Cerebroside and Sulphatide Metabolism**

Disease	Glycolipid abnormality	Deficient enzyme
Metachromatic leukodystrophy	Sulphatide accumulation	Arylsulphatase A (cerebroside sulphatase)
Globoid leuco-dystrophy	Derangement in galactocerebroside	Galactocerebrosidase metabolism

V. BLOOD GROUP SUBSTANCES

The so-called blood group substances are present in erythrocytes, saliva, gastric mucin, cystic fluids and other body secretions. They constitute the different blood group antigens like A, B, H (O) and others. When red cells containing specific type of blood group antigen are mixed with specific antibodies of the serum, agglutination of the cells takes place. Thus, the erythrocytes of type A blood are agglutinated by antibodies (isoagglutinins) found in the serum of type B or H blood; group A serum agglutinates red cells of blood type B and AB.

The first human blood group system (ABH system) was discovered by Landsteiner in 1900. The second blood group system of outstanding clinical significance, the rhesus (Rh) system was discovered after 40 years. In the intervening years two other systems (the MN and P systems) were recognized.

Somewhat paradoxically, the largest amount of information on the chemical basis of blood group specificity has come from chemical and immunological examination of substances that are not derived from the red cells. In humans, substances with A and B activities are present not only in the red cells but also in a water soluble form in the tissue fluid and secretions of most individuals. Five blood group specificities, A, B, H, Le[a] and Le[b] are detectable in human secretions by the use of the corresponding antibodies of human or animal origin. The structures responsible for these specificities are believed to arise from the action of four independent, but closely interrelated gene systems—ABH, Hh, Lele and Sese.

Chemistry of A, B, H and Lewis substances is presented in Figure 7.21. The chemical serological and genetical analyses of ABH and Lewis blood group substances from human secretion, indicate that patterns of synthesis resulting from the sequential action of the product of different blood group genes on a common precursor glycoprotein, lead to the three-dimensional patterns of sugar residues responsible for

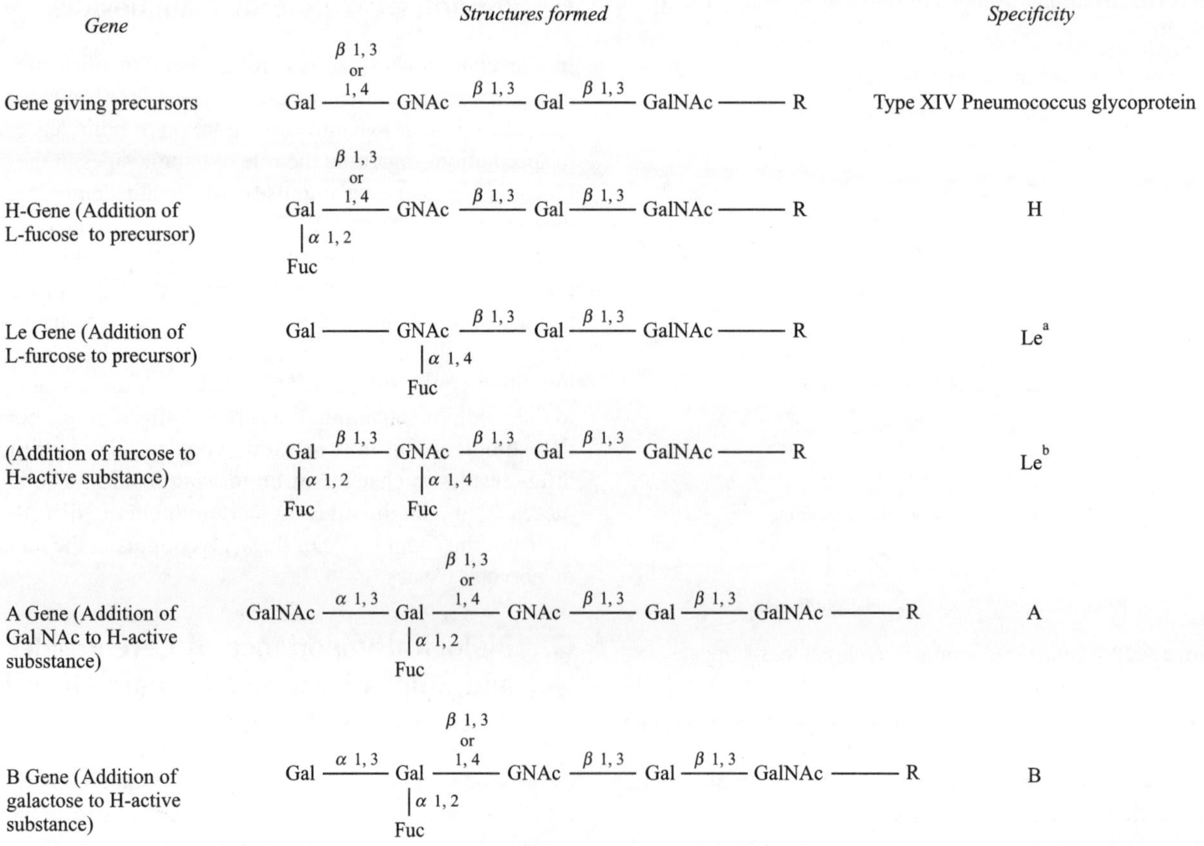

Figure 7.21 Chemistry of blood group substances. Gal = Galactose; GNAc = N-acetyl glucosamine; GalNAc = N-acetyl galactosamine; Fuc = Fucose.

serological specificity. The addition of sugar residue to the end of, or as a branch, on a carbohydrate chain produces a new serological specificity which is primarily determined by the newly added sugar in the chain, but also independent on the nature, sequence, and linkage of those sugars already present in the chain. Thus, H specific structures become part of the A and B specific groupings, when N-acetyl D-galactosamine or D-galactose, respectively, are added to the ends of the chains.

How far the picture for the secreted blood group substances is applicable to the A and B substances on the red cell, is not clear. However, it is almost certain that N-acetyl D-galactosamine and D-galactose respectively, are important serological determinants of the glycolipids A and B substances.

SUGGESTIONS FOR FURTHER READING

Aminoff D. (Ed.) (1970), *Blood and Tissue Antigen,* pp. 461–78, New York, Academic Press.

Aumailley M. and Gayraud B. (1998), Structure and Biological Activity of the Extracellular Matrix, *Mol. Med.,* **76,** 253–265.

Gottschalk (Ed.) (1972), *Glycoprotein—Part A and B,* Elsevier Publishing Company.

Hascall V.C. and Heinegard D.K. (1979), *Structure of Proteoglycans in Gregory Jeanloz Glycoconjugate Research, I.* pp. 341–74, New York, Academic Press.

Harourtz Martin I. and Ward Pigman (Eds.) (1978), *The Glycoconjugates,* Academic Press.

Jaques L.B. (1977), Determination of heparin and related sulphated mucopolysaccharide, *Methods in Biochemical Analysis,* **24,** pp. 203–312.

Kim Y.Z. and Varki A. (1997), Perspectives on the Significance of Altered Glycosylation of Glycoproteins in Cancer, *Glycoconj J.,* **14,** 569–576.

Kornfeld R. and Kornfeld S. (1976), Comparative Aspects of Glycoprotein Structure, *Ann. Rev. Biochem.,* **45,** pp. 217–237.

Montreuil J. (1975), *Pure Appl. Chem.,* **42,** 431–477.

Varma R.S. and Varma R. (Eds.) (1982), *Glycoamino Glycans and Proteoglycans in Physiological and Pathological Process of Body System,* Warren Pa Karger.

8

Eicosanoids (Prostaglandins)

Shail K. Sharma and G.P. Talwar

CONTENTS

During illnesses caused by infectious diseases or other sources of inflammation, a series of brain-mediated responses called the sickness syndrome occurs, which include anorexia, sleepiness, hyperalgesia, and elevated corticosteroid secretion. Much of the sickness syndrome is mediated by *prostaglandins* acting on the brain and can be prevented by nonsteroid anti-inflammatory drugs such as aspirin or ibuprofen that block prostaglandin synthesis.

Prostaglandins were first discovered and isolated from human semen in 1930s independently by Goldblatt and von Euler. It was in 1960s when their biosynthetic relationship to specific essential fatty acids was described. The Nobel Prize for Medicine in 1982 was given to Professors B. Samuelsson, John Vane and Sune Bergstrom for their work on their function. Prostaglandins occur at very low levels in tissues, but they have profound biological activities.

I. NOMENCLATURE AND STRUCTURES OF EICOSANOIDS

Prostaglandins, thromboxanes and leukotrienes are collectively called Eicosanoids. A key feature of all prostaglandins is a five-membered ring encompassing carbons 8 to 12. The thromboxanes are similar but have heterocyclic oxane structures. They are all synthesised by specific enzymes, which confer stereospecificity and chirality on every functional group. In the approved nomenclature each prostaglandin is named using the prefix 'PG' followed by a letter A to K depending on the nature and position of the substituents on the ring (Figure 8.1). Thus PGA to PGE and PGJ have a

keto group in various positions on the ring, and are further distinguished by the presence or absence of double bonds or hydroxyl groups in various positions in the ring. PGF has two hydroxyl groups while PGK has two keto substituents on the ring. PGG and PGH are bicyclic endoperoxides. An oxygen bridge between carbons 6 and 9 distinguishes prostacyclin (PGI). Thromboxane A (TXA) contains an unstable bicyclic oxygenated ring structure, while thromboxane B (TXB) has a stable oxane ring. In addition, all prostaglandins have a hydroxyl group on carbon 15 and a *trans*-double bond at carbon 13 of the alkyl substituent (R_2). Further, a numerical subscript (1 to 3) is used to denote the total number of double bonds in the alkyl substituents, and a Greek subscript (α or β) is used with prostaglandins of the PGF series to describe the stereochemistry of the hydroxyl group on carbon 9 (Figure 8.2).

II. RECEPTORS

Prostaglandins exert their effects by activating rhodopsin-seven transmembrane spanning G-protein-coupled receptors (GPCRs). The receptor subfamily is composed of eight members (DP, EP-1-4, FP, IP and TP), and recently, a ninth prostaglandin receptor was identified, the chemoattractant receptor homologous molecule expressed on Th2 cells. Among the different receptors the IP, DP, EP2 and EP4 receptors increase the intracellular cAMP and have been termed 'relaxant' receptors because they induce smooth muscle relaxation. The TP, FP and EP1 receptors induce calcium mobilization and constitute a 'contractile' receptor group, because they cause smooth muscle contraction.

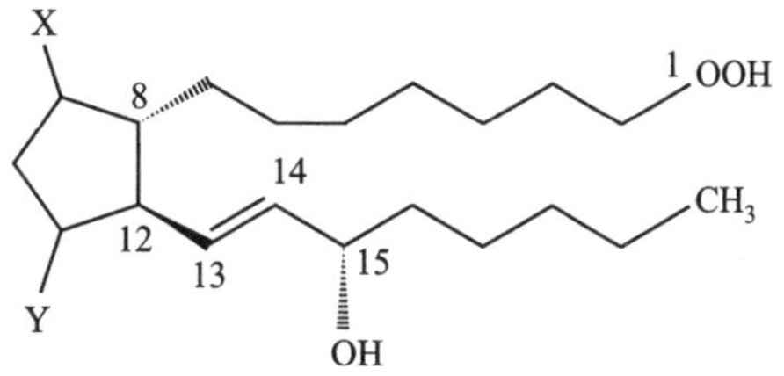

Figure 8.1 Prostaglandins structural features PGA, PGD, PGE, PGF, PGG, PGH, PGI, depending on the functional groups present at X or Y. PGF 1, 2 or 3 depending on the number of double bonds in the linear hydrocarbon chain.

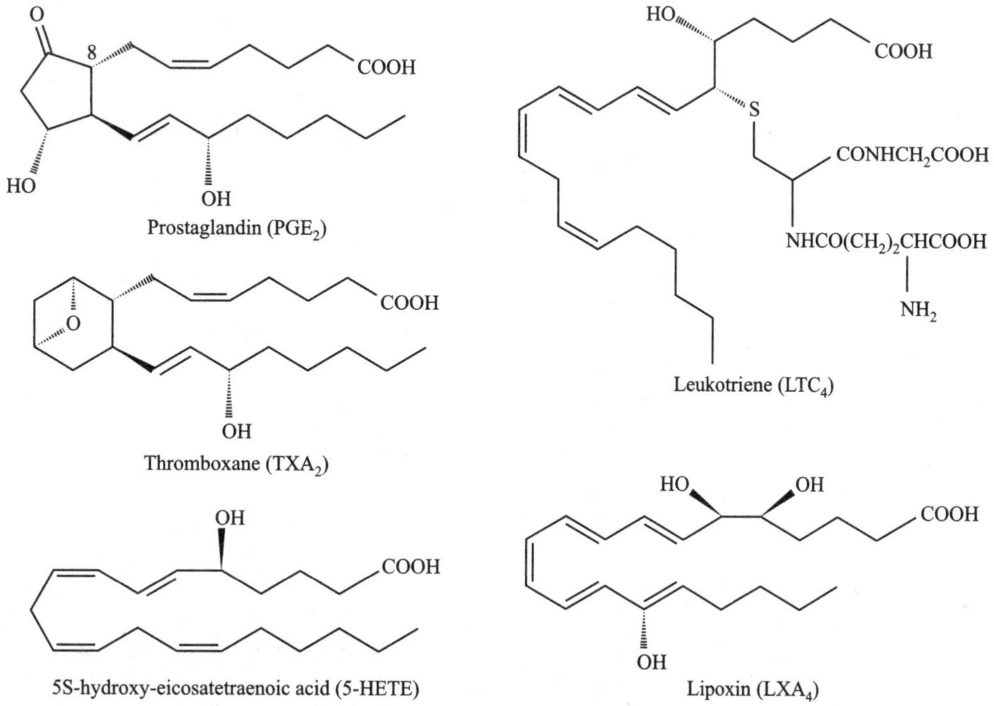

Prostaglandin (PGE$_2$)

Thromboxane (TXA$_2$)

5S-hydroxy-eicosatetraenoic acid (5-HETE)

Leukotriene (LTC$_4$)

Lipoxin (LXA$_4$)

Figure 8.2   Structure of eicosanoids.

III.   FUNCTIONS

Prostaglandins have variety of functions:

Inflammation and Immune System

PGE-2 is the most abundant prostaglandins produced in the body and exhibits versatile biological activities. It is an important mediator of regulation of immune responses, blood pressure, gastrointestinal integrity, and fertility. In inflammation PGE2 is of particular interest, because it is involved in all processes leading to the classical signs of inflammation: redness, swelling and pain. Redness and edema result from increased blood flow into the inflamed tissue through PGE2 –mediated augmentation of arterial dilatation and increased microvascular permeability. Pain results from the action of PGE2 on peripheral sensory neurons and on the central sites within the spinal cord and the brain.

Leukotrienes are a 100 times more potent than histamine. The latter provides a rapid response to an allergen. However in the later stages leukotrienes are principally responsible for inflammation, smooth muscle contraction, constriction of airways, and mucous secretion from the mucosal epithelium.

Cardiovascular Homeostasis

PGI2 is important in regulation of cardiovascular homeostasis. Vascular cells, including endothelial cells, vascular smooth muscle cells, and endothelial progenitor cells are the major source of PGI2.

Reproductive System

Prostaglandins are key regulators of reproductive processes. PGF2α is derived mainly from COX-1 in the female reproductive system, and plays an important role in ovulation, luteolysis, contraction of uterine smooth muscle and initiation of parturition.

Platelets

Platelet aggregation is induced by Thromboxane (TXA2) and prostaglandin endoperoxidases. PGI2 generated in the epithelium, however, reverses platelet aggregation and maintains homeostasis in platelet-vessel wall interaction.

CNS

PGD2 is the major eicosanoid that is synthesized in both CNS and peripheral tissues and appears to function in both an inflammatory and homeostatic capacity. In the brain PGD2 is involved in the regulation of sleep and pain perception.

IV. BIOSYNTHESIS

Cyclo-oxygenase Pathway

Prostaglandins and thromboxanes are formed when arachidonic acid, a 20 carbon unsaturated fatty acid is released from the plasma membrane by phospholipases and metabolized by the sequential actions of prostaglandin G/H synthases, or cyclo-oxygenases and respective synthases. There are 4 principal bioactive prostaglandins synthesized *in vivo*: PGE-2, PGI-2, PGD-2 and PGF-2α. Usually each cell type generates one or two dominant products which act as autocrine or paracrine lipid mediators to maintain local homeostais in the body.

Prostaglandins production depends on the activity of prostaglandin G/Hsynthases, colloquially known as COXs bifunctional enzymes that contain both cyclo-oxygenase and peroxidase activity and which exist as distinct isoforms referred to as COX-1 and COX-2 (Figure 8.3).

COX-1, expressed constitutively in most cells is the dominant source of prostanoids that subserve house keeping functions, such as gasrtric epithelial cytoprotection and homeostasis. CoX-2 induced by inflammatory stimuli, hormones and growth factors is the more important source of prostanoid formation in inflammation and proliferative diseases.

Aspirin suppresses the production of prostaglandins and thromboxanes by irreversible inactivation of cyclo-oxygenase (COX) enzyme (Figure 8.4). Aspirin acts an acetylating agent,

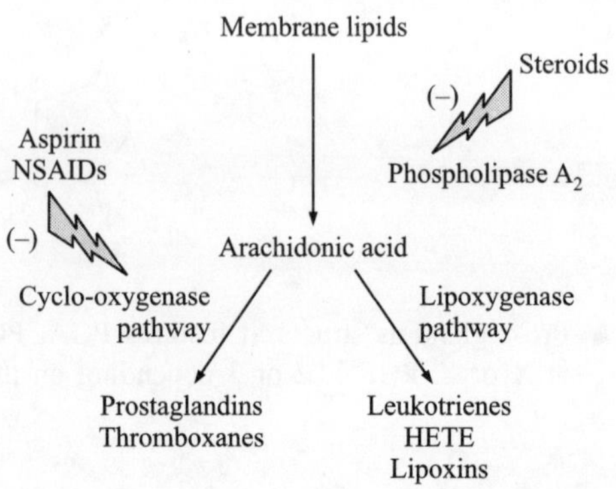

Figure 8.3 Pathways of eicosanoid synthesis.

where an acetyl group is covalently attached to a serine residue in the active site of COX.

Lipoxygenase Pathway

Lipoxygenases are a family of iron-containing enzymes that catalyze the deoxygenation of polyunsaturated fatty acids in the following reaction.

$$\text{Fatty acid} + O_2 = \text{Fatty acid hydroperoxide}$$

Based on the position of the hydroperoxy group three types of lipoxygenases are known. The HPETEs are unstable and are converted to their corresponding hydroxy fatty acids (HETEs).

5 HPETE formed from arachidonic acid is the precursor of four series of leukotrienes. Linoleate gives rise to to the LT3 series, and linolenate to the LT5 series.

Rearrangement of 5HPETE leads to the formation of an unstable 5,6 epoxide –Leukotriene A4 (LTA4) which is transformed to LTB4 or conjugated with glutathione by LTC4 synthase to form LTC4. Subsequent cleavage of glutamic acid and glycine leads to the formation of LTD4 and LTE4, respectively. The slow reacting substance of anaphylaxis (SRS-A) is now recognized to be a mixture of LTC4 and LTD4 (Figure 8.5).

15 HPETE, in leukocytes is converted to lipoxins.

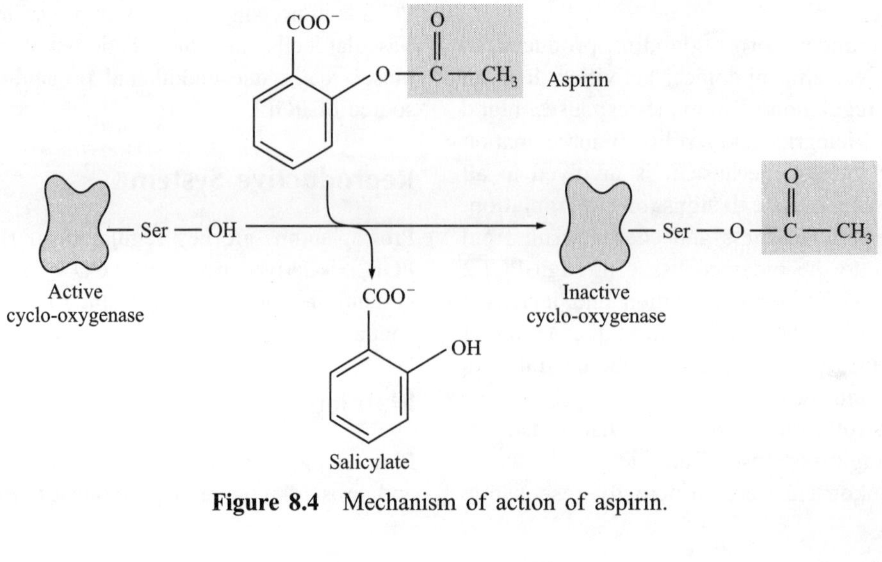

Figure 8.4 Mechanism of action of aspirin.

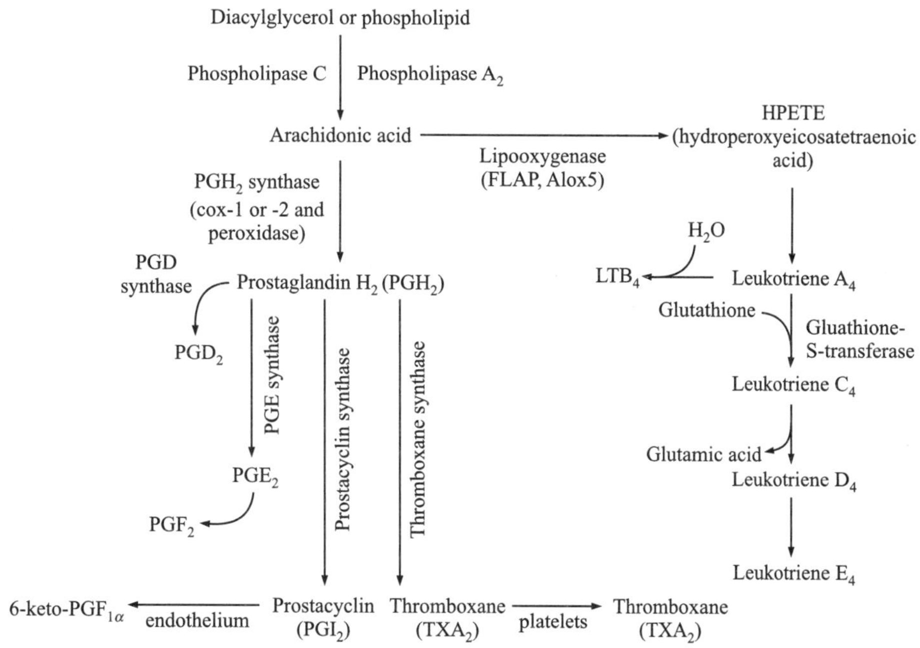

Figure 8.5   Biosynthesis of eicosanoids inactivating cyclo oxygenase blocking the formation of PGE from arachindonic acid (Figure 8.3).

V.   INACTIVATION

Eicosanoids have short half-lives (10 seconds to 5 minutes), so that functions are usually limited to actions on nearby cells. 15-hydroxyprostaglandin dehydrgenase is a key prostaglandin catabolic enzyme catalyzing the oxidation and inactivation of PGE2. Further catabolism of 15-keto-prostaglandin E2 is catalyzed by 15-ketoprostaglandin-Δ(13)-reductase, which also exhibits LTB(4)-12-hydroxydehydrogenase activity these enzymes are expressed in lungs and other tissues.

VI.   THERAPEUTIC USES

The therapeutic uses of the prostaglandins are concentrated in four areas, gynecology, cardiovascular diseases, gastroentereology, and ophthalmology.

Gynecology: Prostaglandins are used in induction of labor and medical termination of pregnancy. PGE-2 analog (dinoprostone) binds EP2 and EP3 receptors. EP3 receptor activation in uterus causes contractions. It is used in cervical ripening and uterine contractions.

Cardiovascular diseases: PGI2 analog (Treprostinil) is used in pulmonary hypertension.

Gastroenterology: PGE-1 analogs (misoprostal) binds to EP3 and EP2 receptors. They are cytoprotective in the gut and are used in suppressing gastric acid secretion.

Ophthalmology: PGF2α analogues (bimatoprost) decrease intraocular pressure and is used to treat glaucoma.

Pulmonary Circulation

PGE1 and PGE2 have vasodilatory effects on the pulmonary circulation. It is mainly used in infants with congenital heart disease. The ductus arteriosus is a normal fetal blood vessel that closes soon after birth. In patent ductus areriosus (PDA) the vessel does not close and remains patent (open). ProstaglandinE1 is responsible for keeping the ductus patent. NSAIDS (indomethacin) infusions are used to close the PDA.

In addition PGI-2 analogues (epoprostenol) are used in inhibition of platelet aggregation, Leukotriene and lipoxygenase inhibitor (Zileuton) cause bronchoconstriction and secretions and are used in asthma.

SUMMARY

Prostaglandins are ubiquitous and have been found in almost all tissue in human beings. They are synthesized at the site of tissue damage or infection, where they cause inflammation, pain and fever as a part of the healing process. Aspirin is a common medication and has two main actions: (1) Anti-inflammatory fever and pain reliever and (2) Anti-platelet (blood coaggluting) effect. Aspirin acts by acetylation of the COX enzyme, thereby stopping in synthesis of prostaglandins. Clinically prostaglandins have been used in child birth or abortion, cyanotic heart defects peptic ulcers, pulmonary hypertension and treatment of glaucoma.

SECTION II

CELLULAR REACTIONS

The dynamics of cellular reactions is transacted by a variety of enzymes. These biocatalysts digest a variety of carbohydrates, fats and proteins and convert these into a variety of compounds. In the process, energy is created and stored in the form of ATP which in turn can be utilized for other reactions, whenever needed. This section familiarizes the readers with mechanics of the body. The significance of cholesterol and various types of lipids in blood is brought out by an eminent cardiologist. Also, this section describes the abnormalities in humans due to inborn errors of metabolism.

9

Enzymes—The Biocatalysts

O.P. Malhotra

CONTENTS

I. INTRODUCTION

A multitude of chemical reactions are known to be brought about by the living cells. Collectively, they are referred to as metabolism. Some reaction sequences lead to breakdown of large molecules and their oxidation to carbon dioxide and water, while others lead to synthesis of such molecules from simpler substances. The individual changes cover a wide variety of organic reactions. For each reaction the cell is equipped with a specific catalyst, an *enzyme*. Together, the enzymes are responsible for an ordered sequence of chemical reactions and control the rates of individual reactions, and therefore, of the overall metabolism. The enzymes themselves are, in turn, under the control of various factors (chemical and physiological) and thus serve as the last but important link in the expression of the manifest effects of hormones and other agents.

In the cell some of the enzymes are arrayed on various cell membranes and other organelles. Some others are present in free soluble form and appear to be floating around in the cytoplasm. Irrespective of its location each enzyme has a specific role to play, namely to catalyse a given chemical reaction.

II. CATALYSIS

The role of a catalyst in the progress of a chemical reaction is illustrated in Figure 9.1. Taking a simple chemical reaction, the conversion of a molecule of A into that of B, the progress of reaction *in the absence of a catalyst* may be represented as curve (1) of Figure 9.1. Although the net reaction A $\rightarrow$ B is a "downhill" process in the sense that energy is released (and, therefore, B occupies a "lower" position on the energy scale), *A is quite stable because of the energy barrier* (Energy of activation) that must be crossed before it is converted into B. This extra energy represents the difference in the energy content of an average molecule of A and that of an "activated" molecule of A, (or the "transition state"). The latter may revert back to A by losing the activation energy to other molecules (by collision), or get converted into B. The rate of reaction is proportional to the population of "activated" molecules. A catalyst alters the nature of transition state and, therefore, the height of the energy barrier between A and B *without affecting the initial and the final states,* namely free A and free B [Figure 9.1; curve (2)]. Since this smaller activation energy is possessed by a larger proportion of the molecules,

101

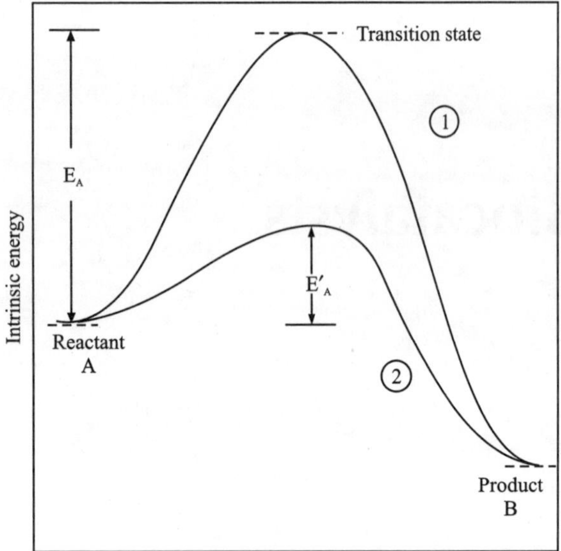

Figure 9.1   Energy diagram for the reaction A → B in the presence and absence of a catalyst. The uncatalysed reaction is shown in curve (1) and the catalysed reaction in curve (2). E_A and E'_A are the energies of activation for the uncatalysed and the catalysed reaction, respectively.

the reaction proceeds at a faster rate. In principle, the enzymes cause catalysis in a similar manner. The enzyme-catalysed reactions are distinguished from other catalysed reactions by the following characteristics.

III. GENERAL CHARACTERISTICS OF ENZYMES

A. Chemical Nature of Enzymes

All enzymes are proteins. The converse is not true, i.e., all proteins are not enzymes. Advances in our understanding of the structure of enzymes and their mechanism are intricately linked with our understanding of the structure of proteins. By now the complete amino acid sequence (primary structure) and three-dimensional structure of several enzymes are known. Consequently, it is now possible to discuss enzyme catalysis in molecular terms. One such example is described in a later section. It can be shown that the integrity of the native three-dimensional structure of the enzyme protein is essential for its catalytic activity. Thus, all such conditions which bring about denaturation of proteins (e.g., extremes of pH, high temperatures, addition of urea etc.) lead to loss of catalytic activity. Many enzymes contain endogenously bound metal ions (like Fe^{2+}, Cu^{2+}, Zn^{2+} etc.). These are referred to as metalloenzymes.

In many cases, the presence of a particular non-proteinous small organic molecule is essential for the enzyme activity. Such a compound is referred to as a *coenzyme*. The protein component, which is inactive by itself, is called *apoenzyme* and the functional complex is called *holoenzyme* or simply, the enzyme [Eq. (9.1)].

$$\text{Apoenzyme} + \text{Coenzyme} \rightleftharpoons \text{Holo-enzyme} \qquad (9.1)$$
$$\text{(Protein)} \qquad \text{(Nonproteinous)}$$

When a coenzyme is very tightly bound to the enzyme protein, it is also referred to as a *prosthetic group*.

B. Enzymes are True Catalysts

A true catalyst affects the rate, but not the equilibrium constant, of a chemical reaction. It is clear from Figure 9.1. The equilibrium constant depends on the energy difference between the reactants and the products. Since this is independent of the pathway connecting them, the equilibrium constant is not affected by a catalyst. Equilibrium constant is also equal to the ratio of the rate constant for the forward and reverse reactions. It follows, therefore, that a catalyst will catalyse the forward and reverse reactions to the same extent.

C. Specificity

Enzymes are highly specific. A given enzyme catalyses a single reaction in one or a few closely related compounds (the "substrates"). The reaction specificity, which defines the type of catalysed reaction, is very stringent. Specificity for substrate is generally relative, in the sense that a number of closely related compounds may function as substrates, but the rate of reaction may vary from one substrate to another. Some enzymes exhibit "absolute specificity" for their substrate, e.g., urease catalyses the hydrolysis of urea only.

D. Formation of Enzyme-Substrate Complex

First step in enzyme catalysis is the formation of a fairly strong complex between the enzyme and its substrate. First suggestions for the complex formation came from kinetic studies. In later years, evidence for complex formation was provided by spectrophotometric experiments in solution and X-ray diffraction studies of crystals of enzyme in the absence and presence of substrates.

Complex formation between an enzyme and its substrate has important consequences for its catalytic efficiency. The catalysed reaction takes place intramolecularly in this complex. Intramolecular catalysis is more efficient than intermolecular catalysis.

In general a substrate molecule is very much smaller than the enzyme molecule. It follows, therefore, that in the enzyme-substrate complex the substrate comes in contact with a small region of the enzyme molecule. Those parts of the enzyme molecule, which come in contact with the substrate in the enzyme-substrate complex or during subsequent chemical reaction, are said to constitute the "Active site" of the enzyme.

E. Catalytic Efficiency

Enzymes are far more efficient than any man-made catalyst. Rate enhancements of the order of 10^{15}, over the rate of uncatalysed reaction, have been reported. The rate of an enzyme-catalysed reaction is expressed as "Turnover number", which is defined as the number of substrate molecules converted into products per enzyme site per second under optimal conditions. Turnover numbers of some enzymes are listed in Table 9.1. In most cases, the number lies in the range 10^2–10^3 per second. Such high rates are remarkable, specially in view of the fact that almost all the enzyme-catalysed reactions proceed under very mild conditions—*neutral aqueous medium at* 30–40°C. In the absence of enzymes, these reactions may not proceed at measurable rates or require rather drastic conditions. Several ideas have been put forward to explain high catalytic efficiency, but these do not adequately explain the quantitative increase in rates brought about by the enzymes.

TABLE 9.1   Turnover Numbers of Some Enzymes

Enzyme	*Turnover number (Molecules of product formed per enzyme site per second)*
Chymotrypsin	10^2–10^3
Trypsin	10^2–10^3
Carboxypeptidase	10^2
Acetylcholinesterase	10^4
Ribonuclease	10^2
Myosin-ATPase	10^2
Urease	10^4
Fumarase	10^3
Enolase	10^2
Carbonic anhydrase	10^5
Catalase	10^7

Rates of enzyme-catalysed reactions are affected by experimental conditions, like temperature, pH and the presence of some specific molecules. Some substances enhance the rates (Activators), and others decrease the rates (Inhibitors). These substances are referred to as "Effectors".

F. Regulation of Enzyme Activity

In living cells, rates of metabolic processes are precisely regulated so that the metabolism may not proceed too fast or too slow. Further, it may become necessary for an organism to increase or decrease these rates in order to meet requirements of a particular situation. Some of these controls operate at the level of enzymes. The enzymes which are subject to such regulation are referred to as *regulatory enzymes:* Structure of regulatory enzymes changes on interaction with the physiological effectors, with consequent changes in catalytic activity.

Some of these general features are discussed in greater detail in the following sections.

IV. CLASSIFICATION AND NOMENCLATURE OF ENZYMES

The specificity of enzymes demands the existence of a multitude of these catalysts in a living cell. Indeed the number of known enzymes is of the order of two thousand. In order to deal with such large numbers, it becomes necessary to evolve a system of classification and nomenclature of enzymes.

In early stages, the enzymes were assigned names by their discoverers in a seemingly arbitrary manner. Such names, as, trypsin, chymotrypsin, diastase and catalase, convey no information about the function of these enzymes, the nature of their substrates, or of the catalysed reaction. Later on, a generic ending "ase" was adopted for adding to the name (full or partial) of the substrate, with or without indicating the nature of chemical reaction. For example, urease, lactase, alcohol dehydrogenase etc. No mention was made, however, of the coenzyme, if any, required for the reaction.

An "official" system of classification and nomenclature of enzymes was proposed in 1961 by the Commission on Enzymes of the International Union of Biochemistry. They divided the enzymes into six major groups according to the type of catalysed reaction (reaction specificity). These classes are shown in Table 9.2. Each class is further divided into sub-classes and sub-subclasses according to the nature of the substrate, its reactive group and coenzyme etc. The same Commission also recommended a system of nomenclature.

TABLE 9.2   Classification of Enzymes

Class Number	*Enzyme Class*	*Catalysed Reaction*
1	Oxido-reductases	Oxidation-reduction $(AH_2 + B \rightleftharpoons A + BH_2)$
2	Transferases	Group transfer $(A–X + B \rightleftharpoons A + B–X)$
3	Hydrolases	Hydrolysis $(A\text{-}B + HOH \rightleftharpoons AH + BOH)$
4	Lyases	Addition-elimination $(A = B + X\text{-}Y \rightleftharpoons AX – BY)$
5	Isomerases	Isomerization $(A \rightleftharpoons A')$
6	Ligases	Joining together ("condensation") of the two different molecules accompanied by expenditure of energy $(A + B + ATP + H_2O \rightarrow A\text{-}B + ADP + Pi)$

The "official" name of an enzyme gives the names of its substrate and coenzyme (if any), specifies the reactive group of the substrate and contains information about the nature of catalysed reaction. Such a name will no doubt distinguish one enzyme from another, but will be too long for ordinary use.

Therefore, other 'trivial' or common names are used more frequently. The trivial name does not convey all the information which is conveyed by a systematic name, but is more handy for everyday use. In many cases, old and familiar names have been retained as "trivial" names. Some examples of systematic and trivial names are given in Table 9.3. In this chapter, we shall be using the trivial names only.

V. ESTIMATION OF ENZYME ACTIVITY

All enzymes are proteins in nature. Therefore, the methods described earlier for protein fractionation (based on differences in solubility, stability, size, electric charge and adsorption properties) are used for the isolation of an enzyme in pure form. The efficiency of any isolation procedure can be judged in terms of the ratio of enzyme activity and total protein (the "specific activity"). This necessitates the estimation of enzyme activity at each purification step. The procedure of activity assay is also of importance in characterizing the enzyme with respect to the effect of various factors, like temperature, pH and the presence of activators and inhibitors.

A measure of enzyme activity, indeed the activity of any catalyst, can be obtained by measuring the rate of catalysed reaction in excess of the rate observed in the absence of the catalyst. The rate of reaction is obtained by monitoring the amount of product formed (or of substrate disappearing) in a unit time. A conveniently observable physical property of the product (or of substrate), *which is directly proportional to its concentration,* is chosen for monitoring the progress of the reaction. On account of convenience and sensitivity, "colorimetric" (or spectrophotometric) methods are most commonly used. For example, p-nitrophenyl acetate has been frequently used as a test substrate for carboxylesterases. The hydrolysis of this colourless substrate yields an intensely yellow coloured product (p-nitrophenol) [Eq. (9.2)].

$$(9.2)$$

The intensity of yellow colour (as measured by absorption of light in the complementary blue-violet spectral region) is proportional to the concentration of p-nitrophenol in the medium. If the enzyme and the above substrate are brought together *at a fixed pH and temperature* and the yellow colour intensity noted at various times (e.g., 30, 60, 90, 120 seconds etc.), we can obtain the *rate* of liberation of the product and, therefore, of the reaction.

In other cases, one of the products of reaction may be made coloured by further reaction with another reagent. For example, hydrolysis of urea with urease gives rise to carbon dioxide and ammonia (Eq. 9.3). The latter product reacts with Nessler's reagent to give a yellow coloured product.

$$NH_2 - CO - NH_2 + H_2O \xrightarrow{\text{Urease}} 2NH_3 + CO_2 \qquad (9.3)$$

In this case, the reaction is allowed to proceed for a fixed period of time and then stopped by inactivating the enzyme. The amount of ammonia is then estimated by adding Nessler's reagent and measuring the resultant yellow colour.

In each case, a blank experiment is run side by side in which the enzyme may be omitted or re-placed by inactivated protein. A proper correction is applied for the "blank" value.

Although colorimetric methods are preferred, wherever possible, other techniques have also been used. These include manometric (estimation of volumes of gases liberated or consumed), titrimetric (estimation of an acid or base liberated during the reaction), and fluorescence measurements. The interested reader is referred to specialized books on enzymes or on analytical techniques.

TABLE 9.3   Nomenclature of Enzymes*

Systematic Name	Catalysed Reaction	Trivial Name
Alcohol: NAD oxidoreductase	$R{-}CH_2OH + NAD^+ \rightleftharpoons RCHO + NADH + H^+$	Alcohol dehydrogenase
β-D-Glucose: oxygen oxido-reductase	β-D-Glucose $+ O_2 \rightleftharpoons$ D-Glucono-δ-lactone $+ H_2O_2$	Glucose oxidase
ATP: D-hexose 6-phospho-transferase	ATP + D-hexose $\rightleftharpoons$ ADP + D-hexose-6-phosphate	Hexokinase
		Carboxyl esterase
Carboxyl ester hydrolase	$R - \overset{\displaystyle O}{\overset{\displaystyle \|}{C}} - OR' + H_2O \rightleftharpoons R - COOH + HOR$	
		Urease
Urea amido hydrolase	$NH_2 - \overset{\displaystyle O}{\overset{\displaystyle \|}{C}} - NH_2 + H_2O \rightleftharpoons 2NH_3 + CO_2$	
	ATP + CH_3COOH + CoA $\rightleftharpoons$	
Acetate: CoA ligase (AMP)	$CH_3{-}\overset{\displaystyle O}{\overset{\displaystyle \|}{C}}{-}CoA + AMP + \text{Pyrophosphate}$	Acetyl–CoA synthetase

*Full names and structures of coenzymes are given in a later section. In this table only the commonly used abbreviated forms (NAD, ATP, ADP, AMP and CoA) are used.

VI. KINETICS OF ENZYME ACTION

We have seen above how the rate of an enzyme-catalysed reaction can be measured. This rate varies with several parameters like substrate concentration, pH of the reaction medium, temperature, concentration of modifiers (activators and inhibitors) of enzyme activity, etc. Various mathematical relationships, derived on simple assumptions, can explain the changes in rate of reaction with various parameters. In this section, we shall discuss mainly the results of the application of these relationships without going into the theoretical derivations. When substrate and enzyme are brought together, there is a lag period (of the order of a millisecond or less) which is followed by a period of linear increase in the concentration of products, i.e., a constant rate of reaction. This constant rate is referred to as the *"initial" steady state rate*. As the substrate is depleted, the rate of formation of products decreases. In the following paragraphs we shall consider the "initial" steady-state rates only.

A. Effect of Substrate Concentration

The simplest observed relationship between the rate of an enzyme-catalysed reaction and the concentration of substrate is illustrated in Figure 9.2. At low substrate concentration, the rate of reaction (v) is directly proportional to the substrate concentration, (i.e., $v \propto [S]$; a first order dependence of v upon [S]). The slope of v versus [S] plot decreases gradually and ultimately the rate becomes independent of substrate concentration ($v \propto [S]^0$; a zero-order dependence of v upon [S]). This type of "saturation" phenomenon is explained on the basis of the following scheme (for a single-substrate system) (Eq. 4).

$$E + S \underset{k_{-1}}{\overset{k_1}{\rightleftharpoons}} E.S. \overset{k_p}{\longrightarrow} E + P \qquad (9.4)$$

In this scheme E, S, E.S and P stand for free enzyme, substrate, enzyme-substrate complex and the product, respectively. Under the conditions of most experiments, *where* $[S] \gg [E]_{total}$ *and where only initial rates are measured*, it can be shown that the rate of reaction (i.e., the rate of formation of products or rate of disappearance of substrate) is given by the Eq. (9.5), known as *Michaelis-Menten rate equation*.

$$v = \frac{k_p [E]_{total}[S]}{K_m + [S]}$$

$$= \frac{V_{max}[S]}{K_m + [S]} \qquad (9.5)$$

where $\qquad K_m = \dfrac{(k_p + k_{-1})}{k_1}$

and $V_{max} = k_p [E]_{total}$

The constant term V_{max} *represents the maximal rate of reaction at very high* [S] (where $[S] \gg K_m$ and, therefore, the

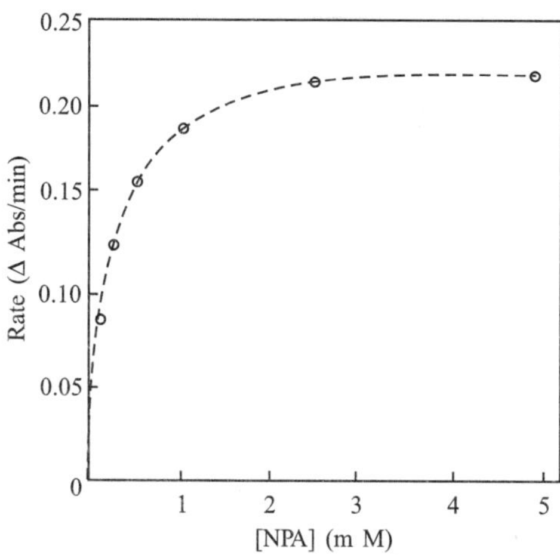

Figure 9.2 Effect of substrate concentration on the rate of hydrolysis of p-nitrophenyl acetate (NPA) catalysed by purified goat intestinal esterase (0.18 mg enzyme per ml in 0.05 M phosphate buffer, pH 7.3 at 30°). Liberation of p-nitrophenol product was monitored spectrophotometrically at 405 nm.

denominator of Eq. (9.5) approaches [S]. Note that V_{max} is independent of [S]. K_m (also called Michaelis constant) is a measure of the overall stability of enzyme-substrate complex,, i.e., affinity of the enzyme for the substrate is inversely related to K_m. Numerically, K_m *is equal to the substrate concentration at which the experimentally observed rate is half of* V_{max}. This can be verified by putting v equal to $V_{max}/2$ in Eq. (9.5).

The two constants, K_m and V_{max} are important for defining an enzyme-substrate system and for comparing different substrates of an enzyme, that is in studies of relative substrate specificity. A convenient method for estimation of K_m and V_{max} was suggested by Lineweaver and Burk using reciprocal of Eq. (9.5). This is done in Eq. (9.6).

$$\frac{1}{v} = \frac{K_m + [S]}{V_{max}[S]}$$

$$= \frac{K_m}{V_{max}} \times \frac{1}{[S]} + \frac{1}{V_{max}} \qquad (9.6)$$

A plot of 1/v versus 1/[S] (a "double-reciprocal plot") should give a straight line with slope equal to K_m/V_{max} and an intercept of $1/V_{max}$ on the 1/v axis (Figure 9.3). Furthermore, the intercept on 1/[S] axis is equal to $-1/K_m$.

In the above discussion, we have considered only a single-substrate system. There are very few enzymes (e.g., isomerases) which conform to such a limitation. In some other cases the reaction may be considered as a pseudo-single substrate reaction in which the concentration of one of the reactants (like water in hydrolysis) remains more or less constant during the course of reaction. In most cases, however, an enzyme catalyses reaction between two substrates (or a substrate and a coenzyme) and the concentrations of both reactants change (two-substrates system). Kinetics of such systems will not be discussed here. A simplified version of such systems is

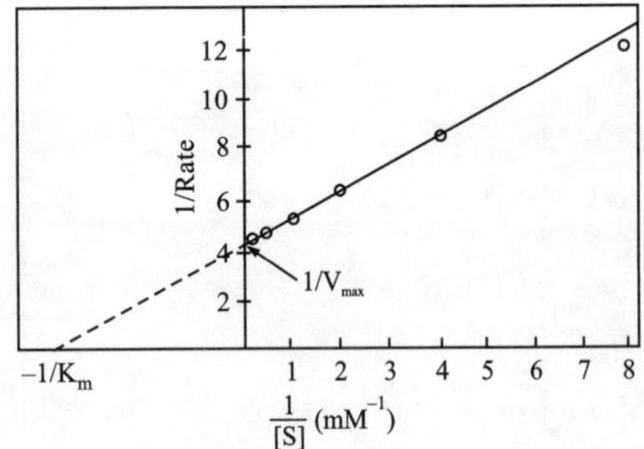

Figure 9.3   Double reciprocal (Lineweaver and Burk) plot of the data of Figure 9.2.

given by Bernhard in his book "Structure and Function of Enzymes."

B.   Effect of Temperature and pH

The velocity of an enzyme-catalysed reaction is affected by such parameters as temperature and pH. As the temperature is raised, the velocity increases, passes through a maximum and then falls at higher temperatures. This is due to the ocurrence of two processes with opposite effects, namely (1) the increase in the rate of catalysed reaction with a rise in temperature, and (2) a steeper rise in the rate of denaturation of the enzyme protein leading to inactivation. The rate of denaturation increases *very steeply* with a rise in temperature so that even a slight rise increases its rate several folds. Similarly, when the temperature is lowered the rate of denaturation decreases sharply. The net result is that at lower temperatures (i.e., under the conditions of normal enzyme assay) denaturation is not significant and may be ignored. Under these conditions, the rate of the catalysed reaction increases with temperature as is the case with any chemical reaction. At higher temperatures, inactivation due to denaturation becomes the dominant factor. As the rate of the catalysed reaction is proportional to the concentration of *active native enzyme,* we observe a *decrease* in the rate of enzyme reaction on raising the temperature. Since the thermal stability of each protein is different from others, some differences may be expected in the so called "optimum temperature" of individual enzymes. The differences are, however, very small.

Qualitatively three types of curves are obtained when the rate of an enzyme-catalysed reaction is plotted against pH of the solution (Figure 9.4). In some cases a bell-shaped curve is obtained (Curve 1) with a well defined (sharp or diffused) pH-optimum. In others a sigmoid curve is observed with the rate being constant at low or high pH's and falling in the other direction (curves 2 and 3). It may be pointed out that the changes shown in Figure 9.4 are *reversible* in the sense that the curves may be retraced by changing the pH in one direction or the other. These are distinct from irreversible inactivation (due to denaturation) of all enzymes at extremes of pH. The results of Figure 9.4 are generally explained on the

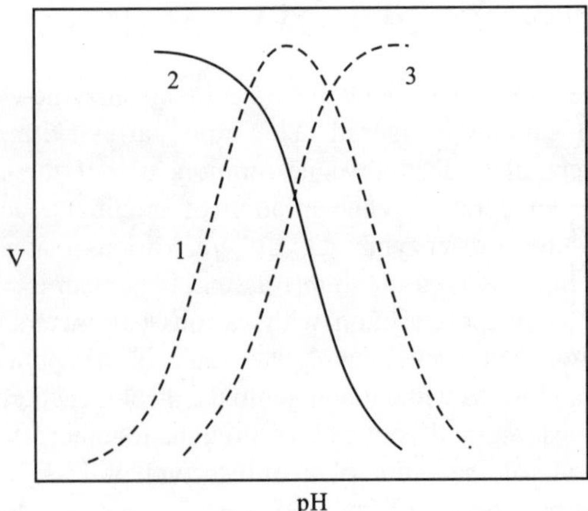

Figure 9.4   Influence of pH on the rate of enzyme catalysed reactions.

basis of acid-base dissociation *in a narrowly defined region of the enzyme molecule* (the "active site"). The dissociation phenomena shown in Eq. (9.7) suffice to explain the bell-shaped curve to Figure 9.4.

$$\underset{\text{inactive}}{EH_2} \xrightleftharpoons{K_1} \underset{\text{active}}{EH} \xrightleftharpoons{K_2} \underset{\text{inactive}}{E} \qquad (9.7)$$

(The relative changes in electric charges are not shown.) EH_2, EH and E represent various acid-base dissociation states of the enzyme molecule, out of which only EH is enzymically active. The acid-base dissociation constants of the two steps are K_1 and K_2, respectively. It is clear that EH_2 (the inactive species) will predominate at lower pH's and E (inactive) at higher pH's. At pH values in between these two extremes, most of the enzyme will exist in the active form EH. In curves 2 and 3 of Figure 9.4 a single dissociation step (K_2 and K_1, respectively) may be involved.

The shape of the rate-pH curve and the position of mid-point on the ascending and descending parts of these curves on the pH-scale provide valuable information about the magnitude of acid-base dissociation constants, K_1 and K_2 of Eq. (9.7). There are only a limited number of types of side-chain groups of proteins which can undergo acid-base dissociation in the pH range of interest here. The dissociation constants of these groups are known and vary within narrow limits. In the enzyme molecule, these may, however, be considerably perturbed (i.e., altered) by the local environments in the enzyme-substrate complex. Dissociation constants (as $pK_a = -\log K_a$) of some of the groups are shown in Table 9.4. Thus the rate-pH curves can provide useful, *albeit* tentative, information about the nature of some of the groups on the enzyme active site.

C.   Inhibition of Enzyme Activity

Conversion of active species EH into inactive species EH_2 involves a *reversible* addition of a proton to the enzyme protein. The resultant decrease in activity may be referred to

TABLE 9.4   Dissociation Constants of Some of the Groups Found in Enzymes

Acidic form of the group	*Amino acid residue*	*pKa* (unperturbed by adjacent groups)*
– COOH	Carboxy terminal	3.0–3.2
– COOH	Aspartate	3.7–4.0
– COOH	Glutamate	4.2–4.5
– Imidazolium$^+$	Histidine	6.7–7.1
–NH$_3^+$	Amino terminal	7.6–8.0
– SH	Cysteine	8.8–9.1
– NH$_3^+$	Lysine	9.3–9.5
Phenolic –OH	Tyrosine	9.7–10.1

*pK$_a$ = – log K$_a$

as "reversible inhibition" of the enzyme by a proton (H$^+$ ion). Many other substances (metal ions, organic compounds) also bring about reversible inhibition and are called inhibitors. The nature of inhibitors and their interaction with the enzyme protein throw considerable light on the catalytic site of individual enzymes and the reaction catalysed by them. The mechanism of action of some drugs at the molecular level can be traced to their inhibitory action against bacterial enzymes. In this way the drug interferes with the bacterial metabolism and, therefore, with their growth and multiplication.

The inhibitors may interact with the free enzyme (thus making it unavailable for interaction with the substrate), with the enzyme-substrate complex (and modify the rate of its conversion into products), or with both. The various interactions are summarized below in Eq. (9.8). The enzyme-substrate interaction (reaction 4) is also included as all these processes proceed simultaneously.

$$E + S \underset{k_{-1}}{\overset{k_1}{\rightleftharpoons}} ES \xrightarrow{K_p} E + P$$

$$E + I \rightleftharpoons EI \quad K_i = \frac{[E][I]}{[EI]}$$

$$ES + I \rightleftharpoons ESI \quad K_{si} = \frac{[ES][I]}{[ESI]}$$

(9.8)

In Eq. (9.8), I represents the inhibitor, and K$_i$ and K$_{si}$ stand for dissociation constants of EI and ESI, respectively. They are referred to as inhibitor constants. Other terms have the same significance as in Eq. (9.4). The species ESI may possibly give rise to products with a different rate constant than k$_p$. A rate equation may be derived which takes into account the above equilibria and also the formation of products from ES and ESI at two different rates. Such an equation is too complex to be of any practical use. However, certain *experimentally observed* limitations of Eq. (9.8) help to simplify the kinetic behaviour and distinguish between different types of inhibitory actions. These are discussed below:

1. *Competitive inhibition*

In some cases, the substrate and inhibitor are mutually exclusive so far as binding to the enzyme is concerned, that is, they cannot coexist on the enzyme active site. *The complex ESI is not formed.* The inhibitor merely *decreases the effective concentration of enzyme* by combining with a fraction of it. It follows that at very high substrate concentration, the inhibitor can be displaced from the enzyme site and the inhibition can be overcome. Indeed in such cases, the inhibition is strong at low [S] and is very weak (if it is observed at all) at high substrate concentrations. A comparison of the rate equation for this type of inhibition (*competitive inhibition,* Eq. (9.9) with Eq. (9.5) shows that only the K$_m$ term is altered, suggesting that only the "apparent affinity" of the enzyme for its substrate is lowered.

$$v = \frac{V_{max}[S]}{K_m\left[1 + \dfrac{[I]}{K_i}\right] + [S]}$$

(9.9)

Typical competitive inhibitors are structural analogues of the substrate. For example, thiourea is a competitive inhibitor in the hydrolysis of urea by urease.

$$\underset{\text{Urea (Substrate)}}{\overset{\displaystyle O \atop \displaystyle \|}{NH_2 - C - NH_2}} \qquad \underset{\text{Thiourea (Competitive inhibitor)}}{\overset{\displaystyle S \atop \displaystyle \|}{NH_2 - C - NH_2}}$$

Similarly, derivatives of sulphanilamide owe their drug action to the similarity of their structure to that of p-amino benzoic acid (PABA), which is a cofactor for some bacterial enzymes.

$$H_2N - \langle\rangle - COOH \qquad H_2N - \langle\rangle - SO_2NH_2$$

p-Amino benzoic acid
(PABA)
(Enzyme cofactor)

Sulphanilamide
(Competitive inhibitor)

Another important class of competitive inhibitors are products of reaction, especially if the substrate does not undergo a profound structural change as a result of the catalysed reaction. Typical kinetic behaviour in the presence of competitive inhibitor is illustrated in the double-reciprocal plot of Figure 9.5. Note that only the slope, but not the intercept on the 1/v axis, is changed by the inhibitor.

2. *Non-competitive inhibition*

In other cases, the binding of substrate or inhibitor does not decrease the affinity of the enzyme for the other. We shall not go into derivation of the rate equation, which is now complicated by the possibility that the complex ESI of Eq. (9.8) may or may not be totally inactive. A *simplified* version of the rate equation for non-competitive inhibition is given in Eq. (9.10).

$$v = \frac{V_{max}[S]}{K_m + [S]} \times \frac{1}{1 + ([I]/K_i)}$$

(9.10)

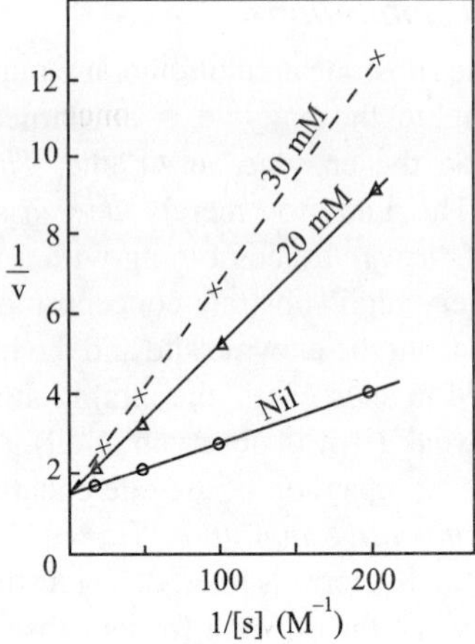

Figure 9.5   Competitive inhibition. Inhibitor concentration is shown on each plot.

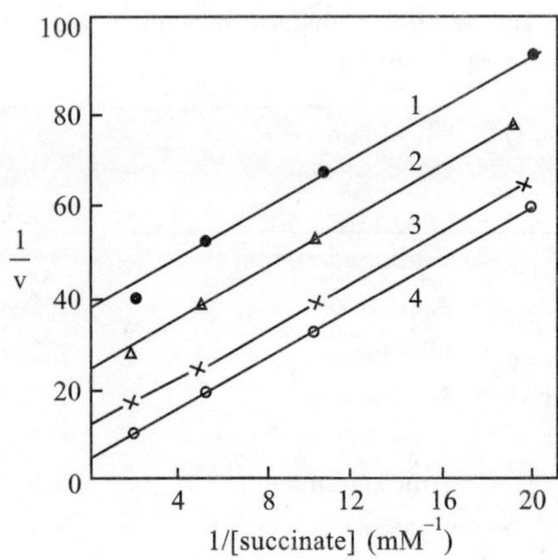

Figure 9.7   Uncompetitive inhibition. Inhibitor concentration is minimum in plot 4 and maximum in plot 1.

It is valid only if (i) $K_{si} = K_i$, that is, the binding of substrate and inhibitor are completely independent of each other; (ii) the complex ESI is completely inactive; and (iii) all equilibria of Eq. (9.8) are established very much faster than the rate of conversion of ES into products. In this case, the extent of inhibition is the same at all substrate concentrations. Results obtained with a non-competitive inhibitor obeying Eq. (9.10) are shown in Figure 9.6. Note the change in the slope as well as intercept of the double-reciprocal plot. The most common example of a non-competitive inhibitor is a proton (H^+ ion). Its size is so small that its binding on the active site does not interfere with the binding of the substrate. Some metal ions (Ag^+, Hg^{2+} etc.) also function as non-competitive inhibitors, particularly when they bind on a site adjacent to (but not overlapping) the substrate binding site.

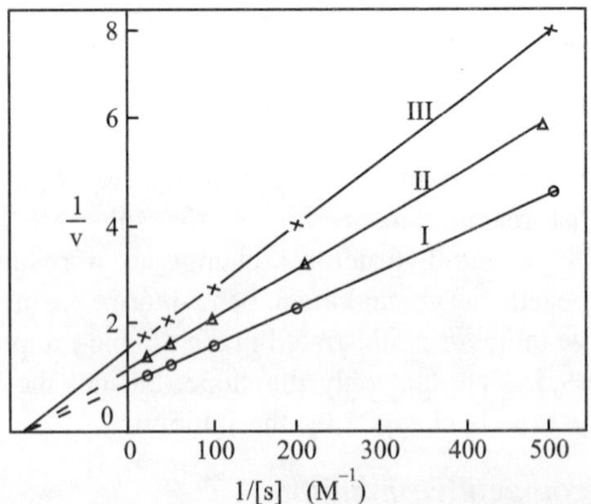

Figure 9.6   Non-competitive inhibition. Plots I, II and III refer to zero, low and high inhibitor concentration.

3. *Uncompetitive inhibition*

In still other cases, the inhibitor does not have any affinity for the free enzyme, but binds to the enzyme-substrate complex. The double-reciprocal plot gives a set of parallel straight lines (Figure 9.7), i.e., the slope remains same but the intercept changes. Uncompetitive inhibition is not so common among

single substrate systems but is frequently observed with reactions involving more than one substrate. In most enzyme reactions, the conversion of ES into products is a multi-step reaction (see Sec. IX). An inhibitor which binds to any enzyme-substrate species from ES to EP will exhibit the kinetic pattern characteristic of uncompetitive inhibition.

From the above discussion of the kinetics of enzyme action, it is clear that each enzyme has a discrete and specific region (or regions) where the substrate binds and where the catalytic reaction takes place. This part of the enzyme is referred to as the "active site", or catalytic site.

VII.   THE ACTIVE SITE

In general, a substrate molecule is very much smaller than the enzyme. Evidently when they combine to form a complex, only a small part of the enzyme molecule (the "active site") comes in contact with the substrate. The active site of an enzyme is said to be constituted by such portions of its molecule which come within *van der Waals contact distance of any atom of the substrate molecule as a result of complex formation, and during subsequent conversion of the complex into products.* It constitutes only a small portion of the enzyme protein. The overall primary structure (amino acid sequence) of a protein defines the specific manner of coiling, folding and superfolding (secondary and tertiary structures) of the polypeptide chain. The latter imparts a unique spatial configuration to the polypeptide backbone and the side-chain groups. Consequently certain amino acid residues which are far removed from each other in the primary sequence are brought close to each other. In such cases where extensive studies have been carried out, it has been shown that the active site of an enzyme is constituted by *side-chain groups of amino-acid residues from different regions of the enzyme protein.* This is illustrated in Figures 9.8 and 9.9 which depict respectively the amino acid residues involved in the active sites of chymotrypsin and ribonuclease. Note that in the case of chymotrypsin, side-chain groups of histidine (residue No. 57), aspartic acid (residue No. 102) and

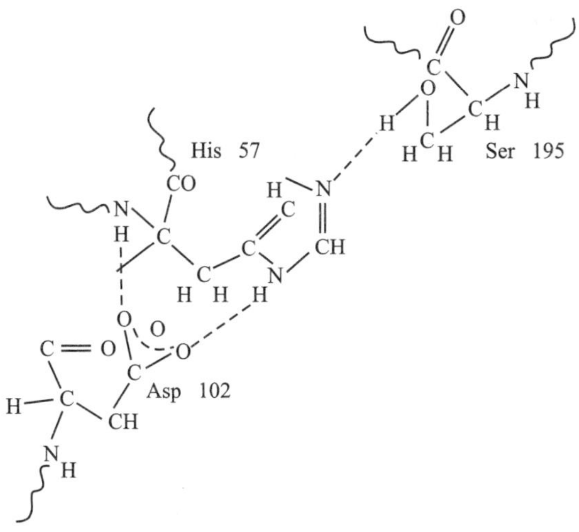

Figure 9.8 Conformation of some amino-acid side chain groups and the "charge relay system" on the active site of alpha-chymotrypsin based upon X-ray diffraction data. Hydrogen bonds are shown as broken lines. The polypeptide backbone is shown as waved lines.

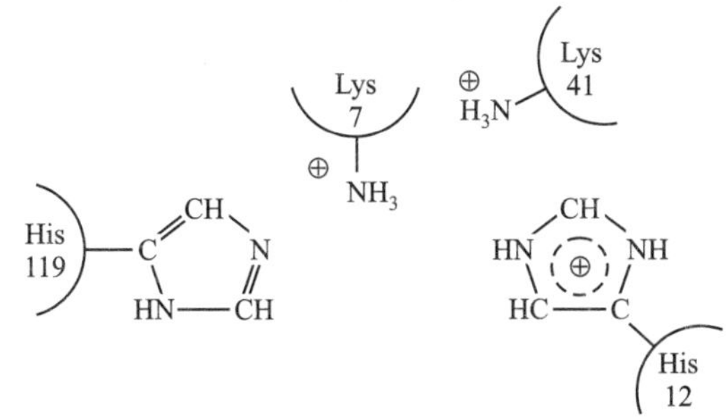

Figure 9.9 Active-site groups of ribonuclease.

serine (residue No. 195) participate in catalysis. Other groups which are involved in the substrate binding (formation of ES complex) are not shown here. In ribonuclease the side-chains involved belong to lysine (residues No. 7 and 41) and histidine (residues No. 12 and 119) residues. The active site groups are brought in specific orientation with respect to one-another by the unique secondary and tertiary structure of the enzyme protein. When these higher order structures are disrupted, e.g., on denaturation, the protein loses enzyme activity. The active site in some cases acquires the requisite conformation as a result of an induced fit caused by the binding of the substrate.

The specificity of an enzyme must be defined by the nature and stereochemical arrangement of groups on its active site. These must, in some way, be complementary to the structure of the substrate in order to permit its specific binding and ensure exclusion of other molecules.

VIII. IRREVERSIBLE INACTIVATION

Once the concept of a discrete active site within the enzyme molecule is appreciated, it is possible to conceive of certain reagents which may be made to react specifically with some critical groups on this site and bring about *inactivation*. If the chemical change in the protein molecule is small, as compared to its overall structure, it may possibly leave the secondary and tertiary structure more or less unaltered. Well-known examples of such reagents are the nerve gases, like di-isopropyl fluoro phosphonate (DFP).

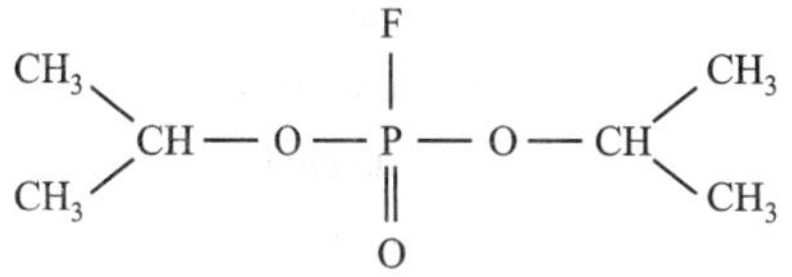

Di-isopropyl fluoro phosphonate (DFP)

These compounds bring about inactivation of several hydrolytic enzymes like acetylcholinesterase, trypsin, chymotrypsin, etc. In each case the reaction is stoichiometric; one mole of the inactivating agent reacts with one mole of the enzyme protein. Also in each case the reagent reacts specifically with a unique serine residue (the active site serine residue) to form a covalent derivative, a di-isopropyl phospho protein.

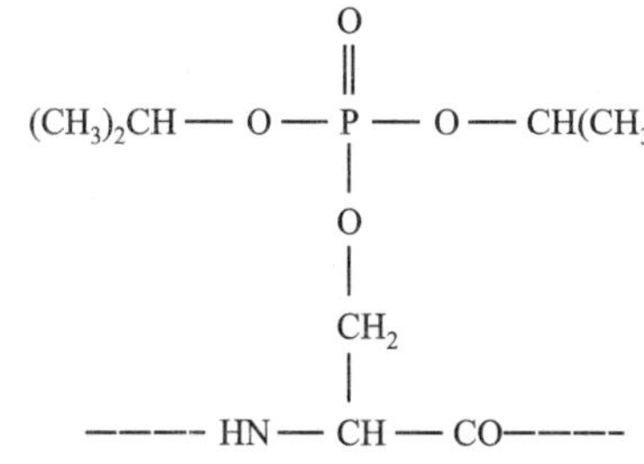

Di-isopropyl phospho (DIP) protein

It is interesting to note that all hydrolases which are *specifically* inactivated by DFP have a uniquely reactive serine residue. The amino acid sequence around this residue is also identical or very similar in these enzymes. This is illustrated in Table 9.5.

The action of inactivators is different from that of inhibitors discussed earlier. In the case of inhibitors the interaction with the enzyme is freely reversible and involves non-covalent forces. The inactivators on the other hand establish covalent linkage(s) on the active site. Reversal of their action requires further chemical treatment.

Non-specific inactivation of enzymes by denaturation has already been discussed together with the effect of temperature on enzyme activity.

IX. GENERAL CONCEPTS OF MECHANISM OF ENZYME ACTION

Identification of the active site groups, their unique orientation and the overall protein structure of several enzymes prompt us to ask the question "How do the enzymes function?" This enquiry includes the specificity and catalytic efficiency of the enzymes. Since the enzymes catalyse a large variety of chemical reactions, the answer to this question (more

TABLE 9.5   Amino-Acid Sequence Near Reactive Serine Residue of Some of the DFP-Sensitive Enzymes

Enzyme	Sequence
Chymotrypsin	–Cys–Met–Gly–Asp–*Ser*–Gly–Gly–Pro–Leu–Val–Cys–
Trypsin	–Cys–Gln–Gly–Asp–*Ser*–Gly–Gly–Pro–Val–Val–Cys
Thrombin	–Gly–Asp–*Ser*–Gly–
Bacterial Phosphatase	–Tyr–Val–Thr–Asp–*Ser*–Ala–Ala–*Ser*–Ala–
Pseudo cholinesterase	–Phe–Gly–Glu–*Ser*–Ala–Gly
Subtilisin	–Thr–*Ser*–Met–Ala

The DFP-sensitive serine residue is shown in italics.

particularly the details thereof) must vary from one enzyme to another depending upon the nature of substrate and the catalysed reaction. However, a study of the action of various enzymes reveals some general features of enzyme catalysis which are widely applicable. We shall discuss these general concepts in this section.

The following generalisations apply to most enzymes:

1. A non-covalent enzyme-substrate complex is formed *prior* to the chemical reaction. The catalytic reaction takes place within this complex. Thus, the catalytic reaction involves intramolecular rather than inter-molecular catalysis.
2. Conversion of the non-covalent enzyme-substrate complex to a non-covalent enzyme-product complex involves the formation of one or more *covalent* enzyme-substrate intermediates. Thus, the catalysed reaction proceeds via a multi-step pathway instead of the conversion of substrate into products in a single chemical step.
3. all enzymes are proteins.

Thus, we may try to answer our question under three heads, namely catalytic consequences of (a) formation of enzyme-substrate complex, (b) multi-step pathway and (c) the use of proteins for catalysis.

A.   Catalytic Consequences of E.S. Formation

One important consequence of the formation of E.S. complex is that the reaction now involves intramolecular catalysis in which the catalytic groups (of the enzyme active site) and the sensitive bond of the substrate molecule are brought together in a specific geometric orientation. Model reactions have shown that this "Proximity effect" leads to a large increase in the rate of reaction. The rate enhancement is higher if the catalytic groups and the substrate groups are held close to each other in a rigid conformation. For example, hydrolysis of esters by carboxylate ion (base) proceeds at a very slow rate (A). The rate is very much faster when the ester and the carboxylate group are part of the same molecule (B). There is a further enhancement (by a factor of 5×10^4) if the two groups are

held in a rigid con-formation (C). The rate of hydrolysis increases by large jumps in going from system A to B to C.

Another important consequence of E.S. formation is "bond distortion" in the substrate molecule. It has been argued that the enzyme has highest affinity for the transition state (cf Figure 9.1) rather than for the reactant or the product. On account of this difference in affinity the substrate undergoes some bond distortion on the formation of E.S. complex, so that its conformation resembles that of the transition state to some extent. Energetically, it is equivalent to taking the substrate part of the way uphill. Consequently, a smaller energy barrier now remains to be crossed on the pathway to the products. Experimental evidence for this postulate has come from two different approaches. Wherever it has been possible to synthesise transition state analogues, these are found to bind to the enzyme with a much higher affinity than the substrates. For example, several glycosidases are inhibited very strongly by the corresponding lactones. The latter possess the half-chair conformation through which the substrate (a glycoside) must pass on the pathway to the products. A more direct evidence for the distortion of substrate came from a study of interaction of lysozyme with its substrate (the mucopolysaccharide of the bacterial cell wall). Studies with synthetic substrates have shown that the sugar ring, the glycosidic bond of which is ruptured during catalysis, is distorted into a half-chair conformation in the enzyme-substrate complex.

Just as the substrate molecule undergoes bond distortion on binding to the enzyme, the latter also exhibits some changes in the relative orientation of active site groups (Induced fit hypothesis). These changes may be small, as found in hexokinase when it binds glucose, or these may be quite large as observed with carboxypeptidase. In the latter case, a specific tyrosine residue moves about 12 Å when a substrate analogue is bound to the enzyme. This distance is equal to about one fourth of the diameter of the enzyme molecule.

B. Catalytic Significance of a Multi-step Pathway

In almost every case, which has been investigated in some detail, it has been shown that the conversion of enzyme-substrate complex into enzyme-product complex involves more than one step. Frequently, a covalent enzyme-substrate compound is formed as an intermediate in the reaction pathway. For example, acylenzyme intermediates are formed in hydrolysis by chymotrypsin, trypsin and various carboxyl-esterases and in the oxidative phosphorylation of glyceraldehyde 3-phosphate by glyceraldehyde 3-phosphate dehydrogenase. Schiff base intermediates are formed in the enzymic reactions of aminoacids using pyridoxal phosphate as coenzyme, and also in the aldolase-catalysed conversion of fructose 1,6-diphosphate into dihydroxyacetone phosphate and glyceraldehyde 3-phosphate. Such a multi-step reaction *may be* more efficient than a single-step reaction, if the activation energy for each individual step is lower than that of the single-step reaction (Figure 9.10). Such a multi-step process is thermo-dynamically feasible in an intra-molecular reaction only. In an inter-molecular reaction, the unfavourable entropy changes will far outweigh the above energetic advantage.

C. Proteins as Biological Catalysts

Functions of enzymes may be discussed under specificity, catalysis and regulation. Specificity requires "recognition" of a compound by the enzyme as its substrate. The recognition takes place not only in the formation of enzyme-substrate complex but also in the subsequent chemical reaction steps. Thus, there are many compounds (substrate analogues or competitive inhibitors) which are bound on the active site with high affinity but do not undergo chemical transformation. Substrate interacts with several groups on the enzyme site, which must be held in a specific and "fairly rigid" geometric arrangement. It is extremely difficult (almost impossible) to hold several groups in a rigid conformation if these are parts of a small molecule. In a macromolecule, like a protein, the groups may be readily brought in specific geometry by folding, coiling and superfolding (secondary and tertiary structures) of the long polypeptide chain and held in that conformation by the cooperative effect of a

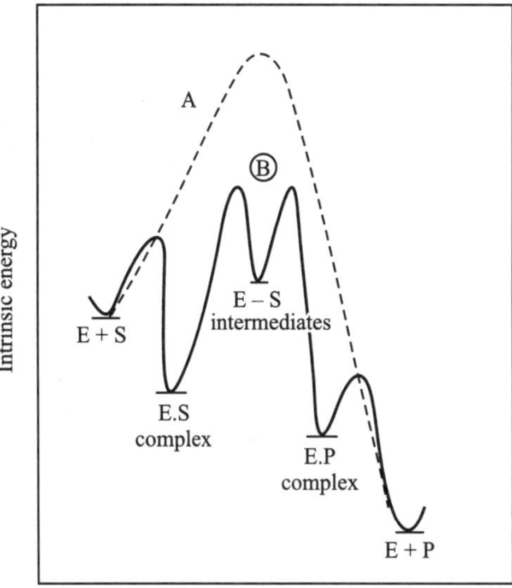

Figure 9.10 Energy diagram for a reaction proceeding via a single step, [curve (A)], and via a number of intermediates [curve (B)].

large number of non-covalent interactions. Individually, these non-covalent interactions are quite weak but their combined effect stabilises the higher order structure of proteins. The same forces also account for the stability of the unique conformation of the groups required for the catalysed chemical reaction. Thus, proteins are more suited for catalytic role than smaller (organic or inorganic) molecules. The proteins are also better suited than the other biologically occurring macromolecules (polysaccharides and nucleic acids) because the former offer a much wider variety of side-chain groups for interaction with the large number of substrates.

In some cases it has been shown that during catalysis the enzyme protein undergoes concomitant (and presumably cyclic) changes in conformation. These conformational changes may assist in catalysis by imposing electronic and/or steric constraints on the sensitive bonds of the substrate molecule.

X. A SPECIFIC EXAMPLE OF MECHANISM OF ENZYME ACTION: CHYMOTRYPSIN

The chemical pathway of catalysed reaction has been elucidated for several enzymes. In this section, we shall describe the mechanism of action of chymotrypsin, which is one of the most extensively studied enzymes.

Chymotrypsin is a proteolytic enzyme which is synthesized in the pancreas as an enzymically inactive precursor, chymotrypsinogen. In small intestines the activation of chymotrypsinogen is initiated by a proteolytic attack of trypsin, which is followed by action of some active chymotrypsin already present in the intestines (Figure 9.11). Two dipeptide fragments are excised from the single polypeptide chain of chymotrypsinogen (245 amino acid residues). The process is accompanied by conformational changes which are vital

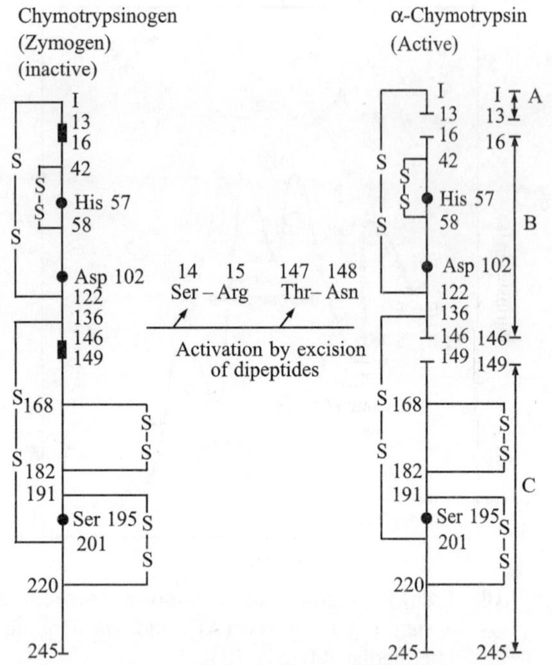

Figure 9.11   Linear representation of the conversion of chymotrypsinogen to chymotrypsin. Note excision of two dipeptide fragments. The chains A, B and C of chymotrypsin are covalently held together by S–S bonds. Groups involved in catalysis are indicated by •.

for the formation of active site of the enzyme. The complete three-dimensional structure of chymotrypsin has been deduced from X-ray crystallographic data.

Chymotrypsin is relatively specific for aromatic aminoacyl peptide bonds (Eq. 9.11).

$$-- NH - CH(CH_2Ar) - CO - NH - CH(CH_2R) - CO -- \xrightarrow{H_2O}$$

$$-- NH - CH(CH_2Ar) - COO^{\ominus} + H_3\overset{\oplus}{N} - CH(CH_2R) - CO -- \tag{9.11}$$

In Eq. (9.11), Ar stands for the aromatic side-chain of a phenylalanine, tyrosine or tryptophan residue. Chymotrypsin exhibits considerable hydrolytic activity on synthetic carboxyl acid esters (like p-nitrophenyl acetate), but amides are poor substrates.

With p-nitrophenyl acetate as the substrate, it was shown that one mole of p-nitrophenol was liberated rapidly for each mole of the enzyme. This was followed by the liberation of p-nitrophenol at a constant steady state rate. This is consistent with a reaction scheme involving a rapid formation of acetyl enzyme intermediate (with the liberation of p-nitrophenol) and

a subsequent slow hydrolysis of the acetylenzyme intermediate. It has been shown that formation of acylenzyme intermediate is common to all the substrates hydrolysed by chymotrypsin (Eq. 9.12).

$$R - \overset{O}{\overset{\|}{C}} - X + E - H \rightleftharpoons (R - \overset{O}{\overset{\|}{C}} - X\ EH)$$
(Substrate)   (Enzyme)   (E.S. complex)

$$(R - \overset{O}{\overset{\|}{C}} - X\ EH) \rightleftharpoons R - \overset{O}{\overset{\|}{C}} - E + X - H$$
(E.S. complex)   (Acyl enzyme intermediate)   (First product)

$$R - \overset{O}{\overset{\|}{C}} - E + H_2O \rightleftharpoons (R - \overset{O}{\overset{\|}{C}} - E.H_2O) \tag{9.12}$$

$$(R - \overset{O}{\overset{\|}{C}} - E.H_2O) \rightleftharpoons (R - COOH.EH)$$
(Enzyme product complex)

$$(R - COOH.EH) \rightleftharpoons R - COOH + E - H$$
(Second product)

Acyl enzyme intermediates have been prepared from radio-actively labelled substrates. Partial degradation and characterisation of radioactive fragments led to identification of the site of acylation as the hydroxyl group of serine-195 residue. The same group is modified in the chemical inactivation of the enzyme by DFP (see Section VIII). Small molecular weight model compounds of serine or the denatured chymotrypsin are neither acylated by any of the substrates, nor do they react with DFP. Other serine residues present in chymotrypsin are also inert against the substrates or DFP.

The extraordinary reactivity of the alcoholic group of serine-195 residue is due to its interaction (in native chymotrypsin) with imidazole group of histidine-57 and carboxylate group of aspartate-102 residues as shown in Figure 9.8. These groups were identified with the use of specific inactivating agents and from an examination of the three-dimensional structure of chymotrypsin.

Formation of acyl enzyme intermediate and its subsequent hydrolysis are believed to proceed as shown in Figure 9.12. The interaction with imidazole group of histidine-57 and carboxylate group of aspartate-102 bestows extraordinary reactivity on the alcoholic group of serine-195 residue. The reactivity of this group now lies between that of an alcohol (RCH_2OH) and an alkoxide ion (RCH_2O^-). This is a good example illustrating how local environments at the enzyme site can drastically modify the chemical reactivities of the participating groups.

Prior to the chemical reactions of Figure 9.12, the substrate has to be bound to the enzyme. X-ray crystallography has revealed the existence of a substrate-binding pocket lying next to the serine-195 residue. This pocket is lined with non-polar side chains and is sufficiently large to accommodate the bulky aromatic side chains. The pocket is stabilised by a very strong

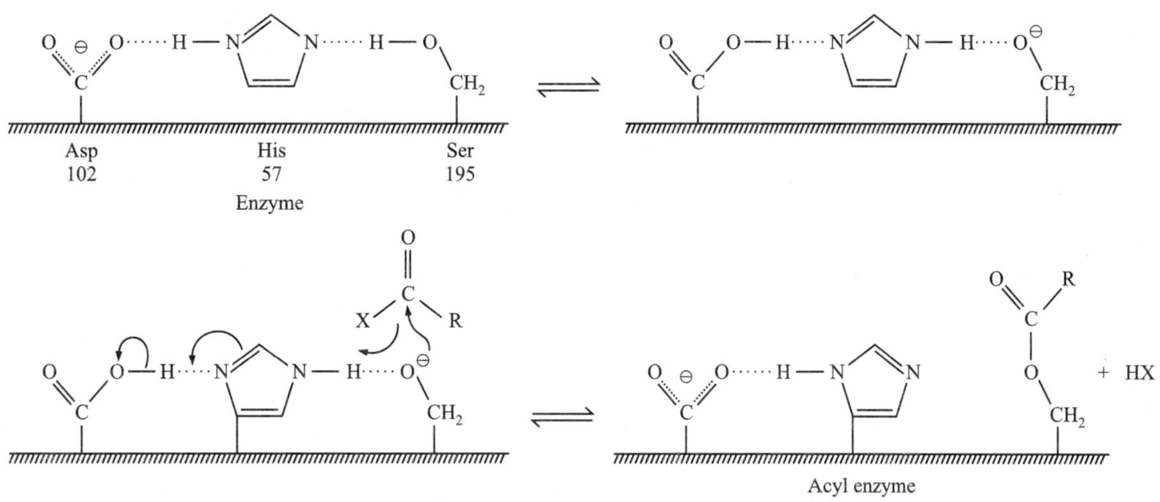

Figure 9.12 Postulated mechanism of action of alpha-chymotrypsin. Note the contribution of serine–CH_2O^- ion in the "pre-transition state" equilibration. The "charge relay system" is disrupted in the acyl enzyme intermediate but is re-established when a water molecule binds in the deacylation step. The deacylation proceeds by a reversal of the electron shifts shown above and the original structure of the enzyme is regenerated.

polar (ion-pair) interaction between free α-amino group of isoleucine-l6 and carboxylate group of aspartate-194, both of which are buried in the interior of the protein molecule. Note that the free α-amino group became available at isoleucine-16 as a consequence of activation of chymotrypsinogen to chymotrypsin. Indeed a comparison of the three-dimensional structures of chymotrypsinogen and chymotrypsin reveals that the characteristic "charge relay system" of the active site (Figure 9.8) is present in both the zymogen as well as the enzyme, but the zymogen lacks the specific substrate binding pocket, without which catalysis cannot take place.

XI. REGULATION OF ENZYME ACTIVITY

In living organisms, metabolism proceeds at precisely regulated rates via a coordination of many highly interdependent biochemical functions (of catabolism, anabolism, transport etc.). Regulation of a metabolic process may be achieved by controlling the availability of substrates (the metabolites) or the rate of enzymecatalysed reactions. Control of substrate availability also depends on regulation of rates of enzyme reactions in a different process. Thus, regulation of the rate of catalysed reaction is an important function of enzymes, in a similar way as catalysis is. The rate of an enzyme-catalysed reaction may be varied by modulation of the enzyme concentration (or level) in a tissue or its activity (i.e., the catalytic efficiency). Both types of controls operate in nature.

A. Regulation of Enzyme Concentration

Enzymes, like other proteins, are synthesised by translation of genetic information contained in DNA. In many cases, a contiguous segment of DNA (an "operon") carries information required for the synthesis of several enzymes, which participate in consecutive reactions in a metabolic process. The expression of the entire operon may be turned on (Enzyme induction) or off (Repression) in response to the physiological environments, like the presence or absence of certain metabolites. This fascinating subject of protein biosynthesis and its regulation is discussed elsewhere in this book.

Level of an enzyme is governed by a balance between the rates of its synthesis and degradation. Nature employs both these opposing processes for regulation of enzyme levels. For example, the level of tryptophan oxygenase increases in response to the presence of tryptophan or glucocorticoid. The tryptophan-induced increase has been shown to be due to inhibition of the degradation of the enzyme (without any effect on the rate of enzyme synthesis), whereas glucocorticoid brings about an increase in the rate of enzyme synthesis.

B. Regulation of Enzyme Activity

Like a capable and efficient chemical engineer, nature has devised (or evolved) a variety of control mechanisms which operate at some critical points in the metabolic chains and thereby regulate the entire metabolism. Thus, the activity of only some of the enzymes (the regulatory enzymes) are modulated in response to physiological changes. This modulation is achieved in a variety of ways, as discussed below:

1. *Feedback inhibition (Allosteric modulation)*

In many biosynthetic processes, the endproduct is found to inhibit the enzyme catalysing the first step of the process. For example, in the hypothetical sequence of Eq. (9.13) the endproduct P of the biosynthetic chain of reactions may act as inhibitor of the enzyme catalysing the first reaction of this chain, namely the conversion of A into B.

$$A \longrightarrow B \longrightarrow C \longrightarrow D \dashrightarrow P \qquad (9.13)$$

P inhibits
this enzyme

The importance of such a modulation for cellular economy is obvious. If too much of P accumulates in the cell, or if P is externally supplied, it is necessary to turn off the synthesis of P. This objective is most economically achieved by inhibition of the first enzyme by P.

An example of an enzyme subjected to this type of regulation is "aspartate transcarbamylase" which catalyses the first step in the synthesis of pyrimidine nucleotides, e.g. cytidine triphosphate (CTP) (Figure 9.13). The latter is a potent inhibitor of aspartate transcarbamylase. The pyrimidine nucleotides are required in the living systems for the synthesis of nucleic acids. The latter have both purine and pyrimidine bases. It is interesting to note here that adenosine triphosphate (ATP) (a purine nucleotide) can reverse the inhibition of aspartate transcarbamylase by CTP (Figure 9.14). Note that neither CTP nor ATP bears any structural resemblance to the substrates of aspartate transcarbamylase.

This lack of structural resemblance has given the special name ("Allosteric" as against "isosteric" for structurally related compounds) to this phenomenon, namely 'Allosteric modulation". CTP and ATP are referred to as allosteric effectors, the former being a negative and the latter a positive allosteric effector. In several cases it has been shown that the allosteric effectors bind to the enzyme at a site different from the substrate binding (active) site. Indeed aspartate transcarbamylase has been shown to be an oligomeric protein composed of two types of subunits. One type (the catalytic subunit) contains the active site, where the substrate binds and catalysis is brought about. The catalytic subunits do not interact with the allosteric effectors. The second type of subunit (the regulatory subunit) do not combine with the substrate but have strong affinity for the allosteric effectors. In the complete enzyme, the two types of subunits are held together by non-covalent interactions (quaternary structure of proteins). Binding of the effector to the regulatory subunit brings about conformational changes in the entire enzyme molecule which alter its catalytic efficiency.

Note that the ("sigmoid") shape of the rate versus substrate concentration curve for aspartate transcarbamylase (Figure 9.14) is different from the rectangular hyperbola of Figure 9.2. The sigmoid curve is indicative of "positive cooperativity" of substrate binding. Such curves are obtained if each enzyme molecule binds more than one substrate molecule and where each succeeding molecule of substrate binds more tightly than those preceding it. Cooperative behaviour is observed in some but not all the regulatory enzymes.

A detailed consideration of "allosteric" is beyond the scope of this chapter. It may, however, be mentioned here that a regulatory enzyme can exist in at least two interconvertible conformational states, one of which has a higher catalytic efficiency than the other. The negative effectors help shift the equilibrium towards the state of lower catalytic efficiency by binding more specifically to this form. The positive effectors and the substrate, on the other hand, bind more specifically to the state of higher catalytic efficiency and thereby shift the equilibrium in its favour.

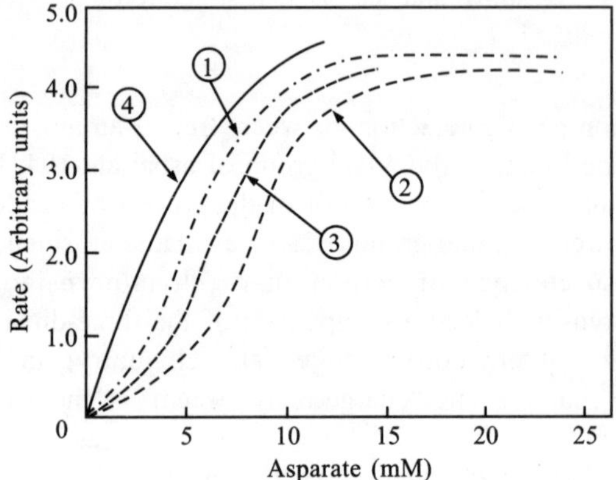

Figure 9.14 Substrate concentration dependence of the rate of aspartate transcarbamylase-catalysed reaction (curve 1) and the effect of CTP (curve 2) and CTP + ATP (curve 3) on the same. Note the "sigmoidal" shape and shifting of the curve along the abscissa in the presence of effectors. Addition of organic mercurial compounds dissociates the regulatory subunits from the catalytic subunits. Under these conditions, the curve is no longer sigmoid but becomes hyperbolic (curve 4).

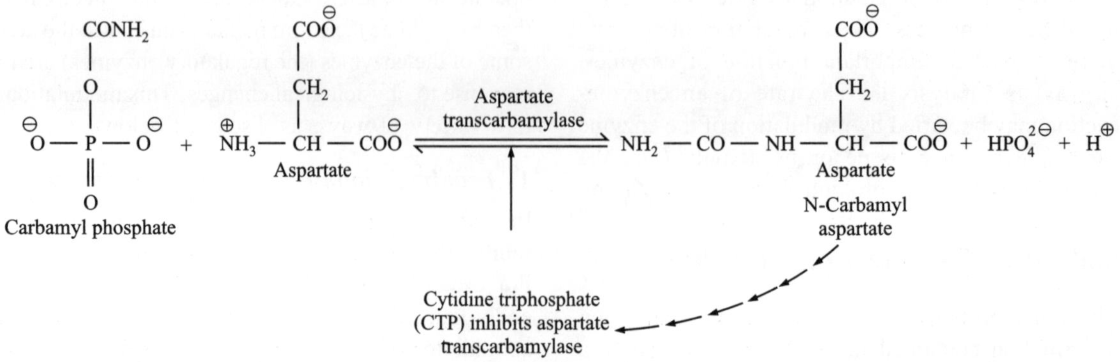

Figure 9.13 The catalysed reaction and feedback inhibition of aspartate transcarbamylase.

2. *Reversible covalent modification of regulatory enzymes*

In this case also, the enzyme exists in two forms of different catalytic efficiency, but their interconversion is brought about by reversible covalent modification of the enzyme protein. In some systems, this kind of regulation reinforces the allosteric controls. A typical example is the enzyme glycogen phosphorylase, which catalyses the first step in the breakdown of glycogen (Eq. 9.14).

$$G_n + Pi \xrightarrow{\text{Glycogen phosphorylase}} G_{n-1} + G\text{--}1\text{--}P$$
$$\text{Glycogen} \quad \text{Phosphate} \qquad\qquad \text{Glucose ion 1--phosphate}$$

Glycogen phosphorylase exists in two forms, namely the more active A and the less active B. Their interconversion involves phosphorylation and dephosphorylation catalysed by two other enzymes phosphorylase kinase and a phosphatase (Eq. 9.l5).

$$\text{Phosphorylase B} + 4ATP \xrightarrow{\text{Phosphorylase kinase}} \text{Phospho-phosphorylase A} + 4ADP$$
$$\text{(less active)} \qquad\qquad\qquad \text{(more active)}$$

$$\text{Phospho-phosphorylase A} \xrightarrow{\text{Phosphatase}} \text{Phosphorylase B} + 4Pi$$

Evidently, the kinase and phosphatase must be under rigorous control, otherwise an uncontrolled coupling of these two enzymes will bring about a wasteful hydrolysis of ATP to $ADP + P_i$. Indeed, phosphorylation of glycogen phosphorylase B constitutes the last step in a cascade of reactions initiated by increase in the level of the hormone, epinephrine (Chapter 24). Note the non-covalent (and pre-sumably allosteric) interaction of 3', 5' cyclic AMP with protein kinase and reversible covalent modification (by phosphorylation/dephosphorylation) of the two succeeding enzymes, phosphorylase kinase and glycogen phosphorylase.

3. *Irreversible covalent modification and activation of enzymes*

This category of enzymes are synthesised as inactive precursors ("zymogens") and later converted into active enzymes at the site of their action. This conversion involves proteolytic cleavage of a small peptide fragment near the N-terminal of the zymogen. Proteolytic enzymes trypsin, chymotrypsin and carboxypeptidase are synthesised as their zymogens in the pancreas and converted into active enzymes in the small intestines. The blood clotting enzyme thrombin circulates harmlessly all the time in the blood in the form of its zymogen, prothrombin. The latter is activated at the site of its necessity by a sequence of reaction. After thrombin has completed its job (of converting fibrinogen into fibrin which aggregates to form the clot), it is inactivated irreversibly by interaction with antithrombin.

XII. MULTI-ENZYME COMPLEXES

We have mentioned above that enzymes of some metabolic chains are synthesised in a "concerted" manner; either all of them are synthesised or none is synthesised. In some cases, the enzymes of a sequence of reactions occur together as an oligomeric complex and are isolated as such. Extra steps have to be incorporated in the working up procedure in order to separate them from each other. There is evidence to show that their combination with each other is not an artefact of the isolation procedure. In all probability, they exist as such complexes inside the cell as well. Several such multi-enzyme complexes are known, as for example fatty acid synthetase, pyruvate dehydrogenase and alpha-ketoglutarate dehydrogenase complexes. Some of these complexes have been dissociated into constituent enzymes, which have been isolated and studied independent of each other. If the isolated constituent enzymes are brought together, or the dissociating agent is removed from the medium, the original complex is reconstituted and can be crystallized as such. Thus the pyruvate dehydro-genase complex is made up of three enzymes, namely, pyruvate decarboxylase (E1), a dehydro-genase-transacetylase (E2) and a dihydrolipoamide dehydrogenase (E3). Each constituent enzyme has its own prosthetic group. These are thiamine pyro-phosphate (TPP), lipoic acid and a flavin (FAD) for E1, E2 and E3, respectively. The individual reactions catalysed by the constituent enzymes and the net reaction are shown below in Eq. (9.16).

$$\text{Pyruvate} + \text{TPP--E1} \rightarrow CO_2 + \text{Hydroxyethyl--TPP--E1}$$

$$\text{Hydroxyethyl --TPP--E1} + \text{Lipoyl E2} \rightarrow \text{S--Acetyl--dihydrolipoyl --E2} + \text{TPP -- E1}$$

$$\text{S--Acetyl--dihydrolipoyl--E2} + \text{CoA} \rightarrow \text{Acetyl--CoA} + \text{Dihydrolipoyl -- E2}$$

$$\text{Dihydrolipoyl--E2} + \text{E3--FAD} \rightarrow \text{Lipoyl--E2} + \text{E3--FADH}_2$$

$$\text{E3--FADH}_2 + \text{NAD}^+ \rightarrow \text{E3--FAD} + \text{NADH} + \text{H}^+$$

$$\text{Sum: Pyruvate} + \text{CoA} + \text{NAD}^+ \xrightarrow{\text{E1, E2, E3}} \text{Acetyl -- CoA} + CO_2 + \text{NADH} + \text{H}^+$$

$$(9.16)$$

Each individual enzyme of a multienzyme complex is regenerated in the succeeding step so that the net reaction (or the sum) is the oxidative decarboxylation of pyruvate in the presence of coenzyme-A (CoA) to give rise to CO_2 and acetyl-CoA. The coenzyme NAD^+ functions as the electron acceptor and is reduced to NADH.

Molecular architecture of pyruvate dehydro-genase complex has been examined and shows that the transacetylase (E2) is located at the core of the large complex. It has been suggested that the long lipoyl lysin side chain of E2 may serve as a flexible "rotating arm" linking the activities of E1, E2 and E3 as depicted in Figure 9.15.

It may be pointed out here that only where the protein: protein interaction is very strong, the concerned enzymes may be isolated as a multi-enzyme complex. It is suspected

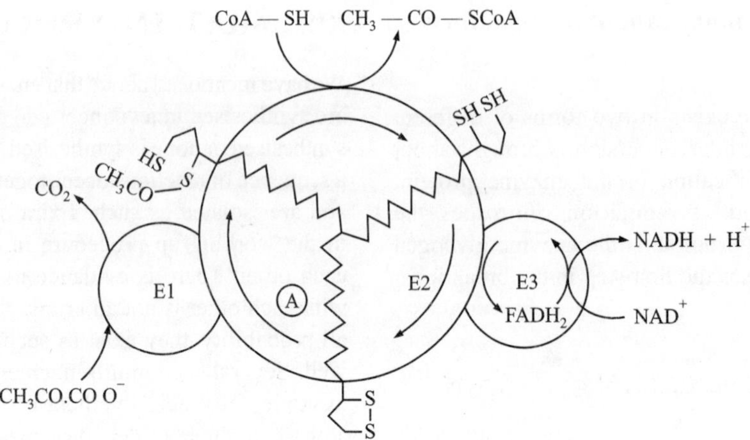

Figure 9.15   Schematic representation of pyruvate dehydrogenase complex, compare with Eq. (9.16).

that in several other cases the enzymes of a metabolic chain may function together as a "complex" *in vivo,* but the nature of their interaction with one another may be such that the complex dissociates during isolation.

XIII.   ISOZYMES

The terms isoenzymes or isozymes refer to different proteins catalysing the same chemical reaction and present in the same species, sometimes even in the same cell. They are detected and separated by electrophoresis and ion-exchange chromatography. A well studied example is that of lactate dehydrogenase. In all, five different farms of lactate dehydrogenases can be isolated from an animal, say man. These are differentiated on the basis of their isoelectric point, stability towards thermal denaturation, K_m value for a given substrate (pyruvate), pH optimum etc. Each lactate dehydrogenase is a tetrameric protein,, i.e., contains four subunits. The five isozymes have been shown to arise from different combinations of two types of subunits, which are commonly referred to as H and M. Thus, the five tetrameric combinations of H and M will be H_4, H_3M, H_2M_2, HM_3 and M_4. It is to be expected that their properties (physical and catalytic) will exhibit a gradual change in going from H_4 to M_4. When any lactate dehydrogenase, other than H_4 or M_4, is allowed to dissociate into subunits and then re-aggregate, it gives rise to all the five forms. Similarly, if a mixture of H_4 and M_4 is subjected to dissociation followed by reassociation, all the five forms are obtained. Different tissues have different proportions of various isozymes. The heart muscle is rich in H_4 and H_3M. Their level in serum increases following damage to the heart tissue in myocardial infarction. This is useful for diagnostic purposes.

Some allosteric enzymes occur as isozymes, which differ in their sensitivity to the allosteric effectors. Phospho-2-keto-3-deoxy heptonate aldolase catalyses the reaction of Eq. (9.17), which is the first step in the biosynthesis of the three aromatic amino acids, namely phenylalanine, tyrosine and tryptophan.

$$\text{D-Erythrose-4-phosphate} + \text{Phosphoenol pyruvate}$$
$$\downarrow$$
$$\text{Deoxyheptulosonic acid 7-phosphate}$$

There are three isozymes of this enzyme. Each of the endproduct aminoacid is a specific allosteric inhibitor for one isozyme only. Thus, all the three aromatic amino acids must be present together to inhibit all the isozymes of this enzyme.

XIV.   IMMOBILIZED ENZYMES

In nature, some enzymes occur in the cytoplasm (the "soluble" enzymes) and others are integral parts of intracellular organelles (the "particulate bound" enzymes). In the latter case, the enzymes are integrated in a membrane. Most of the researches have been carried out on soluble enzymes, because of the ease of their isolation. The particulate enzymes are more difficult to study, because the catalyst and the reactants are in different phases. Some soluble enzymes have been rendered "insoluble" in the laboratory by attaching them covalently to an insoluble matrix like Sepharose (a specific type of polysaccharide), nylon tubes or powder or even glass beads. Such enzyme preparations have been referred to as "insolubilised" or "immobilised" enzymes. These may serve as models for the study of natural membrane bound enzymes. Immobilised enzymes find use in industry because, being insoluble, they can be easily recovered and used again. A familiar medical application of immobilized enzymes is the use of paper strips for a rapid determination of glucose in urine. The strip of paper is coated with two immobilized enzymes (namely glucose oxidase and peroxidase) and an aromatic amine, usually o-tolidine. The chemical changes which take place on contact with urine if it contains any glucose are shown in Eq. (9.18).

$$\text{D-Glucose} + O_2 + H_2O \xrightarrow{\text{Glucose oxidase}} \text{D-Gluconic acid} + H_2O_2$$
(from air)

$$H_2O_2 + \text{o-tolidine} \xrightarrow{\text{Peroxidase}} \text{Oxidation} + H_2O \quad (9.18)$$
product of
o-tolidine

(colourless)   (blue)

The intensity of colour produced is proportional to the amount of glucose present in the urine.

XV. APPLIED ENZYMOLOGY

Enzymes are finding increasing use in industries and in analytical laboratories. For the former, the immobilised enzymes are generally more suited than the soluble form.

Enzymes, being highly specific, are better than the man-made catalysts in preventing side reactions and in stopping the reaction at the desired stage. It is well known that organic cyanogen compounds, also called nitriles, are readily hydrolysed to carboxylic acid ammonium salts and the reaction proceeds via an amide (Eq. 19). However, it is difficult to stop the reaction at the amide stage.

$$R-C \equiv N \xrightarrow{H_2O} R-CO-NH_2 \xrightarrow{H_2O} R-COONH_4 \quad (9.19)$$

With the use of an appropriate enzyme, the reaction can be stopped at the amide stage. An example of this is the commercial process for the preparation of acrylamide from acrylonitrile using the enzyme nitrile hydratase (Eq. 20). Since the enzyme does not catalyse the hydrolysis of amides, the reaction stops here.

$$CH_2 = CH - C \equiv N + H_2O \xrightarrow[\text{hydratase}]{\text{Nitrile}} CH_2 = CH - CONH_2$$

Acrylo-nitrile   Acrylamide

$$(9.20)$$

The sweetener, Aspartame, is synthesised from aspartic acid and phenylalanine-methyl ester using thermolysin as the catalysing enzyme. This reaction provides an example of the synthetic application of an enzyme which normally catalyses hydrolysis of peptides.

$$H_3\overset{+}{N} - CH - CO - NH - CH - COOCH_3$$

Aspartame

An example of the use of enzymes for analytical purposes has already been given above under the immobilised enzymes, namely the paper strips for glucose assay in urine. Immobilised enzymes are also used in "enzyme electrodes". In principle, an enzyme electrode is a combination of any type of electrochemical sensor and a thin layer of immobilised enzyme held close to it. An example is a "Urea electrode". It consists of an ammonium ion electrode (i.e., one which measures ammonium ion concentration in a solution) surrounded by a nylon mesh on which urease has been immobilised. The latter is protected by a layer of dialysis membrane around it. When such an electrode is dipped in a solution of urea, the latter gets hydrolysed by urease to produce ammonium ions and carbon dioxide. The ammonium ion electrode responds to these ions producing an electric potential which is measured with respect to a standard (reference) electrode. The magnitude of electric potential across these electrodes is directly proportional to the concentration of urea in the sample solution.

Several enzymes have been used for medicinal purposes, e.g., the use of pancreatic enzymes in digestive disorders, asparaginase and glutaminase in leukaemia, β-lactamase in penicillin allergy and streptokinase for dissolving blood clots. Several factors limit the scope of such applications. At present, the most successful applications are extra-cellular, e.g., for the removal of toxic materials.

XVI. COENZYMES

The non-proteinous low molecular weight organic compounds required for the activity of some enzymes have been referred to as prosthetic groups or coenzymes, depending upon the affinity of their binding to the enzyme protein. They always participate in the chemical reaction accepting or donating a group, hydrogen atoms, or electrons from or to the substrate. In the overall metabolic changes, another reaction regenerates the coenzymes by transfer of these groups to or from another metabolite. Thus, the coenzymes function as links between different metabolic chains besides participating in individual enzyme reactions. For example, the oxidative phosphorylation of glyceraldehyde-3-phosphate in the glycolytic pathway leads to reduction of the coenzyme nicotinamide adenine dinucleotide. The reduced coenzyme is oxidized to the original form in a subsequent reaction where pyruvate is reduced to lactate. In single enzyme reactions the substrate and coenzyme are *functionally* similar; they are both reaction partners. However, in the overall metabolic sequence the substrate undergoes progressive chemical change while the coenzyme is regenerated. In certain cases, the coenzyme is regenerated in the same reaction, as for example, is the case with reactions requiring pyridoxal phosphate or thiamine pyrophosphate as coenzymes. Further, the binding of a coenzyme to the enzyme generally precedes that of the substrate and the dissociation of the coenzyme-product follows that of substrate-product from the active site.

Many coenzymes are derived from B-vitamins. The latter are compounds which are required by the organisms in small amounts and must be supplied as part of the diet, because they cannot be synthesized by the organism. Their function as coenzymes is indicative of their role in metabolic processes.

A large number of coenzymes are known and they possess varied structures. Some of them are quinones, others are nucleotides and some are large molecules composed of protein and non-proteinous parts (e.g., cytochromes). Some of the coenzymes are listed in Table 9.6 where they are grouped according to the nature of reactions in which they participate. Some of the coenzymes participate in more than one type of reactions. For example, pyridoxal phosphate function as a coenzyme for group transfer, isomerization and elimination reactions of amino acids. The specificity with respect to the catalysed reaction is defined by the enzyme protein. Structural

TABLE 9.6   Classification of Coenzymes

Reaction type	Coenzymes	Corresponding B-vitamin
Oxidation-reduction	Nicotinamide-adenine nucleotides	Nicotinic acid
	Flavin nucleotides	Riboflavin
	Ubiquinone	
	Cytochromes	
	Lipoic acid	
Group-transfer	Adenosine triphosphate (ATP)	Pyridoxine
	Pyridoxal phosphate	Folic acid
	Uridine diphosphate	
	Tetrahydro folic acid	Pantothenic acid
	S-Adenosyl methionine	
	Coenzyme-A	
Isomerization	Pyridoxal phosphate	Pyridoxine
	Uridine diphosphate	
	B_{12}-coenzymes	B_{12}
Addition-elimination	Pyridoxal phosphate	Pyridoxine
	Thiamine pyrophosphate	Thiamine
	Biotin	Biotin

formulae and reactions of some of the common coenzymes are described below.

Two coenzymes containing a nicotinamide moiety participate in oxidation-reduction ("redox") reactions. They are nicotinamide-adenine-dinucleotide (NAD, IA) and its phosphate (NADP, IB).

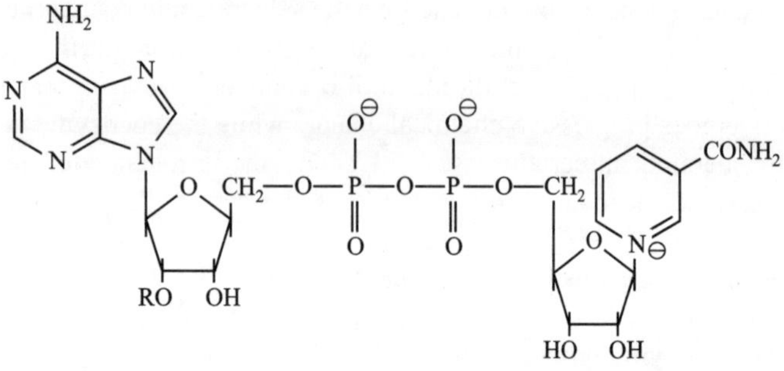

IA $R = H$, $NAD^{\oplus}$ IB $R = PO_{3.1}^{2\ominus}$ NADP

In both these coenzymes, the nicotinamide moiety participates in the redox reaction,, i.e., in the exchange of hydrogen with the substrate (Eq. 9.21).

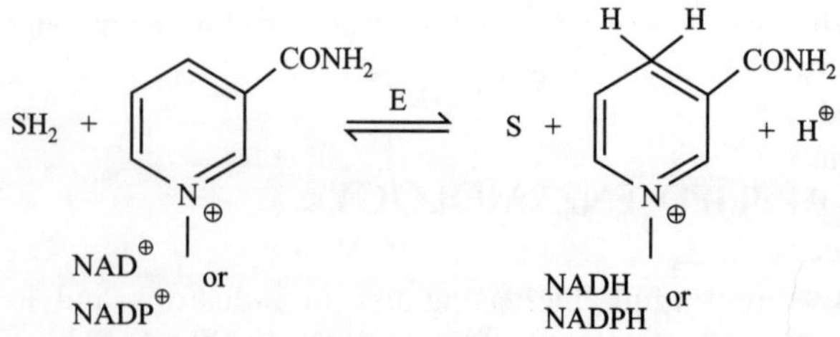

$$SH_2 + \text{(NAD}^{\oplus} \text{ or NADP}^{\oplus}) \underset{}{\overset{E}{\rightleftharpoons}} S + \text{(NADH or NADPH)} + H^{\oplus} \quad (9.21)$$

where SH_2 and S are the reduced and oxidized forms of the substrate and E is the specific enzyme. Enzymes catalysing reactions like 21 are referred to as "oxido-reductases" or as dehydrogenases and are specific for the substrate as well as for the coenzyme (NAD or NADP). Alcohol dehydrogenase and glyceraldehyde-3-phosphate dehydrogenase require NAD as the coenzyme as do most oxido-reductase of degradative metabolic reactions. The reduced form of the coenzyme, NADH, is reoxidized in mitochondria during respiratory oxidative phosphorylation. The reduced phosphorylated coenzyme (NADPH) generally provides reducing power in biosynthetic reactions.

Another pair of coenzymes which participate in oxidation-reduction reactions are derived from riboflavin. These are flavin mono nucleotide (FMN, II) and flavin adenine dinucleotide (FAD, III). In both cases, hydrogen atoms are transferred to (or form) the 1,10 positions of the iso-alloxazine ring of the coenzyme (Eq. 9.9).

Succinate dehydrogenase and fatty acyl coenzyme A dehydrogenases require flavin nucleotides (mostly FAD) for their activity. Reduced flavine coenzymes are reoxidized by cytochromes and cytochrome oxidases.

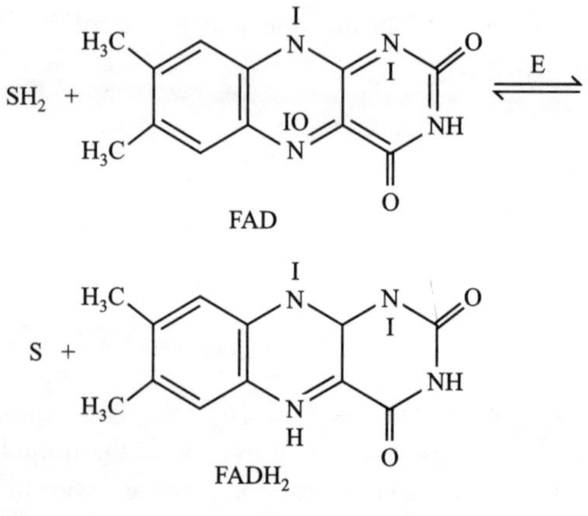

$$SH_2 + FAD \overset{E}{\rightleftharpoons} S + FADH_2 \quad (9.22)$$

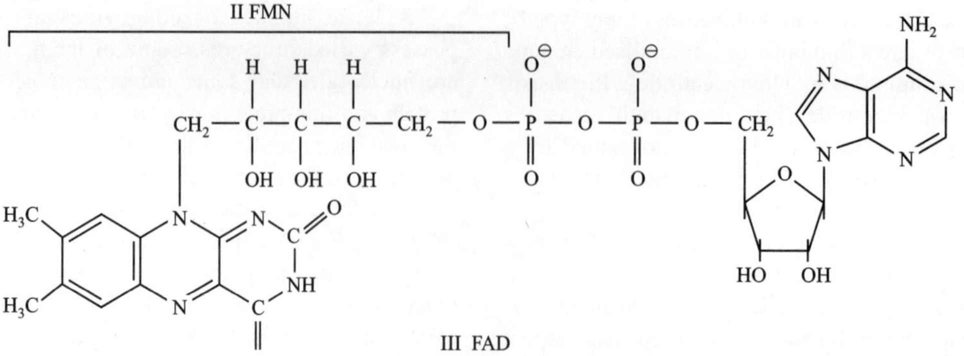

Cytochromes are proteins having porphyrin system as a prosthetic group and are involved in the reoxidation of reduced coenzymes via oxidative phosphorylation. Several different cytochromes (a, b, c, etc.) participate in the process and in the final reaction cytochrome a_3 transfers its electrons to molecular oxygen (Chapter 23). The porphyrin heterocyclic system carries one iron atom by co-ordination (IV). The iron atom is also coordinated to the protein part of cytochrome.

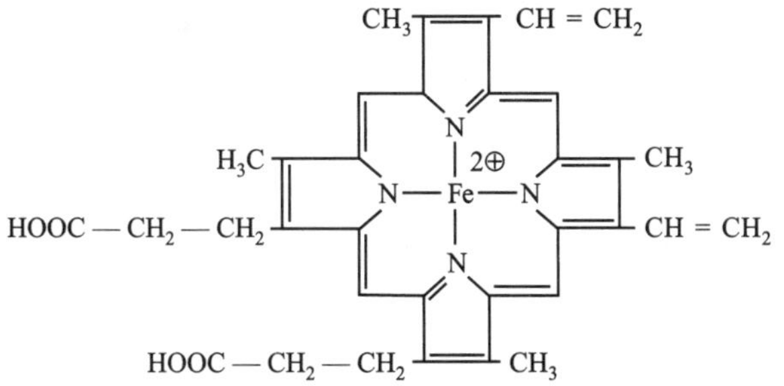

During the oxidation-reduction reaction the valency of iron changes between ferric (Fe^{3+}) and ferrous (Fe^{2+}) states (Eq. 9.23).

$$\text{Reduced Coenzyme Q} + 2\text{Cyt}(Fe^{3+}) \rightarrow \text{Oxidised Coenzyme Q} + 2\text{Cyt}(Fe^{2+}) + 2H^+$$

$$(9.23)$$

The coenzymes of group transfer enzymes show greater structural variety than is the case with coenzymes of oxidation-reduction reactions. Each coenzyme is specific for the transfer of one particular group. Thus adenosine triphosphate (ATP), uridine diphosphate (UDP), and S-adenosyl methionine are involved in transfer of phosphate or adenosyl moiety, glycosyl and methyl groups respectively.

Adenosine triphosphate (ATP, V) is the coenzyme for most phosphate transfer enzymes ("kinases"). It transfers the terminal phosphate group to the substrate giving rise to phosphorylated substrate and adenosine diphosphate (ADP), as is illustrated in reaction 24 catalysed by hexokinase.

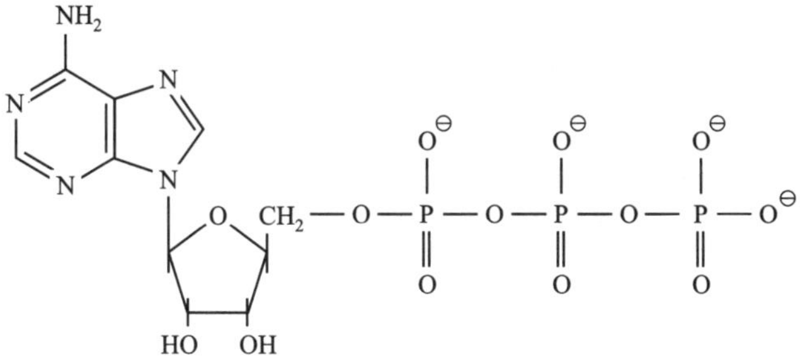

V Adenosine   Triphosphate (ATP)

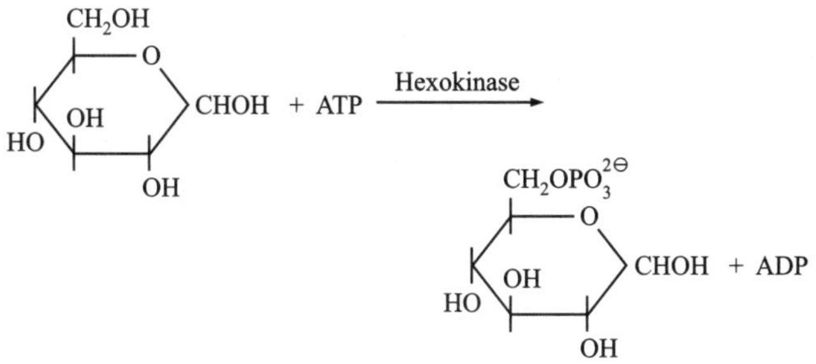

Glucose–6–phosphate

In other reactions ATP transfers its adenosyl or adenosyl mono phosphate (AMP) moiety to suitable acceptors. Besides acting as a group transfer coenzyme, the ADP-ATP pair also functions as a vehicle for capture and transfer of metabolic energy from the point of its liberation to the site of its utilization. It structure (V) containing two pyrophosphate groups is specially suited for this purpose. A pyrophosphate group is an "energy-rich" group.

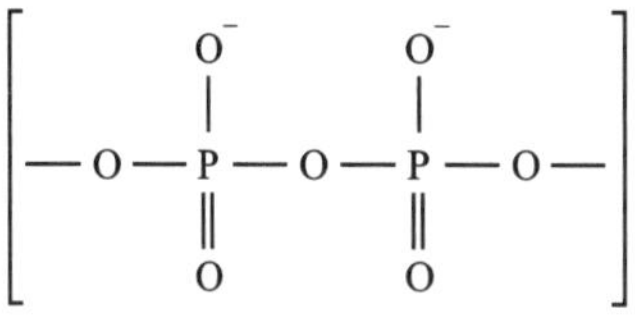

The designation is based on the free energy of hydrolysis at pH 7.0. If the *standard free energy of hydrolysis of a compound at pH 7.0 is 7kcal/mole or more,* the compound is referred to as "energy rich" compound and the bond undergoing hydrolytic rupture as the "energy rich" linkage. The free energy liberated during metabolic breakdown of food materials is captured in the form of synthesis of high energy pyrophosphate bonds of ATP. This energy may be employed for driving energetically unfavourable reactions in biosynthetic processes. Thus, prior to the synthesis of peptides the amino acids are converted into energy rich aminoacyl-AMP ('activation of amino acids") by the reaction of Eq. (9.25).

Similarly, the reaction of Eq. (9.26) "activates" methionine and prepares it for subsequent transfer of methyl group.

S-Adenosyl methionine (VI) is itself a high energy compound and transfers its methyl group to an appropriate acceptor in the presence of specific enzymes as in the reaction 27.

S-Adenosyl homocysteine gets hydrolysed to adenosine and homocysteine. The latter may either undergo degradation or be reconverted into methionine by accepting a methyl group from N^5-methyl tetrahydrofolic acid (VII). Several derivatives of tetrahydrofolic acid are known. These are coenzymes for transfer and metabolism of "one carbon" groups like methyl, formyl (–CHO), etc. Note that in the reactions of S-adenosyl-methionine (VI), the methyl group is transferred in preformed state. Its synthesis takes place somewhere else (involving various derivatives of tetrahydrofolic acid). It is frequently observed that groups or small molecules may be synthesized at one point and then transferred (as building blocks) to suitable acceptors to form bigger and more complex molecules. Formation of oligo and polysaccharides from mono-saccharides (like glucose, fructose etc.) are other examples of the application of this principle. In these reactions *uridine diphosphate* (UDP) acts as a *carrier* of glycosyl groups, in the form of its derivatives like uridine diphosphate glucose (UDPG, VIII). The latter is formed from uridine triphosphate (UTP, IX) and glucose-1-phosphate as in Eq. (9.28).

The glucose moiety of UDPG behaves as an "activated glucosyl" group which can be transferred to a suitable acceptor, like fructose (Eq. 29), or the growing end of an amylose chain.

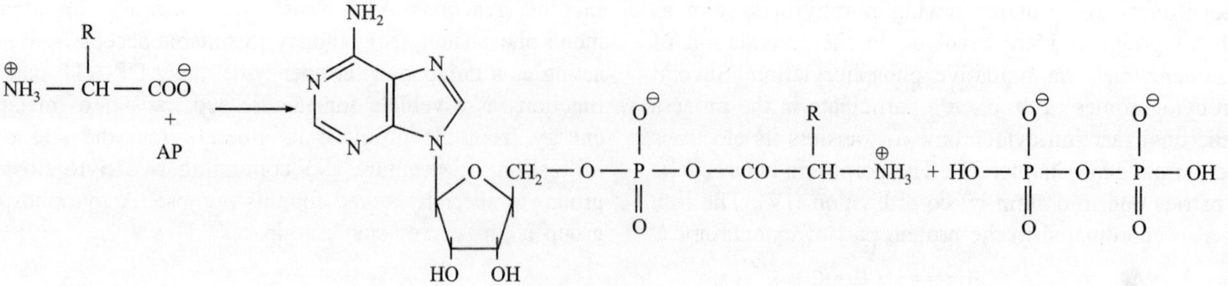

Aminoacyl—AMP

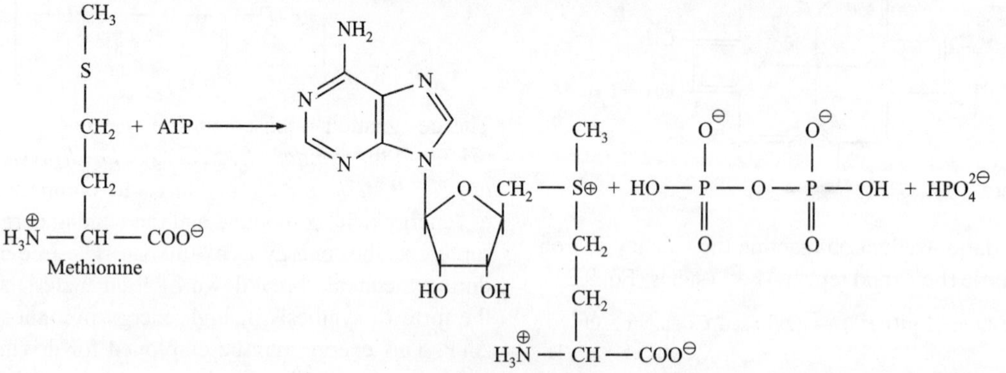

$$(9.25)$$

Methionine

VI   S—Adenosyl methionine

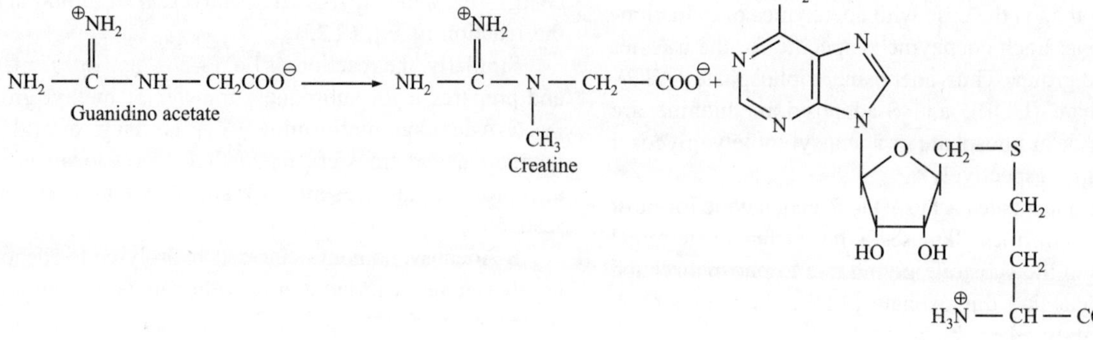

$$(9.26)$$

Guanidino acetate

Creatine

S–Adenosyl–homocystein

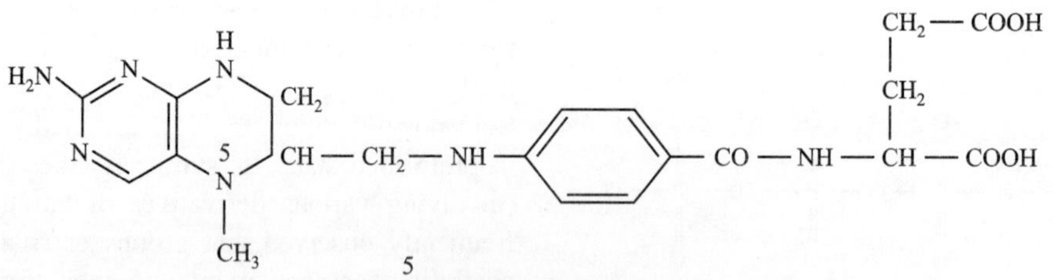

$$(9.27)$$

VII   N–Methyl tetrahydrofolic acid

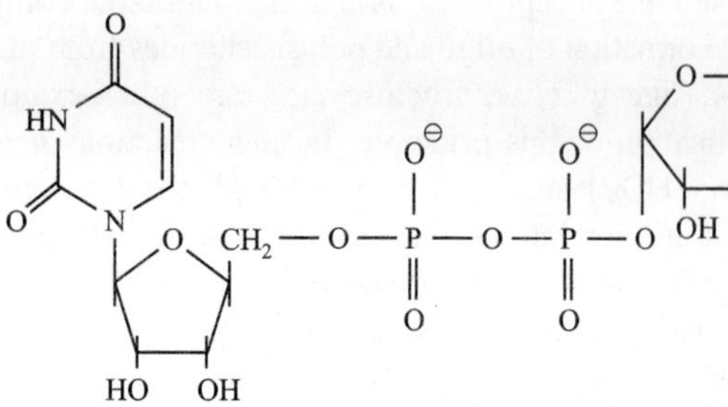

VIII   Uridine diphosphate glucose (UDPG)

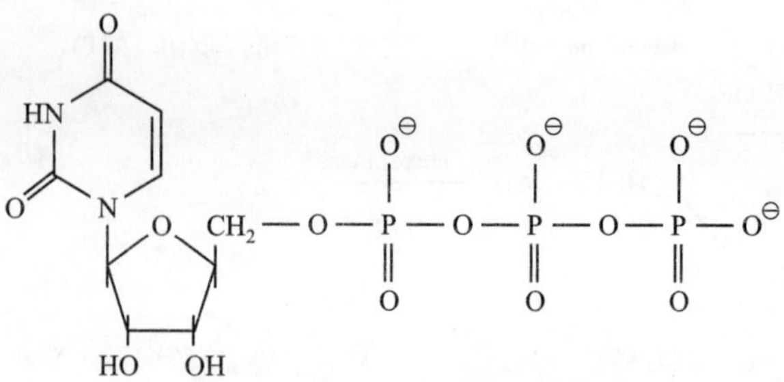

IX Uridine triphosphate (UTP)

$$UTP + G - I - P \rightleftharpoons UDPG + P - P \qquad (9.28)$$

$$\text{UDPG + Fructose} \longrightarrow \text{UDP + Sucrose} \qquad (9.29)$$

UDPG also serves as a source of galactose. The latter monosaccharide is not synthesized as such but is obtained via isomerization of UDPG to UDP galactose (UDPGal) (Eq. 30a). The latter can then transfer the galactosyl moiety to another glucose molecule to form lactose (Eq. 30b).

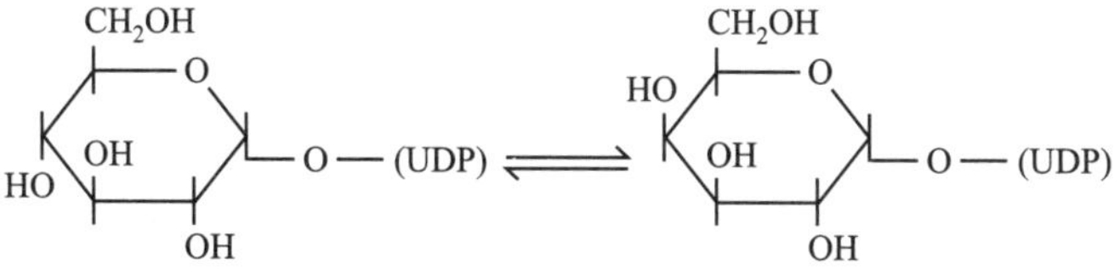

$$(9.30a)$$

$$\text{UDPGal + Glucose} \longrightarrow \text{UDP + Lactose} \qquad (9.30b)$$

Thus we see that depending upon the particular enzyme present *UDP* may act as a *carrier of glycosyl groups* in *transfer or isomerization reactions*. Another coenzyme which participates in more than one type of reaction (group transfer, isomerization, and elimination reactions) is pyridoxal phosphate.

Pyridoxal phosphate (X), derived from the B-vitamin pyridoxine (XI), may be appropriately named as the coenzyme for amino acid metabolism.

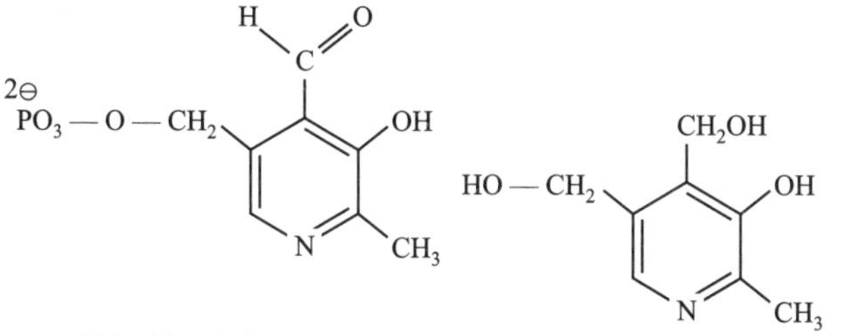

X Pyridoxal phosphate

Xi Pyridoxine

Invariably the first step is the formation of a Schiff's base (XII) as in Eq. (9.31).

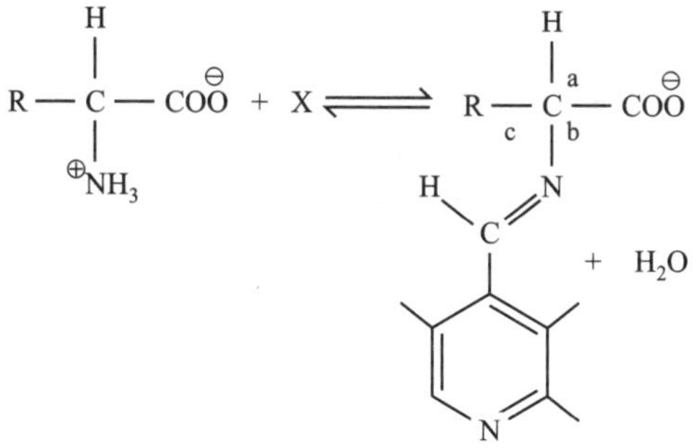

$$(9.31)$$

XII

The peculiar structure of this Schiff's base (namely the presence of alternating single and double bonds leading to the pyridine ring) activates the bonds designated as a, b and c in XII. This is responsible for various isomerization and elimination reactions of this intermediate, leading to different products. One of the functions of the enzyme is to direct the reaction towards one or the other of these pathways. Thus, in the transfer reaction (transmination), XII isomerizes to XIII

(Eq. 9.32a), which gets hydrolysed to a keto-acid (R–CO–COO⁻) and pyridoxamine phosphate (XIV) (32b). The latter then reacts with another keto-acid (R'–CO–COO⁻) and by reversal of reactions 32 and 31 gives rise to a different amino acid (R'–CH–COO⁻). The overall reaction then becomes one of

$$|$$
$$+NH_3$$

transamination (Eq. 9.33).

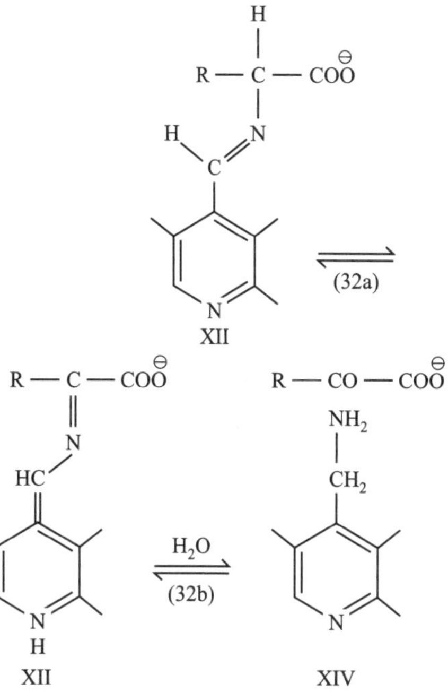

$$(9.32)$$

$$(9.33)$$

Note that in the Schiff's base (XII) the alpha-carbon atom is asymmetric and has L-configuration. On the other hand its isomer XIII has the alpha-carbon atom in a planar configuration (as it is unsaturated) so that its three-dimensional asymmetry has been destroyed. In a reversal of the reaction 32a the hydrogen atom may be attached to this carbon atom either on the same side from which it was removed or on the opposite side. In the latter case the isomeric (D-) form of amino acid will be obtained upon hydrolysis of the Schiff's base (reversal of reaction 31). The overall reaction (9.34) is isomerization of L- and D-amino acids.

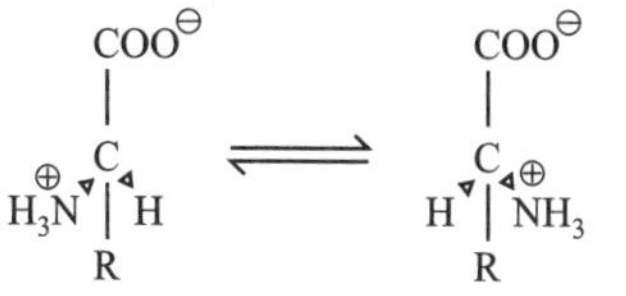

$$(9.34)$$

L-Amino acid   D-Amino acid

Activation of the linkage b of Schiff's base (XII) leads to the loss of carbon dioxide (decarboxylation). The remaining intermediate undergoes isomerization and hydrolysis to give rise to an amine (Eq. 9.35).

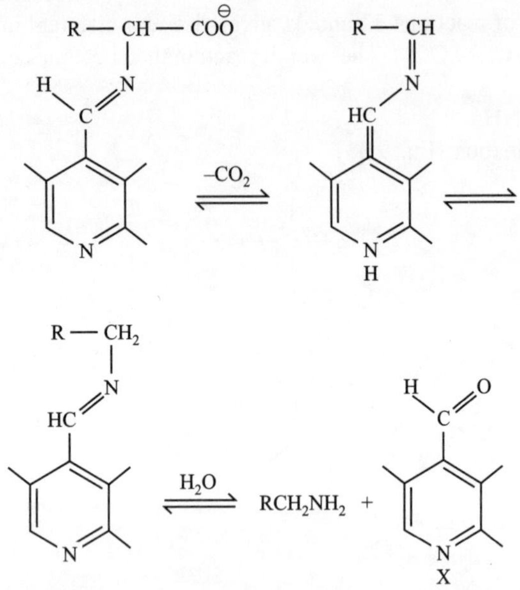

$$R-CH_2-NH_2 \;+\; X \tag{9.35}$$

Summation of reactions 31 and 35 shows that the net reaction (Eq. 9.36) is an elimination (decarboxylation) reaction.

$$R-\overset{\overset{\oplus}{N}H_3}{\underset{|}{C}}H-COO^{\ominus} \;\rightleftharpoons\; R-CH_2-NH_2 \;+\; CO_2 \tag{9.36}$$

Another coenzyme of addition-elimination reactions is thiamine pyrophosphate (TPP, XV) which is derived from thiamine (vitamin B_1).

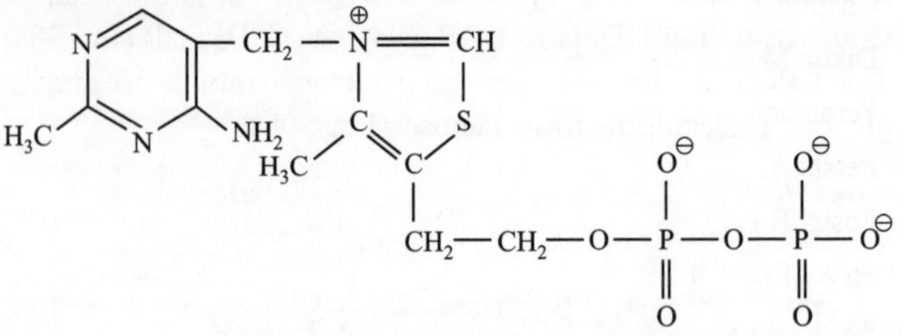

XV Thiamine pyrophosphate (T P P)

The reactive part of XV is the thiazolium ring and its reactions may be illustrated by decarboxylation of pyruvate ($CH_3-CO-COO^-$) to acetaldehyde (CH_3-CHO) which has been shown to proceed as below in Eq. (9.37).

The intermediate (XVI) is referred to as "active aldehyde" and participates in a variety of reactions where an aldehyde function is required. Thus, TPP may also be considered as a carrier of aldehyde group.

Another important coenzyme which serves as a carrier of acyl groups is *coenzyme A* (XVII, CoA). It accepts acyl groups from the points of their formation (as in the breakdown of carbohydrates and fatty acids) and feeds them to other metabolic chains for their ultimate utilization. In each case the acyl group is attached to the terminal SH group to give rise to an acyl-CoA.

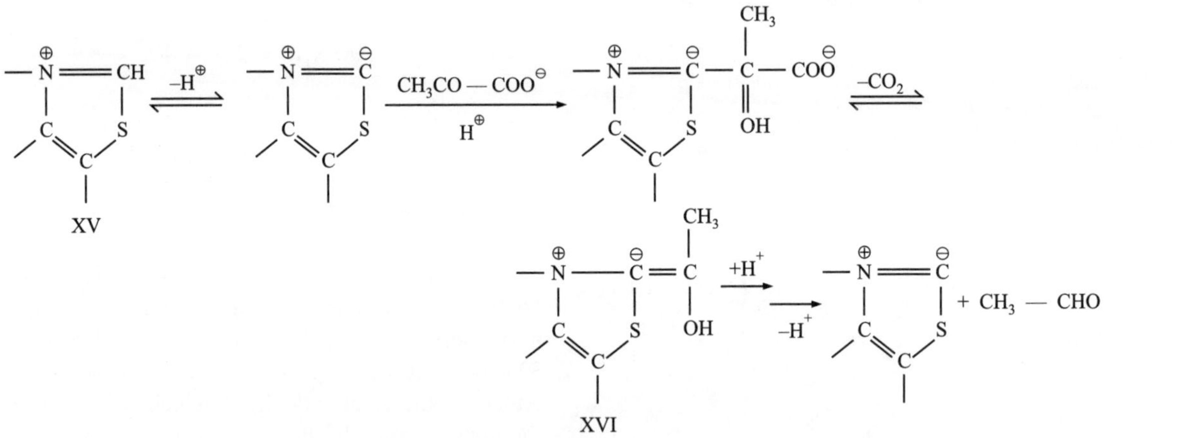

$$\cdots + CH_3-CHO \tag{9.37}$$

XVI

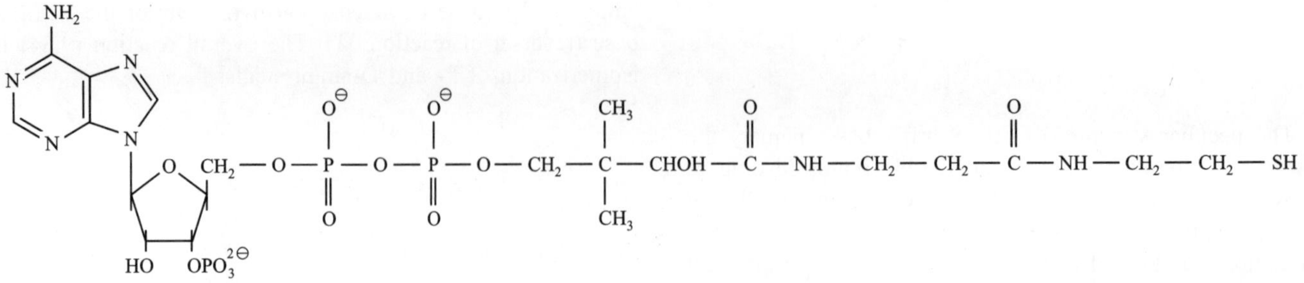

Aminoacyl—AMP

XVII Coenzyme A (CoA)

SUGGESTIONS FOR FURTHER READING

Bover P. (Ed.), (1970): *The Enzymes,* Student Edition, Vols. I and II., Academic Press, New York.

Chaplin M.F. and Bucke C. (1992), *Enzyme Technology,* Cambridge University Press.

Dixon M. and Webb E.C. (1979), *Enzymes,* 3rd ed., Longman, London.

Ferdinand W. (1976), *The Enzyme Molecule,* John Wiley & Sons, New York.

Fersht A. (1985), *Enzyme Structure and Mechanism,* W.H. Freeman & Co., New York.

Foster R.L. (1980), *The Nature of Enzymology,* Croom Helm, London.

Stryer Lubert (1995), Section Enzymes: Basic Concept and Kinetics, *Biochemistry*, 3rd ed., W.H. Freeman and Company, New York.

10

Biological Oxidation

Nibhriti Das

CONTENTS

I. INTRODUCTION

Energy required to maintain the structure and function of living cells is provided from nutrients by anaerobic fragmentation or, more completely, by oxidation of the carbon and hydrogen skeleton of compounds to CO_2 and H_2O. It may be pointed out that the energy in biological systems is essentially generated by the oxidation of hydrogen attached to carbons by molecular oxygen. While oxygen is obtained by inhalation of air by lungs, CO_2 is generated by catabolic degradation of biomolecules. A part of CO_2 is used in cellular reactions, and rest of it is exhaled out. Thus, cellular oxidation is linked to physiological respiration. About half of the energy generated is conserved in the molecule of ATP, the "energy currency of the living cell", discovered by Y. Subba Row (India born) and K. Lohmann (German) independently. Ninety percent of ATP is generated in mitochondria, the powerhouse of an eukaryotic cell, by oxidative phosphorylation.

II. OXIDATION–REDUCTION AND ENERGY CONSERVATION IN BIOLOGICAL SYSTEM

The term oxidation as originally used by Lavoisier meant addition of oxygen atoms to a substance being oxidized or removal of it from a substance being reduced. According to the modern concept, any process in which electrons are removed from a molecule is oxidation and the process in which electrons are added to a molecule is termed as reduction. Electron acceptor is thus an oxidizing agent (oxidant), and an electron donor is a reducing agent (reductant). Such oxidation–reduction reactions can be written as the sum of two separate reactions, one an oxidation reaction and the other a reduction reaction. Thus, oxidation (loss of electrons) of one compound is always accompanied by reduction (gain in electrons) of another substrate.The oxidized and reduced substrates form a 'redox pair'. For example, NAD+/NADH is a redox pair.

A. Redox Potentials

The relative tendency of reductants to donate electrons as compared to hydrogen is termed as oxidation–reduction potential or 'redox' potential. This can quantitatively be specified by a constant E_0 (the standard reduction potential).

By convention, E_0 of hydrogen is taken as zero. A compound having a negative value of E_0 is a better electron donor and those with a positive value are poorer donors of electrons. Therefore, electrons flow from compounds with negative E_0 to those with positive values.

In biological systems, electrons can be transferred in the form of electrons, for example, Fe^{2+}/Fe^{3+}; as hydrogen atoms which contain a single electron and a proton; as a hydride ion, a proton with two electrons or; by combination of a reductant with oxygen. *The term reducing equivalent is used to designate a single electron equivalent participating in an oxidation–reduction reaction irrespective of the mode of electron transfer.* During oxidation–reduction reactions in the metabolic pathways of glucose, fatty acids and amino acids, electrons are accepted by pyridine nucleotides NAD^+, $NADP^+$ or flavin nucleotides FAD and FMN, which get reduced to NADH, NADPH, $FMNH_2$ or $FADH_2$ respectively, These then transfer their high potential electrons to O_2 reducing it to water. In aerobic organisms, in the oxidation of fuel molecules, thus, the ultimate electron acceptor is O_2.

B. Free Energy

Free energy (F_O) is a measure of the energy available for useful work and a change thereof is related logarithmically to the concentrations and also to the difference in redox potentials of the reactants. For a change of 0.2 V in redox potential in a two-electron process, F_O = 9226 calories. This potential drop is required for generation of a "high-energy bond". **Gibbs free energy, G,** expresses the amount of energy capable of doing work during a reaction at constant temperature and pressure.

Free energy change: When a reacting system is not at equilibrium, the tendency to move towards equilibrium represents a driving force, the magnitude of which can be expressed as the free-energy change for the reaction, ΔG. When a reaction proceeds with the release of free energy (that is, when the system changes so as to possess less free energy), energy change ΔG has a negative value and the reaction is said to be exergonic. In endergonic reactions, the system gains free energy and ΔG is positive.

Standard Free energy change: Under standard conditions (298 K/25°C), when reactants and products are initially present at 1 M concentrations or, for gases, at partial pressures of 101.3 kPa, or 1 atm, the force driving the system towards equilibrium is defined as the standard free-energy change, $\Delta G°$.

By this definition, the standard state for reactions that involve hydrogen ions is 1 M or pH 0. Most biochemical reactions, however, occur in well-buffered aqueous solutions near pH 7. Therefore, for biochemical standard state, the defined standard conditions are: pH 7.0 and concentration of water 55.5 M. The standard free energy state at these defined constants is denoted as $\Delta G'0$.

Mathematically, $\Delta G'° = RT \ln K_{eq}$, where R is the gas constant, T is the absolute temperature and K_{eq} is the equilibrium constant of a reaction.

Actual free-energy changes ΔG, however, depend on reactant and product concentrations. The relations between ΔG and $\Delta G'°$ is written as

$$\Delta G = \Delta G'° + RT \ln (C)(D)/(A)(B)$$

where **C & D** and **A & B**, respectively, are concentrations of products and reactants in a reaction. At equilibrium, net energy change is zero and equilibrium constsnt

$$K_{eq} = (C)(D)/(A)(B).$$

Then, $0 = \Delta G'° + RT \ln K_{eq}$ or, $\Delta G'° = RT \ln K_{eq}$.

C. High-energy Bonds

"High-energy bonds" or "energy-rich compounds" are frequently referred to in biochemistry. By implication, their hydrolyses give rise to sufficient free energy of the order of at least 7000 calories/mole at pH 7.0. High–energy bonds is a misnomer since breaking of a bond does not release free energy as such. Since on hydrolysis of these compounds or bonds, products have lower energy levels than the reactants, energy is released. The compounds with this property are regarded as energy–rich compounds. The common types of "energy-rich" compounds are listed in Table 10.1. Although in the literature the terms "energy-rich" or "high energy" are commonly used, and understood for the purpose intended, they should be strictly referred to as compounds with "high group transfer potential". Therefore, it is the acetyl coenzymeA which is energy rich but not the Coenzyme A.

TABLE 10.1   High energy phosphate compounds

Compound	ΔG (KJ.mol^{-1})
Phosphoenolpyruvate	–61.9
1,3-Bisphosphoglycerate	–49.4
Acetyl phosphate	–43.1
Phosphocreatine	–43.1
PPi	–33.5
ATP ($\rightarrow$AMP+PPi	–32.2
ATP ($\rightarrow$ADP+Pi)	–30.5

D.   ATP

Adenosine triphosphate, ATP, gains unique importance in the energy transactions in the living cell in view of its polyanhydride structure.

It has two pyrophosphate bonds, cleavage of each of which leads to the formation of inorganic phosphate and adenosine diphosphate or adenosine monophosphate, each getting stabilized in their resonance hybrid forms, thus, about 7000–8000 calories of energy is released. Many types of cleavages occur at different bonds of the molecule, resulting in either hydrolytic or transfer reactions involving a variety of groups—phosphate, pyrophosphate, adenylic acid and adenosine (Figure 10.1). Singly, or together, these offer a wide armamentarium for the living cell to execute a variety of biosynthetic reactions.

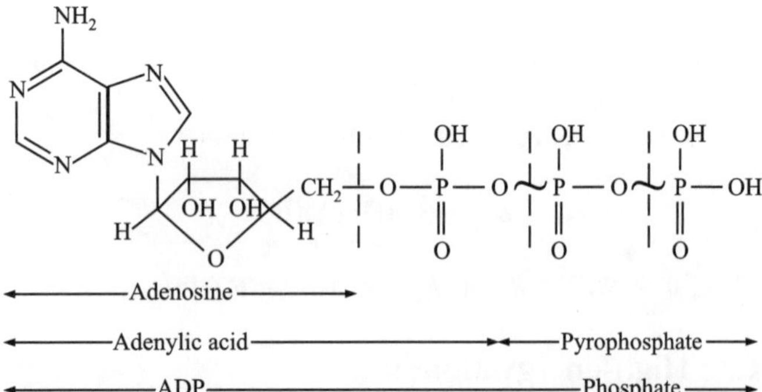

Figure 10.1   Structure of ATP and positions where transfer reactions occur.

ATP has an inseparable companion in Mg^{++} and a chelate of the two is considered to be the active species. Although the hydrolysis of ATP is highly exergonic, ($\Delta G'^{\circ} = -30.5$ KJ/mole), the molecule is kinetically stable at pH 7 because the activation energy for ATP hydrolysis is relatively high. Rapid cleavage of the phosphoanhydride bonds occurs only when the bonds are catalyzed by an enzyme. The actual free-energy change for ATP hydrolysis in living cells is very different from that prescribed for ATP hydrolysis in the standard conditions. Formation of ATP from ADP by phosphate esterification can occur at substrate level using acyl phosphate or enol phosphate intermediates, and more efficiently by oxidative phosphorylation in mitochondria.

III.   MITOCHONDRIA

Mitochondria are the intracellular "power plants" of aerobic cells, which trap the respiratory energy as chemical potential energy by organized enzyme systems. The etymology of the word is of Greek origin with a morphological basis—*Mitos,* a thread and *Chondros,* a grain.

Mitochondria can be seen under a light microscope but only the electron microscope can reveal the details of the internal structure. They occur as spheres or elongated spheres—more like a polymer of repeating structural units. The approximate width or diameter is about 1 micrometer, with varying lengths. The size and shape vary from tissue to tissue. Their locale in the cells is function-oriented; for instance, in the liver they are scattered in the cytoplasm, in the flight muscle they are linearly arranged, and in spermatozoa they are woven into a helical sheath around the midpiece. The number of mitochondria varies in different cells, mostly in the range of 300–3000 per cell, although figures as low as 20 and as high as 500,000 are known.

Mitochondria have characteristic structural features. It is like two bags fitted into each other. The outside shell is generally a smooth membrane. Within this is enclosed another inner membrane which has a varying degree of invaginations of the same membrane continuum, termed cristae (Figure 10.2). Between these two membranes, there is intramembranous space. The design of this architectural oddity seems to provide an increased surface area of the membrane, whereby a large number of the components can be accommodated.

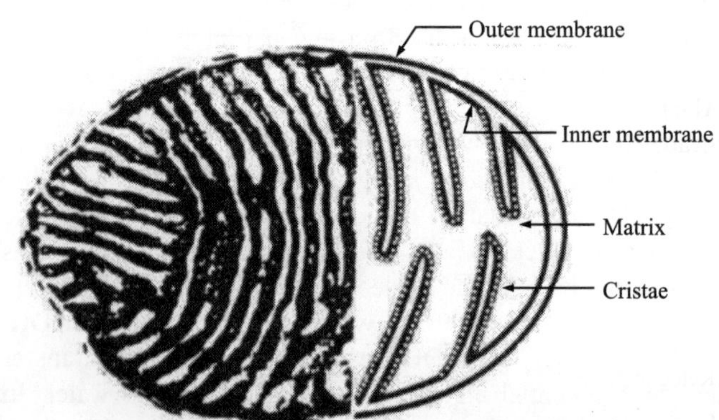

Figure 10.2   Structure of a mitochondrion.

The cytosol inside the inner mitochondrial membrane is called matrix. The matrix contains soluble enzymes including a number of Krebs cycle enzymes. Methods have been developed to separate the inner and outer membranes, as well as derived sub-mitochondrial particles with partial functions. It should be cautioned that artifacts are abundant in this type of work.

The outer membrane is quite permeable to most small molecules and ions because it contains many copies of *mitochondrial porin,* a 30–35 kd poreforming protein also known as voltage-dependent anion channel or VDAC. This channel plays a role in the regulated flux of anionic species such as phosphate, chloride, organic anions, and the adenine nucleotides—across the outer membrane. Some cytoplasmic kinases bind to VDAC, thereby obtaining preferential access to the exported ATP.

The inner membrane is intrinsically impermeable to nearly all ions and polar molecules. A large family of transporters shuttles metabolites such as ATP, pyruvate, and citrate across the inner mitochondrial membrane. Mitochondria contain their own DNA, which encodes a variety of different proteins and RNAs. Human mitochondrial DNA comprises 16,569 bp and encodes 13 respiratory-chain proteins, small and large ribosomal RNAs and tRNAs to translate all codons. However, mitochondria also contain many proteins encoded by nuclear DNA. Cells that contain mitochondria depend on these organelles for oxidative phosphorylation, and the mitochondria in turn depend on the cell for their very existence. The double membrane, circular DNA (with some exceptions), and mitochondrial-specific transcription and translation machinery all point to an *endosymbiotic event* to have occurred whereby a free living organism capable of oxidative phosphorylation was engulfed by another cell. The evidence that modern mitochondria result from a single event comes from examination of the most bacteria-like mitochondrial genome. Mitochodria also contribute to apoptosis.

Mutations in mitochondrial DNA, thyrotoxicosis, other hormonal abnormalities, radiation exposures, toxic chemicals, selected antibiotics, and antinutritients are some of the factors that have adverse effects on mitochondrial structure and functions.

IV. ELECTRON TRANSPORT CHAIN AND OXIDATIVE PHOSPHORYLATION

$NADH^+H^+$ and $FADH_2$ or $FMNH_2$ generated during catabolic degradation of carbohydrates, proteins and fats, transport electrons to molecular oxygen through a number of carriers organized sequentially (Figure 10.3) in four multienzyme complexes (Complex I, II, III, IV) present in the inner mitochondrial membrane. Chain of electron carriers upto molecular oxygen compose electron transport chain or respiratory chain. Coupling of electron transport with ATP formation is known as oxidative phosphorylation.

A. Components and Sequence of the Electron Transport Chain

NAD^+ is the first component which accepts hydrogens from oxidizable substrates in the form of the hydride ion (H^-), a proton having two electrons resulting in neutralization of the charge on the nicotinamide ring. NADH is formed. Electrons from NADH pass through a flavoprotein to a series of iron-sulfur proteins and then to Q.

Flavoproteins, having protein-bound FAD and FMN, are the enzymes which funnel the reducing equivalents from a number of substrates into the main sequence of oxidation. These include the major NADH and succinate dehydrogenases and the other dehydrogenases of substrates such as α-glycerophosphate, choline etc. **Ubiquinone,** a lipid quinone of ubiquitous occurrence in all aerobic cells acts as an electron carrier. It is also called coenzyme Q, signifying its coenzyme activity. The two names are used synonymously. The hydrophobic polyprenyl side chain facilitates interaction with the membrane and the capacity for reversible oxidation-reduction of the quinone moiety permits participation in electron transfer (Figure 10.4). It seems to be situated at the crossroads of a number of dehydrogenases and works as one of the "mobile carriers"—a point at which the $2(H^+ + e)$ transfer is transformed to one of single electron transport sequence. Ubiquinone has additional roles of preferential oxidation of succinate instead of NADH-linked substrates, and generation of heat instead of ATP and also as an antioxidant.

Iron-sulphur proteins are associated with the dehydrogenases. The iron present is not of the heme-type and is bound to s-atoms of elemental sulphur or of cysteine in clusters of 2Fe-2S or 4Fe-4S. They are called "non-heme iron" proteins. The change in their redox state is accompanied by changes

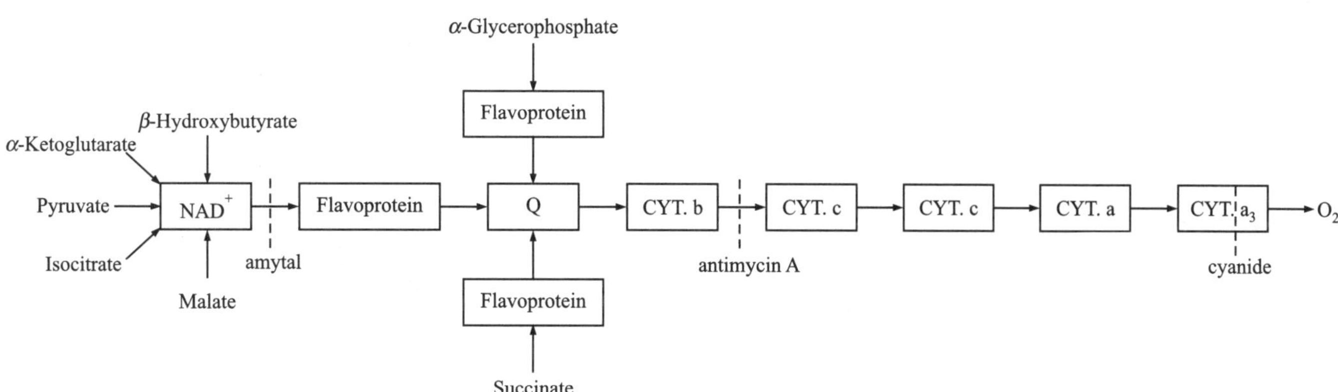

Figure 10.3 Electron transport chain and specific inhibitors.

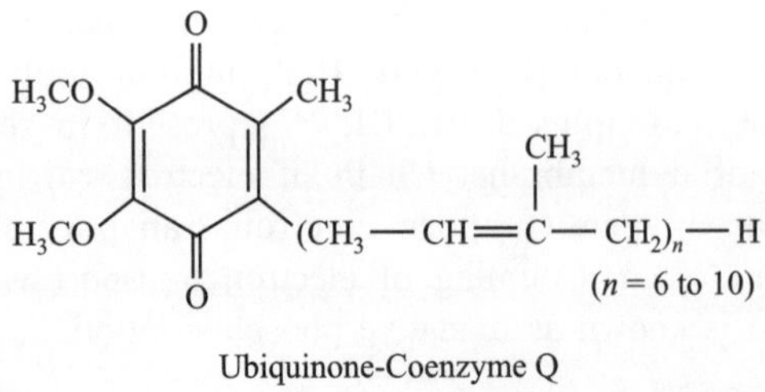

Figure 10.4 Structure of ubiquinone-coenzyme Q.

in the magnetic properties of the molecules, and can be characterized by the electron paramagnetic resonance (EPR) spectra. These are involved in interlinking the main electron transport components.

A number of cytochromes link up the flavoproteins and oxygen and form the "power train" of electron transport down the gradient of lower to higher redox potential. These are hemoproteins with porphyrin-based iron capable of reversibly undergoing transformations between ferrous and ferric states. Spectroscopy has been of immense help in the identification and measurement of this group of electron carriers. In the reduced state they exhibit characteristic bands, known as the Soret bands. The individual cytochromes are distinguished from each other by variable substitution of the porphyrin side chain, the proteins, the redox potentials and the solubility properties. The cytochromes of significance in the mitochondrial oxidations are *b*, c_1, *c*, *a*, and a_3. Of these, cytochrome *a* + a_3, the cytochrome oxidase complex, a copper-containing hemoprotein system, accounts for the reaction with molecular oxygen.

B. The Q-cycle

It is considered that the respiratory chain changes from 2-electron to 1-electron transfer at the Q-b stage. Reduction of cytochrome *b* is oxidant induced which requires transfer of electrons through an Fe-S protein (Reiske protein) to cytochrome *C*. Reduction of Q to QH_2 takes place in two steps—one electron(e) from reduced cytochrome *b* (b_{562}) and another from the dehydrogenase flavoprotein (FP) or its 1-electron oxidation product, flavoprotein radical (FP•). Oxidation of QH_2 by the Reiske protein (Fe-S protein) is needed for reduction of a form of cytochrome *b* (b_{566}) which passes its electron to b_{562}-Ubisemiquinone (Q•) has a dual role—denoting its electron to cytochrome *b* and also receiving the second electron from FP or FP•. These are formulated into "Q-cycle" by which the two electrons from FP are passed on

to cytochromes taking advantage of the mobility of Q and QH_2 in the membrane (Figure 10.5).

Three main considerations that determined the sequence of the components in the respiratory chain were:

1. **Redox potentials:** The sequence could be arranged in the increasing order of positive potential since electrons flow only from low to high potential (Table 10.2).

TABLE 10.2 Standard Redox Potential of selected carriers in the Electron Transport Chain

Redox pair	E_0
NAD+/NADH	–0.32
FMN/FMNH2	–0.22
Cytochromec Fe^{3+}/Fe^{2+}	+0.22
% O_2/H_2O	+0.82

2. **Reaction kinetics:** The components on the "oxygen end" are more rapidly oxidized; relative reaction rates of reduction and oxidation can determine the sequence. For example, in the normal electron transport, first molecule to transfer its electron to molecular oxygen is cytochrome a_3, and hence, this is the first electron carrier to be oxidized followed by cyt.a.cytc, and, so on. In presence of cyanide, electrons from cytochrome a3 will not be transported to O_2 and none of the carriers will be oxidized.

3. **Selective inhibitors:** (a) A specific inhibitor blocking at a point in the chain resulted in the oxidation of the components on the "oxygen end" and reduction of those on the "substrate end". (b) Inhibitors also helped to locate points of entry of some substrates into the cytochrome system (Figure 10.3).

C. Multienzyme Complexes Involved in Electron Transfer from NADH to O_2 Through ETC

The tightly-bound components in each complex appear to support efficient electron transfer. Occurring in clusters with easy mobility in the fluid membrane, these complexes interact with each other and constitute the respiratory chain. Figure 10.6 depicts the complex I, II, III and IV of the Electron Transport chain.

The complex I (NADH dehydrogenase) consists of 26 polypeptides, including three Fe-S proteins, and has FMN.

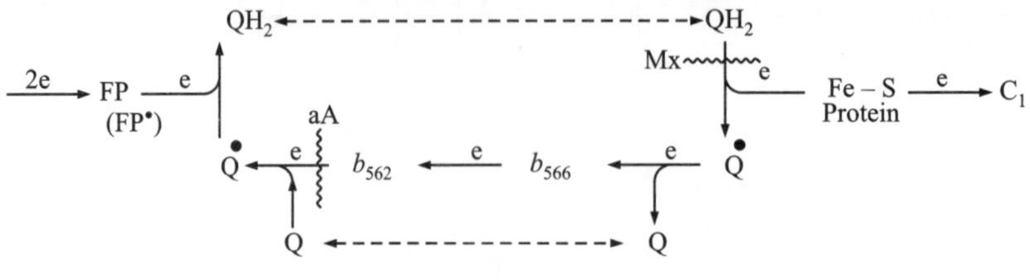

Figure 10.5 The Q-cycle.

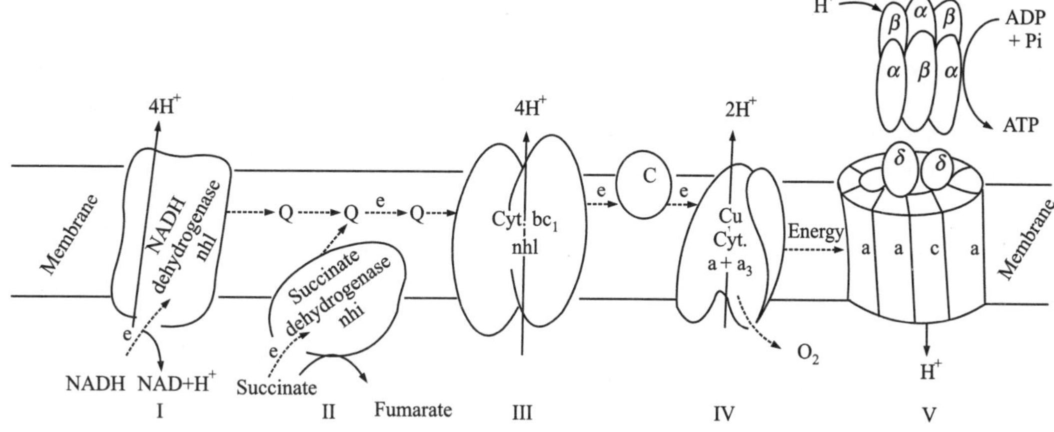

Figure 10.6   The four complexes I, II, III and IV of electron transport system and FO/F1 ATPase (V).

The complex II (succinate dehydrogenase, 97 kDa protein with FAD) also has three Fe-S proteins.

The complex III (bc1 complex) has 9 proteins including cytochrome b (30 kDa), cytochrome c_1 (29 kDa) and an Fe-S protein (Reiske protein).

The complex IV is primarily cytochrome oxidase including both a and a_3 cytochromes. Electron transfer through Complex IV is from cytochrome c to the CuA center, to heme $a,$ to the heme a_3-CuB center, and finally to O_2.

For every four electrons passing through this complex, the enzyme consumes four H^+ from the matrix (N side) in converting O_2 to $2H_2O$. The overall reaction catalyzed by Complex IV is:

4 Cyt c (reduced) + 8H$^+$ (Matrix) + O_2 → 4 Cyt c (Oxidized) + 4H (Intermembrane space) + 2H$_2$O

It should be remembered that these complexes deal with only electron transport and are incapable of oxidative phosphorylation.

IV.   MECHANISMS OF OXIDATIVE PHOSPHORYLATION AND ATP SYNTHESIS

The transfer of a pair of electrons from NADH or FADH2 to molecular oxygen is a highly exergonic process with standard free energy change ($\Delta G'0$) of –220 KJ and –150 KJ per molecule of these two substrates, respectively. ($\Delta G'0$) of ATP synthesis from ADP is 30.5 KJ. Though energetically highly feasible,the mechanism of coupling of electron transport with ATP synthesis remained an enigma for very long time.

A.   Chemical Intermediate and Conformational Intermediate Theory

It was suggested that electron transfer leads to the formation of high energy intermediates that serve as high phosphoryl group transfer compound or to the formation of an activated protein conformation which drives the ATP synthesis. However, presence of any such intermediate could not be evidenced.

B.   Chemiosmotic Mechanism

In 1961, Peter Mitchell, suggested that electron transport and ATP synthesis are coupled by a proton gradient across the inner mitochondrial membrane and not by an activated intermediate or high energy conformation of a protein. The protons are pumped out of the inner mitochondrial membrane into the intermembrane space using the energy discharged by the transport of electrons through the ETC. This generates a gradient of H^+ and the medium becomes acidic. The electrochemical energy inherent in this difference in proton concentration and separation of charge represents a temporary conservation of much of the energy of electron transfer. The energy so conserved is termed as proton-motive force which has two components; a) the chemical potential energy due to the difference in the concentration of the chemical species (H^+) in the two compartments separated by the inner mitochondrial membrane and, electrical potential energy that results from the separation of charge (OH^- inside and the H^+ outside the membrane). The pH outside is 1.4 and the membrane potential is 0,14V. This membrane potential corresponds to 5.2 kcal (21.8 kJ) per mole of protons. The proton gradient is considered to represent energized state which is discharged by pumping the protons back into the matrix through the FO/ F1 ATP synthase, which synthesizes ATP by phosphorylating ADP. Several experimental evidences supported the chemiosmotic hypothesis. However, how does the flow of protons lead to ATP synthesis remained unexplained.

C.   Binding Chain Mechanism for Proton-Driven ATP Synthesis

Experiments showed that about equal amounts of ADP and ATP are bound to the catalytic sites of the enzyme which remains in equilibrium even without the proton gradient. Paul Boyer suggested that the dehydration of ADP + Pi forming ATP occurs at the catalytic subunit of FO/F1-ATPase without energy input by the proton gradient. The tightly bound ATP thus formed is released consequent to energy-induced conformational change of the protein. However, the ATP is not released from the catalytic site without the proton gradient and cosequently no new ATP molecule is formed if proton gradient is disturbed.

D. ATP Synthase and Rotational Catalysis Mechanism of ATP Synthesis

ATP synthase an enzyme complex in the inner mitochondrial membrane causes the synthesis of ATP coupled with election transport. This enzyme complex earlier was known as ATP ase or FO/F1 ATPase because it was discovered as ATPse. It is a large membrane embedded enzyme consisting of the ball like catalytic subunit F1, which protrudes into mitochondrial matrix and a stick like FO subunit that remains embedded into the inner mitochondrial membrane. F1 independently and in isolation from FO has ATPase activity and together with FO has the ATP synthase activity. FO is so named because it binds with oligomycin which inhibits the enzyme.

F1 has three α and three β subunits. Both these bind with nucleotides but the catalytic subunit is β. Each of the three β subunits alternate with three a subunits forming a hexamer. The γ subunit breaks the symmetry of the α_3, β_3 hexamer (Figure 10.7a and b).

FO subunit froms the proton channel consisting of 10–14 c subunits to form the central ring of the channel. Another single 'a' subunit binds to the outer side of the ring. The FO and F1 are connected by the central γ stalk and by an outer column consisting of a, b and ∂ subunits at different stoichiometry. The 'c' subunit and stalk form a moving rotor and rest of the components make a static part, the stator. The ATP synthase uses ADP+Pi as substrate in presence of Mg^{2++} to form ATP.Each of the three β units stay at a comformationally and functionally different form.

One β-subunit remains in T (tight) from with very high affinity of ATP, does not need any extra energy for synthesis of ATP from ADP+Pi; This formed ATP does not leave the catalytic site in this 'T' form. The other p subunit will then be in the L (loose) from which binds with ADP and Pi, does not synthesize ATP and is constrained in releasing ADP bound to it. The third β-submit remains in '0' or open form which

readily accepts and releases nucleotides bound to it and hence may also be in an empty state.

Paul Boyer proposed a **rotational catalysis** mechanism in which the three active sites of F1 take turns catalyzing ATP synthesis.

The conformational changes central to this mechanism are driven by the energy released when the protons enter back to the matrix through FO/F1 ATP synthase.

The streaming of protons through the FO "pore" causes the cylinder of c subunits and the attached γ subunit to rotate about the long axis of γ which is perpendicular to the plane of the membrane. With each rotation of 120°, γ subunit comes into contact with a different β-subunit forcing a conformational change in it. β(T) from gets changed to 'O' form and releases ATP, L form which had ADP and Pi bound to it, becomes T form and synthesizes new ATP which remains tightly bound to it and the O form becomes L form and accepts new ADP and Pi. (Figure 10.8). For the release of ATP, it is mandatory that new ADP and Pi bind to one of the p-subunits. It has been proven experimentally that on rotation of the γ subunit in an opposite direction F1 acts as ATPase releasing ADP and Pi.

E. ATP-ADP Translocase

A new ADP can enter mitochondria only when the preformed ATP leaves the mitochondria. Since ATP and ADP both are highly charged molecules, these proteins are translocated by an antiport, the ATP-ADP translocase abundant in the inner mitochondrial membrane. Inhibition of this transport subsequently inhibits ATP synthesis and cellular respiration. The ATP-ADP translocation is energetically expensive demanding about 1/4th of the energy generated by oxidative phosphorylation.

F. Oxidative Phosphorylation and P/O Ratio

Coupling of ATP synthesis with electron transport is accompanied with consumption of oxygen. The number of

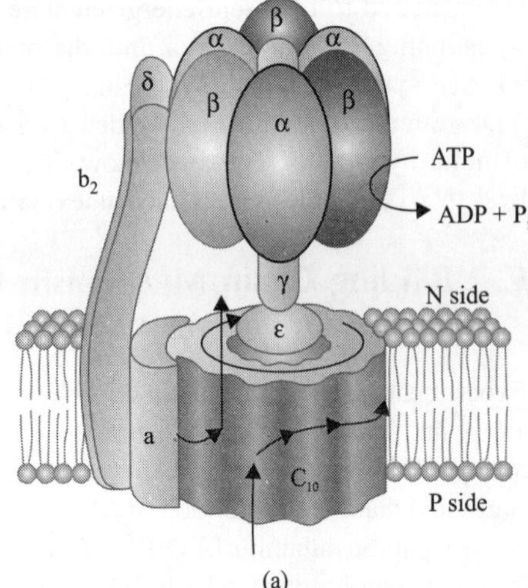

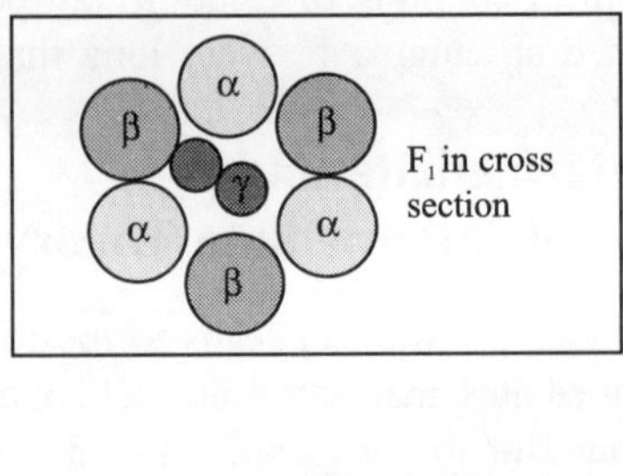

Figure 10.7   Schematic depiction of FO/F1 ATPase and arrangement of α, β hexameraround γ subunit.

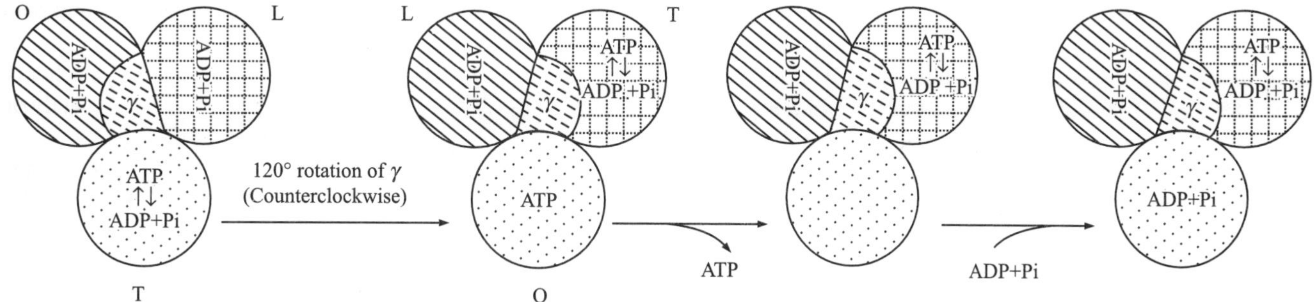

Figure 10.8 Binding chain mechanism and rotational catalysis for proton-driven ATP synthesis.

ATP formed per molecule of oxygen is different for different substrates. This is called P/O ratio.

Earlier experiments, calculations and assumptions suggested that 3 molecules of ATP are formed per molecule of NADH and, 2 molecules of ATP are formed per molecule of $FADH_2$ by oxidative phosphorylation in mitochondria.

Since the acceptance of chemiosmotic hypothesis, the most widely accepted experimental value for number of protons required to drive the synthesis of an ATP molecule is 4, of which 1 is used in transporting Pi, ATP, and ADP across the mitochondrial membrane. 10 protons, 4 each by the complex I and complex III, and two by complex IV are pumped out by energy of electron transport per NADH (Figure 10.6).

Of these 10 protons, 4 protons must flow in to produce 1 ATP, the proton-based P/O ratio is 2.5 (10/4) for NADH as the electron donor and 1.5 (6/4) for succinate.

G. Generation of Free Radicals

Molecular oxygen is the ideal terminal electron acceptor because of its high redox potential. For reduction of one molecule of oxygen, four electrons and four protons are needed. This four-electron reduction of O_2 involves redox centers that carry only one electron at a time, and it must occur without the release of incompletely reduced intermediates such as hydrogen peroxide or hydroxyl free radicals. Despite the fact the cytochrome oxidase does not allow the release of such intermediates, small amounts of hydrogen peroxide and super oxide radicals are formed. Cellular defense mechanisms against free radicals, however, take care of these toxic molecules.

H. Regulation of Oxidative Phosphorylation, the Acceptor Control

During different phases of the cycle of operations of oxidative phosphorylation, mitochondria are said to be in different metabolic "states" as defined by Chance.

State 1: Resting state in the absence of substrate.
State 2: Low respiration in the absence of ADP.
State 3: Active state, complete system.
State 4: Controlled state, ADP fully phosphorylated to ATP.

The significant point to be noted is that active respiration is obtained only when ADP acceptor system is available; it does not start until ADP is added in state 2 and it stops when ADP is exhausted in state 4.

Thus, the rate of respiration and oxygen consumption by mitochondria is limited by the availability of ADP which is also an indicator of the energy needs of a cell. This phenomenon is known as 'Acceptor control' of cellular respiration and oxidative phosphorylation. In some animal tissues, the **acceptor control ratio,** the ratio of the maximal rate of ADP-induced O_2 consumption to the basal rate in the absence of ADP, is at least ten. In fact in perfectly coupled mitochondria there will be complete cessation of oxygen uptake if ADP is exhausted.

The [ATP]/([ADP][Pi] ratio, the energy charge of a cell normally is very high indicating that the ATP-ADP system is almost fully phosphorylated. When energy consuming processes are active, ATP is hydrolyzed to ADP, rate of respiration increases, more of NADH and $FADH_2$ are formed because of increased catabolism of energy fuels,, ADP is phosphorylated to ATP by oxidative phposphorylation till the energy charge comes back to normal. Thus, the system is regulated efficiently so that the amount of ATP hydrolyzed is always replenished by the amount of ATP synthesis governened by strictly regulated metabolism of the energy fuels and acceptor control by ADP.

In tightly coupled mitochondria, the respiration is delicately poised and is controlled by ATP/ADP ratio. Increase in ADP stimulates, while increase in ATP suppresses, the oxygen uptake. This gives a perfect control system. This is logical since the ultimate purpose of oxidation is to provide ATP.

I. Protein Inhibitor, IF1 Prevents ATP Hydrolysis in Ischemic Tissues

In ischemic tissues, as during heart attack and stroke, with the fall of oxygen supply, oxidative phosphorylation and ATP synthesis stops, under these conditions, the ATP synthase tends to operate in reverse, hydrolyzing ATP to pump protons outward. This causes a drop in ATP levels. A small (84 amino acids) protein inhibitor, IF1, which simultaneously binds to two ATP synthase molecules, inhibits their ATPase activity. IF1 is inhibitory only in its dimeric form, which is favored at pH lower than 6.5. In a cell lacking oxygen, glycolysis is the only source of ATP. Its rate increases, with increased lactate production and lowering of pH, this favors IF1 dimerization, leading to inhibition of the ATPase activity of ATP synthase. This prevents ATP hydrolysis and conserves available ATP. When aerobic metabolism resumes, rate of glycolysis and production of pyruvic acid slows, the pH of the cytosol rises, the IF1 dimer is destabilized, and the inhibition of ATP synthase is abolished.

J. Inhibitors of Oxidative Phosphorylation

The sequence of electron carriers and mechanism of oxidative phosphorylation were largely elucidated by the use of respiratory poisons or inhibitors of oxidative phosphorylation or electron transport. Following is a brief account of these inhibitors.

Inhibition of electron transfer

The deadly poisons cyanide and carbon monoxide act by inhibiting the cytochrome oxidase system of mitochondria and thereby cut off the transport of electrons from cytochrome to molecular oxygen, inhibiting the entire process of electron transport and ATP synthesis.

British antilewisite (BAL), 2,3-dimer-captopropanol(DCPP), an agent that neutralizes the arsenical poisons, selectively blocks between cytochrome *b* and *c*. Similar action is exhibited by antimycin A (an antibiotic obtained from Streptomyces).

A fish poison, rotenone, proved to be a specific binder of NADH-flavoprotein and knocks out respiration at very small concentrations.

Myxothiazol an antibiotic from myxobacterium Myxococcus fulvus, Amytal, (a barbiturate drug), DCMU [3-(3,4-dichlorophenyl)-1,1-dimethylurea] a herbiside, inhibit electron transport at the level of coenzyme Q and Fe-s centres. Addition of antimycin *A* blocks oxidation of both Q and *b*.

Inhibition of ATP synthase

Inhibition of ATP synthase is brought about by *Aurovertin, Oligomycin, Venturicidin and* **DCCD** (Dicyclohexyl-carbodiimide) While Aurovertin inhibits F1, oligomycin and venturicin inhibit FO, CFO and DCCD, blocks proton flow through FO and CFO.

Inhibition of ATP-ADP exchange

Atractyloside of plant origin Inhibits adenine nucleotide translocase.

Uncoupling of oxidative phosphorylation

The last class of compounds does not inhibit the electron transport to O_2 but interfere with formation of the proton gradient and therefore, uncouple oxidation with phosphyration. The energy is released as heat.

Chemical uncouplers

Hydrophobic proton carriers like dinitrophenol and FCC [(Trifluorocarbonylcyanide Phenylhydrazone (FCCP)], and K^+ ionophore valinomycin are chemical uncouplers of oxidative phosphorylation.

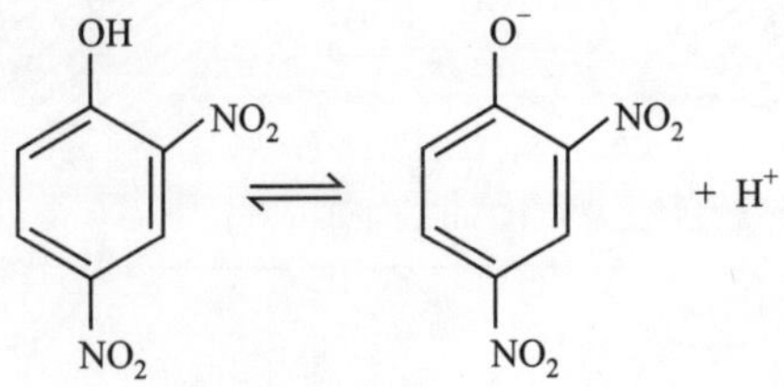

2,4-Dinitrophenol
(DNP)

Uncoupling Proteins (UCP1-5)

UCP1 or thermogenin, a dimer of 33 kd protein that resembles ATP/ADP translocase present in the inner mitochondrial memebrane, creates a proton leak and allows the protons to enter mitochondrial matrix without energy being captured with formation of ATP. Fatty acids activate the action of UCP1. Thermogenin abundant in mitochondria of the brown adipose tissue in human new borns and cold-hibernating animals accounts for non-shivering thermogenesis in them.Thyroxine also serves as physiological uncoupler. UCP2 with 56% homology with UCP1 is found in a wide variety of tissues. UCP3 is presnt in skeletal muscles and brown fat. Role of these UCPs and UCP4 & UCP 5 are not clear.

K. Defects in Oxidative Phosphorylation

Proteins are synthesized in the cytosol and transported into mitochondria. Defects in oxidative phosphorylation are more likely a result of alterations in mtDNA, which has a mutation rate about ten times greater than that of nuclear DNA. Tissues with the greatest ATP requirement (for example, central nervous system, skeletal and heart muscle, kidney, and liver) are most affected by defects in oxidative phosphorylation. Mutations in mtDNA are responsible for several diseases, including some cases of mitochondrial myopathies and Leber hereditary optic neuropathy, a disease in which bilateral loss of central vision occurs as a result of neurorentinal degeneration, including damage to the optic nerve. The mtDNA is maternally inherited because mitochondria from the sperm cell do not enter the fertilized egg.

L. Shuttles for Transport of Reducing Equivalents of NADH from Cytosol to Mitochondria

Since inner mitochondrial membrane does not have a transporter for NADH, In aerobic tissues, reducing equivalents from NADH produced in cytosol are transported to mitochondrial NAD+ or FAD through two different substrate shuttles shown in Figure 10.9(a) and (b).

Glyceraldehyde-3-phosphate shuttle transfers two reducing equivalents of cytosolic NADH to Dihydroxy acetone phosphate by the enzyme glyceraldehydes-3-phosphate dehydrogenase. The product of the reaction glyceraldehydes-3-phosphate crosses inner mitochondrial membrane and is reacted upon by

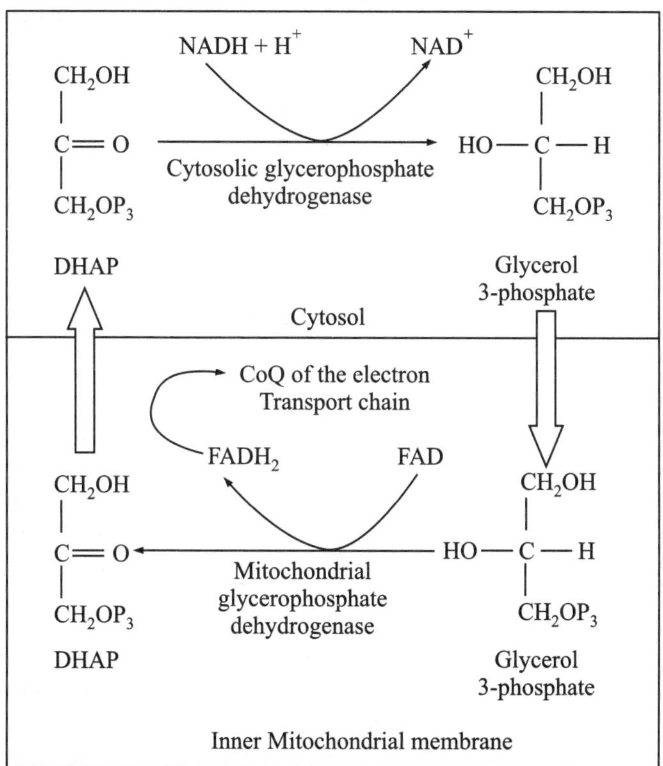

Figure 10.9a Glyceraldehyde-3-phosphate shuttle.

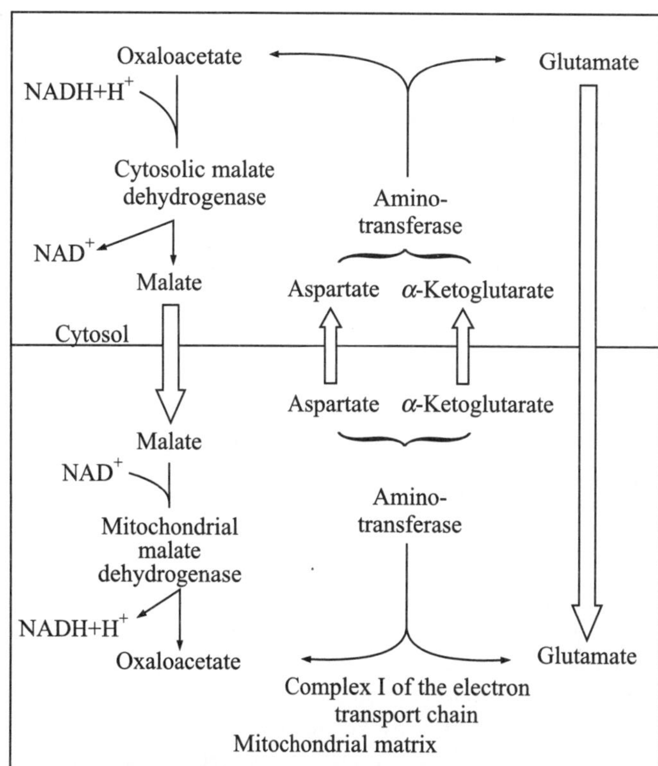

Figure 10.9b Malate-aspartate shuttle.

an isoenzyme of glyceraldehydes-3-phosphate dehydrogenase which gives its reducing equivalents which were transported through it from cytosol to FAD in mitochondria converting it to FADH2. This would generate 1.5 molecules of ATP through oxidative phosphorylation. This shuttle is present in skeletal muscles and brain.

The malate-aspartate shuttle transfers reducing equivalents from cytosolic NADH to oxaloacetate converting it to malate by the enzyme Malate dehydrogenase. Malate permeates through the inner mitochondrial membrane, delivers its reducing equivalents to mitochondrial NAD^+ by an isoenzyme of the enzyme Malate dehydrogenase that converts malate back to oxaloacetate. NADH formed is taken up by complex I. The oxaloacetate which can not traverse through the inner mitochondrial membrane is converted to aspartate by transamination. Glutamate and aspartate from mitochondria come out to cytosol and oxaloacetate is reformed by cytosolic transaminase. This shuttle operates in liver, kidney and heart. This shuttle accounts for 2.5 molecules of ATP per molecule of NADH formed in cytosol.

The total number of ATP molecules that can be generated by a molecule of glucose may either be 30 or 32 depending upon the shuttle that transfers reducing equivalents of NADH to mitochondria.

V. ENERGY-LINKED FUNCTIONS

The function of the membrane-localized energy conservation of mitochondria appears to be ordained not just to make ATP,

although it is the most important of all, but also to carry out mechanical and osmotic work. The following processes are driven directly by the energy of the high energy intermediates, represented by the squiggle (~), or through them, using ATP (Figure 10.10).

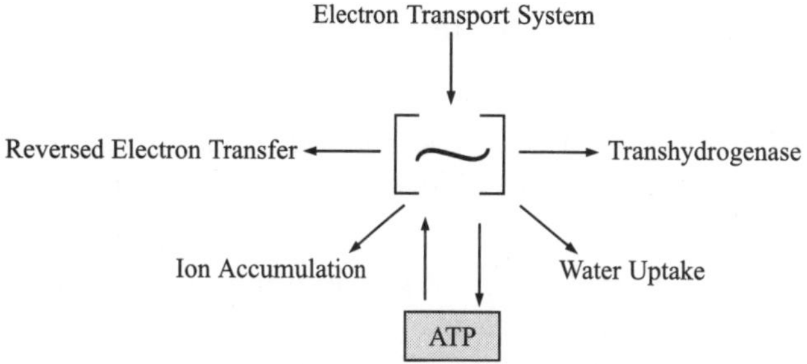

Figure 10.10 Energy-linked functions in mitochondria.

1. Activation of structural changes of the membranes leading to uptake of water—a process called "swelling".
2. Ion uptake and accumulation in coupled translocation.
3. Extrusion of water, also called "contraction", using ATP.
4. Reversal of electron transfer against thermodynamic gradient between NAD^+ and succinate, using ATP.
5. Transhydrogenase, between NADPH and NAD^+, using ATP.

VI. OXIDASES

Quantitatively, terminal electron transport via the cytochrome system accounts for most of the oxygen consumed. In this process the oxygen is fixed in the form of H_2O. Direct utilization of molecular oxygen also occurs in some reactions which are important in detoxication, catabolism, and production of some hormones. These are "true oxidases" and are also called "oxygenases".

A. Cleavage of Aromatic Ring

Oxygenase reaction is employed in the cleavage of aromatic compounds. One of the double bonds is broken, two atoms of the oxygen molecule are added on to the two carbons, and the compound then undergoes further metabolic transformation. Examples of this type are homogentisate oxidase, 3-hydroxyanthranilate oxygenase, and tryptophan pyrrolase. These enzymes are either hemoproteins or require iron in some form for their activity.

B. Mixed-function Oxidases

As the name indicates there are two functions carried out by these enzymes—one of reduction of atomic oxygen to H_2O, and the other, its transfer to an acceptor substrate. These enzymes require both molecular oxygen and a reducing agent, usually NADPH. A variety of important functions are carried out by these enzyme systems and the products are vital to the organism. Deprivation of oxygen is manifested primarily through effects on these enzymes. Since "hydroxyl groups" are added to the substrates they are also known as "hydroxylases". Some examples and their importance are given in Table 10.3.

C. Hydrogen Peroxide Producing Oxidases

These reduce oxygen to H_2O_2 using two reducing equivalents (2H + 2e). Different prosthetic groups are made use of, e.g., Fe, Mo, Cu, NAD^+, $NADP^+$, FMN, and FAD, and in some cases,

combinations of the above. The examples are—glucose oxidase, amine oxidases, xanthine oxidase, and ascorbate oxidase.

Knowledge of the generation of H_2O_2 in cellular oxidations has existed for many years. It has been assumed that H_2O_2 is toxic to cells and the presence of catalase is indicative of a detoxication mechanism. Other radicals of oxygen were recently recognized to be more potent destructive agents of biological material than H_2O_2. Also catalase and other peroxidases utilize H_2O_2 in some cellular oxidation processes leading to several important metabolites. Thus, the generation of H_2O_2 in cellular processes seems to be purposeful and H_2O_2 can not be dismissed as a mere undesirable byproduct. The capacity for generation of H_2O_2 is now found to be widespread in a variety of organisms and in the organelles of the cells. The reduction of oxygen to H_2O by mitochondrial cytochrome oxidase being the predominant oxygen-utilizing reaction had over-shadowed the importance of the quantitatively minor pathways. Under aerobic conditions generation of H_2O_2 by a variety of biomembranes has now been found to be a physiological event interlinked with phenomena such as phagocytosis, transport processes and thermogenesis in some as yet unidentified way. The underlying mechanisms of these processes seem to involve generation and utilization of H_2O_2 in mitochondria, microsomes, peroxisomes or plasma membranes. Cellular H_2O_2 generation is a result of combination of a number of membrane and soluble enzyme systems. In the rat liver it had been estimated that 10% of oxygen uptake accounted for H_2O_2 generation. Invariably, the capacity for generation of H_2O_2 is accompanied by the presence of highly active H_2O_2—scavenging systems, catalase and peroxidases. It is increasingly being realized that this complex metabolic capability has functions to perform in the cellular activities. A number of cellular phenomena are now recognized in which generation or utilization of H_2O_2 seems to play a central role. Among these are the respiratory burst in phagocytosis and in fertilization, maturation of protozoal parasites, hormonal response of cell membranes and transport

TABLE 10.3   Some Hydroxylases and Their Metabolic Importance

Enzyme	Reaction	Cofactor or prosthetic group	Importance
Tyrosinase	hydroxylation of tyrosine to dihydroxy compound followed by oxidation and polymerization to melaninpigments	Copper	Pigmentation, lack of it or masking leads to leucodema
Phenylalanine hydroxylase	phenylalanine $\rightarrow$ tyrosine	tetrahydropterine	lack of it leads to phenylketonuria
Tyrosine hydroxylase	tyrosine $\rightarrow$ DOPA	tetrahydropterine	formation of noradrenaline in adrenals and brain
Tryptophan hydroxylase	tryptophan $\rightarrow$ 5-hydroxytryptophan	tetrahydropterine	rate-limiting oxygen dependent step in the biosynthesis of serotonin
Aromatic hydroxylase (microsomal enzyme)	hydroxylation of aromatic compounds	heme	detoxication
Steroid hydroxylases	specific introduction of OH groups at different positions	cytochrome P-450	production of a number of steroid hormones

processes, cyanide-resistant respiration in plants, initiation of lignification, alternative oxidations and thermo-genesis. Studies on stimuli for generation and utilization of H_2O_2 in a variety of associated cellular phenomena are revealing a significant physiological role for H_2O_2.

D. Peroxidases

Peroxidases utilise H_2O_2 as the oxidant instead of oxygen. These are hemoproteins. The reducing equivalents are transferred from the substrate to H_2O_2, reducing it to H_2O:

$$AH_2 + H_2O_2 \rightarrow A + 2H_2O$$

Catalase is a hydroperoxidase with one molecule of H_2O_2 as substrate and another acting as oxidant:

$$H_2O_2 + H_2O_2 \rightarrow 2H_2O + O_2$$

Catalase also function as a peroxidase physiologically and oxidizes some phenols and alcohols. Oxidation of glutathione by glutathione peroxidase is used extensively by cells to detoxify H_2O_2 generated by oxidases.

SUMMARY

- Energy for all biological processes is derived from carbohydrates and fats, and to a lesser extent from proteins by oxidation of the carbon and hydrogen skeleton of these compounds to CO_2 and H_2O.
- Biological oxidations involve removal of reducing equivqlents (Hydride, hydrogen atoms, electrons as well addition of oxygen).
- About half of the liberated energy is conserved in the form of ATP, the energy currency of the living cell.
- Ninety percent of ATP in an animal cell is generated in mitochondria by oxidative phosphorylation.
- The reducing equivalents from carbohydrates, proteins and fats are transferred to $NADH^+H^+$ and $FADH_2$ or $FMNH_2$. Electrons from these compounds are transported to molecular oxygen through a number of carriers organized sequentially in the order of increasing 'Redox potential' In four multienzyme complexes (ComplexI, II, III, IV) present in the inner mitochondrial membrane. Chain of electron carriers upto molecular oxygen compose electron transport chain or respiratory chain. Coupling of electron transport with ATP formation is known as oxidative phosphorylation. The carrier nearest O_2 (at the end of the chain) gives up its electrons first, the second carrier from the end is oxidized next, and so on.
- The transfer of a pair of electrons from NADH or $FADH_2$ to molecular oxygen is a highly exergonic process.

- The chemiosmotic hypothesis by Mitchell suggests that much of this energy is conserved in the generation of proton motive force. The energy of this proton motive force is coupled with ATP synthesis by the enzyme FO/F1 ATP synthase a multisubunit protein in the mitochondrial inner membrane.
- Paul Boyer suggested a binding chain and rotational catalysis mechanism for ATP synthesis.
- The ratio of ATP synthesized per % O_2 reduced to H_2O (the P/O ratio) is about 2.5 when electrons enter the respiratory chain at Complex I, and 1.5 when electrons enter at CoQ.
- Oxidative phosphorylation generates some free radicals whenever the reduction of oxygen is incomplete.
- NADH equivalents from cytosol moved in the mitochondria by the malate-aspartate shuttle enter the respiratory chain at Complex I and yield a P/0 ratio of 2.5; those moved in by the glycerol 3-phosphate shuttle enter at CoQ and give a P/0 ratio of 1.5.
- The outer mitochodrial membrane is permeable to most small molecules and ions. The inner membrane is intrinsically impermeable to nearly all ions and polar molecules. A large family of transporters shuttles metabolites such as ATP, pyruvate, and citrate across the inner mitochondrial membrane.
- Oxidative phosphorylation is regulated by cellular energy demands. The intracellular [ADP] and the massaction ratio [ATP]/([ADP][Pi]) are measures of a cell's energy status. ATP and ADP concentrations set the rate of electron transfer through the respiratory chain via a series of interlocking controls on respiration, glycolysis, and the citric acid cycle.
- In ischemic (oxygen-deprived) cells, a protein inhibitor blocks ATP hydrolysis by the ATP synthase operating in reverse, preventing a drastic drop in [ATP].
- Site specific inhibitors of electron transport and oxidative phosphorylation helped unravelling the process. These are deadly poisons because they block cellular respiration. The chemical uncouplers like 2,4 Dinitrophenol, ionophores and biological uncouplers of oxidative phosphorylation like thermogenin, fatty acids, thyroid hormones generate heat because of electron transport without ATP synhesis.
- Although, energy conservation is the major task of oxidation-reduction reactions, other functional aspects of these reactions are detoxication reactions by mixed function oxidases, maintenance of free radical homeostasis, reductive biosynthesis and oxidative breakdown of different molecules generating important metabolic intermediates.

SUGGESTIONS FOR FURTHER READING

Boyer P.D. (1997), The ATP Synthase—A Splendid Molecular Machine, *Ann. Rev. Biochem.,* 66, 717–749.

Jeremy M. Berg, John L. and Tymoczko, Lubert Stryer (2010), *Biochemistry,* 7th ed., Freeman and Company.

Khan S. (1997), Rotary Chemiosmotic Machines, *Biochem. Biophys. Acta.,* 322, 86–105.

Mitchell P. (1966), *Biological Reviews,* 41, 445–502.

Nelson David L. and Cox Michael M. (2008), Lehninger *Principles of Biochemistry,* Vth ed., Freeman and Company.

Ramakrishna Kurup C.K. and Ramasarma T. (1989), *Biology Education,* b, 200–217. Senior, A.E. (1988), *Physiological Reviews,* 68, 177–230.

Robert Murray, David Bender, Kathleen M. Botham and Peter J. Kennelly (2012), a Harper's *Illustrated Biochemistry,* 29th ed., McGraw-Hill.

Singhal G.S. and Ramasarma T. (1991), *Trends in Bioenergetics and Biotechnological Processes,* Today & Tomorrow's Printers & Publishers, New Delhi.

Slater E.C. (1987), The Mechanism of the Conservation of Energy of Biological Oxidations, *Eur J. Biochem.,* 166, 489–504.

Wainio W.W. (1970), *Mammalian Mitochondrial Respiratory Chain,* Academic Press, New York.

11

Digestion and Absorption of Carbohydrates

L.M. Srivastava

CONTENTS

Carbohydrates comprise more than 60% of our daily intake of food. In the gastrointestinal tract the polysaccharide, oligosaccharide and disaccharide components of the food are hydrolyzed to hexose sugars before they are absorbed. The polysaccharide component of the diet is mainly starch, which occurs in nature in the form of soluble or insoluble granules, and is resistant to digestion. The cooking of starchy food materials helps in disrupting and partially solubilizing the starch granules and also removes adherent cellulose coating.

I. SALIVARY DIGESTION

In the mouth, saliva is the first secretion that comes in contact with the food. *Salivary amylase* begins the digestion of starch, glycogen and dextrins. The extent of digestion depends on the time of retention of food in the mouth for chewing. Generally the passage of food through the mouth is too rapid to allow any significant action of salivary amylase. If, however, boiled starch is mixed with human saliva at 37°C, the starch is converted to maltose, and lower dextrins in course of time and this reaction can be visualized by staining with iodine.

The optimum pH for the action of salivary amylase is between 6.6 to 6.9. The enzyme is activated by chloride ions. A number of isoenzymes of salivary amylase are known.

II. GASTRIC DIGESTION

The swallowed food material mixed with saliva passes into the stomach and comes in contact with gastric juice. The digestion by salivary amylase continues for a while till the HCl of the stomach penetrates the food material and inactivates the salivary amylase by lowering the pH.

The *gastric secretions* of higher animals *do not contain* any carbohydrate splitting enzymes. However, the rumen or first stomach of the ruminants contains microorganisms that secrete cellulase for the digestion of cellulose. The gastrointestinal tract of man does not contain any enzyme which can break β-1,4 link of glucose units present in cellulose and, therefore, its digestion does not occur. This is the reason that humans cannot obtain energy from fodder or grass, in contrast to ruminants, who can.

III. DIGESTION IN THE SMALL INTESTINE

Once the chyme enters the small intestines, the pancreatic juice and the intestinal enzymes are competent to complete the digestion of carbohydrates.

A. Digestion of Starch by Pancreatic Amylase

Most of the digestion of starch by the action of pancreatic amylase takes place inside the intestinal lumen. Pancreatic amylase, just as the salivary enzyme, is an α-amylase. The substrate for the action of α-amylase is starch amylose and or amylopectin.

The soluble starch or amylose is a linear polymer made up of glucose units joined together via the fourth carbon of one glucose to the first carbon of the neighbouring glucose yielding thereby a α-1,4 linkage. α-amylase (both pancreatic and salivary) acts on the interior α-1,4 link, but will not attack the outermost linkages of the starch molecule. Due to this property the smallest oligosaccharide containing at least 4 glucose molecules joined through α-1,4 linkages, can be acted upon by α-amylase. The digested products of amylose by α-amylase are maltose and maltotriose.

The insoluble starch or amylopectin constitutes a major part of the starch intake. It is usually of a very high molecular weight (1,000,000 or more). Amylopectin is a branched chain polymer. In addition to α-1,4 linkage in the straight chain, there are also α-1,6 branching point linkages. α-amylase hydrolyzes only the interior α-1,4 glucose linkages of the amylopectin molecule. Amylase does not possess specificity for hydrolyzing the α-1,6 branching points. The amylase action is weaker on the straight chain α-1,4 linkages in the neighbourhood of branching points. It is therefore apparent that the action of α-amylases on amylopectin would yield maltose, maltotriose and branched lower dextrins, containing between 6 to 8 glucose units. This was indeed found to be the case by Whelan in 1960 by analysis of the products of hydrolysis of amylopectin under physiological conditions. As α-amylase does not act on the outermost α-1,4 linkage, glucose would normally not be formed by the amylase action. A diagrammatic representation of the action of α-amylase on the starch molecule is given in Figure 11.1.

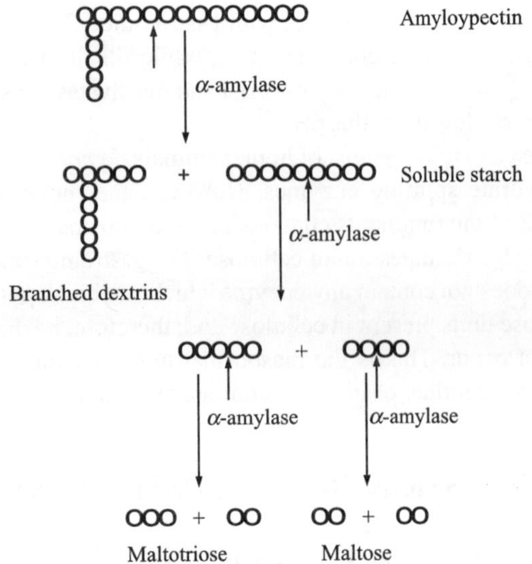

Figure 11.1 Diagrammatic representation of the action of α-amylase. Each circle represents a glucose moiety.

B. Characteristics of Pancreatic Amylase

The optimum pH for the action of pancreatic amylase is 7.1. The enzyme is activated by chloride ions. The enzyme is comprised of a number of isoenzymes. The pancreatic amylase is rather indistinguishable from the salivary amylase. The only known functional difference lies in the fact that while pancreatic amylase can act on native starch, the salivary amylase acts on cooked starch only.

C. Digestion of Disaccharides and Oligosaccharides

The final digestion of dietary disaccharides (sucrose and lactose) and the disaccharides and oligosaccharides obtained by the action of α-amylase on starch takes place in the small intestines. The enzymes responsible for the hydrolysis of these components are localized on the brush border membrane of the intestines. It is now clear that the hydrolysis of disaccharides and oligosaccharides takes place on the outer surface of the permeability barrier of the cell. The enzymes localized on the brush border membrane and helping in the digestion are as follows:

1. **Oligo 1,6-glucosidase (also called α-dextrinase).** This acts on α-dextrins (branched) and hydrolyses them to maltose and glucose. Isomaltose (which contains α-1,6 linkage) is also hydrolyzed by α-dextrinase.
2. **Major disaccharidases.** There are three major disaccharidases—

   (a) *Maltase* which catalyzes the hydrolysis of maltose and maltotriose to glucose. The pH optimum for the enzyme ranges between 5.8 to 6.2.
   (b) *Lactase* catalyzes the conversion of lactose to glucose and galactose; optimum pH for the enzyme is between 5.4 to 6.0, and
   (c) *Sucrase* (*Invertase*) converts sucrose to fructose and glucose. The pH optimum ranges between 5.0 to 7.0.

The individual disaccharidases mentioned above have not been characterized separately in a clearcut manner. Semenza has recently fractionated the extracts of intestinal mucosa by gel filtration and has isolated five distinct fractions having α-specific oligosaccharidases and two β-specific oligosaccharidases. These are as follows:

α-SPECIFIC OLIGOSACCHARIDASES:	
	Substrate attached
Maltase I Maltase II	Maltose (glucose and glucose, α-1,4 linkage)
Maltase III Maltase IV Maltase V	Surcose (glucose and fructose in α-1, β-2 linkage) and Maltose Isomaltose (glucose and glucose in α-1,6 linkage)
β-SPECIFIC OLIGOSACCHARIDASES:	
Lactase I Lactase II	Lactose (glucose and galactose in β-1,4 linkage)

The overall process of the digestion of starch and disaccharides to monosaccharides in the small intestine are depicted in Figure 11.2.

For further details on digestive processes the reader is referred to Chapter 51.

IV. ABSORPTION OF MONOSACCHARIDES

A. General Hypothesis

The final products obtained after the hydrolytic action of amylase followed by the intestinal disaccharidases and oligosaccharidases are the monosaccharides, namely glucose, galactose and fructose. Out of these, glucose accounts for about 80% of the total monosaccharides.

The rates of absorption of various monosaccharides differ considerably, even though the molecular structures of all the hexoses are almost the same and that of pentoses differ only slightly. Cori studied and reported the relative rates of the absorption of monosaccharides in rat intestine. Expressed on the basis of glucose absorption as 100, these are:

Glucose	100	Mannose	19
Galactose	110	Xylose	15
Fructose	43	Arabinose	9

It is apparent that hexose sugars are absorbed at a *faster* rate than pentose sugars. Also, glucose and galactose are absorbed more rapidly than other sugars. Furthermore it was found that hexose sugars can cross the intestinal barrier against an osmotic or concentration gradient. It is therefore obvious that the absorption of these sugars in the gastrointestinal tract cannot be explained by simple passive diffusion. Various hypotheses were put forward to explain the absorption of sugars.

The first hypothesis involving *phosphorylation/dephosphorylation* put forward by Verzar was based on the fact that the selective absorption of glucose and galactose was inhibited by the addition of iodoacetate or phlorizin in experimental systems used for studying absorption. The assumption was that these inhibitors inhibited hexokinase, the enzyme that phosphorylates glucose. This hypothesis proposed the phosphorylation of glucose at the luminal surface of the epithelial cell, the diffusion of glucose-6-phosphate through the cell, followed by a phosphatase-catalyzed dephosphorylation at the serosal border of the cell, and finally a release of glucose, thus formed, into the blood. However, it was subsequently found that iodoacetate and phlorizin were not the specific inhibitors of phosphorylation (hexokinase) reactions. While iodoacetate inhibits any—SH group containing enzyme including those of glycolytic sequence, the phlorizin inhibits glycolysis and oxidation. It, therefore, follows that both iodoacetate and phlorizin inhibit the reactions leading to the generation of ATP. The decrease in the normal absorption of glucose by iodoacetate and phlorizin was more likely due to inhibition of energy yielding processes. In spite of these drawbacks, this hypothesis has not been fully rejected.

B. Current Concept of the Absorption of Glucose and Galactose

The absorption of glucose and galactose is believed to take place by *active* transport. The actively transported sugars are transferred from mucosal to serosal sides against a concentration gradient. Crane *et al.* have reported the minimal structural requirements for sugars to be actively transported by intestinal epithelium. These are the presence of (a) hexose ring and (b) a hydroxyl group at position 2 of these sugars. These are represented as follows:

However it is still uncertain as to what the minimum structural requirements are for the sugars to be actively transported, since L-glucose which differs from D-glucose in the hydroxyl and hydrogen position at carbon 6, and D-xylose which has no sixth carbon, can also be actively transported, though at reduced rates.

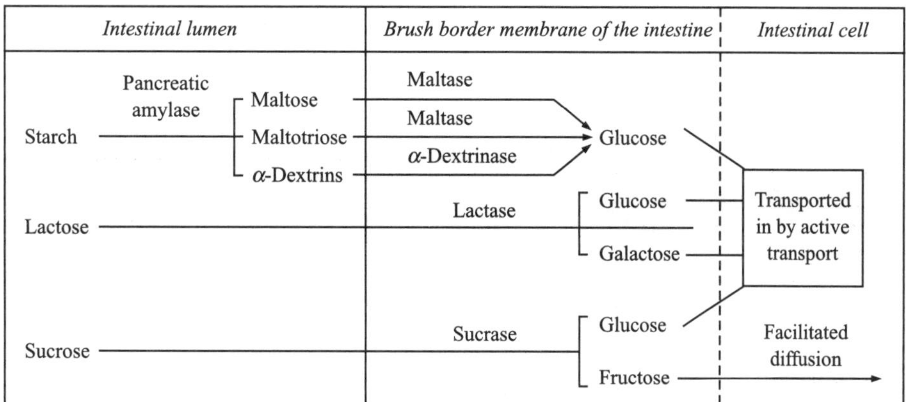

Figure 11.2 Digestion of starch and disaccharides to monosaccharides in small intestine.

For transport to take place, it has been suggested that a specific macromolecule known as *"Carrier"* is present on the outer surface of the brush border membrane of the intestine. The carrier has specific sites for the binding of glucose (and galactose and also Na$^+$). It has been found that the binding of Na$^+$ on the carrier is an absolute requirement for the process, which increases the affinity between glucose and the carrier. The carrier along with sugar and Na$^+$ moves across the brush border cell membrane and liberates glucose into the cytoplasm of the cell which then diffuses into the blood capillaries. The movement of the carrier with sugar and Na$^+$ across the membrane requires energy which is provided by ATP. Na$^+$, which is also liberated into the cell, is again actively transported from the cell. The active transport of sugars thus takes place in two steps:

(i) The movement of sugars against the concentration gradient (active transport) from the brush border membrane into the cell, leading to its accumulation into the cell, and

(ii) The movement of sugars from the cell to blood capillaries by passive diffusion.

The active transport of sugars by carrier mechanism is depicted diagrammatically in Figure 11.3.

C.   Characteristics of the Carrier and the Active Transport

Though the sugar "carrier" in mammalian systems has not been isolated, it is presumably proteinic in nature. The active transport process is rather rapid but it attains a saturation level analogous to enzymatic reactions at high substrate concentrations. In other words when the concentration of sugar (S) is increased on the mucosal border of the intestine, the amount of sugar transported (V) during a fixed time period keeps on increasing as a hyperbolic curve till a maximal rate that persists is obtained. This maximal rate of the transport can be termed as V_{max} and one can calculate $K_{transport}$ (K_t) which would be equal to the concentration of sugars at half the maximal rate of the transport. By knowing the K_t for various sugars one can visualize the affinity of that particular sugar for the carrier. The lower the K_t, the higher the affinity. The above description points towards the similarity in the active transport process by attachment of sugars with carrier to those of the enzyme-substrate reaction. The active transport follows the Michaelis-Menten equation represented as:

$$V = \frac{V_{max}(S)}{K_t + (S)}$$

Active transport requires energy which is provided by the hydrolysis of ATP. Hence, the active transport is completely inhibited by dinitrophenol or by removing oxygen, both of which block the ATP generating system.

It has further been found that both glucose and galactose are transported by the same carrier. When both these sugars are present together, they compete with each other for transport and lend to the hypothesis that they combine to the same carrier.

Some recent experiments have suggested the concept of a polyfunctional nature of the *"carrier"*. This visualizes essentially the involvement of the same carrier with different binding sites for the various groups of the actively transported substances like sugars, neutral, basic and acidic amino acids, and Na$^+$.

D. Absorption of Fructose

Fructose, one of the important sugars obtained after digestion, is transported in a different manner than glucose and galactose. It is known that absorption of fructose is more rapid than pentoses, but slower than glucose and galactose. Furthermore, fructose cannot be accumulated in the cell against concentration gradient, and also, the molecule is comparatively bigger to be carried by simple diffusion process. It is probable that fructose is transported across the brush border membrane by a Na$^+$-independent facilitated diffusion process involving another specific membrane associated protein, possibly glucose transporter (GLUT-5) which is present on the serosal side of the enterocyte and is thought to be responsible for moving glucose (galactose) from the enterocyte to the capillaries.

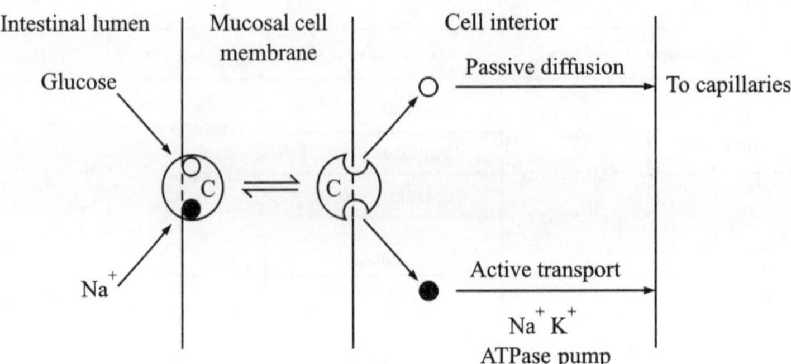

Figure 11.3   Schematic representation of the carrier-mediated transport of glucose. "C" represents the mobile carrier which possesses specific binding sites for glucose and Na$^+$. The carrier along with glucose and Na$^+$ moves from the luminal surface to the cytoplasmic surface of the cell membrane where it liberates both glucose and Na$^+$ into the cell interior. Na$^+$ is then pumped out of the cell actively by the Na$^+$K$^+$-ATPase pump and glucose passively diffuses into the capillaries.

LACTOSE INTOLERANCE

Lactose intolerance is a physiologic change resulting from acquired lactase deficiency. Lactase activity decreases with increasing age in children but this decline in activity is genetically predetermined and demonstrates ethnic variation. Lactase deficiency in the adult black population varies from 45–95%. If symptoms of malabsorption occur after the introduction of milk to adult diets, the diagnosis of acquired lactose deficiency should be considered. A diagnosis is made by challenging the small bowel with lactose and monitoring the rise in plasma glucose. A increase of more than 1.7 mmol/L (20 mg/dL) is diagnostic of lactose deficiency. A rise of 1.1–1.7 mmol/L (20–30 mg/dL) is inconclusive.

SUMMARY AND CONCLUSIONS

This chapter has dealt with the digestion of carbohydrates in the gastrointestinal tract leading to the formation of monosaccharides (Figure 11.4).

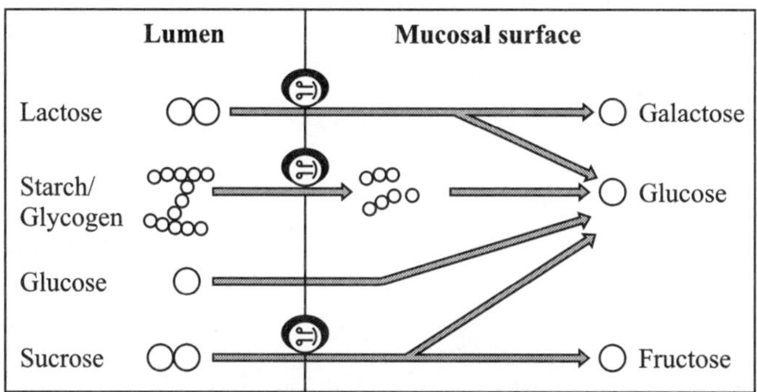

Figure 11.4 Luminal process (digestion).

Glucose, fructose and galactose are the primary monosaccharides resulting from the digestion of dietary carbohydrate. The absorption of these sugars and other minor monosaccharides is via specific carrier-mediated mechanisms (Figure 11.5), all of which demonstrate substrate specificity and stereospecificity, are capable of demonstrating saturation kinetics and can be inhibited specifically. In addition, all monosaccharides can cross the brush border membrane by a simple diffusion process, although this is extremely slow. At least two carrier-mediated transport mechanisms for monosaccharides exist—a Na^+-dependent co-transporter and a Na^+-independent transporter. At the brush border membrane both glucose and galactose are transported by the Na^+-dependent glucose transporter. This membrane linked protein binds with glucose (galactose) and Na^+ at separate sites and transports both into cytosol. The Na^+ is thus transported down its concentration gradient carrying glucose along against its concentration gradient. This transporter mechanism is linked to Na^+-dependent ATPase, which removes Na^+ from the cell in exchange for K^+, with the concomitant hydrolysis of ATP. The transport of glucose (galactose) is thus an indirect active process.

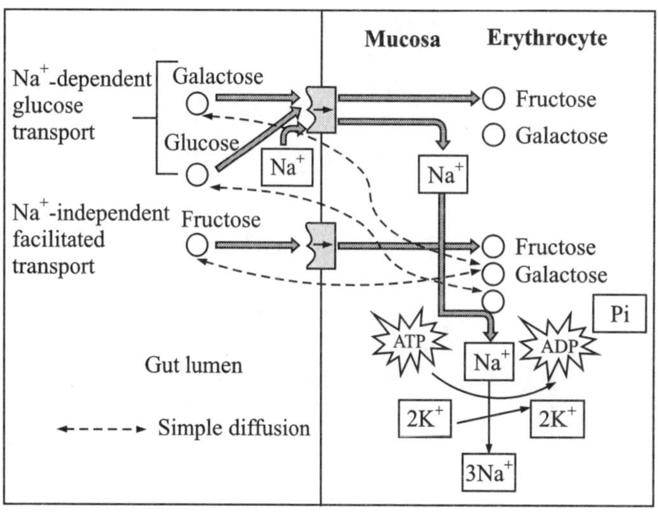

Figure 11.5 Enterocyte processing (absorption).

SUGGESTIONS FOR FURTHER READING

Ao Z., Quezada-Calvillo R., Sim L, Nichols B.L., Rose D.R., Strechi E.E. and Hamakcr B.R. (2007), Evidence of Native Starch Degradation with Human Small Intestinal Maltasc-Glucamylase (Recombinant), *FEBS Letters*, 581:2381–2388.

Bronner F. (1998), Calcium Absorption—Paradigm for Mineral Absorption, *J. of Nutrition*, 128:917–920.

Lentze M.J. (1995), Molecular and Cellular Aspects of Hydrolysis and Absorption, *Amer. Journal of Clinical Nutrition*, 61 (4):9465–9515.

Mwat A.M. (2003), Coeliac Disease—Meeting Point for Genetics, Immunology and Protein Chemistry Lancet, 361:1290–1292.

Quezada-Calvillo R., Robayo C.C. and Nichols B.L. (2006), Carbohydrate Digestion and Absorption, Stipanukon H. (Ed.)—Biochemical, Physiological, Molecular Aspects of Human Nutrition. 2nd ed., Saunders Elsevier, 168–199, Philadelphia.

Quezada-Calvillo, R and Nichols B.L. (2009), In: Digestion and Absorption of Carbohydrates, Pond W.G., Nichols B.L. and Brown, D.L. (Eds.), *Adequate Food for All*: *Culture, Science and Technology for the* 21st *Century*, F L: CRC Press, Boca Raton, 69–87.

Thompson A.B. and Wild G. (1997), Adaptation of Intestinal Nutrient Transport in Health and Disease, Part I Dig. Dis., Sci., 42, 453–469.

Tietz (2006), *Textbook of Clinical Chemistry and Molecular Diagnostics*, 4th ed., Burtis, C.A., Ashwood, E.R. and Browns, D.E. (Eds.), W.B. Saunders, An Imprint of Elsevier, Missouri, USA.

12

Metabolism of Glucose and Other Absorbed Sugars

Nibhriti Das

CONTENTS

I. INTRODUCTION

Dietary carbohydrates are hydrolysed into monosaccharides in the digestive tract and absorbed as simple sugars (Chapter 11). The principal sugar derived from most dietary carbohydrate items is glucose. In this chapter we will discuss the fate of glucose and other absorbed sugars in the body. Carbohydrates serve as the major constituent of our food. Thus, they serve as the main source of energy in the body. The carbon skeleton of a wide range of biomolecules and structural elements are contributed by dietary carbohydrates.

II. TRANSPORT AND DELIVERY TO THE TISSUES

Dietary carbohydrates are hydrolyzed into monosaccharides in the digestive tract which are absorbed as simple sugars. Glucose and other sugars like fructose, galactase are carried by the portal circulation to the liver from intestines. While glucose is channeled to different metabolic pathways in liver and other tissues, rest of the sugars are metabolized in liver and enter the pathways common to glucose metabolism. Therefore, under normal health, glucose is the major sugar in the circulating blood.

From the portal or systemic circulation, glucose is transported to different tissues by glucose transporters which may be Glut1, Glut2, Glut3 and Glut4 varying from tissue to tissue. Glut1 and Glut2 transport glucose to erythrocytes and liver, respectively. Glut3 transports glucose to brain. In brain, liver and red blood cells, glucose is readily taken up by cells from the extracellular compartment. The uptake is dependent on the concentration of glucose. The brain is an organ which is almost exclusively dependent on glucose for its energy needs. The fall of glucose below a threshold level affects the function of this organ markedly. If unchecked, hypoglycemic convulsions or coma can ensue. However, a number of regulatory mechanisms help to maintain the constancy of blood glucose levels. In many tissues, such as muscle and adipose, glucose is transported by Glut4. The expression of Glut4 on the cell surface is insulin dependent. Hence, when the levels of insulin in blood falls low as in fasting, Type I Diabetes mellitus or in Type II Diabetes, where the effect of insulin is impaired, uptake of glucose by these tissues are highly affected. However, insulin dependency of Glut4 mediated glucose uptake is a clever strategy to spare glucose for brain when blood glucose concentrations are low.

III. FATE OF GLUCOSE IN CELLS

In liver, the two-third of glucose arriving from the portal blood is converted to glucose-6-phosphate. The purpose of this phosphorylation is three fold. Firstly, the glucose so entrapped cannot leak out of the cells. Secondly, this is the energetically priming step of glucose metabolism. Glucose-6-phosphate is the starting molecule for the utilization and storage of glucose. Finally, the phosphate entrapped in the molecule of glucose-6-phosphate contributes to the generation of ATP by substrate level phosporylation in the glycolytic pathway. Rest one-third of the incoming glucose is released to blood.

IV. PATHWAYS OF GLUCOSE METABOLISM

After being converted to glucose-6-phosphste, glucose is fluxed to different metabolic pathways. A major fate of glucose in the fed state is its storage as glycogen. The process is known as glycogenesis. During fasting or glucose deficient states, glucose is generated from stored glycogen as glucose-6-phosphate by glycogenolysis. While skeletal muscles utilize glucose so generated for its own energy needs, liver releases glucose to the blood because only liver and kidney have glucose-6-phosphatase to cleave off glucose-6-phosphate to free glucose. Glucose is catabolized by the glycolytic pathway which ends with the formation of pyruvate. Alternatively, it may enter the HMP shunt with the generation of ribose 5 phosphate. Glucose is also converted to UDP glucose and UDP glucuronic acid through glucuronide pathway. At the time of needs, glucose is synthesized de novo from non-carbohydrate sources by gluconeogenesis. All these pathways are tightly regulated.

Blood glucose level is maintained within a limited range at different stages of feeding and fasting. Maintenance of blood glucose levels is vital for life since brain essentially needs glucose for its energy metabolism.

Like brain, erythrocytes and some other tissues can not utilize fatty acids as fuel and hence glucose is indispensible for these tissues.

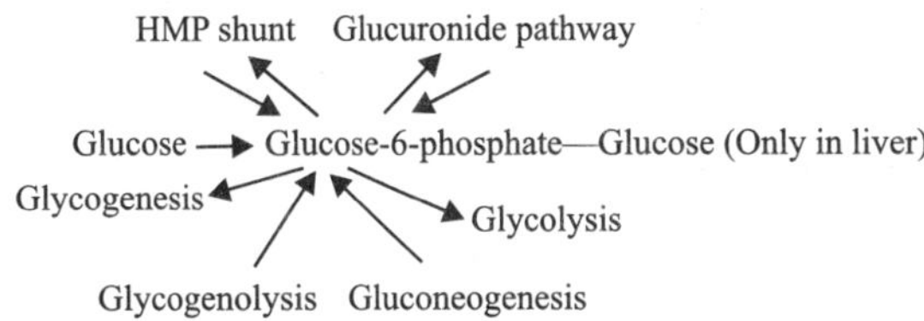

Figure 12.1   Major Pathways of glucose metabolism

A. Storage of Glucose as Glycogen (Glycogenesis)

Glycogenesis (genesis = generation) takes place in liver, muscle and in other tissues of the body. Glycogen is a branched chain polymer of glucose in which, glucose units are condensed to each other through α-1, 4 and α-1, 6 linkages. The bulk of glycogen is stored in liver and muscles.

In glycogenesis, glucose-6-phosphate is first converted to glucose 1 phosphate by the enzyme phosphoglucomutase and then to an activated derivative of glucose, the uridine diphosphoglucose (UDPG) by interaction with UTP. The enzyme is UDPG pyrophosphorylase. From UDPG glucose units are added to the existant glycogen molecule (the primer) with the formation of a linear chain of α-1, 4 glycosidic bonds. The enzyme catalyzing thie reaction is known as glycogen synthase. The formation of branched 1, 6-linkages is catalyzed by another enzyme called the branching enzyme (amylo, α-1–4) to (1–6) transglycosylase which cuts off a chain of 7 glucose residues from the non-reducing end of the elongating α-1–4-glycosidic chain to form the branched chain leaving at least two glucose molecules before and after the branch point of an at least eleven residue long α-1-4-glycosidic chain (Figure 12.2).

B. Breakdown of Glycogen (Glycogenolysis)

At the times of need, glycogen is hydrolysed back to glucose. This proceeds through four enzymes: (i) phosphorylase that breaks glycogen into glucose 1 phosphate, (ii) debranching enzyme, oligo (α-1-6) to (α-1-4) glucan transferase that cleaves the 1-6 bonds, (iii) mutase that converts glucose-1-phosphate into glucose-6-P, and (iv) gucose-6-phosphatase that hydrolyses the phosphate ester into glucose and inorganic phosphate. Glycogen phosphorylase acts repetitively on the non-reducing ends of the glycogen chains and stops at a point four residues away from the branching point. Now debranching enzyme cleaves further glucose residues off the shortened chain. It may be noted that the glycogenesis and glycogenolysis are irreversible pathways that require two different sets of enzymes.

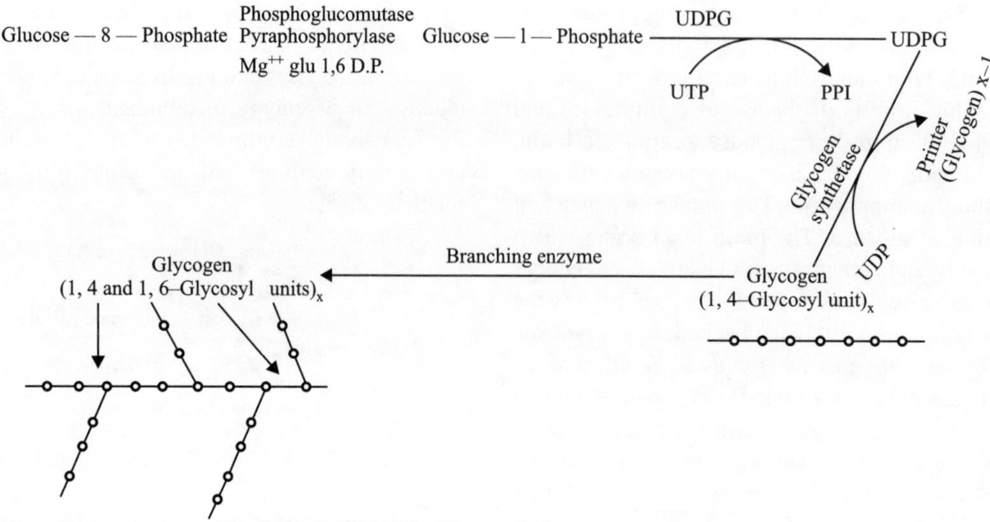

Figure 12.2 Pathway of glycogen synthesis.

C. Regulation of Glycogenesis and Glycogenolysis

Both glycogenesis and glycogenolysis are regulated by hormones by covalent modification of glycogen synthase and glycogen phosphorylase primarily by insulin, glucagon and epinephrine and also by allosteric modulation of these enzymes. Insulin promotes glycogenesis, whereas, glucagon and epinephrine promote glycogenolysis. Liver and skeletal muscles are the major organs where glycogen is stored. Although due to large mass, skeletal muscle stores more glycogen than the liver, the concentration of glycogen is higher in liver than the muscles. The purpose of glycogenesis in liver is to store glycogen for energy economy when blood glucose level is high and to break it down to release glucose to blood during scarcity. Skeletal muscles, however, store and mobilize glycogen to fulfil its own energy needs. Therefore, the regulation of glycogen metabolism in liver and muscle has some differences.

Glycogen synthase is present in an active dephospho and glucose-6-phosphate independent form (a/I form) and, in a much less active phosphorylated glucose-6-phosphate dependent form (b/D form). Conversely, glycogen phosphorylase is active in its phospho form ('a' form) and is inactive in its dephospho form ('b' form). Phosphorylase is normally present in a catalytically inactive form (phosphorylase b). It is converted into the active form (phosphorylase a) by limited phosphorylation of 4 serine residues present in the molecule. Levels of epinephrine and glucagon rise high when the level of blood glucose is low. On binding to the receptors on the plasma membrane, these hormones activate adenylate cyclase through the activation of G-Proteins. Cyclic AMP is generated which activates protein kinase by allosteric modulation. Activated protein kinase phosphorylates and activates phosphorylase kinase. Activated phosphorylase kinase ultimately phosphorylates the enzyme glycogen phosphorylase. Glycogen phosphorylase on activation, breaks down glycogen to glucose-l-phosphate.

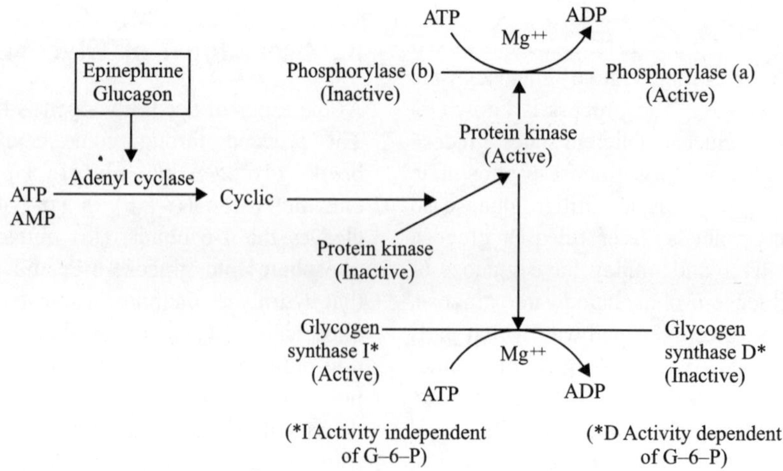

Figure 12.3 Co-ordinated regulation of glycogenesis and glycogenolysis.

Simultaneously, protein kinase phosphorylates glycogen synthase kinase 3 which in turn phosphorylates glycogen synthase and inactivates it. Dephosphorylation and activation of this enzyme is brought about by the enzyme protein phosphatase 1 which is activated by insulin. The level of insulin increases with the rise in blood glucose levels. When blood glucose level is low, consequently, level of insulin is also low. Under these conditions, protein phosphatase 1 remains in an inactive state as it is phosphorylated by protein kinase-A activated by glucagon and epinephrine through cyclic AMP.

Cyclic AMP thus, serves as a composite signal for stimulating glycogenolysis and inhibiting glycogen synthesis. Insulin depletes the levels of cyclic AMP by activating the enzyme phosphodiesterase which converts cyclic AMP to 5′AMP.

While liver is susceptible to the action of glucagon and epinephrine both, the skeletal muscles lack the receptors for glucagon and hence, glycogen mobilization in this tissue is effected only by epinephrine. Levels of epinephrine increases during heavy exercise when glycogen breakdown gets intensive to fulfil immediate energy needs of the exercising muscles.

AMP kinases activated by adiponectin promote glycogenolysis and inhibit glycogenesis. Adiponectin is a hormone released essentially by adipose tissue during fuel scarcity.

In liver, glucose-6-phosphate and glucose activate glycogen synthase and glucose inactivates glycogen phosphorylase by facilitating its dephosphorylation. In muscles, Ca^+ and AMP activate glycogen phosphorylase.

In the event of defective functioning of the enzymes related to glycogen metabolism, abnormal storage of glycogen results. This causes glycogen storage diseases.

V. UTILIZATION OF GLUCOSE FOR ENERGY NEEDS OF THE CELL

One mole of glucose on complete oxidation to CO_2 and H_2O liberates about 700 kilo calories. A substantial amount of this energy is conserved as ATP. Three pathways are involved in the catabolic degradation and oxidation of glucose.

(i) Glycolysis or breakdown of glucose to 3 carbon fragments (pyruvate and lactate). This pathway is operative in anaerobic as well as aerobic conditions and precedes the utilization of glucose by the oxidative pathway of tricarboxylic acid cycle.

(ii) Tricarboxylic acid (TCA) or Krebs citric acid cycle. This is a pathway that can oxidize completely the two carbon fragments formed by the breakdown of carbohydrates, fats and proteins. This aerobic pathway of reactions generates the major portion of energy obtainable from the metabolism of sugars, fats and proteins.

(iii) Alternate direct oxidative pathways—the hexose monophosphates shunt. This pathway is active in the utilization of glucose to variable extent in different tissues. It is important for production of some key metabolites (e.g. pentoses) and for generation of NADPH required for the synthesis of fats and steroids and detoxication of xenobiotics and protection of the cells especially the red blood cells against oxidative damages.

A. Glycolysis/Glycolytic Pathway/Embden-Meyerhoff Pathway

The step-wise degradation of one molecule of glucose to two molecules of pyruvate is known as glycolysis. This is the first metabolic pathway to be discovered and is named after the scientists who elucidated all the enzymatic steps in mammalian cells. In aerobic tissues and, in presence of oxygen and mitochondria, pyruvate is converted to acelyl co-enzyme A which enters TCA cycle. In absence of oxygen or mitochondria, pyruvate is converted to lactate. In erythrocytes, which do not have mitochondria and in exercising skeletal muscle where energy need is high and oxygen falls deficient, the end product of glycolysis is lactate.

Functions of glycolysis

Glycolysis is the central pathway of glucose metabolism, gives quick energy for exercising muscles and other tissues and, is the only source of energy for anaerobic tissues like erythrocytes. Although it is a catabolic pathway, it provides starting molecules for the synthesis of lipids and proteins. It provides feeder to HMP Shunt for skeletal muscles where oxidative phase of HMP shunt is deficient and generates 2,3-bisphosphoglycerate in erythrocytes to regulate the transport of O_2 to tissues by hemoglobin. Glycolysis prepares glucose for complete oxidation by converting it to pyruvate which after getting converted to acetyl coenzyme A gets completely oxidized through TCA cycle. In addition, sugars like galactose and fructose meet their final catabolic fate through glycolysis.

The enzymatic steps of glycolysis

Steps 1–5 constitute the preparatory phase of glycolysis where steps 1 and 3 consume ATP, steps 6–10 constitute the pay-off phase of the pathway as energy is gained through these steps.

Step 1 Conversion of glucose to glucose-6-phosphate

Glucose + ATP → glucose-6-phosphate + ADP

$$\Delta G'^o = -16.7 \text{ kJ/mol}$$

This first step of glucose utilization is energetically irreverstore and is catalyzed by the enzyme hexokinase. Hexokinase activity is present in the soluble fraction of the cytoplasm (cytosol) as well as in the particulate fraction composed of membranes and other organelles. The enzyme transfers the terminal phosphate residue of ATP to the hydroxyl group at aposition 6 of glucose. It requires Mg^{++}. There are four hexokinase isozymes which are different in their tissue distribution and kinetic properties (See regulation of glycolysis). As depicted in a previous section (Figure 12.1), glucose-6-phosphate can enter other pathways of glucose metabolism depending on the need of the cells. This first step of glucose metabolism is highly regulated.

Step 2 Conversion of glucose-6-phosphate to fructose-6-phosphate

$$\text{Glucose-6-phosphate} \longleftrightarrow \text{Fructose-6-phosphate}$$
$$\Delta G'^\circ = 1.7 \text{ kJ/mol}$$

This step is catalyzed by the enzyme phosphohexose isomerase and is freely reversible. The direction of reaction is decided by the concentrations of the reactant and the product. When blood glucose level is high, the reaction proceeds to the right hand direction.

Step 3 Conversion of fructose-6-phosphate to fructosel-6 bis phosphate

$$\text{Fructose-6-phosphate} + \text{ATP} \rightarrow \text{fructose-1,6-bisphosphate} + \text{ADP}$$
$$\Delta G'^\circ = -14.2 \text{ kJ/mol}$$

This is the second irreversible, rate limiting, highly regulated and, committed step of glycolysis. The enzyme is phosphofructokinase I, also known as phosphofructokinase. The large negative free enrgy change of this reaction drags the step 2 to the flow of glycolysis.

Step 4 Formation of trioses and their interconvrsion

$$\text{Fructose l,6-bisphosphate} \longleftrightarrow \text{Glyceraldehydes-3-phosphate} +$$
$$\text{dihydroxyacetonephosphate}$$
$$\Delta G'^\circ = 23.8 \text{ kJ/mol}$$

In this step, triose phosphates are formed by the cleavage of fructose 1,6-bisphosphate by the enzyme aldolase. This step is freely reversible.

Step 5

Trioses are interconvertible by the action of the enzyme Triose phosphate isomerase. Dihydroxy acetone phosphate is the starting molecule for the synthesis of triacylglycerols and phospholipids.

$$\text{glyceraldehydes 3 phosphate} \leftrightarrow \text{dihydroxyacetone phosphate}$$
$$\Delta G'^\circ = 7.5 \text{ kJ/mol}$$

Step 6 Oxidation of glyceraldehydes-3-phosphate

$$\text{Glyceraldehyde-3-phosphate} + \text{NAD}^+ + \text{Pi} \rightarrow$$
$$\text{l,3-bisphosphoglycerate} + \text{NADH} + \text{H}^+$$
$$\Delta G'^\circ = 6.3 \text{ kJ/mol}$$

Glyceraldehyde-3-phosphate is oxidized and phosphorylated to 1,3-bisphosphoglyceric acid. This step is catalysed by an oxidative enzyme glyceraldehydes-3-phosphate dehydrogenase, that utilizes NAD^+ as coenzyme. NAD^+ is reduced to NADH by the removal of 2 atoms of hydrogen from the substrate. The reaction would proceed as long as NAD^+ is available to accept the hydrogen. NAD^+, thus, needs to be regenerated from NADH. In case, oxygen is available (aerobic conditions), Electrons from each molecule of NADH in the form of hydride

ions are transported to the molecular oxygen through the electron transport chain with the regeneration of NAD^+ and generation 2.5 molecules of ATP. Under anaerobic conditions, NAD^+ essentially is regenerated by the conversion of pyruvate to lactate.

Glyceraldehyde-3-phosphate dehydrogenase is an enzyme in which –SH group is present at catalytic site. It is sensitive to inhibition by compounds like iodoacetate and p-chloromercuribenzoate. In the sequence of reactions, glyceraldehyde-3-phosphate first combines with the –SH moiety of the enzyme which is oxidized to thiol ester, while two hydrogens are removed and transferred to NAD^+. In the next step condensation of inorganic phosphate (orthophosphoric acid, H_3PO_4) takes place and the enzyme with the –SH group is liberated. The energy liberated during oxidation is retained in the form of a high energy-sulphur bond, which on phosphorylation becomes a high energy phosphate bond at position 1 of 1,3-bis-phospho-glycerate. Arsenate prevents incorporation of the inorganic phosphate by disallowing the formation of the high energy sulphur intermediate by forming As intermediate of glyceraldehydes-3-phosphate.

Step 7 Formation of ATP

The high energy phosphate of the 1,3-bis-phosphopho glyceric acid is used to generate ATP from ADP by the enzyme phosphoglycerate kinase (PGK). As one molecule of glucose yields 2 triose units, there is a yield of 2 ATP molecules at this step for each molecule of glucose undergoing anaerobic glycolysis. This is an example of *'substrate level phosphorylation'*. A high energy phosphate bond is generated without recourse to the electron transport chain and oxidative phosphorylation. These ATP molecules are not formed in Arsenate poisoning.

$$\text{1,3-bisphosphoglyceric acid} + \text{ADP} \rightarrow \text{3-phosphoglycerate}$$
$$+ \text{ATP}$$
$$\Delta G'^\circ = -18.5 \text{ kJ/mol}$$

Step 8 Conversion of 3-phosphoglycerate to 2-phosphoglycerate

The enzyme is phosphoglycerate mutase. $\Delta G'^\circ = 4.4 \text{ kJ/mol}$

Step 9

Rearrangement of the phosphate group in the phosphoglycerate molecule, followed by an "enolase" reaction in which one molecule of water is removed and phosphoenol pyruvate (PEP) is formed. $\Delta G'^\circ = 7.5 \text{ kJ/mol}$

COOH   COOH   COOH

H—C—OH   Mutase   H—C—O—P   Enolase   H—C—O~℗

H—C—O—P   H—C—OH   (Mg++, H_2O)   CH₂

H   H

3–Phosphoglycerate   2–Phosphoglycerate   Phosphoenol pyruvate (PEP)

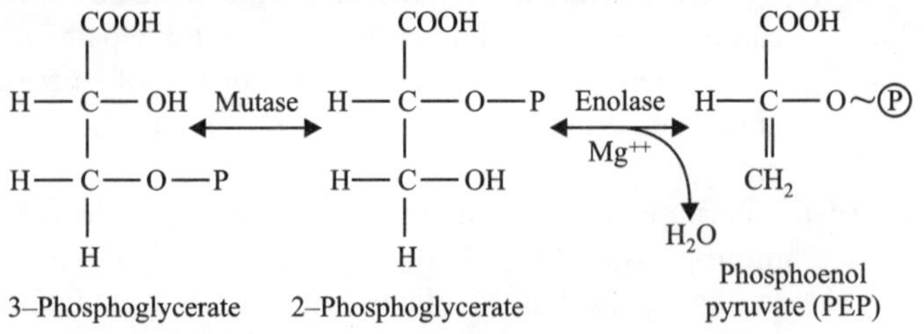

Step 10 Formation of pyruvate from PEP
This is the third irreversible step of glycolysis. PEP is a high energy compound and transfers its phosphate to ADP by an enzyme Pyruvate Kinase (PK). The dephosphorylated product is a Carboxyenol (Enolpyruvic Acid), which tautomerizes into keto-form, namely the pyruvate. This is again a substrate level phosphorylation. $\Delta G'^\circ = -31.4$ kJ/mol

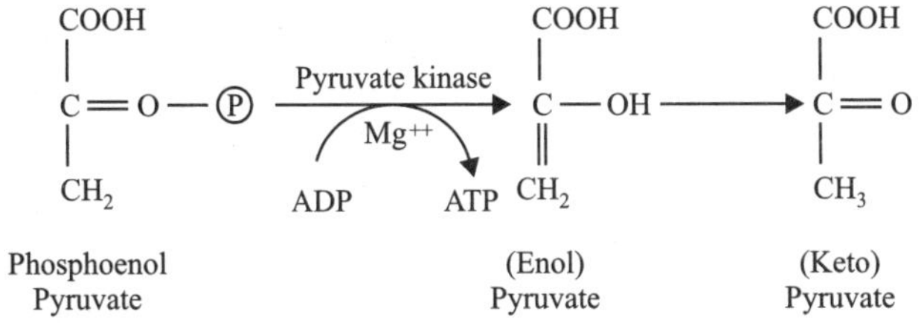

COOH—C=O—(P)—CH$_2$ Phosphoenol Pyruvate → (Pyruvate kinase, Mg^{++}, ADP → ATP) → COOH—C—OH=CH$_2$ (Enol) Pyruvate → COOH—C=O—CH$_3$ (Keto) Pyruvate

B.   Fate of Pyruvate

Pyruvate may get converted to lactate, acetyl coenzyme A, oxaloacetate and alanine. In yeast, pyruvate is converted to ethanol.

Conversion of pyruvate to lactate is an essential step under anaerobic conditions: When oxygen is less available as in exercising skeletal muscles and in tissues where mitochondria is absent like erythrocytes, the major fate of pyruvate is its conversion to lactate. This step is freely reversible and is catalyzed by the enzyme lactate dehydrogenase (LDH). in anaerobic tissues and exercising skeletal muscles glycolysis essentially needs this step for the regeneration of NAD$^+$ for the step catalyzed by glyceraldehyde-3-phosphate dehydrogenase. Lactate formed in these tissues is transported to liver by blood where it is reconverted to pyruvate and is metabolized predominantly through Cori cycle.

Under aerobic conditions pyruvate is converted to the two carbon fragment acetyl-CoA by an oxidative decarboxylation reaction catalysed by a multienzyme complex, the pyruvate dehydrogenase complex. This complex has at least three distinct enzyme activities, namely, pyruvate dehydrogenase, transacetylase and lipoate dehydrogenase which are inseparable as distinct independent enzymes and the intermediate formed by one enzyme is directly passed onto the second enzyme. The complete multienzyme system has a molecular weight of 4 million daltons. A number of coenzymes are utilized in the reaction. These are thiamine (vitamin B1) pyrophosphate (TPP), lipoic acid, coenzyme A, FAD, and NAD$^+$. The sum total is the formation of acetate in an "activated" state, e.g., as acetyl-CoA. NAD$^+$ is reduced to NADH + H$^+$, the hydrogen

and electrons are carried via the electron transport chain in mitochondria yielding 2.5 molecules of ATP. The reaction is essentially irreversible. Acetyl-CoA is an important intermediate and is utilized as a building block in the synthesis of fatty acids, cholesterol and ketone bodies. This enzyme is inhibited by NADH, ATP and acetyl coenzyme-A.

Conversion of pyruvate to oxaloacetate is catalyzed by the enzyme pyruvate carboxylase. Oxaloacetate is the starting molecule for gluconeogenesis and also TCA cycle. Oxaloacetate can also be coverted to aspartate by aspartate transaminase and to malate by malate dehydrogenase. Insulin represses the expression of this enzyme. Acetyl coenzymeA which is generated not only from pyruvate but also from fatty acids especially when blood glucose falls low, is an essential cofactor for this enzyme. Fine regulation of pyruvate dehydrogenase, pyruvate carboxylase along with that of TCA cycle decides the fate of pyruvate in aerobic tissues.

Pyruvate and alanine are interconvertible by the transaminase enzyme alanine aminotransferase.

C.   Lactic Acid Cycle (Cori Cycle)

During intensive exercise, when the rate of supply of oxygen to skeletal muscles is insufficient to generate the required ATP by aerobic pathways, there is an accumulation of lactate as a result of enhanced anaerobic glycolysis. In the resting phase, recovery takes place. About 1/5th of the accumulated lactate is metabolized in the muscles to CO_2 and H_2O through the citric acid cycle after it is converted back to pyruvate. The bulk of lactate however diffuses out of cells and is transported through the blood to the liver, where it is reconverted to glucose and glycogen. The glucose can go to the blood, from which glycogen is reformed in the muscles. There is no direct conversion of lactate to glucose in the muscles.

D.   Glycolysis in Erythrocytes, the Rappaport-Lubering Cycle

In erythrocytes, a bypass from glycolysis produces 2,3-bisphosphoglycerate (2,3 BPG), the most important allosteric modulator of the oxygen carriage and transport by hemoglobin. This molecule is produced in equimolar ratio with hemoglobin and reduces the affinity of hemoglobin for oxygen. This facilitates release of oxygen to tissues when the oxygen tension is low. Erythrocyte contains two specific enzymes, the bisphosphate glycerate mutase which converts 1,3-bisphosphoglycerate to 2,3-bisphosphoglycerate and, another

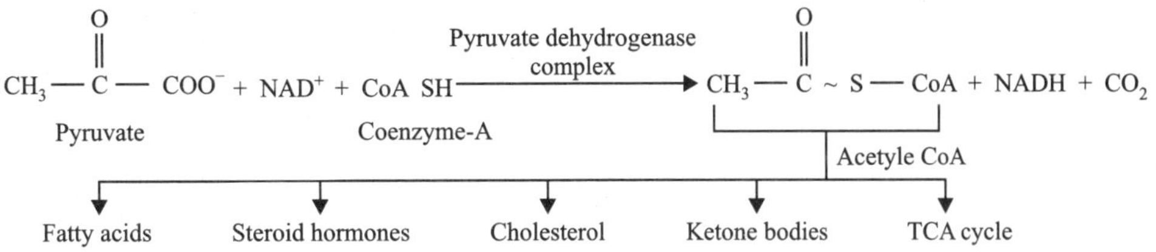

$$CH_3-\overset{O}{\overset{\|}{C}}-COO^- + NAD^+ + CoA\ SH \xrightarrow[\text{complex}]{\text{Pyruvate dehydrogenase}} CH_3-\overset{O}{\overset{\|}{C}} \sim S - CoA + NADH + CO_2$$

Pyruvate — Coenzyme-A — Acetyle CoA

Fatty acids — Steroid hormones — Cholesterol — Ketone bodies — TCA cycle

enzyme, the 2,3-bisphosphoglycerate phosphatase, that converts 2,3-bisphosphoglycerate to 3-phosphoglycerate for the continuance of glycolysis. However, generation of two molecules of ATP is compromized in Rappaport-Lubering cycle. 2,3-bisphosphoglycerate is a competitive inhibitor of bisphosphate glycerate mutase and hence, 2,3-bisphosphoglycerate is formed only in desired amount so the main glycolytic pathway as well as the Rappaport-Lubering cycle operate in desired rate. It is to be remembered that glycolysis is the only pathway for erythrocytes to get ATP needs of these cells.

E. Regulation of Glycolysis

Because glycolytic intermediates feed into several other pathways, the regulation of glycolysis occurs at more than one point. This allows the regulation of several pathways to be coordinated. Glycolysis is regulated at the level of step I catalysed by glucokinase/hexokinase, the third step catalyzed by phosphofructokinase and the last step catalyzed by pyruvate kinase. These enzymes are regulated by allosteric modulation, covalent modulation as well as by transcriptional regulation. In liver, insulin increases the rate of glycolysis, glucagon and epinephrine lowers the same. In skeletal muscles, although insulin increases the rate of glycolysis primarily by increasing the uptake of glucose by the muscle tissue, epinephrine also increases the rate of glycolysis by increasing the rate of glycogenolysis at the time of glucose scarcity or increased energy demands during heavy exercise. The muscle tissue lacks glucagon receptors and hence, glucagon has no effect on muscle tissue.

There are four isenzymes of hexokinase, namely, Hexokinase I, II, III and IV. Hexokinase IV is conventionally known as glucokinase. The enzyme is specific for glucose, gets half saturated at about 10 mM of glucose. The enzyme is predominantly present in liver, kidney and also in pancreas. The enzyme glucokinase is the ideal enzyme for glucose trapping in liver since the enzyme activity keeps on increasing with increasing amount of glucose coming to hepatocytes from the portal blood. The enzyme is activated by glucose and inhibited by Fructose-6-phosphate. This makes sense. The enzyme although is synthesized in endoplasmic reticulum, remains localized in nucleus of the hepatocytes in the inactive state bound to a regulatory protein. Transport of the enzyme to nucleus is mediated by fructose-6-phosphate which binds to it with high affinity. This happens when cell is deprived of glucose and there is a build-up of fructose-6-phosphate by gluconeogenesis. Fructose-6-phosphate thus, prevents the re-utilization of free glucose by hepatocytes so that it can be released to blood without instantly getting converted to glucose-6-phosphate.

When glucose concentration in the hepatocytes are high, glucose enters the nuclear pores, replaces fructose-6-phosphate, brings glucokinase back to cytosol to act on the incoming glucose in the hepatocytes. Although hepatocytes have hexokinase, glucokinase is the enzyme that entraps glucose in this tissue.

Hexokinase (Hexokinase I, II, III) is non-specific for glucose, can convert fructose and galactose to respective 6-phosphates but has very high affinity for glucose. Therefore, fructose and galactose in initial steps are metabolized by different enzymes, and later get merged to the glycolytic pathway. Hexokinase but, not the glucokinase is inhibited by its own product glucose-6-phosphate.

Hexokinase IV and Hexokinase II are transcriptionally upregulated by insulin. This takes care of efficient utilization of glucose by liver and skeletal muscles. Insulin simultaneously increases the density of GLUT4 on the plasma membranes of the skeletal muscles and other extrahepatic tissues like adipose tissue that enhances the uptake of glucose by these tissues when blood glucose level is high in the fed state.

Phosphofructokinase-1 (PFK1) is the key regulatory enzyme in glycolysis since the conversion of F6P to F1,6-bisphosphate is the step committed to glycolysis. The committed step is the one after which the substrate has only one way to go. The enzyme is also the rate-limiting enzyme of glycolysis and is under the allosteric control of ATP, citrate, AMP, ADP, F6P, G6P and F-2,6-bisphosphate (F2, 6BP). ATP serves both as a substrate and inhibitor of this enzyme and exemplifies homotropic allosteric inhibition. Binding of ATP at the catalytic site ensues the reaction. Availability of additional ATP ensures high energy state in the cells and its binding to the allosteric site of the enzyme slows down the reaction. ATP and citrate are negative modulators where as rest all are positive modulators. Fructose-2, 6-bisphosphate is the most effective activator of this enzyme. Negative modulation by ATP and, positive modulation by ADP and AMP have an important impact on determining the pace of glycolysis and on linking it up with the energy needs of the cell. When the cell is utilizing energy, ATP will be broken down to ADP. A decrease in the concentrations of ATP and an increase in ADP and AMP will activate PFK1 to increase the rate of glycolysis. Reverse is true when ATP is high and ADP and AMP concentrations are low. However, the functional role of glycolysis is not limited only to serve as a preparatory to ATP synthesis. Intermediates of glycolysis pathway are important for the synthesis of many other biomolecules such as glyceraldehydes-3-phosphate and dihydroxyacetone phosphate for the synthesis of triglycerides, pyruvate for alanine, acetyl-coenzyme A and oxaloacetate. Acetyl-coenzyme A is used for fatty acid and cholesterol biosynthesis. Thus, glycolysis has important anabolic functions and therefore, inhibition of PFK1 by ATP needs to be overcome at the time when active synthesis of these molecules are going on. Fructose-2,6-bisphosphate is a highly potent activator of glycolysis and can overcome the inhibitory effects of ATP. When fructose-2,6-bisphosphate binds to its allosteric site on PFK1, it increases the enzyme's affinity for its substrate F6P and reduces its affinity for ATP and citrate.

The cellular concentration of fructose-2,6-bisphosphate depends on the activity of phosphofructokinase 2 (PFK2), and fructose-2,6-bisphosphatase (FBPase2). Though these two are catalytically distinct entities, they are part of a single functional protein. PFK2 is covalently modulated by phosphorylation and dephosphorylation. Its kinase function is stimulated when it is in dephospho form and FBPase2 is activated in phospho form. Glucagon and epinephrine lead to phosphorylation of this enzyme through cyclic AMP dependent protein kinase cascade. Phosphorylation leads to reduction in the amounts of F-2,6-bisphosphate because of reduced activity of PFK2 and enhanced activity of F-2,6-bisphosphatase, which results in the diminished rate of glycolysis and enhanced gluconeogenesis in liver because of simultaneous activation of F-1,6-bisphosphatase which is under negative control of F-2,6-bisphosphate. Xylulose-5-phosphate, an intermediate of the HMP shunt keeps PFK2 in the dephosphoform by activating a protein phosphatase (PPA2), and, thus is an activator of PFK2 and increases the concentration of fructose-2,6-bisphosphate (Figure 12.4). Allosterically PFK2 is modulated the same way as PFK1 excepting that ATP does not modulate the activity of this enzyme.

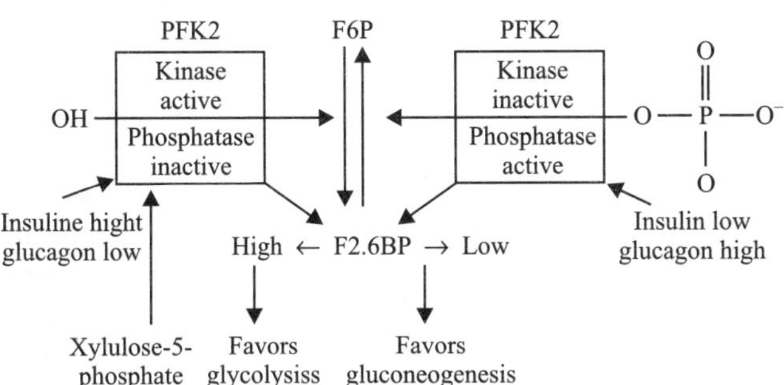

Figure 12.4 Regulation of PFK2 by insulin and glucagon and coordinated regulation of glycolysis and gluconeogenesis.

In skeletal muscles, FBPase is absent and hence, concentrations of F-1,6-bisphosphate is not affected much even if the concentration of F-2,6-bisphosphate is low as may be in the state of active muscle contraction or starvation when levels of epinephrine are high and the kinase activity of PFK2 is low as against the F-2,6-bisphosphatase activity. Further, stimulation of glycogenolysis by epinephrine during emergency need leads to the formation of high amounts of glucose 6 phosphate and then fructose-6 phosphate which is stimulatory to PFK1. Therefore, in skeletal muscles, epinephrine causes acceleration of glycolysis despite the inhibition of PFK2 by cyclic AMP dependent phosphorylation. The ATP concentration in skeletal muscle at rest is higher than liver. Hence on exercise larger amounts of AMP and ADP on PFK1 are formed which further accelerate the rate of glycolysis during active muscle contraction. Thus, even though PFK2 is present in skeletal muscles F-2,6-bisphosphate primarily is required for speeding up glycolysis in liver. Practically glycolysis in liver is not possible without the presence of fructose-2,6-bisphosphate.

Both in liver and muscles, activating role of F-2,6-bisphosphate on PFK1 can be minimized by simultaneous effects of ATP and citrate at high concentrations. High citrate concerntrations increase the inhibitory effects of ATP, and also serve as inter-cellular signal to indicate that the cells needs for energy yielding metabolism and for biosynthetic intermediates are met, and hence, glycolysis can proceed at a slower rate.

Transcriptionally, insulin upregulates both PFKI and PFKII. Xylulose-5- phosphate also upregulates the transcription of glycolytic enzymes by activating carbohydrate response element binding proteins (ChREBP) by dephosphorylating them and localizing them to nucleus to activate the relevant genes. Xylulose-5-phosphate also upregulates the enzymes for the synthesis of fatty acids. Presence of xylulose-5-phosphate indicates active HMP shunt and formation of NADPH, the condition that favors lipogenesis from glucose.

Rate of glycolysis is regulated also at the level of pyruvate kinase which is activated by fructose-1,6-bisphosphate allosterically in a feed forward manner. Further, it is also inhibited by ATP, alanine and acyl coenzyme A which has great implications in gluconeogenesis which takes place under the influence of reciprocal regulation of glycolysis. Whenever the cell has a high concentration of ATP or, whenever ample fuels are already available, glycolysis is inhibited by the slowed down action of pyruvate kinase. This enzyme also exists in an inactive phospho form and active dephospho form which is cyclic AMP dependent and hence is under hormonal control as discussed. The covalent modulation, however, is relevant only to hepatocytes as it is relevant to the co-ordinated regulation of glycolysis and gluconeogenesis.

The regulation of glycolysis, thus, is different in liver and skeletal muscles. Liver has the task of utilizing glucose optimally in all the different pathways and also to produce glucose by gluconeogenesis, where as muscle tissue primarily uses glucose for obtaining energy and partly to synthesize pentoses and to store the surplus glucose as glycogen. The co-ordinated regulation of glycolysis and gluconeogenesis is the specific feature of glucose homeostatsis in liver. Table 12.1 depicts the allosteric modulators of regulatory enzymes of glycolysis.

Table 12.1 Regulatory steps of glycolysis: Allosteric modulation

Enzyme	Positive modulator	Negative modulator
Hexokinase		Glucose 6 phosphate
Glucokinase	Glucose	Fructose 6 phosphate
Phospho Fructo-kinase	F6P, AMP, ADP, F2, 6 BP	ATP, citrate
Pyruvate Kinase	F1, 6 BP	ATP, Alanine, Acelyl, coenzyme A

F. Summary of Glycolysis

1. The pathway involves ten enzymatic steps
2. The 1st and the 3rd step from glucose consume one molecule of ATP each. In the 4th step a molecule of glucose is split off into two triose molecules. This is called as preparatory phase of glycolysis. The following step onwards is the pay-off phase of glycolysis where in the high energy phosphate intermediates 1-3 bisphospho glycerate and later, phosphoenol pyruvate are formed ultimately causing the generation of ATP molecules by substrate level phosphorylation.

3. The 7th step (PGK) and the the last step (PK) generate ATP by substrate level phosphorylation.
4. Net gain in energy directly from the pathway is 2 molecules of ATP.
5. 2 molecules of NADH are produced under aerobic condition. Each one can give 2.5 molecules of ATP. Glycolysis, thus, can account for 7ATP molecules.
6. 1st, 3rd and the last steps are regulatory steps.
7. The third step catalyzed by PFK1 is the committed and rate-limiting step and, hence, is the key regulatory step of glycolysis.
8. Under aerobic conditions/cells glycolysis ends with the formation of pyruvate, under anaerobic condition lactate is formed from pyruvate which is essential for the regeneration of NAD^+ for the continuation of glycolysis. In yeast, ethanol is formed.
9. Glycolytic intermediates are precursors for lipid and amino acid biosynthesis.
10. Regulation of glycolysis is co-ordinated with gluconeogenesis.

G. Understanding Glycolysis: Translational Implications in Chemotherapy and Diagnosis of Cancer

In many types of tumors found in humans and other animals, glucose uptake and glycolysis proceed about 10 times faster than in normal, noncancerous tissues. Most tumor cells grow under hypoxic conditions because, at least initially, they lack the capillary network to supply sufficient oxygen. Cancer cells located more than 100 to 200 μm from the nearest capillaries must depend on glycolysis alone for much of their ATP production. Tumor cells must take up much more glucose than do normal cells. In general, the more aggressive the tumor, the greater is its rate of glycolysis.

This increase in glycolysis is achieved at least in part by increased synthesis of glycolytic enzymes and of the insulin independent plasma membrane transporters GLUT1 and GLUT3 that carry glucose into cells. The hypoxia-inducible transcription factor (HIF-1) is a protein that acts at the level of mRNA synthesis to stimulate the production of at least eight glycolytic enzymes and the glucose transporters when oxygen supply is limited. Inhibitors of glycolysis might target and kill tumors by depleting their supply of ATP. Three inhibitors of hexokinase have shown promise as chemotherapeutic agents: 2-deoxyglycose, lonidamine, and 3-bromopyruvate. The high glycolytic rate in tumor cells also has diagnostic usefulness. The relative rates at which tissues take up glucose which can be detected by positron emission tomography (PET) can be used in some cases to pinpoint the location of tumors.

H. Tricarboxylic Acid (TCA) or Kreb's Citric Acid Cycle

TCA cycle is an amphibolic pathway that represents a series of reactions by which the two carbon unit, acetate, is fully oxidized to CO_2 and H_2O. There are four *oxidative reactions,* three requiring NAD^+ as coenzyme and one requiring FAD. There is also one *substrate level phosphorylation* during the conversion of α-ketoglutarate to succinate. The complete set of reactions is given in Figure 12.5. The enzymes and coenzymes bringing about these reactions and coupling them with formation of ATP bonds, are present in the mitochondria, the *"powerhouse"* of the cell.

In this process, three molecules of NADH and one molecule of $FADH_2$ are generated. Each NADH on reoxidation by electron transport chain in mitochondria provides 2.5 molecules of ATP and 1.5 molecules of ATP are generated from $FADH_2$. GTP readily gets converted to ATP. Thus TCA cycle accounts for 10 molecules of ATP although the cycle per se provides only one molecule of ATP ie from GTP.

Acetyl-CoA condenses with oxaloacetate to yield a 6-carbon tricarboxylic acid, citric acid. The latter is changed into isocitrate, which undergoes oxidative decarboxylation to alpha-ketoglutarate (5 carbon). Another step of decarboxylation and oxidation leads to the 4-carbon dicarboxylic acid, succinate. In these two reactions two carbons are lost as CO_2. Succinate is oxidized to fumarate which on hydroxylation gives malate. Finally malate is oxidized to oxaloacetate which can condense with a fresh molecule of acetyl-CoA to restart the cycle. The *energy yielding reactions* are the ones where dehydrogenation of metabolites followed by oxidation of H_2 to H_2O, takes place. As discussed in Chapter 10, it may be emphasized that in biological systems, energy is generated by the *oxidation of hydrogen* and not of carbon. *The production of* CO_2 *by decarboxylation reaction does not release energy for ATP molecules to be formed.*

Amphibolic role of TCA cycle

TCA cycle ensures complete catabolism of acetyl-coenzyme-A, pyruvate, and amino acids, the catabolic intermediates of carbohydrates, fat and proteins. The intermediates of TCA cycle can enter gluconeogenesis since they all get converted to oxaloacetate and malate. TCA cyle provides succinate for heme synthesis, oxaloacetate and alpha ketoglutarate for amino acid synthesis. It is also linked to urea cycle through fumarate and succinate. Citrate which can diffuse out of mitochondria provides acetyl coenzyme for fatty acid and cholesterol biosynthesis. Thus, TCA cycle is the merging point for all catabolic pathways as well as the starting point for all anabolic pathways and hence, is described as an amphibolic pathway.

Anaplerotic pathways/reactions for TCA cycle

Anaplerotic pathways/reactions are those which provide intermediates to a pathway although they themselves do not constitute the pathway to which they feed in the intermediates. While intermediates of TCA cycle can be removed for various pathways, there are reactions which can provide intermediates for continuance of TCA cycle. Oxaloacetate can come from pyruvate by the reaction of pyruvate carboxylase and from aspartate by its transamination; alpha ketoglutarate can come

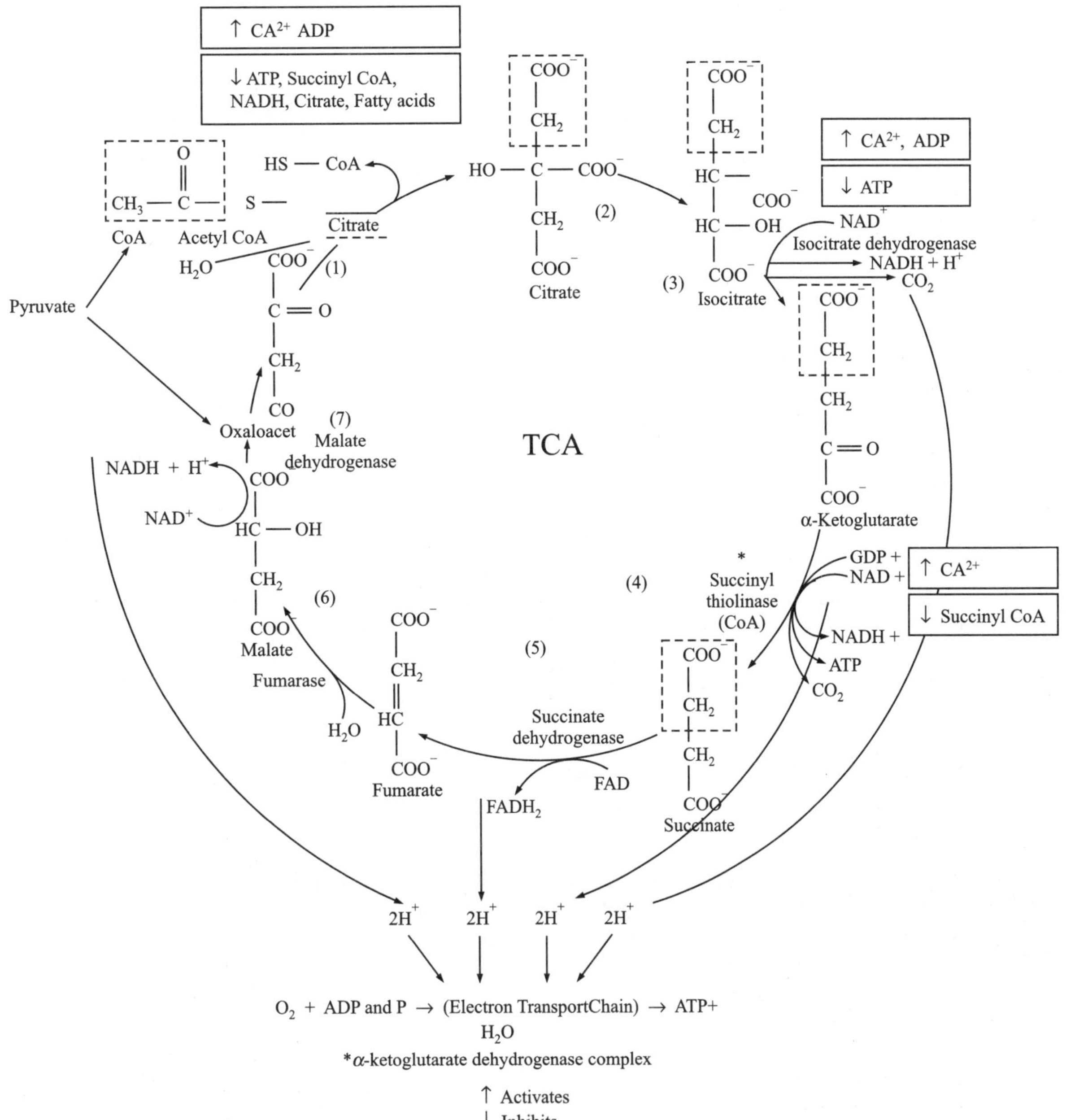

Figure 12.5 TCA cycle and its regulation.

from glutamate, succinate from arginosuccinate. These reactions keep TCA cycle continued under different circumstances. Acetyl coenzyme A comes from pyruvate, fatty acids, ketone bodies and amino acids. Since, pyruvate from glucose catabolism is the main source of oxaloacetate for continuance of the TCA cyle, while acetyl coenzyme A can come both from pyruvate and the fatty acids, it is sometimes stated that "fats burn in the flame of carbohydrates". Fluoroacetate is a suicide inhibitor of TCA cycle.

Regulation of TCA cycle

TCA cycle is regulated by NADH/NAD⁺ ratio, and ATP/ADP ratio, acetyl-CoA, fatty acyl-CoA and feed back by its intermediates at its irreversible steps (Figure 12.5).

I. Energetics of Glucose Catabolism

A total of 30 or 32 molecules of ATP can be formed on complete oxidation of glucose by glycolysis followed by TCA cycle and oxidative phosphorylation. Each molecule of NADH formed in glycolysis can give 2.5 or 1.5 molecules of ATP depending on the shuttle system that transports reducing equivalent from NADH either to NAD or FAD in mitochondria (Chapter 10). See the computation below:

Glycolysis: 2 ATP + 2NADH = 5 or 7ATP
Pyruvate to Acetyl coA: 2NADH = 5ATP
TCA Cycle: 6NADH + 2FAD + 2ATP = 20
Total = 30/32 ATP/molecule of glucose

J. Hexose Monophosphate (HMP) Shunt/ Pentose Phosphate Pathway

Glucose can be metabolised by a direct oxidative path in addition to the channel discussed above (glycolysis followed by TCA cycle). The preponderance of HMP shunt pathway varies from tissue to tissue. In most tissues, it accounts for only 10 to 30% of the total glucose utilized, but in some tissues like the adipose cells, adrenals, gonads, and lactating mammary glands, it may have a major importance. The overall result of this pathway is as follows:

$$C_6H_{12}O_6 + 7H_2O + 12NADP^+ \rightarrow 6CO_2 + 12NADPH + 12H^+ + ADP + Pi$$

The pathway is illustrated in Figure 12.6. G-6-P is diverted into the cycle by the NADP requiring enzyme G-6-P dehydrogenase. In the process NADPH + H$^+$ is formed, which has a role in the formation of fatty acids and steroids. Another step at which this reduced co-enzyme is formed in the cycle is the oxidative decarboxylation of 6-phosphogluconate into the pentose ribulose-5-phosphate. The remaining reactions of the cycle are group transfer reactions catalyzed by transketolase and transaldolase. Tramsketolase reaction needs thiamine pyrophosphate (TPP) as its cofactor. Levels of erythrocyte TPP serve as a marker of Vit. B1 (Thiamine) status in an individual. The direct oxidative pathway is differentiable from the glycolytic cycle by the fact that oxidation takes place early in the cycle and that CO$_2$ which is not a product in E-M pathway is a characteristic by-product of the shunt pathway. It is the carbon at position one of glucose that gets converted into CO$_2$ in this cycle. The contribution of this cycle can therefore be studied by using glucose labelled with radioactive carbon (C^{14}) or heavy isotopic oxygen (O^{18}) at position-1.

Importance of HMP pathway

This pathway is one of the sources of pentoses in human tissues. Another importance of the pathway is to make available the reduced coenzyme NADPH required for the synthesis of fatty acids and steroids. NADPH is essential for the maintenance of gluathione in the reduced state as a cofactor of glutathione reductase. Reduced glutathione is a safeguard against injurious effects of oxygen free radicals. In erythrocytes, glutathione reductase is the main enzyme that maintains the integrity of RBC membranes and also protects hemoglobin from getting converted to met-hemoglobin under the influence of oxidants and peroxides. People deficient in glucose-6-phosphate dehydrogenase which is the only source for the generation of NADPH for erythrocytes, may suffer from hemolytic crisis on exposure to oxidants as it happens with such patients if treated with drugs like premaquin, acetylphenyl hydrazine, etc.

Galactose inhibits glucose-6-phosphate dehydrogenase. It is possible that in galactosaemia, the formation of *cataract* may be due to intervention at this step, as HMP pathway is dominant in the metabolism of the *lens*.

HMP shunt is also an important route for metabolism of glucose in the mammary gland. Insulin may exercise its lipogenic action, at least partially, by stimulation of the shunt pathway enzymes.

VI. FORMATION OF GLUCOSE FROM NON-CARBOHYDRATE SOURCES: Gluconeogenesis

Glucose can be generated in cells from pyruvate, lactate, the glycerol moiety of fats, but not fatty acids, and some amino acids. The pathway is known as gluconeogenesis. There are

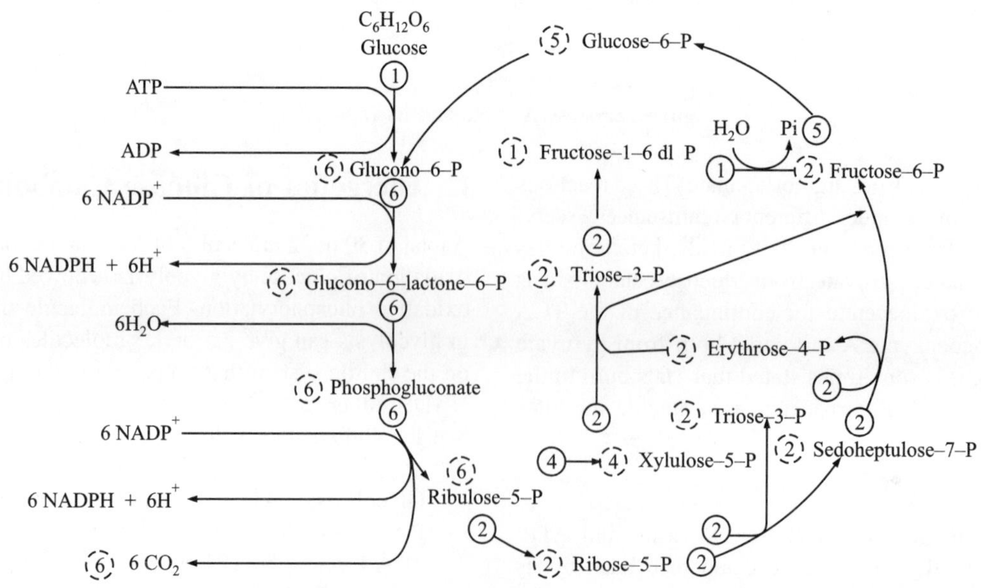

Figure 12.6 Hexose Monophosphate shunt pathway.

three irreversible steps in gluconeogenesis which separates it from glycolysis. The reversible steps and intermediates are common between glycolysis and gluconeogenesis. These two pathways move in opposite directions to each other but one at a time depending on the energy and fuel needs of the cells. Gluconeogenesis takes place predominantly in liver and also in kidneys to maintain blood glucose levels at the time of glucose scarcity. Following is an account of the steps specific for gluconeogenesis. The strategy is to start from the last and irreversible step of glycolysis and end at the formation of glucose-6-phosphate and then its conversion to free glucose.

(i) While pyruvate is the starting molecule for gluconeogenesis, the enzyme pyruvate kinase is irreversible. To by-pass this step, the first step of gluconeogenesis is the conversion of pyruvate to oxaloacetate with the fixation of one molecule of CO_2. This reaction is catalyzed by the enzyme pyruvate carboxylase. This reaction takes place in the mitochondria. The enzyme requires the vitamin biotin as a coenzyme. One molecule of ATP is utilized in the reaction. The oxaloacetate can be converted to phosphoenolpyruvate (PEP) by the enzyme phosphoenol pyruvate carboxykinase (Figure 12.7). Oxaloacetate, however, cannot crossover from the mitochondria to the cytoplasm. The transfer is achieved by the reduction of oxaloacetate to malate, which permeates through the mitochondria to the cytoplasm. It is reoxidized to oxaloacetate in the cytoplasm. These reactions are known as the *malate* or *dicarboxylic acid shuttle.*

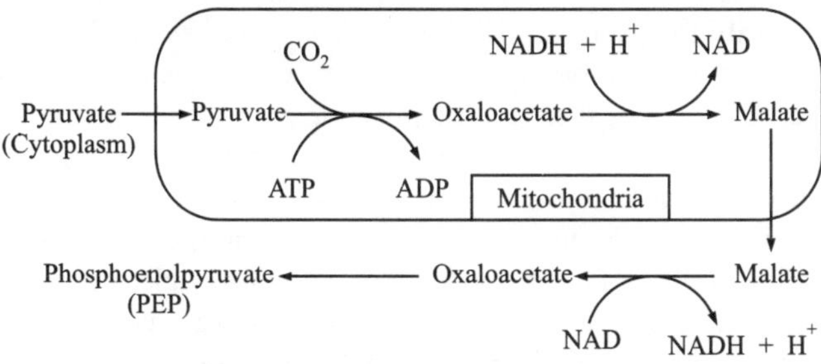

Figure 12.7　Oxaloacetate-malate shuttle.

(ii) PEP is then converted to fructose 1,6-bisphosphate (FDP) by the reversible glycolytic pathway reactions. Its further conversion to fructose-6-phosphate (F-6-P) is brought about by the enzyme Fructose-1,6-bisphosphatase.

(iii) The third nonreversible reaction is the conversion of glucose-6-phosphate (G-6-P) to glucose. This is catalysed by G-6-P phosphatase.

$$\text{Phosphoenolpyruvate} \rightleftharpoons 2PG \rightleftharpoons 3PG \rightleftharpoons \text{G–3–P}$$
$$\Updownarrow$$
$$\text{Glucose} \longleftarrow \text{G6P} \rightleftharpoons \text{F6P} \longleftarrow \text{FDP}$$

A. Futile Cycles in Glycolysis and Gluconeogenesis

These cycles, although energetically futile, are pivotal in the coordinated regulation of glycolysis, gluconeogenesis and blood glucose levels (discussed below).

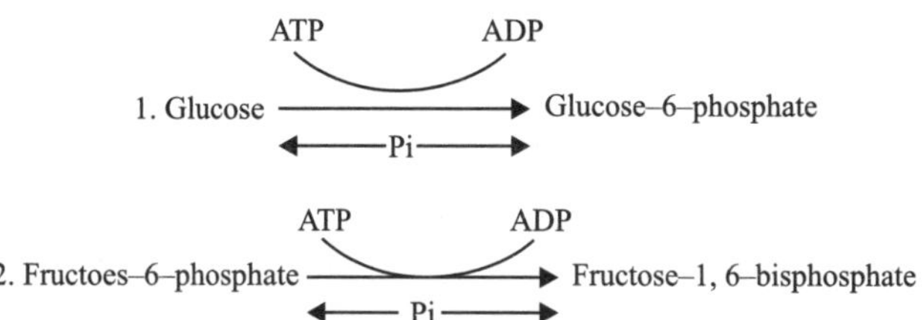

B. Regulation of Gluconeogenesis

The basic strategy in the regulation of gluconeogenesis is the reciprocal regulation of glycolysis since most of the intermediates and enzymes are common between these two pathways. The key allosteric regulation of gluconeogenesis is at the level of phospho-fructokinase I. The co-ordinated regulation of PFKI and FBPase by allosteric modulator is depicted in the Figure 12.8a.

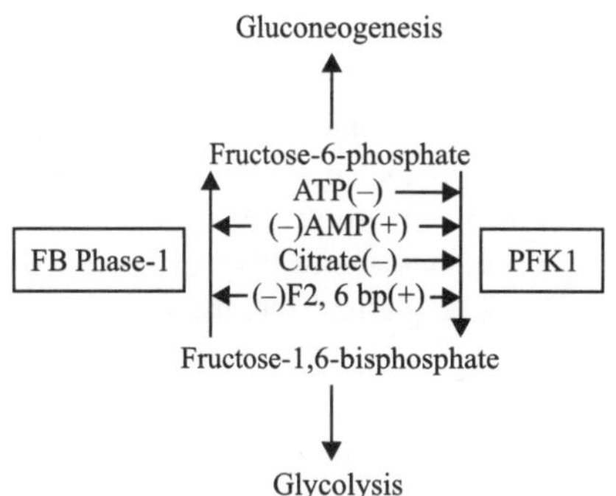

Figure 12.8a　Co-ordinated regulation of glycolysis and gluconeogenesis.

AMP and fructose 2,6-bis phosphate are positive modulators of phospho-fructokinase I and negative modulators of the fructose 1,6-bisphosphatase-1, respectively. The formation of fructose 2,6-bisphosphate is determined by insulin/glucagon ratio and the energy charge. Low blood glucose levels cause increased release of glucagon, epinephrine and glucocorticoids with a fall in the levels of insulin. This inactivates phosphofructokinase II causing a decline in the levels of fructose 2,6-bisphosphate. Consequently, fructose 1,6-bisphosphatase is activated and inhibition of phosphofructokinase take place (Figure 12.8a). This leads to the accumulation of fructose-6-phosphate followed by accumulation of glucose-6-phosphate in the hepatic cells which now enters the endoplasmic reticulum from the cytosol along with the concentration gradient. Due to increased influx of glucose-6-phosphate in the endoplasmic reticulum, glucose-6-phosphatase becomes more active and free glucose is formed.

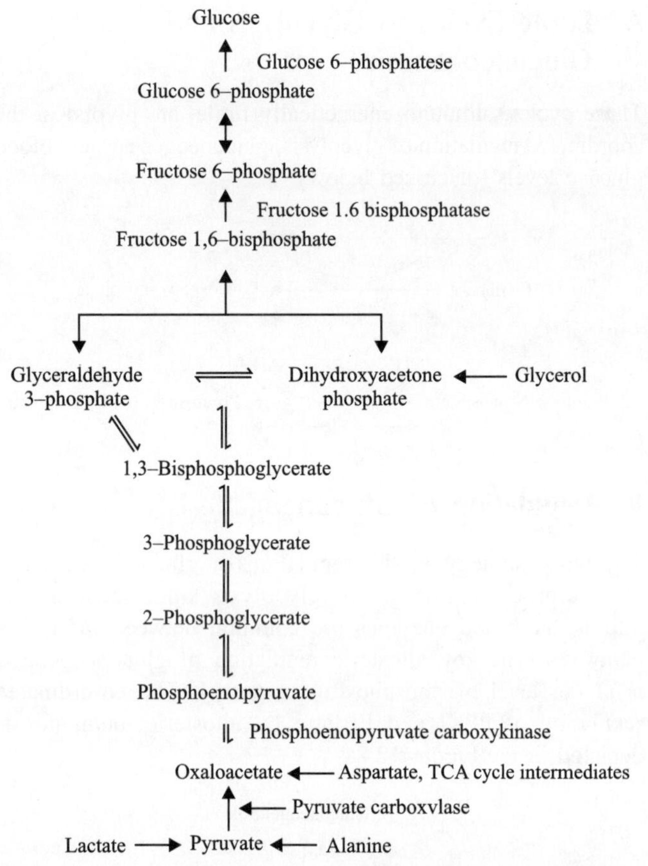

Figure 12.8b The flow chart of gluconeogenesis showing the specific enzymes of gluconeogenesis. The downward arrow depicts flow of glycolysis. Since the conversion of PEP to pyruvate is not reversed by a single enzyme, the step is not shown.

The glucose formed cannot be reconverted to glucose-6-phosphate due to inhibition of hexokinase by glucose-6-phosphate and inhibition of glucokinase by fructose-6-phosphate. Glucokinase which is insulin-dependent high turnover protein, remains at low concentrations because of low levels of insulin.

Pyruvate carboxylase is activated by increased acetyl coenzyme A produced due to enhanced fatty acid oxidation consequent to cyclic AMP induced adipose mobilization caused by low insulin/glucagon ratio and increased levels of epinephrine. Such a hormonal balance is established when blood glucose level is low, and glucose needs to be generated *denovo*. Oxaloacetate formed is flushed out of TCA cycle via malate, as increased NADH/NAD ratio and acyl coenzyme A resulting from oxidation of fatty acids, inhibit the activity of citrate synthase to a great extent. Pyruvate dehydrogenase is in its phospho form and hence less active providing more pyruvate to be acted upon by pyruvate carboxylase which is activated by increasing concentration of acetyl coenzyme A. During gluconeogenesis, the main bulk of pyruvate is built up from alanine through transamination. In addition, induced synthesis of both pyruvate carboxylase and phosphoenol

carboxykinase is increased under the influence of glucagon and glucocorticoids as the gene repression of these enzymes by insulin is no more operative.

Pyruvate kinase is less active due to cyclic AMP mediated phosphorylation and hence, phosphoenol pyruvate (PEP) formed is not converted back to pyruvate, and thus, conversion of PEP to glyceraldehyde-3-phosphate is facilitated. Glyceraldehyde-3-phosphate so formed can neither get back to glycolysis nor to lipogenesis due to low levels of insulin and increased glucagon levels and consequently, is left to enter gluconeogenic pathway.

To conclude, gluconeogenesis is operative in conditions of dietary deprivation (fasting starvation or hormonal imbalances (as in diabetes) low blood glucose levels trigger up the pathways related to gluconeogenesis. Gluconeogenesis is under hormonal control, low blood glucose levels decrease insulin/glucagon ratio and leads to release of glucocorticoids and epinephrine. This slows down the rate of glycolysis simultaneously enhancing gluconeogenesis.

VII. REGULATION OF BLOOD GLUCOSE LEVEL

In human, blood glucose levels are regulated with great efficiency within a very narrow range during intermittant phases of feeding and fasting. In normal individuals the glucose levels in blood samples collected after 12 h of fasting, range between 60–100 mg/dl, which rises to about 150–170 mg/dl on glucose administration (1.75 mg/kg body wt.) by one hour and comes back to normal within 2 to 3 hours time.

This stringent regulation of blood glucose level is essential primarily to ensure availability of glucose to cells and organs like erythrocytes, brain and adrenal medulla which essentially need glucose for their energy metabolism and vital functions. Brain consumes glucose at a rate of 120 mg/day which remains fairly constant until a person goes on prolonged starvation when 75% of brain's energy needs are fulfilled by ketone bodies. Still, brain can not function normally under the stress of hypoglycemia and undergoes coma if blood glucose levels tend to fall below 40 mg/dl. Further glucose serves as the main fuel to meet energy demands of other tissues like skeletal muscle and heart on exercise. At the same time blood glucose levels getting higher than the normal range is detrimental as it causes osmotic imbalance leading to pH and ionic imbalance, haemoconcentration and dehydration of cells. This, along with other metabolic impairments lead to hyperglycemic coma in untreated diabetics.

The blood glucose level is maintained through hormonal control on various pathways of glucose synthesis and utilization and its transport from blood to various tissues. Insulin is the only hormone which promotes each and every avenue of glucose utilization. Thus, insulin lowers the blood glucose level. In contrast, glucagon, epinephrine glucocorticoids are hyperglycemic hormones. All the effects of insulin are counteracted by these hormones to ensure the maintenance of blood glucose levels under the conditions of fasting and

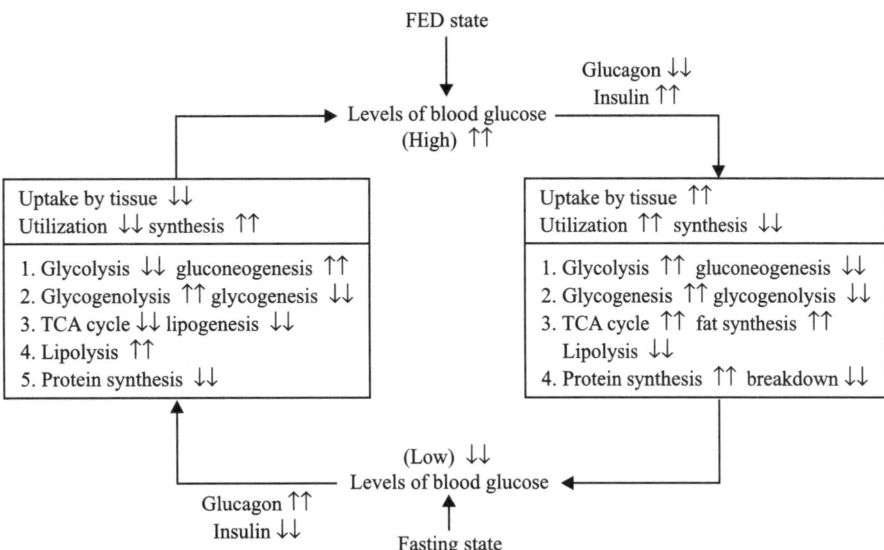

Figure 12.9 Blood glucose regulation: overall dynamics.
↑↑ High/Enhanced ↓↓ Low/ Declined.

starvation when blood glucose levels and insulin tend to fall. Uptake of glucose in skeletal muscles, adipose and many other tissues is insulin dependent. Insulin promotes the number of glucose transporters on the cell surface. Conversely, glucocorticoids and growth hormones demote the same. However, uptake of glucose in liver is not directly insulin dependent.

Brain is totally exempted from this hormonal restriction on glucose uptake and utilization which ensures availability of glucose to brain and maintenance of its energy metabolism even under the conditions of severe hypoglycemia (upto 40 mg/dl).

Liver is the direct source of glucose to blood. Sources of glucose for liver are glucose from diet, glycogenolysis and gluconeogenesis. After the meals, about two-third of the total glucose from intestines carried to the liver enters the hepatic cells freely and is converted to glucose-6-phosphate by glucokinase and rest one-third of the free glucose passes through the liver into the systemic blood. Increased blood glucose levels cause increased release of insulin from the β-cells of pancreas to the blood. Increased levels of insulin induce the synthesis of glucokinase, activates glycogen synthase, phosphofructokinase, pyruvatekinase, and pyruvate dehydrogenase to ensure overall utilization of glucose as metabolic precursor and as a ready source of energy to various tissues, its storage in the form of glycogen in liver and muscles, and as triacylglycerols in adipose tissues.

During intermittant fasting, while blood glucose tends to fall around 60 mg/ml, levels of insulin fall with simultaneous increase in the levels of glucagon. This primarily leads to glycogenolysis in liver and later from muscles due to the release of epinephrine. Since glucose-6-phosphatase is present in liver, not in muscles, only liver can release free glucose to blood. Nevertheless, lactate released form muscles as the end product of glycolysis coupled to glycogenolysis, enters the Cori cycle

and contributes to gluconeogenesis in liver. The glycogen stored in the body is exhausted in 24 hours time. While glycogen is being mobilised, simultaneously triacylglycerols in the adipose also get mobilised under the influence of low insulin and high glucagon and epinephrine levels. Adipose mobilization serves two purposes. On one hand fatty acids released now are utilized by most of the tissues for energy needs sparing glucose for brain, and on the other hand glycerol produced moves to liver due to the deficiency of glycerol kinase in the adipose tissue and is used for gluconeogenesis. Further stress of starvation leads to the release of glucocorticoids which mobilize amino acids from proteins especially from the skeletal muscles (till 72 hr of starvation) to liver, which is utilised for the synthesis of glucose through gluconeogenesis. Under the same hormonal influence the key enzymes for gluconeogenesis from lactate, TCA cycle intermediates, and amino acids, i.e. pyruvate carboxylase and phosphoenol pyruvate carboxykinase get synthesized increasingly as discussed. Phospho-fructokinase I and II become less active with increased activity of fructose 1,6-bisphosphatase due to phosphorylation of PFK II and depletion of fructose 2,6-bisphosphate. Similarly, activity of glucose-6-phosphatase is enhanced and that of glucokinase and hexokinase is at the minimum. Thus, rate of glycolysis is diminished to a great extent, and gluconeogenesis is enhanced. The uptake of glucose in insulin dependent tissues is minimum due to low insulin/glucagon ratio, and therefore, brain gets its needful quota of glucose alongwith other glucose dependent tissues like red blood cells, renal medulla etc. Elevation of blood glucose levels leads to reversal of the hormonal and metabolic profile and blood glucose level is maintained. Figure 12.10 depicts the overall dynamics of blood glucose regulation.

The fuel homosteostatsis at different stages of fasting and starvation is described in Chapter 17.

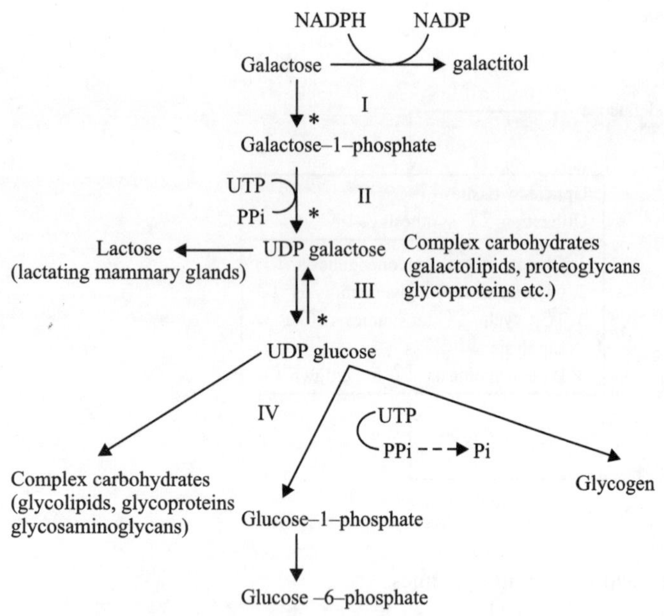

Figure 12.10 Pathways of galactose metabolism.

Blood glucose regulation by kidneys

Kidneys have limited capacity of gluconeogenesis which takes place during late starvation, glutamine being the precursor. During late starvation due to proteolysis of many of the liver enzymes, urea cycle is impaired. Ammonia gets detoxicated by being converted to glutamine which is taken up by kidneys. Kidney is rich in enzyme glutaminase. Glutaminase converts glutamine to glutamic acid which consequently get transaminated, and de-aminated to give alpha keto glutaric acid. Alpha ketoglutaric acid now can proceed to gluconeogenesis after being converted to malate in the TCA cycle. Blood glucose level in excess to renal threshold, i.e., 180 mg/dl is excreted out by kidneys again keeping the further rise in blood glucose levels restricted to a great extent.

Emerging concepts

In totality, regulation of blood glucose levels starts at the point of fine regulation of dietary intakes by hypothalamic orexogenic peptide NPY and the anorexogenic peptide alpha MSH at the Arcuate nucleus under the influence of leptin and adiponectin from the adipose tissue, insulin, and a number of peptides released from gastrointestinal tract. Adiponectin increases insulin sensitivity in promoting uptake of glucose by adipose and skeletal muscles. It acts through AMP kinase, promotes glycolysis and fatty acid oxidation in tissues.

VIII. METABOLISM OF GALACTOSE AND FRUCTOSE

Both these sugars are metabolized in liver and enter the main stream of glucose metabolism.

Galactose

Galactose is phosphorylated as galactose-1-phosphate which gets converted to UDP galactose by the action of galactose-1-phosphate UDP glucose uridyl transferase. UDP-galactose gets converted to UDP glucose by the enzyme UDP galactose 4 epimerase. UDP glucose can either be converted to glucose-1-phosphate by the action of UDP glucose pyrophosphorylase or gets involved in the synthesis of complex carbohydrates and glycogen (Figure 12.10). Hereditary deficiency of galactokinase, gal 1 phosphate uridyl transferase and epimerase are known. The common symptoms in the first two types are refusal to breast milk, vomiting and diarrhea on forced feeding. Galactosemia and galactosuria are common biochemical findings alongwith cataract and mental retardation.

The most benign form of galactosemia is the deficiency of epimerase which is characterized mainly by mental retardation to some extent with no overt symptoms of other types. The severest form of galactosemia is due to the deficiency of galactose-1-phosphate: UDP glucose uridyl transferase. In this type, galactose-1-phosphate accumulates in all tissues and is believed to be responsible directly or indirectly, for the malfunctioning of the gastrointestinal tract causing diarrhea, jaundice and cirrhosis of liver, renal and brain damage leading ultimately to death if the condition is not diagnosed and managed in time. High levels of galactose in patients cause accumulation of galactitol in lens formed by the action of aldose reductase on galactose. Consumption of NADPH in this reaction also is a major cause of injury to lens which leads to cataract. The enzyme aldose reductase has high Km for galactose and hence only the presence of high amounts of galactose leads to formation of significant levels of galactitol. The symptoms of galactosemia are observed right in the perinatal period. Withdrawal of breast milk is adviced. Body's need for galactose for the synthesis of complex carbohydrates and galactolipids can be met by glucose through the action of epimerase.

Fructose

Fructose like galactose, fructose is converted to fructose-1-phosphate by the enzyme fructokinase Fructose-1-phosphate is broken down to glyceraldehyde and dihydroxy acetone phosphate by the action of an enzyme called aldolase B (Figure 12.11). Glyceraldehyde further gets converted to glyceraldehydes-3-phosphate by the action of glyceraldehyde kinase. These intermediates follow the usual path of glucose metabolism. Fructose can be lipogenic since its metabolism escapes regulation of glycolysis by PFK1.

Hereditary deficiency of fructokinase is known. The common symptoms are rejection of fructose in diet, i.e., sucrose, fruits, etc., vomiting and diarrhea on forced feeding. Biochemical

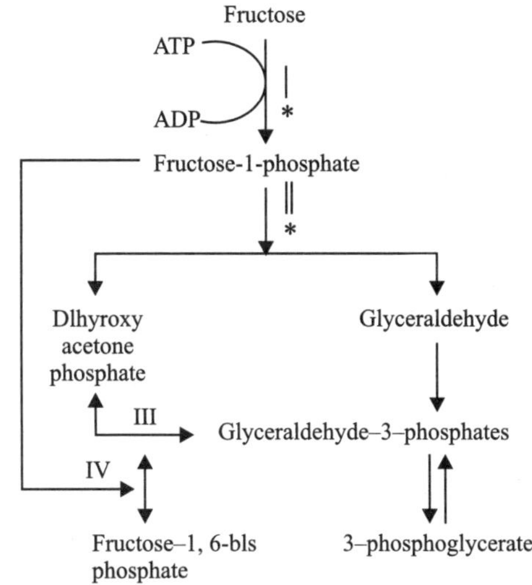

Enzymes I. Fructokinase
    II. Aldolase B
    III. Glyceraldehyde kinase
    IV. Aldolase B

* Deficiency leading to fructosemia

Figure 12.11 Pathways of fructose metabolism.

findings are fructosemia and fructosuria, with no other symptoms.

The more severe form of fructosemia is caused by the deficiency of aldolase B which is characterized by severe hypoglycemia during intermittant fasting and pronounced hypophosphatemia on fructose loading (Figure 12.12). This condition is often referred as to hereditary fructose intolerance.

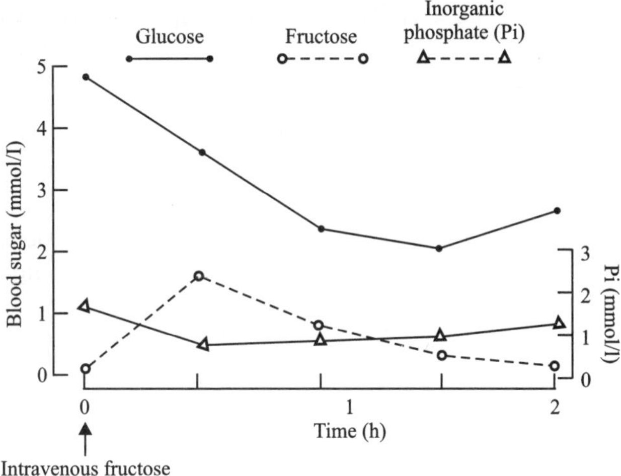

Figure 12.12 Fructose tolerance test.

Deficiency of aldolase B causes the accumulation of fructose-1-phosphate in liver which inhibits aldolase A catalyzed conversion of triose phosphates to F1,6-bisphosphates thereby hampering gluconeogenesis from glycerol and other sources. Continuous trapping of phosphate with no replenishment and recovery of fructose-1-phosphate impairs glycogenolysis. Thus, impairment of gluconeogenesis and glycogenolysis cause

hypoglysemia leading to convulsions and absent reflexes in untreated patients. Further, liver damage is also caused by accumulating fructose phosphate.

Deficiency of fructokinase does not lead to all these complications.

SUMMARY OF THE METABOLISM OF GLUCOSE AND OTHER ABSORBED SUGARS

- Dietary glucose, galactose and fructose are taken from the intestines to liver by portal circulation. Glucose serves as a source of energy for immediate use and is stored as glycogen for future use. Simultaneously, glucose provides carbon skeleton for the synrhesis of different bio molecules.

- Glycogen is synthesized from glucose by the process of glycogenesis. Glycogen is stored predominantly in liver and muscles. Glucose is catabolized to pyruvate through glycolysis. HMP shunt is regarded as an alternative pathway of glucose metabolism that generates NADPH for reductive biosynthesis, detoxication reactions, maintenance of the integrity of erythrocyte membranes against oxidative assaults and, provides pentoses for nucleotide synthesis. It is also a feeder pathway to glycolysis.

- During fasting or starvation, glucose is regenerated by glycogenolysis in liver and muscle tissue. Liver is capable of gluconeogenesis from non-carbohydrate sources. Liver is the only organ that releases glucose to the blood. Limited gluconeogenesis is carried out by kidneys as well.

- All the pathways of glucose metabolism are regulated. Regulation of glycolysis and gluconeogenesis as well as glycogenesis and glycogenolysis are coordinated in a manner that pathways of synthesis and breakdown do not occur at the same time.

- Regulation of these pathways takes place by allosteric modulation, covalent modification and transcriptional regulation of the irreversible steps catalyzed by regulatory enzymes.

- Insulin activates utilization of glucose by maintaining the crucial enzymes in the de-phospho form whereas glucagon activates generation of glucose maintaining the related enzymes in the phospho form which simultaneously keep the enzymes of glucose utilization less active. These two hormones also bring about the transcriptional regulation of these pathways.

- Despite the fact that glycolysis is a catabolic pathway, it provides intermediates for the synthesis of fat and amino acids.

- In erythrocytes, a bypass from glycolysis called Rappaport Lubering pathway generates 2,3-bisphosphoglycerate, a regulator of oxygen carriage by hemoglobin.

- Two molecules of ATP are consumed and four molecules of ATP are formed per molecule of glucose by glycolysis. Two molecules of NADH are also generated. The reducing equivalents enter the mitochondria through shuttle systems to generate ATP by oxidative phosphorylation in aerobic tissues. In anaerobic tissues, NADH is reconverted to NAD^+ by the reaction of lactate dehydrogenase that forms lactate from pyruvate.
- Regeneration of NAD^+ from NADH is essential for continuation of glycolysis.
- Fluoride, Iodoacetate and arsenate are poisons because they inhibit certain glycolytic enzymes.
- Lactate formed in skeletal muscles enter liver for gluconeogenesis or glycogenesis through "Cori Cycle".
- Pyruvate formed in glycolysis is converted to acetyl coenzyme A with the formation of a molecule of NADH. Acetyl coenzyme A enters TCA cycle for complete oxidation to CO_2 along with generation of a molecule of ATP by substrate level phosphorylation. One molecule of $FADH_2$ and three molecules of NADH are also generated in each cycle of TCA cycle. $FADH_2$ and NADH enter the electron transport chain and generate ATP by oxidative phosphorylation.
- Pyruvate is also converted to oxaloacetate, alanine and glucose.

- TCA cycle is an amphibolic pathway meant for complete oxidation of carbohydrates, fats and proteins and prviding intermediates for the biosynthesis of all these molecules. The cycle starts with the formation of citrate by the condensation of oxaloacetic acid and acetyl coenzyme A and ends with the formation of oxaloacetate. Fluoroacetate inhibits TCA cycle.
- Blood glucose level is regulated very stringently by concerted action of insulin, glucagon and several other hormones including those released from gut and adipose tissue. Brain is entirely dependent on glucose for its energy needs. Same is true for erythrocytes. Fatty acids which are major sources of energy for other tissues can not cross blood-brain barrier and RBCs lack mitochondria to oxidize fatty acids.
- Galactose and fructose are metabolized by different enzymes in the initial steps but the intermediates get merged in the glyclytic pathway. These sugars are metabolized mainly in liver and, hence, are not detected in blood under normal health. In born errors of metabolism of these sugars lead to galactosemia/galactosuria or fructosemia/fructosuria along with life-threatening pathological manifestations in infants.

SUGGESTIONS FOR FURTHER READING

Fischer E.H. (1997), Cellular Regulation by Protein Phosphorylation: A Historical Review, *Biofactors,* 6, 367–374.

Fothergill-Gilmore L.A. and Michels P.A. (1993), Evolution of Glycolysis, *Prog. Biophys. Mol. Biol,* 59, 105–235.

Murray Robert K., Benden David A., Botham Kathleen M., Kennelly Petcr J., Rodwell, Victor W. and Weil P. Anthony, Harper's *Illustrated Biochemistry,* 29th ed. (2012), McGraw-Hill, New York.

Nelson David L., Cox, Michael M. (2008), "Lehninger *Principles of Biochemistry*" 5th ed., Freeman and Company.

Sacks D.B. (1999), Metabolism of Carbohydrates Tietz, *Textbook of Clinical Chemistry,* Burtis C.A. and Ashwood E.R. (Eds.), 3rd ed., W.B. Saunders Company, Philadelphia, USA.

13

Lipid Metabolism

K.D. Moudgil

CONTENTS

I. INTRODUCTION

Lipids include heterogeneous chemical compounds that are insoluble in water but soluble in organic solvents such as ether, chloroform, acetone and benzene. Moreover, they have some relation, actual or potential, to fatty acids. The classification, physical properties and chemical structure of various lipids are covered in Chapter 6. The lipids present in the human body are in a dynamic state of metabolism. They are an important source of energy and are an integral part of the cell membrane. In this chapter, emphasis is placed only on the metabolism of various types of lipids.

II. DIGESTION AND ABSORPTION OF LIPIDS

A. Digestion of Lipids

There is little or no digestion of lipids in the mouth or stomach. There is evidence to suggest that a gastric lipase catalyzes limited hydrolysis of fats containing short- or medium-chain fatty acids. The acidic pH of gastric secretion is not conducive to optimum functioning of the gastric lipase.

The main site of digestion of lipids is the small intestine. In the duodenum, food mixes with the bile and pancreatic juice. The bile salts contained in the bile promote emulsification of the lipids, thereby breaking large fat globules into fine droplets. The emulsified fat droplets are further reduced in size to 'micelles'. Micelles are supramolecular aggregates of 0.1–0.5 μm diameter in which the hydrophobic components are localized at the interior of the aggregate, while the hydrophilic components are on the outer surface. The formation of micelles increases the surface area exposed to the enzymes and facilitates the contact between the hydrophilic enzymes and their hydrophobic substrates, the lipids. The catalytic activity of lipases is greatly increased at the lipid–water interface, a phenomenon known as interfacial activation. The main enzyme involved in the digestion of lipids in the small intestine is a lipase (previously known as steapsin) contained in the pancreatic juice. Pancreatic lipase is an α-lipase, which acts specifically at ester linkages at positions 1 and 3 in the triacylglycerols (triglycerides). Another protein, colipase, present in pancreatic juice is essential for action of pancreatic lipase. Colipase anchors the pancreatic lipase close to the surface of the micelle. Pancreatic lipase, colipase and bile salts form a ternary complex, which brings about digestion of lipids. As a result of the action of lipase, exogenous triacylglycerols are primarily converted into fatty acids and 2-monoacylglycerols (β-monoglycerides). The latter is either cleaved by an esterase to form a fatty acid and glycerol, or it is isomerised to 1-monoacylglycerol, which is subsequently acted upon by pancreatic lipase to produce fatty acid and glycerol (Figure 13.1). Several other enzymes contained in pancreatic juice are required for digestion of many dietary lipids other than triacylglycerols. For example, phospholipases and cholesterol esterase act on phospholipids and cholesterol esters, respectively.

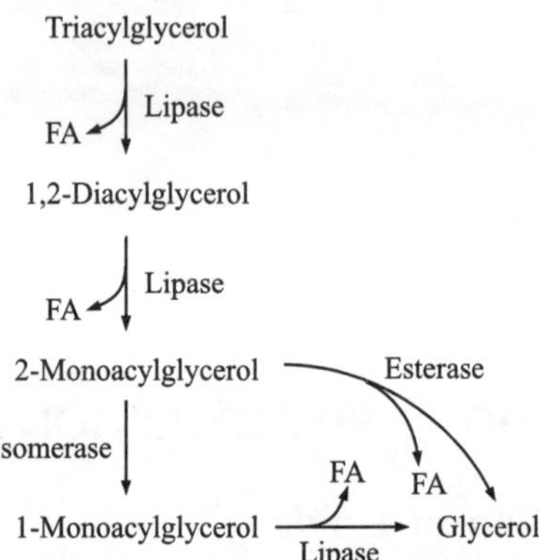

Figure 13.1   Digestion of triacylglycerols (triglycerides) in the small intestine is brought about by pancreatic lipase (FA, long chain fatty acid).

B.   Absorption of Fats and Other Lipids

The main products of digestion of triacylglycerols in the lumen of small intestine are free fatty acids, 2-monoacylglycerol and glycerol. Some triacylglycerols may remain unhydrolysed. The free fatty acids and 2-monoacylglycerols contained in the micelles enter through the microvilli of mucosal epithelial cells of the small intestine by passive diffusion through the cell membrane. The bile salts contained in the micelles are not absorbed by the mucosal epithelial cells. However, the bile salts are absorbed in the lower part of the small intestine. This constitutes the enterohepatic circulation. Inside the mucosal epithelial cells, long chain fatty acids with 12–18 carbon atoms are converted to acyl-CoA derivatives, which undergo reesterification by monoacylglycerols or diacylglycerols to resynthesize the triacylglycerols. The triacylglycerols then interact with apolipoproteins, cholesterol and phospholipid to form chylomicrons and very low density lipoproteins (VLDL), which are discharged into the lymph and then into lacteals, and finally reach the bloodstream via the lymphatic system. On the other hand, glycerol and short to medium chain fatty acids with 6–10 carbon atoms are not reesterified. They are transported bound to albumin to the liver via the portal blood (Figure 13.2). The lipids are transported in the blood mainly in the form of chylomicrons, lipoproteins and unesterified fatty acids.

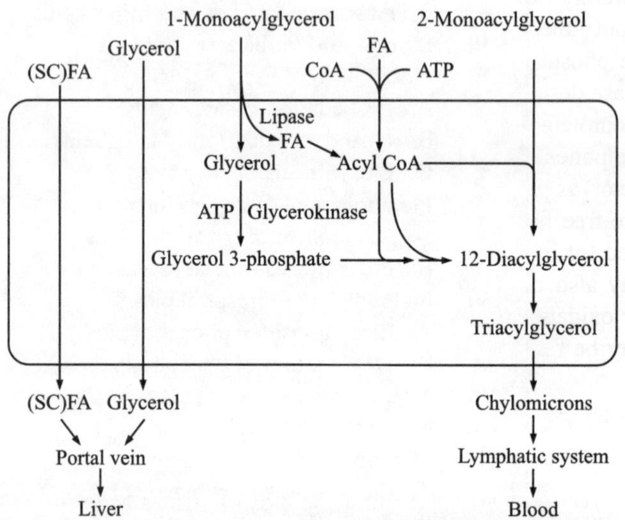

Figure 13.2   The absorption of dietary triacylglycerols in the small intestine (FA, Longchain fatty acid; (SC) FA, Short chain fatty acid).

III.   STORAGE AND MOBILIZATION OF FATS

The lipids are stored as depot fat in the adipose tissue. The adipose tissue lipid comprises mainly of triacylglycerols. These triacylglycerols are continually undergoing hydrolysis (lipolysis) and reesterification by two entirely different biochemical pathways. The relative activities of the two processes determine the magnitude of the free fatty acid pool in the adipose tissue. The latter is reflected in the level of free fatty acids in the circulation.

In the adipose tissue, triacylglycerols can be synthesized from acyl CoA and glycerol 3-phosphate. Acyl CoA is formed from free fatty acid, whereas glycerol-3-phosphate is formed from glucose instead of glycerol. This is so because the activity of the enzyme glycerokinase, which converts glycerol into glycerol-3-phosphate, is low in adipose tissue.

The mobilization of lipids from adipose tissue occurs as free fatty acids. A hormone-sensitive lipase (triacylglycerol lipase) brings about hydrolysis of triacylglycerol to free fatty acids and glycerol. Many hormones stimulate lipolysis by activation of this enzyme, which exists in an active or inactive form. The activity of this enzyme is controlled by phosphorylation–dephosphorylation. Phosphorylation is mediated by a protein kinase, whereas dephosphorylation is brought about by a phosphatase (Figure 13.3). Several hormones such as epinephrine, norepinephrine, serotonin, hydrocortisone, glucagon, thyroxine, adrenocorticotropin (ACTH), thyrotropin (TSH), secretin and luteinizing hormone (LH) stimulate lipolysis. These hormones cause increase in intracellular (cyclic AMP) levels. The cAMP in turn activates the protein kinase responsible for phosphorylation of hormone-sensitive lipase, resulting in activation of this enzyme. On the other hand, insulin, prostaglandin E and nicotinic acid inhibit lipolysis by lowering the intracellular level of cAMP. This effect is brought about either by inhibiting the adenylate cyclase or by stimulating the phosphodiesterase. The hormone-sensitive triacylglycerol lipase described above is different from the hormone-sensitive lipoprotein lipase, which causes hydrolysis of the triacylglycerol component of the lipoproteins. Lipoprotein lipase (or clearing factor) is present in most tissues, including the adipose tissue. The free fatty acids released from triacylglycerols are mainly used for oxidation by cells to generate energy. However, they may also be transported (bound to albumin) to other tissues for oxidation or conversion to other lipids. Alternatively, they may be resynthesized into triacylglycerols.

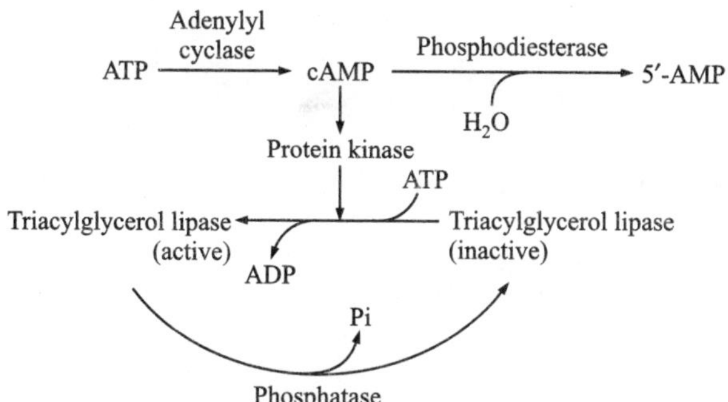

Figure 13.3   The activity of triacylglycerol lipase is regulated via cyclic AMP (cAMP).

IV. METABOLISM OF FATTY ACIDS

The major pathway for the catabolism of fatty acids is the β-oxidation pathway. Other minor routes include the alpha- and omega-oxidation pathways. The biosynthesis of fatty acids occurs from acetyl CoA and the reactions are catalysed by fatty acid synthetase, a multienzyme complex.

A. β-oxidation of Fatty Acids

The catabolism of fatty acids occurs by a stepwise process called β-oxidation. In this process, two carbon units are

sequentially removed, beginning from the carboxy-terminal end. In 1904, Knoop elucidated the principle of β-oxidation. In his study, phenyl derivatives of the fatty acids were prepared by tagging the fatty acid with a benzene ring at the methyl end. These derivatives were fed to dogs and the products excreted in the urine were isolated and characterized. It was observed that the compounds isolated from urine were derivatives of benzoic or phenylacetic acids (Figure 13.4). Benzoic acid and phenylacetic acid are not oxidized by the body. They are conjugated with glycine and excreted in urine as hippuric acid and phenaceturic acid, respectively. Therefore, when benzoic acid or phenylacetic acid were fed to dogs, hippuric acid or phenaceturic acid, respectively, were excreted in the urine. On feeding higher homologs such as phenylpropionic acid or phenylbutyric acid, the product excreted in urine was again hippuric acid or phenaceturic acid, respectively. From these observations, it was concluded that the catabolism of fatty acids involves oxidation of the β-carbon. After the loss of two carbon atoms, the new β-carbon is oxidized.

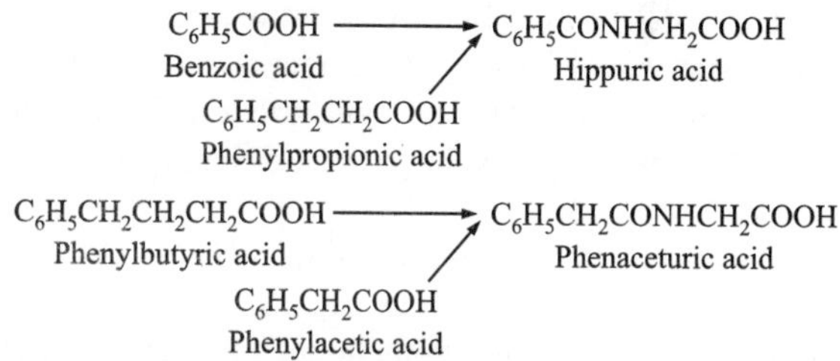

Figure 13.4   The fatty acids are catabolised by β-oxidation. The products of catabolism of phenyl derivatives of benzoic acid or phenylacetic acid, or their higher homologs, are excreted in urine. These products are always the derivatives of benzoic acid or phenylacetic acid.

The enzymes that participate in β-oxidation are present in the mitochondrial matrix. An exception is the NAD-dependent dehydrogenase, which is a component of the inner mitochondrial membrane. A fatty acid undergoing β-oxidation must first be converted into the corresponding active intermediate, the CoA ester. This reaction takes place in the cytoplasm. However, neither the long chain fatty acids nor their CoA derivatives can diffuse across the mitochondrial inner membrane. Their entry into the mitochondrial matrix requires a membrane transport system. Carnitine is a carrier molecule which functions in the transport system.

1. Activation of fatty acids: The fatty acids are converted into their active intermediates, the CoA esters, in a reaction catalysed by acyl CoA synthetases or thiokinases. These enzymes are specific for fatty acids of different chain lengths. Moreover, the reaction requires the participation of ATP or GTP. This is the only reaction in the oxidation of fatty acids that requires energy from nucleotide triphosphates.

Fatty acids with up to 10 and 11 carbon atoms (medium-chain fatty acids) can be converted to the corresponding CoA derivatives by one of the two mitochondrial acyl CoA synthetases (Figure 13.5). The butyryl-CoA synthetase catalyses

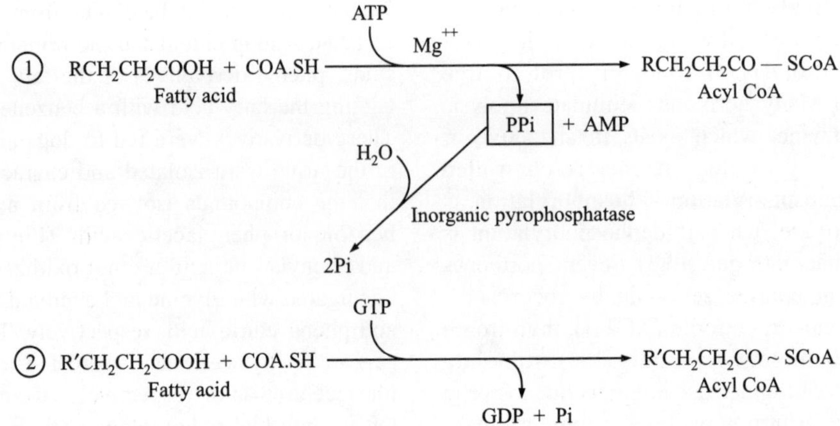

Figure 13.5   A fatty acid is activated prior to β-oxidation. ATP-dependent acyl CoA synthetases catalyse this reaction both within and outside the mitochondria. In addition, a GTP-dependent acyl CoA synthetase is also located inside the mitochondria.

an ATP-dependent reaction, whereas the acyl-CoA synthetase (GDP-forming) requires GTP. On the other hand, the long chain fatty acids are esterified by acyl-CoA synthetase, which is predominantly associated with the microsomal fraction of the cell and the outer membrane of mitochondria. The reaction is similar to that catalysed by butyryl-CoA synthetase.

2. Role of carnitine in β-oxidation. The CoA esters of long chain fatty acids formed outside the mitochondria cannot diffuse across the mitochondrial inner membrane. Their transport is mediated by a carrier molecule known as *L*-carnitine (3-hydroxy-4-trimethyl ammonium butyrate; $(CH_3)_3N^+CH_2CH(OH)CH_2COO^-$). It is synthesized from lysine in the liver and kidney.

The transport of fatty acids across the inner mitochondrial membrane requires the transfer of acyl moiety from acyl CoA to the hydroxyl group of carnitine to form acyl carnitine (Figure 13.6). The transesterification is catalysed by carnitine acyl transferase I and it occurs on the outer surface of the inner mitochondrial membrane. The acylcarnitine is translocated into the mitochondrial matrix. This reaction is coupled with the transport out of carnitine from within the mitochondrial matrix. Carnitine acyl carnitine translocase acts as a membrane exchange transporter. On the inner surface of the inner mitochondrial membrane, another transesterification reaction allows the acyl CoA to be regenerated. However, the CoA now is of intramitochondrial origin. This reaction is catalysed by carnitine acyl transferase II.

In disorders of lipid metabolism associated with deficiency of carnitine or carnitine acyl transferase I, there is impaired oxidation of fatty acids and impaired ketogenesis, which in turn lead to reduced gluconeogenesis. One important clinical manifestation of these impaired biochemical reactions is episodic periods of hypoglycemia. Carnitine deficiency may occur in two forms: (i) *myopathic,* in which there is decreased muscle carnitine but serum carnitine is normal, and (ii) *systemic,* which is characterized by low serum carnitine. Treatment with oral carnitine has proved beneficial in some of these patients. Patients with carnitine acyl transferase II

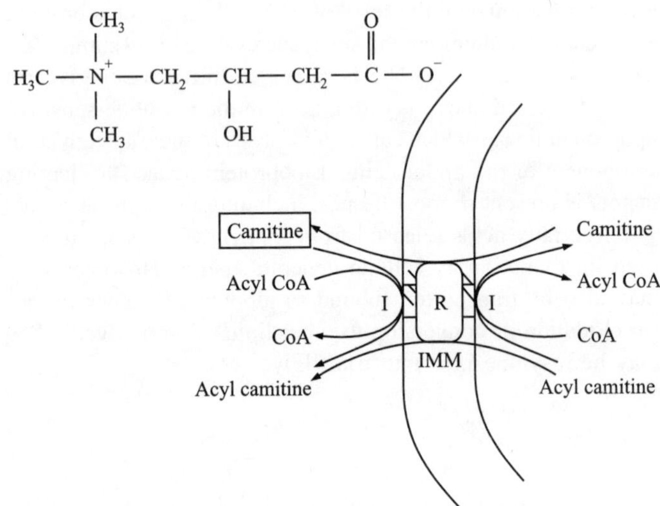

Figure 13.6   The transport of CoA ester of long chain fatty acids across inner mitochondrial membrane is mediated by L-carnitine. I, carnitine acyl transferase I; II, carnitine acyl transferase II; and R, carnitine acyl carnitine translocase.

present with recurrent myoglobinuria. Interestingly, a group of drugs used in the treatment of diabetes mellitus, known as sulfonylureas, mediates their effect on fatty acid oxidation by inhibiting carnitine acyl transferases.

3. Intra-mitochondrial reactions of β-oxidation. Within the mitochondrial matrix, the acyl moiety of acyl CoA molecule undergoes a repetitive series of reactions causing the cleavage of carbon atoms and 2 from the acyl moiety to form another acyl CoA molecule having two carbon atoms less. These sequential reactions are shown in (Figure 13.7).

(a) *Dehydrogenation:* The dehydrogenation of the acyl moiety is catalysed by an enzyme called acyl CoA dehydrogenase. FAD is the prosthetic group of the enzyme, and it functions in conjunction with

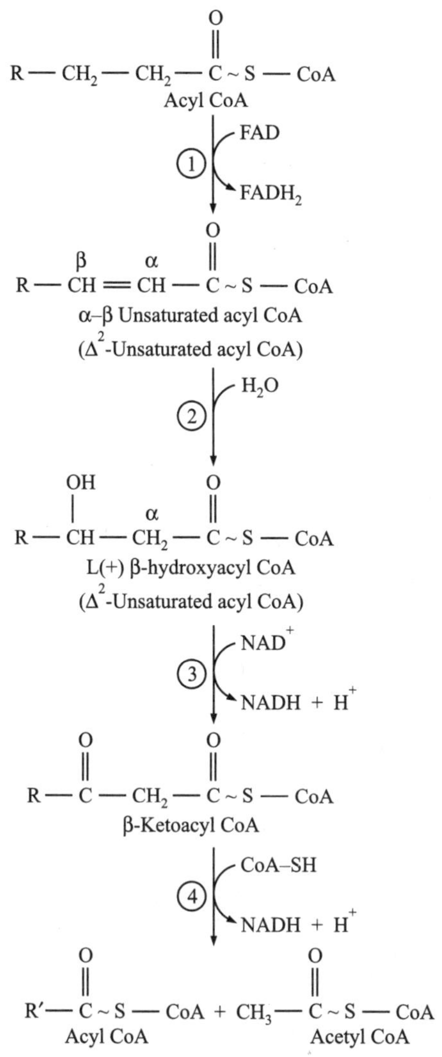

$$R - CH_2 - CH_2 - \overset{\overset{\textstyle O}{\|}}{C} \sim S - CoA$$

Acyl CoA

(1) FAD → FADH$_2$

$$R - \overset{\beta}{C}H = \overset{\alpha}{C}H - \overset{\overset{\textstyle O}{\|}}{C} \sim S - CoA$$

α–β Unsaturated acyl CoA

(Δ^2-Unsaturated acyl CoA)

(2) H$_2$O

$$R - \overset{\overset{\textstyle OH}{|}}{C}H - \overset{\alpha}{C}H_2 - \overset{\overset{\textstyle O}{\|}}{C} \sim S - CoA$$

L(+) β-hydroxyacyl CoA

(Δ^2-Unsaturated acyl CoA)

(3) NAD$^+$ → NADH + H$^+$

$$R - \overset{\overset{\textstyle O}{\|}}{C} - CH_2 - \overset{\overset{\textstyle O}{\|}}{C} \sim S - CoA$$

β-Ketoacyl CoA

(4) CoA–SH → NADH + H$^+$

$$R' - \overset{\overset{\textstyle O}{\|}}{C} \sim S - CoA + CH_3 - \overset{\overset{\textstyle O}{\|}}{C} \sim S - CoA$$

Acyl CoA　　　　　　　　　Acetyl CoA

Figure 13.7　*β*-oxidation of fatty acids. The activated fatty acid is oxidized through a repetitive sequence of four reactions. In each cycle, carbon atoms 1 and 2 of the acyl moiety are cleaved in the form of acetyl CoA. The new acyl CoA thus generated has two less carbon atoms. It again enters the same series of reactions (1 = Acyl CoA dehydrogenase; 2 = Δ^2-Enoyl CoA hydratase or Crotonase; 3 = β-hydroxy-acyl CoA dehydrogenase, and 4 = Acetyl CoA acetyl transferase, also known as *β*-Ketothiolase or thiolase).

another flavoprotein called electron transferring flavoprotein (ETF). Two hydrogen atom, each with an electron, are removed from the acyl moiety. These reducing equivalents are transferred to FAD and then to the electron transport chain (ETC) via the ETF. This reaction results in the formation of α, β-unsaturated acyl CoA (or Δ2-unsaturated acyl CoA).

(b) *Hydration:*　One molecule of water is added to saturate the double bond in the *trans* Δ^2-acyl moiety. The reaction is catalysed by a stereospecific Δ^2-enoyl CoA hydratase (or crotonase). The product of the reaction

is L(+)β-hydroxyacyl CoA (or L(+)-3-hydroxyacyl CoA).

(c) *Oxidation:*　In this reaction, there is oxidation of the secondary alcohol group at carbon 3 (*β*-carbon) of the acyl moiety of *β*-hydroxyacyl CoA formed in the previous reaction. Thus, there is dehydrogenation at the *β*-carbon, and the reducing equivalents are transferred to NAD. This reaction is catalysed by *β*-hydroxyacyl CoA dehydrogenase. This enzyme is localized in the inner mitochondrial membrane unlike other enzymes of *β*-oxidation, which are found in the mitochondrial matrix.

(d) *Cleavage:*　There is a thiolytic cleavage of *β*-ketoacyl moiety at the *β*-carbon in the presence of CoA. The reaction is catalysed by acetyl CoA acetyl transferase (also known as *β*-ketothiolase or simply thiolase). The products of this reaction are acetyl CoA and an acyl CoA molecule. The latter contains two carbon atoms less than the original acyl CoA molecule present at the beginning of the reaction. The new acyl CoA molecule generated in this reaction reenters the oxidative pathway at the dehydrogenation reaction. The reactions of *β*-oxidation now begin again. With each cycle, a two-carbon unit is removed from acyl CoA in the form of acetyl CoA.

The long chain fatty acids with an even number of carbon atoms can be completely degraded to acetyl CoA. Acetyl CoA can enter the citric acid cycle, thus allowing complete oxidation of fatty acids to carbon dioxide and water. However, a saturated, non-branched fatty acid with an odd number of carbon atoms can be degraded up to the point at which propionyl CoA having three carbon atoms is formed. The propionyl CoA is then converted to succinyl CoA, which can enter the citric acid cycle.

Deficiencies of some enzymes involved in *β*-oxidation of polyunsaturated fatty acids may occur in some patients, resulting in impaired *β*-oxidation. For example, deficiency of medium-, long- and short-chain acyl CoA dehydrogenases, and *β*-hydroxy-acyl CoA dehydrogenase, has been reported. Deficiency of medium-chain acyl-CoA dehydrogenase causes impaired *β*-oxidation but enhanced *ω*-oxidation of fatty acids. This disorder is characterized by dicarboxylic aciduria (wherein medium chain dicarboxylic acids are excreted in the urine) and hypoglycemia.

Energetics of fatty acid oxidation

The degradation of caproate (or caproic acid, $C_5H_{11}COOH$), a six-carbon fatty acid, through *β*-oxidation pathway will require two cycles of the reactions of *β*-oxidation. Finally, three molecules of acetyl CoA will be generated. The reducing equivalents from FADH$_2$ and NADH, when fed into the electron transport chain, will lead to synthesis of five high energy phosphate bonds for each of the acetyl CoA formed. Thus, 10 ATP molecules will be formed in this way. In addition, three molecules of acetyl CoA can enter the citric acid cycle, and each round of the citric acid cycle will yield 12 ATP molecules, generating 36 molecules of ATP. Thus, a total of 46 ATP molecules will be generated by complete degradation of caproic acid through β-oxidation. Two high energy phosphate bonds (ATP converted to AMP) are required for initial activation of fatty acid to form acyl CoA. The net yield from the degradation of caproic acid, therefore, consists of 44 (46–2) high energy phosphate bonds. Similarly, palmitic acid ($C_{15}H_{31}COOH$), a 16-carbon fatty acid, will yield 129 high energy phosphate bonds.

Catabolism of branched chain fatty acids

The branched chain fatty acids can also be partially degraded through reactions of the β-oxidation. However, β-oxidation can proceed only up to a branch point at carbon 2 or carbon 3 of the newly formed acyl CoA molecule. Examples of the acyl moieties with a branch point at carbon 2 or carbon 3 are methyl butyryl CoA or isovaleryl CoA, respectively. After this step, alternate pathways are used for further degradation of acyl CoA (Ref. Chapter 16).

Oxidation of unsaturated fatty acids

The metabolism of an unsaturated fatty acid initially parallels that of a saturated fatty acid. The reactions of β-oxidation proceed until a Δ^3- or a Δ^2-*cis*-acyl moiety is formed (Figure 13.8). This depends on the position of the double bond in the molecule. The examples of unsaturated fatty acids leading to Δ^3- or Δ^2-*cis* acyl moieties are palmitoleyl CoA or linoleate CoA, respectively.

In the case of Δ^3-*cis* acyl moiety, first there is isomerization, which converts the *cis* double bond between carbons 3 and 4 into a *trans* double bond between carbons 2 and 3. The resultant acyl moiety is then acted upon by enoyl CoA hydratase, followed by the usual reactions of β-oxidation.

On the other hand, the Δ^2-*cis* acyl moiety is first hydrated by enoyl-CoA hydratase to form D (–)-β-hydroxyacyl moiety instead of the L(+)-β-hydroxy-acyl moiety produced in the course of β-oxidation of saturated fatty acids. The D(–)-β-hydroxyacyl moiety is then acted upon by a racemase to form the L(+)-β-hydroxyacyl moiety, followed by the usual reactions of β-oxidation.

The number of ATP molecules produced by oxidation of unsaturated fatty acids depends on the number of double bonds in the fatty acid molecule. In comparison to the oxidation of saturated fatty acids by β-oxidation, the oxidation of unsaturated fatty acids produces less numbers of ATP molecules. This is because saturated fatty acids have more hydrogen atoms, and thus more reducing equivalents are generated during their oxidation.

B. α- and ω-oxidation of Fatty Acids

1. α-oxidation: This pathway for oxidation of fatty acids has been detected in mammalian brain and liver. There is direct hydroxylation of long chain fatty acids at the α-carbon to generate α-hydroxy fatty acid. The latter undergoes oxidative decarboxylation to eliminate one carbon from the carboxyl end of the molecule (Figure 13.9). The enzymes catalysing these reaction are located in the endoplasmic reticulum. High energy phosphates are not generated in this pathway.

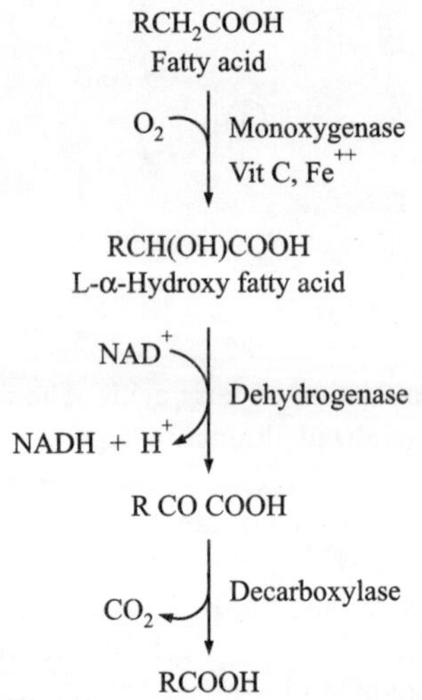

Figure 13.9 α-oxidation of fatty acids.

3,7,11,15-tetramethylhexanoic acid (phytanic acid) derived from phytols present in plant foods contains a methyl group on α-carbon, which blocks the β-oxidation. It can normally be metabolised by an initial reaction of α-oxidation, followed by the usual reactions of β-oxidation. Alpha-oxidation helps in overcoming the metabolic block posed by the methyl group on the α-carbon. In patients with Refsum's disease, there is a genetic defect in the α-oxidation system. Consequently, phytanic acid accumulates in the blood and tissues, leading to neurological abnormalities.

2. ω-oxidation: This pathway has been detected in the liver. There is oxidation of the omega carbon of the carbon chain of

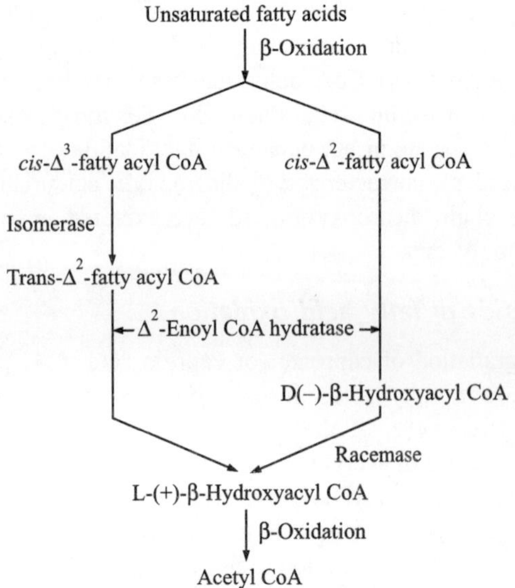

Figure 13.8   Oxidation of unsaturated fatty acids.

the fatty acid involving two reactions (Figure 13.10). The end product is a dicarboxylic acid, which may be further shortened by reactions of β-oxidation. The enzyme system catalyzing ω-oxidation is a monooxygenase system involving cytochrome P450, NADPH, molecular oxygen and a NADPH-cytochrome P450 reductase. It is located in the endoplasmic reticulum.

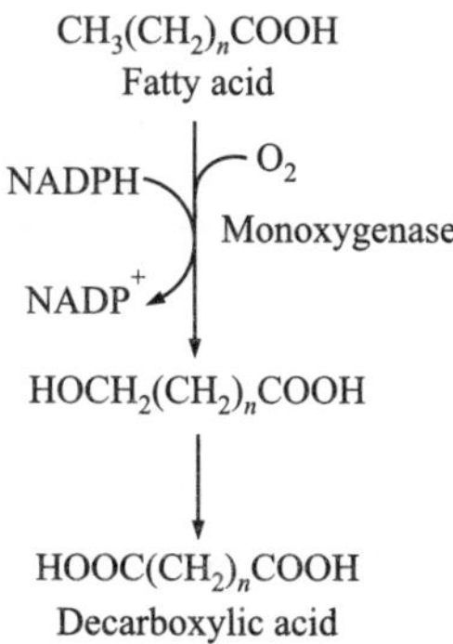

Figure 13.10 ω-oxidation of fatty acids.

C. Biosynthesis of Fatty Acids

The *de novo* synthesis of fatty acids (lipogenesis) occurs by a pathway that is entirely different from the pathway of catabolism of fatty acids by β-oxidation. The enzymes involved in the *de novo* synthesis of fatty acids are located in the extra-mitochondrial portion of the cell. Besides the *de novo* pathway, there is a mechanism for elongation of the carbon chain of an existing fatty acid. There are two active chain elongation systems, one located in the mitochondria and the other in the microsomal fraction of the cell.

(a) de novo synthesis of fatty acids

In mammals, the *de novo* synthesis of fatty acids is catalysed by a multienzyme complex known as fatty acid synthetase (or synthase) complex. It is active in the form of a dimer formed from two identical monomers. The dimer has a molecular weight of about 230,000. Each monomer consists of a polypeptide chain containing seven enzymes and a molecule called acyl carrier protein (ACP), which is located in the terminal part of the chain. Each chain is folded into three domains: (i) domain 1 or the substrate entry and condensation unit, (ii) domain 2 or the reduction unit and (iii) domain 3 or the palmitate release unit. ACP has 4′ phosphopantetheine as its prosthetic group (Figure 13.11). The 4′-phosphopantetheinyl moiety functions as an arm, which allows an acyl moiety bonded to it to be positioned in contact with the enzymes catalyzing the reactions of fatty acid biosynthesis. The sulfhydryl groups of the dimer act in a coordinated manner. They alternate in their function carrying first an acyl group and then a malonyl group. In yeast, the multienzyme complex consists of two polypeptide chains and β, assembled as a subunit $\alpha_6\beta_6$. In bacteria and plants, the enzymes of *de novo* fatty acid biosynthesis are not components of a tightly associated multienzyme complex.

The carbon atoms of the fatty acids are derived from acetyl CoA. The acetyl CoA formed inside the mitochondria

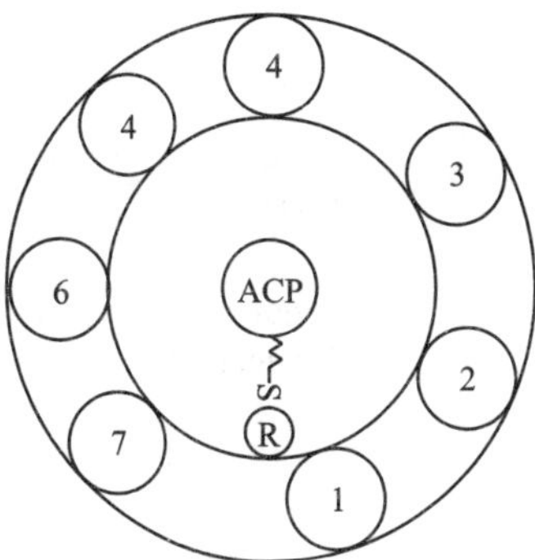

Figure 13.11 Fatty acid synthetase (or synthase) is a multi-enzyme complex, which consists of two identical monomers. Each monomer consists of seven enzymes and a molecule called acyl carrier protein; ACP (1 = Acetyl transferase; 2 = Malonyl transferase; 3 = β-Ketoacyl ACP synthase; 4 = β-Ketoacyl ACP reductase; 5 = β-Hydroxyacyl dehydratase; 6 = Enoyl ACP reductase; 7 = Deacylase; and R = acyl moiety).

cannot directly enter the cytoplasm because the mitochondrial membranes are impermeable to acetyl CoA. The acetyl moiety gains access into the cytoplasm indirectly via citrate, which is synthesized inside the mitochondria from acetyl CoA during the reaction of citric acid cycle. In the cytoplasm, citrate is cleaved into acetyl CoA and oxaloacetate in a reaction catalysed by ATP-citrate lyase (Figure 13.12).

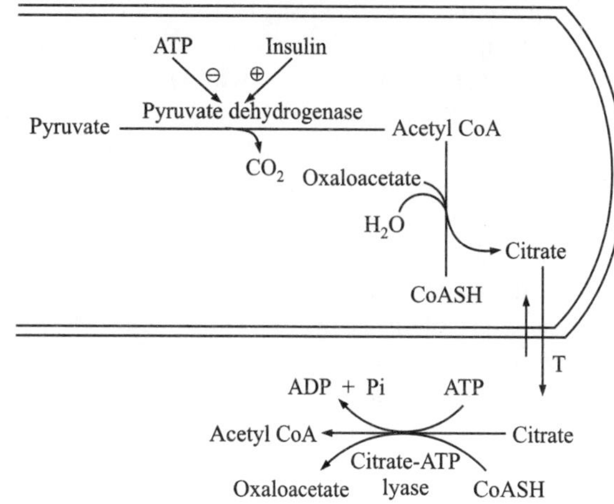

Figure 13.12 The acetyl CoA formed inside the mitochondria gains access into the cytoplasm indirectly via citrate (T = tricarboxylate transporter).

The initiation of fatty acid synthesis from acetyl CoA requires a preliminary activation process. There is ATP-dependent carboxylation of acetyl moiety of acetyl CoA to produce malonyl CoA. The reaction is catalysed by a biotin-containing enzyme, acetyl CoA-carboxylase (Figure 13.13). Bicarbonate is the source of carbon dioxide required in this reaction. This is the rate-limiting reaction in the *de novo* biosynthesis of fatty acids.

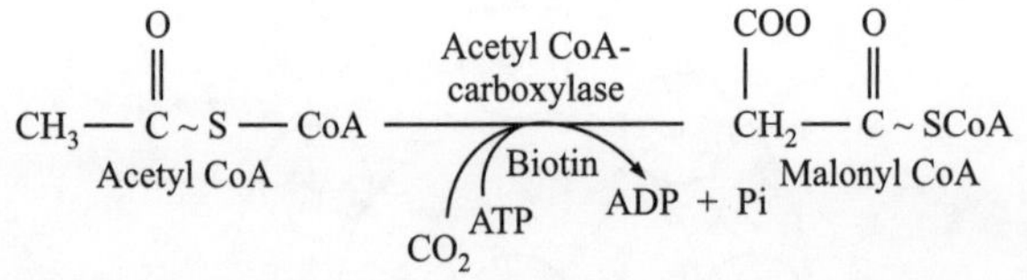

$$CH_3 - C(=O) \sim S - CoA \xrightarrow[\substack{\text{Biotin} \\ ATP \quad CO_2 \quad ADP + Pi}]{\substack{\text{Acetyl CoA-} \\ \text{carboxylase}}} CH_2(COO) - C(=O) \sim SCoA$$

Acetyl CoA → Malonyl CoA

Figure 13.13 The ATP-dependent carboxylation of acetyl CoA is catalysed by acetyl CoA carboxylase, a biotin-containing enzyme. It is a rate-limiting reaction in the *de novo* biosynthesis of fatty acids.

The stepwise biosynthesis of a fatty acid involves a priming reaction, followed by the repetitive cycles of reaction, through which there is addition of malonyl CoA and simultaneous loss of carbon dioxide.

In the priming reaction, the acetyl moiety of a molecule of acetyl CoA is transferred to a sulf-hydryl group of acyl carrier protein (ACP) on one of the monomers of fatty acid synthase complex. This reaction is catalysed by acetyl transferase (Figure 13.14). Similarly, the malonyl moiety from malonyl CoA is transferred to the sulfhydryl group of the other monomer. This reaction is catalysed by malonyl transferase. From this step onwards, the cycles of repetitive reactions begin (Figure 13.15). These reactions are described below:

1. **Condensation:** The acetyl and malonyl moieties undergo a condensation reaction, which forms an acetoacetyl moiety (*β*-ketoacyl ACP). Carbon dioxide is liberated in this process. This reaction is catalysed by *β*-ketoacyl ACP synthase.
2. **Reduction:** The *β*-ketoacyl ACP is then reduced by NADPH-dependent *β*-ketoacyl-ACP reductase. The *β*-keto group is reduced to yield a secondary alcohol group and the product formed is D(–)–*β*-hydroxyacyl ACP, which has the opposite configuration to that of L(+) *β*-hydroxyacyl moiety formed in the process of *β*-oxidation.
3. **Dehydration:** The D(–)*β*-hydroxyacyl moiety is dehydrated to form *α, β* unsaturated acyl ACP. The reaction is catalysed by *β*-Hydroxyacyl dehydratase. The product *α, β* unsaturated acyl ACP contains the *trans* Δ^2-acyl moiety.
4. **Reduction (Saturation):** Finally, the *α, β*-unsaturated acyl ACP undergoes reduction. NADPH-dependent enoyl-ACP reductase catalyses the reaction to yield saturated acyl ACP. With this reaction, one round of the cyclic pathway is completed.

The saturated acyl moiety is now transferred to the sulfhydryl group to which the acetyl moiety was initially attached, and the other sulfhydryl group is made free to react with another incoming malonyl group. Again, the four repetitive steps of condensation, reduction, dehydration and reduction (saturation) are repeated once more. With each round of this cyclic pathway, the acyl chain is elongated by two carbon atoms, which are carbons 1 and 2 of the malonyl moiety (Figure 13.16). This process continues until the carbon chain

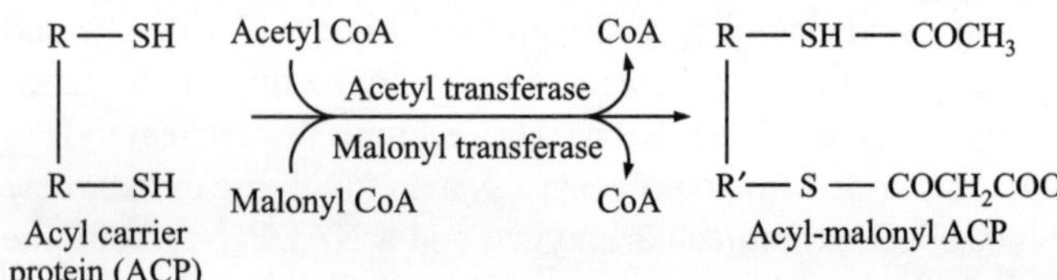

Figure 13.14 The acetyl or malonyl moiety from acetyl CoA or malonyl CoA, respectively, is transferred to a sulfhydryl group of the acyl carrier protein on one of the monomers of fatty acid synthase complex. This is the priming reaction for the biosynthesis of fatty acids.

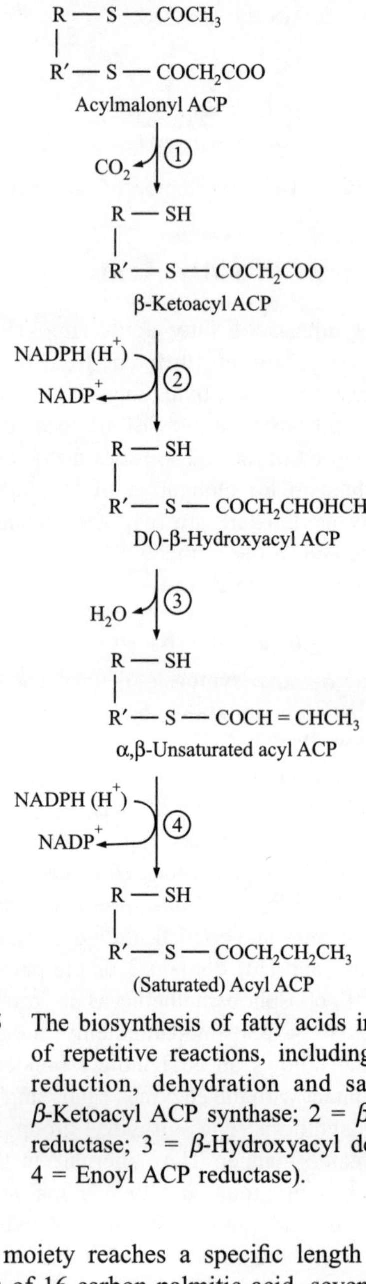

Figure 13.15 The biosynthesis of fatty acids involves a cycle of repetitive reactions, including condensation, reduction, dehydration and saturation. (1 = *β*-Ketoacyl ACP synthase; 2 = *β*-Ketoacyl ACP reductase; 3 = *β*-Hydroxyacyl dehydratase; and 4 = Enoyl ACP reductase).

of the acyl moiety reaches a specific length (for example, for synthesis of 16-carbon palmitic acid, seven rounds of the cyclic pathway of above-mentioned reactions are required). At this point, the acyl moiety is released from the ACP in a hydrolytic reaction, catalysed by a deacylase.

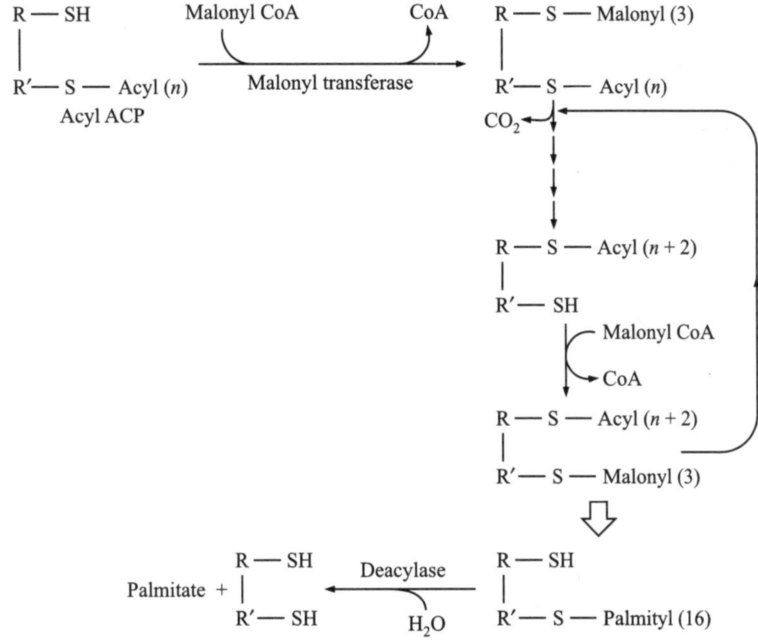

Figure 13.16 During the biosynthesis of fatty acids, with each cycle of repetitive reactions (shown in Figure 13.15), the acyl chain is elongated by two carbon atoms. These carbon atoms are derived from the malonyl moiety. Finally, acyl moiety is released from acyl carrier protein in a reaction catalysed by deacylase.

Unsaturated fatty acids are synthesized by a modification of the metabolic pathway described above. In this process, a β-hydroxyacyl moiety is converted into a *cis* Δ^3-acyl moiety rather than a *trans* Δ^2-acyl moiety. The same enzyme catalyses the formation of both the *cis-* Δ^3-acyl moiety and the *trans* Δ^2-acyl moiety. The *cis* Δ^3-acyl moiety can then undergo the condensation reaction catalysed by β-ketoacyl synthases followed by the remaining three reactions of the biosynthetic pathway.

(b) Systems for chain elongation

Fatty acids having more than 16 carbon atoms are synthesized by reactions that lengthen the carbon chain of the existing fatty acids. Two enzyme systems catalyze the reactions involved in chain elongation. One is associated with mitochondria and the other with the microsomal fraction of the cell.

The microsomal system for chain elongation is the main site for the elongation of existing long chain fatty acids. The substrate for the enzymes are the thioesters of CoA (Figure 13.17). The acetyl donor is malonyl CoA. The reactions are essentially similar to those involved in the *de novo* synthesis of a fatty acid.

The mitochondrial system utilizes acetyl CoA as the source of carbon atoms. The reactions are essentially similar to those of the *de novo* synthesis of a fatty acid. However, this pathway involves NADH instead of NADPH for the reduction of β-ketoacyl moiety. The physiological significance of this pathway is uncertain. It operates when [NADH]/ [NAD$^+$] ratio in the mitochondria is high.

(c) Microsomal desaturase system

The multienzyme system acyl CoA desaturase is located in the endoplasmic reticulum of animal cells, and it catalyses

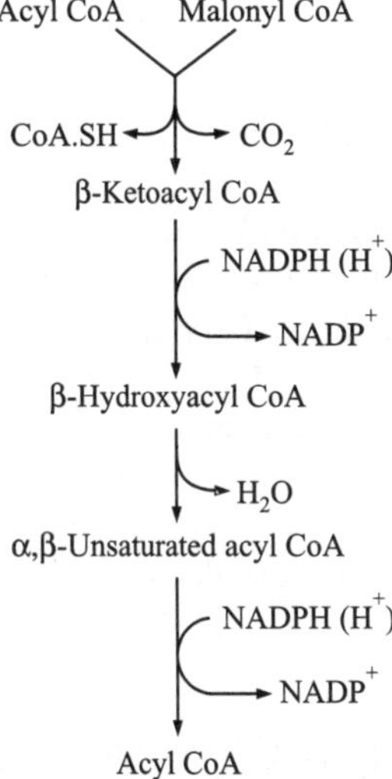

Figure 13.17 The microsomal system for chain elongation.

the desaturation of fatty acids. It is a monoxygenase system involving molecular oxygen, NADH, cytochrome b5 and cytochrome b5 reductase (Figure 13.18). It brings about synthesis of monounsaturated fatty acids from saturated fatty acids by introducing a double bond in the Δ^9 position in the carbon chain of the saturated fatty acid.

$$\text{Stearyl CoA} + \text{O}_2 + \text{NADPH (H}^+) \longrightarrow \text{Oleyl CoA} + \text{H}_2\text{O} + \text{NADP}^+$$

Figure 13.18 The microsomal desaturase system.

The coexistence of the chain elongation system and desaturase system in the endoplasmic reticulum permits the synthesis of non-essential polyunsaturated fatty acids by the animal cell. In a sequence of alternating reactions, there is addition of 2-carbon unit (from malonyl CoA) to the existing carbon chain, followed by desaturation to introduce a double bond.

Regulation of fatty acid biosynthesis (Lipogenesis)

Lipogenesis is controlled by several factors. The nutritional state of the organism is the major factor that controls the rate of fatty acid synthesis. It operates through the concentration of free fatty acids (FFAs) in the plasma. There is an inverse relationship between hepatic synthesis of fatty acids and the concentration of FFA in plasma. The rate-limiting step in fatty acid synthesis is catalysed by acetyl CoA carboxylase. Acetyl CoA carboxylase is activated by citrate but inhibited by long chain fatty acids in the form of acyl CoA molecules. Thus, accumulation of acyl CoA inhibits the synthesis of new fatty acids. Acyl CoA may also prevent the transport of citrate from mitochondria into the cytoplasm by inhibiting the mitochondrial tricarboxylate transporter.

Another key enzyme which controls lipogenesis is pyruvate dehydrogenase. It can be inhibited by acyl CoA. Increased concentration of acetyl CoA and NADH in the mitochondria is inhibitory for pyruvate dehydrogenase.

Glycolysis is an important pathway which supplies acetyl CoA. Therefore, factors which affect the rate of glycolysis will play an important role in the regulation of fatty acid synthesis.

The pancreatic islet cell hormone insulin plays a key role in stimulating fatty acid synthesis. Insulin acts in many different ways. It facilitates the transport of glucose into the cell. Glucose is the source of pyruvate and glycorol-3-phosphate required for fatty acid synthesis. Insulin also activates acetyl CoA carboxylase and pyruvate dehydrogenase. Insulin also acts by decreasing the concentration of long chain fatty acyl CoA, which is an inhibitor of fatty acid synthesis. This effect is brought about by inhibition of lipolysis resulting from the decreased intracellular levels of cAMP.

V. METABOLISM OF TRIACYLGLYCEROLS (TRIGLYCERIDES)

A. Catabolism of Triacylglycerols

The catabolism of triacylglycerols is initiated by hydrolysis of triacylglycerols into their constituent free fatty acids and glycerol. The process of lipolysis is described in detail in section III. The FFAs are taken up by the tissues and subsequently oxidized to produce energy. Alternatively, they can be used for the biosynthesis of triacylglycerols or transported to other tissues for conversion into other lipids. Glycerol can be utilized by tissues that possess the enzyme glycerokinase. Glycerokinase converts glycerol into glycerol-3-phosphate, which can be used either for the biosynthesis of triacylglycerols or for the synthesis of glucose by the process of gluconeogenesis.

B. Biosynthesis of Triacylglycerols

The enzymes involved in the biosynthesis of triacylglycerols are located in the mitochondria and the endoplasmic reticulum of the cell. Synthesis of triacylglycerols requires fatty acids and glycerol. These precursors need to be activated to acyl CoA and glycerol-3-phosphate, respectively, before they can participate in the synthesis of triacylglycerols. Glycerol-3-phosphate can be formed by two pathways (Figure 13.19). In tissues that possess the enzyme glycerokinase (e.g., liver, kidney, intestine, lactating mammary gland), glycerol can be converted to glycerol-3-phosphate. In tissues that lack this enzyme activity, completely or partially (e.g., muscle and adipose tissue), glycerol-3-phosphate is synthesized from dihydroxyacetone phosphate (DHAP). The fatty acids are activated to acyl CoA in a reaction catalyzed by thiokinase (or acyl CoA synthetase).

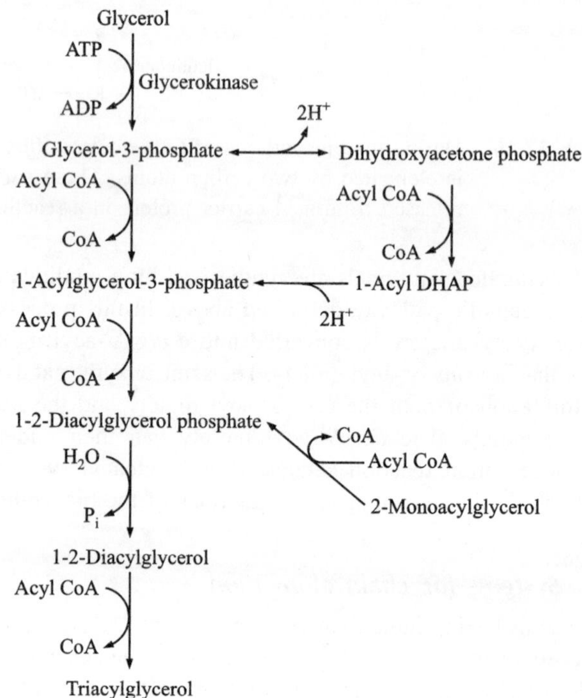

Figure 13.19 Biosynthesis of triacylglycerols (triglycerides).

The formation of a monoacylglycerol phosphate (also known as lysophosphatidate) can take place by two routes. Either glycerol-3-phosphate is acylated or there is acylation of DHAP, followed by reduction of acyl DHAP. The acylation of glycerol-3-phosphate is the pathway more frequently used for the synthesis of monoacylglycerol phosphates. The monoacylglycerol phosphates are acylated to form diacylglycerol phosphates (known as phosphatidic acid or phosphatides). Next, the phosphatide undergoes a hydrolysis reaction in which phosphoryl group is lost and a diacylglycerol is formed. Acylation of the diacylglycerol results in the formation of a triacylglycerol. In the intestinal mucosa, a monoacylglycerol pathway operates in which two acylation reactions lead to the formation of triacylglycerol from monoacylglycerol (Figure 13.19).

VI. METABOLISM OF PHOSPHOLIPIDS

Among the important phospholipids are phosphatidylcholine (lecithin), phosphatidylethanolamine (cephalin), phosphatidylinositol, phosphatidylserine, cardiolipin, lysophospholipids, glycerol ether phospholipids, plasmalogens (phosphatidal ethanolamine) and sphingomyelins. The chemistry of these phospholipids is discussed in Chapter 6.

A. Biosynthesis of Phospholipids

Phosphatidic acid serves as a common precursor for the biosynthesis of triacylglycerols and phospholipids. In mammals, phosphatidic acid initiates two different pathways. In one pathway, phosphatidic acid is converted to a diacylglycerol, which can be converted to phosphatidylcholine, phosphatidylethanolamine and phosphatidylserine (Figure 13.20). In the second pathway, phosphatidic acid is converted to CDP-diacylglycerol, which can be converted to phosphatidylinositol, cardiolipin and in bacteria also phosphatidylserine (Figure 13.20).

The biosynthesis of glycerol ether phospholipids and plasmalogens is initiated by the acylation of dihydroxyacetone phosphate (Figure 13.21).

In sphingomyelins, the alcohol portion is a complex alcohol, sphingosine, instead of glycerol. Sphingomyelins are acyl amides (ceramides) and not acyl esters. The biosynthesis of cardiolipin and sphingomyelins is shown in Figures 13.22 and 13.23, respectively.

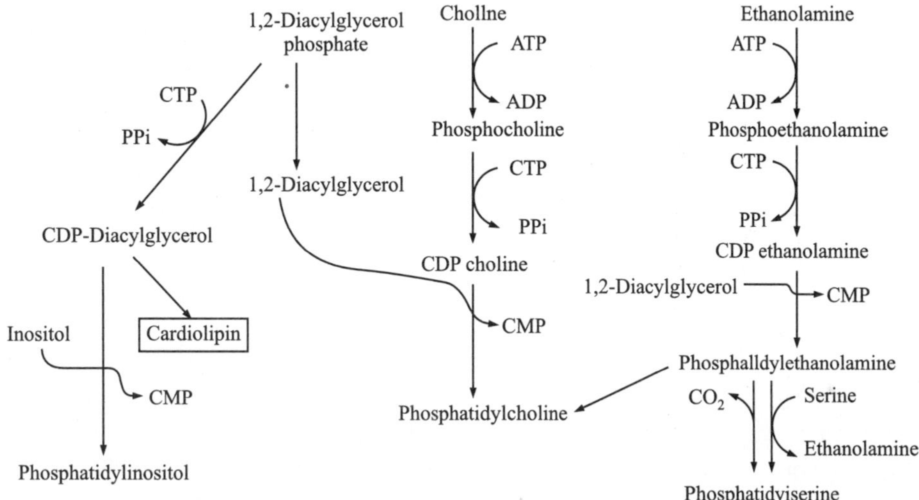

Figure 13.20   Biosynthesis of phospholipids (CTP = cytidine triphosphate; CDP = cytidine diphosphate and CMP = cytidine monophosphate).

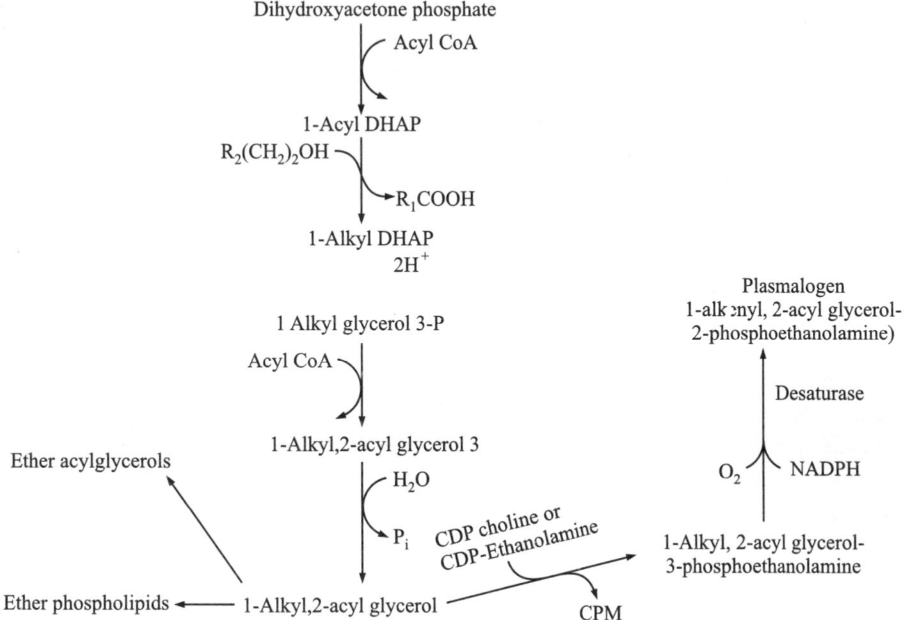

Figure 13.21   Biosynthesis of glycerol ether phospholipids and plasmalogens (CDP = cytidine diphosphate).

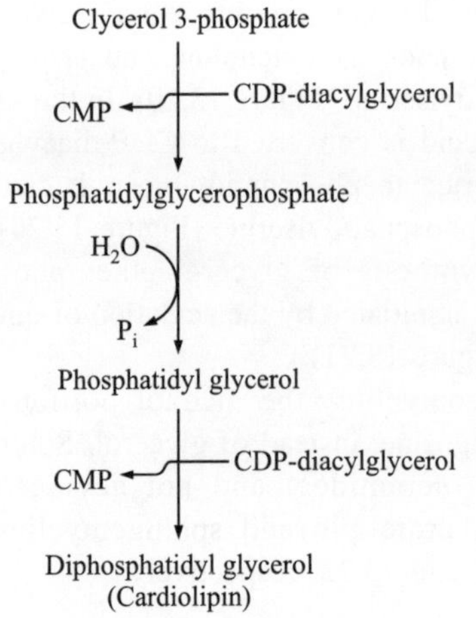

Figure 13.22 Biosynthesis of cardiolipin (CDP = cytidine diphosphate; CMP = cytidine monophosphate).

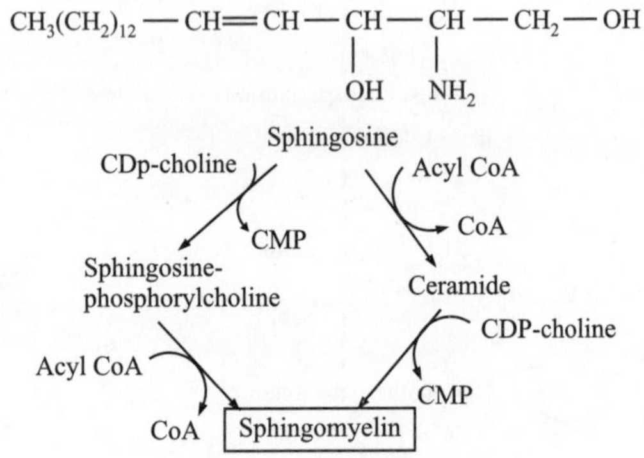

Figure 13.23 Biosynthesis of sphingomyelins (CDP = cytidine diphosphate; CMP = cytidine monophosphate).

B. Degradation of Phospholipids

The phospholipids are actively degraded by hydrolytic cleavage reactions, and each portion of the molecule has a different turnover rate. The enzymes phospholipases allow partial degradation of the phospholipids, followed by resynthesis. A phospholipid like phosphatidylcholine can be degraded by four different phospholipases (Figure 13.24). Presumably, other kinds of phospholipids are catabolized similarly by means of appropriate enzymes. Phospholipase A1 (phosphatidate 1-acylhydrolase) catalyses the hydrolytic cleavage of the ester bond involving the hydroxyl group at carbon 1 of the glycerol moiety of a phospholipid substrate. Phospholipase B (or lysophospholipase) acts similarly to remove the acyl group from carbon 1 in lysophospholipids like lysolecithin. Phospholipase A_2 attacks the ester bond in position 2 of phospholipids. Phospholipase C attacks the ester bond in position 3. Phospholipase D catalyses hydrolytic removal of the nitrogenous base from the phospholipid substrate.

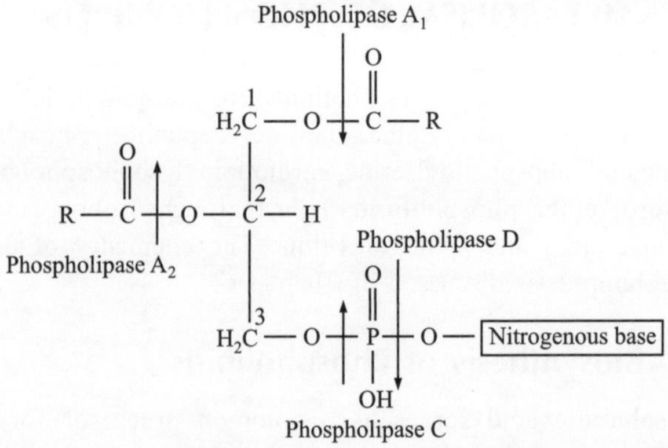

Figure 13.24 Phospholipids are degraded by the action of phospholipases. The sites of action of phospholipases A1, A2, C and D on a phosphatidylcholine molecule are indicated by arrows.

VII. METABOLISM OF GLYCOLIPIDS

Glycolipids (glycosphingolipids or sphingolipids) consist primarily of ceramide (acyl sphingosine) and hexoses. Simple glycolipids such as cerebrosides contain sphingosine, a high molecular weight fatty acid (e.g., lignoceric, cerebronic, nervonic or stearic acid), and galactose. Cerebrosides with sulfated galactose moiety are known as sulfatides or sulfolipids.

Complex glycolipids like gangliosides contain ceramide, a fatty acid, sialic acid (*N*-acetylneuraminic acid; NANA) and three molecules of hexoses.

The biosynthesis of cerebrosides is initiated by a reaction between ceramide and UDP-galactose (Figure 13.25). The acyl moiety of ceramide represents the fatty acid that is to be incorporated into the cerebroside. The cerebrosides react with "active sulfate" (3″-phosphoadenosine-5′-phosphosulfate; PAPS) to form sulfatides (Figure 13.25). For synthesis of gangliosides, activated sugars and *N*-acetylneuraminic acid are added to ceramide in a series of reactions, which is shown in Figure 13.26.

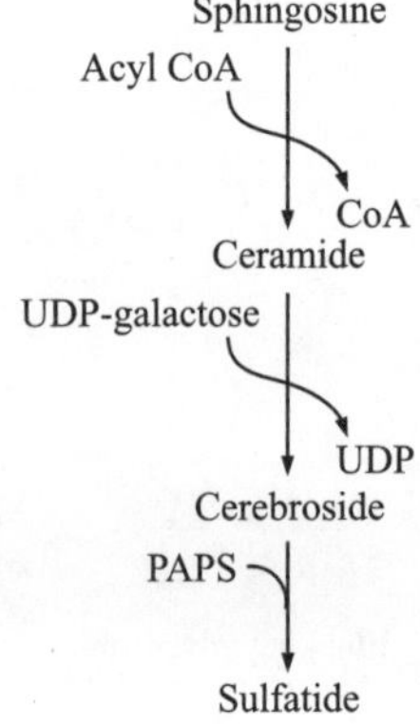

Figure 13.25 Biosynthesis of cerebrosides and sulfatides (UDP = uridine diphosphate; PAPS = 3′-phosphoadenosine-5′-phosphosulfate).

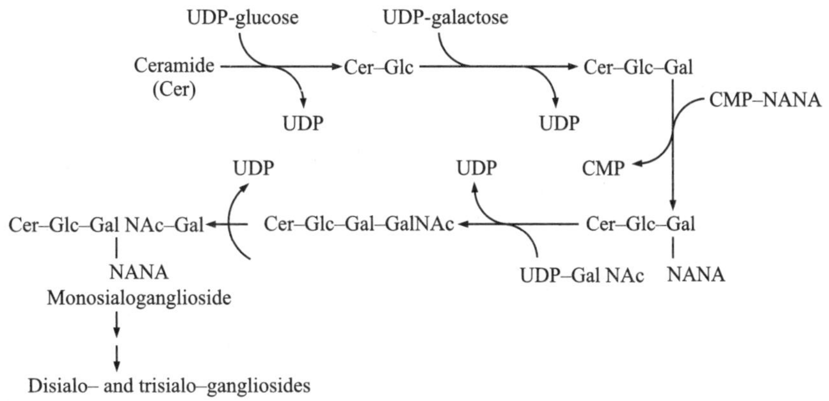

Figure 13.26 Biosynthesis of gangliosides (Cer = Ceramide; Glc = Glucose; Gal = Galactose; NANA = N-acetyl neuraminic acid; GalNAc = N-acetyl galactose, CMP = cytidine monophosphate; and UDP = Uridine diphosphate).

The glycolipids undergo continuous degradation by hydrolytic enzymes generally present in the lysosomes of the cells. The abnormal degradation of glycolipids due to partial or complete deficiency of these enzymes results in many diseases categorised as "lipidoses". These inborn errors of lipid metabolism are discussed in Chapter 18.

VIII. METABOLISM OF KETONE BODIES

Three substances, namely, acetoacetic acid, D(–)-β-hydroxybutyric acid and acetone, are collectively known as "ketone bodies" or "acetone bodies" (Figure 13.27). Ketosis is a condition in which there is an increased level of ketone bodies in the blood (ketonemia) or urine (ketonuria). The increased blood levels of acetoacetic acid and β-hydroxybutyric acid, which are moderately strong acids, result in metabolic acidosis (ketoacidosis). Ketosis results from the production of ketone bodies in excess of the body's ability to utilize them. Ketosis occurs in starvation and severe diabetes mellitus.

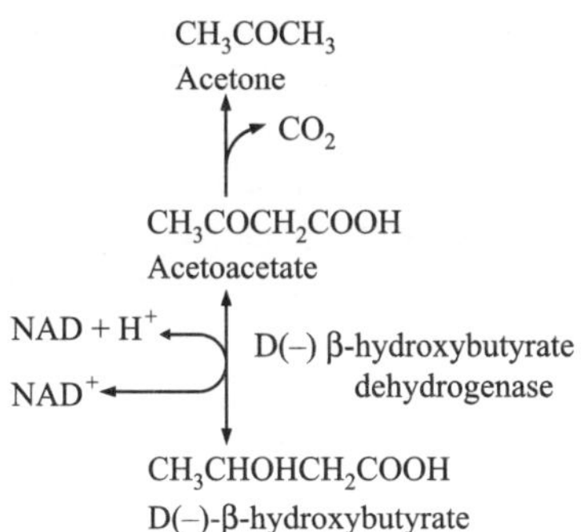

Figure 13.27 The term *ketone bodies* refers to three substances, namely, acetoacetic acid, D(–)-β-hydroxybutyric acid and acetone.

1. Formation of ketone bodies: The ketone bodies are formed from acetyl CoA in the liver. They are normal degradation products of fatty acids. The enzymes involved in the formation

of ketone bodies are located in the mitochondria. There are two pathways for the formation of ketone bodies:

(a) *HMG CoA pathway:* This major pathway involves the synthesis of an intermediate HMG-CoA (β-hydroxy-β-methyl glutaryl CoA) from acetyl CoA (Figure 13.28). HMG CoA is then cleaved to form acetyl CoA and acetoacetate. The latter may be converted to β-hydroxybutyrate.

(b) *Deacylase pathway:* A minor pathway that involves the formation of acetoacetate from acetoacetyl CoA. The reaction is catalysed by a deacylase (Figure 13.28).

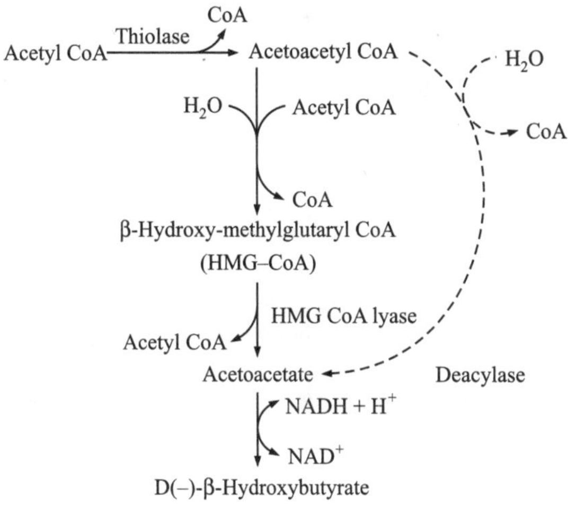

Figure 13.28 Formation of ketone bodies occurs through two pathways. The major pathway involves the synthesis of an intermediate, HMG-CoA. The minor pathway involves a reaction catalysed by deacylase.

2. Utilization of ketone bodies: The ketone bodies formed in the liver are transported to extrahepatic tissues for utilization. Liver has very low activity of enzymes involved in the oxidation of ketone bodies. Hence, practically, ketone bodies are catabolised by the extrahepatic tissues. The acetoacetate can be converted to acetoacetyl CoA by two pathways: (i) pathway that involves succinyl CoA. The reaction is catalysed by succinyl CoA-aceto-acetate-CoA transferase, also known as thiophorase (Figure 13.29), (ii) pathway that involves activation of acetoacetate by ATP (Figure 13.30). The reaction is catalysed by acetoacetic thiokinase. The β-hydroxybutyrate is also converted to acetoacetate, followed by activation of acetoacetate to acetoacetyl CoA. The acetoacetyl CoA is then acted upon by thiolase to form acetyl CoA, which is oxidized to carbon dioxide and water in the citric acid cycle.

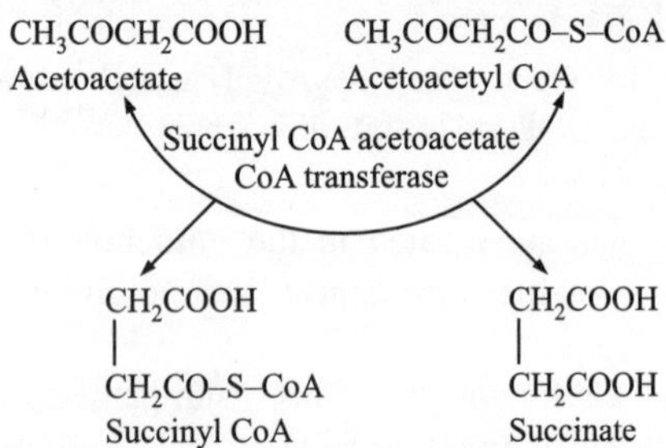

Figure 13.29   Utilization of acetoacetate by conversion to acetoacetyl CoA. Succinyl CoA is required for this reaction.

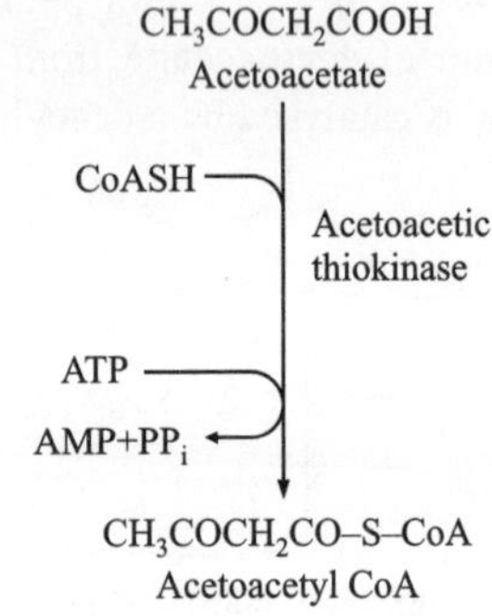

Figure 13.30   Utilization of acetoacetate by activation to acetoacetyl CoA. ATP and coenzyme A are required for this reaction.

Regulation of Ketogenesis

The term *ketogenesis* means the formation of ketone bodies. Free fatty acids are the main substrates for the formation of ketone bodies by the liver. Therefore, factors which increase lipolysis indirectly increase ketogenesis. The increased concentration of FFA indirectly increases oxidation of fatty acids by activation of the enzyme carnitine acyltransferase 1. With increased fatty acid oxidation, more fatty acids are directed towards formation of ketone bodies, whereas less fatty acids are oxidized to produce carbon dioxide. These factors are so balanced that total production of ATP remains constant.

IX. METABOLISM OF CHOLESTEROL

Cholesterol is the most important member of a group of

compounds known as sterols. It is found in foods of animal origin, but is absent (or present in traces) in plant foods. It is an essential component of cell membranes. It is the precursor of various steroid hormones (such as estrogens, progesterone, corticosteroids and androgens), vitamin D and bile acids.

A.   Biosynthesis of Cholesterol

Cholesterol is readily synthesized in the body, particularly in the liver, adrenal cortex, skin, intestine and testis. The enzymes required for the synthesis of cholesterol are located in the endoplasmic reticulum and cytosol of the cell. All the carbon atoms in the cyclopentanoperhydrophenanthrene nucleus and octyl side chain of the cholesterol molecule (Figure 13.31) are derived from acetyl CoA. The formation of cholesterol from acetyl CoA proceeds through formation of the intermediate compounds—mevalonate, squalene and lanosterol. The sequence of reactions involved in the *de novo* synthesis of cholesterol are described below:

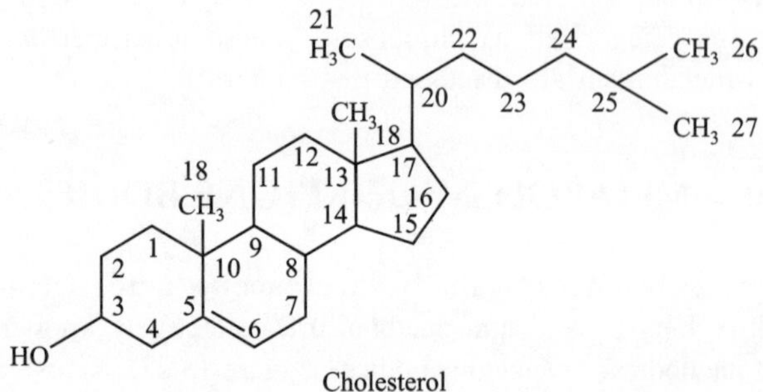

Figure 13.31   The structure of cholesterol. Acetyl CoA is the source of all the carbon atoms in the cyclopentanoperhydrophenanthrene nucleus and octyl side chain of the molecule.

1. Formation of mevalonate: The formation of mevalonate is initiated by the condensation of acetyl moieties derived from two acetyl CoA molecules, resulting in the formation of acetoacetyl CoA (Figure 13.32). This reaction is catalysed by cytoplasmic acetyl CoA acetyl transferase. The acetoacetyl moiety of acetoacetyl CoA then condenses with the acetyl moiety of another acetyl CoA molecule to form HMG CoA (β-hydroxy-β-methyl glutaryl CoA). This reaction is catalysed by cytosolic enzyme, HMG CoA synthetase. Furthermore, HMG CoA is reduced in two stages by microsomal NADPH-dependent HMG CoA reductase to form mevalonic acid (or mevalonate). One molecule of acetyl CoA is released in this reaction.

2. Formation of squalene: The synthesis of 30-carbon molecule of squalene from mevalonate involves the formation of a 5-carbon isoprenoid unit, followed by condensation of the isoprenoid units. Mevalonate is phosphorylated in three ATP-dependent reactions to form mevalonate-3-phospho-5-pyrophosphate, which is an unstable compound (Figure 13.33). It loses carbon dioxide and phosphate to form isopentenyl pyrophosphate, the active 5-carbon isoprenoid unit,

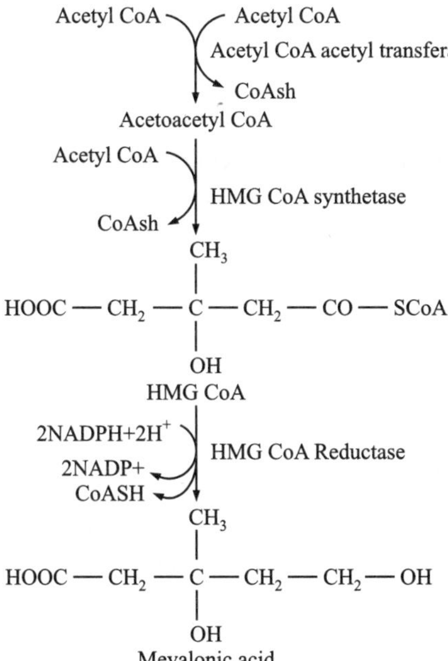

Figure 13.32 Formation of mevalonic acid (mevalonate) from acetyl CoA involves HMG-CoA as an intermediate compound (HMG CoA = β-hydroxy-β-methyl glutaryl CoA).

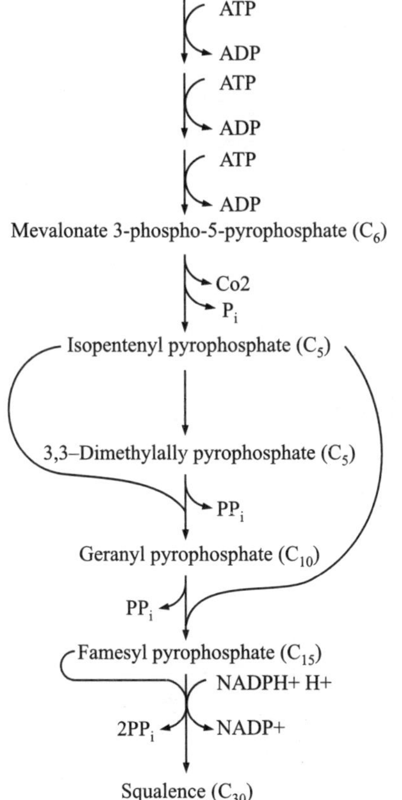

Figure 13.33 Formation of squalene (a 30-carbon compound) from mevalonic acid.

which can be isomerised to 3,3-dimethylallyl pyrophosphate. These two 5-carbon isomers condense in a head-to-tail fashion to produce squalene (30-carbon) through the formation of geranylpyrophosphate (10-carbon) and farnesyl -pyrophosphate (15-carbon).

3. Formation of lanosterol: Squalene is first converted to squalene 2,3-oxide, which undergoes cyclization through a series of reactions to form lanosterol (Figure 13.34). These reactions are catalyzed by squalene epoxidase (or monooxygenase) and squalene oxide cyclase, respectively.

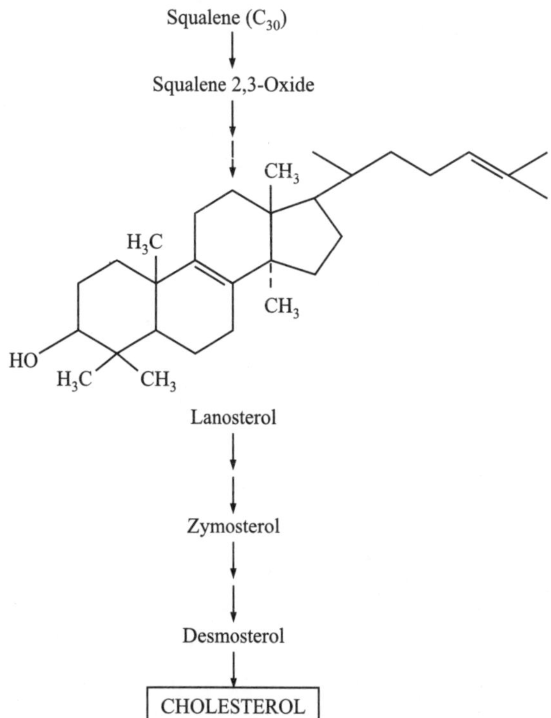

Figure 13.34 Formation of cholesterol molecule from squalene. Lanosterol, zymosterol and desmosterol represent the intermediate compounds formed in this process.

4. Formation of cholesterol: Three methyl groups on carbon atoms 4 and 14 (two on carbon 4 and one on carbon 14) of the lanosterol molecule are oxidized to carboxyl groups, which are eliminated as carbon dioxide by decarboxylation. The product of this reaction is zymosterol (Figure 13.34). The double bond at position 8,9 of zymosterol shifts to position 5,6 to form desmosterol. Finally, the double bonds in the side chain of desmosterol are reduced to form cholesterol.

B.   Regulation of Cholesterol Biosynthesis

Both rate-limiting and regulatory steps in the biosynthesis of cholesterol involve the reaction catalysed by HMG CoA reductase. The activity of HMG-CoA reductase is affected by several factors, which may increase or decrease the enzyme activity. The decrease in enzyme activity is brought about by a negative feedback inhibition by repressing the synthesis of HMG CoA reductase or by inducing the synthesis of enzymes that degrade HMG CoA reductase. These factors may operate singly or in combination. Moreover, HMG CoA reductase may be regulated by phosphorylation–dephosphorylation. Feedback inhibition of HMG CoA reductase in the liver is brought about by cholesterol and bile acids. Bile acids, but not cholesterol, can inhibit the intestinal HMG CoA reductase. Reduced activity of HMG CoA reductase is observed in fasting and following administration of glucagon or glucocorticoids. On the other hand, administration of insulin or thyroxine increases the activity of this enzyme. Mevinolin is a potent competitive inhibitor of HMG CoA reductase. Its structure resembles that of HMG CoA. Mevinolin can effectively block the synthesis of cholesterol. It is useful in reduction of blood cholesterol in patients with familial hypercholesterolemia.

The incorporation of cholesterol into chylomicrons and VLDL, the transport of cholesterol in blood as lipoproteins and the clinical significance of cholesterol metabolism with special reference to cardiovascular system are discussed in Chapter 14.

Interestingly, the modification of proteins at C-terminal cysteine residues by the isoprenoids farnesyl and geranylgeranyl is essential for the biological function of a number of eukaryotic proteins. This highlights another important function of the intermediate isoprenoids.

X. PROSTAGLANDINS

Prostaglandins are derived from prostanoic acid, which is a 20-carbon carboxylic acid consisting of a cyclopentane unit and two aliphatic side chains. There are five series of prostaglandins, designated as PGA, PGB, PGE, PGF and PGH. The prostaglandins are considered to be local hormones. They act by modulating the intracellular cAMP levels and the intracellular flow of calcium. The details of the chemistry and metabolism of prostaglandins are discussed in Chapter 8.

SUGGESTIONS FOR FURTHER READING

Bloch K. and Vance D. (1977), Control Mechanisms in the Synthesis of Saturated Fatty Acids, *Annu. Rev. Biochem.*, 46, 263.

Borén J., Taskinen M.R., and Adiels M. (2012), Kinetic Studies to Investigate Lipoprotein Metabolism, *J. Intern Med.*, 271, 166–73.

Brzozowski A.M., et al. (1991), A Model for Interfacial Activation in Lipases from the Structure of a Fungal Lipase-inhibitor Complex, *Nature*, 351, 491.

Dominiczak M.H. (1997), Lipids: The Story So Far, *Br. J. Cardiol.*, 4, 425–429.

Dominiczak M.H. and Caslake M.J. (2011), Apolipoproteins: Metabolic Role and Clinical Biochemistry Applications, *Ann Clin Biochem.*, 48, 498–515.

Eaton S., Bartlett K. and Poufarazam M. (1996), Mammalian β-Oxidation, *Biochem. J.*, 320, 345–357.

McGarry J.D. and Foster D.W. (1980), Regulation of Hepatic Fatty Acid Oxidation and Ketone Body Production, *Annu. Rev. Biochem.*, 49, 395.

Omer C.A. and Gibbs J.B. (1994), Protein Prenylation in Eukaryotic Microorganisms: Genetics, Biology and Biochemistry, *Mol. Microbiol.*, 11, 219.

Schroepfer G.J. Jr. (1982), Sterol Biosynthesis, *Annu. Rev. Biochem.*, 51, 555.

Scriver C.W., Beaudet A.L. and Sly W.S. et al. (Eds.) (1989), *The Metabolic Basis of Inherited Disease*, 6th ed., McGraw-Hill, New York.

Wakil S.J., Stoops J.K. and Joshi V.C. (1983), Fatty Acid Synthesis and its Regulation, *Annu. Rev. Biochem.*, 52, 537.

14

Significance of Blood Levels of Lipids and Lipoproteins in Health and Disease

Balram Bhargava and Sunil Verma

CONTENTS

I. INTRODUCTION

Plasma lipids are mainly composed of cholesterol, triglycerides, phospholipids and free fatty acids. Owing to the predominance of hydrocarbon chains ($-CH_2-CH_2-CH_2-$) in their structure, they are insoluble in water and are carried in the bloodstream in combination with protein. The lipid–protein complex is called a lipoprotein. Biological lipids serve as (i) storage form of metabolic fuel, (ii) a transport form of metabolic fuel, (iii) a part of outer coat between the body of the organism and the environment and (iv) the structural component of the membrane. They also function in the formation of bile acids, as well as adrenalcortical, androgen and estrogen hormones. Six lipoproteins are classified into six major classes. These are (i) chylomicrons, (ii) very low density lipoproteins (VLDL), (iii) intermediate density lipoproteins (IDL), (iv) low density lipoprotein (LDL), (v) high density lipoprotein (HDL) and (vi) lipoprotein(a). The different chemical and physical characteristics of these lipid-protein aggregates allow classification of hyperlipoproteinemias. Such a classification allows the general finding of elevated cholesterol and triglycerides to be related to a specific class. This in turn gives useful additional clues to the understanding of the underlying metabolic abnormalities and their prognostic and therapeutic implications.

Hyperlipidemia is of concern to physicians for several reasons. It is generally accepted as a coronary risk factor (its presence increases the risk of developing arterial atherosclerotic disease). In some familial lipid metabolic disorders, pancreatitis may occur. Skin rashes and xanthomatas may appear acutely or be chronically present requiring management. Its presence, on the other hand, may indicate that many factors such as unsuitable diet, excessive alcohol intake, stress or disorders causing secondary fat metabolism abnormalities such as hypothyroidism, diabetes mellitus, nephritic syndrome, biliary tract obstructive disease or pancreatitis may be existing unsuspected, requiring evaluation and management Table 14.1.

TABLE 14.1

Factors to consider in patients with hyperlipidemia

Primary hyperlipidemia (Genetic disorders)

LDL (low density lipoprotein) particles
    Familial hypercholesterolemia
    Familial defective apo B100
    Autosomal dominant hypercholesterolemia
    Autosomal recessive hypercholesterolemia
    Abetalipoproteinemia
    Hypobetalipoproteinemia
    Familial sitosterolemia
    Familial lipoprotein (a) hyperlipoproteinemia

Remnant lipoproteins
    Dysbetalipoproteinemia type III
    Hepatic lipase deficiency

Triglyceride-rich lipoproteins
    Lipoprotein lipase deficiency
    Apo C II deficiency
    Apo A V
    Familial hypertriglyceridemia
    Familial combined hyperlipidemia

HDL particles
    APO A1 deficiency
    Tangier disease/familial HDL deficiency
    Familial LCA deficiency syndrome
    CETP deficiency
    Niemann–Pick disease types A and B
    Niemann–Pick disease type C

Secondary hyperlipidemia

Uncontrolled diabetes mellitus
Hypothyroidism
Uraemia
Nephrotic syndrome
Obstructive liver disease
Dysproteinemia (multiple myeloma, lupus erythematosis)

Dietary factors

Caloric intake (recent weight gain)
Content of saturated fats and cholesterol
Alcohol intake

Drugs producing or aggravating hyperlipidemia

Oral contraceptive
Glucocorticoids
Antihypertensives (thiazides and beta blockers)
Retinoic acid derivatives
Testosterone
Immunosuppressive medications (cyclosporine)
Antiviral medications (HIV protease inhibitors)

Some commonly used terms that require familiarization are as follows:

1. *Hyperlipidemia Excess:* cholesterol, triglycerides or both are present in plasma.
2. *Hyperlipoproteinemia (HPL):* One or more of the lipoproteins carrying cholesterol and triglycerides are present in excess.

3. *Secondary hyperlipoproteinemia*: When excess of lipoproteins and their components is as a result of some other disease disorder. Several such diseases have been identified and include uncontrolled diabetes mellitus, hypothyroidism, nephritic syndrome, inappropriate dietary habits, biliary tract obstruction, pancreatitis, etc.
4. *Primary hyperlipoproteinemia*: When the existing elevated serum levels of lipoproteins are not due to one of the many known causes. These are genetically determined defects in lipid or lipoprotein metabolism which are exaggerated by environmental factors through poorly understood mechanisms.

II. LIPOPROTEINS

Lipoproteins are complex macromolecular structures composed of an envelope of phospholipids and free cholesterol, a core of cholesterol ester and triglycerides. Protein moiety of lipoprotein is called apolipoprotein. Lipoproteins are classified according to density of lipoprotein particles as determined by their flotation rate on the centrifugation in 1.063 d sodium chloride solution. The following six fractions are distinguished:

1. *Chylomicrons*: These are lipoproteins of intestinal origin (density <0.95 gm/ml).These consist almost entirely of dietary triglycerides with small amounts of cholesterol, phospholipids and proteins. Chylomicrons are present in the plasma for 1–4 h after a meal. The enzyme lipoprotein lipase (LPL) by its action dissociates triglycerides from the chylomicrons, with the resulting free fatty acid (FFA) transported to adipose tissue, heart and skeletal muscles for utilization. The remnant particles derived from chylomicron following LPL action contain apo E and enter the liver for degradation and reutilization of their core constituents.
2. *Very low density lipoproteins or pre betalipoproteins*: These are synthesized in liver and transport endogenously produced triglycerides (TG) which are derived from the circulating carbohydrates and free fatty acids generally mobilized from the tissue (density <1.006).
3. *Intermediate density lipoprotein*: After hydrolysis of triglycerides that partly depletes VLDL of triglycerides, VLDL particles have relatively more cholesterol. They shed several apolipoproteins (especially the C apolipoproteins) and acquire apo E. The VLDL remnant lipoprotein, called intermediate density lipoprotein (density 1.006–1.019), is taken up by the liver via its apo E moiety or further delipidated by hepatic lipase to form an LDL particle.
4. *Low density lipoprotein or beta lipoprotein:* It is mostly a metabolic product of VLDL and IDL. The density of LDL varies between 1.019 and 1.063. This accounts for most of the cholesterol in the plasma. LDL particles contain predominantly cholesteryl ester packed with

protein moiety apo B100. LDL particle size variation results from changes in the core constituents, with an increase in triglycerides and a relative decrease in cholesteryl ester, leading to smaller, denser LDL particles. LDL has the highest atherogenic potential.

5. *High density lipoprotein*: These are synthesized in the liver and intestine. Their density varies between 1.063 and1.210. HDL particles acquire their components from several sources although these components also are metabolized at different sites. Apolipoprotein A1 is the main protein of HDL. HDL promotes reverse cholesterol transport and can prevent lipoprotein oxidation. HDL levels are inversely related to the risk of atherosclerosis. Increased HDL levels may be a protective factor in the development of vascular disease.

6. *Lipoprotein(a)*: It is synthesized in the liver and its density is between 1.051 and 1.082. Lipoprotein(a) consists of an LDL particle linked covalently with one molecule of apo(a) at its C-terminal end. This addition blocks the receptor apo B site so that Lp(a) is not cleared by this receptor pathway but instead by VLDL receptor. The apo(a) moiety consists of a protein with high degree of homology with plasminogen. There is a weak association between lipoprotein(a) and coronary artery disease.

The protein parts of lipoproteins (named apolipoprotein) play four major roles: (1) assembly and secretion of lipoprotein, (2) structural integrity of lipoprotein, (3) co-activators or inhibitors of enzymes and (4) binding and docking to specific receptors and proteins for cellular uptake of the entire particle or selective uptake of a lipid component. Different lipoproteins and their predominant lipoprotein is given in Table 14.2.

TABLE 14.2 Apolipoproteins and their associated lipoprotein classes

Apoprotein	*Lipoprotein Class*
Apo AI	HDL
Apo AII	HDL
Apo IV	HDL
Apo V	VLDL, HDL
Apo B100	LDL, VLDL
Apo B48	Chylomicrons
Apo C1	Chylomicrons
Apo CII	Chylomicrons, VLDL
Apo CIII	Chylomicrons, VLDL
Apo D	HDL
Apo E	Chylomicrons remnant, IDL
Apo J	HDL
Apo(a)	Lipoprotein(a)
Apo M	HDL
Apo L1–6	HDL

Non-esterified fatty acids (NEFA), free fatty acid: These are derived from adipose tissue and are used by most organs (except some; like brain) as a major source of energy, or are stored after esterification as triglycerides.

Phospholipids

A phospholipid consists of a glycerol molecule linked to two fatty acids. The third carbon of the glycerol moiety carries a phosphate group to which one of the four molecules is linked: choline (phosphatidylcholine or lecithin), ethanolamine (phosphatidylethanolamine), serine (phosphatidylserine) or inositol (phosphatidylinositol). Sphingomyelin is a related phospholipid that has a special function in the plasma membrane in the formation of membrane microdomains.

III. TYPES OF ABNORMAL LIPOPROTEIN PATTERNS

Five major abnormal lipoprotein patterns have been recognized, which may be primary or secondary to the other coexisting secondary disease states (Table 14.3).

TABLE 14.3 Profile of abnormal lipoprotein patterns— Types of Hyperlipoproteinemia and their composition

Type I	Plasma milky
	Chylomicrons +++
	Cholesterol N
	Triglycerides +++
Type II	Plasma clear
	Chylomicrons nil
	Cholesterol ++ or +++
	Triglyceride N or +
	LDL +++
	VLDL N or +
Type III	Plasma opalescent
	Chylomicrons nil
	Cholesterol ++
	Triglycerides ++ or +++
	VLDL ++ to +++
Type IV	Plasma opalescent
	Chylomicrons nil
	Cholesterol +
	Triglycerides +++
	VLDL +++
Type V	Plasma opalescent
	Chylomicrons +
	Cholesterol +
	Triglycerides ++
	VLDL ++

VLDL = Very low density lipoprotein; LDL = Low density lipoprotein N = Normal; + = Slightly increased; ++ = Moderately increased; +++ = Markedly increased

These are as follows:

A. Type I

There is hyperchylomicronemia even in a postabsorptive state (12–14 h after a meal). Plasma is milky and on refrigeration a creamy layer settles at the top while the infranatant fluid is normal. Fasting plasma triglyceride levels are high (>1000 gm/dl), being almost entirely of dietary origin. LDL and VLDL are normal, but cholesterol is only marginally increased. The hypertriglyceridemia results from a markedly reduced or absent LPL activity or, more rarely, the absence of its activator apo CII.

Patients are usually infants or children with recurrent attacks of abdominal pain (probably due to pancreatitis), hepatosplenomegaly, eruptive xanthomas and a creamy serum. Severe hypertriglyceridemia can also be associated with xerostomia, xerophthalmia and behavioral abnormalities. Disordered carbohydrate metabolism is not present. Premature atherosclerosis has not been observed so far although the period of observation is limited and the number of cases is small.

Secondary type I is uncommon but may be seen in uncontrolled diabetes mellitus, pancreatitis and acute alcoholism.

B. Type II (Hyperbetalipoproteinemia)

There is an abnormal increase of low density lipoprotein. This type is subdivided into type IIa and type IIb. In the former type, only LDL is increased. In type IIb, there may also be an increase in VLDL. Characterstically, serum cholesterol is very high although the triglycerides are normal, except in IIb, where these are also high. Cholesterol/triglyceride ratio is always 1.5 or more. Defect in LDL-R gene cause an accumulation of LDL particles in plasma and thus alter the function of the LDL-R protein. To date, there are more than 1000 identified mutations of the LDL-R gene.

Type II hyperlipoproteinemia is the most common cause of the hypercholesterolemia. High serum concentrations of cholesterol need further study in any patient to rule out a secondary cause and require thyroid, hepatic, renal and pancreatic function studies and a survey of nutritional and physical habits, alcohol intake and stress situations. A detailed study usually indicates that the moderate hypercholesterolemia or hypercholesterolemia–hypertriglyceridemia is fundamentally diet induced and a result of a long exposure to a unsatisfactory diet, high in total calories, total and saturated fats, cholesterol and carbohydrate and low in unsaturated fats.

Type II lipoprotein abnormalities are not uncommonly inherited. The disease is then transmitted as an autosomal dominant trait, the hyperlipoproteinemia is being accompanied by xanthomas in palpebral fissure and tuberous xanthomas. *Arcus senilis* is common, and there is accelerated atherosclerosis, especially of coronary arteries. Glucose tolerance is commonly abnormal.

Diagnosis is based on observation of high levels of cholesterol and beta lipoproteins and normal or slightly elevated triglycerides. The lipoprotein abnormalities are present in at least one of the parents and distributed in other relatives.

The very high incidence of atherosclerotic coronary artery disease in familial type II abnormalities gives further support to the importance of high cholesterol and beta lipoproteins in the genesis of atherosclerosis. Data available from human subjects and from experimental animal studies underline the importance of high levels of cholesterols (and of beta lipoproteins) in the causation of coronary heart disease although the exact pathogenic mechanisms are disputed. In humans, except for special situations such as hypothyroidism, nephritic syndrome, etc., associated with high levels of beta lipoproteins and serum cholesterol, diet, specially high animal fat intake, plays a major role in causing this abnormality.

Populations habitually subsisting on low animal fat diets have lower levels of serum cholesterol and also of atherosclerosis. This is seen in people of Italy, Spain, Central America and in many communities even in America and Europe who live on low fat diets (Trappist monks, Seventh day Adventists, etc.).

C. Type III (Dysbetalipoproteinemia or Broad Beta Disease)

The most characteristic feature is a typical pattern of a broad B band between prebeta (VLDL) and beta (LDL) lipoproteins (on lipoprotein agarose gel electrophoresis).The plasma is usually turbid. The lipoprotein profile shows increased cholesterol and triglyceride levels and reduced HDL cholesterol. The ratio of VLDL cholesterol to triglycerides, normally less than 0.7 mmol/l (<0.30 mg/dl), is elevated. Plasma cholesterol varies from 200 to1400 mg/100 ml. Defect results from abnormal apo E, which does not bind to hepatic receptors that recognize apo E as a ligand.

Clinically, patients have early arcus senilis, tuberous xanthomas and palmar striated xanthomas. Advanced and early atherosclerosis of peripheral and coronary vessels is very common. Carbohydrate intolerance is the rule in most patients.

D. Type IV

This has a lipoprotein pattern of endogenous hyperlipemia. Synthesis of triglycerides (mainly hepatic) is increased and exceeds their capacity for their removal from blood. It is an indication of metabolic imbalance and is considered as "carbohydrate-induced hyperlipemia" focusing attention on an associated abnormal glucose tolerance and a family history of diabetes, which is remarkably frequent.

There is a marked increase in the VLDL (hyper-prebeta lipoproteinemia). The triglycerides are increased and also the cholesterol, but to a lesser extent, and the triglyceride/cholesterol ratio is 5:1 or more. Chylomicrons are not present.

Severe cases may have eruptive xanthomas, lipemia retinalis and foam cells in the bone marrow. Patients are prone to abdominal pain with or without chemical evidence of pancreatitis. High serum uric acid is frequent. Obesity and abnormal glucose tolerance are almost universal. Although detection has been made in some children, the disease usually manifests at an older age.

Prebeta lipoproteinemia is associated with higher incidence of coronary artery disease at a relatively younger age, but to lesser extent than that in types II and III.

Secondary type IV disorder is especially common in diabetes mellitus and pancreatitis. The diabetics in this disorder are younger, non-obese, insulin-sensitive, ketosis-prone patients. Hyperlipemia is universal in uncontrolled state, but clears quickly when diabetes is brought under control.

Glycogen storage disease, idiopathic hyperglycemia, hypothyroidism, nephritic syndrome, dysglobulinemia, oral contraceptive intake, gout, Niemann–Pick disease are other disorders associated with this type of lipid abnormality.

E. Type V

Patients are characterized by having chylomicrons and increased amounts of VLDL (prebetalipoproteins) in the postabsorptive state. Cholesterol and triglycerides are elevated but in varying proportions. The overnight refrigerated plasma shows a creamy layer of chylomicrons at the top. The infranatant layer remains latescent after refrigeration.

Symptoms appear in the second or the third decade. Patients are obese and frequently have a family history of diabetes mellitus and obesity. Recurrent abdominal pain often occur and eruptive xanthomas may appear. Hepatosplenomegaly may be present. Accelerated atherosclerosis in patients or their families is not remarkable.

Such a pattern is frequently seen as a familial disorder, but it is also seen in association with uncontrolled insulin-dependent diabetes, chronic pancreatitis or heavy alcohol intake.

IV. FREE FATTY ACIDS (FFA, NEFA)

The free fatty acid content of normal plasma is small (0.3–0.7 meq/l; 8–20 mg/l of the total plasma lipid concentration). It is the most common form in which fat is transported through the plasma compartment. Most of the fatty acids originate from and reassemble into the adipose tissue triglyceride in the blood. FFAs circulate as FFA–albumin complex, which releases the fatty acid at the sites of utilization, such as the liver, myocardium, muscles and other tissues. Excess FFA is converted in the liver to triglycerides. A sensitive hormonal and nervous control exists over the rapidly changing body requirements for the FFA as an energy fuel, depending on the available insulin, catecholamines, adrenocortical hormones, thyroid hormone, glucagon, hormone and sympathetic nervous system activity.

In the state of acute lipid mobilization from adipose tissue such as that after nor-adrenaline administration, thyrotoxicosis and diabetic ketosis, plasma FFA levels are invariably elevated.

V. TRIGLYCERIDES

Calories derived from fats and carbohydrates have to be taken up from the blood. In man, dietary carbohydrates are first assimilated by the liver. A part of this is deposited locally as glycogen and the remainder is transferred into fatty acids, which are incorporated into VLDL and find their way into plasma. The fatty acids of the VLDL triglycerides are then assimilated by adipose tissue. Calories ingested in the form of fats are transported as chylomicrons in the plasma. The triglycerides of both these triglycerides-rich lipoproteins are hydrolyzed by lipoprotein lipase activity in the capillaries and the fatty acids formed are taken up by the adipose tissue and esterified to triglyceride.

The enzyme system lipoprotein lipase (LLA) is responsible for lipolysis of lipoproteins trapped in the capillaries. The lipoproteins are said to require activation in order to be accessible to LLA. Some of the 'C' apolipoproteins probably act as activators or inhibitors. The estimation of postheparin LLA activity in hyperlipoproteinemia in general shows no deficiency, indicating that this does not play an important role as a causative mechanism for hypertriglyceridemia. LLA activity, on the other hand, is related to nutritional or hormonal factors. Fasting decreases and feeding increases LLA in animals. Cyclic AMP plays a role in LLA activity. Agents known to increase cyclic AMP (catecholamines) decrease LLA, while those that decrease cyclic AMP (insulin, nicotinic acid, prostaglandin E) increase LLA in animals. Its relevance to humans has, however, yet to be established. Low LLA in adipose tissue may be a pathogenic mechanism in the rare type I hyperlipoproteinemia. Its importance in other types is not known.

Processes which alter fatty acid incorporation into adipose tissues may be deranged and cause hyperlipoproteinemia with hypertriglyceridemia. Recent information has indicated low average fatty acid incorporation into adipose tissue in all types of hypertriglyceridemias compared to controls, especially in types IV and V hyperlipoproteinemia.

VI. DIAGNOSIS OF HYPERLIPOPROTEINEMIA

Although in past estimation of serum cholesterol has been the mainstay of diagnosis, a complete subtyping requires detailed analysis. This includes estimation of serum cholesterol and serum triglycerides, observation of serum kept overnight at 4°C for chylomicrons and the presence or absence of opalescence in the infranatant layer, electrophoresis of lipoproteins and ultracentrifugation. However, for practical purposes, it is possible to obtain a clear picture in the following ways:

(i) Observation of the serum obtained after overnight fasting and kept at 4°C. This is checked for the presence of the supernatant layer of chylomicrons. The infranatant layer is observed for opalescence, which indicates presence of significant amounts of VLDL.

(ii) Estimation of serum cholesterol, its HDL and LDL fraction and triglycerides in an overnight fasting specimen.

(iii) Electrophoresis of serum lipoproteins may, sometimes, be required.

A number of factors which influence plasma lipids and lipoproteins must be considered in the diagnosis of hyperlipo proteinemia. These include the influence of age, sex, methods of estimating serum lipids and the expected laboratory variation, presence of an acute or chronic disease, specially the former, individual variations, seasonal variation, geographic variation and nutritional status of the individuals. Diagnosis of HPL on the basis of a single sample analysis is undesirable. Estimation of plasma lipids under standard conditions on at least three different occasions should be carried out before accepting a diagnosis of hyperlipoproteinemia.

The merits of classification of systems are the expression hypercholesterolemia and hypertriglyceridemia according to lipoprotein abnormality. However, it gives no idea about the aetiology or pathogenesis of the abnormality or whether it is primary or secondary.

VII. MANAGEMENT OF HYPERLIPOPROTEINEMIA

A. Rationale

As HPL is frequently associated with disease and is an important risk factor in promoting disease such as ischemic heart disease, its correction can decrease the risk of disease and complications. Levels of plasma cholesterols and LDL have constantly been shown to be directly correlated with the risk of coronary artery disease (CAD). Data from Lipid Research Clinics Coronary Primary Prevention Trial and Helsinki Heart Study have provided conclusive evidence that a reduction in LDL-C in hypercholesterolemic men can result in a decrease incidence of coronary events, including fatal and nonfatal myocardial infarction. On the basis of evidences from clinical trials (primary prevention trials such as WOSCOS, AFCAPS, HPS, ALL-HAT LLT, ASCOT-LLA and PROSPER), the 2004 U.S. Adult Treatment Panel (ATP III) modified its guidelines recommending a more aggressive LDL-C target in high risk subjects. A number of secondary prevention trials (4S, CARE, LIPID, HPS, TNT, IDEAL) have shown that LDL-C lowering decreases cardiovascular events. Epidemological studies and other clinical trials are consistent in supporting the observation that for individuals with serum cholesterol levels in the 250–300 mg/dl range, each 1% reduction in serum cholesterol would yield about 2% reduction in the rate of combined morbidity and mortality from coronary artery disease.

B. Therapeutic Approach

In 2006, the AHA/ACC (American Heart Association/American College of Cardiology) updated their secondary prevention guidelines for lipid management. As with the NCEP guide—lines, LDL goal remains at less than 100 mg/dl, and less than 70 mg/dl is added as an optional goal. Table 14.4 gives the updated ATP III guidelines recommendations.

TABLE 14.4 Categories of elevated cholesterol

Category	LDL-C (mg/dl)	HDL-C (mg/dl)	TG (mg/dl)
Low risk/desirable	<130	>40	<150
Intermediate risk	131–160	31–40	151–250
Moderate high risk	161–190	25–30	251–400
High risk	>190	<25	>400

TABLE 14.5

Risk status based on presence of CHD risk factors other than LDL cholesterol

Positive risk factors:

- Age $\geq$ 45 years
- Female $\geq$ 55 years or premature menopause without oestrogen replacement therapy
- Family history of premature CHD (definite myocardial infarction or sudden death before 55 years of age in father or other male first degree relative, or before 65 years of age in mothers or other female first degree relative).
- Current cigarette smoking
- Hypertension
- Low HDL cholesterol ($\leq$35 mg/dl)
- Diabetes mellitus

Negative Risk Factors:
- High HDL Cholesterol (>60 mg/dl)

C. Nonpharmacological Management

Nonpharmacological management consists of diet and exercise program. Dietary management is always initially indicated to reduce serum cholesterol values. A *step one* diet contains 30% total fat, with less than 10% saturated fat, 50–60% carbohydrates, proteins less than 10–20% and cholesterol limited to less than 300 mg per day. If therapeutic goals are not achieved after 3 months of this program, a *step two* diet should be instituted. This contains 30% fat, with less than 7% saturated fat, and 200 mg of cholesterol per day. Total cholesterol can be used as a surrogate monitor for LDL cholesterol during the treatment period and is measured every 6 weeks. In addition to the dietary approach, the patient should be inducted in a regular exercise program. Emphasis is placed on regularity and consistency of exercise rather than extent of the effort expended. In addition to diet and exercise as initial therapeutic alternatives, smoking cessation should be strongly emphasized. Hypertension should also be treated with appropriate antihypertensive agents. Dietary and exercise program should be monitored every 2 months with repeat serum lipid profile. Four to six months should be allowed to optimize the lipid values with diet and exercise before pharmacological intervention is considered.

D. Pharmacological Management

The lipid lowering medications can be classified according to their mode of action on different sites of lipid metabolism. Various groups of lipid lowering medications are available (Table 14.6).

TABLE 14.6 Lipid Lowering Medications

(1) HMG CO-A reductase inhibitors (statins)
    Atorvastatin
    Fluvastatin
    Pravastatin
    Rosuvastatin
    Simvastatin

(2) Bile acid absorption inhibitors
    Cholesteryamine
    Colistipol
    Colesevelam

(3) Cholesterol absorption inhibitors
    Ezetimibe

(4) Fibrates
    Benzafirate
    Fenofirate
    Gemofibrozil

(5) Niacin
    Nicotinic acid

(6) Probucol

(1) HMG CO-A reductase inhibitors

Statins inhibit HMG CO-A reductase enzyme competitively and prevents the formation of mevalonate. This is the rate-limiting step of sterol synthesis. This leads to increased expression of LDL-R and the rate of cholesteryl ester formation declines. These haemostatic adjustments to HMG CO-A reductase inhibition increases LDL cholesterol clearance from plasma and decreases hepatic production of VLDL and LDL. These inhibitors act mainly in the liver. The major indications for this drug is primary severe hypercholesterolemia, as also moderate hypercholesterolemia in patients with coronary heart disease and other risk factors.

2013 ACC/AHA Guidelines on the treatment of blood cholesterol to reduce atherosclerotic cardiovascular risk in adults have given major emphasis on statin treatment. Accordingly, there are 4 major statin benefit groups—(1) with clinical ASCVD (atherosclerotic cardiovascular disease), (2) primary elevations of LDL–C >190 mg/dL, (3) diabetes aged 40 to 75 years with LDL–C 70 to 189 mg/dL and without clinical ASCVD, or (4) without clinical ASCVD or diabetes with LDL–C 70 to 189 mg/dL and estimated 10-year ASCVD risk >7.5%. Second important consideration is that there are 3 types of statin therapy groups depending upon intensity of statin therapy used for primary and secondary prevention. These are high-intensity statin therapy on average lowers LDL–C by approximately ≥50%, moderate-intensity statin therapy lowers LDL–C by approximately 30% to <50%, and lower-intensity statin therapy lowers LDL–C by <30%. High-intensity statin therapy with atorvastatin 40 mg to 80 mg reduced ASCVD risk more than moderate-intensity statin therapy with atorvastatin 10 mg, pravastatin 40 mg, or simvastatin 20 mg to 40 mg bid.

(2) Bile acid absorption inhibitors

Bile acid sequestrants are ion exchange resins that interrupt the enterohepatic circulation of bile acids by inhibiting their reabsorption in the intestine. Currently, they are used mainly in adjunctive therapy in patients with severe hypercholesterolemia due to increased LDL cholesterol (type II).

(3) Cholesterol absorption inhibitors

This class of drugs limits selective uptake of cholesterol and other sterols by intestinal epithelial cells by interfering with the Niemann–Pick C1-like 1 protein 1 (NPC1L1). Ezetimibe is the first of this type of drug. It is indicated for patients with LDL cholesterol levels above target on a maximally tolerated statin dose.

(4) Fibrates

They have a complex mechanism of action which includes the following:

- Enhancement of activity of lipoprotein lipase
- Partial inhibition of synthesis of cholesterol with an increased activity of LDL receptors
- Partial inhibition of synthesis of VLDL triglyceride
- Reduced conversion of cholesterol into bile acids
- Enhanced secretion of cholesterol into bile

Fibric acid derivatives are useful in preventing pancreatitis in patients with severe hypertriglyceridemia and also in patients with moderate hypercholesterolemia.

The mechanism of action of fibrates involves interaction with the nuclear transcription factor PPARα that regulates the transcription of the LPL, apo CII and apo AI genes.

(5) Niacin (Nicotinic acid)

Niacin is a drug that has multiple effects on the plasma lipoproteins. These effects appear to be secondary to its action of inhibiting the hepatic secretion of VLDL and decreased FFA mobilization from periphery. Recent work has identified cell surface receptors for nicotinic acid that belongs to G–protein-coupled heptahelical superfamily. Niacin is particularly effective in increasing HDL cholesterol and lowering triglyceride levels. The effect of niacin on LDL cholesterol is more modest. Niacin and its esters are useful for types II, III, IV and V. hyperlipoproteinemias.

E. Drug Combinations

For patients with severe or mixed hyperlipidemias, combined drug therapy can enhance the response. Bile acid sequestrants combined with niacin are used for heterogeneous familial hyperlipoproteinemia. If the LDL cholesterol level is above the target level on maximally tolerated dose of statin, ezetimibe can be combined with statin. Bile acid binding resins can be used in combination with statins and cholesterol absorption inhibitors in case of severe hypercholesterolemia. Niacin in combination with low dose statin can retard the progression of CAD and decrease adverse cardiac events. Statin with fibrate combination is considered for patients with high risk of coronary heart disease and a residual dyslipidemia after initiation of statin therapy in patients with type II diabetes mellitus.

SUGGESTIONS FOR FURTHER READING

Fredrickson D.S. and Lees S.R. (1965), A System for Phenotyping Lipoproteinemia, *Circulation*, 31, 321, 1965.

Greenland P., Alpert J.S. and Beller G.A. (2010), ACCF/AHA Guidelines for Assessment of Cardiovascular Risk in Asymptomatic Adults, A Report of American College of Cardiology Foundation/American Heart Association Task Force on Practice Guidelines, Circulation, 122:e584–636.

Jacques Genest and Peter Libby (2011), Lipoprotein disorders and cardiovascular disease, in Heart Disease, Braunwald E. (Ed.), 10th ed., Philadelphia, W.B. Saunders Company.

Libby P. (1995), Molecular Basis of the Acute Coronary Syndromes, *Circulation*, 91, 2844–2850.

Rafkind B.M. and Levy R.I. (Eds.) (1997), *Hyperlipidemias: Diagnosis and Treatment*, New York, Grunne and Stratton.

Robert A. Hegele (2009), Plasma Lipoproteins: Genetic Influence and Clinical Implications, *Nature Reviews Genetics*, 10, 109–121.

Robert L. Huang and David J. Maron (2012), Dyslipidemia and Other Cardiac Risk Factors, Hurst's *The Heart*, 13th ed., McGraw-Hill, Inc., New York.

Stone N.J. et al. (2013), ACC/AHA Blood Cholesterol Guidelines, A Report of the American College of Cardiology/American Heart Association Task Force on Practice Guidelines.

Third Report of the National Cholesterol Education Program (NCEP) (2002), Expert Panel on Detection, Evaluation and Treatment of High Blood Cholesterol in Adults, *Adult Treatment Panel III*, final report, Circulation.

WHO Bulletin (1972), Classification of Hyperlipidemia and Hyperlipoproteinemias, *Classification*, 45, 50.

15

Digestion and Absorption of Proteins

L.M. Srivastava

CONTENTS

I. DIGESTION OF PROTEINS

The digestion of proteins in the gastrointestinal tract is carried out by specific proteolytic enzymes present in gastric juice and pancreatic secretions as also by the enzymes of the mucosa of the small intestine. The salivary secretions, on the other hand, do not carry out the digestion of proteins.

A. Proteolytic Enzymes

The proteolytic enzymes may be classified into the following two groups:

1. Endopeptidases

These degrade the larger molecules of the dietary proteins to yield smaller fragments. This includes pepsin, which is formed from a precursor present in gastric juice, and trypsin and chymotrypsin formed from precursors present in pancreatic juice. These three enzymes are grouped under the head *endopeptidases* as these attack the proteins specifically at certain peptide linkages within the molecule.

2. Exopeptidases

These are concerned mainly in completing the process started by the first group of enzymes and eventually liberating the amino acids. This category includes carboxypeptidase of the pancreatic juice, aminopeptidase and dipeptidases of the intestine. Carboxypeptidases and aminopeptidases are grouped

under the head *'exopeptidases'* as these remove amino acids from either the carboxyl end (carboxypeptidase) or the N-terminal end (aminopeptidase) of the protein.

The proteolytic enzymes of gastric juice and pancreatic secretions are initially secreted and stored within the exocrine cells as inactive precursors called *zymogens* which are later converted to the active form of the enzymes in the gut. Their storage as zymogens is presumably to prevent their hydrolyzing action on tissue proteins. In general their activation on emptying into the lumen of the gastro-intestinal tract consists in the removal of a short peptide from the original inactive forms.

B. Gastric Digestion

The proteins in food are present either as native proteins or as cooked denatured proteins. The proteolytic enzyme of the gastric juice is *pepsin*. Pepsin is elaborated and secreted by the chief cells of the gastric mucosa as an inactive precursor *pepsinogen*. After its secretion in the stomach, the inactive pepsinogen is converted to the *active* form of the enzyme pepsin by: (a) the acidity of the gastric juice due to the presence of HCl, and (b) by pepsin itself (autocatalytically). Gastric HCl removes the masking peptide from pepsinogen and the pepsin so obtained hastens the process of conversion at pH 4.6 or below and activates the remaining pepsinogen.

$$\text{Pepsinogen} \xrightarrow[\substack{\text{Pepsin} \\ \text{(autocatalytic)}}]{\text{HCl}\,(\text{H}^+)} \text{Pepsin + masking peptide}$$

Although pepsinogen is stable in neutral solution, pepsin loses its catalytic activity rapidly and completely at neutral pH.

Pepsinogen has a molecular weight around 42,500 and the active enzyme (pepsin) 35,500. In its activation, therefore, a cleavage of peptides with a total molecular weight of around 7500 is involved. As can be seen in Figure 15.1 initially the inhibitor fragment (molecular weight 31,000) is removed along with some miscellaneous peptides whose total molecular weight amounts to approximately 4000.

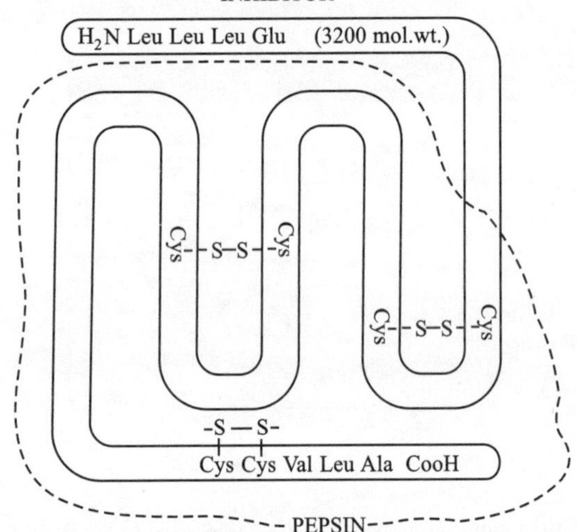

Figure 15.1   The structure of pepsinogen and pepsin indicating the composition of the inhibitor and certain other peptides split off during conversion of pepsinogen to pepsin. (From: Herriott, R.M., Pepsinogen and Pepsin, J. Gen. Physiol. 45, 57, 1962).

Pepsin starts the digestion of many native proteins as well as denatured proteins resulting from cooking of food, and splits them into smaller fragments (polypeptides). Though the action of pepsin is not highly specific, it attacks, preferentially, the peptide linkages formed between L-dicarboxylic acids (glutamic and aspartic) and L-aromatic amino acids (phenylalanine, tyrosine and tryptophan). It also requires that second carboxyl group of dicarboxylic amino acid should be free and there should not be a free amino group (i.e., a basic amino acid) in the neighbourhood of the peptide linkage. It, however, also acts if free carboxyl group of glutamic acid is replaced by sulphydryl group of cysteine. It is not necessary that peptide linkages attacked by pepsin should be adjacent to a free carboxyl or amino group end and, therefore, implies that pepsin acts on the *interior* of the peptide chain and hence it is grouped under the head *"endopeptidase"*.

Pepsin acts mainly at a pH below 3 and the optimum pH ranges between 1.5 to 2.2. As a result of the action of pepsin, the proteins are mainly broken down to proteoses and peptones, i.e., a mixture of polypeptides of different chain lengths.

$$\text{Proteins} \xrightarrow{\text{Pepsin}} \text{Proteoses + Peptones + Amino acids}$$

As the food remains in the stomach for a comparatively short period (2 h), and as pepsin is the only proteolytic enzyme in the gastric secretion, a partial hydrolysis of the proteins is achieved in the stomach. Under optimum conditions only 10–15% of protein may be broken down to amino acids in the stomach.

In infants the stomach also secretes another enzyme, rennin, which causes the coagulation of milk. This enzyme is especially important for the digestion of milk in infants because it prevents the quick removal of milk from the stomach. Rennin converts irreversibly the casein of the milk to para-casein which can then be acted upon by pepsin. This enzyme is apparently absent in the stomachs of adults.

C.   Protein Digestion in the Small Intestine

The partially digested food material from the stomach arrives in the intestine and is at acidic pH. Its pH is brought towards neutrality by the buffers present in pancreatic secretions. The pancreatic juice that is delivered in the duodenum contains four proteolytic zymogens. These are *trypsinogen, chymotrypsinogen, procarboxypeptidase* and *pro-elastase*. In the intestine, *trypsinogen* is activated and converted to *trypsin* by an enzyme called enterokinase (an intestinal enzyme). Trypsin so obtained, activates more of trypsinogen in an autocatalytic manner. Trypsinogen is made up of a single folded polypeptide chain of 229 amino acids while trypsin contains 223 amino acids in the chain—six less than trypsinogen. It is now fully established that during the activation of trypsinogen to trypsin, a hexa-peptide of the sequence Val-(Asp)₄-Lys is split off. This hexapeptide can therefore be termed as the "masking peptide" and the active enzyme trypsin is obtained only after the removal of this masking peptide from the inactive precursor trypsinogen. It is also known that the active catalytic site of trypsin resides at the histidine and serine regions of the molecule.

Chymotrypsinogen is different from trypsinogen in that it is not activated by enterokinase. Its activation is brought about by trypsin. Also, chymotrypsin does not activate chymotrypsinogen in an autocatalytic manner. Chymotrypsinogen of the pancreatic juice are of A and B types. Its conversion to active forms involves the removal of certain peptides such as seryl arginine and threonyl arginine. The chymotrypsinogen molecule has a chain length of 246 amino acids. The active site of chymotrypsin is identical to that of trypsin. It is believed that the conversion of inactive precursor zymogen forms of both trypsin and chymotrypsin to active forms involves, essentially, a change in the conformational state of enzyme protein so as to expose the catalytic site of the enzymes. The activation of chymotrypsinogen in fact serves as a good example for understanding the basic principles of zymogen conversion.

In the activation of chymotrypsinogen to chymotrypsin, the following steps are thought to occur. As is evident in Figure 15.2, the structure of chymotrypsinogen is not favourable for expression of activity of its active site. Both the potential active sites, being embedded in the interior of the molecule, are inaccessible to reaction with the substrate.

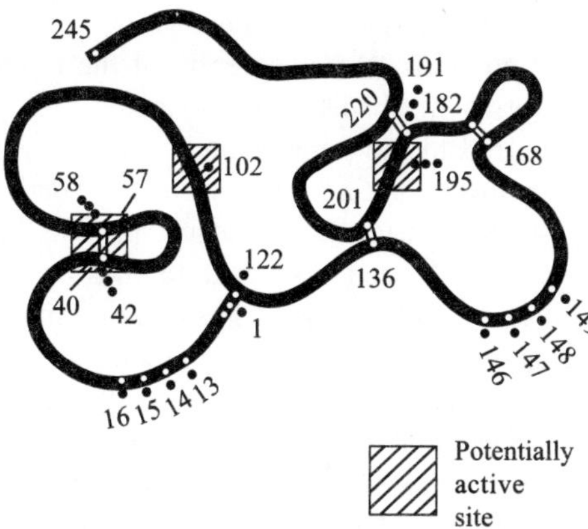

Figure 15.2 Model of structure of Chymotrypsinogen. (Activation of chymotrypsin minimally requires hydrolysis between residues 15 (Arg) and 16 (Ile). Subsequent conformational changes bring residues 195 (Ser), 102 (Asp) and 57 (His) into proximity. Other residues labelled are 14 (Ser), 13 (Leu), 1 (Cys), 40 (His), 42 & 58 (Cys), 146 (Tyr), 147 (Thr), 148 (Asn), 149 (Ala), 191 (Cys) & 245 (Asn). From: Orten, James M. and Neuhaus, Otto. W., *Human Biochemistry*, 9th ed., 1975, the C.V. Mosby Co., St. Louis, Mo., U.S.A.)

Tryptic degradation of a few peptides however change this conformation in the following manner:

(a) The first step is the tryptic cleavage of the peptide bonds between amino acid 16 (Ile) and amino acid 15 (arg) with the formation of π chymotrypsin.

(b) Next is the splitting of the dipeptide Seryl arginine by cleavage of the bond between amino acids 14 and 13. δ chymotrypsin is the result.

(c) Hydrolysis of bond 146–147 and 148–149 from α chymotrypsin.

(d) The N-terminal Ile (amino acid 16) is held by the β carboxylate ion of amino acid 194 (Asp) and this brings His (amino acid 57) into close proximity with Ser (amino acid 195) which potentiates the active site.

(e) Thus three chains are formed, A (1–13), B (16–146) and C (149–246) held together by 5 disulphide bonds.

The *procarboxypeptidase* of the pancreatic juice is converted to the active form *carboxypeptidase* by trypsin. In this process approximately two-third of the original molecule is split off. At least two carboxypeptidases, A and B, are present in pancreatic juice.

Proelastase is also converted to the active form by trypsin.

D. Pancreatic Proteolytic Enzymes

The pancreatic juice is alkaline in nature (pH 8.0) and it brings the pH of the partially digested food material entering the intestine towards the alkaline range for facilitating the action of proteolytic enzymes of pancreatic and intestinal secretions.

The optimum pH for both trypsin and chymotrypsin is in the vicinity of 8–8.5. Trypsin splits the peptide linkages formed by the carboxyl groups of arginine or lysine (basic amino acids). The replacement of these basic amino acid residues by other groups would yield products resistant to trypsin action. It is also necessary for tryptic action that the second amino group of the basic amino acid should be free and unsubstituted while one of the amino group is involved in the peptide linkage. Trypsin further hydrolyzes the peptic digestion products as also the proteins which could not be broken down by pepsin. Trypsin is most active on partially digested and denatured proteins. The end products are amino acids and simpler polypeptides.

Chymotrypsin splits the peptide bonds formed by the carboxyl group of aromatic amino acids, phenylalanine, tyrosine and tryptophan. However, its action is masked by the presence of an amino acid having a free carboxyl group in the immediate vicinity of the peptide linkage. It would therefore imply that chymotrypsin would split peptide linkages between the carboxyl group of aromatic amino acids and amino group of another amino acid, other than glutamic and aspartic acids.

The products obtained after the additive actions of pepsin, trypsin and chymotrypsin are amino acids and smaller polypeptides. The specificities of pepsin, trypsin and chymotrypsin for the peptide linkages are given below:

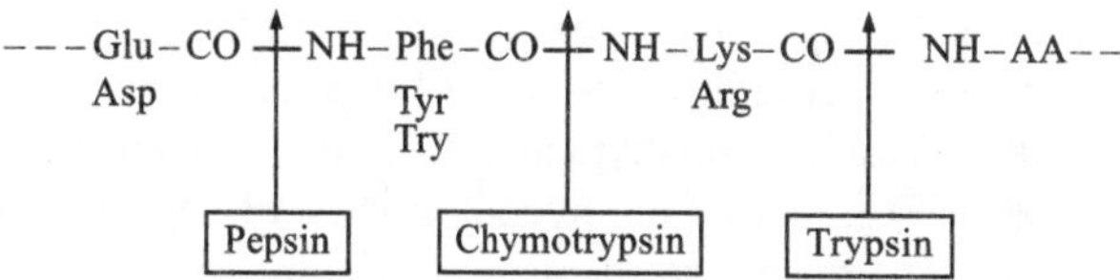

Preferential specificities of pepsin, trypsin and chymotrypsin for peptide linkages

Carboxypeptidase A and B of the pancreatic juice attack an end peptide linkage of the polypeptide having a free carboxyl group. By its action, therefore, this enzyme liberates free amino acids from the polypeptides. Carboxypeptidases are not able to act if there is a free amino group in the vicinity and therefore the enzyme cannot split a dipeptide. The optimum pH of its action is approximately 7.4. *Carboxypeptidase A* contains zinc as an integral component and is essential for the catalytic activity of the enzyme. It splits preferentially the peptide linkage adjacent to C-terminal amino acids containing tyrosine, phenylalanine and tryptophan. *Carboxypeptidase B* acts only on peptide linkages adjacent to arginine or lysine residues.

Elastase acts on elastin as also on several other related proteins and digests them to their constituent amino acids.

E. Small Intestinal Proteolytic Enzymes

The mucosa of the small intestine too contains certain enzymes that hydrolyze peptide linkages. While these enzymes are sometimes found in the intestinal lumen it is now amply clear that their action is mainly intracellular. The hydrolysis of the smaller products of protein digestion (polypeptides and dipeptides) takes place during and following their entry into the mucosal cells of the small intestines.

The intestinal mucosal cells and/or its secretions contain a group of *aminopeptidase* enzymes. The aminopeptidases act on polypeptides having a free amino group, and liberate an amino acid by splitting the peptide linkage adjacent to the free amino group. One of the examples of this group of enzymes demonstrated in intestinal mucosal cells is *leucine amino peptidase* which possesses a wide specificity as regards the N-terminal residues of the polypeptide. It requires Mg^{++} or Mn^{++} for its activity. Aminopeptidases do not act if there is a free carboxyl group in the neighbourhood and therefore this group of enzymes cannot split a dipeptide.

Another group of enzymes present in the intestinal cells are *dipeptidases,* that act on dipeptides specifically. To cite an example, glycylglycine dipeptidase acts on the dipeptide glycylglycine but not tripeptide glycylglycyl glycine. It is thus apparent that dipeptidases require both a free amino group and a free carboxyl group near the peptide linkage for action. Dipeptidases are localized superficially on the brush border membranes of the small intestines.

On the basis of the above description, it is apparent that as many as six enzymes capable of splitting peptide linkages are mainly involved in the stepwise breakdown of native proteins to amino acids in the gastro-intestinal tract. This process results in the hydrolysis of the dietary proteins to free amino acids. Table 15.1 describes these enzymes and their respective functions. Depending on the source of peptidases, the protein digestive process can be divided with gastric, pancreatic and intestinal phases, and by the overall process of Digestion and Absorption of proteins are depicted in Figure 15.3.

II. ABSORPTION OF AMINO ACIDS

The dietary proteins under normal circumstances are almost completely hydrolyzed, as a result of digestive processes, to their component amino acids. On completion of the tremendously efficient absorptive mechanism, amino acids thus produced are then transferred to the portal blood.

The absorption of amino acids from the small intestine is an energy-dependent *active* transport process. The term 'active transport' has been restricted to those processes in which a substance (amino acids in this context) moves across a membrane against a concentration gradient, this movement showing a direct quantitative relationship between the energy utilized and the work performed. The clear demonstration of the movement of amino acids against a concentration gradient was originally made by the use of the everted sac technique of Wilson and Wiseman. In this technique a small segment of the intestine is turned inside out, and after the lumen has been filled with fluid, it is tied at both ends. Movement of the material through the intestinal wall can be measured by assessing the changes in the content of substances on the side of the lumen.

It has been suggested that sugars and amino acids may be absorbed at the same place of the cell structure and a carrier mechanism similar to sugar absorption (Chapter 11) is believed to be involved. A common carrier mechanism for the transport of sugars, Na^+ and amino acids has been suggested. The transport of amino acids that combine with the carrier has been found to be Na^+ dependent, and there are reports about the inhibition of amino acid transport by sugars (mainly glucose and galactose).

Experiments using the everted sac of the intestine have shown that L-isomers (natural) of amino acids are more readily absorbed than D-isomers. L-amino acids are actively

TABLE 15.1 Important Proteolytic Enzymes of the Gastrointestinal Tract

	Active proteolytic enzymes	*Inactive protein precursors (Zymogens)*	*Activators of zymogens*	*Source*	*Functions*
1.	Pepsin	Pepsinogen	HCl and pepsin (autocatalytically)	Stomach (chief cells)	Digests proteins and breaks, preferentially, the peptide linkage between an L-Dicarboxylic acid and an aromatic amino acid
2.	Trypsin	Trypsinogen	Enterokinase and Trypsin (Autocatalytically)	Pancreas	Splits peptide linkage between the carboxyl group of lysine or arginine and of another amino acid
3.	Chymotrypsin	Chymotry-psinogen	Trypsin	Pancreas	Breaks peptide linkage between carboxyl group of aromatic amino acid and the amino group of an amino acid other than glutamic and aspartic acids
4.	Carboxy-peptidase	Procarboxy-peptidase	Trypsin	Pancreas	Removes amino acids from the carboxyl end of a polypeptide with free carboxyl group
5.	Amino-peptidases	–	–	Intestinal mucosa	Removes amino acids from the amino end of a polypeptide with free amino group
6.	Dipeptidases	–	–	Intestinal mucosa	Acts on dipeptides and liberates amino acids

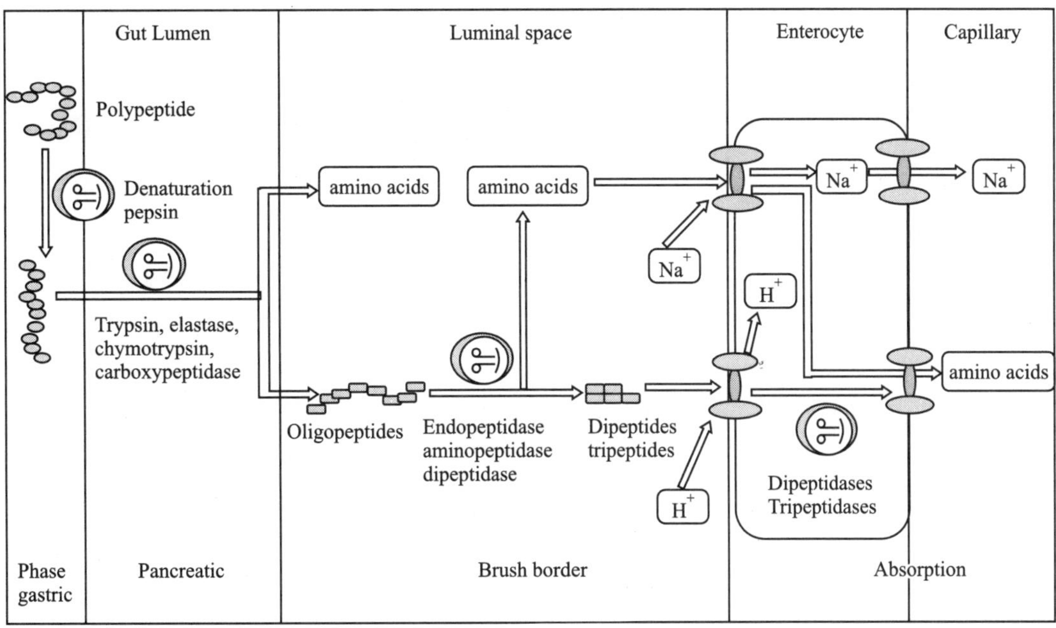

Figure 15.3 Digestion and absorption of protein.

transported and vitamin B_6 (pyridoxal phosphate) is believed to be involved in this transport. The manner in which pyridoxal phosphate is implicated is not clear. The D-amino acids are probably transported by passive diffusion. The neutral and more hydrophilic amino acids are more readily absorbed than basic and hydrophobic amino acids. The following three carriers have been identified for absorption of amino acids:

1. Neutral amino acids are absorbed by a single common transport mechanism; these amino acids compete with each other for transport.
2. Basic amino acids (arginine, lysine and ornithine) are carried by another transport system whose efficiency is 5–10% of the rate of transport of neutral amino acids.
3. The third transport system helps in the absorption of L-proline, hydroxyproline, sarcosine and dimethyl glycine. Proline and hydroxyproline have a stronger affinity for this third carrier system than neutral ones.

There are some observations on the competition between different groups of amino acids (neutral, basic, acidic and N-substituted). It has been found with hamster intestine that D-galactose and L-arginine and their analogues are competitive inhibitors of the active transport of neutral amino acids. On the basis of these observations, a polyfunctional nature for mobile carriers in the brush borders of the cells has been proposed. Such a common polyfunctional carrier may possess different binding sites for different classes of actively transported substances, i.e., sugars, neutral, basic and acidic amino acids and Na^+. The process is considered to be under allosteric control with respect to associated binding sites. To facilitate the process a sodium gradient between lumen and cell is operative and is similar to that described for carbohydrate absorption.

As a result of the active transport from the lumen, once the amino acids accumulate in high concentrations within the cell, they leave by passive diffusion to portal blood.

The absorption of *smaller peptides* from the intestine also takes place to a certain extent. It is also likely that certain individuals are able to absorb some unhydrolyzed proteins and protein fragments of large molecular size. This notion has been derived from the observation that immunological sensitivity to proteins occurs in certain persons when they eat some particular proteins. Proteins are antigenic and are able to elicit an immunological response. On the other hand, the digested products of the proteins, even to the stage of smaller polypeptides, are not antigens.

It has been suggested that the main defect in *nontropical sprue* lies in the small intestinal cell, which permits the absorption of the larger polypeptides obtained by the digestion of gluten, the main protein constituent of wheat. These large polypeptides not only harm the intestine but are also absorbed and stimulate the production of antibodies. These observations prove the feasibility of the absorption of protein components of large molecular size under certain situations.

SUMMARY

This chapter has dealt with the digestion of proteins in the gastrointestinal tract through specific proteolytic enzymes leading to the formation of amino acids. The characteristics of specific proteolytic enzymes involved in gastric, pancreatic and small intestine have been elaborated. In the end, absorption of amino acids from the small intestine through an energy dependent active transport process have been amplified.

SUGGESTIONS FOR FURTHER READING

Kagnoff M.F. (2007), Celiac Disease: Pathogenesis of A Model Immunogenetic Disease, *J. Clin Invest.*, 117:41.

Makhlouf G.M. and Murthy K.S. (1997), Signal Transduction in Gastrointestinal Smooth Muscle, *Cell Signal*, 9:269.

Mowat A.M. (2003), Coeliac Disease—A Meeting Point for Genetics, *Immunology and Protein Chemistry*, *Lancet*, 361:1290.

Thomson A.B. (2001), Small Bowel Review, Normal Physiology, *Dig. Dis. Sci.*, 46:2588.

Wright E.M. (1993), The Intestinal Na$^+$/glucose Cotransporter, *Ann. Rev. Physiol.*, 555.

16

Metabolism of Proteins and Non-protein Nitrogenous Compounds

K.D. Moudgil

CONTENTS

I. INTRODUCTION

Many foodstuffs, whether derived from animal or plant sources contain proteins, the chief form in which nitrogenous substances are ingested. While there is a provision to store carbohydrate and lipid in the body, there is no storage form of protein. However, there is a reserve protein supply which can be drawn from the body tissues themselves. The amino acids derived from exogenous or endogenous proteins serve many essential biological functions. They are metabolized in such a way that the amino nitrogen is converted into ammonia, urea, and uric acid, which are eliminated in urine as nitrogenous end products. The term non-protein nitrogen (NPN) is employed to include the nitrogen from all nitrogenous substances other than proteins. These substances include urea, uric acid, creatinine, creatine, amino acids, ammonia and some peptides.

Nitrogen balance is the relationship between the total amount of nitrogen ingested in the form of nitrogenous substances in the diet and the total amount of nitrogen excreted from the body in the form of nitrogenous end products. Nitrogen balance is an indicator of the overall protein metabolism of the body.

Nitrogen balance is positive if intake exceeds output and negative if output is more than the intake. Nitrogen equilibrium follows when intake and output are equal. A normal healthy adult is ordinarily in a state of nitrogen balance.

II. METABOLISM OF PROTEINS AND NON-PROTEIN NITROGEN

Metabolism of proteins is essentially the metabolism of amino acids. Amino acids derived from the digestion of food proteins are absorbed into the blood from the gastrointestinal tract. In addition, amino acids synthesized in the body and those derived from the breakdown of worn out tissue proteins are added to the body fluids to form the "labile amino acid pool" of the body. This pool contains not only the 20 different kinds of L-α-amino acids contained in proteins, but also other amino acids such as citrulline, ornithine and taurine. This pool supplies amino acids for the synthesis of body proteins and non-protein nitrogenous substances such as nucleotides, creatine, choline and glutathione. The amino acid nitrogen pool of the body is small and turns over rapidly. For example, the daily turnover of the proteins in humans ranges 1–2% of the total body protein.

A. Metabolism of Amino Acid Nitrogen

Amino acids of the body undergo many different types of reactions. Some reactions are common to all or many of the amino acids. Among these are the processes of deamination and transamination. Removal of amino groups from the amino acids is termed *deamination*. The catabolism of amino acids involves deamination and formation of the corresponding keto acids prior to oxidation and the subsequent production of energy. In animals, the amino acid nitrogen is excreted in the form of ammonia, uric acid or urea. In man, the amino group or ammonia is converted to urea and excreted. It may also be "stored" as amide group of glutamine or excreted as NH^{4+} or some other nitrogenous waste products.

The synthesis of amino acids in tissues involves amination of keto acids derived from the metabolism of carbohydrates, proteins and fats. Moreover, amino groups of certain amino acids may be transferred to the keto acids corresponding to other amino acids, thereby effecting amino acid–keto acid interconversion. In this process, amino acid donating the amino group is converted to the corresponding keto acid. This process of amino acid–keto acid interconversion involving two amino acid–keto acid pairs is called *transamination*.

1. Oxidative deamination

(a) *L- and D-Amino acid oxidases:* Many microorganisms and the liver and kidney of animals contain both D- and L-amino acid oxidases. D-Amino acid oxidases (E.C.1.4.3.3) are flavoproteins containing FAD (flavin adenine dinucleotide) and have broad substrate specificity. L-Amino acid oxidases (E.C.1.4.3.2) contain FMN (flavin mononucleotide) instead of FAD. The action of amino acid oxidases may be illustrated as follows:

The amino acid is first dehydrogenated by flavoprotein of the amino acid oxidase, forming an α-imino acid, which is not a stable intermediate. It spontaneously adds water and then decomposes to the corresponding α-keto acid with the loss of α-imino nitrogen as ammonia. Amino acid oxidases are auto-oxidizable flavoproteins. The reduced FAD or FMN is reoxidized directly by molecular oxygen, forming a toxic product hydrogen peroxide (H_2O_2). H_2O_2 is acted upon by catalase widely distributed in tissues, producing oxygen and water.

Although L-Amino acid oxidases of animal tissues deaminate many of the naturally occurring amino acids, the generally low levels of their activity indicates that oxidative deamination involving L-amino acid oxidases is of minor importance in the removal of amino groups of amino acids as ammonia.

The physiological function of D-amino acid oxidases, which are much more active than the L-amino acid oxidases, is not known. D-amino acids are not present in animals and are found only rarely in other naturally occurring materials.

(b) *Glycine oxidase:* Glycine may be oxidatively deaminated by glycine oxidase, which is present in liver and kidney of mammals:

$$H_2C-NH_2 \ \big|\ COOH \quad \xrightarrow[\quad O_2 \quad]{\ H_2O \quad\quad H_2O_2\ } \quad CHO \ \big|\ COOH$$

Glycine$\qquad\qquad\qquad$ NH_3 Glyoxylic acid

Glycine oxidase is a flavoprotein and apparently identical to D-amino acid oxidase.

(c) *L-Glutamate Dehydrogenase (GDH):* L-Glutamic acid is not deaminated by L-amino acid oxidase. A reaction of great biological importance is the reversible oxidative deamination of glutamic acid by L-glutamate dehydrogenase (EC 1.4.1.3), a zinc containing enzyme. It is widely distributed in mammalian tissues. Glutamate dehydrogenase can use NAD^+ or $NADP^+$ as cosubstrate. Liver glutamate dehydrogenase has many allosteric modifiers. ATP, GTP and NADH inhibit the enzyme, while ADP activates it.

Interestingly, L-Glutamic acid is the sole amino acid that undergoes oxidative deamination at a significant rate in mammals. The reaction catalyzed by glutamate dehydrogenase is reversible and functions both in amino acid catabolism and in their biosynthesis. It represents a mechanism by which ammonia is taken up by α-ketoglutaric acid and converted to glutamic acid. The amino group of glutamic acid may then be reversibly transferred to many different amino acids by transamination. Thus, this reaction serves as a mechanism for incorporating ammonia into amino acids and links protein and carbohydrate metabolism through the citric acid cycle. Conversely, by transamination, amino groups of many amino acids are transferred to α-ketoglutaric acid to form glutamic acid and then get disposed of as ammonia via the following reactions.

$$\begin{array}{c} NH_2 \\ | \\ R-C-COOH \\ | \\ H \\ \alpha\text{-Amino acid} \end{array} \quad \underset{\begin{array}{c}O_2 \;\; H_2O_2\\ FP.H_2 \rightleftarrows FP\\ \text{Catalase}\\ 2H_2O_2 \longrightarrow 2H_2O_2 + O_2\end{array}}{\overset{FP \quad FP.H_2}{\rightleftarrows}} \quad 2\left[\begin{array}{c} R-C-COOH \\ || \\ NH \\ \alpha\text{-Amino acid} \end{array}\right]$$

(FP = Flavoprotein enzyme)

$$\xrightarrow[\;NH_3\;]{\;H_2O\;} \quad \begin{array}{c} R-C-COOH \\ || \\ O \\ \alpha\text{-Keto acid} \end{array}$$

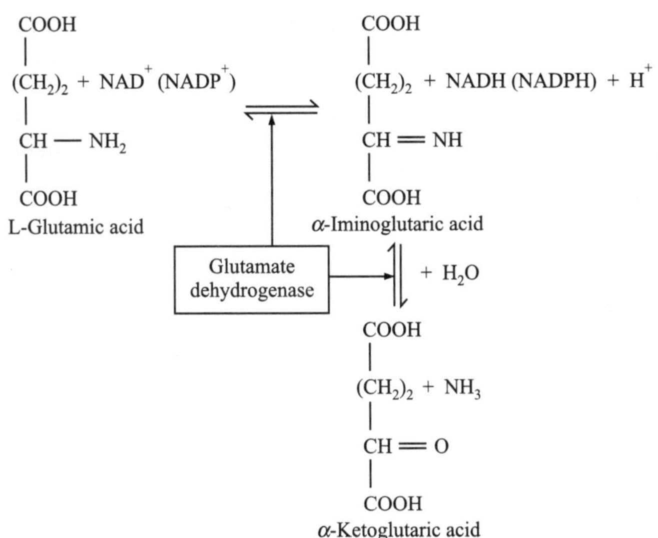

2. Non-oxidative deamination

(a) *Amino acid dehydratases (EC 4.2.1)*: Serine, threonine and homoserine are deaminated by specific dehydratases (serine dehydratase, EC 4.2.1.13; threonine dehydratase, EC 4.2.1.16; and homoserine-dehydratase, EC 4.2.1.15) that catalyze a primary dehydration followed by spontaneous deamination. The dehydratase reaction proceeds via elimination of water from α-amino acid, forming an unsaturated amino acid. The latter rearranges to an α-imino acid, which is spontaneously hydrolyzed to an α-keto acid and ammonia. There is no net gain or loss of water during the dehydratase reaction.

Dehydratases require pyridoxal phosphate (PLP) as a cofactor. Dehydratases for L-α-amino acids are found in the mammalian liver. Many microorganisms contain both D- and L-amino acid dehydratases.

(b) *Amino acid desulphhydrases:* Cysteine desulphhydrase (EC 4.4.1.1) and homocysteine desulphhydrase (EC 4.4.1.2) deaminate cysteine and homocysteine, respectively, by a primary desulph-hydration ($-H_2S$) forming an imino acid. The latter is then spontaneously hydrolyzed to yield a keto acid and ammonia. Pyridoxal phosphate is required as a coenzyme for this reaction.

Desulphhydrases are found in some micro-organisms and in the liver, kidney and pancreas of animals.

(c) *Histidase (EC 4.3.1.3)*: Histidine is deaminated non-oxidatively to urocanic acid.

3. Transamination

Transamination is a process that brings about the transfer of amino group from an α-amino acid to an α-keto analog of another amino acid, resulting in the formation of an α-amino acid (from the corresponding α-keto acid) and an α-keto acid (from the corresponding α-amino acid, which donated its amino group to the keto acid). Thus, transamination affects interconversion of a pair of amino acids and a pair of keto acids. This reaction is catalyzed by a group of enzymes known as transaminases or amino-transferases (EC 2.6.1). The general process of transamination may be represented as follows:

Transaminases are widespread in animal tissues, plants and microorganisms. Liver, kidney, brain and heart have relatively high transaminase activity. In general, the transaminases of animal tissues appear to be specific for L-α-amino acids. Certain bacteria have transaminases specific for both D- and L-amino acids.

Pyridoxal phosphate forms an essential part of the active site of all transaminases. Each transaminase is specific for the specific pair of amino and keto acids as one pair of substrates, but non-specific for the other pair. Most amino acids are substrates for transaminases. Lysine, threonine, proline and hydroxyproline are notable exceptions.

Transamination is a freely reversible process. This permits transminases to function in both amino acid catabolism (by deamination) and biosynthesis of amino acids from the corresponding keto acids (by amination). Alanine transaminase and glutamate transaminase catalyze the transfer of amino groups from most amino acids to form alanine and glutamate from pyruvate and α-ketoglutarate, respectively.

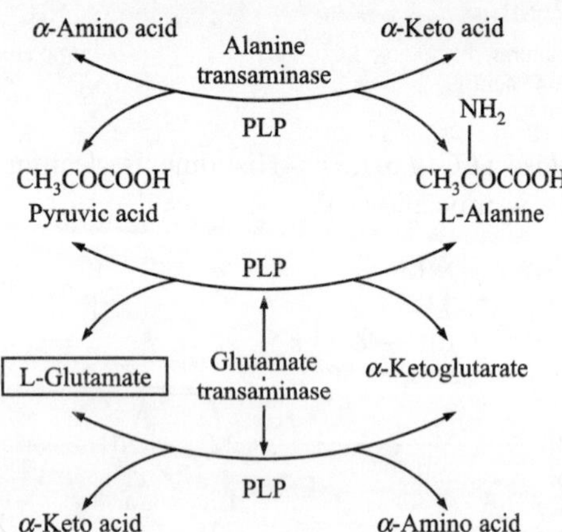

By the action of glutamate transaminase on alanine, all of the amino nitrogen from most of the amino acids can be funneled into glutamate. Oxidative deamination of glutamate by glutamate dehydrogenase then releases ammonia, which can be disposed of by the body.

In some cases, transamination may involve ω-amino acids rather than α-amino acids.

A transamination reaction involving glutamine (or asparagine) and a keto acid requires both transamination and deamidination. In these cases, corresponding α-ketoglutaraemic

and α-ketosuccinamic acids are formed from glutamine and asparagine, respectively. They are then converted to α-keto acids and ammonia by specific ω-amidases.

Mechanism of transamination: Pyridoxal phosphate (PLP) serves as a coenzyme in transamination reactions. In the first phase of the reaction, α-amino group of the amino acid forms a Schiff base with 4-aldehyde group of PLP, which remains bound to the enzyme by a salt bridge. By a series of electron shifts and rearrangements, pyridoxal phosphate becomes pyridoxamine phosphate, and α-amino acid generates the corresponding α-keto acid. In the second phase of the reaction, α-keto acid substrate receives the amino group from pyridoxamine phosphate to generate the corresponding α-amino acid and pyridoxal phosphate (for details, see Chapter 47).

4. Amino acid decarboxylation

Carboxylases (EC 4.1.1) (or decarboxylases) are found in a variety of animal tissues, especially liver, kidney and brain. Pyridoxal phosphate is required as a cofactor in the reactions catalyzed by decarboxylases. Pyridoxamine phosphate is not involved in this reaction. The process involves an intermediate Schiff base as in transamination, but decarboxylation rather than amino group transfer follows. This process is important in the body for the formation of histamine, tyramine, and other amines from amino acids.

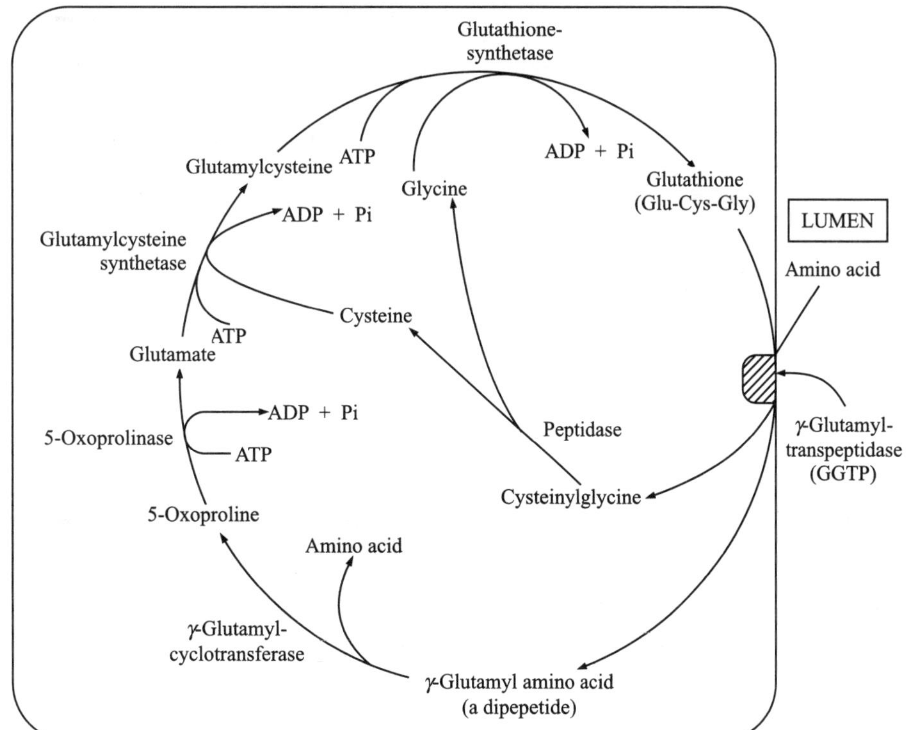

5. Transpeptidation

This is a process involved in the transfer of amino acids or small peptides from one oligopeptide to another peptide or amino acid. The transfer of γ-glutamyl group of glutathione to amino acids and peptides has been observed in various mammalian tissues. A recycling process involving glutathione (γ-glutamyl cycle) is involved in the transport of amino acids across the renal proximal tubular cells. Three molecules of ATP are consumed in each cycle (Figure 16.1).

B. Formation and Disposal of Ammonia

The deamination of amino acids in the body tissues forms large amounts of ammonia. Amino nitrogen from most of the amino acids is funneled into glutamate. By the action of glutamate dehydrogenase on glutamate, ammonia is released. Some of the amino acids (e.g., serine, threonine, cysteine, homocysteine, and histidine) directly liberate ammonia through special deamination processes. Ammonia is also formed in other metabolic processes such as deamination of adenylic acid, catabolism of pyrimidines, and oxidation of amines by amine oxidases. Another source of ammonia is glutamine, which is the chief form in which ammonia is transported in extracellular fluids. Liberation of amide nitrogen of glutamine as ammonia occurs by hydrolytic removal of ammonia in a reaction catalyzed by glutaminase.

In addition to the formation in the tissues, a considerable quantity of ammonia is produced by the gastrointestinal tract, which harbors bacterial flora. Bacterial urease and other bacterial enzymes act on dietary protein and urea present in fluids secreted into the gastrointestinal tract. This ammonia is absorbed from the intestine. Consequently, the concentration of ammonia in portal blood is higher than that in systemic circulation.

Kidneys also produce ammonia and add it to the blood. Considerable amount of ammonia may be excreted in the

Figure 16.1   Transport of amino acids through the γ-glutamyl cycle.

urine. This ammonia is produced by renal tubular cells from intracellular amino acids, especially glutamine.

Ammonia is constantly produced in the body tissues, but it is present in trace amounts in peripheral blood. As even minute quantity of ammonia is toxic to the central nervous system, rapid removal of ammonia from circulation is necessary. This is achieved by conversion of ammonia to much less toxic compounds. This process is called *detoxification*. Some organisms like birds, amphibians and reptiles (*uricotelic* organisms) excrete nitrogen waste products as uric acid. Those organisms (e.g. mammals) that dispose of nitrogenous waste products in the form of urea are referred to as *ureotelic* organisms. Many aquatic animals that excrete ammonia into the surrounding water by diffusion through the skin are called *ammonotelic* organisms.

1. Biosynthesis of urea

In the human system, urea constitutes about 80–90 per cent of the nitrogen excreted. Liver is the principal site of urea formation. Once formed, urea is excreted in the urine. It is essentially non-utilizable. The amount of urea excreted is a direct reflection of the extent of amino acid catabolism, and it varies considerably with dietary protein intake.

Chemical mechanism of urea formation: Urea is synthesized through the urea cycle. The overall process of urea formation was established by Hans Krebs and Kurt Henseleit in 1932. The reactions and intermediates of urea cycle and its relationship with citric acid cycle are represented in Figure 16.2.

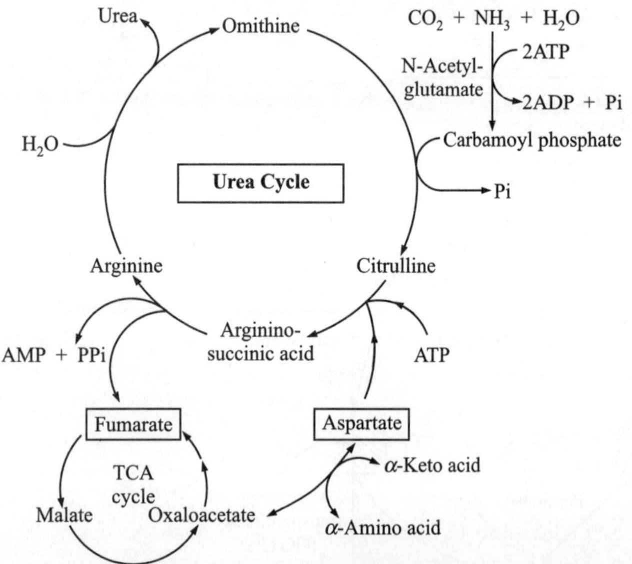

Figure 16.2 Urea cycle is represented xxxxx.

The overall process of biosynthesis of 1 mole of urea requires 3 moles of ATP. One of the nitrogen atoms of urea comes from ammonia, but the other nitrogen atom is derived from aspartate. The carbon atom of urea is contributed by carbon dioxide. Ornithine is the carrier of these nitrogen and carbon atoms. The immediate precursor of urea is arginine. There is no net loss or gain of ornithine, citrulline, argininosuccinate, or arginine. However, NH_3, CO_2, ATP and aspartate are utilized.

(a) *Formation of carbamoyl phosphate:* Initial reaction in the process of urea biosynthesis results in the formation of carbamoyl phosphate from CO_2 and NH_3. Mg^{++}, ATP, and N-acetylglutamic acid are also required for this reaction.

$$CO_2 + H_2O + 2ATP \xrightarrow[\substack{\text{Carbamoyl} \\ \text{phosphate} \\ \text{synthetase}}]{\text{N-acetylglutamate}} H_2N-\overset{\overset{\displaystyle O}{\|}}{C}-O-\overset{\overset{\displaystyle O}{\|}}{\underset{\underset{\displaystyle O^-}{|}}{P}}-O^-$$
$$+ \; 2ADP + Pi$$

This reaction is catalyzed by the liver mito-chondrial enzyme, carbamoyl phosphate synthetase (ammonia), EC 6.3.4.16. N-acetylglutamic acid acts as a positive allosteric effector for carbamoylphosphate synthetase, whereas arginine acts as a specific stimulator for N-acetylglutamate synthesis.

Arginine
Acetyl CoA ⊕ CoA
Glutamate → N-Acetylglutamate
N-Acetylglutamate-synthetase (NAGS)
H₂O

In bacteria, glutamine serves as a substrate for carbamoyl phosphate synthetase.

(b) *Formation of citrulline:* Citrulline is formed from ornithine and carbamoyl phosphate. This reaction is catalyzed by the liver mitochondrial enzyme, L-ornithine transcarbamoylase (EC 2.1.3.3).

$$H_2N-\overset{\overset{\displaystyle O}{\|}}{C}-O-\overset{\overset{\displaystyle O}{\|}}{\underset{\underset{\displaystyle O^-}{|}}{P}}-O^- + \underset{\underset{\displaystyle \text{COOH}}{|}}{\overset{\overset{\displaystyle CH_2NH_2}{|}}{\underset{\displaystyle HC-NH_2}{\overset{\displaystyle (CH_2)_2}{|}}}} \xrightarrow[\substack{\text{transcarba-} \\ \text{moylase}}]{\text{L-Ornithine}} \underset{\underset{\displaystyle \text{COOH}}{|}}{\overset{\overset{\overset{\overset{\displaystyle NH_2}{|}}{C=O}}{\underset{\displaystyle CH_2-NH}{|}}}{\underset{\displaystyle HC-NH_2}{\overset{\displaystyle (CH_2)_2}{|}}}}$$

Carbamoyl phosphate   L-Ornithine   →Pi   L-Citrulline

(c) *Formation of argininosuccinate:* Synthesis of argininosuccinate from aspartate and citrulline requires ATP. The reaction is catalyzed by argininosuccinate synthetase (EC 6.3.4.5), a liver cytosolic enzyme.

$$\text{L-Citrulline} + \text{L-Aspartic acid} + \text{ATP} \xrightarrow[\text{Mg}^{++}]{\text{Argininosuccinic acid synthetase}} \text{AMP} + \text{PPi} + \text{Argininosuccinic acid}$$

(d) *Formation of arginine:*   The cleavage of argininosuccinic acid produces arginine and fumarate. The reaction is catalyzed by a cold-labile enzyme, argininosuccinase (EC 4.3.2.1).

(e) *Formation of urea from arginine:*   Arginine is hydrolyzed to ornithine and urea by the enzyme arginase (EC 3.5.3.1).

$$\text{Argininosuccinic acid} \xrightleftharpoons{\text{Argininosuccinase}} \text{L-Arginine} + \text{Fumaric acid}$$

Ornithine reenters the urea cycle. Arginase is activated by CO^{++} or Mn^{++}, but ornithine and lysine are competitive inhibitors of arginase.

$$\text{L-Arginine} \xrightarrow[\text{H}_2\text{O}]{\text{Arginase}} \text{Omithine} + \text{urea}$$

Synthesis of urea in the liver appears to be a continuous process. In a normal healthy adult, the level of blood urea ranges from 15 to 40 mg/dl. In spite of its continuous production, normal level of blood urea is maintained by rapid and efficient elimination of urea by the kidney.

2. Role of glutamine and glutamic acid in disposal of ammonia

An important pathway of removal of ammonia produced in the tissues is via formation of glutamine and asparagine. Asparagine is formed from aspartic acid by a reaction analogous to that employed in the formation of glutamine from glutamic acid. Glutamine is synthesized through the action of the enzyme glutamine synthetase (GS; EC 6.3.1.2).

$$\text{L-Glutamic acid} + \text{ATP} + \text{NH}_3 \xrightarrow[\text{Mg}^{++}]{\text{Glutamine synthetase}} \text{Glutamine} + \text{ADP} + \text{Pi}$$

This process occurs in a variety of tissues such as the liver, kidney, brain and retina. Blood glutamine is synthesized largely in the liver. Some of the ammonia is also utilized to form glutamic acid in a reaction catalyzed by glutamate dehydrogenase. Glutamic acid undergoes a transamination reaction with α-keto acids forming other amino acids.

Glutamine is involved in many important processes of body metabolism. Glutamine constitutes a large proportion of the available free amino nitrogen of tissues and blood, and of the metabolic nitrogen pool. Glutamine, like glutamic acid, participates in transamination reaction. Glutamine serves as an important storage form of ammonia produced in deamination processes. This helps to avoid an increase in the levels of ammonia in the blood and tissues. Ammonia in the form of glutamine can be utilized for various synthetic processes. Examples of utilization of the amide nitrogen of glutamine include the synthesis of hexosamines, purines, pyrimidines, histidine, and NAD^+. Glutamine and asparagine are also found as constituents of many proteins. In this way, ammonia in the form of glutamine is utilized for the synthesis of various nitrogenous constituents. Thus, ammonia is not allowed to be completely lost in the form of urea.

Glutamine is hydrolyzed by glutaminase (EC 3.5.1.2), which is found in various tissues such as the brain, kidney, liver and retina. Ammonia released by the action of renal glutaminase is utilized for the regulation of acid-base balance and conservation of cations. In conditions of acidosis, the production of ammonia by renal tubular cells increases as a compensatory mechanism to maintain acid-base balance of the body fluids. On the contrary, ammonia production by renal tubular cells is decreased in alkalosis.

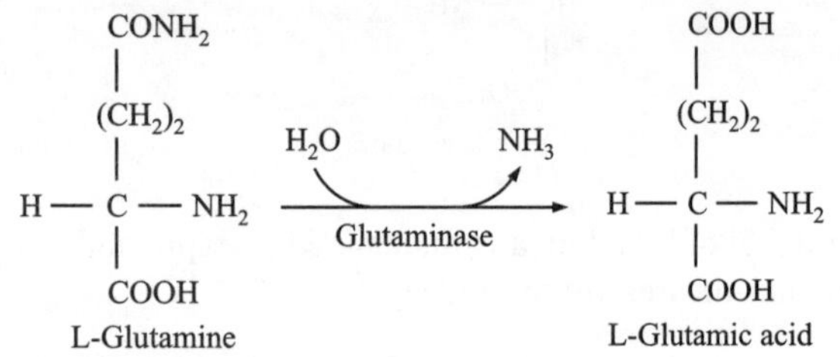

$$
\begin{array}{ccc}
CONH_2 & & COOH \\
| & & | \\
(CH_2)_2 & \xrightarrow[\text{Glutaminase}]{H_2O \quad NH_3} & (CH_2)_2 \\
| & & | \\
H-C-NH_2 & & H-C-NH_2 \\
| & & | \\
COOH & & COOH \\
\text{L-Glutamine} & & \text{L-Glutamic acid}
\end{array}
$$

Glutamine also serves as a source of glutamic acid for the body tissues, particularly the brain, because it can penetrate blood-brain barrier more easily than glutamic acid.

3. Hepatic encephalopathy

Hepatic encephalopathy is a metabolic disorder of the nervous system. It may result from advanced, decompensated liver disease and/or extensive portal-systemic (anatomic or functional) shunting. In these conditions, portal blood is shunted directly into the systemic circulation, thereby bypassing the liver.

The pathogenesis of hepatic encephalopathy remains unclear. No single biochemical or physiological defect has been pinpointed as the specific cause of this condition. It has been suggested that various toxic substances produced in the gut are not metabolized by the liver, and thus accumulate within the brain. Some toxic substances implicated in this process are ammonia and mercaptans, the latter being formed by the degradation of sulphur containing compounds in the gut. Ammonia may affect cerebral tissue by a direct toxic effect and/or due to impairment of functioning of the citric acid cycle. The latter effect may be due to the depletion of citric acid cycle intermediates caused by funelling of α-ketoglutarate towards increased synthesis of glutamate. Glutamate is required by the brain for the removal of excessive ammonia as glutamine. Increased synthesis of glutamine from ammonia also contributes to the amino acid imbalance in the nervous system resulting from changes in the blood-brain barrier in hepatic encephalopathy.

Many other compounds like short chain fatty acids, some (aromatic and branched-chain) amino acids, phenols and false neurotransmitter substances (such as beta-hydroxy-phenylethylamines derived from aromatic and branched-chain amino acids) have been postulated to play a role in the pathogenesis of hepatic encephalopathy. The lack of some

substances normally produced by the liver and essential for normal brain metabolism may also be a contributing factor. It has been suggested that diminished levels of consciousness observed in hepatic encephalopathy might be attributable to increased CNS levels of γ-aminobutyric acid (GABA). Increase in the blood level of ammonia or other nitrogenous substances due to gastrointestinal bleeding, electrolyte disturbances and increased dietary protein can precipitate hepatic encephalopathy.

C. Creatine and Creatinine

Creatine is distributed in many tissues of the body but largest amounts are found in the muscle, which contains about 98 per cent of total body creatine. Most of the creatine in normal tissues (except RBC) exists as creatine phosphate.

1. Biosynthesis of creatine

For the synthesis of creatine, three amino acids namely, arginine, glycine, and methionine (as S-adenosyl-methionine) are required. Chemically, creatine is N-methylguanidoacetic acid. The sarcosine (N-methylglycine) portion of the molecule is derived from glycine, while methyl and guanido groups are derived from methionine and arginine, respectively. The synthesis of creatine occurs in two stages. In the first stage, arginine and glycine react in the presence of an enzyme, arginine-glycine amidinotransferase (EC 2.1.4.1). It catalyzes the transfer of a guanido group from arginine to glycine forming glycocyamine (guanidoacetic acid). This reaction takes place principally in the kidney. Amidinotransferase activity is also found in pancreas, liver and brain.

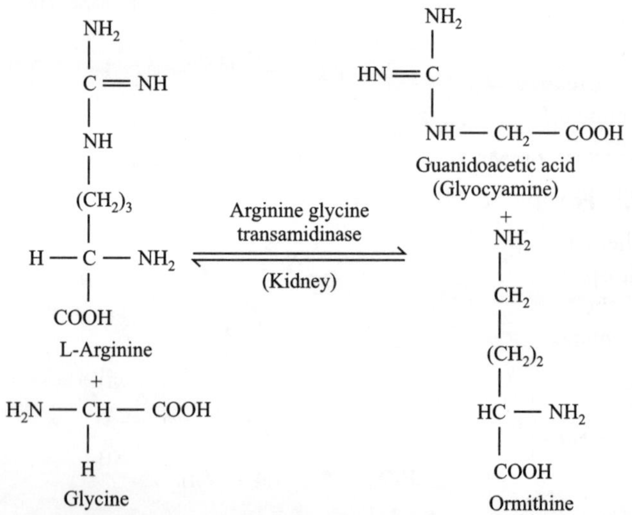

$$
\begin{array}{ccc}
NH_2 & & NH_2 \\
| & & | \\
C = NH & & HN = C \\
| & & | \\
NH & & NH-CH_2-COOH \\
| & & \text{Guanidoacetic acid} \\
(CH_2)_3 & & \text{(Glyocyamine)} \\
| & \xrightarrow[\text{(Kidney)}]{\text{Arginine glycine}} & + \\
H-C-NH_2 & \text{transamidinase} & NH_2 \\
| & & | \\
COOH & & CH_2 \\
\text{L-Arginine} & & | \\
+ & & (CH_2)_2 \\
H_2N-CH-COOH & & | \\
| & & HC-NH_2 \\
H & & | \\
\text{Glycine} & & COOH \\
& & \text{Ormithine}
\end{array}
$$

In the second stage, there is transfer of methyl group from S-adenosylmethionine to guanidoacetic acid, resulting in the production of creatine (methyl-guanidoacetic acid). This reaction catalyzed by guanidoacetate methyl transferase (EC 2.1.1.2) takes place in the liver.

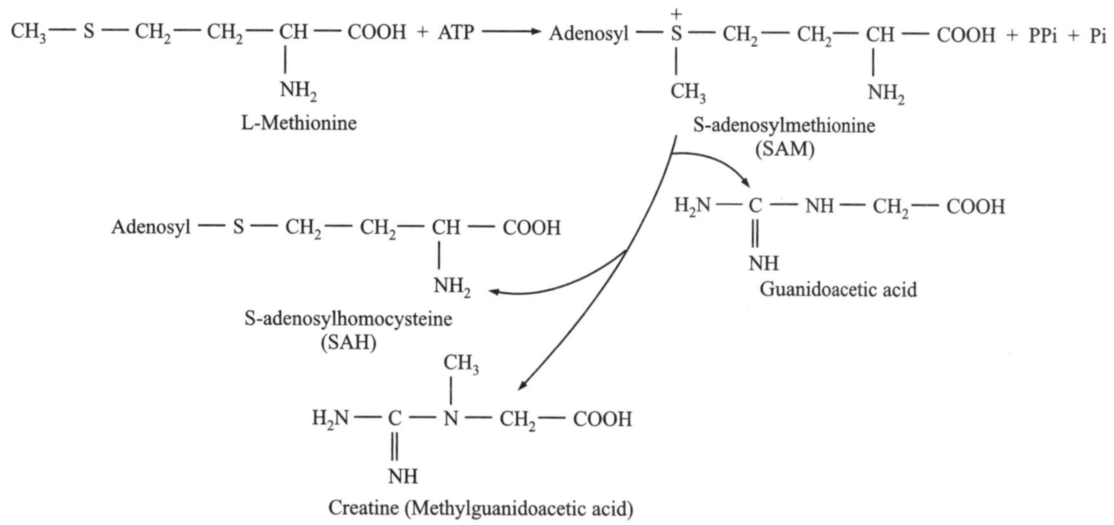

$$CH_3 - S - CH_2 - CH_2 - CH(NH_2) - COOH + ATP \longrightarrow Adenosyl - \overset{+}{S}(CH_3) - CH_2 - CH_2 - CH(NH_2) - COOH + PPi + Pi$$

L-Methionine

S-adenosylmethionine (SAM)

$$Adenosyl - S - CH_2 - CH_2 - CH(NH_2) - COOH$$

S-adenosylhomocysteine (SAH)

$$H_2N - C(=NH) - NH - CH_2 - COOH$$

Guanidoacetic acid

$$H_2N - C(=NH) - N(CH_3) - CH_2 - COOH$$

Creatine (Methylguanidoacetic acid)

Creatine synthesized in the liver is transported to the muscle, where creatine is converted to creatine phosphate by reaction with ATP (Lohmann reaction). This reaction is readily reversible.

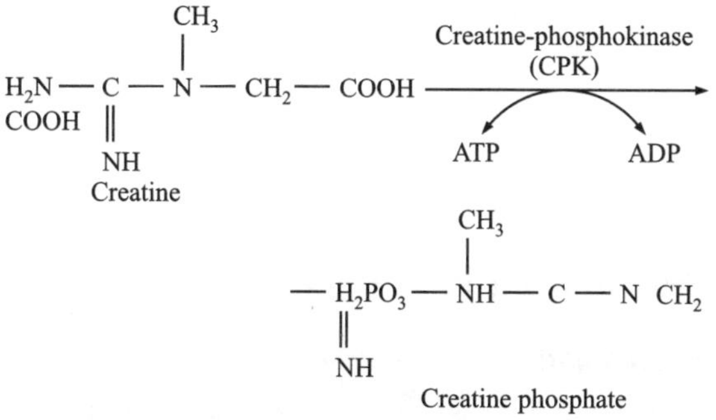

$$H_2N - C(=NH) - N(CH_3) - CH_2 - COOH \xrightarrow[\text{(CPK)}]{\text{Creatine-phosphokinase}} -H_2PO_3 - NH - C(=NH) - N(CH_3) CH_2$$

Creatine

ATP   ADP

Creatine phosphate

Creatine phosphate is one of the high energy phosphate compounds, and its main function is to regenerate ATP during muscular contraction by a reversal of the above reaction.

2. Formation of creatinine

There is a spontaneous conversion of creatine and creatine phosphate into creatinine to the level of about two per cent of the total each day. This is an irreversible, nonenzymatic, spontaneous reaction.

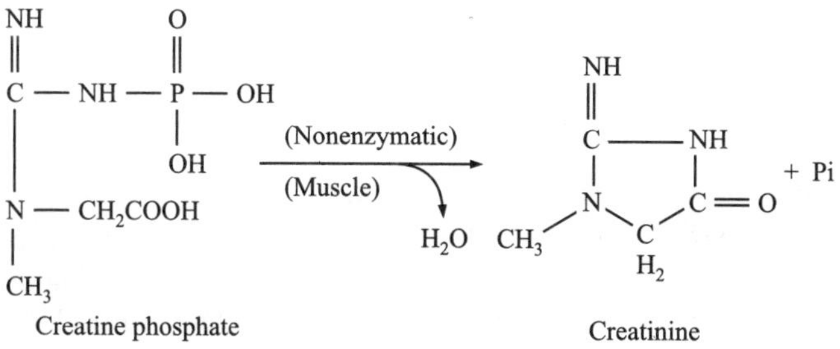

$$\text{Creatine phosphate} \xrightarrow[\text{(Muscle)}]{\text{(Nonenzymatic)}} \text{Creatinine} + Pi$$

This conversion is related to the total muscle mass and thereby to the body weight. It remains almost the same from day to day unless the muscle mass changes. Creatinine is a waste product and is excreted in urine. Therefore, 24-hour excretion of creatinine in urine of a given person is remarkably constant from day to day, and it is proportionate to the muscle mass.

Creatine and creatinine are handled differently by the kidney. Both are filtered at the glomerulus, but whereas there may be some tubular excretion of creatinine, creatine is partly reabsorbed in the tubule. At low concentrations in the plasma, re-absorption of creatine may be so efficient that there is little or no creatine in urine under normal circumstances.

D. Uric Acid

Uric acid forms the principal end product of nitrogen metabolism in birds and scaly reptiles, analogous to the formation of urea in man. In human urine, uric acid occupies a position quantitatively inferior to urea, ammonia and creatinine. Uric acid is the chief end product of purine metabolism in man, the higher apes, and the Dalmatian dog (Coach dog). Other mammals carry the conversion one step further to allantoin through the action of the enzyme uricase (EC 1.7.3.3). In man, uric acid results principally from the destruction of purine rich material obtained from the diet or disintegration of cellular matter of the organism. The enzymatic conversion of purines (adenine and guanine) to uric acid may be summarized as in Figure 16.3.

Uric acid is closely related to the purine bases. According to purine nomenclature, it may be designated as 2, 6, 8-trioxypurine. Uric acid is very sparingly soluble in water, and is insoluble in alcohol and ether. It can exist in tautomeric enol (lactim) and keto (lactam) forms. In its enol form, uric acid acts as a weak dibasic acid, and forms both neutral and acid salts (e.g. sodium, potassium and lithium salts). Neutral salts are much more soluble in water as compared to the acid salts. The two pKa values for uric acid are approximately 5.7 and 9.8. It is present as the monosodium salt (sodium urate) in plasma. In very acidic urine or other body fluids, the proportion of free uric acid is considerable. At pH 5.7, the concentration of uric acid is equal to that of sodium urate. In alkaline medium, sodium urate predominates. The uric acid

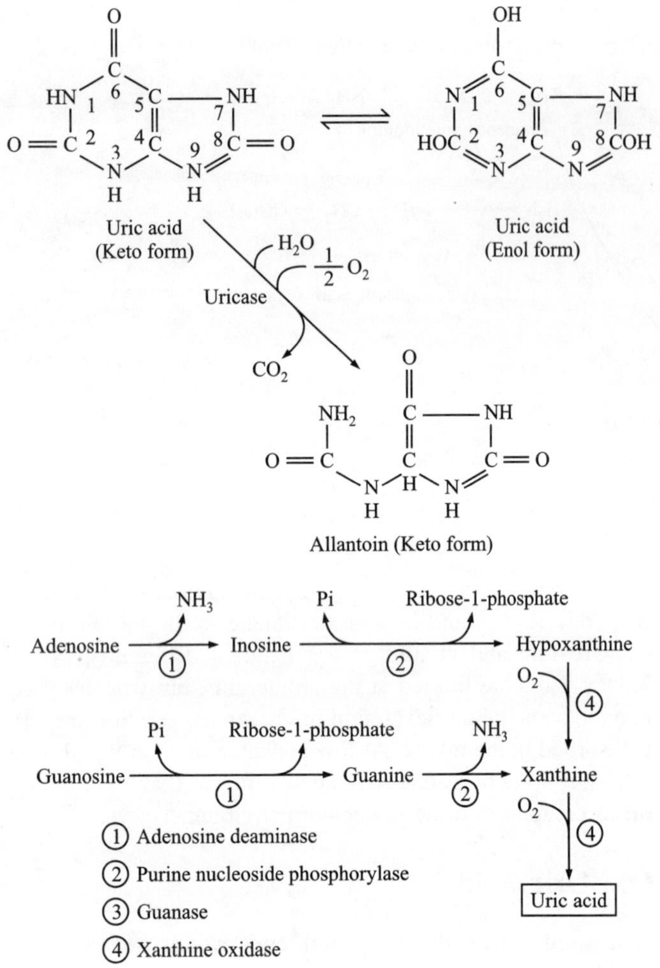
Uric acid
(Keto form)
Uric acid
(Enol form)
Uricase
H_2O
$\frac{1}{2}O_2$
CO_2
Allantoin (Keto form)
Adenosine
Inosine
Hypoxanthine
NH_3
Pi
Ribose-1-phosphate
Guanosine
Guanine
Xanthine
Uric acid
① Adenosine deaminase
② Purine nucleoside phosphorylase
③ Guanase
④ Xanthine oxidase

Figure 16.3 Pathway showing the degradation of purine nucleotides ultimately to uric acid.

content of urine is of importance in relation to the formation of uric acid calculi (stones).

There is turnover of about half the miscible uric acid pool daily. About 600 mg is formed daily and about the same amount is lost. Of the amount lost, about 75 per cent is excreted in urine, and about 25 per cent is excreted into the intestines via gastric juice and bile and subsequently destroyed by the intestinal bacterial flora. Urate filtered at the glomerulus is fully reabsorbed in the proximal tubule. Therefore, urate found in the urine represents that excreted into the distal tubule.

III. METABOLISM OF INDIVIDUAL AMINO ACIDS

A. Nutritionally Essential and Nutritionally Non-essential Amino Acids

All the twenty amino acids are biologically essential for the living systems. However, from nutritional point of view, investigations have revealed that animals can synthesize only some of these twenty amino acids at a rate that is sufficient enough to support the growth and maintenance of health of the animal. These amino acids are called "nutritionally non-essential" (or dispensable) amino acids. Amino acids which can be synthesized, but at a suboptimal rate are called "nutritionally semi-essential (or semi-indispensable)" amino acids. Other amino acids, which have to be present preformed in the diet, are called "nutritionally essential" (or indispensable) amino acids. An individual amino acid may be nutritionally essential for one form of life but non-essential for another. Amino acid requirements of human beings are given in Table 16.1

TABLE 16.1 Nutritive Classification of Amino Acids for Humans

Essential	*Semi-essential*	*Non-essential*
Lysine	Arginine	Alanine
Phenylalanine	Histidine	Glycine
Tryptophan		Serine
Threonine		Tyrosine
Methionine		Cysteine
Leucine		Aspartic acid
Valine		Asparagine
Isoleucine		Glutamic acid
		Glutamine
		Proline
		Hydroxyproline

B. Metabolism of Aromatic Amino Acids

1. Phenylalanine and tyrosine

These two amino acids share a close metabolic relationship. Phenylalanine is an essential amino acid in mammals. Tyrosine can be formed from phenylalanine and thus is not essential, provided the diet contains adequate quantities of phenylalanine. The conversion of phenylalanine to tyrosine is irreversible; tyrosine cannot be converted into phenylalanine in the animal body. Besides entering into the formation of body proteins, these amino acids participate in the synthesis of a variety of physiologically important substances.

(a) *Conversion of phenylalanine to tyrosine:* Phenylalanine is converted to tyrosine by a reaction catalyzed by phenylalanine hydroxylase (EC 1.14.3.1) present in mammalian liver. The phenylalanine hydroxylase complex is a mixed function oxygenase, which has two distinct activities (I and II). The reducing power is supplied by tetrahydrobiopterine coupled with NADPH (Figure 16.4).

(b) *Metabolic breakdown of tyrosine:* A major metabolic pathway of tyrosine results in the formation of acetoacetic acid and fumaric acid (Figure 16.5).

The first enzymatic reaction that converts tyrosine to p-hydroxyphenylpyruvic acid is catalyzed by tyrosine-α-ketoglutarate transaminase (EC 2.6. 1.5). This enzyme requires

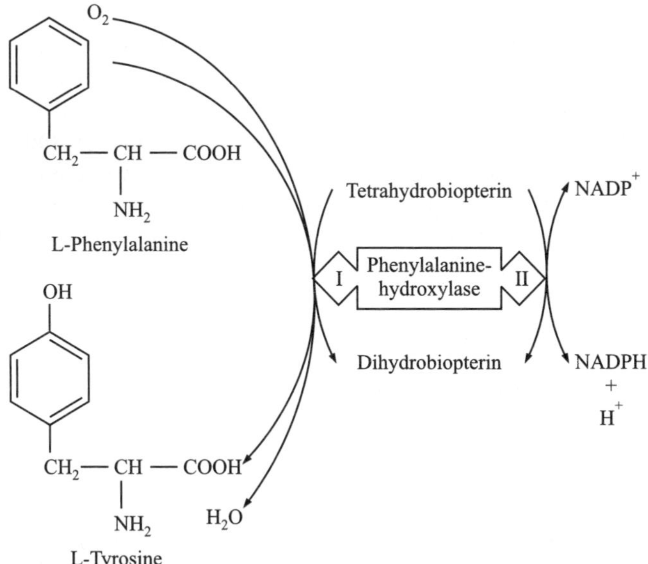

Figure 16.4 Conversion of phenylalanine to tyrosine by the action of phenylalanine hydroxylase.

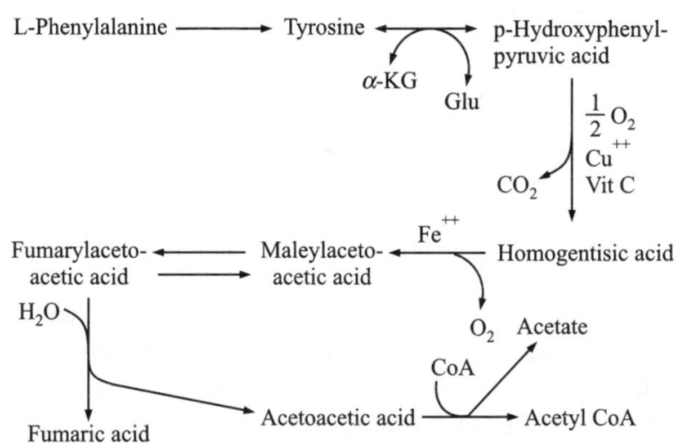

Figure 16.5 The conversion of phenylalanine and tyrosine to fumaric acid and acetoacetic acid.

pyridoxal phosphate as the coenzyme. This is an inducible enzyme present in high amounts in the liver.

p-Hydroxyphenylpyruvic acid is oxidized to homogentisic acid by p-hydroxyphenylpyruvate hydroxylase (EC 1.14.2.2), a copper metalloprotein. The presence of ascorbic acid is also required. The reaction is a complex one involving a shift in the side chain of p-hydroxyphenylpyruvic acid. Finally, homogentisic acid is formed by oxidative decarboxylation of the side chain.

The benzidine ring of homogentisic acid is then oxidatively cleaved forming maleylacetoacetate. The reaction is catalyzed by homogentisate oxidase/oxygenase (EC 1.13.11.5), an iron metalloprotein found in liver and kidney.

Maleylacetoacetate is converted to fumarylacetoacetate by a *'cis'* to *'trans'* isomerization catalyzed by maleylacetoacetate *cis-trans* isomerase (EC 5.2.1.2). It is a sulphhydryl enzyme and requires glutathione as a cofactor.

Fumarylacetoacetate is hydrolyzed by fumarylacetoacetate fumaryl hydrolase (EC 3.7.1.2) to form fumarate and acetoacetate.

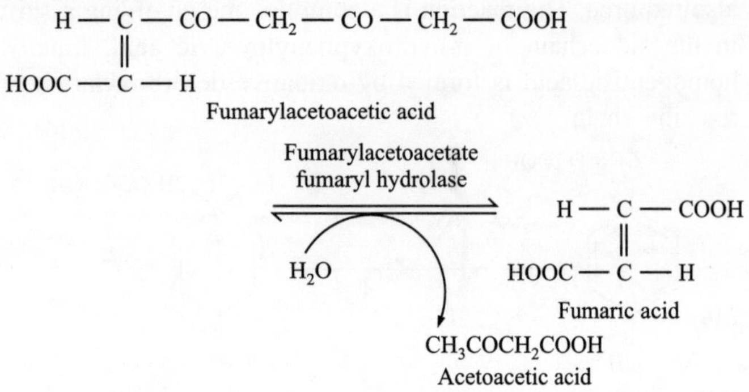

$$H - C - CO - CH_2 - CO - CH_2 - COOH$$
$$HOOC - C - H$$

Fumarylacetoacetic acid

Fumarylacetoacetate
fumaryl hydrolase

H_2O

$$H - C - COOH$$
$$HOOC - C - H$$
Fumaric acid

CH_3COCH_2COOH
Acetoacetic acid

Fumaric acid is convertible through pyruvic acid to glucose (gluconeogenesis) and glycogen. Acetoacetic acid enters the pathway of fatty acid metabolism after its conversion to acetyl CoA. Thus, phenylalanine and tyrosine are both glycogenic and ketogenic.

(c) *Melanin formation:*   Melanin is the pigment of the skin and hair. In the skin, melanin is located in very small granules (melanosomes), which are formed in the melanocytes. Melanosomes are transferred into the keratocytes and thereby get distributed throughout the epidermis by the outward movement of keratinocytes.

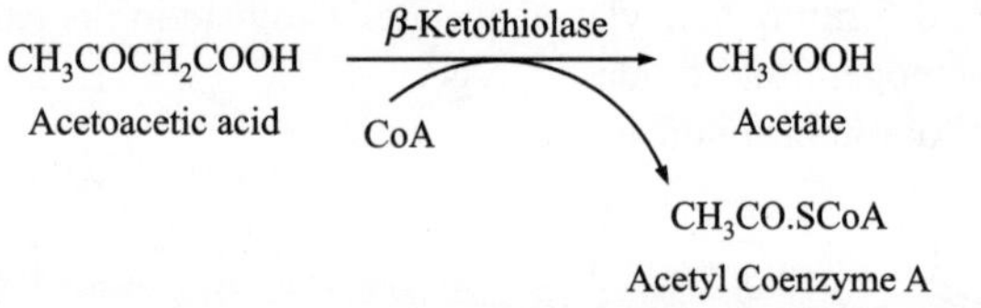

CH_3COCH_2COOH $\xrightarrow{\beta\text{-Ketothiolase}}$ CH_3COOH
Acetoacetic acid CoA Acetate

$CH_3CO.SCoA$
Acetyl Coenzyme A

The early steps of mammalian melanogenesis involve the conversion of tyrosine to β-3, 4-dihydroxyphenylalanine (L-dopa), which is then oxidized to dopaquinone. Both these reactions are catalyzed by a copper containing enzyme, tyrosinase (EC 1.10.3.1), which possesses both tyrosinase and dopa oxidase activities. Further steps in the formation of melanin from dopaquinone are nonenzymatic (Figure 16.6).

Though Figure 16.6 depicts melanin biosynthesis as a polymerization of idole-5, 6-quinone alone, it has been shown that mammalian melanins are heteropolymers containing many different combinations of the precursors.

2. *Tryptophan metabolism*

Tryptophan (α-amino-β-indolepropionic acid) is the only amino acid containing an indole ring. Indole ring cannot be synthesized in mammalian tissues. D-tryptophan can be deaminated to form indolepyruvic acid (a keto acid), which in turn can be aminated to form L-tryptophan, the isomer utilized for protein synthesis. Consequently, D-tryptophan and its α-keto analogue can replace L-tryptophan in the diet. The hydroxy acid, indoleacetic acid, has also been reported to be capable of replacing L-tryptophan in the diet.

D-Tryptophan   L-Tryptophan

NH_3 NH_2
Indolepyruvic acid

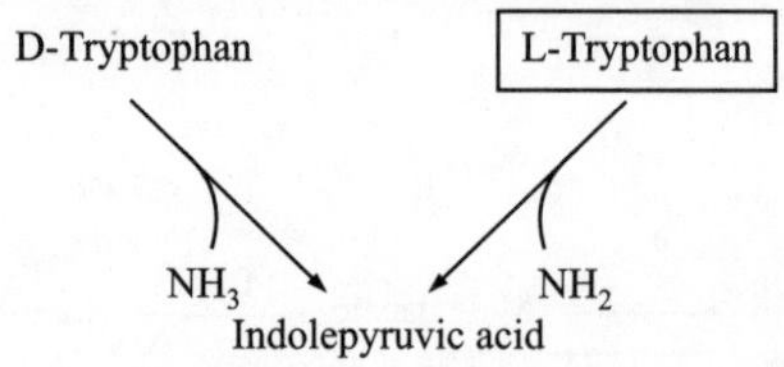

Tryptophan is required in the diet for growth and maintenance of nitrogen equilibrium in man. Tryptophan gives rise to some important biologically active compounds like nicotinic acid and serotonin (5-hydroxytryptamine).

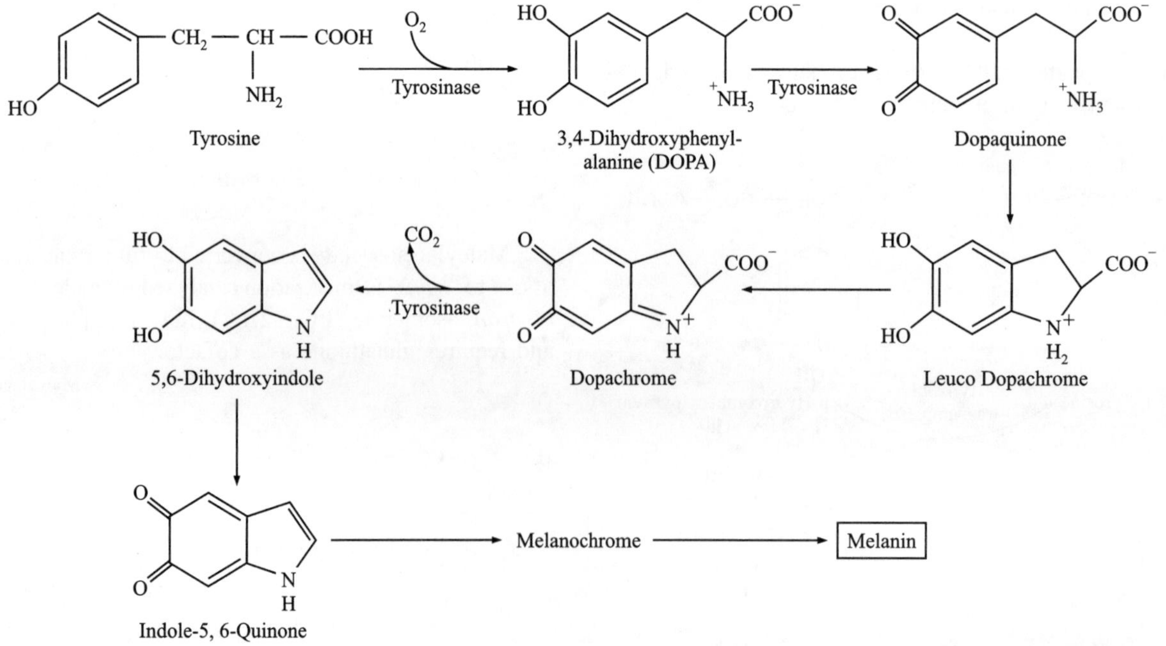

Figure 16.6   Formation of melanin from tyrosine.

(a) *Metabolic breakdown of tryptophan:* The main pathway of degradation of tryptophan proceeds through breakdown of the pyrrole ring, leading to formation of formylkynurenine. Formylkynurenine is then sequentially converted into kynurenine, 3-hydroxykynurenine and 3-hydroxyanthranilic acid (Figure 16.7).

Tryptophan pyrrolase (EC 1.13.1.12) or tryptophan peroxidase is a heme enzyme, which brings about oxidative breakdown of pyrrole ring using molecular oxygen. Kynurenine formylase (EC 3.5.1.9), which converts formylkynurenine to kynurenine and formate, is found in the liver. Hydroxylation of kynurenine is brought about by L-kynurenine hydroxylase (EC 1.14.1.2) present in the liver mitochondria. Kynurenine and 3-hydroxy-kynurenine are cleaved by kynureninase (EC 3.7.1.3) present in the liver and kidney to alanine plus anthranilic acid and alanine plus 3-hydroxy-anthranilic acid, respectively. Kynureninase is a pyridoxal phosphate-dependent enzyme (Figure 16.8).

In pyridoxine deficiency, kynurenine and 3-hydroxy-kynurenine cannot be converted into anthranilic acid and 3-hydroxyanthranilic acid, respectively. Instead, they form kynurenic acid and xanthurenic acid, respectively, which are

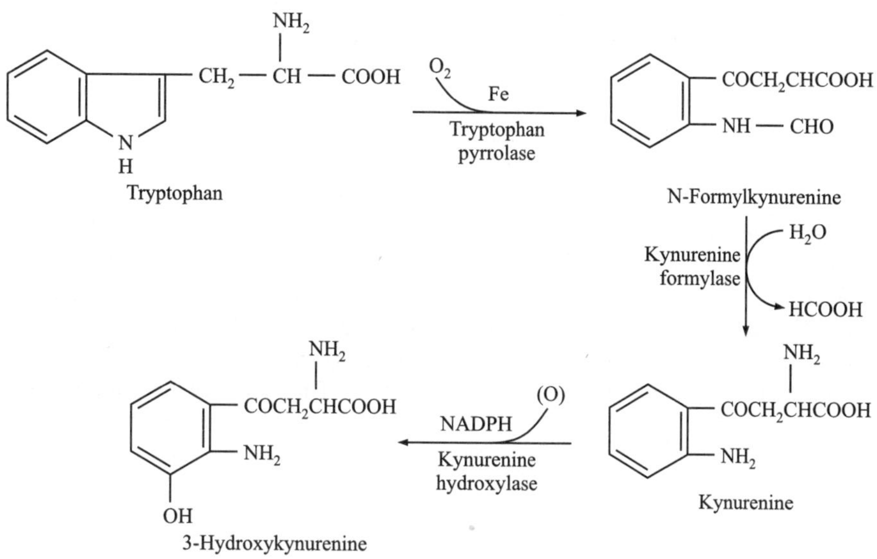

Figure 16.7 The initial steps in the catabolism of tryptophan.

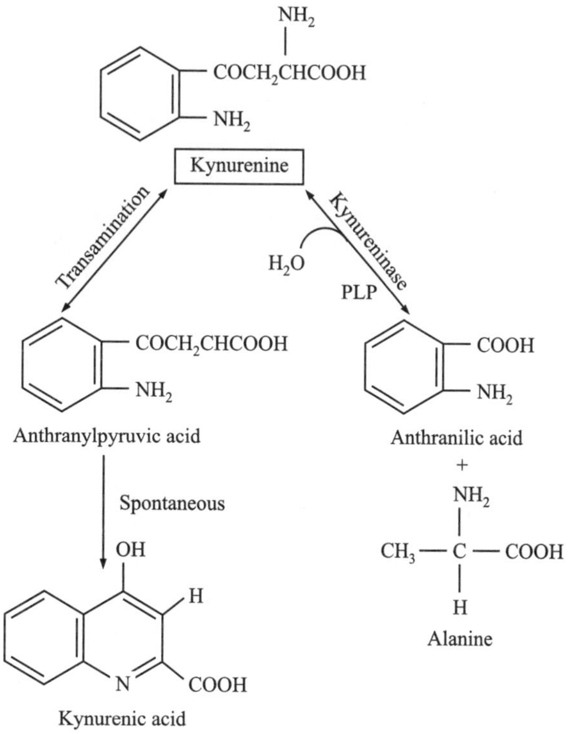

Figure 16.8 The conversion of kynurenine to anthranilic acid and alanine.

not further metabolized. Instead, they are excreted in urine. Estimation of xanthurenic acid levels in urine following tryptophan load in patients suffering from pyridoxine deficiency can reveal the degree of pyridoxine deficiency.

The primary oxidation product of 3-hydroxy-anthranilic acid is α-amino-β-carboxymuconic semialdehyde (2-acroleyl-3-aminofumaric acid), which is an unstable compound. Most of this compound is degraded via α-aminomuconic semialdehyde (2-aminomuconic acid-δ-semi-aldehyde) and α-ketoadipic acid to glutaryl CoA (Figure 16.9). The ultimate end products are acetyl CoA and CO_2. A small proportion of α-amino-β-carboxymuconic semialdehyde forms NAD^+ and $NADP^+$ through quinolinic acid. Quinolinic acid can also be decarboxylated to form nicotinic acid.

(b) *Formation of Nicotinic acid and NAD(P):* The unstable α-amino-β-carboxymuconic semi-aldehyde undergoes spontaneous cyclization forming quinolinic acid, a pyridine dicarboxylic acid. Quinolinic acid may be decarboxylated to form nicotinic acid, or it may react with phosphoribosylpyrophosphate (PRPP) to form quinolinic acid ribonucleotide. Quinolinic acid ribonucleotide is decarboxylated to form niacin ribonucleotide, which further reacts with ATP forming desamido nicotinamide adenine dinucleotide (desamido NAD^+). The latter compound in the presence of ATP and glutamine is converted to nicotinamide adenine

dinucleotide (NAD^+). NAD^+ is phosphorylated to $NADP^+$ by means of NAD-kinase (EC 2.7.1.23) in the presence of ATP (Figure 16.10).

The dietary nicotinic acid is also converted into NAD^+ in the body by a separate enzyme system. However, dietary nicotinic acid can be replaced by intake of sufficient amounts of tryptophan. Man and many other animals are able to meet the tissue requirements of nicotinic acid by converting tryptophan into NAD^+ and $NADP^+$. It has been estimated that a diet containing 600 mg of tryptophan derived from 60 g of mixed protein diet would supply 10 mg equivalents of nicotinic acid (60 mg of tryptophan is equivalent to 1 mg nicotinic acid). The formation of NAD^+ from tryptophan is normally limited by the concentration of α-amino-β-carboxymuconic semialdehyde.

(c) *Formation of Serotonin:* (see Page 216 also Section II).

(d) *Bacterial putrefaction:* Bacteria in the gastrointestinal tract convert tryptophan to a wide variety of products. These compounds may either be excreted in the feces or absorbed, detoxified, and excreted in urine. These excretory substances include indole, indoxyl (3-hydroxyindole), skatole (3-methylindole), skatoxyl (2-hydroxy-3-methylindole), and others. The hydroxy derivatives are excreted in the urine as ethereal sulphates or glucuronides. The potassium salt of indoxylsulphate is called indican (Figure 16.11).

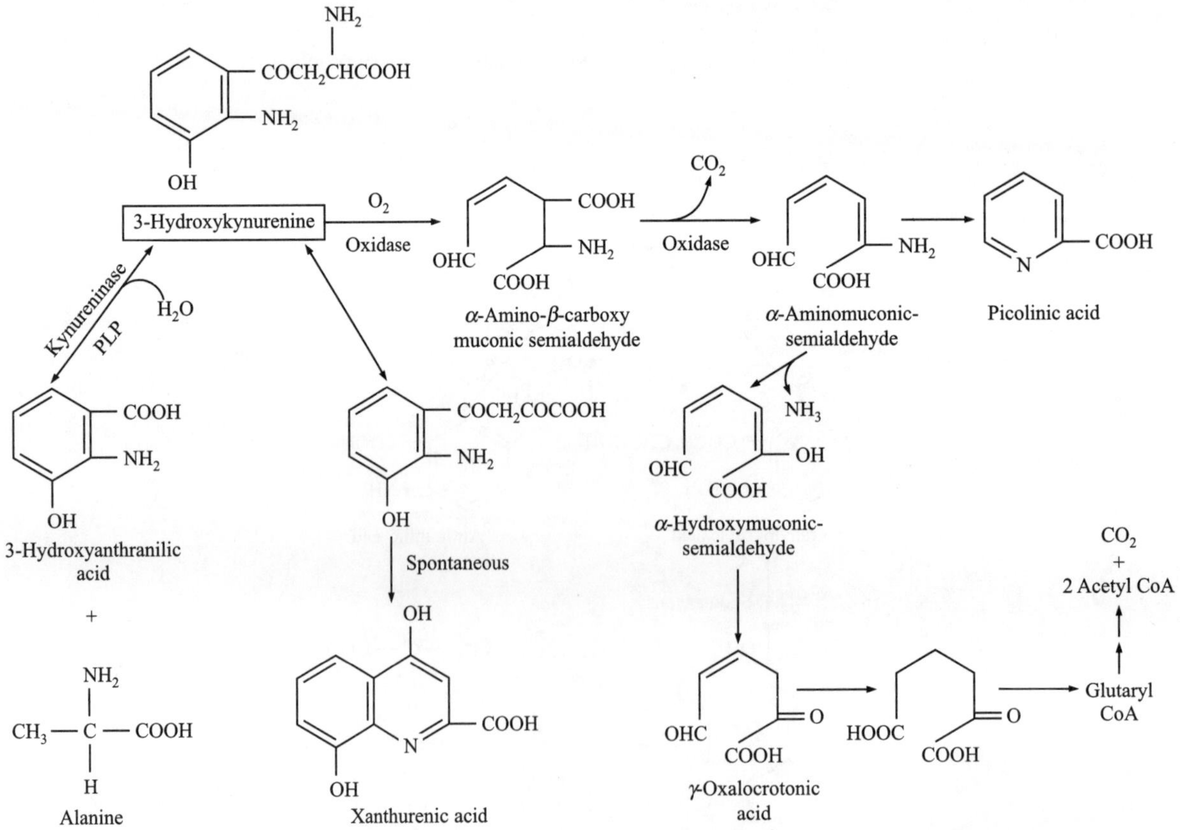

Figure 16.9 Degradation of 3-Hydroxyanthranilic acid to Acetyl CoA via *a*-Ketoadipic acid.

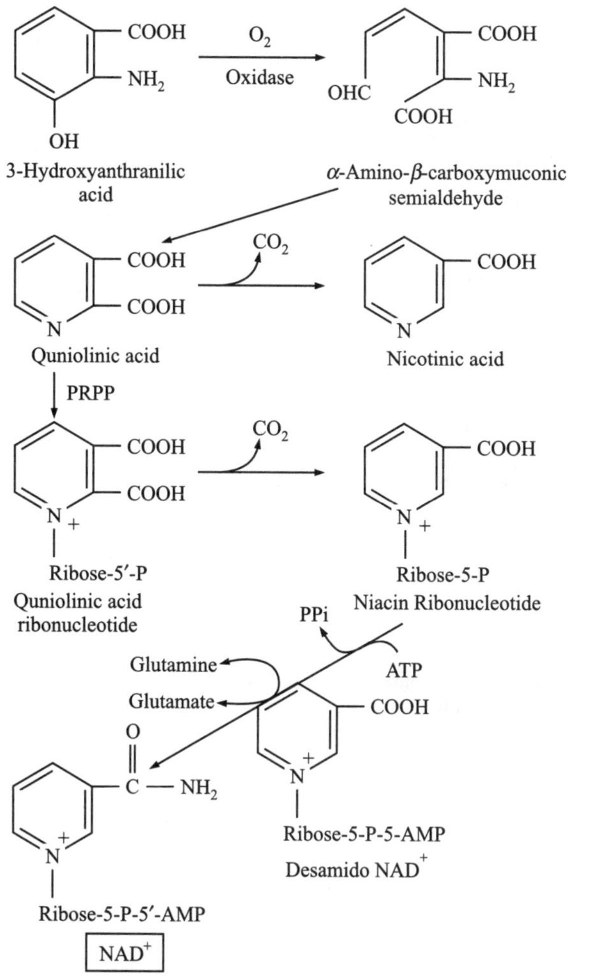

Figure 16.10 Formation of nicotinamide adenine dinucleotide (NAD$^+$) from tryptophan via 3-hydroxy-anthranilic acid.

3. Histidine

Histidine (α-amino-β-imidazolepropionic acid) is characterized among the amino acids by the presence of an imidazole ring.

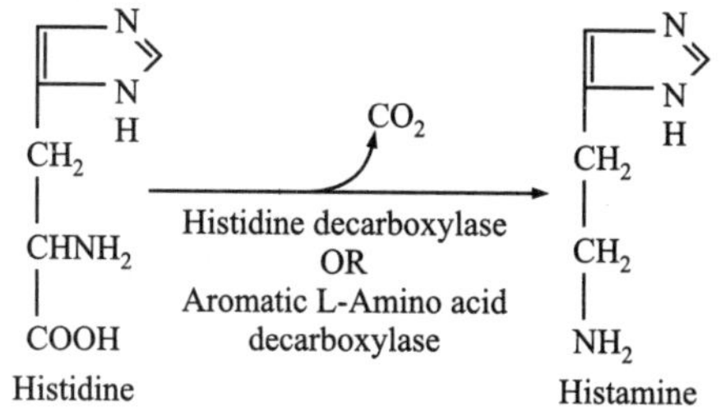

(a) *Metabolic breakdown of Histidine:* The enzyme histidase (EC 4.3.1.3) non-oxidatively deaminates histidine to ammonia and urocanic acid. Urocanic acid is further broken down to glutamic acid, formate and ammonia by the enzyme urocanase (Figure 16.12). Urocanic acid was first isolated from canine urine, hence the name. Later on it was shown to be present in urine of various animals following administration of histidine.

Estimation of formiminoglutamic acid (FIGLU) in urine after a histidine load is helpful for detection of folic acid deficiency (see Chapter 47).

(b) *Formation of Histamine:* Histidine is converted into histamine by the enzyme histidine decarboxylase (EC 4.1.1.22) for which pyridoxal phosphate acts as a coenzyme. Most cells of the body contain histidine decarboxylase.

(c) *Other histidine derivatives:* In the muscle, there are two dipeptides which contain histidine. These are carnosine and anserine. The function of these compounds is not known. Carnosine is a dipeptide of β-alanine and histidine, while

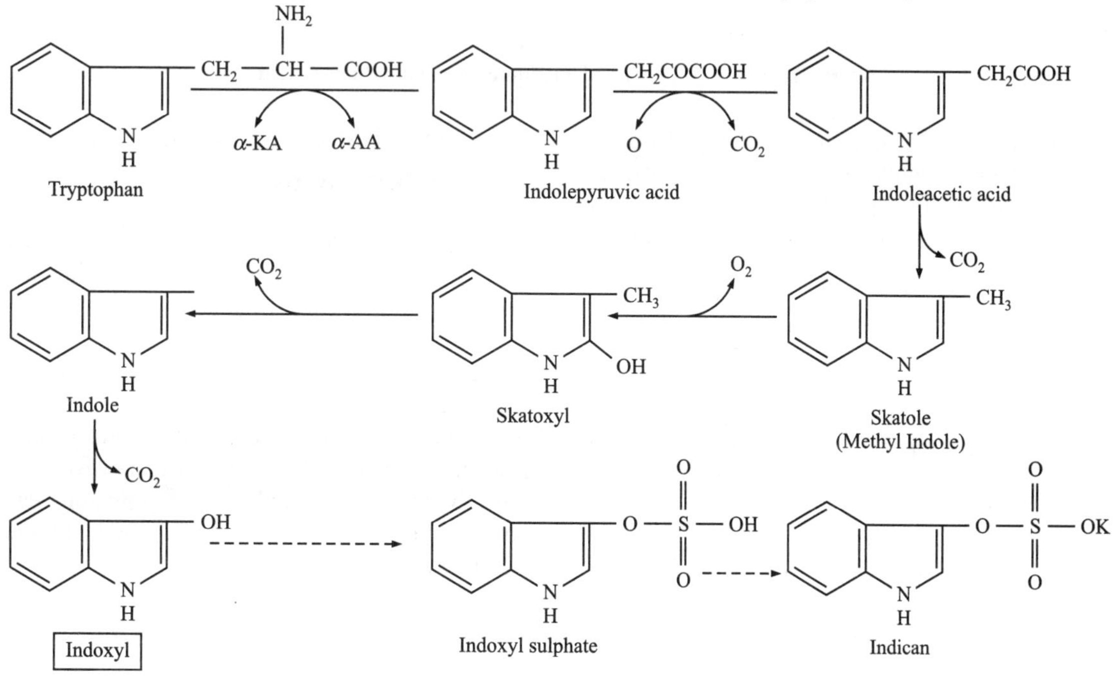

Figure 16.11 Conversion of tryptophan to several indole derivatives.

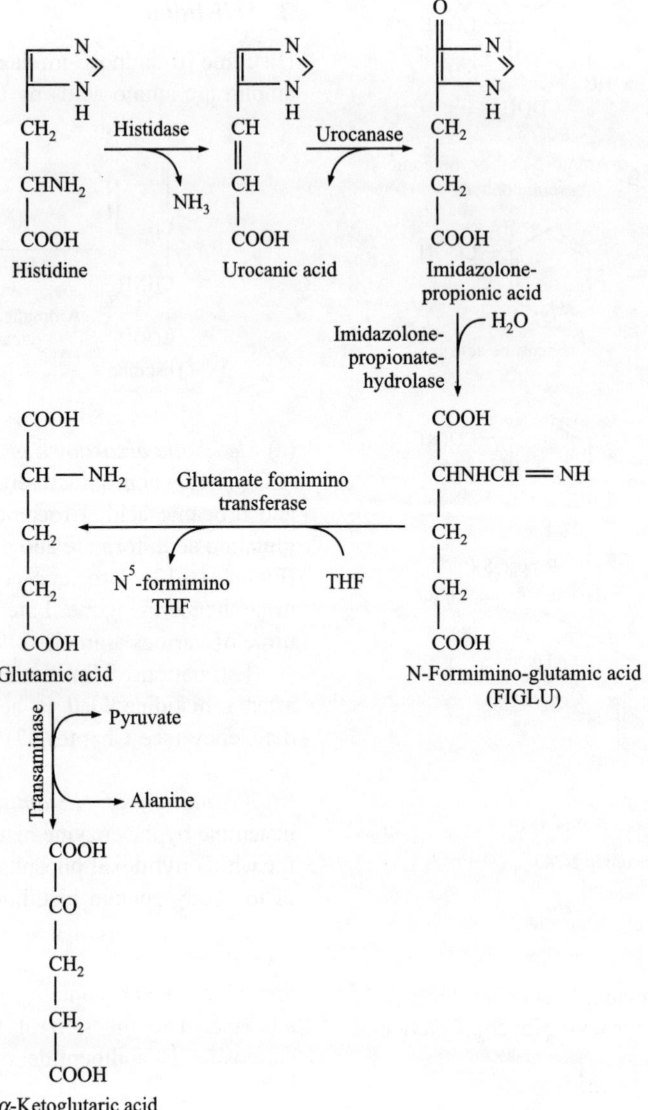

Figure 16.12 Catabolism of histidine to α-ketoglutaric acid.

anserine is a dipeptide of β-alanine and 1-methylhistidine. Ergothioneine, the betaine of thiolhistidine, is present in RBCs. Ergothioneine is derived from the diet, and it is actively taken up by RBCs. It is formed by fungi. It is not synthesized by animals.

C. Metabolism of Sulphur-containing Amino Acids

Most of the sulphur ingested is in the form of sulphur-containing amino acids present in dietary proteins. Of the three sulphur-containing amino acids (methionine, cystine and cysteine), methionine is considered to be an essential amino acid although the methyl group of methionine can be synthesized in the body. The other two amino acids can be synthesized in the body from methionine and are therefore, regarded as non-essential amino acids. Methionine can however, be synthesized in the body by a reversible transamination reaction if the corresponding α-ketoacid (α-keto-γ-methylthiobutyric acid) is available.

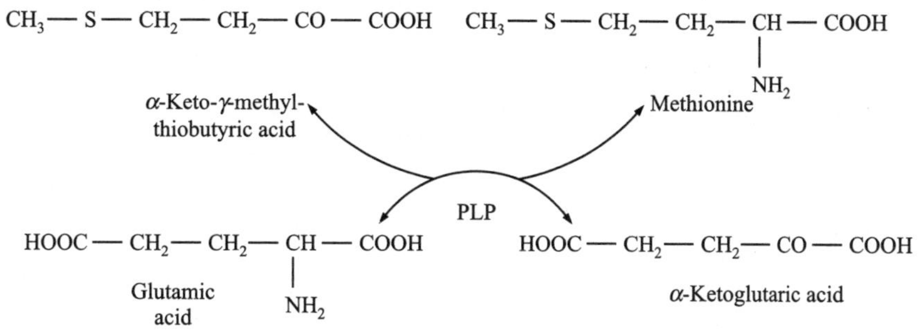

$$CH_3 — S — CH_2 — CH_2 — CO — COOH \qquad CH_3 — S — CH_2 — CH_2 — \underset{\underset{NH_2}{|}}{CH} — COOH$$

α-Keto-γ-methyl-thiobutyric acid → ← Methionine

PLP

$$HOOC — CH_2 — CH_2 — \underset{\underset{NH_2}{|}}{CH} — COOH \qquad HOOC — CH_2 — CH_2 — CO — COOH$$

Glutamic acid — α-Ketoglutaric acid

1. Cysteine and cystine

As cystine can be readily converted to cysteine by various reducing agents, metabolism of cystine proceeds through cysteine.

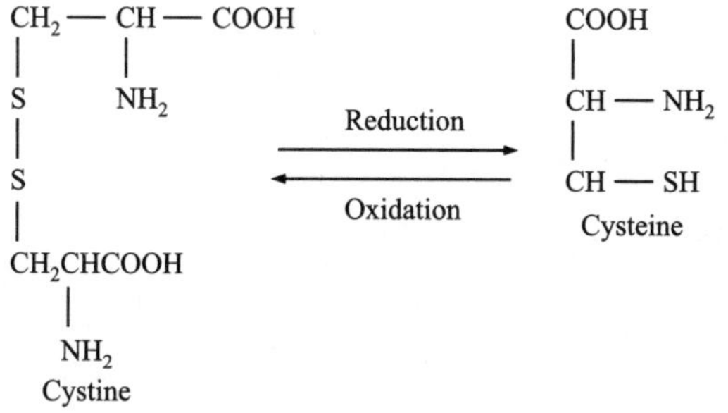

$$\underset{Cystine}{CH_2 — CH — COOH}$$

(Reduction / Oxidation)

Cysteine

L-Cysteine may be synthesized in yeast and certain other microorganisms from L-serine in a reaction catalyzed by cysteine synthase (serine sulphhydrase): EC 4.2.99.8.

$$\text{L-Serine} + H_2S \xrightarrow{\text{Cysteine synthase}} \text{L-Cysteine} + H_2O$$

Cysteine and cystine are formed in the animal organism from methionine.

Cysteine is metabolised by several pathways:

(a) Cysteine can be desulphhydrated to pyruvic acid by the action of cysteine desulphhydrase (EC 4.4.1.1). Pyridoxal phosphate is required in this reaction as a coenzyme.

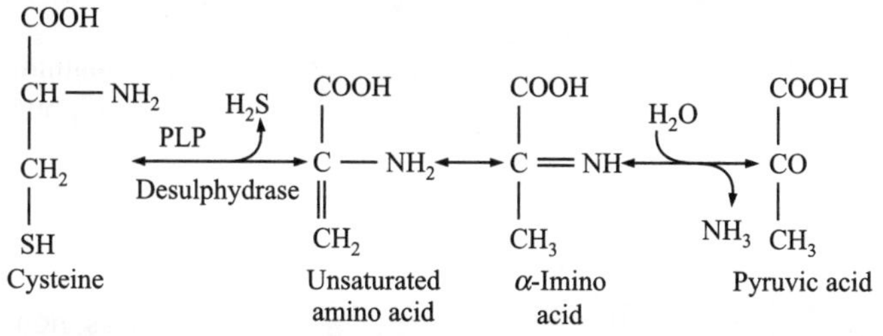

Cysteine → (PLP, Desulphydrase, H_2S) → Unsaturated amino acid → α-Imino acid → (H_2O, NH_3) → Pyruvic acid

(b) Cysteine can form pyruvate via β-mercap-topyruvic acid (β-thiolpyruvic acid). The latter is formed by transamination of cysteine in the presence of α-ketoglutaric acid and cysteineglutamate aminotransferase. β-Mercaptopyruvic acid undergoes desulphuration in the presence of thiolpyruvate *trans*-sulphurase to form pyruvate. The sulphur requires a reduced sulphhydryl compound like reduced glutathione for the formation of H_2S. [The sulphide is converted to sulphite,

and then to sulphate (SO_4^{--}); the latter reaction is catalyzed by sulphite oxidase (EC 1.8.3.1)]. [The sulphate is either excreted in urine or converted to "active sulphate", PAPS (3′-phosphoadenosine 5′-phosphosulphate). PAPS may be utilized for the synthesis of sulphate esters of various alcohols, phenols, steroids, and polysaccharides].

The formation of PAPS takes place as follows [Figure (a)].

(c) Cysteine can form pyruvate via cysteine-sulphinic acid, which is formed by oxidation of cysteine [Figure (b)].

(d) *Trans-sulphuration:* β-Mercaptopyruvic acid formed from cysteine can transfer sulphur to a number of acceptors. The reaction is catalyzed by sulphurtransferases (EC 2.8.1) or *trans*-sulphurases. Sulphocyanate is formed by the transfer of sulphur from β-mercaptopyruvic acid to the cyanide ion (CN–). Thiosulphate is formed by transfer of sulphur from β-mercaptopyruvic acid to the sulphite ion (SO_3^{--}).

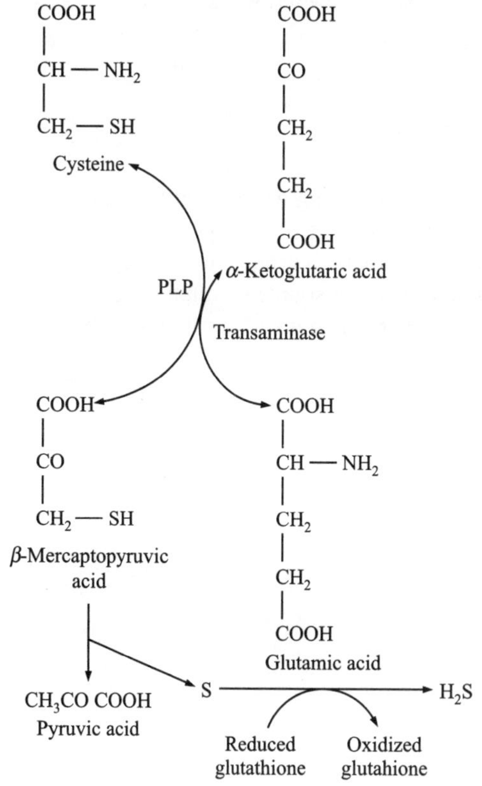

Cysteine → (PLP, α-Ketoglutaric acid, Transaminase) → β-Mercaptopyruvic acid → Pyruvic acid ($CH_3CO\ COOH$) ; Glutamic acid ; S → (Reduced glutathione / Oxidized glutathione) → H_2S

(e) Cysteine can form taurine via cysteine-sulphinic acid (see Taurine in section III).

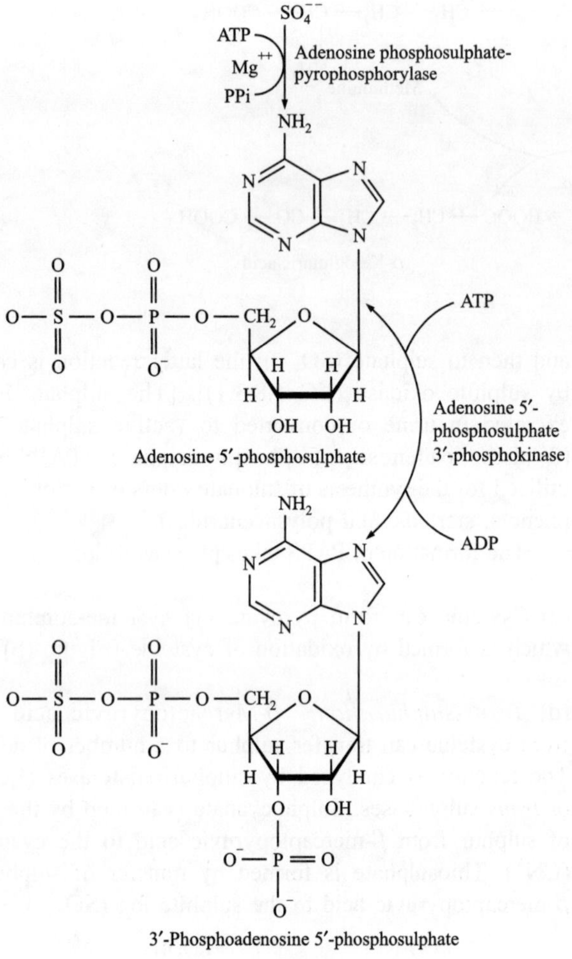

Adenosine phosphosulphate-pyrophosphorylase

Adenosine 5′-phosphosulphate

Adenosine 5′-phosphosulphate 3′-phosphokinase

3′-Phosphoadenosine 5′-phosphosulphate
(PAPS)

(a)

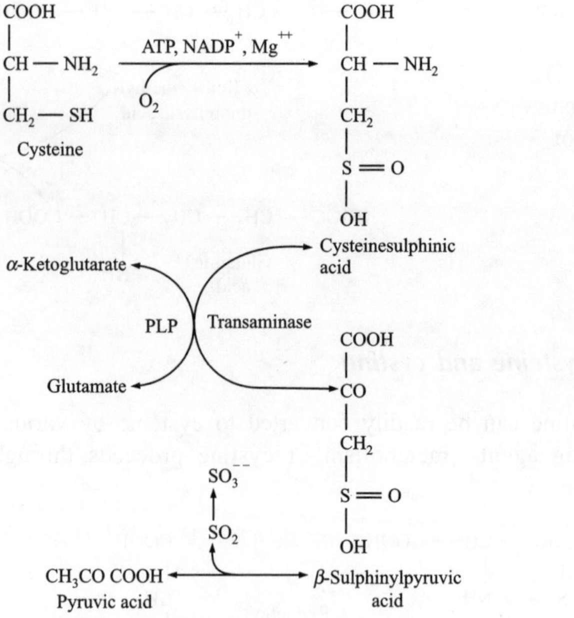

Cysteine

α-Ketoglutarate

Cysteinesulphinic acid

PLP   Transaminase

Glutamate

β-Sulphinylpyruvic acid

$CH_3CO\,COOH$
Pyruvic acid

(b)

β-Mercaptopyruvic acid

Trans-sulphurase

SCN^- + $CH_3COCOOH$
Sulphocyanate   Pyruvic acid

Trans-sulphurase

$S_2O_3^{--}$ + $CH_3COCOOH$
Thiosulphate   Pyruvic acid

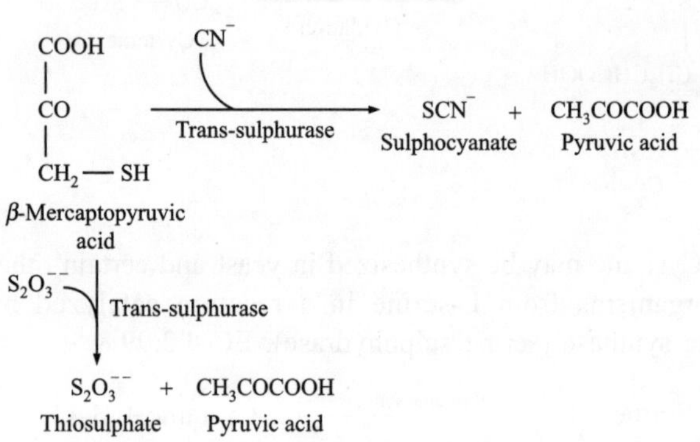

2. Methionine

Of the sulphur-containing amino acids, methionine is peculiar in that it contains a labile methyl group. Methionine is the main source of methyl groups for biological methylations. Methionine is also the source of sulphur contained in cysteine and cystine. Methionine also contributes carbon atoms in the formation of polyamines like spermine and spermidine. In the bacterial system, formylmethionyl tRNA is an initiator of polypeptide chain synthesis.

(a) *Transmethylation:* Methionine, which donates the methyl group to various acceptors, can do so only after an initial activation by reaction with ATP. The actual methyl donor is

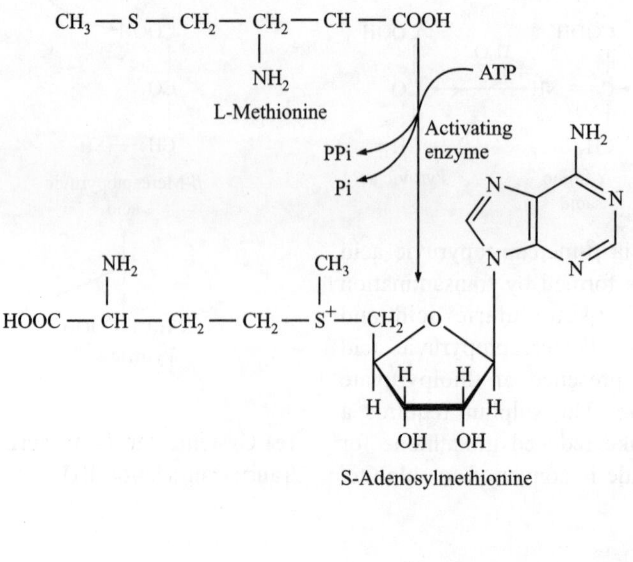

L-Methionine

Activating enzyme

S-Adenosylmethionine

S-adenosyl methionine (SAM). The activating enzyme has been shown to be present in a number of tissues including the brain. The energy in methyl sulphonium bond in S-adenosyl methionine is equal to the energy present in pyrophosphate bond of ATP.

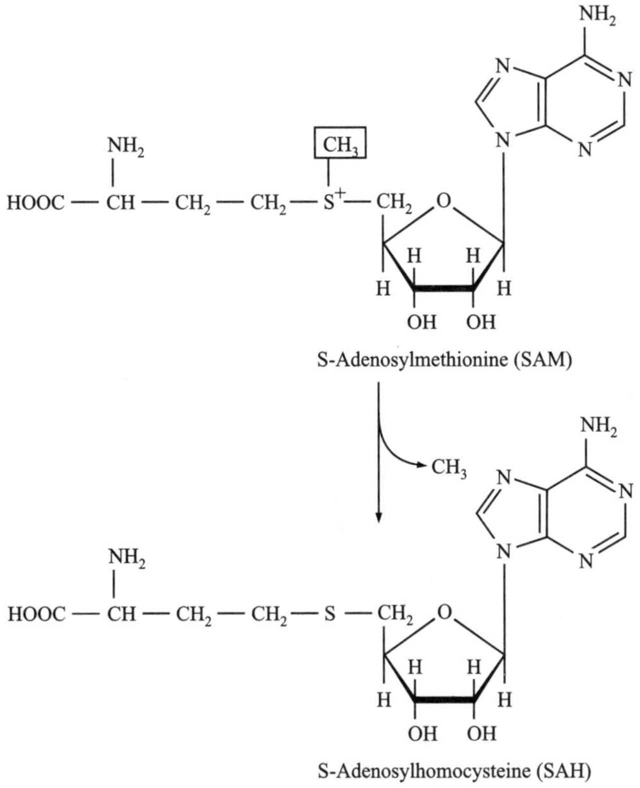

S-Adenosylmethionine (SAM)

S-Adenosylhomocysteine (SAH)

In the process of transmethylation, S-adenosyl methionine (SAM) is demethylated to S-adenosyl homocysteine (SAH). The enzymes involved are the methyltransferases (EC 2.1.1).

Creatine is synthesized in the body by methylation of guanidoacetic acid (see Section I).

Similarly, adrenaline is produced by methylation of noradrenaline.

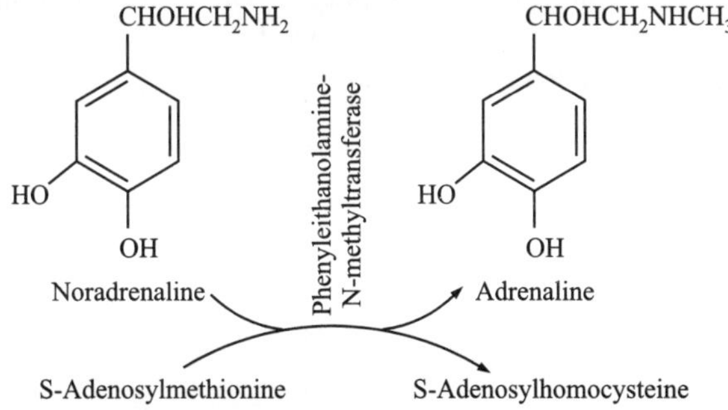

Noradrenaline → Adrenaline (Phenyleithanolamine-N-methyltransferase)
S-Adenosylmethionine → S-Adenosylhomocysteine

These catecholamines are further metabolized and inactivated to some extent by O-methylation. Thus, adrenaline upon O-methylation gives rise to metanephrine, and noradrenaline is converted into normetanephrine. The catechol-O-methyl transferase (EC 2.1.1.6) has been shown to be present in the brain and other mammalian tissues. Therefore,

transmethylation is important for the biosynthesis of adrenaline as well as the inactivation of catecholamines.

Choline, a lipotropic factor and precursor of acetylcholine, can be produced in the body by methylation of dimethyl ethanolamine.

Choline is a constituent of lecithin and other sphingolipids, some of which are constituents of cell membranes. The integrity of cell membranes, therefore, depends upon continuous supply of choline. Most of the phospholipids are constituents of lipoproteins, which are synthesized in the liver. Hence choline, being a constituent of phospholipids, acts as a lipotropic factor. All substances, including methionine, which aid in the production of choline, also act as lipotropic factors. Choline can also donate methyl group after undergoing conversion (by oxidation) to betaine. Betaine can transfer the methyl group to homocysteine to form methionine.

Nicotinic acid in the body is converted into N-methyl nicotinic acid and is excreted in the urine. This process involves loss of methyl groups from the body. As a result of this, methyl group will not be available for synthesis of choline. This will lead to accumulation of lipid in the liver. Thus, nicotinic acid acts as an anti-lipotropic factor.

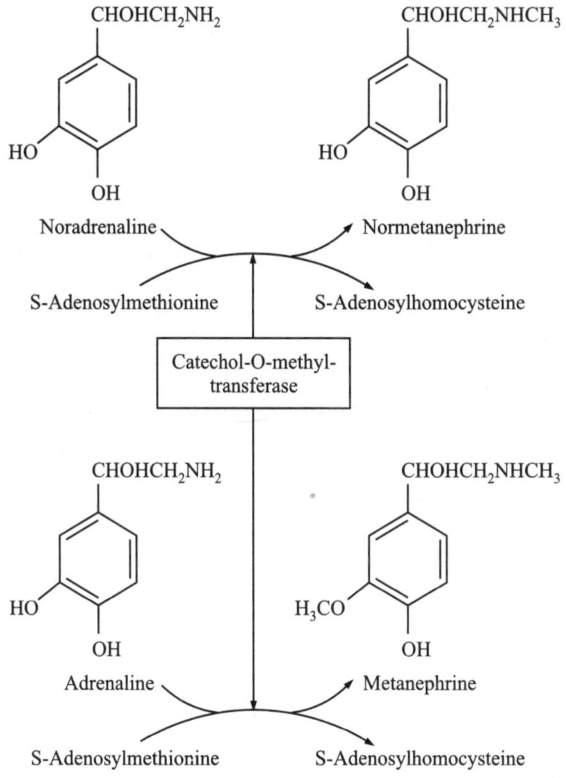

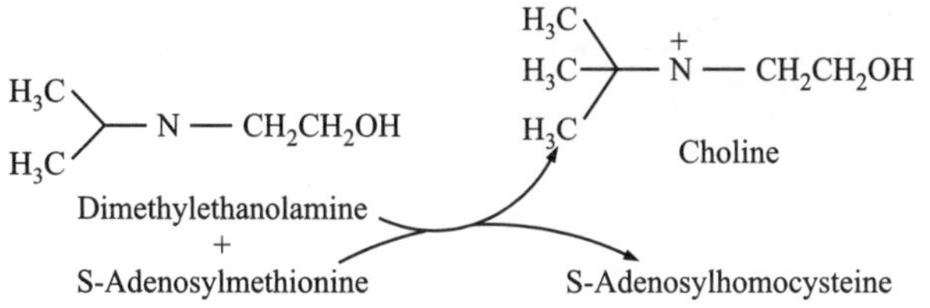

Noradrenaline → Normetanephrine
S-Adenosylmethionine → S-Adenosylhomocysteine
(Catechol-O-methyl-transferase)

Adrenaline → Metanephrine
S-Adenosylmethionine → S-Adenosylhomocysteine

Dimethylethanolamine + S-Adenosylmethionine → Choline + S-Adenosylhomocysteine

In the tissues, methylated nucleic acids are present. Transfer RNA (tRNA) contains a large number of methylated

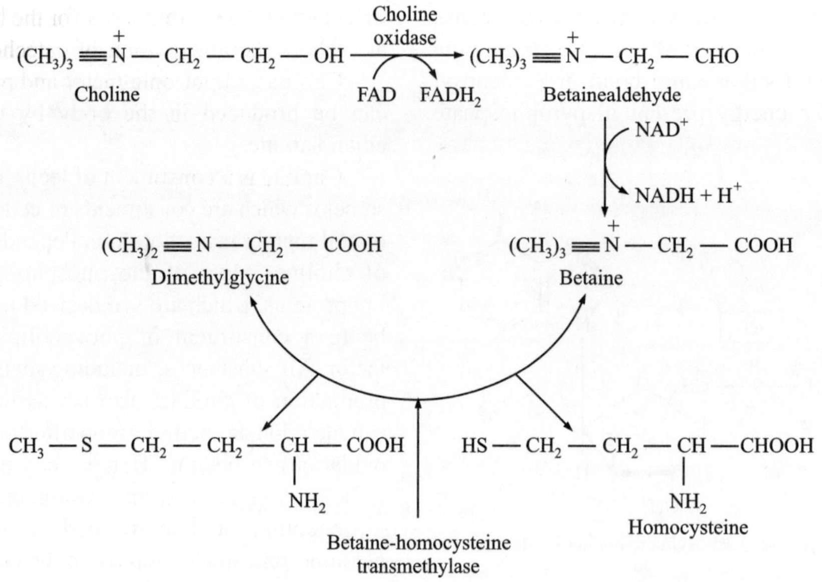

$$(CH_3)_3 \equiv \overset{+}{N} - CH_2 - CH_2 - OH \xrightarrow[\;FAD \quad FADH_2\;]{\text{Choline oxidase}} (CH_3)_3 \equiv \overset{+}{N} - CH_2 - CHO$$

Choline — Betainealdehyde

$$\text{Betainealdehyde} \xrightarrow[NADH + H^+]{NAD^+} (CH_3)_3 \equiv \overset{+}{N} - CH_2 - COOH$$

Betaine

$$(CH_3)_2 \equiv N - CH_2 - COOH$$
Dimethylglycine

$$CH_3 - S - CH_2 - CH_2 - \underset{\underset{NH_2}{|}}{CH} - COOH \qquad HS - CH_2 - CH_2 - \underset{\underset{NH_2}{|}}{CH} - CHOOH$$

Homocysteine

Betaine-homocysteine transmethylase

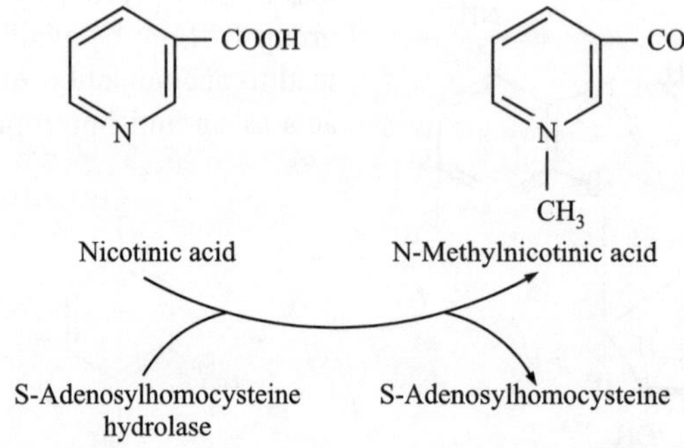

Nicotinic acid → N-Methylnicotinic acid

S-Adenosylhomocysteine hydrolase — S-Adenosylhomocysteine

bases, and it has been shown that methylation occurs after the macromolecule is formed. S-adenosyl methionine acts as the methyl donor. Many enzymes that methylate nucleic acids at the macromolecular level are known. These enzymes are species-specific, and therefore, confer an individuality on the number and distribution of methylated bases in tRNA, rRNA and DNA in each species. It has increasingly been recognized that the methylating ability of tumours towards RNA and DNA is several fold higher compared to that of the normal tissues.

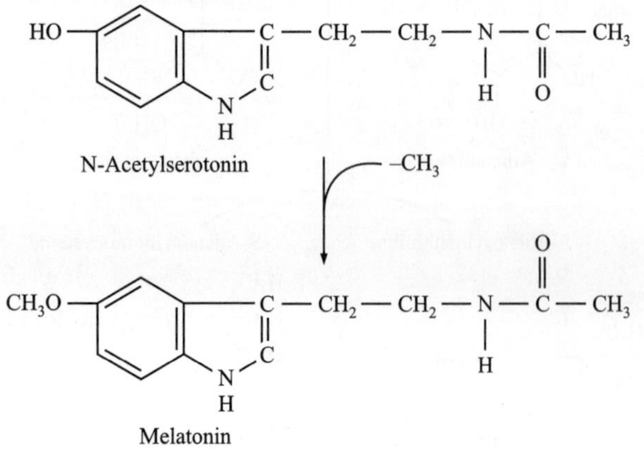

N-Acetylserotonin

$$\xrightarrow{-CH_3}$$

Melatonin
(N-Acetyl-5-methoxyserotonin)

In the past few years, evidence has been accumulating rapidly for the implication of DNA methylation in various biological processes. DNA methylation affects DNA-protein interaction, protects DNA against restriction endonucleases, regulates gene expression in eukaryotes, enhances mutation, and probably affects the structure and replication of DNA.

Transmethylation is involved in the synthesis of melatonin, a hormone of the pineal gland. As a result of demethylation, S-adenosyl-methionine is converted into S-adenosylhomocysteine. In many tissues, S-adenosyl-homocysteine hydrolase acts on S-adenosylhomocysteine to produce adenosine and homocysteine. The equilibrium of the above reaction favours the backward reaction, that is, the synthesis of S-adenosylhomocysteine.

S-Adenosylhomocysteine
$$\xrightarrow[\text{Adenosine}]{H_2O}$$
S-Adenosylhomocysteine hydrolase
L-Homocysteine

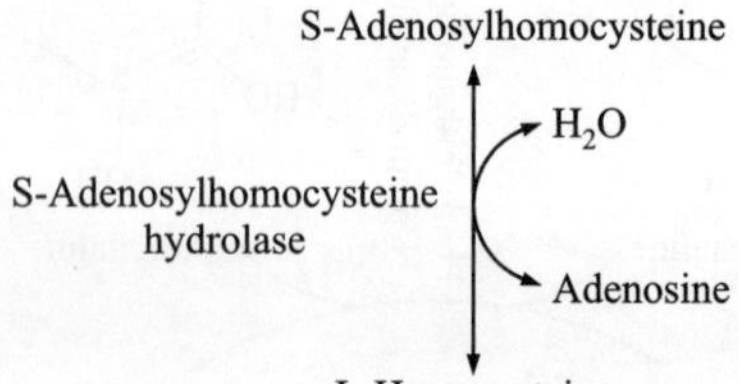

(b) *Trans-sulphuration:* Homocysteine condenses with serine in presence of an enzyme, cystathionine-β-synthase (EC.4.2.1.22), to form cystathionine. The enzyme requires pyridoxal phosphate as a coenzyme. This enzymatic activity is found in mammalian liver. Cystathionine thus formed is cleaved by γ-cystathionase (EC.4.4.1.1) to form cysteine and homoserine. This enzyme also requires pyridoxal phosphate as a cofactor. The intermediate homoserine may be enzyme bound. Homoserine deaminase (EC 4.4.1.1) acts on homoserine to yield NH_3 and α-ketobutyric acid. Thus, the sulphur of homo-cysteine is transferred to cysteine through successive actions of cystathionine synthase and cystathionase. In other words, sulphur of methionine can be utilized to form cysteine in the body. Therefore, the need for cysteine can be fulfilled by dietary methionine. It is interesting to note that brain tissue has all the enzymes involved in the process of trans-sulphuration through cystathionine pathway.

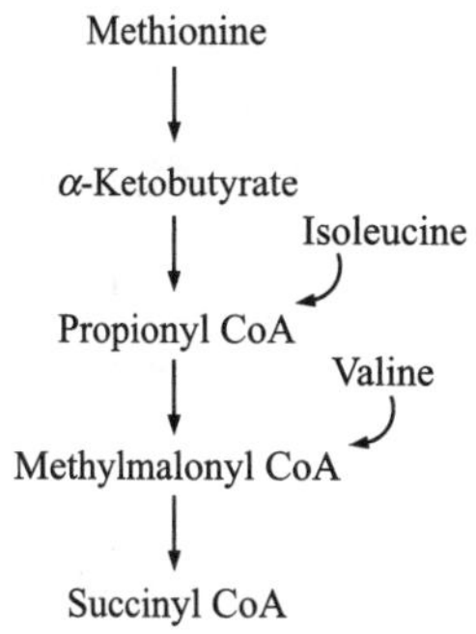

(c) Methionine synthesis: As already indicated, methionine can be formed from homocysteine as a result of transmethylation by choline, betaine, or dimethylthetin.

Synthesis of methionine may also occur from either homocysteine and 5-methyltetrahydropteroyl-L-glutamate, or homocysteine and S-adenosyl methionine. The former reaction is catalyzed by tetrahydropteroyl glutamate methyltransferase (EC 2.1.1.13, or methionine synthase), and the latter by homocysteine methyltransferase (EC 2.1.1.10).

α-Ketobutyric acid may undergo transamination to form α-aminobutyric acid. More important fate of α-ketobutyrate is its conversion to propionyl CoA by oxidative decarboxylation analogous to the oxidative decarboxylation of pyruvate to yield acetyl CoA.

Propionyl CoA is ultimately converted into succinyl CoA through methylmalonyl CoA. Thus, carbon skeletons of methionine and homocysteine enter the citric acid cycle through the formation of succinyl CoA. Therefore, both methionine and homocysteine are glycogenic amino acids.

D. Metabolism of Aliphatic Amino Acids

1. Metabolism of branched chain amino acids (Leucine, Valine, and Isoleucine)

These three amino acids initially share a common metabolic pathway, after which each amino acid follows its own route. Leucine, valine and iso-leucine form the corresponding α-ketoacids by transamination reaction catalyzed probably by a single transaminase. These α-ketoacids are oxidatively decarboxylated in a reaction sequence catalyzed by a single oxidative decarboxylase in man. This reaction is analogous to the reaction catalyzed by pyruvate dehydrogenase complex. The corresponding acyl CoA thioesters produced by oxidative decarboxylation are dehydrogenated to the corresponding α, β-unsaturated acyl CoA thioesters. At least two different dehydrogenases catalyze this reaction, one of them being specific for leucine alone. Further steps in catabolism of individual amino acids are represented in Figures 16.13, 16.14, 16.15 and 16.16.

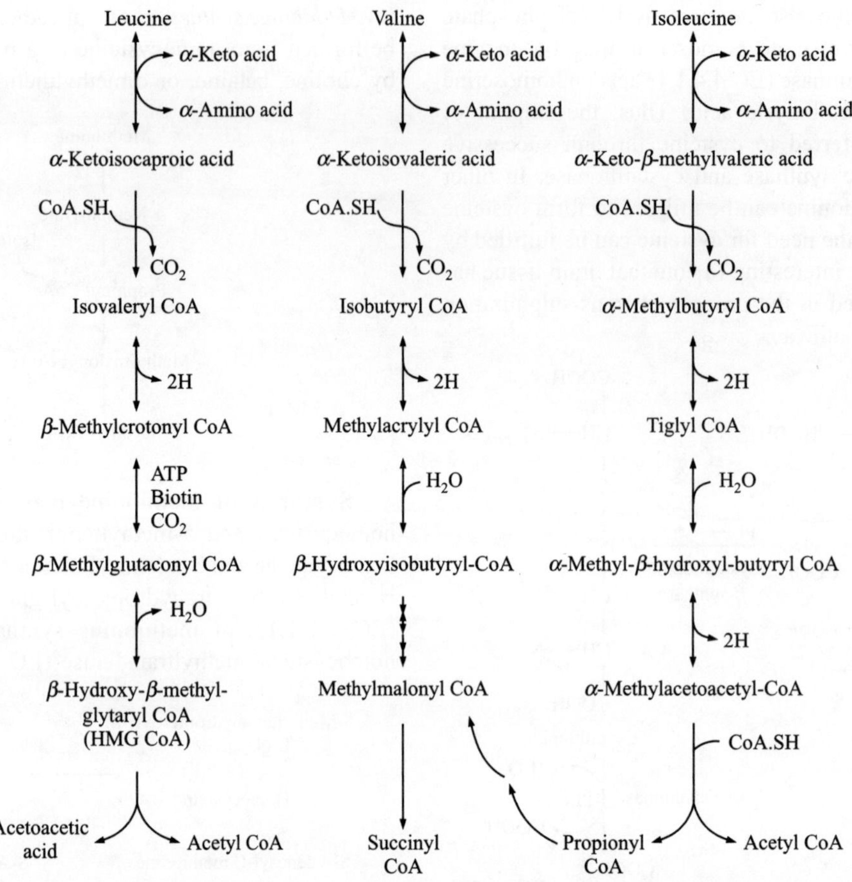

Figure 16.13   Metabolism of branched chain amino acids proceeds through the initial reactions involving transamination, oxidative decarboxylation and oxidation. The details of the succeeding steps are given in Figures 16.14, 16.15, and 16.16.

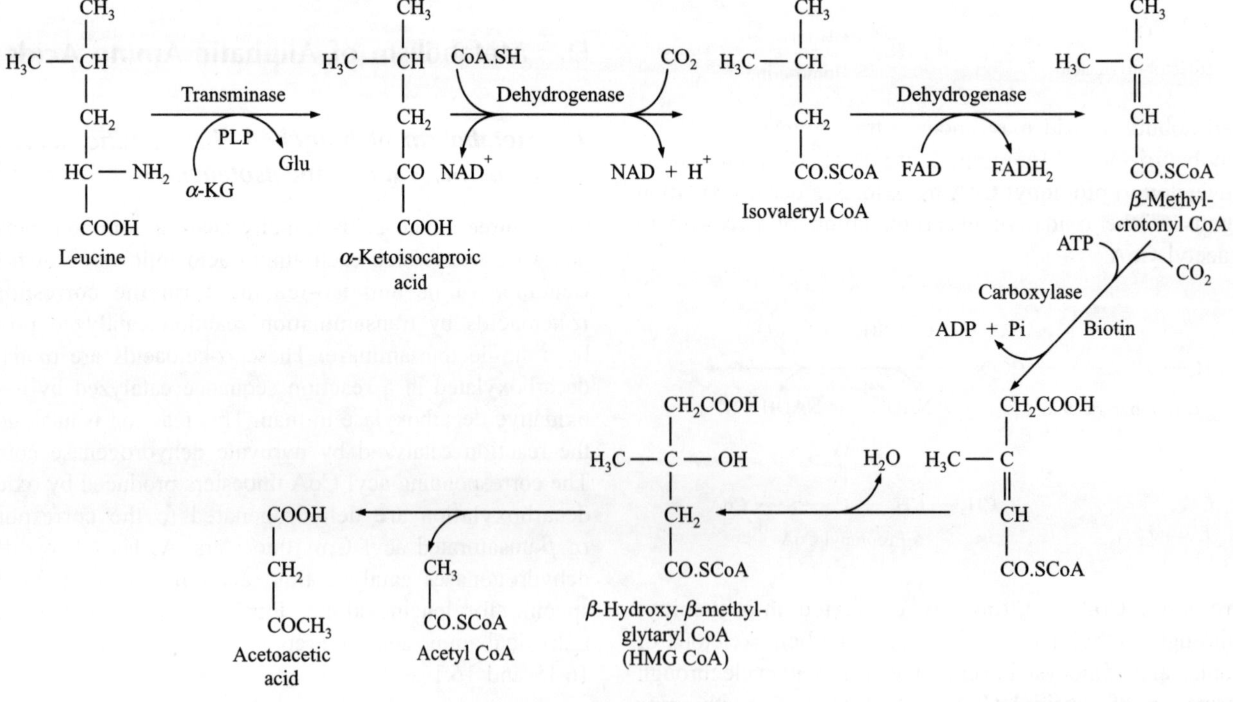

Figure 16.14   Conversion of leucine to acetoacetic acid and acetyl CoA via HMG CoA.

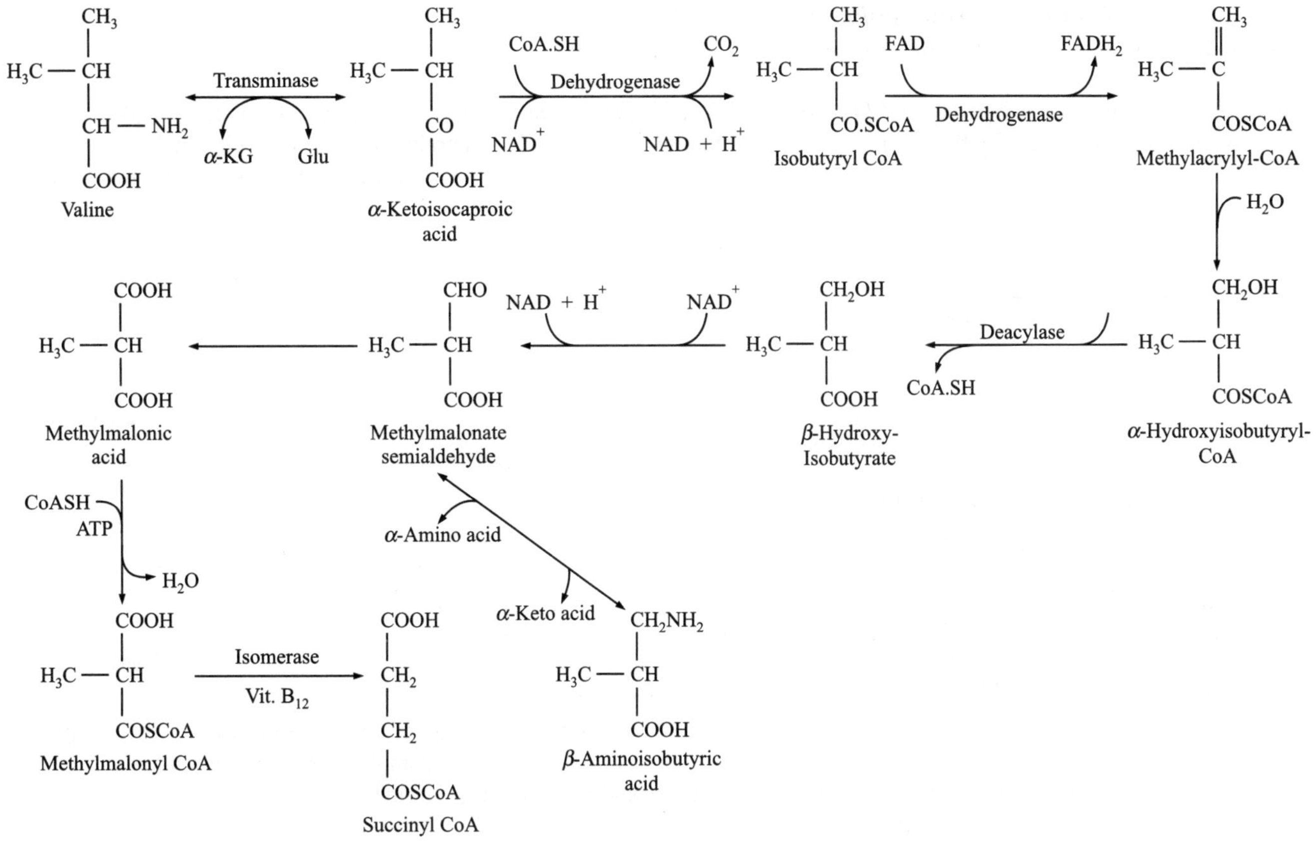

Figure 16.15 Conversion of valine to succinyl CoA via methylmalonyl CoA.

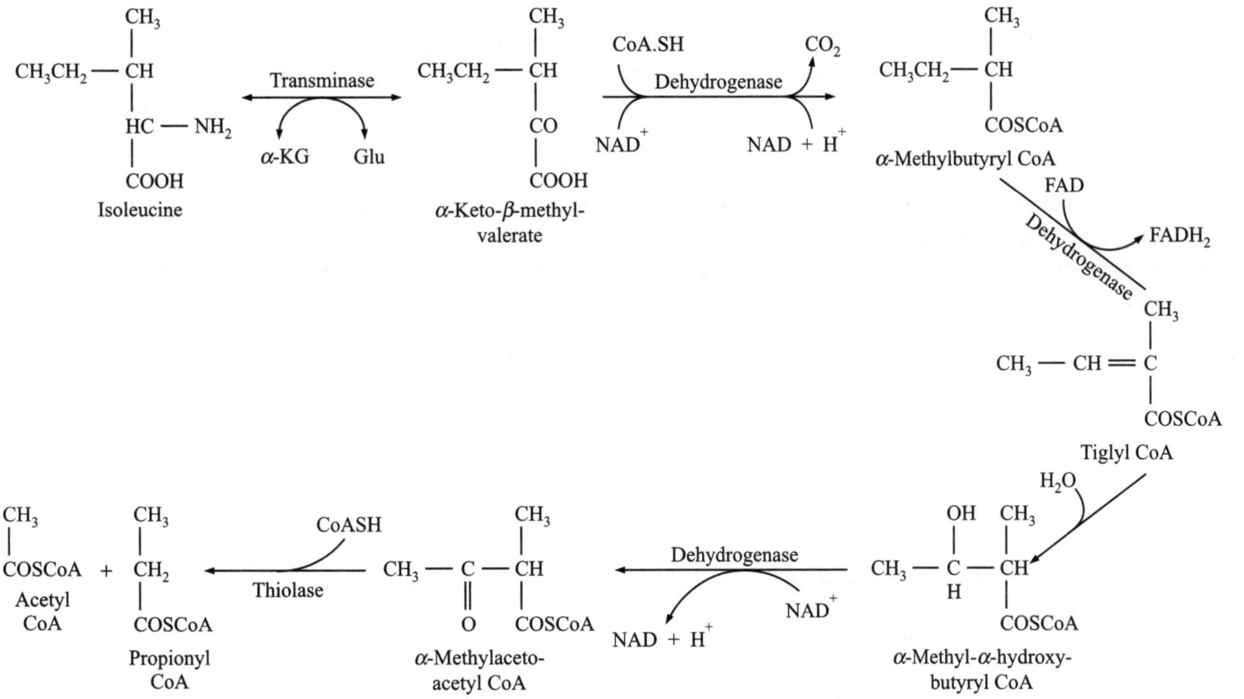

Figure 16.16 Conversion of isoleucine to acetyl CoA and propionyl CoA.

2. Metabolism of other aliphatic amino acids (Glycine, Alanine, Serine, and Threonine)

(a) *Glycine:* Glycine is catabolized to CO_2, NH_3 and N^5, N^{10}-methylene tetrahydrofolate by the glycine cleavage system in the liver.

Glycine and serine can be interconverted by serine-hydroxymethyltransferase (EC 2.1.2.1).

Glycine is glycogenic. It can be converted to serine, which in turn can form pyruvate (see Serine).

There are many biologically important compounds that derive some of their components from glycine. These include: proteins, fumarate (one-carbon pool), glutathione, creatine, heme, purines, hippuric acid, glyoxylic acid, glucose and glycogen.

(b) *Serine:* Serine is mainly catabolized to glycine and N^5, N^{10}-methylene tetrahydrofolate (see Glycine).

Serine can be converted to pyruvate by the action of serine dehydratase (EC 4.2.1.13).

Serine is involved in the synthesis of tissue proteins, cysteine, glycine, ethanolamine, choline, sphingosine, purines and pyrimidine (thymine).

(c) *Alanine:* Alanine can form pyruvate and acetyl CoA.

In the muscle, alanine is synthesized from pyruvate derived from glucose. This alanine is transported to the liver, where it is channelled into gluconeogenesis. The amino groups derived from transamination involving alanine are removed as urea. This series of reactions (glucose-alanine cycle) effects cycling of glucose from liver to muscle, and that of alanine from muscle to liver (Figure 16.17).

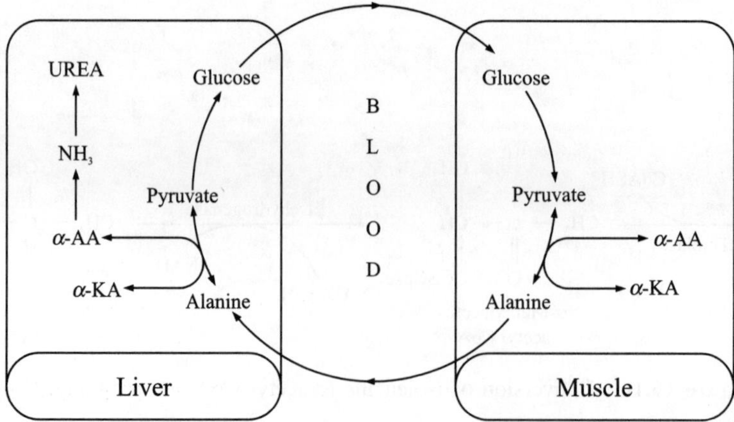

Figure 16.17 The glucose-alanine cycle.

(d) *Threonine:*   Threonine can be metabolized by three different pathways:
   (i) Deamination catalyzed by threonine dehydratase (EC 4.2.1.16).

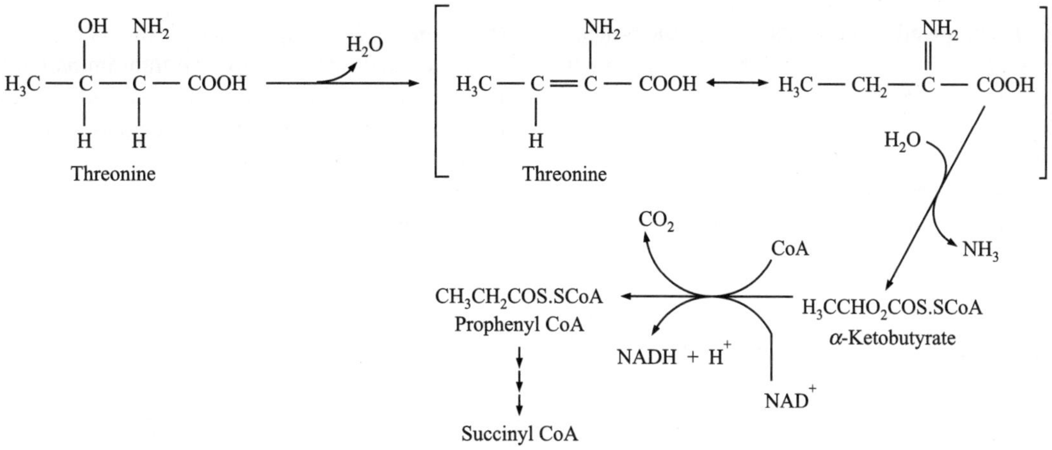

$$
\begin{array}{ccc}
\text{OH}\quad\text{NH}_2 & & \\
\text{H}_3\text{C}-\text{C}-\text{C}-\text{COOH} & \xrightarrow{\;\;\text{H}_2\text{O}\;\;} & \left[\; \text{H}_3\text{C}-\text{C}=\text{C}-\text{COOH} \;\longleftrightarrow\; \text{H}_3\text{C}-\text{CH}_2-\text{C}-\text{COOH} \;\right] \\
\text{H}\quad\text{H} & & \\
\text{Threonine} & & \text{Threonine}
\end{array}
$$

Via CoA / NAD$^+$ / NADH + H$^+$ / CO$_2$ / NH$_3$ / H$_2$O:

$$
\text{H}_3\text{CCHO}_2\text{COS.SCoA (}\alpha\text{-Ketobutyrate)} \longrightarrow \text{CH}_3\text{CH}_2\text{COS.SCoA (Prophenyl CoA)} \longrightarrow \longrightarrow \text{Succinyl CoA}
$$

(ii) Oxidative decarboxylation.

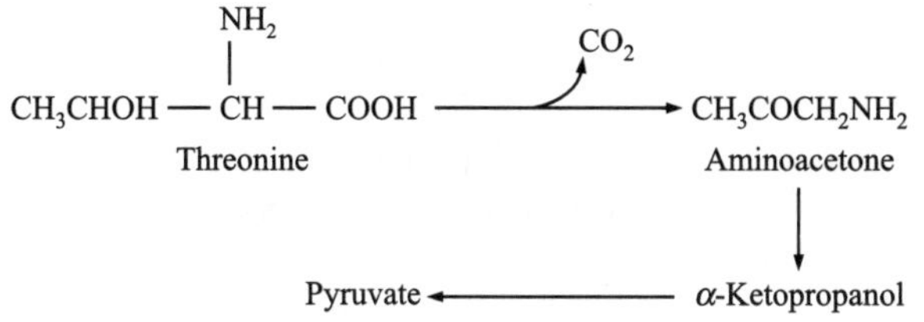

$$
\begin{array}{c}
\text{NH}_2 \\
\text{CH}_3\text{CHOH}-\text{CH}-\text{COOH} \xrightarrow{\;-\text{CO}_2\;} \text{CH}_3\text{COCH}_2\text{NH}_2 \\
\text{Threonine} \qquad\qquad \text{Aminoacetone}
\end{array}
$$

$$
\text{Pyruvate} \longleftarrow \alpha\text{-Ketopropanol}
$$

(iii) Oxidation of β-carbon and splitting of the four carbon chain.

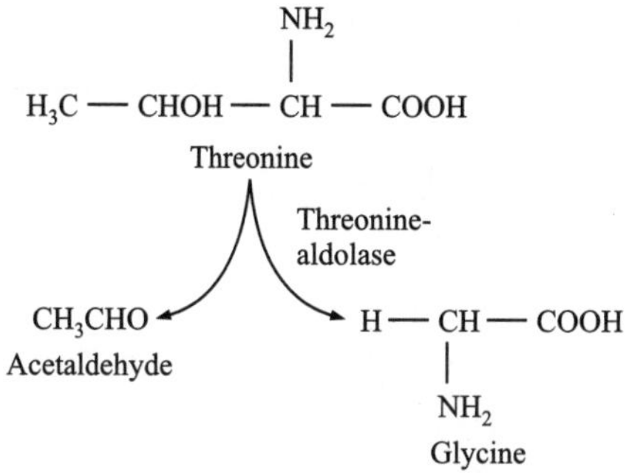

$$
\begin{array}{c}
\text{NH}_2 \\
\text{H}_3\text{C}-\text{CHOH}-\text{CH}-\text{COOH} \\
\text{Threonine}
\end{array}
$$

Threonine-aldolase:

$$
\text{CH}_3\text{CHO (Acetaldehyde)} \longleftarrow \;\;\longrightarrow\; \text{H}-\text{CH}-\text{COOH (Glycine, with NH}_2\text{)}
$$

E.   Metabolism of Acidic and Basic Amino Acids

1.   Acidic amino acids (Glutamic acid and Aspartic acid)

Glutamine and asparagine are deaminated to the corresponding α-amino acids, which in turn are acted upon by the same transaminase to form the corresponding α-keto acids.

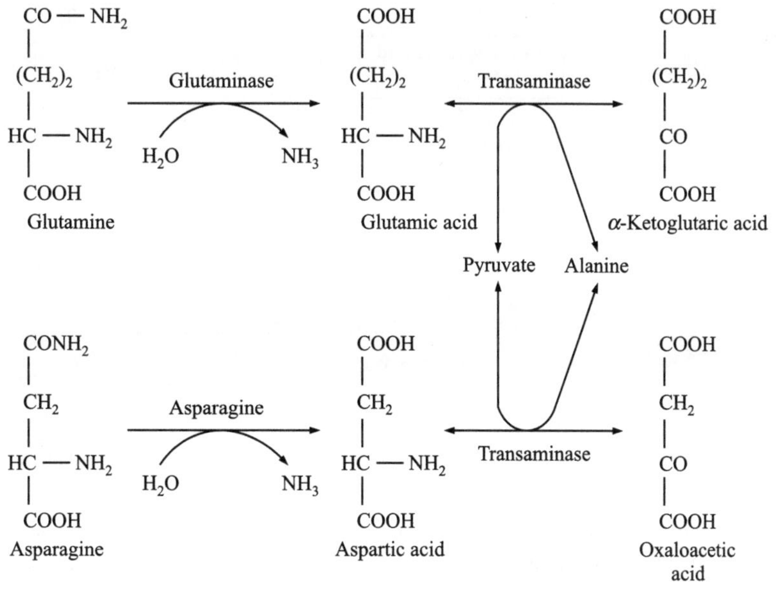

$$
\begin{array}{ccccc}
\text{CO}-\text{NH}_2 & & \text{COOH} & & \text{COOH} \\
(\text{CH}_2)_2 & \xrightarrow[\;\text{H}_2\text{O}\;\;\;\text{NH}_3]{\text{Glutaminase}} & (\text{CH}_2)_2 & \xrightleftharpoons[\;]{\text{Transaminase}} & (\text{CH}_2)_2 \\
\text{HC}-\text{NH}_2 & & \text{HC}-\text{NH}_2 & & \text{CO} \\
\text{COOH} & & \text{COOH} & & \text{COOH} \\
\text{Glutamine} & & \text{Glutamic acid} & & \alpha\text{-Ketoglutaric acid}
\end{array}
$$

Pyruvate / Alanine (Transaminase)

$$
\begin{array}{ccccc}
\text{CONH}_2 & & \text{COOH} & & \text{COOH} \\
\text{CH}_2 & \xrightarrow[\;\text{H}_2\text{O}\;\;\;\text{NH}_3]{\text{Asparagine}} & \text{CH}_2 & \xrightleftharpoons[\text{Transaminase}]{} & \text{CH}_2 \\
\text{HC}-\text{NH}_2 & & \text{HC}-\text{NH}_2 & & \text{CO} \\
\text{COOH} & & \text{COOH} & & \text{COOH} \\
\text{Asparagine} & & \text{Aspartic acid} & & \text{Oxaloacetic acid}
\end{array}
$$

Both glutamic acid and aspartic acid are glycogenic as α-ketoglutarate and oxaloacetate can enter into citric acid cycle, and thereby into the process of gluconeogenesis.

Glutamic acid is involved in the synthesis of tissue proteins, GABA, and glutathione. Aspartate and N-acetylglutamate participate in the urea cycle.

2. Basic amino acids (Arginine and Lysine)

(a) *Arginine:* It is the precursor of urea in the urea cycle. Arginine is converted to putrescine, spermine, and spermidine by intestinal bacteria. Arginine is also used for the synthesis of streptomycin by streptomyces.

(b) *Lysine:* The ε-nitrogen of lysine is removed by a reaction sequence that does not involve transamination. Catabolism of lysine produces glutamic acid, α-ketoadipic acid and glutaryl CoA. In mammals, the exact metabolic fate of glutaryl CoA is not known (Figure 16.18).

Studies using metabolic tracer have suggested some possible pathways for lysine:

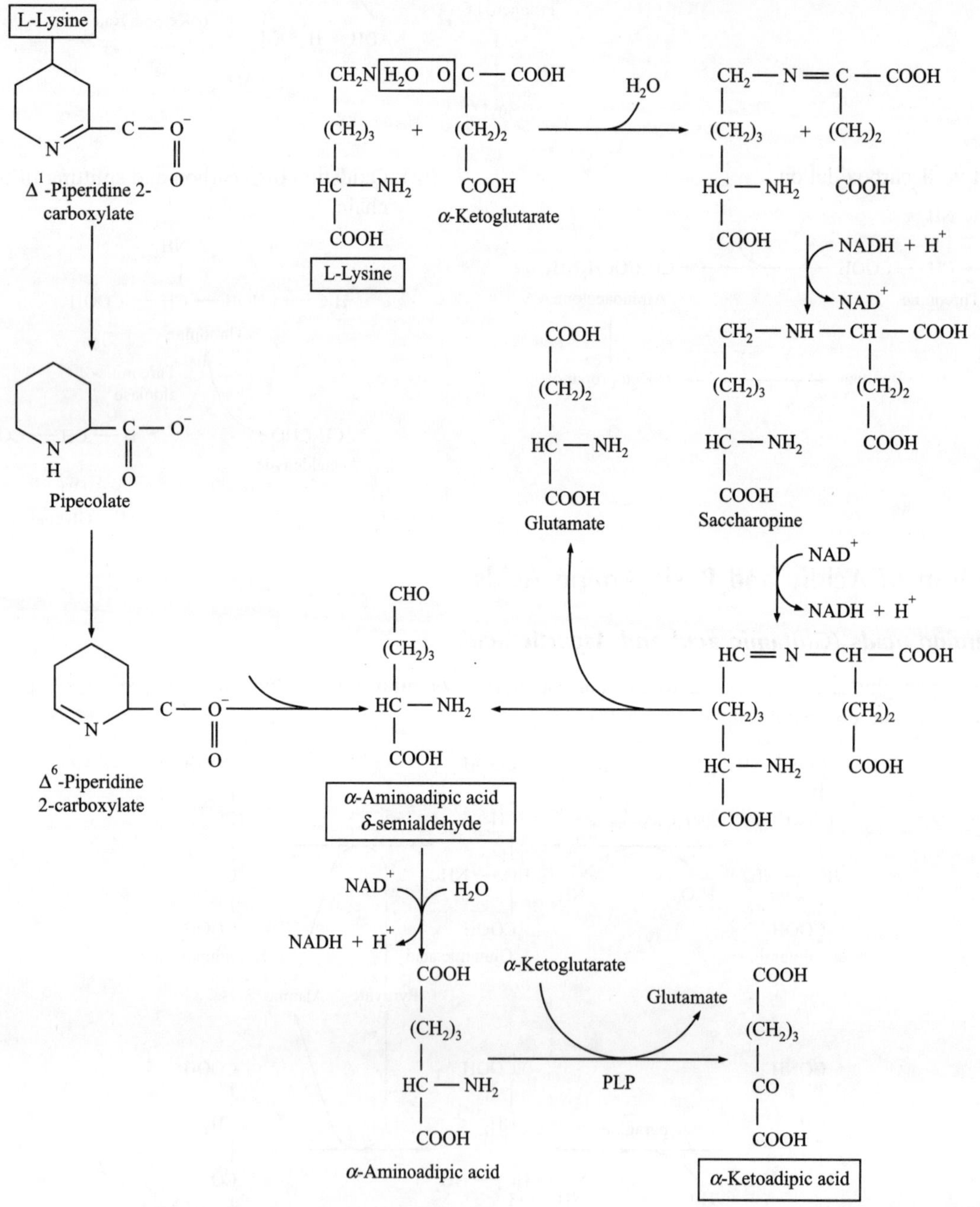

Figure 16.18   Conversion of lysine to α-ketoadipic acid. The major route for the metabolism of lysine proceeds through its sequential conversion to α-ketoadipic acid, glutaconyl CoA, Crotonyl CoA, and finally, acetoacetyl CoA and acetyl CoA. (See also Figure 16.19).

(i) Metabolism of lysine may proceed through formation of cyclic intermediates like L-pipecolic acid, etc. (Figure 16.18).

(ii) Glutaconyl CoA may also be metabolized by two other routes, one leading to acetyl CoA and the other to glutamic acid (Figure 16.19).

F. Metabolism of α-Imino Acids (Proline and Hydroxyproline)

Proline serves as a precursor of hydroxyproline. Hydroxylation of proline is catalyzed by proline hydroxylase (EC 1.14.11.2), a mixed function oxygenase (Figure 16.20).

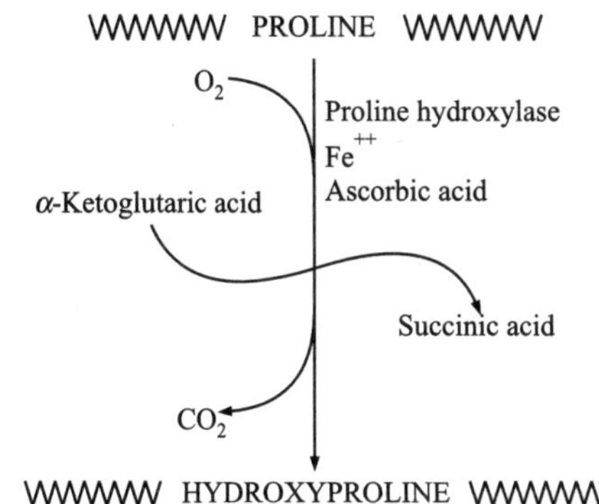

Figure 16.20 Formation of hydroxyproline by hydroxylation of proline.

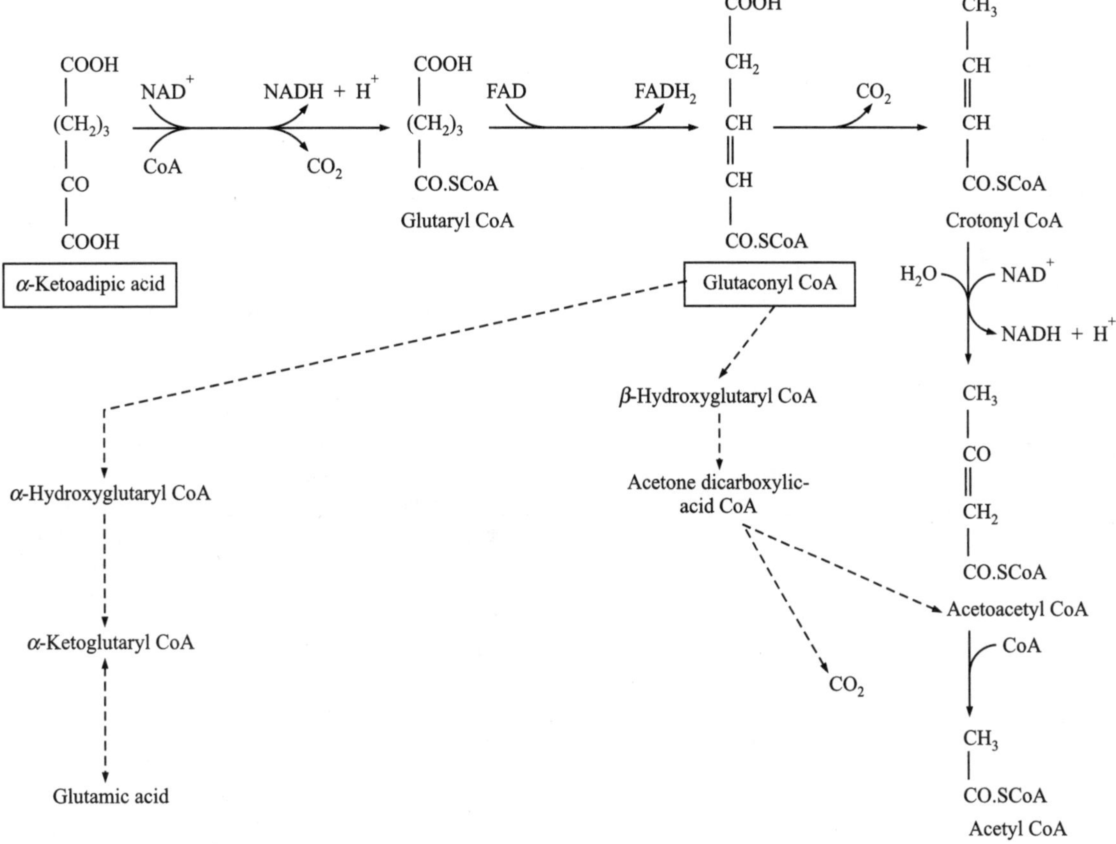

Figure 16.19 Metabolic pathways proposed for the breakdown of α-ketoadipic acid. The major route proceeds via crotonyl CoA.

Proline is catabolized to glutamate and α-ketoglutarate, whereas hydroxyproline is metabolized to yield pyruvate and glyoxylic acid (Figure 16.21).

G. Summary: Catabolism of the Carbon Skeleton of Amino Acids

The breakdown of the carbon skeleton of amino acids produces some important compounds through which metabolism of proteins is linked to metabolism of carbohydrates and fats. A brief summary is presented in Table 16.2.

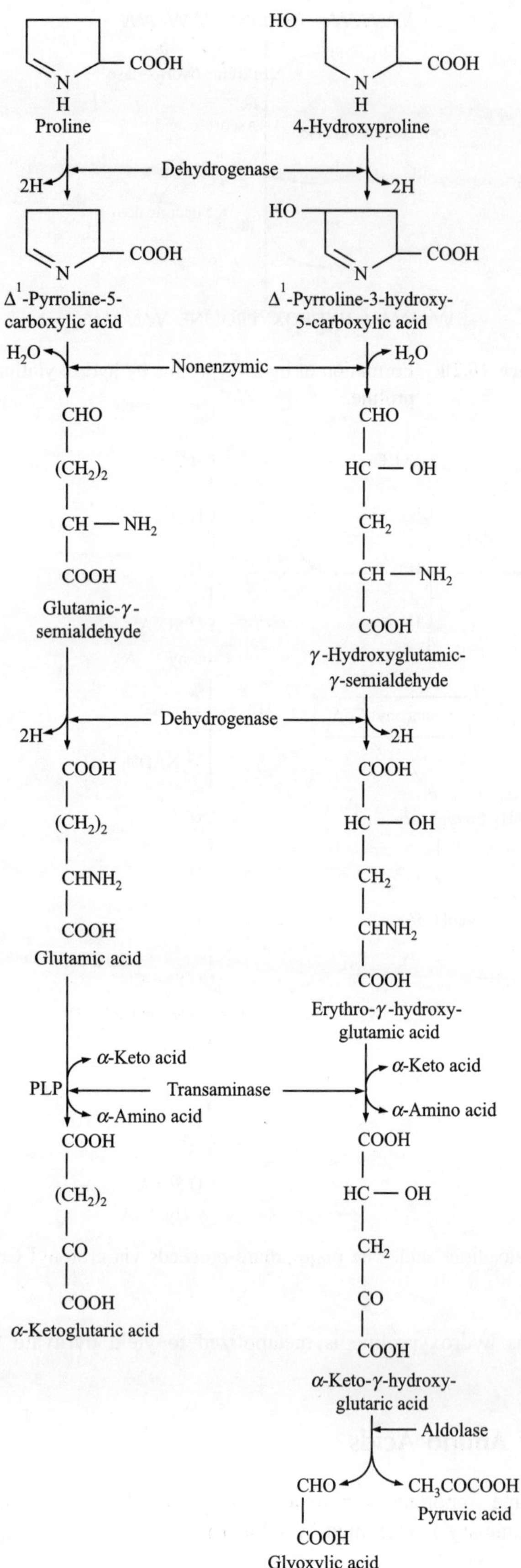

Figure 16.21 Summary of the pathways for the degradation of proline and 4-Hydroxy-L-proline in mammalian tissues.

TABLE 16.2   Catabolism of the Carbon Skeleton of Amino Acids

Group	Amino Acids	Metabolic Product
a.	Glycine, Alanine, Serine, Threonine, Tryptophan, Cysteine Cystine and Hydroxy-proline	Pyruvate
b.	Leucine, Lysine, Phenylalanine, Tyrosine, Tryptophan	Acetoacetyl CoA
c.	Leucine, Isoleucine, Tryptophan, Amino acids in group 'a' (forming pyruvate) and group 'b' (forming acetoacetyl CoA)	Acetyl CoA
d.	Arginine, Glutamine, Histidine and Proline	Glutamate
e.	Glutamic acid, and amino acids in group 'd'	α-Ketoglutarate
f.	Valine, Isoleucine, Methionine, and Threonine	Succinyl CoA
g.	Tyrosine, Phenylalanine	Fumarate
h.	Asparagine	Aspartate
i.	Aspartate, Asparagine	Oxaloacetate

On the basis of the nature of substance formed through the metabolism of individual amino acids, the amino acids may be placed in three groups (Table 16.3).

TABLE 16.3   L-α-Amino Acids

Glycogenic	Ketogenic	Glycogenic + Ketogenic
Glycine, Alanine, Valine, Serine, Threonine, Aspartate, Asparagine, Glutamate, Glutamine, Arginine, Histidine, Cystine, Cysteine, Methionine, Proline, and Hydroxypraline	Leucine	Lysine, Isoleucine, Phenylalanine, Tyrosine, Tryptophan

IV. METABOLISM OF BIOLOGICALLY ACTIVE AMINES DERIVED FROM AMINO ACIDS

A. Catecholamines

(See Section XI)

B. SEROTONIN

5-hydroxytryptamine (serotonin) is found in the brain and other tissues (stomach, intestine, mast cells and platelets, etc.). It is formed by 5-hydroxylation of tryptophan followed by decarboxylation of 5-hydroxytryptophan (Figure 16.22).

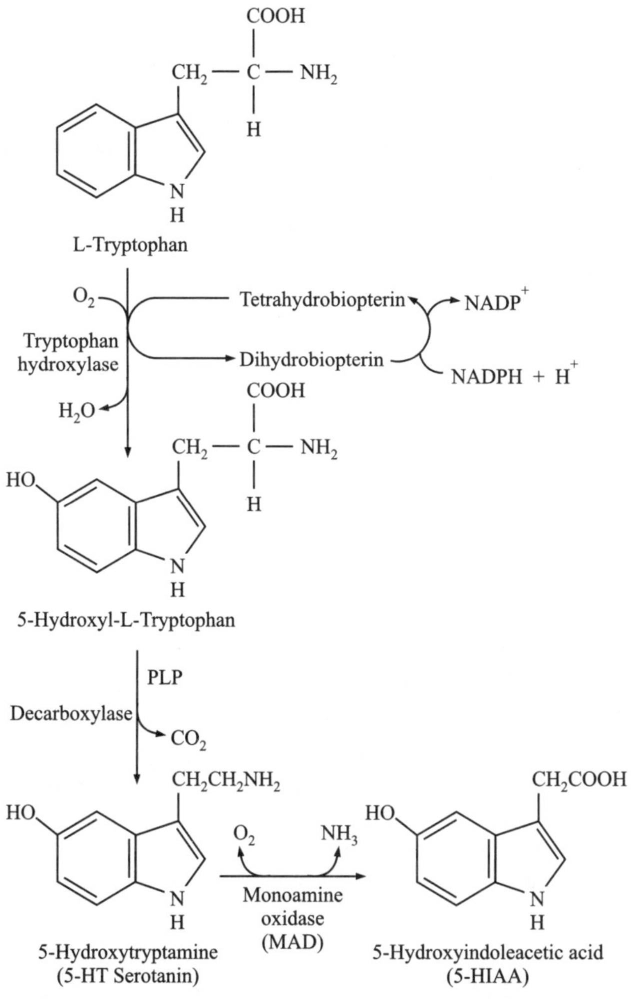

be present in the liver. In the liver, phenylalanine hydroxylase (EC 1.14.16.1) also catalyzes hydroxylation of tryptophan, but the rate is at least 30 times less than hydroxylation of phenylalanine. Tryptophan hydroxylase activity has been shown to be high in the intestine, particularly in the Argentaffin cells. It is also present in mammalian brain. This enzyme requires reduced pteridine as a coenzyme. 5-hydroxy tryptophan decarboxylase (EC 4.1.1.28) is present in several mammalian tissues and requires pyridoxal phosphate as a cofactor.

Metabolism of serotonin: Serotonin is acted upon by mono-amine oxidase (MAO; EC 1.4.3.4), and is ultimately converted into 5-hydroxyindole-acetic acid (5-HIAA). 5-HIAA is excreted in urine. The amount of 5-HIAA in urine is usually a reflection of serotonin metabolism. Patients suffering from malignant carcinoid tumour produce large amounts of serotonin and 5-HIAA. In this condition, 60 per cent of tryptophan metabolism is routed via the serotonin pathway. Normally only 1 per cent of tryptophan is converted to serotonin and its metabolites. Large amounts of 5-HIAA in urine indicate the presence of a carcinoid tumour.

C. Gamma-aminobutyric Acid (GABA)

Gamma aminobutyric acid is formed by alpha-decarboxylation of glutamate. Glutamic acid de-carboxylase (glutamate decarboxylase; EC 4.1.1.15) requires pyridoxal phosphate as a coenzyme. GABA is further metabolized by deamination to succinic-semialdehyde. This deamination is accomplished by a pyridoxal phosphate-dependent enzyme. Succinic semialdehyde may then either be oxidized to succinate or reduced to gamma-hydroxybutyrate (Figure 16.23).

The above pathway provides a route for disposal of GABA. It also bypasses the α-ketogluta-rate dehydrogenase reaction of the citric acid cycle. By this, α-ketoglutarate is converted into succinic acid instead of succinyl CoA.

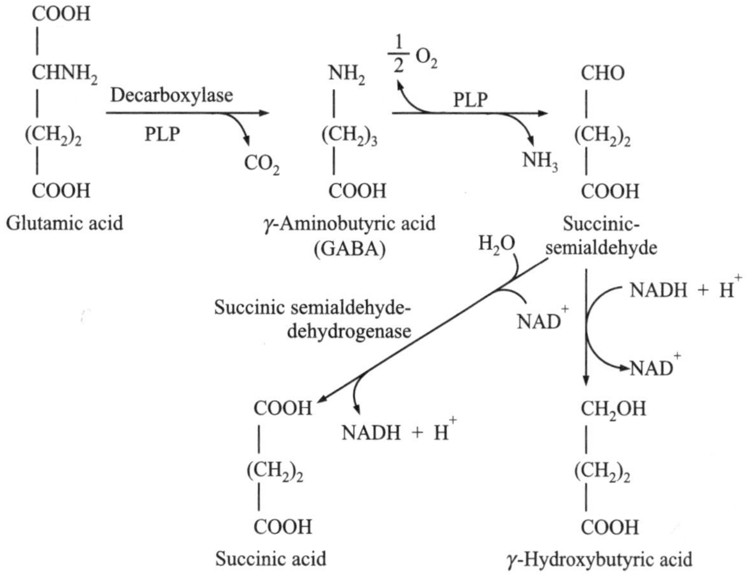

Figure 16.22 Formation and metabolism of serotonin.

Tryptophan hydroxylase (EC 1.14.16.4), the enzyme responsible for hydroxylation of tryptophan, has been shown to

Figure 16.23 Formation and metabolism of γ-aminobutyric acid (GABA).

D. Polyamines (Putrescine, Spermidine, and Spermine)

Putrescine, spermidine, and spermine are small aliphatic amines that are ubiquitous in biological materials. They are present in a wide variety of animals, bacteria, yeasts, and plants.

The synthesis of spermidine proceeds by an interaction of putrescine (1, 4-diaminobutane) and S-adenosyl-methionine (Figure 16.24). In mammals, putrescine is formed by decarboxylation of ornithine. The enzyme L-ornithine decarboxylase (EC 4.1.1.17) has been shown to be present in a number of normal mammalian tissues. In bacteria, putrescine may be formed either by ornithine decarboxylase reaction or by arginine decarboxlase reaction via agmatine. S-adenosyl methionine is decarboxylated to form decarboxylated S-adenosyl methionine [S(5′-adenosyl)-3-methyl-mercaptopropylamine], which donates aminopropyl group to putrescine to form spermidine in the presence of spermidine synthase (aminopropyltransferase; EC 2.5.1.16).

Spermidine so formed is converted into spermine by acceptance of another aminopropyl group from S-(5′-adenosyl)-3-methyl-mercaptopropylamine. The enzyme involved in this reaction is spermine synthase. Putrescine inhibits spermine synthase.

Polyamines have been shown to be growth factors essential for some invertebrates and various types of microorganisms. No evidence exists at the present time for their requirement by higher animals including humans. Several lines of evidence indicate that polyamines, particularly spermidine, are required for cell proliferation, but their precise role in cell proliferation and differentiation is uncertain. These organic polycations stabilize cells, membranes, DNA and ribosomes, etc. It has been suggested that polyamines, by virtue of their charged nature under physiological conditions and their conformational flexibility, serve to stabilize nucleic acids by neutralization of anions (phosphate groups) of nucleic acids, thus reducing their repulsive action. Polyamines stimulate the synthesis of DNA and RNA, protect DNA against depurination, and affect several enzymatic reactions (e.g., methylation involving RNA). Polyamines have been found to influence protein synthesis both at transcriptional and translational levels.

The most active clinical area concerning polyamines has been the use of polyamines as biochemical markers for malignant tumours. Preliminary clinical studies involving the assay of polyamine levels in biological fluids (blood, urine, CSF, etc.) in various kinds of malignant tumours have substantiated the use of polyamines as cancer markers.

E.   Taurine

Taurine (2-aminoethanesulphonic acid), a non-toxic sulphur-containing amino acid, has been found in various organs of many animal species, but it is hardly present in plants. The levels of taurine are abundant in all mammalian tissues, but it is found in very high concentrations in excitable tissues such as the heart, brain, muscle and retinal photoreceptor cells.

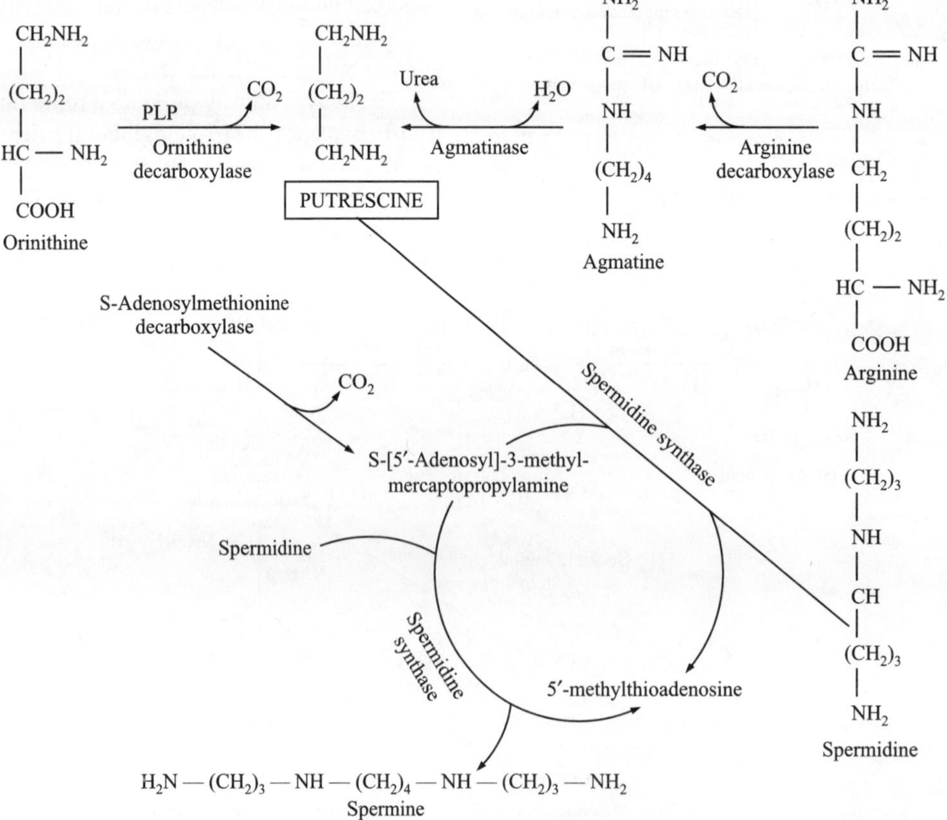

Figure 16.24 Synthesis of putrescine, spermidine and spermine.

Taurine appears to be formed from cysteine-sulphinic acid by two pathways. One pathway involves decarboxylation of cysteinesulphinic acid to hypotaurine and subsequent oxidation of hypotaurine to yield taurine. The cysteinesulphinic acid decarboxylase (EC 4.1.1.29) catalyzing the initial decarboxylation requires pyridoxal phosphate as a coenzyme. Glutathione is also required in the reaction catalyzed by this decarboxylase.

The other pathway involves the oxidation of cysteinesulphinic acid to cysteic acid followed by the decarboxylation of cysteinesulphinic acid and cysteic acid.

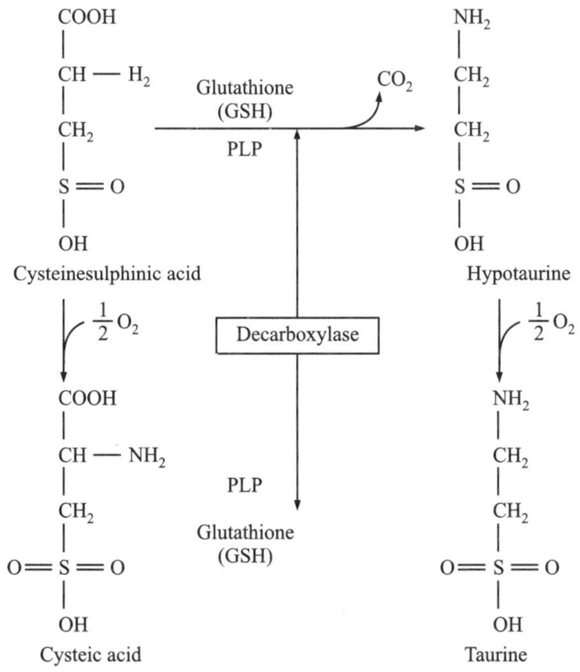

$$
\begin{array}{ccc}
\text{COOH} & & \text{NH}_2 \\
| & & | \\
\text{CH} - \text{H}_2 & \xrightarrow[\text{PLP}]{\text{Glutathione (GSH)}} \; \text{CO}_2 & \text{CH}_2 \\
| & & | \\
\text{CH}_2 & & \text{CH}_2 \\
| & & | \\
\text{S} = \text{O} & & \text{S} = \text{O} \\
| & & | \\
\text{OH} & & \text{OH} \\
\text{Cysteinesulphinic acid} & & \text{Hypotaurine}
\end{array}
$$

Decarboxylase

$$
\begin{array}{ccc}
\text{COOH} & & \text{NH}_2 \\
| & & | \\
\text{CH} - \text{NH}_2 & \xrightarrow[\text{Glutathione (GSH)}]{\text{PLP}} & \text{CH}_2 \\
| & & | \\
\text{CH}_2 & & \text{CH}_2 \\
| & & | \\
\text{O} = \text{S} = \text{O} & & \text{O} = \text{S} = \text{O} \\
| & & | \\
\text{OH} & & \text{OH} \\
\text{Cysteic acid} & & \text{Taurine}
\end{array}
$$

Taurine is degraded by conversion to its deaminated analogue, isethionic acid (2-hydroxyethanesulphonic acid).

$$
\begin{array}{ccc}
\text{NH}_2 & & \text{CH}_2\text{OH} \\
| & & | \\
\text{CH}_2 & \xrightarrow{\text{NH}_3} & \text{CH}_2 \\
| & & | \\
\text{CH}_2 & & \text{O} = \text{S} = \text{O} \\
| & & | \\
\text{O} = \text{S} = \text{O} & & \text{OH} \\
| & & \\
\text{OH} & & \\
\text{Taurine} & & \text{Isethionic acid} \\
& & \text{(2-hydroxyethane-sulphonic acid)}
\end{array}
$$

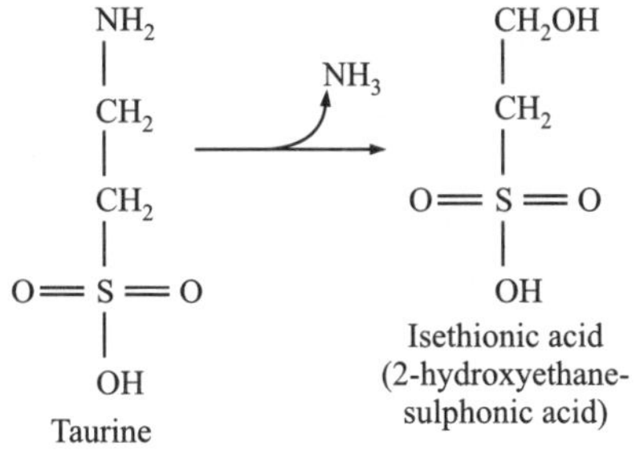

In the liver, taurine combines with cholyl CoA to form taurocholic acid, which is one of the bile acids.

$$\text{Taurine} + \text{Cholyl.SCoA} \longrightarrow \text{Taurocholic acid} + \text{CoA.SH}$$

A physiological function for taurine has not been established, apart from its involvement in bile salt synthesis in some mammals. Taurine exhibits a wide range of physiological and pharmacological properties. Many of the effects of taurine are perhaps related to its ability to modulate calcium binding and movement. There is strong evidence for taurine being a modulator for cation fluxes, particularly calcium, both in central nervous system (CNS) and peripheral organs. Taurine has been found to be essential for muscle contraction. However, neither taurine nor calcium alone govern the contractile response.

Experimental studies have suggested that taurine has a physiological function in the heart as a regulator of ion fluxes, a membrane stabilizer, a modulator of membrane behaviour, an antiarrhythmic agent, or as an osmotic agent.

Taurine has some of the properties of an inhibitory neurotransmitter. Its function as a membrane stabilizer may be correlated with its powerful antiepileptic action in experimental epilepsy. It has been suggested that taurine regulates the glutamate/glutamine balance in the brain, and hence serves an important function in regulation of neuronal excitability. Also, there is a role for taurine in osmotic regulation in mammalian brain. There is evidence to suggest that taurine may be essential for viability of retinal photoreceptor cells.

Alteration in taurine content and taurine transporting ability have been associated with a number of human diseases, including congestive heart failure (CHF), cardiomyopathy, myocardial infarction, epilepsy and ataxia. Retinal degeneration and blindness ensue in cats maintained on a taurine-free diet.

V. AMINO ACID METABOLISM AND GLUCONEOGENESIS

As outlined in the preceding sections, most of the amino acids after deamination give rise to carbon residues, which are either intermediates in citric acid cycle or are converted into these intermediates. The carbon residues of amino acids are thus ultimately oxidized to CO_2 and H_2O. However, these carbon residues of the amino acids could also be converted into glucose and glycogen. This process of conversion of non-carbohydrate substances to glucose is termed as *"gluconeogenesis"*, the biosynthesis of *"new"* sugar (for gluconeogenesis see also Chapter 17).

In this process, not only amino acids but glycerol and certain other organic acids like propionic acid, besides citric acid cycle intermediates, can be converted into carbohydrate (glucose). In nature, metabolic degradation of amino acids is so planned that during the course of their degradation, they give rise to intermediates of carbohydrate metabolism and citric acid cycle. Such a scheme is economical for the cell, as amino acids are ultimately oxidized through a final common pathway, involving the use of fewer enzymes over and over again. A similar pattern is also seen in catabolism of fatty acids which are initially degraded to acetyl CoA, and are finally disposed of through the citric acid cycle (see Chapter 25). The intermediates such as pyruvate, α-ketoglutarate, succinyl CoA, fumarate and malate are, therefore, derived not only from carbohydrate metabolism and citric acid cycle, but are also formed from degradation of amino acids (Figure 16.25).

Gluconeogenesis is essentially a metabolic process developed in the body to ensure a continual supply of glucose to certain tissues such as the central nervous system (CNS),

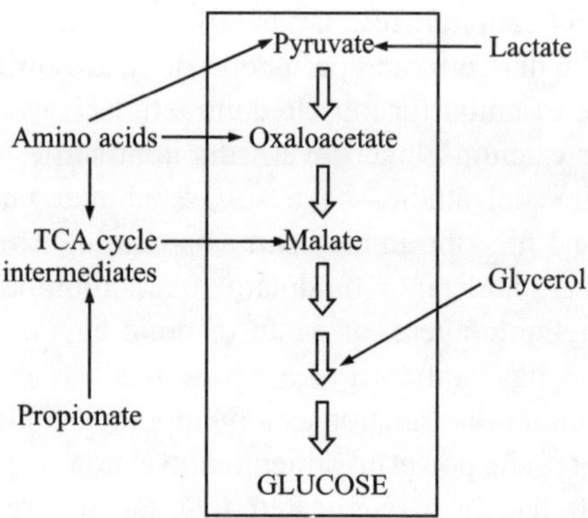

Figure 16.25 Points of entry of some important compounds into the pathways of gluconeogenesis.

adrenal medulla, erythrocytes and skeletal muscle (under anaerobic conditions). These tissues depend exclusively on glucose for their energy requirements and function, when dietary carbohydrate is not available in sufficient amounts. Glucose is also required by adipose tissue and mammary gland, as a source of glycerol and lactose, respectively. In addition, lactate formed from glucose (supplied by the liver) during glycolytic activity in contracting skeletal muscle or erythrocytes, is converted into glucose in the liver (Cori cycle; Figure 16.26). Glycerol produced by adipose tissue may also be converted into glucose by gluconeogenic mechanisms.

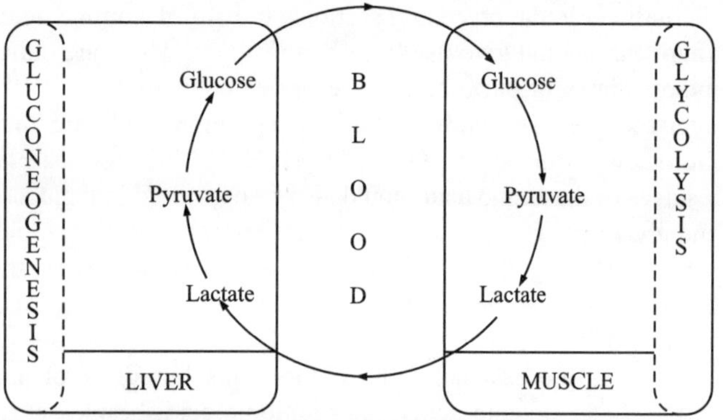

Figure 16.26 The lactic acid (Cori) cycle.

Recent studies have provided for the first time insights into the relative contributions of gluconeogenesis and glycogenolysis in the liver for maintaining blood glucose levels during fasting. During the first 22 hours of fasting, hepatic glycogen content decreased in an almost linear fashion; gluconeogenesis contributed 64% of the total glucose produced during this period. Subsequently, the rate of glycogenolysis decreased over the next 46 hours until gluconeogenesis was responsible for almost all the glucose production. In summary, contrary to the earlier belief in humans, gluconeogenesis provides a significant fraction of glucose produced during fasting, even after a few hours.

In the body, gluconeogenesis occurs mainly in liver and kidney. In these organs, reactions of gluconeogenesis take place in two compartments of the cells (mitochondria and

the cytosol) unlike glycolysis, whose reactions occur only in the cytosol. The pathway of gluconeogenesis is not just reversal of glycolysis, though some reactions are common to both the processes. There are three reactions which are peculiar to gluconeogenesis. These reactions overcome the thermodynamic barriers imposed in the reversal of glycolysis. In glycolysis, reactions catalyzed by pyruvate kinase, phosphofructokinase and hexokinase are essentially irreversible in nature (Figure 16.27).

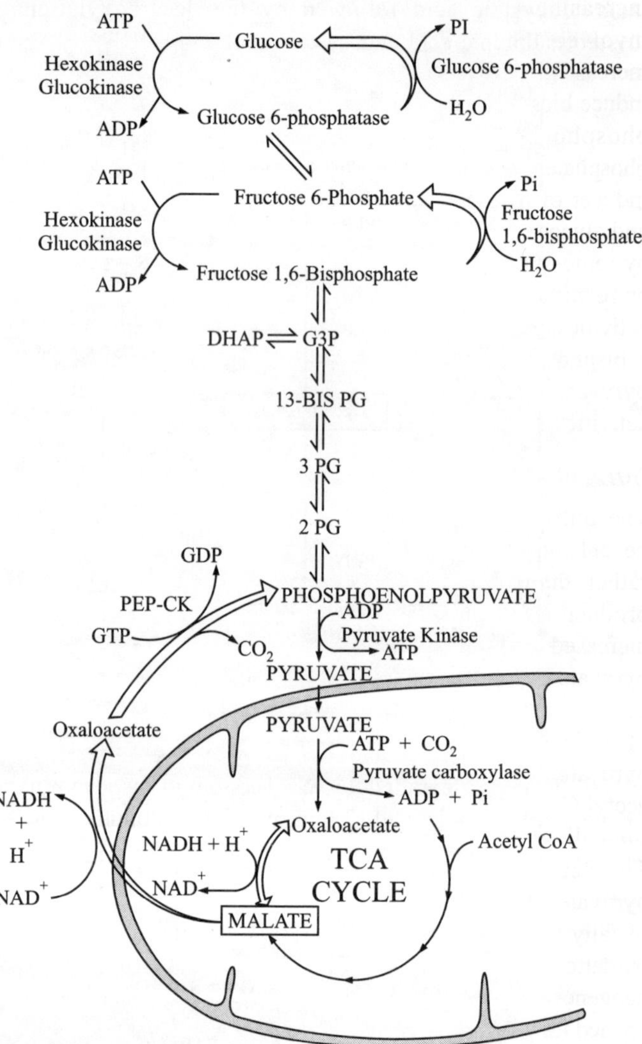

Figure 16.27 Major pathway of gluconeogenesis in the liver.

The first reaction consists of conversion of pyruvate to oxaloacetate in the mitochondria by means of pyruvate carboxylase (EC 6.4.1.1), a biotin-dependent enzyme. The enzyme is activated by acetyl CoA in an allosteric manner.

The oxaloacetate formed in this mitochondrial reaction is then reduced to malate at the expense of NADH by mitochondrial malate dehydrogenase (EC 1.1.1.37). The malate so formed may leave the mitochondria. In the cytosol, malate is reoxidized to oxaloacetate by cytosolic NAD-linked malate dehydrogenase (EC 1.1.1.37). This extramitochondrial oxaloacetate is acted upon by phosphoenolpyruvate-carboxykinase (EC 4.1.1.32) in presence of GTP to yield phosphoenolpyruvate and CO_2.

The energy barriers existing in the reversal of reactions catalyzed by phosphofructokinase and hexokinase are overcome by having two enzymes: fructose 1,6 bisphosphatase (EC 3.1.3.11) and glucose-6-phosphatase (EC 3.1.3.9). Fructose-1, 6 bisphosphatase (Hexose diphosphatase) hydrolyzes fructose 1, 6-bisphosphate to yield fructose-6-phosphate. Glucose 6-phosphate is dephosphorylated to free glucose in a hydrolytic reaction catalyzed by glucose-6-phosphatase.

Many hormones increase gluconeogenesis by specifically increasing the activity of one or more of the enzymes involved. Glucocorticoids, glucagon and catecholamines increase gluconeogenesis. Glucocorticoids have been shown to induce biosynthesis of hepatic and renal pyruvate carboxylase, phosphoenolpyruvate carboxykinase, fructose-1, 6 bis phosphatase and glucose-6-phosphatase. Epinephrine acts as an inducer of pyruvate carboxylase, fructose-1, 6-bisphosphatase and glucose-6-phosphatase. Insulin suppresses gluconeogenesis by repressing the enzymes of gluconeogenesis. In the absence or relative lack of insulin, there is a marked increase in the activities of hepatic and renal glucose-6 phosphatase, fructose-1, 6-bisphosphatase, phosphoenolpyruvate carboxykinase, and pyruvate carboxylase. Administration of insulin brings the activities of these enzymes back to normal levels.

Intracellular regulation

The utilization of pyruvate for formation of glucose would be enhanced by a greater activity of pyruvate carboxylase rather than other enzymes acting on pyruvate. Increased production of acetyl CoA in mitochondria as a result of increased oxidation of fatty acids would increase the activity of pyruvate carboxylase, because acetyl CoA acts as an allosteric activator of pyruvate carboxylase. ADP is an allosteric inhibitor of pyruvate carboxylase. It has also been demonstrated that pyruvate dehydrogenase is inhibited by increased amounts of acetyl CoA, NADH and ATP, and this would further facilitate funnelling of pyruvate towards formation of oxaloacetate. The activation of pyruvate carboxylase and inhibition of pyruvate dehydrogenase by acetyl CoA formed from oxidation of fatty acids helps to explain the sparing action of fatty acid oxidation on oxidation of pyruvate, and stimulation of gluco-neogenesis. The oxaloacetate so formed has several pathways opened for it. Within the mitochondria, oxaloacetate is converted into citrate by citrate synthase. It has been shown that this enzyme is inhibited by succinyl CoA, ATP, NADH, and long chain fatty acyl CoA. Succinyl CoA is the end product of metabolism of amino acids (like valine, methionine, isoleucine and threonine) and propionic acid. In citric acid cycle, it is converted into oxaloacetate. Hence, increased production of succinyl CoA would not only give rise to more oxaloacetate in mitochondria, but would also prevent its conversion to citrate by inhibition of citrate synthase. Oxaloacetate is converted into malate, which diffuses out of the mitochondria. In the cytosol, oxaloacetate is formed from malate. Malate dehydrogenase activity increases under conditions of increased gluconeogenesis. Due to enhanced biosynthesis of phosphoenolpyruvate carboxykinase in gluconeogenic circumstances, oxaloacetate

is converted into phosphoenolpyruvate. The cytosolic supply of GTP is a limiting and controlling factor for this step. Phosphoenolpyruvate is to be prevented from conversion back to pyruvate so that it is routed towards formation of glucose. The pyruvate kinase is shown to be inhibited by ATP, alanine and cAMP. Hence, the availability of more ATP would favour conversion of phosphoenolpyruvate to glucose.

Fructose-1, 6-bisphosphate and AMP act as allosteric activators for phosphofructokinase-1 on one hand, and allosteric inhibitors for fructose 1, 6-bisphosphatase on the other hand. This regulation emphasizes the role of these modulators in facilitating the sequence of reactions towards formation of glucose, when intracellular concentration of fructose 1, 6-bisphosphate and AMP decreases.

The discovery of fructose 2, 6-bisphosphate added a new dimension to understanding the regulation of glycolysis and gluconeogenesis. In the liver, fructose 2, 6-bisphosphate (Fr 2, 6-P2) is formed from fructose 6-phosphate by phospho-fructokinase-2 (PFK-2). The classical phospho-fructokinase catalyzing the formation of fructose 1, 6-bisphosphate has therefore also been re-designated as phosphofructokinase-1 (PFK-1). Fr 2, 6-P2 is hydrolyzed to Fr-6-P by fructose bisphosphatase-2 (FBP-2). Fructose 2, 6-bisphosphate is a potent stimulator of phosphofructokinase-1, but it inhibits fructose 1, 6-bisphosphatase. On the other hand, fructose 6-phosphate accelerates the synthesis of Fr 2, 6-P2 but inhibits its hydrolysis. During fasting, because of low concentration of fructose 2, 6-bisphosphate, PFK-1 is inactive, and the metabolites are channelled towards gluco-neogenesis. In contrast, in the well fed state, fructose 2, 6-bisphosphate is in high amount, which activates PFK-1 but inhibits fructose 1, 6-bisphosphatase (FBP-1). The metabolites then follow the path of glycolysis. Fructose 2, 6-bisphosphate, therefore, acts as a "signal" which signifies that glucose can be utilized, and that gluconeogenesis can be stopped (Figure 16.28).

Both PFK-1 and PFK-2 are stimulated by fructose-6-phosphate, AMP and Pi, and are inhibited by glycerol-3-phosphate, phosphate, phosphoenolpyruvate and citrate. Fructose 2, 6-bisphosphate counteracts the inhibition of PFK-1 by ATP. The regulation of PFK-1 is thus largely dependent upon fructose 2, 6-bisphosphate. The inhibition of

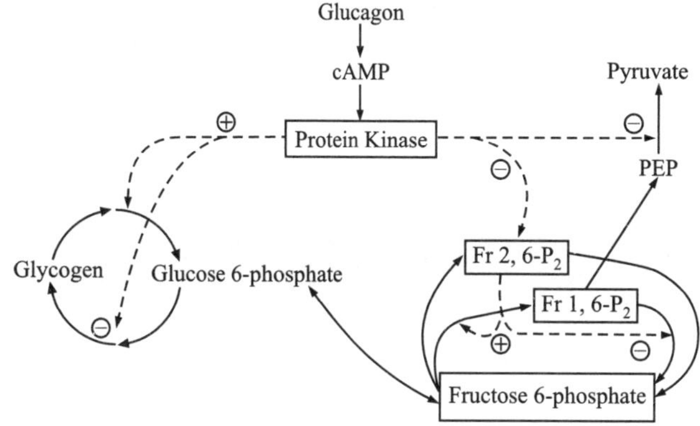

Figure 16.28 Regulation of glycolysis and gluconeogenesis by fructose 2, 6-bisphosphate and glucagon in the liver.

FBP-1 by AMP is potentiated by fructose 2, 6-bisphosphate. Interestingly, PFK-2 and FBP-2 activities reside on a single polypeptide chain. Phosphorylation of this enzyme activates FBP-2 but inhibits PFK-2, whereas dephosphorylation to enzyme has the opposite effect. Recently, it has been shown that unlike in the liver described above, the enzyme catalyzing the synthesis of Fr 2, 6-P2 in the muscle is stimulated by phosphorylation. This explains why epinephrine leads to stimulation of the process of glycolysis in the muscle, whereas the same hormone inhibits glycolysis in the liver.

The effects of glucagon on gluconeogenesis and glycolysis can be explained by the increase in intracellular cyclic AMP brought about by the hormone. Cyclic AMP stimulates a specific protein kinase, which modulates the activity of key regulatory enzymes by phosphorylation-dephos-phorylation. Phosphorylation results in activation of liver phosphorylase kinase, liver phosphorylase and FBP-2, but inactivation of PFK-2 and pyruvate kinase. PFK-1 and FBP-1 are indirectly controlled by the concentration of fructose 2, 6-bisphosphate. Glucagon also enhances transport of pyruvate into mitochondria. The intramitochondrial carboxylation of pyruvate results from the induction of pyruvate carboxylase. Glucagon is also an inducer of glucose-6-phosphatase.

CLINICAL ASPECTS OF SEROTONIN AND GABA

Serotonin has been implicated in the pathophysiology of migraine headaches. Stimulation of a subset of 5-HT receptors can stop an acute migraine attack. Both selective (e.g., triptans) and non-selective (e.g., ergotamine) receptor agonists are available for treatment of migraine. Antagonists acting on another subset of 5-HT receptors are also being used clinically, but as anti-emetic drugs. Such drugs are used to treat vomiting associated with cancer chemotherapy or radiation therapy. Another clinical use of serotonin-based drugs is for the treatment of depression. For example, selective serotonin reuptake inhibitors (SSRIs) (e.g., fluoxetine) are used as anti-depressants. Furthermore, serotonin released by platelets is involved in increasing sensitivity to pain (sensitization) and clotting-associated inflammation. This information is being exploited to further understand the pathophysiology of pain and inflammation and to develop appropriate therapeutic approaches for disorders involving these effector mechanisms.

GABA is a major neurotransmitter in the brain. GABA and its receptors are involved in the pathophysiology of epilepsy. Some of the anti-epileptic drugs act by increasing the availability of GABA and potentiating GABA receptor function. For example, gabapentin and valproic acid can increase the levels of GABA in the brain. Another clinical situation involving GABA is Stiff-person syndrome, which is characterized by hyperexcitation of motor nerves causing stiff muscles. Serum antibodies to glutamate decarboxylase (GAD), the enzyme required for the biosynthesis of GABA, are present in about two-third of patients with Stiff-person syndrome.

SUGGESTIONS FOR FURTHER READING

Behal R.H. et al. (1993), Regulation of the Pyruvate Dehydrogenase Multienzyme Complex, *Annu Rev. Nutrition*, 13, 497.

Bnisilow S.W. and Maestri N.E. (1996), Urea Cycle Disorders: Diagnosis, Pathophysiology and Therapy, *Adv. Pediatr.*, 43, 127–170.

Braissant O. (2010), Current Concepts in the Pathogenesis of Urea Cycle Disorders, *Mol Genet Metab.*, 100 Suppl 1, S3–S12.

Felig P. (1975), Amino Acid Metabolism in Man, *Annu Rev. Biochem.*, 44, 933.

Jones D.P. et al. (1992), Coordinated Multisite Regulation of Cellular Energy Metabolism, *Annu Rev. Nutrition*, 12, 327.

Longo A., Fauci A., Kespar D., Hauser S., Jamesson J. and Loscalzo J. (Eds.) (2012), *Harrisons's Principles of Internal Medicine*, 18th ed., McGraw-Hill, New York, Chapters 14, 366 and 369.

Meister A. (1965), *Biochemistry of the Amino Acids*, 2nd ed., Vols. 1 and 2, Academic Press, Inc.

Morris S.M. Jr. (1992), Regulation of Enzymes of Urea and Arginine Synthesis, *Annu Rev. Nutrition*, 12, 81.

Pilkis S.J. and Granner D.K. (1992), Molecular Physiology of the Regulation of Hepatic Gluconeogenesis and Glycolysis, *Annu Rev. Physiology*, 54, 885.

Ratner S. (1977), A Long View of Nitrogen Metabolism, *Annu Rev. Biochem.*, 46, 1.

Rothman D.L. et al. (1991), Quantitation of Hepatic Gluconeogenesis in Fasting Humans with 13C NMR, *Science*, 254, 573.

Schutz Y. (2011), Protein Turnover, Ureagenesis and Gluconeogenesis, *Int J. Vitam Nutr Res.*, 81:101–107.

Tabor C.W. and Tabor H. (1984), Polyamines, *Annu Rev. Biochem.*, 53, 749.

Tietz (1999), *Textbook of Clinical Chemistry*, 3rd ed., Burtis, C.A. and Ashwood, E.R. (Eds.), WB Saunders Company, Philadelphia, 444–476.

Truffi-Bachi P. and Saari J.C. (1973), Amino Acid Metabolism, *Annu Rev. Biochem.*, 42, 113.

Tyler B. (1978), Regulation of the Assimilation of Nitrogen Compounds, *Annu Rev. Biochem.*, 47, 1127.

Umbarger H.E. (1978), Amino Acid Biosynthesis and Its Regulation, *Annu Rev. Biochem.*, 47, 533.

Wright C.E., Tallan H.H., Lin Y.Y. and Gaull G.E. (1986), Taurine: Biological Update, *Annu Rev. Biochem.*, 55, 427.

17

Interrelation between the Metabolism of Carbohydrates, Fats and Proteins

L.M. Srivastava

CONTENTS

I. INTRODUCTION

The earlier chapters have dealt individually with the metabolism of carbohydrates, fats and proteins in detail. An understanding of these processes is essential to develop a rational approach to the problem of the treatment of disease whenever one encounters abnormalities in the metabolism of such substances. One of the common contributions of the catabolism of these three main dietary components is the release of energy. Energy is generated at various steps of the catabolism leading to a complete oxidation of these substances into CO_2 and H_2O. However, during the intermediary stages of the metabolism of these components, certain compounds (e.g., acetyl-CoA, oxaloacetate) are formed which are common to various pathways. There is a close metabolic inter-relationship between these nutrients in the tissues and there are steps at which their intermediates enter common pathways. The general observation that the animals get fattened on a predominantly carbohydrate diet shows the ease with which the carbohydrates can be converted into fats. Also, the interconversion of nutrients provides a very useful adaptation mechanism for maintaining homeostasis in the body under varying conditions.

II. MAJOR METABOLIC PATHWAYS AND THEIR CONTROL SITES

1. Glycolysis

(a) This sequence of reactions in the cytosol converts glucose into two molecules of pyruvate with concomitant generation of two ATP and two NADH.

(b) The NAD^+ consumed in the reaction catalysed by glyceraldehyde-3-PO_4 dehydrogenase must be regenerated for glycolysis to proceed. Under anaerobic conditions, as in highly active skeletal muscle, this is accomplished by the reduction of pyruvate to lactate. Under aerobic conditions NAD^+ is regenerated by the transfer of electron from NADH of O_2 through the electron transport chain.

(c) Glycolysis serves 2 main purposes:

   (i) it degrades glucose to generate ATP.

   (ii) it provides carbon skeletons for biosynthesis.

(d) Phosphofructokinase is the most important control site. High level of ATP suppresses PFK. This negative effect is enhanced by citrate and reversed by AMP. Rate of glycolysis depends on the need for ATP (i.e., ATP/AMP ratio) and on the need for building blocks (i.e., citrate levels). In liver, the most important regulator of PFK activity is F-2, 6-bisphosphate.

(e) PFK is controlled differently in muscle. Epinephrine increases glycolysis in muscle but decreases glycolysis in liver because of key differences between the isozymes.

2. Citric Acid Cycle

(a) Common pathway for oxidation of fuel molecules — aldehydes, amino acids, fatty acids takes place in the mitochondria.

(b) Most fuels enter the cycle as acetyl-CoA.

(c) The complete oxidation of an acetyl unit generates one GTP, 3 NADH and 1 $FADH_2$.

(d) The four pairs of electron are then transferred to O_2 through the electron transport chain, which results in the formation of a proton gradient that drives the synthesis of eleven ATP. NADH and $FADH_2$ are oxidized only if ADP is simultaneously phosphorylated to ATP. This tight coupling called respiratory control, ensures that the rate of the citric acid cycle matches the need for ATP.

(e) An abundance of ATP also diminishes the activities of three enzymes in the cycle—citrate synthase, isocitrate dehydrogenase and α-ketoglutarate dehydrogenase. The citric acid cycle also has an anabolic role.

(f) It provides intermediates for biosynthesis, such as succinyl-CoA for the formation of porphyrins.

3. Pentose Phosphate Pathway

(a) Two main purposes of this pathway are:

   (i) The generation of NADPH for reductive biosynthesis.

   (ii) Formation of ribose-5-P for the synthesis of nucleotides.

(b) 2 NADPH are generated in the conversion of glucose-6-P into ribose-5-P. The dehydrogenation of glucose-6-phosphate is the committed step in this pathway. This reaction is controlled by the level of $NADP^+$, the electron acceptor.

4. Gluconeogenesis

(a) Synthesis of glucose from the non-carbohydrate precursors such as lactate, glycerol and amino acids by liver and kidneys.

(b) Major entry point is pyruvate, which is carboxylated to oxaloacetate in mitochondria. Oxaloacetate is then decarboxylated and phosphorylated in the cytosol to form phosphoenol pyruvate.

(c) Gluconeogenesis and glycolysis are reciprocally regulated, key control site — Fructose–1, 6, bisphosphatase.

5. Glycogen Synthesis and Degradation

(a) Glycogen is a branched polymer of glucose residues.

(b) The activated intermediate in its synthesis is UDP-glucose which is formed from glucose-1-P and UTP.

(c) Glycogen synthase catalyzes the transfer of glucose from UDP-glucose to the terminal hydroxyl residue of a growing strand.

(d) Glycogen degradation–phosphorylase increases the cleavage of glycogen by orthophosphate to give glucose-1-phosphate.

(e) Glycogen synthesis and degradation are coordinately controlled by a hormone triggered amplifying cascade so that the synthase is inactive when phosphorylase is active and vice versa.

6. Fatty Acid Synthesis and Degradation

(a) Fatty acids are synthesized in the cytosol by the addition of 2-carbon units to a growing chain on acyl carrier protein. Malonyl-CoA, the activated intermediate, is formed by the carboxylation of acetyl-CoA.

(b) Acetyl groups are carried from mitochondria to the cytosol by citrate malate shuttle.

(c) Citrate in the cytosol stimulates acetyl-CoA carboxylase, the enzyme catalyzing the committed step.

(d) When ATP and acetyl-CoA are abundant, the level of citrate increases which accelerates the rate of fatty acid synthesis.

(e) *Fatty acid degradation:* Fatty acids are degraded by a different pathway in a different compartment. They are degraded to acetyl-CoA in the mitochondrial matrix by β-oxidation. Acetyl-CoA can also give rise to ketone bodies. The $FADH_2$ and NADH formed in the β-oxidation transfer their electrons to O_2 through the ETC. Rate of fatty acid degradation is also coupled to the need for ATP.

Metabolic profiles of the major organs

1. Brain: Glucose is the sole fuel for brain except during prolonged starvation. During starvation, ketone bodies partly replace glucose as fuel for the brain. Fatty acids do not serve as fuel for the brain because they are bound to albumin in plasma and do not traverse the blood brain barriers.

2. Muscle: Major fuels are glucose, fatty acids and ketone bodies. Muscle differs from brain in having a large store of glycogen. Muscle lacks glucose-6-phosphatase and so it does not export glucose. It retains glucose for bursts of activity. In actively contracting muscle, the rate of glycolysis far exceeds that of TCA cycle. Much of the pyruvate formed is reduced to lactate which flows to the liver, where it is converted to glucose. These interchanges, known as **Cori cycle,** shift part of the metabolic burden of muscle to the liver (Figure 17.1).

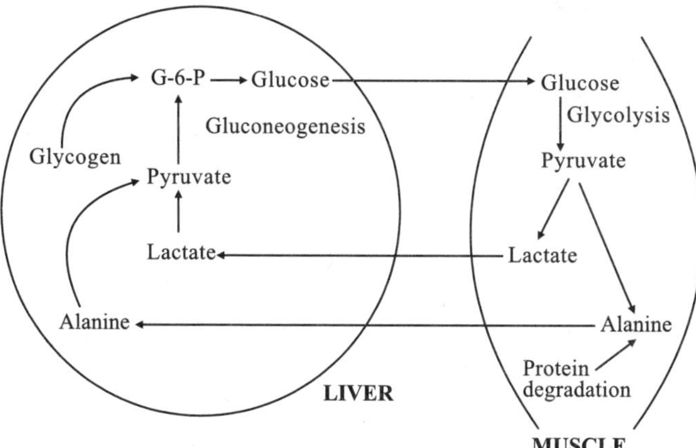

Figure 17.1 Cori cycle.

The metabolic pattern of resting muscle is quite different. Fatty acids are the major fuel. Ketone bodies can also serve as fuel for heart muscle. Heart muscle consumes acetoacetate in preference to glucose.

3. Adipose tissue: It is specialized for the esterification of fatty acids (FA) and for their release from triacylglycerols (TAG) as shown in Figure 17.2.

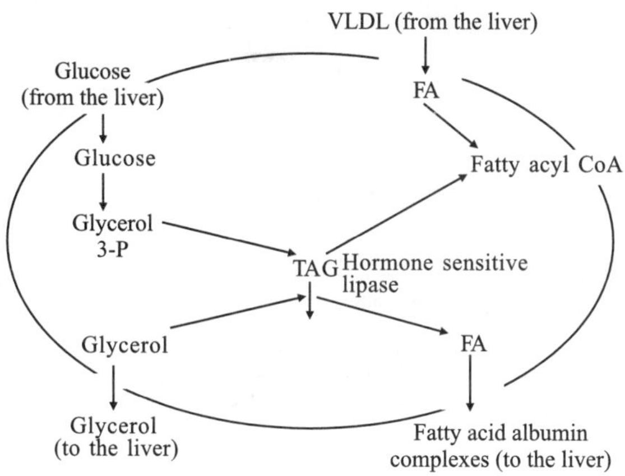

Figure 17.2 Adipose tissue.

4. Liver: It can rapidly mobilize glycogen and carry out gluconeogenesis to meet the glucose needs of other organs. It plays a central role in the regulation of lipid metabolism. When fuels are abundant, FA are synthesised, esterified and sent from the liver to the adipose tissue in the form of VLDL. In the fasting state, however, FA are converted into ketone bodies by the liver. Liver's own fuel are the keto acids derived from the degradation of amino acids which are preferred to glucose. Liver cannot use aceto-acetate as a fuel because it has little of the transferase needed for its activation to acetyl-CoA.

Hormonal regulators of fuel metabolism

1. Insulin: It is secreted by β-cells of the pancreas and is enhanced by glucose and peripheral nervous system. This hormone promotes the dephosphorylation of key interconvertible enzymes. One consequence is to increase glycogen synthesis in both muscle and liver and to suppress gluconeogenesis by the liver, which accelerates glycolysis in the liver, which in turn increases the synthesis of FA. The entry of glucose into the muscles and adipose cells is promoted by insulin. The abundance of fatty acids and glucose in adipose tissue results in the synthesis and storage of triacylglycerols. Insulin promotes the uptake of branched chain amino acids by muscle, which favours a building up of muscle protein. It decreases the intracellular degradation of proteins.

2. Glucagon: It is secreted by α cells of the Islets of Langerhans in response to a low blood sugar level in the fasting state. The main target is liver. It increases glycogen breakdown and decreases glycogen synthesis by triggering the cAMP mediated cascade that leads to the phosphorylation of phosphorylase and glycogen synthase. Glucagon also lowers fatty acid synthesis by diminishing the production of pyruvate and by lowering the activity of acetyl-CoA carboxylase. Glucagon stimulates gluconeogenesis and blocks glycolysis by lowering the level of F-2, 6-bisphosphate. All known actions of glucagon are mediated by protein kinases that are activated by cAMP. The net result is to markedly increase the glucose released by the liver.

3. Epinephrine and norepinephrine: Secreted by adrenal medulla and sympathetic nerve endings in response to low blood glucose level. They increase the mobilization of glycogen and triacyl glycerols by triggering the cAMP mediated cascade. Their glycogenolytic effect is greater in muscle than in liver. They also decrease glucose uptake by muscle. Epinephrine stimulates secretion of glucagon and inhibits insulin secretion. Thus the catecholamines increase the amount of glucose released into the blood by the liver and decreases the utilization of glucose by muscle.

III. TRICARBOXYLIC ACID CYCLE (TCA CYCLE) AS A CENTRAL CORE IN THE INTER-RELATIONSHIPS IN METABOLISM AND INTER-CONVERSIONS OF MAJOR FOODSTUFFS

The TCA cycle, as described in detail earlier (Chapter 12), begins with the condensation of acetyl-CoA and oxaloacetate to produce citrate which then gets isomerized into isocitrate. Through dehydrogenation and decarboxylation processes isocitrate is converted into α-ketoglutarate and then to succinyl-CoA by oxidative degradation. Succinyl-CoA after being converted into succinate is dehydrogenated to form fumarate which is converted into malate and then again to oxaloacetate thereby completing the cycle.

The following three compounds formed in the course of catabolism of major foodstuffs are mainly responsible for the interconversions.

(a) Acetyl-CoA formed from carbohydrates, fats and certain amino acids (e.g., isoleucine).

(b) Oxaloacetate which is formed from pyruvate (and indirectly from carbohydrates) and asparate is another point of entry in the TCA cycle.

(c) α-ketoglutarate, formed from reversible transamination reactions with a number of amino acids.

The interrelationships in the metabolism of these nutrients are depicted in Figure 17.3.

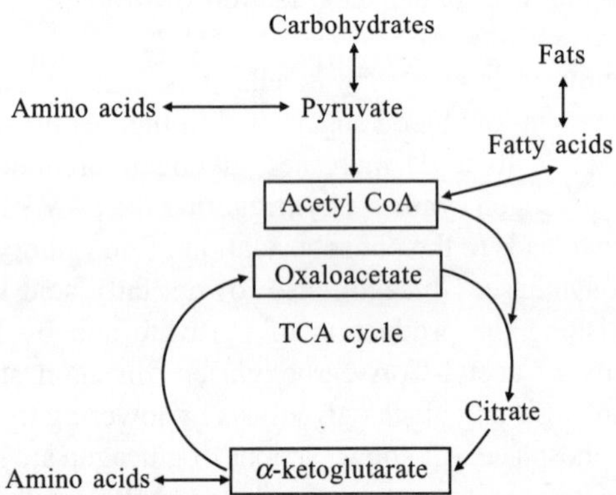

Figure 17.3   Broad links between the metabolism of carbohydrates, fats and proteins.

In the following subsections the salient features of the metabolism of carbohydrates, fats and proteins are described, which are essential in understanding the inter-conversions of major foodstuffs. These inter-linking pathways are shown in the metabolic chart (Figure 17.4).

A.   Metabolism of Carbohydrates

Experiments using radioisotopes have shown that glucose is incorporated into lipids and proteins.

The catabolic processes of carbohydrate metabolism whether starting from glycogen or glucose yield pyruvate. The sugar phosphates of different monosaccharides are interchangeable and enable these sugars to enter the glycolytic pathway at the stage of glucose-6-phosphate. Glucose-6-phosphate is an important metabolite and can be utilized either via the pentose phosphate shunt pathway or the glycolytic sequence (Figure 17.5). The oxidation of glucose via pentose phosphate shunt is of significance for the formation of pentoses, which are required for nucleic acid synthesis. In addition the reduced form of NADP produced in this cycle is utilized as coenzyme for several synthetic processes, for instance, in the synthesis of fatty acids, cholesterol and other steroids.

A substantial portion of glucose is, however, oxidized via the glycolytic pathway and then through the citric acid cycle into CO_2 and H_2O. One of the main reactions of the glycolytic sequence is the breakdown of fructose 1, 6-diphosphate into 2 moles of triose phosphate (one mole of glyceraldehyde-3-phosphate and one mole of dihydroxy acetone phosphate).

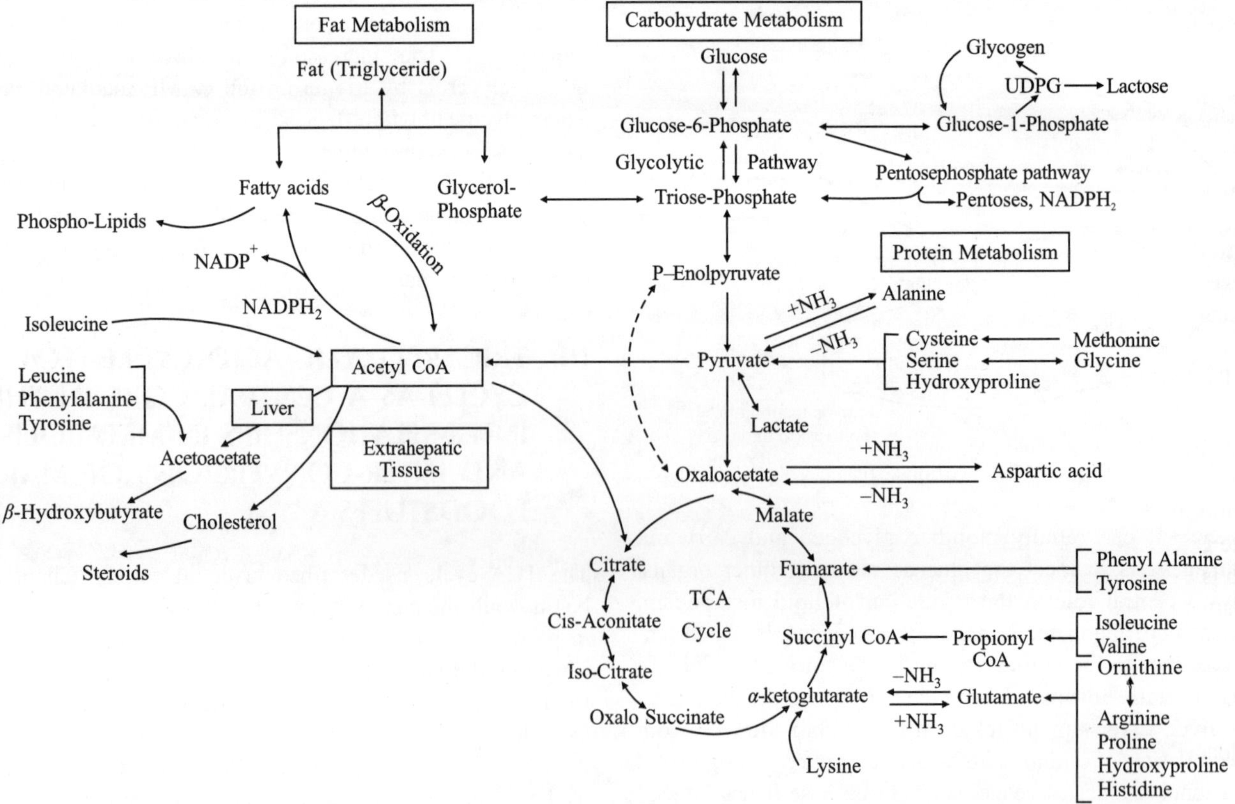

Figure 17.4   Interrelationship between the metabolism of carbohydrates, fats and proteins.

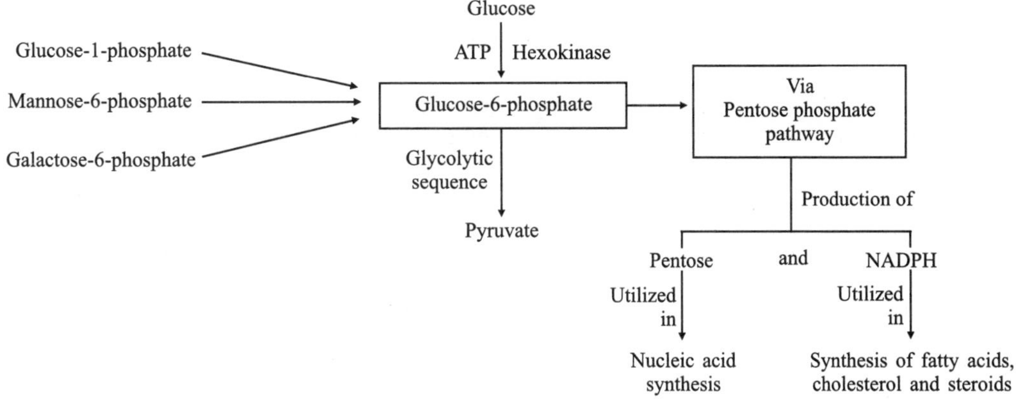

Figure 17.5　Diagram depicting glucose-6-phosphate as an important intermediate of carbohydrate metabolism.

Dihydroxy acetone phosphate can be transformed into glycerophosphate which is necessary for the synthesis of fats and phospholipids.

Triose phosphate, through a series of enzymatic reactions, yields phosphoenol pyruvate and pyruvate. Pyruvate can be transformed into common intermediates of various metabolic pathways via one or more of the following reactions (Figure 17.2).

1. Conversion to acetyl-CoA:　Pyruvate is oxidatively decarboxylated to acetyl-CoA by pyruvate dehydrogenase. This requires NAD^+, CoA, thiamine pyrophosphate, lipoic acid and FAD.

$$\text{Pyruvate} \xrightarrow[\text{Thiamine pyrophosphate}]{\substack{\text{Pyruvate dehydrogenase} \\ NAD^+ : \text{CoA; lipoic acid; FAD}}} \text{Acetyle-CoA}$$

The acetyl-CoA thus formed can find entry into the TCA cycle or be utilized for the synthesis of fatty acids, cholesterol, and steroids. An abundance of acetyl-CoA can also produce the ketone body acetoacetate.

2. Formation of oxaloacetate:　Pyruvate carboxylase in the presence of ATP, biotin and CO_2 converts pyruvate to oxaloacetate.

$$\text{Pyruvate} \underset{-CO_2 \text{ Pyruvate carboxylase}}{\overset{+CO_2}{\rightleftharpoons}} \text{Acetyle-CoA}$$

The oxaloacetate produced by the above reaction can combine with acetyl-CoA and enter into the TCA cycle. Furthermore as the conversion of phosphoenol pyruvate to pyruvate is a *unidirectional* reaction, the pyruvate carboxylase reaction is useful as the first step in the conversion of pyruvate to phosphoenol pyruvate in the gluconeogenic pathway. Oxaloacetate can be converted to phosphoenol pyruvate by an enzyme phosphoenol pyruvate carboxykinase.

$$\text{Oxaloacetate} \xrightarrow[-CO_2]{\substack{\text{Phosphoenol pyruvate} \\ \text{carboxykinase}}} \text{Phosphoenol pyruvate}$$

3. Conversion to malate.　By the action of malic enzyme, pyruvate is converted into malate by carboxylation in the presence of NADPH. H^+

$$\text{Pyruvate} \underset{-CO_2,\, NADP^+}{\overset{\substack{NADPH\, H^+;\, CO_2 \\ \text{Malic enzyme}}}{\rightleftharpoons}} \text{Malate}$$

4. Transformation into lactate:　Pyruvate is reduced under anaerobic conditions by the $NADH + H^+$ to lactate, the reaction being catalyzed by lactate dehydrogenase.

$$\text{Pyruvate} \underset{\substack{NAD^+ \\ \text{Lactate dehydrogenase}}}{\overset{NADH\, H^+}{\rightleftharpoons}} \text{Lactate}$$

5. Formation of amino acids:　The reversible conversion of pyruvate to alanine by amination or transamination provides an excellent way of entry to protein metabolism. Serine and cystine are also known to be converted into pyruvate.

$$\text{Pyruvate} \underset{-NH_3}{\overset{+NH_3}{\rightleftharpoons}} \text{Alanine}$$

A schematic figure of the fate of pyruvic acid is given in Figure 17.6.

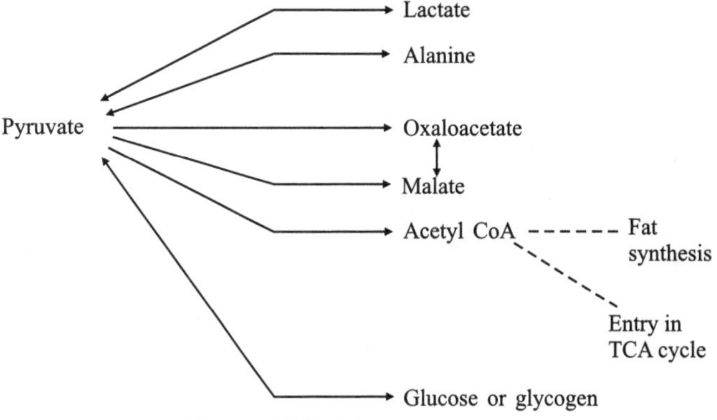

Figure 17.6　Fate of pyruvate.

In glucose synthesis (gluconeogenesis) the starting material is primarily lactate which is formed during anaerobic glycolysis and also from a number of amino acids (Figure 17.4). The key component for conversion is phosphoenol pyruvate, which is obtained from pyruvate through an intermediary product, oxaloacetate.

B.   Metabolism of Fats

The catabolism of fats starts with their hydrolysis into fatty acids and glycerol-phosphate. The fatty acids are further broken down by β-oxidation into acetyl-CoA. It is evident therefore that acetyl-CoA is derived from oxidative decarboxylation of pyruvate as well as from the β-oxidation of long chain fatty acids. The common terminal pathway of various intermediary processes is the TCA cycle. Acetyl-CoA and oxaloacetate condense to form citric acid. Oxaloacetate for the reaction can be obtained either from pyruvate (derived from carbohydrates) or else by deamination of aspartic acid.

Acetyl-CoA is also an important *precursor* for various biosynthetic reactions, the most important of which is the formation of long chain fatty acids. The conversion of fats into carbohydrate synthesis is *limited* because the pyruvate dehydrogenase reaction is essentially non-reversible, which prevents the direct conversion of acetyl-CoA (formed by β-oxidation) to pyruvate. There cannot be a *net* conversion of acetyl-CoA to oxaloacetate via the TCA cycle since one molecule of oxaloacetate is required to condense with acetyl-CoA and only one molecule of oxaloacetate is generated. For such reasons, there cannot be a net conversion of fatty acids having an even number of carbon atoms (which form acetyl-CoA) to glucose or glycogen.

Only the terminal 3-carbon portion of a fatty acid having an odd number of C-atoms is glycogenic as this portion of the molecule would form propionate upon oxidation and enters TCA cycle via succinate. Nonetheless it is possible for labelled carbon atoms of fatty acids to be found ultimately in the glycogen molecule after going through the TCA cycle;

this is essentially because oxaloacetate is an intermediate in both the TCA cycle and in the gluconeogenesis.

Another important synthetic role of acetyl-CoA is in the production of phospholipids, cholesterol, sex hormones and other steroids. In diabetes mellitus, associated with a high rate of fatty acid oxidation because of lack of insulin, acetyl-CoA is converted into acetoacetate and β-hydroxy butyrate (called ketone bodies) in the liver and causes ketosis.

C.   Metabolism of Proteins

Certain amino acids in course of their metabolism produce either the precursors and/or members of the TCA cycle. These can, therefore, be converted into glucose and glycogen via gluconeogenic pathways. Such amino acids are termed glycogenic (or glucogenic) amino acids. The ketogenic amino acids are those which produce acetoacetate and can normally be metabolized via the formation of acetyl-CoA in extra-hepatic tissues. A list of the amino acids which are either glycogenic or ketogenic, on the basis of the product of the metabolism, is given in Table 17.1.

α-ketoglutarate is an important bridge between the carbohydrate and protein metabolism. The transaminases are mainly responsible for the inter-conversion of non-essential amino acids. As is evident from the metabolic chart (Figure 17.4), arginine, proline and histidine enter the TCA cycle indirectly by conversion to glutamate which can give rise to α-ketoglutarate.

Aspartic acid can be converted to oxaloacetate either by transamination or oxidative deamination and find direct entry into the TCA cycle. Alanine enters the TCA cycle by conversion to pyruvate and then to acetyl-CoA.

Some amino acids like valine and isoleucine after transamination of oxidative deamination yield propionyl-CoA which is then carboxylated to succinate. Succinyl CoA can unite with glycine to form δ-aminolevulinate, which is utilized for heme synthesis.

Serine can enter the carbohydrate metabolism through transamination which produces hydroxy-pyruvate. Also,

TABLE 17.1   **List of Glucogenic and Ketogenic Amino Acids and their Breakdown Products**

Glucogenic amino acids		Ketogenic amino acids		Both glucogenic and ketogenic amino acids	
Amino acid	*Product*	*Amino acid*	*Product*	*Amino acid*	*Product*
Alanine, Serine	Pyruvate	Leucine	Acetoacetate	Phenylalanine, Tyrosine	Fumarate
Glycine	Pyruvate succinate				Acetoacetate
Threonine, Valine	Propionate			Isoleucine	Propionyl CoA, Acetyl CoA
Proline, Histidine, Arginine, Lysine	Glutamic acid — Ketoglutarate				
Cystine, Hydroxyproline	Pyruvate				
Aspartic acid	Oxaloacetate				

decarboxylation of serine would produce ethanolamine which can be converted to choline. These two substances are required for phospholipid biosynthesis.

Leucine, during the course of its metabolism produces acetoacetate and is termed a ketogenic amino acid. The metabolism of isoleucine and aromatic amino acids, phenylalanine and tyrosine through deamination and oxidation give rise to fumarate and acetoacetate. These two amino acids are therefore considered as both glucogenic as well as ketogenic.

It is evident from the preceding discussion that the entry of the amino acids by their transformation into the intermediates of the TCA cycle is of immense significance both for their complete degradation as well as for interconversion to carbohydrates and fats. The conversion of carbon skeletons of glucogenic amino acids to fatty acids is possible by conversion to pyruvate and acetyl-CoA or by reversal of non-mitochondrial reactions of the TCA cycle from α-ketoglutarate to citrate followed by the action of ATP-citrate lyase to give acetyl-CoA. The net conversion of amino acids to fats is nevertheless not a significant process except possibly in animals receiving a high protein diet. However, *a net* conversion of fatty acids to glucogenic amino acids is not possible due to non-conversion of acetyl-CoA into pyruvate or phosphoenol pyruvate. Furthermore it is also not feasible to reverse the reactions of the breakdown of ketogenic amino acids (leucine, phenylalanine and tyrosine). That is why these are "essential" amino acids.

Though the transformation of amino acids to carbohydrates is of physiological significance, the conversion of amino acids to fats is not significant. This is best understood under conditions of starvation and diabetes mellitus. In such abnormalities while proteins and amino acids are degraded there is a simultaneous breakdown of fats and, therefore, the only course left for degraded amino acids is to get converted into sugars.

IV. IMPORTANCE OF THE INTEGRATION OF VARIOUS METABOLIC PATHWAYS

The interrelated metabolic pathways (Figure 17.4) help in the understanding and elaboration of the following processes:

A. Maintenance of Constant Blood Glucose Level

(For details the reader is referred to Chapter 12).

B. Regulation of the Intermediary Metabolites of the Metabolism

The integrated pathways of the metabolism would lead to the understanding as to how tissues keep the concentration of its metabolites in equilibrium. Thus, under normal conditions, the reactions of the metabolism of various nutrients proceed in a manner which would allow the maintenance of constant levels of various intermediates. It is only when a certain metabolite is accumulated in an unusual quantity, one finds

that the state of equilibrium of the metabolism is shifted in different directions. One of the examples of such a situation is seen under conditions of severe diabetes mellitus, which in addition to its other effects (Chapter 90), ultimately leads to an increased breakdown of tissue proteins as well, and this process is related to the reactions of the TCA cycle. In diabetes mellitus most of the energy is obtained from the oxidation of fatty acids to acetyl-CoA and, subsequently, through the TCA cycle. For the efficient utilization of acetyl-CoA in the TCA cycle, and adequate amount of oxaloacetate is required. Under normal conditions the requirement for oxaloacetate is met by its production from pyruvic acid, aspartic acid, and glutamic acid, after being converted to α-ketoglutarate. However, in severe diabetes mellitus the formation of oxaloacetate from pyruvate is greatly reduced due to lower rate of glycolysis. Under such circumstances the bulk of oxaloacetate for the operation of the TCA cycle, as well as for the entry into the gluconeogenic pathway is derived mainly from amino acids, thereby depleting the tissue proteins.

Furthermore, during severe diabetes mellitus even the functioning of the TCA cycle is diminished to a marked extent as the higher concentrations of long chain fatty acyl CoA inhibit condensing enzyme—the first enzyme of the TCA cycle which condenses acetyl-CoA and oxaloacetate to citrate. This would lead to only a small utilization of acetyl-CoA in the TCA cycle and would, ultimately, generate less of ATP which could be one of the causes for weakness in diabetic patients. Various ketoacids obtained from amino acids after being converted to oxaloacetate could, however, be transformed to glucose via gluconeogenic pathways in the liver and excreted, while a significant proportion of acetyl-CoA not utilized in the TCA cycle would be converted to ketone bodies. Fatty acyl CoA also inhibits acetyl-CoA carboxylase—the first enzyme of fatty acid synthesis. This is the probable explanation of the depressed lipogenesis under a high fat diet of diabetes mellitus—both of which are associated with increased levels of free fatty acids in the plasma.

C. Metabolic Adaptations during Starvation

(a) By the end of an overnight fast, gluconeogenesis already starts providing glucose from aminoacids. As judged by the output of urinary nitrogen, gluconeogenesis from aminoacids increases during first 3 days and then declines, until the urinary nitrogen is close to the obligatory output at about the thirtieth day of starvation.

(b) Effects pertaining to metabolic fuel.

    (i) *Proteins:* The reduction in the utilization of proteins as a metabolic fuel is essential for longer survival since the protein represents the "fabric" and "enzymatic machinery" of the body and can only be depleted to a certain extent.

    (ii) *Fatty acids:* Gluconeogenesis from amino acids declines as starvation continues because free fatty acids mobilized under conditions of low plasma insulin levels are preferentially oxidized. The mobilization of free fatty acids is unchecked and lasts as long as reserves of triglycerides permit.

(iii) *Ketone bodies:*   The high levels of acetyl-CoA in the liver drive the formation of ketone bodies, which are produced at the maximum rate as early as the third day of starvation. The ketone bodies are readily utilized by cardiac and skeletal muscle as fuel and by the brain when the plasma levels of ketone bodies are high enough.

(iv) *Muscle:*   After about 20 days of starvation, muscle for reasons that are not clear turns wholly to free fatty acids for fuel. Muscle stores the ketone bodies for oxidation by the brain, which then obtains about 60% of its energy from ketone bodies.

D.   Metabolic Adaptations in Diabetes Mellitus

I. There are two types of diabetes:

(i) *Insulin-dependent* (Type I) diabetes is caused by an inability of the body to produce insulin.
   (a) This is brought about by the destruction of the insulin producing β-cells of the pancreatic islets.
   (b) The onset of this disease is often abrupt and occurs before age 40. It is often referred to as juvenile-onset-diabetes.

(ii) *Non-insulin-dependent* (Type II) diabetes is caused by a deficiency or defect in insulin receptors.
   (a) The body is still able to produce insulin when affected by this disorder.
   (b) The onset of this disease often occurs after age 40. Patients are usually overweight and have a family history of diabetes. It is sometimes referred to as adult onset diabetes.

II. Metabolic abnormalities in insulin dependent diabetes: Because insulin production is impaired, the level of insulin is always very low, relative to that of glucagon, regardless of the glucose in the blood. As such, the metabolic profile of the body is very similar to that observed during starvation, despite the presence of an abundance of metabolic fuel.

1. Glucose uptake by tissues is impaired, despite the fact that glucose is being released by the liver.
2. Glycolysis is depressed.
3. Gluconeogenesis is enhanced.
4. Lipolysis of Triacylglycerols is stimulated.
5. Synthesis of fatty acids and triacyglycerol is depressed.
6. Fatty acid oxidation is increased.
7. Glycogen storage gets depleted.
8. Ketone bodies are synthesized in the liver and used as fuel by other tissues.
9. Proteins are degraded and amino acids are used as fuel.
10. The level of glucose, fatty acids and ketone bodies in the blood become very high.
11. Excess glucose leads to glycosylation of haemoglobin and possibly other proteins.

III. Clinical manifestations of untreated diabetes are:

(a) Hyperglycemia—which denotes high levels of glucose in the blood.
(b) Increased urinary output.
(c) Dehydration and electrolyte imbalance from increased output of urine.
(d) Acidosis from high levels of ketone bodies in blood. (Acetoacetate and (β-hydroxybutyrate are acids that cannot be excreted by the lungs and that disrupt the acid-base balance, which lowers pH. This can lead to coma and death if not treated/corrected).
(e) Neuropathy (due to sorbitol accumulation in Schwann cells and subsequent cell damage.
(f) Renal failure
(g) Retinopathy and blindness.

E.   Protein Sparing Effect of Carbohydrates

If an excess of carbohydrates is fed to experimental animals, it is seen that the excretion of nitrogen in the urine is greatly diminished. This would indicate that the animals under these conditions are in positive nitrogen balance. The rationale behind such observations termed as "protein sparing effect of carbohydrate" would be that an excess of carbohydrates leads to an increased formation of the ketoacids of the TCA cycle, which would, in turn, be aminated to give rise to amino acids and subsequently proteins. This would lead to the tying up of nitrogen for useful purposes, and diminish its excretion in urine.

F.   Specific Dynamic Action (SDA)

It is well known that the calorific values of each type of foodstuff, viz. protein, fat, and carbohydrate are 4, 9 and 4 kcal per g respectively. Thus it is expected that the ingestion of, say, 25 g of protein liberate 100 kcal in the body. However, this is not the case; instead, 130 kcal has been found to be the amount of heat produced. Similarly, fats and carbo-hydrates produce 13 per cent and 5 per cent, respectively, more heat than expected. This extra heat produced is known as specific dynamic action (SDA).

For a long time it was thought that this was due to various metabolic activities at the time of digestion e.g., movements of the gut, etc. Even today the precise reason is unknown, but as it has been found experimentally that this phenomenon is not seen in hepatectomized animals, it is believed that this extra heat production stems from deamination.

It is also seen that on a diet comprising of all three principal foodstuffs, the total SDA does not amount to the additive sum of each. Thus, the high specific action of protein is reduced by the others, especially fats.

This phenomenon is of clinical significance in calculating the nutritional requirements of any individual. It is thus customary to add 10 per cent to the total required daily caloric intake to account for the SDA. However, individual types of foodstuffs in the diet will influence this calculation, especially if the fat content is high.

SUMMARY

This chapter has outlined and elucidated the specific steps involved in the interrelationships of the metabolism of carbohydrates, fats, and proteins. The central core in the interrelation between the metabolism of these major food constituents is the tricarboxylic acid cycle and its intermediates. The ease with which the carbohydrates are converted into fats and proteins has been dealt with. Also elaborated upon is the mode of conversion of proteins into fats although the latter has no significance. There is, however, a metabolic block preventing the net conversion of fats into carbohydrates and proteins.

SUGGESTIONS FOR FURTHER READING

Alberts B., Bray D., Lewis J., Raff M., Roberts K. and Watson J.D. (2008), *Molecular Biology of the Cell*, Garland Science, UK.

Banks P., Bartley W. and L.M. Birt (1979), Chapter 45 "The Control of Metabolic Processes in Cells" and Chapter 48—"Tissue Interrelationships" 453–464 in the *Biochemistry of the Tissues*, 2nd ed., John Wiley & Sons, New York, Toronto.

Bender D.A. (2002), *Introduction on Nutrition and Metabolism*, 3rd ed., Taylor & Francis, London, 2002.

Bezkorovainy A. and Rafelson M.E. (1996), Integration of Metabolism, *Concise Biochemistry*, Marcel Dekker Inc., New York, 581–96.

Cooper G.M. (2007), *The Cell*, 4th ed., ASM Press, Washington.

Frayn K.N. (2003), *Metabolic Regulation, A Human Perspective*, 2nd ed., Blackwell Science.

Harris R.A. and Crabb D.W. (1992), Metabolic Interrelationships, *Textbook of Biochemistry with Clinical Correlations*, 3rd ed. (Devlin T.M., Ed.) 576–606; John Wiley & Sons, Inc., New York.

Klover P.J. and Mooney R.A, (2004), Hepatocytes: Critical for Glucose Homeostasis, *International Journal of Biochemistry and Cell Biology*, 36:753.

Mizock B.A. (1995), Alteration in Carbohydrate Metabolism During Stress: A Review of Literature, *Am. J. Med.*, 98, 75–84.

Murray R.K., Garner D.K., Mayes P.A. and Rodwill (2006), Haper's *Illustrated Biochemistry*, 27th ed.; McGraw-Hills, New York.

Nordlie R.C., Foster J.D. and Lange A.J. (1999), Regulation of Glucose Production by the Liver, *Annual Review of Nutrition*, 19:379.

Voet D. and Voet J.G. (2004), *Biochemistry*, 3rd ed., John Wiley & Sons, New York.

18

Inborn Errors of Metabolism

Seema Alam

I. DEFINITION

Single gene mutations can result in alternation in primary protein structure or the amount of protein synthesized which would then impair the function of the protein. The protein could be an enzyme, receptor, transporter, membrane or structural element. *These hereditary biochemical disorders are called inborn errors of metabolism (IEM) or even inherited metabolic disorders.* Most of these genetic mutations may be inconsequential and only a few may cause disease. Following rules would help in clinical diagnosis of an IEM.

1. Presentation of IEMs can be at any age from fetal life to adulthood. Severe forms usually present in the newborn period. But there is a need to recognize which of the newborns or children could be having an IEM.

2. Along with more common conditions, e.g., neonatal sepsis, birth asphyxia, encephalitis, keep a high suspicion of IEMs in a sick neonate. Specially suspect an IEM in case of persistence of unexplained symptoms, after commonly occurring conditions have been ruled out.

3. Most IEM are recessively inherited but some maybe sporadic.

4. Consider treatable IEM initially specially in acutely sick patients.

The manifestations of the IEM are varied but some features could be common. Recognition of these features can help in early identification as well as improving the morbidity and mortality related to IEM.

II. CLASSIFICATIONS OF IEM

As per the different metabolic pathways the IEM can be classified and sub-classified as depicted in Table 18.1 where the common errors of various metabolite pathways are listed. The list is exhaustive and hence from a therapeutic perspective, metabolic disorders can be divided into the following three useful groups.

Group 1: Disorders Causing Intoxication

The symptoms in this group are due to intermediary metabolic block leading to an acute or progressive intoxication from the accumulation of toxic compounds. The IEMs included in this group are that of amino acid catabolism, organic academia (OA), urea cycle defects (UCD), galactosemia, hereditary fructose intolerance (HFI), metal intoxication, and porphyria. The inborn errors of neurotransmitter synthesis and catabolism and the inborn errors of amino acid synthesis can also be included in this group. Almost all disorders in this group do not interfere with the embryo-fetal development and they present with a symptom-free interval and clinical signs of intoxication. Clinical signs of intoxication may have acute or sudden onset or a chronic onset of symptoms. Acute onset could be in the form of diarrhea, vomiting, coma, seizures or liver failure. Chronic onset would present as growth failure. delay in development, cardiomyopathy, lens displacement, etc. Catabolic crisis like fever, intercurrent illness and food intake can provoke intermittent acute metabolic attacks. Nutritional therapy is the backbone of the treatment in this group. It includes approaches to deplete the toxic substrate that accumulates or to replace the crucial metabolic product that is deficient. Strategies to decrease the concentration of toxic substrates or their precursors also involve the administration of a variety of cleansing drugs (carnitine, sodium benzoate, penicillamine) that bind the accumulated metabolites and allow their excretion. Pharmacological doses of vitamins have also shown remarkable efficiency in vitamin-responsive disorders.

Group 2: Disorders Involving Energy Metabolism

This group consists of IEMs with symptoms due to deficiency in energy production or defects in energy utilization in various tissues like liver, brain, muscle, myocardium, etc. Depending

---

TABLE 18.1   Classification of inborn errors of metabolism

I. Disorders of Carbohydrate Metabolism

   (a) Glycogen Storage Diseases (GSD) and Related Disorders
      Hepatic GSD (see table 9)
      Muscle GSD (type V, VII, X, XII, XIII, XIV)
      Cardiac GSD (type I, IIa, IIb, III,)
      Brain GSD
   (b) Disorders of Galactose Metabolism
      (i) Galactose – I Phosphate Uridyltransferese (Galactosemia)
      (ii) Uridine Diphosphate-Galactose-4 Epimerase
      (iii) Galactokinase deficiency
      (iv) Fanconi Bickel Syndrome
   (c) Disorders of the Pentose Phosphate Pathway
      (i) Ribose 5-Phosphate Isomerase Deficiency
      (ii) Transaldolase deficiency
   (d) Disorders of the Fructose Metabolism
      (i) Essential Fructosuria.
      (ii) Hereditary Fructose Intolerance
      (iii) Fructose I, 6-Biphosphatase Deficiency
   (e) Persistent Hyperinsulinemic Hypoglycemia
   (f) Disorders of Glucose transport
      (i) Congenital Glucose/Galactose Malabsorption/SGLTI
      (ii) Renal Glucosuria SGLT2
      (iii) Glucose Transporter Deficiency Syndrome. GLUT I
      (iv) Fanconi Bickel Syndrome. GLUT 2
      (v) Arterial Tortuosity Syndrome GLUT 10

II. Disorders of Mitochondrial Energy Metabolism

   (a) Disorders of Pyruvate Metabolism and Tricarboxylic Acid Cycle**
      (i) Pyruvate Carboxylase deficiency
      (ii) Phosphoenolpyruvate Carboxykinase deficiency
      (iii) Pyruvate Dehydrogenase Complex Deficiency

(Contd.)

TABLE 18.1 Classification of inborn errors of metabolism (*Contd.*)

(b) Disorders of Fatty acid oxidation**
    (i) Fatty acid transport defects
    (ii) Carnitine cycle defects (Carnitine transporter defect, CPT I, CPT II)
    (iii) β Oxidation defects (VLCAD, LCHAD, MCAD, SCAD, SCHAD)
    (iv) Electron transfer defects

(c) Disorders of the Ketogenesis and Ketolysis
    (i) Mitochondrial-3-Hydroxy-3 CoA synthasedeficiency (OH)
    (ii) 3-OH-3-Methylglutatyle CoA lyase deficiency
    (iii) Succinyl CoA-3-Oxoacid CoA transferase deficiency
    (iv) Mitachondrial Acetoacetyl CoA-thiolase deficiency

(d) Defects of Respiratory Chain

(e) Creatine Deficiency Syndrome
    (i) Guanidinoacetate Methyl transferase deficiency
    (ii) Arginine Glycin methyl Amidino transferase deficiency
    (iii) SLC 6A8 deficiency
    (iv) GAMT deficiency

III. Disorders of Amino Acid Metabolism and Transport

(a) Hyperphenylalaninemia
    (i) Phenylalanine hydroxylase deficiency
    (ii) Maternal PKU
    (iii) HPA and Disorders of Biopterin Metabolism

(b) Disorders of Tyrosine metabolism
    (i) Hereditary Tyrosinemia type I.
    (ii) Hereditary Tyrosinemia type II
    (iii) Hereditary Tyrosinemia type III
    (iv) Transient Tyrosinemia
    (v) Alkaptonuria
    (vi) Hawkinsinuria

(c) Branched Chain Organic Acidemia (see Table 11)

(d) Disorders of Urea cycle (see Fig 5)

(e) Disorders of Sulfur Amino Acid
    (i) Homocystinuria
    (ii) Methionine S-Adenosyltransferase deficiency
    (iii) Glycine N-Methyltransferase deficiency
    (iv) S-Adenosyl homocysteine hydrolase deficiency
    (v) Y Cystathionase deficiency

(f) Disorders of Ornithine Metabolism**
    (i) Hyperornithinemia
    (ii) Hyperornithinemia, Hyperammonemia and Homocitrullinuria (HHH) Syndrome.

(g) Cerebral Organic Acid Disorders and Lysine Catabolism disorders**
    (i) Glutaric Aciduria Type I
    (ii) Glutaric Aciduria Type III
    (iii) Canavan Disease

(h) Non Ketotic Hyperglycinemia (NKH)
    (i) Neonatal NKH
    (ii) Late Onset NKH

(i) Disorders of Proline and Serine Metabolism**

(j) Transport Disorders of Amino Acids
    (i) Cystinuria
    (ii) Hartnup Disorder
    (iii) Lysinuric Protein Intolerance

IV. Vitamin Responsive Disorders.

(a) Biotin responsive Disorders
    (i) Holocarboxylase Synthetase deficiency
    (ii) Biotinidase deficiency
    (iii) Biotin responsive basal ganglia disease

(*Contd.*)

TABLE 18.1 Classification of inborn errors of metabolism (*Contd.*)

   (b) Disorders of Cobalamine and Folate Transport and Metabolism**
      (i) Hereditary Intrinsic factor deficiency
      (ii) Defective Transport of Cobalamin by Enterocytes
      (iii) Transcobalamin deficiency
      (iv) Hereditary Folate deficiency
      (v) Cerebral Folate deficiency

V. Neurotransmitter and small Peptide Disorders**
   (a) Disorders of Neurotransmission
      (i) GABA Transaminase deficiency
      (ii) Pyridoxine responsive Epilepsy
      (iii) Pyridoxamine 5′-Phosphate Oxidase
   (b) Disorders in the Metabolism of Glutathione and Imidazole Dipeptide
   (c) Trimethylaminuria and Dimethyle glycine Dehydrogenase deficiency

VI. Disorders of Lipid and Bile acid Metabolism
   (a) Disorder of Exogenous Lipoprotein Metabolism
      (i) Lipoprotein Lipase deficiency
      (ii) Apo C-II Deficiency
   (b) Disorder of Endogenous Lipoprotein Metabolism
      (i) VLDL Overproduction
      (ii) LDL Removal
   (c) Disorder of Disorder of Exogenous & Endogenous Lipoprotein Transport
      (i) Dysbetalipoproteinemia
      (ii) Hepatic lipase deficiency
   (d) Disorder of Reduced LDL-Cholesterol Levels
      (i) Abetalipoproteinemia
      (ii) Hypobetalipoproteinemia
   (e) Disorder of Reverse Cholesterol transport
   (f) Disorder of Cholesterol Synthesis
   (g) Disorder of Bile Synthesis
      (i) 3β-OH-Δ5-C27-Steroid dehydrogenase deficiency
      (ii) Δ4-3-Oxosteroid 5β Reductase deficiency
   (h) Disorders of Phospholipid and Glycosphingolipid Synthesis

VII. Disorders of Nucleic Acid and Heme Metabolism**
   (a) Disorders of Purine and Pyramidine Metabolism
   (b) Disorders of Heme Metabolism
     (X-linked Sideroblastic Anemia, Acute Intermittent Porphyria, 5-Aminolevulinic Acid Dehydratase Porphyria, Congenital Erythropoetic Porphyria, Porphyria Cutanea Tarda)

VIII. Disorders of Metal Transport
   (a) Copper (Wilson's disease, Menkes Disease, Others)
   (b) Iron (Hemochromatosis, Autodegeneration with brian Iron accumulation)
   (c) Zinc (Acrodermatitis Enteropathica)
   (d) Magnesium, Manganese, Selenium

IX. Organelle Related Disorders
   (a) Disorders of Sphingolipid Metabolism, Lysosomes, Peroxisomes, and Golgi and Pre-Golgi systems.
     (Gauchers Disease, Neimann Pick A and B, GM1 Gangliosidosis, GM 2 Gangliosidosis, Krabbe Disease, Metachromatic Leukodystrophy, Fabry disease, Farber disease, Neimann Pick disease)
   (b) Mucopolysaccharidoses and Oligosaccharidoses
   (c) Peroxisomal Disorders
   (d) Congenital disorder of Glycosylation
      (i) Congenital Disorders of Protein N-glycosylation
      (ii) Congenital Disorder of Protein O-glycosylation
   (e) Defects in lipid Glycosylation
   (f) Defects in multiple glycosylation pathways
   (g) Cystinosis (Infantile and ocular cystinosis)

** Only more common disorders listed under each category

upon the location of the defect, this group is further divided into mitochondria or cytoplasm energy defects .

1. Mitochondrial defects are the most severe which include the congenital lactic acidemias (defects of pyruvate transporter, pyruvate carboxylase (PC), pyruvate dehydrogenase (PDH), and the Krebs cycle), and mitochondrial respiratory chain disorders. Most of these are not treatable except coenzyme Q10 synthesis defect, PDH and PC deficiency, fatty acid oxidation and ketone body defects, which are partly treatable.
2. Cytoplasmic energy defects are less severe. Some of the treatable defects in this group are disorders of glycolysis, glycogen metabolism, gluconeogenesis, and hyperinsulinism. Partly treatable disorders of creatine metabolism and the untreatable errors of pentose phosphate pathways are also included in this group.

Although the more commonly occurring glycogen storage defects in this group do not interfere with embryofetal development, mitochondrial respiratory chain and pentose pathway defects can affect embryofetal.

Group 3: Disorders Involving Cellular Organelles

The disorders which involve cellular organelles cause disturbance in the synthesis or the catabolism of complex molecules. In this group the symptoms are more severe, persistent and progressive. These patients do not present intermittently with episodes of catabolic crisis or particular food intake. All lysosomal storage disorders, peroxisomal disorders, disorders of intracellular processes (e.g., α1-antitrypsin, congenital defect of glycosyllation (CDG)) and inborn errors of cholesterol synthesis belong to this group. Enzyme replacement therapy (ERT) has become available for several lysosomal disorders in the last decade. Various cell and organ transplantation strategies have also been developed for certain disorders, some of them are successful, but many others are still experimental or under evaluation. Finally, besides gene therapy, new therapeutic approaches such as chaperon therapy appears promising but currently remain mostly inaccessible in clinical practice, e.g., in Fabry disease a potent α-galactosidase A inhibitor, 1-deoxygalactonojirimycin which is acts as a chaperone to help in appropiate folding of the mutant enzyme thereby improving its stability and transporting it to the lysosomes.

III. NEONATAL MASS SCREENING

Inborn errors of metabolism are individually rare, but many among the now recognized disorders are present during the neonatal period. A proportion of them are treatable, and delay in intervention can lead to permanent damage. The early clinical signs and symptoms of the IEM are nonspecific, hence often misdiagnosed. Many babies with IEM actually become symptomatic before the screening results are available. With the important therapeutic progress, clinical screening for IEM by the first-line neonatologists or pediatricians is mandatory. The goal of the screening should be 'Do not miss a treatable disorder', in particular in the neonatal intensive care unit or emergency room. In UK, phenylketonuria is the only IEM for which neonatal screening is currently done and a trial of screening for medium-chain acyl-CoA dehydrogenase (MCAD) deficiency is in progress. Table 18.2 lists the IEMs which universally screened for in the USA. In Table 18.3, we see a

TABLE 18.2   **Primary disorders recommended for neonatal screening by the National Newborn Genetics Resource Centre**

DISORDERS OF ORGANIC-ACID METABOLISM	DISORDERS OF AMINO-ACID METABOLISM
Isovaleric academia	Phenylketonuria
Glutaric aciduria type I	Maple syrup urine disease
3-Hydroxy-3-methylglutaric aciduria	Homocystinuria
Multiple carboxylase deficiency	Citrullinemia
Methylmalonic acidemia, mutase deficiency form	Argininosuccinic academia
3-Methylcrotonyl-CoA carboxylase deficiency	Tyrosinemia type I
Methylmaionic acidemia, cblA and cblB forms	**HEMOGLOBINOPATHIES**
Propionic academia	Sickle cell anemia
Beta-ketothiolase deficiency	Hemoglobin S-β-thalassemia
DISORDERS OF FATTY ACID METABOLISM	Hemoglobin SC disease
Medium-chain acyl-CoA dehydrogenase deficiency (MCAD)	**OTHER DISORDERS**
Very long-chain acyl-CoA dehydrogenase deficiency (VLCAD)	Congenital hypothyroidism
Long-chain 3-hydroxy acyl-CoA dehydrogenase deficiency (LCHAD)	Biotinidase deficiency
Trifunctional protein deficiency	Congenital adrenal hyperplasia
Carnitine uptake defect	Galactosemia
	Hearing deficiency
	Cystic fibrosis

cblA, cobalamin A defect; cblB, cobalamin B defect; CoA, coenzyme A.
Available at http://genes-r-us.uthscsa.edu/sites/genes-r-us/files/nbsdisorders.pdf

TABLE 18.3 Secondary disorders recommended for neonatal screening by the National Newborn Genetics Resource Centre

ORGANIC ACID METABOLISM DISORDERS	AMINO ACID METABOLISM DISORDERS
Methylmalonic acidemia, Cbl C and Cbl D forms	Hyperphenylalaninemia, benign (not PKU)
2-Methyl 3-hydroxybutyric aciduria	Tyrosinemia type II
Isobutyryl-CoA dehydrogenase deficiency	Tyrosinemia type III
2-Methylbutyryl-CoA dehydrogenase deficiency	Defects of biopterin cofactor biosynthesis
3-Methylglutaconic aciduria	Defects of biopterin cofactor regeneration
Malonic academia	Argininemia
FATTY ACID OXIDATION DISORDERS	Hypermethioninemia
Medium-/short-chain 3-OH acyl-CoA dehydrogenase deficiency	Citrullinemia type II
Short-chain acyl-CoA dehydrogenase deficiency (SCAD)	**HEMOGLOBINOPATHICS**
Medium-chain ketoacyl-CoA thiolase deficiency	Hemoglobin variants (including hemoglobin E)
Glutaric acidemia type 2	**OTHERS**
Carnitine palmitoyltransferase I deficiency	Galactose epimerase deficiency
Carnitine palmitoyltransferase II deficiency	Galactokinase deficiency
Carnitine acylcarnitine translocase deficiency	
Dienoyl-CoA reductase deficiency	

list of IEMs which are selectively screened in population with higher risk. A number of countries have introduced extended newborn screening for inborn errors of metabolism. In contrast to the "classical" OA which typically result in significant CNS damage, Isovaleric academia (IVA) has mild neurologic involvement which is entirely preventable by treatment. This encourages pre-symptomatic diagnosis and reinforces IVA to be qualified for Neonatal blood screening (NBS). A recent appraisal of the neonatal blood screening in the European Union stressed that NBS is a comprehensive program; it should not be focused or even reduced to the part of the screening laboratory. Screening includes confirmation of the positive screening result, decision for or against treatment, choice of appropriate treatment options (e.g., diet, drugs, behavioral measures or mere observation) as well as long-term follow-up of patients and evaluation of outcome. Tandem mass spectrometry screening was estimated to identify 83 affected newborns and cost effectiveness calculation found that NBS would save 1.5 million annually. A large number of IEMs can be identified by this method which requires a few drops of blood on a filter paper, mailed to a specified laboratory. Severe forms of some of these diseases may cause clinical manifestations before the results of the NBS become available. It should also be noted that these methods may identify mild forms of inherited metabolic conditions, some of which may never cause clinical disease in the lifetime of the individual. An example of this is 3-methylcrotonyl CoA carboxylase deficiency, which has been identified in an unexpectedly high frequency in screening programs using tandem mass spectrometry. The majority of these children have remained asymptomatic.

IV. CLINICAL SCREENING

Physicians and other health care providers, who care for children, should familiarize themselves with early manifestations of genetic metabolic disorders, because of the following two reasons:

1. Severe forms of some of these conditions may cause symptoms before the results of neonatal screening studies become available, and
2. The current screening tests identify only small number of IEMs. In the newborn period, the clinical findings are usually nonspecific and similar to those seen in infants with sepsis. In babies where clinical risk factors (Table 18.4) are present should be screened for IEM along with screening for more common disorders. A genetic disorder of metabolism should be considered in the differential diagnosis of a severely ill newborn

TABLE 18.4 Clinical Risk factors suggestive of the Inborn Errors of Metabolism

- History of Consanguinity and/or abortions & neonatal deaths in the family.
- History of sudden infant deaths and psychiatric illnesses
- Very few males in the family could be due to a x-linked disorder
- Recurrence of the clinical presentation in times of catabolic stress
- Recurrent vomiting, diarrhea, failure to thrive, short stature, dysmorphic features, edema,
- Developmental delay, hypotonia and seizures, cataract, unusual odours, rickets, and renal tubulopathy.
- Jaundice, hepatosplenomegaly and hepatic failure, hypoglycemia, lactic acidemia, hyperammonemia and coagulopathy
- Specific Food avoidance: sweet food avoided in Hereditary fructose intolerance and high protein diet in urea cycle defects.

infant, and special studies should be undertaken if the index of suspicion is high. A plan for investigation should be devised before delivery in families known to have a metabolic disease with or without a specific diagnosis, undiagnosed neonatal deaths or unexplained severe illness in childhood. Presence of consanguinity in family history would raise the risk of autosomal recessive disorders with no affected relatives. In X-linked IEM, relatives outside the immediate family may be relevant. Some of these disorders have a high incidence in specific population groups. Ethnicity of the patient helps in the diagnosis since some of the IEMs are more common among some ethnic groups, e.g., French Canadians of Quebec commonly have Tyrosinemia type 1.

Physical examination usually reveals nonspecific findings; most signs are related to the central nervous system. Hepatomegaly is a common finding in a variety of inborn errors of metabolism. Sometimes, body odors may help in the diagnosis, e.g., Maple syrup urine disease (MSUD) have maple syrup body as well as urinary odor. Whilst some babies may have a characteristic odour (e.g., sweaty feet for IVA & glutaric academia type II and boiled cabbage for tyrosinemia), these are uncommon and unfortunately seldom diagnostically useful.

V. MODE AND TIMING OF PRESENTATION

Signs and symptoms such as lethargy, poor feeding, convulsion and vomiting may develop as early as a few hours after birth. Common conditions explaining these clinical presentations should be concomitantly looked for. Lethargy, poor feeding, seizures, and coma may also be seen in infants with hypoglycemia or hypocalcemia. Measurements of blood concentrations of glucose and calcium and response to intravenous injection of glucose or calcium usually establish these diagnoses. Similarly, pyloric stenosis should be ruled out in babies with repeated vomiting.

Inborn errors may present at almost any time and in many different ways, but four particular circumstances are: (A) Antenatally, (B) at birth to young infancy (C) sudden neonatal death and (D) deterioration after a period of wellness.

A. Antenatally

IEMs as mentioned before may interfere with the fetal development, and hence, could be associated with birth defects. The presentation is in three clinical categories where either there is a true malformation (e.g., skeletal malformation or congenital heart defect), dysplasia (polycystic kidneys, liver cysts, cortical cyst) or functional manifestations (hydrops fetalis, IUGR, hepatosplenomegaly and microcephaly). Only the true malformations are irreversible, whereas the dysplasia and functional manifestations are reversible to some extent. Under *Group 1 disorders* of amino and organic acid catabolism causing

intoxication do not cause antenatal manifestations, but untreated maternal phenyleketonuria (PKU) can cause fetal dysplasia. Patients with glutaric acidemia type II can have dysmorphic phenotype (hypertelorism, high forehead, abdominal wall defects with nephromegaly, hypospadias and rocker bottom feet) explained by a energy deficient mechanism similiar to maternal diabetes mellitus. Several of the other OA, such as mevalonic aciduria and 3-OH-isobutyric aciduria have been associated with multiple dysmorphic features. *Group 2 disorders*, especially the respiratory chain defects, cause malformations. PDH deficiency causes dysmorphism resembling that found with fetal alcohol syndrome. The possible explaination is the maternal acetaldehyde may inhibit the fetal PDH.

Group 3 disorders, e.g., lysosomal or perixosomal only cause dysplasia and functional manifestations. Dysmorphism is a phenomenon illustrated clearly by the associated multiple defects in some IEM, e.g., peroxisomal enzymes, including those involved in fatty acid oxidation and plasmalogen synthesis. Among these is Zellweger syndrome, which is associated with hypotonia, epicanthal folds, brushfield spots, large fontanelle, simian creases and renal cysts. The Smith–Lemli–Opitz syndrome is an autosomal recessive disorder where cholesterol biosynthesis defect leads to the dysmorphism of many organ systems.

Isolated malformations may be even more commonly associated with inherited metabolic disorders than are specific malformation patterns. Nonketotic hyperglycemia and PDH deficiency is associated with agenesis of corpus callosum. *Abnormal eye findings* are associated with many IEMs either at presentation or may develop later. Cataracts are commonly seen in disorders of carbohydrate metabolism like galactosemia and heriditary fructose intolerance (HFI). Dislocated lens in homocystinuria and molybdenum co-factor deficiency can be an important diagnostic clue. Careful ophthalmologic assessment can reveal retinal degenerative changes seen in peroxisomal disorders or congenital glaucoma or corneal clouding. Women carrying a fetus with a fatty acid oxidation disorder (FAOD), particularly long-chain 3-hydroxyacyl-CoA dehydrogenase (LCHAD) deficiency, are at risk of HELLP syndrome (hemolysis, elevated liver enzymes and low platelet count), acute fatty liver of pregnancy and hyperemesis gravidarum. Nevertheless, majority of women with these complications have a normal fetus.

B. At Birth to Young Infancy

Most babies with IEM are born at term and initially appear to be well but subsequently deteriorate. The cause of this deterioration could be due to the abnormal metabolites either crossing placenta or catabolism at delivery or oral feeding. The rate of deterioration varies. The initial working diagnosis should be common conditions like sepsis and intracranial hemorrhage (especially in the preterm), duct-dependent heart disease and drug withdrawal. Neonates with IEM may present with nonspecific symptoms as poor suck, lethargic, vomiting, diarrhea, respiratory distress, hypotonia and seizures. Various IEM may present more often with ascites, dysmorphism, seizures or severe hypotonia (Table 18.5). Group 1 disorders causing intoxication usually are born normal and deteriorate after a gap of hours or weeks depending upon the severity of the metabolic error. These babies usually present with lethargy and coma. On the other hand, the group 2 babies with energy defects have a variable and milder presentation. Some of the energy defects may have no symptom free period at birth. Hyperlactatemia with or without metabolic acidosis, cardiac and hepatic involvement are common among energy defects. Patients presenting with neurological abnormalities at this age, particularly in patients with congenital lactic acidosis and early-onset fits, are often misdiagnosed as perinatal asphyxia. Among the group 3 disorders. Most lysosomal diseases appear normal but some of them are dysmorphic or have hydrops at birth. Peroxisomal defects are born with severe dysmorphism and neurological dysfunction. Majority of the babies with IEMs rapidly develop encephalopathy and the challenge is to identify the IEM before the baby becomes too sick or develops neurological sequelae. For this reason, some additional investigations are recommended concurrently with the septic screen, particularly in term babies. These are listed in Table 18.6.

C. Sudden Neonatal Death

Sudden death in a neonate can be due to cardiac arrhythmias in FAOD. Blood spots collected for acylcarnitine profile and skin fibroblast culture can help in knowing the cause of death.

D. Deterioration After Initial Period of Health

Patterns of deterioration for most babies with inborn errors who deteriorate after an initial period of health fall into one or more of the following five categories.

   (i) Neurological deterioration
  (ii) Cardiac disorders
 (iii) Acute parenchymal liver disease
 (iv) Hypoglycaemia
  (v) Acid/base disorders

TABLE 18.6   Investigations for a sick Neonate/Infant

Blood	*Urine*
• pH and blood gas	sugars
• Blood glucose	ketones
• Electrolytes and anion gap	
• Liver function tests	
• Ammonia	
• Guthrie card for amino acids and acylcarnitines	

It is also important to note that other problems may co-exist with inborn errors. Thus, galactosemia predisposes to Gram-negative sepsis, and cerebral or pulmonary haemorrhage may be the terminal event in urea cycle disorders.

Neurological deterioration

This is probably the most common presentation. Ultimately, most patients with IEM become encephalopathic, but to avoid sequelae early assessment and laboratory screening is needed. The rate of deterioration varies considerably while some patients have a brief phase of wellness and others are well for several days before gradual deterioration. Two-thirds of patients with non-ketotic hyperglycinemia develop symptoms within 48 h of birth. Urea cycle disorders and branched-chain OA typically present between 12 and 72 h of age, whereas MSUD usually presents later in the first week. Early signs of encephalopathy are non-specific, such as poor feeding, lethargy, vomiting, abnormal muscle tone or irritability, once again emphasizing the need for vigilance and for strategies that will detect biochemical markers at an early stage to avoid diagnostic delays. Later problems may include drowsiness, fits, hiccups, myoclonus, apneic episodes, marked hypotonia, irritability with cycling movements and coma. The disorders which would present with predominantly encephalopathy or seizures are given in Table 18.7. Among group 1 intoxication disorders, MSUD will present with generalised hypertonia and opisthotonous, whereas organic acidurias have axial hypotonia with limb hypertonia. Energy defects of group 2 have generalised hypotonia with rapid progression of neurological involvement. Lysosomal disorders never present with neurological involvement in neonatal age group but most peroxisomal defects and some congenital disorder of glycosylation present with severe neurogical dysfunction.

TABLE 18.5   Clinical Presentations in IEMs presenting in utero or at birth

Ascites/ Hydrops Fetalis	*Dysmorphism*	*Seizure/Apnoea*	*Hypotonia*
Neonatal hemochromatosis	Peroxisomal biogenesis disorders	Peroxisomal biogenesis disorders	Peroxisomal biogenesis disorders
Lysosomal disorders	Lysosomal disorders	Non ketotic hyperglycemia	Non ketotic hyperglycemia
Erythrocyte enzymopathies	Congenital defect of N-glycosylation	Congenital lactic academia	Congenital lactic acidemia
Congenital defect of N-glycosylation	Disorders of cholesterol synthesis	Pyridoxine dependency	Congenital defect of N-glycosylation
Glycogen storage disease type IV	Dystroglycanopathies	Molybdenum co-factor deficiency	Dystroglycanopathies
Pearson syndrome andother respiratory chain disorders		Folinic acid responsive seizures	

TABLE 18.7 Treatable causes of Neurological deterioration presenting later after birth

Predominant Encephalopathy	Predominant Seizures
• Maple Syrup Urine Disease	• Pyridoxine responsive
• Methylmalonic acidemia	• Hyperinsulinemic hypoglycemia
• Propionic Acidemia	
• Isovaleric Acidemia	• 3 Phosphoglycerate dehydrogenase
• Molybdenum co-factor	
• Deficiency	• Biotin responsive
• Urea Cycle Defects	• GLUT1 deficiency (brain transporter deficiency)

Cardiac disease

Cardiomyopathy is associated with tyrosinemia type 1. Cardiomyopathy and cardiac arrhythmias may be the presentation at birth of some patients with long-chain fatty acid oxidation disorders or respiratory chain defects. The later-onset forms of these disorders tend to have a primary cardiac or skeletal muscle presentation. Neonates with congenital defects of glycosylation sometimes have pericardial effusions or cardiomyopathy. Propionic academia (PA) can be complicated with cardiomyopathy and long QT syndrome. Cardiomegaly is also a feature of congenital hyperinsulinism. Pompe disease is occasionally present during the neonatal period, but most patients develop symptoms later. It has been reported that cardiac involvement is the only manifestation of Fabry disease in some patients. Myocardial abnormalities are characterized mainly by left ventricular (LV) wall thickening without significant cavity dilatation.

Parenchymal liver disease

Inborn errors of metabolism, where hepatomegaly and/ or abnormal liver function form part of the clinical disease, are known as "Metabolic liver disease (MLD)". Dietary exclusions, antioxidants and chelation therapy with immunoglobulins can offer good outcome in some of the treatable MLD. Liver transplantation for MLD shows markedly improved outcome. The presentations of an MLD in an infant or child can be varied (Figure 18.1). Among the five presentations, the first two would present as a sick child. Protocol based age appropriate

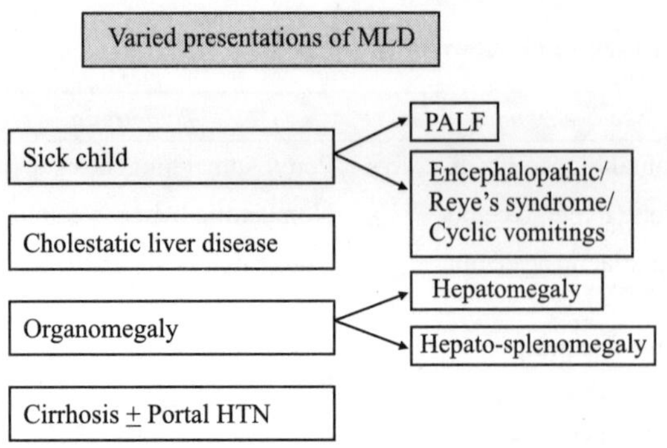

Figure 18.1 Various Presentations of Metabolic Liver Disease (MLD).

approach to the various presentations of MLD developed by our center is depicted in the flow charts [Figure 18.2(a-e)]. This approach makes the management easier to follow and more cost effective.

Common MLDs causing either PALF (including acute decompensation of chronic liver disease/cirrhosis) or encephalopathy with liver dysfunction in various age groups are listed in Table 18.8 and approach to these patients is seen in Figure 18.2 (a, b and e). These patients present with jaundice, coagulopathy and hepatomegaly (sometimes with splenomegaly and ascites). The age of presentation helps in pinpointing the etiology. A newborn presenting with liver failure within 7 days of birth could be Neonatal Hemochromatosis (NH), especially if there is a history of ascites (or hydrops fetalis). Hereditary Fructose Intolerance (HFI) has an association with the age of introduction of sucrose or fructose to the diet. Mitochondrial respiratory chain disorders may present at any age. Galactosemia typically presents in the first or second week of life, and tyrosinaemia may also be present from this age onwards. If HFI or galactosemia is suspected, urine screening tests for mucopolysaccharides and oligosaccharides should be performed. False-positive mucopolysaccharide test results are commonly observed in neonates. Hydrops fetalis can also be seen in Wolman's disease and congenital disorder of glycosyllation.

Alpha-1-antitrypsin deficiency, Niemann-Pick disease type C (NPC) and bile acid synthesis defects usually present after 3 week of age, predominantly with cholestasis. The first signs of cholestasis usually appear before 3 months of age and tend to appear earlier in Progressive familial intrahepatic cholestasis type 2 (PFIC2) than in PFIC1. No confirmed differences in jaundice evolution pattern are seen between PFIC1 and PFIC2. The same difference is also not seen in the development of pruritus, hepatomegaly, or splenomegaly. More than half of the PFIC1 patients present with extrahepatic symptoms during disease evolution, such as diarrhea, pancreatitis and deafness. PFIC2 patients generally have signs of fat soluble vitamin deficiencies and cholelithiasis in 30% of cases. Many PFIC2 children progressed early to severe outcome of liver disease, such as LF and/or HCC. A MDR3 (PFIC3) defect is also likely to be involved in some cases of transient neonatal cholestasis and adult idiopathic cirrhosis. Citrin deficiency type 2 or Citrullinemia can again be present as transient neonatal cholestasis and later as end stage liver disease. Bile acid synthesis defects usually present after 3 week of age. Diagnosis is easier with the protocol based approach [Figure 18.2(c)]. Children presenting with only hepatomegaly could be glycogen storage disorder (GSD) and cholesterol storage disorder [Figure 18.2(d)]. Presence of dolls like face, renal rickets and growth failure would be more prominent in GSD type 1 then type 3. Majority present in infancy but GSD type 1 could even present in neonatal period (Table 18.9). If splenomegaly is prominently present in an infant with liver dysfunction, then lipid storage (NPC, Gaucher disease) or lysosomal disorders (Wolman disease) are more likely. History of prolonged neonatal jaundice and subsequent progressive

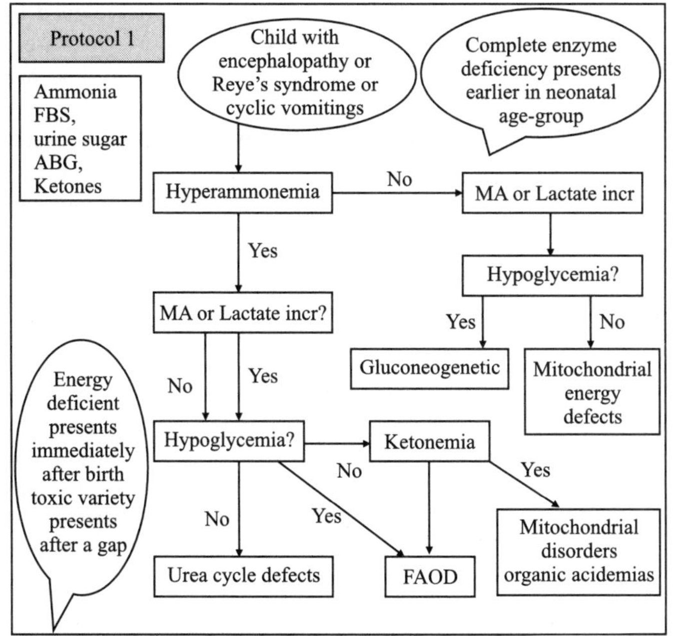

Figure 18.2(a) Approach to a Child with Encephalopathy.

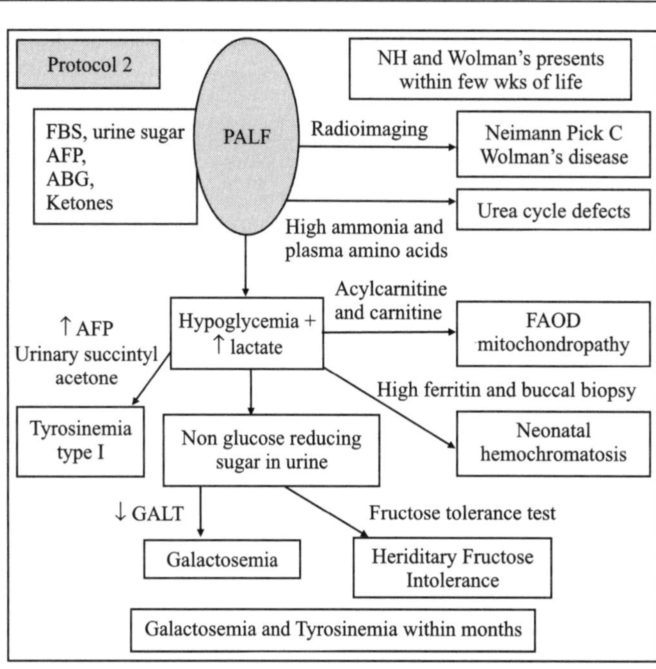

Figure 18.2(b) Approach to Pediatric Acute Liver failure.

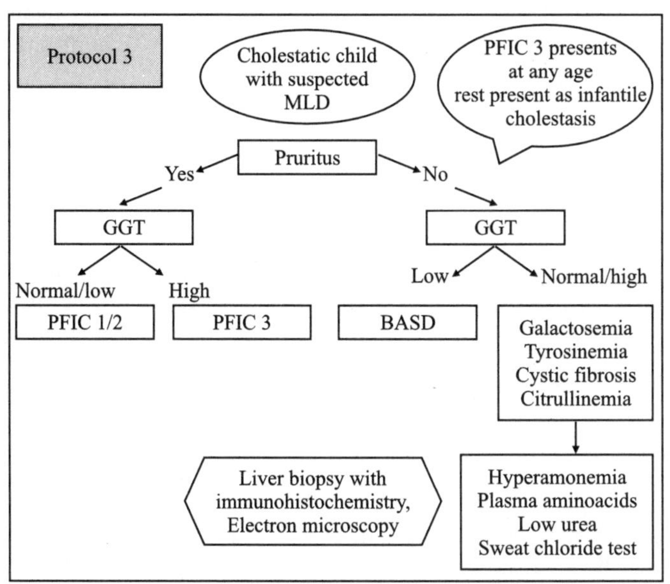

Figure 18.2(c) Approach to Cholestatic infant or child.

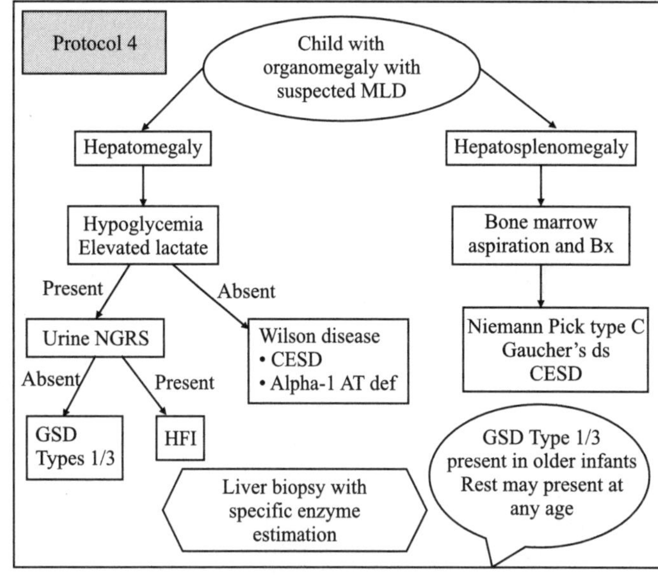

Figure 18.2(d) Approach to a child with Organomegaly.

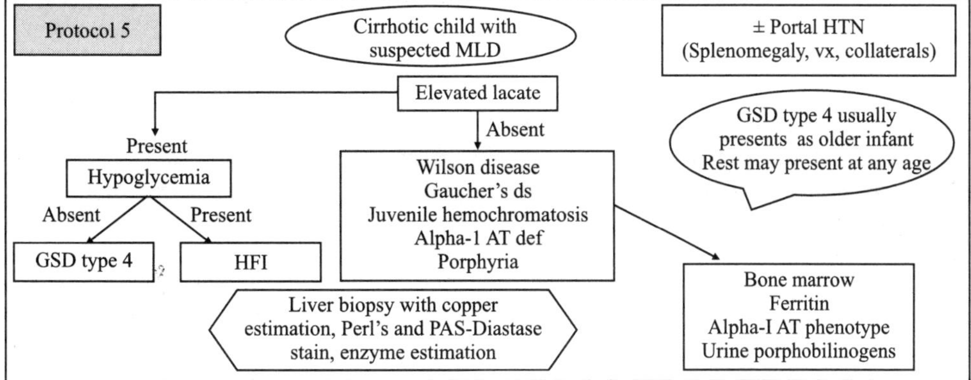

Figure 18.2(e) Approach to a child with Cirrhosis.

TABLE 18.8   Common Metabolic liver diseases in sick neonates, infants and children

	Pediatric acute liver failure and acute decompensation of chronic liver disease	*Encephalopathy with liver dysfunction*
Neonates	Neonatal hemochromatosis, Galactosemia, Tyrosinemia type 1, Mitochondrial cytopathy and Wolman's disease	Organic acedemias, FAOD, gluconeogenetic disorders, mitochondrial cytopathies and UCD
Infants	Tyrosinemia type I, FAOD, Mitochondrial cytopathy, Galactosemia, Hereditary fructose intolerance, UCD and Congenital disorders of glycosylation	Organic acedemias, FAOD, gluconeogenetic disorders, UCD, Congenital disorders of glycosyllation and mitochondrial cytopathies
Children	Wilson's disease, FAOD, mitochondrial cytopathy, hereditary fructose intolerance, UCD and Congenital disorders of glycosylation	UCD, FAOD, Congenital disorders of glycosyllation, gluconeogenetic disorders and mitochondrial cytopathies

TABLE 18.9   Characteristics of Hepatic Glycogen Storage Disorders (GSD)

	GSD 1 a + 1 b	GSD III	GSD IV	GSD VI	GSD IX
Deficient Enzyme	Glucose 6 Phosphatase (G6Pase) & translocase	Liver & muscle debranching enzyme (GDE)	Glycogen Branching enzyme (GBE)	Hepatic glycogen phosphorylase	Phosphorylase Kinase (PHK)
Inheritance	AR/17q21	AR/1p21	AR/ 3p14	AR/14q21-q22 ,	X /PHKA2
Prevalence	1 in 100000	24% of GSD	0.3% of GSD	rare	rare
Hypoglycemia	+++	++	+/–	+/–	++
Doll like facies	++	+	+	+	+
Hepatomegaly	+++	+++	++	++	++
Renal involvement	++	rare	–	–	–
Rickets	+	–	–	–	–
Elevated lactate	++	+/– Fed	–	(hyperketosis)	+
Hyperlipidemia	++	++	–	+	+
Hyperuricemia	++	–	–	–	–
Fibrosis (liver biopsy)	+	+	+++	+	–
Confirmatory Test	G6Pase assay in liver tissue. G6PC gene sequencing	Mutations in exon 3 of AGL gene in c. 1p21.	Histology and GBE deficiency in liver biopsy	Mutation analysis	Mutation of PHK gene
Treatment	Corn Starch diet lipid lowering therapy Growth hormone	Corn starch with high protein diet	Same as GSD III & Liver transplantation	Corn starch diet	Corn starch diet

neurological manifestations like vertical gaze palsy, delayed speech, ataxia, uncontrolled seizures and gelastic cataplexy are suggestive of NPC. Cherry red spot on fundal examination would be suggestive of NPC and Gauchers disease. MLD which can present as cirrhosis and portal hypertension are mentioned in the flow chart [Figure 18.2(e)] depicting the approach to a child with this presentation.

Hypoglycemia

Serum glucose lower than: 40 mg/dL in term and preterm infants, 60 mg/dL in children and 76 mg/dL over age 16 years can be diagnosed as hypoglycaemia. There are many causes of hypoglycemia in neonates, and the majority of neonates with low blood glucose concentrations do not have a primary metabolic or endocrine disorder. Nevertheless, well-grown term babies with severe, prolonged or otherwise unexplained hypoglycemia should be investigated for an underlying metabolic or endocrine. The risk of patients with hypoglycemia suffering from IEM was significantly higher (10% versus 1.8%) than that of the non-hypoglycemia patients in NICU, based on cut-off value of 2.8 mmol/L (50 mg %). Hypoglycemia associated with metabolic acidosis suggests an organic acidemia or defect of gluconeogenesis (GSD 1 or fructose 1, 6-bisphosphatase deficiency). The possibility of a defect of fatty acid oxidation should be considered in patients with hypoglycemia and cholestatic jaundice in the absence of ketonuria. Ketonuria in the form of a positive (not trace) urine dipstick for ketones or which detects acetoacetic acid and acetone suggests organic academia. Definitive test for ketosis would be urine organic acid profile containing excess β-hydroxybutyrate and acetoacetic acid as defined by the norms of the laboratory performing the test. Hypoglycemia is a key feature of gluconeogenesis defects accompanied by hepatomegaly and marked lactic acidosis; latter rapidly resolving with dextrose administration is important diagnostic clue. In gluconeogenesis disorders, both hypoglycemia and lactic acidosis may be intermittently present in the fasting state, e.g., GSD and Fructose 1,6 biphosphatase deficiency. Persistent and severe hypoglycemia should indicate the differential diagnosis of galactosemia, tyrosinemia, HFI, FAOD, OA, and CDG. Hyperammonemia is prominent in FAOD and OA, whereas galactosemia, HFI and CDG would more commonly present with liver failure. HFI should be suspected if there is positive non-glucose reducing substance in urine. Oral fructose tolerance test reveals hypoglycemia with increasing serum uric acid and magnesium.

Acid/base disorders

Acid/base disorders are common in sick neonates; most commonly they have a mixed respiratory and metabolic acidosis. In babies with persistent metabolic acidosis and ketosis but normal tissue perfusion, may suggest an OA or a congenital lactic acidosis. It is important to note that ketosis, if present in neonates, is always abnormal and suggests OA. Mild respiratory alkalosis in the absence of ventilator issues is suggestive of hyperammonemia which is often associated with irritability. Although, generally in OA, the acidosis is severe with high (>20) anion gap, but these acid/base changes are not specific. Patients with OA can be alkalotic, and those with UCD may be acidotic. The latter is particularly likely if the diagnosis has been delayed, since hyperammonemia leads to vasomotor instability and collapse, emphasizing the value of assessing the acid/base status at an early stage of the illness. Hyperammonemia and /or metabolic acidosis should be approached as suggested in the flow chart (Figure 18.3). Similarly, a neonate with hyperammonemia with encephalopathy without liver dysfunction should be approached as per the protocol developed for those with encephalopathy and liver dysfunction [Figure 18.2(a)]. Both these flow charts also decide the approach to neonates with predominant hypoglycaemia. If a neonate presents with congenital lactatemia then further work up should be as per the algorithmic approach depicted in Figure 18.4.

Hyperammonemia

The interpretation of plasma ammonia concentration can give rise to difficulties. Plasma ammonium concentration exceeding

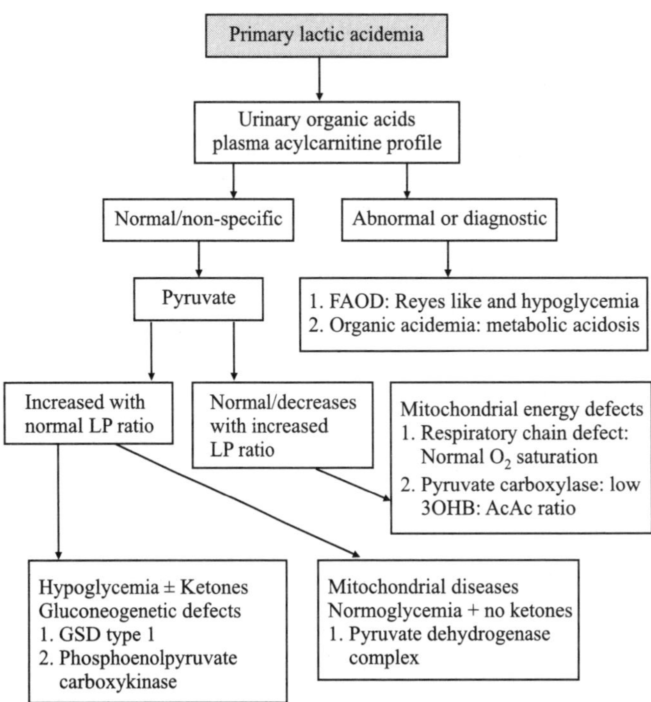

Figure 18.4 Approach to a patient with Congenital Lactic Acidemia.

the reference range for the laboratory performing the test and the age of the affected individual, usually greater than 150 µg/dL in neonates, 70 µg/dL in infants from 1–12 months and 35–50 µg/dL in older children and adults. Values > 200 µg/dL suggest a metabolic disorder and should be investigated urgently. Severe hyperammonaemia (>500 µg/dL) is a very serious complication of many inborn errors, and the outcome is poor, although it may be improving with aggressive treatment. Nevertheless, before proceeding, it is essential to inform the parents about the risk of neurological impairment, the likelihood of further hyperammonemic episodes and the need for complex treatment throughout life.

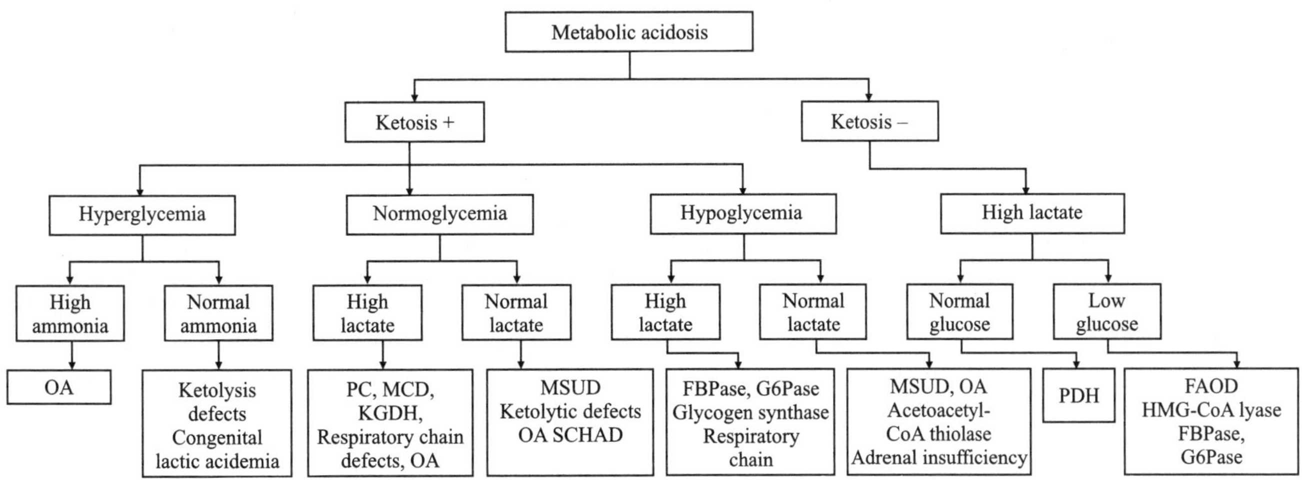

Figure 18.3 Algorithmic approach to a child with Metabolic acidosis.

E.   Late Infancy to Adolescence and Adulthood

In almost half the IEMs the onset of a genetic metabolic condition may occur months or even years after birth. Mutations cause partially nonfunctional genes in these children. Since the oldest diagnosed patients of IEM are in the adulthood, hence the new field of adult metabolic medicine also raises many new therapeutic problems, including the management of pregnancy in affected mothers. Physicians should be aware that *psychomotor retardation/regression, neurological deficits, seizures, myopathy, recurrent vomitings and cardiomyopathy* whether, in a child or an adult, should raise the suspicion of an IEM. These may be episodic or intermittent presentations with acute decompensation. These episodes can be triggered by stress, infection, exercise or excessive intake of protein. The patient may die during one of these acute attacks. The various presentations of IEM are the following categories:

Coma, Strokes, Reyes syndrome, ataxia and psychiatric symptoms

Comatosed state may be associated with predominant *metabolic acidosis*, predominant *hyperammonemia*, predominant *hypoglycemia* and combination of the three states. The approach to such patients should be based on the algorithmic approach mentioned in the neonatal section [Figures 18.1, 18.2(a) and 18.4]. OA with hyperglycemia and ketosis may mimic diabetic coma. Some patients with OA and UCD could be associated with focal neurological signs suggestive of space occupying lesion. Homocystinuria may be associated with cerebrovascular accidents. OA can be associated with extrapyramidal involvement and cerebral hemorrhage. Reyes like syndrome or cyclical vomiting can be presentation of fatty acid oxidation defect and urea cycle. Comatosed child with lactatemia is present in respiratory chain defects. Recurrent ataxia is seen in OA, late onset MSUD, UCD [Ornithine Trans Carbamyalase deficiency (OTC) and Arginosucinate synthetase defiency (ASS)] and can accompany peripheral neuropathy in PDH. UCD, acute intermittent porphyria and homocysteinuria can present as unexplained acute psychosis.

Reyes Like syndrome, dehydration and exercise intolerance

Mitochondrial defects, FAOD, UCD, and OA can cause Reyes like symptoms intermittently. Recurrent dehydration in absence of diarrhea in an older child could be due to OA, cystic fibrosis and FAOD [Figure 18.2(a)]. Exercise intolerance is seen in gluconeogenetic disorders, especially in adolescence and adulthood. Myoglubinuria after intense exercise is seen in FAOD.

Hepatic involvement

Wilson's disease is more common in the 5–15 years age group and 17% of which may present as fulminant hepatic failure. GSD classically presents with dolls facies, hepatomegaly, hypoglycemia, growth failure, lactic acidemia and renal tubulopathy in the infancy or toddlers age group with variations

in the different types (Table 18.9). Some of the GSD may have muscular involvement. Liver symptoms in GSD type III may disappear with puberty. If splenomegaly is prominently present in an infant with liver failure, then lipid storage or lysosomal disorders are more likely. Cholesterol ester storage disorder may present with only hepatomegaly. Most of the well-known lipid storage diseases typically present in the late infancy to childhood. Foamy cells on bone-marrow examination could be an important clue. More definitive diagnosis of lysosomal storage disorders is made by appropriate biochemical studies on leukocytes or cultured skin fibroblasts. Whereas a cholestatic child or adult without pruritis could be bile acid synthetic defect. Majority of cystic fibrosis presents with biliary cirrhosis in second decade of life while just over 1% may present as neonatal cholestasis, where repeated pneumonia can guide to the diagnosis. X-linked UCD, e.g., OTC can present in females as acute liver failure if there is partial deficiency of the enzyme. Among the FAOD, Carnitine palmitoyltransferase I deficiency and MCAD may present with Reye's like syndrome while LCHAD and SCHAD can present as ALF. Respiratory chain defects usually do not present with hepatic involvement in children above 2 years of age. HFI and UCD may show food intolerance (sugar and protein aversion, respectively). GSD type 1 and 3 may have this classic picture associated with or without muscle involvement. GSD type 4 may present with cirrhosis and portal hypertension (Table 18.9). Wilson's disease and HFI can present with cirrhosis later on in life. Approach to these cases should be as per Figure 18.2(b-e).

Cardiac disease

Cardiomyopathy is associated with Pompe disease, tyrosinemia type 1, mucopolysaccharidosis and PA. Cardiomyopathy and cardiac arrhythmias may be the presentation at birth of some patients with LCHAD, respiratory chain defects, hypoparathyroidism, thiamine deficiency dependent state and CDG. Prolonged QTc interval (>440 ms) in 70% of the patients beyond infancy in PA.

F.   Hyperbilirubinemia

Most of the disorders of bilirubin metabolism are autosomal recessive except for Gilbert's syndrome which is autosomal dominant. Persistent elevation of indirect bilirubin beyond the limits of physiologic jaundice, without evidence of hemolysis, suggests the diagnosis of Criggler–Najjar (CN) syndrome. The hyperbilirubinemia in this disorder is related to a partial (CNII) or complete (CNI) deficiency of glucuronyl transferase (enzyme needed for the conjugation of bilirubin to bilirubin diglucuronide). In both CNI and CNII, liver tissue show low bilirubin UDP-glucuronosyltransferase (UDPGA) activity. Hence, patients with these disorders experience a profound block in bilirubin excretion because they lack the ability to conjugate bilirubin. The major differentiating characteristic between the two types is the response in CNII to phenobarbitone or diphenylhydantoin (drugs that stimulate hyperplasia of the ER) with a significant decline in the serum bilirubin level. In patients with this disorder, the standard modalities of phototherapy and exchange transfusion may prevent the development of kernicterus in the neonatal period. Gilbert's syndrome (GS) (OMIM #143500) is chronic or recurrent but mild unconjugated hyperbilirubinemia with normal other liver function tests. Serum unconjugated bilirubin elevation is variable and usually ranges from 1–4 mg/ dL. Frequently, patients are first identified when an elevated serum bilirubin is found on screening blood chemistry or mild jaundice (perhaps only scleral icterus) is noted during a period of fasting or nonspecific viral illness. In Caucasian populations, the homozygous finding of an additional TA repeat in the promoter region, or so-called TATA box [i.e., (TA) 7TAA rather than (TA) 6TAA], of the *UGT1A1* gene has been shown to be a necessary, though not sufficient condition for GS. Asians show mutations involving the exon 1 of the UGT1A1 gene and the commonest mutation is Gly71Arg. Rotor syndrome is marked by mild mixed (conjugated and unconjugated) hyperbilirubinemia which is noticed just after birth or in childhood. Jaundice may be intermittent. Conjunctival icterus may be the only clinical manifestation. Dubin–Johnson syndrome (OMIM #237500) also has elevation of both the conjugated (>50%) and unconjugated serum bilirubin with total level of 1.5–6 mg/dL. Hepatomegaly is seldom seen, liver function tests are normal with no hemolysis. Brown to black liver discoloration is characteristic of DJS. This syndrome is far more common than RS, and jaundice may be worsened by pregnancy and oral contraceptives.

VI. CONFIRMATORY TESTS

Confirmatory tests for common treatable IEM is available in Table 18.10. A more detailed definitive laboratory work up for hepatic GSD (Table 18.9), UCD (Figure 18.5), OA (Table 18.11) and FAOD (Table 18.12) is available in the respective tables. Anemia, thrombocytopenia and neutropenia is present in 82%, 35% and 29% respectively in PA. Absolute neutrophil count (ANC) less than 1500/mm^3 is suggestive of PA. GSD type 1b is also associated with neutropenia and colitis like picture. GSD V can be confirmed with negative staining for myophosphorylase in muscle biosy and identification of common mutation (p.R50X). Pompe disease is the only lysosomal disorder among the glycogenoses. This can be confirmed by the demonstration of lysosomal acid α-glucosidase deficiency in skin fibroblasts. While investigating early onset seizures, plasma &/CSF amino acids, plasma VLCFA, plasma uric acid and blood/CSF lactate are the required tests. For pyridoxine deficiency, a genetic basis has been identified now.

Prenatal Testing

Prenatal diagnosis is possible for galactosemia and some other carbohydrate metabolic defects with measuring enzyme (e.g., GALT in case of galactosemia) activity on chorionic villi cells or by measurement of intermediate metabolite (e.g., galactitol). Known mutation can be detected on DNA from chorionic villi biopsy even at 11 weeks of gestation. Three approaches to prenatal diagnosis may be possible for OA, including measurement of analytes in amniotic fluid or use of cells obtained by amniocentesis to either assay enzyme activity or extract DNA for molecular genetic testing. As all genes for UCD have been mapped so prenatal diagnosis can be best done by mutation analysis in chorionic villi samples. Similarly, succinyle acetone levels in aminiotic fluid and detection of know mutations can help in the prenatal diagnosis of Tyrosinemia as early as 12 weeks of gestation. All enzymes of FAOD are all found in CVS so prenatal diagnosis is possible but mutational analysis is a preferred technique for the molecular defect. In case of mucoplysaccharidosis prenatal diagnosis is possible with chorionic villi samples or in amniotic fluid cells.

VII. MANAGEMENT

For some disorders, supportive care is all that can be offered. Fortunately, treatment is possible for many inborn errors, and the principles are outlined in Table 18.10. General neonatal

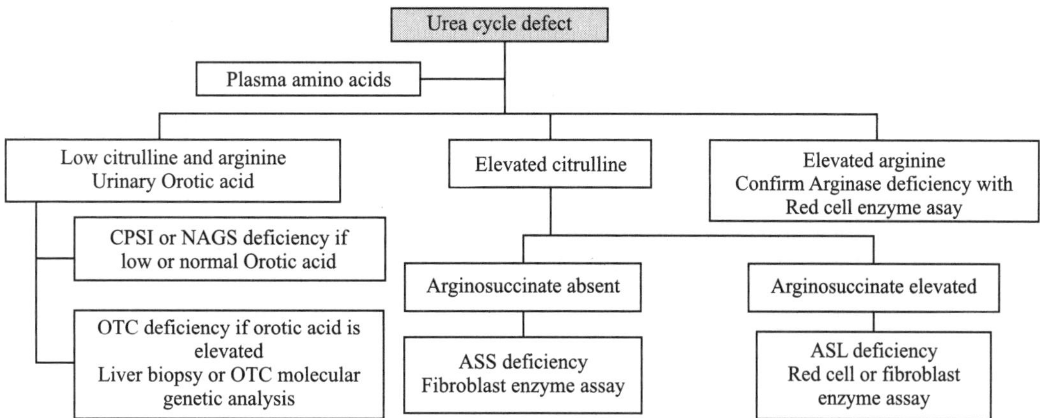

Figure 18.5   Laboratory work up of a suspected Urea cycle defect and the confirmatory tests.
CPSI (Carbamoylphosphate Synthetase I), NAGS (N-acetyl glutamate Synthetase), ASS (Arginosuccinic acid Synthetase), ASL (Arginosuccinic acid Lyase), OTC (Ornithine Transcarbamylase).

TABLE 18.10   Confirmatory tests and Definitive Treatment in the common treatable IEM.

Disorder	Confirmatory Test	Definitive Treatment
Galactosemia	GALT assay Transferase GALT gene mutations	Galactose free diet
Glycogen Storage disorders	Enzyme assay in liver/muscle/cultured fibroblasts	*See* Table 18.9
Hereditary Fructose Intolerance	Fructaldolase B assay in liver tissue	Fructose free diet
Fructose 1,6 Biphosphatase	Molecular analysis of DNA from leukocytes and liver FBP1 mutations in chromosome 9q22.2-q22.3	Fructose free diet Corn starch diet
Phenylalanine hydroxylase deficiency	Hyperphenylalaninemia >1200μmol/L R408W & R243Q mutations in chromosome 12	Phenylalanine restricted Diet and Vitamin B12 supplementation BH4 therapy Liver transplantation
Urea Cycle Defect	Enzyme assay in liver/ cultured fibroblast. *See* Figure 18.5	Ammonia scavengers, protein free diet with essential amino acids supplementation
Organic Acidemia	Enzyme in cultured skin fibroblasts. *See* Table 18.11	Specified formula and Carnitine supplementation Ammonia scavengers, N-carbamoylglutamate Extracorporeal detoxification
Tyrosinemia	Enzyme assay on liver tissue	Low tyrosine and phenylalanine diet Nitisone
Fatty Acid Oxidation Defect	See Table 18.12	Avoid prolonged fasting Breastfeeding in MCAD MCT rich diet in VLCAD & LCHAD Carnitine in carnitine transporter deficiency Bezafibrate in VLCAD
Respiratory chain disorder	Analysis of oxidative phosphorylation complexes I-IV from intact mitochondria isolated from fresh skeletal muscle	Normocaloric and low carbohydrate diet Avoidance of certain drugs Oral Quinone for CoQ10 deficiency Carnitine in carnitine deficiency
Porphyria	Increase urinary porphobilinogen and decreased erythrocyte porphobilinogen deaminase	Intravenous heme arginate or hydroxide and removal of precipitating factors
Wilson Disease	Copper estimation in Liver tissue	D Pencillamine and zinc
Neonatal hemochromatosis	High Serum Ferritin + MRI/oral mucosa biopsy s/o extra hepatic siderosis	Prevention : Maternal IVIG Cocktail therapy (desferroxamine + NAC+ selenium + Vitamin E)
Adult and Juvenile Hemochromatosis	Elevated Ferrirtin. HFE, HJV, TfR2 & SLC40A1 gene mutations	Phlebotomy & Iron Chelators
Lysosomal disorder Mucoplysaccharidosis	Specific enzyme analysis in isolated leukocyte/ cultured fibroblast/lymphoblast Urinary Glycosaminoglycan (GAG) screening tests	Enzyme replacement, substrate reduction therapy and allogenic bone marrow transplant in some cases.
Peroxismal Biogenetic Defect	Elevated VLCFAs, phytanic acid in blood and decreased Plasmogen in erythrocytes	Low phytanate diet and hematopoetic cell transplantation in some.
Congenital defects of Glycosylation (CDG)	Isoelectrofocussing and immunofixation of serum transferrin	Mannose in Mannose phosphate isomerase deficient CDG

TABLE 18.11   Biochemical and Genetic analysis of Common Organic Acidemias.

Disorder	Deficient Enzyme	Diagnostic Analytes by GC/MS and Quantitative Amino Acid Analysis	Gene Symbol
Maple syrup Urine disease (MSUD)	Branched-chain ketoacid dehydrogenase	Branched-chain ketoacids & hydroxyacids in urine Alloisoleucine in plasma	BCKDHA, BCKDHB, DBT
Propionic academia (PA)	Propionyl CoA carboxylase	PA, 3-OH PA, methyl citric acid, propionyl glycine in urine Propionyl carnitine, increased glycine in blood	PCCA, PCCB
Methyle malonemic acidemia (MMA)	Methylmalonyl CoA mutase	MMA in blood and urine PA, 3-OH PA, methyl citrate in urine Acyl carnitines & glycine in blood	MUT, MMAA, MMAB
Isovaleric acidemia (IVA)	Isovaleryl CoA dehydrogenase	3-OH IVA, isovaleryl glycine in urine	IVD

TABLE 18.12   Clinical, Biochemical and Genetic profile of common Fatty Oxidation Disorders

Defect	VLCHAD	MCAD	LCHAD	SCHAD	CPTI
Hypoglycemia and liver dysfunction	+	+	+	+	+
Cardiomyopathy	+	+			+
Rhabdomyolysis	+	+			+
Chronic weakness	+	+			+
Plasma Acylcarnitines	C16:1, C14:2, C14:2 C18:1	C10:1, C8, C6	C18:1 OH, C18-OH, C16:1-OH C16-OHC4-OH		Absent acylcarnitine and High free carnitine

intensive care is essential and acidosis, electrolyte disturbances, dehydration and hypothermia should be corrected, and sepsis monitoring is necessary. Sometimes it is impossible to control the metabolic derangement solely by these means. In some conditions this may be all that is needed (e.g., phototherapy and exchange transfusion in erythrocyte enzymopathies). In other conditions, treatment to minimize catabolism and remove toxic metabolites is required. The outcome of these IEMs is dependent on how soon these measures were started in the patient. Hence to suspect the diagnosis and to start the treatment even before a precise diagnosis is known would be prudent.

A.   Removal of Nutrient

Nutrient that is suspected to have precipitated the illness (e.g., galactose, fructose or protein) should be stopped and a high-energy intake given either orally or intravenously. In PA, restriction of protein transiently for 24–36 h and usage of propiogenic free parenteral amino acids are the therapeutic modalities available. Transition to enteral feedings should be done as soon as they are tolerated. High concentration glucose will require a central line in place. In some cases fat emulsions can be used to increase energy intake. Hyperglycemia due to high rates of glucose infusion may need to be controlled with insulin. Although insulin has an advantage of promoting anabolism, it should be used with caution in neonates. Some patients with metabolic disorders are insulin resistant.

B.   Detoxification

Removal Toxic Metabolites

To avoid neurological damage, duration of exposure to

toxic metabolites such as ammonia or leucine (in MSUD) should be minimized. Quick and effective removal of toxic metabolites often requires some form of dialysis; which will also correct acidosis and fluid overload. Peritoneal dialysis has not been found to be effective in high clearances. Hemodialysis only has higher clearances at high blood flow rates, which require the placement of adequately sized catheters.

Hemodialysis or hemofiltration should be instituted as soon as possible in patients with severe hyperammonemia or rapidly rising concentrations. Ammonia scavenger medications include intravenous sodium benzoate (250 mg/kg) and sodium phenylacetate (250 mg/kg) alone or in combination. These drugs are conjugated to glycine and glutamine, respectively, and the products are excreted, thereby creating an alternative pathway for nitrogen excretion. Evidence available suggests that the drugs are safe and effective in UCD and OA but ammonia scavengers should be used with caution in PA where lowering of glutamine levels may cause hyperammonemia. Oral N-carbamoylglutamate (carglumic acid; 100–250 mg/kg) has also been reported to aid in the detoxification of ammonia during neonatal and acute decompensations. Arginine can also be used for hyperammonemia.

Early-onset fits

Rapid control of fits in neonates has been found to be associated with better outcome. All patients should have a trial of pyridoxine (100 mg), which can be given intravenously or enterally, preferably with EEG monitoring and full resuscitation facilities in case of collapse after the first dose. If there is no response, pyridoxal phosphate (30 mg/kg) should be given enterally followed by folinic acid (3 mg/kg).

C. Megavitamin therapy

Many enzymes require vitamin as co-enzymes where pharmacological doses of precursor vitamins will increase the enzyme activity. Administration of vitamin B12 and biotin is probably justified because vitamin-responsive methylmalonic acidemia (MMA) and holocarboxylase synthetase can present in the pediatric emergency room. The justification for many other vitamins is doubtful, particularly since *vitamin responsive forms of IEMs are milder and do not present in the neonatal age group.*

D. Specific Therapy

After the initial stabilization, early involvement of a specialist center will result in better outcome. Organic acids are excreted in part bound to carnitine. Carnitine concentrations are usually low in OA, particularly during episodes of neonatal decompensation. Carnitine supplementation (100 mg/kg/day IV) may increase the detoxification of organic acids by excretion of carnitine esters in the urine. In IVA, glycine plays an important role in promoting metabolite excretion, and supplements should be given in combination with carnitine. Use of carnitine in FAOD is controversial, since long-chain acylcarnitines may have toxic effects, although this is disputed. Branched-chain free amino acid mixtures should be given as early as possible in MSUD. Tyrosinaemia type 1 sometimes presents with neonatal liver failure, and the use of Nitisinone (NTBC) may be lifesaving. In deficiency of pyruvate dehydrogenase complex (PDHC), phosphorylation of PDHC by pyruvate dehydrogenase kinase (PDK) inactivates the enzyme, whereas dephosphorylation restores PDHC activity. Experimentally, it was found that phenylbutyrate enhances PDHC enzymatic activity by inhibition of PDK.

Enzyme replacement therapy (ERT)

ERT has been approved for 6 lysosomal storage diseases (LSDs) worldwide (Table 18.10). These diseases include Gaucher disease, Fabry disease, mucopolysaccharidosis (MPS) types I, II, and VI, and Pompe disease. The efficacy and safety of ERT for LSDs has been confirmed by extensive clinical trials. ERT can cause adverse events like immune reactions against the infused enzyme and it may have a negative impact on the therapy. Mistargeting of the enzyme due to autophagic build up was seen in a mouse model of Pompe diease. Intractable tissues (brain and bone) is another challenge seen in LSDs.

E. Continuing Management

It is important to recognize that the emergency treatment mentioned above is nutritionally incomplete and, if used for a prolonged period, may exacerbate catabolism, leading to poor metabolic control and frank malnutrition. Balanced diet should be introduced as soon as possible. Breastfeeding of infants with OA is feasible with close monitoring of growth, development

and levels of amino acids, organic acids and ammonia. Failure to thrive in these patients may be related to inadequate protein and energy intake rather than pathology of disease. Inadequate protein intake can also result in decreased resting energy expenditure, cause amino acid deficiency, increased risk of infection, and developmental delay. The detection and management of metabolic decompensations at home are the most crucial part of the chronic management of OA. Home management should be tailored for each patient needing close monitoring with modification of the diet by the metabolic team.

F. Liver Transplantation

With liver transplantation (LT) success reaching >85 % long term survival. Some IEMs do not have liver as a clinical entity but since the enzyme is in the liver hence LT can be a cure for them, e.g., UCD. In some conditions like Tyrosinemia type 1 where the drug NTBC can control the disease but LT is a better option cause it may turn out to be less expensive as against life long medications. Multi-disciplinary approach of the patient with metabolic disease, liver disease and other relevant experts is most appropiate. Living related transplantation from a heterozygote parent is safe and effective. Auxiliary liver transplantation were earlier undertaken for UCD.

G. Future Therapies

Hepatocyte transplantation

Hepatocyte transplantation is moderately successful for liver-based metabolic disease and acute liver failure as a bridge to liver transplantation. However, there are limitations, e.g., unused donor livers (to isolate hepatocytes) are in short supply and often steatotic and of poor quality. Some of the recent advancements in technique are improved cryopreservation protocols, usage of cells labelled with iron oxide (magnetic resonance imaging contrast agents) to track the engraftment of cells and prevention of early cell loss using inhibitors such as low molecular weight dextran sulphate. Alternative sources of hepatocytes like mesenchymal stem cells are currently being investigated particularly for their hepatotropic effects.

Gene therapy

Gene therapy represents an attractive strategy to circumvent the consequences of inherited diseases. Several approaches can be envisaged, including gene addition, gene expression modification such as skipping of a mutated exon, inhibition of a cryptic splice site, or inhibition of the expression of a mutated gene associated with a gain of function. Finally, gene repair is also a potential option. Numerous vectors or methods have been tested with regard to the underlying targeted disease. These include retroviruses, lentiviruses, adeno-associated viruses (AAVs), adenoviruses, transposons and targeted integration. A number of reports in the recent literature provide evidence of efficacy in controlling at least some aspects of disease phenotype. Targeting of the relevant cell population can be an issue, for instance when considering the brain, although stereotaxic injection of vectors or transduced cells can be envisaged. It is difficult to control the level and duration of expression of the transgene product. Transduced cells can be lost spontaneously, or an immune response to viral proteins (AAV) or to the transgene product itself may occur. Expression of the transgene may also cease prematurely. Potential toxicity, as discussed above in the setting of SCID, can also be an issue. However, careful utilization of other nonintegrative vectors in appropriate conditions (i.e. targeting nondividing cells) such as AAV LTR inactivated lentiviruses should provide reasonably safe approaches. Adeno associated viral (AAV) gene delivery has been successful in the treatment of *Pcca* knock-out mice and may represent a future treatment option for PA and MMA patients. In a study, the therapeutic effects of hepatocyte-directed delivery of the Mut gene to mice with a severe form of MMA were examined. It was revealed that a single intrahepatic injection of recombinant adeno-associated virus serotype 8 expressing the Mut gene under the control of the liver-specific thyroxine-binding globulin promoter is sufficient to rescue Mut(-/-) mice from neonatal lethality and provide long-term phenotypic correction.

Chaperone Therapy

A potent α-galactosidase A inhibitor, 1-deoxygalactonojirimycin can act as a chaperone by binding to the active site of the enzyme and help in the appropiate folding of the mutant enzyme. This in turn stabilises the enzyme and transports it to the lysosomes. When orally administered to transgenic mice, it reduces the globotriaosylceramide which is the incriminating metabolite in Fabry disease.

Even with the most aggressive therapy, a number of these children will die. It is essential to appropriately establish a diagnosis. A conventional autopsy seldom provides many clues. Progress with the treatment of inborn errors has been slower than progress on their biochemical and molecular bases. Nevertheless, outcomes are improving with the use of dialysis and drugs to promote the removal of toxic metabolites, coupled to measures to minimize catabolism. Early intervention is crucial, however, if neurological sequelae are to be avoided. This requires constant vigilance and routine measurement of biochemical markers, such as ammonia, at an early stage in patients with suspicious presentations. The management plan outlined here is largely based on personal experience rather than published evidence. There are no clinical trials that document prospectively the presentation and outcome of inborn errors.

SUMMARY

The hereditary biochemical disorders are called inborn errors of metabolism (IEM) or even inherited metabolic disorders. To diagnose IEM certain rules should be followed to make it easy for early diagnosis from the exhaustive list. The IEM can be classified as per the metabolic pathway they involve or more simply they can be classified according the therapeutic approach needed to treat them. There are a proportion of them which are treatable, and delay in intervention can lead to permanent damage. Hence they have been included in the list for Neonatal blood screening. Physicians and other health care providers who care for children should familiarize themselves with early manifestations of genetic metabolic disorders. In the newborn period, the clinical findings are usually nonspecific, hence risk factors should be carefully looked for and high index of suspicion for IEM will help in prevention of irreversible damage. Age appropriate approaches should be adopted to reach the early diagnosis. Early diagnosis of common and treatable IEM should be the priority. Commonly IEM present at birth or shortly after in the neonatal period after a period of wellness. Some of the IEM present in the infancy to adulthood. Moreover some of the IEM can be recognized based on their particular systemic involvement. Metabolic liver disease have various presentations for which appropriate approaches have been developed. Confirmatory tests for common IEM have been depicted in tabulated form. Finally management of IEM includes removal of the possible incriminating nutrient, toxins with the help of drugs & hemofiltration /hemodialysis, some specific therapies available, e.g., enzyme replacement therapy and organ transplantation. Most need liver transplantation. Others would need close monitoring of the patient for toxic accumulation and restricted diet prescribed. Some of the future therapies for IEM, e.g., hepatocyte transplantation, gene therapy and chaperone therapy have been discussed.

SUGGESTIONS FOR FURTHER READING

1. Loeber J.G., Burgard P., Cornel M.C., Rigter T., Weinreich S.S., Rupp K., Hoffmann G.F. and Vittozzi L. (2012), Newborn Screening Programmes in Europe; Arguments and Efforts Regarding Harmonization, Part 1. From Blood Spot to Screening Result. *J. Inherit Metab Dis.*, 35(4):603–11, Epub.

2. Burgard P., Rupp K., Lindner M., Haege G., Rigter T., Weinreich S.S., Loeber J.G. and Taruscio D. (2012), Newborn screening programmes in Europe; arguments and efforts regarding harmonization, Part 2. From Screening Laboratory Results to Treatment, Follow-up and Quality Assurance, *J. Inherit Metab Dis.*, 35(4):613–25. Epub.

3. Fumić K. and Bilić K. (2014), Laboratory Diagnostics of Emergency Conditions Associated with Inborn Errors of Metabolism, *Signa Vitae*, 9 (Suppl. 1):70–72.

4. Ida Vanessa Schwartz, Carolina Fischinger Moura de Souza and Roberto Giugliani, Treatment of Inborn Errors of Metabolism, *Journal de Pediatria* 0021–7557/08/84–04–Suppl/S8.

SECTION III

SPECIALIZED TISSUES AND ORGANS

In the preceding section, the composition of the body has been analysed from the point of view of the basic elements and compounds that go in its making. In this section, we consider the structure and function of some important specialised tissues and organs. There is evidence to suggest that 80–90% of the total genetic information in mammalian cells is dormant and that only about 10% of the genes are expressed for synthesis of proteins and enzymes in a given cell. The set of the genes expressed varies from one tissue to the other leading to the creation of diversity in molecular structure and function. The biochemistry and special features of muscles, bone, skin, connective and epithelial tissues are described. Two organs (heart and kidney) vital for their metabolic contributions have been discussed in an integrated manner. Basic information on the biochemistry of one of the most complex and challenging organs—a pinnacle of evolution and one which has endowed the human being with special capacities, viz. the brain, has been given. Two chapters have been included for the first time in a biochemistry textbook on teeth and gums. A composite chapter describes eye and vision along with molecular basis of cataract and glaucoma, as well as the use of stem cells for restoring vision which is actually in practice in India. The haematopoietic system along with leukemia, as well as lung and its dysfunction have been contributed by experts in these fields.

19

Epithelial Cells

Subimal Roy

CONTENTS

The world epithelium is derived from the term "epithelis" (epi = upon; thele = nipple) which literally means 'layer upon the nipple'. This was originally coined to describe the morphologic features of the margin of the lips which is not identical to the epidermis. Its literal translation, therefore, is a misnomer. In current usage, epithelium represents a layer/layers of cells covering the surface of the body such as skin or lining hollow tubular structures (like gastrointestinal tract, urinary bladder, bronchi etc.), with one free surface facing the lumen. It may also comprise of solid organs, such as endocrine glands and liver.

I. CLASSIFICATION

The following is a simple morphological classification of different types of epithelial tissues.

1. Covering and lining cells—surface epithelium
   (a) Simple—Single layer of cells lining pleural, pericardial and peritoneal surface and also endothelial cells lining the blood vessels.
     — Squamous—flattened cells.
     — Cuboidal—length equal to breadth; lining cells of excretory ducts, renal tubules, thyroid follicles; nucleus usually central and spherical.
     — Columnar—taller than cuboidal epithelium, nucleus usually basal and elongated—cells in gastrointestinal tract.
   — ciliated
   — non-ciliated with microvilli.

   (b) Stratified—more than one layer of cells—Squamous—keratinizing—surface covered by layers of proteinaceous material, formed by surface epithelium, known as keratin as in skin.
     — Nonkeratinizing
     — Cuboidal
     — Columnar
     — Transitional
   (c) Transitional—stratified epithelium having different shapes in different layers—urinary bladder, ureter.
   (d) Pseudostratified—all cells are in contact with basement membrane but not all reach to the apical surface—ducts of salivary gland, respiratory tract epithelium.

2. Glandular—Endocrine—cord or clump type.
   — follicular type
   — Exocrine—sirrple
   — Compound
   Basement membrane—A layer of extracellular matrix protein that forms an interface between cells and underlying tissue.

Functions

Functions of surface epithelium is protection like skin, diffusion, secretion and absorption by intestinal epithelial cells. They have combined functions such as secretion of mucus and digestive juices, absorption of products of digestion and protection from toxic intestinal content by mucus.

The primary function of the glandular cells is to produce a secretion which is poured into the blood by the endocrine, and into the tubular ducts by the exocrine systems, respectively. In this category, the cells may appear to constitute of solid cords or groups devoid of a lumen, as in the case of adrenal cortex or anterior pituitary, or the lumen may be an ultramicroscopic slit between the cells as in the case of liver.

Some morphological classification includes endothelial and mesothelial cells also under the category of epithelium as mentioned above. The justification offered for this classification is that endothelial cells resemble simple squamous cells, and that the mesothelium which lines serous cavities and is normally of insignificant thickness, except when reacting to injury or in a neoplastic state, resembles epithelial cells. These reacting and proliferating cells often constitute glands and even produce a characteristic secretion. In popular pathological usage, however, the two terms are referred to as distinct from epithelium. Malignant tumors arising from epithelium are called carcinomas, those arising from mesothelium are termed mesotheliomas, whereas those arising from endothelium are called sarcomas, a term reserved for malignant tumors arising in the connective tissue.

Embryologically, most epithelium is derived from the ectoderm covering the surface of the embryo, or the endoderm lining the hollow tubes, which originated as an invagination on the surface and is destined to constitute the intestinal canal. In addition, the epithelium of the genito-urinary system and that of the adrenal cortex is derived from the mesoderm, which is also the mother germ layer for the mesothelium and endothelium. The endothelium originates through the flattening of mesenchymal cells, which are the progenitors of connective tissue cells. The common origin of some types of epithelium, mesothelium and endothelium from the mesoderm, explains the morphological kinship of the cells in certain states, and the resemblance of certain types of tumors arising from them. It also explains the dual growth potential of these cells, which are capable of producing tumors that resemble pure "epithelial", "fibroblastic" or sometimes mixed forms.

II. STRUCTURAL FEATURES

It is difficult to generalize the structural features that characterize epithelial cells, owing to the vast variation of secretory products that such cells produce and their specific metabolic requirements, which determine the diversity of ·cell structures. Cells are bound by cell membrane and contain nucleus and cytoplasm. Nucleus is a distinct structure bound by membrane and may have a dense non-membrane round structure known as nucleolus.

Cytoplasm is very variable depending on the type and metabolic activity of the cell. Most epithelial cells contain cytokeratin (intermediate filament) useful to recognize epithelial phenotype by immunohistochemistry in poorly differentiated neoplastic cells. The ultra structure of a cell has already been described in chapter I, in which the different functions of its various components have been elaborated. All the fundamental organelles are present in the epithelial cell; their proportion may be differential depending upon the precise metabolic functions

and the functional stage of the cell. The cells that are actively involved in protein synthesis are rich in rough endoplasmic reticulum. At some stage of the functional cycle of the cell, there is accumulation of a secretory product in particulate form, which can usually, though not invariably, be demonstrated by some special staining technique. Since the primary metabolic function of the epithelial cells is secretory and dominate all other activities, the secretory apparatus of the cell is more conspicuous.

The free surface may be represented by a thin layer of condensed cytoplasm or provided with numerous protoplasmic projections (microvilli in liver cells, striated border in intestinal epithelium, or brush border in the renal tubules). These are distinct from cilia which are much larger, and are motile. Microvilli represent a device by which the cell acquires a reserve of surface area needed for the absorption of metabolic products. The cilia arise from a dense basal body of cell and are motile structure. They help transport particulate matter over the surface, a process which is often aided by slimy mucoid secretions elaborated by these cells. They are seen in epithelial cells of respiratory tract and fallopian tube. Some types of cells secrete a solid substance on the surface which is detachable such as keratin layer on the surface of stratified squamous epithelial cells.

The cells are held together by an intercellular cement substance. Certain types of epithelial cells, e.g., squamous type are held together by small protoplasmic processes called intercellular bridges which link adjacent cell bodies. Another mode of linking is through alternating pegs and grooves which interdigitate. Within the cell, the rough endoplasmic reticulum is well developed, mitochondria are abundant and organized in a definite arrangement and Golgi bodies are grouped often in close proximity to the secretory channel of the cell like the cholangiole of the liver cell. These features are suggestive of the epithelial nature of the cell but do not exclusively characterize them.

The size, shape, structure and degree of chromatin of the nuclei varies widely during the different stages of the secretory cycle. The nuclear membrane is provided with pores through which exchange of metabolites takes place. Intranuclear inclusions such as crystalloids may be periodically encountered.

Histochemically, the basement membrane on which the layer of epithelial cells rests, is rich in carbohydrates and proteins. The intercellular substance is likewise rich in protein but contains smaller quantities of carbohydrates. The free surface cell membrane contains in addition small quantities of lipids.

III. CELL METABOLISM

The function of an epithelial cell is to participate in the metabolism of the body through absorption of substances from outside, elaborate the secretion product and extrude it into the environment. These three steps constituting the secretory cycle are active energy consuming processes. In some cells there is an intermediate step between intracellular elaboration of secretion and its discharge, namely storage. The functional pattern of cells may be one of continuous or rhythmic activity.

In the former, various steps of the cycle are simultaneous, and thus, the functional as well as anatomic status of the cells is identical at all times. There is no change in the organelles during different phases of cell activity.

Endocrine epithelial cells and the mucus-secreting goblet cells of the gastrointestinal mucosa belong to this category.

In other cells, the above three activities of absorption, elaboration of secretion and its extrusion occupy specified periods of time. Thus, examination of a cell at a given time shows considerable structural differences, depending upon the stage of rhythmic activity of the cell. All the cells of the epithelium or gland may be at different stages of activity at a given time; some cells may be in a resting or relatively inactive phase.

The secretion also follows three different patterns:

In **exocrine secretion**, the extrusion of the elaborated product takes place without any noticeable changes in the organelles. The secretion may be continuous or rhythmic. In the latter, the secretory product can be seen as a granule or as a globule at the apical end of the cell which undergoes dissolution and is discharged into the lumen directly or through ducts.

In **holocrine glands**, ingestion, synthesis and intracellular storage of the secretory product take place only once in the lifetime of the cell. The entire contents of the cell are them discharged into the lumen, resulting in the death of the cell, in which all organelles undergo changes. The dead cells are replaced by reserve cells, e.g., Sebaceous gland.

In **apocrine secretion**, the extrusion of the secretory product take place through decapitation of the apical portion of the cell in the form of membrane-bound vesicles. The extruded portion of the cell includes the secretory product, the nucleus, presumably derived from an amitotic division of the parent nucleus which remains at the base. The cell regenerates from the basal portion which includes a large nucleus, a well developed chondrione, and the ergastoplasm.

In endocrine glands secretion is directly discharged into the circulation.

Ingestion

The epithelial cell is delineated by a cell membrane on the luminal, and a basement membrane on the basal side. In the case of the exocrine pancreatic cell these membranes are 50 Å and 150 Å thick, respectively. All exchanges with blood through the extracellular fluid take place through these membranes. These are active energy consuming processes and are not passive, e.g., Na^+ and K^+ ions introduced intravenously appear within the pancreatic cells within three minutes, and their concentration is lower than that in the blood. Absorption of compounds is extremely selective, and the rate of their ingestion is also variable.

IV. INTRACELLULAR ELABORATION

A. Morphology

During intracellular elaboration, the nuclear volume increases and during extrusion it decreases. During intracellular accumulation of secretions certain changes take place in the chromosomes and desoxyribonucleic acid. There is experimental evidence (Altman cited by Brachet & Mirsky, 1964) that during the stage of accumulation of secretion in the cytoplasm, the nucleolus acquires a sheath of Feulgen-positive material which includes two larger clumps—chromocentres. This sheath is connected with similar Feulgen-positive material pressed against the nuclear membrane. As the cells begin secreting, the chromocentres link up with the nuclear membrane and undergo cleavage, making a 'canal' through which the nucleolus, which undergoes swelling in the meantime, communicates with the cytoplasm. The nucleolar substances are thus evacuated into the cytoplasm. This always seems to take place in close connection with the chromosomes. Altman believed that this canal is made up of some portion of the chromosome. Changes of chromosomic spiralization contribute to these phenomena. According to this hypothesis chromatin plays a crucial role in the nuclear activity of the cell during secretion. The nucleolus only acts as a storehouse of chromosomal products which are used up in the cytoplasm during secretory activity.

The area of the basophilic zone of the cytoplasm, representing the ergastoplasm, undergoes marked enlargement after the cell has discharged its secretory product and is beginning to elaborate new granules. This is particularly true of all cells which are active in protein synthesis.

The basophilic area is reduced in size and stainability in the liver cell during fasting, and it enlarges after a protein-rich meal. Similarly, cells regenerating after partial hepatectomy show marked increase in basophilia. Golgi bodies also increase in number and volume during intracellular elaboration of the secretory products which are packaged within membrane bound vesicles. They are carried to the cell surface and are fused with the cell membrane for discharge or with other intracellular structure such as lysosome.

B. Physiology and Biochemistry

Intracellular elaboration consumes most of the energy required for the secretory cycle. The chondriosome (and not the mitochondria) plays a crucial role in the functional activity of the epithelia cell. Most of the cellular oxidation systems and the enzymes involved in Kreb's tricarboxylic acid cycle and oxidative phosphorylation, as well as most of the total energy-rich phosphates are concentrated on the chondriosomes. Glycolysis takes place in the cytoplasm and is not involved in the activity of the pancreatic cell as a source of energy. However, Kreb's cycle does not participate in the secretion phenomena.

Ribonucleic acid is actively involved in protein synthesis. The RNA content is increased during protein synthesis, and decreased during fasting or on a protein deficient diet.

Currently available data indicate that Golgi bodies play a role in the concentration of the secretory products elaborated in other organelles.

V. MUCIN AND ITS ROLE IN INFLAMMATION

Secretion of mucin is characteristic of epithelial cells in several locations, most notably, respiratory passages and gastrointestinal tract. It is common knowledge that the common cold is accompanied by the secretion of excessive fluid from the nasal passage, which is thin and free flowing in the beginning but later as the cold persists, it becomes more and more viscid, and even becomes tenacious with yellowish streaks. The secretion of mucus and its viscidity may have something to do with the inflammatory process which is in fact a defensive mechanism of the body.

Further it has been observed that the respiratory epithelium which is normally made up of pseudostratified ciliated columnar epithelial cell, admixed with a small number of mucus secreting goblet cells acquires a predominantly mucoid character in inflammatory lesions.

Cilia help to propel the mucus-containing entrapped foreign particles or infecting organisms towards larynx and is coughed out, thus act as a protective mechanism. In chronic long standing inflammation of the mucous membrane, groups of mucous secreting goblet cells multiply in number and add to the mucous-secreting apparatus of the epithelium. Of late the thickness of the layer of mucous secreting goblet cells in proportion to the entire thickness of the bronchial mucosa is being expressed as a numerical index relatedto the chronicity of bronchial inflammation.

Mucins are a groups of compounds containing a complex mucopolysaccharide-sialic acid.

They yield sugars and hexosamines which are hydrolysed by dilute acids (chapter 10). In addition, they contain bicarbonates, chlorides, sulphates, phosphates, etc. The viscosity of the mucus is due to a large randomly coiled thread-like molecule and is related to the complexity of the molecule. The staining character and morphology of the epithelial mucus is variable in different situations, organs, and even different cells in the same organ.

Mucoid secretions in pathological situations become increasingly viscous due partly to the DNA derived from the nuclei of dead polymorphs, and bacteria. The yellowish tinge seen in the nasal secretions in a chronic cold is imparted by a pigment secreted by staphylococci andpus.

The precise functions of epithelial mucinare not understood. It has been suggested that in the stomach they have a buffering effect on the acid and pepsin secreted by the gastric mucosa, thus protecting the epithelial coat. It is known that mucoproteins and mucopolysaccharides are not effective as buffers, but it is possible that the bicarbonate which is secreted in considerable quantities in the mucus might have some such although weak, effect.

Further, goblet cells secreting mucin are present throughout the intestinal mucosa. They are characteristically reduced in idiopathic ulcerative colitis. The significance of this change is not understood. In the colon and rectum, it might provide a physical lubricant to ease the passage of solid feaces but this function is not needed in the small intestine.

Florey (1962) has postulated a possible protective action in the respiratory mucosa. As has been stated earlier, the goblet cells undergo marked increase in number and there is a corresponding increase in the amount of mucus secretion in bronchial mucosa in chronic inflammation of bronchi, especially in cigarette smokers. Some viruses get absorbed in the cell membrane by virtue of their affinity for the mucoprotein present in it. The haemagglutinating property of some viruses is due to this mechanism. The absorbed virus on the erythrocyte membrane liberates itself enzymatically by splitting off from the mucoprotein of the red cell membrane, a sialic acid, which probably constitutes a site of attachment of the virus, thus bringing about a dissolution of the cell. The mucins probably shieldthe epithelial cells by providing an alternative sialic acid molecule for the attachment of the virus, thus getting inactivated themselves in the process.

Lamb and Reid's (1968) experiments have shown that in the bronchial epithelium of a rat reacting to chronic irritation by sulphur dioxide, a quantitative as well as qualitative change occurs in the epithelial mucinssecreted. In the mucus secreted by the tracheal mucosa, the AMP is considerable lesser in quantity and susceptible to sialidase. After exposure to sulphurdioxide, the sialdaseresistant AMP extends to the smallest branches of bronchi, and goblet cells normally absent in the latter, make an appearance. The rate of secretion of individual cells increases as indicated by a higher uptake of radioactive sulphate, and an increase in its discharge. Most cells, even though they produce sialomucinpredominantly, also take up some sulphate, indicating a shift from sialo- to sulphomucuns.

VI. EPITHELIAL REGENERATION, HYPERPLASIA AND METAPLASIA

Abnormal Growth

Under normal conditions different types of cells vary considerably in their turnover and potentialities of growth. Some cells like the neurons are (terminally differentiated cell and are) incapable of division and once dead, cannot be replaced. Epithelial cells in some organs are constantly dividing and reviving themselves, as the senile ones drop out and newer ones take their place. The best example of such a renewal is in the squamous epithelium. In this stratified epithelium, made up of many layers, the deepest layer of cells is disposed at right angles to the basement membrane. In these cells, the nuclear cytoplasmic ratio is higher than that in the upper layers of cells and the nucleus is also proportionately hyperchromatic. The function of this basal layer of cells is to act as a stem cells and to regenerate and replace the cells in the topmost layer, which are constantly being shed as dead keratin. Despite the constant regeneration of cells which replace the dead cells the thickness of the eridermis remains more of less constant, because of the perfect equilibrium between the multiplication of cells at the base and death of cells at the surface. To explain this equilibrium, one has to assume that a fixed percentage of the daughter cells arising from multiplication of the stem cells continue to retain the regenerative propriety of their parent (stem) cells, and an equal number move upwards and die. Likewise, the intestinal villus renews itself and is completely replaced in about 1–3 days. The epithelial cells of the endometrium shed with each menstrual cycle and subsequently regenerate. After a partial hepatectomy, the remaining mass of liver cells undergo rapid multiplication which continues till such time that almost the total original mass of cells has been recreated. Thereafter further regeneration of cells stops. This regeneration is therefore strictly controlled and there are delicate in built mechanisms by which the regeneration stops the instant, the dead cells are replaced, and resumes as soon as a gap is recreated. Thus, it appears that growth of cells and tissues is normally controlled and highly regulated. However, they may loose the ability to undergo apoptosis (see below), the pathways that regulate the normal cell cycle from its origin to differentiation and to senescence and may get altered resulting in continuous, uninhibited growth of cells leading to tumor formation. A host of regulatory genes, when inactivated (mutation) or over activated (amplification) can lead to uncontrolled growth of cells resulting in tumor formation.

In certain situations, the turnover rate of cells is increased so that as compared with the normal rate, larger number of cells degenerates and an equal number are replaced. Thus, a new equilibrium is reached and the homeostatic mechanism is reset at the different level. This change is called hyperplasia. Hyperplasia may be physiological or pathological. Examples of physiological hyperplasia are the development of spermatogenic epithelium or the mammary glands with the onset of puberty. The endometrial epithelium undergoes hyperplasiain a certain phase of the menstrual cycle, in each cycle. Hyperplasia may also be a compensatory phenomenon, e.g., after loss or damage of liver and kidney, cells in the remaining tissue increase in number till a stage is reached when enough number of cells is created to make up for the loss, a new equilibrium is established and further regeneration stops. In other situations, the hyperplasia may be pathological as happens in a disease called diffuse or nodular goiter of thyroid in which thyroid epithelium undergoes hyperplasia under certain pathologic stimuli. Other common examples are hyperplasia of endometrium and prostate due to excessive stimulation by oestrogen and androgen respectively. Hyperplasia is believed to be controlled and regulated by specific growth factors acting on the growth fact receptors on the responding cells. Hyperplasia may result from proliferation of the remaining cells or by forming new cells from stem cells.

In another pathological type of regeneration of cells, e.g., neoplasia ("neo" means "new", "plasia" means "growth") an imbalance is created between the generation of new cells and the rate of their apoptosis. When these two processes are uncoordinated, persistent uncontrolled growth of cells lead to the formation of neoplasia. New cells continue to form from the stem cells.

Furthermore, when these cells continue to proliferate and, they do not mature to take over the normal functions of the cells, e.g., production of mucin or keratin or whatever other product they may be secreting and they fail to acquire the morphological characteristics or maturity, they appear "undifferentiated" or "de-differentiated". It is not a retrograde phenomenon, as some of these terms might indicate, in the sense that a keratin producing cells do not de-differentiate after having reached the stage of differentiated maturity, but the stem cells fail to acquire the morphologic features of maturity associated with the production of keratin. By the same analogy, it is understandable that in neoplastic states, there may be no morphological difference between the connective tissue and epithelial cells. When the above state of lack of equilibrium is reached, the stem cells go on multiplying and a new growth or a tumor results, made up of cells often with no distinguishing features indicating their cell of origin. It is important to note that tumor arises from proliferation of a single cell that has undergone genetic change leading to its unregulated proliferation resulting in tumor formation by clonal expansion, thus it is monoclonal in origin. As has been mentioned earlier this happens due to inactivation (mutation) of regulating genes or over activation (amplification) of growth promoting genes (oncogenes). Several growth factors are also known to regulate cell proliferation.

Dysregulation of such factors may also lead to overgrowth and neoplasm. Added factors for the maintenance of neoplasm are angiogenesis and escape from immunity. Thus it appears that all these phenomena of cell proliferation, regeneration and neoplastic transformation are controlled by interactions of varieties of genetic factors. An anomaly in their action may lead to abnormality in these control mechanisms resulting in abnormal growth of cells and their behaviour.

The following points relevant to this chapter are briefly mentioned below:

Stem cells

Most tissues, even in adults, contain some undifferentiated cells that are able to divide and replace the lost cells and thus help to maintain the tissue homeostasis. They have the capacity of prolonged cell renewal. They may be multipotent, capable of forming cells of many organs, such as liver, brain, heart etc. and present in bone marrow. Some stem cells are unipotent forming single type of cells as in skin and in gastrointestinal tract and others may be totipotent and are capable to differentiate into any type of cells, the example being embryonicstem cell.

In recent years stem cells have gained a lot of interest and they are being used to treat to repopulate damaged tissues such as liver, insulin producing pancreatic cells and even to treat damaged tissues like brain, cardiac muscle, previously considered as terminally differentiated permanent cells.

Homeostasis

In normal tissues, cells interact with adjacent cells to maintain its structural (and functional) integrity; the process is controlled by genetically programmed metabolism.

When there is damage to the tissue due to injury or to handle physiological stimuli, a hyperplastic state occurs to reach the stability and as soon as a stable state is established, further growth stops and a state of homeostasis is achieved. It has been suggested by Florey (1962) that suppression of growth occurs through the intercellular exchange of substances, passing on from one cell to the next through the area of contact.

Metaplasia

Metaplasia is defined as replacement of one adult cell type into another cell type. Some common examples of metaplastic change in epithelial cells are metaplasia of columnar epithelium into squamous epithelium in respiratory tract (in smokers) and in mucus secreting glands in the cervix. It also occurs in Barrett's disease where squamous epithelium in esophagus turns into columnar epithelium. Persistent metaplastic change may lead to development of carcinoma. Metaplastic cells are believed to be derived from undifferentiated stem cells, (preseritin normal tissues) induced by cytokines.

Apoptosis

The term apoptosis is derived from Greek word meaning dropping off. It is a form of cell death, also called programmed cells death. It is a regulated intracellular process, caused by degradation of its own nuclear DNA by activation of enzyme. It helps to eliminate unwanted or harmful cells and is essential in normal fetal development, occurring during embryogenesis. It also occurs normally in endometrial cell in menstrual cycle, intestinal epithelium and ovarian follicle in menopause. It occurs in a variety of pathological conditions such as viral infection, radiation injury, due to cytotoxics drugs and cell death in tumors. The process is controlled by molecular events, important part being played by TNF (tissue necrosis factor), Bcl-2 proteins and cytoplasmic endonucleases. Morphologically the cell shrinks with condensation of both nucleus and cytoplasm followed by breakdown of cell into spherical, membrane bound fragments. Unlike necrosis apoptosis does not induce any inflammatory reaction.

SUGGESTIONS FOR FURTHER READING

Elder D.E., Elenitsas R., Murphy G.F., Johnson, B.L. and XnXiawwei, Lever's *Histopathology of the Skin*.

Haul J.C. Sauver's (2005), *Manual of Skin Diseases*, 10th ed., Philadelphia, Wolters Kluwer, Lippincott, Williams and Wilkins. 10th ed.

Kumar V., Abbas A.K. and Fausto N. (2005), *Robbin's and Cotran Pathologic Basis of Diseases*, 7th ed., Elsevier Saunders, Philadelphia.

Young B., Lowe J.S., Stevens A. and Heath J.W. (2010), *Wheater's Functional Histology: A Text and Colour Atlas*, 5th ed., Churchill Livingstone.

20

Connective Tissue

Subimal Roy

<table><tr><td colspan="2">CONTENTS</td></tr></table>

I. COMPOSITION

Connective tissue is of mesodermal origin and consists of cells and intercellular materials that provide structural support for different tissues. Among the cells, the fibroblasts, mast cells, histiocytes and fat cells (lipocytes) are the main types and among the intercellular substances, collagen, elastin and mucopolysaccharides are the important constituents. Besides structural support, it plays important metabolic role in the form of lipid in adipose tissue and helps tissue repair after damage. Connective tissue may appear loose as in skin and lamina propria in gastro-intestinal tract or dense as in tendon and ligament. It also plays an important role in the exchange of various substances (metabolites) between blood vascular system and epithelial and other cells. Collagen forms the major component of connective tissue and is, in a way, one of the most abundant proteins in the human body (see chapter 7). Although the connective tissue is distributed throughout the body, its proportion varies considerably from organ to organ. Liver contains about 4% of collagen fibres, whereas skin contains about 70% and the tendons about 86%. Collagen fibres form about 90% of the organic matrix of bones. Similar variations exist with regard to other components of connective tissue. These compositional variations in different organs and tissues, make for variations in the stress-bearing capacity of connective tissue. The higher the collagen content of a given tissue, the greater is its strength. The tensile strength of collagen fibre is impressive and to break a collagen fibre I mm in diameter, a load of 10 to 40 kg is required.

Elastin and mucopolysaccharides further contribute to tissue strength.

However, in pathological states of collagen diseases, collagen fibres lose their strength very rapidly. Because of the inherent importance of this tissue in health and disease we will discuss these matters in greater detail.

II. CHEMISTRY OF CONNECTIVE TISSUE

Mucopolysaccharides (Ground Substance)

Mucopolysaccharides, which are also known as glycosaminoglycans, are amorphous gel-like substance present in the extracellular and interfibrillar portion of connective tissue. They are polysaccharides containing hexosamine and uronic acid. They are usually formed by the local connective tissue cells and they occur in different concentrations and proportion in different types of connective tissues of the body (see chapter 10). For the synthesis of the substances tabulated previously, glucose is the basic material which is transformed into glucosamine and glucuronic acid to form hyaluronic acid. Similarly chemical changes take place to form other mucopolysaccharide moieties. Thereafter, they form loose complexes with non-collagen proteins present in the extracellular space and effectively carry on their function till they are degraded and excreted in the urine. The various mucopolysaccharides have different turnover rates. Thus, the turnover rate of hyaluronic acid is rapid and its biological half life is only two days. The half life of sulphated mucopolysaccharides present in the skin and cartilage is much longer. After their liberation from the site of production, they circulate in the blood and are excreted through the kidneys. In normal persons, there is a small amount of mucopolysaccharide in the urine. However, in various disorders of connective tissue with an excessive amount of mucopolysaccharide breakdown, such as rheumatoid arthritis, there are excessive amounts of mucopolysaccharides present in the urine.

III. TYPES OF CONNECTIVE TISSUE CELLS AND FIBRES

A. Fibroblasts

The term fibroblast denotes the ability to form fibres, and in the connective tissue all the intercellular material are formed by the fibroblasts either directly or indirectly. It is spindle shaped in the resting stage when it is called a fibrocyte. During the period of active collagen synthesis, it becomes enlarged and its rough endoplasmic reticulum and Golgi apparatus become well developed. Although there was dispute on the origin of these cells, it is now well established that they originate from the proliferation of primitive mesenchymal cells, present locally in the connective tissue. Their main task is to synthesize collagen in a somewhat similar way to the manner in which other active cells synthesize proteins.

Apart from fibroblasts, which are considered as fixed connective tissue cells, lipocytes, chondrocytes and osteocytes are also included in this category.

B. Mast Cells

These are also included as connective tissue cells containing coarse cytoplasmic granules. They are found in supporting tissue, especially in skin, around blood vessels and in gastrointestinal tract. The granules are membrane-bound and are metachromatic; they contain histamine, heparin and cytokine like TNF-a. They are actively involved in immediate hypersensitively (anaphylactic) reaction, activated by IgE resulting in degranulation with release of histamine and other vasoactive mediators.

C. Reticuloendothelial Cells

Reticuloendothelial cells (monocytes, histiocytes and macrophages) are derived from bone marrow, enter into the circulation as monocytes and then migrate and settle in many tissues, especially spleen, lymph node, liver and thymus. Many remain in a resting state (fixed macrophage). An important function is phagocytosis of various particles and microorganisms and when activated, they become large having more abundant cytoplasm. They also process antigenic material before presenting to lymphocytes, immunoreacting cells in response to infection and inflammatory reaction.

D. Collagen and Reticulin fibres

As already stated, collagen is the principal fibrous protein of the connective tissue and its main function is to give tensile strength to this tissue. The chemistry, structure, biosynthesis, degradation and pathobiology of this important protein has been discussed in Chapter 5. Reticulin fibres are also composed of protein collagen but are thinner, forming a delicate network, close to cells such as around smooth muscle cells, fat cells and beneath endothelial cells.

E. Elastic Fibres and Elastin

Elastic fibres are very thin, generally branching with anastomosis forming a network. They are embedded in amorphous material—elastin. Elastin like collagen, is a protein which is distributed in all tissues of the body in different degrees wherever collagen exists, conferring elasticity. It is a yellow fibrillo-membranous structure arranged in the form of a mesh without free ends. It exists in abundance in the aorta, skin, lungs or ligamentum nuchae. The essential physiological characteristic of this substance is that it is elastic like rubber, as the name denotes. Chemically its amino acid composition resembles that of collagen in that the glycine forms about one-third of its total amino acid residues. Otherwise it is different as it has very little hydroxyproline and contains large amounts of amino acids with non-polar side chains such as valine (14.1%).

The peculiarity of its composition makes it relatively resistant to chemical attack. Unfortunatelyl it shows no readily distinguishable features in the electron microscope. It probably consists of peptide chains linked at intervals to form a three-dimensional network. These chains are linked, or randomly coiled, when the tissue is not under tension and because of its peculiar structural arrangement it is extensible like rubber.

Elastin can be identified in the tissues by using specific stains such as orcein, or aldehyde fuchs in stains. The essential physiological role of elastin is to allow, in combination with collagen and mucopolysaccharides, a controlled, reversible framework of the body which undergoes continuous stretch and contraction. Hence the tissues and organs which constantly undergo such changes like the vascular system, lungs, skin and ligamentum nuchae contain a large amount of elastin in the body. It has an important role to play in aorta where the diastolic blood pressure is maintained chiefly by the recoil of the elastin present in the aortic wall. Its biosynthesis, turnover and excretion rate are not very clear, so far. Elastic fibres are degraded by the elastase (different from collagenase) preset in the tissues containing elastin. In the ageing process the elastin fibres are replaced by collagen fibres which leads to the decrease in the elasticity of various tissues.

IV. BIOCHEMISTRY OF COLLAGEN DISEASES

So far we have been discussing the structural and biochemical properties of connective tissues in normal physiological states.

It would be interesting to know the biochemical changes that take place in different pathological states, for a proper understanding of these processes.

A. Ageing

As a person becomes old, a degenerative process sets in the body which causes a gradual alteration in the connective tissue components. With advancing age, although the mucopolysaccharide content of most of the connective tissues decreases, the collagen content increases and tends to be organized in the form of coarser fibres. As a result of these changes, the tissues become Jess hydrated, less elastic and, therefore, less able to resist mechanical forces. There is progressive decline in many normal physiological functions such as proliferative activity and life span of many cells including neurons caused by cellular and molecular damage due to various genetic anomalies and environmental factors. There is reduced synthesis of structural proteins and reduced capacity of repair phenomenon. Some of the common age-related diseases due to structural and functional decline of cells and tissues are senile osteoporosis, osteoarthritis, dementia, cardiac damage, etc.

B. Collagen Diseases

There is a groups of diseases in which there is a diffuse inflammatory change in the connective tissues of the body with localized fibrinoid changes in the certain specific areas.

Among these diseases, rheumatic fever, rheumatoid arthritis, polyarteritis nodosa, lupus erythematosis and scleroderma are important ones. They are believed to be due to deposition of immunocomplex and activation of complements. The characteristic histological features of fibrinoid changes found in these diseases are, the absence of connective tissue cells, and blurring of collagen fibre patterns, which may be entirely replaced by amorphous or structureless eosinophilic material. Biochemical studies of the fibrinoid material found in the connective tissues indicate that there is a decrease in the hydroxyproline content and an increase in the glucosamine content, indicating thereby that there is a decrease in the collagen content and an increase in the mucopolysaccharide content of these tissues. The X-ray diffraction studies show loss of the characteristic collagen pattern which is replaced by a diffuse ring with spacing of about 4-8 Å. The electron microscopic studies show only an amorphous material with broken and disintegrating collagen fibres at the margins. Clinically these collagen diseases tend to become chronic or recurring with marked constitutional symptoms and weight loss. Depending upon the type of collagen disease, local lesions are found in the joints, blood vessels, heart, skin, muscle and the supporting tissues of internal organs. In these organs and tissues, the connective tissue lesions are mostly destructive so that there is a tendency towards the continuous loss of supporting structures leading to atrophic changes. Here it is important to remember that all inflammatory processes found in collagen diseases appear to be completely or almost completely suppressed by cortisone in adequate dosage. However, they show a tendency to recur on the cessation of cortisone treatment.

V. WOUND HEALING

A marked variation in the structure and function of connective tissue takes place during the healing of wounds; hence this problem is briefly discussed here. Following trauma, there occurs an inflammatory response in the injured region with accumulation of extracellular fluid and inflammatory cells. Soon after this the healing process starts, in which there is a marked proliferation of connective tissue cells, specially the fibroblasts in the region, in an attempt to fill up the gap. Very soon these cells engage themselves in the production of material required for healing, such as mucopolysaccharides in the early stage, followed by the production of collagen fibres in the later stage. The details of biosynthesis of collagen have been discussed in Chapter 7, and one can see clearly the formation of collagen during would healing. Gradually as the collagen fibres become well oriented and mature, the quantity of mucopolysaccharides decreases and the fibroblasts also disappear from the region. When the quantity and quality of collagen fibres increase, the tensile strength of the wound also increases, till it reaches its normal level during the course of the next six months or so. This, in brief is the normal sequence of events in wound healing.

There are some conditions in which normal healing of wounds does not take place. In hypoproteinaemia, collagen does not form in enough quantity due to lack of adequate supply of amino acids in the region. This leads to defective would healing with poor tensile strength. This may ultimately lead to break down of the wound in the early stage or development of hernia in the later stages. In vitamin C (ascorbic acid) deficiency an adequate amount of collagen fibres are not formed during would healing, later leading to wound disruption. Although this fact has been known for quite some time, the exact role of the vitamin in this context has become obvious only recently. As already stated earlier, ascorbic acid is an essential cofactor for hydroxylation of the proline into hydroxyproline at the protocollagen stage of collagen formation. In the absence of ascorbic acid, an abnormal collagen-like material of low tensile strength is formed due to the lack of hydroxyproline formation. Due to this, the healing of wounds becomes defective and delayed. With the availability of ascorbic acid, the collagen fibres appear in the wound in normal quality and quantity. This indicates the biochemical role of ascorbic acid in the formation of collagen during healing. Amongst the hormones, cortisone in high dosage has a deleterious effect on wound healing due, possibly, to its gluconeogenic and catabolic activity which causes considerable loss of proteins from the body. However, this deleterious effect of cortisone can be reversed to a great extent by providing anabolic hormone of the testis in sufficient dosage. These are some or' the facts about wound healing in the normal and abnormal state in which all the components of connective tissue play an important role.

Wound healing may also be affected by some local factors such as infection, presence of foreign body and reduced blood supply. Some growth factors and cytokines, epidermal growth factor (EGF), fibroblastic growth factor (FGF), platelet derived growth factor (PDGF), insulin-like growth factor (IGF) and transforming growth factor-P (TGF-p), play important roles in would healing.

SUGGESTIONS FOR FURTHER READING

Asboe-Hansen G. (1963), Connective Tissue, *Annual Review of Physiology*, 25, 41.

Ayer J.P. (1964), Elastic Tissue, *International Review of Connective Tissue Research*, Vol. 2, Hall, D.A. (Ed.), New York: Academic Press, 33.

Brown J.C. and Timpl R. (1995), Chemical and Histochemical Studies of Wound Healing, *New England Journal of Medicine*, 253, 857.

Endres D.B. and Rude R.K. (1999), In: Tietz *Textbook of Clinical Biochemistry*, 3rd ed., Burtis, C.A. and Ashwood E.R. (Eds.) W.B. Saunders, Philadelphia, USA, 1395–1457.

Fricke R. and Hartman F. (Eds.) (1974), *Connective Tissues, Biochemistry and Pathophysiology*, Springer-Verlag, New York.

Glynn L.E., Diseases of Collagen and Related Tissues, *ibid.*, 214.

Grant M.E. and Prockop D.S. (1972), The Biosynthesis of Collagen, *New Eng. J. Medicine*, 286, 194 and 242.

Jackson D.S. and Steve F.S. (1969), Some Biochemical Considerations of Connective Tissue Proteins, *Disease in Biological Vasis of Medicine*, Vol. 3, Bitter, E.E. and Bitter, N. (Eds.), London Academic Press, 249.

Kellgren H.H. (1955), Collagen Diseases, *Symposia of Society for Experimental Biology*, 9, 163.

Kumar V., Abbas A.K. and Fauso N. (2005), *Robbin's and Cotran Pathologic Basis of Diseases,* 7th ed., Elsevier Saunders.

Muir H., Chemistry and Metabolism of Connective Tissue Glycosaminoglycans (Mucopolysaccharides), *ibid.*, 101.

Peacock E. (1967), Dynamic Aspects of Collagen Biology, *Journal of Surgical Research*, 7, 433 and 481.

Verzar F., Ageing of the Collagen Fibre, *Ibid.*, 244.

Young B., Lowe J.S., Stevens, A. and Heath J.W. (2010), *Wheater's Functional Histology: A Text and Colour Atlas*, 5th ed., Churchill Livingstone.

21

Striated and Smooth Muscles

Sudip Sen

<table>
<tr><td colspan="2" style="text-align:center">CONTENTS</td></tr>
<tr><td>I.</td><td>Introduction</td></tr>
<tr><td>II.</td><td>Structural Components of Muscle</td></tr>
<tr><td>III.</td><td>The Muscle Proteins</td></tr>
<tr><td>IV.</td><td>Biochemical Changes during Muscle Contraction</td></tr>
<tr><td>V.</td><td>Regulatory Role of Troponin and Tropomyosin</td></tr>
<tr><td>V.</td><td>Coupling of Excitation and Contraction: role of calcium</td></tr>
<tr><td>VI.</td><td>Diseases Due to Defects in the Ca^{2+} Release Channel</td></tr>
<tr><td></td><td>A. Muscular dystrophies</td></tr>
<tr><td></td><td>B. Energetics of muscle contraction</td></tr>
<tr><td></td><td>C. Cardiac muscle</td></tr>
<tr><td></td><td>D. Smooth muscles</td></tr>
</table>

I. INTRODUCTION

Muscles constitute approximately 40% of the body mass in normal healthy young adults. They also make up the largest single tissue in the human body. These facts are evidence of the extremely important role this tissue plays in converting chemical energy into mechanical energy.

Muscles can be classified as *skeletal, cardiac* and *smooth* on the basis of the function that they perform and their location. Skeletal muscles are usually under voluntary nervous control while cardiac and smooth muscles are under involuntary control. Under the microscope, both skeletal and cardiac muscles appear striated because of organizational alignment, while smooth muscles which lack this alignment, appear non-striated.

II. STRUCTURAL COMPONENTS OF MUSCLE

Muscles are made up of multinucleated muscle fiber cells which in turn are made up of bundles of myofibrils that are arranged in a parallel fashion along the long axis of the muscle fiber (Figure 21.1). The muscle fiber cell is covered by an electrically excitable plasma membrane known as the sarcolemma. The cytoplasm of the muscle fiber cell containing these myofibrils is also known as sarcoplasm. Myofibrils are composed of serially placed, repeating functional unit of the muscle, called sarcomere.

Under electron microscope, the myofibril can be seen to be comprising of alternating dark (anisotropic) and light (isotropic) bands, named as A and I bands respectively (Figure 21.1). The center of the dark A band seems to be less dense and is called the H band while the center of the light I band appears to be more dense and is called the Z line. The area between two adjacent Z lines forms the sarcomere. The myofibril consists of longitudinal thick (diameter of about 16 nm) and thin (diameter of about 7 nm) myofilaments (Figure 21.2). The cross section and the longitudinal section varies according to the state of contraction of the muscle as it portrays the overlapping of the thick and thin filaments. In areas where they overlap, each thin filament lies symmetrically between three thick filaments and each thick filament is surrounded symmetrically by six thin filaments (Figure 21.3). These thick and thin filaments interact via cross bridges present at intervals on the thick filaments. The thick filament is mainly made up of the protein called myosin whereas the thin filament comprises of the proteins actin and regulatory proteins troponin and tropomyosin (Figure 21.4). Each troponin and tropomyosin molecule combines with seven actin monomers to form a functional unit which repeats itself along the length of the actin filament.

III. THE MUSCLE PROTEINS

Myosin contributes 55% while actin makes up 25% of muscle

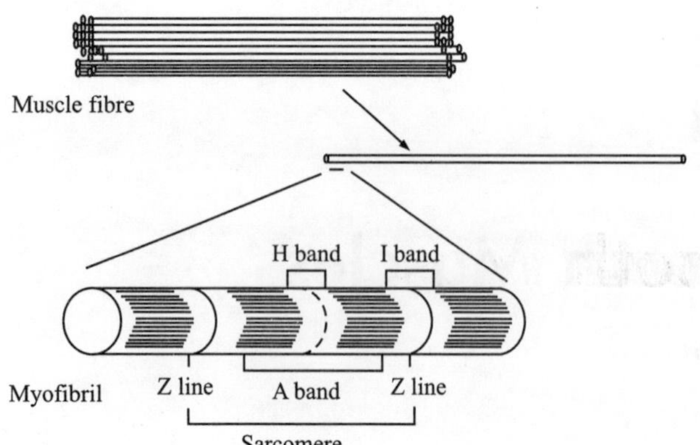

Figure 21.1   Ultrastructural details of skeletal muscle.

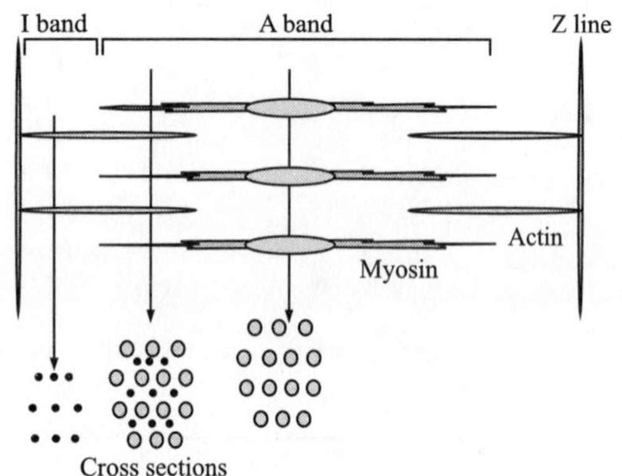

Figure 21.3   Organization of actin and myosin filaments in a sarcomere and cross-sections at different points.

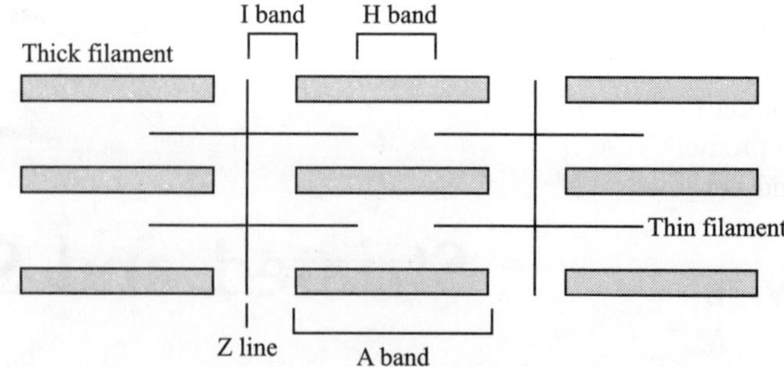

Figure 21.2   Diagrammatic representation of thick and thin filaments during relaxed state.

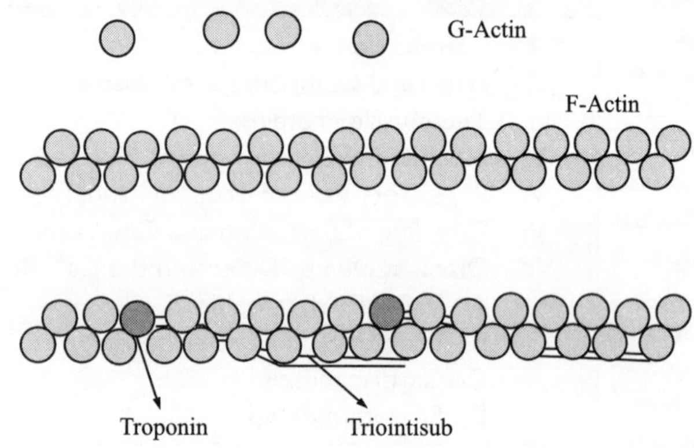

Figure 21.4   Diagrammatic molecular organization of actin filaments; organization of troponin and tropomyosin along actin.

protein by weight. Structural details of actin and myosin have become clearer with X-ray crystallography. Myosin has at least 12 multiple classes in humans and a molecular weight of approximately 460 kDa. Myosin is an asymmetric hexamer made up of one pair of heavy chains and two pairs of light chains. The approximate molecular weights of these heavy and light chains are 200 kDa and 20 kDa, respectively. The light chains can be further subdivided into essential and regulatory light chains. The heavy chain has a globular head attached to a fibrous tail and two such heavy chains have their tails wrapped around each other in the form of an intertwined helix. Skeletal muscle myosin binds actin to form the actomyosin complex, the ATPase activity of which is increased 100 to 200 times compared to the ATPase activity of the unbound form of myosin.

On treating myosin with the enzyme trypsin, two myosin fragments called light and heavy meromyosins are produced. The light meromyosin (LMM) consists of insoluble, aggregated, helical fibers from the tail of the myosin molecule while the heavy meromyosin (HMM) is a soluble part consisting of the globular head as well as a part of the fibrous tail of the myosin molecule (Figure 21.5). HMM exhibits ATPase activity and has the ability to bind to F-actin whereas LMM does not have ATPase activity and does not bind to F-actin. Further digesting the HMM with another enzyme, papain, results in

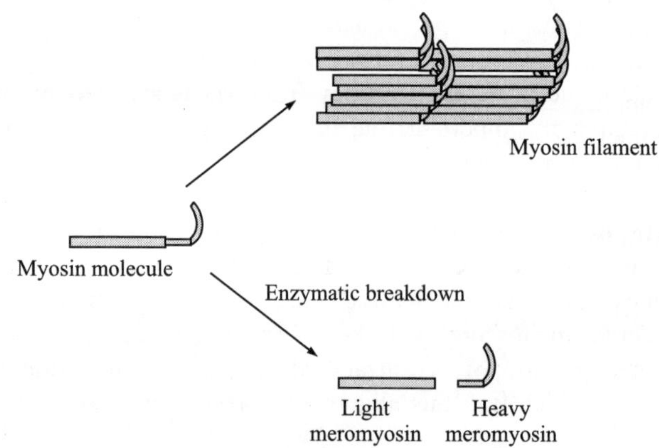

Figure 21.5   Diagrammatic molecular organization of myosin filaments.

the formation of two sub-fragments, S1 and S2. S2 exhibits ATPase activity and essentially comprises of the globular head associated with the light chains and is endowed with actin binding ability, while S1 is the remaining tail portion, and has neither ATPase activity nor the ability to bind to F-actin.

Monomeric globular actin (G-actin) with a molecular weight of 43 kDa, polymerizes non-covalently, to form an insoluble double helical filament called F-actin (Figure 21.4).

This F-actin is one of the constituents of the thin filament and has the ability to bind to myosin to form the actomyosin complex during muscle contraction. F-actin helps by its ability to promote release of ADP and Pi by myosin-ATPase activity and hence catalyses this reaction.

IV. BIOCHEMICAL CHANGES DURING MUSCLE CONTRACTION

The sliding filament cross-bridge model was postulated by Henry Huxley and Andrew Huxley. They observed changes in muscle contraction very minutely and proposed that when muscle contracts, there is no actual change in lengths of the thick or thin filaments. The sarcomere shortens but the actual length of actin or myosin does not change. The contraction or shortening of the sarcomere takes place due to the thick and thin filaments sliding past one another with the help of cross-bridges. The magnitude of shortening is proportionate to the extent of overlap and in turn to the number of cross-bridges that are formed.

It is interesting to note how ATP hydrolysis can lead to muscle shortening, in other words, how chemical energy can be converted into mechanical energy. To comprehend the biochemical reactions behind the sliding filament theory, muscle contraction can be explained as a simple repetitive attachment and detachment of the myosin to the actin to result in repetitive making and breaking of the cross-bridges. The actin and myosin complex formation results in conformational changes in the globular head of myosin. These changes are also affected by the type of nucleotide present along with this complex and result in the power stroke which allows the actin filaments to slide past the myosin filaments. As ATP is hydrolysed to ADP, it supplies the energy needed for the power stroke or muscle contraction. The conformational change occurs in the myosin head as the ADP dissociates from myosin.

The repetitive sequence of events can be represented by the following steps:

(i) Just before muscle contraction begins, the muscle is in a relaxed state. The ATPase activity of myosin hydrolyzes the bound ATP to ADP and Pi but these still remain bound to the myosin in the form of a high energy conformation, the ADP-Pi-myosin complex.

(ii) As excitation initiates muscle contraction, actin binds to the globular head of myosin to form the actin-myosin-ADP-Pi complex.

(iii) The binding of actin promotes release of ADP and Pi from this complex. This release is also associated with a conformational change in the globular myosin head in relation to its tail and is known as the power stroke. The resultant actin-myosin complex is a relatively low-energy state compound.

(iv) Another ATP molecule binds to the myosin head and forms the actin-myosin-ATP complex. As ATP binds to myosin, the latter's affinity for actin diminishes and

hence actin is released from the actin-myosin complex as soon as ATP binds to it. This step signals muscle relaxation and takes us back to the first step where myosin is bound to the ATP.

This biochemical mechanism is also able to explain the phenomenon of rigor mortis that is commonly seen in dead bodies and can help forensic experts determine time of death. After death, as intracellular ATP levels get depleted, ATP is no longer available to bind to the myosin head. This leads to actin not dissociating from the actin-myosin complex and the state of muscle contraction is no longer followed by muscle relaxation. This is manifested by stiffening of the muscles in parts of the body and this process will start as soon as ATP levels in the body start dwindling after death.

V. REGULATORY ROLE OF TROPONIN AND TROPOMYOSIN

Thin filaments in striated muscles comprise of actin as well as troponin and tropomyosin. The troponin complex is almost always found in striated muscles and comprises of the following three polypeptides, troponin T (TpT), troponin I (TpI), and troponin C (TpC). The troponin complex is characteristic of striated muscles. TpT binds to tropomyosin as well as to TpI and TpC. TpI prevents the F-actin-myosin interaction and in addition, binds to TpT and TpC. This action of TpI prevents activation of myosin ATPase that is possible after the F-actin binds to the myosin head. TpC binds to four molecules of calcium. These regulatory proteins prevent spontaneous binding of actin and myosin. Optimal Ca^{2+} concentration promotes formation of actomyosin complexes even in the presence of these regulatory proteins. Calcium specifically binds with the troponin present in actin, altering its conformation and reactivity to various other proteins. There is a significant degree of interaction between the molecular components of the actin-troponin-tropomyosin complex. This complex is also affected by the sarcoplasmic ionic environment, specially to the Ca^{2+} concentration, thus allowing the contractile unit of the muscle to be regulated by membrane dependent ionic fluxes.

V. COUPLING OF EXCITATION AND CONTRACTION: ROLE OF CALCIUM

As filamentous contractile proteins and their interactions are important for muscle contraction, the muscle membrane and its structure also play a very important role in linking neural excitation with the muscle contraction-relaxation cycle. Three important features of the muscle membrane, (1) the sarcolemma, (2) the T system and (3) the sarcoplasmic reticulum are crucial for muscle excitation and relaxation.

The sarcolemma of muscle cells corresponds with the plasma membrane of other cells. A specialized portion of the sarcolemma called the post-synaptic membrane functionally connects the muscle cells with the motor nerve endings. The wave of depolarization that is initiated at the post-synaptic membrane, spreads along the sarcolemma. In association with ionic fluxes that occur during depolarization, free Ca^{2+} ions are released from their binding sites on the inner side of the sarcolemma.

The T-system is a transverse network of tubules that encircle the myofibrils at the Z-line, with intermittent longitudinal connections between adjacent such networks. Sometimes the network extends to the surface of the muscle fibre and merge with the cell membrane thus facilitating communication with the extracellular space. By virtue of these T-system of tubules, excitatory depolarization signals generated at the sarcolemma are conducted deeper into the muscle fibre allowing excitation of all the fibrils that make up the muscle fibre. The T-system is vital to the process and in its absence, mechanical contraction does not occur despite electrical activation being present.

The sarcoplasmic reticulum (SR) surrounds each muscle fibril and runs parallel to its axis. These tubules terminate in the form of an enlarged terminal sac at the end of each sarcomere, near the Z line. The walls of the terminal sacs lie close to the T-tubules, which in turn, lie close to the terminal sacs of the next sarcomere. Ca^{2+} enters the SR with the help of Ca^{2+} ATPase, an active transport system, to allow the muscle attain resting state. Once inside, it is bound to Calsequestrin, a specific Ca^{2+}-binding protein. Ca^{2+} is stored intracellularly in the tubules of the SR and is released following excitation of the sarcolemma with a nerve impulse. Ca^{2+} release channel in the SR, also known as the ryanodine receptor (RYR), releases Ca^{2+} from the SR back into the sarcoplasm. Release of Ca^{2+} from the SR is facilitated by the close proximity of the terminal sac with the T-system, through which the wave of depolarization spreads.

Ca^{2+} plays an important role in regulating muscle contraction and relaxation. The regulation may be actin-based, as in skeletal and cardiac muscles or myosin-based as in smooth muscles. The skeletal muscle system is inhibited at rest with the help of the troponin system and the muscle comes out of this resting state to undergo contraction. The regulatory role of troponin and tropomyosin has already been discussed previously.

VI. DISEASES DUE TO DEFECTS IN THE CA²⁺ RELEASE CHANNEL

Some patients who are exposed to certain anaesthetic agents like Halothane or muscle relaxants like Succinylcholine have a severe reaction called malignant hyperthermia (MH). MH is characterized by rigidity of skeletal muscles, hypermetabolism and high fever. It usually has a genetic predisposition and occurs due to increased levels of cytosolic Ca^{2+} in skeletal muscles. These abnormal levels of cytosolic Ca^{2+} are predominantly due to mutation in the Ca^{2+} release channel. This, in turn, leads to their abnormal functioning, resulting in the Ca^{2+} release channels opening more easily and remaining open

for longer periods, thus allowing release of increased levels of Ca^{2+} into the cytosol. These high levels of cytosolic Ca^{2+} lead to a prolonged state of muscle contraction. MH is a life-threatening condition and needs to be treated immediately by stopping the anaesthetic agent as well as initiating treatment with a skeletal muscle relaxant like Dantrolene. Patients with genetic predisposition to MH should not be treated with agents like Halothane or Succinylcholine; other alternative anaesthetic agents should be used for them.

A. Muscular Dystrophies

Muscular dystrophy is defined as a genetically determined degenerative myopathy. These dystrophies have a few common features: (1) weakness and wasting of groups of striated muscles, (2) progressive deterioration in functional capacity (3) hereditary origin (4) and an intact neuromuscular component. Different muscle groups are involved in different clinical types. For example, pelvic girdle muscles are initially involved in Duchenne muscular dystrophy while facial and pectoral girdle muscles are primarily affected in fascio-scapulo-humoral dystrophy.

Mutations in the gene encoding the muscle protein dystrophin lead to Duchenne muscular dystrophy, Becker muscular dystrophy and some cases of dilated cardiomyopathy. Dystrophin is a muscle protein that is attached to the plasmalemma and is responsible for linking the actin cytoskeleton to the extracellular matrix. This is necessary for the assembly of synaptic junction. Defective dystrophin adversely affect its functions. Other types of muscular dystrophies eg. Limb girdle and congenital muscular dystrophies also exist due to mutations in other structural muscle proteins.

B. Energetics of Muscle Contraction

Skeletal muscles contain two types of fibres: type I (slow twitch fibres) and type II (fast twitch fibres). Type I fibres contain more mitochondria and myoglobin compared to type II as they undergo aerobic metabolism to generate energy to support slow yet prolonged muscle contraction specially for processes like maintenance of posture. On the other hand, type II fibres have fewer mitochondria, lack myoglobin and obtain energy for relatively short duration of muscle contraction through anaerobic glycolysis. Type I fibres predominate in leg muscles of marathon runners while type II fibres are present in sprinters.

ATP needed for muscle contraction and relaxation can be synthesized by different processes like (1) glycolysis (2) oxidative phosphorylation (3) from creatine phosphate and (4) from ADP. A sprinter uses creatine phosphate and anaerobic glycolysis to generate ATP while a marathon runner predominantly depends on oxidative phosphorylation to cater to sustained contraction. These type of muscles also have a high demand for oxygen which is met by presence of greater amounts of myoglobin in them thus bestowing the name, red muscles. Contrarily, muscles with less myoglobin are called white muscles due to the absence of the heme moiety. The lactic acid produced during the anaerobic glycolytic pathway is transported to the liver and used to synthesize glucose by the Cori's cycle.

Liver and skeletal muscles contain large amount of glycogen supplies, however skeletal muscles cannot generate blood glucose directly due to the absence of glucose-6-phosphatase. Glycogen phosphorylase and PFK-I are the enzymes that are two predominant sites of metabolic control during glycogen and glucose metabolism respectively. Glycogen phosphorylase is controlled by Ca^{2+}, epinephrine and AMP while PFK-I is regulated by AMP, Pi and NH_3. In marathon runners, aerobic metabolism is the principal source of ATP and the principal fuel sources are blood glucose and free fatty acids obtained from triglyceride metabolism in adipose tissue, which in turn is controlled by epinephrine.

Though ATP is the main source of chemical energy that is converted into mechanical energy, creatine phosphate is a high energy compound present in the muscles that can be used to generate ATP from ADP. Creatine kinase helps to phosphorylate creatine and also has diagnostic value in acute and chronic muscular diseases.

C. Cardiac Muscle

Cardiac muscle is striated like skeletal muscle and its contraction resembles that of skeletal muscle as it uses the actin-myosin-troponin-tropomyosin system. However, unlike skeletal muscle, cardiac muscle is able to generate its own rhythm of contraction. Cardiac muscle T-tubular system is more developed while the SR is less developed compared to skeletal muscle. Consequently, cardiac muscle needs extracellular Ca^{2+} compared to intracellular Ca^{2+}. Cyclic-AMP has a more important role in the functioning of cardiac muscle.

Ca^{2+} inflow and outflow is regulated through various channels in a regulated manner. Ca^{2+} enters myocytes via Ca^{2+} channels which are selective for Ca^{2+}. There are slow Ca^{2+} channels which are voltage gated and open during depolarization by the spread of cardiac action potential and close when the action potential declines. These slow Ca^{2+} channels are stimulated by cAMP-dependent protein kinases and inhibited by cGMP-dependent protein kinases. Fast Ca^{2+} channels are relatively fewer and possibly have a role in the early phase of increase early phase of increase of cytoplasmic Ca^{2+}. Increased intracytoplasmic levels of Ca^{2+} facilitate opening of Ca^{2+} release channels of the SR and this phenomenon is hence called Ca^{2+} induced Ca^{2+} release.

The primary route of release of Ca^{2+} from the myocytes is the Ca^{2+}-Na^+ exchanger. It is involved in maintaining low levels of free intracellular Ca^{2+} in the myocytes by exchanging with Na^+. This helps in regulating both, muscle excitation as well as relaxation. As discussed previously, $Ca^{2+}ATPase$ also plays a small role in regulating efflux of Ca^{2+} from the cell. Cardiac muscle has a preponderance of ion channels which perform important functions. Abnormalities in these ion channels result in a spectrum of diseases called channelopathies.

D. Smooth Muscles

The smooth muscle is called so because, compared to skeletal or cardiac muscles, it appears smooth, without striations under the microscope. Its also called involuntary muscle because unlike skeletal muscle, its not under voluntary control. Also the smooth muscle contracts slowly but is able to maintain its tone for a longer period. The smallest smooth muscles are found inside blood vessels while the largest smooth muscles are found in the pregnant uterus. The nucleus is located in the central bulging part of the cell. Smooth muscle cells may be surrounded by connective tissue but usually they are present as sheets surrounding hollow organs. These smooth muscles regulate the size of their lumen by their contractions. Smooth muscles, in turn, are controlled by the sympathetic and parasympathetic nervous system.

The contractile mechanism in smooth muscles resembles that of skeletal muscles and are made up of bands of actin and myosin. The intracellular calcium activates the myosin possibly by inactivating the inhibitor troponin. The activated myosin breaks down ATP to release energy. Actin filaments slide on the myosin to produce shortening of the muscle and increase in tone. Carbohydrate metabolism provides the energy needed to resynthesize the ATP. The tone is calcium and energy dependent. Some muscles are slow to contract but maintain their tonicity more economically than faster contracting muscles. Response of skeletal muscles to drugs is more or less uniform throughout the body but this not the case with smooth muscles. Different smooth muscles respond differently to sympathetic or parasympathetic stimulation or drugs.

Smooth muscles generally regulate the volume of the viscus or tube that they surround. After prolonged anaesthesia or abdominal surgery, these muscles have a tendency to lose their tone. This causes retention of fluids or gases in the intestine or urinary retention in the bladder leading to abdominal distension and severe discomfort. Smooth muscle stimulants in the form of parasympathomimetic drugs help in relieving this condition by restoring smooth muscle tone thus alleviating such life threatening conditions. On the other hand, excessive smooth muscle tone may result in colicky pain or bronchoconstriction in the case of bronchial smooth muscles and vasoconstriction in the case of arteries. Smooth muscle relaxants or antispasmodics are used to relieve these potentially life threatening conditions. Uterine smooth muscle tone is physiologically useful during pregnancy and delivery and can be regulated by different drugs to facilitate uneventful and safe delivery of the child.

22

Biochemistry of Bone

Rajesh Malhotra

CONTENTS

I.   Biochemical Basis of Bone Structure
   A. Bone: mineral phase
   B. Organic phase: bone matrix
      1. Bone collagen
      2. Noncollagenous Proteins
      3. Ground substance of bone matrix
      4. Other components
   C. Mineralization of matrix
II. Bone as a Dynamic Tissue
   A. Bone growth and remodelling
   B. Coupling
   C. Bone and mineral homeostasis
III. Biochemistry of the Abnormal Bone
   A. Osteoporosis and other bone remodelling disorders
   B. Scurvy and rickets/osteomalacia
   C. Osteogenesis imperfecta
   D. Osteolathyrism
   E. Fibrous dysplasia

"However basic it is to know the properties of bricks, mortar, steel, wood, glass and soil, in building a house, that knowledge alone can never tell how to get architecture into that pile of stuff."
—Paul Weiss

Bones, despite their grossness in form and hardness in texture, have complex microscopic order, rigorously defined architecture and marvellous functional plasticity. The biochemistry of such a complex tissue can be made easy by a reductionistic approach: that is, merely describe in detail the biochemistry of its components. However, such an approach would provide us with incomplete knowledge, which is unfortunately, characteristic of all the currently available information on bone chemistry. The main aim in this chapter will be to describe the structural chemistry of the components of bone tissue, and highlight its relevance in understanding the function of bones, both normal and abnormal.

I. BIOCHEMICAL BASIS OF BONE STRUCTURE

Unlike most other tissues of the body, the main bulk of skeletal tissue is extracellular. The cellular components of bone, viz. osteoblasts, osteocytes, and osteoclasts are few and buried far between the rigid extracellular substance; though they play a very important role in making bone a functionally dynamic tissue.

Bone cells

Chondrocytes, Osteocytes, osteoblasts and osteoclasts are main types of bone cells, which play a major role in bone formation and bone turnover. Chondrocytes form an integral part of growth plate and produce matrix, which is mineralized later to form bone. Osteoblasts are responsible for formation of new matrix as well as regulation of mineralization of matrix. These osteoblasts are converted into osteocytes, when they are trapped in mineralized bone. These osteocytes act as mechanical sensors and play a role in remodelling of bone according to function. Osteoblasts synthesize more alkaline phosphatase, more type I collagen and more bone sialoprotein than osteocytes, while osteocytes specifically produce sclerostin, a glycoprotein. Sclerostin inhibits osteoblast differentiation. There is sexual dimorphism in the density of osteocytes, as females gain osteoclast lacunar density with increasing age, while males show a decrease in this parameter. This may explain why bone loss in women results in a decrease in trabecular number, while in males there is a thinning of trabeculae.[1, 2]

Osteoclasts are multinucleated giant cells, which are responsible for bone resorption and turnover. Osteoblasts signal them for bone resorption and osteoclasts produce acid and proteolytic enzymes to dissolve the mineralized matrix. This process also stimulates osteoblasts through signaling molecules to produce new bone. This process is also affected by sex hormones, explaining the difference in males and females.[1, 3, 4]

Extracellular bone substance

The extracellular bone substance is composed of two distinct, but functionally interacting phases. These are the inorganic or the mineral phase and the organic or the matrix phase.

A. Bone: Mineral Phase

The mineral phase lends rigidity and hardness to the bone. It consists mainly of calcium phosphate, with small, but significant, amounts of calcium carbonate, and very small quantities of other ionic components such as Mg^{++}, Na^+, K^+, Cl^-, F^- and citrate: X-ray diffraction studies have shown that a proportion of its calcium phosphate exists in crystalline form, having a three-dimensional atomic configuration (crystal lattice) very similar to that of the geological family of calcium phosphate minerals—the apatites. However, the biological apatites (bone apatite), which are truly hydroxy apatites, with a lattice structure $(Ca_{10}^{++}-x(H_3O^+)(PO_4)^6(OH)_2)$, are of poor crystallinity and smaller crystal size. These crystals have calcium: phosphorus ratios, which vary from 1: 3 to 2: 6 and contain, besides calcium phosphates, a large number of other ionic constituents as mentioned above. Some of these constituents cannot be accommodated in the apatite crystal lattice. Hence, the greater the presence of these components within the bone, the higher is the percentage of amorphous material present in bone. These crystals are always oriented parallel to the long fiber axis of the collagenous matrix.[1]

By virtue of their small size, the biological apatite crystals possess the property of 'chemi-sorption' of excess phosphates and other ions, which in turn cause absorption of water and formation of an hydration shell around the mineral crystals. In living bone, there is active exchange of ions across this crystal solution inter-phase. The significant sodium content of bone is believed to be due to cation exchange between Ca^{++} ions of the apatite crystal surface and Na^+ ion of the extracellular fluid, across the hydration shell.

Bone scan is a commonly used diagnostic test, which involves administration of technetium-99m labelled bisphos-phonates such as Tc-99m methylene diphosphonate (MDP) and Tc-99m hydroxy diphosphonate (HDP). These radiolabelled agents bind to hydroxyapatite at sites of active bone formation, which is a nonspecific response of bone seen in various conditions like Stress fractures, infection, tumors etc. Recently FDG-PET (fluorine-18 deoxyglucose positron emission tomography) has shown a diagnostic accuracy of 95–100% in detection of bone infections.

B. Organic Phase: Bone Matrix

The organic phase of bone or bone matrix is predominantly composed of collagen fibres with small amounts of complex protein-polysaccharide moieties as ground substance.

1. Bone collagen

There is predominance of type I collagen in bone. The collagen fibres of bone, which constitute about 95% of its organic matrix, account for 40% of the total body collagen. The fibres of bone collagen increase in diameter and become more closely packed with advancing age. Heterogeneity in terms of size of the collagen fibrils as well as composition and sizes of the crystals deposited on this matrix is important for the mechanical competence of bone.[1]

Bone collagen broadly resembles collagen present in other connective tissues of the body (ref. Chapters 7, 14). Glycine constitutes one-third of the total amino acid content of bone collagen, while alanine, proline, and hydroxyproline together constitute 3/9ths of it. As compared to its occurrence in other connective tissues, hydroxylysine, an unique amino acid, is present in significant proportions, in bone collagen. It is believed to play a very important role in ensuring stable interchain linkage between tropocollagen moieties in bone. Collagen in general, and bone collagen in particular, are relatively poor in aromatic amino acids, as also in essential amino acids such as methionine, cystine and tryptophan. Most of the purified collagen of vertebrates contains 0.6 to 2% carbohydrates, identified as D-glucose and D-galactose. These also are presumed to play an important role in the crosslinkage of collagen sub-units.

The most distinctive physical property of bone collagen is its extreme insolubility in reagents, which easily solubilize collagen from other sources. For example, in mineral solutions, less than 0.5% of the total collagen of bone is soluble, and in dilute acid solutions less than 1% is dissolved. However, as high as 30% of bone collagen can be dissolved by repeated freezing and thawing in 3% acetic acid.

The relative insolubility of bone collagen is believed to be due to the presence of strong inter-chain and intra-chain cross-linkages that form in, and between, the proto-collagen and tropo-collagen units of bone collagen. Apart from the non-covalent form of linkages between sub-units, strong covalent linkages have also been shown to exist in bone collagen. For example, it has been recently shown that amino groups of peptide linked lysine residues undergo rapid enzyme mediated oxidative deamination to form –CHO (aldehyde) residues which interact with amino groups of lysine residues of adjacent polypeptide chains to form pre-dominantly inter-molecular "syndesine" linkage as shown below:

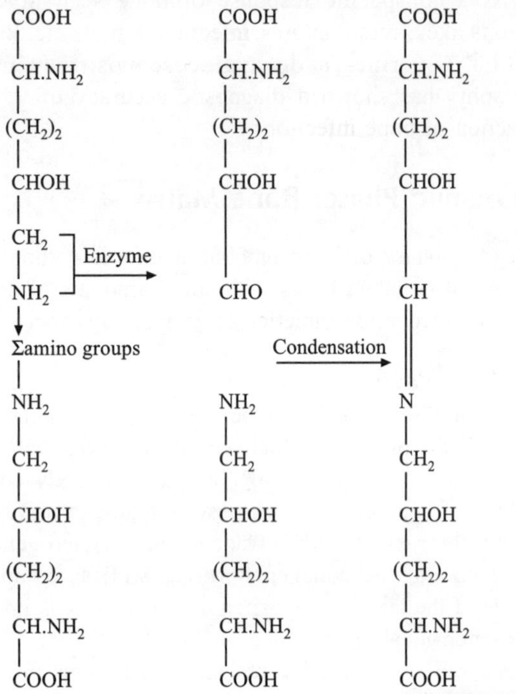

```
COOH              COOH              COOH
 |                 |                 |
CH.NH2            CH.NH2            CH.NH2
 |                 |                 |
(CH2)2            (CH2)2            (CH2)2
 |                 |                 |
CHOH              CHOH              CHOH
 |                 |                 |
CH2 ┐              |                 |
 |  │ Enzyme                        ||
NH2 ┘  ──────►    CHO               CH
 |
 ▼
Σamino groups          Condensation
 |                ──────────────►
NH2               NH2               N
 |                 |                ||
CH2               CH2               CH2
 |                 |                 |
CHOH              CHOH              CHOH
 |                 |                 |
(CH2)2            (CH2)2            (CH2)2
 |                 |                 |
CH.NH2            CH.NH2            CH.NH2
 |                 |                 |
COOH              COOH              COOH
                                    SYNDESINE
```

This 'syndesine' cross linkage, is an extremely rapid, enzyme-mediated, interchain covalent cross-linkage in bone collagen. The enzyme involved needs copper as a cofactor, and is inactivated by certain toxic nitrogenous substances such as beta-hydroxy propiononitrile. There are, perhaps, many more similar covalent intra- and inter-chain cross linkages occurring in bone collagen, which may be responsible for its extreme insolubility. Furthermore, bone collagen, as it gets mineralized, becomes more and more insoluble.

The solubilized bone collagen is mainly composed of alpha chain components (ref. Chapter 5). Basically two types of alpha chain components (alpha 1 and 2) are present in bone collagen. These two chains, though slightly different in amino acid composition, largely resemble in molecular weight. The proportion of these two components in bone collagen differs in comparison to collagen from other connective tissues. For example, it has been shown that the tropocollagen of bone is composed of 2 alpha-1 chains and 1 alpha-2 chain.

Various genetic disorders like osteogenesis imperfect, Marfan syndrome, ehler Danlos syndrome are caused by either defective collagen structure or function.

2. Noncollagenous proteins

There are several proteins in bone matrix other than collagen. Osteocalcin, also called bone gla protein is most abundant of them and has high affinity for hydroxyapatite and calcium. It is commonly used as a bone formation marker[1]. The other protein called matrix-gla protein is mainly found in cartilage and predominantly inhibits hydroxyapatite formation and calcification.[1, 5]

SIBLING proteins (small integrin binding ligand *N*-glycosylated) is another group of proteins present in bone matrix and have signaling functions in addition to interacting with hydroxyapatite and regulating mineralization. This family includes osteopontin (bone sialoprotein 1), dentin matrix protein 1 (DMP1), bone sialoprotein 2 (BSP2), matrix extracellular phosphoglycoprotein (MEPE) and the products of the dspp gene, dentin sialoprotein (DSP) and dentin phosphoprotein (DPP).[1,6]

Small leucine rich proteoglycans (SLRPS) is another protein family, which includes, decorin (the major SLRP produced by osteoblasts), biglycan, osteoadherin, lumican, fibromodulin, aspirin, osteomodulin, Periostin, Tsukushin and mimecan. These protens mainly regulates collagen fibrillogenesis and have an important effect on the mechanical strength of bone.[1,7] They also regulate different aspects of mineralization.

Another group of proteins called 'matricellular proteins' regulate the interactions between the cells and the extracellular matrix. Osteonectin, the matrillins, the thrombospondins, the tenascins, and the galectins are a few of them. Each of these proteins is expressed in higher amounts during development than in adult life, but they are all upregulated during wound repair (callus formation) in the adult.[1] Some other proteins like BAG-75, SPP 24 have also been idenfied and are still under investigation.[1]

3. Ground substance of bone matrix

Apart from collagen, bone matrix contains small quantities of complex protein-polysaccharide moieties as interfibrillar ground substance. They constitute 0.5% of the bone tissue by weight and are extremely complex in their chemical structure and physical properties. The carbohydrate part of these protein polysaccharides contains glycosaminoglycan, sialic acid, etc. while the protein portion contains 40% acidic amino acids. Their average molecular weight is estimated to be about 23,000. The other muco-polysaccharides of bone include chondroitin 4 sulphate, chondroitin 6 sulphate and keratin sulphate.

The exact biochemical role played by these scarce and complex protein-polysaccharides is not clear. They are believed to be of significance in such dynamic functions of the bone as mineralization of matrix, bone resorption, and regulation of bone re-modelling and even in the morphogenesis of bone. They are also believed to function as a well-defined complex molecular chain system providing rigid bridges between collagen fibers.

4. Other components

Several important enzymes, important for regulating intercellular communication and mineralization are also present in extracellular matrix. These include Bone specific alkaline phosphatase, Cathepsin K, Tartrate resistant acid phosphatase, protein kinases etc.[1] Some growth factors such as Bone morphogenetic protein (BMP) and Transforming groth factor –beta are also present in bone matrix.

C. Mineralization of Matrix

Calcifiability is a distinctive characteristic of bone matrix. The biochemical basis for this is not clearly understood. Many hypotheses have been put forward to explain this phenomenon, but none of them, in isolation, seems satisfactory.

A few facts related to the calcification of bone matrix are worth stressing:

(i) Bone mineral crystallites are arranged roughly parallel to the long axis of bone collagen fibrils and are constantly found in relationship with the 640 Å bands of the fibrils. This suggests that specific "structural elements" that occur repeatedly along the collagen fibrils, initiate mineral nucleation and crystal formation (see Figure 22.1).

(ii) Optimal concentrations of the principal ionic components of bone mineral (viz. Ca^{++} and PO_4^- should occur locally in matrix before mineral nucleation and crystallization occur. Bone alkaline phosphatase, perhaps, is partly responsible for this.

(iii) Glycolytic enzymes and ATP are essential for mineralization to occur, *in vivo*.

From the facts recounted above, it appears that mineralization of bone matrix is a complex physico-chemical process which is initiated by the inter-action of "structural elements" in the collagen molecules, and optimal ionic environment, and ATP. The need for ATP suggests that mineral nucleation is, perhaps, an energy requiring process. The detailed mechanism of this process still remains to be elucidated.

Once nucleation of a mineral is initiated, crystal growth proceeds, to form the characteristic apatite crystals all along the collagen fibrils (Figure 22.1) giving rigidity and strength to bone, just as cement and metal give strength and rigidity to the steel scaffolding of reinforced concrete.

II. BONE AS A DYNAMIC TISSUE

Living bone not only grows, but also undergoes constant remodelling to adapt itself to the changing mechanical needs of the organism. Also, being the main reservoir of that functionally vital cation, calcium, it plays a very important role in the calcium homeostasis of the body, in conjunction with the kidneys and the intestines. Furthermore, by such processes as 'chemisorption' and chelation, bone plays a vital role in buffering rapid changes that may occur in the electrolyte composition of the extracellular compartment.

A. Bone Growth and Remodelling

Though we know in considerable detail the chemistry of the structural components of bone tissue, we know hardly anything about the morphogenetic processes that give bones their complex microscopic order and rigorously defined architecture.

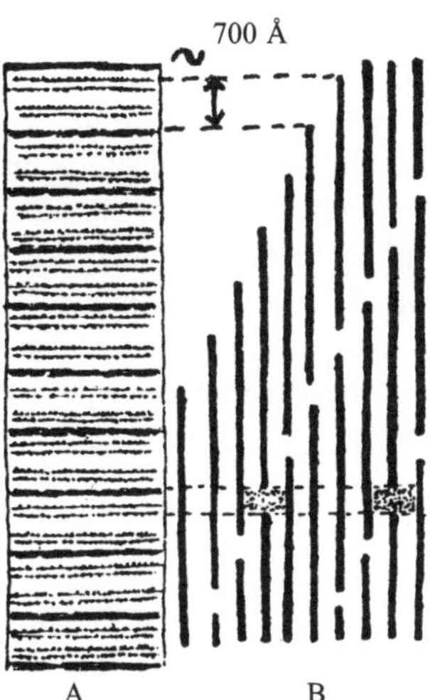

Figure 22.1 A: Depicts schematically the 700 Å axial repeat of bone collagen fibrils as seen by X-ray diffraction. B: Schematic representation of the mode of organisation of individual collagen macro-molecules, to form the fibrils with 700 Å axial repeat: the macromolecules are polymerized head to tail longitudinally and staggered 1/5th of the molecular length. The dotted areas represent the hole zones in the fibril where mineral nucleation occurs.

The innate molecular properties of the individual components, perhaps, bring about their ordered interaction to produce the broad structural design. The cellular elements of bone namely osteoblasts, osteocytes and osteoclasts, play important roles in maintaining and modifying this structure, in accordance with functional demands, by virtue of a variety of humoral and physical influences that impinge on them. For example, the processes of bone resorption, and bone remodelling constantly occur in living bone, and are influenced by mechanical and humoral stimuli. In weight bearing bones like the femur, more bone trabeculae are laid down along the line of transference of force. This is believed to be due to the transduction of mechanical stress to electrical energy (piezoelectricity) by the mineralized collagen fibrils. The piezoelectricity thus produced, stimulates the osteoblasts to lay down more bone trabeculae along the line of transference of force. Conversely, if living bones are not exposed to mechanical stress they show progressive osteoporosis. The rapid osteoporosis (decreased bone mass) that occurs in the bone of young individuals who are suddenly immobilized due to paralytic disease, is believed to be due to the absence of mechanical stress stimulus.

Hormones also play a very important role in the development and maintenance of the structural organization of the bone. Of these the role of thyroxine is of particular significance. In the absence of thyroxine, the growth and maturation of epiphyseal cartilage are affected, causing gross retardation of the growth of long bones. Growth hormone plays a very important role in the anabolic process of bone. This is believed to be mediated by a substance called somatomedin, which is a polypeptide elaborated in the liver under the influence of growth hormone. In the absence of growth hormone and somatomedin, the bones of the body remain small and infantile. Sex hormones play a decisive role in the maturation of bone.

The mechanical and hormonal factors, that influence bone growth, maturation, and remodelling, bring about their effect by acting on the cellular elements of bone, namely, osteoblasts, osteocytes, osteoclasts and the chondrocytes of the epiphyseal cartilage of growing bones. The exact mechanism of action of these agents on these cellular elements remains largely unknown at present.

B. Coupling[9]

Bone remodeling is carried out by a functional and anatomic structure known as the basic multicellular unit (BMU). Bone tissue is continuously remodeled through out life where resorption as well as bone formation go simultaneously through a fine coordinated sequential process between osteoblasts and osteoclasts termed 'Coupling'. Bone resorption signal is usually mediated by osteoblasts under the influence of various hormones or environmental or mechanical stimuli. This signalling involves the release of soluble factors produced by the bone lining cells (**M-CSF** and **osteoprotegerin**) and the interaction of cell surface molecules on the bone lining cells (**TRANCE**) and recruited osteoclasts (**RANK**). Also, the osteoclasts will not bind to bone surfaces covered by unmineralized osteoid, but first require the bone lining cells to degrade the osteoid to expose underlying mineral. Demineralization of bone by the osteoclast releases matrix constituents (such as **BMP** and **TGFβ**) that can stimulate the differentiation and proliferation of osteoblast progenitors or stimulate matrix production by the osteoblasts, and so initiate new bone formation. Osteoclast action is terminated by apoptosis (programmed cell death) and the prevention of new recruitment by osteoprotegerin production by the osteoblasts. Thus each cycle of bone remodeling involves the sequential steps of activation-resorption-formation (**ARF**), carried out by the bone lining cells, osteoclasts and osteoblasts, respectively. Thus bone formation and resorption are tightly coupled. In the young healthy adult resorption and formation are **balanced** such that bone mass remains constant. Pathological states develop when resorption and formation are **unbalanced**, as either excessive bone loss or excessive formation can occur.[9]

Similar to the their diagnostic counterparts, bisphosphonates are common therapeutic agents especially for osteoporosis and attach to hydroxyapatite binding sites on bony surfaces undergoing active resorption. Resorption of bone impregnated with bisphosphonate leads to impairment of ability of osteoclasts to form the ruffled border and to adhere to the bony surfaces leading to inhibition of bone resorption.

C. Bone and Mineral Homeostasis (see also Chapter 43)

Bones contain 95% of body calcium and 75% of body phosphorus. An adult male weighing 70 kg has about 2.5 kg

of crystalline mineral in his bones, and these crystals have an enormous surface area of about 100 acres. A significant portion of this area is exposed to extracellular fluid, and hence is potentially capable of exchanging labile ions with it. As has already been discussed, the hydration shell of these crystals, normally exchange cations and anions such as Na^+, K^+ and HCO_3 with the extracellular fluid. And, depending on the variations in the concentration of these ions in the extracellular fluid, bone can adsorb or release these ions and thus help buffer the extracellular ionic composition. Also, by the hormonally mediated process of bone resorption and accretion, Ca^{++} and $PO4^-$ are dissolved into or absorbed from the extracellular compartment (ref. Chapter 43). Parathormone, vitamin D and calcitonin play very important roles in these processes. Generally speaking, parathormone promotes calcium and phosphorus dissolution from the bone; vitamin D in its biologically active form (ref. Chapter 43) is essential for PTH induced bone mineral mobilization. Calcitonin inhibits this process, and thus indirectly promotes the process of bone formation. Though the actions of these three agents are interrelated, the exact site or mechanism of their action is not known. Recently DeLuca has put forward a hypothesis, according to which, the biologically active form of vitamin D in bone induces the formation of a substance, which brings about the transport of Ca^{++} from the extracellular bone fluid (Figure 22.2) to the intra-cellular fluid of bone cells (osteoclasts or osteocytes). Parathormone and calcitonin compete with each other, to increase (in the case of parathormone) or decrease (in the case of TCT) the permeability of Ca^{++} at the basement membrane sites as indicated in Figure 22.2. Thus it can be said that in vitamin D deficient states the rate-limiting step for Ca^{++} mobilization is at the bone surface of the cell, while PTH and TCT act at the blood surface, to increase or decrease respectively, the permeability of the cell membrane to Ca^{++}.

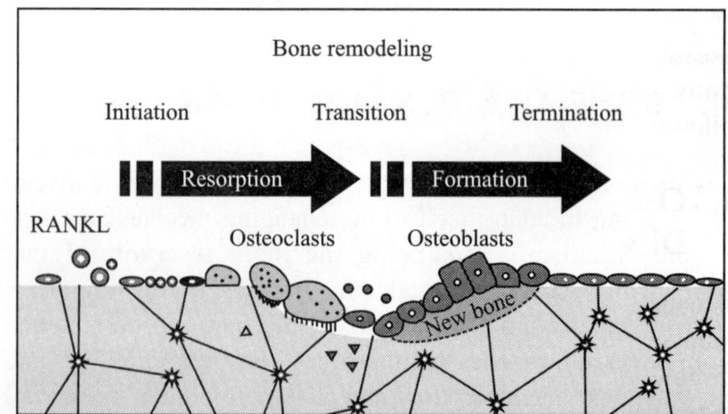

Figure 22.2 Coupling process for bone remodelling.

Whatever the mechanism of action, parathormone is a powerful factor in mineral mobilization from the bone. Because hypocalcaemia stimulates and hypercalcaemia inhibits parathormone secretion, Ca^{++} levels of body fluids are kept constant within a narrow range, either by stimulating actively or inhibiting passively, the mobilization of Ca^{++} from bone.

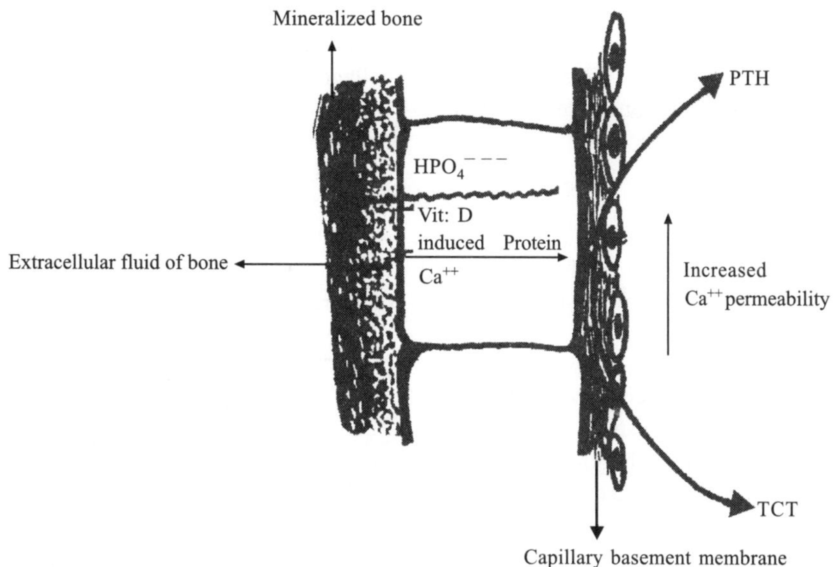

Figure 22.3 Proposed model to explain the inter-dependent action of vitamin D, parathormone (PTH), and thyrocalcitonin (TCT). (After DeLuca, et al.).

III. BIOCHEMISTRY OF THE ABNORMAL BONE

The functional integrity of bone is dependent on the normalcy of its structural components as well as their normal quantitative relationship. Therefore, bone dysfunction would ensue if the structural components were abnormal qualitatively or quantitatively.

There are a large number of bone disorders, both clinical and experimental, which are the result of qualitative abnormality of the structural components of bone. For example, the disease called fluorosis which is widely prevalent in certain areas of North India, is the result of the deposition of an abnormal amount of calcium fluoride in the bones. The cause of this is the presence of excessive fluorides in drinking water. In this disease, the bones become abnormally hard and sclerosed. Many common disorders of bone can be discussed as follows:

A. Osteoporosis and Other Bone Remodelling Disorders

Osteoporosis is a metabolic generalized skeletal disorder charaterized by low bone mass and microstructural deterioration of bone tissue. It is a very common disorder seen in postmenopausal women as well as in geriatric population. This can also occur secondary to various causes like drugs, hormonal imbalance etc. Biochemically, the fat content increases, while mineral and water content decreases[8]. At menopause, deficiency of estrogen leads to increase in number of active osteoclasts resulting in more bone resorption as compared to bone formation. This leads to weaker and more fragile bones with an increased incidence of fractures particularly at hip, spine and wrist.

Hyperparathyroidism is one another bone remodelling disorder involving excessive osteoclastic activity. This increased PTH usually leads to an increasesd bone turnover state characterized by formation of abnormal bone, increased osteoid, and fibrosis. Renal osteodystrophy can also lead to similar state because of secondry hyperparathyroidism associated with it.

Paget's disease is very common in western world and is characterized by increased osteoclast bone resorption with a secondary drastic increase in osteoblast activity leading to an abnormal woven bone formation.

Osteopetrosis is a genetic bone remodelling disorder characterized by by a defect in bone resorption by osteoclasts leading to increased bone density.

B. Scurvy and Rickets/Osteomalacia

Scurvy is a connective tissue disorder that results from ascorbic acid (Vitamin C) deficiency. In this condition there is an impairment of the ability to form normal collagen in bone and other connective tissues. The bones of patients with scurvy are severely osteoporotic and brittle, and often show sub-periosteal haemorrhages, due to poor cohesion between the periosteum and the diaphyses of long bones. There is a complete failure to form osteoid (bone matrix) in the epiphyseal cartilage.

The cause for the failure of formation of collagenous bone matrix is scurvy is not clearly understood. It has been recently shown that human dermal fibroblasts cultured *in vitro* for long, failed to accumulate hydroxyproline or synthesize collagen fibrils. However, when ascorbic acid was added to the culture medium, hydroxyproline accumulation and collagen synthesis reappeared. From this and other related experimental evidences, it is concluded that vitamin C acts as a cofactor for the enzyme catalyzing hydroxylation of proline in protocollagen.

Its deficiency would, therefore, lead to poor formation of collagenous bone matrix, and hence, defective bone.

Rickets is characterized by enhanced bone resorption and increased bone turnover with poor bone mineralization. Osteomalcia is the adult counterpart of Rickets.

C. Osteogenesis Imperfecta

This is a genetically determined disorder of bone, where the latter is poorly formed and brittle. Here the biochemical defect is believed to be due to a genetically determined enzymatic defect in the incorporation of hydroxyproline into the proto-collagen moieties as they are being synthesized in the polyribosomal assemblies of the fibroblasts. A large number of bone disorders, which are the result of abnormalities in the protein polysaccharide ground substance, are described under the broad category of metabolic disorders called mucopolysaccharidoses (see Chapter 32).

D. Osteolathyrism

Osteolathyrism is a disease experimentally produced in rats and chick embryo, by exposing them to beta amino-propiononitrile (BAPN) extracted from the seeds of flowering sweet pea, L-odoratus and also kesari dal. The bone disease thus caused is characterised by thickening and deformation of long bones, misshapen ribs, depressed growth rate of bones, loosening and detachment of tendinous and ligamentous insertions, metaphyseal fibrous defects, detachment of tibial tuberosity, diastasis of sacroiliac joint, etc. Chick embryos treated with this compound showed striking fragility of tissues. Interestingly, an abnormally large percentage of the collagen from lathyritic bones can be extracted with cold saline. The tropocollagen macromolecules from these extracts were comparable in all respects to those extracted from normal bones. This finding indicates the possibility of the defect in lathyritic bone being an impaired polymerization of tropocollagen macromolecules by inter-molecular covalent bonding of the type already described (syndesine linkage). The exact biochemical mechanism by which BAPN brings about the inhibition of inter-molecular collagen bonding is not known. However, it is believed that BAPN and other lathyrogenic substances, inhibit the amino-oxidase enzymes that bring about the oxidative deamination of the Σ- amino groups of peptide-linked lysine residues to the aldehyde (CHO) group. This, being an important initial step in the formation of covalent interchain linkage in collagen fibrils, leads to the formation of a poorly polymerized fragile collagen matrix and hence abnormal bones.

E. Fibrous Dysplasia

This mosaic disorder is associated with sporadic mutations in a G protein gene in osteoblast precursor cells, and results in abnormal differentiation and the excessive proliferation of fibrous tissue in the marrow cavity.[9]

Bone Markers[8]

Various biomarkers of bone formation and bone resorption have been identified, which help in assessment as well as management of various bone pathologies. In some circumstances, it can even act as a surrogate for histological examination. For example markers of bone resorption are excellent indices of disease activity in patients with osteoporosis, Paget disease of bone or bone metastases. Normalization of the test results can be used to help establish the efficacy of treatment.[8] Similarly bone formation markers can be used to assess renal osteodystrophy, osteomalacia, rickets, osteoblastic metastases etc. Total alkaline phosphatase, Bone specific alkaline phosphatase, Osteocalcin, Procollagen type I propeptide are some commonly used bone formation markers. Bone resorption markers can be measured in serum as well as urine and include hydroxyproline, Urinary pyridinoline and deoxypyridinoline, N-telopeptide of collagen type I (NTx), carboxy-terminal cross-linking re- gion of collagen type I (CTx), Serum tartrate-resistant acid phosphatase, Serum cathepsin K, Receptor activator of nuclear factor kappa (RANK), RANK ligand, and its decoy receptor osteoprotegerin.

REFERENCES

1. Orwoll Eric S., John P. Bilezikian and Dirk Vanderschueren (Eds) (2009), *Osteoporosis in Men: The Effects of Gender on Skeletal Health*, Academic Press.

2. Vashishth D., Gibson G.J., Fyhrie D.P., et al. (2005), Sexual Dimorphism and Age Dependence of Osteocyte Lacunar Density for Human Vertebral Cancellous Bone. *Anat. Rec. Part A, Discov. Molec. Cell. Evol. Biol.*, 282:157–162.

3. Novack D.V. and Teitelbaum S.L. (2008), The Osteoclast: Friend or Foe? *Ann. Rev. Pathol.*, 3:457–484.

4. Michael H., Härkönen P.L., Väänänen H.K., et al. (2005), Estrogen and Testosterone Use Different Cellular Pathways to Inhibit Osteoclastogenesis and Bone Resorption. *J. Bone Miner. Res.*, 20:2224–2232.

5. Murshed M., Schinke T., McKee M.D., et al. (2004), Extracellular matrix mineralization is regulated locally; different roles of two gla-containing proteins, *J. Cell Biol.*, 165:625–630.

6. Fedarko N.S., Jain A., Karadag A., et al. (2004), Three Small Integrin Binding Ligand N-linked Glycoproteins (SIBLINGs) Bind and Activate Specific Matrix Metalloproteinases, *FASEB J.*, 18:734–736.

7. Young M.F., Bi Y., Ameye L., et al. (2006), Small Leucine-rich Proteoglycans in The Aging Skeleton, *J. Musculoskelet. Neuron. Interact*, 6:364–365.

8. Singer, Frederick R. and David R. Eyre (2008), Using Biochemical Markers of Bone Turnover in Clinical Practice, *Cleveland Clinic Journal of Medicine,* 75.10:739–750.

9. Roughley P.J., http://www.orthobasicscienceacademy.org/sites/default/files/sample/B1_Cellular_Physiology.pdf

23

Biochemistry of Skin

Somesh Gupta

CONTENTS

I. INTRODUCTION

Skin is the largest organ of the body and constitutes around 16-18% of the normal body weight of an individual. It is around 1.5-4.0 mm thick. It is a dynamic complex organ that constantly renews itself and mediates an array of vital functions by providing a permeability barrier, thermoregulation, sensation, wound healing and physical appearance. It also protects the body against mechanical trauma, ultraviolet radiation and microbial injury.

Skin is comprised of the following interdependent functional layers (Figure 23.1):

1. Epidermis (stratified cellular layer)
2. Dermis (connective tissue layer)
3. Hypodermis (subcutaneous fat layer)

A. Epidermis

The thickness of the epidermis varies from 0.4-1.5 mm[1] and is composed of terminally differentiated stratified squamous epithelium. It consists of keratinocytes, melanocytes, Langerhans cells and Merkel cells. The cells of the epidermis continuously renews itself and forms the derivative structures like the pilosebaceous unit, sweat glands and nails.

1. Keratinocytes

It constitutes around 80% of the epidermal cells[1] and is derived embryonically from the ectoderm. Keratin, an intermediate filament which provides structural support to the cell is the hallmark of keratinocytes[2,3]. There are 34 epithelial and 17 hair keratins[4]. The cells are organized in four layers in all parts of the body except on the palms and soles where there is a fifth layer of stratum lucidum. There are different stages in the maturation of keratinocytes, from the lower-most germinal layer to the end product of horny cells. Normally the cells take around 30 days to move from the germinal layer to the skin surface. This keratinocyte differentiation which occurs as the cells progress upwards through different layers of the epidermis is called as keratinization. Its purpose is to form a terminally differentiated dead keratinocyte[1] which can maintain the permeability barrier function of the skin. The cells undergo various morphological (progressive flattening with increase in size) and metabolic changes with acquisition of specialized organelles as they progress through different layers in the process of keratinization.

The various layers of the epidermis from below upward are (Figure 23.2):

(a) *Basal cell layer (Stratum germinativum):* It contains cells with different proliferation potential such as stem cells,[5,6] transit amplifying cells and the postmitotic cells.[1] In the skin, the long lived stem cells are present in the stratum germinativum and in the bulge of the hair follicle.[7–10] The basal layer is formed by mitotically active, columnar or cuboidal shaped keratinocytes. They are arranged in a single cell layer and are attached to the basement membrane zone by hemidesmosomes. The cells are attached to each other through desmosomes. The cells contain keratin filaments type K5 and K14.

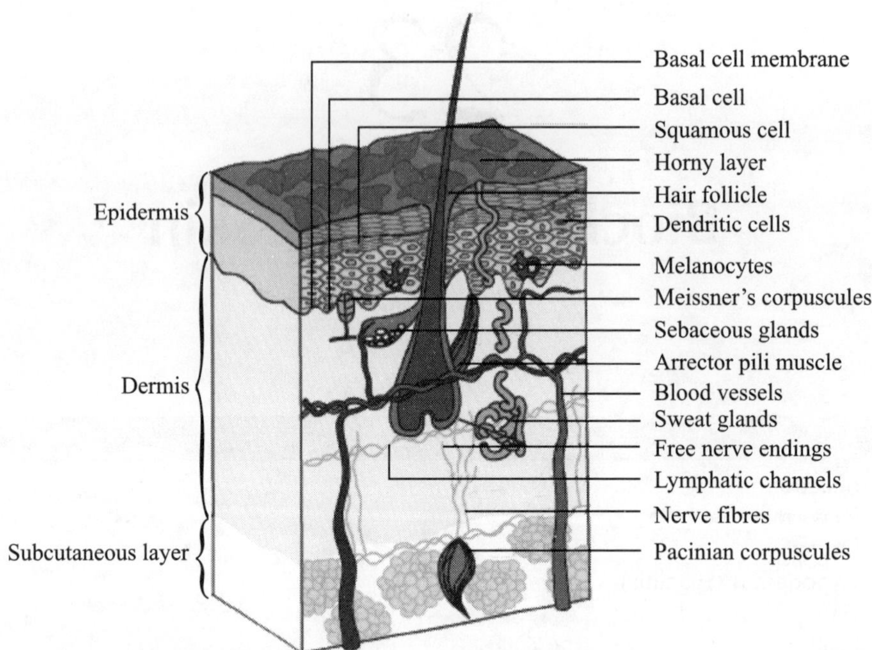

Figure 23.1 Structure of skin.

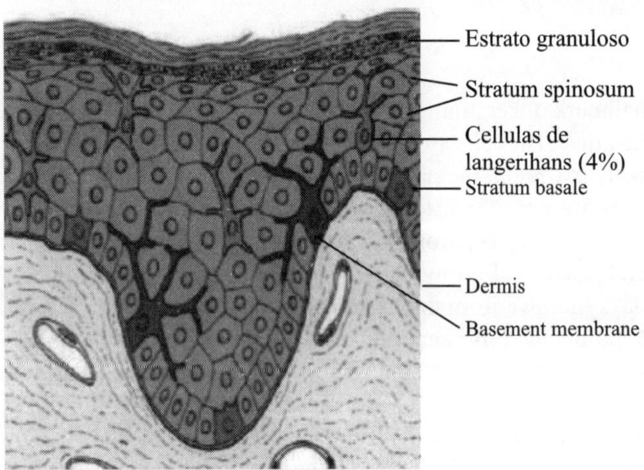

Figure 23.2 Structure of the epidermis.

(b) *Prickle cell layer (Stratum spinosum):* It is named so because of the spine like appearance of the cell margins due to desmosomal attachments which helps in adhesion and protection from mechanical stress.[11] It constitutes most of the thickness of the living epidermis. The shape of the cells is polyhedral or polygonal. K1 and K10 keratins are present in this layer. The cell contains specialized secretory organelle called as *lamellar granules,* which provides the precursors of the intercellular lipids in the stratum corneum which helps to maintain the barrier function.

(c) *Granular cell layer (Stratum granulosum):* It is normally 2–3 cells thick but may become hyperplastic in states associated with excessive formation of mature keratin. The cells are rhomboidal and flattened. The layer is called granular because of the appearance of

a new organelle called as keratinohyaline granule. The granules are composed of profillagrin, keratin filaments and loricrin all of which help in the formation of the cornified cell envelope required to maintain the barrier function of the cell. In the process of differentiation from granular cells into the corneocytes, all the cellular organelles are destroyed except the keratin filaments and the fillagrin matrix. It is at this transition that the contents of the lamellar granules are released to provide the intercellular lipids.

(d) *Stratum corneum:* It is the most superficial layer and is composed of dead, non nucleated, flattened, fully keratinized relatively dehydrated cells. It contains K5 and K14 keratins. This layer is primarily responsible for providing the protection from mechanical trauma, barrier to water loss and to prevent the permeation of soluble substances from the environment.[12,13] The barrier is constituted by the formation of two-compartment model in which the protein rich keratinocytes are surrounded by extracellular lipid. These non-viable cells are constantly, albeit invisibly being shed off the surface of the skin. It takes approximately 14 days for a cell to move from the basal layer and enter the stratum corneum and another 14 days to move through the stratum corneum to finally desquamate.

(e) *Stratum lucidum:* It is only found on the palms and soles, interposed between the stratum granulosum and stratum corneum. It is composed of clear, dying, anucleated flattened epidermal cells.

2. Melanocytes

Melanocytes are derived from the neural crest. They are

pigment producing dendritic cells which are present in the basal cell layer of the epidermis and the hair follicles in the skin.[14] The melanocytes produce a pigment called melanin, which protects the skin against the harmful ultraviolet (UV) radiations from the sun. The melanocytes transfer the melanin to the neighbouring keratinocytes by phagocytosis[15] and this functional unit has been designated as "epidermal melanin unit". Melanin synthesis is discussed in detail later in this under the section 'Biochemistry of skin'.

Control of pigmentation: The behaviour of melanocyte in the skin is influenced by signals from the keratinocytes, autocrine signals and external factors like UV radiation. UV light induces melanin synthesis by a number of processes such as increase in the number of functioning melanocytes, increase melanocyte dendricity,[16,17] increase in the number of melanosomes (both in the melanocytes and keratinocytes), increase in keratinocyte derived factors and by an increase in the tyrosinase activity. The various stimulators of melanogenesis are melanocyte stimulating hormone (MSH) and adrenocorticotrophic hormone (ACTH). Both are derived from propiomelanocortin which is synthesized by the pituitary gland and keratinocytes. MSH is believed to cause the dispersion of melanosomes and increase their melanogenic activity. The α-MSH also helps in repairing DNA damage by UV radiation in melanocytes, reduces the UV induced hydrogen peroxide formation and maintains melanosomal pH. Other stimulators of melanin synthesis are endothelin-1,[18,19] steel factor, inflammatory mediators like prostaglandins and leukotrienes, neurotrophins, basic fibroblast growth factor, nitric oxide and catecholamines. The various melanogenic inhibitors include sphingolipids and bone morphogenic protein BMP-4. Vitiligo is caused by an autoimmune mechanism resulting in depletion of melanocytes.

3. *Other cells of the epidermis*

Merkel cells are slow-adapting mechanoreceptors which are located at the sites of high tactile sensitivity[1] such as digits, nail bed, genitalia and lips.[20] They are derived from the neural crest. They are located on the basal layer of the epidermis and in the outer root sheath of the hair follicle and specialise in the perception of light touch.

Langerhans cells are bone marrow derived antigen-processing and presenting cells of the epidermis which contain the characteristic Birbeck granule.[21]

B. Dermis

Dermis is derived from the mesoderm. In contrasts to the cellular epidermis, dermis is largely a fibrous structure[22,23] composed of collagen, elastin and reticulin fibres that lends elasticity and resilience to the skin as well as supports the blood vessels, nerves and lymphatics. The non-fibrous component of the dermis is formed by the 'ground substance' which contains glycoproteins, proteoglycans and glycoaminoglycans.[24] The cellular component is formed by fibroblasts, macrophages, mast cells and cells of immune system. Dermis is divided into the upper papillary dermis and the deeper reticular dermis. It functions to protect the body against mechanical trauma, binds water, helps in thermoregulation and provides receptors for sensory stimulation. The skin appendages like the pilosebaceous unit and the sweat glands are also located in the dermis. The sebaceous glands secrete the oily secretion called 'sebum' that serves the function of retaining the normal state of hydration of the skin. The eccrine sweat glands are closely coiled structures which are located at the dermo-hypodermal junction. The apocrine glands are located chiefly in the axilla, pubic and anal region.

C. Hypodermis (Subcutis)

It is composed of adipose tissue and functions to insulate the body, protects the skin and provides energy reserve.

D. Hair

Hair is a keratinized structure which is only found in mammals. It is formed by downward projection of epidermal cells called hair matrix which surrounds a dermal invagination called papilla, the two together forming a unit called hair bulb. The canal containing the hair shaft in the dermis is called hair follicle. The number of hair follicles throughout adult life remains constant for a particular region. Hair shows a cyclic growth with periods of growth (anagen), involution (catagen) and rest (telogen).[25] The duration of each phase differs in different region of the body and the length of the hair is determined by the length of the anagen which varies from 2–8 years for scalp hair. In the scalp around 90%–93% of follicles are in anagen and rest are in telogen.[26]

The hair shaft is lodged in the pilosebaceous canal lined by an external root sheath. Its inner core called 'medulla' is surrounded from within outwards by cortex, cuticle, inner root sheath (companion layer, Huxley's layer, Henle's layer and cuticle), outer root sheath and the connective tissue sheath (Figure 23.3). The upper part of the follicle which is permanent consists of the infundibulum and the isthmus. The lower part of the follicle which regenerates with each hair cycle consists of suprabulbar and bulbar region. Attached to the hair follicle and opening into it are sebaceous glands and apocrine glands. A smooth muscle forms a sling around the sebaceous glands and attaches to the dermo-epidermal junction. The toughness of the hair is probably attributable to the strong covalent bonds contributed by sulphur-containing amino acids cystine and methionine. Glycogen is demonstrable in high concentration in various parts of hair follicle.

Disorders of both hair loss (alopecia) and excess hair growth (hirsutism and hypertrichosis) are known. Hormones, particularly the sex hormones, affect the hair growth in males and females. Androgens are responsible for the growth of terminal hair on the chest, upper lips and beard and may be seen in women in association with virilising tumours of ovary or adrenals.

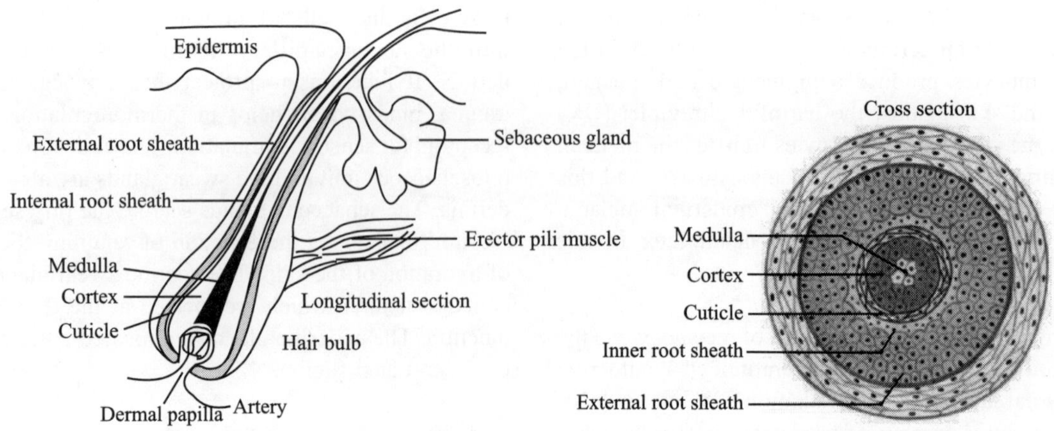

Figure 23.3   Structure of hair.

E.   Nail Development

The development of nail begins during the 9th embryonic week. The nail is formed as a rectangular area from the epidermis from the tip of the digit.[27] The nail apparatus consists of the nail plate, proximal nail fold, nail matrix, nail bed and hyponychium (Figure 23.4). In transverse section the nail plate is formed by 3 parts—dorsal nail plate, intermediate nail plate and the ventral nail plate.[28] The dorsal and the intermediate parts of the nail plate are formed by the keratinisation of the epithelium of the proximal and distal nail matrix, respectively.

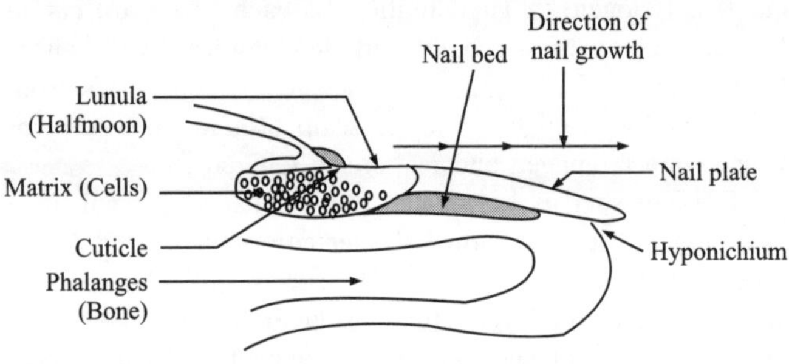

Figure 23.4   Structure of nail.

The ventral part of the nail plate is formed by the nail bed. Nail plate is surrounded proximally and laterally by the proximal and lateral nail folds, respectively. The proximal nail fold has a dorsal and a ventral part. Cuticle is the horny layer of the proximal nail fold. The firm attachment of the cuticle to the superficial nail plate prevents the separation of the nail fold from the nail plate. Nail matrix consists of a proximal and distal part. Keratinization of the nail matrix occurs in the absence of granular layer and along a diagonal axis. Hyponychium separates the nail plate from the nail bed at the tip of the digit. Lunula is the half moon shape, white, opaque area present on the fingernails[1]. The isthmus or the onychocorneal band is present on the fingernails as a distal transverse white band, and represent the most distal point of firm attachment of the nail bed to the nail plate.[29,30]

The onychodermal band is a pink band that separates the onychocorneal band from the white free edge of the nail plate. Extending from the isthmus to the distal margin of the lunula is the nail bed.

The nail serves the function of giving aesthetic beauty to the digits and protects the distal phalanges. Fingernails help in tactile discrimination, help in picking up small objects and in scratching. Toenails help in pedal biomechanics. Fingernails grow at a rate of 3.5 mm/month in adults, which is faster as compared to 1.5 mm/month growth rate of toenails. Time taken for complete replacement of fingernails is around 6 months whereas for toenails it is around 12–18 months.

II.   BIOCHEMISTRY OF SKIN

The dermis

It is composed of a network of connective tissue, predominantly collagen fibrils providing support and elastic tissue providing flexibility, embedded in a mucopolysaccharide gel.[31]

Skin Appendages

Skin plays a role in the body homeostasis, therefore is well-equipped with secretory (release of chemicals from cells for physiological function) and excretory (elimination of waste products of metabolism) capacity. Sweat glands can be sweat secreted with strong odour (apocrine) or with a faint odour (eccrine).

Eccrine Sweat

Contains mainly water (99.0–99.5%) and has a pH of around 5. It also contains electrolytes NaCl, K^+ and HCO_3^-, and other simple molecules-lactate, urea, ammonia, amino acids (serine, ornithine, citruline, aspartic acid) and minerals. Mineral composition varies with the individual according to their acclimatization to heat, exercise and sweating, the duration of sweating, and the composition of minerals in the body.

Apocrine sweat

More viscous, with milky consistency due to high content of fatty acids, cholesterol, squalene, triglycerides, androgens, ammonia, sugars.

Mineral Composition of Sweat

Sodium	0.9 g/l
Potassium	0.2 g/l
Calcium	15 mg/l
Magnesium	1.3 mg/l
Microelements	
Zinc	0.4 mg/l
Copper	0.3 – 0.8 mg/l
Iron	1 mg/l
Chromium	0.1 mg/l
Nickel	0.05 mg/l
Lead	0.05 mg/l

Sebaceous Glands

Glands secrete an oily/waxy matter, called **sebum**, to lubricate the skin and hair and help to maintain the surface pH at around 5. It is composed of 25% wax monoesters, 41% triglycerides, 16% free fatty acids, 1% squalene, small amount cholesterol esters, farnesol (a hydrocarbon that gives the Liebermann-Burchard test) and cholesterol.

Dermal Proteins

Collagen constitutes about 90% of total dermal proteins with predominance of type I (85–90%). Others are type III (8–11%), minor type V (2–4%), (papillary dermis, matrix around vessels, nerves) and type VI associated with fibrils and interfibrillar spaces (responsible for fine structure in early prenatal development of skin). Rest 10% includes elastin, proteoglycans and glycoproteins etc.

A. The Epidermis

The stratum basale

The keratinocytes of the stratum basale are attached to the basement membrane (dermo-epidermal membrane) by hemidesmosomes, which act rather like proteinaceous anchors for these lowest layer cells. Within the stratum basale and the adjacent cell layer, the stratum spinosum, keratinocytes are connected through desmosomes, again highly specialised proteinaceous cellular bridges. Cells in this layer are maturing/aging keratinocytes, melanocytes, Merkel cells (receptor cells).

Melanins

Melanocytes synthesise the pigment melanin from tyrosine. Melanins are polymorphous and multifunctional polymers of eumelanin, pheomelanin, mixed melanins (a combination of the two); and neuromelanin. Mammalian cells produce black-brown **eumelanin** and yellow-redish **pheomelanin.** Eumelanin: highly heterogenous polymer consisting of DHI and DHICA units in reduced or oxidized states. Pheomelanin: mainly sulfur-containing benzothiazine derivatives. Melanin granules formed in the melanocytes tend to be a mixture of these two forms. Melanin absorbs UV light at a wavelength of 280-320 nm and is free-radical scavenger. Both eumelanin and pheomelanin play important protective role in binding to cations, anions, drugs, chemicals, etc. The melanin granules accumulate above the nuclei of keratinocytes and absorb harmful UV-R before it can reach the nucleus and damage the DNA. Three enzymes in melanosomes which are absolutely required for different melanin type synthesis: Tyrosinase (TYR) – responsible for critical step of melanogenesis (tyrosine hydroxylation); Tyrosinase-related protein 1 (TYR1) and DOPAchrome tautomerase (Table 23.1).

TABLE 23.1 Melanin synthesis

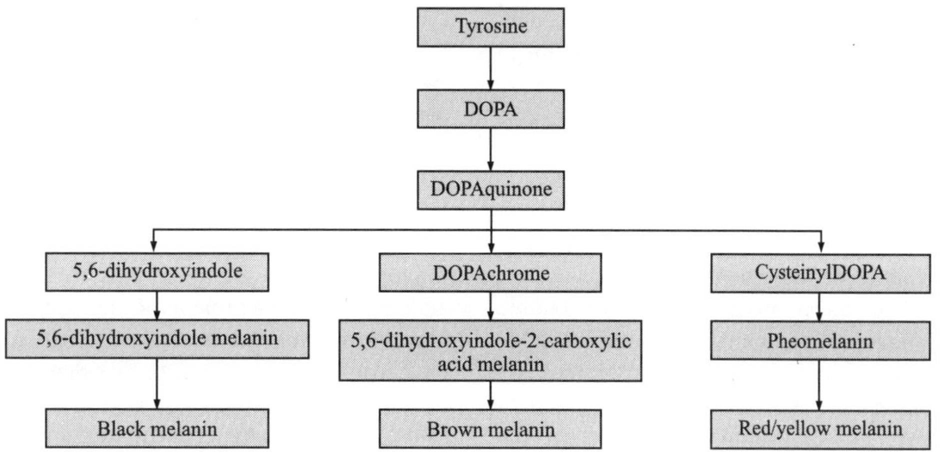

DOPA- 3,4-dihydroxyphenylalanine

(DHI = 5,6-dihydroxyindole; DHICA = 5,6-dihydroxyindole-2-carboxylic acid)

The stratum spinosum

Within this layer the keratinocytes begin to differentiate and synthesise keratins that aggregate to form tonofilaments. Desmosomes connecting the cell membranes of adjacent keratinocytes are formed from condensations of the tonofilaments. It includes 8–10 sheets of keratinocytes with limited dividing capacity and Langerhan's cells.

Langerhans cells

They reside in suprabasal layer and are attracted to keratinocytes by E-cadherin receptor. Their motion is regulated by specific integrin receptor and by α–TNF. UV B stimulates synthesis and release of TNF-α by keratinocytes which in turn modifies the behavior and morphology of Langerhans cells, decreases their total number.

The stratum granulosum: It contains enzymes that begin degradation of the viable cell components such as the nuclei and organelles. Keratohyalin granules mature the keratins within the cell. It includes 3–5 sheets of non-dividing keratinocytes producing keratino-hyalin.

The stratum lucidum: Occasionally, droplets of an oily substance may be seen in this cell layer, possibly arising from the disintegration of lysosomes.

The stratum corneum: This thin membrane, consisting of dead, anucleate, keratinized cells embedded in a lipid matrix, allows for survival of terrestrial animals without desiccation. The stratum corneum has been represented as a 'brick and mortar' model[32] in which the keratinised cells are embedded in a mortar of lipid bilayers. The barrier nature of the stratum corneum depends critically on its unique constituents; 75–80% is protein, 5–15% is lipid with 5–10% unidentified on a dry weight basis.[31] The protein is located primarily within the keratinocytes and is predominantly alpha-keratin (around 70%) with some beta-keratin (approximately 10%) and a proteinaceous cell envelope (around 5%). Enzymes and other proteins account for approximately 15% of the protein component. Crosslinked layer of protein called the cornified envelope protects the body from dehydration, abrasion and infection. The cell envelope protein is highly insoluble and is very resistant to chemical attack. This outer keratinocyte protein has a key role in structuring and ordering the intercellular lipid lamellae of the stratum corneum; the keratinocyte is bound to a lipid envelope through glutamate moieties of the protein envelope.

The envelope is constructed from a variety of soluble (e.g., involucrin and small proline-rich proteins) and insoluble (e.g., loricrin, periplakin, envoplakin) proteins[33] that are glued together by isopeptide bonds formed by the action of transglutaminases.[34] Type I transglutaminase (TG1), an enzyme that catalyzes the formation of ε-(γ-glutamyl) lysine bonds, is the key protein responsible for generation of the crosslinks. Protein, tazarotene-induced gene 3 (TIG3), is expressed in the differentiated layers of the epidermis and its expression is associated with transglutaminase activation and cornified envelope formation.

Lipid content of the stratum corneum includes ceramides, fatty acids, cholesterol, cholesterol sulfate and sterol/wax esters. The stratum corneum lipids are arranged in multiple bilayers, but in contrast to all other lipid bilayers in the body, phospholipids are largely absent. However, eight classes of uniquely structured ceramides (designated ceramide 1 to 5, 6.1, 6.2 and 8) are present in the lipid matrix which, together with the fatty acids, cholesterol and cholesterol sulfates; provide the amphiphilic properties necessary to form lipid bilayers. Natural moisturizing factor (NMF) is a highly efficient humectant synthesized and located within the stratum corneum and helps to maintain suppleness. S. corneum includes 15–30 sheets of non-viable, but biochemicaly active corneocytes.

Transglutaminases in epidermis

Transglutaminases are calcium-activated enzymes that catalyze isopeptide bond formation.[35] Type I transglutaminase[36], TG3[37] and TG5[38] are present in the keratinocytes in surface epithelia. Calcium is considered a necessary cofactor for TG1 activation and TG1 activity is increased at high intracellular calcium concentrations.

Keratins

Keratinocytes contain filaments of the keratin intermediate filament (KIF) family. Hair, nails, horny layers of the skin are formed from keratin cytoskeleton of dead cells. There are two primary groups of keratins, the α-keratins and the β-keratins. α-keratins occur in mammals, β-keratins in birds, reptiles. Both form are right handed helical structure. They are of 2 types: type I – acidic keratins; type II – basic keratins.

Composition and structure of Keratin

Human skin contains ~20 genetically different keratins. Long stretches α-helix is interrupted by short non-helical segments. The most abundant amino acid are glycine and alanine, cysteine can account for up to 24%. Contact between 2 α-helices are formed by hydrophobic amino acid side chain on 1 edge of each helix. Intra- and intermolecular hydrogen bonds and disufide bridges occur at all keratins.

The epidermal permeability barrier

Barrier function in human epidermis depends on transglutaminase-mediated cross-linking of structural proteins and lipids (biological glues). Bonds formed by transglutaminase exhibit high resistance to proteolytic degradation. Proteins are than highly resistent to mechanical perturbation and proteolysis. The quality of the S. corneum barrier depends on the presence of equimolar concentration of ceramides, cholesterol and fatty acids. Changes in the concentration of any of these can affect barrier quality.

Fatty Acids in Epidermis

Together with the acids of the triglycerides, the non esterified fatty acids comprise approximately 70% of the total fatty acid content of human surface lipid (the remainder occurs in mono and diglycerides, wax esters and sterol esters). Arachidonic acid and 20-carbon PUFA can be metabolized by either cyclooxygenase or lipoxygenase pathways.

Epidermal enzyme systems

Most phase 1 (e.g. oxidation, reduction, hydrolysis) and phase 2 (e.g., methylation, glucuronidation) reactions can occur within the skin, though these tend to be at <10% of the specific activities found in the liver.[39] However, esterases tend to have relatively high activities within skin and, considering that there is a large skin surface area, the metabolism of some drugs can be significant.

Skin Metabolism

Primary source for energy production in epidermis is glucose from circulation which diffuses into keratinocytes without effect of insulin. Large proportion of glucose is catabolized into lactate (even in presence of oxygen). Citric acid cycle does operate in epidermis but it is inefficient due to wide fluctuation of temperature and blood flow in skin. 20% of glucose is metabolized by pentose-phosphate pathway (PPP): production of NADPH and pentose for both FA synthesis and nucleic acids.

Secondary source of energy: **Fatty acids** derived from both epidermal stores and exogenous sources (when glucose flow is limited, then FA are metabolized). **Glycogen:** small amount under physiological conditions, however, elevation in all manner of injury of epidermis or during hair growth in follicle.

Lipid metabolism: Synthesized from both glucose catabolism, from AA and circulatingfatty acids. Lipogenesis is going on in all layers of epidermis e.g. sebum synthesis in sebaceous glands.

Degradation: Generally with lipases yielding FA for neutral lipids in outermost layers of epidermis.

SUMMARY

Skin is dynamic organ made up of epidermis, dermis and hypodermis. The appendages present in the skin are the pilosebaceous unit, sweat gland and nails. It seems that in the skin, endogenous glycogen accounts for a small percentage of respiratory activity and that lipids and proteins are far more important sources of energy. Skin plays a balancing role of protecting the host from the external environment and also allows interaction with it. It serves various vital functions of protection, thermoregulation, sensation, wound regeneration and physical appearance.

REFERENCES

1. Chu D.H. (2012), Development and Structure of Skin Goldsmith, Katz S.I., Gilchrest B.A., Paller A.S. and Leffell D.J., Wolff K. (Eds.). *Fitzpatrick's Dermatology in General Medicine*, 8th ed., McGraw-Hill, 58–75.

2. Freedberg I.M., et al. (2001), Keratins and the Keratinocyte Activation Cycle, *J. Invest Dermato*, 116(5):633–640.

3. Fuchs E. (1995), Keratins and the Skin, *Annu Rev. Cell Dev. Biol.*, 11:123–153.

4. Schweizer J., et al. (2006), New Consensus Nomenclature for Mammalian Keratins, *J. Cell Biol.*, 174(2):169–174.

5. Barrandon Y. and Green H. (1987), Three Clonal Types of Keratinocyte with Different Capacities for Multiplication, *Proc Natl Acad Sci.*, USA, 84(8):2302–2306.

6. Blanpain C. and Fuchs E. (2006), Epidermal Stem Cells of the Skin, *Annu Rev Cell Dev Biol.*, 22:339–373.

7. Potten C.S. (1974), The Epidermal Proliferative Unit: The Possible Role of The Central Basal Cell, *Cell Tissue Kinet*, 7(1):77–88.

8. Cotsarelis G., Sun T.T. and Lavker R.M. (1990), Label-retaining Cells Reside in the Bulge Area of Pilosebaceous Unit: Implications for Follicular Stem Cells, Hair Cycle and Skin Carcinogenesis, *Cell*, 61(7):1329–1337.

9. Ito M., et al. (2005), Stem Cells in the Hair Follicle Bulge Contribute to Wound Repair But Not to Homeostasis of the Epidermis, *Nat Med.*, 11(12):1351–1354.

10. Waters J.M., Richardson G.D. and Jahoda C.A. (2007), Hair Follicle Stem Cells. Semin: *Cell Dev Biol.*, 18(2):245–254.

11. Yin T. and Green K.J. (2004), Regulation of Desmosome Assembly and Adhesion. Semin. *Cell Dev Biol.*, 15(6):665–677.

12. Segre J.A. (2006), Epidermal Barrier Formation and Recovery in Skin Disorders, *J. Clin Invest*, 116(5):1150–1158.

13. Elias P.M. (2005), Stratum Corneum Defensive Functions: An Integrated View, *J. Invest Dermatol*, 125(2):183–200.

14. Passeron T., Mantoux F. and Ortonne J.P. (2005), Genetic Disorders of Pigmentation, *Clin Dermatol*, 23(1):56–67.

15. Haass N.K. and Herlyn M. (2005), Normal Human Melanocyte Homeostasis as a Paradig for Understanding Melanoma, *J. Investig Dermatol Symp Proc.*, 10(2):153–163.

16. Scott G. (2002), Rac and Rho. The Story Behind Melanocyte Dendrite Formation, *Pigment Cell Res.*, 15(5):322–330.

17. Friedmann P.S. and Gilchrest B.A. (1987), Ultraviolet Radiation Directly Induces Pigment Production by Cultured Human Melanocytes, *J. Cell Physiol.*, 133(1):88–94.

18. Yada Y., Higuchi K. and Imokawa G. (1991), Effects of Endothelins on Signal Transduction and Proliferation in Human Melanocytes, *J. Biol Chem.*, 266(27):18352–18357.

19. Imokawa G., Yada Y. and Miyagishi M. (1992), Endothelins Secreted from Human Keratinocytes are Intrinsic Mitogens for Human Melanocytes, *J. Biol Chem.*, 267(34):24675–24680.

20. Halata Z., Grim M. and Bauman K.I. (2003), Friedrich Sigmund Merkel and His "Merkel cell", Morphology, Development, and Physiology: Review and New Results, *Anat Rec A Discov Mol Cell Evol Biol.*, 271(1):225–239.

21. Mutyambizi K., Berger C.L. and Edelson R.L. (2009), The Balance Between Immunity and Tolerance: The Role of Langerhans Cells, *Cell Mol Life Sci.*, 66(5):831–840.

22. Burgeson R.E. and Nimni M.E. (1992), Collagen Types. Molecular Structure and Tissue Distribution, *Clin Orthop Relat Res.*, 282:250–272.

23. Christiano A.M. and Uitto J. (1994), Molecular Pathology of The Elastic Fibers, *J. Invest Dermatol*, 103:53S-57S.

24. Kielty C.M. and Shuttleworth C.A. (1997), Microfibrillar Elements of The Dermal Matrix. *Microsc Res. Tech.*, 38(4):413–427.

25. Stenn K.S. and Paus R. (2001), Controls of Hair Follicle Cycling, *Physiol Rev.*, 81:449 [PMID: 11152763].

26. Whiting D.A. (1996), Chronic Telogen Effluvium: Increased Scalp Hair Shedding in Middle-aged Women [see comments]. *J. Am Acad Dermatol*, 35:899.[PMID: 8959948].

27. Zaias N. (1990), *The Nail in Health and Disease*, 2nd ed., Norwalk C.T., Appleton and Lange.

28. Dawber R.P.R. (1980), The Ultrastructure and Growth of Human Nails., *Arch Dermatol Res*, 269:197.

29. Sonnex T.S., Griffiths W.A. and Nicol W.J. (1991), The Nature and Significance of the Transverse White Band of Human Nail, *Sem Dermatol*,10:12.

30. Perrin C. (2008), The 2 Clinical Subbands of The Distal Nail Unit and The Nail Isthmus, Anatomical Explanation and New Physiological Observations in Relation to The Nail Growth. *Am J. Dermatopathol*, 30:216.

31. Wilkes G.L., Brown I.A. and Wildnauer R.H. (1973), The Biomechanical Properties of Skin, C.R.C. *Crit Rev Bioeng*, 453–495.

32. Elias P.M. (1981), Epidermal Lipids, Membranes and Keratinization, *Int J. Dermatol.*, 20:1–9.

33. Nemes Z. and Steinert P.M. (1999), Bricks and Mortar of The Epidermal Barrier, *Exp Mol Med.*, 31:5–19.

34. Jarnik M., Simon M.N. and Steven A.C. (1998), Cornified Cell Envelope Assembly: A Model Based on Electron Microscopic Determinations of Thickness and Projected Density, *J. Cell Sci.*, 111(Part 8):1051–1060.

35. Grenard P., Bates M.K. and Aeschlimann D. (2001), Evolution of Transglutaminase Genes: Identification of A Transglutaminase Gene Cluster on Human Chromosome 15q15, Structure of The Gene Encoding Transglutaminase X and A Novel Gene Family Member, Transglutaminase Z., *J. Biol Chem.*, 276:33066–33078.

36. Polakowska R., Herting E. and Goldsmith L.A. (1991), Isolation of cDNA for Human Epidermal Type I Transglutaminase, *J. Invest Dermatol.*, 96:285–288.

37. Kim I.G., Lee S.C., Lee J.H., Yang J.M., Chung S.I. and Steinert P.M. (1994), Structure and Organization of The Human Transglutaminase 3 Gene: Evolutionary Relationship to The Transglutaminase Family, *J. Invest Dermatol.*, 103:137–142.

38. Candi E., Paradisi A., Terrinoni A., Pietroni V., Oddi S., Cadot B., Jogini V., Meiyappan M., Clardy J., Finazzi-Agro' A. and Melino G. (2004), Transglutaminase 5 is Regulated by Guanine/Adenine Nucleotides, *Biochem J.*, 381:313–319.

39. Hotchkiss S.A.M. (1998), Dermal Metabolism, In: Roberts M.S., Walters K.A. (Eds.), Dermal Absorption and Toxicity Assessment, New York: Marcel Dekker, Chapter 2.

24

Brain

Shail K. Sharma

CONTENTS

I. Neurotransmitters and Drugs
II. The Ultimate Purpose of Life is Pleasure
III. Cognition
IV. Neuroplasticity
V. Growth Factors
VI. Neurogenesis
   Summary

"Every man can, if he so desires, can become the sculptor of his own brain"

—Santiago Ramony Cajal

Brain is amazingly complex machinery. It receives signals, which range in affecting our sensory perceptions of sight, sound, taste, smell and touch to our memories and emotions. How does it perform all these tasks?

It does so through its specialized composition, inlaid with billions of nerve cells, termed neurons. Upon receiving a neural signal, the neuron converts it to an electrical signal and triggers the release of chemicals at special locations called synapses. These chemicals are called neurotransmitters. They act as second messengers and via the generation of electrical signals act in relaying the information from cell to cell.

There are approximately 100 trillion synaptic connections in the brain and its goal is to maintain these connections or restore them back if they were lost in brain injury or trauma. It keeps forming synaptic connections all the time. A single neuron may be interknitted to as many as 200,000 other neurons, via synaptic junctions throughout the body, and can transmit neural signals at speeds of about 100 meters per second. Within a neuron, signals are transmitted by a change of membrane voltage—a variation in the difference in electrical charge between the inside and outside of the cell termed *action potential*. This enables brain to respond to the ever changing events rapidly. Thus human brain can do amazing things; and we humans can do amazing things because of it. We can accept light along the visible wave length reflected from objects in our environment into our, eyes, our retinal neurons convert the light into electrical impulses, the language brain uses to communicate with. As this electricity passes along a neural network, it is sequentially processed and perceived into the form of a 3D-picture of the object in the visual cortical regions of the brain.

I. NEUROTRANSMITTERS AND DRUGS

Otto Loewi and Sir Henry Dale were awarded the Nobel prize in 1936 for showing that neural signals are typically transmitted across synapses by chemicals called *neurotransmitters*. Beneath every thought, dream or action lies a remarkable network of chemical reactions brought about by neurotransmitters that are in constant flux throughout the brain, synthesized and released by the billions of neurons a human brain possesses. In fact these are the chemicals which bring order to human existence.

Neurotransmitters act on the target neurons by binding to specific receptor proteins that convey the signal to the cell and bring about a physiological response. For example, *glutamate* is the excitatory neurotransmitter—that is it causes the hosting neuron to get excited and release another mix of neurotransmitters. N-methyl-D-aspartate (NMDARs) subtypes of glutamate receptors is the predominant molecular device for controlling synaptic plasticity and memory function. Calcium flux through NMDARs is a cellular mechanism for learning and memory. Glutamate crops up in foods either alone or in its flavor enhancing form—monosodium glutamate. Visual transduction cascade involves the molecular machinery of the Ca^{2+}-modulated ROS-GC membrane guanylate cyclase, which generates cyclic GMP and acts as a second messenger of the light signal.

Another neurotransmitter γ-aminobutyric acid (*GABA*) dampens and inhibits nervous activity. Drugs such as benzodiazepine (Alprax) are activators of GABA receptors and are used to help people get more sleep or to lessen anxiety. *Serotonin* is associated with mood—a persons overall state of mind. A lack or deficiency of serotonin in the brain is associated with depression. Drugs such as fluoxetine (Prozac), which cause an increase in the overall levels of serotonin in the brain are commonly prescribed to treat depression. Certain drugs like *MDMA* (ecstasy) 3, 4-methylenedioxymethemphetamine, and *LSD* (Lysergic acid diethylamide) can also stimulate different serotonin receptors, leading to altered or extreme moods.

II. THE ULTIMATE PURPOSE OF LIFE IS PLEASURE

The reward system (feeling of pleasure) of the brain refers to a network where *dopamine* travels around to tell the brain that something good is happening. Of all the neurotransmitters in brain dopamine is most associated with pleasure. Every addictive substance known today affects dopamine release in what is known as the brains *'reward pathway'*. Drugs such as *cocaine* and *amphetamines* lead to sharp, temporary rise in dopamine in the brain. *Cannabinoids* (natural highs) and *opioids* (poppy derived) are commonly used drugs of abuse as they produce euphoria, relaxation and stress reduction.

Opiates and *heroin* deaden pain and create euphoria but also suppress the body's immune system to dangerously low levels. These drugs are addictive and the craving for the drug is what brings about depression and the degeneration of the body systems.

There is experimental evidence that endogenous chemicals *"endorphins"* and other reward chemicals are released during exercise which promote healthy function of the brain and contribute to the natural flow of neurotransmitters.

Meditation, prayer, mindfullness activities are some of the good ways to heal the brain because they shutdown the stress circulatory system. Exercise is one activity we know triggers positive chemicals called *endorphins* which create euphoria and have analgesic activity.

III. COGNITION

Cognitive control is defined by a set of neural processes that allow us to interact with our complex environment in a goal directed manner. Cognition is the mental process of knowing, including aspects such as awareness, perception, reasoning and judgement. Humans regularly challenge these control processes when attempting to simultaneously accomplish multiple goals (multitasking). Multitasking occurs in the frontal lobe that plans, executes and multitasks networks. Regardless of age

our brain has the ability to make new neurons and construct new neural pathways throughout our life. When we engage in new experiences or think in novel ways, new pathways are forged. Everytime we think a specific thought, a specific pathway of neurons fires up, neurotransmitters are released, and synapses are subtly altered. With repetition of the same event this pathway is strengthened. *Understanding how the brain can refashion its own connections is the key to unlocking the double power of positive thinking.*

Hippocampus is the part of the brain critical for memory. The beginnings of memory come into the hippocampus, which organizes them and then goes out and stores them elsewhere in the brain and when cued, then goes out and brings them back. In Alzheimer's disease hippocampus is one of the first regions of the brain to suffer damage, memory loss and disorientation. The *cerebral cortex* can create work-arounds to help us with the memory recalls, e.g., pictures, smell, etc. Everything streams through the *amygdala*—which is the part of the *limbic system*—amygdala decides whether 'incoming' is trustworthy.

IV. NEUROPLASTICITY

Adaptability and plasticity of the brain are the key to human development, cognition and evolution. Plasticity or neuroplasticity, is the lifelong ability of the brain to reorganize neural pathways based on new experiences. As we learn we acquire new knowledge and skills through instructions or experience. In order to learn or memorize a fact or skill, there must be persistent functional changes in the brain that represent the new knowledge.

Data from both animal and human studies shows that plasticity plays a central role in the normal development of neural systems allowing for adaptation to both exogenous and endogenous inputs. The capacity for reorganization and change is a critical feature of neural development, particularly in the postnatal period. With development neural systems stabilize and optimal patterns of functioning are achieved. Stabilization reduces but does not eliminate, the capacity of the system to adapt.

Right now there are no treatments that block the progression of neurodegenerative diseases such as Alzheimer's, Parkinsons and Glaucoma. These and many other diseases involve synaptic loss. If we can find a way to rebuild and restore these connections between neurons, we might be able to halt the progression of these diseases and even reverse them.

V. GROWTH FACTORS

Emerging evidence suggests that *neurotropic* factors that promote the survival and differentiation of developing neurons may also protect mature neurons from neuronal atrophy in the degenerating human brain. The use of neurotrophic factors as therapeutic agents is a novel approach at restoring and maintaining neural function in the CNS. Members of the *nerve growth factors* (*NGF, BDNF and NT-3*) and *trk receptors* (trk receptors are a family of tyrosine Kinases that regulate synaptic strength and plasticity in the nervous system) as well as additional growth factors (*GDNF, TGF-α and IGF-1*) may play important role in the neurodegenerative disorders of the human brain.

V. NEUROGENESIS

Our brains are incredibly elastic. A few decades ago it was thought that brain has essentially no regenerative capacity and was unable to produce new neurons after development. At the time such thinking made sense since brain is a complex organ with millions of intricate connections and if new neurons were added the stability necessary for long term storage of memory and experiences seemed impossible. On this backdrop the discovery of neural stem cells in the adult was a paradigm shift. Although the main idea of brain stability holds true in large part, we have come to accept that neural stem cells in the CNS proliferate and give rise to new neurons throughout life. Today we can optimistically think about CNS repair. Recent discoveries have helped to generate a robust field of *"neural progenitor cell biology"* with relevance to CNS development, pathogenesis and the search for novel neurological therapies. Information on neural stem cells is dovetailing with studies of the ESC and induced pluripotent stem cells We have the technology to make patient matched motor neurons for spinal cord injury or retinal cells to restore vision and identifying factors that inhibit self renewal and it is imperative to do so. Establishing sites of stem cells in vivo is important because the strategy chosen to encourage CNS repair will be quite different based on whether we need to direct progenitor cell migration from remote zones such as *dentate gyrus* of the *hippocampus* or the *subventricular zone* of the *lateral ventricles* or we can activate the local progenitor cells. In animal experiments it has been demonstrated that new borne cells move out of the neurognic zones towards sites of ischemic injury attracted by chemokines such as CXCL12 and BDNF. In animal experiments after cortical injury or infusion of growth factors such as BDNF, CNTF and Shh, new neurons appear in the parenchyma of cerebral cortex, adult striatum, septum, thalamus and hypothalamus. The hope is that with time methods will be developed to have a precise control over this regenerative potential, to direct cells to locations of cell loss or cell injury, to replace appropriate cell populations and to recreate cell functionality. One day there will be a new franchise *"New Brain Inc."* that caters to promote brain enhancement. It turns out that excercise, playing with toys, avoiding stress etc are all proneurogenic. Neurogenesis is not just a constitutive phenomenon, it exhibits both plasticity and elasticity, and with this the spark of purpose.

SUMMARY

For years the doctrine of neuroscientists has been that the brain is a non-repairable :machine: break a part and you loose that function permanently. But more and more evidence is turning up to show that brain can rewire itself, even in the face of catastrophic trauma: essentially the functions of the brain can be strengthened just like a weak muscle.

Nervous system is the most provocative and complex organ, there is wealth of knowledge about cell types and disposition, the function of specific parts and the interaction between the areas. Identifying stem cells in a particular region of the brain which give the feeling of well being, pleasure, emotions, memory , motor functions and visual processing has incumbent implications.

REFERENCES

Brain basics: Know your brain_NINDS_National Institutes of Health, NIH Publications No. 11–3440a las updated 2014.

www.ninds.nih.gov/disorders/brain_basics/know_your_brai.htm

25

Opioid Peptides
Enkephalins Endorphins and Dynorphins

Shail K. Sharma

CONTENTS

I. Introduction
II. Endogenous Opioid Peptides
III. Multiple Opioid Receptors
   Summary

I. INTRODUCTION

Morphine has been used in medicine for centuries to alleviate pain. An optimal exploitation of its beneficial effect has been beset by the wide-spread abuse of the drug arising from its euphoric properties. In addition, the physical dependence on the drug is evidenced by a characteristic stereotyped withdrawal syndrome. In 1973, a major breakthrough came, when specific high affinity binding sites for opiates were identified in brain and in certain peripheral tissues. The fact that these binding proteins (receptors) were found to be clustered in areas of the brain which were earlier mapped by anatomists and physiologists as the emotional centres and the pathways associated with pain, added to the credibility of the above observations.

The question as to why should the animal cells synthesize receptor proteins so highly specific for a drug that is foreign to the body has been a puzzle to the investigators for many years. Is it just incidental? Or, does it indicate the existence of some hitherto unknown factors responsible for higher levels of consciousness eliciting pain and pleasures? This led to an extensive search by scientists for endogenous (natural) ligands possessing a high affinity towards putative opiate receptor proteins present in the brain. The discovery of two endogenous opioid penta-peptides (opioid is the term used for range of gene products with affinities for morphine and its congeners) receptors, (opiate is the term used for opium-derived alkaloids), by Hughes and Kosterlitz in 1975 has been a major landmark in the neurosciences. These were named enkephalins-leucine-enkephalin and methionine-enkephalin (Figure 25.1). The enkephalins bind to opiate (morphine, heroin and other addicting opiate drugs) receptors in the brain and have been shown to be the physiological agonists. Pain relief through

Figure 25.1

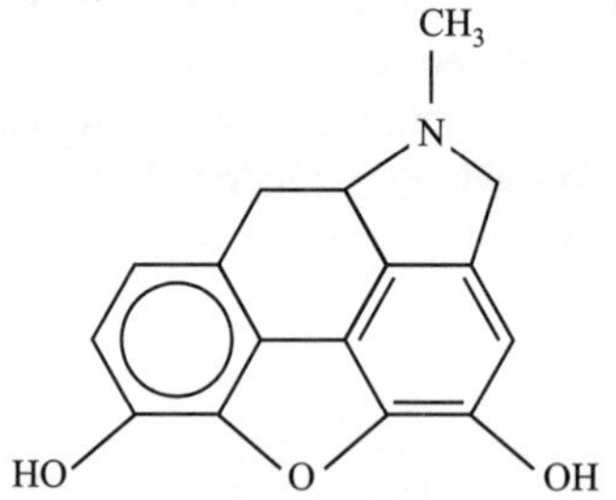

the use of acupuncture, some type of yogic exercises, and phenomenon such as "stress induced analgesia" (deadening of pain sensations), may be mediated by these endogenous opioid peptides. Enkephalins thus represent one of the body's own mechanisms for control of pain.

II. ENDOGENOUS OPIOID PEPTIDES

Three distinct families of peptides have been identified thus far: the *Enkephalins,* the *Endorphins* and the *Dynorphins*. Each is derived from a distinct precursor polypeptide. These precursors are *preproopiomelanocortin* (POMC), *preproenkephalin* and *preprodynorphin* (Figure 25.2).

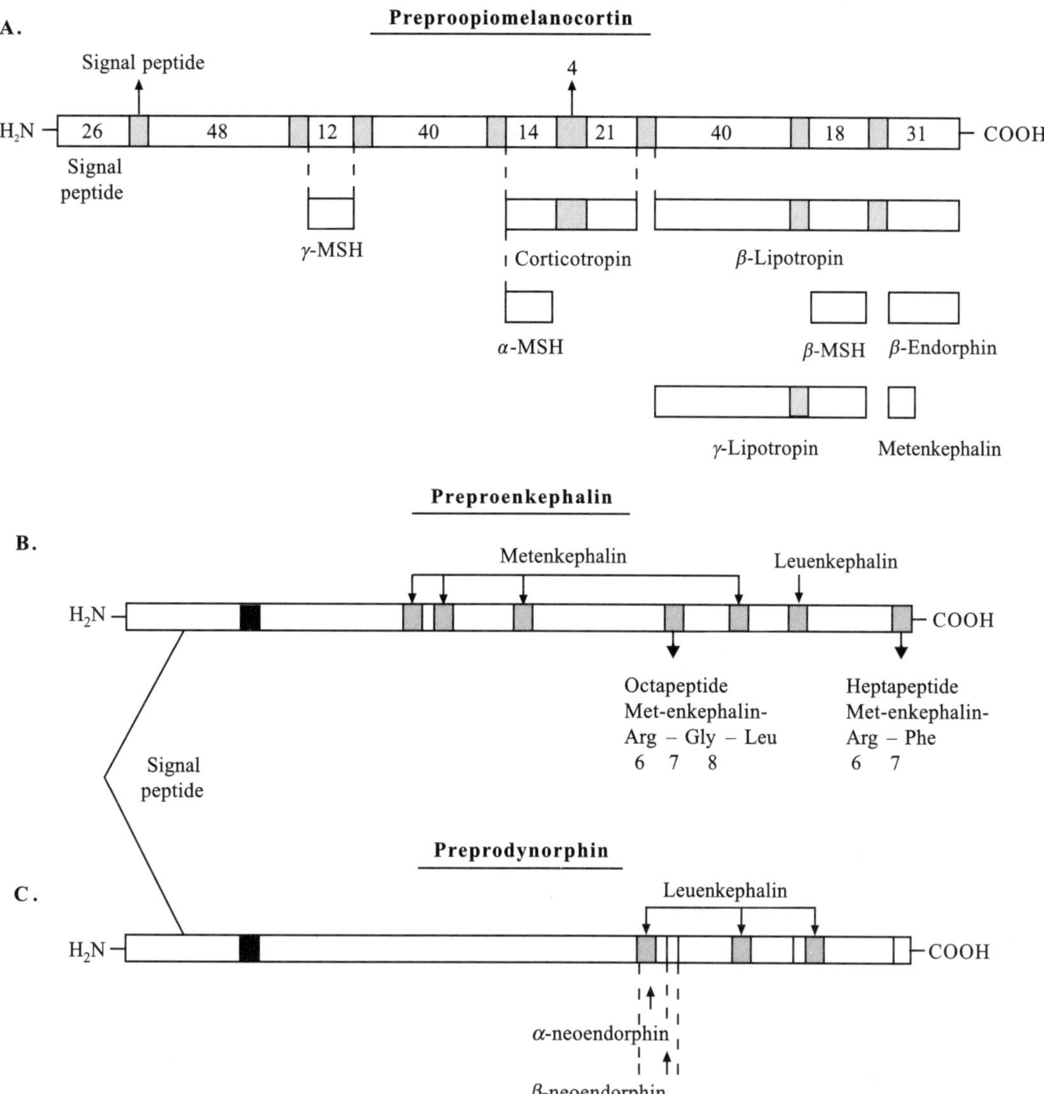

Figure 25.2

POMC is a precursor for corticotropin (ACTH), β-lipotropin (β-LPH), three melanocyte-stimulating hormones (MSH), endorphin and one metenkphalin. Two other preproenkephalin and preprodynorphin are precursors of enkephalins and dynorphins. When several different polypeptides are cleaved from common precursor, the cleavage pattern can vary to yield peptides depending on the cell types. Thus, in the anterior and intermediate lobes of the pituitary gland, the precursor (POMC) is cleaved to yield corticotropin and β-lipotropin. In the intermediate lobe of the pituitary these are further cleaved to yield the γ-MSH and α-MSH; γ-lipotropin; and β-endorphin (Figure 25.2A).

The peptides from prodynorphin and pro-enkephalin are distributed widely throughout the central nervous system (CNS). Proenkephalin peptides are particularly present in areas of the CNS that are related to the perception of pain. The peptides from proenkephalin are also present in adrenal medulla and exocrine glands of the stomach and intestine.

III. MULTIPLE OPIOID RECEPTORS

There is evidence for three major categories of opioid receptors in the CNS, designated μ (mu), κ (kappa), and δ (delta); and subtypes of each have been identified. Molecular cloning and expression of each of the three receptors has been achieved. With the development of selective agonists and antagonists, it has become obvious that each type of opioid receptor has unique properties. Morphine is the prototype agonist of the μ-opioid receptor, enkephalins bind with high affinity to the δ-receptor and ketocyclazocines to the κ receptors. Analgesia is thought to involve activation of μ receptors (largely at the supraspinal sites), κ receptors (principally within the spinal cord) and δ receptors (at both spinal and supra-spinal sites). All the three receptors (μ, δ, and κ) are coupled through G_i to the adenylate cyclase, and at least some of the cellular effects of opiates are manifested by decreasing the intracellular concentration of cyclic AMP.

SUMMARY

Enkephalins, endorphins and dynorphins are the endogenous peptides produced from the larger proteins. They bind to distinct opioid receptors and have been implicated as the endogenous pain killers.

REFERENCE

Brownstein M.J. (1993), A Brief History of Opiates, Opioid Peptides and Opioid Receptors, *Proc. Natl. Acad. Sci.* USA, 90, 5391–5393.

26

Neurotransmitters in Health and Disease

Rajarshi Kar

I. HISTORY OF NEUROTRANSMITTERS

Till the beginning of 20th century, neurologists had conflicting opinion on how nerve impulses are transmitted. Most believed that our nervous system is an electric circuit with continuous flow of current and the synapses acting as switches in between, whereas a few of them argued that chemicals are involved in connecting one neuron with the other. This dispute was settled by Otto Loewi, a professor of pharmacology in Vienna, Austria, with the help of a simple but path breaking experiment. He got the idea of the experiment in his dream and rushed to his laboratory to test his theory as soon as he woke up. He made two frog heart preparations in buffered saline (Ringer's Solution), one with vagus nerve intact (A) and the other without the vagus nerve (B) (Figure 26.1). Stimulation of vagus nerve was known to slow heart rate. As expected when the vagus nerve supplying heart A was stimulated the heart rate slowed down. Interestingly, when he transferred the saline water from heart A to heart B, the rate at which B was beating also slowed down as though its vagus nerve, which has already been severed, has been stimulated. So, it was proved beyond doubt that a chemical was released when vagus nerve was stimulated and this

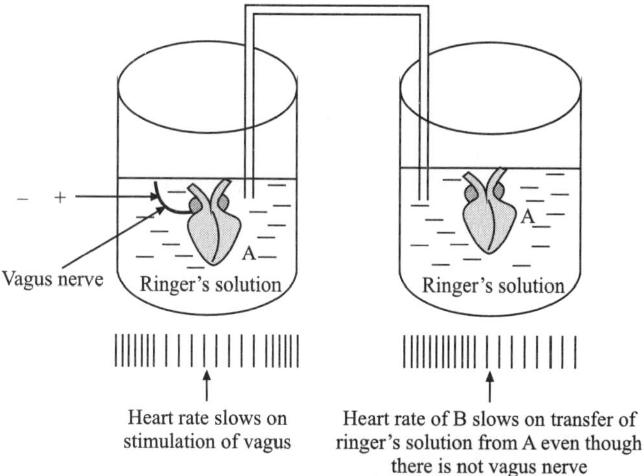

Figure 26.1 Experiment by Otto Loewi proving that neurotransmission is mediated by chemicals.

chemical was actually responsible for slowing the heart rate. Loewi called this substance Vagusstoff (vagal substance) which was later proved to be acetylcholine. Similarly he also showed that the stimulatory affect of sympathetic nerve could also be transferred to the second heart (B) by another chemical released in the ringer solution of heart A. He named it acceleransstoff (accelerating substance) which is now established to be noepinephrine. Sir Henry Dale later showed that such transmission also exists in peripheral nerves at neuromuscular junctions. Loewi and Dale were awarded the Nobel Prize in 1936 for their work on chemical transmission of nerves.

II.   OBJECTIVES

After going through this chapter the reader should be able to:

- Define, characterize and classify neurotransmitters.
- State the distribution and function of major neurotransmitters; their synthesis, secretion and termination of action.
- Classify receptors of the neurotransmitters and describe their mechanism of signal transduction.
- Correlate dysfunction of the neurotransmitters with various neuropsychiatric disorders.

III.   INTRODUCTION

The major role of the nervous system is to integrate external and internal stimuli and send signals to the end organs so that the organism can respond accordingly. This requires effective transmission of signals from one neuron to another and end organs, like muscles, which is carried out by chemicals known as **neurotransmitters**. They are synthesized in neurons and stored in pre-synaptic vesicles. On stimulation of the neuron these are released into the synaptic cleft so that they bind to receptors present on post synaptic neuron or end organs and transmit the signal (Figure 26.2). After the signal is transmitted the neurotransmitters are rapidly degraded to terminate the action. Some chemicals released from neurons do not have an independent effect on the post synaptic neuron. They can only modulate the effect of other neurotransmitters and are known as **neuromodulators**. Proper understanding of

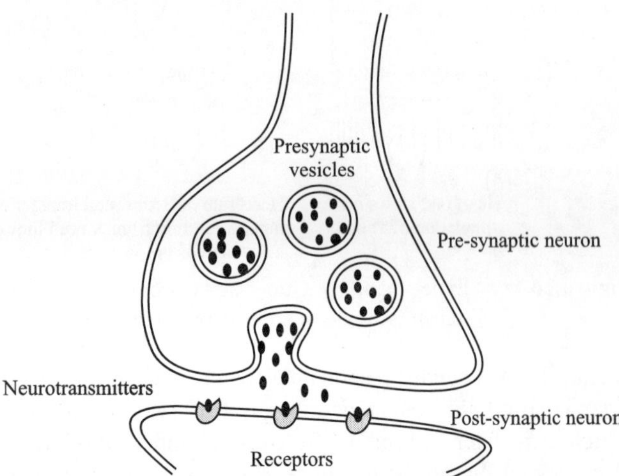

Figure 26.2   Schematic diagram showing neurotransmission.

the role of neurotransmitters and neuromodulators in health and pathophysiology of diseases is important for designing effective and specific therapy for neurological and psychiatric conditions.

For a chemical to be called a neurotransmitter:

- The chemical must be produced by the pre-synaptic neuron and stored in pre-synaptic vesicles.
- It must be released on stimulation of the pre-synaptic neuron in sufficient quantity.
- Receptors for the transmitter must be present on the post-synaptic neuron.
- Experimentally applying the chemical to target cells will have similar effect as stimulating the pre-synaptic neuron.

IV.   CLASSIFICATION OF NEUROTRANSMITTERS

Neurotransmitters, on the basis of their chemical nature, can be broadly classified into three major categories; **small molecule neurotransmitters**, **large molecule neurotransmitters**, and **gaseous neurotransmitters**

- Small molecule neurotransmitters
  - Acetylcholine
  - Amines
    Catecholamines (Dopamine, epinephrine and norepinephrine), 5-Hydroxytryptamine and Histamine
  - Amino acids
    Glutamate, Aspartate, GABA, glycine
  - Purines
    ATP and Adenosine
- Large molecule neurotransmitters
  - Peptides
    Endorphins, substance P, neuropeptide Y
- Gaseous neurotransmitters
  - nitric oxide and carbon monoxide

A.   Receptors for Neurotransmitters

Most neurotransmitters, with the exception of the gaseous ones, cannot cross the lipid bilayer of target cells. Therefore, to act on the target cells they need the presence of receptors on the cell membrane. The receptors for neurotransmitters are present on the post-synaptic neuron, target organs like muscles, Gastro intestinal tract and also on pre-synaptic neuron. One neurotransmitter can act through multiple receptor types and thus have varied action at different locations. For example, norepinephrine can bind to five different receptors; α_1, α_2, β_1, β_2 and β_3.

The receptors can be broadly classified structurally and functionally into two types; *ligand gated channels (ionotropic receptors) and G-protein coupled receptors (metabotropic receptors)* (Figure 26.3). The ligand gated receptors are ion channels which open when a ligand binds to the receptor for

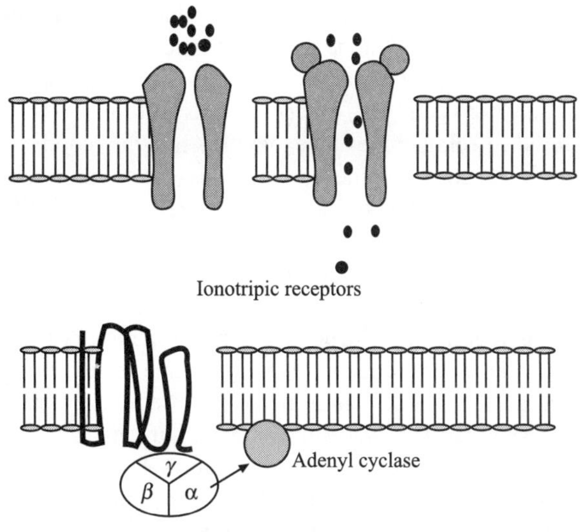

Ionotripic receptors

Adenyl cyclase

Metabotropic receptors

Figure 26.3   Receptors for neurotransmitters.

a brief period of time (few milliseconds). The metabotropic receptors are G-protein coupled receptors which activates a second messenger when a ligand binds to it. The second messenger can in turn modulate an ion channel. Almost all neurotransmitters act through both the type of receptors, the ionotropic for a rapid response and the metabotropic for a delayed and more sustained effect. Prolonged exposure of these receptors to their ligands causes their **desensitization** making the cells unresponsive to the neurotransmitter.

The receptors on the pre-synaptic neurons are of two types—**autoreceptors** (the neurotransmitter acts on the pre-synaptic receptor of the same nerve from which it is released and causes feedback inhibition) and **heteroreceptors** (the ligand is a neurotransmitter or neuromodulator other than the one secreted by the nerve on which the pre-synaptic receptor is located) Figure 26.4.

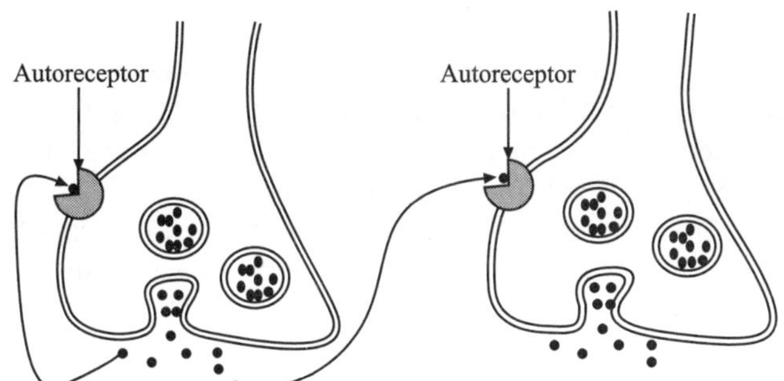

Figure 26.4   Pre synaptic receptors: autoreceptors and hetero-receptors.

B.   Small Molecule Neurotransmitters

Acetylcholine

Acetylcholine is the neurotransmitter of parasympathetic neuroeffector junction, neuromuscular junctions, autonomic ganglia and central nervous system. All the nerves coming out of the CNS, cranial nerves, motor neurons and preganglionc neurons have acetylcholine as the neurotransmitter. In the CNS, cell bodies of cholinergic nerves are present in sepatal nuclei and nucleus basalis of the basal forebrain and the post mesencephalic cholinergic complex. Their axons projects into widespread areas of neocortex, hippocampus and thalamus and is involved in regulation of sleep wake states, learning and memory.

Synthesis of acetylcholine and termination of action

Acetylcholine is synthesized from acetyl-CoA and choline. The enzyme involved is *choline acetyltransferase*. This enzyme is present in high concentration in cholinergic neurons. Immunochemistry with antibodies specific for this enzyme can help identify cholinergic neurons and synapses. Choline required for synthesis is taken up by Na^+ **dependent choline transporter** from extracellular space. After synthesis, acetylcholine is concentrated in pre-synaptic vesicles by a **vesicle-associated transporter**. When a nerve impulse reaches the synaptic junction there is influx of Ca^{2+}, resulting in fusion of the vesicle with pre-synaptic membrane and release of acetylcholine into the synaptic cleft. Acetylcholine then diffuses across the cleft and binds to the receptors present on the postsynaptic neuron or effector cells (Figure 26.5).

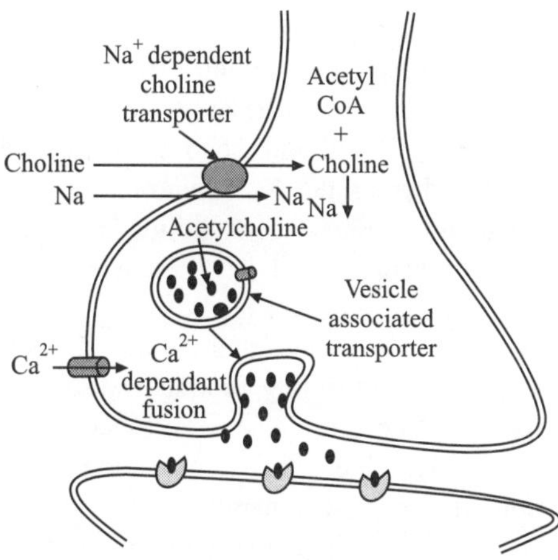

Figure 26.5   Synaptic transmission in cholinergic neurons.

Botulinum toxin blocks acetylcholine release, whereas black widow spider venom causes excessive release of acetylcholine, both of which is fatal.

Acetylcholine has to be rapidly degraded to terminate the action of cholinergic neurons and repolarize the post synaptic neurons or target cells. After transmission acetylcholine is rapidly hydrolysed to acetic acid and choline by the enzyme acetyl cholinesterase. Cholinesterase is present at high concentration at cholinergic nerve terminals. Organophosphate and carbamates inhibit cholinesterase enzyme and prevents breakdown of acetylcholine. This causes persistent action of Acetyl choline at the synaptic cleft in cases of poisoning with these insecticides. If left untreated it may cause death of the patient. Atropine is effective in treating these cases as atropine competes for the receptors of acetylcholine and antagonizes its action.

Receptors

Acetylcholine receptors are classified into two types based on the agonists (other than acetylcholine) which stimulate them; those which are activated by muscarine, the poison present in toad's stool, are **muscarinic receptors** and those activated by nicotine **are nicotinic receptors.** Muscarinic receptors are transmembrane proteins, spanning the membrane seven times and act via G-protein (**metabotropic**). There are five types of Muscarinic receptors M1 to M5. M1, M4 and M5 receptors are present in the central nervous system, M2 in heart and M3 in smooth muscles and glands. These receptors are linked through G-protein to adenylcyclase, K+ channel and/or phospholipase C.

The muscarinic receptors are found in all effector cells stimulated by postganglionic neurons of the parasympathetic system as well as those stimulated by the postganglionic cholinergic neurons of the sympathetic system. Activation of muscarinic receptors result in miosis (constriction of pupil), salivation, lacrimation, nausea and vomiting, slowing of heart rate (bradycardia), increased gastro-intestinal motility, micturition, increased bronchial secretion and bronchoconstriction and decrease in blood pressure.

The nicotinic receptors are present in motor end plates of skeletal muscles (N_M) and in the synapses between the preganglionic and postganglionic neurons of both the sympathetic and parasympathetic system (N_N). The N_M receptors are made of 5 subunits; two α-subunit, one each β, γ and δ subunit, whereas the N_N receptors are made of only of α and β subunits. When acetylcholine binds to **nicotinic acetylcholine receptors** on skeletal muscle fibers (N_M), it opens ligand-gated sodium channels in the cell membrane (**ionotropic**). Sodium ions then enter the muscle cell, initiating a sequence of steps that finally produce muscle contraction. The subunits of N_N receptors also form a Na^+ channel which opens when acetylcholine binds to the subunits and causes depolarization of the post synaptic neuron. Stimulation of nicotinic receptors at sympathetic ganglia results in increase in heart rate and blood pressure and dilation of pupil.

Damage to cholinergic neurons in the central nervous system is associated with loss of memory. In **Alzheimer's disease,** which is characterized by progressive loss of memory, there is significant decrease in acetylcholine levels in the CNS. In **myasthenia gravis,** autoantibodies against nicotinic receptors blocks action of acetylcholine at neuromuscular junction, resulting in severe muscle weakness.

Catecholamines

The major catecholamines are **norepinephrine, epinephrine** and **dopamine** and the neurons secreting them as neurotransmitters are noradrenergic, adrenergic and dopaminergic neurons, respectively. The major site for synthesis and secretion of catecholamines is the adrenal medulla, which is actually a modified ganglion. The cholinergic pre-ganglionic neurons innervate the secretory cells which has ATP bound catecholamines stored in them. When the acetylcholine released from the preganglionic neurons binds to the receptors present on the secretory cells Ca^{2+} enters the cells and causes release of the stored catecholamines by exocytosis into the blood stream. The released catecholamines have various effects on the nervous system, cardiovascular system and also exert metabolic effects. Besides adrenergic, noradrenergic and dopaminergic neurons are also present in the central nervous system and are part of the autonomic nervous system as well.

Norepinephrine is the neurotransmitter in most sympathetic post ganglionic nerve endings. Norepinephrine secreting neurons are also present in the central nervous system. They originate in the locus ceruleus and innervate various areas of cortex, hypothalamus, cerebellum and spinal cord. The noradrenergic neurons also regulate secretion of anterior pituitary hormones and inhibit secretion of posterior pituitary hormones; vasopressin and oxytocin. Decrease in norepinephrine levels in synapses in brain has been associated with depression.

Adrenergic neurons secreting **epinephrine** are present in the central nervous system though their function has not yet been elucidated. They originate in the medulla and project into thalamus, periaqueductal gray and spinal cord.

Dopaminergic neurons have their cell bodies in autonomic ganglia and in some parts of the central nervous system. They have been discussed in detail later in the chapter

Synthesis of catecholamines and termination of action

The catecholamines—dopamine, norepinephrine, and epinephrine—are synthesized from tyrosine in the chromaffin cells of the adrenal medulla and those neurons in which they are the neurotransmitters. Epinephrine constitutes 80% of catecholamines synthesized in medulla and is usually not synthesized in extramedullary tissues. In contrast, most of the norepinephrine and dopamine are synthesized in the noradrenergic and dopaminergic neurons respectively.

The biosynthetic pathway and the enzymes involved are illustrated in Figure 26.6.

Tyrosine is the immediate precursor of catecholamines, and **tyrosine hydroxylase** is the rate-limiting enzyme in catecholamine biosynthesis. It converts L-tyrosine to L-dihydroxyphenylalanine (**L-dopa**). Catecholamines cannot cross the blood-brain barrier; hence, in the brain they must be synthesized locally. In certain central nervous system diseases (e.g., Parkinson's disease), there is a local deficiency of dopamine synthesis. L-Dopa, the precursor of dopamine, readily crosses the blood-brain barrier and so is an important agent in the treatment of Parkinson's disease. DOPA decarboxylase converts L-dopa to 3,4-dihydroxyphenylethylamine (**dopamine**). Compounds that resemble L-dopa, such as α-methyldopa, are competitive inhibitors of this reaction. α-Methyldopa is effective in treating some kinds of hypertension. Dopamine is then converted to **norepinephrine** by dopamine β-hydroxylase. Norepinephrine is the only neurotransmitter that is synthesized in the vesicle. PNMT (**phenylethanolamine-N-methyltransferase**) catalyzes the N-methylation of norepinephrine to form **epinephrine**

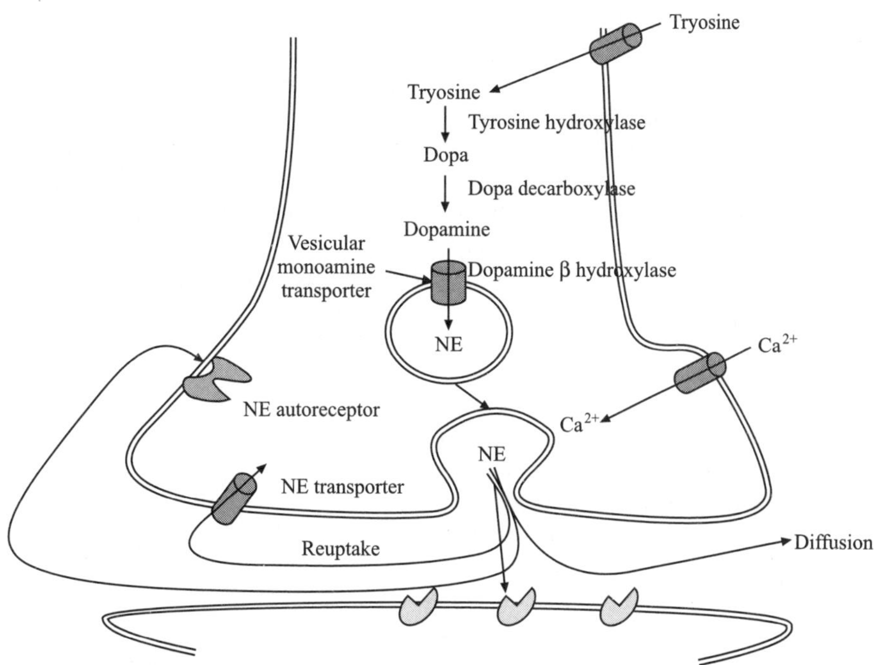

Figure 26.6 Synaptic transmission in adrenergic neurons neurons.

in the epinephrine-forming cells of the adrenal medulla and adrenergic neurons. In these cells norepinephrine comes out of the vesicle and is converted to epinephrine in the cytoplasm

After secretion of catecholamines by the neurons their action is terminated by three mechanisms. First, reuptake by the adrenergic nerve endings, second, diffusion into blood and surrounding tissues and third, metabolism by the enzyme *Monoamine oxidase* (MAO), found in nerve endings, or by *catechol-O-methyl transferase* (COMT) which is distributed in all tissues.

Receptors for catecholamines

Epinephrine and norepinephrine acts through α and β adrenergic receptors. Subtypes of α and β receptors have been identified which are differentially distributed in various tissues. The distribution of adrenergic receptors and their responses are described in Table 26.1. All adrenergic receptors are metabotropic (G-protein coupled). α_1 is linked to Gq and activates IP3/DAG pathway and stimulates postsynaptic neurons. α_2 receptors are associated with inhibitory G protein (Gi). It inhibits adenyl cyclase, causes hyperpolarization by opening G-protein coupled K^+ channel and inhibit neuronal Ca^{2+} channels. Therefore α_2 receptors cause inhibition of postsynaptic neurons. They are also present in pre-synaptic neurons and provide feedback inhibition and prevent further release of norepinephrine. β receptors are linked to stimulatory G-protein and causes activation of adenyl cyclase.

Dopamine

In some neurons catecholamine synthesis stops at Dopamine as rest of the enzymes are absent in these neurons. These are dopaminergic neurons and when stimulated they secrete

TABLE 26.1

Receptor	Tissue	Responses
α_1	Vascular smooth muscle, genitourinary smooth muscle, liver, intestinal smooth muscle Heart	Contraction Contraction Glycogenolysis, gluconeogenesis Relaxation Increased contractile force
α_2	Pancreatic islets, Pre-synaptic nerve terminals vascular smooth muscles Platelets,	Inhibits insulin release Feedback inhibition Contraction Aggregation
β_1	Heart Juxta glomerular cells	Increased force and rate of contraction Increased rennin secretion
β_2	Smooth muscle (vascular, bronchial, Gastro Intestinal, and Genito Urinary) skeletal muscle, and liver	Relaxation glycogenolysis
β_3	Adipose tissue	Increased lipolysis

dopamine into the synaptic cleft. Termination of action is by reuptake through Na^+ co-transport and metabolism by MAO and COMT.

Dopaminergic fibers arise from two main brainstem areas; one is substantia nigra, which projects to striatum (**nigrostriatal system**) and the other ventral tegmental area projecting to the neocortex and subcortical areas (**mesocortical**

system). The nigrostriatal fibers are involved in fine tuning of motor movements. Damage to these fibers is responsible for **Parkinson's disease**. The mesocortical system is essential for cognitive functions and is thought to be involved in motivational and emotional response. This system is probably involved in the pathogenesis of schizophrenia, a disease characterized by hallucination, delusion, lack of emotion and motivation, and apathy. Addictive drugs also act by stimulating the mesocortical system.

Dopamine Receptors: They are G-protein coupled receptors (metabotropic) and are classified into two groups depending on the type of G-protein coupled to the receptor. D_1—like receptors are associated with stimulatory G-protein (Gs) whereas D_2 like receptors activate inhibitory G-protein (Gi). There are five members of dopamine receptors; D_1 and D_5 belongs to D_1 like family and D_2, D_3 and D_4 belongs to D_2—like family. D_2 receptors have been implicated in the pathogenesis of schizophrenia.

Serotonin

It is a neurotransmitter secreted by serotonergic neurons in the CNS and enterochromaffin cells of GIT. It is also found in high concentration in the platelets. In the CNS, cell bodies of serotonergic neurons are concentrated in the midline raphe nuceus of the brain stem. From there the axons project to a widespread area of the brain including hypothalamus, limbic system, neocortex, cerebellum and spinal cord.

Synthesis and Termination of Action

5-HT is synthesized from tryptophan in serotonergic neurons and enterochromaffin cells. Platelets cannot synthesize serotonin but concentrate it by a 5-HT transporter. The same transporter is involved in reuptake of serotonin by the nerve endings to terminate the action of serotonin at the synapses. **Tryptophan** is converted to **5-hydroxytryptophan** by *tryptophan hydroxylase*. 5-hydroxytryptophan is then decarboxylated by the enzyme *5-hydroxytryptophan decarboxylase* to form **5-hydroxytryptamine** or **serotonin**. (Figure 26.7) 5-HT is metabolized by Monoamine oxidase (MAO) to 5-hydroxy indole acetaldehyde. The aldehyde is then converted to 5-hydroxyindoleacetic acid (5 HIAA) by aldehyde dehydrogenase. 5 HIAA is then actively transported out of the brain.

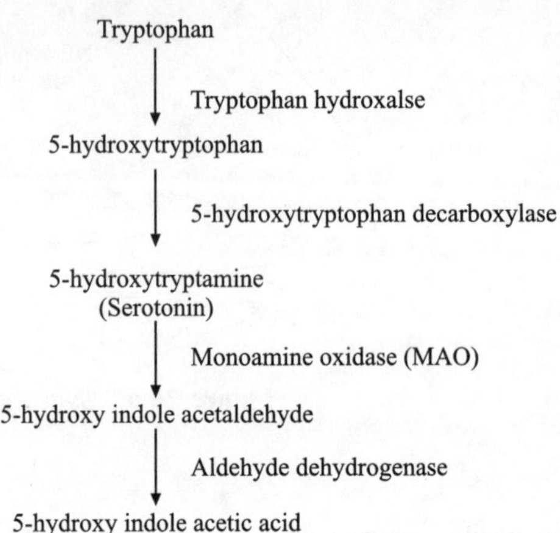

Figure 26.7 Synthesis and termination of action of serotonin.

Receptors

There are 14 subtypes of 5-HT receptors. The distribution and function of these receptors are described in Table 26.2. Most of the receptors act by activating adenyl cyclase. $5HT_{2A}$, $5HT_{2B}$ and $5HT_{2C}$ receptors activate phospholipase C (PLC)

TABLE 26.2 Distribution and function of 5-HT receptors

Receptor subtype	Second messenger	Distribution	Function
5HT1A		Hippocampus	Autoreceptor
5HT1B		Substantia Nigra	Autoreceptor
5HT1D	Inhibits adenyl cyclase	Cranial blood vessels	Vasoconstriction
5HT1E		Cortex	
5HT1F		Cortex	
5HT2A		Platelet	Platelet aggregation
		Smooth muscle	Contraction
	Activates IP3/DAG pathway	Cerebrl cortex	Neuronal activation
5HT2B		Stomach fundus	Contraction
5HT2C			
5HT3	Ligand operated ion channel	Area prostema	Vomiting response
5HT4	Activates adenyl cyclase	striatum, substantia nigra and olfactory tubercle	Excitatory transmission
5HT5A	Inhibits adenyl cyclase	cerebral cortex, hippocampus, granule cells of the cerebellum,	Unknown
5HT5B	Not known	hippocampus, habenula and the dorsal raphe nucleus	Unknown
5HT6	Activates adenyl cyclase	striatum, nucleus accumbens, olfactory tubercle, hippocampus and cerebral cortex	Unknown
5HT7	Activates adenyl cyclase	cortex, septum, thalamus and perioheral neurons	Causes smooth muscle relaxation in periphery

and so the second messengers are IP3 and DAG. 5HT3 is the only receptor which is linked to ligand operated ion channel.

Serotonin plays an important role in modulating various functions of the nervous system, sleep-wake cycle, cognition, sensory and motor response, appetite, mood and sexual behavior, and thermoregulation. The serotonergic fibres are also responsible for controlling aggression and impulsivity. Besides the above CNS affects, serotonin also increases gastric motility, vascular smooth muscle contraction and platelet aggregation. 5HT3 receptors present in area prostema, along with those present in the Gastro intestinal tract, mediate the vomiting response. Ondansetron which is an antagonist for 5HT3 receptors is used to prevent vomiting, especially in patients receiving chemotherapy for the treatment of cancer. Serotonin levels in brain are important in preventing anxiety and depression. Drugs which inhibit reuptake of serotonin are used to treat anxiety states and depression.

Histamine

Most of the histamine is synthesized in mast cells and basophils and is involved in inflammatory and hypersensitivity reactions. Apart from its central role in the mediation of allergic reactions, gastric acid secretion and inflammation in the periphery, histamine serves an important function as a neurotransmitter in the central nervous system. The histaminergic neurons originate from the tuberomamillary nucleus of the posterior hypothalamus and send projections to most parts of the brain and spinal cord. Histamine consists of an imidazole ring and amino group. It is formed by decarboxylation of **histidine** by the enzyme *histidine decarboxylase.*

The central histamine system is involved in many brain functions such as arousal, control of pituitary hormone secretion, suppression of eating and cognitive functions. The effects of neuronal histamine are mediated via G-protein-coupled H_1-H_4 receptors. H_1 receptors activate phospholipase C and H_2 receptors act via stimulatory G-protein to increase cAMP levels. H_3 receptors are mostly presynaptic and decrease release of histamine as well as other neurotransmitters. The newly discovered H_4 receptors are probably involved in regulating cells of the immune system.

Research shows that histamine probably plays a role in regulating sleep-wake cycle, blood pressure, pain threshold, addiction and sexual behavior. It also regulates secretion of anterior pituitary hormones. The prominent role of histamine as a wake-promoting substance has drawn interest to treat sleep-wake disorders, especially narcolepsy, via modulation of H_3 receptor function. Histaminergic system has been implicated in the pathogenesis of neurological and psychiatric diseases. Brain histamine levels are decreased in Alzheimer's disease patients, whereas abnormally high histamine concentrations are found in the brains of Parkinson's disease and schizophrenic patients. Low histamine levels are associated with convulsions and seizures. The release of histamine is altered in response to different types of brain injury, e.g., increased release of histamine in an ischemic brain trauma might have a role in the recovery from neuronal damage.

V. AMINO ACID NEUROTRANSMITTERS

The most widely distributed amino acid neurotransmitters are glutamate, aspartate, GABA and glycine. Glutamate and aspartate are excitatory whereas GABA and glycine are inhibitory neurotransmitters.

A. Glutamate and Aspartate

These amino acids are so widely distributed in the central nervous system that it was initially difficult to establish them as neurotransmitters. Both act via the same receptors, the glutmate receptors, and have excitatory effect on the post synaptic neurons. Glutamate alone is responsible for approximately 80% of excitatory neurotransmission in the CNS.

In the glutaminergic neurons glutamate is available from two sources; one is from α-ketoglutarate of TCA cycle, which is converted to glutamate by the enzyme **transaminase.** The other source is from neighboring glial cells which take up the released glutamate and convert it to glutamine by **glutamine synthase.** The glutamine is then transported back to the neuron where it is converted back to glutamate by **glutaminase.** Glutamate is concentrated in the synaptic vesicles by specific transporters. Glutamate action is terminated by reuptake into neighboring glial cells and the glutaminergic neurons secreting it (Figure 26.8).

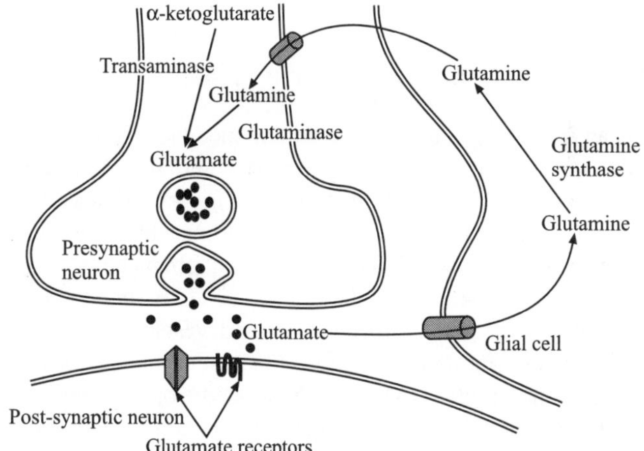

Figure 26.8   Synaptic transmission in Glutaminergic neurons.

The glutamate receptors are broadly classified into **ligand gated ion channel** (ionotropic) receptors or **G-protein coupled receptors** (metabotropic). Ionotropic receptors are always excitatory whereas the metabotropic receptors can be either excitatory or inhibitory. The ionotropic receptors are further classified on the basis of type of agonists that activate them. The NMDA receptors are activated by N-methyl-D-aspartate and the non-NMDA receptors which are again of two types, activated by AMPA (α-amino-3-hydroxy-5-methyl-4-isoxazole propionic acid) and Kainic acid.

The NMDA receptors require binding of both glutamate and glycine to respond. The receptors are ion channels which are blocked by extracellular Mg^{2+} and opens only when there is partial depolarization due to activation of AMPA or kainate receptors. The NMDA, AMPA and kainate receptors are co-localized on the post synaptic neurons and act together to bring about appropriate signal transduction. AMPA and Kainate receptors are in general concerned with point to point transmission and results in quick depolarization of post synaptic neurons, whereas NMDA receptor is mostly involved in fine tuning these responses.

The metabotropic glutamate receptors activate the phospholipase C pathway or inhibit adenyl cyclase via inhibitory G-protein to increase IP3/DAG or decrease cAMP, respectively. There are eight subtypes of metabotropic glutamate receptors present at pre-synaptic or post symaptic neurons. The receptors present pre-synaptically **(autoreceptors)** regulate release of glutamate.

High concentration of glutamate can cause excessive activation of NMDA or AMPA/kainate receptors. The activation of these receptors results in massive influx of Ca^{2+} and death of neurons. **Ischemia** or **hypoglycemia** results in massive release of glutamate and decrease in reuptake. This causes continued action of glutamate and neurotoxicity. Excessive glutamate is also responsible for neurodegenerative disorders like **Amyotropic lateral sclerosis (ALS)**, **Parkinson's disease** and **Alzheimer's disase**. Though gltuaminergic neurons are distributed widely in the CNS, drugs which modulate them are still to be developed. NMDA receptor antagonists are being tried intraspinally or extradurally to relieve chronic pain.

B. GABA (γ-amino Butyric Acid)

It is an inhibitory neurotransmitter and mediates the inhibitory actions of local interneurons in the brain and also presynaptic inhibition in spinal cord. It acts as neurotransmitter at the synaptic junction between small interneurons and nerve bodies of cerebellar cortex, olfactory bulb and hippocampus. The entire output from cellebellar cortex is GABAergic, which sends inhibitory signals to cerbellar and vestibular nuclei. It also mediates inhibitory impulses between caudate nucleus and substantia nigra. GABA is synthesized from glutamate by glutamate decarboxylase. Termination of action is by reuptake via Na^+–Cl^- transporter and then it is converted to succinic semiladehyde by transaminase and then to succinate.

GABA acts by three different receptor type A, B and C (Figure 26.9); $GABA_A$ is a ligand gated Cl^- channel, $GABA_B$ receptors is coupled to inhibitory G-protein (Gi) and inhibits adenyl cyclase enzyme, also activates K^+ channel and reduce Ca^{2+} conductance. $GABA_C$ is transmitter gated Cl^- channel and is more potent than $GABA_A$. They are found in neurons of retina, spinal cord and pituitary.

GABA receptors are the site of action of CNS depressants, like benzodiazepines, barbiturates, ethanol and volatile anesthetic agents like halothane. Barbiturates are better anticonvulsant because they not only potentiate GABAA receptors but also antagonize AMPA receptors.

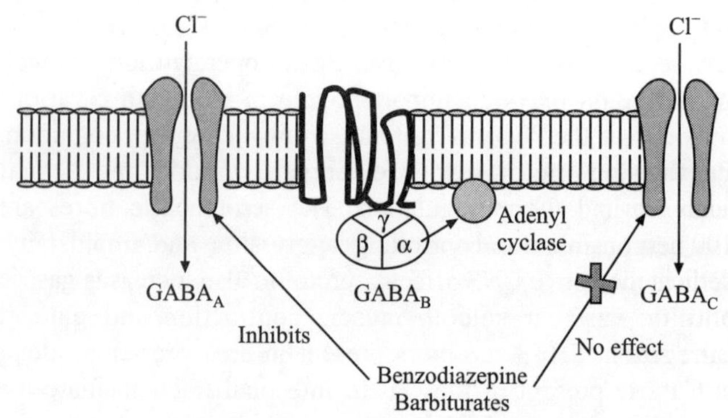

Figure 26.9 GABA receptors.

C. Glycine

Glycine is the major inhibitory neurotransmitter in the brainstem, spinal cord, cerebellum and retina. it participates in a variety of motor and sensory functions. Glycine acts as cotransmitter with glutamate to activate NMDA receptors as described earlier. Thus, glycine is both inhibitory and excitatory in the CNS.

Glycine is formed from serine by the enzyme serine hydroxymethyltransferase (SHMT). Glycine, like GABA, is released from nerve endings in a Ca^{2+}-dependent fashion. The actions of glycine are terminated primarily by reuptake via Na^+/Cl^--dependent, high-affinity glycine transporters, similar to GABA transporters. The specific uptake of glycine has been demonstrated in the brainstem and spinal cord in regions where there are also high densities of inhibitory glycine receptors.

Glycine receptors are ionotropic Cl^- channels with pentameric structure composed of 3α and 2β subunits. Glycine on binding to these receptors increases Cl^- conductance and hperpolarizes the post-synaptic membrane.

Glycine is antagonized by strychnine which is a deadly poison. The convulsion seen in strychnine poisoning emphasizes the importance of inhibitory neurons in maintaining proper motor function.

D. Purines

ATP acts as both neurotransmitter and co-transmitter and is practically present in all synaptic vesicles. It is often released with other neurotransmitters like Norepinephrine, serotonin, glutamate, dopamine and GABA. ATP is metabolized to adenosine which also acts as neurotransmitter. ATP has both ionotropic (P2X) and metabotropic (P2Y) receptors. P2X are cationic channels permeable to Na^+, K^+ and Ca^{2+} ions. They are present in dorsal horns and are probably involved in sensory response. P2X antagonists are being developed to treat chronic pain. P2Y are G-protein coupled receptors which activate K^+ currents, modulates NMDA receptors and voltage gated Ca^{2+} channels. P2Y is present more on astrocytes. Adenosine receptors are present pre-synaptically and act to inhibit synaptic transmission by inhibiting influx of Ca^{2+}.

VI. LARGE MOLECULE NEUROTRANSMITTERS

A. Neuropeptides

Peptides acting as neurotransmitters are usually 3-40 amino acids long. They have receptors throughout the CNS. More than 100 neuropeptides have been identified. The peptides may have dedicated neurons where they are the only transmitters or more often they are co-released with other small molecule transmitters. Like small transmitters they are packaged in vesicles, release is Ca^{2+} dependent and binds to receptor. However there are some differences; unlike small molecule transmitters, which are synthesized at nerve endings, peptides are synthesized at cell bodies of the neurons. They are synthesized as pro-peptides and then broken down in the vesicle to the active form. The vesicles where they are stored are denser and larger than that of small molecule transmitters. They diffuse out after release and so proximity of post synaptic neuron is not necessary and their action is terminated by proteolysis. We will restrict our discussion to three important peptide neurotransmitter, Opioid peptides, substance P and neuropeptide Y.

Opioid peptides are synthesized from large precursor molecules, like proopiomelanocortin found in anterior pituitary. The endogenous opioids are enkephalins, dynorphin and endorphin. They are widely distributed in the CNS and they inhibit neurons involved in pain perception. Opioid receptors are of three classes; μ, κ and δ. They are G-protein coupled receptors and all inhibit adenyl cyclase. Activation of μ receptors increases K^+ conductance and that of κ and δ receptors closes Ca^{2+} channels.

Substance P is an 11 amino acid long peptide. It is secreted by neurons present in brain, GI tract and the dorsal root ganglia. It is probably the neurotransmitter between primary sensory neuron and the spinal interneurons in the dorsal horn of the spinal column and is involved in pain transmission. It is a co-transmitter with dopamine in the nigro-striatal system and is also found in high concentration in the hypothalamus. It is responsible for axon reflex and increased peristalsis in the gastro intestinal tract. It acts through a G-protein coupled receptor.

Neuropeptide Y is a 36-amino acid peptide that acts as a neurotransmitter in the brain and in the autonomic nervous system. It is mainly produced by neurons of the sympathetic nervous system and serves as a strong vasoconstrictor. In the brain neuropeptides Y secreting neurons are found in various locations including the hypothalamus, and are thought to increase food intake and fat stores, reduce anxiety and stress, pain perception, and affect circadian rhythm. Like Substance P, neuropeptide Y receptors are G-protein coupled and mobilize Ca^{2+} and inhibit adenyl cyclase. There are eight neuropeptide Y receptors (Y_1-Y_8). Y_1 and Y_5 antagonists are being tried as anti-obesity drugs.

B. Gaseous Neurotransmitters

Gases like Nitric oxide (NO) and carbon monoxide (CO) have been shown to act as neurotransmitters at synapses between inhibitory motor neurons of enteric motor nerves and GI smooth muscle cells. They also serve as transmitters at various regions of the CNS. They are synthesized when there is depolarization of the nerves where they act as transmitters. Increased Ca^{2+} after depolarization activates enzymes involved in synthesis of NO and CO. As they easily diffuse through membranes, they do not have receptors. They directly bind and activate enzymes. NO modify activity of NMDA receptors and the Na^+ K^+ ATPase pumps by nitrosylating them. It also activates guanyl cyclase and increases cGMP, which is responsible smooth muscle relaxation in arteries.

SUMMARY

- Neurotransmitters and neuromodulators are divided into three major categories: small-molecule transmitters, large-molecule transmitters and gaseous neurotransmitters.
- Neurotransmitters are synthesized by enzymes specifically expressed in the neurons which secrete them and stored in synaptic vesicles. They are released into synaptic cleft from vesicles on activation of the neuron. They bind to pre or post synaptic receptors and receptors present on target cells. Their action is terminated by reuptake or through metabolism by specific enzymes.
- Receptors of neurotransmitters are of two types; ionotropic or metabotropic. Ionotropic receptors are fast acting whereas the metabotropic receptors are slow acting. Ionotropic receptors are ligand gated ion channels whereas the metabotropic receptors are G-protein coupled.
- Major neurotransmitters include acetylcholine, catecholamines, serotonin, histamine, glutamate, GABA, and glycine, and opioids. ATP, NO, CO can also act as neurotransmitters or neuromodulators.
- Deregulation of neurotransmitters have been implicated in the pathogenesis of neuropsychiatric disorders, like schizophrenia, Parkinson's disease, depression, anxiety, ALS, Azheimer's disease and host of other diseases. Various drugs modulating the receptors of these neurotransmitters are used therapeutically in the above mentioned diseases.

27

Neurobiology of Sleep and Sleep Disorders

B.S. Shankarnarayan Rao, P.N. Ravindra and M. Kutty Bindu

CONTENTS

I. Definition
II. Classification of Sleep Stages
III. Sleep Architecture (Hynpogram)
IV. Neural Substrates of Wake and Sleep
V. REM sleep
VI. Importance of Sleep as an Adaptation to Environment
VII. An Overview on Sleep disorders
VIII. Obstructive Sleep Apnea (OSA)

Sleep is a distinct behavioral state essential for health and survival. We spend almost one third of our life time in sleep. Until 1950s, Sleep has been considered as an enigma without much information on the brain mechanisms associated with sleep. With the land mark discoveries of Moruzi and Magoun in 1949(1), a paradigm shift occurred in our understanding of sleep wake behavior. Since then, sleep is no more considered as a passive, idling state of mind, but a specific physiological, dynamic process, initiated and maintained by specific neural mechanisms. Later studies have demonstrated the distinct brain mechanisms associated with wake and sleep states.

I. DEFINITION

Mammalian sleep is defined as a physiological state of rapidly reversible period of immobility associated with characteristic recumbent posture, reduced motor activity and increased response threshold for external sensory stimulation with no awareness about the outside world.

II. CLASSIFICATION OF SLEEP STAGES

Sleep states and complexities have been evolved over the years. Lower vertebrates like fish, amphibians and reptiles show only the NREM (Non-Rapid Eye movement) sleep states. Whereas, REM (Rapid eye movement) sleep has been evolved later in phylogeny and is present in birds and mammals. Thus human sleep consists of two distinct sleep states; the Non-Rapid Eye movement sleep (NREM) and the Rapid eye movement sleep (REM). Whole night Polysomnography techniques (poly = multiple, somno = sleep, graphy = recording, PSG) which comprises of Electroencephalogram (EEG), electro occulo gram (EOG) and electromyogram (EMG). A conventional 10–20 system of electrode placement on the scalp been accepted internationally to have a uniform pattern of polysomnography recording with EEG (scalp recording of cerebral cortical activity), EMG (electrical activity of skeletal, i.e., chin and tibilias anterior muscles) and EOG (movement of ocular muscles recorded from External Canthi) (2). The behavioral changes occurring in these three electrode leads are given in Table 27.1 below. According to the recent updated Manuel of American Society for Sleep Medicine (2), NREM sleep is subdivided into three stages (stage N1, stage N2 and stage N3). EEG during REM sleep is desynchronized, but EMG activity is completely reduced. REM sleep is also called as 'Paradoxical sleep' because it is difficult to wake up a person even though the EEG resembles that of wake state. The threshold to wake up is more during NREM–N3 stage.

III. SLEEP ARCHITECTURE (HYPNOGRAM)

The representation of sleep stages recorded with whole night PSG recordings is termed as 'hypnogram' (Figure 27.1). Human sleep architecture has a distinct morphology. The hypnogram of a normal healthy adult human subject demonstrates certain characteristics; we spent about 80% of our total sleep in NREM (N1 = 10%, N2 = 50%, N3 = 10–15%) and 20% in REM.

TABLE 27.1 **PSG details of each stage of sleep. Sleep stages depicted in brackets (S1, S2, S3 and S4) are based on the Reckschaffen and Kales (1968) (3) classification of sleep stages that are corresponding to N1, N2 and N3 according to American Academy of sleep Medicine (AASM) classification (2).**

Behaviour	*EEG*	*EMG*	*EOG*
Wake	Desynchronized (low voltage, mixed frequency waves) and alpha waves (8–12 Hz)	Highly active	Eye blinks present
NREM-N1(S1)	Alpha waves decreases	Active but less than wake	Eye ball rolling
NREM-N2 (S2)	Spindle waves (12–14 Hz), K complex	Less active	Eye movements reduced
NREM-N3 (S3 and S4 or SWS or deep sleep)	Synchronized - Theta (4–8 Hz) and delta (1–4 Hz) waves	Significantly reduced	Eye movements reduced
REM Sleep	Desynchronized - wake like	Atonia /Hypotonia	Phasic eye movements called Rapid (sharp deflections) of eye movements) super imposed over the Tonic phase (No eye movements)

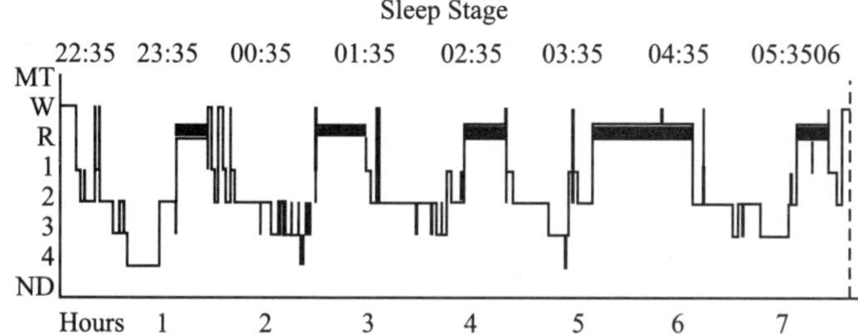

Figure 27.1 Normal hypnogram of whole night sleep in a normal adult. Y axis represents Sleep stages according to Reckshaffen and Klaes (1968) classification and X axis duration in hours. W-wake, NREM sleep stages – S1, S2, S3 and S4, REM – rapid eye movement sleep. Note that slow wave sleep (SWS or S3 and S4) are predominant during first half of night and REM sleep increases in second half of the night. (*hynogram obtained from thesis of Dr. Ravindra P.N, sleep research laboratory, Dept of Neurophysiology, National Institute of Mental Health and Neurosciences, Bangalore*).

The N3 stage (or SWS) occur predominantly during the first half of the night. As the night progress, the REM sleep predominates during the second half of the night. Typically, sleep initiates with stage N1 and progress to N2 and deepens to N3, later, from N3 via N2 the first episode of REM occurs. This cycle (starting from NREM to end of REM stage) is called 'sleep cycle' and normally healthy adults will have 4–5 sleep cycles during the night sleep.

There are basic parameters that need to be kept in mind while observing the hypnogram.

1. *Sleep onset latency (SOL):* Duration from switch off light to onset of any first sleep stage. Normal latency is about 10 minutes. SOL is important in diagnosing insomnia.

2. *REM onset latency (ROL):* Duration taken from sleep onset to the initiation of first REM episode.

Normally between 90–120 minutes. ROL is important in diagnosing Narcolepsy.

Sleep undergoes age associated changes. NREM sleep is more vulnerable to age than REM sleep. N3 is more vulnerable to age and starts declining from second decade of life and by fourth or fifth decade of life, it is decreased considerably to 4–5% of total sleep or disappears completely. However, the REM sleep is preserved up till 70 years and starts declining later. Additionally, aging is associated with increased frequency of intermittent awakenings and sleep onset latency. All these changes are associated with reduced sleep quality which is the most common complaint of old age. **Regular exercise, yoga, meditation practices help to mitigate the age associated changes and help to restore the sleep quality and organization (4,5).**

IV. NEURAL SUBSTRATES OF WAKE AND SLEEP

Until 1950s, the passive theory of sleep existed. Sleep has been considered as a passive idling state, occurring between active wake states (maintained by incoming sensory afferents to cortex). When brain gets fatigued and with the cessation of incoming sensory inputs, sleep ensues. However, the golden era in sleep science had begun since 1950. Lesioning and stimulation studies in cats by Moruzi and Magoun (1949) (1) have discovered that wake state is maintained by specific neural projections, the ascending reticular activating system (ARAS), originate from the brain stem reticular formation and project to various subcortical and cortical structures. Later, with the discovery of immunohistochemical methods, the specific neuromodulatory sub systems of the ARAS has been discovered. These are the cholinergic tegmental system (PPT-LDT the pedunculo-pontine tegmentum and latero dorsal tegmentum) the aminergic group of projections (the median dorsal raphae consisting of the 5HT-serotonergic system, the locus ceruleus nor adrenergic system and the histaminergic projections of tuburomammillary nucleus of posterior hypothalamus) and the cholinergic projections from basal forbrain structures (6). Apart from these wake promoting ARAS projection system, the Orexin/hypocretin system of lateral hypothalamus gives excitatory inputs to the wake promoting ARAS. This orexinergic input is reported to be very important to heighten the wake activity of ARAS to ensure a stable wake state and prevent untimely transition of wake-sleep cycle (7). In addition, the enhanced dopaminergic activity associated with wake enhances the motivational and reward seeking behavior.

The studies of Viennese neurologists Von Economo (8) has provided convincing evidences on the existence of sleep promoting centers of hypothalamus. However, studies on the sleep promoting substrates of ventral lateral pre optric area (VLPO) has provided us with more insights on sleep wake mechanisms. The VLPO consists of two distinct cell populations of GABA and Galanin, which are distributed as medial clusters of cells, the VLPO core (cVLPO), and the more dorsolateral extended VLPO (eVLPO) cells. Studies have clearly demonstrated the role of cVLPO in mediating the NREM sleep states and the eVLPO in maintaining the REM sleep states (9,10). The neurons of cVLPO project extensively to the TMN of posterior hypothalamus (11) and the eVLPO to the dorsal and medial raphae system (12). In addition, the wake promoting monoaminergic and histaminergic neurons project densely to VLPO (13). The VLPO and ARAS wake promoting neurons are thus reciprocally innervated and are mutually inhibitory to each other.

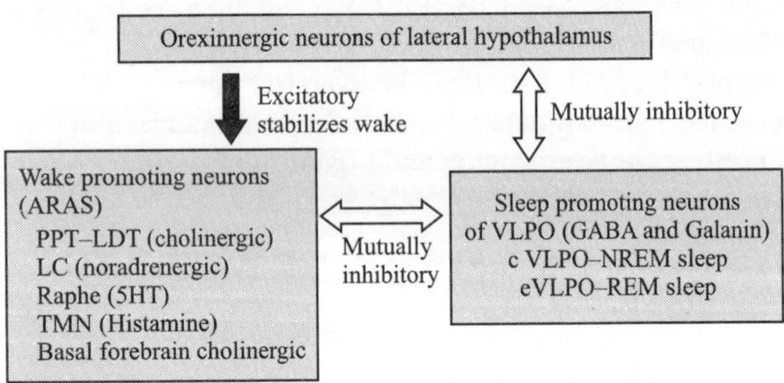

Figure 27.2   Depicts the basic understanding of neurobiology of sleep. Wake and sleep promoting neurons mutually inhibit each other during their respective behavioral state. Orexinergic neurons send excitatory input to wake promoting neurons there by stabilizing the wake state and has mutually inhibitory to VLPO.

V. REM SLEEP

REM sleep was described more than fifty years ago (14). Jouvet and co-workers demonstrated the brain stem neural substrates of REM sleep in early 1980s in cats. They found that two major groups of brain stem neurons, the 'REM On' pontine cholinergic and 'REM off' non cholinergic mesopontine aminergic neurons (6) are reciprocally innervated and are inhibitory to each other. Recent studies (15) using rats however have delineated in detail the role of non cholinergic

and non-monoaminergic 'REM sleep generators' and their interactions in initiation and maintenance of REM sleep and also for REM-NREM alternations. Thus, besides the cholinergic and aminergic neuronal modulations (earlier considered as 'REM on' and 'REM off') of REM sleep, additional 'REM generator' systems of the mesopontine tegmentum, the GABAergic and glutamatergic 'REM on' neurons of Sublaterodorsal tegmentum (SLD) and the GABA ergic 'REM off' neurons of vetrolateral periaqueductal grey matter (vlPAG) and lateral pontine tegmentum (LPT). Anterograde tracer injection studies suggest that, the putative 'REM on' regions such as SLD and the 'REM off' regions, the vlPAG-LPT are mutually inhibitory in nature. These brainstem GABAergic neurons interact with forebrain REM on (GABAergic eVLPO) and REM off (orexinnergic Lateral hypothalamus) group of cells. This mutual interaction between brainstem GABAergic and forebrain GABAergic and orexingeric neurons form a 'flip-flop' model of REM sleep generation (16). The newly discovered REM generators and modulators of mesopontine system and their interactions provide convincing evidence of the existence of independent pathways to mediate the REM characteristics such as muscle atonia, EEG desynchrony and Rapid eye movement during REM sleep.

The circadian and homeostatic regulatory role of sleep wake states also important besides the neural regulations of sleep wake behavior (Figure 27.3). Thus there are multiple wake and sleep promoting centers in the brain and it is the synchronized, coordinated activities help to maintain the different behavioral states. Table 27.3 provides the overall summary of events occurring during sleep wake states.

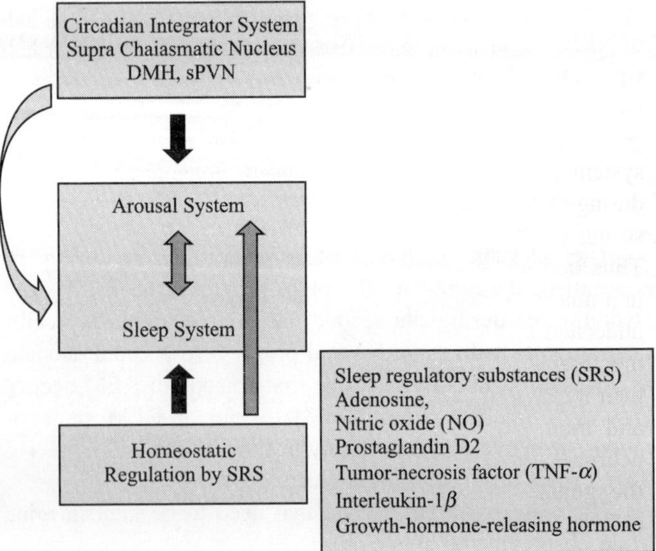

Figure 27.3   Schematic representation of Sleep-wake regulatory mechanism. Black solid arrows represent excitatory projections and unfilled arrows represent the inhibitory influence

The Figure 27.3 provides the interactions among sleep wake states, and the circadian and homeostatic regulatory

TABLE 27.3 Summary of events occurring during various sleep stages

	Wake	*NREM sleep*	*REM sleep*
EEG	Fast low voltage (desynchronized)	Slow high voltage (synchronized)	Fast low voltage (desynchronized
EMG	++	+	nil
EOG	Eye blinks, saccadic eye movements etc	Eye rolling (N1)	Rapid eye movement
LDT-PPT	+	Reduced	++
LC/Raphe/ TMN	++	+	nil
VLPO	nil	++	nil
eVLPO	nil	nil	++
orexin	++	nil	nil

functions on the different behavioral states. During wake, sleep regulatory substances such as adenosine gets accumulated in various areas of the brain more so in the basal forebrain. Adenosine accumulation increases the homeostatic drive for sleep and the brain response to this drive by increasing the amount of SWS and slow wave activity (SWA), a measure of sleep homeostasis. Over the years many sleep regulatory substances have been identified.

Neurophysiology of EEG Desynchronization And Synchronization Associated with Wake And Sleep: EEG during wake and sleep exhibits desynchronized and synchronized patterns respectively. The two distinct synchronised wave patterns – spindles and delta waves characterizes various stages of NREM sleep (N1,N2,N3). Delta waves (0.5–4Hz) forms the signature of deep sleep (stage N3) and spindles (7–14Hz) the N2 stage. Thalamic neurons are of paramount importance in generating these waves. During wake, the ARAS wake active system activates the thalamus and also the cortex. However during sleep, the ARAS activity reduces and hence there is no strong excitatory inputs to activate the thalamus and cortex. Thus during NREM sleep, the thalamus and cortex behaves in a much closed oscillatory network mode of activity. Deeper understanding of thalamic neurons suggests that, it is the passive membrane properties of the thalamic neurons; the richness of their ionic conductance, intrinsic electrophysiological properties and modulation by various neurotransmitters makes these neuronal group to have many functional capabilities including the genesis of spindle waves, delta waves, K complex. (17,18). The transition of desynchronized EEG pattern during wake or REM sleep towards synchronized pattern during NREM is associated with progressive hyperpolarization of thalamocortical cells (TC cells). These TC cells generate a repetitive burst discharges on top of slow depolarizing potentials and abolition of these slow depolarizing spikes brings about burst of fast action potentials. Presence of transient Ca^{2+} waves are responsible for variations in the firing mode of

TC cells. Voltage clamp analysis has revealed the presence of both low threshold or transient and high threshold Ca^{2+} currents in TC cells. Low threshold channels show activation (at membrane potential – 65 mv millivolt) at faster rate and gets inactivated slowly. Therefore, when the membrane potential gets depolarized from hyperpolarized state, initially low threshold Ca^{2+} current gets activated and later inactivates slowly generating slow Ca^{2+} spikes. In turn, these Ca^{2+} spikes make the membrane potentials to be more positive (–55 mV) thereby generating bursts of fast action potentials. This alternating firing mode of thalamocortical cells is associated with variable responsiveness to the peripheral stimuli. Hence, during EEG synchronization (NREM stage – N3 or deep sleep) with progressive hyperpolarisation of thalamic neurons there will be greater attenuation of responsiveness to the peripheral stimuli. Thalamic reticular nucleus (RE cell) rich in GABA ergic neurons encapsulates the thalamic relay neurons (TC) and forms an interpose between thalamus and cerebral cortex. These reticular neurons receive axon collaterals from thalamocortical and corticothalamic neurons, thus, forming a important part of the reciprocal link between thalamus and cortex (Figure 27.4). Reticular cells also have low threshold Ca^{2+} channels which are unique when compared to their counterparts in thalamic cells. These channels generate low threshold Ca^{2+} spikes upon depolarization even at membrane potential at –65 mV. This unique property generates synchronized thalamocortical rhythm, the spindle waves (18).

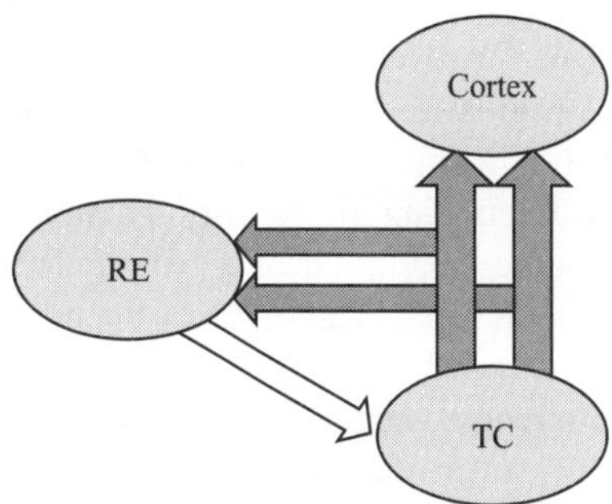

Figure 27.4 Diagrammatic representation of Thalamocortical network. Thalamocortical cells (TC-cell) and cortex has reciprocal innervations and are excitatory, project their collaterals to reticular (RE) nucleus of thalamus. Thus, RE cells form an interpose between thalamocortical and cortico thalamic interaction. TC cell possess low and high threshold calcium channels whose electrical discharges are transmitted to cortex. RE cell rich in GABA modulates the TC cells and inactivates low threshold calcium channels thereby brining about rebound calcium spikes. These calcium spikes (spindle waves) transmitted to cortex which is hallmark of N2. Further, progressive hyperpolarisation of TC brings about more synchronized waves (theta and delta waves) which are signature of deeper stages (N3) during NREM sleep. Unfilled arrow indicates inhibitory input and the thick arrows are excitatory.

VI. IMPORTANCE OF SLEEP AS AN ADAPTATION TO ENVIRONMENT

During sleep, awareness about the outside world is lost and brain engages in an offline mode of functioning. There are now emerging evidences to show that various activities like cellular energetic like glycogen, electron transport, astrocyte to neuron lactate shuttle, uncoupling of proteins, clock transcription proteins gets modulated during sleep wake cycle (19). It is proposed that these metabolic pathways **help the brain to transit from a catabolic wake state to an anabolic sleep state (19)**. If sleep be the only way to conserve energy, then nature would have exclusively provided only this behavioral state. However, animals generally enter into hypothermic state either by torpor or hibernation as a mode of energy conservation which forms the basis of their survival. Therefore, energy conservation may not be the exclusive function of sleep, and sub serve other important functions. Increase in SWS is observed in animals after arousal from these hypothermic conditions. It is suggested that increase in SWA after hibernation or torpor may not be the result of sleep debt, but an essential and most needed state for neural repair and regeneration, and to maintain synaptic homeostasis. Extremely low temperatures are known to limit synaptic transmission, loss of synapses or change in their structure thereby reducing the synaptic efficiency. Post hibernation increase in SWS helps to restore the neural connections between various regions of the brain. Therefore, SWS is important for adaptive neuronal plasticity which plays an important role for survival along with energy conservation.

The tight regulation of synaptic strength is the basis to bring about efficient plasticity changes and is of great importance in functioning and energy dynamics of the brain. It has been hypothesized that quality of wakefulness and sleep determines the total synaptic strength of the brain (20). During wake, synaptic strength increases, reaches maximum before sleep and the strength decreases during sleep which reaches its baseline level when sleep is terminated. Thereby, synaptic homeostasis is maintained and is correlated with SWA. During wake, the process of learning and interaction with environment increases the strength of existing synapse and also helps in establishing new connections. Whereas, during sleep, a sleep dependent down scaling of neurons play an important role in learning and memory, neuronal development and growth (21). Therefore, it appears that SWA during sleep is the most crucial to bring about greater beneficial effects ranging from cellular to behavioral level.

Presently, understanding of sleep is undergoing a great paradigm shift. The classical EEG changes during various sleep stages represent the overall global cortical changes. If we consider brain as a large network, wake and sleep are considered as network properties. There are now emerging evidences that SWA can be local use dependent (22). Enhanced SWA has been studied in cortical columns following learning a specific task (23). This local sleep (SWA), induced by learning task brings about local plastic changes which aids in enhancing the performance. Further, these down scaling events globally or locally play an important role in increasing the signal to noise ratio in neuronal circuits (21). These energy efficient events during sleep brings about great beneficial effect on brain per se therefore the cognition and performance.

Conclusion

Sleep either understood as a global or local phenomenon, it doesn't answer where sleep is located in brain?. In spite of various insults such as large brain lesions, stroke, sleep is not completely abolished. Thus sleep is a self organizing fundamental property of viable neurons and glia and an auto regulated phenomenon (24). Therefore, the two aspects of sleep, one as a global behavioral phenomenon and another as a local use dependent phenomenon helps understanding sleep from behavioral to cellular level. The fact that nature has retained the phenomenon called 'sleep' at all levels from system to cells emphasises that it is an important behavioral state and should sub serve the functions that cannot be undertaken during waking state. In spite of all these evidences, to pinpoint the specific function of sleep is challenging and exciting.

VII. AN OVERVIEW ON SLEEP DISORDERS

Complaints associated with sleep forms one of the major symptoms either associated with medical, psychiatric or neurological illness or primarily due to disorder of sleep itself. The International classification of sleep disorders classifies sleep disorders as follows (25).

1. Dyssomnia
2. Parasomnia
3. Sleep disorders related with mental, neurologic or other medical disorders
4. Proposed sleep disorders

Eventhough more than 80 different sleep disorders have been classified by the International classification of sleep disorders, only few common disorders are explained in brief in present section.

Dyssomnias – are the disorders that produce either difficulty in initiating or maintain sleep or excessive sleepiness. The most common complaint is disturbed sleep in the night or impaired wakefulness. Dysomnias are further classified based on the causative factors as

(a) Intrinsic sleep disorders
(b) Extrinsic sleep disorders
(c) Circadian rhythm sleep disorders

Intrinsic sleep disorders develop within the body or arise form causes within the body, e.g. Psycho physiologic Insomnia, Narcolepsy, Obstructive sleep apnea syndromes etc.

Extrinsic sleep disorders originate or develop from the causes outside the body. For example, inadequate sleep hygiene, environmental sleep disorders, food allergy sleep disorder, alcohol dependent sleep disorder etc.

Circadian rhythm sleep disorders develop due to major misalignment in circadian rhythm and sleep-wake behaviour, e.g., Jet lag syndromes, irregular sleep wake pattern, delayed sleep phase syndrome, advanced sleep phase syndrome etc.

Insomnia: Insomnia is inadequate night sleep *in spite of getting an adequate opportunity to sleep*. The nocturnal and day time symptoms are as follows:

Nocturnal symptoms are:

Long sleep latency

Sleep maintenance difficulty

Short sleep duration

Difficulty returning to sleep

Day time symptoms are:

Not feeling rested after night sleep

Irritability

Anxiety

Reduced cognition

Fatigue

Somatic aches and pains

Increased sleepiness

Insomnia is classified as transient (symptoms less than 2 months), intermittent (repetitive episode of transient) and chronic (symptoms persisting for 6 months). Irrespective of causative factor for insomnia, sleep hygiene forms the most important aspect in its management along with other specific modalities like pharmacotherapy, cognitive behavioral therapy and others. Meticulously by following the sleep hygiene one can get adequate refreshing sleep. The important aspects of sleep hygiene are:

1. Maintain same time to bed and arise in morning. However, irrespective of sleep time try to keep constant wake up time.
2. Avoid heavy physical exercise, caffeinated beverages after evening.
3. Use bed room only for sleep and physical intimacy with spouse.
4. Avoid non sleep activities on bed like reading, video games, laptops on bed.
5. Make bed room completely dark.
6. Avoid day time sleep, however, a short nap at noon is advisable.
7. Avoid tobacco and alcohol consumptions.
8. Avoid work that demands high level of concentration shortly before sleep time.

VIII. OBSTRUCTIVE SLEEP APNEA (OSA)

As per the International classification of sleep disorders (25) OSA is characterized by repetitive episodes of upper airway obstruction that occur during sleep, usually associated with a reduction in blood oxygen saturation. People with OSA commonly present with loud 'snoring' or brief gasps which alternates with episodes of silence that lasts for 20–30 seconds. Snoring can be sometimes very loud that it can disturb the sleep of the partner or other members of the family. Most common cause of OSA is overweight/obesity. 10% increase in body weight increases the severity of OSA by 6 times. Symptoms of OSA can worsen with alcohol intake, weight gain, hypothyroidism, micrognathia, tonsillar and adenoid hypertrophy, drugs like benzodiazepines, opioids, baclofen. Persistent OSA for longer duration of time will be most important risk factor in developing metabolic syndromes, hypertension, arrhythmia, atherosclerosis, congestive heart failure, stroke etc. OSA can be diagnosed with polysomnography where in the recordings of abdominal, chest respiratory movement, nasal airflow and oxygen saturation is carried out along with EEG, EMG and EOG. If PSG demonstrates cessation of breathing movement and airflow at least for 10 sec associated with oxygen desaturation which is generally followed by EEG signs of arousal is considered to be apenic event. If the airflow is reduced by 50% with desaturation, then these episodes are termed as hypopnea. The number of apneic and hyponeic episodes are calculated for the duration of recording and their ratio i.e apnea hypopnea index (AHI) is calculated. AHI is the yardstick to categorize OSA into grades of severity. Life style modification, continuous positive airway pressure (CPAP), treating underlying the medical cause are the main modes of treatment.

Circadian sleep disorder

The major feature in circadian sleep disorder is the major misalignment between individuals sleep pattern and the sleep pattern that is desired. The underlying problem is that the individual cannot sleep when sleep is desired, needed or expected. Therefore, as a result sleep occurs at inappropriate times and wake periods at undesired times. Thereby, patient complains of insomnia and excessive sleepiness. The salient features of common circadian sleep disorders are described below.

Jet lag (Time zone change syndrome)

Consist of varying degrees of difficulties in initiating or maintaining sleep, excessive sleepiness, decrements in subjective daytime alertness and performance, and somatic symptoms (largely related to gastrointestinal function) following rapid travel across multiple time zones (25). Severity depends on the number of time zones travelled, direction of travel (symptoms lasts long for eastward flight). Diagnostic criteria for jet lag is as follows.

1. Patient has primary complaint of insomnia or excessive sleepiness.
2. There is a disruption of normal circadian sleep-wake cycle.
3. The symptoms began within 1 or 2 days after air travel across two time zones.
4. At least two of the following symptoms should be present.
   (a) Decreased daytime performance
   (b) Altered appetite or gastrointntestinal function

(c) An increase in nocturnal awakening to urinate

(d) General malaise

Shift work sleep disorder

Consist of insomnia and excessive sleepiness that occur as transient phenomenon in relation to work schedules.

Parasomnia: These are the clinical disorders not abnormalities of process of sleep or wake perse, but, rather, an undesirable physical phenomenon that occurs predominantly during sleep (25). Parasomnias are disorders of arousal, partial arousal and sleep stage transition. During parasomnia predominant feature will be the activation of central nervous system, autonomic nervous system and skeletal muscle, e.g., sleep walking, sleep terror, confusional arousal, sleep talking, nocturnal leg cramps etc. Parasomnia associated with REM sleep are night mares, sleep paralysis, REM sleep behavioural disorders etc. The details of parasomnia are out of preview of present chapter.

Other medical and mental disorders could be associated with sleep disorders. The most common mental disorders associated with sleep disorders are psychoses, mood disorders, anxiety disorders, panic disorders and alcoholism. Neurological conditions like Parkinsonism, cerebral degenerative disorders, dementia, sleep related epilepsy, sleep related headaches etc., and are associated with sleep disorders. Medical conditions like sleeping sickness, nocturnal cardiac ischemia, COPD, sleep related asthma, peptic ulcer disease, fibromyalgia etc are also associated with sleep disorders. Therefore, understanding the sleep associated symptoms should also be emphasized in general medical and specialty health care.

REFERENCES/SUGGESTIONS FOR FURTHER READING

1. Moruzzi G. and Magoun H.W. (1949), Brainstem Reticular Formation and Activation of the EEG. *Electroencephalogra, Clin. Neurophysiol*, 1:455–473.

2. The AASM Manual for the Scoring of Sleep and Associated Events (2007), Rules Terminology and Technical Specifications, *American Academy of Sleep Medicine Manual for Scoring Sleep.*

3. Rechtsaffen A.K.A. (1968), Techniques and Scoring System for Sleep Stages of Human Subjects, A Manual of Standardised Terminology, Brain Informaton Service/Brain Research Institute, UCLA, Los Angeles, CA.

4. Nagendra R.P., Maruthai N., and Kutty B.M. (2012), Meditation and Its Regulatory Role on Sleep, *Front Neurol.*, 3:54.

5. Ravindra P.N., Sulekha S., Sathyaprabha T.N., Pradhan N., Raju T.R. and Bindu M. Kutty (2010), Practitioners of Vipassana Meditation Exhibit Enhanced Slow Wave Sleep And REM Sleep States Across Different Age Groups, *Sleep and Biological Rhythm*, 8:34–41.

6. Pace-Schott E.F., Hobson A.J. (2002), The Neurobiology of Sleep: Genetics, Cellular Physiology and Sub Cortical Networks. *Nature Reviews Neuroscience*, 3: 591–604.

7. Dean Wagner, et al. (2000), Distribution of hypocretin-containing neurons in the lateral hypothalamus and C-fos-immunoreactive neurons in the VLPO. *Sleep Research*, 3(1):35–42.

8. Von Economo C. (1930), Sleep as A Problem of Localization, *Journal of Nervous and Mental Disease*, 71:249–259.

9. Clifford B. Saper, Thomas C. Chou and Thomas E. Scammell (2001), The Sleep Switch: Hypothalamic Control of Sleep and Wakefulness, *Trends in Neuroscience,* Dec.Vol. 24 (12):726–731.

10. Clifford B. Saper, Thomas E. Scammell and Jun Lu. (2005), Hypothalamic Regulation of Sleep and Circadian Rhythms *Nature,* Vol. 237, 27 Oct, 1257–1263.

11. Sherin J.E., et al. (1998), Innervation of Histaminergic Tuberomammillary Neurons by GABAergic and Galaninergic Neurons in The Ventro Lateral Preoptic Nucleus of Rat. *J. Neurosci*, 18: 4705–4721.

12. Bjorkum A.A., Ha Q.H. and Saper C.B. (1999), Afferents to and Efferent From The Raphae Nuclei to the Ventrolateral Preoptic Nucleus, *Soc. Neuroscience. Abstract*, 25:625.

13. Chaou T.C., Bjorkum A.A., Gaus S.E., Lu J., Scammell T.E., and Saper C.B. (2002), Afferents to Vento Lateral Preoptic Nucleus, *J. Neuroscience*, 22(3): 977–90.

14. Aserinsky E. and Kleitman (1953), Regularly Occurring Periods of Eye Motility, and Concomitant Phenomena, During Sleep, *Science*, 118:273–274.

15. Pace-Schott E.F. and Hobson A.J. (2008), Basic Mechanisms of Sleep, New Evidence on the Neuroanatomy and Neuromodulation of the NREM–REM Cycle, In: Neuro-psychopharmacology, 5th Generation of Progress, American College of Neuropsychopharmacology, 5034–A Thoroughbred Lane, Nashville, TN 37027.

16. Muller P.M., Saper C.B. and Jun Lu (2007), The pontine REM switch: Past and Present, *J. Physiology*, 3:735–741

17. David A., McCormick and Thierry Bal. (1997), Sleep and Arousal: Thalamocortical Mechanisms, *Ann Rev Neurosci.*, 20:185–215.

18. Steriade M., McCormick D.A. and Sejnowski T.J. (1993), Thalamocortical Oscilations in the Sleeping and Aroused Brain. *Science*, 262: 679–685.

19. Scharf M.T., Naidoo N., Zimmerman J.E. and Pack A.I. (2008), 'The Energy Hypothesis of Sleep Revisited' *Prog Neurobiol.*, 86(3):264–80.

20. Tononi G. and Cirelli C. (2003), Sleep and Synaptic Homeostasis: A Hypothesis, *Brain Res Bull.*,15;62(2):143–5023.

21. Tononi G. and Cirelli C. (2006), Sleep Function and Synaptic Homeostasis, *Sleep Med Rev.*,10(1):49–62.

22. Huber R., Ghilardi M.F., Massimini M. and Tononi G. (2004), Local Sleep and Learning, *Nature*, 1;430(6995):78–81.

23. James M. Krueger and Guilio Tononi (2011), Local Use Dependent Sleep: Synthesis the New Paradigm, *Curr Top Med Chem.,* September 1,11(19):2490–2492.

24. Velayudhan Mohan Kumar (2012), Sleep is an Autoregulatory Global Phenomenon, *Frontiers in Neurology,* June, 3(94):1–2.

25. The International Classification of Sleep Disorders (2001), *Revised Diagnostic and Coding Manual.*

28

PRIONS and STATINS

Shail K. Sharma

I. PRIONS

The word *prions* derived from *protein and infection*, was coined in 1982 by Stanley B Pruisner. Prions are aberrantly folded proteins that self propagate through recruitment of their normally folded counterparts. They accumulate in certain transmissible diseases in the central nervous system (CNS) such as *scrapie* (disease of sheep involving CNS and *Cruetzfeldt-Jakob disease* (CJD – human form of prion disease). Three Nobel prizes have been awarded for the seminal work performed in the prion field. Daniel Carleton Gajdusek received the award in 1976 for the discovery of the slow virus that is now known as prion disease *"kuru"*. Stanley B Prusiner subsequently received the award in 1997 for the discovery of prion as the sheep scrapie agent and introduction of the *"Protein only hypothesis"* which stated that misfolded proteins are the infectious agents that cause prion disease. Finally Kurtwuthrich received the award in 2002 for using NMR to unravel the structure of prions.

Prion protein is found throughout the body even in healthy people and animals. Its normal function is in transmembrane transport or signalling. The normal form of the protein is called PrP(C) [C refers to cellular] while the infectious form is referred to as PrP(Sc) [Sc refers to scrapie]. PrP(C) protein is encoded by the highly conserved single copy gene PRNP and is post-translationally refolded into partially resistant β sheet enriched conformation PrP(Sc). The human prion disease variant CJD is believed to be caused by a prion which typically infects cattle and is transmitted through infected meat. Infection can also occur from prions in manure in crop fields.

Prion diseases are classified into sporadic, familial and acquired by infection (Table 28.1). PrP(C) conversion into pathogenic PrP(Sc) is the underlying prion disease. The human cellular prion protein is a glycosylphosphatidylinositol anchored membrane glycoprotein with 2 N glycosylation sites; a disulfide bond and a glycolipid anchor in specialized cholesterol rich domains (calveoli) of the cell plasma membrane.

TABLE 28.1   Etiologic classification of human prion diseases

Etiology	Mechanism	Diseases
Sporadic	Somatic mutation or spontaneous conversion of PrP(C) to PrP(Sc)	Cruetzfeldt- Jakob disease. Fatal insomnia
Hereditary	Germ line mutations in the PRNP gene.	Familial Creutzfeldt-Jakob disease Fatal familial insomnia Gertssman Straussler Scheinkers disease.
Acquired	Infection through ritualistic cannabilism. Infection with bovine prion. Infection with prion tainted human Growth homone, gonadotropins etc.	Kuru variant Creutzfeldt-Jakob disease Iatrogenic Creutzfeldt-Jakob disease

305

In normal conformation PrP(C) comprises a C-terminal globular domain which is rich in α-helices. In contrast The infectious PrP(Sc) exists as insoluble aggregates of different sizes with predominantly β-sheet structure. Aggregation of the abnormal isoforms of PrP(Sc) form highly structured amyloid fibres which accumulate to form plaques. This disrupts the normal tissue structure which is characterized by *"holes"* in the tissue with resultant *spongy architecture* due to vacuole formation in the neurons. All known prion diseases collectively called *"transmissible spongiform encephalopathies"* *(TSEs)* are a family of rare progressive neurodegenerative disorders that affect both humans and animals. They have long incubation periods and characteristic spongy changes associated with neuronal loss.

With advancing age the brain becomes increasingly susceptible to neurodegenerative diseases, most of which are characterized by misfolding and errant aggregation of specific proteins. Prion propagation is thought to occur by a template assisted process in which PrP(Sc) acts as a template onto which PrP(C) is refolded into the infectious conformation . The induction of aggregation involves a crystallization like seeding mechanism by which specific protein is structurally corrupted by its misfolded conformer. Once formed, proteopathic seeds spread from one locale to another via cellular uptake, transport and release. The precise mechanism of PrP(C) conversion to PrP(Sc) is not completely understood but available evidence supports a *seeding/nucleation* model in which infectious PrP(Sc) as an oligomer acts as a seed to bind PrP(C) and catalyzes its conversion into the growing polymer. This is called *protein misfolding cyclic amplification (PMCA).*

Misfolded protein aggregates are associated not only with prion disease, but also with a large group of diseases known as protein misfolding disorders which include Alzheimers, Parkinsons, Huntingtons and over 20 different human maladies.

In animal studies (mutant mice) it is suggested that misfolded prion protein causes the buildup of *voltage gated* Ca^{2+} *channels (VGCCs)* inside the cell, interfering with normal neuronal communication and causing motor dysfunction. Mutant PrP binds VGGCs and prevents them from being delivered to the cell surface Mutations in the PRNP gene can lead to PrP protein aggregation and can cause a variety of neurological diseases in humans that are commonly linked to motor dysfunctions. VGCCs are required at the surface of synapses for normal release of glutamate. Glutamate is the principal excitatory neurotransmitter in brain.

Therapeutics for prion diseases: Currently no effective treatment is available for prions and many other neurodegenerative diseases. The most obvious practical approach continues to be to discover drugs that could cross the blood brain barrier and will block conversion of PrP(C) to PrP(Sc).

II. STATINS

Statins, originally discovered in molds, are the most effective medications to decrease the low-density cholesterol (LDL-C)

which is the key participant in atherogenesis. Statins accomplish this by inhibiting the enzyme 3-hydroxy-3-methylglutaryl coenzyme A (HMG-CoA) reductase a rate-limiting enzyme in cholesterol biosynthesis, which converts HMGCoA to mevalonate. Inhibition of HMG CoA reductase stimulates in the liver the LDL receptors, resulting in an increased clearance of LDL particles from the bloodstream and by this mechanism has proved to be the most widely used medications in reducing the risk in cardiovascular diseases. Mevalonate is also the precursor of many biologically active compounds.

The first statin, Mevastatin was discovered in 1970's in the fermentation broth of *Penicillineum citrinium* during the search of potential new suppressors of lipid metabolism. Lovastatin the first clinically used statin was discovered in Aspergillus terreus in 1978 and was approved for treatment in 1987.

Statins are divided into two groups: hydrophilic (pravastatin and rosuvastatin) and lipophilic (simvastatin, lovastatin, fluvastatin, atrovastatin and pitavastatin). Their structures are shown in Figures 28.1 and 28.2.

Lovastatin is a natural fungal metabolite, simvastatin and pravastatin are synthesized from lovastatin by chemical modifications and the remaining statins are synthetic compounds. The gene SLCO1B1 *(soluble carrier organic anion transporter family 1B1)* codes for an organic anion-transporting polypeptide that is involved in the absorption of statins. Statin induced cholesterol depletion results in

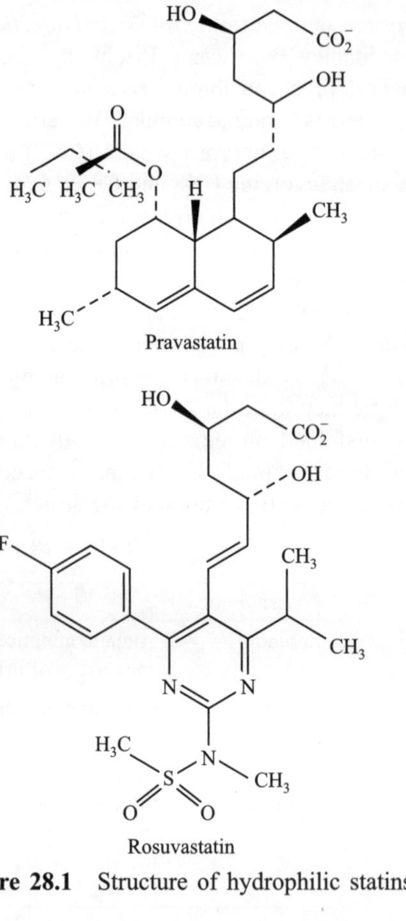

Figure 28.1 Structure of hydrophilic statins.

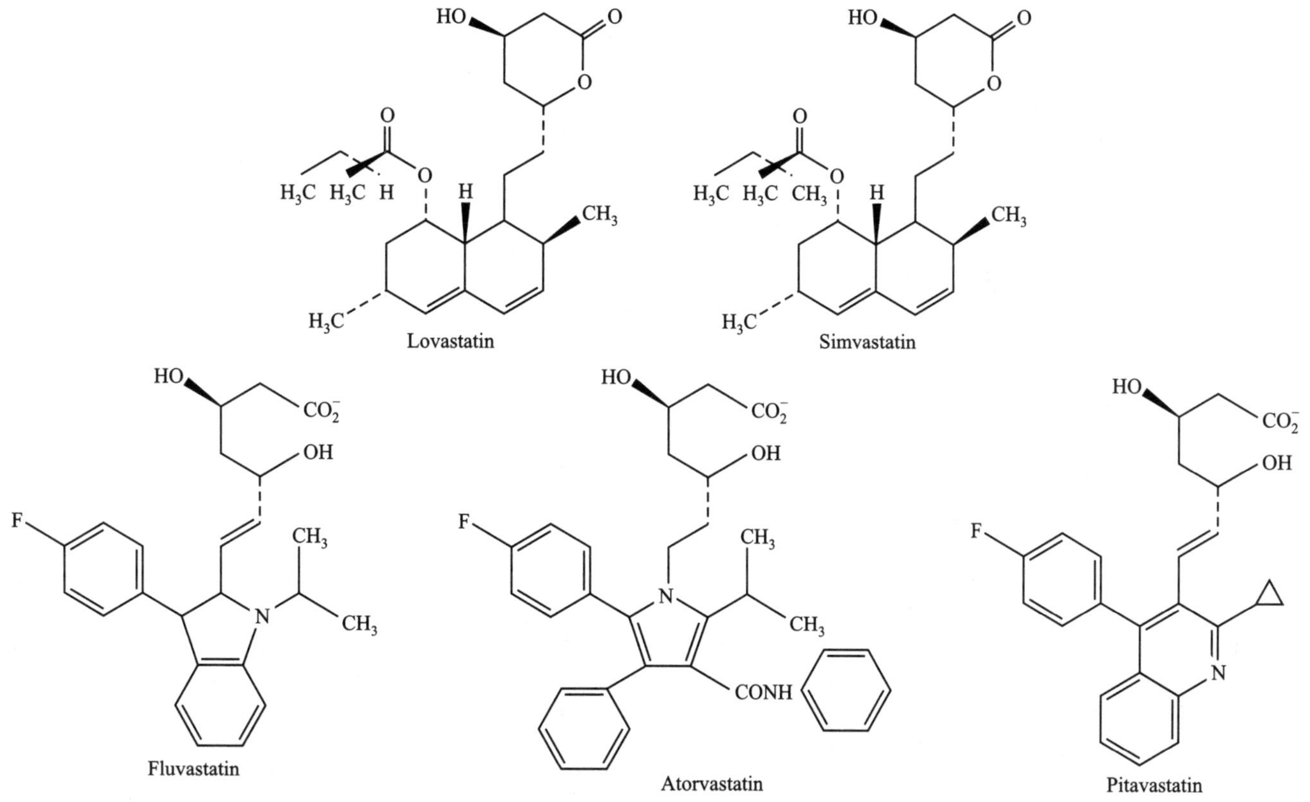

Figure 28.2 Structure of lipophilic statins.

the activation of *sterol regulatory element-binding protein 2*-the sterol-sensitive transcription factor that up-regulates the expression of LDL receptors. Since LDL is mainly metabolized in the liver, hydrophilic statins effectively reduce plasma LDL cholesterol level and LDL receptors are the major mechanism of LDL clearance from the circulation. In contrast, lipophilic statins easily permeate plasma membranes and are effective in both hepatic and peripheral tissues.

A. Pleiotropic Effects of Statins

Beneficial effects of statins extend beyond their cholesterol lowering effects. Apart from cholesterol, mevalonate is also the substrate for the synthesis of *nonsteroid isoprenoids* (Figure 28.3), which include *farnesylpyrophosphate* and *geranylgeranylpyrophosphate* which serve as lipid donors for the post-translational modification (*prenylation*) of proteins

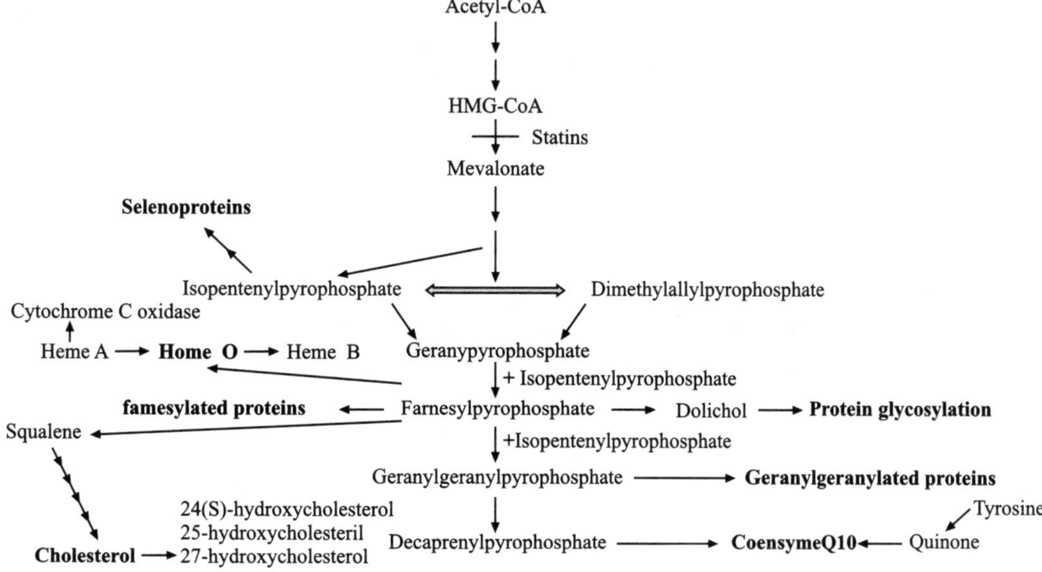

Figure 28.3 Mevalonate cascade

that possess a characteristic C-terminal motif. The *prenylation* reaction is catalyzed by *prenyltransferases*. The lipid prenyl group facilitates to anchor the proteins in cell membranes and mediates protein-protein interactions. A variety of important intracellular proteins undergo prenylation, including almost all members of *small GTPase superfamilies as well as heterotrimeric G protein subunits and nuclear lamins*.

B. Statins and Rho/Rho Kinase (ROCK)

Isoprenylated proteins have key roles in membrane attachement and are likely to be involved in the pathogenesis and progression of *atherosclerosis and Alzheimers* disease. Inhibition of isoprenoid synthesis by statins, leads to the inhibition of intracellular signalling molecules Rho, Rac and Cdc42. ROCKs are the downstream effectors of the small *glutamyltranspeptidase* (GTPase) Rho. These proteins are post-translationally modified through *isoprenylation* which is inhibited by statins. By inhibiting mevalonate synthesis, statins prevent membrane targeting of Rho and its subsequent activation of ROCKs. Cdc42 another member of the Rho family alongwith Rho and Rac regulates the formation of actin to produce stress fibres.

The Rho GTPases form a subgroup of the Ras superfamily of 20–30 KD GTP-binding proteins that have been shown to regulate a wide spectrum of cellular functions. These proteins are ubiquitously expressed across species from yeast to man. Like all members of the Ras superfamily. the rho GTPases function as molecular switches cycling between an inactive GDP-bound state and an active GTP-bound state. Rho proteins, which are geranylgeranylated, regulate vesicle transport and assembly, cytoskeleton reorganization, activate phagocytic and non-phagocytic NADPH oxidases and Rho-dependent kinases, which regulate smooth muscle contractility and growth. Rho GTPases play crucial roles in the regulation of cytoskeletal organization in response to extracellular growth factors in membrane trafficking, transcriptional regulation and cell growth and development (Figure 28.4).

The major Rac signalling pathway includes remodeling of the actin cytoskeleton and generation of reactive oxygen species (ROS). Ras proteins, which are farnesylated, are involved in signal transduction by growth factor receptors and activate mitogen-activated protein kinase cascade. Inhibition of protein farnesylation and/or geranylgeranylation is responsible for lipid-independent or *"pleitropic"* effects of statins such as amelioration of oxidative stress, improvement of endothelial nitric oxide (NO) production, inhibition of adhesion protein expression and leukocyte migration, decreased synthesis of pro-inflammatory cyto and chemokines, inhibition of cell hypertrophy/proliferation (including anti-hypertrophic effect on vascular smooth muscle cells crucial for growth of atherosclerotic plaques).

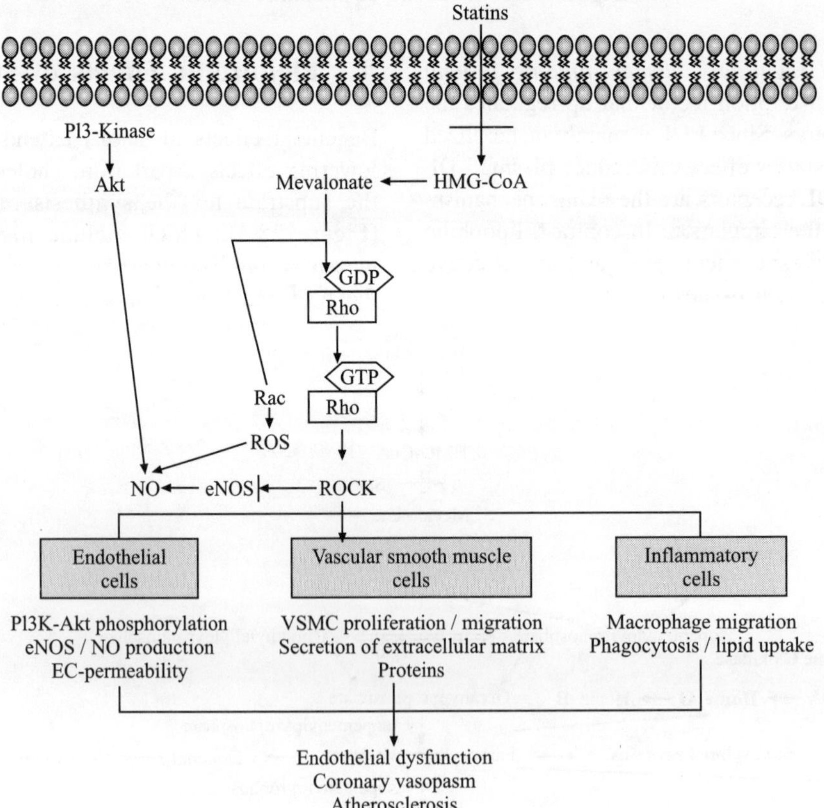

Figure 28.4   Regulation of the GTPase cycle and upregulation of eNOS by statins. Inhibition of Rho isoprenylation and subsequently ROCK actively prevent membrane targetting of Rho. The inhibition of the Rho/ROCK pathway may mediate some of the pleiotropic effects of statins on the vascular wall and endothelial cells. PI3K, phosphatidylinositol-3 kinase; ROCK, Rho kinase; eNOS, endothelial nitric oxide synthase; GDP, guanosine diphosphate; GTP, guanosine triphosphate.

C. Cellular Effects of Statins

Endothelium derived nitric oxide (NO) is an important regulator of endothelial function. Through Rho/ROCK signaling statins could lead to increased endothelial NO synthase (eNOS) expression. Statins promote the activation of serine-threonine protein kinase Akt. in endothelial cells, leading to eNOS phosphorylation and angiogenesis. Cardiac hypertrophy is mediated in part by myocardial oxidative stress. Rac 1 is required for NADPH oxidase activity, and statins may inhibit cardiac hypertrophy through an antioxidant mechanism involving inhibition of geranylgeranylation of proteins. One of the main contributors to this protective effects of statins is the increase in NO, resulting in increased vasodilation and facilitation of regional blood flow under hypoxic conditions. Statins can also attenuate cytokine-mediated vascular smooth muscle cells proliferation.

D. Statins in Atherosclerosis

The causes of atherosclerosis appear to be lipid retention, oxidation and modification which provoke chronic inflammation at susceptible sites in the walls of all major conduit arteries. Initial fatty streaks evolve into fibrous plaques, some of which develop into forms that are vulnerable to rupture, causing *thrombosis* or *stenosis*. Statins have been found to modulate immune activation and to exert anti-inflammatory effects on the vascular wall by decreasing the number of inflammatory cells in atherosclerotic plaques. The mechanism is due to the immunomodulatory ability of statins to decrease the expression of endothelial adhesion molecules such as intercellular adhesion molecule-1, vascular cell adhesion molecule –1 and E-selectin. Statins favorably alter plaque size, their cellular and chemical composition and biological activities centered on inflammation and cholesterol metabolism. Inhibition of prenylation of the small GTP-binding proteins Ras and Rho, and to the disruption, or depletion, of cholesterol rich membrane micro-domains (membrane rafts). Through these pathways statins modulate immune responses by altering cytokine levels and by affecting the function of cells involved in both innate and adaptive responses. Anti-inflammatory and immunosupressory properties of statins provide the rationale for their potential application in conditions in which inflammation and immune response represent key pathogenic mechanisms, such as *rheumatoid arthritis* and *systemic lupus erythematosus*.

E. Statins in Alzheimers Disease

Alzheimers disease (AD) is characterized by a progressive impairment of memory and other cognitive functions. Many experimental studies suggest that hypercholesterolemia accelerates the production of amyloid-β-peptide by shifting amyloid precursor protein metabolism from alpha to beta clevage products, by β- and γ-secretases. These secretases are embedded in a membrane structure known as lipid raft, with high cholesterol content. Another type of evidence linking cholesterol and AD is that several lipoproteins involved in cholesterol metabolism are also related to amyloid deposition One of these lipoproteins is apolipoprotein (Apo) E, which is one of the strongest genetic risk factors known for AD. Recent attention has been directed to ApoJ (clusterin) another lipoprotein associated with AD and also to the LDL receptor which is related to the synaptogenic impairment found in AD. The mechanism by which hypercholesterolemia leads to an increased neuronal content (blood-brain barrier effectively prevents the entry of circulating cholesterol) is not known. However 27-hydroxycholesterol (27-OHC) which is predominantly presnt in circulation can cross the blood brain barrier and is known to increase Aβ levels in experimental studies. Another metabolite 24-hydroxy cholesterol (24-OHC) is brain- specific and is the main mechanism for eliminating cholesterol from the brain. Plasma levels of 24-OHC are reduced in advanced AD and correlate with the degree of brain atrophy and neuronal loss. Although a potential role of cholesterol in the pathogenic process of AD has been established through different experimental models, statin therapy is yet to show a clear general effect on the treatment or prevention of AD.

Recently additional pleiotropic effects of statins on "cellular senescence" have been reported in different cell types, suggesting that the anti-aging effects of statins may be linked to their ability to inhibit telomere shortening.

F. Adverse Effects of Statins

Statins are generally well-tolerated, adverse effects may occur in some cases. These effects are due to impaired protein prenylation resulting in deficiency of coenzyme Q10 which is involved in mitochondrial transport and antioxidant protection, abnormal protein glycosylation due to dolichol shortage, or deficiency of selenoproteins. *Myopathy* is the most frequent side effects of statins. In the reduced form coenzyme Q10 is the most important lipophilic antioxidant, preventing the generation of free radicals as well as oxidative modification of proteins, lipids and DNA.

With the data available from basic studies regarding inhibition of isoprenylation and Rho/ROCK pathways and their subsequent effects on the endothelial cells, vascular smooth muscle cells, and inflammatory cells, it may be possible to use statins for their pleiotropic effects.

SUMMARY

Prions are misfolded protein molecules and propagate when they enter the healthy organism. All known prion diseases affect the brain and are fatal.

Statins are the most common drugs used in the prevention of cardiovascular diseases. They act by inhibiting cholesterol biosynthesis. Beneficial effects of statins extend beyond their cholesterol lowering effects. These are restoring endothelial functions, enhancing the stability of atherosclerotic plaques, decreasing oxidative stress and vascular inflammation.

REFERENCES

Aguzzi A. and Falsig J. (2012), Prion Propagation, Toxicity and Degradation, *Nature Neurosci*, 15, 936–939

Blum A. and Shanburek R. (2009), The Pleiotropic Effects of Statins on Endothelial Functions, Vascular Inflammation, Immunomodulation and Thrombogenesis, *Atherosclerosis*, 203, 325–330.

Qian, Z. and Liao J.K. (2010), Pleiotropic Effects of Statins: Basic Research and Clinical Perspectives, *Circ. J.*, 74, 818–826.

29

Leptin

Uma Sinha and Indranil Sinharoy

CONTENTS

I. LEPTIN

Leptin is a single-chain proteohormone with a molecular mass of 16 kDa produced by differentiated adipocytes in proportion to the body fat stores. It is the product of ob gene in obesity-mice. It acts through its receptors (ObR) in the brain to regulate satiety, energy balance and neuroendocrine function. It also regulates several peripherally mediated actions like regulation of growth, reproduction, tissue remodeling, autonomic activities and immune functions.

Discovery of Leptin

Leptin in humans is encoded by a gene located in chromosome 7q31.3 just similar to that in animal models. Leptin was originally discovered through positional cloning of the ob/ob mice, a mouse model of obesity at Jackson Laboratories in 1994. It was termed leptin from the Greek word "leptos" meaning 'thin' when the discovery of this adipocyte secreted protein-cytokine generated excitement that the answer to obesity had probably been found. In mice, the products of ob gene and db gene are the main candidates for the ligand leptin and its receptor respectively. Mice with ob mutation have a deficiency of leptin and those with db mutation are deficient in the hypothalamic receptors for leptin. Genetically mediated states of leptin deficiency or resistance result in hyperphagia, obesity, preferential storage of calories as adipose tissue, infertility, susceptibility to diabetes, dyslipidemia, hypometabolism, somatic growth impairment and elevated blood glucocorticoid concentration. In humans, leptin is translated as a 167 amino acid product with an amino terminal secretory signal sequence

of 21 amino acids which is subsequently removed from the main peptide. Therefore, leptin circulates in human blood as a protein of 146 amino acid residues. Recombinant human leptin produced in *E. coli* is a single, non-glycosylated, polypeptide chain with a four-helix bundle with one very short strand segment and two relatively long interconnected loops containing 147 amino acids and having a molecular mass of about 16 kDa.

II. BIOLOGY OF LEPTIN

For last 15 years much has been known about leptin biology. Leptin is secreted mainly by white adipose tissues in a pulsatile fashion. Its circulatory levels are positively correlated with the amount of stored body fat and can change with acute changes in calorie intake. Its production has also been demonstrated in other tissues such as the fundus of stomach, skeletal muscles, liver, ovaries and the placenta. Leptin has a significant circadian variation with higher levels in the early morning, peak level being achieved during 02:00 hrs at night when serum concentration of leptin remains 30–100% high. It has been seen that serum leptin level remains stable for at least five freeze-thaw cycles. Samples are stable for about two months when stored at 4°C and for at least two years when stored at –20°C.

In adults, usually there is no change in serum leptin level or its receptor level with age. Women usually exhibit higher serum leptin with lower levels of soluble leptin receptors in comparison with the male counterparts.

BMI (body mass index) is the best predictor of circulating leptin concentration. Still there is a great variation in leptin concentration at any given index of body fat mass. Only 5% of the obese population may have 'relative' leptin deficiency, who could be benefited by leptin therapy. Distribution of body fat, visceral fat, basal glucose level and ethnicity do not seem to influence circulatory leptin concentration.

Leptin production is primarily regulated by insulin induced change in adipocyte metabolism. Short term change in energy balance can change leptin concentration temporarily. After 24 hours of fasting, leptin concentration decreases up to 30% of its basal level. Massive refeeding for more than 12 hours again shows increase in leptin level by 50% of initial basal value. Thus leptin may serve as a sensor of short term change in energy expenditure. Cold exposure in rodents decreases the expression of leptin mRNA acting through the β_3-adrenergic receptors of adipocytes. Significance of these β_3-adrenoceptors in the control of leptin production in human beings is still uncertain, though, it is evident that β_3-adrenoceptors have a role in lipolysis in human omental and subcutaneous adipose tissues.

Several variables that appear to regulate serum leptin levels in different animal and human models have been identified.

Candidates shown to have leptin reducing properties:

- Androgens
- β_3-adrenoceptors
- Fasting
- Cold exposure
- Long term exercise
- Somatostatin
- Cyclic AMP
- Thiazolidinedions
- Smoking
- Free fatty acid
- IGF-1

Candidates shown to increase serum leptin levels:

- Obesity
- Overfeeding
- Impaired renal function
- Insulin
- Glucose
- Glucocorticoids
- TNF-α
- Oestrogen
- IL-1
- Endotoxins
- Alcohol

III. LEPTIN RECEPTORS (OBR)

Leptin mediates its effects by binding to specific leptin receptors (ObR) expressed in brain and peripheral tissues. Leptin receptors belong to the cytokine class I receptor family, found ubiquitously in the body such as in liver, heart, kidney, lungs, gut, testes, ovaries, spleen, pancreas and adipose tissues.

Leptin receptors are the products of a single Lepr gene that contains 17 common exons and several alternatively spliced 3' exons. In mice six distinct isoforms of leptin receptors have been identified (Ob-Ra, Ob-Rb, Ob-Rc, Ob-Rd, Ob-Re and Ob-Rf). All isoforms are divided into three classes—short, long and secreted forms. Short form receptors (Ob-Ra, Rc, Rd, Rf in mice) and long form receptor (Ob-Rb in mice) only differ in sequence due to the alternative splicing of 3' exons. The short leptin receptor isoform Ob-Ra is called leptin transporter and it is thought to play an important role in transporting leptin across the blood-brain barrier. High concentration of short isoforms of Ob-Ra and Ob-Rc in choroid plexus and brain microvessels suggests their role in blood-brain barrier transport. Secreted forms contain only extracellular domains to bind circulating leptin probably to regulate the concentration of free leptin in serum. Interestingly, regulated ectodomain shedding of membrane spanning leptin receptors can generate secreted or soluble isoforms. Ob-Re is the soluble form of the transmembrane leptin receptor. Ob-Rb is the long form isoform containing an intracellular signaling domain and it is crucial for leptin activity. It consists of 1162 amino acids. Ob-Rb is strongly expressed in hypothalamus, mainly in the arcuate nucleus, ventromedial, dorsomedial, lateral hypothalamic nuclei and paraventricular nucleus.

Circulating leptin crosses blood-brain barrier and binds to the centrally located Ob-Rb receptors to activate several signal transduction pathways including Janus-Kinase Transducer and Activator of Transcription 3 (JAK-STAT 3) for the regulation of energy balance and Phosphatidylinositol 3-Kinase (PI 3K) for regulation of food intake and glucose metabolism. Other pathways including Mitogen-activated protein kinase (MAPK), 5'Adenosine Monophosphate-activated Protein Kinase (AMPK) and the Mammalian Target of Rapamycin (mTOR), thought to be downstream of leptin signaling are under investigation. Effects of leptin on immune system and vasculature are mediated by direct influence on hematopoietic cells containing Ob-Rb receptors. Leptin also modulates glucose metabolism independent of its action on adipose tissues partly centrally and partly by regulating pancreatic β-cells and insulin sensitive tissues.

IV. FUNCTIONS OF LEPTIN

The functional axis of leptin is a complex one. Leptin appears to act in endocrine, paracrine and autocrine fashion. Leptin performs a wide range of physiological roles like growth factor like activity, regulation of energy expenditure, a permissive factor for puberty, signaling metabolic activities, modulation of feto-maternal metabolism and interaction with other hormones such as insulin, glucagon, insulin-like growth factors (IGF), growth hormone and glucocorticoids. It is now well established that leptin has got both central as well as peripheral mechanisms of action.

Leptin has been found to interconnect three important physiological axis—(1) adipocyte metabolism, (2) hypothalamic energy homeostasis and (3) reproductive system. Leptin acts as a principal lipostat that controls fuel utilization, mobilization and expenditure. Leptin suppresses neuropeptide Y (NPY, a strong appetite stimulator) production and secretion by arcuate nucleus. Thus it *causes increased energy expenditure, decreased appetite and weight loss*. It also acts peripherally on skeletal muscles, liver, pancreas and adipose tissues mainly via 5′-AMP activated protein kinase to decrease anabolic pathways and increase catabolic pathways (glucose and lipid utilization). The expression of leptin by syncytiotrophoblasts has supported the role of leptin in fetal nutrient transfer.

It has been proposed that leptin may improve insulin resistance through several mechanisms such as (1) activating insulin signaling pathways including skeletal muscle phosphatidylinositol 3-kinase and AMP-activated protein kinase; (2) preventing 'lipotoxicity': decreasing intrahepatic and intramyocellular fat by activating fatty acid oxidation in skeletal muscles; (3) decreasing caloric intake; (4) decreasing body weight and fat mass. It has also been suggested that leptin can improve hepatic insulin sensitivity irrespective of its effect on peripheral insulin sensitivity and independent of reduction in visceral adipose tissue.

Functions of leptin at a glance (in animal and human models):

A. Centrally Mediated Function (Hypothalamus)

- Neuroendocrine function—Leptin decreases appetite and food intake and promotes energy expenditure. Leptin via Ob-Rb receptor acts on two different population of arcuate neurons. One population synthesizes orexigenic (appetite stimulating) hormones like neuropeptide Y (NPY) and agouti-related peptide (AgRP) and the other population produces anorectic (decreases appetite) neuropeptides mainly pro-opiomelanocortin (POMC) and its product α-melanocyte stimulating hormone (α-MSH). Leptin stimulates the synthesis and secretion of POMC/α-MSH to produce anorexia and reciprocally suppresses the levels of NPY/AgRP.
- Leptin inhibits reward seeking behavior associated with feeding acting mainly through mesolimbic dopaminergic pathways.
- Leptin interacts with other neurohormones like neurotensin, galanin, IGF, growth hormone.
- Synaptic plasticity—leptin can rapidly modify synaptic connections between neurons balancing both excitatory and inhibitory synapses. A good number of genes positively regulated by leptin play vital roles in local protein synthesis to maintain the function, remodeling and coordination of synapses. Leptin also acts as a neurotrophin in the developing brain after birth.

B. Peripheral Actions of Leptin

- Metabolism—Leptin influences insulin-mediated metabolic pathways and thus modulates glucose homeostasis and lipid metabolism. Some of the actions of insulin appear to be synergistic to leptin activity like increased peripheral uptake of free fatty acids and glucose. But leptin antagonizes insulin in other activities like it inhibits insulin-mediated suppression of fatty acid oxidation. Thus insulin is anabolic in nature and stores fuel and aminoacids whereas leptin appears to mobilize and utilize fatty acids. Leptin also inhibits insulin-mediated long term glucose storage and oxidation.
- Pancreas—pancreatic β-cells are rich in leptin receptors but how leptin influences insulin secretion is still controversial.
- Adipose tissue—experiments on mice models have shown that leptin causes inhibition of insulin binding, insulin-mediated glucose transport, glycogen synthesis and lipogenesis.
- Skeletal muscle—Stimulation of free fatty acid oxidation, attenuation of insulin mediated antioxidation and lipogenesis are some of the important aspects of leptin activity.
- Uterus—placental leptin may have physiological effects like angiogenesis, growth and immunomodulation. Leptin helps in fetal growth and metabolism. Leptin levels are increased in pregnancy specially in 2nd and 3rd trimester.
- Reproduction—human ovary and prostate express mRNA for leptin receptors. Studies have shown that a significant increase in serum leptin levels occurred during controlled ovarian hyperstimulation suggesting the role of leptin in follicular growth and maturation. Correction of leptin deficiency in ob/ob mice models activated reproductive axis and restored fertility in both sexes.
- Blood—leptin leads to proliferation and differentiation of hematopoietic precursors. Vascular endothelial cells express long forms of leptin receptors in both rodents and humans. In dose-dependent manner leptin stimulates angiogenesis and endothelial growth to accelerate wound repair. But experimentally in higher concentrations leptin induces toxic effects like capillary leakage and avascular zones.

V. LEPTIN RESISTANCE

The concept of *leptin resistance is quite analogous to the syndrome of insulin resistance. Elevated circulatory leptin levels proportionate to the adiposity and not a deficiency of leptin underlie most cases of obesity*. Failure of elevated leptin levels to suppress feeding and prevent weight gain in obese persons defines the state of leptin resistance in obesity. This is most likely to be the result of leptin desensitization of the leptin induced signals. Leptin resistance or tolerance to its weight reducing effects can occur at two distinct levels:

I. Leptin-induced leptin resistance—chronically raised leptin levels in obesity decrease the transport of leptin across the blood-brain barrier. One evolutionary hypothesis behind this peripheral leptin resistance states that hypertriglyceridemia associated with obesity inhibits leptin transport into brain.

II. Attenuated leptin receptor activation and or signaling – cellular leptin resistance is caused by induction of inhibitors of leptin action such as suppressor of cytokine signaling 3 (SOCS-3) and leptin stimulated phosphorylation of Tyr 985 on Ob-Rb receptors.

VI. LEPTIN—CLINICAL UTILITY

A number of leptin deficiency syndromes that are treatable with leptin replacement have been identified. *Complete leptin deficiency from mutations in leptin gene, as found in some rare genetic conditions is associated with infantile morbid obesity and endocrine dysfunction including insulin resistance and hypogonadotropic hypogonadism.* Severe lipodystrophy, both genetic and acquired (as in HIV associated lipodystrophy syndrome) is another hypoleptinemic state that is characterized by selective loss of adipose tissue from subcutaneous spaces mainly from face, arms and legs , hypertriglyceridemia, severe insulin resistance and even overt diabetes mellitus. Recombinant human leptin replacement therapy at physiologic replacement dose (0.04–0.08 mg/kg sc. daily) in both the situations has improved insulin sensitivity, glucose tolerance, levels of fasting glucose and HbA1c, hypertriglyceridemia, transaminitis and changes in body composition (weight loss with decreased adipose tissue and lean mass) and thus the need for insulin or oral hypoglycemic agents could be lessened with the help of leptin. In the management of HIV lipodystrophy and metabolic syndrome recombinant human leptin (metreleptin) has recently been extensively studied in the context of many open-label, clinical trials. Only hypoleptinemic HIV infected patients are the actual beneficiaries and are selected for recombinant leptin treatment trials to combat metabolic syndrome. As serum leptin level varies in general population depending on age, sex, feeding status and circadian rhythms, hypoleptinemia is usually considered, when serum leptin level is < 3 ng/ml in men and < 4 ng/ml in women.

Pathological states associated with low serum leptin level (hypoleptinemia) provide enough scope of research on the clinical utility of leptin replacement. Recent studies have shown some new areas of therapeutic application of leptin such as in some neuroendocrine disorders like hypothalamic amenorrhea (HA) and anorexia nervosa. *Hypothalamic amenorrhea is associated with low leptin levels where leptin replacement therapy has been found to restore reproductive function in women suffering from HA, some of whom had not menstruated for years.* Among the women with anorexia nervosa and patients with cachexia resulting from cancer or severe chronic infection with prominent immune abnormalities, recombinant leptin administration was able to suppress these immune abnormalities at lower doses than are required to suppress food intake or body weight.

VII. LEPTIN VERSUS GHRELIN

Leptin and ghrelin are complementary yet antagonistic to each other in their activity. *Ghrelin is a gut hormone primarily involved in the short-term as well as long term regulation of appetite, food intake and energy metabolism.* Ghrelin is a peptide synthesized as a pre-prohormone predominantly in the epithelial cells of fundus of stomach. Placenta, pituitary, hypothalamus and kidney are some of the other sources of ghrelin. Ghrelin was discovered in 1999 as a novel growth hormone releasing acetylated peptide containing 28 amino acids. The term ghrelin came from the proto-Indo-European origin of the word "ghre" meaning 'grow'. Ghrelin is the endogenous ligand for the receptors known as growth hormone secretagogue receptors (GHS-R). These receptors have been identified in pituitary, hypothalamus, adipose tissues and also in heart and blood vessels.

Ghrelin activates GHS-R on the anterior pituitary to stimulate growth hormone secretion. It is an orexigenic peptide that antagonizes leptin action through activation of hypothalamic NPY/Y1 receptor pathway. Vagal afferent neurons and brainstem nuclei connected to hypothalamus are the important mediators of the central signaling pathway through which both ghrelin and leptin act. *Ghrelin also stimulate lactotroph and corticotroph secretion from pituitary.* Metabolically ghrelin is adipogenic and inhibits fat oxidation. It is a strong gastrokinetic agent that *increases gastric acid secretion*, motility, gastric emptying and mucosal turnover. Ghrelin has recently been known for its cardioprotective role to augment cardiac output and decrease blood pressure. Its diverse action opens the possibility of its clinical application in near future in certain conditions like growth hormone deficiency, eating disorders, gastrointestinal disorders and states of chronic wasting and cachexia.

Gastric ghrelin production is regulated by several nutritional and hormonal factors. Fasting and low protein diets tend to increase ghrelin levels whereas somatostatin, IL-1β, growth hormone, high calorie diet and vagal stimulation seem to inhibit ghrelin expression. Pre-prandial increase and post-prandial fall in plasma ghrelin levels occur in adults. Ghrelin concentration is found to be high in negative energy states like low calorie diets, anorexia nervosa and cancer cachexia. Fasting plasma ghrelin concentrations in obesity are significantly lower which may lead to a blunted GH secretion in obesity and are inversely correlated with body mass index, percent body fat, fasting insulin and leptin concentrations and plasminogen activator 1 (PAI-1) levels that remain elevated in insulin-resistant subjects. The lack of postprandial suppression of plasma ghrelin in obese population may contribute to increased food consumption in them. This low ghrelin secretion in common obesity may be a physiological adaptation to long-term positive energy balance although a particular subset of obesity as in Prader-Willi syndrome may be associated with high levels of ghrelin.

TABLE 29.1 Leptin versus ghrelin—at a glance

	Leptin	*Ghrelin*
Primary source	Adipose tissue	Stomach
Energy balance	Mediates body weight loss	Adipogenic, increases body weight
Appetite	Anorexigenic, suppresses appetite	Orexigenic, stimulates appetite
Food intake	Decreased	Increased
Fat oxidation	Increased	Decreased
Gut motility	Decreased	Increased motility, gastric emptying
Metabolic role	Catabolic	Anabolic
Plasma level	Decreased after prolonged fasting	Increased in fasting state
Obesity	Circulatory leptin level is high	Circulatory ghrelin level is low

SUMMARY

- Leptin is a hormone secreted from adipose tissues in proportion to fat stores. It regulates energy homeostasis, neuroendocrine function, and metabolism.
- Adequate leptin levels mediate energy expenditure in the processes of reproduction, tissue remodeling, and growth, apart from the regulation of the autonomic nervous system, neuroendocrine functions and the immune system.
- Failure of elevated leptin levels to suppress food intake and promote weight loss in the common forms of obesity defines a state known as leptin resistance, thus limiting the scope of leptin therapy in obesity.
- Lack of leptin signaling due to either mutation of leptin (e.g., ob/ob mice) or of the leptin receptor (ObR) (e.g., db/db mice) in rodents and humans results in complete or severe leptin deficiency states with a typical phenotype of congenital lipodystrophy and lipoatrophy (including hypothyroidism, growth failure, infertility and immune deficiency) in addition to obesity.
- Relative leptin deficiency is a prevalent state which is acquired secondarily due to conditions like anorexia nervosa, hypothalamic amenorrhoea and HIV associated lipodystrophy syndrome.
- Recombinant human leptin (metreleptin) in physiological replacement doses is an upcoming potential therapy for leptin deficient states (complete/relative) to normalize neuroendocrine and metabolic defects.
- Ghrelin is an orexigenic and adipogenic gut-hormone which antagonizes leptin action in the regulation of food intake and energy metabolism.

SUGGESTIONS FOR FURTHER READING

Kelesidis T., Kelesidis I., Chou S. and Mantzoros C.S. (2010), Narrative Review: The Role of Leptin in Human Physiology: Emerging Clinical Applications, *Ann Intern Med.*, 152(2):93–100.

Margetic S., Gazzola C., Pegg G.G. and Hill R.A. (2002), Leptin: A Review of Its Peripheral Actions and Interactions, *Int J. Obes Relat Metab Disord.*, 26:1407–33.

Meier U. and Gressner A.M. (2004), Endocrine Regulation of Energy Metabolism: Review of Pathobiochemical and Clinical Chemical Aspects of Leptin, Ghrelin, Adiponectin, and Resistin, Clin Chem., 50(9):1511–25.

Oswal A. and Yeo, G. (2010), Leptin and the Control of Body Weight: A Review of Its Diverse Central Targets, Signaling Mechanisms, and Role in the Pathogenesis of Obesity, *Obesity.*, 18:221–9.

Sinha U., Sinharay K., Sengupta N. and Mukhopadhyay P. (2012), Benefits of Leptin Therapy in HIV Patients, Indian Journal of Endocrinology and Metabolism, 16(9):637–43.

30

The Haematopoietic System
Blood Coagulation—Haemostasis

Susan Turi Blasko and G. Blasko

I. INTRODUCTION

Haemostasis is that property of the circulatory system whereby circulating blood cells and plasma are maintained within the blood vessels and does not loose its fluid physical condition. Based on the classical works of Virchow and Rokitansky this property is attributable to the exceptionally well-coordinated function of three systems, which tightly interact at multiple common metabolic steps and pathways. These are as follows:

1. Blood coagulation
2. Platelets
3. The vessel wall

Haemostatic balance is the end-result of many other subordinated, well regulated-in itself biochemical reactions, including the systems of coagulation-fibrinolysis, aggregation and disaggregation of platelets, etc. Considering all these phenomena, Stormorken suggested use of the term "thrombo-haemorrhagic balance"; which upon being upset, leads to thromboembolic diseases on one hand and more or less serious bleeding on the other hand.

History

Prior to the last century, little was understood about the mechanism of haemostasis. A bleeding disorder associated with scurvy had been described by James Lind, in 1753. Fontane observed an increased tendency for coagulation after the bites of certain snakes, whereas haemophilia—being an inherited bleeding disorder—has been studied only at the end of the 18th century. Buchanan was the first to elaborate on the concept of blood coagulation describing it as a fermentative process, similar to the one seen when milk curdles.

The scheme proposed by Morawitz in 1905 remained unchanged for a long period of time, however he considered separately the formation of fibrin and that of thrombin from thrombokinase in his work. Morawitz also described an inhibitor present in the normal plasma, which he termed 'antithrombin'. The importance of platelets in haemostasis was recognised first in 1882 by Hyem but the first platelet defect was described by Glanzmann only in 1918. The molecular nature of clotting factors, modes of their interaction were elucidated only in the past two decades as a result of wide-range research procedures, whereas the exact role of metabolic processes in the vascular endothelium is being recognized only recently.

In response to vascular injury they are sequentially activated to serine proteases in a linked series of reactions, with each serine protease catalyzing the subsequent zymogen (inactive proenzyme)—protease transition. Finally thrombin converts fibrinogen to fibrin and fibrin is cross-linked. This process resembles a "waterfall" (Macfarlane, 1964) or "cascade" (Davie and Ratnoff, 1964). The clotting factor activated next in the sequence is produced at a much higher molar concentration than the previous one and thus massive amounts of thrombin can be generated. This amplification-system works only with limited efficacy, however based on certain theoretical calculations it has been concluded that activation of a single molecule of Factor XI—disregarding activities of the protease inhibitors—gives rise to the formation of 200×106 fibrin molecules.

The cascade mechanism of blood coagulation is summarized briefly in Figure 30.1, whereas factors involved in this process are listed in Table 30.1. All the active enzymes of this system contain a serine-residue at their active centre and hence they are termed serine proteinases. All the active factors split their substrate protein (the zymogen) via a limited proteolysis, i.e., they hydrolyze only few peptide bonds (mainly by arginine) at very well-determined specific sites throughout the polypeptide chain. A sequence homology has also been established between the proteinases of blood coagulation.

The first seconds of the development of thrombosis are very complex phenomenos of which the starting points are

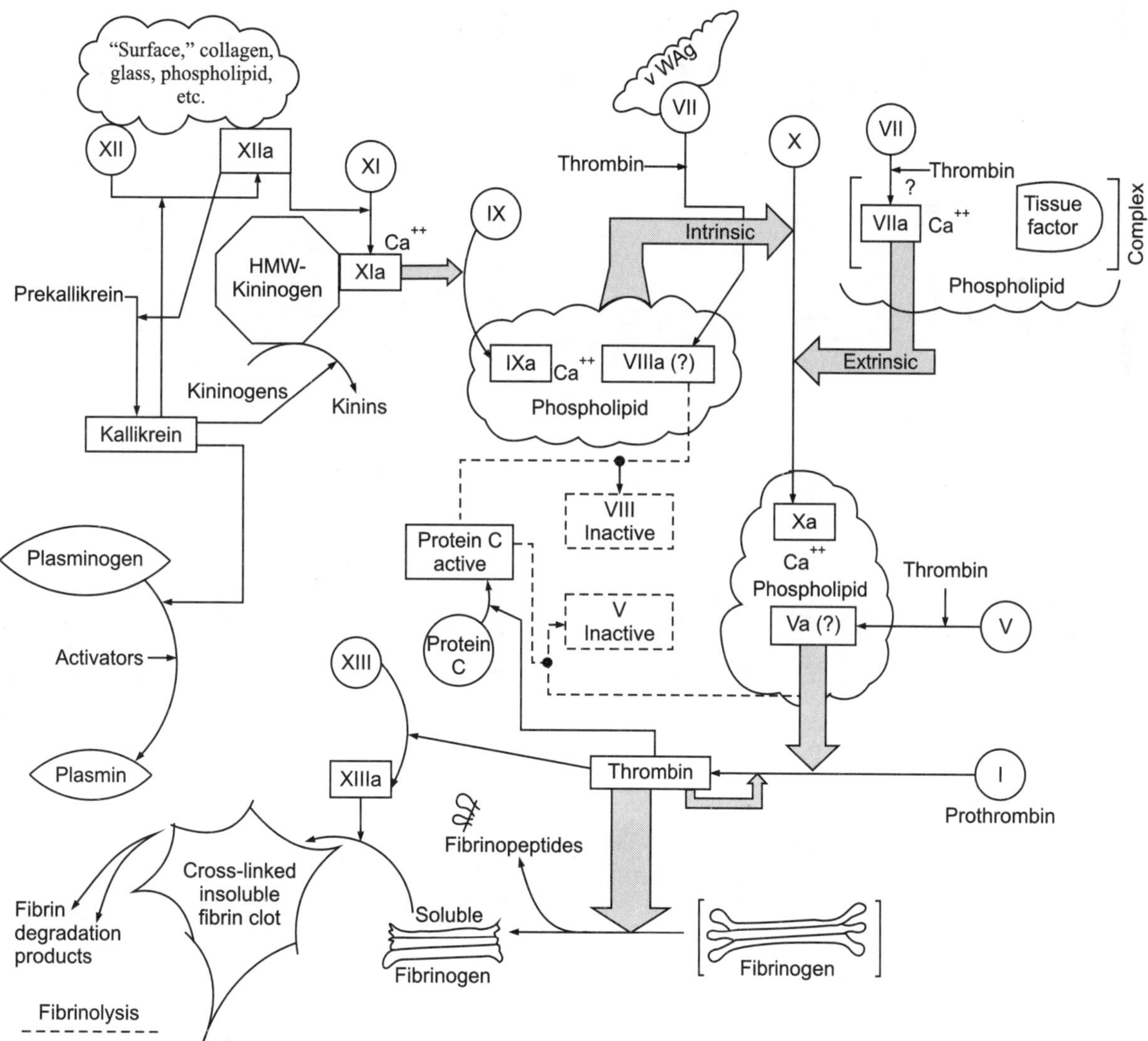

Figure 30.1 The coagulation cascade precursor zymogens are represented by circles, active factors by squares. Activation processes are shown by dashed arrows and some inhibitory reactions by dotted arrows.

TABLE 30.1

	Synonyms	Site of synthesis	Molecular mass (kDa)	Function characteristics	Protein concentration	Plasma concentration	Chromosomal organization	Gene
Factor XII	Hageman factor	Hepatocyte	80	Protease zymogen		30 µg/ml	5	14 exons
Factor XI		Hepatocyte	160	Protease zymogen	Homodimeric	5 µg/ml	4	15 exons
Factor IX		Hepatocyte	56	Protease zymogen	Vitamin K-dependent	5 µg/ml	X	8 exons
Factor VIII	Antihemophilic factor	Hepatocyte	330	Cofactor	Cofactor to factor IXa	0.1 µg/ml	X	26 exons
Factor VII		Hepatocyte	50	Protease zymogen	Vitamin K-dependent	0.5 µg/ml	13	9 exons; adjacent to factor X gene
Tissue factor (factor III)	Tissue thromboplastin	Many cell types	37	Cofactor	Transmembrane protein	0	1	6 exons
Factor X	Stuart factor	Hepatocyte	56	Protease zymogen	Vitamin K-dependent	10 µg/ml	13	8 exons; adjacent to factor VII gene
Factor V	Proaccelerin	Hepatocyte, megakaryocyte	330	Cofactor	Cofactor to factor Xa	10 µg/ml	1	Similar to factor VIII
Prothrombin (factor II)		Hepatocyte	72	Protease zymogen	Vitamin K-dependent	100 µg/ml	11	14 exons
Fibrinogen (factor I)		Hepatocyte	340	Structural	Dimeric, with each monomer having three subchains	200–400 mg/dl	4	Separate genes for α, β and γ chains (5, 8, and 9 exons, respectively)
Factor XIII	Fibrin-stabilizing factor; Laki-Lorand factor	Hepatocyte	320	Transamidation	Tetrameric (a2b2)	10 µg/ml	1(b) 6(a)	a and b genes on different chromosomes
von Willebrand factor		Endothelial cell, (multimers megakaryocyte	220 up to 20×10^6)	Platelet adhesion	Carries factor VIII in plasma	10 µg/ml	12	52 exons; pseudogene on chromosome 22
Protein C		Hepatocyte	62	Regulatory (protease zymogen)	Vitamin K-dependent	5 µg/ml	2	8 exons
Protein S		Hepatocyte	80	Regulatory	Vitamin K-dependent	25 µg/ml	3	15 exons; pseudogene on chromosome 3
Thrombomodulin		Endothelial cell	75–105	Regulatory	Expressed on endothelial surface	0	20	No introns
Antithrombin		Hepatocyte	60	Regulatory	Forms complex with heparin	150 µg/ml	1	6 exons
Tissue factor pathway inhibitor	Lipoprotein associated coagulation inhibitor; extrinsic pathway	Endothelial cell	33	Regulatory	Circulates in plasma; disulfidelinked to apolipoprotein A-II	115 µg/ml	2	9 exons

not well understood yet, however the fortcoming processes are much better known.

The vascular endothelium takes part very actively in the maintenance of normal, non-turbulent flow through the following processes: It releases NO, a dissolved gas being potent platelet inhibitor and vasodilator, prostacyclin (PGI$_2$), it expresses thrombomoduline, which is one receptor of thrombin. The rupture of the atherosclerotic plaque or the damage of the vessel, the subendothelium becames free (collagen and von Willebrand factor) and it is the most powerful inducer of the thrombin-induced platelet aggregation. The platelets become active, the tissue factor releases and the developed active thrombin aggregates platelets and converts fibrinogen to fibrin.

The visible result of clotting is the clot, which forms from the circulating 2 μM of prothrombin to enough thrombin. Interestingly, nobody could demonstrate circulating thrombin even in the cases of serious thromboembolic disorders, which only means, the serious thrombotic disease er cause by an enzyme circulating below $<10^{10}$ M concentraticn.

As it can be seen from Figure 30.1 there are three key-steps in the cascade activation mechanism:

1. Formation of the activated Factor XII/XIIa/—the contact phase
2. Formation of the activated Factor X/Xa/
3. The formation of thrombin and fibrin

The other side of the balance is the fourth:

4. The fibrinolytic system

The coagulation-fibrinolytic system is strongly linked with other enzyme systems and plasma proteins, besides it is also related to platelets and the reticuloendothelial system. Briefly,

1. Activation of the contact phase of coagulation results in a simultaneous activation of the kallikrein-kinin system.
2. The source of phospholipid, which is an absolute requirement for the activation of Factor X and prothrombin, is the platelets (platelet factor 3).
3. The main physiological inhibitor of the coagulation system, antithrombin-III, can inhibit C$_1$ complement.
4. Thrombin is one of the most important physiological inducers of platelet aggregation.
5. Thrombin is capable of converting complement components C$_3$ and C$_5$ to fragment very similar to those formed by the action of highly specific complement enzymes.
6. Thrombin binds to endothelial cells and induces the release of prostacyclin, a species of natural prostaglandins which inhibit platelet aggregation very actively.
7. Reticuloendothelial system is responsible for elimination of the complexes by means of the active coagulation enzymes and proteinase inhibitors.
8. The endothelial cells are now considered to be a new, separate organ, having a large, active surface and possessing a variety of biological, biochemical actions.

(See part VI. for the details!) The roles of endothelium in the regulation of coagulation are diverse as follows:

- triggers the intrinsic phase of coagulation, i.e., factors XII–XI and kallikrein.
- tissue factor is being expressed on the surface of the endothelium.
- binds Protein C and Protein S in quiet state (some cytokines and TNF are inhibit the expression of thrombomodulin).
- secretes t-PA (tissue plasminogen activator—see fibrinolysis).
- cumulates heparin and low molecular weight heparins, and releases them.
- secretes the tissue factor pathway inhibitor (TFPI).
- produces PGI$_2$ and NO, which are potent inhibitors of platelet aggregation.
- produces ADPases and eicosanoids.
- secretes continuously von Willebrand factor, collagen, thrombospondin, vitronectin abluminally.
- the endothelial cells are activated similarly to that of the platelets and they bind IXa, and Xa factors, produce Factors III and V, PDGF, PAF and integrins.
- participates in the cancer procoagulant activity.

II. BLOOD COAGULATION

A. The Contact Phase of Blood Coagulation— Contact Factors

Factors included under this heading are Factor XII (Hageman factor), Factor XI (PTA), prekallikrein (Fletcher factor), high molecular weight kininogen (HMW-kininogen or Fitzgerald factor) and the Factor XII-dependent plasminogen proactivator (See Table 30.2).

TABLE 30.2   Proteins of the Contact Activation System

Protein	Structure	Molecular weight of zymogen and its activated form		Human plasma concentration
Factor XII	Single chain	80,000	80,000	30 μg/ml
Factor XI	Two identical di-sulphide linked chains	160,000	144,000	6 μg/ml
Prekallikrein (Fletcher factor)		85,000	–	50 μg/ml
HMW-kininogen (Fitzgerald, Flaujeac, Williams factor)	Single chain	110,000	–	70 μg/ml

Molecular interactions during contact activation

Exposure of human plasma to negatively charged, so-called "foreign" surfaces like glass, kaolin, etc., leads to initiation of the contact activation system, which is in turn accompanied by triggering of the kinin-forming and fibrinolytic pathways. Factor XII gets attached to these surfaces and this process induces a conformational change in Factor XII.

This results in the cleavage of a single chain native molecule within a disulphide loop, generating a two-chained active enzyme designated as Factor XIIa, which is a serine proteinase*.

Functions of Factor XIIa:

(a) it activates Factor XI to form Factor XIa, which participates in the next step in the intrinsic pathway of coagulation
(b) it is a potent activator of prekalikrein; it triggers the kininforming system
(c) it activates some of the plasminogen activators thus initiating the fibrinolytic system
(d) it cleaves Factor VII resulting in the formation of Factor VIIa, which has a significant role to play in the activation of the extrinsic pathway of coagulation
(e) it activates prorenin to renin
(f) it contributes to the activation of complement C1S; furthermore it was demonstrated that the magnitude of activation of the alternative complement pathway by zymosan is dependent upon the presence of Factor XII. It also contributes to the formation of the classical C3 convertase.

Conversion of Factor XII to XIIa is accelerated by the presence of high molecular weight kininogen (HMW-kininogen). The question, what initiates the activation of XII is still unanswered (The lack of XII causes thrombosis, the first patient, Mr Hagemann died from pulmonary embolism!). Surface-bound Hageman factor is 500 times more susceptible to proteolytic activation than soluble Factor XII and it also activates Factor XI in surface-bound form. The precursor of Factor XI is surface-bound, too. It is suggested, XIIa may activate trace amounts of VII to VIIa, but its aktivity is not important until tissue factor is not present. It is suggested that HMW-kininogen links both Factor XI and prekallikrein to the exposed surface, where they undergo a cleavage and activation by the surface-bound XIIa. Schematic representation of this process is shown in Figure 30.2.

Factor XIIa-induced activation of Factor XI is accomplished by cleavage of an internal ArgIleu peptide bond in each chain of zymogen, resulting in a dimeric molecule having two active centres, as seen from Figure 30.3.

Physiological activation of factor XII

The mechanism of activation mentioned above is true in cases of *in vitro* studies in the presence of artificial surfaces,

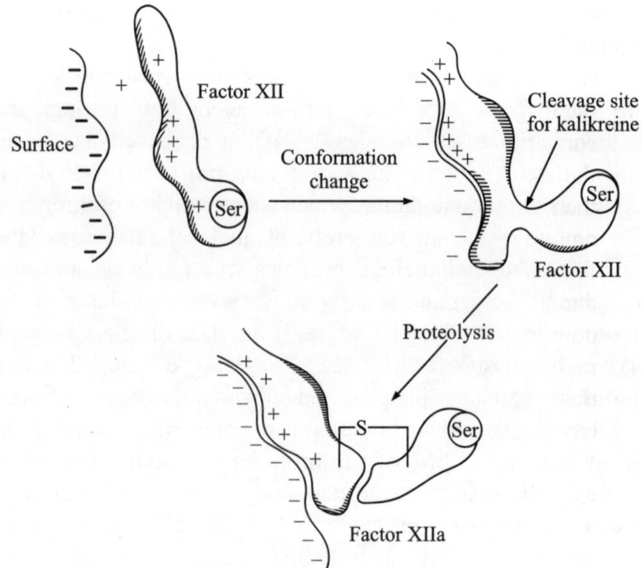

Figure 30.2 The scheme of contact activation.

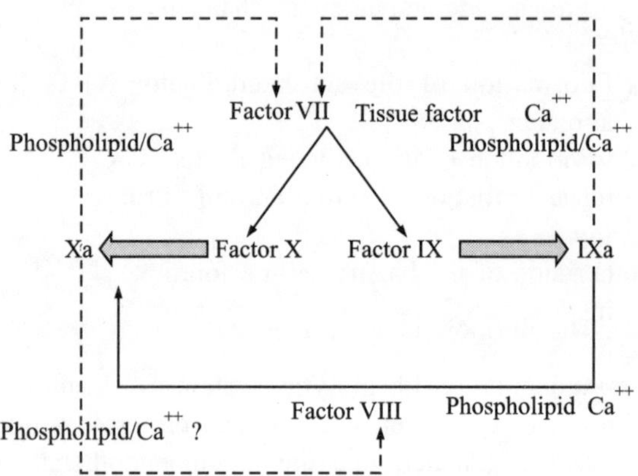

Figure 30.3 Interactions between Factors IX and X in the extrinsic pathway. Dotted lines represent feedback activations

but much less information is available regarding the exact mechanism of activation of the intrinsic clotting system under physiological conditions. It has been demonstrated that certain collagen preparations, activated platelets may have an important influence on the activation of Factor XII, within the circulation. There is evidence suggesting the existence of a platelet dependent mode of activation of Factor XI to XIa in the absence of Factor XII. A lipid constituent of platelets, which differs from the phospholipid-termed platelet factor 3 is implicated to be responsible for this reaction. Since endothelial cells play a key-role in maintaining the integrity of the circulation, evidences have been found that the damaged endothelium contributes to activation of the contact system too. The endothelial activator of Factor XII has been associated with the 100,000 g pellet containing membrane fragments.

---

*Activated forms of the clotting factors are indicated by a lower case 'a' following the Roman numerals.

The components of the contact system have numerous interactions:

- the initiation and amplification of this system on artificial surfaces
- the interaction with biological surfaces
- the interaction of the kininogen with cells of the vascular compartment (e.g. platelets have 1000 binding site/cell for kininogens, the same value is 50,000 site/granulocyte and 1 million site/endothelial cell)
- domains of the plasma kininogens participate in cell membrane expression
- the different domains of kininogens have different biochemical and physiological functions:, i.e., domain 1 binds calcium, domain 2 is cysteine protease inhibitor with anticalpain and anti-inflammatory functions, domain 3 inhibits thrombin binding (antithrombotic), domain 4 regulates endothelium, domain 5 is a heparin binding and kaoilin binding part possessing antithrombotic, and finally domain 6 binds cells and has a procoagulant activity.

Inherited deficiencies of the contact factors

(a) Factor XII deficiency was discovered by Ratnoff and Colopy in 1955 in a patient named Hageman, who had a prolonged clotting time. This deficiency, together with deficiencies of Fletcher factor, Fitzgerald factor however does not lead to severe bleeding (Hageman died of pulmonary embolism). All are inherited as autosomal recessive traits. The gene is located on chromosome 5 and several variants have been described. The individuals do not experience haemorrhagic complications, even major surgery can be performed.

(b) The main activator of Factor XI is thrombin, but the role of Factor XII is not excluded. Factor XI deficiency is also inherited as an autosomal recessive trait affecting mainly Jews originating from the Askenazi descent. It is very rare in members of other origins, i.e., the Oriental or Sephardic Jews. It causes an increased bleeding tendency, menorrhagia in homozygous patients. The diagnosis can be proved only by detailed laboratory investigations. This disease results in decreased thrombin generation, but the decrease is never so severe as in the deficiencies of Factors VIII, IX and X. This deficiency should be differentiated from that one caused by an acquired anti-Factor XI antibody (mainly in SLE and immunological diseases).

(c) deficiencies of Prekallikrein and High-Molecular-Weight Kininogen. The prekallikrein gene is located on chromosome 4q and the patients do not have bleeding. It is a proactivator of plasminogen as well. Some postulate its role in impaired fibrinolysis. The HMW-kininogen (also called Fletcher, Williams or Flaujeac factor) deficiency is inherited by autosomal recessive pattern and the patients are asymptomatic.

B. Formation of the Activated Factor X (Xa)

The enzymes participating in the coagulation cascade can work most efficiently in the form of complexes. Their binding to each other is regulated by well-directed cofactors, phospholipids, ions and the proenzymes. In the presence of every participant, the velocity of the reactions is increased by several magnitudes. These complexes are necessary for the formation of Xa and thrombin.

The assembled complex of factor VIII and Factor IX on a cell membrane (or on any phospholipid membranes, i.e., platelet factor 3) efficiently cleaves Factor X zymogen: this complex is called TENase. The assembly of TENase complex requires the exposure of Factor VIII receptors on the platelet membrane, activation and delivery of Factor VIII and activation and delivery of Factor IX. The TENase complex is inactive in the basal state. Cloning and purification of these factors led to an increased knowledge regarding the interaction between these proteins and with their regulatory proteins.

Factor X can be activated via two pathways: when coagulation is initiated by interaction of the contact factors followed by activation of Factor IX by Factor XIa,—the process is known as the "intrinsic" pathway of coagulation. The term "extrinsic coagulation pathway" defines a sequence of events initiated by the interaction of tissue factor (Factor III) with some plasma proteins. Tissue factor is ubiquitous in human tissues and its contribution to the activation process leads to rapid coagulation through a more simple way. The sharp discrimination between extrinsic and intrinsic pathways became useless, since it was demonstrated that activated Factor VII (from the extrinsic pathway) is able to activate Factor IX: this is the so-called Josso-loop. The activation mechanisms have been studied in flowing blood, too. First we shall discuss the intrinsic way of activation Factor X as a consequence of the already described contact activation and then proceed onto the extrinsic—tissue factor—pathway of blood coagulation.

The frequent bleeding episodes in haemophilias, i.e., in deficiencies of Factor VIII and IX, clearly indicate that there is a requirement for TENase complex but shed limited light upon the biochemical niche of the TENase complex. While episodes of bleeding seem painfully frequent to haemophiliacs, they are infrequent compared to the spontaneous or nearly continuous bleeding that occurs in people with severe deficiencies of all the Vitamin K-dependent coagulation enzymes induced by large doses of coumarins. This difference suggests that the TENase complex is required infrequently in comparison with other parts of the coagulation mechanism. Furthermore, bleeding from mucous membranes bleeding, frequent in subjects with platelet and PROTHROMBINase complex disorders, and petechiae, frequent in platelet disorders, are infrequent in haemophilia. This difference indicates that bleeding from small vascular defects may be controlled without the TENase complex.

Intrinsic activation of factor X

The following proteins are implicated in this process: Factor IX (antihaemophilic globulin B, Christmas factor etc.), Factor XIa, Factor VIII (antihaemophilic globulin A, phospholipid and calcium ions) (see Table 30.3).

TABLE 30.3   Proteins Involved in the Intrinsic Way of Activation of Factor X

Protein	Structure	Molecular weight of zymogen and its activated form		Human plasma concentration
Factor XIa	See details in contact phase single chain	55,000	46,500	3–5 µg/ml
Factor IX	17% carbohydrate	160,000	144,000	6 µg/ml
Factor VIII/von Willebrand factor		860,000*	–	5–10 µg/ml
a. VIII C (procoagulant)	It can be split by thrombin resulting in a 30-fold increase in the procoagulant activity	85 –93,000	–	
b. vWF (von Willebrand factor)		200,000 and above	–	?
Factor X	Two chains light: heavy: 10% carbohydrate	56,000 16,500 39,300	44,000 – –	6–8 µg/ml

*Recently it was demonstrated that purified Factor VIII/vWF exists as series of multimers with molecular weight extending up to 20,000,000.

It should be noted that the activity of Factor VIII which means a procoagulant activity is closely related to von Willebrand Factor. New terminologies commonly use the terms Factor VIII/von Willebrand antigen. The molecular complex has two separate activities: the procoagulant part of Factor VIII is involved in the activation process of Factor X, while the other part of the complex is important for the aggregation of platelets. The assembly of the TENase complex upon platelet membranes in flowing blood starts with diffusion of von Willebrand factor and bound Factor VIII to the exposed collagen in a vascular defect, changing their conformation (step 1). Platelets from flowing blood bind to the immobilized von Willebrand factor (step 2) and are activated by contact with collagen (step 3). Activated platelets express Factor VIII receptor and bind von Willebrand factor and fibrinogen from blood, leading to platelet aggregation (step 4). Small amounts

of thrombin, formed at the platelet surface, activate Factor VIII (step 5), which forms Factor VIIIa and diffuses a short distance through the boundary layer and binds to exposed platelet receptors (step 6). Factor IXa diffuses from plasma and through the boundary layer, above the vascular defect to bind to platelet-bound Factor VIIIa and form the TENase complex. Active Factor IX is present in plasma as a result of an extrinsic activation and may be formed at the site of the vascular defect as a consequence of tissue factor exposure. Congenital disorders of these proteins differ genetically, i.e., deficiency of procoagulant fraction results in the classical, X-chromosome-linked haemophilia A, whereas deficiency of von Willebrand antigen causes a platelet aggregation disorder termed von Willebrand disease which is inherited autosomally. Under physiological conditions these proteins circulate in the form of high molecular weight complexes despite the difference in their origin. The exact structure of the complex (asymmetric complex or identical subunits etc.) is under discussion.

Molecular Interactions during Intrinsic Activation of Factor X: The activation of Factor IX by Factor XIa proceeds in two steps: first Factor IX is cleaved between an Arg-Ala bond giving rise to a two-chained intermediate. The intermediate has no enzymatic activity whatsoever. In the second step an Arg-Val bond in the heavy chain is hydrolyzed resulting in the formation of an activation peptide and Factor IXa. These reactions require calcium ions.

It is possible to split Factor IX by various other serine-proteinase like Factor Xa, thrombin, kallikrein, but these proteolytic reactions do not lead to the formation of active Factor IXa, whereas the protease of Russell's viper venom activates it by only a single-step reaction by splitting the Arg-Val bond. This indicates the great specificity of this particular reaction.

The following step in the activation cascade is the binding of Factor IXa to phospholipid via Ca^{++}-bridges, with the help of its N-terminal γ-carboxy-glutamic acid (Gla) residues. Factor VIII also binds to the phospholipid micelle. The zymogen Factor X is also (it can also be bound through Gla-residues) present. If all these above criteria are fulfilled, a single specific Arg-Ileu peptide bond is cleaved very rapidly in the heavy chain of Factor X, resulting in the liberation of an activation peptide, with a molecular weight of 11,000. The assembly and function of the TENase complex, Factor VIIIa-Factor IXa, which activates Factor X to Xa is also a membrane-bound process and shows remarkable homology to the PROTHROMBINase complex. The assembly of the complex on membrane surface increases the rate of Factor Xa generation approximately 10^9-fold more than that of Factor Xa alone. The association of Factor VIIIa to Factor IXa on the surface stabilizes the cofactor and prolongs the functional lifetime of the intrinsic TENase. Factor VIII plays a modulatory role in this multicomponent molecular complex, but recent evidences have shown, that its function could be improved by limited proteolysis, thus converting it to VIIIa. This protein does not exhibit proteolytic function, however Davie and his associates postulated that the procoagulant activity of Factor VIII could be inhibited by disopropylfluorophosphate and therefore it could be a serine proteinase, too. Factor IXa splits and activates Factor X: in the presence of optimal concentrations of all the components mentioned above, however, the rate of activation increases many thousand-folds. The phospholipid required for activation is supplied either by tissues or platelets (platelet factor 3) under physiological circumstances. Platelet phospholipid usually becomes available following the activation of platelets by thrombin, i.e., after coagulation has been initiated.

Extrinsic activation of factor X

Extrinsic activation of Factor X proceeds in the presence of Factor VII (proconvertin), tissue factor (so-called Factor III) and Ca^{++} ions on the phospholipid surface (Table 30.4).

TABLE 30.4   Proteins of the Extrinsic Activation of Factor X

Protein	Structure	Molecular weight of zymogen and its activated form	Human plasma concentration
Factor VII	Single chain	53,000 –	500 µg/ml
Factor VIIa	Two-chained	52,000	
Tissue factor	?	56,000 –	n.a.
		Factor X See details in Table 30.3	

The carboxy-terminal region of Factor VII zymogen is largely homologous to that of all the other serine-proteinases. Ten glutamic acid residues near the amino terminus of Factor VII are post-translationally modified to Glas. It contains an enzymatically active centre. Several proteases are capable of splitting Factor VII, converting it to a more active – VIIa – form, resulting in the release of an activation peptide and formation of a two-chained molecule. These are as follows: chymotrypsin, Factor XIIa, Factor Xa, thrombin, kallikrein, and the tissue factor—Factor VII complex. The half-life of Factor VIIa is 2.5 hours, much longer than that of other activated coagulation factors. In individuals on stable warfarin therapy, the plasma level of Factor VII activity varies between 12 and 40%, whereas the level of Factor VII antigen is about 50%. The extrinsic TENase complex dysplays different requirements from those of the other procoagulant complexes in terms of generation of its serine protease component.

Tissue factor has been identified as a cell membrane component and it has been demonstrated in various tissues. Its main sources are: the brain, lungs, liver and big vessels. Tissue factor deficiencies have not been described in the literature. The human tissue factor gene resides on the short arm of chromosome 1, it is 12.4 kbase in length and contains six exons and 5 introns. The molecular polymorphism occurs in exon 6 and intron E. The control mechanism of tissue factor production is unknown as yet and we have no additional data regarding any other of its physiological roles apart from its function in blood clotting. It is a lipoprotein in nature as the brain extract-devoid of its lipid content is inactive. The molecular structure has been clarified, showing some homologies with the citokine/interferon receptor family proteins. Its extracellular region is depicted as two domains that form a V-shaped, ligand-binding through. The transmembrane domain transverses the plasma membrane and is followed by a short, cytoplasmic tail. The homolog-sequences are highly conservative. Activity of the reconstituted lipoprotein is highly dependent upon the experimental conditions and procedures. When reconstitution was attempted by using purified phospholipids—phosphatidylethanolamine and phosphatidylcholine were seen to be most effective (however they do not play any role in other coagulation reactions), whereas phosphatydil-serine failed to restore tissue factor activity (though it is active in other coagulation reactions, which require phospholipids). Platelets do not, but endothelial cells and monocytes do contain tissue factor. Tissue factor – VII a complexes formed from the low levels of circulating Factor VIIa also be may responsible for activating Factor VII bound to tissue factor and for enhancing extrinsic TENase activity. The pathway of physiologic Factor VIIa generation is unknown. The importance of the extrinsic TENase trigger of coagulation is well documented. If extrinsic TENase complex could provide significant amount of Factor Xa in vivo, the severe bleeding disorders associated with the deficiencies of Factors VIIIa and IXa (hemophilia A and B) would not exist. Activation of of Factor X by the extrinsic TENase might, in theory, offset the lack of factor X activation by the intrinsic TENase complex in hemophilias A and B and therefore provide sufficient Factor Xa to allow for coagulation at high concentrations achieved by giving recombinant Factor VII, but normal physiological levels are inadequate to reverse the documented pathology of hemophilia A and B.

The proposed mechanism of two-dimensional transfer of enzyme intermediates between membrane-dependent complexes proceeds as follows: Factor IXa is transferred by the membrane surface from the extrinsic TENase complex (TF-VIIa) for Factor VIIIa bound at a separate site on the membrane. The extrinsic TENase and the intrinsic TENase complexes participate in the activation of Factor X to transfer of Factor X to Factor Va in the same lipid-dependent manner. Prothrombin (Factor II)–membrane interactions are also important in efficient generation of thrombin. Diffusion of Factor IXa and Factor Xa enzymes along the lipid surface accelerates the rate of complex assembly and protects free enzymes from inactivation by circulating inhibitors.

These cells may be induced to express tissue factor by a large number of inducers (endotoxin, interleukin-1, tumor necrosis factor, thrombin, phorbol esters, activated complement, immune complexes, tuftsin, cholesterol, lipoproteins, but by adherence, cytotoxic agents, allogeneic stimulation, tumor cell exposure, etc.), whereas its expression could be enhanced by T-lymphocyte collaboration, platelets, CD11b/CD18 receptor occupancy, PAF, calmodulin inhibitors etc. and it could effectively be inhibited by steroids, heparin, dipyridamole, PGE1, cyclosporin and retinoic acid.

The tissue factor generation ex vivo proceeds in inflammatory diseases, SLE, trauma, unstable angina, infections, diabetes, and cancer.

Molecular Interactions in Extrinsic Activation of Factor X: Factor VII binds to the phospholipid surface via its Gla residues alongwith Factor X and tissue factor (Gla residues are not described in tissue factor). Proteolytic activation of Factor VII to yield Factor VIIa can be accomplished by the action of Factor XIIa, but recently it has been proved that the complex formed by Factor VII zymogen and tissue factor could also activate Factor IX, converting it to Factor IXa. This is the Josso-loop. Factor VIIa is shown bound to tissue factor and Factor X is shown in the process of being activated by the Factor VIIa-tissue factor complex. The gamma-carboxyglutamic (Gla) acid, the epidermal growth factor (EGF) and catalytic domains of Factor VIIa and Factor X are depicted as globular structures, as is the activation peptide of Factor X. Calcium ion binding by the Gla domain in Factor VIIa is required for Factor VIIa binding to tissue factor, and the EGF-domain in Factor VIIa plays an important role in its interaction with tissue factor. Calcium ion binding of the Gla-domain of Factor X is required for its association at the phospholipid surface with Factor VIIa—tissue factor (Figure 30.3).

The tissue factor (TF) is a 45 kD protein, which can be found on all external membrane, which do not have a direct contact to the blood (i.e. fibroblasts, smooth muscle cells). The tissue factor apoprotein contains three domains: an aminoterminal extracellular domain, a transmembrane and a cytoplasmatic C-terminal domain. It is a member of Class II cytokine receptor superfamily and also contains the receptors for α-, β-, γ-interferon and interleukin-10. The extracellular part shows two fibronectin III domains linked end-to-end with a very hydrophobic interface resulting in a very rigid structure. In case of the damage of the vessel, blood flows out and comes in contact with the TF bearing cell surfaces. Those cells which are in direct contact with blood do not express TF, however the monocytes and endothelial cells do express it. In cases of Gram-negative bacterial sepsis the bacterial lipopolysaccharides bind to endothel (or monocytes) and induce rapidly the mRNA synthesis of TF, which appears of the surface of the cells. It results in generalized thrombin formation, (In case thrombin, it is not eliminated rapidly, it leads to death. In the USA, yearly it is a 170.000 septic patients die). If the formation of the clot as lesser extent, it starts fibrinolysis and then thrombosis and bleeding occur simultaneously. This situation is called diffuse intravascular coagulation (DIC). Tissue factor may appear in the circulation in case of large tissue damage (dead fetus syndrome, giving birth, orthopedic operations, malignancies, etc.). It is important to know that in the presence of the other necessary cofactors, tissue factor increases the velocity of activation by 10 million times (10^7). Its Kd for VIIa is < 1 nM.

While the intrinsic and extrinsic pathways are usually described as separate and independent entities, many findings points to a tight interaction between them. This phenomenon explains the lack of bleeding in patients suffering from contact factor deficiency.

Inherited Deficiencies Affecting the Middle Phase of the Coagulation Cascade

(a) *Haemophilia A* (the classical haemophilia): It is the least common of all these disorders and occurs in all ethnic groups without geographical limitations. Its incidence is about 5 in 100,000. It results from the failure to synthesize Factor VIIIC partially or totally or from synthesis of an abnormal molecular variant of VIIIC protein. Severity of the disease is parallel to the degree of deficiency; moderately affected patients have VIIIC levels of 2–5 units/100 ml (normal values vary between 50 and 200 units/100 ml; mean: 100 U/100 ml). The hemophilic patients have markedly delayed thrombin generation despite significant tissue injuries. There is serious bleeding at levels below 0.5 units (100 ml) but the severity and incidence of these bleeding episodes vary and are related to stress situations and emotional disturbances, too. The inheritance is X-chromosome-linked, on a locus closely related to glucose-6-phosphate dehydrogenase. Affected males carry the mutant allele on their single X-chromosome (hemizygous) whereas heterozygous females carry it on one of their X-chromosomes hence, they are not affected. True homozygosity is possible but its calculated frequency would be about 3 in 1 billion. All daughters of a haemophilic male will be carriers but all his sons will be normal and will not transmit the deficiency. Detection of the carriers, prenatal diagnosis, preoperative arrange-ments and the treatment of such patients pose special problems for the haematological centres to be dealt with. The factor VIII replacement therapy has an extended literature for clinicians, but recently the human experiments with gene therapy have also been started. To reach a 1–2% factor level by the new treatment nodalities is of utmost importance: many clinical trials are in progress. However there are fundamental problems to be solved regarding long-term transgene expression, to find the ideal target tissue and many safety issues.

(b) *Haemophilia B* (Christmas disease): It is the deficiency of Factor IX which manifests either due to a failure in the synthesis or formation of different types of apparent Factor IX variants—including an abnormal active site, abnormalities in calcium binding and disorders of the activation process itself. One third of the haemophilia B patients have a de novo mutation of the gene positioned on the long arm of the X chromosome. There are many genetic variants (listed on World Wide Web: www.kcl.ac.uk./lip/petergreen/haemBdatabase.html.

The inheritance is X-linked. Haemophilia B is not less severe as compared to haemophilia A but Factor IX requirements for haemostasis are much less than that of Factor VIII—considering normal physiological conditions. Detection of the carriers, diagnostic and therapeutic problems are all similar to that of classical haemophilia.

(c) *Factor X deficiency* (Stuart-Prower disease): The first propositi who had a proven autosomal recessively inherited lack of synthesis of Factor X were Mr. Stuart and Miss. Prower. The abnormality has also been found in an Italian valley, Friuli. This is however a molecular variant of an immunologically reactive Factor X, as proved by laboratory investigations. The gene is located on chromosome 13 at 13q34 position close to the Factor VII gene. Its variants are found on http: archive.uwcm. ac.uk/uwcm/mg/hgm0.htmk. The clinical manifestations are as severe as in Favtor VIII deficiency including haemarthros, soft-tissue haemorrhages, retroperitoneal and CNS bleedings etc.

(d) *Factor VII deficiency:* It is inherited as autosomal recessive but homozygous patients are not severely affected; the symptoms are not related closely to Factor VII levels. Some molecular variants (e.g., Factor VIIPadua) have also been described (see http. europium.csc.mrc.ac.uk). This deficiency should be distinguished from the acquired deficiencies (caused by liver damage, vitamin K deficiency). The therapy is replacement in order to reach 10% of the normal level which is sufficient.

C. The Prothrombinase Complex and the Formation of Thrombin

The formation of thrombin is initiated by the contact of tissue factor to the blood. TF results in an increase of formation of VIIa by several magnitudes and the activated Xa finally activates prothrombin to thrombin. These reactions are very rapid in forming of complexes and are irreversible.

Prothrombin is physiologically activated to thrombin by the serine proteinase activated Factor X in association with Factor V, on a phospholipid surface, the presence of calcium ions. Interactions between Factor Va (the modulator protein cleaved by trace amounts of thrombin), and membrane sites as well as Factor Va, Factor Xa and additional membrane sites lead to Prothrombinase assembly. Two designated domains of prothrombin that interact with site and the heavy chain of Factor Va, respectively.

Factors included under this heading are listed in Table 30.5.

TABLE 30.5

Protein	Structure	Molecular weight of zymogen and its activated form		Human plasma concentration
Factor Xa	See details in Table 30.3			180 nmol/l
Prothrombin	Single chain 8% carbohydrate	72,000		100–200 μg/ml
Thrombin	Two chains	–	38,000	
Factor V	Single chain 20% carbohydrate	350,000	94,000 and 74,000	30 nmol/l

Molecular Interactions During the Activation of Prothrombin

Prothrombin is a single-chain glycoprotein containing nearly 600 amino acid residues. Its amino-terminal region contains γ-carboxy-glutamic acid residues, which are responsible for the binding to phospholipid. The formation of thrombin from prothrombin by Factor Xa takes place through the cleavage of two peptide bonds in the prothrombin molecule. The first split gives rise to two parts of approximately equal size, namely: to a thrombin precursor designated Prethrombin 2 and a fragment termed Fragment 1.2. These two polypeptide chains are associated via noncovalent forces during the second proteolytic splitting which then transforms Prethrombin 2 to thrombin (Figure 30.3). The prothrombinase complex assembly proceeds through initial formation of Factor Va-lipid and Factor Xa-lipid complexes. The individual binary protein-lipid complexes then combine on the membrane surface through translational or rotational diffusion to in the prothrombinase complex.

Although Factor Xa is the sole enzymatic component necessary for conversion of prothrombin to thrombin the rate of this reaction increases by 3 magnitudes if the other components, i.e., Factor Va, phospholipid, calcium are present at optimal concentrations, too.

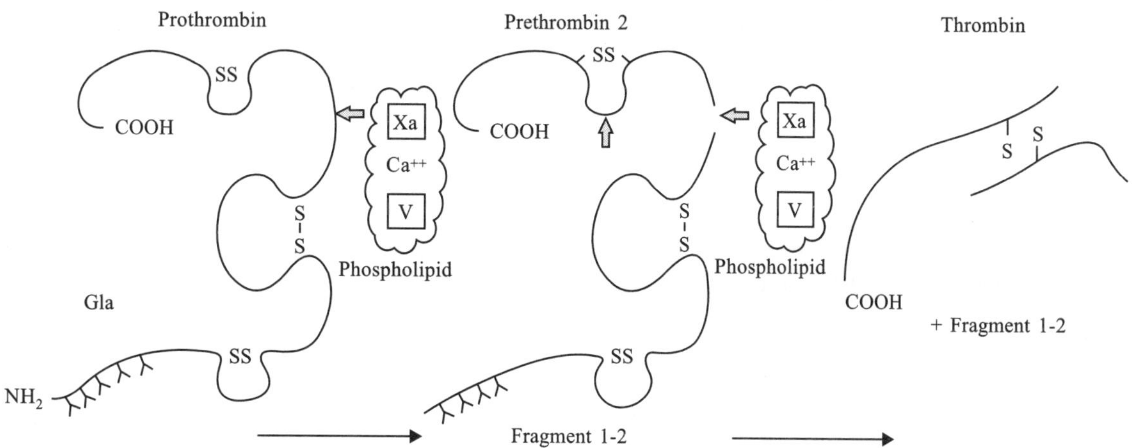

Figure 30.4 Activation of prothrombin.

Thrombin can also be formed by means of autocatalysis from prothrombin, i.e., after the first traces of active thrombin are formed it cleaves a peptide from prothrombin which is shorter than the one cleaved by Factor Xa—called Fragment 1—and the remaining prothrombin 1 is then converted rapidly to prothrombin 2 by Factor Xa. The latter reaction results in the formation of an additional peptide called Fragment 2. This Fragment 2 contains the Factor V—binding portion of the prothrombin molecule. According to this thrombin can control its own formation from prothrombin. Besides the above mentioned reactions thrombin also cleaves Factor V, forming Factor Va in a two-step proteolysis, resulting in an increase of the procoagulant activity of Factor V. This process however exposes Factor V, rendering it more susceptible for proteolytic inactivation.

The role of phospholipids in the coagulation cascade

Phospholipids and cell membranes originating from platelets play a very important role in the activation of coagulation enzymes, i.e., the activation of Factor X via both the extrinsic and intrinsic pathways. The activation of prothrombin also requires phospholipid. This is the surface onto which components of the activation system can bind. The absorption or binding of the components to the surface results in a local increase in concentrations of the constituents: the participating components are present in a compartment thus being separated from the circulating proteins. Vitamin K-dependent coagulation factors (namely-Factor II, VII, IX, X; see details later) bind to all membranes having negatively charged lipids without any structural specificity, i.e., phosphatidyl-serine, phosphatidyl-inositol, cardiolipin, phosphatidyl-glycerol are equally satisfactory. They are all zwitterionic lipids being asymmetrically distributed between the two bilayers of the plasma membrane. There is a great difference regarding coagulation-promoting activities of external and internal lipid surfaces. Using erythrocyte membranes it was demonstrated that the right-side-out membranes (which is the normal condition for circulating erythrocytes) were ineffective, whereas the inside-out red cell membranes were very effective in shortening the recalcification times. Interaction of the coagulation factor(s) with the phospholipid molecule is represented in Figure 30.5 in a schematic form.

The Ca-bridge is formed between the carboxylate group of γ-carboxyglutamic acid residues and the carboxylate or phosphate groups of the negatively charged phospholipid molecule. A variable number of such bridges can result depending on the number of Gla residues available and the calcium ion concentration. The formation of γ-carboxyglutamic acid residues is discussed under the heading of "Biosynthesis of the coagulation factors".

Activation of prothrombin on the surface of platelets: Briefly summarizing the available data it has been shown that platelets possess receptors for binding Factor Xa and prothrombin. It was also revealed that the rate of reaction of prothrombin activation takes places at a much higher pace

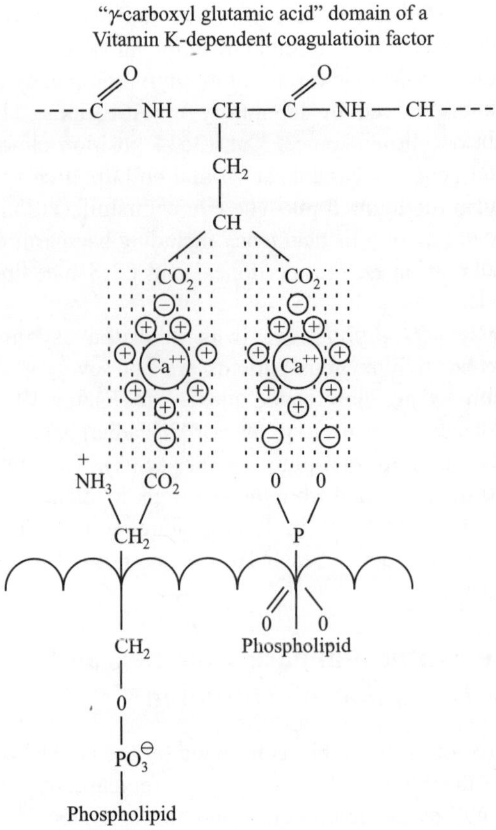

Figure 30.5 Schematic representation of the interaction of a phospholipid molecule with a Vitamin K-dependent coagulation factor. (after Jackson C.M.)

when platelet membrane is used as phospholipid source rather than the synthetic phospholipids. Platelets absorb the factors necessary for prothrombin activation and the phospholipid extruded from the platelets (during the aggregation process) constitutes the surface required.

Regulation of prothrombin activation

The mode of regulation of plasma prothrombin levels is not known. Prothrombin is synthesized by the liver and all factors influencing protein synthesizing capacity of this organ affect prothrombin concentration, too.

The requirement for Factor V and Factor Va opens another possibility for regulation: Following an activation process, Protein C—in its new form C'—rapidly inactivates Factor Va. Activation of Protein C by thrombin is slow, but it is enhanced by the presence of a cofactor derived from the endothelial cells. Protein C protects factor Xa from inactivation but only in a complex form—alongwith Factor VIII, phospholipid and calcium.

Activation of prothrombin is also regulated by inactivation of the product, i.e., thrombin by the plasma inhibitor system. Antithrombin inactivates not only thrombin but Factors Xa, IXa, XIa, XIIa as well. Therefore it exerts a great regulatory effect on the prothrombin activation process. Antithrombin (earlier called antithrombin-III due to historical reasons), a 64 kDa protein is synthesized in the liver and circulates at 4 μM concentration in the blood stream. It is quite specific for thrombin and Xa, but is not able (or very slowly) to inactivate serin proteases being bound in a complex. Thrombin recognizes antithrombin as its substrate (binds to a peptide bond besides an Arg), but the intermedier does not hydrolyse further and the enzyme remains in a complex with the inhibitor. The velocity of inactivation is quite slow: (90% is inactivates within 10 minutes of the total amount of thrombin appears in the circulation). The anticoagulant heparin is a polysaccharide which increases the rate of this complex formation by 3500-fold. After the inactivation of the enzyme in the complex, heparin is released from the complex and is able to form new terner complex: therefore heparin was the first example of non-protein biocatalysts.

There are two ways of thrombin formation via feed-back regulation:

(a) Thrombin splits Factor V and VIII to FVa and FVIIIa (the sign "a" might be misleading because these modulator proteins does not exert any enzymatic activity, (but after the splitting they underway a conformational change and become more active in procoagulant activities). Thrombin activates Factor XI as well. These reactions result in positive feed-backs(+). Thrombin binds to thrombomodulin (TM) and splits Protein C (PC) into its active form (aPC): it is an active proteolytic enzyme and degrades Factor Va and VIIIa. These inactive cofactors mean negative feed-backs (–). This degrading activity of aPC is further enhaced by the presence of Protein S (PS).

(b) The free thrombin is able to activate Factors V, VII, IX and XIII and platelets and convert fibrinogen to fibrin but acts very weakly on PC: these phenomena increase the efficacy of coagulation. Thrombomodulin (TM) being bound to the membrane of endothelial cells binds thrombin and activates only PC, because the other substrates of it are protected. The aPC inhibits further formation of thrombin.

Therefore thrombin, depending on its binding and conformation may exert both thrombotic and antithrombotic activities.

Recent summary of the coagulation cascade until the formation of thrombin

Briefly summarizing: Clotting is initiated by Factor VIIa—tissue factor complexes formed on the surfaces of cells in injured tissues. These complexes activate Factor X (to Xa) and Factor IX (to IXa). The result is the generation of a small amount of thrombin. The system is amplified by the assembly and activation of coagulation factors and cofactors on the surfaces of platelets that have been activated by stimuli such as thrombin itself. The propagation of the coagulation system is achieved by the assembly of the TENase complex and PROTHROMBINase complex on activated platelet surfaces, leading to a thrombin-stimulated fibrin formation.

Functions of thrombin

Thrombin has multiple control functions on the process of haemostasis besides its clotting activity. A brief list of its functions follows here:

(a) In the circulating plasma:

- It converts fibrinogen to fibrin.
- Activates Factor XIII to Factor XIIIa (important for fibrin stabilization).
- Proteolytically activates Factor V, Factor VIII.
- Rapidly activates Factor Va, Factor VIIIa.
- It activates Protein C to its activated form (Protein Ca).
- It cleaves complements C3 and C5, forming the appropriate fragments.
- Causes a platelet-mediated activation of the complement system.
- Forms complexes with proteinase inhibitors, i.e., antithrombin-III, α_2-macroglobulin and α_1-proteinase inhibitor.

(b) Its effects on blood cells:

- Thrombin binds to platelets and stimulates platelet aggregation resulting in the release of certain platelet-specific substances (Ca^{++}, platelet factor 3, platelet factor 4, platelet mitogen, etc.).
- Thrombin stimulates platelet thromboxane A_2 synthesis.
- Thrombin incorporates into fibrin gels deposited on the surfaces of erythrocytes.

(c) Its action on the vessel wall:

- It binds to endothelial cells with high affinity and stimulates prostacyclin synthesis.
- Activates integrins for leukocyte adhesion.
- Stimulates the secretion of endothelin, PGDF, nitric oxide (NO), from the endothelial cells.
- Cells exhibiting thrombin-bound thrombomodulin are stimulated to produce tissue plasminogen activator (t-PA).
- It modulates leukocyte chemotaxis and proliferation, modified cytokine production.
- It is mitogen for smooth muscle cells.
- It modulates neurite growth regulation.
- It binds to fibroblasts thus initiating cell division and hence it also plays a role in the process of wound healing.

Thrombin shows different affinity to platelets, endothelial cells, plasma proteins, etc.: Kd for endothelial cells is about 10^{-10} M, for platelets 1.5×10^{-9} M, for antithrombin 10^{-10} M respectively. Therefore, at the lowest concentration, thrombin activates the endothel-bound Protein C, then increasing the concentration, it aggregates platelets, and finally, after a further increase it starts to get involved in coagulation. Interestingly, at the lowest concentrations, thrombin is an anticoagulant. Under physiological conditions thrombin cannot be detected in the circulation. Long ago Copley forwarded a theory stating that "formation and inactivation of thrombin takes place simultaneously and constantly." The so-called "compartmentalization" of definite phases of the coagulation process results in a great variety of local procoagulant protein and modulator concentrations and therefore we cannot answer the question—which factors or what conditions regulate the actual activity of thrombin. It seems probable that thrombin acts primarily at cellular level and because of the various inactivating (consumptive) processes only subsequently after increasing its plasma concentration sufficiently—it acts at the fluid-phase level in blood.

Inherited deficiencies of prothrombin activation

(a) Prothrombin deficiency: It is a very uncommon disorder. The number of the affected families is cca 30 throughout the world. In cases of true deficiencies a reduced level of prothrombin is demonstrated by immunological and functional methods. The haemorrhagic complications of dysprothrombinaemias (i.e., the molecular varians) are related to the degree of abnormality. When the thrombin generation is defective or thrombin is dysfunctional, the normal fibrin formation is impaired. Only homozygous patients suffer from increased bleeding tendency (the complete absence of prothrombin is incompatible with life); the heterozygous ones are symptomless. A great number of molecular variants of prothrombin (dysprothrombinaemias) have been described— each exhibiting different functional abnormalities. These proteins show altered activities towards various prothrombin activators, i.e., factor Xa, snake venoms, etc. (Prothrombin Barcelona, Prothrombin San Juan, Prothrombin Brussels, etc.). The differential diagnosis is essential: acquired hypoprothrombinaemia is usually caused by warfarin administration. Acquired antibodies have also been described.

(b) Factor V deficiency: Earlier it was described as para-haemophilia by Owren in 1947, i.e., deficiency of the labile factor. It is autosomally inherited, symptomatic patients are usually homo-zygous and consanguinity has also been noted. Its gene is located on chromosome 1. Symptoms are related to bruising, epistaxis, oral haemorrhage and excessive bleeding following a tooth extraction. Deficiency of Factor V occurs more frequently in combination with Factor VIII than can be accounted for by chance. (Type I FMCFD: familial multiple coagulation factor deficiency).

TABLE 30.6   Types of FMCFDs

Type	Defects in factors
I	V and VIII
II	VIII and IX
III	II, VII, IX, X, Protein C Protein S
IV	VII and VIII
V	VIII, IX and XI
VI	IX and XI

Antibodies against Factor V are known, but no purified concentrated preparations are available for replacement (fresh plasma is used).

(c) Protein C deficiency: The evidence of this factor and its deficiency were described only recently. Protein C inactivates Factor Va and Factor VIIIa rapidly: deficiency of this factor manifests by an increased thrombotic tendency. As Protein C and its cofactor Protein S are Vitamin K-dependent inhibitors of coagulation, their deficiencies are frequent causes of thromboembolic diseases, i.e., thrombophilias (see inhibitors of coagulation). They act on the surface of the endothelial cells with interaction of thrombomodulin (their terner complex exerts the biological activity): the deficiency of thrombomodulin(s) has recently been described.

D. Formation of Fibrin

The most important final reaction in the mechanism of coagulation is the thrombin catalysed conversion of fibrinogen to fibrin, whereby an insoluble polymer network is formed from a soluble protein, with great precision, at the site of the wound.

Fibrinogen (molecular weight: 340 kDa) is a glycoprotein, having a plasma concentration of approximately 3 mg/ml (10 μM) with a half-life of 3 days. A highly schematic model of fibrinogen is presented in Figure 30.8, after Doolittle. The heap of N-terminals is called region E, the heap of N-terminals is called region D.

Fibrinogen is a dimer comprised of three dissimilar pairs of polypeptide chains, namely Aα, Bβ and γ, consisting of 610-, 461- and 411 amino acids, respectively. It contains 58 cysteine residues, all of which participate in the formation of disulphide bonds. Four carbohydrate clusters bound to the two β-chains and two γ-chains make up the molecule complete. Its primary structure has been elucidated almost completely (Doolittle et al., and Henschen et al.). The disulphide arrangement is such that six chains are likely to be gathered in the form of a horsetail— then the tuft apparently bifurcates into two bundles, of three strands each. Fibrinogen therefore has a very unique structure, and is a fragile molecule which easily undergoes denaturation.

Molecular interactions during the formation of fibrin

(a) Proteolytic phase: Thrombin cleaves four Arg-Gly bonds in the Aα and Bβ chains, thereby liberating a pair of peptides termed fibrinopeptide A and B. The γ-chain is not split. Fibrinopeptide A is released faster than fibrinopeptide B. This reaction—catalyzed by thrombin—is highly specific and the presence of calcium ions has a stimulating effect on its rate. Liberation of fibrinopeptides is necessary for the ensuing gel formation. Other enzymes (for instance the Russell's viper snake venom—Reptilase) which liberate only the A peptide do not lead to gel formation. The remaining molecule is termed fibrin monomer (after liberation of the fibrinopeptides) which is badly soluble in water and it enters into the next phase which is that of polymerization.

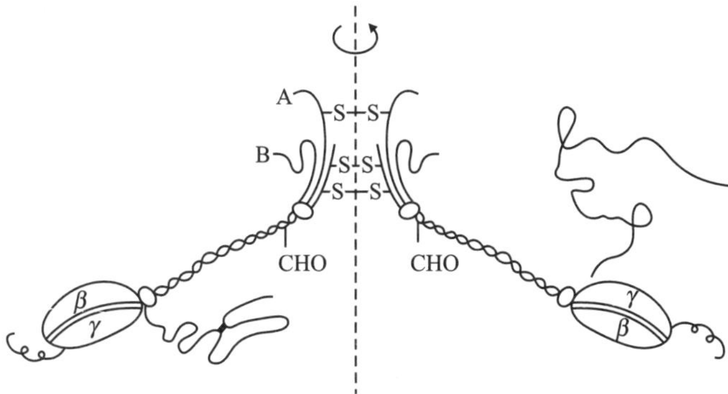

Figure 30.6 Schematic model of fibrinogen. A: Aα chain, B: Bβ chain of fibrinogen. CHO: carbohydrate moieties. (after Doolittle et al.)

The place of release of fibrinopeptides is the so-called E domain. Then the fibrin monomer molecules align in a staggered half-overlapping orientation. The molecule is held together by the D–E contacts to form a two-stranded protofibril. The protofibrils also polymerize laterally to form thicker fibres, which branch to form a three-dimensional gel. In the presence of calcium, plasma Factor XIII and thrombin the fibrin polymers are crosslinked (see point *c* below).

(b) Polymerization: Polymerization of fibrin monomers is a non-enzymatic self-assembly process involving end-to-end and lateral associations of the monomers under physiological conditions (i.e. ionic strength of the plasma). The polymer can also be formed in the absence of calcium and it is soluble in 6 M urea. It forms a semi-solid mass; the propagating fibres get entangled in the cellular elements of blood—including red and white cells but especially platelets—which are the preferred elements for interaction—reactions. The nodular structure of fibrinogen and the given scheme of polymerization limit inter-molecular contacts between "nodules" during interaction within the molecule. This fibrin network is not stable. The stability commences via formation of covalent bonds between the monomer fibres.

(c) Stabilization of fibrin: Soluble fibrin is stabilized in the presence of another plasma enzyme—activated Factor XIII (fibrinoligase, Laki-Lorand factor). Laki-Lorand factor consists of two pairs of non-identical subunits in an inactive tetramer form, that can be activated by thrombin in the presence of calcium ions resulting in the formation of Factor XIIIa. The activation of plasma Factor XIII and the formation of cross-linked fibrin are processes that are very closely related to each other. Platelets also contain Factor XIII having only *a* subunits with an identical amino acid composition to that of plasma Factor XIII. When Factor XIIIa is formed it stitches together neighbouring molecules in the fibrin polymer. Stabilized structure is ready when the XIIIa catalyses the formation of a peptide bond between the Lys and Gln of two β-chains. Fibrin polymers enhance XIIIa activation by thrombin. Fibrinogen and fibrin promote the calcium-dependent dissociation of thrombin-

cleaved Factor XIII A'_2B_2 to yield the active enzyme, Factor XIIIa, which has an active-site Cys314 residues exposed for catalytic crosslinking of fibrin polymers. All molecular forms of Factor XIII will interact and bind to fibrinogen and fibrin. Other proteins are also cross-linked by Factor XIII, i.e., α_2-antiplasmin cross-links to the α-chain of fibrinogen, resulting in a formation of a very effective plasmin inhibitor physiologically, or fibronectin, which is a major plasma protein substrate for Factor XIIIa, or collagen, platelet cytoskeletal proteins, vitronectin, etc. If von Willebrand Factor is cross-linked to fibrin, clotting is slowed down. Cross-linkage is due to the formation of amide bonds between the side-chains of suitably disposed lysines and glutamines. The result of such a union is a (γ-glutamyl)-lysine cross-link. These intermolecular linkages form mainly between the γ-chains of fibrin. Subsequently, multiple cross-links between α-chains can also result. Six of these bonds may be incorporated by each fibrin unit in succession giving rise to covalently stabilized fibrin which is mechanically more strong and insoluble. The mechanism of fibrin stabilization is shown in Figure 30.7.

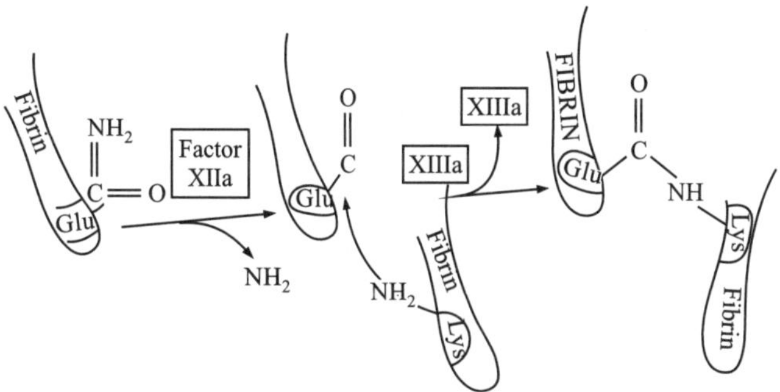

Figure 30.7 The cross-linking of fibrin.

2 Hereditary deficiencies of fibrin formation

(a) Disorders of fibrinogen synthesis can be divided into two groups:

    (i) deficient synthesis (afibrinogenaemia, hypofibrino-genaemia), and

    (ii) defective synthesis (dysfibrinogenaemia)

Afibrinogenaemia is a rare autosomally inherited trait (cca. 160 cases were described). Chromosome 4 contains all the 3 genes of the chains of fibrinogen. There is a coordinate synthesis and an ordered assembly of the 3 chains into a homodimeric molecule. The disease manifests only in homozygous patients causing severe bleeding in the first year of life. History of consanguinity is often present in the parents. It is very interesting to note that total absence of fibrinogen in the plasma of such patients is still compatible with life. It is not known however what type of a molecule substitutes for fibrinogen.

Inherited dysfibrinogenaemias have a similar mode (autosomal dominant with high penetrance) of inheritance but they can further be divided into subgroups according to the biochemical abnormalities present:

Fibrinogen Detroit, Fibrinogen Metz, Fibrinogen Giessen —show an impaired release of fibrinopeptide where no fibrinopeptide B is liberated. These conditions are associated with some changes in the primary structure. About several hundreds of named fibrinogen variants have been described with delayed polymerization, e.g., Fibrinogen Paris, Fibrinogen Amsterdam, Fibrinogen Nagoya, Fibrinogen Montreal. See the website www.geht.org. In the background of acquired deficiencies there may be hepatomas, renal cell carcinoma and autoimmune diseases present.

(b) Factor XIII deficiency: It is an extremely rare disease, just over 100 patients have been described so far. It is carried as an autosomal trait with high incidence of consanguinity in the affected families. It presents with excessive bleeding from the umbilical stump during the first few days of life in the homozygous individuals. All the coagulation tests are normal but the plasma clot is soluble in urea. Three types of deficiencies have been described: Type I, a combined deficiency of subunits A and B, type II in which the A subunit is defective or absent, and Type III in which B subunit is defective or absent.

The role of Factor XIII in wound healing is essential, because fibrin provides the provisional extracellular matrix for it. Cross-linked fibrin is important for fibroblast growth, cross-linked fibro-nectin in tissue remodeling. Remarkable clinical responses to infusion of Factor XIII have been reported in ulcerative colitis, Crohn's disease, scleroderma, Henoch Schönlein purpura.

3 Inhibitors against coagulation factors

(a) Factor VIII inhibitors without haemophilia: they are rare and occur in association with SLE, asthma, pregnancy and cancer. The antibodies are autoantibodies of the IgG class. The diagnosis of the IgG has a specific algorithm in specialized centers.

(b) Factor VIII inhibitors with haemophilia. From clinical point of view this is the most important group, because it develops in 25% of the treated haemophiliacs mainly, where the supplemental products are changed during the lifetime of the affected patients. The haemophila centers usually check their patients yearly with the Bethesda assay. The treatment is highly dependent on the titer of the antibody and needs special laboratory and drugs.

(c) von Willebrand factor inhibitors: may be congenital or acquired (i.e., in hypothyroidism, neoplasms, myeloproliferative disorders, etc.).

(d) antibodies developed for other coagulation factors, i.e., V, VII, IX, X, XI, prothrombin, thrombin, fibrinogen/ fibrin and Factor XIII are very rare and their common immunological origin needs specialized centers for diagnosis and treatments.

E. Biosynthesis of the Factors of Blood Coagulation

Liver is said to be the site of synthesis of Factors XII, XI, prekallikrein and high-molecular-weight-kininogen, however,

the perfusion experiments performed did not confirm it. This suggestion has been based only on finding reduced levels of these proteins in hepatocellular diseases. As far as the rest of the blood coagulation factors are concerned—with the exception of Factor VIII—the liver is the site of their biosynthesis.

Factor V is synthesized partly by the hepatocytes and also by the reticuloendothelial cells of the liver.

Factor VIII and von Willebrand antigen are synthesized by the endothelial cells whereas the procoagulant part of this molecule does not seem to be produced by the same cells as von Willebrand factor; no conclusive evidence has been reached so far as to the exact type of cell responsible for its synthesis. Common to the endothelial cells of liver—spleen, various macrophages, reticuloendothelial cells, vascular endothelial cells, fibroblasts—are all involved in the synthesis of this protein.

Vitamin K-dependent Factors

Liver is the main source of prothrombin, Factor VII, IX, X, Protein C and Protein S. The polypeptide chains and carbohydrate moieties of these factors are produced by the hepatocytes in the same general way as other plasma proteins or fibrinogen. These factors contain various numbers of a previously undescribed amino acid, γ-carboxy-glutamic acid (Gla) on their N-terminal portion. The step of carboxylation takes place in the microsomal fraction of the hepatocytes and this reaction is Vitamin K-dependent. Coumarin derivatives are proved ant-agonists of Vitamin K since a long time. Blood coagulation factors synthesized during coumarin treatment have been called PIVKA (protein *i*nduced in *V*itamin *K* *a*bsence or antagonist). These proteins differ from their normally synthesized forms in the following: they can be activated but extremely slowly, they do not bind calcium ions. Detailed investigations have proved that the N-terminal glutamic acids are carboxylated in the presence of carbon dioxide, the dihydronaphtoquinone form of Vitamin K and Vitamin K-dependent carboxylase. Vitamin K gets converted to Vitamin K-2, 3-epoxide in this reaction. The mechanism of action of Vitamin K-antagonists is through inhibition of regeneration of Vitamin K from its 2,3-epoxide form. Thus these drugs prevent carboxylation reaction and result in the formation of an abnormal "acarboxy"- form of clotting factor. The acarboxy-forms of factors cannot participate in the coagulation process because of their inability to bind to phospholipids and cell membranes. Phospholipid-binding property is of particular importance in cases of prothrombin, Factor X, Factors VII and IX however Gla has been found in other tissues (bone, atheromatous plaque, etc.) too.

Biosynthesis of fibrinogen

Fibrinogen is synthesized by the hepatocytes under the control of inflammatory cytokines of the inter-leukin-6 family (Il-4, Il-10 and Il-13), which together are called hepatocyte stimulating factors. The three non-identical chains of fibrinogen appear to be products of three discrete genes. These three chains are formed by independent translations but the exact mechanism regarding how these gene products get assembled into bundles and packed together as dimers still remains unknown. It is possible that the nascent chains are actually assembled before they are released from their respective ribosomes. Doolittle et al. state that the driving force for this assembly could certainly reside in the strong tendency to form coiled coils, coupled with an unusual stability-attributable to the disulphide "rings". A hereditary component to plasma fibrinogen concentration has also been shown. Several epidemiological studies have shown a link between plasma fibrinogen concentration and a risk of arterial diseases, such as myocardial infarction or stroke. Fibrinogen is an "acute phase protein" its concentration may increase in inflammatory diseases, autoimmune diseases, malignancies, atherosclerosis etc.

Various types of liver diseases are commonly associated with clotting abnormalities. Pathogenesis of these disorders is highly complex, it includes decreased synthesis of various factors, synthesis of structurally abnormal proteins, failure of the hepatic clearing mechanism, abnormalities of fibrinolysis and diffuse intravascular coagulation. Liver malignancies are also associated with the failure in synthesizing coagulation factors.

Dysfibrinogenaemias

The qualitative abnormalities of fibrinogen are inherited in an autosomal dominant manner. The dysfibrinogenaemias may cause bleeding as well as history of thromboembolic diseases. Interestingly the fibrin formed from fibrinogen$_{Chapel Hill III}$ is resistant to the cleavage of plasmin. Fibrinogen$_{Dusart}$ is also resistant against fibrinolysis and maintains a familial thrombotic risk.

III. FIBRINOLYSIS

A. The Fibrinolytic System

Fibrinolysis is an enzymatic process by which water-insoluble fibrin is broken down into soluble fibrin degradation products (FDP). Fibrinolytic system regulates the formation, activity and inactivation of plasmin, the key-enzyme of this system which dissolves blood clots thereby facilitating restoration of vessel patency and reestablishment of blood flow. The fibrin degradation and the activation of fibrinolytic proenzymes proceeds hand-in-hand: thrombolysis is a better wording.

Plasmin is activated from its inactive precursor plasminogen by the action of specific plasminogen activators. The process is shown in Figure 30.8.

Fibrinolytic system consists of the following proteins: plasminogen, plasminogen activators, fibrinogen (fibrin degradation products) and the inhibitors of the fibrinolytic system (Table 30.7).

Molecular Interactions During Plasminogen Activation

Activators of the fibrinolytic system are present in body fluids and tissues and are produced by specific strains of haemolytic

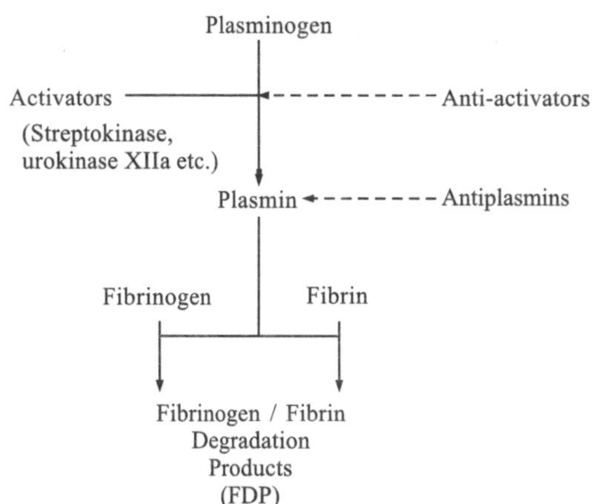

Figure 30.8 The activation of plasminogen.

streptococci and staphylococci (streptokinase, staphylokinase. A stoichiometric complex is formed between the zymogen plasminogen and the activator(s) which then becomes a direct activator. It has been thought that release of a so-called "preactivation peptide" (molecular weight of 8,000) is required for subsequent splitting of the 560 Arg-561 Val bond, resulting in the formation of the two-chained plasmin molecule. Glu-plasminogen (which has an N-terminal Glu) can be activated more slowly than Lys-plasminogen (having Lys on N-terminal): the composition of this form of plasminogen is Glu-plasminogen minus the "preactivation peptide". It is suggested that activation of fibrinolysis can be enhanced "autocatalytically' in such a way that the earliest plasmin formed would convert the residual predominant amounts of Glu-plasminogen to the Lys-form thus increasing the rate of Lys-plasmin formation (see Figure 30.12). The native form of plasminogen is activated slowly because the N-terminal part of the protein inhibits the binding of activators to arg561-Val562 (the hydrolysis of this bond results in a conformational change by which the active center become free on the smaller chain). Irrespective of the origin, when this "covering" of N-terminal stops, plasminogen is activated very rapidly. The activators might be exogenous or endogenous. The activators work through "positive feed-back" the ready-made plasmin may activate both activators (urokinase-type: uPA and tissue-type: tPA)

Fibrin network also interacts with active plasmin formation, i.e., active plasmin is formed not only from the adsorbed plasminogen but also from the plasminogen found within the fibrin network. Thus lysis of the fibrin network takes place at multiple points simultaneously.

Plasminogen activation may occur via 3 different pathways:

1. The contact system dependent pathway (kininogens and factor XII, etc.), i.e., intrinsic activators.
2. The tissue-type plasminogen activator (t-PA) and urokinase-dependent ways, i.e., extrinsic activators.
3. Finally the pharmacologic activator-dependent pathway (exogenous pathway), containing strepto-kinase, staphylokinase, and all the artificial hybrid drugs.

TABLE 30.7   **Proteins of the Fibrinolytic System**

Protein	Structure	Mol. wt.	Human plasma conc. (origin)
plasminogen	single-chain, 794 amino acids, 5 kringle domain	92 kDa	20 µg/ml (2 µmol/1)
plasmin	two-chain the kringle-structures are involved in fibrin binding	60 kDa	
Plasminogen activators:			
t-PA	275 + 252 amino acids, 72 kDa 2 kringles	half-life = 6 min 0.1 nmol/1	
u-PA	single-chain	54 kDa	half-life = 3 min from human embryo andkidney cells
streptokinase	single-chain, 415 amino acids	47 kDa	β-haemolytic Streptococci
elastase	218 amino acids, 2 asp linked CHO chain 25.9	kDA	degrades fibrin, but splits Factors VII, VIII, IX, XI, XII, von Willebrand etc
Inhibitors of fibrinolysis:			
α_1-antitrypsin		53 kDa	24.5 µmol/1
α_2-antiplasmin		67 kDa	0.9 µmol/1
α_2-macroglobulin		725 kDa	3.5 µmol/1
antithrombin		58 kDa	2.6 µmol/1
C1 esterase inhibitor		105 kDa	1.7 µmol/1
activated Protein C inhibitor		51 kDa	0.05 µmol/1
Plasminogen activator inhibitors (PAIs):			
PAI-1	from endothelial, vascular smooth muscle, hepatoma, placental cell, fibroblasts, granulosa cells, platelets, plasma	52 kDa	5–20 ng/ml
PAI-2	placental histiocytes, macrophages, PMN cells, tumor cells urine, plasma	60 kDa	not known
PAI-3	protease nexin	51 kDa	5 ng/ml
	fibroblasts, heart muscle cells, kidney epithelial cells thrombin-activable fibrinolysis inhibitor (TAFI)	47 kDa	not determined
	92 amino acid single-chain, Zn-containing center	60 kDa	?

The intrinsic activation with HMW-kininogen, Factor XIIa, Factor XI is not fully understood and they have a minor importance to the contribution of the total fibrinolytic capacity.

The extrinsic activation proceeds with at least two different types of activators, namely t-PA and the urokinase-type plasminogen activators (u-PA) including urokinase and single-chain urokinase-type plasminogen activators (scu-PA). They differ in the rate, binding and specificity of the reaction, but all cleave plasminogen and convert it to plasmin. Native t-PA is synthesized predominantly in the endothelial cells, and is released by the stimuli of bradykinin, thrombin, histamine, interleukin-1, epinephrine, LMWHs, gonadotropins, exercise, shear stress and venous occlusion. Its binding to fibrin is more tight than to fibrinogen. After fibrin-binding it manifests its full activity: fibrin enhances the catalytic efficiency to t-PA several hundred fold. Scu-PA (single-chain urokinase type plasminogen activator) has lower affinity to fibrin than of t-PA, but it has a fibrin selectivity to fibrin-bound plasminogen.

The exogenous activators: Streptokinase is a product of Group C beta-haemolytic streptococci. In itself it is not an enzyme, it acts indirectly to activate the fibrinolysis. The steps of activation are briefly as follows:

1. streptokinase + plasminogen $\Rightarrow$ streptokinase-plasminogen complex
2. streptokinase-plasminogen complex + plasminogen $\Rightarrow$ plasmin
3. streptokinase-plasminogen complexes $\Rightarrow$ streptokinase-plasmin complex

The problem with streptokinase and urokinase is that they randomly activate plasminogen (not only at the place of thrombosis—as it is in the case of tPA; therefore plasmin is formed even at places where fibrin does not exist (fibrinogenolysis). To exert its theraputic effect cca. 50% of the total amount of plasminogen is activated (at this point it exhausts the level of α-2-antiplasmin) and the fibrinogen degradation products inhbit thrombin very effectively –similarly to the fibrin degradation products—therefore the bleeding risk is much higher.

Staphylokinase acts similar to that of streptokinase. It is a recent new promise in fibrinolytic treatment.

Activation of plasminogen by streptokinase and urokinase is of important therapeutical value in thrombolytic therapy. Receptors for plasminogen activators, t-PA and u-PA and for plasminogen have been detected on the surface of a variety of cells: most cells can express receptors for both types of plasminogen activators and for plasminogen. These receptors may participate in the regulation of the activity of fibrinolysis by colocalizing fibrinolytic enzymes to the cell surfaces and by protecting cell-associated plasmin or plasminogen activators from serpin inhibitors.

Fibrinogen (fibrin)-plasmin interaction

Plasmin-degradation of fibrinogen, free fibrin monomer and soluble fibrin monomer complexes all proceed at a similar rate and the degradation products obtained are also similar. The degradation of cross-linked fibrin however proceeds at a slower rate.

Peptides having a molecular weight of about 40,000 are released from the C-terminal ends of Aα chains (resulting in the formation of an intermediate fragment X) and simultaneously (but more slowly) a peptide-containing fibrinopeptide B- is removed from the N-terminal end of Bβ chains. Thereafter all three chains present at one side of the fibrinogen dimer are split resulting in the formation of a single Fragment D molecule. The remaining larger Fragment Y is rapidly cleaved to a second Fragment D and one Fragment E. Thus, total fragmentation of a fibrinogen results in two D fragments and one E fragment. Digestion of the cross-linked fibrin by plasmin (i.e. in case of thrombi—*in vivo*) results in a single D-dimer fragment and two D-fragments bound by γ-γ-cross-links. Fragment E and a variety of cross-linked and non-cross-linked peptides are also released during lysis of the cross-linked fibrin.

Actions of FDP: Fragments Y, D and E form soluble complexes but do not clot. They are powerful inhibitors of thrombin inhibiting the conversion of fibrinogen to fibrin. Platelet function gets inhibited at high concentrations of FDP. Determination of the presence of fibrinogen and fibrin degradation products has great clinical significance in cases of laboratory control during thrombolytic therapy and in the diagnosis of diffuse intravascular coagulation (DIC).

B. Inhibitors of Fibrinolysis

Fibrinolytic system can be inhibited directly by inactivating plasmin and also by inhibiting plasminogen activation.

1. Inhibitors of plasmin: α_2-antiplasmin is the most potent plasmin inactivator. It is a single-chain glycoprotein with a molecular weight of 70,000. It forms a stable 1:1 stoichiometric complex with plasmin resulting in loss of the proteolytic and esterolytic activities of plasmin. Its plasma concentration is approximately 1 μM. Plasmin is capable to act only after its binding to fibrin, this complex protects it from inactivation. The second-order rate constants show, that tPA is works better in the presence of fibrin. Elastase is also highly protected against α_2-antiplasmin in the presence of fibrin (the rate is 2000-times slower!)

It is a general principle in biochemistry that enzymes in the presence of their substrates are more protected against their inhibitor.

Inhibitor levels show a temporary decrease during thrombolytic therapy and they also decrease in severe liver diseases and DIC. If α_2-macroglobulin is saturated with the enzyme *in vivo* the excess plasmin is neutralized by α_2-macroglobulin. Besides these there are three other plasma protease inhibitors, i.e., α_1-antitrypsin, antithrombin and C$_1$ inactivator which play a less important limited role in the neutralization of plasmin.

Thrombin-activable fibrinolysis inhibitor (TAFI): It is a carboxypeptidase-B enzyme and splits such lysine residues from fibrin, which are responsible for the cofactor function of fibrin in binding to tPA-plasminogen complex. Clinical observations show that an increased activity of TAFI increases the risk of thrombosis. The system is interesting because TAFI is synthesized from procarboxypeptidase-B, which is also activated by the thrombin-thrombomodulin complex. At present we suggest the central role of TAFI in the modulation of function of plasmin(ogen) and cell migration.

2. Inhibitors of plasminogen activation: We have only preliminary information regarding the physiological inhibitors of plasminogen activation; the heparin-antithrombin III complex, human C$_1$ inactivator and α_2-macroglobulin have been described as being inhibitors of the activation of plasminogen.

PAI-1, a single-chain, 379 amino acid glyco-protein is the major inhibitor of plasminogen activation *in vivo*. Its concentration within the endothelial cells is 5000-fold higher than in plasma: its higher plasma level is correlated with an increased risk of cardiovascular disease. It forms an equimolar complex with tPA or uPA in which the enzyme looses its proteolytic activity.

PAI-2, may be glycosylated or not, but its affinity for u-PA is higher than for t-PA. This serpin does not inhibit plasmin. Its exact function is not clear, because its substrates are extracellular enzymes and the primary locus for PAI-2 action is intracellular.

PAI-3 inhibits two-chain urokinase, t-PA and thrombin and it is very similar to the inhibitor of activated Protein C. The physiological role of PAI-3/APCI has yet to be determined.

Human plasma contains antibodies directed against streptokinase which are produced as the result of a previous infection by β-haemolytic streptococci. 95% of the healthy population possesses such antibodies in a very wide range of concentration. Following thrombolytic treatment with streptokinase the antistreptokinase titer rapidly increases to 50–100 fold remaining high for 6 months during which period a repeated treatment is unreasonable. 6-aminohexanoic acid (EACA), p-aminomethyl-benzoic acid (PAMBA) and tranexamic acid are powerful inhibitors of plasminogen activation and they are used as haemostatic drugs.

3. Compartmentalisation of fibrinolysis: It may be considered as a regulatory mechanism of fibrinolysis that compartments of the system—which flow in the bloodstream— form termporary compartments in which the enzymatic, kinetic reactions cannot be described or explained by the classical enzymology rules applicable to reactions taking place in aueous solutions. The reactions take place between solid and fluid phases, where the concentrations and rates of reaction may change by magnitudes and result in new biological aspects. These compartments in the fibrinolysis are: (1) the thrombus itself, (2) the surface of the endothelium and (3) the surroundings of activated polymorphonuclear leukocytes (PNM). We may find numerous molecular and cellular components in the thrombi; activated platelets, phospholipids, myosin, red blood cells, PNM-cells, albumin, immunglobulins and endogenous protease inhibitors. Most of these modulate fibrinolysis. The reaction rate of activation of plasminogen increases immediately by one magnitude in the presence of fibrin. The formed plasmin is capable of activating prourokinase in its environment; the appearance of uPA further increases the activation of plasminogen and the participants are unable to get diluted in this compartment. The plasmin starts to split fibrin immediately and this enzyme being bound to the other molecules is protected against the inhibitors and its inactivation is slower by several magnitudes compared to that of the circulating enzyme. Therefore the efficiency of fibrinolysis is high. How long? Until thrombus exists! Similar phenomenon occurs on the surface of the endothelial cells as well. Basic changes happen in the environment of the activated PNM cells: Elastase coming into fluid phase is inactivated immediately in the presence of the high amount of α-1-protease inhibitor, whereas the fibrin-bound form will be inactivated more slowly. The PNM cells secrete myeloperoxidases which inactivate

α_1-protease inhibitor in this compartment (oxidize a Met in the inhibitor). Elastase breaks down not only fibrin, but many members of the clotting system as well, enhancing the efficiency of the fibrinolytic system.

IV. INHIBITORS OF HAEMOSTASIS, HYPERCOAGULABLE STATES, ANTICOAGULANTS

Thrombophilia is a tendency, proneness to thrombosis, but inherited thrombophilia is applied to individuals with genetic defects that predispose them to thromboembolism. The fact that minor injuries or contact activation do not lead to clotting of the total circulating blood mass—as a result of cascade-wise amplification of enzyme activation—is attributable to the powerful neutralizing action of inhibitors on the activated enzymes. The following inhibitors seem to be responsible mostly for the inactivation of thrombin (and of Factor Xa and other serine-proteinases):

(a) antithrombin is responsible for about 70% of the total thrombin-inhibitory capacity in human plasma (earlier it was named AT-III due to historical reasons), whereas
(b) α_2-macroglobulin and
(c) α_1-proteinase inhibitors are responsible for the remaining inhibitory potency. Inhibitors are listed in Table 30.8.
(d) tissue factor pathway inhibitor (TFPI) is novel, very early suggested but recently clarified, very potent inhibitor of the Factor VIIa-tissue factor complex but also inhibits the active site of Factor Xa. Its synonyms are: lipoprotein-associated coagulation inhibitor (LACI), anticonvertin, etc.

These inhibitors—with the exception of α_2- macroglobulin— form 1:1 stoichiometric enzymatically inactive complexes with

TABLE 30.8   Proteinase Inhibitors in Human Plasma

Inhibitor	Molecular weight	Concentration	Inhibitory function in plasma
Antithrombin	65,000	25 mg% (4.5 µM)	thrombin, Xa, IXa, XIa, XIIa, plasmin, trypsin, C1 complement
α_2-macroglobulin	725,000	260 mg% (3 µM)	thrombin, trypsin, endopeptidases, ficin, bromelain, cathepsin, papain, thermolysin, collagenase
α_1-protease inhibitor (α_1-antitrypsin)	54,000	290 mg% (42 µM)	trypsin, chymotrypsin and to a less extent thrombin
tissue factor-pathway inhibitor	32,000– 40,000	100 µg/ml 2.5 nmol/l	Factor VIIa Factor Xa tissue factor
heparin cofactor II	65,600	?	see AT-III, but its activity is enhanced by dermatan sulfate
α_1-antichymotrypsin	69,000	50 mg% (74 µM)	chymotrypsin
C_1-inactivator	104,000	25 mg% (2.5 µM)	C1s, kallikrein, Factor XIIa, XIa, plasmin
Inter-α-trypsin inhibitor	160,000	50 mg% (2.8 µM)	trypsin
α_2-antiplasmin	70,000	7 mg% (1 µM)	plasmin

the enzymes. They react at varying rates with different enzymes and based on this they can be classified as rapid and slow inhibitors, i.e., antiplasmin is a rapid inhibitor of plasmin but slow for others whereas antithrombin is a slow inhibitor but it turns to a rapid one in the presence of heparin.

A. Molecular Interactions via Inactivation Reactions

α_1-Proteinase inhibitor

A complex formation between this inhibitor and thrombin leads to the cleavage of a Lys-Thr bond in the inhibitor resulting in the release of a small peptide with a molecular weight of 8,000. The reactive center is formed by a Met at position 358 which is situated on an exposed loop of the inhibitor forming a "bait" for the target enzymes. This loop is under considerable tension, i.e., in a metastable form. The 358Met-3359Ser bond fits to the active site pocket of the serine protease precisely. Thereafter a stable complex is formed between the "activated inhibitor" and the enzyme. It is suggested that there is formation of an ester-bond between the OH-group of serine in the active center of the enzyme and a carboxyl group of the inhibitor. Few mutations, (as α-1-antitrypsin$_{Pittsburgh}$) were described with bleeding disorders. Its real role in physiological regulation of coagulation is questionable.

α_2-macroglobulin

This inhibitor is a dimer of dimers, i.e., consists of 4 chains of m.w. 180 kDA each. Active sites of the enzymes are necessary for linkage to the inhibitor but enzymatic activities are not completely inhibited by the binding. The inhibitor can form a complex with many enzyme molecules at a time.Its concentration is 3 µM. Inhibition of thrombin by α2-macroglobulin results in the loss of its activity toward fibrinogen substrate only whereas activity of the enzyme remains preserved towards the small molecular weight substrates; this suggests that the large substrate molecules are excluded by steric hindrance. Binding of the enzyme(s) and the inhibitor is initiated by a limited proteolysis within the inhibitor resulting in such a conformational change so that the enzyme is trapped irreversibly within the inhibitor. Covalent binding of α_2M to proteinases is not necessary for irreversible inhibition. Few genetic polymorphisms were described in Japan and Mexico. The following types of enzymes are targets for α_2M: serine proteases (trypsin, chymotrypsin, plasmin, thrombin kallikrein, urokinase, PNL elastase) neutral proteinases from S. aureus. P. vulgaris and arvin (a coagulant enzyme of Malayan pit viper), thiol proteases (cathepsin B1, papain, ficin, bromelain) carboxy-peptidase (cathepsin D) metal-proteinases (thermolysin, collagenases, aminotransferases. Exopeptidases and nonproteolytic hydrolases do not react with α_2M. Its total absence by inheritance is not compatible with life. Its plasma level is high in diabetes, carcinoma, nephrotic syndrome, COPD etc. whereas it is low after CABG, in M. Kahler, preeclampsia and rheumatoid arthritis.

Tissue factor pathway inhibitor (TFPI)

TFPI inhibits Factor Xa by forming a 1:1 stoichiometric complex, which does not require Ca^{++} ions. In contrast to antithrombin this inhibition is enhanced by phospholipids and Factor Va, but heparin also has a similar, but not exactly clarified action. The proposed mechanism for the Factor Xa-dependent inhibition of Factor VIIa-tissue factor by TFPI involves the formation of a *quaternary* Factor Xa-TFPI-Factor VIIa-tissue factor complex. Most of the TFPI circulating in plasma is bound to LDL lipoproteins. It is synthesized mainly in endothelial cells, but unlike t-PA or von Willebrand factor, TFPI levels in plasma are not increased by infusion of DDAVP or venous occlusion. More than 90% of the cell-associated TFPI is an endothelial cell line stimulated with TNFα. There is a broad range of concentrations in normal individuals (mean : 2.5 nM). Heparin and LMWHs increase it by 3-fold. Its synthesis and excretion are regulated by thrombin via the PAR-1 receptor (protase activable receptor-1). The TFPI inhibited clotting became a major pathway for maintaining blood fluidity. TFPI-induced feed-back inhibition of Factor VIIa-tissue factor can explain the clinical need for both the extrinsic and intrinsic coagulation pathways and is consistent with the *in vitro* results showing the deficient Factor Xa production in haemophilic plasma induced to clot by small amounts of tissue factor. In normal haemostasis, Factor VIIa-tissue factor complex is responsible for the initial Factor Xa generation. Cell-associated TFPI appears to be important in the clearence and elimination of Factor Xa and in down-regulation of cell surface FVIIa-TF activity. Its total absence is associated with intrauterine demise, whereasthe heterozygous deficiencies cause severe thrombotic complications.

Antithrombin

This inhibitor is a single-chain glycoprotein of 425 amino acids. It inactivates thrombin irreversibly by forming a complex in which both the proteins are inactivated. The reactive site of antithrombin is positioned near the carboxy-terminal end: the available evidences indicate that the serine active centre of coagulation enzymes binds to this reactive site of antithrombin. The interaction induces a partial insertion of the amino terminal region of the reactive site into the core of the inhibitor which arrests cleavage of the reactive site bond and there-by the coagulation enzymes are trapped in a stable complex with antithrombin. Heparin has a high affinity to antithrombin with two binding sites (thrombin has a heparin-binding site as well). In the presence of heparin all the complex formations are highly accelerated (90% of thrombin disappears within seconds). The biological half-life is 48 hours.

Serine protease inhibitors, i.e., Serpins, similarly to thrombin, are multifunctional inhibitors. Besides inhibiting the active coagulation factors, they modify the complement cascade, fibrinolysis, protein folding, extracellular matrix remodeling, cell migration and differentiation, prohormone conversion and intracellular proteolysis, blood pressure, tumor progression, hormone transportation and viral and parasitic pathogenicity. It also inactivates all the active serine proteinase clotting enzymes of blood coagulation. It is synthesized in the liver and its decreased level in plasma (or decreased activity) results in a hypercoagulable state manifesting as various forms of thromboembolic diseases. Inacti-vation of each enzyme proceeds as an easily measurable lingering process when only antithrombin is present whereas addition of trace amounts of heparin greatly increases the rate of this reaction. The defects of antithrombin lead to high risk of thrombosis: the molecular defects may occur at the thrombin-binding (Type IIa)-, as well as at the heparin-bindig sites (Type IIb). The type I is the low level of antithrombin-antigen. Besides cases of decreased synthesis of the inhibitor, families having normal protein levels associated with impaired biological functions (abnormal antithrombin-III molecules are synthesized, e.g., antithrombin-IIIBudapest) have also been reported. These proteins may differ in heparin-binding and in their electrophoretic mobility in heparinized agarose. The inheritance is autosomal dominant in certain families.

Heparin cofactor II

This protein–in contrast to antithrombin does not form a complex with Factor Xa or other serine proteases but needs heparin for its action.

Protein C deficiency

This deficiency was described by Griffin in 1981 and is one of the most frequent thrombophilia-disorders (prevalence 1–4% in thrombotic outpatients). According to the Leiden Thrombophilia Study, the heterozygous form increases 7-fold the risk of thrombosis. Its clinical importance araouse after the discovery of warfarin-induced skin necrosis (purpura fulminans, a high mortality disease). The half-life time of Protein C is 7 hours and its level fell down within few hours to zero if loading dose of warfarin was used; therefore the remaining Factor Va and VIIIa caused thrombosis in the capillaries of life-important organs. When the loading dosage was changed to gradually increasing warfarin dosage (bridging regimen) this severe initial complication of warfarin treatment disappeared.

This mechanism was enforced with the discovery of congenital Protein C deficiency in neonatal purpura fulminans. Its low level in disseminated intravascular coagulation (DIC) is also important. More than 200 different mutations have been described in the Protein C gene.

Protein S deficiency

It is an autosomal recessive disorder and was reported in 10% of familial thrombophilias. Physiologically, 60% of Protein S is complexed to a complement component, i.e., C4b-bindig protein and 40% is free as a cofactor for thrombomoduline-Protein C complex to accelerate the splitting of Factor Va and VIIIa. In Type I it has low total and free antigen level, in Type II the free level is normal, but the activity is low and in Type II the total antigen is normal, whereas the free and the activity are low.

Protein Z inhibitor (PZ-I) and Protein Z cofactor

This system has been discovered recently and its mechanism is similar to TFPI . PZI, however is able to inactivate Factors XI and Xa and its cofactor increases its efficacy. The cofactor and the inhibitors circulate only in a complex, however we do not know the importance of the cofactor alone. It seems that its biological role is to avoid the redundant clotting in the vessels.

Activated Protein C Resistance

Dahlbäck described this novel mechanism of familial thrombophilia in 1993 in Leiden. Patients with family histories of thrombosis have a poor response to activated Protein C (APC) in aPPT assay. Later it was found to be the most fequent cause of familial thrombosis. Its background was shown to be a defect in Factor V involving a mutation of Arg 506 to Gln506, which resulted in a resistance against break down of Factor Va by Protein C. Therefore it is Factor V_{Leiden}. The homozygous form has a prevalence of 3 % and a 17-fold increased risk, the heterozygous form has 10 % prevalence and a 7-times risk elevation. This mutation originates presumably 15000 years before between Tiger and Euphrat rivers. The risk of myocardial infacton is also higher. Obstetrics complications, preeclampsia, placental abruption, stillbirth are associated with Factor V_{Leiden}.

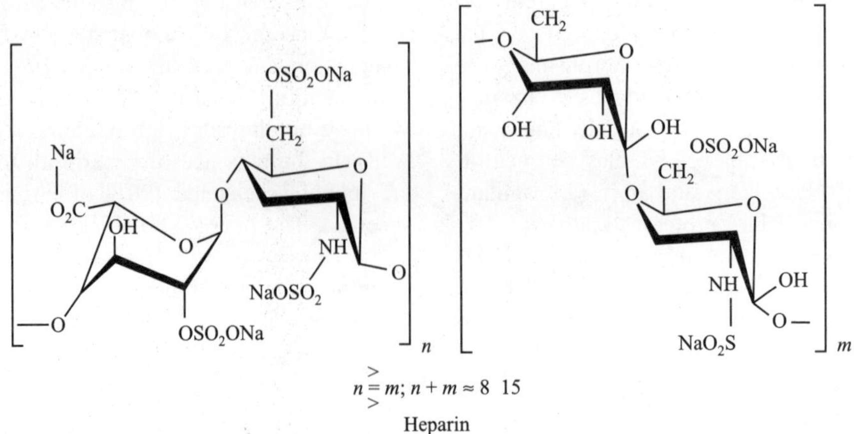

$$n \geq m;\; n+m \approx 8\text{–}15$$

Heparin

Figure 30.9 The chemical structure of a tetrasacchharide unit of heparin.

The thromboprophylaxis, treatment of symptomatic thrombosis and pulmonary embolism of Leiden mutants has already been elaborated and published in guidelines.

Prothrombin gene mutation

The sequencing of the prothrombin gene showed a quite frequent substitution of G to A at a nucleotide 20210 in the 3′ untranslated region of prothrombin gene which was associated with an elevated level of prothrombin as a consequence of increased efficiency of prothrombin mRNA 3′-end formation. The average prevalence is 2% in Southern Europe but is is rarely identified in Asia and Africa. PCR methods are used for the detection and the thromboprophylactic measures are identical to the other thrombophilias.

B. Anticoagulants

Heparin

Heparin plays a key-role in the regulation of inactivation of the active coagulation enzymes. Heparin is a natural strong-anionic glycosamino-glucuronan sulphate ester composed of tetra-saccharide units the exact sequences of which are yet to be determined. The tetrasaccharide structure comprises L-iduronic acid-N-acetyl D-glucosamine- 6-sulphate-D-glucuronic acid-N-sulphated D-glucosamine-6-sulphate (see Figure 30.8).

Heparins and LMWHs (see later) are indirect anticoagulants, because an internal factor (antithrombin) is necessary to exert their effects. The molecular weight of active heparin may vary between 6,000 and 25,000. The minimal anticoagulant unit is a pentasaccharide, which was recently synthesized chemically, whereas the other, low-molecular weight preparates show different antiXa-, and low thrombin-inhibiting activities, their average molecular mass ranges from 4 to 7 kDa. They vary in activity but have a much better bioavailability than unfractionated heparin. Recently definite sequences have been demonstrated in the polysaccharide "recognizing" different proteins, i.e., Factor Xa-binding site and antithrombin-binding sites have been isolated. Proteins participating in the inactivation reaction also possess specific binding site(s) for heparin. Different theories have been forwarded regarding the mechanism of action of heparin. The inhibitor-fitting theory suggests that heparin activates antithrombin-III whereas the enzyme-fitting hypothesis assumes the same for thrombin and for other enzymes. Taking all available evidences into consideration it seems most likely that most efficient inhibition of thrombin (or other enzymes) occurs when both antithrombin-III and thrombin get bound to the same heparin molecule thus forming a ternary complex. Since binding of heparin to the thrombin-antithrombin complex is less stable than that to thrombin or antithrombin-III alone, heparin is released from the complex. Taking into account other characteristics of heparin (binding nature, reaction rate enhancement, Michaelis-Menten kinetics, thermodynamic data, etc) heparin is a unique specimen of a non-protein biocatalyst. Its plasma concentrations have been determined only in pathological conditions suggesting that its normal concentration has to be lower than 10–8 M. Therefore it can play only a modulatory role *in vivo* in maintaining the thrombo-haemorrhagic balance. Therapeutically it is widely used in the treatment of cases of manifest thrombosis. It is given in low dosages for prophylaxis of venous thrombosis and anticoagulation in heart-lung machines etc.

Besides anticoagulation, heparin exerts a wide variety of effects on different enzymes. **Heparin:**

activates: lipoprotein lipase, brain tyrosine hydroxylase, pepsinogen, skin esterase, DNA-polymerase, etc.

inhibits: hyaluronidases, myosin ATP-ase, β-glucuronidase, alkaline and acid phosphatases, catalase, pyruvate kinase, lysozyme, glutathione, reductase, pepsin, cathepsin B1, reverse tran-scriptase, etc.It inhibits angiogenesis and the secretion of β-amyloid precursor, TNFα via a CD11-dependent mechanism.

liberates: lipoprotein lipase from the adipose tissue and liver, diamine oxidase from the intestines, acid ribonuclease from lysosomes, etc. (and cells) it increases the negative electric potential of vessel walls, inhibits proliferation of arterial smooth muscle and endothelial cells, mobilizes eosinophil cells, activates macrophages, increases migration of B-lymphocytes, produces lymphocytosis, etc.—it exerts a protective effect on an animal as a whole—against toxic bases, certain drugs, toxins. It also liberates TFPI and prostacyclin. Heparin inhibits sensitivity reactions by inhibiting the generation of amplification convertase; it inhibits anaphylaxis, antigen-antibody reactions in LE-test, in Arthus phenomenon, in Coombs' test, etc. (These data were collected by Jaques, 1980).

modulates the synthesis and release of various growth factors and the expression of adhesion-molecules.

Commercial heparin and heparinoids have been reported to change the activities of as many as 50 enzymes (stabilization, inhibition, destabilization, activation of the proenzyme, release of the enzyme, formation of adducts, etc.). Heparins are taken up rapidly by the endothelium and the reticuloendo-thelial system thus providing a protection against thrombosis. Artificial heparin—copolymers may play an important role in avoiding thrombogenicity caused by vascular prostheses. Unfractionated heparin does not get absorbed orally, its half-life time is in average 90 min, which increases with the dose. Bioavailability is 25%.

The low molecular weight heparins (LMWHs) cumulate to a less extent in the endothelial cells than unfractionated heparin, but have a much more marked effect in releasing TFPI from these cells: this is a new, unique, clinically very effective anti-coagulant mechanism of LMWHs. Their pharmaco-kinetic profile also differs from unfractionated heparin showing a prolonged half-life of 100–200 minutes after sc. injection. Their half-life does not increase with increasing dosage.Their availability is 98%. Besides the prevention and treatment of thrombosis, there are new indications for LMWHs: as angiogenesis inhibition, against developing tumor metastases, regulation of lypolysis.treatment of DIC, etc.

Novel derivatives of heparin and LMWHs are: *danaparoid* (heparan sulphate+dermatane sulphate and chondroitin sulphate) for the treatment of heparin-induced thrombocytopenia (HIT) *fundaparinux*, a synthetic pentasaccharide containing the most important anticoagulant sequence of heparin, *idraparinux* (the long-lasting variety of fundaparinux and finally *idrabiotaparinux*, a biotinylated form the latter, the effect of which can easily be suspended by avidin.

Oral anticoagulants

A veterinary research has proved that it was the agent dicoumarol which caused the severe haemorrhagic disease of cattle—"sweet clover disease"—in the early 1920s. All oral anticoagulants, i.e., dicoumarins, 4-oxycoumarins and indanediones (Figure 30.10) prepared in tablet form have a common mechanism of action. They prevent carboxylation of the glutamic acid residues of Vitamin K-dependent coagulation factors. Carboxylation of glutamic acid results in formation

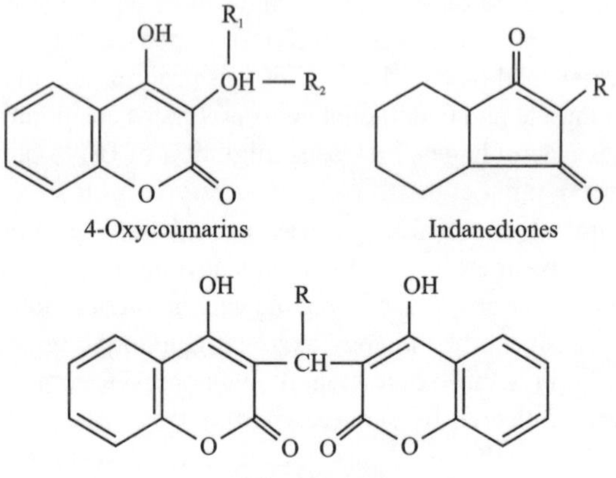

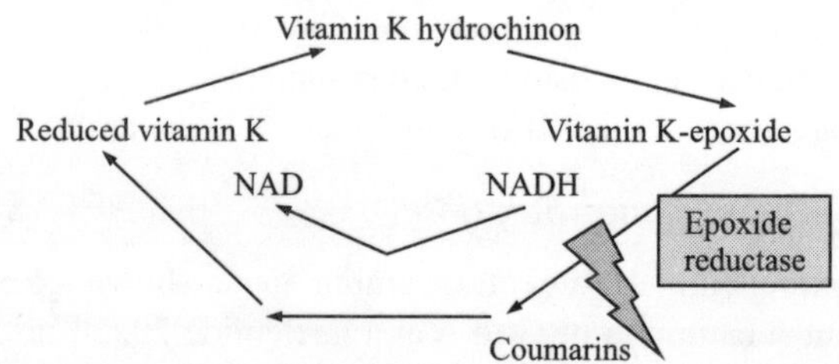

Figure 30.11   The target point of coumarins is the Vitamin K epoxid reductase.

the displacement of few per cent from proteins results in a high increase of the free form, and severe bleeding complications. The dosage should be individually adjusted with the control of International Normalized Ratio (INR) of the prothrombin time. The special importance of food and drug interactions complicate this therapy.

Novel oral anticoagulants: These new compounds are direct inhibitors of thrombin (dabigatran etexilate) and Factor Xa (rivaroxaban, apixaban, betrixaban), being highyly specific. Their usage does not require regular laboratory control and they are almost free of serious interactions. At the same time they do not have antidotes.

Figure 30.10   Basic structures of anticoagulant drugs.

of the unique amino acid, γ-carboxyglutamate (Gla). These residues allow Factors II, VII, X, IX and Protein C and S to bind calcium during the coagulation cascade on the surface of phospholipids. Calcium binding is essential to coordinate the activator complex formation (TENase, PROTHROMBINase). γ-carboxylation is achieved by a microsomal enzyme system which requires reduced vitamin K as a cofactor. Coumarins are competitive inhibitors of the Vitamin K-epoxide reductase. Its action interrupts the conversion of Vitamin K-epoxide to the reduced form of Vitamin K. The acarboxy-factor(s) Proteins Induced in Vitamin K Absence (PIVKA-proteins and factors) do not bind to phospholipid surface.

Main clinical indications for administration of oral anticoagulant therapy are as follows: venous thrombosis and pulmonary embolism, for the prophylaxis of venous thrombosis, heart valve prosthesis, cardiac arrhythmias, myocardial infarction and some cerebrovascular occlusive diseases. The treatment must regularly and frequently be monitored by laboratory tests.

The coumarins have excellent bioavailability but it is important to know that only the not-bound, free coumarin exerts its anticoagulant effect, which is 2–4 % of the orally given dose. The others bind to plasma proteins, mainly albumin. Therefore

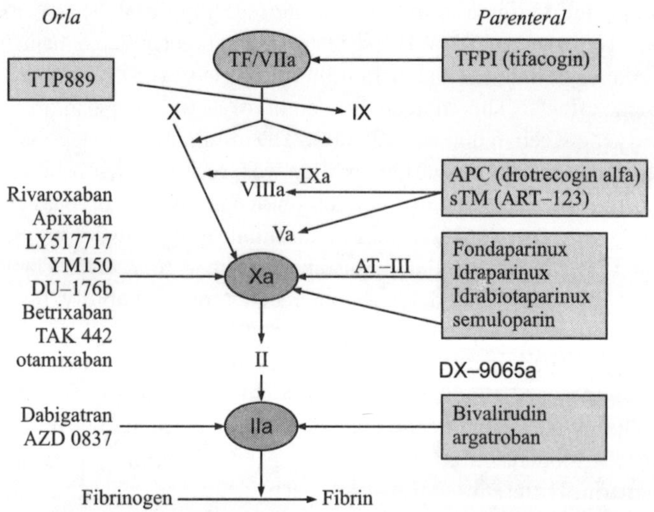

Figure 30.12   An overview of the target points of the novel anticoagulants.

V.   PLATELETS

The main physiological function of platelets is to adhere to damaged vessel walls and form aggregates. This contributes to the formation of a haemostatic plug and cessation of bleeding. On the other hand platelets release certain materials such as ADP, serotonin, calcium and specific platelet factors. The liberated ADP causes further aggregation whereas secretion of specific platelet proteins, constituents, etc. by platelets can influence or modify various functions of other cells, e.g., they contribute to tissue repair, wound healing and inflammation.

The secreted platelet constituents are listed in Table 30.9. Their life-time in the circulation is 9–10 days.

They have 3 main functions:

- adhesion to a "foreign" surface and aggregation (interaction to each other)
- it serves as a phospholipid surface for the acceleration of the activation of TENas and PROTHROMBINase
- they are able to secrete biologically active molecules (the basic prerequisite of it is the function of the contractile system)

Platelets contain the following microfilament-related cytoskeletal proteins (the functions are in parentheses):

actin, 42.5 kDa (it forms microfilaments, cyto- and membrane skeleton), *myosin* (contributes in the contraction of actin filaments), *actin-binding protein* (right-angle crosslinking of actin filaments, it binds glycoprotein Ib), *gelsolin* (serves act in filaments, promotes filament assembly, modulates PIP hydrolysis), *α-actinin* (boundles actin filaments), *profilin* (promotes formation of ATP-actin complex), *β4-thymosin, caldesmon* (regulates myosin binding), *talin* (membrane attachment), *tropomyosin* (stabilizes F-actin), *vinculin* (membrane attachment), *spectrin* (it is a part of membrane skeleton).

A.　Platelet Aggregation

Aggregation of platelets is a nonspecific response of the cells to various stimuli. These are as follows: primarily collagen, thrombin, PAF, furthermore ADP, epinephrine and arachidonic acid. However aggregation is also induced following exposure of platelets to glass beads, phospholipids, viruses, antigen antibody complexes (immune complexes), snake venoms, endotoxins, plant lectins, Factor Xa, ristocetin, etc.

Summarizing the activation of platelets briefly: The shape of the resting platelet is maintained by its microtubule coil; immediately beneath the plasma membrane is a membrane skeleton that is composed of two interlocking networks, one that is spectrin-rich like the erythrocytes, and one that is composed predominantly by actin. Actin-binding protein bound to actin is also able to bind glycoprotein Ib-IX complex. The change in platelet shape that follows exposure of the platelet to soluble agonists (as ADP, serotonin, thromboxane A₂ etc.) or to a surface is due to a net increase in the actin filament content of the cell. Phosphorylation of myosin leads to its incorporation into the cytoskeletal network and provides the contractive force that causes the platelet's granules to move forwards the centre of the cell. The α-granules fuse with the invaginations of the open canalicular system. Talin binds to actin and vinculin beneath the plasma membrane, as well as the cytoplasmic tail of one or more members of the intergrin family, thereby providing a connection between the cytoskeleton and externally bound molecules, such as fibrinogen and Factor Va.

During the first phase of aggregation there is a change in the shape of the platelets whereby pseudopods and small aggregated clumps are formed. This process is reversible. The second phase of aggregation is associated with the release of ADP together with serotonin and other constituents; at the same time various oxidative metabolites of arachidonic acid are formed which promote primary aggregation: these are cyclic prostaglandin endo-peroxides and thromboxane A₂ (TXA₂).

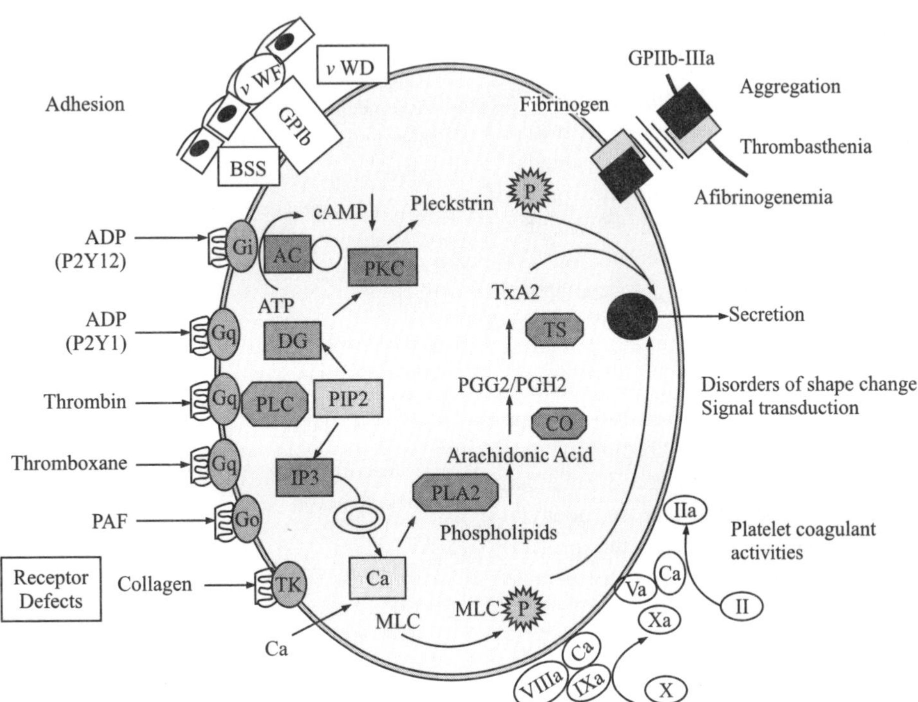

Figure 30.13　The scheme of platelet functions and defects.

Aggregation of platelets requires certain prerequisites: an optimal 1–2 mmole/1 concentration of calcium ions, presence of fibrinogen and von Willebrand antigen, a neutral pH and an intact glycolytic and oxidative phosphorylation pathway in them.

The nucleotides contained in the platelets are present in two major compartments:

(i) the metabolic compartment in the mitochondria and cytosol (40%).

(ii) the non-metabolic or storage pool which contains 60% of the total nucleotides stored together with other constituents in the granules.

Receptors of platelets:

1. *Glycoprotein receptors:* (GPIb-IX-V). This is an adhesion receptor, its main part and functional unit is the GPIb. This receptor makes it possible, and allows the platelet bind to collagen via the von Willebrand receptor, in case of vessel damage. This leads to restructuring of the platelet membrane and thereafter the collagen receptors of platelets bind collagen the and platelet becomes active. The mutation in this molecular complex is the Bernard-Soulier syndrome

2. *Receptors for collagen:* 3 of them are functionally important.

   (a) GP-VI receptor:it belongs to the Ig receptor superfamily which binds non-covalently to the □ chain of Fc receptor. This receptor initiates those signalling processes (through phosphorylation of Tyr), which result finally in the activation of phospholipase C and increasing the Ca^{++} content of the cytosol.

   (b) Fibrinogen/vWF receptor (integrin $\alpha IIb\beta_2$ or (GPIIb/IIIa). Each platelet contains 60-80.000 such a receptor. It binds fibrinogen and makes possible to bind platelets via fibrinogen bridge.

   (c) alfa2beta1 receptor: It is an integrin protein. It binds to collagen and decelerates the speed of the flow. The binding results in activation of phospholipase C.

3. *Protease-activated receptors (PARs):* 4 structures are known, i.e., PAR1-4: commonly in case of a split in N-terminal leads to a conformational change forming the thrombin receptor.

4. *Purinergic receptors:* There are two receptors are on the platelets: $P2Y_1$ and $P2Y_{12}$. They bind ADP (which may be released from red blood cells) and through the G-protein system it inhibits adenylate cyclase and low cAMP level accelerates the formation of thromboxane A_2 and activation of the contractile system. These receptors accelerate the activation of platelets and each platelet has 800 purinergic receptors. Thienopyridine-type of inhibitors such as clopidogrel, prasugrel, etc. are effective inhibitors of $P2Y_{12}$ and exhibit a clinically useful antithrombotic effect.

Other types of characterisation of platelet receptors

A. *Platelet activation receptors:* ADP-, epinephrine-, serotonin, TXA_2-, vasopressin, platelet activating factor (PAF)-, thrombin-, and collagen receptors have been identified, however, the ADP-, collagen- and thrombin-receptors are of utmost importance. Besides these there are receptors for aggregated IgG, immune complexes and selected monoclonal antibodies (e.g., GP IIb/IIIa, platelet CD9 and GP IV) which require the binding site of platelet Fc receptors.

B. *Platelet inhibitor receptors:* agents, that stimulate adenylate cyclase and increase intracellular level of cAMP. They are PGEI, PGD2 and PGI2 receptors.

C. *Platelet adhesion receptors:* they are membrane glycoproteins playing critical role in normal platelet adhesion to components of blood vessel wall and platelet interactions with each other.

TABLE 30.9   Platelet Granule Contents

Alpha-granules	*Dense granules*
fibrinogen	ADP
von Willebrand factor	ATP
platelet factor 4	Ca^{++}
β-thromboglobuline	pyrophosphate
IgG	serotonin
albumin	pp60c–src
platelet derived growth factor	granulophysin
fibronectin	P-selectin
P-selectin (CD-62, PADGEM)	LIMP, CD-63
thrombospondin	
Factor XIII	
Factor V	
Glycoprotein IIb/IIIa	*Lysosomes:*
transforming growth factor-β	β-N-acetylglucosaminidase
osteonectin	beta-galactosidase
histidin-rich glycoprotein	β-glucuronidase
vitronectin	cathepsin
HMW-kininogen	acid phosphatase
C1-esterase inhibitor	arylsulfatase
protease nexin II	heparitinase
Protein S	aryl phosphatase
PAI-1	β-glycerophosphatase
platelet basic protein	α-mannosidase
connective tissue-activating peptide	α-arabinosidase
α_2-macroglobulin	β-N-acetyl-galactosaminidase
α_2-antiplasmin	α-galactosidase
α_1-antitrypsin	α-fucosidase
epidermal growth factor	β-glucosidase
IgA	α-glucosidase
Factor XIa inhibitor	LAMP-1, LAMP-2
GMP-33	LIMP-CD63
platelet-derived endothelial cell growth factor	

D. *Adhesion receptors of the integrin family:* These alpha/beta heterodimer protein complexes are involved in cell-matrix and cell-cell interactions. Platelets contain 5 different integrins. These receptors include: the fibrinogen receptor which is identical to the GP IIb/IIIa and is the most abundant receptor on the platelet membrane (50000 receptor per cell). The binding of this receptor to fibrinogen, the subsequent protein phosphorylation and its interaction with the cytoskeletal proteins is under extensive research. The receptors for vitronectin, laminin, GP Ic-IIa, the collagen receptor (which is identical of GP Ia/IIa) which have currently been discovered.

E. *Adhesion receptors of the leucine-rich glycoprotein family:* it includes the receptor of von Willebrand Factor, identical to GP Ib/IX, and binds vWF and thrombin. This glycoprotein complex constitutes the major attachment site between the platelet membrane skeleton and the plasma membrane: the interaction is mediated by actin-binding protein. GP V is involved in Bernard-Soulier syndrome but its function is not known.

F. *Adhesion receptors of the selectine gene family:* These include P-selectin, platelet activation-dependent granule to external membrane (PADGEM) also designated as CD62.

G. *The adhesion receptors of the immuno-globulin gene family include PECAM-1:* it contributes to the interaction of endothelial cells and platelets.

Intracellular Regulation of Platelet Aggregation

There are many complex regulatory systems within the platelets to maintain their normal metabolic activity and regulate their reactivity to stimuli.

The mechanism of platelet activation: Platelets are equipped with specific plasma membrane receptors that organize these various stimuli and transform them into biologic responses. This transformation occurs to large extent through classical mechanisms of transmembranous stimulus-response coupling that result in the generation of specific intracellular second messengers. The primary activation pathway of platelets has two second messengers derived from hydrolysis of inositol phospholipids, i.e., inositol-trisphosphate (IP$_3$) and sn-1,2-diaglycerol (DAG). There are different mechanisms for strong and weak agonists. A strong agonist (thrombin, collagen) binds to a specific receptor and through a Gp-mediated signalling process, activates PIP2-specific phospholipase C. This results in second messengers that cause the elevation of Ca^{++} and the activation of protein kinase C, which together stimulate the release of membrane-bound arachidonic acid and granule constituents (as ADP). The so-called weak agonist binds to its specific receptor and through a process that depends on Na$^+$/H$^+$ exchange, activates phospholipase A$_2$ (see b). This also results in the release of arachidonic acid which metabolizes to prostaglandin endoperoxides that then feedback to activate phospholipase C directly.

The shear stress also induces platelet adhesion, release and aggregation. It may induce aggregation independent of adhesion, depending on vWF, GP Ib and GP IIb/IIIa but not upon plasma fibrinogen.

Role of prostaglandins in platelet aggregation: Arachidonic acid metabolism of platelets and the vascular endothelium have been proved to be in tight interaction. Arachidonic acid can be transformed into biologically active molecules very rapidly in the cells of both systems. Free arachidonic acid is liberated from the platelets and arterial phospho-lipids due to the action of phospholipase A2 or the sequential actions of phospholipase and diglyceride lipase. The metabolism of arachidonic acid in human platelets and endothelial cells is shown in Figure 30.13.

The enzyme cyclooxygenase (COX·) converts arachidonic acid to prostaglandin endoperoxide PGG$_2$ which in turn converts it to PGH$_2$ by a peroxidase. The endoperoxides may either be converted into stable prostaglandins—PGE$_2$, PGD$_2$ (either enzymatically by an isomerase or non-enzymatically) and PGF$_2\alpha$ (enzymatically by a reductase) or to labile TXA$_2$ by thromboxane synthetase (which rapidly breaks down to form the stable TXB$_2$). 12L, 5, 8, 10-heptadecatrienoic acid (HHT) and malondialdehyde can also be formed from endoperoxides. In platelets roughly equal amounts of 12L, 5, 8, 10, 14-eicosatetraenoic acid (HETE), HHT and TXB$_2$ appear to be formed from the endoperoxides but when compared to other tissues, TXB$_2$ is relatively the main product. COX-1 is an constitutive enzyme ubiquitous, whereas COX-2 is an inducable one. It does not expressed in normal cells, but after certain stimuli such as cytokines, growth factors etc. the inflammatory processes may get modulated.

Physiological properties of prostaglandins involved in platelet arachidonic acid metabolism:

1. *Prostaglandin endoperoxides:* PGG$_2$ and PGH$_2$ are very unstable but are powerful inducers of platelet aggregation (more potent than ADP). Aggregation is potentiated by PGE$_2$ but is inhibited by PGD$_2$. Plasma albumin markedly accelerates transformation of PGH$_2$ into a mixture of PGD$_2$, PGE$_2$, and PGF$_2\alpha$. Endoperoxides cause constriction of aortic tissue and respiratory smooth muscle.

2. *PGD$_2$:* It is an inhibitor of platelet aggregation and is involved in inflammatory reactions.

3. *Thromboxane A$_2$ (TXA$_2$):* It is identical with the previously described "rabbit aorta contracting substance". It has a short half-life in aqueous solution, at pH 7, 4 (30 sec) and is hydrolyzed rapidly to TXB$_2$. It causes contraction of vascular and respiratory smooth muscles; it has a Ca-ionophoric effect: it promotes calcium influx into the cells and induces platelet aggregation. Aspirin and indomethacine inhibit prostaglandin biosynthesis of platelets. Aspirin acetylates a serine residue in the active centre of COX-1 irreversibly but other non-steroidal anti-inflammatory drugs also inhibit it. Its maximal antiplatelet effect

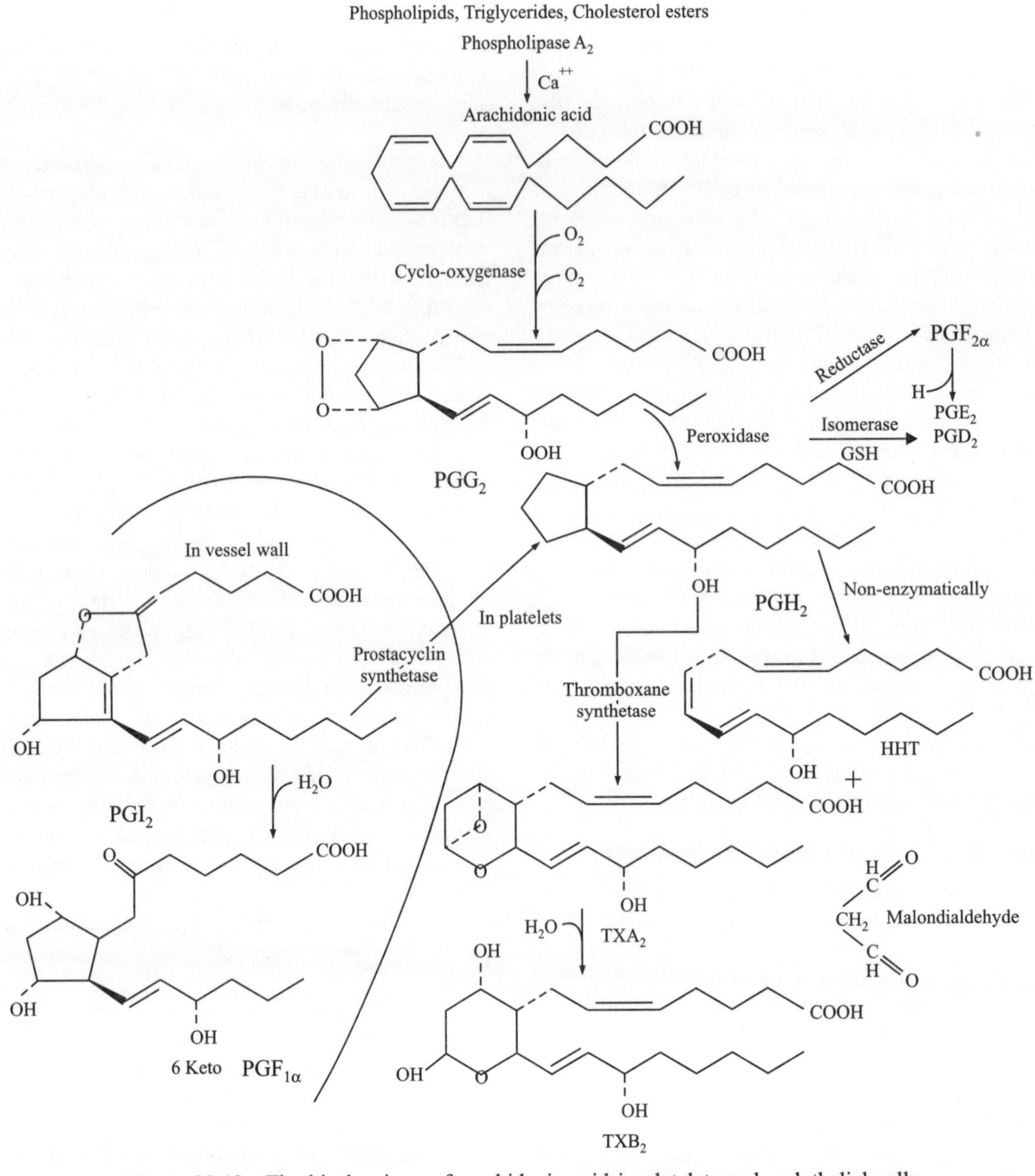

Figure 30.13   The biochemistry of arachidonic acid in platelets and endothelial cells.

is exerted within 1-7 minutes, whereas its antipyretic effect develops after 2-12 hours. The "salicylated" platelets circulate until these platelets are removed by the RES. The turnover of COX-1 in the endothel is more rapid compared to that in platelets: it is possible to choose a low enough dose of aspirin to inhibit platelets but without inhibition of endothelial cell in prostacyclin production!

One third of the arachidonic acid liberated from platelet membranes is metabolized via the lipoxygenase pathway. Arachidonic acid may be converted to 12L-hydroperoxy-5,8,10,14-eicosatetraenoic acid (HPETE) and then by a peroxidase can be reduced to HETE (Figure 30.14).

The physiological role of lipoxygenase pathway in platelets is not yet clarified however HETE shows chemotactic activity towards white cells. HPETE has an inhibitory action on thromboxane synthetase activity.

Role of cyclic nucleotides and calcium: Platelets contain a membrane-bound adenylate-cyclase and cytoplasmic nucleotide phospho-diesterase. Platelet cAMP is in a state of constant turnover its level being regulated by the activity of synthesis on one hand and by the subsequent hydrolysis on the other hand:

$$\text{ATP} \xrightarrow[\text{Mg}^{++}]{\text{adenylate cyclase}} \text{cAMP} \xrightarrow[\text{Mg}^{++}]{\text{phosphodiesterase}} \text{AMP}$$

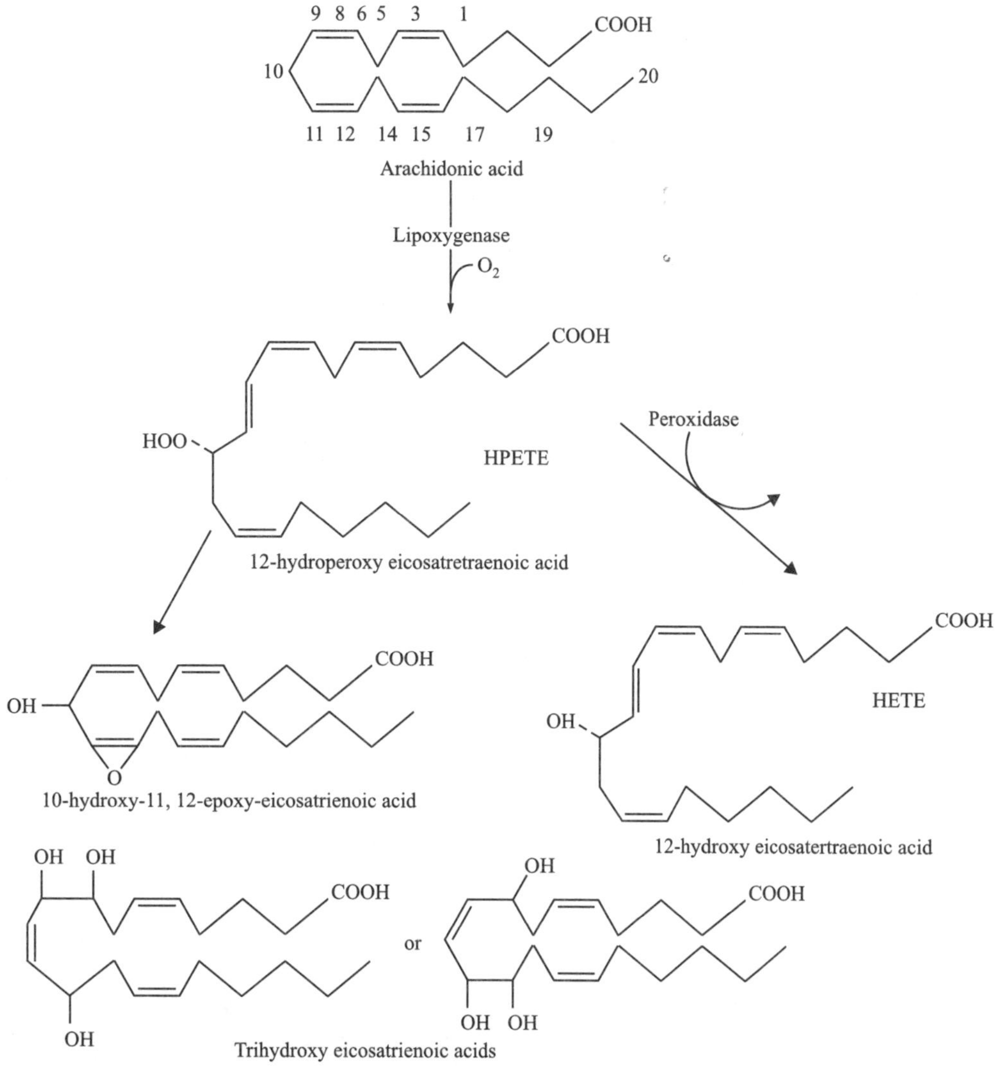
9 8 6 5 3 1
COOH
10
20
11 12 14 15 17 19
Arachidonic acid
Lipoxygenase
O2
COOH
HOO
HPETE
12-hydroperoxy eicosatretraenoic acid
Peroxidase
COOH
HETE
OH
12-hydroxy eicosatertraenoic acid
COOH
OH
O
10-hydroxy-11, 12-epoxy-eicosatrienoic acid
OH OH
COOH
OH
or
OH
COOH
OH OH
Trihydroxy eicosatrienoic acids

Figure 30.14 The lipoxygenase pathway. The scheme of the major pathways of platelet aggregation including the most important target points of the inhibitors of platelet aggregation.

cAMP plays a key-role in aggregation: if the relatively low basal levels of cAMP get elevated (either by activation of adenylate cyclase or via inhibition of phosphodiesterase) platelets exhibit marked resistance against the aggregating stimuli. Adenylate cyclase is localised on the internal surface of the cell membrane and it is regulated by action of specific receptors many of which are known to be found on the external surface of the same membrane (signal-transmission system). Three fractions of cyclic nucleotide phosphodiesterases have been described mainly in the cytoplasm, although a small but important fraction is present bound to membranes. Inhibition of phosphodiesterases by various drugs, e.g., methylxanthines, papaverine, etc. has clinical significance in inhibiting platelet aggregation.

Platelets also contain guanylate cyclase which produces cGMP. This enzyme has higher activity in platelets than in other tissues and cGMP levels increase with marked sensitivity in response to the effects of various aggregating agents (ADP, thrombin, collagen).

As to the effects of calcium it functions pri-marily as an intracellular transmitter of information from the surface membrane. A cAMP-dependent protein kinase has been described in platelets to phosphorylate a protein which is involved in the regulation of intracellular Ca^{++}. This sequester Ca^{++} in its phosphorylated form whereas its dephosphorylated form cannot retain Ca^{++} and thus Ca^{++} is liberated initiating the release reaction. Fluxes of intracellular calcium are important in the regulation of platelet release. Mobilization of calcium leads to changes in the shape, causes alterations in the cell membrane, initiates the contractile activity, causes changes in the microtubules and micro-filaments of the platelets and liberation of platelet factor 3—thus exerting an effect on secretory processes, etc.

The increased intraplatelet-cAMP concentration activates a protein kinase, which is phosphorylating the so-called "40 kDa protein". The phosphorylated form of 40 kDa protein increases the uptake of Ca^{++} into the storage pool, therefore the high level of cAMP inhibits the aggregation of platelets independently from the species of inducers. This protein is the missing link between cAMP and platelet inhibition.

Numerous feed-back mechanisms are known to exist between cAMP and calcium. Phospholipase A_2 which releases arachidonic acid from platelet phospholipids requires Ca^{++} and it is activated by thrombin and collagen. cAMP inhibits this activation. cAMP prevents mobilization of calcium ions within the cell. Calcium inhibits adenylate cyclase and this activity can be antagonized by cAMP. cAMP inhibits TXA_2 synthesis and the function of the contractile system as well. Calmodulin a calcium-binding protein with multiple functions is found in the platelets and appears to regulate phospholipase A_2, calcium-dependent membrane ATP-ase (Ca-pump), myosin light-chain kinase (contractile system), calcium-dependent protein kinase, etc. and it is thought to have an interaction with phosphodiesterase and adenylate cyclase, too. Prostacyclin (PGI_2) activates adenylate cyclase via the G_5-protein, therefore by the elevation of intracellular cAMP the activation of the platelets will be effectively inhibited.

Connections between the adenylate cyclase system and prostaglandins: Proaggregatory cyclic endoperoxides and TXA_2 inhibit adenylate cyclase; whereas PGE_1, PGI_2—formed by the vessel wall—activate it resulting in an increased level of intracellular cAMP. cAMP has a regulatory function in platelets which is dependent upon the balance of these substances exerting antagonistic effects on cAMP levels. The exact role of high cAMP concentration in the cells has not yet been clarified fully: according to some elevated cAMP levels antagonize cyclooxygenase and block the formation of endoperoxides. Others point out the importance of inhibitory action of cAMP on phospholipase A_2 but recent evidences reinforce this concept only in respect of thrombin-induced platelet aggregation. One of the most attractive hypotheses in an attempt to integrate actions of prostaglandins, endoperoxides, TXA_2 and cAMP involves mobilization of the intracellular calcium. Endo-peroxides and TXA_2 might act on adenylate cyclase indirectly causing the release of calcium from storage pools. Mobilized calcium would then lower the adenylate cyclase levels, enhancing aggregation. Phospholipase A_2 also requires calcium. However elevated cAMP concentrations prevent calcium fluxes and the contraction of thrombosthenin.

Collagen, thrombin or adrenalin (or other stimuli) interact with receptors on the platelet mem-brane releasing the amounts of calcium necessary for the action of phospholipase A_2. Phospholipase A_2 liberates arachidonic acid from membrane phospholipids which in turn is converted rapidly to endoperoxides and TXA_2. TXA_2 binds to calcium acting as a physiological Ca-ionophor and the Ca^{++} released stimulates release of the stored constituents from the granules and inhibits adenylate cyclase. Thus release and aggregation proceed. Substances inhibiting aggregation increase intracellular cAMP levels (either by activation of adenylate cyclase or inhibition of phosphodiesterase). cAMP also inhibits phospholipase A_2 and cyclooxygenase thus inhibiting TXA_2 formation.

Platelet activating factor (PAF): A number of immunological mechanisms have been described that induce release reaction in platelets. A soluble factor released from the leucocytes which is capable of aggregating platelets and inducing histamine release has been described and termed platelet activating factor (PAF). PAF is a 1-lyso-glycerophospholipid (1-0-alkyl-2-acetyl-sn-glycerol-3-phosphorylcholine). Considerable variations exists in reported PAF receptor ligand-binding characteristics (the number of receptors is 160–20000 site per platelet): PAF binding is enhanced by Mg^{++} and inhibited by GTP, being characteristic of ligand-receptor interaction coupled to pertussis toxin-sensitive G proteins. The putative PAF receptor of 180 kDa has been isolated. It is released from the leucocytes by specific antigens and it activates platelets independent of arachidonic acid meta-bolism. It causes bronchoconstriction, hypotension but these effects cannot be antagonized by aspirin or antihistaminic agents. Current evidences suggest that PAF is probably an important mediator in inflammation and platelet aggregation and may represent such a mechanism of aggregation which is an alternative to the TXA_2- or ADP-dependent system.

B. Interaction of Platelets and the Vessel Wall

Normally platelets do not adhere to an intact endothelium, however if the endothelium is damaged or denuded, platelets readily adhere to the exposed subendothelium. Numerous factors including mechanical forces, cholesterol, viruses, antigen-antibody complexes, some products of smoking, etc. have all been implicated as being responsible for causing injury to the vessel wall. Platelets interacting with the subendothelial collagen fibres, basement membranes, release the contents of their granules via generation of prostaglandin endoperoxides and TXA_2. The mass of aggregated platelets forms a haemostatic plug and as a consequence to this, the pathway of coagulation gets activated either through activation of Factor XII or availability of platelet factor 3 or Factor V which if adsorbed onto platelets can bind Factor X, etc. At the same time several factors tend to limit thrombus formation:, i.e., thrombin promotes liberation of PGI_2, circulating blood may dislodge the aggregated platelets (the rheological problems are subjects of a separate study), the injured wall produces a plasminogen activator, etc. These processes are in equilibrium in the circulating blood.

Arachidonic acid metabolism of the vessel wall

As it can be seen from Figure 30.15 endothelial cells metabolize arachidonic acid via a common pathway for platelets till the step of cyclic prostaglandin endoperoxide formation. Endoperoxides are converted to PGI_2 by the help of prostacyclin synthetase. It is a labile but predominant metabolite of all vascular tissues. Its half-life is about 30 sec at neutral pH and it rapidly breaks down to 6-keto-$PGF_1\alpha$ non-enzymatically. It is a potent vasodilator and is the most powerful inhibitor of platelet aggregation known so far. Vane and Moncada suggested existence of a balance between PGI_2-synthesizing activity of the vessel wall and TXA$_2$ synthesis of the platelets and according to them this balance plays an important role in pro- and disaggregation of platelets in thearteries.

Endothelial cyclooxygenase can also be inhi-bited by acetylsalicylic acid and indomethacin, however, the turnover of the enzyme in these cells is more rapid. This finding has led to the introduction of the mini-dose aspirin treatment, i.e., blocking of TXA$_2$ synthesis in platelets without influencing PGI$_2$ production by the vessels.

Some lipid peroxides like 15-hydroxy-arachidonic acid block the conversion of endoperoxides to PGI$_2$: this particular one is a powerful inhibitor of prostacyclin synthetase. Those parts of the vascular endo-thelium which contain lipid peroxides incorporated into the cells do not produce PGI$_2$; this finding is of importance in cases of atherosclerosis emphasizing the role of arachidonic acid metabolism.

Since PGI$_2$ is synthesized continuously—mainly in the lungs—one expects it to be a circulating hormone; however on the contrary its stable metabolite is not always demonstrable. Recently the authors considered 6-keto-PGE$_1$—a labile derivative of PGI$_2$—responsible for the long lasting effects of PGI$_2$. PGI$_2$ exerts its activity via the activation of adenylate cyclase whereas its stable breakdown product 6-keto-PGF$_1\alpha$ is ten orders of magnitude less potent in this respect. It is also a powerful vasodepressor agent. It causes relaxation of the coronary arteries and hypotension in several animals and man. It is suggested that PGI$_2$ is a very important regulator of the vascular tone (assuming that it is a circulating hormone).

Clinical importance of the prostacyclinthrom-boxane system has been established by the finding that plasma concentrations of TXB$_2$ were elevated in cases of Prinzmetal's angina, severe sclerosis of the coronary vessels, in diabetes mellitus, smoking, etc. whereas demonstrable levels of 6-keto-PGF$_1\alpha$ were below normal. The low incidence of coronary heart diseases and life style in eskimos—whose diet is rich in eicosapentaenoic and docosahexanoic acid because of the traditional hunting-fishing habits—can be explained by the increased PGI$_3$ production which also inhibits platelet aggregation and elevates cAMP levels in platelets, whereas the formed TXA$_3$ does not have any effect.

C. Mechanism of Action of Drugs Inhibiting Platelet Aggregation

According to the biochemical processes and path-ways mentioned above, the following theoretical possibilities exist for the drug-induced inhibition of platelet function.

(a) suppression of the energy metabolism in platelets (it is not used clinically)

(b) inhibition of phospholipase A$_2$ to prevent arachidonic acid liberation (e.g. quinacrine—the known antimalarial agent)

(c) inhibition of cyclooxygenase (acetylsalicylic acid, indomethacin, other NSAIDs, etc.)

(d) inhibition of thromboxane synthesis (Dazoxiben)

(e) activation of adenylate cyclase (adenosine analogues, PGE$_1$, PGI$_2$, PGD$_2$, etc.)

(f) inhibition of phosphodiesterase (methyl-xanthines, papaverine, etc.)

(g) antagonizing the receptors on the membrane (e.g., 13-azaprostanoic acid)

(h) stabilization of the membrane (e.g. chlorpromazin)

(i) inhibition of protein kinase to prevent the release of bound calcium

(j) interference with calcium movement within the cells (TMB-8)

(k) thromboxane receptor antagonists (Sulotroban)

(l) serotonin receptor antagonists (Ketanserin)

(m) fibrinogen receptor antagonists (monoclonal antibodies against GP IIb/IIIa, i.e., abciximab, 7E$_3$, RGD-peptides "fibans": tirofiban, orbofiban, xemilofiban etc.)

(n) guanyl cyclase activators (endothelium-derived relaxing factor, i.e., nitric oxide, organic nitrates)

(o) von Willebrand factor receptor antagonists (aurin tricarboxylic acid)

(p) purinerg receptor antagonists (ticlopidin, clopidogrel, prasugrel....)

(q) others, ω-3 fatty acids etc.)

Bhatt and Topol summarized the potentially effective target-molecules, target receptors in 2003, which may result in clinically useful inhibitors of platelet functions: These are as follows: inhibitors of P$_2$Y$_2$, P$_2$Y$_{12}$ —occasionally concommittantly(!),- inhibitors acting on GPIb-V-IX-von Willebrand complex, inhibitors of P-selectins, RANTES, CD-40 – CD40 ligand system, the CD39 ADPase, inhibitors of PAR receptors, pepducins and the arp2/3 complex, binding to the platelet leptin receptor, and blocking the growth arrest specific gen 6 (Gas 6).

D. Inherited Disorders of Platelet Function

These disorders can be classified according to the site of the genetic abnormality whether the plasma membrane or the intracellular components are involved. Only the most important ones are mentioned below:

1. *Membrane abnormalities:*

   (a) *Bernard-Soulier syndrome*—Platelets contain less amounts of sialic acid and there is a glycoprotein deficiency in the membrane. Only homozygous individuals are clinically affected.

   (b) *Thrombasthenia (Glanzmann's disease)*—Platelets do not react to any of the known aggregation stimuli. Two closely associated membrane components are absent, the contractile system is defective.

   (c) *Platelet factor 3 deficiency*—It is extremely rare and is associated with coagulation defects.

TABLE 30.10   **The participation rate of the defects of clotting proteins and platelets in different, but frequently occuring thrombosis (After Bick)**

Defect	DVT/PE (%)	Cerebrovasc. thr. (%)	TIA (transit.ischemic attack)(%)	Acute myocardial infarction (AMI) %
antiphospholipid sy.	29	65	28.5	18
APC resistance	18	3	14	?
sticky platelet sy.	14	19	30	18
prothrombin 20210A	7	7	?	6
PS defect	6	?	?	7
PC defect	5.5	?	?	12
antithrombin defect	4	3	?	5
decrase in tPA	?	1	11	16
increase of PAI-1	?	?	?	11

2. *Intracellular defects:*
   (a) *Storage-pool deficiency*—This is deficiency of dense-bodies of the platelets. Such platelets show aggregation abnormalities to various stimuli. Storage pools of the cells are empty, they do not contain ADP, serotonin, etc. in granules. Several syndromes are known to be associated with storage—pool disease, e.g., Chediak-Higashi, Wiscott-Aldrich, Hermansky-Pudlak, etc.
   (b) *Defects of thromboxane synthesis*—Two inherited deficiencies are listed, i.e., cyclooxygenase deficiency and thromboxane synthetase deficiency. These diseases have been described only recently on the basis of failure to produce the aspirin-response in these conditions.
3. *von Willebrand disease:* It does not belong closely to the group of diseases mentioned above because a defect in a plasma protein is responsible for the abnormal platelet function. It is inherited as an autosomal dominant trait. Homozygous patients are severely affected; besides purpura, large haemophilia-like bleedings occur. Factor VIII-related antigen is absent; or it may be abnormal in its biochemical characteristics. Ristocetin-induced aggregation cannot be elicited whereas other agents are effective in inducing it. von Willebrand disease is heterogenous, but the most common inherited bleeding disorder with at least 20 distinct clinical subtypes described, but these can be divided into two main categories based on whether it is a quantitative (type I and III) or qualitative (type II and the other types). Its prevalence in Sweden is 1 in 10000. 70% of them belong to Type I, whereas the von Willebrand antigen: ristocetin cofactor; Factor VIII are pro-portionally decreased but the multimer distribution of the factor is normal, whereas in Type II of the concentrations of the factors mentioned are low and is associated with the absence of large and intermediate size of multimers with prominent satellite bands. (It is a missense mutation.) The definitive diagnosis occasionally becomes difficult.

Acquired platelet dysfunctions can be associated with various diseases. The most common and important ones are as follows: uraemia, myelo-proliferative diseases, acute leukaemias, liver diseases, scurvy, immunologically damaged platelets, drug abuse, etc.

VI.   ENDOTHELIUM

The complete surface of the endothelial cell covers a 1000 m^2 area and it is an important barrier between the cells of the body and the blood stream. The exchange of materials, food, excrements etc. is going through these cells. Morphologically they are similar, but functionally they may differ depending upon their placement (big vs. small vessel or vessels in different organs).

Antithrombotic characteristics of the endothelium.: The most important question in the haemostasis is, why clotting is not triggered (or kicked-off) on the surface of endotheliel cells (EC)? If the monolayer of EC discontinues, the clotting starts and there is no fibrinolysis. In case of EC damage due to any cause like in atherosclerosis, this would ultimately result in a leading cause of death. It is of utmost basic importance to know the role of EC in haemostasis.

The thrombomodulin-thrombin-Protein C –Protein C receptor system ensures, if the clotting had already started and the formed thrombin activated Factors V and VIII by positive feed-back, then the activated Protein C can inactivate (split) the Factors Va and VIIIa preventing the formation of PROTHROMBINase. Thrombin binding to thrombomodulin, activates Protein C to Protein Ca very effectively. Thrombomodulin (TM) is located on the surface of EC. It has many regions or domains: as the lectin-like, the hydrophobic, the EGF-like, the Ser/Thr rich, transmembrane and finally the cytoplasmic regions. It consists of 557 amino acids.

The EGF-region is responsible for binding PC, thrombin and Ca^{++}, whereas the Ser/Thr-rich region increases the affinity of thrombin to TM. The splitting down of TM and its appearance in the circulation is the sign of the damage of endothelium. The thrombin-TM complex activates TAFI as well (see there) and TAFIa inhibits fibrinolysis.

It has a role in embryogenesis as well: its total absence is lethal. The cause is the apoptosis of throphoblasts

The Protein C receptor binds PC and PCa with equal affinity. This receptor does not exert any activity in coagulation, but in its presence the activity of TM and thrombin is enhanced in the process of activation of PCa. This has a role insepsis.

EC binds heparan sulphate as well modulating the activity of surplus thrombin.

EC may protect at the abluminal site against thrombin: in the reaction Protease nexin I has a primary role.

EC secretes TFPI which protects the cascade from starting the clotting without an effective trigger.

In case of activation of platelets in the environment of EC, prostacyclin inhibits the aggregation and the ectoADPase exerts a similar effect.

Finally, if sufficient thrombin had formed to initiate the coagulation cascade, the healthy EC starts to secrete tPA and other plasminogen activators.

In case of damage of this system described at any points, the risk of thrombosis increases and ultimately thrombosis develops.

The matrix mettalloproteases are synthesized in a pre-form in the EC. The formation of the active enzyme is due to plasmin extracellularly and furin intracellularly. The metalloproteases contain a Zn^{++} in the active center and play an extended role in the sterile inflammatory processes and atherosclerosis.

The formation of NO (nitrogen monoxide): Citrullin is formed from arginin in two steps and inbetween NO is released. It is a dissolved gas at a concentration of 100 nM with a half-life of 10-15 seconds, exerts a very powerful inhibitory effect on the aggregation of platelets. The reaction is catalysed by NO-synthase (NOS) and is stimulated by calmodulin-Ca^{++} complex. NOS may be constitutive and inducible as well.

ECs are synthese the endothelins (ET). ET-1 is synthesized by every cell at a different extent. It is a 21-member polypeptide and having a powerful vasoconstrictive effect. It effects through G-proteins in smooth muscle cells, enhances the Ca^{++} level in the cytosol and causes a long-lasting hypertension.

The well-studied renin-angiotensin and kallikrein-kinin systems are described elsewhere: but it is clear, the EC system is a real compartment of molecules/systems of different biological effects, where different events with opposite effects do the fine-tuning. The endothelium has a tissue-specific network in the circulation, in which every phenomenon may be modulated essentially. Only few examples are listed here: thrombomodulin occurs mainly in the peripheral vessels and does not exists in CNS-vessels; the deficiency of Protein C causes mainly venous, but not arterial thrombosis; the deficiency of thrombomodulin results in a 30-fold increase of fibrin deposits in the lungs; 5-fold in the heart; 17-fold in the spleen; but no fibrin deposits are there in the brain and kidneys. At the same time in the deficiency of tPA fibrin deposits are formed in the brain and kidneys and mainly in the liver!

Interactions of the cellular elements of the blood with the endothelium: Under normal conditions there are no, or minimal and transient cellular interactions, but in case of damage and activation, remarkable changes occur.

(a) *binding of platelets to endothelium:* thrombin-activated platelets after plaque rupture aggregated to the damaged EC, but the contraregulatory functions of ectoADPase, PGI_2 production inbibit the further platelet activation. The free subendothelial conponents following the EC damage activate platelets as well: they may interact with von Willebrand factor, collagen, fibronectin. laminin, thrombospondin etc. furthermore the increase of shear-rate alone may activate platelets.

(b) *Interactions with PNM leukocytes and red blood cells:* these interactions occur mainly in immune reactions, e.g., leukocytes are migrating, spreading onto EC (the selectins of EC recognize carbohydrate ligands of leukocytes). Thrombin stimulates expression of P-selectin on ECs, which binds leukocytes via sialmucin ligands.

(c) *The differentiation and EC and angiogenesis:* it is a separate section of biochemistry but it is in a tight relation with the clotting and formation of metastases. Plasminogen and elastase together inhibit mitosis. Pretreatment of rat with plasminogen kringles inhibits tumor growth. Heparin(s) have antiangiogenetic activities and sometimes prolong survival in malignancies.

(d) *Formation of tumor metastases:* tumor cells produce tPA, the formed plasmin activates matrix-prometalloproteinases, which digest collagen, elastin and break up extracellular matrix which is a soil for the migrating tumor cells. However their fixation needs fibrin, which will be the results of the action of tumor procoagulants (an SH protease, which activates Factor X directly).

(e) *Atherosclerosis:* This disease of the developed countries starts with plaque formation (the details are mentioned elsewhere in this book). The rupture of the fibrous cap of the plaque results in the formation of TF and thrombin, followed by a rapid aggregation of platelets. Therefore the so-called atherothrombosis is the new clinical entity for summarizing all the processes regarding the formation of plaques and the results of the rupture, e.g., myocardial infarction, stroke and peripheral aretery disease.

SUGGESTIONS FOR FURTHER READING

Apostolakis S. Lip GYH., Lane D.A. and Shantsila E. (2010), *The Quest for New Anticoagulants: From Clinical Development to Clinical Practice,* Cardiovascular Therapeutics 9, 1–11.

Bick R.L. and Haas S. (2003), Thromboprophylaxis and Thrombosis in Medical, Surgical, Trauma and Obstetric/Gynecologic Patients. *Hematol. Oncol. Clin. N. Am.* 17, 217–258.

Bloom A.L. and Thomas D.P. (1987), *Haemostasis and Thrombosis,* Churchill Livingstone, London, New York.

Collen P., Wiman B., and Verstraete M. (1979), *Physiological Inhibitors of Blood Coagulation and Fibrinolysis,* Elsevier—North Holland, New York.

Davie E.W., Fujikawa K., Kusachi K. and Kisiel W. (1979), The Role of Serin Proteases in The Blood Coagulation Cascade, *Advances in Enzymology,* 49, 277.

Fareed J., Messmore H.L., Fenton II J.W. and Brinkhous K.M. (1981), *Perspectives in Haemostasis,* Pergamon Press, New York.

Fenton J.W. II (1981), Thrombin Specificity, *Annals of the New York Academy of Sciences,* 370, 469–495.

Horellou M.H. and Legrand C. (1996), *Haemostasis: State of the Art,* Karger. Med. Sci. Publ., Basel.

Jaques L.B. (1980), Heparins—Anionic Polyelectrolyte Drugs, *Pharmacological Reviews,* 31, 99–166.

Lewis P.J. and O'Grady J. (1981), *Clinical Pharmacology of Prostacyclin,* Raven Press, New York.

Loscalzo J. and Schhafer A. (1994), *Thrombosis and Haemorrhage,* Blackwell Sci. Publ., Oxford.

Magnusson S., Petersen T.E., Sottrup-Jensen L., and Claeys H. (1975), In: *Proteases and Biological Control,* Reich, E., Rifkin, D.B. and Shaw, E. (Eds.), Cold Spring Harbor Conference on Cell Proliferation.

Mandl J. and Machovich R. (2001), *Pathobiochemistry,* Medicina Budapest.

Philp R.B. (1981), *Methods of Testing Proposed Antithrombotic Drugs,* CRC Press Inc. Boca Raton, F1, USA.

Poller L. and Ludlam C.A. (Eds.) (1997), *Recent Advances in Blood Coagulation,* Edinburgh, Churchill Livingstone.

Ratnoff O.D. and Forbes C.D. (Eds.) (1996), *Disorder of Hemostasis,* 3rd ed., Philadelphia: Saunders.

Segatchian M.J., Samama M.M. and Hecker S.P. (1996), *Hypercoagulable States,* CRC Press Boca Raton, USA.

Thomson J.M. (1980), *Blood Coagulation and Haemostasis,* Churchill Livingstone, London, New York.

Vane J.R. and Bergström S. (1979), *Prostacyclin,* Raven Press, New York.

31

Hematopoiesis and Leukemia

Lalit Kumar and Gaurav Prakash

CONTENTS

I. HEMATOPOIESIS

Blood cells and bone marrow are one of the most dynamic tissues in human body. Incessant activity of bone marrow is vital for renewal and maintenance of blood cells throughout the life of a human being. Synthesis of various mature blood cells from the pluripotent stem cells in bone marrow is known as haematopoiesis. The bone marrow has an enormous production capacity; it is estimated that 10^{10} erythrocytes and 10^8 to 10^9 leukocytes are produced per hour in the steady state. This production capacity may increase many folds during the periods of increased demand like blood loss, infection, and hypoxia. Hematopoietic stem cells (HSC) are pluripotent and have the capacity to differentiate into the cells of all hematopoietic lineages like erythrocytes, platelets, neutrophils, eosinophils, basophils, monocytes, lymphocytes, natural killer cells, and dendritic cells.

A. Bone Marrow: The Niche for Haematopoiesis

Bone marrow resides in the medullary cavity of the bones and is supported by trabecular meshwork and osseous vasculature. A functioning bone marrow has two components- cellular component and supportive microenvironment. Cellular component mainly comprise of hematopoietic stem cells (HSC) and progenitor cells of various lineage. These stem cells have a unique characteristic of proliferation and self-renewal simultaneously. Under the influence of various stimuli and growth factors, stem cells undergo rapid proliferation and give rise to progenitors of various lineages which would eventually differentiate into mature blood cells. During this process a small fraction of daughter cells grow as HSC in order to replenish their pool.

Supportive microenvironment of bone marrow consist of the surrounding stroma, bone trabeculae working as mechanical scaffolding, capillaries and blood sinusoids. Stromal cells are primarily comprised of fibroblastoid cells, endothelial cells, and macrophages. These cells provide support to HSC not only by direct cell to cell contact but also through paracrine effect by releasing various cytokines and hematopoietic growth factors. Bone marrow also has specialised circulatory system in which perforating arteries from cortical bone give rise to radial capillaries which eventually convert into BM sinusoids and end up in venous system. In BM sinusoids, blood is in direct contact with cellular component of BM so that nutrients are directly supplied for dividing stem cells and egress of newly formed cell is also collected to be released in systemic circulation.

"

B. Cell lineage

On the basis of morphology and immune-phenotype all blood cells can be divided into 4 lineages:

1. Lymphoid lineage—T and B Lymphocytes, plasma cells and memory cells
2. Myeloid lineage—granulocytes (Neutrophils, eosinophils, basophils) and monocyte/macrophages
3. Erythroid lineage—Red blood cells
4. Megakaryocytic lineage—platelets.

Cells of myeloid, erythroid and megakaryocytic lineage originates from a common myeloid precursor cell and all lymphoid lineage cells originates from common lymphoid precursor under the influence of specific growth factors. They are often collectively referred to as colony forming cells (CFC) or units (CFU). Interleukin-3 (IL-3) and erythropoietin (EPO) drive the proliferation and maturation of primitive erythroid burst-forming units (BFU-E), Stem Cell Factor, granulocyte colony-stimulating factor (G-CSF), and granulocyte-macrophage colony-stimulating factor (GM-CSF) drive the development of granulocyte and macrophage colony-forming units (CFU-GM). Thrombopoietin (TPO) helps in proliferation of megakaryocytes and platelet production.

Bone marrow plasticity is a unique feature due to which hematopoietic stem cells have capacity to differentiate in to various tissues normally derived from other embryonic germ layers like neurons, muscles and cardiac myocytes. This property has expanded therapeutic use of hematopoietic stem cells beyond haematological disorder to various neurodegenerative and myopathic disorders. Homeostasis of hematopoietic system is of utmost importance for sustenance of normal life. Reduction in BM function, production of dysmorphic blood cells and malignant transformation of BM stem cells are three major classes of BM dysfunction of which, malignant disorders of bone marrow are known as leukemia.

II. LEUKEMIA

A. Introduction

The term 'leukemia' was coined by Rudolf Virchow, the renowned German pathologist, in 1856. As a pioneer in the use of the light microscope in pathology, Virchow was the first to describe the abnormal excess of white blood cells in patients with the clinical syndrome described by Velpeau and Bennett. As Virchow was uncertain of the etiology of the white blood cell excess, he used the purely descriptive term 'leukemia' (Greek: "white blood") to refer to the condition. Finally, in 1900, the myeloblast, which is the malignant cell in AML, was characterized by Otto Naegeli, who divided the leukemia into myeloid and lymphocytic.

Thus, leukemias are disorders of white blood cells characterized by uncontrolled clonal proliferation of malignant cells of different lineages with a reduced capacity to differentiate into more mature cellular elements. As a result, there is an accumulation of leukemic blasts or immature forms in the bone marrow, peripheral blood, and occasionally in other tissues, with a variable reduction in the production of normal red blood cells, platelets, and mature granulocytes. The increased production of malignant cells, along with a reduction in these mature elements, result in a variety of systemic consequences including anemia, bleeding, and an increased risk of infection.

Classification

Broadly, leukemia can be classified according to the cell line of origin like myeloid leukemia (originating from myeloid precursors) or lymphoid leukemia (originating from lymphoid precursors). Infrequently, complex leukemic cells are identified which bears features of multiple cell lines (e.g., lymphoid, myeloid or NK cells) and are labelled as mixed phenotypic or multi-lineage leukemia. Such subtypes are relatively uncommon and they respond poorly to the standard leukemia therapy. According to their clinical course and natural history, leukemia can be divided into:

1. Acute leukemia and
2. Chronic leukemia.

Clinically, patients of acute leukemia present with a shorter history spanning over few days or weeks and rapid deterioration of their health. While patients of chronic leukemia, usually present in relatively preserved general health with a long history spanning over many weeks to months. Basis of these differences lies at cellular level. In acute leukemia, main cellular event is the failure of differentiation and maturation of hematopoietic precursor cell (Blasts). As these cells continue to divide without differentiation, these accumulate in bone marrow and eventually spill over in the peripheral blood. High proliferative index of blasts leads to competitive inhibition of hematopoiesis in bone marrow and eventually rapid development of features of bone marrow failure. On the other hand, in chronic leukemia, genetic changes impart power of uncontrolled proliferation to cells at relatively later stage of maturation. As a result there is accumulation of malignant cells which are more mature and have low proliferative index than blasts. Due to slow proliferation chronic leukemia cells have low metabolic demand and affect normal hematopoiesis after a prolonged subclinical phase.

B. Clinical Presentation of Acute Leukemia

Rapid accumulation of blasts cells in the bone marrow, peripheral blood, and occasionally in other tissues attributes to signs and symptoms of acute leukemia. Compromised hematopoiesis affects all three lineages resulting in anemia, leukopenia and thrombocytopenia. As a result, progressive pallor, easy fatigability, fever (infections) and non-traumatic bleeding from multiple sites are the common presenting features. Patients of Acute lymphoblastic leukemia often have infiltration of lympho-reticular system giving rise to lymphadenopathy or splenomegaly of certain degree which is less common in acute myeloid leukemia. Infiltration of specific tissues like gums, skin and other soft tissues may occur in AML, particularly of monocytic origin.

Tumor lysis syndrome

Blasts are aggressive but highly fragile cells with a prominent nucleus rich in nitrogen rich DNA. Uncontrolled and rapid proliferation of blasts can lead to lysis of these cells, either at initiation of therapy or sometimes spontaneously. This can lead to biochemical abnormalities, e.g., Hyperuricemia, hyperkalemia, hypocalcaemia and renal failure and is called as "tumor lysis syndrome". Anticipation of tumor lysis, adequate hydration and prophylactic use of xanthine oxidase inhibitor, allopurinol are the mainstay of prevention of this complication.

Leukostasis: When peripheral white blood cell count is very high (higher than 1,00,000 cells/cmm) can lead to leukostasis; white cell plugs are formed in the micro-vasculature and clinically patient presents with headache, dizziness, and respiratory or neurological distress. Around 10–20% patients have features suggestive of leukostasis at the time of presentation. Extra corporeal removal of blast cells (called leukapharesis) is the main treatment for this complication.

Epidemiology

ALL is primarily a disease of childhood, with an incidence of 2.8 cases per 100, 000. The peak incidence occurs between two and five years of age, and it is more common among boys than girls. AML accounts for less than 10 percent of acute leukemia in children less than 10 years of age. In adults, AML is the most common acute leukemia and accounts for 80 percent of cases. In the west, the annual incidence of acute leukemia is 3 to 5 cases per 100,000. In India, the incidence of acute leukemia is 1.2 to 4.6 per 100,000 (ICMR). In the West, the median age at diagnosis is 66 years. According to recent SEAR data, approximately 12.6% were diagnosed under age 35; 18.4% between 35 and 55; and 69.1% were above 55 years of age. In India and other Asian countries, the median age is 29 to 32 years. This difference could be attributed to the earlier age of onset in India, difference in population demographics (in India, more than 50% of population is below the age of 25 years and more than 65% are below the age of 35 years) and possibly under reporting in many elderly patients.

C. Etiology of Acute Leukemia

Leukemogenesis is a multistep process in which the most important step is acquisition of a critical change in DNA of the cell from which leukemia originates. Many risk factors have been incriminated for such change and include both genetic predisposition and environmental factors, e.g., exposure to chemicals, radiation, tobacco, or chemotherapy drugs. Genetic abnormalities (e.g., Down syndrome, Fanconi's

anemia, Bloom's syndrome, familial RUNX1 mutations, Li Fraumani syndrome, Diamond blackfan syndrome), and other benign (e.g., paroxysmal nocturnal hemoglobinuria, aplastic anemia) and malignant (e.g., myelodysplastic syndrome and myeloproliferative disorders) have been associated with increased risk of acute leukemia of myeloid lineage. However the specific mutational event(s) required for this progression are not currently well defined. Most of these inherited disorders are associated with defective DNA repair mechanism. As a result hematopoietic cells fail to repair randomly acquired mutations which eventually accumulates and leads to leukemogenesis. Cohort of Atomic bomb survivors and cancer survivors who received therapeutic radiation have shown increased incidence of AML. Ionizing radiation and chemotherapeutic drugs like alkylating agents cause damage DNA, usually by inducing single/double strand breaks that may cause the mutations, deletions, or translocations required for hematopoietic stem cell transformation. Exposure to benzene and certain chemotherapy drugs is associated with increased risk of leukemia. The drugs most commonly implicated are topoisomerase inhibitors like etoposide (VP-16), anthracycline antibiotics, and nitrosoures, e.g., methyl CCNU.

There is evidence suggesting virus mediated leukemogenesis. Incorporation of oncogene containing provirus near a critical promoter site leads to activation of oncogene and eventually to malignancies. Human T cell Leukemia and lymphoma virus mediated adult T-cell Leukemia is one of the example of infection related leukemogenesis. EBV is implicated in oncogenesis of a subset of Burkitt's lymphoma/leukemia, 30–40% of Hodgkin's lymphoma and primary CNS lymphoma in immune-compromised patients.

Classification of AML

Acute myeloid leukemia is classified on the basis of morphology of the blast cells and on their cytochemical features.

According to French-American-British (FAB) classification system AML is classified into 8 subtypes according to their extent of differentiation and cell type involved (Table 31.1).

Recently WHO classification of AML has included molecular features into classifications and four sub-classes (Table 31.2).

For diagnosis of AML, a marrow blast count of $\geq 20\%$ is required, except for AML with recurrent genetic abnormalities t(15; 17), t(8; 21), inv (16) or t(16; 16) and some cases of erythroleukemia.

[c] >20% blood or marrow blast and any of the following: previous history of myelodysplastic syndrome (MDS), or myelodysplastic/myeloproliferative neoplasm (MDS/MPN); myelodysplasia-related cytogenetic abnormality (see below); multi-lineage dysplasia; and absence of both prior cytotoxic therapy for unrelated disease and aforementioned recurring genetic abnormalities; cytogenetic abnormalities sufficient to diagnose AML with myelodysplasia-related changes are:

- Complex karyotype (defined as three or more chromosomal abnormalities

TABLE 31.1   FAB Classification of AML

AML subtupe	Name	Frequency	Common cytochemical properties
M0	Undifferentiated acute leukemia	5%	MPO–
M1	AML, without maturation	15%	MPO+, NSE+
M2	AML with Maturation	25%	MPO+++, NSE+
M3	Acute promyelocytic leukemia	10%	MPO+++, NSE+
M4	Acute myelomonocytic leukemia	20%	MPO±, NSE+++ with fluoride inhibition
M4E0	Acute myelomonocytic leukemia with eosinophilia	5%	
M5	Acute monoblastic leukemia	10%	MPO–, NSE+++ without fluoride inhibition
M6	Acute erythroid leukemia	5%	MPO±, NSE–, PAS+++
M7	Acute megakaryocytic leukemia	5%	MPO–

Abbreviations: MPO-myeloperoxidase, NSE-nonspecific esterase, PAS-Periodic acid Schiff

Unbalanced changes: 7 or del (7q), -5 or del (5q), i(17q), or t(17p), 13(13q), del(11q), del(12p), del(9q)

Balanced changes: t(11; 16); t(3; 21); t(11; 3); t(2; 11), t(5; 12); t(5; 7); t(5; 10); t(3; 5).

[d]Cytotoxic agent implicated in therapy-related hematologic neoplasms: alkylating agents; ionizing radiatlion therapy; agent targeting topoisomerase II; others.

[e]BCR-ABL1-Positive leukemia may present as mixed phenotype acute leukemia, but should be treated as BCR-ABL1 positive acute lymphoblastic leukemia.

D.   Evaluation of a Case of Acute Leukemia

All suspected cases of acute leukemia should be given urgent attention due to the life threatening nature of their illness. Initial investigations include–medical history, physical examination, performance status, analysis of co-morbidities, hemogram with detailed peripheral blood film examination, renal and liver function tests, serum electrolytes, coagulogram, uric acid and a chest radiograph. Bone marrow aspiration is required to confirm the diagnosis. At the time of BM evaluation, apart from smear, samples should be collected for flowcytometric analysis (for immunophenotyping) and karyotyping for diagnostics and prognostic purposes. In flow-cytometric analysis of blast cells, various cell surface or cytoplasmic antigens are studied in order to ascertain lineage of origin of blast cells. CD 13, CD33, myeloperoxidase, CD 117 and HLA DR for AML;

TABLE 31.2   WHO classification of acute myeloid leukemia (AML) and related precursor neoplasms and acute leukemia of ambiguous lineage (adapted from reference 2)

Acute myeloid leukemia (AML) with recurrent genetic abnormalities

AML with t (8; 21)(q22; q22); RUNXI-RUNX1T1

AML with inv(16)(p13.1q22) or t(16; 16)(p13.1; q22); CBFB-MYH11

Acute promyelocytic leukemia (APL) with t(15; 17)(q22; q 12); PML-RARA

AML with t (9; 11)(p22; q23); MLLT3-MLL

AML with t (6; 9)(p23; q34); DEK-NUP214

AML WITH INV (3)(q21q26.2) or t(3; 3)(q21; q26.2); RPN1-EV11

AML (megakaryoblastic) with t (1; 22)(q13; q13); RBM15-MKL1

Provisional enity: AML with mutated NPM1

Provisional enity: AML with mutated CEBPA

AML with myelodysplasia-related changes[c]

Therapy-related myeloid neoplasm[d]

AML, not otherwise specified (NOS)

AML with minimal differentiation

AML without maturation

AML with maturation

Acute Myelomonocytic leukemia

Acute monoblastic/monocytic leukemia

Acute erythroid leukemia

Pure erythroid leukemia

Erythroleukemia, eryhroid/myeloid

Acute megakaryoblastic leukemia

Acute basophilic leukemia

Acute panmyelosis with myelofibrosis (acute myelofibrosis; acute myelosclerosis)

Myeloid sarcoma (extra medullary myeloid tumor; granulocytic sarcoma; chloroma)

Myeloid proliferation related to Down syndrome

Transient abnormal myelopoiesis (transient myeloproliferative disorder)

Myeloid leukemia associated with Down syndrome

Blasticplasmacytoid dendritic cell neoplasm

Acute leukemia of ambiguous lineage

Acute undifferentiated leukemia

Mixed phenotype acute leukemia with t(9; 22)(q34; q11.2); BCR-ABL1[e]

Mixed phenotype acute leukemia with t(v; 11q23); MLL rearranged

Mixed phenotype acute leukemia, B/myeloid, NOS

Mixed phenotype acute leukemia, T/myeloid, NOS

Provisional entity: Natural-killer (NK) cell lymphoblastic leukemia/lymphoma

CD10, CD 19, CD 79a, for B ALL, CD3, CD7, CD 45RO for T cell ALL are commonly used immunologic markers in acute leukemia. Conventional karyotyping is of particular significance in patients with AML for their prognostication into favourable, intermediate and unfavourable groups (Table 31.3). Since samples for cytogenetics occasionally fail to produce adequate numbers of metaphase cells, fluorescence in-situ hybridization (FISH) for the more common AML cytogenetic anomalies can be performed, including monosomies or deletions involving chromosomes 5 and 7, trisomy 8, 11q23 rearrangements, inv(16) or t(16; 16), and t(8; 21). Mutation analysis for FLT3-ITD, CEBPA, and NPM1 mutations is optional and should be performed, if available.

As ALL is uncommon form of acute leukemia in adults and elderly, care should be taken to differentiate it from lymphoid blast crisis of CML due to important therapeutic and prognostic differences in both the conditions. It is advisable to test for bcr-abl traslocation by PCR in all cases of adult ALL with suspicion of blast crisis (Tables 31.4 and 31.5).

All febrile patients need detailed clinical examination with particular attention to occult sites like oral cavity/throat, axilla and perineum. Blood cultures, urine culture and swab culture from any suspicious infected site should be sent.

E. Treatment of Acute Myelogenous Leukemia

Once the diagnosis of acute leukemia is established, induction chemotherapy is given with the goal of rapidly restoring normal bone marrow function which is defined as complete remission (CR).The latter is defined as resolution of disease related sign

TABLE 31.3 **Overall survival in AML according to cytogenetic abnormalities (Age Group: 16 to 59 years, Median age 43 years): Data source MRC /NCRI AML Trials (adapted from reference 7)**

	Cytogenetic abnormality	No of pts.	Overall survival at 10 years (%)
Favorable	t(15; 17)	759	73
	t(8; 21)	401	61
	Inv (16)(t(16; 16)	266	55
Intermediate	t(9; 11)	56	39
	t(6; 9)	34	26
Adverse	Inv (3)(t(3; 3)	65	3
	t(9; 22)	47	14
	Other t(11q23)	61	21
	AML with MDS related changes	797	16

TABLE 31.4 **WHO Classification of Acute Lymphoblastic Leukemia (adapted from reference 2 and 3)**

B lymphoblastic Leukemia

B lymphoblastic leukemia, NOS

B lymphoblastic leukemia with recurrent genetic abnormalities

B lymphoblastic leukemia with t (9; 22)(q34; q11.2); BCR-ABL 1

B lymphoblastic leukemia with t (V; 11q23); MLL rearranged

B lymphoblastic leukemia with t (12; 21)(P13; q22); TEL-AML1 (ETV6-RUNX1)

B lymphoblastic leukemia with hyperdiploidy

B lymphoblastic leukemia with hypodiploidy

B lymphoblastic leukemia with t (5; 14) (q31; q23); IL3-IGH

B lymphoblastic leukemia with t (1; 19) (q23; p13.3); TCF3-PBX1

T lymphoblastic leukemia

TABLE 31.5 **Main recurrent genetic abnormalities in B-lineage ALL and their relation to immunophenotypic features and prognosis (adapted from reference 6)**

Cytogenetics/FISH abnormality	Molecular abnormality	Prevalence		Characteristic immunophenotype	Prognosis
		Children	Adults		
t(9; 22)(q34; q11)	BCR-ABL1	3%	25%	None	Unfavorable
t(4; 11)(q21; q23)	MLL-AFF1	8%	10%	CD10$^{-, N}$ G2$^+$, CD15$^+$, CD65$^+$	Unfavorable in infants
t(11; 19)(q23; q13.3)	MLL-ELL				
t(12; 21)(p13; q22)	ETV6-RUNX1	22%	2%	Frequently CD13$^+$ and/or CD33$^+$	Favorable
Hyperdiploidy 51-65 Chromosomes		25%	7%	CD45 dim	Favorable
Hypodiploidy < 45 Chromosomes		1%	2%	None	Unfavorable
t(1; 19)(q23; q13.3)	TCF3-PBX1	5%	3%	Cytoplasmic u$^+$	Favorable
t(x; 14)(p22; q23) or	IGH-CRLF2	5%		None	
t(Y; 14)(P22.3q22.33) or del X or del (Y)(p11.32p11.32)					
	IKZFI deletions or mutation	15%		None	Unfavorable

and symptoms, absence of circulating blasts, reduction of BM blasts to <5% and recovery of peripheral blood cell counts. Most commonly used induction therapy is combination of 3 days of anthracycline given by short infusion with 7 days of continuous infusion of cytosine arabinoside (Ara-C), commonly known as 3 + 7 regimen. Daunorubicin is the most commonly used anthracycline, other similar drugs like idarubicin, doxorubicin and mitoxantrone have also been used. Approximately 60–70% patients achieve complete remission with initial therapy while remaining would require a repeat course of chemotherapy for the same. Attempts have been made to improve the results obtained by standard 3+7 regimen by adding a third drug like etoposide (VP-16) or 6-thioguanine; or by changing the schedule of drug infusion, or by increasing the dose of anthracycline. In most cases these variations of 3+7 regimen have produced comparable results only. Severe myelosuppression in the following 2–3 weeks of induction is the most critical phase and requires close observation and prompt therapy for any signs of infection as well as blood and platelet support. Response assessment is done at 4 weeks of initiation of induction treatment. Patient attaining CR would proceed for post, remission therapy which is based on their risk of relapse.

F. Post Remission Therapy (Consolidation)

Post remission therapy includes 3–4 cycles of high dose cytosine arabinoside (15-18 G/m2 given in divided doses over 5 days). Patients with *favorable* cytogenetics benefit most. Patients in *unfavorable group* have higher risk of relapse. Once remission is achieved, search for HLA-identical sibling or matched unrelated donor must be started and if a suitable match is available then allogeneic hemopoietic stem cell transplant (HSCT) should be considered. Patient with normal cytogenetic studies (CN-AML) are classified as *intermediate risk*. Decision of post remission therapy in this group depends on FLT3-ITD (Farnesyl like Transferase 3-Internal Tandem Duplication) mutation status. Patients of CN-AML having FLT3ITD mutation have clinical behaviour like unfavourable AML and are offered Allogeneic HSCT as post remission therapy. Those without FLT3ITD mutation are treated with standard high dose Ara-C consolidationtherapy like in favorable group.Patients with high WBC count at diagnosis (≥50, 000/cmm) or those who fail to achieve remission after one cycle of induction chemotherapy are at high risk of relapse. Such patients benefit with use of allogeneic stem cell transplantation.

Autologous stem cell transplantation is an alternative post-remission therapy in intermediate-risk AML patients. However, in this subgroup, superiority of autologous transplant over conventional consolidation is not yet proven. Benefit of maintenance therapy with the aim of improving outcome is controversial in AML.

G. Acute Promyelocytic Leukemia

AML-M3 or acute promyelocytic leukemia is a distinct subtype of AML that occurs as a result of specific cytogenetic abnormality t(15; 17) resulting in fusion of promyelocytic leukaemia (PML) and retinoic acid receptor-α (RARα) gene products, which leads to maturation arrest of developing myeloid cell at promyelocyte stage and leads to accumulation of promyelocytes and blasts in bone marrow and peripheral blood. Management of Acute promyelocytic leukemia differs from the other AML subtypes: for induction ATRA (all Trans' retinoic acid, a congener of vitamin A) and daunomycin is used. ATRA binds to retinoic acid receptors and leads to differentiation of abnormal promyelocytes. This is followed by consolidation 1-2 cycles (ATRA and daunomycin) followed by maintenance therapy using 6-mercaptopurine and methotrexate for 1.5 years. Long term cure rates are 80 to 90% in APML. Patients who relapse can be effectively salvaged using Arsenic tri-oxide, another effective agent in this disease. Multiple studies are currently undergoing for evaluation of arsenic trioxide as first line therapy. Patients of APML have higher risk of bleeding diathesis—usually attributed to thrombocytopenia and/or disseminated intravascular coagulation (DIC), which is believed to result from release of a pro-coagulant factor from the promyelocyte granules.

H. Treatment for Acute Lymphoblastic Leukemia

Treatment of ALL includes remission induction, consolidation phase, CNS prophylaxis (in view of high risk of CNS relapse) and maintenance therapy for 1.5 to 2 years.

Adult ALL differ in many aspects from the childhood ALL and the main differences are in incidence rates, bcr-abl translocation positivity rates, response to chemotherapy and long term survival. Around 20% cases of adult acute leukemia are ALL of which one-fourth carry bcr-abl gene rearrangements. In contrast to the 90-95% survival of a standard childhood ALL case, long term survival in adults is only 30–40%.

Induction chemotherapy

Corticosteroids, vincristine and anthracycline (daunomycin or Adriamycin) and L-asparginaseare the main drugs administered over 4 weeks. Complete response rates are achieved in >90% patients of childhood ALL and around 70–75% patients of adult ALL. For adults—generally 3 drug regimen (corticosteroids, vincristine and anthracycline) is used in view of higher toxicity of L-asparginase. In patients with bcr-abl positive ALL, tyrosine kinase inhibitor drug, imatinib mesylate is started along with induction therapy and is continued till treatment completion. However, these patients have low event free survival and should be considered for allogeneic stem cell transplant, in case a HLA-matched donor is available.

Consolidation

Principle of consolidation regimen is to administer non cross resistant drugs in order to further reduce the residual leukemic clones in the body at microscopic level and targeting cells with primary resistance to the drugs used in induction phase. Commonly used drugs in this phase are cyclophosphamide, cytosine arabinoside, 6-mercaptopurine and/or etoposide (VP-16).

CNS directed therapy

The aim of CNS directed therapy is to destroy residual leukemic cells sequestered in the CNS, which remain protected from therapeutic concentration of systemically administered drugs by blood–brain barrier. In most protocols, CNS directed therapy is given along with consolidation phase of ALL. Whole brain radiation, weekly intrathecal chemotherapy for four weeks, and high dose chemotherapy like methotrexate are few ways to address the issue of CNS involvement. Every patient of ALL undergoes base-line CSF examination for malignant cells. Patients without CNS involvement are candidate for prophylactic cranial irradiation with dose ranging from 12–18Gy along with weekly intrathecal methotrexate. In instances when CSF is positive for malignant cells, therapeutic dose of whole brain radiation is used along with thrice a week triple drug intrathecal therapy comprising, methotrexate, cytosine arabinoside and hydrocortisone. High dose methotrexate and cytosine arabinoside are used in certain protocols for CNS targeted therapy. Patients presenting with high white blood cell count (>1, 00, 000/cmm), T-cell immunophenotye, and testicular involvement are at increased risk of CNS involvement.

Repeat induction and maintenance

Following consolidation and CNS prophylaxis induction, prolonged maintenance for 24–30 months is given in an attempt to eradicate or continually suppress any residual leukaemic cell clone. Anitimetabolite drugs, e.g., 6-mercaptopurine (6-MP) and methotrexate are the mainstay of maintenance therapy. Many protocols use a repeat course of induction regimen after the completion of induction and consolidation treatment with the aim of eradicating minimal residual disease in the bone marrow.

Supportive care

Chemotherapy for acute leukemia is a highly specialised treatment regimen and should ideally be administered in unit equipped with facilities for administration of such complex regimen and manages its after effects. Good intravenous access is helpful to administer chemotherapy, antibiotics and in certain cases parenteral nutrition. Packed red cells and other blood component like platelets and plasma are required to replenish chemotherapy induced cytopenia. Single donor apheresis platelets are useful in preventing transfusion related allo-immunisation. In bone marrow transplant candidates, leuko-depleted packed cells or transfusion through leukocyte filters is advisable. Prevention of infection is of paramount importance during therapy of acute leukemia, risk of infection is significantly higher during induction therapy. Exogenous infection can be controlled to a certain extent by limiting contact of patient with fomites and confining patient in controlled areas. Prophylactic antibiotics against Pneumocystis jirovecii and herpes zoster are advisable during intensive phase of therapy. Maintaining good hygiene, proper diet/nutrition, education of parents is important component of acute leukemia care.

SUMMARY

There has been a major progress in the understanding of the biology of acute leukemia in the past two decades. With currently available methods and experience it is possible to make the diagnosis of ALL and AML quickly. Further characterization – immune-phenotype, and cytogenetic studies have helped to identify a subgroup of patients at high risk for relapse with standard treatment protocols; such patients benefit with more aggressive approach. Identification of genetic abnormalities among AML patients with normal cytogenetics (Intermediate group) have helped to identify patients at higher risk for relapse and can be considered for consolidation with allogeneic stem cell transplantation. Current challenge is to incorporate estimation of minimal residual disease in routine monitoring, to establish optimal time points for MRD sampling and to develop appropriate methods to monitor patients for MRD who lack a suitable marker.

SUGGESIONS FOR FURTHER READINGS

1. Bhutani M., Kumar L. and Kochupillai V. (2002), Lympho-hemopoietic Malignancies in India. *Med Oncology*, 19:141–150

2. Dohner H., Estey E.H., Amadori S., et al. (2010), Diagnosis and Management of Acute Myeloid Leukemia in Adults: Recommendations from an International Expert Panel, on Behalf of the European Leukemianet. *Blood*, 115:453–74.

3. Vardiman J.W., Thiele J. and Arber D.A. (2009), The 2008 Revision of the World Health Organization (WHO) Classification of Myeloid Neoplasms and Acute Leukemia: Rationale and Important Changes, *Blood*, 114:937–51.

4. Cheson B.D., Bennett J.M. and Kopecky K.J. (2003), Revised Recommendations of the International Working Group for Diagnosis, Standardization of Response Criteria, Treatment Outcomes, and Reporting Standards for Therapeutic Trials in Acute Myeloid Leukemia, *J. Clin. Oncol.*, 21(24):4642–4649.

5. Jagarlamudi R., Kumar L. and Kochupillai V. (2000), Infections in Acute Leukemia: An Analysis of 240 Febrile Episodes, *Med. Oncol.*, 17:111–6.

6. Coustan Smith E. and Campana D. (2010), Acute Lymphoblastic Leukemia, In: *Diagnostic Techniques in Hematological Malignancies*, Wendy N. Erber (Ed.), Cambridge University Press, 127–145.

7. Grimwade David (2010), Acute Myeloid Leukemia, In: *Diagnostic Techniques in Hematological Malignancies,* Wendy N. Erber (Ed.), Cambridge University Press, 146–169.

32

Anemia

Renu Saxena and Mrinalini Kotru

CONTENTS

Globally, anemia affects 1.62 billion people, which corresponds to 24.8% of population, according to the WHO global database on anemia. 1993–2005. It is one of commonest laboratory findings found in hospital admissions. It is detected by examination of blood for hemogram. This is a frequent investigation that is universally advised for a patient with a severe or a chronic illness. Hemogram includes estimation of hemoglobin, total leucocyte count, platelet count and red cell morphologic assessment. Hemoglobin gives an indication of anemia. Anemia can occur alone or may be associated with abnormalities of other cell lines. Presence of anemia, changes in the leucocytes and platelets gives some important clues to the diagnosis of underlying disease and help decide on important treatment related decisions Even though anemia is frequently seen it is *not a disease but a sign of a disease*. It is thus important to develop a systematic approach towards diagnosis of anemia.

I.   DEFINITION OF ANEMIA

Anemia is functionally defined as an insufficient RBC mass to adequately deliver oxygen to peripheral tissues. However, objectively it is measured by estimation of hemoglobin concentration (Hb) expressed in g/l commonly. Hematocrit is also used to measure anemia. The cut off thresholds for defining anemia vary with the gender as well as the age of the patient. Following table shows hemoglobin thresholds used to define anemia.

Age or Gender Group	Hb Threshold value (g/l)
Children (0.5–4.99 yrs)	110
Children (5–11.99 yrs)	115
Children (12–14.99 yrs)	120
Non- pregnant women (≥15 yrs)	120
Pregnant Women	110
Adult men (≥15 yrs)	130

II.   CLINICAL PRESENTATION

Clinical presentation of anemia may be due to anemia primarily or due to the underlying disease causing it. Severity of clinical presentation however, depends on the degree of anemia as well as the rate of fall of hemoglobin. A very chronic fall of hemoglobin may be better tolerated compared to an acute fall in minor amount of hemoglobin. The associated co-morbid conditions also determine the response to anemia.

Patients with anemia usually present with easy fatigability, dyspnea on exertion and palpitations. On examination, they typically have pallor, which can be observed in conjunctiva, palms and the tongue. Signs of cardiovascular compensation may be apparent in the form of tachycardia, postural hypotension, systolic murmurs and pedal edema. Brittle nails, platynychia, koilonychia, thinning and early greying of hair may be observed. Pernicious anemia may also be associated with glossitis, atrophy of papillae and parasthesias.

The history and the examination should be directed towards looking for the cause of anemias like hemolytic anemia (history of jaundice, gall stones). An attempt should be made to identify whether anemia is isolated or is associated with involvement of other blood cell lines. History of bleeding or recurrent infections should thus be elicited. Examination should specifically include the lympho-reticular organs namely liver, spleen and peripheral lymphnodes.

III. PATHOPHYSIOLOGY OF ANEMIA

Anemia occurs due to fall of red cell mass or hemoglobin content or both. This can happen in isolation or may be accompanied by alterations in the other cell lines. When all the three cell lines, i.e., red cells, white blood cells and platelets are reduced it is called **pancytopenia.** It could be due to hypersplenism or primary bone marrow pathology like aplastic anemia, leukemia, lymphoma and other hematological as well as non-hematological malignancies. Bone marrow examination is mandatory in these settings to look for causes of pancytopenia in marrow.

Anemia can also be **accompanied by leucocytosis** where peripheral smear examination becomes mandatory. Differential leucocyte count will show whether leucocytosis is due to neutrophilia, eosinophilia, lymphocytosis or monocytosis. It will also pick up abnormal cells or increased blasts. This would give a clue to the probable cause of anemia. A good clinical correlation along with a bone marrow aspirate or biopsy would then be required. This may just be a spectrum of an infective etiology or be an ominous disease like acute leukemia. Hence, anemia may represent a variety of diseases ranging from a nutritional deficiency to a malignancy.

There are numerous reasons for anemia occurring in isolation. It can range from stem cell/progenitor cell defect to peripheral destruction or loss of red cells. Due to the diverse and contrasting nature of etiopathogenesis anemias can be pathophysiogically divided into Hypoproliferative, i.e., anemia due to decreased production and Hyperproliferative anemia, i.e., anemia due to increased destruction of red cells. Heart of this division lies in the estimation of *reticulocyte count.*

Reticulocytes are immature red cells, typically composing about 1–2% of the total red cells. They are released from the bone marrow and circulate for about a day in the blood stream before developing into mature red blood cells. They are called reticulocytes because of a reticular pattern on supra vital staining. This is because of ribosomal RNA present in them. Assessment of reticulocyte count is reflective of the bone marrows capacity of distinguishing between a hypoproliferative

or a hyperproliferative marrow. It is useful in distinguishing anemia due to decreased production or increased destruction. Therefore, should be done in all cases for work up of anemias.

A. Hypoproliferative Anemia

This includes anemia resulting from decreased red cell production .This can be due to failure of Pluripotential stem cell or erythroid progenitor cell . It could also result from functional impairment in the stem cells and the progenitor cell or nutritional deficiencies or anemia of chronic disease or Thalassemias.

Low reticulocyte count represents the cases of hypoproliferative anemias. However, occasionally, reticulocyte count may be in normal range or mildly increased, leading to misinterpretation. In such cases, reticulocyte production index (RPI) and corrected reticulocyte count are reflective of the true hypoproliferative state of the marrow and therefore, should be done in addition to reticulocyte count in a suspected case of hypoproliferative anemia. RPI is measured by the following formula:

Corrected reticulocyte count =

$$\frac{\text{Reticulocyte count} \times \text{Patients hematocrit}}{\text{Normal hematocrit}}$$

$$\text{RPI} = \frac{\text{Corrected reticulocyte count}}{\text{Reticulocyte maturation time}}$$

Hematocrit (%)	Reticulocyte survival (days) = Maturation correction
36–45	1.0
26–35	1.5
16–25	2.0
15 and below	2.5

RPI < 1% is representative of the hypoproliferative status of bone marrow.

RPI > 2% is representative of the hyperproliferative status of bone marrow.

Causes of Hypoproliferative Anemias

  (a) Pleuripotent stem cell failure
    1. Aplastic anemia
    2. Myelodysplastic Syndrome and acute leukemias
    3. Marrow infiltration
    4. Post-chemotherapy
    5. Hereditary bone marrow failure syndromes
  (b) Erythroid progenitor cell failure
    1. Pure red cell aplasia (Both Acquired and Hereditary)
    2. Endocrine disorders
    3. Acquired sideroblastic anemias
    4. Congenital dyserythropoitic anemias
  (c) Functional impairment
    1. Iron deficiency anemia
    2. Anemia of chronic disease
    3. Megaloblastic anemias
    4. Sideroblastic anemias

5. Thalassemia
6. Anemia of chronic disease
7. Anemia due to chemical injury

B. Hyperproliferative Anemias

Hyperproliferative anemias represent a type of anemia where marrow retains the ability to proliferate and also step up production in response to reduced red cell mass. This can be secondary to peripheral red cell destruction (called Hemolytic anemia) or blood loss or also due to redistribution of red cells.

Hemolytic anemia is occurs due to reduced red cell life span. This can be measured by signs of increased RBC breakdown and increased bone marrow red cell production. Increased unconjugated bilirubin, S. LDH, and increased reticulocyte count with or without presence of splenomegaly represent classic cases of hemolytic anemia. Hemolytic anemia can be because of red cell destruction inside the blood vessels (intravascular hemolysis) or in the reticuloendothelial system chiefly spleen (extravascular hemolysis). Distinguishing between the two is often helpful as it helps in narrowing down the likely possibilities. Increased Plasma hemoglobin, absent S. Haptoglobin and presence of urine hemosiderin are indicators of intravascular hemolysis. There are very limited causes of intravascular hemolysis like Paroxysmal Nocturnal hemoglobinuria, Certain types of Autoimmune hemolytic anemia , blackwater fever, micro as well as macroangiopathic hemolytic anemias, toxic and chemical injuries, G6PD deficiency, transfusion reactions etc. Extravascular hemolysis is often associated with splenomegaly and shows low haptoglobin levels instead of absent levels. Common causes of acquired extravascular hemolysis being Autoimmune hemolytic anemia (Direct Coombs test-Positive). Incubated Osmotic fragility test (IOFT) is another test which is done in the workup of hemolytic anemias accomapanied by splenomegaly and helps estabilish the diagnosis of hereditary spherocytosis. (Figure 32.4).

Anemia can also result due to blood loss and blood redistribution.

Causes of hyperproliferative anemias/ hemolytic anemias:

(a) Intrinsic Red Cell defects
  1. Red Cell Membrane defects
     - hereditary spherocytosis, elliptocytosis, pyropoikilocytosis
     - heriditary stomatocytosis, abetolipoproteinemia
     - McLeod syndrome
     - Spur cell anemia
  2. Red Cell Ezymopathies
     - Kreb's pathway defects: commonly Pyruvate Kinase
     - HMP pathway defects: commonly G6PD deficiency
     - other rarer; 5'nucleotidase
  3. Hemoglobinopathies
     - Sickle cell anemia

     - Unstable hemoglobinopathies and others
  4. Some porphyrias
  5. Paroxysmal Nocturnal Hemoglobinuria
(b) Extrinsic Red Cell defects
  1. Antibody mediated
     - Autoimmune Hemolytic anemia
     - Alloimmune Hemolytic anemia (e.g. transfusion reactions)
  2. Mechanical
     - Microangiopathic hemolytic anemia
     - DIC, TTP, HUS, vasculitis
     - Macroangiopathic hemolytic anemia
     - March hemoglobinuria, artificial heart valves
  3. Miscellaneous
     - Chemicals (copper, arsenic, snake venoms, etc.)
     - Physical injury (heat, oxygen, radiation)
     - Hypersplenism

III. RED CELL MORPHOLOGY IN ANEMIA

Red cell morphology is a vital tool for ascertaining underlying etiology. Normal Red cell size is 8–10 fl. Mean Corpuscular Volume (MCV) represents the red cell size on a hemogram. Normal MCV is 80–100 fl. However, Red cell size can also be assessed by careful examination of peripheral smear. Normal red cells are normocytic normochromic (Figure 32.1). They are equivalent to the size of nuclei of small lymphocyte. In anemia this can either reduce or increase or remain within range.

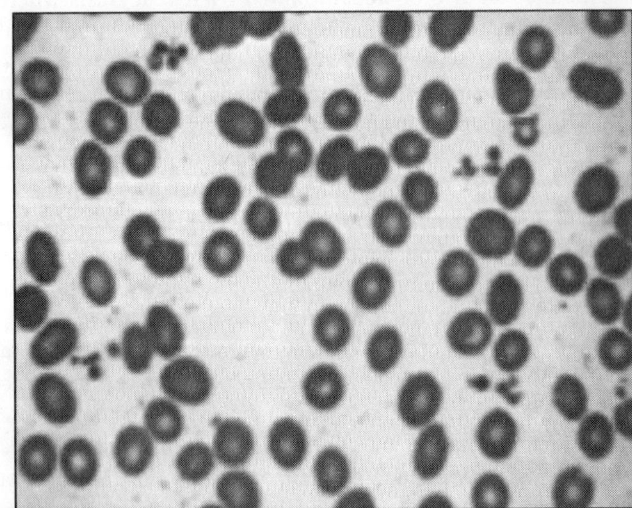

Figure 32.1 Normocytic Normochromic RBC.
(see Plate 3 for colour figure)

The cells which are smaller than this are called microcytes (Figure 32.2) and cells which are larger than this are called macrocytes (Figure 32.3). Apart of the size the other parameters may also be altered. Hypochromia is a morphological finding which accompanies microcytosis and is said when there is increase of central pallor (>1/3rd of the normal red cell radius). There is also variation in cell size called anisocytosis.

Poikilocytosis, i.e., variation in cell shape is a common finding in anemias and is quiet severe especially in certain hemoglobinopathies.

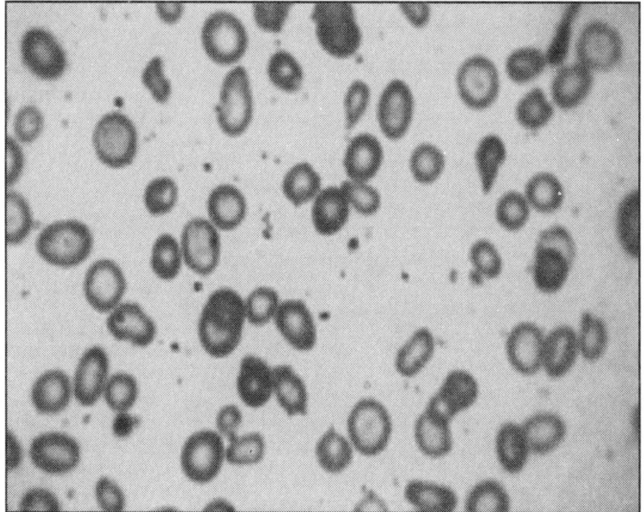

Figure 32.2 Microcytic hypochromic red cells. (*see Plate 3 for colour figure*)

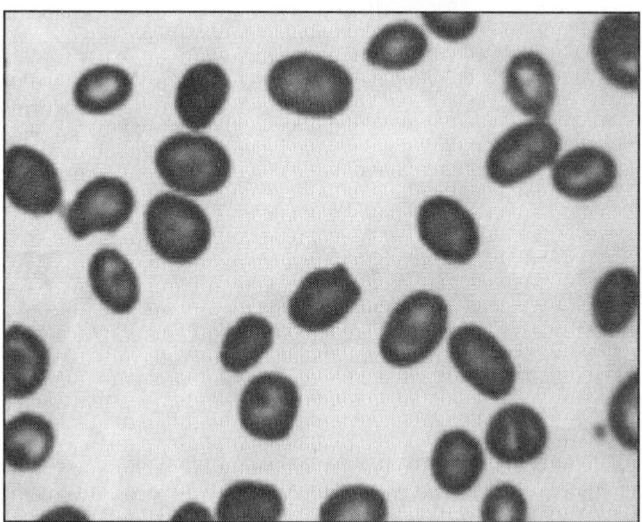

Figure 32.3 Macrocytic red cells. (*see Plate 3 for colour figure*)

Mixed population of cells can also be seen. This variation in cell size in not random but depends on the etiology behind it. For example iron deficiency anemia leads to a microcyic hypochromic red cells and vitamin B_{12} deficiency leads to macrocytic red cells. It is because of this cause and effect relationship which has been exploited to classify anemias morphologically.

Morphologically anemias can be classified as Microcytic Hypochromic, Normocytic normochromic and Macrocytic anemia

A. Microcytic Hypochromic Anemia

Anemia is microcytic hypochromic when MCV is < 80 fl. Most of the cases falling in this category are due to deficient hemoglobin synthesis. Iron deficiency anemia is the most common cause of microcytic hypochromic anemia and it is worthwhile to give a therapeutic trail with iron in absence of further investigations. Anemia of Chronic Disorders and beta thalassemia trait are also common and need to be differentiated from each other by Serum iron studies, Serum Ferritin and Hb HPLC. Serum iron and transferrin saturation are low in iron deficiency anemia and anemia of chronic disease. However, S. Ferritin is low in iron deficiency and high in anemia of chronic disease. Hb HPLC is diagnostic of Beta thalassemia trait (Figure 32.4).

Microcytic hypochromic red cells

Microcytic hypochromic red cells can be due to the following reasons:

Defects in iron metabolism: This can happen due to nutritional deficiency of Iron, Anemia of Chronic Disease and rarely due to hereditary diseases like atransferrinemia.

Defects in globin chain synthesis: α and β-Thalassaemia, Heamoglobin E&C syndromes and unstable hemoglobin characteristically present as microcytic hypochromic anemia.

Defects in Heam metabolism: This encompasses a rare group of disorders which include hereditary and acquired sideroblastic anemias and lead intoxication.

B. Macrocytic Anemia

Macrocytic anemia is defined when MCV is >100 fl. Macrocytosis is generally due to megaloblastosis. Megaloblasts are characteristically present in megaloblastic anemia. These are characterized by large size and more immature appearing nucleus compared to cytoplasmic stage of maturation (nucleocytoplasmic asynchrony). This morphologic change results due to impaired DNA synthesis because of either Vitamin B12 or Folate deficiency or inherited or drug or toxin induced impairement of DNA synthesis. The presence of megaloblast in the bone marrow is pathogonomic of Megaloblastic anemia. The common causes being Folic acid deficiency secondary to poor nurtition and increased physiologic demands. Autoimune gastritis, i.e., pernicious anemia is an important cause of megaloblastic anemia in the West, but is not common in India.

Causes of megaloblastic anemia

- Vitamin B12 Deficiency
  - Dietary deficiency
  - Lack of intrinsic factor
  - Pernicious anemia
  - Gastric surgery
  - Malabsorption syndrome
- Folate Deficiency
  - Dietary deficiency
  - Increased requirements (Pregnancy, Infancy, chronic hemolytic anemia)

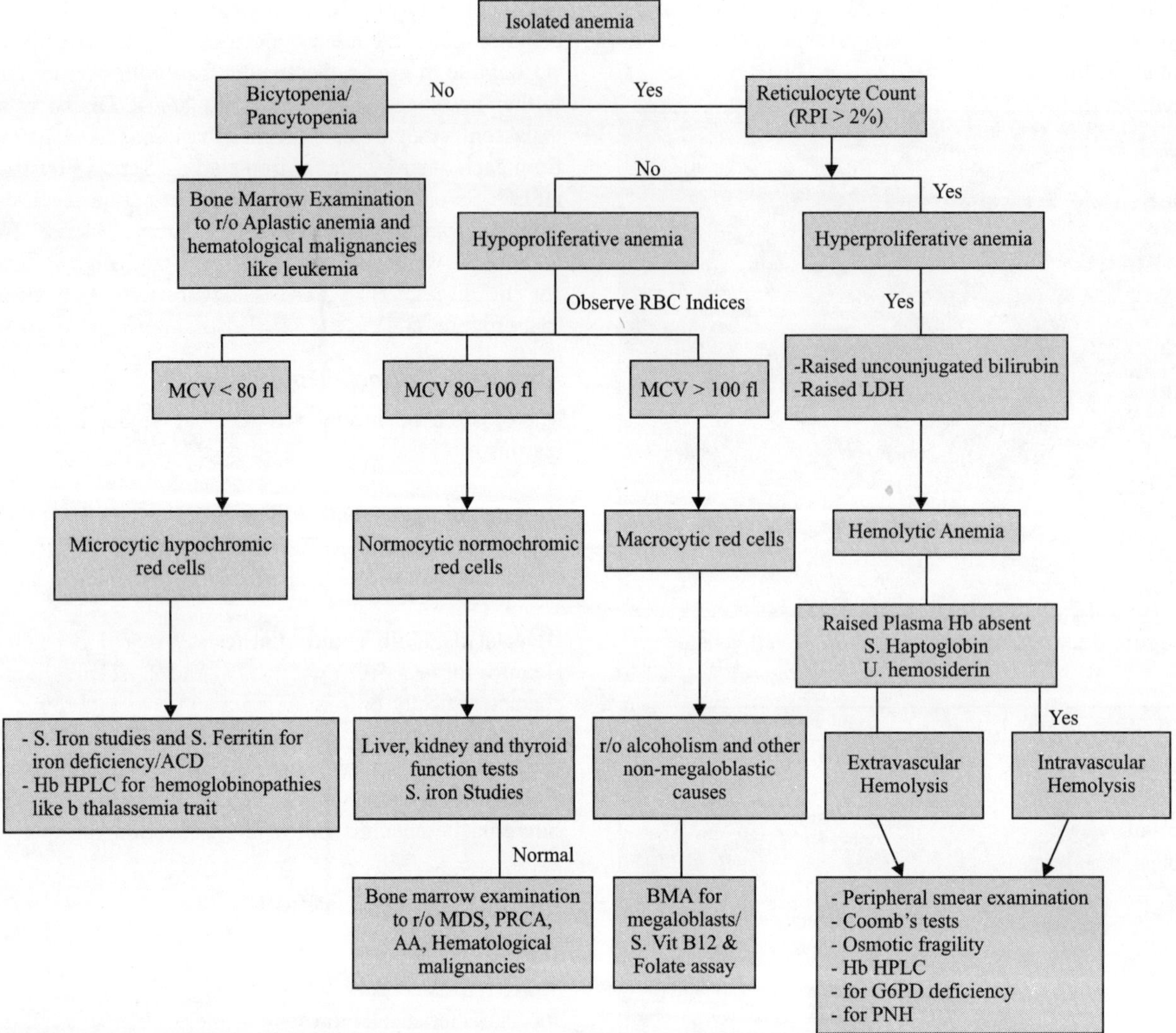

Figure 32.4 Approach to laboratory diagnosis of anemia

- Alcoholism, drug-induced folate deficiency
- Intestinal & jejuna resection
- Combined Folate and Vitamin B12 Defciency
  - Tropical sprue
  - Gluten-sensitive enteropathy
- Inherited disorders of DNA synthesis
  - Homocytinuria & methylmalonic aciduria
  - Orotic aciduria and other rare defects
- Drug and Toxin induced disorders of DNA Synthesis
  - Folate antagonists
  - Purine antagonists
  - Pyrimidine antagonists
  - Alkylating agents
  - Antibiotics like trimethoprim
- Erythroleukemia

However, there are several Non-megaloblastic causes of macrocytic anemia. History of alcoholism and examination of blood for liver function becomes important in such cases. Most commonly mild degree of macrocytosis is due to increased reticulocyte count which is seen hemolytic anemia or secondary to hemorrhage. However, in hypoproliferative settings following causes need to be excluded.

Causes of non-megaloblastic macrocytic anemia

- Alcoholism
- Liver disease
- Myelodysplastic
- Myelophthisic anemia
- Aplastic anemia
- Acquired sideroblastic anemia
- Congentital dyserythropoitic anemia
- Hypothyroidism

C. Normocytic Normochromic Anemia

When anemia is associated with normal RBC size (MCV-80–100fl), it is called Normocytic Normochromic anemia. However, normal MCV may also represent a mixed population of microcytic hypochromic cells and macrocytic red cells. A raised Red cell distribution width (RDW) will be informative and suggest that the red cell population under study has a wide range of sizes. Hemolytic anemias may also present as normocytic normochromic anemia.

Causes of normocytic normocytic anemia

- **Decreased erythropoietin secretion**
  - Anemia of renal insufficiency
  - Anemia of liver disease
  - Anemia of endocrine deficiency
  - Protein-calorie malnutrition
  - Anemia of chronic disorders
- **Anemia with impaired marrow response**
  - Bone marrow hypoplasia
  - Pure Red blood cell aplasia (PRCA)
  - Transient erythroblastopenia of childhood
  - Transient aplastic crises associated with hemolysis
  - Aplastic anemia (pancytopenia); Myelodysplastic syndromes
  - Bone Marrow infiltrative disorders

IV. MANAGEMENT ISSUES

Depending on the underlying etiology, respective therapies may be instituted to treat the underlying causes. However when Hb falls to less than 7g/dl or there is congestive heart failure due to anemia packed red cells may be transfused.

V. CONCLUSION

Anemia is not a disease per se but sign of an underlying disease. It is caused by a variety of diverse illness, which may or may not overt by themselves. Hence, it requires a logical and systematic approach towards diagnosis which encompasses both clinical and laboratory data to arrive at a conclusion. The examination of peripheral smear is not overstated as it provides very important clues to the underlying diagnosis. The examination of bone marrow examination is important in many cases of anemia and should be done at the earliest in clinically indicated cases.

SUGGESTIONS FOR FURTHER READINGS

1. *Wintrobe' Clinical Hematology*, 16th ed., 2009 Lippincott Williams and Wilikans.

2. WHO Global Database on Anemia, 1995–2003.

3. Robbins and Cotran, *Pathologic Basis of Disease*, 8th ed., Saunders Elsevier.

33

Heart
Normal Function and Dysfunction

Manish Bansal, Karen M. Modesto, Jagat Narula and Partho P. Sengupta

CONTENTS

I. Myocardial Structure and Function
   A. Normal myocardial architecture
   B. Electrical activation sequence as a determinant of the myocardial motion
   C. Sequence of myocardial motion during the cardiac cycle
II. Physiological Factors Affecting Myocardial Contraction
III. Alterations in Myocardial Motion in Disease States
   A. Predominant subendocardial dysfunction
   B. Predominant subepicardial dysfunction
   C. Transmural dysfunction
IV Technique for Assessment of Myocardial Motion
V. Dynamics of Blood Flow in the Cardiovascular System
   A. Significance in disease states
   B. Techniques for assessment of blood flow patterns
Summary

Human heart, with its unique anatomy and highly sophisticated functioning, has fascinated, and at the same time intrigued, researchers and clinicians for centuries. Though it had appeared deceptively simple initially, understanding the normal functioning of the heart, from genetic and molecular level to macroscopic level, has actually been quite challenging. Nevertheless, the decades of research involving animal and human experimentations and utilizing an array of imaging techniques, have allowed remarkable advancements in our understanding of the myocardial architecture and function. A lot has been learnt about the arrangement of myofibers within the cardiac walls, their motion during the cardiac cycle, translation of myocardial motion into physiological function, movement of the blood within the heart, etc. and their implications in normal health and diseases. This chapter describes the current understanding of the myocardial architecture and function, with its relevance to the clinical practice.

I. MYOCARDIAL STRUCTURE AND FUNCTION

A. Normal Myocardial Architecture (Figure 33.1)

The human myocardium has a complex, multilayered muscle fiber architecture. Myofibers in different layers are arranged in spirally oriented bundles which start from the apex and spread towards the base forming counter-directional helices with the direction of the fibers changing progressively from subendocardium to subepicardium. For the purpose of description, a right handed helix is the one in which the fingers of the open right hand point in the direction of the muscle fibers whereas thumb points towards the long-axis of the helix. Following this description, the fibers in the subendocardial layer are arranged in a right-handed helical configuration and are

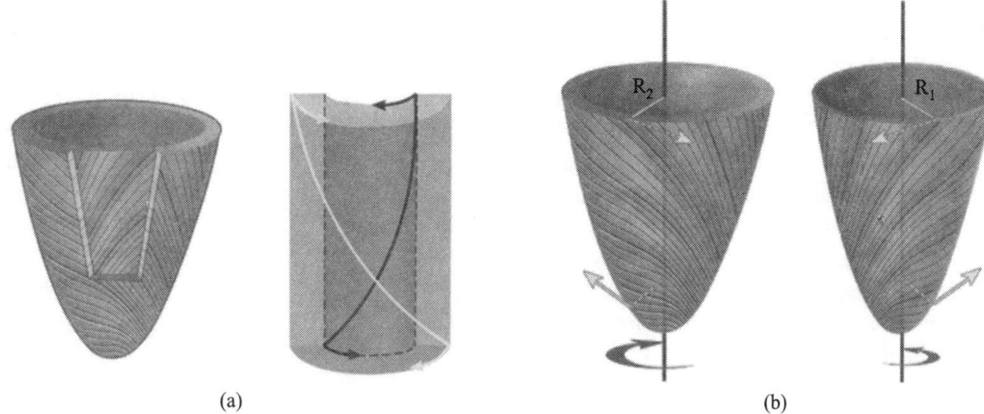

(a)   (b)

Figure 33.1 Myofiber architecture of left ventricle and models [(a) right-Ingels et al., and (b) Taber et al.] for understanding left ventricular twist dynamics. Myofibers in the left ventricular myocardium are arranged in counter-directional helices with left-handed orientation in the subepicardium which changes smoothly to a right-handed helix in the subendocardium [(a), left]. Arrows (a) depict the circumferential components of force that results from force development in each fiber direction. The radii (R_1 for subendocardium and R_2 for the subepicardium) are the lever arms, which convert these circumferential components of force into torque about the long axis of the cylinder. As the subepicardial fibers have a longer arm of moment than the subendocardial fibers, the force generated by these fibers dominates. Figure illustration done by Rob Flewell. From Sengupta PP, Tajik AJ, Chandrasekaran K, Khandheria BK. Twist mechanics of the left ventricle: Principles and application. *J Am Coll Cardiol Img.*, 2008; 1(3):366–376, with permission.

aligned almost parallel to the long-axis of the left ventricle, forming an angle of approximately +85 with its circumference. In contrast, the fibers in the subepicardial layer are arranged in a left-handed configuration, at an angle of roughly –85 with the short-axis of the ventricle. In the mid myocardial layers, the fibers are predominantly in horizontal direction and are parallel to the circumference of the left ventricle. During the cardiac cycle, these different layers contract in different directions and slide over each other to generate shear forces which provide the framework for normal cardiac functioning. However, despite this multilayered arrangement, the entire left ventricular myocardium behaves as a singly functional unit in which contraction of myofibers in one layer cannot happen without reciprocal changes in the other layers.

As is well recognized now, the counter-directional helical arrangement of myofibers is the most critical element of myocardial anatomy and is fundamental to the normal cardiac functioning. Such a myocardial architecture provides the mechanism for translating unidirectional shortening or lengthening of the myofibers in to multi-directional deformation of the ventricle wherein longitudinal shortening occurs simultaneously with shortening in the circumferential direction and thickening in the radial direction. In addition, as described below, owing to the helical configuration of the muscle fibers, longitudinal contraction of the myocardium results in rotation of the left ventricular apex and base in opposite directions, producing twist or torsion of the left ventricle. The twist of the left ventricle greatly enhances systolic function of the ventricle by facilitating ejection in a manner analogous to wringing of a towel. Furthermore, the deformation of the myocardial matrix during systole results in storage of the mechanical energy which is responsible for rapid untwist and recoil during early

diastole, generating diastolic suction required for early rapid filling of the ventricle. Thus, the unique myocardial architecture interlinks systolic and diastolic functions of the ventricle, with major implications in health and disease.

The helical myofiber arrangement is also responsible for another, important mechanistic property of the left ventricle. With the onset of systole, as the earliest activated segments of the left ventricular myocardium begin to contract, the remaining segments get stretched. This early systolic myocardial stretch of the myocardial segments invokes a phenomenon known as "stretch activation response" that allows muscle fibers to adjust the force of shortening according to the preload. In addition, simultaneous contraction and stretch of the different myocardial segments helps reduce endocardial stress and myocardial oxygen demand by allowing uniform distribution of myocardial stress across the wall. Finally, stretch of the late activated segments during the isovolumic contraction phase permits development of strong contractile force despite fixed cavity volume and primes the ventricle for subsequent forceful ejection.

B. Electrical Activation Sequence as a Determinant of the Myocardial Motion

The myocardium has a highly specialized network of fibers which brings about ultrafast depolarization of the myocardium. The presence of intercalated discs with gap junctions and minimum branching facilitates rapid conduction of current. However, despite a very rapid spread of electrical current within the left ventricular myocardium, a gradient of electrical activation still develops. In a study on seven isolated normal human hearts from individuals who died of non-cardiac causes, it was shown that the electrical current initially reached the endocardial aspect of left ventricle at three points simultaneously: (1) an area high on anterior paraseptal wall just below the mitral valve, (2) a central area in the left side of the interventricular septum, and (3) the posterior paraseptal area at approximately one third of the distance from the apex to the base. These areas quickly became confluent, and almost the entire endocardial layer of the LV was activated, with the exception of the posterobasal area, in approximately 30 milliseconds. The epicardial aspect of the left ventricle was activated later. The posterobasal region of the left ventricle was the last one to get activated. The more recent *in vivo* studies utilizing electrocardiographic imaging, a noninvasive cardiac electrical imaging modality, have confirmed these observations. The lateral-apical region of the left ventricle gets activated initially and the activation front then gradually travels toward the posterolateral base in an apex to base direction.

Unlike ventricular depolarization, the sequence of ventricular repolarization remains less well characterized at present. While some of the animal studies have shown that the repolarization has opposite direction-to-activation sequence, other animal and human studies utilizing electroanatomical mapping have reported contrary findings. These studies demonstrated that the repolarization sequence followed the sequence of activation in most subjects. Similarly, controversy also exists about the duration of action potential of different myocardial regions and about the transmural sequence of repolarization.

C. Sequence of Myocardial Motion during the Cardiac Cycle (Figures 33.2–33.4)

The normal myocardial motion is a systematic, sequential, well-orchestrated process which represents the net result of interplay between electrical activation of different myocardial regions, contractile forces developing in different myocardial layers, shear forces between the myocardial layers and the loading conditions. From a functional stand-point, the cardiac cycle can be divided in to four major phases-isovolumic contraction, ejection phase, isovolumic relaxation and diastolic filling phase.

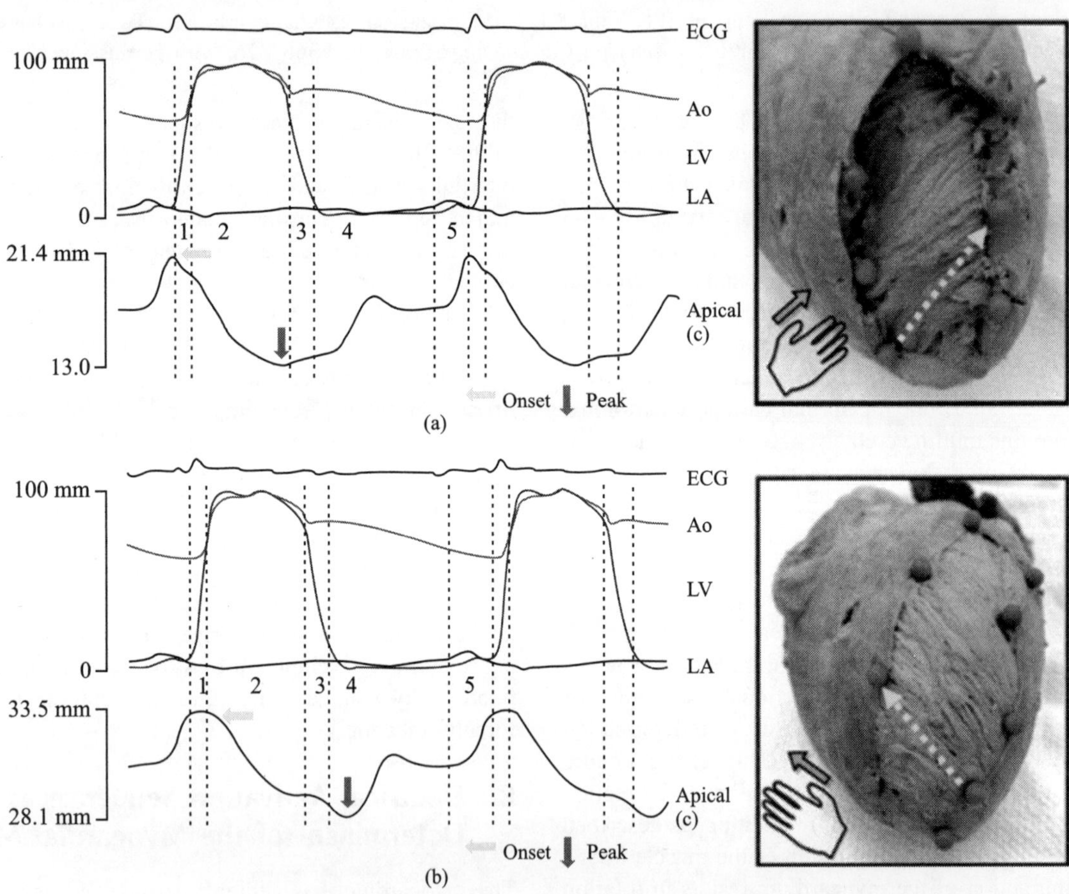

Figure 33.2 Transmural sequence of deformation in left ventricle using sonomicrometry (black tracing). (a) During isovolumic contraction, shortening is initiated along subendocardial (right-handed helical) fiber direction. (b) Onset of shortening in subepicardial (left-handed helical) fiberdirection is delayed and coincides with onset of left ventricular ejection. Tissue specimens (right) show corresponding subendocardial and subepicardial fiber arrangements, position of sonomicrometry crystals (orange markers), and pairs of crystals used to determine deformation along fiber direction (yellow arrows).Phase 1, pre-ejection; 2, ejection; 3, isovolumic relaxation; 4, early diastole; 5, late diastole. From Sengupta PP, Krishnamoorthy VK, Korinek J, et al. Left ventricular form and function revisited: Applied translational science to cardiovascular ultrasound imaging. *J Am Soc Echocardiogr.*, 2007; 20(5):539–551, with permission. (*see Plate 3 for colour figure*)

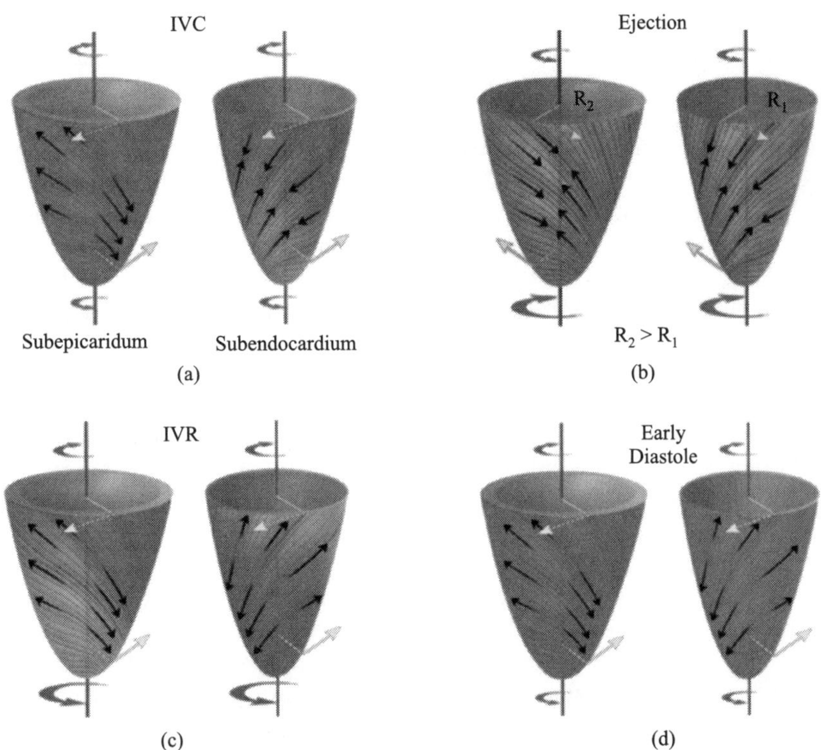

Figure 33.3 Schematic depiction of the sequence of left ventricular twist mechanics. During isovolumic contraction (IVC) (a), the subendocardial myofibers (right-handed helix) shorten with stretching of the subepicardial myofibers (left-handed helix), producing a brief clockwise rotation of the apex and a counterclockwise rotation of the left ventricular base. During ejection (b), the subendocardial and subepicardial layers shorten simultaneously, with shortening strains near the apex exceeding those base. The larger arm of moment of the subepicardial fibers dominates the direction of twist, causing rotation of the apex and base in counterclockwise and clockwise directions, respectively. During isovolumic relaxation (IVR) (c), subepicardium lengthens from the base towards the apex and the subendocardium from the apex towards the base. The subsequent period of diastole is characterized by relaxation in both layers with minimum untwisting (d). Figure illustration done by Rob Flewell. Reprinted From Sengupta PP, Tajik AJ, Chandrasekaran K, Khandheria BK. Twist mechanics of the left ventricle: Principles and application. *J Am Coll Cardiol Img.*, 2008; 1(3):366–376, with permission.

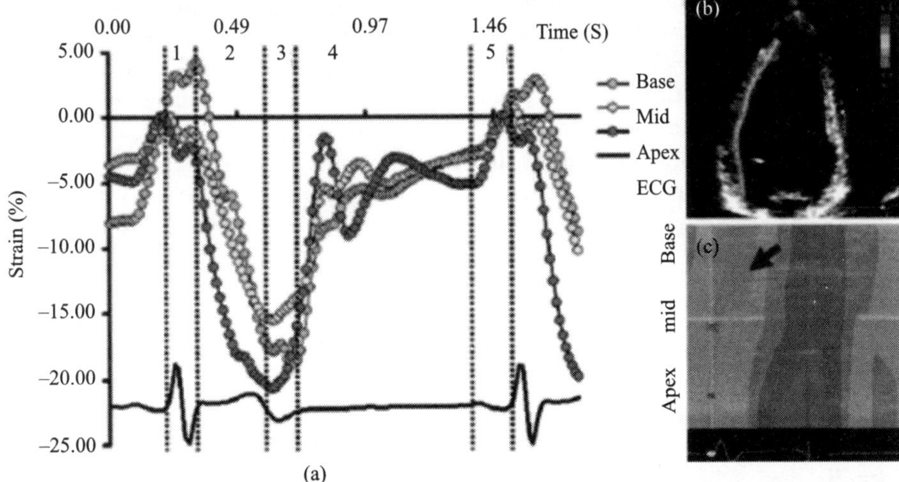

Figure 33.4 Speckle-tracking echocardiography to demonstrate longitudinal deformation in the lateral wall of the left ventricle in a healthy woman. Segmental deformation traces (a), region of speckle tracking in the lateral wall of LV (b) and color M-mode profile of longitudinal deformation (c) are shown. Note that both shortening and relaxation are delayed in the basal segments. In addition, during ejection, shortening strains at apex are higher than that at base. Phase 1, preejection; 2, ejection; 3, isovolumic relaxation; 4, early diastole; 5, late diastole. From Sengupta PP, Krishnamoorthy VK, Korinek J, et al. Left ventricular form and function revisited: Applied translational science to cardiovascular ultrasound imaging. *J Am Soc Echocardiogr.*, 2007; 20(5):539–551, with permission. (*see Plate 4 for colour figure*)

Isovolumic contraction

As described above, the earliest region of left ventricular myocardium to get depolarized is the subendocardium in the apicolateral segment. Consequently, the subendocardial fibers near the left ventricular apex are the earliest to contract during the isovolumic contraction phase. Contraction of these fibers results in clockwise rotation (when viewed from the apical side) of the apex, owing to the right-handed configuration of the subendocardial helix. At the same time, asynchrony in the electrical activation of different myocardial segments ensures that the early subendocardial shortening is accompanied by variable stretching of the other, relatively late activated myocardial segments. Thus, the subendocardial fibers in the posterobasal regions of left ventricle, the subepicardial fibers and the papillary muscles get paradoxically stretched during the isovolumic contraction phase. The stretch of the papillary muscles can be attributed to the increasing intraventricular pressure in early systole that leads to the closing of the atrioventricular valves and increasing tension on the chordae. These reciprocal waves of contraction and stretch help prepare the left ventricle for the subsequent ejection phase in a manner akin to loading of a gun.

Ejection phase

As the electrical depolarization front spreads to the subepicardial layers, more and more subepicardial fibers get recruited in the contractile process generating a powerful contractile wave, which spreads from apex to base. Meanwhile, the contraction of the subendocardial fibers also continues in the apex-to-base direction. Thus, the contractile waves in both the subepicardial and subendocardial layers originate at the left ventricular apex and spread along the ventricular wall to progressively involve the mid and basal segments. Experimental studies utilizing high-speed digital video acquisition system and surface markers have shown that this wave of contraction takes approximately 23 milliseconds to spread across the ventricle.

The absolute magnitude of the contractile force generated in the subendocardial layers is greater than the subepicardial layers. However, subepicardial fibers, being situated on the outer side, have larger moment of arm and hence generate greater amount of torque for the same degree of contraction as compared to the subendocardial fibers. Accordingly, the subepicardial fibers dominate, resulting in rotation of the left ventricular apex in counterclockwise direction owing to the left-handed helical configuration of the subepicardial fibers. The left ventricular base, at the same time, gets rotated in the opposite (i.e. clockwise) direction, producing twist or torsion of the ventricle.

Since the subendocardial and subepicardial layers are arranged largely parallel to the long-axis of the left ventricle, contraction of these fibers contributes to the longitudinal shortening of the ventricle, while also producing circumferential shortening and twist of the ventricle. The mid-myocardial fibers, which are arranged parallel to the circumference of the left ventricle, predominantly cause circumferential shortening.

At any given point of time, the magnitude of circumferential contraction exceeds that of longitudinal shortening.

Isovolumic relaxation and post-systolic shortening

With the onset of myocyte repolarization, the myofibers begin to relax and the force of ejection starts to decrease progressively. When the intraventricular pressure falls below the aortic pressure, the aortic valve closes and the phase of isovolumic relaxation begins. Similar to the myocardial contraction phase, this early phase of relaxation is also characterized by regional asynchrony which results in creation of lengthening-shortening gradients within the ventricle that allow diastolic restoration to be initiated despite fixed volume of the left ventricle. In the apical region, relaxation begins in the subendocardial layers but the fibers in the subepicardium continue to shorten for a brief period even after aortic valve closure. The continued shortening of the muscle fibers beyond aortic valve closure is known as post-systolic shortening. It is a normal physiological phenomenon and is believed to play an important role in active diastolic restoration of the left ventricle. However, when unduly prolonged, as in ischemic heart disease, post-systolic shortening becomes an undesirable feature and starts to impede normal cardiac functioning. These early diastolic changes in the apical region are accompanied by reciprocal changes at the base with onset of relaxation in the subepicardial region and ongoing shortening along the subendocardial fibers. Thus, unlike systole in which both the subepicardial and subendocardial helices contract in the same direction (from apex to base), during diastole the two helices relax in opposite directions (apex to base for subendocardium and base to apex for subepicardium).

The relaxation of myofibers during early diastole results in rapid untwist and recoil of the left ventricle. The apex undergoes a brief twist in counterclockwise direction resulting from continued shortening of the subepicardial fibers, followed by rapid recoil in the clockwise direction. The twist of the left ventricular base is opposite to that of the apex but is significantly lower in magnitude. It is important to note that most of the untwisting occurs during isovolumic relaxation only and is completed well before the end of isovolumic relaxation. This rapid untwist creates a suction force which facilitates early rapid filling of the left ventricle when the mitral valve opens.

II. PHYSIOLOGICAL FACTORS AFFECTING MYOCARDIAL CONTRACTION

Several physiological variables affect left ventricular myocardial motion and twist, the most important being the loading conditions. Left ventricular twist increases with preload and decreases with after load, with after load producing more marked effect than preload. Thus, an increase in the left ventricular end-diastolic volume (a measure of preload) will increase the magnitude of twist if the end-systolic volume is held constant. Converse is true for changes in the left ventricular end-systolic volume (a measure of after load).

Myocardial contractility, as expected, has profound effect on left ventricular deformation. Measures that increase left ventricular contractility, such as inotropic stimulation, augment left ventricular twist whereas negative inotropic agents reduce the magnitude of twist.

Exercise affects left ventricular twist through its effects on both loading conditions and myocardial contractility. Short-term isotonic exercise leads to increase in left ventricular preload, fall in after load and increase in myocardial contractility—all the effects that increase left ventricular twist. Long-term exercise training, however, leads to adaptation of cardiovascular system such that the resting left ventricular twist decreases but there is more marked torsional reserve available for recruitment as the circulatory demands increase during vigorous exercise.

Age also affects left ventricular deformation and twist. During the childhood, counter-clockwise apical rotation remains constant but the basal rotation increases with age. The base rotates in clockwise direction during infancy which becomes neutral during early childhood and reverses to adult pattern by adolescence. Accordingly, left ventricular twist, which is the sum total of apical and basal rotation, increases with age from infancy to adulthood. With further advancement of age during the adult life, the subendocardial function gets progressively compromised as a consequence of the age-related degenerative changes taking place in the subendocardium. The relatively unopposed subepicardial function leads to progressive increase in apical rotation and left ventricular twist with age. However, this increase in the left ventricular twist at rest is accompanied by attenuation of torsional reserves during exercise. At the same time, the early diastolic untwist velocity is delayed and may even be reduced as a manifestation of reduced elasticity of myocardial wall associated with ageing.

III. ALTERATIONS IN MYOCARDIAL MOTION IN DISEASE STATES (Figure 33.5)

The recent insights gained in to the myocardial mechanics have facilitated improved understanding of the pathogenesis and clinical manifestations of cardiac dysfunction developing in various cardiac and non-cardiac disorders. For example, it is now becoming increasingly apparent that systolic and diastolic heart failure—the two major forms of heart failure and traditionally believed to be two distinct entities—are not two entirely different disorders but are interrelated and likely represent a continuum of myocardial dysfunction. Recognizing this, a new scheme of classifying heart failure

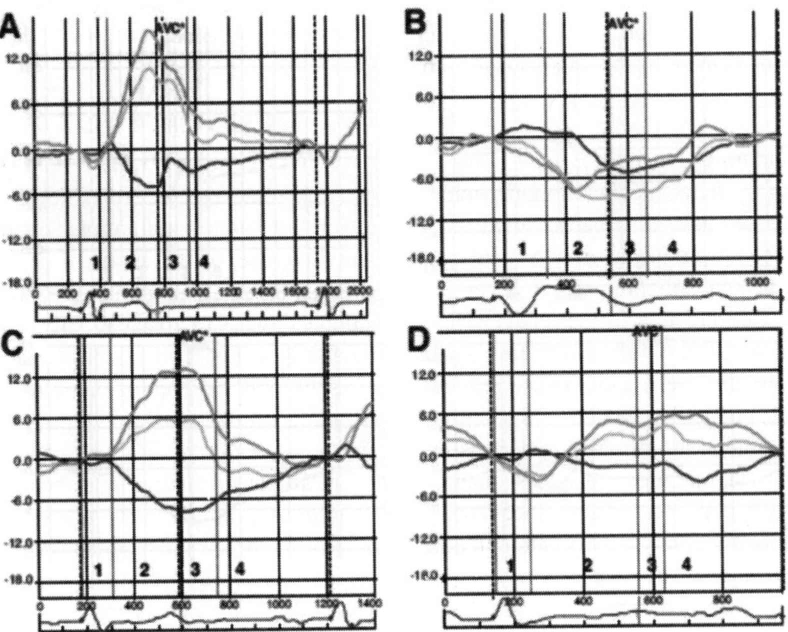

Figure 33.5 Left ventricular twist in health and disease. Rotation of the left ventricular apex, the base, and the net twist angle (shown in yellow, brown and orange colors, respectively) are assessed by speckle-tracking echocardiography in a normal subject (A), a patient with dilated cardiomyopathy with systolic heart failure (B), a patient with cardiac amyloidosis presenting as heart failure with normal ejection fraction (diastolic heart failure) (C), and a patient with constrictive pericarditis (D). Net ventricular twist is negative in dilated cardiomyopathy because of complete reversal of the left ventricular apical rotation (B). In contrast, a patient with amyloid cardiomyopathy shows relatively preserved magnitude of net left ventricular twist angle but the onset of untwisting is delayed after aortic valve closure (C). The patient with constrictive pericarditis (D) shows reduced magnitude of net ventricular twist and marked delay in the onset of untwisting. Phase 1, pre-ejection; 2, ejection; 3, isovolumic relaxation; 4, early diastole. AVC- aortic valve closure. From Sengupta PP, Tajik AJ, Chandrasekaran K, Khandheria BK. Twist mechanics of the left ventricle: Principles and application. *J Am Coll Cardiol Img.*, 2008; 1(3):366–376, with permission. (*see Plate 4 for colour figure*)

has been proposed which categorizes cardiac disorders based on myocardial layer-specific involvement. Such an approach groups together disorders that have similar underlying pathophysiological alterations and are therefore likely to have similar clinical presentation, therapeutic options and outcomes. According to this scheme, heart failure can be classified into three broad subgroups—heart failure with predominant subendocardial dysfunction, heart failure with predominant subepicardial dysfunction and heart failure with transmural dysfunction.

A. Predominant Subendocardial Dysfunction

In general, subendocardial layer of the myocardium is the most vulnerable to the pathological disturbances involving the myocardium. Relatively poor coronary supply to the subendocardial region is one of the important reasons underlying the greater vulnerability of the subendocardial fibers to the disease processes. Thus, in most of the cardiac disorders, evidence of myocardial injury and fibrosis are first seen in the subendocardial layers as demonstrated by magnetic resonance imaging and autopsy studies. These observations are further supported by the experimental studies demonstrating transmural gradients in changes at the molecular level also. For example, in heart failure patients, beta-receptor density is significantly reduced in the subendocardial layer and not in the subepicardial layer. The use of long-term beta-blocker therapy restores beta-receptor density in the subendocardium, which is accompanied by measureable improvements in myocardial function and clinical outcomes. Similarly, the levels of sarcoplasmic reticulum Ca adenosine triphosphatase (ATPase) and the activity of cyclic adenosine monophosphate (cAMP) and phosphodiesterases have also been shown to be reduced primarily in the subendocardial layers.

Since the myofibers in the subendocardial layer are arranged almost parallel to the long-axis of the left ventricle, the longitudinal contraction of the left ventricle is the first to get compromised in disease states. Thus, reduced longitudinal deformation of the left ventricle is a common early manifestation in ischemic heart disease, hypertension, diabetes, cardiomyopathies, valvular heart diseases such as aortic stenosis, mitral regurgitation, etc, or any other disease affecting the left ventricle globally. At this stage, circumferential contraction and the twist mechanics are largely preserved or may even be accentuated due to the lack of the opposing effect from the diseased subendocardial fibers as well as the compensatory hypertrophy taking place in the subepicardial layers. Thus, the overall left ventricular ejection remains preserved at this stage.

The early subendocardial dysfunction also leads to delayed untwist of the ventricle, although the peak untwist velocity may remain normal initially. However, with progressive myocardial involvement, even peak untwist velocity is also reduced. These changes in diastolic function result in increase in left ventricular diastolic pressure which clinically manifests as diastolic heart failure.

B. Predominant Subepicardial Dysfunction

Certain disorders that affect the heart from outside may involve subepicardial layers preferentially. Constrictive pericarditis is a classical example of this. Involvement of the subepicardium by the pericardial pathology results in marked reduction in the circumferential function and the twist mechanics of the left ventricle. The longitudinal function, which is determined primarily by the subendocardial fibers, remains unaffected and tries to maintain the overall ejection fraction within the normal range. In addition, as the normal twist is critical to the maintenance of normal diastolic function, loss of twist also compromises the diastolic function. The untwisting velocity is delayed and reduced with progressive increase in left ventricular diastolic pressures.

C. Transmural Dysfunction

In any disease affecting the myocardium, as the severity of the disease process worsens, the involvement of myofibers becomes increasingly transmural. Involvement of both the subendocardial and subepicardial fibers leads to marked reduction in the left ventricular long-axis function which is accompanied by impairment of circumferential shortening and twist also. Consequently, the systolic function gets progressively compromised and the left ventricular ejection fraction falls with the emergence of systolic heart failure.

IV. TECHNIQUE FOR ASSESSMENT OF MYOCARDIAL MOTION

Over the years, explorations on myocardial mechanics have entailed use of an array of techniques such as implanted radiopaque markers, biplane cineangiography, sonomicrometry, optical devices, gyroscopic sensors, cardiac magnetic resonance imaging, speckle tracking echocardiography etc. However, at present, only cardiac magnetic resonance imaging and speckle tracking echocardiography appear suitable for regular clinical use.

Cardiac magnetic resonance imaging is the current gold-standard for non-invasive assessment of cardiac deformation. It permits three-dimensional evaluation of cardiac deformation without resorting to any mathematical assumptions. The two most common magnetic resonance methods used for measuring myocardial motion are tagging and phase contrast velocity mapping. Tagging is usually a time-consuming technique and can be performed manually or semi-automatically, although semiautomatic techniques generally require some extent of manual correction. In contrast, tissue phase mapping directly encodes the velocity of myocardial motion into the magnetic resonance signal and offers high spatial resolution (1–3 mm). However, as both the above magnetic resonance methods require breath-holding, the temporal resolution is limited by the length of the breath-hold period to 30 to 80 milliseconds. This limitation has been addressed by the development of a respiratory-gated, free-breathing method for tissue phase mapping that allows measurement with a temporal resolution comparable to tissue Doppler imaging.

Speckle-tracking echocardiography is a relatively newer technique which permits relatively angle-dependent assessment of myocardial deformation from gray-scale echocardiographic images. On a gray-scale echocardiographic image, myocardium consists of a series of echogenic speckles which form due to constructive and destructive interference of the ultrasound beams within the tissue. The automated software identifies these speckles and tracks them frame-by-frame to gather information about the myocardial motion during the cardiac cycle. From this information, global and regional myocardial deformation in any direction can be easily resolved. Compared with magnetic resonance imaging, speckle-tracking echocardiography is relatively simple, less time consuming, less expensive and has excellent temporal resolution. In addition, it has much wider availability and can be used even in patients with metallic implants. For these reasons, speckle-tracking echocardiography has fast emerged as the method of choice for the assessment of myocardial deformation *in vivo*. However, it has its own limitations with inter-vendor variability, dependence on gray-scale image quality and variability due to through-plane motion being the most important ones.

V. DYNAMICS OF BLOOD FLOW IN THE CARDIOVASCULAR SYSTEM

Much like the myocardial mechanics, blood flow within the cardiovascular system also exemplifies exceptional mechanical efficiency of the cardiovascular system. The dynamics of the blood flow through the cardiac chambers are such that the most efficient energy utilization is achieved with minimum possible energy dissipation. One of the most salient features of the blood flow that helps achieve this objective is the formation of spiral rings or vortices. Although Leonardo da Vinci was the first to observe formation of vortices within the sinuses of Valsalva and also explained their functional significance, much of the current knowledge about vortex formation has been gained only during the past few decades.

It was long believed that the blood flow within the cardiovascular system was predominantly laminar and unidirectional and that the laminar flow was the most efficient way to prevent energy wastage. However, such an understanding could not explain how the moving blood stream could be smoothly steered from inflow to outflow segments within the cardiac chambers while at the same time resulting in minimum energy loss. The more recent animal and human experiments have provided these explanations. It is now learnt that at multiple points within the cardiovascular system the blood stream becomes non-laminar, multi-directional and assumes vortical shape. As the blood stream moves through the cardiac chambers, multiple fluid layers are created that slide against each other and against the myocardial wall and tend to curl or spin to produce spiral streamlines, known as vortices.

These vortices are formed as a consequence of the opposing influences of the pressure-head driving the forward flow on one hand and the slowing-down effect of the viscosity and the shear stress between the fluid layers and the containing boundaries on the other hand. The phenomenon is the greatest at places where the blood-stream separates from the wall, as typically occurs when blood enters heart cavity from a vein or through a valve or when a relatively narrow vascular segment transitions in to a dilated segment (e.g., carotid sinus or at the site of vascular stenoses).

The formation of these vortices provides a framework for maintenance of circulation under a wide range of physiological and non-physiological conditions by allowing a delicate balance between energy conservation, conversion and dissipation. For example, because of the vortical flow pattern, the blood entering from the left atrium in to the left ventricle is very swiftly and efficiently directed towards the left ventricular outflow for ejection (Figure 33.6). As a result, the kinetic energy carried by the incoming blood in to the left ventricle is not completely lost during the diastole itself but is channelized for use during the subsequent ejection. It fact, it has been recently estimated that roughly $16 \pm 8\%$ of the kinetic energy of diastolic mitral flow is conserved to the end of diastole and carried over into the next beat. In addition, the formation of vortices also assists diastolic filling and contributes to the closure of the mitral valve. During diastole, the larger vortical swirl breaks down into smaller eddies which continue to exert lateral pressure on the left ventricular wall, facilitating further diastolic filling. At the end of diastole, the reversed direction of blood flow in the left ventricle exerts pressure on the mitral leaflets and assists in closure of the mitral valve. The eccentric orientation of the mitral valve in normal ventricles favors formation of vortices by directing the blood towards the left ventricular wall which leads to swirling of the blood.

Vortex formation is not restricted to left ventricle alone but occurs in all cardiac chambers and great vessels including right atrium, right ventricle, left atrium and aorta (Figures 33.6 and 33.7). In the right atrium, flows coming from superior and inferior vena cava swirl into a vortex that empties into the inlet of the tricuspid valve. In the right ventricle, the tricuspid inflow is converted in to vortical streams beneath the tricuspid leaflets which get reoriented toward the outflow region. However, unlike the left ventricle, the change in the direction between inflow and outflow is less acute in the right ventricle. In contrast, the flow pattern in the left atrium is quite distinct with two temporally discrete vortices. The first vortical flow begins to develop immediately before ventricular systole, increases in magnitude during systole and disappears with the onset of early diastole when the flow from all four pulmonary veins is directly propagated in to the left ventricle. The second left atrial vortex develops during mid-diastole and disappears with the onset of atrial contraction. Transient reversal of flow into the pulmonary veins can be recorded during this phase.

The vortex formation has been demonstrated in the aorta also (Figure 33.7). Following ejection, vortices form within the coronary sinuses of Valsalva that provide a mechanism for

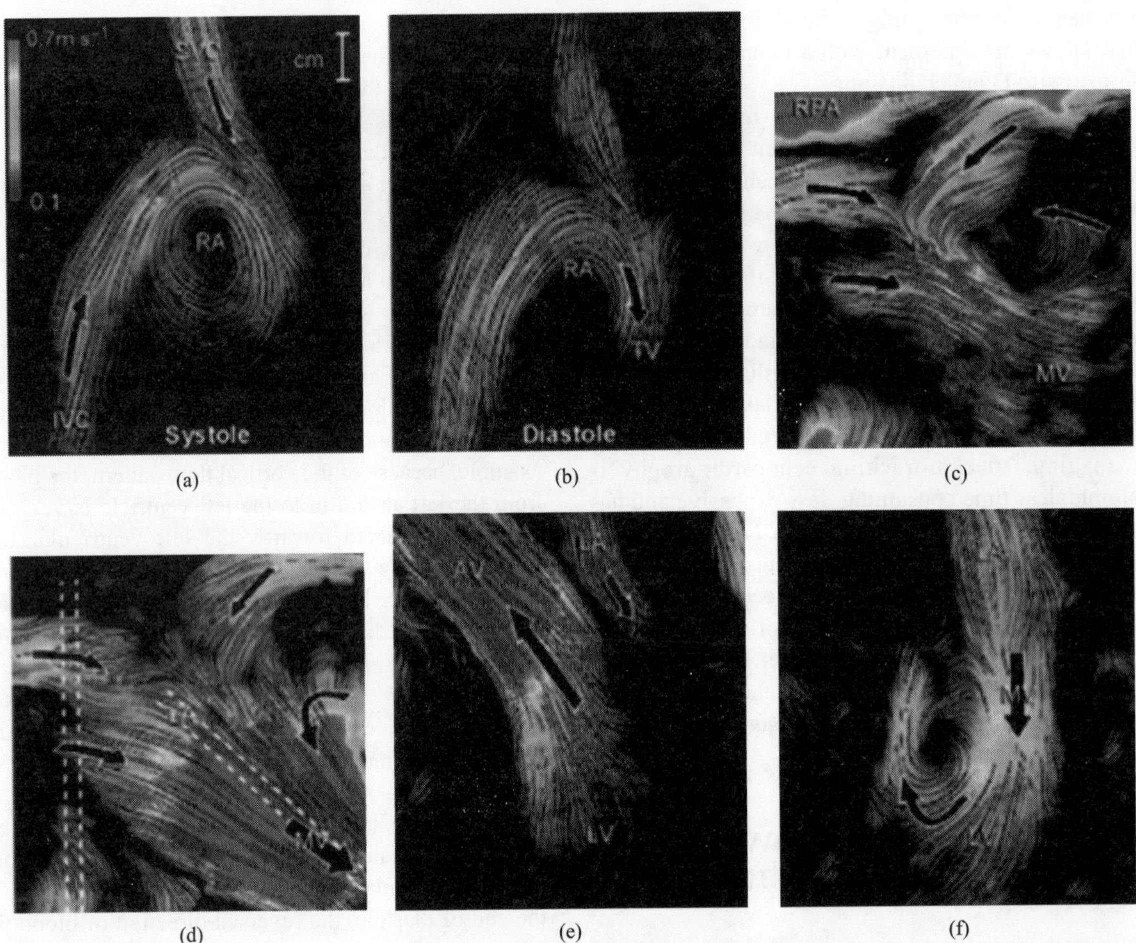

Figure 33.6 Intracardiac flow sequence in a beating heart. Colored streamlines computed from magnetic resonance velocity acquisitions show local speed as indicated by the color scale. (a) The right atrial flow has been viewed in ventricular systole in a sagital plane from the subject's right side. Blood entering from superior and inferior caval veins rotates and forms a well-defined vortex in the expanding chamber. (b) In early ventricular diastole, blood is redirected through the open tricuspid valve. (c) In the left atrium, in ventricular systole, blood enters from the upper and lower pulmonary veins on each side, indicated by arrows in this coronal plane viewed from the front (lower veins lie out of the plane posteriorly).Rotating streams of blood flow are redirected toward the mitral valve. (d) In early ventricular diastole, there is further inflow from veins while blood passes through the open mitral valve to the left ventricle. Vertical and oblique dotted lines indicate planes orthogonal to this panel, represented by panels A, B, E, and F. (e) In systole, streamlines pass from the left ventricle through the aortic valve. (f) In early diastole, flow from the left atrium is redirected from the inflow to the outflow through the formation of a large anterior vortex. In panels E and F only, the color scale is modified to reach red at $1\ ms^{-1}$. AV, aortic valve; IVC, inferior vena cava; LA, left atrium; MV, mitral valve; RA, right atrium; RPA, right pulmonary artery; SVC, superiorvena cava; TV, tricuspid valve. From Kilner PJ, Yang GZ, Wilkes AJ, et al. Asymmetric redirection of flow through the heart. *Nature*, 2000; 404:759, with permission. (*see Plate 4 for colour figure*)

closure of the aortic valve at the end of systole. The central jet however continues in the main aortic arch, towards the anterior right wall of the ascending aorta from where it gets reflected postero-laterally towards the inner curvature, creating opposing helices. A right-handed helix is formed along the left wall and a left-handed helix along the right wall. At the same time, retrograde flow occurs along the inner curvature between these two helices. In young individuals (<50 years), the blood flows from the aortic valve to the mid-descending aorta within one systole but takes two-to-three beats in older individuals. With the exception of young individuals, the direction of blood flow in most individuals is triphasic, with forward flow during early systole that reverses in mid and late systole and goes forward again in diastole.

A. Significance in Disease States

It is clearly apparent from the above discussion that the formation of the vortices is crucial to ensure optimum systolic and diastolic performance of the left ventricle. An inability to produce normal vortical flow pattern therefore contributes to the development and progression of various cardiac disorders. For example, in patients with dilated cardiomyopathy, the direction of flow of blood from left atrium to left ventricle is not favorable

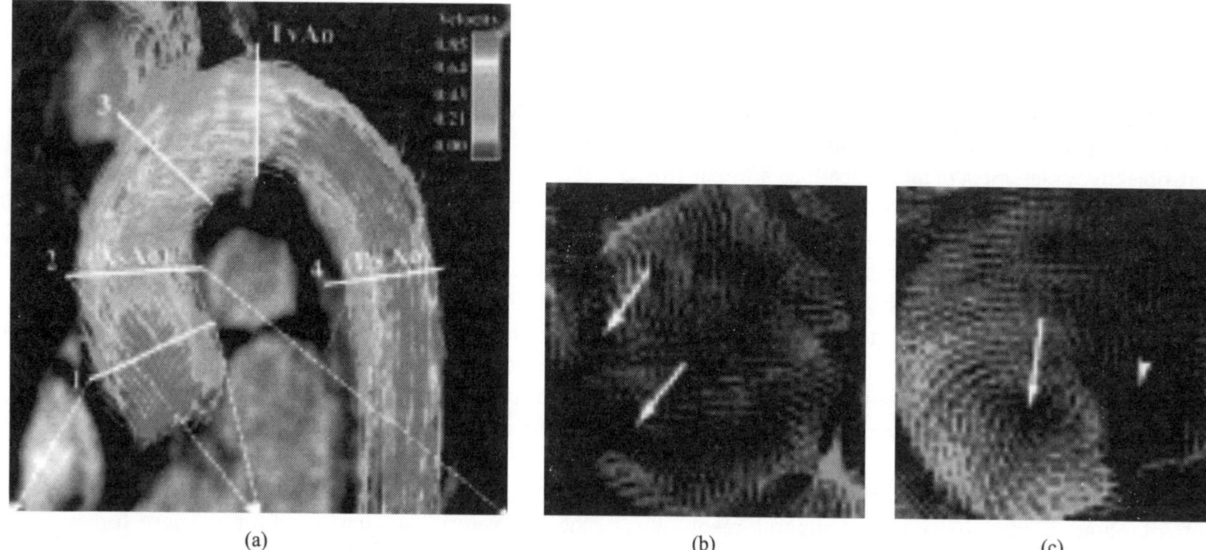

(a)   (b)   (c)

Figure 33.7 Pattern of blood flow in the thoracic aorta discerned by MR phase contrast velocity mapping. Streamlines in a healthy volunteer are shown at 15% of the cardiac cycle. Velocities in all images are measured in meters per second. (a) A right-handed helix can be seen on the surface of the streamlines, as well as a slowing of velocities at the aortic arch (b) Cross-sectional plane with velocity vectors placed orthonormal to the aorta at plane 1 shows two opposing helices, one clockwise and one counterclockwise along the posterior wall of the aorta (arrows). (c) At plane 2, the right-handed helix on the left (arrow) is larger than the left-handed helix on the right (arrowhead). All planes are viewed from a perspective distal to the plane, with the inner curvature at the bottom of the image. For example, the larger vortex (arrow) would be graded as occurring at a 9 o'clock angle, rotating in a counterclockwise direction, encompassing 50% of the aortic lumen, and with its center 50% of the way between the center of the lumen and the vessel wall. AsAo, ascending aorta; DsAo, descending aorta; TvAo, transverse aorta. From Hope TA, Markl M, Wigstrom L, et al. Comparison of flow patterns in ascending aortic aneurysms and volunteers using four-dimensional magnetic resonance velocity mapping. *J Magn Reson Imaging*, 2007; 26:1473, with permission. (*see Plate 5 for colour figure*)

for formation of vortices and leads to greater energy loss as the blood is redirected towards the left ventricular outflow tract. Furthermore, the vortices formed in such patients have reduced propagation velocity and have longer stagnation time, thereby promoting thrombus formation. Abnormal flow patterns are seen in patients with regional wall motion abnormalities also and can lead to formation of clots.

In patients with hypertrophic cardiomyopathy, abnormal flow patterns provide a mechanism for left ventricular outflow tract obstruction. In normal hearts, the outflow jet is directed close to the septum wherein the large anterior vortex tries to keep the mitral leaflets closer to the posterior wall. However, in hypertrophic cardiomyopathy, this process is reversed as a result of abnormal mitral valve-papillary muscle geometry. The abnormal vortices thus formed create drag forces drawing the mitral leaflets in to the left ventricular outflow tract and producing blood flow obstruction.

The analysis of blood flow pattern is also helpful in evaluation of valvular diseases, guiding surgical repair of valve regurgitation, assessing myocardial dyssynchrony and in optimization of resynchronizing devices. Similarly, the application of these blood-flow visualization techniques to atria may provide insights in to atrial function/ dysfunction in various congenital and acquired cardiac illnesses and may help in stratifying risk of clot formation in these disease states.

Finally, the knowledge of the abnormalities of blood-flow patterns in the great arteries may be helpful in quantifying the risk of developing aortic dilatation, aortic dissection, pulmonary thromboembolism, pulmonary hypertension, etc. In the aorta, the presence of spiral flow helps prevent vascular damage by reducing the lateral pressure exerted by the blood stream. Indeed, loss of spiral flow in the abdominal aorta has been shown to be associated with increasing severity of renal artery stenosis and more rapid deterioration of renal function. On the other hand, the presence of retrograde flow from complex plaques in descending thoracic aorta may result in development of cerebral embolization leading to strokes.

B. Techniques for Assessment of Blood Flow Patterns

Recent advancements in cardiac magnetic resonance imaging and echocardiographic techniques have permitted in-vivo visualization and characterization of the blood flow patterns, which otherwise had remained a major challenge so far.

On magnetic resonance imaging, velocity-encoded, cine, phase-contrast technique is the most commonly used technique for intracardiac blood flow visualization. Using ECG and respiratory gating, the complete time-resolved, three-dimensional velocity field can be measured over a volume that covers the complete heart or large vessels. From this data, three-dimensional streamlines and path-lines can be constructed for visual, qualitative assessment of cardiac flow. At the same time, a number of quantitative parameters can also be derived to allow more objective assessment of the intracardiac flow patterns.

On echocardiography, particle imaging velocimetry is the most promising technique presently available for characterizing the intracardiac blood flow patterns (Figure 33.8). The technique utilizes ultrasound contrast agents to opacify the blood cavity and the movement of the particles within the field is then tracked frame by frame, in a manner analogous to speckle-tracking echocardiography. The displacement data thus obtained is converted to derive spatial distribution of flow velocities. With this technique, the flow can be measured at a temporal resolution as high as 4 ms and an effective spatial resolution of about 4 mm. This high temporal resolution allows comprehensive analysis of the flow sequences through different phases of the cardiac cycle, including the brief intervals of isovolumic contraction and relaxation. However, there are certain limitations with this technique at present. The high velocities are usually underestimated, depending on the acquisition frame rate. In addition, so far, the technique has been developed mostly for two-dimensional flow tracking and for a sector width size of 45° only. Complete three-dimensional visualization of flow patterns is presently not possible with this technique.

SUMMARY

Human heart has complex myocardial anatomy which is characterized by arrangement of muscle fibers in the form of counter-directional helices. This helical arrangement of the myofibers provides the framework for multi-directional deformation of the left ventricle that takes place during the cardiac cycle. During systole, the left ventricle undergoes shortening in longitudinal and circumferential directions, thickens in radial direction and, at the same time, also undergoes twist. These different components of myocardial deformation contribute to maintenance of the normal systolic performance of the ventricle. At the same time, deformation of the myocardial matrix during systole stores potential energy which is released during early diastole to produce rapid recoil of the ventricle, generating suction forces driving the left ventricular diastolic filling. These forces initiate and maintain movement of blood, which is facilitated further by formation of vortical streamlines as the blood moves within the cardiac chambers. Any abnormality of the myocardial motion or blood flow patterns compromises cardiac function and provides the pathophysiological basis for the development of cardiac disorders. The newer magnetic resonance imaging and echocardiographic techniques allow non-invasive assessment of myocardial function and flow patterns and have the potential to bring about a paradigm shift in our approach to the clinical assessment of cardiac structure and function.

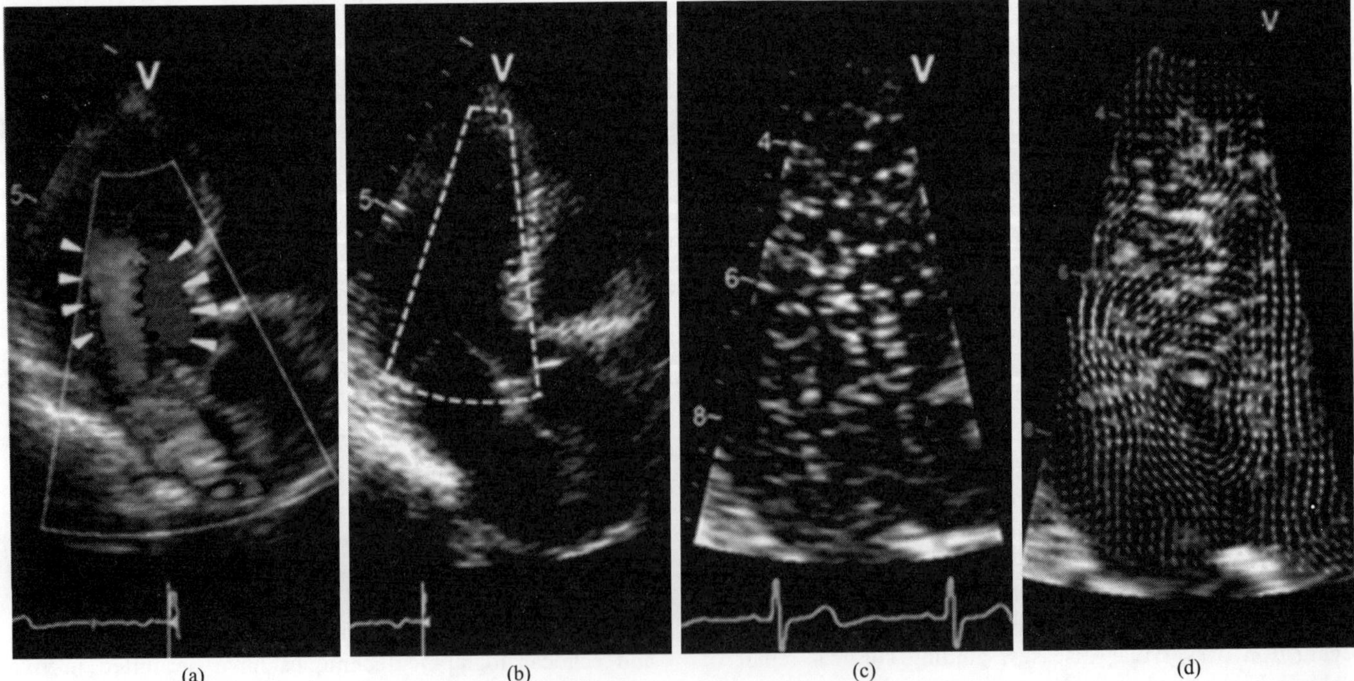

(a)   (b)   (c)   (d)

Figure 33.8 Echocardiographic-contrast particle imaging velocimetry. (a) Color Doppler imaging showing the bidirectional flow during early diastole. (b) Delineation of the region of interest in the submitral and left ventricular outflow region. (c) High-temporal resolution contrast echocardiography is performed for the region of interest. (d) Contrast particle imaging velocimetry shows the flow field with velocity vectors. Note the presence of a clockwise vortex (arrows). In press, *European Heart Journal—Cardiovascular Imaging*, with permission. (*see Plate 5 for colour figure*)

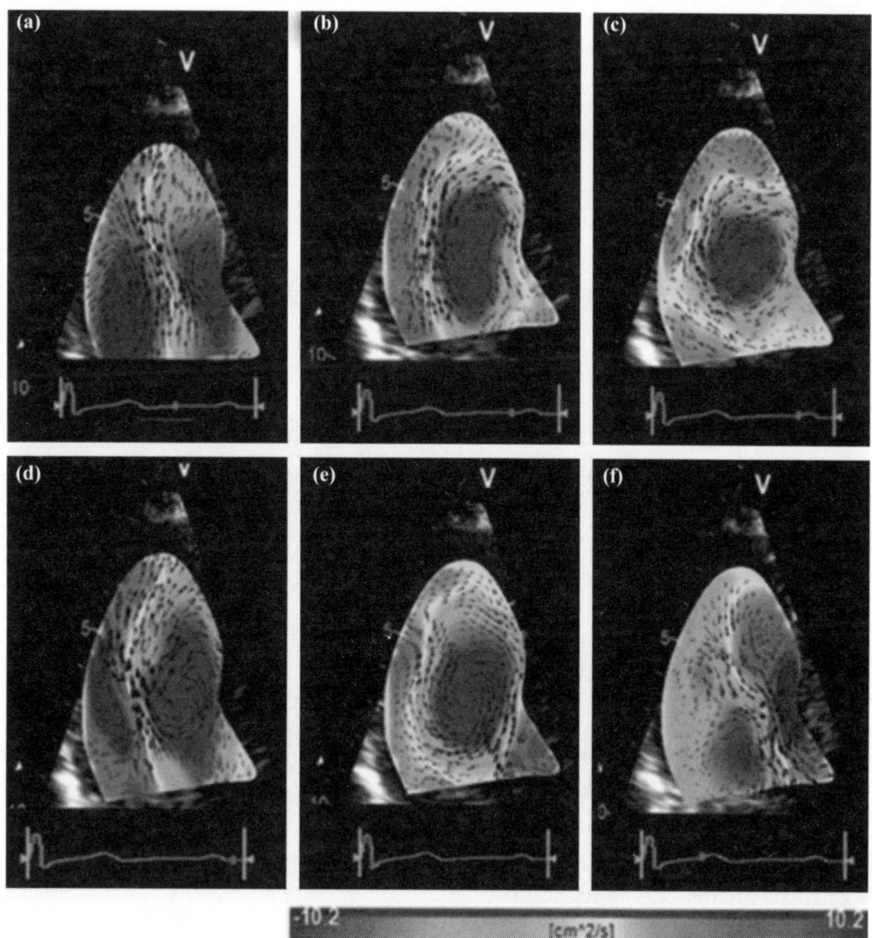

Figure 33.9 The left ventricular flow sequence during phases of cardiac cycles in a non-heart failure control subject. Left ventricular intracavitary planar vorticity maps of a non-heart failure control subject (ejection fractum 61%) during phases of cardiac cycle are shown in (a)–(f). Note that the presence of initial clockwise (blue) and counter-clockwise (red) vortices during early-diastolic filling (a). The counterclockwise flow (red) becomes weaker (b) resulting into predominantly clockwise (blue) vortex during the phase of diastasis (c). The atrial flow in late diastole is characterized by an increase in the strength of the clockwise vortex (d), which persists further during isovolumic contraction (e). Ejection is characterized by accelerating flow towards the left ventricular outflow surrounded by clockwise and counter clockiwise shear layers (f). In press, European Heart Journal-Cardiovascular Imaging, with permission. (*see Plate 5 for colour figure*)

SUGGESTIONS FOR FURTHER READING

1. Sengupta P.P., Korinek J. and Belohlavek M. (2006), Left Ventricular Structure and Function: Basic Science for Cardiac Imaging, *J. Am Coll Cardiol.*, 48(10):1988–2001.

2. Sengupta P.P., Khandheria B.K. and Narula J. (2008), Twist and Untwist Mechanics of the Left Ventricle, *Heart Fail Clin.*, 4(3):315–324.

3. Sengupta P.P., Tajik A.J., Chandrasekaran K. and Khandheria B.K. (2008), Twist Mechanics of the Left Ventricle: Principles and Application, *J. Am Coll Cardiol Img.*, 1(3):366–376.

4. Sengupta P.P., Krishnamoorthy V.K., and Korinek J. (2007), Left Ventricular Form and Function Revisited: Applied Translational Science to Cardiovascular Ultrasound Imaging, *J. Am Soc Echocardiogr.*, 20(5):539–551.

5. Sengupta P.P. and Narula J. (2008), Reclassifying Heart Failure: Predominantly Subendocardial, Subepicardial and Transmural, *Heart Fail Clin.*, 4(3):379–382.

6. Mor-Avi V., Lang R.M. and Badano L.P. (2011), Current and Evolving Echocardiographic Techniques for the Quantitative Evaluation of Cardiac Mechanics: Ase/eae Consensus Statement on Methodology and Indications Endorsed by the Japanese Society of Echocardiography, *J. Am Soc Echocardiogr.*, 24(3):277–313.

7. Sengupta P.P., Pedrizzetti G. and Kilner P.J. (2012), Emerging Trends in CV Flow Visualization, *J. Am Coll Cardiol Img.*, 5(3):305–316.

8. Sengupta P.P., Burke R., Khandheria B.K. and Belohlavek M. (2008), Following the Flow in Chambers, *Heart Fail Clin.*, 4(3):325–332.

34

Kidney
Normal Functions and Dysfunctions

Sanjay Kumar Agarwal and Akansha Agarwal

CONTENTS

I. INTRODUCTION

Claude Bernard coined the term "La fixit de milieu interior". It means that in spite of an ever-changing external environment, the internal environment remains constant. This maintenance of the internal environment was termed by WB Cannon as 'homeostasis'. In order to maintain normal cellular functions in the body, homeostasis needs to be maintained in terms of fluid volume, pH, temperature, ionic balance, osmotic pressure and concentration of different organic constituents in the biological fluids. All the organs and tissues of the body perform functions to help a constant milieu interior. The internal and external environment can interact through the major organ systems such as the lungs, kidneys, skin and gastrointestinal tract to achieve homeostasis.

The interstitial fluid constitutes the internal environment in which the cells are bathed. Cells derive nutrition from this medium and pass metabolic waste products into it. By products of metabolism such as carbon dioxide from oxidation of carbohydrates, nitrogenous wastes like urea and creatinine from protein metabolism, uric acid from purine metabolism, minerals like sulphur and phosphorus from

protein metabolism and excess quantities of water and electrolytes (sodium, potassium, calcium and chloride) are exchanged between the cell and and the extracellular fluid. The intracellular fluid composition needs to be tightly controlled at all times for functioning of the intracellular enzymes. The extracellular and/or the interstitial compartment acts as a buffer between the external environment and the intracellular compartment. Either the excess or deficit in the interstitial fluid gets corrected and rebalanced with the external environment via the lungs, kidneys, skin and GI tract. Lungs control concentrations of oxygen and carbon dioxide and kidneys maintain optimum chemical composition of body fluids.

One of the most important functions of the kidneys is to maintain the optimal chemical composition of body fluid by excreting substances from blood into urine. The kidneys rid the body of waste material that are either ingested or produced by metabolism. Another essential function is to control the volume of body fluids. The kidneys perform these functions by the following sequential processes: (a) filtration of the blood plasma by the glomeruli, (b) selective reabsorption of substances by the tubule, (c) secretion of substances from the tubules into the tubular lumen for excretion into the urine, (d) exchange of hydrogen ions and production of ammonia for conservation of base. Apart from excretory function, the kidneys play a role in maintaining blood pressure, pH regulation, calcium homeostasis and erythropoiesis. Thus the major functions of the kidneys include

(a) Excretion of metabolic waste products and foreign chemicals
(b) Regulation of water and electrolyte balances
(c) Regulation of body fluid osmolality and electrolyte concentrations
(d) Regulation of acid-base balance
(e) Regulation of arterial pressure
(f) Secretion, metabolism and excretion of hormones
(g) Gluconeogenesis

II. ANATOMY

The kidneys are bean shaped, paired retroperitoneal organs situated on the posterior abdominal wall, one on each side of the vertebral column. Each kidney weighs around 125–170 g in the adult male and 115–155 g in the adult female. The kidney has two major regions—an outer cortex and an inner medulla. The medulla is divided into multiple cone-shaped masses of tissue called renal pyramids. The pyramids terminate in the papilla, which project into the funnel shaped continuation of the upper end of the ureter, called the pelvis.

Each kidney has approximately 1 million functional units called a nephron (Figure 34.1).

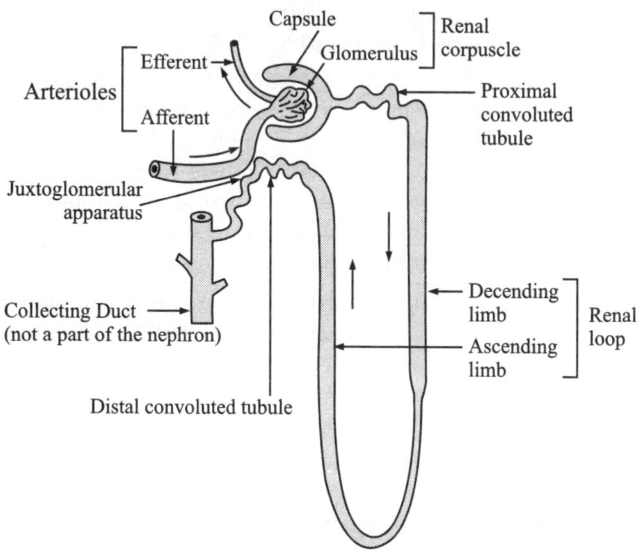

Figure 34.1 Morphology of a nephron.

A. Morphology of a Nephron

Each nephron consists of the Malphigian corpuscle, proximal convoluted tubule, loop of Henle, and the distal convoluted tubule. The distal convoluted tubule joins an arched collecting tubule that in turn enters a medullary pyramid and terminates into the renal pelvis. The renal cortex comprises glomeruli, proximal and distal convoluted tubules, ascending and descending segments of the thick loop of Henle and the collecting tubules. The medullary pyramids are composed of collecting tubules, descending and ascending segments of the thin loop of Henle and small caliber blood vessels.

Although the basic structure of the nephrons remains the same, there are some differences on how deep the nephron is located within the kidney. Those nephrons which have glomeruli located in the outer cortex are called cortical nephrons; they have short loops of Henle which penetrate only a short distance into the medulla. Around 20–30% nephrons have glomeruli which lie deep in the cortex near the medulla and are called juxtamedullary nephrons. They have long loops of Henle that dip deeply into the medulla.

B. Histology and Electron Microscopic Picture of Different Parts of the Nephron

1. Malphigian corpuscles are composed of a tuft of capillaries (glomerulus) enveloped by the Bowman's capsule having a visceral and parietal layer lined by squamous epithelium. The glomerulus is composed of a network of branching and anastomosing glomerular capillaries with high hydrostatic pressure. The afferent arterioles divide into 2–5 basal branches, which in turn divide into smaller branches giving rise to the capillary plexus. Within the corpuscle, blood and the filtrate are separated by a layer of endothelial cells with pores of 0.06 μ diameter, a basement membrane and a layer of epithelial cells comprising the visceral layer of Bowman's capsule.

The visceral layer is formed by the podocytes with numerous elongated thin pedicles, which interdigitate and are attached to the surface of the capillaries. The pedicles have small free spaces in between each other. These spaces are called slit-pores with a minimum width of 70–100 Å. The pedicles and the slit pores with the basement membrane and the endothelial cell layer comprise the filtration barrier (Figure 34.2).

2. Proximal convoluted tubule (PCT) originates at the Malphigian corpuscle and is confined to the renal cortex. It is lined by strongly eosinophilic, large columnar cells with a brush border. Each cell has around 150 long and smaller brush border extensions per square micron. The cells contain round or oval nuclei and numerous mitochondria in the basal aspect.

3. Loop of Henle: The proximal convoluted tubule, as it enters the medullary ray, is continued as the descending limb of the loop of Henle.

(a) **Descending limb:** The cells have light cytoplasm because of few mitochondria and a few sub-microscopic ribonucleoprotein particles. The cell surface shows scattered brush border extensions, which are shorter and coarser than that in the PCT. These features indicates a less active resorption occurring in this part of the tubule compared to the PCT. Gradually, the columnar

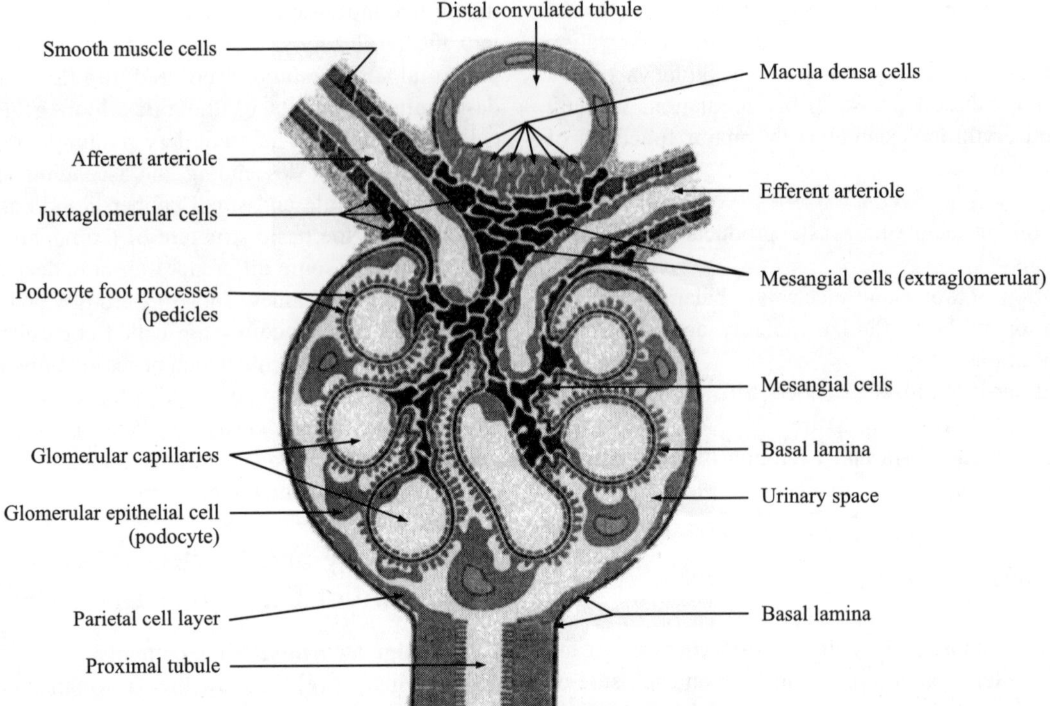

Figure 34.2   Structure of the glomerulus.

cells of the descending limb are replaced by squamous epithelium.

(b) **Thin segment:** This is where the descending limb bends and becomes continuous with the ascending limb of the loop of Henle. The cytoplasm of the squamous cells of the thin segment contains few mitochondria.

(c) **Ascending limb:** It is lined by cuboidal or low columnar epithelium. The cells increase in height and width as the nephron leaves the hair pin turn and approaches the distal convolution. There is simultaneous increase in the number of mitochondria, which eventually become quite densely packed.

The loop of Henle is closely related with the vasa recta and the collecting tubules throughout its course. This is part of the counter-current multiplier system which is essential for the production of hypertonic urine.

4. Distal convoluted tubule: The thick ascending limb enters the medullary ray, passes into the cortex and becomes continuous with the distal convoluted tubule (DCT). The DCT is shorter and less coiled than the PCT. The cells are more cuboidal, less eosinophilic, smaller and don't have a brush border like the cells of the PCT. Microvilli are present on the cell surface. A large number of microvesicles are found in the luminal aspect of most cells. The nucleus occupies the main part of the cell and the mitochondria are few and spherical. Relatively few sub-microscopic ribonucleoprotein particles are present. The DCT finally joins the collecting tubule which enters the medullary ray to join the straight tubule. The latter then passes into the pyramid. When the distal convolution approaches the

vascular pole of the glomerulus, the afferent arteriole and the DCT are in close contact with each other for a short distance. At this point, the cells of the tubule are more closely aggregated than in the other regions. Collections of these cells are known as the *macula densa*, because the cells are narrow and their nuclei are closely opposed to each other. The cells of the macula densa cannot be well differentiated in their ultra structure from the cells of the DCT. In the same region, the wall of the afferent arteriole is thickened by the presence of epithelioid cells which contain granules. These granule-containing cells form the *juxtaglomerular complex*, which contain renin. The efferent arterioles in this region may also contain similar granules. The highest concentration of renin is found in the superficial layers of the renal cortex. The structural organization of the juxtaglomerular apparatus suggests a regulatory function. The macula densa senses the change in composition (sodium and chloride concentration) of the distal urine and this information is used to adjust the tone of the glomerular arterioles, thereby producing a change in glomerular blood flow and filtration rate. This is known as the '*tubular glomerular feedback*' mechanism. This system also determines the amount of renin that is released into the circulation, thereby acquiring great systemic relevance.

5. Cortical collecting tubule establishes continuity between the DCT and the collecting tubule. It has two types of cells—intercalated cells and principal cells. The intercalated or dark cell has a large number of spherical mitochondria clustered above the nucleus. The luminal surface has numerous microvilli and the apical cytoplasm has numerous microvesicles. There are also a lot of ribonucleoprotein particles. The nucleus is large, occupying two-thirds of the cell. The principal or the light cell has a light cytoplasm with few mitochondria, scattered ribonucleoprotein particles, and absent microvilli and vesicles.

6. Collecting tubule: In the medulla, the straight collecting tubules unite to form the papillary ducts of Bellini. These open onto the renal papilla in the area cribrosa. The cells of the collecting tubules are cuboidal or low columnar without any brush border and possess an agranular, faintly eosinophilic cytoplasm. There are few small, scattered mitochondria. The luminal surface has a variable number of very short microvilli and the basal plasma membrane shows short and narrow infolding. The nuclei are large, and occupy a significant portion of the cell.

7. Papillary duct is lined by columnar cells with basal nuclei. There are few short, coarse villi on the luminal surface. The basal cell surface is smooth, and the lateral surface is thrown into short, irregular folds. Numerous desmosomes are present along the lateral cell border, securing a firm intercellular attachment. There are few, widely scattered mitochondria. Dense lipid granules are present in the cytoplasm.

C. Correlation of Ultra Structure with Function

1. Proximal convoluted tubule is adapted for bulk reabsorption by active process. The microvilli (brush border) on the apical surface, provides a large absorptive area; the basolateral membrane is thrown into folds that similarly enhance surface area. The cells are rich in mitochondria (concentrated near the basolateral membrane) and lysosomal vacuoles, and the tight junctions between adjacent cells are relatively leaky.

The proximal tubule is responsible for the bulk of sodium, potassium, chloride and bicarbonate reabsorption, and almost complete reabsorption of glucose, amino acids, and low-molecular-weight proteins. Most other filtered solutes are also reabsorbed to some extent in the proximal tubule (e.g., ~60% of calcium, ~80% of phosphate, ~50% of urea). The proximal tubule is highly permeable to water and most filtered water (~65%) is also reabsorbed at this site.

Thus, no quantitatively significant osmotic gradient can be established.

2. Loop of Henle: Very little active reabsorption takes place in the descending limb and the thin segment, because of the few brush border extensions and mitochondria, limited oxidative enzymes and lack of basal infolding. The structure of the ascending limb helps in the reabsorption of small amounts of fluid, due to the presence of few short microvilli. The abundance of microvesicles in the apical part of the cells with a possible movement towards the surface presents evidence for active secretion. Ammonia is synthesized and secreted in this part of the nephron and acidification of urine also takes place here. Enzymes necessary for deamination of amino acids are abundantly found in the mitochondria of the ascending limb.

The thick ascending limb continues reabsorption of solutes (Na^+, Cl^-, K^+, Ca^{2+}, Mg^{2+}) and is responsible for the kidney's ability to generate a concentrated or dilute urine.

3. Distal convoluted tubule: Functionally, it is responsible for the isotonicity of the tubular fluid. This is mediated by active uptake of water from the tubular fluid (ADH dependent) with the help of the mitochondrial enzymes present in this region. The presence of numerous microvesicles is also evidence of ammonia synthesis.

4. Collecting tubule and papillary duct: The principal cell of the cortical collecting duct is responsible for sodium reabsorption and potassium secretion. The other cells in the late distal tubule and cortical collecting duct, the intercalated cells, are responsible for secretion of hydrogen (by α-intercalated cells) or bicarbonate (by β-intercalated cells) into the final urine. In the medullary collecting duct, there is a gradual transition in the epithelium. There are fewer and fewer intercalated cells while the "principal cells" are modified; although they reabsorb sodium, they have no apical potassium channels and therefore do not secrete K+.

D. Renal Blood Supply

Blood flow to the two kidneys is about 22% of the cardiac output (1100 ml/min). Thus an amount of blood equal to the total blood volume passes through the renal circulation in 4-5 minutes time. As such, 660 ml of the plasma passes through the kidneys per minute, of which 120 ml of fluid gets filtered through the glomerular membrane. The renal artery enters the kidney through the hilum and then branches progressively to form the interlobar arteries, arcuate arteries, interlobular arteries, and afferent arterioles, which lead to the glomerular capillaries, where glomerular filtration of the plasma takes place. The distal ends of the capillaries of each glomerulus coalesce to form the efferent arteriole, which leads to a secondary capillary network, the peritubular capillaries, which surround the renal tubules.

The renal circulation is unique as it has two capillary beds—glomerular and peritubular. These are arranged in series and are separated by the efferent arterioles which help regulate the hydrostatic pressure in both capillary beds. High hydrostatic pressure in the glomerular capillaries (~60mmHg) causes rapid fluid filtration, whereas lower pressure in the peritubular capillaries (~13 mmHg) permits rapid fluid reabsorption. By adjusting the resistances of the afferent and efferent arterioles, the kidneys can regulate the hydrostatic pressures in both glomerular and peritubular capillaries, thereby changing the rate of glomerular filtration and/or tubular reabsorption in response to body homeostatic demands.

The peritubular capillaries empty into the venous system, which run parallel to the arteriolar system and progressively form the interlobular vein, arcuate vein, interlobar vein and renal vein, which leaves the kidney beside the renal artery and ureter (Figure 34.3).

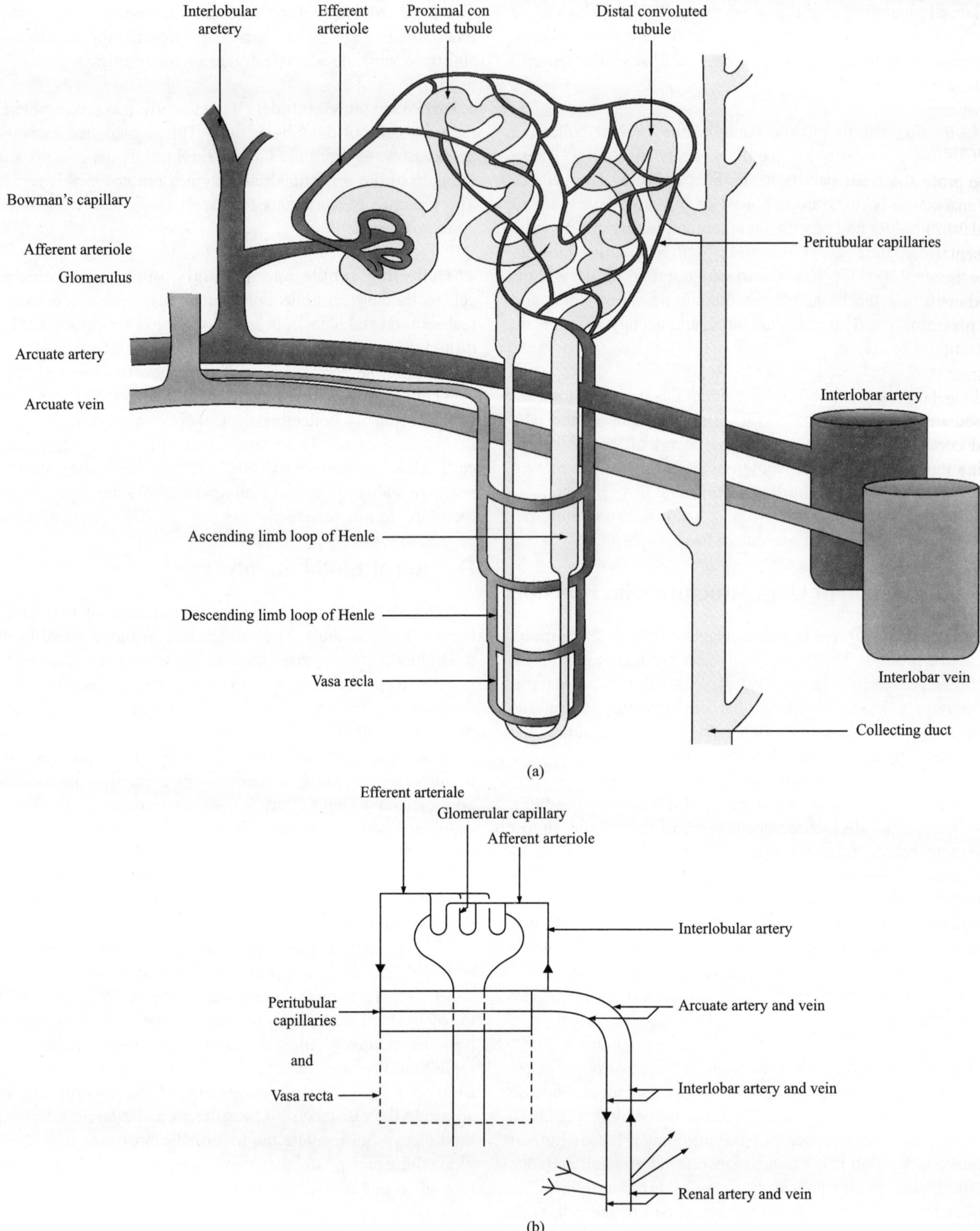

Figure 34.3 Blood supply of the nephron. (*see Plate 6 for colour figure*)

III. FUNCTIONS OF KIDNEY

As mentioned above, of the many functions performed by the kidney, the formation and excretion of urine is probably the most important.

Formation of urine

The process of urine formation will be better understood with an analysis of the composition of plasma and urine (Table 34.1 and 34.2). It will be evident that protein and glucose are totally absent from urine and other substances like urea, uric acid, creatinine and ammonia are present in very high concentrations. Sodium and calcium are present almost in same proportions in plasma and urine. As blood comes to the glomerulus, it gets filtered through the Bowman's capsule. As the cell-free filtrate passes down the renal tubules, the composition of the fluid gets altered. Certain substances are reabsorbed into the bloodstream to maintain optimum concentration in the blood, and certain others that need to be excreted are added to the urine through the process of secretion.

A. Glomerular Filtration

The composition of glomerular filtrate obtained by micro puncture is similar to plasma without proteins; the concentrations of sodium, chloride, potassium, glucose and urea are almost identical to those in plasma. Substances which have larger molecular size than the glomerular membrane pores cannot pass through the barrier, and hence do not appear in urine. Solutions of polysaccharide dextran containing molecules of different sizes, can be prepared and allowed to pass through the glomerular membrane. Molecules of 3 nm diameter are as freely permeable as water, whereas those above 14 nm in diameter are impermeable. Substances with diameter 3-14 nm have variable clearance. It is clear from this that the glomerular membrane has a variable pore size.

Normally, plasma proteins are not allowed through the glomerular membrane. However, small quantities (100 mg/day) can be detected in urine normally. This shows that such protein which passes through the glomerular membrane is reabsorbed by the tubules. The presence of abnormal protein in urine (proteinuria) is a common finding in renal disease. More proteins leak through the glomerulus due to defective filtration in the diseased state. Some degree of proteinuria also occurs in conditions where the hydrostatic pressure in the glomerulus is increased, such as congestive cardiac failure. Orthostatic proteinuria, a physiological condition, occurs in a small proportion of young people while maintaining an erect posture, and disappears on lying down. This is proposed to be due to increased pressure in the renal veins. The permeability of the filtering membrane is increased in many abnormal conditions, such as inadequate blood supply (as in circulatory failure), anoxia and action of toxic agents including drugs and certain bacterial poisons. The normal epithelium retains serum globulin

TABLE 34.1   Composition of plasma, glomerular filtrate and urine

Constituent	Concentration in plasma (mg/100ml)	Total in 170 litres of glomerular filtrate	Total excreted in 24 hours (urine)	Total reabsorbed in tubules
Water	–	170 L	1.5 L	168.5 L
Protein	8 g	none	none	--
Glucose	100	170 g	none	170 g
HCO_3^-	150	255 g	0.1 g	255 g
Na^+	330	560 g	5 g	555 g
Cl^-	365	620 g	9 g	611 g
K^+	17	29 g	2.2 g	26.8 g
Phosphate	3	5.1 g	1.2 g	3.9 g
Calcium	10	17 g	0.2 g	16.8 g
Urea	30	51 g	30 g	21 g
Uric acid	0.3	–	30 g	–
HSO_4^-	2	3.4 g	2.7 g	0.7 g

TABLE 34.2   Composition of plasma and urine

Constituent	Concentration in plasma (g/100ml)	Concentration in urine (g/100ml)	Ratio Urinary conc / Plasma conc
Urea	0.03	2	60
Uric acid	0.003	0.045	15
Creatinine	0.001	0.1	100
Ammonia	0.0001	0.05	500

(mol. wt. 170,000) and serum albumin (mol. wt 70,000), but lets through substances of smaller molecular weight eg. injected hemoglobin (mol. wt 68,000), egg albumin, Bence Jones protein or gelatin (mol. wt 35,000). In disease, the membrane permeability increases; serum albumin escapes and appear in the urine (albuminuria). With graver injury to the Malphigian corpuscles, both red and white blood cells may pass through, or frank hematuria may occur.

The driving force or the energy for filtration in the glomerulus comes from the hydrostatic pressure of the blood. As the renal artery is a direct branch of the abdominal aorta, the net pressure exerted in the glomerular capillaries is ~75 mmHg and is higher than in the capillaries elsewhere. It is opposed by the osmotic pressure exerted by the plasma proteins (~30 mmHg), interstitial pressure (~10 mmHg) and a renal intratubular pressure (~10 mmHg). Thus, the effective force under which filtration takes place is ~ 25 mmHg; termed the effective filtration pressure. As the opposing force is 50 mmHg, the hydrostatic pressure of blood must be more than this to effect filtration in the glomerulus. If the blood pressure is lower than the opposing force, the formation of urine will be suppressed. This phenomenon is noted in situation of shock where systolic blood pressure falls. Obstruction to filtration may also occur when interstitial pressure increases, as in inflammatory processes. Increased resistance to flow in tubular systems of the excretory channels, such as obstruction in collecting tubules, ureters or urethra, will affect formation of urine to a greater extent.

B. Tubular Handling of Substances

The concentrations of the constituents of glomerular filtrate are very different from that of urine. As mentioned earlier, composition of glomerular filtrate is similar to plasma minus the protein. The alterations in the composition of the filtrate occur as it passes through the tubules. It may be either reabsorption or secretion. Substances like glucose, amino acids, chloride and water, which are essential for the body but filtered from the glomerulus, are reabsorbed. Substances like urea, creatinine and uric acid, which are to be eliminated, are excreted into the urine. Some of these substances are also secreted to a certain extent.

1. Threshold substances

Substances which are reabsorbed completely by the tubules when their concentration in plasma remains within normal range, and appear in urine when normal levels are exceeded, are known as threshold substances. Substances which are reabsorbed slowly or not at all are low threshold substances eg. urea, creatinine, uric acid. Those substances which are essential for the body and are completely reabsorbed are known as high threshold substances eg. glucose, amino acids. A correlation between the ultra structures of different segments of the nephron with their function has already been described. It has been stated that the presence of the brush border and abundant mitochondria in the PCT help in active reabsorption of various substances. Water, however, is passively reabsorbed.

In the proximal tubule, two-thirds of the water and sodium are reabsorbed. Sodium is reabsorbed actively and water passively. As equivalent amounts of sodium and water are reabsorbed, the tubular fluid remains iso-osmotic with plasma. The other substances are reabsorbed in unequal proportions and at different situations. The first part of the proximal tubule reabsorbs amino acids, glucose and small amounts of protein which are filtered at the glomerulus. To a large extent, reabsorption of phosphate also occurs in this area. Sodium chloride and bicarbonate are reabsorbed uniformly along the length of the proximal and distal tubule, whereas potassium and uric acid are reabsorbed in the proximal, but secreted into the distal tubule.

2. Glucose

Glucose filtered in the glomerulus is completely reabsorbed in the PCT by phosphorylation. At blood sugar concentration of 100 mg/dl, 120 mg of glucose will be filtered per minute through the glomerulus as the glomerular filtration rate (GFR) is 120 ml/min. This amount of glucose is completely reabsorbed from the filtrate. As the glucose concentration in the filtrate is increased by raising the blood sugar levels, the tubules continue absorbing glucose from the filtrate till a limit is reached when excess glucose cannot be absorbed completely and consequently will spill over into the urine. This is because the tubules can absorb glucose only up to a certain limit depending upon the plasma concentration of glucose and glomerular filtration rate. The critical level of blood sugar at which saturation of absorption by the tubular cells takes place is known to be 180 mg/dl and is termed as *renal threshold* for glucose. With a blood sugar level of 180 mg/dl and GFR of 120 ml/min, 216 mg of glucose is filtered per minute and that saturates tubular absorption in man. The threshold concentration is not fixed but varies inversely with the GFR. A patient of diabetes mellitus with dehydration and GFR of 30 ml/min will need a blood sugar level of 720 mg/dl to have 216 mg glucose in his glomerular filtrate per minute. Hence sugar will appear in urine only above a blood sugar level of 720 mg/dl. At levels below this, despite blood sugars remaining very high there will be no glycosuria. This is a paradox of hyperglycemia without glycosuria.

Another situation arises when the concentration of glucose in blood is high enough to saturate the reabsorptive processes in all the tubules. In such circumstances, the rate of reabsorption becomes maximum and then constant. This is known as the maximal tubular reabsorptive capacity for glucose (TmG). This measures the total mass of functioning tubular cells and has been determined to be 350 mg/min. The phenomenon of tubular reabsorption maximum can be explained on the basis of a carrier mechanism. The carrier combines with glucose in the inner side of the tubular cell and transports it to the outer side where glucose separates from the carrier, which is then free to transport more glucose. Like any other transport mechanism, this mechanism can carry only a limited number of glucose molecules at a given time. At its maximum, it becomes saturated.

If more glucose is filtered than can be reabsorbed in the tubules, the excess glucose which escapes reabsorption will be excreted in the urine. Under circumstances where blood sugar remains high, if the GFR is reduced, as occurs in hypertensive cases, less glucose is filtered per minute through the glomerulus and as this is less than TmG, is completely reabsorbed. Hence a paradoxical situation arises where there is no sugar in urine even if the blood sugar remains high. The reabsorptive capacity for glucose is raised above normal in hyperthyroidism and in diabetes mellitus. The reabsorption of glucose also changes during pregnancy. The reabsorption maxima is lowered; hence sugar appears in urine while blood sugar level remains normal.

3. Water

Though the glomerular filtrate is formed at the rate of 120 ml/min, urine is passed at a rate of only 1 ml/min. Rest of the water is reabsorbed from the tubules. Absorption of water occurs mostly in the proximal convoluted tubule, descending loop of Henle, distal convoluted and collecting tubules. While water is freely permeable in the proximal tubule and descending loop of Henle, reabsorption of water in the distal convoluted and collecting tubules is under the influence of ADH. Reabsorption of water can be classified into two categories: (a) Obligatory reabsorption, and (b) Facultative reabsorption.

(a) Obligatory reabsorption: Approximately 80% of the volume of glomerular filtrate is reabsorbed from the PCT. In spite of this, the fluid remains iso-osmotic with the original filtrate. This is because of passive reabsorption of water as a solvent for the actively reabsorbed solutes, such as sodium chloride and glucose. Thus "obligatory reabsorption" means the portion of water which is reabsorbed secondary to the reabsorption of solutes, independent of the water requirement of the body. In case reabsorption of water is hampered at this site, there will be diuresis. This can happen if excess amounts of solutes are excreted, as in diabetes mellitus. With the use of diuretics (e.g., thiazides, loop diuretics), reabsorption of sodium chloride is impaired and hence more water is lost with unabsorbed sodium chloride.

(b) Facultative reabsorption: The GFR in man is 120ml/min, though only 1/8th of this gains access to the distal convoluted tubule, i.e., 16 ml/min. The usual rate of urine formation is 1 ml/min. Thus reabsorption of 15 ml water per minute takes place in the distal and collecting tubules under the influence of anti-diuretic hormone (ADH). The secretion of ADH from the posterior pituitary can be temporarily abolished by drinking water (water diuresis). The rate of urine formation then approaches 16 ml/min, indicating that in these circumstances, little, if any, water is being reabsorbed by the distal and collecting tubular segments. Similarly, damage to the hypothalamus interferes with ADH formation and urine flow may reach 16 ml/min or about 24 liters per day, as occurs in diabetes insipidus.

In water diuresis, the epithelial wall of the distal and collecting tubule is impermeable to water in the absence of ADH. ADH increases the permeability of the tubules to water. The facultative reabsorption which occurs under the influence of ADH in the DCT, renders urine isotonic. Further, reabsorption of water in the collecting tubule renders urine hypertonic. In man, maximal hypertonicity of the urine leaving the collecting tubules is 1.4 osmoles/liter, i.e., almost 5 times the osmolar concentration of plasma (0.3 osmoles/liter). The normal osmolarity of plasma is largely attributable to the presence of electrolytes. It is maintained by the kidney by varying the volume as well as the osmolar concentration of urine. For example, an excess of water which tends to dilute the plasma and thus reduce osmotic pressure, produces a renal response which results in excretion of an increased volume of hypo-osmolar urine. Thus when the plasma osmotic pressure rises due to dehydration, more water is retained by decrease in urine volume. Variation on the osmolarity of plasma is detected by the osmoreceptors located in the anterior hypothalamus, which in turn either decreases or increases the secretion of ADH so as to increase or decrease diuresis according to the need of the body. Thus, increased water intake leads to dilution of plasma and decrease in its osmolarity. This leads to suppression of ADH causing increased urine volume and loss of water until the osmotic pressure is normalized. The reverse occurs if the osmolarity of plasma rises. With water deprivation, plasma becomes hypertonic and this stimulates ADH secretion. More water is reabsorbed as a result and urine volume is decreased.

ADH (vasopressin) controls the reabsorption of water in the distal and collecting tubules. It is a nonapeptide synthesized in specialized neurons of the supraoptic and paraventricular nuclei. It is transported from these nuclei to the posterior pituitary and released in response to increases in plasma osmolality and decreases in blood pressure as described above. Osmoreceptors are found in the hypothalamus, and there is also input to this region from arterial baroreceptors and atrial stretch receptors. The actions of vasopressin are mediated by three receptor subtypes: V1a, V1b, and V2 receptors. V1a receptors are found in vascular smooth muscle and are coupled to the phosphoinositol pathway; they cause an increase in intracellular calcium, resulting in contraction. V1b receptors are found in the anterior pituitary, where vasopressin modulates adrenocorticotropic hormone release. V2 receptors are found in the basolateral membrane of principal cells in the late distal tubule and the whole length of the collecting duct; they are coupled by a Gs protein to cyclic adenosine monophosphate generation, which ultimately leads to the insertion of water channels *(aquaporins)* into the apical membrane of this otherwise water-impermeable segment. The apical insertion of aquaporins allows reabsorption of water in the distal and collecting tubules.

A number of drugs act to suppress ADH secretion and this increases urine flow. Certain stresses like that of surgery or severe trauma as well as some anesthetic drugs can cause excessive ADH secretion. In the immediate post-operative period, ADH secretion remains high leading to oliguria and excessive water retention. Failure to recognize this has led to serious over-hydration of patients in an attempt to correct oliguria.

Besides absorption of water at different sites in the renal tubule, there is readjustment in the electrolyte concentrations in the loop of Henle leading to concentration of urine. Wirz (1956) postulated that the tubular filtrate becomes progressively more concentrated during its passage through the descending loop of Henle. Water without solute is reabsorbed from the descending limb, and sodium chloride without water is actively reabsorbed in the ascending limb into the surrounding interstitial fluid. This active reabsorption of sodium provides the driving force for the operation of this scheme. The presence of abundant mitochondria in the ascending limb also helps the process. Thus, in the descending limb urine gets concentrated due to reabsorption of water and in the ascending limb it gets diluted due to reabsorption of sodium chloride. This process of concentration and dilution in opposite segments of a loop, where flow of urine in one is opposite in direction to the other, has been designated as the counter-current multiplier system of Wirz.

The hair pin bend of the loop of Henle and close apposition of the two limbs help the counter-current exchange mechanism. As the water is reabsorbed in the descending limb, the concentration of sodium chloride rises and reaches a maximum at the bend of the loop. In the ascending limb as sodium is actively reabsorbed, its concentration gradually falls with the ascent of urine in the limb. The urine becomes progressively hypoosmotic and enters the DCT, where reabsorption of water takes place under the influence of ADH. Absorbed sodium makes the interstitial fluid hypertonic and from this the electrolyte is again passively reabsorbed into the descending limb, raising its concentration in ascending order from the proximal tubular end, towards the bend of the loop. A portion of sodium ion from the interstitial fluid is also absorbed into the blood vessels.

Hyperosmotic interstitium on the other hand extracts water from the water permeable descending limb. Thus, the descending limb of the loop contains fluid which is hypertonic as compared to plasma. As the fluid rounds the bend of the loop, the descending and ascending limbs contain fluid which is momentarily iso-osmotic. The osmotic pressure gradient between the loops is abolished due to the subsequent active secretion of sodium from the ascending limb, which sets up a new osmotic gradient, whereupon the tubular fluid descending to the tip becomes more hypertonic. A counter-current multiplier system is set up.

The counter-current multiplier system has various advantages. It prevents excessive diffusion of either sodium or urea from the tubules into the interstitial fluid without expenditure of energy because of the equivalent concentrations of both the substances in the interstitium and the ascending limb. A concentration gradient in the interstitium increasing from the cortico-medullary junction to the tip of the loop is established due to this system. This mechanism further helps to concentrate the urine by passive absorption of water from the water-permeable collecting ducts. The vasa recta, serving as counter-current exchangers, promote the overall efficiency by carrying the water absorbed from the urine through the ascending capillaries. The tubular fluid and the capillary blood also become progressively more concentrated as they approach the tip of the loops and subsequently become more dilute as they ascend towards the cortex. (Figure 34.4)

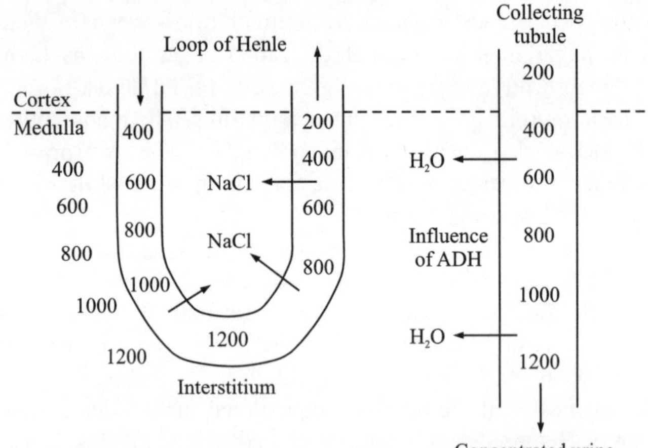

Figure 34.4 Counter-current exchange multiplier. (Numbers indicate osmolality in milliosmoles/L)

4. Ions

(a) Sodium: The sodium ion concentration does not alter throughout the proximal tubule segment; 560 g of sodium is filtered per day and since 7/8ths of water has been reabsorbed, therefore 490 g sodium must have been reabsorbed. The difference is that the sodium ion is reabsorbed actively while water is reabsorbed passively. It has been observed that with intravenous infusion of mannitol, an inert sugar which is filtered but not reabsorbed, a large osmotic effect is exerted in the proximal tubule and reabsorption of water is limited to a great extent. Hence, a large volume of urine is voided (osmotic diuresis). Only 1/3rd of water instead of 7/8ths is reabsorbed from the glomerular filtrate. Nevertheless, 75% of the sodium is reabsorbed and only 25% escapes in the urine.

Further, sodium is not reabsorbed alone; it is accompanied by anions, majorly chloride ion. Thus, reabsorption of chloride is passive and is secondary to reabsorption of sodium. When 620 g of chloride ion is filtered, 500 g is reabsorbed in the proximal tubule and 110 g in the distal tubule per day, leaving 10g to be voided in the urine.

70 g of sodium ion is presented to the loops of Henle and the distal tubules. As the average loss in the urine is normally 5 g/day, a large part is reabsorbed in these segments. In salt deprivation, the urinary concentration of sodium falls below the plasma level. In such circumstances, at least part of this reabsorption must be active. The distal reabsorption of sodium in under the influence of aldosterone. (Figure 34.5)

(b) Potassium: The normal clearance of potassium is about 100 ml/min. The glomerular filtrate contains about 5 mEq/L of

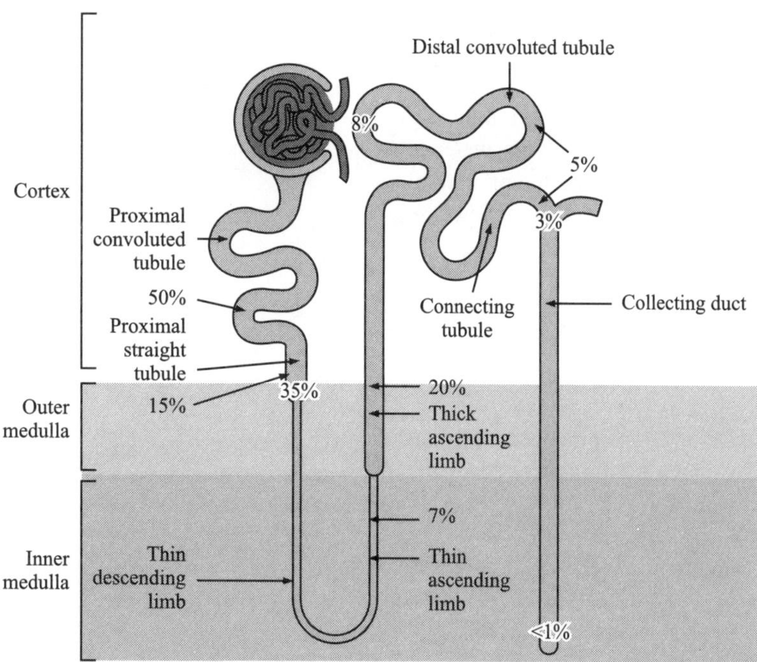

Figure 34.5 Renal sodium handling. (*see Plate 6 for colour figure*)

potassium ion. Clearance of potassium exceeds the GFR. This indicated that potassium is excreted both by filtration and by tubular secretion. Normally, the entire potassium ion which is filtered is reabsorbed in the PCT. The potassium which appears in the urine is added by the process of secretion in the DCT. Secretion of potassium is closely associated with hydrogen ion exchange and acid-base equilibrium. The capacity of the kidney to excrete potassium is so great that hyperkalemia does not normally occur, even after ingestion or intravenous injection of relatively large quantities of potassium. When kidney function is impaired, as in dehydration, potassium should not be administered until normal urination is resumed.

In chronic kidney disease, where glomerular filtration is reduced, blood urea is elevated but serum potassium remains normal or slightly increased. This is because urea is excreted only by filtration, but potassium is excreted by both filtration and secretion. Tubular secretion of potassium is seriously affected only in the later stages of kidney failure. Reabsorption of sodium and potassium ions is under the hormonal control of adrenal corticoids, particularly aldosterone. It stimulates reabsorption of sodium and urinary loss of potassium. The reverse takes place in Addison's disease, a condition of adrenocortical insufficiency. There occurs acute loss of sodium ion and retention of potassium in this disease, leading to further complications of dehydration and circulatory failure.

In congestive heart failure, there is retention of sodium due to reduction in GFR and increased venous pressure in the renal circuit. There is also some evidence of increased aldosterone secretion from the zona glomerulosa. In nephrotic syndrome also there may be increased aldosterone secretion. Aldosterone is secreted from the zona glomerulosa under the influence of angiotensin II. The rate of formation and liberation of renin (and hence formation of angiotensin II), is dependent upon the mean pressure and pulse pressure of the arterial blood. Low blood pressure as sensed through the juxta-glomerular apparatus increases renin output, and this increases the formation of angiotensin II, which in turn stimulates aldosterone secretion.

Other factors affecting urinary sodium excretion are – renal hemodynamics, compositional factors, neural and systemic hemodynamic factors. Brain natriuretic peptide also influences urinary sodium excretion in response to changes in volume status.

(c) Bicarbonate: The pH of the proximal tubular fluid remains same as that of the filtrate despite reabsorption of 7/8ths of the sodium and water. Bicarbonate concentration also remains fairly constant as the tubular pCO_2 does not differ widely from that of venous blood in any part of the nephron. This shows that 7/8ths of the bicarbonate must have been reabsorbed. The details will be discussed later.

(d) Phosphate: glomerular filtrate contains around 1mEg/L phosphate. Hence, 170mEq phosphate are filtered daily, of which 40mEq/l are excreted in urine. Phosphate reabsorption is restricted to the proximal tubule. As there is no change in pH of the proximal fluid, it is supposed that HPO_4 and H_2PO_4 are reabsorbed in the same ratio (4:1) as these ions exist in the plasma and glomerular filtrate.

5. Urea

Urea, the end product of protein metabolism, is expected to be excreted completely. However, only 20–60% of urea filtered at the glomerulus is excreted in the urine. The rest is reabsorbed by the renal tubules to be filtered again. The reabsorption of urea is a passive process and is secondary to the reabsorption of water. About 50% of the filtered urea is reabsorbed in the proximal tubules. In the loop of Henle, there is no significant change in urea concentration. As much urea returns to the tubular fluid in the descending limb as leaves it in the first portion of the ascending limb. It is in the collecting tubule that the largest amount of back diffusion occurs, if free water is greatly reabsorbed. Part of this urea on reaching the interstitial fluid is returned to blood and a part is carried back into the thick portion of the loop of Henle and the distal tubule.

IV. ROLE OF KIDNEY IN ACID BASE EQUILIBRIUM

The pH of blood remains constant at 7.4 in spite of the fact that different acidic metabolites are being continuously produced and added to it. The acids produced are

(a) From carbohydrate oxidation: lactic acid, pyruvic acid and CO_2

(b) From oxidation of fat: Ketoacids (acetoacetic acid and β-hydroxy butyric acid) and CO_2

(c) From protein oxidation: Sulphuric acid, phosphoric acid and CO_2

Different organs and tissues of the body try to combat any change in pH of blood through their individual action and by means of their coordinated function between one another (Figure 34.6).

The main systems which are working in this regard are:

A. Blood – buffers of plasma and red blood cells
B. Kidney – excretion of fixed acids and conservation of base
C. Lungs – elimination of CO_2
D. Intestine – Dilution

Buffers present in blood are mainly engaged in reacting with the incoming acids and converting them into their corresponding salts and weak acids. Thus the pH of blood is not affected. The main buffer systems are enumerated below:

A. Plasma
  (a) Bicarbonate   (b) Phosphate
  (c) Plasma proteins

B. RBC
  (a) Hemoglobin   (b) Bicarbonate
  (c) Phosphate   (d) Protein

While bicarbonate buffer contains a volatile aid (H_2CO_3) and operated via the lungs, phosphate buffer deals with fixed acids and functions through the kidney. The kidneys control acid-base balance by excreting either acidic or basic urine. Excreting an acidic urine reduces the amount of acid in extracellular fluid, whereas excreting a basic urine removes base from the extracellular fluids. (Figure 34.7)

The overall mechanism by which kidneys excrete acidic or basic urine is as follows: Large numbers of bicarbonate ions are

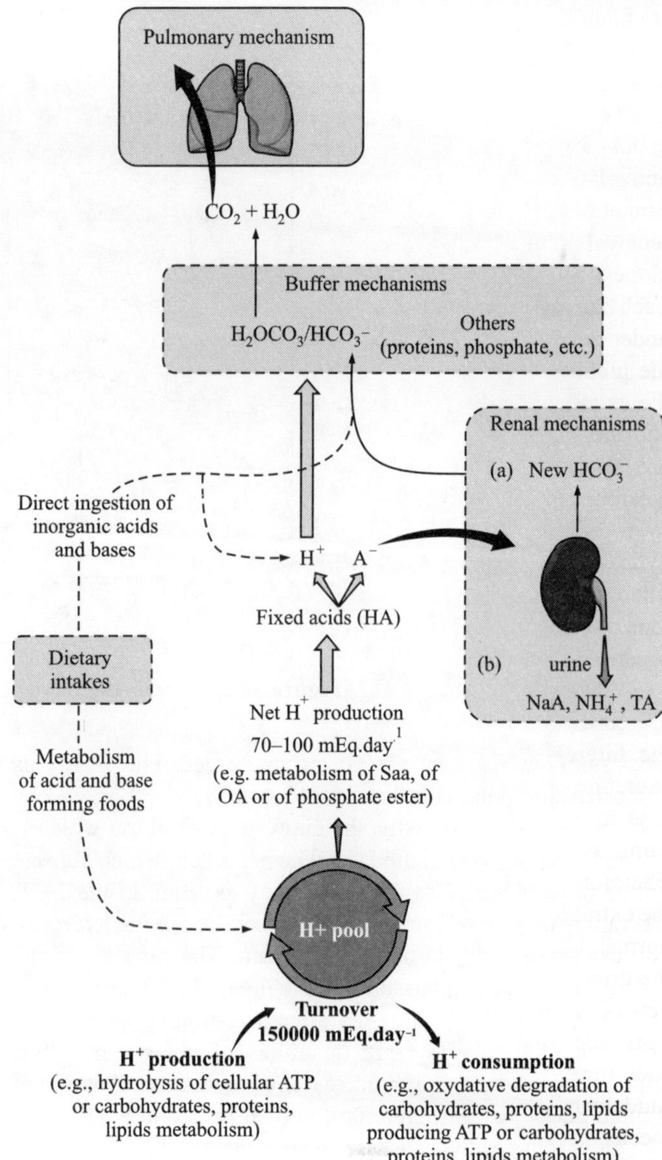

Figure 34.6 Organ systems in acid-base homeostasis.

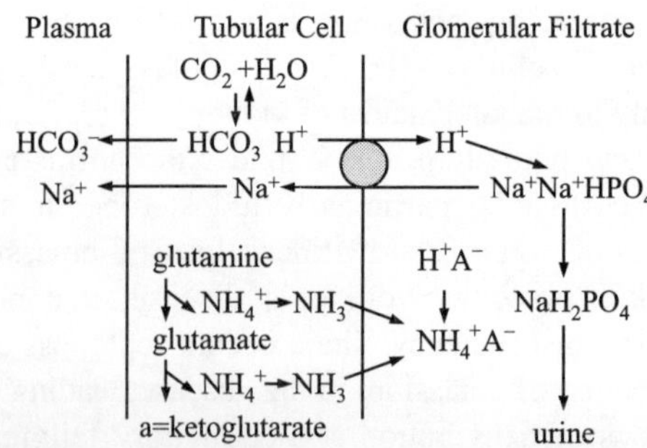

Figure 34.7 Buffer systems in the renal tubular cell.

filtered continuously into the tubules, and if they are excreted into the urine, this will remove base from the blood. Large numbers of hydrogen ions are also secreted into the tubular lumen by the tubular epithelial cells, thus removing acid ions the blood. If more hydrogen ions are secreted than bicarbonate ions are filtered, there will be a net loss of acid from the extracellular fluid. Conversely, if more bicarbonate ions are filtered than hydrogen ions are secreted, there will be net loss of base.

The body produces 80 mEq of nonvolatile acids per day, mainly form the metabolism of proteins. These acids are called nonvolatile because they are not carbonic acid, and therefore cannot be excreted from the lungs. The primary mechanism of removal of these acids from the body is by renal excretion. The kidneys must also prevent the loss of bicarbonate in the urine. Each day the kidneys filter about 4320 mEq of bicarbonate, and under normal conditions, almost all of this is reabsorbed from the tubules, thereby conserving the primary buffer system of the extracellular fluids. Both, the reabsorption of bicarbonate and the excretion of hydrogen ions are accomplished by the process of hydrogen ion secretion in the tubule. Because the bicarbonate ion must react with a secreted hydrogen ion to form carbonic acid before it can be reabsorbed, 4320 mEq of hydrogen ions must be secreted each day just to reabsorb the filtered bicarbonate. Then an additional 80 mEq of hydrogen ions must be secreted to rid the body of the nonvolatile acids produced each day.

When there is a reduction of the extracellular fluid hydrogen ion concentration (alkalosis), the kidneys fail to reabsorb all the filtered bicarbonate, thereby increasing the bicarbonate excretion. Because bicarbonate ions normally buffer hydrogen ions in the extracellular fluid, this loss of bicarbonate is the same as adding a hydrogen ion to the extracellular fluid. Therefore, in alkalosis, the removal of bicarbonate ions raises the extracellular fluid hydrogen ion concentration back towards normal. In acidosis, the kidneys do not excrete bicarbonate into the urine but reabsorb all the filtered bicarbonate and produce new bicarbonate, which is added back to the extracellular fluid. This reduces the extracellular fluid hydrogen ion concentration back towards normal. Thus, the kidneys regulate extracellular fluid hydrogen ion concentration through three fundamental mechanisms: (a) secretion of hydrogen ions, (b) reabsorption of filtered bicarbonate ions, and (c) production of new bicarbonate ions.

In the phosphate buffer as disodium hydrogen phosphate (Na_2HPO_4) passes through the renal tubules, one sodium ion is exchanged for hydrogen ion. As a result, sodium dihydrogen phosphate (NaH_2PO_4) formed in the tubule is excreted in urine and sodium ion is reabsorbed into blood. Each molecule of disodium hydrogen phosphate is excreted as sodium dihydrogen phosphate. The ratio of $Na_2HPO_4/NaH_2PO_4 = 4:1$. In every buffer unit there are 9 units of sodium ion ($4 \times 2 = 8$ in Na_2HPO_4 and one in NaH_2PO_4) in 5 moles of both Na_2HPO_4 and NaH_2PO_4. From 9 units of sodium, 5 are excreted as NaH_2PO_4 and 4 are conserved. The 5 units of sodium ions excreted are replenished from the diet.

Mechanism of conservation of sodium

As has already been stated, sodium is reabsorbed in the PCT, ascending loop of Henle and DCT. While reabsorption is obligatory in the PCT and active in the ascending limb of loop of Henle, it is controlled by aldosterone in the DCT. For conservation of sodium, hydrogen ion is exchanged in the tubular cells. The hydrogen ion comes from the ionization of carbonic acid formed from the combination of carbon dioxide and water in the presence of enzyme carbonic anhydrase. In the PCT the exchange of hydrogen ion proceeds first against sodium of sodium bicarbonate, as it is abundant. Sodium bicarbonate is formed inside the tubular cell and is reabsorbed into the blood and the carbonic acid that is formed inside the tubule decomposes into carbon dioxide and water. Carbon dioxide diffuses back into the tubular cells and is again utilized in the formation of carbonic acid. There occurs no change in the pH of urine in this part of the tubule after reabsorption of sodium bicarbonate.

At normal plasma levels of bicarbonate, about half the amount that is filtered at the glomerulus is reabsorbed in the PCT and the remainder in the distal tubule. In the distal tubule, secretion of hydrogen ion is much more than the bicarbonate ion available; hence after bicarbonate ions are used up, hydrogen phosphate is used as hydrogen ion acceptor. As fluid passes slower through the distal tubules than through the proximal tubules, the equilibrium regarding exchange is achieved. The exchange of sodium ion for hydrogen ion converts disodium hydrogen phosphate to sodium dihydrogen phosphate with consequent increase in acidity of urine and diminution in urinary pH.

In the DCT cells, potassium can also be exchanged for sodium. Normally both hydrogen and potassium ions can be exchanged for sodium. Under the circumstances when intracellular hydrogen ion concentration is high and that of potassium is low, as occurs in potassium deficient states, the distal tubular cells can secrete more hydrogen ion. There is excessive reabsorption of bicarbonate accompanied with excessive secretion of hydrogen ion causing urine to be more acidic. Administration of a quantity of sodium bicarbonate sufficiently lowers both intracellular and extracellular hydrogen ion concentration. As the hydrogen ion concentration of the renal tubular cell falls, more of potassium is exchanged for sodium instead of hydrogen. Hence in alkalosis more potassium will be excreted. Even slight hyperventilation (producing respiratory alkalosis) will raise potassium excretion. Administration of excess potassium salts produces a decline in hydrogen ion concentration within the cells and an increase (acidosis) in extracellular fluid accompanied by the excretion of alkaline urine, as more potassium ion is excreted in the urine instead of hydrogen. Conversely, potassium depletion is associated with the development of an alkalosis in the extracellular fluid an increase in hydrogen ion concentration within the cell. This is followed by increased excretion of hydrogen ion in urine and concurrent highly acidic urine even in the face of raised bicarbonate content in the plasma (paradoxic aciduria). These latter circumstances occur in patients on prolonged cortisone therapy or hypercortisolism. It also occurs in the case of those post-operative patients who are maintained with potassium-free fluid infusion.

The diuretic acetazolamide enhances excretion of potassium ion by reducing hydrogen ion concentration inside tubular cells by inhibiting the carbonic anhydrase enzyme.

The rate of renal exchange of potassium for sodium is regulated by the hormone aldosterone.

The third mechanism of elimination of hydrogen ions from the tubular cells involves ammonia generation. The distal tubular cells produce ammonia by the process of deamination of glutamine by glutaminase. Though 60% of the ammonia is derived from glutamine, the remainder comes from the α-amino nitrogen of other circulating amino acids (glycine, alanine, leucine and aspartic acid) by the action of amino acid oxidase. Dissolved ammonia is a proton acceptor, absorbing hydrogen ion from an aqueous environment to become ammonium ion.

Cells are permeable to ammonia but not to ammonium ion. Hence ammonia can diffuse from the tubular cell into the tubular fluid where it captures hydrogen ion as outlined above. The higher the concentration of hydrogen ion, the greater is the formation of non-diffusible ammonium ion which is consequently excreted. The acidic salts of ammonium ions are formed which are excreted in the urine. This mechanism is specifically helpful for conserving bases from salts like sodium chloride. Simple exchange of sodium ion with hydrogen ion will produce a strong acid like hydrochloric acid, but on the other hand, ammonium ion will produce ammonium chloride, which is comparatively a weak acid. It is also known that in acidosis and in conditions with low urine pH, more ammonia is formed in the renal tubular cells and vice versa.

Acidification and ammonia formation go hand in hand. The formation of ammonia in response to ingestion of acid is not abrupt. It takes 2–3 days before maximum quantity of ammonium ion is excreted. Similarly, after acidosis has receded, ammonia formation persists for a while. It is also known that the total amount of ammonium ion excreted in 24 hours is larger than that of titrable acidity. Daily excretion of ammonium ion at pH 6 of urine is 30 mEq and that of titrable acid is 25 mEq.

Influence of conservation of base or bicarbonate reabsorption

As increase in carbon dioxide tension will increase intra-cellular hydrogen ion concentration through accelerated formation of carbonic acid. Raised hydrogen ion concentration leads to bicarbonate reabsorption. Decreased carbon dioxide tension would act conversely. This mechanism operates in the case of respiratory acidosis or alkalosis. In respiratory acidosis, compensation is achieved by an increased formation of bicarbonate. In respiratory alkalosis, more bicarbonate is eliminated due to decreased reabsorption.

Both the mechanisms, reabsorption of sodium against excretion of hydrogen ion, and the ammonia mechanism, operate in collaboration with each other and with the respiratory mechanism of carbon dioxide elimination to maintain normal blood pH.

In compensatory acidosis, there is not only hyperventilation to wash out carbon dioxide but also increased reabsorption of sodium and increased ammonia formation in the tubular cells. As a result, highly acidic urine is formed. Conversely, in compensatory alkalosis, the body tries to retain carbon dioxide by depressing respiration. The kidney also accommodates by decreasing bicarbonate reabsorption and ammonia formation. On the whole, there is an attempt to retain hydrogen and eliminate base.

V. ROLE OF KIDNEY IN BLOOD PRESSURE CONTROL

The kidneys play a dominant role in long-term regulation of blood pressure by excreting variable amounts of sodium and water. When the extracellular fluid volume increases, the blood volume and arterial pressure rise. The rising pressure in turn has a direct effect to cause the kidneys to excrete the excess extracellular fluid, thus returning the blood pressure towards normal. Increase in the total body salt content is more likely to increase the arterial pressure than water content. This is because pure water is normally excreted by the kidneys almost as rapidly as it is ingested, but salt is not excreted so easily. As it accumulates in the body, salt indirectly increases the extracellular fluid volume. When there is excess salt in the body, the plasma osmolality increases, and this in turn stimulates the thirst center, making the person drink more water to dilute the extracellular salt to a normal concentration. The increase in osmolality also stimulates the hypothalamic-posterior pituitary axis to secrete more ADH. ADH causes the kidneys to reabsorb increased quantities of water, thereby increasing the extracellular volume.

Thus it is clear that the kidneys, by regulating salt and water excretion in the body, play a very important role in blood pressure control. Aside from controlling the blood pressure through changes in the extracellular fluid volume, the kidneys also have another powerful mechanism for controlling blood pressure—the renin-angiotensin system. Renin is synthesized and stored in an inactive form called prorenin in the juxtaglomerular cells of the kidneys. When the blood pressure falls, intrinsic reactions in the kidneys themselves cause the prorenin molecules to spilt and release renin. Most of the renin enters the systemic circulation, and a small amount remains in the local fluid of the kidney to initiate intrarenal functions.

Renin acts on a plasma protein called angiotensinogen to release angiotensin I. Angiotensin I has mild vasoconstrictor properties. Within a few seconds after formation of angiotensin I, it is converted to angiotensin II. This conversion occurs in the lungs, and is catalyzed by the angiotensin-converting enzyme. Angiotensin II is an extremely potent vasoconstrictor. It causes intense, rapid arteriolar vasoconstriction, thereby increasing the peripheral resistance and hence the blood pressure. It also causes mild venoconstriction which increases the venous return to the heart, and helps the heart to pump against the increasing pressure. Another mechanism by which angiotensin increases the arterial pressure is by acting directly on the kidneys to decrease the excretion of salt and water. It causes renal arteriolar constriction and stimulates aldosterone secretion. This slowly increases the extracellular fluid volume, which in turn increases the blood pressure. This long-term effect, acting through the extracellular fluid mechanism, is even more powerful than the acute vasoconstrictor mechanism.

VI. ROLE OF KIDNEY IN MINERAL METABOLISM

The kidneys play an important role in metabolism of mineral such as calcium, phosphate and magnesium.

(a) Calcium: The kidneys produce the active form of vitamin D, 1,25-dihydroxy vitamin D_3 (calcitriol) by hydroxylating this vitamin at the number '1' position. Calcitriol is essential for normal calcium deposition in bone and calcium reabsorption by the gastrointestinal tract.

The kidneys play a major role in the minute-by-minute regulation of calcium homeostasis; whereas the intestine and the skeleton ensure homeostasis in the mid and long term. The kidneys adjust blood calcium levels by modulating tubular calcium reabsorption in response to the body's needs. In the proximal tubule, most of the calcium is reabsorbed by convective flow (as for sodium and water). In the distal segments of the tubule, the transport mechanisms are more complex. States of excess volume delivery to the kidney, such as a high-sodium diet, diminish the concentration gradient between proximal tubule and peritubular capillary, reducing calcium absorption and increasing calcium in the urine. On the other hand, volume depletion states increase salt, water, and calcium reabsorption in the proximal tubule, exacerbating states of hypercalcemia. In the thick ascending loop, the transport of calcium is primarily passive by the paracellular route, depending on the electrical gradient.

Numerous factors control the glomerular filtration and tubular reabsorption of calcium. Elevated renal blood flow and glomerular filtration pressure lead to an increase in filtered load. Hypercalcemia also increases ultrafilterable calcium, whereas hypocalcemia decreases it. Parathormone (PTH), a hormone secreted by the parathyroid gland, increases calcium reabsorption in the distal nephron. However, PTH also induces hypercalcemia, and because of the increase in serum calcium, the excretion of filtered calcium is elevated overall. Metabolic and respiratory acidosis lead to hypercalciuria, respiratory acidosis through an increase in plasma calcium and metabolic acidosis through calcium release from bone and an inhibitory effect on tubular calcium reabsorption. Conversely, alkali ingestion reduces renal excretion of calcium. The enhancing effect of phosphate depletion on urinary calcium elimination can partly occur through changes in PTH and calcitriol secretion. Dietary factors modify urinary excretion of calcium, mostly by their effects on intestinal calcium absorption.

(b) Phosphate: Phosphate is freely filtered in the glomerulus and reabsorbed primarily in the proximal tubule of the kidney. Normally, the daily amount of phosphate excreted in urine equals that absorbed in the intestine, usually comprising 5% to 20% of the ultrafiltered phosphate load.

After passage through the glomerulus, part of the filtered phosphate load is recovered by the tubule, depending on the body's needs. The major fraction of phosphate is reabsorbed in the proximal convoluted tubule.

(c) Magnesium: Magnesium is mainly eliminated by the kidney. Losses through intestinal secretion and sweat are negligible under normal conditions. 80% of total plasma magnesium is filtered at the glomerulus, but only 5% of this filtered load is excreted in the urine. The major portion of filtered magnesium is reabsorbed by the renal tubules – 25% in the proximal tubule, 65% in the thick ascending loop of Henle, and 5% in the distal convoluted tubule. Magnesium transport in the thick ascending limb is primarily passive, through the paracellular route. An electrical, lumen-positive gradient is induced by sodium chloride reabsorption that creates the driving force required for the reabsorption of divalent cations such as magnesium and calcium. In the distal convoluted tubule and the connecting tubule, magnesium is reabsorbed through the transcellular route against an uphill electrochemical gradient.

Tubular magnesium transport is modulated by serum magnesium and calcium and extracellular fluid volume. An increase of plasma magnesium and calcium concentration results in a decrease in magnesium transport. Extracellular volume expansion produces a decrease in proximal tubular magnesium reabsorption, in parallel with that of sodium and calcium.

VII. RENAL FUNCTION IN THE INFANT

The presence of a large number of lobulations in the newborn kidney is indicative of immaturity. Unlike other essential organs like the heart and lungs, the kidney does not take up new functions immediately after birth. The organ matures rather slowly. The persistence of undifferentiated nephrogenic zone cells in premature babies gives evidence that new nephrons are formed in extrauterine life.

Clearance measurements in newborn infants show lower values than those in older children or adults, on the basis of surface area and body weight. By the first year of extrauterine life, the glomerular filtration gradually improves to almost adult levels on the basis of surface area. The infant has limited ability to concentrate urine. Hence, there is a tendency to a fixed urinary concentration of the organic excretory substances and electrolytes. Therefore, improved excretion of urea or creatinine can be accomplished only by increase in urinary output. Thus nitrogenous retention occurs even in mild dehydration and this can be corrected by parenteral fluid infusion. The osmotic content of the infant's urine is almost the same as plasma compared to adults in whom it is 3–4 times that of plasma.

Adjustment of body content of sodium and chloride of dehydrated infants is difficult. If glucose solutions are infused, sodium loss in the urine may be alarmingly high. On the other hand, infusion of normal saline may produce edema and/or hyponatremia. This is why frequent sodium and chloride analyses are necessary during the treatment of dehydration in infants.

VIII. KIDNEY IN THE ELDERLY

Aging is associated with a gradual decline in renal function associated with a progressive loss of nephrons, with glomerular and tubulointerstitial scarring. These changes begin in the fourth decade of life and accelerate between the fifth and sixth decades. There is a progressive decrease in GFR after the age of 40 years, approximately 10 ml per decade.

Aging is also associated with impaired sodium excretion of a salt load as well as defective conservation in the setting of sodium restriction. Proximal sodium reabsorption is increased in aging, whereas distal sodium reabsorption may be reduced. Water handling is also impaired in the aging kidney. Both concentration and dilution capacity are affected, and nocturia is common. There is a reduced maximal urinary osmolality and thirst response to hyperosmolality, which may predispose to dehydration and hypernatremia. The increase in urine osmolarity in response to antidiuretic hormone (vasopressin) is blunted compared with younger subjects. There is slower excretion of a water load, leading to an increased predisposition to hyponatremia.

Potassium excretion is impaired, probably because of a decrease in functioning tubules. Hyperkalemia occurs more frequently in elderly subjects treated with drugs that interfere with potassium excretion (such as potassium-sparing diuretics). Other factors contributing to the predisposition for hyperkalemia include decreased GFR and lower basal levels of aldosterone. Hypokalemia is also common because of renal or extrarenal losses.

Acid-base disturbances may occur in the elderly due to impaired distal tubular acidification. Aging patients do not excrete an acid load as effectively as younger subjects do. Also, assessment of renal disease in the elderly is difficult because serum creatinine is a less reliable indicator of renal function in the aging population. This is because creatinine generation reflects muscle mass, and muscle mass decreases with aging.

Thus, we can see that almost all aspects of renal function are impaired in the elderly. This has significant clinical implications. Given the limitations of glomerular and tubular function in elderly people, fluid and electrolyte management needs careful attention. Adverse renal effects of commonly used drugs such as NSAIDS and diuretics should also be kept in mind when treating an elderly patient.

IX. RENAL FUNCTION TESTS

Various renal function tests can be assessed in a group and are mentioned in brief below and described in detail in other sections.

Tests for excretory function

- Tests for excretory function include measurement of serum levels of nitrogenous waste products that are usually excreted by the kidney—urea, blood urea nitrogen and creatinine.
- The glomerular filtration rate (GFR) is difficult to measure directly. However, it can be estimated by measuring the urinary clearance of certain filtration markers. The most widely used marker is creatinine. Creatinine clearance can be measured from the urinary and serum creatinine concentrations
- *Estimated GFR:* GFR can also be estimated by various equations such as the Cockcroft Gault formula and the MDRD (Modified Diet in Renal Disease) equation. The normal GFR is approximately 130 ml/min per 1.73 m^2 for men and 120 ml/min per 1.73 m^2 for women, and is affected by various factors age, sex, body size, physical activity, diet, drugs, and physiologic states such as pregnancy.

Markers for tubular damage

- Low-molecular-weight plasma proteins are readily filtered by the glomerulus and subsequently reabsorbed by the proximal tubule. Thus, normally only small amounts of the filtered proteins appear in the urine. The urinary excretion of these proteins rises when proximal tubular reabsorption is impaired. Because there is no distal tubular reabsorption, measurement of urinary low-molecular-weight proteins (e.g., β_2-microglobulin) acts as a marker of proximal tubular damage.
- In addition to low molecular weight protein in urine, renal glycosuria, PO_4 fractional excretion are also some of the tests of tubular function done in clinical practice.

Urine examination

This is one of the basic tests to evaluate kidney and urinary tract disease. Urine is examined for:
- Physical properties – colour, turbidity, specific gravity.
- Chemical properties – pH, glucose, protein, ketones.
- Microscopy – red and white blood cells, casts, crystals, bacteria.

Imaging

Common imaging modalities useful in assessment of renal disease are:
- Plain radiograph – detection of renal size, radio-opaque renal calculi, etc.
- Ultrasound – one of the most commonly used tests to assess kidney location, contour and size, and to look for evidence of obstruction in the urinary tract.
- Pyelography – intravenous, retrograde or antegrade. Although the use of pyelography has significantly decreased now a days after availability of ultrasound and CT scan.
- CT and MRI – used to assess renal masses, renal calculi
- Nuclear scans – DTPA scan to assess renal function and GFR, DMSA scan to look for structural abnormalities in the kidney such as cortical scarring.

Kidney biopsy

Kidney biopsy assesses histopathological changes in kidney which are useful to make many structural diagnoses of kidney diseases. It is usually done in patients with nephrotic syndrome, proteinuria, hematuria, systemic disease with renal manifestation, unexplained renal dysfunction to know the exact nature and pathology of the disease.

35

Role of Kidney in Hypertension and Haematopoiesis

Rajendra Prasad Mathur

CONTENTS

I. ROLE OF KIDNEY IN HYPERTENSION

A. Introduction

The kidney facilitates the initial development and subsequent maintenance of primary (essential) hypertension and in turn suffers its consequences. Thus hypertension is both a cause and effect of kidney disease.

Approximately 90-95% of patients suffering from hypertension have primary (essential) hypertension, characterized by a sustained systolic pressure >140 mmHg and a diastolic BP of >90 mmHg. The diagnosis of essential hypertension depends on excluding the known (secondary) causes of hypertension.

Guyton and Coleman proposed around 40 years back, that an increase in arterial pressure produces a sustained increase in sodium and water excretion (renal pressure natriuresis and diueresis) leading to long term control of arterial pressure by regulating blood volume. Therefore, the role of kidney is central in the pathogenesis of hypertension. Impairment of excretory ability of the kidney in the presence of increase in blood volume, cardiac output and microvascular injury causes initial rise in BP. Subsequently, the hypertension is sustained by increased systemic vascular resistence and activation of renin-angiotensin-aldosterone system (RAAS). Endothelial dysfunction due to decreased production of nitric oxide and increased production of endothelin may also contribute to increased vascular tone.

However, the initial abnormality of the kidney need not be intrinsic in the development of hypertension. The other factors predisposing to the development of hypertension include: genetic (often polygenic), congenital low nephron number, environmental effects such as high fructose diet, hyperurecemia, high sodium low potassium diets. Obesity or metabolic syndrome with hyperglycemia, hyperinsulinemia may also cause endothelial dysfunction as a result of damage mediated by oxygen free radicals and decrease bioavailability of nitric oxide. Altered kidney function may also be the result of extra renal disturbances such as increase sympathetic nervous system activity or excessive formation of antinatriureteric hormones, e.g., aldosterone.

B. The Renal Body Fluid Volume Concept for long term regulation of Blood Pressure

According to this concept for the long-term regulation of arterial pressure, the extracellular fluid volume is determined by the balance between intake and excretion of salt and water by the kidneys. Intake of salt and water in excess of output leads to rise in extracellular fluid volume and BP. This results in increase in sodium excretion by the kidneys, decrease in extracellular fluid volume thus reducing venous return and cardiac output until BP returns to normal and fluid intake and output are balanced. On the other hand, if sodium output exceeds intake and extracellular fluid volume and BP fall below normal, the kidneys retain sodium and water till BP is restored to normal set-point. This set-point for long term BP control is the arterial pressure at which sodium and water intake and output are at equilibrium.

The key factor of this concept of salt and water regulation is pressure natriuresis/diueresis which is based on the effect of increase arterial pressure to raise sodium and water excretion by the kidneys. Various hormonal and neural control mechanisms can amplify or blunt the pressure natriuresis response. Maintenance of normal BP inspite of elevation of sodium intake is the result of decreased formation of antinatreuretic hormones and/or increased formation of natriuretic factors thus enhancing the effectiveness of pressure natriuresis mechanism to maintain sodium balance with little or no increase in arterial pressure. Conversely, excessive activity of antinatreuretic hormones and/or decreased formation of natriuretic factors can reduce the effectiveness of pressure natriuresis requiring greater increase in arterial pressure to maintain sodium and water balance resetting the set-point for long term control of BP (Figure 35.1).

Primary increase in total peripheral vascular resistance or cardiac output do not result in long term increase in arterial pressure and hypertension occurs only when some factor impairs the excretory function of the kidneys and shifts the relation between sodium excretion and arterial pressure towards higher level.

C. Maintenance of Sodium Balance in Hypertension: Role of Renal Pressure Natriuresis

In hypertensive states, rise in arterial pressure is a compensatory mechanism to maintain sodium balance in presence of a sodium retaining defect such as aldosterone excess or increased levels of angiotensin II. In such situations with excess levels of sodium retaining hormones, the pressure natriuresis mechanism is important to maintain the sodium balance.

D. Renal Pressure Natriuresis: Relationship with Renal Perfusion Pressure

Alteration in sodium excretion by kidneys in response to changes in renal perfusion pressure is mediated by changes in tubular reabsorption of sodium. A decrease in renal perfusion pressure leads to an increase in reabsorption of sodium by tubules resulting in reduced sodium excretion. On the other hand, increase in renal perfusion pressure leads to reduced sodium absorption and enhanced sodium excretion, the phenomenon called Renal Pressure Natriuresis.

This pressure natriuresis is mediated by intrarenal haemodynamic factors. Raised renal perfusion results in increase in medullary blood flow and renal interstitial hydrostatic pressure (RIHP) due to nitric oxide induced reduction in medullary vascular resistence (Figure 35.2). Preventing RIHP to increase in response to increase in renal perfusion pressure attenuates pressure natriuresis. Conversely, direct increase in RIHP results in marked reduction in sodium reabsorption in proximal tubules and increased sodium excretion.

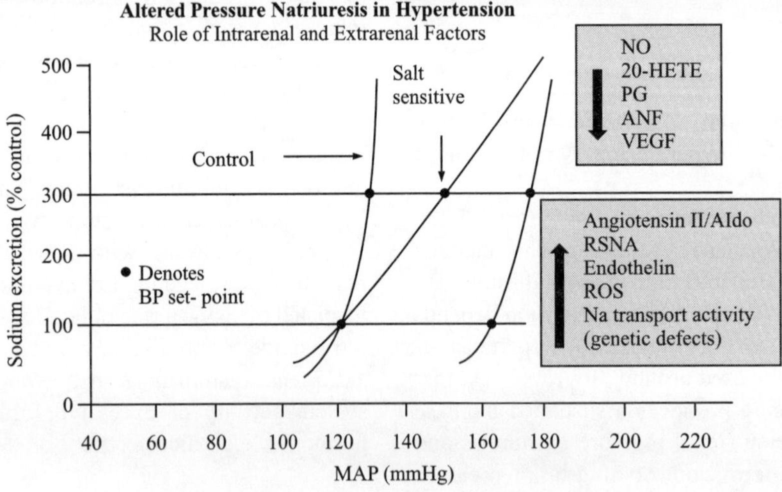

Figure 35.1 Steady-state relationships between arterial pressure and urinary sodium excretion and sodium intake for control subjects with normal kidneys and for subjects with a rightward hypertensive shift in the pressure-natriuresis relationship. A shift in the pressure natriuresis relationship can occur as a result of intrarenal abnormalities such as enhanced formation of angiotensin II (ANGII), reactive oxygen species (ROS), inflammatory cytokines and endothelin (via ET A receptor activation), or decreased synthesis of nitric oxide (NO), natriuretic prostanoids (PG), vascular endothelial growth factor (VEGF), or genetic defects that enhance renal sodium transport mechanisms. In other instances, the altered kidney function is caused by extrarenal disturbances, such as increased sympathetic nervous system (SNS) activity or excessive formation of antinatriuretic hormones such as aldosterone (Aldo) or decreased atrial natriuretic peptide (ANP).

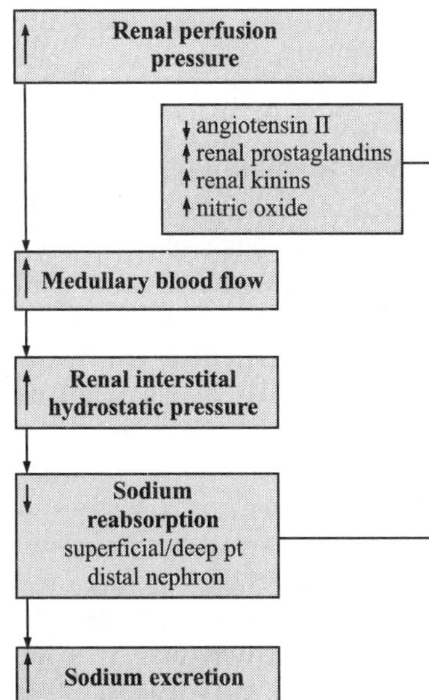

Figure 35.2 Mechanisms whereby increases in renal perfusion pressure enhance sodium excretion.

E. Impaired Renal Pressure Natriuresis in Hypertension

The kidney plays a central role in the genesis and maintenance of hypertension. This has been proved by renal cross-transplantation studies. Transplantation of a kidney from a hypertensive donor to a normotensive individual leads to sustained rise in BP in several genetic models of hypertension. The study by Curtis et al has shown that when hypertensive patients received kidneys from normotensive donors, their BP returned to normal levels. Several experimental studies have proved that abnormal kidney function is the primary factor in genesis of hypertension. Various experimental and clinical situations causing altered renal haemodynamics or tubular absorption, leading to reduction in excretion of sodium and water by kidneys lead to hypertension. Thus, the impaired renal pressure natriuresis is the abnormality that is found in almost all cases of primary or idiopathic hypertension.

The impaired renal pressure natriuresis may be the result of intrarenal factors such as excessive synthesis of angiotensin II, inflammatory cytokines, reactive oxygen species and endothelin or reduced synthesis of natriuretic prostanoids, nitric oxide and genetic defects affecting renal sodium transport. The extrarenal factors causing impaired sodium excretion include increased synthesis of antinatriuretic hormones like aldosterone or enhanced sympathetic nervous system activity.

Role of intrarenal factors

Renin-angiotesin-aldosterone system (RAAS): RAAS plays a crucial role in the pathogenesis of various forms of hypertension including essential hypertension and renovascular hypertension. RAAS exerts a significant effect on renal haemodynamics and sodium reabsorption through Angiotensin II (ANG II) via ANG II type I (AT 1) receptor (Figure 35.3).

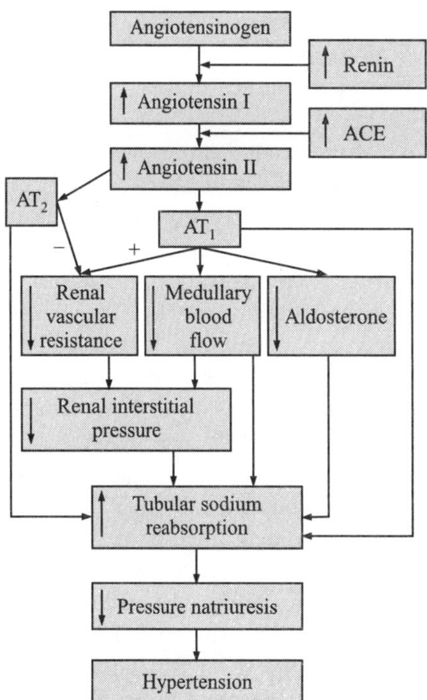

Figure 35.3 Renal mechanism whereby activation of the renin-angiotensin system reduces pressure natriuresis relationship and leads to hypertension.

ANG II also acts via AT 2 receptors to cause renal vasodilatation and natriuresis. RAAS (via AT 1 receptor) regulates sodium balance and a relatively normal BP as sodium intake is increased from low to high level. With high sodium intake ANG II levels are suppressed leading to enhanced sodium excretion. On the other hand if sodium intake is restricted ANG II levels rise and sodium excretion is reduced. Inability of kidneys to suppress RAAS in response to increase in sodium intake is an important mechanism for salt sensitive hypertension.

ANG II reduces Renal Pressure Natriuresis leading to hypertension as a result of its stimulating effect on tubular sodium transport to cause enhanced sodium reabsorption. The other effect of RAAS is to reduce peritubular capillary and renal interstitial hydrostatic pressure thereby affecting the tubular reabsorption of sodium. ANG II induced constriction of efferent arterioles of the juxtamedullary nephrons causes reduced medullary blood flow, which in turn increases sodium reabsorption by increasing medullary interstitial tonicity. ANG II also has a direct effect on proximal tubular transport of sodium due to activation of Na/H exchange.

Inflammatory cytokines: Several animal models of hypertensive animals have shown accumulation of inflammatory cells including macrophages and T cells in the kidneys. Plasma level of proinflammatory cytokines is associated with raised BP in experimental animal models as well as human hypertension. Study by Lee and coworkers has revealed that hypertension produced by ANG II excess may be partly dependant on the presence of interleukin 6 (IL-6). It was shown that during 2 weeks of ANG II infusion, the mice with IL-6 lock out had significantly lower BP than wild-type mice. However, the importance of inflammatory cytokines in human hypertension is yet to be proved by further studies.

Endothelin system: Endothelin-1 (ET-1) receptor binding sites are present throughout the body with largest numbers of receptors in kidney. ET-1 on activation of endothelin type A (ET_A) receptors can elicit antinatriuretc, prohypertensive response leading to renal vasoconstriction. Coversely, ET-1 acting via endothelin type B (ET_B) receptors cause antihypertensive and natriuretic response. Activation of renal ET_B receptor increases synthesis of nitric oxide and prostaglandin E2 and suppression of RAAS thereby, leading to decrease in sodium retention and lowering of BP.

In experimental hypertension, especially salt sensitive animal models, the significant role of ET-1 has been shown, however, its role in the pathogenesis of human hypertension remains to be proved. Krum et. al. reported that Bosentan, a nonselective ET receptor antagonist, reduces BP in patients with essential hypertension. Bosentan has been reported to block the effects of ANG II on BP and renal blood flow in rats. This indicates that a part of these ANG II actions may be mediated by endothelin.

Nitric oxide: Nitric Oxide (NO) is the primary endogenous vasodilator. Although the role of NO in the genesis and regulation is yet to be proven, several studies have demonstrated its effects on BP and renal haemodynamics. In healthy human subjects, inhibition of NO synthetase by N-monomethyl-L-arginine (L-NAME) results in acute increase in BP, peripheral vascular resistence and fractional excretion of sodium. These effects are partly mediated by interactions between NO and RAAS. NO is an important mediator of renal blood flow and renal microcirculation.

Reduction in NO synthesis leads to reduction in renal sodium excretion by a direct increase in renal vascular resistence or by enhancing the renal vascular responsiveness to vasoconstrictors like ANG II or norepinephrine (Figure 35.4). In this situation, reduction in renal medullary blood flow is an important mechanism leading to a hypertensive shift in renal

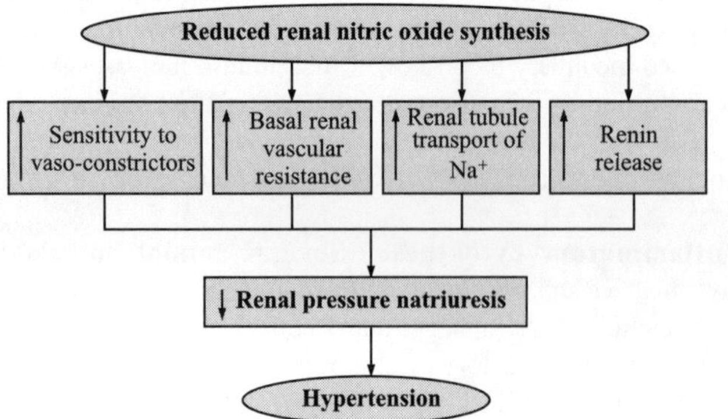

Figure 35.4   The renal effector mechanisms whereby reductions in NO synthesis decrease pressure natriuresis and increase blood pressure.

pressure natriuresis. NO also affects sodium reabsorption by its direct effect on tubular transport or its indirect effect on alteration of renal interstitial hydrostatic pressure.

Arachidonic acid metabolites: Arachidonic Acid is metabolized by cycloxygenase into prostaglandin (PG) G2 and subsequently to PGH2, which is further metabolized to PGs and thromboxane. Out of the many PGs produced by the kidney, PGE2 is the major renal PG mainly synthesized in medulla, which controls sodium excretion. Other arachidonic acid metabolites such as 20-HETE, prostacyclin and thromboxane may also affect renal pressure natriuresis and regulation of BP.

Non specific cyclooxygenase inhibitors do not affect sodium excretion or BP responses to chronic dietary sodium alterations indicating that renal PGs may not be playing an important role in sodium excretion during chronic dietary sodium alterations.

The role of renal PGs is important in the clinical situations with enhanced activity of RAAS. They protect the preglomerular vessels with ANG II induced vasoconstriction. If this protective mechanism is impaired, it would result in altered renal haemodynamics, reduced sodium excretion and hypertension.

COX-1 and COX-2 are the two distinct cycloxygenases. COX-2 inhibition reduces urinary sodium excretion resulting in hypertension. Nonsteroidal anti-inflammatory drugs (NSAIDs) in general and COX-2 inhibitors in particular lead to exacerbation of pre- existing hypertension, reduced renal blood flow and GFR. Increased COX-2 expression is seen in animal models associated with compromised renal haemodynamics and progressive kidney injury. Hence, NSAIDs may cause acute kidney injury in clinical states with compromised renal perfusion.

The other metabolites of arachidonic acid produced by cytochrome P450 (CYP) monooxygenase are 20-HETE and EETs. 20-HETE constricts renal arterioles and may have a role to play in autoregulation of renal blood flow and tubuloglomerular feed back mechanism. These metabolites also inhibit sodium reabsorption in the proximal renal tubule and thick ascending limb of loop of Henle. The renal production of these metabolites is altered in genetic and experimental models of hypertension.

Studies in humans suggest that impaired renal production of 20-HETE contributes to reduced renal pressure natriuresis in human hypertension especially associated with obesity.

Renal oxidative stress (ROS):   ROS produced by migrating inflammatory cells or vascular cells leads to endothelial dysfunction, impaired pressure natriuresis and hypertension by increasing renal vascular resistence, enhancing tubuloglomerulo feedback and increased renal tubular reabsorption of sodium thereby, reducing GFR. However, the importance of ROS in human hypertension is not yet established, and needs further studies.

Vascular endothelial growth factor (VEGF):   Although VEGF or VEGF-A is a known regulator of angiogenesis and arteriogenesis, various studies have also suggested its role in BP regulation. VEGF belongs to a family of secreted glycoproteins including VEGF-B, C, D and placenta growth factor (PlGF). Inhibitors of VEGF receptor signaling increase BP and reduce GFR as a result of endothelial dysfunction, reduced production of endothelium derives NO and PG and increased production of vasoconstrictors, e.g., thromboxane and endothelin. These changes ultimately lead to reduced renal blood flow and GFR and impair the kidney's ability to excrete sodium and water (Figure 35.5). How the macrophage depletion or inhibition of VEGF-C signaling leads to alteration in renal pressure natriuresis needs further study.

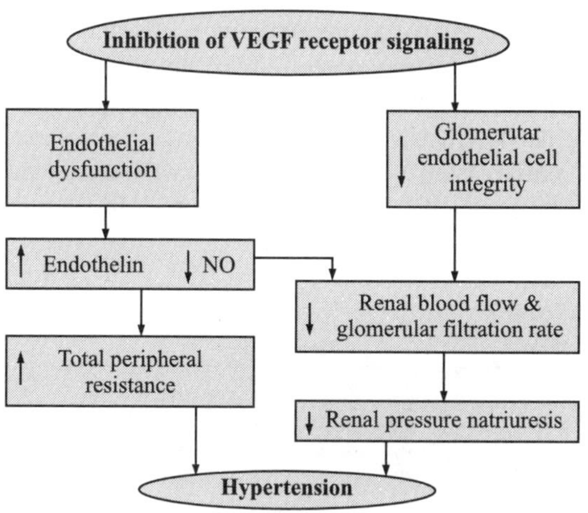

Figure 35.5 Potential mechanisms whereby inhibitors of vascular endothelial growth factor (VEGF) receptor signaling raise blood pressure.

Role of extrarenal factors

Aldosterone: Aldosterone increases sodium reabsorption in the the distal tubules and cortical collecting ducts, thereby, altering the renal pressure natriuresis relationship. The sodium retaining effect of aldosterone is the result of its binding to the intracellular mineralocorticoid receptor and activation of transcription by target genes which stimulate sodium-potassium ATPase pump on the basolateral epithelial membrane and activation of amiloride-sensitive sodium channels on the luminal side of the epithelial membrane.

Increased sodium intake suppresses aldosterone levels allowing sodium excretion by the kidneys. On the other hand if sodium intake is reduced aldosterone levels increase causing reduced sodium excretion. Inability to reduce aldosterone production in response to increased sodium intake is an important mechanism for salt sensitive hypertension in humans. Excessive activation of mineralocorticoid receptor by aldosterone is seen in primary hyperaldosteronism, renovascular hypertension, resistant hypertension and obesity related hypertension.

Sympathetic nervous system: The renal sympathetic nervous system (SNS) can reduce renal pressure natriuresis leading to chronic hypertension. Various studies have revealed increased SNS acivity in early hypertension and in the normotensive offsprings of hypertensive parents among whom a large number are likely to develop hypertension in future. Renal nerves act on the tubules to increase sodium reabsorption and also cause increased renal vascular resistance, reduced medullary blood flow and renal interstitial pressure. Increase in renal sympathetic nerve activity can enhance tubular sodium reabsorption by activating RAAS. Animal studies have demonstrated that renal denervation delays or attenuates the development of hypertension. Obesity related experimental hypertension has been linked to enhanced SNS activity.

The role of renal nerves in human hypertension is yet to be confirmed. The recent observation of radio frequency ablation of renal nerves in cases with resistant hypertension leading to significant reduction in BP is a compelling evidence in favour of importance of renal nerves in human hypertension.

Atrial natriuretic peptide: Atrial Natriuretic Peptide (ANP) is a 28 amino acid peptide released by atrial cardiocytes in response to stretch. ANP exerts an antihypertensive and natriuretic effect via its receptors (NPR-A). ANP enhances sodium excretion by intra-renal and extra-renal mechanisms. Deficiency in ANP production or a defect in its receptor leads to hypertension and reduced pressure natriuresis by increased tubular reabsorption of sodium either directly by enhancing active tubular transport of sodium or indirectly by reduction in medullary blood flow, increased renin release and activation of RAAS.

ANP decreases release of aldosterone by adrenal zona glomerulosa cells and reduces renin secretion and ANG II levels, thus contributing to long term effect on sodium balance and arterial pressure regulation.

Plasma levels of ANP increase in the states of blood volume expansion, saline loading and chronic increased intake of dietary salt. Several studies have suggested that genetic deficiencies in ANP or natriuretic receptor activity could play a role in salt sensitive hypertension.

II. ROLE OF KIDNEY IN HAEMATOPOIESIS

A. Introduction

The blood contains several different types of cells. Each of these cell types is quite distinct in appearance and has a specific biological function. Although there is extreme structural and functional difference among the various cells of the blood still, all the blood cells are the progeny of a single type of cell known as the Hematopoietic Stem Cell (HSC). The processes involved in the production of various cells of the blood from the HSCs are collectively called hematopoiesis. Hematopoiesis includes HSC self-renewal, HSC commitment to specific lineages, and maturation of lineage-committed progenitors into functional blood cells. During normal steady state, HSCs reside mainly in the marrow cavity, but under certain conditions HSCs can migrate and colonize other organs like liver and spleen in a process called extramedullary hematopoiesis.

B. Regulation of Haematopoiesis

The regulation of hematopoiesis is a complex process. Some regulatory factors influence overall hematopoiesis by affecting very early progenitor cells: HSCs and their progeny that have not yet undergone commitment to a single cell lineage. There are several growth factors each capable of supporting a particular spectrum of hematopoietic cell type. Specific regulatory growth factors play key roles in stimulating the production of cells in each lineage.

C. Growth Factors for Erythroid Cells

Erythropoietin (EPO), is the physiologic regulator of erythrocyte production which keeps the red cell mass within narrow limits. The major physiologic stimulus for RBC production is tissue hypoxia, primarily in the peritubular cells, located in the renal cortex. The production of EPO is regulated by the O_2 activity in the vicinity of these specialized EPO-producing cells in the kidney. By sensing O_2 activity, they essentially measure the oxygen delivery capacity of the blood, and adjust EPO production to achieve the number of erythrocytes needed for normal tissue O_2 tension. EPO acts on committed erythroid progenitors to support the later phases of erythroid differentiation. The regulation is achieved by EPO's action to modulate apoptosis of these progenitors (Figure 35.6).

The liver also contains specialized cells that can produce EPO in an oxygen-dependent pathway however, the contribution of the liver to total EPO production is much less than that of the kidney.

In the specialized kidney and liver cells, the transcription of the EPO gene is controlled by an oxygen-dependent transcription factor known as Hypoxia-Inducible Factor (HIF), that interacts with DNA sequences corresponding to the 3′ untranslated sequence of the mRNA and also with sequences in the EPO promoter region. HIF ubiquitination

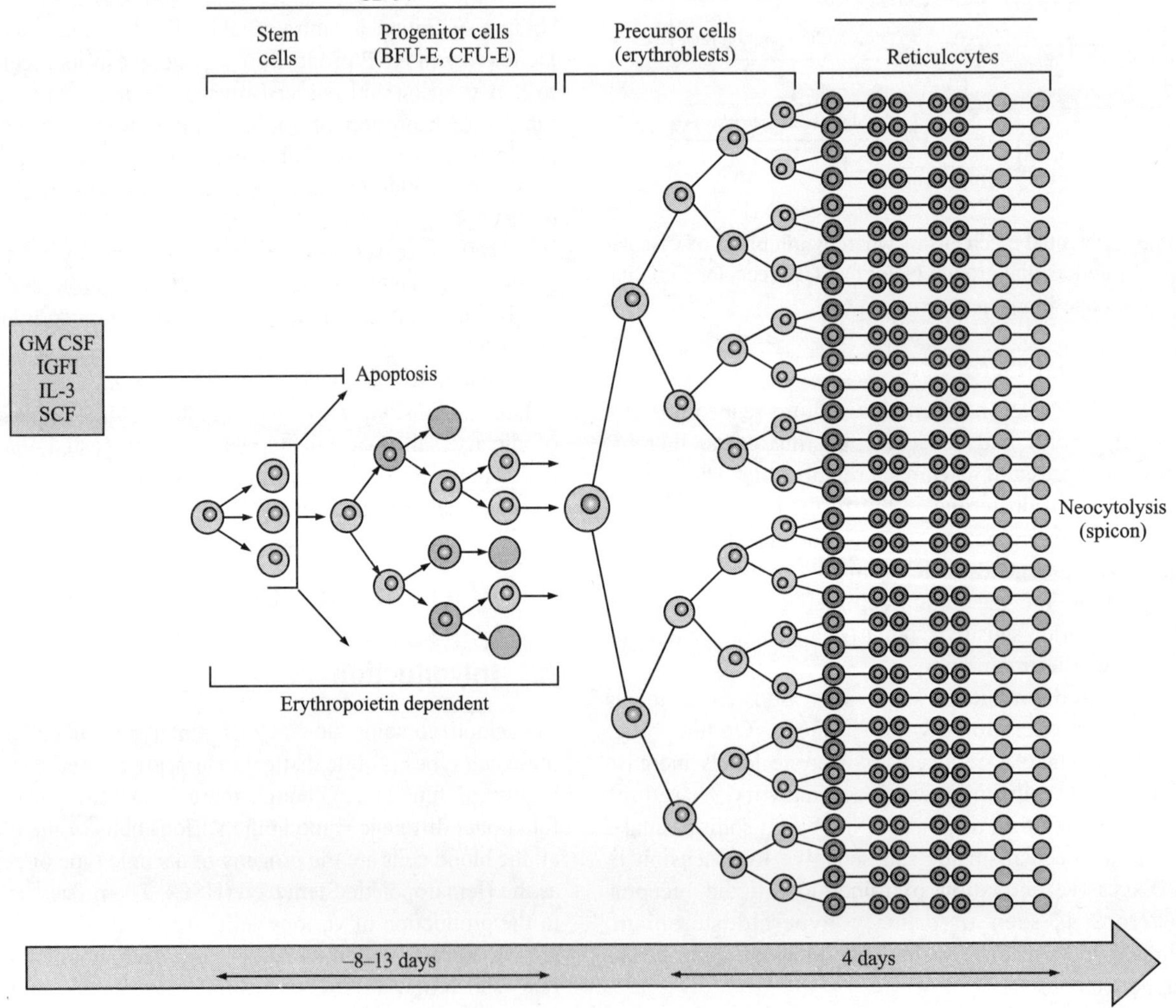

Figure 35.6 Normal erythropoiesis. The early stages of erythropoiesis, which require 8–13 days, involve the differentiation of the multipotent stem cell into the burst forming unit-erythroid (BFU-E) and colony forming unit-erythroid (CFU-E). These precursors require erythropoietin (EPO) and other growth factors to prevent apoptosis. The later stages of erythropoiesis, which require 4 days, involve the differentiation of erythroblasts to reticulocytes. Reticulocytes require EPO to avoid neocytolysis (premature destruction in the spleen).

and subsequent proteasomal degradation are dependent upon the hydroxylation of two specific prolines and an asparagine of HIF by nonheme, iron-containing hydroxylases that use molecular oxygen as a substrate for the reactions. With normoxia, HIF is rapidly hydroxylated and degraded. With hypoxia, HIF is not hydroxylated and degraded, but it forms part of a transcription complex that binds the 3′ enhancer sequence and induces EPO gene transcription. In addition, tissue specificity of expression in the kidney requires specific, cis-acting DNA sequences. EPO is secreted rapidly into the circulation, and in the bone marrow it binds to EPO receptors on erythroid progenitor cells in the CFU-E through early erythroblast stages. The EPO–EPO receptor interaction not only triggers signal transduction, but also leads to endocytosis and degradation of both the EPO and EPO receptor. This erythroid progenitor-mediated consumption of EPO appears to be a major determinant of the metabolic fate of EPO both in vitro and in vivo.

Regulation of apoptosis in late-stage progenitors, is an important function of EPO. EPO increases transcription of the bcl-x gene during late stages of erythroid differentiation, and also stimulates the Akt/protein B kinase pathway. In a model of erythropoiesis based on EPO preventing apoptosis in late-stage erythroid progenitors it has been demonstrated that the level of EPO ultimately controls erythrocyte production by regulating the number of dependent progenitors that survive or die.

The other growth factors involved in erythroid cell development are Kit Ligand (KL), IGF-1, and Granulocyte-Macrophage Colony-Stimulating Factor-1 (GM-CSF-1).

SUMMARY

Kidneys play a central role in the long term regulation of body fluid volume and arterial pressure. Renal pressure natriuresis is the main mechanism of renal-body fluid feedback control system to maintain long term BP regulation. Renal pressure natriuresis is abnormal in all forms of experimental and clinical hypertension. Impaired renal pressure natriuresis leading to chronic hypertension can be the result of factors leading to reduced GFR and/or increased renal tubular absorption of sodium. A shift of Renal pressure natriuresis may be due to intrarenal factors like enhanced formation of ANG II, ROS and endothelin or decreased synthesis of NO or natriuretic prostaglandins or it may be the result of extrarenal factors such as enhanced activity of SNS or excessive antinatriuretic hormones like aldosterone.

The kidney plays an important role in haematopoiesis by secretion of Erythropoietin by the peritubular cells of renal cortex. EPO acts on committed erythroid progenitors to support the later phases of erythroid differentiation.

SUGGESTIONS FOR FURTHER READING

1. Granger J.P. and George E.M. (2013), Role of the Kidneys in Hypertension, In: *Schrier's Diseases of the Kidney,* Vol. I, 9th ed., Lippincott Williams and Wilkins, 1086–1108.

2. Guyton A.C. and Coleman T.G. (1967), Long-term Regulation of the Circulation: Inter-relationship with Body Fluid Volumes. In: Reeve E. and Guyton A.C., (Eds.), *Physical Basis of Circulatory Transport: Regulation and Exchange*, Philadelphia: Saunders, 179–201.

3. Krum H., Viskoper R.J. and Lacourciere Y. (1998), The Effect of An Endothelin-receptor Antagonist, Bosentan, on Blood Pressure in Patients with Essential Hypertension, *N. Engl J. Med.*, 338:784–790.

4. Lee D.L., Sturgis L.C. and Labazi H. (2006), Angiotensin II Hypertension is Attenuated in Interleukin-6 Knockout Mice, *Am J. Physiol Heart Circ Physiol.*, 290:H935–H940.

5. Kassab S., Kato T. and Wilkins F.C. (1995), Renal Denervation Attenuates the Sodium Retention and Hypertension Associated with Obesity, *Hypertension*, 25:893–897.

6. Esler M.D., Krum H. and Sobotka P.A. (2010), Renal Sympathetic Denervation in Patients with Treatment-resistant Hypertension (The Simplicity HTN-2 Trial): A Randomised Controlled Trial, *The Lancet*, 376 (9756):1903–1909.

7. Cowley A.W. and Roman R.J. (1996), The Role of The Kidney in Hypertension, *JAMA*, 275:1581–1589.

8. Garvin J.L. and Ortiz P.A. (2003), The Role of Reactive Oxygen Species in the Regulation of Tubular Function, *Acta Physiol Scand*, 179: 225–232.

9. Cheng H.F. and Harris R.C. (2004), Cyclooxygenases, The Kidney, and Hypertension, *Hypertension*, 43: 525–530.

10. Knox F.G. and Granger J.P. (1992), Control of Sodium Excretion: An Integrative Approach, In: Windhager E. (Ed.), *Handbook of Renal Physiology*, New York: Oxford University Press, 927–967.

11. Granger J.P. and Alexander B.T. (2000), Abnormal Pressure Natriuresis in Hypertension: Role of Nitric Oxide. *Acta Physiologica Scandinavia*, 168:161–168.

12. Goddard J., Kohan D. and Pollock D. (2008), Role of Endothelin-1 in Clinical Hypertension: 20 years on, *Hypertension*, 52:452–459.

13. Curtis J.J., Luke R.G. and Dustan H.P. (1983), Remission of Hypertension After Renal Transplantation, *N. Engl J. Med.*, 309:1009–1015.

14. Hall J.E., Guyton A.C., and Brands M.W. (1996), Pressure-volume Regulation in Hypertension, *Kidney Int.*, 49 (55):S35–S41.

15. Navar L.G., Paul R.V. and Caramines P.K. (1986), Intrarenal Mechanisms Mediating Pressure Natriuresis: Role of Angiotensin and Prostaglandins, *Fed Proc.*, 45:2885–2891.

16. Koury M.J., Mahmud N. and Rhodes M.M. (2009), Origin and Development of Blood Cells, In: *Wintrobe's Clinical Haematology*, 12th ed., Lippincott Williams and Wilkins, 79–105.

36

Dental Hard Tissues and Pulp

O.P. Kharbanda, Jitendra Sharan and Sonal Punjani

CONTENTS

I. INTRODUCTION

Tooth is a hard calcified structure of human body which encases a core of soft pulp within. Dental pulp that extends from crown to root tip, is immediately surrounded by dentin. Tooth has a crown portion which has an outer most layer of enamel and a root portion that is lined with cementum (Figure 36.1). Tooth has a very interesting phylogenic history and they have been considered as the dermal skeletal in a wide range of jawed vertebrates (Dnoghue et al., 2002). Teeth develop from both the ectoderm covering the facial processes and the underlying mesenchymal cells which are derived from the cranial neural crest cells.

Humans possess two sets or generations of teeth (Diphyodont), primary or deciduous and permanent teeth. Deciduous dentition has 20 teeth located in two jaws (dental formula 212/212). First primary tooth erupts in neonate's mouth around 6 months of age, and by the two years of age, 20 deciduous teeth have erupted in occlusion. Rarely teeth may erupt in mouth at the time of birth or soon after birth. These are called neonatal teeth. Deciduous teeth start shedding with the eruption of permanent successors around 6 years of

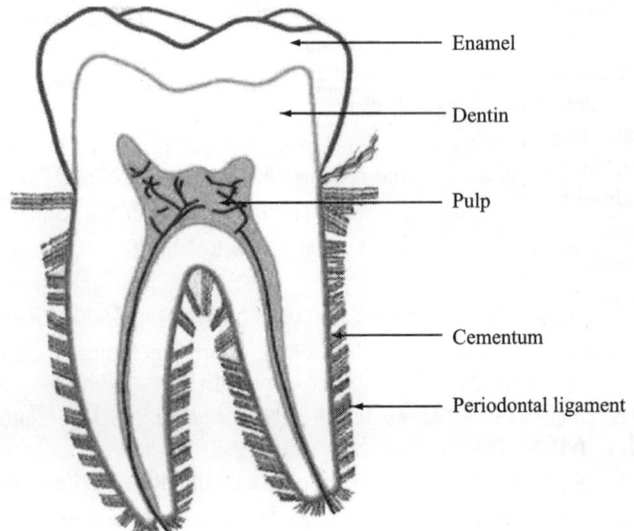

Figure 36.1 Components of tooth; enamel, dentin, cementum and pulp. Enamel, dentin and cementum are hard tissue component of tooth; on the other hand pulp is the only soft tissue component of tooth. (*see Plate 6 for colour figure*)

Numbering system	
Universal numbering system	In this system for the permanent maxillary dentition teeth are numbered through 1 to 16 beginning with upper right 3rd molars. The mandibular teeth are numbered through 17 to 32 beginning with lower left 3rd molar.

1	2	3	4	5	6	7	8	9	10	11	12	13	14	15	16
32	31	30	29	28	27	26	25	24	23	22	21	20	19	18	17

The Universal system notation for primary dentition utilizes upper case alphabets.

A	B	C	D	E	F	G	H	I	J
T	S	R	Q	P	O	N	M	L	K

Federation Dentaire Internationale (FDI) system

This is a two digit system that has been adopted by WHO. In this system the 1st digit indicates the quadrant and the 2nd digit indicates the tooth within the quadrant. 1 to 4 and 5 to 8 of the 1st digit indicates permanent and primary dentition respectively. 1 to 8 and 1 to 5 of the 2nd digit indicates permanent and primary teeth respectively.

Primary teeth —

55	54	53	52	51	61	62	63	64	65
85	84	83	82	81	71	72	73	74	75

Permanent teeth —

18	17	16	15	14	13	12	11	21	22	23	24	25	26	27	29
48	47	46	45	44	43	42	41	31	32	33	34	35	36	27	38

Palmer system

Primary dentition —

E	D	C	B	A	A	B	C	D	E
E	D	C	B	A	A	B	C	D	E

Permanent dentition—

1	2	3	4	5	6	7	8	9	10	11	12	13	14	15	16
32	31	30	29	28	27	26	25	24	23	22	21	20	19	18	17

age and the process goes on till 12–14 years. Dental formula for permanent dentition is 2123/2123.

Dentition in human beings can be categorized in three or four sub groups depending on the type of dentition. In deciduous dentition these are incisors, canines and molars; and for permanent dentition these are categorized into four groups, i.e., incisors, canines, premolars and molars. Numbering systems have been proposed to identify the tooth and dentition. Most commonly used systems are Universal numbering system, Palmer system and FDI system. Palmers system has traditionally been used conventionally by most dentists in clinical practice. Identification of tooth/teeth using computers necessitated development of a two digit identification system. The system was jointly developed by Federation Dentaire Internationale and World Health organization in the year 1971.

II. MORPHOGENESIS OF DENTAL STRUCTURES

Development of teeth and supporting structures is a result of a complex series of interactions between epithelium and the underlying mesenchymal tissues. At about 6th week of intra uterine life the ectodermal covering of oral cavity is 2–3 epithelial cells thick layer. This epithelial cell layer proliferates at the future location of alveolar process and forms the dental lamina (Figure 36.2a). The maxillary and mandibular dental lamina gives rise to 52 tooth buds. Out of this, 20 are primary teeth, which arise between the 6th and 8th week and remaining 32 forms the permanent teeth, which appear at later periods. Permanent dentition develops from successional tooth buds of their deciduous predecessors (this occur in utero at 5th month of age for the central incisor and 10 months of age for premolars). Distal growth of dental lamina will give rise to the 1st permanent molars at 4th prenatal months and 2nd permanent molar at 4th year of age (Tencate et al., 2005).

Tooth development is a continuous process and developmental stages are termed according to the shape of epithelial component of the tooth, i.e., dental lamina, bud, cap and bell stage (Tencate et al., 2005) (Figure 36.2).

In the bell stage of tooth development, enamel is secreted by the ameloblasts at a rate of approximately 4 micrometer/day. Studies have been shown that, for enamel to be secreted by ameloblasts, a layer of dentin is very much essential (i.e., laid down dentin acts as an inducer for ameloblasts to secrete enamel). At the outset, this involves differentiation of the cells of dental apparatus at the tips of the cusp outline which form the epithelial outline of the tooth, followed by deposition of dentin and enamel.

Amelogenesis is divided into three main functional stages (presecretory, secretory and maturational). Ameloblasts secrete

Developmental stage	Histological features
Dental lamina	Thickening of oral epithelium with non distinguishable tooth sites.
Bud	Epithelial cells of dental lamina depict rounded, localized growth. This stage is also considered as initiation/ proliferative stage, as initial proliferation of oral epithelial cells and adjacent mesenchyme will start. Developing mesenchymal cells surround the bud and form an ectomesenchymal condensation.
Cap	As the development proceeds, epithelial bud acquires a concave surface and develops a cap like appearance. Dental mesenchyme which is surrounded by the cap shaped enamel organ is dental papilla. Cells which surround the enamel organ and dental papilla divide and form dental sac, from which cementum, periodontal ligament and adjacent alveolar bone arise.
Bell	Further development of enamel organ and dental papilla results in an increase in size, which in turn changes the appearance of tooth germ from cap to bell. In this stage, shape of future tooth crown is defined and ameloblasts are differentiated from precursors, which secrets enamel. Root formation is also initiated at this stage with elongation of cervical loop, which gives rise to epithelial root sheath and epithelial diaphragm. These structures will finally lead to the development of root.

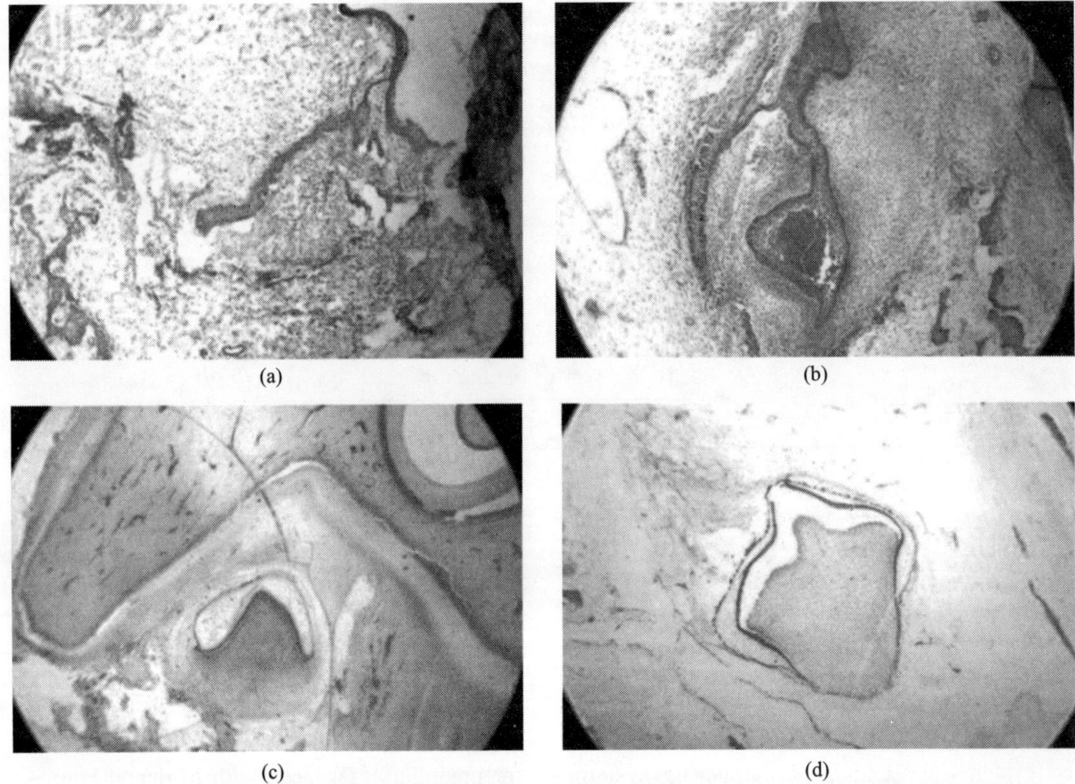

Figure 36.2 Various stages of morphogenesis of dental structures under microscope. (a) Bud stage; showing developing tooth bud (20X). (b) Cap stage; showing cap shape enamel epithelium and various structures of enamel organ (20X). (c) Bell stage; showing changes in shape of enamel organ and formation of dental hard tissues (20X). (d) Advanced bell stage; showing further development of enamel organ and deposition of enamel and dentin (20X). (*see Plate 7 for colour figure*)

enamel (amelogenesis), which is a two stage procedure. In the first stage partially mineralized enamel is formed. Once full width of enamel is deposited, it is followed by second stage in which influx of inorganic component will increase significantly along with the removal of organic components and water. Ameloblasts secrete matrix proteins and these are responsible for creating and maintaining an extracellular environment which favors mineral deposition. Ameloblasts secret enamel and are lost once the tooth has erupted in the oral cavity. Thus, making enamel a non vital structure with no reparative capacity (Nanci 2005).

At 3rd to 4th months of intra uterine life, deciduous upper and lower incisors show beginning of the calcification, followed by upper and lower laterals (4th month in utero) and upper and lower first molars (4th month in utero). Once 3/4th of the root is formed, dental eruption begins. Eruption of tooth can be divided in two easily definable aspects:

Intraosseous eruption: It involves resorption of bone or, roots of deciduous tooth (in case of eruption of permanent tooth) and translocation of the tooth within the bone.

Supra-osseous eruption: This stage involves movement of teeth in the occlusal direction once the part of the crown has pierced the oral mucosa and is above the alveolar crest.

Eruption of tooth is a result of combination of various integrated factors, like bone remodeling, root elongation/growth, vascular pressure and periodontal ligament traction (Kardos et al., 1979; Marks et al., 1996; Sutton et al., 1985). Studies have clearly shown that the eruptive forces reside in the dental follicle-periodontal ligament complex.

In normal circumstances, root resorption of primary tooth (primary upper and lower incisors) starts at 4 years of age. This is triggered by eruption of secondary/permanent tooth which produces pressure on the primary tooth and alveolar bone. Thus, expressing various bio-active molecules and inflammatory mediators. This will initiate resorption of root of primary tooth and alveolar bone thereby enhancing the eruption of permanent tooth. In permanent dentition upper and lower permanent first molars will erupt in the oral cavity followed by lower and upper central incisors. By the age of 12 years, all the teeth of permanent dentition will erupt in the maxillary and mandibular arches except 3rd molars, which will erupt between by age of 18 to 25 years.

III. ANATOMY, HISTOLOGY AND BIOCHEMISTRY

A. Enamel

Enamel is one of the most important structures of tooth, both from function as well as esthetic point of view. It is derived from oral epithelium and forms the protective covering over the exposed surface of crown.

Thickness of enamel varies from tooth subtypes and maximum thickness of enamel is seen on permanent molars. It behaves as a semi permeable membrane, which allows selective passage of different molecules. Enamel is considered as the hardest tissue of human body, which is due to high inorganic content and their crystalline arrangement (Figure 36.3). It

consists of more than 96% of inorganic material and remaining 4% of organic material and water (Ehrlich et al., 2009). Calcium hydroxyapatite is the major inorganic component of teeth. Studies have also shown that hydroxyapatite is the best possible inorganic structure of human dental enamel (Torbrog et al., 1948; Wilson et al., 1999). General formula for inorganic component of enamel is $Ca_5(PO_4)_3X_2$, i.e., apatite; where X can be replaced by OH, Cl and F also magnesium, sodium and potassium may replace calcium in the crystal lattice. Hydroxyapatite crystals are arranged to form enamel rods or prism.

Organic content of enamel is comprised of unique proteins and lipids. Enamel proteins can be categorized into two groups: amelogenins and nonamelogenins. Amelogenins which are heterogeneous group of low molecular weight proteins are major constituent of enamel proteins (up to 90%). They are rich in proline, histidine, glutamine and leucine; therefore they are hydrophobic in nature. Whereas non-amelogenins include enamelin, ameloblastin and tuftelin which constitute up to 10% of the enamel protein. They are rich in glycine, serine and aspartic acid and have high molecular weight.

Organizational units of human enamel are rods/prisms, rod sheath and cementing inter-prismatic substance. In cross section, enamel rods appear as hexagonal or prism like. Course of enamel rods are oblique and wavy, this makes its length more than the width (Figure 36.4). A single enamel prism is developed from one ameloblast. The boundary between rod and interrod enamel contains an organic material known as the rod sheath and has increased content of organic material and water (Garnett et al., 1990).

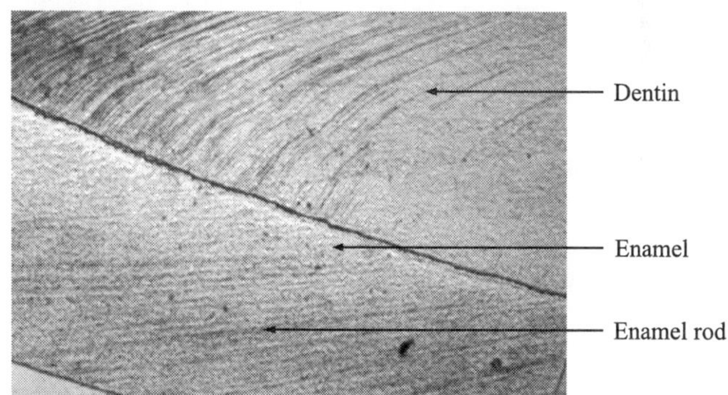

Figure 36.4 Longitudinal ground section of tooth showing enamel rod. Enamel rods are obliquely placed and wavy in orientation (4X). (*see Plate 7 for colour figure*)

B. Dentin

Dentin is a calcified dental hard tissue which forms the bulk of dental hard tissue of tooth. It extends from crown to root of a tooth where it is encircled by enamel and cementum, respectively. It is laid down earlier than enamel thus determining the shape of the crown and number and size of the roots. Hardness of dentin is intermediate between enamel and bone. Dentin is formed by odontoblasts and their projections are present within the tubules, while cell bodies lie on the pulpal

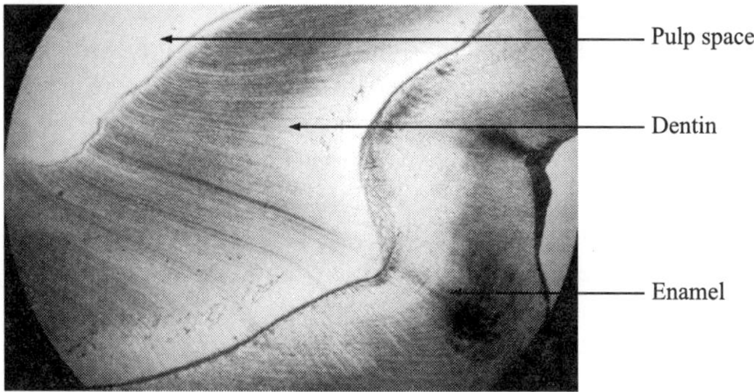

Figure 36.3 Longitudinal ground section of tooth under microscope showing various dental hard tissue structures; from right to left are enamel, dentin and pulp space (4X). (*see Plate 7 for colour figure*)

layer. Thus, unlike enamel, dentine is considered as a living tissue.

It is viscoelastic in nature and withstands slight distortion and deformation. It is deposited as a layer of unmineralized structure (i.e., predentin) and consists mainly of collagen and non collagen proteins. Predentin transforms itself to dentin with addition of various non-collagenous matrix proteins and minerals.

Mature human dentin is made up of approximate 70% inorganic material, 20% organic material and remaining is water by weight (Tjaderhane et al., 2009). Inorganic component consists of hydroxyapatite and it also contains small amount of phosphates, carbonates and sulfates.

Organic components of dentin are complex and consist of collagenous fibrils embedded in ground substance. Collagen fibers form approximately 30% of the organic component. Type 1 collagen is principle collagen of human dentin along with minor amounts of type lll and V (Acil et al., 2007). Type 1 collagen acts as a scaffold for a large proportion of mineral in dentin. Major constituents of ground substances are proteoglycans: chondroitin sulphates, decorin and biglycan, glycoproteins-dentin sialoprotein, osteonectin, osteopontein, phosphoprotein-dentin phosphoprotein and phospholipids. Dentin sialoprotein and dentin phosphoprotein are specific to dentin. Bone morphogenic proteins, insulin like growth factor, fibroblast growth factor and transforming growth factors are also present in the matrix of dentin and play an important role in dentin mineralization.

Dentin is categorized into three groups (Figure 36.5):

Primary dentin: Major type of dentin which forms the bulk of tooth.

Secondary dentin: It develops after the root formation is complete. Deposition of this type of dentin continues over an extended period of time with slower rate.

Tertiary dentin (reactive/reparative dentin): It is deposited in response to stimuli such as attrition, caries and dental restoration.

Figure 36.5 Longitudinal ground section of tooth under microscope showing different types of dentin; from top to center are primary, secondary and reparative dentin (4X). (*see Plate 7 for colour figure*)

C. Pulp

Pulp is a soft connective tissue component of tooth, which supports the dentin and occupies the core of the tooth. Pulp extends from crown to the apical foramina of the root and later it continues with the periapical connective tissues. Pulp has four different distinct zones when seen histologically from dentin towards the core/ center of the pulp: Odontoblastic zone, cell free zone, cell rich zone and pulp core (Figure 36.6).

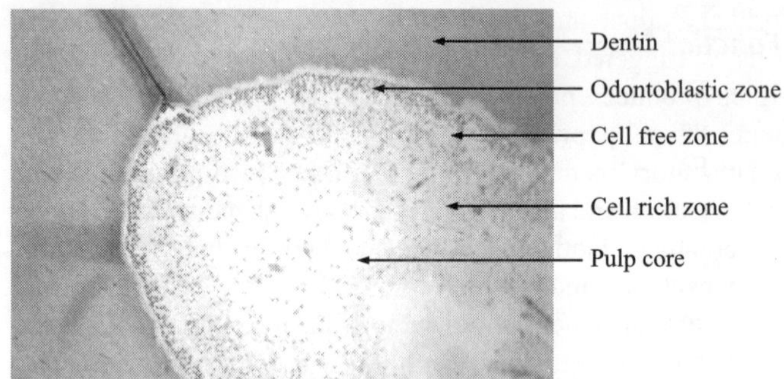

Figure 36.6 Transverse section of tooth under microscope showing dentin at the periphery and pulp at the centre. Pulp tissue show various zones viz odontoblastic zone, cell free zone, cell rich zone and pulp core (10 X). (*see Plate 7 for colour figure*)

Pulp is a loose connective tissue of mesodermal origin and can be subdivided into cellular component and matrix/ ground substance (Lindhe 1985). Fibroblasts are most important and numerous cell type in the pulp. They secrete collagen fibers and play an important role in pulpal inflammation. Odontoblasts are the second most prominent cells of pulp, which are usually seen at the roof of the pulp chamber with cell bodies in the pulp. Undifferentiated mesenchymal cells of pulp are totipotent in nature and considered as primary cells. Histiocytes/macrophage, dendritic cells, lymphocytes, plasma cells and eosinophils form the defense cells of pulp.

Intercellular matrix consists of acid mucopolyssachrides and protein polysaccharides. During tooth formation chondroitin A, B and hyaluronic acid are present in the ground substance in sufficient amount which gradually decreases as the tooth matures. Glycoproteins are also present in the matrix and its amount decreases as the age advances. Tissue fluid pressure of the pulp is contributed by glucosaminoglycans, which is hydrophilic and forms a gel. Syndecan and versican contributes to the major bulk of proteoglycans content of the pulp. Syndecan acts as an adhesion promoter for fibroblasts and collagen fibers. Tenascin and fibronection are cell adhesion promoters and are essential for the vitality of the cellular components. Integrin and laminin form the cell surface adhesion receptors, which promote the cellular adhesion with syndecan. Fibroblasts synthesize and secrete FGF2 and VEGF, which play an important role in the inflammatory reaction of the pulp and healing. It also secretes type I (major collagen of pulp tissue) and III collagen fibers, which are embedded in the ground substance. IL 8 is a chemotactic factor released by the odontoblasts. It also releases nitric oxide synthetase along with pulpal endothelial cells, which is important for control of hydrostatic and blood pressure in the pulp tissues. Dendritic cells and macrophages express CD14 and 68 and they present the antigen to the T cell. Pulp stem cells secrete TGF beta1 and BMP2, which are involved in differentiation and proliferation of endothelial cells. Dentonin is a complex peptide which is derived from matrix extracellular phosphoglycoprotein and has been shown to stimulate and transform the pulp stem cells into various cell lines (Liu et al., 2004; Six et al., 2007).

Functions of the pulp

- Induction – it interacts with oral epithelial cells, leads to its differentiation into dental lamina and enamel organ.
- Formative – odontoblasts cells which are present in the periphery of the pulp chamber secretes dentin, which in turn induces secretion of enamel from ameloblasts.
- Nutritive – as pulp has blood vessels and lymphatic channel, it nourishes the dentin through odontoblasts.
- Protective – nerves in the pulp have sensory innervation for various stimuli, it also initiates and maintains reflexes for circulatory control in the pulp tissue.
- Reparative – pulp tissues have remarkable property for repair. It lays down reparative dentin in response to various noxious stimuli.

D. Cementum

Refer Chapter 37 "Periodontium".

IV. ODONTOGENSIS AND MOLECULAR CONTROL

Teeth and its supporting structures develop as a result of a complex series of interactions between epithelium and underlying neural crest derived mesenchymal tissue (Thesleff et al., 1997). Sequential, reciprocal and inductive interactions and signaling plays a vital role and determines various outcomes in odontogensis. Three types of signaling observed during odontogensis are (Peisco et al., 2002):

Short range signaling: Direct cell to cell contact and their interactions. It is important for cell to cell communication and formation of differentiated tissue layers within the developing enamel organ and papilla.

Mid range signaling: Requires diffusion of the signaling molecules to the cells in the vicinity and is important during Odontogensis. This also includes the interaction between the cells and the matrix and vice versa, which is a must for morphogenic activity.

Long range signaling: They have generalized influence on the development of body including development of teeth, bones and muscles to functional unit.

Individual tooth germs develop as an anatomically distinct unit. Local concentrations of signaling molecules and positional signals have an effect on the rate of cell division which ultimately determines the positioning of teeth and tooth types. Neural crest cells are important for initiation and development of craniofacial structures. During odontogensis short range signaling is mediated by cell surface molecules and is important for cell differentiation. Various signaling molecules have been expressed during development of tooth and the most important ones are sonic hedgehog (SHH), Wnt, fibroblast growth factor (FGF) and transforming growth factor beta (TGF-beta) which also includes BMP4 (Seppo et al., 1993). These signaling factors interact with their specific receptors resulting in specific gene expression.

Homeobox transcription factors are most important genes for patterning of dental tissues. Lef 1, Pitx2/Otlx2, Barx1, Lhx6, Lhx7, Pax9, Msx and Dlx are important transcription factors which are involved in initiation of tooth development, tooth morphogenesis and dental patterning (Peters et al., 1999; Thesleff et al., 2000; Thesleff et al., 1997).

Wnt family of signaling molecules are involved in early developmental processes of craniofacial region which includes neural crest migration and establishment of dental arches. Once the first branchial arch is established, homeobox transcription factor Pitx2 is expressed in oral epithelium. This is followed by establishment of dental epithelial thickening with expression of FGF8. FGF8 induces the expression of transcription factor Pax9 in the mesenchymal tissue, which is essential for further development of tooth analog. BMP4 has an antagonistic effect on tooth primodia as far as effect of Pax9 is concerned, thus BMP4 will inhibit the tooth development in the dental arch. Syndecan binds tenascin, a matrix glycoprotein abundant in dental mesenchyme, suggesting involvement of cell-matrix interactions. Syndecan also binds growth factors, and its association with cell proliferation in the dental mesenchyme suggests roles in the regulation of cell number in the condensing cells. Inductive interactions between the epithelial and mesenchymal tissues regulate tooth development at all stages (Thesleff et al., 1996) (Figure 36.7).

FGF8 induces expression of Lhx6 and Lhx7 in the dental mesenchymal tissue during bud stage of odontogensis. Transcription factor Lef1 expression is first observed in dental epithelium followed by mesenchyme. Lef1 is an important component of Wnt signaling pathway. Interaction of tenascin, which is an extracellular matrix molecule with cell surface molecule syndecan, facilitates condensation of mesenchymal tissue in the bud stage (Ehrismann et al 1986). Various genes which are activated by transcription factors Lef1 and Pax9 are essential for progression of odontogensis from bud to cap stage.

Experimental studies have shown that instructions for pattern of dental hard tissues and odontogensis are initiated by dental epithelial tissue and are tightly regulated by the mesenchyme. Before the initiation of tooth germ homeobox genes like Msx1/2, Dlax1 and Barx1 are expressed. Expression of Barx1 is restricted to the molar field along withDlx1/2 and is important for molar patterning. On the other hand, Msx1 is

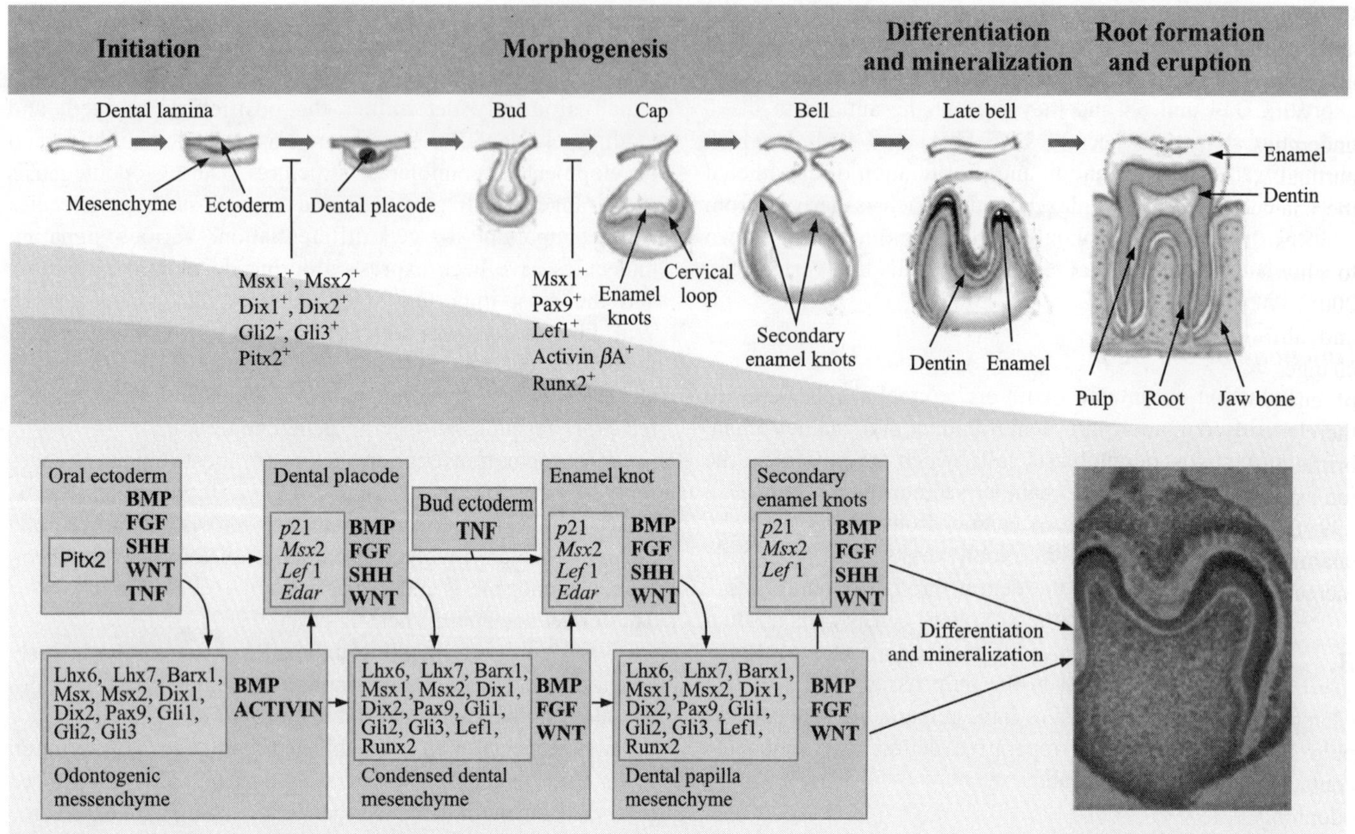

Figure 36.7 Schematic diagram of signaling pathways and molecular control during various stages of odontogensis (Adapted from *J Cell Sci.*, 2003, 116, 1647–1648)

expressed in the incisor fields and leads to the development of incisors (Thesleff et al., 1996).

SHH functions as an early patterning molecule for the incisor area by acting prior to the establishment of epithelium thickening, by regulating the expression of Pax transcription factors. BMP4 is expressed in the epithelium which is present in the overlying incisor field, which induces expression of Msx1; on the other hand FGF8 is expressed over the molar region and induces expression of Barx1 (Thesleff et al., 1996).

Primary enamel knot is a transient structure during odontogensis and produces various signaling molecules; important ones are SHH, BMPs 1/2/4/7, FGFs4/9 and Wnt10a/10b (Thesleff et al., 2003). These signals affect epithelium and mesenchyme leading to the growth of future cusps, followed by apoptosis of cells in the enamel knots. Secondary enamel knots will appear over the tips of cusps of multicuspid teeth and they control differentiation and proliferation around the developing cusps much like the primary enamel knot.

Extracellular matrix plays an im ortant role in initiation, development and proliferation of various stages of tooth. Basement membrane is an unique extracellular matrix component which is deposited between the cells and tissues. Various molecules like FGF, which are present in basement membrane diffuse easily through it and play an important inductive and regulative role in end stages of

odontogensis. In the final stages of tooth development FGF signaling along with sulfated proteoglycans are very important in Wnt signaling.

V. TEETH AND AGING

In general dental hard and soft tissues show morphological and histological measurable changes due to aging. Aging is also observed in dental hard tissue under the influence of environmental insults or diseases such as caries, extensive restorative procedures, and trauma (Pashley et al., 2002). Pulp is a loose mesenchymal tissue, which also shows degenerative changes over time.

A. Enamel

Once the enamel is laid down by ameloblasts, there is no further deposition, as these cells undergo apoptosis once the tooth has erupted in the oral cavity. Recently erupted teeth are covered with rod ends and perikymata which begin to disappear with age, followed by generalized loss of rod ends and flattening of perikymata. With advances in age, enamel becomes discolored, which is attributed to the deposition of organic material from the environment and also by deepening of dentin color owing to thinning of overlying enamel layer.

Enamel does undergo changes subsequent to various physical and chemical processes like incisal, cuspal and interproximal attrition (Ketterl 1983), abrasion, erosion and dental caries.

With aging composition of surface layer of enamel undergoes an ionic exchange with the oral environment. In particular, there is a marked increase in fluoride content of the enamel (Nanci 2005).

Resistance to decay of enamel may be increased due to change in the crystal lattice structure of enamel (Nanci 2005). With increasing age the tendency of enamel to crack and abfraction also increases and is attributed mainly to changes in composition of mineral and organic components of enamel. Calcification of enamel progressively increases thereby decreasing the permeability and making enamel more brittle and translucent. Enamel crystals changes occur due to ion exchange, remineralization and demineralization (Ferguson 1999). These result in formation of wear facets, decrease in enamel mass and density which become pronounced with increase in age (Nanci 2005).

B. Age Changes in Dentin

Changes in dentin are not viewed alone but in unison with pulp as pulp-dentin complex, as pulp is present in intimate contact with dentin. Environmental insults result in injury to odontoblasts leading to inflammation and if the extent of insult is severe, it will cause death of pulp. These cells are replaced by new odontoblasts, which are derived from the undifferentiated mesenchyme. Newly formed odontoblasts deposit the hard tissue termed as reparative dentin or tertiary dentin, which is characterized by fewer and more twisted dentinal tubules. This dentin is also referred as irregular secondary dentin owing to the irregular nature of the dentinal tubules. The mineral content of reparative dentin is higher than the normal dentin. Also, dentin phosphophoryn which is present in other forms of dentin is not present in reparative dentin (Kumar 2009).

Dentin is exposed to the environment with loss of covering enamel, leading to secondary stimulation of odontoblasts cells resulting in deposition of a layer of highly calcified material on the walls of the tubules, i.e., sclerotic dentin. In this type of dentin the tubules are completely occluded by deposition of calcified material which is similar to peritubular dentin. Mineral density of this dentin is greater than the normal dentin making it hard with reduced fracture toughness (Ferguson 1999; Kumar 2009) (Figure 36.8).

Dead tracts are another aspect of age changes associated with aging of dentin. This develops subsequent to complete retraction of the odontoblastic processes from the tubule and through death of the odontoblast, which makes the dentinal tubules empty. These air filled tubules appear black "dead tracts" with transmitted light (Nanci 2005) (Figure 36.9).

Biochemically, changes are witnessed in dentin that give an idea of chronology of dentin; marked biochemical changes take place in dentin owing to its high organic content. Racemisation of L to D form of aspartic acid, an amino acid component of collagen takes place with age. Moreover, deoxypyridinoline, a

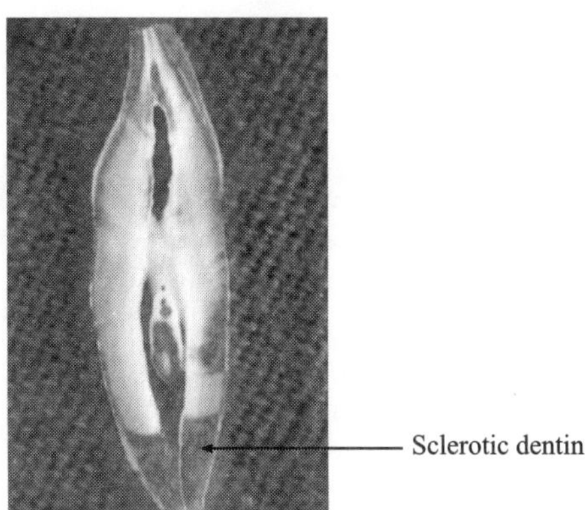

Figure 36.8 Longitudinal section of tooth showing sclerotic dentin. Sclerotic dentin has high inorganic content which makes it transparent in thin section. Thin section of tooth is kept on geometric pattern to show the transparency of sclerotic dentin. (*see Plate 8 for colour figure*)

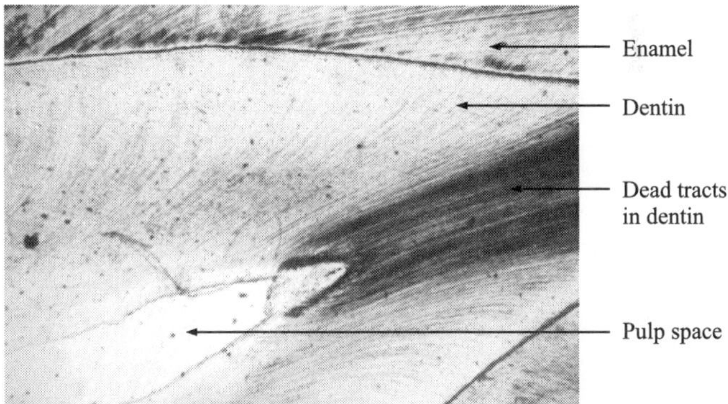

Figure 36.9 Longitudinal ground section of tooth showing dead tracts, which appear dark in transmitted light due to empty dentinal tubules secondary to death of odontoblasts (4X). (*see Plate 8 for colour figure*)

non-reducible collagen cross link in dentin increases with age (Heras et al., 1999). Enzyme Gelatinase A has been implicated in the remodeling and degradation of the dental basement membrane during human tooth morphogenesis. The levels of this enzyme show a gradual increase in aging dentin (Heras et al., 2000).

C. Age Changes in Pulp

Pulp shows changes in response to its environment. Since it is in close proximity to dentin, it shows changes in its internal structure and the surrounding hard tissue. Aging pulp demonstrates an increase in the fibrous component (mainly collagen) and a simultaneous decrease in the number of cells. Increased fibrosis is attributed not only to the increase in collagen, but to a persistence of connective tissue sheath in an increasingly narrowed pulp chamber.

With advancing age, pulp space decreases in size by the continuous deposition of dentinal matrix by odontoblasts resulting in narrowing of the root canal and occasionally obliteration of it. Fibroblasts and odontoblasts show a considerable decrease in number and also cells appear relatively inactive. The cells exhibit a decrease in size and number of cytoplasmic organelles. In the aging pulp, fibroblasts are characterized by less perinuclear cytoplasm and possess long, thin cytoplasmic processes. Moreover, the intracellular organelles like mitochondria and endoplasmic reticulum are reduced in size and number (Kumar 2009).

Bernick et al., had observed a decrease in the number of blood vessels and nerves supplying the pulp with increasing age. There is a progressive reduction in the size of the pulpal chamber and progressive deposition of calcific masses that originate in the root pulp and progress into the coronal pulp. As a result of calcification there is decrease in the number of blood vessels and nerves in the coronoal pulp. Aging pulp exhibits degeneration of myelinated and unmyelinated axons with simultaneous reduction in sensitivity (Nanci 2005). However, the connective tissue sheath of the blood vessels and nerves persist even after their degeneration and give the fibrotic appearance to the pulp (Bernick et al., 1975).

Another age change seen in pulp is the occurrence of irregular areas of calcification called pulp stones or denticles (Figure 36.10). They consist of concentric layers of mineralized tissues around blood thrombi, collagen fibers or dying cells. Pulp stones at times are attached to dentinal wall and get surrounded by secondary dentin, called as attached pulp stones. Free pulp stones are surrounded by soft tissue. Milch (1965) postulated that with increased cross linkage of collagen during advancing age there is an enhanced tendency for these fibers to become mineralized (Nanci 2005).

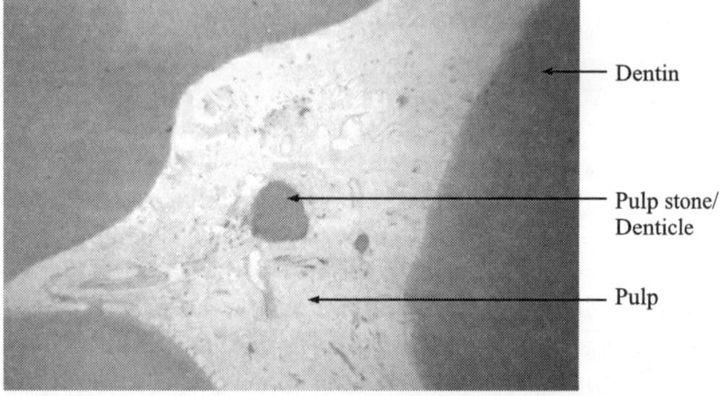

Figure 36.10 Transverse section of tooth under microscope showing pulp stone/denticle at the centre of pulp (10X). (*see Plate 8 for colour figure*)

In aging pulp the active cells demonstrate fewer organelles associated with synthesis and secretion. This further reduces the capacity of pulp to combat and recover from injury (Pashley et al., 2002). Aged pulp also show decrease in connective tissue matrix mainly mucopolyssachrides. Biochemical changes are seen in pulp which includes an up regulation of growth factors like CTGF, FGFs, and TGFB 1. These proteins are secreted extracellularly and take part in cellular functions including control of cell growth, cell proliferation, cell differentiation and apoptosis.

Concentration of apoptosis regulators like Bax and Bcl2 increases in aging pulp (Yang et al., 2010). Also, older dental pulp shows a rise in type IX and type X collagen. These changes signify that the vitality of the pulp decreases with age. This is attributed to the low activity of genes coding for transcription regulators and a high expression of genes involved in apoptotic process (Tranasi et al., 2009).

Age changes of pulpal tissues get accelerated in response to environmental stimuli like caries or attrition of enamel. In case of pulpal injury, the age of pulp determines its reparative potential. With age the potential for differentiation of new odontoblasts from the mesenchymal cells is greatly reduced (Nanci 2005).

D. Age Changes in Cementum

Refer Chapter 37 "Periodontium".

VI. MORPHOLOGIC VARIANTS OF DENTAL HARD TISSUES

A multitude of developmental alterations of teeth can occur due to various factors. These alterations can be primary or can manifest secondary to the environmental influences (Neville 2009). The coronal portion of tooth is made of enamel, dentin and pulp. Change in any of these structures results in a subsequent change in the external appearance of tooth (Watts et al., 2001). Traditionally, morphologic change occurring as a result of discoloration can be categorized as intrinsic or extrinsic. A further category of internalized stain or discoloration is also being considered.

Intrinsic stain/ discoloration	Extrinsic stain/ discoloration	Internalized stain/ discoloration
1. Amelogenesis imperfecta	1. Metallic 2. Non-metallic	1. Developmental defects
2. Dentinogenesis imperfecta		2. Acquired defects
3. Dentin dysplasia		(a) Tooth wear and gingival recession
4. Dentin Dysplasia		(b) Dental caries
5. Fluorosis/Enamel hypoplasia Tetracycline staining		(c) Restorative materials
6. Alkaptonuria		(Watts 2001)
7. Congenital erythropoietic porphyria		
8. Congenital hyperbilirubinaemia		
9. Pulpal hemorrhagic products		

Amelogenesis imperfecta (AI)

AI represents a group of conditions, genomic in origin, which affects the structure and clinical appearance of the enamel of all or nearly all the teeth in a more or less equal manner and which may be associated with morphologic or biochemical changes elsewhere in the body (Aldred et al., 2003).

Amelogenesis imperfecta exists in at least 14 different hereditary subtypes with a wide range of inheritance and clinical manifestations (Neville 2009). According to Witkop and Sauk, the clinical features of three major types of Amelogenesis imperfecta can be broadly established as:

1. *Hypoplastic type:* The enamel has not formed to its full thickness on newly erupted developing teeth.
2. *Hypo calcified type:* The enamel is so soft that it can be removed by a prophylaxis instrument.
3. *Hypo maturation type:* The enamel can be pierced by an explorer point under firm pressure and can be lost by chipping away from the underlying normal-appearing dentin (Figure 36.11).

Clinical appearance varies in different forms of AI. Generally, all teeth of both the dentitions are affected to some extent. If discoloration is present on the coronal portion of the teeth, it ranges from yellow to dark brown (Figure 36.11). Enamel may be totally absent and in cases where it is present, it can have a chalky appearance or a cheesy consistency or can be hard. At times, enamel present is smooth with presence of vertical grooves. Occlusal surfaces of teeth are severely abraded resulting in open contacts.

Mutations in five genes AMELX, ENAM, MMP-20, KLK4, DLX3 have been implicated for Amelogenesis Imperfecta. These genes code for various proteins like amelogenin, enamelin, enamelyisn and kallikrein-4 (Neville 2009). Biochemically, enamel formation is considered to be a highly organized and unusual structure strictly in control of ameloblasts by a number of organic matrix molecules like enamelin and amelogenins. At a micro structural level in amelogenesis imperfecta, ground sections show a very thin enamel layer composed of laminations of irregularly arranged enamel prisms (Chanmougnanada et al., 2012).

In hypocalcified amelogenesis imperfecta, enamel is found to contain albumin as one of the major protein fraction. Hence, indicating that hypo calcified variant is caused by a disturbance in matrix protein degradation during the maturation phase (Takaqi et al., 1996).

Dentinogenesis imperfecta

Dentinogenesis Imperfecta is an autosomal dominant disease which affects dentin mineralization. Dentinogenesis imperfecta is broadly classified into three clinical entities:

1. *Dentinogenesis Imperfecta Type I:* The dentin mineralization defects are also associated with osteogenesis imperfecta of bone, a mixed connective tissue disorder of type I collagen.
2. *Dentinogenesis imperfecta Type II (Hereditary opalescent dentin):* Associated with dentin mineralization defects but no collagen defects have been identified. This is the main condition related to the teeth alone.
3. *Dentinogenesis Imperfecta Type III (Brandywine isolate hereditary opalescent dentin):* A complex heterogeneous disorder in which both collagen and mineralization are affected (Figure 36.12).

Dentin Sialophosphoprotein (DSPP), is the gene implicated for Dentinogenesis imperfecta. This gene codes for two major noncollagenous proteins found in human dentin, dentin sialoprotein and repeat containing dentin phosphoprotein. DSPP mutation has also been correlated to Dentin dysplasia.

Genes implicated for this disease are located on the long arm of human chromosome 4q12-q21. Dentinogenesis imperfecta affects both the primary and the permanent dentition. Affected teeth show a characteristically gray to brownish blue discoloration (Figure 36.12). Dentin formed is poorly mineralized and is rapidly attrited, affecting the shape of the crown. The crown of the affected teeth is bulbous and there is associated narrowing of the pulp chamber owing to the deposition of the defective mineralized dentin matrix with increased water content.

DSPP mutation has a negative effect on the dentin matrix assembly/mineralization, because odontoblasts are unable to make and secrete normal DSPP and/or type I collagen in an organized manner and also in sufficient quantities to make the rapidly forming and mineralizing matrix (McKnight et al., 2008).

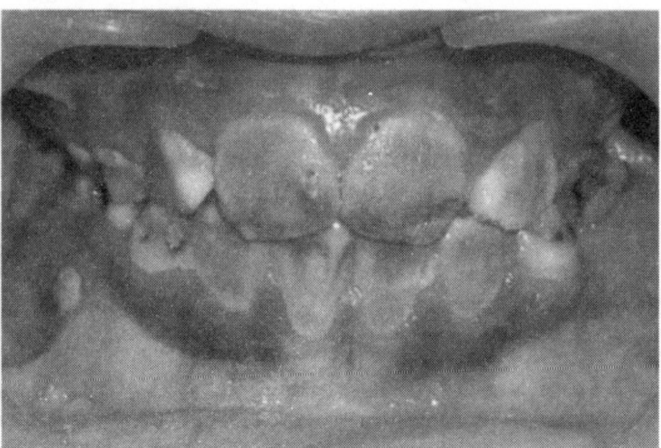

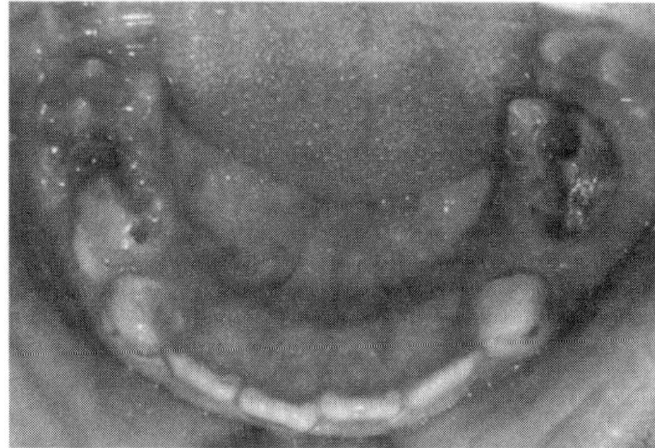

Figure 36.11 Clinical photograph of a child in mixed dentition with Amelogenesis Imperfecta; showing yellowish brown discoloration of the dentition with soft and rough teeth surfaces. (*see Plate 8 for colour figure*)

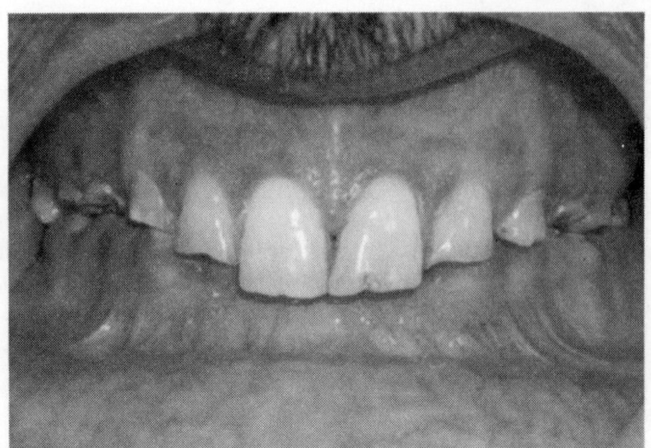
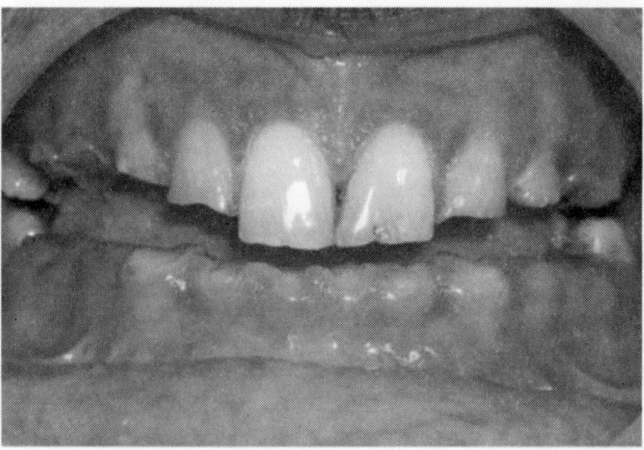

Figure 36.12 Clinical photograph of an adult with permanent dentition showing Dentinogenesis Imperfecta showing mild discoloration of teeth with loss of crown structure due to rapid wear. (*see Plate 8 for colour figure*)

At the micro structural level there is a presence of irregular dentinal tubules with irregular polarization of odontoblasts with mineralization defects in dentinogenesis imperfecta.

Dentinal tubules in dentinogenesis imperfect are irregular in size and direction when compared to the normal dentinal tubules. Few areas of mantle dentin are nearly atubular with less differentiation between mantle and circumpulpal dentin (Thotakura et al., 2000; Watts et al., 2001).

Dentin dysplasia

Dentin dysplasia is dominantly inherited, nonsyndromic disease of dentin. There are two variants of dentin dysplasia:

Type I dentin dysplasia: Both the primary and the secondary dentition are of normal shape and form but may have an amber translucency. However, the pulp is obliterated.

Type II dentin dysplasia: Exhibits thistle shaped pulp chamber with numerous pulp stones. Teeth may show brown discoloration of teeth.

As in dentinogenesis imperfecta, dentin dysplasia is also correlated with DSPP genetic mutations. Both the diseases show strikingly similar features. However, one major difference between the two is that the visible portion of the crown of the secondary teeth appears relatively normal in dentin dysplasia.

At the cellular level, in dentin dysplasia, odontoblasts are not able to secrete normal type I collagen in an organized manner and also there is an associated defective dentin mineralization. This is because of defects in DSPP gene that codes for two noncollagenous proteins in dentin, dentin sialoprotein and dentin phosphoprotein (Watts et al., 2001; McKnight et al., 2008).

Fluorosis

Dean, in the year 1932 described an association between fluoride intake and its effect on enamel. This can occur endemically from the naturally occurring water supplies or from fluoride delivered from the mouthrinses, tablets and toothpastes as supplements. Clinically, enamel may manifest as areas of flecking to diffuse opacities, referred as "mottling" (Figure 36.13). The color of the enamel may range from white to dark brown to black.

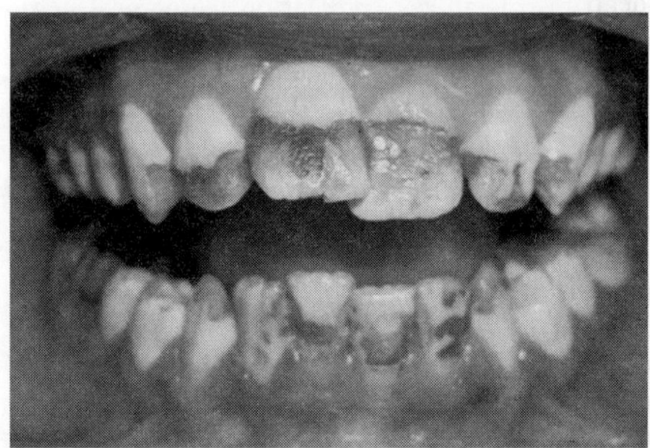

Figure 36.13 Clinical photograph of permanent dentition showing fluorosis, i.e., "mottled dental enamel"; affected tooth is characterized by white spot, brown stains, pitting or mottling of enamel. (*see Plate 8 for colour figure*)

Fluoride causes fluorosis in concentrations greater than 1ppm in drinking water.

Biochemically, fluoride enhances the chemical reaction resulting in precipitation of calcium phosphate. Normally an equilibrium exists in the oral cavity between calcium and phosphate ions in the solution phase, saliva and in the solid phase, enamel. Fluoride shifts this equilibrium in favor of the solid phase.

It results in stabilization of hydroxyapatite lattice. Hydroxyapatite has inherent voids due to missing hydroxyl groups and fluoride fills up these voids. Owing to the replacement of missing hydroxyl ions and subsequent formation of fluorapatite there is definitive decrease in solubility of enamel thus, making enamel resistant to caries.

$$Ca_{10}(PO_4)OH_2 + 2F^- \longrightarrow Ca_{16}(PO_4)F_2 + 2OH^-$$

The conversion of hydroxyapatite to fluorapatite is in the range of 3–5%; hence, the use of the term fluorhydroxapatite is advocated.

When a localized region of enamel loses its mineral content, the enamel may be remineralized if the offending agent is removed. The remineralization process is enhanced in the presence of fluoride. The severity of fluorosis is related to age and dose. Fluorosis affects both the primary and the secondary dentition.

Enamel hypoplasia

Enamel hypoplasia is a defect in which an incomplete or defective organic enamel matrix of the teeth is formed in the embryonic stage of the tooth. It can be hereditary or environment induced. In the environmental type, an ectodermal disturbance occurred during the embryonic development of the enamel and the mesodermal components are normal. It involves both deciduous and permanent teeth and affects only enamel. Environmental type enamel hypoplasia is caused by the environmental factors that cause damage to the enamel cells. Either deciduous or permanent teeth are involved and sometimes a single tooth is involved; both the enamel and the dentin are involved in varying degrees. Enamel shows various morphological and chemical alterations depending on the etiological factors.

Tetracycline staining

Tetracycline administered for systemic diseases during odontogensis is associated with unsightly stains on teeth. Tetracycline may also disturb enamel formation resulting in malformed hypoplastic teeth.

Teeth affected by tetracycline exhibit a yellowish or brown gray discoloration which after eruption reduces in intensity. Tetracycline fluoresces under ultraviolet light and imparts bright yellow color to the tooth. Light exposure changes the color to brown and thus the anterior teeth are more likely to be affected subsequent to light exposure. Severity of tooth discoloration is affected by the age at the time of administration, duration of administration and dosage of the drug administered. Dentin is heavily stained than enamel (Figure 36.14).

Most likely explanation for the incorporation of tetracycline into the teeth is through the chelation of tetracycline with calcium ions in the molecular structure. This takes place at the time of calcification; as a result the effect on the teeth is seen during the last trimester of pregnancy in childhood.

Tetracycline has an unique ability to form complexes with calcium ions on the surface of hydroxyapatite crystals within dental tissues.

Since tetracycline has the ability to cross placental barrier, it is recommended to avoid the drug from 29 weeks in utero till full term. Also, since permanent teeth continue to develop until 12 years of age, tetracycline administration should be avoided in children below the age of 12 and in also in breast feeding and expectant mothers (Venkateshwarlu et al., 2011; Watts et al., 1998).

Other common pre-eruptive causes of intrinsic discolorations are congenital Erythropoietic Prophyria, congenital Hyperbilirubinaemia, Alkaptonuria. These are metabolic disorders, which cause deposition of various metabolites and end products in enamel or dentin or both. Thus, causing discoloration of tooth (yellowish brown to brown). On the other hand, discoloration of tooth due to pulpal hemorrhagic products is a post eruptive cause of intrinsic discoloration. Acute trauma to an erupted tooth results in pulpal hemorrhage, which gives a reddish tinge to the tooth. This further changes to gray-brown as the pulpal tissue undergoes necrosis. Heme group is released after the hemolysis of red blood cells, which reacts with putrefying pulp tissue to form black iron sulfide (Marin et al., 1997), which get deposited in the dentin.

SUMMARY

Teeth form an integral part of oral cavity and are one of the hardest calcified structures present in the human body. Initiation of tooth germ, its growth and development, eruption in the oral cavity and survival, all are mediated by signaling and interactions through several biochemical mediators. Enamel, dentin and cementum are calcified structures of teeth. While enamel is acellular structure, dentine, and cementum have some cellular components and pulp is rich in vascular and cellular components. Tooth enamel is exposed to a variety of physical and chemical stimuli therefore start undergoing alterations in its morphology and chemical composition as soon as it erupts in the oral cavity.

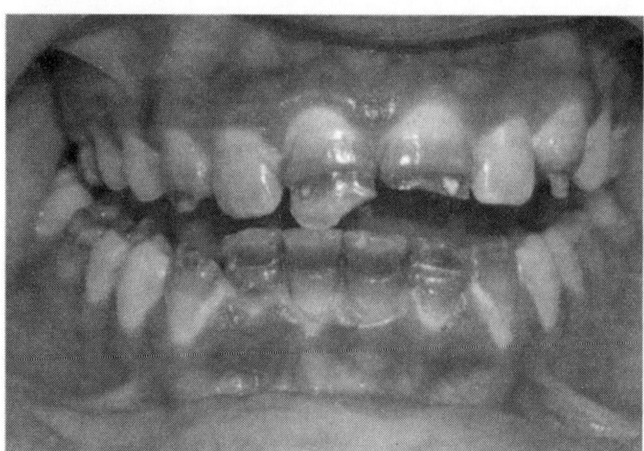

Figure 36.14 Clinical photograph of permanent dentition showing tetracycline related diffuse brownish discoloration. A. Mild discoloration of teeth B. Sever diffuse discoloration of teeth. (*see Plate 8 for colour figure*)

Acknowledgement

Figures 36.2, 36.3, 36.4, 36.5, 36.6, 36.8, 36.9 and 36.10, Department of Oral and Waxillofacial Pathology, Manipal College of Dental Sciences, Mangalore, India.

SUGGESTIONS FOR FURTHER READING

Acil Y., Mobasseri A.E., Warnke P.H., Terheyden H., Wiltfang J. and Springer I. (2005), Detection of Mature Collagen in Human Dental Enamel, *Calcif Tissue Int.*, 76, 2, 121–126.

Aldred M.J., Crawford P.J.M. and Savarirayan R. (2003), Amelogenesis Imperfecta—A Classification and Catalogue for the 21st Century, *Oral Dis.* 9, 19–23.

Bernick S. and Nedelman C. (1975), Effect of Agining on the Human Pulp, *J. Endod.* 1, 3, 88–94.

Chanmougananda S.C., Ashokan K.A., Sxhokan S.C., Bojan A.B. and Ganesh R.M. (2012), Literature Review of Amelogenesis Imperfecta with Case Report, *J. Indian Aca Oral Med Radiol.*, 23 (1), 83–87.

Donoghue P.C.J. and Sansom I.J. (2002), Origin and Evolution of Vertebrate Skeletonization, *Microsc. Res. Techn.* 9, 352–372.

Ehrlich H., Koutsoukos P.G., Demadis K.D. and Pokrovsky O.S. (2009), Principles of Demineralization: Modern Strategies for the Isolation of Organic Framework, Part ll. *Decalcification Micron.*, 40, 169–193.

Ferguson D.B. (1999), *Oral Bioscience*, Churchill Livingstone: China, 291–300.

Garnett J. and Dieppe P. (1990), The Effects of Serum and Human Albumin on Calcium Hydroxyapatite Crystal Growth, *Biochem J.* 266, 863–868.

Heras S.M., Valenzuela A. and Villanueva E. (1999), Deoxypyridine Cross Links in Human Dentin and Estimation of Age, *Int J. Legal Med.*, 112, 222–226.

Heras S.M., Valenzuela A. and Overall C.M. (2000), Gelatinase in Human Dentin as a New Biochemical Marker for Age Estimation, *J. Forensic Sci.*, 45, 4, 807–811.

Kardos T.B. and Simpson L.O. (1979), A Theoretical Consideration of the Periodontal Membrane as a Collagenous Thixotropic System and its Relationship to Tooth Eruption, *J. Perio Res.*, 14(5), 444–451.

Ketterl W. (1983), Age-induced Changes in the Teeth and their Attachment Apparatus, *Int. Dent. J.*, 33(3), 262–71.

Kumar G.S. (2009), *Orban's Oral Histology and Embryology*, 12th ed., Elsevier: India, 45–136.

Lindhe A. (1985), The Extracellular Matrix of Dental Pulp and Dentin, *J. Dent Res.* Apr., 64 Spec. No. 523–529.

Liu H., Li W., Goo C., Kumagai Y., Blacher R.W. and Den Besten P.K. (2004), Dentonin, a Fragment of MEPE, Enhanced Dental Pulp Stem Cell Proliferation, *J. Dent. Res.*, 83, 6, 496–499.

Marin P.D., Bartold P.M. and Heithersay G.S. (1997), Tooth Discoloration by Blood: An in Vitro Histochemical Study, *Endod Dent Traumatol.*, 13, 132–138.

Marks S.C. and Schroeder H.E. (1996), Tooth Eruption: Theories and Facts, *Anat Rec.*, 245, 374–393.

McKnight D.A., Simmer J.P., Hart P.S., Harl T.C. and Fisher L.W. (2008), Overlapping DSPP Mutations Cause Dentin Dysplasia and Dentinogenesis Imperfecta, *J. Dent. Res.*, 87, 12, 1108–1111.

Nanci A. Enamel (2005), Composition, Formation, and Structure, In: Nanci A. (Ed.), *Tencate's Oral Histology Development, Structure and Function*, 6th ed., *Mosby*: India, 145–191.

Nanci A. (2005), Dentin-Pulp Complex. In: Nanci A. (Ed.), Tencate's *Oral Histology Development, Structure and Function*, 6th ed., *Mosby*: India, 192–239.

Neville B.W., Damm D.D., Allen C.M. and Bouquets J.E. (2009), *Oral and Maxillofacial Pathology*, 3rd ed., Elsevier, St Louis, Missouri, 54–99.

Pashley D.H., Walton R.E. and Slavkin H.C. (2002), Histology and Physiology of the Dental Pulp, In: Ingle J.I., Bakland L.K. (Eds.), *Endodontics*, 5th ed., B.C. Decker, Hamilton, 25–62.

Peisco N.P. and Avery J.K. (2002), Development of Teeth: Crown Formation, In: Avery (Ed.), *Oral Development and Histology*, 3rd ed., Thieme, Germany, 72–107.

Peters H. and Balling R. (1999), Teeth: Where and How to Make Them, *Trends Gen.*, 15, 2, 59–65.

Schofield J.D. and Weightman B. (1978), New Knowledge of Connective Tissue Aging, *J. Clin. Pathol.*, 31, 174–190.

Six N., Septier D., Chaussain M.C., Blacher R., DenBesten P. and Goldberg M. (2007), Dentonin, A MEPE Fragment, Initiates Pulp Healing Responses to Injury, *J. Dent. Res.*, 86, 8, 780–785.

Suttin P.R.N. and Graz H.R. (1985), The Blood Vessel Theory of Tooth Eruption and Migration, *Med. Hypo.*, 18(3), 289–295.

Takaqi Y., Fujita H., Katano H., Shimokawa H. and Kuroda T. (1998), Immunohistochemical and Biochemical Characteristics of Enamel Proteins in Hypocalcified Amelogenesis Imperfecta, *Oral Surgery, Oral Medicine, Oral Radiology and Endodontology*, 85(4), 424–430.

Tencate A.R., Sharpe P.T., Roy S. and Nanci A. (2005), Development of the Tooth and Its Supporting Structures, In: Nanci A (Ed.). Tencate's *Oral Histology Development, Structure and Function*, 6th ed., *Mosby*: India, 79–110.

Thesleff I. (2003), Epithelial Mesenchymal Signaling Regulating Tooth Morphogenesis, *J. Cell Sci.*, 116, 1647–1648.

Thesleff I. (2000), Genetic Basis of Tooth Development and Dental Defects, *Acta Odont Scand.*, 58, 191–194.

Thesleff I. and Sharpe P.T. (1997), Signaling Networks Regulating Dental Development, *Mech. Dev.*, 67(2), 111–123.

Thesleff I., Vaahtokari A., Vainio S. and Jowett A. (1996), Molecular Mechanism of Cell and Tissue Interactions During Early Tooth Development, *Anat Rec.*, 245(2), 151–161

Thotakura S.R., Mah T., Srinivassan R., Takagi Y., Veis A. and George A. (2000), The Non-collagenous Dentin Matrix Proteins are Involved in Dentinogenesis Imperfecta Type II (DGI-II), *J. Dent Res.*, 79(3), 835–839.

Tjaderhane L., Caeeiho M.R., Breschi L., Tay F.R. and Pashley D.H. (2009), *Dentin Basic Structure and Composition—An Overview*, Endodontic Topic, 20(1), 3–29.

Torborg J.A. and Moller A. (1948), Determination of Size and Shape of Apatite Particles in Different Dental Enamels and Dentin by the X-ray Powder Method, *J. Dent. Res.*, 27, 524–531.

Tranasi M., Sberna M.T. and Zizzari V. (2009), Microarray Evaluation of Age Related Changes in Human Dental Pulp, *J. Endod.*, 35(9), 1211–1217.

Venkateshwarlu M. and Naga S.R. (2011), Tetracycline Induced Tooth Discoloration, *Ind. J. Dent. Adv.*, 3(1), 457–462.

Watts A. Addy M. (2001), Tooth Discoloration and Staining: A Review of the Literature, *British Dent J.,* 190, 309–316.

Wilson R.M., Elliott J.C. and Doker S.E.P. (1999), Rietveld Refinement of the Crystallographic Structure of Human Dental Apatites, *Am Min.*, 84, 1406–1414.

Witkop C.J. Jr. (1988), Amelogenesis Imperfecta, Dentinogenesis Imperfecta and Dentin Dysplasia Revisited: Problems in Classification, *J. Oral Pathol.*, 17, 547–553.

Yang H., Zhu Y.T., Cheng R., Shao M.Y., Fu Z.S., Cheng L., Wang F.M. and Hu T. (2010), Lipopolysaccharide Induced Dental Pulp Cell Apoptosis and the Expression of Bax and Bcl2 in vitro, *Braz J. Med. Biol. Res.*, 43(11), 1027–1033.

37

Periodontium

O.P. Kharbanda and Jitendra Sharan

CONTENTS

I. Anatomy and Histology
   A. Gingiva
   C. Cementum
            B. Periodontal ligament
            D. Alveolar bone

II. Gingival Crevicular Fluid (GCF)
   A. Origin
   C. Function
            B. Composition
            D. Clinical significance

III. Periodontium and Aging
   A. Gingiva
   C. Cementum
   E. GCF
            B. Periodontal ligament
            D. Alveolar bone

IV. Periodontium and its Biochemistry
   A. In health
   B. In disease

V. Orthodontic Tooth Movement and Periodontium
   A. Biochemical changes in periodontal structure during and after orthodontic tooth movement

VI. Periodontium in Root Resorption
   A Mechanism and biochemistry of orthodontic tooth movement
   B. GCF markers of root resorption

VII. Periodontium and Implants
   A. Prosthetic dental implants
   B. Orthodontic mini-implants

The periodontium consists of the biological structures around the tooth. These are: alveolar bone, cementum, periodontal ligaments and gingiva all of which are essential for support and vital to the very existence of the dental structures.

I. ANATOMY AND HISTOLOGY

A. Gingiva

The oral tissue around the tooth is called gingiva, connective tissue of which is covered with epithelium. Healthy gingiva is pale pink in color and firm with a knife edge scalloped appearance around the teeth. It is differentiated into free and attached gingiva. Dark pigmentation of gingiva is seen in African population and is rare in Indian population. This pigmentation of gingiva is considered normal and not pathological (Figure 37.1).

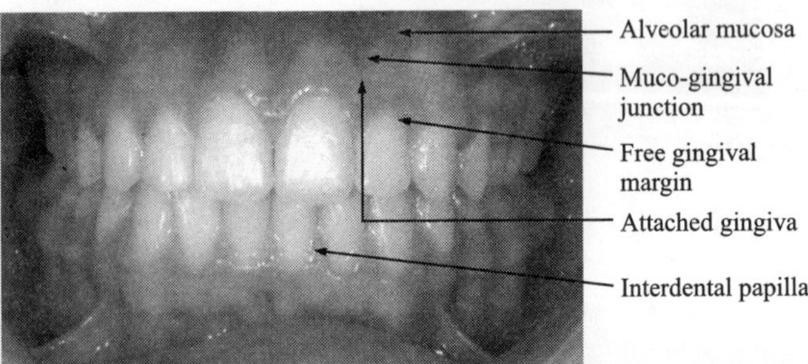

Figure 37.1 Clinical photograph of a healthy mouth showing healthy gingiva, clinical landmarks includes the free gingiva at the cervical margin of the teeth; the interdental papilla; the muco-gingival junction and alveolar mucosa. (*see Plate 8 for colour figure*)

Gingiva makes an attachment on tooth crown through gingival epithelial apparatus. The margin of the gingiva is free and makes a sulcus with tooth surface called gingival sulcus.

Gingival tissues merge apically with alveolar mucosa. This part of the gingiva is called Junctional epithelium.

Gingival groove is a shallow line or depression on the gingival surface at the junction of the free and attached gingivae.

In healthy conditions and fully erupted tooth, the free gingiva is in close contact with the enamel surface and its margin is located 0.5–2 mm coronal to the cementoenamel junction. The mean histologic depth of gingival sulcus is 1.8 mm in healthy conditions, with a variation of 0 to 6 mm.

The attached gingiva is comparatively immobile in relation to the underlying tissue; this is due to its firm attachment to the underlying alveolar bone and cementum by connective tissue fibers. Width of attached gingiva differs in different areas of mouth.

Gingiva gets its blood supply mainly through branches of the alveolar arteries. Lymphatic capillaries originate in the connective tissue papillae and stroma near or adjacent to the attachment epithelium. Gingiva is well innervated with different types of nerve endings.

Microscopic features

The principle cell type of gingiva epithelium is keratinocyte and non keratinocytes (Langerhans cells, Merkel cells and melanocytes). Keratinized epithelium is made up of four cell layers (Figure 37.2):

- Stratum basale
- Stratum spinosum
- Stratum granulosum
- Stratum corneum

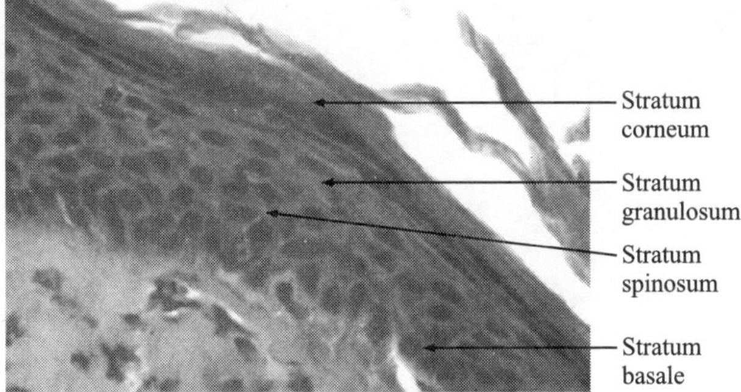

Figure 37.2 Horizontal section of gingival tissue under microscope showing various cell strata from top to bottom, Stratum basale, Stratum spinosum, Stratum granulosum and Stratum corneum (40 X). (*see Plate 9 for colour figure*)

As the cells move from stratum basale to the stratum corneum, they express keratohyaline granules; this will cause chemical and morphological changes of the cell shape and size. Cells will get flattened with an increasing tonofilaments and intercellular junction. Other proteins like keratolinin, flaggrin and involucrin are also expressed during the maturation process of the cell. This will lead to formation of a keratinized squama, a dead cell filled with densely packed protein contained with a rigid cell membrane.

Metabolic features of gingival epithelium

Oral epithelium: Oral epithelium contains many types of keratins and their distribution is very specific. Like keratins K1, K2 and K10 are specific to the epidermal type differentiation are expressed in high intensity in ortho-keratinised and less in the para-keratinized areas.

K6 and K16 are expressed in highly proliferative epithelium, K5 and K14 in stratification specific cytokeratins and para-keratinized areas expresses K19. Concentration of glycogen is inversely related to the degree of keratinization and inflammation.

Junctional epithelium: Length of Junctional epithelium is in the range of 0.25 to 1.35 mm. It expresses keratin K19, K5 and K14. Activity of the glycolytic enzyme is low and it lacks acid phosphatase activity. The Junctional epithelium is attached to the tooth surface by means of an internal basal lamina and to the gingival connective tissue by an external basal lamina. Histochemical studies have shown the presence of neutral polysaccharides in the zone of epithelial attachment.

Gingival fiber: Gingiva contains collagen fibers bundles known as gingival fibers and consist of mainly type I collagen. They are arranged in three groups (Figure 37.3):

- *Dento-gingival group:* Present on the facial, lingual and interproximal surfaces. They are attached in the cementum beneath the epithelium at the base of the gingival sulcus.
- *Circular group:* They traverse through the connective tissue of the marginal and interdental gingiva and encircle the tooth in collar like fashion.
- *Transeptal group:* They are present in the interproximal area and extend between the cementum of the adjacent teeth.

B. Periodontal Ligament

The periodontal ligament is composed of a complex vascular and highly cellular connective tissue that surrounds the root and connects it to the inner wall of the alveolar bone through sharpey's fibers. Sharpey's fibers are collagen fibers which are embedded into cementum on one side of the periodontal space and into alveolar bone on the other side. Average width of the periodontal ligament space is about 0.2 mm, even though considerable variation exists (Figure 37.4).

Periodontal ligament houses several types of cells:

Synthetic cells	Resorptive cells	Progenitor cells	Defense cells
Fibroblasts	Osteoclasts		Mast cells
Osteoblasts	Fibroblasts		Macrophages
Cementoblats	Cementoclasts		Eosinophils

Periodontal ligament fibers mainly consist of collagen, which are arranged in bundles and follow a wavy pattern. The

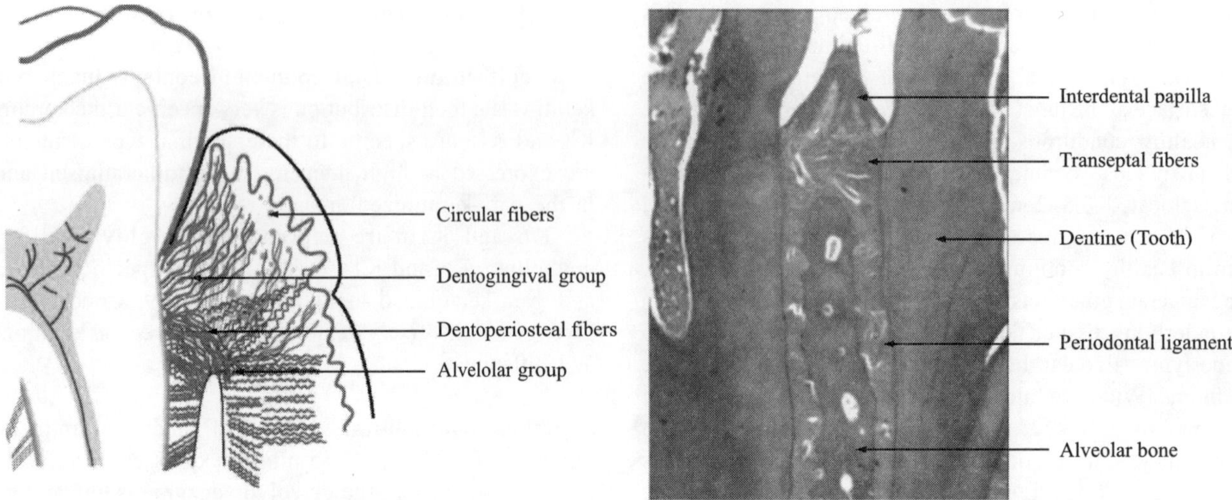

Figure 37.3 Gingival group of fibers A. Gingival fibers are attached to the gingival tissues and are arranged in three groups, i.e., dentogingival group of fibers, circular group of fibers and transeptal group of fibers. B. Histological section of the periodontal ligament between teeth in vertical plane showing arrangement of trans-spetal fibers (4 X). (*see Plate 9 for colour figure*)

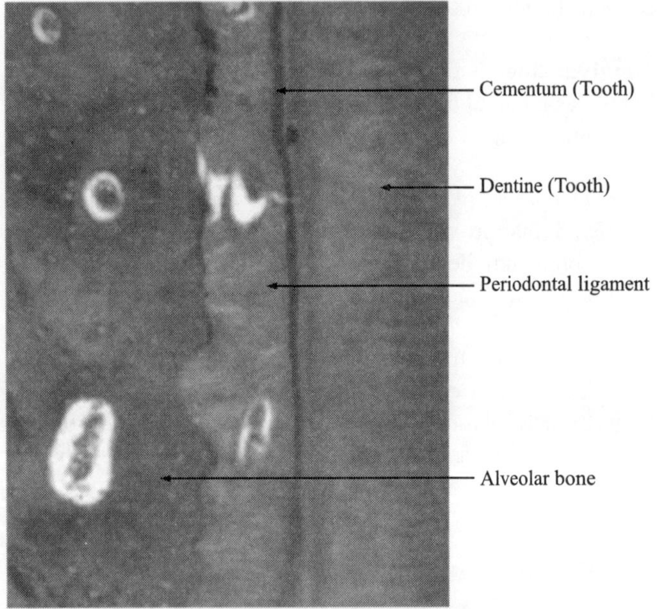

Figure 37.4 Histological section of periodontal ligament in vertical plane showing cementum, periodontal ligament and alveolar bone (10x). (*see Plate 9 for colour figure*)

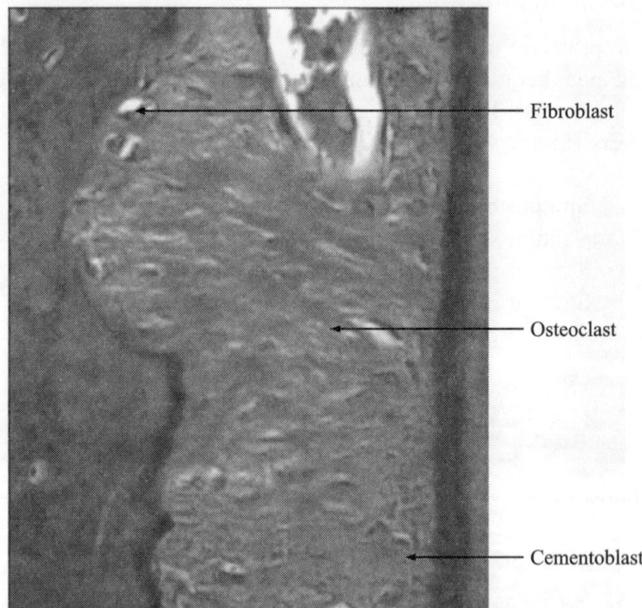

Figure 37.5 Histological section of periodontal ligament in vertical plane from left to right showing cementum, periodontal ligament and alveolar bone. Periodontal ligament showing osteoblast, fibroblast and cementoblast in the histological section (40X). (*see Plate 9 for colour figure*)

principle fibers of periodontal ligaments are attached to the tooth in a unique arrangement and can be grouped into six types on the basis of their orientation pattern (Figures 37.5 and 37.6).

The periodontal ligament fibers remain uncalcified over a period of time; this is due to the interaction of various molecules like "osteopontin" Msx2, bone sialoprotein, osteoprotegrin, Matrix 'Gla' protein and glycosaminoglycans.

On radiographs periodontal ligament space appears as a radiolucent line between the tooth and the alveolar bone (Figure 37.7).

Periodontal ligament is highly vascularised through alveolar arteries of maxilla and mandible and venous channels accompanying their arterial counterpart. Lymphatic drainage of the PDL is towards and into the adjacent alveolar bone. Periodontal ligament is supplied by the branches of superior and inferior alveolar nerve and there are four types of neural terminal ends in the periodontal ligament:

- Free nerve ends
- Ruffini's corpuscles
- Meissner's corpuscles
- Spindle type of nerve ending

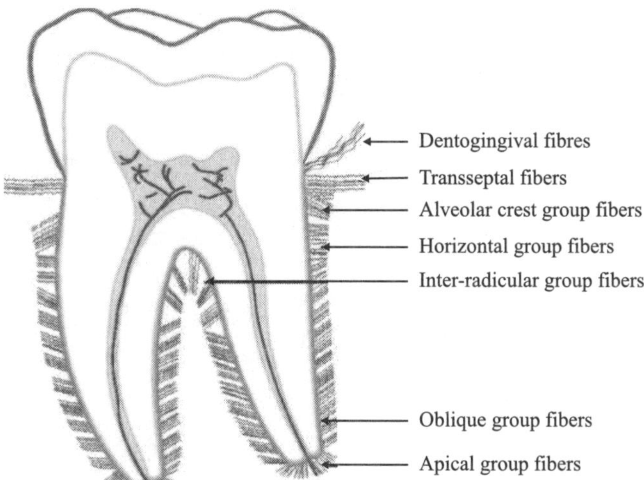

Figure 37.6 Arrangement of principle fibers of periodontal ligament. Alveolar crest fibers, Horizontal group fibers, Oblique group fibers, Inter radicular group fibers, Apical group fibers.

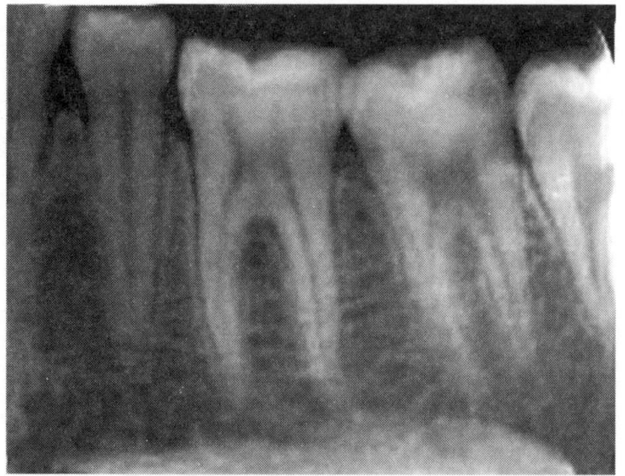

Figure 37.7 Intra Oral Peri-Apical Radiograph (IOPA) showing periodontal ligament space (radio-opaque/dark area between the alveolar bone and the tooth).

Functions of periodontal ligament:

- It provides attachment of the teeth to the alveolar bone.
- Transmits the forces from tooth to the alveolar bone.
- It maintain the gingival tissues in their normal relationship to the teeth.
- Function in tooth eruption.
- Reparative function (bone and cementum).
- It has nerve fibers which allow the passage of proprioceptive signals to higher center.
- It provides soft tissue casing to protect the nerves and vessels from mechanical forces.

C. Cementum

Cementum is a mineralized, avascular mesenchymal tissue which covers the root of the tooth. Thickness of cementum varies among different teeth and in the same teeth in different regions and at age. It is thinnest at the cemento-enamel junction and thickest towards the apex of the root. Main cells of the cementum are cementoblasts, which secretes the cementum and get embedded in its matrix and known as cementocytes. They lay down the organic matrix of cementum which later gets mineralized by minerals from oral fluids. Cementum is of two types (Figure 37.8):

Acellular cementum: Covers the cervical third of the tooth, it is formed before the tooth has reached to the occlusal table. It consist mainly of sharpey's fiber, most of the fibers are inserted into the cementum at the right angle and deeply penetrated into cementum.

Cellular cementum: It is formed once the tooth has reached the occlusal table. Most of the time it is irregular and contains cell in canaliculi and is less calcified than the acellular one.

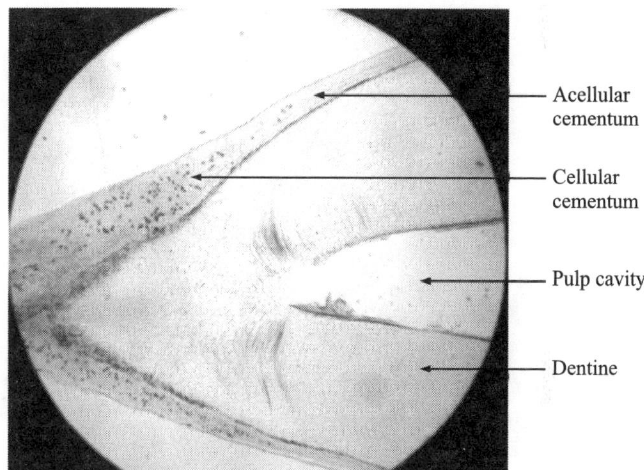

Figure 37.8 Ground glass section of apical third of root under microscope showing cellular and acellular cementum (4X). (*see Plate 10 for colour figure*)

Cementum on dry weight basis in fully formed human permanent teeth contains approximately 45–50% of inorganic substance (mainly in the form of hydroxyapatite) and 50–55% of organic substance. Organic substance of cementum is similar to bone and contains mainly type I collagen, whereas noncollagenous proteins (osteopontin, heparin sulphate, chondroitin sulphate, hyaluronic acid, keratan sulfate, lumican) have important role in matrix deposition.

Functions of cementum

- It provides a means of attachment of periodontal fibers to the root.
- It protects the underlying dentinal tubules from exposure to oral fluids and bacteria
- The new cementum around the apex of the root compensates for tooth wear and the occlusal surface and provides a means of continuous eruption of the tooth.

D. Alveolar Bone

(Latin, alveus = trough or cavity, used by Vesalius to describe the bone of the tooth socket)

Tooth is encased inside the alveolar process of the maxillary and mandibular bone. These bones are formed by membranous ossification of the mesenchymal tissues.

The level of alveolar bone around each tooth is surprisingly constant at 2 mm below the level of the enamel (the cemento-enamel junction). Alveolar bones are always in a constant state of flux. Remodeling of the alveolar bone is dependent on various factor like functional loading of the tooth, systemic condition and local factors. This will affect the density, contour and height of the alveolar bone.

Alveolar bone consists of (Figure 37.9):

- External cortical plate
- Inner socket wall (alveolar bone proper)
- And in between these two is cancellous bone

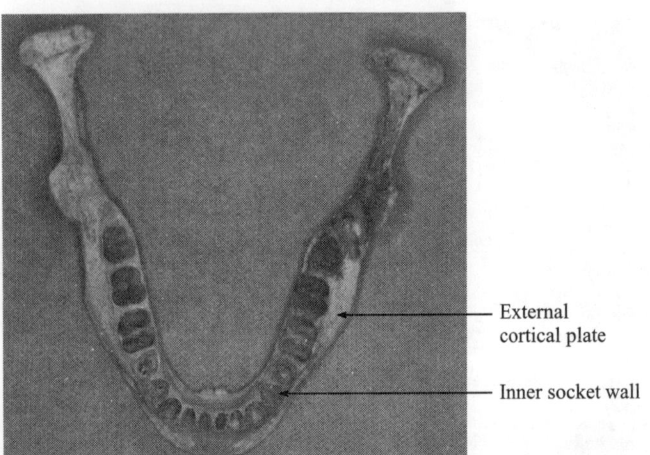

Figure 37.9 Mandibular alveolar bone showing (A) External cortical plate (B) Inner socket wall (C) Cancellous bone (present between external cortical plate and inner socket wall)

Dry weight of bone consists of 1/3rd of organic and 2/3rd of inorganic matter and similar organic contents as cementum. Inorganic components consist mainly of hydroxyapatite crystal and small amount of calcium phosphate. Cells, collagen and glycoproteins form the organic component of the bone. Bone cells consist of osteoblasts, osteocytes, osteoclasts, osteoprogenator cells. Collagen form the main bulk of the organic component of the bone (mainly of type I, and less amount of type III, type V and type XII) along with non collagen proteins like osteocalcin, osteopontein, bone sialoprotein, osteonectin, proteoglycans, bone morphogenic proteins.

Histology of alveolar bone

Circumferential and concentric lamellae form the cortical plates and lamellar dura of the alveolar bone (Figures 37.10 and 37.11). The intervening trabeculae have the same basic histology, but as a result of resorption, they have been remodelled into a honeycomb like system of trabeculae. The spaces between the trabeculae are occupied by red marrow (haemopoetic tissue) in the young, but this is replaced in the adult by fatty tissue.

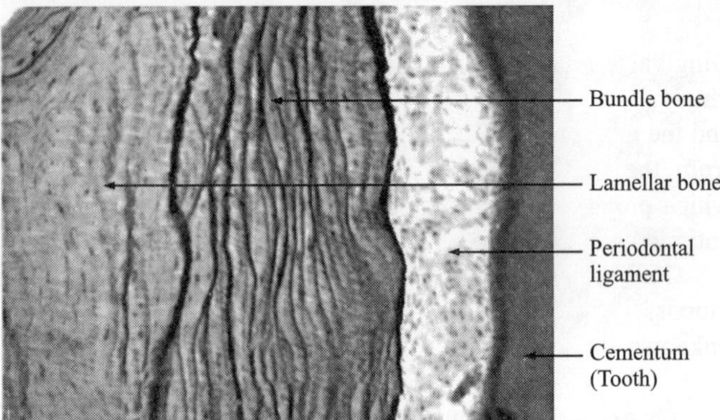

Figure 37.10 Histological section of alveolar bone in vertical plane showing bundle bone and lamellar bone (20 X). (*see Plate 10 for colour figure*)

The tooth socket is a sieve like structure; it provides attachment to the Sharpey's fibers of the periodontal ligament. These sieve like structures are traversed by blood capillaries which form the network in the PDL space. On an x-ray a lamina dura will appear as a radiopaque line surrounding the root. The radiolucent space between root and bone is occupied by periodontal ligament. Mineral content of the laminar dura is more than the inner cancellous bone, which makes the former more bright cancellous bone.

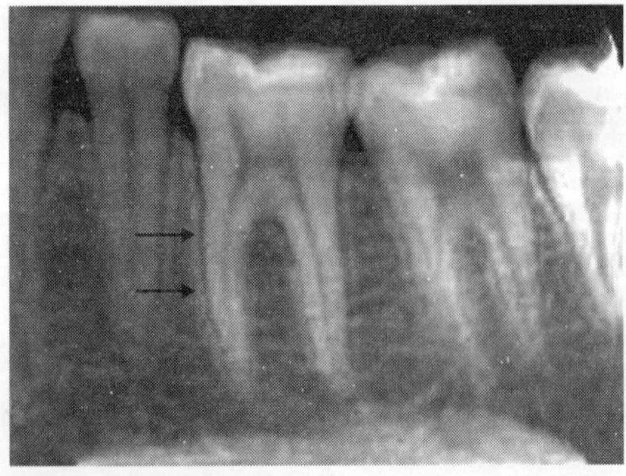

Figure 37.11 Intra Oral Periapical Radiograph (IOPA) of inner socket wall crib (Radiolucent line around the root of the tooth).

Functions

- Protection—it forms and protects the sockets for the teeth.
- Attachment—it provides the attachment to the periodontal ligament fibers, which are the principle fibers. These fibers which enter the bone and regarded as sharpey's fibers.
- Support—it supports the tooth roots on the facial and on the palatal/lingual sides.

- Shock absorber—it helps to absorb the masticatory forces by disseminating the forces to underlying tissue.

II. GINGIVAL CREVICULAR FLUID (GCF)

Gingival crevicular fluid is present in minute quantity in gingival crevice, i.e., in between the free marginal gingiva and the tooth surface. It is an exudate which cleanses material from the crevice; it also contains sticky plasma proteins which promote adhesion of the epithelial attachment and has antimicrobial properties.

GCF is produced by the sulcular epithelium of the oral mucosa. The exact mechanism of production of GCF is unknown, although it is know that gingival fluid:

- is not a salivary product, since the high protein content is not seen in saliva and an outward flow of the fluid has been shown.
- is not a glandular secretion, since no glands can be demonstrated in the gingiva.
- is not a straight vascular transudate, as this would have an ionic composition similar to plasma.

Alfano (1974) was first to propose the mechanism of production of crevicular fluid. According to him a proinflammatory flow of gingival fluid may be osmotically mediated. When the gingiva is healthy, very small amount of subgingival plaque will give rise to limited quantities of macromolecular byproducts which gets removed by absorbing to the surface of desquamating epithelial cells or, through phagocytosis. When more macromolecules are present, they will diffuse to the basement membrane, this will create an osmotic gradient and flow of gingival fluid is created. This osmotically modulated fluid is not an inflammatory exudate.

Gingival fluid production is modulated by the passage of fluid from capillaries into the tissue and removal of interstitial fluid by the lymphatics of the gingiva. When the production of fluid from capillaries is greater than the lymphatic uptake, GCF is formed.

Challacombe (1980) measured the volume of gingival cervical fluid in humans individual. The mean crevicular fluid volume for molar teeth ranges from 0.43 to 1.56 micro lit and in anterior teeth, the volume was between 0.24 to 0.43 micro lit per tooth. In healthy mouth approximately 0.5 to 2.4 milliliter of crevicular fluid is secreted per day. Studies done on circadian pattern of secretion of crevicular fluid by various researchers have confirmed that there is a gradual increase in the flow of GCF from 6 am to 10 pm and a decrease afterwards.

Due to dynamic nature of GCF secretion its contents and composition may vary. Glucose concentration in GCF is 3 to 4 times greater than the serum, while the total protein content of GCF is much less than that of serum.

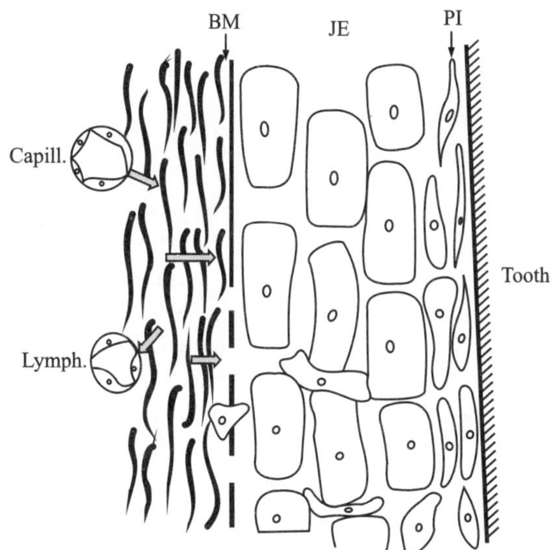

Figure 37.12 Schematic representation of mechanism of gingival fluid formation. in the absence of inflammation, macromolecules originated from plaque (Pl) will diffuse inter-cellularily and reach the basement membrane (BM) where they create an osmotic gradient and attract interstitial fluid towards the sulcus.

A. Functions of GCF

- Protective effects
  - Clearance of cells and potentially dangerous bacterial molecules
  - Antibacterial action of immunoglobulins
  - It dilutes and washes away harmful enzymes
- Negative effects
  - Tartar formation is induced by alkaline phosphatase
  - Proteolytic enzymes are harmful for the gingival sulcus and the other gingival tissues,

Cellular and humoral activity in GCF:
- Analysis of GCF provides the information about cellular and humoral responses in both health and disease.
- IL-1β, TNFα, IL-1 alpha and IL-1 beta are known to increase the binding of polymorphnuclecytes, monocytes/macrophages to endothelial cells, stimulate the production of PGE-2 and release of lysosomal enzymes and simulates bone resorption.
- Presence of Interferon gamma in GCF may have a protective role in periodontal disease because of its ability to inhibit bone resorption activity of IL-1 beta.
- Some of GCF components which are involved in the initiation of the periodontal disease:
  - Cathepsin-K
  - Interleukin-4 and interferon-gamma
  - Leptin
  - Osteocalcin
  - Alkaline phosphatase

An increase in concentration of these substances can be detected very early in the GCF before the actual clinical manifestations of inflammation.

B. Clinical Significance

- Amount of GCF is greater when inflammation is present and proportional to the severity of the inflammation is more.
- GCF production increases due to mastication of coarse foods, tooth brushing and gingival massage, ovulation, hormonal contraceptives, smoking and periodontal disease.
- Female sex hormone increases GCF flow, by enhancing the vascular permeability.
- Drugs in GCF
  - Drugs which are excreted through the GCF may be used advantageously in the periodontal therapy.
  - Tetracyclines are excreted through GCF. Minocycline is another antibiotic which is detected in the GCF with clinically significant concentration.

It is now an established fact that the constituents of GCF are reflection of systemic as well as local condition.

- The amount of fluid recoverable from gingival crevices is small and it can be only used for sensitive immune assays, which permits the analysis of the specificity of the antibodies.

C. Orthodontic Treatment and GCF

Orthodontic tooth movement which is used to correct abnormal tooth position is controlled by inflammatory responses to the applied orthodontic force/s. Prolonged application of mechanical forces that exceed the bioelastic limits of tooth-supporting structures induces tooth movement. During orthodontic treatment, the applied forces produce a distortion of the alveolar bone (bone bending), periodontal ligament extracellular matrix, resulting in alterations in cellular shape and cytoskeletal configuration. This will cause synthesis of tissue-degrading enzymes, acids, and inflammatory mediators. It also induces cellular proliferation and differentiation which promotes wound healing and tissue remodeling. These changes may modify both the GCF flow rate and its components.

Most of the studies which relates the GCF markers with orthodontic tooth movement can be categorized in four group:

1. Markers that are metabolic products of para-dental remodeling
2. Markers that are inflammatory mediators
3. Markers that are enzyme and enzyme inhibitors
4. Markers that are correlated to roots resorption

D. GCF and Ageing

The amount of GCF increases with age; this is directly related to the amount of gingival inflammation.

III. PERIODONTIUM AND AGEING

Anatomic and functional changes have been reported in the periodontium with aging; this will cause alteration in the periodontium dynamics.

A. Gingiva

Junction epithelium migrates from its normal position to a more apical position on the root surface, causing gingival recession; this decreases the level of attached gingiva. In the edentulous gingivae marked loss of nerves and terminal endings has been observed.

Changes in the structural elements of the epithelium and connective tissue take place with age. The basal membrane of epithelium becomes loose and shows more micro-filaments in the cytoplasm. Granular layer cells show increase in keratohyaline granules in cytoplasm, the cells become large and their typical form is lost. Basal, spinous and granular layer cells show decrease in the cytoplasmic organells. Mitochondria acquire the appearance of electron translucent cavities. More changes occur in intercellular inter-relations like widening of intercellular spaces. The ultra structural rearrangements of epitheliocytes in the human gingiva demonstrate certain disturbances in keratinization processes, in mechanical firmness, as well as in barrier function of the epithelial layer.

The gingival connective tissue becomes thick, coarse and dense. Along with this there is increase rate of conversion of soluble collagen to insoluble collagen, which is caused by interlinking of the soluble collagen causing increase mechanical strength. Experimentally it has been shown that there is increased collagen content in the gingiva of older animals despite of lower rate of collagen synthesis with age.

The role of reactive oxygen species (ROS) and gingival fibroblastic cell dysfunction was pointed by Sohal et al. They have mentioned that ROS is linked to the periodontal aging and related diseases. The effect of aging related increased in ROS might lead to the damage of the lipid peroxidant and DNA damage of the fibroblast.

B. Periodontal Ligament

Periodontal ligament cells have great regenerative capacity, which are involved in the maintenance of the local structure as per the requirement. The fibers and the cellular components of PDL decrease and the organization of it become more irregular with age. With ageing there is significant reduction in chemotaxis, mobility and proliferation of the periodontal ligament cells. Reduction in osteoblasts chemotaxis and lower rate of osteoclasts differentiation in the cells of older age group has been observed. Periodontal ligament derived from the elderly people showed an increase in synthesis and secretion of plasminogen activator, IL-8 and IL-6. These products will affect the PMN migration, proliferation and migration of epithelial cells as well as fibroblasts.

C. Cementum

Formation of the cementum is a continuous process which occurs throughout the life, this leads to increase thickness of cementum. Acellular cemental deposition is seen at apical 1/3rd of the root throughout life, this is to compensate for occlusal wear of tooth. Cementum accumulates in large quantities along the interradicular surfaces of the tooth and continual re-apposition of new layers of cementum represents the aging of the tooth.

Cemental resorption and apposition increases with age and may be responsible for an increased irregularity of the cemental surface with advanced age.

D. Alveolar Bone

Bone formation steadily reduces with age after adulthood, leading to significant reduction in the bone mass this is due to decrease in osteoblastic precursor proliferation; it might also be due to decrease in synthesis and secretion of essential bone matrix protein. Gottlieb and Orban have mentioned that with advancing age alveolar bone resorption occurs and may be related to gingival recession. Alveolar process resorption is known to occur without the former loss of dentition, this will reduce the height of the alveolar bone. The extracellular bone matrix around the osteoblasts plays important role in bone metabolism; and studies have shown that with advancing age there is dysfunction of this extracellular matrix of bone.

IV. PERIODONTIUM AND ITS BIOCHEMISTRY

A. In Health

Gingiva

Gingival epithelium varies in their degree of keratinization, number of cells and presence of rete pegs. There are no fibrous protein components of the epithelial extracellular matrix and the non-fibrous epithelial components. They synthesize and secrete sulfated molecules that contribute to intercellular cementing substance of gingival epithelium. Hyaluronic acid, decorin, syndecan CD 44 has been observed in the intercellular spaces of the gingival epithelium.

Fibroblasts form approximately one tenth of the gingival connective tissue volume. These cells are responsible for producing connective tissue elements in health and disease. Collagen is the most abundant biochemical components in the gingival connective tissue and it forms approximately 60% of the total tissue protein. Gingival collagen fibers are made up of a heterotypic mixture of collagen types. These fibers are arranged in two patterns, dense bundles of thick fibers or in a loose pattern of short fibers mixed with a fine reticular network. They contain mainly Type l and lll collagen. Integrins like alpha2, beta1 and alpha3 and beta1 are present on basal keratinocytes in the epithelium.

Gingiva contains several noncollagenous proteins. Fibronection is one of the non collagenous proteins and is localized throughout the connective tissue. Elastin is a minor component of the gingival connective tissue, along with osteonectin.

The major glycosaminoglycans of the gingival connective tissue are dermatan sulfate followed by heparin sulfate and hyaluronic acid and chondroitin sulphate. Proteoglycans species identified in the gingiva include decorin, biglycan, verisican and syndecan. CD44, is a cell surface proteoglycans located on fibroblasts, and at the cell to cell contact areas at basal and spinous areas of the oral epithelium.

Periodontal ligament

In mature and completely developed periodontal ligament, collagen fibers are organized into distinct structural forms. The major collagen of the periodontal ligament extracellular matrix is Type l, with presence of Type lll, V and Vl in small quantity. Type lll collagen is predominantly localized with major fibrils, while Type V collagen has a limited distribution. Type Vl collagen is present in the periodontal ligament as fine micro fibrils appearing to interconnected cross striation. Type Xll collagen is believed to involved in the three dimensional organization of the extracellular matrix, this collagen is restricted to mature tissues and is not expressed in developing one, but is expressed during tooth movement.

The periodontal ligament contains fibronectin and small quantities of many other noncollagenous proteins like tenascin, which is present in between densely packed collagen fibrils. Elastin is present within the periodontal ligament as a meshwork made up of elastic lamellae. Osteonectin and other nonmatrix components such as vitronectin is also present Integrin subunits locate in periodontal ligament include alpha 5, and beta 1.

The glycosaminoglycan component which is present in the periodontal ligament includes hyaluronic acid, heparan sulfate, dermatan sulfate and chondroitin sulphate. The periodontal ligament glucosaminoglycan has molecular weight in the range of 18–20 kd.

Cementum

Biochemically it is made up of an inorganic matrix and an organic matrix which is mainly composed of Type l and lll collagen. Sharpey's fiber which is embedded in bone as well as cementum is made up of Type l collagen. The organic matrix of the cementum also contains bone sialoprotein, osteopontein, tenasin, fibronectin and proteoglycans.

Bone sialoprotein has cell adhesion properties, appears to be associated with mineralization during the cementogenesis. Osteopontin is another adhesion protein which is associated with acellular cementum. The cementum sequesters several growth factors which include Fibroblast growth factor-1 and 2, bone morphogenic protein, Tumor growth factor- beta, and Insulin like growth factor-I like molecule and growth factor receptors. Cementoblasts of the cellular cementum appear to produce above mentioned proteins along with TFF-beta and IGF-I.

Alveolar bone

The major organic constituent of the alveolar bone matrix is Type I collagen. Type I collagen is co-distributed with Type III collagen. Type I collagen is mainly expressed by osteoblasts and by osteocytes in areas where alveolar bone is undergoing remodeling fibronectin, its receptor and trenascin is also expressed in expressed in periosteal and endosteal layers. Alveolar bone also expresses bone specific noncollagenous proteins like, osteocalcin, bone sialoprotein, osteopontein and osteonectin.

In the extracellular matrix of the alveolar bone, fibronectin, bone sialoprotein, osteopontein and several other cell adhesion molecules are present along with fibroblast growth factor-1 and 2, platelet derived growth factor, IGF-I, TGF-beta and BMP.

B. In Disease

Periodontal disease is one of the most prevalent chronic diseases of mankind. The term periodontal disease refers to all diseases of the periodontium that can affect the tissues around and supporting the teeth.

The disease which affects the periodontium can be braodly categorized into two groups:
- Inherited (Due to defects in gene structure of proteins)
- Acquired (This group includes inflammatory disease that affect the periodontium directly, and those that are secondary to other disease)

When the inflammation is localized or, restricted to the gum tissue it is knowm as gingivitis. In the presence of certain genetic and organic factors or chronically poor oral hygiene, this process may extend to the supporting structure of the tooth (like periodontal ligament, bone), which results in an irreversible damage to the periodontium (i.e, periodontitis).

Periodontitis presents the classic clinical features of chronic inflammation with the infiltration of mononuclear cells, their conversion to macrophage, and antibody production by the plasma cells. In the situation where lesion is prolonged, focal area of fibrosis and granulation tissue will develop.

One of the most striking features of the inflammatory periodontal disease is the destruction of gingival connective tissue. When the inflammation of gingiva has just started, the tissue destruction begins within the peri-vascular extracellular matrix. Gingivitis causes matrix degradation due to the activity of matrix metalloproteinases, which is produced by the macrophages and fibroblasts. A quantitative and qualitative change in the gingival collagen of patients with periodontal disease has been observed. In gingiva, the collagen becomes more soluble; Type 1 and Type III collagens are lost at the site of inflammation. The ratios of collagen types are also altered; the amount of Type V collagen increases and its amount may exceed that of the Type III. Noncollagenous gingival proteins are also destroyed along with the collagens. This destruction is carried out by matrix metalloproteinases released by the neeutrophils and macrophages. The breakdown

of the matrix is due to an imbalance between the activated matrix metalloproteinases and their endogenous inhibitors (Reynolds et al., 1997).

The gingival proteoglycans are also frequently affected by the inflammatory process. In the inflamed gingival tissues, the amount of dermatan sulfate decreases while, the concentration of chondroitin sulfate increases. In the inflamed gingival connective tissue, there is degradation of proteoglycans core proteins and hyaluronic acid is frequently seen (Bartold et al., 1986). Inflammatory cells also synthesize proteoglycans; especially proteoglycans containing chondroitin sulfate, this will increases the chondroitin sulfate concentration at the site of inflammation (Bartold et al., 1986).

Periodontal ligament manifests topographic changes in the distribution of various glycoseaminoglycan; the major change is increase in the presence of chondroitin sulfate (Kirkham et al., 1992). Changes in the hard tissue matrices of bone and cementum differ to a great extent due to their different anatomic locations. Cementum might get altered due to its exposure to the oral and periodontal environment, in which there is a loss of collagenous attachment and both its organic and inorganic components are changes.

The GCF contains the breakdown down products of the various periodontal tissue matrices, this provide an easy way to access the status of the periodontal structure in the beginning stage itself.

V. ORTHODONTIC TOOTH MOVEMENT AND PERIODONTIUM

Orthodontic forces are used to correct the abnormal tooth positions. Orthodontic tooth movement is result of a biologic response to interference in the physiologic equilibrium of the dentofacial complex by an externally applied force. Forces from the orthodontic appliances get transferred to the tooth and then to the periodontal ligament and finally to the bone. There are three mechanisms of tooth movement:

1. The application of pressure and tension to the periodontal ligament (PDL): Tooth movement can be explained in a most simplified way through the pressure-tension hypothesis. This hypothesis explained that, on the pressure side, the PDL displays disorganization and diminution of fiber production. Here, cell replication decreases due to vascular constriction. On the tension side, stimulation produced by stretching of PDL fiber bundles results in an increase in cell replication. This enhanced proliferative activity leads eventually to an increase in fiber production.

According to this theory the orthodontic force induce the PDL progenitor cells differentiation into compression-associated osteoclasts and tension-associated osteoblasts, causing bone resorption and apposition, respectively. It is suggested that width changes in the PDL cause changes in cell population and increases in cellular activity.

2. Bending of the alveolar bone: Alveolar bone bending plays a pivotal role in orthodontic tooth movement. When an orthodontic force is applied to the tooth, forces delivered to the tooth are transmitted to all tissues near force application. These forces bend bone, tooth, and the solid structures of the PDL. Bone was found to be more elastic than the other tissues and to bend far more readily in response to force application. The active biologic processes that follow bone bending involve bone turnover and renewal of cellular and inorganic fractions. These processes are accelerated while the bone is held in the deformed position.

The applied mechanical forces will generate the bioelectrical signals, i.e., electric potentials in the stressed tissues. It has been observed that concave side of orthodontically treated bone is electronegative and favors osteoblastic activity, whereas the areas of positivity or electrical neutrality—convex surfaces—showed elevated osteoclastic activity.

3. Fluid dynamic theory: This concept is also known as blood flow theory of tooth movement, it occurs as a result of alterations in fluid dynamics in periodontal ligament. Periodontal space is a confined space & passage of fluid in and out of this space is limited which will create a hydrodynamic condition which resembles a hydraulic mechanism and a shock absorber.

On application of orthodontic force blood vessels get trapped between principal fibers and stenosis results. Vessels above stenosis get dilated which results in formation of aneurysm. These can act as minute springs between tooth and bone, help in replenishing squeeze film, dissipate the applied force. Also there is alteration in the chemical environment at the site of vascular stenosis.

The orthodontic loading of periodontal tissue will provide a favorable microenvironment for tissue deposition and resorption. Removal of orthodontic forces causes significant changes in the number and density of the periodontal vasculature. During early stages of change in the periodontal ligament, a rapid change in local blood circulation is observed. There is increase in the number of blood vessels and the blood flow to the tension side compared to the compression side is observed. Studies have confirmed that there is increase in the Type I collagen and fibronectin. These changes suggest increase metabolic activity of the PDL.

Once the tooth has moved to correct position, retention phase of treatment begins. In this phase the bone around the tooth is still in the phase of remodeling. First bone which is laid down around the tooth after orthodontic tooth movement is woven bone, which has irregular arrangement of collagen and less mineral content. This immature (woven) bone will transform into the lamellar or mature bone over a period of 6–7 months after tooth movement. During this the bony trabeculae will get organized and the mineral content will increase.

After orthodontic tooth movement the periodontal mechanoreceptors, although having some of their response properties altered during orthodontic force application, were able to recover and adapt to the newly acquired intraoral condition after removal of the orthodontic force.

VI. PERIODONTIUM IN ROOT RESORPTION

Root resorption is breakdown or destruction, and subsequent loss, of the root structure of a tooth. This is caused by living body cells attacking part of the tooth, i.e., root cementum and dentine.

The main etiological factors associated with root resorption can be categorized as:

1. *Physiological:* In this pattern the craters of root resorption are seen on root surfaces.
2. *Pathological:* Common cause for this pattern of root shortening is localized inflammation, tumor and cyst, localized trauma which exert the pressure on the root and releases inflammatory mediators.
3. *Iatrogenic:* Root resorption is common undesirable and unpredictable side effect of orthodontic treatment. Various factors like magnitude of applied orthodontic force, the duration of force application, the type of orthodontic appliance along with individual variations play a major role as an etiological factor for root resorption. It has been also observed that genetic factors also predispose the root resorption in some cases.

A. Mechanism of Root Resorption

Periodontal vasculatures usually get compressed by local factors (like tumor, cyst, and orthodontic force) and develop a stage of transient ischemia. This will cause release of various chemical mediators of inflammation. The inflammatory products along with the cells (Macrophage, Monocyte) will cause shortening of the root.

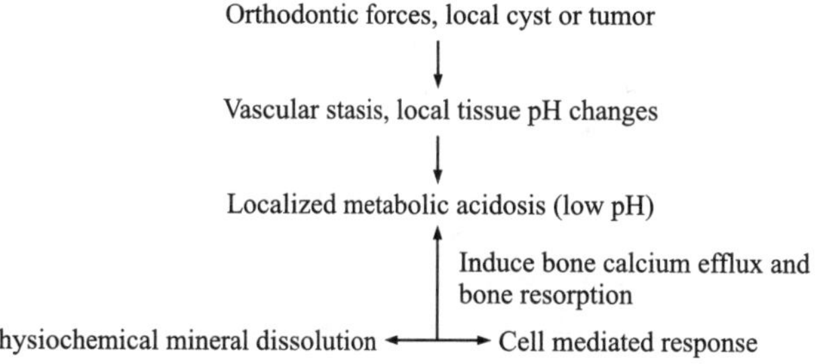

The loss of tooth structure generally proceeds in a sequential manner to the loss of cementum, prior to the dissolution of dentine. During orthodontic tooth movement, the focal areas of cementum are resorbed which are subsequently repaired if stimulus causing the resorption is withdrawn. Therefore, the proteins of cementum may not be indicative of permanent loss of root surface. On the other hand, once the resorption has reached to the dentine, the dentine protein will be removed from the surface and will appear in GCF. Very small area of dentine which has been resorbed undergoes repair, but moderate to severe dentine resorption hardly undergoes repair.

The healing process of resorbed craters is believed to start when the applied orthodontic force is discontinued or, reduced below a certain level or, the pathological factor is removed.

VII. PERIODONTIUM AND IMPLANTS

(Prosthetic dental implants and Orthodontic mini-implants/ Temporary anchorage devices)

A dental implant is a metallic screw like structure which is placed inside the alveolar bone. Prosthodontic implant is used to replace the missing tooth for improving the occlusal table and improving the occlusion. Prosthodontic implant system was designed and developed to provide the osseointegration of the implant material with surrounding bone for replacing missing teeth.

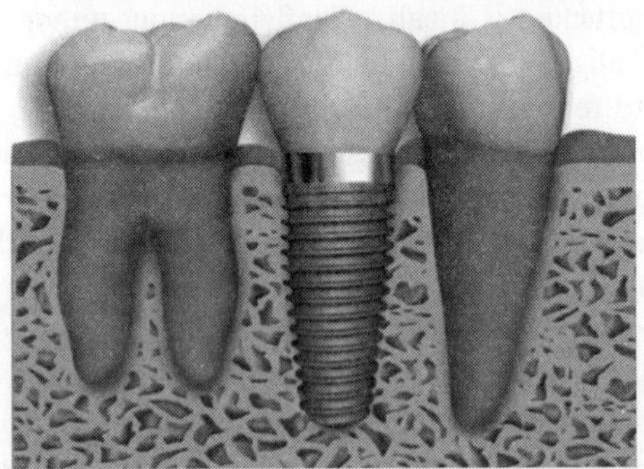

Figure 37.13   Prosthetic dental implant.
(see Plate 10 for colour figure)

Most of prosthodontic implants are self tapping type and the portion which is going to be in contact with the alveolar bone is coated with bone growth conducing material so that bone can grow on the surface of implant and enhance osseointegration. The endosseous implants and the bone relationship can be one of two types:

- Osseointegration – In this bone is in intimate but not ultrastrcutre contact with the implant
- Fibrosseous integration – Here soft tissue like fibers, are interposed between the two surfaces.

The concept of osseointegration was purposed by Branemark et al. Osseointegration is defined as a direct structural and functional connection between ordered living bone and the surface of a load carrying implant.

Prosthetic implants are made up of pure titanium or, titanium alloys. The clinical success of endosseous implants is associated with the formation and maintenance of bone implant inter-phase. At the prosthetic implant interphase, the presence of osteoprotegrin and alpha-2-HS-glucoprotein clearly demonstrate the contribution of osteoblastic matrix protein in the formation of the interface. This electron dense interfacial transition zone represent a glycoprotein rich, collagen free region of extracellular matrix that resembles cemental lines of bones. Osteopontin is a bone protein which is present in abundance at the intact bone implant surface of both titanium and hydroxyapatite implants.

Osseointegration of metallic implants involves the formation of an acellular, amorphous interface which is interposed between the implant surface and vital tissue. The bone formation around the prosthetic implant is enhanced by means of hydroxylapatite coating and making the surface rough by grit blasting. Bone formation will takes place toward the implant surface, which is shown by H3 Proline autoradiography. There is gradual mineralization which is directed toward the implant surface. Active bone formation will proceeds from the surgical margins of bone towards the implant surface. Macrophages adhere to both the implant and hydroxyapatite coating of the implant. The surface of hydroxyapatite and hydroxyapatite coated implants is resorbed over a period of time. Adell et al. (1981) and Albrektsson et al. (1986) have mentioned that the mean crestal bone loss around the prosthetic implant is 0.9–1.6 mm during the first year of implant placement and its function, which reduces to 0.05–0.13 mm per subsequent year.

Orthodontic mini implant/miniscrew is a device that is temporarily fixed to the bone for the purpose of enhancing orthodontic anchorage either by supporting the teeth of the reactive unit or, by obviating the need for the reactive unit altogether, and which is subsequently removed after use. Orthodontic mini-implants are made up of titanium or titanium alloy or surgical grade IV stainless steel. They are used for short time for various orthodontic or, orthopedic purposes and once it has served the purpose it is removed.

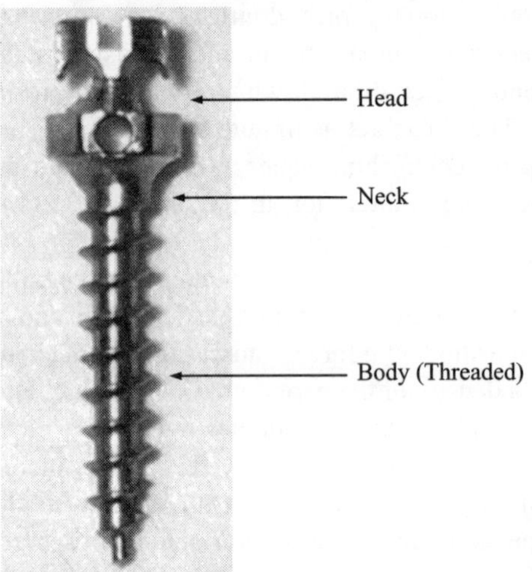

Figure 37.14   Orthodontic mini-screw. Orthodontic miniscrew consists of three parts like head, neck and body.

This device is placed inside the bone. Bone consists of periosteum, cortical bone, trabecular bone and bone marrow. Mini-implant comes in contact with the dental hard and soft tissue (gingiva) once it is placed. The surfaces of implant surface milled and smooth so as to discourage osseointegration.

Clinical studies have shown that miniscrew have success rate of more than 80%. Major reason for this failure is gingival inflammation around the orthodontic mini-implant. Healthy peri-implant tissue plays an important role as a barrier to pathological micro-organism. In most of the clinical situation the gingival seal is not formed between the gingival collar of the implant and the soft tissue thus providing the place for bacteria and fluid to get retained which produces the inflammation.

Therefore peri-implant crevicular fluid can be used to detect the level of pro-inflammatory products to monitor the health of the tissues surrounding the mini-implant. It has been proved that the elevated level of IL-1 beta in mini-implant crevicular fluid should be used as a marker for monitoring the health status of dental implants.

Studies have been shown there is increase in the level of IL-2, 6, 8 and chondroitin sulphate levels around the orthodontic mini-implants during orthodontic tooth movements, depicting the bone remodeling around the mini-implant.

SUMMARY

The periodontium is the investing and supporting tissues of the tooth. It is subjected to morphological and functional variations as well as changes associated with age. Biochemical nature of the periodontium changes in health and disease. Inflammation is the central pathological feature of the periodontal disease. An orthodontic tooth movement is also a controlled inflammatory process which alters the biochemical nature of the tooth supporting structures (root resorption, gingival recession etc). The soft and hard tissues which surround the dental implants have many similar features and some important differences with the periodontal tissues.

Acknowledgement

Figures 37.2, 37.3b, 37.4, 37.5, 37.8, 37.10, Department of Oral and Waxillofacial Pathology, Manipal College of Dental Sciences, Mangalore, India.

SUGGESTIONS FOR FURTHER READING

Abiko Y., Shimizu N., Yamaguchi M., Suzuki H. and Takiguchi H. (1998), Effect of Ageing on Functional Changes of Periodontal Tissue Cells, *Ann Periodont.*, 350–369.

Alfano M.C. (1974), The Origin of Gingival Fluid, *J. Theor Biol.*, 47,126–136.

Bartold P.M., Walsh L.J. and Narayan S. (2000), Molecular and Cell Biology of the Gingiva. *Perio.*, 24, 28–55.

Borden S.M., Golub L.M. and Kleinberg I. (1977), The Effect of Age and Sex on the Relationship Between Crevicular Fluid Flow and Gingival Inflammation in Humans. *J. Periodontol Res.,* 12(3), 160–165.

Branemark P.I., Hansson B.O., Adell R. and Breine U. (1977), *Osseointegrated Implants in the Treatment of the Edentulous Jaw.*, Almquist and Wiksell Stockholm, 132.

Cabrini R.L. and Caranza F.A. (1966), Histochemistry of Periodontal Tissue, A Review of Literature, *Int. Dent. J.*, 16, 476.

Carlsson L.V., Albrettsson T. and Berman C. (1989), Bone Response to Plasma Cleaned Titanium Implant. *Int J. Oral Maxillofac Implants*, 4, 199–204.

Cimasoni G. (1983), Crevicular Fluid Update, 45–102, Karger, Basel.

Del F.M., Francetti L., Bulfamante G., Cribium M.G. and Weinstein R.L. (2001), Fluid Dynamics of Gingival Tissues in Transition from Physiological Condition in Inflammation, *J. Periodontol.*, 72, 1, 65–73.

Ebersole J.L. (2003), Humoral Immune Responses in Gingival Crevicular Fluid: Local and Systemic Implication, *Perio.,* 31, 135–166.

Embrey G. (1991), An Update on the Biology of the Periodontal Ligament, *Euro J. Ortho.*, 12, 77–80.

Goodson J.M. (2003), Gingival Crevicular Fluid Flow, *Perio.*, 31, 43–54.

Hamamci N., Filiz A.K., Ersin U. and Beran Y. (2011), Identification of Interleukin 2,6 and 8 Levels Around Miniscrews During Orthodontic Tooth Movement, *Advanced Access.*

Imai I., Suttichai K., Prachya K., Sirwan O.C. and Boonsiva B. (2010), Chondroitin Sulphate (WF6 Epitope) Levels in Peri–Miniscrew Implant Crevicular Fluid During Orthodontic Loading, *Eur J. Orthod.*, 32, 60–65.

Karthikeyan B.V. and Praddep A.R. (2007), Leptin Levels in Gingival Crevicular Fluid in Periodontal Health and Disease, *J. Periodontal Res.,* 42(4), 300–304.

Kovalev E.V. and Gusev S.A. (1988), Age-related Changes in the Ultrastructural Organization of the Human Gingival Epithelium, *Arkh Anat Gistol Embriol.*, 94(4), 44–47.

Kumar G.S. (2006), Orban's *Oral Histology and Embryology,* 12th ed., Elsevier.

Kuroda S., Sugawara Y., Deguchi T., Kyung H.M. and Yamamoto T.T. (2007), Clinical Use of Miniscrew Implants as an Orthodontic Anchorage: Success Rates and Postoperative Discomfort, *Am J. Orthod Dentofac Orthop.*, 131, 9–15.

Lamster I.B. (1992), The Host Response in Gingival Crevicular Fluid: Potential Applications in Periodontal Clinical Trials, *J. Periodontol.*, 63(12):1117–1123.

Masada M.P., Person R., Kenney J.S., Lee S.W., Page R.C. and Alison A.C. (1990), Measurement of Interleukin–1 Alpha and 1 Beta in Gingival Crevicular Fluid: Implication for the Pathogenesis of Periodontal Disease, *J. Periodont Res.*, 25, 3, 156–163.

Melsen B. (2001), Tissue Reaction to Orthodontic Tooth Movement—A New Paradigm, *Eur J. Ortho.*, 23, 671–681.

Monga N., Kharbanda O.P. and Duggal R. (2011), A. Clinical and Biochemical Study of *Peri–implantitis Around Temporary Anchorage Device (TAD)*, Masters Thesis. CDER, AIIMS. Nov. 85

Myriam A., Cecilia E.C. and Karina M.S. (2009), Enzymatic Profile of Gingival Crevicular Fluid in Association with Periodontal Status, Lab Medicine, 40, 5, 277–280.

Nanci A. and Bosshardt A.A. (2006), Structure of Periodontal Tissues in Health and Disease, *Perio 2000*, 40, 11–28.

Nanci A. (2008), Ten Cate's Oral Histology—Development, Structure and Function, 7th ed., Elsevier.

Newman M.G., Takei H., Klokkevold P.R. and Carranza F.A. (2012), Carranza's Clinical Periodontology, 11th ed., Elsevier.

Perinetti G., Paolantonio M., D'Attilio M. and D'Archivio D. (2003), Aspartate Aminotransferase Activity in Gingival Crevicular Fluid During Orthodontic Treatment, A Controlled Short-term Longitudinal Study, *J. Periodontol.*, 74(2), 145–152.

Perinetti G., Serra E., Paolantonio M., Bruè C., Meo S.D. and Filippi M.R. (2005), Lactate Dehydrogenase Activity in Human Gingival Crevicular Fluid During Orthodontic Treatment: A Controlled, Short-term Longitudinal Study, *J. Periodontol.*, 76(3), 411–7.

Perinetti G., Paolantonio M., Femminella B., Serra E. and Spoto G. (2008), Gingival Crevicular Fluid Alkaline Phosphatase Activity Reflects Periodontal Healing/Recurrent Inflammation Phases in Chronic Periodontitis Patients, *J. Periodontol.*, 79(7), 1200–1207.

Sampathnarayan A. and Bartold P.M. (1998), Biology of the Periodontal Connective Tissues, Quintessence Publishing.

Storey E. (1973), The Nature of Tooth Movement, *Am J. Ortho*, 63, 292–315.

Tuncer B.B., Ozmeric N., Tuncer C. and Teoman (2005), Levels of IL–8 During Tooth Movement, *Angle Ortho.*, 75, 631–636.

Vander Velden U. (1984), Effect of Age on the Periodontium, *J. Clin Periodontol*, 11, 281–294.

Verna C., Zaffe D. and Sicillani G. (1999), Histolmorphometric Study of Bone Reaction During Orthodontic Tooth Movement in Rats, *Bone*, 24, 4, 371–379.

38

Biochemistry of the Eye

D. Balasubramanyam

CONTENTS

I. INTRODUCTION: THE OUTER SURFACE OF THE EYE

The mammalian eye is a well-protected organ, whose main function is to offer vision to the individual. The anatomy of the various parts of the eye is presented in Figure 38.1. The entire eyeball is placed securely inside the socket in the skull. It is protected in the front or anterior by the eyelid and lashes, and is kept clean and sterile by constant blinking (about 20 times a minute), during which time it is kept moist and lubricated by the tear fluid. We describe here the various components and functions of the ocular surface, which are shown in Figure 38.2.

The lacrimal gland (the word lacrima means tears), and the Meibomian gland are both present in the lid, and together they generate the tear fluid. The normal tear contains 6–10 mg/ml of as many as 500 proteins and peptides. Particular mention must be made of hydrolases and protease inhibitors, immune response agents, and antibiotic proteins and peptides such as lysozyme, lactoferrin and histatin, all of which help the ocular surface remain clean and contamination-free. The Meibomian gland produces lipids, particularly fatty acids, triglycerides, cholesteryl esters, steroids, and wax. As much as 75% of the Meibomian gland fluid is wax monoesters and sterol esters. These allow the tear fluid to form a film adhering to the ocular surface, keeping it scratch-free during blinking. Tables 1 and 2 list some of the major constituents of the tear fluid. The major functions of the tear fluid components are to (a) maintain a smooth scratch-free surface, (b) lubricate the surface, (c) provide nutrients and transporting dissolved oxygen, (d) regulate the pH and salt concentrations, (e) stop pathogen attack and (f) remove foreign material from the cornea and the conjunctiva.

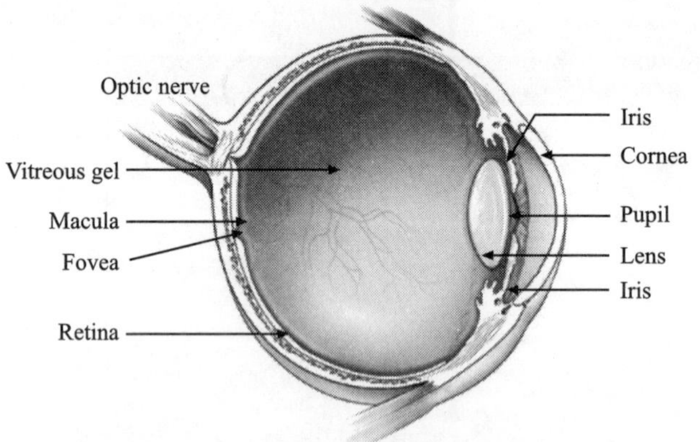

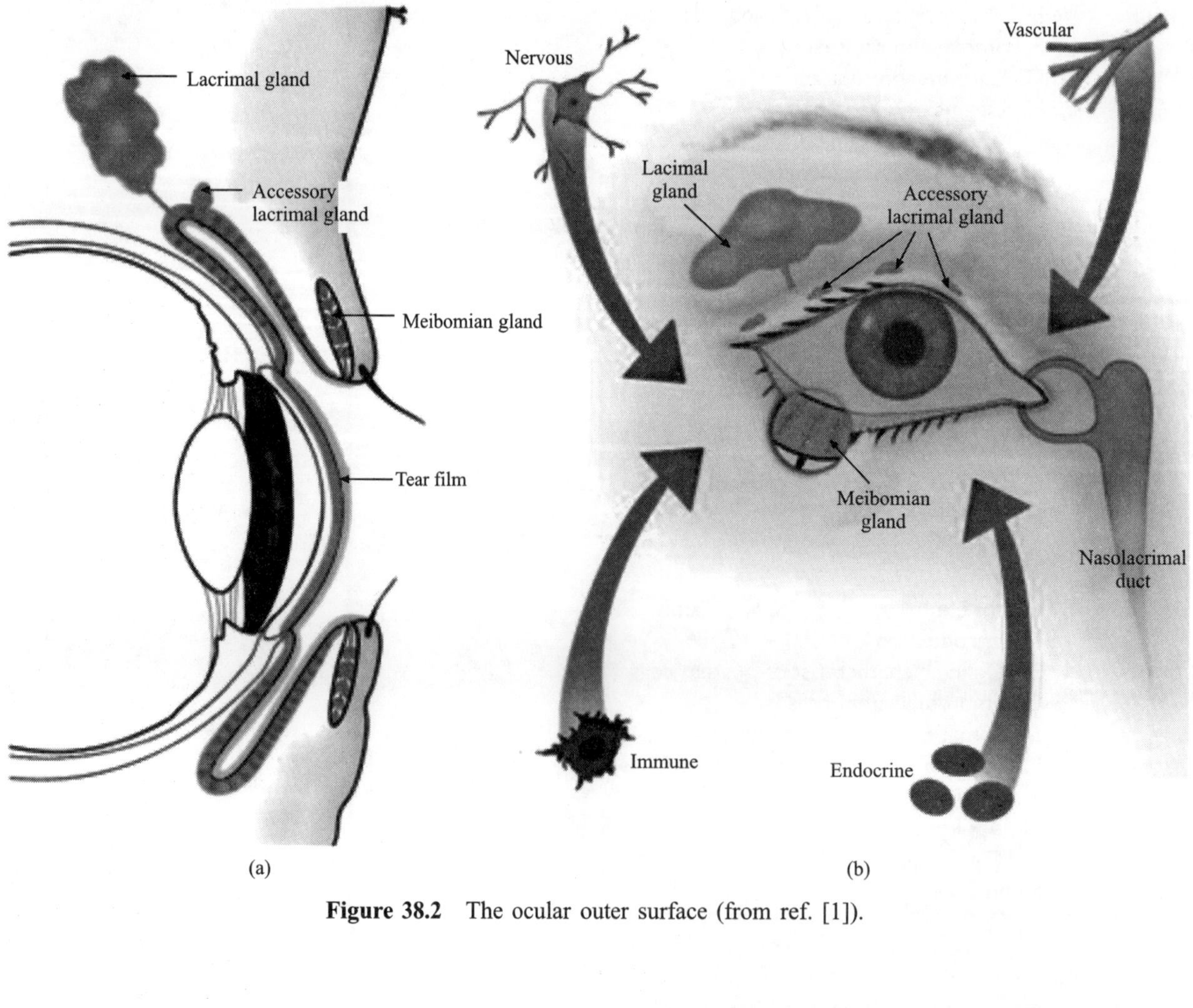

Figure 38.1 The human eye and its parts (from NEI.NIH.gov).

Figure 38.2 The ocular outer surface (from ref. [1]).

Unlike many other parts of the body, and indeed even the back of the eye, the anterior portion of the eye does not have access to blood for nutrition; in fact it cannot, because if blood vessels were to populate the anterior portion, it would affect and compromise the transparency and transmission of light from the external world into the retina. Hence the blood vessels are tucked away from the visual path of the eye. Blood vessels are however found in the back of the eye, behind the retinal screen, not affecting light transmission and focusing. Avascularity is thus a feature in this portion of the eye, hence the necessity for what is referred to as the blood-aqueous barrier. It also means that the anterior part of the eye must obtain its nutrition by other means. The tear fluid offers a small contribution towards this. As Tables 38.1 and 38.2 show, it has nutritional components feeding the front surface of the eye. However these are quite low in comparison to the plasma, yet important.

TABLE 38.1   Major Proteins and Lipids in the Tear Fluid

Lipids*	Major Proteins**
Phosphatidylethanolamines (PEs)	Albumin
Plasmalogen PEs	Transferrin
Lyso phosphatidyl cholines (Lyso-PCs)	Matrix metalloproteases (MMPs)
Sphingomyelins	Apolipopotein
Triglycerides	Immunoglobulins: IgA, IgG, IgM
Cholesteryl Oleate	Lactotransferrin
Ceramides	Lysozyme C
	Cathepsins
	Aquaporin 5
	Cystatins
	Lipocalin

Notes: * From ref.[2].; ** from ref.[3].

TABLE 38.2   Comparison of the compositions of human tears and plasma

Material	Tears	Plasma
Water	98.2%	94%
Total solids	1.8%	6%
Na^+	142 mEq/L	135-140 mEq/L
K^+	15-30 mEq/L	5 mEq/L
Cl^-	120–135 mEq/L	100 mEq/L
HCO_3^-	26 mEq/L	24 mEq/L
Ca^{++}	2.3 mg/100 ml	- -
Total proteins	0.6–2.0 g/100 ml	6.8 g/100 ml
Amino acids	8 mg/100 ml	- -
Glucose	3–10 mg/100 ml	80–90 mg/100 ml
Urea	0.04 mg/100 ml	20–40 mg/100 ml

Tables compiled from Ref. [4].

The conjunctiva is a thin membranous layer covering the sclera (or the white of the eye) containing a thin transparent layer of cells, some of which (called the goblet cells) secrete the mucin that helps in stabilizing the tear film. The conjunctiva is rich with nerve endings that keep the front of the eye healthy but also in its sensitivity to touch, rubbing and foreign bodies. Blood supply to the conjunctiva comes from various arteries of the eye. The conjunctiva thus acts as a barrier against infection and foreign bodies, and provides stability and homeostasis of the tear film components, and allowing the rotation of the globe.

The conjunctiva ends in the cornea via the limbus (see Figure 38.3), which is rich in stem cells which form the corneal epithelial cells. These are housed in a pocket termed the palisades of Vogt, which also has the pigment melanin protecting the limbus from UV radiation this area. Its component cells provide what is referred to as the niche or the micro-environment for the stem cells of the limbus.

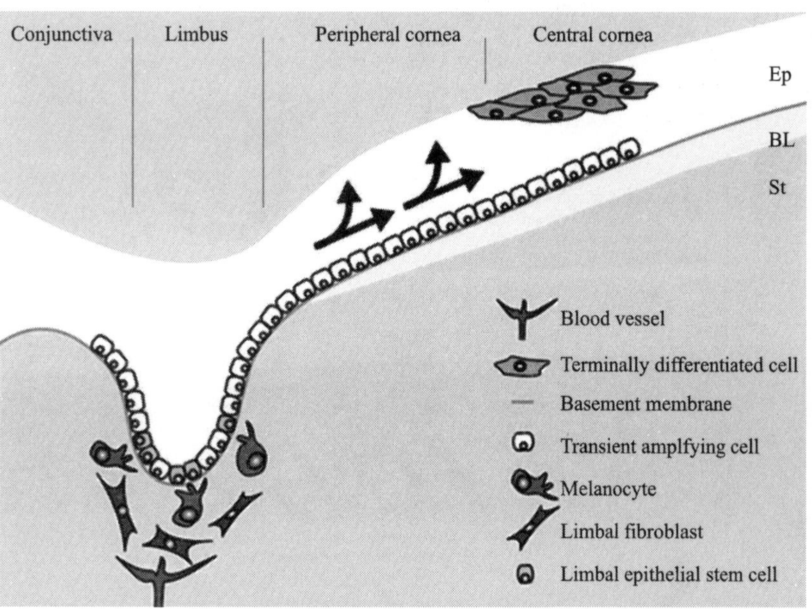

Figure 38.3   Limbus sandwiched between the cornea and the conjunctiva (wikipedia.org).

Thus, one can see that the front surface of the eye facing the external world is well protected, thanks to the presence of tears and the tear film clinging to the outer surface, the epithelial cells covering the sclera and conjunctiva, and the proteins and peptides present within. They not only protect by forming a barrier that prevents microbes and foreign bodies from attaching to the surface, but also attack the invader and stop its proliferation, and offer adaptive immunity. The most common bacterial species infecting the ocular surface are *Pseudomonas aeruginosa, Serratia marcescens, Staphylococcus aureus, Staphylococcus epidermidis*, and *Streptococcus pneumoniae*. Herpes simplex viruses (HSV1 and HSV2) are the common viral pathogens seen on the eye, while the fungal infectious agents are *Aspergillus, Candida, and Fusarium*.

While components of the tear fluid such as lysozyme, secretory phospholipase A2 (slpA2), cationic antimicrobial peptide (CAMP) and human neutrophil peptides HNP 1, 2, 3 are antimicrobial, the secretory immunoglobulin IgA removes the organisms at the entry level itself. The collectin class of lectins in the tear fluid binds to the carbohydrate components of the invading microbe and to receptors in the phagocytic cells. The surfactant proteins SP-A and SP-D inhibit the pathogen attachment using the oligopolysaccharide- dependent mechanism. And the mucins bind to and trap the invader and get rid of it during blinking.

Besides these components of the tear fluid, the ocular surface cells themselves offer the second line of defense. Tight junctions between epithelial cells stop the entry of the pathogen.

In addition, the polarity of the cells, maintained by differences in the composition and distribution of surface molecules between the apical and basolateral surface do not allow the pathogen to settle down. In addition, ocular surface epithelial cells are known to produce a variety of cytokines, notably IL1, IL6, IL8, and TNF-alpha, which regulate inflammation and immunity.

In addition, the surface is studded with pathogen-recognizing receptors, termed toll like receptors (TLR). This is a class of proteins that play a key role in the innate immune system; they are single, membrane-spanning, non-catalytic receptors which recognize structurally conserved molecules derived from microbes. Once these microbes have breached physical barriers such as the skin, they are recognized by TLRs, which activate immune cell responses. (They get their name, since the proteins are similar to a protein called toll in the fruitfly drosophila). To date, ten such TLRs have been reported on corneal and conjunctival epithelial cells. Figure 38.4 is an illustrative diagram of the distribution of some of the TLRs their ligands and signaling pathways. Furthermore, the turnover of the epithelial cells by constant expulsion and replacement offers yet another layer of protection against infection. Each time the eyelid blinks, epithelial cells are expelled and replaced. Defense and protection are thus molecular, antibacterial, surface-active, immunological, via pathogen recognition tight junctions, slough and regeneration through turnover.

Most of the antimicrobial peptides on the ocular surface are small in size (less than 50 kDa), usually cationic and also amphipathic in nature. The two major categories of

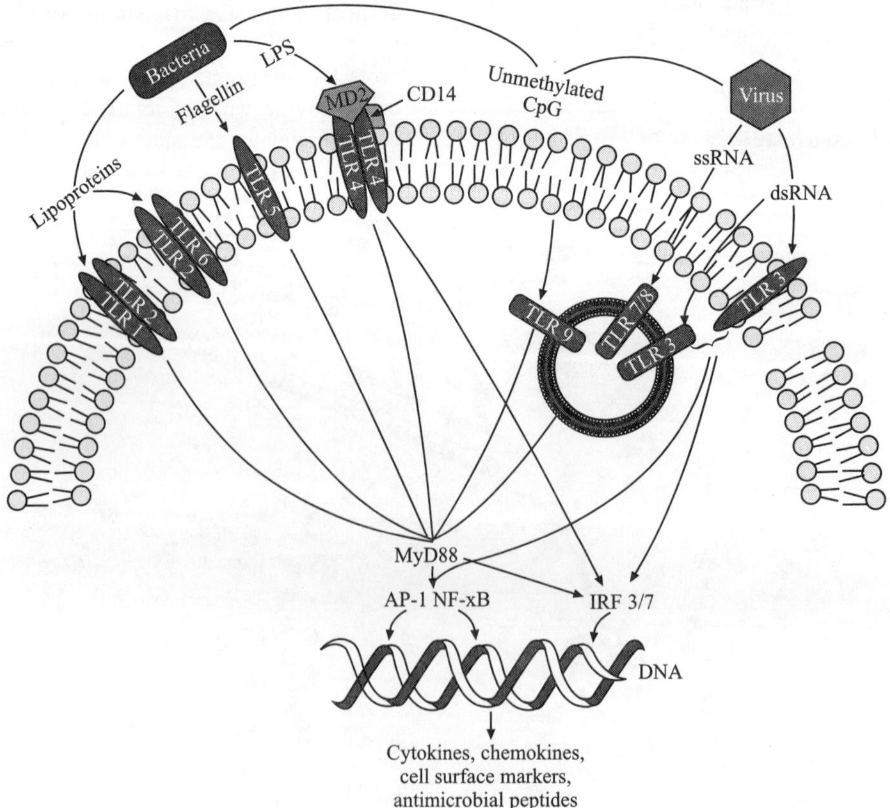

Figure 38.4   Defense offered by toll like receptors on the ocular surface (from [5]). (*see Plate 10 for colour figure*)

antimicrobial peptides (AMPs) are the defensins and the cathelicidins. The mechanism of action of some of these antimicrobial peptides is of molecular interest, since they offer a direct, almost mechanical, insight into the process. The review by Kim Brogden (6) summarizes the three major modes by which these peptides attack and destroy the invading pathogen (Figure 38.5). These modes are referred to as (a) the barrel stave model, wherein the peptide aggregates in the membrane of the pathogen microbe in such a fashion that the hydrophobic regions of the peptides are inserted facing, and interacting with the nonpolar lipid chains of the membrane making a barrel, while the hydrophilic regions of this aggregate generate the pore through which water and electrolytes can enter the interior of the microbial cell, and bursting it open through osmotic pressure, or (b) the carpet model, wherein the peptide molecules line up on top of, or parallel to the bilayer arrangement of the microbial membrane, thus disrupting it, and (c) the toroidal model, in which the peptide continuously bends the lipid monolayers and thus disrupts the architecture of the microbial cell and kill it. Another comprehensive review of the role of AMPs at the ocular surface is by A.M. McDermott [7].

A. The Cornea

The cornea is the transparent dome-shaped disc in the front of the eye, sandwiched between the conjunctiva, with the annular ring called the limbus surrounding it. While the sclera is white in colour and opaque, this central window component called the cornea is transparent, allowing all visible light in the 350 – 750 nm wavelength range enter the eye. It is estimated that the cornea provides almost 70% of the focusing power of the eye, with the lens behind it contributing the remainder 30%. The refractive index of the human cornea (expressed as n'), along with the tear film that covers it in the anterior, has been estimated to be 1.377. Given that the refractive index of air (n) is 1.00, the refractive power D of the anterior part of the cornea, expressed in diopter units (diopter is the inverse of the focal length, given in meters) would be

$$D = \frac{n' - n}{r} = \frac{1.377 - 1.000}{0.0078} = 48.33$$

where r = the radius of curvature of the anterior cornea ($r = 0.0078$ meters). In the posterior, the cornea is more curved ($r = 0.0065$ meters) and is bathed in the aqueous humor (whose refractive index is 1.336), we notice the refractive power of the posterior cornea would be

$$D = \frac{1.336 - 1.377}{0.0065} = -6.31$$

Together the total refractive power of the cornea is $48.33 - 6.31 = 42.0$ diopters. Given that the total diopter value of the human eye is 60 D, we note that the cornea accounts for 70% of the focusing/refractive power of the eye [8].

This calculation also shows why we are not able to see sharply under water, or when our eyes are filled with tears. The additional layers of water alter the refractive power of the cornea. However, if a mask or goggles gripping the eye are worn while under water, keeping the cornea-air interface intact, clear vision becomes possible under water. Likewise, when a person has high refractive errors (e.g., myopia), it is possible to alter the curvature of his cornea surgically in a precise manner, and thus attempt to avoid the use of corrective spectacles. Procedures of sculpting the cornea in this manner, such as laser-assisted *in situ* keratomileusis (LASIK) or photorefractive keratectomy (PRK), help in this process.

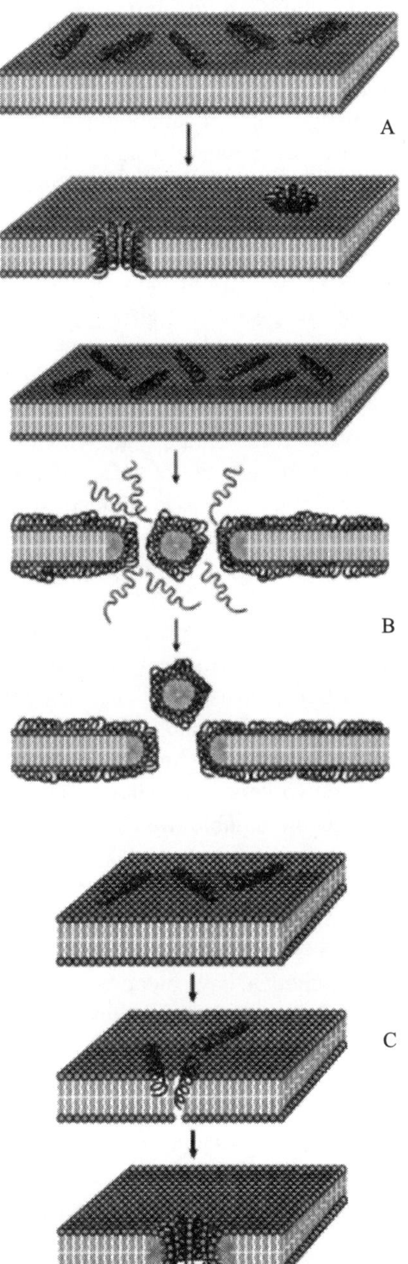

Figure 38.5 Three ways in which the antimicrobial peptide disrupts the bacterial membrane and kills the pathogen. Top (A): the barrel-stave model, Middle (B): the carpet model and bottom (C): the toroidal model (taken from ref. 6).

Figure 38.6 shows that the cornea is arranged as a five-layer sandwich. On the outermost side, facing the world is the epithelium. This is a multi-layer arrangement with squamous cells on the top, followed by mucous cells (also called winged cells) which give it strength and some level of rigidity. And at the bottom or deepest are the basal cells. This setup together is about 250 μm in thickness and is filled with thousands of nerve endings, making the cornea very sensitive to pair when rubbed or scratched. The epithelium is the protective shield of the eye, allowing oxygen to be absorbed and transported inside, blocking any foreign material from entering and since it is covered on the outside with tear film, is also sterile and fights microbial pathogens. The epithelial cells are constantly sloughed and regenerated. The source of regeneration is the limbus, the outer O- ring around the cornea, separating it from the conjunctiva. In this sense, the corneal epithelium is akin to the skin epidermis.

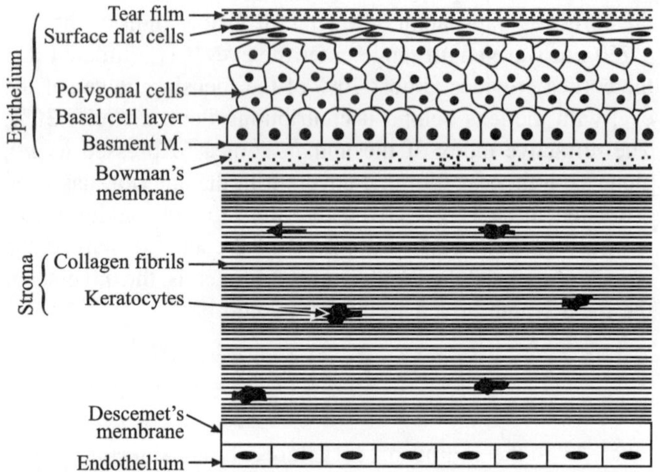

Figure 38.6   The layers of the cornea (taken from [9]).

Beneath the epithelial layer, and separating it from the stroma is a transparent 10 μm thick layer called the Bowman's layer (named after Sir William Bowman), made use of layers of collagen. Beneath it is the stroma, the major structural component of the cornea, made predominantly of collagen and the proteoglycan called keratan sulfate proteoglycan, which are together arranged in a unique geometrical arrangement that makes the cornea a perfectly transparent, protein-rich tissue.

Posterior to the stroma is a 3–10 μm thick sheet called the Descemet's membrane (named after the French scientist Descemets) which serves as a barrier against infection and insult. And finally at the posterior end, as the 5th or bottom layer of the cornea, is the thin layer of endothelial cells, constituting the corneal endothelium. While the entire cornea is bathed in water (from the tear film in the front and the aqueous humor in the back), it is the job of the endothelium to act as the pump which maintains an appropriate level of water, keeping the cornea well hydrated and transparent. Any defect in the pumping would lead to excess water, which would swell the stroma and disturb the transparency. Likewise, if the water content is below a critical amount, the cornea can

'dry up' again affecting the light transmission. The corneal endothelium thus functions to keep the cornea clear and wet.

In essence then, corneal transparency is maintained by the leaking of the aqueous humor into the tissue and pumping out back to the aqueous to the aqueous humor. Unfortunately however, the corneal endothelial cells do not regenerate as well as the epithelial cells do, and it is not clear whether there are endothelial stem cells. Recent experiments suggest that modulating the biochemistry of the endothelial cells using what is referred to as Rock–1 inhibitor might offer a way of maintaining the optimal number and level of the endothelium.

The number density of cells in the corneal endothelium is a vital parameter. During the procedure and technique of corneal transplantation, the staff in the eye bank measure the number of endothelial cells per square mm of the cornea, using a specular microscope; a donor cornea which has a density below 2000 cells/mm^2 is not usable for transplantation to a recipient, since its pumping action is compromised and the transplanted cornea might lose its transparency.

From the above, we would have noted that the cornea, a 5 mm thick tissue, is a beautifully structured protein assembly, largely made of collagens and proteoglycans besides a whole list of thousands of proteins, which is perfectly transparent in the visible region of the spectrum (350–800 nm). While its neighbours, the sclera and limbus, are vascular (filled with blood vessels for nutrition), the cornea which is right in the middle of the structure stays as a perfectly transparent disc with no blood vessels or vasculature. Indeed it needs to be a vascular, since of it were not, its optical transmission would be compromised.

The biochemical question then is how does a structure which is sandwiched on either side by a vascular tissue remain completely blood-free or avascular? The answer to this comes from a comprehensive paper published by two Indian-origin scientists, BK Ambati and JK Ambati, in collaboration with a large group of coworkers [10]. They show that the cornea expresses the molecule soluble Vascular Endothelial Growth Factor Receptor–1 (sVEGFR–1, also known as sflt–1). This molecule binds to, and suppresses the action of VEGF-A, a potent stimulator of blood vessel growth (angiogenesis). This is particularly striking since the cornea's neighbours, namely, the limbus and conjunctiva, have blood vessels. The VEGFR–1 in the cornea binds to, and inactivates the VEGF-A, thus blocking the growth of blood vessels here. They were able to prove this by three ways; (a) by treating corneas with a specific antibody and blocking the action of VEGFR–1, (b) by knocking down the expression of the gene for the receptor using siRNAs, and (c) by deleting the gene for VEGFR–1 using a specific enzyme (pCre); in all these cases of suppressing the action of the rector allowed the corneas to vascularize. They further showed that the levels of VEGFR–1 are also high in other animals such as dolphins, whales, and elephants (all of which have avascular corneas), suggesting that VEGFR–1 (or sflt–1) is evolutionarily conserved. Again, this is the reason why corneal transplantation can be done using the tissue from any donor and to any needy recipient, without the need for any blood group matching. The cornea is thus an immunologically privileged tissue.

Proteins in the human cornea

A recent mass special analysis of the human corneal proteome (the entire set of proteins present in the tissue) by the Danish group [11] has identified a total of 3250 proteins. Of these a major portion (2737) is in the corneal epithelium (the outer 50 micron-thick part), 1679 in the stroma (450 microns thick) and 880 proteins in 20 micron-thick corneal endothelium. Thus the human cornea is very rich in its protein components. And these proteins (a) offer the tissue its structural integrity and stability, (b) help in maintaining its perfect transparency and focusing of incoming light in the visual range of wavelength 400–750 nm by their architectural arrangement, (c) protection against incoming ultraviolet radiation (below 320 nm), (d) antibacterial action and (e) even immunological protection. One special protein component helps in keeping the tissue vascular, so that no blood vessels (which will hamper optical transparency) are formed in the tissue.

The proteins in the cornea are distributed mainly in three components. The epithelium is rich with members of keratin family. Type I keratin makes about 20% and type II 25% They after the epithelium its structural strength and also "freshness". The corneal epithelial cells are shed continuously but replenished from the limbus, which is the annular ring-like structure surrounding the cornea, separating it from the conjunctiva. In this sense, the corneal epithelium is remarkably similar to the skin. The corneal epithelium is renewed approximately every 12–15 days so that damage is handled through such turnover.

Besides the keratins, the epithelium is rich is enzymes such as aldehyde dehydrogenase, alpha-enolase, protein kinases and glyceraldehydes–3-phosphate dehydrogenase. Interestingly but expectedly the cornea (and its epithelium) contains proteins that offer protection and defense response against infections and physical trauma or damages, while immunoglobulins and complement factors are found more in the stroma, the cytokeratins and in particular cytokeratin 6A are seen to possess antibacterial activity. The glycine-rich C-terminal region of this protein is implicated in such activity. Smaller length synthetic analogs of this keratin-derived antimicrobial peptides (KDAMPs) have been shown to possess rapid bactericidal activity. It has recently been realized [12] that (a) knockdown of cytokeratin 6A using siRNA reduced the bacterial activity of corneal epithelial cells and (b) synthetic peptides containing the glycine- rich sequence derived form cytokeratin 6A (such as those containing the sequence GGLSSVGGGS) exhibit powerful antibacterial activity against pathogens such as *S. aureus, S. epidermidis, P. aeruginosa* and *E. coli.* Thus, apart from their roles as structural components, some of the keratins present in the cornea offer innate immunity through their antimicrobial activity.

The corneal stroma, the major structural component of the tissue, is rich is collagen. Almost 70% of the dry weight of the cornea is due to its collagens. All types of collagen termed types I, II, III, IV, and V are present and the role of each type in contributing to the structure and function of the cornea and its stroma is still a subject of contemporary interest. All these collagens have the classical tripeptide repeat motif gly-pro-x (where X is often hydroxyproline). This special sequence allows the molecular to come together in three chains (each called the α-chain) wound together in the famous Ramachandran Triple Helix. While the gly residues are within the helical interior, the side chains of pro and hypro are exposed to the environment, and interact with molecules there. Endogenous collagen molecules display a variety of properties. The main one is to assemble together to form fibrils, while other are to form networks, anchor the fibrils, connect with the members and so forth. Typically each fibril, organized in the triple helical mode, has a repeat run of 67 nm, sequentially denoted as a, b, c, d and e in electron micrograph pictures [13]. Of the five types mentioned above, type V associates with type I (and also type III) and regulates the diameter of the fibrils formed in the cornea. Type II collagen is particularly, found in the developing cornea while type III helps during wound healing.

As we turn to other macromolecules in the stroma, we find enzymes to constitute less than 2%, but the other major classes of molecules are the proteoglylans. Of these keratin sulfate proteoglycan (KSPG, also called lumican) constitutes more than 50% of this family of glycosaminoglycans, while dermatan sulfate preoteoglycan (DSPG, also called decorin) and versican (a chondroitin sulfate poteolgycan, CSPG) occur in minor amounts. The chemical structures of two of these, namely, dermatan sulfate and keratin sulfate, are illustrated in Figure 38.7.

When a glycosaminoglycan chain is covalently linked to a protein core, the result is a proteoglycan. Generally these complexes are made up of one or more glycosaminoglycans attached to a protein core. Figure 38.8 shows such a representative complex, taken from [14]. And Figure 38.9 shows a three-dimensional reconstruction of the interactions between collagen fibrils and proteoglycan filaments visualized by electron tomography of the mouse corneal stroma [15].

An interesting feature is encountered when we turn to the enzymes and other soluble proteins in the cornea. Aldehyde dehydrogenases (ALDH3A1 and ALDH1A1) comprise a major portion (3–50%) of the soluble proteins in the mammalian cornea, predominantly the epithelium and stroma; yet they do not function as enzymes alone but do double duty as antioxidant protective agents. As light falls on the cornea, the proteins present within absorb light in the UVB region 280–320 nm and below, and upon so doing, might undergo photochemical damage leading to reactive oxygen species (ROS) such as hydroxyl radical (OH°), peroxide radical (O_2^{-2}°) and singlet oxygen ($^1O_2$). These ROS generate the oxidative damage of other endogenous molecules (DNA, RNA, proteins, lipids and sugars), and do so often in a chain reaction. Interestingly enough the enzyme ALDH3A1 is seen to effectively act as antioxidants by metabolizing toxic aldehydes, quenching the ROS, as well as by directly absorbing UV radiation [16–18]. The other enzyme ALDH1A1, present more in the rabbit cornea,

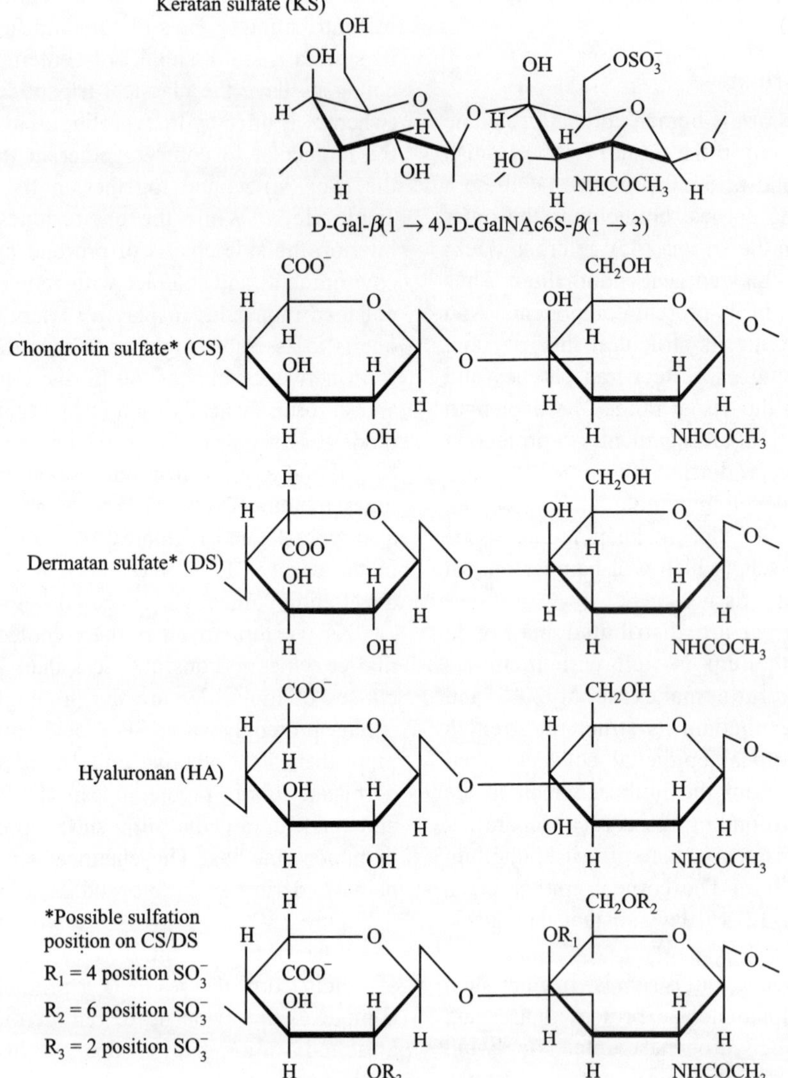

Figure 38.7 Chemical structures of various glycosaminoglycans.

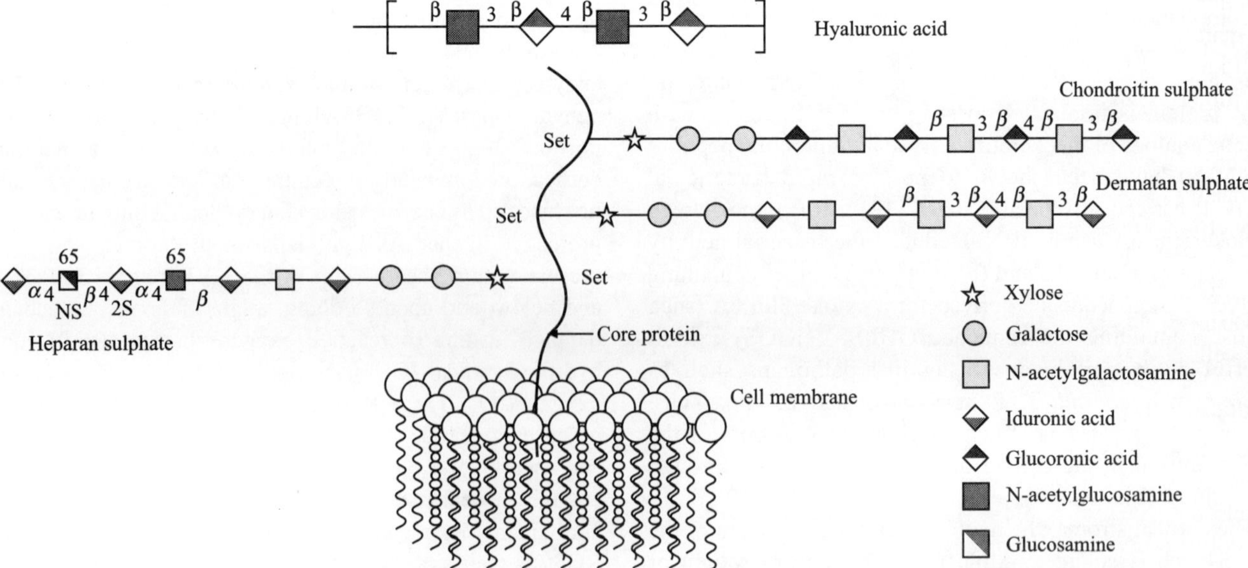

Figure 38.8 Representation of a glycosaminoglycan linked to a protein core in proteoglycans [taken from 14]. (*see Plate 10 for colour figure*)

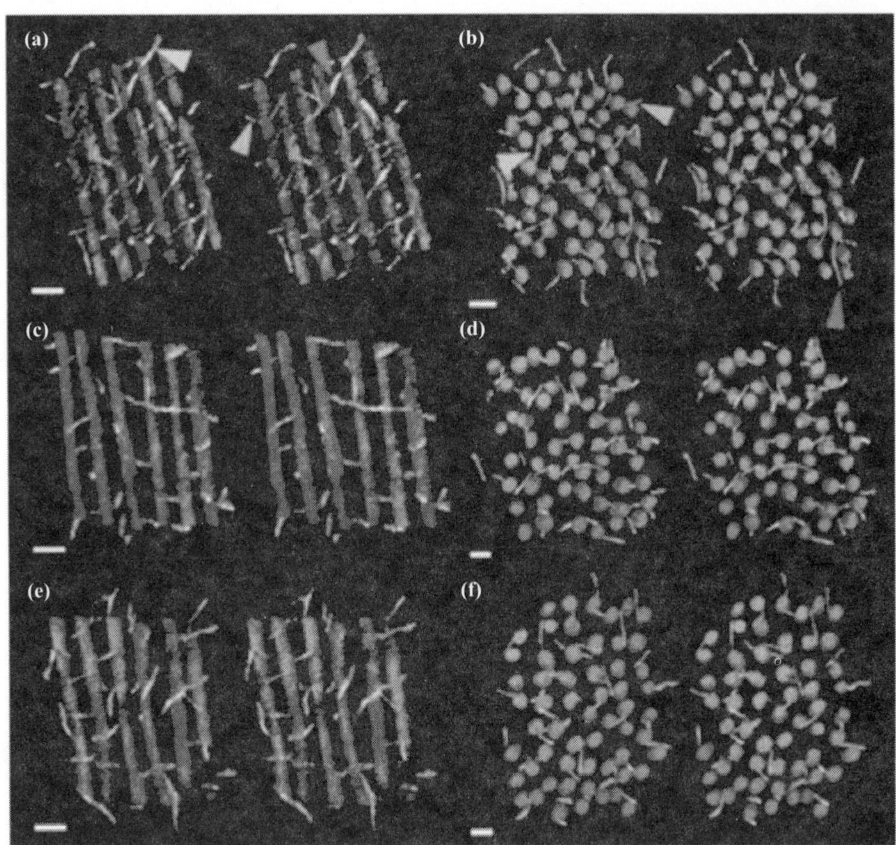

Figure 38.9 Three-dimensional representation of the collagen-proteoglycan complexes in the corneal anterior (taken from [15]). (*see Plate 11 for colour figure*)

behaves similarly as an antioxidant. Similar protective action has been reported with other enzymes present in the cornea, e.g., transketolase (TKT) and isocitrate dehydrogenase. (ICD) And the role of serum albumin (present up to 10–13% in the water-soluble protein content in the human cornea too has been suggested to be antioxidant and protective. Interestingly, these molecules (ALDH, TKT, albumin) perform additional protective functions (besides their classical roles), and act as a second (or parallel) line of defense for the cornea, along with the well known non-enzymatic antioxidant systems there, namely, ferritin, glutathione and its redox cycle system, ascorbate (vitamin C) and α-tocopherol (vitamin E) [18]. Another important enzyme in the human cornea is corneal glutathione S-transferase (GST). This enzyme degrades xenobiotic agents and protects cellular membrane integrity, and also mediates a pathway alternative to ALDH for aldehyde detoxification. Thus GST and ALDH play an important role in the defense mechanism against cytotoxic aldehydes that are generated in the cells after membrane lipid peroxidation.

Metabolism and energy production in the cornea

The epithelium is fed its energy through the glucose in the aqueous humor, using which the epithelial cells store high levels of glycogen. Glucose metabolism here is largely by glycolysis and the HMP shunt pathway. The cornea also receives, absorbs and utilizes significant amounts of oxygen, both in the dissolved form and from air. Thus, when a person wears contact lenses, it must be ensure that the material used does not hinder the flow (permeability and transport) of oxygen from outside to the epithelium. Failing which, not only does aerobic metabolism get hindered but the front surface of the eye reddens due to stress as well (contact lens associated redness of the eye, CLARE). Thankfully, such high-oxygen-permeable contact lenses are readily available in the market and should be used.

It has been estimated that the corneal epithelium receives its oxygen when it is exposed to a partial oxygen pressure of 155 mm Hg in the tears, in the open eye condition, which drops to 55 mm Hg when closed. And under open eye condition, the amount of O_2 consumed by the corneal epithelium is estimated to be about 3.5 μL per cm^2 per hour. Under such aerobic conditions, glycolysis goes through the tricarboxylic acid cycle. But under hypoxic conditions, pyruvate might convert to lactate, which can increase the acidity and cause epithelial edema, thinning and erosion. The importance of oxygen to the cornea is thus evident.

The cornea also utilizes the alternate hexose monophosphate (HMP) shunt pathway, which generators NADPH and pentoses. In this anabolic, rather than catabolic, process ribose 5-phosphate (R5P) is produced, used in the synthesis of RNA and DNA, as well as for fatty acid synthesis. In addition glucose is also seen to enter the sorbitol pathway, which is not desirable since accumulation of sorbitol can cause osmotic damage to the cells.

The need for oxygen permeability is felt even more in the corneal endothelium, which pumps fluid across the stroma and epithelium, constantly hydrating the entire tissue. This consumes metabolic energy which is 5-6 times higher than that of the epithelium. Thus aerobic metabolism is vital and glycolysis is estimated to account for over 90% of the conversion of glucose–6-phosphate to pyruvate, and the tricarboxylic and cycle converts the resultant pyruvate into ATP. Deficit of oxygen to the endothelium can result in a shift to anerobic metabolism, acidification due to the conversion of pyruvate to lactate and thus endothelium dysfunction.

Corneal dystrophy

The term dystrophy refers to wasting away due to lack of nutrition. While the corneal epithelial cells and thus the epithelium itself is continuously being shed and regenerated, the stroma and endothelium are relatively longer lived. Thus defects or inadequate supply of nutrients there lead to dystrophy, opacification of the cornea and compromised vision. Since the cornea is avascular and thus immunologically "privileged" as a tissue, transplantation of a donor cornea (from a deceased person, with the permission of the family) can restore vision to the affected individual.

For a video display of how the defective cornea is removed by surgery and a donor cornea is transplanted, please access the website www.youtube.com/watch?v=rp9eUsLksgQ

However, since the corneal endothelium does not renew itself as well as the other cells, defects in the endothelium pose a problem even after transplantation. Such defects in the supply of nutrients to the endothelium occur in individuals due to genetic defects (mutations), and a number of such genetic defects have been identified and studied (19).

Four different types of corneal dystrophies have been described. They are: (a) Bowman layer dystrophies; (b) stromal dystrophies; (c) epithelial and sub-epithelial dystrophies; and (d) Descemet membrane and endothelial dystrophies. Most of these dystrophies occur in an autosomal dominant (AD) manner and present themselves as blister- like lesions or band-shaped irregularities in the cornea. Some of these, such as posterior polymorphous corneal dystrophy (PPC) and congenital hereditary endothelial dystrophy (CHED) are Mendelian disorders, while others such as Fuchs endothelial late onset corneal dystrophy (FECD-late onset) are more complex. Another form called X-linked corneal endothelial dystrophy (XECD) occurs more in men than women, and presents as "moon-like craters" on the endothelium of all affected. While gene therapy has been attempted in some monogenic dystrophies, repeated corneal transplantation (called penetrating keratoplasty) has so far been the standard treatment. Some researches such as Professor May Griffith have been attempting to make artificial corneas by making transparent sheets made of recombinant collagen III on which she grows cells. These have been tried on animals with some success and human trials are on the way (20).

Stem cell therapy for the cornea

Compared to genetic defects causing dysfunctions of the cornea, accidents such as chemical or thermal burns cause damage to the tissue, blinding the individual. In many of these cases, the recent procedure called cultivated limbal epithelial transplantation (or CLET) is a successful one which regenerates a stable ocular surface, suitable to restore vision after the surgery, particularly after a keratoplasty is followed. CLET is based on the fact that the limbus stores stem cells which move to the cornea and differentiate into corneal cells (Figure 38.3). In the procedure, the corneal surgeon first takes a tiny amount of limbal tissue from the patient by biopsy, and cultivates the limbal stem cells on suitable scaffold, e.g., a sheet of human amniotic membrane, thus differentiating the limbal cells into the corneal cells. Now he removes all scar tissue from the cornea of the patient and puts in the amniotic membrane on which the corneal epithelium has been generated and sutures or glues the membrane. In a few days the wound is healed and a stable ocular surface is regenerated. Details of the CLET procedure are given in Sub-chapter 'D' on stem cells for ophthalmology, in later pages.

However, a typical CLET procedure can be watched on the video by accessing the website:
www.youtube.com/watch?v=NlvYrTdbycs

B. The Aqueous Humor

The eye, particularly the front part which receives and focuses the incoming light using the cornea and the lens, needs to be colourless, transparent and scatter-free. Therefore it cannot have blood as the nutrient. Instead, nutrition here is offered by the fluid referred to as the aqueous humor. This colourless fluid not only offers nutrition to these avascular structures but also removes waste material coming out of metabolism, and stabilizes the shape of the eye by offering an optimum intraocular pressure of about 15 mmHg above normal atmospheric pressure. It also permits inflammatory cells and mediators to circulate in the eye in pathological conditions, as well as drugs to be distributed to different parts of the eye. The aqueous humor is generated by the ciliary body, enters the posterior chamber and exits through the uveo-scleral pathway Figure 38.10 describes the ciliary body, where it is produced, the trabecular meshwork through which it flows and the Schlemm's canal (which is the channel in the limbus) from where it goes via 25 or 30 collector canals to the episcleral veins through the collector channels.

The aqueous humor is thus steadily pumped in and out, at an optimum siphoning rate and pressure. Any error that affects this steady flow rate and pressure will affect the structure and dynamics leading to either a high intraocular pressure (IOP), which can press and stress the optic nerve connecting the eyeball to the brain, leading to a slow atrophy of the nerve. The ultimate result is loss of vision through the condition known as glaucoma. On the other hand, too low an IOP, due to ineffective production of aqueous humor too is harmful and

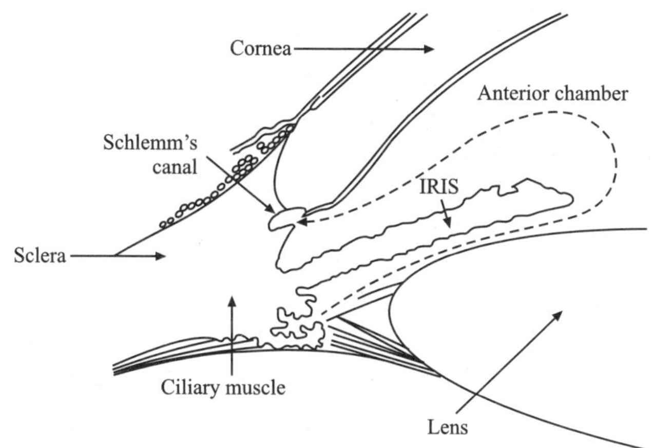

Figure 38.10 Normal outflow of the aqueous humor (taken from [21]).

will not provide nutrition to the needy parts of the eye. Thus, an optimal rate and amount of production, flow rate and siphoning out is vital, failing which vision can be compromised through the disorder referred to as glaucoma affecting the retina. The average rate of aqueous humor turnover is about 2.4 ± 0.6 µl/min in healthy individuals. It shows a day-night variation, with the high rate of 3.0 µ/min, about 2.4 µl/min during the day and dropping to 1.5 µl/min at night.

The formation of aqueous humor occurs through passive diffusion, ultrafiltration and active transport. Lipid soluble substances diffuse through the cell membranes. Ultrafiltration of water-soluble substances occurs from the capillaries of the ciliary stroma. Together the diffusible and ultra-filtered molecules arrive at the posterior chamber of the aqueous humor. But the major source of aqueous humor formation is through active secretion from the pigmented epithelial cells. The membrane spanning proteins called aquaporins, in particular aquaporin AQP1 and AQP4 contribute to aqueous humor formation in a major manner. Likewise, the enzyme carbonic anhydrase, found in the pigmented and non-pigmented ciliary epithelia, contributes to the transport of bicarbonate ions ($HCO_3{}^-$) which influences fluid transport through pH regulation.

The major constituents of aqueous humor are organic and inorganic ions and electrolytes, small molecules such as carbohydrates, amino acids, urea, glutathione, proteins and O_2 and CO_2. A detailed set of components of the fluid is listed in Tables 38.3–38.5. It can be seen from them that while the electrolytes and small molecular weight components are comparable in abundance to those seen in the plasma, the protein levels here are considerably lower than in the plasma. Much of the proteins seen in aqueous humor are glycoproteins and also some specific immunoglobulins such as IgG. Some glycosaminoglycans, notably hyaluronic acid and also chondroitin sulfate, are seen, as constituents of the extracellular matrix. Note that the aqueous humor is richer in antioxidant molecules (ascorbate, glutathione) and in enzymes (collagenase) that help maintain extracellular matrix in proper condition. It is also rich in receptors for transferrin, some growth factors, endothelin and indoleamine 2,3-dioxygenase.

TABLE 38.3 **Biologically active substances in aqueous humor and plasma**

Components	Aqueous Humor (mg.ml^{-1})	Plasma (mg.ml^{-1})
Prostaglandins	2	–
Cyclic AMP	8	–
Catecholamine		
Noradrenalin	0.8–1.14	0.311
Adrenalin	0 - 0.13	0.097
Dopamine	0.12	0.037

TABLE 38.4 **Proteins composition of aqueous humor in comparison to plasma**

Components	Aqueous Humor (µg.ml^{-1})	Plasma (mg.ml^{-1})
Protein (total)	12.4 ± 2.0	7000
Albumin	5.5-6.5	3000
Transferrin	1.3–1.7	–
Prealbumin	0.3–0.4	–
Fibronectin	0.25	29
Immunoglobulins		
IgG	3.0	1270
IgE	< 0.75	16–218

TABLE 38.5 **Electrolytes and low molecular weight solutes in human aqueous humor and plasma**

Components	Aqueous Humor (mM)	Plasma (mM)
Na^+	142	130–145
K^+	4	3.5-5.0
Ca^{2+}	1.2	2.0–2.6
Mg^{2+}	1	0.7–1.1
Cl^-	131	92–125
HCO_3^-	20	24–30
Ascorbate	1.1	0.04–0.06
Lactate	4.5	0.5-0.8
Citrate	0.1	0.1
Glucose	2.7–3.9	5.6–6.4
Urea	4.1	3.3–6.3
Glutathione	0.001–0.01	-
H_2O_2	0.024–0.069	-
Amino acids (total)	0.17	0.12

Tables 38.3, 38.4 and 38.5 are taken from [22].

As mentioned above, a steady non-obstructive and dynamic flow of aqueous humor is vital not only for the nutrition of the anterior parts of the eye, but also to keep the eyeball (globe) in proper shape and optical property. How and when does resistance to such an even flow occur? About 75% of the resistance to the outflow seems to be at the trabecular meshwork level and 25% beyond the Schlemm's canal. In order to correct for this resistance, eye surgeons operate on these two sites, using procedures defined as trabeculotomy and trabeculectomy. Also, just as stents are used by heart surgeons to regularize blood flow, glaucoma surgeons insert a value in order to regularize aqueous humor flow.

As mentioned above, a steady non-obstructive and dynamic flow of aqueous humor is vital not only for the nutrition of the anterior parts of the eye, but also to keep the eyeball (globe) in proper shape and optical property. How and when does resistance to such an even flow occur? About 75% of the resistance to the outflow seems to be at the trabecular meshwork level and 25% beyond the Schlemm's canal. In order to correct for this resistance, eye surgeons operate on these two sites, using procedures defined as trabeculotomy and trabeculectomy. Also, just as stents are used by heart surgeons to regulate blood flow, glaucoma surgeons insert a valve, or glaucoma drainage implant, or a shunt in order to regulate aqueous humor flow. Pharmacological approaches to lower the intraocular pressure in high IOP patients include the cholinergic drugs such as the alkaloid pilocarpine, the hormone epinephrine (which stimulate alpha and beta receptors), beta blockers such as timolol, carbonic anhydrase inhibitors such as acetazolamide, prostaglandins such as latanoprost (a phenyl-substituted PGF2α–isopropylester), or specific combinations of these as advised by the glaucoma specialist. The later Sub-chapter 'C' describes the biochemistry of glaucoma at some length.

What molecular events and mechanisms are responsible for such obstruction of the flow is not clear yet, but one culprit appears to be the glycosaminoglycans of the extracelluar matrix in the trabecular meshwork, which might cause edema and swelling due to hydration and also deposits obstructing the flow. Treatment for this involves the use of corticosteroids. An excellent review of aqueous humor and its dynamics is the one by Goel et al [23].

C.   The Iris and The Pupil

The iris is similar to the aperture control in a camera, adjusting the amount of light passing through the lens and falling on the retina. It is an impressive and colourful tissue that is aptly named after Iris, the Greek goddess of the rainbow. It has a disc-like structure that can open wide or close into a pinpoint. The opening is referred to as the pupil. An excellent review of the genetics and molecular analysis of the iris is given by Davis-Silberman and Ashery-Padan (24). Figure 38.11 illustrates a typical human iris. The iris is made up of three layers. The innermost layer is referred to as the iris pigmented epithelium or IPE. Above the IPE are the iridial muscles, above which lies the iris stroma. The iris stroma has cells and connective fibers that form a meshwork containing blood vessels and nerves. Both the iris and the IPE contain the coloured pigment melanin. It is the amount and distribution sets of colours such as blue,

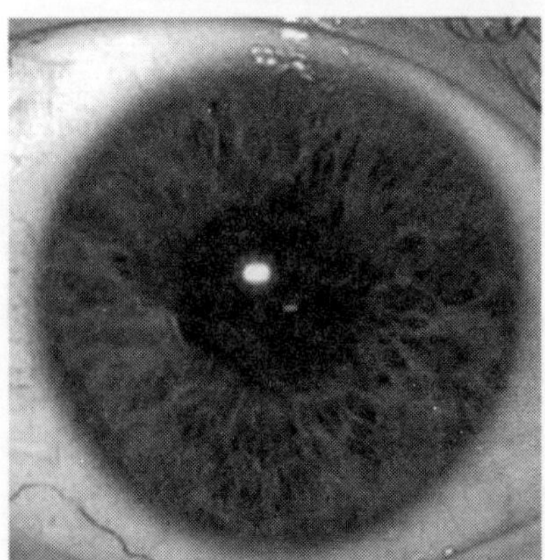

Figure 38.11   The iris (taken from commons.wikimedia.org). (*see Plate 11 for colour figure*)

brown, black (or none at all in melanin-negative albinos). In people with blue eyes, the pigment cells are mostly in the IPE while the pigment is also found in the stromal cells in brown and black eyes. And the distribution and pattern of the geometric arrangement of the cells, blood vessels and nerves in the iris are specific to each individual, just as finger prints are. It is this individuality that has led to iris scans as a means of identifying people.

The iris is attached at its base or root to the ciliary body (CB) and to the cornea-sclera junction. This part of the eye is referred to as the irido-corneal angle. It is through the iridocorneal angle that the aqueous humor is secreted out of the eye via the trabecular meshwork and Schlemm's canal. Thus, any error (genetic or metabolic) in the iris can affect the efficiency of drainage of the aqueous humor and increase the intraocular pressure, leading to glaucoma. One such example is the pigment-dispersion syndrome, where deposits of the pigment block the trabecular meshwork. As a result, people with this syndrome suffer from glaucoma, degeneration of the optic nerve and hence loss of sight.

The development of the iris itself is the result of a coordinated expression of a variety of genes. This coordination involves the patterning of the optic cup and the formation of the iris and ciliary body, followed by migration of the cells into appropriate locations. Figure 38.12 lists the genes involved in iris development.

Given this large number of genes involved in the development and action of the iris, it is not surprising that mutations in one or several of these can affect vision. One major defect of this kind is the disorder known as aniridia, which is associated with partial or total absence of the iris itself. This abnormality leads to glaucoma, cataract and corneal disorder. But the most prominent genetic basis of aniridia is mutation in the gene for the transcription factor PAX–6. PAX–6 belongs to the class of paired box protein family of transcription factors. These proteins attach to specific sites of DNA and control the activity of the genes. In the case of the eye, the relevant protein is also called oculorhombin. This transcription factor is essential not only for the development of the iris and thus vision, but also for the olfactory system,

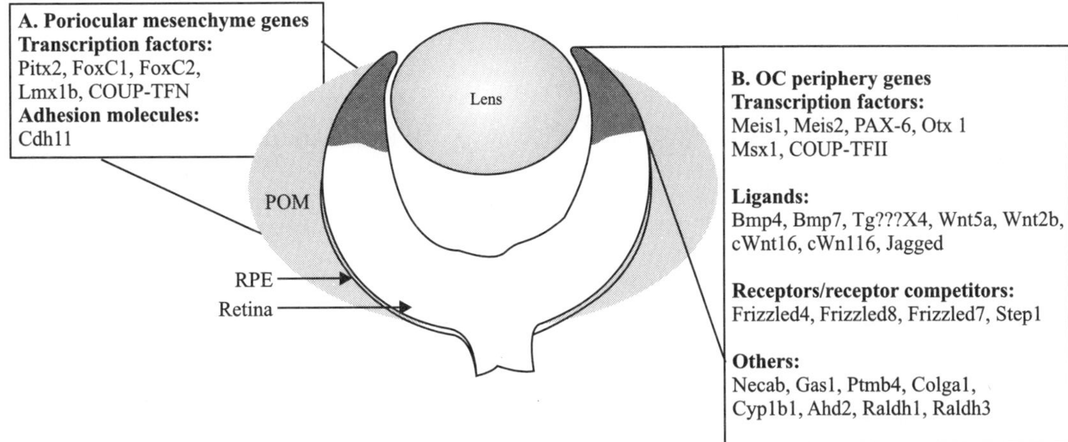

Figure 38.12   Some of the major genes involved in iris development (taken from [24]).

the pancreas, and the central nervous system. It appears that PAX–6 is involved in regulating the expression of molecules essential for iris development.

The second major mutation base disorder of interest is the Axenfeld-Rieger syndrome or ARS. This involves the malformation of the anterior seqgment of the eye, particularly the iris, pupil and the cornea. The associated gene here is referred to as PITX2 (which is located in chromosome 11, 11p 13), coding for the transcription factor referred to as Paired-like homeodomain transcription factor 2, also called the pituitary homeobox 2. This transcription factor is known to regulate the protein procollagen lysyl hydroxylase, and is responsible for the establishment of the left-right axis, the asymmetric development of the heart, lungs and the twisting of the guts. Pitx–2 is also known to play a role in myogenesis or the development of the muscles. The gene is located in chromosome 4 (4q25).

Beside PITX–2, Axenfeld-Riege, syndrome is also associated with mutations in the other transcription factor FOXC1, which is known as fork head/winged helix type transcription factor. It is involved in the development of many embryonic tissues and in the case of the eye, the periocular mesenchyme. Mutations in FOXC1 lead to severely eccentric pupils, corneal opacity, under- or undeveloped Schlemm's canal and trabecular meshwork.

Two other regulatory pathways of importance to the development and functioning of the iris and anterior sequent of the eye are through the Bone Morphogenic Proteins (BMPs) and Wnt signaling pathway. The BMPs are growth factors that play a key role during embryonic development. Of the various sets of BMPs, BMP4 and BMP7 appear relevant to the development of the ciliary body and the iris. In mice, when the level of BMP4 is reduced, hypoplasia of the iris was noticed, leading to irregular and eccentric pupils, as also abnormality in the iridocorneal angle. And in other animal experiments, when the inhibitor noggin was expressed in mice, thus completely blocking the expression of BMP4 and BMP7 expression, the mice were found to have completely lost the ciliary body. However, the situation in humans appears a little different since BMP4 is expressed here in the trabecular meshwork.

One particular point of interest in the iris is the remarkable properly of IPE cells to trans-differentiate to other cell types. These cells thus appear to maintain stem cell properties, or at least progenitor properties. Chick IPE cells have been able to produce lentoids or lens cell aggregates, and also form neurospheres that can differentiate into retinal cell types. Subretinal transplant of IPE cells appears to increase photoreceptor cell survival and reduce choroidal neovascularization. How useful and applicable this would be needs to be confirmed.

The importance of the iris in biometrics and personal identification

The technology here involves pattern recognition of the geometric details and texture that comprise the blood vessels and nerve connection profile of each individual. No two irises are like, not of identical twins, nor even of the right and left eye of the same individual. Since the recognition program uses 240 points of reference as a basis of match, compared to 60 in fingerprints, iris biometric identification is more trustworthy. Further, unlike fingerprints which could be compromised due to loss by damage, overuse or other factors, the iris pattern stays the same since the age of 10 months to life time. A frame from a video capture of an iris scan is digitized into a 512 byte file and stored on a computer database. Unlike fingerprinting, no physical contact is needed (nor ink that sticks to the fingers even after wash), since the image can be recorded a foot away from the eye, and spectacles or contact lenses do not interfere. Indeed even blind people, as long as they have an iris present to scan, can be identified. Since 1987, when two ophthalmology professors Leonard Flom and Aran Safir patented the idea of using iris recognition and requested the computer expert John Daugman to write the recognition algorithm for computer analysis, the idea had grown into an everyday identification and biosecurity device. Today, It has a false match rate of 10^{-11} (one in a hundred billionth error rate), and even if some medical and surgical procedures affect the colour and shape of the iris, the texture stays undisturbed for at least as much as 30 years (unlike fingerprints which can be lost or disfigured due to manual labour or excessive use). Figure 38.13 summarizes the main steps involved in iris scanning and identification.

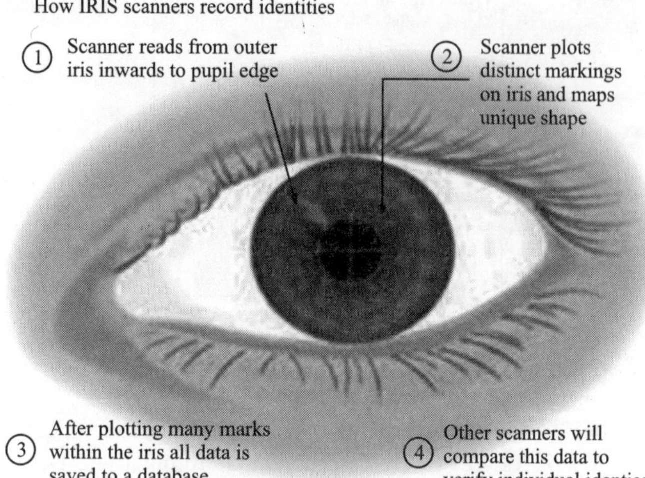

Figure 38.13 How iris scanners record identities (taken from news.bbc.co.uk). (*see Plate 11 for colour figure*)

A detailed tutorial description of iris scanning as a personal identification is given by the University of Cambridge, UK in the you tube video, which can be seen by accessing the site: http://www.youtube.com/watch?v=pbFFHkP9j4c&list=TLbHp8aBn9dEE

D. The Lens

Right behind the iris, attached vertically from the top and bottom by the ciliary muscles hangs the eye lens, a transparent, protein-packed gel. It is encapsulated by a 25-30 μm thick transparent capsule, the lens capsule. The lens capsule is a viscoelastic sheet with a refractive index of 1.40, made up of extracellular matrix proteins (notably collagen IV and laminin), a sulfated glycoprotein termed nidugen or entactin, as well as the familiar proteoglycans such as heparan sulfate proteoglycans (HSPG) fibronectin, osteonectin (also called SPARC) and several growth factors. The capsule protects the lens and keeps it in position as the lens changes shape while allowing the diffusion of small molecules such as water, O_2, CO_2, salt glucose, small peptides and proteins. A comprehensive review of the lens capsule is provided by Danyush and Duncan (25).

Packed inside the capsule is the egg-shaped eye lens, flatter in the front side facing the world, more conical in the rear, with radii of curvature about 10 mm and −6 mm, respectively. The axial thickness of the adult human lens is about 4 mm while its equatorial diameter is about 10 mm. Figure 38.14 shows the structure of the lens.

The anterior surface is decked with a monolayer of globular lens epithelial cells, which grow, metabolize, divide and differentiate in the usual classical way. As they proliferate, they move sideways towards the lens equator, where they start differentiating to produce long, thin fiber cells. The signal for the differentiation most likely comes from the posterior part of the eye. But something remarkable happens upon differentiation

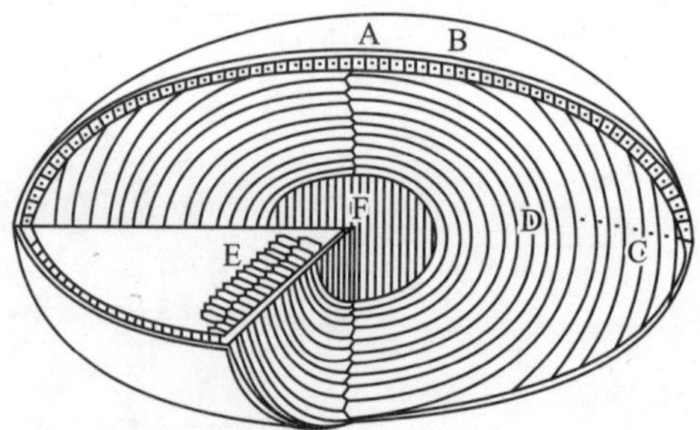

Figure 38.14 The human eye lens; A: capsule, B: epithelial cells, C: equatorial region, D: fiber cells removing their organelles denoted by the black dots, E: fiber cells with their characteristic hexagonal shape, and F: nuclear region (from [26]).

into lens fiber cells. These latter get rid of their nuclei and other organelles—in effect any particle that tends to scatter light. Thus the lens fiber cells are metabolically inert; they are made to last for life, with little or no innate biochemistry or indeed biological activity. (In effect then, lens fiber cells are membrane-enclosed bags filled with proteins and small molecules). They are in essence cells that are 'dead', where only chemistry occurs. There is no metabolism, division, differentiation, or discharge of waste or unwanted material. While the epithelial cells are active, generating ATP essentially from glucose and nutrients that come from the aqueous humor, both aerobic and anerobic pathways operate, roughly in equal proportions, metabolic energy for the whole lens (including epithelial and fiber cells) is in a major fashion due to anerobic glycolysis.

The fiber cells connect up with one another to make long thin concentric shells or layers, much in the shape of an onion (Figure 38.14). The earliest formed fiber cells thus accumulate in the deep inside of the lens, in what is referred to as the embryonic or fetal nuclear region, while as the lens grows with age, the latter fiber cell layers form the cortex. While the lens of a newborn baby is about 30 mg (wet weight), it grows rapidly to 150 mg by the time the baby is 4 years old, and about 250 mg at 60 years of age.

Each lens fiber cells is connected to its neighbor through a protein called major intrinsic polypeptide (MIP 26, denoting its mass as 26kDa), also called aquaporin 0, which transports water between cells and also regulates the volume. Another group of proteins called connexins (three of them, termed CX43, CX46 and CX50, again the numbers denoting the molecular weights), belonging to the gap junction protein family, are involved in transporting nutrients and other small molecules between cells. Aquaporin 0 (AQP0) is the major membrane protein of the lens fiber cells which regulates water permeation across the fiber cell membrane. This allows for keeping the right osmotic balance in the lens. The structure of bovine aquaporin 0 has been determined by X-ray crystallography, and reveals a remarkably beautiful organization of the monomer molecule (each about 270 amino acid residues long) into tetramers as shown in Figure 38.15. Each monomer is folded in such a fashion that a 28 Å long cylindrical channel (ranging in diameter 1.99–2.50 Å) through which water is transported. The structure is such that it also allows for cell-cell adhesion using AQP0.

While AQP0 helps in water transport, there are the gap junction proteins, connexins CX43, CX46 and CX50, which transport nutrients across fiber cells. Their functional architecture too is equally fascinating. Gap junctions refer to the contacts between cells using a cluster of inter-cellular channels through which molecules are exchanged in a tunnel-like fashion without involving intercellular space. An excellent review of gap junctions and the connexins has been published by Sohl and Willecke (28). Connexins too form multimeric complexes, actually hexamers, making what is referred to as a hemi-channel. One hemi-channel hexamer in cell 1 docks up with another hemi-channel in the neighboring cell 2, to make a full channel, as shown in Figure 38.16.

Each connexin molecule is folded in a manner that it has four trans-membrane domains (M_1-M_4) and two extracellular domains (E_1 and E_2). While the trans-membrane domains are largely hydrophobic in order to traverse the nonpolar membrane region, the E_1 and E_2 domains have two 'loop' motifs which hang out but are held together by disulfide bonds, as shown in Figure 38.16. While two hemichannels from adjacent cells dock, a full channel is formed. The connexins that go to make

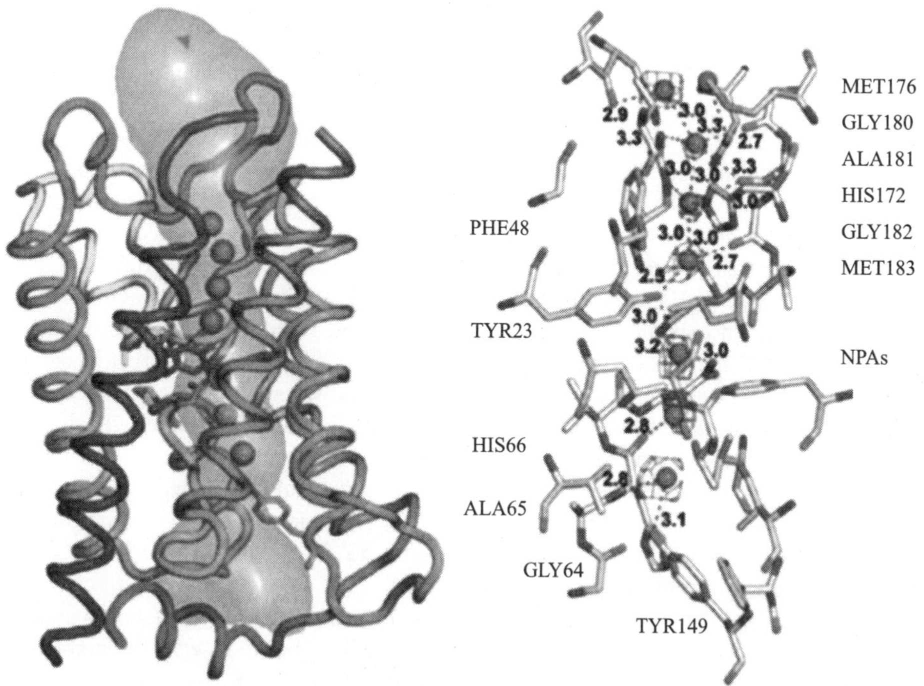

Figure 38.15 The crystal structure of Aquaporin 0, highlighting some of the residues that help generate the water pore/channel (taken from [27]). (*see Plate 12 for colour figure*)

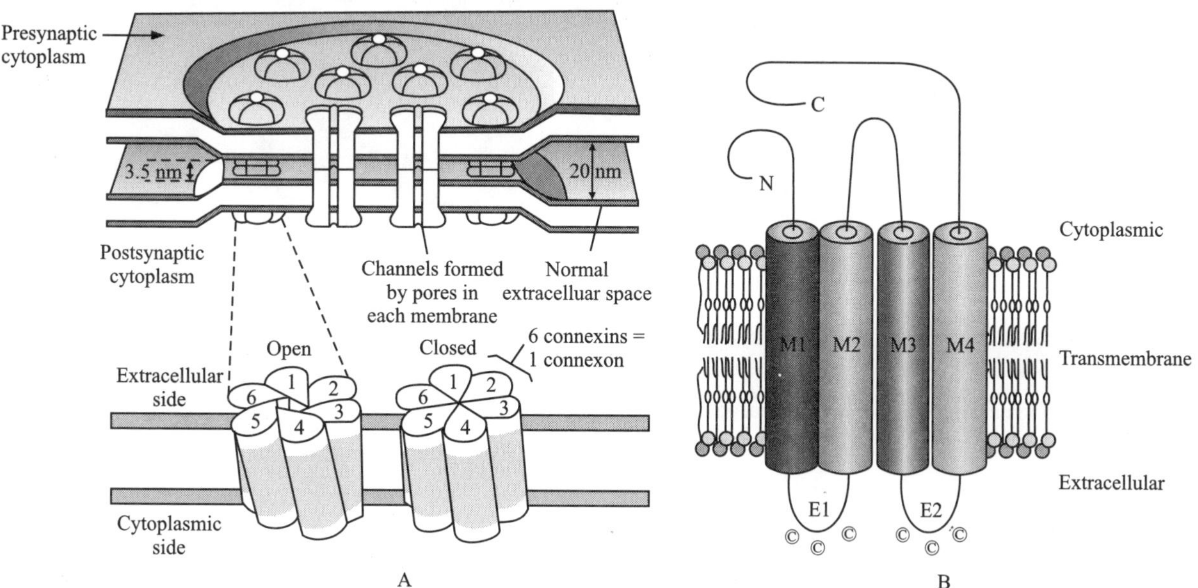

Figure 38.16 Gap junction proteins and their organization in membranes (from Ref. [28]).

the channel act together by changing their conformation in synchrony in order to open and close the hemichannels. The channel closes when each subunit slides against one another and rotates in a screw-like manner, while opening is done by the reverse. Of the three connexins found in the lens, CX43 is essentially in the epithelial cells while CX46 and CX50 are more abundant in the fiber cells.

Crystallins

The family of crystallins is the most abundant water-soluble ones in the lens, together constituting about 90% of all water soluble proteins therein. Indeed, it is estimated that the crystallins alone constitute about 35% of the net weight of the human lens. They come in three types, the α-crystallins (αA and αB), each of molecular weight about 20 kDa, multimerize into large units as many as 40-mers with a molecular weight of 800 kDa. These are of ancient origin and belong to the small heat shock protein (SHSP) family, and actually do display stress response effects. Interestingly, though originally found in the lens, they have been identified to be present elsewhere in the eye (particularly in the retina), where they display (particularly αB) stress response properties. Together αA and αB crystallins account for 40% of the total crystallin content of the human lens. Their function in the lens is thought to be twofold: act as a structural material and also as chaperone-like molecules which bind to misfolded (and precipitated/insoluble) proteins (such as the $\beta\gamma$-crystallins) and help fold them back to their native, water-soluble conformations. Addition of α-crystallin to insoluble aggregates of $\beta\gamma$- crystallins brings them back into solution (29). While the crystal structure of α-crystallin is yet to be determined, chimera map fitting, based on available cryo-electron microscopy and crystal structures, shows the molecule to be aggregated in the form of a donut (30).

molecule contains a N-terminal domain comprising two such motifs and a C-terminal domain with two motifs. The two domains are interlinked in the γ-crystallins through a short penta-peptide linker, while the linker region in the β-crystallins is far longer. The short linker makes the C-terminal domain in γ-crystallins to bind to the N-terminal domain and fold intramolecularly into a monomeric molecule (about 20 kDa), but since the linker peptide is longer in β-crystallins, this allows for inter-molecular interactions leading to dimers and multimers in them (mass 40–120 kDa). Human γ-crystallins come in three (highly homologous) types: γC, γD and γS, while β-crystallins come in 7 types, 3 acidic and 4 basic forms. The characteristic Greek key motifs and inter-domain interactions lead the $\beta\gamma$-crystallins to be folded into compact, globular, highly stable molecules that do not denature even at 8M urea or upon heating to 70°C. The crystal structure of human γD-crystallin, determined by Basak el al. (31) is shown in Figure 38.18. Notice the two domains in the molecule, joined by a linker region in the bottom. Each domain has two Greek key folds, and each fold has four interlocking beta sheet 'runs'. (This interlocking geometry is reminiscent of the interlocking patterns used in Greek art, hence the name Greek key). And since each of these molecules is folded tight and in a globular fashion, with little or no nonpolar side chains exposed, they are able to be packed at very high concentration in a small compact volume, with just short-range interactions between one another. Since the linking region between the N-terminal and C-terminal domains is a short sequence of just a few residues in gamma crystallins, the C-terminal domain tends to fold over the N-terminal domain intramolecularly, leading to a very compact structure. In beta crystallins, the linker region is much longer, which allows inter-molecular interactions between each monomer, leading to dimers and multimers in these cases, yet compactly packed. This feature plus their globular shape, each "ball" no bigger than 10–20 nm in size, is seen to be the basis behind the transparency of the lens, with no scattering-size particle present within (32).

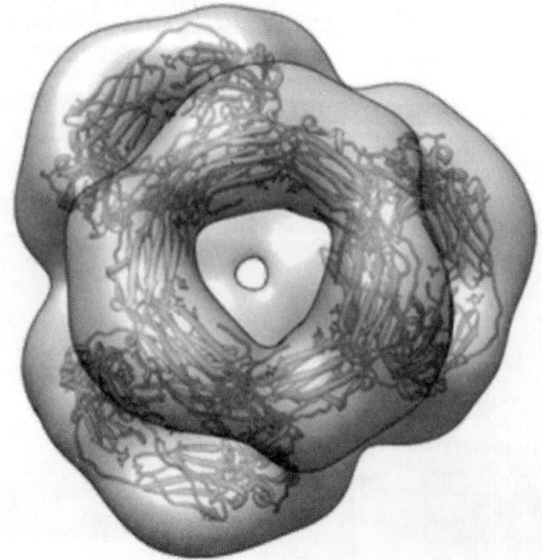

Figure 38.17 The modeled multimeric structure of α-crystallin (from ref. [30]). (*see Plate 11 for colour figure*)

The $\beta\gamma$-crystallins are found more abundantly in the nuclear and cortical regions of the lens. They too have evolved from archaean sources and possess a typical chain conformation comprising what is referred to as the Greek key motif. Each

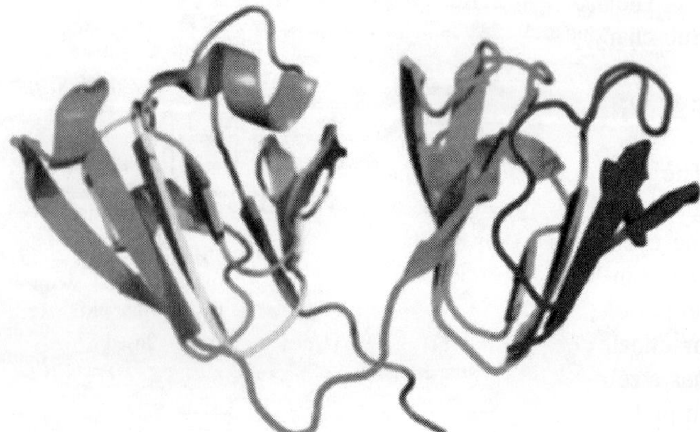

Figure 38.18 The Greek key fold of human γD-crystallin (from Ref. [31]). (*see Plate 12 for colour figure*)

It also appears that there is a relationship between the structural integrity of the Greek key topology (which keeps the proteins compact and stable) and the central lens transparency. Mutations in the $\beta\gamma$-crystallin genes, which distort the topology are associated with nuclear cataract while those not affecting the topology lead to peripheral cataract. It appears that distortion of the topology exposes many otherwise buried residues to the surface, causing intermolecular interactions and protein aggregation. When such aggregates fall out of solution and also scatter incoming light, lens transparency is compromised, particularly in the central nuclear region of the lens, which is rich in $\beta\gamma$-crystallins (33).

Unlike α-crystallin, which has a double-role in the lens-namely structural component and chaperone-like action, it is not clear whether the $\beta\gamma$-crystallins have any roles besides offering the lens a stable gel-like packing. But increasing evidence is being gathered that they might bind to and sequester free calcium ions (Ca^{2+}) and act as calcium depots in the lens. The level of total calcium in the lens is in the order of 0.5-1.0 mM, yet the amount of free Ca^{2+} ions is in the μM range; this means that there must be molecules that can sequester calcium ions. Free Ca^{2+} has the tendency to activate proteolytic enzymes such as calpain (present in the lens) and nucleases, which is not a desirable thing since it could lead to protein degradation and lens dysfunction. Work done at CCMB Hyderabad shows the $\beta\gamma$-crystallins to bind free Ca^{2+} ions, with a modest affinity; but given the high concentrations of these proteins in the lens (300–400 mg/ml), this offers significant sequestration ability (34) and a protection mechanism in the lens. In addition to their roles in the lens as structural elements and chaperone-like activity, α-α-crystallins are also expressed outside the lens, particularly in the retina. Here they act as stress response molecules, and as promoters of neural cell survival (35). And just as α-crystallins are known to be active in the retina, $\beta\gamma$-crystallins too appear to be expressed outside the lens, and seen to play a role in the regulation of ciliary neurotropic factor (CNTF), which is involved in devascularization (or vascular regression) of the anterior part of the eye (36, 37).

The biochemistry of cataract is described in the subsequent Sub-chapter II.

E. Vitreous Humor and the Proteoglycans

The vitreous humor, called simply as vitreous (meaning glass-like), is a clear transparent jelly-like material that fills the eyeball between the lens capsule in the anterior and the retina in the posterior. It occupies a volume of about 4.5 ml and or close to 70% of that of the eyeball. It is avascular, with practically no cells or any other light-scattering material, and has a refractive index of 1.40. It is thought to keep the retina in place, since if it liquefies, the chances of the retina getting detached from the choroid is high. Figure 38.19 describes the anatomy of the vitreous in same detail.

Its cell content is less than 1%, and the cells are mostly phagocytes which help in removing undesired cellular material from the visual field. It is rich in proteoglycans, which are complexes of glycosaminoglycans with collagens (and

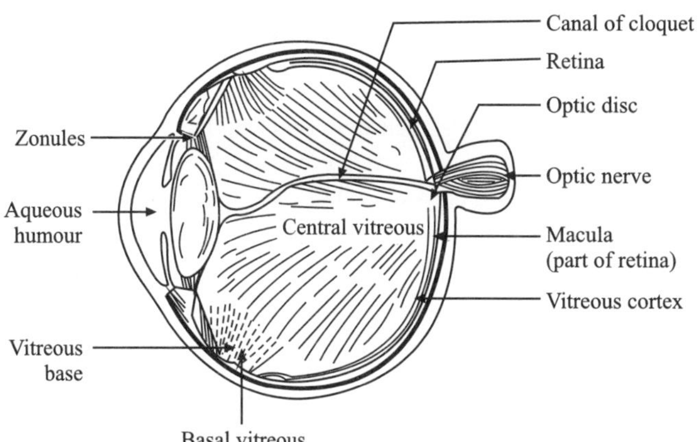

Figure 38.19 The anatomical features of the human vitreous humor (from [38]).

related proteins), which help in keeping the vitreous glass-like, transparent and stable. As can be seen in the figure, its proteoglycans are attached more firmly in the front to the vitreous base (close to the ciliary) than in the posterior to the vitreous cortex (in the retinal region). Thus when it loses its gelatinous structure with age or any pathological reasons, it tends to liquefy from the rear, and this could cause the detachment of the retina, affecting vision mildly. It is believed to be synthesized from the ciliary body (non-pigmented ciliary epithelium), although some of its collagen content is derived from the fetal neural retina.

The vitreous has a variety of physiological roles. It plays a role in the development of the eye itself, through its gelatinous structure and spatial distribution, which help in coordinating the various steps in the growth and development of the eye. It certainly protects the eye during any mechanical trauma. It is also suspected to have components which inhibit angiogenesis, and the role of its phagocytes has been mentioned above.

Much of the structural specialty and integrity of the vitreous gel is due to its rich proteoglycan content. As we saw in the case of the corneal stroma, the macromolecular architecture of the proteoglycans offers the vitreous its shape, stability and transparency. Thus it is valuable to digress a little here and describe the structures, variety, interactions, supramolecular structures of the proteoglycans and their roles in the biochemistry of the eye.

Proteoglycans are comprised of three components. One is the family of linear polysaccharides termed glycosaminoglycans or GAGs. The second are the collagens while the third component is non-collagenous proteins. Taking GAGs first, these are long, linear polysaccharides whose basic building units are disaccharides, which are repeated to various lengths. One component of the disaccharide is a hexose sugar such as glucose or galactose which may or may not carry a carboxyl acid group, e.g. glucose or glucuronic acid, iduronic acid. This unit is invariably attached to a hexosamine, usually N-acetyl glucosamine or N-acetyl galactosamine (see Figure 38.8, shown earlier in the section on cornea).

The nature of this repeating dimeric unit, its length and amount is tissue-specific. And invariably, the repeating unit is also sulfated (carries a SO_3^- group), making it highly anionic charged. In the vitreous, the GAGs found are hyaluronan (D-glucuronic acid linked to N-acetyl D-glucosamine), chondroitin sulfate (D-glucuronic acid attached to N- acetyl-D-galactosamine (which is sulfated in either position 4 or 6) and heparan sulfate (D-glucoronic acid plus N-acetyl D-glucosamine, sulfated at position 4 or 6. GAGs are invariably long chains which are viscous and mucous, offering lubricating properties. And they are attached covalently as branches to a protein core (usually collagen or elastin). They are invariably found as part of the extracellular matrix. Because of their sugar content and charges, they are very hydrophilic, absorbing water like sponge. When squeezed, the water is expelled and they are forced to occupy a small volume. When the compression is relieved, they rehydrate and fill back to their original volume. Because of their long chains carrying ionic charges, they invariably adopt rigid rod- like macromolecular shapes. Hyaluronan, for example, has a linear left-handed threefold helical fold, which can laterally aggregate to from a three dimensional network. Upon hydration it can bloom in size thousand-fold, to get back to a tiny stick when dehydrated. GAGs are thus useful for water retention, lubrication and volume filling.

The second important component of proteoglycans is the family of collagens. As many as 19 different collagens are known, encoded by 30 genes. While their amino acid sequences vary, they all have the signature tripeptide repeat sequence (Gly-Pro-X) where X can be any amino acid but more often than not hydroxyproline (or hypro). This special repeat motif causes the collagen molecule to adopt a three-strand interlinked triple helix, first discovered by Prof. G.N. Ramachandran at the University of Madras. Many collagens (e.g., type IX) have sequences where the tripeptide motif is missing (the so called non-collagenous sequence) which allows the molecule to fold on itself at places, like a long chain which has rod-like regions folded compactly using the flexible regions. Other collagens (e.g., type II) have non-collagenous sequences only at the start (N-terminus) and the end (C-terminus), called telopeptide sequences, which become useful in intermolecular crosslinking.

Collagen molecules line up together in the extracellular matrix (and in the cornea and the vitreous), forming fibrils. The vitreous humor has both fibrillar and non-fibrillar collagens (the latter called fibrillins), plus a member of the leucin-rich-repeat proteins (LRR proteins), notably opticin. Figure 38.20 illustrates how individual collagen molecules are assembled to form fibrils and fibers.

The third components are the non-collageneous proteins, whose role in proteoglycans is not as clear as with collagens. But it is know that opticin (monomer molecular weight 45 kDa), which is more likely thought to adopt a beta-sheet dimeric conformation, is known to bind to heparan and chondroitin sulfates, thus helping the vitreous gel-structure, acting as a molecular glue in the process of the vitreous adhering to the retina.

A proteoglycan is thus a combine of GAGs attached to a protein core, as shown in Figure 38.8, shown earlier. The core protein, collagen here, binds to the GAG generally using its serine side chain. Together, the proteoglycans ensures a regular, fibrillar structure which is geometrically well-laid out in the vitreous (and the corneal stroma), with appropriate spacings in a manner that no scattering of light occurs. The role of the interaction between GAGs and collagens in organizing the macromolecular conformation hand and spacing in these two instances thus offers optical transparency on are hand and structural stability and strength on the other. The role of proteoglycans in other instances, e.g., synovial fluid, is different; there it acts as a lubricant. In tendons or cartilage, they offer strength and elasticity. Some other GAGs bind to antithrombin III, and inhibit blood clotting. GAGs and proteoglycans are thus multi-task assemblies, with each component suited specifically for one end use.

The vitreous does not get regenerated easily once it is removed for any physiological or surgical purpose. Thus when a surgeon removes the vitreous for any retinal surgery, he substitutes it with silicone oil, or synthetic polymers with properties as close to the natural material as possible, as a stop-gap arrangement.

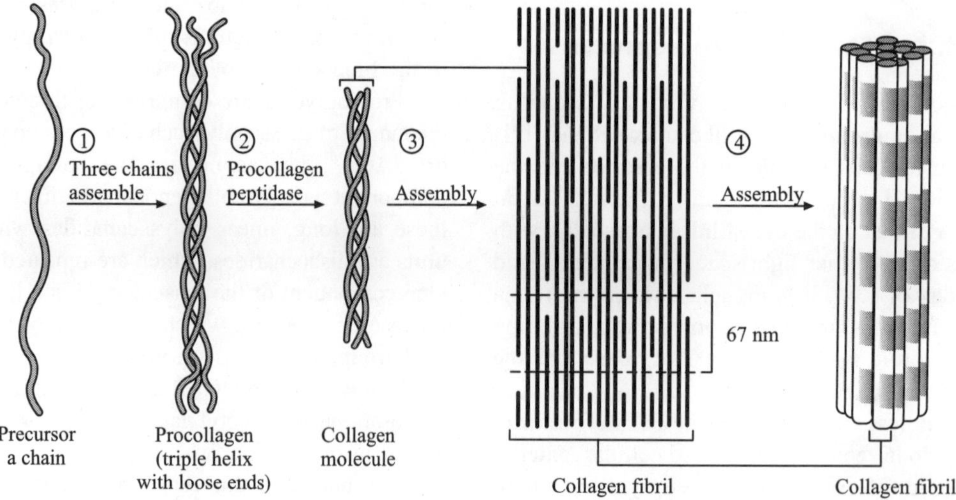

Figure 38.20 Assembly of individual collagen molecules to form fibrils and fibers (taken from ref [38]).

F. The Retina

The retina is actually to be regarded as part of the brain, since it is essentially a collection of neurons and synapses. The human retina is a ten-layered sheet placed right behind the vitreous humor, and is the screen where incoming light is focused and processed, to be sent to the brain in the form of electric signals. In other words, the retina is a photo-electric device. A cupped sheet of circumference about 50 mm and thickness 0.5 mm, it is composed of ten specific layers, each saving a specific purpose for this light transduction. Figure 38.21 shows these layers. Closest to the vitreous is a basement membrane (also called the inner limiting membrane), behind which is a nerve fiber layer made up from Muller cells. Next layer is the ganglion cell layer which contains the retinal ganglion cells (the word ganglion literally means a mass of nerve cells), after which is the inner plexiform layer, containing the synapse or connections between the bipolar cell axons and the dendrites of ganglion and amacrine cells. Layer 5 is the inner nuclear layer which houses the bipolar cells. Next in series are the outer plexiform layer (plexiform meaning network) containing the projections of the rod cells and cone cells. These rod and cone shaped cells actually absorb light (and hence also called photoreceptor cells) and form the next, outer nuclear layer, and the external limiting membrane. Finally, at the posterior are the retinal pigment epithelial cells (RPE Cells).

We thus see that the retina is essentially a multilayer sheet, populated by neurons or nerve cells that conduct the electrical signals produced upon light absorption through the optic nerve to the brain-in effect a printed circuit board where each component has a specific role to play in this visual transduction. What role does each type of cells in the retina have? Muller cells are glial cells serving as support for the neurons of the retina, stretching radically across the tissue. The word glia means glue, thus the Muller cells act as the glue, which offer the architectural base to hold the retinal neurons in place and supplying the necessary nutrients. They do not carry any electrical impulse, but provide homeostatic regulation. When the retina is injured, Muller cells are seen to de-differentiate to produce neural progenitor cells; they are thus doubly useful.

The ganglion cells are located near the inner surface and carry the final electrical output of the retina. Indeed, they are the only cells in the retina that have long axons, and thus able to send across nerve impulses. These cells receive electric signals from bipolar and amacrine cells and transmit the signal through their axons to the brain. Thus any defect in the ganglion cell function would compromise vision. Such defects or damages are caused when, for example, the intra-ocular pressure in the eye is increased leading the stress-related changes in the ganglion cell function, and the end effect is the disorder called glaucoma.

The inner nuclear and plexiform layers contain amacrine cells, with the term amacrine meaning "no axon". They play the role of inter-neuron connections in the retina, and they direct the pathway photoreceptor-bipolar-ganglion cells, by integrating, modulating and timing the photo-electric message. They adjust the image brightness and by integrating the timing of sequential neurons, help in detecting motion.

Bipolar cells are neurons located in the inner nuclear layer and help in passing on the information from the outer plexiform layer cells to the inner retina, i.e., amacrine and ganglion cells. They receive signals from the photoreceptor cells and help us see well under both bright and dim light conditions.

Horizontal cells, coming in three types (likely responding to the three types of cone cells), are laterally placed as a long horizontal strip in the inner nuclear, or bipolar layer, connecting with the photoreceptors. They provide the pathways for both local and long-range interactions between photoreceptors, offering feedback signals which help adjust the photoreceptor output. This feedback adjustment helps in allowing the eye to see well under the bright and dim light conditions.

Next in the outer plexiform layer are the photoreceptor cells, which come in two types. The first are rod shaped cells, simply called rods (Figure 38.22). The normal human retina is

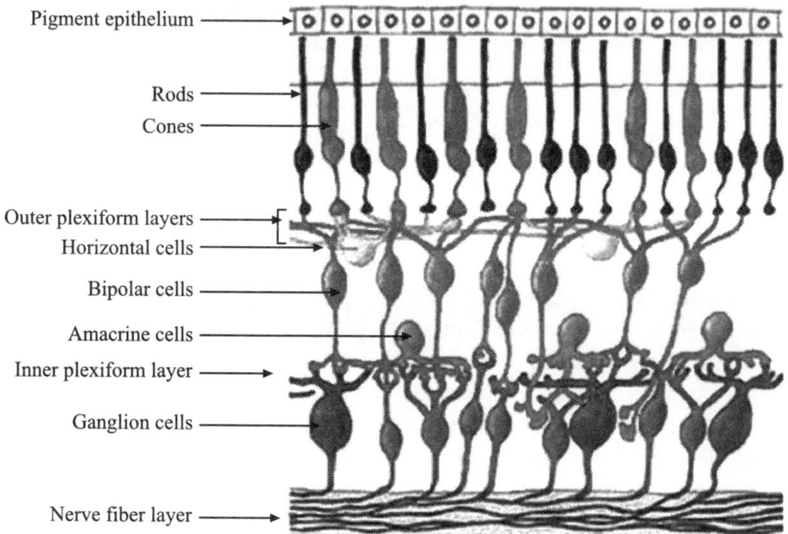

Figure 38.21 The cellular organization of the retina (taken from www.arn.org). (*see Plate 12 for colour figure*)

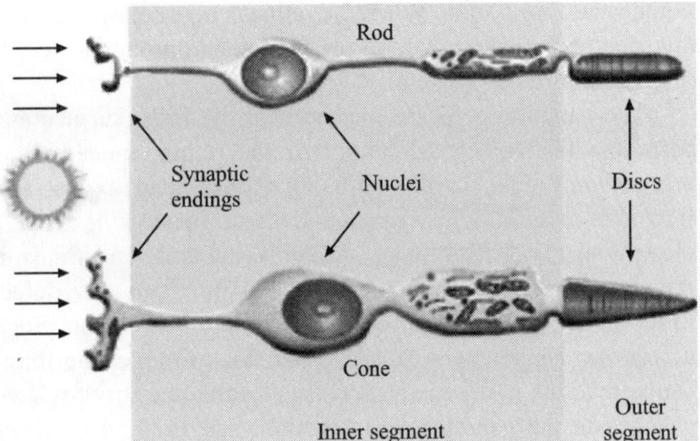

Figure 38.22   Rod cells and cone cells in the retina (taken from askabiologist.asu.edu). (*see Plate 13 for colour figure*)

rich with about 120 millions of them. The top part of these cells, called the rod outer segment or ROS is stacked with membrane enclosed disc-like structures containing the protein rhodopsin, at a concentration estimated to be as high as 25,000 molecules per μm^{-2}. These discs are placed on top of one another like a stack of coins. Besides the usual nutrients, ROS contains a very high concentration (in the millimolar range) of cyclic guanosine monophosphate or cGMP. The inner segment of the rod is rich in mitochondria, providing metabolic energy. The cell nucleus is in the lower end, and each cell ends with a synaptic terminal through which it connects to other cells. It is through this synapse that the electrical impulse created in the rod cells is transmitted. Figure 38.26 also shows the cone cells, which are packed essentially similar to the rods, except for their shape. In addition, while rod cells respond to the entire visible range of wavelengths, but generate only black and white images, cone cells come in three types; cells responsive to blue, to green and to red light. While there are as many as 120 million rod cells, there are only 7 million cones in the retina, and are maximally populated in the fovea. While rods are very sensitive to light and help us see well at night, cones are less sensitive but provide colour vision. Colour in the entire range of 400–750 mm (VIBGYOR range) is perceived by the combined response of the three colour-specific cones (red, green and blue). Thus if even one set of them is inactive or missing (by inheritance or mutation), colour blindness of one form or the other can arise. As long as the rods are active, even if the cones are defective, vision is by and large preserved to a large extent.

Finally, at the back end of the retina, attached to the choroid, is a layer of retinal pigment epithelial (RPE) cells. These are vital for the health, metabolism and functioning of the retina, since they act as the cargo—a two way shuttle of cells—which bring in nutrients, glucose, retinal, fatty acids and other small molecules from the blood vessels of the choroid and deliver them to the photoreceptor cells, and in the reverse mode, helping in the phagocytosis of photoreceptor cells and clearing away other debris (metabolic endproducts) from the retina to the blood. RPE is also believed to secrete a variety of growth factors that help in retinal cell health.

Interestingly, note that the light actually has to travel almost 0.5 mm of the retina before being absorbed by the rod cells and the cone cells. The electricity generated by the photoreceptors in the posterior of the retina is transmitted through the horizontal cells to the amacrine and bipolar cells in the front. They in turn transmit the signal to the ganglion cells. From the ganglion cells, the signal goes to the visual cortex of the brain through the optic nerve. Thus, the main process in the retina follows the scheme: photoreception-transmission to bipolar cells-transmission to the ganglion cells and transmission of the signal to the brain along the optic nerve. Notice too that in effect the retina is arranged in an inverted fashion–light reception at the posterior and signal transmission to the brain from the anterior! This inversion seems to have happened since the eye is an outgrowth of the brain; also the process of light to electricity conversion requires a lot of energy, which needs to be supplied through the blood stream. Had the vessels been placed in the front of the eye, transparency would have been a problem.

The process of vision is transduction of the incoming and focused light that falls on the retina into electrical signals which are interpreted by the brain. The molecular process by which this light to electrical signals is mediated by a series of chemical and biochemical processes known as the visual cycle. The main player in this act is the membrane bound protein rhodopsin. This molecule, consisting of 348 amino acid residues is actually called opsin to which the aldehyde chromophore called retinal (a derivative of vitamin A) is covalently bound, leading to the composite molecule rhodopsin. The residue lys296 in the seventh trans-membrane helix segment is covalently attached through its terminal amino group to the aldehyde moiety of the retinal by a Schiff base linkage. The retinal is bound here in its 11-cis isomeric form. This covalent attachment of the highly conjugated retinal molecule results in the retinal-bound rhodopsin displaying an absorption maximum at around 495 mm, and thus be visibly coloured, hence it is often also referred to as 'visual purple'. Figure 38.23 shows the chemical structures of the aldehyde in both the 11-*cis* form and the all-trans form. Note that the molecule has 5 double bonds in the long side chain, besides the double bond between C5 and C6. Thanks to this highly conjugated and delocalized electronic structure of 11-*cis*-retinal, visible light in the 400–700 nm wavelength region is absorbed by the molecule. Upon absorption, the electron is sent to the excited state, from where it causes the isomerization and change in the shape of the chromophore to the all-*trans* form. This light mediated isomerization occurs extremely fast (within a few picoseconds, 10^{-12} s); in the dark and from the ground state, the isomerization is extremely slow, occurring once in 1000 years $(3 \times 10^{10}$ s). This contrast suggest that the energy required to

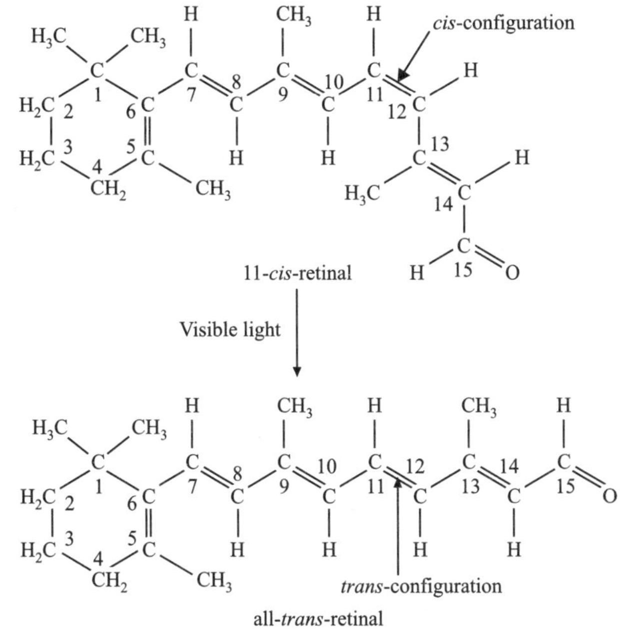

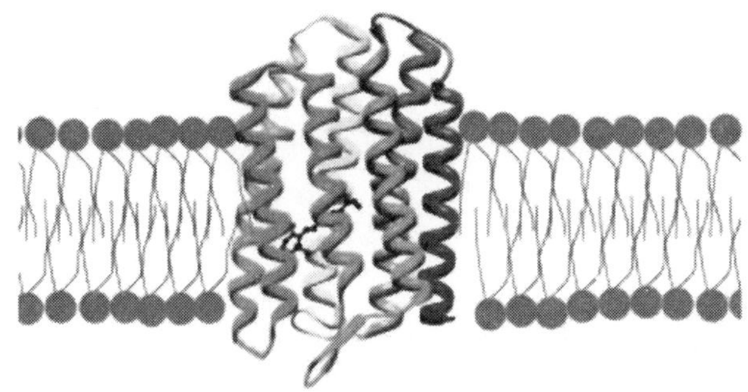

Figure 38.23 Structures of 11-*cis*-retinal and all-*trans* retinal (from wwwchristian.npage.de).

Figure 38.24(a) Representation of rhodopsin molecule bound to *cis*-retinal; note the seven *trans*-membrane helical runs, and the packing of the cofactor 11-*cis*-retinal in the interior of the molecule (taken from wordpress.mrreid.org). (*see Plate 13 for colour figure*)

'activate' the stable *cis* isomer to form the trans isomer is very high, and is supplied upon absorbing photons in the visible region of the spectrum. This excitation "breaks" the double bond between carbon atoms C11 and C12 of this cis isomer, making for a single bond which allows free rotation around this bond. This feature called photo-isomerization leads to a conformational change, resulting in the isomer all-*trans*-retinal. The originally bent molecule 11-*cis*-retinal now straightens out as a rod in the all-*trans* form (the shape change is that of a bent rod in the shape of the number 7 straightening out into a stiff rod with a longer end-to-end distance between say carbon atom 4 and the terminal oxygen atom). In this isomerization, the 11-*cis* isomer rotates about 180 degrees about the C11-C12 double bond and also about 140 degrees about the C12-C13 bond. The result of this isomerization of the protein-bound retinal is to cause a conformational change in the protein as well; the retinal in its cis form is able to be accommodated in a compact manner by the protein; when its shape changes to a straightened form, the protein has to readjust its own conformation to accommodate the all-*trans*-form of the cofactor. The detailed 3-dimensional structure of rhodopsin has been determined. Since it is a highly hydrophobic molecule, insoluble in water, the structure had to be determined by dissolving it in surfactant micelle. Figure 38.24a shows how the molecule is a membrane-bound protein, highly α-helical in conformation and how the molecule is packed in the cell membrane, spanning it using seven trans-membrane helical runs. The figure also shows how the cofactor 11-*cis*-retinal is compactly packed into the molecule within the membrane.

Figure 38.24b shows the molecule in a different perspective, and points out the important post-translational changes that help in an anchoring the protein in the discs of the ROS. Much of these details have come from the work of Dr Hargobind Khorana. His group showed that (a) the disulfide bond formed between cys 110 and cys 187 positions the N-terminal intra-discal segment in place (b) thio-esterification of the two adjacent cys residues C322 and C323 by palmitic acid (palmiotoylation) anchors the C-terminal domain in the membrane, and (c) glycosylation (with the hexasaccharide Man_3-Glc-Nac_3), positions the rhodopsin molecule appropriately in the intra-discal space, enabling it to interact optimally with the protein transducin, which is the subsequent step in the phototransduction pathway.

The first step in the vision process is the absorption of light by the retinal-bound rhodopsin molecule in the photoreceptor cells. The light-mediated change of shape of the cofactor rod poses problem for the compact accommodation of the retinal in the protein structure and as a result the protein changes its conformation in a series of steps, each step with a different absorption maximum value. The steps involved in the process, as a function of the photo-isomerization are as given below.

Rhodopsin, purple in colour (λ_{max} 498 nm) → bathorhodopsin, yellow (543 nm)
→ lumirhodopsin (497 nm) → metarhodopsin I (487 nm)
→ metarhodopsin II, colourless (380 nm).

Note how as the molecule adjusts its structure its absorption maximum also changes. The final product, called metarhodopsin II, absorbs not in the visible region but in the ultraviolet (380 nm) and is therefore colourless. The 'visual purple' has been bleached into a colourless conformational isomer. Figure 38.25 describes how the conformational change affects the packing of the retinal in the protein.

The first step in visual transduction has been completed upon the conversion of rhodopsin (abbreviated as R) to metarhodopsin II (abbreviated as R*), which is now ready to initiate the second step. In order to do so, it now binds to a protein called transducin. Transducin is a heterotrimeric protein, with the subunit Tα (molecular weight 39 kDa) and T$\beta\gamma$ (mol wt 36 kDa and 10 kDa, respectively). The subunit Tα binds in its functional form to GDP (hence the name G-protein to

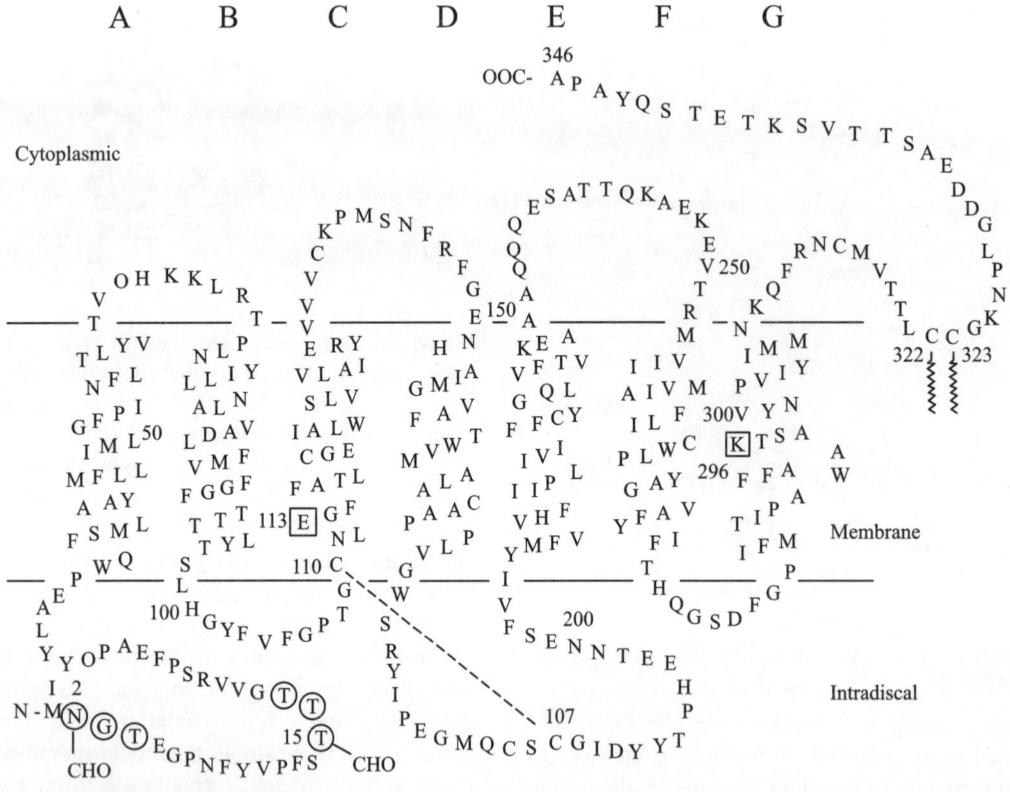

Figure 38.24(b)   A schematic secondary structural detail of rhodopsin. The disulfide bond between C110-C187 locates the N-terminal domain in the intradiscal zone, as also the glycosylation (shown as CHO) at N2 and N15. Palmitoylation at C322 and C323 insert the C-terminal segment in the membrane. A to G show the seven membrane-embedded helical segments (taken from [39]).

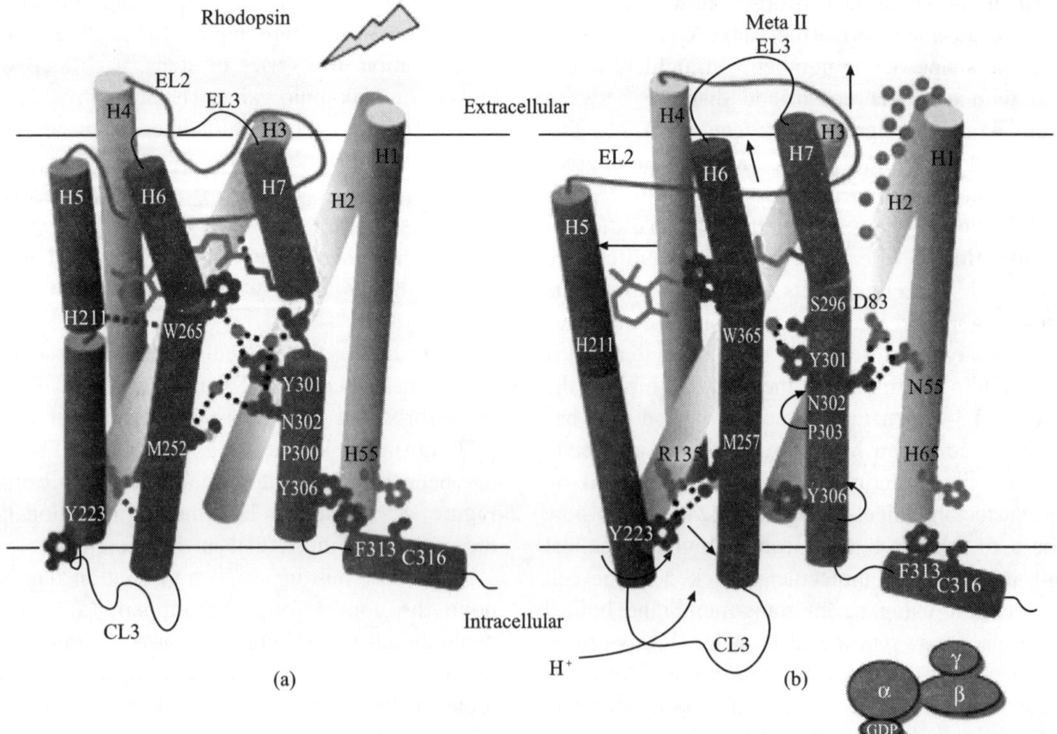

Figure 38.25   Picture (a) is the representation of the protein in the dark, bound to *cis*-retinal and picture (b) after isomerization of the retinal has occurred (taken from ref [40]).

transducin), while the heterodimer T$\beta\gamma$ is involved in binding to metarhodopsin II (R*). It is interesting to note that transducin binds selectively to metarhodopsin II and not to metarhodopsin I or other isoforms. The roles of the cytoplasmic loops 2, 3 and 4 of the membrane-bound rhodopsin conformation appear important in this specificity of binding. It is also worth noting that the heterodimer T$\beta\gamma$ acts in the combined form; individual subunits Tβ or Tγ do not appear to do so. This is a general feature of the G-protein family of molecules.

Now comes the third step in phototransduction. Upon R* binding to transducin, an exchange reaction occurs wherein the GDP bound to the subunit Tα is released and substituted by GTP from the cytoplasm. Since R* has a greater affinity for Tα-GDP than for Tα-GTP, it leaves the latter, to recycle rapidly and activate more transducin molecules, almost in an enzyme-like catalytic fashion. Next, once the Tα is loaded with GTP binding, it too is released from the T$\beta\gamma$ subunits and goes to activate the next step. The T$\beta\gamma$ subunits are now ready to recycle and attach to freshly available Tα-GDP and resume binding to R*. This recycling pathway has been described in detail and is illustrated in Figure 38.26.

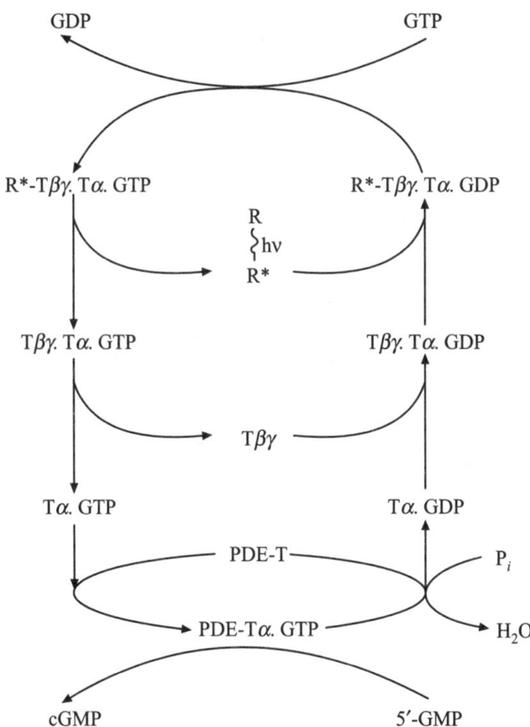

Figure 38.26 The cyclic cGMP cascade of molecular recycling in vision (from [41]).

Now is the next, crucial fourth step involving the Tα subunit of transducin. This subunit has three functional domains. One of them is involved in binding to R*, in coordination with its partner subunits T$\beta\gamma$, as discussed above. A second site in the molecule is for binding to GDP, and the third functional domain is to bind and activate the enzyme known as cGMP phosphodiesterase in the photoreceptor cells, termed as PDE6. This enzyme is the last step in the cascade, since it hydrolyzes the cGMP molecules present in abundant amounts

in the photoreceptor cell cytoplasm into linear GMP. PDE6 is a multimeric protein, with a catalytic dimer of α and β subunits (P$\alpha\beta$) and two inhibitory γ-subunits (Pγ) which are tightly bound to the (P$\alpha\beta$) subunit in the native state. In order for it to be activated to hydrolyze the phosphodiester bond, this tight binding must be released so that (P$\alpha\beta$) can catalyze the hydrolysis. This happens when the Tα subunit of transducin attached to GTP (i.e., Tα–GTP) binds to PDE6 and releases the (Pγ) bound to the (P$\alpha\beta$) catalytic dimer of the enzyme.

An easily understandable and explanatory video on the process of visual phototransduction can be had by accessing the website www.youtube.com/watch?v=KosDT4z6NBc

The molecule cGMP has an important role in keeping the electric potential status of cells. In general, cells have a lower electric potential (more negative charge) than the medium they are in, and are thus (negatively) polarized. What cGMP does is to open the sodium ion (Na$^+$) channels in the plasma membrane of the cell and allow Na$^+$ ions to come in and balance the polarization. But when transducin activates the enzyme PDE6, the level of cGMP falls, leading to a blockage of Na$^+$ entry and since charge neutralization does not happen, the cells are hyperpolarized. It is this large potential difference which is then transmitted as a signal from the photoreceptor cells to the horizontal and bipolar cells. The main steps in the conversion of absorbed light energy into electrical energy have been achieved.

Visual transduction as a G-protein cascade

The process described above is an excellent example of what has been known as the G-Protein cascade of events. G-proteins are a family of proteins that bind to guanosine nucleotides such as cGMP, GMP, GDP or GTP and are involved in transmitting signals from a variety of stimuli outside a cell into the cell. Olfaction is one such example. G-proteins are more often than not heterotrimers, with an α subunit and a binary complex of β and γ subunits. They are located inside cells and activated by molecules called G-Protein coupled receptors (or GPCR). Upon binding to the GPCR, a G-protein triggers an enzyme which helps in signal transmission through the release of a 'second messenger'. The classic example is that of adenyl cyclase which produces cAMP, for working out the details of which Drs. Gilman and Rodbell were awarded the Nobel Prize in physiology or medicine in 1994. Thus, as can be seen, metarhodopsin II (or R*) is the GPCR while transducin is the G-Protein which activates PDE6, causing it to hydrolyze the cGMP molecules in the cell. This has been highlighted in the paper by Lamb and Pugh [42].

Bleach and recycle pathway

Now comes the recycling of the molecules. Once the phototransduction step is done, R* moves to a site in the ROS where it is acted upon by the enzyme known as All-Trans-Retinol Dehydrogenase (ATRDH). This is a 360 amino acid-long protein that uses NAD$^+$ and Zn as cofactors for its action, which is to reduce the all-*trans*-retinal to its alcohol. The product all-*trans*-retinol is then captured by the chaperone protein Interphotoreceptor Retinoid Binding Protein (IRBP), a 135 kDa glycoprotein, which transports the molecule from the rod cell to the retinal pigment epithelium (RPE). Once there, the all-*trans*-retinol is volleyed to and bound by yet another chaperone protein called Cellular Retinol Binding Protein (CRBP). This is a protein of the lipocalin family found in several tissues in the body, and is known to be a specific carrier of retinol. Here at the retinal pigment epithelium, the alcohol is transferred to an enzyme called Lecithin Retinol Acyl Transferase (LRAT, a 230 residue-long esterase) which esterifies the all-*trans*-retinol. It is at this stage of the ester that isomerization of the all trans form into the 11-*cis* form, as well as de-esterification occurs through the mediation of the protein called RPE65 (molecular weight 65 kDa). The isomerized and de-esterified product, 11-*cis*-retinol is now ready to be oxidized (dehydrogenated) using the enzyme RDH5 (11-*cis*-Retinol Dehydrogenase) to produce 11-*cis*-retinal which is transferred from the RPE back to the rod cells and cone cells, completing the visual transduction recycling pathway. It is interesting to note that while the steps involving light absorption and transduction into electric signal take place very fast, within a millisecond, the regeneration process takes place much slower, in a matter of several minutes. Figure 38.27 illustrates the entire visual transduction and recycling pathway in a concise manner.

SUMMARY

The eye ball is placed securely within the orbit of the eye. It is covered in the front by the eye lids which have a row of tiny hairs at the end, which help in protecting the eye from dust and similar air-carried particles.

Next is the tear film which not only lubricates the eye during blinking, but also provides a host of molecules that go to fight infections and provides some level of immunity.

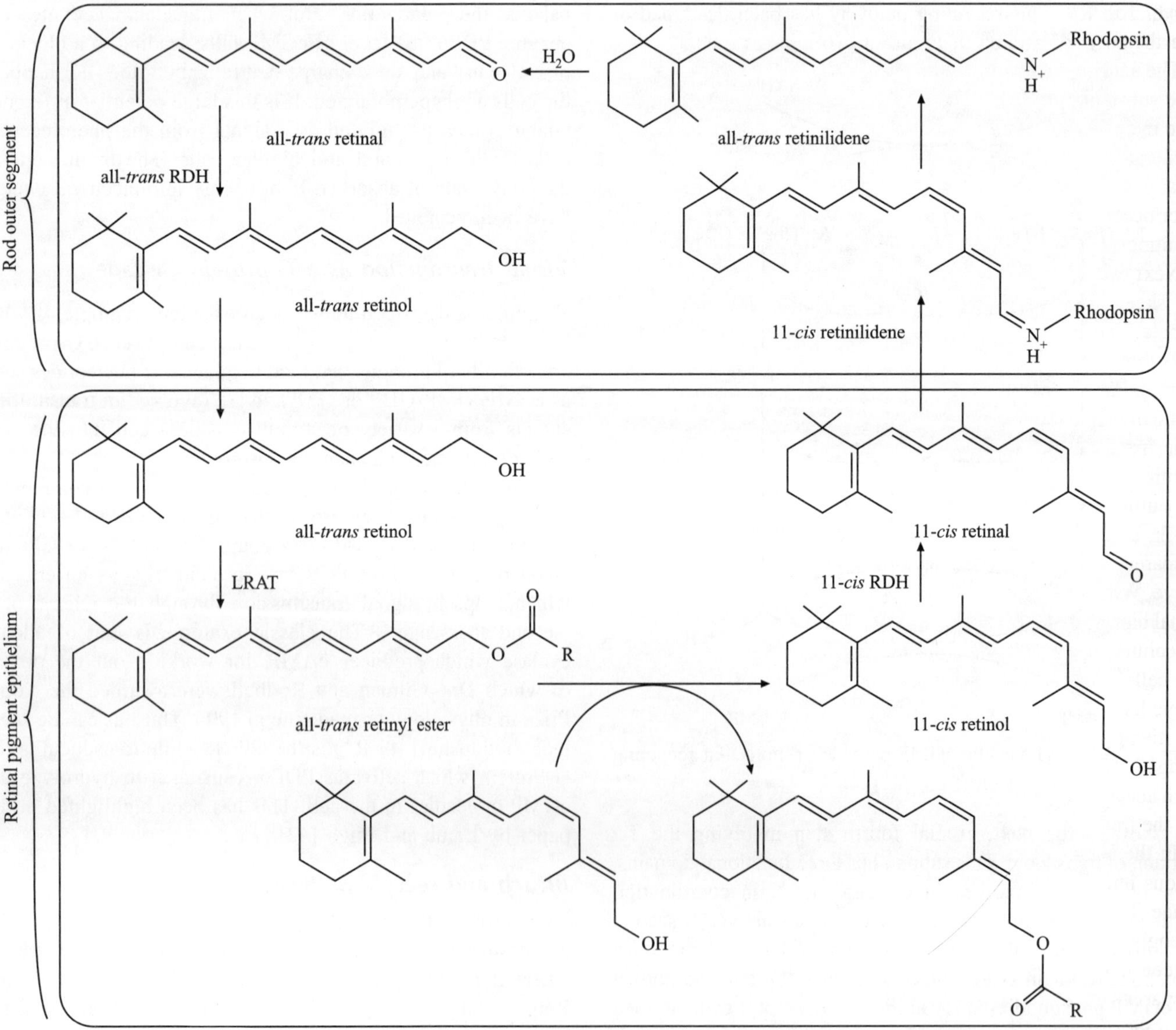

Figure 38.27   The visual phototransduction recycling pathway (from wikimedia.org).

The front surface of the globe is the white surface called sclera which clears in the middle into a transparent disc called the cornea; the two are separated by an annular ring called the limbus which contains adult stem cells which, upon differentiation produces the corneal epithelial cells. The cornea happens to provide as much as 70% of the focusing power of the eye and is free of any blood vessels. It is this avascularity which allows us to transplant a donor cornea into a patient who might need corneal replacement.

As many as 2700 proteins are present in the cornea, many of which provide very important functions, and the major structural proteins that go to make the cornea are collagen and proteoglycans such as dermatan sulfate, keratin sulfate, and hyaluronic acid. It is the proper intermolecular placement of these molecules and interactions between these proteoglycans and the collagen which causes the perfect transparency of the corneal tissue. The cornea has three layers and each layer is separated by a membranous surface. Corneal epithelium faces the outside while the corneal endothelium faces the inside of the eye. Together the endothelium and aqueous humor inside make the cornea well hydrated and transparent.

The aqueous humor has a very large number of biologically active substances which provide nourishment to the front part of the eye, and also removes any metabolic waste material. It is generated in the ciliary body and exits through the uveo-scleral pathway. This production and siphoning of the aqueous humor occurs at a intra-ocular pressure of 10–20 mmHg higher than atmospheric pressure.

Next we have the iris which is a protein-made fabric of a sheet or a disc which can open and close,using the sphincter muscles, thus alter the size of the aperture (the pupil) through which light enters the lens. The colouring of the iris is largely because of the pigment content and thus is genetically dependent. Also the intricate patterns that one sees in the iris sheet are individual-specific, thus allowing bionic identifications of individuals such as fingerprints do.

Further inside the eye after the iris is the lens which is a thick, elastic and plastic protein-containing gel, completely transparent, which thus forms the a second refracting surface in the eye. While the top or the epithelial part of the lens contains normal biological cells, these cells lose all their nuclear material and cellular organelles, upon differentiation, to produce lens fiber cells which are stitched together to produce the bulk of the eye lens. Because the fiber cells do not have any biological activities-metabolic, differentiation or turn-over, they get damaged over a time and it is this time or age-related changes which accumulate (and not get cleared off and renewed) which alter the transparency of the lens thus causing cataract.

In the back of the lens is a jelly-like substance called the vitreous humor which supports and keeps in place the retinal surface. The major components of the vitreous humor are again proteoglycans and collagen. Here too the specific interaction between the proteoglycans and the proteins which leads to a transparent thick and viscous fluid called the vitreous humor.

In the back of the eye is retina which is really where the sensation of vision occurs. The retina is a ten-layer tissue containing as many as ten different types of cells. Towards the back part of this ten layer sandwich are the photoreceptor cells called rods and cones. These cells are rich in a light-sensitive protein called rhodopsin. The light-capturing part here is the molecule called retinal, which is covalently bound to the protein.

Upon capturing visible light, the retinal molecule undergoes a shape change (11-*cis* into all-*trans* conformation), which triggers the opening of the sodium channels in the plasma membrane of the photoreceptor cells, allowing external sodium to come in and cause hyperpolarization. This step of photo-electric transduction results in an electric signal, which is passed on to the ganglion cells in the front of the retina. These ganglion cells pass the electric signal through their axons to the visual cortex of the brain. This is the process known as visual photo transduction, and an excellent example of what is known as the G-protein cascade of events. Following each such phototransduction, recycling of the photo pigment molecules also occurs, using another set other proteins. Thus we have not only phototransduction but also a bleaching and recycling pathway so that the next event of light capture and visual transduction is ready to go.

II. BIOCHEMISTRY OF CATARACT

A. Introduction

The eye lens collects the light coming through the cornea and helps focus it on to the retina. In order to do so efficiently the lens (a) must be totally transparent and not scatter the incoming light out of focus, (b) should be colourless (since if it were not, it might transmit some wavelengths better than others, or display colour-bias), and (c) should be able to readily change shape and thus its refractive index (so as to focus near and far objects equally well).

A healthy lens indeed is able to do all these tasks well, thanks to it being essentially a plastic and elastic tissue. It is plastic since it changes shape readily (thanks to the action of the ciliary muscles which hold it in place), and elastic because once the force changing its shape is released, it bounces back into original shape. The lens is able to do this thanks to it being a highly water-filled tissue, arranged as a multi layered gel of cells, which are rich in globular proteins, mainly the crystallins as its structural components, as we had discussed in Sub-chapter A above.

The lens however is a biologically inert tissue. Most of it (all fiber cells) is devoid of DNA and cellular organelles, and thus of metabolism and regeneration as well. As the human lens grows in time as the individual ages, this inability to metabolize, remove damaged molecules (waste removal) and replace them with fresh ones becomes a problem, affecting the structure and hence the function of the lens. The very first property to be affected as the individual (and hence the individual's lens) ages is its gel-like nature. The lens becomes harder with time, largely due to the slow loss of water content.

As the lens hardens, its ability to focus with equal ease is compromised. This condition, known as presbyopia, is a very common affliction, and the individual needs "reading glasses" to focus better on near objects; and a more convenient solution to this is the use of bifocal lenses or spectacles.

Note that though presbyopia has occurred, the transparency of the lens has not been compromised in any major manner. However, as the individual ages, other physical and chemical changes occur to the lens which can affect its transparency in a manner that can compromise vision itself. When the lens loses its transparency, it is no longer able to focus the incoming light on the retina and individual can no longer be able to see well. Use of spectacles does not help because they only affect the focal length and not transparency.

This loss of transparency, or clouding and opacification of the eye lens is defined as cataract, and can occur due to a variety of physical and chemical changes in the lens. Cataract can occur in either of the two lenses (unilateral) or both of them together (bilateral). Figure 38.28 shows tow such examples; panel A is a unilateral cataract while panel B illustrates a bilateral cataract. When cataract is diagnosed, the ophthalmologist reaches out through the corneal edge of the eye, reaches out to the cataract-affected lens, removes it and replaces it with a synthetic polymer-based artificial lens (called intraocular lens or IOL) and restores sight. Cataract surgery is thus a simple 20 minute procedure, which is quite successful and widely practiced.

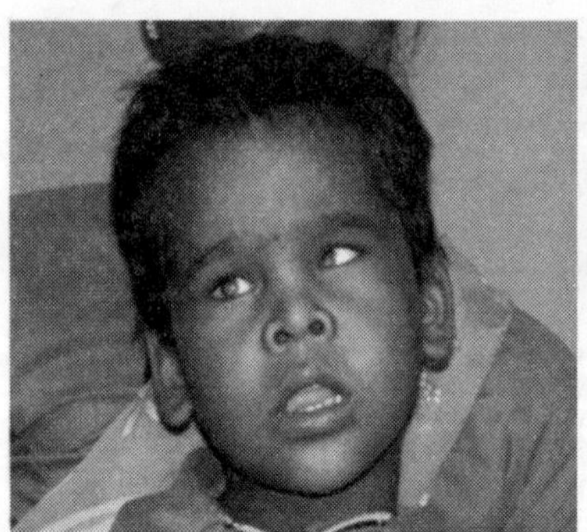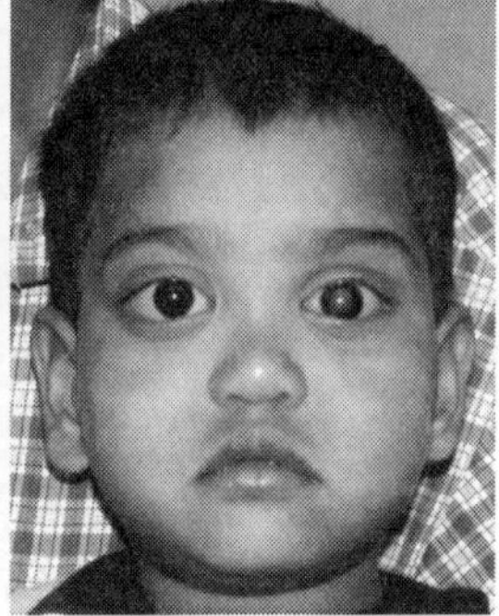

Figure 38.28   Cataracts. Left panel is bilateral and right panel is unilateral. (*see Plate 13 for colour figure*)

Cataract is in effect an age-related disorder, affecting a very large number of people. It usually starts by the time the individual is in his/her fifties and sixties. But who will be affected, and at what age, is still largely based on metabolic factors that go to affect the molecules and cells that constitute the individual lens. A host of factors is cataractogenic and includes continuous intense light exposure (such as the intense sparks of light when a welder is at work during welding), or photochemical, oxidative damage to the lens components, osmotic imbalance due to abnormal hydration of the lens, accumulation of sugar molecular in the lens (due to diabetes) and the consequent chemical reactions between them and the lens proteins, degradation of the constituent proteins due to a variety of factors leading to their precipitation, smoking and associated life style habits, and genetic factors such as mutation in the genes coding for lens proteins. Let us look at some of these factors and how they are cataractogenic.

B.   Osmotic Swelling, and how it affects Lens Structure and Function

As mentioned above, the lens is a water-bathed tissue, comprising about 70-80% water. The lens volume is a balance of two opposing forces; one is the permeability of the lens cell membrane to electrolytes such as Na^+ and K^+, and small molecules (the lens membrane is not permeable to proteins and other macromolecules). The other is a pump in the lens cells that keeps on removing K^+ ions from the interior and takes in Na^+ ions from the outside. This exit of K^+ and entry of Na^+ is vital for the health of the lens, and is controlled by the action of the enzyme Na-K ATPase which acts as the cation pump. This pumping is in proper equilibrium with the normal "leaking" or passive diffusion of Na^+ into the lens and K^+ out of it. When this equilibrium between the pumping and leaking or "pump-leak" is disturbed, a change in the osmotic balance occurs and the hydration of the lens becomes abnormal, highly swollen. This "Donnan" swelling leads to a compromise in the transparency of the lens (osmotic swelling or Donnan swelling occurs due to a high concentration of bound, non-diffusing ions located inside the structure).Thus agents or factors that affect the ability of efficiency of Na-K ATPase and the "pump-leak" equilibrium are cataractogenic [43]. One typical example is the accumulation of sugar molecules such as glucose, galactose or xylose in the lens as happens in diabetic patients.

C.   Polyol Pathway and Aldose Reductase

While what has been described is a simple, physical effect of Donnan swelling compromising the lens, further damage can also happen due to associated biochemical reactions. As the ketose sugar (such as glucose or galactose) accumulates in the lens, it is reduced from its keto form into the alcohol form by the endogenous enzyme called aldose reductase (AR) .The accumulating alcohol, a polyol since it has several hydroxyl groups in its structure; Figure 38.29 is not metabolized, nor is it able to penetrate the membrane and leave the cell. These accumulate leads to an influx of water from outside, again leading to osmotic stress, which is cataractogenic. Of the three sugars that accumulate, the worst is xylose, then galactose and then glucose, and AR is most active against these sugars in the same order.

Figure 38.29   The polyol pathway.

The polyol pathway (shown in Figure 38.29) not only leads to hydration and swelling but also to biochemical ill-effects. Glucose is reduced by AR to sorbitol, and in the process the cofactor NADPH is oxidized to $NADP^+$. The sorbitol so formed is next oxidized by the enzyme sorbitol dehydrogenase (SD, or more generally polyol dehydrogenase or PD) to fructose, and in the process the cofactor NAD^+ is reduced to NADH. Under normal circumstances, the fructose so formed is acted upon by the enzyme hexokinase and is taken through the glycolysis pathway to produce ATP as energy. But as more and more glucose accumulates in the lens (due to e.g., diabetes), this high concentration simply overloads the glycolysis pathway (through the law of mass action) and sorbitol accumulates in the tissue. To add to this further, the levels of the cofactors NADPH and NAD^+ also decrease. These two molecules are vital for several redox reactions in the cell, leading to the synthesis of important molecules such as glutathione, nitric oxide (NO) and myoinositol. The paucity of these protective molecules not only affects the lens but even the nervous system, leading to not only sugar-mediated cataract but also retinopathy (affecting the retina) neuropathy (nerves) and nephropathy (affecting the kidney function).

It is for this reason that researchers have been working on compounds and substances that act as inhibitors of AR, or aldose reductase inhibitors, and thus attempt to prevent the formation of sorbitol. Some drugs such as Epalrestat are synthetic molecules made by organic and medicinal chemists while others are constituents of naturally occurring spices and food stuff such as green tea, spinach, cumin seeds, fennel seeds and cinnamon. Some examples of molecules isolated from them are quercetin, curcumin and the catechins.

Called aldose reductase inhibitors, they are largely of the flavonoid family of molecules though some synthetic molecules have also been tried with success. One such popular inhibitor is the acid called Epalrestat (chemical structure shown in Figure 38.30), also called by the names such as Eparel 50, Aldonyl, Aldorin etc. Among the naturally occurring molecules used as AR inhibitors are the flavonoid (basic structural backbone in Figure 38.30a) such as quercetin (found in sweet potato, caper berry, dill and red onion) and its derivatives, the catechins of green and black tea. As Steke [44] points out, the AR inhibitors can be classified on the basis on how they bind to the enzyme active site. One type binds to the catalytic site through its acidic group [e.g., epalrestat (Figure 38.30e)] while the other through its hydantoin moiety [e.g., sorbinil (Figure 38.30c)]. These are called the carboxylate type and hydantoin type inhibitors. The third class of inhibitors is illustrated by flavonoids such as quercetin or epigallocatechin gallate. These bind to the catalytic site of the enzyme using the 7-hydroxyl group and the catechol moiety in the B ring, in particular the 3'4' dihydroxy moiety strongly [as seen in quercetin and epigallocatechin gallate (Figure 38.30b)] and inhibit AR activity. Indeed, the presence of the aromatic ring containing multiple hydroxyls group also offers the molecule high antioxidant activity. It is this multiple ability of the plant flavonoids (and their synthetic derivatives) that makes them attractive not only as anti-cataract agents but useful in the prevention and treatment of ocular disorders as well [45].

D. Diabetes and Cataract

Apart from these osmotic and AR-related effects, an even more worrisome set of biochemical changes that occur in the eye lens is due to the accumulation of sugars, particularly glucose, as in the case of diabetes. Apart from the osmotic imbalance, this can create chemical modifications in the lens proteins brought about by the chemical reaction of the sugar molecule with some of the free amino groups in the side chains of lys, arg, gln and asn residues of the protein chain, both enzyme-catalyzed and non-enzymatic. When an enzyme catalyses the reaction, it is termed glycosylation whereas the non-enzymatic reaction is termed glycation.

Termed the Maillard reaction, it is the chemical reaction of the carbonyl group of glucose with the amino groups in the protein side chain to produce a glycosylamine. This then undergoes what is termed an Amadori rearrangement, forming ketosamines, which leads to what is known as the group of Advanced Glycation Endproducts or AGEs (Figure 38.31). Advanced glycation alters the arrangement of the collagens in the membranes, altering their stability. In addition AGEs that have dicarbonyl compounds cause aggregation of the lens proteins—both of which cause cataract. Table 38.6 lists a partial list of AGEs found in the diabetic human lenses, some of which, because of their absorption bands occur in the near ultraviolet and visible region of the spectrum, impart colour to the lens. This colouration is somewhat akin to the 'browning' or caramelization seen during cooking seen dishes in the kitchen. Figure 38.32 shows how glycation of lens proteins has led to both browning and high molecular weight covalent aggregates (which, due to their low solubility in water, scatter light and thus opacify the lens).

In addition to these effects, some of the AGE molecules listed in Table 38.6 are also photodynamic in nature, i.e., after

TABLE 38.6   **A partial list of advanced glycation end products (AGE) found in diabetic human eye lenses, compiled largely from [46]**

1-(5-Ammonio-5-carboxyopentyl)-3-oxido-4-(hydroxymethyl) pyridinium

2-Ammonio-6-(3-oxidopyridinium-1-yl) hexanoate (OP-lysine)

Arg-hydroxy-triosidine

Argpyrimidine

N^ε-Carboxyethyl-lysine

N^ε-Carboxymethyl-lysine

Glyoxal-lysine dimer (GOLD)

3 Isomers of hydroimidazole

Lys-hydroxy-triosidine

Methylglyoxal-lysine dimer (MOLD)

Oxalate monoalkylamide

Pentosidine

Tetrahydropyrimidine

Vesperlysine A

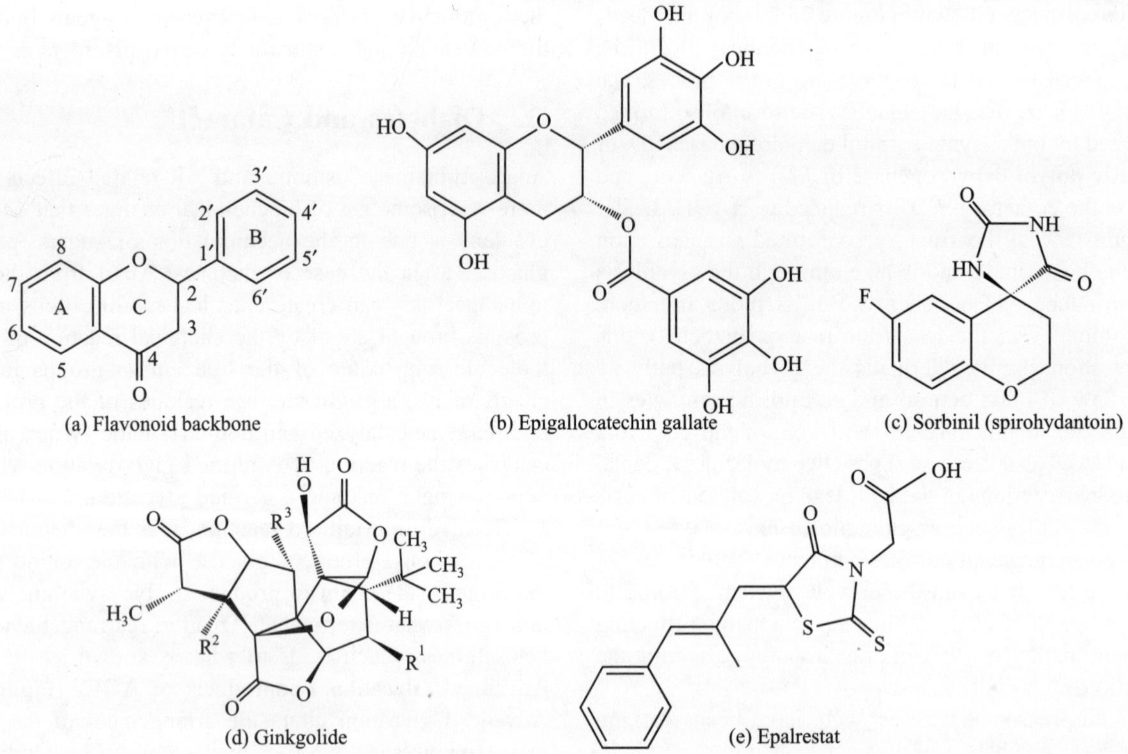

(a) Flavonoid backbone

(b) Epigallocatechin gallate

(c) Sorbinil (spirohydantoin)

(d) Ginkgolide

(e) Epalrestat

Figure 38.30 A to F: chemical structures of some ARIs.

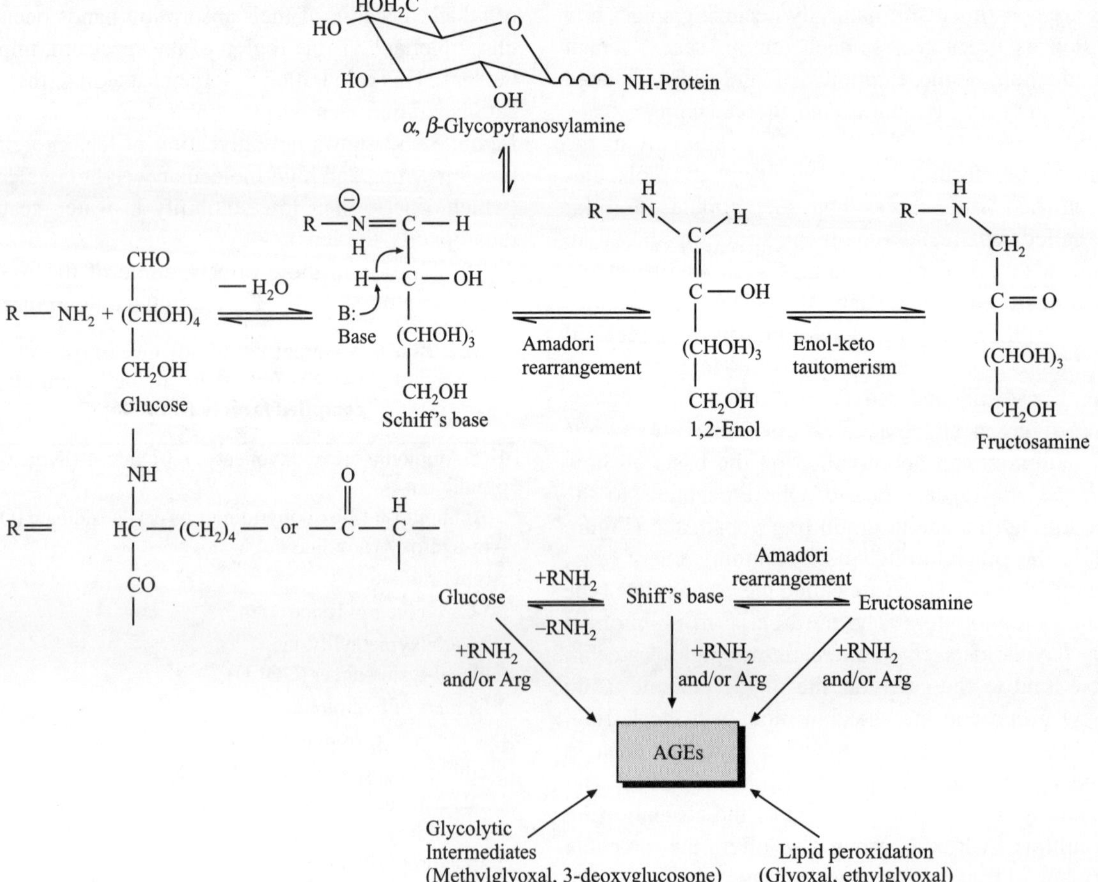

Figure 38.31 Production of Advance Glycation Endproducts (AGEs) generated by the Amadori rearrangement.

the absorb light they generate free radicals and related reactive species in the excited state, which react with the proteins and damage their chemical structure. Their role in cataractogenesis is discussed next.

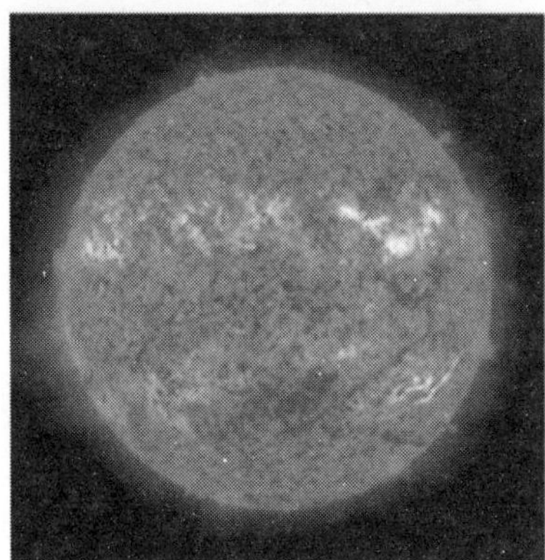

Figure 38.32 Browning of the diabetic cataract lens, and protein aggregates found in that lens. (*see Plate 13 for colour figure*)

E. Photochemical Damage and Cataract

Sunlight reaching the earth is made up of visible (750–400 nm), ultraviolet A radiation (UV-A; 400–320 nm) and UV-B (320–280 nm), besides infrared radiation. The lens has no chromophores which absorb in the visible region of the spectrum (and is thus colourless and fully transmits visible light). Turning to UV, proteins have the following chromophores: peptide bonds that absorb below 220 nm in the UV-C, cys residues (235 nm), his (240 nm), phe (265 nm), tyr (275 nm) and trp (280 nm region). Given that the cornea blocks wavelengths below 295 nm, the major chromophores inherently present in the lens, and of photochemical interest are trp (and to some extent tyr). In addition are the compounds that accumulate in the lens due to metabolic events such as the AGE molecules listed in Table 38.1, which contain chromophores that absorb in the near-UV and visible region. Turning to the innate chromophores first, since the concentrations of proteins in the eye (cornea and lens) is very high, the absorption spectrum is huge and has its wings spread well beyond 330 nm or so in the UV-A region (the red edge). Since the incident light on the lens does have UV-A component, photochemical damage due to absorption of the radiation followed by chemical reactions from the excited state does occur both in the cornea and the lens. The proteins present in the cornea are damaged by this near-UV photochemical damage, but since corneal cells are regularly shed and repopulated, this damage is reduced or handled with efficiency in the cornea. But the lens being largely a no-turn-over system, damage here can accumulate, despite the counter action mediated by endogenous antioxidants. One result of this is the photo-oxidation of the Trp residues by opening the indole ring and generate molecules like N-formylkynurenin (NFK), kynurenine, kynurenic acid, xanthurenic acid and others, which not only absorb in the near-UV and visible region (thus generating colour in the normally colourless lens) but also

can generate photo-oxidative products. Some of these are pro-oxidants (they generate oxidative damage upon light absorption, e.g., NFK, kynurenic acid) while some others are benign or even antioxidants (kynurenine, 3-hydroxykynurenine [46]). The end result of such oxidative damage is invariably precipitation of the soluble proteins due to structural changes, covalent and noncovalent modifications and aggregation of the proteins, to yield scatting particles disturbing transparency. Table 38.7 lists many molecules that have been isolated from aged human lenses and characterized using analytical spectroscopy, while Figure 38.33 shows the chemical structure of tryptophan and several of its derivatives mentioned above.

TABLE 38.7 Classification of the roles of molecules that accumulate in aged and cataract human eye lenses*

Sensitizer Pro-oxidant role	Antioxidant role	Role uncertain
N-Formylkynurenine	Kynurenine	(4-Hydroxy 3-glycine)-quinoline
Kynurenic acid	3HK	Ditryosine
Anthranilic acid	3HKG	ABHG
Quinaldic acid	β-Carbolines	DHQ- carboxylic acid
Xanthurenic acid		Glutathionyl-3-HKG
XA8OG		Compounds that accumulate in diabetic lenses

*Compiled from Ref. [46]. Abbreviations: XAOG: xanthurenic acid O-glucoside; 3HK: 3-hydroxykynurenine; 3HKG: 3-hydroxykynurenine glucoside; ABHG: 4-(2-amino-3-hydroxyphenyl)-4-oxobutanoic acid; DHQ-carboxylic acid: 4,6-dihydroxyquinoline-carboxylic acid.

Several of them act as sensitizers, i.e., absorb light in the visible and near-UV light and generate Reactive Oxygen Species (ROS) and Reactive Nitrogen Species (RNS) such as singlet oxygen ($^1O_2$), hydroxyl radical (^oOH), superoxide anion radical (O_2^{-o}), nitrous oxide (NO) and peroxynitrite anion ($OONO^-$). And these ROS and RNS cause covalent damage to the endogenous molecules in the lens, making it cataract-prone. Some of them are more photodynamically active in this manner then others. Kynurenic acid is far more effective as a sensitizer than NFK; on the other hand photoproducts like kynurenine, 3 hydroxykynurenine and the β-carbolines actually quench the ROS and RNS, acting thus as antioxidants or protective agents (just as ascorbic acid and α-tocopherol are). And yet other accumulants in the ageing lens, such as kynurenine, 3 hydroxykynurenine and the β-carbolines actually quench the ROS and RNS, acting thus as antioxidants or protective agents (just as vitamin C or ascorbic acid and vitamin E or a-tocopherol are). And the role of yet other accumulants in the ageing lens, such as dityrosine, is not clear-whether they are just benign photofilters or have any other properties is yet to be found out. But this discussion highlights the fact that the very act of collecting, refracting and focusing incoming light in the seemingly safe region of the spectrum (visible and near

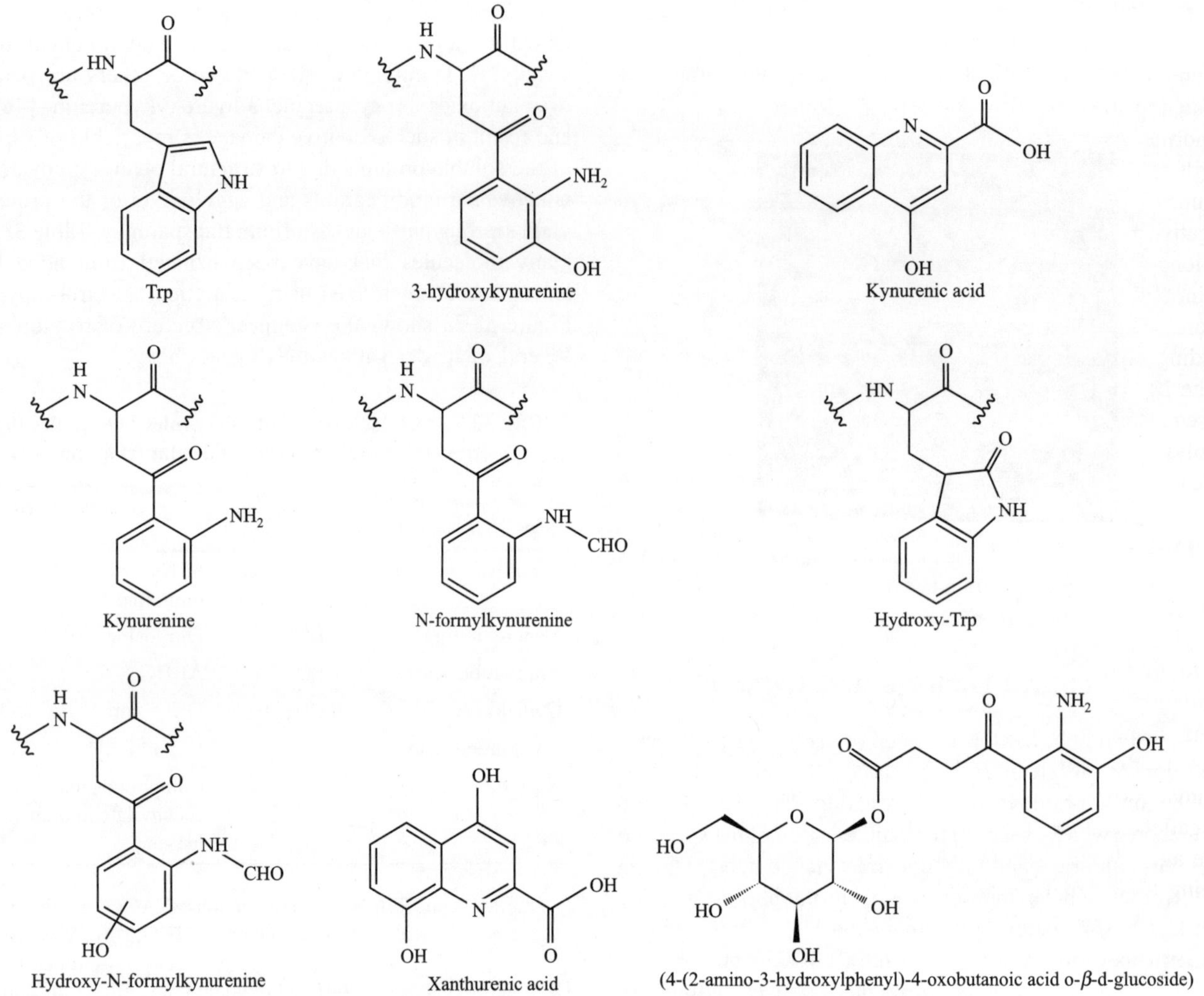

Figure 38.33   Tryptophan and some of its photoproducts seen in the lens.

ultraviolet) could lead to some irreversible changes in the lens chemistry and consequent function with time. And since many of the photoproducts are fluorescent, absorbing in the near-UV and emitting in the visible region of the spectrum, the colour of human lenses slowly changes with age from totally colourless at birth and childhood to yellow and brown with age.

While the lens has endogenous defense systems in terms of antioxidants, enzymes and α-crystallin to counter the effects of photo-, metabolic and pathology-related results, note that it is biochemically inert. Hence as these protective agents deplete with time and age, the colour and the transparency of the lens do get compromised with age. In addition, as the nuclear region of the lens gets more and more packed with time, it becomes harder, and less hydrated. This leads to a gradual decrease in the ability of the lens to change its shape. Since the lens is not able to "accommodate" to more spheroidal and increase its refractive index, the need for reading glasses or the use of bifocal glasses (the top part of the spectacles with one focal length or diopter value and the bottom half with a different diopter value) becomes necessary for such people with presbyopia. Lens manufacturers are now able to fuse the two halves in the same glass and offer a set of 'progressive' eye glasses.

F.   Smoke and Cataract

One of the risk factors for lens opacification and cataractogenesis is smoking and inhalation of fuel smoke (and its condensates). Epidemiological analysis has shown that people who smoke regularly have a higher risk of cataract than those who do not. The 'odds ratio' for cataract in regular smokers in comparison to those who do not is 1.66, and current smokers have a higher risk for being afflicted with cataract than those who have quit smoking [47, 48]. As the smoke is inhaled, a portion of it does reach the eye and the lens, causing chemical damage. Here too, the mechanism behind this risk is oxidative in nature, since smoke contains polycyclic hydrocarbons (phenanthrene, 1,2- benzanthracene) and free radicals. While the latter directly attack and damage the molecules in the eye and the lens, the aromatics cause photodynamic damage in the way discussed above [49]. In addition, tobacco leaves contain significant amounts of the metals Cd and Fe, which generate free radicals and reactive oxygen species (ROS) and reactive nitrogen species (RNS) which cause molecular damage [50]. As one quits smoking, the accumulation of these damaging substances decrease with time and the odds ration reduces.

G. Calcium and Cataract

The concentration of calcium in the aqueous humor is estimated to be about 1.2–1.7 mM; yet, the amount of calcium in a normal healthy is estimated to be in the 200 μM range. Though calcium ions are needed for normal lens development and function, studies have shown that excess Ca^{2+} ions have a negative effect on the lens, and associated with cataract. Ca^{2+} ions are known to activate proteolytic enzymes such as calpain, and also some nucleases. How the intracellular Ca^{2+} ions are sequestered and kept out of activating such protein degrading enzymes has been studied and it has been found that the $\beta\gamma$-crystallin family of lens proteins has binding sites that sequester free Ca^{2+} ions, thus protecting the les from proteolysis [51]. Mutations in these proteins are thus expected to release Ca^{2+} and lead to lens opacification.

H. Deamidation and Cataract

The reaction that removes the side chain amine groups in asparagines and glutamine is referred to as deamidation, and leads to glutamate and aspartate, respectively. Modification by deamidation introduces negative charges in the protein chain, causing changes in the protein tertiary structure and, in turn, affects its structural stability and even insoluble aggregates. Deamidation of βB2-crystallin has been found to increase its aggregation tendency [52]. Similar results have been reported on the age-dependent deamidation of αA- and αB-crystallins, affecting their structural and functional properties.

I. Genetics: Congenital Cataracts

Cataract can also occur at birth, as a congenital disorder. More often than not, this is due to genetic errors (though infection of the fetus *in utero*, usually through measles, is also a cause). As many as 50 genetic loci of 28 genes have been identified as related to congenital cataract [53, 54]. Of these, the most frequent mutations are encountered in the genes for the crystallins (α, β, and γ), gap junction proteins (CX46, CX50) and some enzymes. Genetic changes affect the structure of the coded protein and very likely its function. The same protein molecule may affect more than one reaction or pathway. In that case, the mutation might be associated with more than one phenotype or pathological syndrome. For example the frameshift mutation Thr203Asn fsX47, is a single base insertion in the codon 203, leading to a frame shift and translation of 46 "nonsense" or aberrant amino acids followed by truncation of the chain. Analysis of the mutant protein function revealed that the molecule disrupts the trafficking of proteins in the lens cells, causing cataract [55]. Likewise, the two mutations E134G and T138R, reported in the protein aquaporin-0 (also called membrane intrinsic protein) act by interfering with water channel activity across lens membranes [56].

Congenital cataract occurs in two major types. One is termed nuclear cataract, which is seen as opacification of the central nuclear region of the infant lens. This is a serious affliction since it blocks the central visual axis. Because of this, the development of the eye of the growing infant is affected, and the surgeon needs to intervene at the earliest (days or weeks after birth). The other type is peripheral or cortical cataract where, since the central visual axis is not affected, surgery can wait until even 4-5 years. Interestingly, a significant number of cases of congenital nuclear cataracts are seen to be associated with mutations in the $\beta\gamma$-crystallins, which are found in high amount in the nuclear region of the lens. Analysis of the mutation and the three dimensional structure of the mutant proteins in solution has suggested that when the mutation distorts even one of the four Greek key structural topology of the molecule, the cataract seen is a nuclear one, whereas if the mutation does not affect this structural motif in any manner, peripheral cataracts seem to result. Thus, loss of the conformational motif in $\beta\gamma$-crystallins upon mutation leads to loss of solubility and precipitation to produce nuclear cataract, blocking vision at the central axis [57].

J. Anti-cataract Agents and their Effectiveness

Since the basic biochemistry of cataract is reasonably well understood, efforts have gone on to produce and use anti-cataract agents. Since the lens is not easily reached due to its anatomical positioning, direct access to it is not possible, unlike the cornea. Thus, anti-cataract agents are necessarily administered orally. Using them as eye drops, as has been attempted, is not an efficient mode since the substance will have to cross too many barriers before reaching the lens (tears, cornea, aqueous, humor, iris and lens capsule); and it needs to reach the interior of the lens past the epithelium, to counteract the damage done in the cortical and nuclear regions of the lens (where no turnover occurs). Even oral administration is not efficient, since the amount that reaches the target tissue will only be a minuscule part of what has been administered. Anti-cataract therapy is thus not an instant or short-term one but a long-term process. (However, the use of antioxidants and other such defense substances will be useful for the retina which is easier reached through the blood stream). It is for this reason that physicians recommend the intake in the daily diet of spinach, citrus, fruits and the other plant products rich in antioxidants and micronutrients for a variety of degenerative metabolic insults including cataract. And eye doctors recommend the use of sun glasses (goggles) when venturing in bright sunlight, as protection against photo-induced damage to the lens proteins and enzymes. People involved in occupations such as welding face flashes of intense light which can literally "burn" off the eye-cornea, lens and the retina—are thus advised, indeed forced, to use photo-shields as protection.

Since the lens is not able to get rid of its waste products and regenerate new cells in most of its structure, cataract or opacification of the lens is a time dependent or age-related disorder. By the time an individual turns 70, his/ her lens would have lost its transparency by as much as 80% or more and a cataract surgery becomes essential in order to restore proper vision. Indeed cataract is the most prevalent age-related disease. People who are otherwise in perfect health would still need to undergo cataract surgery. In this procedure, the eye surgeon reaches the lens through a small incision at the edge of the cornea, uses an ultrasonic probe and breaks the lens into minute pieces and aspirates out the emulsified material (the procedure is called phacoemulsification). Next, the surgeon inserts an artificial polymer-based lens inside the now emptied capsule, and the patient's sight is restored for good.

For a video demonstration of how a cataract surgery is done in the above manner, please access www.youtube.com/watch?v=Go82c4f1emc

Conclusion

Cataract is an age-related disease that affects just about everyone. In India, over 10 million people are affected by cataract (but sadly only 50% of them get surgical care). The reasons for this high number are not hard to seek: daily exposure to intense sunlight, inadequate nutrition, plus inaccessibility of cataract surgeons in the rural areas. At least the last reason is in part taken care of by the program of conducting rural eye camps where free cataract surgeries are done. Thanks to this step, India leads the developing world in fighting and winning over cataract blindness. India does over 5000 cataract surgeries per million population. In Africa this cataract surgical rate is less than 2000 and in China it is even lower. It would be ideal if, rather than temporary and occasional eye camps, cataract hospitals are established on a permanent and sustained basis across rural India. Encouraging steps along these lines have been taken by non-profit, non-government eye care centers in some states, notably in Andhra Pradesh and Tamil Nadu.

SUMMARY

The lens is a plastic and elastic tissue which is essentially a protein-filled jelly like transparent globe. This transparency allows incoming light to be focused appropriately in to the retina.

The plasticity of the lens allows its easy change of shape, and thus its focal length, thus allowing us to visualize near objects as well as objects far away. When the muscles which caused the change of lens shape relax, the lens regains its original shape, thanks to its elasticity.

The lens is metabolically active only in its top layer, or the epithelium. The cells here, called the lens epithelial cells, are similar to other cells in the body in their action. They further differentiate to produce fiber cells which constitute over 95% of the lens cell population. These fiber cells are devoid of all organelles and nuclei and thus metabolically inactive—no biosynthesis, no metabolism, no waste disposal or turnover. Thus, all changes that happen because of light or chemical based oxidation and degradation, and due to diabetic and sugar accumulation—all these cause progressive loss of lens transparency causing cataract.

The major protein components of the eye lens are of the crystalline family of soluble proteins. Of these, alpha crystallin acts as a stress response protein and also shows some chaperone-like activity, helping insoluble proteins get back into solution. The beta-gamma crystallins belong to one family of tightly and compactly packed globules, with a highly stable structure, packed together as globules, with only short range interactions. This is the basis of the transparency of eye lens. As the lens loses its transparency, cataract occurs and the ophthalmologist reaches the lens through a small incision in the cornea, and either pulls it out physically or emulsifies it and aspirates out the emulsified liquid; he then implants an intra ocular lens made of synthetic polymers in its place, thus restoring vision and curing cataract.

Cataract is an age-dependent protein disorder disease, and affects most people upon ageing, because of the metabolic inertness of the lens. However it is believed that continual intake of antioxidant and anti diabetic molecules might reduce the damage to the lens constituents and delay or postpone the onset of cataract.

III. BIOCHEMISTRY AND GENETICS OF GLAUCOMA

A. Introduction

Glaucoma is a heterogeneous group of eye diseases characterised by progressive optic neuropathy that involves cupping and structural degeneration of the optic nerve and deterioration of retinal ganglion cells and its axons ultimately leading to visual field loss [58]. It is the second major cause of irreversible blindness in the world next to cataract [59]. Since the peripheral vision is lost first and the disease is typically pain free, it goes unrecognized until substantial degree of vision loss occurs. Hence, glaucoma is called 'the sneak thief of sight'. Although rise in intraocular pressure is a major risk factor, there is no set threshold for intraocular pressure that causes glaucoma. The patient loses peripheral vision first (**tunnel vision**) followed by central vision. If left untreated, it leads to permanent damage of optic nerve head and loss of visual field and permanent blindness. Other important risk factors include family history, aging, high myopia, and primary vascular dysregulation and low blood pressure [58].

This sub-chapter will focus on the aqueous humor production and outflow, present knowledge of various forms of glaucoma, the associated genes known till date and also the molecular aspects of some of the glaucoma-associated proteins and their mutants.

Aqueous humor outflow pathway

As we had discussed in Sub-chapter A above, the anterior chamber (the fluid filled space between the cornea and the iris) of the eye is filled with a liquid called aqueous humor. Aqueous humor is a clear, slightly alkaline liquid (pH 7.1–7.3) that nourishes the anterior chamber of the eye. Its chemical composition is basically derived from the ciliary epithelium by active secretion and from plasma by passive diffusion. The major components of the aqueous humor are organic and inorganic ions, carbohydrates, glutathione, urea, amino acids and proteins, oxygen, carbon dioxide and water. Aqueous humor is slightly hypertonic to plasma in a number of mammalian species. It resembles blood plasma in composition but unlike plasma it contains less glucose and protein but more ascorbic acid and lactic acid.

After being produced by the ciliary body, the aqueous humor passes through the pupil into the anterior chamber. From there it leaves the eye by passive flow by two pathways. In conventional pathway, aqueous humor passes through a sieve like region called trabecular meshwork and then across Schlemm's canal in the anterior chamber angle and finally draining collector channels, aqueous veins and episcleral veins. The non-conventional pathway or uveoscleral outflow pathway consists of uveal meshwork and anterior face of the ciliary muscle. Needless to say, aqueous humor provides nourishment including oxygen to the anterior regions of the eye which are avascular and also removes waste products. The fine balance between its production and drainage maintains the intraocular pressure of the eye within its physiological range (10–22 mm Hg). In certain pathological states, the aqueous humor gets accumulated in the eye, leading to the raise in intraocular pressure (IOP).

The mechanism of elevated IOP in glaucoma is either impaired outflow of aqueous humor because of the abnormalities in the drainage system of the eye (open angle) or because of impaired access of aqueous humor to the drainage system (closed angle) (Figure 38.10 in Sub-chapter A). This causes damage to the optic nerve and visual impairment. The loss of vision is because of degenerative changes in the retina, particularly retinal ganglion cells—the sensory cues of the eye.

Optic nerve head cupping and glaucomatous neurodegeneration

The optic nerve head (optic disc) consists of axons from approximately one million ganglion cells that have their cell bodies in the retina. The axons converge on the optic disc to form the optic nerve. The fibres exit the eye through lamina cribrosa and synapse in the lateral geniculate nucleus of the brain. The convergence of the axons in the optic disc form a central depression called optic cup. The neuroretinal rim of the optic nerve head is pink in colour and surrounds the cup. Glaucomatous progression is characterized by decrease in the width of the neuroretinal rim with concomitant enlargement of the cup. As axons within the nerve die (largely through apoptosis) and the plates of the lamina cribrosa sclerae (a specialized area of sclera) collapse due to IOP or ischemia, loss of optic nerve tissue produces a characteristic excavation or "cupping" of the optic nerve head (Figure 38.34). Thinning or notching of the disk rim, or disk hemorrhages might also be seen. When a vertical cup-to-disk ratio of 0.6 or greater is seen, glaucoma should be suspected. The cupping of the optic nerve head is directly related to the loss of one's peripheral

visual field leading to tunnel vision. With peripheral vision loss, a person can see in front of him- or herself but has lost the vision to the side which is called tunnel vision (Figure 38.35).

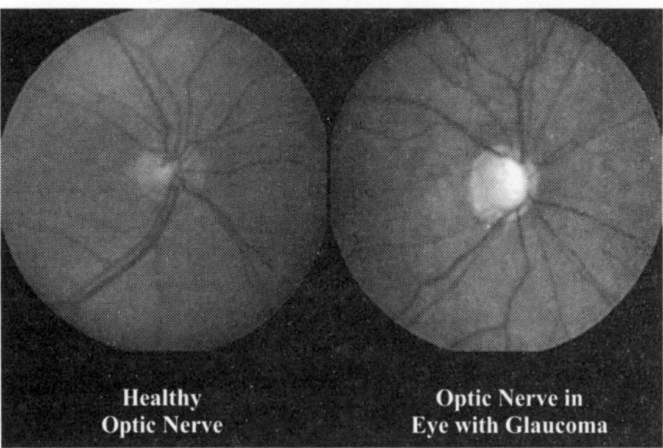

Figure 38.34 Optic nerve head in healthy and glaucomatous eye (from www.glaucoma.org) (*see Plate 13 for colour figure*)

World as seen by a healthy individual

World as seen by a glaucoma patient

Figure 38.35 Tunnel vision, experienced by a glaucoma patient (right panel).

B. Factors Contributing to Glaucomatous Neurodegeneration

Apart from rise in IOP, few other factors play a major role in glaucomatous neurodegeneration. They include ischemia, glutamate excitotoxicity, aberrant immunity and alteration in glial cells [58] (Figure 38.36). The relative significance of each specific factor varies between patients depending on the magnitude of IOP elevation, oxidative stress and individual life style [60].

(a) **Ischemia:** Elevated IOP induces ischemia, which results in blockage of axoplasmic flow and obstruction in normal flow of growth factors leading to RGC death. Optic nerve blood flow is compromised in glaucomatous patients. Ischemia enhances oxygen free radical, glutamate release, and nitric oxide production, which increase the levels of Ca^{2+} that show detrimental effect on RGC cells.

(b) **Excitotoxicity:** Glutamate is an essential excitatory amino acid in the central nervous system and also in the retina. It has been shown in several neurological diseases that excessive glutamate induces over-excitation of neurons, which ultimately leads to the death of neuronal cells. It has recently been shown that glutamate excitotoxicity plays an important role in the pathogenesis of glaucoma. The evidence comes from the fact that (i) vitreal glutamate levels are elevated in experimental and clinical glaucoma, (ii) ischemia (loss of blood supply) plays a role in glaucoma, (iii) neuroprotection based animal studies and (iv) secondary degeneration of RGCs after optic nerve injury.

(c) **Autoimmunity:** In addition, factors such as autoimmunity [61] and excessive activation of glial cells (which are normally helpful in limiting nerve cell damage) [62] are also responsible for disabling retinal ganglion cells (RGC), leading to glaucoma.

An easy to understand video that describes the disease glaucoma can be accessed at the site www.youtube.com/watch?v=TIkYzWr44b0

C. Genetics and Biochemistry of Glaucoma

Glaucoma is a genetically heterogeneous disease and involves multiple factors. While smaller proportion follows monogenic inheritance, majority of the cases involve multiple factors. Although autosomal dominant and recessive glaucomas exist, the disease is much more complex in many individuals, hence, making it difficult to understand the molecular genetics. The most common forms of glaucoma do not exhibit Mendelian inheritance but the immediate relatives of the proband are at higher risk than general population. Multiple genetic and environmental factors play a major role in its etiology. Molecular geneticists have identified at least 20 chromosomal loci till date, which are linked with various forms of glaucoma but only 5 genes are identified. The 4 known POAG genes include *Myocilin/TIGR, Optineurin (OPTN), WDR36* and *ASB10. Cytochrome P450-1B1 (CYP1B1) and latent transforming growth factor-beta-binding protein 2 LTBP2)* are the candidate genes for congenital glaucoma [63, 64].

Glaucoma can be classified based on different criteria like etiology (primary and secondary), nature of the iridocorneal angle (primary open angle (POAG) and primary angle closure glaucoma (PACG) and the time of onset (infantile, juvenile and adult onset). In general, glaucoma can be classified into three major categories: (i) Primary Open Angle Glaucoma (POAG) (ii) Primary Congenital Glaucoma (PCG) and (iii) Primary Angle Closure Glaucoma (PACG).

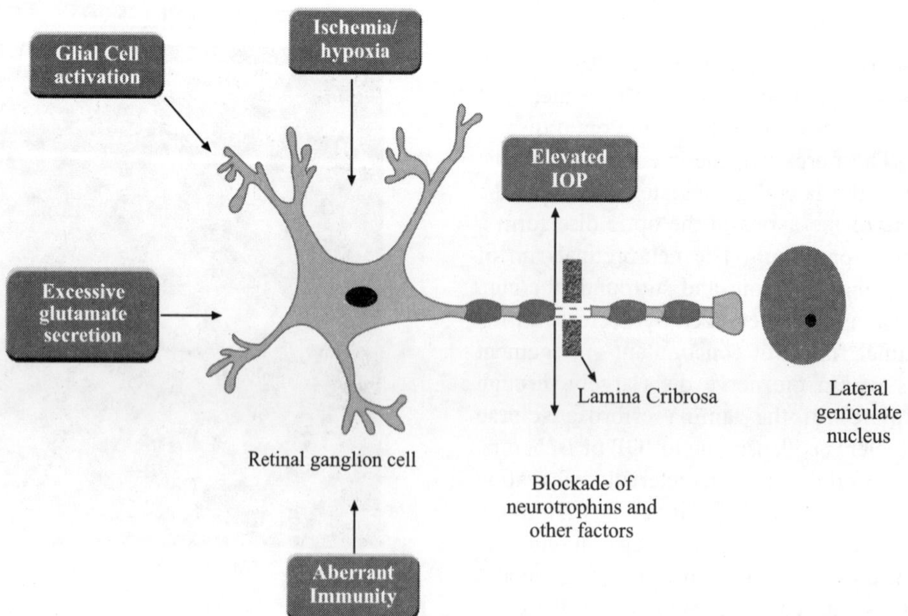

Figure 38.36   Factors contributing to glaucomatous neural degeneration.

D. Primary Open Angle Glaucoma (POAG)

POAG is the most common type of glaucoma. As the name POAG suggests, here the angle between the cornea and iris is unimpeded, yet there is obstruction in the outflow of aqueous humor. Based on the age of onset, POAG is divided into juvenile onset POAG (JOAG) and adult onset POAG. JOAG occurs at the age of 5-35 and adult onset POAG occurs at an older age. POAG patients have IOP consistently above 22 mm Hg and hence called high tension glaucoma (HTG). Approximately, one-third (33%) of POAG patients have IOP within normal range (<22mm Hg) and is called normal tension glaucoma (NTG) or low-pressure glaucoma (LPG). But the ultimate theme in both HTG and NTG is retinal ganglion cell death and visual field loss whose mechanism is still unclear. At least 20 genetic loci have been mapped for POAG through linkage analysis but only 4 genes are known as mentioned earlier, viz., *myocilin (MYOC/GLC1A), optineurin (OPTN/GLC1E), WD repeat domain36 (WDR36/GLC1G)* and very recently ankyrin repeats and suppressor of cytokine signaling box containing protein 10 *(ASB10/GLC1F)*. Apart from these, association studies have identified at least 27 genes susceptible to POAG including *TANK-binding kinase-1 (TBK1), cytochrome P450-1B1 (CYP1B1),* and *neurotrophin-4 (NTF-4)*. The molecular insights into *MYOC, OPTN and WDR36* which are well known POAG genes are discussed below.

Myocilin (MYOC)

In 1997, a gene called *myocilin/TIGR* (trabecular meshwork induced glucocorticoid response), located in the chromosomal position 1q23-25, encoding for the protein myocilin was identified as the first candidate for POAG [8]. Mutations in this *myoc* gene are associated with approximately 3% of adult onset POAG and a major proportion of juvenile open angle glaucoma (JOAG).

Molecular structure and functions: The *MYOC* gene consists of three exons which code for a 504 amino acid-long secretory glycoprotein with a predicted molecular weight of 55kDa. The protein consists of an amino terminal signal sequence, a myosin-like domain, a leucine zipper motif within two coiled-coil domains and what is called the c-terminal olfactomedin domain [65]. The line diagram of the *MYOC* gene is shown in Figure 38.37. This olfactomedin domain is highly conserved among mammals and also among diverse species suggesting the importance of this domain in structure-function relationship. The

significance of this domain can also be corroborated by the fact that majority of the mutations studied till date are concentrated in this region which is encoded by a single exon. Studies have predicted the existence of four potential glycosylation sites and ten putative phosphorylation sites [66].

Three-dimensional protein structure prediction studies revealed that the olfactomedin domain of myocilin has a six-bladed (which is numbered 1–6) β-propeller structure. This model is consistent with some earlier studies showing that olfactomedin domain is globular and contains mainly β-sheets. Menaa et al. [66] have predicted the three dimensional structure of myocilin and suggested it may have a role in methylation process.

Myocilin is expressed in both ocular (iris, cornea, ciliary body, trabecular meshwork, optic nerve, lamina cribrosa, aqueous humor and vitreous humor) and non-ocular tissues (skeletal muscle, heart, brain). Despite extensive efforts, the physiological function of myocilin and its role in glaucoma remain elusive. Myocilin is normally secreted into the aqueous humor of the eye. It is believed that misfolding, abnormal aggregation, and defects in secretion play an important role in the involvement of myocilin in glaucoma. Immuno-histochemical studies have revealed that myocilin is found in trabecular meshwork cells, trabecular beams and the connective tissue of the juxtacanalicular part of the human trabecular meshwork *in situ*. Cell biology studies have shown that myocilin is associated with organelles of the secretory pathway such as the endoplasmic reticulum and the Golgi apparatus prior to its secretion into the extracellular environment. It is also seen to be associated with mitochondria and microtubules, actin fibers and proteins involved in signalling and metabolism.

Disease association: POAG-associated mutations affecting the myocilin gene exhibit an autosomal dominant inheritance, suggesting that it is the presence of the mutant protein rather than the absence of wild type protein that causes glaucoma. Since its discovery in 1997, substantial efforts have been made to determine the function of myocilin in normal healthy eye. Though the physiological role of the protein is not yet known, considerable progress was made in understanding the pathophysiology of MYOC associated glaucoma. One of the first hypotheses regarding the association of MYOC and POAG is based on the fact that excessive myocilin protein levels in trabecular meshwork either because of increased protein production or compromised degradation leads to increased resistance in aqueous humor outflow which in turn causes elevated IOP [67]. Till date, more than 70 POAG causing

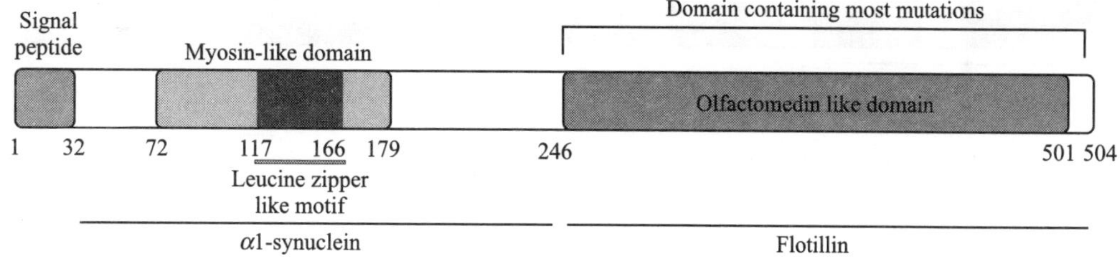

Figure 38.37 Line diagram of Myocilin with various domains and region of binding with two of the known interacting partners. (*see Plate 13 for colour figure*)

mutations are recorded by various teams all over the world. These mutations have a strong genotype-phenotype correlation. Majority of the mutations having amino acid substitution are not secreted and are accumulated in the ER as homo or heterodimers. Indeed these mutants render the protein insoluble and would interfere with its secretion, dimerization and its interaction with extracellular matrix (ECM) components of the trabecular meshwork [68]. Proper oligomerization is essential for the wild type protein to function. The mutants prevent this oligomerization and as a result there is increased production of misfolded protein which instead of getting secreted gets accumulated in the ER. Accumulation of mutant protein in ER leads to ER stress and unfolded protein response. Hence, the association of MYOC with POAG is thought to be what is called a "gain-of-function".

Optineurin (OPTN)

In 2002, Rezaie et al. identified *OPTN* as a candidate gene for POAG, particularly NTG [69]. Later on, mutations in *OPTN* are also known to be causative of amyotrophic lateral sclerosis (ALS) and also known to be genetic risk factors for Paget's disease of bone. Optineurin is found to be localized to pathological structures in ALS, neurofibrillary tangles and dystrophic neuritis in Alzheimer's disease, Lewy bodies and Lewy neuritis in Parkinson's disease, ballooned neurons in Creutzfeldt–Jakob disease, glial cytoplasmic inclusions in multiple system atrophy, and Pick bodies in Pick disease [70].

Molecular structure and functions: *OPTN* gene contains 16 exons out of which first three are non-coding. The remaining thirteen exons code for a 577 amino acid protein which is 67kDa [70]. Alternative splicing at the 5′-UTR generates four isoforms and all have them have the same reading frame. The protein optineurin contains several putative domains including multiple coiled-coil motifs, leucine zipper, NEMO-like domain, ubiquitin binding UBAN domain (UBD), LC3-interacting region and a C-terminal zinc finger domain (Figure 38.38). Optineurin is neither N- nor O-glycosylated but it is known to get phosphorylated at serine 177 adjacent to LC3 interacting region [70].

Optineurin is a cytosolic and endogenous protein, processed through the ubiquitin-proteasome pathway [71]. Optineurin is known to form hexamers and also as a supermolecular complex along with Rab8, myosinVI, transferrin receptor (TfR). It is also known to increase myocilin mRNA levels which is a POAG gene itself and this regulation is achieved primarily through control of the mRNA stability.

Optineurin is expressed in both ocular and non-ocular tissues. In the eye its expression is seen in the non-pigmented ciliary epithelium, trabecular meshwork and in the retina at significantly high levels. Endogenous optineurin is localized to cytoplasm with some population in the Golgi [72]. Though the structure of optineurin is not known, it is one of the glaucoma candidates whose physiological role is well studied. It is a multifunctional protein interacting with proteins belonging to different classes having a role in NF-κB signalling (CYLD, A20, RIP1, TAX1, TAX1BP1), vesicular trafficking (Rab8, TBC1D17, TfR, Huntingtin, MyosinVI), autophagy (LC3, TBK1), cell division (plk1, MYPT1), antiviral signalling (TBK1, TRAF3, TAX1, TAX1BP1) and cell survival (neurotrophin3 & CTNF) (Figure 38.39). The functional significance of most of these interactions is well deciphered [70]. Also, the pathological mechanism of few important mutations like E50K, M98K and H486R is also revealed which is discussed later in this section.

Apart from its role in NF-κB signalling, optineurin interacts with many proteins involved in vesicle trafficking like Rab8, TfR and Myosin VI and huntingtin. Optineurin acts as an adaptor protein and helps in recruitment of various proteins during vesicular trafficking [69]. It is found to regulate endocytic trafficking of TfR to the juxta-nuclear region and to mediate the interaction of TBC1D17 with Rab8 and regulate the Rab8- mediated TfR trafficking [73, 74].

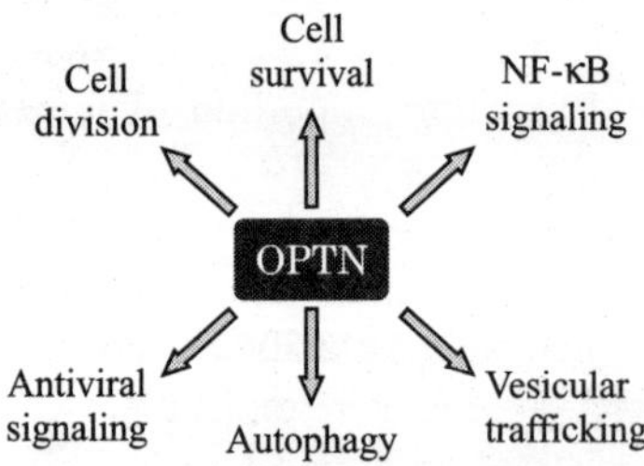

Figure 38.39 Optineurin—a multifunctional protein involved in multiple pathways.

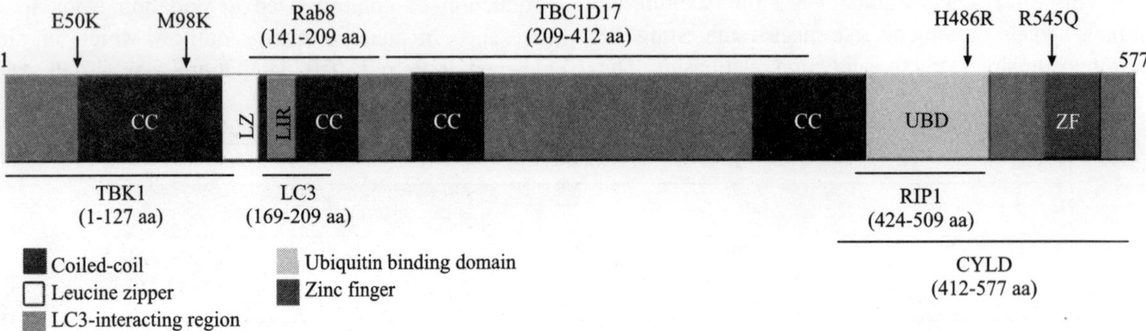

Figure 38.38 Line diagram of optineurin with various domains, interacting proteins and some glaucoma associated mutations. (*see Plate 13 for colour figure*)

Optineurin is recently discovered to be an autophagy receptor by binding to the protein called LC3 through its LC3 interacting region (sequence 169-209). Upon phosphorylation at Ser177 by TANK binding kinase1 (TBK1), the LC3 binding efficiency of optineurin increases which helps in autophagic clearance. TBK1 phosphorylates optineurin at ser177 and colocalizes with optineurin on protein aggregates and helps in their clearance through autophagy-lysosome pathway [75].

Disease association: In the initial study, mutations in OPTN were reported in 16.7% of families with hereditary POAG, with most of them having NTG. Also, 13.6% of risk-associated alterations were observed in both familial and sporadic individuals. Out of 16.7% of mutations, the major proportion of patients, i.e., 13.5% of them carried a missense mutation, E50K, where glutamic acid at position 50 was mutated to lysine.

Cell biology studies have revealed that this E50K mutant caused cell death selectively in retinal ganglion cell (RGC-5) and not in other non-ocular cell lines. This cell death was mediated by oxidative stress and treatment with antioxidants rescued cells from the E50K-mutant induced cell death [76]. In addition, the role of E50K in NTG was further corroborated by a study where E50K transgenic mice showed severe retinal ganglion cell loss, reduction of retinal thickness and excavation of optic nerve head [70]. This mutant also causes defective endocytic trafficking and recycling of TfR by causing inactivation of Rab8 mediated by its GTPase activating protein, TBC1D17 [73, 74]. A functional UBD is required for the proper trafficking of transferrin to the juxtanuclear region of the cell [74].

M98K, which is described as a risk associated factor for NTG in Asian populations, is said to enable a gain of function where it shows enhanced autophagosome formation and also enhanced the autophagic degradation of TfR. TfR plays an important role in maintaining iron homeostasis. This decrease in TfR levels in turn causes death of retinal ganglion cells specifically. Moreover a functional UBD and LC3-interacting region are required for the M98K mediated TfR degradation and cell death in RGC-5 cells. The study also reveals TfR as the first protein cargo for optineurin delivered for autophagic degradation [77].

WDR36

The third candidate gene in POAG, WDR36 mapped onto chromosome *5q22.1 (GLC1G)* [78] is comprised of 23 exons and codes for a 951 amino acid residue protein. Though the gene has been mentioned as a causative of POAG, genetic studies by various groups have shown that *WDR36* may act as a modifier of POAG, and/or as a causative gene for POAG in certain populations. The encoded protein consists of at least four known motifs: a guanine nucleotide binding protein (or G)-beta WD40 repeat, an AMP-dependent synthetase and ligase, a cytochrome cd1-nitrite reductase-like (C-terminal heme d1) and an Utp21-specific WD40 associated putative domain. These domains are involved in protein-protein interactions.

In human and mouse tissues, *WDR36* is widely expressed, and it was also found to be upregulated during human T-cell proliferation. RT-PCR analysis has revealed WDR36 mRNA expression in lens, iris, sclera, ciliary muscles, ciliary body, trabecular meshwork, retina and optic nerve [78]. Endogenous WDR36 localized to the plasma membrane and it acts as a scaffold protein tethering a G-Protein coupled receptor-thromboxane A2 receptor (TPβ), Gαq and PLCβ in a signalling complex [79]. The disease-associated variants (L25P, N355S, I604V, D658G and M671V) of WDR36 affected its ability to modulate Gαq-mediated signalling by TPβ. Molecular modelling by Chi et al. [80] has shown the WDR structure to have two distinct domains. As seen in Figure 38.40, each individual domain is a seven-bladed β-propeller, and each blade is arranged as four anti-parallel β-strands starting from the center of the propeller; these blades are stabilized by hydrogen bonds between the main chain atoms of β-strands. They have also constructed the molecular models of three different mutations of WDR36, namely, D606G, Del605-607 (deletion in the 7th beta propeller of the second domain) and Del601-640 (removal of the entire 7th beta propeller region). Such mutations destabilized the propeller structure, thus destabilizing the molecule; these predictions are in accord with actual experiments involving transgenic mice with these

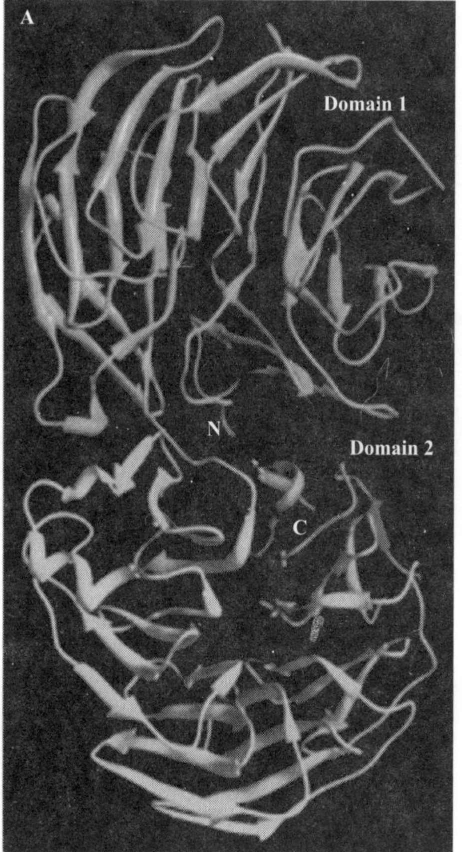

Figure 38.40 3D structure predicted for WDR36 (taken from [80]). (*see Plate 14 for colour figure*)

mutations showed loss of retinal ganglion cells and retinal degeneration. The β-propeller structures found in WD40 repeat-containing proteins appear to act as platforms for the stable or reversible association of binding partners.

E. Primary Congenital Glaucoma (PCG)

PCG is commonly called infantile or congenital glaucoma and is an autosomal recessive disorder. PCG is an ocular developmental anamoly characterised by an abnormal development of trabecular meshwork and anterior chamber angle, elevation of IOP, enlargement of the globe (buphthalmos) and opacification of the cornea. PCG is a rare eye disorder which accounts for 0.01–0.04% of total blindness. The disease is usually manifested at birth or early childhood (before 3 years of age). The incidence of PCG is most common in communities where consanguinity is high [81].

Currently, four chromosomal loci have been implicated in PCG on *GLC3A* at 2p21-22, *GLC3B* at 1p36.2-36.1, and *GLC3C* at 14q24.3 and *GLC3D* at 14q24. The candidate genes include *CYP1B1* on *GLC3A* and *latent transforming growth factor-beta-binding protein 2* (*LTBP2*) on *GLC3D*. Also, myocilin (MYOC) which is mainly implicated in POAG may be involved in congenital glaucoma through a digenic inheritance along with CYP1B1.

(i)   Cytochrome P4501B1 (CYP1B1)

Identification as a candidate gene for PCG: CYP1B1 (Cytochrome P4501B1) was the first gene of the CYP450 superfamily whose mutations are accounted for autosomal recessive primary congenital glaucoma [82]. It was later also reported as a modifier locus for POAG that together with MYOC mutation expedite the disease progression from adult onset to a juvenile form in a digenic mode of inheritance.

Structure and functions: The human CYP1B1 consists of 3 exons coding for a 543 amino acid-long membrane-bound protein. It contains a hydrophobic amino terminal region of about 50 amino acids which serves as a membrane anchor, proline rich hinge region and cytosolic globular domain consisting of a conserved core structure. Included in these structures are four helix bundles (helices D, I, and L and the antiparallel helix E), helices J and K, β-sheets 1 and 2, the heme-binding region, and the "meander" region just N-terminal

of the heme-binding domain (Figure 38.41). Most of these regions which are located at the C-terminal half of the molecule are expected to be involved in the heme binding and proper folding of the molecule. The protein cytochrome *P*450 1B1 (CYP1B1) is a heme-thiolate monooxygenase that is involved in the NADPH-dependent phase I monooxygenation of a variety of substrates, including fatty acids, steroids, and xenobiotics. It is also involved in the metabolism of carcinogens apart from the synthesis of steroid hormones, arachidonic acid, retinoic acid, and melatonin which regulate growth and differentiation of tissues (Figure 38.42).

It is through these signal transduction pathways it may be regulating ocular differentiation. Although CYP1B1 is expressed in normal tissues, its expression is higher in tumor cells compared to its normal healthy counterparts. CYP1B1 is constitutively expressed, because its mRNA has been detected in many human tissues, especially human kidney, prostate, mammary, ovary, and uterus. In the eye, the protein is expressed in the cornea, ciliary body, iris, and retina but not in the trabecular meshwork. Interestingly, CYP1B1 protein levels are higher in the fetal eye compared to that of adult suggesting that it might be crucial for metabolizing an important substrate that plays a key role in the development and maturation of ocular tissues.

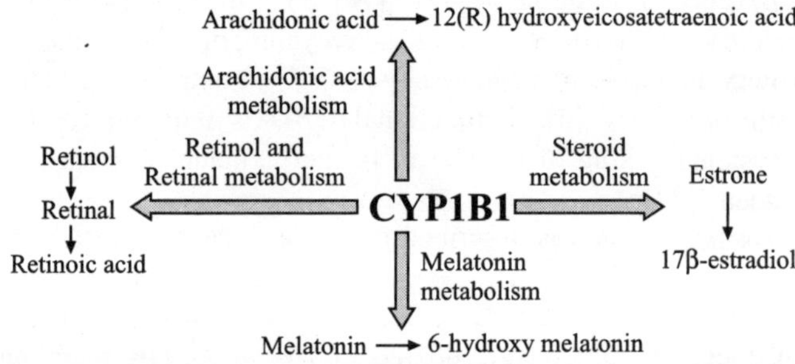

Figure 38.42   CYP1B1 is involved in a variety of metabolic activities.

Disease association: Till date more than 70 mutations harboured in the CYP1B1 gene have been linked to PCG [83]. One-third of the mutations in CYP1B1 are either deletions or insertions suggesting that CYP1B1 is relatively susceptible to recombination [84]. There seems to be no particular hotspot for the location of CYP1B1 mutations because these mutations span the entire gene. It is an autosomal recessive trait with

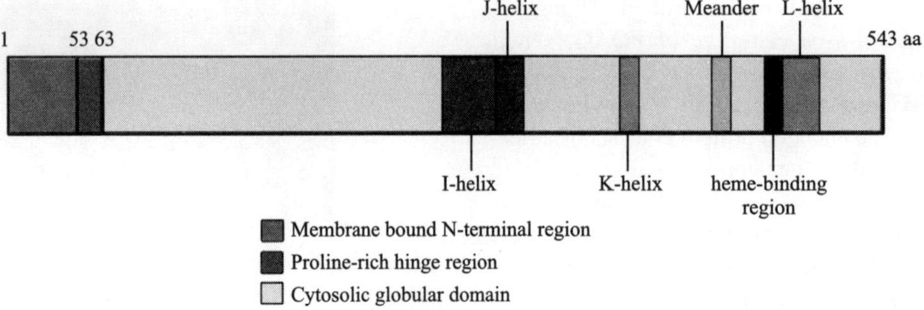

Figure 38.41   Line diagram of CYP1B1 with various conserved regions. (*see Plate 14 for colour figure*)

incomplete penetrance. Mutations in this gene have been observed in different populations, accounting for 20-100% of all primary congenital glaucoma (PCG) patients [85, 86]. A majority of the CYP1B1 mutations are missense mutations affecting the amino acid residues located either in the hinge region or the conserved core structures in the cytosolic region. These mutations, therefore, are expected to interfere with fundamental properties of the protein such as folding, heme binding, substrate accommodation, and interaction with the redox partner. *Insilico* studies performed with a few CYP1B1 mutants have shown that these mutations caused several structural defects which in turn impact functional domains. Molecular docking studies by Achary and Nagarajaram [89] showed that CYP1B1 mutants showed altered protein–ligand interactions compared with the wild-type as a consequence of changes in the geometry of the substrate-binding region and in the position of heme relative to the active site. An important difference in ligand–protein interactions between the wild-type and mutants is the presence of stacking interaction with phenyl residues in the wild-type, which is either completely absent or considerably weaker in mutants.

(ii) Latent transforming growth factor-beta-binding protein 2 LTBP2

LTBP2 gene contains 35 exons coding for 1821 amino acid long protein. It contains many putative domains like 20 epidermal growth factor (EGF)-like domains, 4 transforming growth factor beta binding protein (TB)-like modules each containing 8 cysteine residues, and an amino-terminal signal peptide. Sixteen of the EGF domains have calcium-binding motifs and it has been suggested that these adopt a rod-like molecular arrangement upon calcium binding in order to present a specific surface for protein-protein interactions. The cysteine-containing TB domains provide a degree of conformational flexibility to LTBP-fibrillin superfamily. The secretory signal targets the protein to the extracellular matrix, where it has cell-adhesive properties, a functional role as a docking molecule for elastic fiber assembly, and a structural role as a component of fibrillin-rich microfibrils in connective tissues [88].

Though mutations in LTBP2 are reported as causative of PCG, the precise mechanism by which these mutations cause the disease is not known.

F. Primary Angle Closure Glaucoma (PACG)

In PACG, the iridocorneal angle (angle between iris and cornea) is closed, and as a result there is a blockage for the aqueous humor to flow through trabecular meshwork in to the venal system. As a result IOP rises and causes optic nerve head cupping and blindness. ACG is more common in people who have shallower anterior chambers and narrower angles, like eg., patients with hypermetropia (far sightedness). It is more common in people with Asian racial background. ACG can be either chronic or acute. Unlike POAG, patients with ACG may have intermittent symptoms like eye ache with cloudy vision, nausea, headache, vomiting etc. Patients with ACG can

be relieved by performing a surgery called laser iridotomy where a hole is made in the iris to allow the normal outflow of AH through the exit channels. Genome wide association studies have shown three susceptibility loci but no genes are known yet [89].

G. Diagnosis and Treatment Strategies for Glaucoma

An easy to follow video that explains treatment procedures that ophthalmologists use for some of these forms of glaucoma can be had by accessing <www.youtube.com/watch?v=X9NgArAMAQ8>

Early diagnosis is the key to successful management of glaucoma. Since glaucoma occurs without any signs or symptoms, regular eye-check ups are a must particularly for people above the age of 40. The diagnosis of glaucoma no longer simply relies on the presence of pressure within the eye. It requires that there be optic nerve damage or a strong suggestion of damage, which can be clearly seen during a dilated eye examination of the optic nerve. In general, the hallmark sign of this condition is a loss of peripheral vision. Glaucoma evaluation has several components. Apart from measuring the IOP by tonometry, the health of the optic nerve is also examined by ophthalmoscopy and visual field test. The treatment strategies for glaucoma usually include medicated eye drops which lower the IOP by either decreasing aqueous humor production or increase its outflow. In extreme cases when the medications do not improve the condition, surgeries are performed like trabeculoplasty where high energy laser beam is used to shrink one part of the trabecular meshwork so that the other half stretches and allows drainage of aqueous humor or conventional surgery like trabeculectomy. In this surgery trabecular meshwork is excised from the patient completely so that the aqueous humor can easily leave the eye through the hole. In a few cases like secondary glaucoma or for children with glaucoma drainage implants are inserted for effective drainage of aqueous humor. While the above surgical techniques are applied for open angle glaucoma, angle closure glaucoma needs a different surgery called laser iridotomy.

The primary goal of glaucoma treatment is to preserve vision. However, current knowledge of the factors causing optic nerve damage and visual loss is limited. Proven existing therapies focus mainly on IOP reduction although elevated IOP is no longer a diagnostic feature for glaucoma. So, a multidrug approach will be promising which includes agents targeted towards lowering IOP as well as an agent directed at preserving and protecting the optic nerve from glaucoma.

H. Complete Therapy

Glaucoma is said to be a neurodegenerative disease and shares common features with few other neurodegenerative disorders like parkinson's. Current therapeutic agents under investigation include neuroprotectants, which target the disease progress manifested by death of retinal ganglion cells, axonal loss and irreversible loss of vision. The use of neuro-protective agents in conjunction with IOP-reducing therapy is defined as complete therapy for glaucoma management. Till date, no such medications have been approved. Drugs which protect RGC degeneration belong to different groups, which include NMDA antagonists such as memantine and dexanabinol, inhibitors of inducible nitric oxide synthase (NOS-2), neurotrophic factors such as nerve growth factors (NGF), brain derived neurotrophic factor (BDNF), neurotropin 3 and others.

Conclusion

Glaucoma being a multifactorial disease, single treatment strategy is not possible for the patients. Moreover, the disease is still understudied in terms of the clinical progress made for patients particularly carrying NTG. The only treatment mode involves drugs or surgeries for lowering IOP. Neuroprotective stratagies might be helpful to patients particularly belonging to NTG category. There is also a need to search for compounds which show maximum advantage, with minimum or no side effects. One such possibility is the use of natural anti-oxidants, which can also work through different and multiple mechanisms, such as oxidative stress induced or excitotoxicity induced mechanisms of RGC degeneration in glaucoma.

SUMMARY

Glaucoma refers to a group of eye diseases that lead to progressive optic neuropathy or the progressive disabling of the optic nerve, leading to progressive blindness. This is the second major cause of irreversible blindness in the world and as of today there is no cure for this. The disease is associated with difficulties or errors in the inflow or outflow of aqueous humor, the fluid that nourishes the front part of the eye.

While the normal intraocular pressure in the eye is generally between 10–22 mm Hg above atmospheric pressure, in many cases of glaucoma the intraocular pressure increases highly and it is this which leads to slow atrophy of the optic nerve. Optic nerve is cupped in the back of the eye in a disc and one measures the cup to disc ratio, which in essence measures the number of axons from the retinal ganglion cells that lead to the visual cortex.

A variety of factors contribute to glaucoma. Some of these are ischemia, excite-toxicity, mutations in a few and biochemical defects, Glaucoma is classified in various groups such as primary open angle glaucoma, primary angle closure glaucoma, normal tension glaucoma and primary congenital glaucoma. Some genes associated with glaucoma conditions are myocilin, optineurin, WDR 36 and Cytochrome P450 1b1. Most recently several more genes are being implicated with glaucomatic conditions; one of them is referred to as LTB2.

Early diagnosis of glaucoma is important for successful management of the condition. Several drugs such as prostaglandins and NMDA antagonists such as memantine and dexanabinol are used to manage glaucoma; surgically the ophthalmologist also places a small valve to help regulate aqueous humor flow.

IV. USE OF STEM CELLS IN RESTORING VISION

A. Introduction

Cells are the basic unit of structure and function in all livings organisms. With the advent of microscopes the first living cell was reported by Anton van Leeuwenhoek, who in 1674, described algae spirogyra as animalcules or "small animals". However the term "cell" was coined much earlier by Robert Hooke who was reminded of the walled compartments that monks lived in while looking at the tiny pores in sections of a cork used to stopper bottles. Further observations over the years led to the development of the cell theory that explains the basic relationship between cells and living things. The cell theory states that:

- All living things or organisms are made of cells and their products.
- New cells are created by old cells dividing into two.
- Cells are the basic building units of life.

These tenets of the cell theory are applicable to all organisms ranging from a single celled bacterium and yeast to complex multicellular organisms like humans.

There are two basic categories of cells that make up a human body. These are the germ cells and somatic cells. Germ cells produce gametes, i.e., sperms and eggs and carry the genetic information that is passed on from one generation to the next. These cells are immortal since they can continue to produce more of themselves, a trait made possible by the activity of an enzyme called telomerase. This enzyme extends the telomeres of the chromosome, preventing chromosome fusions and other negative effects of shortened telomeres. On the contrary, most somatic cells have limited proliferative capacity (~30–50 cell divisions) defined by the Hayflick limit. The stem cells, similar to germ line cells, have active telomerase enzyme activity that allows them indefinite proliferative capacity. Thus stem cells can be described simply as cells that can make more copies of themselves indefinitely, a trait called self-renewal and also differentiate into mature cells with specialized functions. The general defining characteristics of stem cells are described below.

B. Defining Characteristics of Stem Cells

Stem cells are responsible for cellular replacement and tissue regeneration. The following are some of the unique properties of the stem cells which allow them to achieve this task.

1. Self-renewal: This term simply denotes the ability of the stem cells to make copies or clones of themselves so as to maintain a constant reserve of stem cells. Stem cells can generate their clones either through asymmetric or symmetric cell division. Asymmetry refers to only the final fate of the daughter cells. Here one of the daughter cell retains all the properties of the parent stem cell thus replenishing the stem cell pool while the other daughter cell takes the path of differentiation. Thus in this case the asymmetry induces otherwise similar daughter cells to behave differently. Conversely, when a stem cell undergoes symmetric cell division, either both daughter cells retain their stemness or both take the path of differentiation. In other words these cells are self-renewing only half the time. The differentiating daughter cell can either be a precursor or a progenitor cell. If it is a precursor cell, it directly undergoes terminal differentiation (e.g., germ cells) and in the case of a progenitor, the daughter cell will further amplify itself and then differentiate into many cell types within the tissue.

2. Quiescence (which indicates low mitotic activity): Although stem cells are generally endowed with high proliferative potential, under steady-state conditions, they exhibit extremely low rates of proliferation. This low mitotic activity of stem cells is described as quiescence and this property allows them to have a long life span equivalent to the life of the organism. This is thought to be a protective mechanism against incurring mutations during cell division.

3. Error-free Proliferation: This is a very important requirement of stem cells. This is because any genetic changes at the level of the stem cells is going to be passed on to all subsequent clones (self-renewal) leading to the differentiated cells being abnormal and possibly dysfunctional. To minimize any error made during mitosis, several protective mechanisms have been developed. First, stem cells unlike the transient amplifying cells are relatively quiescent during the state of steady growth. The transient amplifying cells typically undergo periodic cell divisions in order tomaintain the population numbers during homeostasis. Mutations or genetic changes at the level of transient amplifying cells therefore will affect only a limited generation of cells since all other cells except stem cells have limited life cycle.

4. Potency: To establish a cell as a stem cell, an assessment of its plasticity or potency becomes essential. And the fertilized egg or the zygote formed by the fusion of the germ cells (each of which carries half the genetic information from the two parents), can be considered as the most primitive of stem cells. This is because the zygote can not only divide to produce all cell types required to create a complete organism, but it can also produce all the necessary support systems (extra-embryonic tissue like placenta) that are required to help the embryo develop. Therefore a zygote can be said to have totipotency. Once the zygote starts differentiating, the first step being the differentiation into the embryonic (inner cell mass) and extraembryonic (trophectoderm) components of a blastocyst, the potency of these cells diminishes to that of pluripotency. This means that although the inner mass cells can give rise to all cells of every organ, they cannot produce the trophectoderm necessary for implantation. Next in the hierarchy are the multipotent stem cells that can differentiate into cells of several but not all lineages. These types are referred to as 'adult' stem cells since these stem cells demonstrate multipotency, meaning that that they exhibit the potential to generate a variety of differentiated cells of different.

Having described the salient features of a stem cell, it begs the question of whether the zygote can be classified as a true stem cell. These cells have the ability to create an entire organism but do not self-renew. The same can be asked of the embryonic stem cells which are derived from cells only transiently found during embryonic development. There is no concrete evidence so far for the presence of the embryonic stem cells or a supporting niche for these cells in the adult human. This is in contrast to the plants wherein a population of pluripotent stem cells are maintained throught life and can be reprogrammed to create a whole new plant.

Thus it needs to be remembered that all populations of stem cells cannot meet all of the above described criteria. The definition of a stem cell can be context dependent, organism dependent and is driven by a need for uniformity in understanding. However, the above criteria serve as a good guideline for identifying a cell as a "stem cell".

C. Embryonic Stem Cells

The clinical relevance of stem cells became apparent with the demonstration that cells derived from the inner mass (embryonic stem cells) can be isolated and expanded in the laboratory using the essential support systems such as mitotically inhibited mouse fibroblast cells and specific growth factors (LIF and FGF). These pluripotent cells have the ability to differentiate into cells of all three germ layers namely, ectoderm, mesoderm and endoderm. This means that in theory, these cells from the embryo can be used to make any type of cell found in the body, and therefore can potentially be used to repair any dysfunctional tissue/organ or even regenerate new organs for transplantation. These cells are called embryonic stem cells (abbreviated as ES cells or ESCs). That ES cells can be cultured was first demonstrated by two independent groups; one led by Martin Evans [90] and the other by Gail Martin [91] using inner mass cells taken from mice. It was nearly 2 decades later that James Thomson developed a technique to isolate and culture the human ES cells in the laboratory [92].

Although the potential for clinical application of ES cells is infinite, it has been fraught with controversies. The main concern being that of ethics; Being derivatives of fertilized eggs (that do not get used for in-vitro fertilization) the critics of this technique equate the use of ES cells obtained from the inner mass to the murder of a fetus. There are other scientific issues related to the use of ES cells such as the possibility of immune rejection, teratoma formation, scale up of production, banking and spontaneous mutations that have been reported to occur in these cells with their long term culture. While these scientific challenges will likely be overcome, the ethical concern unfortunately has posed a great impediment to furthering research on and with ES cells.

D. Somatic Cell Nuclear Transfer

The problems faced with ES cells forced scientists to search for or develop alternatives. Induced pluripotent cells (iPS) and adult stem cells have captured the imagination of researchers since both these sources of stem cells are free of ethical concerns. Shinya Yamanaka and John Gurdon won the Nobel prize in 2012 for their discovery that specialized or terminally differentiated cells can be reprogrammed to a pluripotent stem cell state capable of giving rise to cells of all three germ layers. Gurdon was the first to show in frogs in the 1960s that the genetic memory of a stem cell is not altered by its differentiation into a specialized cell type [93]. To prove this, he physically removed the nucleus from a terminally differentiated skin cell of a frog and placed it within an enucleated frog egg cell. When induced to proliferate, these eggs could produce tadpoles that were genetically similar to the donor of the nucleus. This technique called somatic cell nuclear transfer (SCNT) was the basis for the creation of the famous sheep called Dolly [94]. To create Dolly the nucleus of a Scottish blackface ewe egg was replaced with the nucleus taken from a somatic cell of a Finn dorset ewe and induced to divide using an external stimulus. Dolly a Finn dorset sheep was born in 1996 and lived until 2003. This confirmed that complete reprogramming of the genetic information occurs such that a somatic cell nucleus converts into a pluripotent cell capable of producing an organism similar to the gene donor. Figure 38.43 describes the various steps used in SCNT and the birth of Dolly.

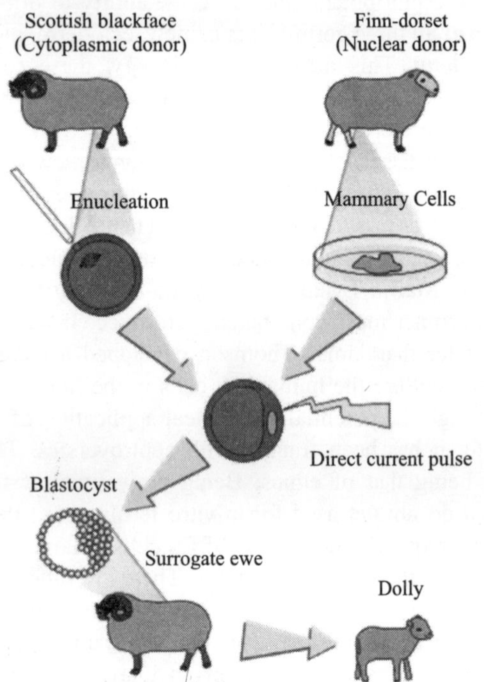

Figure 38.43 The cloning of the famous sheep "Dolly" (from en.wikipedia.org).

Note here that, contrary to the ES cells derived from IVF clinics, fertilization in SCNT does not use sperms since, unlike the germ cells, somatic cells are diploid and therefore already have the necessary copy numbers of the chromosomes. Instead the egg is 'tricked' into dividing by simulating the entry and fusing of sperm cell either by using electrical impulses or by altering the intracellular calcium levels.

The success of Dolly made big news as can be expected but also induced fear, fear that humans may end up playing god by using SCNT technique to create humans. Cloning humans is certainly a possibility albeit an unlikely one because the morality and sensitivity of humans will prevent their cloning at least in the near future.

The cloning efficiency using this technique is quite poor (1 success in 200 attempts) and so the question then was, can we directly induce a specialized cell to revert to a pluripotent state given that the genetic information is retained intact in the hierarchy of differentiated cells? This would mean that there would be no need for eggs or for transferring nuclei to egg cells.

E. Induced Pluripotent Stem Cells (iPSCs)

The answer to this question is 'yes' and the successful generation of such induced pluripotent cells was reported by Yamanaka's group in 2006 [95]. Based on the experience with culturing ES cells in the laboratory, Shinya Yamanaka set out to identify the factors that help the ES cells retain their pluripotent status in culture and see if the use of these factors could help a fully differentiated somatic cell to de-differentiate and take on a pluripotent state. This discovery came after laborious screening of 24 candidate genes in various permutations and combinations until the 4 essential genes, known as the Yamanaka factors, required for inducing this transformation were identified. The cells reprogrammed using these 4 genes (Oct4, Sox2, Klf and c-Myc) were named as induced pluripotent stem cells (iPS). The iPSCs therefore simply are terminally differentiated somatic cells that have been genetically reprogrammed to take up an embryonic stem cell–like state. Figure 38.44 summarizes the creation of iPSCs and their differentiation to various terminally differentiated cells.

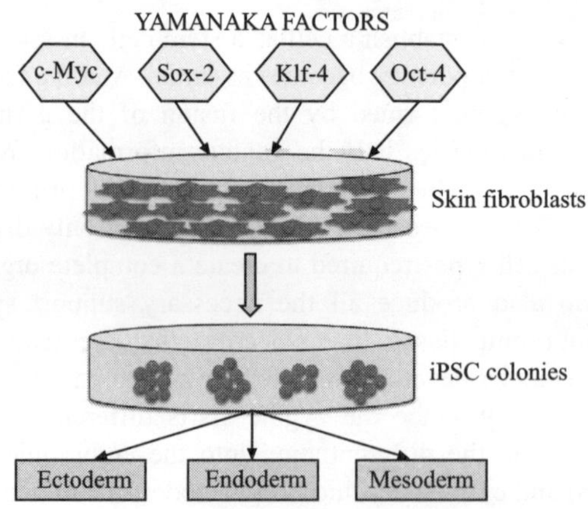

Figure 38.44 The creation of induced pluripotent stem cells using the key transcription factors [96].

The major advantage with iPS cells is that they do not carry the same ethical concerns as the ES cells and will not only allow us to generate autologous (patients own) cells for transplantation but also create *in vitro* models of diseases and use these models for drug testing. Technically these cells carry the same problems as the ES cells (teratoma formation and gene alterations); thus, there is a long way to go in terms of our understanding of these cells and their true capacity before they can be used for clinical therapy.

F.　Adult Stem Cells

Although ES cells and iPS cells have high potential for therapeutics, it has not been possible yet to use them widely in transplantations for reasons described above. When compared to these cells, adult stem cells have proven to be the less controversial and more productive alternative in terms of their current use in clinical applications. Adult stem cells are undifferentiated cells found in a fully developed organism amongst differentiated cells in a given tissue or organ. In keeping with the general definition of stem cells, these cells exhibit the tendency to self-renew and differentiate into all or some of the specialized cells that populate the organ. It appears that the primary role of these stem cells is to maintain homeostasis and repair the tissue when injured. For example, the skin epithelium or the ocular surface epithelium is replaced regularly by stem cells located in specific niches within the respective organs.

While there is some evidence for the presence of pluripotent stem cells (e.g., in the hair follicle), the differentiation potential of adult stem cells is certainly limited when compared to ES or iPS cells. Adult stem cells are also referred to as somatic stem cells to distinguish these from the germ line stem cells. Most adult stem cells appear to be multipotent exhibiting the potential to differentiate into cells of other lineages. For example, cells from the bone marrow (mesodermal derivative) can give rise to brain cells that are ectodermal in origin. But caution has to be exercised while interpreting these data since there are two possible explanations for these observations. One, the stem cells of interest are truly plastic and can produce cells from different origins and two, the tissue harbors more than one type of stem cells that then give rise to cells of different types.

Hematopoietic and mesenchymal stem cells

As with the pluripotent stem cells, the discovery of adult stem cells has generated a great deal of excitement. With many more adult tissues containing a reserve of stem cells within them, their possible use for therapy has become an exciting possibility, and in some instances a reality. At present adult stem cells have been identified in organs including bone marrow, peripheral blood, blood vessels, skeletal muscle, skin, eye, teeth, heart, gut, liver, ovarian epithelium and testis, and the list keeps growing with each passing year. Research has even identified stem cells in the brain once thought to be a tissue with no plasticity or ability to regenerate.

But the first evidence for the presence of cells with stem-like properties in adults was obtained from bone marrow as early as the 1950s. Interestingly, bone marrow transplantation was being offered as treatment, especially for repopulating the blood cells after radiation exposure or chemotherapy in leukemia patients, even before the existence of stem cells was formally established. Further studies on the bone marrow showed that it housed two types of stem cells. The first known as hematopoietic stem cells (HSCs) can differentiate into all red blood cells (B lymphocytes, T lymphocytes, natural killer cells, neutrophils, basophils) [97]. Figure 38.45 shows the various branching that HSCs are capable of.

The other population of stem cells called the mesenchymal stem cells (MSCs) gives rise to tissues required to support the formation of blood like bone, cartilage, fat and other connective tissue cells. Bone marrow transplantation is the first and best known success story of adult stem cell transplantation in humans. The second best success story of using adult

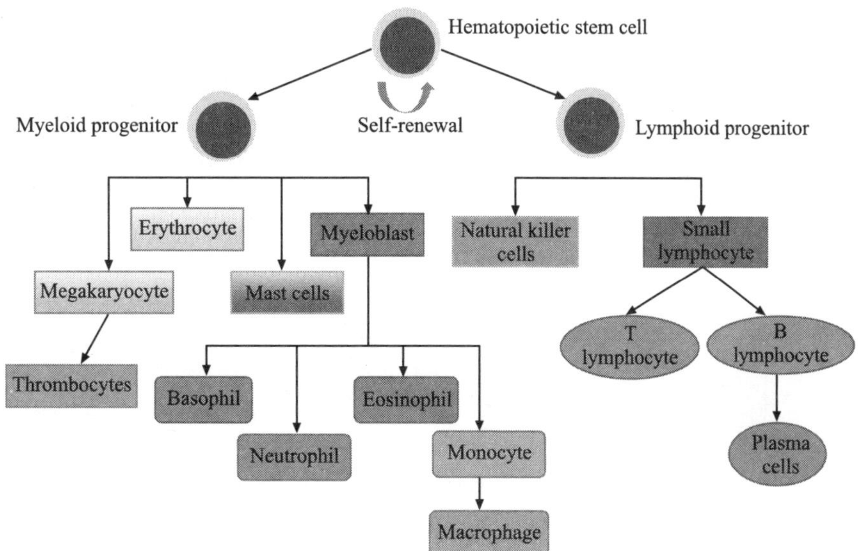

Figure 38.45　Differentiation of hematopoietic stem cells into the different cellular components of blood.

stem cells for regenerative purposes is the transplantation of keratinocytes for regenerating skin in burn victims. Howard Green demonstrated that the skin has a reserve of stem cells that could be cultured in the laboratory for regenerative purposes in the 1970s [98]. What this has allowed us to do is twofold: one, it has allowed us to dramatically reduce the amount of starting tissue material required for transplantation since by expanding cells in the laboratory a large number of cells can be obtained from a small amount of material (skin biopsy) and two because only a small amount of starting material is required this has enabled us to perform autologous transplantations meaning that the patients own healthy cells can be use for transplantation thus reducing the chances of rejection by a large factor. Use of cultured cells for regenerating the epidermis of the skin is being widely used in the clinic today.

This was the first clinical use of MSCs to regenerate skin. The other popular use of MSCs is in generating heart muscle cells or cardiomyocytes. Since the discovery of this differentiation and demonstration in a murine model [99], this has been used on the operation table on humans by heart surgeons, who engraft bone marrow derived MSCs into the myocardium to generate cardiomyocytes.

G. Application of Stem Cells in Ophthalmology

There are several conditions of the eye that may benefit from cell-based therapeutic interventions to help prevent or restore vision. Some of these include glaucoma, macular degeneration (age-related or juvenile), corneal endothelial dystrophies, limbal stem cell deficiency and retinal disorders.

Ocular surface regeneration using stem cells

One of the success stories of stem cell therapy in the eye is the replacement of corneal epithelial stem cells in a condition called limbal stem cell deficiency.

The corneal epithelium is a constantly renewing stratified structure similar to the skin epithelial cells (Figure 38.46). Regeneration is a multistep process involving cell division, migration, maturation and finally death of the superficial squamous cells, which are removed through desquamation. Only the basal cells adjacent to the basement membrane can proliferate and the dividing cells slowly move up to the surface of the cornea. This cyclical replacement of the epithelium for maintenance of its health has been elegantly described by the X, Y, Z hypothesis of Thoft and Friend [100], and shown in Figure 38.47.

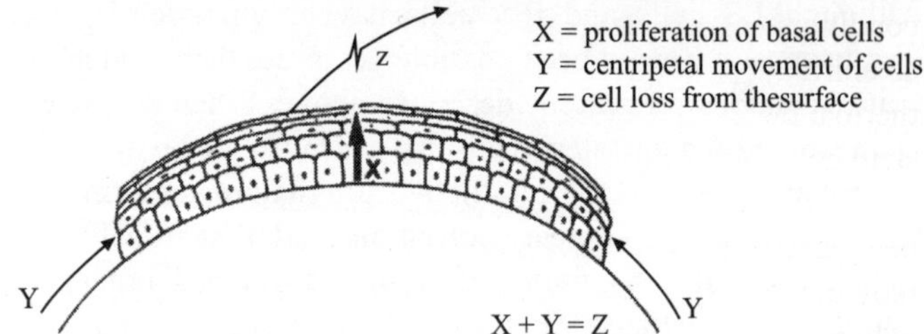

Figure 38.47 X,Y,Z hypothesis describing the balance of corneal epithelial cell loss and regeneration during homeostasis [100].

In Figure 38.47, X describes the proliferation of the basal epithelial cells, Y describes the centripetal movement of the cells from the periphery to the center and Z describes the loss of cells from corneal surface. Thus the loss of surface epithelium needs to be balanced by lateral migration and proliferation of basal cells. Any imbalance in this equation would lead to an unstable ocular surface marked by repeated erosions and epithelial defects. For the constant repopulation and more importantly the healing of an epithelial defect, a reserve of stem cells is located in the limbal region specifically in the basal regionof papillary structures called the palisades of Vogt (shown in Figure 38.3).

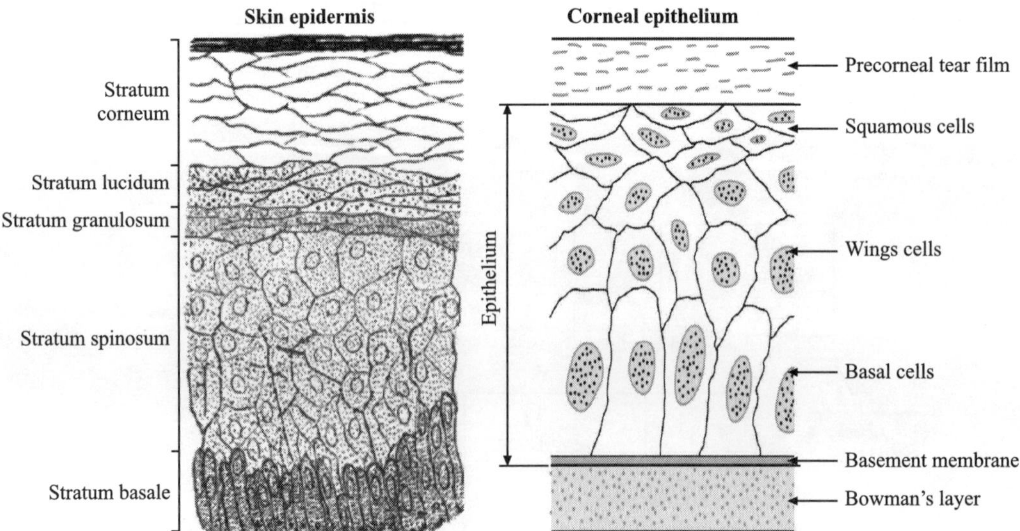

Figure 38.46 Similarity between the skin epidermis and corneal epithelium (from the fullwiki.org). Both have several layers of cells with the topmost layer being the squamous epithelium that gets sloughed off to be replaced by the layers of cells below. (*see Plate 14 for colour figure*)

In the case of corneal epithelium, the limbus acts as a niche, a term used to describe the microenvironment that supports the stem cells and protects them from several intrinsic and environmental assaults. The niche provides the necessary nutrients and growth factors, structural support and acts as a signaling hub to provide information about the integrity of the tissue. The limbal niche is located close to blood vessels from where the necessary growth factors and nutrients are drawn which will otherwise not be available in the avascular cornea. The limbus also acts as a border between the avascular cornea and the vascularized neighboring tissue the conjunctiva to prevent any mix up of cells between these two tissues. Thus any condition that affects the structural integrity of the limbus such as aniridia (developmental absence of iris tissue), chemical or thermal burns, andrepeat surgeries can alter its support for the stem cells resulting in a deficiency of the stem cell population.

In the limbus, the identity and location of stem cells has been determined based on their ability to form holoclones (colonies of undifferentiated cells capable of self-renewal) in culture, replicative quiescence and the expression of putative stem cell markers such as, p63, ABCG2 and c-EBP. The estimated percentage of limbal stem cells is between 1–5% of the total limbal cell population. Unlike certain mammals [101], the central cornea in humans is devoid of stem cells and is made up of a bulk of transient amplifying cells and the progressively differentiated cells constituting post-mitotic and terminally differentiated epithelial cells. This makes the presence of a healthy population of stem cells in the limbus a necessity to maintain the homeostasis of the corneal epithelium, the loss of which is clinically termed as limbal stem cell deficiency (LSCD). The presence of LSCD is marked by the growth of the vascularized conjunctiva over the transparent corneal surface and recurrent epithelial erosions resulting in a painful decrease in vision.

The method of treatment for LSCD depends primarily on whether the condition is partial or total. In the case of partial LSCD, where the deficiency involves only a few sectors of the cornea, the first treatment option has been transplanting human amniotic membrane (hAM) or mechanical debridement of the encroaching conjunctiva especially when the central cornea is spared. Several studies have reported good success in treating partial LSCD with repeated mechanical debridement of the conjunctiva until the limbal barrier was restored [102]. Engraftment of hAM was found to reduce ocular inflammation allowing for the restoration of a stable ocular surface and improvement in vision. Thus in most cases of mild to moderate LSCD, engraftment of hAM or debridement of the encroaching conjunctiva coupled with close follow up of the patient has been the preferred choice of treatment.

Unlike partial LSCD, in total LSCD, whether unilateral or bilateral, the treatment is surgical. It was in the late 1980s that the first human limbal transplantation using autologous tissue taken from the healthy unaffected eye was performed in cases of unilateral LSCD. Since then, several variations in allogeneic and autologous limbal transplantation with good clinical success have been devised.

Most notable of these advances being the culture of the epithelial cells in the laboratory for transplantation a technique termed as cultured limbal epithelial transplantation or CLET. This was possible because unlike the hematopoietic stem cells, the limbal epithelial cells lend themselves to be expanded in vitro quite easily. Using the same technique as that of skin keratinocytes, cultured limbal epithelial cells were first transplanted in 2 patients leading to complete regeneration of a stable ocular surface [103]. Since then, CLET has been improved such that the culture and transplantation of the limbal stem cells has been made xenobiotic (animal product) free. For example, the use of autologous serum in the place of bovine serum, and the use of human amniotic membrane or fibrin as the carrier material for delivering the cells to the ocular surface have removed all animal derived products from the equation and made the transplantation technique very safe for use in humans [104, 105].

Suspension culture vs explants culture of limbal stem cells

There are two prevalent culture techniques used for the culture of the limbal cells for clinical transplantation. One technique removes or isolates the epithelial cells from the limbus using enzymes that chop off the intercellular connections and the connections between the cells and the substrate. The isolated cells are then expanded on mitotically inactive cells that are called feeders, which provide the limbal cells with the necessary structural, and biochemical support needed to maintain the stem cell population. The commonly used feeder cells are the mouse-derived fibroblasts called the NIH3T3-J2 cells that are FDA approved for use in patients. This method of culturing the isolated epithelial cells is known as suspension culture technique. The other popular method called the explant technique differs from the suspension method in that it does not separate the epithelial cells from the limbal niche but keeps the natural limbal structure intact. Since the limbus has an inherent population of fibroblast cells, this culture system does not require additional support in the form of feeder cells.

And technically this is a much easier culture system since all that needs to be done is cut up the limbal biopsy into small pieces and place them on the human amniotic membrane for expansion which can then be directly transferred onto the ocular surface not requiring any intermediate steps to prepare the cells for transplantation. So the amniotic membrane does a double act of providing the necessary support for the cells to grow and function as the degradable carrier membrane for cell transfer. The absence of feeder cells and the use of autologous serum makes this culture system completely xenofree thus reducing worry about introducing any animal pathogens in the patient. In short the salient feature of culturing cells in the laboratory irrespective of the technique used is the ability to get a large number of cells from a small biopsy (~2x2 mm). This means that there are no deficiencies induced in the donor eye, and small biopsy allows for repeat CLET procedure in failed cases again without adversely affecting the health of the donor eye [106].

A typical CLET procedure can be watched on the video by accessing the website: www.youtube.com/watch?v=NlvYrTdbycs

There are other variants of the CLET procedure like the co-culture of limbal epithelial cells with conjunctival cells used in severe cases of LSCD where there is extensivedamageto the neighboring conjunctiva as well [107]. Cells of the conjunctiva are necessary to produce the mucin and lipid layers of the tears that in turn help to keep the corneal epithelial cells healthy, infection free and nourished. Thus it was shown that the replacement of both conjunctival and corneal cells in severe cases allows for better success of the transplantation than when only the epithelium is replaced. The CLET procedure has also been useful in treating patients with bilateral LSCD by taking the biopsy from living related donor (allogeneic) for transplantation. In situations where even allogeneic CLET is not possible or not recommended alternate sources of tissues have been explored for the generation of the corneal epithelium. One commonly targeted source is the oral mucosa from where a small biopsy is taken, as is from the limbus, and cultured on the amniotic membrane for transplantation [108]. When cultured in the laboratory, these cells retain their mucosal phenotype marked by the expression of specific cytokeratins (CK4, 15, and 13) and stem cell marker p75. Interestingly these cells express the cornea specific CK3 but not its partner 12 the absence of which affects the integrity of the epithelial cells. Transplantation of the cultured cells in humans led to a complete regeneration of a stable ocular surface with restoration of vision in many cases but the main drawback with these cells has been the vascular supply that they attract to the otherwise avascular cornea leading to eventual failure of graft and loss of vision. One possible way to address this issue would be to use this treatment in conjunction with topical medications that can prevent blood vessel growth.

What is interesting is that the clinical or surgical outcome of reversing the stem cell deficiency is comparable between the suspension and explant culture systems. In recent studies, with the largest patient population of 160 and 200, it was reported that successful ocular surface regeneration could be achieved and maintained in over 68% of the patients 2 to 10 years after they underwent the procedure [109, 110]. This means that both culture techniques are able to sufficiently support the limbal stem cell population to regenerate the deficient ocular surface. However, a lot in terms of the basic understanding of the biology is still left unanswered. For example, are we truly expanding the stem cell population in the laboratory using the current system of culture? What are the components of the limbal niche that help these stem cells survive? And are the same components present during culture? And the biggest challenge has been our inability to track the cells in humans to really find out what happens to the stem cells after they are transplanted onto the surface. Do the stem cells eventually reach the limbal region to repopulate it ?which would define the long-term success of stem cell transplantation.

As with all systems, CLET has disadvantages that restrict limbal stem cell transplantation to a few centers around the world. The main disadvantage is the need for clean room facilities for processing amniotic membranes and also for culturing the cells for transplantation. Also theman hours involved in cell culture, the need to check for pathogens in the cultured cells (to ensure that none have been introduced during culture) and the use of specific growth factors and nutrient media have all increased the cost per transplantation.

From CLET to SLET

A recent development in the treatment of LSCD is a technique called SLET or simple limbal epithelial transplantation. This technique was developed primarily to minimize the drawbacks while maximizing the benefits of CLET. This is a one step surgical procedure wherein the limbal biopsy is obtained from the healthy eye, chopped into small pieces, stuck to the hAM using fibrin glue and then placed on the prepared recipient cornea. This procedure is similar to CLET in that only a small biopsy is taken but differs in that the cells are cultured in situ using the eye as the incubator and the tears as the nutrient supplement. A stable, epithelialized and avascular ocular surface was observed within 6 weeks in the small clinical trial that was conducted on 6 patients [111]. And follow up of the patients up to 11 months showed that there was considerable improvement in the visual acuity of all patients with no complications either to the donor or recipient eye. Figure 38.48 shows how the steps of laboratory culture and transfer of the cultured sheet (the two steps shown in the extreme right of Figure 38.48) are done away with in SLET.

Of course the true value of this procedure can be assessed only when longterm follow-up of at least a few years becomes available. But this advancement in the treatment of LSCD has certainly reduced the cost of the procedure, the wait time and hopefully the treatment for LSCD becomes easily accessible to surgeons and patients worldwide.

Apart from LSCD, the use of stem cells for replacing lost photoreceptor cells in retinal degenerative disease has been tried in humans. What is most interesting about this trial is that the cells are derived from human ES cells, the first evidence for use of ES derived cells in humans. The result of this trial has been very promising and the details are discussed below.

ESCs in the treatment of Stargardt's Disease of the Retina

The retinal tissue has a far more complex structure and function when compared to the cornea. (Please see the detailed description of the retina in Sub-chapter A above). The retina is comprised of 10 layers starting with the photoreceptor cells (two types called rods and cones) which are specialized neurons that can detect photons of light and convert them into electrical signals, through the process described as photo-transduction. From the photoreceptors the electrical signals are relayed via the intermediate bipolar cells to the ganglion cells which finally carry the information to the visual cortex in the brain. The photoreceptor cells lie in close association with a monolayer of pigmented epithelial cells (RPE) that play a crucial role in the maintenance of the photoreceptor function and health.

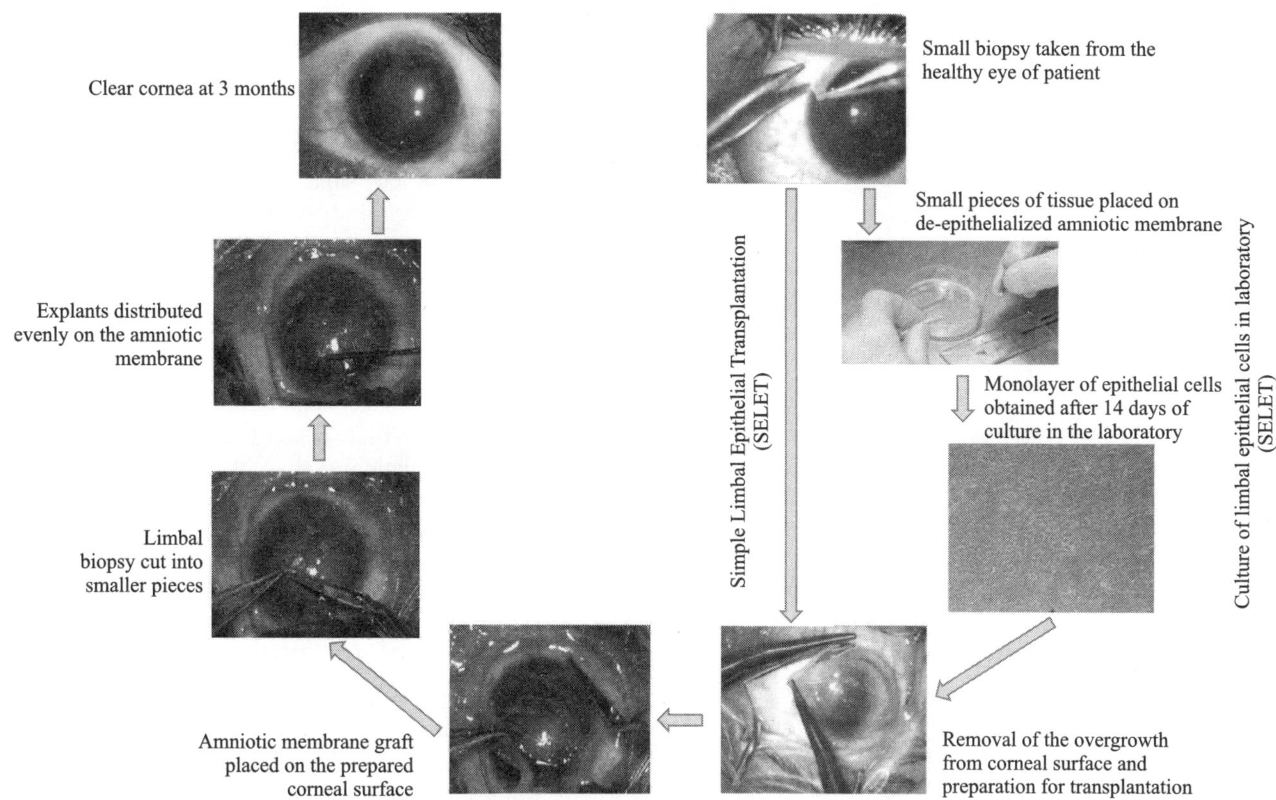

Figure 38.48   Comparison of CLET and SLET techniques of limbal epithelial transplantation. (*see Plate 14 for colour figure*)

Several diseases of the retina either affect the RPE cellsand/or the photoreceptor cells leading to the disruption of phototransduction cascade. Most of the retinal degenerative diseases (such as retinitis pigmentosa, macular degeneration, or Leber's congenital amaurosis) have underlying mutations in one or many genes, making targeted treatment difficult. There are currently two approaches for treating these degenerative diseases. One approach is cell-based and the other is gene-based. In the former the idea is to replace all the defunct cells with normal functioning healthy cells (stem cells or differentiated cells). The later involves correcting the mutated gene in the affected cells. In the example below, therapy using RPE cells derived from ES cells has been done in humans in an attempt to see if there is integration of cells with the defective retina and to assess the safety of ES cell use in humans. This is the first study to use ES derived cells in humans since their discovery a decade ago and the results are quite promising.

Stargardt's disease was discovered in 1909 by the ophthalmologist Karl Stargardt. This disease has an early onset and is marked by the degeneration of photoreceptor cells especially in the macular region thus affecting the central vision. The cause of retinal degeneration has been identified to be due to mutations in a gene called ABCA4 and the defective genes are inherited in an autosomal recessive pattern [112]. Similar macular degenerative disease also occurs in old age but presents a more complex gene mutation pattern. ABCA4 belongs to a family of genes called ATP-binding cassette transporters (ABC). This gene transcribes a large retina-specific protein with two transmembrane domains (TMD), two extracellular domains

(ECD), and two nucleotide-binding domains (NBD). Also known as the rim protein, the ABCA4 protein is located in the outer segment of the photoreceptor cells. The mutations cause the production of a dysfunctional protein that cannot perform energy transport to and from photoreceptor cells in the retina. The photoreceptor cells then degenerate, causing vision loss.

The eye is an ideal candidate for transplantations because it is an immune-privileged site protected by blood–ocular barriers, and is characterized by antigen-specific inhibition of both the cellular and humoral immune responses. Studies in a rat models of retinal degenerative disease had shown that subretinal transplantation of ESC-derived RPE resulted in extensive photoreceptor rescue and improvement in vision without evidence of untoward pathological effects or tumor formation. In 2012, a group transplanted RPE cells derived from human ES cells into patients with Stargardt's disease and age related macular degeneration [113]. Most recently, they generated RPE cells using human embryonic stem cells, and treated patients with age-related macular degeneration, and Stargardt's with success [114]. Cells for transplantation were generated from the ES cell line MA09. Using their differentiation protocol this group was able to get over 99% pure RPE cells and approximately 50,000 cells were injected in the sub-retinal space. Specific regions in the retina, where the photoreceptors were not completely lost to the disease, were chosen to receive the injectionsin order to maximize the integration of the cultured cells. After 4 months of procedure, the group reported that the cells remained viable and had integrated with the host retina without any signs of abnormal proliferation. Most importantly, there was no immune rejection or teratoma formation in these patients, which are two of the major concerns with the use of ES cells. Of course this study has been conducted in only 2 patients per group, but has certainly given hope for future use of ES cells in restoring cells lost to disease.

There is still a long way to go before cells, be it from the ES, iPS or adult stem cells, can be offered as a regular treatment option for many of the eye diseases. With retina, unlike the cornea, just replacing the dysfunctional cells would be insufficient to restore vision. Reconnecting the regenerated or replaced neurons to the brain has to be achieved for vision to be restored and this has proven to be a major challenge. In diseases where photoreceptors are affected, formation of synaptic connections by the replaced cells with bipolar cells is necessary for transmitting information.

Diseases where the retinal ganglion cells (RGCs) are affected, as in the case of like glaucoma, establishing neural connections becomes a bigger problem. This is because the optic nerve (made of axons of RGCs), which carries information to the brain, is considered to be a part of the central nervous system (CNS). The fibers are covered with myelin produced by oligodendrocytes instead of Schwann cells found in the peripheral nervous system (PNS). Regenerating the connections in CNS neurons is far more challenging when compared to neurons of the PNS. But, a lot of research is on in order to overcome these challenges.

Stem cells for other ocular diseases

There are several other ocular conditions that can potentially benefit from cell transplantations and avenues to replace cells in these tissues being explored. One such disease is glaucoma and the other being dystrophies of the corneal endothelium.

Glaucoma

Glaucoma is a group of disorders with multi-factorial etiology presenting with a characteristic increase in intraocular pressure and associated optic neuropathy. If left untreated it can lead to gradual but irreversible loss of vision. (For a detailed discussion of glaucoma, please refer to Sub-chapter A). This disease, especially the type primary open angle glaucoma, is thought to occur because of dysfunction of cells located at the trabecular meshwork. Cells in this meshwork function to regulate the amount of fluid (aqueous humor), which fills the region between the back surface of the cornea and the lens. Aqueous humor is secreted by cells adjacent to the lens and exits the eye through the sieve like trabecular meshwork located at the angle where the corneal endothelium meets the iris. This blood plasma-derived liquid provides the avascular corneal cells with the much needed nutrients. It also functions to maintain the pressure within the eye a constant. In old age, some individuals develop a blockage at the level of the trabecular meshwork resulting in a decrease in the drainage of aqueous from the eye which leads to an increase of intraocular pressure much like the increase in blood pressure in hypertension.

The increased pressure within the eye in turn compresses the axons of the ganglion cells at the optic disc leading to their gradual death that manifests as progressive loss of vision starting from the periphery. Stem cell or cell therapy can potentially help at two levels in this condition. One- cell replacement can be envisioned at the level of the trabecular meshwork to replace the dysfunctional cells so that aqueous drainage can be regulated in order to control the rise in intraocular pressure, and two- restoring or preventing the loss of retinal ganglion cells (RGCs) can be achieved with the help of cell transplantation. However, RGC transplantation might be more difficult because these are cells that need to form the necessary neuronal connections in order to carry information to the brain. Research has begun to explore both these areas and recently a population of stem like cells were found to be located in the trabecular meshwork. Trials to repopulate the meshwork tissue and to replace the ganglion cells in animals have shown promising results and it is possible that we might have a treatment option for this disease soon.

Corneal endothelial dysfunction

The corneal endothelium (Figure 38.49) is a monolayer of cells that function to keep the cornea transparent; a prerequisite for clear vision. The cells achieve this by keeping excess water out of the stroma using their ion pumps (e.g., Na^+/K^+ ATPase) and transporters. When these cells become dysfunctional as a result of genetic disorders, age and after surgeries that increase the rate of cell death, it leads to corneal decompensation. Cloudy cornea due to edema means the vision is poor or in severe cases completely lost. Poor vision due to endothelial dysfunction affects millions worldwide and the treatment for this problem at present is to either replace only the endothelial cell layer (surgery called endothelial keratoplasty) or to replace the whole cornea (penetrating keratoplasty). The success with these surgical techniques is excellent (~60%)

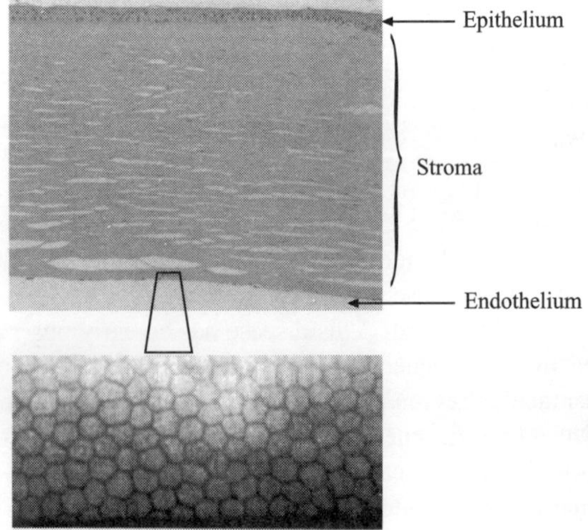

Figure 38.49   Layers of the cornea showing the epithelium, stroma and endothelium with a magnified image of the hexagonal shaped monolayer of corneal endothelium. (*see Plate 15 for colour figure*)

despite the donor tissue being allogeneic which can partly be attributed to the avascular nature of the cornea.

The bottleneck in treating endothelial dysfunction is the limited availability of healthy donor corneas for transplantation. Attempts to culture these cells in the laboratory have not been very successful as these cells have very poor proliferative capacity both in vivo and in vitro. Therefore a lot of research is now being focused on finding an alternate source of cells for deriving the corneal endothelium with stem cells being the obvious target. Research looking at iPS, ES and adult stem cells as possible sources has just begun but the goal should be easily attainable given that it is just a monolayer of cells and that surgical expertise for delivering the cells already exist.

This covers almost all the major layers of the eye and it can be seen that research in the field of ocular regeneration is quite advanced. It will soon be possible to replace almost every cellular layer of the eye but the challenge does not end here. Some are asking the question if we can do complex hand and other organ transplantations, can we not transplant an eyeball or better still can we not build one?

H. Can We Grow a Whole Eye? The Baby Steps

At present we can only build parts of the eyeball such as the cornea. And the first steps towards creating a whole eyeball have already been taken. Recently it was shown that the optic cup from ES cells can be grown in culture [115]. The eye is one of the first organs that begins to form in an embryo and retina is the first structure to form within the eye occurring as early as the 5th week post-fertilization. It forms with the invagination of the optic vesicle to form the optic cup, the inner surface of which differentiates later into the retinal layers. This group demonstrated the formation of an optic cup with stratified neural retina complete with RPE cells. The formation of the optic cup started around day 5 from embryoid bodies and by day 9 there was differentiation of the retina into its different layers. The time course of optic cup formation was comparable to what occurs during normal development of the embryo. What is most interesting about this study is the fact that the retina self-formed meaning that there was no need for the neighboring structures to aid their

formation in culture. This is the opposite of what occurs during the normal development of the embryo where inputs from the lens and surface ectoderm are crucial for the formation of the retina. Thus what this study shows is an intrinsic ability of the cells to self-organize without the need for external inputs. Whether this means that we can soon make retinas for potential transplantations remains to be seen.

SUMMARY

The eye does possess stem cells, particularly in the limbal region between the sclera and the cornea. The definition of a stem cell is that it can (a) self renew and (b) it can also differentiate under suitable conditions to produce other types of cells.

Limbal stem cells are efficient in doing so, and in the eye they are responsible for the generation of cells in the corneal epithelial layer. These limbal stem cells can be isolated and expanded either in suspension culture or explant culture on a membrane or a suitable scaffold in the laboratory to generate the corneal epithelial sheet, which can then be transplanted on to a patient who suffers from corneal damage. This procedure known as cultivated limbal epithelial transplantation or CLET has now become standard translational research practice and is accepted as clinical practice in India.

The retina also appears to have stem cells although they are yet to be isolated and defined fully.

However it is now possible to take any somatic cell from the body and induce it to become a stem cell by adding specific transcription factors know as Yamanaka factors. In such a procedure, we induce pluripotency to this cell, and it is now called an induced pluripotent stem cell (iPSC). It is now possible to take such cells from a patient (say his skin cells) and make iPSCs out of them, and then to convert them into chosen retinal cells such retinal pigment epithelial cells or photoreceptor cells.

While CLET has become a reality for taking care of corneal damage [116], the use of stem cells in therapeutically treating diseases or damage to the retina is still in the development stage.

REFERENCES

1. Gipson I. (2007), The Ocular Surface: The Challenge to Enable and Protect Vision: The Friedenwald Lecture, *Invest Ophthalmol Vis Sci.*, 48:4391–98.

2. Rantamaki A.H., Seppanen-Laakso T., Oresic M.J., Auhiainen M. and Holopainen J.H. (2011), Human Tear Fluid Lipidome: From Composition to Function, PLoS One 6(5): E19553. doi: 10.1371/journal.pone.0019553.

3. de Souza G.A., Godoy L.M.F. and Mann M. (2006), Identification of 491 Proteins in the Tear Fluid Proteome Reveals a Large Number of Proteases and Protease Inhibitors, *Genome Biology,* 7:R72. doi: 10.1186/gb–2006–7–8–r72.

4. Khurana A.K. (2012), *Comprehensive Ophthalmology*, 5th ed., New Age International, India.

5. Dartt D. (2010), *Encyclopedia of the Eye*, Academic Press, NY, USA.

6. Brogden K. (2005), Antimicrobial Peptides: Pore Formers or Metabolic Inhibitors in Bacteria, *Nat Rev in Microbiol.*, 3(3):238–250.

7. McDermott A.M. (2009), The Role of Antimicrobial Peptides at the Ocular Surface, *Ophthalmic Res*, 41:60–75. doi: 10:1159/000187622.

8. Wyszecki G. and Stiles W.S. (1982), Color Science: Concepts and Methods, Quantitative Data and Formulae, 2nd ed., John Wiley and Sons, New York.

9. http://simple-med.blogspot.in/search?q=histological+structure+of+the+cornea

10. Ambati B.K., Nozaki M., Singh N., Takeda A. and Jani P.D. (2006), Corneal Avascularity is Due to Soluble VEGF Receptor–1, *Nature*, 443:993–997.

11. Dyrlund T.F., Poulsen E.T., Scavenius C., Nicolajsen C.L. and Thogersen I.B. (2012), Human Cornea Proteome: Identification and Quantication of the Proteins of the Three Main Layers Including Epithelium, Stroma and Endothelium, *J. Proteome Res.,*11:4231–39.

12. Tam C., Mun J.J., Evans D.J. and Fleiszing S.M.J. (2012), Cytokeratins Mediate Epithelial Innate Defense Through Their Antimicrobial Properties, *J. Clin Invest*, 122:3665–3667.

13. Chapman J.A. (1974), The Staining Pattern of Collagen Fibrils, *Connect. Tissue Res.*, 2:137–150.

14. Gandhi N.S. and Mancera R.L. (2008), The Structure of Glycosaminoglycans and their Interactions with Proteins, *Chem. Biol. Drug Des.,* 72:455–482.

15. Knupp C., Pinali C., Lewis P.N., Parfitt G.J. and Young R.D. (2009), The Architecture of the Cornea and Structural Basis of its Transparency, *Adv. Protein Chem. Structural Biol.*, 78:25–49.

16. Uma L., Hariharan J., Sharma Y. and Balasubramanian D. (1996), Effect of UVB Radiation on Corneal Aldehyde Dehydrogenase, *Curr. Eye Res.*, 19:685–690.

17. Uma L., Hariharan J., Sharma Y. and Balasubramanian D. (1996), Corneal Aldehyde Dehydrogenase Displays Antioxidant Properties, *Exp. Eye Res.*, 63:117–120.

18. Lassen N., Black W.J., Estey T. and Vasiliou V. (2008), The Role of Corneal Crystallins in the Cellular Defense Mechanisms Against Oxidative Stress, *Semin. Cell Dev. Biol*, 19:100–112.

19. Kannabiran C. (2009), Genetics of Corneal Endothelial Dystrophies, *J. Genet.*, 88:487–491.

20. Griffith M., Hakim M., Shimmura S., Watsky M.A. and Li F. (2002), Artificial Human Corneas: Scaffolds for Transplantation and Host Regeneration, Cornea, 21:S54–S61.

21. Medeiros F.A. and Weinreb R.N. (2002), *Medical Backgrounders*, Drugs Today, 38:563.

22. Shahidullah M., Hassan Al-Malki W. and Delamere N.A. (2011), Mechanism of Aqueous Humor Secretion, Its Regulation and Relevance to Glaucoma, In: *Glaucoma—Basic and Clinical Concepts*, Shimon Rumelt (Ed.), ISBN: 978–953–307–591–4.

23. Goel M., Picciani R.G., Lee R.K. and Bhattacharya S.K. (2010), Aqueous Humor Dynamics: A Review, *Open Ophthalmol J.*, 3; 4:52–9. doi: 10.2174/1874364101004010052.

24. Davis-Silberman N. and Ashery-Padan R. (2008), Iris Development in Vertebrates: Genetic and Molecular Considerations, *Brain Res.*, 1192:17–28.

25. Danyush B.P. and Duncan M.K. (2009), The Lens Capsule, *Exp Eye Res.*, 88:151–164.

26. Perng M.D., Zhang Q. and Quinlan R.A. (2009), Insights into the Beaded Filament of the Eye Lens, *Exp Eye Res.*, 313:2180–2188.

27. Harries W.E.C., Akhavan D., Miercke L.J.W., Khademi S. and Stound R.M. (2004), The Channel Architecture of Aquaporin O at a 2.2 A Resolution, *Proc Natl Acad Sci.*, USA, 101:14045–14050.

28. Sohl G. and Willecke K. (2004), Gap Junctions and the Connexin Protein Family, *Cardiovas Res.*, 62:228–232.

29. Horwitz J. (1992), Alpha-crystallin can Function as a Molecular Chaperone, *Proc Natl Acad Sci.*, USA, 89:10449–53.

30. http://www.cgl.ucsf.edu/chimera/data/3dem-june2011/fitdemo_long.html

31. Basak A., Bateman O., Slingsby C., Pande A., Asherie N., Ogun O., Benedek G.B. and Pande J. (2003), High-resolution X-ray Crystal Structures of Human Gamma D Crystallin (1.25 A) and the R58H Mutant (1.15 A) Associated with Aculeiform Cataract, *J. Mol Biol.*, 328(5):1137–47.

32. Delaye M. and Tardieu A. (1983), Short-range Order of Crystallin Proteins Accounts for Eye Lens Transparency, *Nature,* 302:415–417.

33. Vendra V.P., Agarwal G., Chandani S., Talla V., Srinivasan N. and Balasubramanian D. (2013), The Structural Integrity of the Greek Key Supersecondary Topology in the Bg-crystallins is Vital for Central Lens Transparency, PLoS One.

34. Aravind P., Mishra A., Suman S.K., Jobby M.K., Sankaranarayanan R. and Sharma Y. (2009), The Betagamma-crystallin Superfamily Contains A Universal Motif for Binding Calcium, *Biochemistry*, 29:12180–90. doi: 10.1021/bi9017076.

35. Fort P.E. and Lampi K.J. (2011), New Focus on Alpha-crystallins in Retinal Neurodegenerative Diseases, *Exp. Eye Res*, 92:98–103.

36. Sinha D., Valapala M., Bhutto I., Patek B., Zhang C., Hose S., Yang F., Cano M., Stark W.J., Lutty G.A., Zigler J.S. and Wawrousek E.F. (2012), bA3/A1–crystallin is Required for Proper Astrocyte Template Formation and Vascular Remodeling in the Retina, *Transgenic Res.*, 21(5):1033–42. doi: 10.1007/s11248–012–9608–0. Epub

37. Prokosch V., Schallenberg M. and Thanos S. (2013), Crystallins are Regulated Biomarkers for Monitoring Topical Therapy of Glaucomatous Optic Neuropathy, PLoS One, 8(2):e49730. doi: 10.1371/journal.pone.0049730. Epub 2013 Feb 26.

38. Bishop P.N. (2000), Structural Macromolecules and Supramolecular Organization of the Vitreous Gel, *Prog. Retin., Eye Res.*, 19:323–344.

39. Kaushal S., Ridge K.D. and Khorana H.G. (1994), Structure and Function in Rhodopsin: The Role of Asparagine-linked Glycosylation, *Proc Natl Acad Sci.*, USA, 91:4024–8.

40. Hornak V., Ahuja S., EIlers M., Goncalves J.A, Sheves M., Reeves P.J. and Smith S.O. (2010), Light Activation of Rhodopsin: Insights from Molecular Dynamics Simulations Guided by Solid-state NMR Distance Restraints, *J. Mol. Biol.*, 396:510. doi: 10.1016/j.jmb.2009.12.003.

41. Fung B.K. (1983), Characterization of Transduction from Bovine Retinal Rod Outer Segments. I. Separation and Reconstitution of the Subunits, *J. Biol. Chem.*, 258:10495–502.

42. Lamb T.D. and Pugh E.N. Jr. (2006), Phototransduction, Dark Adaptation, and Rhodopsin Regeneration: The Proctor Lecture, *Invest Ophthalmol Vis Sci.*, 47:5137–52.

43. Kinoshita J.H. (1974), Mechanisms Initiating Cataract Formation: Proctor Lecture, *Invest Ophthalmol.*, 13:713–724.

44. Steke M. (2011), Natural Flavonoids as Potential Multifunctional Agents in Prevention of Diabetic Cataract, *Interdiscip Toxicol*, 4:69–77. doi: 10.2478/v/10102–011.0013.y

45. Majumdar S. and Srirangam R. (2010), Potential of the Bioflavonoids in the Prevention/Treatment of Ocular Disorders, *J. Pharm Pharmacol*, 62:951–965.

46. Balasubramanian D. (2005), Photodynamics of Cataract: An Update of Endogenous Chromophores and Antioxidants, *Photochem Photobiol*, 81:498–501.

47. Krishnaiah S., Vilas K., Shamanna B.R., Rao G.N., Thomas R. and Balasubramanian D. (2005), Smoking and Its Association with Cataract: Results of the Andhra Pradesh Eye Diseases Study from India, *Invest Ophthalmol Vis. Sci.*, 46:58–65.

48. Ye J., He J., Wang C., Wu H., Shi X., Zhang H., Xie J. and Lee S.Y. (2012), Smoking and Risk of Age-Related Cataract: A Meta-analysis, *Invest Ophthalmol Vis. Sci.*, 53:3885–3895.

49. Shalini V.K., Luthra M., Srinivas L., Rao S.H., Basti S., Reddy M. and Balasubramanian D. (1194), Oxidative Damage to the Eye Lens Caused by Cigarette Smoke and Fuel Smoke Condensates, *Indian J. Biochem Biophys.*, 31:261–266.

50. Ramakrishnan S., Sulochana K.N., Selvaraj T., Abdul Rahim A., Lakshmi M. and Arunagiri K. (1995), Smoking of Beedies and Cataract: Cadmium and Vitamin C in the Lens and Blood, *Br J. Ophthalmol.*, 79:202–206.

51. Mishra A., Suman S.K., Srivastava S.S., Sankaranarayanan R. and Sharma Y. (2012), Decoding the Molecular Design Principles Underlying Ca(2+) Binding to βγ-crystallin Motifs., *J. Mol Biol.*, 415:75–91. doi: 10.1016/j.jmb.2011.10.037.

52. Takata T., Smith J.P. Arbogast B., David L.L. and Lampi K.J. (2010), Solvent Accessibility of BetaB2–Crystallin and Local Structural Changes Due to Deamidation at the Dimer Interface, *Exp Eye Res.*, 91:336–346.

53. Shiels A. and Hejtmancik J.F. (2007), Genetic Origins of Cataract, *Arch Ophthalmol.*, 125:165–173.

54. Shiels A., Bennett T.M. and Hejtmancik J.F. (2010), *Cat-Map: Putting Cataract on the Map.* Mol Vis., 16:2007–2015.

55. Chalasani M.L.S., Muppirala M., Ponnam S.P.G., Kannabiran C. and Swarup G. (2013), A Cataract-causing Connexin 50 Mutant is Mislocalized to the ER Due to Loss of the Fourth Transmembrane Domain and Cytoplasmic Domain, *FEBS Open Bio.*, 3:22–29.

56. Francis P., Chung J.J., Yasui M., Berry V., Moore A., Wyatt M.K., Wistow G., Bhattacharya S.S. and Agre P. (2000), Functional Impairment of Lens Aquaporin in Two Families with Dominantly Inherited Cataracts, *Hum Mol Genet.*, 9:2329–34.

57. Vendra V.P, Agarwal G., Chandani S., Talla V., Srinivasan N. and Balasubramanian D. (2013), The Structural Integrity of the Greek Key Supersecondary Topology in the βγ-crystallins is Vital for Central Lens Transparency, PLoS One, 8(8):e70336. doi: 10.1371/journal.pone.0070336.

58. Weinreb R.N. and Khaw P.T. (2004), *Primary Open-angle Glaucoma*, Lancet 363:1711–1720.

59. Quigley H.A. and Broman A.T. (2006), The Number of People with Glaucoma Worldwide in 2010 and 2020, *Br J. Ophthalmol.*, 90:262–267.

60. John S.W. (2005), Mechanistic Insights into Glaucoma Provided by Experimental Genetics; The Cogan Lecture, *Invest Ophthalmol Vis Sci.*, 46:2649–2661.

61. Grus F.H., Joachim S.C., Hoffmann E.M. and Pfeiffer N. (2004), Complex Autoantibody Repertoires in Patients with Glaucoma, *Mol Vis.* 10:132–137.

62. Tezel G. and Wax M.B. (2000), Increased Production of Tumor Necrosis Factor-alpha by Glial Cells Exposed to Simulated Ischemia or Elevated Hydrostatic Pressure Induces Apoptosis in Cocultured Retinal Ganglion Cells, *J. Neurosci*, 20:8693–8700.

63. Fuse N. (2010), Genetic Bases for Glaucoma, *Tohoku J. Exp. Med.*, 221:1–10.

64. Fingert J.H. (2011), Primary Open-angle Glaucoma Genes, *Eye* (Lond) 25:587–595.

65. Stone E.M., Fingert J.H., Alward W.L., Nguyen T.D., Polansky J.R., Sunden S.L., Nishimura D., Clark A.F., Nystuen A., Nichols B.E., Mackey D.A., Ritch R., Kalenak J.W. Craven E.R. and Sheffield V.C. (1997), Identification of A Gene that Causes Primary Open Angle Glaucoma, *Science,* 275:668–670.

66. Menaa F., Braghini C.A., Vasconcellos J.P., Menaa B., Costa V.P., Figueirido V.P. and Melo M.B. (2011), Keeping An Eye on Myocilin: A Complex Molecule Associated with Primary Open-angle Glaucoma Susceptibility, *Molecules* 16:5402–5421.

67. Nguyen T.D., Chen P., Huang W.D., Chen H., Johnson D. and Polansky, J.R. (1998), Gene Structure and Properties of TIGR, An Olfactomedin-related Glycoprotein Cloned from Glucocorticoid-induced Trabecular Meshwork Cells. *J. Biol Chem.*, 273:6341–6350.

68. Zhou Z. and Vollrath D. (1999), A Cellular Assay Distinguishes Normal and Mutant TIGR/Myocilin Protein, *Hum Mol Genet* 8:2221–2228.

69. Rezaie T., Child A., Hitchings R., Brice G., Miller L., Coca-Prados M., Héon E., Krupin T., Ritch R., Kreutzer D., Crick R.P. and Sarfarazi M. (2002), Adult-onset Primary Open-angle Glaucoma Caused by Mutations in Optineurin, *Science,* 295:1077–1079.

70. Ying H. and Yue B.Y. (2012), Cellular and Molecular Biology of Optineurin, *Int. Rev. Cell Mol. Biol.*, 294:223–258.

71. Shen X., Ying H., Qiu Y., Park J.S. Shyam R., Iwata T. and Yue B.Y. (2011), Processing of Optineurin in Neuronal Cells, *J. Biol Chem.*, 286:3618–3629.

72. Chalasani M.L., Balasubramanian D. and Swarup G. (2008), Focus on Molecules: Optineurin, *Exp. Eye Res.,* 87:1–2.

73. Vaibhava V., Nagabhushana A., Chalasani M.L., Sudhakar C., Kumari A. and Swarup G. (2012), Optineurin Mediates A Negative Regulation of Rab8 by the GTPase-activating Protein TBC1D17, *J. Cell Sci.*, 125:5026–5039.

74. Nagabhushana A., Chalasani M.L., Jain N., Radha V., Rangaraj N., Balasubramanian D. and Swarup G. (2010), Regulation of Endocytic Trafficking of Transferrin Receptor by Optineurin and Its Impairment by A Glaucoma-associated Mutant, *BMC Cell Biol.*, 11:4.

75. Korac J., Schaeffer V., Kovacevic I., Clement A.M., Jungblut B., Behl C., Terzic J. and Dikic L. (2013), Ubiquitin-independent Function of Optineurin in Autophagic Clearance of Protein Aggregates, *J. Cell Sci.*, 126:580–592.

76. Chalasani M.L., Radha V., Gupta V., Agarwal N., Balasubramanian D. and Swarup G. (2007), A Glaucoma-associated Mutant of Optineurin Selectively Induces Death of Retinal Ganglion Cells which is Inhibited by Antioxidants, *Invest Ophthalmol Vis. Sci.,* 48:1607–1614.

77. Sirohi K., Chalasani M.L., Sudhakar C., Kumari A., Radha V. and Swarup G. (2013), M98K-OPTN Induces Transferrin Receptor Degradation and RAB12–mediated Autophagic Death in Retinal Ganglion Cells, *Autophagy* 9:510–527.

78. Monemi S., Spaeth G., DaSilva A., Popinchalk S., Ilitchev E., Liebmann J., Ritch R., Héon E., Crick R.P., Child A. and Sarfarazi M. (2005), Identification of a Novel Adult-onset Primary Open-angle Glaucoma (POAG) Gene on 5q22.1, *Hum Mol Genet*, 14:725–733.

79. Cartier A., Parent A., Labrecque P., Laroche G. and Parent J.L. (2011), WDR36 Acts as a Scaffold Protein Tethering a G-protein-coupled Receptor, Galphaq and Phospholipase Cbeta in a Signalling Complex, *J. Cell Sci.,* 124:3292–3304.

80. Chi Z.L., Yasumoto F., Sergeev Y., Minami M., Obazawa M., Kimura I., Takada Y. and Iwata T. (2010), Mutant WDR36 Directly Affects Axon Growth of Retinal Ganglion Cells Leading to Progressive Retinal Degeneration in Mice., *Hum Mol Genet.*, 19:3806–3815.

81. Stoilov I., Akarsu A.N. and Sarfarazi M. (1997), Identification of Three Different Truncating Mutations in Cytochrome P4501B1 (CYP1B1), as the Principal Cause of Primary Congenital Glaucoma (Buphthalmos) in Families Linked to the GLC3A Locus on Chromosome 2p21, *Hum. Mol. Genet.*, 6:641–647.

82. Mandal A.K. and Chakrabarti S. (2011), Update on Congenital Glaucoma, *Indian J. Ophthalmol.*, 59:S148–157.

83. Kaur K., Mandal A.K. and Chakrabarti S. (2011), Primary Congenital Glaucoma and the Involvement of CYP1B1, *Middle East Afr. J. Ophthalmol.* 18:7–16.

84. Vasiliou V. and Gonzalez F.J. (2008), Role of CYP1B1 in Glaucoma, *Annu. Rev. Pharmacol. Toxicol,* 48:333–358.

85. Reddy A.B., Kaur K., Mandal A.K., Panicker S.G., Thomas R., Balasubramanian D. and Chakrabarti S. (2004), Mutation Spectrum of the CYP1B1 Gene in Indian Primary Congenital Glaucoma Patients, *Mol. Vis.* 10:696–702.

86. Mashima Y., Suzuki Y., Sergeev Y., Ohtake Y., Tanino T., Kimura I., Miyata H., Aihara M., Tanihara H., Inatani M., Azuma N., Iwata T. and Araie M. (2001), Novel Cytochrome P4501B1 (CYP1B1), Gene Mutations in Japanese Patients with Primary Congenital Glaucoma, *Invest. Ophthalmol. Vis. Sci.*, 42:2211–2216.

87. Achary M.S. and Nagarajaram H.A. (2008), Comparative Docking Studies of CYP1b1 and Its PCG-associated Mutant Forms, *J. Biosci.*, 33:699–713.

88. Ali M., McKibbin M., Booth A., Parry D.A., Jain P., Riazuddin S.A., Hejtmancik J.F., Khan S.N., Firasat S., Shires M., Gilmour D.F., Towns K., Murphy A.L., Azmanov D., Tournev I., Cherninkova S., Jafri H., Raashid Y., Toomes C., Craig J., Mackey D.A., Kalaydjieva L., Riazuddin S. and Inglehearn C.F. (2009), Null Mutations in LTBP2 Cause Primary Congenital Glaucoma, *Am. J. Hum. Genet.,* 84:664–671.

89. Vithana E.N., Khor C.C., Qiao C., Nongpiur M.E., George R., Chen L.J., Do T., Abu-Amero K., Huang C.K., Low S., Tajudin L.S., Perera S.A., Cheng C.Y., Xu L., Jia H., Ho C.L., Sim K.S., Wu R.Y., Tham C.C., Chew P.T., Su D.H., Oen F.T., Sarangapani S., Soumittra N., Osman E.A., Wong H.T., Tang G., Fan S., Meng H., Huong D.T., Wang H., Feng B., Baskaran M., Shantha B., Ramprasad V.L., Kumaramanickavel G., Iyengar S.K., How A.C., Lee K.Y., Sivakumaran T.A., Yong V.H., Ting S.M., Li Y., Wang Y.X., Tay W.T., Sim X., Lavanya R., Cornes B.K., Zheng Y.F., Wong T.T., Loon S.C., Yong V.K., Waseem N., Yaakub A., Chia K.S., Allingham R.R., Hauser M.A., Lam D.S., Hibberd M.L., Bhattacharya S.S., Zhang M., Teo Y.Y., Tan D.T., Jonas J.B., Tai E.S., Saw S.M., Hon do N., Al-Obeidan S.A., Liu J., Chau T.N., Simmons C.P., Bei J.X., Zeng Y.X., Foster P.J., Vijaya L., Wong T.Y., Pang C.P., Wang N. and Aung T. (2012), Genome-wide Association Analyses Identify Three New Susceptibility Loci for Primary Angle Closure Glaucoma, *Nat Genet.*, 44:1142–1146.

90. Evans M.J. and Kaufman M.H. (1991), Establishment in Culture of Pluripotential Cells from Mouse Embryos, *Nature*, 292(5819):154–6.

91. Martin G.R. (1981), Isolation of a Pluripotent Cell Line from Early Mouse Embryos Cultured in Medium Conditioned by Teratocarcinoma Stem Cells. Proceedings of the National Academy of Sciences of the United States of America, 78(12):7634–8.

92. Thomson J.A., Itskovitz-Eldor J., Shapiro S.S., Waknitz M.A., Swiergiel J.J., Marshall V.S. and Jones J.M. Embryonic Stem Cell Lines Derived from Human Blastocysts, *Science*, 282(5391):1145.

93. Gurdon J.B. (1962), The Developmental Capacity of Nuclei Taken from Intestinal Epithelium Cells of Feeding Tadpoles. *Journal of Embryology and Experimental Morphology*, 10:622–40.

94. Wilmut I., Schnieke A.E., McWhir J., Kind A.J. and Campbell K.H. (1997), Viable Offspring Derived from Fetal and Adult Mammalian Cells, *Nature*, 385(6619):810–3.

95. Takahashi K. and Yamanaka S. (2006), Induction of Pluripotent Stem Cells from Mouse Embryonic and Adult Fibroblast Cultures by Defined Factors, *Cell*, 126(4):663–76.

96. Nasir A. and MacLellan W.R. Induced Pluripotent Stem Cells for Regenerative Cardiovascular Therapies and Biomedical Discovery, Advanced Drug Delivery Reviews, 63(4–5):324–30.

97. Till J.E. and McCulloch E.A. (1980), Hemopoietic Stem Cell Differentiation, *Biochimica et Biophysica Acta*, 605(4):431–59.

98. Rheinwald J.G. and Green H. (1975), Serial Cultivation of Strains of Human Epidermal Keratinocytes: The Formation of Keratinizing Colonies from Single Cells, *Cell*, 6(3):331–43.

99. Toma C., Pittenger M.F., Cahill K.S., Byrne B.J. and Kessler P.D. (2002), Human Mesenchymal Stem Cells Differentiate to a Cardiomyocyte Phenotype in the Adult Murine Heart, *Circulation*, 105(1):93–8.

100. Thoft R.A. and Friend J. (1983), The X,Y,Z Hypothesis of Corneal Epithelial Maintenance, *Investigative Ophthalmology and Visual Science*, 24(10):1442–3.

101. Majo F., Rochat A., Nicolas M., Jaoude G.A. and Barrandon Y. (2008), Oligopotent Stem Cells are Distributed Throughout the Mammalian Ocular Surface, *Nature*, 456(7219):250–4.

102. Sangwan V.S., Matalia H.P., Vemuganti G.K. and Rao G.N. (2004), Amniotic Membrane Transplantation for Reconstruction of Corneal Epithelial Surface in Cases of Partial Limbal Stem Cell Deficiency, *Indian Journal of Ophthalmology*, 52(4):281–5.

103. Rama P., Bonini S., Lambiase A., Golisano O., Paterna P., De Luca M. and Pellegrini G. (2001), Autologous Fibrin-cultured Limbal Stem Cells Permanently Restore the Corneal Surface of Patients with Total Limbal Stem Cell Deficiency, *Transplantation*, 72(9):1478–85.

104. Koizumi N., Inatomi T., Quantock A.J., Fullwood N.J., Dota A. and Kinoshita S. (2000), Amniotic Membrane as a Substrate for Cultivating Limbal Corneal Epithelial Cells for Autologous Transplantation in Rabbits, *Cornea*, 19(1):65–71.

105. Fatima A., Sangwan V.S., Iftekhar G., Reddy P., Matalia H., Balasubramanian D. and Vemuganti G.K. (2006), Technique of Cultivating Limbal Derived Corneal Epithelium on Human Amniotic Membrane for Clinical Transplantation, *Journal of Postgraduate Medicine*, 52(4):257–61.

106. Basu S., Ali H. and Sangwan V.S. (1012), Clinical Outcomes of Repeat Autologous Cultivated Limbal Epithelial Transplantation for Ocular Surface Burns, *American Journal of Ophthalmology*, 153(4):643–50, 50 E1–2.

107. Sangwan V.S., Vemuganti G.K., Singh S. and Balasubramanian D. (2003), Successful Reconstruction of Damaged Ocular Outer Surface in Humans Using Limbal and Conjictival Stem Cell Culture Methods, *Bioscience Reports*, 23(4):169–74.

108. Kinoshita S., Koizumi N. and Nakamura T. (2004), Transplantable Cultivated Mucosal Epithelial Sheet for Ocular Surface Reconstruction, *Experimental Eye Research*, 78(3):483–91.

109. Rama P., Matuska S., Paganoni G., Spinelli A., De Luca M. and Pellegrini G. (2010), Limbal Stem-cell Therapy and Long-term Corneal Regeneration, *The New England Journal of Medicine*, 363(2):147–55.

110. Sangwan V.S., Basu S., Vemuganti G.K., Sejpal K., Subramaniam S.V., Bandyopadhyay S., Krishnaiah S., Gaddipati S., Tiwari S. and Balasubramanian D. (2011), Clinical Outcomes of Xeno-free Autologous Cultivated Limbal Epithelial Transplantation: A 10-year Study, *The British Journal of Ophthalmology*, 95(11):1525–9.

111. Sangwan V.S., Basu S., MacNeil S. and Balasubramanian D. (2012), Simple Limbal Epithelial Transplantation (SLET): A Novel Surgical Technique for the Treatment of Unilateral Limbal Stem Cell Deficiency, *The British Journal of Ophthalmology*, 96(7):931–4.

112. Briggs C.E., Rucinski D., Rosenfeld P.J., Hirose T., Berson E.L. and Dryja T.P. (2001), Mutations in ABCR (ABCA4), in Patients with Stargardt Macular Degeneration or Cone-rod Degeneration, *Investigative Ophthalmology and Visual Science*, 42(10):2229–36.

113. Schwartz S.D., Hubschman J.P., Heilwell G., Franco-Cardenas V., Pan C.K., Ostrick R.M., Mickunas E., Gay R., Klimanskaya I. and Lanza R. (2012), Embryonic Stem Cell Trials for Macular Degeneration: A Preliminary Report, *Lancet,* 379(9817):713–20.

114. Schwartz S.D., Regillo C.D., Lam B.L., Eliott D., Rosenfeld P.J. and Gregori N.Z. (2014), Human Embryonic Stem Cell-derived Retinal Pigment Epithelium in Patients with Age-related Macular Degeneration and Stargardt's Macular Dystrophy: Follow-up of Two Open Label Phase 1/2 Studies, Lancet, Pii: S0140–6736(14)61376–3. doi: 10.1016/S0140–6736(14)61376–3. [Epub Ahead of Print]

115. Sasai Y., Eiraku M. and Suga H. (2012), *In vitro* Organogenesis in Three Dimensions: Self-organizing Stem Ceils, *Development*, 139(22):4111–21.

116. Ramachandran C., Basu S., Sangwan V.S. and Balasubramanian D. (2014), Concise Review: The Coming of Age of Stem Cell Treatment for Corneal Surface Damage, *Stem Cells Transl. Med.*, (10):1160–8. doi: 10.5966/sctm.2014–0064. Epub.

39

Lung and Lung Dysfunction

Surendra K. Sharma, Ragesh R. and Animesh Sharma

CONTENTS

Respiration is the physiological process by which an organism is able to extract oxygen from the air in the environment which is then carried attached to haemoglobin in the blood to various tissues of our body. Body tissues in turn produce carbon dioxide as a product of metabolism which is then transported back in a similar manner from the tissues to the lungs and finally is exhaled out. In humans, the upper and lower respiratory tract, along with bronchi and their further divisions carry air in and out of lungs and this part of the respiratory tract does not participate in the process of gas exchange which is a function of the gas exchange units of the lungs and capillary network.

I. ANATOMY

The respiratory tract can be divided anatomically at the level of the larynx into the upper respiratory tract and the lower respiratory tract. The upper respiratory tract consists of the nostrils, nasal sinuses, pharynx and epiglottis. The larynx can be considered as a part of the upper or the lower tracts. The lower respiratory tract consists of the trachea, bronchi, bronchiolesand the lungs.

Physiologically, the respiratory tract can be subdivided into three parts, i.e., the conducting airways, the transitional airways and the lung tissue. The conducting airways, spanning from the trachea to the terminal bronchioles, participate solely in the process of air flow while not engaging in gas exchange. The transitional airways consist of respiratory bronchioles and since these consist of regions with and without alveoli, these carry out the dual function of air flow and gas exchange. The lung tissue is made up of alveolar ducts and alveolar sacs. Table 39.1 describes physiological division of airways.

TABLE 39.1 Physiological division of airways

Conducting airways	Trachea, Bronchi, Bronchioles, Terminal bronchioles
Transitional airways	Respiratory bronchioles
Respiratory zone	Alveolar ducts, Alveolar sacs

A. Parts of the Upper Respiratory Tract

Nose

Air enters the respiratory tract through the nose which combines the functions of a filter and conditioner by warming

"

and humidifying the air that enters it. The region involved in respiration is lined with pseudostratified ciliated columnar epithelium or respiratory epithelium. The region involved in olfaction is lined with a specialized type of pseudostratified columnar epithelium containing receptors for perception of sense of smell.

Paranasal sinuses

The paranasal sinuses are cavities of varying numbers containing air named after the bones of the skull in which these are located: frontal, maxillary, ethmoid and sphenoid. These are also lined with pseudostratified ciliated columnar epithelium and are known to assist in humidification and warming of air along with various other functions. The frontal, maxillary and anterior ethmoidal sinuses drain into the middle meatus. The posterior ethmoidal and sphenoidal sinuses drain into the superior meatus and the spheno-ethmoid recess respectively.

Pharynx

The pharynx, which is a conduit for food and air, extends from the base of the skull upto larynx and is divided into three parts: the nasopharynx, the oropharynx and the laryngopharynx. While the nasopharynx is lined with the respiratory epithelium, the oropharynx and laryngopharynx are lined with stratified squamous epithelium. The nasopharynx is the area behind the nose, extending from the base of the skull to the upper surface of the soft palate. The middle ear is connected to the nasopharynx through the eustachian tubes and plays a fundamental role in equalisation of pressures, by opening up during swallowing or yawning. The oropharynx is the area extending from uvula to the level of the hyoid bone. It facilitates the process of swallowing while at the same time preventing aspiration by

coordinating the activity of the pharyngeal muscles and the epiglottis. The laryngopharynx extends from the epiglottis to the level where the pharynx divides into the oesophagus and larynx.

Larynx

The larynx is a musculocartilaginous structure that performs the functions of phonation, respiration and the protection of airways. It consists of three paired cartilages: the arytenoids, the corniculate and the cuneiform, three unpaired cartilages: the epiglottis, the thyroid and the cricoid and two muscular folds: the upper false cords and the lower true cords. Superior laryngeal nerve provides sensory supply to the larynx while all the intrinsic muscles of the larynx are supplied by the recurrent laryngeal branch of the vagus nerve except for the cricothyroid which in turn is supplied by the superior laryngeal nerve.

B. The Lower Respiratory Tract

After passing through the larynx the inspired air reaches the trachea, which is partly intrathoracic and partly extrathoracic. It extends from the level of the sixth cervical vertebra to the fifth thoracic vertebra, where it bifurcates into two bronchi, one entering each lung. In order to prevent collapse, the trachea and major bronchi have cartilaginous rings that are deficient posteriorly. The right and left main bronchi further subdivide into lobar, segmental and sub-segmental bronchi. The segmental bronchi supply the bronchopulmonary segments (illustrated by Figure 39.1) which are functionally independent pulmonary units. Table 39.2 describes the subdivisions of the bronchus.

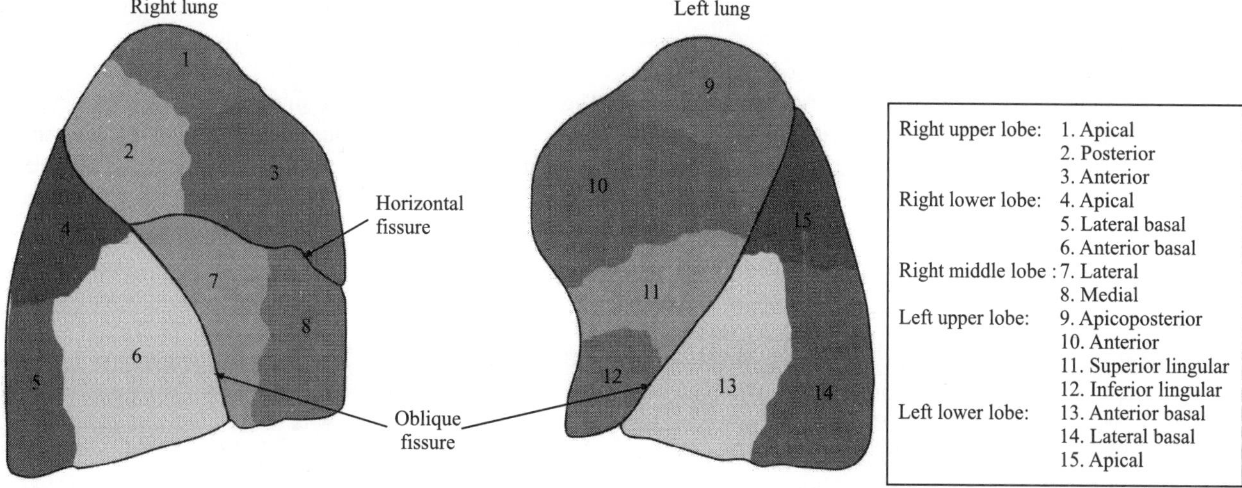

Figure 39.1   Shows the lateral view of right and left lung with the different bronchopulmonary segments. The left lung is divided into the upper and lower lobes by the oblique fissure. The right lung has two fissures the horizontal fissure and oblique fissure which divides the right into the upper lobe, middle lobe and the lower lobe. The bronchopulmonary segments have been labelled in the diagram numerically.

TABLE 39.2   Division of the bronchus

Bronchus	Lobar bronchi	Segmental bronchi
Right main	Right upper	Apical, anterior, posterior
	Right middle	medial and lateral
	Right lower	Apical, anterior, lateral, posterior, and medial basal
Left main	Left upper	Apicoposterior and anterior segmental branches, superior lingular and inferior lingular
	Left lower	Apical, anterior, lateral and posterior basal

Since the right main bronchus is more in line with the axis of the trachea, foreign bodies are frequently found lodged on the right side more frequently. The airway divisions that contain cartilage are called bronchi whereas the smaller divisions lacking cartilage or mucus glands are called bronchioles. These successively branch into the terminal bronchioles and then respiratory bronchioles, finally terminating in the alveolar ducts which lead into the clustered alveolar sacs. As compared to peripheral airways (< 2 mm diameter), central airways offer 90% of airway resistance because of their small total cross-sectional area. Each lung contains about 300 million alveoli. A rich capillary network derived from the pulmonary artery surrounds the alveolar sacs. The alveoli are the actual site of gas exchange as it is here that the blood-gas barrier finally thins out enough to allow simple diffusion of respiratory gases. This barrier consists of a layer of epithelium supported by an extracellular matrix of elastin and collagenalong with the capillary endothelium. The volume of gas exchange is to the tune of 250 ml/min of oxygen alone. The epithelium consists of three types of cells: Type I-squamous cells, Type II-surfactant secreting cells and Type III-macrophages.

C. The Lung

The anatomical organisation of the lung parenchyma is as follows. The alveolar duct, alveolar sacs, associated blood vessels and the lymphatic's constitute the primary lobule. The secondary lobule is the smallest discrete portion of the lung containing several primary lobules and is well defined by connective tissue septae. It comprises of three to five terminal bronchioles and the respiratory tissue distal to it. The connective tissue septae contain large and medium sized veins and major lymphatic channels. The Kerley-B lines on chest radiographs seen in heart failure represent these channels becoming prominent. Kerley-B lines are seen in the lung bases and appear as horizontal lines perpendicular to the pleural surface. There are two other types of Kerley lines which are found in heart failure: Kerley-A and Kerley-C. Kerley-A lines extend from the lung periphery to the hila, representing distended lymphatic channels while Kerley-C lines appear as reticular shadows seen in the base of lung.

D. Pulmonary Circulation

The lungs have a dual blood supply, being perfused by the pulmonary arteries that carry deoxygenated blood from the heart and primarily have the function of gas exchange as well as by the bronchial arteries carrying oxygenated blood arising from the aorta, which provide nutrition for the parenchyma. The alveoli are perfused by a capillary network formed by the branches of the pulmonary artery. The superior and inferior pulmonary veins derive tributaries from the pulmonary capillaries and venules and finally drain into the left atrium.

E. Lymphatic Drainage

The lymphatic vessels draining the lungs are derived from the superficial subpleural and the deep lymphatic plexuses. The superficial lymphatics drain the lung parenchyma and the visceral pleura. The deep lymphatic plexus drains the bronchial and the peribronchial area. All these lymphatics drains successively into the intrapulmonary nodes and the bronchopulmonary or hilar nodes. The inferior lobe of both lungs drains into the central inferior tracheobronchial or carinal nodes, which primarily drain to the right superior tracheobronchial lymph nodes. However, the other lobes of both lungs drain primarily to the ipsilateral superior tracheobronchial lymph nodes.

The point of clinical interest is that enlargement of the para-aortic lymph nodes belonging to the superior tracheo-bronchial group frequently results in left recurrent laryngeal nerve palsy, as these are located along the inferior border of the arch of aorta. The final destination of the lymphaticsis the thoracic duct on the left side and the lymphatic duct on the right side which in turn drain into their respective innominate veins.

F. Nerve Supply

The lung derives its innervation from the anterior and posterior pulmonary plexuses which are supplied by the branches of the vagus nerves and the upper six thoracic sympathetic ganglia. Vagal stimulation results in increased parasympathetic activity leading to bronchoconstriction and increased mucus secretion. Sympathetic stimulation results in bronchodilation, reduction in mucus secretion and vasoconstriction.

G. The Pleura

The pleural space is the potential space between the visceral and the parietalpleura. The visceral pleura envelops the lung and the parietal pleura lines the chest wall, diaphragm and the mediastinum. Both the layers join at the hila of the lung. The pleural space contains pleural fluid (normally less than 20 ml) formed at the parietal layer and absorbed at the visceral layer that facilitates movement of the visceral over the parietal layer. The intrapleural pressure is normally below the atomspheric pressure and becomes further sub-atmospheric with inspiratory effort causing the underlying lung to expand. For example the intrapleural pressure at the base of lung at the beginning of inspiration is around –2.5 mm Hg. With the contraction of the inspiratory muscles it becomes –6 mm Hg and the underlying lung expands. With stronger inspiratory effort the intrapleural pressure can further decrease, causing greater expansion of the lung. The parietal pleura is pain sensitive while the visceral pleura is pain insensitive.

II. PHYSIOLOGY

A. Normal Respiration

Respiration involves the following steps: pulmonary ventilation, diffusion of oxygen and CO_2 between the alveoli and the blood, perfusion of alveoli, transport of O_2 and CO_2 by the blood, exchange between the cells and the blood. The lungs are crucial for the initial three steps. Inspiration is an active process carried out by the diaphragm (predominantly an inspiratory muscle), the external intercostals and parasternal intercartilagenous muscles. When the inspiratory muscles contract it causes decrease in the intrapleural pressure, causing expansion of the lung and lowering of airway pressures. This results in air to flow into the lungs. Under resting conditions, in normal subjects, expiration is a passive process. The accessory muscles including the sternocleidomastoid, the scalene and the alaenasi become active in dyspnoeic patients. Muscles of the anterior abdominal wall sometimes aid the passive process of expiration.

Ventilation is greater at the lung bases when compared to the apex. Similarly perfusion of blood is also higher at the lung bases. But these differences are steeper for perfusion, thereby leading to higher ventilation/perfusion ratio at the lung apices. This predisposes the upper lobes to be affected by tuberculosis, because the causative organism *Mycobacterium tuberculosis* thrives at higher oxygen concentration. Diffusion of gases occurs at the alveolo-capillary membrane, and is governed by the Fick's law which states that diffusion is a passive process influenced by the concentration gradient, the distance the molecules have to travel and the surface area available. The membrane provides the moist medium, a mere two-cell thick distance that needs to be traversed and the large surface area essential for the diffusion of gases.

The process of diffusion depends on the concentration gradient, which in the case of gases is represented in terms of partial pressure. The partial pressure of a gas is the pressure exerted by it independently in a mixture of gases. The partial pressure of oxygen in the alveoli is 104 mm Hg, that in the pulmonary capillaries 40 mm Hg and that in the tissue cells is 40 mm Hg. The partial pressure of carbon dioxide in the alveoli is 40 mm Hg, that in the pulmonary capillaries 45 mm Hg and in the tissue cells 45 mm Hg. Because of these differences in partial pressures, according to Fick's Law, gases diffuse from the region of their higher concentration to the region of their lower concentration. Thus CO_2 is transported from the tissues

via the venous blood to the lungs to be exhaled out, while O_2 is transported from the alveoli to the tissues via the arterial blood. Once the oxygen diffuses into blood, it is transported bound to haemoglobin (98.5%) and dissolved in plasma (1.5%). Carbon dioxide is transported as bicarbonate ions (60%), bound to haemoglobin (30%) and dissolved in plasma (10%).

B. Oxygen-Haemoglobin Dissociation Curve

Oxygen-haemoglobin dissociation curve is a sigmoid shaped curve representing the relationship between the partial pressure of arterial oxygen (PaO_2) and the oxygen saturation of haemoglobin. The P_{50} is the PaO_2 at which haemoglobin is 50% saturated, which in a healthy individual corresponds to approximately 26 mm Hg. It is also evident that 90% saturation of haemoglobin corresponds to a PaO_2 of about 60 mm Hg. At this point the curve becomes flatter meaning thereby that the haemoglobin saturation would not significantly increase relative to the increase in the PaO_2. The conditions which result in the oxygen-haemoglobin dissociation curve shifting to right or left include changes in pH, temperature, concentration of 2,3-diphosphoglycerate (2,3 DPG) and CO_2 concentration. The Bohr effect describes the decrease in affinity of haemoglobin to O_2 with decrease in the pH. The oxygen-haemoglobin dissociation curve shifts to the right thereby leading to release of O_2. With exercise 2,3 DPG concentration and tissue temperature increases, shifting the curve to the right leading to increased unloading of O_2 at the exercising muscles. The Figure 39.2 illustrates the oxygen-haemoglobin dissociation curve along with the factors influencing left/right shift of the curve. Pulse oximetry is a non-invasive technique to measure the oxygen saturation the blood.

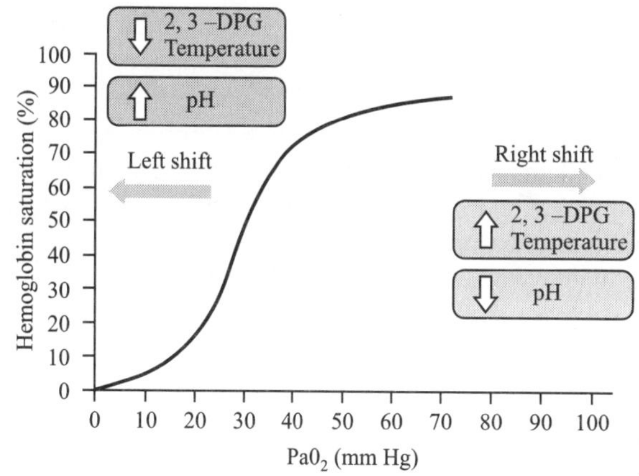

Figure 39.2 Shows the oxygen-haemoglobin dissociation curve along with the factors influencing shift of the curve to left or to the right.

C. Alveolar-Arterial Oxygen Pressure Difference

Alveolar-arterial oxygen pressure difference or gradient is the difference between the partial pressure pressures of the alveolar and arterial oxygen. It takes into account the fact that alveolar and therefore arterial oxygen tension changes depending on the alveolar ventilation.

It is calculated using a simplified form of alveolar gas equation

$$P(A–a)O_2 = PAO_2 – PaO_2$$

where, $PAO_2 = FIO_2 \times (PB – PH_2O) – PaCO_2/R$

PAO_2 = partial pressure of alveolar oxygen

PaO_2 = partial pressure of arterial oxygen

FIO_2 = fractional concentration of inspired oxygen (0.21 when breathing room air at sea level)

PB = barometric pressure (760 mm Hg at sea level)

PH_2O = water vapour pressure (47 mm Hg when air is sully saturated at 37°C

R = respiratory quotient (the ratio of carbon dioxide production to oxygen consumption assumed to be 0.8 under steady state conditions). The respiratory quotient depends on the substrates being metabolized as fuel source.

Substituting the above values,

$$PAO_2 = 150 – 1.25 \times PaCO_2$$

In healthy young persons aged 20 years, the $P(A–a)O_2$ is normally 4 to 15 mmHg and the value increases with increasing age. When hypoxaemia is purely due to a low inspired oxygen or alveolar hypoventilation, $P(A–a)O_2$ is normal. If $P(A–a)O_2$ and $PaCO_2$ are both elevated, additional mechanisms such as ventilation-perfusion mismatch or shunt are likely to be contributing to hypoxia.

D. Control of Respiration

The rate as well as depth of respiration are controlled by an involuntary central control mechanism located in the medulla as well as voluntarily by the cerebral cortex. In the medullary pre-botzinger complex there are interconnected neurons that fire rhythmically during inspiration or expiration. These lead to inspiration and expiration via their connections to the motor neurons of the respiratory muscles. Pontine pneumotaxic centre and the vagal afferents influence the neurons in the pre-botzinger complex. Changes in PaO_2, CO_2 and blood pH influence the pre-botzinger complex via the carotid body, aortic body and the brainstem chemoreceptors. Increases in CO_2, decrease in PaO_2 and pH cause increased ventilation.

III. PULMONARY DYSFUNCTION

Any type of lung dysfunction will produce symptoms and signs according to the localisation of the dysfunction. Lung dysfunction can occur due to several causes and include infective, malignant, idiopathic, immunological and systemic diseases. Infective etiology is illustrated by community acquired pneumonia, malignancy by carcinoma lung, idiopathic by idiopathic pulmonary fibrosis, immunological like Good pasture syndrome and systemic diseases like sepsis. The cardinal respiratory symptoms are dyspnea, cough, haemoptysis, and

chest pain. Table 39.3 lists the common causes of these symptoms. These symptoms in combination with detailed history including that of smoking and occupational exposure and description of other symptoms, physical examination and judicious use of relevant investigations would help in establishing the diagnosis of the aetiology of the dysfunction.

TABLE 39.3 Common respiratory symptoms and their causes

Symptom	Causes
Dyspnea	Asthma
	Chronic obstructive pulmonary disease
	Idiopathic pulmonary fibrosis (IPF)
	Pulmonary artery hypertension
	Kyphoscoliosis
	Diaphragmatic palsy
	Non cardiogenic pulmonary edema
	Cardiogenic pulmonary edema
	Anemia
	Deconditioning
Cough	Respiratory tract infection
	Aspiration
	Asthma
	Chronic eosinophilic bronchitis
	Chronic obstructive pulmonary disease
	Idiopathic pulmonary fibrosis (IPF)
	Pulmonary tuberculosis
	Bronchiectasis
	Lung cancer
	Gastroesophageal reflux
	Post nasal drainage
	Drugs like angiotensin converting enzyme inhibitors
Haemoptysis	Acute or chronic bronchitis
	Pulmonary tuberculosis
	Pneumonia
	Bronchiectasis
	Foreign bodies
	Airway trauma
	Lung cancer
	Goodpasture's syndrome
	Wegener's granulomatosis
	Pulmonary arteriovenous malformations
	Mitral stenosis
Chest pain	Empyema
	Pleuritis
	Lung cancer
	Pulmonary thromboembolism
	Ischemic heart disease
	Pericarditis
	Aortic stenosis
	Esophageal spasm
	Costochondritis
	Herpes zoster

The lung maybe affected at the level of its airways like bronchial asthma and chronic obstructive lung disease. The lung parenchyma maybe affected in diseases like diffuse parenchymal lung diseases and pneumonia. The vasculature of the lung maybe affected in diseases such as pulmonary vasculitis, pulmonary artery hypertension, hypertension and pulmonary thromboembolism. Lung pleura is affected in pleural effusion due to causes like tuberculosis and pneumothorax due to trauma.

A. Dyspnea

Dyspnea or breathlessness is a subjective feeling of breathing discomfort. Dyspnea while in lying down position is orthopnea, occurring usually in congestive heart failure or marked obesity. Patients with platypnea havedyspnea in upright position, with relief in supine position, usually caused by hepatopulmonary syndrome. Diseases of the lung like bronchial asthma, chronic obstructive lung disease (COLD), bronchiectasis and idiopathic pulmonary fibrosis (IPF) can cause dyspnea. The non-respiratory causes of dyspnea are cardiovascular disorders and anaemia. The severity of dyspnea is graded by modified Medical Research Council's (MRC, UK) dyspnea scale as given in Table 39.4.

TABLE 39.4 **Modified Medical Research Council Dyspnoea Scale (mMRC) grading of exertional breathlessness**

mMRC Grade	Symptom
mMRC Grade 0	I only get breathless with strenuous exercise
mMRC Grade 1	I get short of breath when hurrying on the level or walking up a slight hill
mMRC Grade 2	I walk slower than people of the same age on the level because of breathlessness, or I have to stop for breath when walking on my own pace on the level
mMRC Grade 3	I stop for breath after walking about 100 meters or after a few minutes on the level
mMRC Grade 4	I am too breathless to leave the house or I am breathless when dressing or undressing

Source: The Global Initiative for Chronic Obstructive Lung Disease (GOLD) guidelines 2011

B. Cough

Cough can be classified according to its chronicity, with cough of more than 8 weeks being classified as chronic cough, cough less than 3 weeks acute and the cough of duration in between 3-8 weeks as subacute. Cough can be caused by various pulmonary and non-pulmonary disorders. Bronchial asthma, COPD, lung cancer, pulmonary tuberculosis, tracheobronchitis and interstitial lung disease may cause cough. Figure 39.3 illustrates the cough reflex in a simplified manner. Weak cough reflex due to condition like diaphragmatic weakness can lead to retention of respiratory secretion predisposing to lower respiratory tract infection. Cough suppressants like codeine and dextromethorphan act by suppressing the cough centre in

the brainstem. Non-pulmonary causes include cardiovascular disorders, adverse effects of medications and gastro-esophageal reflux disease.

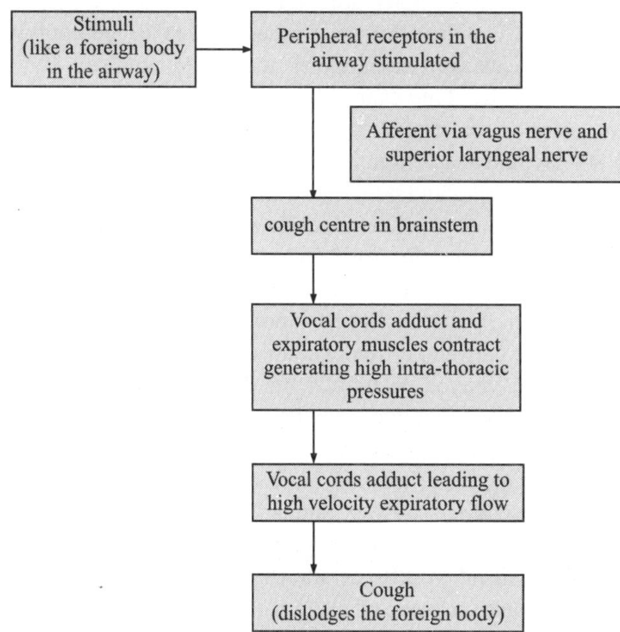

Figure 39.3 Illustrates the cough reflex in a simplified manner.

Hemoptysis

Coughing out of blood is haemoptysis. Massive haemoptysis would be coughing of more than 200–600 ml of blood in 24 hours which can potentially be life threatening. A patient with massive haemoptysis needs to be admitted in a hospital for emergency treatment and further evaluation with CT scan (computed tomography of the chest) and bronchoscopy may be needed. Figure 39.4 shows the CT scan of a patient with haemoptysis due to a cavitary lesion of the lung. The localisation of the bleed can be anywhere distally from

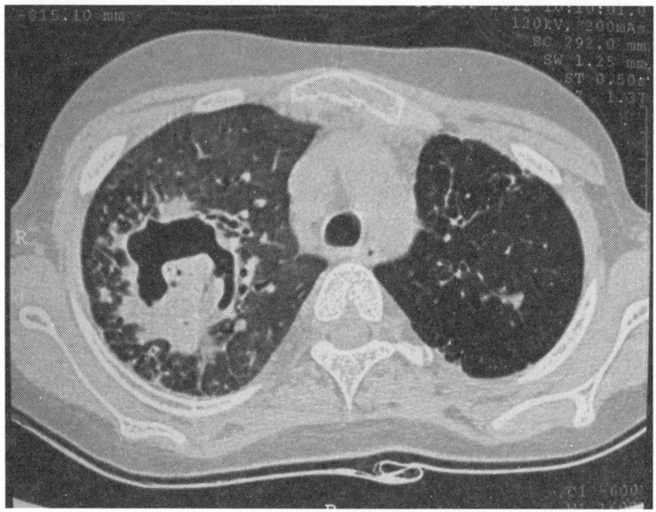

Figure 39.4 Shows the CT scan of a patient with haemoptysis due to a cavitary lesion of the lung.

the alveoli to proximally the larynx. Pathologies affecting the bronchial arteries tend to produce more bleeding when compared to bleeding from the pulmonary arterial system since the former is at a higher pressure being part of the systemic circulation. Besides various cardiac causes of hemoptysis, pulmonary causes of haemoptysis include tuberculosis and its various sequelae, lung cancer, bronchiectasis, lung abscess and acute venous thromboembolism.

Chest pain is a common clinical symptom. In the lung only the parietal pleura and pulmonary vasculature is pain sensitive. Respiratory causes include pleuritic chest pain due to pleuritic irritation for example from pneumonia and pneumothorax, pulmonary embolism leading to pulmonary infarction and pulmonary artery hypertension. It is very relevant to differentiate between different causes of chest pain since many of the causes maybe potentially life threatening. The different non pulmonary aetiologies may be stable angina pectoris from coronary artery disease, acute myocardial infarction, pericardial disease like acute pericarditis, gastroesophageal reflux disease, musculoskeletal causes like traumatic injury to the chest wall. Investigation like electrocardiogram (ECG) and a chest X-ray would be helpful to differentiate cardiac causes of chest pain from pulmonary causes of chest pain.

History of exposure to smoking both passive and active is very relevant to the patient with respiratory disorder. Diseases of the lung like carcinoma lung, COPD and non respiratory disorders like coronary artery disease and Buerger's disease have direct casual relationship with smoking. Occupational history and history of exposure to chemicals is important as many illnesses like asbestosis (exposure to asbestos), silicosis (exposure to silica), coal workers pneumoconiosis (coal exposure) and byssinosis (cotton dust exposure). Oxidative stress plays a role in several respiratory disorders.

For example, reactive oxygen species are involved in the pathogenesis of bronchial asthma. Genetic defects like in cystic fibrosis and alpha-1-antitrypsin deficiency cause respiratory dysfunction. Mutations in BMP-2 receptors (Bone Morphogenetic Protein receptors) are implicated in familial pulmonary artery hypertension. Family history is a strong risk factor for developing bronchial asthma. The immune system plays a significant role in the pulmonary dysfunction apart from providing defence against pathogens. For example in hypersensitivity pneumonitis the T_H1 cells respond to the inhalation of the antigens by the delayed hypersensitivity reaction associated with release of IL-12 and interferon gamma. Disorders of the lungs may affect the airways, the lung parenchyma, the vasculature or the pleura. Table 39.5 provides a summary of the common pulmonary disorders. A brief description of the common pulmonary disorders is given below.

Bronchial asthma

Bronchial asthma is a chronic airway inflammation characterised by hyper-responsive airways leading to reversible airflow obstruction. Asthma can develop at any age. The patient generally present with history of recurrent episodic breathlessness, cough and wheeze with nocturnal worsening. Allergens, various food products, air pollutants and hyperventilation may trigger an acute attack of asthma in a patient with bronchial asthma. In asthmatic airways T_H2 cells dominate and cause increased IL-4, IL-5 release and IgE synthesis leading to the characteristic eosinophilic airway inflammation. After the allergen enters the airways these bind to the allergen specific IgE attached on the mast cell. This leads to mast cell activation and release of mediators of bronchoconstriction like cysteinyl leukotrienes, histamine, various chemokines and different cytokines.

TABLE 39.5 Summary of the different types of interstitial lung diseases

Known causes	*Unknown causes*
Inhaled inorganic dust	Idiopathic pulmonary fibrosis
• Asbestosis (exposure to asbestos)	Sarcoidosis
• Silicosis (exposure to silica)	Nonspecific interstitial pneumonia
• Coal worker's pneumoconiosis (exposure to coal dust)	Cryptogenic organizing pneumonia
• Berylliosis (exposure to beryllium)	Pulmonary alveolar proteinosis
• Stannosis (exposure to tin)	Connective tissue disease
Inhaled organic dusts (Hypersensitivity pneumonitis)	• Systemic lupus erythematosus
• Farmer's lung (exposure to moldy hay)	• Systemic sclerosis
• Bird fancier's disease (exposure to bird droppings)	• Rheumatoid arthritis
• Bagassosis (exposure to moldy bagasse)	Lymphangioleiomyomatosis
• Compost lung (exposure to compost containing Aspergillus)	
• Mushroom worker's lung (exposure to mushroom compost)	
Drug-induced pulmonary toxicity	
• Amiodarone	
• Gold	
Smoking-related	
• Desquamative interstitial pneumonia	
• Langerhans cell granulomatosis	
Radiation induced ILD	
Aspiration pneumonia related ILD	

Histamine and leukotrienes cause bronchoconstriction and increased mucus secretion. Figure 39.5 gives a conceptual overview on the pathophysiology of bronchial asthma. Due to the airway inflammation the oxidative stress is high which can negatively impact treatment. Spirometry in asthma shows reversibility of airflow obstruction after administration of a rapid acting inhalational bronchodilator. Treatment is with inhalational and oral corticosteroids, short acting and long acting β_2 agonists (a bronchodilator), antileukotrieneagents, mast cell stabilizers and anti-IgE antibody like omalizumab. Figure 39.6 shows a metered dose inhaler, a device frequently used in asthma to provide inhalational drug therapy with bronchodilator like salbutamol or steroids like budesonide.

Chronic obstructive pulmonary disease

GOLD (Global Initiative for Chronic Obstructive Lung Disease) defined COPD, as a common preventable and treatable disease, characterized by persistent airflow limitation that is usually progressive and associated with an enhanced chronic inflammatory response in the airways and the lung to noxious particles or gases. The major risk factors for COPD include smoking both active and passive, air pollution which can be indoor (like exposure to chulhas) as well as outdoor and exposure to occupational dusts. Inflammation in response to cigarette smoke in the airways as well in the lung parenchyma leads to increased oxidative stress and activity of proteinases like matrix metalloproteinase and serine proteinase. This leads to airway remodeling, decreased parenchymal elasticity and destruction of the alveolar walls Patients have symptoms of cough, dyspnea and expectoration. Unlike in bronchial asthma the airflow obstruction in COPD is not reversible as measured by spirometry. There is controversy over the pathogenesis of COPD. The Dutch hypothesis suggests that COPD and bronchial asthma are different expressions of a single disease. On the contrary the British hypothesis suggests that COPD and bronchial asthma are two entirely

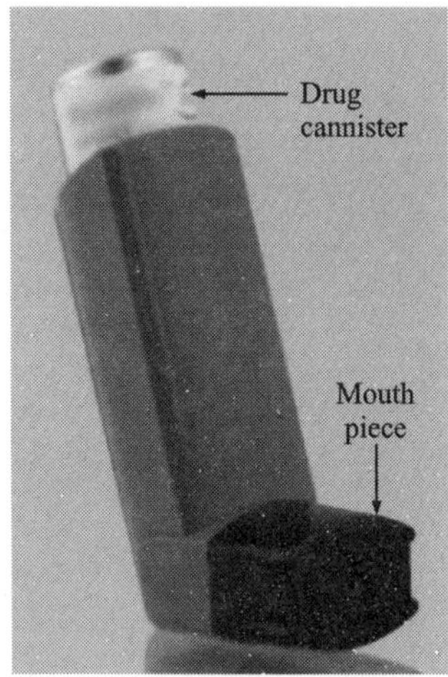

Figure 39.6 Shows a metered dose inhaler (MDI), a device frequently used in asthma and COPD to provide inhalational drug therapy with bronchodilator like salbutamol or steroids like budesonide.

different disease entities. The controversy is still unsettled as of now. Treatment is smoking cessation, inhalational anti-muscarinic agents, inhalational β_2-agonists like salbutamol and ipratropium bromide, glucocorticoids, oxygen inhalation, and surgical options like lung volume reduction surgery and lung transplantation. Figure 39.7 shows a nebulizer, a device frequently used in COPD to provide inhalational drug therapy with bronchodilator like ipratropium bromide or salbutamol.

Obstructive sleep apnea

With the advent of technological era and an already highlighted sedentary lifestyle Obstructive Sleep Apnea (OSA) has become a common disease. International classification of sleep disorders defines OSA as an intrinsic sleep disturbance characterized by the appearance of repeated episodes of upper airway obstruction (apneas) occurring during sleep usually associated with a

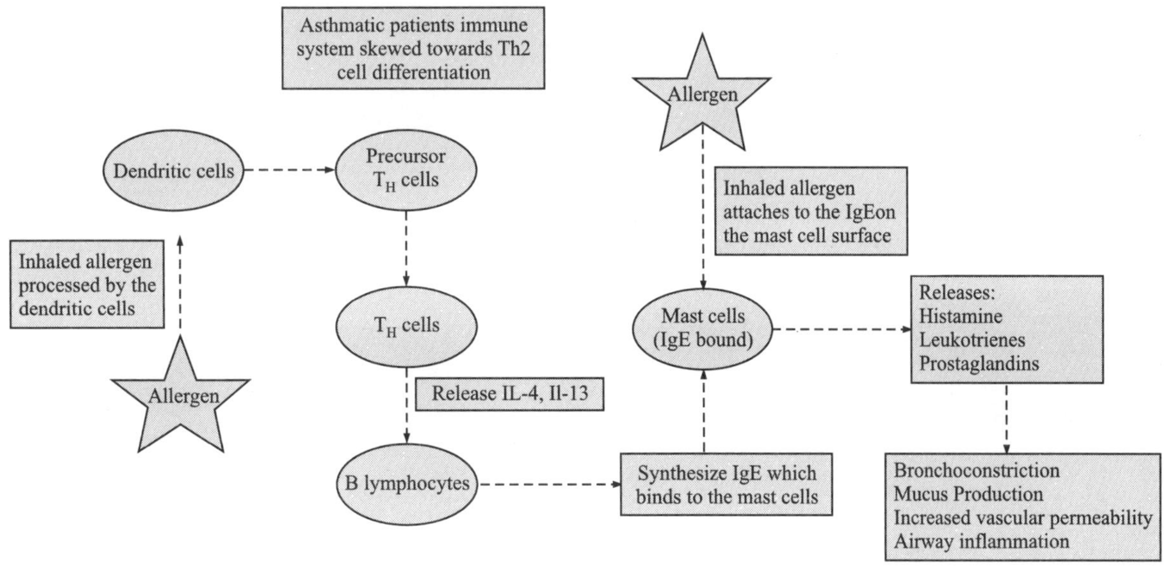

Figure 39.5 Gives a conceptual overview on the pathophysiology of bronchial asthma.

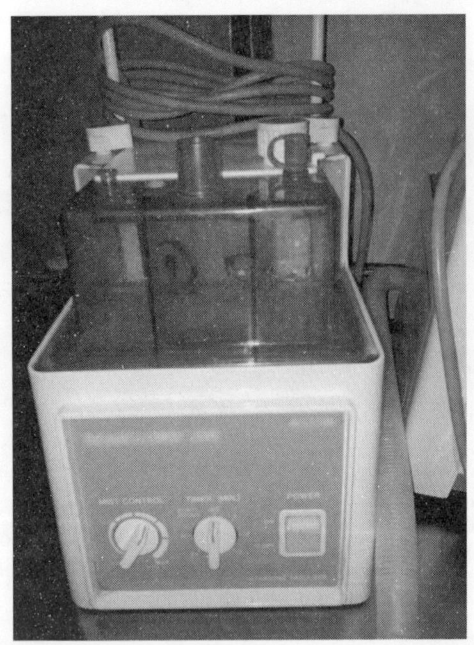

Figure 39.7 Shows a nebulizer, a device frequently used in COPD to provide inhalational drug therapy with bronchodilator like ipratropium bromide or salbutamol. (*see Plate 15 for colour figure*)

reduction in blood oxygen saturation. The symptoms consist of snoring while sleeping and excessive day time sleepiness. Obesity, craniofacial abnormalities, male gender and positive family history are the major risk factors in developing OSA. OSA can lead to hypertension, coronary artery disease, heart failure, cardiac arrhythmias, stroke, neurocognitive deficits, increased insulin resistance and metabolic syndrome. Diagnosis is by using overnight polysomnography. Figure 39.8 shows the polysomnography reading of a patient. Treatment consists of weight loss, continuous positive airway pressure (CPAP) therapy, bilevel positive airway pressure (BiPAP), weight reduction, bariatric surgery for obesity if body mass index is more than 35 kg/m^2.

Interstitial lung disease

The disease starts in the interstitial tissue of the lung and subsequently may involve adjacent lung parenchyma. Various examples of common interstitial lung disorders include idiopathic pulmonary fibrosis (usual interstitial pneumonia [UIP]), sarcoidosis, various occupational lung disorders, hypersensitivity pneumonitis and interstitial lung disorders associated with connective tissue disorders like rheumatoid arthritis. In the interstitial lung diseases there is variable combination of inflammation, parenchymal injury and

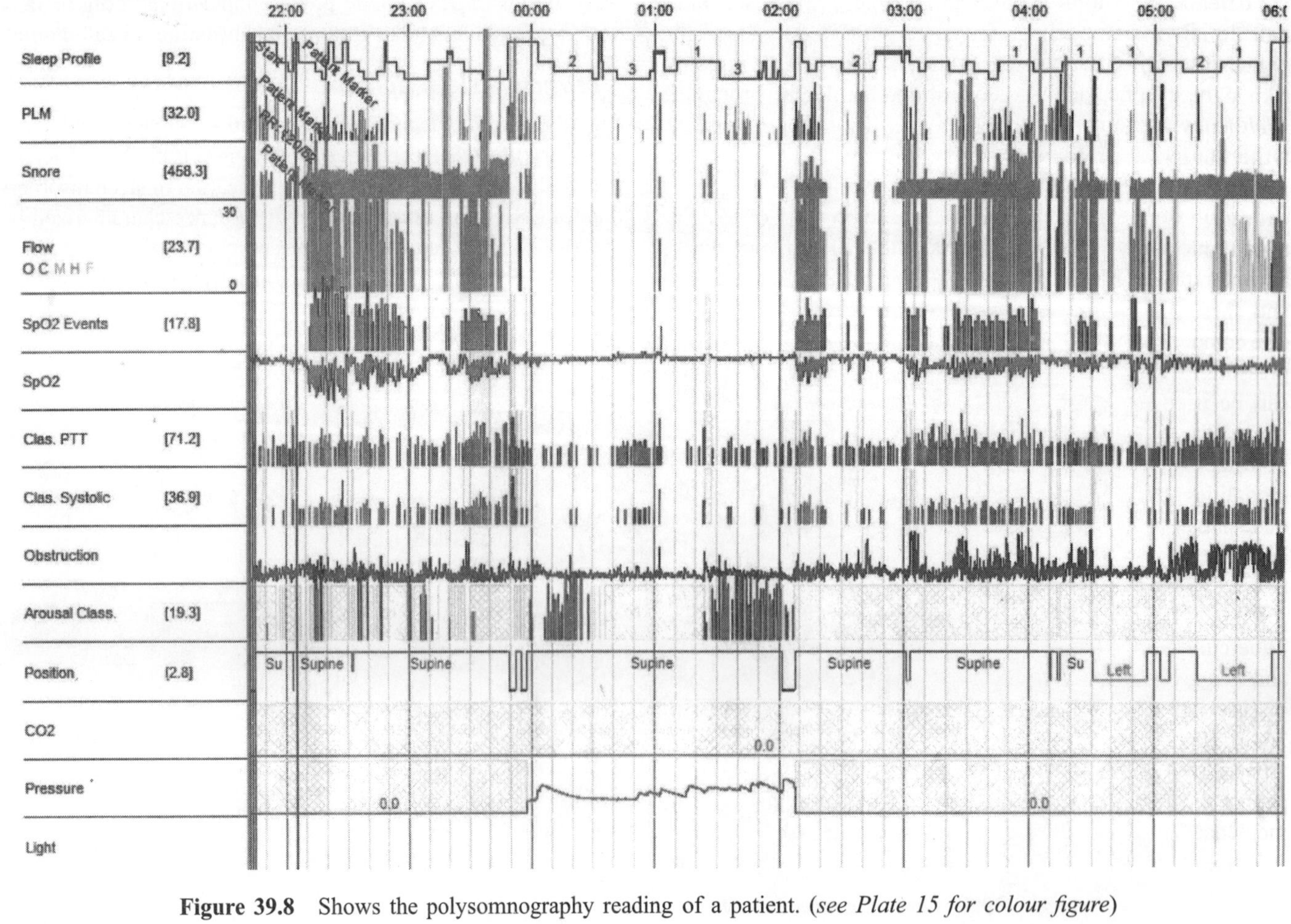

Figure 39.8 Shows the polysomnography reading of a patient. (*see Plate 15 for colour figure*)

fibrosis. Sarcoidosis is a multisystem disease which affects predominantly the lung. T_H1 response to an unknown antigen leads to increased IL-2, interferon gamma and TNF secretion by T-Helper cells as well as the macrophages. This leads to formation of non-caseating granulomas in the affected organs such as lung and other extrapulmonary tissues. Serum angiotensin converting enzyme (ACE) maybe elevated in patients with sarcoidosis. Table 39.5 summarizes the different types of interstitial lung diseases usually seen.

Pneumonia

It is a disorder of the lung parenchyma where the etiological agent can be a bacteria, virus or fungus. Pneumonia can be community acquired or maybe acquired from the hospital known as hospital acquired pneumonia. Hospital-acquired pneumonia (HAP) is pneumonia that occurs 48 hours or more after admission to hospital, but was not incubating at the time of admission. Ventilator-associated pneumonia (VAP) is a type of HAP (hospital acquired pneumonia) that develops more than 48 hours after endotracheal intubation. Healthcare-associated pneumonia (HCAP) is a pneumonia that occurs in a non-hospitalized patient with extensive healthcare contact, like that of a patient who received intravenous antibiotic therapy within the last 30 days. HAP, VAP and HCAP are usually associated with multidrug resistant organisms. Pneumonia results when the pathogens proliferate in the lung parenchyma after bypassing the defense mechanism of the respiratory tract like coughing, hair lining the nose and alveolar macrophages. Common organisms causing pneumonia include bacterial microorganisms like *Streptococcuspneumoniae, Haemophilusinfluenzae, Klebsiellapneumoniae, Mycoplasma pneumonia* and *Legionella*. Other organisms like adenovirus, influenza viruses and *Pneumocystis jirovicii* etc can also cause pneumonia. The causative pathogens in hospital acquired pneumonia are usually multi drug resistant and typically challenging to treat. Symptoms include fever, cough with expectoration, dyspnea and chest pain. Different scoring systems like the CURB-65) (the components being Confusion, Urea, Respiratory rate, Blood pressure, Age >65 years) or pneumonia severity index (PSI) can be used to decide on the location of treatment varying from ICU, non-ICU ward or outpatient setting. Diagnosis can be established by history, physical examination and imaging modalities like chest X-ray. Figure 39.9 shows the chest X-ray of a patient with pneumonia. Treatment includes antibiotics and supportive care like inhalational O_2 therapy.

Tuberculosis

Tuberculosis is a common disease in the developing nations and has a bountiful contribution towards the large burden of disease in such countries. According to the World TB report 2013 released by WHO, there were 8.6 million new cases of tuberculosis worldwide in the year 2012. Around 1.3 million people died from this infectious disease in the same year. The incidence of TB in the year 2012 in the WHO South East Asia Region (SEAR) was 187 cases per 1,00,000 population. Table 39.6 provides a summary of the disease burden in

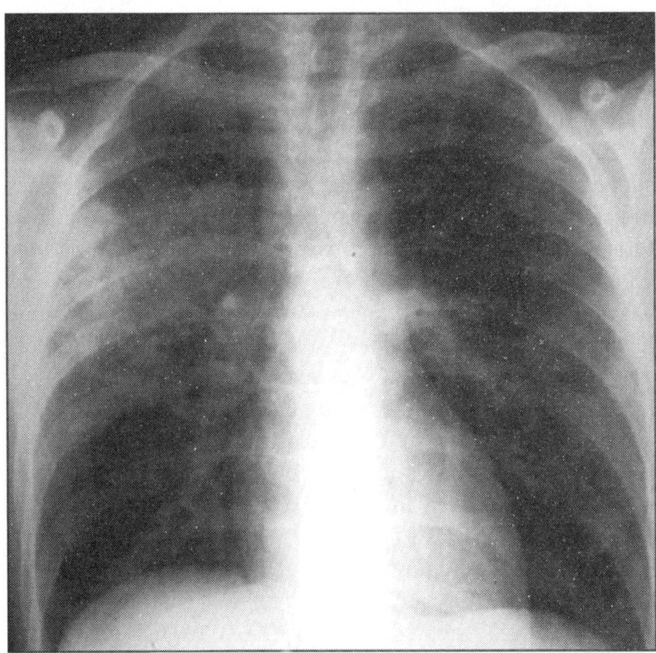

Figure 39.9 Shows the chest X-ray of a patient with pneumonia in the right lung.

India for the year 2012. It is caused by *Mycobacterium tuberculosis* which is an acid fast bacillus. This organism can affect any organ of the body but most commonly infects the lung. Pulmonary tuberculosis is spread via droplet nuclei from the affected patient by coughing or simply talking. The risk of developing active pulmonary tuberculosis after being infected by the *Mycobacterium tuberculosis* is high in patients who are malnourished, immunocompromised like patients with diabetes mellitus or HIV/AIDS. The body's immune response towards *M. tuberculosis* is outlined by the macrophage activation and delayed type hypersensitivity via the $T_H I$ cells trying to curb the infection thereby leading to the formation of granulomas with caseous necrosis.

TABLE 39.6 Summary of the TB burden in India for the year 2012

TB incidence	176 per 1,00,000 population
Prevalence (includes HIV+TB)	230 per 1,00,000 population
New smear-positive cases	53%
New smear-negative	27%
New extrapulmonary	20%
Percentage of new TB cases with MDR-TB	2.2%
Mortality (excludes HIV+TB)	22 per 1,00,000 population
Mortality (HIV+TB only)	3.4 per 1,00,000 population

Source: Global tuberculosis report 2013 (eighteenth Global Report on Tuberculosis (TB) published by World Health Organization)

Patient may develop lung cavities and pleural effusion which can be commonly observed on chest imaging. The bacilli may disseminate though blood and cause cause extrapulmonary tuberculosis like tuberculous lymphadenitis, pleural effusion due to TB, central nervous system tuberculosis. Figure 39.10 shows the computed tomography (CT) of a patient with TB of the mediastinal lymph nodes. Sputum smears of the patient stained with Ziehl-Neelsen stain are examined for the presence of acid fast bacilli. Sputum should also be inoculated for culture to isolate the *Mycobacterium tuberculosis* in medium like the LJ medium or using BACTEC cultures. Patients suffering with TB and having pleural effusion have been found to have high levels of adenosine deaminase in their pleural fluid. Treatment is complex and tedious with combination chemotherapy using antituberculous drugs like rifampicin, isoniazid, pyrazinamide and ethambutol owing to the propensity of the bacilli to become resistant towards drugs during monotherapy. *Mycobacterium tuberculosis* has been found to become resistant towards the treatment and has lead to the emergence of MDR-TB (multi drug resistant-tuberculosis). MDR-TB is defined as resistance to *at least* isoniazid and rifampicin with or without resistance to other drugs. Another type of TB on the rise these days is the XDR-TB (extensively drug resistant TB) which is resistant to isoniazid and rifampicin, along with resistance to any one of the fluoroquinolones and one of three injectable second-line drugs (Amikacin, Capreomycin or Kanamycin). MDR-TB and XDR-TB are pure laboratory diagnosis. Drug susceptibility testing can be done using the traditional culture methods which is the gold standard and by using molecular methods. The two commonly used molecular methods for drug susceptibility tests are Line probe assay and the GeneXpert MTB test. Line probe assay is a PCR based rapid screening method recommended for smear positive (Positive for AFB) to detect INH and rifampicin resistance. GeneXpert is a rapid real time PCR based test for detection of MTB and rifampicin resistance which can give results in less than 2 hours. Smear positive, smear negative and extrapulmonary tissue can tested with GeneXpert. The patients with TB have to be followed up for few years even after cure. This is because they may have recurrence due to relapse or re-infection. Re-infection and relapse can be differentiated on the basis of restriction fragment length polymorphism (RFLP). In relapse the RFLP patterns of the current and past *Mycobacterium tuberculosis* isolates may be identical. In many individuals infection with *Mycobacterium tuberculosis* may not cause active disease, that is the bacilli may remain quiescent. But these patients are at risk of developing active disease. Latent TB infection can bedetected with the help of tuberculin skin testing (TST) or interferon gamma release assay (IGRA). QuantiFERON-TB Gold In-Tube (ELISA based test) and T-SPOT. TB assay are two types of IGRAs used commonly. Early secreted antigenic target 6 (ESAT-6) is one of the antigens used for these tests.

Lung cancer

Lung cancer over the past few decades has steadily made its way on the list of the top three cancers worldwide affecting both genders and having a high mortality rate. Risk factors

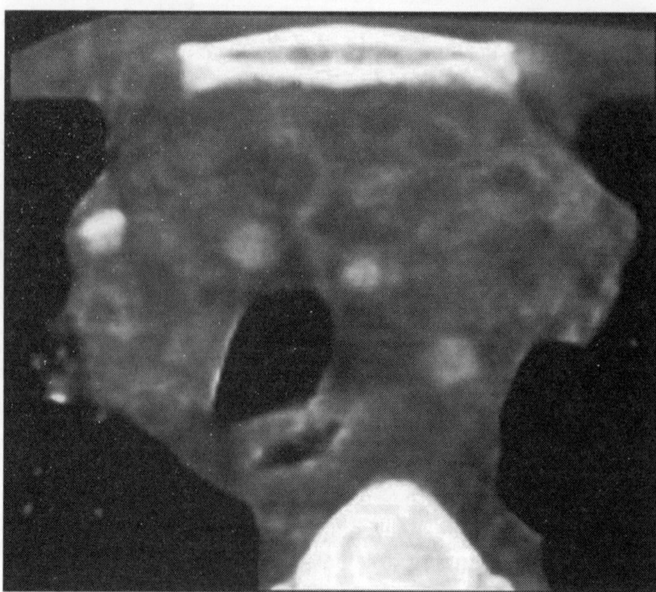

Figure 39.10 Shows the CT of a patient with TB of the mediastinal lymph nodes with features suggestive of necrosis.

include cigarette smoking (both active and passive), exposure to asbestos, radiation exposure and a positive family history of lung cancer. Cigarette smoke contains polycyclic hydrocarbons and nitrosamines which can lead to mutations in the genes. Mutations in protooncogenes like the EGFR and in the tumor suppressor genes like the P53 are involved in the pathogenesis of carcinoma lung. Table 39.7 gives an overview on the genes involved in lung cancer. Small cell lung cancer (SCLC) and the non-small cell lung cancer (NSCLC which includes adenocarcinoma, squamous cell carcinoma, and large cell carcinoma) are the 2 major types of lung cancer. Small cell lung cancer are notoriously known to secrete adrenocorticotrophic hormone (ACTH) and arginine vasopressin (AVP) leading to paraneoplastic syndromes like Cushing's syndrome or syndrome of inappropriate secretion of antidiuretic hormone (SIADH) respectively. Squamous cell carcinoma may secrete PTH-rp

TABLE 39.7 Overview on the genes involved in lung cancer

Oncogenes	*Tumor-Suppressor Genes*
EGFR	TP53
MYC	TP63
BCL-2	RB1
ALK	STK11

EGFR-Epidermal growth factor receptor
MYC-Myelocytomatosis oncogene
BCL-2 -B-cell CLL/lymphoma 2
ALK-Anaplastic lymphoma receptor tyrosine kinase
TP53-Tumor protein p53
TP63-Tumor protein p63
RB1- Retinoblastoma protein 1
STK11-Serine/threonine kinase 11

(parathyroid hormone-related peptide) leading to hypercalcemia. Anti–voltage-gated calcium channel antibodies may develop in carcinoma lung which can lead to Eaton-Lambert syndrome, leading to muscle weakness. Standard protocols of treatment include chemoradiation or surgery depending on the subtype and the stage of the cancer. Figures 39.11 and 39.12 show the computed tomography (CT) and chest X-ray of 2 different patients with carcinoma lung.

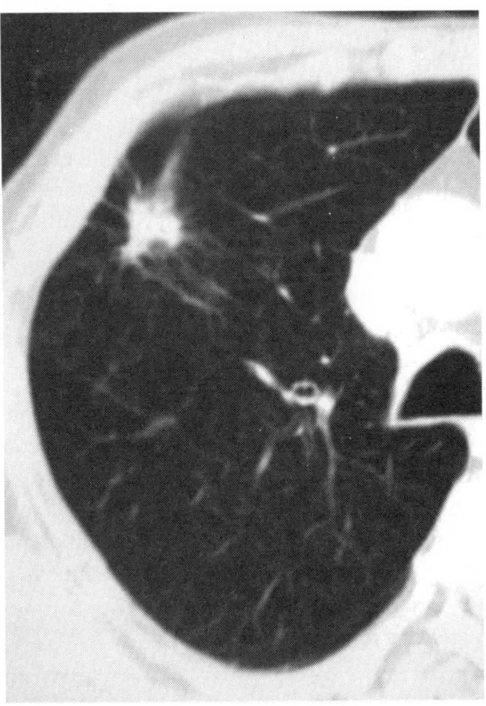

Figure 39.11 Shows the chest CT of a patient with carcinoma lung with speculated margins.

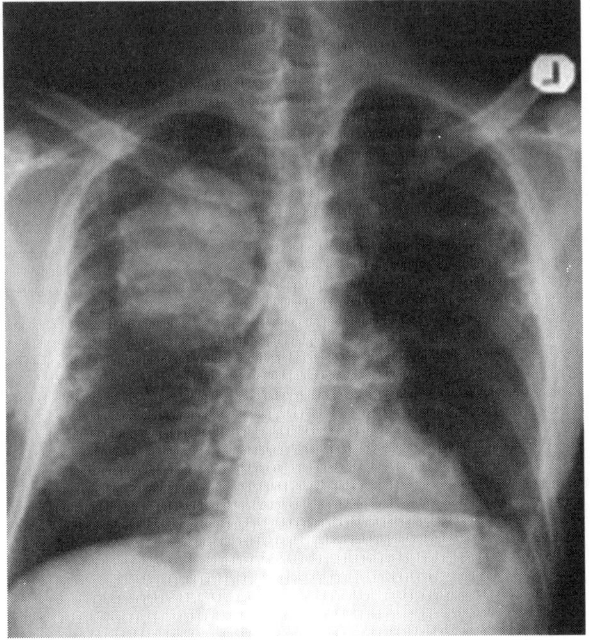

Figure 39.12 Shows the chest X-ray of a patient carcinoma lung with the tumour in the right lung.

Pleural effusion

Accumulation of excessive fluid in the pleural space results in pleural effusion. Pleural effusion can be either exudative or transudative. Pleural effusion is a not a disease per se, but a manifestation of many disorders. Causes for transudative pleural effusion are congestive cardiac failure, liver cirrhosis and other conditions with fluid overload. Exudative pleural effusion is caused by pneumonia, tuberculosis, and malignancy etc. In order to diagnose and treat an underlying pleural disorder it is imminent to differentiate between the transudative and exudative nature of the pleural fluid. Generally lactate dehydrogenase levels and protein levels are found to be elevated in the exudative pleural fluid. Adenosine deaminase (ADA) is helpful in differentiating pleural effusion due to tuberculosis from other causes. ADA value below 40 U/L has high negative predictive value for tuberculous pleural effusion. Amylase in the pleural fluid may be elevated in cases of pleural effusion due to acute pancreatitis. In patients with pleural effusion due to right heart failure, N-terminal pro-brain natriuretic peptide (NT-proBNP) may be elevated in the pleural fluid as well as in the blood. Pleural fluid may look milky in chylothorax, which can be due to malignancy. In chylothorax triglycerides are usually elevated above 110 mg/dL and chylomicrons may also be seen while in chyliform (pseudochylous) effusions usually seen in chronic pleural effusions due to tuberculosis and rheumatoid arthritis, cholesterol content is >200 mg/dl. In pleural effusion due to *Aspergillus* the pleural fluid may look blackish. Pleural fluid glucose levels maybe very low in pleural effusion due to rheumatoid arthritis or malignancy. Table 39.8 gives an overview on the biochemical differences between exudative and transudative effusion. Figure 39.13 shows a chest X-ray of a patient with pleural effusion.

TABLE 39.8 **Overview on the biochemical differences between exudative and transudative effusion**

	Transudative pleural effusion	*Exudative pleural effusion*
Protein	< 3 g/ dl	> 3 g/ dl
Pleural fluid protein/ serum protein ratio	< 0.5	> 0.5
Pleural fluid LDH/serum LDH ratio	< 0.6	> 0.6
Pleural fluid cholesterol	< 45 mg/dl	> 45 mg/dl

LDH = lactate dehydrogenase

Various investigations help in assessing the aetiology and extent of the pulmonary dysfunction. Imaging modalities like chest X-ray, CT scan, MRI and ultrasonography, CT angiography (CTA), CT pulmonary angiography (CTPA), ventilation and perfusion scan, are useful in delineating the abnormality. Overnight polysomnography (PSG) helps to diagnose obstructive sleep apnea (OSA). Pulmonary function testing and plethysmography help in characterizing the dysfunction into obstructive lung disease or restrictive lung disease by assessing the flow rates and the lung volume. Obstructive diseases like bronchial asthma are characterized

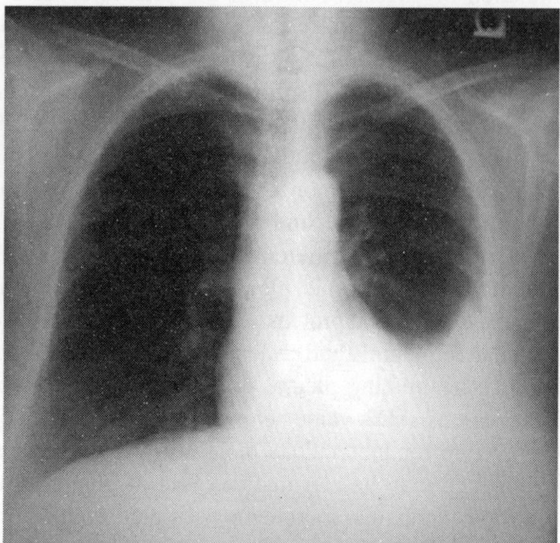

Figure 39.13   Shows the chest X-ray of a patient with a left sided pleural effusion.

by decreased flow rate. Restrictive lung diseases like idiopathic pulmonary fibrosis are characterized by normal airflow rate but decreased total lung volume. Arterial blood gas analysis can help in assessing the impact of the lung dysfunction on oxygenation and carbon dioxide elimination. Bronchoscopy can be used to visualise the bronchus and obtain biopsy from the bronchus.

High altitude pulmonary edema

At high altitudes the barometric pressure is lower when compared to the sea level pressures. This leads to lower partial pressure of oxygen in the alveoli at locations like Himalayas. In susceptible individuals this leads to patchy vasoconstriction of the pulmonary vasculature, thereby exposing capillaries to high intravascular pressure. This leads to a form of non-cardiogenic pulmonary edema known as high altitude pulmonary edema (HAPE). The patient may present with symptoms like cough and dyspnea. Prevention is by adequate acclimatisation before ascending heights. Calcium channel blocker nifedipine and inhaled beta$_2$agonists can also be used for prophylaxis. Treatment is by bringing the patient to lower altitude, giving supplemental oxygen and hyperbaric therapy.

SUMMARY

The understanding of the normal anatomy and physiology of the lungs is critical to the study of its dysfunction. Lung dysfunctions produce different symptoms and signs which with the help of relevant investigations help in diagnosing its aetiology. Table 39.9 is a summary of the common pulmonary disorders, their symptoms, diagnosis and treatment.

TABLE 39.9   Summary of the common pulmonary disorders

Disease	Symptoms	Investigation	Treatment
Pulmonary tuberculosis	Fever, weight loss, cough, expectoration, Haemoptysis	Chest X-ray (CXR), sputum for acid fast bacilli (AFB) staining, sputum culture for TB, GeneXpert	Antituberculosis drugs (combination chemotherapy)
Pneumonia	Fever, dyspnoea , pleuritic chest pain, expectoration, cough	Chest X-ray, computed tomography scan (CT Scan), blood culture, sputum gram staining, sputum culture and sensitivity	Antibiotics, oxygen and supportive treatment
Bronchial asthma	Dyspnoea, cough Wheeze	Spirometry with assessment of reversibility	Inhalation therapy • Bronchodilators • Corticosteroids oxygen
Bronchiectasis	Cough with purulent expectoration	Chest X-ray, high resolution computed tomography scan (HRCT)	Chest physiotherapy, cough expectorants and antibiotics
Chronic obstructive pulmonary disease (COPD)	Cough, dyspnoea	Chest X-ray, spirometry, arterial blood gas analysis (ABG)	Inhaled bronchodilators and inhaled steroid, oxygen inhalation, smoking cessation
Idiopathic pulmonary fibrosis (IPF)	Dyspnoea, cough	Chest X-ray, high resolution computed tomography scan, ABG	Supportive care Oxygen
Sarcoidosis	Cough, dyspnoea	Chest X-ray, computed tomography scan of the chest, transbronchial lung biopsy, serum ACE levels	Steroids
Pulmonary thrombombolism	Chest pain, dyspnoea	Chest X-ray, contrast enhanced computed tomography scan, pulmonary angiography, D-dimer, electrocardiogram, echocardiography, ventilation and perfusion scans	Anticoagulation, thrombolysis

(*Contd.*)

TABLE 39.9   Summary of the common pulmonary disorders (*Contd.*)

Disease	Symptoms	Investigation	Treatment
Pulmonary hypertension	Dyspnoea, chest pain	Chest X-ray, echocardiography, pulmonary artery catherization	Calcium channel blockers, bosentan, sildenafil, epoprostenol
Obstructive sleep apnoea (OSA)	Snoring, excessive day time sleepiness	Polysomnography (PSG)	Continuous positive airway pressure (CPAP) therapy, bilevel positive airway pressure (BiPAP), weight reduction, bariatric surgery for obesity if body mass index is more than 35 kg/ m2
Lung cancer	Cough, weight loss, haemoptysis	Chest X-ray, contrast enhanced computed tomography of the chest, bronchoscopy guided biopsy, sputum cytology	Chemotherapy, surgery depending on the stage of the surgery, radiotherapy

SUGGESTIONS FOR FURTHER READING

1. Longo, Farici, Kasper, Hauser, Janson and Loscalzo (2011), *Harrison's Principles of Internal Medicine*, 18th ed., McGraw-Hill.

2. Barreti Kim E., Barman Susan M., Boitano Scott and Brooks Heddwen (20120, *Ganong's Review of Medical Physiology*, 24th ed., McGraw-Hill, LANGE Basic Science.

3. John E. Hall (2010), *Guyton and Hall Textbook of Medical Physiology*, 12th ed., Elsevier Saunders.

4. Snell Richard S. (2011), *Clinical Anatomy by Regions*, 9th Ed., Lippincott Williams and Wilkins.

5. Moore Keith L. (2013), *Clinically Oriented Anatomy*, 7th ed., Lippincott Williams and Wilkins.

SECTION IV

FOODS AND NUTRITION

Whether we live to eat or eat to live, food is a central concern. The caloric needs of the body at different stages of life and in different states of activity have been outlined as also the principles on which such deductions are based. With the definition of the caloric content of a gram of carbohydrate, fat and protein, the basis is available to relate the amount of various foods to the caloric intake. In Chapter 41, tables are given to furnish at a glance the composition of a wide variety of Indian dietary items in terms of proteins, carbohydrates, fats, minerals and vitamins. A special chapter has been included on 'milk and milk products' because of its particular importance as a source of animal proteins in a vegetarian diet. In view of the fact that nutritional deprivation in fetal and early childhood can have serious effects on brain development and the immune system, a chapter devoted to the nutritional needs in pregnancy and infants is included. Protein and caloric malnutrition is widespread in the country—hence the coverage of Kwashiorkor and marosmus. Last, but not the least, is a chapter on the special diets required by the sick and convalescents, a question which is constantly asked but the answer is seldom provided in the textbooks.

40

Energy Needs of the Body

K.K. Deepak

CONTENTS

I. BASIC CONCEPTS OF ENERGY EXPENDITURE

Biological work is of two general categories: (i) external work indicative of movements of external objects brought about by the contraction of skeletal muscles, and (ii) internal work comprising all other forms of biological activity happening inside the body, including the skeletal muscle activity not actually moving external objects. All biological activity involves the expenditure of energy which is liberated by the breakdown of organic molecules. According to the first law of thermodynamics, energy can neither be created nor destroyed. It only changes its form, the total energy content of any system remaining the same. Thus, energy liberated inside the body during the breakdown of an organic molecule can either appear as heat or be consumed for performing work.

All internal work done by the tissues too, ultimately gets transformed into heat except, of course, during of growth. For example, cardiac contraction performs mechanical work by pumping blood for circulation into various tissues, but the ultimate fate of such mechanical energy is heat which is generated by resistance (friction) to flow offered by the blood vessels. Similarly, internal work is performed during secretion of HCl by the stomach, and $NaHCO_3$ by the pancreas, but

when finally hydrogen and bicarbonate ions react in the small intestine, heat is produced. Again, when plasma proteins are synthesized, internal work is done. But the proteins are then, sooner or later, catabolized with the inevitable production of heat. However all body constituents are being constantly built up and broken down, and during the periods of net synthesis of protein, fat, etc., energy is also being stored in the form of molecular bonds and does not appear as heat. Thus, the total energy liberated by the catabolism of organic molecules may be transformed into body heat, appear as external work during activity or be stored in the body in the form of molecules, the latter occurring during growth of net fat deposition in obesity. Therefore, at any point of energy balance in a living organism:

$$\text{Total energy produced} = \text{heat produced} + \text{external work} + \text{energy stored}$$

It is worthwhile remembering that, in the animal cell in general, the energy used for work is first incorporated into ATP molecules which are subsequently broken down to supply the immediate energy source for work. For example, almost half the energy released when a nutrient like glucose is metabolized, appears immediately as heat. Some part of it goes to form ATP which can subsequently be broken down, and the energy thus released is expended to perform work. Ultimately, as mentioned above, all energy that performs internal or external work is completely converted into heat (Figure 40.1)

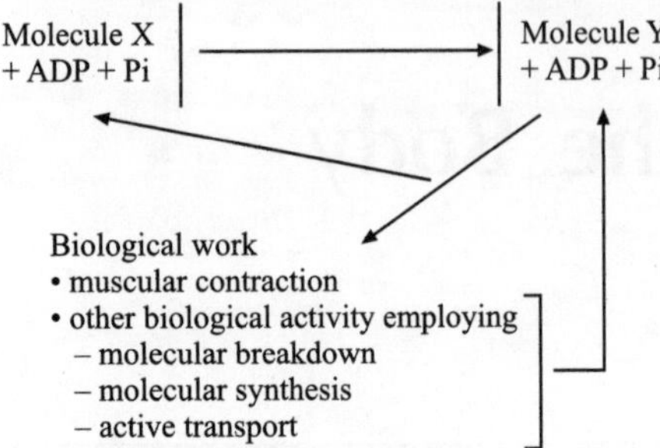

Figure 40.1   General scheme of energy liberation and utilization in the living body.

A.   Unit of Energy

Since all forms of energy in a biological system tend ultimately to appear in the form of heat, it is understandable that units of heat should find a universal acceptance for expressing the amount of energy used or liberate. The amount of heat necessary to raise the temperature of 1 g of water from 14.5°C to 15.5°C is called a gram-calorie or simply a calorie. The principal unit employed by physiologists however is a *kilocalorie* (kcal) which is also written as Calorie (note that it starts as capital C).

B.   Measurement of Energy Expenditure

Total energy expenditure per unit time is called the metabolic rate. It can be measured directly or indirectly. The direct method is rather difficult to perform and requires an immense amount of technical skill. It is interesting to know how direct method was used in past. First used by Atwater, an American physiologist, in 1900, it consisted of a chamber in which a person could live and work for several days allowing at the same time the measurement of his total output of heat (Figure 40.2). The energy expenditure, thus, measured can be related to net energy lost in urine and faeces. Atwater's experiments, measuring energy intake and energy output, lasted a number of days and he was able to demonstrate consistently a fair amount of agreement between the input and the output. Although nobody uses the human calorimeter these days on account of the difficulties of technique, Atwater's experiment was the first of its kind which demonstrated that the human body behaved like any engine running on combustion of fuels.

The energy production, however, is quite accurately related to oxygen utilization. The calculation is based on the principle that when 1 liter of oxygen is utilized in the oxidation of organic nutrients, approximately 4.8 kcal of heat is liberated. The measurement of oxygen consumption which is a relatively simple technique is now universally employed to estimate the metabolic rate. This is *indirect calorimetry*. The following three methods are the ones most commonly used:

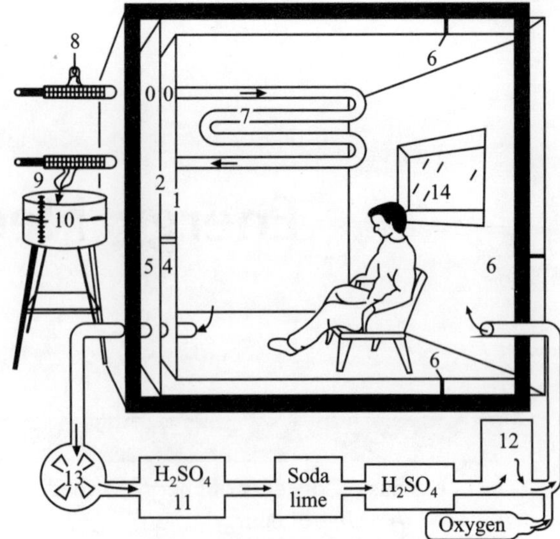

Figure 40.2   Schematic diagram of Atwater-Benedict Human Calorimeter. The structural elements are represented as numbers 1–6. The water flows in the tubes at (7) and its temperature is measured at the inlet (8) and the outlet (9). The air leaves the calorimeter aided by a fan, water is removed from it by absorption to sulphuric acid and CO_2 by soda lime, and then goes back to the chamber with O_2 added at measured rate (10–12). The calorimeter possesses and efficient thermal insulation. The heat gained by flowing water is a fairly accurate indication of the external heat produced by the subject. Activities of the subject can be observed through the double glass in the window (14).

1. The Benedict-Roth Spirometer Method (Figures 40.3a and b).

This is a closed-circuit breathing apparatus which is filled with oxygen and has a capacity of about 6 liters. Oxygen is contained in a metal drum which floats on a water seal. The person whose O_2 consumption is to be measured, breathes in oxygen through an inspiratory valve and breathes out into the drum through an expiratory valve and a sodalime canister, so that the CO_2 produced is absorbed. As the O_2 is used up, the drum sinks and its movement is recorded on a moving paper mounted on a kymograph; from this, the rate of oxygen consumption cab be read. The apparatus is accurate and simple to use. It has the disadvantage that it can be used only when the person is at rest or doing very light exercise.

2. The Douglas Bag method

This is a canvas or plastic bag with variable capacity, usually 100, 200 or 300 liters. The subject breathes through a mouth piece which contains inspiratory and expiratory valves. Room air is breathed in, but breathing out is into the Douglas bag so that all the expired air is collected in it. Douglas bag so that all the expired air is collected in it. (Figure 40.4). The bag is then emptied through a gas meter and a sample of the

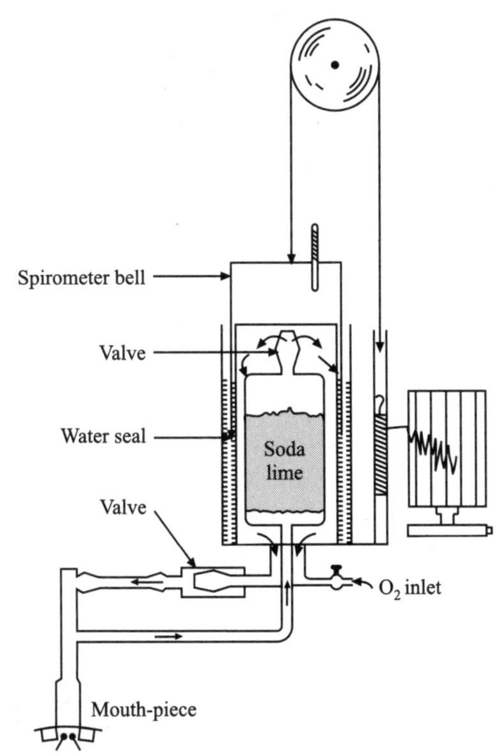

Figure 40.3a Schematic diagram of the Benedict-Roth Spirometer used for measuring oxygen consumption.

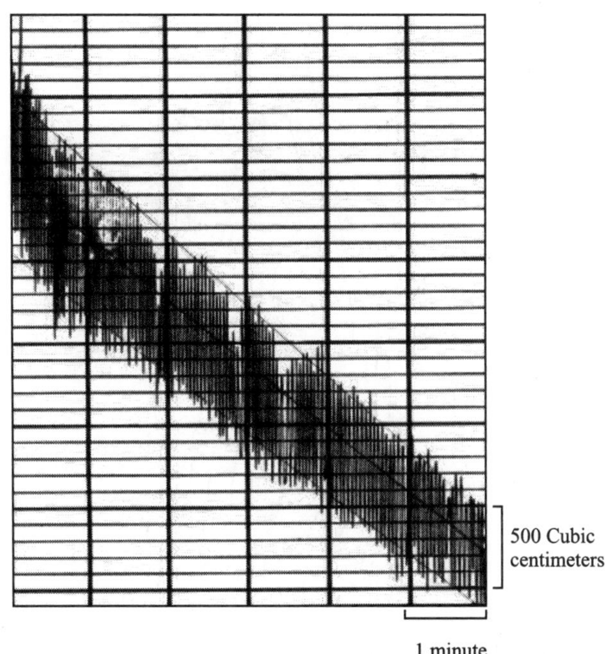

Figure 40.3b A typical spirometer record for measuring oxygen consumption taken in laboratory in the Department of Physiology. Steps for calculating metabolic rate are given below. (*Courtsey:* Dr. Kumar Sarvottam, MD, Sr. Demonstrator, Department of Physiology, AIIMS, New Delhi).

A. Calculation of metabolic rate

Date: 11.01.13 Time: 11.30 AM Barometric Pressure: 748

Age of the subject : 30 years, Male; Weight $= 67$ kg, Temperature $= 32°$ C, Height $= 165$ cm

O_2 Consumption at ATPS in 6 minutes $= 2300$ ml

O_2 consumption at ATPS in 60 minutes $= 23000$ ml $= 23$ litres/60 min (or 383 ml/min)

O_2 consumption in STPD in 60 minutes $= 23 \times 0.945$ litres

$= 21.735$ litres/hour (or 362 ml/min)

Now,

Since 1 litre of O_2 yields $= 4.86$ kcal

So, 21.735 litres yield $= 4.86 \times 21.735$ kcal

$= 105.6321$ kcal

Body surface area for the given height and weight $= 1.74$ m^2

So, Basal metabolic rate $= \dfrac{105.6321 \text{ cal/m}^2/\text{hr}}{1.74}$

$= 60.70$ kcal/m^2/hr

B. Temperature Corrections

Degree C	Subtract
10, 11, 12	4%
13, 14, 15	5%
16, 17, 18	6%
19, 20, 21	7%
22, 23, 24, 25	8%
26, 27, 28	9%
29, 30, 31, 32	10%

C. Barometric Corrections

For each 10 mm, below 760 subtract 1%

For each 10 mm, above 760 add 1 %

Figure 40.4 Person breathing out into the Dougles bag.

expired air is taken for analysis of O_2 and CO_2 content from which the rates of oxygen utilization and CO_2 production can be calculated. The method has the advantage that both O_2 consumption and CO_2 production can be measured at varying grades of activity or muscular exercise.

3. Doubly Labeled Method

It is indirect method of calorimetry which was discovered about 3 decades ago. In this method a loading dose of water labeled with isotopes (both for Hydrogen, H; and oxygen, O) is given. The elimination rates of H and O are monitored. As evident from usage of these in metabolic pathways, the H is eliminated as water and O is eliminated as both water and CO_2. The difference between the elimination rates provides the value of CO_2 production. From CO_2 one can easily calculate the energy production. This is precise and stable method to measure energy production for several weeks in unstrained humans.

II. BASAL METABOLIC RATE (B.M.R.)

A. Definition

The metabolic rate determined when the subject is at, as complete a mental and physical resting state as possible, in a room with comfortable temperature, and 10–12 hours after the last meal, is called the basal metabolic rate. It essentially implies the minimum energy required to keep the body going. However, one cannot be very exact in such an estimation for the metabolic rate during sleep is generally less than the B.M.R. and, as will be clear subsequently, there are large number of factors of nutritional status, geographical as well as individual variations, which affect the B.M.R. Nevertheless, it is a fair indication of the body's energy need at rest.

The B.M.R. of an average Indian man is 1750–1900 kcal/day. In terms of oxygen consumption it would amount to about 15 litres/hr. Large sized animals have higher B.M.R., but the B.M.R. per unit body weight is higher in the smaller animals. The variable that correlates most with the B.M.R. is the surface area of the body. The B.M.R. for human is around 1000 kcal/m² body surface area. can be calculated by the following formula.

$$S = 0.007184 \times W^{0.425} \times H^{0.725}$$

[where, S = surface area in sq. meters; W = body weight in kg; and H = height in cm.]

Nomograms (Figure 40.5) constructed from this formula are available for easy estimation of the body surface are (Several programs are available on internet to calculate this). In clinical practice the B.M.R. is usually expressed as percentage of increase or decrease from a reference level of standard normal values. Thus a value of + 50% will mean that the B.M.R. is 50% more than the standard for his age and sex. The reference values for height and weight for Indians have been provided by Indian Council of Medical Research (ICMR) in 2010 (Table 40.1).

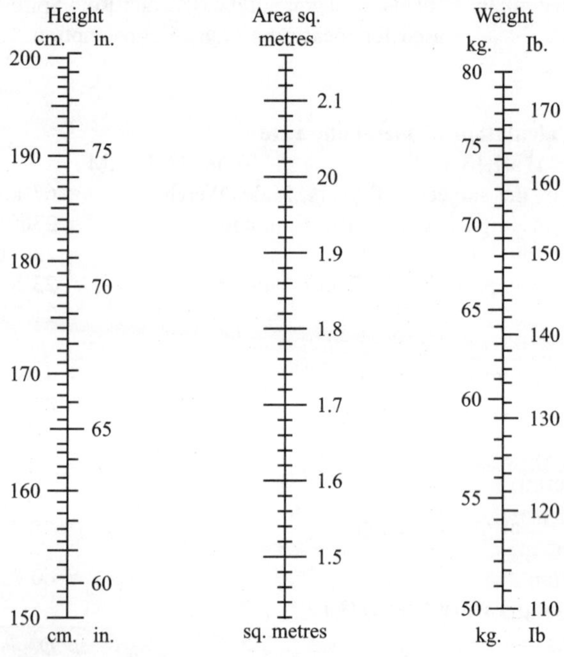

Figure 40.5 Nomograms for the calculation of body surface area.

B. Factors Influencing B.M.R.

There are many factors that affect he B.M.R. These include body temperature, age, sex, race, emotional state, climate and circulating levels of hormones like catecholamines (epinephrine and norepinephrine) and those secreted by the thyroid gland. Indians and Chinese seem to have a lower B.M.R. than the Europeans. This may as well be due to dietary differences between these races. Table 40.2 provides predictive equation's

TABLE 40.1   Indian Council of Medical Research (ICMR) definition of Reference Indian

Parameters	Reference Indian man	Reference Indian woman
Age range	18-29 years	18-29 years, non-pregnant non-lactating (NPNL)
Weight	60 kg	55 kg
Height	1.73 m	1.61 m
Body Mass Index	20.3	21.2
Physical activity details	Physically fit for active work. On each working day engaged in 8 hours of occupation which involves moderate activity.	Physically fit for active work. On each working day engaged in 8 hours of occupation which usually involves moderate activity.
Activities during non-work periods (hours spend at home)	Spends 8 hours in bed, 4–6 hours in sitting and moving about, 2 hours in walking and in active recreation or household duties.	Spends 8 hours in bed, 4–6 hours in sitting and moving about, 2 hours in walking and in active recreation or household duties.
Disease state	Free from disease	Free from disease

Source: ICMR document on nutrient requirements and recommended dietary allowances for Indians, 2010; reproduced with permission from ICMR.

TABLE 40.2   Equations for predicting BMR proposed by FAO/WHO/UNU and those proposed for Indians

Gender	Age	Prediction Equation to compute BMR		Correlation coefficient	Standard Deviation
		Proposed by FAO/WHO/UNU consultation (1985)	Proposed by ICMR e × pert group for Indians (1989)		
Male	18–30	$15.1 \times$ B.W. (kg) $+ 692.2$	$14.5 \times$ B.W.(kg) $+ 645$	0.65	151
	30–60	$11.5 \times$ B.W. (kg) $+ 873$	$10.9 \times$ B.W.(kg) $+ 833$	0.60	164
	> 60	$11.7 \times$ B.W. (kg) $+ 587.7$	$12.6 \times$ B.W.(kg) $+ 463$	0.79	146
Female	18–30	$14.8 \times$ B.W. (kg) $+ 486.6$	$14.0 \times$ B.W.(kg) $+ 471$	0.72	121
	30–60	$8.1 \times$ B.W. (kg) $+ 845.6$	$8.3 \times$ B.W.(kg) $+ 788$	0.70	108
	> 60	$9.1 \times$ B.W. (kg) $+ 658.5$	$10.0 \times$ B.W.(kg) $+ 565$	0.74	108

Source: ICMR document on nutrient requirements and recommended dietary allowances for Indians, 2010; reproduced with permission from ICMR. [BMR of Indian is 5% lower than the international values (FAO/WHO/UNU) B.W. (body weight), kg (kilogram); FAO (Food and Agricultural Organization), WHO (World Health Organization); UNU (United Nations University).]

to calculate the B.M.R. for Indians. An increase in body temperature as a result of fever increases the B.M.R. by 14–15% per degree C which, evidently, is due to the increased rate of metabolic reactions of the body.

Standards of basal metabolic rate were first arrived at in Europe and North America. Later it was discovered that these figures were higher by 10% in comparison to the B.M.R. of individuals living in tropical climates, e.g., Calcutta or Singapore. It is difficult to account for the change in B.M.R. values resulting from climatic variations. The thyroid hormones stimulate oxidation and heat production within the cells. In experimental animals removal of these glands produces a marked fall in the basal level of oxygen consumption. In the disease myxoedema, which is due to atrophic changes in the thyroid gland, B.M.R. is very low. Similarly, over-activity of the thyroid gland in hyperthyroidism markedly increases the B.M.R. Although the term "basal conditions" signifies complete mental and physical rest, it is impossible to assess these objectively. Anxiety and tension may not show on the face but they do produce an increased tensing of the muscles, and release of norepinephrine even though the subject is seemingly quiet. Both these factors tend to increase the metabolic rate. Therefore, when metabolic rate is determined at rest it is called resting metabolic rate (RMR). The RMR serves as an important metabolic parameter in scientific research.

C.   Thermic Effect of Feeding (TEF)

Metabolic rate is increased immediately after ingestion of food and remains so for 6 hours or more. This can easily be demonstrated by measuring oxygen consumption of a resting man before and after eating, and is known as thermic effect of feeding (earlier called as specific dynamic action, S.D.A. of foods). After most meals it usually amounts to an increase of 10–15%. It is known, however, that ingestion of proteins causes maximum increase, which is about 30%, carbohydrates increase it by 6%, and fats by 4%. This entire extra energy expenditure is not because of requirements of digestion and absorption of ingested food, which accounts for only a small proportion of energy. Besides, intravenous injection of amino acids produces almost the same TEF effect as oral ingestion of the same material. Experiments have demonstrated that TEF does not occur in an animal whose liver has been removed. It thus seems that the cause of the TEF of proteins is related to the process of deamination of the amino acids which occurs in the liver after their absorption. However, it is likely that the TEF of fats is due to direct stimulation of metabolism by the liberated fatty acids, and that of carbohydrates may be due to energy requirements for the synthesis of glycogen. Be that as it may, it does imply that the amount of energy available on ingestion of food is, in effect, less than the actual caloric values of these foods. Prolonged fasting on the other hand produces a decrease in the metabolic rate. The reason for this is the reduction of body mass that occurs in starvation. It is an adaptive mechanism which helps the body in times of scarcity.

D. Resting Metabolic Rates of Different Tissues

It is possible to measure the metabolic rates of different organs by measuring the rate of blood flow and the oxygen content of their arterial and venous blood. Data of oxygen consumption of various organs of a resting individual, indicates that it is the brain, which weights less than 2.5% of the body weight, that consumes almost 20% of the energy required by a resting individual. It is interesting to note that this amount does not appreciably increase however hard the mental work one may be engaged in. By contrast the oxygen consumption of skeletal muscles may increase several fold on exercise. It is also interesting that the kidneys do more work than the heart.

III. ENERGY REQUIREMENTS UNDER VARIOUS CONDITIONS

A. Effect of Pregnancy

A pregnant mother requires to maintain the metabolism of the foetus, builds up a reserve of fat and additional maternal tissue. By the end of a pregnancy the mother gains approximately 1 kg protein, mostly in the uterus and products of conception, and 4 kg of fat in her own fat depots. Energy will be required for the metabolism of the foetus and placenta, metabolism of additional uterine and breast tissues and for the extra work of heart and respiration. According to one estimate an extra amount of about 70,000 kcal spread evenly at an almost constant rate over the last 2/3rd period of pregnancy is generally required. An average woman at rest, thus, will need an extra amount of approximately 350 kcal/day.

B. Lactation

A lactating mother must also consume extra food to meet the energy cost of lactation. The efficiency of milk production is probably 80%, so that for the production of milk having an energy value of 80 kcal, the mother must expend 100 kcal, thus requiring the same amount of energy in her diet. A healthy baby may require 850 ml of milk daily. At an energy value of 68 kcal/100 ml of milk this will mean the provision of 578 kcal/day. A lactating mother will thus require an extra diet of approximately 600 kcal/day to meet the demands of lactation. However, the fat reserves which are carefully laid down during pregnancy are also handy for this energy requirement.

C. Energy Expenditure in Various Activities

The energy requirements are currently compiled by factorial method. Each activity is assigned a factor value over and above the BMR. This is called physical activity ratio (PAR). The PAR is expressed as the ratio of the energy cost of an individual activity per minute to the cost of the basal metabolic rate (BMR) per minute.

$$\text{Physical Activity Ratio (PAR)} = \frac{\text{Energy cost of an activity per minute}}{\text{Energy cost of basal metabolism per minute}}$$

By doing several experiments the PAR values for various daily activities have been estimated. PAR values for activities performed in a day can be aggregated over that period to yield the Physical Activity Level (PAL), which is the ratio of the energy expenditure for 24 hours and the BMR over 24 hours. For example, a person spending 8 hours in sleep (with a PAR of 1), 8 hours in domestic and leisure activity (with an average PAR of 2) and 8 hours at work (with an average PAR of 3), would have a total PAR-hour value = $(8 \times 1) + (8 \times 2) + (8 \times 3)$ = 48 PAR-hours.

$$\text{The PAL (for the day)} = \frac{\text{Total PAR-hours}}{\text{Total time}} = \frac{48}{24} = 2.0$$

Table 40.3 classifies the physical activity in terms PAL values as derived from the data of rates of O_2 consumption. The maximum amount of work that a physically fit person can perform, however, is higher than the very heavy activity indicated in this table. Maximum oxygen consumption up to 4–5 liters/min have been recorded in world class athletes, which essentially implies an expenditure of energy amounting to about 19–24 kcal/min. However, middle-aged or elderly man, or a person whose respiratory or cardiovascular functions are a bit compromised, can do light work and enjoy light exercise if he can increase his resting metabolic rate by about four times, say up to 1 liter/min of oxygen consumption or 5 kcal/min of energy expenditure. It is not easy to designate various daily activities in terms of sedentary, moderate or heavy activity as classified in Table 40.3 and 40.4. For example, work on a farm may involve moderate work, though some of it may be heavy, while some may be just light. Similarly, sports for some adults may be moderate and enjoyable physical exercise, but for a champion of a class like Saina Nehwal is, or for a sprinter like Milkha Singh, it is always a heavy and strenuous work on the courts or tracks. Nevertheless Table 40.4 gives

TABLE 40.3   Physical Activity Levels (PAL) values for different categories of work for Indian

Level of activity	ICMR 2009	FAO/WHO/UNU
Sedentary Work	1.53	1.40–1.69
Moderate Work	1.8	1.70–1.99
Heavy Work	2.3	2.0–2.40

Source: ICMR document on nutrient requirements and recommended dietary allowances for Indians, 2010; reproduced with permission from ICMR [For abbreviations please see Table 40.2]

TABLE 40.4   Energy requirements of Indians for different category of work levels

Gender	Category	Body weights	Requirement (kcal/d)[1]
Man	Sedentary work	60	2320
	Moderate work	60	2730
	Heavy work	60	3490
Woman	Sedentary work	55	1900
	Moderate work	55	2230
	Heavy work	55	2850
	Pregnant woman	55 + GWG[2]	+350
	Lactation	55 + WG[3]	+600
			+520

[1] Rounded off to the nearest 10 kcal/d

[2] GWG – Gestational Weight Gain. Energy need in pregnancy should be adjusted for actual bodyweight, observed weight gain and activity pattern for the population.

[3] WG – Gestational Weight gain remaining after delivery

[Reproduced with permission from ICMR]

a fair estimate of the energy expenditure required for various types of routine activities.

Knowledge about energy expenditure in sitting, standing and walking is important as we all spend a good part of our time during the day sitting and standing. Evidently energy will be needed for maintaining different body postures. Passmore during 1950s conducted experiments in which oxygen consumption of individuals sitting in their homes while watching television, reading or listening to radios was measured. It was observed that the mean values of energy expenditure were up by about 20% while sitting and 50% while standing when compared with values of resting metabolic rates. Mental tension accompanied by fidgeting did not raise the energy expenditure to any appreciable extent. Passmore's studies show that walking at a moderate speed of 5 kilometers/hour increases oxygen consumption about four fold indicating energy expenditure of about 5 kcal/min. Since most of the work done while walking is against gravity, the energy cost of walking is proportional to body weight. The speed of walking also matters and there is a fine correlation between the speed of walking and energy expenditure in a man of given weight.

D.   Daily Rates of Energy Expenditure

It now remains for us to know how a plausible figure for the daily energy requirement of an individual can be arrived at. The energy requirement will vary from individual to individual depending upon weight, sex, environmental temperature and other factors. Still, fairly accurate formulations can be made on the basis of a Indian "reference man" and a "reference woman". The definition of an Indian reference man and woman has been defined by ICMR. Table 40.5 provides value for different levels of energy expenditure.

TABLE 40.5   Computation of energy expenditure of reference Indian person from physical activity ratio (PAR)

Main daily activities	Duration (h)	Physical activity ratio (PAR) values for daily activities and energy expenditure		
		Sedentary	Moderate active	Heavily active
Sleep	8	1.0	1.0	1.0
Occupational activity	8	1.5	2.3	3.8
Non-occupational activity	8	2.1	2.1	2.1
Mean PAR values		**1.53**	**1.80**	**2.30**
Non-occupational activity details				
Personal care	1	2.3	2.3	2.3
Eating	1	1.5	1.5	1.5
Commuting to work by bus or by vehicle or by walk	1	2.0	2.0	2.0
General household or other activities	2	2.5	2.5	2.5
Walking at various speeds without load	1	3.2	3.2	3.2
Light leisure activity	2	1.4	1.4	1.4
Mean non-occupational activity	8	2.1	2.1	2.1
Computation of energy requirement (PAR value × BMR)				
Men: body wt. (60 kg), BMR (1515 kcal)	–	2318	2727	3485
Women: body wt (55 kg), BMR (1241 kcal)	–	1899	2234	2854

[Reproduced with permission from ICMR]

IV. REGULATION OF ENERGY BALANCE

With these basic concepts of energy expenditure and metabolic rates as a foundation, we can consider the regulation of energy balance much in the same way as any other balance, i.e. in terms of input and output. Thus in a steady state, the total caloric expenditure of an individual will be equal to total body caloric fuel input i.e.,

Food energy intake + internal heat produced + external energy + energy stored

In the normal course some energy is lost in urine, faeces, and sloughed hair and skin, but this energy loss is negligible. In certain diseases, however, the urinary loss or organic molecules may be significantly large so as to disturb the energy balance, e.g., diabetes mellitus. The energy stored is reflected as gain in body weight. In a situation when the energy value of food intake is equal to internal heat production and external work, body weight does not change. If, however, the energy obtained from food is more than the demands of internal heat production and external work, body weight is increased. Excessive eating by adults having sedentary occupations is one of the important causes of obesity. On the other hand, in negative energy balance, when food intake is less than the demands of internal heat production and external work, which is the bane of populations living in scarcity areas, body weight will decrease.

In most adults body weight remains fairly constant over long periods of time which means that all the three factors, i.e. food intake, internal body heat, and external work, are quite precisely controlled for the regulation of energy balance. Control mechanisms for the control of energy intake (food intake) are well investigated and described in textbooks of physiology and are perhaps, the dominant controls in the regulation of energy balance. Control mechanisms of heat production are primarily aimed at regulating the body temperature, rather than the total caloric balance. For example, if a person is exposed to cold, his body shivers to produce more heat even though he may be starving and incurring a negative energy balance. Controls of external work are not automatic and therefore do not exercise a regulatory influence on energy balance unless consciously willed by the subject. For example, a person when he becomes fat, does not automatically feel impelled to indulge in jogging or playing tennis. More often it is quite the contrary! Thus although internal heat production and external work are important factors for total body energy balance, these do not have automatic physiological controls for energy regulation. It is essentially, the control on food intake which provides a regulation of energy. Thus, it is when the individual is exposed to cold or takes up exercise, thereby expending more energy, that he automatically increases his food intake.

If volunteers whose normal daily caloric intake is around 3000 kcal are restricted to a calorie-deficient diet say of 1500 kcal/day, they first lose weight very rapidly but ultimately get stabilized at a weight which is about 25% less than their average normal weight. In other words caloric balance is reestablished in spite of the continued decrease in the caloric intake. This is due to reduction in their B.M.R. which, in turn, is due to the decreased total body mass and also the decreased metabolism of liver and other organs. It is interesting, though understandable, that these individuals also become apathetic and reluctant to indulge in much physical activity. Thus it is clear that although the control of food intake is the major physiological mechanism for the regulation of energy balance, there are physiological adaptations which allow regulation of energy balance in the face of scarcity leading to chronic malnutrition but only at the price of reduced amount of external work and efficiency.

41

Nutritive and Calorific Value of Food

N.C. Sharma

CONTENTS

I. INTRODUCTION

Food provides energy and materials for the repair of wear and tear of tissues and also for the synthesis of compounds of physiological importance. It is therefore important to provide the right quality and quantity of food to an individual. The quality of a food can be judged from the nutritive values of different foods which in turn can be assessed from the composition of food in respect of the contents of carbohydrates, proteins, fats, minerals, and vitamins. In the subsequent pages the composition of common foods will be given in tables. This will tell us which food is rich in particular *proximate* principles and/or particular mineral and vitamin.

The biological values of carbohydrates, proteins, fats, minerals and vitamins have been described in the appropriate chapters but proteins and fats need special mention.

II. PROTEINS

Proteins are practically the sole source of nitrogen for the body. The nutritive values of different proteins are not the same. Proteins from animal sources are almost completely absorbed, whereas 10% to even 40% of vegetable proteins may not be absorbed. Besides the difference in absorbability there is a difference in the amino acid content. With the exception of gelatin, a protein which is deficient in tryptophan and methionine, animal proteins contain all the essential amino acids, but the vegetable proteins are deficient in some of the essential amino acids e.g., leucine, isoleucine, lysine, valine, phenylalanine, tryptophan, methionine, threonine and

histidine. Actually speaking the requirement of proteins should be expressed in terms of individual amino acids but as yet we do not know the requirement of the correct proportion of the different amino acids. We also do not know the amino acid content of all the food proteins. Also the nature of the linkages of the amino acids in the protein molecule may be resistant to the highly specific digestive enzymes and hence the amino acids involved may not be liberated on digestion and thus lost to the system. Hence, evaluation of protein according to the amino acid content has its limitations. Even so it is advisable to take mixed proteins of which about half should be animal protein. But in a poor country like ours a larger section of the people depends primarily on vegetable protein for nitrogen supply because animal protein is expensive and hence out of reach of the common man. However a mixture of properly selected vegetable proteins will ensure a supply of all the essential amino acids.

III. FATS

The nutritive value of fat depends largely on two points—(1) the highest caloric value in contrast to carbohydrates and proteins, and (2) the valuable adherents like the fat-soluble vitamins specially vitamins A and D. There is little difference between animal and vegetable fat in respect of energy value but animal fat is considered to be biologically superior because of the presence of the valuable adherents.

The significance of essential fatty acids—linoleic acid, linolenic acid, and arachidonic acid present in fat deserves special mention. These acids are either not produced, or are produced in insufficient amount in the tissue. These essential fatty acids are polyunsaturated acids. It has been experimentally shown that young rats on diets deficient in these acids cease to grow in 10 to 12 weeks and develop an eczematous condition of the skin and a scaly tail. Recently these acids have been reported to be involved in atherosclerosis since a diet containing these acids tends to decrease serum cholesterol levels. It is now established that essential fatty acids are required by human infants but there are no definite indications as yet about its requirement by adult men.

All living forms require many inorganic elements for normal processes of life. They are called mineral elements. Of these, particular care has to be taken for the supply of calcium, phosphorous, iron and iodine. Many others, which are required in smaller quantities (except sodium chloride), are usually available from the mixed diet in such amounts that we do not suffer any deficiency symptoms.

The nutritive value of foods in respect of vitamins has been dealt with in detail in Section V.

It may be noted that certain signs and symptoms may appear as a result of the deficiency of one or more of the following in the diet—carbohydrates, fats, proteins, minerals and vitamins. But it must be appreciated that none of these work in isolation and there is real interrelationship for the normal functioning of the body. Hence the nutritive value of a mixed diet will serve a real purpose for the body only when the necessary ingredients are available in adequate amounts and forms.

IV. GENETICALLY MODIFIED FOODS

Genetically modified (GM) foods are foods derived from plants, animals or organisms whose genes are modified in the laboratory by recombinant DNA (rDNA) technologies. rDNA or genetic engineering technologies have been dealt with in details in Chapter 55. GM Foods have the potential to solve many of the hunger and malnutrition problems of the world and help protect and preserve the environment by increasing yield and reducing dependence upon chemical pesticides and herbicides. GM Foods were first introduced on the market in 1996 and by now have been eaten by/millions of people worldwide with no adverse health effects (National Research Council, Committee on Identifying and Assessing Unintended Effects of Genetically Engineered Foods on Human Health, 2004; Royal Society of Medicine, 2008). The first commercially grown GM whole food crop was a tomato called "FlavrSavr" which was modified to ripen without softening by Calgene, a California based Company. GM Foods currently on the market are aimed at increased production, self life and nutritive value by inserting genes to introduce pest resistance, herbicide tolerance, disease resistance, cold tolerance, draught tolerance, salinity tolerance or increased nutritional components. At present, several GM Foods are commercialized. Table 41.1 gives a brief account, of some of the available GM Foods and the associated genetic modifications. In India GM cotton (BT cotton) only has been grown since 2002.

V. PHYTOCHEMICALS

Most bioactive food constituents are derived from plants and are called phytochemicals. A large number of them are redox active molecules and, therefore, defined as antioxidants. Vitamin A/carotenoids, vitamin C, vitamin E, selenium, flavonoids/polyphenols, lycopene, lignan and lutein containing foods foods constitute rich sources of antioxidants. Food antioxidants can eliminate in vivo free radicals and other active oxygen and nitrogen speeies responsible for many chronic diseases and bring about beneficial health effects in antioxidant defense, longevity, cell maintenance and DNA repair. Table 41.2 gives total contents of antioxidants in some of the popular foods.

TABLE 41.1 Genetically Modified Foods

S.No.	Name of food crop	Genetic Modifications	Foods
1.	Corn	Pesticide resistance	Flour, Oil, Syrup
2.	Cotton	Inserted gene for toxin production from Bacillus thuringiensis (BT)	Cottonseed oil
3.	Flax	Herbicide resistance	Seeds, Flax oil
4.	Honey	Canadian honey comes from bees Collecting nectar from GM canola plants	Honey
5.	Meat	Meat and dairy products come usually from animals fed GM foods	Meat and Dairy Products
6.	Papaya	Virus resistance	Fruit
7.	Rapeseed	Pesticide resistance and improved to be free of toxic erucic and glucosinolates. In Canada "double zero" rapeseed was renamed Canola (Canadian oil) to differentiate it from non-edible rapeseed.	Canola oil,
8.	Rice	Genetically modified to contain high amounts of vitamin A	Rice
9.	Soybean	Herbicide resistance	Beverages, Tofu, Oil, Flour
10.	Sugar cane	Pesticide resistance	Sugar
11.	Tomato	Polygalacturonase is suppressed retarding fruit softening after harvesting	Tomato

TABLE 41.2   Antioxidant Contents of Some Common Foods mmol/100 g)

S.No.	Food	Amount	S.No.	Food	Amount
	Legumes, grains and nuts			**Fruits and vegetables**	
1.	Barley	1.0	17.	Amla (Indian gooseberry dried)	261.5
2.	Bread (whole meal)	0.5	18.	Apple	0.4
3.	Bread (whole meal flour)	1.4	19.	Apricot (dried)	3.5
4.	Maize (whole flour)	0.8	20.	Chilli (red and green)	2.4
5.	Millet	1.3	21.	Dates (dried)	1.7
6.	Peanut (roasted)	2.0	22.	Mango (dried)	1.7
7.	Pistachio	1.7	23.	Okra	4.2
8.	Sunflower seed	6.4	24.	Oranges	0.6
9.	Walnut	21.9	25.	Plums (dried)	3.2
	Spices			**Beverages**	
10.	Cinnamon (dried, ground)	77.0	26.	Black tea (prepared)	1.0
11.	Clove (dried, whole ground)	277.3	27.	Coffee (prepared, filtered and boiled)	2.5
12.	Ginger (dried)	20.3	28.	Grape juice	1.2
13.	Mint leaves (dried)	116.9	29.	Green tea (prepared)	1.5
14.	Nutmeg (dried, ground)	26.4	30.	Orange juice	0.6
15.	Saffron (dried, ground)	44.5	31.	Tomato juice	0.5
16.	Thyme (dried, ground)	56.3			

VI. MEASUREMENT OF CALORIC VALUES

The body requires energy for its internal and external work. This is provided by the oxidation, in the tissues, of the three proximate principles of food, e.g., carbohydrates, fats, and proteins. The foodstuffs contain varying amounts of carbohydrates, fats, and proteins and, therefore, the energy obtained from different foods must vary.

The energy value is most conveniently measured in calories. One calorie denotes the amount of heat required to raise the temperature of one gram of water by 1°C. This quantity of heat is very small, hence, in nutrition it is customary to use the large calorie written as capital 'C', which is 1,000 time the small calorie. In other words it is a kilocalorie.

The caloric value of foods depends on the amount of carbohydrates, fats, and proteins in them. This can be determined by two methods—direct and indirect.

A. Direct Method

The caloric value of a foodstuff can be determined by measuring the heat produced when a given amount is completely burnt in oxygen. It is done in a *bomb calorimeter* where the oxygen is put in under considerable pressure. Since it requires a calorimeter of robust construction, it has been called a bomb calorimeter.

The one commonly used for the purpose is the Atwater bomb calorimeter. It consists of heavy steel bomb A, with a platinum or gold plated copper lining. It has a cover which is held tightly by a strong screw-collar. A weighed amount of the sample is placed in B. The bomb is charged with an oxygen valve. The valve is then closed and the bomb is immersed in a weighed amount of water W. The burning of the sample is set off by an electric spark and the heat liberated is measured by the rise in temperature of the surrounding water by means of a differential thermometer T which can read up to one-thousandth of a degree. Deduction of the heat arising out of accessory combustions is made in order to obtain the heat liberated in calories from the combustion of the actual sample.

B. Indirect Method

The caloric value can also be determined indirectly by burning the food in oxygen in an oxycalorimeter. The volume of oxygen required to burn the food sample is measured and the caloric value can then be calculated by means of factors previously determined in the bomb calorimeter.

The energy obtained as a result of complete combustion is the potential energy but the energy liberated in the body is not the same, and this is called the physiological energy. Since carbohydrates and fats contain carbon, hydrogen, and oxygen, they can be completely burned to CO_2 and water and hence the potential energy is the same as the physiological energy. However, in the case of the proteins, the nitrogen is eliminated as urea, etc., so the physiological energy is less than the potential energy.

The average heat released from foodstuffs in the bomb calorimeter is:

Carbohydrates	4.1	Calories per gram
Fats	9.45	Calories per gram
Proteins	5.65	Calories per gram

But when these foodstuffs burn in the body, the heat released is:

Carbohydrates	4.1	Calories per gram
Fats	9.45	Calories per gram
Proteins	4.35	Calories per gram

Since there is some loss during digestion (approximately—carbohydrates 2%, fats 5%, and proteins 8%) the subsequent physiological values, are:

Carbohydrates	4	Calories per gram
Fats	9	Calories per gram
Proteins	4	Calories per gram

Naturally, therefore, foods will vary in their caloric values. Some foods can be classed as *high caloric* value ones, and some as *low caloric* value ones. Those which are rich in fat or low in water content, are high caloric value foods, e.g., all fatty foods and those with low water content such as cheese, dried fruits, dried legumes, etc. Fresh fruits and green leafy vegetables are low caloric foods because of their high water and cellulose content. Cereals and starchy vegetables are intermediate in caloric value.

VII. COMPOSITION AND NUTRITIVE VALUE OF INDIAN FOODS

The information of the composition and the nutritive value of foods presented in the following tables is collected from the work done in Indian laboratories. A common serial number is assigned to a food item in Tables 41.4 and 41.5. The absence of a particular serial number and values indicates that data for these items is not available.

English names of the foods are given in these tables. For scienific names and names in other Indian languages the Special Report Series No. 42 on "The nutritive value of Indian foods and the planning of satisfactory diets", published by the Indian Council of Medical Research, New Delhi, may be referred to.

In order to enable the conversion of the figures to Indian measures or to the avoirdupois system of weights, the following relationship may be used:

1 seer (weight)	= 2 lb	= 907.2 g
1 lb (Avoirdupois)	= 8 chhataks	= 453.6 g
2.2 lb	= 1 kg	= 1000.0 g
1 chhatak	= 2 oz	= 56.8 g
3.5 oz (Avoirdupois)	= 8.62 tolas	= 100.0 g

TABLE 41.3   Essential Fatty Acid Content of Certain Common Fats and Oils

S.No.	Fat	Poly-unsaturated fatty acids %	
		Linoleic acid	Linolenic acid
1.	Coconut	2.6	–
2.	Cotton seed	30.4	–
3.	Egg yolk	21.7	2.9
4.	Flax	58.0	14.0
5.	Groundnut	27.4	–
6.	Linseed	46.7	60.9
7.	Margarine	4.8	–
8.	Milk fat (buffalo)	2.8	–
9.	Milk fat (cow)	3.8	–
10.	Milk fat (goat)	1.5	–
11.	Mustard	13.0	8.6
12.	Niger	53.5	–
13.	Olive	15.5	–
14.	Pig depot fat	15.6	–
15.	Rapeseed	29.0	–
16.	Safflower	78.0	–
17.	Sesame	40.4	–
18.	Sheep depot fat	5.0	–
19.	Soyabean	50.0	7.0
20.	Vanaspati (hydrogenated groundnut oil)	4.9	–

ψ Omega-6 fatty acid; *Omega-3 fatty acid

TABLE 41.4   Proximate Principles, Energy, Minerals and Vitamins (Per 100 g of Edible Portion)

Sl. No.	Name of foodstuff	Edible Portion %	Protein	Fat	Carbohydrates	Minerals Total	Calcium mg	Magnesium mg	Total Iron mg	Total Phosphorus mg	Vitamin A I.U.	Thiamine mg	Riboflavin mg	Nicotinic acid mg	Vitamin C mg
	Cereals and Grain Products														
1.	Bajra	84	11.6	5.0	67.5	2.3	42	125	14.3	269	220	0.33	0.16	3.2	0
2.	Barley	100	11.5	1.3	69.6	1.2	26	127	3.0	215	79	0.37	0.28	1.8	0
3.	Maize (dry)	100	11.1	3.6	66.2	1.5	10	144	2.0	348	1502	0.42	0.10	1.4	0
4.	Rice (raw, milled)	100	6.8	0.5	78.2	0.6	10	48	3.1	160	0	0.09	0.03	1.9	0
5.	Wheat (whole)	100	12.1	1.7	71.2	2.7	41	139	4.9	306	108	0.45	0.12	5.0	0
6.	Wheat flour (whole)	100	11.8	1.5	69.4	1.5	48	55	11.5	423	49	0.49	0.29	4.3	0
7.	Wheat flour (refined)	100	11.0	0.9	73.9	0.6	23	42	2.5	121	43	0.12	0.07	0.9	0
8.	Wheat (suji)	–	10.4	0.8	74.8	16	–	–	1.6	102	–	–	76	2.4	–
	Pulses and Legumes														
9.	Bengal Gram (whole)	100	17.1	5.3	60.9	3.0	202	168	10.2	312	316	0.30	0.51	2.1	3
10.	Green Gram (whole)	100	24.0	1.3	56.7	3.5	124	171	7.3	326	158	0.47	0.39	2.1	1
11.	Green Gram (dhal)	100	24.5	1.2	59.9	3.5	75	189	8.5	405	83	0.72	0.15	2.4	0
12.	Kesari Dhal	100	28.2	0.6	56.6	2.3	90	92	6.1	317	200	0.39	0.41	2.2	0
13.	Lentil	100	25.1	0.7	59.0	2.1	69	94	4.8	293	450	0.45	0.49	1.5	0
14.	Peas (dry)	100	19.7	1.1	56.5	2.2	75	124	5.1	298	66	0.47	0.38	1.9	0
15.	"Rajmah"	–	22.9	1.3	60.6	3.2	260	–	5.8	410	–	–	–	–	–
16.	Soya Bean	–	43.2	19.5	20.9	4.6	240	–	11.5	690	710	0.73	0.76	2.4	–
	Vegetables														
17.	Bathua leaves	–	3.7	0.4	2.9	2.6	150	–	6	45	–	–	–	–	–
18.	Cabbage	88	1.8	0.1	4.6	0.6	39	10	0.8	44	2000	0.06	0.03	0.4	124
19.	Cauliflower Greens	–	5.9	1.3	7.6	3.2	626	–	40.0	107	–	–	–	–	–
20.	Coriander leaves (tender)	70	3.3	0.6	7.5	2.3	184	64	18.5	62	11530	0.05	0.06	0.8	135
21.	Lettuce (leaves)	66	2.1	0.3	2.5	1.2	50	30	2.4	28	1650	0.09	0.13	0.05	10
22.	Mint (tender leaves)	45	4.8	0.6	5.8	1.9	200	–	15.6	62	–	–	–	–	–

(Contd.)

TABLE 41.4   **Proximate Principles, Energy, Minerals and Vitamins (Per 100 g of Edible Portion)** *(Contd.)*

Sl. No.	Name of foodstuff	Edible Portion %	Protein	Fat	Carbohydrates	Minerals Total	Calcium mg	Magnesium mg	Total Iron mg	Total Phosphorus mg	Vitamin A I.U.	Thiamine mg	Riboflavin mg	Nicotinic acid mg	Vitamin C mg
23.	Mustard leaves	–	4.0	0.6	3.2	1.6	155	–	16.3	26	4370	0.03	–	–	33
24.	Onion tops	–	1.2	0.8	5.3	1.0	78	–	–	14	–	–	–	–	–
25.	Spinach leaves	87	2.0	0.7	2.9	1.7	73	84	10.9	21	9300	0.03	0.07	0.5	28
26.	Beetroot	85	1.7	0.1	8.8	0.8	200	9	1.0	55	0	0.04	0.09	0.4	88
27.	Carrot	95	0.9	0.2	10.6	1.1	80	14	2.2	30	3150	0.04	0.02	0.6	3
28.	Onion	–	1.8	0.1	12.6	0.6	40	–	1.2	60	–	–	–	–	–
29.	Potato	100	1.6	0.1	22.6	0.6	10	20	0.7	40	40	0.10	0.01	1.2	17
30.	Radish (white)	99	0.7	0.1	3.4	0.6	50	–	0.4	22	5	0.06	0.02	0.5	15
31.	Sweet potato (shakarkand)	–	1.2	0.3	28.2	1.0	20	–	0.08	50	10	0.08	0.04	0.7	24
32.	Tapioca (Simla Alu)	–	0.7	0.2	38.7	1.0	50	–	0.9	40	–	0.05	0.10	0.3	25
33.	Turnip	65	0.5	0.2	6.2	0.6	30	–	0.4	40	0	0.04	0.04	0.5	43
34.	Bittergourd	97	1.6	0.2	4.2	0.8	20	17	1.8	70	210	0.07	0.09	0.5	88
35.	Brinjal	91	1.4	0.3	4.0	0.3	18	16	0.9	47	124	0.04	0.11	0.9	12
36.	Cauliflower	70	2.6	0.4	4.0	1.0	33	20	1.5	57	51	0.04	0.10	1.0	56
37.	Cucumber	83	0.4	0.1	2.5	0.3	10	11	1.5	25	0	0.03	0.01	0.2	7
38.	French Beans	94	1.7	0.1	4.5	0.5	50	29	1.7	28	221	0.08	0.06	0.3	14
39.	Giant chillies	97	1.3	0.3	4.3	0.7	16	–	1.5	32	596	0.29	0.05	0.1	80
40.	Ladies' fingers (Okra)	84	1.9	0.2	6.4	0.7	66	43	1.5	56	88	0.07	0.10	0.6	13
41.	Mango (green without seed)	72	0.7	0.1	10.1	0.4	10	21	5.4	19	150	0.04	0.01	0.2	3
42.	Mogra (green)	–	1.6	0.4	4.5	0.6	98	–	2.5	34	1090	0.04	0.03	0.4	74
43.	Mushroom (paddy straw)	88	3.8	0.7	5.1	1.4	100	–	1.5	100	0	0.14	0.61	2.4	12
44.	Peas	53	7.2	0.1	15.8	0.8	20	34	1.5	139	139	0.25	0.01	0.8	9
45.	Pumpkin	79	1.4	0.1	4.6	0.6	10	14	0.7	30	84	0.06	0.04	0.5	2
46.	Radish	100	0.5	0.1	3.3	0.7	20	–	1.0	20	8	0.02	0.03	1.4	21
47.	Tinda	99	1.4	0.2	3.4	0.5	25	14	0.9	24	23	0.04	0.08	0.3	18
	Nuts and Oilseeds														
48.	Almond	–	20.8	58.9	10.5	2.9	230	–	4.5	2.4	0	0.24	0.15	2.5	0
49.	Cashewnut	–	21.2	46.9	22.3	2.4	50	–	5.0	2.0	100	0.63	0.19	2.1	0
50.	Coconut (dry)	–	6.8	62.3	18.4	1.6	400	–	2.7	–	0	0.08	0.06	0.6	7

(Contd.)

TABLE 41.4   Proximate Principles, Energy, Minerals and Vitamins (Per 100 g of Edible Portion) (*Contd.*)

Sl. No.	Name of foodstuff	Edible Portion %	Protein	Fat	Carbohydrates	Minerals Total	Calcium mg	Magnesium mg	Total Iron mg	Total Phosphorus mg	Vitamin A I.U.	Thiamine mg	Riboflavin mg	Nicotinic acid mg	Vitamin C mg
51.	Gingelly seeds	100	18.3	43.3	25.0	5.2	1450	–	10.5	570	100	1.01	0.06	4.4	0
52.	Groundnut	–	26.7	40.1	20.3	1.9	50	–	1.6	390	63	0.90	0.30	14.1	0
53.	Linseeds	99	20.3	37.1	28.8	2.4	170	–	2.7	370	50	0.23	0.07	0.5	0
54.	Mustard seeds	–	22.0	39.7	23.8	4.2	490	–	17.9	700	270	0.65	0.07	4.0	0
55.	Pistachio nut	–	19.8	53.5	16.2	2.8	140	–	13.7	430	240	0.67	0.03	1.4	0
56.	Sunflower seeds (kernel only)	52	19.8	52.1	17.9	3.7	280	–	5.0	670	0	0.86	0.03	5.0	1
57.	Walnut	45	15.6	64.5	11.0	1.8	100	–	4.8	380	10	0.45	0.05	1.6	0
Condiments and Spices															
58.	Chillies (green)	90	2.9	0.6	3.0	1.0	30	24	1.2	80	292	0.19	0.39	0.9	111
59.	Cardamom	–	10.2	2.2	42.1	5.4	130	–	5.0	160	0	0.22	0.17	0.8	0
60.	Cloves (dry)	100	5.2	8.9	46.0	5.2	740	–	4.9	100	422	0.08	0.13	0	0
61.	Coriander	–	14.1	16.1	21.6	4.4	630	–	17.9	393	1570	0.22	0.35	1.1	0
62.	Cumin seeds	–	18.7	15.0	36.6	5.8	1080	–	31.0	511	870	0.55	0.36	2.6	3
63.	Garlic (dry)	85	6.3	0.1	29.0	1.0	30	–	1.3	310	0	0.16	0.23	0.4	13
64.	Ginger (fresh)	–	2.3	0.9	12.3	1.2	20	–	2.6	60	67	0.06	0.03	0.6	6
65.	Pepper (dry)	–	11.5	6.8	49.5	4.4	460	–	16.8	198	1800	0.09	0.14	1.4	–
66.	Turmeric	–	6.3	5.1	69.4	3.5	150	–	18.6	282	50	0.03	0	2.3	0
Fruits and Fruit Products															
67.	Apple	90	0.3	0.1	13.3	0.3	0.9	7	1.0	20	0	0.12	0.03	0.2	2
68.	Apricots (fresh)	86	1.0	0.3	11.6	0.7	20	–	2.2	25	3600	0.04	0.13	0.6	6
69.	Bael fruit	64	1.8	0.3	31.8	1.7	85	–	0.6	50	93	0.13	1.19	1.11	8
70.	Banana (various)	64–83	0.7–1.5	0.1–0.8	18–37	0.7–1.0	10–26	–	0.3–2.1	50	100	0.02	0.02	0.3	6
71.	Ber	–	2.0	0.3	17.6	1.3	4	–	1.8	9	35	0.02	0.05	0.7	76
72.	Cherries (red)	88	1.1	0.5	13.8	0.8	24	–	1.3	25	0	0.08	0.08	0.03	7
73.	Dates, Indian	–	1.2	0.4	33.8	1.7	22	–	–	38	–	–	–	–	–
74.	Figs	99	1.3	0.2	7.6	0.6	60	–	1.2	30	270	0.06	0.05	0.6	5
75.	Grapes (pale green)	–	0.5	0.3	16.5	0.6	20	–	1.5	30	–	–	–	–	–
76.	Guava (Country)	100	0.9	0.3	11.2	0.7	10	8	1.4	28	0.03	0.03	0.3	0.3	212

(*Contd.*)

TABLE 41.4 Proximate Principles, Energy, Minerals and Vitamins (Per 100 g of Edible Portion) (*Contd.*)

Sl. No.	Name of foodstuff	Edible Portion %	Protein	Fat	Carbohydrates	Minerals Total	Calcium mg	Magnesium mg	Total Iron mg	Total Phosphorus mg	Vitamin A I.U.	Thiamine mg	Riboflavin mg	Nicotinic acid mg	Vitamin C mg
77.	Indian prune (ripe and fresh)	–	0.5	0.3	12.8	0.6	10	–	–	18	–	–	–	–	–
78.	Jackfruit	30	1.9	0.1	19.8	0.9	20	27	0.05	41	292	0.03	0.13	0.4	7
79.	Jamun (big)	–	0.4	0.2	10.9	0.3	8	–	–	3	–	–	–	–	–
80.	Lemon (sour)	97	0.6	0.7	8.2	0.8	100	–	2.4	20	0	–	–	0.1	26
81.	Lemon (sweet)	79	0.7	0.3	7.3	0.5	30	–	0.7	20	0	–	–	0	45
82.	Lichies (pulp)	68	1.1	0.2	13.6	0.5	10	10	6.7	35	0	0.02	0.06	0.4	31
83.	Loquat	76	0.6	0.3	9.6	0.5	30	–	1.3	20	933	–	–	0	0
84.	Malta (sweet lime)	67	0.7	0.2	7.8	0.4	30	–	1.0	20	0	–	–	–	0.54
85.	Mango ripe (various)	58–79	0.3–1.1	0.1–0.8	12–23	0.3–0.7	10–42	22–27	0.3–3.2	30	2600	0.13	0.2	6.0	80
86.	Melon, water	78	0.2	0.2	3.3	0.3	11	13	7.9	12	0	0.02	0.04	0.1	1
87.	Melon, white	80	0.6	0.1	5.4	0.6	65	31	1.3	20	450	0.11	0.08	0.5	32
88.	Orange (Nagpur)	70	0.6	0.2	8.9	0.3	20	–	0.5	20	0.15	–	–	–	–
89.	Papaya, ripe	75	0.6	0.1	7.2	0.5	17	11	0.5	13	1100	0.04	0.25	0.2	57
90.	Peaches	88	1.2	0.3	10.5	0.8	15	21	2.4	41	0	0.02	0.03	0.5	6
91.	Pears (Kashmiri)	–	0.2	0.3	14.9	0.4	20	–	1.0	20	–	–	–	–	–
92.	Phalsa	61	1.3	0.9	14.7	1.1	129	72	3.1	39	699	–	–	0.3	22
93.	Pineapple	60	0.4	0.1	10.8	0.4	20	20	1.2	9	30	0.20	0.12	0.1	39
94.	Plums (various)	89–95	0.5–0.7	0.2–1.0	9–12	0.3–0.7	10–20	–	0.3–1.7	10–25	150–1000	0.03–0.12	0.03–0.18	0.1–0.6	0–13
95.	Pomegranate (plain)	68	1.6	0.1	14.6	0.7	10	12	0.3	70	0	–	–	0	10
96.	Pomeloe	–	0.6	0.1	10.2	0.5	30	–	0.3	30	200	0.03	0.03	0.2	20
97.	Raisins (preserved)	–	2.0	0.2	76.4	2.0	100	20	4.0	80	0	0.06	0.10	0.5	0
98.	Rasberry	–	1.1	0.6	11.6	0.9	10	–	2.3	110	2080	–	–	0.8	30
99.	Sapota, Cheekoo (various)	82–83	0.5–1.1	0.6–1.2	19–26	0.4–0.5	20–30	–	0.1–6.7	10–72	0–500	0.01–0.02	0.01–0.04	0.1–0.8	6–10

(*Contd.*)

TABLE 41.4   Proximate Principles, Energy, Minerals and Vitamins (Per 100 g of Edible Portion) (*Contd.*)

Sl. No.	Name of foodstuff	Edible Portion %	Protein	Fat	Carbohydrates	Minerals Total	Calcium mg	Magnesium mg	Total Iron mg	Total Phosphorus mg	Vitamin A I.U.	Thiamine mg	Riboflavin mg	Nicotinic acid mg	Vitamin C mg
100.	Sharifa (Punjab)	–	1.7	0.6	23.6	1.1	10	–	2.9	60	0	–	–	0.08	58
101.	Strawberry	96	0.7	0.2	9.8	0.4	30	–	1.8	30	30	0.03	0.02	0.2	52
102.	Tomato ripe	100	0.9	0.2	3.6	0.5	48	12	0.4	20	585	0.12	0.06	0.4	27
	Fishes, Eggs, Meat and their Products														
103.	Fishes (various)	43–100	9–25	1–20	0–9	1–5	100–1600	–	1–115	175–740	0–20	0.05–0.20	0.01–0.55	0.4–4.8	3–32
104.	Beef Muscle	–	22.6	2.6	–	1.0	10	–	0.8	190	60	0.15	0.04	6.4	2
105.	Duck	–	21.6	4.8	–	1.2	4	–	–	235	–	–	–	–	–
106.	Egg (hen)	–	13.3	13.3	–	1.0	60	–	2.1	220	1200	0.10	0.18	0.1	0
107.	Fowl	–	25.9	0.6	–	1.3	25	–	–	245	–	–	–	–	–
108.	Goat meat muscle	–	21.4	3.6	–	1.1	12	–	–	193	–	–	–	–	–
109.	Grey quail	–	21.9	1.7	–	1.4	22	–	–	282	–	–	–	–	–
110.	Liver (goat)	–	20.0	3.0	2.3	1.3	17	–	–	279	–	–	–	–	–
111.	Mutton (muscle)	–	18.5	13.3	–	1.3	150	–	2.5	150	31	0.18	0.27	6.8	–
112.	Pigeon	–	23.3	4.9	–	1.4	12	–	1.8	290	–	0.10	0.20	5.6	–
113.	Pork (muscle)	–	18.7	4.4	–	1.0	30	–	2.3	200	0	0.54	0.09	2.8	2
	Fat and Edible Oils														
114.	Butter	100	–	81.0	–	2.5	–	–	–	–	2500	–	–	–	–
115.	Ghee (buffalo)	100	–	100.0	–	–	–	–	–	–	2000	–	–	–	–
116.	Vegetable cooking oil	100	–	100.0	–	–	–	–	–	–	–	–	–	–	–
117.	Vanaspati	100	–	100.0	–	–	–	–	–	–	2500	–	–	–	–
	Milk and Milk Products														
118.	Milk (cow)	100	3.2	4.1	4.4	0.8	149	–	2.3	96	150	0.05	0.18	0.1	2
119.	Milk (buffalo)	100	4.3	8.8	5.1	–	210	–	0.2	130	160	0.04	0.10	0.1	3
120.	Milk (goat)	100	3.3	4.5	4.4	0.8	130	–	–	–	60	0.05	0.12	0.2	2
121.	Milk (human)	100	1.1	3.4	7.5	0.1	34	–	–	12	70	0.02	0.02	–	3
122.	Curds	100	3.1	4.0	2.9	0.8	149	–	0.3	93	102	0.05	0.16	0.1	1
123.	Skimmed milk (liquid)	100	2.5	0.1	4.6	0.7	120	–	0.2	90	–	–	–	0.1	1
124.	Channa (cow)	100	18.3	20.8	2.3	2.6	208	–	–	138	366	0.07	0.02	–	3

(*Contd.*)

TABLE 41.4 Proximate Principles, Energy, Minerals and Vitamins (Per 100 g of Edible Portion) (*Contd.*)

Sl. No.	Name of foodstuff	Edible Portion %	Protein	Fat	Carbohydrates	Minerals Total	Calcium mg	Magnesium mg	Total Iron mg	Total Phosphorus mg	Vitamin A I.U.	Thiamine mg	Riboflavin mg	Nicotinic acid mg	Vitamin C mg
125.	Cheese	100	14.6	31.2	20.5	3.1	650	–	5.8	420	273	–	–	–	–
126.	Khoa (buffalo)	100	24.1	25.1	6.3	4.2	790	–	2.1	520	497	0.24	0.41	0.4	–
127.	Skimmed milk powder (cow)	100	38.0	0.1	51.0	6.8	1370	–	1.4	1000	0	0.45	1.64	1.0	5
128.	Whole milk powder (cow)	100	25.8	26.7	38.0	6.0	950	–	0.6	730	1400	0.31	1.36	0.8	4
	Miscellaneous Foodstuffs														
129.	Arecanut	–	4.9	4.4	47.2	1.0	50	–	1.5	130	–	–	–	–	–
130.	Betel leaves	–	3.1	0.8	6.1	2.3	230	–	7.0	40	9600	0.07	0.03	0.7	5
131.	Cane sugar (commercial)	–	0.1	–	99.4	0.1	12	–	–	1	–	–	–	–	–
132.	Coconut water (5 month old nut)	100	0.1	0.1	5.9	<0.1	30	–	0.2	<10	0.01	0.01	0.1	2	–
133.	Coconut kernet	100	4.5	41.6	13.0	1.0	10	–	1.7	240	–	–	–	–	–
134.	Honey	–	0.3	0	79.5	0.2	5	–	0.9	16	0	0	0.04	0.2	4
135.	Jaggery	–	0.4	0.1	95.0	0.6	80	–	11.1	140	280	0.02	–	1.0	0
136.	Neera	–	0.4	–	10.9	0.5	Trace	–	<0.1	140	–	–	–	–	–
137.	Pappad	–	18.8	0.3	52.4	8.2	80	–	17.2	300	–	–	–	–	–
138.	Sugarcane juice	–	0.1	0.2	9.1	0.4	10	–	1.1	10	10	–	0.04	–	–
139.	Toddy	–	0.1	<0.1	3.5	0.2	40	–	1.3	10	–	–	–	–	–
140.	Yeast (dried)	–	35.7	1.8	46.3	8.4	160	–	21.5	2090	–	3.20	–	27.0	–

Note: The folic acid content in micrograms per 100g of edible portion for some food items is as follows: Bengal gram—125; Green gram (whole)—145; Kesari Dhal—100; Lentil—107; Peas (dry)—51; Mango—(green without seed)—7; Jackfruit—3; Lichie (pulp)—9.

TABLE 41.5   Nutritive Value of the Protein in Various Foods

Sl. No.	Name of the foodstuff	Biological efficiency value*	Protein efficiency ratio**	Sl. No.	Name of the foodstuffs	Biological value*	Protein efficiency ratio**
	Cereals and Products				**Nuts and Oilseeds**		
1.	Bajra	83	1.1	22.	Almond	51	–
2.	Maize	60	1.0	23.	Cashewnut	72	–
3.	Rice (raw, milled)	80	1.7	24.	Coconut	77	1.0
4.	Wheat (whole)	66	1.3	25.	Gingelly	67	1.0
				26.	Groundnut, raw	57	–
	Pulses and Legumes				**Fish**		
5.	Bengal Gram dhal (raw)	61	1.1	27.	Katla	78	1.8
6.	Black Gram	63	1.0	28.	Mrigal	72	1.7
7.	Green Gram	54	0.8	29.	Prawn	67	–
8.	Khesari dhal	47	0.3	30.	Shark	62	–
9.	Lentil	49	0.5	31.	Singhi	89	1.5
10.	Peas	59	1.1	32.	Sundried Hilsa	74	1.7
11.	Soyabean	60	0.9				
	Leafy Vegetables				**Meat and Eggs**		
12.	Amaranth	72	–	33.	Beef muscle	69	–
13.	Bathua leaves	79	–	34.	Egg, hen	95	3.8
14.	Chawli	73	–	35.	Egg, yolk, hen's	–	3.2
15.	Mustard leaves	84	–	36.	Egg white, duck's	61	2.4
16.	Radish tops	77	–	37.	Mutton	60	–
17.	Spinach	87	–				
	Roots and Tubers				**Milk and Milk Products**		
18.	Potato	68	1.9	38.	Cow's milk	82	2.0
19.	Sweet potato	72	1.5	39.	Buffalo milk	67	2.0
				40.	Curds	67	2.1
				41.	Goat's milk	68	1.0
				42.	Khoa (cow)	69	2.3
	Other Vegetables				**Miscellaneous Foods**		
20.	Brinjal	71	–	43.	Yeast (food)	32	1.7
21.	Ladies fingers (Okra)	82	–				

*Biological value is a percentage of the absorbed nitrogen retained when a particular food is the only source of nitrogen.
**Protein efficiency ratio is defined as the gain in the weight in gram per gram of protein consumed.

SUGGESTIONS FOR FURTHER READING

Carlsen, et al. (2010), The Total Antioxidant Content of More than 3100 Foods, Beverages, Spices, Herbs and Supplements Used Worldwide, *Nutrition Jour.*, 9, 3.

Ekici K. and Sancak Y.C. (2011), A Perspective on Genetically Modified Food Crops, *African Jour. Agri. Res.*, 6, 1639–1642.

Suzie K.J., Ma K.C. and Drake, M.W. (2008), Genetically Modified Plants and Human Health, *Jour. Royal Soc. Med.*, 101, 290–298.

42

Milk and Milk Products

A.K. Srivastava, (Late) N.C. Ganguli, Sunita Grover, Virender K. Batish and A.K. Singh

CONTENTS

I. INTRODUCTION

Milk is the sole food of the off springs on their birth. It has been immaculately tailored to meet the nutritional requirements of the baby. The nutrients in milk are so well designed and structured that no other food can substitute for it. The biospecificity of the mammary gland is chiefly attributed to the nutritional excellence of the milk. It is defined as "a white or yellowish fluid consisting of small fat globules suspended in water and secreted by the mammary glands to fulfil the nutritional requirements of the newborn in terms of energy, essential amino-acids, fatty acids, vitamins, minerals, carbohydrates etc." Milk and milk products make significant contribution towards providing nutrients to adolescents and adults. Milk is considered as Mother Nature's perfect food. Apart from providing valuable nutrients, milk also possesses wide array of life promoting components. It consists of about 10,000 compounds in varied concentrations. With the advent of nutraceuticals and functional foods. the difference between diet and medicine is slowly diminishing. Researchers across the world are looking for biomolecules derived from natural resources which also possess disease preventing and combating ability.

The convergence of major consumer food and nutrition trends with significant scientific breakthroughs in biotechnology and medical sciences will have a profound impact on the therapeutics in 21st century. Consumers are beginning to understand the powerful influence of naturally occurring components particularly in diet on health and well being. Increasing scientific evidence confirms that specific components in diet might tend off certain chronic diseases. It has revitalized the interests not only in consumers, but also among researchers and processors to isolate bioactive components and utilize them to formulate products, which are "Natural," "Therapeutic" and "Nutritional." as well. In this regard, an exceedingly high value is placed on milk components and since these molecules are safe and do not cause any toxicity, these may become the ultimate choice for developing novel food formulations. Ingredient suppliers are also keeping a close eye on trends impacting health conscious consumer's purchase power along with their liking and preference for novel formulations. Marketing ingredients for pharmaceuticals require equal emphasis on its physiological functionality apart from other advantage associated with it. Dairy milk ingredients including non fat dry milk, whey proteins and milk minerals, a variety of new product formulations meant for specialized group of peoples by adding nutritional value, providing functional benefits and improving sensory acceptability.

New products meant for specialized group of people utilizing the advantage of dairy ingredients include popular sports drinks, meal replacement beverages, nutritional bars, infant formula. Consumers have a natural fondness for the taste of dairy products and consider dairy ingredients as good stuff. Milk comprises proteins, fats, lactose, minerals and vitamins. Of these, the first three are the most dominant members whereas the other two fall under a minority group. Minor milk constituents like whey proteins; peptides, lactose and lactose derivatives have been thoroughly investigated by researchers for their nutritional and therapeutic characteristics and are promising ingredients for pharma as well as food industry. The components of milk are discussed in the following section.

II. COMPOSITION

A. Milk

General composition

A closer look at the composition of milk from three different species, namely, cow, buffalo, and human, shows that human milk is quite different from cow milk in composition and even more so from buffalo milk (Table 42.1 and Figure 42.1). Being a bio-fluid, its control in chemical composition is at its site of synthesis. At parturition, the mammary gland starts synthesis and secretion of milk. In the first few days, milk varies significantly in composition from the ultimate stable fluid milk. It is distinguished by the name *colostrum*. Colostrum contains more solids than the regular milk. It is richer in protein and fat, but not in lactose. Colostrum has bioactive components, i.e., growth factors and antibodies called immunoglobulins such as IgA, IgG, IgM that prevent newborn from infections/diseases. The antibodies in colostrum provide passive immunity to the newborn, while growth factors stimulate the development of the gut. Several types of colostrum preparations are available in market, e.g. Lactobin[R]N from New Zealand which contains cytokines, growth factors, lactioferrin and immunoglobulins. Generally, colostrum is available in dried form such as Naturade symbiotic colostrums plus, APS Biogroup colostrums products (whole colostrum, skim colostrum, immulox prodcuts). Dr Reddy's lab at Hyderabad has launched New Life Colostrum in collaboration with Symbiotics of New Zealand in 2005. The product is available in a novel chewable form. New Life Colostrum prevents infections among children and is a useful supplement for children suffering from recurrent infections.

Factors controlling composition of milk

There are several factors that control the composition of milk since fresh milk drawn from the udder varies in composition.

(a) *Species and breed variation:* A major determinant of milk composition is the animal species responsible for its synthesis. Buffalo milk is richer in all constituents, except carotene and water, than cow milk. An Indian cow gives more fat in milk than the foreign breeds. Sometimes, variation in composition among individual cows of one breed may be greater than that among breeds. Goat and sheep milk fats have low contents of butyric acid but high of caproic, caprylic and capric acids. While, human milk is a good reservoir of lactose, reindeer's milk is a repository of fat.

(b) *Stage of lactation:* The period of milk secretion, commonly known as lactation, is the most important physiological variable

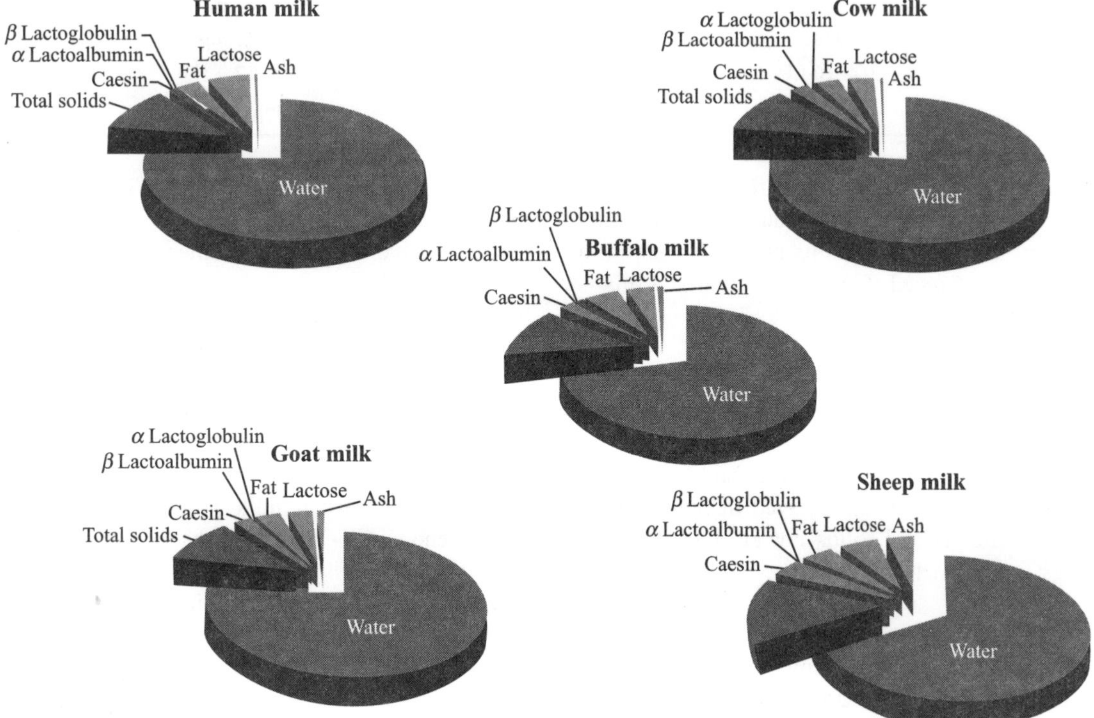

Figure 42.1 Average composition of human, cow, buffalo milk, goat and sheep milks. (*see Plate 16 for colour figure*)

TABLE 42.1   Average composition of Human, Cow, Buffalo Milk, Goat and Sheep Milks

Composition	Human Milk %	Cow's Milk %	Buffalo's Milk %	Goat's Milk %	Sheep's Milk %
Water	87.6	87.2	83.6	87.0	81.0
Total Solids	12.4	12.8	16.4	12.3	19.3
Casein	0.4	2.7	2.8	2.5	4.6
α-Lactalbumin	0.4	0.4	0.5	–	–
β-Lactoglobulin	0.2	0.2	0.3	–	–
Total proteins	1.1	3.3	3.6	3.6	6.0
Fat	3.8	3.8	6.5	4.1	7.0
Lactose	7.0	4.8	5.1	4.1	4.8
Ash	0.21	0.71	0.71	0.82	0.96
Sodium	0.015	0.058	–	0.050	0.044
Potassium	0.055	0.138	–	0.204	0.138
Calcium	0.034	0.126	0.16	0.134	0.194
Magnesium	0.004	0.013	0.01	0.014	0.018
Iron, mg	0.03	0.05	0.12	0.05	0.10
Chlorine	0.043	0.100	0.06	0.15	0.27
Phosphorous	0.016	0.099	0.10	0.111	0.158
Sulphur	0.014	0.030	–	0.028	0.029
Calories/100 ml:	71	69	94	70	105
Vitamins/100 ml:					
Vitamin A	53 µg	34 µg	69 µg	57 µg	44 µg
Carotenoids	27 µg	38 µg	0	–	–
Thiamine	16 µg	42 µg	50 µg	48 µg	65 µg
Riboflavin	43 µg	157 µg	107 µg	138 µg	356 µg
Niacin	172 µg	85 µg	171 µg	277 µg	419 µg
Pyridoxine	11 µg	48 µg	325 µg	46 µg	60 µg
Folic acid	0.18 µg	0.23 µg	0.55 µg	0.2 ug	0.2 ug
B12	0.18 µg	0.56 µg	0.40 µg	0.07 µg	0.71 µg
Vitamin C	4.30 mg	1.80 mg	2.54 mg	1.31 mg	4.22 mg
Vitamin D	0.4–10 I.U.	0.3–4.0 I.U.	–	0.12 I.U.	0.18–0.88 I.U.
Vitamin E	–	184 µg	197 µg	70 µg	120 µg

and has three distinct phases, namely, early, mid and late lactation. Milk composition is more stable during mid lactation when the mammary factory maintains its production unit at a steady speed. Its function recedes in late lactation and this is reflected in milk composition.

(c) *Seasonal variation:*   Milk composition has been found to be fluctuating during winter, summer and rainy season. In winter, milk is normally richer in solids. This phenomenon has been linked with the intake of feeds and the fodder available during this season. Generally, it is observed in regions with temperate climate. Milk fat and solids not fat are highest in winter and lowest in summer.

(d) *Number of milkings:*   Milk synthesis in a mammary gland is a continuous process for lactation. However, once the milk cistern is full, the gland operates its safety valves to slow down the rates of synthesis. As a result, a compositional difference occurs in milk if a cow is milked once, twice or three times a day. The milk of the first stripping will show difference from the mid and last strippings. The former is poor in fat whereas the latter have a higher fat content.

(e) *Genetic factors:*   In the same species, there can be variations in the composition of milk. It is well established that milk from an indigenous breed of cow is richer in fat than that of an exotic breed. Milk proteins are likely to be more influenced in cross bred cows since their synthesis in the mammary gland is a gene-controlled phenomenon. Appearance or absence of certain milk protein variants in the milk of phenomenon. Appearance or absence of certain milk protein variants in the milk of the progeny of the same breed, is an inherited character. Even within the same breed, individual animals can exhibit variation in components of casein or β-lactoglobulin.

(f) *Lactogenic variations:*   The other important agent is the dairyman who can conveniently alter milk to suit his compositional formula more commonly with water. Fat can also be easily withdrawn from milk. Milk powder (milk solids) can be dispersed in water by mechanical processing. To make milk easily available at a cheaper price, it is defatted to different levels. Milk composition can therefore, be suitably changed to varying fat levels without altering the solids-non-fat (SNF). These types of milk come under the broad category of standardized milk. Suitable names have been coined for these types of milk like toned milk (with 3% fat and 8.5% SNF) and double toned milk (with 1.5% fat and 9.0% SNF).

(g) *Nutrition:* Animal feed/nutrition has a direct bearing on the fat composition in milk. Mostly compositional variations in milk is attributed to dietary manipulations which are related to changes in ratios of acetate, butyrate and propionate therein. Low protein diet causes a decrease in protein content and high protein diet causes non protein N content to increase in milk. Underfed cows may reduce milk fat and protein percentage. Forage and concentrate ration also affect fat due to lower production of acetate and butyrate by ruminal micro-organisms. Incorporation of dietary fat in feed may alter the profile of milk fatty acids. Milk fat differs from other sources of animal fat since it contains relatively more amount of short chain fatty acids than long chain fatty acids. Feed consisting of canola, sunflower or flaxseed has been found to reduce the concentrations of short (C4-C10) and medium chain (C16:0) fatty acids and increase the concentrations of C18:0 and C18:1 fatty acids.

(h) *Diseases:* Infection of udders also influences the composition of milk. Severe inflammation of udders leading to mastitis due to pathogenic bacteria can result into alteration in milk composition. There is increase in the number of somatic cells in milk which is generally used as an indicator test for the detection of mastitic milk. The concentration of proteins, fats, solids not fat, lactose, whey proteins also get reduced and sodium and chloride are increased in the mastitic milk.

Detailed Composition

The various components of milk are discussed in detail below:

(a) *The proteins:* The major milk proteins unique to milk are caseins and whey proteins (α–Lactalbumin and β–Lactoglobulin) which are synthesized by mammary gland. The other relatively less abundant proteins include milk enzymes, proteins involved in transporting nutrients and disease resistance, immunoglobulins, growth factors, etc.

Casein accounts for about 80% of the total proteins in milk and the value of others is given in Table 42.1.

The contribution by proteins towards the weight of milk is less than 4.0%. Although their quantity is low, these are next to egg proteins in their biological value. This is due to the presence of all the "essential" amino acids in these proteins.

(i) Casein: Casein accounts for approximately 80% of the total protein and is highly digestible and possess all the essential amino-acids. These are phosphoproteins with molecular weight more than 20 kDa and can be precipitated at pH 4.6 using acid or by rennet (chymosin). Casein, in milk, is dispersed as aggregates called micelles. The chemical composition of the micelles is such that these remain as floating entities in milk. The floating tendency of casein micelles along with their uniform distribution in milk is the reason why milk looks white. Like blood is red because of its haemoglobin, milk is white on account of its casein micelles. The micellar population is normally of stable nature and it can only be disturbed through drastic treatment, e.g., with acids, salts and by application of centrifugal force. If milk is centrifuged at 100,000 g, all casein micelles will settle and milk loses its whiteness. The casein proteins are composed of four major members knows as αs–casein, β-casein, γ-casein and k-casein. Distinct proportions of these members bridged together by calcium and phosphate linkage constitute the complex casein. Of these members, αs-casein and k-casein have the major responsibility for maintaining the casein micelle in a floating state. The k-casein is the 'body guard' of αs-casein and protects it from its contact with calcium, which would result in the formation of insoluble precipitates of αs-casein. The k-casein being resistant to calcium attack, acts as a cementing agent around αs-casein. Through this protective role of k-casein, all other members of the casein group remain dispersed in milk.

Under an electron microscope, the casein micelle is found to range between 50–250 milli-microns. It is a mixture of casein particles of myriad sizes in its native state. These particles normally occur in isolation in milk. However, certain abnormal situations produce an aggregation of the particles and such an agglomeration is an alarming phenomenon resulting in a partition between the solid and the liquid counterparts of the fluid. Such a process results in a clot.

(ii) Whey proteins: The other class of proteins which has importance next to casein, resides in the milk serum, more popularly called milk whey. The whey or milk serum proteins comprise of approximately 20% of the original proteins of milk ranging from 4 to 7 g/lt of which 3.7 g is β–Lactoglobulin, 0.6 g is α–Lactalbumin, 0.3 g is Bovine Serum Albumin, and 1.4 g is proteose–peptone fractions (Marshall, 1982). In addition, it contains other proteins such as lactoferrin, immunoglobulins, ceruloplasmin, and milk enzymes such as lysozyme, lipase, and xanthine oxidase, which are present in low concentrations. In fact, whey proteins exist in two classes, named major and minor proteins on the basis of their concentration. Two of the major categories known as α–lactalbumin and β–lactoglobulin, contribute 0.6% of 'milkspace' and account for about 18% of the milk proteins. One of them cannot tolerate a salt-free water medium and hence precipitates like a globulin. On this basis, the water soluble protein is named α–lactalbumin and insoluble protein as β–lactoglobulin. The former controls the operation of the lactose factory whereas the latter contributes to the cooked flavour of milk through its donation of reduced sulphur groups (–SH) during heat treatment of milk.

Under the minor proteins come γ–globulins, proteose-peptone, and enzymes. γ–globulin is a fraction which appears in milk from blood. Proteose-peptone is a heat-resistant fraction, unlike other enzymes which are heat-labile. Milk has several enzymes (about 20) having different roles. One of these, lactase synthetase, can synthesize lactose of milk. Another enzyme alkaline phosphatase has gained significance for its "coinactivation", along with pathogenic bacteria present in milk, on heat treatment. Hence, its activity in milk is used as a yard-stick of "pasteurization".

(b) *The fats:* The cream portion of milk is primarily due to the presence of fat. This family occupies about 4% of the total milkspace in cow milk. Milk fat comprises primarily of triglycerides; the mono and diglycerides occur in trace quantity. The fatty acids present in milk triglycerides are normally of 10 different types and occur in amounts of more than 1% of the total. These fatty acids differ in their chain length. The beauty of milk fat is that, unlike other body fats, it has selectively accommodated short chain fatty acids like butyric acid and caproic acid. These are responsible for contributing the typical fat flavour to milk. Fat globules remain as minute droplets, their location in milk depends very much on the temperature and physical treatment. Agitation like churning, heating and cooling will force these droplets to assemble together to form the creamy layer at the top. On the other hand, a physical process like homogenization will disperse the fats more finely and prevent cream formation.

Milk fat is enveloped in a membrane which is derived from the outer membrane of the lactating mammary cell during secretion. The fat globule membrane is responsible for maintaining its droplet structure. The droplets are minute in size, having a diameter of 3 to 4 microns. In addition to triglycerides, the fat droplets also contain phospholipids, cholesterol, vitamins A, D and E, and carotene (the pigment which imparts yellow colour to butter is present in cow milk and absent in buffalo milk).

(c) *The carbohydrates:* Milk contains a unique sugar, lactose. It is responsible for imparting sweetness to milk although it is one-fifth as sweet as sucrose. Lactose is a disaccharide composed of one molecule of glucose linked to galactose. The latter has a special nutritional importance for the developing brain.

Lactose serves two major nutritional functions: (a) supplies galactose as a structural unit to the growing infant, and (b) gets metabolically transformed to lactic acid in the intestine which is harmful to mischievous bacteria, and provides a salubrious environment for the growth of benevolent microbes.

(d) *The minority group:* Some of the minor constituents are of major significance. These are milk enzymes, vitamins and minerals. While the first one controls, the fate of other constituents in milk, the second and third group maintain the prestige of milk as a health giving tonic.

(i) Milk enzymes: Among the milk enzymes alkaline phosphatase, lipase and xanthine oxidase are dominating. The enzymes maintain their catalytic activities till milk is either boiled or pasteurized. For example, lipase has the innate property of degrading milk fat. Milk fat as a result gives rise to fatty acids which in turn change the flavour profile and contribute to the development of rancidity in milk. Alkaline phosphatase is profitably utilised as a marker for pasteurization efficiency. This enzyme is inactivated at a temperature which is enough to kill pathogenic bacteria of milk. Hence, pasteurized milk exhibiting negative phosphatase test acquires the 'gate-pass' for its sale and consumption.

(ii) Vitamins: The discovery of vitamins stemmed from milk feeding experiments with rats about fifty years ago by Professor F.G. Hopkins in Cambridge, U.K. Milk thereby made significant contribution to the early development of knowledge on vitamins. Milk is a unique food having at least 10 to 11 vitamins present in different proportion. For meeting the body demand for certain vitamins, milk can serve as the sole food and can be looked upon as a 'broad spectrum vitamin package'. Vitamins appear in milk primarily from the feeds and fodder the animal digests and are partly also contributed by the microflora of the rumen and digestive tract. Both fat soluble and water soluble vitamins are present. Their amounts are given in Table 42.1. Certain differential phenomena exist in the appearance of vitamin A in cow and buffalo milk. β-carotene, the precursor for vitamin A, is not solely transformed to the latter in case of cow milk. As a result, the unprocessed β–carotene appears in cow milk and imparts a yellow colour. This is not the case with buffalo milk where a total conversion of β–carotene to vitamin A takes place and thereby β–carotene cannot leave the mammary glands. That is why buffalo milk has more vitamin A, no β–carotene and is whiter in look unlike cow milk.

(e) *Salts in milk:* Milk can be considered as a repository of certain minerals. The most important are calcium, phosphorous and magnesium. Milk salts include sodium, potassium, chloride, citrate, sulphate and bicarbonates (Table 42.1). Other mineral elements in milk particularly copper and iron contribute to undesirable activity by catalyzing the oxidation of milk lipids resulting in development of "off" flavour.

The stability and heat tolerance of milk is dependent on the calcium and phosphate ions. An alteration in their levels either by dilution or by concentration is likely to disturb the organization of milk components, e.g., casein. Buffalo milk is more sensitive to such derangements. Calcium and phosphorous, exist in milk in two molecular states, namely, free form and bound form. It is the bound form which has a more prominent role in keeping the casein particles uniformly distributed. A cleavage or attack on the linkage of these minerals with casein will result in aggregation of the particles into a clot or gel.

B. Milk Products

Milk products are primarily made for certain reasons, namely, (a) to conserve surplus milk, (b) to preserve the milk nutrients in a more convenient form, and (c) to serve as a concentrated milk food. In India, about 49% of the total milk produced is converted into one milk product, namely, 'ghee'. Another 20% is used for preparing other milk products.

Milk products can be categorized under (a) fermented products, (b) condensed products, (c) dried products, (d) cheese and cheese products, and (e) sterilized products. Milk can be separated into distinct components by physicochemical methods for the preparation of products like butter, casein, lactose, and whey products. The composition of these foods is given in Table 42.2.

C. Indigenous Milk Products

Most of the indigenous milk products are prepared either though severe heat treatment, by fermentation process, or by preserving the product in high sugar concentration. These products were involved mainly to save the excess fluid milk for its preservation at room temperature. Hence, the primary objective in such processing technology is to remove water from the product or develop high acidity so that no detrimental microorganisms can develop to spoil the food quality of milk. The products which are prepared by subjecting milk to high heat treatment, lose their nutritional quality. It is possible, therefore, to classify these products into two major classes, (a) heat induced products, and (b) acidulated/fermented products.

III. NUTRITIVE VALUE OF MILK

Milk very often has been regarded as the most balanced and complete food available in nature and is consumed by large groups of people in the world. Historically, milk has been an integral part of our diet since time immemorial because of its high nutritional value that makes it advantageous for the nutrition of the young and ,adults and for infants in particular. It is apparent from the list of nutrients present in milk (Table 42.1) that most of our body requirements can be met by this fluid. It is also clear that there are subtle differences in the levels of these nutrients present in cow, buffalo and human milk. This is because the synthesis and secretion of milk in a particular species is primarily occasioned to meet the nutritional demands of the respective offspring.

The composition of milk and its nutritional quality are dependent on various factors, some of which have already been discussed above. When milk is exposed to boiling, some of the heat labile vitamins are lost. That is why pasteurized milk has more nutritive value than boiled milk. This is due to the fact that the time and temperature which are applied in pasteurization causes low damage to the valuable vitamins present in milk.

The major milk proteins, namely casein, α-lactalbumin and β–lactoglobulin have been often referred to as "high class proteins". Milk proteins have a high biological value (79 to 84 as measured in rats) and in this contest are only next to egg proteins.

Milk also has a symbiotic relationship with other food products when ingested in combination. For instance, the biological value of crude protein rated at 52 increases to 67 when supplemented with cheese. Similarly, potato protein with the biological values of 71 is enhanced to 86 when supplemented with skim milk (having a biological value of 89). In a similar manner, it has also been established that the nutritive value of milk as a whole is higher than the values for each constituent individually.

The high biological value of any protein, in turn, is dependent on its 'amino acid composition' and to the extent that it resembles the composition of human proteins. It is the essential amino acids which contribute to the rating of biological value. It is not only the presence of the whole spectrum of essential amino acids, but also their relative concentration, which are responsible for the high biological value of milk proteins. In this respect, the proteins from both cow and buffalo milk are fairly similar. However, the digestibility of proteins from buffalo milk was recorded to be at a slightly lower rate compared to the corresponding proteins from cow milk.

Milk fat is considered as unique because it contains glycerides with short chain fatty acids. Significant quantity of unsaturated fatty acids (essential fatty acids) are present in milk which are necessary for our growth. The presence of short chain fatty acids not only renders milk more digestible but also contributes towards the typical milky flavour.

The milk carbohydrate, lactose, is unique for its nutritive value. Lactose is considered as indispensable for the infant because galactose one of the two monosacharides can be made available only from lactose on hydrolysis. Furthermore, intake of lactose causes the proliferation of benevolent microorganisms in the intestinal tract due to its selective fermentation, giving rise to lactic acid. These microorganisms in turn synthesize certain vitamins of the B group which are ultimately absorbed by the body. Hence, the consumption of milk as a source of lactose is of considerable biological significance for infants and at the same time helps in digestion in the young and the old. Fermentation of lactose to lactic acid by lactic acid bacteria in milk enables the transformation of milk to curd which is another popular milk product with better digestibility.

Milk contains about two dozen minerals. The level of key minerals in milk is shown in Table 42.1. Milk has a high

TABLE 42.2   Composition of Some Dairy Products

Name	Proximate composition (g per 100 g)					Minerals (mg/100 g)			Calories		Vitamins (per 100 g)						
	water	fat	protein	Lactose	ash	calcium	phosphorous	iron	per oz	Per 100 g	A IU	B1 mg	B2 mg	Nicotinic acid mg	B12 (µg)	C mg	D IU
Cow milk (whole)	86.5	4.5	·3.5	4.8	0.7	122	92	0.1	22	80	150	0.04	0.17	0.1	6	1	4
Buffalo milk (whole)	83.1	7	4	5.2	0.7	150	100	0.1	29	103	240	0.04	0.14	0.1	6	1	7
Toned milk (3.5% fat)	87.1	3.5	3.7	5	0.7	122	93	0.1	19	66	115	0.04	0.15	0.1	6	1	3
Cow milk (skimmed)	90.5	0.1	3.6	5	0.8	130	96	0.1	10	35	Trace	0.04	0.18	0.1	6	1	0
Buffalo milk (skimmed)	89.5	0.1	4.2	5.4	0.8	160	105	0.1	11	39	Trace	0.04	0.18	0.1	6	1	0
Butter (salted)	16	80	1	0.5	2.5	20	16	Trace	203	726	3300	Trace	0.01	0.1	0.05	0	92
Skimmed milk (powder)	3.5	1	35.6	32	7.9	1300	30	0.6	101	362	40	0.35	1.96	1.1	36	7	Trace
Ghee	0.3	99.5	0.1	0	0.1	0	0	Trace	251	895	3800	0	0	0	0	0	99
Cheese (Cheddar)	37	32.2	25	2.1	3.7	725	495	1	111	398	1400	0.02	0.42	Trace	66	0	35
Cheese (Cheddar processed)	40	29	23.2	2	4.9	675	787	0.9	104	370	1300	0.02	0.41	Trace	66	0	30
Cheese (Surti)	66	20.5	9.3	2.3	1.9	250	200	0.2	65	231	830	0.01	0.2	0.1	4	0.6	20
Ice-cream (plain)	62.1	12.5	4	20.6*	0.8	125	100	0.1	58	207	520	0.04	0.19	0.1	6	1	12

* With added sugar

content of calcium, phosphorous and potassium. It is, however, deficient in iron, copper and magnesium. Buffalo milk is richer in calcium as compared to cow milk. As a result, buffalo milk exhibits higher curd tension and therefore is not preferred for infant feeding without proper alteration of its composition.

From the compositional data indicated in Table 42.1, it will be noticed that 1 kg of cow milk a day can furnish an average man approximately all the needed fat, calcium, phosphorous, riboflavin, about 1/3rd of the needed proteins and vitamin A, 1/4th of the needed energy and various other constituents. The protein in 1 kg of milk is roughly equivalent to that of 125 g of meat or fish. The energy equivalent is about 3/4 kg of potatoes or 350 g of meat.

Table 42.3 gives the detailed picture on the availability of nutrients in milk, and daily necessary consumption. It is quite clear that milk alone can serve as an excellent source for calcium, phosphorous, riboflavin, as a good source for protein, vitamins A and thiamine for both adults and children.

Milk when converted to certain milk products, can serve as a good repository of nutrients. Products like cheese can meet the nutritional demand for protein, fat, minerals and vitamins. The quantities of certain nutritive components in milk and milk products are shown in Table 42.2.

IV. THERAPEUTIC VALUE OF MILK

Nutraceutical preparations derived from milk and colostrum possess immense therapeutic potential, besides being highly effective, easily bioaccessible and safe and hence can find broader applications as prophylactics as well as therapeutics for infants, adults and elderly. Milk and colostrum being richer in proteins and peptides can develop innate immunity and lead to development of Gut Associated Lymphoid Tissue (GALT). Milk contains many components associated with health benefits. Therapeutic effects of milk proteins particularly whey proteins have been well documented. Bioactive peptides and whey proteins derived from milk have been the subject of intensive studies and include opioid peptides, antihypertensive peptides, anti-thrombotic peptides, immunomodulating peptides, anti-microbial peptides, mineral binding peptides etc. which possess potential physiological and bio-functional activities. The opioid peptides originate from three precursor proteins: proopiomelanocortin (endorphins), proenkephalin (enkephalin) and prodynorphin (dynorphins). These peptides have the same N terminal sequence, Tyr – Gly – Gly – Phe. Opioid peptides that are derived from a variety of precursor proteins are called "atypical", since they carry various amino acids sequence at their N terminal regions; only the N terminal tyrosine is conserved. The N terminal sequence of "atypical" opioid peptides is Tyr – X – Phe or Tyr – X_1 – X_2 – Phe. The tyrosine residue at the N terminal and the presence of another aromatic amino acid at the third or fourth position form an important structural motif that fits into the binding site of the opioid receptors. Whey proteins contain opioid – like sequences, namely α-Lactalbumin (both bovine and human), f (50–53) and β-Lactoglobulin (bovine), f (102–105), in their primary structure. These peptides have been termed α- and β-Lactorphins. Proteolysis of α-Lactalbumin with pepsin produces α-Lactorphin, and while digestion of β-Lactoglobulin with

TABLE 42.3 Nutritive Components and Required Intake of Milk

Nutrient	Amount in 1 qt milk	Amount Per 100 kcal portion	Requirements per day		Amount of milk needed to meet total requirement	
			Adults	Children	Adults	Children
Energy	650 kcal		3000 kcal	2500 kcal	4.6	3.8
Protein	33 g	5.0 g	70 g	70 g	2.1	2.1
Calcium	1.12 g	0.17 g	0.8 g	1.2 g	0.71	1.1
Phosphorous	0.94 g	0.14 g	0.9 g	1.2 g	1.0	1.3
Iron	2.26 mg	0.35 g	12 mg	12 mg	5.3	5.3
Copper	0.26 mg	0.04 g	1.0 mg	1.0 mg	4.0	4.0
Iodine	0.04–0.07 mg		0.05 mg	0.15 mg		
Vitamin A	500–1000 I.U. Winter 2000–3000 I.U. Pasture	75–460 I.U.	5000 I.U.	5000 I.U.	1.7–10.0	1.7–10.0
Vitamin D	5–15 I.U.	0.75–2.25 I.U.		400–450 I.U.		30-90
Thiamine	0.35–0.40 mg	0.06 mg	2.0 mg	1.2 mg	5.0	3.0
Riboflavin	1.5 mg	0.23 mg	2.5 mg	1.8 mg	1.7	1.2
Niacin	0.2–1.2 mg	0.03–1.8 mg	20 mg	12 mg	15–100	10–60
Panthothenic acid	2.9 mg					
Ascorbic acid	20 mg (fresh milk)	0.75 mg	75 mg	75 mg	3.7–15	3.7–15

pepsin and then with trypsin, or with trypsin and chymotrypsin, yields β-Lactorphin. Bovine blood serum albumin f (399–404) named serrophin, also displays opioid activity.

Bioactive peptides particularly antihypertensive peptides with angiotensin-converting enzyme activity have been well studied and used even in commercial preparations. Angiotensin, a blood polypeptide exists in two forms, the physiologically inactive angiotensin–I and the active angiotensin–II. The inactive form is converted into active form by angiotensin–I converting enzyme (ACE), which is a key enzyme in the regulation of peripheral blood pressure. ACE plays a major physiological role in the regulation of local levels of several endogenous bioactive peptides. Two ACE – inhibitory tri peptides, Val–Pro–Pro and Ile–Pro–Pro, obtained during the fermentation of milk proteins have been widely used in controlling hypertension. Several peptides have been isolated from whey proteins digested with proteinase–k, among them is a peptide corresponding to β-Lactoglobulin f (78–80) (β-Lactosin), which showed the highest ACE inhibitory activity.

Milk has also been found to contain antithrombotic peptides, obtained largely through enzymatic hydrolysis of Kappa-casein. Kappa-casein is structurally similar to the gamma chain in human fibrinogen. Milk is also a rich source of antimicrobial proteins and peptides, e.g., lactoferrin which has antimicrobial and antiviral activity. Milk protein–derived peptides are known to have an effect on the cells of the immune system, as well as on downstream immunological responses and cellular functions.

HAMLET (human α-lactalbumin made lethal to tumor cells) is a complex of apo α-lactalbumin and oleic acid (C18:l) formed in casein after low pH treatment of human milk which kills tumor cells while healthy cells are spared (Svensson et al., 2000). Structural variation of α-lactalbumins does not preclude the formation of HAMLET-like complexes and that natural HAMLET formation in casein was unique to human milk, which also showed the highest oleic acid content (Pettersson et al., 2006)

Oligosaccharides, short chain carbohydrates, present in human milk provide health benefits by enhancing the proliferation of Bifidobacteria in the large intestine besides preventing pathogen infection. The Oligosaccharide content of milk varies greatly between and within species and individuals. These complex oligosaccharides are also present in goat and bovine milk. Oligosaccharides also exhibit an anti-inflammatory effect. Human milk oligosaccharides (HMO) result in proliferation of *Bifidobacterium infantis*. These bacteria lead to modulation of the immune system and restrict pathogen access to preferred colonization sites among other benefits.

Studies have shown that omega 3 fatty acid—docosahexaenoic acid (DHA) and long-chain poly unsaturated fatty acids that are present in human milk enhance neurodevelopment in infants. Serpero et al. (2012) have recently characterized neurotrophic factors such as Activin A; calcium binding protein such as S100B and heat shock protein from milk, known to be involved in oxidative stress response (namely hemeoxygenase-1, HO-1 or Heat shock Protein 32, HSP32 present in human milk. Among milk lipids, conjugated linoleic acid (CLA) has been demonstrated to show its effects against cancer, hypertension, atherosclerosis and diabetes as well as to improve immune function.

Camel milk has medicinal properties especially for dropsy, jaundice and conditions affecting the lungs and spleen, malaria and gastro-intestinal disorders, pneumonia as well as tuberculosis (Tezera, 1998; Alemayehu, 2001; Ilse, 2004). In a study conducted in India by Mal et al (2000), camel milk was found to act as an adjuvant nutritional supplement in human tuberculosis patients. It is a common practice in Africa, Asia, and the Middle East to drink camel milk for treating diabetes mellitus (Yagil et al. 1994). Large concentration of insulin is present in camel milk probably the protein bears a lot of similarity to human insulin. However, research is required to be carried out to purify the protein and characterize it further for commercial exploitation in the management of DM-2 (Zagorski, 1998).

Goat milk has higher buffering capacity, easier to digest, less allergic than cow milk and has host of other potential medicinal properties. It has been found to be effective in controlling diarrohea, gastro-intestinal disorders, constipation and respiratory tract infections, etc. However, human milk contains a large number of antigens that interacts with gut mucosa of infant and protects the infant from gastro-intestinal infections. Human milk enhances the immunity and there are decreased cases of mortality and morbidity in breast fed children compared to formula fed children. Human milk possesses antimicrobial properties due to immunoglobulins, proteins (lysozyme, lactoferrin, haptocorrin) and oligosaccharides as well as glycoproteins. The human milk oligosaccharides selectively increase the population of Bifidobacteria in the infant gut. All these attributes make human milk the obvious choice for maintaining the health and growth of all the age groups of human population

V. HUMAN MILK AND ITS VIRTUES

There is a teleologic slogan "human milk is for the baby, cow's milk is for the calf". Let us open this issue to examine its validity. The composition of milk in different species is always 'bio-tailored' to meet the nutritional demands of its offspring. That is why reindeer milk has a high fat (33%) and protein (10%) content; buffalo milk has higher fat (7%), protein (5%) and calcium (0.2%) relative to cow milk; and human milk has higher levels of lactose (7%), low protein (1.2%) and fat (3.5%) content. Hence, the ideal situation from the point of view of the nutritional demands of the offspring, is to feed it with its mother's milk.

Let us now look more closely at, the compositional profiles of milk from these species. The data in Table 42.1, provides adequate evidence that human milk is distinct in composition from cow milk, and more so from buffalo milk. (i) Unlike cow or buffalo milk, it lack in αs-casein and β–lactoglobulin, two principal components of milk proteins. The particle size of casein, is significantly smaller in human milk. It is, therefore, not the low concentration of protein in human milk which matters, but it is the different chemical make up of these proteins. (ii) It has more unsaturated fatty acids. (iii) It has more oligosaccharides and high lactose. (iv) It has low concentration of calcium, unlike, for example, buffalo milk. (v) It has very high lysozyme content, an enzyme responsible for bactericidal action.

One has to, therefore, pay distinct attention while considering cow/buffalo milk as a baby food, since the baby is supposed to thrive on human milk in its infancy. One of the solutions would be to dilute cow milk with an equal volume of water so as to equalize the protein content and add cream and sugar to restore the composition of these constituents to a figure as close as possible to the human milk. This is one of the ways in which cow milk has been utilized for infants. Industry has also put out a number of baby food preparations where the aim has been to approach the human milk as far as possible in terms of general composition of major constituents.

Nutritional requirement of an infant

Like any other growing animal, a baby is dependent on protein, fat, carbohydrates, minerals, and vitamins. But what is more important in such biological development is the specific nutritional demand of an infant. Feeding of mother's milk has certain unique advantages, e.g., (a) quick stomach emptying time, (b) enhanced lysozyme activity in stool, (c) remarkable increase in the number of bifidus bacilli in the intestinal microflora, (d) less burdening of the kidney due to low levels of minerals, and (e) easy digestion and absorption efficiency attributable to the small casein particle size.

VI. HUMANIZATION OF MILK: A NEW SOLUTION

While preparing baby food, one has to identify the nutrients specific to mother's milk, which are more or less in market milk. High level of mucopolysaccharides, lysozyme, unsaturated fatty acids, and lactose and low level of protein, fat, and minerals, as in mother's milk, should be the objectives in humanizing cow/buffalo milk. Hence the problem for a more satisfying degree of conversion is not merely a standardization of fat and SNF, but a constitutional change of cow/buffalo milk.

For substitution, a baby food has to be formulated from cow/buffalo milk so as to be almost a duplication of human milk. This can be accomplished by enzymatic and chemical processing. Cow/buffalo milk is fortified with mucopolysaccharides, vegetable oil, selective degradation of casein particles, and withdrawal of minerals by electrodialysis. In other words, humanization of milk by the above approach would provide a better nutritional food for the infant.

VII. DESIGNER MILK

Although, milk is often considered as nearly the most perfect and ideal food as discussed above to meet our nutritional requirements, there is still considerable scope to improve its functional properties to suit to the needs of the consumers by introducing appropriate modifications in its composition. The attention is now focused on adding more value to milk and studying its health implications. With the recent developments in the Biotechnological techniques, e.g., Genetic Engineering, rDNA technology, Protein Engineering and advances in animal cloning and transgenic techniques, it is now possible to alter milk composition at will for better manufacturing/technological properties to add variety to our traditional dairy products and also to produce variants of milk to cater to the needs of specific consumers from health and nutritional perspectives. With the advent of modern gene transfer and expression methodologies, new opportunities have been created for the modification of animal production traits including milk production with altered composition. Animal udder can virtually be used now as an efficient biological vat or Bioreactor for the production of homologous and heterologous proteins, sugars and fats. Transgenic animals which constitute a useful experimental tool for assessing the ability and effect of transgenic mammary gland specific expression are mainly concerned with either producing biologically important and active proteins such as pharmaceuticals in milk of transgenic animals or to alter the intrinsic properties and composition of milk itself by genetically adding a new or modified protein for better manufacturing properties for dairy industry. Our growing understanding of the lactation process in the ruminants at molecular level and continual innovations in dairy processing have presented exciting opportunities for genetic manipulations that are not possible through traditional, nutritional, and classical genetic approaches.

Biotechnology based strategies for altering the properties of milk

Advances in biotechnology and genetic engineering have led to exploring new initiatives that were hitherto not even thought possible in the field of dairying. It is now firmly established that a new generation of value-added products can be produced and harvested from milk and milk products. While until recently, emphasis has been on breeding large animals to produce more milk, the current interest of animal scientists is now on producing designer milk by expressing homologous/heterologous proteins and introducing appropriate alterations in the major milk constituents in the milch animals through animal cloning and transgenic technology. By a thorough understanding of the biochemistry, genetic traits and changes in the cows diet that affect milk synthesis and composition, ways and means to manipulate milk composition to suit specific needs can now be explored judicially. By combining the two approaches of nutritional and genetic interventions, researchers are now hoping to develop **'designer milk'** tailored to consumer preferences or rich in specific milk components that have implications in health as well as milk processing. The current interest in the modification of milk composition include the healthful and therapeutic aspects of milk and milk products. To realize the full potential of these advantages, it would be desirable to have the opportunity to alter milk composition in several ways. For diet and human health measures, the actions that would be beneficial include:

(a) generate a greater proportion of unsaturated fatty acids (USFA) in milk fat
(b) reduce lactose content in milk in order to cater to persons suffering from lactose intolerance and
(c) remove β-lactoglobulin (β-l g) from milk.

From a technological stand point, there exist vast opportunities in:

(a) alteration of primary structure of casein to improve technological properties of milk
(b) production of high-protein milk
(c) engineering milk meant for cheese manufacturing that leads to accelerated curd clotting time
(d) increased yield and/or more protein recovery
(e) milk containing nutraceuticals and
(f) replacement for infant formula.

Five basic areas that might be highly useful for introducing desired alterations in milk are listed in Table 42.4. Within these broad areas, a wide variety of modifications to milk can be exploited. These include; adding extra copies of an existing gene (αsl, κ- and β-casein), down regulating the expression of a gene (α-lactalbumin), adding new genes such as those encoding human lysozyme or lactoferrin, removal of a gene (β-casein, β-lactoglobulin or acetyl-CoA carboxylase), and adding a mutated gene (αsl, κ- and β-casein) etc. In this direction, preliminary research has already been carried out using transgenic mice as model systems in the first four of these categories.

TABLE 42.4 **Potential areas for introducing desired manipulations in milk targeting its key components for value addition**

Altering the proteins to change the manufacturing properties of milk
Changing the amino acid composition of milk to improve human nutrition
Increasing the overall protein content of milk
Altering the type and amount of fatty acids in milk
Increasing the antimicrobial activity of milk

VIII. MILK GENOMICS

The composition of milk can have a significant impact on nutritional and technological properties of milk and milk products. As previously described, there are several factors which affect the composition of milk. Another very important factor that can play a vital role in introducing desired or undesired changes in the milk composition at molecular level and influence the yield and functionality of milk constituents is the milk genomics. It is a great challenge to determine the biological functions of milk components when consumed. The era of genomics, nutrigenomics (Figure 42.2), personalized medicine have opened new avenues to look for the influence of each component of diet on health. Nutrigenomics can be explored to decipher the function of each and every component of milk on individual's genome. Milk can be used as a model to show how

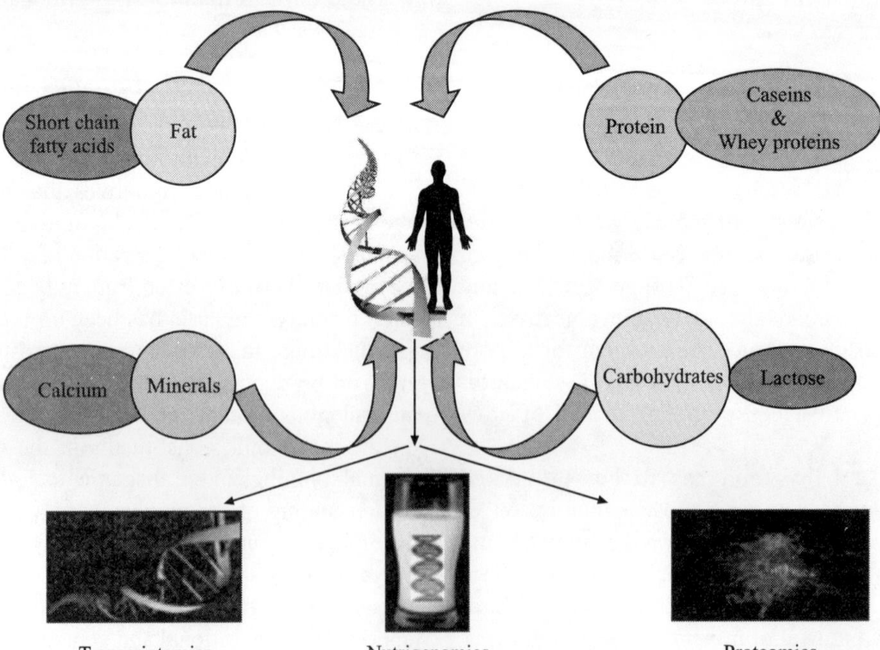

Figure 42.2 Role of Nutrigenomics, Proteomics and Transcriptomics to establish linkage between the milk constituents and human health. *(see Plate 16 for colour figure)*

diet influences health. This would be a step towards designing functional/customized/personlized dairy products for enhancing the human health. Functional genomics will be used to annotate the function of all the essential and non-essential nutrients present in milk. It is important to discover functional ingredients using molecular biology approaches. The composition of milk of different species points towards the different nutritional needs of neonates of various species. However, functionality of each component of milk may differ which depends on genetic variations in human and the environment.

A number of milk components have been identified using proteomics and metabolomics. Proteomics has been used to study milk proteome during lactation. Differentially expressed proteins in lactating and non-lactating animals, under different stages of lactation can be used to differentiate samples since these can be used as biomarkers. Studies are needed to identify the exact function of different proteins, differences in their nutritional value and their physiological impact. During pregnancy, changes occur in the cellular composition of mammary glands for producing milk. The genes responsible have been identified which can be altered according to the requirement.

Lot of work on milk genomics and proteomics is currently in progress to unravel the function of various genes and identify new proteins. The application of high throughput gene expression microarray analysis in lactation research is producing a large data on the expression of genes in the mammary tissue. Comparison of expression data from different species will lead to better understanding of the function of genes in the mammary gland which may ultimately lead to enhanced lactation in dairy cattle. This is an emerging area of considerable commercial interests from human health perspective and has bright future prospects in designing novel strategies to tailor milk based on the genomic profile of the individuals for treating and preventing diseases.

SUMMARY

Milk is a 'biological package' of nutrients harmonized together in adequate proportion. Due to the high biological value of proteins, presence of a number of short chain and unsaturated fatty acids and the unique carbohydrate lactose in milk, our food habits and portfolio should contain milk and milk products. Milk as a source of infant food occupies a pivotal position in the nutritional market. Every biological species synthesizing milk, tailors at the genetic level, the amount and nature of milk constituents, to suit the nutritional requirements of their offspring. Likewise, human milk possesses additional virtues compared to cow and buffalo milk as an infant food. A possible 'reorganization' of nutrients in cow/buffalo milk has been proposed to 'humanize' such milk for its nutritional consonance with human milk. Besides this, it is now possible to add lot of variety to milk by altering its composition at molecular level by applying advanced biotechnological tools for product diversification and to produce customized milk tailored for specific target populations and consumers to boost their overall efficiency, productivity and health. Furthermore, nutrigenomics in future can play a vital role in reshaping the dairy sector through production of individual's need based dairy products from human health perspectives. This would be a step towards designing functional/customized/ personlized dairy products for enhancing the human health.

SUGGESTIONS FOR FURTHE READING

Ganguli N.C. (1968), Research on Milk Products in India, *Indian Journal of Vet. Science,* Vol. 38, 1.

Ganguli N.C. (1980), Numerous Nutrients in Milk, *Arogya,* Vol. 6, 1.

Hariman A.M. and Dryden L.P. (1965), *Vitamins in Milk and Milk Products,* American Dairy Science Association.

Jenness R. and Patton S. (1959), *Principals of Dairy Chemistry,* John Wiley and Sons, Inc., New York.

Mal G., Suchitra S.D., Jain, V.K., Singhvi N.M. and Sahani, M.S. (2000), Role of Camel Milk as an Adjuvant Nutritional Supplement in Human Tuberculosis Patients, *Livestock International,* 4:7–14.

Pettersson J., Mossberg A.K. and Svanborg C. (2006), α-Lactalbumin Species Variation, HAMLET Formation, and Tumor Cell Death, Biochemical and Biophysical Research Communications, 345:260–270.

Rusoff L.L. (1970), Milk: Its Nutritional Value at a Low Cost for People of all Ages, *Journal of Dairy Science,* Vol. 53, 1296.

Serpero L.D., Alessandro F. and Gazzolo D. (2012), Human Milk and Formulae: Neurotrophic and New Biological Factors. Early Human Development, 88 Suppl 1:S9–12.

Srinivasan M.R. and Anantakrishnan C.P. (1964), *Milk Products of India,* Indian Council of Agricultural Research, New Delhi.

Svensson M., Hakansson A., Mossberg A.K., Linse S. and Svanborg C. (2000), Conversion of Alpha-lactalbumin to a Protein Inducing Apoptosis, *Proc. Natl. Acad Sci.* USA, 97:4221–4226.

Veldee M.S. (1999), In: Tietz *Textbook of Clinical Chemistry,* 3rd ed., Burtis, C.A. and Ashwood, E.R. (Eds.), W.B. Saunders Co., Philadelphia, USA, 1359–1394.

Yagil R. (1994), The Camel in Todays World, A Hand Book on Camel Management. Germany-Israel Fund for Research and International Development and Deutsche Welthungerhilfe, Bonn, 74.

Zagorski O., Maman A., Yafee A., Meisles A., Creveld C.V. and Yagil R. (1998), Insulin in Milk —A Comparative Study, *International J. Animal Science,* 13:241–244.

43

Nutrition During Pregnancy and Problems of Infant Nutrition

Neerja Bhatla, Shweta Rai and Divya Awasthi

CONTENTS

I. INTRODUCTION

Pregnant and lactating women form one of the most vulnerable segments of the population from a nutritional point of view. Although it may be difficult at times to ascertain the extent to which diet, as distinct from other environmental factors, influences the outcome of pregnancy, a good nutritious diet appears to be related to adequate normal weight of the newborn and consequent chances of survival. There is no doubt that nutritional needs increase in pregnancy. The fetus is an obligate parasite on the mother and so limited variations in nutrient intake of the mother may still be compatible with the health of the fetus. However, the symbiosis may break down with disastrous results in cases of severe nutritional compromise. Maternal under-nutrition is associated with low birth weight and all its attendant adverse consequences. Too early, too close, too many and too late pregnancies certainly have an adverse impact on the nutritional status of pregnant women.

While under-nutrition continues to be a major problem as in earlier decades, the current decade has witnessed also the progressive rise of over-nutrition in women in the reproductive age group belonging to affluent populations in both urban and rural areas. It has, therefore, become even more imperative to assess the nutritional status of pregnant women and give them appropriate advice and care.

A. Weight Gain During Pregnancy

Generally the weight gain during pregnancy is about 9 to 12 kg, most of which is gained in the second half of pregnancy. The Institute of Medicine has made recommendations of gestational weight gain (GWG) based on pre-pregnancy Body Mass Index (BMI) expressed as kg/m^2 (Table 43.1). There is evidence that maternal weight gain had a positive correlation with birth weight, and women with the greatest risk for delivering an infant weighing less than 2500 g are those with weight gain less than 6 kg.

B. Recommended Dietary Allowance (RDA)

The recommended dietary allowance is the daily dietary nutrient intake level that is sufficient to meet the nutrient requirement of nearly all healthy individuals in a particular life stage and gender group. An expert group from ICMR (2010) recommended dietary allowances for women who are pregnant or lactating, taking into account the individual variability in Indian women and the nutrient bioavailability from their habitual diet (Table 43.2).

i. Calories

Pregnancy is a state of increased basal metabolic rate (BMR) even in a sedentary female. An additional 80,000 kcal of

TABLE 43.1   **Recommended Ranges of Total Weight Gain for Pregnant Women by Pre-pregnancy Body Mass Index (BMI) for Singleton Gestation (Institute of Medicine 2009)**

Category	BMI* (kg/m2)	Recommended Total Weight Gain Range (kg)	Rates of Weight Gain 2nd and 3rd Trimester Mean (range) in kg/week
Low	< 18.5	12.5 – 18	0.51 (0.44–0.58)
Normal	18.5–24.9	11.5 – 16	0.42 (0.35–0.50)
High	25.0–29.9	7 – 11.5	0.28 (0.23–0.33)
Obese	> 30.0	5 – 9	0.22 (0.17–0.27)

*Institute of Medicine (2009) has adopted WHO BMI categories for its recommendations.

TABLE 43.2   **Summary of Recommended Dietary Allowance (RDA) for various nutrients in Pregnant and Lactating Women in India (2010)**

		Pregnant woman	Lactating Woman	
			First six months	After six months
Net energy (kcal/d)	Sedentary	2250	2500	2420
	Moderate work	2580	2830	2750
	Heavy work	3200	3450	3370
Protein (g/d)	Sedentary	78	74	68
	Moderate work	78	74	68
	Heavy work	78	74	68
Visible fat (g/d)	Sedentary	30	30	30
	Moderate work	30	30	30
	Heavy work	30	30	30
Fat-soluble vitamins				
Vitamin A (µg/d)	Retinol	800	950	
	B-carotene	6400	7600	
Vitamin Da (µg/d)		5	5	
Vitamin Ea (mg/d)		15	19	
Vitamin Kb (µg/d)		90	90	
Water-soluble vitamins				
Vitamin C (mg/d)		60	80	
Thiaminec (mg/d)		1.2–1.6	1.3–1.7	1.2–1.6
Riboflavinc (mg/d)		1.4–2.0	1.5–2.1	1.4–2.0
Niacinc (mg/d)		14–18	16–20	315–17
Vitamin B6 (mg/d)		2.5	2.5	
Folate (µg/d)		500	300	
Vitamin B12 (µg/d)		1.2	1.5	
Minerals				
Calcium (mg/dl)		1200	1200	
Iron (mg/dl)		35	21	
Zinc (mg/dl)		12	12	
Iodined (µg/d)		250	250	
Magnesium (mg/d)		310	310	

ICMR (2010). Nutrient requirement and recommended dietary allowances for Indians, A report of expert group of the ICMR

(a)   RDA of Food and Nutrition board of Institute of Medicine (2008)
(b)   Recommendations measured as Adequate Intake
(c)   Range varies between sedentary, moderately working and heavy working ladies.
(d)   Recommended dietary intake, Geneva: WHO, 2007

energy is required which are accumulated primarily in the last 20 weeks. Before 2002 it was recommended to increase the energy consumption by 300 kcal/day in second and third trimester. In 2002, Institute of Medicine revised the Dietary Reference Intake (DRI) recommendations for energy intake during pregnancy. The new recommendations advise no additional calories for the first trimester, add 340 kcal for the second trimester and add 452 kcal for the third trimester. Also it has been recently advised that additional needs should be tailored based on mother's pre–pregnancy BMI. Of the total calories, 50–55% should be supplied by carbohydrates, 15–20% by proteins and 20–30% should come from fat.

ii. Carbohydrates

The rapid growth of fetus and the increased metabolic rate require ample amount of energy in the form of glucose, which is made readily available by breaking the carbohydrates. The Recommended Dietary Allowance (RDA) for carbohydrate during pregnancy is 175 g/day. Low carbohydrate diet is dangerous during pregnancy and can put the baby at risk of poor growth. A mild restriction may be recommended in women who are diabetic, particularly if they are also obese.

iii. Proteins

Proteins provide the basic 'building blocks' necessary for formation of enzymes, antibodies, muscle and collagen. The mother must consume adequate protein to meet the demands of her growing fetus in addition to her own increased needs for growth of the uterus, as well as development of breasts. To accommodate for increasing demand, the mother's body adapts to conserve protein. The interplay of various hormones gives anabolic signals, which make her body retain nitrogen for protein synthesis. The 2002 DRI for pregnant women recommends 1.1 g/kg/d of body weight or additional 25 g/d to meet the demands of pregnancy.

iv. Fats

Lipids are essential for formation of cell membranes and hormones and are necessary for proper eye and brain development. Fat is also a source of concentrated calories and may be beneficial to women at risk of energy malnutrition. Moderation is essential since excessive fat can cause undesired weight gain. There is no separate DRI/RDA for fat intake in pregnancy and the recommendations remains 20% to 35% of total calories. Sources that provide essential fatty acids and choline should be emphasized. Supplementation of omega 6 and omega 3 amino acids may be required. The 2002 DRI/RDA recommends for adequate intake of 13 g/d of omega-6 and 1.4 g/d of omega-3 and the ratio of the two should not exceed 5:1.

v. Minerals

Generally a diet that results in adequate weight gain will cover all mineral requirements with the exception of iron, calcium and iodine. These three are very important for both maternal and fetal health and are discussed in more detail in the section of nutritional supplements (*see below*). Deficiency of some minerals like zinc and selenium has been associated with fetal abnormalities like dwarfism, hypogonadism and cardiomyopathy; most other mineral deficiencies have not been found to be significantly related to pregnant state.

vi. Vitamins

The increased requirements for vitamins during pregnancy are usually supplied by a healthy well-balanced diet that provides adequate calories and protein. The only exception to this is folic acid. This is the only vitamin supplement that is recommended prophylactically during pregnancy. Its usefulness is discussed in the section covering nutritional supplements during pregnancy.

vii. Fibre

Fibre is a very important component of the prenatal diet. Although it does not affect the development of fetus, it significantly increases the comfort of mother by helping to reduce constipation, which is generally a concern as progesterone decreases bowel motility. Adequate intake in pregnant women is 28 g/d.

C. Nutritional supplements

i. Iron supplementation

Pregnant women are particularly at risk for iron deficiency and iron-deficiency anemia because of increased iron needs during pregnancy superimposed on a background of rampant anemia. Approximately 1000 mg of iron are needed during pregnancy, 500 mg are used to support the expanding maternal hemoglobin mass and 300 mg for the development of the fetus and placenta. On an average, the daily iron needs are between 6 and 7 mg as opposed to 1 mg/day in normal physiologic conditions. During the last 6 to 8 weeks of pregnancy, the need increases to up to 10 mg/day. Although iron absorption is substantially increased during pregnancy and is adequate in healthy iron-replete women it fails to meet the requirements in iron-deplete pregnant women. In women who enter pregnancy with low iron stores, iron supplements often fail to prevent iron deficiency.

The prevalence of iron-deficiency anemia in pregnant women is estimated to be between 35 and 75% (average 56%) in developing countries including India, whereas in industrialized countries the average prevalence is 18%. It is not documented that iron supplementation has any substantial effect on birth weight or various complications in pregnancy. However, supplementation corrects the iron stores and biochemical parameters of iron deficiency including hemoglobin (Hb) concentration and maintains the maternal iron stores in the puerperium. Recent literature also suggests that iron supply to the pregnant women may have beneficial effects on the iron content of neonates in the first year of life. However oral iron has been associated with gastric irritation and altered bowel habit (i.e., constipation or diarrhoea).

Thus iron supplementation for prophylaxis during pregnancy remains a controversial issue and should be individualized. In an iron-depleted population like ours, daily iron supplementation with iron preparations giving 100 mg elemental iron should be offered routinely to all pregnant women after the first trimester. Parenteral iron may be required in some women who are unable to tolerate oral preparations or have malabsorption syndromes.

In a non-anemic population, as is seen in the West, supplementation with even lower doses of elemental iron (30 mg) is as effective as daily supplementation, with the advantage of better tolerance and compliance due to fewer side effects.

ii. Calium

Calcium intake is crucial during pregnancy and lactation because of the potential adverse effect on maternal bone health if maternal calcium stores are depleted. There is often a transient lowered bone mineral density and increased rate of bone resorption, with the greatest consequence during the third trimester and throughout lactation due to increase demand. Also, calcium supplementation in pregnancy has been associated with a reduced risk of pregnancy-induced hypertension, but this effect is only seen in persons with a low basal calcium intake. There are also reports in literature that calcium supplementation in pregnancy protects against low-birth weight in newborns.

It is recommended that a daily consumption of 1000 mg calcium is required for pregnant and lactating women, being the amount contained in one litre of milk. Thus, proper calcium consumption can be attained by diet with healthy nourishment including 3-4 snacks of milk or milk-derived products such as yogurt and cheese. Other calcium rich foods include beans, especially soya bean, and fish.

However the average consumption of calcium even in developed countries is about 800 mg in young women. Therefore calcium supplementation in pregnancy should be encouraged, especially during the second and third trimester of pregnancy and during lactation.

iii. Vitamin D

During the last trimester, the fetus begins to calcify the skeleton, thereby increasing maternal demand for calcium. This demand is met by increased production of 1,25(OH)2D by the mother's kidneys and placenta. Circulating concentrations of 1,25(OH)2D gradually increase during the first and second trimesters, owing to an increase in vitamin D-binding protein concentrations in the maternal circulation. However, the free levels of 1,25(OH)2D, which are responsible for enhancing intestinal calcium absorption, are only increased during the third trimester. Pregnant women are at high risk for vitamin D deficiency, which increases the risk of preeclampsia and cesarean section. Its deficiency in pregnancy has been associated with neonatal rickets (rarely) and reduced bone mineral accrual in childhood. Emerging evidences suggests that vitamin D supplementation may be offered routinely to pregnant women after first trimester. Daily doses of 600 IU do not prevent vitamin D deficiency in pregnant women. Their daily regimen should include a prenatal vitamin containing at least 1000 IU vitamin D.

iv. Iodine

According to WHO, iodine deficiency at critical stages in fetal life and early childhood remains the single most important and preventable cause of mental retardation globally. Iodine requirement is increased during pregnancy due to 1) increased production of thyroxine (T4) by mother to maintain her euthyroid state, 2) transfer of iodine in form of thyroid hormone to fetus and 3) increased renal excretion secondary to increased renal clearance. WHO recommends women who are pregnant or breastfeeding should take a daily oral iodine supplement so that the total daily intake is 250 µg. The current salt iodisation guidelines (= 15 ppm of iodine at consumer level) is designed to deliver only 150 µg/L of iodine per day. This level is inadequate for pregnant women, whose dietary requirements are much greater. Thus iodine supplementation in pregnancy above the amount of iodine supplied by salt fortification needs to be explored. Therefore fortification of other food items, e.g., oil is being tried.

v. Folic acid

Periconceptional folate supplementation was found to substantially reduce the prevalence of neural tube defects (relative risk 0.28). There was a reduction both where the mother had not had a previously affected fetus or infant (relative risk 0.07) and when the mother had given birth to a previously affected infant (OR 0.31). Recently it is also reported that maternal folic acid supplementation was associated with decreased risk of other congenital anomalies, including cardiovascular defects and limb defects and some pediatric cancers including leukaemia, pediatric brain tumours and neuroblastoma. It is estimated that only one-third of women take folic acid supplements before conception. As folic acid is needed at the time of embryogenesis and many women do not plan pregnancy, folic acid-fortified foods have been advocated and have been found to be effective in achieving beneficial levels of red-cell folate. However, increasing intake through foods naturally containing folates has not been found to be effective.

It is recommended that folic acid supplementation should be at a dose of 400 micrograms/day before conception and up to 12 weeks of gestation.

D. Other Nutritional Concerns during Pregnancy

Some foods can cause harm to the developing fetus. All meats should be thoroughly cooked to avoid exposure to toxoplasmosis, salmonella, and other harmful bacteria. Tobacco smoke, drug use, and alcohol consumption should be stopped. Caffeinated beverages (soda, coffee) should be restricted to the minimum. Raw papaya in first trimester is likely to increase the risk of abortion.

II. PROBLEMS OF INFANT NUTRITION

A. Breast Feeding

What is natural is often the best. The inherent advantage of breast milk is not only in an ideal composition most suited for the human infant's stomach and needs, but also in the emotional satisfaction provided to both the mother and her offspring. Breast milk is economical, readily available at the right temperature, sterile and fresh. For all these reasons, it is also a convenient method of feeding. It is bacteriostatic, partly because of a high content of lactoferrin. Neonatal tetany is rare in breast-fed infants. Recent evidence points to its beneficial effect in the prevention of infantile eczema and other forms of atopy in susceptible children. For the first few days, colostrum is secreted which is particularly rich in antibodies and immunoglobulins and helps develop the infant's resistance to infections.

Most failures of breast-feeding are in quantity of milk available and not in quality, whatever folklore might say. The common comment that breast milk looks 'thin and watery' is without logic or biological sense. Good antenatal care of the breasts, physiological support during pregnancy and perinatal period, complete emptying of the breast after each feed, avoidance of fatigue, and optimum nutrition would ensure adequate amount and flow of milk.

Breast feeding may be difficult in a few situations and might require temporary suspension – very weak infants, those with cleft palate or lip, mastitis, cracked inverted nipples, and acute or chronic illness in the mother. Some drugs taken by the lactating mother can be present in the milk in sufficient quantities to affect the infant. These include chemotherapy drugs, reserpine, laxatives, iodides, aspirin, barbiturates and sulfonamides. HIV positive mothers are also advised to avoid breast-feeding.

B. Weaning and Supplemental Feeding

After four months of age, the infant's need exceed the quantity and variety of nutrients present in breast milk. Supplements with orange juice, cereals, mashed and whipped fruits and vegetables, and egg yolk are indicated, one new food item being introduced in the diet every few days. One or two feeds with a bottle or a cup may be necessary to suit the mother's needs of rest and work at home or outside it. Cow's or goat's fresh milk is satisfactory; most proprietary preparations require handling to prepare, with consequent increase in chances of contamination, and are costly. The choice of a particular type of milk is entirely empirical since the young infant is almost certainly unable to distinguish one brand from another. Weaning should be completed by about 9 months of age.

Undernutrition

The velocity of growth and development in the early years of life is many times higher than it in late childhood. The intense anabolic activity, cell division and maturation cannot proceed at an optimum pace when there is deficiency of one or more essential nutrients. It is now confirmed that a period of significant under-nutrition in early life would retard physical growth and development, which on re-feeding would soon catch up. But this catch-up growth is less likely to be complete and will require a longer period if nutritional deficiency has retarded growth at a stage of life when normal growth velocity is very rapid.

Animal studies have shown conclusively that early malnutrition at a time when growth of brain is maximum can lead to severe reduction in size, cell content, DNA and lipids with resultant effects on function. Several studies have shown that malnutrition in the first few months after birth is associated with reduction in head circumference and brain size. Depending upon the severity, timing and duration of under nutrition, brain damage may be permanent or reversible.

It is being recognized increasingly that the foundations of atherosclerosis are laid in infancy and childhood. Low birth weight is associated with increased rates of cardiovascular disease, hypertension and type 2 diabetes in adulthood. The fetal origins hypothesis states that these effects may be a consequence of 'fetal programming', whereby a stimulus or insult at a critical, sensitive period in development hs a permanent effect on physiology, structure or metabolism of tissues and organs.

C. Over-nutrition in the Young

Excessive intake, together with physical inactivity, leads to obesity. The common etiopathogenetic causes are over-enthusiastic parents, emotional and psychological problems: parent–child or peer conflicts, constitutional and familial predisposition. Endocrine and metabolic problems are rare causes. A chubby, fat child often becomes an overweight obese adult, probably because the hypothalamic hunger-satiety balance has been set at an abnormally high level of dietary intake. Management should be based on dietary restriction, physical activity, social participation and helpful parental attitudes.

Over-supplementation with vitamin concentrates can cause toxic symptoms. Hypervitaminosis A manifests as nausea and vomiting, lethargy, bulging of the anterior fontanelle, hepatosplenomegaly, pruritus, hyperostosis and periostetis. Overdosage with vitamin D causes vomiting, convulsions, coma, abdominal pain, bone pain, urinary urgency, often with polyuria, hypercalciuria, nephrocalcinosis, and metastatic calcification. Large doses of water-soluble vitamin K to the newborn, especially the one with low birth weight, can result in hemolytic anemia, jaundice and kernicterus with its dreadful sequelae.

Effect of nutrition on early growth and development

The velocity of growth and development in the early years of life is many times higher than it is in late childhood. The intense anabolic activity, cell division and maturation cannot proceed at an optimum pace when there is a deficiency of one or more essential nutrients. "Catch up" growth is less likely to be complete or will require a prolonged period, if nutritional deficiency has retarded growth at a stage of life when the normal growth velocity is very rapid. For development of immunity mechanisms and central nervous system, such a critical period is during intra-uterine life and early infancy. For physical growth, the velocity is high during fetal life, especially the first and third trimesters of gestation, infancy and again in the prepubertal period.

1. Physical growth and maturation: There is an association between adult stature and socio-economic status. This may not be apparent in one generation but over succeeding ones. The relationship between socioeconomic status and growth is not, of course, solely through nutrition. Other adverse extrinsic causes are important factors for impairment of development. It has also been suggested that where means of subsistence have been severely compromised for generations, genetically small strains may be naturally selected as an advantageous type since they are more capable of reaching maturity and leaving descendants. It is possible that children with a genetic make up which promotes rapid growth on adequate diet would suffer more from malnutrition than those with modest demands.

The ultimate height achieved depends upon the rate of growth of long bones and the timing of epiphyseal fusion which would stop further increment in growth. The latter is sex linked. But the completeness of the recovery process would depend upon the time at which restriction took place, its extent and the efficiency and duration of rehabilitation. A lifetime of starvation would reduce a man's stature, which would then have an effect on the subsequent generations; thus a vicious cycle is set up.

2. Brain size, composition and function: In the case of several organs, the growth in size and chemical and functional development occur simultaneously. The nervous system is different; replacement of its water content by lipids and nucleic acids may involve little or no change in weight, and therefore weight and size are not necessarily good measures of brain development. Under-nutrition may lead to dissociation between weight, chemical composition and biological function.

Animal studies have shown conclusively that early malnutrition at a time when growth of the brain is maximal, can lead to severe reduction in its size, cell content, DNA and lipids, with resultant effects on its function. Depending upon the severity, timing and duration of under-nutrition, brain damage may be permanent or reversible. Rats and pigs starved during fetal life or during suckling manifest a reduction cell number, lower content of DNA cholesterol, cerebrosides and phospholipids. The spinal cord does not exhibit the same growth characteristics as the brain, it does not have an early growth spurt and it grows both in thickness and in length. Under-nutrition retards growth in the length of the spinal cord but still permits growth in thickness that is appropriate for the chronological age. Moreover, the concentration of the lipids and DNA is less severely altered.

Several studies have shown that malnutrition in early life, especially during the intra-uterine period and the first few months after birth, is associated with reduction in head circumference and brain size. The extent of catch up varies with the period when the growth was impaired, its severity, and the quality of rehabilitation. The mildly undernourished, if given an adequate diet for a sufficient period of rehabilitation, would make a complete recovery. The severely malnourished show the least potential for recovery, which would take prolonged period and may not be complete. In such cases, the reduced head circumference, brain size and dysfunction are seemingly permanent. Obviously, the majority of children lie between these two extremes and are likely to make a near complete recovery on careful dietary management. Although gross function tests and developmental quotient are often reported to be normal in such children on follow-up, finer tests of comprehension, association, cognition, integration, and logic may show continuing deficits for as long as 5-7 years after malnutrition had been corrected.

3. Development of immunocompetence: In man, immunity mechanisms evolve and mature in fetal life, beginning in the sixth week of intrauterine life. It is recognized that the human newborn is capable of mounting a fairly adequate immune response in the event of an infection. A severe restriction of nutrients during pregnancy, directly or through placental insufficiency, can adversely affect the development of immunocompetence of the fetus. Intrauterine growth restriction results in a low birth weight infant who has increased susceptibility to infections which contribute to high rates of morbidity and mortality. Low birth weight infants who are small for gestational age have several defects of immune response, including impaired cell mediated immunity, low levels of IgG and polymorph dysfunction. As with physical growth, it is possible that interference with the development of immunity mechanisms in utero may have a far greater and more lasting effect compared with post-natal malnutrition, in terms of susceptibility to infection, development of neoplasia and autoimmune disease, and even longevity.

Postnatal nutritional deficiency is associated with a high frequency of infections with bacteria, viruses and fungi. There seems to be a synergistic interaction between malnutrition and infection, each potentiating the other. A variety of defects in immunocompetence in malnourished children have been described. There is significant involution of the thymus and other lymphatic tissues, especially the lymphocyte population. The pathogenesis of such morphological changes is not understood clearly, stress, hormonal changes, infection, failure of cell proliferation, defective mobilization, activation of latent viruses and endotoxemia may all be contributory factors. The serum immunoglobulins of undernourished children are normal, low or high, the latter being the result of concomitant or past infections. Antibody responses are generally adequate, though secretory antibody production may be deficient. There is an impairment of delayed hypersensitivity and cell mediated immunity, which correlates with the selective reduction in rosette forming thymus dependant Iymphocytes. Complement in component C3 in normal or low. A reduced level is often seen in children with infection and it may be due to decrease synthesis or increased consumption and catabolism. A severe impairment in intracellular bacterial killing and metabolic function of polymorphs has been noted in association with protein-calorie deficiency and iron under-nutrition. After nutritional rehabilitation, most defects in immunocompetence are reversible.

SUMMARY

An adequate intake of all the necessary nutrients is essential, especially for growth and development and during periods of stress such as pregnancy and lactation. While prevention of nutritional deficiency is highly desirable, avoidance of excesses is equally relevant. A well-nourished individual less readily succumbs to infection and a non-obese person is less liable to develop degenerative vascular disorders.

REFERENCES

1. Institute of Medicine, Weight Gain During Pregnancy: Reexamining the Guidelines. Washington, DC: The National Academies Press, 2009.

2. ICMR, Nutrient Requirement and Recommended Dietry Allowances for Indians, A Report of the Expert Group of ICMR, 2010.

44

Nutrition in the Management of Liver Diseases

Jaya Benjamin

CONTENTS

This chapter deals with the aspects of management of diseases of the liver through modifications in the nutritional requirements or with nutritional therapy oral, enteral or parenteral.

I. CHRONIC LIVER DISEASE (CLD)

A. Malnutrition in CLD

Liver is the metabolic hub of the body that orchestrates many physiological processes essential for a well nourished state of the body. The functional reserve of the liver is so great that 80–90% of hepatocytes have to be damaged before manifesting as any physiological insult. Destruction of large proportion of hepatocytes results in liver failure/hepatic insufficiency. Chronic hepatic insufficiency is seen in end stage cirrhosis where most of the liver is replaced by scar tissue.

The reported prevalence of malnutrition in CLD ranges from 30–50% depending upon the criteria used for the assessment of nutritional status. In a well compensated disease protein energy malnutrition (PEM) may be seen in 20% of patients however it may be up to 60–90% in cases of severe hepatic insufficiency. Malnutrition is characterized by both reduced visceral as well as somatic proteins manifesting as a decreased muscle mass or body cell mass accompanied by muscle weakness and an impaired quality of life. A number of factors contribute to malnutrition in patients with CLD (Table 44.1). The impact of malnutrition is profound on these patients. Severe depletion of muscle mass and fat are the independent predictors of survival. The severity of malnutrition also correlates with the severity of liver disease, prevalence of complications and both long- and short-term survival. Malnutrition also results in poor outcome of liver transplant, poor quality of life, impaired immune functions, decreased muscle strength and increased fatigue. Hence a timely diagnosis of malnutrition is very important. Various methods are available for the assessment of nutritional status in patients with liver disease (Table 44.2); however it is very difficult to pick any one method for the assessment of nutritional status. The value of traditional methods is debated as they are confounded by changes in metabolism, body composition, and immune functions occurring in liver disease per se independent of nutritional status. Nevertheless, anthropometry, handgrip strength, subjective global assessment (SGA), and assessment

TABLE 44.1   Causes of Malnutrition in CLD

Decreased food intake	↓appetite
	Salt restricted diets
	Disguesia
	Ghrelin concentration
	Iatrogenic
	Mental depression
	Medications
	Ascites and gastroparesis causing early satiety
Hypermetabolism	↑REE coupled by ↑requirements
Altered metabolism	↓glycogen stores
	↓splanchnic glucose production
	↑gluconeogenesis
Decreased absorption	Small intestinal bacterial overgrowth
	↓intestinal motility
	Fat malabsorption
Increased losses	Protein losses during paracentesis
Effect of drugs	Neomycin – causes villous blunting and interferes with lipid metabolism
	Lactulose – ↓absorption
	Cholestryamine – causes steatorrhea
	Diuretics – cause K, Mg, Zn depletion
Complications	Sepsis, SBP, UGI bleed, cause ↑protein catabolism
Alcohol	Causes vitamin and mineral deficiency in particular
Procedures	Upper GI endoscopy causes starvation

TABLE 44. 2   Methods of Nutritional Assessment in CLD

Method	Comments
A. Energy Balance	
Diet intake	Spontaneous or usual diet intake
Energy expenditure	Measured by indirect calorimetry Predicted by Harris Benedict equation
B. Body composition	
Anthropometry	Triceps skin fold (TSF) and mid arm muscle circumference MAMC most common
	Compared with 5th or 10th percentile of reference population Local standards are recommended Convenient and economical, but high intraobserver and interobserver variability
Body mass index (BMI)	Fallacious in case of ascites Calculate BMI from estimated dry weight or define cut-offs for diagnosis of malnutrition according to presence of ascites
Creatinine Height Index	Very common method to assess whole body muscle mass Renal dysfunction and liver disease itself can effect results
Muscle strength (Handgrip dynamometry)	Simple, quick, inexpensive, non-invasive Muscle weakness correlates with severity of malnutrition in cirrhosis
Bioelectrical Impedance Analysis (BIA)	Relatively inexpensive, quick, reproducible, non-invasive Lacks credibility in ascites and odema Valid tool for nutritional assessment in non-ascitic cirrhotics
Dual Energy X-ray Absorptiometry (DEXA)	Precise, safe, reproducible Validity questionable in ascites and odema
Isotopic dilation and in-vivo neutron activation (IVNAA)	Ideal methods but facility are rarely available for routine practice
Biological Parameters (Albumin, C reactive protein, Prealbumin)	Affected by liver disease per se, excess alcohol intake and inflammatory state
Immune markers (delayed cutaneous hypersensitivity)	Affected by hpersplenism, alcohol abuse etc.
Composite Scores Subjective Global Assessment (SGA) Royal Free Hospital-Subjective global assessment (RFH-SGA)	Convenient, good interobserver reproducibility but may not be useful for serial measurements

of body composition by DEXA or bioelectrical impedance analysis (BIA) are the most commonly used reproducible, convenient and economical methods.

B.   Compensated CLD

The nutritional requirements of stable patients are not much different from normal individuals. However some differences do occur on the basis of presence of malnutrition, risk of encephalopathy and presence of specific vitamin deficiencies like in patients with alcoholic liver disease. Our understanding about nutritional management has changed over the last five decades.

Energy

In general stable patients tolerate normal calorie provision irrespective of the aetiology however in cases with cholestatic liver diseases reduction of calories contributed from fat is recommended without reduction of total calories provided. Similarly in patients with non-alcoholic fatty liver disease (NAFLD) or non-alcoholic steatohepatitis (NASH) with an associated metabolic syndrome benefit from reduction of total calories in order to manage body weight is recommended (discussed later). There have been conflicting data in terms of the calorie requirement of compensated or stable patients. Some studies recommended higher calorie intake than normal in view of the hypercatabolic state in cirrhosis. Hence a provision of 25–35 kcal/kg of energy is appropriate but for malnourished patients 35–40 kcal/kg may be needed to promote anabolism.

Protein

Patients with CLD often have reduced synthesis of proteins by the liver not only albumin but other proteins like clotting factors is a common feature in these patients. They also exhibit decreased urea synthesis and increased production of aromatic amino acids. In spite of these differences the net protein turnover in patients with compensated liver disease is not significantly different from that of normals. Despite the metabolic alterations patients with stable cirrhosis tolerate normal protein intake without an increased risk of encephalopathy. Thus in patients with compensated liver disease with normal renal functions without any pre-existing encephalopathy a protein intake of about 1.0 to 1.2 g/kg is needed. However if the patient is malnourished with a good renal function may require 1.5 to 2 g/kg body weight. However in stable patients with deranged renal function slight protein restriction is needed. This protein should be provided from high biological value sources. Proteins rich in BCAA may be useful (discussed later).

Stable cirrhotics often have fat soluble and water soluble vitamins, zinc, and selenium deficiency hence it should be supplemented in them. In cases with a history of alcohol consumption thiamine and folate supplementation should be considered. These stable patients should be counselled to meet the adequate requirements with normal diet avoiding high salt foods even in the absence of ascites. Small frequent meals ~4–7 meals/day with one late evening snack are recommended. In case of malnutrition when oral diet is not able to meet the requirements additional oral nutritional supplements (ONS) with specialized formulas should be given. However in patients with less advanced disease additional food combined with nutritional counselling is adequate. Fluid restriction should be done only in the presence of hyponatremia (<120 mmol/L).

C. Decompensated CLD

Decompensated cirrhotics with ascites should be managed with a low salt diet. In case of moderate ascites, sodium intake should be limited to <1000 mg (i.e., 1/4 tsp of salt) per day. The protein requirement increases to 1.2–1.5 g/kg/day. In case of impaired kidney function a modest reduction in protein intake can be done on a temporary basis until renal functions improve. Fluid intake is restricted only in case of hyponatremia and reduced urine output.

Protein controversy

Many factors contribute to HE like the degree of hepatocellular failure, portosystemic shunting and exogenous factors such as sepsis and variceal bleeding. The pathogenesis of HE is complex and uncertain. In most cases it is precipitated by factors found in addition to impaired hepatic functions like

azotemia (30%), sedatives and analgesics (30%), GI bleeding (20%), metabolic acidosis (10%) and dietary protein (9%). Impaired detoxification of ammonia absorbed from the gut is one of the important factors and dietary proteins are believed to contribute to this ammonia.

Hence protein restriction in patients with HE has been the cornerstone of treatment in the 1950's. However the efficacy of protein restriction has not been proven by randomized trials. On the contrary there is a rise in the aromatic amino acids due to endogenous protein breakdown thus contributing five times the amount of amino acids from normal dietary intake. Hence suppression of catabolism is a more effective strategy than dietary protein restriction. Skeletal muscle metabolises ammonia in patients with CLD, hence a loss of lean body mass depletes this 'ammonia sink' and increases the ammonia load to the brain thereby worsening HE. Concomitantly malnutrition and resultant protein breakdown in these patients may further deteriorate HE through decreased peripheral ammonia detoxification, increased release of aromatic amino acids, impaired liver functions and increased susceptibility for infections; all of these are the known precipitants of HE. Protein restriction of <40 g/day aggravates malnutrition and worsens HE. Protein intake of 1.06 g/kg/day has been associated with improvement in HE in patients with severe liver disease with PEM. Proteins are also very important for hepatocyte survival and functions.

In case of acute hepatic encephalopathy (overt HE) temporary protein restriction of 0.8 g/kg/d is done only for a few days until the cause of HE is diagnosed and eliminated. However normal intake of protein (1–1.2 g/kg/day) is resumed as soon as possible. Dairy (casein and whey) and vegetable proteins are preferred. Casein based diets are 22% BCAA along with an appropriate BCAA to AAA ratio of 2:1 in addition to a low methionine content A higher calorie formula with a 35 kcal/kg/day late evening snack has been found to promote protein anabolism.

In chronic hepatic encephalopathy patients show variation in the tolerance to dietary protein depending on its source. Long term protein restrictions may worsen the PEM in patients with chronic encephalopathy. A protein intake of 0.8 g/kg/day has been related with improvement in nutritional status. Patients have also shown to tolerate up to 100 g of vegetable protein without, deterioration of mental status. Randomized trials have demonstrated that diets containing 1.2 g of protein can be safely administered to patients with liver cirrhosis suffering from episodic encephalopathy and even transient protein restriction does not confer any benefit to these patients. In case of malnutrition higher amounts of protein can be sought by carefully titrating the maximal tolerable amount without deterioration of mental status. Vegetable and dairy protein show better tolerance than meat protein.

Largely uncontrolled observations have lead to the restriction of protein intake in HE and this practice is very popular with medical practitioners and nutritionists. The negative effects of protein restriction are clear. Recent trials have shown that in patients with acute hepatic encephalopathy low protein diet exacerbates protein breakdown without inducing any clinical benefit.

Cirrhotics show better tolerance to vegetable based proteins rather than meat based proteins. Vegetable proteins have a high fiber content which enhances the colonic motility and increases the intestinal transit time thereby promoting nitrogen clearance from the gut. They also reduce the colonic pH and prevent absorption of ammonia. Vegetable proteins also have better carbohydrate to protein ratio thereby promoting better anabolic response. Though vegetarian diets may cause abdominal bloating and flatulence; when combined with milk based proteins this effect can be reduced.

D. Branch Chain Amino Acids (BCAA)

In chronic liver diseases there is an imbalance in the BCAA and aromatic amino acids (AAA). The normal ratio of BCAA/AAA is 3 to 3.5; however in cirrhotics this ratio is very low. Decreased capacity of the liver to metabolize AAA leads to increase in AAA. On the contrary hyperinsullinemia, hyperglycemia, starvation and most importantly hyperammonaemia all lead to decreased levels of BCAA. BCAA have a three pronged action in cirrhosis (1) BCAA promotes the synthesis of glutamine thereby detoxifies the ammonia and eventually reduces the flux of AAA to the brain leading to improvement and prevention of HE. (2) The stimulatory effects of leucine on protein synthesis and hepatic stellate cells results in tissue repair and (3) BCAA also help in prevention and treatment of cachexia and muscle protein catabolism.

In a long term multicenter study by Marchesini et al. in 2003 found significantly low death rates and low progression rates of HE along with fewer hospital admissions, better appetite and health related quality of life in patients receiving BCAA versus an equicaloric maltodextrin and equicaloric equinitrogenous lactalbumin diet. A Cochrane review in 2004 states significant increase in number of patients improving from HE after supplementation with either enteral or parenteral BCAA. A metaanalysis by Muto et al. in 2005 showed that patients with cirrhosis who receive BCAA are more likely to recover from HE than those who do not receive this supplementation. BCAA are found to be more beneficial when given as a late evening snack. BCAA are highly recommended for patients awaiting liver transplantation. Milk proteins both casein and whey proteins are a good source of BCAA. Hence a milk/curd based diet forms the foundation of vegetarian diets in cirrhotics. Recently in a meta-analysis by Gluud et al., BCAA have been found to improve the manifestations of hepatic encephalopathy without any effect on overall survival.

E. Nutrition in Critically ill Patients with CLD

Critically ill patients with CLD include those with acute exacerbation of chronic hepatic failure, hepatorenal syndrome, sepsis, spontaneous bacterial peritonitits (SBP), acute variceal bleed or patients with acute liver failure. The mortality rates ranges from 40–90% in patients with CLD admitted to ICU.

Management of these patients begins with supportive measures including clinical assessment, fluid resuscitation, assessment of respiratory and haemodynamic status. In conjugation with an appropriate nutrition regimen to achieve anabolism, proper management of electrolytes balance, glycemic control and fluid maintenance is a vital parts of the treatment.

Early enteral nutrition (EN) is a standard care for all ICU patients and CLD is not an exception. EN reduces septic complications by maintaining gut barrier functions and immune competence. Patients with cirrhosis should achieve a energy intake of 35–40 kcal/kg/day and protein intake of 1.2–1.5 g/kg/day. If oral intake of diet is inadequate then additional oral nutritional supplements or tube feeding should be initiated. Due to profound hemodynamic instability, massive ascites and fluid retention, often enteral feedings are not tolerated by these patients. Under such conditions parenteral nutrition should be considered. Tube feeding is recommended even in the presence of esophageal varices, however in our clinical practice we wait for 24 hours following endoscopic procedure for bleed and then place a tube and start feeding. Nasogastric tubes are used more commonly as nasoenteric tubes often need endoscopic procedures. Nonetheless nasodueodenal and nasojejunal tubes are better than nasogastric tubes. Since most of the patients have fluid retention hence the total volume of the enteral formula is restricted, thus under such a situation a high energetic density (1.5 kcal/ml) feed with low sodium content is recommended.

Clinical trials do not suggest an increased risk of variceal bleeding due to insertion of feeding tube although while feeding patients with advanced HE the risk of aspiration during tube feeding must be weighed against the potential complications of parenteral nutrition. In patients with an esophageal upper gastrointestinal stricture there may be an indication for insertion of percutaneous feeding jejunostomy (PEJ) or gastrostomy (PEG) tubes, but presence of ascites, impaired coagulation profile and portosystemic collaterals are significant risk factors. A whole protein formula is generally recommended as per the ESPEN guidelines. As discussed earlier a BCAA rich formula should be used in patients with hepatic encephalopathy.

PN is the second choice of nutritional support as it is associated with gut mucosal atrophy, overfeeding, hyperglycaemia, adverse effects on immune functions, increased risk of infectious complications and increased mortality in critically ill patients. However in patients intolerant to enteral feeds it is mandatory. In moderate to severely malnourished cirrhotics immediate commencement of PN should be considered. A temporary abstinence from enteral/oral diet does not indicate use of PN, however IV glucose at 2–3 g/kg/d should be used. When fasting periods last for >72 hours then TPN is required. Carbohydrates should be given in the form of glucose to cover 50–60% of non-protein energy requirements. In case of hyperglycemia insulin infusion should be used. Lipids should provide around 40–50% of nonprotein energy requirements. The fat emulsions should be lower in *n*-6 unsaturated fatty acid content in order to have less stimulatory effects on proinflammatory cytokines and less suppressive effects on the leukocytes and immune cells.

Amino acids should be administered in amounts of 1.2 and 1.5 g/kg/day in patients without malnutrition and severe malnutrition respectively. Liver adapted formulas with increased amounts of BCAA and low AAA are recommended in these patients. Substrate utilization is better when glucose and lipids are infused simultaneously compared to individual nutrients alone. TPN mixtures containing amino acids, dextrose, lipids, electrolytes, vitamins and minerals, are available for infusion through central catheters. The high osmotic strength of PN admixtures requires the infusion of large amount of fluids that may be excessive for patients with ascites, hence EN is the preferred route.

In one study Plauth et al. have shown that small intestine metabolism contributes to post feeding hyperammonemia and worsens HE in cirrhosis in patients with transjugular intrahepatic portosystemic shunting (TIPS). Hence in such cases PN may be superior to EN. Cabre et al. in 2005 have suggested an increased absorption of dietary fats through the portal route in cirrhosis with a concomitant inflow of fat in the liver along with impairment in release of hepatic VLDL. This could facilitate the hepatic inflammation, fibrogenesis and steatosis, hence providing fat intravenously may be beneficial than enteral route in such a situation.

Acute liver failure

Acute liver failure is a serious condition with substantial loss of hepatic parenchymal function, characterized by profound metabolic dysfunction almost invariably complicated by multiple organ failure. Unfortunately there is very limited data from clinical trials on which the rationale for nutritional therapy or metabolic interventions can be based. Data only from descriptive physiology and experimental studies is available. Like any other critically ill patient, if oral nutrition is not likely to resume in 5–7 days (irrespective of the current nutritional status) enteral or parenteral nutrition is initiated. The primary goals of the nutrition therapy in ALF are to stabilize the metabolic and vital functions and treat brain odema. Hence adequate provision of energy with glucose and lipids at the same time maintaining euglycemia and also ensuring adequate protein synthesis by providing proteins and amino acids is the main focus of nutritional therapy.

ALF is a state of extreme catabolism due to hepatic impairment and co-resistant sepsis, in spite of a generally preserved nutritional state at presentation. The REE is increased by 1.2 to 1.3 fold, i.e., nearly 18–30% more than the healthy individual. Due to impaired gluconeogenesis, depleted hepatic glycogen and hyperinsulinemia, hypoglycaemia is a well known problem in ALF which has received appropriate attention in clinical practice. However at later stages secondary to sepsis hyperglycemia related to insulin resistance is observed

The reported rate of glucose infusion ranges from 0.6 to 10 g/kg/day. Hence a sufficient glucose provision (2–3 g/kg/day) is mandatory for prophylaxis and treatment of hypoglycaemia. There are no studies in the literature that have formally addressed strict glycemic control in ALF. In a survey by Schutz et al. found that only 39% of the participating centres aimed for blood glucose levels below 180 mg/dl. Hyperglycemia can also be problematic in ALF hence aiming at euglycemia had benefit over survival and morbidity. In view of a more widely prevalent hypoglycaemia care should be taken with insulin regimens.

For the hepatocytes oxidation of fatty acids and ketogenesis are the main energy yielding processes, hence adequate provision of lipids is a plausible therapeutic objective. However in patients with ALF due to loss of parenchymal mass hepatic utilization of fatty acids decreased. Moreover there is mobilization of mesenteric fat. Unlike the septic patients the splancnic viscera of ALF patients does not take up rather release free fatty acids. Like glucose there is no systematic data on the role of lipids and lipid monitoring in this context. However in most cases exogenous lipids are well tolerated. Lipids can be given at a rate of 0.8–1.2 g/kg/day simultaneously with glucose. Lipid utilization should be monitored by measuring plasma triglyceride levels and aim to maintain levels no higher than 4–5mmol/L. In most centres PN in ALF comprises of physical mixtures of LCT and MCT emulsions.

Unlike sepsis or other catabolic states, in ALF the released amino acids from the peripheral tissues are not taken up by the splancnic bed due to the loss of hepatocytes. Resulting in three to four fold rise in the plasma amino acid levels (hyperaminoacidemia). The ammonia released from the intestine cannot be extracted sufficiently by the failing liver in spite of ammonia detoxification by skeletal muscle thus resulting in hyperammonemia. The levels of BCAA may be decreased or normal and that of AAA is increased. Due to hyperaminoacidemia, hyperammonimeia, cerebral odema and encephalopathy, amino acids are often omitted from infusions due to the fear of aggravating the above conditions. But it is prudent to adjust the provision of amino acids (mainly as BCAA) according to ammonia levels. In hyperacute liver failure amino acid administration is not mandatory but in acute and sub-acute liver failure 0.8–1.2 g/kg/day of protein in PN and EN should be given to support protein synthesis. In general there is uncertainty about the use of amino acid in PN solutions. Almost more than half of the centers used amino acid solutions in ALF for hyperacute, acute and subacute liver failure also. Confusion regarding the use of amino acids arises because there are no accepted means of monitoring amino acid utilization apart from plasma ammonia levels.

In actual clinical practice nutritional therapy in ALF patients shows wide variation. There is also a lack of evidence regarding the optimal amount and composition of substrate. In hyperacute ALF, nutrition therapy is not important until transplant or recovery. Nearly 76% of hepatology centers use PN; EN has become the alternative and is very safe however both EN and PN require adequate metabolic monitoring.

Sepsis in liver failure

When systemic inflammatory response is triggered by infection it leads to sepsis. Sepsis is the most common cause of ICU mortality. A patient with liver disease is in a immunocompromised state and has a high risk of developing infections, sepsis and septic shock eventually. Resuscitation protocol of septic patients with liver disease is a logical step. Albumin is most commonly used to maintain intravascular volume. In such patients muscle wasting secondary to liver disease adds insult to injury. Dysphagia and gastroparesis is common in these patients. Presence of HE increases the risk of aspiration during enteral feeding, hence an early post pyloric feeding tube is recommended under fluoroscopic guidance. A high caloric high protein formula is recommended. Enteral formulas containing eicosapentaeoic acid (EPA) and γ-linolenic acid have shown to reduce mortality by 19.4% in septic patients. EPA has been shown to attenuate muscle atrophy by down regulation of the increased expression of the ubiquitin proeasome preolytic pathway. There is a high chance of cholestasis, steatosis and hyperglycemia in patients with sepsis and liver failure. Although adequate protein supplementation is essential but may be complicated by hyperammonaemia, azotemia, and deranged kidney function. A short term accommodation may be achieved by limiting protein intake in the predialysis phase.

Hypoglycemia commonly develops in patients with liver disease stressed with infection, however, hyperglycemia may develop rapidly with adequate resuscitation. Thus appropriate insulin therapy is essential to achieve euglycemia. Under such conditions enteral formulas that contain soluble fibers and slow digestible carbohydrates is mandatory.

II. NON-ALCOHOLIC FATTY LIVER DISEASE (NAFLD)

This disease was first described as "nutritional liver disease" in the 1960s. NAFLD is defined as >5% of liver weight consisting of fat in the absence of significant amount of alcohol consumption. It is considered as the hepatic component of metabolic syndrome as in majority of patients NAFLD is associated with metabolic risk factors like obesity, DM, dyslipidemia. The liver fat accumulation (pure fatty liver or steatosis) may be associated with variable degree of necroinflammation and fibrosis (NASH) and later may progress to advanced fibrosis and cirrhosis and finally to HCC.

Though only a minority of patients with NAFLD develop cirrhosis and ESLD it is the large majority of the masses that have NAFLD makes it a significant health care concern. By stable isotope technique it has been ascertained that in the presence of steatosis 59% of the triglycerides present in the liver arise from recirculation from adipose tissue 26% from de novo synthesis from dietary carbohydrates and 15% from dietary lipids. Lipotoxicity is thought to play a major role in the pathogenesis, modification of diet and exercise are the best therapeutic option. In this chapter we focus on the nutritional interventions that can potentially make a difference in the management of the disease.

A. Weight Loss

Patients with NAFLD have been found to have higher energy intake as compared to controls even in the Indian context as shown by Sathiaraj et al. in 2011. There is no established medical therapy available for NAFLD patients. Lifestyle modifications including both dietary modifications and increased physical activity is the most valid therapy for treatment of patients with NAFLD.

Studies have also documented histological improvement in steatosis and inflammation following weight loss in NAFLD along with improvement in transaminase levels and hepatic steatosis. Huang et al. found that even with an average weight loss of 2.9 kg in patients with biopsy proven NASH nearly 60% of patients showed histological improvements. The efficacies of low fat diets versus low calorie diets with low carbohydrates alone in producing weight loss have been equivocal. Nonetheless, unanimously all studies support the fact that even small degree of weight loss of around 5–10% show clear benefit. However rapid weight loss of around >1.6 kg/week may have an opposite effect on worsening the inflammation in patients with NASH by increasing the delivery of fat to liver from visceral fat breakdown. Prolonged fasting, excessively hypocaloric diets (<500 kcal/day) and jejunoileal bypass surgery aggravates steatohepatitis or may even result in cirrhosis and liver failure. Hence even slight weight loss may prove beneficial for patients but in patients with NAFLD particularly those with advanced liver disease slow and controlled weight loss over time is the goal. Though as low as 5% weight loss may have beneficial effects on NAFLD an initial 10% weight loss is recommended and supported by the National Heart Lung and Blood Institute, National Institute of Diabetes and Digestive and Kidney Disease clinical guidelines. Later 0.5 kg–1 kg/week is recommended rate of weight loss to avoid rapid weight fluctuations. Maintaining the weight loss is very important as regaining weight within subsequent months and years is commonly seen in these patients. There could be many physical, economic, and other barriers for patients to maintain weight loss. Large fluctuations in weight can exacerbate liver injury and dysfunction. Several factors are associated with maintenance of weight loss, i.e., moderate intensity exercise ~ 60 min/day, eating breakfast daily, increased emotional support and lower intake of soft drinks.

Thus it is important to follow these points and set realistic goals of weight loss through sustainable changes in lifestyle resulting in natural weight loss over time rather than weight loss itself through specific dietary interventions.

Apart from weight loss through caloric restriction, modification of the composition of diets also influences the development of NAFLD as the manipulation of the micronutrient and macronutrient content effects HTN, level of inflammation, serum lipids and IR independent of weight loss. In this chapter we review the effect and contribution of manipulation in each macro and micro nutrient on NAFLD.

Overall growing patterns towards a fast food diets have paralleled the increase in the incidence of obesity, DM, hypertension, dyslipidemia and in general NAFLD. These patterns include high composition of fructose and soft drinks, high meat and saturated fat and cholesterol intake and low consumption of fish & ω3FAor PUFA, fiber and vitamins.

B. Carbohydrates

The percentage of carbohydrate in drinks and glycemic index (GI) of foods influence the development of NAFLD. A carbohydrate rich diet promotes the development of fatty liver via increased *de novo* synthesis of fatty acids. High carbohydrates lead to increase in circulatory insulin concentrations leading to elevated fasting triglycerides (TG) levels. In a study, patients consuming carbohydrates as >54% of energy compared to <35% had a 6.5 fold increased risk of hepatic inflammation. In another study, patients receiving lower percentage of calories from carbohydrates with equal protein and equal calorie deficit had lower ALT levels as compared to high carbohydrate group, despite equal weight loss. In a yet another study patients with NAFLD on a carbohydrate restricted diet (<20g/d) had drastic reduction in hepatic TG levels (55%) as compared to control group on a low calorie diet only. However weight loss was comparable in both the groups.

Simple carbohydrates

Over the last few decades the consumption of simple carbohydrates and fructose has increased and parallel the increase in NAFLD, obesity and diabetes. Food industry has also replaced sucrose with high fructose corn syrup (HFCS) (containing 42–90% fructose). HFCS is used in the manufacturing of soft drinks; increased consumption of HFCS as soft drinks is linked with conditions like metabolic syndrome and increased liver enzymes. In a study the consumption of HFCS was found to be two fold higher in patients with NAFLD as compared to patients with other types of CLD.

Unlike glucose, fructose stimulates de novo fatty acid synthesis directly and promotes weight gain. Fructose is less responsive to insulin compared to glucose and sucrose and is more lipogenic. Fructose bypasses the rate limiting step in glycolysis, catabolised by phosphofructokinase and this results in lack of control over fructose metabolism. Unlike glucose, fructose does not stimulate leptin secretion effectively bypassing normal satiety signals that are paramount for regulation of food intake and body weight. Moreover fructose also up regulates the transcription factors like SREBP-1e and ChREBP which are involved in the lipogenesis thus promoting de novo lipogenesis thereby increased intrahepatic triglyceride (IHTG) synthesis.

Glycemic index

Valtuena et al. in 2006 found a close relationship between the degree of hepatosteatosis and dietary glycemic index independent of total energy or carbohydrate intake. Also considering the effect of GI on other comorbidities associated with NAFLD it may be an important aspect of diet to be considered while dietary counseling of patients with NAFLD. A low GI in conjugation with exercise does reduce post parandial hyperinsulinemia. High consumption of high glycemic index foods causes rapid and intensive accumulation of glucose this exceeds the glycogen synthesis capacity of the liver and their hepatic denovo lipogenesis occurs. Conversely low GI foods only slightly raise the blood glucose which results in a low insulin response. Nonetheless GI is a gross measurement whose direct influence on NAFLD remains unknown.

C. Fats

The total amount of fat in the diet irrespective of the total calorie and its chemical properties are important in patients with fatty liver. Increased intake is linked with insulin resistance and impaired postprandial lipid metabolism and increased the IHTG. In one study patients with NASH were found to ingest high percentage of energy from fat (37%) hence high fat diet is an independent factor for the progression of NAFLD. A reduction in the net amount of fat from 27% to 19% results in reduction of AST and ALT levels in NAFLD patients.

Saturated fats

Saturated fatty acids (SFA) have adverse effect on lipids and glucose homeostasis and worsen the metabolic syndrome and NAFLD. SFA have been found to cause hepatocyte damage by increasing the endoplasmic reticulum (ER) stress. It has been suggested by two studies that patients with fatty liver consume more SFA and less PUFA, fiber, and antioxidants, hence a diet low in SFA, i.e., ~ <10% SFA is beneficial however a further reduced consumption of SFA (i.e. <6%) has deleterious effects on the plasma lipid levels and may increase TG and decrease the HDL levels along with the decrease in LDL-C levels. Experimental studies suggest that high dietary SFA worsens IR, NAFLD and cardio vascular diseases in rats.

Mono-unsaturated fatty acids (MUFA)

Increased MUFA intake has been shown to have a positive effect on the lipid profile where it decreases the LDL-C levels and total cholesterol levels without reducing the HDL-C levels. MUFA also improves insulin resistance and if replaced for SFA in the diet it also offsets the proinflammatory effect of SFA and improves hepatic steatosis. Walnuts, hazelnuts, olives are rich in MUFA. Assy et al. in 2009 found that olive oil rich diet decreases the accumulation of TG in liver and also improves the postprandial TG levels and glucose response in patients with insulin resistance.

Poly unsaturated fatty acids (PUFA)

PUFA comprise the n-3 and n-6 fatty acids. In patients with NAFLD however the ratio of n-6 to n-3 is important. The n-3 PUFA (EPA and DHA) act as ligands of the peroxisome proliferators activated receptor α (PPAR α) and down regulate the fatty acid synthesis and thereby reduce the hepatic inflammation and oxidative stress. Low circulating levels of *n*-3 PUFA are associated with higher de novo lipogenesis.

Increased hepatic uptake of circulating free fatty acids and decreased fatty acid oxidation both can worsen hepatic steatosis. In experimental studies in animals with insulin resistance consumption of n-3 PUFA had favourable effect on regulating the plasma lipid levels and cardiovascular diseases, immune function and insulin. Moreover even in patients with metabolic syndrome and CVD, increased n-3 PUFA consumption improved dyslipidemia. n-3 PUFA enriched diets have been found to reduce the hepatic TG content and development of steatosis and improved the biochemical parameters (ALT & GGT) and imaging findings in patients with NAFLD. A higher consumption of fish rich in n-3 PUFA or walnuts may reduce the risk of NAFLD and improve dyslipidemia.

Trans fats

Trans fats are early found in natural foods but formed during the production of partially hydrated oils. Increased consumption of trans fatty acids (TFA) have been shown to be associated with obesity, insulin resistance and CAD. Animal experiments have documented the role of TFA in the pathogenesis of NAFLD and NASH, however lacks corroboration from studies in NAFLD, however they should be avoided in patients with NAFLD.

D. Protein

Deficiency of protein and malnutrition are known to predispose individuals to NASH however there is no conclusive evidence on a definitive effect of dietary proteins on NAFLD. Protein intake expressed as the percentage of energy has not been reported to be altered in NAFLD. However Zelbersagi et al. found a higher consumption of meat protein in patients with NAFLD. Moreover, higher red meat intake has also been associated with insulin resistance, increased diabetes and increased risk of CVD. The negative effect of protein deficiency is known but the upper limit of protein intake remains unclear.

E. Vitamins

Increased oxidative stress on injury is a well accepted pathophysiological feature in NASH. Hence the role of vitamins like vitamin E and C have been studied as antioxidants therapies in patients with NAFLD. Vitamin E suppresses lipid peroxidation and oxidative stress thereby improving inflammation and fibrosis in patients with NASH. Interestingly patients with NASH have been found to have lower intakes of antioxidants.

Many randomized trials have demonstrated improvement in liver histology scores and fibrosis scores with independent supplementation of vitamin E or in combination with vitamin C in patients with NASH. However certain high dose (>400 IU/day) trial's have shown an increase in all cause mortality therefore clinicians should be cautious when prescribing these vitamins to these patients. Study done by Nobili et al. did not show any additive effects of vitamin E and vitamin C therapy on fibrosis as compared to weight reduction by diet and physical exercise alone. Nevertheless patients with NASH have also been found to consume less vitamin C and E as compared to controls. More conclusive evidence is needed on the role of vitamins in NAFLD.

There have been many studies on the role of different types of diets on metabolic syndrome (MS) and its implications on NAFLD. There is no consensus as to what diet and lifestyle approach is ideal for patients with NAFLD and NASH, however inclusion of n-3fatty acids, high MUFA, fruits, vegetables, low glycemic index (GI) foods, high fiber, decreased fat, and simple carbs and sweetened drinks may universally be recommended in NAFLD patients.

III. NUTRITION IN ALCOHOLIC LIVER DISEASE

The prevalence of malnutrition/protein calorie malnutrition in patients with alcoholic liver disease is reported to be around 20–60% and nearly 100% in hospitalized patients with acute alcoholic hepatitis. Unlike the earlier belief, evidence from studies suggests no difference in the prevalence of malnutrition between patients with alcoholic and non-alcoholic cirrhosis.

Alcoholics have a dual burden; alcoholics suffer from primary malnutrition when alcohol replaces the other nutrients in the diet resulting in a net decrease in the overall nutrient intake. Secondary malnutrition occurs when alcoholics in spite of adequate intake of nutrients suffer from the toxic effect of alcohol and its interference in the digestion and absorption of these nutrients. The spectrum of nutritional status among patients with alcoholic liver disease spans from severe loss of muscle mass on one end to morbid obesity on the other. This wide range of nutritional status depends on the proportion of calories from alcohol ('how much they drink') and the composition of the diet ('what they eat') When only 16–23% of total calories come from alcohol then alcohol replaces only the carbohydrates in the diet but when >30% of calories come from alcohol it replaces fats and protein as well along with the calories and vitamins. To top it all the high caloric content of alcohol (7 kcal/g) along with high fat and/or calorie diet and sedentary life leads to the development of truncal obesity. A coexistent metabolic syndrome (MS) along with alcoholic liver disease leads to quick progression of fibrosis and more severe disease. Nevertheless these obese patients are not resistant to nutrient deficiencies associated with ALD.

Several mechanisms contribute to malnutrition in ALD like decreased caloric intake, decreased intestinal absorption and digestion and decreased processing and storage of nutrients coupled with a hypercatabolic state. The nutritional therapy for patients with ALD depends on the severity of liver disease, presence of factors like pancreatic insufficiency leading to malnutrition and presence of obesity. Obese patients with less severe disease should be referred for weight reduction by dietary restrictions and regular exercise. There is limited data on the outpatient nutritional therapy in patients with ALD. Most of the data on nutritional therapy pertains to inpatients with alcoholic hepatitis. In a study on patients with acute alcoholic hepatitis the mortality rates were more than 80% when voluntary oral intake was less than 1000 calories per day and almost no mortality in patients with an oral intake >3000 calories/day. Cabre et al. in a multicentre study compared the effect of 40 mg prednisone daily with a liver specific formula containing 2000 Kcal/day by NG tube; it was found that 1 year mortality rate was significantly lower in the enterally fed group mainly due to reduced infectious complications. Another study has demonstrated the benefit of tube feeding compared to regular diets in ALD. Hence in ALD and AH there is a potential need for tube feeding as nearly 67% of patients have been found to consume less than recommended calories in the hospital in spite of aggressive dietary and nursing support. Hence NS is vital for patients with ALD. American College of Gastroenterology (ACG) and American Association of Study of the Liver (AASLD) guidelines recommend 1.2–1.5 g/kg of protein and 35–40 kcal/kg of body weight per day in patients in ALD. Though the role of BCAA is still debatable, studies have shown benefit of BCAA supplementation on the health status and overall complications in cirrhosis. The nutritional recommendations for patients with alcoholic liver disease are no different from the general recommendations for CLD. However there is special emphasis on micronutrients and vitamins.

Zinc is the second most abundant trace elements after iron in the body. Alcoholics are often deficient in zinc manifesting as anorexia, skin lesions, neurosensory deficits, immune dysfunctions. Alcoholics often receive zinc supplementation in the form of zinc sulphate.

Thiamine deficiency is the most important cause of tissue damage in alcoholics and occurs due to impaired absorption and poor intake. Wernicke-Korsakoff syndrome (WKS), a severe mental disturbance is caused by thiamine deficiency in alcoholics. It is generally reversible and large doses of thiamine are used for its treatment. Most commonly thiamine deficiency is associated with peripheral neuropathy along with deficiency of other vitamins. Methionine deficiency occurs as a consequence of folate, betaine and choline deficiency. This along with impaired activity of S-adenosyl methyltransferase in cirrhosis leads to low levels of S-adenosylmethionine (SAM) which is essential for maintaining cell membrane integrity and glutathione synthesis. SAM administration has been found to increase glutathione concentration in ALD patients and prevents oxidative injury by maintaining mitochondrial structure. Pyridoxine deficiency occurs due to acetaldehyde (generated from alcohol metabolism) causing competitive decrease in albumin binding leaving the unbound vitamin to be lost in urine. Niacin deficiency or pallegra has been associated with chronic alcoholism. Alcoholic pallegra encephalopathy (APE) is less common than WKS and is thought to be under diagnosed. Hence niacin supplementation is also very important with thiamine and pyridoxine in patients with alcoholic encephalitis. Vitamin A deficiency (low serum levels and abnormal dark adaptation) is seen in approximately 50% of alcoholic cirrhotics. Hepatic vitamin A stores may become severely depleted in ALD even if liver damage is moderate and serum vitamin A and retinol binding protein levels are still within normal limits. Patients with alcoholic liver disease have low vitamin E levels as well. Vitamin E administration may be useful as it decreases the vulnerability of liver by alcohol but even normal vitamin E levels cannot prevent development of liver disease especially fibrosis.

Magnesium deficiency is also a common feature in alcoholics. The reasons are decreased intake and/or accumulation of saturated fatty acids on cell membrane. It is associated with muscle cramps and its supplementation improves muscle cramps in patients with alcoholic cirrhosis. Dworkin et al. found decreased levels of selenium in patients with ALD. Selenium levels were related to prothrombin activity and nutritional status. In a German study mild to moderate ALD in males consuming a normal diet (with alcohol as added calories) selenium status was depressed.

IV. NUTRITION IN LIVER CANCER

Liver cancer is the 5th most common malignancy in the world which brings about alteration in the metabolism of the host leading to malnutrition and chacexia thus influencing the overall morbidity and mortality. Due to the inherent nature of cancer per se, in particular HCC the role of nutrition has been better studied and explained in the pathogenesis and prevention of cancer rather than its treatment.

Dietary factors may contribute up to 30% of all malignancies in western countries and diet is the second preventable cause after tobacco of all causes. Alcohol consumption remains the main diet related risk factor for the development of liver cancer. Mycotoxin aflatoxin is an important dietary risk factor for the development of liver cancer. Mycotoxin aflatoxin is a common and unavoidable contaminant of food. This toxin is produced by the action of fungi during production, harvest, storage and food processing. The risk of HCC is based on the cumulative lifetime dose of mycotoxin. Aflatoxin may cause deletion of the p53 tumour suppressing genes and also bring about activation of dominant oncogene. A very strong association has been reported between aflatoxin and chronic hepatitis and in the development of HCC. Utmost attention to nutrition in terms of cancer has been given to the preventive aspect of cancer through dietary modifications. Among the various nutrients antioxidants like vitamin C, E, and selenium has been the focus of research in this area along with many non-nutritive chemicals like β carotene, epigallic acid, curcumin, lycopene, coenzyme Q10, epigallocatechin gallate, and N acetyle cysteine etc. These compounds contain antioxidant activity when consumed in the diet. The polyphenols in tea and coffee and other non-nutritive agents mentioned above are involved in radical scavenging of reactive oxygen and nitrogen species. Their mechanisms of actions are complex and are determined by the structure of the compound, redox status of the environment and interations with other agents. Though preliminary data so far from experimental epidemiology studies have suggested a preventive role of antioxidants; however the same has not been proved by randomized controlled trials. In a recent Cochrane review, antioxidants like β carotene, vitamin A, C, E and selenium have not proven beneficial for the prevention of gastrointestinal cancers. Paradoxically antioxidants have been associated with an increase in all cause mortality. The potential role of tea and coffee on liver cancer prevention has already been extensively studied (discussed in detail in Chapter 45 of this book). Compounds like lycopene have been studied extensively. Lycopene is a red coloured carotene and carotenoid pigment and phytochemical found in tomatoes and other red fruits and vegetables. High lycopene intake and high serum levels of lycopene have been associated with a decreased risk of HCC.

Because of conflicting results from these studies it is difficult to make recommendations to increase the intake of antioxidants in order to decrease the risk of HCC. Diet therapy for liver cancer be addressed from two angles (1) malnutrition and chachexia and (2) using diet as the therapy to directly treat HCC. Chechexia is a common feature seen in patients with HCC, which has an independent effect on the morbidity and mortality of these patients. Malnutrition undoubtedly is a risk factor for poor outcome following surgical resection as well as increased toxicity following chemo and radiotherapy. Chachexia is caused by a multiplicity of factors like hypermetabolism, acute phase protein response failure of anabolism and inadequate intake of food. Timely recognition of PCM is important, hence due attention should be given to the food intake pattern, food aversions, changes in taste, pain/discomfort while eating etc. For those patients who fail to meet the needs adequately by oral diet, EN or PN is imperative.

American studies have shown that n-3 PUFA may significantly retard growth of tumors and improve the efficacy of many chemotherapeutic drugs. Whereas n-6 FA may potentially increase tumor development. A large metaanalysis of 38 cancer incidence studies found little evidence in favour of supplementation of n-3 FA in reducing risk of cancer incidence. Hence AICR and other organizations recommend a balance between n-6 and n-3 FA keeping the dietary fat <=30% of total energy intake.

Oxidative stress has been proposed to be involved in the etiology and progression of cancer including liver cancer. On one hand an antioxidant therapy along with a low fat, high fiber diet with lifestyle modifications has been shown to markedly improve the efficacy of standard and experimental cancer therapies. But on the other hand supplemented antioxidant can interfere with the oxidative breakdown of cellular DNA and cell membrane necessary for the chemotherapeutic agents to work. Additional supplementation of antioxidants may prevent the spontaneous apoptotic breakdown of tumor cells. The beneficial effects of supplementation of different antioxidants like vitamin E, C, β carotene and selenium as an adjunct in the treatment of cancer has been offset by the contradictory data in support of detrimental effects of high dosage of these antioxidants. Like vitamin E may act as a pro-oxidant in cigarette smokers particularly if diet is high in *n*-6 fatty acids. Similarly, consuming excessive amounts

of vitamin C may transform its beneficial properties into a harmful pro-oxidant nature that may interfere with standard therapy. β-carotene supplementation have not been shown to confer any benefit for the prevention of cancer and actually may cause harm in certain subgroups, thus it is not routinely advised for cancer patients.

In conclusion there is no convincing evidence in favour of antioxidants in amounts over and above the amounts obtained from normal diets rich in fruits and vegetables. Specifically in HCC only selenium supplementation has shown clear benefit. Hence the American Institute of Cancer Research (AICR) does not recommend over supplementation of antioxidants, a diet containing 5 or more servings of fruits and vegetables and the amount of antioxidants obtained in a reliable daily multivitamin pill at the level comparable with RDAs is good enough. Hence the bottom line is to follow a balanced diet avoiding over indulgence in calories, sugars and fats; focusing on adequate amounts of fruits and vegetables in the prevention and even treatment of hepatocellular cancer.

V. NUTRITION IN CHOLESTATIC LIVER DISEASE

Diseases like PBC, PSC and autoimmune cholangiopathy represent the cholestatic liver disease characterized by increased bilirubin, often have associated malnutrition. PCM is not a very common finding in patients with cholestatic liver disease without cirrhosis. However they may develop deficiencies of vitamin and minerals especially when bilirubin is >5 mg/dl. Severe PCM occurs in cholestatic diseases only when the disease is long standing for more than one or two decades. Cholestasis due to morphological, clinical or functional causes results in diminished flow of bile into the duodenum; insufficient to form micelles resulting in a decreased fat absorption. Failure of emulsification of fat leads to steatorrhea and decreased absorption of fat soluble vitamins and its resultant complications. Fat malabsorption clinically manifests as steatorrhea with a fecal fat excretion of >10 gm/24 hrs (normal value <9 gm/72 hrs). Drugs like cholestryamines used to treat pruritis may worsen steatorrhea. Steatorrhea leads to weight loss, weakness, diarrhea, along with bone disorders, night blindness and follicular hyperkeratosis.

These patients require a high calorie intake ranging from 25–35 kcal/day and protein up to 1–1.5 gm/day without sodium restriction. However in steatorrhea patients should be given ~15–20% of calorie from fat. The total fat is decreased to 30–40 gm/day with additional fat in the form of medium chain triglycerides (MCTs) can be added to the diet to increase the calorie intake, starting from 1 teaspoon/meal and gradually increasing the dose. MCTs are C5 or C10 fatty acids that are absorbed directly into the portal circulation as free fatty acids without undergoing micelles formation. Hence patients may be asked to cook food in coconut oil as it is a rich source of MCTs.

Chronic cholestasis may also cause hyperlipidemia in the form of hypertriglyceridemia and total cholesterol. These patients may also have skin xanthomas but only when the serum lipids exceed 1000 mg/dL. Xanthomatous neuropathy also occurs when excess lipids deposit in the peripheral nerves. Usually pharmacological therapy for hyperlipidemia is suggested only in patients with multiple risk factors for coronary artery disease but not otherwise.

Hepatic osteodystrophy (osteopenia and osteomalacia) is a common finding in patients with cholestatic liver disease. A number of factors may lead to hepatic osteodystrophy in these patients like poor dietary intake of both calcium and vitamin D, fat malabsoption and the related deficiency of fat soluble vitamins including vitamin D, cholestasis may also cause an impaired hydroxylation to form 25OH vitamin D, and lack of photoisomerization (due to insufficient sunlight exposure). Cholestasis may also indirectly influence calcium absorption; as calcium would bind to the free fatty acids in the gut (steatorrhea) thereby reduce the total amount of absorbable calcium. Hence in patients with chronic cholestatic liver disease dual energy X-ray absoptiometry (DEXA) should be routinely performed to screen for hepaticosteodystropy and also have a periodic assessment of vitamin D status. Adequate sun exposure, increased intake of calcium and phosphorus in the diet should be encouraged. In case of vitamin D deficiency, supplementation with 400–800 IU/day of vitamin D is recommended. In general calcium should also be supplemented along with vitamin D. In case of patients with steatorrhea, supplementation of MCTs would help in the absorption of calcium along with calcium supplementation given between meals. Calcium usually binds with oxalates in the intestine, but in case of fatty acid malabsorption calcium forms soap with fatty acids causing an increase in oxalate absorption resulting in a risk of kidney stone formation. Hence patients with severe PBC and PSC may be advised a low oxalate diet.

Vitamin E deficiency may be seen in steatorrhic patients with cholestatic disease. Its deficiency may be associated with psychomotor dysfunction, ataxia, opthalmoplegia, myopathy, and pigmented retinopathy. However the efficacy of vitamin E supplementation in adults has not been established. Patients with cholestatic liver disease may also have prolonged prothrombin time due to either impaired liver function or may be due to vitamin K deficiency; therefore patients with a severe disease may require intramuscular injections of vitamin K. Similarly, these patients should be screened for vitamin A deficiency, i.e., serum retinol levels and dark adaptation tests. If found deficient, the diet of a non-alcoholics should be supplemented with 5000–15000 IU of vitamin A daily.

SUGGESTIONS FOR FURTHER READING

1. Delich P.C., Siepler J.K. and Parker P. (2007), Liver Disease, In: The ASPEN Nutrition Support Core Curriculum: A Case Based Approach—The Adult Patient, Gottschlich M.M. (Eds.), ASPEN, USA.

2. Plauth M., Cabré E., Riggio O., Assis-Camilo M., Pirlich M., Kondrup J.; DGEM (German Society for Nutritional Medicine), Ferenci P., Holm E., Vom Dahl S., Müller M.J. and Nolte W.; ESPEN (European Society for Parenteral and Enteral Nutrition), (2006), ESPEN Guidelines on Enteral Nutrition: Liver Disease, *Clin Nutr.* 25(2):285–94.

3. Plauth M., Cabré E., Campillo B., Kondrup J., Marchesini G., Schütz T., Shenkin A. and Wendon J. (2009) ESPEN, ESPEN Guidelines on Parenteral Nutrition: Hepatology, *Clin. Nutr.* 28(4):436–44.

4. O'Brien A. and Williams R. (2008), Nutrition in End-stage Liver Disease: Principles and Practice, *Gastroenterology*, 134(6):1729–40.

5. Prakash R. and Mullen K.D. (2010), Mechanisms, Diagnosis and Management of Hepatic Encephalopathy, *Nat Rev Gastroenterol Hepatol.*, 7(9):515–25.

6. Marchesini G., Bianchi G., Merli M., Amodio P., Panella C., Loguercio C., Rossi Fanelli F. and Abbiati R. (2003), Italian BCAA Study Group, Nutritional Supplementation with Branched-chain Amino Acids in Advanced Cirrhosis: A Double-blind, Randomized Trial, *Gastroenterology*, 124(7):1792–801.

7. Gluud L.L., Dam G., Borre M., Les I., Cordoba J., Marchesini G., Aagaard N.K., Risum N. and Vilstrup H. (2013), Oral Branched-chain Amino Acids have a Beneficial Effect on Manifestations of Hepatic Encephalopathy in a Systematic Review with Meta-analyses of Randomized Controlled Trials, *J. Nutr.*, 143(8):1263–8

8. Als-Nielsen B., Koretz R.L., Kjaergard L.L. and Gluud C. (2003), Branched-chain Amino Acids for Hepatic Encephalopathy, *Cochrane Database Syst Rev.*, (2):CD001939.

9. Muto Y., Sato S., Watanabe A., Moriwaki H., Suzuki K., Kato A., Kato M., Nakamura T., Higuchi K., Nishiguchi S. and Kumada H. (2005), Long-term Survival Study Group, Effects of Oral Branched-chain Amino Acid Granules on Event-free Survival in Patients with Liver Cirrhosis, *Clin. Gastroenterol Hepatol*, 3(7):705–13.

10. Cabral C.M. and Burns D.L. (2011), Low-protein Diets for Hepatic Encephalopathy Debunked: Let Them Eat Steak, *Nutr. Clin. Pract.*, 26(2):155–9.

11. Milke García M.P. (2011), Nutritional Support in the Treatment of Chronic Hepatic Encephalopathy, *Ann Hepatol*, 10 Suppl 2:S45–9.

12. Pontes-Arruda A., Aragão A.M. and Albuquerque J.D. (2006), Effects of Enteral Feeding with Eicosapentaenoic Acid, Gamma-linolenic Acid, and Antioxidants in Mechanically Ventilated Patients with Severe Sepsis and Septic Shock, *Crit. Care Med.*, 34(9):2325–33.

13. Schütz T., Bechstein W.O., Neuhaus P., Lochs H. and Plauth M. (2004), Clinical Practice of Nutrition in Acute Liver Failure—a European Survey, *Clin. Nutr.*, 23(5):975–82.

14. McCarthy E.M. and Rinella M.E. (2012), J. Acad Nutr Diet, The Role of Diet and Nutrient Composition in Nonalcoholic Fatty Liver Disease, 112(3):401–9.

15. Colak Y., Tuncer I., Senates E., Ozturk O., Doganay L. and Yilmaz Y. (2012), Nonalcoholic Fatty Liver Disease: A Nutritional Approach. *Metab Syndr Relat Disord.*, 10(3):161–6.

16. Sathiaraj E., Chutke M., Reddy M.Y., Pratap N., Rao P.N., Reddy D.N. and Raghunath M. (2011), A. Case-control Study on Nutritional Risk Factors in Non-alcoholic Fatty Liver Disease in Indian Population., *Eur J. Clin Nutr.*, 65(4):533–7.

17. Singal A.K. and Charlton M.R. (2012), Nutrition in Alcoholic Liver Disease, *Clin. Liver Dis.*, 16(4):805–26.

18. Sarin S.K., Dhingra N., Bansal A., Malhotra S. and Guptan R.C. (1997), Dietary and Nutritional Abnormalities in Alcoholic Liver Disease: A Comparison with Chronic Alcoholics without Liver Disease, *Am J. Gastroenterol.*, 92(5):777–83.

19. Mendenhall C., Roselle G.A., Gartside P. and Moritz T. (1995), Relationship of Protein Calorie Malnutrition to Alcoholic Liver Disease: A Reexamination of Data from two Veterans Administration Cooperative Studies, *Alcohol Clin. Exp. Res.*, 19(3):635–41.

20. Cabré E., Rodríguez-Iglesias P., Caballería J., Quer J.C., Sánchez-Lombraña J.L., Parés A., Papo M., Planas R. and Gassull M.A. (2000), Short- and Long-term Outcome of Severe Alcohol-induced Hepatitis Treated with Steroids or Enteral Nutrition: A Multicenter Randomized Trial, *Hepatology*, 32(1):36–42.

21. Russell R.M. (1980), Vitamin A and Zinc Metabolism in Alcoholism, *Am J. Clin Nutr.*, 33(12):2741–9.

22. Dworkin B.M., Rosenthal W.S., Stahl R.E. and Panesar N.K. (1988), Decreased Hepatic Selenium Content in Alcoholic Cirrhosis, *Dig. Dis. Sci.*, 33(10):1213–7.

23. Assy N., Nassar F., Nasser G. and Grosovski M. (2009), Olive Oil Consumption and Non-alcoholic Fatty Liver Disease, *World J. Gastroenterol.*, 15(15):1809–15.

24. Zelber-Sagi S., Nitzan-Kaluski D., Goldsmith R., Webb M., Blendis L., Halpern Z., Oren R. (2007), Long Term Nutritional Intake and the Risk for Non-alcoholic Fatty Liver Disease (NAFLD): A Population Based Study, *J. Hepatol.*, 47(5):711–7

25. Nobili V., Manco M., Devito R., Ciampalini P., Piemonte F. and Marcellini M. (2006), Effect of Vitamin E on Aminotransferase Levels and Insulin Resistance in Children with Non-alcoholic Fatty Liver Disease, *Aliment Pharmacol Ther.*, 24(11–12):1553–61.

26. Labriola D. and Livingston R. (1999), Possible Interactions Between Dietary Antioxidants and Chemotherapy, *Oncology* (Williston Park), 13(7):1003–8

27. Kong Q. and Lillehei K.O. (1998), Antioxidant Inhibitors for Cancer Therapy. *Med Hypotheses*, 51(5):405–9.

28. LaRusso N.F., Shneider B.L., Black D., Gores G.J., James S.P., Doo E., and Hoofnagle J.H. (2006), Primary Sclerosing Cholangitis: Summary of a Workshop, *Hepatology*, 44(3):746–64.

29. Mustafa Alnounou and Santiago J. Munoz (2006), Nutrition Concerns of the Patient with Primary Biliary Cirrhosis or Primary Sclerosing Cholangitis, *Practical Gastroenterology*, 92–100.

45

Beverages
Tea and Coffee

Y.K. Joshi and Jaya Benjamin

CONTENTS

I. Tea
  A. Introduction
  B. Antioxidant activity of tea
  C. Anti-inflammatory activity of tea
  D. Potential health benefits of tea consumption
  E. Potential health risks of tea consumption
II. Coffee
  A. Introduction
  B. Potential health benefits of coffee consumption
  C. Potential health risks of coffee consumption
  Summary

Tea and coffee are the most popular beverages in the world. They have been consumed by humans since ages, but only recently their effects on health have been recognized and studied in great detail. This chapter covers the chemical components in tea and coffee along with their proposed mechanisms of action in the body and also a review about their beneficial and harmful effects on human health.

I. TEA

A. Introduction

Tea is the second most popularly consumed beverage in the world, widely known for its refreshing and stimulating effects. Nearly three billion kilograms of tea is produced and consumed every year. The history of tea timeline dates back from 2737 BC when the Chinese emperor named Shen Nung accidentally discovered this beverage when the leaves of the plant named *Camellia sinensis* blew into his cup of hot water. Nearly 78% of the tea is produced worldwide as black tea (BT), 20% as green tea (GT) and 2% as Oolong tea (OT). China and India are the largest producers of tea in the world. As the Chinese proverb goes "drinking a daily cup of tea will surely starve the apothecary", the health benefits of tea have been known since its origin; however it is only in the past thirty years that the scientific research of this beverage had been in progress.

The health benefits of tea are related to the bioactive compounds called the flavanoids, which have anti-inflammatory, antioxidant and antimutagenic properties. Flavanoids are aromatic compounds having the benzopyrane skeleton. Catechins are the major flavanoids in tea; these include epicatechin (EC), epigallocatechin (EGC), epicatechin-3-gallate (ECG) and epigallocatechin-3-gallate (EGCG), catechin gallate, gallocatechin, gallocatechin gallate, epigallocatechin digallate, methylepicatechin and methyl EGC. EGCG is the most abundant, accounting for 50–75% of the total catechins. Green tea (GT) contains higher amounts of catechins as compared to black and oolong tea. Tea also contains flavanols, including quercetin, kaempferol, myricetin and their glycosides. Other compounds that are present in tea are gallic acid, quinic esters of gallic, coumaric and caffeic acids together with the purine alkaloids theobromine and caffeine, proanthocyanidins and trace levels of flavones. Tea beverage prepared by 3 minutes brewing of 1 gram leaf in 100 ml of water would typically contain 250–350 mg of tea solids, comprising of 30–42% catechins and 3–6% caffeine.

The four basic varieties of tea are produced from the same plant through different processing techniques. White tea (WT) is prepared when leaves are picked and harvested before the tea buds fully open, while still covered by fine white hair. The processing of WT and GT involves steaming the leaves quickly soon after harvesting to inactivate the enzymes that prevent the oxidation of polyphenols, and then drying. In case of BT the leaves are rolled, disrupting the cellular compartment and bringing the phenolic compounds into contact with the polyphenol oxidases, thus the young tea leaves undergo oxidation for 90–120 minutes before drying. This process is called 'fermentation', wherein the flavan-3–ols are converted to complex condensation products like theaflavin and thearubigins. BT derives its unique taste and red-orange color from the Benzotropolone ring structure present in the flavins. Oolong tea is manufactured with a shorter fermentation period than BT and has the taste and color somewhere between GT and BT.

B. Antioxidant Activity of Tea

The antioxidant property of tea polyphenols is the main factor associated with health benefits of tea. The polyphenols in tea prevent the formation of reactive oxygen species (ROS) and also possess strong metal ion chelation properties. The chemical structures of polyphenols in tea are responsible for its strong antioxidant capacity which allows electron delocalization, conferring high reactivity to quench free radicals. The B ring of catechins consists of vicinal dihydroxy or trihydroxy groups which can chelate metal ions and prevent the generation of free radicals. The radicals quenching ability of green tea is usually higher than black tea. EGCG is most effective among all the other catechins against the ROS. Specific structural characteristics of the green tea polyphenols like pyrogallol type B ring and/or galloyl group, electron withdrawing substitutes

(carbonyl and ketal carbon) and/or intramolecular hydrogen bonding are responsible to prevent auto-oxidation reaction.

Tea preparations can trap ROS like superoxide radicals, singlet oxygen, hydroxyl radical, peroxyl radical, nitric oxide, nitrogen dioxide and peroxynitrite and prevent the damage to lipid membranes, protein and nucleic acids. These antioxidant properties of tea contribute to the prevention of atherosclerosis and cardiovascular diseases. It is also believed that these tea polyphenols can stimulate the antioxidant responsive elements (AREs) present in the promoter regions of many antioxidant genes and thereby stimulate antioxidant transcription and detoxify system through ARE. The antioxidant capacity of tea polyphenols against peroxyl radicles has been found to be much higher than vegetables like garlic, spinach and Brussels sprouts. A single dose of tea improves plasma antioxidant capacity of healthy adults within 30–60 minutes after ingestion. After ingestion of tea though the plasma antioxidant capacity peaks in about 1–2 hours but it subsides shortly after. A repeated consumption of tea or its extracts for 1–4 weeks has been found to decrease biomarkers of oxidative status. The evidence on this aspect of tea is rather conflicting. There are reports in the literature suggesting an augmentation of the total antioxidant capacity of plasma after consumption of tea (Table 45.1), at the same time some reports (Table 45.2) fail to demonstrate the antioxidant potential of tea and its extracts.

Similarly there has been conflicting data regarding the role of tea consumption on the markers of lipid oxidation. Only a few studies have shown a positive effect of tea consumption on reducing lipid peroxidation, as most of the intervention studies did not show any effect of tea consumption on markers of lipid oxidation in healthy subjects.

TABLE 45.1   Studies showing the evidence of antioxidant capacity of tea

Year	Author	Result
1994	Serafini et al.	First evidence of how GT and BT boost plasma antioxidant capacity in humans
1996	Serafini et al.	After ingestion of both GT and BT the total antioxidant capacity (TAC) increased significantly after a single consumption of 300 ml of either GT and BT. No difference seen between GT and BT.
1997	Van Let Hof et al.	Significant increase in plasma TAC seen after consumption of 6 cups of GT/day for 4 weeks. No significant effect seen with BT.
1998	Pietta et al.	Increase in plasma TAC, EGE-gallate and EC-gallatae after GT consumption. 6–8 hours urine showed substantial amounts of final catechin metabolites
1999	Benzie et al.	Increase in TAC 40 minutes after ingestion of tea in healthy volunteers along with significant increases in TAC in urine
2000	Leenen et al.	Single dose of GT/BT increased plasma TAC and plasma catechins in 24 volunteers within 3—60 minutes after ingestion. Rise in plasma catechin concentration were more after consumption of GT compared to BT.
2000	Langley Evans	Increase in plasma TAC after consumption of BT in a crossover randomized study design
2000	Sung et al.	A positive dose response relation was seen in TAC after consumption of 300 and 450 ml of GT
2002	Young et al.	Intervention of GTE for 3 wks increased plasma TAC in a mixed group of smokers and non-smokers
2007	Kyle et al.	Significant increase in the plasma TAC and concentration of phenols and catechins was seen 80 minutes after consumption of BT in a acute intervention study
2010	Pecorari et al.	GTE at a concentration of 2g/L significantly increased plasma antioxidant potential and plasma reducing power (FRAP) at 1 hour of consumption in 15 healthy volunteers

TABLE 45.2 Studies not showing the evidence of antioxidant capacity of tea

Year	Author	Results
1996	Maxwell et al.	No significant change in TAC in healthy subjects after ingestion of BT
1998	McAnlis et al.	No change in plasma TAC was found
2000	Hodgson et al.	No significant difference on TAC was seen after ingestion of BT or GT in healthy subjects
2001	Deiffy et al.	No effect on plasma TAC was seen in 60 patients with coronary artery disease
2002	Kimura et al.	Acute ingestion of tea polyphenols did not change the plasma TAC but increased the EGCG concentration. A long term supplementation did not have any affect on TAC and EGCG concentration
2005	Henning et al.	After ingestion of either purified EGCG or GT extracts no change in plasma TAC and EGCG was seen
2010	Li L et al.	GTE (up to 200 mg/d) alone or in combination with lutein failed to increase the overall antioxidant activity of lipid peroxidation of subjects
2010	Muller et al.	The markers of oxidative stress (8–iso-prostaglandin-F(2a); DNA strand breaks) were not affected by consumption of 600 ml white tea or GT in healthy non-smokers

C. Anti-inflammatory Activity of Tea

Unchecked inflammatory response causes irreparable damage to the tissue and gives way to a number of diseases like obesity, cardiovascular diseases, diabetes and cancer. The anti-inflammatory activities of tea polyphenols may be due to the stimulation of a negative feed back mechanism of the inflammatory process such as suppression of leukocyte adhesion to endothelium and subsequent transmigration through inhibition of transcriptional factors mediated production of cytokines and adhesion molecules. The anti-inflammatory activities of tea also include inhibition of proinflammatory enzymes modulation of signal transduction, inhibition of production of NO and expression of iNOS messenger ribonucleic acid (mRNA) by macrophages; decreased expression of different proinflammatory cytokines/chemokines and enhanced production of anti-inflammatory cytokines like IL-10. There is inconclusive evidence on the effect of tea polyphenols on the inflammatory markers. In a few studies consumption of BT and GT had no effect on IL6. IL1β, TNFα and CRP. However in a few studies in patients with coronary disease there was an increase in the plasma catechin levels but no reduction in the CRP levels. Similarly in patients with diabetes BT and GT administration did not influence the level of CRP, IL6, TNFα, fibrinogen, intercellular adhesion molecule (ICAM) and vascular adhesion molecule (VCAM).

D. Potential Health Benefits of Tea Consumption

Tea and its constituents have been found useful in a few disease conditions.

Cancer

Many experimental studies have studied the role of tea polyphenols and tea on cancer prevention. Various mechanisms have been proposed in this aspect. These include inhibition of MAP kinases and P13 K/AKT pathway, inhibition of NFkB and AP-1 mediated transcription growth factor mediated signaling, along with inhibition of aberrant arachidonic acid metabolism and other activities. All these mechanisms may finally result in inhibition of tumor cell growth, induction of apoptosis or inhibition of angiogenesis.

The epidemiological studies support the role of tea in decreasing the risk of cancer, likewise the *in vitro* and *in vivo* experimental models also generate convincing data. In spite of lack of information on the exact mechanism of action of tea catechins, the anti-cancer activity of GT has been demonstrated in a number of cancer models like lung, breast, skin, oesophagus, stomach, liver, intestine, colon, prostate etc. Nevertheless the concentration of catechins found to be effective in the in vitro studies is 10–100 times higher than that concentration achieved in plasma or tissue. Moreover the results obtained from these experimental studies may not be directly extrapolated to humans. Several mechanisms are likely to be involved in these experimental studies out of which some of the proposed mechanisms may not be relevant to cancer prevention in humans unless these mechanisms are demonstrated in human tissue by identifying the direct targets of action of tea catechins. Hence there is an undoubtful need of clinical trials on this aspect.

Gastrointestinal tract cancer: The positive role of EGCG, GT extracts and BT on prevention of stomach and duodenal carcinogenesis in rats and intestinal carcinogenesis in mice have been reported earlier. Mu et al. found 81% and 16% decrease in risk of development of stomach cancer and 39% and 31% decrease in the risk of oesophageal cancer in alcoholics and cigarette smokers respectively. In spite of the strong evidence from experimental data in support of GT and BT as a potential chemo preventive agent in colorectal cancer, the epidemiological data from 25 studies across 11 countries was found to be insufficient to conclude on the protective effect of either BT and GT against colorectal cancer in humans. Similarly the metaanalysis of 14 epidemiological studies on 6123 patients with gastric cancer and 13,4006 controls did not show any association of GT with gastric cancer prevention **even** with highest consumption level of >5 cups/day. On the contrary another metaanalysis of recent cohort studies in 2009 suggested that the highest given GT consumption was shown to significantly increase the risk of stomach cancer.

Breast Cancer: EGCG have been found to suppress Wnt signaling in invasive breast cancer cells. Similarly, treatment of EGCG rich GT polyphenols in mice resulted in reduced tumor size along with inhibited metastasis to lungs and improved survival after treatment. GT consumption has been found to be associated with a reduced risk of breast cancer in Chinese, Japanese and Filipino American women. Reportedly women who consume 3 or more cups of tea per day were 37% less likely to develop breast cancer compared to those who don't drink tea. Seely et al. in 2005 found in their mataanalysis, that consumption of >5 cups of GT/day showed a insignificant trend towards prevention of breast cancer however GT consumption may possibly prevent breast cancer recurrence in early stages of the cancer. Sun et al. in 2006 also showed a lower risk of breast cancer with GT and a minor effect of BT consumption in a metaanalysis which included 13 articles in 8 countries. Increased GT consumption (>3 cups/day) is found to be inversely associated with the risk of breast cancer recurrence; however the association of GT consumption and breast cancer incidence remains unclear.

Liver cancer: Positive effects of tea catechins tea extracts have been seen in terms of decreased incidence of NNK induced liver tumors and hepatic preneoplastic foci formation in rats. GT consumption may also reduce the risk of cancer caused due to exposure of pollutants like pentachlorophenol. Mu et al. reported that regular consumption of GT reduces the risk of development of liver cancer by 78% among alcoholic and 43% among cigarette smokers. A recent metaanalysis investigated the association between tea consumption and risk of primary liver cancer from 6 case control and 7 cohort studies; consumption of tea showed preventive effects on the development of primary liver cancer.

Ovarian cancer: In a metaanalysis of 5 cohort studies, consumption of tea was found to be significantly and inversely associated with risk of ovarian cancer. Contrary to this another metaanalysis Zhou et al. did not find any association between consumption of tea and reduction in the risk of ovarian cancer. Recently in 2010, Nagle et al. found that consumption of >4 cups of tea per day (either BT, GT, Herbal tea) is protective for ovarian cancer. A population based Swedish study suggested that tea consumption is associated with a decreased risk of epithelium ovarian cancer in a dose response manner. Butler et al. in 2011 reported an inverse association between GT intake and risk of ovarian cancer; however no association was seen with BT in this metaanalysis. In a recently published cohort study 330849 women were studied and tea composition was quantified by FFQ methods. A total of 1244 women developed ovarian cancer during a median follow-up of 11.7 years;

however no association was seen between tea consumption and ovarian cancer.

Prostate cancer: Interest in the role of tea consumption and prostate cancer developed with the observation of significantly lower incidence of prostate cancer among Japanese and Chinese populations who have a high intake of green tea. GT polyphenols significantly inhibit the development and metastasis of prostate cancer in mice. EGCG induces apoptosis, cell growth inhibition and cell cycle dysregulation in cell cultures using human prostate cancer cells. Treatment of human prostate cancer cells with EGCG and COX2 inhibitors resulted in inhibition of cell growth, apoptosis and inhibition of NF-kB. In a large cohort study it was found that drinking >5 cups of GT per day reduces the risk of developing advanced cancer compared to those having 1 cup/day. In a recently published meta-analysis in 2011, GT was found to have a protective effect on prostrate cancer in Asian population; however no such effect was seen with BT consumption.

Cardiovascular diseases

Formation of atherosclerotic plaque is a cardinal feature in cardiovascular diseases (CVD). These fatty plaques are formed by the initial deposition of LDL on the wall of the arteries, followed by oxidation of LDL, thus stimulating the inflammatory cascade and further leading to the accumulation of macrophages on the oxidized lipids thereby resulting in the formation of atherosclerotic plaques itself. Increased oxidative stress, inflammation platelet aggregation, impaired endothelial functions are some of the anomalies associated with CVD. The tea falvan-3ols or polyphenols are beneficial in CVD through their vasculoprotective, antioxidative, anti-thrombolytic, anti-inflammatory and lipid lowering properties. GT extracts have been found to decrease the LDL significantly and also been preserve and improve arterial compliance and endothelial function.

In a large cohort study consumption of >10 cups of GT/day was found to decrease the relative risk of death from CVD in Japanese population, Earlier an inverse relation between flavanol intake and CHD has been reported in a large cohort study of 800 men. Sesso et al. have reported a 44% reduction in cardiovascular risk in individuals drinking more than one cup of tea per day. Similarly intake of catechin has been inversely correlated with the CHD associated mortality in a large cohort study of 12763 men under 25 years of follow-up from 7 different countries.

Consumption of tea has been found to lower the blood pressure and improve the serum lipid profile. In a recent metaanalysis by Zheng et al. in 2011, administration of GT beverages or extracts resulted in a significant reduction of serum TG and LDL cholesterol concentration.

Consumption of BT along with normal diet has been found to decrease the independent cardiovascular risk factor and improve overall antioxidant status in humans in a randomized controlled trial by Bahorun et al. in 2012. However Wang et al. in their meta-analysis of 18 epidemiological studies showed no protective role of BT and an uncertain association of GT consumption with reduced risk of CAD.

E. Potential Health Risks of Tea Consumption

Hepatotoxicity of GT and its extracts has been reported in a few case reports, systemic reviews and animal experiments. Anecdotal evidence from a few case reports in 2009 suggest that consumption of oral green tea derivatives against hair loss reportedly caused acute hepatitis; while cessation of the nutritional additive showed rapid and sustained recovery. Similarly a recent Danish case report showed a case of toxic hepatitis following consumption of 4–6 cups of GT per day for 6 months. Mazzanti et al., in 2009 suggest a causal relationship between GT and liver damage by analyzing 34 cases of hepatitis published between 1999 and 2008.

Significant liver injury has been reported after the intake of tea extracts of *Camellia sinensis* or other herbal products in a systematic review of the published studies and case reports from 1990 to 2010 on liver injury by dietary supplements. Administration of high concentrations of GTE (upto 1000 mg/kg) in mice has shown to potentiate acetaminophen (APAP) induced hepatotoxicity when GTE is administered after the APAP dose thus highlighting strong drug–diet interactions. Similarly a single high dose of EGCG (1500 mg/kg) increased plasma ALT levels by 138 fold and reduced survival by 85%. Under all these circumstances hepatotoxicity was associated with oxidative stress including increased hepatic peroxidation and depletion of glutathione (a potent antioxidant).

In 2008 a systematic review was carried out by the US pharmacopeia on 216 case reports on green tea products. Of the 34 reports on liver damage 27 reports were categorized as possible causality and 7 of 34 as probable causality due to GTE. Consumption of GT concentrated extracts on empty stomach is more likely to have adverse effects than in the fed state. Based on this safety review though the Dietary Supplement Information Expert Committee (DSIEC) remains concerned about the safety issues concerning GT extracts, yet the requirement of cautionary labeling on the GTE monograph was deferred. In a Spanish study in 2008, of all the DILI cases associated with the use of herbal remedies and dietary supplements; *Camellia sinensis* was found to be the main causative herb followed by other substances.

Consumption of tea and coffee can cause heart burn and has been associated with frequency of GERD in Indian population. In an Icelandic study on healthy subjects regular consumption of tea and coffee have been shown to exert adverse effects on the function of lower esophageal sphincteric.

High levels of tea consumption have been found to be closely associated with persistent complaints of poor sleep in the elderly subjects. Although there have been also reports of rapid increase in alertness and self reported improvements in mood after consumption of tea. This could be because of the amino acids in tea (Thiamine) acting as a neurotransmitter.

Tea consumption has been speculated to have a detrimental effect on oral health. However in a review; Gardner et al. did not find any detrimental effect of tea on adult dental health. Nevertheless a recent study in 2012 found that EGCG and salivary protein interactions might be both protective and detrimental with respect to oral health.

Recently the high concentrations of EGCG have been found to have cytotoxic and genotoxic properties in mammalian cells. These effects of bioflavonoids are believed to be mediated by their action on topoisomerase II.

In a case report published in 2010 drew attention on the harmful aspect of GT. The case of a 38 year old woman who showed hemolytic purpura and marked thrombocytopenia due to consumption of GT preparations for two months to loose weight. Her condition improved soon after stopping the intake of GT preparations.

II. COFFEE

A. Introduction

Legendary stories suggest that coffee was fortuitously discovered way back in the 9th century by an Ethiopian goat herder named Kaldhi. He noticed the lively behavior of his goats after eating the berries of the coffee tree. The earliest drinking of coffee dates back to the middle of 15th century, later it reached the Middle East consequently spreading to Europe, Asia and America. The word 'coffee' is adapted from the Dutch word Koffie which in turn is derived from the Arabic word qahhwat-al-bun meaning the 'wine of the bean'. Coffee is extracted from the berries of the plant belonging to the family Rubiaceae of about 85 species of the shrub of the genus Coffea. The three main species—*Coffea canephora, C. arabica* and *C. liberica* have been used successfully for commercial use. Over 2.5 billion cups (30 ml/cup) of coffee is consumed every day worldwide, making coffee the second most popular drink after water in the world. It has become the second largest traded commodity in the world after petroleum. Coffee is a dark aromatic non-alcoholic brew having stimulating effects on the mental and physical activities.

The chemical composition of coffee is complex comprising of thousand different chemicals, mainly holocellulose (fiber), hemi-cellulose, starch, sucrose, oils, proteins, chlorogenic acid, trigonelline and caffeine. This composition varies according to the variety of coffee, climatic conditions and practice of agriculture, processing and storage. Processing of coffee involves the steps of harvesting, drying, grading, roasting, grinding and brewing. Both dry and wet methods of processing coffee may be used. Decaffeination and filtration may also be carried out to remove certain components like caffeine and lipids.

Humans have been consuming coffee for over 1000 years now. Consumption of coffee is a double edged sword for human health. Several diseases have been alleged to be caused by coffee; nevertheless it has been found to have antioxidant activity useful in a number of diseases. Talking of the health effects of coffee the attention goes on the key chemical compound in coffee like caffeine, chlorogenic acid, kahweol and cafestol and other micronutrients. Caffeine is a purine alkaloid which is rapidly absorbed in the stomach and small intestine and distributed to all the tissues including the brain. Liver is the primary site of its metabolism. Caffeine exerts its biological effects through the antagonism of the A_{2A} subtypes of adenosine receptors. The caffeine content of 100 ml of coffee may range from 27 mg in brewed to 185 mg in moccachino shot of coffee. Chlorogenic acids are important dietary phenols. Coffee is the richest source of chlorogenic and cinnamic acids (cafeic acid). A cup of coffee (200 ml) may have 70–350 mg of chlorogenic acid and 35–175 mg of caffeic acid. The chlorogenic acids are readily absorbed in the intestine and metabolized by the colonic microflora. Chlorogenic acid is hydrolyzed to caffeic and quinic acids. Cafestol and Kahweol are the cholesterol raising diterpenes found in the oil derived from coffee. Their individual concentration can range from 0.1–7 mg/ml. higher levels of the compounds are found in unfiltered coffee such as expresso, French press or boiled Turkish/Greek coffee. Filtered and instant coffee have low cafestol and kahweol. The mechanisms of these diterpenes on lipoprotein metabolism are not clear but their consumption is associated with increased LDL cholesterol and also colorectal cancer. However the anti-tumor mechanisms of these compounds are poorly understood.

B. Potential Health Benefits of Coffee Consumption

Diabetes mellitus

Coffee consumption has been associated with a reduced risk of diabetes mellitus and metabolic syndrome. The exact mechanism of actions still remain to be unrevealed, however a few possible ways in which coffee may exert a positive effect on DM have been suggested. The components of coffee may act as antioxidants and enhance the insulin sensitivity. Caffeine increases the uncoupling protein expression and also lipid oxidation which further leads to decreased glucose storage capacity thereby affecting the glucose metabolism and reducing the extent of DM. Chlorogenic acid (CGA) exerts an antioxidant effect and inhibits glucose-6–phosphatase activity. CGA also stimulates the transport of glucose into skeletal muscle via the activation of AMPK. Decaffeinated coffee can also delay the intestinal absorption of glucose and enhance the glucose induced insulin secretion and action of insulin.

In most of the cohort studies significant inverse correlation have been seen between coffee intake and the risk of type 2 DM. Most of these studies are on Dutch and Scandinavian population who are amongst the greatest consumers of coffee. Nonetheless some studies also fail to demonstrate any such effects of coffee on DM. Huxley et al. in 2009 reported in their metaanalysis that after adjustment of confounders a log linear relationship of diabetes was seen with coffee in a way

that an additional cup of coffee was associated with a 7% reduction in risk of DM (RR 0.93; CI 0.91–0.95). However recommendation of increased coffee intake for general public cannot be based on these studies. More long term randomization intervention studies are needed with the explanation of a detailed mechanism of action.

Parkinson's disease

Parkinson's disease (PD) is a brain disorder characterized by degeneration of dopaminergic neurons in the substantial nigra. An inverse association between coffee consumption and PD has been reported in many epidemiological studies; where the effectiveness of caffeine and coffee was found to be gender dependent. The risk of development of PD was 3–5 times more in men who did not drink coffee than those who drank at least 28 oz daily. However in women no such inverse relationship was seen. Estrogen replacement therapy (ERT), received very commonly by the postmenopausal women seems to have an effect on mechanism of action of coffee in PD. The exact mechanism of estrogen is yet not clear but ERT has been found to inhibit CYP1A2 – an important caffeine metabolizing enzyme found in the liver. Coffee seems to induce the expression of mRNA and enzymes alleviating the negative effects of free radicals on neurodegeneration, thus coffee seems to boost the antioxidant action of the immune system. Caffeine acts as an antagonist of the A1 and A_{2A} receptor and reduces the dopaminergic neurotoxicity.

Caffeine supplementation in mice models of 1–methyl-4phenyl-1,2,3,6–tetrahydropyridine (MPTP) induced parkinsonism have been found to induce losses of striatal dopamine and dopamine transporter. Presently the epidemiological studies support the beneficial effects of coffee in PD, however its mechanisms of action needs to be understood before recommending its increased consumption to prevent PD.

Liver injury

The effect of coffee on liver pathology ranging from abnormal liver biochemistry to cirrhosis and hepatocellular carcinoma has been studied in several epidemiological studies. Large observational studies have found significant inverse relationship between coffee intake and elevated transaminases. High coffee intake was also associated with lower ALT levels in patients at a high risk of liver disease, i.e., patients with a history of alcohol intake, insulin resistance, diabetes, viral hepatitis, and iron overload. A significant inverse correlation between coffee and liver disease progression has also been seen in a large cohort of 766 patients with hepatitis C. A similar inverse relationship is also seen in alcohol related cirrhosis. Molloy et al. in 2012 have reported a significant reduction in the risk of fibrosis among patients with NASH. In another study published in 2011, consumption of coffee (>3 cups/day) has been found as an independent predictor of improved virologic response to peginterferon plus ribavirin in patients with hepatitis C. In a Japanese study coffee drinking has been shown to reduce the risk of developing HCC by 50%. Several case control studies have supported the role of coffee on reducing the risk of HCC in patients with cirrhosis. A metaanalysis of total 10 studies showed an overall risk reduction of developing HCC for coffee drinkers vs. non-coffee drinkers RR:0.57 (95% CI 0.49–0.67). The most commonly studied compounds in coffee in relation to liver disease are caffeine, diterpenes (cafestol and kahwoel) and chlorogenic acid.

Several mechanisms of actions of these compounds have been proposed. The compounds in coffee seem to have an anti-inflammatory action; a reduction in serum transaminases and inflammatory cytokine gene expression was seen in mice. However cell culture and animal studies suggest an antioxidant action of coffee. In animal models coffee exerts a protective effect on histological and clinical fibrosis through the suppression of fibrogenic cytokines; collagen I and stellate cell activation. Caffeine blocks the stimulation of connective tissue growth factor expression by inducing the degradation of the profibrotic SMAD2 gene and hence weakening the signaling of transforming growth factors. In HCC coffee is found beneficial as it inhibits the activity of phase 1 activating enzymes associated with carcinogene activation. Coffee may also block the early mutageneic events by inducing phase II detoxifying enzymes and antioxidants. C&K in coffee also conjugate aflatoxin B1 an important carcinogen implicated in HCC. C&K may also inhibit cytochrome P450 enzyme pathway in carcinogen activation. The chlorogenic acid in coffee seems to inhibit iron absorption and alter gut microflora, thereby protecting the liver.

Alertness, mood and performance

Coffee has been generally found to have effects on increased alertness, reduced fatigue, and improvements in measures of reaction time, vigilance, memory and mood. Caffeine in coffee has been shown to increase alertness especially in situations where arousal is low (night shift workers and early in the mornings). It also seems to increase vigilance in daytime as assessed by both visual and auditory tests. Consumption of caffeine in the dose of 250 mg twice per day showed better auditory vigilance tat than the placebo group. High levels of coffee consumption are associated with improved performance in reaction time, verbal memory and visuospatial reasoning. Chlorogenic acid in coffee has been reported to reduce anxiety related behavior, improve spatial learning and memory in many animal studies. On the contrary there are reports in the literature to suggest that caffeine can produce anxiety or exacerbate anxiety in adults with persistent anxiety disorders. In a recently published report in 2012, the anxiogenic effect of caffeine has been demonstrated.

C. Potential Health Risks of Coffee Consumption

Cardiovascular health

It is well established that acute consumption of dietary caffeine raises the blood pressure in normotensive individuals; however this pressure effect is more striking in hypertensive individuals. Coffee or caffeine seems to enhance the arterial stiffness leading to hypertension. A recently published metaanalysis in 2011 suggests that caffeine intake does produce an acute rise in blood pressure in hypertensive individuals however there is a lack of association of long term coffee consumption with increased BP. Similarly the diterpenes C&K in coffee bring a rise in the serum lipids and also enhance the risk of cardiovascular diseases (CVD). A metaanalysis of 14 RCTs showed that boiled coffee increased the serum total and LDL cholesterol while filtered coffee had insignificant effects on serum lipids. Nevertheless the normal intake of coffee of ~3–4 cups/day cause a small increase in both LDL and HDL cholesterol which alone is not sufficient to cause coronary heart disease. A recent metaanalysis of 12 studies published in 2012 suggests that intake of unfiltered coffee significantly increases the total cholesterol, LDL and TG levels. Coffee consumption may trigger an episode of coronary event and also increase the infarct size in a selected group of patients. High coffee intakes are associated with an increased risk of CHD and myocardial infarction (MI). High intake of coffee consumption (≥5 cups/day) causes 40–60% high chances of CHD than those who did not drink coffee. Excessive coffee intake of ≥8 cups/day may aggravate cardiac arythemias and also raise the plasma homocysteine levels. Caffeine also causes an acute increase in the arterial wave reflection which may increase the pulsatile load of the heart.

On the contrary these studies have been counter challenged by other studies which do not support the association between coffee consumption and the risk of CHD.

The most recently published metaanalysis by Wu et al. 2009, Larsson et al. 2011, and Mostofsky et al. 2012, have paradoxically found an inverse association between coffee consumption and risk of heart failure or stroke. It has been suggested that the strong association reported between high coffee consumption and risk of CHD in case control studies does not hold true in long term follow-up cohort studies. The debate continues with the positive aspects of coffee on reducing the risk of CHD. Chlorogenic acid improves the antioxidant status of the body and reduces the LDL oxidation. Lopez Gaarcia et al. in 2008 found an inverse association between markers of inflammation and coffee consumption hence coffee seems to provide protection against endothelial dysfunction. Coffee consumption may also reduce the activation of platelet and plasma C reactive protein in healthy men thus sustaining the cardiovascular health.

Osteoporosis and bone health

Many metabolic studies have examined the effect of coffee consumption and caffeine on calcium homeostasis. Caffeine intake has been associated with increased urinary calcium excretion in both men as well as women. The caffeine induced hyperclaceurea was not affected by oestrogen status. Caffeine does seem to have a negative effect on the calcium balance either through increased urinary calcium excretion or decreased intestinal absorption. The effect of caffeine on calcium loss is associated with total amount of coffee or caffeine intake, as in one study coffee intake of >1000 ml/day increased the calcium loss while an intake of 150–300 ml/day has little impact on calcium balance. Another study found an association between loss of BMD and caffeine intake but only in women who had a low dietary intake of calcium. Similarly Connor et al. found that women who did not drink one glass of milk per day exhibited a coffee associated decrease in BMD. Women with excessive coffee intake (~9 cups/day) have an increased risk of fracture. Prospective cohort studies in the US have found a positive association between risk of hip fracture and caffeine consumption in women.

However the study done by Travers Gustfson et al. in 1995 did not show an effect of caffeine consumption on increased risk of bone trauma. Most of the case control studies did not find an association between caffeine consumption and risk of hip fracture. It is not known whether studies that have shown increased risk of fracture with coffee intake have adjusted for the dietary calcium intakes or not, as the effect of caffeine on bone metabolism is a complicated matter which is influenced by a number of confounders like calcium intake, age, cigarette smoking, alcohol consumption etc. It has been suggested by Dew et al. in their review that the deleterious effects of caffeine on bone health seem to be overstated and a reassessment of the role of caffeinated beverages needs to be done. In another crossectional study heavy habitual consumption of caffeinated beverages was not associated with deterioration in bone mineral content or BMD in young women. In a recent report of cell culture studies in 2013 it has been reported that caffeine at low doses has a stimulatory effect on the differentiation of adipose-derived stem cells (ADSCs) to osteoblasts; however at high doses this effect is reversed.

Reproductive health

During the last three decades a substantial amount of evidence has accumulated relating to the effect of coffee on reproductive health. Caffeine the main compound in coffee gets readily absorbed from the gastrointestinal tract, crosses the placenta and gets distributed to all foetal tissues including the central nervous system. Since the enzymes required for the oxidation of methylated xanthenes derived from coffee are absent in the foetus the exposure of the foetus to caffeine is increased as the half life of caffeine is increased in the fetus.

There have been contradictory data originating from the studies on the aspect of coffee consumption and spontaneous abortions. There are certain methodological issues with epidemiological studies including measurement of the actual caffeine consumption of the participants, caffeine content of coffee, and elucidation of other risk factors for miscarriage like nausea, smoking etc. A large prospective cohort study including 3135 pregnant women found that coffee consumption may have contributed to the risk of spontaneous abortions. Similarly a retrospective case control study on 226 Saudi Arabian women found an increased risk of miscarriage with the consumption of >150 mg of caffeine/day. In a retrospective cohort study in 1996 the increased risk of spontaneous abortions was associated only with the highest amount of coffee and tea consumption (3 or more cups/day ~ ≥300 mg caffeine/day).

A metaanalysis combining the data of epidemiological studies though showed a positive association of spontaneous abortions with a consumption of >150 mg caffeine/day, however this metaanalysis failed to adjust for the major confounders like maternal age, smoking, alcohol intake etc. Most studies show an association between coffee consumption and risk of spontaneous abortions at a level of intake of caffeine ≥300 mg/day.

Nonetheless there have been many studies which failed to show any association between coffee and increased risk of spontaneous abortions. Christian et al. in 2001 found that pregnant women not smoking or taking alcohol and who consumed moderate amounts of caffeine are unlikely to develop any reproductive problems. The evidence from some epidemiological studies have failed to demonstrate a causal association between caffeine consumption and risk of spontaneous abortions.

Caffeine consumption seems to influence the foetal growth also. The mechanisms of action of caffeine are through its effect on increasing the levels of cAMP through the inhibition of phosphodiesterases; cAMP interferes with foetal cell growth and development. Caffeine may also cause hypoxia in the foetal tissue by blocking the adenosine receptors involved in maintaining the balance between the availability and use of tissue oxygen. Coffee may also increase maternal epinephrine concentrations and decreased intervillous placental blood flow. The data from the epidemiological studies on this aspect is rather conflicting. Some studies do suggest an association between caffeine consumption and low birth weight or intrauterine growth retardation. A few reports have found the detrimental effects on the foetus (decreased birth weight and small head circumferences) at the level of caffeine intake of ≥300 mg/day even after adjustment of confounding factors like maternal nicotine use. In a large cohort study on 40455 pregnancies, it was found that coffee consumption at ≥10 cups/day was associated with low birth weight and at 5–9 cups/day, it was associated with lower birth weight for gestational age even after adjusting for confounders like maternal age, smoking, and alcohol use. A metaanalysis examining 7 original studies on 64268 pregnancies reported a significantly increased risk (by 50%) for low birth weight babies in pregnant women consuming >150 mg caffeine/day; however this metaanalysis failed to adjust for the confounders. A case control study in 2003 suggested a higher intake of caffeine in 3rd trimester in mothers of small for gestational age infants as compared to mothers of non-SGA infants. These results persisted even after adjusting for confounders.

Nevertheless there have been some reports in the literature which did not find any association between caffeine consumption and birth weight or foetal growth retardation even at a level of 300–400 mg/day during pregnancy after controlling for confounders.

Thus majority of data from epidemiological studies does not support the causative role of coffee or caffeine consumption on premature delivery and congenital malformations. A recent experimental report in 2012, suggests that maternal caffeine consumption impairs gonadal development in male offspring rats. A case control study by Schliep et al. in 2012 found an inverse association of caffeine intake ($\geq$200 mg/day) with free estradiol concentration in white women while a positive correlation was seen among Asian women. Similarly another study by Al-Saleh et al. in 2010 did not find an association between coffee consumption and pregnancy rates and various performance parameters of in-vito fertilization, yet this study suggests that caffeine can reach the follicular fluid and has a potential of causing harmful effects on the reproductive health.

SUMMARY

The refreshing and stimulatory effects of tea and coffee on the human body are known for centuries, making them the two most popular beverages in the world. Both tea and coffee are a conglomeration of multitude of complex chemical compounds, however the most active ingredients of tea are the polyphenols mainly EGCG and that of coffee are caffeine and chlorogenic acids. The nature and the concentration of these active ingredients depends on the different types of these two beverages.

Many potential health benefits of tea and coffee are reported in the literature like prevention of cancer and cardiovascular diseases with the consumption of tea and the hepatoprotective, and anti-diabetic effects of coffee. However there is a flip side of this story where both tea and coffee have been reported to be harmful under certain situations. Tea may be hepatotoxic in nature while coffee can have detrimental effects on bone and cardiovascular health.

There is a plethora of research on these beverages ranging from epidemiological to cell line studies. The cellular and in-vitro studies elucidate the molecular aspects and the various mechanisms of action, while the epidemiological studies give the direct evidence of these compounds on human health. Notwithstanding the expansion of scientific literature the data is still equivocal and unconvincing. There are a lot of methodological issues with the epidemiological studies; while the concentrations reached in the cellular studies are much higher than those obtained through normal diet.

It rather sounds sensational for every day products like tea and coffee to have such therapeutic or detrimental effects however there is clear cut need for 'good' evidence mainly in the form of well planned randomized controlled trials. These beverages are freely marketed and very popular among masses, hence before endorsing their increased consumption substantial amount of convincing evidence needs to be gathered. Presently the '*mantra*' to good health is consumption of these beverages in moderation.

SUGGESTIONS FOR FURTHER READING

1. Kanwar J., Taskeen M., Mohammad I., Huo C., Chan T.H. and Dou Q.P. (2012), *Recent Advances on Tea Polyphenols*, Front Biosci (Elite Ed.), 4:111–31.

2. Serafini M., Del Rio D., Yao D.N., Bettuzzi S. and Peluso I. (2011), Health Benefits of Tea, In: Benzie IFF, Wachtel-Galor S. (Ed.), *Herbal Medicine: Biomolecular and Clinical Aspects*, 2nd ed., Boca Raton (FL): CRC Press, Chapter 12.

3. Khan N. and Mukhtar H. (2007), Tea Polyphenols for Health Promotion, *Life Sci.*, 81(7):519–33.

4. Yang C.S., Lambert J.D., Ju J., Lu G. and Sang S. (2007), Tea and Cancer Prevention: Molecular Mechanisms and Human Relevance, *Toxicol Appl Pharmacol.*, 224(3):265–73.

5. Schönthal A.H. (2011), Adverse Effects of Concentrated Green Tea Extracts, *Mol Nutr Food Res.*, 55(6):874–85.

6. Hu M.L. (2011), Dietary Polyphenols as Antioxidants and Anticancer Agents: More Questions than Answers, *Chang Gung Med. J.*, 34(5): 449–60.

7. Chen D., Wan S.B., Yang H., Yuan J., Chan T.H. and Dou Q.P. (2011), EGCG, Green Tea Polyphenols and Their Synthetic Analogs and Prodrugs for Human Cancer Prevention and Treatment, *Adv. Clin. Chem.*, 53:155–77.

8. Mazzanti G., Menniti-Ippolito F., Moro P.A., Cassetti F., Raschetti R., Santuccio C. and Mastrangelo S. (2009), Hepatotoxicity from Green Tea: A Review of the Literature and Two Unpublished Cases, *Eur J. Clin Pharmacol.*, 65(4):331–41.

9. Higdon J.V. and Frei B. (2006), Coffee and Health: A Review of Recent Human Research, *Crit. Rev. Food Sci. Nutr.*, 46(2):101–23.

10. Nawrot P., Jordan S., Eastwood J., Rotstein J., Hugenholtz A. and Feeley M. (2003), Effects of Caffeine on Human Health, *Food Addit. Contam.*, 20(1):1–30.

11. Butt M.S. and Sultan M.T. (2011), Coffee and Its Consumption: Benefits and Risks, *Crit. Rev. Food Sci. Nutr.*, 51(4):363–73.

12. George S.E., Ramalakshmi K. and Mohan Rao L.J. (2008), A Perception on Health Benefits of Coffee, *Crit. Rev. Food Sci. Nutr.*, 48(5): 464–86.

SECTION V

VITAMINS AND MINERALS

Although carbohydrates, fats and proteins form the predominant portion of the diet, quantitatively minor components such as vitamins and minerals have nonetheless a major role in normal cell function. These are not synthesized by the body but are essential, and have to be supplied in the diet for health. Vitamins are conventionally divided into two categories—fat soluble and the water soluble. All fat soluble vitamins except Vitamin D are covererd in Chaper 46. Vitamin D along with two relevant hormones—parathyroid hormone and calcitonin have been more appropriately dealt along with the calcium and phosphorus metabolism on account of their inherent relationship. The manner in which different vitamins exercise their effects is not known in all cases, but the role of the B group of vitamins as coenzymes is well characterized. Vitamin C has gained additional attention from the reports by the Nobel Laureate, Linus Pauling on the beneficial effects of massive doses of this vitamin as prophylaxis against flu and other infections. Amongst the minerals, calcium, phosphorus and iron have been considered in separate chapters because of their structural role in bones and red blood cells, respectively. Calcium also has a key role in neural excitability and muscular contraction. Its translocation on or across the membrane of cells may provide a signal for elicitation of the effects of some hormones, drugs etc. Iron is a component of several enzymes involved in the oxidative chain. The role of numerous other minerals required by the body has been outlined in the terminal chapter of the Section. Iodine is discussed along with the thyroid.

46

Chemistry and Biological Role of Vitamins A, E, K and Coenzyme Q

K.D. Moudgil

CONTENTS

I. INTRODUCTION

The importance of certain dietary factors in curing a number of nutritional diseases was known long before the concept of 'vitamin' evolved. However, only after the epoch-making discoveries by McCollum, Funk, Hopkins and others it was established that each of the nutritional diseases like nyctalopia, beri-beri, rickets and scurvy is caused by a specific vitamin deficiency.

The biochemical mode of action of most of the water-soluble vitamins has been thoroughly elucidated in terms of their cofactor roles in specific enzyme systems, but that of the fat-soluble vitamins is, as yet, only partly understood. Extensive studies have been carried out on experimental animals, but our knowledge based on human subjects is still limited. Recent studies have revealed much greater role for them than that known earlier like involvement of vitamin A (retinoic acid) in gene expression, antioxidant role of vitamin E, and γ-carboxylation of glutamic acid in proteins of blood clotting and bone metabolism by vitamin K.

In this chapter, the importance of fat-soluble vitamins A, E and K and coenzyme Q (CoQ) (ubiquinone) is discussed. (Vitamin D is dealt with separately elsewhere in the book.)

II. VITAMIN A

The term vitamin A covers all compounds with biological activity of retinol. Many compounds which include natural forms of vitamin A and synthetic analogs of retinol irrespective of their biological activity are called retinoids. The three forms, namely retinal (vitamin A aldehyde), retinol (vitamin A alcohol) and retinoic acid have functions of their own.

A. Sources of Vitamin A

Vitamin A is of animal origin and is largely found in marine fish liver oil, milk, butter, cheese, egg yolk and in the intestines. It is largely present in liver oil in the form of palmitate esters. The well known physiological functions for which vitamin A or retinol is required are growth, cell differentiation, vision and reproduction. In vitamin A deficiency, growth is retarded resulting in decline in weight, vision is impaired resulting in complete blindness, stones are formed in bladder and kidney and reproduction is considerably hindered, leading often to resorption of fetuses in female animals.

Vitamin A_2 is largely found in fresh water fish liver oil (from river and lake fish) and differs from vitamin A in containing an extra conjugated double bond in β-ionone ring (Figure 46.1). In rats, it is 40 per cent as biologically active as vitamin A.

Various forms of vitamin A including the provitamin A are isoprenoids (five carbon compounds with alternate double bonds, see below).

Retinal (vitamin A, retinaldehyde) has CHO in place of CH_2OH of retinol, whereas retinoic acid has COOH in place of CH_2OH of retinol. Retinyl esters are esters of retinol with fatty acids like acetate (retinyl acetate), palmitate (retinyl palmitate) etc.

The corresponding forms of vitamin A_2 are 3-dehydro retinol, -retinal and -retinoic acid.

Vitamin A is also available in plant kingdom in the form of provitamin A carotenoid pigments (α-, β- and γ-carotene and cryptoxanthin), the most common being β-carotene in carrots.

B. Carotenoids

The carotenoids are the most important of the various classes of pigments in living organisms. They are ubiquitous in nature as they occur in plants, flowers, insects, bacteria, birds and other animals.

Carotenoids are precursors of vitamin A. However, not all the carotenoids are completely absorbed, and after absorption there is considerable selectivity in the formation of vitamin A depending upon the structural configuration of the carotenoid. On weight basis, one sixth of β-carotene and one twelfth of other carotenoids from the diet are absorbed for conversion to retinol in the intestinal enterocytes. Dietary fat is required for the absorption of carotenoids.

The most important functions of carotenoids are: (a) they act as protective agents in order to prevent cells from undergoing damage due to photodynamic action, (b) they can act as accessory pigments in photosynthetic organisms by way of transferring radiant energy to the actual pigments in photosynthesis, and (c) β-carotene serves as an antioxidant by stabilizing organic peroxide free radicals by virtue of which it has anticancer activity.

On account of a large number of conjugated double bonds in carotenoids and vitamins A, care has to be taken during isolation, saponification and chromatography to avoid light and high temperatures in order to prevent isomerism. Usually γ-tocopherol is added as an antioxidant for maintaining stability of vitamin A.

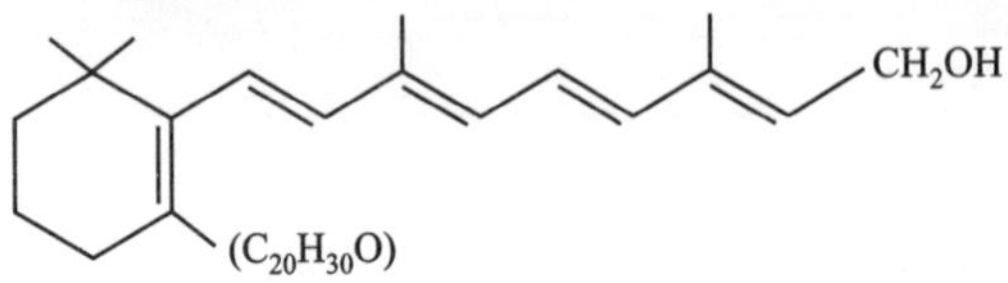

$(C_{20}H_{30}O)$

Vitamin A (retinol)
(mol. wt. 286)

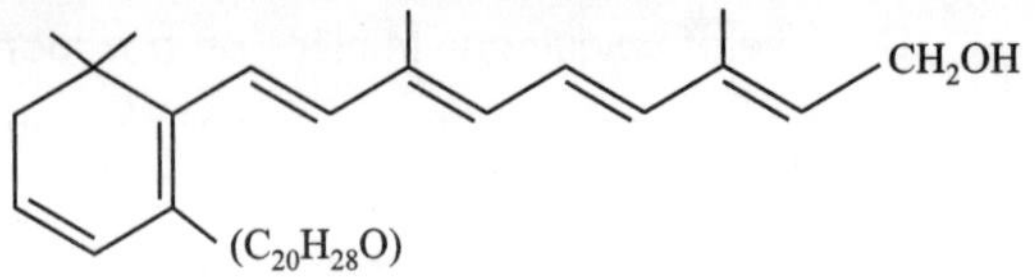

$(C_{20}H_{28}O)$

Vitamin A_2 (3-dehydroretinol)
(mol. wt. 284)

Figure 46.1

1. Structure-function relationship: Carotenoids are yellow to red pigments composed of isoprene units linked so that the two methyl ($-CH_3$) groups nearest the centre of the molecule are in position 1:6, whilst all the other lateral methyl groups are in positions 1:5 as shown in Figure 46.2.

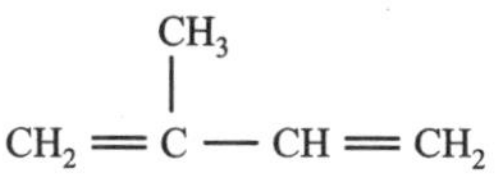

$$CH_2 = C - CH = CH_2$$

with CH_3 attached to C.

Isoprene unit

Many carotenoids have been identified. β-carotene, which has two unmodified and intact β-ionone rings similar to those in vitamin A in rings A and B, has maximum biological activity with respect to vitamin A formation. But α- and γ-carotenes, cryptoxanthin and β-carotene monoepoxide in green grass, and apahanin echinenone in sea urchins, which have in ring A one β-ionone ring as in vitamin A but have in ring B either a modified or oxygenated carotenoid, have only half the biological activity of β-carotene. Other carotenoids having modified or oxycarotenoids in both rings A and B such as lutein and zeaxanthin in green grass, rhodoxanthin in seed coats, and astaxanthin and astacin in lobster body and eyes do not show vitamin A activity.

Carotenoids and vitamins A exist naturally in different isomeric forms on account of the large number of double bonds in the molecules. The *all trans* isomer and the *cis*-isomer of vitamins A1 and A2 show differences in biological activity.

2. Conversion of β-carotene to vitamin A: The first evidence for a site other than the liver was established when large amounts of vitamin A were observed on the lining of the intestines of various types of fish. Several investigations unequivocally showed that the intestinal wall is the main site of conversion of provitamin (β-carotene) to vitamin A.

Mechanism of cleavage of carotenoids: Dietary β-carotene may be cleaved either centrally giving two molecules of retinal or peripherally by stepwise degradation from one end of the conjugated carotene chain adjacent to one β-ion one ring forming only one molecule of retinal. In the latter case, β-carotene was only half as active as vitamin A on weight basis according to many nutritional studies. More recent studies also seem to favour the view that β-carotene may be cleaved both centrally as well as peripherally by two separate enzymes.

A soluble enzyme from the intestine and liver of rabbit, guinea pig, hog and rat, which cleaves β-carotene into two molecules of vitamin A aldehyde (retinal), has been isolated. This enzyme is called β-carotene dioxygenase (β-carotene 15, 15′-oxygenase). Molecular oxygen and bile salts are required for this reaction. Retinal formed is reduced to retinol

by retinaldehyde reductase for which NADPH (NADH) is required.

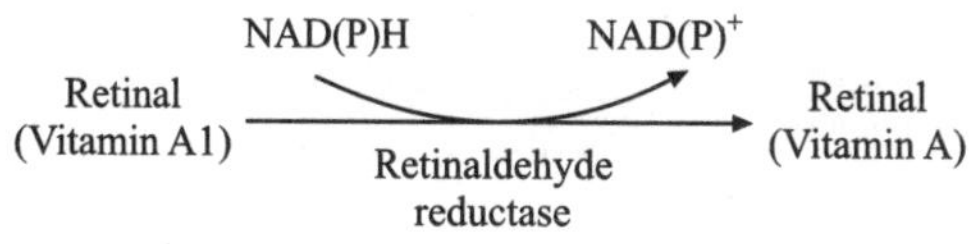

$$\text{Retinal (Vitamin A1)} \xrightarrow[\text{Retinaldehyde reductase}]{\text{NAD(P)H} \quad \text{NAD(P)}^+} \text{Retinal (Vitamin A)}$$

However, retinal bound to the intestinal cellular retinal binding protein type II (CRBP II, see under C) may be reduced by a membrane-bound microsomal enzyme.

A small amount of retinaldehyde is oxidized to retinoic acid (*all trans*) by a NAD+/FAD-requiring enzyme.

$$\text{Retinaldehyde} \xrightarrow{\text{NAD}^+\text{FAD}} \text{Retinoic acid (all trans)}$$

9-*cis* and 13-*cis* retinoic acid are also formed.

The conversion of β-carotene to vitamin A occurs by stepwise and central cleavage as shown in Figure 46.3. Carotenoids are generally estimated by colorimetric methods.

Carotene and thyroid metabolism: It has been shown that thyroxine administration stimulates absorption of carotene from the intestine and its subsequent enhanced storage to vitamin A in the liver, whereas anti-thyroid drugs, such as thiouracil and thiourea, considerably decrease the absorption of carotene from the intestinal tract.

C. Absorption and Transport

Recent studies have shed light on the complex mechanism of absorption of vitamin A and carotenes from food for which dietary fat is also required and on its transport to the site of action in the target cells (Figure 46.4). Esters of vitamin A (retinol) in animal foods are not well absorbed. They are first hydrolysed to retinol in the intestine by the pancreatic or brush border esterase.

The carotenes of vegetables, as mentioned earlier, are cleaved to retinal by the intestinal dioxygenase and then reduced to retinol. The free retinol is absorbed through the cell membrane of the intestines both by facilitated and passive diffusion depending on the concentration into enterocytes. Here it is reesterified with long chain fatty acids especially palmitate either by lecithin: retinol acyltransferase (LRAT) or by acyl CoA: retinol acyltransferase (ARAT) under normal or excess load of retinol, respectively. The retinyl ester (RE) is incorporated into chylomicrons and released into lymph to enter into blood. However, some vitamin A ester is also absorbed directly into blood circulation. After the chylomicrons

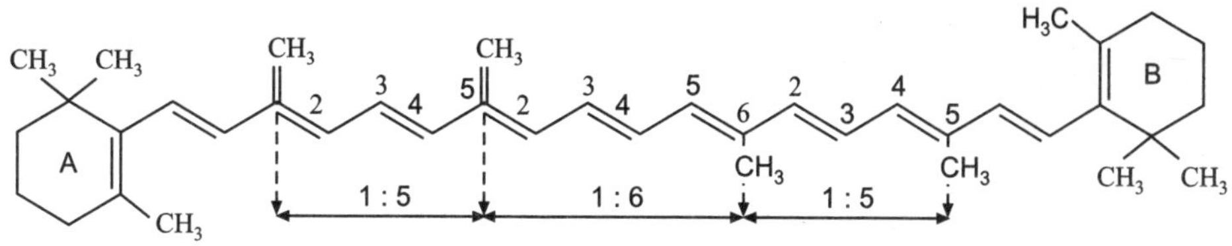

Figure 46.2

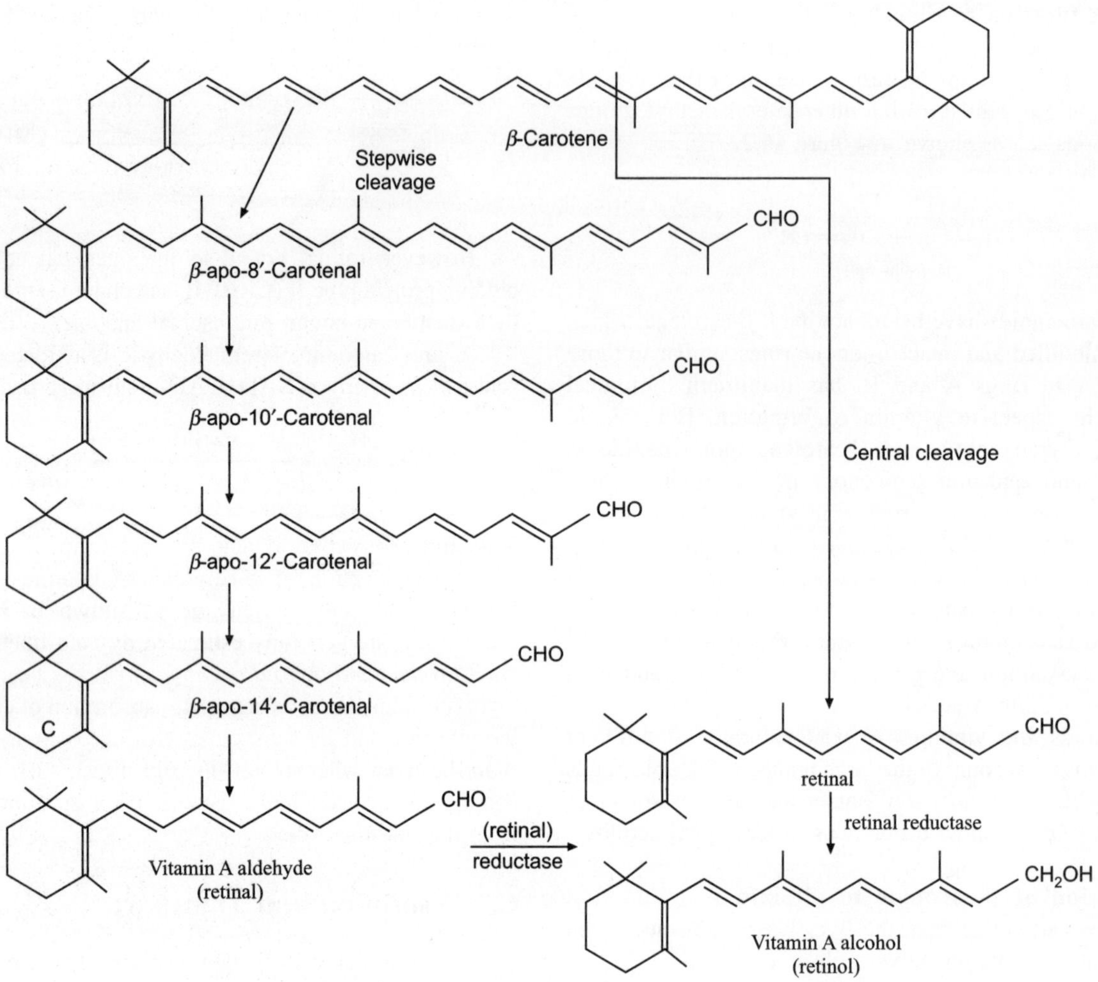

β-Carotene
Stepwise cleavage
CHO
β-apo-8′-Carotenal
CHO
β-apo-10′-Carotenal
Central cleavage
CHO
β-apo-12′-Carotenal
CHO
β-apo-14′-Carotenal
C
CHO
retinal
retinal reductase
CHO
Vitamin A aldehyde
(retinal)
(retinal)
reductase
CH_2OH
Vitamin A alcohol
(retinol)

Figure 46.3

Intestine
Liver
HC
SC
R-CRBP I
RE (Storage)
RE → R
Diet
Vitamin A ester
R
RE
Lymph
(Chylomicron)
–RE
R
RBP
Carotenes
R
RE
Blood
R-RBP
Acyl T
CHYREM-RE
Target cell
LR
Nucleus
RA-CRBP
R-RBP-transthyretin
RA
Chromatin
CRBP
RA
Kidney
RBP
R
CRBP
RBP-RCP
R-CRABP
R
Specific effects

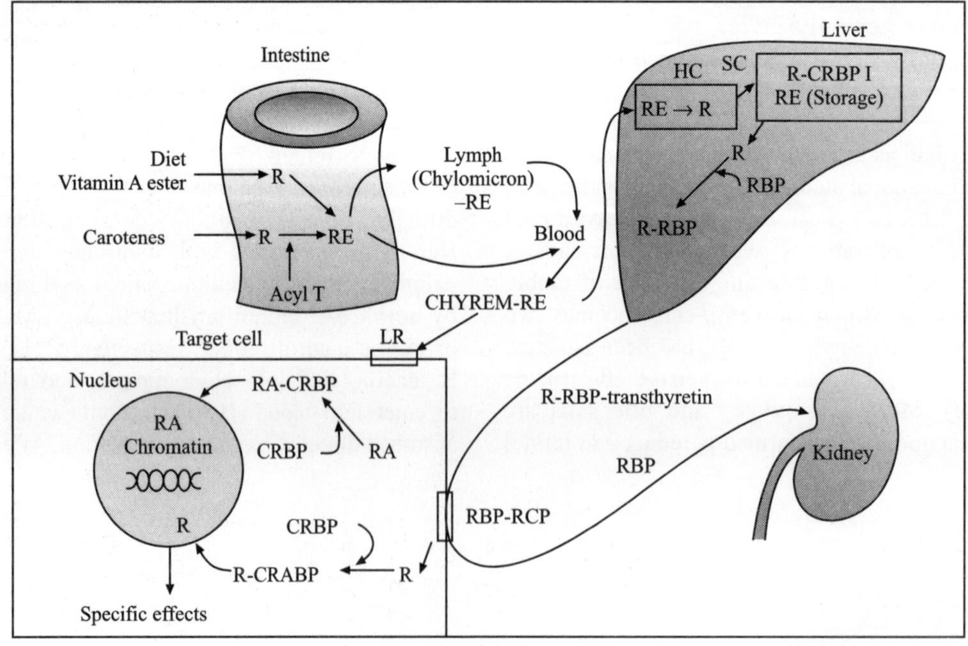

Figure 46.4

are metabolised, chylomicron remnants which contain retinyl ester are formed.

Storage of vitamin A in liver: The vitamin A ester is taken up from the chylomicron remnants by the parenchymal cells (hepatocytes) of the liver probably through low density lipoprotein (LDL) receptor and its related protein. It is then immediately hydrolysed by retinyl ester hydrolase. The free retinol is bound to plasma retinol binding protein (RBP) present in hepatocytes in higher concentrations. After processing in endosomes, endoplasmic reticulum and Golgi complex, it is rapidly transported (as RBP-retinol) to perisinusoidal stellate cells (lipocytes) which take up retinol for storage. In these cells, retinol is bound to cellular retinol binding protein type I (CRBP I), which like intestinal CRBP II gives retinol for reesterification by lecithin: retinol acyltransferase (LRAT) for ultimate storage as palmityl ester. It is worth noting that the binding of retinol to CRBPs in various tissues ensures that retinol is delivered to the proper enzyme. LRAT is in stellate cells of liver. Of the total vitamin A and its ester in the body, 50–80% is stored in liver stellate cells as retinyl ester. The gene for CRBP I is under the control of retinoic acid or its receptor.

Release of liver vitamin A and transport in plasma: Release of retinol from the liver occurs after the hydrolysis of vitamin A ester. The transport of retinol to the target tissues proceeds in a manner analogous to that of steroids. The liver synthesizes a protein (not known exactly whether in hepatocytes or stellate cells) called retinol binding protein (RBP molecular weight about 2,000). Free retinol is bound to apo RBP and processed in the Golgi apparatus for secretion. RBP accumulates in vitamin A deficiency. In the blood, vitamin A is transported as the retinol-RBP complex which has α1-mobility in electrophoresis. This binds with prealbumin transthyretin (TTR) to form a circulating ternary complex (normal level equivalent to 30–60 µg retinol/dl) which is transferred to various tissues. RBP is the only retinol-specific binding protein in plasma. But there are other retinoid-binding proteins as well. An inter photoreceptor retinoid-binding protein (IRBP) seems to function in the intercellular transport of retinoids during the visual cycle. A protein related to RBP carries retinol to the placenta in animals. On the target cells, there are specific receptors for this circulating complex. While retinol enters the cell, RBP does not; the latter is degraded in the kidney. Thus, there is a complex mechanism for absorption and transport to deliver *all trans* retinol to the plasma membrane of the target cell.

Uptake of retinol and retinoic acid by extrahepatic tissues: Free retinol in plasma is taken up by various tissues. However, a cell surface receptor or transport protein is involved in the uptake of RBP-bound retinol. Retinoic acid present in smaller amounts (5 to 10 nM) is bound to albumin in blood and appears to be spontaneously transferred from plasma to cells. Within the cell, retinol complexes with a specific cytoplasmic receptor protein, the cellular retinol-binding protein (CRBP). Retinoic acid or retinal cannot bind to CRBP. However, the

cells have a separate protein called retinoic acid-binding protein (CRABP), which binds only retinoic acid but not retinol or retinal. The CRBP and CRABP of the target cells differ from RBP of the storage organ, the liver. The retinol-CRBP complex in the cytoplasm is transported to the nucleus where retinol (but not CRBP) is bound to specific proteins in the chromatin and brings about the changes attributed to it. CRABP also functions in a similar way with retinoic acid. This is shown in Figure 46.4. Another protein called cellular retinal-binding protein (CRALBP) is found in retinal pigment and plays a role in visual cycle.

Two cytoplasmic retinol-binding proteins CRBP (liver) and CRBP II (intestinal) and two cytoplasmic retinoic acid-binding proteins CRABP I and CRABP II have been well characterised. CRBP and CRBP II have amino acid sequence identity of 56%. The two CRABPs have 72% amino acid identity. Functionally, the cellular retinol- and retinoic acid-binding proteins are involved in modulating intracellular retinoid metabolism.

The plasma retinol concentration is maintained at about 2 µM. In tissues, retinal and retinol are interconvertible by NAD^+ or $NADP^+$ requiring dehydrogenases or reductases. But retinoic acid, which is formed by oxidation of retinol, cannot be converted back to retinal or retinol. Retinoic acid can only perform its role in growth and differentiation but is unable to serve the functions of retinal in vision or retinol in reproduction.

Retinoyl phosphate obtained by phosphorylation of retinoic acid carries oligosaccharides and takes part in the synthesis of glycoproteins required for growth and cellular differentiation.

D. Structure-function Relationship of Vitamin A

It has been shown conclusively by the brilliant work of Wald and Morton that retinal plays a pivotal role in vision, whereas retinoic acid is important for systemic growth and cellular differentiation of animals. However, rats maintained on retinoic acid become blind though they grow properly and appear normal. Rats maintained on retinoic acid diet during reproduction conceive but their fetuses are resorbed. Resorption could be prevented by giving retinol. Vitamin A is also required for the development of the chick embryo, whereas retinoic acid does not serve this function. Thus, a structural alteration of retinol to retinoic acid has a deleterious effect on vision and reproduction though it effectively sustains the systemic function (growth). Thus the studies seem to indicate that the three retinoids, namely retinol, retinal and retinoic acid have different functions. Retinol appears to function as a hormone (similar to steroid hormones in its transport and mechanism of action), and maintains all the physiological functions of vitamin A such as vision, reproduction, growth and differentiation. As a component of visual pigment rhodopsin, retinal (the isomer 11-*cis* retinal) plays a role in vision (see under F). Retinoic acid supports normal growth rate and differentiation and also helps in the synthesis of glycoproteins but has no role in vision or reproduction. The above relationships are summarized in (Figure 46.5).

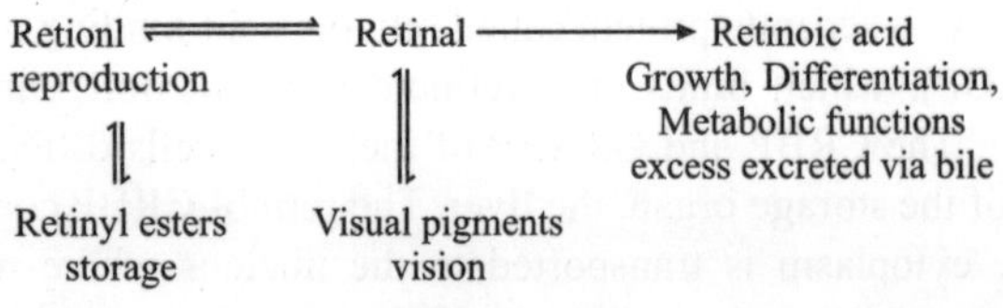

Figure 46.5

Detailed studies have been carried out to get an insight into the structure-function relationship of vitamin A.

Biological studies show that the epoxide of retinol, under any form, has a deleterious effect on growth, vision and reproduction. Thus the conjugate double bond in the β-ionone ring in retinol molecule is as important as the other four conjugated double bonds in the side chain for biological activity of retinol.

A retinal containing protein called bacteriorhodopsin is of importance in the metabolism of halobacteria.

E. Retinoic Acid and Gene Expression

Retinoic acid is necessary for growth. It also induces cell differentiation. It acts on embryonic tissues, epithelium, connective tissues (bone, cartilage) and hematopoietic tissue. Retinoic acid brings about cellular responses by diffusing into the nucleus, binding to specific nuclear receptors and regulating gene transcription. Retinoic acid nuclear receptors have been identified.

Retinoic acid receptors: These receptors resemble in some respects and have homology to nuclear receptors of the thyroid hormone and steroid hormones. All belong to the same superfamily of nuclear receptors. However, retinoic acid receptors (RARs) have their own unique features. The DNA binding domain of all these nuclear receptors have a carboxy terminal region (region C) containing 8 cysteine residues, which are involved in the chelation of zinc ions (Zn^{2+}) called the zinc binding fingers. Three types of retinoic acid receptors RAR-α, RAR-β and RAR-γ can be isolated from many vertebrates. All of them have similar structure and are almost identical in the C region of the DNA binding domain. RAR-α binds with high affinity to 13-*cis*-retinoic acid, all-*trans* retinoic acid and other retinoids having a role in cell differentiation. But retinal and retinol are bound to RAR-α with much lower affinity. It is expressed in tissues like cerebellum, adrenals, testis and leukemic cells (promyelocytic leukemia). RAR-β seems to be expressed in kidneys, prostate and cerebral cortex but not much in leukemic cells. RAR-γ appears to be expressed in the skin.

The crystal structure of the ligand binding domain of RAR-γ has been elucidated at 2Å resolution. Out of the several helices, helix 12 is folded around the ligand retinoic acid. Lysine 236, serine 289 and arginine 278 of RAR are involved in retinoic acid binding.

Besides RARs, nuclear receptors (X type) called RXRs appear to mediate the action of retinoids. RXR-α and RXR-β have tissue specificity. RXR-α is expressed in the liver and intestinal villi and RXR-β may be specific to the central nervous system. RXR-γ has also been identified.

Retinoic acid responsive element (*RARE*): Retinoic acid nuclear receptor interacts with a nearby hormone response element or enhancer and acts as a transcription factor promoting gene expression. The RAREs are responsive to the three nuclear receptors RAR-α, RAR-β and RAR-γ. It appears that retinoic acid brings about gene expression by binding to RARs that are already bound to the responsive elements (Figure 46.6).

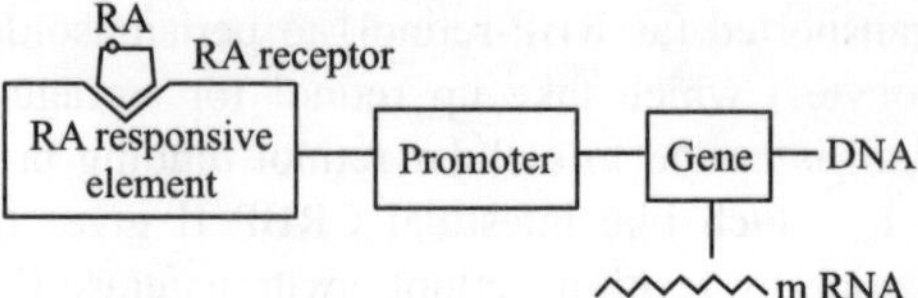

Figure 46.6

In view of the close similarity (high homology) of the nuclear receptors (C region) for thyroid hormone and retinoic acid, RAR-α also can bind to thyroid hormone responsive element (palindromic sequence TGACCGGTCA) and thereby control growth hormone production along with thyroid hormone.

F. Vitamin A and Vision

Visual perception is the result of several definite processes, e.g., light perception with differences of light intensity and colour perception from light of different wavelengths.

There are two types of cells in the retina, the rods and the cones. Rods are primarily concerned with simple light perception, which would be of great use to an animal in dark vision, also called scotopic vision. Cones are useful for bright light vision, called photopic vision as well as for colour vision.

Birds are known to sleep at dusk and get active at day break. Their retina contains predominantly cone cells. The owl is a notable exception as its retina shows a predominance of rod cells. An owl is most active at night-time, but is inactive during the day. Mammals possess both rods and cones. Cats, rats, mice, dogs, and foxes largely have rod cells in their retina. Man has a predominance of cone cells along with a fair proportion of rod cells.

An analysis of the nerve fibers to retinal excitation by means of an electroretinogram explains the mechanism of vision as follows: 'When a nerve, which is connected to a muscle, is stimulated either mechanically or electrically, the muscle contracts owing to the conduction of the stimulus along the nerve fiber. The conduction of the impulse along the nerve is accompanied by a wave of change in electrical potential, and if electrodes attached to a suitable amplifying and recording device are attached to the nerve, then the change in electrical potential accompanying the passage of the impulse can be recorded. Conduction along the nerve fibers connected with the retinal rods and cones has been demonstrated. When the intensity of light was increased, the frequency of electrical charges also increased, showing that stimulation of retina by light is directly related to conduction of impulses along the optic nerve'.

The pink-colored substance isolated from the rod cells of the frog and called *visual purple* or *rhodopsin* bleaches when it is exposed to light with the formation of *all-trans* retinal. Visual purple is found in the rod cells of the retina but never in the cone cells. Similarly, another typical pigment called iodopsin is always found in the cone cells but never in the rod cells. In ordinary day light, the retina contains relatively little rhodopsin because the pigment gets bleached the moment the light falls on it. However, it is constantly re-formed by the retinal enzymes.

Effective night vision is dependent on an adequate store of vitamin A in the liver. With the aid of enzymes and a carrier-protein in the blood, a continuously good supply of vitamin A is sustained in the retina.

1. The visual cycle: A special isomeric form of vitamin A called 11-*cis* retinal is required. Retinal can be prepared in vitro by the oxidation of retinol with manganese dioxide. Rhodopsin is a conjugated protein whose prosthetic group is a vitamin A derivative, 11-*cis* retinal (Figure 46.7). It is easy to prepare rhodopsin in the laboratory by taking vitamin A, crystalline alcohol dehydrogenase from horse liver, yeast cozymase and opsin from retina. If the mixture of the above four substances is placed in the dark, it will synthesize rhodopsin. When rhodopsin is bleached, the *all trans* retinal isomer emerges and comes out from its binding site on the protein, which then undergoes a change in its conformation. It induces a calcium ion channel in the rod cell membrane which initiates the nervous impulse passing on to the optic nerve and an image is formed in the brain (Figure 46.8).

Porphyropsin, whose prosthetic group is 11-*cis*-3-dehydroretinal, is the rod pigment in the eyes of certain animals, e.g., fresh water fish. Similar to the rod pigments named rhodopsin and porphyropsin, there also are the two analogous cone pigments, iodopsin and cyanopsin, which are also dependent on vitamin A1 and A2, respectively. The composition and spectroscopic properties of the four visual pigments are summarized as follows:

Rod opsin + 11-*cis* retinal rhodopsin (500 nm)
Rod opsin +11-*cis* 3-dehydroretinal porphyropsin (520 nm)
Cone opsin +11-*cis* retinal iodopsin (560 nm)
Cone opsin +11-*cis* 3-dehydroretinal cyanopsin (620 nm)

These four visual pigments are studied in detail after isolation from the retina of specific species of animals. Their presence could also be shown by action spectra by connecting electrodes to the nerve region of the animal.

2. Colour vision: In the cones, a similar mechanism operates to give rise to colour vision. Each individual cone contains one of three pigments with absorption maxima at 430, 540 and 575 nm. Each is an opsin combined with the same prosthetic group, viz. 11-*cis* retinal. The differences in their absorption maxima are due to the differences in the protein structure, and such differences stimulate colour discrimination. Colour blindness is primarily due to the absence of one of the three pigments.

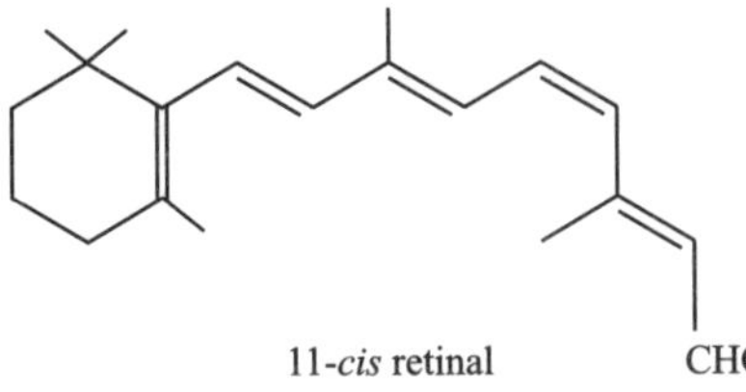

G. Effects of Vitamin A Deficiency

Vitamin A deficiency occurs in persons with poor diet that lacks vegetables (β-carotene). When there is depletion of vitamin A in the liver, the earliest sign is night blindness, that is impaired dark adaptation. Night blindness or nyctalopia is due to inadequate amount of vitamin A required for the regeneration of rhodopsin. The individual is unable to see in the dim light or to adapt to a decrease in light intensity. Rods, which are involved in dark vision, need vitamin A for this function (Figure 46.8). Night blindness is associated with cirrhosis of liver. It may also occur in diabetics if they are on large amounts of green or yellow vegetables without insulin, because they cannot convert carotene into retinol. Night blindness is readily cured by vitamin A. Night blindness is more common in males than in females.

Xerophthalmia occurs within 40 days in rats given a diet devoid of vitamin A. Furthermore, epithelial lesions and keratinization of tissues is observed in the respiratory, intestinal and urinary tracts, salivary glands and the genital system. In addition, teratogenic effects leading to congenital deformities are seen in rats whose mothers are kept deficient in vitamin A over a prolonged period. Bitot's spots in the eye and typical skin lesions called Darrier's disease are often observed in children and adults. Occurrence of xerophthalmia and keratomalacia are the characteristic manifestations of vitamin A deficiency. Xerophthalmia is characterized by dryness of the eyes, and the lacrimal glands get keratinized and stop secreting tears. Xerophthalmia is characterized by the following three stages depending upon the severity of vitamin A deficiency.

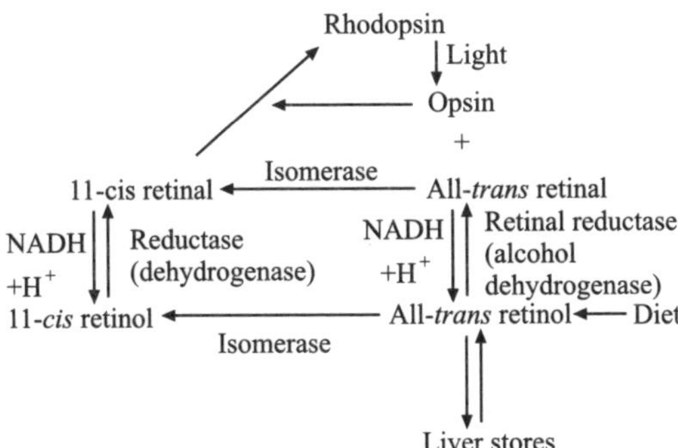

Figure 46.7

Figure 46.8

(a) Xerosis—manifested by the dryness of the conjunctiva leading to a wrinkled and thickened appearance of the eyes.

(b) Bitot's spots are often associated with vitamin A deficiency and xerosis can extend to the cornea.

(c) Keratomalacia—the cornea becomes softened and may extend to iris and lens. In fact, keratomalacia occurs mainly due to secondary infection as the bacteria are not washed away due to drying up of tear-secreting glands. Hyperkeratotic papules of the hair follicles are also characteristic of vitamin A deficiency. However, in severe cases, if also accompanied by the deficiency of B complex vitamins, there is formation of 'toad-like' skin. While the adults store vitamin A in the liver, the infants do not possess adequate stores of vitamin A. Xerophthalmia thus affects large number of children but rarely adults. Xerophthalmia, if not treated in time, leads to total blindness. If the children also have Kwashiorkor (see under I), they become more susceptible to infections. There may be renal calculi due to cornification of kidney medulla.

Growth retardation is another early manifestation of vitamin A deficiency. There is also a failure of skeletal growth as a result of defective synthesis of chondroitin sulphate and excessive lysosomal sulphatase activity. While the development of bony structure is retarded, the growth of nervous tissues continues causing lesions in the central nervous system.

Atrophy of the germinal epithelium results in sterility of the male. In the females, there may be normal ovulation and implantation, but normal offspring are rarely born in vitamin A-deficient mothers due to cornification of cells and defective placentas.

H. Hypervitaminosis A

The excessive consumption of polar bear liver by man and dogs can cause severe illness due to hypervitaminosis A. Injudicious administration of massive doses of vitamin A to children leads to an excessive production as well as sudden rise of the cerebrospinal fluid pressure. Prolonged overdosage with vitamin A in young children leads to chronic type of hydrocephalus, and external cortical thickening of both tibia and fibula. In elderly persons, usually a stooping posture, pain in movement, and a bulge in the right temporal area is observed. The major manifestation of hypervitaminosis A in man and in experimental animals is an increased fragility of bones with frequent fractures. The symptoms of hypervitaminosis A usually occur if RBP is saturated and excess unbound retinol is present.

Vitamin A contents of liver and blood in normal human subjects and in disease conditions: The amount of vitamin A in the liver of normal subjects is about 220 I.U./g whereas it is higher-up to 300 I.U./g in persons suffering from thyroid diseases and in diabetes. On the other hand, individuals with other diseases show a remarkable decrease in the liver reserves

of vitamin A, with the lowest values in enteric fever, urinary infections, chronic nephritis, tuberculosis and cancer. The average concentration of vitamin A in blood plasma is 170 I.U./100 ml in normal controls, whereas in some other diseases, and particularly during fever, there is a remarkable lowering in vitamin A levels in the plasma. However, an intake of ethyl alcohol by dogs or by human beings causes mobilization of vitamin A stored in tissues and both the blood level and the dark adaptation test show a remarkable increase in vitamin A.

I. Kwashiorkor and Vitamin A Metabolism

During kwashiorkor—a protein deficiency syndrome—the mobilization of vitamin A from the liver of rats is remarkably reduced, resulting in higher vitamin A storage. However, the vitamin A levels in the blood are reduced markedly and there are signs of xerophthalmia. Furthermore, enzymes of vitamin A metabolism, viz., retinal oxidase, retinal reductase, and retinyl palmitate hydrolase show a considerable decrease during kwashiorkor. All these effects are reversed if the rats are restored to normal protein-rich diet showing that the damage is not irreversible.

Like other polyprenols, vitamin A (a tetraprenol) forms a monophosphoric ester named retinol phosphate, which can transfer a sugar like mannose or galactose to glycoproteins of the membranes or glycolipids. A role for vitamin A in glycoprotein synthesis has been implicated.

J. Determination

Biological activity of vitamin A is expressed as International Units (I.U.). One I.U. is equal to 0.3 µg of vitamin A or 0.344 µg of vitamin A acetate. Vitamin A can be estimated by rat growth method. But colorimetric and fluorometric methods are preferred for convenience. Crystalline vitamin A is the reference standard.

Vitamin A requirements: The recommended daily dietary intake for Indians by Indian Council of Medical Research 1984 is indicated in Table 46.1.

TABLE 46.1

		Vitamin A (µg)	Carotene (µg)
Infants	birth–6 months	–	–
Infants	7–11 months	300	1200
Children	1–3 years	250	1000
	4–6 years	300	1200
	7–9 years	400	1600
	10–12 years	600	2400
Adolescent boys and girls	13–18 years	750	3000
Adults (men and women)		750	3000
Lactating women		1150	4600

K. Vitamin A and Cancer

There appears to be increased risk of cancer with low intake of vitamin A. But in adenocarcinomas, the reverse is reported. Increased susceptibility to cancer could be due to several reasons such as altered metabolism of carcinogens or altered immunocompetency. Further, in vitamin A deficiency animals, many tissues show altered states of differentiation and rates of proliferation (especially in the lung, kidney and bladder) as evidenced by increased DNA synthesis. Under such circumstances carcinogens could be more potent.

III. VITAMIN E

As early as 1920, it was found that rats fed only on cow's milk did not bear young. Their fertility was restored by wheat germ oil called vitamin E. It was given the name tocopherol (Greek, tokos, child birth; pheros, to bear; ol, derivative of an alcohol). The physiological effects of vitamin E are mediated by a series of naturally occurring compounds, which are chemically related.

A. Structure of Tocopherols

A number of different compounds have been isolated, of which the three important ones are α-, β- and γ-tocopherols (Figure 46.9). Biologically, the most active form is α-tocopherol (called RRR-α tocopherol).

They are derivatives of tocol, 2-methyl, 2-(4', 8', 12')-trimethyltridecyl-chroman-6-ol. The 6'-hydroxychromane α-tocopherol is 5, 7, 8, tri-methyltocol. β-tocopherol is 5, 8, dimethyl tocol, while γ-tocopherol is 7, 8 dimethyl tocol (Table 46.2).

TABLE 46.2

	Tocopherols	R_1	R_2	R_3
(i)	α-	CH_3	CH_3	CH_3
(ii)	β-	CH_3	H	CH_3
(iii)	γ-	H	CH_3	CH_3
(iv)	δ-	H	H	CH_3
(v)	ε-	CH_3	H	H
(vi)	ζ-	CH_3	CH_3	H
	Tocol	H	H	H

B. Occurrence

The best natural sources of vitamin E are vegetable oils such as wheat germ oil, safflower oil and sunflower oil, which contain high amounts of α-tocopherol. Soyabean oil and corn oil contain mostly γ-tocopherol. Cottonseed oil, rice germ oil, palm oil and other seed germ oils contain varying amounts of α- and γ-tocopherols. Lettuce, alfalfa, etc., contain considerable amounts of vitamin E, whereas fruits contain negligible amounts of the vitamin. Animal oils are devoid of vitamin E. Fish liver oils which are predominantly rich in the vitamins A and D, are poor in vitamin E. The livers of horse and cattle contain high amounts, as they are herbivorous, but no vitamin E is stored in the liver of the rat. A few other compounds of unknown constitution are even stronger antioxidants than vitamin E and protect the vitamin against oxidation.

The tocopherols are effective antioxidants, the γ-isomer being more effective than the β-isomer, which in turn is more effective than the α-isomer. Thus, the antioxidant power of vitamin E is the inverse of its activity.

The normal serum values range from 0.5 to 1.0 mg per 100 ml. However, the plasma vitamin E level does not exactly reflect the vitamin E intake. The reason is that higher intake of vitamin E does not result in a proportionate increase in plasma vitamin E concentration.

C. Determination

Vitamin E content can be determined either by the biological method (occurrence of pregnancy and prevention of resorption of fetuses in virgin female rats) or by spectroscopic or colorimetric method. Synthetic α-tocopherol acetate has been adopted as the reference standard for vitamin E. The international unit for vitamin E is defined as the specific activity of 1.0 mg of the standard preparation. This quantity is the average amount, which when administered orally, prevents resorption gestation in rats devoid of vitamin E.

D. Deficiency

Symptoms of vitamin E deficiency vary considerably in different species of animals. The most prominent effects are on the reproductive, muscular and nervous systems and on erythrocytes. The fertility in both males and females is affected. In vitamin E-deficient male rats, immotility of spermatozoa and degeneration of the germinal epithelium occur which can not be corrected by the administration of vitamin E. Vitamin

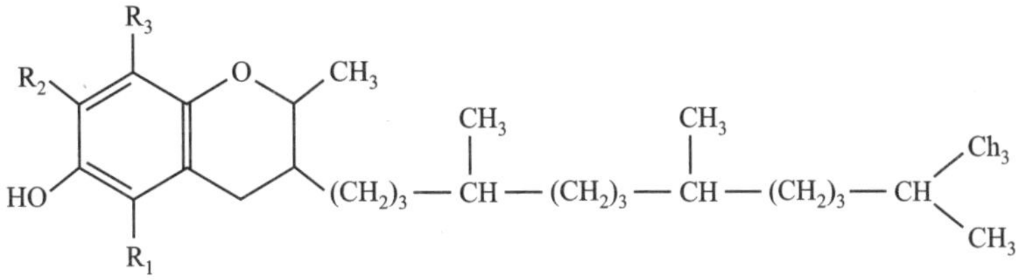

Figure 46.9

E-deficient female rats have healthy ova, normal placenta or uterus and become pregnant. However, in a short period of time, the fetuses die and resorption occurs. This can be prevented if vitamin E is given to such rats before the sixth day of gestation.

Muscular dystrophy is another equally important consequence of vitamin E deficiency. Skeletal, cardiac and smooth muscles and peripheral vascular system need vitamin E for their optimal functioning. Skeletal and cardiac muscle of vitamin E-deficient animals consume more oxygen due to peroxidation of unsaturated fatty acids. Some characteristic features are the inability of the skeletal muscle to use creatine which is thereby excreted in urine, degeneration of renal tubular epithelium, and depigmentation of teeth. These changes ultimately result in muscular dystrophy, paralysis of hind legs and creatinuria. In chicks, damage to the nervous system occurs due to impairment of the blood vessels of the brain. In humans, prolonged vitamin E deficiency affects nervous system. In rats, if there is deficiency of both vitamin E and proteins, hepatic necrosis is seen which could be prevented not only by tocopherol but also by cysteine, methionine or selenium. Muscular dystrophy has been observed in other animals also, viz. turkey, lamb and cow, etc.

In view of its antioxidant properties, vitamin E is added in the diet to prevent oxidation of vitamin A and other unsaturated fats. This sparing effect of vitamin E on vitamin A and carotene enables the latter two compounds to display better potency. Selenium and ascorbic acid enhance the antioxidant effects of vitamin E.

There is no conclusive evidence to show that vitamin E is required for human fertility although there are claims about the beneficial role of tocopherol in sterility, habitual abortions and in abnormalities of premature infants. The deficiency symptoms of vitamin E are observed only in persons with chronic impairment of fat absorption.

Vitamin E deficiency is observed in premature infants with hemolytic anemia, low birth weight babies on nutrients high in polyunsaturated fatty acids (PUFA) but low in vitamin E, malabsorption of vitamin E from diet, defects in plasma lipoprotein transport and delivery to target cells, and secondary to fat malabsorption (steatorrhea), pancreatic in-sufficiency, hepatobiliary tract diseases, abetalipoproteinemia.

Another effect of vitamin E deficiency is on the fragility of erythrocytes. Serum α-tocopherol levels are low in some nutritional anemias (macrocytic) which are cured by vitamin E. α-Tocopherol maintains the integrity of the membranes of the erythrocytes and prevents their hemolysis by preventing the oxidation of unsaturated fatty acids of the red cell membrane. Because of its role in the synthesis of heme and nucleic acid metabolism, vitamin E deficiency affects activities of some of the enzymes involved. In rats, decreased formation of delta aminolevulinic acid by inhibiting its synthetase and consequently inhibition of the synthesis of heme are also observed in vitamin E deficiency. The enzyme delta aminolevulinic acid dehydratase involved in porphobilinogen synthesis is also markedly affected in the liver. Certain catabolic enzymes like DNase, RNase, aryl sulphatase and cathepsins

are increased. Vitamin E is also claimed to be useful in the circulatory disturbances of old men.

E. Absorption, Transport and Delivery to Target Tissues

Vitamin E is utilized by oral route but not well by the parenteral route. Vitamin E is readily absorbed from the intestines in a way similar to that of other lipids using processes involved in their digestion and absorption. Pancreatic esterases hydrolyse tocopheryl esters in the diet. Bile acids are also required for vitamin E absorption along with other lipids like cholesterol, triglycerides, phospholipids and other fat soluble vitamins into the lymphatic system (thoracic duct). Diseases affecting pancreatic juice like pancreatitis, or bile acids like biliary obstruction, result in poor absorption and consequently deficiency of vitamin E.

Through the lymphatic system, vitamin E enters circulation and is transported in chylomicrons. Lipoprotein lipase of the lining of the capillaries seems to play a part in the transfer of vitamin E along with fatty acids from chylomicrons to the tissues like adipose tissue, muscle, brain and skin. When the chylomicrons are catabolized, different forms of vitamin E are carried in chylomicron remnants and delivered to the liver. Liver differentially handles diverse forms of vitamin E (α, β, γ and δ-tocopherols). It preferentially transfers RRR-α tocopherol to VLDL. Further catabolism of VLDL results in the transfer of α-tocopherol to HDL and other lipoproteins. α-Tocopherol acquired by all the lipoproteins is delivered to different tissues. It appears that there is no separate transport protein for vitamin E. The transport and delivery of vitamin E is linked to that of other lipids, especially fatty acids in lipoproteins (Figure 46.10).

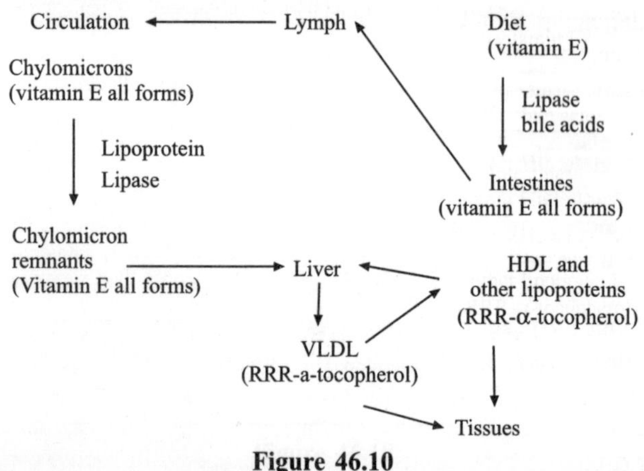

Figure 46.10

α-Tocopherol transfer protein: Liver plays a central role in recognising different stereochemical configurations (α-, β-, γ- and δ-forms) of vitamin E with the help of a protein called α-tocopherol transfer protein. This protein specifically transfers RRR α-tocopherol to VLDL and other lipoproteins for onward delivery to other peripheral tissues. The human and rat α-tocopherol transfer proteins have been purified.

They have 94% homology. Its gene in humans is localized in chromosome 8. Patients with a genetic defect of this protein cannot differentiate between different forms of vitamin E. Such patients show low plasma vitamin E, neurological abnormalities and ataxia. These symptoms can be prevented by supplementation of vitamin E. Liver diseases and chronic alcohol intake might lead to decreased vitamin E in the liver.

α-Tocopherol from plasma lipoproteins is recycled back to the liver. Vitamin E can be delivered to tissues through LDL receptor as well. There is no specific storage organ for α-tocopherol. But adipose tissue contains a major amount of vitamin E. It is also concentrated in sites such as plasma membranes. The turnover of vitamin E is fast in plasma, red blood cells and liver but low in muscle, testes, brain and spinal cord.

Measurable amounts of vitamin E are found in the blood. The blood of female rats contains approximately 0.5 µg and that of male rats 0.6 µg of tocopherols per 100 ml of the serum. The blood levels could be increased up to 100 µg by oral administration of a high dose of vitamin E. Upon an intake of vitamin E esters, the free vitamin E appears in the blood. The fetus can absorb sufficient vitamin E for its needs through the placenta if the mother is fed an adequate diet. Storage in adults occurs in the adipose tissue.

No significant excretion of vitamin E is observed as long as normal doses of the vitamin E are given. The intake of an excess amount of vitamin E causes the excretion of a certain amount in the feces, but only traces are found in the urine. α-Tocopherol is oxidized to tocopherylquinone, then reduced to its hydroquinone and excreted into bile after conjugation with glucuronate. It is further degraded in the kidneys to tocopheronic acid and excreted in urine.

A minimum of α-tocopherol to polyunsaturated fatty acid (PUFA) ratio of 0.6 is essential in the diet to prevent vitamin E deficiency syndromes in man. Whereas whole cow milk and human milk have a ratio of 1.0, synthetic diets prepared from skimmed milk and vegetable oils have a ratio of only 0.2–0.3.

F. Biochemical Functions

An understanding of the precise metabolic role of vitamin E is complicated by the fact that several chemically unrelated compounds, viz. synthetic antioxidants, ubiquinone, selenium, sulphur-containing amino acids reverse the avitaminosis E syndromes and that there are many deficiency diseases in different species of animals on diets lacking in vitamin E. However, some metabolic roles have been very well documented for α-tocopherol. It is a very potent fat soluble antioxidant and has a specific role in the metabolism of selenium.

1. Antioxidant role: Vitamin E acts as an antioxidant to protect lipids like unsaturated fatty acids (PUFA) of phospholipids of biological membranes and lipoproteins of plasma against damage by oxidation. The organic peroxide radical (ROO°) is converted to the corresponding hydroperoxide (ROOH) by the hydroxyl group of vitamin E which itself is

transformed into tocopheroxyl radical (VitEO°) as shown in Figure 46.11. α-Tocopherol is regenerated by reduction of the tocopheroxyl radical by a hydrogen donor like vitamin C, thiols, glutathione, etc. Thus, vitamin E plays an important role as an antioxidant in the body defense mechanism to prevent peroxidation of lipids and the damage by lipid peroxidation products. For performing this function, vitamin E must be absorbed from the intestinal lumen and transported to the target cell membranes in adequate amounts.

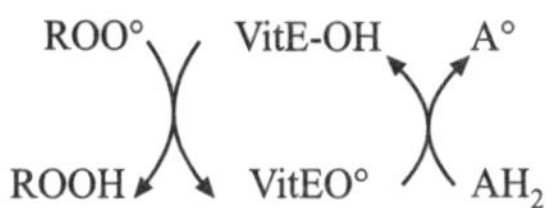

Figure 46.11

Supplementation of Vitamin E in diseases and aging: Peroxidation products are found in increased amounts in many diseases like atherosclerosis, diabetes mellitus, some cancers, and during aging. Therefore, the protective role of vitamin E, either alone or in combination with other antioxidants like vitamins A and C, against free radical damage has been investigated.

Coronary artery diseases: According to one study lower levels of vitamin E in the blood seem to be related to the risk of angina pectoris. Normally, vitamin E, which is associated with LDL along with other antioxidants, forms the body defense system to balance the free radicals and protect LDL. Formation of increased peroxidation products within LDL itself due to oxidative stress by free radicals leads to the depletion of vitamin E within LDL and, in general, other antioxidants and increases the risk of atherosclerosis. Supplementation of vitamin E along with other antioxidants like ascorbic acid, which regenerate α-tocopherol from α-tocopheroxyl radical (as explained above), seems to decrease the susceptibility of LDL to oxidation by increasing α-tocopherol in LDL. However, vitamin E in much higher doses has been shown to act as a prooxidant and increase lipid peroxidation.

Diabetes and aging: Results of limited scale studies indicated that supplementation with vitamin E (400 mg to 1 g) prevented vascular complications in diabetes by increasing thromboxane A2 levels and decreasing platelet aggregation as well as enhanced cell-mediated immunity in elderly persons over 60 years. Prolonged deficiency of vitamin E brings about neurological changes also. There are suggestions that early supplementation with vitamin E could be of use in senile cataract.

Parkinson's disease and Alzheimer's disease: Vitamins A, C and E given together in physiological doses that could be obtained from a well balanced diet seem to be beneficial in the above-mentioned conditions with increased levels of peroxidation products. However, therapeutic role of supplementation of vitamin E above the recommended dietary requirements (15–30 mg), either alone or in combination with other vitamins, is controversial and is the basis of many ongoing studies.

The antioxidant-responsive diseases which can be prevented partly or wholly either by vitamin E or synthetic antioxidants such as methylene blue, thiodiphenylamine, rosaniline, malachite green, methyl violet, 2:6-dichlorophenol indophenol, etc., include *in vitro* hemolysis of vitamin E-deficient erythrocytes, damage to membrane systems, dental depigmentation, brown discolouration of the uterus, mitochondrial swelling, muscular dystrophy and reproductive failure. Vitamin E is indispensable for fertility in rats. The tocopherols are also added as antioxidants in vitamin A-rich oils to prevent oxidation.

2. Interrelationship of vitamin E and selenium: In the chick, muscular dystrophy can be prevented by cystine or methionine in the diet. The primary cause of necrotic liver degeneration in rats, and muscular dystrophy in turkeys, lambs, and cows is due to deficiency of selenium in the diet. Furthermore, vitamin E is not well absorbed by the selenium-deficient chick.

Recent studies have revealed a close functional inter-relationship between vitamin E and selenium. It has been shown that glutathione peroxidase is a selenium-containing enzyme. In infants, it has been shown that vitamin E deficiency is coincident with low glutathione peroxidase levels. In adults, on diets sufficient in selenium as well as vitamin E, the activity of this enzyme is normal. Some peroxides, which are still formed even in the presence of vitamin E, are destroyed by glutathione reductase. Thus, initially vitamin E and subsequently selenium and glutathione reductase constitute an effective defence system against the peroxidative damage and maintain the integrity of the cells and its organelles, especially the membranes. Selenium has a sparing effect on vitamin E and reduces its requirement by preventing the peroxidation of polyunsaturated fatty acids and also by helping digestion and absorption of lipids including vitamin E. In a similar way vitamin E also has a sparing effect on selenium and prevents its loss from the body. When vitamin E acts as an antioxidant, both its chroman ring and the side chain are oxidized and excreted in the bile after conjugation with 2 molecules of glucuronic acid.

The requirements for vitamin E as recommended by the ICMR group are 5 μg for infants (human milk has some vitamin E), 10–25 μg for children and 10 μg for adults with low intake of polyunsaturated fatty acids (PUFA) but the need increasing up to 30 μg with higher intake of PUFA.

3. Action on membranes.   Vitamin E is effective in preventing the vitamin A-induced hemolysis of rabbit erythrocytes by preventing the breakage of the cell membrane *in vitro,* as well as during vitamin E deficiency as observed in *in vivo* experiments. These observations are also valid at the level of intestinal cell membranes explaining the poor absorbability of amino acids during vitamin E deficiency. Vitamin E prevents lipid peroxidation, which is toxic to membranes.

Vitamin E and amino acid metabolism: During avitaminosis, the creatine content of the striated muscles is greatly reduced in rabbits and in rats, and the urinary excretion of creatine is considerably increased while the excretion of creatine remains unchanged. A marked reduction of the creatine content in the urine is observed following the administration of vitamin E. These effects could be explained from the point of view of the poor absorbability of amino acids from the intestinal cell membranes during vitamin E deficiency. However, no such changes have been found in the urine of humans suffering from progressive muscular dystrophy, and any other neuromuscular disease.

4. Vitamin E at cellular level: The mode of physiological action of vitamin E involves its ability to direct certain activities of the cell nucleus. The vitamin function is intimately concerned with the process of cell maturation and differentiation with mitosis. Vitamin E promotes growth as demonstrated in *in vitro* experiments using tissue cultures of liver, spleen, heart cells. The considerable decrease in the weight of the testis even when calculated in relation to the reduced body weight is, however, specific for this vitamin. Vitamin E affects growth of prematurely born children and heals skin wounds, thus supporting the concept that vitamin E influences such tissues where cellular proliferation and differentiation proceed at a rapid speed.

Vitamin E-depleted rats show histopathological changes in the testis. The chromatin of the spermatozoa undergoes a typical change, a lysis, and the nuclei become crescentic. Upon extended development of the deficiency disease, spermatocytes and spermatozoa fuse into giant cells with several nuclei. Besides chromatolysis, there appears to be an interference in the formation of chromatin. Similar signs are also observed in the fetus. The earliest sign of vitamin E deficiency in the embryo is observed in the hematopoietic tissues, where the growth is greatly retarded. The mesodermal tissues where rapid cellular activity prevails are also affected. The nervous and degenerative lesions in the muscles, which are signs of vitamin E deficiency, seem to be related to the cerebral cortex, a tissue of high cellular activity.

IV.   VITAMIN K

This vitamin was discovered by the Danish scientist Dam who observed that a hemorrhagic disease in chicks fed a synthetic diet could be prevented by a factor isolated from alfalfa leaves and fish meal. This was called Koagulation vitamin.

A.   Structure

Vitamin K1 and K2 represent two naturally occurring vitamins K. Chemically, they are derivatives of naphthoquinone. The product from alfalfa is vitamin K1 or phylloquinone. It is 2-methyl-3-phytyl-1, 4-naphthoquinone. The products from fish meal and a number of microorganisms are vitamin K2 or menaquinones. They are derivatives of 2 methyl-3-n (isoprenyl)-1, 4-naphthoquinone where n is the number of isoprenyl units which vary. Out of many synthetic analogs, menadione (iii) is water soluble.

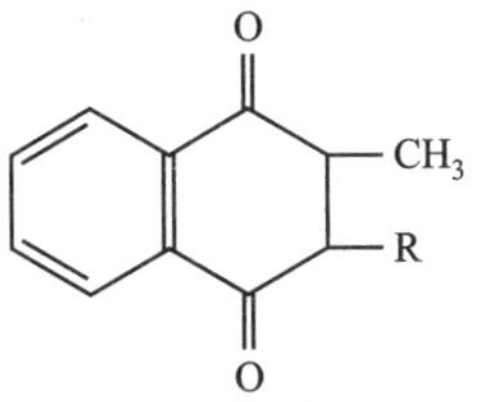

![Figure 46.12 chemical structure]

(i) Vitamin K_1:

$R = CH_2CH = C.CH_2[CH_2.CH_2.CH.CH_2]_2. CH_2. CH_2. CH.CH$

(ii) Vitamin K_2:

$R = [CH_2.CH = C.CH_2]_n$. where n varies from 4 to 12

(iii) Menadione: R = H
Synthetic (2-methyl-1, 4-naphthoquinone)

Figure 46.12

B. Occurrence

Vitamin K is synthesized by plants and by certain microorganisms. In plants, the site of synthesis is in the green leaves. The tops of carrots contain a considerable amount of vitamin K, while the roots contain a negligible quantity. Phylloquinone (vitamin K_1) is in highest amount in green leafy vegetables (1000–8000 µg/kg). Fruits and other vegetables contain 10–500 µg/kg. Dairy products contain only 3–70 µg/kg depending on fat content. Grains have only 0.5–70 µg/kg. Vitamin K synthesis is achieved and influenced in the presence of sunlight. Peas grown in the dark contain small amounts of vitamin K, while control plants raised in light contain considerably higher quantities of the vitamin. The outer leaves of cabbage contain about four times more vitamin K than the inner leaves. The green leaves of plants predominantly contain vitamin K_1, whereas the microorganisms, especially bacteria, contain vitamin K_2. The microorganisms in the intestinal tract contain large quantities of vitamin K, which is synthesized in them from shikimate and α-keto-glutarate. Most putrefied animal and plant materials contain high amounts of vitamin K due to bacterial growth. Animal products contain very little vitamin K, although dairy products like yogurt, milk, cheese contain about 300 µg/kg, and eggs contain some amounts. Hog liver is the richest animal source of vitamin K.

The human digestive tract contains much more vitamin K than is actually required to prevent delayed blood clotting, i.e. reduction in prothrombin time. Menaquinones consisting of different number of isoprenoid units usually 4 to 12 (designated MK-4 to MK-12) are detected in humans. Major amounts are in the gut and liver. Liver, which synthesizes most of the vitamin K-dependent proteins, contains mostly (90%) menaquinones (MK-9 to MK-11) and small amounts (10%) of phylloquinones. Menaquinones are mostly in mitochondria. In serum, very small amounts of vitamin K are found, out of which menaquinones are more than phylloquinones.

When menadione is tested, an activity of about four times that of vitamin K1 per weight unit is observed. This enhanced activity of menadione prompted the use of this compound in clinical practice. Its activity is almost equal to that of vitamin K1 on a molar basis.

C. Determination

The K-group of vitamins are thermostable and show typical absorption characteristics in the UV region with maxima at 243, 248, 261, 270 and 328 nm.

For many years the standard method for determination of vitamin K was either by the time taken for the blood of vitamin K-deficient chicks to clot (blood clotting time) or by prothrombin time which is a measure of prothrombin activity. Recently modified method is based on estimation of serum γ-carboxylated prothrombin. Another γ-carboxylated protein secreted in bones called osteocalcin is released into circulation. Serum level of osteocalcin or under-carboxylated osteocalcin is estimated because it is considered to be a more sensitive index of vitamin K status of an individual.

D. Absorption

Bile salts which help in the digestion of fatty acids and other fat soluble vitamins are necessary for the absorption of vitamin K from the upper part of the small intestine. Therefore, phylloquinone solubilized in fat is better absorbed from intestines than as such. Phylloquinone is incorporated into chylomicrons for transportation to liver and other tissues. Any condition which causes fat malabsorption like obstruction of bile duct or insufficient secretion of bile salts, pancreatic dysfunction and steatorrhea reduces the absorption of vitamin K and leads to its deficiency. Menadione and its water-soluble derivatives do not need the presence of bile salts for their absorption. It is not known how exactly vitamin K is transported and cleared but phylloquinone is transported in chylomicrons.

E. Deficiency

In vitamin K deficiency, there is delayed blood clotting and increased prothrombin time as a result of which profuse bleeding occurs even from minor injuries. Vitamin K deficiency is found only in such animals in which the vitamin K synthesized is not sufficient to meet the requirements. In newly hatched chicks not supplemented with vitamin K, there is gradual diminution of prothrombin in blood. Administration of vitamin K corrects this deficiency and normal blood clotting is restored. In rats, males are more susceptible than females to the effect of vitamin K deficiency.

In humans, the synthesis of vitamin K by the intestinal flora meets the requirements of this vitamin. But sterilisation of the large intestine by the prolonged use of sulfonamides and antibiotics or diarrhoeal diseases like sprue, ulcerative colitis and conditions which reduce fat absorption may lead to vitamin K deficiency. Many antibiotics like penicillin, cephalosporin, aminoglycoside, chloramphenicol, amphotericin B, and erythromycin are reported to cause vitamin K deficiency and hypoprothrombinemia (vitamin K-responsive type). Antibiotics containing N-methylthiotetrazole side chain or methyl-thiadiazolethiol side chain can inhibit hepatic vitamin K epoxide reductase, and like coumarins, affect the synthesis of coagulation factors and may be weak anticoagulants. In the newborn infants, the intestines are sterile and their prothrombin level is low. Therefore, they may show signs of vitamin K deficiency and bleeding, which is called hemorrhagic disease of the new born (HDN). This condition persists until the bacterial flora are established in the infant's intestines, usually a week, and vitamin K is synthesized. Hemorrhagic disease of the new born is one of the causes of infant mortality. In cases of intracranial hemorrhage in infants, there may be death, or in cases of survival, brain injury occurs leading to mental and neurological disorders. Treatment with vitamin K of mothers before parturition or the infants soon after birth is helpful in preventing the harmful effects of vitamin K deficiency.

F. Biochemical Functions

Vitamin K is not part of the prothrombin molecule. Recent work at molecular level has indicated that vitamin K plays a very significant role in the formation of functional prothrombin in liver. For this reason vitamin K cannot correct the deficiency symptoms in hepatectomised animals. The mechanism of blood clotting involves a series of steps (details in Chapter 19) in which many factors take part. Essentially it involves the conversion by the enzyme thrombin of soluble plasma fibrinogen into insoluble fibrin. In this process a peptide is removed by proteolysis. But thrombin itself is formed from its inactive precursor prothrombin by the action of thromboplastin (thrombokinase) in presence of calcium only when required. These steps are shown in Figure 46.13 in a simplified form, although a number of other factors are involved.

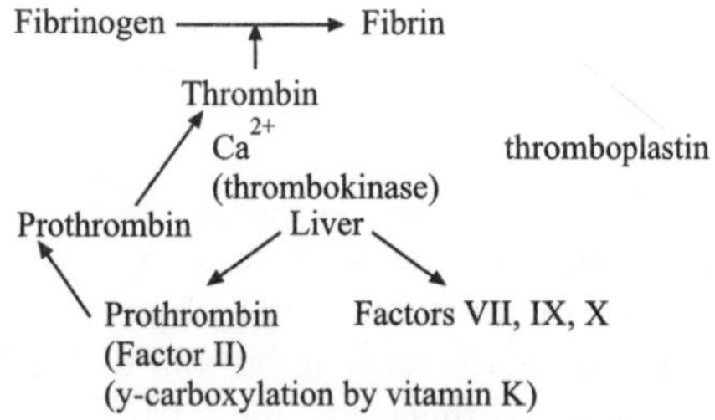

Figure 46.13

Vitamin K is necessary for the maintenance of normal levels of not only prothrombin (factor II) but also blood clotting factors VII (proconvertin, serum prothrombin conversion accelerator), IX (Christmas factor, plasma thromboplastin component) and X (Stuart Prower factor). All these four blood clotting factors, II, VII, IX and X, are synthesized in the liver as inactive precursors and the conversion to their active forms requires vitamin K.

Vitamin K and γ-carboxylation of glutamic acid: Previously it was believed that vitamin K was necessary at transcriptional level for the induction of mRNA for prothrombin. The discovery in 1974 that bovine prothrombin contains a previously unrecognised amino acid γ-carboxyglutamic acid has revolutionised the role of vitamin K in blood coagulation. More recent work has revealed that vitamin K plays a hitherto unexpected role of carboxylation at γ-position of glutamic acid residues (10 in number) on the prothrombin molecule already synthesized, i.e., the effect is after translation. Similar post-translational modification by carboxylation of glutamic acid (Glu) to γ-carboxyglutamic acid (Gla) (Figure 46.14) occurs even with the other three blood clotting factors VII, IX and X. These previously unknown Gla residues chelate calcium ion in the presence of phospholipid in a specific interaction which is necessary for clotting.

The vitamin K-dependent carboxylase or γ-glutamyl carboxylase appears to be a specialised microsomal electron transport system linked to CO_2 fixation. This reaction does not require ATP but utilizes energy due to reoxidation of reduced vitamin K (KH_2) for which NADH is required. The carbonyl group next to methyl group in the vitamin K molecule is involved in the carboxylation reaction.

Two mechanisms have been proposed for the oxidation of KH_2 to vitamin K and regeneration of KH_2 during CO_2 fixation of Glu-peptide.

According to the first mechanism oxygen taken up yields H_2O_2 and vitamin K (Figure 46.15). NADH regenerates reduced vitamin K for carboxylation. In the second mechanism, the reduced vitamin K and oxygen form water and vitamin K epoxide. The latter is converted back to vitamin K by a reductase for which dithiothreitol or other sulphhydryl compound is necessary. Since reduced vitamin K is regenerated, it is called vitamin K cycle.

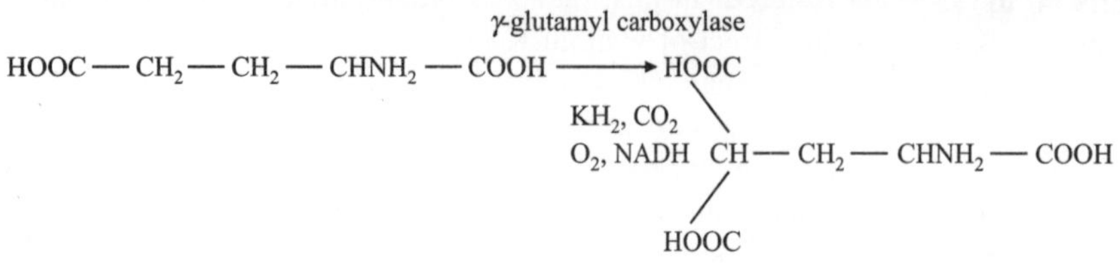

Figure 46.14

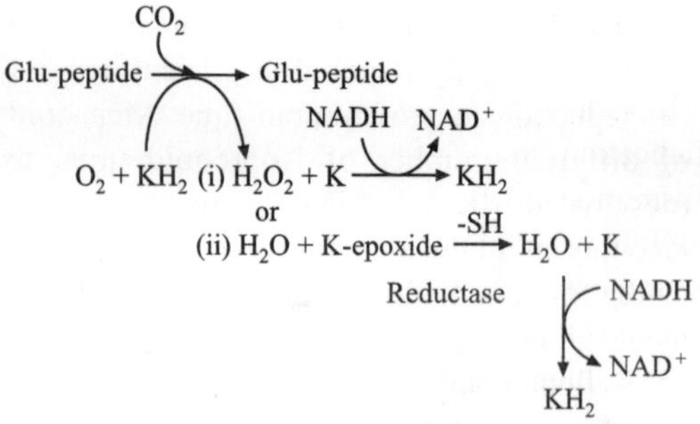

Figure 46.15

Besides the above blood clotting factors, a number of other proteins (with functions not clearly established) in plasma (C, S and Z proteins), bone (osteocalcin or bone Gla protein BGP and matrix Gla protein MGP), kidney, lung and spleen also have γ-carboxyglutamic acid for which vitamin K is necessary.

Role in bone metabolism: Substantial amounts of vitamin K are found in bone. Osteoblasts in bone produce three proteins which undergo post-translational γ-carboxylation of glutamic acid (Gla proteins) for which vitamin K is a cofactor. These proteins (osteocalcin, also called bone Gla protein (BGP), matrix Gla protein (MGP) and protein S) are secreted in such a way that main part of each protein remains bound to the hydroxyapatite matrix in bone. Out of these three Gla proteins, osteocalcin is exclusively found in osteoblasts and odontoblasts. But about 20% of it is released into blood circulation. The MGP and protein S are produced in other tissues also. MGP is formed in many soft tissues, while protein S is produced in hepatocytes, megakaryocytes and endothelial cells. About 80% of the body store of Gla proteins is in bones. These three bone Gla proteins have some similarities with Gla proteins of blood coagulation.

The three bone Gla proteins are secretory proteins with a common intracellular precursor having a hydrophobic leader sequence which helps in translocation of these proteins across the endoplasmic reticulum. These proteins also have an 18-amino acid residue which recognises the vitamin K-dependent γ-carboxylase. These recognition sequences have highly conserved amino acid residues-16 (Phe) and -10 (Ala/Gly). The Gla domain of all Gla proteins has the sequence Gla-x-x-x-Gla-x-Cys. The α-helical Gla domain of osteocalcin is bound to hydroxyapatite matrix (mineralized), while its carboxy terminal part is attracted to osteoclast cells involved in resorption of bone. It is therefore believed that osteocalcin regulates the mineralization and remodeling of bones.

In spite of a high content of hydrophilic amino acids, matrix Gla protein is an insoluble low molecular weight (9.6 kDa) protein. While most of the secreted MGP is in circulation, it is present in higher amounts in bone and cartilage. It is suggested that MGP may protect against tissue calcification and takes part in extracellular matrix involved in cell adhesion. Protein S is a single chain-protein with 11 Gla residues. It acts as an inhibitor of blood coagulation by acting as a cofactor for activated protein C. It is involved in metabolism and is a constituent of organic bone matrix.

Vitamin K supplementation and deficiency: There are indications that vitamin K supplementation stimulates bone formation and reduces resorption as revealed by *in vitro* and *in vivo* studies in humans. MK4 (menaquinone with four isoprenoid units in side chain) reduced prostaglandin E_2, which is a bone-resorbing agent. Gla proteins play an important role in blood clotting and bone metabolism. In older persons with less vitamin K intake, blood coagulation is not affected much, but bone formation is affected. So bone tissue may be more prone to vitamin K deficiency than liver which synthesizes (in hepatocytes) Gla proteins involved in blood clotting. There is under-carboxylation of osteocalcin. In view of this, some investigators believe that instead of impaired blood coagulation tests, circulating osteocalcin in under-carboxylated form is a more sensitive marker of vitamin K status of an individual. Both dietary phylloquinones from green leafy vegetables and dairy products and synthetic vitamin K (menaquinone) supplements are important sources of vitamin K. As stated above, triacylglycerols help in the absorption of phylloquinones. According to some investigators, plasma triglycerides could be important determinants of circulating phylloquinones.

One population study suggests that low vitamin K intake may also be associated with atherosclerotic calcification because the Gla proteins osteocalcin and MGP are also found in atherosclerotic plaque.

The recommended daily dose of vitamin K is 35–40 μg for infants, 65–120 μg for children, depending on their age and 100–200 μg for adolescents and adults. There is a possibility of hypervitaminosis K with menadione in children leading to hemolysis and hyperbilirubinemia.

As a quinone, vitamin K has been expected to play a role in the electron transport chain like coenzyme Q. There is no evidence in support of such a function for vitamin K in mammalian systems. However, the work of Weber and Brodie in 1957 indicated a role for vitamin K2 (K9H) in the electron transport of *Mycobacterium phlei*. In the respiratory chain of mycobacteria, vitamin K occupies a position before cytochromes and after flavin. Vitamin K is necessary for malate oxidation and coupled phosphorylation requiring FAD but not NAD^+ as a coenzyme. Thus, vitamin K appears to have a role in the electron transport of mycobacteria.

G. Antagonists

Dicumarol (Figure 46.16) and its synthetic derivative warfarin are anti-coagulants that function as vitamin K

Figure 46.16

antagonists. These drugs inhibit vitamin K epoxide reductase. Therefore, vitamin K can be used as an antidote for dicumarol toxicity.

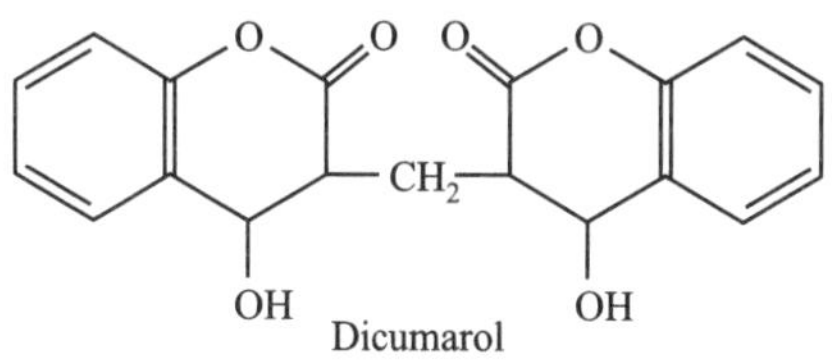

V. UBIQUINONE (COENZYME Q)

Ubiquinone was discovered by Morton in 1955, and Crane isolated it in 1957.

Ubiquinones are found in most of the aerobic organisms and in animals, plants and microorganisms. Generally, ubiquinone is associated in the membrane fraction of the cells. Rat liver mitochondria contains about 40–50 per cent of ubiquinone. Some anaerobic organisms and Gram-positive bacteria are devoid of this lipid quinone. However, Gram-negative bacteria contain ubiquinone. It is observed that the tissues having a high respiratory rate contain large amounts of ubiquinone, which is consistent with their role in electron transport.

Ubiquinones have been considered to have vitamin function because of the ability of these compounds to protect against vitamin E deficiency in some species of animals.

A. Structure

Ubiquinone has a structural relationship with fat-soluble vitamins—with vitamin A, in having a polyprenoid side chain; with vitamin E, in being able to form a cyclized chromenol derivative; and with vitamin K, in having a quinone capable of oxidation-reduction (Figure 46.17).

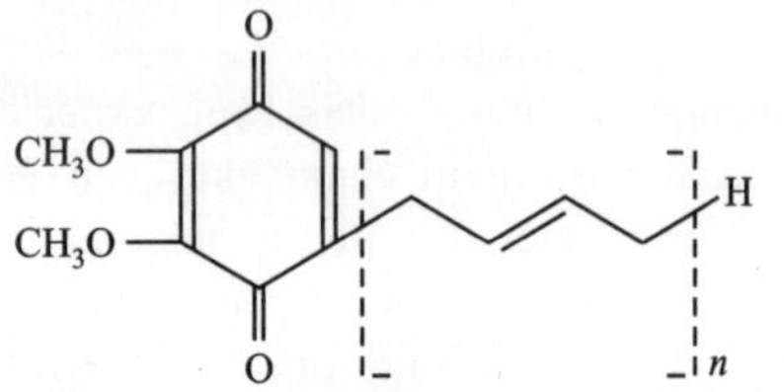

Ubiquinone-*n*

Figure 46.17

Ubiquinone-n: The predominant form found in higher animals is ubiquinone-10 (Q-10), with the notable exception of Q-9 in the rat, which is present to about 85 per cent. Ubiquinone is present in all parts of the plant; plant seeds contain large amounts of it. Ubiquinone has been found in human blood, urine, feces and other tissues.

Ubiquinone treatment has a potential curative effect at least in some types of dystrophy in the monkey and man, and ubiquinone-4 shows a similar effect in dystrophic mice.

B. Estimation of Ubiquinone

Spectroscopic method: Most of the ubiquinones show a strong characteristic absorption band at 275 nm. However, they have varying molecular extinction coefficients, depending on the chain length. The method of estimation consists of finding the change in absorption at 275 nm on addition of a small quantity of sodium borohydride solution to an ethanolic solution of ubiquinone.

For crystalline ubiquinone-10 in ethanol, E1%1cm values at 275 nm are 165 and 23 for oxidized and reduced spectra, respectively.

The oxidation-reduction potential of ubiquinone measured against the benzoquinone-hydroquinone system (E0 = 720 mV) is 542 mV.

C. Ubiquinone and Vitamin A Deficiency

The original observation by Moore, Morton and coworkers on a compound with absorption peak at 275 nm was made in the livers of rats given a vitamin A-deficient diet. The progressive increase in the liver ubiquinone content had been suggested as an indication of vitamin A deficiency syndrome in the rat. The increase in ubiquinone became much more rapid during acute avitaminosis A. As chickens and guinea pigs did not show these effects, the relationship was mainly characteristic of the rat and its liver tissue. The increase in the ubiquinone content in the liver can be prevented by supplying retinal or its different analogs, viz., retinylacetate, retinal, retinoic acid, 3-dehydroretinal, etc. It is likely that the increased liver ubiquinone in vitamin A deficiency resulted from impaired catabolism and not due to an increase in synthesis.

D. Role of Ubiquinone in Electron Transport

On account of the capacity of the quinones to undergo oxidation-reduction, they are of prime importance in electron transport activities.

In beef heart mitochondria, the concentration of ubiquinone is 4 times that of cytochromes a, b and c, and c1, about 6 times that of flavin. One of the important criteria to be satisfied for a component to be an obligatory member of the electron transport system is that it should undergo oxidation-reduction at a rate commensurate with the overall enzymatic activity. Although ubiquinone is shown to undergo oxidation-reduction during electron transport, from rate considerations its position as a main component is assigned at the converging point of a number of flavoproteins and before the cytochrome chain.

In the chemiosmotic hypothesis of oxidative phosphorylation, which is widely accepted, the proton motive ubiquinone cycle has a prominent role. Ubiquinone is not bound to any protein but it is present as a small mobile pool in the lipid phase of the membrane. It accepts electrons from one group of enzymes like NADH dehydrogenase (Complex I), succinate dehydrogenase (Complex II) and also substrates like acyl CoA, glycerophosphate, choline and sarcosine, each having its own flavin and iron sulfur proteins. Thus, reduction of ubiquinone is the common entry point of electrons from all the above-mentioned substrates. It then delivers electrons to cytochromes b and c (Complex III). Therefore, it serves as a mobile carrier between the two sides as shown in Figure 46.18. During its oxidation and reduction in the respiratory chain it also accounts for the release and uptake of protons necessary for the events under the chemiosmotic hypothesis.

The participation of ubiquinone in oxidative phosphorylation is covered in Chapter 23. Ubiquinone is synthesized from a number of precursor molecules in the inner mitochondrial membrane. The isoprenyl side chain is obtained from mevalonate, ring from tyrosine, the hydroxyl groups from molecular oxygen and the methyl groups from S-adenosylmethionine.

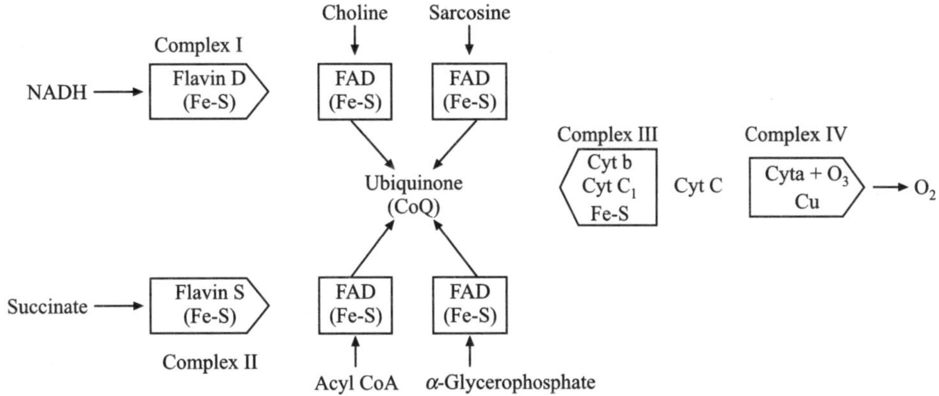

Flavin D = NADH flavo protein
Favin S = Succinate flavoprotein
Complexes I to IV are the four complexes of the respiratory chain
Fe-S are iron sulfur centres as integral part of the respective proteins

Figure 46.18

SUGGESTIONS FOR FURTHER READING

Blomhoff R., Green M.H. and Norum K.R. (1992), Vitamin A: Physiological and Biochemical Processing, *Annu. Rev. Nutr.*, 12, 37–57.

Blomhoff R., Green M.H., Berg T. and Norum K.R. (1990), Transport and Storage of Vitamin A., Science, 240, 399–404.

Brodie A.F. and Gutnick D. (1972), In: *Electron and Coupled Energy Transfer in Biological Systems*, King and Klingenberg, Marcel Decker (Eds.), Vol. 1, Part B., 599.

Clagett-Dame M. and DeLuca H.F. (2002), The Role of Vitamin A in Mammalian Reproduction and Embryonic Development, *Annu. Rev. Nutr.*, 22, 347–81.

De Luca L.M. (1972), In: *Vitamins and Hormones*, P.L. Monson, E. Diczfalusy, J. Glover, and R.E. Olson (Eds.), Acad. Press, New York, 40, 105, 144.

Dowd P., Ham S.W., Naganthan S. and Hershline R. (1995), The Mechanism of Action of Vitamin K, *Annu. Rev. Nutr.*, 15, 419–440.

Ganguly J., Rao M.R.S., Murthy S.K. and Sarada K. (1980), In: *Vitamins and Hormones*, P.L. Monson, E. Diczfalusy, J. Glover and R.E. Olson (Eds.), Acad. Press, London, 38, 1–54.

Genova M.L. and Lenaz G. (2011), New Developments on the Functions of Coenzyme Q in Mitochondria, *Biofactors*, 37, 330–54.

Hill D.L. and Grubbs C.J. (1992), Retinoids and Cancer Prevention, *Annu. Rev. Nutr.*, 12, 161–181.

Isler O. (1971), *Carotenoids*, Birkhauser Verlag, Basel.

Joshi Y.B. and Praticò D. (2012), Vitamin E in Aging, Dementia and Alzheimer's Disease, *Biofactors*, 38, 90–7

Krinsky N.I. (1993), Actions of Carotenoids in Biological Systems, *Annu. Rev. Nutr.*, 13, 561–587.

Lakshmanan M.R. and Cama H.R. (1970), *International Encyclopaedia of Food and Nutrition*, R.A. Morton (Ed.), Pergamon Press, London, 409.

Li E. and Norris A.W. (1996), Structure/Function of Cytoplasmic Vitamin A Binding Proteins, *Annu. Rev. Nutr.*, 16, 205–234.

Murthy P.S. (1978), In: *World Review of Nutrition and Dietetics*, 31, 210–215.

Olson, R.E. and Rudney H. (1983), In: *Vitamins and Hormones*, Editor-in-Chief G.D. Aurbach, D.B. McCormick (Ed.), Academic Press, New York, 40, 1–43.

Ong D.E. and Chytil F. (1983), In: *Vitamins and Hormones*, Editor-in-Chief G.D. Aurbach, D.B. McCormick (Ed.), Academic Press, New York, 40, 105–144.

Petkovich M. (1992), Regulation of Gene Expression by Vitamin A: The Role of Nuclear Retinoic Acid Receptors, *Annu. Rev. Nutr.*, 2, 443–471.

Shearer M.J., Fu, X. and Booth S.L. (2012), Vitamin K Nutrition, Metabolism, and Requirements: Current Concepts and Future Research, *Adv. Nutr.*, 3, 182–95.

Traber M.G. and Sies H. (1996), Vitamin E in Humans: Demand and Delivery, *Annu. Rev. Biochem.*, 16, 321–347.

Vermeer C., Jie. K.S.G. and Knapen M.H.J. (1995), Role of Vitamin K in Bone Metabolism, *Annu. Rev. Nutr.*, 15, 1–52.

Weatherman R.V., Fletterick R.J. and Scanlan T.G. (1999), Nuclear-receptor Ligands and Ligand-binding Domains, *Annu. Rev. Biochem.*, 68, 559–581.

47

B-Complex Group of Vitamins

K.D. Moudgil and S.H. Venkatesha

CONTENTS

I. INTRODUCTION

The group of B-complex vitamins includes many compounds which are all water soluble. These vitamins are not related chemically but they have common dietary sources and complex physiological interrelationships. All these vitamins function in cells as coenzymes. In order to function as a coenzyme, a vitamin often has to undergo structural modifications. The biochemical importance of these vitamins therefore, resolves into a study of coenzymes derived from them. The members of the vitamin B-complex important for human nutrition include thiamin, riboflavin, nicotinic acid, pantothenic acid, pyridoxine, biotin, vitamin B_{12} and folic acid. p-Aminobenzoic acid (PABA) is not a true vitamin for any mammalian species, but it is required by certain bacteria for the synthesis of folic acid.

II. FOLIC ACID

Synonyms: Pterolyglutamic acid; Vitamin M; Folacin.

A. Chemical Structure

Chemically, folic acid belongs to a group of compounds called pteridines. It has three component structural units: pteridine, p-aminobenzoic acid (PABA) and glutamic acid (Figure 47.1).

The majority of folates in nature are polyglutamates. The number of glutamate residues can vary from one to seven in different folates. These glutamate residues form peptide linkages involving γ-carboxyl group of one glutamate residue and amino group of the other. This linkage differs from the peptide bonds in proteins, which involve α-carboxyl groups. The polyglutamate side chain of folate is involved in the attachment of the coenzyme to apoenzyme, whereas the pteridine ring is involved in the transfer of one-carbon moieties.

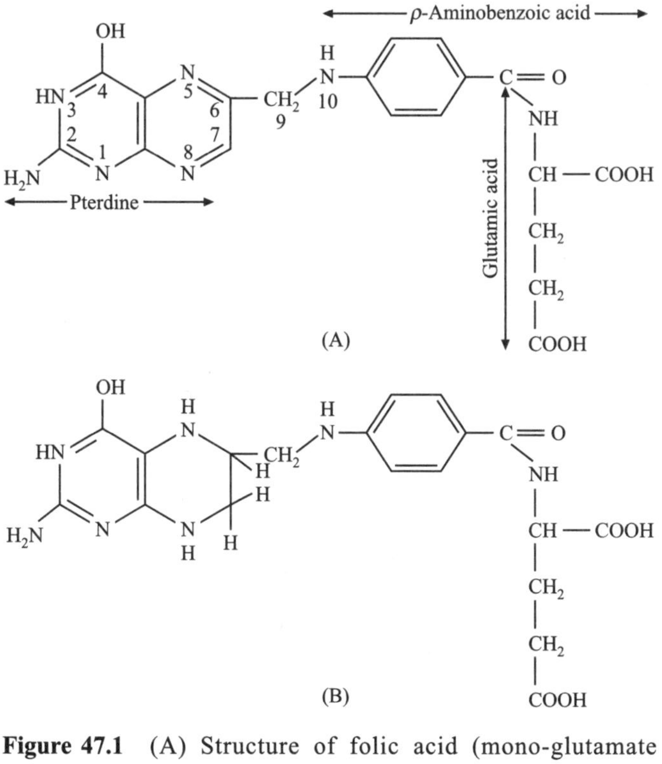

B. Physiological Role

PABA (p-aminobenzoic acid) always functions as a part of folate molecule in the human system. In bacteria, enzymes are available that can synthesize folate from pteridine, PABA and gluamate. As man cannot synthesize folic acid, the entire folic acid molecule must be supplied in diet as a vitamin. The fact that PABA is used in synthesis of folate by bacteria but not by man is a feature that is exploited usefully in sulphonamide therepy.

Before folate can function physiologically, it must be reduced in the pteridine moiety. This reduction takes place in two steps.

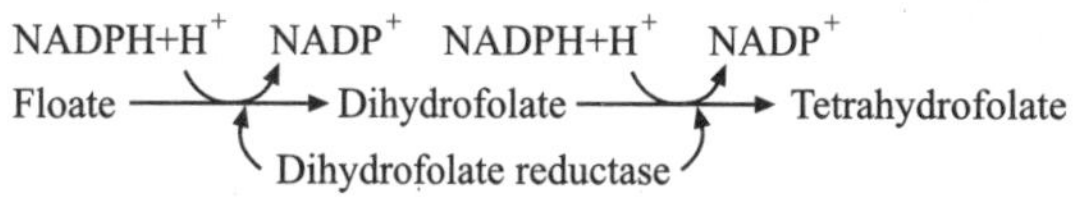

$$\text{Floate} \xrightarrow[\text{NADPH+H}^+ \quad \text{NADP}^+]{} \text{Dihydrofolate} \xrightarrow[\text{Dihydrofolate reductase}]{\text{NADPH+H}^+ \quad \text{NADP}^+} \text{Tetrahydrofolate}$$

It is tetrahydrofolate or THF (Figure 47.1) and its various derivatives, which constitute the functional coenzymes of folic acid. 5-methyl THF, a mono-glutamate, is the transport form of folate in plasma and cerebrospinal fluid (CSF), etc.

1. One-carbon metabolism

THF plays an important role in one-carbon metabolism (a one-carbon compound is an organic molecule that contains only a single carbon atom, e.g., CH_3OH, $HCHO$, and $HCOOH$). Enzymes containing folate coenzymes catalyze transfer of a one-carbon moiety ($-CH_3$, $-CHO$, $-HCOO$, $-CH_2OH$, etc.) between various substrates. Folate coenzymes also participate in enzymatic reactions concerned with oxidation or reduction of one-carbon compounds.

It is important at this stage to recall various oxidation states of carbon. Carbon dioxide is recognized as the highest state of oxidation of carbon, while methane is its most reduced form. There are various states of oxidation in between. Hence, CH_4, CH_3OH, $HCHO$, $HCOOH$, and CO_2 may be regarded as representing increasing states of oxidation of carbon.

Methane is not important in the human system, and CO_2 is handled mainly by coenzyme forms of thiamin, pyridoxine and biotin rather than by tetrahydrofolate. Folate coenzymes are most intimately concerned with the interconversions of formic acid, formaldehyde and methanol. In such interconversions, these compounds are covalently held on to THF through one or both of the folate nitrogens (N-5 and N-10), examples of which are shown in Figure 47.2.

It should be stressed at this point that the nomenclature used for the THF derivatives of various one-carbon groups is not clearly suggestive of the oxidation state of that group.

Figure 47.1 (A) Structure of folic acid (mono-glutamate derivative). Note the numbering of atoms of folic acid. (B) Structure of tetrahydrofolate (monoglutamate derivative).

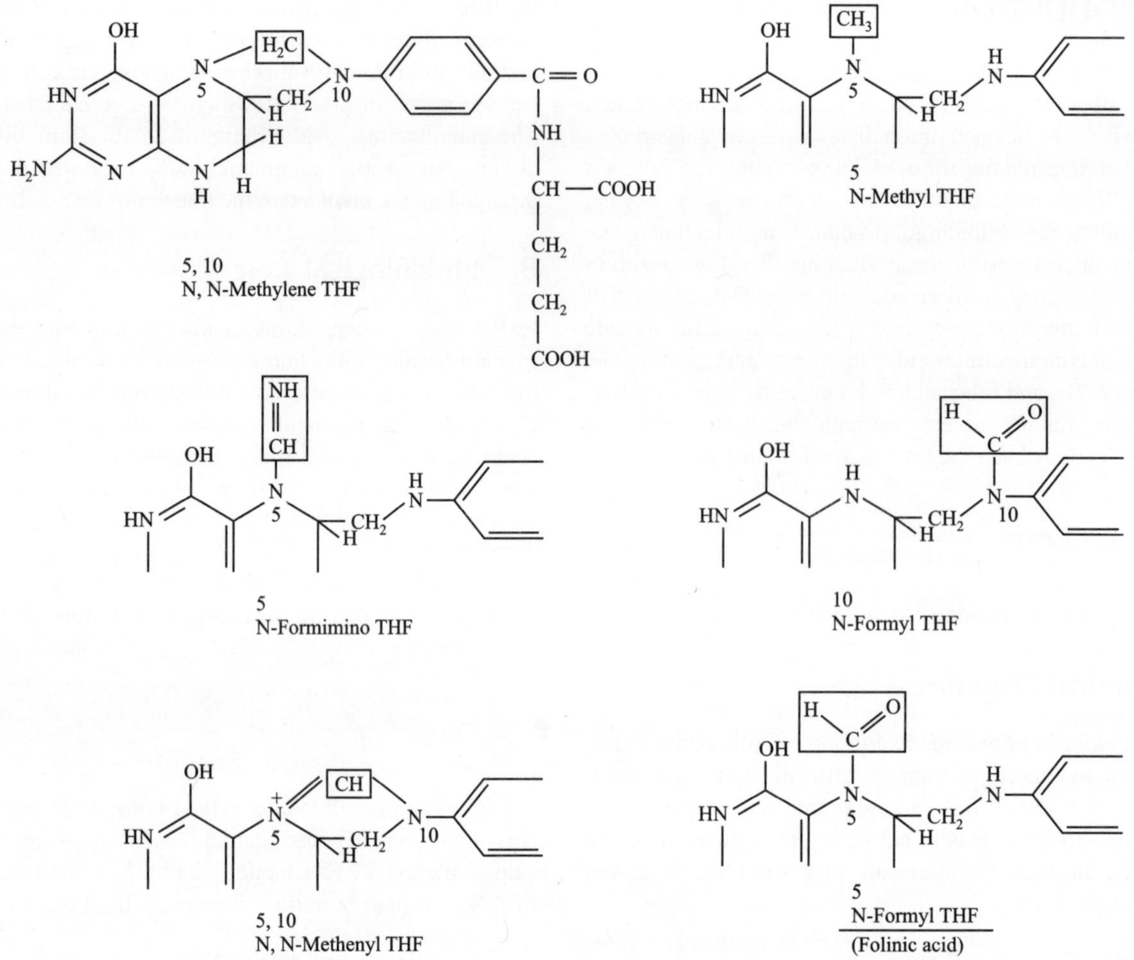

Figure 47.2 Some derivatives of tetrahydrofolic acid. Full structure is given only for 5, 10-methylene THF. In all other formulae only partial structures of THF are shown. The groups attached to THF are enclosed in the box.

Table 47.1 lists the parent compounds of different groups on the left, and the corresponding names and abbreviated formulae of their THF derivatives on the right. Thus, the conversion of methylene THF to formyl-THF is equivalent to the oxidation of formaldehyde to formic acid.

The interaction of serine with THF in the presence of serine hydroxymethyltransferase leads to the formation of 5, 10-methylene THF. 5, 10-methylene THF may either be

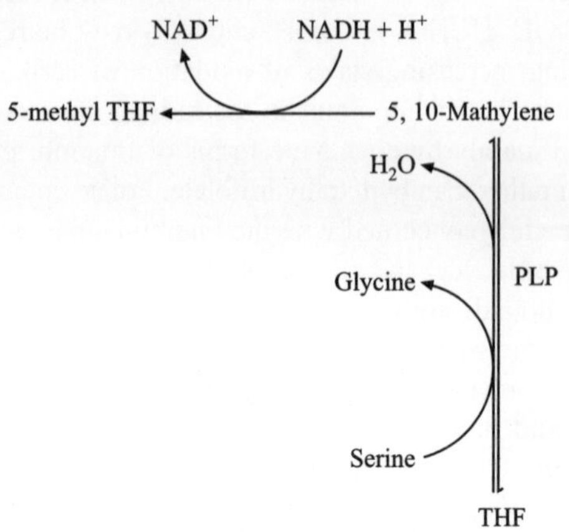

TABLE 47.1

Oxidized	
↑ Formic acid →	Formyl THF
(HCOOH)	(CHOTHF)
↑	Methylene THF
Formaldehyde	(5, 10-CH$_2$THF)
(HCHO)	
Methanol	Methyl THF
↓ (CH$_3$OH)	(CH$_3$THF)
Reduced	

oxidized or reduced by appropriate enzymes. Reduction of 5, 10-methylene gives rise to 5-methyl THF, and its oxidation results in the formation of 10-formyl THF via an important intermediate, 5, 10-methenyl THF.

2. Role of THF coenzymes

(a) *Conversion of homocysteine to methionine:* Methylation of L-homocysteine requires 5-methyl THF and methylcobalamin.

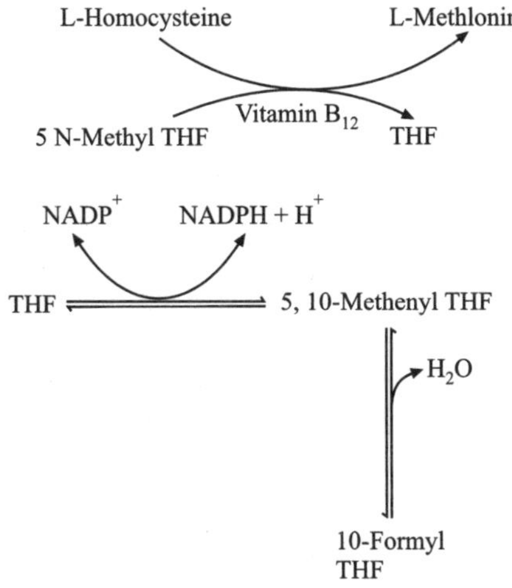

(b) *Biosynthesis of purines and thymidylate:* THF coenzymes participate in the biosynthesis of purines and thymidylate. Two of the carbon atoms in the purine nucleus are derived from formate, and both of these are donated by THF derivatives: 10-formyl THF donates C-2, and 5, 10-methenyl THF donates C-8 (Figure 47.3). The conversion of 2-deoxyuridylate (dUMP) to thymidylate is catalyzed by thymidylate synthetase. The methyl group transferred to the uracil moiety of dUMP is donated by 5, 10 methylene THF (Figure 47.4). This carbon atom is transferred to the pyrimidine ring at the oxidation level of formaldehyde and is reduced to methyl by the pteridine ring of the folate coenzyme; the result is the formation of dihydrofolate. To function again as a cofactor, dihydrofolate must first be reduced to THF by dihydrofolate reductase.

(c) 10-formyl THF acts as a source of the formyl group on N-formylmethionine-t-RNA, which initiates synthesis of peptide chains on ribosomes in microorganisms.

$$10\text{-Formyl THF} + \text{Methionyl} + \text{tTRNA}_F$$
$$\longrightarrow \text{THF} + \text{N} - \text{Formylmethionyl} - \text{tRNA}_F$$

(d) *Conversion of serine to glycine:* The interconversion of glycine and serine is catalyzed by serine hydroxy-methyltransferase, a pyridoxal phosphate-containing enzyme.

The role of this enzyme in forming 5, 10-methylene THF has been mentioned above. The same enzyme along with 5, 10-methylene THF is involved in serine-glycine interconversion.

$$\text{TFH} \qquad \text{Methylene THF}$$
$$\text{Serine} \longleftrightarrow \text{Glycine}$$
$$\text{Serine hydroxymethyltransferase}$$

(e) *Histidine degradation:* THF is also known to play a role in the catabolism of certain biological compounds. Histidine is a good example. Histidine breakdown through the urocanate pathway leads to formation of N-formimino-L-glutamate (FIGLU) and subsequently to L-glutamate; the formimino group is accepted by THF to form 5-formimino THF. This is subsequently converted to 5, 10-methenyl THF and finally to 10-formyl THF.

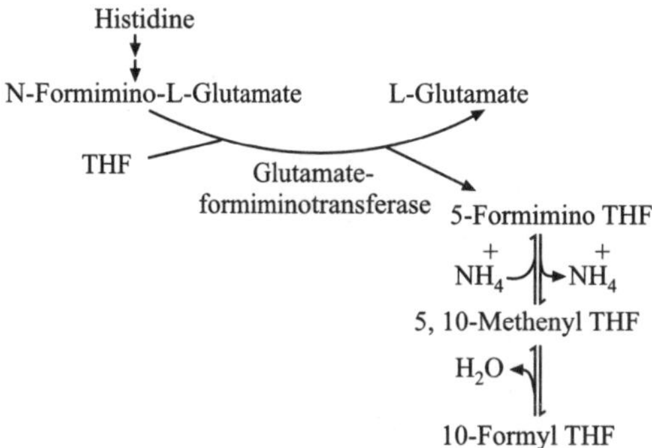

A deficiency of THF introduces a blockage in histidine breakdown because formimino-L-glutamate cannot be metabolized further. In severe folate deficiency, therefore, large quantities of formimino-glutamate (FIGLU) may appear in urine after an oral load of histidine. This may be used as a biochemical parameter of folate deficiency. Normally up to 18 mg of FIGLU are excreted in 8 hours after an oral dose of 15 g of histidine.

(f) *Utilization or generation of formate:* This reversible reaction utilizes THF and 10-formyl THF. Formyltetrahydro-pteroylglutamate synthetase is involved in the utilization of formate. The reverse reaction is brought about by 10-formyl-tetrahydropteroylglutamate deacylase.

Figure 47.3 Origin of the ring atoms of the purine nucleus.

Figure 47.4 Conversion of deoxyuridylate into thymidylate.

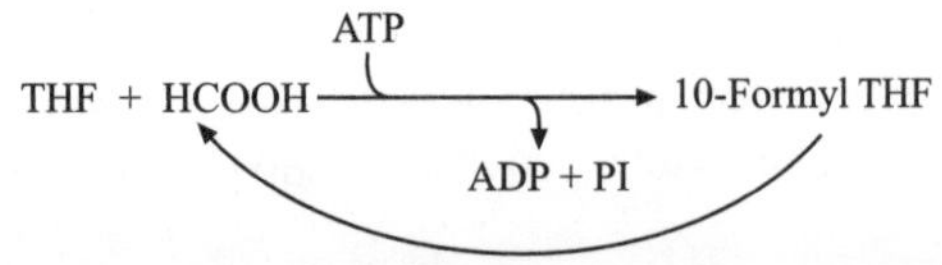

$$\text{THF} + \text{HCOOH} \xrightarrow[\text{ADP} + \text{PI}]{\text{ATP}} \text{10-Formyl THF}$$

(g) *Glycine catabolism:*   Glycine is catabolized by the glycine cleavage system in the liver as follows:

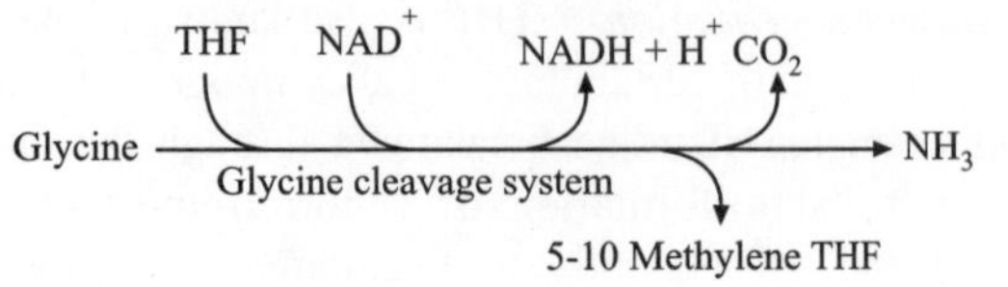

$$\text{Glycine} \xrightarrow[\text{Glycine cleavage system}]{\text{THF} \quad \text{NAD}^+ \quad \text{NADH} + \text{H}^+ \text{CO}_2} \text{NH}_3$$

5-10 Methylene THF

C. p-Aminobenzoic Acid (PABA) and Sulphonamide Therapy

Certain bacteria must meet their folate requirement by synthesizing it from pteridine, PABA, and glutamate. Sulphanilamide (Figure 47.5) and sulphonamides (synthetic aromatic sulphanilamide derivatives) are structural analogs and competitive antagonists of PABA, and thus prevent normal bacterial utilization of PABA for the synthesis of folic acid. More specifically, sulphonamides are competitive inhibitors of the bacterial enzyme responsible for the incorporation of PABA into dihydropteroic acid, the immediate precursor of folic acid. Such microorganisms are sensitive to sulphon-amide therapy due to inhibition of their growth caused by folate stravation. Those microorganisms that either do not require folate or can utilize pre-formed folic acid escape from the bacteriostatic action of sulphonamides. Sulphonamides do not affect mammalian cells since they require preformed folic acid and cannot synthesize it.

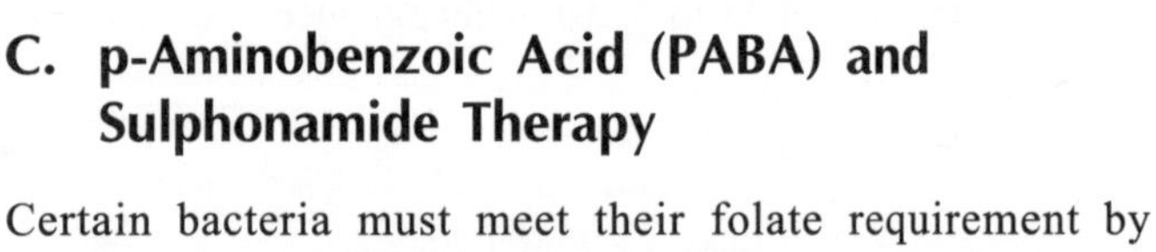

$$\text{H}_2\text{N} - \boxed{\text{COOH}} \qquad \text{H}_2\text{N} - \boxed{\text{SO}_2\text{NH}_2}$$

(A)   (B)

Figure 47.5   Structure of p-aminobenzoic acid (A), and sulphanilamide (B). Note similarities in the two structures. The groups enclosed in the box are the only differentiating features in these molecules.

D.   Folate Deficiency

Folate deficiency along with deficiency of B_{12} is very widespread amongst the Indian population, affecting chiefly the vegetarians. In folate deficiency, the intracellular concentration of THF coenzymes is reduced. As a result, the activity of the enzymes that require these coenzymes is decreased. Thymidylate synthetase is the enzyme most seriously affected, and this results in an inadequate production of thymidylic acid. Due to the shortage of this nucleotide, DNA synthesis cannot proceed and cell division is arrested. A tissue whose cells divide very rapidly is naturally the one to suffer most from folate deficiency. Such tissues are intestinal mucosa and bone marrow. Consequently, the symptoms that predominate in folate deficiency are those related to gastrointestinal and haemopoietic systems. These symptoms may be very marked if folate deficiency has been induced by administration of methotrexate or other folate antimetabolites. Varying degrees of diarrhoea (sometimes haemorrhagic), vomiting and abdominal pain may be seen. Polymorphs show a hypersegmentation of their nuclei. Megaloblastic anaemia is a prominent feature.

It must be mentioned here that megaloblastic anaemia is not restricted to folate deficiency; it can also be caused by a deficiency of vitamin B_{12}. The megaloblastic anaemia that results from folate deficiency cannot be distinguished from that caused by a deficiency of vitamin B_{12} because of the final common pathway of the major intracellular metabolic functions of the two vitamins. However, folate deficiency is rarely, if ever, associated with neurological abnormalities. Due to very limited stores of folate in the body, the appearance of anaemia following deprivation of folate is much more rapid than that caused by deprivation of vitamin B_{12}. Folic acid therapy causes total remission of the anaemia if it is due to folate deficiency. Folic acid, particularly in large doses, can correct the megaloblastic anaemia of vitamin B_{12} deficiency without affecting the neurological abnormalities. The neurological manifestations may even be aggravated by folic acid therapy.

Folinic acid (leucovorin; citrovorum factor) is useful as a therapeutic agent. The principal indication for the use of folinic acid is to circumvent the action of inhibitors of dihydrofolate reductase. It is not indicated for use in the treatment of folic acid deficiency.

Folate deficiency can be assessed by the following parameters: folate absorption from intestine studied by comparison of blood folate levels after oral and intravenous doses of folate; measurement of erythrocyte and serum folate concentration; histidine loading test (measurement of urine 'FIGLU' excretion after histidine load); and study of peripheral blood smear and bone marrow smear (macrocytic anaemia, hypersegmented polymorphonuclear leucocytes, and megaloblastic bone marrow).

E.   Folate Antagonists

The beneficial effect of methotrexate (4-amino-10-methylfolic acid) in certain leukaemias is due to the fact that it is a folate antimetabolite. Pyrimethamine and trimethoprim are also antifolate compounds and inhibit the multiplication of certain protozoan parasites, including the malarial parasite, plasmodium. Pyrimethamine is used as an antimalarial drug. Methotrexate, trimethoprim and pyrimethamine, along with various other folate antimetabolites, are structural analogs of folic acid (Figure 47.6). Methotrexate differs from folate in only two groups. The structural relationship of pyrimethamine and trimethoprim to folate is partial, but is effective enough for them to function as antifolate agents. Because of their structural resemblance to folic acid, these anti-metabolites occupy the catalytic site on dihydrofolate reductase, the enzyme that reduces dihydrofolate to THF. When dihydrofolate reductase is blocked by an anti-metabolite, dihydrofolate cannot be reduced to THF, and an acute intracellular deficiency of THF and its various coenzymes is produced. The folate coenzyme become trapped as dihydrofolate polyglutamates, which cannot function metabolically. One-carbon transfer reactions crucial for *de novo* synthesis of purine nucleotides and thymidylate cease, with subsequent interruption of the synthesis of DNA and RNA along with other vital metabolic reactions. Although the synthesis of RNA and proteins is also inhibited by the lack of THF coenzymes, the thymidylate block is considered to be the most important site of cytotoxic action leading to "thymineless" cell death.

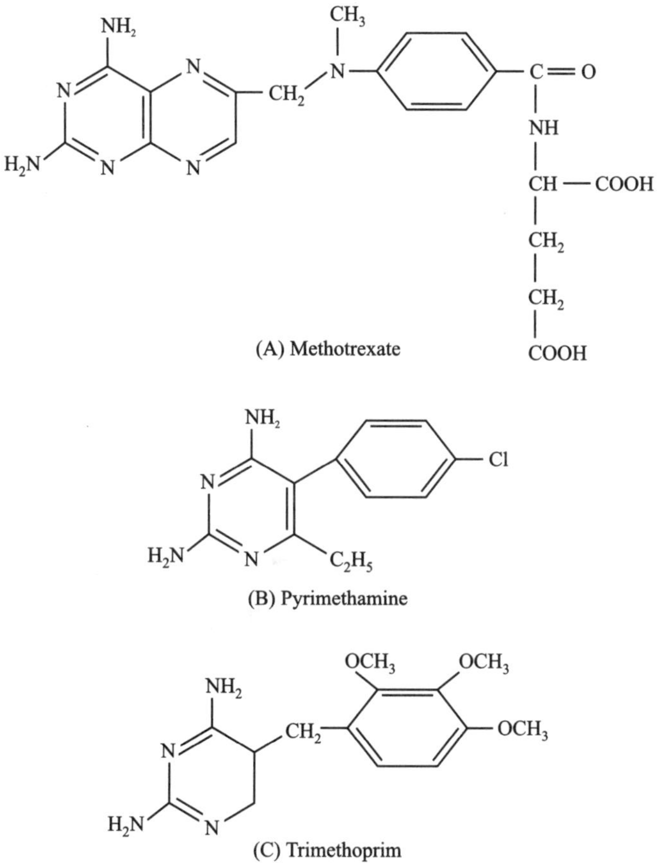

Figure 47.6 Structures of folate antimetabolites: (A) Methotrexate; (B) Pyrimethamine, and (C) Trimethoprim.

Combinations of trimethoprim and sulphonamides act by inhibiting two essential sequential steps in the pathway of THF synthesis. The utilization of PABA in the synthesis of dihydrofolic acid is inhibited by sulphonamides; and the reduction of dihydrofolic acid to tetrahydrofolic acid is inhibited by trimethoprim, which is a potent and selective inhibitor of microbial dihydrofolate reductase. Such combinations, e.g., bactrim (tri-methoprim-sulphamethoxazole) are bactericidal and are valuable in the treatment of some bacterial infections.

III. VITAMIN B_{12}

Synonyms: Cyanocobalamin; cobamide; Extrinsic Factor of Castle; anti-pernicious anaemia factor.

Vitamin B_{12} is abundant in the liver, kidney, brain and other foods of animal origin. It is markedly deficient in cereals, green vegetables, and potatoes. This explains as to why a predominantly vegetarian group often shows symptoms of nutritional B_{12} deficiency. Vitamin B_{12} in the diet is bound to R-proteins. (The term R-binders or R-proteins has been used to describe certain B_{12} binding proteins in food, tissues and body fluids. The letter 'R' refers to their more rapid electrophoretic mobility under certain conditions). In the stomach, dietary cobalamin is released, which then binds to gastric R binders. The vitamin is released from R-proteins in the duodenum by proteolytic enzymes and acid, and combines immediately with intrinsic factor (IF).

Intrinsic factor, also known as gastric intrinsic factor (GIF), is a glycoprotein with a molecular weight of about 50,000, produced by the gastric parietal cells. Vitamin B_{12} in bile is also attached to intrinsic factor. At neutral pH, in the presence of Ca^{++}, vitamin B_{12}-IF complex interacts with a specific receptor on the brush border of the ileal mucosal cells. Intrinsic factor is destroyed inside the cell but vitamin B_{12} is transferred to another transport protein, transcobalamin II (TC II). B_{12}-TC II complex is then secreted into the blood, from which it is taken up by the liver and other tissues. Most circulating B_{12} is bound to trans-cobalamine I (TC I). Methylcobalamin is the main form of B_{12} in plasma.

Vitamin B_{12}-TC II complex interacts with specific cell surface receptors in various tissues, and is endocytosed. Free cobalamin is released as hydroxycobalamin in the cytoplasm, where it may be converted to methylcobalamin. Hydroxycobalamin entering the mitochondria is converted into 5′-deoxyadenosylcobalamin. The turnover of vitamin B_{12} in man involves the loss of B_{12} in bile, urine, and desquamated cells. Also, there is enterohepatic circulation of B_{12}.

A. Chemistry

The chemistry of B_{12} was a challenge to the chemists for a long time. Finally, Dorothy Hodgkin showed that the basic structure consists of a planar corrin ring system of four pyrrole rings that surround a cobalt atom, and are bonded to it through their nitrogens (Figure 47.7A). The corrin ring is attached at right angles to a nucleotide portion –5, 6-dimethylbenzimidazole, which in turn is joined to ribosephosphate. In vitamin B_{12} as commonly isolated, the cobalt atom bears cyanide group (as an artifact of isolation), hence the term cyanocobalamin. The cyanide group can be replaced by a number of other groups inside the cell. Cyanocobalamin as such does not have a physiological role. Cyanocobalamin undergoes chemical modifications to form functional coenzymes (Figure 47.7). These modifications involve removal of the cyanide group and its replacement by:

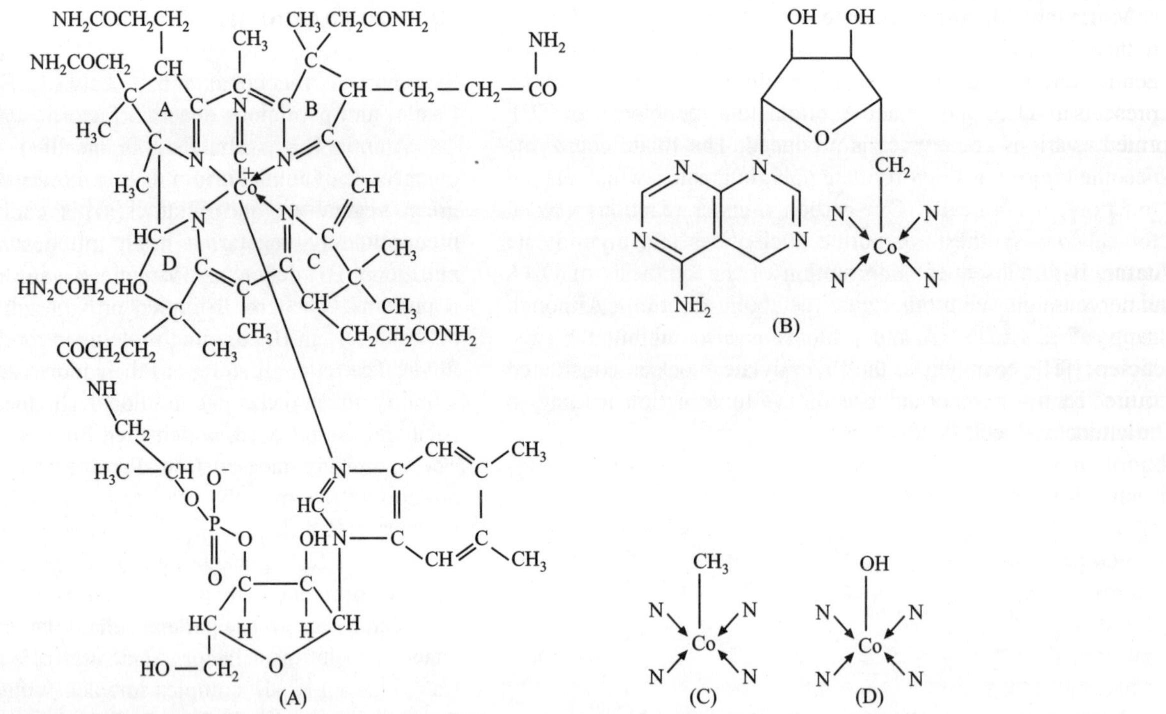

Figure 47.7   Structures of vitamin B_{12} derivatives: (A) Cyanocobalamin, (B) 5′-deoxyadenosylcobalamin-partial structure, (C) Methylcobalamin-partial structure, and (D) Hydroxycobalamin-partial structure.

(i) 5′-deoxyadenosyl group; the resulting compound is called 5′-deoxyadenosylcobalamin (ado-B_{12}).

(ii) methyl group; the product is referred to as methyl-cobalamin (methyl-B_{12}).

(iii) hydroxyl group; the product is named as hydroxy-cobalamin.

Ado-B_{12} and methyl-B_{12} serve as coenzymes for B_{12}-dependent enzymes in the animal system.

B.   Functional Role

Many B_{12}-containing enzymes are known in bacteria. However, only two enzymes have been shown to be present in animal tissues. Each of these enzymes requires a different type of B_{12} coenzyme.

1. Methylmalonyl CoA isomerase: Methyl-malonyl CoA is formed in the body from propionyl CoA. Methylmalonyl CoA is converted by methylmalonyl CoA-isomerase (which contains 5′-deoxyadenosylcobalamin as a coenzyme) into its isomer, succinyl CoA. Succinyl CoA is readily metabolized in the citric acid (TCA) cycle. Jointly with propionyl CoA carboxylase, methyl-malonyl CoA isomerase provides an elegant route for utilization of propionic acid in the body. In man, the metabolism of valine, isoleucine, methionine and threonine leads to production of methylmalonyl CoA, either directly or via propionic acid. Methyl-malonyl CoA is converted into succinyl CoA which is metabolized further through TCA cycle. In B_{12} deficiency, conversion of methylmalonyl CoA to succinyl CoA is seriously affected. This results in accumulation of methylmalonyl CoA in the body. Methylmalonyl CoA is eliminated from the body in urine, but not before its coenzyme has been stripped off. Urine examination shows high levels of methyl-malonic acid (MMA). This condition, clinically described as methylmalonic aciduria, is an indi-cation of B_{12} deficiency.

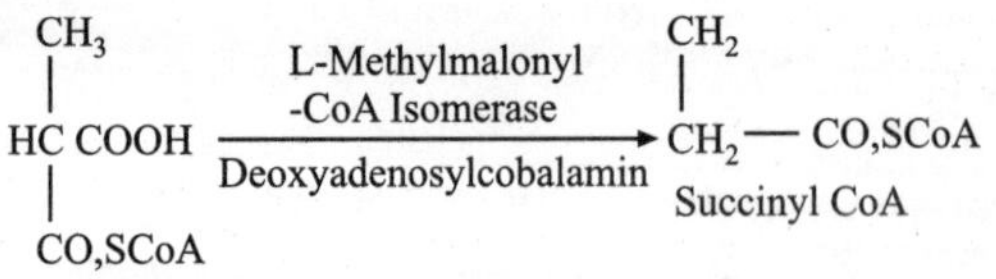

2. S-Adenosylhomocysteinemethyltransferase: This enzyme contains methylcobalamin as a coenzyme and the reaction catalysed by it also requires the participation of a folic acid coenzyme.

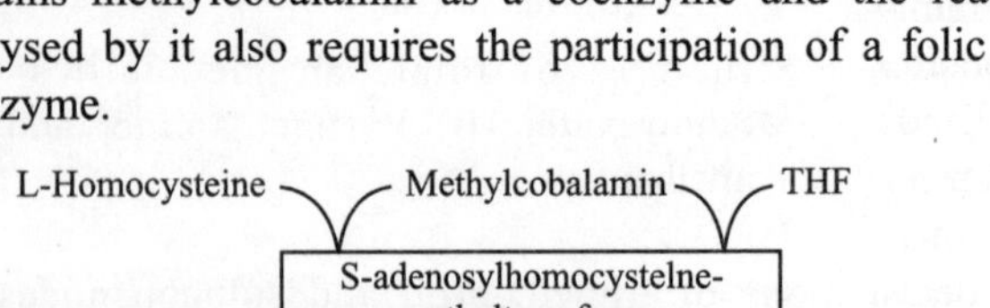

The methyl group is in fact donated by 5-methyl THF to L-homocysteine, but it is funneled through B_{12}. B_{12} captures the methyl group from 5-methyl THF and passes it on to L-homocysteine. In B_{12} deficiency, this reaction is interrupted, leading to accumulation of folate in the form of 5-methyl THF. This constitutes the "methylfolate trap".

Methionine is formed through this route in such low amounts that from nutritional point of view it still remains an essential amino acid. Methylation of homocysteine probably represents a useful means of utilizing L-homocysteine, which is formed as a byproduct of certain other biochemical reactions.

C. Deficiency of B_{12}

Vitamin B_{12} deficiency predominantly affects the haemopoietic and nervous systems. In vitamin B_{12} deficiency, folate becomes "trapped" as methyltetra-hydrofolate, causing functional deficiency of other vital intracellular forms of THF coenzymes required for synthesis of purines and pyrimidine (thymidylate). The latter phenomenon is the most likely cause for haematological abnormalities observed in B_{12} deficiency. Erythropoiesis is arrested towards the terminal stage. Consequently, mature RBCs cannot be formed, and immature precursors escape into the blood stream. These immature cells are larger than normal erythrocytes, and a condition of megaloblastic anaemia results. Administration of B_{12} leads to a rapid improvement in the condition. Folic acid, particularly in large doses, can correct the haematological abnormalities of B_{12} deficiency.

Another important feature of B_{12} deficiency is the irreversible damage to the nervous system. 5′-deoxyadenosylcobalamin is required as a coenzyme for the enzyme methylmalonyl-CoA isomerase, which converts L-methylmalonyl CoA to succinyl CoA. Since utilization of propionic acid by some animals proceeds via conversion of methylmalonyl CoA to succinyl CoA, vitamin B_{12} is involved in both lipid and carbohydrate metabolism.

In B_{12} deficiency, the lack of adenosylcobalamin leads to a large increase in the tissue levels of methylmalonyl CoA and its precursor, propionyl CoA. In such a situation, abnormal nonphysiological fatty acids containing odd number of carbon atoms are produced as a result of competition between acetyl CoA and accumulated methylmalonyl CoA in the biosynthetic pathway of fatty acids. These fatty acids are incorporated into myelin. This has been suggested as an explanation for the neurological damage in vitamin B_{12} deficiency. The central nervous system shows demyelination and cell death especially affecting cerebral cortex and fibres of the dorsal column and pyramidal tract of the spinal cord (sub-acute combined degeneration; SACD). Signs and symptoms include paraethesia of the hands and feet, and diminution or loss of sensation of vibration and position resulting in unsteadiness. In the later stages, there is loss of memory, confusion, loss of central vision, delusion, hallucination, or even frank psychosis. Unlike the condition of anaemia described above, the neurological symptoms are relieved only by B_{12} and not by folic acid. The neurological manifestations may in fact be aggravated by folic acid therapy.

Deficiency of vitamin B_{12} can be assessed by the following parameters: Measurement of vitamin B_{12} in serum; measurement of urinary methylmalonic acid (MMA); histidine loading test (measurement of urinary 'FIGLU' excretion after histidine load), peripheral blood and bone marrow morphology; tests

of gastric function (e.g., basal acid output, maximal acid output, and Schilling test); and a therapeutic trial of vitamin B_{12}. In Schilling test, the patient is given radioactive vitamin B_{12} by mouth, followed shortly by an intramuscular injection of unlabeled vitamin B_{12}. The proportion of the administered radioactive B_{12} in the urine excreted during the next 24 hours reflects the intestinal absorption of vitamin B_{12}. Normal persons excrete more than 10 per cent of the oral dose and patients with defective B_{12} absorption excrete less than that. Schilling test performed after the oral administration of labeled cobalamin bound to intrinsic factor can help to delineate the mechanism of abnormal absorption of B_{12}.

IV. THIAMIN

Synonyms: Thiamine; Vitamin B_1; anti-beri-beri substance; aneurine.

Thiamin is present in large quantities in meat and the outer coating of various grains. It is for this reason that whole wheat flour and unpolished rice have a better nutritive value than "refined" foods, such as white flour and polished rice. Following ingestion, thiamin undergoes a direct enzymatic conversion to 4-thiamin pyrophosphate (TPP; cocarboxylase). The pyrophosphate group is derived from the two terminal phosphates of ATP. The reaction is catalyzed by thiamin pyrophosphotransferase.

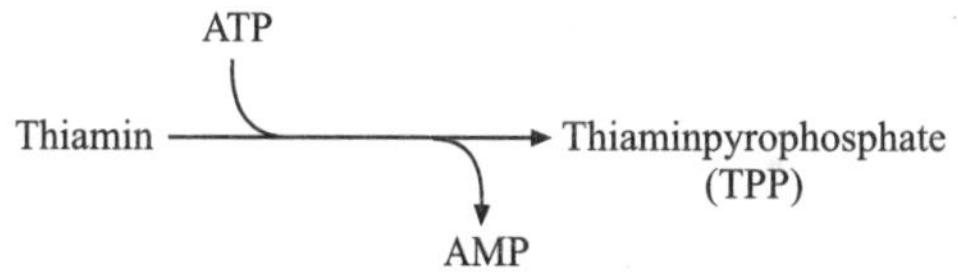

$$\text{Thiamin} \xrightarrow[\text{AMP}]{\text{ATP}} \text{Thiaminpyrophosphate (TPP)}$$

A. Structure

Chemically, this vitamin or its coenzyme consists of a substituted pyrimidine ring (2, 5-Dimethyl-6-aminopyrimidine) connected to a substituted thiazole ring (4-Methyl-5-hydroxyethylthiazole) by means of a methylene bridge (Figure 47.8). Antimetabolites

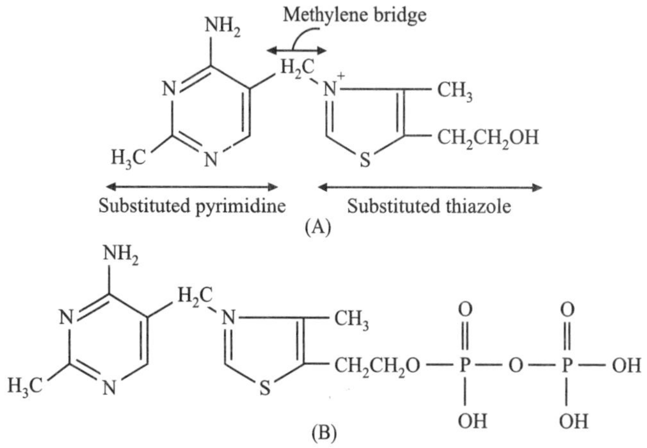

Figure 47.8 Structures of thiamin (A) and thiamin pyrophosphate (B).

to thiamin have been synthesized; the most important of these are neopyrithiamin (pyrithiamin) and oxythiamin. Some thiaminase enzymes can inactivate thiamin by breaking the two rings apart. Thiaminases are found in some foods and in microorganisms in the gut.

B. Physiological Role

Thiamin functions as a coenzyme, thiamin pyrophosphate (TPP). In discussing the functions of TPP, we may consider two groups of enzymatic reactions transferring an activated aldehyde unit. The enzymes catalyzing these reactions require TPP as the coenzyme.

(a) The first group of enzymes is comprised of enzyme systems that bring about oxidative decarboxylation of α-keto acids (TPP is also important in some non-oxidative decarboxylation reactions in bacteria). In these reactions, there is a cleavage of covalent linkage between two carbon atoms. This function of TPP may be considered opposite to that of biotin, which helps in forming a covalent linkage between two carbon atoms.

1. The classical example of this group of enzyme systems is pyruvate decarboxylase, a component of the pyruvate dehydrogenase complex. The overall reaction catalyzed by the pyruvate dehydrogenase complex of enzymes is the breakdown of pyruvate through a series of reactions into acetyl CoA and carbon dioxide. The step in which CO_2 is split off from pyruvate is handled by pyruvate decarboxylase. In order to accomplish this, pyruvate first binds to the thiazole nucleus at the carbon atom between nitrogen and sulphur (the electronic chemistry of this carbon atom between two powerful electrophiles, nitrogen and sulphur, makes such a binding particularly easy). This is followed by the decarboxylation step as shown in Figure 47.9. On binding to the thiazole nucleus, the keto group of pyruvate is converted to an α-hydroxyl group. The product resulting from the binding of pyruvate to TPP is, therefore, referred to as lactyl TPP (hydroxyethyl TPP) and not pyruvyl TPP. Lactyl TPP transfers the acetaldehyde moiety to lipoamide, forming acetyl lipoamide. The latter transfers the acetyl group to coenzyme A. The oxidized form of lipoamide is regenerated hy dihydrolipoyl dehydro-genase, which contains flavin adenine dinucleotide (FAD) as the prosthetic group.

2. Another analogous biochemical reaction that requires TPP is the decarboxylation step involved in the degradation of α-ketoglutarate into succinyl CoA and CO_2. This enzymatic system is distinct from the previous one (pyruvate dehydrogenase complex) and is referred to as the α-ketoglutarate dehydrogenase complex. However, the mode of action of the two systems is strikingly similar. Here, it is the α-ketoglutarate decarboxylase (analogous to pyruvate

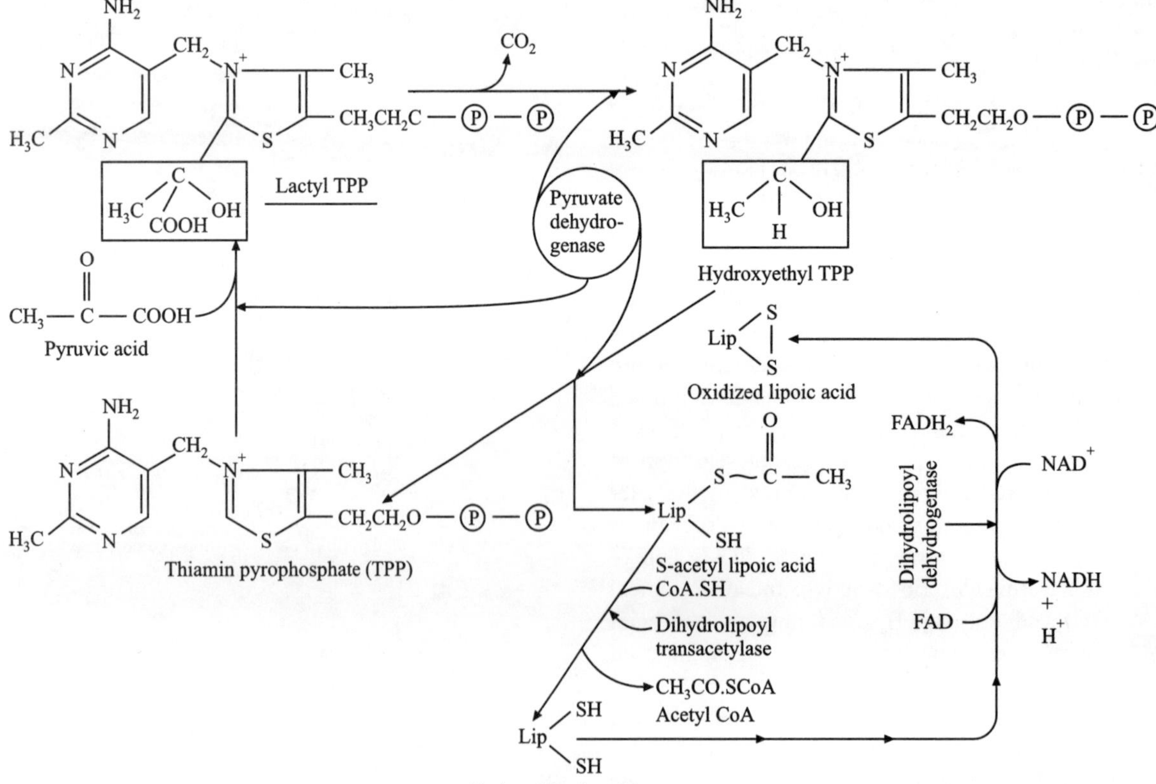

Figure 47.9 Role of thiamin pyrophosphate in the conversion of pyruvate to carbon dioxide and acetyl CoA. These reactions are catalyzed by enzymes of the pyruvate dehydrogenase complex. P-P refers to pyrophosphate groups.

decarboxylase) that is directly responsible for the cleavage of carboxyl group as CO_2.

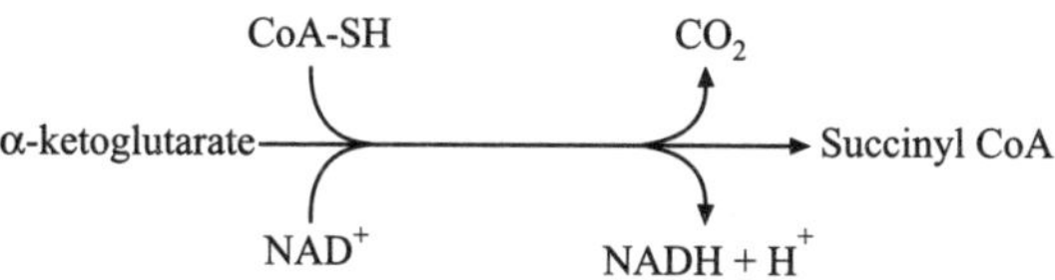

3. Other biochemical reactions that require TPP include decarboxylation steps involved in the degradation of α-ketobutyrate and α-keto derivatives of branched chain amino acids (valine, leucine and isoleucine; see Chapter 16).

In all the above-mentioned reactions, the substrates are always α-keto carboxylic acids, and in each case the carboxyl group is cleaved off as CO_2. Consequently, the product always has one carbon atom less than the substrate.

(b) The second group of enzymes that use TPP as a coenzyme is illustrated by transketolase. This enzyme splits off a ketol group ($-CO-CH_2OH$; glycoaldehyde group) from a donor compound, e.g., xylulose-5-phosphate. This ketol group is then passed on to an acceptor molecule, e.g., ribose-5-phosphate or erythrose-4-phosphate. The resulting products are shown below.

This is an example of a direct transfer of the ketol group from one compound to another (hence the term "transketolase") in contrast to the simple decarboxylation catalyzed by the first group of enzymes. However, as in the latter case, the substrate must have a keto group on the bond being cleaved. It is interesting to compare this cleavage with that brought about by another enzyme of the hexose monophosphate shunt, transaldolase. This enzyme does not contain TPP and the bond which is broken is one carbon atom away from the keto group. The above-mentioned reactions emphasize the role of thiamin (as TPP) in carbohydrate metabolism.

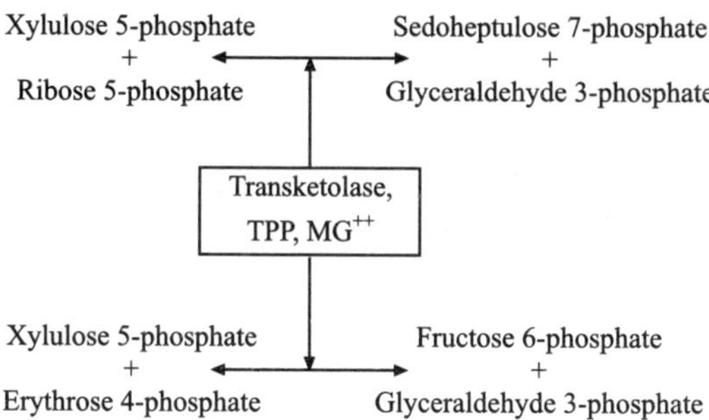

The requirement of thiamin is related to metabolic rate and is greatest when carbohydrate is the source of energy. This is of practical significance for patients who are maintained on parenteral alimentation, who receive practically all their calories in the form of glucose. Thiamin requirements of such patients are quite high. There may be significant loss of thiamin from the body of patients undergoing diuretic therapy or dialysis, and during diarrhea. The absorption of thiamin is defective in situation of malabsorption, alcoholism and folate deficiency.

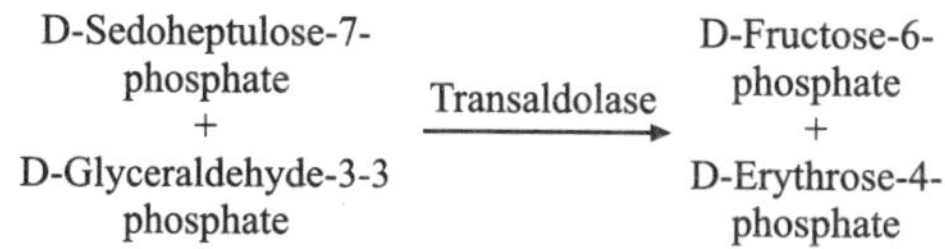

C. Deficiency of Thiamin

Deficiency of thiamin leads to the disorder called beriberi. The major symptoms are related to the nervous system ('dry' beriberi) and/or the cardiovascular system ('wet' beriberi). In 'dry' beriberi there is peripheral neuritis with sensory disturbances in the extremities. There is muscular weakness, which may progress to complete paralysis. Usually there is no edema, hence named 'dry' beriberi. In 'wet' beriberi, there is edema due to cardiac failure.

Wernicke-korsakoff syndrome or 'cerebral' beriberi is also caused by thiamin deficiency. However, it can result from any disease in which nutrition is severely affected. The clinical features are those of encephalopathy (ophthalmoplegia, nystagmus and cerebellar ataxia) along with psychosis.

Thiamin deficiency seriously affects the nervous system because of the central role that thiamin coenzyme plays in pyruvate metabolism. In the absence of TPP, the breakdown of pyruvate into acetyl CoA (which enters the TCA cycle), the immediate source of energy for all cells, is adversely affected. The nervous system, which is primarily dependent on glucose metabolism for its energy requirements, is the one to be most severely affected. In addition, thiamin possibly plays a specific role in neuronal function, and therefore, its deficiency may lead to neurological manifestations.

Finally, there is a condition of thiamin deficiency seen in chronic alcoholics. The symptoms are those of polyneuritis with motor and sensory defects. Because of the high caloric value of alcohol, the thiamin requirements of these individuals are well above the normal recommended amount. Hence, at the usual level of vitamin intake, which may be adequate for most normal individuals, a chronic alcoholic may reveal symptoms of mild beriberi. It has been suggested that susceptibility to neurological manifestations in chronic alcoholics may be genetically determined.

Interestingly, many thiamin-responsive inborn errors of metabolism have been described. For example, megaloblastic anemia, lactic acidosis, branched-chain ketoaciduria, etc. These patients respond to pharmacologic doses of thiamin.

Thiamin deficiency can be assessed by the following parameters: measurement of blood levels of thiamin, pyruvate, α-ketoglutarate, lactate and glyoxylate; urinary excretion of thiamin and an acetylated thiamin metabolite, and urinary methylglyoxal; thiamin loading test; measurement of erythrocyte transketolase activity; and assessment of clinical response to thiamin administration. For measurement of transketolase activity, patient's haemolysed RBCs are incubated with pentose-5-phosphate. The transketolase activity is measured in the presence as well as the absence of the coenzyme, TPP. The increase in activity in the presence of added TPP is expressed as a percentage known as 'TPP effect'.

V. BIOTIN

Synonyms: Vitamin H; anti-egg white injury factor.

A. Structure

The biotin molecule consists of a ring structure (fused imidazole and thiophene rings) with a valeric acid side chain, which holds the molecule onto the apoenzyme (Figure 47.10). The carboxyl group of the side chain forms an amide linkage with the ε-amino group of a lysyl residue of the apoenzyme. The enzyme holocarboxylase synthetase catalyzes the attachment of biotin to the proper lysyl residue of the apoenzyme, carboxylase. Schematically, biotin may be regarded as being attached to its apoenzyme through a long 'lever' containing nine carbon atoms and one nitrogen atom. This offers the molecule considerable flexibility in motion and this feature is of help to it in its role in fatty acid biosynthesis. The active form of biotin is the 'd' isomer.

B. Role of Biotin in CO_2 Fixation

Biotin plays a very important role as a coenzyme for some enzymes that fix carbon dioxide into various substrates. In the first step of this fixation process, a molecule of CO_2 is captured by biotin of the enzyme. The resulting product is carboxybiotin enzyme in which the carboxyl group is covalently attached to a nitrogen (N1) of the biotin molecule. The energy required for the formation of this bond is provided by ATP. In the second step of the carboxylation reaction, the activated carboxyl group is transferred to the substrate and a free biotinyl enzyme is regenerated.

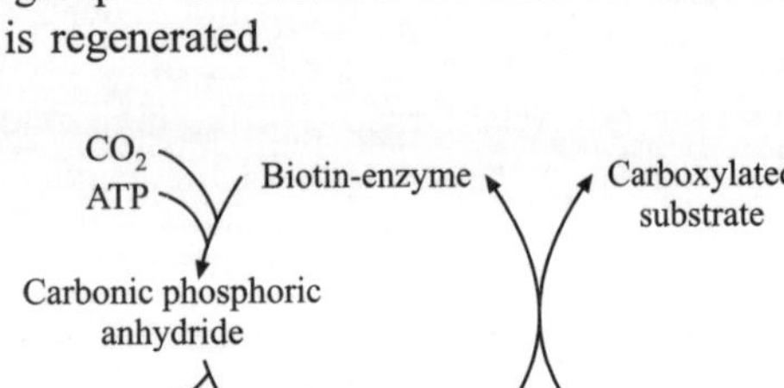

There are many examples of CO_2 fixation reactions brought about by biotinyl enzymes; however, a few are listed below because of their special importance:

1. *Acetyl CoA carboxylase:*   This enzyme converts acetyl CoA to malonyl CoA. This is an important reaction in biosynthesis of fatty acids.

$$CH_3CO — SCoA + ATP + CO_2$$

Acetyl CoA

Acetyl CoA-Carboxylase; Biotin

$$COOH$$
$$|$$
$$CH_2—CO\text{-}SCoA + ATP + CO_2$$

Malonyl CoA

2. *Propionyl CoA carboxylase:*   This converts propionyl CoA to methylmalonyl CoA.

$$CH_3CH_2CO.SCoA + CO_2 + H_2O$$

Propionyl CoA

ATP

Biotin

Propionyl CoA carboxylase

$$ADP + Pi$$

$$COOH$$
$$|$$
$$CH_3CHCO.SCoA$$

D-Methylmalonyl CoA

Methylmalonyl CoA is further converted into succinyl CoA in a reaction catalyzed by methylmalonyl CoA isomerase. This reaction is a useful means of utilizing propionyl CoA resulting from the breakdown of odd chain fatty acids through β-oxidation. Since propionyl CoA cannot be further metabolized by β-oxidation, the tissues adopt a different means of utilizing this molecule. In sheep and other ruminants, the importance of propionyl CoA-carboxylase is much greater. It deals with the copious amounts of propionic acid produced by the ruminal bacteria and thereafter, absorbed into the blood stream of the animal.

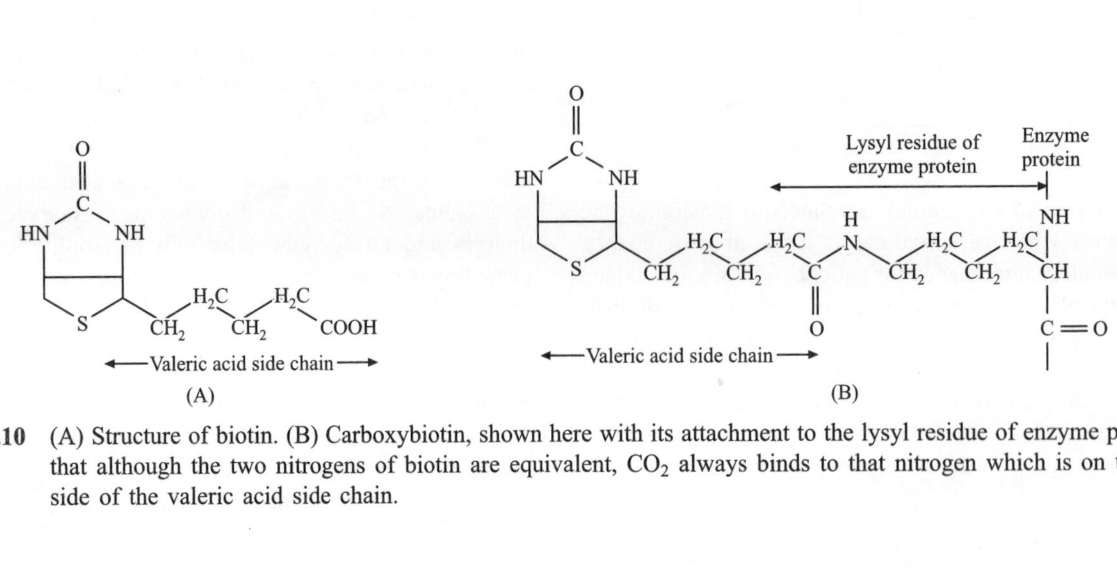

Figure 47.10   (A) Structure of biotin. (B) Carboxybiotin, shown here with its attachment to the lysyl residue of enzyme protein. Note that although the two nitrogens of biotin are equivalent, CO_2 always binds to that nitrogen which is on the opposite side of the valeric acid side chain.

3. *Pyruvate carboxylase:* The product of this reaction is oxaloacetate.

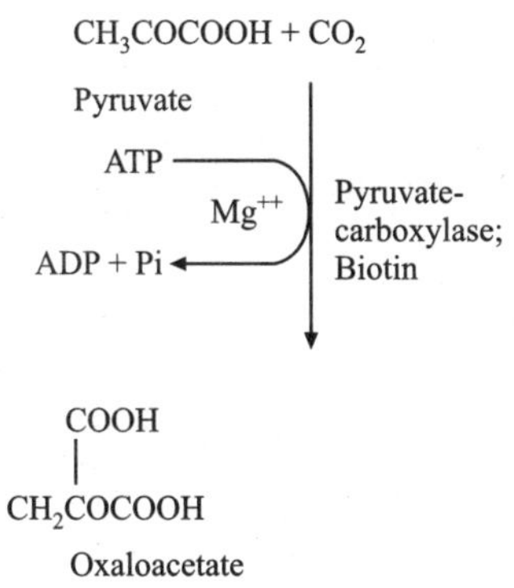

$$CH_3COCOOH + CO_2$$

Pyruvate

ATP $\longrightarrow$

Mg^{++} — Pyruvate-carboxylase; Biotin

ADP + Pi $\longleftarrow$

$$\begin{array}{c} COOH \\ | \\ CH_2COCOOH \end{array}$$

Oxaloacetate

This reaction provides an important means of maintaining a balance between the biosynthesis and breakdown of sugars.

In all of the above-mentioned reactions, it is interesting to note that the carboxylation of substrates takes place on the carbon atom alpha to the carbonyl group (α-carboxylation reaction). Carboxylation at the α-carbon atom is facilitated because this carbon atom is covalently linked to coenzyme A (in examples 1 and 2 above), or to a carboxyl group (in example 3 above), both of which have electron-withdrawing effects.

As a cofactor for the three above-mentioned enzymes, biotin plays an important role in both carbohydrate and lipid metabolism.

C. Biotin Deficiency

Symptoms of biotin deficiency are not very common, but they can be induced in individuals that are fed with large quantities of avidin, a protein found in raw egg white. Avidin has the property of complexing very strongly with biotin. The resulting complex cannot be absorbed in the gastrointestinal system. Therefore, biotin present in the diet is not made available to the individual. Feeding on large amounts of raw egg white will lead to a condition of biotin deficiency. A number of compounds antagonize the actions of biotin. Among these are biotin sulphone, desthiobiotin and certain imidazolidone carboxylic acids.

When biotin deficiency is induced, the signs and symptoms include dermatitis, atrophic glossitis, hypersthesia, muscle pain, lassitude, anorexia, slight anaemia and changes in Electrocardiogram. These disappear on administration of biotin.

VI. VITAMIN B$_6$

Synonyms: Pyridoxine; pyridoxal; pyridoxamine

A. Structure

Vitamin B$_6$ represents a family of three related pyridine derivatives: pyridoxine, pyridoxal and pyridoxamine. The chemical differences between these compounds are confined to the side group on carbon-4 of the molecule (Figure 47.11A). Pyridoxine is a primary alcohol, pyridoxal is the corresponding aldehyde, and pyridoxamine contains an aminoethyl group in this position.

The main supply of B$_6$ compounds in food is in the form of pyridoxine. This is readily converted into pyridoxal and pyridoxamine in the body. These compounds are phosphorylated in a pyridoxal kinase-catalyzed reaction utilizing ATP. The phosphate is esterified with the alcohol at position 5 of the pyridine ring. Pyridoxal phosphate (PLP) and pyridoxamine phosphate serve as the coenzyme forms of B$_6$ (Figure 47.11B). Pyridoxal phosphate binds to its apoenzyme via a Schiff base (between its 4-aldehyde group and an ε-amino group of a lysine residue of the enzyme protein) and an ionic bond between

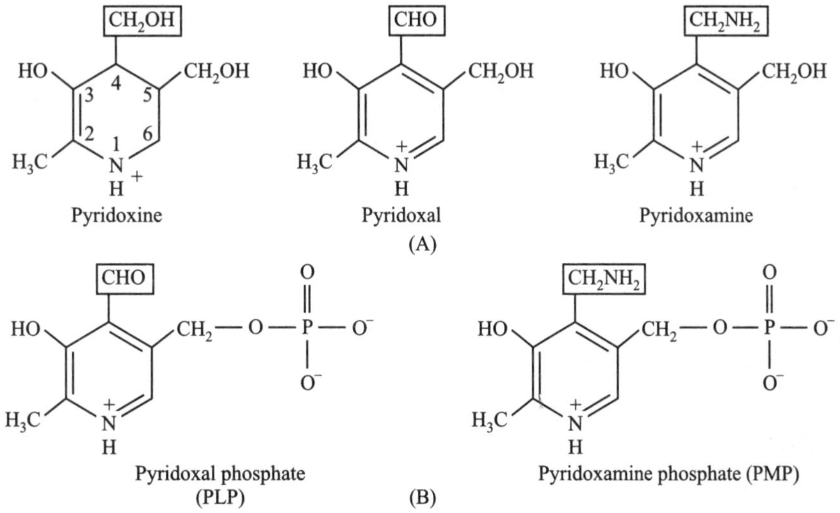

Figure 47.11 (A) Structures of B$_6$ vitamins. The side groups on positions 2, 3 and 5 are the same in all three compounds; however, the substituent on C-4 (enclosed in the box) is different in each case. The latter substituent is also the active site of the molecule. (B) Structures of B6 coenzymes.

its phosphate and the enzyme. The incoming amino group of the substrate displaces the ε-amino group of the lysyl residue, forming a new Schiff base. However, coenzyme remains bound to the enzyme by the salt bridge.

B. Functions

By far, the most important role played by these coenzymes is in amino acid metabolism. A deficiency of this vitamin leads to disorders of protein metabolism. Pyridoxal phosphate is a coenzyme for a variety of enzymes including: Transferases (e.g., aminotransferases and hydroxy-methyl transferase), Hydrolases, Lyases (e.g., carboxylyases hydrolyases, carbon-sulphur lyases, and aldehyde lyases), and Isomerases (e.g., racemases). Some examples of the reactions catalyzed by the above-mentioned categories of enzymes are discussed below.

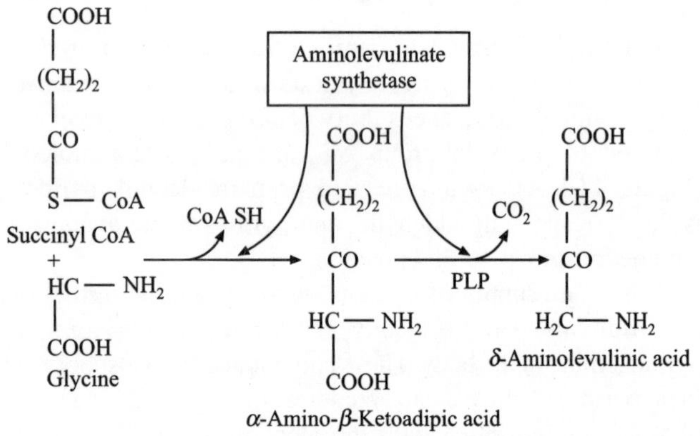

α-Amino-β-Ketoadipic acid

1. Transamination reaction is catalyzed by amino transferases (transaminases) which employ PLP as

a coenzyme. An attempt is made in Figure 47.12 to illustrate the molecular mechanism of transamination reaction involving B6 coenzymes. Pyridoxal phosphate and an amino acid combine to form a Schiff base (an aldimine; a Schiff base is formed when an aldehyde reacts with an amine). This Schiff base then isomerises to a Ketimine. Finally, there is hydrolytic cleavage of this intermediate (Ketimine) to pyri-doxamine phosphate and an α-ketoacid. If the reaction is followed in the reverse direction, an α-ketoacid could be converted into its corresponding α-amino acid. It should be clear that when an α-amino acid is converted into α-ketoacid in a transmination reaction, pyridoxal phosphate is converted to pyridoxamine phosphate, and vice versa.

2. Glycine can be converted to serine in a reaction catalyzed by serine hydroxymethyl-transferase, a pyridoxal phosphate-dependent enzyme (see Folic acid).

3. Succinyl CoA: glycine C-succinyl transferase (decarboxylating) or δ-Aminolevulinate synthase is also a pyridoxal phosphate-dependent enzyme. It catalyses the following reaction:
the product δ-aminolevulinic acid is the starting unit for haem biosynthesis, and the importance of vitamin B6 in maintaining blood haemoglobin level can be readily appreciated.

4. 3-Hydroxykynurenine, one of the intermediates of L-tryptophan catabolism, is acted upon by kynureninase (a hydrolase), which uses pyridoxal phosphate as a coenzyme. Alanine is split off, leaving 3-hydroxyanthranilic acid.

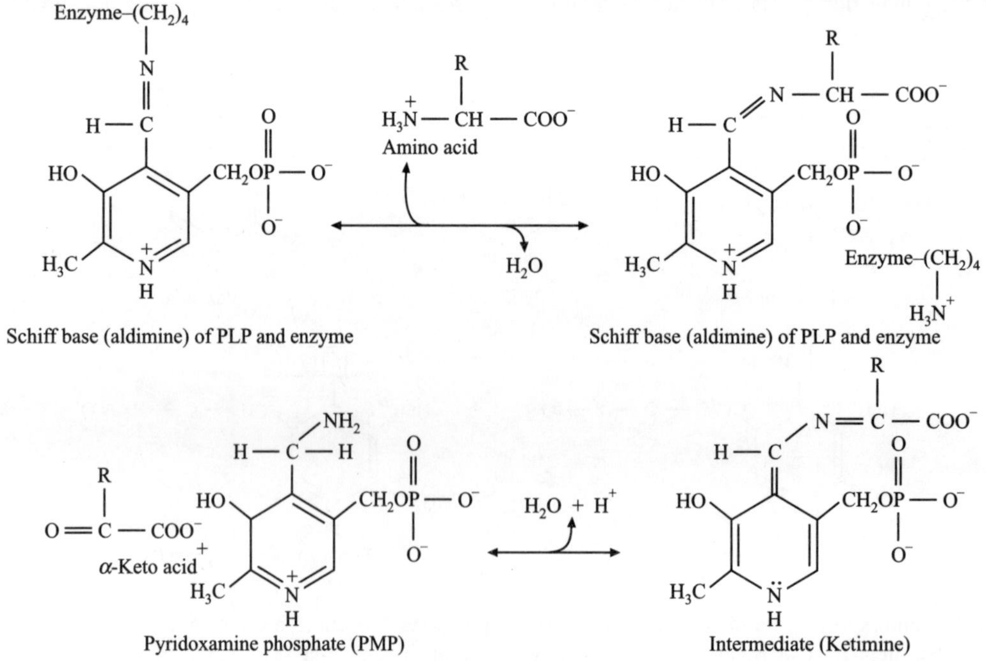

Figure 47.12 Molecular mechanisms involved in the transamination reaction.

3-Hydroxy-Kynurenine $\xrightarrow[\text{PLP}]{\text{Kynureninase}}$ L-Alanine + 3-Hydroxy-anthranilic acid

5. Pyridoxal phosphate serves as a coenzyme in decarboxylation reactions of amino acids catalyzed by decarboxylases (carboxylyases). As in transamination reaction, there is formation of the intermediate Schiff base and subsequent shift of the double bonds. There are several amino acid decarboxylases in mammalian tissues, e.g., histidine decarboxylase, ornithine decarboxylase, aromatic amino acid decarboxylase and glutamic acid decarboxylase.

Glutamic acid $\xrightarrow[\text{PLP}]{\text{Decarboxylase}}$ $H_2N—CH_2—CH_2—CH_2—COOH$ + CO_2
γ-Aminobutyric acid (GABA)

6. Serine and Threonine (both containing a hydroxyl group in the side chain) can be deaminated by dehydratases (hydrolyases) to yield pyruvate and α-ketobutyrate respectively. There is dehydration (hence the name of the enzyme) followed by rehydration. The net result is deamination of amino acid. Pyridoxal phosphate is required as a coenzyme for dehydratases.

7. Homocysteine (derived from methionine) condenses with serine in the presence of cystathionine-β-synthase (a hydro-lyase) to form cystathionine. Hydrolytic cleavage of cystathionine by γ-cystathionase or cystathionine-γ-lyase (a carbon-sulphur lyase) results in the formation of homo- serine and cysteine. The net effect is conversion of homocysteine to homoserine, and of serine to cysteine. Both the enzymes mentioned above contain pyridoxal phosphate as their coenzyme.

8. The removal of sulphur from homocysteine is catalyzed by homocysteine desulphhydrase (a carbon-sulphur lyase), which requires pyridoxal phosphate as the coenzyme.

$$\text{L-Homocysteine} + H_2O \xrightarrow{\text{Homocysteine desulphhydrase}} \text{2-Oxobutyrate} + NH_3 + H_2S$$

9. Pyridoxal phosphate is also required for enzymes (aldehydelyases) catalyzing aldol cleavage of amino acids, e.g., threonine aldolase.

L-Threonine $\xrightarrow[\text{PLP}]{\text{Threonine aldolase}}$ Glycine + Acetaldehyde

10. L-amino acid can be converted to D-amino acid, and vice versa, by amino acid racemase (an isomerase). Pyridoxal phosphate is required as a coenzyme in this reaction. Racemases acting on amino acids have been described in bacteria and other microorganisms, but not in mammals.

C. Deficiency of B$_6$

In a situation of B_6 deficiency, the B_6-containing enzymes function poorly. The normal ability of the body to synthesize any urgently needed amino acid from other amino acids is compromised, because transamination cannot take place in the absence of B_6 coenzymes. The cells are unable to meet the continuous demand for biogenic amines (e.g., histamine, serotonin, epinephrine, and norepinephrine, etc.) and γ-aminobutyric acid (GABA) because all these compounds are formed by pyridoxal phosphate-dependent decarboxylases. These substances affect neural activity significantly. Neurological symptoms are therefore, quite commonly observed in B_6 deficiency. Anaemia (hypochromic, microcytic) is also observed since haem synthesis is adversely affected in B_6 deficiency.

A serious deficiency of B_6 will also affect tryptophan metabolism since 3-hydroxykynurenine cannot be metabolized further. This provides a good parameter to diagnose B_6 deficiency. Since 3-hydroxykynurenine is not broken further by normal pathways, a part of it is converted into xanthurenic acid. Both xanthurenic acid and its parent compound are then excreted in urine. The quantity of xanthurenic acid excreted increases with the amount of L-tryptophan ingested and this indicates a serious deficiency of B_6. It has been shown that urine collected within 24 hours of ingestion of 2 g of tryptophan contains more than 30 mg of xanthurenic acid.

Such interpretations must be made with great caution in a pregnant subject because pregnancy leads to changes in many physiological and biochemical norms.

Deficiency syndrome: Clinical features of B_6 deficiency are related to the skin, nervous system, and erythropoiesis. However, there is no clearly defined specific deficiency disease in man. The earliest observations on B_6 deficiency were made in rats, who showed a peculiar dermatitis on their tails, around their eyes, ears, and mouth—a condition called *acrodynia*. This condition was cured by pyridoxine. In man, seborrhea-like skin lesions around the eyes, nose and mouth, accompanied by glossitis and stomatitis can be observed in experimentally-induced B_6 deficiency.

In children, B_6 deficiency leads to predominantly neurological symptoms such as hyperirritability and convulsions, but anaemia is not very marked. In adults, microcytic hypochromic anaemia may occur, though rarely. In man, a peripheral neuritis associated with synovial swelling and tenderness, especially of the carpal synovia (carpal tunnel disease) has been attributed to B_6 deficiency. The syndrome is reversible upon treatment with high dosages of the vitamin.

There are many genetic conditions involving abnormal vitamin B_6 metabolism, which respond to vitamin B_6 therapy: (a) In some children γ-aminobutyric acid (GABA) is synthesized at a decreased rate due to a defective glutamic acid decarboxylase (an apoenzyme with a decreased binding affinity for pyridoxal phosphate). The patients develop convulsions and brain damage, and die in infancy, if not treated early by B_6 supplementation. (b) Some patients with xanthurenic-aciduria or cystathioninuria respond to vitamin B_6 therapy. These patients have mutant apoenzymes (kynureninase or γ-cystathionase) which have defective interaction with pyridoxal phosphate. (c) Some patients with homocystinuria respond to B_6 therapy because pyridoxal phosphate increases the activity of the deficient enzyme, cystathionine synthetase. (d) Some patients have pyridoxine-responsive anaemia which is not due to nutritional deficiency of pyridoxine. In these patients, there is probably a deficiency of an unknown enzyme involved in hemoglobin synthesis, or a defect in utilization of pyridoxine due to defective pyridoxine kinase (which converts pyridoxine to pyridoxal phosphate).

Vitamin B_6 deficiency can be assessed by the following parameters: Tryptophan load test (measurement of urinary excretion of xanthurenic acid after tryptophan load); measurement of urinary homocysteine and cystathionine after a methionine load; measurement of urinary excretion of B_6 or its metabolite, 4-pyridoxic acid; serum serine/glycine ratio: erythrocyte glutamate oxaloacetate transaminase (EGOT) activity; and *in vitro* effect of pyridoxal phosphate on EGOT activity.

D. Effect of Drugs on Activity of the B_6 Coenzymes

4-deoxypyridoxine (Figure 47.13) is one of the most active antimetabolite to pyridoxine. 4-deoxypyridoxine-5-

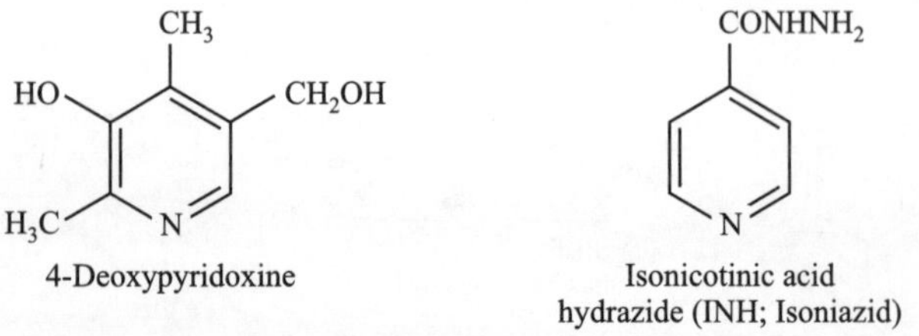

Figure 47.13 Antimetabolites to pyridoxal phosphate.

phosphate, formed from 4-deoxypyridoxine in the body, is a competitive inhibitor of several pyridoxal phosphate-dependent enzymes.

Isonicotinic acid hydrazide (Isoniazid; Figure 47.13) combines with pyridoxal or pyridoxal phosphate to form hydrazones, which are potent inhibitors of pyridoxal kinase. Consequently, these compounds inhibit the formation of B_6 coenzymes. The pyridoxal hydrazone is rapidly excreted in urine, causing vitamin B_6 deficiency. It is only in the group of "slow acetylators" of isoniazid that isoniazid forms hydrazone of pyridoxal. The hydrazones may in addition act directly as convulsants. Cycloserine is also an antagonist of vitamin B_6. It probably forms a complex with pyridoxal phosphate, and this complex competes with pyridoxal phosphate for apoenzymes. Penicillamine acts as a B_6 antagonist by forming a thiazolidine derivative with pyridoxal phosphate.

Vitamin B_6 enhances the peripheral decarboxylation of levodopa. Consequently, less levodopa is available for conversion to dopamine in the basal ganglia. This reduces the efficacy of levodopa in the treatment of Parkinson's disease. Vitamin B_6 and multivitamin preparations should thus be avoided in patients receiving levodopa.

Slight vitamin B_6 deficiency may be seen in many women taking oral contraceptive steroids and in some women during pregnancy. Many estrogenic steroids induce increased synthesis of tryptophan pyrrolase resulting in increased production of kynurenine and 3-hydroxykynurenine. Increased activity of several pyridoxal phosphate-dependent enzymes (kynureninase and other enzymes involved in conversion of tryptophan to niacin) produces a strain on the availability of vitamin B_6 elsewhere in the body and thus leads to B_6 deficiency.

VII. RIBOFLAVIN

Synonyms: Riboflavine, Vitamin B_2, Lactoflavine.

A. Structure

Riboflavin (Figure 47.14A) has a three-ring cyclic structure, isoalloxazine, bearing two methyl groups, one at C-6 and the other at C-7 (6,7-dimethyl-isoalloxazine), along with a ribitol side chain at N-9 (ribitol is the alcohol corresponding to ribose sugar, in which the terminal CHO group is reduced to CH_2OH).

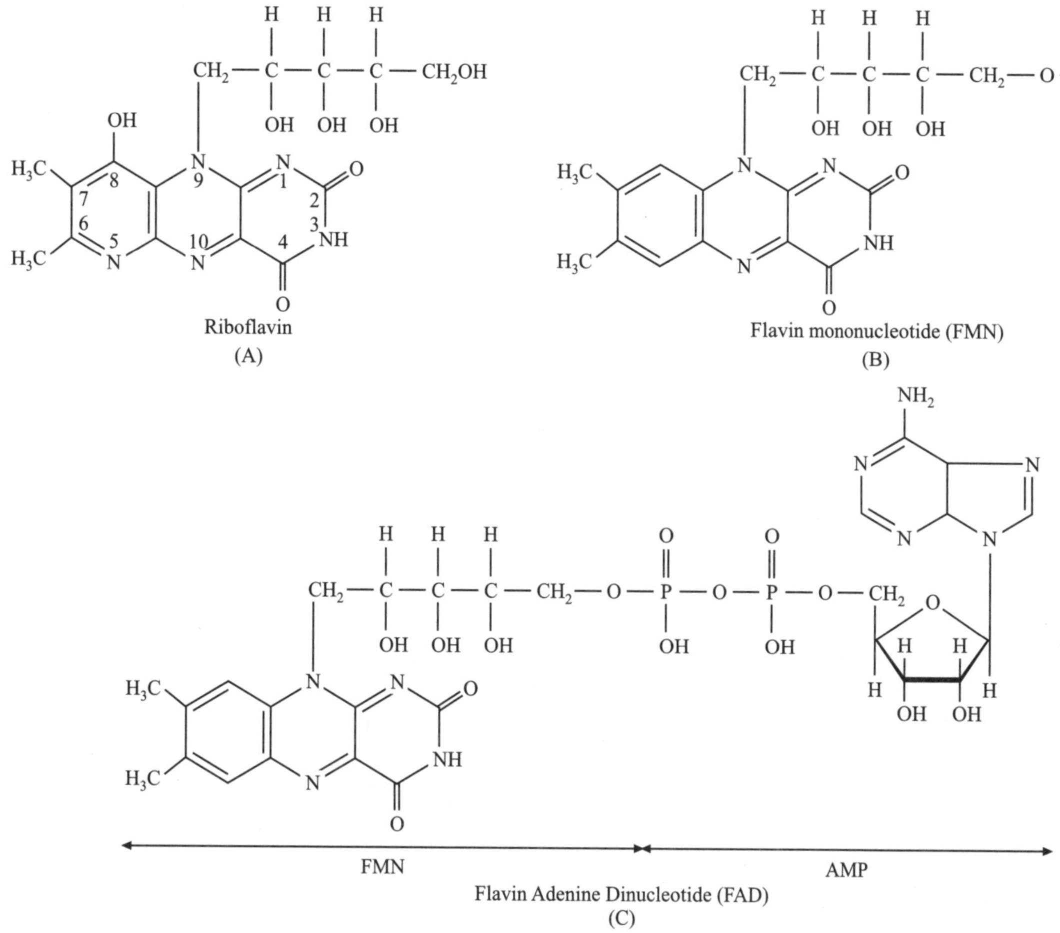

Figure 47.14 Structures of riboflavin (A), FMN (B) and FAD (C). Note the numbering of atoms of the ring system. Riboflavin has no phosphate group; FMN has one phosphate group on the ribitol side chain; and FAD has two phosphate groups (besides a ribose and adenine) on the ribitol side chain.

B. Coenzymes of Riboflavin and Their Functional Role

Like most other members of the B-complex group of vitamins, riboflavin needs to be converted into coenzyme forms before it can function in the biological system. Two main coenzyme derivatives of this vitamin are known: Flavin mononucleotide (FMN, in which an orthophosphate molecule is esterified to the ribitol side chain; Figure 47.14B) and Flavin adenine dinucleotide (FAD, in which the FMN is further linked through its phosphate to a nucleotide, AMP, thus creating an acid anhydride linkage; Figure 47.14C). During absorption from the intestine, there is ATP-dependent phosphorylation of riboflavin to FMN by mucosal flavokinase. FAD is formed by the transfer of an AMP moiety from another ATP molecule to FMN.

$$\text{Riboflavin} + \text{ATP} \rightarrow \text{FMN} + \text{ADP}$$
$$\text{MN} + \text{ATP} \rightarrow \text{FAD} + \text{PPi}$$

The Flavin enzymes contain either FMN or FAD and each enzyme has a high specificity for either of the two coenzymes. Proteins containing flavin compounds are in general referred to as flavoproteins. They often have a yellowish tinge because of the intense yellow colour of the constituent riboflavin coenzymes.

Functionally, the flavin enzymes deal with oxidation-reduction mechanisms (Note that enzymes containing NAD and NADP are similarly concerned with oxidation-reduction processes). In the reduction process, two hydrogen atoms are accepted by each molecule of FMN or FAD which is reduced to $FMNH_2$ or $FADH_2$, respectively. The sites at which these hydrogen atoms are accepted are the same in both the coenzymes. These are the two nitrogens (N-1 and N-10) of isoalloxazine nucleus as illustrated in Figure 47.15.

A few examples of reactions involving flavoprotein enzymes are given below to illustrate the features of the flavin system. There are numerous other examples which are discussed elsewhere in this book.

1. *Amino acid oxidation:*   One reaction by which amino acids are catabolized is catalyzed by amino acid oxidases. These enzyme are specific for either D-or-L-amino acids. This group of enzymes usually contains FAD, although there are instances where L-amino acid oxidases contain FMN (e.g., L-amino acid oxidase of rat kidney). In the course of oxidation of

Figure 47.15 Reduction of isoalloxazine ring in FMN and FAD.

amino acid by the amino acid oxidase, flavin is reduced. The reduced flavin is directly reoxidized by molecular oxygen and hydrogen peroxide is released as a byproduct. The oxidized flavin is then available for another cycle of catalytic activity.

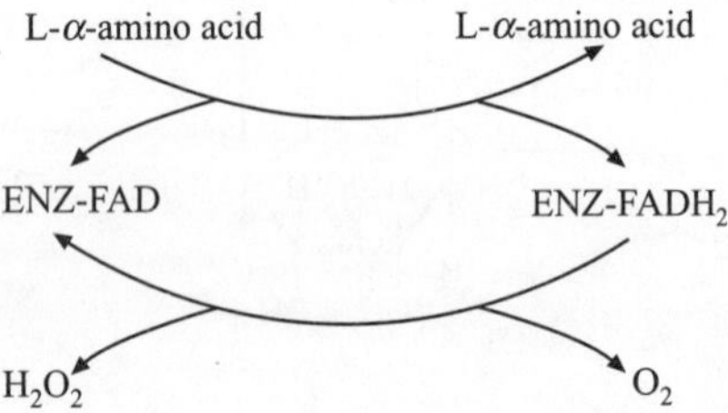

2. *Dihydrolipoate dehydrogenase:* There are other flavin enzymes involved in dehydrogenation reactions in which NAD or NADP rather than molecular oxygen is the eventual acceptor of hydrogens. An example of this is the dihydrolipoate dehydrogenase, which is a component of pyruvate dehydrogenase complex and other related enzyme systems. Each molecule of the enzyme contains two molecules of FAD. The reaction may be schematically depicted as follows:

3. *Flavin dehydrogenases of the electron transport chain (ETC):* These enzymes play an important role in the transport of electrons in the electron transport chain. NADH-dehydrogenase and the dehydrogenases acting on succinate, glycerol 3-phosphate, acyl-CoA, pyruvate and α-ketoglutarate are flavoprotein enzymes. NADH dehydrogenase contains FMN but the rest contain FAD (see Chapter 23).

C. Riboflavin Deficiency

Natural deficiency of riboflavin in man is rather uncommon, partly because riboflavin is normally synthesized by the intestinal flora. Riboflavin deficiency usually accompanies other deficiency diseases such as beriberi, pellagra and kwashiorkor. Phototherapy for neonatal jaundice causes transient and probably harmless riboflavin deficiency. However, when deficiency does occur, the symptoms are confined to the skin and mucosa. There is no clearly defined specific deficiency disease. Sore throat, angular stomatitis, glossitis, cheilosis (red denuded lips), seborrheic dermatitis, normochromic normocytic anaemia, and neuropathy have been observed. In some patients, corneal vascularization and cataract have been described.

Riboflavin deficiency can be assessed by measurement of urinary excretion of riboflavin and erythrocyte glutathione reductase activity (as FAD is the coenzyme of glutathione reductase present in RBC). For the latter test, haemolysed erythrocytes are incubated with glutathione with and without the addition of FAD. An 'FAD effect' is measured as the activation coefficient.

VIII. NIACIN

Synonyms: Nicotinic acid; Niacinamide; P.P. (Pellagra-preventing) factor of Goldberger.

A. Coenzymes

Niacin is a pyridine derivative. After absorption from the gastrointestinal tract, niacin (Figure 47.16A) is converted to its coenzyme forms: nicotinamide adenine dinucleotide (NAD) and nicotinamide adenine dinucleotide phosphate (NADP). The older names for NAD and NADP were diphosphopyridine nucleotide (DPN) or coenzyme I and triphospho-pyridine nucleotide (TPN) or coenzyme II, respectively. When the coenzyme is formed, the vitamin appears not as nicotinic acid but as nicotinamide, i.e., its carboxyl group is amidated (Figure 47.16A).

B. Chemistry

NAD consists of two distinct fragments: nicotin-amide mononucleotide and 5′-AMP linked by an acid anhydride bond (Figure 47.16B). Similarly, NADP contains two fragments: nicotinamide mononucleotide and 2, 5′-ADP (Figure 47.16B). In other words, the only difference between NAD and NADP is the presence of an additional phosphoryl group on C-2 on one of the ribose units of the latter.

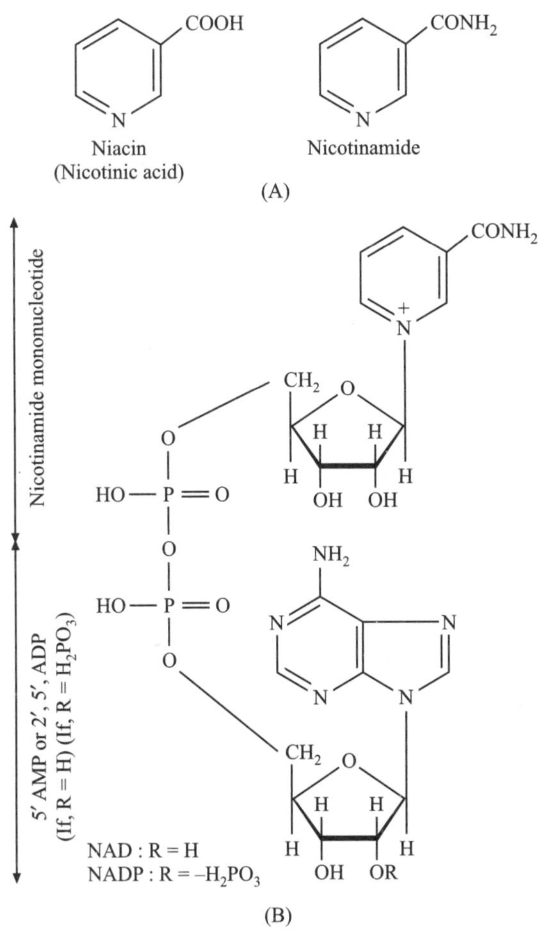

Figure 47.16 (A) Structures of niacin (nicotinic acid) and nicotinamide; (B) Structures of NAD^+ and $NADP^+$.

C. Role in Oxidation-Reduction Reactions

Both NAD and NADP function as coenzymes for enzymes that participate in oxidation-reduction reactions in cells. Moreover, NADH is primarily oxidized by the respiratory chain to generate ATP, whereas NADPH is used almost exclusively for the reductive biosynthesis (Riboflavin coenzymes are similarly important in oxidation-reduction reactions). In general, NAD and NADP are loosely bound to their apoenzymes so that they can be readily separated. A flavin coenzyme, on the other hand, is more tightly bound and is somewhat difficult to separate from its apoenzyme.

A brief discussion on the mechanism of oxidation-reduction process involving NAD or NADP will be of interest (Figure 47.17). The reactive site in both cases is always the carbon atom 4 of the nicotinamide ring, and rest of the molecule serves to bind the coenzyme to the apoenzyme. In the oxidized form, nitrogen of the nicotinamide residue, which is always in a quaternary state, forms an extra fourth bond, and thus it acquires a positive charge. It is for this reason that in depicting bio-chemical reactions,the oxidized form of the coenzyme is usually written as NAD^+ or $NADP^+$. In the process of reduction, NAD^+ or $NADP^+$ accepts a hydride ion rather than a hydrogen

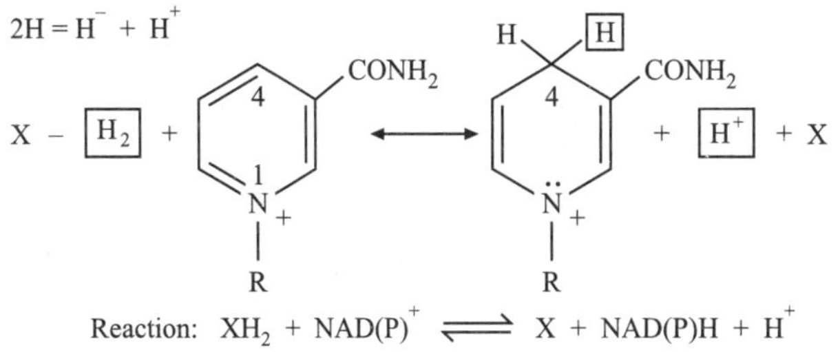

$$2H = H^- + H^+$$

$$Reaction:\ XH_2 + NAD(P)^+ \rightleftharpoons X + NAD(P)H + H^+$$

Figure 47.17 Reduction of NAD or NADP. Only partial structure of the coenzyme is shown. Only one hydrogen and one electron from XH_2 are accepted by the coenzyme. The remaining proton is released into the surrounding medium.

atom [a hydride ion (H–) is a hydrogen atom carrying an extra electron; it can dissociate into a hydrogen atom plus an electron, i.e. $H^- \rightarrow H + e^-$]. When a compound XH_2 is to be oxidized by an NAD-containing enzyme, there are two hydrogen atoms that would need to be removed by NAD. These two hydrogen atoms are equivalent to one hydride ion plus one proton:

It is this single hydride ion that is accepted by the nicotinamide moiety, while the proton is released into the surrounding medium. The electron of the hydride ion neutralises the positive charge on the nitrogen (which reverts to a valency of three), while the hydrogen atom binds to the carbon atom (which is in para position to the nitrogen atom) in either the A or B position (according to the stereospecificity determined by the particular dehydrogenase catalyzing the reaction). The reaction may be depicted as follows:

$$XH_2 + NAD^+ \longrightarrow X + NADH + H^+$$

From a study of the biochemical reactions of the body, it is obvious that NAD and NADP play vital roles in the life processes. The body therefore needs a substantial supply of the parent vitamin. Part of this requirement is met by L-tryptophan because this amino acid can be converted to a small extent to niacin. However, this is an inadequate source since L-tryptophan is itself an essential amino acid. The major supply of niacin is still through the diet unless the intake of L-tryptophan is very large. This is why niacin deficiency can sometimes be countered by foods that are very poor in this vitamin, but are extremely rich in L-tryptophan. In man, about 1 mg of niacin is formed from 60 mg of dietary tryptophan.

D. Deficiency

Deficiency of niacin leads to a clinical condition called "pellagra". Pellagra is characterised by signs and symptoms referable especially to skin (an erythematous eruption resembling sunburn involving areas exposed to sunlight), gastrointestinal tract (stomatitis, glossitis, and diarrhoea) and central nervous system (headache, dizziness insomnia, depression, impairment of memory and dementia). This triad is frequently referred to as dermatitis, diarrhoea, and dementia, or the "three Ds". Decreased conversion of tryptophan into serotonin may be responsible for some of the clinical manifestations related to the nervous system.

In the past, pellagra has mainly been seen in the tropical and subtropical countries amongst people whose staple diet is maize (American corn) because maize is deficient in both niacin and tryptophan. It has also been associated with the use of millets, particularly Sorghum Vulgare (jowar, guinea corn). However, the relation between the intake of the above cereals and the development of pellagra is not straightforward; the disorder may be due to an imbalance in dietary amino acids or to a complex vitamin deficiency.

There are three non-nutritional causes of pellagra, where dietary intake of tryptophan and niacin is adequate but there is disturbance of tryptophan metabolism leading to reduced capacity for synthesis of nicotinamide from tryptophan. These are:

(a) *Isoniazid therapy for tuberculosis*, which leads to vitamin B_6 deficiency. Vitamin B_6 deficiency causes reduced synthesis of nicotinamide from tryptophan due to blockage of tryptophan metabolic pathway at the level of kynureninase, a pyridoxal phosphate-dependent enzyme. Isoniazid also inhibits the utilization of dietary nicotinamide by inhibiting the enzyme nicotinamide deaminase, which catalyzes an essential step in the utilization of nicotinamide.

(b) *In Hartnup disease,* there is failure of a specific intestinal absorption—and renal tubular reabsorption—system for the neutral amino acids. Consequently, little tryptophan is available for the synthesis of nicotinamide. Moreover, due to metabolism of tryptophan by the intestinal flora, some abnormal metabolites (including indole and indoleacrylic acid) are produced, which inhibit tryptophan pyrrolase and possibly other enzymes involved in tryptophan metabolism.

(c) *In carcinoid syndrome,* a major proportion of the available tryptophan is utilized for the synthesis of serotonin by the tumour. Consequently, little tryptophan is available for the synthesis of nicotin-amide in advanced cases.

Niacin deficiency can be assessed by measurement of urinary N1–Methylnicotinamide, which is the most important metabolite of nicotinic acid. Plasma tryptophan and the levels of NAD and NADP in the red cells are also decreased.

IX. PANTOTHENIC ACID

Synonyms: Chick antidermatitis factor; vitamin B_x.

A. Structure

The structure of pantothenic acid (Figure 47.18A) is quite simple. However, structure of the coenzyme built from it (coenzyme A) is rather complex (Figure 47.18B). Pantothenic acid contains β-alanine and D-pantoic acid in amide linkage. The biological activity is characteristic of only the 'd'isomer.

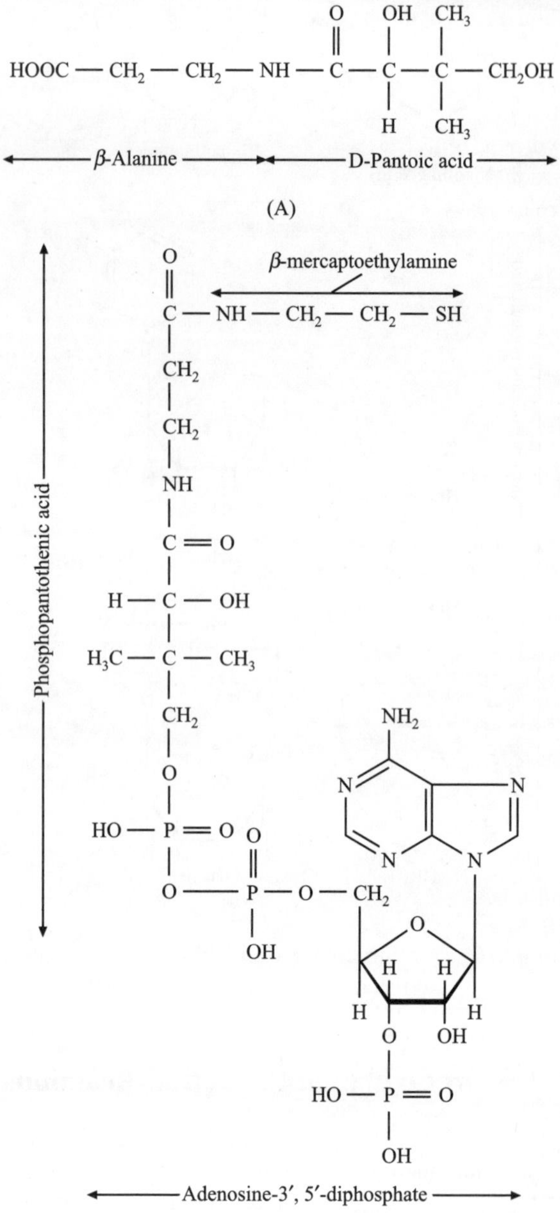

Figure 47.18   (A) Structure of D-pantothenic acid; (B) Structure of coenzyme A.

The coenzyme A built from pantothenic acid phosphate has adenosine 3′, 5′-diphosphate and β-mercapto-ethylamine in addition. The sulphydryl group on the latter is most important since this is the active site. Free (reduced) coenzyme A is often abbreviated as CoA.SH.

B. Functions

Pantothenic acid functions by entering into the formation of coenzyme A. Coenzyme A is a component of fatty acid synthetase complex. Therefore, panthothenic acid deficiency would seriously affect the biosynthesis of fatty acids. Coenzyme A is important as a carrier of acyl groups. Its crucial role in the TCA cycle would be apparent when it is recalled that the conversion of pyruvate to acetyl CoA is not possible without this coenzyme. CoA is widely distributed in the body. Many reactions involving coenzyme A, which are related to acyl group transfer, are important in the oxidative metabolism of carbohydrates, gluconeogenesis, synthesis and degradation of fatty acids, and the synthesis of sterols, steroid hormones and porphyrins.

C. Deficiency

Deficiency of pantothenic acid is not easy to produce because this vitamin is widely distributed amongst most forms of life. There is also a ready supply in the gut, where it is synthesized by the bacterial flora. Experimental human deficiency has produced fatigue, headache, sleep disturbances, nausea, vomiting, abdominal cramps, and paraesthesias in the extremities, muscle cramps, and impaired coordination.

A "burning foot" syndrome associated with paraesthesia in the legs and seen in malnourished person may be a form of pantothenic acid deficiency, since there is a favourable response to the vitamin.

SUGGESTIONS FOR FURTHER READING

Barker, B.B. and Bender, D.A. (Eds.) (1980), Vitamins in Medicine, 4th ed., Vol. 1, William Heinemann Medical Books Ltd., London.

Benkovic S.J. (1980), On the Mechanism of Action of Folate—and Biopterin—Requiring Enzymes, *Annu. Rev. Biochem.*, 49, 227.

Hayashi H., et al. (1990), Recent Topics in Pyridoxal 5′-Phosphate Enzyme Studies, *Annu. Rev. Biochem.*, 59, 87.

Kapadia C.R. (1995), Vitamin B12 in Health and Disease: Part I–Inherited Disorders of Function, Absorption and Transport, *Gastroenterologist*, 3, 329–344.

Knowles J.R. (1989), The Mechanism of Biotin-dependent Enzymes, *Annu. Rev. Biochem.*, 58, 195.

Lamers Y. (2011), Indicators and Methods for Folate, Vitamin B-12, and Vitamin B-6 Status Assessment in Humans, *Curr. Opin. Clin. Nutr. Metab. Care.*, 14:445–54.

McCormick D.B. and Wright L.D. (1979), Vitamins and Coenzymes, In: Colowick, S.P. and Kalpan, N.O., *Methods in Enzymology*, Vol. 62, Part D., New York, 1979, Academic Press, Inc.; Vol. 66, Part E; Vol. 67, Part F., 1980, Academic Press Inc.

McCormick D.B. and Greene H.L. (1999), Vitamins, In: Tietz *Textbook of Clinical Chemistry*, 3rd ed., Burtis, C.A. and Ashwood, E.R. (Eds.), W.B. Saunders Co., Philadelphia, USA, 999–1028.

Metz J. (1992), Cobalamin Deficiency and the Pathogenesis of Nervous System Disease, *Annu. Rev. Nutrition*, 12, 59.

Mildvan A.S. (1974), Mechanism of Enzyme Action, *Annu Rev. Biochem.*, 43, 357.

Morgan S.L. and Baggott J.E. (2010), Folate Supplementation During Methotrexate Therapy for Rheumatoid Arthritis, *Clin Exp. Rheumatol.*, 28:S102–9.

Walsh C. (1979), *Enzymatic Reaction Mechanisms, Freeman.*

W.H. Sebrell, Jr. and R.S. Harris (Eds.) (1967), The Vitamins, 2nd ed., Vols. II, IV and V., Academic Press Inc., New York.

Wilson J.D., Braundwald E., Isselbacher K.J., Petersdorf R.G., Martin J.B., Fauci A.S. and R.K. Root (Eds.), (1991), Harrison's *Principles of Internal Medicine*, 12th ed., McGraw Hill, Inc., New York.

Wood H.G. and Barden R.E. (1977), Biotin Enzymes, *Annu Rev. Biochem.*, 46, 385.

48

Vitamin C

K.D. Moudgil

CONTENTS

I. Structure and Biologically Active Forms
II. Metabolism
III. Functional Roles
IV. Sources and Supplemental Forms
V. Recommended Dietary Allowance
VI. Deficiency
VII. Prevention of Chronic Diseases
VIII. Therapeutic Use
IX. Adverse Effects

Vitamin C, also known as ascorbic acid, is an essential nutrient for humans. Unlike most plants and animals, humans cannot synthesize vitamin C. The deficiency of vitamin C may lead to scurvy in humans. Albert Szent-Györgyi was awarded the Nobel Prize in Medicine in 1937 for his work that led to the identification of vitamin C as the antiscorbutic factor. High dose vitamin C supplementation has been suggested to offer health benefits against common cold and cancer.

I. STRUCTURE AND BIOLOGICALLY ACTIVE FORMS

Vitamin C refers to the L-enantiomer of ascorbic acid (L-ascorbic acid) and its oxidized forms. Ascorbic acid is a six-carbon lactone with molecular formula $C_6H_8O_6$. Its structural formula closely resembles that of carbohydrates. It is a white crystalline solid and its melting point is about 190°C. It is highly water soluble and sour in taste. It is optically active

and has the selective absorption maxima at 245 nm. Humans lack the enzyme, L-gulono-1,4-lactone oxidase (GULO), which catalyzes the final step of vitamin C synthesis from glucose by converting L-gulonolactone into L-ascorbic acid. Vitamin C occurs in two forms, ascorbic acid (reduced form) and dehydroascorbic acid (DHA) (oxidized form). Both forms are biologically active. Ascorbate refers to the anionic form of ascorbic acid. It can prevent other compounds/molecules being oxidized by donating electrons and in this process ascorbate undergoes reversible oxidation to DHA. This is the most important chemical property of vitamin C and it forms the basis of its known anti-oxidative and other physiological actions (Figure 48.1).

II. METABOLISM

Small intestine is the major site for vitamin C absorption. Vitamin C in its reduced form (ascorbate) constitutes the

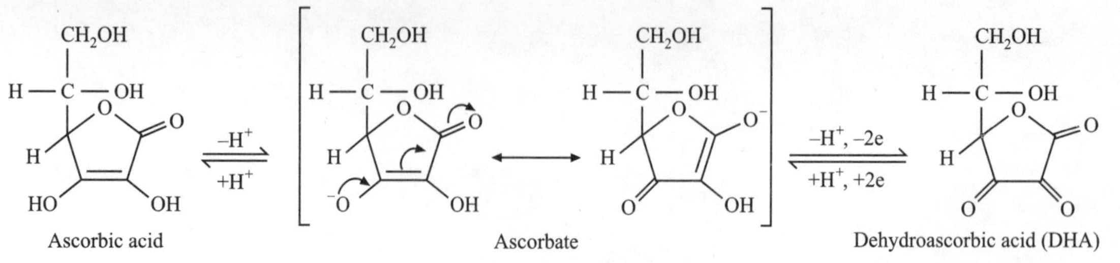

Figure 48.1

majority (80-90%) of this vitamin in food. Absorption occurs by both active transport system at low concentrations and by diffusion at higher concentrations. Ascorbate is absorbed by a Na^+-dependent process involving sodium-dependent vitamin C transporters (SVCT), whereas, the absorption of DHA is a Na^+-independent process and it involves glucose transporters (GLUT). The majority of DHA is immediately reduced to ascorbate upon crossing the serosal membrane by glutathione peroxidase. Recycling between DHA and ascorbate is a prominent feature of vitamin C metabolism and appears to aid in maintaining antioxidant reserves in the body. The coexpression of GLUT1 with stomatin in erythrocyte membranes allows uptake of DHA. It constitutes a compensatory mechanism in those mammals who are unable to synthesize vitamin C. Ascorbic acid is widely distributed throughout the tissues. A portion of excess ascorbic acid in the body is directly excreted in urine unchanged, whereas the rest is metabolized to 2,3-diketogulonic acid and oxalate and then excreted in urine.

III. FUNCTIONAL ROLES

A. Metabolic Reactions

Ascorbate as a cofactor in hydroxylation reactions: Ascorbate plays a vital role in various hydroxylation reactions. Hydroxylation is catalyzed by Fe(II)-dependent nonheme oxygenases which belong to the family of α-ketoglutarate-dependent dioxygenases. Examples of the former include prolyl-, asparaginyl- and lysyl hydroxylases. Ascorbate is not consumed in the reaction catalyzed by ascorbate-dependent dioxygenases. Instead, ascorbate maintains nonheme iron in dioxygenase enzymes in the active Fe(II) form. The role of ascorbate in the hydroxylation of proline and lysine residues during the formation and stabilization of collagen molecule has been studied extensively (Figure 48.2).

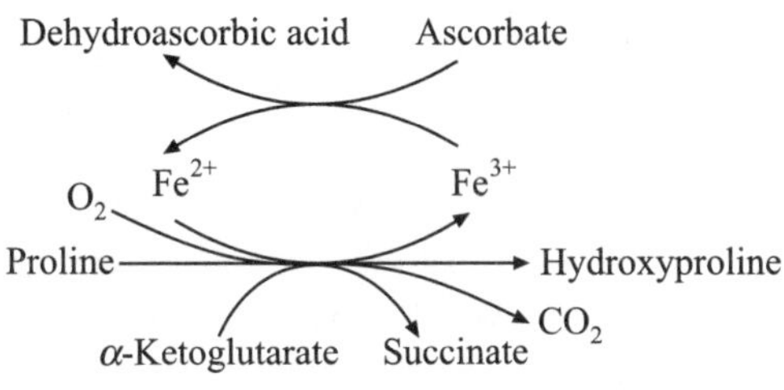

Figure 48.2

Ascorbate-dependent hydroxylation also plays a role in gene transcription mediated by hypoxia-inducible factor (HIF-1), which is expressed in all nucleated cells under hypoxic conditions. Specifically, the hydroxylation of proline and asparagine residue of HIF-1 facilitates its binding to the Von Hippel-Lindau tumor suppressor, which causes ubiquitin-dependent degradation of HIF-1. Ascorbate thus helps to maintain normoxia. HIFs regulate a variety of genes including those controlling growth and apoptosis, energy metabolism and matrix and barrier functions.

Vitamin C as an antioxidant and free radical scavenger: Reactive oxygen species (ROS) and free radicals are derived from normal metabolic processes in the human body as well as external sources such as environmental pollutants. ROS reacts with biological molecules such as DNA, proteins and lipids causing structural and functional damage leading to pathological conditions including neurodegenerative diseases and cancer. Ascorbate is a good electron donor and it reduces compounds with unpaired electrons, for example, superoxide (O_2^-), peroxyl radicals (O_2^{-2}), and hydroxyl radicals ($\cdot OH$). Another mechanism of reduction is by donating hydrogen to free radicals. During the process, ascorbate itself becomes an ascorbate radical. But the ascorbate radical is very stable because of its resonance structure. Moreover, ascorbate is readily regenerated by NADH- or NADPH-dependent reductase (Figure 48.3). Through these actions, ascorbate reduces oxidative damage.

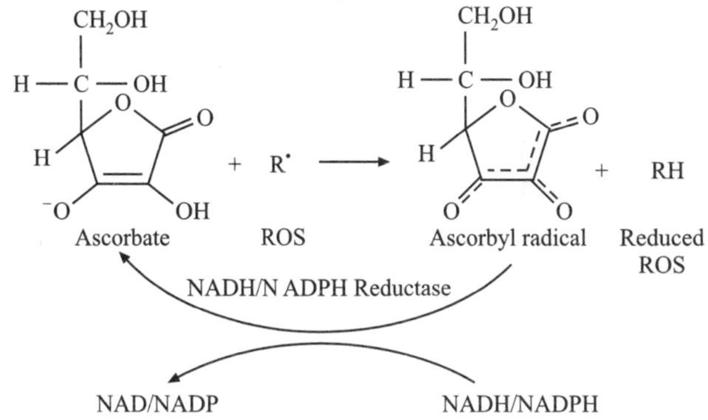

Figure 48.3

Vitamin C scavenges free radicals both in the plasma and within cells. For cellular membranes and other sub-cellular compartments, it plays an indirect antioxidant role by recycling α-tocopherol. Many of the ROS-producing enzymes (e.g., electron transport chain, xanthine oxidase, myeloperoxidase and NADPH oxidase) are embedded in lipid bilayers. This renders lipids preferred targets of oxygen radicals. Increased oxidative stress causes peroxidation of membrane lipids resulting in membrane damage. The most important antioxidant in the cell membrane is α–tocopherol, which scavenges carbon-based peroxide free radicals of the membrane. The interaction of radical oxidants with α-tocopherol produces α-tocopheroxyl radical. This α-tocopheroxyl radical can be regenerated back to α-tocopherol by ascorbate. Thus, vitamin C can protect against membrane peroxidation.

Pro-oxidant activity: Paradoxically, vitamin C may also manifest pro-oxidant activity under certain situations. As a pro-oxidant, vitamin C promotes the formation of ROS such as hydrogen peroxide and hydroxyl radicals. ROS interacts with cellular proteins and sub-cellular components resulting in their oxidative breakdown. This leads to reduced viability of cancer cells without harming normal healthy cells.

Role of ascorbate in the redox homeostasis of certain organelles: Ascorbate has an essential role in stabilizing the mitochondrial redox balance and decreasing ROS formation. Oxidative respiration in mitochondria is a major source of the intracellular production of ROS. DHA represents the oxidized form of vitamin C that is transported into mitochondria across the membrane via glucose transporter (GLUT). DHA is very unstable. DHA can be reduced back to ascorbate, which then helps in maintaining redox homeostasis. Electron transport chain complex III can donate electrons for the reduction of DHA and this in turn stabilizes the mitochondrial redox balance by decreasing the reduced state of electron chain. This also reduces the electron leakage of the electron transfer chain and subsequent ROS formation (Figure 48.4). DHA also can be reduced back to ascorbate by DHA reductase and reduced glutathione (GSH). The oxidized glutathione formed is reduced back by the action of glutathione reductase at the expense of NADPH. Finally, Lipoic acid (LA)-containing enzyme complexes can also reduce DHA. The main effect of reduction of DHA is on restoring the redox homeostasis.

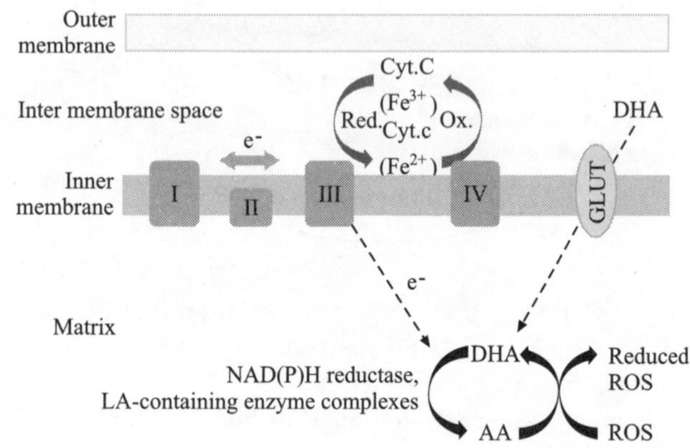

Figure 48.4   Vitamin C protects mitochondrial membrane depolarization by decreasing reduced state of the electron transfer chain and scavenging ROS. Dehydroascorbic acid (DHA) is transported into mitochondria via the glucose transporter GLUT. DHA is reduced to ascorbic acid (AA) at the site of complex III by donating electron (e-). DHA can also be reduced to AA by NAD(P)H reductase or lipoic acid (LA)-containing enzyme complexes. AA quenches reactive oxygen species (ROS). Cyt.c = Cytochrome c; Red.= reduction; Oxd.= oxidation; Fe = iron.

The endoplasmic reticulum (ER) possesses powerful electron transfer chains. Ascorbate plays a role in maintaining the ER redox potential similar to that in the mitochondria. Various ROS-generating reactions (e.g., oxidative protein folding) occur in this compartment and many luminal enzymes of the secretory pathway use ascorbate as a cofactor.

Ascorbate and deaminative cleavage of glypicans: Glypicans represent a family of glycosylphosphatidylinositol-anchored, cell-surface heparin sulphate proteoglycans. The deaminative degradation of glypican-1 heparan sulphate is defective in patients with Niemann-Pick type C1 disease, a degenerative disorder affecting the nervous system. Cysteines in glypican-1 can be nitrosylated by nitric oxide in a copper-dependent reaction. Exposure of glypican-1 to ascorbate releases nitric oxide from the intrinsic S-nitroso groups of the glypican-1 core protein. The nitric oxide thus released mediates the deaminative cleavage of heparin sulphate by heparanase at sites where glucosamines possess a free amino group.

Other functions: (i) Ascorbate plays a role in maturation of red blood cells (RBCs). Folic acid is necessary for the synthesis and maturation of RBCs, and ascorbate prevents the oxidation of the active form of folic acid, namely tetrahydrofolic acid (THFA). (ii) Vitamin C is involved in the production of carnitine, which is necessary for the generation of energy from fatty acids in cardiac and skeletal muscle cells. (iii) In addition, it is necessary for iron absorption, and plays a role in the conversion of cholesterol to bile acids. Vitamin C may also affect prostaglandin metabolism. (iv) Ascorbate appears to be involved in histamine metabolism and allergic reactions. In the presence of copper ions, ascorbate prevents the accumulation of histamine and assists in its degradation and elimination. (v) Evidence gathered from different studies has shown an inverse relationship between vitamin C intake, either through regular diet or via vitamin C supplementation, and the blood levels of lead. It has been proposed that vitamin c might act at the level of intestinal absorption and/or urinary excretion of lead.

B.   Host Immunity

Vitamin C may help in preventing a microbial infection and controlling an ongoing infection. This beneficial effect of vitamin C can be attributed to the ability of this vitamin to enhance immune response and thereby facilitate the induction of effective host immunity against infectious agents such as viruses and bacteria. Vitamin C can contribute to enhanced immune defense against pathogens via facilitating the production of immunoglobulins (e.g., IgG, IgM); enhancing the chemotactic, phagocytic and pathogen-killing activities of leukocytes; increasing the production of interferons which have anti-viral activity; inducing the generation of lymphocytes through the action of certain prostaglandins; and reducing the levels of stress hormones that otherwise can suppress immune system. Phagocytic cells consume vitamin C, and extra vitamin C uptake may prevent damage to phagocytes themselves by neutralizing oxidants. For above reasons, vitamin C has been reported to be beneficial in patients with Chédiak-Higashi syndrome. Supplementation with vitamin C as an adjunct has been reported to be beneficial in recovery from tuberculosis and streptococcal infections. The role of vitamin C in preventing "common cold" has been of particular interest over the past few decades and it is discussed below in more detail.

C.   Anti-cancer Activity

The proposed beneficial effect of vitamin C against cancer is attributable to its many actions. Vitamin C can intervene at different steps of carcinogenesis. Vitamin C is an antioxidant. It can protect against oxidative damage in normal tissues along with scavenging of free radicals leading to the inhibition of formation of carcinogens. Further, the pro-oxidant activity of vitamin C renders cancer cells more susceptible to cell death than normal cells.

Most tumor cells are unable to transport ascorbic acid into their interior. Therefore, cells obtain vitamin C in the form of DHA. In addition, tumor cells are relatively deficient in anti-oxidant enzymes compared to normal cells. Therefore, tumor cells produce high amount of ROS in the presence of vitamin C. This in part explains the selective cytotoxic action of vitamin C on tumor cells while sparing normal cells. Above physiological concentrations, vitamin C may act as a prodrug, which can form ascorbate radicals and hydrogen peroxide. DHA also has anti-tumor activity. DHA reacts with homocysteine thiolactone, a constituent of normal cells, forming 3-mercaptoproponaldehyde, which is cytotoxic. Vitamin C, along with vitamins A and E, can influence gene expression. It has been speculated that vitamin C may also contribute to protection against cancer via strengthening the extracellular matrix through increased collagen synthesis. In addition, vitamin C can inhibit angiogenesis in cancer cells.

IV. SOURCES AND SUPPLEMENTAL FORMS

Fresh citrus and other fruits and green vegetables are good sources of vitamin C (L-ascorbic acid). Some of the common food items in this regard include broccoli, bell pepper, Brussels sprouts, cantaloupe, cauliflower, grapefruit, kiwifruit, lemon, lychee, orange, papaya, parsley, potato, strawberry, and tomato. For an easy reference, five servings (2.5 cups) of fruits and vegetables provide about 200 mg of vitamin C. Other sources of vitamin C are milk and animal products (e.g., fish, liver, and kidney). Contrary to the general belief, vitamin C is not totally lost during cooking or storage of food. The vitamin is partially preserved following commonly used methods of food processing such as boiling, steaming, freezing and dehydration. Natural and synthetic forms of vitamin C are chemically and functionally identical. Ascorbic acid, sodium ascorbate, and calcium ascorbate are the commonly used forms of vitamin C supplements. In regard to vitamin C supplementation, the vitamin is unstable. At high concentrations, its absorption from the intestine is poor, and it is rapidly excreted in the urine. Calcium ascorbate is the main component of Ester-C®, which in addition contains a few other components. Ascorbyl palmitate is a vitamin C ester in which vitamin C is esterified to palmitic acid, a saturated fatty acid, yielding a fat-soluble vitamin C. This form of vitamin C is used in skin creams for its use as a topical anti-oxidant.

V. RECOMMENDED DIETARY ALLOWANCE

The recommended dietary allowance (RDA) for vitamin C (Table 48.1) varies from country to country. It is aimed at the

TABLE 48.1 Recommended dietary allowance (RDA) for vitamin C

Group	Age (month (mo) or year (yr))	Ascorbic acid (mg/d)	
		Males	*Females*
Infants	0–6 mo	40	40
	7–12 mo	50	50
Children	1–3 yr	15	15
	4–8 yr	25	25
	9–13 yr	45	45
Adolescents	14–18 yr	75	65
Adults	19 yr and above		
	Non-smokers	90	75
	Smokers	125	110
	Pregnant	N/A	85
	Lactating	N/A	90

N/A= not applicable. RDA in USA as per the National Academy of Sciences, USA.

Source: The Linus Pauling Institute, Oregon State University, Oregon, USA.

prevention of deficiency disease. The requirement of vitamin C is higher in pregnant individuals and those under stress such as infection or trauma. Smokers and persons undergoing dialysis for kidney diseases also need higher daily amounts of vitamin C. The requirement for vitamin C in smokers is higher to control extra oxidative stress owing to toxic components in cigarette smoke compared to non-smokers. The tolerable upper level (UL) of intake of vitamin C from diet for adults is 2 g/day.

VI. DEFICIENCY

Vitamin C deficiency may result in a clinical condition known as "Scurvy". In the past, scurvy was common among sailors who were away from home and frequently did not have access to fresh fruits and vegetables. The availability of synthetic vitamin C in 1930s subsequently allowed easy supplementation of this vitamin in the diet of various populations vulnerable to vitamin C deficiency. Unlike in the past, scurvy is no more a global health problem. Presently, scurvy is limited to subsets of individuals. These groups include poor and elderly persons not fully cared for, chronic alcoholics, patients with kidney diseases who are undergoing dialysis, infants maintained on food formulas lacking vitamin C, and persons consuming an unbalanced diet.

Scurvy

Collagen formation is most adversely affected in situations of vitamin C deficiency. Impaired collagen formation affects the connective tissue in different tissues, including blood vessels. Consequently, the skin, the blood vessels, and the skeletal system are affected the most by the pathological events associated with scurvy. Inadequate connective tissue support to the blood vessels results in bleeding tendencies. Patients with scurvy suffer from fatigue, weight loss, inflammation and bleeding of gums, dry and pigmented skin with signs of bleeding into the skin (petechiae, perifollicular hemorrhages, etc.), impaired wound healing, and bleeding into the joints (causing swollen and painful joints). The patients complain of aches and pains in the muscles, bones and joints of extremities. Additional complications may occur following bleeding into the pericardium, the peritoneal cavity, the brain, eye, or other tissues. Hematuria and melena may occur. In children, inadequate bone formation may lead to impaired bone growth. In addition, the sternum may sink inwards coupled with a raised prominent rib cage margin, "scorbutic rosary". Anemia may accompany severe deficiency of vitamin C.

Vitamin C deficiency can be assessed by measuring the concentrations of vitamin C in the plasma and white blood cells (WBCs). The concentration of vitamin C in leukocytes is much higher than that in plasma. It reflects the tissue concentration of the vitamin and is a preferred test. If needed, a vitamin C loading or saturation test can be performed. In this test, a defined amount of ascorbic acid is administered orally or intravenously, and then periodic blood and urine samples are collected for measuring the concentration of vitamin C. Depending on the results, patients can be assigned to different categories.

Treatment of patients with scurvy with exogenous vitamin C results in rapid improvement of symptoms.

VII. PREVENTION OF CHRONIC DISEASES

The results of several studies point to the disease-preventive effect of this vitamin in certain chronic diseases. Notable among these are cardiovascular diseases, cancer, gout and cataract.

A. Cardiovascular Diseases

Many studies have highlighted the association between reduced intake of vitamin C and increased risk of cardiovascular diseases. Two of the main conditions addressed in these studies are coronary artery disease and stroke. In some of these studies, the reported association was with dietary intake of vitamin C in the form of fruits and vegetables, while in others it was with supplemental vitamin C. However, clinical trials on the effects of vitamin C on prevention of heart disease or stroke have yielded equivocal results.

B. Cancer prevention

The relationship between increased intake of fresh fruits and vegetables and reduced risk of cancer has been suggested by the results of many studies conducted in different settings. The projected benefit of vitamin C against cancer prevention is reinforced by the guidelines for consumption of fresh fruits and vegetables in a balanced diet as suggested by several national and international health agencies. However, studies on vitamin C and cancer prevention were based on different types of cancer including those of the gastrointestinal tract, lung, cervix, and breast. Therefore, a direct comparison of results was often difficult. In addition, in some studies, vitamin C was used in combination with vitamin A, β-carotene, and vitamin E, making it difficult to pinpoint the individual contribution of vitamin C to the observed effects. Similarly, in studies where fresh fruits and vegetables were the main source of vitamin C, it was difficult to dissociate the effects of vitamin C per se from those of other natural ingredients in these foods. Furthermore, many of these studies were not well-controlled. Other caveats were the use of different dosages of vitamin C supplements in different studies and the variations in the nutritional and socio-economic status as well as other life-style differences (e.g., smoking) of the study populations. Also, the effect of vitamin C in cancer prevention might be indirect in some cases. For example, the bacterium *Helicobacter pylori* is now known to be the cause of gastric ulcers. It also is a predisposing factor for subsequent gastric cancer. It has been proposed that vitamin C supplementation might be a beneficial adjunct to the treatment of *Helicobacter pylori* infection with the aim of reducing the risk of gastric cancer. Additional large-scale, prospective, and controlled studies are required to establish the role of vitamin C in prevention of cancer.

C. Gout and Cataract

Studies in patients with gout, a form of arthritis linked with increased blood level of uric acid, have consistently shown a correlation between higher intake of vitamin C and reduced levels of serum uric acid as well as reduced risk of gout. Similarly, an inverse correlation has been reported between the levels of vitamin C in the blood (or the lens of the eye) and the severity of cataract, a disease in which the lens of the eye becomes opaque. However, the correlation is relatively much less convincing in cataract than in gout.

VIII. THERAPEUTIC USE

Several studies have suggested but not proven unequivocally the therapeutic utility of vitamin C in diseases such as cardiovascular diseases, cancer, and common cold.

A. Cardiovascular Diseases

Inadequate relaxation of blood vessels coupled with reduced blood supply is one of the major contributors to the disease process in myocardial infarction (heart attack) and stroke. In this context, the therapeutic effects of vitamin C against cardiovascular disease are attributable mostly to its ability to induce effective vasodilatation of blood vessels and to reduce blood pressure. Vitamin C has also been shown to reduce systolic blood pressure in patients with hypertension who otherwise were on conventional anti-hypertensive medications.

In other studies, an inverse correlation was found between vitamin C blood levels and prevalence of hypertension. Also, the effect on systolic pressure was much more marked than on diastolic pressure. Similarly, reduced levels of serum vitamin C were found to correlate with increased risk of developing stroke in hypertensive and over-weight men. The use of vitamin C supplements combined with reduction in body weight might reduce the risk for stroke.

Vitamin C also has been shown to be effective in reducing the cardiovascular complications of diabetes mellitus, but only in a subset of patients. In other diabetic patients, vitamin C either had no effect or worsened the symptoms. The precise reasons for the differential responses of subsets of diabetic patients are not yet clear. Studies on genetic differences, which form the rationale behind "personalized medicine", might shed light on the influence of such differences on the outcome of vitamin C treatment in diabetes mellitus and other diseases. The above studies were based on 500 mg/day or less dose of vitamin C unlike mega dose therapy described below for cancer and common cold. The beneficial therapeutic effects of vitamin C against cardiovascular diseases need to be rigorously examined in double blind and controlled clinical trials.

B. Mega-dose Vitamin C Therapy for Cancer and Common Cold

Linus Pauling developed and propagated the concept of "orthomolecular medicine", that is the treatment of diseases by employing naturally occurring endogenous substances (e.g., vitamins and minerals) that play an important role in the maintenance of health and well being. This approach was termed "orthomolecular" implying the use of "right molecules". Examples of the use of endogenous substances as therapeutic agents are insulin for diabetes mellitus and vitamin B12 for methylmalonic acedemia. The counterpart of this is withholding the intake of potentially harmful dietary substances, for example, in individuals with inherited metabolic disorders such as phenylketonuria. The idea of high dose vitamin C intake stemmed from Pauling's reasoning that the liver of some adult mammals (e.g., goat) that are endowed with the ability to synthesize vitamin C produces and releases into blood approximately 10 g or more of vitamin C per day. The findings regarding the role of high dose vitamin C for protection against cancer and common common cold are summarized below.

Cancer: Linus Pauling and colleagues proposed in 1970s that high dose (e.g., 10 g/day) vitamin C therapy was effective in increasing both the survival as well as the quality of life of patients with terminal cancer. The treatment regimen consisted of intravenous injection followed by oral administration of vitamin C (10 g/d). However, subsequent trials by other investigators using the same dose of vitamin C (10 g/day) but administered only orally in cancer patients failed to replicate the success observed in the earlier studies by Pauling et al. A retroactive analysis of these trials revealed that a difference in the route of administration of vitamin C might have been one of the factors contributing to the disparate outcomes in these studies. It has been proposed that the bioavailability of vitamin C and consequent effects on cancer cells might be significantly different when the same vitamin is given intravenously versus orally. Regardless, re-analysis of the earlier studies which had been conducted by Pauling and colleagues revealed that the comparison groups lacked appropriate controls and randomization. In addition, there is evidence from in vitro as well as animal studies that vitamin C might interfere with effectiveness of chemotherapy in some cases. Thus, the proposed mega dose therapy of vitamin C in cancer needs to be further evaluated using well-controlled clinical trial protocols. In addition, it is recommended that vitamin C treatment, if instituted, should be an adjunct to, rather than a substitute for, conventional anti-cancer treatment .

Common cold: Linus Pauling's proposal in late 1960s to use high dose (over 1 gram/day) of vitamin C to prevent common cold generated great interest in this vitamin in the general public and the professionals alike. Several studies have been conducted in the past three decades to examine the effect of vitamin C supplementation on the incidence, severity and duration of common cold. Vitamin C was administered either before (prevention) or after (treatment) the onset of clinical symptoms of common cold. The daily dose of vitamin C tested in these studies ranged from 0.25-4 g/day. The overall results of different studies are that there was a slight reduction in the duration and/or incidence of common cold when vitamin C was used in preventive regimen, but not when used in therapeutic regimen. Furthermore, the beneficial effects against cold were most marked in groups of individuals (e.g., soldiers and athletes), who were under stress owing to physical activities or extreme cold environment compared to physically unstressed individuals. Also, relatively higher dose of vitamin C (e.g., 0.5g/day) effectively reduced the incidence but not the severity or duration of colds when compared with lower dose (e.g., 0.05g/day), which was ineffective for the effects under study. A few other studies showed at best a slight reduction in the severity of cold. Taking together the results of over 30 experimental clinical studies, Pauling's claims of benefit of high dose vitamin C for cold were not validated. Additional well-controlled clinical trials are required to validate the significance of high dose vitamin C in affording protection against common cold.

C. Other Applications of Vitamin C.

(i)   Vitamin C plays a vital role in collagen biosynthesis and extracellular matrix formation, and it has been found to be beneficial in patients with osteogenesis imperfecta.

(ii) Under conditions of experimentally-induced stress and in response to histamine, rats produce high amounts of vitamin C. It has been suggested that vitamin C may help in countering stress involving histamine release. In fact, it has been suggested that this anti-histamine effect of vitamin C might account for the mild beneficial effect of this vitamin against common cold in some cases.

(iii) Vitamin C has been used in combination with certain anti-psychotic medications such as haloperidol to enhance the efficacy of that drug in the treatment of acute phencyclidine psychosis.

(iv) Vitamin C has also been used in the treatment of severe burn injury as an adjunct to conventional treatment.

IX. ADVERSE EFFECTS

Vitamin C is believed to be free from adverse effects in a wide dose range per day. The tolerable upper level (UL) of intake of vitamin C from diet for adults is 2 g/day. A higher dose may lead to diarrhea and other gastrointestinal symptoms (e.g., nausea, abdominal cramps) in susceptible individuals, particularly following rapid consumption of vitamin C. However, these symptoms are easily reversible upon dose reduction or discontinuation of the vitamin.

A. Drug Metabolism and Drug Interactions

Certain drugs can lower vitamin C levels and thereby necessitate an increase in the requirement for vitamin C during that period. For example, aspirin and contraceptives containing estrogen can lower vitamin C levels in the blood and leukocytes. Certain proton pump inhibitors (e.g., Omeprazole) used as antacids may reduce the bioavailability of vitamin C. Another type of drug interaction of vitamin C involves a reduction in the efficacy of a pharmacological agent, for example, Warfarin (Coumadin), an anti-coagulant. In a susceptible individual, the dose of Warfarin may have to be increased to overcome the interference by vitamin C. The vitamin may also interfere with Bortezomib used for the treatment of multiple myeloma. In addition, vitamin C can increase the metabolic clearance of Antipyrine.

In experimental systems, vitamin C can alter the activity of the enzymes of the hepatic mixed function oxidase system and impair the metabolism of many drugs. Also, vitamin C can interfere with the interpretation of laboratory results, for example, the levels of serum bilirubin or creatinine, and tests for occult blood. The use of vitamin C in combination with other anti-oxidants has been implicated but not proven in reducing the efficacy of certain cholesterol-lowering drugs such as statins, which are inhibitors of 3-hydroxy-3-methyl-glutaryl-CoA (HMG-CoA) reductase.

A few possible contraindications of high dose vitamin C intake include adverse effects in individuals suffering from iron overload conditions because vitamin C can increase iron absorption, and in those with glucose-6-phosphate dehydrogenase deficiency as vitamin C may aggravate hemolysis.

In the case of high dose vitamin C therapy, a sudden reduction in the dose of vitamin C may result in a "rebound effect" because of continued activity of enzymes that convert vitamin C into other metabolites. Accordingly, individuals taking high dose vitamin C are advised to gradually decrease the dose over several days.

B. Kidney Stones

There is some concern about increased risk of kidney stone formation in individuals given high dose vitamin C supplements. High dose vitamin C intake is associated with increased oxalate excretion in the urine, which in turn represents a predisposing factor for kidney stone formation or aggravation of a pre-existing condition. Although the cause-and-effect relationship between high dose vitamin C consumption and kidney stone formation has not been validated scientifically, it is recommended that patients with an existing oxalate kidney stone, or those predisposed to its formation, should refrain from taking high doses of vitamin C.

C. Other Effects

Some unsubstantiated claims have been made for the association of high dose vitamin C intake with reduced vitamin B_{12} absorption causing vitamin B_{12} deficiency leading to pernicious anemia; reduced absorption of copper leading to copper deficiency; dental enamel erosion; and induction of abortion.

SUGGESTIONS FOR FURTHER READING

1. Hoyumpa A.M. (1986), Mechanisms of Vitamin Deficiencies in Alcoholism, *Alcohol Clin. Exp. Res.*, 10:573.

2. Rudman D. and Williams P.J. (1984), Nutrient Deficiencies During Total Parenteral Nutrition, *Nutr. Rev.*, 43:1.

3. Boxer L.A., et al. (1971), Correction of Leukocyte Function in Chédiak-Higashi Syndrome by Ascorbate, *N. Engl. J. Med.*, 295:1041.

4. Levine M. (1986), New Concepts in the Biology and Biochemistry of Ascorbic Acid, *N. Engl. J. Med.*, 314:892.

5. Montel-Hagen A., et al. (2008), Erythrocyte Glut1 Triggers Dehydroascorbic Acid Uptake in Mammals Unable to Synthesize Vitamin C. Cell, 132:1039.

6. Reuler J.B., et al. (1985), Adult Scurvy. JAMA, 253:805.

7. Chalmers T.C. (1975), Effects of Ascorbic Acid on the Common Cold, *Am J. Med.* 58:532.

8. Shin H.B., et al. (1976), Ascorbic Acid-induced Uricosuria: A Consequence of Megavitamin Therapy, *Ann. Intern. Med.*, 84:385.

9. Mamede A.C., et al. (2011), The Role of Vitamins in Cancer: A Review, *Nutrition and Cancer*, 63:479.

10. Hemila and Herman (1995), Vitamin C and the Common Cold: A Retrospective Analysis of Chalmer's Review, *J. Am Coll. Nutr.*, 14:116.

11. Douglas R.M., et al. (2007), Cochrane Database Syst. Rev., 18:CD000980.

12. Hemila H. (1997), Vitamin C Supplementation and the Common Cold—was Linus Pauling Right or Wrong? *Int J. Vitam. Nutr. Res.*, 67:329.

13. Linus Pauling Institute, Oregon State University, Oregon, USA; website:<http://lpi.oregonstate.edu/infocenter/vitamins/vitaminC/>

14. Pauling L. (1976), Vitamin C and the Common Cold, San Francisco: WH Freeman.

15. Pauling L. (1976), Vitamin C, the Common Cold and the Flu. San Francisco: WH Freeman.

16. Pauling L. (1986), How to Live Longer and Feel Better, New York: WH Freeman.

49

Iron Metabolism

S.K. Sood and Usha Rusia

CONTENTS

Iron is one of the most plentiful elements on earth, yet its deficiency is probably one of the commonest nutritional disorders. Iron is present in all organisms and in all cells. It does not exist in the free state, but is always present in organic combination, usually with proteins. It is a transition metal having more than one relatively stable state—Fe^{++} (ferrous) and Fe^{+++} (ferric)—capable of undergoing a reversible valency change on reduction by organic substrate and reoxidation by O_2. It thus serves as an oxygen and electron carrier and is incorporated into redox enzymes and substances charged with the function of oxygen transport such as haemoglobin and cytochromes. The prosthetic group with which iron is attached confers varied functions to this metal. Some of the important iron-containing compounds of biological interest are listed in Table 49.1.

I. BIOLOGICALLY ACTIVE COMPOUNDS THAT CONTAIN IRON

A. Haemic Compounds

In these compounds the protoporphyrin is combined with iron to form haem (divalent iron) and haematin (trivalent iron).

1. **Haemoglobin:** Iron in haemoglobin is by far the largest compartment and constitutes about 70% of the total body iron. The haemoglobin molecule consists of four haem groups (iron-protoporphyrin complexes) attached to four polypeptide chains of a globin (Chapter 19). Iron in haemoglobin is in divalent form. Haemoglobin contains 0.34% iron (3.4 mg iron per g haemoglobin).

TABLE 49.1 Biologically Active Compounds Containing Iron

Compound	Function
A. Haem iron	
Haemoglobin	O_2 carrier
Myoglobin	
Cytochromes	Electron carrier
Peroxidase and catalase	Peroxide breakdown
B. Non-haem iron	
Transferrin	Transport of iron
Lactoferrin	Transport of iron
Ferritin and haemosiderin	Storage of iron
Flavoproteins, dehydrogenases and oxidases	Oxidizing enzymes

In iron deficiency the amount of haemoglobin is reduced but its iron content remains unaltered.

2. **Myoglobin:** Unlike haemoglobin, myoglobin has a single haem group attached to a long polypeptide chain. Its molecular weight is approximately 16,700. Its function seems to be to provide an oxygen reserve for quick utilization by the muscle during contraction. In man, myoglobin iron is not mobilized during iron deficiency for synthesis of haemoglobin. In severe iron deficiency in animals, however, the myoglobin content of the muscle is markedly reduced.

3. **Cytochromes:** These are enzymes of cellular respiration and are located in the mitochondria. Their function is transmission of electrons. This is achieved by reversible change in iron from divalent to trivalent form. Anoxia, enzyme depletion or its inhibition by compounds such as cyanide, results in defective energy production, and accumulation of intermediate metabolites and cell death.

4. **Catalases and peroxidases:** The iron in these is always in the trivalent form. These enzymes act on peroxides (e.g. H_2O_2) to liberate oxygen.

B. Non-haemic Compounds

1. **Transferrin (siderophilin):** Physiological function of transferrin is to transport iron to erythron and to other tissues. Transferrin is synthesized chiefly in the liver but lesser amounts are made in other tissues including CNS, ovary, testis and $T4^+$ subsets. Its molecular weight is 83,000. It consists of a single polypeptide chain and has two binding sites, one at each end and each capable of binding one atom of Fe^{+++}. On paper electrophoresis it migrates like $\beta1$ globulin. It has strong affinity for iron at the pH of blood. There are several genetically determined molecular variants of transferrin, but these show no functional differences. Normal plasma level in man is 240 mg per 100 ml and this can bind about 350 μg of iron. The levels of

transferrin and the iron it carries are subject to great variation in health and disease and show an inverse relationship to iron stores so that with depletion of iron stores, the transferrin levels are increased while in iron excess the levels are decreased.

2. **Lactoferrin:** Lactoferrin is an iron binding protein of milk. Besides milk it is present in neutrophils and other secretions. Its iron binding properties are similar to those of transferrin. Its precise function is unknown. It possibly protects the infant from infection, as its high affinity for iron restricts iron supply to the microbe.

3. **Ferritin and haemosiderin:** About 15% of iron is stored in the body tissues bound to the protein apoferritin in the form of ferritin and haemosiderin. Ferritin is a brown, water soluble and crystallizable protein containing about 20–24% of iron. It is maximally synthesised by hepatocytes which can also take up non-transferrin bound iron. Its molecular weight is 460,000. The protein shell consists of two distinct ferritin subunits designated H (for heavy or heart) and L (for light or liver). These are functionally different as H subunits contain a ferroxidase centre and are able to acquire iron more rapidly, while ferritn rich in L chains is more stable and resistant to denaturation. Electron microscopic examination has shown that ferritin consists predominantly of octachedral particles. Due to superimposition of apices these appear as tetrads. Most of the iron in the ferritin molecule is in the form of ferric oxyhydroxide. Besides its function as a storage protein, ferritin is also thought to have a role in metal detoxification.

Like many other proteins, ferritin obtained from different species and even tissues show significant heterogeneity of size and charge. Within a given tissue many 'isoferritins' can be demonstrated by isoelectric focussing procedures. Clinical significance of these variants remain uncertain.

On the other hand, haemosiderin is water insoluble, golden brown, granular complex protein containing about 25–30% iron. Electron microscopically haemosiderin contains similar tetrads as are present in ferritin. Haemosiderin is probably a complex aggregate of partially denatured and partially deproteinized ferritin. In bone marrow smears and tissue sections, haemosiderin gives a positive Prussian blue reaction. In the past, ferritn was considered exclusively to be water soluble and Prussian blue negative, while hemosiderin was waster insoluble and Prussian blue positive. However these distinctions are not reliable as ferritin can be stained with Prussian blue and can be insoluble.

Ferritin is present in most cells of the body and in the plasma, whereas haemosiderin is found predominantly in the cells of the reticuloendothelial system. Iron is readily available for synthesis of haemoglobin both from ferritin and haemosiderin whenever a need arises, although mobilization from hemosiderin occurs relatively slowly compared to the rapid iron mobilization from ferritn.

4. **Miscellaneous compounds:** There are several iron containing enzymes in nature which perform important biological functions. Iron is required as a cofactor for action of several other enzymes. The steps controlled by these enzymes such as: (i) reduction of ribonucleotides to deoxy-ribonucleotides by the enzyme ribonucleotide reductase, (ii) conversion of citrate to isocitrate in the Kreb cycle by mitochondrial aconitase and iron containing enzyme 4Fe–4S, and (iii) detoxification by cytochrome p 450, are essential for cell survival (Table 49.2).

TABLE 49.2   Miscellaneous iron containing enzymes and compounds

Compound	Biological function	Remarks
Siderochromes, Sideramines	growth factor	for certain microorganism
Sideromycines	antibiotic	
Ferroxidase	conversion of ferrous to ferric iron	Oxidase activity of plasma is due to ceruloplasmin. This is enhanced by ferrous ions which also act as substrate to the enzyme
Ferredoxin	electron transfer	takes part in photosynthesis in green plants and respiratory chain or bacteria
Aconitase	interconversion of citric, isocitric and aconitic acids	requires iron for its action
Flavin containing enzymes	electron transfer	the better known compounds are succinate dehydrogenase, cytochrome C reductase, choline dehydrogenase and α-glycerophosphate

II. IRON METABOLISM

A. Distribution

The total amount of iron in the body is distributed as follows:

Haemoglobin	1.5–3 g
Storage iron	0.6–1.5 g
Myoglobin and enzyme iron	0.3 g
Plasma iron	3–4 mg
Total	3–5 g

Physiologically body iron may be considered to be in two main compartments, viz. functional and storage iron. The functional component consists largely of the iron in circulating haemoglobin and smaller amounts in body tissue myoglobin and haem and non-haem enzymes. The storage compartment exists as ferritin and haemosiderin mainly in liver, spleen and bone marrow. Plasma iron is also in exchange with interstitial and intra-cellular compartments. The iron in these compartments is generally referred as "labile iron pool" and is estimated to be in the order of 80–90 mg. Here the iron may stay briefly on the cell membrane before its incorporation into haem or storage compounds (Figure 49.1). Indians have lower iron stores and higher incidence of iron deficiency anemia than other population groups.

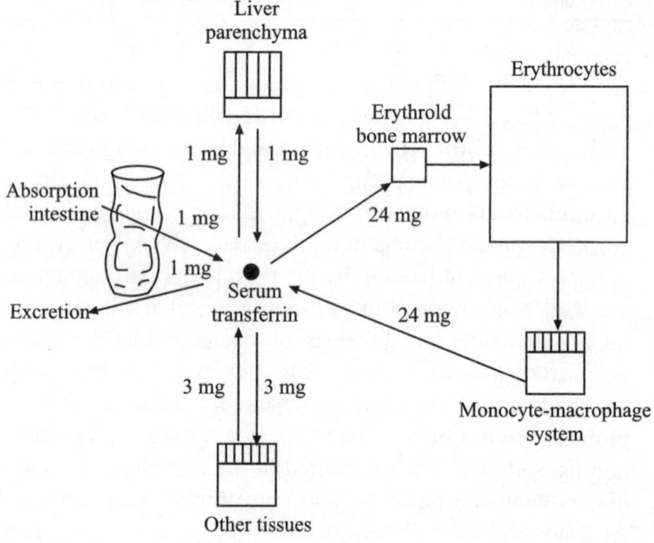

Figure 49.1   Iron metabolism in man. The importance and sizes of the different iron pools is depicted by the relative sizes of the boxes. Hatched area represents storage iron.

B. Absorption

Under physiological conditions iron balance is regulated through control of its absorption. Absorbed iron is highly conserved by the organism as no specific mechanism exists for its excretion except for obligatory losses through exfoliation of gastrointestinal mucosal cells and smaller losses through bile, urine and skin. These losses of 0.5–2 mg per day are balanced by absorption of an equivalent amount from diet. Absorption mainly occurs in the duodenum and the proximal jejunum where key proteins of iron absorption are expressed maximally. The amount absorbed depends principally on the body needs. As a means to circumvent iron starvation and toxicity, cells have acquired specific regulatory controls during evolution which allows them to adopt to changes in extracellular iron levels and to internal physiological changes like cell proliferation and differentiation. Dietary iron absorption across the apical membrane is regulated by diet (dietary regulator), iron stores

(stores regulator) as well as factors such as hypoxia, iron deficiency and hemolytic anemia. It is also influenced greatly by a third regulator, the erythropoetic regulator which is independent of the cellular iron levels or daily iron requirements and seems to be a more potent stimulator than iron deficiency.

Under normal physiological conditions, a small fraction of the iron in the lumen enters the mucosal cell and a portion of this is transported across the cell to the circulation and the rest is held in the cell as ferritin. In the normal regenerative cycle of the intestinal epithelium, the mucosal cell travels to the tip of the villus and is desquamated. The iron trapped as ferritin in the enterocytes is thus discarded. In iron deficiency, the fraction of iron entering the intestinal epithelial cell is greater and almost all of this is transported to the circulation, practically none being retained as ferritin. Conversely, in iron replete subjects more of the absorbed iron is retained as ferritin is greater and the delivery of iron to the plasma is reduced.

Depending upon the body needs, the intestinal mucosa regulates the absorption of iron; an increase in absorption occurs in children, adolescent girls, during pregnancy, and also when erythropoiesis is stimulated by anoxia of high altitude, haemolytic anaemia and bleeding. A decrease in absorption occurs when iron reserves are high or erythropoiesis is depressed.

Absorption also depends on the form in which iron is taken, the dietary composition and intra-luminal factors. Ferrous iron is better absorbed than ferric and simultaneous administration of reducing substances such as ascorbic acid promotes its absorption. Dietary iron is present as haem or non-haem iron. The acidic pH of the gastric juice stabilizes dietary non-haem iron as Fe^{+++} preventing its precipitation. At least three pathways have been observed for the transport of iron into the cell, as there are separate uptake pathways for heme and non-heme ferrous and ferric iron. The separate methods of transport of ferric and ferrous iron have been documented by competitive inhibition studies. .

Most heme iron is derived from hemoglobin, myoglobin and other heme proteins in foods of animal origin. Heme iron is enzymatically digested free of globin in the intestinal lumen and enters the enterocytes as an intact metalloprotein. Within the enterocyte, heme is degraded by heme oxygenase and the liberated iron then follows the same pathways as those used by non-heme iron. Most dietary iron is in the ferric state however acidic pH of the stomach allows its chelation with mucin, ascorbate, histidine and fructose, so that when it enters the duodenum it is in a soluble chelated form. Most of this dietary iron is converted into ferrous iron and is absorbed into the cell by DMT-1 (divalent metal iron transporter) pathway (previously also called Nramp-2 and DCT-1 pathway). The DMT-1 protein also transports other non ferrous metals of nutritional importance such as manganese, besides ferrous iron. DMT-1 pathway is regulated by body iron stores and may also be susceptible to regulation by dietary iron. Only some ferric iron enters the enterocyte in combination with mobilferrin integrin (IMP pathway). But this transport mechanism has not been fully characterised.

Absorption of iron is better from foods derived from animal sources which is relatively unaffected by the overall composition of the diet. However other dietary constituents have a profound effect on the absorption of non-heme iron. Phytates, oxalates and phosphates present in the food form insoluble compounds with iron and therefore the iron in the food stuffs rich in these compounds is not readily absorbed. Tea and eggs also inhibit iron absorption. Protein deficiency is associated with decrease in absorption. Gastric acidity facilitates ionization of iron and hence promotes its absorption

C. Mucosal Uptake of Iron

Iron absorption from dietary sources by the villous enterocytes differs from cellular iron uptake. Firstly, the villous enterocytes do not synthesise apotransferrin and secondly they lack transferrin receptors. Therefore while at the cellular level the uptake is through the process of transferrin/TFR endocytosis, in the villous entertocytes iron uptake is through divalent metal transporter (DMT1) which functions in non transferrin bound iron uptake. As iron cannot pass through the lipid bilayer of the cells it requires this specific carrier protein (DMT1).

Iron absorption involves atleast two distinct steps (a) mucosal uptake mediated via DMT1 and (b) transfer across the basolateral membrane into the portal circulation. It is essential for the iron to be in the ferrous form for its uptake by DMT1. A duodenal cytochrome B (DCYTB) like ferrireductase is probably involved in reducing dietary Fe^{+++} to Fe^{++}. It has been shown that both DMT1 and DCYTB increase appreciably in iron deficiency anemia (IDA) to facilitate iron absorption.

Once the iron is absorbed at the mucosa it may take two routes; (a) some remains within the cell as ferritin while (b) the remainder is transported across the basolateral membrane. Thus when demands are high iron preferably passes into the portal circulation but in conditions of iron excess it is sequestered within the mucosal cells as ferritin and is lost with the shedding of the senescent enterocytes.

The export of iron across the mucosal cell requires ferroportin as the transporter. Ferroportin (also known as IREG-1, MTP-1) is expressed in enterocytes, has features of iron responsive mRNA and mediates the efflux of iron from cells. Ferroportin is also required in the maternoembryonic iron transport and in the export of iron from macrophages. The ferrous iron must be oxidized to ferric form to enable it to bind to apotransferrin. This is facilitated by hephaestin, a membrane bound ferroxidase, which closely resembles ceruloplasmin and is involved in efflux of iron from macrophages and hepatocytes (Figure 49.2a).

D. Transport and Intracellular Iron Uptake

Iron is transported bound to the iron-transport protein, transferrin. This transferrin-iron complex (Fe-TF) then binds to transferrin receptors (TFR) which are ubiquitously present on surface of all cells. The Fe-TF-TFR complex so formed is endocytosed. The release of iron within the cytosol from this endocytosed Fe-TF-TFR complex depends upon several interactive factors. One important factor is the acidification of the endocytic vesicle. An acidic pH favours dissociation of iron from the Fe-TF-TFR complex. The iron so released emerges in the cytosol bound ATP and is delivered to the mitochondrial iron receptors, while the TF-TFR is recycled (Figure 49.2b). Within the mitochondria majority of the iron is incorporated into heme and a small amount of iron is diverted to ferritin. Another related molecule TFR-2 which shares 45% homology with TFR also binds transferrin but with lower affinity. Patients carrying mutations in TFR-2 gene develop an iron overload clinically indistinguishable from familial hemochromatosis (Type 1).

It has been proposed that the amount of intracellular iron uptake is determined by the number of TFR the cell expresses, the number being inversely proportional to the amount of iron availability. A decrease in iron requirement diminishes the cell receptors while an increase in its requirement increases the receptor number, such as occurs in iron deficiency anemia.

Regulation of iron homeostasis

In the past it was suggested from mammalian models that mucosal iron absorption is regulated or modulated by the multipotent precursor stem cells in the crypts of Leiberkuhn. The cells in the crypts express transferrin receptor (TFR) and endocytic mechanism impart information about body iron stores based on plasma iron levels. Thus the crypt cells are programmed with information of the iron requirement of the body when they move up to the apex of the villi and accordingly modulate the iron absorption. However the discovery of hepcidin, an iron regulatory hormone, has added greater insight into the mechanism of iron homeostasis and the above hypothesis has been abandoned though not totally

Hepcidin was first discovered in human urine. The word hepcidin is derived from three words: *hep*atic bacteri*cid*al prote*in*. Its role as a regulator of intestinal iron absorption and in release of iron from macrophages has been established. It is a 25 amino acid peptide, synthesized predominantly in the liver and excreted in the urine. It acts by inhibiting iron efflux through ferroportin from basolateral membrane of the enterocytes as well as from hepatocytes and macrophages. Hepcidin directly binds to ferroportin causing it to be internalized and enter a lysosomal compartment to be degraded thereby preventing iron export into the plasma (Figure 49.3). Blocking of iron export by hepcidin would result in rise of intracellular iron and suppression of synthesis of DMT-1 inhibiting further iron uptake at the mucosal surface. Hepcidin thus acts as a negative regulator of intestinal iron absorption. Its synthesis is increased in iron overload, anemia of inflammation due both to infectious and non-infectious causes and sepsis. IL-6 is a potent stimulator of hepcidin synthesis and its infusion causes marked increase in hepcidin levels resulting in immediate hypoferremia which is Nature's way aimed at eliminating microbial infection. Inappropriately low levels of hepcidin are central to the pathogenesis of Familial hemochromatosis due to mutation in genes including HFE, TFR-2 and Hemojuvelin (HFE-2).

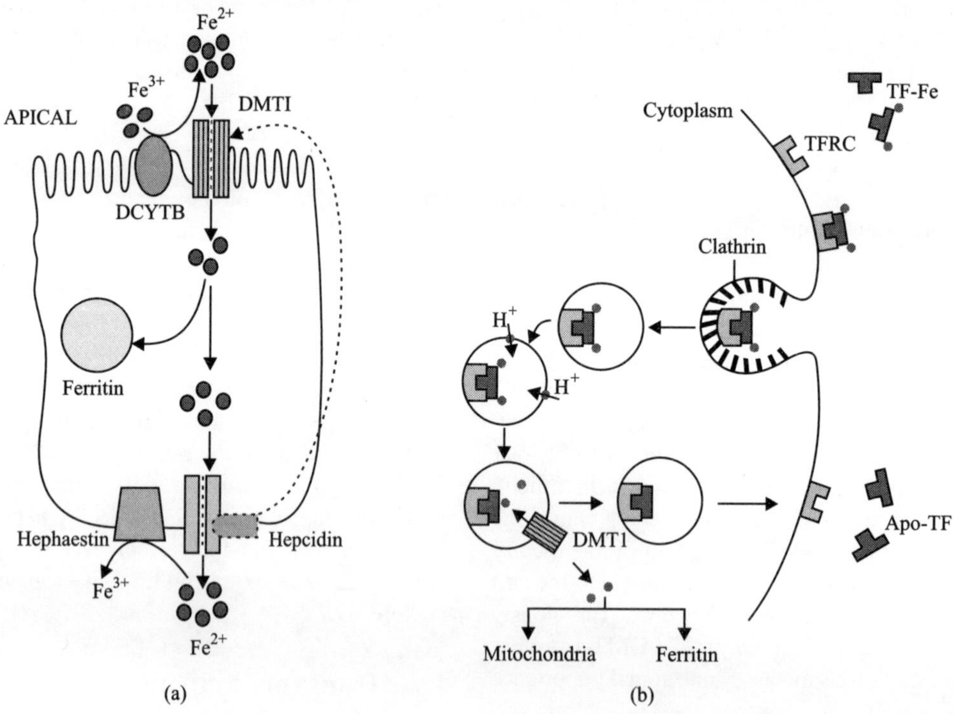

Figure 49.2 (a) Non Heme iron absorption by the absorptive enterocyte through DMT-1 and its export at the basolateral membrane by ferroportin within the entrocyte. Some iron is stored as ferritin which is lost when it is shed. Note the negative feedback of iron absorption (broken line) through the action of Hepcidin. (b) Intracellular iron uptake mediated by Fe-TF-TFR. The receptor complex enters the cell through clathrin coated pits. The iron is released from TF-TFR complex at low pH and transported by DMT-I to mitochondria. TF-TFR returns to the cell surface to participate in further rounds of iron delivery.

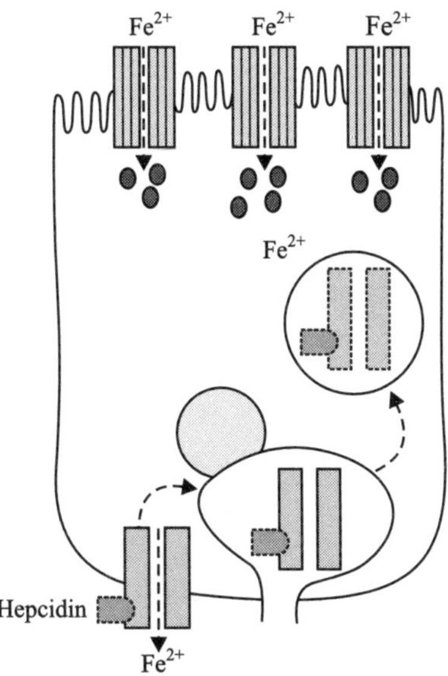

Figure 49.3 Control of iron export from basolateral membrane of entrocytes. In iron excess Hepcidin secreted by hepatocytes interacts with ferroportin, causing it to be endocytosed and degraded, thereby preventing iron export into plasma.

Molecular regulation of iron proteins: As indicated transferrin receptors and ferritin each play important role in iron metabolism in the cell. Before iron is incorporated into haemoglobin, porphyrins must be synthesized. This is accomplished within the mitochondria through the enzyme δ-amino-levulinic acid (δ-ALA) synthase.

The mechanism which coordinates protoporphyrin synthesis with the iron supply depends upon the iron-responsive elements (IRE) in the messenger RNA for apo-ferritin, transferrin receptors and δ-ALA synthase and a protein factor that binds to IRE. This factor initially called the IRE-BP (iron-responsive element binding protein) has been renamed as iron regulatory protein (IRP). There are two iron dependent regulatory proteins namely IRP-1 and IRP-2. These are capable of binding to mRNA of transferrin receptors as well as mRNA of ferritin. The manner in which IRP appears to regulate the synthesis of apo-ferritin and transferrin receptors is schematically shown in Figure 49.4. It is proposed that with a decrease in intracellular iron, the IRP binds to IRE of m-RNAs of apo-ferritin, TFR and δ-ALA synthase, thereby repressing the translation of apo-ferritin m-RNA and simultaneously increasing the translation and stability of m-RNA for the transferrin receptors and δ-ALA synthase. This increases the number of molecule

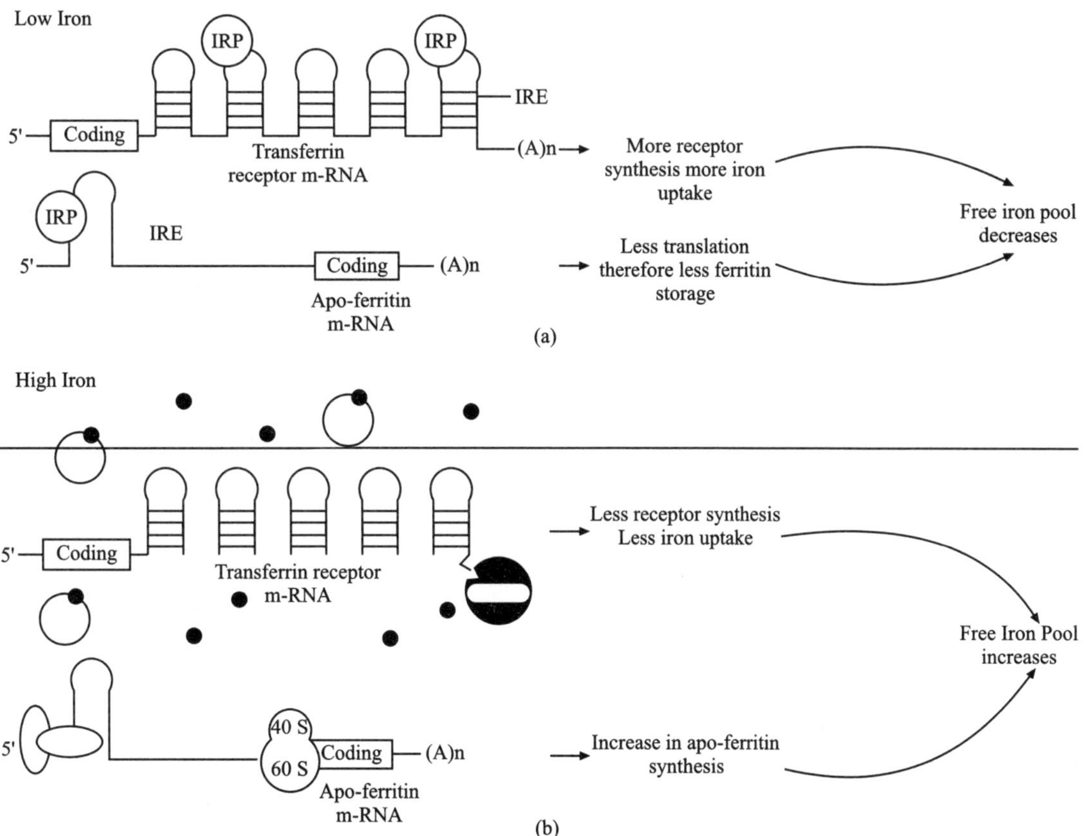

Figure 49.4 Coordinated regulation of ferritin and transferrin receptor synthesis at the mRNA level in condition of low and high cytosolic iron availability. The IRP senses low iron levels and binds to IRE of mRNA of both transferrin receptor and apoferritin. The effect is to repress the translation of apo-ferritin mRNA thereby reducing the amount of apo-ferritin formed and stabilizing and increasing the translation of mRNA for transferrin receptor so as to increase the expression of transferrin receptors on cell surface. High cytosolic iron leads to the opposite effect.

of TFRs on cell membrane while decreasing the ferritin trapping of iron that enters the cell. Conversely, when there is abundance of iron in the cytoplasm the IRP is displaced from the IRE of both m-RNA. This results in derepression of apo-ferritin synthesis and destabilisation and degradation of transferrin receptor m-RNA causing an increase in ferritin within the cytosol and decrease in the number of transferrin receptors on cell surface.

E. Excretion

Physiological excretion of iron is minimal. The normal routes of excretion are urine, bile, faeces, cellular desquamation, and sweat. Daily excretion in an adult male is estimated to be upto 1 mg. In women of reproductive age additional loss through menstruation averages to 1 mg per day. These physiological losses do not seem to contribute much to pathogenesis of IDA. It is the pathological blood loss which is the most important aetiological factor in production of iron deficiency.

F. Storage

Stores of iron are maintained chiefly in the liver, spleen and bone marrow in the form of ferritin and haemosiderin which can be measured by histochemical techniques. Evaluation of iron stores in bone marrow or liver is an important tool for diagnosing deficiency, adequacy or excess of iron.

The foetus acquires most of its iron during the last trimester. In a premature newborn, the blood volume and iron reserves are inadequate. Soon after birth, physiological haemolysis occurs. It releases iron which is deposited in the stores. Milk being a poor source of iron, enough iron is not ingested to meet requirements of growth and increase in blood volume. In 3–6 months the infant utilizes its reserves at birth and develops iron deficiency unless the diet is supplemented with iron.

It has been shown that the iron stores in Indians are much less than in Western populations. Indians, therefore, are more vulnerable to the development of IDA. Women have lower stores than men, and therefore, develop anaemia much more frequently than men especially during reproductive age group.

Iron stores are increased in haemochromatosis, severe haemolytic anaemias, aplastic anaemia and in persons receiving multiple blood transfusions, prolonged oral or parenteral iron therapy.

In the bone marrow of normal subjects 40 to 50% erythroblasts contain ferritin granules which give a positive Prussian blue reaction. These cells are described as 'sideroblasts'. These are markedly reduced in IDA but are increased in megaloblastic anaemia, haemolytic anaemia, sideroblastic and refractory anaemias.

F. Kinetics

The plasma iron pool is only 3–4 mg, constituting about 0.1% of the total body iron. However, it is a very dynamic compartment and almost all the internal exchange of iron from one compartment to another occurs through plasma. Approximately 35 mg of iron enters and leaves the plasma iron pool each day. About 80% of this is incorporated into the red cells as haemoglobin. The same amount is contributed to the plasma from the breakdown of red cells (catabolism of haemoglobin). Minor exchanges of plasma iron occur with the storage and tissue iron. The understanding of iron kinetics has been achieved, through use of radioactive isotopes of iron. Intravenously injected radioiron (Fe^{59}), bound to transferrin is cleared from the circulation at a constant rate so that the fall in plasma radioactivity reaches to one half of its initial value. ($T\frac{1}{2}$) in about 60 to 120 minutes. The rate of clearance is markedly altered in several clinical disorders. The clearance is faster in IDA, polycythaemia vera, thalassaemia (conditions in which erythropoiesis is acclerated). The iron clearance is slower when erythropoiesis is depressed as is witnessed in aplastic anaemia. The plasma iron turn over rate (PITR) which, in general, is a measure of erythropoiesis can be calculated by the formula:

$$PITR = \frac{0.693 \times \text{Serum iron} \times \text{Plasma volume}}{Fe^{59}T^{1/2}}$$

If the radioactivity of the red cells is determined by serial measurements every day after injection of Fe^{59} normally a plateau is reached in 10–14 days. From the red cells mass, the amount of radioactivity injected and the radioactivity present in the red cells, one can measure the fraction of iron incorporated into haemoglobin. The iron utilization is then expressed as a percentage of the administered dose. In a normal situation, about 70 to 85% of the administered dose is incorporated into red cells in 10 to 14 days. The radioiron utilization is increased in IDA and polycythaemia vera, and is reduced in thalassaemia major, megaloblastic anaemia, anaemia of renal disease, infection, and aplastic anaemia. In certain conditions, e.g., thalassaemia major, erythropoiesis is accelerated (as evidenced by rapid iron clearance and greater initial uptake of radioactivity by the bone marrow), but the iron utilization is decreased. This is called ineffective erythropoiesis and is explained on the basis of either inadequate synthesis of haemoglobin, decreased rate of cell proliferation or intramedullary haemolysis. On the other hand in aplastic anaemia there is failure of erythropoiesis, iron clearance is slower, bone marrow uptake is poor and iron utilization is markedly depressed.

Iron balance: Iron absorption and excretion maintain a balance. A negative iron balance is set up when body requirements exceed its availability. If prolonged it results in progressive iron deficiency and anaemia. Excessive assimilation of iron results in iron overload.

III. REQUIREMENTS

Daily iron requirements in different groups are tabulated below. These are increased during periods of active growth.

Group		Daily requirement (mg)*
Infants		0.5
Children		1.0
Adolescent: Boys		1.8
Girls		2.4
Menstruating women		2.8
Pregnant women	1st half of pregnancy	0.8
	2nd half of pregnancy	3.0
Lactating women		2.4
Normal men and non-menstruating women		0.9

*If 25% of the calories are provided from foodstuffs of animal origin, the iron absorption is about 20%. If only 10% of the calories are from animal foods, the absorption of iron is 10%. In population groups whose intake of foodstuff of animal origin is negligible, there is need to ingest much larger amounts of iron to get the prescribed daily requirements.

IV. LABORATORY MEASUREMENTS OF IRON STATUS

In the plasma, the iron is bound to transferrin which is only partially saturated. The unsaturated transferrin is measured as unsaturated iron binding capacity (UIBC) of plasma or serum. The total iron binding capacity (TIBC) corresponds to total iron that the transferrin can bind when fully saturated and is the sum of plasma/serum iron (PI/SI) and UIBC. The transferrin saturation (TS) is derived by the formula SI/TIBC multiplied by 100. The normal levels are as follows:

S Iron 70–180 µg per 100 ml (12.5–32.2 µmol/1)
TIBC 300–400 µg per 100 ml (53.7–71.6 µmol/1)
TS   20–60%

SI is subjected to diurnal variation, the levels being maximum in the morning and falling progressively through the day.

The levels of SI, TIBC and TS are subjected to great pathophysiological variation. SI is decreased in iron deficiency anaemia (IDA), pregnancy, infection and protein deficiency. It is elevated in haemochromatosis and in some cases of haemolytic anaemias and megaloblastic anaemia. The TIBC is increased in IDA and pregnancy. It is decreased in haemochromatosis, megaloblastic anaemia, haemolytic anaemia, protein deficiency and infection. SI below 50 µg per 100 ml and TS below 15% is highly suggestive of IDA. In haemochromatosis the TS is close to 100 per cent.

Small amounts of ferritin about 100 µg/ml are present in the plasma. Plasma ferritin levels correlate well with the body iron stores and are reduced in iron deficiency anaemia and elevated in situations of iron overload. Serum ferritin values of <12 ng/ml is a definitive indicator of iron deficiency anaemia. Direct histological examination of bone marrow aspirate or bone

biopsy is a well established technique for evaluating storage iron. This being an invasive procedure has been replaced by serum ferritin estimation which shows excellent correlation with bone marrow iron in uncomplicated iron deficiency.

Serum transferrin receptors (TFR) has provided a new laboratory measurement of erythropoiesis. It can be measured by immunological methods. Serum TFR are significantly elevated in iron deficiency and haemolytic anaemia and are a sensitive index of early tissue iron deficiency. Ratio of serum TFR to log of the ferritin value (STFR/log ferritin) has been shown to be a better discriminant of iron deficiency anemia and anemia of chronic inflammation. When the ratio is >1.5 iron deficiency anemia is usually present while a ratio of <1.5 suggests impaired iron utilization due to anemia of chronic disorders.

Free erythrocyte protoporphyrin is another important parameter reflecting iron deficiency as it increases with the non-availabilty of haem-iron to bind with protoporphyrin to form haemoglobin.

A newer approach to identify iron deficiency takes advantage of the measurement of reticulocyte hemoglobin content by advanced automated Hematology analyzers.

V. IRON DEFICIENCY

Iron deficiency is the commonest cause of nutritional anaemia and is prevalent all over the world.

A. Causes of Iron Deficiency

1. **Dietary deficiency:** As estimated from chemical analysis, iron content in the Indian diet is comparable to the Western diet. The excessive amount of phytates in cereals, consumed in abundance in India, may be responsible for non-absorbability of this iron. The World Health Organisation recommends a higher daily intake of iron for those whose diet is chiefly derived from vegetable sources (see under Requirements).
2. **Lack of absorption:** This may be seen in malabsorptive syndromes.
3. **Increased demand:** This occurs during rapid growth in infancy and pregnancy.
4. **Poor stores at birth:** These are found in prematurity and twin pregnancy.
5. **Pathological blood loss:** This is the most important cause of iron deficiency; with loss of 1 g of haemoglobin 3.4 mg of iron is lost. In India, hook-worm infestation is the most important single factor responsible for blood loss. Other sources of blood loss are bleeding piles, peptic ulcer, hiatus hernia, cancer of gastrointestinal tract, chronic aspirin ingestion, oesophageal varices, ulcerative colitis, epistaxis, haematuria, haemoptysis and uterine haemorrhages. Professional blood donors in this country, very often develop iron deficiency anaemia.

B. Sequence of Events in Development of Iron Deficiency

Iron deficiency follows a negative iron balance which implies that iron loss or iron needs are more than the iron supply. The iron is gradually mobilized from the stores which eventually get exhausted and little stainable iron is demonstrable in the liver or reticuloendothelial cells in the bone marrow. This is the stage of latent iron deficiency as there is no anaemia at this time. It is associated with an increase in iron absorption, a fall in serum ferritin below 10 ng/ml and a rise in free erythrocyte protoporphyrin. If the negative iron balance continues, the iron binding capacity rises, followed by a fall in serum iron and transferrin saturation. Sideroblast count gradually falls to less than 10%. A serum iron level of less than 50 μg/100 ml and transferrin saturation of less than 15% restrict the supply of iron to the bone marrow and haemoglobin synthesis is retarded. This results in anaemia which is initially normocytic normochromic, and later microcytic hypochromic. The mean corpuscular volume (MCV), mean corpuscular haemoglobin (MCH), and mean corpuscular haemoglobin concentration (MCHC) are markedly reduced. Ferrokinetic studies show that iron is cleared at a faster rate from the plasma and almost 100% is incorporated into red cells. Bone marrow shows normoblastic micronormoblastic erythroid hyperplasia.

As iron deficiency progresses, the tissue iron respiratory enzymes are also reduced. Patients complain of lassitude, weakness, fatigue, dyspnoea on exertion and palpitation. The skin and mucosa show pallor. Finger nails become thin and flattened. In severe cases nails become concave or spoon-shaped (koilonychia).

C. Non-haematological Effects of Iron Deficiency

Effects of iron deficiency with or without anaemia have been studied in experimental animals and in man.

Iron deficiency causes reduction in myoglobin, cytochromec, flavin containing enzymes, monoamine oxidase, α1-glycerophosphate dehydrogenase and other enzymes in different tissues. Iron deficiency results in decreased work capacity and productivity and behavioural changes. Other changes of iron deficiency include impaired lymphocyte transformation, decrease in macrophage inhibition factor, reduced bactericidal activity of neutrophils, impaired DNA synthesis, increased blood and urine catecholamines, elevated levels of thyroxine (T4) and reduced concentration of tri-iodothyronine (T3).

D. Treatment of Iron Deficiency Anaemia

The treatment of iron deficiency anaemia is very rewarding. Anaemia responds to oral iron therapy. The commonly used preparations are ferrous sulphate, ferrous fumarate and ferrous gluconate. All of them are equally effective and the side effects of all three drugs are similar. In some cases where there is evidence of intolerance or when the absorption is defective, parenteral therapy is indicated. The commonly used compounds are iron dextran (which can be administered both intramuscularly and intravenously), iron sorbitex (which is given intramuscularly) and saccharated iron oxide (given intravenously). After haemoglobin returns to normal some extra iron is often given to build up the stores.

VI. IRON OVERLOAD

A. Primary

Hemochromatosis may occur as a primary disorder or secondary to excessive entry of exogenous iron into the body. Primary haemochromatosis type 1 is an autosomal recessive disorder and is characterized by an excessive intestinal absorption of iron and progressive iron loading of parenchyma of organs. The haemochromatosis gene (HFE) has been localised on the short arm of chromosome 6 in close proximity to HLA-A locus. Individuals homozygous for HFE gene mutation present at about the 4th to 5th decade with accumulation of large amounts of iron (up to 40 g). Excessive amounts of iron are deposited in the parenchymal organs like the liver, pancreas, and heart. The disease is manifested by pigmentation of skin, cirrhosis of liver, pancreatic damage with diabetes mellitus, endocrine disturbances and ultimately heart failure. The bronzing of the skin is due to excessive melanin deposition though some iron is also deposited in the corium. The serum iron is markedly elevated, and the transferrin saturation may reach 80–100%. Besides HFE gene mutation, hemochromatosis type 2a,2b 3,4 are due to mutations in Hemojuvelin gene (HJV), Hepcidin gene, TFR-2 gene and Ferroportin gene respectively resulting in iron overload which is clinically similar to hemochromatosis type 1.

B. Secondary

1. **Haemosiderosis due to excessive ingestion of iron:** The two classical examples of iron overload due to increased intake are haemochromatosis in African Bantus and a few cases of alcoholic cirrhosis.
2. **Post-transfusional haemosiderosis:** This occurs in individuals who receive repeated blood transfusion for refractory anaemias.

In secondary haemochromatosis, in the early stages the iron is deposited mainly in the reticulo-endothelial cell in the liver and the spleen as against the characteristic parenchymal deposition in idiopathic haemochromatosis.

C. Miscellaneous

Haemosiderosis of lungs without involvement of other organs may occur in idiopathic pulmonary haemosiderosis. Kidney tubules may contain large amount of iron when there is intravascular haemolysis, e.g., paroxysmal nocturnal haemoglobinuria. Siderosis may also occur in refractory anaemias, sideroblastic anaemias, haemolytic anaemias, congenital atransferraemia and congenital iron overload.

VII. IRON POISONING

Fatalities have been reported in children following accidental swallowing of large doses of oral iron preparations.

The toxicity is characterized by vomiting, abdominal pain and melaena followed in severe cases by hepatocellular damage, metabolic acidosis, coma, and death.

Powerful iron chelators have been developed. These have strong affinity for iron. The most commonly known is desferrioxamine. The drug is very useful in treatment of acute iron poisoning. Desferrioxamine (DFX) is still the most widely used iron chelator. It works best when administered as continuous infusion. Recently an orally effective iron chelator 1, 2 dimethyl 3 hydroxypyrid-4-one (L1) has been found to be highly specific for ferric iron and is able to remove iron directly from ferritin.

SUGGESTIONS FOR FURTHER READING

Bothwell T.H., Charlton R.W., Cook J.D. and Finch C.A. (1979), *Iron Metabolism in Man,* Blackwell, Oxford.

Cook J.D., Skikne B.S. and Baynes R.D. (1993), Serum Transferrin Receptors, *Ann. Rev. Med.,* 94, 63–74.

Finch C.A. (1982), *Seminars in Haematology,* Vol. 19, Clinical Aspects of Iron Deficiency and Excess, Grune and Stratton, New York.

Food and Agriculture Organization (1988), Requirements of Vitamin A., Iron, Folate and Vitamin B12, Report of a Joint FAO/WHO Expert Group, FAO, Rome.

Ganz T. (2007), Molecular Control of Iron Transporter, *J. Am Soc Nephrol.,* 18, 394–400.

Hallberg L., Harwerth H.G. and Vannotti A. (1970), *Iron Deficiency, Pathologenesis, Clinical Aspects Therapy*, Colloquia Geigy, Academic Press, London and New York.

Handin R.I., Lux S.E. and Stossel T.P. (Eds.) (1995), *Blood: Principles and Practice of Haematology,* J.B. Lippencott Company, Philadelphia.

Hershko, C. (2005), Iron Diseases, Best Practice and Research Clinical Haematology, Vol. 18, No. 2. Elsevier Science Ltd. Amsterdam, Netherlands.

Sood S.K. (1984), Nutritional Anaemia in the Developing World—Challenges in its Management and Control, In: *Progress in Clinical Medicine,* Vol. V, M.M.S. Ahuja (Ed.), Arnold Heinmann, India, 251.

Worwood M. (2002), Iron Metabolism, *Best Practice and Research Clinical Haematology*, Vol. 15, No. 2, Elsevier Science Ltd. Amsterdam, Netherlands.

50

Metabolism of Trace Elements

B.L. Jailkhani

CONTENTS

I. INTRODUCTION

The importance of minerals in animal and human nutrition has been known for over a century. The role of more abundantly occurring elements like Ca, P, Fe, Na, K, etc. has been discussed in other sections. Here, we will consider the so-called "trace elements". The "trace elements" are called thus because they occur in very small amounts. Advanced analytical techniques like mass spectrography and polarography have made the quantitation of these elements possible with precision and accuracy. One can now measure simultaneously twenty to thirty elements in a given biological sample. A number of trace elements in a wide range of concentrations are present in plant and animal tissues. About ten of these are known at present to be essential for higher animals; some are possibly essential, while others have as yet not been discovered to have any vital role (Table 50.1). The requirements, doses and tolerance levels of essential elements are decided on the basis of effects on growth, health, fertility and other relevant criteria. The values are commonly expressed as parts per million (ppm) of the dry tissue mass or the diet intake. It should, however, be noted that the availability of a mineral depends

TABLE 50.1* Commonly Occurring Trace Elements in Animals and Human Body

Category	Elements
The dietary Essentials	Fe, Zn, Cu, Mn, I, Mo, Mg, Co, Se, Cr
Suspected Essentials	Ni, F, Br, As, Va, Cd, Ba, Sr,
Non-Essentials	Al, Sb, Hg, Ge, Si, Rb, Ag, Au, Pb, Bi

*See also Table 1.1.

on its geographical location, chemical form, its absorption from GIT, and the presence of such materials in diet which may influence these.

II. MODE OF ACTION

Most of the trace elements function in body as catalysts in a variety of enzyme systems. Two types of roles are recognizable. One, as activators of enzymes, in which the metal-protein complex is rather weak and easily dissociable by dialysis. The loss of enzyme activity that results on metal removal is recovered easily by addition of the original or another metal. The other important biological active forms of the trace elements are the metalloenzymes. Table 50.2 lists some important metalloenzymes. In metalloenzymes the association of the metal with the protein is more specific and very firm. The number of metal ions per molecule of enzyme is fixed. The complex is not easily dissociable by dialysis. However, the removal of the metal by drastic treatment, e.g., with metal chelators results in an irreversible loss of enzyme activity.

TABLE 50.2   Some Metallo-Enzymes and Metallo-Proteins*

Metal	*Enzyme/Protein*
Cu	Phenol oxidase, tyrosinase, monoamine oxidase, uricase, ceruloplasmin; galactose oxidase, haemocyanin, erythrocuprein, milk copper protein
Fe	Bacterial ferredoxin, haemoglobin, myoglobin, transferrin, ferritin, cytochrome C reductase, succinic dehydrogenase, choline dehydrogenase, dihydroorotic reductase, xanthine oxidase, Catalase
Zn	Protease, carboxypeptidase, carbonic anhydrase, alkaline phosphatase, ethanol dehydrogenase, glutamic dehydrogenase, metallothionein, leucocyte zinc protein, superoxide dismutase
Mn	NADH nitroreductase; nitrite reductase
Mo	NADPH nitrite reductase, xanthine oxidase, aldehyde dehydrogenase.
Co	Transmethylases, glutamic isomerase, diol dehydrase

*Compiled from various sources.

Besides acting as enzyme activators and metalloenzymes, many metals, especially divalent ions, act by activation of substrate, e.g., Mg^{++} and Mn^{++} in the activation of ATP. Metal ions may also be essential for biological activity of hormones like insulin and for the structural integrity of other proteins, e.g., iron in heme proteins.

III.   INDIVIDUAL ELEMENTS

A.   Zinc

The total zinc content of the human body is 1.4–2.3 g, about half that of iron. It is widely distributed in body tissues. Prostate (102 µg/g), liver, muscle and kidney (50–58 µg/g) have the highest levels. In adrenals, brain, testis, lungs and spleen the concentration range is 12–24 µg. Serum contains 0.8 to 1.2 µg/ml, nails, hair and teeth contain unexchangeable though appreciable amounts of zinc. In liver and mammary gland the metal is distributed in nucleus, mitochondria and the soluble fraction. About 18% of tissue zinc is firmly bound with macromolecules like metalloenzymes and nucleic acids. This cannot be easily removed by dialysis or with EDTA.

The remainder which is exchangeable, is bound to imidazole or –SH groups of proteins.

Zinc is absorbed from small intestine, predominantly in duodenum. It is equally well absorbed as oxide, carbonate and free metal, but poorly as sulphite, or from mixed oxides of Zn. Fe and Mn, Copper, calcium and cadmium interfere with the absorption of zinc. Zinc and copper competitively block each other's absorption. Absorbed or injected zinc combines with plasma proteins, through which it gets distributed to body tissues. The pancreas, liver, kidney and spleen show a rapid uptake and turnover rate of retained zinc. In contrast the uptake is slower by bone and CNS, but the metal remains bound for a longer time.

Excretion of zinc occurs mainly in faeces, which consists mostly of unabsorbed zinc and a small amount from endogenous origin. Endogenous zinc is primarily secreted into small intestine via pancreatic juice. Very small amounts are secreted through bile or appear in urine. The normal urinary excretion of zinc is low, about 0.1–0.7 mg/day, after a daily intake of 10–15 mg. These levels rise very high in nephrosis, post alcoholic hepatic cirrhosis, in hypertension, in starvation and following administration of EDTA. Significant quantities of zinc may be lost in sweat, especially in hot weather.

Zinc deficiency syndrome as demonstrated in animal and man is characterised by growth retardation or failure, lesions of skin, its appendages and other epithelial tissue, and by impaired reproductive development and function. In animals and birds, zinc deficiency has been shown to cause severe defects in bone development. The metal plays a vital role in wound healing. In reproduction zinc has a profound role. Its deficiency affects spermatogenesis and development of accessory sex organs in the male, and in the female, every aspect from cyclicity to parturition is affected. Severe zinc deficiency during pregnancy is harmful for both the mother and the fetus. Zinc therapy was found beneficial in certain cases of atherosclerosis, a condition of diseased blood vessels affecting arterial perfusion.

Recently, zinc deficiency has been recognised as an underlying factor in persistent diarrhoea in children and in impaired immunity. In zinc deficient animals the activity of brush border enzymes is decreased with impaired Na^+ and H_2O transport. Deficiency of Zn and its supplementation in Age Related Mascular Degeneration (ARMD) is a matter of interest in recent years. Excessive ingestion may cause its toxicity.

In immune system, zinc deficiency affects cell mediated immunity, with decreased number of T-Cells, and abnormal T-helper and/or suppressor cell function, depressed delayed type hypersensitivity, *in vitro* response to mitogens, deficiency of natural killer cells and antibody dependent cytotoxicity. Zinc therapy (supplementation) restores CD3 and CD4 cells; the latter being important in antibody formation.

A fundamental role for zinc in nucleic acid metabolism has been indicated for microorganisms, plants and animals. The synthesis of nucleotide phosphates and their incorporation into nucleic acids of liver, testis, spleen, kidney and thymus was found impaired in zinc deficiency. In these cases the content of RNA and protein was low, while the activity of RNase was high. The DNA synthetic enzymes and RNA polymerases from several prokaryote and eukaryote cells are zinc metalloproteins. A recently recognised role of zinc is in DNA stability and function, as an integral part of DNA binding protein the "Zinc fingers". In a comprehensive study, Prasad and colleagues have shown decreased activities of several zinc containing and zinc dependent enzymes in kidney, testis, liver and bones of zinc deficient animals. Amongst these were the enzymes LDH, ADH, MDH and alkaline phosphatase. The lowered activities of enzymes could be due to direct effects of zinc deprivation or because of decreased synthesis due to transcriptional impairment.

B. Copper

The total body content of copper in the human is about 60–100 mg, which is next only to iron and zinc. In the newborn the values per gram tissue are higher than in adults. The serum levels are slightly higher in female (1.23 µg/ml) than in the male (1.1 µg/ml). During pregnancy the levels rise steadily and reach a peak of 2.7µg/ml at delivery. In man, about 32% of orally taken copper is absorbed, mainly from duodenum. The net absorption depends on the form of copper, the acidity of GIT and the presence of other factors in diet. Phytate, high ascorbic acid, zinc, Mo, Cd, Ag, and Hg generally impair the absorption and alter the tissue distribution of copper. Similarly, certain aminoacids and ligands which complex with copper, reduce its rate of absorption. Not much is known about the mechanism of copper absorption. It is postulated that in chicks a Cu-binding protein present in duodenal mucosa may play a role.

The element is distributed to the tissues through plasma, where it is loosely associated with albumin. The key organ in the metabolism of copper is liver. In it, the element is stored in soluble fraction (52%), nuclei (15%), mitochondria (12.4%) and microsomes (14%). The stored copper is released for incorporation into copper containing proteins and enzymes of cells, e.g., ceruloplasmin in liver and erythrocuprein in bone marrow. A high proportion of injected copper appears in faeces, consisting of unabsorbed copper and that excreted through bile. Smaller amounts are removed via urine. Negligible amounts are lost in menstrual flow.

A number of disorders have been associated with dietary deficiency of copper. These include depressed growth, bone disorders, anaemias, neonatal ataxia, impaired reproductive function, gastro-intestinal disorders, and cardiovascular defects leading to heart failure. Anaemias in Cu-deficient conditions are associated with impaired release of iron into plasma from duodenal mucosa, the reticulo-endothelial system and liver parenchyma. Transfer of iron from tissues to plasma requires oxidation of ferrous, which is catalysed by Cu-containing protein ceruloplasmin. In addition, copper deficiency impairs the incorporation of iron into haemoglobin in normoblasts.

In the etiology of bone disorders, neonatal ataxia, and cardiovascular dysfunction, the primary biochemical lesion is a drastic reduction in the activity of cytochrome oxidase and amine oxidase. A preferential depletion of brain stores of copper, compared to other tissues, leads to a deficiency of cytochrome oxidase in motor neurones and to neonatal ataxia. This nervous disorder is characterized, in animals, by incoordination of movement, demyelination of nerve fibres, and histopathological lesions in cerebral regions, brain stem and spinal cord.

The cardiac disorders in Cu-deficiency involve atrophy of myocardium with replacement fibrosis. Basically, there is derangement of elastic tissue of aorta, coronary and pulmonary arteries. Invariably the major blood vessels rupture and death ensues. The degeneration of elastic membranes is caused by lack of sufficient cross linking in elastin molecules. The cross-bridges are formed by condensation of lysine residues through ε-amino groups. This requires the Cu-containing enzyme amine oxidase, whose formation is impaired in copper deficiency. Cu/Zn imbalance is suggested in human hypertension.

Evidence of copper deficiency syndrome has been obtained in malnourished human infants, during rehabilitation phase on high calorie but low copper diet. They developed hypocuperemia, anaemia, marked neutropenia and "scurvy-like" bone changes. These cases responded promptly to copper supplementation. Hypocuperemia occurs in Kwashiorkor, where it arises due to low protein intake, in nephrosis and Wilson's disease, both of which are associated with increased urinary excretion of copper. In Wilson's disease the copper deposition in basal ganglia, liver and cornea has been noticed. Decreased urinary excretion, impaired Cu absorption and decrease in hepatic Cu content with normal RBC Copper is observed in Menkes, an X-linked genetic copper deficiency disease. Hypercuperemia is noticed in most acute and chronic infections, in leukemias, Hodgkin's disease, collagen disorders, and in myocardial infarction. In animals administration of ACTH, hydrocortisone, or conditions causing stress lead to increased levels of copper (ceruloplasmin).

C. Magnesium

The essentiality of magnesium for normal growth and development was discovered in 1926. The adult human body contains about 21 to 28 g of magnesium, about half of which is present in bones, in complex salts with Ca and P. The levels in other tissues in meq/1 are, 20 in liver and muscle, 17 in kidney, 13 in brain and 6 in RBC. The levels in serum are 1.8–2.1 meq/l, i.e. 2–2.5 mg/100 ml and in cerebrospinal fluid, 2.4–3 meq/l. Magnesium, in the form of its soluble salts is readily absorbed from small intestine. A daily intake of 200–250 mg for adults and about 150 mg for infants is adequate. During pregnancy the requirement is increased to 400 mg/day. The magnesium of the stool represents mostly the unabsorbed element. Of the ingested element about a third is excreted in urine.

Magnesium deficiency is characterized by general growth retardation, neuromuscular irritability, generalized seizures, Cardiac arrhythmias, behavioral disturbances and in chronic deficiency by death. In magnesium deficiency the basic metabolic rate is increased. Large amounts of calcium aggravate magnesium deficiency. High protein intake may cause deficiency symptoms even on an adequate Mg diet. Generally magnesium deficiency symptoms are similar to hypocalcerdemic tetany.

In diabetic patients the pretreatment serum Mg levles are very high, about 9 meq/l which rapidly fall to 0.6 meq/l on insulin therapy. During acidosis, which occurs on insulin withdrawal, large amounts of Mg are excreted. In chronic alcoholics with delirium tremens, and in condition of alcoholic cirrhosis, the serum magnesium levels are depressed. Magnesium therapy proved beneficial in patients of Mg-deficiency tetany and in certain cases of hypocalcemic vitamin D deficiency rickets.

The vital role of Mg in cellular metabolism is exercised more by way of enzyme activation, i.e. as metal-enzyme complexes rather than as metallo-enzymes. The metal seems to activate all enzyme systems which catalyze the transfer of phosphate from ATP (GTP) to a substrate, or from a phosphorylated compound to ADP. Since ATP is required in activation of acetate, formate, sulphate, in oxidative phosphorylation, in muscular contraction, in group transfers and in active transport of nutrients across cellular membranes, the important role of Mg is implicit in all these functions. In enzyme reactions involving ATP, the actual substrate is suspected to be ATP-Mg complex. In these systems the enzyme activities are optimum when Mg: ATP ratio is one or more than one. Mg may also interact directly with the enzyme protein and stimulate its activity. This role of Mg may not be specific, it can be replaced by divalent ions like Mn^{++}.

The list of enzymes activated by Mg is very large. These include alkaline phosphatase, the enzymes of carbohydrate metabolism like hexo-kinase, fructokinase, phosphofructokinase, phospho-pyruvic and acetyl transphosphorylases, and the enzymes which phosphorylate vitamins and their cofactors. The metal is also required for activation of adenyl cyclase systems and for the activity of cyclic AMP dependent kinases like those involved in phosphorylation of nucleotides and in glycogen metabolism. Mg^{++} activated RNA-polymerases and membrane-associated ATPases are widely distributed in mammalian tissues. Reduced number of antibody forming cells is reported in Mg deficiency, in animals and in culture studies. Alternate complement pathway activation and T-cell mediated cytolysis requires magnesium ion.

D. Manganese

The estimated values of manganese in an adult man of 70 kg are 12 to 20 mg. In the tissue on wet weight basis it is maximum in liver (1.68 ppm), kidney (0.98 ppm) and pancreas (1.21 ppm). In brain, testis, ovary and heart; the levels range from 0.2–0.4 ppm. Within the cells the element is mostly concentrated in mitochondria and nuclei. The capacity of liver to accumulate Mn is rather limited compared to its ability to concentrate copper and zinc. In human blood Mn is bound to β-globulins. In cells about 80% of the metal is complexed with proteins and nucleic acids, only 20% is dialysable.

About 3–4% of orally taken Mn is normally absorbed, which is rather quickly excreted, via bile, in the faeces. When the hepatic route is blocked or Mn overload occurs, the excretion mainly takes place via pancreatic juice. Excess dietary Ca and P reduce Mn availability. Exogenous corticoids, or their endogenous activation by ACTH causes redistribution of Mn in the body. A shift in the partition of the metal from liver to other parts occurs. Daily adequate dietary intake of Mn is about 0.6 mg. Mn intoxication due to occupational hazard with signs like Parkinsons disease are reported.

Experimental Mn deficiency has been developed in several animals but not so convincingly in man. The characteristic deficiency manifestations are impaired growth, skeletal abnormalities, disturbed or repressed reproductive function and neonatal ataxia. The defects in bone formation in Mn deficiency are not identical to those caused by deficiency of other elements. These are related to the role of Mn in the synthesis of organic matrix of cartilage, in which a drastic reduction in chondroitin sulphate occurs. In the biosynthesis of chondroitin sulphate, Mn is required by the enzymes involved in the formation of polysaccharides and its cross linkage with the protein.

The impairment of reproductive function in female rat depends on the severity of deficiency. These range from neonatal ataxia and infant mortality to infertility due to disruption of estrus cycle and ovulation. The ataxia in the offspring of Mn deficient animals is rather peculiar. No histological lesions develop in the brain or spinal cord, nor do any abnormal or defective enzyme patterns result, as is the case in ataxia caused by copper deficiency. Furthermore, Mn deficiency neonatal ataxia is irreversible. This ataxia is believed to be identical with a specific congenital ataxia caused by a mutant gene. In the manganese-deficient male rat, sterility and absence of libido are associated with degeneration of seminal tubules, lack of spermatogenesis and accumulation of degenerating cells in the epididymis.

The most important Mn containing metalloenzyme is pyruvate carboxylase. Mn otherwise also influences carbohydrate metabolism by affecting the peripheral utilization of glucose and its conversion to mucopolysaccharides. Pancreatic hypoplasia, with a pronounced decrease in β-cells and a diabetes-like glucose tolerance test are features of Mn deficiency syndrome. Mn has a lipotropic action on liver, stimulates the hepatic synthesis of cholesterol and fatty acid in rat and is an essential factor for conversion of mevalonic acid to squalene. The metal is required for the optimal activity of enzymes, e.g., Arginase, RNA-polymerase, hepatic and intestinal phosphatases, choline esterase and various carboxylases. Recently it was shown that Mn and other divalent ions are involved in the transport of nucleotides, and in maintenance of the structural stability of RNA and DNA.

E. Cobalt

It has been known for over 50 years that cobalt is present in plant and animal tissues, though in very low amounts. The first indication that cobalt is an essential element came in 1935 from studies on the "Wasting disease" in Australian sheep. The functional role of cobalt is due entirely to its presence in the antipernicious anaemic factor or vitamin B_{12} of which about 4% is cobalt. The element is concentrated mostly in liver, kidney and bones. The presence of cobalt in forms other than vitamin B_{12} has been reported in sheep, dogs and chickens. However, little is known about the nature of the physiological role of these forms. In animals orally taken cobalt is mainly excreted, unabsorbed, in faeces and in small amounts in urine. In man, the pattern is characterized by high absorption and predominant excretion in urine. Reference human serum value of Co is 0.18 mg/ml.

A dietary deficiency of cobalt, *per se* has not been produced in humans. Its significance and role in human nutrition, as far as is known, is confined to its presence in vitamin B_{12}. Sporadically, cobalt has been used as a nonspecific stimulant of erythropoiesis in the treatment of nephritic and infectious anaemias. Cobalt also seems to augment the therapeutic effects of iron in iron-deficiency cases. However, the amounts of Co required are too large (20–30 mg) and serious toxic effects like thyroid hyperplasia, myxedema and congestive heart failure in infants may develop.

F. Nickel

Completely convincing evidence that nickel, is an essential nutrient for plants and animals has not been obtained so far. It is however distributed widely, though in small amounts in animal tissues. Nickel is poorly absorbed from normal diet and is excreted in faeces. Ingested or injected Ni is rapidly eliminated in urine, about 60–70% in two days. Spectrographic studies have shown that the body does not significantly retain Ni nor does the element accumulate in any tissue with age, except in lungs. The normal plasma levels of Ni are 0.1–0.6 μg/g blood. In a number of patients with myocardial infarction the values were significantly higher, (0.5–2 μg). Interestingly, the nickel content of the hair of human female is higher (3.96 1 ppm) than that of the male (0.97 0.14 ppm). This sex difference is shared by Ni with Cu and Co.

The precise biological role of Ni is not known. The metal is a potent activator of several enzymes *in vitro*. These include arginase, carboxylases, acetyl CoA-synthetase, and trypsin; the activity of acid phosphatase is inhibited. Nickel may play a role in pigmentation in fish, birds, insects and animals and in structural organization of RNA. The metal has been consistently found present in RNA from varieties of organisms in concentrations much above those present in original source material. Increased incidence of lung and nasal carcinomas is reported due to industrial excessive exposure to Ni.

G. Molybdenum

It was first discovered in 1935 that molybdenum is an essential growth factor for *Azotobacter* and for nitrogen fixing bacteria, *Aspergillus niger.* It is present in low concentrations in all animal and plant tissues. The essential role of molybdenum in animal biology was evident from the discovery that the flavoprotein enzyme, xanthine oxidase, is a molybdenum metalloenzyme. Direct evidence that molybdenum is an essential nutritional factor for animals was obtained later with the use of purified diets.

The molybdenum content of animal tissues is rather low, comparable to that of Mn. Liver and kidney accumulate more Mo than other organs. The element is readily and rapidly absorbed from most diets. Its absorption, retention and route of excretion is profoundly affected by sulphate. At higher sulphate intake the content of Mo in all tissues is low, both when dietary molybdenum is high or low. Sulphate increases the urinary excretion of Mo without affecting urine volume.

Excess molybdenum intake causes molybdenosis syndrome with toxic effects. These effects include growth retardation, loss of body weight, anaemias, diarrhea and skeletal deformities. The extent to which these toxic manifestations develop depends on Mo intake relative to that of Cu and sulphate. Copper and to some extent methionine and cystine are effective in alleviating Mo toxicity in rats. Conversely, high levels of Mo impair Cu absorption and utilization of tissue Cu. In the presence of sulphate, Mo helps in the formation of insoluble copper sulphide. In hypermolybdenosis the activity of sulphide oxidase is acutely depressed, resulting in deposition of sulphide in tissues like liver. The activity of hepatic alkaline phosphatase is reduced in Mo toxicity but that of kidney and intestine is increased.

Though xanthine oxidase, the known Mo-containing enzyme, is present in human tissues, few direct studies have been conducted on the status of molybdenum in human health and disease. The mean dietary intake for the adult is estimated to be 75 to 200 μg/day. Most of this appears in urine. An adequate protein intake seems to be essential for the excretion of Mo. At a high protein consumption (2.5 g/kg) the retention of metal was low, but was increased five fold at a lower protein intake (0.7 g/kg). Epidemiological studies in Hungary and New Zealand and experiments with mice suggest a beneficial effect of Mo on incidence and severity of dental caries. In this disease Mo can augment the well established effect of F.

H. Chromium

Nearly seventy years ago chromium was first discovered to have growth promoting effects in plants and animals. The metal stimulates the synthesis of biotin in *Aerobacter aerogenes*. In the human, it is distributed widely in low and variable concentrations. The levels in most adult tissues are 0.02–0.04 ppm; the total body content is less than 6 mg. Unlike most trace elements the tissue levels of chromium decline with age. Its blood levels are 0.009–0.055 μg/g, i.e., an average of 0.027 μg/g. In plasma the metal is largely bound to transferring fraction of β-globulins. The affinity of Cr^{+++} for transferrin approaches that of Fe^{+++} and the two ions compete for the metal binding sites. Within the tissues Cr is associated with RNA in a non-dialysable form. A chromium containing nucleo-protein has been isolated from beef liver.

The absorption, retention and excretion of chromium depends on its ionic form. Orally taken trivalent Cr is poorly absorbed to the extent of 1% or less. The unabsorbed metal is excreted as insoluble complex in faeces. Cr^{6+} is better absorbed than Cr^{3+} and readily passes through erythrocyte membrane and binds to haemoglobin. Varieties of techniques have been developed in recent years to label cells and plasma proteins with radioactive Cr to determine their life span or survival time.

Chromium deficiency is characterized by impaired growth and life span, and by disturbances in glucose, lipid and protein metabolism. In rats, chromium intake of less than 0.1 ppm/day causes corneal lesions. Chromium supplementation prevents deterioration but does not cure the developed defects. Chromium is implicated in glucose uptake by tissues as a cofactor with insulin at the cellular level, presumably as ternary complex between insulin, chromium and membrane sites as "glucose tolerance factor". An impaired glucose tolerance test develops in chromium deficiency in rat and man. In patients with mild diabetes, a significant improvement in glucose tolerance test occurred during oral chromium supplementation. Chromium cannot *per se* alleviate the diabetic condition, but only acts in association with insulin. In the rat cells, chromium stimulates uptake of glucose and its utilization for lipogenesis, CO_2 formation and glycogen synthesis, only in the presence of insulin.

Chromium plays a role in cholesterol homeostasis. In general the element suppresses serum cholesterol levels and their tendency to increase with age. Diabetes in man and animals is associated with increased vascular lesions, deposition of stainable lipids and fluorescent material in aorta. In chromium-supplemented rats the incidence of these lesions was reduced. Synthesis of cholesterol and fatty acids from acetate in the liver of chromium-deficient rats is stimulated by trivalent chromium. Chromium has a stimulatory effect on insulin-mediated tissue uptake of several aminoacids and their incorporation into proteins.

The threshold concentration of chromium for toxicity in rats is rather high, about 1 mg/100 g compared to normal requirements which are 0.5–1 µg/100 g. Hexavalent chromium is more toxic than trivalent chromium. Daily dietary estimated intake of Cr in an adult human is about 0.05 to 0.2 mg.

I. Selenium

The biological significance of selenium was first recognized more for its toxic effects in animals, a condition generally called selenosis. The physiological role of selenium was evident from its prevention of liver necrosis in the rat and of exudative diathesis in chicks. These conditions are also responsive to vitamin E. Subsequently it was shown that a certain type of muscular dystrophy in lambs and calves manifested selenium deficiency which could be corrected by selenium therapy.

No clear evidence has been obtained so far on selenium deficiency or selenium toxicity in man. The element occurs in most animal tissues but is particularly high in kidney cortex, pancreas, pituitary and liver. In an autopsy examination in human, the levels of selenium per g tissue were found to be 0.44 µg in liver, 0.27 µg in skin, and 0.37 µg in muscle. From normal diets or added soluble salts selenium is rapidly absorbed mainly in duodenum. In blood, the element is apparently associated with plasma proteins. Within the cells it is found as selenocystine and selenomethionine, replacing sulphur in protein bound and non-protein forms. Selenium is excreted in faeces, urine and via exhalation.

Selenium and vitamin E play important roles in tissue respiration. Selenium may be involved in the biosynthesis of coenzyme Q (ubiquinone). Coenzyme Q is structurally related to vitamin E (α-tocopherol) and is involved in the respiratory chain. The interrelationship between vitamin E and selenium is not clear. However, tissue levels of vitamin E are reduced in selenium deficiency states. Selenium may act as a nonspecific antioxidant, providing protection against peroxidation in tissues and membranes. Vitamin E acts at the level of cell membranes, preventing the formation of toxic lipid peroxides from polyunsaturated fatty acids. Blood and plasma Se levels reflect dietary intake. Selenomethionine is major Se constituent of wheat. Se level in healthy control are reported as about 7.5 µg/dl and daily estimated intake as 0.05 to 0.2 mg. Excessive intake induces hair loss, irritability and dermatitis. Se is integral component of Glutathione Peroxidase (Gpx), an enzyme which uses glutathione and peroxides and is a component of antioxidant system. In Se deficiency Gpx decreases. In Gpx Se is present as Selenocysteine at the active site, where sulphur of cysteine is replaced by Se. Se is also required for pancreatic function which helps lipid absorption including vitamin E, a vitamin with antioxidant property. Se has synergestic antioxidant effect with vitamin E. Deficiency is reported to cause cardiomyopathy, degeneration of skeletal muscle and liver necrosis. Both humoral and cell mediated immunity was depressed in Se deficient animal and supplementation of Zn increased the enhancement of vaccine induced immunity in mice.

J. Cadmium

Cadmium, which is related to zinc, was found in high amounts in the kidney. An important role for Cd in mammalian cells was postulated by Valle and colleagues. They isolated from equine renal cortex a protein containing 5.9% Cd and 2% zinc and called it metallothionein. A similar complex was found in human kidney and liver, with proportions of Zn and Cd which could account for all the Cd in these organs. No true cadmium-metalloenzymes have been discovered, though *in vitro,* Cd activates or depresses the activity of several enzymes.

The estimated total body content of Cd is 30 mg, 10 mg of which is in kidney and 4 mg in liver. In the new born the element is nearly absent in these organs, but accumulates with age. Other tissues have normally very low levels of Cd. About 1–2% of supplemented Cd is absorbed through intestines in man. The metabolism of Cd is greatly influenced by relative uptakes of Cu and Zn. A mutual antagonism exists between Cd and Zn. In the presence of Cd the requirements for Cu and Fe are increased. Excess Cd intake causes effects similar to iron deficiency, such as hyperplastic bone marrow, and hypochromic microcytic anaemias, which may be associated with raised levels of plasma transferrin. Recent studies suggest that Cd may be a causal factor for hypertension in humans. In such patients the plasma and urinary levels of Cd may rise 40–50 times above normal. Even small doses of Cd, e.g., 0.02–0.04 mM/kg causes haemorrhagic necrosis in testis. Substances like Se, Zn, estrogens and some thiol compounds can protect the testis against Cd toxicity.

Cd is non essential and accumulates in human tissues with increase in age. Infants absorb more Cd than adults.

Cd decreases bioavailability of Fe. Decrease in both humoral (B-lymphocytes mediated) and cell mediated immunity are reported in conditions of Cd exposure and prolonged feeding. Cd levels of smokers are higher than non smokers. Increased incidence of prostatic carcinoma is reported due to prolonged exposure to Cd. Cd has long half life in the body.

K. Strontium

^{90}Sr is a potentially hazardous radioactive by product of nuclear fission. The problem of the retention and movement of Sr in the body has therefore attracted considerable attention. Strontium is poorly absorbed from normal human diets. Bone tissue appear to have a stronger affinity for the element, where it is present to the extent of 0.02%. Strontium has not been shown to be essential for the growth of animals and plants. Some studies however suggest that the element may help in the calcification of bones and teeth, and in the prevention of dental caries.

L. Fluorine

Fluorine has not been shown to be essential for animals and humans. In very small amounts the mineral plays a significant role in the development of teeth and prevention of dental caries. However, even a marginal excess causes mottling of the enamel of teeth of dental fluorosis. Fluorine toxicity, or fluorosis, in animals is accompanied by loss of appetite and body weight, gastroenteritis, muscular weakness, and respiratory and cardiac

failure. Fluoride ion affects the activity of a large number of enzymes including alkaline phosphatase, glucose-6-phosphate dehydrogenases and fatty acid oxidase. There may, however, be endemic areas where the fluorine content of water or soil may constitute a potential hazard.

In India, where fluorosis has been suggested to be an important health problem, are the States of Andhra Pradesh, Maharashtra, Tamil Nadu, Gujarat, Rajasthan, Madya Pradesh, Bihar, U.P., Punjab and Haryana. The role of fluoride in preventing osteoporosis by virtue of its inhibiting effect on bone resorption is also reported. Low calcium intake may precipitate fluorosis. Fluoride gets incorporated into hydroxyapatite to form fluoroapatite to make teeth harder and resistant to acids.

Fluoride inhibits glycolysis by inhibiting enzyme enolase and is used as preservative (NaF) *in vitro* while collecting blood sample for glucose analysis. In this reaction fluoride first reacts with phosphate to make fluorophosphate ion (inhibiting agent) which binds to Mg. Mg is essential for enolase activity. Fluoroacetate is a poison and is used as pesticide for rodent control; it inhibits enzyme aconitase and thus inhibits Kreb's-cycle. Fluorocitrate which is formed is inhibitor of phosphofructokinase, hence no ATP is formed.

M. Arsenic

The total body content of arsenic is estimated as 15 to 20 mg, with highest concentrations in skin, nails and hair. The hair content of arsenic rises rapidly with excessive intake, and is a helpful index in the diagnosis of arsenic poisoning. The average estimated dietary intake of As in Canada, USA and UK is about 0.5 to 4.2 mg/day. Burning coal and metal smelting especially copper smelting are sources of air borne As pollution. Tobacco smoking also contributes to As pollution. Long term exposure is associated with hyperpigmentation, skin cancer, leukopenia, portal hypertension, peripheral neuropathy, there is also impairment of interferon and antibody production. The biochemical lesion in As toxicity is decreased production of ATP (Substrate level) during glycolyis. Because of its structural and chemical similarities with phosphate (Pi), during glycolysis, 1-arseno-3-phosphoglycerate is formed (instead of 1,3-diphosphoglycerate), which spontaneously hydrolyses to produce heat and 3-phosphoglycerate, without ATP production.

SUGGESTIONS FOR FURTHER READING

Abdulla M., Vohora S.B, and Athar M. (Eds.) (1995), *Trace and Toxic Elements in Nutrition and Health,* Wiley Eastern Ltd., New Age International Publishers Ltd., New Delhi.

Chandra R.K. (Ed.) (1985), Trace Elements in Nutrition of Children, *Nestle Nutrition Workshop Series,* Vol. 8, Raven Press, New York.

Evans S. (1998), *Metals and Immune System,* Chapman and Hall, New York.

Hambidge K.M., Casey C.E. and Krebs N.F. (1988), Zinc, In: *Trace Elements in Human and Animal Nutrition,* 5th ed., Mertz, W. (Ed.), Academic Press, Florida, Vol. 2, 1–137.

Keen C.L., Lonnerdal B. and Hurley L.S. (1984), Manganese, In: *Biochemistry of the Essential Ultratrace Elements,* Frieden, E. (Ed.), New York, Plenum, 89–132.

Klug A. and Rhodes D. (1987), Zinc Fingers: A Novel Protein Motif for Nucleic acid Recognition, *Trends Biochem Science,* 464–469.

Milne D.B. (1999), Trace Elements, In: Tietz *Textbook of Clinical Chemistry,* Burtis, C.A. and Ashwood, E.R. (Eds.), 3rd ed., W.B. Saunders Company, Philadelphia, USA, 1029–1055.

Prasad A.S. (Ed.) (1982), *Clinical, Biochemical and Nutritional Aspects of Trace Elements,* New York, Alan R. Liss.

Prasad A.S., Cavdar B., Brewer G. and Aggett P.J. (Eds.) (1983), *Zinc Deficiency in Human Subjects,* New York, Alan R. Liss.

Spallthotz J.E., Boylan L.M. and Larsen H.S. (1990), Advances in Understanding Selenium's Role in the Immune System, *Ann. N.Y. Acad. Sci.,* 587, 123–139.

Teotia S.P.S., Teotia M. and Singh R.K. (1981), Hydro-geochemical Aspects of Endemic Skeletal Fluorosis. in *India–An Epidemiologic Study, Fluoride,* 14, 69–74.

SECTION VI

GASTROINTESTINAL TRACT AND HEPATOBILIARY SYSTEM

The gastrointestinal system is the focus of attention in this section. The first chapter covers comprehensively the physiology and biochemistry of the system. This is followed by three chapters describing the manner in which normal and abnormal functioning of some of the important organs of this system can be established. Liver and biliary functions have been discussed in Chapters 56, 57 and 59 of its sub-section B. Due emphasis has been laid on these aspects in view of the widespread disorders of the G.I. tract encountered in a tropical country like India. Malabsorption has been dealt with in a separate chapter. An important chapter on gut microbiome has been included. Finally, it has been considered appropriate to include a chapter on viruses causing hepatitis for its obvious relevance.

51

Gastrointestinal System

K.N. Sharma

CONTENTS

The gastrointestinal system is composed of a variety of structural elements and is capable of executing a multitude of diverse functions. It is concerned principally with the transfer of food from external sources to internal environment for subsequent distribution to the cells of the body. Since most of the foods taken orally consist of large masses of high molecular weight substances such as proteins and polysaccharides which cannot cross the cell membrane, the gastrointestinal system provides physical and chemical changes in the consumed foodstuffs so that complex and insoluble food substances are transformed into simpler, soluble and diffusible forms. This breakdown processing of food called *digestion,* is brought about by digestive juices secreted into the alimentary tract and are shown in Tables 51.1 and 51.2. The final products of digestion such as monosaccharides and amino acids cross the cell membrane of the intestinal tract and enter blood by a process called *absorption.* As the process of digestion and absorption is going on, the foodstuffs progress along the digestive tract aided by its movements, and resulting finally in the elimination of unabsorbed remains from the anal end.

TABLE 51.1 Enzymes of the Gastrointestinal Tract

Source	Enzyme	Inactive precursor	Chemistry	Activators	Substrate	Optim. pH	Site of action	End products
1. Salivary glands	Salivary amylase (ptyalin)		Protein: mol. wt. 50,000*	Chloride ions	Cooked starch	6.6 to 6.8	1-4 gluco-sidic bonds	Maltose, Isomaltose
2. Stomach	Pepsins I, II A & B, III	Pepsino-gens	Proteins: mol. wt. 42,000*	HCl initially later auto-catalytic	Protein	1.6 to 3.2	Endo-peptidase; amino group of aromatic a.a.	Polypeptides
3. Exocrine pancreas	Trypsin	Trypsino-gen	Protein: mol. wt. 25,000*	Entero-kinase auto-catalytic at pH 7.9	Proteins, polypep-tides	8.0	Endo-peptidase: carboxyl group of basic a.a.	Poly-peptides dipeptides
	Chymo trypsins	Chymo-tryp-sinogens	Protein: mol. wt. 25,700*	Trypsin	Proteins, polypep-tides Caesinogen	8.0	Endopep-tidase carboxyl group of aromatic a.a.	Polypeptides dipeptides Calcium caesinate
	Carboxy-peptidases (A & B)	Pro-carboxy-peptidases	Protein: mol. wt. 30,000*	Trypsin	Polypep-tides		Carboxyl terminal bond	C-terminal a.a. & poly-peptide
	Pancreatic lipase			Bile salts	Neutral fats	8.0		Glycerides, fatty acids
	Pancreatic amylase		Protein: mol. wt. 50,000*	Chloride ions	Starch	6.7 to 7.1	1-4 gluco-sidic linkage	Disacchar-ides
	Nucleases Phospho-lipase A	Prophos-pholipase			Nucleic acids Lecithin			Polynucleotides Lysolecithin
4. Duodenal mucosa	Entero-peptidase (enterokinase)				Trypsino-gen	5.2 to 6.0	Cleaves terminal hexapeptide	Trypsin
5. Intestinal mucosa	Amino-peptidase				Polypep-tides	8.0	Exopeptidase N terminal bond	N terminal a.a. & poly-peptide a.a.
	Dipeptidases Tripeptidases Disaccharidases				Dipeptides Tripeptides Disaccharides			a.a., dipeptides Monosaccharides
	Maltase				Maltose		1-4 glucoside linkage	Glucose
	Lactase				Lactose		1-4 galacto-side linkage	Glucose, galactose
	Sucrase (invertase)				Sucrose		Glucoside linkage	Glucose, fructose
	Isomaltase				—limit dextrins		1-6 glucosidic linkage\	Glucose
	Intestinal lipase				Monogly-cerides			Glycerol & fatty acids
	Nucleases (nucleotidases, nucleosidases)				Nucleic acids			Purine, pyrimidine bases, pentoses

*Approximate mol. wt.   a.a. = amino acid.

TABLE 51.2   Daily Secretions of the Gastrointestinal Tract in Man

Secretion	Volume (ml)
Saliva	500–1500
Gastric juice	1200–2000
Pancreatic juice	800–1200
Bile	500–1000
Brunner's glands secretions	(?) 50
Succus entericus	1000–3000
Large intestinal secretions	60

The three major aspects of digestive, propulsive and absorptive functions of the gastrointestinal tract are highly integrated series of activities whose co-ordination depends on nervous and hormonal control. The alimentary tract receives extensive para-sympathetic and sympathetic innervation, in addition to the elaborate nervous organisation of intramural plexus distributed in the gut wall (Figure 51.1). A considerable number of afferent nerve fibres, arising from different layers of the gut wall either terminate in the intramural plexus or travel via vagus and sympathetic nerves to various parts of the central nervous system. These afferent endings are excited by the physical and chemical properties of food present in the lumen as well as by the tonic and phasic contraction state of the gut wall, and reflexly bring about control and regulation of many gastrointestinal functions. Also, the hormones of the alimentary tract (Table 51.3) help to regulate the volume and character of the secretions as well as influence gut motility. There is a remarkable margin of safety provided by duplication of function by different parts of the alimentary tract. This may at least partially offset the derangement but not lead to the total abolition of digestive and absorptive functions initiated by disease.

In the following sections gastrointestinal motility will be described followed by a similar description of gastrointestinal secretions beginning from the oral cavity. Absorption has been dealt with elsewhere in the book (Chapter 49) and will not be described here.

I. MOVEMENTS OF THE GASTROINTESTINAL TRACT

The gastrointestinal tract is adapted to exhibit various types of movements, whose function is primarily concerned in propelling the ingested food to the areas specialized to carry out digestive and absorptive, function. These movements are also involved in the process of mechanical break up of food with the digestive juices, and expulsion of the unwanted residue.

A. Mechanical Aspects of Food Ingestion

Mastication and deglutition (swallowing) are the two major activities concerned with mechanical aspects of food ingestion.

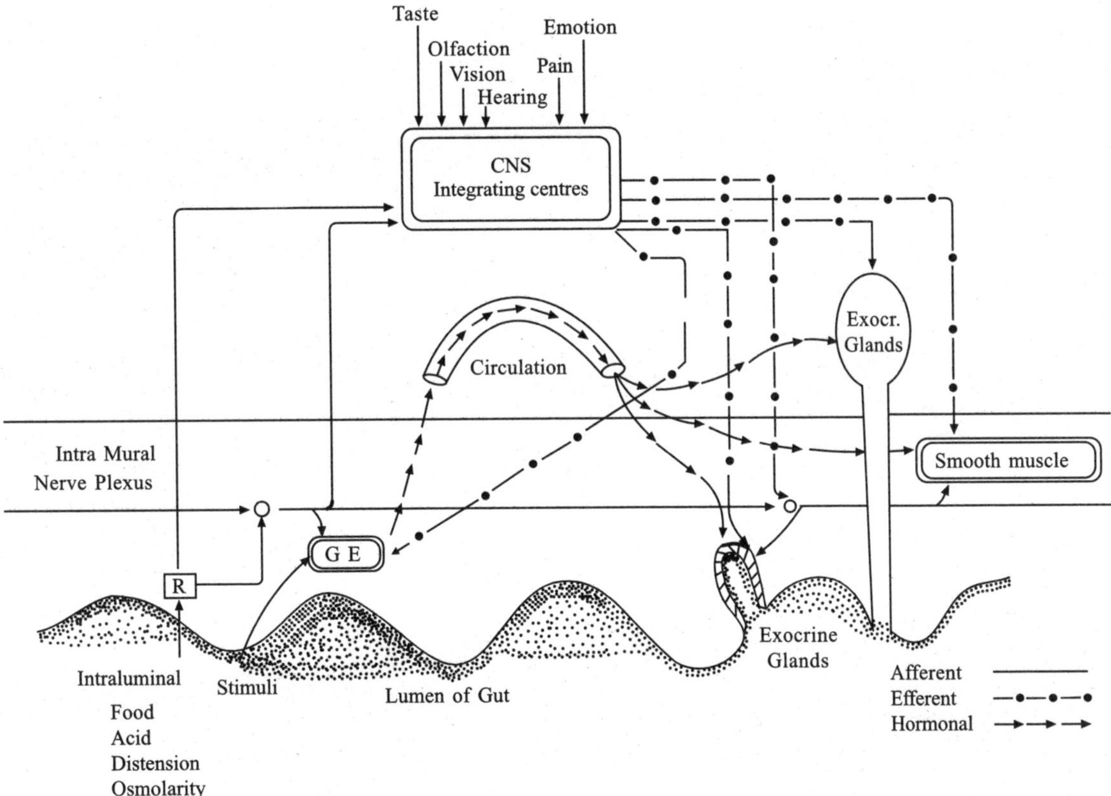

Figure 51.1   Schematic drawing to show neural and hormonal control of gastrointestinal secretions and motility. GE: gastrointestinal endocrine glands; R: receptor; CNS: central nervous system.

TABLE 51.3 Hormones of the Gastrointestinal Tract

Hormone	Source	Chemistry	Physiological stimulus for release	Actions
1. Gastrins	Antral mucosa and duodenal bulb	Polypeptides consisting of G 14, G 17, G 34, and a big gastrin with 45 amino residue (in both sulfated and non-sulfated form)	Protein breakdown Products Distension of stomach Vagal stimulation acid Calcium Epinephrine	Gastric acid, pepsins and intrinsic factor stimulation Increases growth of gastric mucosa Stimulation of gastric motility Closure of gastro-oesophagal junction Stimulates glucagon secretion
2. Cholecystokinin Pancreozymin (CCK-PZ or CCK)	Mucosa of the upper small intestine and nerves of the distal part of ileum and colon	Polypeptides with 39 amino acids. Also present in various forms (structure resembles gastrins)	Secretin Products of protein digestion Soaps and fats	Secretion of enzyme rich pancreatic juice Trophic action on pancreas Contraction of gall bladder Relaxation of sphincter of Cdi Inhibits gastric emptying and prevents regurgitation of duodenal contents into the stomach Stimulates glucagon secretion
3. Secretin (First hormone to be discovered in the body by Bayliss and Starling in 1902)	Cells present in the deep seated glands, from the upper intestine	Linear polypeptide structure similar to glucagon, glicentin (GLI) VIP, G.I.P.	Acid chyme in duodenum Vagal stimulation Products or protein digestion	Copious bicarbonate rich pancreatic juice secretion Increases bile secretion (cholerectics) Decreases gastric acid secretion Causes contraction of pyloric sphincter Augments cholecystokinin action
4. Somatostatin	Gastrointestinal mucosa	14 amino acid residue with disulphide bridge	—	Inhibits gastric secretion
5. Bombesin	-do-	14 amino acid residue	—	Increases gastric secretion
6. G.I.P. (Gastric Inhibitory peptide)	Duodenal and jejunal mucosa	43 amino acid residue	—	Inhibits gastric secretion and motility Stimulates insulin secretion
7. V.I.P. (Vasoactive Intestinal Peptide)	Gland calls and nerves or the G.I. tract	28 amino acid residue	—	Stimulates intestinal secretion of water and electrolytes Inhibits gastric acid secretion Potentiates the action of acetyl choline on salivary glands
8. Enterogastrone	—	—	Fats and fatty acids	Inhibition or gastric acid secretion and motility
9. Substance P	Endocrine cells of the G.I. tract	11 amino acid residue	—	Stimulates intestinal motility
10. Motilin	Duodenal mucosa	22 amino acid residue	—	Stimulates gastric acid secretion
11. Glucagon (Intestinal glucagon)	'A' cells of the mucosa or stomach and duodenum	29 amino acid residue	—	Plays a role in hyperglycemia of diabetes Functions in G.I.T. are not known
12. Glicentin (Glucagon like immunoreactivity G.L.I.)	-do-	Larger molecule than glucagon	—	Exact mode of action in G.I.T. is not known
13. Villikinin*	—	—	Acid chyme in the intestine	Movements of the villi
14. Enterocrinin*	—	—	Mechanical distension of small intestine and presence of food	Increases secretion of succus entericus

*These hormones have not been isolated so far.

1. Mastication or the chewing of food occurs in the mouth and the teeth present there seem to be particularly designed to suit this purpose. The incisors provide a strong cutting action, while the posteriorly placed molars grind the ingested food. The movement of jaws to bring about apposition of the two rows of teeth is made possible by contraction of the muscles of mastication. These muscles are innervated by the trigeminal nerve. Stimulation of the pontine reticular formation, and hypothalamic and limbic areas produces reflex mastication by influencing the mastication centre in the hind-brain. The normal masticatory reflex is initiated by the presence of food in the mouth. The alternate up and down movements of the jaw are maintained by stretch reflexes of the masticatory muscles. Mastication involves not only the up and down but also the side to side movements of the jaws along with the movement of the tongue. Although they form a part of the conscious activity of a person, they tend to become automatic, reflex activity always remaining under voluntary control.

Mastication is responsible for the mechanical break up of food, especially its undigestible cellulose content. It also helps in mixing the saliva with food and thus exposing a greater area of the food to the digestive enzyme.

2. Deglutition or swallowing is a well co-ordinated complex mechanism, which permits the bolus of food to enter the oesophagus and allows it to be carried down into the stomach. The entire process occupies only a few seconds and is divided into three stages: (a) *buccal,* (b) *pharyngeal,* and (c) *oesophageal.*

(a) *The buccal stage* is under voluntary control. With the closure of the mouth, the bolus of food collects on the upper surface of the mouth, the bolus of food collects on the upper surface of the tongue and is compressed backwards against the oropharyngeal isthmus.

(b) *The pharyngeal stage* is involuntary and is controlled by neurons present in the medulla and lower pons, constituting the *deglutition or swallowing centre.* The afferents for this reflex travel in sensory division of V and IX nerve while the efferents go via V, IX, X, XI and XII cranial nerves. The soft palate is elevated and touches the posterior wall of the pharynx, closing off the nasopharynx in the process. There is an involuntary cessation of respiration, while the larynx rises and the bolus passes over the back of the tongue. The backward movement of tongue towards pharyngeal wall, bends the epiglottis to cover the laryngeal opening. Simultaneous contraction of the muscles in the neighbourhood prevents food from entering the larynx and the trachea. With the arrival of food, the cricopharyngeus muscle with a sphincter like action at the oral end of the oesophagus, relaxes for a second allowing the food bolus to enter it. The larynx and the epiglottis regain their original shape while the cricopharyngeus contracts preventing air from entering. This marks the end of the pharyngeal stage.

(c) *The oesophageal stage:* The food in the oesophagus is carried down by its propulsive movements. These are wave like contractions travelling at 4 cm/sec and developing pressures from 20 to 50 mm Hg. They are initiated reflexly by afferents travelling in the V, IX and X cranial nerves; the efferents being present in the vagus. The vagal innervation is not necessary, for high vagotomy does not abolish propulsive contractions in the oesophagus. These movements are probably initiated and controlled by the myenteric plexus. The propulsive wave of contraction in the oesophagus is preceded by a relaxation wave, which prepares the stomach for the reception of food—*receptive relaxation*—and also causes relaxation of the cardio-oesophageal junction permitting the bolus of food to enter the stomach. This functional sphincter at the cardio-oesophageal junction also prevents the reflux of food into the oesophagus.

B. Gastric Movements

The receptive relaxation coincident with the arrival of food in the stomach, is in keeping with its reservoir function. In addition, the stomach also exhibits mixing and propulsive movements that fragments food into smaller particles and mix the chyme with gastric secretions. Gastric motility also serves to empty gastric contents into the duodenum at a controlled rate. The waves of peristaltic movements originate as weak contractions in the body of the stomach and proceed towards the pyloric antrum where they become very strong. Usually, one to three waves are seen at a time. The peristaltic movements in the upper two-thirds of the stomach are weak and many of these, originating in the cardia, do not pass along unless they are very strong. These movements mix the food with the gastric secretions and help break up the stomach contents into several portions. The peristaltic contractions in the muscular pyloric antrum are very strong and they help to 'milk' the chyme into the duodenum. These latter contractions of the distal stomach act as a peristaltic pump which moves semi-solid or liquid food from the stomach into the duodenum. Further, these antral contractions move only small amounts of food into the duodenum, whereas the bulk of the contents pushed along are squeezed backwards into the stomach helping the mixing and maceration of gastric contents. These peristaltic movements constitute the *mixing and propulsive movements of the stomach.*

The peristaltic rhythm of the stomach is autonomous and is dependent on the intrinsic plexuses, mainly the myenteric plexus. The role of these autonomic plexuses are constantly modulated and modified by the vagus: the importance of the latter being demonstrated by decreased motility following vagotomy.

Gastric Emptying and its Control (Figure 51.2): The working of the peristaltic gastro-duodenal pump which regulates the rate of emptying by the stomach is influenced by the volume and quality of the gastric contents.

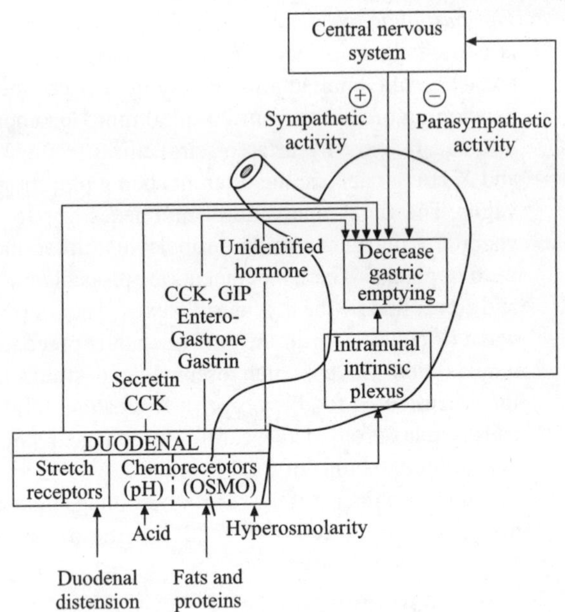

Figure 51.2 Neural and hormonal control of gastric emptying.

The effect of *volume* of gastric contents may be mediated by changes in tension, since the distension by the contents would influence the latter; and this in turn increases the peristaltic rhythm and activity. These tension changes could act directly on the muscle or through neural reflexes, the afferents and efferents of which are probably in the vagus: vagotomy annuls the influence of volume changes on the gastroduodenal peristalsis.

The *contents* of the stomach also effect the rate of gastric emptying. These are primarily feedback mechanisms initiated by the nature of the chyme delivered to the duodenum. (a) *Osmolarity* of the contents influences the gastro-duodenal pump. The peristaltic movements causing emptying are inhibited by hyperosmolar solutions while solutions of low osmolarity bring about rapid emptying. It is believed that there are osmoreceptors in the duodenum which act by inhibiting the peristaltic pump. With hyperosmolar solutions, the osmoreceptors shrink in size and actively inhibit, while hypo-osmolar solutions swell up the osmoreceptors and there is a withdrawal of a normal inhibition that is present under isotonic states. (b) *Acid chyme* in the duodenum reduces gastric motility and the peristaltic pump. Further, lowering of the pH increases this inhibition, suggesting the presence of pH receptors. Their effects are either mediated by inhibitory reflexes passing through the intrinsic plexuses or may be mediated by the release of secretin and pancreozymin-cholecystokinin. (c) *Products of digestion,* especially that of *fat* and proteins, also influence gastric emptying. The presence of fatty acids in the duodenum and jejunum activates neural mechanisms that decrease the rate of gastric emptying but the response is probably caused mainly by release of cholecystokinin from the duodenum and jejunum. Cholecystokinin stimulates contractions of the gastric antrum and constriction of the pyloric sphincter. The net effect of cholecystokinin is to decrease the rate of gastric emptying.

The presence of fatty acids in the duodenum and jejunum also elicits the release of another hormone, gastric inhibitory peptide (GIP), that decreases the rate of gastric emptying. (d) *Duodenal distension* also seems to influence gastric emptying. Duodenal stretch receptors are activated by distension and they reduce the activity of the gastro-duodenal pump by decreasing the electrical and motor activity of the pyloric antrum. (e) *The central nervous system* also directly influences both gastric motility and the emptying rate. These effects are without doubt mediated by the autonomic nervous system.

C. Small Intestinal Movements

The movements of the small intestine mix chyme with digestive secretions, bring fresh chyme into contact with the absorptive surface of the microvilli, and propel chyme toward the colon. The small intestine, particularly the duodenum and jejunum, is the site of most digestion and absorption. The jejunum and ileum are characterized by movements which are concerned with mixing of their contents and propelling them onwards. Movements termed *segmentation contractions* are responsible for the mixing and churning of the chyme. When the small intestine is examined flouroscopically, it is seen to divide into a number of approximately equal distended segments, by contraction of the muscular walls. A few seconds later, the distended segment contracts at the middle, while the contracted portion relaxes and distends. This alternate contraction and relaxation of the same area of the intestines constantly breaks up the contents into segments, and helps mixing up of the food. These rhythmic contractions are excited by distension of the intestine produced by the chyme and are due to the intrinsic muscle activity (myogenic).

The propulsion of food down the intestines is achieved by wave-like movements called *peristalsis*. These are due to contractions of the circular muscle of the gut behind the bolus in intestines and which is propelled into a preceding area of relaxation of intestinal muscle. According to Alvarez, true peristaltic movements are fast waves travelling at 2 to 25 cm/minute along the gut and occur 1 to 12 times per minute. These propelling movements are again initiated by mechanical stimulation of bowel and are dependent on the integrity of the myenteric plexus. Both these movements of the intestines decrease from the jejunum downwards to the ileum showing a gradient of activity. Vagal stimulation increases both the segmentation and the peristaltic movements, while the sympathetics inhibit them. In addition to these two types of movements, the intestinal smooth muscle also exhibits a constant *tonic contraction* on which the other movements appear to be superimposed.

The chyme moved down by the peristaltic activity is held up for a short time at the ileocolic sphincter, a thickening of the circular muscle fibres at the point where the ileum opens into the caecum. When the wave of peristalsis arrives here the sphincter opens, and permits the contents to enter the large intestine, and then quickly closes to prevent regurgitation back into the ileum.

The intestinal villi also exhibit some movements which include rapid contraction and shortening of the finger-like projection followed by a slow relaxation back to its original form. These contractions are irregular and occur approximately six times per minute, quite independent of the neighbouring villi. These movements are dependent on the intact submucosal plexus. Vagal stimulation increases movements of the *villi*, while sympathetic activity leaves the villi short, pale and motionless. It appears, a hormone released by the duodenal mucosa to the acid chyme–*villikinin*—also stimulates movements of the villi.

Intestinal Reflexes

(i) The myenteric reflex is the response of the intestine to stretch. When a bolus of material is placed in the small intestine, the intestinal wall is stretched leading to its contraction oral to the bolus and relaxation aboral to the bolus. This response, originally called the law of the intestine, helps to propel the bolus in an aboral direction, like a peristaltic wave.

(ii) The intestointestinal reflex is the relaxation of the smooth muscle of the small intestine in response to overdistension of one segment of the intestine. This reflex requires an intact extrinsic innervation.

(iii) The ileogastric reflex is the reflex decrease in gastric motility brought about by distension of the ileum.

(iv) The gastroileal reflex is the increased motility of the terminal part of the ileum and the accelerated movement of material through the ileocecal sphincter in response to distension of the stomach. The hormone gastrin may play a role in this response.

D. Movements of the Large Intestine

The chyme that enters the large intestine slowly moves and fills the colon. This filling movement appears to be apparently passive since very little peristaltic activity is observed in the large intestine.

The more important motor activity of the large intestine, especially that of the distal colon and rectum, is in the evacuation of the residue through the anal opening. This is initiated by a *mass peristaltic movement* which is responsible for the bulk of the faeces being transferred from the colon to the rectum. The filling of the rectum brings about a sensation and a desire to evacuate which under suitable conditions culminates in the actual expulsion of unwanted residue by a process called defaecation.

Reflex control of colon activity

1. Colocolonic reflex is the reflex relaxation of other parts of the colon following distension of another part of the colon. It is mediated partly by the sympathetic fibres that supply the colon.

2. Gastrocolic reflex is the increase in motility and frequency of mass movements brought about reflexly after the entry of a meal into the stomach. It is most pronounced in infancy and childhood.

Defaecation is a complex act, controlled by the nervous system, and involving not only the contraction of the distal colon and the rectum but also a simultaneous relaxation of the sphincters that guard the anal opening. This is accompanied by contraction and elevation of the muscles of the pelvic floor and also contraction of the abdominal muscles and diaphragm. The entire mechanism is a reflex, and the afferents and efferents travel in both the pelvic splanchnic and pudendal nerves. Although the act of defaecation is a reflex, with its centre in the spinal cord, it is actively inhibited by the higher centres. The release of central inhibition required for defaecation to take place, is the reason why among human beings defaecation depends very much on personal habits and available circumstances.

II. SALIVARY SECRETION

Saliva is the first of the digestive secretions encountered by the food during its passage in the gastrointestinal tract. In man and most mammals, it is the composite secretion of three pairs of salivary glands: the *parotid*, the *submandibular* and the *sublingual*. Numerous smaller glands scattered within the oral and pharyngeal cavities and secreting mucus, also contribute to a smaller extent. The parotid gland opens through the Stenson's duct at the level of the second upper molar tooth. This duct can be most easily cannulated in animals for the collection of saliva and has been extensively used for studies on conditioned reflexes. The submandibular (often called the submaxillary) gland opens by Wharton's duct into the mouth, while the sublingual glands open by means of 10 to 20 short ducts on to the floor of the mouth.

A. Histological Structure

The salivary glands are typically compound tubular or *racemose* glands (like a bunch of grapes). The secreting unit called the *acinus*, is made up of a single row of glandular cells arranged around a central cavity into which the secretions are poured. A number of adjacent acini join together by means of short, *intercalated ducts* which then join to form larger duct systems.

The glandular cells of the acini are of two kinds: *serous* and *mucous*. The serous cells possess a large centrally placed vesicular nucleus, and small (dark) granules of the enzyme precursor in the cytoplasm. The *mucous* cells have a small, dark staining nucleus which appears compressed to the lower pole of the cell by large translucent granules in the cytoplasm. These are mucinogen granules, the precursor of mucin. The acini of the parotid gland are entirely serous and they secrete a clear watery juice, rich in enzyme content. The other two glands are *mixed* glands, having both serous and mucous acini. The submandibular gland is predominantly serous and the sublingual, chiefly mucous. The intercalated ducts which appear specialised to transport inorganic ions and water (in both directions), modify the primary secretions of the acini.

B.   Nervous Innervation

The salivary glands are innervated by both sympathetic and parasympathetic nerves (Figure 51.3). The sympathetic supply arises from dorsal segments of spinal cord and relays in cervical ganglia. Postganglionic fibres from these ganglia then travel along external carotid artey to reach the salivary glands. The parasympathetic fibres have their origin in the superior and inferior salivary nuclei situated in the medulla. The fibres reach by way of the chorda tympani branch of the facial nerve and later through the lingual branch of the trigeminal nerve to the submandibular and sublingual glands. Innervation to the parotid gland is by glossopharyngeal nerve, the postganglionic fibres of which travel in the auriculotemporal branch of the trigeminal nerve.

Parasympathetic nerve stimulation and acetyl-choline administration causes salivary secretion which is blocked by atropine. The parasympathetics also appear to be vasodilator to the blood vessels of the salivary glands. This vasodilation, however, is atropine resistant and is considered to be mediated by the release of an enzyme *kallikrein* from the active gland cells which act upon an α_2 globulin in the interstitial fluid to form a vasodilator peptide called *bradykinin*. Sympathetic stimulation also causes salivary secretion, but the effects are less marked and variable.

C.   Composition of Saliva

The precise composition of human saliva is difficult to determine as it comes from a number of different sources and is influenced by a variety of factors not easily definable accurately (Table 51.4). Usually, saliva is a dilute, colourless and opalescent fluid with specific gravity ranging between 1002 to 1010 (average 1003). In man, the daily volume secreted varies from 500–1500 ml, large amounts being secreted during meal times. Submandibular glands contribute almost 70% of the volume secreted, while parotid and sublingual glands contribute about 25% and 5% respectively. Depending on its exposure to air and bacterial action, the pH of saliva varies from 5.75 to 7.05 (the pH being directly dependent on the CO_2 content of arterial blood). Saliva is normally hypotonic, its osmolarity being half to three-fourths that of plasma. Table 51.4 shows the average composition and other characteristics of mixed saliva.

The composition of saliva varies with the rate of secretion and the type of stimulus initiating it. Sodium concentration in saliva varies from 5 to 100 mEq/L, being highly dependent upon its flow rate. The potassium concentration is relatively higher than in plasma, mixed saliva usually containing 8–20 mEq/L. Salivary calcium concentration may reach upto 3–4 mEq/L and like sodium increases with the flow rate.

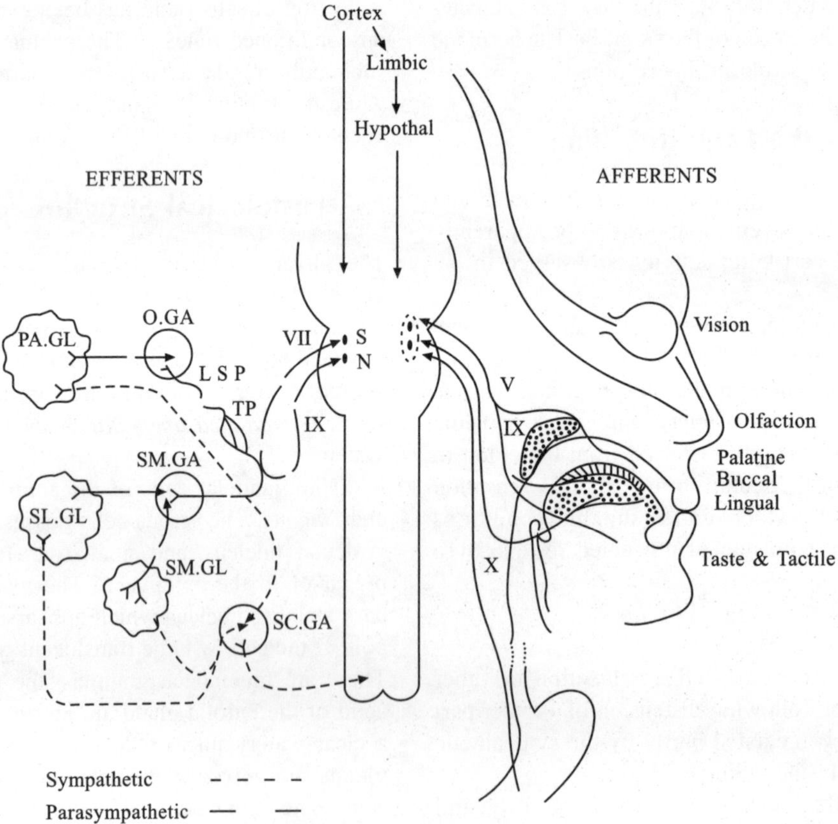

Figure 51.3   Neural regulation of salivary secretion. V, VII, IX and X a're the cranial nerves. SN: salivary nucleus; PA.GL: Parotid gland; SL.GL: sublingual gland; SM.GL.: submandibular gland; O.GA: otic ganglia; SM.GA: submaxillary ganglia; SC.GA: superior cervical ganglia; LSP: lesser superficial petrosal nerve; TP: tympanic plexus.

TABLE 51.4 Physical Characteristics and Composition of Human Saliva

Volume/day	500–1500 ml
pH	5.75–7.05
Specific gravity	1.002–1.010
Osmolarity	Hypotonic to plasma
Water	99.5%
Inorganic constituents	0.2%
Cations:	
sodium	8.5–24 mEq/L
potassium	12.5–16 mEq/L
calcium	2.3–5.5 mEq/L
Anions:	8.5–17.5 mEq/L
chloride	7.5–21 mEq/L
phosphorus (inorganic)	6–60 mEq/L
bicarbonate	0.3%
Organic constituents	
Ptyalin (α-amylase)	
Mucin	
Urea, uric acid, creatinine	
Amino acids	
ABO agglutinogens	
Kallikrein	
Lysozyme	

The salivary chloride concentration is always below that of plasma and may vary from 5–70 mEq/L. Generally its concentration is linearly related to the flow rate. The bicarbonate concentration is directly influenced by the partial pressure of CO_2 in arterial blood. The phosphate concentration is about twice that of plasma, and appears to be independent of the flow rate.

The chief organic constituents of saliva are mucin—a glycoprotein which is responsible for the viscosity and lubricating properties of saliva, and an enzyme ptyalin, which is an *α-amylase* and catalyses the breakdown of starch to maltose. This enzyme has an optimum pH of 6.8 and requires chloride ions for its activation. The other organic materials in saliva are made up of small amounts of other enzymes like carbonic anhydrase, free amino acids, urea, uric acid and creatinine. It contains a bacteriolytic enzyme called *lysozyme,* which is mucoprotein in nature. Saliva also contains soluble polysaccharides like ABO specific blood group (agglutinogens) substances, enzymes like kallikrein, and a protein component that acts as a nerve growth factor (NGF).

D. Formation of Saliva

Saliva is the active secretory product of the acinar cells which is modified during its passage along the ducts. The primary secretion of the acinar cells contain K^+, Cl^- and HCO_3^- ions, and the secretion appears to be iso-osmotic with plasma. Chloride ion is actively transported from the interstitial fluid into the acinar cell and drags along with it cations, mainly potassium, to maintain electrical neutrality. The osmotic force generated takes in water. The resultant increase in hydrostatic pressure within the cell causes ions and water to escape into the lumen, though ionic 'pumps' may be implicated. Bicarbonate is derived both from the metabolism of the cells and its transfer from plasma. The primary secretion contains little sodium. The active nature of salivary secretion is substantiated by secretory pressures which exceed systolic arterial pressure and increased oxygen utilization during secretion, ruling out the possibility of its being an ultrafiltrate of plasma. Experiments using radioactive isotopes of various ions suggest that the primary acinar secretion is modified during its passage along the intercalated ducts by the absorption of sodium and the secretion of potassium. Since potassium is reabsorbed more rapidly than the entry of sodium, the final secretion becomes hypotonic to plasma. Substances like urea are added by passive diffusion and others like iodide are actively secreted. It stands to reason that the dependence of most of the ionic constituents on the salivary flow rate is secondary to activity of the cells lining the ducts.

E. Regulation of Salivary Secretion

The human salivary glands produce only a few milliliters of saliva per hour at rest. When food is ingested, the chemical nature of food as well as its textural and other physical qualities, stimulate oral receptors sensitive to both chemical and mechanical stimuli. The afferent impulses reach salivary centres in the medulla and reflexly bring about salivary secretions. Figure 51.3 illustrates the pathways for salivary secretion. Depending on the type of food, the quality and the quantity of salivary secretion varies. Chewing of insipid paraffin wax produces upto 250 ml of saliva per hour. With dry food the saliva is thin and watery whereas meat induces a flow of thick saliva rich in mucin. A high carbohydrate diet may increase amylase fraction while a high protein diet increases mucin secretion and thus enhancing its buffering capacity. This relationship between the type of food and the composition of saliva is more definite in animals though little is known in man. In addition to stimulation of the taste buds in the mouth as well as other chemo- and mechanoreceptors in the oropharyngeal cavity, reflex salivary secretion may be caused by stimulation of oesophagus, e.g. by passage of a rubber tube, or reflux of acid gastric contents into the lower oesophagus (oesophago-salivary reflex).

The salivary secretions can also be stimulated or inhibited by impulses from higher regions of central nervous system. It is now claimed that taste and small areas of cerebral cortex or amygdala may, through the hypothalamus, influence salivary centres and bring about increase or decrease in salivary secretion. Thus the secretion is induced not only by *innate* or *unconditioned* reflex initiated by materials placed in the mouth, but also by *conditioned reflex* originating in structures outside the oral cavity. Pavlov in his classic experiments on salivary secretion in dogs, showed that ringing of a bell which by itself did not induce salivary secretion, could after appropriate pairing with food reinforcement, acquire the properties of a conditioned stimulus and produce salivary secretion. *Conditioned reflex salivation* can readily be demonstrated in animals; though it is difficult to condition man to salivate. The oft quoted example of "mouth watering" at the thought or sight of food is probably an awareness of the saliva already present in the mouth.

F. Functions of Saliva

Unlike most other secretions of the gastrointestinal tract which are primarily digestive in nature, saliva subserves a number of functions of which *digestion* appears to be of minor significance.

(i) Perhaps the most important function of saliva is its ability to moisten the food and facilitate swallowing by its *lubricating action*. Moistening of the oral mucosa by saliva aids chewing, dissolving, and mixing of the food with saliva in the mouth.

(ii) Ptyalin—the α-amylase, converts cooked starch into the disaccharide (maltose). The rapid ingestion of food leaves little time for digestion of starch to occur in the mouth. The slow mixing of food in the stomach allows sufficient time for the enzyme to act within the bolus before the low pH of the stomach inactivates the enzyme completely. Thus salivary digestion occurs mainly in the stomach.

(iii) Constant flow of saliva helps keep the mouth and the teeth clean. This *cleansing* action is further aided by the bactericidal effect of lysozymes present in saliva.

(iv) Saliva keeps the mucous membrane of mouth and lips moist and thus helps in *articulation* and *speech*.

(v) Being a good *solvent* for several substances in food, saliva helps in the perception of *taste*. Taste is a chemical sense and dissolving of food is a prerequisite to stimulation of taste buds.

(vi) Certain substances like lead, mercury and iodides as well as some alkaloids like morphine and the viruses of rabies and poliomyelitis are *excreted* into saliva. The habit of spitting saliva, not uncommon in India, provides an excretory function for saliva!

(vii) Saliva subserves a part of the regulatory mechanism concerned with *water balance* of the body. With adequate water content, there is constant and continuous flow of saliva. During conditions like dehydration, salivary secretion is suppressed and the dryness of the oral and pharyngeal mucosa arouses the *sensation of thirst*.

(viii) In animals like dog, salivation associated with panting is a means for the disposal of excess heat, and may be concerned with *heat regulation* of the body.

III. GASTRIC SECRETION

After mastication and deglutition, the food reaches the stomach, though still in large lumps. Here it is temporarily stored and acted upon by gastric juice to form a homogeneous semifluid mass, the *chyme*. Perhaps the most important function of stomach in man is to store the ingested food and control its further passage into the intestine where most of the digestion and absorption takes place.

The shape of the full stomach may vary and it is conveniently divided into an upper part called the *fundus*, the main portion of the stomach, the *body*, and the more distal area, the *pyloric antrum*. The oesophagus communicates at the proximal end called the *cardia*, and the duodenum is separated from the antrum by the *pylorus*.

A. Histological Structure

The mucosa of the stomach consists of numerous branched tubular glands arranged close to each other at right angles to the surface. Several of these branched tubular glands open into a common depression in the mucosal surface, called the *gastric pit* or *foveola*. The blind end of these glands reach the muscularis mucosa to form the *base* of the gland. The body of the gland forms its main tubular part and is joined by the neck to the *isthmus*, which opens into the gastric pit. The surface epithelium of the gastric mucosa consists of a single layer of tall columnar cells that secrete mucus.

The secreting cells present in the tubular glands are mainly of three types: *mucous cells, parietal (or oxyntic) cells, and chief (or peptic cells)*. In man, the pyloric area is mainly made up of glands with mucus secreting cells while the rest of the stomach has varying proportions of these three types of cells. Mucus cells are mainly confined to the neck region of the gland and secrete soluble mucus, which is different from the mucus of the surface epithelial cells. Parietal cells, disposed peripherally and giving nodular appearance to the gland, are the sources of HCl and can be easily distinguished by their staining characteristics. The body of the gland is composed principally of *chief cells* or *pepsinogen cells*. Special stains reveal few, scattered, *argentaffin* cells in the basal region.

B. Nervous Innervation

Both parasympathetic and sympathetic fibres innervate gastric glands. The parasympathetic efferents arise in the dorsal nucleus of the vagus nerve (in the floor of the IV ventricle) and travel in the right and left vagi. These preganglionic fibres synapse in the intramural plexuses and reach glandular cells by short postganglionic fibres. The sympathetic supply originates from the lateral horns of thoracic spinal segments (T5–10). These are preganglionic fibres and synapse either in the sympathetic ganglia or (to a larger extent) in the coeliac plexus. The postganglionic fibres accompany the arterial supply to the stomach. Both the vagus and the sympathetic nerves carry afferents from the stomach.

C. Composition of Gastric Juice

Gastric juice is the secretory product of the three types of glandular elements, and the mucus laden epithelial cells lining the stomach. Its major constituents are: water, hydrochloric acid, inorganic ions and organic matter like enzymes, mucus and mucoproteins, and certain specialized substances like intrinsic factor and the blood group ag-glutinogens. Gastric juice is sometimes considered a mixture of *parietal cells secretions,* i.e. HCl and secretion from other glandular structures, i.e. the nonparietal component. Details of volume, physical characteristics and composition of human gastric juice are depicted in Table 51.5. Both the quality and the amount of secretion can vary and depends on the rate of flow and the source of secreting elements.

TABLE 51.5   **Physical Characteristics and Composition of Gastric Juice**

Volume/day	1200–2000 ml
pH	0.9–1.2
Specific gravity	1.006–1.009
Osmolarity	Same as plasma
Water	99.45%
Inorganic constituents	0.15%
(at maximal secretory rates)	
Cations:	
Hydrogen ions	150 mEq/L
Potassium ions	10–20 mEq/L
Sodium ions	?
Anions:	
Chloride ions	
total	168 mEq/L
neutral	10–20 mEq/L
Organic constituents	0.40%
Pepsinogens I, II (A & B), III	
Gelatinase	
Elastase	
Gastric lipase (? regurgitated from	
dudodenum	
Mucin	
ABO agglutinogens	
Intrinsic factor	

1. Hydrochloric acid formation. HCl is secreted from the parietal cell-containing body and funds of the stomach and not from the pyloric antrum which is devoid of these cells. Electron microscopy of parietal cells reveals a very well developed intracellular canalicular system which shows characteristic changes during their activity. It is now well recognized that the acid formation requires large amounts of energy which is provided by enzymes, flavoproteins and cytochrome oxidases, present in substantial quantities in these cells. Carbonic anhydrase, an enzyme involved in the secretory mechanism of HCl, is also present in high concentrations in these cells. A number of suggestions have been advanced to explain the precise mechanism of HCl formation. Figure 51.4, summarizes the present concept of the mechanism of HCl formation. H^+ and HCO_3^- are derived from CO_2 produced by the metabolism of the parietal cell. H^+ is secreted into the lumen of the secretory canaliculus by the H^+, K^+ – ATPase in

exchange for K^+. Cl^- flows from the cytosol to the lumen of the canaliculus through an electrogenic Cl^- channel.

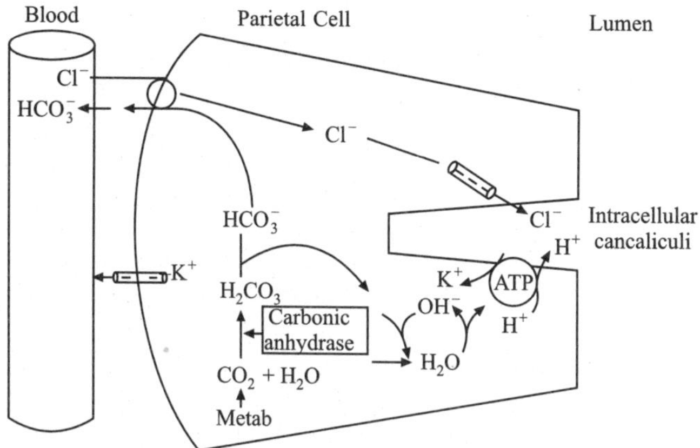

Figure 51.4 Sketch of parietal cell in stomach, showing formation of hydrochloric acid.

The secretion of H^+ into the secretory canaliculus causes the pH of the parietal cell to become alkaline. If the alkalinity could not be dissipated, H^+ secretion would cease. To dissipate the excess base, HCO_3^- and OH^- leave the cell at the basolateral membrane via the anion exchanger in exchange for Cl^-. HCO_3^- and OH^- leave the cell down their electrochemical potential gradient. The HCO_3^- that enters the blood stream is presumably responsible for the 'alkaline tide' that follows a meal. The active uptake of Cl^- at the basolateral membrane is essential for its passive flow through the Cl^- channel, in the membrane of the secretory canaliculus.

Since pure parietal or nonparietal secretions have not been separately collected, their (inorganic) *ionic* composition can only be computed. The H^+ ion concentration reaches 150 mEq/L at maximum secretory rates and the Cl^- ion concentration of the parietal cell component is 10–20 mEq/L higher than that of H^+ ions, the excess anions being balanced by the cation K^+. The ions of the nonparietal part, chiefly the anion chloride, are secreted along with the cations like sodium. The nonparietal cell chloride constitutes the *neutral chloride* and its concentration in gastric juice is an inverse function of the acidity of the secretions.

2. Enzymes: The enzymes of gastric juice are mainly the *pepsins.* They are a complex of four chemically distinct enzymes, pepsins I, II A and B, and III, the physiological significance of these various forms being largely unknown. They are produced by the peptic or chief cells and stored as *pepsinogens.* The entire stomach secretes pepsinogen I whereas pepsinogens II and III are produced only by the mucosa of the body and the fundus. These pepsinogens are transformed autocatalytically into pepsin at acidic pH. These pepsins are active proteolytic enzymes with optimum pH of action ranging from 1.5 to 3.2. At pH 5 and above, the pepsins are inactivated. Pepsins attack peptide linkages in which the amino group is contributed by aromatic amino acids like phenylalanine and tyrosine. This proteolytic activity of gastric juice is not indispensable to life, since the main end-products are only short peptide chains preparatory to further breakdown to individual amino acids by other digestive juices (pancreatic and intestinal). Gastric pepsins also help in clotting milk by converting caesinogen to caesin which later forms insoluble calcium caesinate.

Gastric juice also contains a significant amount of a weak lipolytic enzyme (*gastric lipase*) which is a tributyrase with an optimum pH 4.0 and 5.0. The exact cell of origin in the tubular glands is not known; and many consider this as regurgitated from the duodenal contents. Smaller amounts of less significant enzymes like lysozyme, gelatinase and urease, are also present in gastric juice.

3. Gastric *mucus* is made up of mucopolysaccharides and mucoproteins, and is of two types: (a) *visible mucus*, secreted by the surface epithelialcells, is thick and viscous and is present even in the resting, empty stomach; (b) *soluble mucus* is secreted by the tubular glands of pyloric, cardiac and fundic areas and is clear and transparent with a pH of 7.0 to 7.5. It has a high buffering capacity and the mucoitin-sulphuric acid in it has antipeptic activity (acts as an *antipepsin*). The surface epithelial cells also secrete watery fluid with Na^+ and Cl^- concentrations similar to plasma but with higher K^+ and HCO_3^- concentrations than in plasma. The high HCO_3^- makes the visible mucus alkaline. When food is eaten the rates of secretion of visible mucus and of HCO_3^- by the surface eipthelial cells increase.

The mucus forms a gel on the luminal surface of the mucosa. The gel protects the mucosa against damage by HCl and pepsin. The mucus and alkaline secretions are part of the gastric mucosal barrier that prevents damage to the mucosa by gastric contents.

4. *Intrinsic factor* is indispensable for the absorption of vitamin B_{12} (cyanocobalamin). It is present in the mucus fraction of the gastric juice and is probably a mucopolypeptide or a mucoprotein. In man, the fundus and cardiac portions of the stomach produce the intrinsic factor. The pepsinogen secreting cells may be its cell of origin, though (gastric intrinsic factor) antibody studies suggest that the factor originates in the parietal cell region of the glandular mucosa. The controversy remains unresolved and deficiency of this factor is associated with glandular atrophy in both these regions of the mucosa causing a megaloblastic (pernicious) anaemia. Intrinsic factor is released in response to the same stimuli that evoke secretion of gastric acid from the parietal cells.

The intrinsic factor—B_{12} complex is highly resistant to digestion. Receptors in the mucosa of the ileum bind the complex, and the B_{12} is taken by the ileal mucosal epithelial cells.

5. Gastric mucous secretions also contain the *ABO blood group agglutinogens* in 80% of the population.

D.   Regulation of Gastric Secretion

Secretions of the stomach are regulated by both neural and humoral (hormonal) mechanisms. Neural control mechanisms are mediated by the vagus nerve supplying secretory fibres directly, or by means of the intramural plexus to all three glandular elements—mucus, parietal and chief cells. The humoral mechanisms involved in the regulation of gastric secretion are either stimulatory or inhibitory and mainly influence the acid output, though to some extent pepsin secretion is also influenced. Contribution to the understanding of these regulatory mechanisms has a long romantic history in which Alexis St. Martin, a patient of Dr. William Beaumont, forms an important milestone in the early years of the 19th century. He acquired a gastric fistula as a result of a gunshot wound which did not heal, and this afforded an opportunity for making observations directly on the human stomach. Since then a number of elegant experimental methods have been developed to prepare several types of gastric pouches: innervated, partially or completely denervated, or transplanted varieties. A portion of the stomach is separated from the main cavity and communicates with the outside through an opening in the abdominal wall. Gastric juice uncontaminated with food can thus be collected. In some cases oesophageal opening is made so that food after chewing and swallowing passes outside through this opening and does not reach the stomach. This procedure is called *sham feeding*.

The sight, smell or thought of appetizing food can initiate secretion of gastric juice. When food is ingested, its presence in the mouth, the act of mastication and the taste sensation it evokes, causes gastric secretion. Gastric secretion can also occur under conditions of *sham feeding* when food in the mouth bypasses its normal course to the stomach. In animals conditioned stimuli paired with food presentation, stimulation of the anterior hypo-thalamus or limbic areas and the dorsal nucleus of the vagus, all have similar stimulatory effects on gastric secretion. These can be simulated by vagal stimulation or administration of cholinergic drugs (like acetylcholine), and are abolished by cutting both vagi or blocking it by anticholinergic drugs like atropine. These initial stages of gastric secretion which occur even before the food enters the stomach, are mediated by the vagus nerve and constitute the *cephalic or psychic phase*.

When food enters the stomach, further stimulation of gastric secretion takes place. This is both neurally and hormonally mediated and is called the *gastric phase of secretion*. Gastric distensions, caused by food in the stomach, induce reflex stimulation of secretion. The afferents and efferents of this reflex lie in the vagus nerve (vagovagal reflex). Simultaneously, the distension by food and particularly its quality (e.g., protein contents) can cause secretion of acid (also pepsin) in a completely denervated or transplanted pouch of the stomach. This secretion is mediated either by *secretagogue* substances or by the release of a hormonal substance (*gastrin*) from the mucosa of the pyloric antrum which is carried by circulation and influences primarily the secretion of acid. Figure 51.5 shows the pathways involved in the regulation of gastric secretion, and indicate that the quantity as well as the products of digestion influence reflexly the gastric secretions through neural and humoral mechanisms.

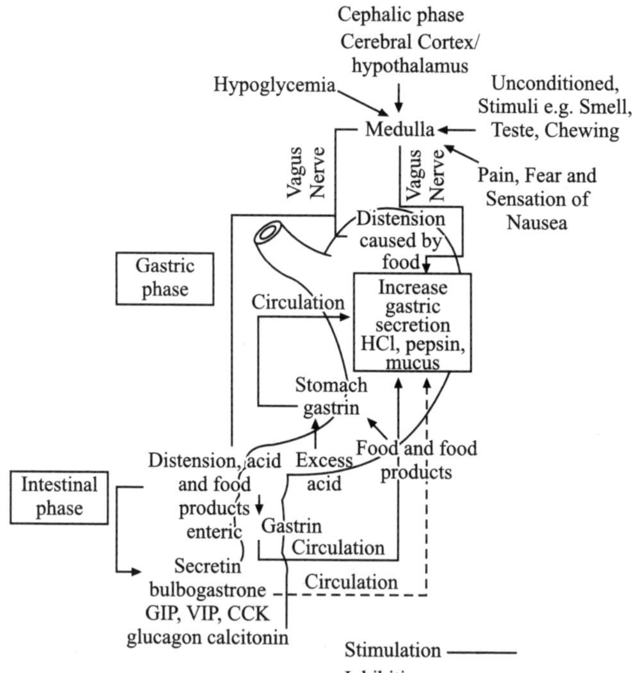

Figure 51.5 Factors and pathways involved in the regulation of gastric secretion.

The antral mucosal hormone called *gastrin*, discovered by Edkins, has been isolated in several species and is a polypeptide. The active centre of the molecule is a C-terminal tetrapeptide blocked by an amide group and possessing all the physiological actions of the entire molecule. These include stimulation of pepsin and intrinsic factor in addition to its primary acid secretory effect. Intraluminal stimuli (meat extracts, distension) in the antrum may not directly influence the gastrin secreting cell to release the hormone. Experimental evidences strongly support the possibility that these stimuli act on surface receptors which influence gastrin release mechanisms through a cholinergic link. It is also claimed that histamine, a potent acid secretory stimulant, may be the final local chemostimulator of the parietal cell, mediating even the effects of the hormone gastrin. Calcium ions and epinephrine also increase gastrin secretion.

In the gastric phase of secretion, the inter-dependence of nervous and hormonal factors on acid secretion are well established. Vagal stimulation causes gastrin release and vagal denervation diminishes the quantum of hormone released for the same intraluminal stimuli. Similarly, the vagus potentiates the acid secreting effect of the same dose of gastrin. Thus, the release of gastrin to intraluminal stimuli and the action of the released gastrin on the parietal cell are both potentiated by the intact vagus nerve.

The secretory glands of the stomach are also influenced by food products and distension of the upper small intestine. This is evident even if food bypasses the stomach following anastomosis of the oesophagus with the duodenum; and is probably mediated by a gastrin like hormone released by the duodenal mucosa and constituting the *intestinal phase* of secretion. The intestinal phase is characterized by a long latent period and low secretory activity as compared to cephalic or gastric phase, but the secretion continues for a prolonged period of 8 to 10 hours.

Certain inhibitory mechanisms, nervous and humoral, also operate to influence gastric secretion, particularly the acid output. Depressing emotions, pain, fear and sensation of nausea inhibit secretion by neural mechanisms. The inhibitory humoral mechanism is mediated by the hormones enterogastrone, GIP, VIP, secretin, glucagon and calcitonin, released by excess fat or acid in the duodenum. Excess of acid in the antrum has a 'breaking' effect on the release of gastrin.

No doubt gastric secretion is conventionally described under three phases, it really constitutes a closely correlated physiological mechanism, each phase representing a manifestation of a single complex: a coordinated neurohumoral control of a continuous gastric secretory response to feeding.

E. Functions of the Stomach

(i) The primary purpose of the stomach—a saccular dilatation in the gastrointestinal tract, is to act as a reservoir for the ingested food. This is particularly relevant to the populations of developing countries, who are used to large, bulky, infrequent feeds with plenty of starch and roughage.

(ii) In addition to storing food, the stomach acts to regulate the entry of semi-digested food into the upper intestines where most of the digestion and absorption occurs.

(iii) The movements of the stomach help to mechanically mix the food with the secretions and also breakup and macerate the stomach contents.

(iv) Special types of reflex movements like vomiting may help to eliminate harmful substances from the gut.

(v) The digestive functions of stomach are mainly proteolytic in nature and considerably differ in different species. In man these digestive functions are not indispensable to life.

(vi) Intrinsic factor produced by the gastric mucosa is necessary for the absorption of vitamin B_{12} which is required for proper haemopoiesis.

(vii) The low pH of gastric juice (due to the HCl) inhibits bacterial activity. It also helps in the digestive functions of the stomach.

(viii) There is negligible absorption in the stomach and small amounts of water and alcohol may be absorbed.

(ix) The stomach contains a variety of receptors which are considered to play an important role in the peripheral mechanism of hunger and thirst.

IV. SECRETIONS INTO THE DUODENUM

The semi-digested food, reduced to *'chyme'* in the stomach, enters the upper small intestine gradually. Depending on the quality (and also the quantity) of food, it is "milked" into the duodenum in small spurts in a complex but well organized manner. The acid chyme now confronts the secretions poured into the duodenum, viz. the potent digestive enzymes and neutralizing bicarbonate of *pancreatic juice*, alkaline, hydrotropic and emulsifying *bile*, and the mucosa protecting *secretions of the Brunner's glands* of the duodenum. For purposes of clarity each of these will be considered separately.

A. Pancreatic Juice

Pancreatic juice is an alkaline secretion of the exocrine part of the pancreas and contains enzymes which act on all the three principal types of food, i.e. proteins, carbohydrates and fats.

The exocrine glands of the pancreas are racemose glands, and are quite similar to the salivary glands. The secreting units called *acini*, are made of spherical or ovoid cells filled with *zymogen* granules, arranged polygonally around the central cavity. The number of granules vary with the activity of the gland, being more abundant in the apical (luminal) border of the cells. A characteristic feature of the pancreatic acinus, is the infiltration of the terminal duct cells into the acinus to form *centroacinar* cells. The secretions are carried by the pancreatic duct which opens into the duodenum, guarded by the sphincter Oddi.

The secretory units of the exocrine pancreas have both parasympathetic and sympathetic innervation. The former are vagal fibres, most of them passing through the coeliac plexus without relaying there, to reach the intrinsic ganglia in the pancreas. Postganglionic unmyelinated fibres then run a short course to innervate the acinar cells and the smooth muscles of the ducts. Sympathetic fibres in the splanchnic nerves synapse in the coeliac and other related ganglia; postganglionic fibres then travel along the pancreatic blood vessels.

1. Composition of pancreatic juice: Pancreatic juice is a clear, colourless, alkaline fluid with strong odour. Table 51.6 shows the composition of pancreatic juice in humans.

Sodium bicarbonate is the chief electrolyte in pancreatic juice and is responsible for its strong, alkaline nature, capable of neutralising acid gastric juice. The bicarbonate concentration increases with rapid flow rates, while chloride varies reciprocally to that of the former, such that the sum of their concentrations (in milliequivalents) is an approximate constant. The organic matter in pancreatic juice are largely the *proteolytic, lipolytic* and *amylolytic* enzymes.

Trypsin is a proteolytic enzyme secreted as an inactive precursor trypsinogen. Trypsinogen is a single polypeptide chain with molecular weight of 25,000 and activated by a duodenal mucosal enzyme *enterokinase* or *enteropeptidase* which cleaves hydrolytically a terminal hexapeptide. Formation of sufficient trypsin activates the rest of the precursor autocatalytically. Trypsin acts on peptide bonds where carboxyl groups are

TABLE 51.6   Physical Characteristics and Composition of Pancreatic Juice

Volume/day	800–1200 ml
pH	8.0–8.3
Specific gravity	1.010–1.018
Osmolarity	Same as plasma
Water	97.6%
Inorganic constituents	0.6%
Cations: Sodium, potassium, calcium, magnesium, zinc	
Anions: Bicarbonate, chloride, phosphates, sulphates	
Organic constituents	1.8%
Enzymes: trypsinogen, chymotrypsinogen, pro-carboxypeptidase, elastase pancreatic lipase, phospholipase pancreatic amylase ribonuclease, deoxyribonuclease	
Albumin and globulin (in traces)	

contributed by arginine or lysine present deep inside the polypeptide chain (*endopeptidase*) at an optimum pH of 8.0. Certain inorganic ions, chiefly calcium, help activation and also accelerate the action of trypsin. *Chymotrypsin*, secreted as chymotrypsinogen, is also a single polypeptide of molecular weight approximately 25,700. It is activated by trypsin, following cleavage of the terminal 15 amino acids. An endopeptidase like trypsin, it acts preferentially on peptide bonds where carboxyl groups belong to aromatic amino acids like tyrosine and phenylalanine, at an optimum pH of 8.0. Unlike trypsin, chymotrypsin coagulates milk. *Carboxypeptidase* is the other proteolytic enzyme of the pancreatic juice, secreted as pro-carboxypeptidase and activated by trypsin. Being an *exopeptidase*, it cleaves off terminal amino acids with free carboxyl groups. The *nucleases*, ribonuclease and deoxyribonuclease, which break up nucleic acid to nucleotides, are also present in the pancreatic juice.

Pancreatic lipase is a powerful lipolytic enzyme which hydrolyses triglycerides to lower glycerides and free fatty acids. This hydrolysis occurs only in the presence of bile salts, though high concentrations of bile salts and salts of heavy metals inhibit their action. Certain *phospholipases* are present as inactive precursor prophospholipases, and convert lecithin to lysolecithin.

Pancreatic amylase is an α-amylase (similar to ptyalin) of molecular weight 45,000 and requires chloride ions for activation. It hydrolyses starch to maltose by splitting the 1–4 glycosidic linkages at an optimum pH of 6.5 to 7.2. It also digests glycogen.

2. Formation of pancreatic juice: The separate cellular origins of the ionic and enzyme components of pancreatic secretions are almost certain. The centroacinar cells provide water and bicarbonate while the acinar cells secrete the enzymes. Each component may be influenced by distinct regulatory mechanisms.

Bicarbonate ions are secreted by centroacinar cells in higher concentrations than present in plasma, while Na^+ and K^+ concentrations are identical to that in the plasma. Since chloride concentration in the centroacinar secretion is lower than in the plasma, it is exchanged for bicarbonate ions to maintain iso-osmolarity of the composite secretion. With increase in secretory rate, the available time for this exchange during the passage in the ducts diminishes considerably and hence the chloride concentration in the final secretion falls. This bicarbonate-chloride shift is responsible for the reciprocal relationship between the concentration of these two ions in the secretion.

The enzymes are synthesized and stored as zymogen granules in the acinar cells. Under adequate stimuli, these zymogen granules are discharged into the central cavity by a process called *emiocytosis* (reverse pinocytosis).

3. Regulation of pancreatic secretion: Pancreatic secretion is regulated both by neural and hormonal mechanisms (Figure 51.6), the latter being more dominant. Vagal stimulation or injection of cholinergic drugs produce small amounts of viscous secretions rich in enzymes, and the effect is blocked by atropine. Small amount of secretion may take place reflexly during cephalic phase of gastric secretion and is dependent on the integrity of vagal fibres. The effects of sympathetics on the secretion are less potent and variable.

The humoral mechanisms are more important regulators of the exocrine pancreas. Bayliss and Starling are credited to be the famous first who identified in 1902 a gastrointestinal hormone and showed that the hormone increases the pancreatic secretion. Two hormones are now known which influence pancreatic secretion. *Secretin*, a poly-peptide of molecular weight 3200–3500 is released from the duodenal mucosa by intraluminal acid stimuli. It mainly acts on centroacinar cells and brings about the secretion of large volumes of thin, watery juice rich in bicarbonate (Figure 51.6). Unlike gastrin, vagal innervation does not seem to appreciably potentiate the secretin release or its action. The other hormone *pancreozymin* (CCK-PZ), described by Harper and Raper in 1943, is also a polypeptide and is released from the duodenal mucosa. The same stimuli responsible for secretin release seem to be effective for pancreozymin release also. As shown in Figure 51.6, pancreozymin causes the secretion of the pancreatic juice, rich in enzymes, and is coincident with the depletion of zymogen granules in the acinar cells.

Following a meal, the secretion of pancreatic juice is reflexly initiated by vagal fibres. As acid food enters the duodenum, the secretion is increased by the hormonal mechanisms. These secretions diminish quite rapidly as the food leaves the upper small intestine. Normally, pancreatic secretions appear to be conditioned to the diet of the animal. High carbohydrate diet produces secretions rich in amylase whereas high protein food produces more proteolytic enzymes.

B. Bile

Bile is the other major secretion that mixes with, and helps in, the digestion of the chyme entering the duodenum. Being alkaline, it also helps in neutralizing the acidic nature of chyme. Though formed in the liver, biliary secretions are temporarily stored in the gall bladder and are intermittently poured into the duodenum on the arrival of food there. For details refer to Chapter 56.

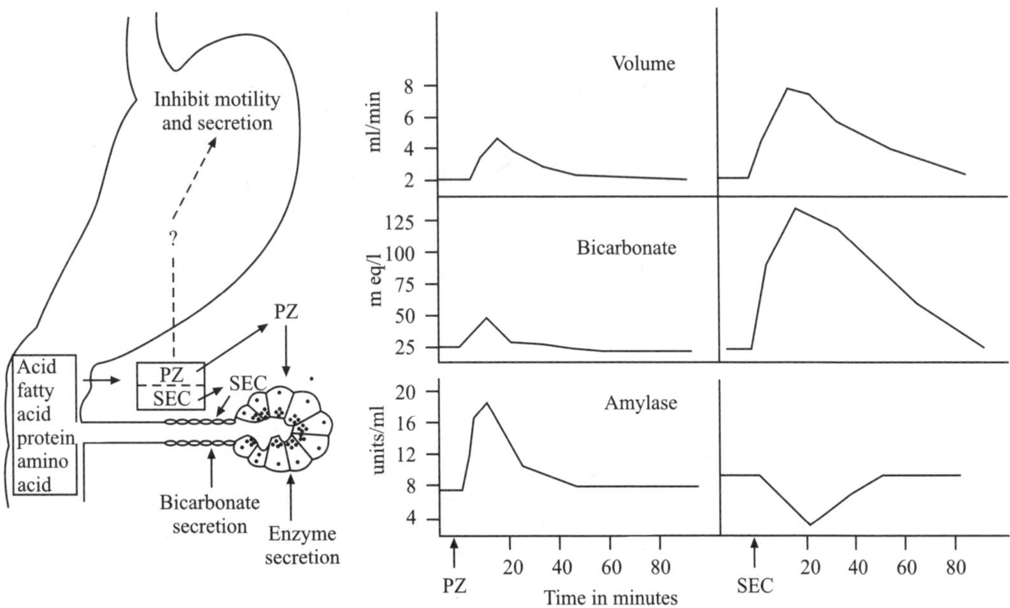

Figure 51.6 *Left half* of the figure shows the factors influencing the pancreatic secretion. *Right half* shows the effects of pancreozymin (PZ) and secretin (Sec) on the volume and bicarbonate and amylase contents on the pancreatic secretion.

The parenchyma of the liver is made up of single cell thick walls which are arranged in a honey comb fashion. The cavities within the honey comb structure called *lacunae*, contain blood sinusoids. At the junction between two cells, the membrane is grooved to enclose a narrow channel, the biliary *canaliculus*. The cytoplasm of the parenchymal cell bordering the canalicular groove is electron dense and the membrane shows microvilli. These canaliculi join together and form a network of small bile ducts, which unite to form larger hepatic ducts.

The two main hepatic ducts join together to form the common bile duct which along with the pancreatic duct open at the ampulla of Vater. The common bile duct is also joined by the cystic duct coming from the gall bladder.

1. Composition.   Bile is a golden-yellow (or greenish-yellow) fluid, alkaline in nature and bitter in taste. The composition of bile formed in the liver varies from that stored in the gall bladder and is shown in Table 51.7. Both hepatic and gall bladder bile are made up of *bile salts, bile pigments, cholesterol, neutral fats, phospholipids, inorganic ions,* and *water.*

TABLE 51.7   Physical, Characteristics and Composition of Human Bile

	Hepatic bile	*Gall-bladder bile*
Volume/day		500–1000 ml
pH	8.0–8.6	7.0–7.6
Specific gravity	1.010–1.011	1.026–1.032
Osmolarity	Same as plasma	Same as plasma
Water	96.0–98.0%	88.0–90.0%
Total solids	2–4%	10–12%
Bile salts	0.2–2%	6.0%
Bile pigments	0.02–0.07%	2.5%
Cholesterol	0.1–0.3%	0.4%
Phospholipids	0.2–0.8%	
Inorganic salts	1.0%	0.8%

Bile salts are sodium and potassium salts of the glycocholic and taurocholic acids. These bile acids are formed on conjugation of amino acids glycine and taurine with cholic acid which is related to, and probably derived from, cholesterol. The glycocholates are more abundant in human bile. *Bile pigments* are the breakdown products of haemoglobin, released by the haemolysis and destruction of old and effete red blood cells. The main pigments are bilirubin and biliverdin and are responsible for the colour of the bile. In human bile (adult) the former is abundant, whereas in herbivores and in young infants the latter predominates. In addition to these two main pigments, reduction products of bilirubin like bilinogens are also found in smaller amounts. These bilinogens are colourless unless oxidised, and are re-excreted after absorption from the gut.

Lecithin and free cholesterol form a small percentage of total solids in bile. Normal plasma proteins have been demonstrated in bile, and along with mucin, muco-proteins, fats and free fatty acids, constitute the rest of the organic, constituents. The inorganic fraction includes sodium, potassium, chloride and bicarbonate.

2. Formation and storage of bile.   There is ample evidence to show that liver cells are capable of pushing water into the canaliculi even if the pressures in the duct system are higher than that of the blood sinusoids. This secretion is temperature dependent, and requires oxygen and energy. Since water is actively secreted and not filtered, the ions probably move freely along with water, and their concentrations are similar to that in the blood plasma. This leads hepatic bile to be isoosmotic with the plasma.

The organic constituents of bile are also actively secreted by the hepatic cells. About 0.7 g of bile salts are secreted into the bile per day. The part of it is probably synthesized by conjugation of the amino acid with cholic acid in the hepatic parenchyma. Most of the bile salts are re-excreted after being absorbed from the gut. The balance between synthesized and re-excreted bile salts is delicately maintained by a homeostatic mechanism. The bile pigments are also secreted by the hepatic cells after being conjugated with glucuronic acid. The water soluble bilirubin glucuronides, 5–29 g of which are secreted per day, are converted in the gut to colourless bilinogens. A part of this is reabsorbed and then resecreted by the liver into the bile. Thus both the bile salts and the bile pigments participate in this cycle of *entero-hepatic circulation* constituted by their secretion in bile, reabsorption from gut, and again their secretion into the bile.

The origin of phospholipids (mainly lecithin) and cholesterol in bile is not known. It is, however, agreed that the cholesterol concentration in bile is independent of blood cholesterol levels. Certain substances, not normally present in the body, e.g., dyes like bromsulphalein (BSP), and detoxified products of normal body constituents, e.g. steroid hormones, are also secreted by the hepatic parenchymal cells with the possible exception of inorganic ions which freely diffuse across the membrane.

The bile formed in the liver is stored in the gall bladder prior to its secretion into the duodenum. During storage, the gall bladder mucosa acts on the bile and changes its composition considerably. It becomes concentrated, mainly by absorption of water and inorganic ions and the addition of thick mucus secreted by the epithelial cells of the mucosa. Due to absorption of bicarbonate ions, the gall bladder bile is less alkaline (pH 7.0).

3. Regulation of biliary secretion.   The rate at which bile enters the duodenum is determined by two factors: (1) the rate of formation of bile by the liver, and (2) the evacuation of bile from the gall bladder. The *rate of bile secretion by the hepatic cells* is influenced by several substances. Substances which increase both the volume and solid content of bile from the liver are called *choleretics. Hydrocholeretics* are the substances which increase only the volume without influencing the output of bile solids. Vagal stimulation seems to have a mild choleretic effect and serves as the final pathway for emotional and other cephalic factors.

Bile salts act as powerful choleretics and participate in sustaining the continued increase in bile production following a meal. Hydrochloric acid and acid chyme in the duodenum act as moderately good choleretics. Their effect is mediated by the release of secretin which increases the volume flow of bile. Gastrin also seems to have this hydrocholeretic effect.

Evacuation of bile from the gall bladder seems to be regulated both by nervous and humoral mechanisms. These mechanisms contract the gall bladder and simultaneously relax the sphincter of Oddi, in an integrated fashion. Substances which bring about the contractions of the gall bladder and evacuation of bile into the duodenum are called *cholagogues*.

The neural effects on gall bladder contraction and sphincter relaxation may be due to extrinsic nerves or intrinsic nervous plexuses. The extrinsic neural pathways are mediated by the vagus and the sympathetic—both of which show weak contractions of the gall bladder with relaxation of the sphincter. Short reflexes initiated by food in the stomach, duodenum, and intestines through the intrinsic plexuses, may also initiate emptying of the gall bladder. Nevertheless, in man, the neural control of the gall bladder is of little importance.

The presence of acidic chyme or fats in the duodenum can bring about rapid and complete evacuation of a denervated or transplanted gall bladder. This release of bile into the duodenum is mediated by a hormone, *cholecystokinin* (CCK-PZ). Fats, peptide fragments and acid in the duodenum release pancreozymin (CCK-PZ). Release of this hormone following intraluminal stimuli are mediated by local nervous reflexes (Figure 51.7). Thus the release of pancreatic secretions and the evacuation of bile into the duodenum take place simultaneously.

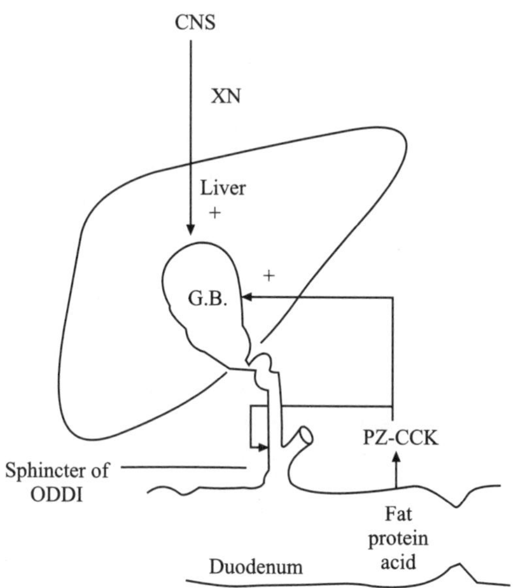

Figure 51.7   Shows the factors and the pathways involved in the evacuation of gall bladder. G.B.: gall bladder; PZ-CCK: Pancreozymin-cholecystokinin.

C.   Secretions of the Brunner's Glands

Brunner's glands produce small amounts of mucoid secretions which are protective to the duodenal mucosa. These glands are structurally specialized mucus glands similar to the glands in the pylorus. They are long, tortuous, tubular glands, frequently branched and characteristically penetrating the muscularis mucosa to form submucosal glands. The ducts of these glands open into the *intestinal crypts*. Brunner's glands are typically located between the pylorus and the ampulla of Vater, and are not found beyond the duodeno-jejunal junction. Other glands, similar to the intestinal glands, are also found in the duodenum. Brunner's glands secrete a small amount of viscid fluid, containing a large amount of mucus. In animals where the amount secreted has been determined, it varies from 0.5 to 3.4 ml/hour. In man, probably the amount secreted is 50 ml/per day. The secretion has a pH of 8.0 to 8.9 and appears to show weak proteolytic activity.

Secretion of these duodenal submucosal glands is regulated by nervous and hormonal mechanisms specific to them; and having little effect on other intestinal glands. Vagal stimulation and cholinergics stimulate secretion while sympathetics have no influence. Denervated duodenal pouches secrete to intestinal stimuli, suggesting the existence of a hormonal mechanism. It was earlier believed that secretin is probably the hormone mediating this effect. Later it was suggested that a specific hormone, *duocrinin,* released by the duodenal mucosa, was possibly responsible for the secretion of Brunner's glands but this is disputed now.

D.   Functions of Duodenal Secretions

The primary function of the three secretions that confront the chyme in the duodenum, is to *neutralize* its acidic nature and to protect the intestinal mucosa. In addition, each of the secretions has specific functions.

The *pancreatic* juice is responsible for major digestive activity of the gastrointestinal tract. The potent proteolytic, lipolytic and amylolytic enzymes break up the three principal food materials, prior to their final digestion by the intestinal secretions. The bicarbonate content of pancreatic juice not only neutralises the acid chyme, but also participates in maintaining the acid-base balance of the body.

Bile salts are the most important constituents of *bile* for digestive functions. (a) They combine with insoluble substances like cholesterol and fat soluble vitamins to form water soluble complexes called choleic acids. This is termed the *hydrotropic* property of bile. (b) The lipolytic enzymes of pancreatic juice and succus entericus (lipase) are activated by the bile salts. (c) The surface tension reducing properties of bile salts help emulsification of fats in the chyme. The fats are reduced to fine droplets providing a greater surface area for the lipase activity. The emulsion formed is stabilized by the bile salts. The emulsifying property of bile salts is also responsible for the easy absorption of digested fats in the small intestines. (d) Circulating bile salts act as powerful *choleretics* and promote continuous secretion of bile by the hepatic parenchymal cells aided by enterohepatic circulation.

A fairly large group of substances of varied nature are *excreted* in the bile, e.g., salts of heavy metals like copper or iron, enzymes like alkaline phosphatase, and even pathogenic bacteria like the typhoid bacillus. Bile exerts a mild laxative effect on the gut by *stimulating peristalsis* in the small bowel. It also has an *antiseptic action* and inhibits growth of certain bacteria. Absence of bile increases putrefaction in the large intestines.

The *secretions of the Brunner's glands* are protective to the duodenal mucosa and they prevent the potent proteolytic enzymes from acting on the mucosal epithelium. The high mucus content of these secretions also help in the lubrication of the intestinal contents.

V. INTESTINAL SECRETION

The duodenal chyme is already in the advanced stages of digestion as a result of the actions of the pancreatic juice. It, however, still needs to be acted upon by the digestive enzymes of small intestine so that the products could be finally reduced to absorbable forms.

A. Histological Structure

Since the small intestine is geared primarily for absorption, the intestinal mucosa is adapted to provide an extremely large surface area. This is achieved by the finger like projections of the mucosa called *villi* which in turn have a brush border of epithelial cells having *microvilli*. The numerous tubular glands called the *crypts of Lieberkuhn*, are present at the base of the villi. They do not penetrate the muscularis mucosa and are lined by a single layer of columnar cells inter-spersed with goblet cells that secrete mucus. Enormous mitotic activity and rapid turnover of the intestinal mucosal cells is in keeping with the suggested function of these glands, i.e., to continuously and constantly replace the epithelial cells lining the intestinal mucosa. In addition, *argentaffine or enterochromaffin* cells, which synthesize serotonin, and special cells with large acidophilic nuclei called *Paneth cells*, are also present. The combined secretion of these intestinal glands is called *Succus Entericus*.

The study of intestinal secretion has been made possible as a result of various methods used for collecting uncontaminated juice. These techniques include preparations of Thiry fistula, Thiry-Vella loop, and denervated and transplanted loops of the small intestine. In humans, it is possible to collect intestinal secretions by using multi-lumen intestinal tube (Miller-Abbott tube).

B. Composition

The small intestinal juice is a colourless, straw coloured fluid. It is cloudy in appearance due to the shed epithelial cells and flecks of mucus. Due to rapid exchange of fluid in both directions of the intestinal mucosa, it has been difficult to determine the actual amount secreted. It is estimated that daily secretions may range between 1000 to 3000 ml. Succus entericus is composed of water, inorganic salts and organic material. Details of its composition are shown in Table 51.8.

The organic matter consists mainly of enzymes, cellular debris and mucus. A number of the enzymes are located

TABLE 51.8 Physical Characteristics and Composition of Succus Entericus

Volume/day	1000–3000 ml
pH	7.0–8.5
Specific gravity	1.010
Water	98.4%
Inorganic constituents	1.0%
Cations:	
sodium, potassium,	
calcium, magnesium	
Anions:	
bicarbonate, chloride,	
phosphate	
Organic constituents	0.6%
Enzyme:	
Aminopeptidase,	
dipeptidase, entero-	
peptidase	
Lipase, phosphatase*.	
esterases	
Sucrase, invertase*	
maltase*, lactase*	
Nuclease, nucleotidase*	
Mucin	

*Intracellular enzymes present in shed mucosal cells.

intracellularly (within the shed epithelial cells). It is doubtful whether secretion in the true sense occurs in the small intestine, since it is likely that the free enzymes present in the intraluminal contents are probably released by the dissolution of the exfoliated and shed epithelial cells. Further, the concentration of the enzymes in the secretion appear small and insignificant compared to pancreatic and other secretions, implying that their presence within the lumen is not of much consequence. May be, complete breakdown of all food substances is not essential, and that part of the digestive process occurs intracellularly in the intestinal wall.

Peptidases are the most important group of proteolytic enzymes in succus entericus. They bring about the final breakdown of polypeptides to amino acids. *Di-* and *tripeptidases* break up di- and tripeptides respectively, while *aminopeptidases* act on peptide linkages of terminal amino acids possessing free amino groups. *Enteropeptidase* or *enterokinase* activates trypsinogen. Its release in succus entericus is triggered by the presence of pancreatic juice in the duodenum.

Lipase is seen in the intestinal juice along with a large group of amylolytic enzymes mainly the *disaccharides* like lactase (α-galactosidase), sucrose (invertase), maltase (α-glucosidase) and isomaltose (oligo-1-6-glucosidase). Most of the latter group of enzymes are located intracellularly.

Regulation of secretion. The local stimulatory effect of chyme seems to be important in bringing about the release of succus entericus. Mechanical stretching or distension of the bowel induces secretion while the chemical composition of chyme probably determines the amount secreted. Stimulation of the parasympathetic nerves induce the secretion of juice containing chiefly enzymes. The sympathetic nerves, on the other hand, have a basic inhibitory effect. The glands are released from this tonic sympathetic influence only when stimulated by the parasympathetics and/or the hormones.

Nasset and his associates (1953) showed that transplanted loops of intestine secrete when chyme reaches the intestines. Extracts of intestinal and upper colonic mucosa when injected stimulated secretion and these extracts were presumed to contain a hormone called *enterocrinin*. Further experimental work suggests the possibility that they are in fact a mixture of two hormones, one influencing water and inorganic ion secretion while the other evokes enzyme secretion. The physiological role of enterocrinin is not as yet certain, though it is believed that the chyme in the upper intestines release enterocrinin and thereby 'prime' the lower intestinal glands to maximally secrete to local mechanical or chemical stimulation.

C. Secretions of the Large Intestine

The small amount of unabsorbed chyme entering the large intestine, is reduced substantially to a semi-solid or solid residue mainly as a result of water absorption from the gut.

The large intestines consist of colon and the caecum, and appear histologically similar in many respects to the small intestine. Though the mucosa is smooth with no villi, they have crypts and a number of closely packed glands composed largely of mucus secreting goblet cells. Rapid replacement of the mucus epithelium is suggested by the marked mitotic activity seen in the base of the epithelium. Within 3 days the mucosal cells are completely replaced.

It has been difficult to collect colonic secretions, and when collected, it is a small amount of viscous and opalescent watery fluid with flecks of mucus. The high concentration of sodium bicarbonate, up to 80–90 mEq/L in the large intestinal secretions makes it alkaline (pH 8.0). The organic matter consists primarily of large amounts of mucus. Colon exhibits an enormous capacity to secrete mucus especially when it is irritated (mechanically or in the presence of bacterial infections). It appears as though all the enzymes of the succus entericus (except entero-peptidase) are present in much smaller amounts in these secretions.

Stimulation of the parasympathetic nerves causes secretion of a mucoid fluid at much higher rates than at rest. Stimulation of parts of the large intestine can also reflexly bring about secretion. Sympathetic effects seem to be variable. It appears that the predominant neural control leads to excess secretion of colonic mucus under conditions of psychosomatic disorders.

D. Faeces and Its Composition

About 500 ml of chyme that enters the large intestine, is reduced to a semi-solid or solid *faeces* by the end of 24–36 hours. The average quantity of faeces produced per day is around 100 g, usually ranging from 80–200 g and depending on the nature of the diet. The consumption of carbohydrate rich cereals, common in our part of the world, results in larger faecal bulk. Faeces has a surface pH of 7.0 to 7.5, and its colour is due to oxidised pigment sterobilin which is derived from bile pigment metabolism. The characteristic odour of faeces is due to the presence of indole and skatole as also due to gases produced by bacterial fermentation.

Water forms the bulk of faeces, and constitutes 60–80% of its moist weight. Depending on the duration of the faeces retained in the colon as well as its cellulose content, the water concentration varies. About 350 ml of water is usually reabsorbed mainly in the ascending colon and the caecum. The faeces of healthy persons is made up chiefly by residues of gastrointestinal secretions, like mucus, desquamated and shed epithelial cells, leukocytes and large quantities of bacteria; and may account for the fair amount of faeces passed by starving individuals. Part of the faecal mass is contributed by actual undigested food, mainly cellulose and roughage. Approximately 10 to 20% of faecal solids are inorganic ions, mainly calcium and phosphates. The large amount of nitrogenous material present in faeces, is derived from non-dietary sources: the bacteria, digestive enzymes, and the desquamated cells. Similarly, faecal fat is also mostly attributable to bacterial origin. In pancreatic disease and conditions like malabsorption syndrome, the faecal fat content rises above the normal level. Several sterols are also present in faeces. Except for the cellulose and roughage content of food, neither the constitution of the diet nor the quantity of water ingested, influences the composition and bulk of faeces very much.

E. Bacterial Flora of the Gut

The gastrointestinal tract of a new born infant is free of bacteria. Within a few days following birth, various types of organisms invade the intestine and the gut is not bacteriologically sterile any more. The stomach has limited bacterial flora since the gastric hydrochloric acid appears to be bacteriostatic. The antibacterial mechanisms of the small intestine also tend to discourage the growth and multiplication of bacteria, but the large intestine has an enormous amount of bacteria, which mainly include anaerobic forms like bacteriodes, coliform bacilli and streptococci, among others. These bacteria of the human colon can be considered as the normal inhabitants of the human colon.

The bacterial flora of the intestines seem to be useful to the host animal in many ways. They may digest the cellulose content of fruits and vegetables or synthesize amino acids thereby enhancing the nutritional status of the host. Though not relevant to man, the microbial flora in ruminants synthesize amino acids from nitrogenous wastes like urea and supply from a quarter to a third of the host's protein requirement. The bacteria also synthesize vitamins. In man, a substantial part of the daily requirement of vitamin K is provided by the intestinal bacteria. Since the gut of the new-born infant is sterile, vitamin K deficiency may be manifested as spontaneous haemorrhage at that stage. This disorder disappears gradually as bacteria invade the infant's gut and produce sufficient amounts of vitamin K. In modern paediatric practice vitamin K is routinely administered to the newborn infant to tide over the period during which the intestinal flora are becoming established. Large amounts of B group vitamins, also synthesized in the large intestine of man, are not absorbed. Colonic bacteria produce amines and phenols like indole and skatole (which are responsible for odour of faeces) by the decarboxylation and deamination of proteins and amino acids. Intestinal bacteria also help in the conversion of the bile pigments (bilirubin and biliverdin) to the bilinogens.

The diet of the host animal seems to influence the nature and the metabolic activity of the microorganisms in the alimentary tract. It is also being suggested that the type and amount of bacterial growth in the intestines influences the vital economy of the host animal. The bacteria of the intestines not only contribute essential nutrients but also exert beneficial effects on the development of the gastrointestinal tract and protect the host against pathogenic and noxious organisms. The normal microbial flora of the gut is thus a delicately balanced ecological system; and disturbing it may have profound effects on the host. Indiscriminate use of antibiotics and drastic changes in diet may upset this ecology and endanger the health of the host.

VI. PATHOPHYSIOLOGY OF THE GASTROINTESTINAL TRACT

A. Esophagus

1. *Dysphagia* is the difficulty in swallowing manifested as a brief feeling of pressure or the feeling that "something has got stuck". Frequent causes are esophagitis, achalasia, benign and malignant tumors, malformation and sclerodermia.
2. *Achalasia* is due to a neuromuscular disorder of the caudal esophagus in which the lower-esophageal sphincter fails to relax normally as food approaches. As a result, food passage from the esophagus into the stomach is greatly impeded leading to the dilatation of the esophagus. The condition is caused by malfunction of the Auerbach's plexus.
3. *Heart-burn* is an inability of the lower esophageal sphincter to close adequately after food has entered the stomach. Hydrochloric acid from the stomach contents can irritate the esophageal wall, resulting in a burning sensation.

B. Stomach

1. *Peptic ulcers* develop in areas of the gastrointestinal tract exposed to acid gastric juice. Accordingly, peptic ulcers are found in the lower part of the esophagus, in the stomach (gastric ulcer) and in the duodenum (duodenal ulcer). Most peptic ulcers are duodenal. Factors responsible for the formation of peptic ulcer include hypersecretion of acid gastric juice and pepsin and hyposecretion of mucus.
2. *Zollinger-Ellison syndrome* results from a gastrin-producing islet–cell tumor. The raised gastrin secretion elicits an excessive hydrochloric acid and pepsinogen production leading to the production of peptic ulcers, diarrhoea with hypokalemia, and steatorrhea.
3. *The dumping syndrome* develops due to a rapid uncontrolled emptying of the stomach contents into the jejunum following gastric resection. It leads to a sudden overstretching of the jejunum and withdrawl of large amounts of fluid from the blood plasma into the hyperosmolar chyme. As a con-sequence of this, acute attacks of palpitations, sweating, dizziness, fullness, retching and/or vomiting occur.

Further complications include iron deficiency anemia due to decreased iron absorption, megalo-bastic anaemias resulting from lack of production of intrinsic factor, steatorrhea as a consequence of insufficient mixing of the food with the pancreatic secretion, protein deficiency due to reduced protein absorption and osteopathy due to a chronic calcium deficiency.

C. Small Intestine

1. *Lactose intolerance* is due to the failure of the small intestine to produce lactase which is essential for the digestion of lactose leading to milk intolerance.
2. *Malabsorption* due to damage to the small intestinal mucosa as occurs in sprue leading to diarrhoea and deficiency states.
3. *Ileus* is the incapacity of the intestine to transport its contents further. It could be mechanical ileus due to the obstruction of the intestinal lumen or it could be paralytic ileus due to muscular insufficiency.

D. Colon and Rectum

1. *Diarrhoea* is the frequent emptying of watery or mushy faeces. The pathophysiology could be any one or a combination of the following mechanisms:
   osmotic diarrhoea due to reduced absorption of osmotically active substances from the intestinal lumen.
   secretory diarrhoea due to the intensified secretion of electrolytes and water into the intestinal lumen.
   diarrhoea due to the increased permeability of the intestinal mucosa.
   diarrhoea due to disturbed intestinal motility.
2. *Constipation is the condition of delayed emptying of dry and hard faeces due to a disturbed emptying reflex or a delayed intestinal passage.*

52

Tests for Assessment of Gastric Function

(Late) P.N. Chuttani and N. Sharma

CONTENTS

I. PHYSIOLOGICAL ASPECTS

Biochemically important specific secretory products of the gastric juice are:

(i) Hydrochloric acid secreted by the parietal cells.
(ii) Enzymes
    (a) Mainly pepsinogen secreted by the chief cells, and activated to pepsin by the acid, which also provides the acid medium required by this enzyme for its action on proteins.
    (b) Rennin, which clots milk by converting the milk protein caseinogen to casein.
    (c) A weak lipase is also present.
(iii) The haemopoietic factor, the absence of which is responsible for the development of pernicious anemia.
(iv) Mucus.

Gastric juice also contains amino acids, histamine and other amines, urea, ammonia, and neutral inorganic salts such as chloride.

Gastric secretion has three phases. The first of these is the psychic or cephalic phase, related to a nervous mechanism and mediated through the vagus nerve. This is evoked by pleasurable sensations accompanying thought, sight, smell and taste of palatable food. Inhibition of secretion may result from psychic influences, such as worry and anxiety and sight or smell of disagreeable food. The second is the gastric phase, where mechanical and chemical stimulation of the gastric mucosa in the pyloric antrum causes the release of a substance called gastrin, which, transmitted by the blood stream, acts as a gastric secretagogue. There exists evidence that a local nervous mechanism (vagus terminals in the stomach wall), is involved in the production and liberation of this hormone. The third is the intestinal phase, evoked by certain products of gastric digestion when they enter the duodenum, since these act as chemical excitants to gastric secretion. This holds good for water, meat extracts, saponin, soaps and magnesium sulphate. This phase is thus due to secretagogues in the food absorbed in the intestine involving the vagus terminals in the wall of the stomach or intestine. The intestinal phase is inhibited by fat (before absorption). The acidity, volume, and especially peptic activity of the secretion, is lowered. The factor responsible for this effect has been termed enterogastrone. Other factors include insulin, which stimulates gastric secretion through the vagus mechanism. The volume, acidity and peptic activity of the gastric juice are all in-creased. Mecholyl, pilocarpine and nicotine increase the volume considerably but the acidity to a small degree. Alcohol is a powerful stimulant of acidity and volume, while acids and atropine are secretory depressants. Histamine is a powerful stimulant of acid secretion. Different foodstuffs influence the secretion in different ways. Meat produces a gastric juice of high acid and moderate pepsin content, bread, a juice of low acid and high pepsin content, while milk produces a juice of moderate acid and low pepsin content. Fat depresses peptic activity more than acidity or volume.

Residuum is the name given to the contents of the fasting stomach. Its composition can elicit some information, which by itself is not important.

Amount, i.e., the volume usually ranges between 20 and 100 ml (average 50 ml). An increase in volume may be due to hypersecretion, retention, or regurgitation from the duodenum. However, levels of other components like its acid content, etc. would determine the condition conclusively.

Colour of the freshly secreted fluid should be whitish to colourless, but 55% of normals have a residuum that is either yellow or green in colour due to regurgitation of bile from the duodenum. Increased amount of bile results from obstruction or ileal stasis. Dark brown, dark red or black colour indicates stale blood and bright red colour would be due to freshly exuded blood which may also be due to trauma caused by intubation. Blood in the residuum may also be due to lesions such as carcinoma of the stomach, portal cirrhosis, peptic ulcer, acute gastritis and bleeding disorders due to any cause.

Consistency should normally be rather fluid with a small amount of ropy mucus which is derived from the nasopharynx. Increased amount of sediment indicates retention and increased quantities of mucus point to catarrhal inflammation of the stomach.

Organic acids such as lactic, butyric, and other fatty acids may be found in the residuum. It is believed that lactic acid may be secreted by the gastric mucosa in carcinoma of the stomach. However, a majority of observers believe that organic acids result from stagnation of gastric contents with consequent bacterial action and fermentation. Since the latter situation is inhibited by high free acidity, gastric retention and hypo-chlorhydria usually accompany the presence of lactic acid and the combination of all three points to carcinoma of the stomach.

Enzymes present are usually pepsin, trypsin (regurgitated from the duodenum) and very seldom rennin. The absence of pepsin, concomitant with achlorhydria, constitutes true achylia encountered in pernicious anemia.

Free and total acidity is expressed as the number of millilitres of tenth-normal sodium hydroxide needed to neutralize 100 ml of gastric contents ("clinical units" or mEq/litre). Free acidity ranges from 0 to 30 with the average at 18.5 while total acidity which includes acidity due to free hydrochloric acid, hydrochloric acid combined to protein, acid salts (phosphates), and organic acids such as lactic acid, ranges from 10–50 units or mEq/litre, the average being 30. Free HCl is practically never found when total acidity is below 10. Absence of both free HCl and total acidity, particularly if pepsin is also absent, is very significant. Free HCl values above 30 or total above 50 should be considered abnormal and indicative of the state of hyperacidity.

II. INVESTIGATION OF GASTRIC SECRETORY ACTIVITY

Principles underlying the evaluation of gastric secretory activity are based upon the observations that:

(i) hypersecretion or hyperacidity occurs commonly in duodenal ulcer.
(ii) hyposecretion or hypoacidity is a common feature of gastric cancer, and
(iii) pernicious anemia is invariably accompanied by achlorhydria.

Hypersecretion may often be manifested by an abnormally high concentration of free HCl or total chloride in the gastric residuum, but a normal level does not exclude the possibility of hypersecretory response to stimulation, just as the absence of free HCl does not necessarily indicate achlorhydria or hyposecretion. It will consequently be evident that more examination of the residuum will not provide much significant information about differential diagnosis in cases of pyloric obstruction.

A. Tubeless Gastric Analysis

This originated from the work of Segal and his co-workers and is based on the principle that when a quininium resin indicator is given by mouth, hydrogen ions from the hydrochloric acid in the stomach can liberate quinine ions (QH + cation) at pH values less than 3. Quinine hydrochloride is formed and is absorbed in the intestine and begins to be excreted in the urine about 15 minutes after the resin was administered. The quinine can be extracted from the urine and determined by measuring the fluorescence in ultraviolet light. This forms the basis of the "Diagnex test". After studying several resins, they introduced Diagnex Blue which is prepared by reacting a carbacrylic cation exchange resin with Azure A (3-amino-7-dimethyl-amino-phenazathionium chloride). It is a safe indicator. The hydrogen ions of the resin are exchanged with Azure A ion, a reaction which is reversed in the stomach.

B. The Diagnex Blue Test

The test is carried out in the fasting state. The bladder is emptied and the urine discarded. The tablets of caffeine sodium benzoate are given orally with a glass of water. Urine is passed an hour later and keep as the control specimen. The Diagnex blue granules are then administered orally suspended in a quarter of a glass of water. The bladder is emptied after two hours which gives the test urine in which Azure A can be excreted either in its blue form or as a conjugated colourless form, or as a mixture of the two. Any colourless form present may be converted into the blue form by acidifying, boiling and standing for two hours. Analysis consists of making up the urine to standard volume and observing the colour (if indicated after hydrolysis) in a simple comparator with controls made by adding ascorbic acid to reduce the dye to the colourless form. The comparator supplied with each box of test units has two standards representing a colour intensity of 0.6 mg and 0.3 mg of Azure A respectively. If the colour intensity of the test urine is equal to or exceeds the 0.6 mg standard, free HCl is present in the stomach. If the colour is between that of the two standards it indicates hypochlorhydria, and if it is less than that of the 0.3 mg standard it is presumptive evidence for achlorhydria.

Vitamin preparations should not be taken on the day preceding the test. The presence of pyloric obstruction, impaired intestinal absorption, renal disease and obstruction to the flow of urine may invalidate the test and produce false negatives. False positives may result if inorganic ions are being given. A negative test cannot be regarded as sufficient evidence of achlorhydria but should be followed by an augmented histamine test.

III. STIMULATION METHODS

Test Meals

Formerly a number of different test meals (e.g., Ewald, Riegel, Boas and Heckman) were employed to study the secretory and motor activity of the stomach. They are currently not used.

Since normal individuals exhibit an extremely wide variation in their response to test meals, stronger stimulation procedures are resorted to where the agents used are pentagastrin (pentavlon), alcohol, caffeine, histamine, and insulin.

1. Alcohol stimulation. This is carried out after collecting the basal secretion by giving a dose of 50 ml of 7 per cent ethyl alcohol, instilled into the stomach through the tube. Samples are aspirated for analysis at intervals of ten to twenty minutes for one and a half to two hours.

2. Caffeine stimulation. After collecting the basal secretion at intervals of ten to fifteen minutes for one hour, 500 mg caffeine sodium benzoate in 200 ml of warm water is instilled into the stomach. The entire stomach contents are aspirated after thirty minutes. Samples are then withdrawn at intervals of ten minutes for ninety minutes. Volume and acidity are measured in all samples. In normal subjects the secretion reaches a maximum at about thirty minutes and returns to the

basal level at sixty to eighty minutes. The total HCl secretion is less than 300 units in over 90% of normal individuals and in less than ten per cent of subjects with duodenal ulcer, who also exhibit a prolonged curve of secretion.

3. Histamine stimulation. This is the most potent procedure, producing as it does, a juice of maximal acidity and relatively low pepsin content. After collecting basal samples at intervals of fifteen minutes for sixty minutes, histamine base is injected subcutaneously in a dosage of 0.01 mg/kg body weight. A total dose of 0.3 ml or a 1 : 1000 solution of histamine phosphate may be used instead. The samples are aspirated again at intervals of fifteen minutes for sixty to ninety minutes.

A histamine analog (Histalog: 3-beta-amino ethylopyrazole dihydrochloride) may be utilized instead of histamine as it is a more potent stimulant, has fewer side effects and may be given orally as well as subcutaneously. In normal subjects, the free HCl increases promptly, reaching a maximum at twenty to forty minutes and then it falls. The volume of the juice varies correspondingly. Maximum response in normals is 30 to 160 units (30–160 millimoles HCl, 0.1–0.6 per cent HCl). Because this method produces a maximal response, it must be employed before clinching the diagnosis of achlorhydria.

A double histamine test has been proposed for demonstrating the capacity of the stomach for maintaining a maximal secretion over prolonged periods of time and involves a second injection given sixty to ninety minutes after the first, when the initial curve is beginning to fall. In normal subjects, a second rise and fall occurs, similar to the first curve in shape and magnitude. In subjects with hypersecretion (e.g., duodenal ulcer) the peak of acidity after the first stimulation is maintained at a constant level during the double test period.

The "Augmented histamine test" (Kay) represents an approach to an estimation of the active parietal cell mass and involves continuous aspiration of the gastric contents one hour before and forty five minutes after administration of histamine or histalog usually in fifteen minute samples. It is the most reliable test for the detection of complete achlorhydria.

Method: The patient fasts overnight. A tube is passed so that the tip is in the most dependent part of the stomach and the contents are completely removed. Gastric juice is then collected over one hour, the stomach being emptied every five to ten minutes by hand suction using a syringe (an automatic pump is less satisfactory since the end of the tube may be blocked by mucus or adherence of the mucous membrane). Forty minutes after the start of the "basal collection", 100 mg of anthisan is given intramuscularly. At the end of an hour, histamine is given subcutaneously in a dose of 0.04 mg of histamine acid phosphate per kilogram of body weight. Gastric juice is collected for 1 hour in four 15 minutes samples and all specimens are analyzed for their acid content.

Procedure for analysis: The volumes of both specimens are measured. The acid content of each hourly specimen is determined by titration with standard alkali to pH 7.0–7.4. Measure 5 ml of gastric juice and add 2 drops of bromothymol blue indicator solution. A blue colour indicates that no acid is present. If the colour is yellow, titrate against 0.05 N-sodium hydroxide (50 mEq/L) from a 10 ml burette to a blue-green end point, or the titration can be carried out using a pH meter to an end point at pH 7.4. If the volume of the fluid is too small, measure the pH with short range hydrion-indicator.

Calculations:

$$\text{Gastric acid} = \text{Titre} \times \frac{50}{5} = \text{Titre} \times 10 \ (\text{mEq per litre})$$

Rate of gastric acid production (mEq per hour)

$$= \frac{\text{Titre} \times 10 \times \text{Volume in ml}}{1000}$$

(Titre—Volume of titrating solution used)

4. Insulin stimulation. Hypoglycemia causes stimulation of the vagus centres in the brain; its one manifestation being increased gastric secretion. This procedure is used chiefly to ascertain the effectiveness of vagotomy in patients with ulcer. After collecting the basal secretion at intervals of fifteen minutes for sixty minutes, blood is drawn for determination of sugar and 20 units of regular insulin are injected intravenously. The samples are withdrawn at fifteen minute intervals for two hours and blood is drawn at thirty to forty-five minutes (or whenever hypoglycemic symptoms appear, e.g., salivation or perspiration), and again at the end of the test period. Glucose should be at hand for intra-venous administration and sweetened orange juice should be given at the conclusion of the test.

Insulin test: As originally suggested by the Hollander, gastric stimulation was assumed if there was a rise in the concentration of free acid of 10 mEq/L above the base line if the resting contents were an acid, or of 20 mEq/L above the mean basal secretion if this contained acid. A negative result was discounted if the blood sugar did not fall to 50 mg% if gastric secretion could not be demonstrated in response to histamine. It is now suggested that a rise in an acid concentration of 20 mEq/L over the basal level during the first hour after insulin is the most certain indication of incomplete vagotomy, the significance of zero rise in the second hour is less certain.

The stimulation procedures are followed by subsequent titration of the acid concentration of the gastric juice which involves the differential titration of a mixture of strong and weak acids. The universally employed procedure of titrating the free hydrochloric acid present with Topfers reagent as the indicator has the disadvantage of lack of a sharp colour change and a rapid fading of the colour at the end point. It has consequently been substituted by thymol blue, which also has the advantage of having two colour changes, i.e., red to yellow at pH 1.2 to 2.8 and yellow to blue at pH 8.0 to 9.5 and thus, also, eliminates the need of utilizing of combined acid (phenolphthalein). The first colour change would thus give the free acid measure and the second, the total. But since the change from red to yellow occurs at a more acid pH than with Topfers reagent (red to yellow from pH 2.9 to 4.0) small amounts of free acid may not be detected.

(a) Sahli's method for estimating free hydrochloric acid: This method requires more time but gives more accurate results because of a sharper end point. It is based on the liberation of iodine from the reagent employed in the presence of free hydrochloric acid. The iodine is titrated with sodium thiosulfate, using a starch indicator.

1. Prepare Sahli's reagent, which is a mixture of equal parts of a 48 per cent aqueous solution of potassium iodide and an 8 per cent aqueous solution of potassium iodate.
2. Prepare a 0.01 N solution of sodium thiosulfate.
3. Prepare a 1 per cent aqueous solution of starch.
4. Place a 1 ml of the strained sample and 10 ml of distilled water in a porcelain evaporating dish.
5. Add 1 ml of Sahli's reagent, mix and allow to stand for 5 minutes.
6. Titrate with 0.01 N solution of sodium thiosulfate until only a faint yellow colour of the liberated iodine remains.
7. Add 0.5 ml of the soluble starch solution. The mixture turns blue. Continue the titration until the blue colour disappears.
8. The total number of ml of 0.01 N sodium thiosulfate used in the titration of 1 ml of gastric juice is equivalent to the number of ml of 0.01 N sodium hydroxide necessary to neutralize the free hydrochloric acid in 1 ml of gastric contents. This value, multiplied by 10, represents the number of ml of 0.1 N sodium hydroxide necessary to neutralize 100 ml of stomach contents.

(b) Milliequivalents per litre and milligrams per 100 ml: Milliequivalents express with greater clarity the acid base status of a medium, since they imply the number of charged particle rather than the weight. Since the reaction of a medium depends upon the number of charged units, it will be evident that their weight is of no consequence.

To convert mg per 100 ml into milliequivalents per litre it is only necessary to multiply by 10 and divide by the equivalent weight, thus:

$$\text{milliequivalents per litre} = \frac{\text{mg per 100 ml} \times 10}{\text{equivalent weight}}$$

For the chief inorganic ions the equivalent weights are: Sodium 23, Potassium 39, Calcium 40/2 = 20, Magnesium 24/2 = 12, Chloride 35.5, Phosphate (as phosphorus) 30/1.8 = 17.2, Sulphate (as sulphur) 32/2 = 16, Bicarbonate 61.

If mg per 100 ml sodium chloride and sodium bicarbonate are to be converted into milliequivalents per litre the factors are 58.5 and 84 respectively.

The factor 1.8 used for phosphate is derived from the proportion of dihydrogen and monohydrogen phosphate present at the pH of body fluids.

To convert alkali reserve expressed as volumes CO_2 per cent, into milliequivalents per litre, divide by 2.24. Since one molecular weight in grams of bicarbonate gives 22.4 litres of carbon dioxide, 1 mEq. gives 22.4 ml. So to convert volumes per cent to mEq. per litre, multiply by 10 and divide by 22.4, i.e., divide by 2.24. Proteins are anions at the pH of plasma. To convert grams of proteins per 100 ml into mEq. per litre Van Slyke used a factor of 2.43, so that milliequivalents per litre of protein = grams protein per 100 ml × 2.43.

IV. MEASUREMENT OF OTHER GASTRIC COMPONENTS

The gastric rediuum as well as the fractions aspirated or the vomitus, could be subjected to the following tests for normal and abnormal constituents as giving additional information for help in the diagnosis.

A. Total Chloride

In the event of excessive regurgitation of duodenal contents into the stomach or excessive alkaline secretion of the pyloric glands, a large proportion of the secreted HCl may be neutralized so that neither the curve of free HCl nor that of total acidity can be regarded as indicating the true state of gastric secretory activity. The curve of total chloride is a true index of the true state of gastric acid secretory activity, since it includes in addition to free HCl and HCl combined with protein, salts of HCl which are neutral in reaction. The chloride curve may continue to rise for a variable period after the acid curve has begun to fall, thus indicating secretion of HCl which is being simultaneously neutralized. Determination of total chloride is particularly valuable in the differentiation between true and false achlorhydria. In true achlorhydria no free HCl appears in the gastric contents during the digestive cycle and the chloride curve is also low. In false achlorhydria, a normal amount of HCl may be secreted but is not present in the free state due to neutralization, and the curve of chloride is essentially normal. Total chloride content of normal fasting juice varies from 160 to 550 mg/100 ml, i.e., 45–155 mEq/1; with histamine stimulation it may reach a maximum of about 600 mg/100 ml or 172 mEq/l.

Total chloride varies within narrower limits than the free acidity.

B. Tests for Occult Blood

Tests for occult blood in amounts or forms not observable on inspection and thus referred to as occult blood are important in the diagnosis of lesions of the alimentary tract. The benzidine test was the most commonly used test until the carcinogenic nature of benzidine was revealed. It was then replaced by the orthotoluidine and orthoanisidine test—the chemicals however were found to have the same defect. Hence the pre-cautions of avoiding inhaling the material or handling it have to be taken when performing these tests. These tests depend upon the pseudoperoxidase activity of haemoglobin and its iron containing derivatives. The active group containing iron in the haemoglobin can transfer oxygen from hydrogen peroxide to oxidizable substances. The oxygen liberated by hydrogen peroxide is received by benzidine or orthotoluidine giving blue compounds. Degradation products of meat and chlorophyll can give a positive response, so that, regard to the recent diet of the patient has to be made. Inorganic iron does not give the orthotoluidine reaction. Orthotoluidine test is more sensitive, and, as with the benzidine test, the sensitivity varies with the concentration used which should be about 3% and not 10%. Higher concentrations give a positive response for even minute traces which may be present in normal fluids.

1. **Benzidine test:** To 3 ml of a saturated solution of benzidine in glacial acetic acid add 2 ml of the solution to be tested and 1 ml of 3% hydrogen peroxide. A blue green colour indicates a positive response (excess of hydrogen peroxide interferes with the reaction).
2. **Orthotoluidine test:** Boil the liquid to be tested with the addition of a little water to destroy the peroxidases that may be present. Cool and mix equal volumes of the orthotoluidine and hydrogen peroxide in another test tube and add to the boiled cooled fluid drop by drop until an amount equal in volume has been added. If a blue or greenish-blue colour is produced the test is positive.

Use the 4% solution first and if positive use the 1–2% solution and if positive with this also then the 0.4% solution and report results as negative or weakly positive, positive or strongly positive, respectively, as the case may be.

Meat free diet and avoidance of liver would avoid false positives.

Reagent
(a) *Orthotoluidine 4 per cent in glacial acetic acid:* Add orthotoluidine to the acid in a flask and shake to dissolve. Dilute 30 ml and 10 ml to 100 ml to give 1.2% and 0.4% solutions.
(b) *Hydrogen peroxide 10 volumes per cent:* Prepare at frequent intervals from 100 volumes per cent by diluting 1 to 10 with water.

C. Gunzberg's Test

Technique. Mix 1 to 2 drops of gastric contents with 1 to 2 drops of the reagent in a small evaporating dish and evaporate to dryness on a waterbath. If free HCl is present a reddish colour is produced. Other mineral acids give the test too but organic acids do not.

Gunzberg's Reagent. Dissolve 2 grams of phloroglucinol and 1 gram vanillin in 100 ml of 95 per cent ethanol. Prepare freshly.

D. Tests for Lactic Acid

1. **Ferric chloride test:** Extract 5 ml of gastric contents with 20 ml of ether. To 5 ml of ether extract add 20 ml distilled water and 2 drops of 10% ferric chloride solution. Shake the solution well. Lactic acid gives a yellowish green colour.
2. **MacLeans test:** To 5 ml of filtered gastric contents add a few drops of MacLeans solution (100 ml of saturated solution of mercuric chloride and 1.5 ml of hydrochloric acid in which 5 g of ferric chloride are dissolved). A reddish colour is given by lactic acid. Compare with a blank carried out on normal gastric controls.
3. **Uffelmann's test:** Mix equal volumes of Uffelmann's reagent and filtered gastric contents. A yellow or greenish colour is given by the lactic acid.

Uffelmann's Reagent. Add 10% ferric chloride solution drop by drop to 1% solution of phenol in water until a bluish colour is produced. The solution should be prepared just before use.

E. Tests for Bile (Pigments)

1. **Schlesingers test:** This iodine test is the oldest one employed in which tincture of iodine drops added to 1 or 2 volumes of the fluid are layered in a test tube over a solution of 1 gram of zinc acetate in ethanol. A green ring at the junction of the layers or a greenish yellow fluorescence on mixing, are indicative of the presence of bile pigments.
2. **Fouchet's test:** It is convenient to perform and fairly conclusive. Add a few ml of 10% barium chloride solution to the fluid after adding a drop or two of sulphuric acid. Filter and drain well and spread out the filter paper on another dry paper and add a drop or two of Fouchets reagent (25 g of trichlor acetic acid in 50 ml of water and to it add 10 ml of 10% ferric chloride and make up to 100 ml with water). A greenish blue colour is an indication of a positive response.

F. Peptic Activity

Pepsin is present in normal gastric juice, often along with trypsin which is present due to regurgitation from the duodenum. In the absence of free hydrochloric acid, the determination of pepsin is of importance for distinguishing between achlorhydria and achylia.

A number of procedures are outlined using different substrates for the quantitation of the concentration of rennin, pepsin and trypsin in gastric contents (Hawk and Oser, pp. 484–886, 1965).

1. A simple test for Rennin. Add 5 drops of gastric juice to 5 ml of fresh milk in a test tube and place in an incubator or water bath at 40°C for 15 minutes. If the milk coagulates, rennin is present.

2. Determination of the peptic activity of gastric contents. Mett's earlier method is fully described by Hawk et al. A convenient method published by Hunt uses the digestion of dried serum and later methods use Fibrin (Hawk and Oser, 1965).

Reagent
   (a) Trichloracetic acid, 0.35 N solution: This is approximately 6 per cent w/v. standardize to 0.35 N.
   (b) Sodium hydroxide, 0.25 N
   (c) Hydrochloric acid, pH 2.1.
   (d) *Folin and Ciocalteu reagent:* Dissolve 100 g of sodium tungstate ($Na_2WO_4.2H_2O$) and 25 g of sodium molybdate ($Na_2MoO_4.2H_2O$) in about 700 ml of water in a 2-litre round bottomed flask. Add 50 ml of syrupy 85% phosphoric acid. Reflux for ten hours. Use all glass apparatus if possible, otherwise wrap the stopper in tin foil. Add 150 g of lithium sulphate, 50 ml of water and a few drops of bromine. Boil without a condenser for about a quarter of an hour to remove the excess of bromine. Cool, make up to a litre, and filter. The solution should have no greenish tint. This is an improved phenol reagent which reduces the possibility of turbidity due to insoluble urates.
   Dilute this solution 1 in 3 for use, that is, to 1 volume of phenol reagent add 2 volumes of water.
   (e) *Standard phenol solution:* Five mg per 100 ml solution.
   (f) *Substrate:* Dissolve 5.6 g of dried citrated plasma or dried serum in 100 ml of distilled water with sufficient hydrochloric acid to give an acidity to pH 2.1 as determined by a glass electrode at 18° to 22°C. Stir the solution and filter through cotton wool.

Technique: For each determination place four tubes, (1) and (2) containing 5 ml of substrate, and (3) and (4) with 10 ml of the trichloracetic acid, in the water bath at 37°C and add 1 ml of the mixture to each of tubes (1) and (4). Incubate for fifteen minutes, at the end of which time tube (1) is mixed with tube (3) and tube (2) with tube (4), by pouring backwards and forwards. Allow to stand in the bath for about four minutes. (1) + (3) gives the test, (2) + (4) the control.

Filter: Fifteen to thirty minutes after the beginning of filtration, pipette 2 ml of the filtrates into 50 ml stoppered cylinders containing 20 ml of 0.25 N sodium hydroxide. Mix by gentle rotation, and add 1 ml of Folin and Ciocalteu reagent, and again mix by gentle rotation. Read the colour in thirty minutes, using a red filter or transmission at 680 millimicrons.

Calculation: The difference between the test and the control, gives a measure of the peptic activity. As standard, mix 2 ml of a freshly prepared phenol solution, containing 5 mg per 100 ml, with 20 ml 0.25 N sodium hydroxide, and add immediately 1 ml of Folin and Ciocalteu reagent. Read in the colorimeter after five to ten minutes. Put up a blank omitting the phenol. Gastric juice is said to have 50 units to peptic activity per ml when the colour produced in the above estimation equals that of this standard.

3. Hypoacidity. It is usually encountered in chronic gastritis, carcinoma of the stomach, gastric and other neuroses, secondary anemia, chronic debilitating disease, tuberculosis, gastric ulcer, hyperthyroidism and in about 20% of normals. Three-fourths of normal pregnant women are reported to have subnormal secretion of free hydrochloric acid and pepsin. This may play a part in the development of iron deficiency during pregnancy.

4. Hyperacidity. It does not imply that the gastric glands are secreting a juice of abnormality high acidity, often, the normal maximum may not be exceeded.

Hyperchlorhydria. This is said to occur when the maximum free acidity exceeds 45 mEq per litre (range 40–60).

Achlorhydria. It denotes absence of secretion of hydrochloric acid though enzymes such as pepsin may still be present. However, when both are absent it is called *achylia gastrica*. The most sensitive measurement of gastric acidity is pH determination. Achlorhydria is said to be present when gastric juice does not become more acid than pH 6 after maximal histamine stimulation (0.04 mg/kg body weight). Achlorhydria may be found in some apparently normal people, increasing with age to about 60 to 65 years and in wasting febrile diseases, tuberculosis, Addisons disease and later stages of malignant conditions, ulcerative colitis and fevers. In microcytic anemia due to iron deficiency, there is a considerable increase in the incidence of achlorhydria present in about 80 per cent of cases (Wintrobe) and a high incidence in cancer of the stomach. The combined acid gets greatly increased ranging from 30 to 80 mEq/l and may be even higher.

Chaudhuri, et al. made a comparative study between the modified non-isotopic saline load test of Goldstein and Boyle with the recently introduced radio-isotopic method designed to measure gastric-emptying time. This radio-isotopic method employs a gamma camera with a computer. A liquid meal of isotonic saline is used with or without an ideal radiopharmaceutical, namely ^{99m}-Tc-diethylene triamine penta-acetic acid (DTPA). Results indicate that gastric-emptying time T½ (8.8 ± 3.5 min) obtained by saline load test was shorter than that obtained by the radio-isotopic method (12 ± 3 min). This is probably a discrepancy due to the inherent error in the complete aspiration of the fluid in the non-isotopic method, thus giving a false faster emptying time. More variations were observed in T½ value in the same individual in the aspiration method than in the isotopic method for assessing gastric retention.

Purification of gastrin accomplished by Gregory and Tracy has made possible the routine radioimmunoassay for this hormone. This is very helpful in the diagnosis of the Zollinger-Ellison syndrome in which serum gastrin levels are considerably elevated as compared to values obtained in normal subjects.

Conflicting opinions exist regarding the role of cyclic GMP as a mediator in gastric acid secretion. Rise in levels of cyclic GMP occurred following pentagastrin treatment as contrasted to cyclic AMP levels which did not rise.

SUGGESTIONS FOR FURTHER READING

Babkin B.P. (1950), *Secretory Mechanism of the Digestive Glands*, 2nd ed., Paul. B. Hoeber, Inc., New York.

Cantrow Abraham and Trumper Max (1962), *Clinical Biochemistry*, 6th ed., W.B. Saunders Company, Philadelphia, London, 462–475.

Code C.G. (1956), *Histamine and Gastric Secretion*, J & A Churchill Ltd., London.

Dahms B.B. and Qualman S. (1997), *Gastroenterology Diseases*, Farmington, C.T., Karger AG.

Grossman M.I. (1950), Gastrointestinal Hormones, *Physiol. Rev.*, 30, 33.

Hawk P.B. and Oser B.L. (1965), *Hawks Physiological Chemistry*, 14th ed., McGraw-Hill, Inc., New York.

Henderson A.R. and Rinker A.D. (1999), Gastric, Pancreatic and Intestinal Function, In: Tietz *Textbook of Clinical Biochemistry*, Burtis C.A. and Ashwood, E.R. (Eds.), 3rd ed., W.B. Saunders Company, Philadelphia, USA, 1271–1327.

King Earl J. and Wootton I.D.P. (1964), Micro-analysis in Medical Biochemistry, 4th ed., J & A Churchill Ltd., 215–217.

Kolmer Spaudling, John A., Robinson Earle H. and Howard W. (1969), Approved Laboratory Technic, Appleton-Century-Crofts, Meredith Publishing Company, New York, 221.

Levin, E., Kirsner J.B. and Palmer W.L. (1951), A Simple Measure of Gastric Secretion in Man, *Gastroenterology*, 19, 88.

Marks I.N. and Shay H. (1960), Augmented Histamine Test, Evald Test-meal and Diagnex Test, *Am. J. Digest*, Dis. 5, 1.

Varley Harlod (1967), *Practical Clinical Biochemistry*, 4th ed., William Heinemann Medical Books Ltd. (London) and Inter Science Books Inc. (New York), Tests of Gastric Function, 327–347.

53

Evaluation of Exocrine Pancreatic Function

Anoop Saraya and Namrata Singh

CONTENTS

I. NORMAL PANCREATIC PHYSIOLOGY

An understanding of normal pancreatic function is necessary to comprehend the rationale for different pancreatic function tests. The adult pancreas is a transversely oriented retroperitoneal organ extending from the "C" loop of the duodenum to the hilum of the spleen. Pancreas is a complex lobulated organ with distinct endocrine and exocrine elements. The pancreas combines critical exocrine and endocrine functions and has complex regulatory mechanisms. The endocrine portion constitutes only 1% to 2% of the pancreas and is composed of about 1 million cell clusters, the islets of Langerhans; these cells secrete insulin, glucagon, and somatostatin. The exocrine pancreas is composed of acinar cells that produce the digestive enzymes, and the ductules and ducts that convey them to the duodenum.

The acinar cells produce mostly proenzyme forms of digestive enzymes and store them in membrane-bound zymogen granules. Digestive enzymes are primarily secreted in inactive or zymogen form (all proteases in particular) by pancreatic acinar cells, amylase and lipase are exceptions and are secreted in an active form. The major proteolytic, amylolytic, lipolytic, and nuclease digestive enzymes are:

Proenzymes: Cationic trypsinogen, Anionic trypsinogen, Mesotrypsinogen, Chymotrypsinogen (A, B), Kallireinogen, Procarboxypeptidase A (1, 2), Procarboxypeptidase B (1, 2), Prophospholipase, Proelastase.

Enzymes: Amylase, Carboxylesterase, Sterol esterase, Lipase, DNase, RNase

Some of the enzymes are present in more than one form (e.g., cationic trypsinogen, anionic trypsinogen, and mesotrypsinogen). Enzymes that could digest the pancreas are stored in the pancreas and secreted into the pancreatic duct as inactive precursor forms. The epithelial cells lining the ducts are also active participants in pancreatic secretion: cuboidal epithelial cells lining the smaller ductules secrete bicarbonate-rich fluid, while the columnar epithelial cells lining the larger ducts produce mucin. The secretions of the exocrine pancreas, approximately 1 L/day, consist of water, bicarbonate and digestive enzymes. These secretions are transported to the duodenum through a series of anastomosing ducts. The proenzymes remain largely inactive until they reach the duodenum; there, enteropeptidase (a brush-border enzyme) cleaves trypsinogen into active trypsin. Trypsin is then able to activate the other proteases to their active form. Both vagal and hormonal pathways mediate the secretion of enzymes by acinar cells. Vagal input initiates pancreatic secretion in the cephalic phase of eating. Continued prandial and post-prandial pancreatic enzyme secretion is mediated by cholecystokinin (pancreozymin), acting primarily through vagal afferents. Secretion of bicarbonate is under the control of the hormone secretin, released in response to duodenal acidification.

II. EXOCRINE PANCREATIC FUNCTION TEST IN CHRONIC PANCREATITIS

Various tests have been devised to measure the secretory function of the pancreas in order to diagnose disorders such as chronic pancreatitis and pancreatic cancer. The function tests fall into two general categories: direct and indirect. Direct tests of pancreatic secretory function measure the output of enzymes or bicarbonate from the pancreas after intravenous administration of a secretagogue or a combination of secretagogues. Indirect tests of pancreatic secretory function measure the released enzymes indirectly (through its action on a substrate or its level in blood or stool) include the measurement of pancreatic enzymes in duodenal samples after nutrient ingestion; the measurement of products of digestive enzyme action on ingested substrates; the measurement of pancreatic enzymes in the stool; and the measurement of the plasma concentration of hormones or other markers that are altered in pancreatic insufficiency states.

Direct test: Direct tests provide a gold standard for measurement of pancreatic function. Stimulation of secretion has been described most commonly with secretin, CCK, or the two combined. The combination provides the complete information about acinar and ductular cell secretions. In fact, it is accepted that the direct tests are more sensitive than current imaging techniques in the diagnosis of early pancreatic disease.

A. Secretin-cholecystokinin (CCK) Test

The direct function tests are based on the principle that maximal water, bicarbonate, and enzyme secretion are related to the functional mass of the pancreas. Historically, the secretin test (intravenous administration of secretin, with volume and bicarbonate measurement) first provided information about the function of the pancreas in various clinical settings. Administration of CCK and the measurement of digestive enzyme secretion also have been used successfully to demonstrate pancreatic insufficiency. Because the combination of secretin and CCK administration provides stimulation of both functional units of the exocrine pancreas, this combination is most commonly used. After an overnight fast, a double lumen tube under fluoroscopic control, is placed with the tip at the ligament of Treitz. The gastric and duodenal juice is aspirated continuously. Continuous intravenous infusion of secretin and CCK is given over 90 min. Sampling of duodenal juice in 10-min aliquots over the last 60 min of hormone infusion is done. Followed by quantification of volume, bicarbonate concentration and amylase, lipase and protease activities. The secretin-CCK test allows classification of the severity of exocrine pancreatic dysfunction into:

Normal: Normal output of enzymes and bicarbonates.

Mild dysfunction: Secretion of enzymes and bicarbonates ≥75% of the lower limit of normal

Moderate dysfunction: Secretion of enzymes and bicarbonates 30–75% of the lower limit of normal

Severe dysfunction: Secretion of enzymes and bicarbonates <30% of the lower limit of normal

The sensitivity and specificity of this test for the diagnosis of Chronic Pancreatitis is 90% and 94%, respectively.

Some experts have suggested that the pancreas has such reserve that 30% to 50% damage to the gland is necessary before direct pancreatic function tests yield reliably positive results. Despite their theoretical advantages, direct pancreatic function tests have a number of limitations. First, they have not been well standardized across institutions offering the test. Second, they are available at only a very few referral centers and so are not available to the majority of clinicians seeing patients with chronic pancreatitis. Third, it can be difficult for patients to tolerate unsedated placement of an oroduodenal tube for the hour or more required for the test. Fourth, accurate measurement of bicarbonate concentrations or enzyme output may be challenging. False-positive results have been reported in patients who have undergone Billroth II gastrectomy, in patients with diabetes, celiac disease, and cirrhosis, and in patients recovering from a recent attack of acute pancreatitis.

A direct pancreatic function test is most useful in patients with presumed chronic pancreatitis in whom easily identifiable structural and functional abnormalities have not been demonstrated on more widely available diagnostic modalities such as CT (e.g., a patient with small-duct disease). This type of test is most useful in ruling out chronic pancreatitis in patients who present with a chronic abdominal pain syndrome suggestive of chronic pancreatitis, saving these patients the label of chronic pancreatitis with its negative repercussions and the risk of such diagnostic tests as ERCP.

B. Intraductal Secretin Test

There are variations of direct pancreatic function tests that may be easier for patients to tolerate (by sedating them) and might be able to be made more widely available. One proposed variation is to collect pancreatic secretions at the time of ERCP by placement of a catheter directly in the pancreatic duct (the so-called intraductal secretin test). This test typically samples pancreatic output for only 15 minutes, to minimize the likelihood of ERCP-induced pancreatitis. It is not standardized and does not appear to be as accurate as standard direct pancreatic function testing, probably because of the rather brief collection time.

C. Endoscopic Test

An adaptation of the direct secretory test to upper endoscopy also has been described. At the time of endoscopy, either secretin, CCK, or the combination is administered intravenously, and pancreatic secretions are collected via the endoscope positioned in the duodenum. This variation attempts to bypass the difficulties limiting the widespread application of standard direct pancreatic function tests, such as passage of the collection tube in unanesthetized patients. Then the bicarbonate concentration and/or pancreatic enzyme activity in samples of duodenal juice obtained and measured. The initial descriptions of this test used a 60-minute collection with timed aspirates of duodenal fluid every 15 minutes. A short version of the test is based on the collection of duodenal juice for only 10 min. The peak bicarbonate concentration and lipolytic activity in duodenal juice collected is lower in patients of chronic pancreatitis compared with patients with normal pancreas. Although the endoscopic pancreatic function test is a promising procedure, it is far from being the current standard.

Indirect tests: The desire to develop indirect tests of pancreatic function is an outgrowth of the complexity, discomfort, and limited availability of direct pancreatic function testing. Indirect tests can generally measure pancreatic enzymes in blood or stool.

D. Fecal Chymotrypsin Activity

Fecal chymotrypsin is a simple test based on enzymatic quantification of of chymotrypsin activity in an isolated small stool sample. This enzyme can be measured by employing N-acetyl-L-tyrosine-ethylester as the substrate for chymotrypsin. Fecal chymotrypsin concentration is significantly lower than normal in patients with pancreatic insufficiency. A cut-off of 7 U/g of stool is generally accepted as an abnormal test. Patients with fecal chymotrypsin activity of less than 7U/g of stool are considered having pancreas exocrine pancreatic

insufficiency. However, chymotrypsin is variably inactivated during intestinal passage in such a way that fecal chymotrypsin activity does not accurately reflect pancreatic secretion of the enzyme. In addition, false-positive results have been reported in other malabsorptive conditions (celiac disease, Crohn's disease), in diarrheal diseases in which the stool is diluted, and in severe malnutrition. Because the test is usually normal in patients without steatorrhea, it is reliably positive only in advanced chronic pancreatitis. These measurements are about 85% specific and only 57% sensitive in advanced pancreatic dysfunction and relatively insensitive in mild to moderate disease. Orally administered pancreatic enzymes therapy for pancreatic insufficiency interact with the determination of chymotrypsin in stool. So this enzyme therapy should be discontinued 48 hrs preceeding stool sample collection.

E. Fecal Elastase Concentration

Compared with chymotrypsin, pancreatic elastase is highly stable during gastrointestinal transit and the fecal concentration of the enzyme correlates with the amount secreted by pancreas. A random stool sample is required for estimation of elastase by the enzyme linked immunoassay. Pancreatic elastase 1 can be measured in the stool using a monoclonal antibody against human elastase 1. Since the methodology used for its estimation is based on human specific monoclonal antibodies, oral enzyme supplementation does not interfere with the test. Levels more than 200 µg per gram of stool are considered normal. The test is reasonably accurate in more advanced chronic pancreatitis with 70% sensitivity and 85% specificity. Fecal elastase can be low in other diseases causing diarrhea, such as short bowel syndrome and small bowel bacterial overgrowth. This test, like other indirect tests, is not sensitive in detecting mild to moderate pancreatic disease.

F. Fecal Fat Quantification

Maldigestion of fat occurs after 90% of pancreatic lipase secretory capacity is lost. The simplest evaluation of pancreatic lipase action is the measurement of fecal fat excretion during a 72-hour collection of stool. The acid steatocrit method, which includes digestion, extaction and estimation and estimation of fat is used for the estimation of fecal fat. Although theoretically quite simple, the test is difficult to perform in practice. The patient must follow a diet containing 100 g/day fat for at least three days before the test, and complete collection of the sample is difficult to achieve. In health, less than 7 g of fat (7% of the ingested dose) should be present in stool. Measuring fecal fat requires that the dietary content of fat be known exactly, which is impossible unless the patient is housed in a general clinical research center. A qualitative analysis of fecal fat can also be performed with a Sudan III stain of a random specimen of stool. The finding of more than six globules per high-power field is considered a positive result, but as with fecal fat excretion, the patient must be ingesting adequate fat to allow measurable steatorrhea. Sudan

III staining of stool is positive only in patients with substantial steatorrhea. Because steatorrhea occurs only with advanced pancreatic disease, measurement of fecal fat is not useful in the diagnosis of mild or moderate disease, nor is steatorrhea specific for pancreatic disease. It should be noted that fecal fat quantification is a nonspecific pancreatic function test since any other cause of maldigestion (i.e., obstructive jaundice) or malabsorption (i.e., sprue, chrons disease) may also induce abnormal fecal fat excretion.

G. Serum Trypsinogen

Serum trypsinogen (often called serum trypsin) can be measured in blood and provides a rough estimation of pancreatic function. Very low levels of serum trypsinogen (<20 ng/mL) can be seen in patients with advanced chronic pancreatitis with steatorrhea. Serum trypsin is not decreased in patients with other forms of steatorrhea, but low levels of serum trypsinogen may be seen in patients with pancreatic ductal obstruction, including malignant obstruction.

H. Lundh Test Meal

The subject ingests a 300-mL liquid test meal composed of dried milk, vegetable oil, and dextrose (6% fat, 5% protein, and 15% carbohydrate). After ingestion of this meal, samples are aspirated from the intubated duodenum at intervals for measurement of digestive enzyme concentration. Usually only trypsin activity is measured; however, the additional determination of lipase or amylase may improve test sensitivity. The test is not valid in patients with mucosal disease (e.g., celiac disease) or altered gastroduodenal anatomy (e.g., following a vagotomy and drainage procedure or a Billroth II gastrectomy). Secretin-CCK test is more sensitive compared to Lundh test meal in detecting mild forms of pancreatic disease, whereas in more advanced disease the tests are comparable. This test is not recommended for clinical use since it has insufficient diagnostic accuracy.

I. *N*-benzoyl-L-tyrosyl-*p*-aminobenzoic Acid (NBT-PABA)

The bentiromide test is an indirect tubeless test in which the synthetic peptide *N*-benzoyl-l-tyrosyl-*p*-aminobenzoic acid (NBT-PABA) is specifically cleaved by chymotrypsin to NBT (*N*-benzoyl-l-tyrosyl) and para-aminobenzoic acid (PABA). PABA is then absorbed in the intestine, conjugated in the liver, and excreted in the urine. The PABA metabolite can be measured in either serum or urine. Prior gastric surgery, small bowel disease, liver disease, and renal insufficiency may interfere with the measurements. So may the use of several drugs (acetaminophen, phenacetin, chloramphenicol, benzocaine, lidocaine, procaine, sulfonamides, sulfonylureas, and thiazides) and prior ingestion of certain foods (prunes and cranberries).

A broad range of sensitivities has been reported for the NBT-PABA test. In patients with severe pancreatic insufficiency and malabsorption, the sensitivity is 80% to 90%. In those with mild to moderate impairment, test sensitivity is as low as 40%.

J. Fluorescein Dilaurate (Pancreolauryl) Test

Another test similar to the NBT-PABA test is the fluorescein dilaurate (pancreolauryl) test. The principle underlying this test is the same as that for the NBT-PABA test. Fluorescein dilaurate is an ester, poorly soluble in water, that is hydrolyzed by pancreatic carboxylesterase into lauric acid and free water-soluble fluorescein. The fluorescein is readily absorbed into the intestine, partly conjugated in the liver, and excreted in the urine. Fluorescein dilaurate is given in the middle of a breakfast meal. Urine is collected for 10 hours after breakfast, and the fluorescein excreted in the urine is measured. Like the PABA metabolite, fluorescein can also be measured in the serum. To evaluate the subject's absorption, conjugation, and excretion, the test is repeated two to three days later with free fluorescein. The recovery rate on both days is expressed as a ratio. Like the NBT-PABA test, the pancreolauryl test is highly sensitive and specific for advanced pancreatic disease and less so for mild or moderate disease.

K. ^{13}C-substrate Breath Tests

Several substrates,mainly ^{13}C-labelled, have been used to evaluate exocrine pancreatic function by means of breath tests. In these tests, the labeled substrate is given orally together with the test meal. After intraduodenal hydrolysis of the substrate by specific pancreatic enzymes, ^{13}C-marked metabolites are released, absorbed from the gut, and metabolized within the liver. As a consequence of hepatic metabolism, $^{13}CO_2$ is released and thereafter eliminated with the expired air. The amount of $^{13}CO_2$ expired, which indirectly reflects exocrine pancreatic function, can be measured by means of mass spectrometry or infrared analysis.

Most substrates used in breath tests,among them mixed ^{13}C-triglyceride, cholesteryl ^{13}C-octanoate, ^{13}C-hioleinand ^{13}C-triolein, are hydrolysed by pancreatic lipase. In this way pancreatic function breath tests should be seen as fat digestion tests and thus considered as an alternative to fecal fat quantification.

The breath test is a simple, noninvasive and accurate method for diagnosis of exocrine pancreatic insufficiency. The utility of the test is not only limited to the diagnosis of exocrine pancreatic insufficiency but can also be extended to control the efficacy of oral enzyme therapy in these patients.

L. "Enhanced Imaging" Pancreatic Function Tests (S-MRCP)

In this technique, MRI with secretin stimulation to increase the flow and volume in the pancreatic duct, has been used to diagnose pancreatic disorders. The filling of the duodenum can besemi-quantitated to assess for chronic pancreatitis. The limitation of this test is that it measures volume of pancreatic flow rather than bicarbonate concentration. MR images are acquired over at most 30 min, which is often insufficient length of time during secretin stimulation and which may lead to reduced sem.

M. Other Tests

Several other indirect tubeless tests like dual-label Schilling test and plasma measurements of pancreatic polypeptide and amino acids have been established to improve sensitivity for identifying milder forms of exocrine pancreatic dysfunction. These tests have not shown to have greater sensitivity than the indirect tubeless tests. In addition, many of them require radioactive isotopes or expensive equipment, making their usefulness less desirable.

III. EXOCRINE PANCREATIC FUNCTION TEST IN ACUTE PANCREATITIS

A. Serum Amylase

In healthy persons, the pancreas accounts for 40% to 45% of serum amylase activity, and the salivary glands account for the rest. Simple analytic techniques can separate pancreatic and salivary amylases. Because pancreatic diseases increase serum pancreatic (P) isoamylase, measurement of P-isoamylase can improve diagnostic accuracy. However, this test is rarely used.

The total serum amylase test is most frequently done to diagnose acute pancreatitis because it can be measured quickly and cheaply. It rises within 6 to 12 hours of onset and is cleared fairly rapidly from the blood (half-life, 10 hours). Probably less than 25% of serum amylase is removed by the kidneys. The serum amylase is usually increased on the first day of symptoms, and it remains elevated for three to five days in uncomplicated attacks. Sensitivity of serum amylase is greater than 85% but it may be normal or slightly elevated during mild pancreatitis, acute on chronic pancreatitis or during severe fatal pancreatitis. It may also be normal during recovery from acute pancreatitis. Serum amylase levels are known to be spuriously low during pancreatitis due to hypertriglyceridemia because of a possible amylase inhibitor associated with high triglyceride levels. In such a case, if serum is tested after serial dilutions, amylase level can be shown to be high.

Hyperamylasemia is not specific for pancreatitis because it occurs in many conditions other than acute pancreatitis. In fact, one half of all patients with an elevated serum amylase may not have pancreatic disease. In acute pancreatitis, the serum amylase concentration is usually more than two to three times the upper limit of normal; it is usually less than this with other causes of hyperamylasemia. However, this level is not an absolute discriminator. Thus, an increased serum amylase level supports rather than confirms the diagnosis of

acute pancreatitis. In addition, there are some individuals who have persistent hyperamylasemia without clinical symptoms. This has been reported due to macroamylasemia or pancreatic hyperamylasemia on a familial basis. Nonpancreatic diseases that cause hyperamylasemia include pathologic processes in organs (e.g., salivary glands, fallopian tubes) that normally produce amylase. Furthermore, mass lesions such as papillary cystadenocarcinoma of the ovary, benign ovarian cyst, and carcinoma of the lung, cause hyperamylasemia because they produce and secrete salivary-type isoamylase. Transmural leakage of pancreatic-type isoamylase and peritoneal absorption probably explain hyperamylasemia in intestinal infarction and in perforated viscus. Renal failure increases serum amylase up to four to five times the upper limit of normal due to decreased renal clearance of this enzyme. Patients on hemodialysis tend to have higher serum amylase levels than those on peritoneal dialysis. In patients with chronic kidney disease, there is no clear correlation between the creatinine clearance and serum levels of amylase, and about one third of patients with marked renal insufficiency have normal pancreatic enzyme levels.

Chronic elevations of serum amylase (without amylasuria) occur in macroamylasemia. In this condition, normal serum amylase is bound to an immunoglobulin or abnormal serum protein to form a complex that is too large to be filtered by renal glomeruli and thus has a prolonged serum half-life. Macroamylasemia may complicate the diagnosis of pancreatic disease, but it has no other clinical consequence. The urinary amylase-to-creatinine clearance ratio (ACCR) increases from approximately 3% to approximately 10% in acute pancreatitis. However, even moderate renal insufficiency interferes with the accuracy and specificity of the ACCR. Other than to diagnose macroamylasemia, which has a low ACCR, urinary amylase and the ACCR are not used clinically. Macroamylasemia can also be measured directly using serum. Deliberate contamination of urine with saliva, as in Munchausen's syndrome, can increase the urine amylase, with the serum amylase being normal. This situation can be excluded by measuring salivary amylase in the urine.

B. Serum Lipase

The sensitivity of serum lipase for the diagnosis of acute pancreatitis is similar to that of serum amylase and is between 85% and 100%. Lipase may have greater specificity for pancreatitis than amylase as serum lipase is normal when serum amylase is elevated as in salivary gland dysfunction, tumors, gynecologic conditions, and macroamylasemia. Serum lipase always is elevated on the first day of illness and remains elevated longer than does the serum amylase. Consequently some suggest combining lipase with amylase as a test for acute pancreatitis. However, we and others have found that combining enzymes does not improve diagnostic accuracy.

Lipase has problems of specificity similar to estimation of amylase. Patient with renal insufficiency may have elevation up to two times of normal lipase levels witihout having pancreatitis.

In abdominal conditions mimicking acute pancreatitis lipase levels may rise up to three times normal likely by absorption through inflamed or perforated intestine and small bowel obstruction may also have similar rise in lipase levels. Some studies have reported that lipase is equally sensitive and more specific than serum amylase estimations and should be preferred but other studies find no similar advantage.

Other pancreatic enzymes

During acute pancreatic inflammation, pancreatic digestive enzymes other than amylase and lipase leak into the systemic circulation and have been used to diagnose acute pancreatitis. They include PLA_2, trypsin/typsinogen, carboxylester lipase, carboxypeptidase A, colipase, elastase, and ribonuclease. None, alone or in combination, are better than serum amylase or lipase, and most are not available on a routine basis.

IV. CLINICAL USE OF PANCREATIC FUNCTION TESTS

It is obvious that a wide variety of pancreatic function tests are available to clinicians. Whilst clinicians might consider using these tests to diagnose chronic pancreatitis, to determine the severity of pancreatic insufficiency or to follow patients over time to determine the progress of chronic pancreatitis. The diagnosis is suspected on the basis of compatible signs and symptoms and most often confirmed by imaging studies or pancreatic function tests. In general the blood tests are more useful for the diagnosis of acute pancreatitis disease, while the tests on stool and pancreatic juice are for that of chronic diseases of the pancreas.

V. CONDITIONS THAT ALTER PANCREATIC EXOCRINE SECRETION AND MAY REQUIRE ENZYME THERAPY

Chronic pancreatitis is the most frequent and relevant indication for pancreatin supplementation therapy. In an unselected group of patients with chronic pancreatitis 80–90% show some degree of pancreatic exocrine insufficiency but in most patients, severe pancreatic exocrine insufficiency with overt steatorrhoea is a late event in the course of the disease. Once overt malabsorption has occurred, patients definitely need pancreatic enzyme supplementation therapy. Still, even today, refractory steatorrhoea despite high doses of enzymes is a frequent problem in enzyme supplementation therapy. A therapeutic trial may also be indicated in chronic pancreatitis patients without overt malabsorption but with abdominal symptoms which cannot be attributed to the inflammatory process itself but may be due to regulatory disturbances.

During the first few weeks following onset of *acute pancreatitis*, pancreatic exocrine function appears to be regularly impaired and a high proportion of patients with necrotising disease may show severe exocrine insufficiency.

Milder forms of pancreatic exocrine insufficiency usually recover step by step but particularly patients with expanded necrosis may need enzyme therapy for a prolonged period of time.

Severe pancreatic exocrine dysfunction and malabsorption occur in 80–90% of patients with *pancreatic cancer*. Site and degree of ductal obstruction appear to be major determinants of residual pancreatic exocrine secretion in these patients. In view of the poor prognosis of these patients and the impact of nutritional status on quality of life and survival indication for enzyme therapy should be administered generously.

Depending on the genetic background *cystic fibrosis* is associated with pancreatic exocrine insufficiency in the vast majority of patients. Patients presenting with steatorrhoea and/or other symptoms of malabsorption definitely need pancreatic enzyme supplementation therapy. Enzyme supplementation has been shown to markedly improve nutrient digestion and absorption though it is rarely normalised. This may in part be due to the fact that in cystic fibrosis further disturbances of gastrointestinal secretory and motor functions—that is, alterations of gastrointestinal pH, motility, and transit—additionally compromise assimilation of nutrients.

Pancreatic resections naturally decrease pancreatic secretory capacity and it depends on the original disease process and on the type and extent of resection whether patients develop maldigestion and malabsorption. In patients with previously normal pancreatic function, even a 90–95% resection of the pancreas is generally tolerated without clinical signs of severe pancreatic exocrine insufficiency. By contrast, in patients with pre-existing chronic pancreatitis, the percentage of patients with steatorrhoea is markedly increased by less extensive distal resections. Thus, pancreatic exocrine function needs to be monitored carefully postoperatively, particularly in patients who are at increased risk of developing severe pancreatic exocrine insufficiency. Naturally, high doses of enzymes with every meal are needed following total pancreatectomy. Moreover, all patients with overt malabsorption following limited pancreatic resections need enzyme therapy.

Partial or total gastrectomy and *short bowel syndrome* may also be associated with intraluminal lack of pancreatic enzymes postprandially, mainly because of postcibal asynchrony—that is, dissociation between intestinal delivery of nutrients and discharge of digestive secretions. In patients operated on because of malignant disease, extensive denervation of the pancreas including vagotomy contributes to pancreatic dysfunction. Thus, pancreatic exocrine function should be measured in patients with symptoms of malabsorption following gastric or extensive small bowel resections. Indirect tests may be preferred because they reflect digestive action of endogenous enzymes under the patient's pathological conditions. Enzyme supplementation therapy should be initiated in those patients with documented exocrine insufficiency.

Mostly mild to moderate pancreatic exocrine insufficiency is observed in up to 40% of patients with *coeliac disease*. Impairment of pancreatic exocrine function is mainly due to decreased release of stimulatory mediators by the diseased upper intestinal mucosa and to postcibal asynchrony of secretory and motor functions. It remains unclear whether there may be direct pancreatic insufficiency in a subset of patients. At least in children with coeliac disease, low faecal enzyme concentrations might be used to select patients who would probably benefit from enzyme treatment during the first few weeks of dietary therapy.

Diabetes mellitus is associated with morphological and functional impairment of the exocrine pancreas. Mostly mild to moderate pancreatic exocrine insufficiency is observed in about 25–80% of patients with IDDM and in 15–73% with NIDDM. Severe pancreatic exocrine insufficiency associated with overt steatorrhoea which obviously necessitates enzyme therapy affects only a minority of these patients. However, patients with diabetes mellitus frequently suffer from a wide range of abdominal symptoms which markedly contribute to impairment of quality of life and might in part be caused by a distal shift of nutrient digestion and absorption as observed in mild to moderate pancreatic exocrine insufficiency. In such patients, a therapeutic trial with enzyme preparations may be justified but necessitates close control of blood glucose.

In ***Crohn's disease*** decreased pancreatic exocrine insufficiency as assessed by routine pancreatic function tests occurs in 5–15% of patients. Mostly, mild to moderate exocrine insufficiency is observed which would not be expected to cause clinical symptoms—that is, no steatorrhoea in subjects with an otherwise healthy digestive system. It might be speculated, however, that in Crohn's disease patients with reduced absorptive capacity due to small intestinal resections, extensive inflammation, or scars and/or motility disturbances, lesser degrees of pancreatic insufficiency might be enough to induce overt malabsorption. Accordingly, selected patients might benefit from enzyme supplementation. Moreover, a distal shift of nutrient digestion and absorption and consecutive regulatory disturbances may contribute to symptoms and might respond to enzyme treatment.

In ***Zollinger-Ellison syndrome*** intraluminal enzymes are destroyed by excessive amounts of gastric acid which leads to steatorrhoea in about 5–10% of patients. Thus, these patients need high doses of proton pump inhibitors and/or surgical therapy but usually do not benefit from enzyme supplementation.

VI. ENZYME SUPPLEMENTATION THERAPY

Modern pancreatic enzyme preparations consist of pH sensitive pancreatin microspheres. In order to reduce steatorrhoea to less than 15 g of fat per day, a minimal dose of 25,000–50,000 IU of lipase per meal is required. In order to ensure release of enzymes throughout the digestive period it is recommended that for small meals or snacks, one tablet/capsule should be swallowed with the start of the meal, and that for major meals a second dosage should be administered during the meal.

Acid resistant coating of the microspheres protects exogenous enzymes from acidic denaturation in the stomach but allows disintegration and release of enzymes at about pH 5.5–6 which is achieved physiologically in the postprandial period. The small size of microspheres enable simultaneous delivery of enzymes and meal nutrients to the duodenum because larger indigestible particles (exceeding about 2 mm in diameter) are retained within the stomach until the end of the digestive period. Microsphere preparations have been shown to be superior to equivalent doses of unprotected pancreatin powder and to acid resistant monolithic capsules or tablets. However, due to the physicochemical properties of the coating the microspheres may be somewhat retained in the stomach postprandially (in the lipophilic phase of the meal) and—more importantly—release of enzymes within the intestinal lumen does not occur instantly but is retarded particularly if intestinal pH is below the pH threshold cited above due to loss of bicarbonate secretion. Thus, nutrient digestion and absorption are shifted from the duodenum to the more distal small intestine. These appear to be the main reasons why lipid digestion and absorption cannot be normalised in a subset of patients with severe pancreatic exocrine insufficiency despite administration of high doses of modern enzyme preparations.

Following (total) gastrectomy and in patients with gastric acid suppression therapy due to other reasons, unprotected pancreatin powder should be preferred. On the other hand, application of proton pump inhibitors together with enzyme supplementation may increase fat absorption in patients who do not respond adequately to standard enzyme treatment. Moreover, failure of enzyme therapy to adequately improve steatorrhoea and other symptoms of exocrine insufficiency may be due to low patient compliance, intestinal infections or other disorders associated with malabsorption. These require specific diagnostic work up and therapeutic intervention.

SUMMARY

A wide variety of tests are available nowadays for evaluation of exocrine pancreatic function. Which pancreatic function test should be used depends on the clinical question and the characteristics and availability of the test. The exocrine pancreas has a very large functional reserve. Maldigestion and malabsorption do not occur until the functional capacity of digestive enzyme secretion is reduced to 5% to 10% of normal. Thus, many tests relying on the conversion of an ingested substrate by digestive enzymes to a measurable product will be insensitive in pancreatic disease unless moderate to severe pancreatic insufficiency is present. Therefore, the measurement of duodenal digestive enzymes after the intravenous administration of pancreatic secretagogues provides the greatest sensitivity and specificity. The major drawbacks to the direct tests are the requirements for duodenal intubation and the fact that very few centers are proficient in performing the studies properly. Improved imaging techniques for diagnosing pancreatic disease have greatly decreased the use of the tests. On certain occasions, however, pancreatic function tests are necessary for supporting the diagnoses of pancreatic disease. The non invasive tests tend to perform poorly in patients with early, mild disease.

SUGGESTIONS FOR FURTHER READING

Chowdhury R.S. and Forsmark C.E. (2003), Review Article: Pancreatic Function Testing. *Aliment Pharmacol Ther*, 17:733–750.

John G. Lieb II and Peter V. Draganov (2008), Pancreatic Function Testing: Here to Stay for the 21st Century, *World J. Gastroenterol*, 14(20): 3149–3158

Juan Enrique Dominguez-Munoz (2004), Pancreatic Function Tests for Diagnosing and Staging of Chronic Pancreatitis, Cystic Fibrosis and Exocrine Pancreatic Insufficiency of Other Etiologies: Which Tests are Necessary and How should they be Performed in Clinical Routine, In: *Clinical Pancreatology for Practising Gastroenterologists and Surgeons*. Juan Enrique Dominguez-Munoz (Ed.), Wiley-Blackwell, 259–26.

Keller J. and Layer P. (2005), Human Pancreatic Exocrine Response to Nutrients in Health and Disease. Gut, 54 (Suppl VI): Vi1–vi28.

Stephen J. Pandol (2010), Pancreatic Secretion, In: Sleisenger and Fordtran's *Gastrointestinal* and *Liver Disease*, Pathophysiology/Diagnosis/Management. Edited by William O. Tschumy Jr., Lawrence S. Friedman and Lawrence J. Brandt, Elsevier.

Tandon R.K. and Tandon B.N. (2002), *Evaluation of Exocrine Pancreatic Function*, In: *Textbook of Biochemistry* and *Human Biology*. G.P. Talwar and L.M. Srivastava (Eds.), PHI Learning Private Limited, 623–633.

54

Malabsorption Syndromes
Procedures for Studying Absorption

Govind K. Makharia and Hanish Sharma

CONTENTS

I. INTRODUCTION

The gastrointestinal (GI) tract consists of a hollow organ extending from mouth to the anus and the associated glandular organs (i.e. salivary glands, liver, gallbladder and pancreas) that empty their contents into the GI tract. The assimilation of a meal involves major physiological processes in the GI tract such as motility, secretion, digestion, absorption, and elimination. Normal food is a complex mixture of polymers of carbohydrate, fat, proteins, and fibers in addition to micronutrients including vitamins. All these polymers cannot be absorbed as such and has to be broken down in small absorbable sub-units. The main function of the gastrointestinal tract is digestion and absorption of the nutrient components of the food. The processes of digestion and absorption are complex. Food is taken into the mouth as large particles containing macromolecules that are not absorbable. The breaking down of food into absorbable subunits occurs by grinding and mixing of the food (motility) with various secretions containing enzymes and water that enter the GI tract. The enzymes convert the macromolecules into absorbable molecules in a process termed digestion. The products of digestion, as well as the secretions from the upper parts of the GI tract, are then transported across the epithelium to enter the blood or lymph by a process termed absorption. Secretions and luminal contents are moved from the mouth to the anus and eliminated by GI motility. The coordination of GI function is regulated in a synchronized way to maximize digestion and absorption by means of multiple control mechanisms.

The liver and pancreas are anatomically coupled to the GI tract through the biliary and pancreatic ductal systems, respectively. Further the liver receives blood circulation directly from the GI tract through the portal venous circulation. The liver has numerous vital functions, including bile formation and secretion; metabolism of carbohydrate, proteins and fats; synthesis of proteins; detoxification; and immune surveillance. The exocrine pancreas secrets the major digestive enzymes and bicarbonate, whereas the endocrine pancreas seecrets hormones that plays a central role in the uptake, storage and release of nutrients in the liver, as well as in other tissues such as skeletal muscle and adipose tissue. Defects in any of the abovementioned functions can lead to disturbances in the absorptive function of the intestine.

II. OVERVIEW OF INTESTINAL DIGESTION AND ABSORPTION

The role of the stomach and pancreas in digestion of food has been elaborated in another chapter in the book. Basically the stomach and pancreatic secretions in addition to bile ensure activation of proteolytic and lipolytic enzymes which breaks down ingested food into oligopeptides, fatty acids and simple sugars. The intestinal epithelium is responsible for further breakdown to monosaccharides, disaccharides, amino acids and fatty acids leading to absorption across the intestinal epithelium. There is active receptor mediated absorption of monosaccharides, disaccharides, amino acids, fatty acids and divalent cations. Water is absorbed by passive transport due to osmotic drag.

The basic absorptive unit of the intestinal surface is organized as a crypt-villous unit (Figure 54.1). The luminal surface of the small intestine is thrown into circular folds called duodenal folds. These folds are covered by millions of small projections called villi, which increase the active surface exponentially. These villi are covered predominantly with mature, absorptive enterocytes, along with occasional mucus-secreting goblet cells. Enterocytes have a limited life of 3-5 days, then they are replaced by new cells and are shed into the lumen. Crypts are tubular invaginations of the epithelium around the villi, lined largely with younger epithelial cells, which are involved primarily in secretion. At the base of the crypts are stem cells, which continually divide and provide the source of all the epithelial cells in the crypts and on the villi. The crypt–villous unit also follows a definite pattern of absorptive and secretory functions distributed from the base of the crypt to the tip of the villous. Virtually all nutrients, including all amino acids and sugars, enter the body across the epithelium covering small intestinal villi. As shown in Figure 54.1, each villus contains a capillary bed and a lymphatic channel. After crossing the epithelium, most of these molecules diffuse into a capillary network inside the villus, and hence into systemic blood. Some molecules, fats in particular, are transported not into capillaries, but rather into the lymphatic vessels, which drain from the intestine and finally flows into blood.

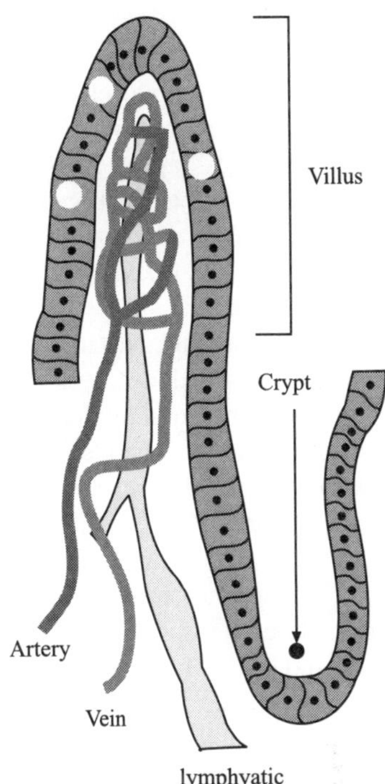

Figure 54.1 Crypt villous unit. (*see Plate 17 for colour figure*)

A. Physiology of Digestion of Carbohydrate

The primary carbohydrates in the human diet are primarily starches and simple sugars. Simple sugars in the diet include disaccharides such as sucrose and lactose, and monosaccharides such as glucose and fructose. Starches are composed of polysaccharides and the two major polysaccharides in the starch are amylose and amylopectin. While disaccharides consist of two monosaccharide units, oligosaccharides contain three to nine monosaccharide units. Polysaccharides consist of more than nine monosaccharide units joined in long linear or branched chains, as occurs in starches. The human diet also consists of non-digestible sugars such as stachyose and raffinose, which are present in legumes, and sugar alcohols such as sorbitol, xylitol, mannitol, lactilol, and maltitol.

The hydrolytic enzyme α-amylase initiates the digestion of starch in the lumen of small intestine. Two different iso-enzymatic forms of α-amylase, salivary and pancreatic, mediate the luminal phase of starch digestion. The terminal products of luminal digestion of amylopectin are maltose, maltotriose, and α-limit dextrins, which are a series of oligosaccharides containing four or more glucose molecules linked by α-(1,6)-glycosidic bonds. The optimal enzymatic activity of salivary α-amylase is limited to the narrow pH range of 6.6–6.8. Therefore, although salivary α-amylase initiates starch digestion, its contribution is minor because it is rapidly inactivated in the acidic environment of the stomach. Starch digestion is mediated largely by pancreatic amylase, which is secreted into the intestinal lumen in amounts exceeding that required for the hydrolysis of starch.

The terminal digestion of dietary carbohydrates is mediated by hydrolytic enzymes namely sucrase–isomaltase, lactase, and trehalase which are located on the apical brush border membrane of epithelial cells of intestinal mucosa. These saccharidases are responsible for the hydrolysis of the products of luminal starch digestion (i.e., maltose, maltotriose, and α-limit dextrins) and of ingested disaccharidases (i.e., lactose and sucrose). The enzymatic action of the brush border saccharidases results in the production of monosaccharides primarily glucose, galactose, and fructose that can be transported across the apical membrane.

The absorption of glucose and galactose through the apical membrane of enterocytes is mediated by a sodium-dependent active transport mechanism (sodium/glucose co-transporter, SGLT-1) in which each monosaccharide is transported along with two sodium ions. SGLT-1 is capable of mediating the apical membrane transport of glucose or galactose into the enterocyte. (Figure 54.2) SGLT-1 and glucose absorption also play a major role in intestinal water transport. The human SGLT-1 transporter can co-transport approximately 264 water molecules for every molecule of glucose absorbed through the enterocytes. The ingestion of an average carbohydrate diet results in transportation of upto 5–6 litre of water through the enterocytes by SGLT1 co-transport mechanism in the human intestine. Increased intestinal water transport coupled to glucose absorption is the rationale for adding glucose to oral rehydration solutions used to treat dehydration. Glucose and galactose are transported out of the enterocyte into the portal circulation by the GLUT-2 transporter located on the basolateral membrane. The fructose is transported across the apical intestinal brush border membrane by a sodium-independent facilitated diffusion mechanism involving the GLUT-5 transporter.

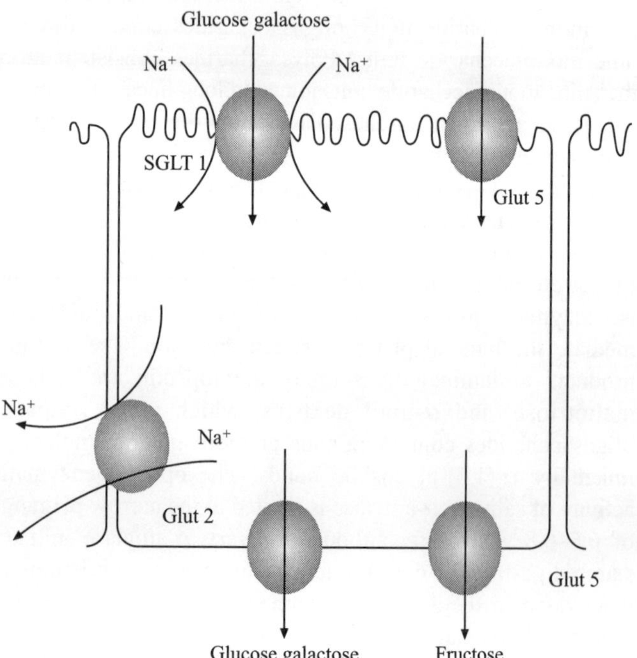

Figure 54.2 Absorption of glucose, galactose and fructose through the enterocytes.

B. Physiology of Digestion of Protein

Dietary amino acids are ingested predominantly in the form of proteins rather than in the form of free amino acids or small peptides. Protein digestion begins in the lumen of the stomach with the action of proteases named gastric pepsins, which are secreted by the chief cells present in the stomach. Pepsins are acidic proteases that get activated in the acidic pH, hence their activity is restricted to the stomach only. Pepsins are endoproteases that hydrolyze internal peptide bonds in the proteins and liberate large peptide fragments. Pepsins act preferentially on peptide bonds formed by the aromatic amino acids such as phenylalanine and tyrosine, and branched-chain amino acid such as leucine.

The intestinal phase of digestion of protein consists of three phases, which are classified depending upon the exact location of the digestion in the intestine. The three phases are the luminal phase, the brush border phase, and the intracellular phase. The luminal phase of protein digestion in the small intestine begins in the duodenum where the pancreatic juice enters the small intestine and is mediated by pancreatic proteases and peptidases. The pancreas secretes three endoproteases (trypsinogen, chymotrypsinogen, and proelastase) and two exopeptidases (procarboxypeptidase A and procarboxypeptidase B). In order to protect the auto-digestion of pancreas, each of these enzymes is synthesized and secreted as an inactive precursor. Activation of these inactive enzymes is initiated in the duodenum by enteropeptidase, an enzyme secreted in the small intestine, especially duodenum. Pancreatic endoproteases and exopeptidases function in a complementary manner to carry out the efficient digestion of proteins. The action of these pancreatic enzymes on proteins in the intestinal lumen produces a mixture of oligopeptides and free amino acids. The amino acids thus generated are absorbed as such by the enterocytes. The brush border above the mucosal surface of enterocytes consists of high levels of amino-peptidases, which mostly are exopeptidases, which are capable of hydrolyzing the terminal peptide bonds on the amino terminal of oligopeptides. In addition to several exo-peptidases, the intestinal brush border membrane also contains endopeptidases and dipeptidases.

The absorption of free amino acids into enterocytes across the brush border membrane is mediated by a number of distinct amino acid transport systems. Peptides consisting of two or three amino acids are absorbed intact across the brush border membrane by a specific peptide transport system. The intestinal brush border membrane possesses a specific transport system that accepts dipeptides and tripeptides as substrates consisting of neutral amino acids, anionic amino acids, and cationic amino acids. Free amino acids in the cytoplasm of the enterocyte enter into various metabolic pathways, such as degradation, conversion into other amino acids, and incorporation into cellular proteins; they also enter the portal circulation by way of specific amino acid transport systems through the basolateral membrane. Glutamine, glutamate, aspartate, and arginine are preferentially used by the enterocyte as metabolic fuel.

C. Physiology of Digestion of Fat

Dietary fat is ingested in the form of triglycerides with three long chain fatty acids on a glycerol backbone. The properties of the fatty acids that determine their solubility and mobility include the length of the carbon chain and presence of additional reactive groups such as phosphorylation. Based on the length of the carbon chain in the fatty acid, they are classified as short chain fatty acids (having <6 carbon chain), medium chain fatty acids (having 6-12 carbon chains) and long chain fatty acids (>12 carbon chains).

The digestion of dietary lipids in humans begins in the stomach. The initial step involves the enzyme gastric lipase, which is secreted by the gastric mucosa. Gastric lipase has its highest activity in the acidic environment of the stomach and breaks down fat into diacylglycerol and fatty acids. Milk fat, the primary source of nourishment to neonates, consists mainly of medium-chain triglycerides, which are efficiently hydrolyzed by gastric lipases.

The released fatty acids along with ingested proteins, aid in stimulating the release of cholecystokinin from the duodenal mucosa. The cholecystokinin then stimulates the release of pancreatic secretions and causes gallbladder contraction with relaxation of the sphincter of Oddi, which leads to expulsion of bile salts in the duodenum. Bile salts helps in the emulsification of fat and micellar formation. An emulsion is a suspension of fat in water that allows exposure of larger surface area of dietary lipids to the pancreatic lipase.

The important enzymes involved in fat digestion in the intestine are secreted from pancreatic acini and carried through the pancreatic duct into duodenum. The pancreatic lipolytic enzymes are pancreatic lipases, phopsholipases, and cholesterol esterase. Pancreatic lipase is responsible for most of the digestion of triglycerides in the upper part of the intestinal lumen. Pancreatic lipase is active at the interface between the oil and aqueous phases. Pancreatic lipase is abundant in pancreatic juice accounting for 2%–3% of the total protein secreted in the intestine. Pancreatic lipase acts on the triglyceride molecule to release 2-monoacylglycerol (2-monoacylglycerol) and free fatty acids, which are absorbed by the small intestine. Digestion and absorption of fat is very efficient and most humans can absorb 90% to 95% of the ingested fat. Fatty-acid transporters are localized on the brush border membranes, and their expression is highest in the jejunum, less abundant in the duodenum, and least in the ileum. Various absorbed fat molecules migrate from the site of absorption to the endoplasmic reticulum, where biosynthesis of complex lipids occurs. They are packaged as chylomicrons and very low density lipoproteins (Figure 54.3).

The absorption of medium chain triglycerides (MCTs) resembles more close to that of carbohydrates than that of fats. MCTs are more water-soluble, do not require micellar digestion and are transported rapidly across the enterocytes. These properties make MCTs as the preferred fat in clinical conditions where normal fat digestion is disrupted. Some of these conditions are insufficiency of the pancreas (where secretion of pancreatic enzymes are compromised), obstructive diseases of biliary system (therefore bile is not secreted in the duodenum which affects micelle formation which are required for the digestion of long chain triglycerides) and intestinal lymphangiectasias (where the lymphatic system is blocked affecting transport of long chain triglycerides).

D. Physiology of Absorption of Vitamin B$_{12}$

The main source of vitamin B$_{12}$ is animals and dairy products, therefore, pure vegans suffer from vitamin B$_{12}$ deficiency. Vitamin B$_{12}$ is released from the food by gastric acid and pepsin. In the stomach, vitamin B$_{12}$ combines with intrinsic factor, which is secreted by the gastric parietal cells. The vitamin B$_{12}$-intrinsic factor complex is then absorbed in the ileum. Vitamin B$_{12}$ also binds to R-proteins in intestinal secretions the

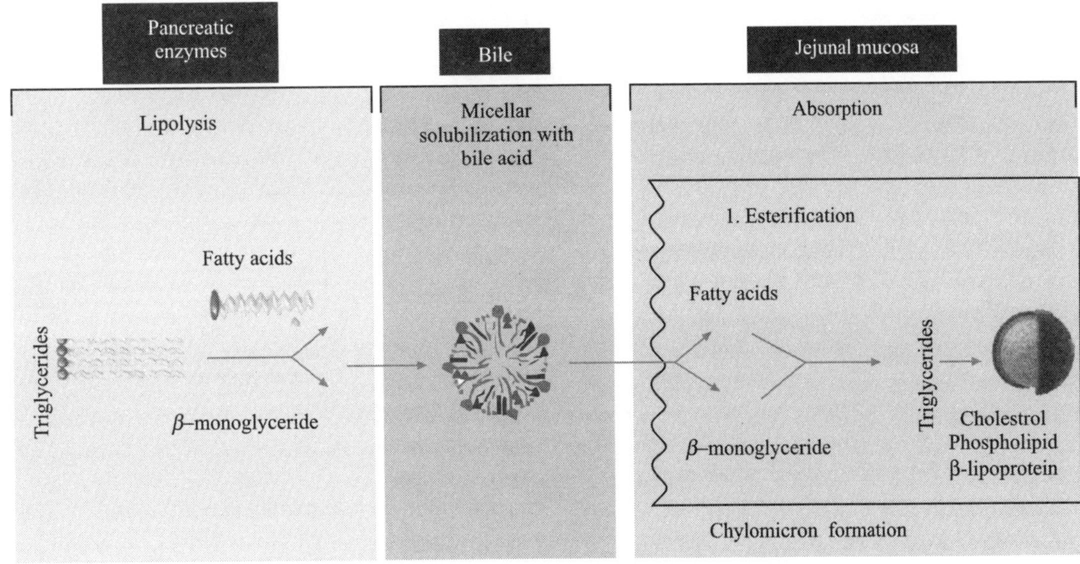

Figure 54.3 Digestion and absorption of fat. (*see Plate 17 for colour figure*)

R-protein-B_{12} complex cannot be absorbed by the ileum. The pancreatic proteases cleave this bond making B_{12} available to bind with intrinsic factor and enabling subsequent absorption.

Therefore diseases affecting secretion of intrinsic factor (pernicious anemia, gastrectomy), diseases of the pancreas and diseases of terminal ileum can lead to malabsorption of vitamin B_{12}, and thus, its deficiency. Deficiency of vitamin B_{12} leads to anemia and neurological symptoms.

E. Physiology of Absorption of Folic Acid

Folate is ingested as polyglutamates, which is hydrolyzed to monoglutamates by the enzyme folate diconjugase in the small intestine brush border membrane. Folate is then avidly taken up by the enterocytes by folate transporters. Absorption of folic acid occurs throughout the small intestine, more in the proximal small intestine.

F. Physiology of Absorption of Iron

Iron is essential for multiple metabolic functions. It helps in synthesis of heme, a component of red blood cells, which carries oxygen. The absorption of iron occurs in the proximal small intestine. Inadequate absorption is a significant cause of iron deficiency.

Site-specific absorption of nutrients in the GI tract

Absorption of nutrients occurs at the specific sites in the small intestine as shown in Figure 54.4.

III. MALDIGESTION AND MALABSORPTION

As discussed in the physiology of carbohydrate, fat and proteins, they are first digested in the intestinal lumen by enzymes secreted by stomach, pancreas and the gallbladder. The

final phase of digestion occurs in the small intestinal mucosa from where they are absorbed too. If there is impairment of any of the many steps involved in the complex process of nutrient digestion and absorption, intestinal malabsorption may ensue. Any impairment in the intraluminal digestion is called maldigestion, which occurs mainly due to diseases of the pancreas, and hepato-biliary diseases. On the other hand, malabsorption refers to impaired absorption of nutrients through the small intestinal mucosal layer. It can result from the disease affecting small intestinal mucosa. If the abnormality involves a single step in the absorptive process, as in primary lactase deficiency, or if the disease process is limited to the very proximal small intestine, selective malabsorption of only a single nutrient (iron or folate) may occur. However, generalized malabsorption of multiple dietary nutrient develops when the disease process is extensive.

Impaired intestinal lymphatic drainage because of obstruction of the intestinal lymphatic system leads to rupture of the intestinal lymphatics. Rupture of lymphatics leads to loss of lymph which is a high source of proteins. This protein is lost in the stool, a condition called protein losing enteropathy.

IV. DISEASES CAUSING MALABSORPTION SYNDROMES

As discussed earlier, the causes of malabsorption are numerous but a few of them are the most common. Furthermore the causes of malabsorption vary in different geographical regions.

V. MANIFESTATIONS OF MALABSORPTION

Depending on the nature of the disease process causing malabsorption and its extent, gastrointestinal symptoms may range from severe to subtle. In some, there may not be any

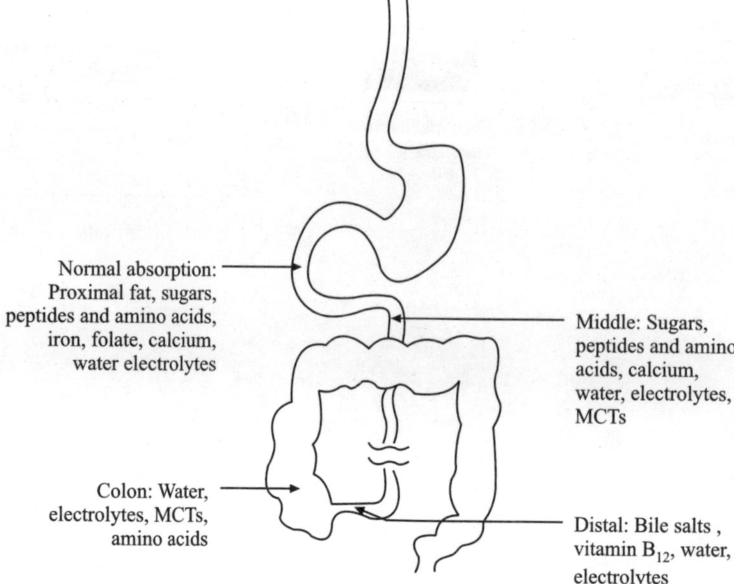

Figure 54.4 Absoption of different nutrients at different sites of the small intestine.

TABLE 54.1 Classification of causes of intestinal malabsorption

Conditions with predominant impaired intraluminal digestion
- Pancreatic disease: chronic pancreatitis
- Hepatobiliary disease
- Postgastrectomy malabsorption

Conditions with predominant impaired mucosal digestion, uptake, and transport
- Celiac disease
- Tropical sprue
- Parasitic infestations
- Crohn's disease
- Immunoproliferative small intestinal disease (IPSID) and lymphoma Abetalipoproteinemia
- Eosinophilic gastroenteritis

Conditions with impaired post mucosal transport
- Primary intestinal lymphangiectasia
- Secondary intestinal lymphagiectasia

TABLE 54.2 Features seen in vitamin deficiency states

Vitamins	*Manifestations*
Vitamin A	Xerophthalmia, nightblindness, bitot spots, follicular hyperkeratosis immune dysfunction
Thiamine	Beriberi: neuropathy, muscle weakness and wasting, cardiomegaly, edema, confabulation
Riboflavin	Angular stomatitis, Cheilosis
Niacin	Pellagra: pigmented rash of sun-exposed areas, bright red tongue, diarrhea, apathy, memory loss, disorientation
Vitamin B_6	Seborrhea, glossitis, convulsions, neuropathy, depression, confusion
Foliac acid	Megaloblastic anemia, atrophic glossitis
Vitamin B_{12}	Megaloblastic anemia, loss of vibratory and position sense, dementia
Vitamin C	Scurvy, poor wound healing
Vitamin D	Rickets: skeletal deformation, rachitic rosary, bowed legs; osteomalacia
Vitamin E	Peripheral neuropathy, spinocerebellar ataxia
Vitamin K	Elevated prothrombin time, bleeding

symptoms despite presence of malabsorption. Malabsorption of carbohydrate, protein and fat leads to passage of undigested food in the stool that leads to increase in both the frequency and the volume of the stool. Undigested carbohydrate is broken down in the colon by the colonic bacteria, which forms hydrogen and carbon dioxide. Methanogenic bacteria in the intestine convert hydrogen and carbon dioxide into methane. Therefore, patients with malabsorption have feeling of fullness in the abdomen (bloating) and passes a lot of gas as flatus, which may be, foul smelling. Malabsorption of carbohydrate leads to loss of energy therefore these patients feels very weak, and gets tired easily. Malabsorption of fat leads to passage of undigested fat in the stool that leads to greasy stool (steatorrhea) and low fat mass in the body. Failure of absorption of nutrients leads to failure to gain in the weight in case of growing children or can lead to loss of weight in adults.

Malabsorption of iron, folate or vitamin B_{12} leads to deficiency of these nutrients. All these are required for heme synthesis. Therefore, anemia is a very common finding in patients with malabsorption. Malabsorption of calcium and vitamin D (both required for bone formation) leads to deficiency of calcium and vitamin D. This leads to weakening of bones called osteopenia and osteoporosis. Weak bones predispose to bone fractures even with minor trauma. Malabsorption of proteins leads to deficiency of both somatic proteins and visceral proteins. Low somatic proteins mean low muscle mass and low visceral protein means low circulating serum proteins. One of the visceral proteins is albumin, which exerts oncotic pressure and keeps fluid drawn in the vascular compartment. Low serum albumin leads to low oncotic pressure, that leads to leakage of fluid from capillaries into the interstitial space which causes edema (swelling of feet). Edema is more common in diseases where protein is lost from the intestine in a disease called protein-losing enteropathy caused by lymphatic obstruction or extensive mucosal ulcerations. Vitamins and micronutrients are very useful in many metabolic processes and their deficiency can lead to many features, a summary of which is shown in Table 54.2.

VI. INVESTIGATIONS FOR ASSESSING DIGESTION AND ABSORPTION

While there are clinical features, as described above, which suggest presence of malabsorption, one needs to confirm presence of malabsorption by investigations. The basic principle of investigation to test absorptive function include:

- Investigations to confirm normal absorption
- Investigations to confirm mucosal malabsorption
- Investigations for assessment of absorption of fat
- Investigations for assessment of absorption of carbohydrate
- Investigation for assessment of absorption of proteins
- Intestinal permeability and its assessment

A. Investigations to Confirm Normal Absorption

A few general investigations may indicate absorptive function such as level of hemoglobin (suggests absorption of iron, folate and vitamin B_{12}), serum calcium or serum albumin, serum folate and vitamin B_{12} levels.

All that is not absorbed in the intestine is passed through the GI tract as stool. While stool tests and collection are cumbersome, it provides good information regarding absorption of carbohydrate, proteins and fat in a patient with malabsorption. Some of these are done frequently others are rarely required as same information can be drawn from non-stool based investigations.

Weight of the stool

Stool weight is a good surrogate marker of malabsorption as all that is not absorbed in the intestine is passed through the GI tract as stool. While stool weight is 200 g/day in the western countries, the stool weight of most Indians is higher than 200 g/day (up to 400 g/day) as diet of most Indians contain lot of fibers (such as leafy vegetables, salads). Weighing of stool is not done in most cases, however visual inspection by the patient or the health care provider gives an idea about the bulkiness of the stool.

pH of the stool

Malabsorption of carbohydrate in the small intestine leads to passage of partially digested carbohydrate in the large intestine where they are acted upon by colonic bacterial flora. The action of bacterial flora on the partially digested carbohydrate leads to production of short chain fatty acids and lactic acid which are passed along with the malabsorbed stool making stool acidic. The pH of stool <5.3 indicates predominant carbohydrate malabsorption. The pH of stool of patients having pan-malabsorption of carbohydrate, fat and protein malabsorption ranges between 6.0–7.5.

B. Investigations for Mucosal Absorption

Investigations to assess absorptive function of proximal intestinal mucosa.

D-xylose test

The D-xylose test is used widely to assess the absorptive capacity of the proximal small intestine. D-xylose is a pentose monosaccharide that is incompletely absorbed in the small intestine. The mechanism of D-xylose absorption appears to be different from those of other monosaccharides, which are absorbed by a sodium-hexose co-transporter. D-xylose is absorbed by facilitated diffusion, but the dose used for absorptive studies crosses the mucosa largely by passive diffusion. The D-xylose tolerance test therefore reflects mucosal surface area rather than specific carbohydrate transport. Of the 50 percent of D-xylose absorbed by the small intestine, approximately half is metabolized, and the remainder is excreted in the urine.

The test is simple to perform. After an overnight fast, a 25 g dose of D-xylose is given orally, and the patient is encouraged to drink to maintain a good urinary output. Urine is collected for the next five hours. Instead of 25 g of D-xylose, many laboratories including ours, use 5 g of D-xylose. One can also measure levels of D-xylose in the venous sample one hour after ingestion of D-xylose and serum xylose concentration below 20 mg/dl is suggestive of impaired intestinal absorption.

Interpretation: Urinary excretion of 20% of ingested D-xylose suggests intact mucosal absorption. Therefore, excretion of 5 g (with 25 g D-xylose test), 1 g (with 5 g D-xylose test) over five hours is considered abnormal and suggests defects in the mucosal absorption.

False-positive test occurs in the following situations:

- If the total amount of D-xylose is not ingested (vomiting, spillage)
- If there is delay in the passage of the D-xylose through the stomach to the intestine (for example motility disorder of the stomach),
- If there is bacterial overgrowth in the intestine: D-xylose is eaten away by excessive bacteria in the intestine
- If excretion of the absorbed D-xylose is affected as in states of kidney failure (chronic or acute kidney diseases) (serum values of D-xylose should be normal in this situation)
- If the volume of distribution of a molecule is large, such as ascites (fluid in the peritoneal cavity), and heart failure

This test appears to be the best noninvasive method to detect small intestinal mucosal dysfunction. In the presence of steatorrhea, a normal D-xylose absorption test suggests pancreatic exocrine insufficiency.

Investigations to assess absorptive function of distal small intestinal mucosa

Distal small intestine absorbs specifically vitamin B_{12} and bile acids. Therefore assessment of absorptive function of the distal small intestine can be done by assessing the absorption of vitamin B_{12} and bile acids.

Test to assess absorption of vitamin B_{12}: The traditional test to assess the absorption of vitamin B_{12} is Schilling test. Vitamin B_{12} is absorbed by specialized receptors in the terminal ileum. Absorption depends upon the presence of intrinsic factor. Part one of the Schilling test involves administration of radiolabeled B_{12} alone; part two involves the addition of intrinsic factor, and part three involves administration of antibiotics prior to part two.

In this test, patient is given [58]Co-labelled cobalamin orally. In order to saturate the receptors of the B12 in the intestine, non-radioactive cobalamin is administered intramuscularly simultaneously. The B_{12} carriers are allowed to saturate with the extra dose of non-radioactive B_{12}, so that radioactive B_{12} absorbed by the intestine is excreted in the urine. After that, urine is collected for 24 hours and in that urine level of radioactive B_{12} is measured. If the excretion of radioactive B_{12} is less than 7–10% of the administered dose, it suggests vitamin B_{12} malabsorption. Now the next step is to find out the site of vitamin B12 malabsorption. Is it gastric disease (where intrinsic factor is synthesized) or ileal disease (where vitamin B12 is absorbed). To resolve this, part II of Schilling test is performed. In this stage, radiolabelled vitamin B_{12} is administered with intrinsic factor. In diseases of the stomach where secretion of intrinsic factor is affected, it gets corrected by simultaneous administration of intrinsic factor. If the excretion of vitamin B_{12} is still low, it suggests ileal disease.

Schilling test is presently not performed at most centers because of non availability of human intrinsic factor. Furthermore, malabsorption of vitamin B_{12} and folic acid can be assessed by measuring their levels in the serum.

Assessment of absorption of bile acid absorption

Bile acids are absorbed in the distal small intestine including terminal ileum. Terminal ileal disease caused by tuberculosis, Crohn's disease or intestinal resection can lead to malabsorption of bile acids. Bile acid malabsorption can be assessed by measurement of fecal bile acid output, measurement of turnover of radiolabelled bile acids, and indirectly by a therapeutic trial of bile acid binding resins.

Measurement of fecal bile acid output: This method involves measurement of the quantity of endogenous bile acids excreted in the stool. The measurement can be performed by either an enzymatic method or gas chromatography.

^{14}C-Glycocholate bile acid absorption test: This test involves quantifying the fecal recovery of ^{14}C glycocholate in stool over 48–72 hours after ingestion of an oral load of this marker. The fecal recovery of ^{14}C glycocholate determines the rate of intestinal bile acid absorption.

^{75}Selenium labeled homotaurocholic acid test (^{75}Se-HCAT): This test involves ingestion of this synthetic analogue of naturally conjugated taurocholic acid. This analogue enters the entero-hepatic circulation of bile acids. The retained fraction is assessed by a gamma camera seven days after oral administration and values less than 15% are suggestive of bile acid malabsorption. This test can also be used to test the functional integrity of terminal ileum.

Current status: Most of these tests are not available and they are not even well standardized.

Therapeutic trial of bile acid binding resins (Cholestyramine): Due to nonavailability of other tests and the cumbersome methods of measurement, a therapeutic trial of cholestyramine is given. Chlestyramine binds with unabsorbed bile acids and increase their excretion in the stool. Excessive bile acids in the stool leads to diarrhea. Cessation of diarrhea within 3 days of starting cholestyramine suggests bile acid malabsoprtion.

C. Investigation to Assess Absorption of Fat

Fat absorption is a complex process involving integrated action of pancreas, biliary system and small intestinal mucosa. It is not surprising, therefore, that measures of fat absorption are used as global indicators of digestive and absorptive functions. Although excessive fat in the stool may look greasy and oily, simple inspection is a poor guide to fecal fat content and even stool with a high fat content may have a normal appearance. While measurement of fecal fat is a good test for confirming malabsorption, it cannot reliably distinguish between different causes of fat malabsorption.

There are two kinds of tests for fat malabsorption: qualitative estimation and quantitative estimation.

Qualitative estimation of fat in the stool

Sudan stain test for stool fat: Sudan test for fat globules is a screening test for fat excretion in the stool and it is a semi-quantitative test. The steps of Sudan stain test include: (a) a small stool sample is dissolved in water or saline, (b) glacial acetic acid is added to hydrolyze the insoluble salts of fatty acid (c) a few drops of alcoholic solution of Sudan III is added (d) the sample is spread on a microscopic slide, and heated twice to boil and (e) the slide is examined under light microscope. A count of 100 fat globules with a diameter of less than 4 mm per high power field is considered as normal fat excretion. Larger globules or increase in number of fat globules in the stool suggests excessive malabsorbed fat in the stool.

If properly performed, a Sudan stain on a spot sample of stool can detect more than 90 percent of patients with clinically significant steatorrhea.

Quantitative stool collection and analysis

The excreted fat in the stool is composed of neutral fats and phospholipids. The principle of quantification of stool fat is based on the principles of hydrolysis of fatty acid esters and measurement of the fatty acids that are released. The gold standard for fecal fat analysis is the Van de Kramer method, which is the quantitative titration of fatty acid equivalents, and the results are expressed as fecal fat in grams per 24 hours.

Collection of stool: The excretion of fat in the stool varies from day to day and for quantification of which requires collection of total stool passed per day. Since fat excretion varies from day to day, it is advised that fat excretion is measured over three days stool excretion and an average of the three days is regarded as stool fat excretion per day. Collection of stool for three days is quiet cumbersome for patients. Hence, many laboratories including ours, collect stool for 24 hours (1 day) and estimation of fat in the stool is done.

For conduct of this test, patients are advised to ingest 75–100 g of additional fat (butter) per day for six consecutive days to ensure 100 g of fat intake per day (fat ingestion of in a normal diet varies from 20–30g/day). During the first three days (day 1, 2, 3), the stool fat excretion reaches equilibrium. Stool is not collected on the first three days. For the next three days (day 4, 5, 6), patients are advised for collection of total stool in three different containers (one for each days stool). Alternatively, as discussed earlier, using the same protocol, stool is collected only on one day (day 4) (24 hour stool fat estimation). Patients are asked to keep a record of all food intake on these six days.

The stool samples are submitted in the laboratory where stool is weighed (per day volume) and homogenized before analysis.

Principle of estimation of fat in the stool: Stool is saponified with concentrated potassium hydroxide in ethanol. Fatty acids are allowed to be liberated by addition of hydrochloric acid to the alkaline solution. Ethanol is then added and the fatty acids are extracted with petroleum ether. In an aliquot sample of the petroleum ether layer, the fatty acids are titrated with alkali using thymol blue as an indicator. This method of estimation of total fat content in the stool takes about 35 minutes.

Procedure: About 5 g of homogenized stool is taken in a 150 ml Erlenmeyer flask. After addition of 10 ml of 33% alkali and 40 ml of ethanol containing 0.4 percent amyl alcohol, the mixture is boiled for 20 minutes under a reflux condenser, and then thoroughly cooled. Seventeen ml of 25% HCl is added, after which the mixture is again cooled. Exactly 50 ml of petroleum ether is then added, and the flask is closed with a rubber stopper and shaken vigorously for 1 minute. After complete separation, 25 ml of the petroleum ether layer is transferred into a small Erlenmeyer flask using a pipette. After addition of a piece of filter paper, the petroleum ether is evaporated and 10 ml. of neutral ethanol is added. The fatty acids are titrated with 0.1 N NaOH from a micro burette, with thymol blue as indicator, until the yellow color begins to change.

Calculation: The calculation is carried out assuming an average molecular weight of 284 for fatty acids.

Hence,

$$\frac{A \times 284 \times 1.04 \times 2 \times 100}{10,000\,Q} = \frac{5.907\,A}{Q} = \frac{\text{Fatty acids in gm}}{\text{per 100 g stool}}$$

(In which A = mililiter of 0.1 N alkali required for titration, Q = the g of stool taken for analysis)

The factor 1.04 is used as the petroleum ether layer increase 1 percent in volume when shaken with alcoholic hydrochloric acid and because 3 percent of the amount of fatty acids remains in solution in the acid alcoholic layer.

Semi-quantitative fat analysis—Acid steatocrit test: The stool fat excretion can also be assessed semi-quantitatively using acid steatocrit test. A 500 mg stool sample is diluted with distilled water in 1:3 ratio and is homogenized. $HCLO_4$ is added so as to separate out the fatty layer. Acid steatocrit is measured by the ratios of fatty layer and solid layer. The acid steatocrit percentage is calculated using the formula:

$$\text{Acid steatocrit\%} = \frac{\text{FL}}{(\text{FL} + \text{SL})} \times 100$$

where, FL = Fatty layer and SL = Solid layer.

The stool fat content is quantified as follows:
Stool fat = –0.43 + [0.45 (acid steatocrit %)] g/24 hrs.
The normal value of stool acid steatocrit is <31%.

Normal values for fat excretion: Generally 93% of the ingested fat is absorbed. Therefore, in the quantitative test as discussed above where 100 g of fat is ingested per day, a stool fat excretion of 7 g/day is considered normal. Patients with pancreatic disorders have fat excretion generally more than 14 g/day, whereas patients with small intestinal diseases have the stool fat excretion between 7–14 g/day.

Normal values in Indian population: The fat excretion in normal Indian healthy people in a study from Bangalore including 600 apparently healthy adults were estimated using the acid steatocrit method on a random stool sample. The average daily fat excretion in stool of normal healthy people in India was found to be 8.72 g, which was higher than the average daily fat excretion in people from the West where the average daily fat excretion is 7 g.

Breath tests for fat absorption: [14]C triolein breath test: Oral ingestion of triolein labeled with radioactive ^{14}C releases $^{14}CO_2$ after digestion of triolein, which is exhaled in the breath air. In case of fat malabsorption there is impaired pulmonary excretion of $^{14}CO_2$. While this test does not require cumbersome collection of stool and has acceptable sensitivity and specificity; this test is rarely done.

D. Investigations for Absorption of Carbohydrates

Carbohydrate malabsorption is either due to global malabsorption because of diffuse diseases of the small intestine or due to specific isolated enzyme deficiency such as lactase enzyme deficiency. The carbohydrate malabsorption leads to incomplete digestion of the carbohydrates. The malabsorbed carbohydrates are passed through the colon where they are acted upon by the colonic bacterial flora. The action of colonic bacterial flora on the malabsorbed carbohydrates leads to production of short chain fatty acids, and production of hydrogen and methane. The malabsorbed carbohydrate increases the osmolality in the lumen of the intestine, which further draws water and sodium. All of these leads to increase in the volume of the intraluminal contents and passed in the stool (increase in the volume of the stool). The tests for carbohydrate malabsorption thus use the abovementioned principle such as hydrogen excretion in the breath and measurement of specific nutrients after ingestion of a test dose.

Lactose tolerance test

Lactose malabsorption leads to transit of lactose to the colon where it is fermented rapidly to organic acids and gases including hydrogen. A loading dose of 25 g of lactose is given and an increase in breath hydrogen of 20 ppm above baseline within 4 hours suggests lactose malabsorption. There are two forms of the lactose tolerance test. First, the serum test consists of a oral test dose of 50 g of lactose with subsequent measurements of blood glucose levels at 0, 60 and 120 minutes. An increase of blood glucose less than 20 mg/dL and development of symptoms (e.g. diarrhea, abdominal cramping, and flatulence) is diagnostic of lactose intolerance.

Lactose hydrogen breath test

Nitrogen, oxygen, carbon dioxide, hydrogen and methane account for more that 99% of expelled intestinal gas. In humans, hydrogen and methane are exclusively produced by intestinal bacteria in the large intestine. In addition, hydrogen and methane can also be formed in the small intestine especially once there is bacterial overgrowth in the small intestine. Approximately 80% of hydrogen and methane is expelled through the flatus; and the rest 20% is exhaled by the lungs which can be measured in the exhaled breath using gas chromatography. Hydrogen is produced by bacterial fermentation of the saccharides in the intestinal lumen.

An increase in breath hydrogen of more than 20 ppm is diagnostic of lactose intolerance. Studies have shown that simply looking at hydrogen values at 0 and 120 minutes after the lactose challenge is sufficient to determine lactose malabsorption. Other investigators have noted that gastrointestinal symptoms after a lactose challenge are more strongly associated with hydrogen excretion than with glucose concentration. False positive results occur with the breath hydrogen test when patients have bacterial overgrowth. The other problem that occurs with the execution of this diagnostic test is that it can be fundamentally difficult to perform.

Test for fructose absorption: Fructose malabsorption has been identified as a potential cause of gastrointestinal symptoms. 25 g dose of fructose is given and a positive test is defined as a rise in breath hydrogen or methane excretion of > 20 ppm or a rise of > 5 ppm on consecutive breath samples as a positive test.

Stool pH: Patients with carbohydrate malabsorption has acidic stool as discussed earlier.

Tests for protein absorption: In the clinical setting, test for protein malabsorption is generally not performed. The surrogate markers of protein malabsorption are the status of somatic protein (muscle mass of the body) and visceral protein (albumin level in the serum). There are many conditions of the GI tract such as ulcerations in the intestine that leads to leakage of proteins in the intestinal lumen and then in the stool. This condition is called protein-losing enteropathy. Rupture of lymphatics, in a condition called intestinal lymphangiectasia, can also lead to protein-losing enteropathy. The protein losing enteropathy can be detected by detection of passage of radioactive albumin injected intravenously (^{99m}Tc-human serum albumin scintigraphy). The radioactive albumin in the lumen is detected by use of gamma camera. Enteral protein loss can also be assessed by measurement of the alpha-1 anti-trypsin clearance.

Assessment of intestinal permeability: Transport of molecules across the intestinal epithelium takes place through two major routes, i.e., transcellular and paracellular. Transport of solutes across the transcellular route (through the cells) is mediated by carriers and channels. Transport of molecules through paracellular route across the intestinal epithelium occurs by the process of diffusion and does not require carriers. Assessment of intestinal permeability is done to assess the overall function of transport through the intestinal epithelial paracellular route. Intact intestinal epithelial barrier is essential for preventing penetration of the intestinal mucosa by the noxious substances. The disruption of this barrier system results in "leaky" intestine leading to increase in intestinal permeability.

Intestinal permeability mainly concerns the permeation of molecules with a molecular weight greater than 150 Da. There are series of aqueous pores along the crypt villous axis of the small intestine. While the tip of the villous contains relatively small (< 6 Å) pores and they are present in abundance, the crypts contain much larger pores (50–60 Å) but they are in low abundance. Base of the villous contain intermediate size (10–15 Å) pores. Most of the probes used to measure intestinal permeability are water soluble, which cannot penetrate the lipid cell membrane of enterocytes and thus use the paracellular route through the tight junctions. The smaller probes can easily pass through the small, more numerous and more accessible tight junctions of the villous tips, whereas the larger probes have to make use of the larger and less accessible and less numerous pores at the crypt base.

Non-invasive technique for measurement of intestinal permeability

Intestinal permeability is measured with the help of non-metabolizable probe molecules that pass across the mucosal barrier and are excreted in the urine after being absorbed into the systemic circulation. Quantitation of the probe in a timed urine collection provides a measure of the fraction of the ingested probe that had penetrated the mucosal barrier. Only a small proportion of this large probe should get through the intestinal mucosa, reach the circulation, get filtered by the kidney and get measured in the urine over a defined point of time. If there is an abnormality in the paracellular transport; the passage of the probe is enhanced and thus more is excreted in the urine, which suggests abnormality in the intestinal permeability. Intestinal permeability can be assessed using a variety of marker probes such as lactulose, mannitol, rhamnose and cellobiose, polyethylene glycol (PEG) 400, PEG 1000, [51]CrEDTA, and [99m]TcDTPA.

Single sugar test: Earlier a single probe (such as lactulose, cellobiose, PEG and ^{51}Cr-EDTA) was used to measure intestinal permeability. The urinary excretion of single dose of a test probe is dependent on a number of non-mucosal factors (such as gastric emptying, intestinal transit, renal clearance and incomplete urine recovery) other than mucosal integrity which not only reduce the sensitivity and specificity of test procedure, but also poses a problem in the interpretation of the data.

Differential sugar test: To overcome the mucosal and non-mucosal factors, use of two sugars or a combination of a sugar and non-sugar probe for assessment of intestinal permeability is used. It involves the simultaneous oral administration of a disaccharide and a monosaccharide and measurement of their excretion in urine over a defined period of time. Intestinal permeability is measured as the ratio of percentage of excretion of disaccharides/monosaccharides in the urine.

The individual variations due to the non-mucosal factors (such as gastric emptying, intestinal transit, renal clearance and incomplete urine recovery) are circumvented when the urinary recovery is expressed as a ratio (lactulose: mannitol ratio or LMR) since both of them are affected equally by these factors, except for the route of permeation. Any variation in premucosal and post mucosal determinants of the sugars affects overall permeation of the test probe equally, so that urinary excretion ratio (percentage of orally administered test dose) is not affected.

E. Method of Assessment of Intestinal Permeability

After an overnight fasting, subjects undergoing assessment of intestinal permeability are asked to ingest sugar probes (5 g of lactulose, 2 g of mannitol in 100 ml water for lactulose mannitol ratio) after voiding their first urine (which is collected). No food or drink is allowed until the completion of the test. Water intake is permitted after one hour of the ingestion of the test solution. All the subjects should be asked to drink a minimum of 1000 ml of water during urine collection time to ensure sufficient urine output. All the urine passed in the subsequent 5 hours should be collected into a 2-liter plastic container containing 1 ml of chlorohexidine (20 mg/ml) as a preservative. The volume of the urine is measured and a small volume (10–100 ml) is preserved and stored in –80°C until analysis.

The concentration of one or more sugar probe present in the urine can be measured using enzyme assay, high performance liquid chromatography (HPLC), ion-exchange chromatography in combination with mass spectrometry and liquid chromatography with mass spectrometry. If the PEG has been used as a probe, then the concentration of the PEG can be assessed using capillary column gas chromatography.

Lactulose and mannitol ratio is the most commonly used test for assessment of intestinal permeability and the most reliable method of measurement of concentration of lactulose and mannitol in the urine is HPLC. HPLC is a chromatographic technique used to separate a mixture of compounds with the purpose of identifying, quantifying and purifying the individual components of the mixture. Small amount of urine sample is injected into a HPLC column which is filled with solid particles (called stationary phase) and the sample mixture is separated into compounds as it interacts with the column particles. These separated components are detected at the exit of this column by a flow-through device (detector) that measures their concentration. An output from this detector is called a "liquid chromatogram". The separation of the components in the sample depends upon many parameters including chemical structure, molecular weight; but the most common parameter

for the identification of individual compounds in the sample is its retention time (the time it takes to elute from the column after injection). Measurement of amount of solute in urine by HPLC is done by extrapolating the total area of known concentration of solute with that of area of solute found in the in the sample.

F. Interpretation of Intestinal Permeability

Urinary excretion of disaccharides and monosaccharides and ratio of their excretion is a basis of measurement of intestinal permeability. After the measurement of the concentration of the probes in the urine; the results are expressed as the ratio of percentage excretion of the ingested dose of lactulose and mannitol in the urine [lactulose mannitol ratio (LMR) = % lactulose/% mannitol].

The lactulose and mannitol ration is a useful, simple, non-invasive and a reliable test for estimation of intestinal permeability. In this test, while mannitol uses for its passage predominantly smaller and more numerous tight junctions at the villous tips; lactulose being a larger molecule passes through larger pores in the crypts. Under normal physiological condition, disaccharide due to its large size is restricted from passing across the villous tip whereas monosaccharides can do so with relative freedom. The size of the probe and the pore size in the intestinal mucosa have a bearing in the abnormalities in the permeability in different disease conditions.

G. Imaging Studies

The discussion in the previous section pertained mainly to the function and dysfunction of the digestive and absorptive processes. In the diseases of the GI tract, hepatobiliary system and the pancreas, it is also important to look at the anatomical and pathological status of the particular organ in order to know the disease process. Structure of these organs can be seen and studied by various imaging and histological techniques.

Information on the intestinal lumen, intestinal wall and the surrounding structures including lymph nodes is required for a complete evaluation of the intestine. One also needs to know about the length, site, multiplicity and activity of the stricture(s), if present. The assessment of intestinal lumen has traditionally been done by conventional barium studies including barium meal follow through and small-bowel enteroclysis. While enteroclysis provides information on the size and site of intestinal strictures, it fails to provide information on the activity of the strictures and surrounding structures. Much interest has been focused on computerized tomography (CT) enteroclysis and magnetic resonance (MR) enteroclysis in recent times to overcome the individual deficiencies of CT/MRI (no distention of the small intestine) and conventional enteroclysis (no extraluminal information). Both the CT enteroclysis and MR enteroclysis are reported to be highly accurate in depicting mucosal abnormalities and extra-intestinal complications in patients with diseases of the intestine. Therefore, they are combined with a cross sectional imaging technique such as CT enteroclysis or more often MR enteroclysis. The hepatobiliary system can be imaged using ultrasound, abdominal computed tomography (CT), abdominal magnetic resonance imaging (MRI), and in some cases endoscopic retrograde cholangio-pancreatography (ERCP), magnetic resonance cholangiopancreatigraphy (MRCP), and endoscopic ultrasonography

H. Histological Evaluation of the Intestinal Mucosa

Because intestinal malabsorption may occur in a number of diseases in which mucosal involvement occurs, multiple biopsy specimens are obtained from several sites in the duodenum or proximal jejunum. These biopsies are examined by pathologists for crypt villous architecture and inflammation.

Approach to a patient with malabsorption

Talking with patient (typically called history taking) and a thorough examination leads to many clues and based on that a diagnosis is suspected. Subsequently, a few investigations are done for confirmation of malabsorption and for detection of cause of malabsorption. All the tests discussed above are not used in every patient. A detailed discussion on the approach to a patient with malabsorption is out of the scope of this book.

VII. CONCLUSIONS

The process of digestion and absorption is complex, yet efficient. There are a number of tests for assessment of absorption of carbohydrate, fat and proteins. (Table 54.3)

TABLE 54.3   Tests for assessing absorption

Tests for mucosal malabsorption:
- D-xylose test

Tests for assessment of fat malabsorption
- Qualitative fecal fat analysis—Sudan stain test
- Semiquantitative fecal fat analysis—Acid steatocrit test
- Quantitative fecal fat analysis
- Breath tests for fat malabsorption

Tests for carbohydrate malabsorption
- Stool pH
- Lactose hydrogen breath test
- Fructose hydrogen breath test
- Measurement of fecal carbohydrates

Tests of protein malabsorption
- Fecal clearance of radiolabelled albumin
- Fecal clearance of alpha 1 antitrypsin

SUGGESTIONS FOR FURTHER READING

Arrieta M.C., Bristiz L. and Meddings J.B. (2006), Alterations in Intestinal Permeability, Gut, 55:1512–1520.

Craig R.M. (1999), Ehrenpreis ED, D-xylose Testing, *J. Clin Gastroenterol*, 29:143–50.

Khouri M.R., Huang G. and Shiau Y.F. (1989), Sudan Stain of Fecal Fat: New Insight into an Old Test, *Gastroenterology*, 96:421–7.

Ramakrishna B.S., Venkataraman S. and Mukhopadhya A. (2006), Tropical Malabsorption, *Postgrad Med J.,* 82:779–87.

Saad R.J. and Chey W.D. (2007), Breath Tests for Gastrointestinal Disease: The Real Deal or Just a Lot of Hot Air? *Gastroenterology*, 133:1763–6.

Simko V. (1981), Fecal Fat Microscopy, Acceptable Predictive Value in Screening for Starorrhea, *Am J. Gastroententerol*, 75:204–208.

Van De Kamer J.H., Huinink H.T.B. and Weyers H.A. (1949), Rapid Method for Determination of Fat in Feces, *J. Biol Chem.*, 177:349–55.

55

The "Uncultured" but Significant World of the Human Gut Microbiome
New Metagenomic Insights

G.B. Nair, Sourav Sen Gupta and Bhabatosh Das

CONTENTS

I.   Introduction—"Human"—The Super-Organism
II. The Gut Metagenome—Our Microbial Defence Partners
   A. The human gut as an ecosystem
   B. Implications for concepts of health and disease
   C. The "core"
III. Gut Metagenomics
IV. Metagenomic Insights: Some Significant Metagenomic Studies of The Human Gut
V.   Bioinformatic Challenges
VI. Conclusions

I. INTRODUCTION—"HUMAN"—THE SUPER-ORGANISM

Bacteria make up most of the flora in the colon and 60% of the dry mass of faeces (Guarner, F., J.R. Malagelada, 2003). The large bowel contains about 1×10^{13} to 1×10^{14} microbes, weighing around 1.5 kg (Gill et al., 2006). Our adult bodies harbour ~10 times more microbial than human cells (Bäckhed et al., 2005). Their genomes (the microbiome) endow us with physiologic capacities that we have not had to evolve on our own and thus are both a manifestation of who we are genetically and metabolically, and a reflection of our state of well-being. Our distal gut is the highest density natural bacterial ecosystem known, the most comprehensively surveyed to date, and the most highly represented in pure culture. It contains more bacterial cells than all of our other microbial communities combined.

The total number of genes in the various species represented in our indigenous microbial communities likely exceeds the number of our human genes by at least two orders of magnitude (Bäckhed et al., 2005). If the average genome size and gene density of the species represented in the gut microbiota are similar to those in *Escherichia coli* (5 megabase pairs genome, 4000 genes), the microbiome would have a complexity of 2.5 giga basepairs and contain 2 million genes (Hooper et al., 2002). Thus, it seems appropriate to consider ourselves as a composite of many species—human, bacterial, and archaeal—and our genome as an amalgam of human genes and the genes of our microbial 'selves'. (Turnbaugh et al., 2006). The trillions of microbes residing in the human gut help us digest our foods, break down toxic compounds that we ingest, and play an important role in keeping pathogenic organisms in check. Somewhere between 15,000–36,000 different species belonging to at least 1,800 genera live in the gut (Frank et al., 2007). However, it is probable that 99% of the bacteria come from about 30 or 40 species (Beaugerie, L. and J.C. Petit, 2004). Fungi and protozoa also make up a part of the gut flora, but little is known about their activities. The colon has the greatest numbers of bacteria and the most different species, and the activity of these bacteria make the colon the most metabolically active organ in the body. The acid in the stomach, as well as bile and pancreatic secretions, hinder colonization of most bacteria in the stomach and proximal small intestine (O'Hara, A.M., F. Shanahan, 2006). The first part of the colon is mostly responsible for fermenting carbohydrates, while the latter part mostly breaks down proteins and amino acids. Bacterial growth is rapid in the cecum and ascending colon, which has a low pH, and slow in the descending colon, which has an almost neutral pH (Guarner, F. and J.R. Malagelada, 2003). The body maintains the proper balance and locations of species by altering pH, the activity of the immune system, and peristalsis. Over 99% of the bacteria in the gut are anaerobes, but in the cecum, aerobic bacteria reach high densities. Members of the anaerobic genera such as *Bacteroides*, *Eubacterium*, *Bifidobacterium*, *Ruminococcus*, *Clostridium* and *Fecalibacterium* have been found to constitute the majority of the adult human intestinal microbiota (Eckburg et al., 2005) (Rappé, M.S. and S.J. Giovannoni, 2003). Microorganisms in the gut perform a host of useful functions, such as fermenting unused energy substrates, training the immune system, preventing growth of harmful species, regulating the development of the gut, producing vitamins for the host (such as biotin and vitamin K), and producing hormones to direct the host to store fats. Populations of species vary widely among different individuals but stay fairly constant within an individual over time, even though some alterations may occur with changes in lifestyle, diet and age (O'Hara, A.M. and F. Shanahan, 2006).

II. THE GUT METAGENOME—OUR MICROBIAL DEFENCE PARTNERS

A. The Human Gut as an Ecosystem

The wet, anoxic and energy rich human gut serves home to several of the 100 trillion of our microbial partners. The human gut from a microbial point of view is an ecosystem hosting multiple interacting spieces with a complex interplay between food, host cells and microbes. The human gut ecosystem is very dynamic and consists of various bacterial populations that are either permanent gut residents, i.e., autochthonous components, or transient inhabitants, i.e., allochthonous members, introduced from the environment. The gut flora is responsible for successful and efficient absorption of nutrients. Bacterial co-operation occurs during the breakdown of complex carbohydrates, which originate from the diet or host. We can consider our gut microbiota to be an efficient natural bioreactor, programmed to break down food and supply us with the extracted energy and nutrients. This bioreactor is stable, i.e., it is resistant to chaotic blooms of subpopulations (or pathogens) that could be disruptive. Functional redundancy encoded in genomes from widely divergent bacterial lineages would provide the host with insurance against disruption of the food web caused by loss of keystone species from selective sweeps (e.g., by phage attacks). Ecologic principles predict that host-driven ("top-down") selection for functional redundancy would result in a community composed of widely divergent microbial lineages (divisions) whose genomes contain functionally similar suites of genes (Ley et al., 2006). Another prediction is the widespread occurrence of, and abundant mechanisms for, horizontal gene transfer. In contrast, competition between members of the microbiota would exert "bottom-up" selection pressure that results in specialized genomes with functionally distinct suites of genes (metabolic traits). Once established,

these lineage-specific traits can be maintained by barriers to homologous recombination (Majewski et al., 2000). The mammalian gastrointestinal tract microbial ecosystem shares a homeostatic relationship with the host's immune system and both microbe and mammal benefits from this symbiotic relationship as expected in co-evolving systems (Juengst, E.T., 2009). The gut microbiota is an effector and a reporter of many aspects of our normal physiology. The human gastrointestinal mucosa is uniquely adapted to the proximity of a massive load of potentially antigenic microbes. Specialized mucosal immune cells continually sample intestinal microbes and initiate local immune responses without inducing a systemic response (Macpherson et al., 2005) by inhibiting nuclear factor (NF) - κ B activation. Commensal flora block adhesion of pathogenic bacteria, modulates the host immune response via upregulation of phagocytic activity, and stimulates the production of nonspecific and specific secretory IgA. All these activities ensure immune homeostasis within the gut associated lymphoid tissue [GALT]. Commensal microbes such as bifidobacteria not only maintain the mucosal barrier; they also facilitate interkingdom signaling among different microbes across it. Bifidobacteria also ferment oligosaccharides, often liberated from more complex polysaccharides by the Bacteroides, to produce short-chain fatty acids such as acetate and lactate, which in turn are converted into butyrate, the main energy source for the colonic mucosa, by other dominant members of the gut microflora (Kinross et al., 2008).

Comparisons of germ free and colonized animals have shown that the microbiota helps regulate energy balance, both by extracting calories from otherwise inaccessible components of our diet and by controlling host genes that promote storage of the extracted energy in adipocytes. *Bacteroides thetaiotaomicron*, a dominant member of the normal distal intestinal microbiota, hydrolyzes otherwise indigestible dietary polysaccharides and supplies humans with 10% to 15% of their caloric requirement (Xu et al., 2003). *Lactobacillus* species are responsible for a significant proportion of bile acid deconjugation, a process that efficiently reduces lipid absorption in the gut (Narushima et al., 2006).

The microbiota directs myriad biotransformation, ranging from synthesis of essential vitamins to the metabolism of the xenobiotics that we ingest and the lipids that we produce. The microbiota modulates the maturation and activity of the innate and adaptive immune system: an immune system educated to allow the host to tolerate a great degree of microbial diversity provides a selective advantage since this diversity ensures the stable functioning of a microbiota in the face of environmental stresses. Although four phyla—*Firmicutes*, *Bacteroidetes*, *Actinobacteria*, and *Proteobacteria* dominate the human gut microbiota and are relatively constant across individuals, the makeup of a person's microbiota at the species and strain levels can be as individual as a fingerprint; we may share as little as 1% of the same species (Eckburg et al., 2005). Each host thus has a unique, genetically determined response to both aggressive and protective bacterial species.

B. Implications for Concepts of Health and Disease

The gut microbiota has been invoked as a factor that determines susceptibility to diseases ranging from gastrointestinal and other malignancies, atopic disorders (asthma), infectious diarrheas, various immunopathologic states including inflammatory bowel diseases and previously unassociated systemic conditions such as diabetes and obesity (Tsukumo et al., 2009). It is also considered to be a key contributor to individual variations in the bioavailability of orally administered drugs.

Development of mechanisms to define, at regularly scheduled intervals, how the gut microbiota and microbiome are changing in humans living in distinct geographic regions of the planet, under varied socio-economic conditions will provide an opportunity to monitor our micro-evolution during a time of increasing travel and great climatic change. New tools and metrics for identifying, forecasting and responding to national and world-wide changes in disease susceptibility will form the path for new age science.

Metagenomic studies to examine the correlations between the human gut microbiome and human health have commenced. The central question to be answered is whether changes in the microbiome can be correlated with disease state, and whether such changes are the cause of the disease or are caused by the disease, using metagenomic tools.

The complex and dynamic communities of microbes that are present on and within the human body (the human microbiota), especially that of the gut are thought to profoundly influence human health in a variety of ways, through effects on human physiology, nutrition, immunity and development. Disruptions in community composition, structure or dynamics may reflect, trigger, or influence the course of various disease states. Studies on humans and vertebrate animal models have generated evidence that this is the case for some specific diseases like obesity and Crohn's disease and suggest that further studies in this area may be vital for the understanding, prevention and treatment of many human diseases, as well as maintenance of homeostasis. However, studies of the human gut microbiome are truly in their infancy, and the range and magnitude of the effects of the collective microbiota upon health and disease are not known. Future studies will lead to disease prevention strategies, personalized healthcare regimens, and the development of novel therapeutic interventions.

Human gut microbial communities are also influenced by other, less predictable factors such as interactions between individual resident members of the microbiota (Zoetendal et al., 2006), as well as transient members such as pathogens and probiotics. The concept of "bioecologic" control of the gut has recently been proposed as a novel method of improving human health. This approach encompasses the use of "functional foods"—probiotics, prebiotics that are nondigestible food ingredients that benefit the host by selectively stimulating the growth or activity of bacteria that can improve the host's health, and synbiotics—a combination of both. There is now considerable evidence to support the efficacy of this approach, with numerous studies suggesting that functional foods improve outcome in irritable bowel syndrome, inflammatory bowel disease, and antibiotic-associated diarrhea in the critically ill (Kinross et al., 2008).

C. The "Core"

Following birth every human (baby) develops a complex and active intestinal ecosystem starting from an assumed sterile environment. In just a matter of a couple of days the gastrointestinal track of a newborn becomes home to a large bacterial community, whose total number quickly exceeds the total number of cells of its host. *Bacteroidetes* and *Firmicutes* dominate the human bacterial "flora," but there is considerable variation in the composition of indigenous bacterial communities among individual humans. Development of the gut flora have been hypothesized as a succession in which a gastrointestinal community inhabited by diverse, opportunistic colonizers gradually are succeeded by microbes typical of the adult gut, which presumably are of greater adaptive fitness. Host selection plays a very significant role in the composition of the microbiota, so that microbiomes differ greatly among individuals. Generally individuals with similar diets have more closely related gut microbial communities. Diet is a pivotal variable in influencing the composition of the intestinal microbiota. Studies on chemically well defined diet components have proven a clear correlation between diet and the presence of specific bacterial groups. It has been shown that a diet rich in inulin and related fibers promote an increase in *bifidobacteria*, whereas the intake of dietary sulphate favours several genera of sulphate-reducing bacteria over methanogenic Archaea. Diet affects not only the microbiota composition, but more significantly the metabolic activities of the microorganisms which are also influenced by factors relating to the gut environment, i.e., local conditions of pH, oxygen and hydrogen, metabolite concentration and gut transit time.

A very important question is whether there is a "normal" "core microbiome." The idea that there is a "normal" flora may be simplistic. Instead, there may be a spectrum of normalcy and several models of health. 'Core' human gut microbiome refers to a set of features shared across all or the vast majority of gut microbiomes (e.g., genes and/or metabolic capabilities). Recent studies at Center for Genome Sciences, Washington University School of Medicine, St Louis, USA, suggests that a core microbiome can be found at the gene level, despite large variation in community membership, and that variations from the core are associated with obesity, providing answers to longstanding questions like "Is there a core microbiome shared between humans, and should this core be defined in terms of organisms, genes, or functional characteristics? And finally, are there genes and/or metabolic pathways in the human microbiome that can be identified as being associated with obesity?" (Turnbaugh, P.J. and J.I. Gordon, 2009). Obesity was found associated with phylum-level changes in the microbiota, reduced bacterial diversity, and altered representation of bacterial genes and metabolic pathways, including those involved in nutrient harvest (Turnbaugh et al., 2006), emphasizing the importance of early environmental exposures in shaping the human gut microbiota and that a diversity of organismal assemblages can nonetheless yield a core microbiome at a functional level, and show that deviations from this core are associated with different physiological states (obese compared with lean) (Turnbaugh et al., 2009) (Tschöp et al., 2009). Despite marked interpersonal variation in species assemblages, there is an identifiable core gut microbiome composed of genes encoding various signaling and metabolic pathways. This convergence of diverging species assemblages onto common functional states (Hamady, M. and R. Knight, 2009) is a feature of macro-ecosystems and implies considerable functional redundancy between different taxa comprising gut communities.

III. GUT METAGENOMICS

Phylogenetic analysis of small-subunit rRNA genes (SSU rDNA) present in all cellular organisms serves as the gold standard for molecular identification of microbial species. These genes can be isolated directly from a specimen (for instance, a faecal sample) by PCR (Polymerase chain reaction) with broadly specific (universal) rDNA primers. Microbes in the specimen then can be identified by comparison of the rDNA gene sequences to previously analyzed sequences, including those from well characterized isolates. Based on the sequence similarity of the newly sequenced genes to those in various databases, microorganisms can be more or less precisely identified. The human gastrointestinal tract is particularly well sampled and so species or genus-level assignments generally are possible (Frank, D.N. and N.R. Pace, 2008). But 16S data do not provide any information on functional features of the microbial community. In the most previous studies, gene libraries were constructed by 'shotgun' cloning of bulk environmental DNA, rather than specific genes like the 16S or antibiotic resistance genes and these libraries were sequenced and subjected to intensive computational analysis. But the degree of gene coverage is largely dependent on sequencing depth and complexity of the communities and biases are also introduced through the cloning and PCR step. Employing next-generation DNA sequencers based on massively parallel sequencing technologies by which the cloning step is eliminated and sequence quantity is increased by orders-of-magnitude compared with that of conventional Sanger sequencers have been able to successfully resolve this problem.

Inability to culture most members of the human gut microbiome have led to the quest of development and application of newer culture independent techniques like metagenomics to gain insights into this complex world of gut microbiome. Metagenomics is a rapidly evolving field that emerged from rapid advances in DNA sequencing methods, with a focus on the use of culture-independent ways to study the structure, function and dynamic operation of microbial

communities (Kinross et al., 2008). It is a technique of isolating and sequencing community DNA directly from an environment that have made it feasible to analyze complex mixtures of entire genomes with reasonable coverage and extends to the profiling of gene expression at the level of RNA (metatranscriptomics), and protein (metaproteomics), as well as community metabolism (metabolomics) (Zoetendal et al., 2008). The proportion of the protein coding regions in bacterial genomes can be as high as 80%, so that most of the metagenomic sequences obtained contain at least partial gene regions directly related to function. Metagenomic, PCR-free identification of microbial diversity of the dominant microorganisms of the human gut revealed domination by very few phyla when compared with other complex ecosystems such as soils and oceans, but nonetheless highly diverse and complex at the level of 'phylotypes'. The limited diversification of the intestinal microbiota may reflect the relatively short time of existence of the human intestine as a habitat, i.e., 100 million years for mammals, as opposed to over 3.85 billion years for the ocean. Metagenomic studies of the gut are considered to represent the method that produces results of 'gold standard' quality.

10 to 50% of microbial population of the human gut have been reported unculturable using rDNA based molecular analysis techniques, thus direct metagenomic studies of collective DNA of community members bypassing the need for isolation and laboratory cultivation of individual species have provided avenues in deciphering the interactions of the human gut and its microbiota. As we get to know our microbiota better, we will improve our ability to culture and study them in the laboratory environment.

IV. METAGENOMIC INSIGHTS: SOME SIGNIFICANT METAGENOMIC STUDIES OF THE HUMAN GUT

The first comprehensive molecular survey of the gut microbiota was published by a group composed of members of the Relman laboratory (Stanford University School of Medicine, USA) and TIGR (The Institute for Genomic Research, Maryland, USA) (Eckburg et al., 2005). This study produced 13,335 16S rRNA gene sequences from mucosal biopsy samples harvested from the proximal to the distal colons of three healthy individuals, plus one stool sample from each person. The result is the largest database of 16S rRNA sequences from a single study of any ecosystem. Three hundred ninety-five bacterial and one archaeal phylogenetic types ('phylotypes') were identified based on the criterion that ≥99% sequence identity was required for any pair of sequences to be assigned to unique phylotypes. The number of individual sequences representing each phylotype is a measure of abundance. Thus, this study provided the most complete view to date of microbial composition ("who's there") and diversity ("who's there and in what numbers") in the distal human gut. In contrast to microbial communities in natural environments where <1% of phylotypes are represented by laboratory isolates, at least 22% of the 395 phylotypes have an available cultured representative. Results from this study also show that human GI tract is predominantly a bacterial ecosystem. Cell densities in the colon (10^{11}–10^{12}/ml contents) are the highest recorded for any known ecosystem. The vast majority of phylotypes belong to two divisions (superkingdoms) of bacteria—the *Bacteroidetes* (48%) and the *Firmicutes* (51%). The remaining phylotypes are distributed among the *Proteobacteria*, *Verrucomicrobia*, *Fusobacteria*, *Cyanobacteria*, *Spirochaetes* and *Vadin*BE97.

Early studies of two healthy adults to understand ways in which commensal and pathogenic microbes adapt themselves to the human habitat, revealed that their faecal microbial metagenomes were enriched in genes involved in energy metabolism, including production of short-chain fatty acids, which provide energy to the intestine. The study also reported that the gut microbiome contains gene sets for significantly enriched metabolism of glycans, amino acids, and xenobiotics; methanogenesis; and 2-methyl-d-erythritol-4-phosphate pathway-mediated biosynthesis of vitamins and isoprenoids (Gill et al., 2006).

In another collaborative study at the Relman laboratory utilising deep 16S rRNA pyrosequencing, it was shown that a short course of ciprofloxacin in healthy individuals has a significant impact on gut microbiota and affects about one third of all taxa found in the gut. While most taxa rebound to their pretreatment level after four weeks, some taxa failed to return for six months. Interestingly, these subjects suffered no apparent ill effects from the antibiotic administration, suggesting that there may be functional redundancy in the gut microbiota (Dethlefsen et al., 2008).

Shotgun metaproteomic studies of the human distal gut microbiota carried out at the OakRidge National Laboratory, USA, used a novel approach of non-targeted, shotgun mass spectrometry-based whole community proteomics to directly identify microbial proteins in faecal samples to gain information about the genes expressed and about key microbial functions in the human gut, proteins for translation, energy production and carbohydrate metabolism, human proteins, including antimicrobial peptides and several unknown proteins, were identified, revealing previously undescribed microbial pathways or host immune responses and a novel complex interplay between the human host and its associated microbes.

Functional metagenomic studies for characterization of key functional members of the human gut microbiome revealed that these key highly functionally active members most influence host metabolism and hence health by influencing numerous host pathways.

The human stomach, which was thought to be a relatively sterile and a simple ecosystem, when metagenomically analysed has revealed unexpected gastric bacterial diversity. Despite the different chemical environments, gastric and intestinal microbiotas have been found to overlap significantly at the phylum level. Gastric biopsies were found to predominantly contain members of the phyla Proteobacteria (majority being *Helicobacter pylori*), Firmicutes, and Bacteroidetes, while representatives of the phyla Actinobacteria, Fusobacteria, TM7,

Deferribacteres, and Deinococcus/Thermus typically occurred in diminishing proportions (Bik et al., 2006).

Studies to demonstrate profound alterations in the gastrointestinal microbiota of patients with Crohn's disease and ulcerative colitis found particular, members of the phyla *Firmicutes* and *Bacteroidetes* typically are underrepresented in inflammatory bowel disease patients compared with controls.

Metagenomic approaches have at best been able to uncover only about 50% of the predicted species-level diversity (Frank et al., 2007) among all sampled metagenomes including the gastrointestinal track. Lot remains to be explored for gastrointestinal microbial diversity as accumulation of predicted new taxa have not reached a plateau yet.

Scientific interest in the human microbiome is so intense that the US National Institutes of Health recently committed $115 million to launching the Human Microbiome Project, which aims to characterize the microbial components of the human genetic and metabolic landscape using metagenomic and other molecular techniques (Kinross et al., 2008). It is an international effort where the mission is to generate resources enabling comprehensive characterization of the human microbiota from different body sites including the gastrointestinal tract in several hundreds of healthy and disease afflicted subjects and analysis of its role in human health and disease (Turnbaugh et al., 2007). Part of this project aims to identify a "core human microbiome," a set of microbial genes, common to all humans, and completely sequence nearly 1000 human commensal reference microbial genomes from these sites. It also aims at developing standards in storing and managing metagenomics data and metadata to facilitate smooth analysis. A Data Analysis and Coordination Center (http://www.hmpdacc.org/) has been tasked to manage the enormous amount of data that will be generated by Human Microbiome Project (Hsiao, W.W. and C.M. Fraser-Liggett. 2009).

V. BIOINFORMATIC CHALLENGES

Sequence information must be interpreted (annotated) in terms of genes, proteins and the functions they perform. Advances in high throughput sequencing have made it feasible to analyze complex mixtures of entire genomes with reasonable coverage. Annotation of genomic sequences, by comparative sequence analysis, reveals the types of genes present in a specimen. The resulting information describes the collective genetic content—the metagenome—of the community, from which physiological and metabolic lifestyles can be inferred (Frank, D.N. and N.R. Pace, 2008). One of the major challenges in the emerging field of metagenomics is computational analysis of metagenomic data. New tools for multivariate analyses are being developed to utilise metagenomic data to develop testable hypotheses about the functions of members of complex microbial communities.

New massively parallel sequencing technologies can sequence up to one billion bases in a single day at low costs but produce read lengths as short as 35–40 nucleotides, posing challenges for genome assembly and annotation. 454 Life Sciences/Roche, Solexa/Illumina and Applied Biosystems (SOLiD technology) are already into business with next generation sequencers and a competing technology from Helicos is also due to appear soon. Next generation sequencing technologies greatly increase sequencing throughput as millions of DNA fragments are laid out on a single chip and sequencing of all these fragments are done in parallel (Pop, M. and S.L. Salzberg, 2008). This promises an overwhelming sweep of raw sequences as researchers across the globe are beginning to use these technologies to sequence various metagenomes. The short read lengths and absence of paired ends make it difficult for assembly software to completely resolve repeat regions, therefore resulting in fragmented assemblies. The sequence reads in a metagenomic experiment might originate from hundreds or even thousands of different species, presenting a much greater assembly challenge than a single genome sequencing project. A simple BLAST-based strategy for each read using tblastn to translate the sequence in all six frames and search a protein database for matches have been used in the Sargasso Sea project (https://research.venterinstitute.org/sargasso/), which sampled the bacterial population of a region of the Atlantic ocean. Since then many other modified bioinformatic tools like the MEGAN, CARMA and the MetaGene system have been used to analyse metagenomic data from a number of other projects to assemble short read sequence data.

Various new annotation pipelines for metagenomic projects like in the Human Microbiome Project are being developed for efficient assembly of individual sequences into longer "contigs" and their efficient annotation. These "pipelines" leads from the raw sequence to database that integrates all that is known about the homologs of a given putative gene product and has features to allow "community-based" curation. The function of up to 75% of the genes can be deduced in this way, which is similar to the proportion of genes in individual genomes that is possible to annotate automatically. The automated annotation is very consistent, which is of great importance for comparing different metagenomes.

Finally, open access release of both software and data will accelerate efforts to develop better less error prone bioinformatic tools to analyse metagenomic data by allowing the scientific community to join forces in addressing the challenges and promises of the new sequencing technologies.

VI. CONCLUSIONS

This new science of metagenomics will not replace any of the existing fields of biology but will supplement them with a new and powerful foundation of information and deeper insight. Microbiota could potentially form the basis of a 21st century pharmacopeia. Microbes have the capacity to synthesize many novel chemical entities that sustain mutually beneficial relationships with us. Through the identification of natural products, exploration of the effects on host signalling, examination of metabolic pathways and gene manipulation, we

have the opportunity to identify and develop many possible new biologically active compounds. It is also feasible that intestinal flora will be used not only as a novel method of drug delivery but also as a bioreactor, using new cloning and biosynthetic expression strategies for the controlled secretion of biologically active molecules and vaccines. A new class of pharmaceutical compounds derived from previously uncultured commensal microorganisms in appropriate *in vitro* or animal models may be produced to develop long-term host immunotherapy for diseases such as Inflammatory Bowel Disease. Patterns of gut microbiota, as well as specific bacterial species that are now found to be associated with certain diseases will also constitute a new type of drug target.

Researchers are now also considering questions like clinical and health applications such as probiotic use, "microbiome transplants" and effects of the microbiome on behaviour and potential forensic uses of microbiome profiles.

Metagenomic approaches are beginning to provide us with a greater understanding of biological function and the complex host-microfloral metabolic cross-talk of humans at a "superorganism" level. These tools permit in-depth and non-invasive analysis of intestinal ecology. The initial evidence provided by such an approach suggests that the mammalian-microbiome symbiosis is vital to human health and fundamental to numerous pathologic processes (Cani, P.D. and N.M. Delzenne, 2007) (Kinross et al., 2008). Metagenomic insights have most notably sighted that global changes in the gastrointestinal microbiota composition may accompany a variety of disease states. The notion that the intestine is instrumental in the development of many systemic conditions not previously associated with the gut implies that the role of the gastroenterologist is becoming increasingly important. Metagenomic analysis has significant advantages. It allows a high throughput approach to process samples, which contrasts significantly with the time-consuming and labour-intensive classical culturing techniques. Moreover, metagenomic studies do not require immediate processing, as samples can be frozen and subjected to analysis later, while DNA can easily be transported between laboratories. Among the many hurdles of metagenomic gut studies and metagenomic studies all together is the unavailability of "reference" microbial genome sequences, which in turn will aid in the analysis of the complex metagenomic sequence data from microbial populations resident in the human gut. The extensive characterization of the human gut microbiome will create a technological and data rich research resource that will enable in-depth study of its variation in relation to any number of relevant variables (e.g., genotype, disease, age, nutrition, medication and other environmental factors) and its influence on health and disease. As high throughput sequencing becomes cheaper, metagenomics as a routine quantitative method for diversity analysis will be within the reach of most laboratories though it is rather complicated with respect to the standardization of extraction of genetic material to cover every member of a given bacterial community. In fact, the results achieved by metagenomic studies may be significantly influenced by the efficacy of the protocols used for directly extracting DNA from environmental samples (e.g., faecal or human tissues) as well as by the specificity of PCR primers and associated PCR conditions (Hattori, M. and T.D. Taylor, 2009). Metadata outlining the microbial diversity of the human gut should nevertheless be interpreted with caution because the large majority of published metagenomic data are based on faecal samples rather than intestinal biopsies. Thus, the identified phylotypes do not necessarily represent the actual diversity of the intestinal microbiota. In fact, the faecal microbial community (or at least the microbial DNA present in faeces) represents all bacterial species living in the gastrointestinal track and so metagenomic analysis of faecal samples seems to overestimate the diversity of the human gut microbiota, a notion which was corroborated by the finding that the microbiota of the proximal regions of the human gastrointestinal track is different from that present in other parts of the intestine (Rajilić-Stojanović et al., 2007). Despite this, faecal samples are often chosen to investigate the microbial composition of the intestinal microflora because from a practical, clinical and ethical perspective they are easy to collect. Another limitation of the available metagenomic data is that the composition of the predominant bacterial community is host-specific and that the number of genes within a given community is so large that the number of analyzed fragments derived from a single sample must be enormous to obtain a statistically relevant representation of the complete microbiome. Furthermore, not all elements of a metagenomic library, even if such a library were to be complete, are sequenced and bacterial species that are only minor components of the microbiota may appear to be absent. Therefore, considerable caution must be applied in the design, analysis and interpretation of the results obtained by metagenomic studies.

It is anticipated that exploration of newer and culture independent ways to understand human gut microbiota will provide insights that will enable the development of new approaches to monitor health status, improve understanding of the etiology of disease, and stimulate the development of new preventive and therapeutic strategies, through the maintenance or re-establishment of a healthy microbiota, cataloguing both the presence and prevalence of specific organisms and their gene content, determining the physiological capabilities of each organism in the community by elaborating their genetic makeup and to be able to interpret the functional significance of changes in community structure. Moreover metagenomic approaches such as the total microbiota genome sequencing might be important in order to identify new phylotypes with no previous cultured relatives. The analysis of such genome sequences together with functional genomic approaches will be crucial in order to highlight sequence redundancy and thus may provide very valuable information on both, strategies for survival under gastrointestinal track conditions and processes behind the mutualistic relationship established between certain microorganisms and the host. In fact, symbionts may dedicate part of their genomes to functions that are beneficial for both bacterium and host. Thus, the identification of such functions will aid in the development of novel probiotics as it is important that any newly developed probiotic bacterium must be genetically adapted to survive in the intestinal ecological niche. Such requirements will provide these bacteria with genetically determined capabilities to interact with both the human host as well as with the other indigenous components of the intestinal microflora, thus allowing them to exert their health-promoting effects. Greater depth of knowledge on the autochthonous component of the intestinal microbiota will provide the possibility to screen or develop important characteristics within novel probiotic microbes, including the capacity to be highly stable in the human intestine and especially have reliable or enhanced probiotic effects (Ventura et al., 2009). Furthermore as our understanding of the ecology and rich diversity of the human gut microbiota develops, members isolated from intestinal biopsies (autochthonous component of the intestinal microbiota), beside faecal isolates, will contribute to the next generation of probiotic microbes. Knowledge on evolution of species and strain diversity in the intestine can thus open up new possibilities that are expected to lead to the development of a next generation of designer probiotics compatible with the genetic background of the targeted host and specific gastrointestinal disorders keeping in mind individual commensal microbiome enabling customised disease prevention strategies and treatment regimes (Preidis, G.A. and J. Versalovic, 2009).

REFERENCES

Bäckhed F., Ley R.E., Sonnenburg J.L., Peterson D.A. and Gordon J.I. (2005), Host-bacterial Mutualism in the Human Intestine. *Science*, 307(5717):1915–20, Review.

Beaugerie L. and Petit J.C. (2004), Microbial-gut Interactions in Health and Disease, Antibiotic-associated Diarrhoea. *Best. Pract. Res. Clin. Gastroenterol*, 18(2):337–52, Review.

Bik E.M., Eckburg P.B., Gill S.R., Nelson K.E., Purdom E.A., Francois F., Perez-Perez G., Blaser M.J. and Relman D.A. (2006), Molecular Analysis of the Bacterial Microbiota in the Human Stomach, *Proc. Natl. Acad. Sci. USA*. 103(3):732–7.

Cani P.D. and Delzenne N.M. (2007), Gut Microflora as a Target for Energy and Metabolic Homeostasis, *Curr. Opin. Clin. Nutr. Metab. Care.*, 10(6):729–34, Review.

Dethlefsen L., Huse S., Sogin M.L. and Relman D.A. (2008), The Pervasive Effects of an Antibiotic on the Human Gut Microbiota, as Revealed by Deep 16S rRNA Sequencing, *PLoS Biol*. 18; 6(11):E280.

Eckburg P.B., Bik E.M., Bernstein C.N., Purdom E., Dethlefsen L., Sargent M., Gill S.R., Nelson K.E. and Relman D.A. (2005), Diversity of the Human Intestinal Microbial Flora. *Science*, 308(5728):1635–8.

Frank D.N., St Amand A.L., Feldman R.A., Boedeker E.C., Harpaz N. and Pace N.R. (2007), Molecular-phylogenetic Characterization of Microbial Community Imbalances in Human Inflammatory Bowel Diseases, *Proc. Natl. Acad. Sci.* USA, 104(34):13780–5.

Frank D.N. and Pace N.R. (2008), Gastrointestinal Microbiology Enters the Metagenomics Era, *Curr. Opin. Gastroenterol,* 24(1):4–10, Review.

Gill S.R., Pop M., Deboy R.T., Eckburg P.B., Turnbaugh P.J., Samuel B.S., Gordon J.I., Relman D.A., Fraser-Liggett C.M. and Nelson K.E. (2006), Metagenomic Analysis of the Human Distal Gut Microbiome, *Science*, 312(5778):1355–9.

Guarner F. and Malagelada J.R. (2003), Gut Flora in Health and Disease, *Lancet*, 361(9356):512–9, Review

Hamady M. and Knight R. (2009), Microbial Community Profiling for Human Microbiome Projects: Tools, Techniques and Challenges, *Genome Res.* 19(7):1141–52, Review.

Hattori M. and Taylor T.D. (2009), The Human Intestinal Microbiome: A New Frontier of Human Biology, *DNA Res.*16(1):1–12, Review.

Hooper L.V., Midtvedt T. and Gordon J.I. (2002), How Host-microbial Interactions Shape the Nutrient Environment of the Mammalian Intestine, *Annu. Rev. Nutr.*, 22:283–307, Review.

Hsiao W.W. and Fraser-Liggett C.M. (2009), Human Microbiome Project-paving the Way to a Better Understanding of Ourselves and Our Microbes, *Drug Discov. Today*,14(7–8):331–3.

Juengst E.T. (2009), Metagenomic Metaphors: New Images of the Human from 'Translational' Genomic Research, 129–145, In: Martin A.M., Drenthen, F.W., Jozef Keulartz and James Proctor (Eds.), *New Visions of Nature—Complexity and Authenticity*, Part-3 Springer Netherlands, Netherlands.

Kinross J.M., Von Roon A.C., Holmes E., Darzi A. and Nicholson J.K. (2008), The Human Gut Microbiome: Implications for Future Health Care, *Curr. Gastroenterol. Rep.* 10(4):396–403, Review.

Ley R.E., Peterson D.A. and Gordon J.I. (2006), Ecological and Evolutionary Forces Shaping Microbial Diversity in the Human Intestine, *Cell.*,124(4):837–48, Review.

Macpherson A.J., Geuking M.B. and McCoy K.D. (2005), Immune Responses that Adapt the Intestinal Mucosa to Commensal Intestinal Bacteria, *Immunology*, 115:153–162.

Majewski J., Zawadzki P., Pickerill P., Cohan F.M. and Dowson C.G. (2000), Barriers to Genetic Exchange Between Bacterial Species: *Streptococcus pneumoniae* Transformation, *J. Bacteriol.*, 182(4):1016–1023.

Narushima S., Itoha K., Miyamoto Y., Park S.H., Nagata K., Kuruma K. and Uchida K. (2006), Deoxycholic Acid Formation in Gnotobiotic Mice Associated with Human Intestinal Bacteria, *Lipids*, 41(9):835–43.

O'Hara A.M. and Shanahan F. (2006), The Gut Flora as a Forgotten Organ, *EMBO Rep.*, 7(7):688–93, Review.

Petrosino J.F., Highlander S., Luna R.A., Gibbs R.A. and Versalovic J. (2009), Metagenomic Pyrosequencing and Microbial Identification, *Clin. Chem.*, 55(5):856–66, Review.

Pop M. and Salzberg S.L. (2008), Bioinformatics Challenges of New Sequencing Technology, *Trends Genet.*, 24(3):142–9, Review.

Preidis G.A. and Versalovic J. (2009), Targeting the Human Microbiome with Antibiotics, Probiotics, and Prebiotics: Gastroenterology Enters the Metagenomics Era, *Gastroenterology*,136(6):2015–31, Review.

Rajilić-Stojanović M., Smidt H. and De Vos W.M. (2007), Diversity of the Human Gastrointestinal Tract Microbiota Revisited, *Environ. Microbiol.*, 9(9):2125–36, Review.

Rappé M.S. and Giovannoni S.J. (2003), The Uncultured Microbial Majority, *Annu. Rev. Microbiol.*, 57:369–94, Review.

Tschöp M.H., Hugenholtz P. and Karp C.L. (2009), Getting to the Core of the Gut Microbiome, *Nat. Biotechnol*, 27(4):344–6.

Tsukumo D.M., Carvalho B.M., Carvalho-Filho M.A. and Saad M.J. (2009), Translational Research into Gut Microbiota: New Horizons in Obesity Treatment, *Arq. Bras. Endocrinol. Metabol.*, 53(2):139–44.

Turnbaugh P.J., Ley R.E., Mahowald M.A., Magrini V., Mardis E.R. and Gordon J.I. (2006), An Obesity-associated Gut Microbiome with Increased Capacity for Energy Harvest, *Nature*, 444(7122):1027–31.

Turnbaugh P.J., Ley R.E., Hamady M., Fraser-Liggett C.M., Knight R. and Gordon J.I. (2007), The Human Microbiome Project, *Nature*, 449(7164):804–10.

Turnbaugh P.J. and Gordon J.I. (2009), The Core Gut Microbiome, Energy Balance and Obesity, *J. Physiol.*, 587(Pt 17):4153–8.

Turnbaugh P.J., Hamady M., Yatsunenko T., Cantarel B.L., Duncan A., Ley R.E., Sogin M.L., Jones W.J., Roe B.A., Affourtit J.P., Egholm M., Henrissat B., Heath A.C., Knight R. and Gordon J.I. (2009), A Core Gut Microbiome in Obese and Lean Twins, *Nature*, 457(7228):480–4.

Ventura M., Turroni F., Canchaya C., Vaughan E.E., O'Toole P.W. and Van Sinderen D. (2009), Microbial Diversity in the Human Intestine and Novel Insights from Metagenomics, *Front. Biosci.*, 14:3214–21, Review.

Xu J., Bjursell M.K., Himrod J., Deng S., Carmichael L.K., Chiang H.C., Hooper L.V. and Gordon J.I. (2003), A Genomic View of the Human-Bacteroides Thetaiotaomicron Symbiosis, *Science*, 299(5615):2074–6.

Zoetendal E.G., Rajilic-Stojanovic M. and De Vos W.M. (2008), High-throughput Diversity and Functionality Analysis of the Gastrointestinal Tract Microbiota, *Gut.*, 57(11):1605–15, Review.

Zoetendal E.G., Vaughan E.E. and De Vos W.M. (2006), A Microbial World Within us, *Mol. Microbiol*, 59(6):1639–50, Review.

56

The Biliary System
Normal Physiology and Microanatomy

Shiv Kumar Sarin and Cyriac Abby Philips

CONTENTS

I. ANATOMIC ORGANIZATION

The biliary system is an integral part of the human body that includes the gall bladder and the intrahepatic and extrahepatic ductal system and the peribiliary glands (Figure 56.1).

A. Gall Bladder and Cystic Duct

The gall bladder is a small sac-like pyriform organ with a wall thickness of less than 3 mm, that lies at the inferior part of the liver in relation to the right lobe. The gall bladder is divided into the fundus, body and neck which is continuous with the cystic duct; the duct which joins it to the hepatic ducts. It lies obliquely in the antero-posterior direction and medially, such that the fundus that is the most anterior portion, lies below the level and away from the midline as compared to the neck region.

B. The Right and Left Hepatic Ducts

The right and left hepatic ducts are 3 to 4 mm in diameter and lie within the hepatoduodenal ligament. They join to form the common hepatic duct at the porta hepatis (hilum of the liver) within 1 cm of their exit from the liver. The left hepatic

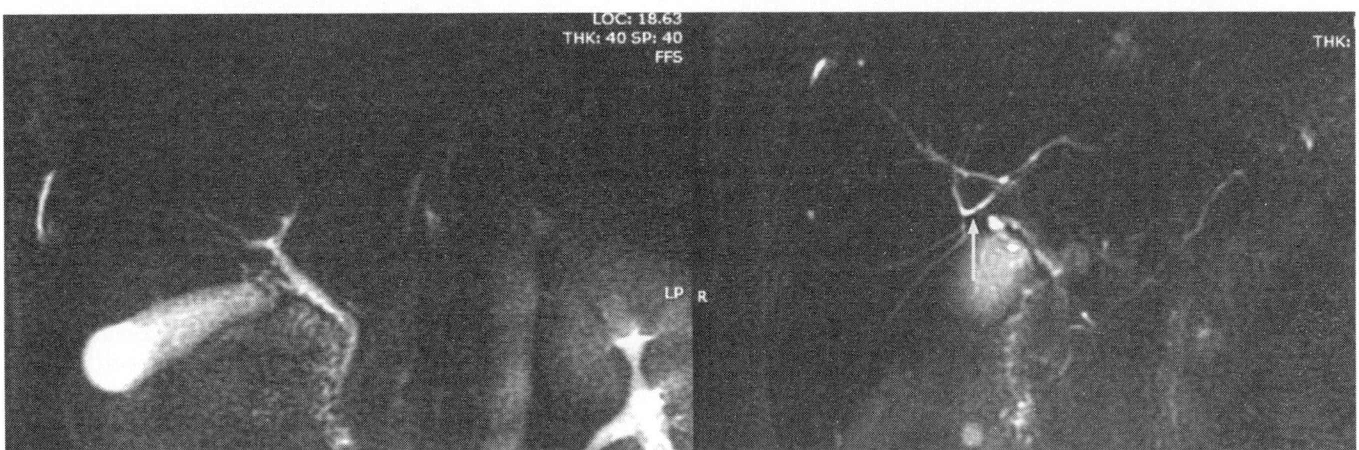

Figure 56.1 The biliary tree or the biliary system (Magnetic resonance cholangiopancreatography picture). The second image shows the third to fifth order branching of the biliary tree also (arrow). (*see Plate 17 for colour figure*)

duct drains the three segments of the left lobe: II, III and IV. Segment IV is drained by mediosuperior and medioinferior branches. The right hepatic duct is mostly an extrahepatic structure, and drains the four subsegments of the right lobe: V, VI, VII, and VIII. The biliary drainage of the caudate lobe (segment I) enters both the right and the left hepatic duct systems in 80% of individuals.

C. The Common Hepatic Duct

The common hepatic duct is formed by the union of the left and right hepatic ducts 1 cm outside the hepatic hilum in the transverse fissure of the liver. This common duct runs medially, downwards in the gastro-hepatic omentum lateral to the hepatic artery and in front of the portal vein and cystic artery, thereafter joining the cystic duct to form the common bile duct. The common hepatic duct is 2 to 8 mm in diameter and is 1–5 cm long.

D. The Common Bile Duct

This is formed by the union of the cystic duct and the common hepatic duct. The common bile duct from its origin, passes downwards and left and then slightly forward. It then again passes downwards and to the right and to the back – thereby displaying a double curvature, with a convexity directed towards the left and another forwards. It is classically divided into 3 portions. The supra-pancreatic division, the pancreatic or middle division, and the inferior or duodenal division which pierces the walls of the second part of the duodenum to open into it along with the pancreatic duct at the ampulla of Vater.

II. MICROSCOPIC ORGANIZATION

A. The Gall Bladder

The gall bladder consists of the mucosa, which is composed of simple columnar epithelium which throws numerous folds into

the lumen to facilitate increment in absorptive and concentrative functions. The mucosa of the gall bladder secretes mucous and not bile. The inner mucosal layer has multiple polygonal compartments which give it a 'honeycomb' appearance on histological examination (Figures 56.3a, 56.3b, 56.3c). The

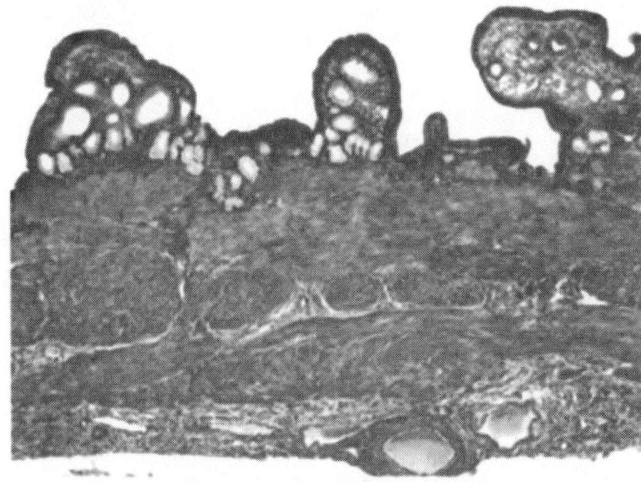

Figure 56.2 Histopathology of the Gall Bladder epithelium. H&E stain 100X. (*see Plate 17 for colour figure*)

Rokitasnky Aschoff sinuses are mucosal outpouchings that go through the muscularis layer into the subserosal tissue. They are common occurrences in a normal gall bladder.

B. The Ductal System and the Peribiliary Glands

The intrahepatic bile ducts are divided into large and small segments. The large type encompasses the right and left hepatic bile ducts and their first to third branches called the segmental area bile ducts. The small intrahepatic bile ducts, the branches of the large intrahepatic bile duct are of two types—the septal and the interlobular both of which are microscopic (Figure 56.4).

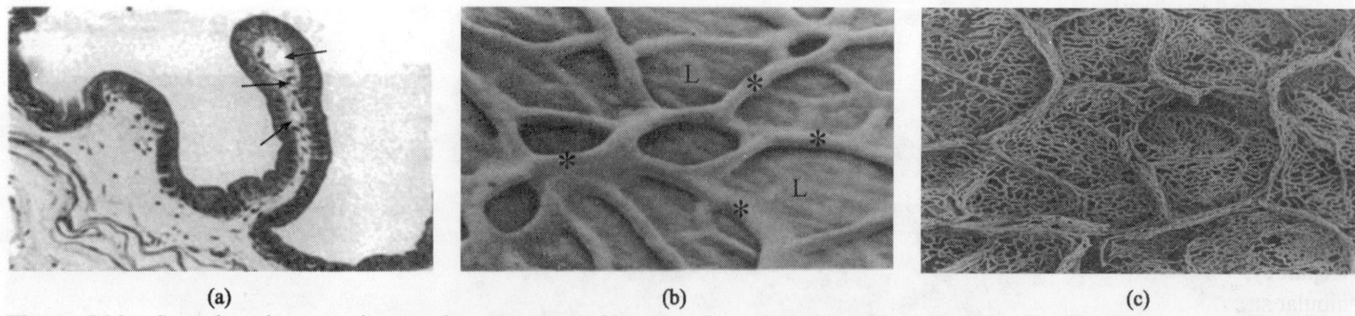

Figure 56.3 Scanning electron microscopic appearance of inner surface of gall bladder epithelium showing numerous capillaries and epithelial folds with rich vascular network (Vascular Corrosion Cast Imaging). *From: Alberto Caggiati et al: Scanning electron microscopy of the rabbit gallbladder mucosal microvasculature.* Arrows—thin capillaries, Asterix—Epithelial folds and L—lacunae

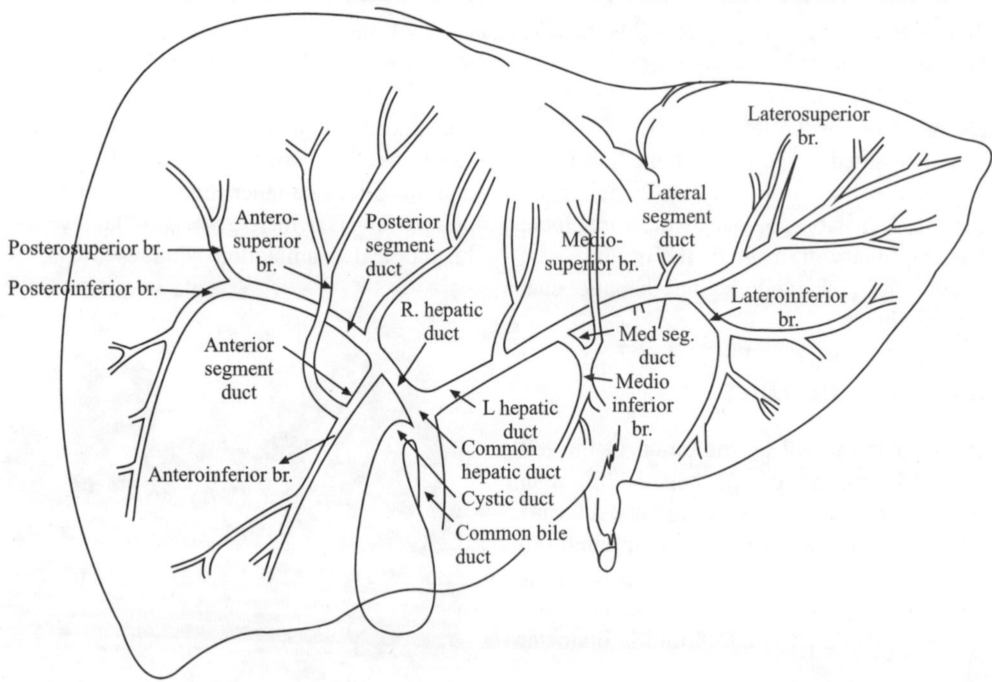

Figure 56.4 The intrahepatic organization of the biliary system.

Bile canaliculi are formed from the membrane of adjacent parenchymal cells, separated from the perisinusoidal space by junctions. The bile flow is from the canaliculi through the Canals of Herring into the interlobular bile ducts that are evident in the portal pedicles. Here the bile ducts are found above and veins and arteries beneath.

The biliary cells have a basement membrane. The Canal of Hering is the bridge between the anatomic and physiological link between the intralobular canalicular system and the biliary tree. The Canal of Hering continues into a channel lined by the cholangiocytes—the ductule. Ductules may have an intralobular course and an intraportal course depending on how much they traverse the limiting plate. The ductules link the smallest interlobular bile ducts. The peribiliary glands are seen within in the fibromuscular walls of the extrahepatic bile ducts and also along the large intrahepatic bile ducts. These glands are again of three types—the intramural glands, non branching tubular glands and extramural ramified glands. The unique expression pattern of PDX1 and HES1 and increased expression of endodermal S/P cell markers in the peribiliary glands may be involved in biliary pathophysiologies.

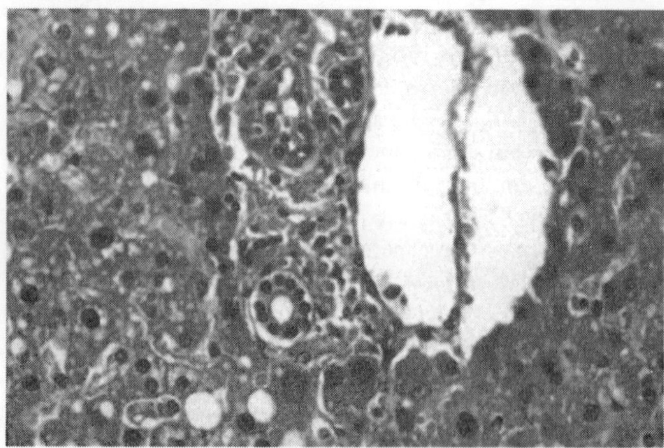

Figure 56.5 The bile duct on histopathology examination. H&E staining at 100X. (*see Plate 17 for colour figure*)

To summarize, the bile duct system is categorized according to the original classification proposed by Ludwig; based on diameter. It consists of terminal cholangioles (diameter <15 mm), interlobular ducts (15–100 mm), septal ducts (100–300 mm), area ducts (300–400 mm), segmental ducts (400–800) and hepatic ducts (>800 mm) (Figure 56.6a and b).

The biliary apparatus is a convergent system of canals that begins in the canaliculi ending in the common bile duct. Hepatocyte forms the primary bile. Biliary canaliculi are mild tubular structures with very surface to volume ratio that favours osmotic gradient mediated bile flow. Cholangiocytes secrete 30% of the bile volume and help to modify the canalicular bile by secretory and reabsorption processes. The hepatocyte exhibits structural and functional polarity through different domains. The basolateral membrane of hepatocytes is involved in secretions and regulation of bile flow. The Golgi complex and network of microtubules and microfilaments help in formation of bile. The vesicles formed from the Golgi complex are vehicles for substances that are excreted in bile and form plasma proteins including transporters that form the structural network of apical or basolateral membranes. The novelly synthesized apical ABC (ATP-binding cassette) transporters are transferred from Golgi apparatus to the canalicular membrane. The hepatocytes and the biliary system in close relation to the blood vasculature form a functional unit, known as the **acinus.** The blood flow that is in opposite to the bile flow generates concentrative gradients along the sinusoids. This forms the periportal zone or zone I (more oxygenated), intermediate zone or zone II and centrilobular zone or zone III (least oxygenated) (Figure 56.7a, b, c and d).

Figure 56.6a The microanatomy of small and large ducts and their contrasting functional delineation. *From: Alpini, McGill and LaRusso. The Pathobiology of Biliary Epithelia. Hepatology, Vol. 35, No. 5, 2002.*

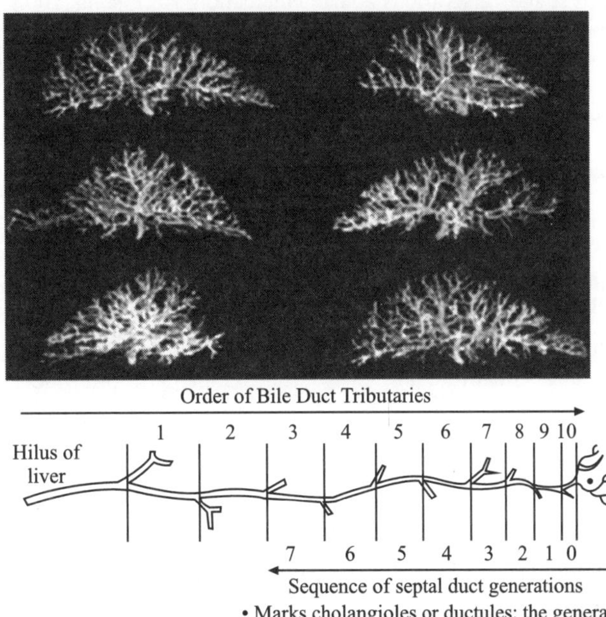

Figure 56.6b Computer-generated cholangiographic views of a normal biliary tree, based on contiguous 2–mm slices imaged with electron-beam computed tomography. Schematic representation of branching orders and duct generations of the biliary tree. The numbering of duct orders increases against the direction of the bile flow, whereas the numbering of duct generations increases in the direction of the bile flow. Ducts draining the surgical segments of Coineau are of the third to fifth order. *From: Ludwig et al. Hepatology Vol. 27, No. 4, 1998 893–899.*

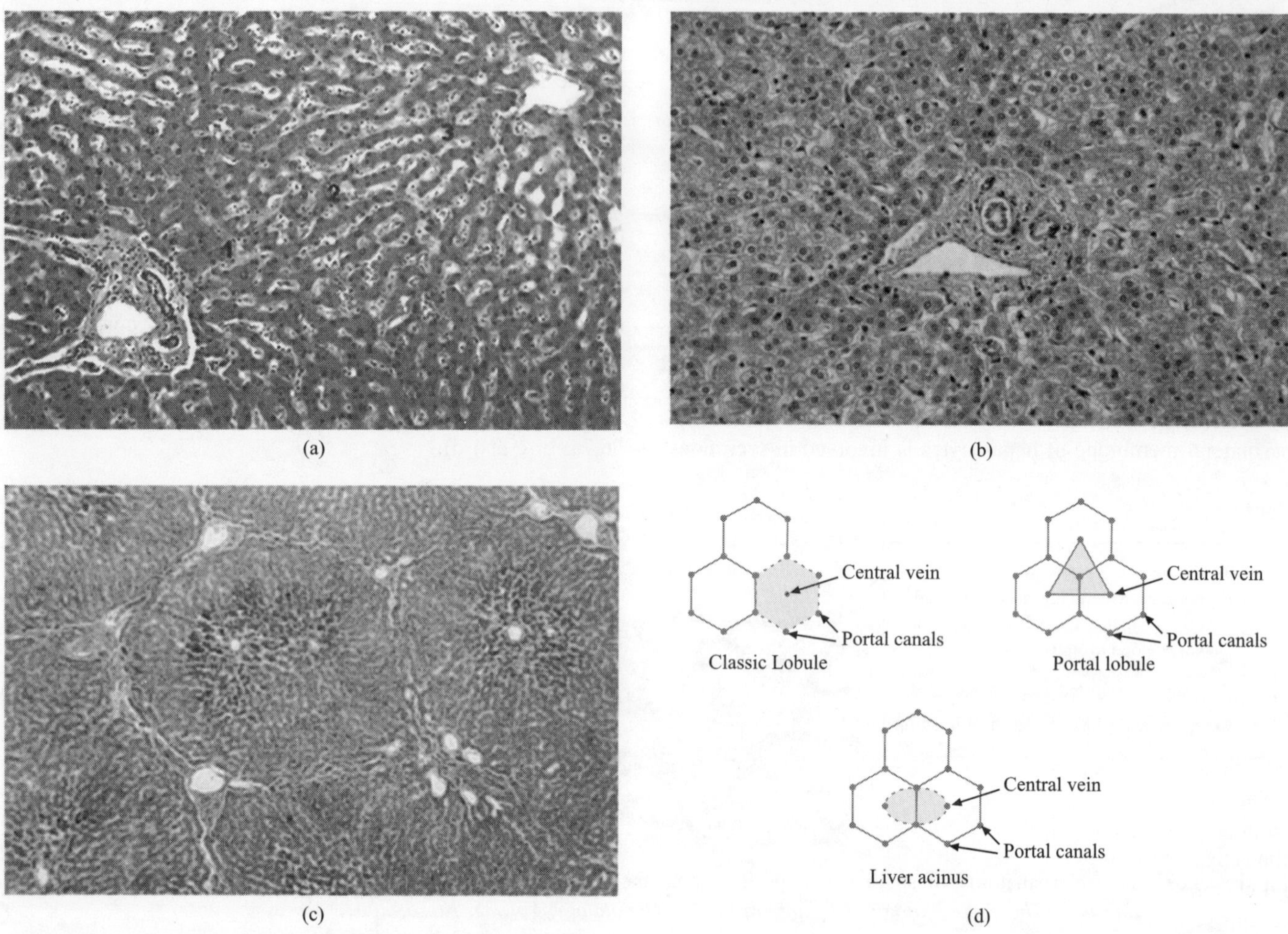

Figure 56.7   (a) The acinar unit of the liver showing various zonal demarcation. H&E stain, 100X, (b) The portal triad, (c) The lobules of the liver. H&E staining at 100X and (d) The functional unit models of the liver. (*see Plate 18 for colour figure*)

III. MOLECULAR ORGANIZATION OF THE BILIARY SYSTEM

A. The Hepatocyte

Hepatocytes are the parenchymal cells of the liver. They are majorly concerned with the synthetic and metabolic functions of the liver. They are large polygonal cells with complex rhomboid form with many surfaces and constitute almost 80% of the liver parenchyma. A large proportion of the hepatocyte surface faces other hepatocytes and the rest face the sinusoidal space throwing microvilli into the space of Disse. Tight junctions, intermediate junctions and desmosomes form adhesion complexes that form a permeability barrier between the space of Disse and the bile canaliculi. The bile canaliculi form a thin continuous stretch of extracellular space that moves along the lengths of the hepatic plates and connects the portal ends with the bile ducts eventually. The canalicular membrane is modified to secrete bile. Hepatocytes are rich in mitochondria, lysosomes, peroxisomes, Golgi complexes, microtubules and filaments and numerous rough and smooth endoplasmic reticulum.

B. The Cholangiocyte (Biliary Epithelial Cell)

The cholangiocytes (Figure 56.8a) are the main cells that line the biliary tract. The tract begins with the Canals of Herring (that are points where liver epithelial stem cells reside and they differentiate into cholangiocytes or hepatocytes) (Figure 56.8b) and progressively merges into the interlobular, septal and then the major ducts which then combine to form the extrahepatic bile ducts that eventually drain bile into the gall bladder and the duodenum. Four to five cholangiocytes line the small ductules. These cholangiocytes are cuboidal in shape, has a basement membrane, form tight junctions with the adjacent cells and have microvilli projecting into the bile duct lumen. The cholangiocytes, in addition also have primary cilia on their apical cell membrane; the functions of which are many.

The major function of the cholangiocyte is ductal bile production initiated by a series of secretory and absorptive processes. The primary cilium of the cholangiocyte is a solitary long organelle that extends into the lumen.

The major characteristics of the primary cilium include:
1. It arises from the mature mother centriole of a pair of centrioles.

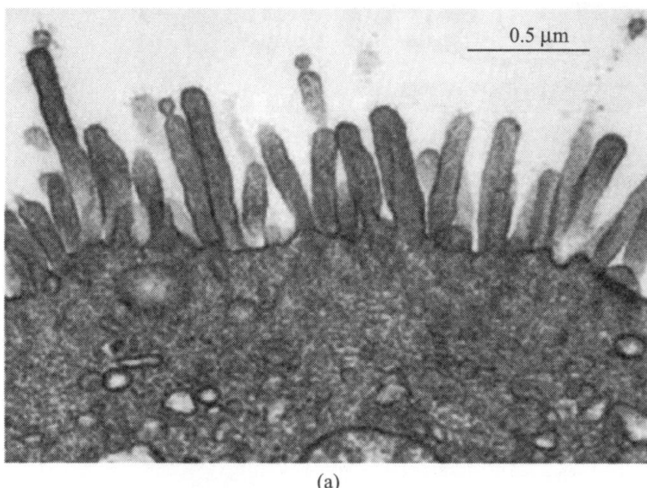
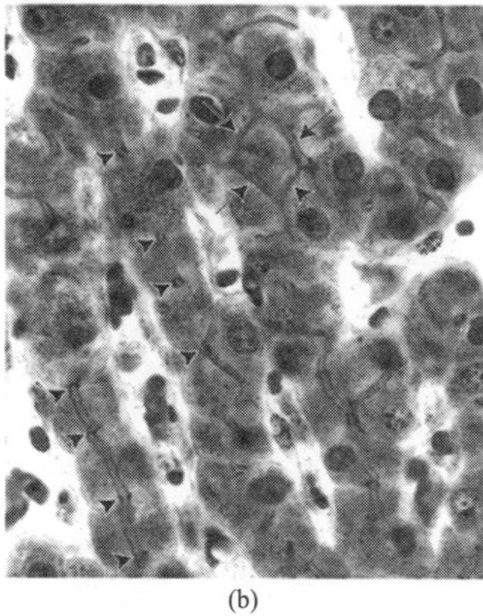

(a)　(b)

Figure 56.8　(a) Electron micrograph of the luminal border of a cholangiocyte showing microvilli, (b) The bile canaliculi system criss-crossing across between the hepatocytes (arrows). (*see Plate 18 for colour figure*)

2. Consists of a basal body and a '9+0' axoneme with nine outer doublet microtubules.
3. Lacks the central pair of microtubules and dynein arms
4. They are functionally heterogeneous.

The size of the cilia depends upon the size of the cholangiocyte anywhere from 2 to 8 micrometer being bigger in the bigger ducts. The main generally accepted function of the cilia is that of sensory. They sense mechanostimuli, osmostimuli and chemostimuli. They do so by their strategic placement at the apical plasma membrane whereby they detect changes in bile flow, composition and osmolality and converting them into intracellular signals. The calcium and cAMP signaling pathways are the major players in the intracellular regulatory mechanism by which the extracellular or the ciliary signal affect ductal bile formation.

Release of calcium ions from the intracellular stores in response to activation of G protein coupled receptors/Tyrosine kinase receptors on the cholangiocyte membrane or influx of the extracellular calcium ions into the cell through dedicated channels on the plasma membrane results in the activation process. Various factors like type of neurotransmitters, growth factors, hormones etcetera produce specific patterns of calcium ion activatioin; for example acetylcholine induced 'wavy' activation from apex to base and oscillatory pattern of activation produced by adenosine triphosphate. The increase in calcium ions in cholangiocytes results in production of inositol triphosphate (IP3) which acts through the IP3R (receptor) seen in the endoplasmic reticulum. This results in calcium release into the cytoplasm from the endoplasmic reticulum. The predominantly expressed isoform of IP3 receptor type in cholangiocyte in the isoform III.

The purinergic receptor family which includes the P2Y and P2X receptor families are also associated with the calcium ion and cAMP based signaling on the apical plasma membrane of the cholangiocyte. Out of these receptors only $P2Y_{12}$ is associated with the cholangiocyte cilia and works through the cAMP pathway. The other channels through which calcium ion activation occurs include the PC1 and PC2 (Polycystin group) and the TRV4 (transient receptor potential vanilloid 4) channel. PC1 and PC2 form the mechanosensory complex in primary cilia. The cilia respond to changes in bile flow thorugh transient increases in intracellular calcium ion concentrations. TRPV4 is the most sensitive osmolality change sensing channel on the cholangiocyte (Figure 56.9).

The cAMP pathway consists of major components that include adenyl cyclase, phosphodiesterase and serine threonine protein kinase A along with the exchange proteins and A kinase anchoring proteins. The cAMP signaling in cholangiocytes are by multiple mechanisms. Adenyl cyclase acts through Gs, Go, Gi and Gq proteins resulting in increase or decrease in cAMP levels. The cAMP activities are also under control of the phosphodiesterase receptor family. The cAMP signaling plays a major role in mechano and chemosensory functioning of the primary cilium. The calcium signaling in cholangiocytes are activated by extracellular phenomenon such as bile flow and bile osmolality, whereas the cAMP signaling is inhibited by the bile flow and biliary nucleotides.

The cholangiocytes form only 1% of the total number of cells in the liver parenchyma. Cholangiocytes together with the peribilliary plexus of capillaries form a metabolic unit that manages and modifies the content of water and solute in the canalicular bile.

1. Hormone/peptide receptors (basolateral domain of cholangiocytes)—modulate ductal choleresis OR inhibit basal and secretin-stimulated choleresis.
2. The choleretic effect of secretin is mediated by increases in cAMP, activation of CFTR with subsequent ductal bicarbonate secretion. [Aquaporin water channels (regulated by secretin) play an important role]
3. Concomitant with cAMP-dependent chloride secretion, physiologic concentrations of ATP in bile activate P2u receptors, which mobilize intracellular calcium stores

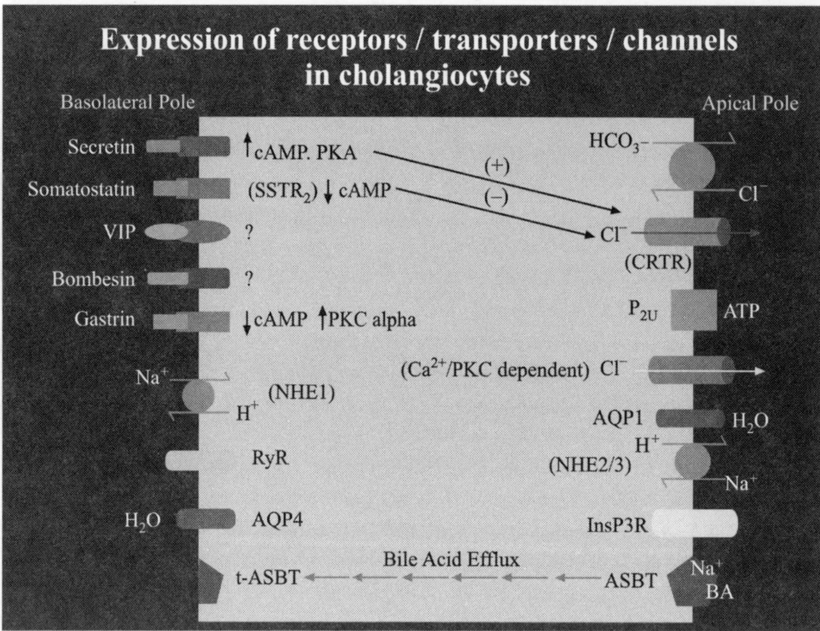

Figure 56.9   Representation of localization of some of the membrane transporter proteins expressed in cholangiocytes. (*see Plate 18 for colour figure*)

activate calcium dependent apical membrane chloride channels, which seem to play an important role in bile secretion.

4. The NHE1 isoform regulates secretin-stimulated ductal secretion.

5. Type I InsP3R is expressed apically whereas type II is distributed in the basolateral surface. Type III InsP3R is expressed at the apical pole of cholangiocytes where its activation initiates a wave of calcium release from apical to basolateral that is independent of extracellular calcium concentration.

6. While the apically located ABAT allows the entry of bile salts into cholangiocytes, the truncated form of ABAT favour handling and elimination of bile salts from the basolateral membrane of cholangiocytes.

C. The Molecular Antigenic Structure of the Biliary Tree

Each anatomic components of the biliary system have specific antigenic components that are unique to each other. The cholangiocytes lining the large bile ducts contain columnar epithelial cells and contain and secrete mucus; whereas mucin is not detectable in the interlobular bile ducts and ductules. The cholangiocytes that line the intrahepatic large bile ducts express the MUC3 and MUC6 whereas the small intrahepatic bile ducts do not. The MUC1, MUC2 and MUC5 is seen more commonly in disease conditions like hepatolithiasis and not in normal livers. Cytokeratin expression CK7 and CK19 are seen in cholangiocytes of the biliary tree and also in peribiliary glands, while EpCAM is seen expressed in bile ductules. CK8 and CK18 are expressed in hepatocytes and also some cholangiocytes of the biliary tree.

IV. THE FUNCTIONAL ORGANIZATION OF BILIARY SYSTEM

A. The Hepatocyte Couplet Model

The hepatocyte couplet (Figure 56.10) is a primary secretory unit for electrophysiological studies of biliary secretory functions. It is made up of two adjacent hepatocytes that surround a lumen or a vaccoule that represents the functional equivalent of bile canaliculus in the intact liver.

B. Cholangiocyte Level Functional Aspects

The cholangiocytes in the small biliary ducts differ very much from the ones present in the large biliary ducts. The functional components of the biliary system (at the cholangiocyte level) include – bile formation, proliferation, injury repair, fibrosis, angiogenesis and regulation of blood flow which have been further broken down into their respective descriptions as below (Figure 56.11).

C. The Bile and Biliary Lipids

Bile mostly consists of water with organic and inorganic substances (Figure 56.12) which are in equilibrium in between states. The concentration of sodium, potassium, calcium and bicarbonate are higher than that of plasma concentration. The biliary chloride is usually lower than that in plasma. Bile acids are in the concentrations of 2 to 45 mmmol/L and the pigments in the range of 50 to 200 mg/100 ml. The phospholipid concentrations range between 25 to 810 mg/100 ml and cholesterol is about 60 to 320 mg/100 ml. The ratio on an average, of phospholipids to bile acids is 0.3 and cholesterol

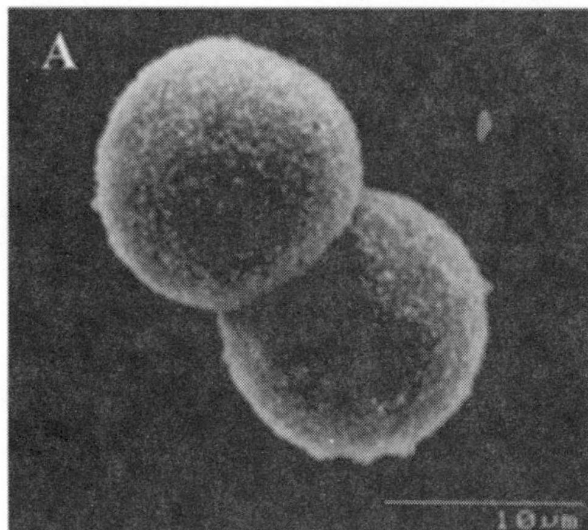
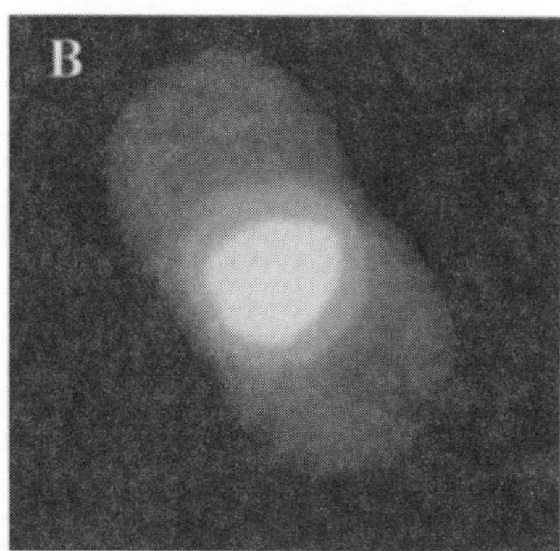

Figure 56.10 The hepatocyte couplet. *From: Gautam A et al. Quantitative assessment of canalicular bile formation in isolated hepatocyte couplets using microscopic optical planimetry. The Journal of Clinical Investigation.1989 February; 83(2)565.*

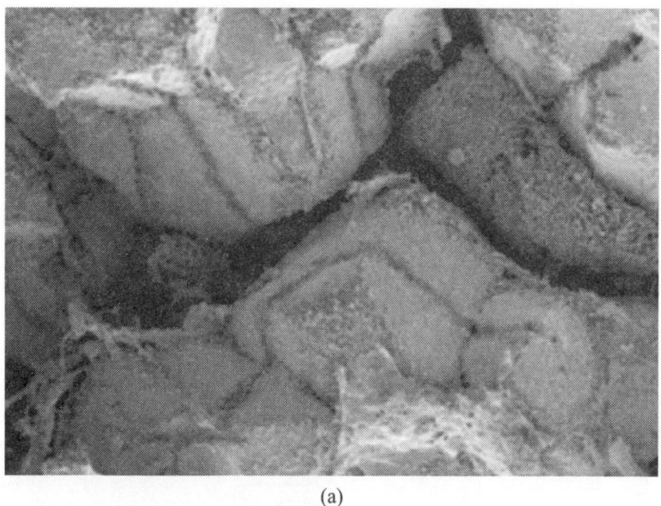
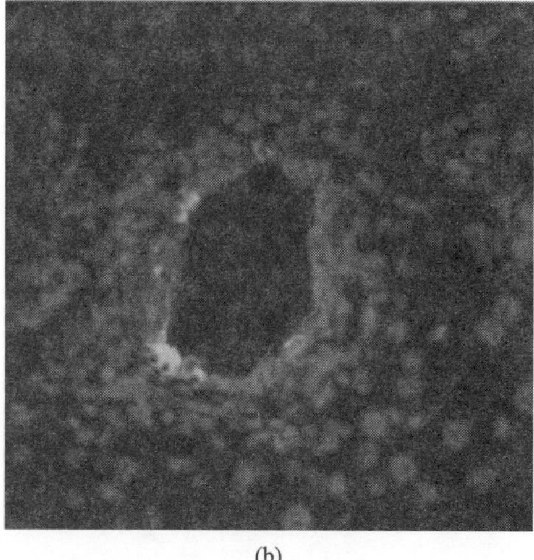

(a)   (b)

Figure 56.11 (a) Electron micrograph of biliary canaliculi draning into larger ductule, (b) Cholangiocytes lining the biliary canalicular region. (Immunofluroscent Study). (*see Plate 19 for colour figure*)

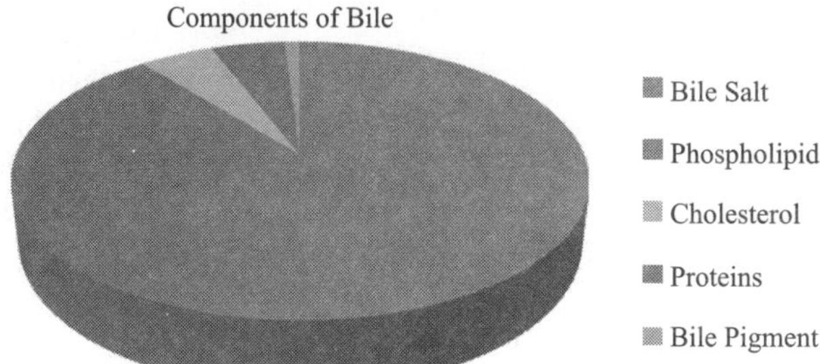

Figure 56.12 Components of the bile. (*see Plate 19 for colour figure*)

to bile acids is 0.07. The mean basal bile flow is around 620 ml/day. The bile acid dependent canalicular fraction is around 220 ml/day. The bile acids have choleretic activity where in water is incorporated into the bile in realization with the bile acid concentration which then regulates the flow. The canalicular bile independent of the osmotic force of the bile acids is called the independent fraction and is around 235 ml/day.

Biliary lipid secretion (Figures 56.13a, b, c, and d) is driven by the presence of bile salts. There is a striking contrast between the composition of phospholipids in bile and in the canalicular membrane. Biliary phospholipids consists almost exclusively of phosphatidylcholine and canalicular membrane contains sphingomyelin, phosphatidyl ethanolamine and phosphatidyl lysine. Biliary phospholipids predominantly contain palminate fatty acid. MDR2 (in murines) and MDR3 (in humans) glycoporteins are essential for biliary lipid secretion. It belongs to the family of P glycoproteins which belong to the super family of ABC transporters. The majority of the phospholipids of biliary origin is derived from pre-existing microsomal and extrahepatic pools. Within the hepatocyte, the phopholipid molecules are remodeled with regards to their fatty acyl chains. The cytosolic phospholipid transfer portien functions in the supply of phospholipids to the inner canalicular leaflet. Phospholipid and cholesterol secretion are tightly coupled. In the absence of MDR3 activity, there is no secretion of cholesterol. In the heterozygous state, there is

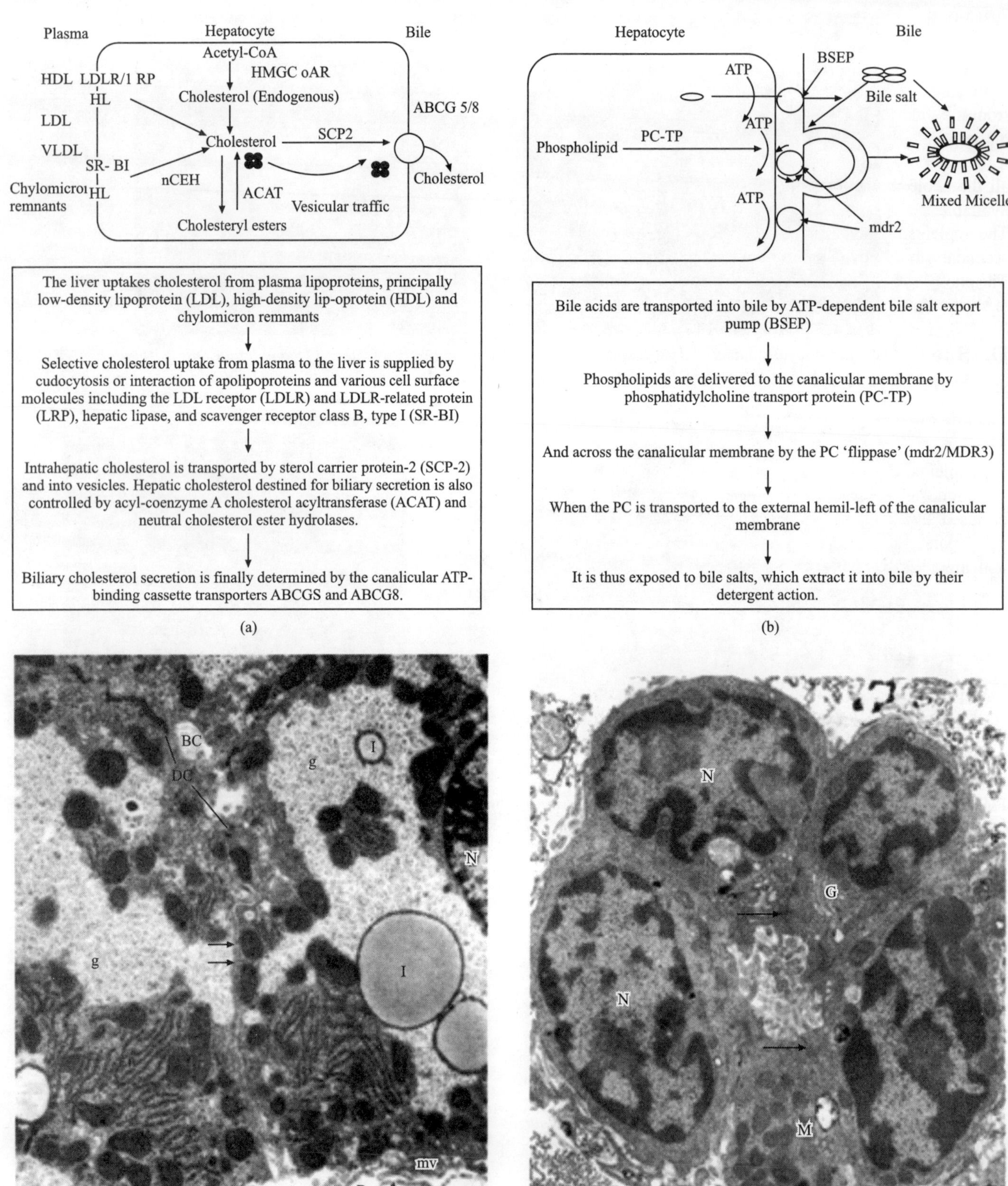

The liver uptakes cholesterol from plasma lipoproteins, principally low-density lipoprotein (LDL), high-density lip-oprotein (HDL) and chylomicron remmants

Selective cholesterol uptake from plasma to the liver is supplied by cudocytosis or interaction of apolipoproteins and various cell surface molecules including the LDL receptor (LDLR) and LDLR-related protein (LRP), hepatic lipase, and scavenger receptor class B, type I (SR-BI)

Intrahepatic cholesterol is transported by sterol carrier protein-2 (SCP-2) and into vesicles. Hepatic cholesterol destined for biliary secretion is also controlled by acyl-coenzyme A cholesterol acyltransferase (ACAT) and neutral cholesterol ester hydrolases.

Biliary cholesterol secretion is finally determined by the canalicular ATP-binding cassette transporters ABCGS and ABCG8.

(a)

Bile acids are transported into bile by ATP-dependent bile salt export pump (BSEP)

Phospholipids are delivered to the canalicular membrane by phosphatidylcholine transport protein (PC-TP)

And across the canalicular membrane by the PC 'flippase' (mdr2/MDR3)

When the PC is transported to the external hemil-left of the canalicular membrane

It is thus exposed to bile salts, which extract it into bile by their detergent action.

(b)

(c)

(d)

Figure 56.13 (a) Phospholipid transport into bile, (b) Cholesterol transport into bile, (c) Transmission electron microscopy of the biliary system. BC – bile canaliculus, OC – occluding junctions, sD-space of Disse, mv-microvilli, Si-sinusoid, SC-sinusoid lining cells, l-lipid droplet, g-glycogen, m-mitochondria, rER-rough endoplasmic reticulum, N-nucleus and (d) Micrograph of an isolated intrahepatic bile duct. N-nucleus, M-mitochondria, L-lumen, Microcilli project into L, Arrows – tight junctions.

40% of phospholipid secretion decrement, but the cholesterol secretion is not hampered. A proper transporter gene for puritan cholesterol secretion has not been delineated yet. There has been an advocacy that spontaneous flip-flop of cholesterol from the inner to the outer leaflet is rapid enough to generate the required cholesterol flux. Also, the lipids from the inner leaflet laterally diffuse through the tight junctions. As a result, all the cholesterol present in the entire plasma membrane is available for extraction from the canalicular outer leaflet. The replenishment of the free cholesterol pool is possible secondary to de novo synthesis in the endoplasmic reticulum. The phospholipid and cholesterol secretion in the bleary system is shown below.

D. Bile Production and Bile Component Functions (Figure 56.14, Table 56.1)

The bile flow is related directly to the osmotic forces, on the amount of bile salts secreted into the canaliculi. Different experimental models have been used to study the biliary secretion. These include, chronic bile fistula, isolated and perfused liver, hepatocyte couplets, isolated bile duct units and separation of the membranes from basolateral and apical domains. Bile is isotonic and consists of inorganic electrolytes,

phospholipids, cholesterol, bile acids and pigments, albumin and IgA and glycoproteins, mainly mucin, detoxified xenobiotics and some hormones. The formation of bile is initiated at the hepatocyte interface which is then secreted into the bile canaliculi. Canalicular bile comprises of two fractions. The bile salt dependent fraction, mediated by the ATP dependent bile salt export pump (BSEP) and the bile salt independent fraction that is formed secondary to the active canalicular secretion of principally glutathione organic anion related ATP dependednt multispecific organic anion transporter, MRP2. Canalicular bile regulation is dependent of the supply of bile salts, vagal stimuli and hormonal manipulation, and most importantly the rate at which bile alts re circulate from the intestine—the enterohepatic circulation. Secretin hormone (and glucagon) is the most potent stimulant of ductular bicarbonate secretion and ductal bile formation. Somatostatin is a potent inhibitor of the same. The ductal secretion is in main part due to the increase in cAMP concentrations, which leads to phosphorylation of cystic fibrosis transmembrane conductance regulator (CFTR). This results in chloride and bicarbonate efflux resulting in sodium absorption. The CFTR is a cAMP activated slow conductance chloride channel located in the apical region of plasma membrane on the cholangiocyte. The efflux secondary to activation of this channel results in the export of bicarbonate

TABLE 56.1 Various transporters at the hepatocyte and cholangiocyte level

Hepatocyte	Bile Acid-Dependent Bile Flow	
NTCP (SLC10A1)	Basolateral membrane	Na^+-dependent bile acid and xenobiotic uptake
OATP1B1 (SLCO1B1)	Basolateral membrane	Na^+-independent bile acid and xenobiotic uptake
OATP1B3 (SLCO1B3)	Basolateral membrane	Na^+-independent bile acid and xenobiotic uptake
Na^+, K^+ATPase	Basolateral membrane	Secretion of 2 Na^+ m exchange for 3 K^+
BSEP (ABCB11)	Canalicular membrane	ATP-dependent bile acid export
MDR3 (ABCB4)	Canalicular membrane	ATP-dependent phosphatidylcholine export
A8CG5	Canalicular membrane	ATP-dependent sterol export
ABCG8	Canalicular membrane	ATP-dependent sterol export
NPC1L1	Canalicular membrane	Sterol import
FIC1 (ATP881)	Canalicular membrane	ATP-dependent aminophospholipid flipping
	Bile Acid-Independent Bile Flow	
MRP2 (ABCC2)	Canalicular membrane	ATP-dependent transport of glucuronide glutathione and sulfate conjugates
OATP(1B1 1B3 2B1)	Basolateral membrane	Na^+-independent transport of organic anions, cations and neutral steroids
	Sinusoidal Bile Acid Export	
MRP3 (ABCC3)	Basolateral membrane	ATP-dependent export of bile acids and glucuronide conjugates
MRP4 (ABCC4)	Basolateral membrane	ATP-dependent export of glutathione and bile acids
$OST\alpha$-$OST\beta$	Basolateral membrane	Bile acid export
Cholangiocyte	Ductular Secretion	
Aquaporin 1 (AQP1)	Apical membrane	Water transport
Aquaporin 4 (AQP4)	Basolateral membrane	Water transport
AE2 (SLC4A2)	Apical membrane	HCO_3^+ secretion in exchange for Cl^-
CFTR (ABCC7)	Apical membrane	Cl^- secretion
ASBT (SLC10A2)	Apical membrane	Bile acid uptake (cholehepatic shunt)
Heal Enterocyle		
ASBT (SLC10A2)	Apical membrane	Na^+-dependent bile acid uptake
NPC1L1	Apical membrane	Sterol import
OSTc-OSTp	Basolateral membrane	Bile acid export
MRP3 (ABCC3)	Basolateral membrane	Bile acid export

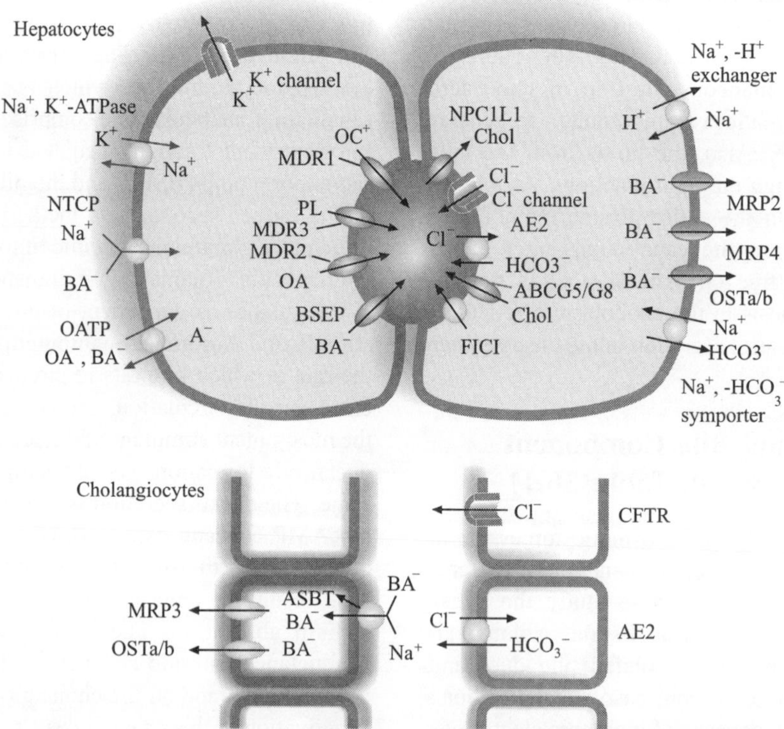

Figure 56.14 Hepatocyte and cholangiocyte transporters important for bile acid secretion. At the sinusoidal membrane of hepatocytes, the Na^+-taurocholate cotransporting polypeptide (NTCP; gene symbol *SLC10A1*) mediates the uptake of conjugated bile acids. Sodium-dependent uptake of bile acids through the NTCP is driven by an inwardly directed sodium gradient generated by the Na^+,K^+-ATPase. The Na^+-independent bile acid uptake is mediated by the organic anion-transporting polypeptides OATP1B1 *(SLCO1B1)* and OATP1B3 *(SLCO1B3)*. The sinusoidal membrane also contains a sodium-hydrogen exchanger and a sodium-bicarbonate cotransporter (symporter). At the canalicular membrane, bile acids are secreted via the bile salt export pump (BSEP; *ABCB11*), and sulfated or glucuronidated bile acids are secreted via the multidrug resistance–associated protein 2 (MRP2; *ABCC2*). The canalicular membrane also expresses ATP-dependent export pumps that transport phospholipid (multidrug resistance protein 3, MDR3; *ABCB4*) and cholesterol and drug metabolites (MDR1; *ABCB1*) into bile. The canalicular membrane expresses the chloride-bicarbonate anion exchanger isoform 2 (AE2) that secretes biocarbonate. FIC1 *(ATP8B1)* is a P-type ATPase mutated in progressive familial intrahepatic cholestasis type 1. Within the cholangiocytes of the large bile ducts, conjugated bile acids are absorbed by the apical Na^+-dependent bile acid transporter (ASBT; *SLC10A2*). Bile acid may then exit at the basolateral surface into the hepatic arterial circulation via the heteromeric transporter OSTα-OSTβ or an ATP-dependent carrier, MRP3 *(ABCC3)*.

into the bile duct lumen. The chloride bicarbonate exchanger has been localized to the canalicular membrane of hepatocytes as well as the apical membrane of the small and medium sized bile ducts. Ductal bicarbonate secretion is also stimulated by bombesin and vasoactive intestinal peptide, which also increases the activity of the chloride bicarbonate channels. The ductal bicarbonate secretion is in contrast, inhibited by somatostatin and gastrin, more potently by somatostatin through the SSTR2 receptors. This occurs almost exclusively on the large bile ductal cells; wherein basal and secretin stimulated bile flow is inhibited. Gastrin does not affect the bile flow. Gastrin affects the bicarbonate secretion by way of decreasing secretin induced cAMP levels and expression of secretin receptor in cholangiocytes.

The cholangiocytes are also involved in reabsorbtion of bile components like the bile salts (for their cholehepatic circulation, leading to regulation of bile flow) glutathione (leading to detoxification processes in liver and glucose. The bile salts are reabsorbed from the secreted bile by means of an apical sodium channel on the cholangiocytes that then secrete them into the peribilliary plexuses through the two important transporters ASBT and MRP3 (also known an phosphatidyl choline flippase). This ability of the cholangiocytes to take up bile acids/bile salts represents a therapeutic strategy where in ursodeoxycholic acid is used for chloride channel stimulation leading to CFTR mediated ATP release and subsequent clinical outcome.

Bile acids and bile salts are major functional components of the biliary secretion. They are divided into primary bile and acids which have 25–27 carbons atoms and are mostly in conjugated form with either glycine or taurine. Conjugation increases the water solubility. Bile salts or bile acids have a structure that is derived from a saturated tetracyclic hydrocarbon perhydrocyclopentophenathrene structure, the steroid nucleus. This nucleus consists of three six member rings and five membering curved or flat structure. Nearly all primary bile acids have a 7 alpha configuration with ursodeoxycholic acid being the notable exception. The most important physiological property of bile salts is lipid transport by solubilization and excretion of cholesterol into the intestinal tract. This is due to the amphipathic nature of bile salts in which there is hydrophilic side (alpha face) and a hydrophobic side (beta face).

The primary bile acids are cholic acid and chenodeoxycholic acid and the secondary bile acids include lithocholic acid and the secondary bile acids include deoxycholic acid and lithocholic acid. The secretion of bile acids into the canaliculi of the biliary system generates an osmotic pressure that accounts for the bile acid dependent raction of bile flow. These stimulate biliary lipid secretion and are able to form mixed micelles with the biliary phospholipids which allows solubilization in bile of cholesterol and other lipophilic compounds. These micelles also yield in the emulsion of dietary fat and liposoluble vitamins in the gut, thereby facilitating their absorption. The bile acids are also potent antimicrobial agents and prevent overgrowth of bacterial in the intestinal system. The bile acids also act as signaling molecules through the farnesoid X receptos and the membrane receptor TGR5. They do this through paracrine and endocrine modalities. Bile acids are also involved in adaptive responses to cholestasis and other injurious insults. The bile acids also control general energy related metabolism and more precisely in hepatic glucose handling.

E. Physiology of Bile Acid Homeostasis

Primary bile acids are synthesized in the hepatocytes from cholesterol; the rate limiting enzyme involved being 7 alpha hydroxylase. Via cytochrome P4507A1. Chenodeoxycholic acid and cholic acid are the abundant primary bile acids in humans. Secondary bile acids are lithocholic acid and deoxycholic acid, formed by 7 alpha dehydroxylation of primary bile acids. Bile acids undergo conjugation with glycine or taurine in the liver by amidation processes. The bile acids formed in the hepatocytes are transporter vectorially from sinusoidal blood across the basolateral membrane and across the canalicular membrane into the bile. Bile acids are taken up into hepatocytes by uptake protens sodium taurocholate cotransporting polypeptide (NTCP) and organic anion transporting polypeptides (OATPs). NTCP takes part in sodium dependent bile acid uptake whereas sodium independent transport is mediated by OATPs. The efficiency of hepatic uptake is dependent on the bile acid structure. Trihydroxy bile acids are taken up most and unconjugated bile acids are taken up least.

At the canalicular membrane, bile acids are excreted predominatly through the bile salt export pump in an ATP dependent manner. The multidrug resistance protein 2 (MRP2) is the main mediator for bile salt independent bile flow through canalicular excretion of reduced glutathione. The osmotic forces that are generated by acid secretion together with the

coordinated contraction of actin filaments that surround the canliculus generate the pressure necessary to pump the bile forward in the biliary tree. The cholangiocytes in the bile ducts express ion and organic anion transporters on the apical region (apical sodium dependent bile salt transporter (ASBT, OATP1A2) and the basolateral organic solute transporter (OST) that modify the bile before it passes into larger bile ducts. The bile acids are then stored in the gall bladder from where they are expelled into duodenum in response to hormone signals. The hepatocytes also efflux bile acids across the basolateral membrane into sinusoidal blood through MRP3, MRP4 and recently identified heteromeric OST. MRP4 plays an important role in basolateralefflux of bile acids in humans. OST transports glycine and taurine conjugated bile acids by facilitated diffusion. In the intestinal lumen the taurine/glycine conjugated bile acids are deconjugated and reabsorbed by the enterocytes in the terminal ileum or by passive diffusion. The bile acids then enter the mesenteric circulation through basolateral transport proteins such as MRP3 and OST and return to the hepatocytes. Bile acids not absorbed from the intestine are eliminated in feces.

F. Bile Modification Functions of the Gall Bladder

Around a liter of hepatic bile is produced in a 24 hour period. The bile produced is stored in the gall bladder and then released accordingly, in a timely manner in relation to meals and hormonal stimulation. The spincheter of Oddi keeps the common bile duct closed during the inter digestive period at the ampulla of Vater. This leads to bypass of formed bile into the gall bladder from the liver. This stored bile undergoes concentrational modification within the gall bladder. Concentration occurs by way of active mucosal absorption of water and electrolytes. The major differences between gall bladder bile and hepatic bile are shown in the table below.

The gall bladder bile is then emptied into the duodenum mediated by the release of CCK from the mucosa of the proximal intestine, stimulated by free fatty acids and aromatic amino acids and increased vagal tone.

G. The Secretory/Absorptive and Transport Functions of the Biliary System (Figure 56.15a, b)

The primary bile flows through the bile ducts on its route to the duodenum and the composition is continuously modified along its way by the intrahepatic biliary epithelium that reabsorbs fluids, amino acids, glucose and bile acids all the while secreting water, electrolytes and immunoglobulin A. The ultimate secretion and alkalinization of the bile is dependent on the net flux of chloride ions and bicarbonate into the lumen which induces the secretion of water which then regulates the pH of bile. The facilitated membrane transport in cholangiocytes is provided by the sodium/potassium/ATP-ase which actively extrudes sodium from the cell and together with potassium

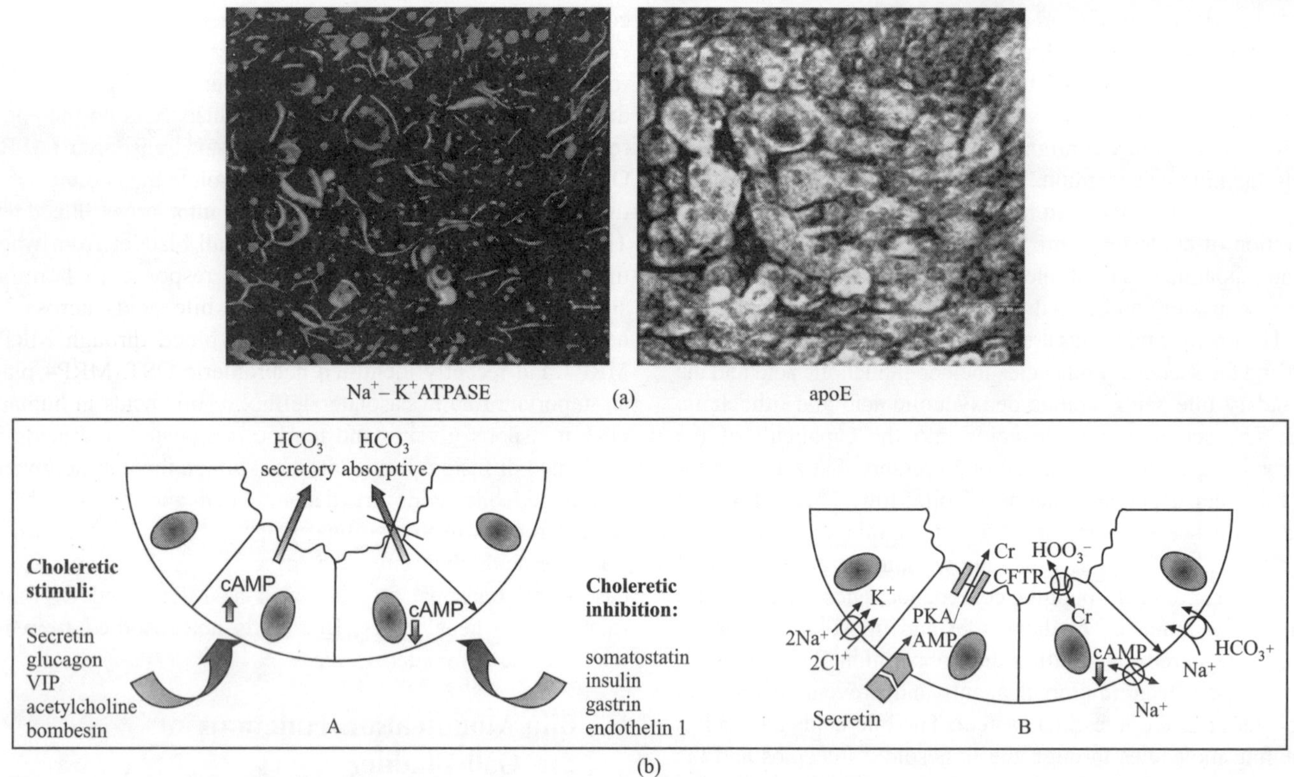

Figure 56.15 (a) The basolateral membrane transporters of the cholangiocyte (b) Cholangiocyte modification of bile flow and alkalinity by exerting both secretory and absorptive activities. **A:** Cholangiocyte possesses numerous different molecules located on their apical (luminal) and basolateral domain finely regulated by a balance of prosecretory and antisecretory **B:** In particular, a relevant role in normal cholangiocyte physiology is played by secretin that increases intracellular cAMP levels and consequently activates CFTR through PKA phosphorylation: the consequent Cl$^-$ efflux into the lumen generates the driving force for HCO$_3^-$ secretion by AE2 exchanger into bile. *Modified from: Strazzabosco et al. Functional anatomy of normal bile ducts. Anat Rec (Hoboken) 2008. Jun; 291(6):653–60.* (*see Plate 19 for colour figure*)

channels, maintains the transmembrane potential. At the basolateral side, the sodium gradient regulates the sodium and hydrogen exchanger isoform 1 (NHE1) and the sodium bicarbonate symporter or the sodium dependent chloride bicarbonate exchanger (NCHE) which mediate reabsorption of bicarbonate necessary for extrusion of acid. At this time, the sodium-potassium-chloride co transporter (NKCC1) the major determinant of fluid secretion, mediates the chloride uptake into the cell. The apical side of the cell regulates the chloride efflux, mainly mediated by the cyclic adenosine monophosphate (cAMP) activated slow conductance regulator (CFTR). The opening of the chloride channels in the apical membrane leads to an efflux of chloride and the generation of an osmotic gradient which induces the release of water into the lumen by way of the aquaporin receptors (1 and 2). The chloride gradient then regulates the sodium independent chloride-bicarbonate exchanger (AE2) which then pushes out bicarbonate into the bile, providing for the biliary alkalinization. The sodium dependent glucose transporter (SGLT1) –1, the glutamate transporter and the ileal bile acid transporter (iBAT) which are expressed on the apical side of the cholangiocyte mediate the reabsorption of the biliary constituents such as glucose and glutathione breakdown products and conjugated bile

acids. This step is crucial because the bile acids can stimulate proliferation of the biliary epithelial cells. The biliary bile acids are then secreted into the peribiliary plexus through the t-ASBT, a truncated isoform of the apical sodium dependent bile acid transporter or through the multidrug resistant protein 3 (MRP3), a p glycoprotein. This is the cholehepatic circulation of the bile acids and is very important for the overall regulation of bile secretion. The secretory function of the bile ducts is closely and finely regulated by the rapid hormone mediated signaling. The amount of fluid and bicarbonate secreted is regulated by the secretin, glucagon, vasoactive intestinal peptide, acetylcholine and bombesin – which are secretory stimulants and also by the anti secretory stimulants like the somatostatin and endothelin stimuli. The hormonal pathways ultimately act through the adenylcyclases that regulate the intracellular level of the second messenger cAMP, converting ATP to cAMP. Secretin is the main choleretic hormone that increases the cAMP to protein kinase A (PKA) ratio. This leads to activation of the CFTR and stimulation of chloride and bicarbonate efflux and inhibits the sodium hydrogen exchanger dependent sodium absorption. Cholinergic pathways also regulate bile secretion through the cholinergic pathways that act through beta adrenergic agonism, via the cAMP and PKA pathways (Figure 56.16).

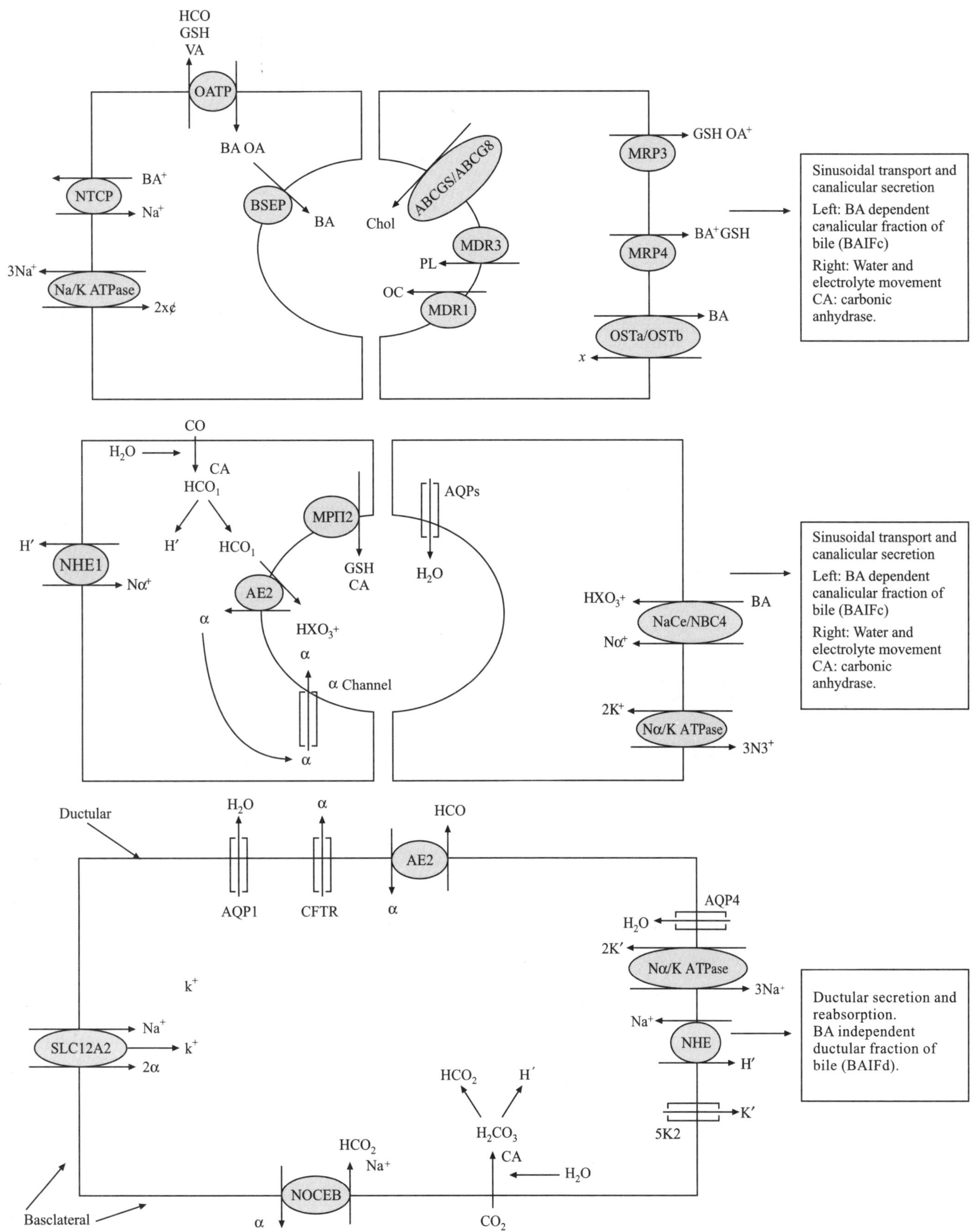

Figure 56.16 Physiology of bile secretion in a nutshell. *Modified from: Alejandro Esteller. World J Gastroenterol 2008 October 7; 14(37): 5641-5649.*

The cholangiocytes are capable of active reabsorption of glucose, bile acids, conjugated bilirubin, glutamic acid, low molecular weight organic ions that eventually leads to modification of the final bile constitution. This ability of the cholangiocytes to take up bile acids can be used as an alternative to therapy to stimulate ion secretion by cholangiocytes – by way of use of ursodeoxycholicacid – the stimulation of chloride secretion by means of CFTR and ATP mediated release. After secretion of bile salt from the hepatocytes, they are transported across the basolateral membrane to the canalicular region. Canalicular bile salt secretion is the rate limiting step in bile formation. Canalicular membrane has both ATP dependent and ATP independent transport systems. ATP dependent transport are members of the ATP binding cassette (ABC) superfamily. These transport biliary constituents across the canalicular membrane by hydrolysis of ATP, there by generating energy for the process. Majority of the ABC proteins belong to the multidrug resistance protein (MRP) family. The canalicular membrane also contains a bile salt export pump (BSEP) for monovalent bile salts and a conjugate export pump (MRP2) for divalent bile salts; the GSH pump for bile salt independent canalicular bile flow, a multidrug export pump (MDR1) for heavy amphipathic organic cations and a phospholipid flippase (MDR3) for phosphatidylcholine. Cholangiocytes play an important role in normal bile secretion. Large and medium sized bile ducts contain many transport systems for secretory and absorptive functions. In cholangiocytes, bile salts are taken up by the ISBT which is also expressed in the terminal ileum. After uptake, bile salts effluxed through the basolateral membrane of the cholangiocytes into the peribiliary plexus. This is a sodium independent process. This is achieved by MRP3 at the basolateral memebrane of the cholangiocytes and gall bladder epithelium.

H. Immune Functions of the Biliary System (Figure 56.17)

The biliary tree has the potential to be contaminated from the surrounding intestinal components in the form of lipopolysaccahrides and bacterial DNA and other products. The biliary tract contains defence mechanisms that include physical, chemical and immunological agents that protect and secure the environment. The physical agents include bile flow and biliary mucus; the chemical agents include bile salts and acids; immunological agents include secretory IgA. There is also rich expression of toll like receptors and adaptor molecules present intracellularly. The biliary system also secretes antibiotic peptides and inflammatory cytokines, lactoferrin, lysozymes and B-defensins along with cathelicidin. The human B defensins are of two types, type 1 predominates constitutively in biliary epithelium and type 2 predominates in large intrahepatic biles ducts and extrahepatic bile ducts. They are richly expressed in

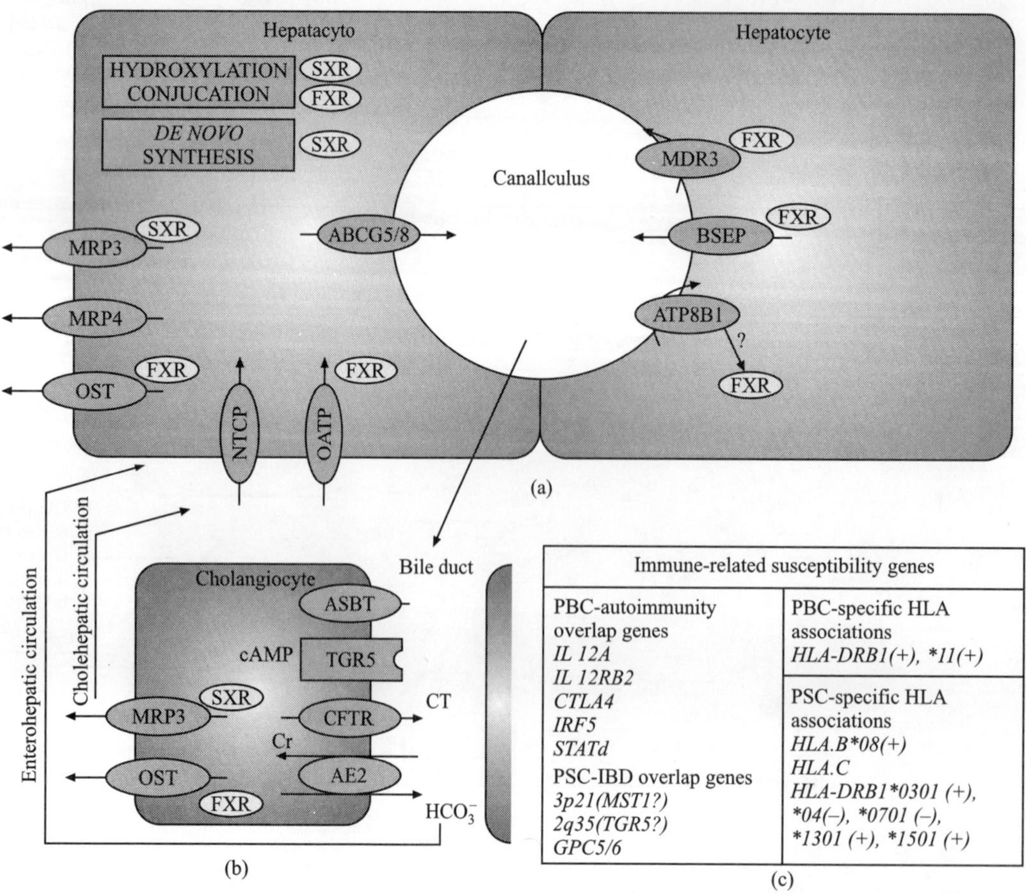

Immune-related susceptibility genes	
PBC-autoimmunity overlap genes *IL 12A* *IL 12RB2* *CTLA4* *IRF5* *STATd* PSC-IBD overlap genes *3p21(MST1?)* *2q35(TGR5?)* *GPC5/6*	PBC-specific HLA associations *HLA-DRB1(+), *11(+)* PSC-specific HLA associations *HLA.B*08(+)* *HLA.C* *HLA-DRB1*0301 (+),* *04(–), *0701 (–),* *1301 (+), *1501 (+)*

Figure 56.17 Biliary immunity related genetic susceptibility genes.

condition of extrahepatic biliary obstruction, hepatolithiasis and chronic inflammatory autoimmune conditions such as primary biliary cirrhosis and primary sclerosing cholangitis. Mucosal repair in the biliary tract is undertaken by the Trefoil factor family of receptors which are expressed at the apical surface. Biliary epithelial lymphocytes which are mostly positive for CD57.

IgA mediated immunity in the biliary system is important and its mainin bile includes protection of the biliary tree and liver against invading organisms, by preventing their adherence and/or toxins; clearance of harmful antigens from the circulation by formation of immune complexes and impeding a systemic inflammatory response. IgA in bile also provides protection against specific pathogens that may selectively infect the biliary tree and thereby prevent its spread into the systemic circulation.

Cholangiocytes also express and secrete chemokines and cytokines important for innate immune defenses. They secrete IL-8 and monocyte chemotactic protein-1 (MCP-1) which helps in recruiting neutrophils and monocytes to the biliary tract which ultimately prevents ascending biliary infections and helps in ameliorating pathological inflammatory responses in diseases like primary biliary cirrhosis and primary sclerosing cholangitis.

I. Innervative and Vascular Functions of the Biliary System

The biliary epithelium is supplied by the terminal branches of the hepatic artery and constitutes the complex peribiliary plexus (PBP). The canals of Hering and part of the bile ductules are not directly vascularized. The blood flow is opposite to that of bile flow.

The intra and extra hepatic biliary epithelium is innervated by autonomic nerves that originate from the celiac plexus and the vagus nerve. These nerves form the hepatic plexus surrounding the portal vein and bile duct. Anterior and posterior hepatic plexus nerve fibers penetrate the liver parenchyma following branches of the hepatic artery, portal vein and biliary tree. Classical neurotransmitters like adrenaline, noradrenaline and acetylcholine along with numerous aother neuropeptides like neuropeptide Y, calcitonin gene related peptide, somatostatin, vasoactive intestinal polypeptide, bombesin and enkephalin are seen to function and act in the biliary tree.

Neural innervations are sparse in the intrahepatic biliary epithelium in contrast to rich nervous system innervation seen in the extrahepatic and peribiliary regions. The cholangiocyte possess activity for M3 subtype of receptor for acetylcholine. This parasympathetic innervations aids in functioning such as secretin induced choleresis, cell proliferation and apoptosis, all of which are regulated by the intracellular cAMP levels. Likewise, the sympathetic activity in the cholangiocyte regulates hepatic proliferation. These are mainly expressed due to activity within the B1 and B2. The serotoninergic activity of the cholangiocyte through the 1A and 1B receptors inhibits cholangiocyte growth and choleretic activity. The neuropeptide

Y receptors present almost exclusively in extrahepatic bile ducts take part in modulation of the bile flow. Cholangiocytes produce and respond to certain vascular molecules, specially vascular endothelial growth factor (VEGF). Cholangiocytes also promote angiogenesis, a remodeling process in which newly formed blood vessels sprout to sustain the maturation and stabilization of a previously formed vascular network. This function is aided through action of Angiopoetins.

Platelet derived growth factor (PDGF) which encompasses types A to D are important in fibrogenic potentiation in the wake of bile duct injury. This also adds to the cirrhotic acceleration that is mediated by injured cholangiocytes. Along with the VEGF and PDGF pathways, Notch and Hedgehog signaling pathways also add to the deleterious functions of the cholangiocytes. These are dealt with in detail in the section on dysfunction of the biliary system. In short, bile duct epithelial cell signaling with VEGF promotion helps in arterial vasculogenesis during normal development and neoangiogensis in diseases of the biliary system. Cholangiocytes also have the potential to act as autocrine modulators, a feature wherein they provide protective mechanism against strong ischemic insults in the form of intense vasculogenic potentiation. VEGF, PDGF, Notch and Hedgehog activity has also been shown to be of fibrogenic potential in patients with chronic liver disease, leading to progression to cirrhosis and cancerogenosis.

J. Cholangiocyte Proliferation and Hormonal Function in Health and Disease

Three types of cholangiocyte proliferation have been described.

The type 1 or typical cholangiocyte proliferation is a hyperplastic reaction involving an increase in number of intrahepatic bile ducts that are confined to the portal space. These proliferating cholangiocytes form well differentiated tubular structures that show a well-defined lumen. Experimentally, type 1 cholangiocyte proliferation is seen in extrahepatic bile duct ligation models and partial hepatectomy or in chronic treatment with L-proline and after administration of carbon tetrachloride. In human livers, the type 1 proliferation is seen in acute obstructive cholestasis and early stages of chronic cholestasis. This is due to stretching of the existing bile ducts localized in the portal tracts.

Type 2 or atypical proliferation classically occurs after administration of carbon tetrachloride in rodents and after massive hepatic necrosis seen in humans, in alcoholic liver disease, focal nodular hyperplasia and in primary biliary cirrhosis and primary sclerosing cholangitis. The proliferation of cholangiocytes extends from the portal areas to the periportal area and adjacent liver parenchyma. This leads to formation of irregular and tortuous ductular structures which does not have a well defined lumen and is associated with infiltration by neutrophils. These newly formed ducts are not functionally efficient. In type 2 proliferation these is presence of transitional cells with characteristics (phenotypical) of both hepatocytes and cholangiocytes, thereby making this a metaplasia model of proliferative injury.

Type 3 proliferations or the oval cell proliferation is seen in early stages of carcinogenesis and is characterized by formation of tubular irregular and disorganized structures without well defined lumen and which protrude inside the hepatic lobule with altered hepatic parenchymal structures. These oval cells are located at the level of canals of Hering. Sex steroid hormones are able to influence the development and course of chronic liver disease. Cholangiocytes are estrogen sensitive cells and express both ER-alpha and ER-beta receptors in contrast to hepatocytes that express only ER-alpha receptors. ER receptor expression is decreased in alcoholic liver disease, primary biliary cirrhosis and primary sclerosing cholangitis. This decrement in ER expression leads to vanishing bile duct syndrome in late stages. On the other hand, in patient of cholangiocarcinoma, the ER-alpha and beta expression are both elevated. Testosterone plays an autocrine role in sustaining cholangiocyte proliferation during cholestasis. The expression of androgen receptors was low in normal patients and high in patients with primary biliary cirrhosis. Cholangiocytes also express the short form of prolactin receptor. The proliferative effect of prolactin is associated with activation of IP3/Calcium dependent and MAPK pathways and also via phosphorylation of the JAK2/STAT5 signalling medium. The insulin like growth factor-1 synthesized in the liver has a complex interaction with the estrogen receptor and together are involved in the modulation of proliferation, apoptosis and cell differentiation promotion processes of tissue repair in the liver and biliary tree. Secretin hormone regulates biliary secretion. Secretin stimulates ductal secretion in large cholangiocytes, as they are the only ones that express secretin receptors. This interaction leads to activation of adenylyl cyclise and consequent increase in synthesis of cAMP and its action on PKA. Phosphorylation of PKA leads to opening of chloride channel, CFTR and subsequent activation of ion pump Cl/HCO_3 with secretion of bicarbonate and for osmotic gradient formation, water. In pathology, where there is small cholangiocyte hyperplasia, there is increase in secretin receptor expression and secretin stimulated cAMP and its subsequent downstream actions. In the case of pathology to the large bile ducts, there is decreased proliferation and downstream regulation of secretin and subsequent actions, with finally, decreased bicarbonate secretion in bile. Secretin is shown to inhibit cholangiocarcinoma growth both in in vitro and in vivo models. Gastrin, interacts with the receptor cholecystokinin B (CCK-B) located in the basolateral membrane of cholangiocytes and inhibits secretin stimulated cAMP levels and bile secretion. It also inhibits cholangiocyte hyperplasia. Cholangiocytes also express somatostatin receptor subtype SSTR2. Somatostatin inhibits secretin induced ductal hypercholeresis and exocytosis by interacting with SSTR2 by downregulation of cAMP levels; it also inhibits large cholangiocyte proliferation. Large and small cholangiocytes express histamine HI to H4 receptors. Histamine by way of its action in HIHR increased small bile duct mass, whereas the through H2HR it increases the large bile duct mass. Histamine increases cholangiocarcinoma growth by autocrine mechanism and blockage of histamine precursor, histidine decarboxylase decreases cholangiocarcinoma growth. Cholangiocytes express melatonin MT1 and MT2 receptors and melatonin is shown to reduce the oxidative damage and hepatic proliferation and thereby ameliorates liver fibrosis. Cholinergic, adrenergic and dopaminergic nerve fibres modulate cholangiocyte function. Proliferation is positively modulated by the parasympathetic and sympathetic innervation and negatively by serotoninergic systems. Acetylcholine via M3 subtype of receptor promote proliferative and secretory activity of cholangiocyte. Alpha 1 and 2 adrenergic activation reduces cholangiocyte apoptosis and increase cholangiocyte proliferation. High levels of sensory neuropeptide alpha-CGRP promote biliary hyperplasia in cholestatic states. Neurokinins and substance P also promotes cholangiocyte proliferation.

REFERENCES

1. Vijayan V. and Tan CE. (1999), Development of the Human Intrahepatic Biliary System. *Ann. Acad. Med. Singapore*, 28(1):105–8.

2. Han Y., Glaser S., Meng F., Francis H., Marzioni M., McDaniel K., Alvaro D., Venter J., Carpino G., Onori P., Gaudio E., Alpini G. and Franchitto A. (2013), Recent Advances in the Morphological and Functional Heterogeneity of the Biliary Epithelium, *Exp Biol Med* (Maywood), 238(5):549–65.

3. McCuskey R.S. (1968), Dynamic Microscopic Anatomy of the Fetal Liver, *Anat Rec.*, 161(3):267–79

4. Roskams T. and Desmet V. (2008), Embryology of Extra- and Intrahepatic Bile Ducts, the Ductal Plate. *Anat Rec.*, (Hoboken), 291(6): 628–35.

5. Nakanuma Y., Hoso M., Sanzen T. and Sasaki M. (1997), Microstructure and Development of the Normal and Pathologic Biliary Tract in Humans, Including Blood Supply, *Microsc Res Tech.*, 38(6):552–70

6. Terada T., Kitamura Y. and Nakanuma Y. (1997), Normal and Abnormal Development of the Human Intrahepatic Biliary System: A Review, *Tohoku J Exp Med.*, 181(1):19–32.

7. Shiojiri N. (1997), Development and Differentiation of Bile Ducts in the Mammalian Liver, *Microsc Res. Tech.*, 39(4):328–35.

8. Kawarada Y., Das B.C. and Taoka H. (2000), Anatomy of the Hepatic Hilar Area: The Plate System, *J. Hepatobiliary Pancreat Surg.*, 7(6):580–6.

9. Strazzabosco M. and Fabris L. (2012), Development of the Bile Ducts: Essentials for the Clinical Hepatologist, *J. Hepatol.*, 56(5):1159–70.

10. Gaudio E., Franchitto A., Pannarale L., Carpino G., Alpini G., Francis H., Glaser S., Alvaro D. and Onori P. (2006), Cholangiocytes and Blood Supply, *World J. Gastroenterol.*, 12(22):3546–52.

11. Alvaro D. (1999), Biliary Epithelium: A New Chapter in Cell Biology, *Ital J. Gastroenterol Hepatol.*, 31(1):78–83.

12. Glaser S., Francis H., Demorrow S., Lesage G., Fava G., Marzioni M., Venter J. and Alpini G., Heterogeneity of the Intrahepatic Biliary Epithelium, *World J. Gastroenterol.*, 12(22):3523–36.

13. Tietz P.S., Chen X.M., Gong A.Y., Huebert R.C., Masyuk A., Masyuk T., Splinter P.L. and LaRusso N.F. (2002), Experimental Models to Study Cholangiocyte Biology, *World J. Gastroenterol.*, 8(1):1–4.

14. Boyer J.L. (1996), Bile Duct Epithelium: Frontiers in Transport Physiology, *Am J. Physiol.*, 270(1 Pt 1):G1–5.

15. Alvaro D. and Mancino M.G. (2008), New Insights on the Molecular and Cell Biology of Human Cholangiopathies. *Mol Aspects Med.*, 29(1–2):50–7.

16. Firrincieli D., Zuniga S., Poupon R., Housset C. and Chignard N. (2011), Role of Nuclear Receptors in the Biliary Epithelium, *Dig Dis.* 29(1):52–7.

17. Elsing C., Kassner A., Hübner C., Bühli H. and Stremmel W. (1996), Absorptive and Secretory Mechanisms in Biliary Epithelial Cells, *J. Hepatol.*, 24 Suppl 1:121–7.

18. Jezequel A.M., Benedetti A., Marucci L. and Orlandi F. (2001), Morphological Features and Modulation of the Intrahepatic Biliary Epithelium, *Ital J. Anat Embryol.*, 106(2 Suppl 1):363–9.

19. Strazzabosco M. and Fabris L. (2008), Functional Anatomy of Normal Bile Ducts, *Anat Rec* (Hoboken), 291(6):653–60.

20. Kwiatkowski A.P., McGill J.M. (1997), Electrolyte Transport in Biliary Epithelia. *J. Lab Clin Med.*, 130(1):8–13.

21. Kanno N., LeSage G., Glaser S., Alvaro D. and Alpini G. (2000), Functional Heterogeneity of the Intrahepatic Biliary Epithelium, *Hepatology*, 31(3):555–61.

22. Reynoso-Paz S., Coppel R.L., Mackay I.R., Bass N.M., Ansari A.A. and Gershwin M.E. (1999), The Immunobiology of Bile and Biliary Epithelium, Hepatology, 30(2):351–7.

23. Sternlieb I. (1972), Special Article: Functional Implications of Human Portal and Bile Ductular Ultrastructure. *Gastroenterology*, 63(2):321–7.

24. Strazzabosco M., Zsembery A. and Fabris L. (1996), Electrolyte Transport in Bile Ductular Epithelial Cells, *J. Hepatol.*, 24 Suppl 1:78–87.

25. Esteller A. (2008), Physiology of Bile Secretion, *World J. Gastroenterol.*, 14(37):5641–9.

26. Monte M.J., Marin J.J., Antelo A. and Vazquez-Tato J. (2009), Bile Acids: Chemistry, Physiology, and Pathophysiology, *World J. Gastroenterol.*, 15(7):804–16.

27. Cowen A.E., Campbell C.B. (1997), Bile Salt Metabolism: I. The Physiology of Bile Salts, *Aust N Z J Med.*, 7(6):579–86

28. Wolkoff A.W. and Cohen D.E. (2003), Bile Acid Regulation of Hepatic Physiology: I. Hepatocyte Transport of Bile Acids, *Am J. Physiol Gastrointest Liver Physiol.*, 284(2):G175–9.

29. Chiang J.Y. (2003), Bile Acid Regulation of Hepatic Physiology: III. Bile Acids and Nuclear Receptors, *Am J. Physiol Gastrointest. Liver Physiol.*, 284(3):G349–56.

30. Fuchs M. (2003), Bile Acid Regulation of Hepatic Physiology: III. Regulation of Bile Acid Synthesis: Past Progress and Future Challenges. *Am J. Physiol Gastrointest Liver Physiol.*, 284(4):G551–7.

31. Jansen P.L. and Müller M. (2000), Genetic Cholestasis: Lessons from the Molecular Physiology of Bile Formation, *Can J. Gastroenterol.*, 14(3):233–8.

32. Trauner M. and Boyer J.L. (2003), Bile Salt Transporters: Molecular Characterization, Function and Regulation. *Physiol Rev.*, 83(2):633–71.

33. Suchy F.J. and Ananthanarayanan M. (2006), Bile Salt Excretory Pump: Biology and Pathobiology, *J. Pediatr Gastroenterol Nutr.*, 43 Suppl 1:S10–6.

34. Meier P.J. and Stieger B. (2002), Bile Salt Transporters. *Annu Rev Physiol.*, 64:635–61.

35. Tischoff I., Wittekind C. and Tannapfel A. (2006), Role of Epigenetic Alterations in Cholangiocarcinoma, *J. Hepatobiliary Pancreat Surg.*, 13(4):274–9.

36. Akita H., Suzuki H., Ito K., Kinoshita S., Sato N., Takikawa H. and Sugiyama Y. (2001), Characterization of Bile Acid Transport Mediated by Multidrug Resistance Associated Protein 2 and Bile Salt Export Pump, *Biochim Biophys Acta.*, 1511(1):7–16.

37. St-Pierre M.V., Kullak-Ublick G.A., Hagenbuch B. and Meier P.J. (2001), Transport of Bile Acids in Hepatic and Non-hepatic Tissues, *J. Exp Biol.*, 204(10):1673–86.

38. Boyer J.L. (2008), Bile Canicular Secretion—Tales From Vienna and Yale, *Wien Med Wochenschr.*, 158(19–20):534–8.

39. Coleman R. and Roma M.G. (2000), Hepatocyte Couplets, *Biochem Soc. Trans.*, 28(2):136–40.

40. Boyer J.L. (2013), Bile Formation and Secretion. *Compr Physiol.*, 3(3):1035–78.

41. Syal G., Fausther M. and Dranoff J.A. (2012), Advances in Cholangiocyte Immunobiology, *Am J. Physiol Gastrointest Liver Physiol.*, 303(10):G1077–86.

42. Ghonem N.S., Ananthanarayanan M., Soroka C.J. and Boyer J.L. (2013), PPARα Activates Human MDR3/ABCB4 Transcription and Increases Rat Biliary Phosphatidylcholine Secretion, *Hepatology*.

43. Treyer A. and Müsch A. (2013), Hepatocyte Polarity, *Compr Physiol.*, 3(1):243–87.

57

The Biliary System
Functions and Dysfunctions

Shiv Kumar Sarin and Cyriac Abby Philips

CONTENTS

I. GALL BLADDER DYSFUNCTION AND SPHINCTER OF ODDI DYSFUNCTION

Gall bladder, along with the sphincter of Oddi modulates the flow of bile from the biliary tree into the duodenum. This occurs as a result of simultaneous contraction of gall bladder and relaxation of the sphincter. During the fasting state, the gall bladder stores the bile produced from the biliary tree and concentrates and modifies it. In the fasting state, the gall bladder periodically contracts to export bile into the duodenum, an action mediated by the motilin receptors which acts through the vagal cholinergic nerves and synchronized with the migratory motor complexes. During eating, local gastroduodenal reflexes and cephalic reflexes by way of mainly cholecystokinin hormonal action on the gall bladder through the cholinergic neural pathways. Concurrently, the non adrenergic and non cholinergic nerves produce synchronous relaxation of the sphincter of Oddi by way of activity through the vasoactive intestinal peptide and nitric oxide. The neural supply to the biliary tract is also through norepinephrine, vagal

694

afferent nerves, substance P. Derangements of these components produce clinical syndrome of recurrent/intermittent abdominal pain, elevations in liver and pancreatic enzymes (self limited), ductal dilatation and episodes of pancreatic inflammation.

Gall bladder dysfunction is characterized when the gall bladder empties insufficiently with biliary symptoms and in the absence of demonstrable structural abnormalities. The reasons are supposedly obscure, but theories of functional obstruction at the transitional zone of gall bladder and cystic duct, inflammation related amplification of sensory firing, primary hypersensitivity leading to neuroplastic changes in the neuronal configuration responsible for visceral hypersensitivity.

Sphincter of Oddi dysfunction is usually seen in patients who are cholecystectomized.

Patients suffering from sphincter dysfunction can be divided into three groups. The first or Type 1 includes patients with biliary type pain with deranged liver function tests documented on two or more occasions, delayed contrast drainage that is more than 45 mins and a common bile duct diameter more than or equal to 12 mm. The second group of patient, i.e., Type 2 dysfunction, present with biliary type pain and either one/two of deranged liver function test/delayed contrast drainage/positive ERCP findings. The third group of patients, Type 3; have only biliary pain and none of the other findings. The pathophysiology of sphincter of Oddi dysfunction has been suggested to be secondary to impediment in bile flow because of spastic contractions secondary to ischemic insults and hypersensitivity of the papilla or the common bile duct. Evaluation methods in biliary tract functional derangements include quantitative hepatobiliary scintigraphy which is a noninvasive scintigraphic method where the patency of the biliary system and the rate of bile flow can be measured. The tracers used include 99 m Technitium EHIDA and its other forms like Mebrofenin and DISIDA. The level of contrast excretion in the common bile duct and the duodenum peaks at 45 and 60 minutes respectively. Gall bladder phase imaging measures the gall bladder ejection fraction by placing the region of interest over the gall bladder and another over the liver and then generating a time activity curve for the gall bladder where by the ejection fraction is calculated using set formulas. This phase imaging is done under the secretory effect of cholecystokinin. A gall bladder ejection fraction of above 35% is considered healthy, if the dose of cholecystokinin used is 10ng/kg administered within 10 minutes. In patient who undergo ejection fraction studies to delineate the functioning, some experience biliary type pain while during the test. This can be abolished by administration of nitroglycerine before the cholecystokinin infusion. In patients of sphincter dysfunction, the presence of tracer retention 120 minutes in the prominent bile ducts on static images was suggestive of abnormal biliary functioning/drainage. Manometric studies asses the biliary function and sphincter of Oddi dysfunction more expertly than tracer studies. The sphincter basal pressure relative to the common bile duct pressure was the best feature to suggest a sphincter dysfunction and presence of biliary symptoms. Sphincter manometry is done during ERCP a triple lumen catheter which is placed into the common bile duct and the area of the sphincter and reading taken through use of external transducers and via perfusion from a hydraulic capillary infusion pump. In normal subjects the baseline pressure is approximately in the range of 5 to 15 mm of Hg with a phasic wave amplitude of 100 to 140 mm of Hg and wave duration and frequency of waves of to 6 seconds and 3 to 6 waves per minute respectively. Sphincter dysfunction is defined when there is a baseline pressure of more than 40 mm of Hg, phasic amplitude of above 350 mm of Hg and frequency of more than 7 per minute with 50% or more reversal of wave direction.

Ultrasonography guided gall bladder function is sometimes preferred in view of no exposure to radiation. In this method, the gall bladder volume can also be estimated, using the Dodd Formula in which the three axes values of the gall bladder organ are measured and multiple by pi value and divided by 6. Thereafter gall bladder ejection fraction is also calculated accordingly measuring the gall bladder volume (maximum) and gall bladder volume minimum and their differential ratio. Pain evaluation studies using endoscopic manipulation of the papilla, pain inductance with intra ductal contrast injections and duodenal specific visceral hypersensivity to pain are upcoming methods to delineate sphincter dysfunction.

Symptomatic patients with gall bladder dysfunction and patients post cholecystectomy with sphincter dysfunction may improve with cholecystectomy and endoscopic sphincterotomy respectively.

II. CHOLELITHIASIS, CHOLEDOCHOLITHIASIS AND CHOLECYSTITIS

Cholelithiasis (Figure 57.1) is a very prevalent biliary system disease with a great burden on the health care system. It has been proposed to have multifactorial etiologies and can be associated with many chronic disorders. The main factor for development of cholelithiasis is impaired metabolism of cholesterol, bilirubin and bile acids leading to stone formation in the hepatic duct, common bile duct or gall bladder. Cholelithiasis is more common in women than in men and age and other factors such as obesity, rapid weight loss, glucose intolerance, insulin resistance, high dietary glycemic load, alcoholism, dyslipidemia, drugs and super physiological states like pregnancy predispose to gall stone formation.

The factors that contribute to bile cholesterol supersaturation increases with age and among women. The age related increase in dyslipoproteinemia resulting in a linear increase in cholesterol excretion into the bile and also added by the sysnthesis of bile acids, secondary to the decreased activity in the enzyme, 7 alpha hydroxylase results in supersaturation and gall stone formation. The pregnant X receptor, a xenobiotic receptor and its loss leads to lithogenicity in bile. This leads to decrease in concentration of bile salts and phospholipids and increases the formation of cholesterol crystals. The reduced expression of the 7 alpha hydroxylase related gene CYP7A1 results in

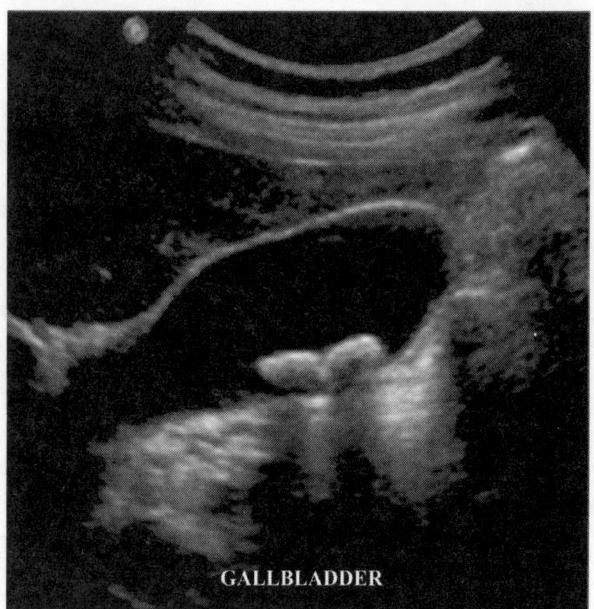

Figure 57.1 Ultrasonography of the abdomen showing presence of gall stone.

the induction of fibroblast growth factor and lithogenicity. Dysfunction of the gall bladder due to decreased hemoperfusion with increase in age and also sclerotic changes leads to gall bladder dysmotility and dysfunction leading to lithogenic bile formation. This is increased in the presence of infection and inflammation. There are many evidences that cholelithiasis many genetically driven. The risk is around four times higher in people who have a related member with gall stone disease.

There is evidence that it is an autosomal dominant disease with susceptible gene loci identified as in the apolipoprotein B mRNA editing protwin (APOBEC1) and the peroxisome poroliferator activated receptor gamma (PPARG). It was also detected that the Nr1h4 gene encoding the nuclear bile salt receptor, Farnesoid X receptor has lithogenic candidacy. There is also linkage to the hepatocanalicular cholesterol transporter ATP binding cassette G5 and G8 (ABCG5/8) which regulate biliary cholesterol absorption and excretion as players in lithogenic bile formation. These are involved in increase in hepatobiliary cholesterol secretion leading to the supersaturation of bile. The T400K polymorphism maybe independently associated with gall stone formation in males. The LITH genes are mainly located on chormosomes 3, 4, 9 and 11. Increase in expression of HMG CoA reductase enzyme is also suggested to play a role in lithogenicity of the bile. Genetic polymorphism in apolipoprotein genes like APOA1-75G/A has been shown to predispose to gall stone disease. The polymorphism in the low density lipoprotein receptor related protein (LRPAP1) might have an association with gall stone formation. Mucin, an important component of bile has been shown play a major factor in stone formation. Mucin associated genes (MUC), MUC2, MUC5AC, MUC5B and MUC6 are recognized to play important part in biliary stone formation. The epidermal growth factor, EGFR has shown to

regulate this aspect of stone formation. The single nucleotide polymorphisms at MUC1 and MUC2 has been associated with stone formation in men, but not in women. Large fluctuations in weight and more weight cycles regulate stone formation in susceptible individuals. The beta adrenergic receptor (ADRB3) is a transmembrane receptor highly expressed in adipose tissue that predispose to stone formation, as it is also highly expressed in gall bladder epithelium, involved in gall bladder contraction. The ADBR3 Trp64Arg polymorphism has been shown to promote stone formation. Weight loss promotes elevated mucin level and calcium in cystic bile thereby giving rise to sludge and stones in the gall bladder. A low fibre diet slows the transit of intestinal components and promotes the increased formation and absorption of secondary bile acids thereby increasing the lithogenic potential of bile. Long term parenteral nutrition promotes gall bladder dyskinesis and results in lithogenicity of the bile. In patient of cirrhosis of the liver, the etiological agents hepatitis B and C have shown to increase gall stone formation. Primary biliary cirrhosis patients are also predisposed to formation of lithogenic bile. Immune resistance associated with polymorphism of genes encoding receptors in the adipocytes, mainly the retinoid X receptor and peroxisome proliferator activated receptor promotes occurrence of cholelithiasis. Several drugs such as estrogens, immunosuppressant medications, octreotide, clofibrates and nicotinic acid in the long term use, predispose to gall stone formation. Long term corticosteroid therapy predisposes to dyslipidemia and in the long run cause cholelithiasis. Elevated levels of cholesterol leads to change in bile acid to cholesterol ratio and leads to lithogenicity in the bile. Ceftriaxone therapy especially in pediatric age group predisposes to biliary sludge formation and lithogenic potential, when used in doses above 2 g per day. Genetic polymorphism is genes that encode steroid biosynthesis has been shown to promote lithogenicity in bile. Association between short alleles for both c1092/3607(CA) 5-27 and c172(CAG)5-32 repeat polymorphisms of the estrogen receptor beta and androgen receptor was also found recently.

The most important pronucleators in gall stone formation is mucin glycoprotein gel. Mucins contain oligosaccharide side chains attached to serine or threonine residues of the apomucin backbone by O-glycosidic linkage. They are high molecular weight glycoproteins. The bile mucin has two main domains, one which is rich in serine, threonine and proline which contains the mainly covalently bound carbohydrates and another nonglycosylated domain rich in serine. In the presence of lithogenic bile, the mucin secretion increases and the bile undergoes modification contributing to sludge formation. Gene-gene interaction between MUC1 and MUC5AC expression is additive to mucinogenic modification of biliary bile. Bilirubin is frequently found in center of cholesterol stones secondary to cholesterol crystallization in the protein pigment complex. Cholesterol precipitation is a constant process in the gall bladder and the contractility of the gall bladder clears this out normally. In dysfunctional states, the gall bladder filling and emptying leads to non clearance of this cholesterol and mucin and leads to nidus formation and stone precipitation. Gall bladder emptying is dysfunctional in pregnancy, total parenteral nutrition, sudden weight loss, starvation, small intestinal diseases and with age. Cholecystokinin receptor A gene polymorphism increases the rate of cholelithiasis due impaired motility. Gall stone disease is also affected by increase in expression of the gene encoding synthesis of the type II receptor to pituitary polypeptide that activates adenyl cyclase in gall bladder epithelium, leading to impaired motility. Drugs that can cause impaired gall bladder motility include somatostatin, methylscopalamine and also morphine (which causes cholecystokinin like action on gall bladder, but produces spasmodic contractions of the sphincter of Oddi). Gall bladder cholesterosis produces wall muscle dyskinesia by accumulation in the membranes of myocytes and weakening of the wall. Impaired enterohepatic circulation of bile acids as seen in small intestinal diseases (celiac disease, Crohn's disease) and post intestinal resection surgeries also leads to gall stone formation.

The gall stone components are mostly unesterified cholesterol, unconjugated bilirubin, calcium bilirubinate salts, fatty acids, carbonates, phosphates and mucin glycoproteins. The three main types of stones are cholesterol stones, majority, 75% of all stones, pigment stones and mixed stones (Figure 57.2). The cholesterol gallstones comprise mainly of cholesterol monohydrate (at least 50% of stone). Color cathodoluminescence scanning electron microscopy is used to characterize the components of the gall stones. Pigment gall stones are those that contain less than 30% of cholesterol. These include black and brown pigment stones. The brown stones are mainly

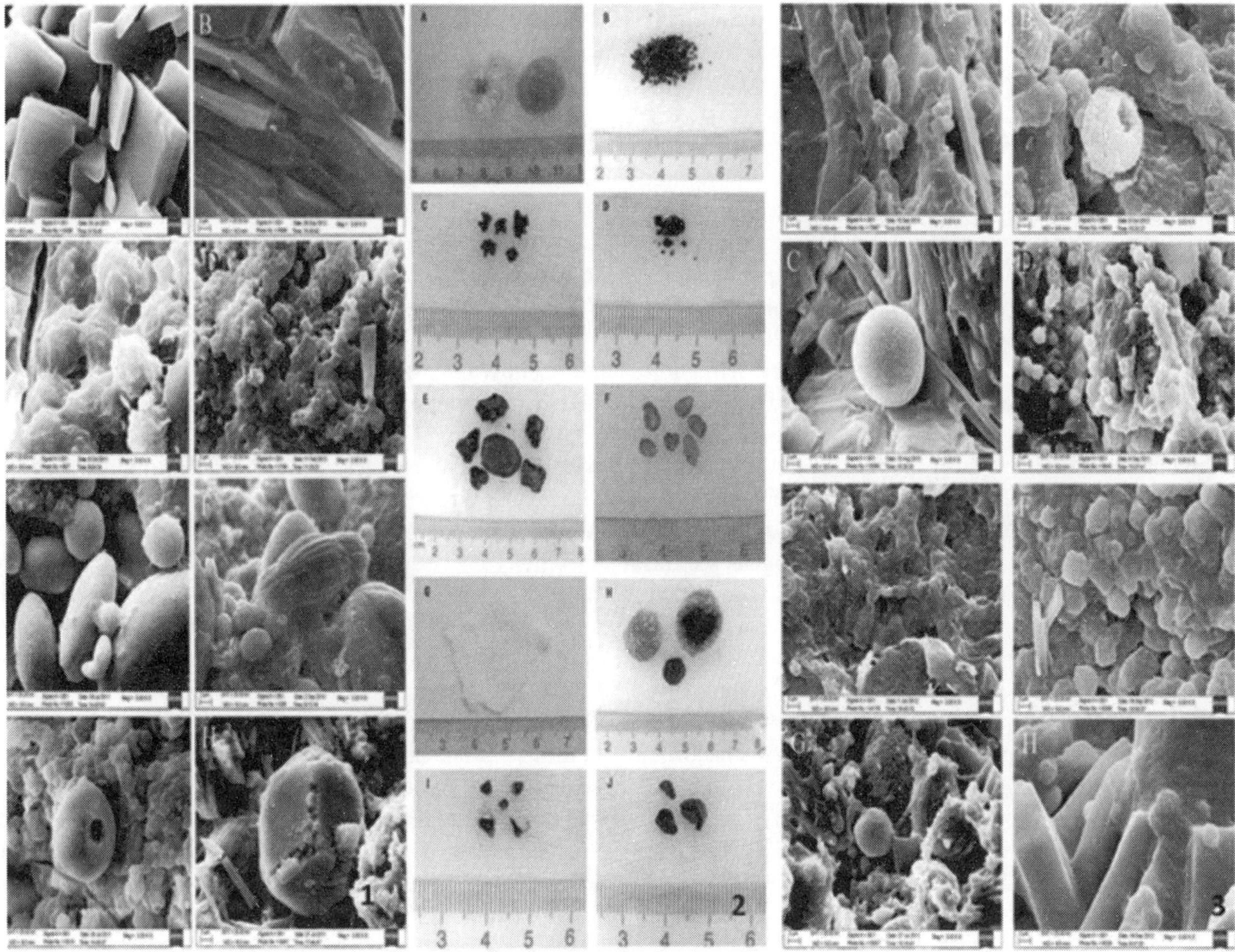

Figure 57.2   **Panel 1**: A, B. Cholesterol stone. A. Plate-like cholesterol crystals. B. Lamellar cholesterol crystals. C, D. Pigment stone. C. Clumping-like bilirubinate particles. D. Irregular bilirubinate particles. E – H. Calcium carbonate stone. E. Bulbiform and ellipsoid calcium carbonate crystals. F. Fusiform calcium carbonate crystal. G. Hawthorn-like calcium carbonate crystal. H. Cuboid calcium carbonate crystal. **Panel 2:** The appearance of each type of gallbladder stone. A. Cholesterol stone B. Pigment stone C. Calcium carbonate stone D. Phosphate stone E. Calcium stearate stone F. Protein stone G. Cystine stone H. Cholesterol-bilirubinate mixed stone I. Bilirubinate-calcium carbonate mixed stone J. Bilirubinate-phosphate mixed stone. **Panel 3:** Microstructure of mixed stones (original magnification ×3000). A. Cholesterol- bilirubinate mixed crystals. B. Bilirubinate-calcium carbonate mixed crystals. C. Cholesterol-calcium carbonate mixed crystals. D. Bilirubinate-phosphate mixed particles. E. Bilirubinate-calcium stearate mixed particles. F. Cholesterol-phosphate mixed crystals. G. Cholesterol- bilirubinate-calcium carbonate mixed crystals. H. Cholesterol-bilirubinate-phosphate mixed crystals. *From: Qiao T, Ma R-h, Luo X-b, Yang L-q, et al. (2013) The Systematic Classification of Gallbladder Stones. PLoS ONE 8(10): e74887. (see Plate 19 for colour figure)*

located in the bile ducts and they contain calcium bilirubinate and cholesterol and calcium palmitate/stearate. Supersaturation of bile with unconjugated bilirubin results in agglomeration. This type of gall stone is typically seen in chronic hemolytic conditions. The variant of TATA-Box in the promoter region of the UDP-glucoronosyltransferase 1A1 (UGT1A1) gene is associated with the development of cholelithiasis. Intrahepatic stones contain high levels of free bile acids that are formed secondary to bacterial infection, which results in deconjugation of glycine and taurine, leading to gall stone formation. There is presence of high levels of free saturated fatty acids in the stone as well as the involvement of phospholipases particularly phospholipase A1 which break down the phospholipids. Pure calcium stones are composed of calcium carbonate which are very rare in adults. Single mixed gall stone displays protein cholesterol composition in the core and the multiple mixed gall stone demonstrates protein bilirubin composition in the core. Cholesterol nucleation is the initial stage that leads to formation of cholesterol gall stones. This nucleation occurs from the liquid-crystalline phase or mesophase after the aggregation and fusion of the cholesterol rich unilamellar vescicles. Polarizing ligh microscopy aids in visualization of the crystal formation process and development of lithogenic bile. In addition to mucin, the proteins that accelerate precipitation include N aminopeptidases, immunogloblulins and phsopholipases. The components that prevent nucleation include apolipoprotein A1 and A2, aspirin and other NSAIDS. Intrahepatic stone formation is mostly related to infective processes as described above. The neck of the gall bladder contains the biggest bacterial load in the gall bladder structure and this is partly due to the presence of Rokitansky Aschoff sinuses (mucosal folded invaginations into the lumen, normally seen in this area). The major bacteria such as *E. coli*, Klebsiella and other enteric pathogens are the ones involved mainly in stone formation and the ones that produce beta glucoronidase and mucus only give rise to pigment or mixed stones and never pure cholesterol stones. The gall stone can grow at the rate of 3–5 mm per year. In summary, the formation of cholelithiasis is multifactorial in nature and depends on genetic makeup of the susceptible individual, the presence of inflammation, gall bladder dysmotility, metabolic dysregulations, supra physiological states and environmental and secondary factors that all culminate to produce a nidus from which the stone development begins.

Choledocholithiasis is a gall stone related complication in which secondarily the stone can pass from the gall bladder through the cystic duct to the common bile duct, thereby complicating and adversely affecting the outcome of gall stone disease. Stones can also form primarily in the common bile duct. Bile stasis, biliary system infections, chronic inflammation, sludge formation and other factors that predispose to gall bladder stone formation also predispose to choledocholithiasis. In around 7% of the cases, choledocholithiasis is seen during cholecystectomy. Approximately 25 to 50% of choledocholithiasis with become symptomatic in the long run necessitating interventions. Fever, pain and jaundice (the Charcot triad) develop in the presence of cholangitis. The

migrated stones to the common bile duct can be identified by their constituent as cholesterol or pigment. The brown stones usually develop de novo in choledocholithiasis. Bacteria can be cultures from the surface of the cholesterol stones, but not from the inside because they do not help in nidus formation and nucleation property during cholelithiasis. Pigment stones are more common in patient with cirrhosis and ileal disease.

Cholecystitis can be acute or chronic. 90% of acure cholecystitis is caused due to stones in the cystic duct. The rest 10% occur in the absence of gall stones and is known as acalculous cholecystitis. Obstruction of the cystic duct leads to distension of the gall bladder which leads to diminished flow of blood and lymphatics of the gall bladder which further leads to mucosal ischemia and necrosis with subsequent clinical course. In acalculous cholecystitis, there is injury to the gall bladder epithelium secondary to bile stasis which leads to bile indiced inflammation and injury. During prolonged fasting periods or in critical care scenario, there is diminished activity of cholecystokinin thereby promoting more bile accumulation and stagnation related inflammatory injures. Endotoxemia is also a suspected reason for cholecystitis. Endotoxemia leads to necrosis, hemorrhage, fibrin deposition and mucosal loss leading to gall bladder stasis and insult. Ultrasonography is 90 to 95% sensitive for cholecystitis and around 80% specific. For the diagnosis of gall stones, it is 95% sensitive and specific, where in the gall stone is more than 2 mm in diameter. The ultrasonography findings suggestive of acute cholecystitis include the presence of pericholecystic fluid, gall bladder wall thickening more than 4mm (in the abscene of ascites and hyalbuminemia) and sonographic Murphy's sign [Tokyo Guideline]. With computed tomography and magnetic resonance imaging for predicting acute cholecystitis the sensitivity and specificity is greater than 95%. Hepatobiliary scintigraphy has been found to be sensitive in 99–100% of the time (with a specificity of 85 to 95%) in coming to a diagnosis of acute cholecystitis. In a standard study, the gall bladder, common bile duct and small bowel fills within 30 to 45 minutes. If the gall bladder is not visualized, intravenous morphine administration can improve accuracy of the scintigraphic scanning by increasing the resistance to flow through the sphincter of Oddi, resulting in gall bladder filling if the cystic duct is patent. The features of chronic cholecystitis include the presence of fibrosis, flattening of mucosa and chronic inflammatory mixed cell population and morphologically a chronic contracted gall bladder.

III. CHOLANGITIS

The presence of biliary epithelial cell damage associated with inflammatory cell infiltrate is seen in cholangitis. Cholangitis can occur in the presence or absence of suppuration. These might be associated with ductal or periductal fibrosis. When this process affects the bliliary tree as a whole, the term cholangitis is used. In suppurative cholangitis, the intense neutrophilic inflammatory reaction in seen within the wall and also around the lumen of the ducts. Sometimes, this is associated with presence of clustering of inflammatory cells forming microabscesses. The presence of endotoxemia and chemokine and cytokine storm secondary to microbial invasion is the suggested destructive process in this condition. In non suppurative cholangitis, the inflammation occurs in the presence of granulomatosis, fibrosis, lymphoid infiltration or pleomorphic destructive processes. Granulomatous cholangitis is destructive in nature and is seen classically in primary biliary cirrhosis, but can also be seen in tubercular, sarcoid and drug induced cholangitis. The interlobular ducts are most affected in these conditions. Lymphoid aggregates are also seen in these cases and go hand in hand with granulomatosis in certain cases, but can also be seen as a separate entity. Pleomorphic cholangitis is seen with mixed inflammatory infiltrates and the presence of fibrosis or sclerosis of the biliary tree. These could result in obliterative cholangiopathy which is classically seen in primary sclerosing cholangitis. In these types, the cholangiocyte cell loss is characteristic. Ischemic cholangitis is another entity seen in patients with hypotensive episodes and certain conditions of the Biliary tree that would result in fibrosclerosis of the bile ductules resulting in vasculogenic compromise and eventually biliary epithelial cell loss. There is marked deposition of collagen in the walls of the bile ducts and ductules and an increase in number of c-Kit receptor positive mast cells that secrete fibrogenic agents like fibroblast growth factor, histamine, tumor necrosis factor alpha and other factors like transforming growth factor beta and platelet derived growth factors which are well expressed and activated and secreted by the biliary epithelial cells also. These sclerosed and fibrosed ductules and ducts express vimentin as part of final mesenchymal transformation and also result in marked attenuation of the peribiliary vascular plexus.

Bile duct loss or ductopenia intervenes as the final result which is secondary to the imbalance between the death and dropout of the biliary epithelial cells and regenerating biliary cells. Immunostaining for biliary cytokines (CK7 and CK19) maybe absent in these cases. Ductopenia is diagnosed microscopically when there is a lack of more than 50% of interlobular bile ducts from the portal tracts visualized. This leads to advanced fibrosis and chronic cholestasis.

IV. MUCOBILIA

Impaction of mucin in the biliary duct lumen leads to formation of mucus lakes leading to blockade. This maybe severe enough to cause extravasation and drainage of mucus from the ampulla of Vater. Mucin producing intraductal tumors can lead to this condition. Benign conditions such as primary sclerosing cholangitis and hepatolithiasis can lead to such a state and most of these are associated with occult neoplastic lesion underlying the primary lesion. Abudant mucin can be seen in the intra as well as extrahepatic biliary tree. The mucin is rich in albumin and electrolytes and drainage of this could lead to severe electrolyte disturbances. Hepatolithiasis can also

result secondary to this entity and there is always a risk of developing cholangiocarcinoma in such cases. Primary tumors that produce mucobilia are extremely rare and include biliary papillomatosis, mucin producing cholangiocarcinoma and cystic mucinous neoplasms of the pancreas. Intraductal papillary mucinous neoplasm (IPMN) is a separate entity that can lead to mucobilia. It is of two types, IPMN-P (pancreatic origin) and IPMN-B (biliary origin). These tumors consist of frond like invaginations lined by columnar epithelium surrounding a fibrovascular core. Cystic dilatations of the proximal ducts are commonly encountered. The pattern on cytokeratin 20 in seen on immunohistochemistry. Clinical presentation is that of obstructive jaundice, right sided abdominal pain, abdominal mass and deranged liver function tests. Imaging studies like ultrasonography, CT imaging and ERCP are very sensitive in diagnosing these conditions. ERCP or percutaneous cholangiography provides direct evidence of mucin and demonstrates multiple elongated linear, ovoid or amorphous filling defects in dialated bile ducts. Fine needle aspiration cytology is usually contraindicated in these cases as they may result in inoculation of infections into the presumable sterile environment of the biliary system. On macroscopic analysis, there is prominent intraductal papillapry proliferation and mucin hypersecretion. The neoplastic subtypes in this condition include the pancreatobiliary type in which columnar cells with eosinophilic cytoplasm and rounds nuclei predominate; the intestinal type where in stratified tall columnar cells with goblet cells resembling intestinal adenoma or carcinoma predominate and the gastric subtype in which columnar cells with abundant intracytoplasmic mucin predominate and finally the rare oncocytic type which has eosinophilic cytoplasm and round nuclei representing a pancreatobiliary type variant, but more aggressive. IPMN-B lesions are classified according to the Yeh as shown in Table 57.1.

Surgical management is standard of care. Partial hepatectomy with common bile duct exploration/excision with or without cholecystectomy and portal lymphadenopathy is done is fit and selected group of patients. Palliative procedures include drainage and stenting.

V. HEMOBILIA

Hemobilia is defined as when there is an abnormal communication between a small caliber splanchnic vessel with the bile duct leading to hemorrhage into the bile duct causing upper gastrointestinal bleeding/melena. Hemobilia is clinically characterized by 'Quinke's Triad' which consist of upper gastrointestinal bleeding, biliary colic and obstructive jaundice. Blood containing material is visualized on ultrasonography as echogenic, nonshadowing and sometimes merge with the liver parenchyma. Sudden transitions from hyperechoic to hypoechoic material in the bile ducts also suggest hemobilia. Management modality consists of promoting homeostasis and removal of biliary obstruction. Angiographic embolization is particularly very effective in managing hemobilia. Pre

TABLE 57.1   Classification of IPMN

	Description
Histopathological classification	
Type 1	Low-grade dysplasia
Type 2	High-grade dysplasia
Type 3	Microinvasive adenocarcinoma (carcinoma in situ)
Type 4	Stromal invasive adenocarcinoma
Cholangiographic classification	
IA	Hepatolithiasis with stricture
B	Fusiform bile ducts and amorphous filling defects
1C	Disproportionate biliary dilatation
IIA	Intrahepatic polyploidy/cystic neoplasia
IIB	Intrahepatic polyploidy/cystic neoplasia extending into extrahepatic bile ducts
IIIA	Type 1 and 2 with concurrent operable malignancy
1MB	Type 1 and 2 with concurrent inoperable malignancy

(*From:* Yeh TS, Tseng JH, Chiu CT, Liu NJ, Chen TC, Jan YY, et al., Cholangiographic spectrum of intraductal papillary mucinous neoplasm of the bile ducts. *Ann Surg.* 2006; 244: 248–53.)

procedure, the portal vein patency must be confirmed so as to avoid inadvertent hepatic parenchymal ischemia. Hemobilia is in all likelihood secondary to trauma or aneurysmal rupture resulting in sometimes, a catastrophic scenario.

VI. THE DUCTULAR REACTION IN BILIARY SYSTEM DYSFUNCTION

Ductular reactions (Figure 57.3) are reactive processes that arise in pathological biliary system dysfunction, mainly at the interface of the portal or septal and parenchymal compartments in liver. Ductular reactions occur in many acute and chronic liver and biliary disorders and are a main component in the pathway of hepatic stem cell and progenitor cell activity in liver regeneration, hepatic fibrosis and neoplastic processes affecting the liver. Ductular reactions have varied appearances and are important in delineating biliary versus hepatic pathology and acute versus chronic pathology. The epithelial cells that are involved in the reactive process are varied and include intermediate hepatobiliary cells, activated cells of the canals of Hering, ductular cells, marrow derived precursors, biliary metaplastic hepatocytes and mesenchymal cells.

Normal liver do not have ductular reactions. The canals of Hering and ductular units (the hepatocyte canalicular system linking the canals of Hering) inter communicate across the

limiting plate. This limiting plate transformation starts in the periportal parenchymal region as the first small branch of the biliary tree, crosses the limiting plate and becomes the interlobular bile duct. Biliary obstruction is the prototype of ductular reactions. This shows multiplication of small ductules at the periphery of edematous portal stroma, with variable nuclear size and the absence of bile. Ductular reactions of the obstructive type show markedly expanded portal tracts; and in the presence of bile concretions in the dilated ductular lumen, termed as cholangitis lenta, a super imposed insult such as sepsis usually is associated. In the long run, the edema decreases and extracellular matrix of the ductular reaction becomes dense with collagen deposition, leading to obstructive biliary cirrhosis. Dense fibrous stroma is classically seen only in primary biliary cirrhosis and the pattern seen in extrahepatic biliary atresia and primary sclerosisng cholangitis is that of chronic dense stroma along with obstructive type of ductular reaction due to presence of smaller and larger duct involvement. The most striking type of ductular reaction is seen in acute (Fulminant) hepatic failure. Here there is presence of sparse fibrosis or inflammation, but massive ductular reactions, in the presence of more hepatocyte like cells than that of biliary type epithelial cells. A 'Star-Burst' pattern of ductular reaction in which there is expansion of the ductular reaction into the hepatic parenchyma is seen in the fibrosing Cholestatic variant of hepatitis B and C. In chronic viral hepatitis, the pattern is that of compacted stromal parenchymal interface ductular reaction which appears in the later stages of the disease. In autoimmune hepatitis, the pattern of ductular reaction is varied, in which hepatocytic rosettes usually predominate. These rosettes are mostly of the intermediate cell type. In fatty liver disease, the ductular reaction is inconspicuous and is more prominent with steatohepatitis. In vascular diseases of the liver producing ischemia, the reaction is centrilobular.

The epithelial cells that take part in ductular reaction are called intermediate hepatobiliary cells. They also represent a population of bipotent progeny of hepatobiliary stem cells that can give rise to both hepatocyte and cholangiocytes.

The region of dispersion of these cells is mostly parenchymal border for hepatocyte like intermediate cells and the portal with stromal border for the cholangiocyte like intermediate cells.

Immunostaining is the key investigation that can differentiate the different types of epithelial hepatobiliary cells in ductular reaction. Epithelial membrane antigen (EMA) is suggestive of activity of mature cholangiocytes and neural cell adhesion molecule (NCAM) or CD56 and CD10 is seen in the hepatocellular counterpart.

The only known factor specific for progenitor cells found on the intermediate cells is TWEAK. The pathways involved in transformation of intermediate cells into bipotency to form regenerative hepatobiliary cells involve the Wnt, Hedgehog signaling, Notch, TGF-bone morphogenic pathway and JAK STAT pathway.

Ductular reactions result in parenchymal regeneration (in association with potent peribiliary hepatocytes and post natal

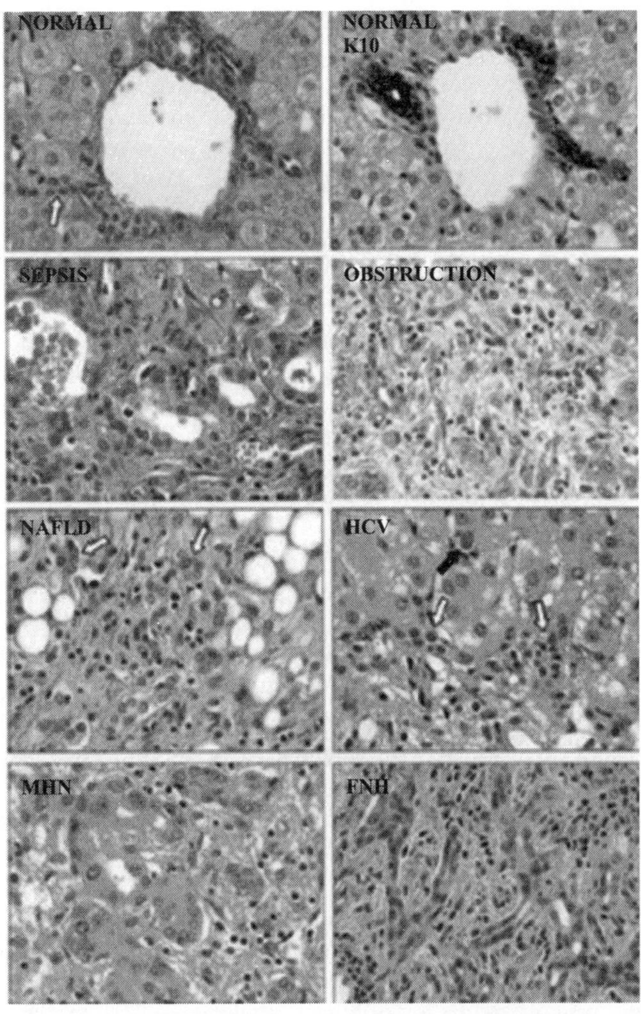

Various Types of Ductular Reactions (DR)

Left to Right, from Top to Bottom

1. Normal portal tract with canal of Hering (arrow)
2. Normal portal tract with K19 staining to show normal limiting plate extension
3. DR in sepsis – little stroma or inflammation, dilated ductule, inspissated bile seen
4. DR in acute large duct obstruction – abundant edematous stroma, neutrophilia with hepatocyte cholate stasis
5. DR in NAFLD – dense scar, intermediate hepatobiliary cells, steatotic changes andscant mononuclear infiltrate (arrows)
6. DR in HCV – dense compact scar, mononluclear cells, intermediate cells, acidophil bodies amidst necroinflammatory changes (arrows)
7. DR in Massive Hepatic Necrosis – intermediate cells with prominent hepatocytic features, less compacted stroma
8. DR in Focal Nodular Hyperplasia – intermediate cells with more cholangiocyte features with well defined ductal lumen and variable stroma

Figure 57.3 Ductular reactions in various pathological conditions. *Modified from:* Gouw, Clouston, and Theise. Ductular reactions in human liver – Diversity at interface. *Hepatology*, Vol. 54, No. 5, 2011. (*see Plate 20 for colour figure*)

hepatoblast population), thereby mediating repair. On the other hand, ductular reactions can also result in fibrogenesis as a strong predisposing and continuiung injury factor drives the cells to promote profibrogenic development in lieu with replicative senescence and oxidative stress, therby shifting the paradigm to fibrogenic injury and promotion of chronicity of the disease.

An important component to this fibrosis is the epithelial to mesenchymal transistion seen within the intermediate cells that take part in the ductular reaction.

A third outcome seen in ductular reaction is that of neoplastic transformation, where the stem cell/progenitor cells undergo malignant transformation and contribute to development of tumor microenvironment. This can be classically attributed to the mixed Hepatocellular-cholangiocarcinoma of the liver. In a patient of end stage liver disease, the presence of diminishment of ductular reactive cells, paves way for hepatocellular transformation. Thus ductular reaction is a diverse play between the molecular, cellular and tissue level environments seen in the hepatobiliary system which has varied outcomes depending upon the microenvironmental characteristic.

VII. THE DUCTAL PLATE MALFORMATIONS

A. Introduction

Jørgensen first described and coined the term ductal plate malformations. The congenital disease associated with ductal plate malformations include congenital hepatic fibrosis, autosomal dominant and recessive polycystic kidney diseases, Caroli's disease and syndrome and the von Meyenberg complexes and its associations. Ductal plate malformations (DPM) arise due to persistence of embryonic biliary tissue structures after birth. Embryonic structures include biliary cell clusters or duct like structures with luminal elongation and varied morphology. These form part of many developmental dysfunction of the biliary system. The main pathology in these conditions is the lack of remodeling of the ductal plate during fetal period. Hepatoblasts develop into biliary precursor cells during the embryonic period (Figure 57.4). These precursor cells form the ductal plate, which is a single layered sleeve of cells located in the vicinity of the portal mesenchyme. Primitive ductal structures lead to tubulogenesis. These are lined on the

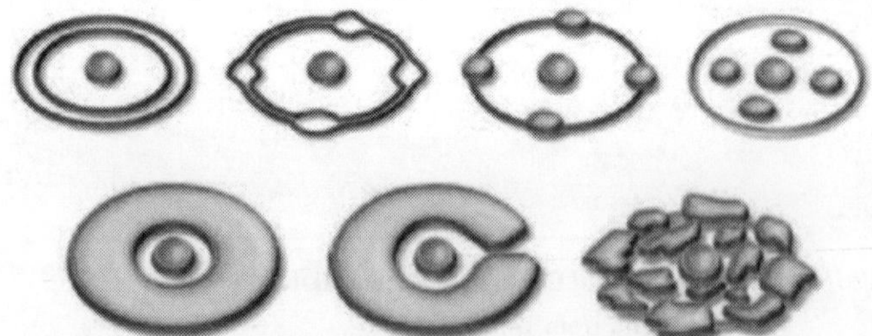

Figure 57.4 The above figure shows the development of biliary tract. The ductal plate is shown in purple and consists of double layer of flattened cuboidal cells; the portal vein branches are shown in blue. In figure below shows ductal plate malformation. Complete lack of remodeling of ductal plate is shown, resulting in either single dilated duct or interrupted circle of ectatic bile ducts. *From:* Santiago et al. Congenital Cystic Lesions of the Biliary Tree. *AJR,* 2012; 198:825–835. (*see Plate 20 for colour figure*)

portal side by the ductal plate cells, asymmetrically and on the parenchymal side near the hepatoblast like cells. Once the primitive ductal structures mature enough into ducts, the portal and the parenchymal sides differentiate into mature cholangiocytes, thereby symmetrically lining the duct lumina. These ducts are further integrated into the portal mesenchyme and the plate cells not involved in tubulogenesis involute.

Hepatocyte nuclear factors 6 and 1b are involved in these processes. There is morphogenetic abnormality in HNF-6 deficiency leading to biliary tree perturbations and abnormal expansion of ducts occurred in HNF-1b deficiency. Dysregulation of differentiation of biliary precursor cells, maturation of primary ductal plate cells and duct expansion along the developments hence form the crux of ductal plate malformations. The main morphologic end point in these ductal plate malformations were the absence of cilia. This resulted in malpositioning of the basal body and thus leading to abnormality in biliary cell polarity in which the apicobasal polarity was much more affected than others. This leads to persistence or lack of remodeling of the embryonic ductal plate normally observed during the biliary ductal development. The cells that help in the biliary system development are the hepatoblasts that have bipotentiality, which are located near the portal veins. These form a band of potential cholangiocytes which form the ductal plate, encircling the portal veins. The remodeling process of this ductal plate starts at the oldest cell population near the portal veins, at the hilum and then move towards the periphery of the liver along the portal system. The ductal cells that do not contribute to this process then involute. If this involuting cell population do not receive proper signaling or are insensitive to proper cell signaling, then they lead to the development of malformed ductal plates. Thus there is a fine coordinated process that ultimately connects the completed intrahepatic biliary system to the extrahepatic ductal system, which functions at the molecular level on certain specific

functions, dictated by epithelial and mesenchymal cells. The classification of ductal plate malformations is according to distinct defects in biliary tubulogenesis wherein ductal plate malformations can be classified on the basis of three distinct defects

- Differentiation of biliary precursor
- Maturation of PDS
- Duct expansion

In this classification, as described above, there is delineation in the structure that develops surrounding a formed lumen as either primitive ductal structure or a mature duct. The primitive ductal structure can of two types of cells—one with the presence or absence thereof markers of expression like SRY-related HMG box transcription factor 9 (SOX9), hepatocyte nuclear factor 4 (HNF4) and trandforming growth factor receptor type II (TbRII). The primitive ductal structure is asymmetrical, where in the cells on the portal side express SOX9 and the cells at the parenchymal side express the other two mentioned. A mature duct is on the other hand, symmetrical and expresses SOX9. As described in the study, the prior hilar intrahepatic biliary ductal formation and these theories were clubbed together to highlight defects at different steps of bile duct tubulogenesis. The absence of hepatocyte nuclear factor 6 produced early defect in biliary cell differentiation and the loss of HNF1b produced a defect in the primary ductal system maturation. A third type of defect was observed in the absence of cystin-1, where an abnormal expansion of ducts was seen even when the biliary cells differentiated normally. Deficiency of another protein known as Cystin-1 does not lead to defects in differentiation of biliary tubulogenesis, but produces ductal plate malformations seen in patients with autosomal recessive polycystic kidney disease. The polarity of the cholangiocytes also plays a very important role in the formation of ductal plate structures. The apical basal polarity was the most affected in case of HNF1b and HNF6 deficient patients. This results in absence of osteopontin expression leading to abnormal localization of centrioles and Golgi apparatus leading to ductal malformations. The deficiency of HNF6 and HNF1b has differential modification in the tight junction marker, zonula occludens–1 (ZO-1) and therefore leads to further changes in apical pole from basolateral pole of cholangiocytes. The loss of HNF6 results in normal apical-lateral localization of ZO-1 as against low levels of ZO-1 with improper apical localization on the portal side in HNF1b deficiency. The ZO-1 in HNF-1b mutations is irregularly expressed in dysplastic ducts. Absence of Cystin-1 did not influence apical polarity of cholangiocytes. These mutations also led to disruption in functioning and distribution of centrioles leading to ciliary dysfunction. Further animal models and human studies are required to completely delineate the pathology behind ductal plate malformations so that future directions can be made to prognosticate and for therapeutic application in this condition.

B. Congenital Hepatic Fibrosis and Caroli's Disease (Syndrome) (Figure 57.5)

This is a developmental disorder of the portobiliary system in which there is defective remodeling of the ductal plate leading to abnormal branching of the intrahepatic portal veins and progressive fibrosis of portal tracts. This may be associated with macroscopic cystic dilatation of the intrahepatic bile ducts. Most patients present with a large liver disease with well preserved liver function and portal hypertension with splenomegaly, hypersplenism and presence of esophageal and gastric varices. These diseases are also clubbed under primary ciliopathies with associated renal diseases called the hepato-renal fibrocystic diseases. The syndromes associated with congenital hepatic fibrosis are mostly autosomal recessive in nature. Congenital hepatic fibrosis is a histopathological diagnosis in its entirety. The three main characteristic of this disorder include

- Defective remodeling of the ductal plate
- Abnormal branching of the intrahepatic portal veins
- Progressive fibrosis of the portal tracts

Congenital hepatic fibrosis progressively leads to portal hypertension which leads to various complications related to increase in portal pressure. The pathogenesis of portal hypertension in this situation is purely due to increase in vascular resistance to portal blood flow in the liver secondary to increase in stiffness and fibrosis of the liver parenchyma and associated congenital vascular abnormalities. Hepatic encephalopathy and ascites is less commonly seen as a complication of portal hypertension in this condition than seen in cirrhosis. These patient are also at an increased risk for developing cholangitis, cholangiocarcinoma and hepatolithiasis. The liver histology in such patients reveal the presence of:

- Abundant abnormally formed bile ducts in portal tracts caused by an excess of embryonic bile duct structure
- Abnormal branching of portal veins
- Periportal fibrosis without inflammation
- Portal to portal bridging fibrosis
- Multiple bile duct hamartomas within the dense stroma (also known as von Meyenberg Complex)
- Inspissated bile in the lumen of some ducts

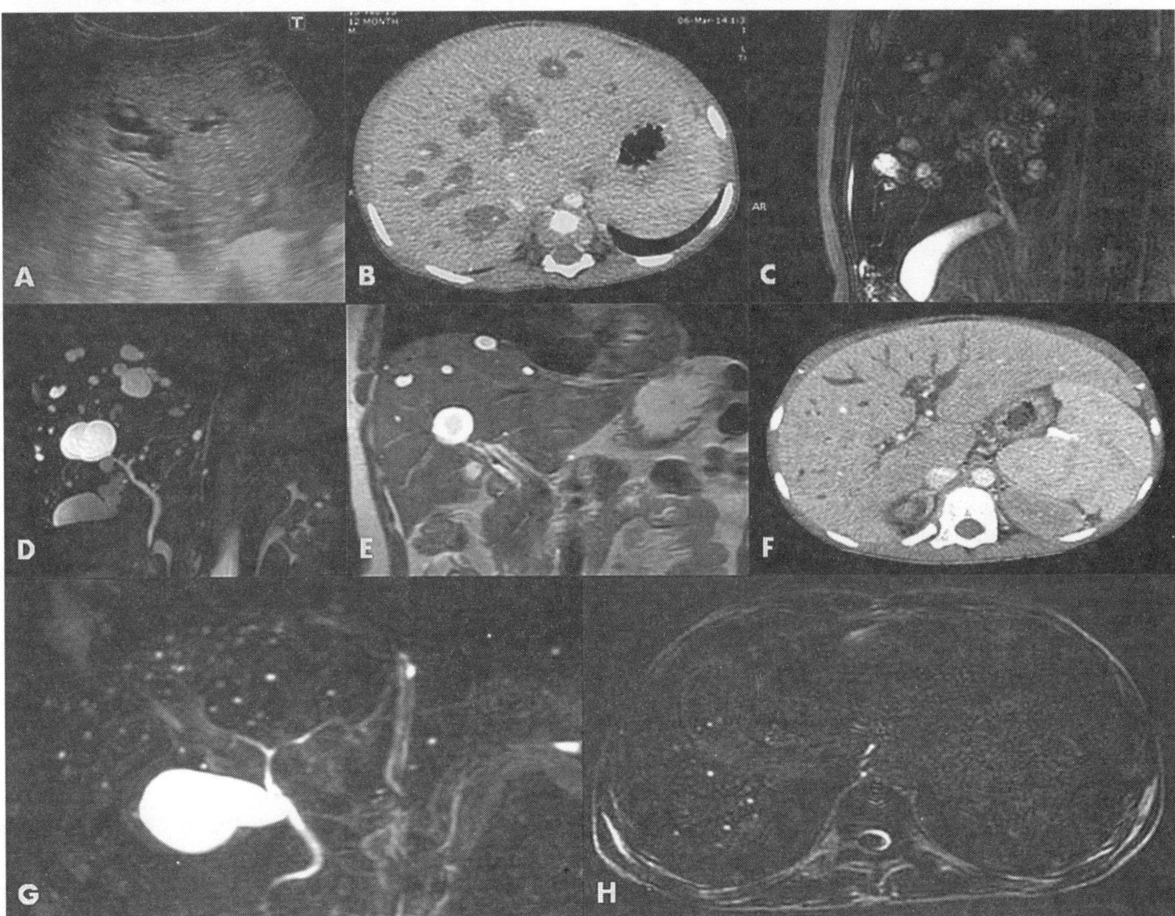

Figure 57.5　A, B and C—Ultrasonography, cross-sectional computed tomography and 2-D magnetic resonance imaging of Caroli's disease respectively; D and E—3D magnetic resonance imaging and abdominal cross sectional computed imaging of polycystic liver disease respectively; F—Caroli's syndrome (Caroli's disease with congenital hepatic fibrosis—note the large liver, splenomegaly, small abdominal collateral and associated biliary tract change); G and H—2D and cross sectional magnetic resonance imaging of von Meyenberg complexes.

An association of congenital hepatic fibrosis is the renal-hepatic-pancreatic dysplasia (RHPD) is characterized by cystic dysplastic kidneys, DPM of the liver, and fibrocystic dysplasia of the in association with partial or total situs inversus. NPHP3 is the first gene known to be associated with renal-hepatic-pancreatic dysplasia an association with congenital hepatic fibrosis. Other genetic factors purely associated with hepatic fibrosis of the congenital variety have not been delineated yet.

As described before, the ductal plate is a cylindrical layer of cells that surround a branch of portal vein and is also the embryonic precursor of intrahepatic bile ducts. The interlobular and intralobular bile ducts develop from this precursor ductal plate. Remodelling of the ductal plate starts at 12th week of gestation and is completed around 20th week. Arrest of maturation and remodeling leads to persistence of these immature remnants which stimulates portal fibrous tissue formation and periportal fibrosis that contributes to disease manifestations. The ductal plate malformation leads to a "pollard window" abnormality of the portal vein in which too many small and closely branched portal veins are seen around a thrombosed portal vein. Histopathologically this is seen as immature ductal paltes surrounding several hypoplastic and obliterated portal vein branches. Depending on the stage of arrest, either the small interlobular bile ducts—leading to congenital hepatic fibrosis or the mediumsized intrahepatic bile ducts—leading to Caroli's disease maybe involved.

The hepatic stelate cell is the main participant of the hepatic fibrotic process and also is inevitable to disease progression. This, in conjuction with the transforming growth factor beta (TGF-B) with increased expression, derived from the Kupffer cells has been shown to promote the disease. Other factors that add to fibrotic potentiality is the overexpression of osteopontin, matrix metalloproteinase under activity, plasmin over activity and laminin over expression. A liver biopsy is essential in the diagnosis and differential diagnosis of CHF, as the presence of small bile duct dilatation and proliferation would rule out other metabolic disorders of the liver.

Caroli's disease is characterized by the presence of non obstructive saccular or usiform dilatation of larger intrahepatic bile ducts. The cystic lesions are in continuity with the biliary tree which can be demonstrated with the help a number of imaging modalities. Histologiccally, the main features include non obstructive localized dilatation of the bile ducts, intraluminal bulbar protrusions of the ductal wall and intraductal vascular tracts containing patent portal venous and hepatic arterial channels that traverse the true lumen and terminate within the lumen. The 99 Technitium Sulphur scan and isotope scan shows a cold area in case of Caroli's disease that is classical of this condition. The presence of the "central dot sign" on imaging, notably on computed tomography and also on ultrasonography is pathognomonic of Caroli's disease. This is because of the fibrovascular bundles containing portal vein radical and a branch of hepatic artery bridging the saccule appear as central dots or linear streaks. Diffuse forms of Caroli's disease can be treated with liver transplantation and localized form with resection. The patient clinically presents with recurrent episodes of cholangitis. Caroli's disease in association with congenital hepatic fibrosis is termed Caroli's Syndrome.

C. The von Meyenburg Complexes

Multiple cystic lesions in the liver can represent a spectrum of diseases from simple cysts to complex parasitic diseases to the rare microbiliary hamartomas, termed the von Meyenburg complexes. There is predominance of multiple small, less than 1.5 cm nodular, cystic lesions which are mostly derived from malformed ductal plates, that affects the smallest intrahepatic biliary ductules. Typically the patient is asymptomatic and diagnosis is mostly incidental with characteristic findings in histology and imaging. This condition is related to autosomal dominant polycystic kidney disease, Caroli's disease and congenital hepatic fibrosis. Histology showd the presence of dilated bile ducts lined by a single layer of regular cuboidal epithelium embedded in a collagenous stroma. The computed tomographic findings consists of lesions with multiple irregularly delineated hyperintense cystic nodules not communicating with the biliary tree (as against Caroli's disease). Contrast enhancement is important to differentiate von Meyenburg complexes from Caroli's disease. Periodic follow up of these patients is advised as these lesions have malignant transformation potential. This malignant transformation is secondary to loss of heterogenosity in loci of oncogenes associated with these diseases, which maybe part of the ductal plate malformation and its consequences in the long term.

D. Polycystic Liver Disease

The polycystic liver diseases include those genetic diseases that primarily affect the bile duct and renal tubule epithelia. The autosomal dominant kidney disease is associated with this liver disease entity in about 50% of the time. Isolated polycystic liver disease is rarer and not associated with other phenotypes. The genes affected include the PKD1 and PKD2 which encode the polycystin-1 and polycystin-2 respectively. The surface complex polycystin-1/2 helps the cell in receiving signals from the surrounding environment which also includes the mechanical primary ciliary signals. Polycystin-1 is a mechanical receiver and the polycystin-2 is a calcium channel that causes intracellular signaling. Isolated autosomal dominant polycystic liver disease results from a defect in the PRKCSH, a hepatocysteine protein gene which codes for proteins that are required for endoplasmic reticulum maturation in N glycosylation of cell surface proteins. Defects in PKHD1 encoding the fibrocystin protein results in polycystic liver disease associations with Caroli's disease and congenital hepatic fibrosis. The fibrocystin protein dysregulation results in aberancy in primary ciliary mechanisms leading to cystogenesis in these conditions. Epidermal growth factor activation and probably estrogen excess results in excessive proliferation of the biliary epithelial cysts.

Polycystin and fibrocystin are predominantly expressed in the primary cilia of cholangiocytes. Ductal plate malformations lead to morphologically aberrant development of biliary epithelium that retains immature, ductal remnants which further leads to formation of multiple biliary microhamartomas that progressively dilate to macroscopic cysts, scattered throughout the liver parenchyma. As discussed previously the isolated polycystic liver disease is caused by mutations in PRKCSH. This gene codes for a protein kinase C substrate 80 K-H also called as hepatocystin or in the SEC63 gene. This gene encodes for a component of the molecular machinery regulating translocation and folding of newly synthesized membrane glycoproteins. These proteins are expressed in the endoplasmic reticulum and not in the cilia.

VIII. CHOLEDOCHAL CYSTS

Choledochal cysts are rare conditions where in abnormal cystic dilatations of the biliary tree prevail. It is of five types, in which the type IA shows marked cystic dilatation of the entire extrahepatic biliary tree with sparing of the intrahepatic ducts. In type Ib, there is focal, segmental dilatation of the extrahepatic bile duct. In type IC cysts, there is smooth fusiform dilatations of the entire extrahepatic bile duct extending from the pancreaticobiliary junction to the intrahepatic biliary tree. The type II cysts are discrete diverticulous extensions of the extrahepatic duct with a narrow stalk connection to common bile duct. Type III cysts are called choledochocele and consist of dilatation of the distal common bile duct that is confined to the wall of the duodenum and bulge commonly into the duodenal wall. The outer lining of the cyst is always lined by duodenal mucosa, the inner lining can be either be duodenal or biliary epithelium. The choledochoceles are further divided into five subtypes based on the relation to ampulla of Vater and the pancreatic duct. Type IVA cysts consists of multiple intrahepatic and extrahepatic dilatations. Todani et al. further classified Type IVA cysts into cystic-cystic, cystic-fusiform or irregular types. Type IVB cysts refer to multiple dilatations of the extrahepatic biliary tree only (bunch of grapes appearance, radiologically). The occurrence of these types of cysts are as follows: 50%–80% are type I, 2% type II, 1.4%–4.5% type III, 15%–35% type IV and 20% type V. Alonso and colleagues initially classified choledochal cysts which was further modified by Todani and colleagues (Figure 57.6)

The pathogenesis of cyst formation has been advocated as the Babbitt's Theory of Cysts caused by an abnormal pancreaticobiliary duct junction (APBDJ) wherein the pancreatic duct and the common bile duct meet outside the ampulla of Vater and form a long common channel, which allows mixing of the pancreatic and biliary juices leading to inflammation and deterioration of the biliary ductal wall, leading to dilatation. Chronic pancreatocobiliary reflux leads to dysplasia. Studies have shown that the levels of trypsinogen and phospholipase A2 levels in the cyst bile is elevated. The tyrpsinogen was then activated to trypsin in the presence of ectopically generated enterokinase enzyme that was secretd by

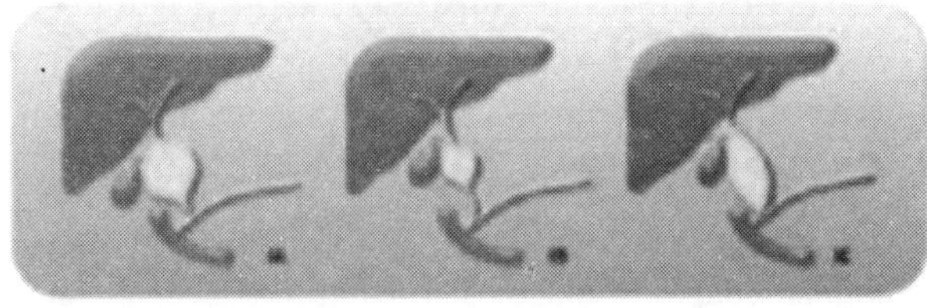

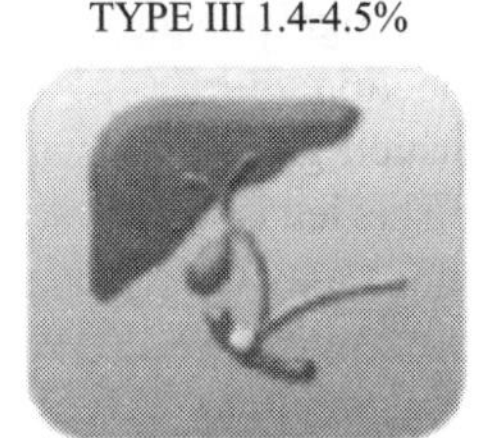

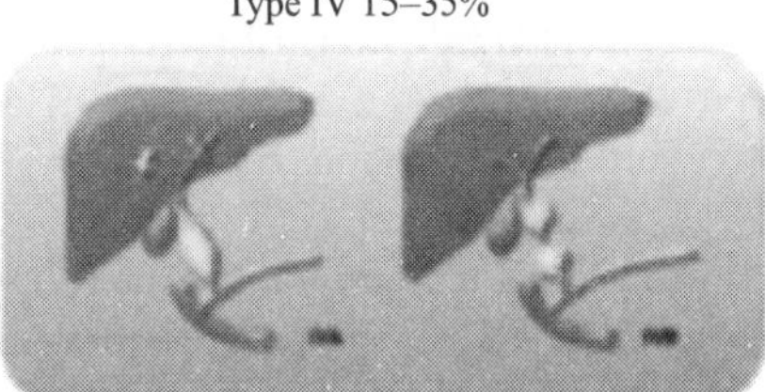

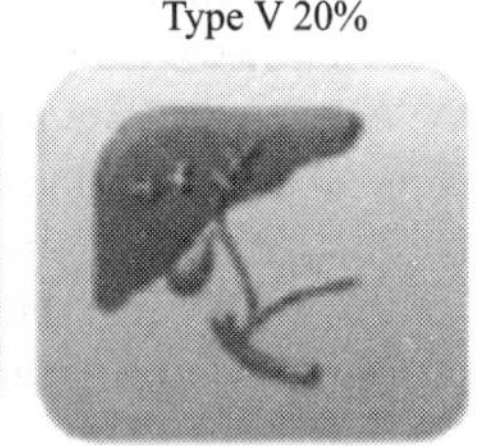

Figure 57.6 Todani Classification of Choledochal cysts. TypeIA is marked cystic dilatation of entire extrahepatic bile duct. IB is focal segamental dilatation, usually dital end; IC is smooth fusiform dilatation of entire extrahepatic bile duct; II is discrete diverticulum of extrahepatic bile duct; III dilatation of distal common bile duct confined to wall of duodenum; IVA multiple sites of dilatation both intra and extra hepatic; V multiple sites of saccular or cystic dilatation of only intrahepatic biliary tree (Caroli's disease or communicating cavernous ectasia). (*see Plate 20 for colour figure*)

the dysplastic epithelium. Further activation of phospholipase A2 hydrolyzes the epithelial lecithin to lyso-lecithin leading to inflammation and bile wall breakdown. The presence of immature neonatal pancreatic acini has also been referred to as a potential factor for development of choledochal cyst, antenatally. The congenital theory of choledochal cyst formation states that there an embryologic overproliferation of epithelial cells resulting in dilatation durin the developmental cannulation period. Congenital cysts are almost always found to be round in shape and acquired ones, being fusiform. Choledochal cysts are associated with other abnormalities like colonic atresia, duodenal atresia, imperforate anus, pancreatic AV malformations, multiseptate gall bladder, cardiac defects, fical nodular hyperplasia of liver and familial adenomatous polyposis. Type V choledochal cyst or the Carolis disease has a compeletly different pathogenesis as compared with other types and has been described in the section on ductal plate malformations.

IX. BILIARY SYSTEM DYSFUNCTION IN THE LIGHT OF IMMUNE DYSREGULATION

A. Introduction

Molecular mechanisms of biliary innate immunity fall heavily on the expression of PAMP-recognizing receptors and intracellular adaptor molecules. The toll-like receptor family is the most abundant cell surface receptors recognizing PAMP's.

The recognition and response to lipopolysaccharides occurs in conjunction with TLR4 and its accessory proteins MD-2 and CD14 which go on to trigger the transduction of intracellular signals. This leads to activation of myeloid differentiation factor 8 (MyD88) and the IL-1 receptor associated kinase, leading to the activation of nuclear factor-kB and then to synthesis of proinflammatory cytokines and antibodies. This reaction is mostly mediated by the bacterial components. In contrast, the double stranded RNA are recognized by TLR3, Interferon inducible helicase retinoic acid induced protein I (RIG-I) and melanoma differentiation associated gene-5 (MDA-5). The stimulation results in activation of transcription factor interferon regulatory factor 3 (IRF3) as well as NF-kB. NOD 1 and NOD 2 are involved in the intracellular recognition of microbes through specific interactions with derivatives of pathogen specific peptidoglycans. TLR's are expressed abundantly in the human cholangiocytes and biliary epithelium. Expression and activation of TLR's has been demonstrated during bacterial, viral, parasitic and malignant transformations affecting the biliary tree. This is documentation enough that the human biliary epithelial cells posses functional PAMP-recognizing receptors and an innate immune system against viruses and bacteria. In response to PAMP activation, the biliary cells secrete polymeric immunoglobulin A and produce several antibiotics against bacteria, particularly lactoferrin, lysozyme and defensins. The defensins are the major determinants of innate immunity. The defensins are of two types, alpha and beta defensins. Human beta defensins are produced by cholangiocytes also and protect against mucosal infections. The common defensins produced by the biliary epithelial cells include the hBD1 and hBD2 (Figure 57.7). Apart from protecting against bacterial invasion, the cholangiocytes also protect against viral attack because they posses TLR3, RIG-1 and MDA-5 receptors which recognize dsRNA viruses. Cholangiocytes participate in innate immunity directly without the help of immunocompetent cells. Cholangiocytes also produce several inflammatory cytokines and chemokines such as IL-6, TNF-alpha, IL-8, fractalkine, monocyte chemotactic protein-1 and CXCL16. IL-6 increases DNA content in cholangiocytes, IL-8 increases neutrophilic

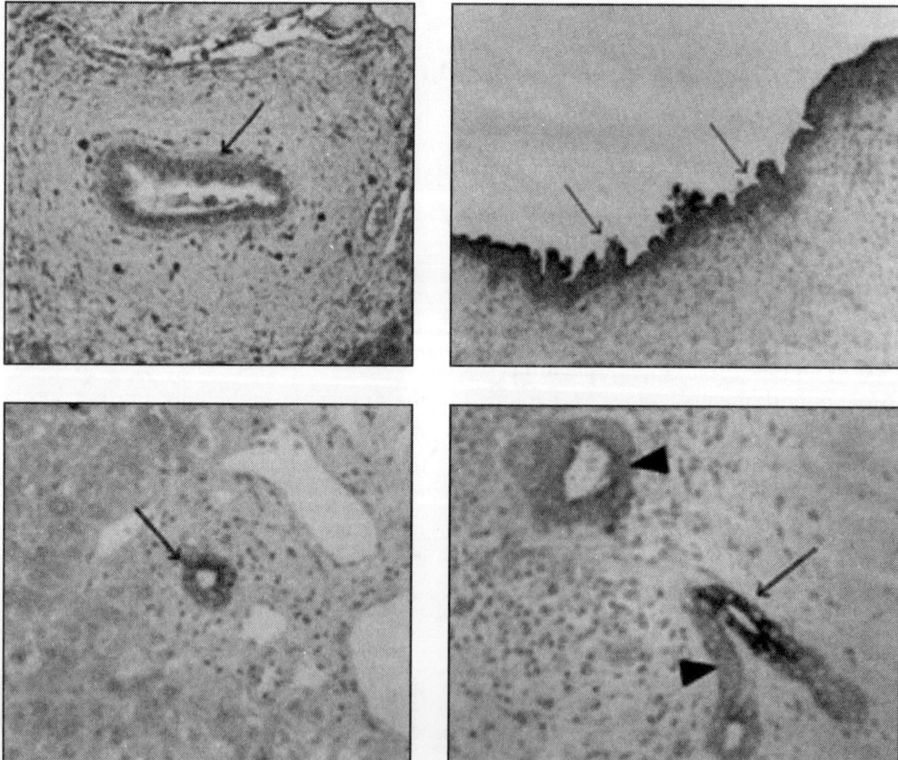

Figure 57.7   Immunohistochemical staining (above panel) staining for human beta defensin (hBD-1) and hBD-2) in which the first box shows normal liver where in septal bile ducts stain positive and the second box shows features of extrahepatic obstruction with biliary epithelium of large bile ducts showing strong expression of hBD-2. In the panel below, the immunohistochemistry for peroxisome proliferator activated receptor gamma is shown. Here the first box shows normal liver, the receptor is expressed in the cytoplasm of bile ducts (arrow); in the second box, the damaged ducts in primary biliary cirrhosis show reduced expression of the receptor. *From: Kenichi Harada and Yasuni Nakanuma : Biliary Innate Immunity: Function and Modulation Mediators of Inflammation.Volume 2010, Article ID 373878.*

activity in leading to bile ductular proliferation, ductular cholestasis and ductular epithelial injury (as in cholangitis lenta). The chemokines take part in recruitment of T cells, macrophages, Kupffer cells, hepatic stellate cells and their activation. Cholangiocytes function as antigen presenting cells and control inflammatory processes. The biliary innate immune responses are also involved in functional regulation of tight junctions in cholangiocytes. Th-1 related pathogenesis is cause for most of the biliary system immune dysfunctions, namely the upregulation of TLR4 and TLR9, which is classically seen in primary biliary cirrhosis and primary sclerosing cholangitis. Micro RNA mediated post transcriptional pathways are critical for cholangiocyte regulation of biliary system infections/immune dysregulation. Even though the luminal cholangiocytes are continuously exposed to bacterial contents in the bile, there is a lack of response to PAMP's especially lipopolysaccharides in the long run. This is known an endotoxin tolerance and is a method by which the biliary tree prevents septic shock syndromes, thereby maintaining homeostasis in organs. On the contrary, the cholangiocytes on the luminal side are not resistant to virus derived dsDNA, thereby, causing immune dysregulation.

B. Primary Biliary Cirrhosis (PBC)

PBC (Figures 57.8a, b) involves the selective destructive loss of interlobular bile ducts which is a chronic non suppurative destructive cholangitis. The factors leading to this phenomenon are many and include recurrent genitourinary infections, bacterial and viral factors in bile and molecular mimicry of the pyruvate dehydrogenase complex, common to bacteria and humans (PDCE2) which is a major epitope of the antimitochondrial antibody which defines PBC. The biliary epithelial cells produce translocation of the PDCE2 complex into apoptotic bodies, thereby forming apoptopes, thereby leading to triad formation defined by the presence of apoptope in cholangiocytes, antimitochondrial antibodies and activated macrophages that cause inflammatory cytokine production within the biliary system. The dysregulation of microorganisms with their environmental counterparts produce the cholangiopathy that promote the pathogenesis of this disease. The cholangiocytes in PBC express TNF-alpha and IL-6 excessively. Th1 dominant cytokine dysregulation and inflammation produce the bile duct injury seen in PBC (the histopathological correlates as described in Ludwigs Classification).

The Th1 cytokine IFN-gama upregulates the expression of TLR's, sensitize the susceptibility to PAMPs in cholangiocytes thereby impairing regulation of biliary innate immunity. Loss of peroxisome proliferator activated receptor gamma activity in cholangiocytes cause homeostatic aberrations leading to loss of maintenance of biliary innate immunity, upregulation of TLR4 and TLR9 in cholangiocytes and TLR3 and IFN in portal tracts. Th17 cells are also involved in the chronic inflammatory process of PBC. IL-17 is produced by the Th17 cells under the influence of IL6, IL1beta and IL23. In PBC, the IL17 positive mononuclear cells are scattered at the interface areas showing interface hepatitis at these areas. The Th17 cytokine rich peribiliary tissue sustains th cholangiopathy in PBC. The biliary innate immunity is responsible for induction and maintanence of Th17 cells in the periductal area in PBC and also formation of dendritic cells and tissue macrophages. Langerin positive Langerhans cells are also seen within the damaged biliary ducts and ductules. These are scattered around the periductal areas and are also responsible for the maintanence of non suppurative inflammation in the biliary tree. Thus PBC is an autoimmune mediated disease where in there is female predominance, clinical homogeneity, production of multilineage immune response to mitochondrial autoantigens, inflammatory non suppurative destruction of small bile ducts and in late stages, fibrosis and cirrhosis. The serological response prototype includes antomitochondrial antibodies. The multilineage immune response is directed at the E2 component of 2-oco-dehydrogenase pathway, mainly the PDC-E2; there is an autoreactive T cell precursor response for both CD4 and CD8 cells; the formation of apoptope of biliary cells containing PDC-E2 which produces a proinflammatory response and the

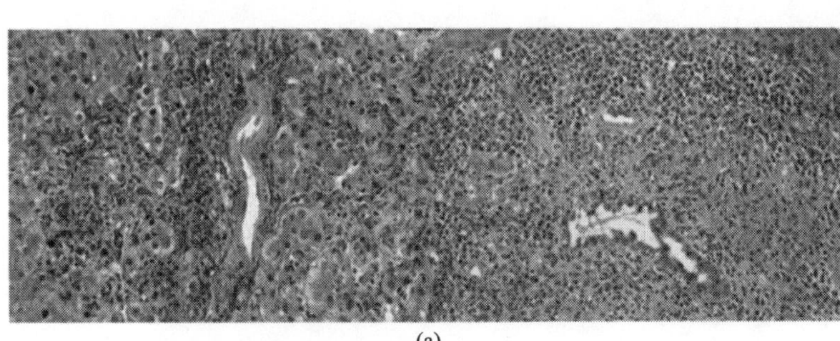

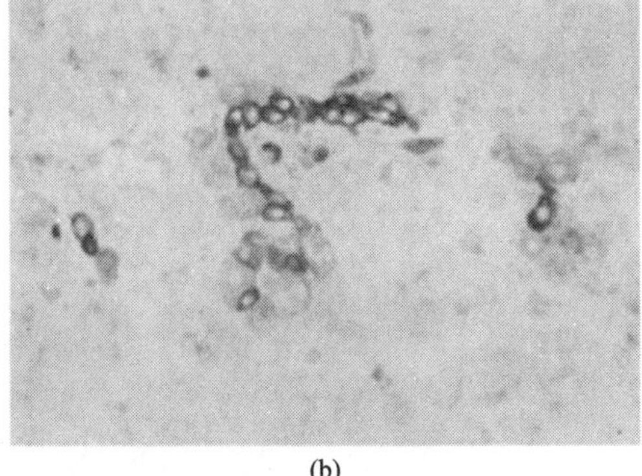

(a)   (b)

Figure 57.8 (a) Histopathology of Primary Biliary Cirrhosis showing florid duct lesions with chronic inflammatory infiltrates, lymphocytic infiltration and ductal sclerosis and desctruction, (b) Ductular reaction in a case of primary biliary cirrhosis, immunohistochemically stained for chromogranin A (on frozen tissue). Note the morphology midway between cholangiocyte and hepatocyte (Mayer's hematoxylin counterstain, original magnification 250X). (*see Plate 20 for colour figure*)

loss of tolerance to PDC-E2 and the critical role of the IL-12 signaling pathway – all of which culminates into the clinical entity that PBC is.

C. Biliary Atresia (BA)

This disease is characterized by a progressive sclerosing obstruction of the extrahepatic bile ducts. It has been associated with type 3 reoviral infection (Rotavirus type C). The presence of enhanced apoptosis in biliary epithelial cells lining the extrahepatic biliary system is a major feature of this disease (Figure 57.9). TNF-alpha related apoptosis inducing ligand (TRAIL) and CD 95-Fas ligands mediate this occurrence in cholangiocytes. This is partly due to the epitheliotropism shown by the dsRNA viruses and its innate immune dysregulatory effects in the bile ducts. The cholangiocytes of the extrahepatic bile ducts shows an increase in TLR3 expression and activation of NF-kB and IRF-3. Obstructive cholangipathy is maintained in the biliary system leading to BA because of lack of innate immune tolerance to dsRNA in the cholangiocytes of

the extrahepatic biliary system. Epithelial to mesenchymal transition of the cholangiocytes eventually take place leading to histogenesis of the sclerosing lesion. This further leads t periductal fibrosis and portal fibrosis leading to biliary atresia. The normal epithelial features are lost, cell to cell adhesion properties are attenuated and mesenchymal phenotypic changes occur where in cholangiocytes acquire fibroblastic spindle morphology and cytoskeletal reorganization. There is increased expression of mesenchymal markers vimentin and S100A4 or the fibroblast specific protein-1 and upregulation of transition factor Snail takes place leading to fibroblastic growth factor increment. The basic faulty technique in BA is the incomplete induction of apoptosis within the cholangiocytes caused by the dysregulated biliary innate immune system and further development of sclerosing pathology.

The onset of BA is restricted to the neonatal period and the target injury is restricted to the biliary system. There are classically two types of BA explained. The embryonic type and the perinatal type. In both phenotypes, the patients undergo inflammatory and fibrosing process triggered by an unknown

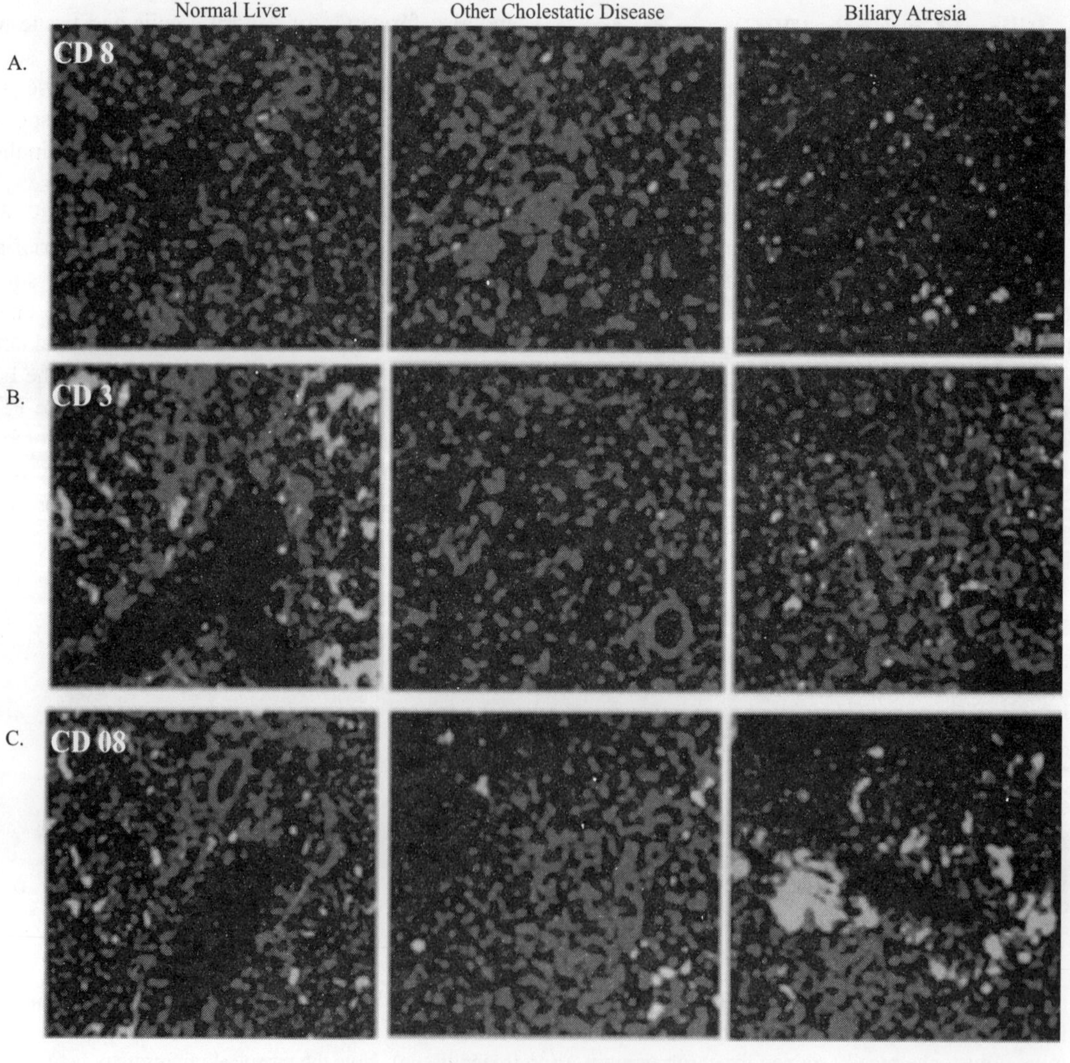

Figure 57.9   *(Contd.)*

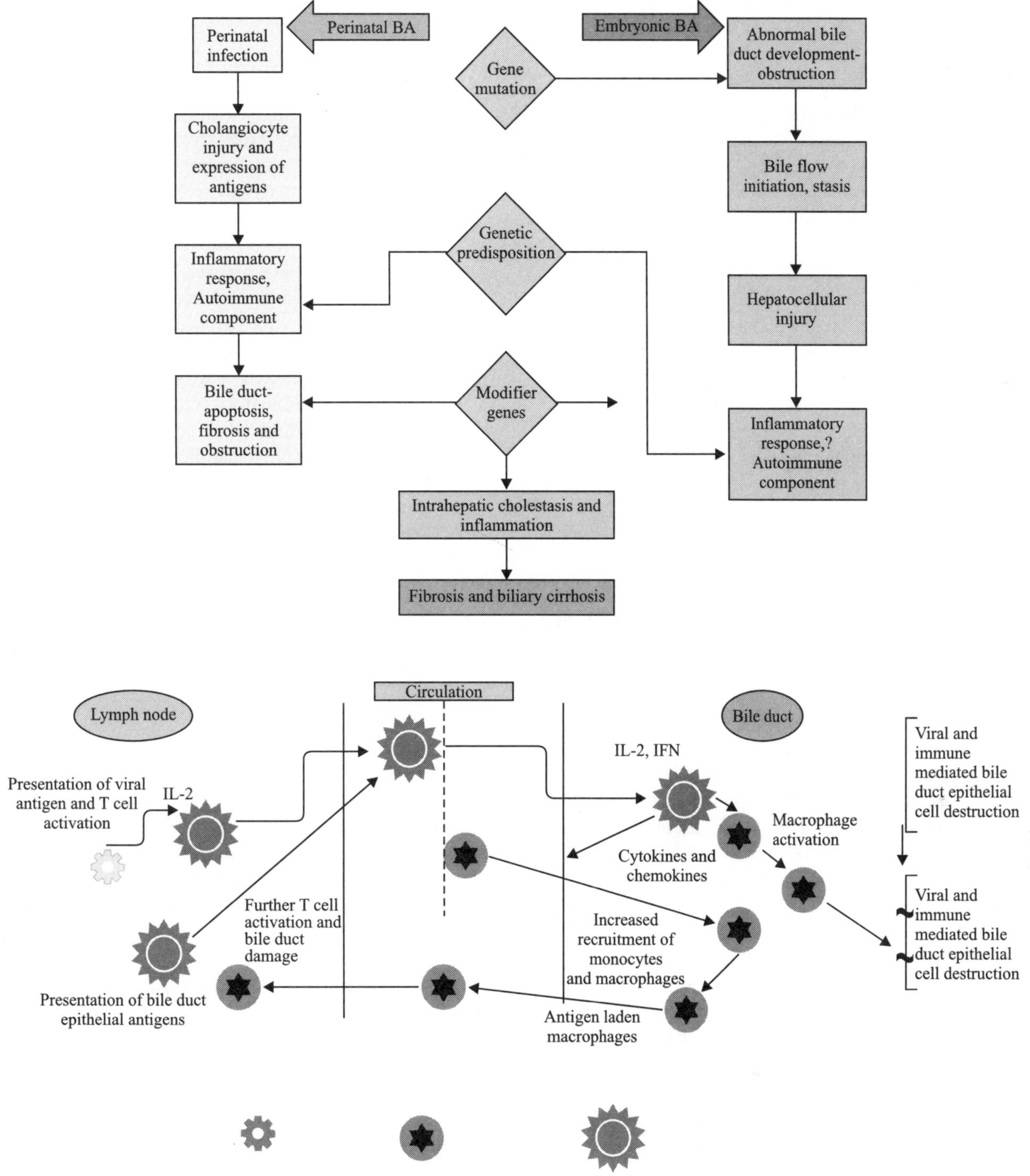

Figure 57.9 Etipathogenesis of Biliary Atresia (BA) – LEFT: Perinatal BA develops in the event of perinatal insult like in the presence of a cholangiotropic virus, trigerring bile duct epithelial injury. Inflammatory processes leads to immune response that is dysregulated leading to apoptosis and necrosis of extrahepatic bile duct epithelium leading to fibro-obliteration of lumen and bile duct obstruction. Embryonic BA results from mutations in genes regulating normal bile duct formation/differentiation secondarily inducing inflammatory or aberrant immune response within common bile duct and liver parenchyma. The end result is one of intrahepatic cholestasis and portal tract fibrosis leading to biliary cirrhosis. RIGHT: Fluroscent immunohistochemistry revealing an increase in CD8+ T cells, CD3+/CD4+ T cells and increase Kupffer cells in the portal tracts in biliary atresia, much more than that seen in normal liver and cholestatic liver disease of other causes. *From:* Cara L Mack and Ronald L Sokol – Unravelling the etiology and pathogenesis of biliary atresia. *Pediatric Research* (2005) **57**, 87R–94R. (*see Plate 21 for colour figure*)

entity. Five mechanisms have been proposed that culminates into the formation of BA.

1. Defect in morphogenesis of the biliary tract—inappropriate persistence or lack of remodelling of the embryonic ductal plate – the ductal plate malformation has been proposed to cause BA. Evaluation of neonates after birth has revealed the presence of duct remnants proximally and distally to the early cystic dilatations of the biliary tree in BA. The main associated dysmorphgenetic lesions includepoly or asplenia, cardiovascular defects, abdominal situs inversus, intestinal malrotation and portal vein and hepatic artery malformations. These have been related to the recessive insertional mutation of the inversion gene shown in murine models. The identification of loss of function in CFC1 gene encoding the CRYPTIC protein in patients of BA has also been advocated. Yet another gene that participates in the expression of BA is the Jag-1 gene associated with single nucleotide polymorphisms. These factors show that genetic aberrations indeed add to the pathobiology of BA.

2. Defect in fetal/prenatal circulation—impaired blood flow through the hepatic artery in the early developmental period has been advocated an an inciting factor for injury to the biliary system leading to biliary atresia. This has been shown in the form of arterial hyperplasia and hypertrophy as described in affected childrens liver specimens.

3. Environmental toxin exposure—Toxin exposure leading to BA development has been shown to be feasible due to time-clustering of cases. These are mostly derived from animal studies and no specific agent could be delineated, which could incite or promote BA.

4. Viral infection – the isolation of viruses in the livers of affected neonates lead to the belief that there is potential infective role in development of BA. Varied viral types have been documented in the liver specimens in many series. No study has ever proven the etiological role of a single agent in the development of BA. Among all viruses implicated, the reovirus type 3 and the rotavirus type C has been the most consistently seen and studied. There is a high prevalence of IgG and IgM immunophenotypes of these viruses derived from affected neonates. The use if virus specific amplification by reverse transcription polymerase chain reaction has been a boon in further demonstrating these infective causative agents. The prototype murine model of reoviral etiological association is seen in the "oily fur syndrome" mice models. The administration of rhesus rotavirus type A in mice have produced disease models that resemble BA. There is a segmental or continuous obstruction of the extrahepatic duct lumen leading to the formation of BA phenotypes in these models.

5. Immunologic/inflammatory dysregulation—In neonate livers of BA the presence of cholangiocyte pyknosis and necrosis have been associated with infiltration of mononuclear cells into the walls of the interlobular bile ducts and also lymphocytic infiltration into portal tracts, the duct walls at porta hepatis and common bile duct remnants. Phenotypically, the lymphocytes infiltrating portal tracts are CD4+ rather than CD8+ T cells. These cells express T helper lymphocyte activation and interleukin 2 receptors. This further leads to differentiation of the T-helper cells into a proinflammatory phenotype of Th1 type. The cholangiocytes which normally express MHC class I, but not class II antigens aberrantly express HLA-DR, a major MHC class II molecule and acts as antigen presenting cells. The infiltration of the portal tracts by CD14+ Kupffer cells and lymphocytes further adds to this proinflammatory and immune dysregulatory component which then through expression of IL-8 promotes IFN-gamma productions and Th1 differentiation of lymphocytes leading to ductal injury and fibrosis. Recent evidence has also shown that the extrahepatic biliary system and ventral pancreas share a common progenitor cell. Out of these, the Sox 17, a protein involved in formation of endodermal organs, whose expression is dysregulated by mutagenesis has been shown to be associated with BA. The Lgr4 gene (family of leucine rich repeat containing G protein coupled receptors) has also been vindicated in BA development through mutational changes, in murine models.

Thus BA is an infantile disorder characterized by the complete obstruction of a portion or entire length of the extrahepatic biliary tree by a fibro-inflammatory process that disrupts physiological bile flow, which is brought about by multifactorial dysregulatory changes seen both at the histological, molecular and environmental levels.

D. Primary Sclerosing Cholangitis (PSC)

PSC is a chronic cholestatic liver disease with fibroobliterative sclerosis of intra- and/or extrahepatic bile ducts, eventually leading to biliary cirrhosis (Figure 57.10). The identification of certain HLA variants, notable HLA DR3 and HLA B8 in PSC patients have emphasized on the genetic role of development of this entity. The genome wide association studies in PSC has shown strong associations with a subset of HLA and non-HLA genes involved in bile homeostasis associated with regulatory inflammatory pathways that define PSC pathobiology. Earlier studies identified the association of HLA DR3(DRB1*0301) and HLA B8(HLA-B*0801) haplotypes as susceptibility markers in PSC. This also led to further studies which showed that the HLA-A1 allele, HLA-C7, MHC class I chain related A (MICA) and TNF-alpha promoter A allele were also identified as being susceptibility loci for PSC. Negative associations have been found for HLA DRB1 and DQA1. Killer immunoglobulin like receptors (KIR) has been implicated in pathogenesis of PSC.

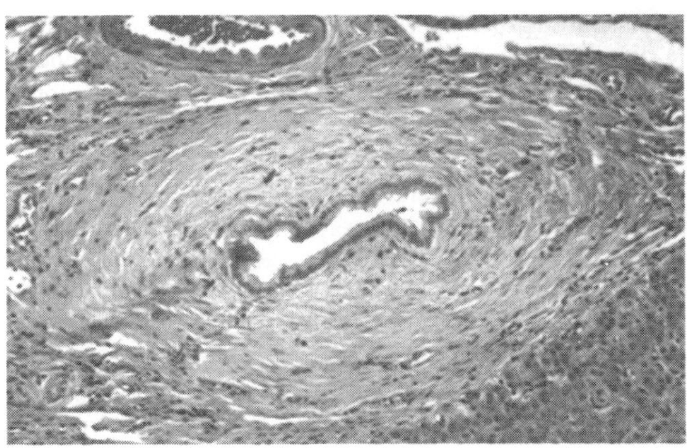

Figure 57.10 Histopathology of Primary Sclerosing Cholangitis showing classical 'onion skin fibrosis'. (*see Plate 22 for colour figure*)

It functions as natural killer cell receptor which bind to HLA class I molecules and influence susceptibility to autoimmune diseases – produced mainly secondary to imbalances between the two. 80% of PSC patients harbor ulcerative colitis, an association secondary to commonness in susceptibility genes. The intracellular adhesion molecule-1 (ICAM-1, CD54) gene polymorphisms have been implicated in the susceptibility to PSC. There is expression of this molecule on proliferating bile ducts in late stages of PSC. The G protein coupled bile acid receptor 1 TGR5 has also been implicated in PSC pathogenesis. It serves a dual function in biliary bicarbonate secretion and anti inflammatory effects in macrophages of the biliary system. The multi drug resistance gene 1 (MDR1) provides for PSC susceptibility. It is a membrane protein mediating efflux of xenobiotics and toxins and whose gene variants maybe associated with disease progression in PSC and ulcerative colitis. The association of PSC and ulcerative colitis has led to many models of injury that explained several pathogenetic roles leading to development of PSC. The first of this hypothesis, known as the "leaky gut hypothesis" stated that translocation of bacteria or its components entering the portal venous system increased intestinal permeability resulting from an inflamed gut wall and induction of inflammatory processes concentrated within the portal fields leading to PSC pathobiology. The bacterial antigens functioned as molecular mimicry agents to trigger a dysregulated immune response in the biliary tree. The second hypothesis, known as the "gut lymphocyte homing hypothesis" stated that the CCR9+ Alpha4beta7+ memory T lymphocytes primed in the inflamed gut persists as long living memory T cells that undergo enterohepatic circulation that triggers portal inflammation and aberrant expression of adhesion molecules in the liver and intestines. The adhesion molecules whose dysregulation leads to this phenomenon include the ICAM-1 and the vascular adhesion protein-1 (VAP-1). The genetic modification of bile composition was advocated to induce sclerosing cholangitis and biliary fibrosis in animal models. The disruption of MDR2 (ABCB4 gene) encoding the canalicular phospholipid flippase lead to development of cholangitis and the typical 'onion skin' type of periductal fibrosis that is the hallmark of PSC in humans. This disruption leads to defective biliary phospholipid secretion resulting in an increased concentration of free non micellar bile acid and subsequent inflammatory changes. Even though there are many models and hypothesis that try to answer the pathobiology and molecular mechanisms in PSC, none of them are proven until now and further studies are ongoing to deeply describe this entity as a whole, phenotypicaly and genotypicaly.

E. The IgG4 Related Cholangiopathy

Different from PSC, the IgG4 related sclerosing cholangitis (IgG4-SC) is a unique systemic inflammatory condition in which there is tumourous swelling of affected organs and high serum igG4 levels. The prototype of this disease is autoimmune pancreatitis. The disease can involve many organ systems like salivary glands, lacrimal glands and retroperitoneum and kidneys. Concerning the biliary tree, IgG4-SC can present as a diffuse sclerosing cholangitis or hilar pseudotumorous mass. IgG4 cholangiopathy is almost always associated with large duct lesions. The characteristic lesion in IgG4-SC is extensive infiltration by IgG4+ plasma cells on immunostaining, diffuse lymphoplasmacytic infiltration, storiform fibrosis, obliterative phlebitis and eosinophilic infiltration. Obliterative phlebitis is a finding characteristic of IgG4 related disease affecting any organ system. The Th2 type immune response is active in IgG4 related diseases. There is a higher ratio of IL-4/IFN-gamma and IL-5/IFN-gamma and IL-13/IFN-gamma in this group of diseases, which is seen differently from other autoimmune related diseases. Lymphocytes with IL-4 expression were most commonly seen on in situ hybridization techniques. Peripheral blood mononuclear cells in IgG-SC produce predominantly tH2 type of cytokines such as IL-4, IL-5, IL-10 and IL-13 after T cell stimulation and activation. The number of regulatory T cells is increased in affected tissues and in blood. The expression of A regulatory T cell specific transcriptional factor, the Foxp3 (forkhead box P3) mRNA was higher in this group of disease. The regulatory cytokines TGF-beta and IL-10 are also over expressed in IgG-SC. The number of Foxp3+ cells was significantly correlated with the number of IgG4+ plasma cells in IgG4-SC. The number of naïve regulatory T cells is decreased and that of CD4+CD25 regulatory T cells are high. The hyporeactive naïve regulatory T cells are mostly involved in the pathogenesis of IgG4 related disease and hyper reaction of CD4+CD25 high regulatory promotes disease progression. Some studies have shown the involvement of Helicobacter pylori in the pathogenesis of IgG4 related diseases. The molecular mimicry between human CA-II and alphacarbonic anhydrase of Helicobacter was found to be one of the triggering agents and was also the presence of antibodies against plasminogen binding protein of Helicobacter. The amino acid sequence of palsminogen binding protein exhibited homology with the ubiquitin protein ligase E3 component recognon 2, an enzyme expressed in pancreatic acinar cells. Helicobacter theory hold good for mostly autoimmune IgG4 related pancreatitis and not other IgG4 related diseases as of now.

X. DRUG INDUCED CHOLESTASIS

Drug induced liver injury can be of three types—hepatocellular, cholestatic or mixed. Dysregulation of the bile acid homeostasis leads to drug induced cholestasis and its clinical components. The drug induced liver injury spectrum ranges from acute bland cholestasis, acute cholestasis with hepatitis, acute drug induced cholestasis with bile duct injury and the chronic drug induced cholangiopathies or the vanishing bile duct syndromes. Bile acids induce biliary lipid secretion and solubilize cholesterol in bile, promoting cholesterol elimination. The bile acids are potent activators of nuclear receptors like the farnesoid X receptor and pregnane X receptor that play an important role in lipid homeostasis (Figure 57.11).

Bile acids become cytotoxic when abnormal accumulation occurs within hepatocytes. Canalicular membrane defects, alteration in fluidity of the bile, impaired contraction of actin filaments in the pericanalicular region along with bile duct patency abnormalities decreases bile flow and propagation. The more hydrophobic a bile acid is, the more toxic it is. The bile acid toxicity is highest in LCA >CDCA>DCA>CA>UDCA. Accumulation of bile acids in hepatocytes leads to mitochondrial damage and necrosis/apoptosis. Drug induced cholestasis is due to disruption of bile acid homeostasis by direct inhibition of bile acid transport or by indirect ways. Drug mediated functional changes in hepatic bile acid transporters can lead to intracellular accumulation of harmful bile acids and hepatocyte damage. Intracellular accumulation of bile acids depends on the basolateral and canalicular processes. Systemic concentrations of the drug promote results in uptake transporter inhibition and the intracellular drug concentration is responsible for inhibitory effects on acid efflux. There are several transport proteins identified as potential loci for drug induced cholestasis. These include basolateral uptake transporters, NTCP and OATP, canalicular efflux transporters, BSEP, MRP2 and MDR3 and basolateral efflux transporters MRP3 and MRP4. Bile salt excretory protein (BSEP - member of the ATP binding cassette gene superfamily) mediates the rate limiting step in bile formation across the canalicular membrane. Defects of the BSEP coding or gene regulation leads to inherited and acquired cholestatic disorders such as progressive familial intrahepatic cholestasis type 2 (PFIC 2), benign recurrent intrahepatic cholestasis type 2 (BRIC 2) and intrahepatic cholestasis of pregnancy. Troglitazone, bosentan, rifampin, cyclosporine and gilbenclamide inhibit BSEP mediated bile acid secretion leading to bile acid mediated hepatocyte injury. Insect and animal models have shown that BSEP inhibition is one of the causes for drug induced cholestatic injury. Most of the BSEP inhibitors directly cis-inhibit BSEP while the estradiol and progesterone metabolites trans-inhibit BSEP after secretion into bile canaliculus by Mrp2. MDR3 an ATP dependent phospholipid flippase translocates phosphatidyl choline from the inner to the outer compartment of the canalicular membranes whereby the canalicular phospholipids are solubilized by canalicular bile salts to form mixed micelles, protecting cholangiocytes from the detergent action of bile

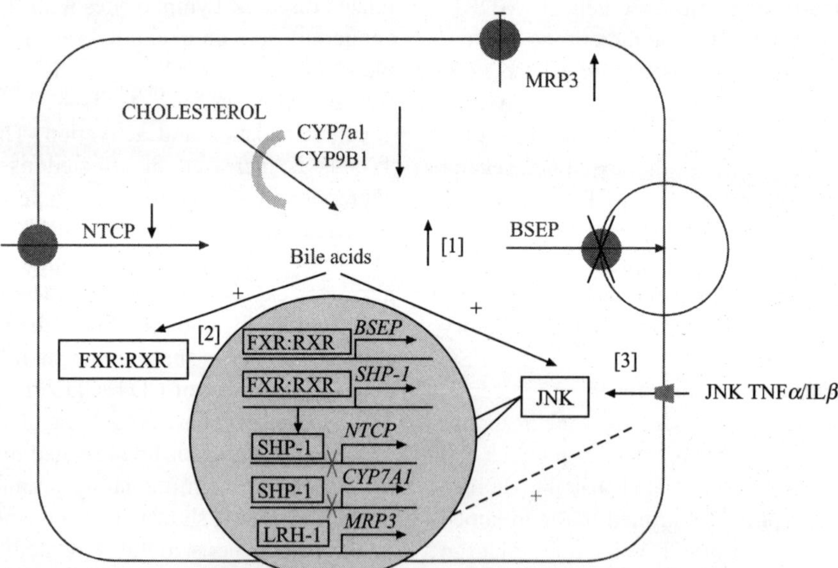

Figure 57.11   Gene regulation by bile salts. Bile salts are taken up in the liver by NTCP and secreted into bile by canalicular BSEP. In cholestasis BSEP activity is reduced the intracellular bile salt concentration increases (1). Bile salts serve as ligands for FXR, which forms a heterodimer with RXR and translocates to the nucleus (2). The heterodimer activates the transcription of the BSEP and SHP-1 genes. SHP-1 antagonizes the expression of the bile acid biosynthetic enzymes CYP7A1 and CYP8B1 and the transporter NTCP. In addition, Kupffer cells produce TNFa and interleukin-1b during cholestasis and, via the c-Jun N terminal kinase-dependent (JNK) pathway, they reduce the expression of NTCP and CYP7A1 (3). Recent evidence indicates that Lrh-1 (liver receptor homolog-1)-mediated Mrp3 transcription is enhanced via a TNF-a signalling pathway. *Modified from:* Jansen PLM, Sturm E. Genetic cholestasis, causes and consequences for hepatobiliary transport. *Liver International* 2003: 23: 315–322.

salts. Mutations at the MDR3results in impaired biliary excretion of phosphatidylcholine and cause PFIC3 a severe pediatric liver disease requiring liver transplantation for curative potential. Drugs like itraconazole inhibited the MDR3 mediated efflux of phophatidylcholine and not BSEP related canalicular secretion. Verapamil and cyclosporine are also two other drugs that act via inhibition of the MDR3 promoting phospholipid accumulation related cholestatic injury. The basolateral efflux proteins, MRP3 and MRP4 are up regulated in cholestatic conditions to compensate for impaired biliary excretion. This compensatory mechanism prevents hepatic bile acid accumulation and helps in renal elimination of bile acids. In the event of impaired functioning of MRP3 and MRP4, toxic bile acids accumulate in the hepatocytes. Troglitazone, a major BSEP inhibitor also inhibits MRP4 mediated dehydroepiandrosterone transport and produces potent inhibition at the canalicular and basolateral efflux of bile acids adding to the bile acid toxicity. Thus MRP4 is also a potent pathway of inhibition in drug indiced cholestasis. Uptake of bile acids are dependent on both basolateral and canalicular processes. Bosentan is a potent inhibitor of human BSEP and NTCP. It has been shown that bosentan is a more potent inhibitor of NTCP mediated sodium dependent TCA uptake, thereby proving that uptake as well as secretion is a closed control process and insults that affect both these processes simultaneously led to more toxic effects than either of them singly, due to the compensatory actions of the other.

The other factor that involves cholestasis is the bile acid concentration in the enterocytes. Bile acids activate the intestinal farnesoid X receptors leading to liberation of fibroblast growth factor 19 by the enterocytes. This migrates to the liver and activates the FGF receptor 4 leading to signal inhibition of hepatic bile acid synthesis. In conditions of ASBT deficiency or bile acid sequestration, the FXR activity is impaired and bile acid secretion is not inhibited at the hepatic level leading to more accumulation of toxic bile acids within hepatocytes. Multiple other nuclear receptors like the liver receptor homologue 1 (LRH1), the constitutive androstane receptor (CAR) and liver X receptor (LXR) also tightly controls the regulation of bile acid homeostasis and genetic aberration among these also culminates into cholestatic diseases. Drugs that act as nuclear receptor activators, like rifampicin and dexamethasone can increase the clearance of other drugs and form toxic metabolites that can cause liver injury. Anticholestatic compounds like UDCA, Phenobarbital and rifampicin are nuclear receptor agonists and are proposed treatments for cholestatic liver disease because they repress bile acid uptake and synthesis and promote bile acid excretion by activation of canalicular bile acid transporters. For diseases like PBC and PSC, the stimulation of canalicular bile acid excretion could worsen liver injury. In such cases, FXR antagonists work better in the sense, they act as a negative regulator of the basolateral bile acid transport through MRP4 protein whose increased expression normally protects the liver accumulation of toxic bile acids and increases renal bile acid secretion.

XI. ESTROGEN AND INFLAMMATION INDUCED CHOLESTASIS (FIGURE 57.12)

The causes for cholestasis secondary to changes in transport protein expression or localization in contrary to inhibition due to drug treatment or cholestatic agents are very few. The prototype of this example is Estrogen Related Cholestasis and the C17 alkylated steroid induced cholestasis. There agents

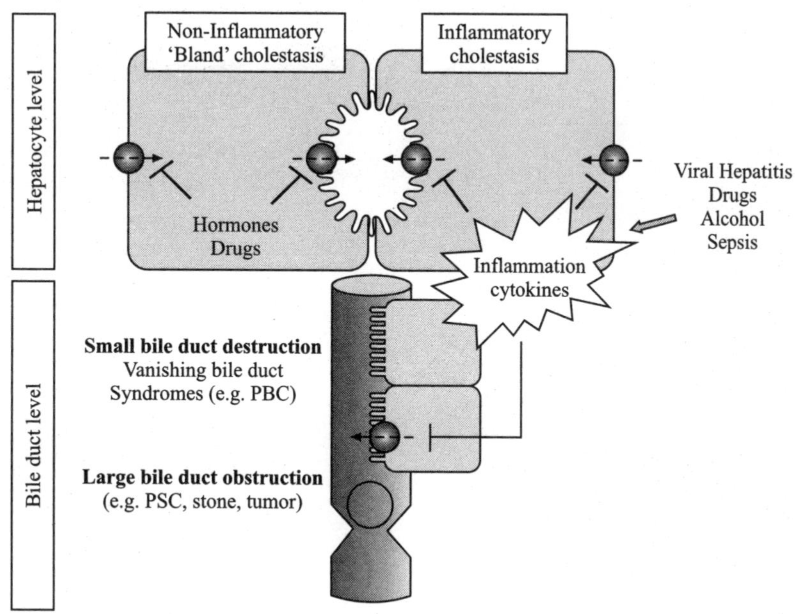

Figure 57.12 The inflammatory and noninflammatory mechanisms of cholestasis. *From: Mechanisms of cholestasis, Zollner and Trauner; Clin Liver Dis 12 (2008) 1–26.*

produce a clinical picture similar to intrahepatic cholestasis. Experimental evidence of this is seen in rodent models were 17 D glucuronide or the synthetic estrogen ethinylestradiol has been used. There occurs trans inhibition of BSEP mediated bile acid transport and internalization of BSEP and MRP2 that results in impairment of excretory function by redusing the amount of protein in the canalicular membrane, leading to cholestasis. The activation of classical calcium dependent protein kinase C and phosphoinositol 3 kinase signaling pathways result in internalization of BSEP and MRP2 (Figures 57.13a and b).

In patients with extrahepatic infections and inflammatory processes, the cytokines or the endotoxins lead to profound reductions in bile flow. The prototype of this inflammation induced cholestasis is sepsis associated cholestasis (Tables 57.2). The lipopolysaccharide endotoxin from the gram negative bacterial cell wall is a potent inflammatory agent. The endotoxins are cleared from the biliary system mainly by the Kupffer cells, by producing anti inflammatory cytokines and nitric oxide. The inflammatory cytokines reduce expression of the bile acid uptake proteins NTCP and OATP and decrease expression of the canalicular bile acid efflux pumps BSEP and MRP2 and downregulate the Phase I and II metabolic reactive enzymes thereby promoting cholestasis. Another key factor is the translational regulation resulting in reduced mRNA transcription and hence protein synthesis. This signal transduction occurs through phosphorylation or decreased

binding of nuclear transcription factors resulting in reduced amount and function of these nuclear receptors. The increase in severity of inflammation is associated with decreased hepatic mRNA expression of MRP2, MDR1 and OATP1B1. The severity factors do not affect the expression of MRP3, as it was demonstrated in rodent models of acute hepatitis.

TABLE 57.2

Mechanisms of Cholestasis of Sepsis
Decreased basolateral transport of bile acids
Inhibition of basolateral membrane Na-K-ATPase activity
Decreased basolateral membrane fluidity
Down-regulation of transporters
Decreased NTCP function
Decreased canalicular transport of bile acids
Down-regulation of transporters
Decreased BSEP function
Decreased MRP2 function

Liver Test Abnormalities in Sepsis
Conjugated hyperbilirubinemia: total bilirubin ranging from 2 to 10 mg/dL
Elevated alkaline phosphatase: rarely more than 2-3 times upper limit of normal

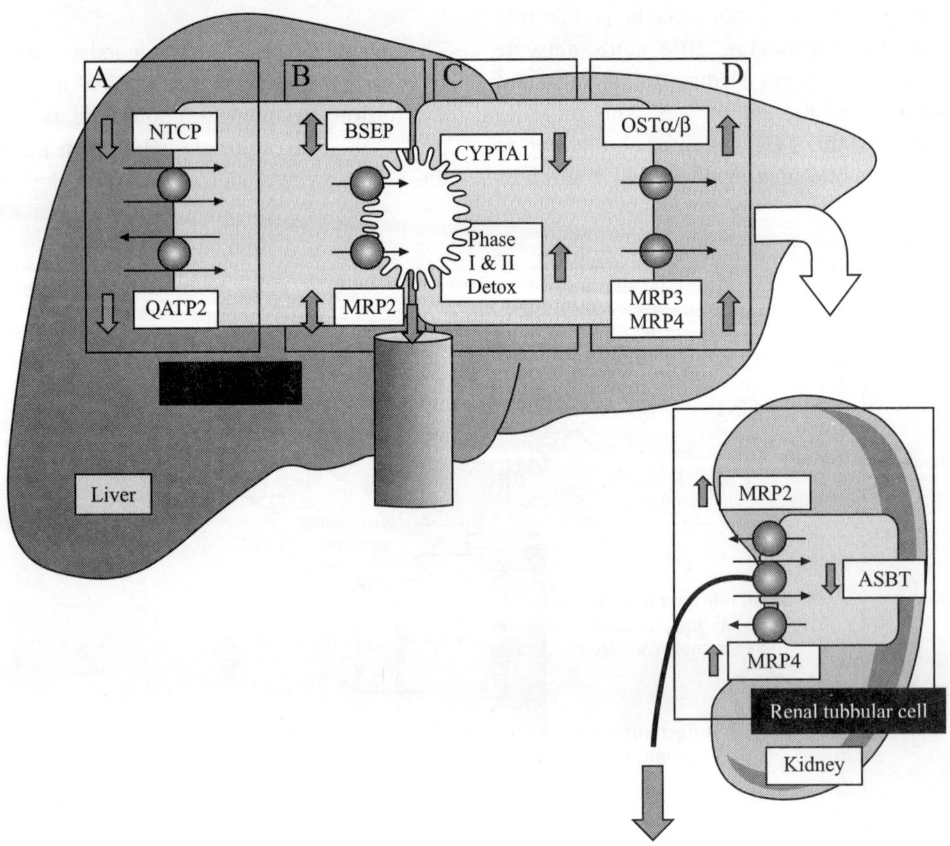

Figure 57.13a   Mechanisms of secondary biliary acid clearance in the event of cholestasis – liver related cholestatic injury leading to renal over expression of transport proteins. *From: Mechanisms of cholestasis, Zollner and Trauner; Clin Liver Dis 12 (2008) 1–26.*

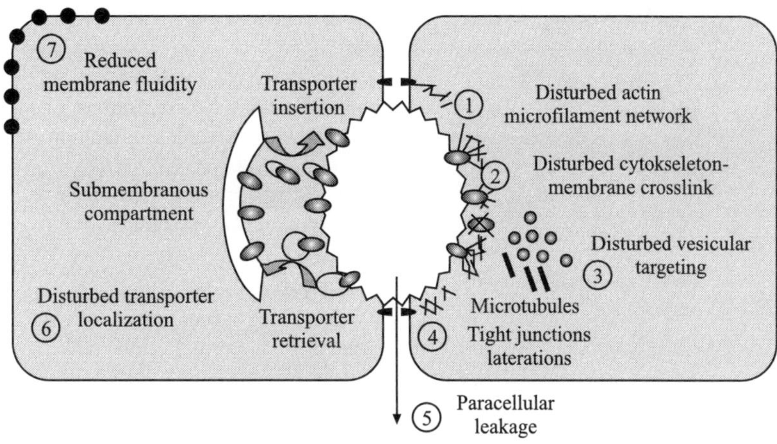

Figure 57.13b Cytoskeletal and other hepatocellular changes in cholestasis. Cholestasis is associated with profound alterations of the cytoskeleton of hepatocytes. *From: Mechanisms of cholestasis, Zollner and Trauner; Clin Liver Dis 12 (2008) 1–26.*

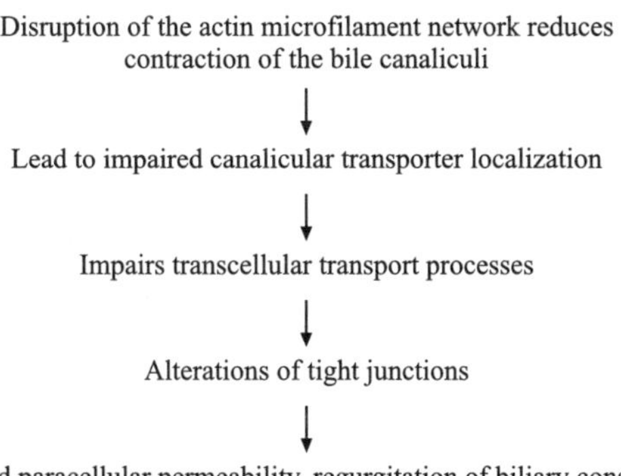

XII. HEREDITARY/GENETIC CHOLESTATIC DYSFUNCTION

A number of diseases can be caused by the defects of ABC transporter proteins and have varied presentations, even though the crux of the pathology is cholestasis. A brief review of these is shown in the Figure 57.14, Table 57.3.

The genetic diseases of transporter protein defects include progressive familial intrahepatic cholestasis (PFIC), benign recurrent intrahepatic cholestasis (BRIC), cholestasis of pregnancy, intrahepatolithiasis, cystic fibrosis, adrenoleukodystrophy and Dubin Johnsosn syndrome.

PFIC constitutes a group of autosomal recessive diseases which is characterized by cholestasis that begins in infancy. It is of three types, I, II and III. In PFIC type I or Byler disease, the cholestatic episodes leads to permanent cholestasis with fibrosis and eventually cirrhosis and liver failure within the first two decades of life. The children are small for age with recurrent episodes of diarrhea and pancreatitis. The larger bile ducts are usually normal and the liver histology shows features of bland canalicular cholestasis without much bile duct proliferation, inflammation or fibrosis. There is paucity of canalicular microvilli and a thickened pericanalicualr network of microfilaments. The coarse granular bile in the

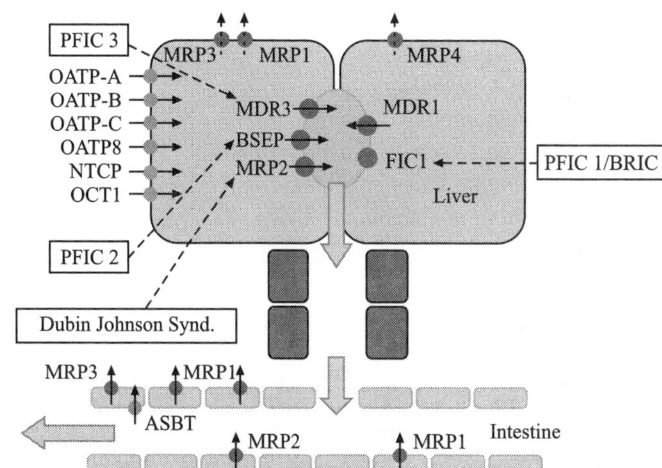

Figure 57.14 Human hepatobiliary transport proteins involved in bile formation, secretion and reabsorption. Transporter proteins located in the basolateral membrane are responsible for hepatic uptake of bile salts (NTCP, OATPs), bulky organic anions, uncharged compounds (OATPs) and cations (OATPs, OCT1). Transporter proteins located in the canalicular membrane are responsible for the biliary secretion of bile salts, phosphatidylcholine, cholesterol and glutathione and the excretion of drugs and toxins. *From: Jansen PLM, Sturm E. Genetic cholestasis, causes and consequences for hepatobiliary transport. Liver International 2003: 23: 315–322.*

canaliculi is called the 'Byler Bile'. The serum gamma glutamyl transpeptidase activity is not high and the primary bile salt levels, mostly the chenodeoxycholic acid levels are increased. Serum cholesterol is normal usually. Recurrent familial intrahepatic cholestasis (Summerskil Syndrome) is a benign disorder, that unlike PFIC1 does not progress to cirrhosis and liver failure. The serum gamma glutamyl transpeptidase is not elevated. The F1C1 gene locus on chromosome 18 is

TABLE 57.3 Diseases comprising of genetic cholestasis

Disease	Chromosome	Gene	Phenotype	Therapy
PFIC type 1	18q21	FIC 1(ATP8B1) P-type ATPase, acts as an aminophospholipid translocator	First recurrent, later permanent cholestasis, bile duct proliferation is a late phenomenon. Diarrhea, pancreatitis, pruritus, short stature. Coarse granular bile on EM. Normal gamma-Gl	Ursodeoxycholic acid, bile diversion, liver transplantation
Benign recurrent intrahepatic cholestasis	15q21	FIC1 (ATP8B1)	Recurrent episodes of cholestasis with severe pruritus, steatorrhea and weight loss. Normal gamma-Gl	Cholestyramine and/or ritampicine as symptomatic antipruitius therapy
PFIC type 2	2q24	BSEP (ABCB11), bile salt export pump	Neonatal hepatitis, progressive cholestasis, pruritus, short stature, bile duct proliferation is a late phenomenon, lobular and portal fibrosis BSEP protein absent. Amorphous bile on EM. Normal oarrtma-GT	Ursodeoxycholic acid bile diversion, liver transplantation
PFIC type 3	7q21	PGY3(ABCB4, MDR 3), P-glycoprotein 3	Cholestasis, portal hypertension, extensive bile duct proliferation and periportal fibrosis. MDR3 is not expressed. Elevated gamma GT	Ursodeoxycholic acid, liver transplantation
Intrahepatic cholestasis of pregnancy	e.g. 7q21	e.g. MDR3	Cholestasis in third trimester of pregnancy. High gamma-GT in case ot MDR3 defect; low gamma-GT cases may be caused by genetic defects of other transporter proteins. High incidence of fetal loss.	Ursodeoxycholic acid causes symptomatic relief in the mother and decreases fetal loss
Aagenaes stbdrome	15q	LCS1, LCS2	Episodic cholestasis, lymphedema, normal gamma-GT	Liver transplantation but persistence of lymphedema
Familial hypercholanemia	9q12-q13 9q22-q32	TJP2/20-2 BAAT	Elevated bile acids, severe pruritus, fat malabsorption, failure to thrive, rickets, vitamin K coagulopathy	Liver transplantation
Bile acid synthesis defects	e.g. 8q2.3	3β-$\Delta5$-C27- hydroxysteroid oxidorcductase: $\Delta4$-3-oxosteroid-5β reductase, 3β-hydroxy C27 steroid dehydrogenase/isomerase; oxysterol 7α-hydroxylase; 24-25-dihydroxy-cholanoic cleavage enzyme	Intrahepitic cholestasis, neonatal giant cell hepatitis. Normal or elevated gamma-GT, low or elevated serum total bile acids	Ursodeoxycholic acid chenodcoxychotic acid or cholic acid alone or in combination, depending on subtype

found to be involved in BRIC. PFIC and BRIC (types 1) are genetically related even though their expressions are phenotypically different.

PFIC type 2 is mapped to chromosome 2q24 and the affected gene is the ABCB11 (BSEP). Antibiodies have been demonstrated against BSEP in the canalicular domain of the hepatocyte plasma membrane and the liver specimens of PFIC type 2 patients do not stain for canalicular BSEP proteins in the presence of antibiodies. The serum gamma glutamyltranspeptidase is normal in this subset of patients also and bile duct proliferation is absent on histology. The PFIC type 2 disease starts as a non specific giant cell hepatitis which is difficult to be differentiated from neonatal giant cell hepatitis. The patients are frequently jaundiced and the disease rapidly progresses to persistent and progressive Cholestasis requiring

liver transplantation within the first decade, unlike the patients of PFIC type 1. The liver histology evaluation shows more features of inflammation with giant cell transformation and lobular and portal fibrosis. The bile in the patients of PFIC type 2 is amorphous or filamentous unlike Byler Bile.

PFIC type 3 presents later in life than in patients with PFIC type 1 and 2. The liver failure also occurs at a later age and jaundice is less apparent in the earlier stages of the disease. The serum levels of gamma glutamyltranspeptidase activity is markedly elevated and the liver histology shows extensive bile duct proliferation and portal and periportal fibrosis. The MDR3 gene (ABCB4) defect has been described to be associated with disease. MDR3 is a phospholipid flippase translocating phospholipids from the inner to the outer leaflet of the canalicular membrane where they can be extracted by bile acids. Phospholipids in bile are required for the formation of mixed micelles with bile acids and cholesterol to protect the bile duct epithelium from the detergent properties of bile acids. Reduced or absent phospholipid excretion into bile in PFIC-3 causes bile duct injury in these patients. These patients often respond to UDCA therapy and those patients in whom there is a complete defect, needs to be transplanted. Patients with mild mutations of the MDR3 gene respond to UDCA therapy while, for most other patients of PFIC, external biliary diversion is a procedure of choice or in end stage disease, transplantation is curative.

PFIC 4 is not a transporter defect disorder, but is one due to dysfunction of the bile acid synthetic pathway. Jacquemin et al reported the occurrence of patients in whom there is a defect in the 3-hydroxyl group steroid dehydrogenase activity. These patients had histologic features of cholestasis with mild fibrosis and serum analysis showed normal levels of 3-hydroxyl group steroids with normal gamma glutamyl transpeptidase activity and elevated transaminases. Clinically, there was no pruritis. Another example of transporter mutation is the biliary excretory function affection in Dubin Johnson syndrome. This syndrome is caused by the mutations of canalicular bilirubin export pump, MRP2/ABCC2 leading to reduced biliary excretion of endogenous and exogenous compounds. But these patients are non cholestatic, do not have pruritis and have normal serum bile acid levels. Mutations in sitosterol and cholesterol transporter ABCG5/G8 causing Sitosterolemia also do not cause cholestatic injury or pattern of symptoms.

XIII. INTRAHEPATIC CHOLESTASIS OF PREGNANCY (ICP)

ICP occurs in the third trimester and is characterized by pruritis and elevated serum bile acid levels which resolve after delivery. ICP can lead to increased incidence of fetal distress, premature birth and stillbirth. ICP patient are of two types, based on gammaglutamyltranspeptidase levels. One group have elevated gammaglutamyl transpeptidase levels (70%) while the other group has normal levels (30%). The intense cholestatic effects of estrogen leads to culmination of disease symptoms as seen in the third trimester, wherein these hormone levels are the highest. There is decrease in bile acid independent bile flow which is due to marked endocytic internalization of MRP2 from the canalicular membrane into the intracellular membranes leading to decrease in MRP2 transport activity and protein expression. There is also inhibition of the biliary secretion of glutathione leading to more toxic bile acid effects. The effect of this on the NTCP and to a lesser degree, on the anion transporting proteins, OATP reduces the membrane fluidity leading to impairment in the hepatobiliary transport of bile salts. All these models of injury and pathogenesis have been mostly studied in rats and need to be ascertained in humans. It has also been proposed that levels of sulphated progesterone metabolites increase in ICP and adds to the injury that is ongoing. These progesterone metabolites trans inhibit

BSEP. Patients prone to ICP are prone to cholesterol gallstone development and also to chronic liver disease in the future. Only a minority of cases of ICP are caused by transporter defects.

XIV. INBORN ERRORS OF BILE ACID METABOLISM

Bile acids (Figure 57.15) belong to the steroid group and are classified as acidic sterols. Principle bile acid formation in humans is in the liver. These bile acids have hydroxyl groups substituted in the nucleus at carbon positions C-3, C-7 and C-12. Bile acids are primary or secondary. The primary bile acids are conjugated to the amino acids glycine and taurine. Bile acids form the major pathway for removal of cholesterol from the body, provide an excretory route for endogenous and exogenous toxic substance removal and facilitate detergent action for absorption of fats and fat soluble vitamins from the intestine. The normal bile acid pool is 2 to 4 grams in an adult human. Daily less than 5% of the pool is lost in stool. A fraction of the hepatic bile acids is concerted to secondary bile acid, lithocholic and deoxycholic acids, wherein they undergo deconjugation, enterohepatc circulation and then reconjugation in the liver. Bile acids are also needed for promotion of bile flow normally. The disorders of bile acid synthesis can be primary or secondary.

The primary disorders include the enzyme defects such as

(a) Cholesterol 7α hydroxylase deficiency
(b) 3β hydroxyl C27 steroid oxidoreductase deficiency
(c) D4-3-oxosteroid 5β reductase deficiency
(d) Oxysterol 7α hydroxylase deficiency
(e) 27 hydroxylase deficiency (cerebrotendinous xanthomatosis)
(f) 2-methyl-acyl-CoA racemase deficiency
(g) Trihydroxycholestanoic acid CoA oxidase deficiency
(h) Deficiency of bile acid CoA ligase
(i) Side chain oxidation defect in 25 hydroxylation pathway for bile acids

The secondary bile acid synthesis disorders include those defects that impact primary bile acid synthesis like

(a) Cerebrohepatorenal syndrome of Zellewegers
(b) Smith Lemli Opitz syndrome due to deficiency of D7 desaturase

In bile acid synthesis disorders, there is markedly reduced or complete lack of cholic and chenodeoxycholic acids in the serum, bile and urine and elevated concentrations of atypical bile acids and sterols that lead to toxic effects. Mass spectrometric techniques are currently employed to detect these abnormal bile acids. In cerebrotendinous xanthomatosis, these presence of progressive neurologic dysfunction, dementia, ataxia, cataracts and xanthomata in the brain and tendons of affected infants along with neonatal cholestasis. There is

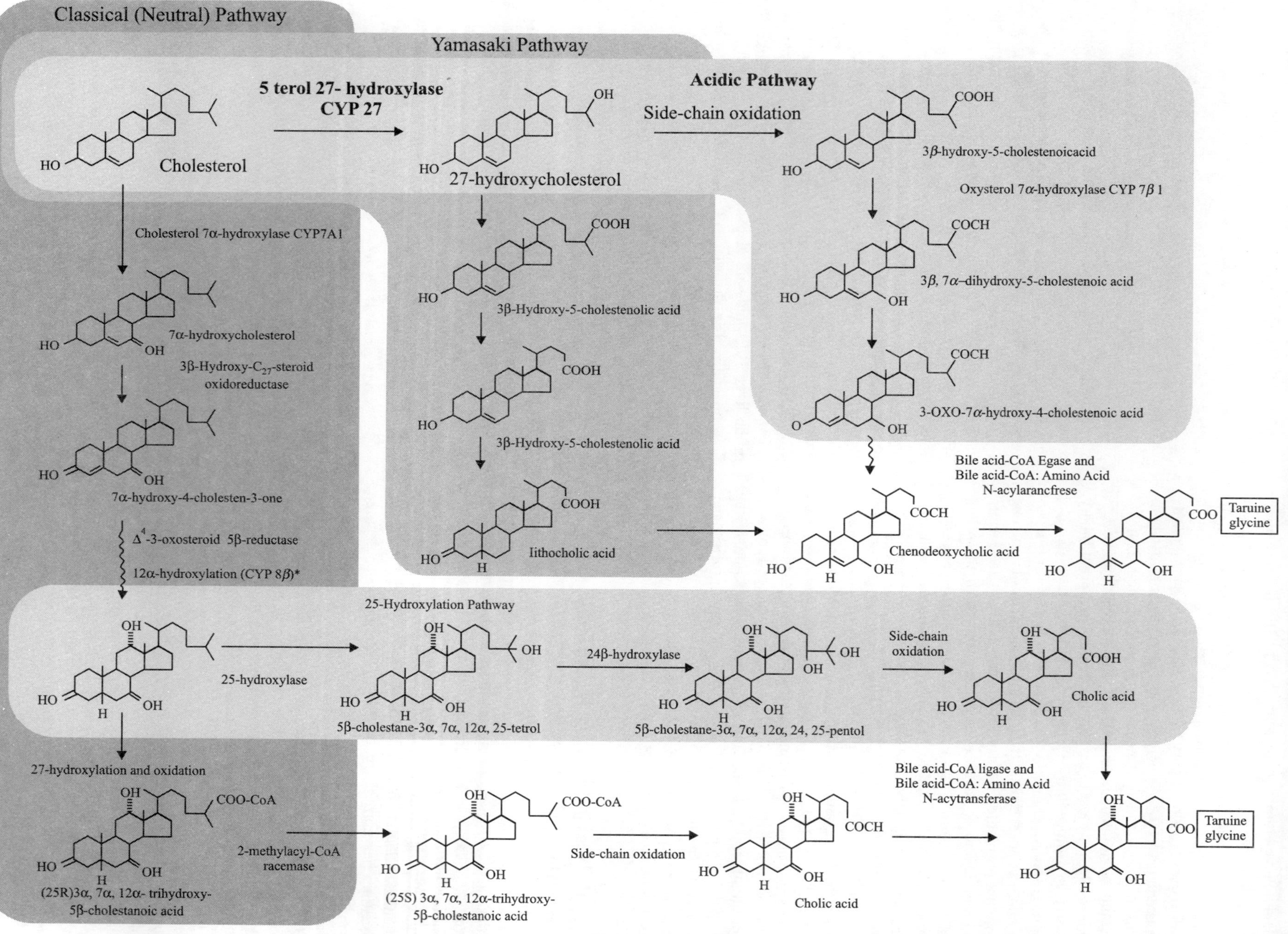

Figure 57.15 The bile acid metabolism pathway. *Modified from:* Kevin E Bove; *The Childhood Liver Disease Research Network.*

significantly reduced primary bile acids, elevation in biliary, urine and fecal excretion of alcohol glucoronides, low plasma cholesterol and elevations in cholestenol. Point mutations of the gene on chromosome 2 leading to inactivation of sterol 27 hydroxylase enzyme leads to this disorder. The presence of elevated cholestanol is a unique feature of this disease.

The first metabolic defect in bile acid synthesis to be described in the 3β-hydroxy-C27-steroid oxidoreductase deficiency. This is the most common among the bile acid synthesis disorders. Most of the neonates afflicted by this disease present with elevated serum transaminases, conjugated hyperbilirubinemia and normal serum gamma glutamyltranspeptidase levels. They have hepatomegaly with our without splenomegaly, steatorrhea, vitamin malabsorption and absence of pruritis. Giant cell cholesatic picture is seen on liver histology. Definitive diagnosis depends upon mass spectrometric studies and electrospray techniques. Molecular technical evolution has led to the discovery of the gene related to this disease – HSD3B7 encoding the specific enzyme that is absent in these group of patients. In patients with D4-3-oxosteroid 5β-reductase deficiency, the clinical picture is similar to that seen in 3β-hydroxy-C27-steroid oxidoreductase deficiency, but with the exception that gamma glutamyl transpeptidase levels are elevated in the former. The disease presentation is much more severe with rapid progression to cirrhosis and liver failure. The liver histology in these patients reveal the presence of acinar disarray with giant cell and pseudoacinar transformation and hepatocellular and canalicular bile stasis and extramedullary hematopoesis. The bile canaliculi are small and slit like with absent microvilli. Oxysterol 7α-hydroxylase deficiency has been described in only one infant and presented with progressive liver failure which worsened with oral UDCA therapy. Unlike the other two enzyme deficiencies described above, this particular disorder does not respond t primary bile acid therapy and is more aggressive in its clinical course. 2-methylacyl-coa racemase deficiency lead to both bile acid and fatty acid pathway defects. These patients can present in adulthood or in infancy and mostly with symptoms of sensory motor neuropathy, fat soluble vitamin deficiency and cholestasis. The mass spectrometric analysis in these patients reveal similar features seen in peroxisome related disorders. The primary bile cholic acid therapy has been found to beeffective in normalizing liver enzymes and addition of phytanic and pristanic acids in the long term is found to be useful in this diease group. THCA-CoA oxidase deficiency primarily presents with neurological symptoms and less of cholestatic symptoms. Ataxia is the main clinical feature of this syndrome and the disease onset occurring at a median age of 3 years. Patients of bile acid CoA ligase deficiency and defective amidation present mostly with fat and fat soluble vitamin malabsorption along with neonatal cholestasis. Side-chain oxidation defect in the alternate 25-hydroxylation pathway presents with severe intrahepatic cholestasis and giant cell hepatitis. These patients have high levels of alcohol glucoronides in the bile, serum and urine and reduced primary bile acids in the serum and bile,

similar to the ones seen in cerebrotendinous xanthomatosis. The patients with cholesterol 7α-hydroxylase deficiency have mutations in the CYP7A1 gene leading to abnormal serum lipids without liver dysfunction. The clinical phenotype is of markedly elevated total and low density lipoprotein cholesterol level and premature gall stones with peripheral arterial and coronary heart disease. The elevated lipids are unresponsive to therapy with HMG-CoA reductase inhibitor (Statins).

XV. BILIARY SYSTEM DYSFUNCTION: THE NEOPLASMS

A. Gall Bladder Carcinoma (GBC)

Gall bladder cancer is seen among women more than men and is most frequently found in North India, Chile, Pakistan and Eucador. The various risk factors for GBC include gall stone disease, obesity and metabolic syndrome, environmental toxins and chemical pollutants, female sex, biliary infections – chronic (especially Salmonella) and recurrent and anatomical aberrations like porcelain gall bladder and anomalous pancreaticobiliary communications.

GBC histogenesis can be explained on the basis of two models (Figure 57.16)—the metaplasia dysplasia sequence and the adenoma-dysplasia sequence as delineated by Roa et al. shown in Figure.

In the first model, exophytic lesions are not formed and the precursor lesions are detected mostly incidentally during microscopic evaluation of gall bladder specimens taken post cholecystectomy for gall stone disease/cholecystitis. The main cause for such incidental lesions are chronic inflammatory processes. Pyloric type and intestinal type of metaplasia are detected usually, of which intestinal metaplasia is more sinister. Metaplsia progresses to dysplasia and then to invasive carcinoma. The adenoma to dysplasia and carcinoma sequence was described by Kozuka et al.; they noted that the progression to carcinoma correlated when the size of the adenoma was more than 12 mm leading to carcinomatous transformation and in those with >30 mm, leading to invasive carcinoma. Median age of patients with invasive carcinoma was around 60 years. The most frequent histological type was that of adenocarcinoma if which, more than two thirds were of poorly differentiated type. The molecular and genetic alterations that occur during the malignant transformation is through a complex interaction between oncogenes, tumor suppressor genes and DNA repair genes. These genetic alterations occur through loss of heterozygosity (LOH), mutation and methylation. Microsatellite instability and DNA repair inactivation also form part of this transformation. K-ras aberration is one of the most frequently seen genetic modifications in neoplasms of glandular origin. These mutations occur most commonly at codon 12 and correspond to transition of guanine to adenine and substitution of aspartic acid by glycine. In patient's wit anatomic aberrancy, like that in anomalous pancreaticobiliary union, the frequency of K-ras mutations is much higher. K-ras

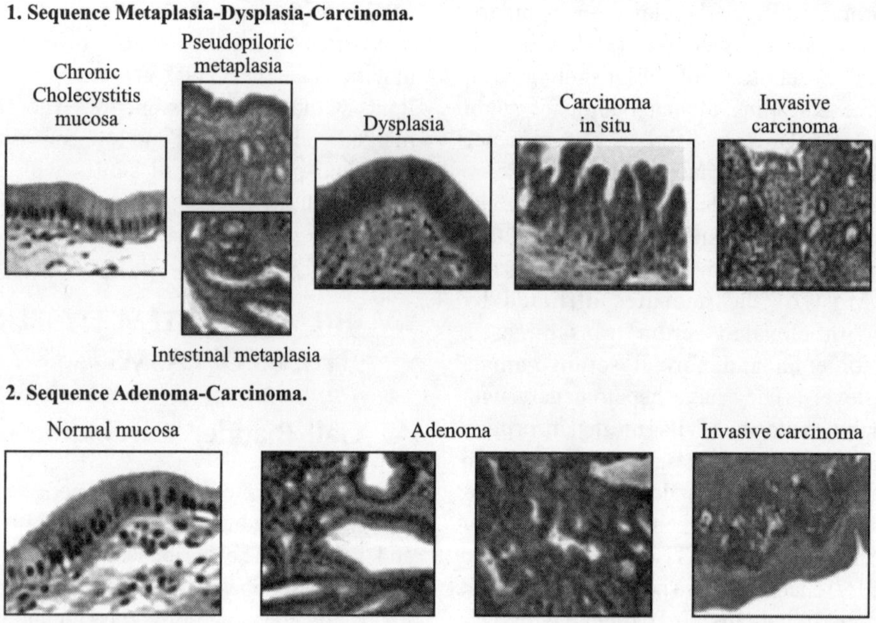

Figure 57.16 Pathogenesis of gall bladder cancer. From : Carlos J, Roa JC. Preneoplastic lesions of a gallbladder from morphological and molecular points of view. In: Litchfield JE, editor. *New Research on Precancerous Conditions.* New York: Nova Science Publishers; 2006. (*see Plate 22 for colour figure*)

normally takes art in signalling pathways and has GTPase activity for cell growth and differentiation.

The CDKN1A or the p21 cyclin dependent kinase inhibitor is seen in about 33% of GBC. p21 expresses its carcinogenic potential through interaction with other pathways like the p27 and p53. C-erbB2 oncogene on chromosome 17 codes for tyrosine kinase in the plasma membrane. The amplification/overexpression of this gene is associated with a worse prognosis and aggressiveness of GBC. Cell cycle progression promotion by cyclinD1 and cyclinE plays a major role in progression of precursor lesions to carcinomatous transformation. LOH of p53 tumour suppressor gene has been shown to promote carcinogenesis in GBC. p53 has a role in maintaining the integrity of the genome and regulated cell cycle by DNA synthesis and repair, promoting healthy cell differentiation and apoptosis. Point mutations—transitions and transversions, most commonly in exons 5–8 have been seen in p53 aberrant GBC. The p16 or CDKN2A tumor suppressor gene located at chromosome 9 dysregulation by means of homozygous deletion, hypermethylation of promoter and mutations result in carcinogenesis. Mutations of this gene are most commonly seen in GBC. The progression of cells from G1 to S phase occurs via phosphorylation of the Rb gene protein product by cyclin dependent kinase complexes. This functional loss results in carcinogenicity. The fragile histidine triad gene on chromosome 3 has been identified as part of carcinogenesis in GBC. Microsatellite instability (MSI) are short repetitive sequences of nucleotide bases that are seen to span the entire genome especially in areas of spontaneous mutations. The DNA mismatch repair system consists of main genetic components, the MLH1, MSH2, PMS2 and MSH6. When there is a deactivation of these genes by way of methylation

or mutation, replication errors, as shown by presence of MSI occur. The majority of GBC transformation (even though small in this setting of cancer) occurs due to hypermethylation of MLH1 region leading to its downregulation and unchecked proliferation potential. Cadherin-catenin complex is a major part of the cell adhesion system. The adenomatous transformation into carcinomatous lesions seen in GBC is commonly linked to altered beta catenin expression wherein, exon 3 of this gene complex is affected. Cadherins maintain normal architecture of the epithelium by forming homotpic unions among the mucosal cells. Among the cadherins, E cadherin is the most studied. In GBC, loss of E-cadherin has been found to be a common association. The other adhesion molecules whose aberrant expression result in GB carcinogenesis include ICAM-1/CD54, CD44 located on chromosome 11 the expression of which is prognostic for metastatic spread, CD99 or Mic2 a transmembrane glycoprotein whose overexpression is seen in sarcomas, but loss of expression is associated with GBC. Among genetic polymorphisms and susceptibility, many groups have found that by using CA242 and CA125 together as susceptibility biomarkers, discrimination of GBC from cholelithiasis was more apparent. Heparanase and HIF-1 activity are also frequently increased in GBC and is associated with a poor survival. Chronic inflammation in the gall bladder can activate oncogenes and promote carcinogenesis in GB. Mitochondrial DNA mutations in which D310 mutation at mitochondrial DNA displacement loop is a relatively frequent and early event in sequential pathogenesis of GBC secondary to chronic inflammatory processes. Nitric oxide (NO) has a double effect in GBC. It has been found that NO interferes with energy metabolism in tumor cells, inhibits protein synthesis, damages DNA directly or indirectly and induces apoptosis in tumour cells. The iNOS fraction of NO can also induce unchecked cellular proliferation, when they are formed in chronically inflamed cells. This type of iNOS can damage the cell, break DNA and cause mutagenesis thereby promoting oncogenesis and angiogenesis in the affected cells. In times of chronic inflammation, the cytokines promote the activation of cyclooxigenase-2 (COX-2) enzyme which then promotes the overexpression of prostaglandins, which are arachidonic acid derivatives that mediate and promote tumorigenesis by altering cell proliferation, differentialtion and adherence and by modification of vascular response. The induction of COX-2 is an important event in formation of GBC, through formation of low grade dysplasia to high grade dysplasia. Vascular endothelial growth factor (VEGF) induces a mitogenic effect on endothelial cells when there is underlying ischemia or inflammation. This causes vascular permeability to increase and leads to tumoral neoangiogenesis. This also has a role in hematogenous or lymphomatous spread of tumoral dissemination, especially that of GBC. Last but not the least, the human telomere or hTERT activity, wherein telomerase enzyme catalyzes the addition of telomere repeats (TTAGGG) to telomeres occur in neoplastic cells (as well as normally in germline cells) in an aberrant fashion. The human telomerase has three sub units—the RNA component known as hTR, associated protein TEP1 and catalytic subunit hTERT (a reverse transcriptase). The ectopic expression of hTERT fraction is excessively found in GBC. These complex molecular mechanisms promote biliary dysfunction in the GB epithelium, promoting neoplastic transformation.

B. Cholangiocarcinoma

Cholangiocarcinoma (CC) arises from the biliary epithelial cells and are divided into intrahepatic and extrahepatic types. CC has poor prognosis mainly due to the fact that early diagnosis is a difficult process in these cases. The lack of specific symptoms and the absence of tools for early diagnosis make this set of neoplasms very morbid and with a high mortality rate due to unresectability at diagnosis. The development of CC is attributed to the accumulation of epigenetic alterations in regulatory genes in biliary epithelial cells that lead to activation of oncogenes and dysregulation of tumour suppressor genes. The malignant cholangiocytes have uncontrolled growth with high capacity for tissue invasiveness and metastases. Cholangiocarcinogenesis develops in thr background of chronic inflammation of bile ducts and cholangiocyte damage associated with biliary obstruction. They can develop in the normal or cirrhotic liver, the former especially in the perihilar regions. Primary sclerosing cholangitis is one of the most solid documentation as a risk factor for CC. The other risk factors include biliary parasitic infestations like Opthisthorchis viverrini, Clonorchis sinensis and Schistosoma japonica; hepatolithiasis, Caroli's disease, congenital choledochal cysts, bilioenteric surgeries, anomalous pancreatocibiliary union, age more than 65 years, bile duct adenomas, papillomatosis, cirrhosis of the liver, smoking, diabetes and chemical exposure to thorotrast, vinyl chloride and

HIV and hepatitis C virus infections are also documented. The chronic inflammatory cytokine presence triggers and maintains the process of cholangiocarcinogenesis. On this background pf chronic inflammation, there is promotion of neoplastic processes brought about by protooncogene dysregulation, DNA mismatch and repair gene processes, deactivation of tumor suppressor genes, neoangiogenesis and apoptosis, cell growth proliferation and invasiveness. Like in other gastrointestinal malignancies. K-ras, p53, p14ARF, p16INK4a and beta catenin gene mutations play a role in development of CC. NKG2D, the natural killer group 2, member D cell receptor expressed by natural killer cells and T lymphocytes plays an important role in tumour surveillance through cell mediated cytotoxicity. Single nucleotide polymorphisms of the NKG2D gene is associated with risk of CC in patients of primary sclerosing cholangitis. Another gene product, the activation induced cytidine deaminase (AID) a member of the DNA/RNA editing enzyme family is associated with increase in CC in primary sclerosing cholangitis patients. Dysregulated expression of the AID gene led to somatic mutations in tumour related genes like p53, c-myc and INK4A/p16 sequence.

IL-6 leads an important aspect in pathogenesis of CC. IL-6 has a mitogenic effect which increases in conditions where chronic inflammation predominates. IL-6 acts through autocrine and paracrine mechanisms through aberrant stimulation of intracellular pathways involved in survival and growth of malignant cholangiocytes, like the p44/p42 and p38 MAP kinases. This stimulation decreases the expression of p21, a cell cycle controller protein and leads to growth independent of normal signalling. IL-6 also upregulates myeloid leukemia cell- or Mcl-1 a major component of antiapoptotic Bcl-2 family of proteins, through increase in activation of STAT-3, promoting resistance to apoptosis. IL-6 increases malignant cholangiocyte resistance to tumour necrosis factor related apoptosis inducing ligand (TRAIL).

The Janus kinase/signal transducer and activator or transcription (JAK/STAT) pathway is a key signalling mechanism in the development of CC. This is the fundamental behind the use of Sorafenib, a multikinase inhibitor that induces STAT3 dephosphorylation by stimulating phosphataseactivity; thereby promoting TRAIL mediated apoptosis and subsequent cancer cell death. Transforming growth factor- beta (TGF-beta) is expressed by cholangiocytes during cholestatic processes. TGF-beta inhibits proliferation of human CC cells by way of modulation of p21 cyclin dependent kinase inhibitor. Mutations in TGF-beta produce an inhibitory effect of TGF-beta effect on the malignant cells by altering the intracellular signalling mediators together with intracellular overexpression of cyclin D1. The inhibition also leads to deposition of fibrotic tissue, expressed in abundance by biliary malignant cells. DP4/Smad is a tumour suppressor gene and comes downstream in TGF-beta signalling. Smad4 along with another tumor suppressor gene, PTEN inhibits proliferative activity of CC. It has also been shown that the degree of Smad4 loss correlated well with the TNM staging of CC. ErbB-2 is a protooncogene receptor tyrosine kinase that is over expressed in biliary malignancies. It stimulates the proliferation of CC cells and also overproduces COX-2 which interacts with IL-6 leading to mitogenic effects. Cyclooxygenase is responsible for generation of prostaglandins, in the wake of inflammation. Cyclooxygenase type 1 is normally present in cells and cyclooxygenase 2 is the inducible isoform. The inducible form is stimulated by lipopolysaccharides, cytokines and other inflammatory molecules leading to its excessive generation of prostaglandins, forming increment in malignant transformation of biliary epithelial cells into CC. Selective cyclooxygenase-2 inhibitors prevent CC progression by means of inducing apoptosis. But this inhibitory effect and apoptosis induction of cyclooxygenase 2 antagonists comes only at high doses, with a price of extreme side effects and is clinically not feasible. Cyclooxygenase 2 inhibition also activates cyclin dependent kinase inhibitors which acts to arrest the cell cycle at G1/S phase. Inducible nitric oxide synthetase (iNOS) is expressed heavily in the background of inflammation. The production of nitric oxide leads to inhibition of DNA repair, promotes DNA damage accumulation and hence mutations, stimulates cyclooxygenase 2 expression and modulates. This carcinogenic effect is through Notch-1 signalling which is constitutively hyperexpressed in premalignant conditions. Bcl-2 belong to the superfamily of antiapoptotic proteins. Its overexpression in malignant cells of CC demonstrates the carcinogenic action of this protein. This anti apoptotic activity is via reduction of caspase 3 activation, preventing

cytochrome c release from mitochondria. As described above, the TRAIL protein indices apoptosis only in malignant cells without having effect on normal cells. CC cells are resistant to TRAIL because of the presence of myeloid cell leukemia protein-1 (Mcl-1) expression in them. Small interfering mRNA or stable transfection of Mcl-1 results in increased sensitivity of tumour cells to TRAIL dependent apoptosis, a potent therapeutic model. Deoxycholic acid, in the event of bile stasis increases Mcl-1 expression by blocking the protein degradation through activation of EGFR/Raf-1 pathway. The tumoral cells proliferate in abundance around rich supply of vasculature. The bproliferation of tumoral blood vessels enhanced in the presence of over expression of VEGF or the vascularendothelial growth factor. This protein is activated by TGF-beta or beta-catenin. 17 beta estradiol stimulates CC cell growth. Stimulation of estrogen receptors on tumoral cells also activates VEGF in abundance leading to more proliferation. Blockade of estrogen/estrogen receptors through Fas/APO-1 (CD95) signalling induction results in decrement in tumor growth and proliferation. Since biliary malignancies express insulin like growth factor in abundance, their measurement can help differentiate between extrahepatic biliary tumors and carcinoma of the pancreas. Thus, the emergence of CC is a complex process in which multiple proteins, genetic modulation and environmental factors come together to promote cholangiocarcinogenesis from normal biliary epithelial cells in the wake of chronic inflammation.

REFERENCES

1. Baillie J. (2010), Sphincter of Oddi Dysfunction, *Curr. Gastroenterol Rep.*, 12(2):130–4.

2. Chuang S.C., Hsi E. and Lee K.T. (2013), Genetics of Gallstone Disease, *Adv Clin Chem.*, 60:143–85

3. Chuang S.C., Hsi E. and Lee K.T. (2012), Mucin Genes in Gallstone Disease, *Clin Chim Acta.*, 413(19–20):1466–71.

4. Sookoian S. and Pirola C.J. (2012), Genetic Determinants of Acquired Cholestasis: A Systems Biology Approach, Front Biosci Landmark (Ed.), 17:206–20.

5. Pauli-Magnus C., Meier P.J. and Stieger B. (2010), Genetic Determinants of Drug-induced Cholestasis and Intrahepatic Cholestasis of Pregnancy, *Semin Liver Dis.*, 30(2):147–59.

6. Pauli-Magnus C. and Meier P.J. (2005), Hepatocellular Transporters and Cholestasis, *J. Clin Gastroenterol.*, 39(4 Suppl 2):S103–10.

7. Jansen P.L., Müller M. and Kuipers F. (2000), Bile Formation and Cholestasis, *Ned Tijdschr Geneeskd.*, 144(50):2384–91.

8. Le Bail B. (2012), Pathology: A Pictorial Review, A Selected Atlas of Paediatric Liver Pathology, *Clin Res Hepatol Gastroenterol.*, 36(3): 248–52.

9. Akita H., Suzuki H., Ito K., Kinoshita S., Sato N., Takikawa H. and Sugiyama Y. (2001), Characterization of Bile Acid Transport Mediated by Multidrug Resistance Associated Protein 2 and Bile Salt Export Pump, *Biochim Biophys Acta*, 1511(1):7–16.

10. Kullak-Ublick G.A. and Meier P.J. (2000), Mechanisms of Cholestasis, *Clin Liver Dis.*, 4(2):357–85.

11. Park S.M. (2012), The Crucial Role of Cholangiocytes in Cholangiopathies, *Gut Liver*, 6(3):295–304.

12. Priester S., Wise C. and Glaser S.S. (2010), Involvement of Cholangiocyte Proliferation in Biliary Fibrosis. *World J. Gastrointest Pathophysiol.*, 1(2):30–7.

13. Nicolaou M., Andress E.J., Zolnerciks J.K., Dixon P.H., Williamson C. and Linton K.J. (2012), Canalicular ABC Transporters and Liver Disease, *J. Pathol.*, 226(2):300–15.

14. Kubitz R., Dröge C., Stindt J., Weissenberger K. and Häussinger D. (2012), The Bile Salt Export Pump (BSEP) in Health and Disease, *Clin Res Hepatol Gastroenterol.*, 36(6):536–53.

15. Hirschfield G.M. (2013), Genetic Determinants of Cholestasis. *Clin Liver Dis.*, 17(2):147–59.

16. Marin J.J. (2008), How We have Learned about the Complexity of Physiology, Pathobiology and Pharmacology of Bile Acids and Biliary Secretion, *World J. Gastroenterol.*, 14(37):5617–9

17. Igarashi S., Sato Y., Ren X.S., Harada K., Sasaki M. and Nakanuma Y. (2013), Participation of Peribiliary Glands in Biliary Tract Pathophysiologies, *World J. Hepatol.*, 5(8):425–32.

18. Nakanuma Y., Hoso M., Sanzen T. and Sasaki M. (1997), Microstructure and Development of the Normal and Pathologic Biliary Tract in Humans, Including Blood Supply, *Microsc Res Tech.*, 38(6):552–70.

19. Stieger B. (2011), The Role of the Sodium-taurocholate Cotransporting Polypeptide (NTCP) and of the Bile Salt Export Pump (BSEP) in Physiology and Pathophysiology of Bile Formation, *Handb Exp Pharmacol.*, (201):205–59.

20. Alpini G., McGill J.M. and Larusso N.F. (2002), The Pathobiology of Biliary Epithelia. Hepatology, 35(5):1256–68.

21. Alvaro D., Mancino M.G., Glaser S., Gaudio E., Marzioni M., Francis H. and Alpini G. (2007), Proliferating Cholangiocytes: A Neuroendocrine Compartment in the Diseased Liver, *Gastroenterology*, 132(1):415–31.

22. Elferink R.P. and Groen A.K. (1999), The Mechanism of Biliary Lipid Secretion and its Defects, *Gastroenterol Clin North Am.*, 28(1):59–74.

23. Adams D.H. and Afford S.C. (2002), The Role of Cholangiocytes in the Development of Chronic Inflammatory Liver Disease, *Front Biosci.*, 7:e276–85.

24. Strazzabosco M., Spirlí C. and Okolicsanyi L. (2000), Pathophysiology of the Intrahepatic Biliary Epithelium, *J. Gastroenterol Hepatol.*, 15(3):244–53.

25. Danzer C. and Mattner J. (2013), Impact of Microbes on Autoimmune Diseases, *Arch Immunol Ther Exp.* (Warsz). 61(3):175–86.

26. Adams D.B. (2013), Biliary Dyskinesia: Does It Exist? If So, How Do We Diagnose It? Is Laparoscopic Cholecystectomy Effective or a Sham Operation? *J. Gastrointest Surg.*, 17(9):1550–2.

27. Desmet V.J. (1998), Ludwig Symposium on Biliary Disorders—part I, Pathogenesis of Ductal Plate Abnormalities, *Mayo Clin Proc.*, 73(1):80–9

28. Erlinger S. (2011), Ductal Plate Malformations: A Morphogenetic Classification Based on Genetic Defects, *Clin Res Hepatol Gastroenterol.*, 35(10):604–6.

29. Hirschfield G.M., Heathcote E.J. and Gershwin M.E. (2010), Pathogenesis of Cholestatic Liver Disease and Therapeutic Approaches, *Gastroenterology*, 139(5):1481–96.

30. Nakanuma Y., Tsuneyama K. and Harada K. (2001), Pathology and Pathogenesis of Intrahepatic Bile Duct Loss, *J. Hepatobiliary Pancreat Surg.*, 8(4):303–15.

31. Omenetti A., Bass L.M., Anders R.A., Clemente M.G., Francis H., Guy C.D., McCall S., Choi S.S., Alpini G., Schwarz K.B., Diehl A.M. and Whitington P.F. (2011), Hedgehog Activity, Epithelial-mesenchymal Transitions and Biliary Dysmorphogenesis in Biliary Atresia, *Hepatology*, 53(4):1246–58.

32. Alvaro D., Mancino M.G., Onori P., Franchitto A., Alpini G., Francis H., Glaser S. and Gaudio E. (2006), Estrogens and the Pathophysiology of the Biliary Tree, *World J. Gastroenterol*, 12(22):3537–45

33. Kock K., Ferslew B.C., Netterberg I., Yang K., Urban T.J., Swaan P.W., Stewart P.W. and Brouwer K.L. (2013), Risk Factors for Development of Cholestatic Drug-induced Liver Injury: Inhibition of Hepatic Basolateral Bile Acid Transporters MRP3 and MRP4, *Drug Metab Dispos.*

34. Strubbe B., Geerts A., Van Vlierberghe H. and Colle I. (2012), Progressive Familial Intrahepatic Cholestasis and Benign Recurrent Intrahepatic Cholestasis: A Review, *Acta Gastroenterol Belg.*, 75(4):405–10.

35. Munoz-Garrido P., García-Fernández de Barrena M., Hijona E., Carracedo M., Marín J.J., Bujanda L. and Banales J.M. (2012), MicroRNAs in Biliary Diseases, *World J. Gastroenterol.*, 18(43):6189–96.

36. Patel T. (2013), New Insights into the Molecular Pathogenesis of Intrahepatic Cholangiocarcinoma, *J. Gastroenterol.*, [Epub Ahead of Print].

37. Li Y., Zhang J. and Ma H. (2013), Chronic Inflammation and Gallbladder Cancer, *Cancer Lett.*

38. Ehlken H. and Schramm C. (2013), Primary Sclerosing Cholangitis and Cholangiocarcinoma: Pathogenesis and Modes of Diagnostics, *Dig. Dis.*, 31(1):118–25.

FURTHER READING

1. Monga, P.S. Satdarshan (2011), Molecular Pathology of Liver Diseases, Springer Publications.

2. Irwin M. Arias (2009), The Liver: Biology and Pathobiology, John Wiley and Sons Limited.

3. Dufour J.F., Pierre Alain Clavien et al., Springer Publications.

4. Lodish H. et al., Molecular Cell Biology, W.H. Freeman Publishing.

5. Geoffrey M. Cooper and Robert E. Hausmann (2013), The Cell: A Molecular Approach, Sinauer Asssociates Inc.

58

Hepatitis Viruses and Viral Hepatitis

Shahid Jameel and Rakesh Aggarwal

CONTENTS

Viral hepatitis is an important global cause of morbidity and mortality. Five different hepatotropic viruses, called hepatitis viruses A to E, which predominantly affect the liver, are responsible for the disease. Besides these, the liver can also be infected with viruses such as cytomegalovirus, dengue virus, Epstein-Barr virus, and others that affect several organs in a systemic manner. Despite similarities in the disease they cause, hepatitis viruses A to E differ widely in their biology, taxonomy and modes of replication. These viruses infect their susceptible human (and in some cases animal) hosts by either the enteral or parenteral route. The two enterally transmitted hepatitis viruses, i.e., hepatitis A virus (HAV) and hepatitis E virus (HEV) are generally endemic to tropical areas where they are feco-orally transmitted, with contaminated drinking water and improperly cooked food being the primary sources of infection. The three parenterally transmitted viruses, i.e., hepatitis B virus (HBV), hepatitis C virus (HCV) and hepatitis D virus (HDV) have a global presence, with contaminated needles, untested (or improperly tested) blood, unprotected sex and materno-fetal transmission being the major causes of infection. While all hepatitis viruses cause an acute infection followed by disease of varying severity, the enteral viruses generally persist in an immunocompetent host for just a few weeks, but the parenteral viruses can lead to significant rates of chronic infection and long-term sequelae. The estimated global burden of hepatitis viruses is summarized in Table 58.1.

This chapter focuses on the causative agent, epidemiology, pathogenesis, disease, diagnosis, prevention and treatment aspects of each of the five types of viral hepatitis. Table 58.2 provides a comparison of the salient differences between these viruses and their epidemiological and clinical features.

TABLE 58.1 Estimated global burden of the hepatitis viruses

Virus	Number infected	Number chronic infections
Hepatitis A	5 billion	0
Hepatitis B	4 billion	350 million
Hepatitis C	200 million	170 million
Hepatitis D	17 million	15 million
Hepatitis E	2 billion	0

Table 58.3 provides a summary of serological investigations that are used to determine the causative agent for viral hepatitis.

I. HEPATITIS A

Though hepatitis A virus (HAV) infection is distributed globally, its epidemiological characteristics vary with socio-economic development. In developing countries with poor sanitation and living conditions, transmission rates are high and most infections occur in early childhood. Since the primary infection confers strong life-long protection, infection and disease among adults is uncommon in these regions. In contrast, in developed countries, viral transmission during childhood is much less frequent, and the cases arise from travel to endemic areas, person-to-person spread or common-source contamination of food. Humans are the only host, and hence the only source of HAV.

A. Virology

Hepatitis A virus belongs to the *Picornaviridae* family, to which also belong other important human pathogens such as polioviruses and the common cold rhinoviruses. It is stable at low pH and at 60°C for up to one hour, but is destroyed by autoclaving, boiling for at least 5 minutes, ultraviolet radiation, formalin, β-propiolactone, iodine or chlorinated compounds.

The virus particle of 27–34 nm is made up of four proteins and a single-stranded, positive sense RNA of ~7.5 kb (Figure 58.1). The RNA genome is linear and includes, starting from its 5′ end—a noncoding region (NCR) of about 750 nucleotides, linked at its 5′ end to the VPg viral protein; an open reading frame that encodes the viral proteins; and a short 3′ NCR terminating in a poly(A) tract. The open reading frame is divided into 3 regions, of which P1 encodes the four viral structural proteins VP1 to VP4, and P2 and P3 encode the nonstructural proteins with biochemical activities that support viral genome replication and protein processing.

TABLE 58.2 Virological, epidemiological and clinical features of hepatitis viruses

Feature	Virus				
	HAV	*HBV*	*HCV*	*HDV*	*HEV*
Nucleic acid	RNA	DNA	RNA	RNA	RNA
Virion size (nanometer)	28	42	50	35–37	32–34
Incubation period (days)	14–28	45–180 (usually 60–90)	15–150	30–180	15–60
Usual routes of transmission	Fecal–oral	Parenteral Mother–to–infant Sexual	Parenteral Mother–to–infant Sexual	Parenteral Sexual	Fecal–oral
Clinical syndromes	Acute viral hepatitis Acute liver failure (infrequent)	Acute viral hepatitis Chronic hepatitis, including liver cirrhosis and liver cancer Acute liver failure (infrequent)	Chronic hepatitis, including liver cirrhosis and liver cancer Acute viral hepatitis (infrequent)	HBV–HDV co–infection, presenting as acute viral hepatitis HDV super–infection on chronic HBV infection (chronic hepatitis, acute on chronic liver disease)	Acute viral hepatitis Acute liver failure (infrequent) Chronic hepatitis (very rare)
Chronicity	No	Yes; may lead to cirrhosis and liver cancer	Yes; may lead to cirrhosis and liver cancer	Yes, with chronic HBV infection	Rare, in transplant recipients and immunosuppressed persons
Treatment	None	Oral nucleoside (or nucleotide) analogues, or interferons	Combination of interferon and ribavirin	Interferons	None for acute hepatitis
Vaccine	Yes	Yes	Not available	Yes (HBV vaccine)	Developed but not yet commercially available

TABLE 58.3   Summary of serological investigations commonly done for assessment of exposure to or infection with hepatitis viruses

Agent	Test	Interpretation
HAV	IgM anti-HAV	Presence in a patient with acute hepatitis indicates recent HAV infection.
	IgG anti-HAV	Indicates prior exposure and immunity to HAV. Useful for sero-epidemiological studies.
HBV	HBsAg	Indicates HBV infection, either acute or chronic; detection indicates presence of virus in the body.
	Anti-HBc IgM	Indicates acute (recent) HBV infection.
	Anti-HBs	Indicates immunity against HBV acquired after either natural infection or vaccination.
	HBeAg	Indicate replicative status of HBV and high infectivity of the infected individual.
	Anti-HBe	Indicate non-replicative status of HBV and low infectivity of the infected individual (in some persons, present despite high viral replication – implies infection with an HBeAg-negative viral mutant).
	HBV DNA	Quantitative estimations show number of viral genome copies/ml of serum. Higher value reflects active viral replication and high infectivity.
	HBV genotype	Genotype of infecting virus; may have clinical implication in relation to treatment.
HCV	Anti-HCV	Presence indicates either current infection with or prior exposure to HCV. Occasionally false positive.
	HCV RNA	Presence indicates current infection. Quantitation of HCV RNA useful in deciding treatment and in followup of patients receiving treatment.
	HCV genotype	Helps determine treatment duration.
HEV	IgM anti-HEV	Presence in a patient with acute hepatitis indicates recent HEV infection.
	IgG anti-HEV	Indicates prior exposure to HEV. Useful for sero-epidemiological studies.

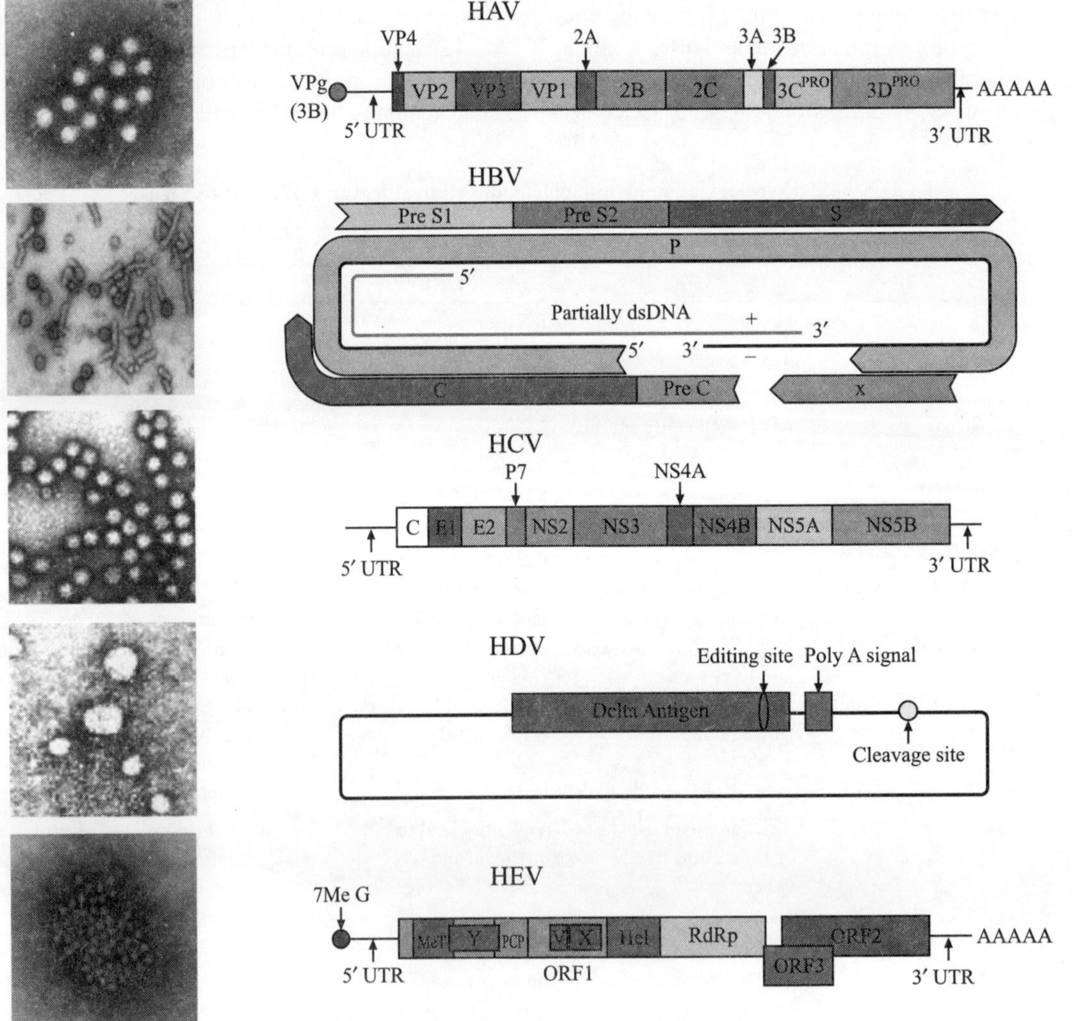

Figure 58.1   Hepatitis viruses and their genomes. A summary of all five hepatitis viruses is presented, showing electron microscopic images (left) and illustrations of the viral genomes (right). The genomes are not drawn to scale. See text for details. (*see Plate 23 for colour figure*)

Following HAV entry into liver cells, the genomic RNA is translated into a polyprotein, which is processed into 11 different proteins through the actions of the viral 3C protease and unidentified cellular proteases (Figure 58.1). A highly structured RNA element within the 5′ NCR, called the internal ribosome entry site (IRES), aids RNA translation by binding to ribosomes. The viral 2C (helicase) and 3D (RNA polymerase) proteins replicate the genomic plus-stranded RNA through an amplification cycle that involves an antigenomic minus-stranded RNA intermediate. Once enough genomic RNA is made, it is packaged by the capsid proteins to form new virus particles. These are secreted across the apical surface of hepatocytes into biliary canaliculus, passing into the bile and small intestine, from where these are excreted in the stool of an infected person.

The virus can be cultured *in vitro* on various primary and continuous cell lines of primate origin, which include African green monkey kidney (AGMK) cells, human fibroblasts and MRC5 human diploid lung cells. Many HAV strains have been characterized, of which HM175 has been adapted to different cell types to yield a range of viral variants. Though at least seven HAV genotypes are reported, they correspond to a single serotype, enabling a single vaccine or natural infection to protect against all strains.

B. Epidemiology

Hepatitis A virus is excreted in stools of infected persons and has a predominantly fecal-oral transmission. Interpersonal contact with infected family members and consumption of contaminated food or water are important routes of acquisition of HAV infection. Several types of contaminated foods including salads, milk, fruits, etc. have been implicated in causation of outbreaks of hepatitis A. Such contamination usually occurs either in agricultural farms or through contact with infected persons. Seafood such as shellfish and mollusks contaminated through human sewage pose particular risk since these are known to concentrate the virus.

The epidemiology of hepatitis A varies widely around the world. Based on differences in rates of transmission, three main epidemiological patterns are observed. In developing countries (e.g., India), HAV infection is highly endemic and most (>90%) persons in the population are exposed to it in early childhood. Infection leads to life-long immunity, and thus HAV infection among adults is rare. Since childhood HAV infections are usually asymptomatic, illness due to HAV is uncommon in these areas though infection is ubiquitous. In comparison, in highly developed countries (e.g., Scandinavia), HAV infection has been virtually eradicated. Thus, though the residents of these countries are not immune to HAV, there is no exposure to HAV, and hence both infection and illness due to HAV are uncommon. In areas with intermediate infection rates, a proportion of children grow up without acquiring HAV infection; exposure to HAV of these non-immune adolescents and adults leads to hepatitis A disease. It is in these areas that HAV poses the greatest disease burden. Importantly, epidemiologic patterns may vary in different segments of a population (for instance, in different socio-economic strata). Recent data suggest that an increasing proportion of Indian children from higher socio-economic groups are non-immune and thus at risk of hepatitis A in adulthood.

A. Pathogenesis and Disease

The incubation period of hepatitis A varies from 15 to 30 days. The virus enters the intestinal mucosa, and reaches the liver through portal circulation, possibly with a phase of replication in the enterocytes. It multiplies in the hepatocytes and is shed into the bile, whereby it reaches the intestine and is excreted in feces. The death of liver cells appears to be related not to viral replication but to killing by HAV-specific, cytotoxic T lymphocytes. This is associated with a marked increase in transaminases.

IgM antibodies to HAV can be detected early in the course of clinical illness, i.e., during the phase of transaminase elevation, and gradually disappear over a period of several months. IgG anti-HAV antibodies, which appear during convalescence, persist at a detectable level throughout life and confer protection against reinfection.

The viremia and viral excretion are generally short. However, a few patients may have a relapsing course lasting for several months with prolonged viremia and viral excretion.

The most commonly recognized clinical illness is acute hepatitis. The illness usually begins with abrupt onset of a variable combination of fever, malaise, fatigue, abdominal discomfort, loss of appetite, aversion to food and smoking, nausea and vomiting. This is followed in a few days by appearance of dark urine and jaundice of variable intensity. Some patients may report clay stools and itching. Fever is mild to moderate, not associated with chills and usually improves as the jaundice appears. The jaundice gradually improves over a few weeks. Clinical examination is usually unremarkable except for icterus and a slightly enlarged, smooth and mildly tender liver; spleen tip may be palpable in some patients. Presence of ascites, a palpable gallbladder or of a firm, irregular liver should suggest alternative diagnoses.

The illness usually lasts less than 8 weeks. In some patients, itching is prominent and the course is prolonged (maximum 6 months) though most patients ultimately recover. This entity is referred to as cholestatic hepatitis. Rarely, liver injury may be more severe and leads to fulminant hepatic failure, which is heralded by the appearance of irritability, insomnia and altered behavior, and may progress to frank coma. The mortality rates in such patients may reach 60% to 90% despite optimum medical care.

The likelihood and severity of symptoms in HAV infection is related to age. In children, most HAV infections (70% below the age of 6 years) are asymptomatic. Even among symptomatic children, most have a 'flu-like' illness without any jaundice. On the other hand, in adolescents and adults, symptomatic infection and jaundice are more frequent.

Rarely, other organ systems may be affected during acute hepatitis A, with manifestations such as Guillain-Barre syndrome, acute pancreatitis, cholecystitis, aplastic anemia, renal failure, etc. HAV infection can occur in persons with pre-existing chronic liver disease (cirrhosis); in such persons, the disease is more often severe and prolonged, and is associated with increased complications and mortality.

Some patients may have a prolonged or relapsing course. However, chronic infection with HAV, or association with chronic hepatitis, cirrhosis of the liver or increased risk of hepatocellular carcinoma are not known.

D. Diagnosis

Similar to other forms of acute hepatitis, hepatitis A is characterized biochemically by marked elevation of serum alanine aminotransferase (ALT) and aspartate aminotransferase (AST), and a variable rise in serum bilirubin levels. Specific diagnosis of HAV infection is based on detection of IgM anti-HAV antibody. During acute hepatitis A, viral particles or viral RNA may be detected in stool specimens; however, techniques for these are too cumbersome for routine use.

E. Treatment

No specific treatment is available for HAV infection; in fact, none is needed since the illness is mild and self-limiting. Dietary modifications such as restriction of fat and proteins, and liberal consumption of sugars and carbohydrates have no therapeutic role and may in fact jeopardize the patient's nutrition. Avoidance of alcohol is prudent. Similarly, bed rest and restriction of daily activities are not advisable except in seriously ill patients. Anti-emetic drugs may be used if needed. Patients with severe pruritus may benefit from antihistaminic agents or cholestyramine.

Admission to hospital is required only for patients with concurrent serious medical problems, those with severe vomiting that interferes with adequate fluid intake or those with incipient liver failure. Patients with severe disease need close observation. FHF is managed with lactulose, oral non-absorbable antibiotics and measures aimed at reduction of cerebral edema (intravenous infusion of mannitol). Corticosteroids have no role in the management of hepatitis A. Liver transplantation may be considered in patients with FHF. Elective surgery should be avoided during acute hepatitis, as it may precipitate FHF.

F. Prevention and Control

Improvements in hygiene that include provision of clean drinking water, proper disposal of human waste and better living conditions reduce the opportunities for HAV infection. Hand washing, and avoidance of raw or inadequately cooked foods or foods with potential for contamination with water (e.g., salads and iced drinks) help reduce the risk of infection.

Passive immunization with immunoglobulins prepared from the plasma of persons who have recovered from hepatitis A is efficacious either before or shortly after exposure to HAV.

When administered prior to exposure, intramuscular (IM) doses of 0.02 mL/kg and 0.06 mL/kg provide protection for about 3 and 5 months, respectively. The only accepted indication for passive pre-exposure prophylaxis of hepatitis A is travel to endemic areas, the usual recommended dose being 0.02 mL/kg IM within the preceding 2 weeks. Post-exposure prophylaxis depends on the time interval between exposure and immunoglobulin administration, reaching 85–95% with a single dose of 0.02 mL/kg IM given within 2 weeks of HAV exposure; later administration reduces efficacy, but may still attenuate disease severity. Post-exposure prophylaxis is recommended for close personal contacts such as household and sexual partners, regardless of age, but is not recommended for casual contacts of hepatitis A patients (i.e., at school or work).

Two different hepatitis A vaccines are currently available; these include Havrix from GlaxoSmithKline and VAQTA from Merck. Both contain formalin-inactivated attenuated strains of HAV, are highly immunogenic and safe. For each vaccine, two IM doses separated by at least 4 weeks are recommended. These vaccines offer protection against HAV infection when given before, and even a few days after exposure, and protection lasts for 10–20 years. Indications for the use of vaccines are similar to those for immunoglobulins. However, these have advantages of providing active, long-term protection, without interference with other childhood immunizations or the risk of transmission of blood-borne infections. Their use is recommended in persons with pre-existing chronic liver disease, particularly hepatitis C, in whom HAV infection may be associated with a particularly severe disease.

A combined vaccine called Twinrix from Glaxo SmithKline is also available for protection against both hepatitis A and B in persons 18 years or older. Though its use reduces the number of injections required for protection against two types of viral hepatitis, it does not fit well into the childhood immunization schedule. The current high cost of HAV and combination vaccines remains a major challenge for use in endemic areas, which also have limited resources.

II. HEPATITIS B

The hepatitis B virus (HBV) chronically infects about 350 million people worldwide causing disease ranging from acute hepatitis to chronic hepatitis, cirrhosis and hepatocellular carcinoma (HCC). It is responsible for nearly 1 million deaths annually. Chronic HBV infection with cirrhotic liver has been considered as the single most important factor associated with the development of HCC.

A. Virology

Hepatitis B virus belongs to the *Hepadnaviridae* family, which also includes other HBV-like viruses that infect animals such as woodchucks, ground squirrels and ducks. A distinctive feature of this family is the DNA genome that goes through a reverse transcription step in its replication. The intact virus particles of 42 nm (also called Dane particles) are found in the sera of infected persons together with empty spherical or tubular particles of 20–22 nm (Figure 58.1).

The HBV genome is a partially double-stranded DNA of about 3200 bases, which is held in a circular conformation due to overlapping ends. It contains four overlapping open reading frames (ORFs) – S (surface), C (core), P (polymerase) and X (Figure 58.1). The S ORF codes for the HBV surface envelope proteins, which include the major hepatitis B surface antigen (HBsAg) and two extended versions called the pre-S1 and pre-S2 proteins. The C ORF encodes the hepatitis B core antigen (HBcAg) that makes up the viral capsid. Alternative translation initiation also gives rise to the hepatitis B e-antigen (HBeAg) whose function remains undefined, but its presence in serum is taken as a marker of active viral replication and persistence. The P ORF encodes the viral polymerase, which is functionally divided into three domains—the terminal protein (Tp), which is involved in encapsidation and initiation of minus strand synthesis; the reverse transcriptase (RT) domain, which catalyzes genome synthesis; and the ribonuclease H (RNase H) domain, which degrades pregenomic RNA and facilitates replication. The X ORF encodes a protein of about 16 kDa (HBxAg), with multiple roles in modulating the infected cells that include signal transduction, transcriptional activation and modulation of the cell cycle. The HBxAg is also suggested to contribute to HBV-mediated transformation of liver cells and the development of hepatocellular carcinoma (HCC).

The mechanisms of viral entry are not fully understood. Several candidate receptors for HBV have been described and polymerized human serum albumin is shown to facilitate the entry process. The life cycle of HBV is complex but unique. Although not a retrovirus, HBV still uses reverse transcription as an important step in its replication process. After gaining entry into cells, the viral genomic DNA is converted into fully double-stranded covalently closed circular DNA (cccDNA) in the nucleus. This serves as a template for transcription of four viral mRNAs with the help of host RNA polymerase, which are transported to the cytoplasm. The largest of these is the pregenomic RNA, which is (a) translated into the viral capsid and polymerase proteins, and (b) reverse transcribed by the viral DNA polymerase into a DNA copy [L(–) strand]. As the viral polymerase copies the L(–) strand to generate the S(+) strand, the viral particles assemble at the hepatocyte surface and bud off.

HBV strains were initially classified into 4 subtypes—*adw, adr, ayw* and *ayr* based on the main antigenic determinant 'a' and two pairs of mutually exclusive serotype determinants *d/y* and *w/r* in the surface antigen (HBsAg). These subtypes can be further classified into 9 serotypes (ayw1, ayw2, ayw3, ayw4, ayr, adw2, adw4, adrq$^+$ and adrq$^-$). Now based on phylogenetic analysis of the complete viral genome sequences, HBV strains are classified into ten genotypes named A through J that show a distinct geographical distribution pattern. For example, genotypes B and C are most prevalent in East Asian countries where vertical transmission is common whereas genotypes A and D are generally found in Africa, Europe, Middle-East and Indian subcontinent and which relate to horizontal transmission.

B. Epidemiology

Hepatitis B is prevalent worldwide with nearly 350 million chronic carriers. The rate of infection varies widely, and countries are classified based on chronic hepatitis B carrier rates as having low (<2%), intermediate (2–8%) and high (8%) endemicity. With an average carrier rate of ~4%, India is in the intermediate zone.

HBV is present in variable quantities in the blood of patients with hepatitis B. It is transmitted parenterally, through transfusion of contaminated blood or blood products and use of contaminated instruments and needles—maternal-infant transmission and sexual routes. Risk factors associated with increased risk of HBV infection include: transfusion of unscreened blood and blood products (including persons with hemophilia, thalassemia, etc), injection drug use with shared injection equipment, long-term dialysis, healthcare workers, sexual promiscuity, and residence in a household with a chronic HBV carrier.

In high-endemicity areas, most HBV infections occur around birth or in early childhood, often from HBV-infected mothers. The risk of mother-to-child transmission is much higher (90%) if the mother is also positive for HBeAg (reflects high viral load) than if she is negative for HBeAg (10–20%).

C. Pathogenesis and Disease

The outcome of HBV infection depends heavily on age at the time of acquisition of infection. Nearly 90% of exposed infants develop chronic HBV infection; this proportion declines with increasing age to reach 3–5% by 5 years of age, a risk that continues into adulthood. In addition, HBV infection in children is often asymptomatic, whereas nearly 30% of adults develop acute hepatitis. As with other forms of acute hepatitis, a small proportion of patients with acute hepatitis B progress to FHF.

Acute HBV infection manifests after an incubation period of 6 weeks to 6 months with an illness similar to other forms of acute viral hepatitis. In some patients, acute liver failure develops, which is likely to be related to an enhanced immune response.

Chronic hepatitis B can be subdivided into four phases. The initial 'immune tolerant' phase is characterized by lack of host immune response, a high rate of viral replication (very high serum HBV DNA), detectable HBeAg, normal ALT and lack of inflammation on liver biopsy. This is followed by the 'immune active' or 'immune clearance' phase, where killing of liver cells by the host immune response leads to a reduction in rate of viral replication (medium to high HBV DNA), high ALT levels and evidence of active hepatitis on liver biopsy. In the third 'inactive' or 'low replication' phase, viral replication is absent or minimal, HBeAg is negative and ALT returns to normal, though HBsAg continues to be positive. In some persons who have reached the inactive phase, one or more bouts of reactivation of HBV replication may occur with findings similar to the 'immune active' phase. In some patients, a viral mutant that cannot produce HBeAg emerges; despite negative HBeAg they may have liver injury. A small proportion of infected persons develop 'seroconversion'; this phase is characterized by disappearance of HBsAg and appearance of protective anti-HBs antibodies.

Chronic HBV infection can progress in about one-fourth of affected persons to liver cirrhosis or hepatocellular carcinoma.

D. Diagnosis

Several HBV proteins induce specific antibodies in infected people and are thus of diagnostic importance; these include: the viral envelope protein (detectable as hepatitis B surface antigen; HBsAg), the viral nucleocapsid protein (hepatitis B core antigen; HBcAg) and a soluble nucleocapsid protein (hepatitis B e antigen; HBeAg).

Detection of HBsAg in serum is the most widely used marker of HBV infection. In contrast, anti-HBs antibodies indicate recovery from infection and protection against future HBV infection. HBcAg cannot be detected in the serum. Anti-HBc antibodies belonging to IgM isotype are a marker of recent (acute) HBV infection whereas IgG anti-HBc antibodies are a marker of past exposure to HBV – whether cleared or persistent. HBeAg is a marker of active viral replication and its presence is associated with high levels of circulating HBV DNA.

E. Treatment

Most patients with acute hepatitis B improve spontaneously and do not need any specific treatment. Antiviral therapy may have a role in patients with severe or fulminant hepatitis B; however, further studies are needed on this subject.

In patients with chronic hepatitis B, the goals of treatment include clinical and histological improvement, prevention of serious complications such as liver cirrhosis, hepatic decompensation and liver cancer, improvement of quality of life and prolongation of survival. However, most clinical trials have relied largely on surrogate endpoints – serological (loss of HBeAg, seroconversion from HBeAg positivity to anti-HBe positivity, HBsAg loss), virological (reduction in HBV DNA level or its undetectability in serum), biochemical (normalization of liver enzymes) and histological (reduction of necroinflammation and fibrosis).

No treatment is indicated for patients in immune tolerant and inactive carrier states. Patients in immune active phase, i.e., those with substantial viremia and significant liver inflammation (as evidenced by raised ALT or liver biopsy) may benefit from treatment. Two different approaches—(1) interferon and its pegylated analogues, which boost the host immune response, and (2) oral nucleotide or nucleoside analogues (NAs), which suppress viral replication, are available. Each approach has certain advantages and disadvantages. Interferon treatment has a short, finite duration and better HBsAg seroconversion rate than NAs; however, it requires injections and has more

adverse effects. Moreover, interferons are contraindicated in patients with underlying autoimmune disease, thyroid dysfunction, pancytopenia, and liver cirrhosis. In contrast, NAs are administered orally and have minimal side effects; however, the duration of treatment with these is indefinite (several years, possibly life-long) and drug resistance may occur with time. Interferons are costlier than NA; however, life-long NA treatment may well translate to a higher per patient cost. The treatment thus needs to be individualized, after considering factors such as the clinical condition of the patient, viral load, co-morbid conditions, financial constraints, likelihood of compliance to therapy, and the patient's preference.

Treatment with NAs is particularly useful in patients with decompensated cirrhosis (jaundice, ascites, variceal bleed, and/or hepatic encephalopathy), with improvement in liver function, reduction in mortality, and movement of some patients off the liver transplant list. The choice of NA in such patients is based on viral load, cost considerations, potential for drug resistance, etc.

Several international professional organizations have issued guidelines for the treatment of HBV infection, including indications, contraindications, duration, follow-up and prognosis. These guidelines show minor differences. The treatment protocols that these lay down are complex and hence referral to physicians experienced in the management of such patients is useful.

It is important to recognize that both the currently available forms of therapy clear HBsAg in only <10% of those successfully treated.

F. Prevention and Control

Screening of donated blood, plasma, virus inactivation in plasma-derived products and implementation of infection control practices are effective means to prevent HBV transmission. However, the single most effective prevention measure is routine immunization for infants, especially those born to HBsAg carrier mothers, and high-risk individuals. Universal infant vaccination will be the key to successful elimination and subsequent eradication of hepatitis B.

An effective HBV vaccine has been available since the early 1980s and is already part of the universal child immunization programmes in various countries. This was the first recombinant vaccine licensed for human use and includes HBsAg produced in engineered yeast cells, which is then purified and combined with alum. The vaccine is given as doses of 20 µg protein at 0, 1 and 3–6 months by intramuscular injection. It has a remarkable safety record with over 1 billion doses used since its introduction in the market. Booster doses are recommended every 5 years but there is no data to support this in immunocompetent individuals who have completed the 3–dose primary vaccination. In individuals who are immunocompromised booster doses may be necessary. Whether boosters are needed beyond 20 years of hepatitis B vaccination in children and adults needs to be studied.

While each country or region assesses its needs and chooses a strategy depending upon epidemiology, disease burden, economic implications and systemic readiness, it is clear that immunization will be the key to controlling hepatitis B. The seroprevalence of HBsAg in India ranges from 2% to 10%, placing it in the group of intermediate to high endemicity for HBV infection. Assuming an average HBsAg carrier rate of 4%, the total number of HBV carriers in India is estimated to be about 40 million or about 15% of the total world pool of HBV carriers. The Indian Academy of Pediatrics (IAP) also endorses the recommendations of the World Health Organization on HBV vaccination at birth whether the mother is a carrier or not. In India, hepatitis B immunization programme has been carried out in selected states with the vaccine given at birth (if possible) and at 6, 10 and 14 weeks of age.

III. HEPATITIS C

Hepatitis C virus (HCV) is estimated to have infected almost 200 million people. Of those infected with HCV, a proportion of individuals have an acute infection, which is cleared spontaneously; however, in the large majority, the virus persists causing a chronic (>6 months) infection. Persistent infection with HCV is associated with the development of chronic hepatitis, hepatic steatosis, cirrhosis and HCC.

A. Virology

Hepatitis C virus is an enveloped virus, which is a member of the *Flaviviridiae* family that also consists of other human pathogenic viruses such as the yellow fever, dengue and Japanese encephalitis viruses. The genome consists of a single-stranded positive sense RNA molecule of ~9600 bases that encodes a large open reading frame, which is flanked by highly structured 5′ and 3′ untranslated regions (UTR) (Figure 58.1). The 5′ UTR harbours the Internal Ribosome Entry Site (IRES) that mediates cap-independent translation of HCV RNA. The 3′ UTR includes a polyU/UC tract of variable length and a conserved 3′ X tail region that helps in RNA replication.

The viral proteins are translated as a single polyprotein of about 3000 residues, which is cleaved into 10 proteins with the help of both host cell and viral proteases to generate the structural and non-structural proteins (Figure 58.1). The polyprotein is functionally divided into three segments— (1) the amino-terminal region, which includes the structural proteins (core, E1, E2); (2) a central region including two proteins (p7, NS2) that are not part of the viral structure or required for RNA replication, but are important for new virus formation and release; and (3) the carboxy-terminal region that includes the non-structural proteins (NS3, NS4A, NS4B, NS5A, NS5B), which are required for RNA replication. The NS2 protein together with the amino-terminal region of the NS3 protein constitutes the NS2–NS3 proteinase that catalyzes a single autocatalytic cleavage to release NS2 and NS3 proteins. Later, the N-terminal region of NS3 serves as a serine protease

and cleaves at different sites within the polyprotein to release other non-structural proteins. The C-terminal part of NS3 binds to HCV RNA and serves as a RNA helicase. The NS4A protein acts as a cofactor of NS3 activity, while the NS4B protein interacts with the viral replicase. The NS5A protein has been shown to be involved in HCV pathogenesis causing endoplasmic reticulum and oxidative stress, and resistance to IFN therapy. The NS5B protein is a RNA dependent RNA polymerase (RdRp) that replicates the viral RNA.

HCV enters its target cells in a highly coordinated process involving components of the virus particle and several cellular receptors that include CD81, SR-B1 (the scavenger receptor class B type 1), claudin 1 (CLDN1) and occludin (OCLN). The understanding of HCV RNA replication received a major boost after the availability of the replicon systems in 1999. Since then, several advances, including the development of full-length infectious HCV genomes and more permissive clones of existing hepatic cell lines, have enhanced the *in vitro* culture of HCV. This is proving invaluable in understanding the biology of HCV as well as in screening potential natural and synthetic compounds for their anti-HCV activity.

There are six reported genotypes of HCV, which differ in their virulence and susceptibility to antiviral therapy. Genotypes 1, 2 and 3 are the major types observed in Japan, Western Europe and North America. Type 4 has been found in Central and Northern Africa and in the Middle East. Type 5 has been described in South Africa and type 6 in Southeast Asia. In India, genotype 3 is predominant, accounting for about three-fourths of cases; most of the remaining cases are due to the genotype 1 virus.

B. Epidemiology

HCV is transmitted through parenteral routes similar to those for HBV. Transmission through percutaneous routes, such as transfusion of HCV-contaminated blood or blood products, unsafe injections, and hemodialysis, is the most efficient mode of transmission of HCV. In most countries around the world, the risk of transmission of HCV with blood transfusions has come down drastically in recent years following the introduction of screening of blood and blood products for HCV infection.

In comparison with HBV, maternal-infant and sexual routes are relatively inefficient modes for transmitting HCV. Only about 5% of infants born to HCV-infected mothers get infected; the risk is higher if the mother also has human immunodeficiency virus infection. Though persons with multiple sexual partners have an increased risk of HCV infection, the risk of transmission of HCV in stable monogamous sexual relationship is quite low. The risk of acquiring HCV infection following needlestick injury is also quite low.

A large proportion (55–85%) of those infected with HCV develop chronic infection (lasting >6 months), which is known to progress in the long-term to chronic hepatitis, cirrhosis and liver cancer. The risk of development of liver cirrhosis increases with duration of HCV infection, and is higher among men, those who acquire infection at an older age, and persons with substantial alcohol intake or HBV or HIV coinfection. It is believed that about 20% of those infected with HCV will develop liver cirrhosis over 20–25 years, though in community-based studies the absolute risk has been much lower.

C. Pathogenesis and Disease

After entering the host, HCV multiplies in the liver cells, and induces humoral and cellular immune responses. Anti-HCV antibodies develop within a few weeks of the occurrence of HCV infection. These antibodies appear not to confer protection against reinfection. The cellular immune responses include both innate and adaptive responses. These determine the outcome of HCV infection – whether self-limited or persistent. A strong and broadly reactive cellular immune response is associated with early viral clearance, whereas a limited response is associated with persistent HCV infection.

HCV is a non-cytopathic virus and liver injury is mediated by the host immune response, in particular the cytotoxic T lymphocytes. The resulting inflammation leads to activation of hepatic stellate cells, culminating in progressive liver fibrosis and liver cirrhosis. The consequent regenerative process is believed to be responsible for an increased risk of liver cancer.

The incubation period of HCV infection varies widely—from 2 weeks to 6 months. HCV infection is most often not associated with any symptoms during the initial months though laboratory tests often show mild liver enzyme elevation. Some of those infected develop an acute hepatitis like illness with nausea, vomiting, general malaise and jaundice, which are indistinguishable from those in other forms of acute viral hepatitis. FHF due to HCV infection is extremely rare.

Chronic hepatitis C is also often asymptomatic or associated with non-specific symptoms such as malaise, fatigue and right upper abdominal discomfort. Diagnosis is thus usually made following incidental detection of abnormal results for biochemical liver function tests or anti-HCV antibody testing during evaluation for other diseases, general health checkup, blood donation, etc. Alternatively, HCV infection may come to light for the first time when a person has developed liver cirrhosis, decompensated liver disease or liver cancer.

D. Diagnosis

Diagnosis of hepatitis C involves detection of anti-HCV antibodies and/or HCV RNA. Initial diagnosis is usually based on detection of anti-HCV antibodies using enzyme immunoassays. The currently available assays for anti-HCV are highly sensitive and specific. However, the predictive value of a positive test depends on pre-test probability of HCV infection in the person tested. Thus, a positive anti-HCV test result in a healthy person (for instance a blood donor with normal liver function tests) needs confirmation by another assay. Immunoblot assays have been used for this purpose in the past; however, currently, HCV RNA testing is usually preferred. A negative HCV RNA test in this situation indicates that the anti-HCV antibody test was either a false positive test result or related to a resolved HCV infection.

Detection of HCV RNA in clinical specimens is done using amplification based assays (reverse transcription and polymerase chain reaction). Both qualitative and quantitative assays are available; the former assays have now largely been replaced by real-time quantitative assays since these provide additional information on viral load, which is useful for treatment.

Patients with HCV infection often undergo tests to assess the effect of infection on liver structure and function. Biochemical tests show raised levels of alanine aminotransferase and aspartate aminotransferase. In addition, patients with liver cirrhosis may show evidence of decreased liver synthetic (reduced serum albumin and abnormal results on tests of coagulation) and excretory (high serum bilirubin) capacity. Liver biopsy shows evidence of inflammation and fibrosis; several schemes for scoring the severity of liver injury have been devised; these are particularly useful in research studies on natural history and response to treatment.

Distinction between acute and chronic HCV infection is difficult, except when the test for anti-HCV or HCV RNA becomes positive in a person who was previously known to be negative for the particular marker.

E. Treatment

Treatment in patients with HCV infection is aimed at eradication of the virus in an attempt to prevent progression of liver damage, improve the quality of life and survival of the patient, and reducing the risk of transmission to others.

The current standard of care for HCV infection is combination treatment with pegylated interferon (either alpha 2a or alpha 2b) and Ribavirin for 6–12 months. This treatment is associated with sustained virological response (SVR), defined as absence of detectable HCV RNA in blood at the end of treatment and at 6 months after cessation of treatment, in nearly half the patients. The treatment is associated with frequent side effects and is contraindicated in persons with advanced liver disease.

Several host and viral factors have been shown to influence the response to this treatment. Persons infected with genotype 2 or 3 HCV attain SVR rates of 70–80% even with a shorter duration of treatment (6 months). Other predictors of good response to treatment include lower HCV RNA load, younger age, female gender, higher ALT levels, normal body weight (compared to overweight), Caucasian race (compared to African), insulin sensitivity, good compliance to treatment and favorable host interleukin-28B genotype.

Two directly acting anti-viral drugs (Boceprevir and Telaprevir), which inhibit the NS3/4A serine protease enzyme of genotype 1 HCV, were approved for treatment of hepatitis C in several countries in the past year. These drugs inhibit viral replication. Addition of one of these drugs to pegylated interferon and Ribavirin greatly improves SVR rates in patients infected with such virus, whether treatment-naïve or previously failed on interferon-Ribavirin combination therapy. In several countries, these drugs are now included in initial treatment regimens for genotype 1 HCV infection. However, high cost, frequent adverse events, and lack of proven efficacy for HCV genotypes other than genotype 1 limit the use of these drugs. Further, they carry a high risk of development of drug-resistant mutants, if compliance is poor or the treatment is not closely monitored.

Additional directly acting anti-viral drugs that include nucleoside and non-nucleoside polymerase inhibitors, NS4B/NS5A protease inhibitors, immunomodulators and drugs that interfere with viral lipid metabolism are currently in clinical trials. In future, this may allow the development of interferon-free, oral drug regimens, making the treatment of hepatitis C easier.

Anti-viral therapy in patients with acute HCV infection may reduce the rate of progression to chronic HCV; however, optimum dose, duration and timing for starting therapy remain unclear.

F. Prevention and Control

There is no vaccine to prevent hepatitis C. Vaccine development efforts are impeded by the extensive heterogeneity of HCV strains that circulate worldwide and the ability of the virus to mutate rapidly. At the same time, there is a large reservoir of HCV-infected individuals who can serve as a source of transmission to others. Thus, effective public health programmes are the only tools available to prevent new infections. These include ensuring a safe blood supply, implementing good infection control practices, and limiting high-risk drug and sexual behaviors. In resource poor regions, the major source of infection is from improperly screened blood. Implementing donor screening and testing policies will be the centerpiece of prevention efforts, as also safe injection practices and the safe use and disposal of medical equipment.

IV. HEPATITIS D

The first evidence for the existence of hepatitis delta virus (HDV) was presented in mid-1970s when a previously unrecognized nuclear antigen was shown in hepatocytes of patients with chronic hepatitis B. Like other hepatotropic viruses, HDV also causes liver inflammation and produces symptoms similar to the other acute viral hepatitis diseases, just probably more severe.

A. Virology

Hepatitis D virus, also called the delta agent, is a defective virus that requires HBV to complete its life cycle. The 36 nm particles are roughly spherical with an envelope made up of HBsAg and the nucleocapsid made up of ~70 copies of the delta antigen. The HDV genome exists as a negative sense, single-stranded, closed circular RNA genome of ~1.7 kb—the smallest of known human pathogens (Figure 58.1). As the nucleotide sequence is 70% self-complementary, the viral genome forms a partially double stranded RNA structure, a part of which can function as a ribozyme. This activity is required during viral replication to produce unit length copies of the genome from longer RNA concatamers.

The HDV genome encodes only two highly basic proteins called the small delta antigen (HDAg-S or p24) and large delta antigen (HDAg-L or p27), which can bind nucleic acids. These two proteins are produced from a single open reading frame and have 195 common amino acids; the HDAg-L protein (215 amino acids) carries additional amino acids at the C-terminus. Despite sharing 90% identical sequences, the two proteins play different roles during the course of viral infection. While HDAg-S is produced during the early stages of infection and is essential for viral replication, HDAg-L is produced late in infection and functions as an inhibitor of viral replication but facilitator of viral assembly.

HDV resembles viroids in many ways. Viral replication occurs in the nucleus of primary hepatocytes with the help of host cell machinery using a double-rolling circle mechanism. The host RNA polymerase II is used for replication—a rare usage of DNA-dependent RNA polymerase as an RNA-dependent RNA polymerase. The ribozyme activity associated with viral genome is utilized for cleavage and self-ligation of unit lengths of genome. New virions are produced only in the presence of HBsAg, which acts as the envelope protein. Therefore, HDV can cause productive infection only in persons with HBV infection.

B. Epidemiology

As replication of HDV is closely linked to that of HBV, its transmission follows the same routes as HBV, i.e., mainly through parenteral exposure. However, the global distribution pattern of HDV differs from that of HBV, being highly endemic in Mediterranean countries, the Middle East, Central Africa, and northern parts of South America.

Two distinct patterns of HDV infection are recognized, depending on the relative times of acquisition of HDV and HBV infection, namely coinfection and superinfection. In the former, HBV and HDV are acquired simultaneously, whereas in the latter, a person who is already infected with HBV acquires HDV infection.

C. Pathogenesis and Disease

HBV-HDV coinfection often presents as an acute hepatitis like illness, which is more severe than with HBV alone. Most patients with such coinfection clear both the viruses, but some progress to chronic HBV and HDV dual infection.

HDV superinfection can occur any time during the course of chronic hepatitis B. It may thus present as an illness resembling acute hepatitis in a previously asymptomatic HBV carrier or as worsening of pre-existing chronic hepatitis B. HDV superinfection in such patients often becomes chronic, and persons with chronic HBV and HDV dual infection have a more rapid progression of chronic liver disease than those with HBV infection alone.

D. Diagnosis

Available tests for HDV infection include detection of IgM or IgG anti-HDV, HDV antigen and HDV RNA.

IgM anti-HDV appears transiently in persons with self-limited HDV infection but may persist for a variable duration in those with chronic infection. IgG antibodies appear late in acute hepatitis, and persist in chronic HDV infection. HDV antigen can be detected during acute infection. Detection of HDV RNA using a PCR-based assay is a highly sensitive and specific test for both acute and chronic HDV infection.

Distinction between coinfection and superinfection is based on presence or absence of IgM anti-HBc, a marker of recent HBV infection.

E. Treatment

Treatment of HDV infection is aimed at eradication of both HBV and HDV. Interferon alpha or pegylated interferon alpha, in higher dose and longer duration than those used for HBV infection alone, have shown some efficacy, though results continue to be suboptimal. Addition of nucleoside analogues directed against HBV appears to provide no additional benefit.

F. Prevention and Control

No specific measures are available for the prevention of HDV infection in those already infected with HBV. For those without HBV infection, measures aimed at preventing HBV infection, including hepatitis B vaccine, are highly effective in preventing HDV infection as well.

V. HEPATITIS E

A large epidemic of waterborne hepatitis in New Delhi, India during 1955–56 was the first reported outbreak of hepatitis E. Stored sera of patients from this epidemic and another outbreak in Kashmir (1978–79) showed the absence of serological markers for hepatitis A and hepatitis B, suggesting the existence of a new viral agent, then called the enterically-transmitted non-A, non-B hepatitis virus. The first sequence for a new virus called hepatitis E virus (HEV) became available in 1990. Subsequently the sequences of several geographically distinct isolates of human HEV and related viruses that infect pigs, rabbits, rodents and chicken have been described. These viruses are now classified in the new *Hepeviridae* family.

A. Virology

Hepatitis E virus is a 27–34 nm nonenveloped virus, which includes a capsid made from a single protein and a single-stranded, linear, positive sense RNA genome of about 7.2 kb (Figure 58.1). The genome is capped at its 5′ end and includes a short 5′NCR of about 25–30 nucleotides, a protein-coding region with three open reading frames called orf1, orf2 and orf3, and a short 3′NCR terminating in a poly(A) sequence (Figure 58.1). The ORF1 non-structural polyprotein contains methyltransferase, RNA helicase, RNA-dependent RNA polymerase and possibly also protease activities that aid in viral genome replication. The ORF2 protein sequesters the RNA genome and forms the viral capsid. The ORF3 protein is required for virus egress from infected cells and optimizes the host cell environment for viral replication through its interaction with various cellular proteins and intracellular pathways; it is required for virus infection *in vivo*, but not for replication in cultured cells.

At least four HEV genotypes are recognized, which include human and animal HEV isolates. Genotype 1 includes Asian and African human strains, genotype 2 includes the single Mexican strain and few isolates from Africa, genotype 3 includes human and swine HEV strains from industrialized countries, and genotype 4 includes human and swine HEV strains from Asia. Viruses of genotypes 1 and 2 are associated with outbreaks due to efficient human-to-human feco-oral transmission, while genotype 3 and 4 viruses are maintained in animal species and show inefficient cross-species transmission to humans. All four genotypes of HEV belong to a single serotype, which is good news for cross-strain protection.

The entry of HEV into target cells has not been characterized and the proposed replication strategy is based on analogy to similar viruses. Following entry and uncoating, the orf1 on genomic RNA is translated into the non-structural polyprotein. This also includes the viral replicase, which replicates the positive-stranded RNA genome through a negative-stranded RNA intermediate. A subgenomic positive-stranded RNA is also produced, which is translated into the ORF2 (and possibly also the ORF3) proteins. The assembly and egress of virions has not been fully characterized. Only limited success has been achieved in propagating HEV *in vitro* in cultured cells and in primary cynomolgus hepatocytes. There are no small animal models of HEV infection. Nonhuman primates can be experimentally infected. While these display rise in liver enzymes, jaundice is not observed. Similarly, no clinical liver injury is observed in pigs infected with genotype 3 HEV.

B. Epidemiology

Serological data indicate that HEV infection occurs all over the world. Clinical disease is particularly common in developing countries in Asia, Africa and Central America. Two distinct epidemiological patterns have been reported—in areas where disease is highly endemic and where it is not—that differ in routes of transmission, affected population groups and disease characteristics.

In hyperendemic areas, the infection is transmitted mostly through the fecal-oral route, usually through water supplies contaminated with feces of persons infected with HEV.

Transmission through contaminated food possibly occurs. Materno-fetal and transfusion-related transmission has been documented but the contribution to overall disease burden may be small. Person-to-person transmission of HEV appears to be uncommon. Also, there is little evidence of animal-to-human transmission in most endemic areas.

In high-endemicity areas, the infection causes disease outbreaks as well as sporadic cases. Incubation period varies from 15 to 70 days.

In low-endemicity areas (United States, western Europe, and developed countries of Asia-Pacific), only occasional cases with locally acquired HEV infection are observed. Zoonotic transmission seems to be the main route of transmission. The virus circulates among several animal species, in particular pigs, from where it appears to spread to humans, either through close contact or consumption of meat.

C. Pathogenesis and Disease

In disease-endemic areas, HEV infections are usually asymptomatic. Most commonly encountered clinical presentation is as acute hepatitis resembling that described with HAV infection; the illness improves spontaneously over a few weeks. In both outbreak and sporadic setting, young adults (15–35 years old) are more commonly affected. Some patients progress to acute liver failure, a serious illness with potentially fatal outcome. Some patients progress to acute liver failure. Pregnant women appear to be particularly prone to develop clinical disease and serious form of disease. Also, the case fatality rate is much higher among pregnant women. The reason for this specific predilection among pregnant women remains unknown, though immunological or hormonal factors may play a role. HEV infection in a person with pre-existing liver disease can present as acute-on-chronic liver failure. Chronic HEV infection has not been reported in these areas.

In contrast, in low endemicity areas, most cases occur among middle-aged to elderly persons, often with coexisting diseases, and generally have milder liver injury. In addition, chronic HEV infection has been reported, primarily among immunosuppressed persons, such as solid organ transplant recipients who are receiving immunosuppressive agents, persons with human immunodeficiency virus infection and those with hematological malignancies. All cases of chronic hepatitis E have been related to genotype 3 HEV. Chronic hepatitis E can progress to liver cirrhosis.

Both IgM and IgG anti-HEV antibodies develop early in the course. The IgM antibodies can be detected up to about 6 months. IgG antibodies persist longer than IgM; however, data on the longer-term durability of these antibodies are unclear.

D. Diagnosis

Detection of IgM anti-HEV antibodies is a marker of recent infection with HEV and is used for diagnosis of acute hepatitis E. In contrast, presence of IgG anti-HEV indicates exposure to HEV either recent or remote, and is used for sero-epidemiological studies.

Available assays for these antibodies are far from perfect, and both false positive and false negative results are known. In particular, in low-endemicity areas, detection of HEV RNA is necessary for a reliable diagnosis of current HEV infection.

E. Treatment

Acute hepatitis E is usually self-limited and treatment is usually not necessary. Chronic HEV infection in immunosuppressed persons has been treated successfully in some small case series using either interferon or Ribavirin.

F. Prevention and Control

As in the case of hepatitis A, proper treatment and safe disposal of human excreta, provision of safe drinking water and improvement in personal hygiene are the mainstays of hepatitis E prevention. Sanitary food handling practices, and avoiding consumption of undercooked or uncooked meat and vegetables should be recommended. In epidemic settings, boiling and chlorination of water supplies may be useful. Unlike in hepatitis A, the administration of immunoglobulins has not been shown to protect humans against hepatitis E.

Truncated recombinant versions of the ORF2 protein form highly immunogenic virus-like particles (VLPs) when these are produced in insect cells in culture or in *E. coli*. Two vaccines containing the insect or bacterial cell versions of the recombinant VLPs have undergone safety and efficacy studies in humans. Study subjects receiving 20 μg of alum-adjuvanted recombinant HEV protein given as three doses at 0, 1 and 6 months, showed >95% protection. One of these vaccines, Hecolin, developed at Xiamen University in China received regulatory approval in December 2011 and is now ready for the market. More data are needed on vaccine efficacy among pregnant women and children, and in special groups such as persons with chronic liver disease. Widespread use in endemic regions will depend on cost considerations, the duration of protection afforded by the vaccines and their ability to interrupt the transmission of infection.

SUMMARY

Viral hepatitis is a heterogeneous disease condition, caused most often by infection with one of the five currently known viruses; a few cases are possibly caused by viruses that have not yet been discovered. The known causative agents vary widely in their structure, organization, viral replication strategy and in epidemiological features; however, the clinical illnesses caused by these are often similar. Available serological assays allow specific diagnosis of various infections. Specific treatment is indicated currently only for selected cases with HBV or HCV infection. Specific vaccines are available and are being widely used for HAV and HBV. HDV infection is prevented by HBV vaccine, obviating the need for a separate vaccine. A vaccine has recently become available for HEV infection, though further data are needed before it can be widely used.

Acknowledgement

The authors thank Mr. Imran Ahmed for preparation of the figure illustrating hepatitis viruses and their genomes.

REFERENCES

Craig A.S. and Schaffner W. (2004), Prevention of Hepatitis A with the Hepatitis A Vaccine, *N. Engl J. Med.,* 350:476–81

HAV: Martin A. and Lemon S.M., Hepatitis A Virus: From Discovery to Vaccines (2006), *Hepatology*, 43:S164–172;

HBV: Dienstag J.L. (2008), Hepatitis B Virus Infection, *N. Engl J. Med.,* 359:1486–500.

HCV: Rosen H.R. (2011), Clinical Practice, Chronic Hepatitis C Infection, *N. Engl J. Med.,* 364:2429–38.

HDV: Hughes S.A., Wedemeyer H. and Harrison P.M. (2011), Hepatitis Delta Virus, Lancet, 378:73–85.

HEV: Aggarwal R. and Jameel S. (2011), *Hepatitis E. Hepatology,* 54:2218–26.

Liaw Y.F. and Chu C.M. (2009), Hepatitis B Virus Infection, *Lancet*, 373:582–92.

59

Fatty Liver and its Implications

Shiv Kumar Sarin and Ramesh Kumar

CONTENTS

I. INTRODUCTION

Fatty liver is characterized by increased accumulation of fat, especially triglycerides in the liver cells. In the spectrum of fatty liver disease, there is hepatic steatosis with or without varying degree of hepatic inflammation, hepatocyte injury and fibrosis. Broadly, fatty liver has been classified as alcoholic or nonalcoholic fatty liver. The nonalcoholic fatty liver disease (NAFLD) is the most common cause of fatty liver, and is commonly associated with obesity hyperlipidemia, and diabetes mellitus (DM). NAFLD may also be due to several other etiologies as depicted in Table 59.1. NAFLD is a major cause of chronic liver disease in developed as well as developing countries. Studies report NAFLD prevalence rates of 10% to 20% in lean population, 60% to 74% among the obese and over 90% in the morbidly obese. The prevalence of NAFLD in Indian population ranges from 5 to 28%, which is comparable to the West. There is a strong association between NAFLD and other potentially life threatening diseases and clinical conditions. NAFLD is associated with increased mortality from cardiovascular diseases, malignancy and end stage liver disease. NAFLD constitutes a substantial proportion of patients with cryptogenic cirrhosis. NAFLD is associated with severity of other chronic hepatitis. NAFLD is associated with several extra-hepatic disorders, such as colorectal neoplasm, breast cancer, hypothyroidism, polycystic ovary syndrome, and gall stone disease. The fatty liver has poor ability to regenerate after volume loss which may lead to development of liver failure and increased mortality after extended liver resection. Alcohol, another common cause of fatty liver, is the most commonly used intoxicating substance in India with 32% prevalence rate among adult males. The most common manifestation of alcoholic liver disease is fatty liver. The clinical course, complications, and outcome of alcoholic fatty liver are different from those of NAFLD, and a detailed discussion on which is out of scope for this chapter. Hence, this chapter will be focused on implications of NAFLD only.

II. MECHANISTIC PATHOGENESIS

The pathogenesis of NAFLD is still poorly understood. It may be a consequence of multi-hit process (Figure 59.1). According to this, the first hit is macrovesicular steatosis which is believed to be triggered by insulin resistance. The second hit involves adipo-cytokines alteration, oxidative stress and mitochondrial dysfunction leading to hepatocyte injury, inflammation and fibrosis which result in development of non alcolic steatohepatitis (NASH), cirrhosis, and hepatocellular carcinoma (HCC). Hepatic iron, leptin, anti-oxidant deficiencies,

737

TABLE 59.1   Etiology of non-alcoholic fatty liver Disease

Etiology of Non-alcoholic Fatty Liver Disease	
Insulin resistance	(a) Metabolic Syndrome
	Obesity
	Diabetes Mellitus
	Hypertriglyceridemia
	Hypertension
	(b) Lipoatrophy
	(c) Mauriac Syndrome
Inborn error of metabolism	(a) Abetalipoproteinemia
	(b) Hypobetalipoproteinemia
	(c) Andersen's disease
	(d) Weber-Christian syndrome
	(e) Glycogen Storage Disease
Drugs	(a) Amiodarone
	(b) Methotrexate
	(c) Tamoxifen
	(d) Steroids
	(e) Tetracycline
	(f) L-Asparaginase
	(g) Estrogen
Severe weight loss	(a) Jejunoileal bypass
	(b) Gastric bypassa
	(c) Severe starvation
Others	(a) Total parental Nutrition
	(b) Refeeding Syndrome
	(c) Toxins: dichlorethylene, ethyl bromide
	(d) Inflammatory Bowel Disease
	(e) Severe Anemia

and endogenous alcohol production have all been suggested as potential factors involved in NAFLD pathogenesis. Some reports have suggested role of small intestinal bacterial overgrowth, endotoxin from gut-derived bacteria and lipopolysaccharide in the pathogenesis of NAFLD. The peripheral insulin resistance might contribute to steatosis by decreasing insulin suppression of lipolysis and thus increasing delivery of free fatty acids to the liver. The environmental factors and heritability appear to influence the development of NAFLD. An increased intake of dietary fat, soft drinks, and meat has been associated with an increased risk of NAFLD, independently of age, gender, adiposity and total calories. Several inherited disorders of lipid metabolism are associated with hepatic steatosis. In addition, polymorphisms in genes affecting lipid metabolism, oxidative stress, insulin resistance and immune regulation have been identified as predisposing factors to the development of hepatic steatosis and the development of progressive liver injury.

III.   CLINICAL PROFILES AND DIAGNOSIS

Patients with NAFLD are generally asymptomatic, although some report fatigue or right upper quadrant abdominal discomfort which may be mistaken for irritable bowel syndrome or gallbladder disease. Hepatomegaly is commonly noted on physical examination. These patients may have features of metabolic syndrome, and health problems directly relating to excess adiposity, such as arthritis, obstructive sleep apnoea, dyspepsia, gastroesophageal reflux disease, dyspnoea, hypertension, social stigmatization, and depression. Patients who have advanced liver disease may present with muscle wasting, jaundice, gastrointestinal bleeding, or ascites. NAFLD is the most common cause of mild to moderate and asymptomatic elevation of plasma aminotransferases. The ratio of aspartate aminotransferase to alanine aminotransferase is usually less than 1 in the presence of steatosis, but this ratio is usually more than 1 in alcoholic fatty liver and advanced parenchymal fibrosis. Serum alkaline phosphatase and gamma-glutamyltransferase are often above normal ranges. Fatty liver can be diagnosed using liver ultrasonography,

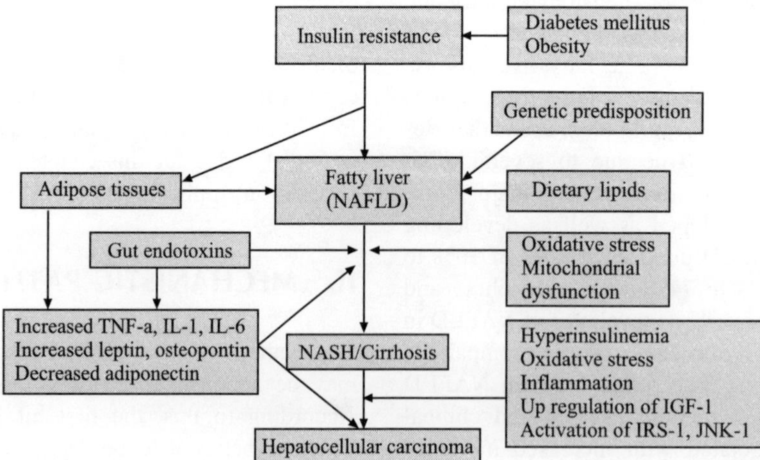

Figure 59.1   The mechanistic pathogenesis of NAFLD and its progression to NASH, cirrhosis, and hepatocellular carcinoma. It appears to involve a complex interplay between genetic predisposition and the mechanisms described earlier. *Abbreviations:* FFA—free fatty acids, TNF—tumour necrotic factor, IL—interleukin, ER—endoplasmic reticulum, NASH—nonalcoholic steatohepatitis, IGF-1—insulin-like growth factor-1, IRS1—insulin receptor substrate-1, JNK-1-c—Jun amino-terminal kinase 1.

computed tomography (CT) scan, magnetic resonance imaging (MRI), MR spectroscopy (MRS), and controlled *attenuation parameter using Fibroscan.* Ultrasound is the most commonly used imaging modality, and it has acceptable sensitivity and specificity. On ultrasound, liver looks enlarged and has diffusely increased echogenicity. The normal echogenicity of the liver is determined by comparing the liver echogenicity with that of the cortex of the kidney. Hepatic echogenicity is usually equal to or greater than that of the renal cortex. In severe fatty liver, acoustic penetration may be decreased resulting in indistinctness of blood vessels and the diaphragm. The presence of >33% fat on liver biopsy has been shown to be optimal for detecting fatty liver on radiological imaging. As the amount of fat dropped below this, the sensitivity of these tests diminishes. Non-contrast CT is more sensitive to detect and quantify steatosis. However, MRI with chemical shift sequence detects microscopic fat better than CT. Liver fat can be most reliably detected and quantified by proton MRS. While all of these imaging methods can detect a significant accumulation of fat, none can differentiate steatosis from more advanced stages of NAFLD. Histological examination of the liver is the only way to make this distinction. However, liver biopsy itself has several problems. Liver biopsy is associated with sampling error, inter and intra-observer variability, expense, invasiveness and risk of complications. Furthermore, fat infiltration of the liver can have a nonuniform distribution in some cases which may lead to inaccurate and misleading information on liver biopsy.

IV. IMPLICATIONS OF FATTY LIVER: AN OVER VIEW

NAFLD was once dismissed as a harmless condition, and was ignored as a significant clinical entity until several reports documented the development of liver failure in some patients following surgical jejunoileal bypass for morbid obesity. The liver histology in such patients was identical to that seen in alcoholic hepatitis. Subsequently, emerging evidences demonstrated strong associations between NAFLD and several other life threatening diseases and clinical conditions (Figure 59.2). A significant proportion of these patients develop progressive liver injury, and some patients may develop cirrhosis and hepatocellular carcinoma (HCC). NAFLD is associated with increased mortality from cardiovascular diseases, malignancy and hepatic complications. NAFLD has also been found to be associated with extrahepatic disorders such as colorectal neoplasm, breast cancer, hypothyroidism, polycystic ovary syndrome, gall stone disease, renal dysfunctions. These patients may have health problems directly relating to excess adiposity, which may include arthritis, obstructive sleep apnoea, dyspepsia, gastroesophageal reflux disease, hypetension, and adverse pregnancy outcome. Unrecognized NAFLD constitutes a substantial proportion of patients with cryptogenic cirrhosis. NAFLD is associated with severity of other chronic hepatitis, such as chronic hepatitis C virus infection, alcoholic liver disease, and hemochromatosis. The fatty liver has poor ability to regenerate after volume loss which may lead to development of liver failure and increased mortality after extended liver resection. Also, transplantation of steatotic liver results in increased rate of poor graft function, primary nonfunction of graft, and poorer outcome. There is high frequency of recurrence of fatty liver disease in patients transplanted for NASH.

V. NATURAL HISTORY, DISEASE PROGRESSION AND OVERALL MORTALITY

The natural history of NAFLD is poorly known because of lack of long-term follow-up, no standardized definitions, preponderance of data from tertiary care centers, and presence of confounding factors. Population-based studies to determine the

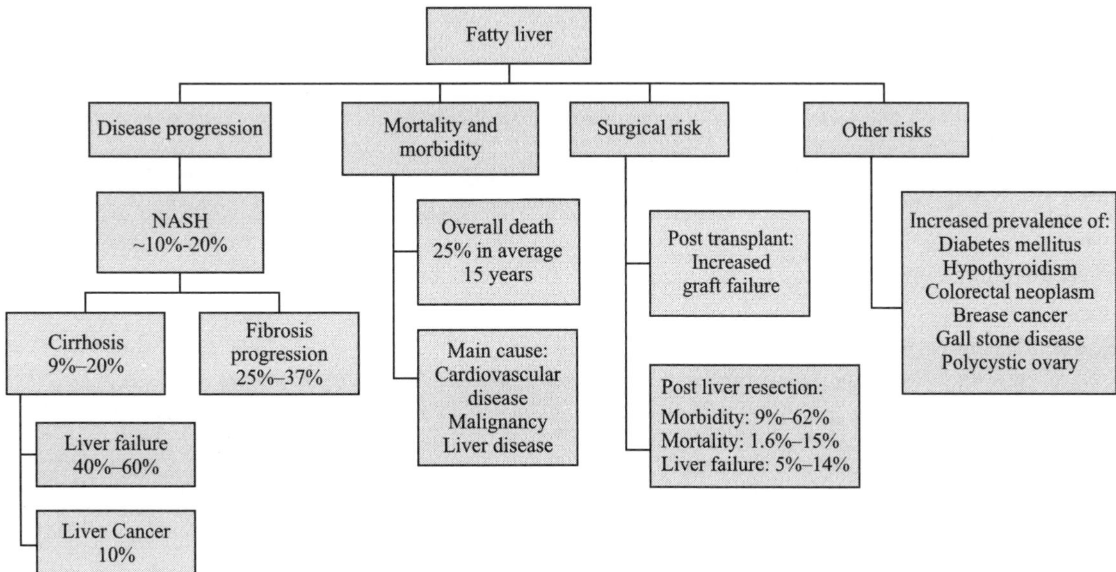

Figure 59.2 Fatty liver: natural history, mortality, and implications in surgical and other diseases.

long-term outcome of such patients is limited. The progression of liver disease in NAFLD varies across the different stages of NAFLD. While progressive liver disease is quite common in patients with NASH, it is not clearly known what proportion of patients with simple steatosis (SS) progress to advanced liver disease. Unconfirmed studies suggest that about 20% of patients with SS progressed to NASH in 3 years. It appears that about 9% to 20% of NASH progresses to cirrhosis in about 15 years. NASH may progress insidiously to cirrhosis, and NASH is now believed to be major cause of cryptogenic cirrhosis. A systematic review of ten longitudinal studies comprising two hundred twenty one ($n = 221$) biopsy proved NASH patients who had undergone two liver biopsies at median interval of 3.7 years revealed an overall progression of fibrosis in 37% patients. While 13% patients had 2 stages progression in fibrosis, 24% had one stage increase in fibrosis. Notably, even regression of fibrosis was noted in about 21% of patients.

Several large community-based studies have found increased mortality in NAFLD patients compared to the expected mortality of the general population of same age and sex (Table 59.2). A Danish cohort study which followed 7372 individuals with fatty liver, including 1770 with NAFLD, found that the mortality was 2.6-fold higher among patients with NAFLD as compared with the Danish general population. The stage of NAFLD is important determinant of long term prognosis. Patients with SS have not been shown to have higher liver related mortality. A recent study has demonstrated a liver-related mortality of 17.5% for the NASH cohort compared with only 2.7% for the non-NASH cohort with a median follow-up of 18.5 years. A meta-analysis of 7 longitudinal studies with follow-up ranging from 7.3 to 24 years has revealed an overall mortality 1.5 times higher in NAFLD compared to the general population. The most common cause of death in NAFLD was malignancy (28% of all death), closely followed by cardiovascular disease (25% of all death) and liver disease (13% of all deaths). While liver-related mortality is higher in NASH compared to SS (Odds ratio: 5.71), cardiovascular mortality, interestingly, does not differ between SS and NASH.

VI. CARDIOVASCULAR DISEASE RISK

Cardiovascular disease (CVD) is among the most important causes of morbidity and mortality in patients with NAFLD. The increased cardiovascular risks are independent of the risk conferred by traditional risk factors. NAFLD is associated with higher Framingham risk score in general population. A systematic review and meta-analysis of cross-sectional studies ($n = 3497$ subjects) has confirmed that NAFLD is strongly associated with increased carotid-artery intimal medial thickness and an increased prevalence of carotid atherosclerotic plaques. A population-based study has suggested premature atherosclerosis in patients with NAFLD. Overall death from CVD in NAFLD ranges from 13% to 30%. A recent systematic review of 8 longitudinal studies with follow up ranging from 3.3 years to 24 years has found that pooled odds ratio for CVD mortality in NAFLD compared to general population was 2.16.

VII. RISK OF MALIGNANCY

Patients with NAFLD have increased risk of both hepatic and extra-hepatic malignancy. A large study by the American Cancer Society demonstrated increased death rates from all types of cancers among obese patients. Compared to patients with normal BMI, the relative risk of mortality from HCC is 1.68 times higher in women and 4 times higher in men with BMI > 35 kg/m^2. There are multiple established risk factors for the development of HCC in NAFLD including obesity, diabetes, and iron deposition. HCC is the third greatest cause of liver-related deaths in NAFLD patients, responsible for 0.5% of overall deaths. About 4% to 27% of NASH patients develop HCC after onset of cirrhosis. Longitudinal studies over 19.5 years have reported the prevalence of HCC in NAFLD to be 0%–0.5%, and in NASH to be 0%–2.8%. NAFLD is a risk factor of HCC even without cirrhosis, however, the risk appears to be minimal as shown in a recent systematic review of 61 studies which revealed that cumulative HCC mortality

TABLE 59.2   Long-term mortality in patients with different stages of nonalcoholic fatty liver disease

Study	Diagnosis*	n	Follow-up (years)	Liver related death	Overall death
Matteoni et al 1999	NAFLD	98	8.3	9 (9%)	48 (49%)
Evans et al. 2002	NASH	26	8.7	0	04 (15%)
Dam-Larsen et al 2004	Simple steatosis	109	16.7	01 (0.9%)	27 (24.8%)
Adams et al. 2005	NAFLD	420	7.6	07 (1.7%)	53 (12.6%)
Ekstedt et al. 2006	NAFLD	129	13.7	2 (1.6%)	26 (20.2%)
Sanyal et al. 2006	NASH cirrhosis	152	10	22 (14.55)	29 (19.1%)
Rafiq et al. 2009	NAFLD	131	18.5	12 (9.2%)	78 (59.5%)
Soderberg et al 2010	NAFLD	118	21	09 (7.6%)	47 (39.85)
Bhala N et al. 2011	NAFLD with advanced fibrosis	264	7.1	14 (5.6%) †	33 (13.4%) †

*NAFLD include any stage of NAFLD at baseline
†Death or liver transplantation

in NAFLD cohorts with few or no cases of cirrhosis was only of 0%-3% over 20 years. In contrast, studies from Japan have reported a considerable higher proportion of NASH related HCC patients had non-cirrhotic liver. The typical non-cirrhotic NASH patient who presents with HCC appears to be older, male, and has one or more features of the metabolic syndrome.

Recent evidence demonstrated an association between NAFLD and several extra-hepatic malignancies. Metabolic syndrome has been associated with an increased risk for colorectal cancer in several epidemiologic studies, and colorectal adenoma, a premalignant lesion of colorectal cancer, is closely associated with obesity and NAFLD. In a colonoscopy screening studies, NAFLD patients had a significantly higher prevalence of colorectal adenomas (34.7% vs. 21.5%) and advanced neoplasm (18.6% vs. 5.5%) than healthy controls. NAFLD has an independent association with colorectal adenomas after adjustment for age, sex, body mass index and glucose intolerance. The postulated mechanism involves Insulin and the insulin like growth factor which are involved in cell growth. A recent case-control study has found a significant association between NAFLD and breast cancer. Similar to HCC, there is a roust association between pancreatic cancer and obesity. Likewise NAFLD, nonalcoholic fatty pancreas disease may promote development of chronic pancreatitis and pancreatic cancer. Obesity, diabetes and gallstones are known risk factors for gallbladder cancer

VIII. IMPLICATIONS IN OTHER TYPE OF CHRONIC HEPATITIS

There is unfavorable interaction between fatty liver and other types of chronic hepatitis. NAFLD often add to disease progression in patients with liver disease of various other etiologies. The hepatic steatosis is commonly found in patients with chronic hepatitis C (HCV) virus infection, occurring in about one half of infected patients. Several studies have identified fatty liver as an independent predictor of progressive liver fibrosis in patients with chronic hepatitis C. Hepatic steatosis also conveys an independent risk for HCC development in HCV infected patients. Fatty liver is associated with a poor response to antiviral therapy. The excess body weight has been recognized as a risk factor for the development of cirrhosis in alcoholic liver disease. In subjects with heavy alcohol consumption, excess body weight markedly increases the risk of acute alcoholic hepatitis and cirrhosis. Similarly, in patients with hemochromatosis, the odds of having liver fibrosis increases when steatosis is present.

IX. IMPLICATIONS IN NON-HEPATIC DISEASES

NAFLD is strongly associated with type 2 DM. Also, NAFLD has been identified as an early and independent risk factor of future development of type 2 DM. The majority of patients with type 2 DM have NAFLD, and that as many as 50% or more may have NASH. Both NAFLD and DM may serve as a progression factor for the each other. Patients' with type 2 DM and NASH have more severe hepatic insulin resistance and progressive liver disease. In a cohort of 337 diabetic patients, the presence of NAFLD carried a 2.3-fold increased risk (95% CI 0.9–5.9) of dying from malignancy, after adjusting for age, smoking habits, obesity, diabetes duration, hyperlipidemia, earlier malignancy, and CVD. A higher prevalence of hypothyroidism has been demonstrated in patients with NAFLD compared to controls (21% vs. 9.5%). Patients with hypothyroidism are more likely to have NASH which may suggest additive effect of two diseases on liver disease. Obstructive sleep apnea (OSA) is well known to be associated with hypoxia, insulin resistance and glucose intolerance, and it is now recognized that there is increased risk of NAFLD with OSA. Recent findings suggest that women with NAFLD may be risk factor for development of polycystic ovary syndrome, and vice versa. A systematic review has shown that the prevalence of both polycystic ovary syndrome and non-alcoholic fatty liver disease rises proportionally to the degree of insulin resistance and increases in the mass of adipose tissue.

X. IMPLICATIONS IN HEPATIC SURGERY

The liver is unique in its ability to repair itself after suffering loss of tissue mass from surgical resection. This regenerative capacity has made it possible for surgeons to remove large portions of liver without permanent impairment of function. Liver resection remains the treatment of choice with significant chance of long-term survival for patients with primary or secondary malignant liver tumors and some of benign hepatic lesions. However, sometimes extended liver resection may lead to development of progressive liver failure. Liver failure may be only transient if the liver has the ability to regenerate, but is prolonged when regeneration is impaired. Tolerance to ischemia, reperfusion, remnant liver volume, and postoperative regenerative capacity of hepatocytes are crucial for a good postoperative recovery of liver function and volume after resection. Fatty hepatocytes have reduced tolerance against ischemic injury and the poor ability to regenerate after major hepatic resection, and this predisposes NAFLD patients at higher risk of post liver resection liver failure which is associated with very high mortality, up to 60% to 90%.

A retrospective review, of 135 patients who had undergone major hepatic resection from 1990 to 1993 at Department of Surgery Mayo Clinic, has reported an increased postoperative mortality, morbidity, and longer operative time in the presence of hepatic steatosis. Furthermore, 14% of patients with steatotic liver had acute liver failure versus 4% in patients with normal liver. Another study involving a cohort of 478 liver resection patients showed that steatosis was an independent risk factor for postoperative complications (8% in patients with steatosis versus 2% in patients with normal liver). While simple steatosis may not been associated with a significant increase in mortality, patients with NASH are at higher risk for liver failure and death after major hepatectomy. A recent analysis on outcome after hepatectomy for colorectal liver metastases in 406 patients has demonstrated that patients with steatohepatitis had a significantly higher 90-day posthepatectomy mortality rate than did patients who did not have steatohepatitis (14.7% vs. 1.6%).

XI. IMPLICATIONS IN LIVER TRANSPLANTATION

Fatty liver is detrimental to graft function both in cadaveric and living donor liver transplantation. Because of preponderance of NAFLD in the general population, a preoperative identification of steatosis is important prior to liver transplantation (LT). Studies have revealed a negative impact of steatosis on both recipient and early allograft survival. In a large series of LT patients, those receiving livers with up to 30% of macrovesicular steatosis had a higher rates of primary graft non-function (5.1% vs. 1.8%) and worse 2-yr patient (77% vs. 91%) and graft (70% vs. 82%) survival as compared to patients receiving graft without steatosis. Another large, retrospective study has found an increased initial poor graft function, increased graft loss rates, and higher 3-month patient mortality among recipients of livers with moderate and severe steatosis (>30%). However, grafts containing less than 30% macrosteatosis may be safe for LT. Also, microvesicular steatosis does not appear to affect graft survival. Presence of NAFLD has been found to be an independent predictor of post transplant metabolic syndrome. Many studies have found recurrence of fatty liver disease in patients transplanted for NASH. Fatty change occurs in 60-100% of patients within a few months of transplantation, and approximately 10–40% develops NASH. Approximately 10% of patients progress to bridging fibrosis or cirrhosis by 10 years post-transplant.

XII. PRINCIPLE OF TREATMENT FOR FATTY LIVER DISEASE

The appropriate treatment of fatty liver is not known. The underlying cause should be identified as some of them may be related to alcohol, drugs, toxin, and systemic diseases. For NAFLD, current management strategies are directed at treating the individual components of the metabolic syndrome. Current therapies for NAFLD can be divided into those directed at the abnormal metabolic components and those directed primarily at the liver. Weight loss through diet and exercise has been shown to result in improved insulin resistance and improvement of markers of liver inflammation. For patients with simple steatosis, in view of benign prognosis, these strategies are all that it is required. However, patients who develop obesity over several years often find difficulty with effort at lifestyle modification. For morbidly obese patients (BMI > 35), bariatric

surgery is now a viable option for achieving sustained weight loss. Bariatric surgery induces weight loss, reduces the number of risk factors for fatty liver and reduces the severity of features of metabolic syndrome. In obese NASH patients who underwent a laparoscopic adjustable gastric band procedure, regression of NASH was shown in paired liver biopsies. The reductions in cardiovascular risk have also been observed. Patients with more advanced NAFLD require long-term follow-up and surveillance for complications including esophageal varices and HCC. These patients will be candidates for emerging pharmacological therapies. The rationale for NAFLD therapies is based on a growing understanding of disease pathogenesis with a particular focus on reducing insulin resistance and oxidative stress. There is no universally accepted pharmacologic

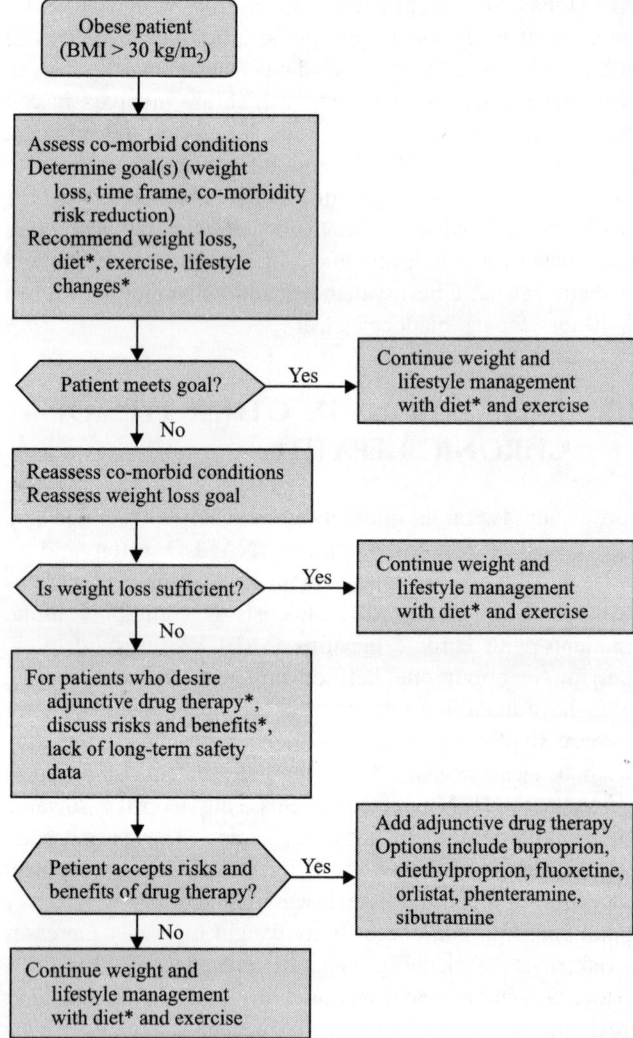

Figure 59.3 Algorithm for managing obesity (BMI = body mass index). *From:* Snow V., Barry P., Fitterman N, Qaseem A., Weiss K., Clinical Efficacy Assessment Subcommittee for the American College of Physicians. Pharmacoloic and Surgical Management of Obesity in Primary Care: A Clinical Practice Guideline from the Americal College of Physicians. *Ann Intern Med.* 2005; 142:525–531.

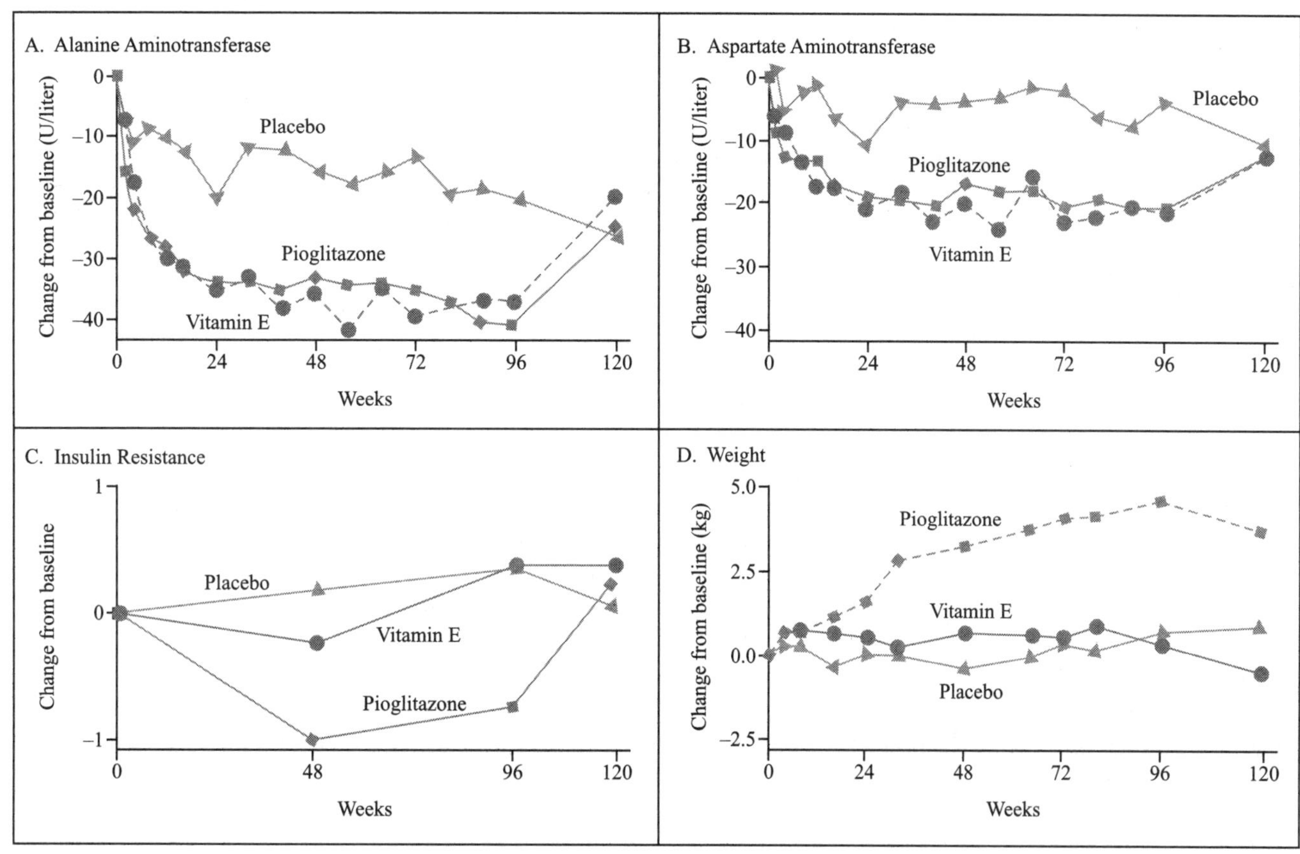

Figure 59.4   The PIVENS Trial on vitamin E versus pioglitazone versus placebo in nonalcoholic steatohepatitis. *From:* Sanyal A.J., Chalasani N., Kowdley KV, et al., Pioglitazone, vitamin E, or placebo for nonalcoholic steatohepatitis. *The New England Journal of Medicine.* 2010; 362(18):1675–1685. (*see Plate 23 for colour figure*)

treatment for NASH. Multiple pharmacologic interventions have been attempted with variable success; these include vitamin E, orlistat, ursodeoxycholic acid, pentoxifilline, metformin, pioglitazones, and lipid-lowering agents. In a recent large study, PIVENS trial vitamin E therapy compared with placebo was found to be associated with a significantly higher rate of improvement in NASH.

SUMMARY AND CONCLUSION

Fatty liver is increasing in worldwide populations and is commonly seen in conjunction with obesity; alcohol use, dyslipidemia, and the presence of type 2 DM. Nonalcoholic fatty liver disease (NAFLD), once dismissed as an innocuous condition, is now considered as the commonest cause of chronic liver disease worldwide. Different imaging procedures are available for the detection of fatty liver, including ultrasonography, computed tomography scan, and magnetic

resonance imaging; however, liver biopsy still remains the gold standard for accurate diagnosis as well as staging of fatty liver disease. There is a strong association between NAFLD and several potentially life threatening diseases. NAFLD may progress to cirrhosis and HCC. Several large community-based studies have found increased mortality in NAFLD patients compared to the expected mortality of the general population of same age and sex. Cardiovascular disease is an important cause of morbidity and mortality in patients with NAFLD, and account for up to 30% of overall death. NAFLD is associated with severity of other chronic hepatitis and several other non-hepatic disorders. The fatty liver is more susceptible to ischemia/reperfusion injury, and has poor ability to regenerate after volume loss. Thus, extended liver resection may lead to development of progressive liver failure and increased mortality. Also, use of grafts with steatotic liver for LT results in increased rates of poor graft function, and poorer outcome. Therefore, surgeons should take steatosis into account when planning the liver resection or transplantation.

SUGGESTIONS FOR FURTHER READING

1. Angulo P. (2002), Nonalcoholic Fatty Liver Disease, *N. Engl. J. Med.,* 346:1221–1231

2. Ascha M.S., Hanouneh I.A., Lopez R., Tamimi T.A., Feldstein A.F. and Zein N.N. (2010), The Incidence and Risk Factors of Hepatocellular Carcinoma in Patients with Nonalcoholic Steatohepatitis, *Hepatology*, 51:1972–8.

3. Adams L.A., Lymp J.F., St Sauver J., Sanderson S.O., Lindor K.D., Feldstein A. and Angulo P. (2005), The Natural History of Nonalcoholic Fatty Liver Disease: A Population-based Cohort Study, *Gastroenterology*, 129:113–21.

4. Wong V.W., Wong G.L., Tsang S.W., et al. (2011), High Prevalence of Colorectal Neoplasm in Patients with Non-alcoholic Steatohepatitis, *Gut*, 60:829–36.

5. Wong V.W., Wong G.L., Choi P.C., et al. (2010), Disease Progression of Non-alcoholic Fatty Liver Disease: A Prospective Study with Paired Liver Biopsies at 3 years, *Gut*, 59:969–74.

6. Clark J.M. and Diehl A.M. (2003), Non-alcoholic Fatty Liver Disease: An Underrecognized Cause of Cryptogenic Cirrhosis, *JAMA*, 289:3000.

7. Bugianesi E., Leone N., Vanni E., et al. (2002), Expanding the Natural History of Non-alcoholic Steatohepatitis: From Cryptogenic Cirrhosis to Hepatocellular Carcinoma, *Gastroenterology*, 123:134–140.

8. Jepsen P., Vilstrup H., Mellemkjaer L., et al. (2003), Prognosis of Patients with a Diagnosis of Fatty Liver—A Registry-based Cohort Study, *Hepatogastroenterology*, 50:2101–4.

9. Matteoni C.A., Younossi Z.M., Gramlich T., et al. (1999), Non-alcoholic Fatty Liver Disease: A Spectrum of Clinical and Pathological Severity, *Gastroenterology*, 116:1413–1419.

10. Dam-Larsen S., Franzmann M., Andersen I.B., et al. (2004), Long-term Prognosis of Fatty Liver: Risk of Chronic Liver Disease and Death, *Gut*, 53:750–755.

11. Sanyal A.J., Banas C., Sargent C., et al. (2006), Similarities and Differences in Outcomes of Cirrhosis Due to Non-alcoholic Steatohepatitis and Hepatitis C., *Hepatology*, 43:682–9.

12. Musso G., Gambino R., Cassader M., et al. (2011), Meta-analysis: Natural History of Non-alcoholic Fatty Liver Disease (NAFLD) and Diagnostic Accuracy of Non-invasive Tests for Liver Disease Severity, *Ann Med.*, 43:617–49.

13. Sookoian S. and Pirola C.J. (2008), Non-alcoholic Fatty Liver Disease is Strongly Associated with Carotid Atherosclerosis: A Systematic Review, *J. Hepatol*, 49:600–607.

14. Calle E.E., Rodriguez C., Walker-Thurmond K., et al. (2003), Overweight, Obesity, and Mortality from Cancer in a Prospectively Studied Cohort of U.S. Adults, *N. Engl J. Med.*, 348:1625–1638.

15. Siegel A.B. and Zhu A.X. (2009), Metabolic Syndrome and Hepatocellular Carcinoma: Two Growing Epidemics with a Potential Link, *Cancer*, 115:5651–5661.

16. Starley B.Q., Calcagno C.J. and Harrison S.A. (2010), Nonalcoholic Fatty Liver Disease and Hepatocellular Carcinoma: A Weighty Connection, *Hepatology*, 51:1820–1832.

17. Ahmed R.L., Schmitz K.H., Anderson K.E., et al. (2006), The Metabolic Syndrome and Risk of Incident Colorectal Cancer, *Cancer*, 107:28–36.

18. Lonardo A., Adinolfi L.E., Loria P., et al. (2004), Steatosis and Hepatitis C Virus: Mechanisms and Significance for Hepatic and Extrahepatic Disease, *Gastroenterology*, 126:586–597.

19. Ohata K., Hamasaki K., Toriyama K., et al. (2003), Hepatic Steatosis is a Risk Factor for Hepatocellular Carcinoma in Patients with Chronic Hepatitis C Virus Infection, *Cancer*, 97:3036–3043.

20. Musso G., Olivetti C., Cassader M., et al. (2012), Obstructive Sleep Apnea-hypopnea Syndrome and Non-alcoholic Fatty Liver Disease: Emerging Evidence and Mechanisms, *Semin Liver Dis.*, 32:49–64.

21. Redaelli C.A., Wagner M., Krahenbuhl L., et al. (2002), Liver Surgery in the Era of Tissue-preserving Resections: Early and Late Outcome in Patients with Primary and Secondary Hepatic Tumors, Redaelli et al., *World J. Surg.*, 26:1126 –1132.

22. Behrns K.E., Tsiotos G.G., DeSouza N.F., et al. (1998), Hepatic Steatosis as a Potential Risk Factor for Major Hepatic Resection, *J. Gastrointest Surg.*, 2:292–298.

23. Sato N., Eguchi H., Inoue A., et al. (1986), Hepatic Microcirculation in Zucker Fatty Rats, *Adv Exp Med Biol.*, 200:477–483.

24. Adam R., Reynes M., Johann M., et al. (1991), The Outcome of Steatotic Grafts in Liver Transplantation, *Transplant Proc.*, 23:1538–1540.

25. Soejima Y., Shimada M., Suehiro T., Kishikawa K., Yoshizumi T., Hashimoto K., et al. (2003), Use of Steatotic Graft in Living-donor Liver Transplantation, *Transplantation*, 76:344–8.

26. Malik S.M., DeVera M.E., Fontes P., et al. (2009), Outcome After Liver Transplantation for NASH Cirrhosis, *Am J. Transplant*, 9:782–93.

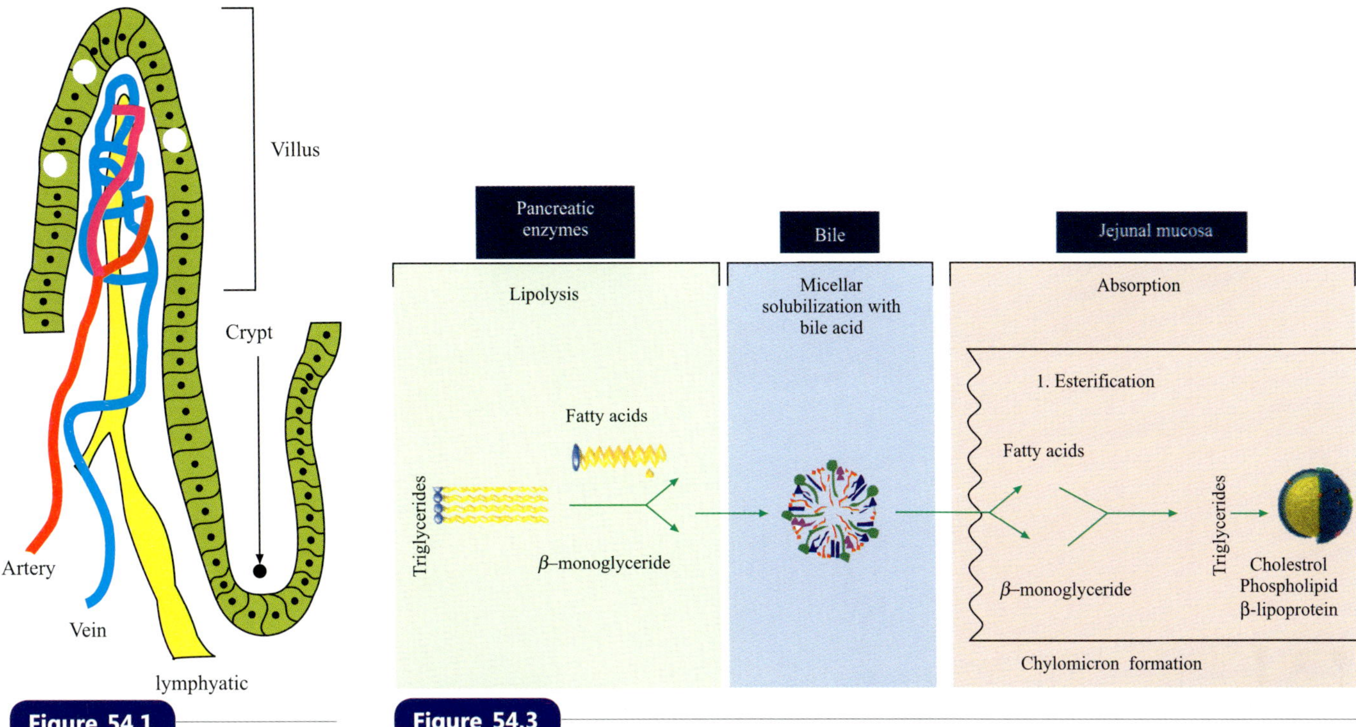

Figure 54.1

Figure 54.3

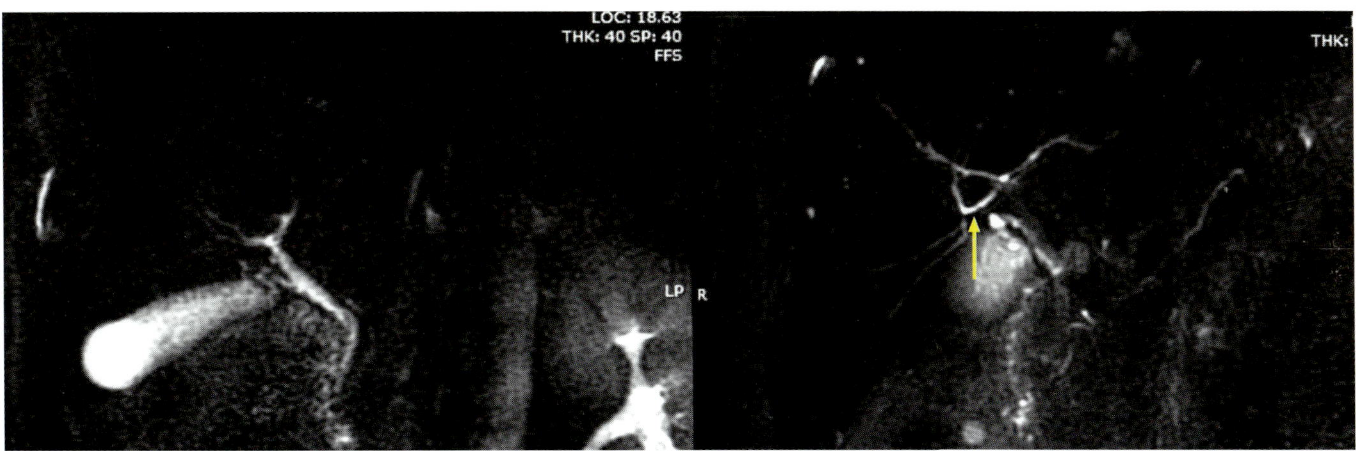

Figure 56.1

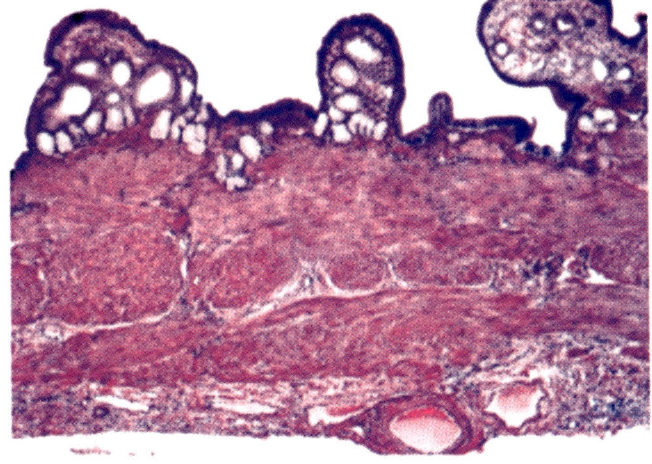

Figure 56.2

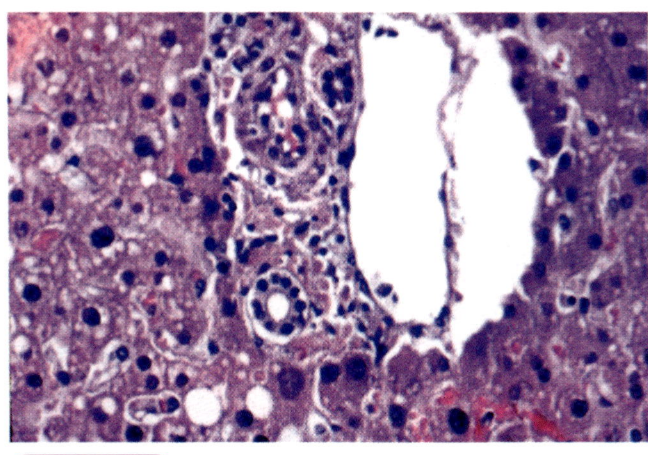

Figure 56.5

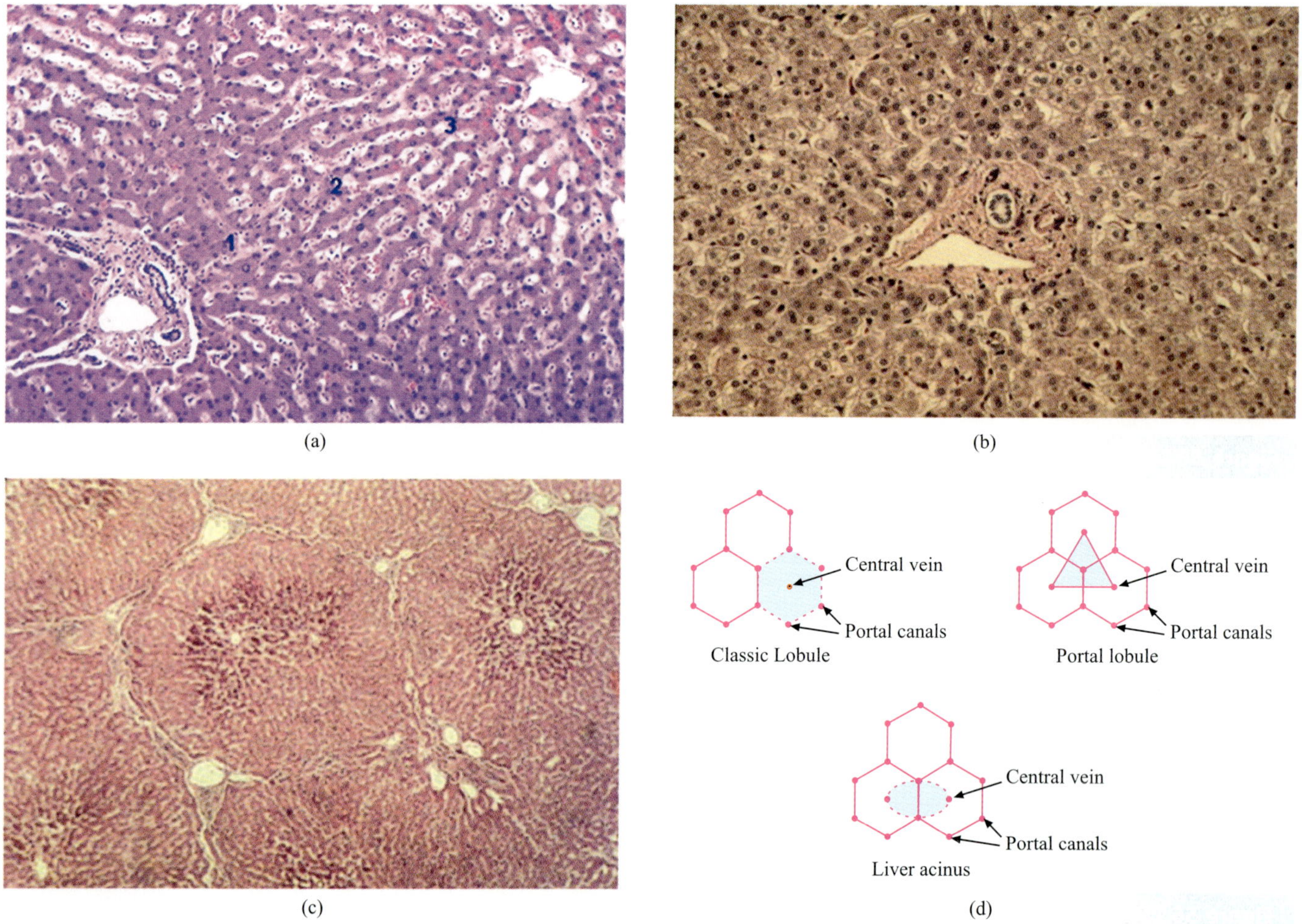

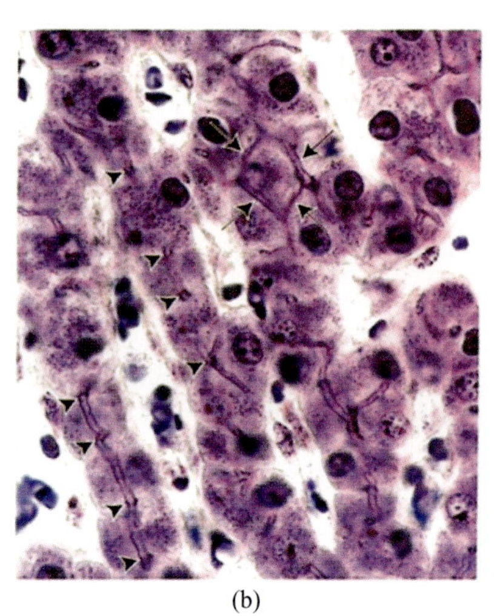

(a)

(b)

(c)

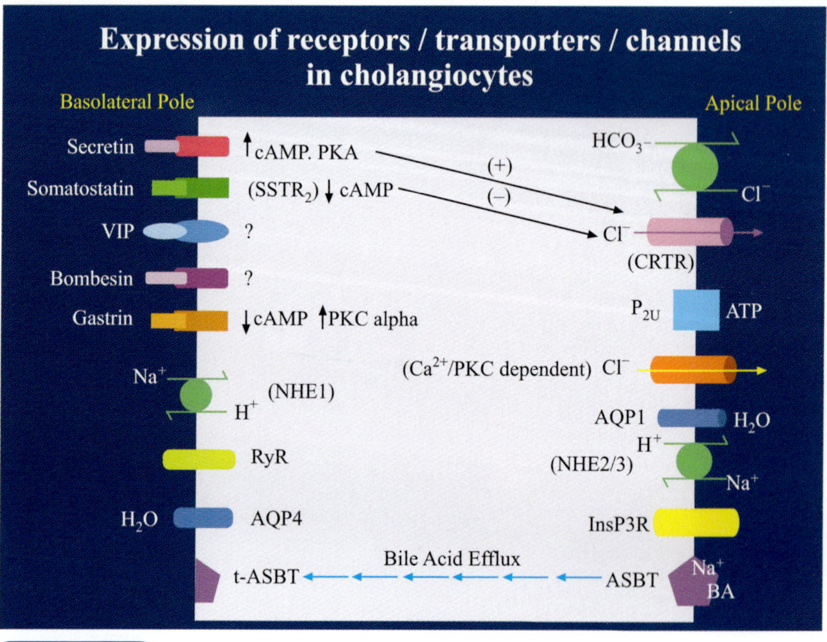

(d)

Figure 56.7

(b)

Figure 56.8

Figure 56.9

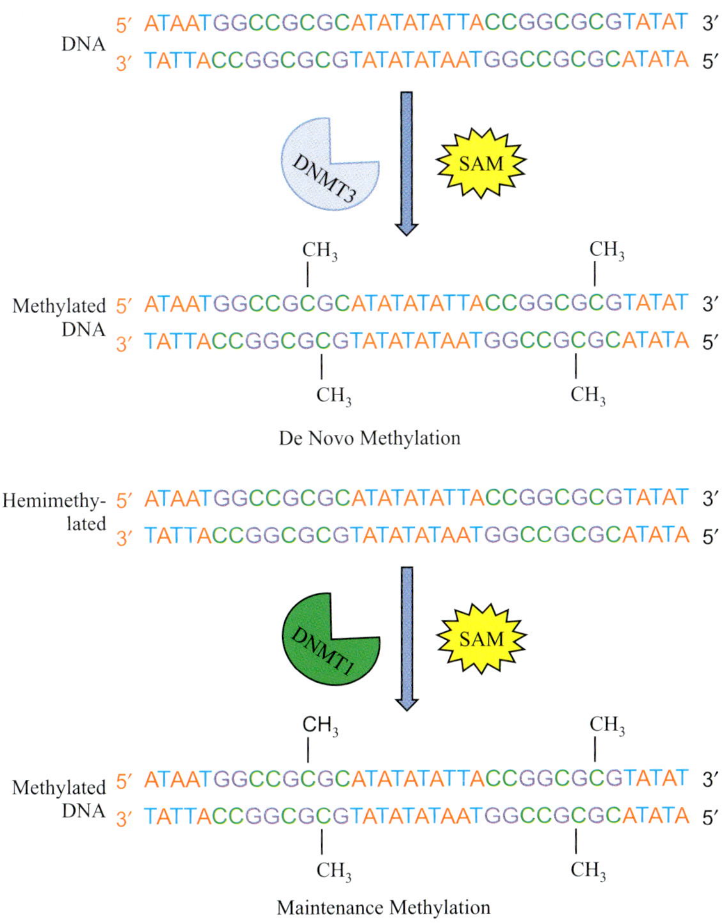

Figure 70.2

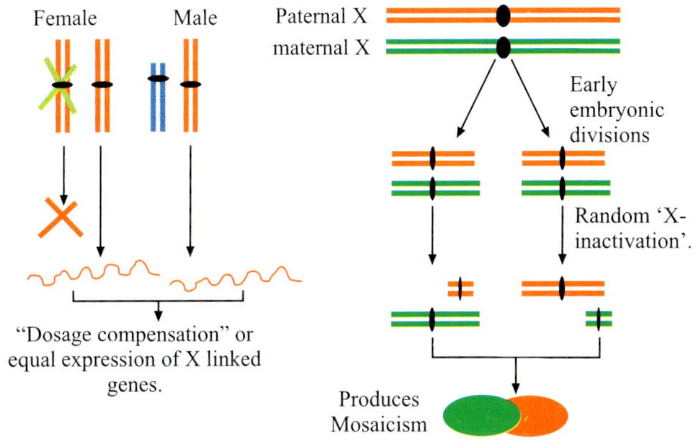

Figure 70.3

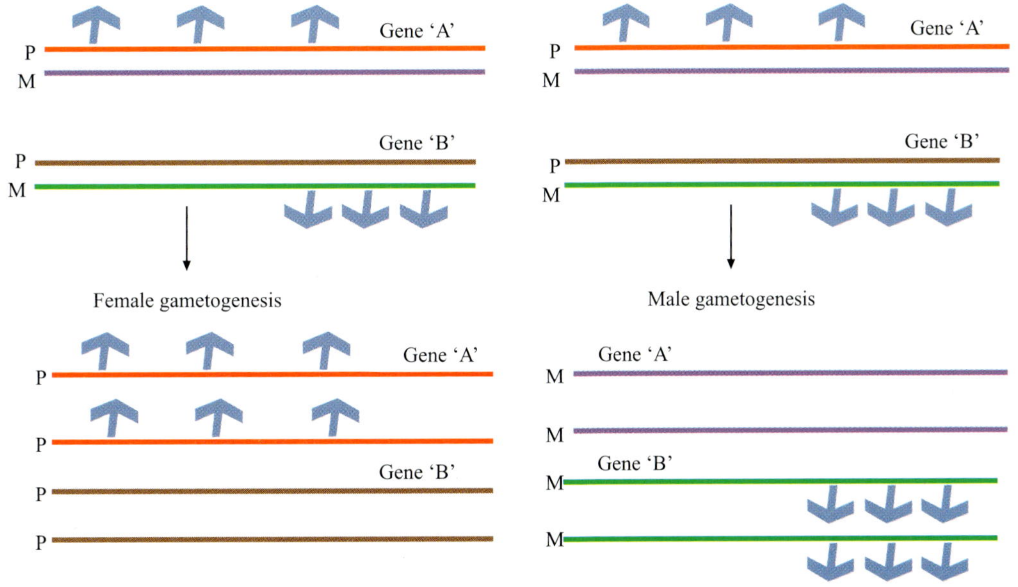

Figure 70.5

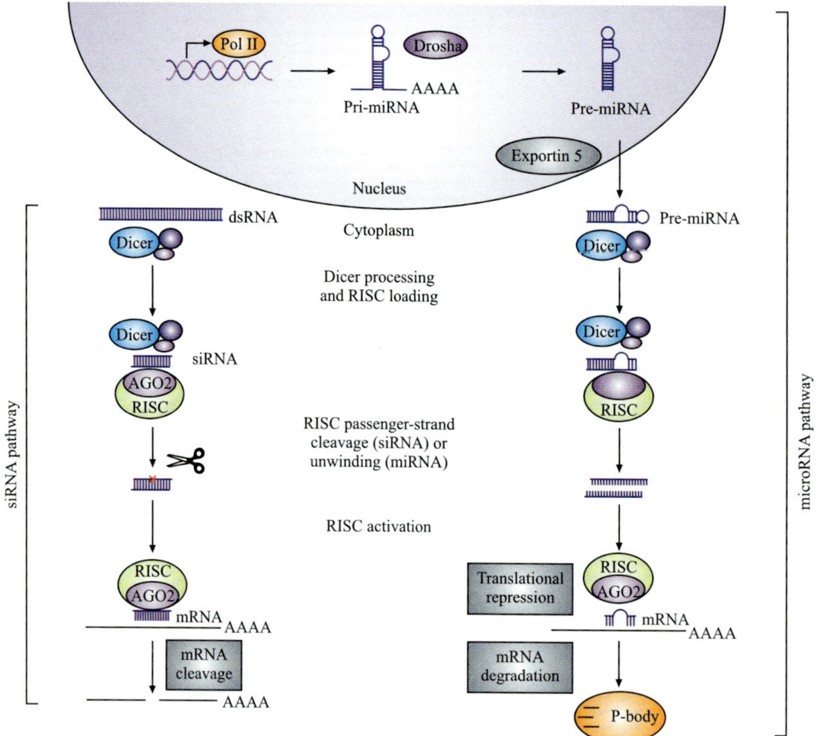

Figure 71.1

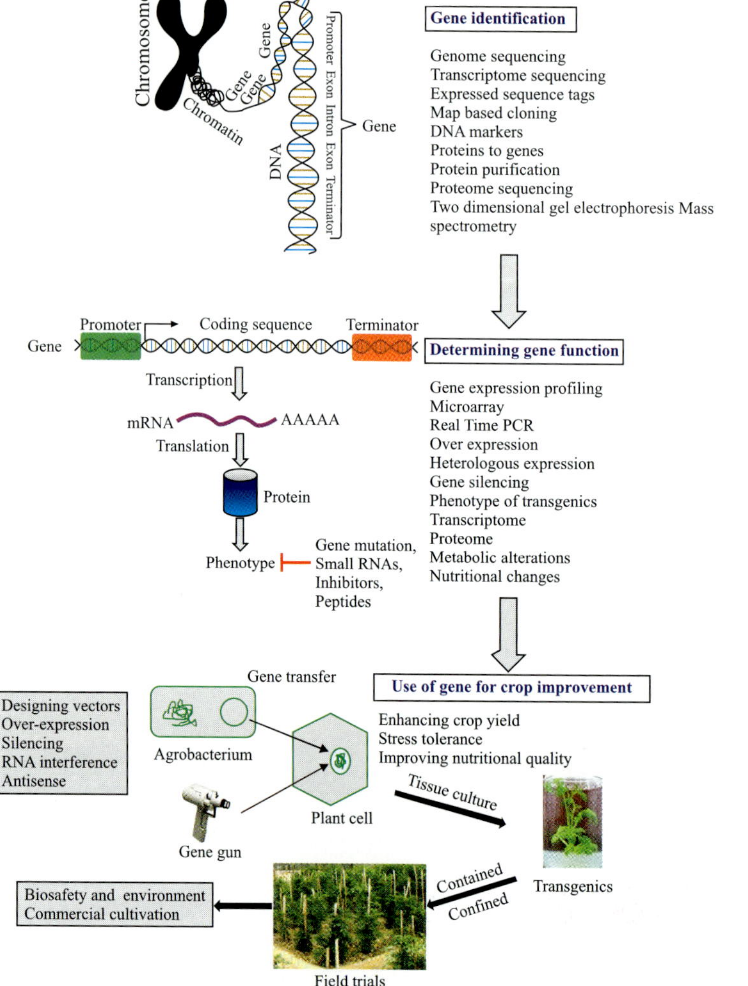

Figure 72.2

SECTION VII

PROBIOTICS

The human body harbours "friendly" microorganisms, termed as Probiotics in various tissues. They play pre-dominant role in gastrointestinal tract and vagina. A "healthy" vagina has a pH of about 4.5 owing to the essential presence of Probiotic Latobcilli.

60

Probiotics
Role in Health and Intestinal Diseases

Neerja Hajela

CONTENTS

I. INTRODUCTION

The human body is home to *far more microbial cells* than human cells and is inhabited by a vast number of bacteria, archaea, viruses and unicellular eukaryotes. The collection of microorganisms that live in peaceful coexistence with their host has been referred to as the microbiota, microflora or normal flora and the genes that they encode is known as the microbiome (Kunz C et al., 2009; Morelli L, 2008; Neish AS et al., 2009).

The past decade has been phenomenal in providing tremendous insights into the previously unsuspected enormous diversity of the microbiota and it has only recently been estimated that we harbor at least *100 trillion microbial cells* that encode 100 fold more unique genes than their own genome. (Ley et al., 2006). This microbiota colonizes virtually every surface of the human body that is exposed to the external environment. Microbes are found on our skin, genitourinary, gastrointestinal and respiratory tracts (Chiller K et al., 2001; Hull MW et al., 2007; Neish AS et al., 2009; Verstraelen H et al., 2008). By far the most heavily colonized organ is the gastro intestinal tract (GIT), the colon itself is estimated to contain over 70% of all the microbes in the human body (Ley RE et al., 2006; Whitman WB et al., 1998).

II. THE INTESTINAL MICROBIOTA

Illuminating work from the Meta Hit and Human Microbiome projects have thrown light on the fact that the gut harbors the largest consortium of microorganisms (Nelson KE et al., 2010) which has the ability to function as a "virtual" organ system (Forsythe P et al., 2010). Advanced high throughput sequencing tools have galvanized the study of the molecular cross talk between commensal organisms and the host and it is now understood that the gut microbiota aids host nutrition and maintains homeostasis. This forgotten organ is also implicated in *epithelial cell turnover, gastrointestinal motility, breaking down dietary toxins and carcinogens, synthesizing micronutrients, fermenting indigestible food substances, assisting in the absorption of certain electrolytes and trace minerals, and affecting the growth and differentiation of enterocytes and colonocytes through the production of short-chain fatty acids. The gut microflora also plays an important role in augmenting the host's innate and acquired immune response by interacting with the immune system and providing signals to promote the maturation of the immune cells* and the normal development of the immune functions (Chow et al., 2010).

The composition and activity of the microbiota has a profound influence on health and a mutualistic relationship between the beneficial symbionts and commensals is very important for the maintenance of wellbeing; alterations in the balance of the intestinal microflora results in dysbiosis and ultimately in clinical disease (Round et al., 2009). Our understanding of human physiology has thus transformed and created a new paradigm in the way we think and regulate health and disease. The focus is shifting from individual pathogens to an ecological approach that considers the microbial community as a whole where disease prone microbiota pattern can be re modeled to a more robust, resilient and disease free-state.

There is no consensus on what constitutes an ideal intestinal microbiota but a healthy microbiota is one that is *predominantly saccharolytic*, containing large numbers of *Lactobacillus* and *Bifidobacteria* so that their activities predominate over the activities of the potentially harmful microbes. The end products from asaccharolytic fermentation may be considered benign or even positive. Metabonomic studies have demonstrated that probiotics can modulate the gut microbiome and metabolism of short chain fatty acids, amino acids, bile acids and plasma lipo proteins demonstrating the diversity of symbiotic co-metabolic connections between the gut microbial content and the host.

A. Early Colonization and Evolving Development of the Microbiota

The human infant is born with a sterile gut but is rapidly colonized by microorganisms from the mother and immediate environment shortly after birth (Ley RE et al., 2006). Various factors influence the development of microbiota in early infancy such as the route of delivery, gestational age, use of antibiotics in the perinatal period, especially in the neonatal intensive care unit setting. *Increased colonization with Bifidobacteria* and not Lactobacilli is *associated with vaginal births as compared to caesarean section deliveries* (Chen J et al, 2007). By contrast caesarean section births are associated with increased colonization by Klebsiella, Enterobacter and Clostridia (Conroy ME et al., 2009) which are organisms prevalent in hospital settings. Diet and host genetics also have a profound influence on the microbiota's subsequent development. Breast Feeding is associated with an increased presence of Bifidobacteria whereas formula fed babies have a more complex adult like microbiota with increased presence of both Clostridia and Bacteroides (Conroy ME et al., 2009).

In humans the microbiota evolves during the different stages of life from infancy to adulthood to old age (Mariat D et al., 2009). Early environmental exposure to microbes is fundamental in shaping an individual's microbiota (Turnbaugh PJ et al., 2009) and although the diversity of microorganisms is very low in the infant gut, it climbs through early development. A detailed time series of a single infant demonstrates that the phylogenetic diversity increases gradually over time. (Figure 60.1).

B. Composition of the Intestinal Microbiota in the Gastro Intestinal tract

The highly complex collection of microorganisms varies considerably in composition and numbers along the length of the gastrointestinal tract (Figure 60.2). Bacterial numbers are lowest in the stomach and duodenum due to the low pH (1–3) and fast transit time, with numbers typically around 10^3 colony-forming units (cfu) per ml of the contents (Holzapfel et al., 1998). The bacterial numbers increase as we traverse to the jejunum and ileum that has a relatively higher pH although

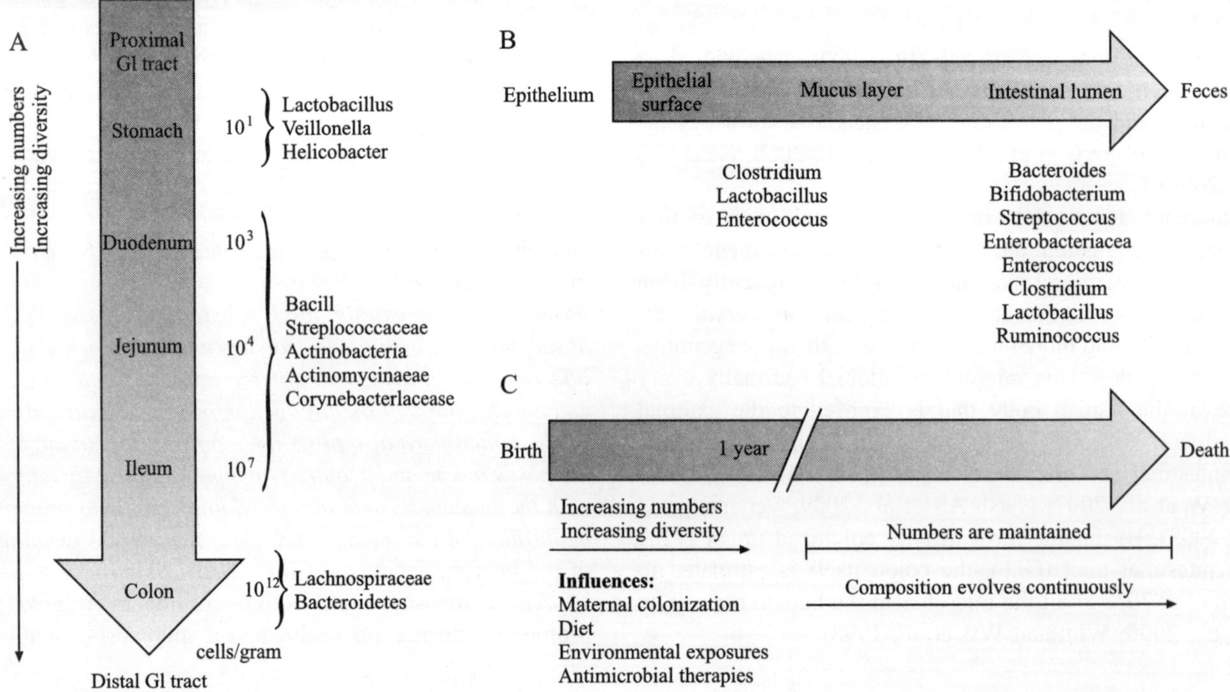

Figure 60.1 Spatial and temporal aspects of intestinal microbiota composition (Sekirov I et al, 2010).

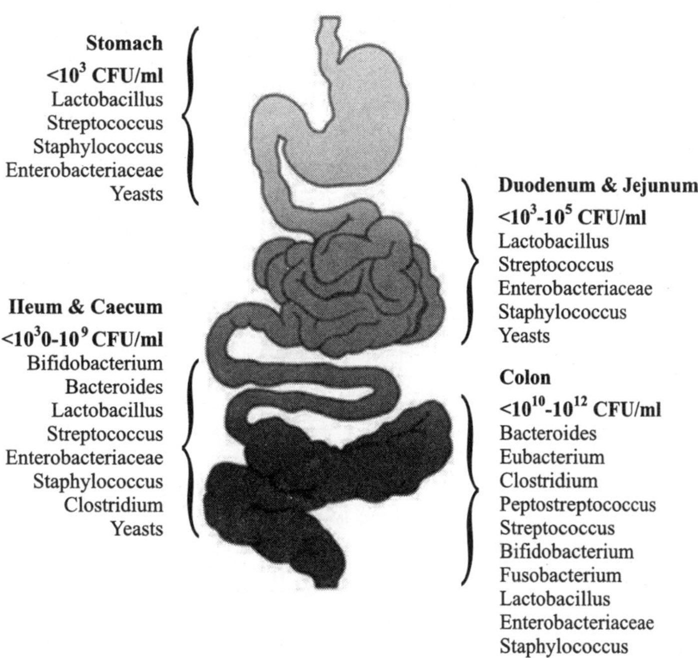

Figure 60.2   Distribution of dominant bacterial population groups in the various sections of the human gastrointestinal tract (BlautM et al, 2003).

diversity is limited by rapid transit time and digestive secretions such as bile acids and pancreatic juices. The main inhabitants are mainly acid tolerant bacteria: Lactobacilli, Streptococci and Enterobacteria. In the large intestine, the levels can be as high as 10^{12} cfu/g due to the slower transit, favorable pH, substrate availability and oxygen levels with a predominance of Bacteroides, Eubactrium and Bifidobacterium.

C. Factors Affecting the Gut Microflora (Dysbiosis)

Although the microbiota is generally stable within individuals over time, the composition of the intestinal flora fluctuates over time and can be negatively altered due to external pertubations such as poor diet, lifestyle habits, stress, indiscriminate use of antibiotics and other medication (Figure 60.3). In later years of life, there is a gradual shift in numbers and an increase in diversity of the microbial population which may be considered unhealthy: in the elderly Bifidobacteria numbers drop while Clostridia and Enterobacteria increase.

III. PROBIOTICS

The probiotic definition has been widely debated, however FAO/WHO have defined Probiotics as "live micro-organisms which when administered in adequate amounts confer a health benefit on the host" (FAO/WHO, 2002). It appears that to provide a potential benefit, probiotics must survive the harsh conditions of the intestinal passage to influence the microflora of the targeted location in the body, which is in most cases the intestinal tract. The desired characteristics of a good probiotic are as follows:

- Non pathogenic and Non toxic and specified by Genus and strain
- Exert a beneficial effect when consumed
- Has the capacity to survive and metabolize in the gut

Probiotic benefits are strain specific, therefore it is important that each probiotic is supported by its own dossier of scientific evidence with relevant studies conducted using a similar intervention as that recommended for daily consumption. There is no minimal recommendation for the probiotic count. Experts differ in their recommendation, most advise a daily intake of 10^8 cells per day.

A. Probiotic Organisms

With the understanding and universal adaption of the probiotic definition given by FAO and WHO, many different species of bacteria as well as yeasts have been used as probiotics. The most common are Bifidobacteria and Lactobacilli; the latter being lactic acid producing bacteria, which have been used for centuries in fermented foods. The yeast *Saccharomyces cerevisiae* and some *E. coli* and *Bacillus* species are also used as probiotics. The term probiotic has however been reserved for live microbes that impart a health benefit in controlled human studies.

Although, probiotics are microorganisms that are not associated with pathogenicity, safety of the probiotic strain should be demonstrated by human and other studies. The careful selection of specific organisms based on desired clinical outcome is imperative in defining a probiotic strain.

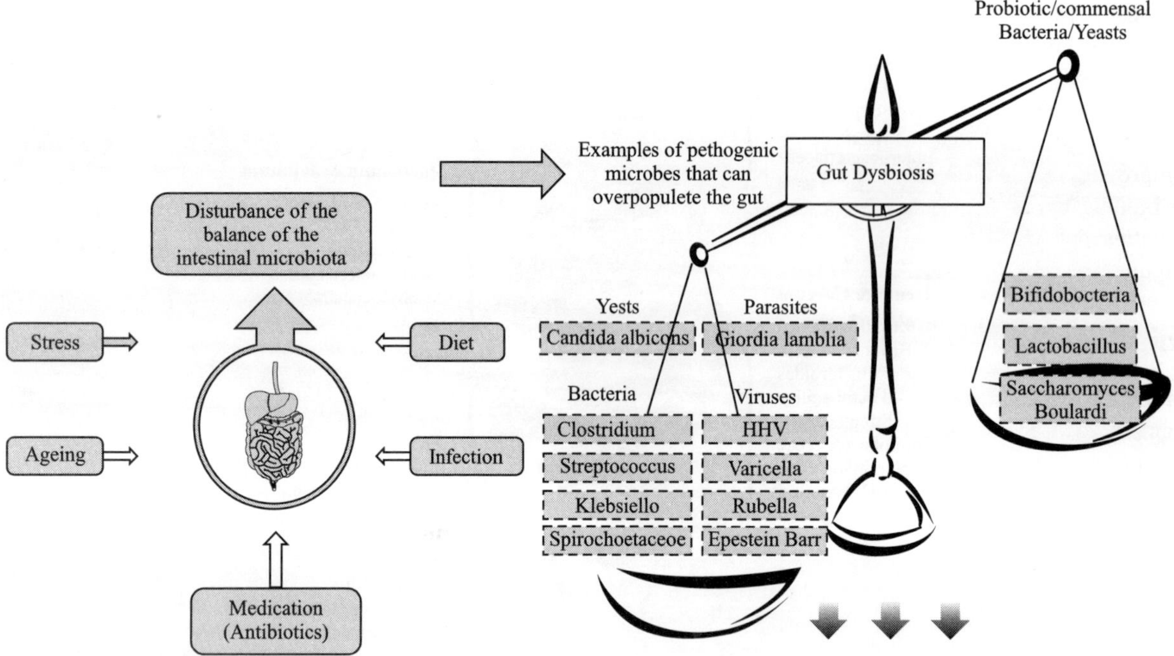

Figure 60.3   Dysbiosis of Intestinal Microbiota (PNAS March 15, 2011 Suppl.).

B.   Probiotic Mechanisms

The precise mechanism influencing the cross talk between the microbe and the host remains unclear but there is growing evidence to suggest that probiotics have multiple and diverse influences on the host. They act through diverse mechanisms that affect the composition and function of the commensal gut microbiota. There are three general classes of probiotic mechanisms: *Direct Antagonism, Exclusion and Immune modulation.*

Direct antagonism

Probiotics can antagonize pathogenic bacteria by producing antibacterial substances called *bacteriocins* and *defensins* that have antimicrobial activities. The inhibitory activities of bacteriocins varies; some inhibit other lactobacilli or taxonomically related Gram-positive bacteria and some are active against a much wider range of Gram-positive and Gram-negative bacteria as well as yeasts and molds (Nemcova R, 1997). Mice pre-treated with *Lactobacillus salvarius*, strain UCC118 are protected from *Listeria* infection, as evidenced by reduced pathogen translocation to the liver and spleen in a manner that is dependent on the production of a bacteriocin (Corr SC et al., 2007). *Escherichia coli* preparation induces the production of human beta defensin 2 from colonic epithelial cells that has potent antimicrobial activity *in vitro* (Mondel M et al., 2009). Several *in vitro* studies have also reported down regulation of virulence factors in pathogens exposed to probiotics or their cell free supernatants.

Immunomodulation

The central role of gut microbiota in the development of immunity is not surprising, considering that the intestinal mucosa represents the largest surface area in contact with the antigens of the external environment (Rakoff-Nahoum S et al., 2008). The intestinal microbiota influences both arms of the immune system: the innate or natural immunity (the non-specific, immediate response) and the adaptive (specific) immune response. Probiotics elicit a variety of responses from immune cells in vitro and in vivo, through mostly unknown mechanisms (Forsythe P et al., 2010). The immunomodulatory effects of probiotics can be species (Christensesn HR et al., 2002) and strain specific (Lin YP et al., 2008; Thomas CM et al. 2010) and involve multiple mammalian signaling pathways that affect immune cell phenotypes. Some probiotic Lactobacillus strains *increase production of tumor necrosis factor* through activation of the transcription factors. Differential immune regulation might prime the immune system to limit infections, inflammation and pathogen-mediated damage. Bacterial probiotic strains also act by functionally modulating dendritic cells and T cell functions through glycoproteins present on their cell surface. *L. acidophilus*, strain NCFM, has a surface-layer protein A, which, when mutated, results in loss of the ability of the probiotic to bind to the dendritic cell receptor (Konstantinov SR et al., 2008). Probiotic bacteria have multifunctional activities, exemplified by the mechanism of action of *Lactobacillus casei* strain Shirota (LcS) whose ability to augment the host's immune system has been extensively examined (Shida K et al., 2011).

Exclusion

Exclusion is a catch all term for probiotic mechanisms that make the GI environment less hospitable for pathogens.

The mechanisms include altering the resident microbiota by producing a physiologically restrictive environment that makes the GI environment less favorable for the growth of pathogens. Probiotics affect this by *reducing the luminal pH*, redox potential and by hydrogen sulfide production. Other mechanisms include improving epithelial barrier function, interfering with pathogen binding by down regulating specific host receptors and stimulating the production of mucin (Meddings J et al., 2008) (Figure 60.4).

C. Clinical Evidence

Given that the intestinal tract is the largest reservoir of microbes in the human body, it is not surprising that the use of probiotic organisms in disease has been investigated extensively in intestinal disorders.

Acute Infectious Diarrhoea

Infectious diarrhoea occurs more commonly in the developing rather than industrialized countries (Guerrant RL et al., 1990). In developing countries the mortality rates due to infectious diarrhoea is highest in children younger than five years where as in the developed countries it is seen in the elderly (Savarino SJ et al., 1993).

Oral rehydration is the mainstay of therapy for acute diarrhoea and should continue to be fostered, encouraged and supported. However, despite its proven efficacy, oral rehydration therapy remains underused (Szajewska H et al., 2000) because it does not reduce the frequency of bowel movements and fluid loss, nor does it shorten the duration of the illness. The role of probiotics in the form of a ubiquitous, simple, safe intervention has been evaluated for the prevention and treatment of acute infectious diarrhoea.

A systematic review that included 12 randomized, controlled trials (RCTs) in the Cochrane database (majority from affluent countries) concluded that *probiotics reduced the mean duration of acute diarrhoea in children* by 29.2 hours in a fixed-effects model and by 30.48 hours in a random-effects model (Allen SJ et al., 2004). Two meta-analyses that evaluated similar studies found statistically significant but modest reduction of diarrhoea duration (Huang JS et al., 2002; Szajewska H et al., 2001).

A more recent systematic review of 63 studies (56 of these studies recruited infants and young children) that included 8014 participants concluded that used alongside rehydration therapy, probiotics appear to be safe and have clear beneficial effect in shortening the duration and reducing stool frequency in acute infectious diarrhoea.

A meta-analysis of acute pediatric diarrhoea concluded that there was significant data for probiotic based reduction of diarrhoea duration, treatment failure and prevention (McFarland LV et al., 2006). Multiple studies using different probiotic strains of Lactobacilli have also shown benefit in reducing the duration of rotaviral shedding; an observation that has favorable epidemiologic implication (Guarino A. et al., 1997; Henker J. et al., 2007).

Two community based trials have evaluated the role of probiotics in diarrhoea prevention; children in Peru had 13% fewer diarrheal episodes after 15 months of *L. rhamnosus* (Oberhelman RA et al., 1999) whereas diarrhoea frequency was reduced by 14% among children in India who received daily doses of *L. casei* strain Shirota for 12 weeks, with a 12-week follow-up period (Sur D et al., 2011).

Thus far, the beneficial effects of probiotics in acute infectious diarrhoea appear to be moderate (~1 day reduction in duration of diarrhoea), strain dependent (Canani RB et al., 2007), dose dependent (greater for doses of $> 10^{10}$–10^{11} CFU/d), significant in viral gastroenteritis and not as significant in bacterial diarrhoea or invasive diarrhoea and more evident when treatment with probiotics is initiated early in the course of the disease and more evident in the developed countries (Szajewska H. et al., 2005).

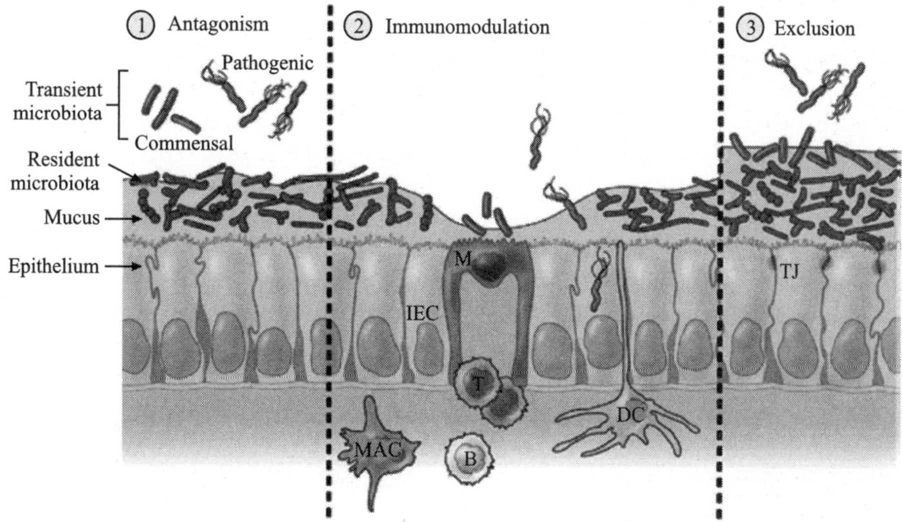

Figure 60.4 Mechanism of Probiotic Action (Preidis GA et al, 2011).

The use of probiotics for infectious diarrhoea in children is however, an accepted therapy in Europe. The European Society for Paediatric Gastroenterology, Hepatology and Nutrition and European Society of Paediatric Infectious Diseases expert Working Group have stated that selected probiotics with proven clinical efficacy and in appropriate dosage according to the strain and the population may be used as an adjunct for the management of children with acute gastroenteritis being given rehydration therapy (Guarino A et al., 2008). Other probiotic strains may also be used provided their efficacy is documented in high quality RCT's (or in Meta-analysis).

Persistent diarrhoea

Persistent diarrhoea (diarrhoea lasting for more than 14 days) accounts for one third of all diarrhoea related deaths in developing countries (Black RE et al., 1993). It is an important cause of morbidity and mortality in children under five years old especially in developing countries. The illness is associated with malnutrition, malabsorption and growth failure and is widely prevalent in developing countries where 8% of the acute diarrhoeas result in persistent diarrhoea (Walker-Smith JA et al., 1990; Bernaola AG et al., 2010). The cause of persistent diarrhoea is not completely understood but is likely to be complex thereby making the management of this condition difficult.

A meta-analysis of four trials that included 464 participants showed that probiotics reduced the duration of persistent diarrhoea. Stool frequency was reduced with probiotics in two trials. One trial reported a shorter hospital stay, which was significant, but the numbers were small. No adverse effects were reported (Black RE et al., 1993).

Two RCT's evaluated probiotics for children with persistent diarrhoea in developing countries and reported dramatic reductions in diarrhoea duration — 4.8 and 3.9 days in Argentina (Gaon D et al., 2003) and India (Basu S et al., 2009) respectively.

Antibiotic associated diarrhoea

The incidence of Antibiotic Associated Diarrhoea (AAD) ranges between 1% to 44% and symptoms vary from mild episodes that resolve when antibiotics are stopped or lead to serious complications such as toxic mega colon, bowel perforation and death. Risk is increased with extremes of age, comorbidity, use of oral broad spectrum antibiotics, antibiotic administration for prolonged duration, previous antibiotic associated diarrhoeas and hospitalization (Butler CC et al., 2012). Potentially probiotics maintain or restore gut microecology during or after antibiotic treatment through receptor competition, competition for nutrients, inhibition of epithelial and mucosal adherence of pathogens, introduction of lower colonic pH favouring the growth of non-pathogenic species, stimulation of immunity or production of antimicrobial substances (Rolfe RD et al., 2000; Cremonini F et al., 2002).

A meta-analysis of 34 masked, randomized, placebo controlled trials of which only one was community based from a developing country revealed that probiotics significantly reduced the incidence of antibiotic associated diarrhoea by 52% (95% CI 35 to 65%) (Sazawal S et al., 2006). Several systematic reviews with or without meta-analysis documented that most of the tested probiotics have been shown to be effective in reducing the risk of AAD in the general population (Cremonini F et al., 2002; D'Souza AL et al., 2002; Hawrelak JA et al., 2005). Evidence from three recent systematic reviews of RCT's also suggests that probiotics reduce the risk of AAD in children (Szajewska H et al., 2006; Johnston BC et al., 2006; Johnston BC et al., 2007).

A very recent meta-analysis that included 82 RCT's revealed that the pooled relative risk of meta-analysis of 63 RCT's which included 11811 participants indicated a statistically significant association of probiotic administration with reduction in AAD. Majority of the trials used Lactobacillus based interventions alone or in combination with other genera but strains were poorly documented. Therefore additional research is needed to determine which probiotics are associated with the greatest efficacy and the antibiotic against which the probiotic would be efficacious (Hempel S et al., 2012).

Six placebo controlled trials are in progress to examine the effect of probiotics in preventing antibiotic associated diarrhoea in hospitalized patients which will provide information on the role of probiotics to prevent antibiotic associated diarrhoea in a wider range of hospitalized patients and may be large enough to provide information on which subgroups of patients are at the greatest risk and are most likely to benefit (Butler CC et al., 2012).

Clostridium difficile Diarrhoea

More recently the role of probiotics has extended to the prevention of *Clostridium difficile* diarrhoea. (McFarland LV et al., 2008). While the organism can remain latent in many patients, those receiving antibiotics have an increased risk of developing diarrhoea and pseudomembranous colitis. *C. difficile* associated disease (CDAD) is responsible for around 10–20% of all cases of AAD (Bartlett JG et al., 2002) and it can occur up to eight weeks after antibiotic therapy (Gerding et al., 1986). The three risk factors include antibiotic use, increasing age and hospitalization.

C. difficile is contagious and spreads easily from patient to patient, thereby representing a very serious problem for health care providers. In the more recent years there has been an emergence of hyper virulent strains that have led to an increased incidence of CDAD, elevated severity, higher relapse rates, increased mortality, and greater resistance to fluoroquinolone antibiotics (Cartman ST et al., 2010).

There is weak or inconclusive evidence from two systematic reviews for the effectiveness of probiotics in prevention and treatment of *C. difficile* diarrhoea. The authors of the first review concluded that available evidence (all in adults) does not support the administration of probiotics with antibiotics to prevent or treat *C. difficile* diarrhoea (Dendukuri N et al., 2005). The conclusions from the second systematic review (with meta-analysis) do support probiotic use although the meta-analysis has been criticized for combining the results of one study on prevention of *C. difficile* diarrhoea with the results from five other studies with *C. difficile* treatment and pooling data on different probiotics, different conditions and different patient characteristics (Lewis S et al., 2007; Dendukuri N et al., 2007).

However, a very recent meta-analysis showed that administration of probiotics led to a statistically significant relative risk reduction of CDAD by 71%. A systematic review and meta-analysis of twenty trials that included 3818 participants concluded that probiotics reduced the incidence of CDAD by 66%. In a population with 5% incidence of antibiotic associated CDAD (median control group risk), probiotic prophylaxis would prevent 33 episodes per 1000 persons. Of probiotic treated patients, 9.3% experienced adverse events, compared with 12.6% of control patients (Johnston BC et al., 2012)

Traveller's diarrhoea

More than 50% of travellers are affected by Traveller's diarrhoea (TD) especially those travelling to developing countries. In most cases bacterial pathogens are the causative agents and symptoms include abdominal cramps, nausea and diarrhoea. A significant reduction of 11.8% in the incidence of diarrhoea was seen in healthy Finnish adults who travelled to Turkey and took probiotics as compared to placebo (Oksanen PJ et al., 1990). Two double blind placebo controlled RCT's conclude that capsules or sachets containing *L rhamnosus* GG might prevent TD. In yet another two studies it was observed that the probiotic yeast *S. boulardii* imparted a significant benefit to tourists travelling to warmer climates (Kollaritsch et al., 1989). A mixture of probiotic strains was useful for travellers who visited Egypt by reducing the incidence of TD from 71% in the placebo group to 43% in the treatment group (Black FT et al., 1989).

The pooled relative risk in a recent meta-analysis has indicated that some probiotics can significantly prevent TD (McFarland LV et al., 2007). However the heterogeneity of the included studies with respect to the probiotic strain, dose and duration of treatment, travel destination and unidentified causal agents of TD still hamper a more specific use of probiotics. In addition, very few studies report on the stability of the probiotic products, compliance to treatment, and/or recovery of the strain from fecal samples, and such factors may be ultimately important, especially during travel.

Constipation

Constipation is a common condition affecting children and adults. Differences in the intestinal microflora between constipated and healthy adults have been observed. In constipated children, the number of Bifidobacteria decreased whereas the number of non pathogenic *E.coli,* Bacteroides and total number of organisms increased. It has also been reported that constipation predominant IBS patients showed increased amounts of *Veillonella*spp. (Malinen E et al., 2005).

Research on the relationship between constipation and nutrition has focussed on fibre and non-digestible oligosaccharides. The effect of probiotics on various forms of constipation has been explored only in a few studies and that too without controls. Constipations never the less has a significant impact on the quality of life, and the beneficial effects of probiotics in the treatment of constipation appear to be promising.

A recent meta-analysis of 5 RCT's with a total of 377 subjects of which 266 were adults and 111 were children concluded that *Bifidobacteriumlactis* DN – 173010, *Lactobacillus casei* strain Shirota and *Escherichia coli* Nissile 1917 had a favourable effect defecation frequency and stool consistency in adults and *L. casei rhamnosus* Lcr 35 had a positive impact in children (Chmielewska A et al., 2010).

In a double blind placebo controlled trial conducted on 70 adults who were randomized to receive a probiotic fermented milk drink containing *Lactobacillus casei* strain Shirota or a sensorially identical placebo, 89% in the probiotic group and 56% in the treatment group reported a positive impact of the probiotic (Koebnick C et al., 2003).

The beneficial effects of probiotics on constipation may be explained due to various mechanisms. Changes in the composition of the intestinal flora may result in changes in metabolites of bacterial fermentation resulting in enhanced motility and shortening of intestinal transit time.

Irritable bowel syndrome

Irritable Bowel Syndrome is a heterogeneous condition often diagnosed by exclusion of other underlying diseases, including IBD. Symptoms include abdominal pain, bloating and altered bowel habits. An altered intestinal microbiota has been observed in IBS sufferers with higher facultative anaerobes and lower Bifidobacteria and Lactobacilli. Disease risk factors include those that disturb the microbiota such as antibiotics, gastrointestinal surgery and infection.

A meta-analysis by McFarland and Dublin (2008) reviewed 20 IBS trials and reported that probiotic use was associated with improvement of IBS symptoms compared to placebo. Moayyedi P et al., 2010 reviewed 19 RCT's involving 1650 patients with IBS and concluded that probiotics were better than placebo. Clarke et al., 2012, reviewed 42 RCTs to determine the effect of lactic acid bacteria on IBS symptoms. Thirty-four of these trials reported benefit in at least one of the end points studied.

The precise manner in which the probiotic strains maintain the gut epithelial barrier function and prevent intestinal dysbiosis remains to be determined. Given the heterogeneity of IBS, research that focuses on the use of probiotics for specific IBS subgroups and for specific IBS related symptoms is probably the best way forward.

Inflammatory bowel disorders

There is increasing evidence that an altered gut microbiota has an important role in the pathogenesis of IBD. The number of different commensal bacteria is altered in IBD patients with increased Bacteroides, adherent or invasive *Escherichia coli* and Enterococci and reduced Bifidobacteria and Lactobacillus species. For example, the abundance of an anti-inflammatory commensal organism, *F. prausnitzii*is decreased in patients with Crohn's disease compared with healthy individuals (Sokol H et al., 2008). Absence of this commensal organism is also associated with an increased risk of disease recurrence following surgical resection in patients. In addition adherent invasive *E coli* (AIEC) strains are identified with increased frequency in patients with ileal Crohn's disease compared with healthy individuals (Darfeuille-Michaud A et al., 2004). A recent study in twins revealed that a reduced abundance of the commensal bacterium *Fecalibacteriumprausnitzii* and an increased abundance of *E coli* species was associated with an ilealcrohn's disease phenotype altered gut permeability, mucosal inflammation and ulcerations are characteristic features of IBD.

The rationale for using probiotics in Inflammatory Bowel Disease stems from study of dysbiosis in the Intestinal Microbiota in Ulcerative Colitis, Crohn's Disease and Pouchitis detected using conventional anaerobic culture or molecular probes. The potential for probiotics to modulate the microbiota provide beneficial immunomodulatory effects and restore epithelial barrier defects suggests that a probiotic strategy might prove a viable future treatment option for patients with IBD. Investigations have focussed mostly on remission maintenance, rather than treatment of active disease, with evidence strongest for pouchitis > UC > CD.

In a UC trial, lasting one year remission was improved by combining normal therapy with an intervention of *B breve* Yakult, *B bifidum* Yakult and *L acidophilus* Yakult YIT0168 (Ishikawa et al., 2003).

A meta-analysis of 286 papers was reviewed of which 13 randomized controlled studies met the selection criteria. Seven studies evaluated the remission rate, eight studies estimated the recurrence rate; two studies evaluated both remission and recurrence rates. The study showed that the recurrence rate of ulcerative colitis patients who received probiotics was 0.69 in comparison to non-probiotics group. The group who received *Bifidobacteriumbifidum* treatment had a recurrence rate of 0.25 as compared to non-probiotics group (Sang LX et al., 2010).

An open-label preliminary trial was conducted on 10 patients with active ulcerative colitis (UC) who received *Lactobacillus casei* strain Shirota (LcS), (8 x 10^{10}cfu/day) in addition to conventional therapy daily for 8 weeks. The trial revealed that LcS effectively treats UC at least in part through the inhibition of interleukin-6 signaling (Mitsuyama K et al., 2008).

Similar to the results of studies in adults, children with IBD have found to have a higher number of mucosa associated facultative anaerobic and aerobic bacteria in the intestine and colon compared with healthy individuals (Conte MP et al., 2006).

Necrotizing enterocolitis (NEC)

Necrotizing Enterocolitis is a severe enterocolitis that affects preterm infants and has both serious morbidity and mortality rate. Although the precise mechanism underlying the etiology of this disease remains unknown, altered microbial colonization,

formula feeding and neonatal stress are each thought to be involved in its pathogenesis (Morowitz MJ et al., 2010). A retrospective study of very low birth weight infants showed that prolonged antibiotic treatment in the first few days of life increases the risk of developing necrotizing enterocolitis and subsequent death (Cotton CM et al., 2009). This finding highlights the role of an altered gut microbiota in disease pathogenesis.

At the present time, NEC is associated with 30% mortality, despite extensive medical and surgical efforts, and with severe and costly sequelae if the patient survives. The immature intestine of preterm infants is especially prone to inflammation and loss of epithelial integrity (Nanthakumar NN et al., 2000).

Scientific evidence suggests that reduced bacterial translocation, amelioration of intestinal barrier defects and decreased apoptosis in the face of exogenous stressors are underlying mechanisms of action of probiotics that prevent the occurrence of NEC in high risk premature infants. Premature newborns are susceptible to infectious complications, including bacteremia, sepsis and meningitis. Preterm babies have delayed gut colonization with pathogens compared with full-term newborns (Morowitz MJ et al., 2010).The systematic review reported in 2007 identified 7 RCT's and found that most of the investigated probiotics might reduce the risk of NEC in preterm neonates with < 33 wk gestation (Deshpande G et al., 2007).

Helicobacter pylori

It has been established that this pathogen is associated with chronic gastritis, peptic ulcers and gastric cancer. Numerous studies have shown that the ability of various probiotic formulations to decrease side effects of anti-Helicobacter treatments such as AAD, epigatsric pain, discomfort and flatulence. *L. rhamnosus* GG was able to reduce the occurrence of eradication treatment side effects in 2 separate studies. The results for *L. rhamnosus* GG were confirmed later in a study comparing *L. rhamnosus* GG, *Saccharomyces boulardii* and a combination of *L. acidophilus* and *B. lactis* (Cremonini F et al., 2002).

Probiotics have been investigated either to reduce the risk of the disease or as a therapy. In a small trial, consumption of *Lactobacillus casei* strain Shirota for three weeks resulted in *H-pylori* inhibition in 64% of the patients compared to 33% in an untreated control group (Cats et al., 2003). Probiotics as an alternative to antibiotics have also been the focus of several trials. Also, regular consumption of probiotic products with a specified probiotic strain as an alternative to antibiotics may have some potential in suppression of *H. pylori* infection and gastric inflammation.

CONCLUSION

As the gut microbiota appears to contribute to nearly every aspect of the host's growth and development, it is not surprising that a tremendous array of diseases and dysfunctions have been associated with an imbalance in either composition, numbers or habitat of the gut microbiota.

However many more questions still remain to be answered, and fuel ongoing research in these areas seeking to tease out the many subtle dysregulations in the interactions between the host and the microbiota and how they impact the disease process.

The first step to realizing the full potential of probiotics is to define the specific microbial genes, small molecules, and host-microbe interactions that mediate their beneficial functions. Basic scientists must identify, isolate, and characterize bacterial fermentation products, immuno-modulatory factors, antimicrobial agents, and cell-wall components that produce discrete physiological effects through specific host interactions. These types of studies are required to improve the understanding of probiotic function (Preidis GA et al., 2011).

Prospective RCTs should aim to reduce short-term pathologies associated with acute diarrhoea, prevent long-term morbidities from recurrent or persistent infections, and increase vaccine efficacy. Specific strain and dose recommendations should be made, with the long-term goals of improving survival, growth, and development during childhood. With trials in multiple geographic locations and ethnic groups, patterns will emerge to guide selection of specific microbial-based therapies for specific regions of the world (Ng SC et al., 2009).

REFERENCES

1. Kunz C., Kuntz S. and Rudloff S. (2009), Intestinal Flora, *Adv Exp Med Biol*, 639:67–79.

2. Morelli L. (2008), Postnatal Development of Intestinal Microflora as Influenced by Infant Nutrition, *J. Nutr.*, 138:S1791–S1795.

3. Neish A.S. (2009), Microbes in Gastrointestinal Health and Disease, *Gastroenterology*, 136:65–80.

4. Ley R.E., Peterson D.A. and Gordon J.I. (2006), Ecological and Evolutionary Forces Shaping Microbial Diversity in the Human Intestine, *Cell*, 124:837–848.

5. Chiller K., Selkin B.A. and Murakawa G.I. (2001), Skin Microflora and Bacterial Infections of the Skin, *J. Invest Dermatol Symp Proc.*, 6:170–174.

6. Hull M.W. and Chow A.W. (2007), Indigenous Microflora and Innate Immunity of the Head and Neck, *Infect Dis Clin North Am.*, 21: 265–282.

7. Verstraelen H. (2008), Cutting Edge: The Vaginal Microflora and Bacterial Vaginosis, *Verh K Acad Geneeskd Belg.*, 70:147–174.

8. Ley R.E., Peterson D.A. and Gordon J.I. (2006), Ecological and Evolutionary Forces Shaping Microbial Diversity in the Human Intestine, *Cell*, 124:837–848.

9. Whitman W.B., Coleman D.C. and Wiebe W.J. (1998), Prokaryotes: The Unseen Majority, *Proc Natl Acad Sci.,* USA, 95:6578–6583.

10. Nelson K.E., Weinstock G.M., Highlander S.K., et al. (2010), A Catalog of Reference Genomes from the Human Microbiome, *Science*, 328:994–999.

11. Forsythe P., Sudo N., Dinan T., Taylor V.H. and Bienenstock J. (2010), Mood and Gut Feelings, *Brain Behav Immun*, 24:9–16.

12. Sekirov I., Russell S.L., Antunes L.C.M. (2010), Finlay BB: Gut Microbiota in Health and Disease, *Physiol Rev.*, 90:859–904.

13. Round J.L. and Mazmanian S.K. (2009), The Gut Microbiota Shapes Intestinal Immune Responses During Health and Disease, *Nat Rev Immunol.*, 9:313–323.

14. Chen J., Cai W. and Feng Y. (2007), Development of Intestinal Bifidobacteria and Lactobacilli in Breast-fed Neonates, *Clin Nutr.*, 26: 559–566.

15. Conroy M.E., Shi H.N. and Walker W.A. (2009), The Long-term Health Effects of Neonatal Microbial Flora, *Curr Opin Allergy Clin Immunol.*, 9:197–201.

16. Mariat D., et al. (2009), The Firmicutes/Bacteroidetes Ratio of the Human Microbiota Changes with Age, *BMC Microbiol.*, 9:123.

17. Turnbaugh P.J. and Gordon J.I. (2009), The Coregut Microbiome, Energy Balance and Obesity, *J. Physiol.*, 587:4153–4158.

18. Gibson G.R. and Collins M.D. (1999), Concept of Balanced Colonic Microbiota, Probiotics and Synbiotics, In: Hanson L.A. and Yolken R.H. (Eds.) Probiotics, Other Nutritional Factors and Intestinal Microflora, Vol. 42:139–152. Lippin-cott-Raven Publishers, Philadelphia, PA USA.

19. Holzapfel W.H., Haberer P., Snel J., Schillinger U. and Huis in't Veld J.H. (1998), Overview of Gut Flora and Probiotics, *Inter J. Food Microbiol.*, 41:85–101.

20. Blaut M. (2003), Influence of Food Components on Intestinal Microbiota Composition.

21. Joint FAO/WHO Working Group (2002), Guidelines for the Evaluation of Probiotics in Food: Report of a Joint FAO/WHO Working Group on Drafting Guidelines for the Evaluation of Probiotics in Food, London, Ontario, Canada.

22. Nemcova R. (1997), Criteria for Selection of Lactobacilli for Probiotic Use, *Vet Med* (Praha), 42:19–27.

23. Corr S.C., Li Y., Riedel C.U. and O'Toole P.W., et al. (2007), Bacteriocin Production as a Mechanism for the Antiinfective Activity of *Lactobacillus Salivarius* UCC118, *Proc Natl Acad Sci.,* USA, 104(18):7617–21.

24. Rakoff-Nahoum S. and Medzhitov R. (2008), Innate Immune Recognition of the Indigenous Microbial Flora, *Mucosal Immunol,* Suppl 1:S10–S14.

25. Forsythe P. and Bienenstock J. (1996), Immunomodulation by Commensal and Probiotic Bacteria, *Immunol Invest,* 64:5225–5232.

26. Christensen H.R., Frokiaer H. and Pestka J.J. (2002), Lactobacilli Differentially Modulate Expression of Cytokines and Maturation Surface Markers in Murine Dendritic Cells, *J. Immunol,* 168:171–178.

27. Lin Y.P., Thibodeaux C.H., Pena J.A., et al. (2008), Probiotic *Lactobacillus Reuteri* Suppress Proinflammatory Cytokines via c-Jun, *Inflamm Bowel Dis.,* 14:1068–1083.

28. Thomas C.M. and Versalovic J. (2010), Probiotics-host Communication: Modulation of Signaling Pathways in the Intestine, *Gut Microbes,* 1:1–16.

29. Konstantinov S.R., Smidt H., de Vos W.M., et al. (2008), S. Layer Protein A of *Lactobacillus Acidophilus* NCFM Regulates Immature Dendritic Cell and T Cell Functions, *Proc Natl Acad Sci.,* USA, 105(49):19474–9.

30. Shida K., Nanno M. and Nagata S., Flexible Cytokine Production by Macrophages and T Cells in Response to Probiotic Bacteria, *Gut Microbes,* 2(2):109–114.

31. Mondel M., Schroeder B.O., Zimmermann K., et al. (2009), Probiotic E. Coli Treatment Mediates Antimicrobial Human Beta-defensin Synthesis and Fecal Excretion in Humans, *Mucosal Immunol.,* 2 (2):166–72.

32. Meddings J. (2008), The Significance of the Gut Barrier in Disease, *Gut,* 57:438–440.

33. Preidis G.A., Hill C., Guerrant R.L., Ramakrishna B.S., Tannock G.W., Versaovic J. (2011), Probiotics, Enteric and Diarrheal Diseases, and Global Health, *Gastroenterology,* 140:8–14.

34. Guerrant R.L., Hughes J.M., Lima N.L. and Crane J. (1998), Diarrhea in Developed and Developing Countries: Magnitude, Special Settings, and Etiologies, *Rev Infect Dis.,* 12 Suppl 1:S41–50.

35. Savarino S.J. and Bourgeois A.L. (1993), Diarrhoeal Disease: Current Concepts and Future Challenges, Epidemiology of Diarrhoeal Diseases in Developed Countries, *Trans R Soc Trop Med Hyg.,* 87 Suppl 3:7–11.

36. Szajewska H., Hoekstra J.H. and Sandhu B. (2000), Management of Acute Gastroenteritis in Europe and the Impact of the New Recommendations: A Multicenter Study, The Working Group on Acute Diarrhoea of the European Society for Paediatric Gastroenterology, Hepatology, and Nutrition, *J. Pediatr Gastroenterol Nutr.,* 30(5):522–7.

37. Allen S.J., Okoko B., Martinez E., Gregorio G. and Dans L.F. (2004), Probiotics for Treating Infectious Diarrhoea, *Cochrane Database Syst Rev.,* 2:CD003048.

38. Huang J.S., Bousvaros A., Lee J.W., Diaz A. and Davidson E.J. (2002), Efficacy of Probiotic Use in Acute Diarrhea in Children: A Meta-analysis, *Dig Dis Sci.,* 47(11):2625–34.

39. Szajewska H. and Mrukowicz J.Z. (2001), Probiotics in the Treatment and Prevention of Acute Infectious Diarrhea in Infants and Children: A Systematic Review of Published Randomized, Double-blind, Placebo-controlled Trials, *J. Pediatr Gastroenterol Nutr.,* 33 Suppl 2:S17–25.

40. McFarland L.V. (2006), Meta-analysis of Probiotics for the Prevention of Antibiotic Associated Diarrhea and the Treatment of Clostridium difficile Disease, *Am J. Gastroenterol.,* 101(4):812–22.

41. Guarino A., Canani R.B., Spagnuolo M.I., Albano F. and Di Benedetto L. (1997), Oral Bacterial Therapy Reduces the Duration of Symptoms and of Viral Excretion in Children with Mild Diarrhea, *J. Pediatr Gastroenterol Nutr.,* 25 (5):516–9.

42. Henker J., Laass M., Blokhin B.M., Bolbot Y.K., Maydannik V.G., Elze M., et al. (2007),The Probiotic *Escherichia coli* Strain Nissle 1917 (EcN) Stops Acute Diarrhoea in Infants and Toddlers, *Eur J. Pediatr.,* 166 (4):311–8.

43. Oberhelman R.A., Gilman R.H., Sheen P., Taylor D.N., Black R.E., Cabrera L., et al. (1999), A Placebo-controlled Trial of Lactobacillus GG to Prevent Diarrhea in Undernourished Peruvian Children, *J. Pediatr,* 134(1):15–20.

44. Sur D., Manna B., Niyogi S.K., Ramamurthy T., Palit A., Nomoto K., et al. (2011), Role of Probiotic in Preventing Acute Diarrhoea in Children: A Community-based, Randomized, Double-blind Placebo-controlled Field Trial in an Urban Slum, *Epidemiol Infect,* 139(6): 919–26.

45. Canani R.B., Cirillo P., Terrin G., Cesarano L., Spagnuolo M.I., De Vincenzo A., et al. (2007), Probiotics for Treatment of Acute Diarrhoea in Children: Randomised Clinical Trial of Five Different Preparations, *BMJ,* 18;335 (7615):340.

46. Szajewska H. and Mrukowicz J.Z. (2005), Use of Probiotics in Children with Acute Diarrhea, *Paediatr Drugs,* 7 (2):111–22.

47. Guarino A., Albano F., Ashkenazi S., Gendrel D., Hoekstra J.H., Shamir R., et al. (2008), European Society for Paediatric Gastroenterology, Hepatology, and Nutrition/European Society for Paediatric Infectious Diseases Evidence-based Guidelines for the Management of Acute Gastroenteritis in Children in Europe, *J. Pediatr Gastroenterol Nutr.,* 46 Suppl 2:S81–122.

48. Black R.E. (1993), Persistent Diarrhea in Children of Developing Countries, *Pediatr Infect Dis J.,* 12(9):751–61; Discussion 62–4.

49. Walker-Smith J.A. (1990), Management of Infantile Gastroenteritis, *Arch Dis Child.,* 65(9):917–8.

50. Bernaola Aponte G., BadaMancilla C.A., Carreazo Pariasca N.Y., Rojas Galarza R.A. (2010), Probiotics for Treating Persistent Diarrhoea in Children, *Cochrane Database Syst Rev.*, (11):CD007401.

51. Gaon D., Garcia H., Winter L., Rodriguez N., Quintas R., Gonzalez S.N., et al. (2003), Effect of Lactobacillus Strains and *Saccharomyces boulardii* on Persistent Diarrhea in Children, *Medicina* (B Aires), 63(4):293–8.

52. Basu S., Paul D.K., Ganguly S., Chatterjee M., and Chandra P.K. (2009), Efficacy of High-dose *Lactobacillus Rhamnosus* GG in Controlling Acute Watery Diarrhea in Indian Children: A Randomized Controlled Trial, *J. Clin Gastroenterol.*, 43(3):208–13.

53. Butler C.C., Duncan D. and Hood K. (2012), Does Taking Probiotics Routinely with Antibiotics Prevent Antibiotic Associated Diarrhoea? *BMJ*, 344:e682.

54. Rolfe R.D. (2000), The Role of Probiotic Cultures in the Control of Gastrointestinal Health, *J. Nutr.*, 130(2S Suppl):396S–402S.

55. Cremonini F., Di Caro S., Nista E.C., Bartolozzi F., Capelli G., Gasbarrini G., et al. (2002), *Meta-analysis*: The Effect of Probiotic Administration on Antibiotic-associated Diarrhoea, *Aliment Pharmacol Ther.*, 16(8):1461–7.

56. Sazawal S., Hiremath G., Dhingra U., Malik P., Deb S. and Black R.E. (2006), Efficacy of Probiotics in Prevention of Acute Diarrhoea: A Meta-analysis of Masked, Randomised, Placebo-controlled Trials, *Lancet Infect Dis.*, 6(6):374–82.

57. D'Souza A.L., Rajkumar C., Cooke J. and Bulpitt C.J. (2002), Probiotics in Prevention of Antibiotic Associated Diarrhoea: Meta-analysis, *BMJ*, 8;324 (7350):1361.

58. Hawrelak J.A., Whitten D.L. and Myers S.P. (2005), Is *Lactobacillus rhamnosus* GG Effective in Preventing the Onset of Antibiotic-associated Diarrhoea: A Systematic Review, *Digestion*, 72(1):51–6.

59. Szajewska H., Ruszczynski M. and Radzikowski A. (2006), Probiotics in the Prevention of Antibiotic-associated Diarrhea in Children: A Meta-analysis of Randomized Controlled Trials, *J. Pediatr.*, 149(3):367–72.

60. Johnston B.C., Supina A.L. and Vohra S. (2006), Probiotics for Pediatric Antibiotic-associated Diarrhea: A Meta-analysis of Randomized Placebo-controlled Trials, *CMAJ*, 15;175(4):377–83.

61. Johnston B.C., Supina A.L., Ospina M. and Vohra S. (2007), Probiotics for the Prevention of Pediatric Antibiotic-associated Diarrhea, *Cochrane Database Syst Rev.*, (2):CD004827.

62. Hempel S., Newberry S.J., Maher A.R., Wang Z., Miles J.N., Shanman R., et al. (2012), Probiotics for the Prevention and Treatment of Antibiotic-associated Diarrhea: A Systematic Review and Meta-analysis, *JAMA*, 9;307(18):1959–69.

63. McFarland L.V. (2008), Update on the Changing Epidemiology of Clostridium difficile-associated Disease, *Nat Clin Pract Gastroenterol Hepatol.*, 5(1):40–8.

64. Bartlett J.G. (2002), Clostridium difficile-associated Enteric Disease, *Curr Infect Dis Rep.*, 4(6):477–83.

65. Gerding D.N, Olson M.M., Peterson L.R., Teasley D.G., Gebhard R.L., Schwartz M.L., et al. (1986), Clostridium difficile-associated Diarrhea and Colitis in Adults. A Prospective Case-controlled Epidemiologic Study, *Arch Intern Med.*, 146(1):95–100.

66. Cartman S.T., Heap J.T., Kuehne S.A., Cockayne A. and Minton N.P. (2010), The Emergence of 'hypervirulence' in Clostridium difficile, *Int J. Med Microbiol.*, 300(6):387–95.

67. Dendukuri N., Costa V., McGregor M. and Brophy J.M. (2005), Probiotic Therapy for the Prevention and Treatment of Clostridium difficile-associated Diarrhea: A Systematic Review, *CMAJ*, 173(2):167–70.

68. Lewis S. (2006), Response to the Article: McFarland LV: Meta-analysis of Probiotics for the Prevention of Antibiotic-associated Diarrhea and the Treatment of Clostridium difficile Disease, *Am J. Gastroenterol.*, 101:812–22. *Am J. Gastroenterol.*, 102(1):201–2.

69. Dendukuri N. and Brophy J. (2007), Inappropriate Use of Meta-analysis to Estimate Efficacy of Probiotics, *Am J. Gastroenterol.*, 102(1):201; Author Reply 2–4.

70. Johnston B.C., Ma S.S., Goldenberg J.Z., Thorlund K., Vandvik P.O., Loeb M., et al. (2012), Probiotics for the Prevention of Clostridium difficile-associated Diarrhea: A Systematic Review and Meta-analysis, *Ann Intern Med.*, 157 (12):878–88.

71. Oksanen P.J., Salminen S., Saxelin M., Hamalainen P., Ihantola-Vormisto A., Muurasniemi-Isoviita L., et al. (1990), Prevention of Travellers'Diarrhoea by Lactobacillus GG, *Ann Med.*, 22(1):53–6.

72. Kollaritsch H. (1989), Traveller's Diarrhea Among Austrian Tourists in Warm Climate Countries: I. Epidemiology, *Eur J. Epidemiol.*, 5(1):74–81.

73. Kollaritsch H., Holst H., Grobara P. and Wiedermann G. (1993), [Prevention of Traveler's Diarrhea with *Saccharomyces boulardii*, Results of a Placebo Controlled Double-blind Study]. *Fortschr Med.*, 111(9):152–6.

74. Black F.T., Andersen P.L., Orskov J., Orskov F., Gaarslev K., and Laulund S. (1989), Prophylactic Efficacy of Lactobacilli on Traveler's Diarrhea, In: PD Dr. Med. Robert Steffen, Dr. James Haworth, Professor David J. Bradley, (Eds.), *Travel Medicine*, Springer Berlin Heidelberg, 333–5.

75. McFarland L.V. (2007), Meta-analysis of Probiotics for the Prevention of Traveler's Diarrhea, *Travel Med Infect Dis.*, 5 (2):97–105.

76. Malinen E., Rinttilä T., Kajander K., et al. (2005), Analysis of the Fecal Microbiota of Irritable Bowel Syndrome Patients and Healthy Controls with Realtime PCR, *Am J. Gastroenterol.*, 100:373–382.

77. Chmielewska A. and Szajewska H. (2010), Systematic Review of Randomized Controlled Trials: Probiotics for Functional Constipation, *World J. Gastroenterol.*, 16:69–75.

78. Koebnick C., Wagner I., Leitzmann P., Stern U. and Zunft H.J.F. (2003), Probiotic Beverage Containing *Lactobacillus casei* Shirota Improves Gastrointestinal Symptoms in Patients with Chronic Constipation, *Can J. Gastroenterol.*, 17 (11):1–5.

79. McFarland L.V., and Dublin S. (2008), Meta-analysis of Probiotics for the Treatment of Irritable Bowel Syndrome, *World J. Gastroenterol.*, 14:2650–2661.

80. Moayyedi P., Ford A.C., Talley N.J., Cremonini F., Foxx-Orenstein A.E., Brandt L.J., and Quigley E.M.M. (2010), The Efficacy of Probiotics in the Treatment of Irritable Bowel Syndrome: A Systematic Review, *Gut*, 59:325–332.

81. Clarke G., Cryan J.F., Dinan T.G., et al. (2012), Review Article: Probiotics for the Treatment of Irritable Bowel Syndrome-focus on Lactic Acid Bacteria, *Aliment Pharmacol Ther.*, 35:403–13.

82. Barrett J.S., Canale K.E.K., Gearry R.B., Irving P.M. and Gibson P.R. (2008), Probiotic Effects on Intestinal Fermentation Patterns in Patients with Irritable Bowel Syndrome, *World J. Gastroenterol.*, 14:5020–5024.

83. Sokol H., et al. (2008), *Faecalibacterium prausnitzii* is an Anti-inflammatory Commensal Bacterium Identified by Gut Microbiota Analysis of Crohn's Disease Patients, *Proc Natl Acad Sci.,* USA, 105:16731–16736.

84. Darfeuille-Michaud A., et al. (2004), High Prevalence of Adherent-invasive *Escherichia coli* Associated with Ileal Mucosa in Crohn's Disease, *Gastroenterology*, 127:412–421.

85. Conte M.P., et al. (2006), Gut-associated Bacterial Microbiota in Pediatric Patients with Inflammatory Bowel Disease, *Gut*, 55:1760–1767.

86. Sang L.X., Chang B., Zhang W.L., Wu X.M., Li X.H. and Jiang M. (2010), Remission Induction and Maintenance Effect of Probiotics on Ulcerative Colitis: A Meta-analysis, *World J. Gastroenterol.*, 16:1908–1915.

87. Mitsuyama K., Matsumoto S., Yamsaki H., Masuda J., Kuwaki K., Takedatsu H., Nagaoka M., Andoh A., Tsurata O. and Sata M. (2008), Beneficial Effects of Lactobacillus Casei in Ulcerative Colitis: A Pilot Study, *J. Clin Biochem Nutr.*, 43:78–81.

88. Morowitz M.J., Poroyko V., Capalan M., Alverdy J. and Liu D.C. (2010), Redefining the Role of Intestinal Microbes in the Pathogenesis of Necrotizing Enterocolitis, *Pediatrics*, 125:777–785.

89. Cotton C.M., et al. (2009), Prolonged Duration of Initial Empirical Antibiotic Treatment Is Associated with Increased Rates of Necrotizing Enterocolitis and Death for Extremely Low Birth Weight Infants, *Pediatrics*, 123:58–66.

90. Nanthakumar N.N., Fusunyan R.D., Sanderson I., et al. (2000), Inflammation in the Developing Human Intestine: A Possible Pathophysiologic Contribution to Necrotizing Enterocolitis, *Proc Natl Acad Sci.,* USA, 97:6043–8.

91. Deshpande G., Rao S. and Patole S. (2007), Probiotics for Prevention of Necrotizing Enterocolitis in Preterm Neonates with Very Low Birthweight: A Systematic Review of Randomized Controlled Trials, *Lancet*, 369:1614–20.

92. Cremonini F., Di C.S., Covino M., Armuzzi A., Gabrielli M., Santarelli L., Nista E.C., Cammarota G., Gasbarrini G., Gasbarrini G. and Gasbarrini A. (2002), Effect of Different Probiotic Preparations on Anti-Helicobacter Pylori Therapy-related Side Effects: A Parallel Group, Triple Blind, Placebo-controlled Study, *Am J. Gastroenterol.*, 97:2744–9.

93. Cats A., Kuipers E.J., Bosschaert M.A.R., Pot R.G.J., et al. (2003), Effect of Frequent Consumption of A Lactobacillus Casei-containing Milk Drink in Helicobacter Pylori-colonized Subjects, *Aliment Pharmacol Ther.*, 17:429–435.

94. Ng S.C., Hart A.L., Kamm M.A., et al. (2009), Mechanisms of Action of Probiotics: Recent Advances, *Inflamm Bowel Dis.*, 15(2):300–10.

61

Probiotics in the Female Reproductive Tract

G.P. Talwar

CONTENTS

I. INTRODUCTION

An astonishing observation is that ***the pH of healthy vagina is in the acidic range of around 4.5, whereas the pH*** of the human body is highly regulated to be around 7.4. This observation was initially made by A. Doderline, in 1894; *Zentralbl Gynakol,* 18: 10-4. It was ascribed to the presence of living bacteria on the mucus lining of the vagina. In course of evolution, Lactobacilli and bifidobacteria have co-survived on the lining of cavities; gastrointestinal tract (G I) and vagina, amongst others. Although there is communication between the oral and the vaginal tract via the rectofecal route, the species residing in the genital tract are not identical to those in the gastrointestinal tract. A few reports are available on the species of Lactobacilli resident in the vagina of women from USA, Canada, Brazil, Sweden and Turkey, besides the first study done by us in India in 2009. Many species are common. However, their preponderance differs in many countries. In India, the species isolated from vagina of 80 healthy women attending the Family Planning and Post Natal Clinics in Delhi are given in Table 61.1.

It will be seen that majority of women in Delhi, who volunteered, harbored Lactobacilli species classifiable in category IV. Amongst these *L. reuteri, L. fermentum* and *L. salivarius* were frequent. On the other hand, in Western countries *L. crispatus, L. jensenii* are more frequent. The identification and characterization of species was done not only on the basis of their being Gram-positive, Catalase negative but also on the basis of their genus and species specific PCR (Polymerase Chain Reaction) profiles confirmed by sequencing of 16 sRNA.

TABLE 61.1   Lactobacillus species isolated from healthy vagina of women (n = 80)

Group	Species	Number of isolates identified (%)
Group IV: 64 (80%)	*L. reuteri*	26 (3.25%)
	L. fermentum	20 (25%)
	L. salivarius	13 (16.25%)
	L. plantarum	5 (6.25%)
Group II: 11 (13.75%)	*L. crispatus*	4 (5%)
	L. jensenii	3 (3.75%)
	L. gasseri	2 (2.5%)
	L. acidophilus	2 (2.5%)
Group III: 5 (6.25%)	*L. casei*	0 (0)
	L. paracasei	1 (1.25%)
	L. rhamnosus	4 (5%)
Group I: 0	*L. delbruckii*	0 (0)

From: Garg et al., 2010

II. PROPERTIES OF LACTOBACILLI

The probiotic Lactobacilli *benefit their host by coping with and preventing infections.* This is achieved by:

(a) Their making and secreting D & L-lactic acid, which keeps the pH in the acidic range, discouraging the growth of many infectious micro-organisms which do not grow at this acidic pH.

(b) Being catalase negative, they make H_2O_2 and many secrete the same. A cloud or envelope of H_2O

surrounding the resident Lactobacillus acts as a local 'antiseptic'.

(c) Many strains of Lactobacilli carry Arginine deaminase, which prevents the formation of malodorous derivatives of this amino acid, thereby preventing foul odor, occurring in vagina lacking such organisms.

(d) Hydrophobicity: Strains differ in their hydrophobicity, which aids their sticking to and colonize onto the vaginal mucosa.

(e) Bacteriocins: Lactobacilli make and secret peptides and proteins which have anti-microbial action.

III. SELECTION OF PROBIOTICS FOR VAGINAL USE

Vaginal infections occur more frequently in women, whose vaginal pH > 5, indicating thereby that either the resident Lactobacilli have declined/disappeared or replaced by strains, not producing adequately D & L-lactic acid required to maintain the vaginal pH in the acidic range. Both manifestations are seen. In a study that we conducted at the All India Institute of Medical Sciences and at Sir Gangaram Hospital, in over 200 women suffering from recurring episodes of vaginosis with vaginal pH above 5; 60% of such women were observed to be devoid of resident Probiotics. However, from 40% of the rest, one or more strains of Lactobacilli were isolated via culture of the vaginal swabs. The question that arose was that if Lactobacilli are indeed present, why do these women have vaginal pH > 5. Along with Kavita Garg Bansal, my Ph.D. student and with the help of Prof. Kriplani (AIIMS) and Prof. Ganguli (SGRH), who enabled our collecting the strains isolated from women suffering from vaginosis. We determined that the capacity of Lactobacilli strains isolated from women with pH > 5, to make D-Lactic acid was much lower than the D-Lactic acid made by Lactobacilli strains isolated from women with healthy vagina: 3.94 ± 0.72 mM/L as compared to 8.04 ± 1.07 mM/L respectively. This implied to our amazement that each *species of Lactobacillus has strains differing in their metabolic properties*. In selection of desirable Probiotics for replenishment, it is not only necessary to select the species but also strains of that species possessing meritorious properties.

From amongst the 80 strains that we isolated from healthy vagina, 3 strains were selected on the basis of their ability to make large quantity of D-Lactic acid, high hydrophobicity, H_2O_2 secretion, Arginine deaminase activity and production of Bacteriocins having anti-microbial properties. The strains are: *Lactobacillus fermentum* # TRF 36; *Lactobacillus salivarius* # TRF 30 and *Lactobacillus gasserei* # TRF 8 (Figure 61.1). These Probiotics individually and in combination are in clinical trials to determine whether they can colonize in the vagina, restore the pH of the vagina to acidic range and prevent vaginosis in women suffering from recurring episodes of infections.

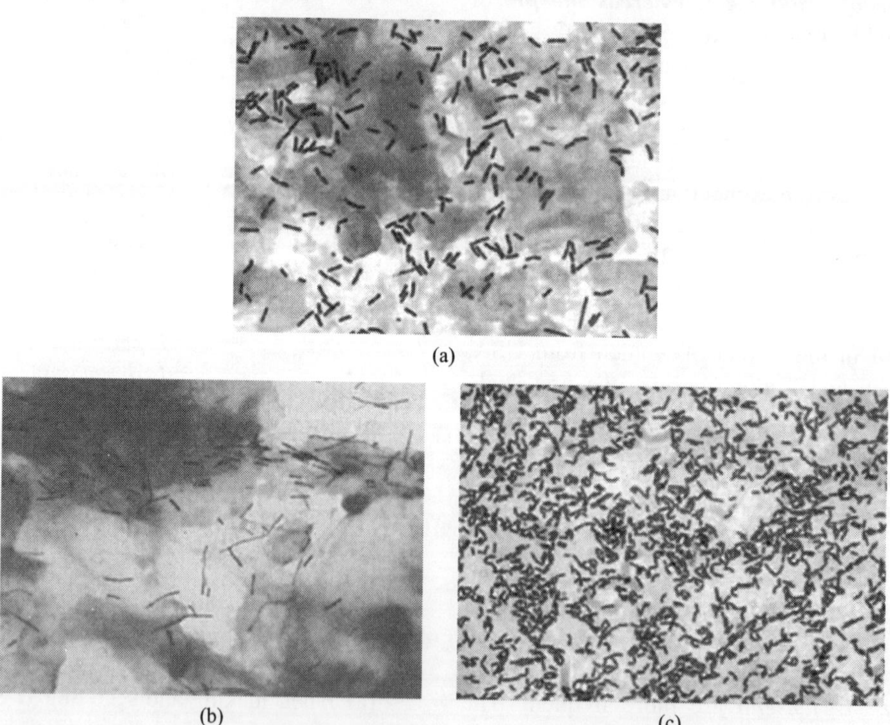

(a)

(b)   (c)

Figure 61.1   (a) *L. fermentum* TRF # 36, (b) and *L. gasseri* TRF # 8 adhered on vaginal epithelial cells of subjects APB and A. S respectively 3 days after administration as seen under 1000 X in light microscope. (c) *L. salivarius* TRF#30, Gram strain from culture. (*see Plate 24 for colour figure*)

IV. AVAILABILITY TO PUBLIC

These 3 Lactobacilli strains have been licensed to Microbax, a company in Hyderabad, who has grown these organisms on a mass scale, lyophilized these along with stabilizers and packaged these in veg cellulose capsules to make these available to the Public. *To our knowledge these will be the first Probiotics from India for utility in Reproductive Health of Women.*

REFERENCES AND SUGGESTIONS FOR FURTHER READING

Garg Kavita Bansal, Ganguli Indrani, Das Ram and Talwar G.P. (2009), Spectrum of Lactobacillus Species Present in Healthy Vagina of Indian Women, *The Indian Journal of Medical Research,* 129(6):652–657.

Garg K.B., Ganguli I., Kriplani A., Lohiya N.K., Thukar J. and Talwar G.P. (2010), Metabolic Properties of Lactobacilli in Women Experiencing Recessing Episodes of Bacterial Vaginosis with Vaginal PH$\geq$5, *Eur. J. Clin. Microbiol. Infect. Dis.*, 29, 123–125.

Vallor A.C., Antonio M.A.D., Hawes S.E. and Hillier S.L. (2001), Factors Associated with Acquisition of, or Persistent Colonization by Vaginal Lactobacilli: Role of Hydrogen Peroxide Production, *J. Infct Dis.*, 184:1431–6.

Vasquez A., Jakobsson T., Ahrne S., Forsum U. and Molin G. (2002), Vaginal *Lactobacillus* Flora of Healthy Swedish Women, *J. Clin Microbial.*, 40:2746–9.

62

The Enigma of Probiotics in Life

Shvetank Sharma

<table><tr><td colspan="2">CONTENTS</td></tr></table>

I. HISTORY OF PROBIOTICS

Humans have been using "probiotics" since ages, even when the term was not even coined. The use of fermented foods and dairy products date back to few thousands of years. There are various scriptures and hieroglyphs across the globe that would attest to this. Passed on as traditional wisdom, manufacturing of curd or yogurt has been practised in almost all the countries, with or without the knowledge of fermenting microbial species being involved in the process. The use of microbes colonizing human body were first tested in recorded history, by Henry Tissier (from Pasteur Institute) in 1899 (Tissier, 1900). Tissier determined that supplementation with *Bifidobacteria* species was beneficial in treating infant diarrhoea. *Bifidobacteria* is the first colonizers of the gut in new-borns. A few years later, in 1906, a Russian scientist from Pasteur Institute, Ilya Ilyich Metchnikoff postulated that lactic acid bacteria were capable of providing health benefits thus promoting longevity. He tested upon himself "sour milk" fermented with what he called as "Bulgarian Bacillus" (Metchnikoff 1907). He suggested that this bacterium was able to replace the toxin producing microbes such as *Clostridium sp.* in the gut. Metchinkoff shared the Nobel prize with Paul Ehrlich in 1908, for his work.

During World War-I there was an outbreak of shigellosis. In 1917, before Sir Alexander Fleming's discovery of penicillin, German professor Alfred Nissle isolated a non-pathogenic *Escherichia coli* strain from the feces of a soldier who did not develop enterocolitis during this outbreak (Nissle, 1918). In 1930, a Japanese scientist Minoru Shirota, discovered that a stronger strain of *Lactobacillus* (later named as *L. casei Shirota*) was able to replace the pathogenic bacterial strains in the intestine. This led him to formulate the well-known probiotic drink Yakult.

Lilly and Stillwell first introduced the term "probiotics" in 1965. The term was coined in contrast to existing term like antibiotics (Lilly and Stillwell, 1965). Probiotics were defined as microbially derived factors that stimulate the growth of other organisms. Marcel Roberfroid in 1995 coined the term "Prebiotics" and defined them as non-digestible food ingredients that promote the growth or activity of the bacteria in the digestive tract. Roberfroid redefined prebiotics in 2007 as "a selectively fermented ingredient that allows specific changes, both in the composition and/or activity in the gastrointestinal microflora that confers benefits upon host well-being and health" (Roberfroid, 2007). Another term "Synbiotics" is also now in use and is defined as a product that contains both probiotic and prebiotic components.

II. BIOLOGY OF PROBIOTICS

A. Distribution

The human microbiome inhabits the body both externally as well as internally. The microbes colonize the whole body at variable compositions. Areas of microbial colonization range from scalp, retroauricular crease, congunctiva, anterior nares, buccal mucosa, hard palate, saliva, palatine tonsils and antecubital fossa to duodenum, ileum, colon, vagina and skin in general. In 2010, the Human Microbiome Project (HMP) of National Institute of Health, USA, identified 178 genomes that inhabit the human body (Huttenhower, 2012). It is estimated that the number of microbial cells outnumber the human body cells by 10 to 1. While the project is in the final stages of completion, it is projected that there would be more than 900 genomes that would be defined by the end of it. While this megaproject will be helpful in determining the diveristy and distribution of human microbial diversity, there have been other studies that have investigating some very interesting facts about the species distribution. For example, the investigation by Fierer et al. (2008) from the University of Colorado, USA, has determined that the "microbial signature" on the hands of each individual is unique (Fierer, Hamady et al., 2008). Not only that, there is a distinction between the two hands of the same person. As determined by the group only 13% of phylotypes were common between the two hands. As exciting as it may sound, the results need to be taken with caution. The microbial diversity in temperate areas of the world is much limited in comparison to tropical areas, which are known diversity hotspots. Social and cultural differences on how we interact with our environment have a role to play in the kind of microbes that we associate with ourselves. An interesting example of this comes from communities in the Ladakh region of India. Wirth et al., (2004) studied the distribution of *Helicobacter pylori* in Buddhist and Muslim population living in these high altitude areas. The study resulted in identification of two distinct genetic population of *H. pylori* existing between the two ethnic groups. It was established that this bacterium is transmitted within families, with the Buddhist population having the Ladakhi gene signature of the bacterium and the Muslim having an Indo-European signature (Wirth, Wang et al., 2004).

Amongst the various species inhabiting the human host, *Acinetobacter calcoaceticus*, *Burkholderia cepacia*, *Pseudomonas pseudoalcaligenes*, *Peptostreptococcus spp* are distributed across the body and found at various "locales". Distribution of major groups of microbial taxa is depicted in Figure 62.1. Table 62.1 presents a list of some commonly associated species with their habitats across the human body.

B. Species Characteristics

As is evident from Figure 62.1 and Table 62.1, there are a number of species that inhabit different sites on the human body while there are speceis that are restricted to certain areas

TABLE 62.1 List of human body sites and some representative associated microbial species inhabiting these regions

Localization	Species
Hair follicle	*Staphylococcus aureus*
Ear (external)	*Corynebacterium, Staphylococcus aureus, Staphylococcus epidermidis*
Eye	*Chlamydia trachomatis, Moraxella* sp, *Staphylococcus aureus, Staphylococcus epidermidis*
Mucosa	*Chlamydia trachomatis, Hemophilus influenzae, Staphylococcus aureus, Staphylococcus epidermidis*
Mouth	*Actinomyces naeslundii, Arachnea propionica, Bacteroides gingivalis, Campylobacter sputorum, Fusobacterium nucleatum, Leptotrichia buccalis, Mycoplasma* spp, *Porphyromonas gingivalis, Streptococcus* sp, *Treponema denticola*
Saliva	*Lactobacillus* sp, *Mycobacterium chelonae, Neisseria* spp,
Pharynx	*Acinetobacter* spp, *Bacteroides pneumosintes, Candida albicans, Haemophilus paraphrophilus, Neisseria elongata, Peptostreptococcus* spp, *Streptococcus viridans*
Ileum	*Achromobacter* spp., *Aeromonas* spp, *Clostridium sordellii, Enterococcus faecalis, Flavobacterium* spp, *Mycobacterium* spp, *Macoplasma* spp, *Pseudomonas aeruginosa, Staphylococcus aureus, Vibrio* spp.
Cecum	*Runimococcus* sp.
Colon	*Achromobacter* sp, *Acidamoncoccus fermentans, Aeromonas* spp, *Alcaligenes faecalis, Bacillus* sp, *Bifidobacterium* sp, *Clostridium difficile, Flavobacterium* sp, *Mycobacterium* sp, *Mycoplasma* sp, *Propionibacterium* sp, *Staphylococcus aureus, Sarcina* sp, *Vibrio* sp.
Nasopharynx	*Acinetobacter* sp, *Campylobacter* sp, *Moraxella* sp, *Neisseria* sp, *Selenomonas* sp, *Candida albicans.*
Urinogenital (both male and female)	*Bacteroides* spp, *Bifidobacterium* sp, *Candida albicans, members of Enterobacteriaceae family, Neisseria gonorrhoeae, Mycoplasma hominis, Mobiluncus curtisii*

only. The landmark study published by the HMP Consortiun in the journal Nature in 2012, demonstrated that the species composition, diversity and abundance of microbial communities inhabiting various sites on human body was not only variable between sites, but also differed between various individuals for the same site. Each of the body habitats were determined to be colonized by a given set of taxa, which was variable from one indivdual to another in the study. For example, based on the metagenomic investigations of the 16s rRNA, it was determined that *Propionibacterium acnes* and *Corynebacterium accolens*

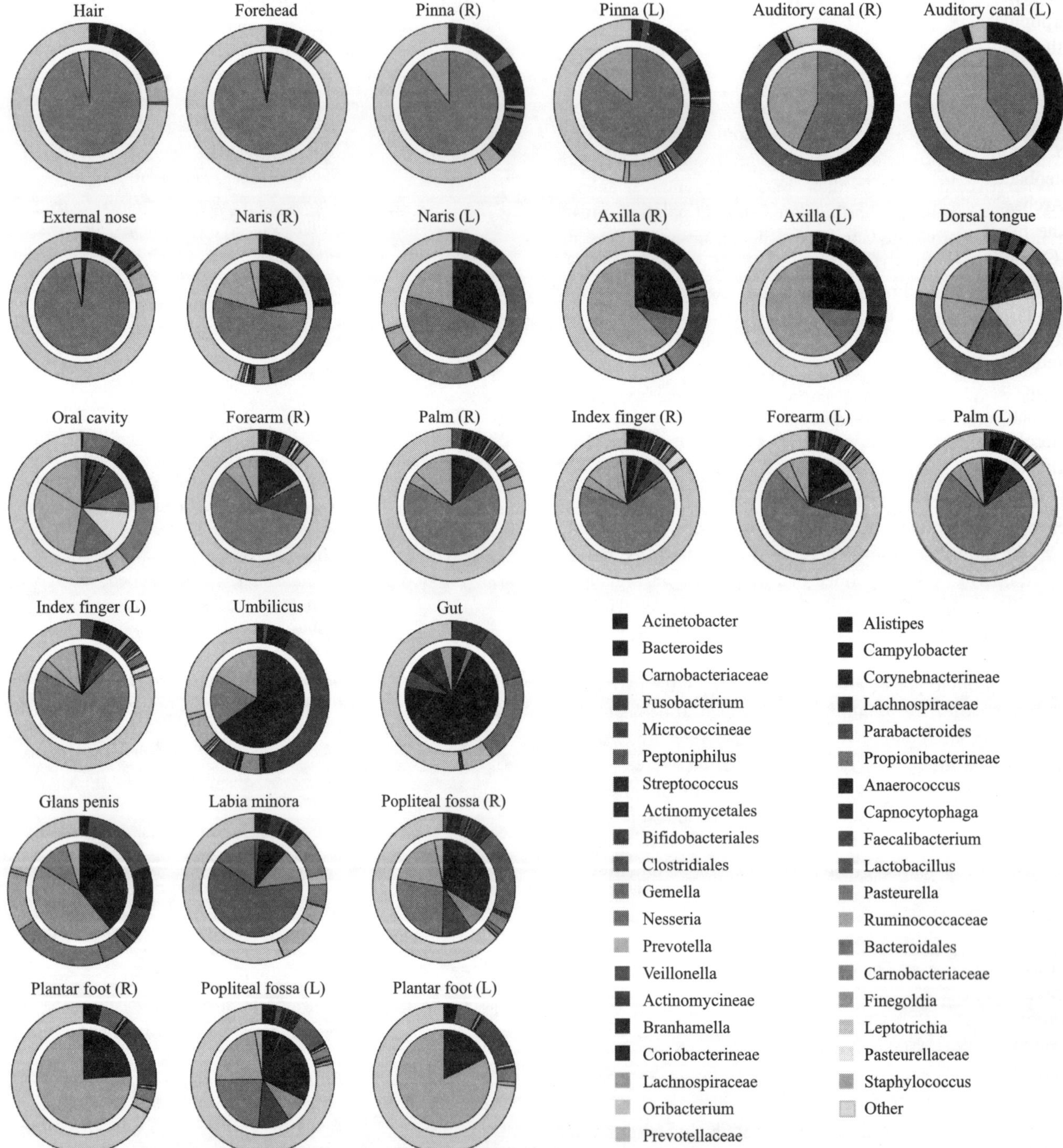

Figure 62.1 Distribution of microbial taxa on different regions of the bioscope. The outer ring depicts the proportion of taxa that were not differentially present in the sampled individuals, and the inner graph shows representation of those taxa that were present in all the sampled individuals. (*Adapted from Costello et al. Science, 2009*) (*see Plate 24 for colour figure*)

were the most abundant species in the anterior nares area; in contrast in the retroauricular crease, besides *P. acnes* another species *C. koppenstedtii* were most abundant (even though *C. accolens* was also present in the community at both the sites) (Huttenhower, 2012). In continuation to the work of the HMP, Faust et al. (2012) analyzed the ecological interactions among bacteria in the human microbiome. They determined

that there were a few *"hub" microbes or signature microbes that determined the structure of the community in that location.* Also, species belonging to genus *Streptococcus* or *Bacteroides* functioned as connectors between multiple related body sites. Positive associations could result due to nutritional requirements of the species, co-aggregation, co-colonization or even sharing signaling pathways in the given environments. In conjunction the negative forces like toxins, immunomodulation or simply an overpopulation of the niche by a species may lead to change in the status of mutualism, commensalism, amensalism, or predator-prey relationships within the community (Raes and Bork, 2008; Faust, Sathirapongsasuti et al., 2012).

The microbiota of the human gut can be grouped into five phyla—Firmicutes, Bacteroidetes, Actinobacteria, Proteobacteria and Verrucomicrobia. Another phyla belonging to Archaea, Euryarchaeota, is also a dominant form present in the human gut. The less common groups are constituted by Cyanobacteria, Fusobacteria, Lentisphaerae, Spirochaetes and TM7. The Firmicutes are majorly represented by species like *Clostridium, Lactobacillus, Ruminococcus*. A number of representatives of this groups have been employed as ingredient in commercially available probiotics. Species from *Eubacterium, Faecalibacterium* and *Roseburia* also contribute significantly to the representation from this group. The phyla Bacteroidetes is represented by *Bacteroides, Prevotella* and *Xylanibacter* degrade a variety of complex glycans. *Collinsella* and *Bifidobacterium* (various strains of from these are common ingredients in probiotics) represent the Actinobacteria phylum in the gut. Common found Proteobacteria are represented by *Escherichia* and *Desulfovibrio* (members of this group have sulphate-reducing properties). *Verrucomicrobia* was recently discovered and includes *Akkermansia* (specialize in mucus degradation). Euryarchaeota member *Methanobrevibacter* is a major player in intestinal methanogenesis (Tremaroli and Backhed, 2012).

C. Function

With all these variations in the distribution and composition of the microbial communities, the composition of the probiotic needs to be carefully designed. There have been multiple investigations on the composition of probiotics and their efficacies.

Probiotics have been packaged into different kinds of products like foods, drugs, and more commonly dietary supplements. *Lactobacillus* and *Bifidobacterium* are the most common species represented in various probiotics. Yeast (*Saccharomyces cerevisiae*) and some *E. coli* or *Bacillus* species are also now commonly available in different formulations of probiotics. *Lactobacillus* species and other lactic acid-modifying species, which have been used for thousands of years for fermentation, have been demonstrated in experimental models to provide health benefits. Fermentation of food besides providing characteristic taste profiles also lowers the pH, which prevents colonization by potential pathogens.

As far as microbial colonization is concerned, the human gut is akin to tropical rainforest or intertidal zone on the human biogeographical plane. The diversity of species, the interactions thereof, the microenvironmental variation in the digestive tract (starting from the bucal cavity throught the ileum and colon up till the rectum), alterations in the gut microenvironment sometimes on a diurnal basis and other times due to sudden stress of antibiotics, make the gut a highly dynamic ecological landscape or "bioscape". The species inhabiting this "bioscape" interact between themselves while acting on the modified microenvironment provided by various food molecules. The gut carries thousands of species and all the species do not function identically or on all the molecules crossing the gut.

Carbohydrate processing

The microbial species assist in modifications of major food components like carbohydrates and proteins or even bile acids to be utilized by the body. Certain carbohydrates that escape the upper GI tract digestion due to presence of poly- or oligosaccharide chains, are connected with chemical bonds. Such carobhydrates (like cellulose, hemicelluloses etc.) are modified by the gut microbiota. The main products of this digestion turn out to be gases like CO_2, acids like acetic acid, butyric acid, formic acid, propionic acid, lactate etc. Acetate and propionate are taken up by the liver and used as substrates for lipogenesis and gluconeogenesis. The combined effect of various microbial species working on the carbohydrates and their by-products is formation of short-chain fatty acids (SCFA). These short chain faty acids have been demonstrated to show anti-inflammatory properties mediated through the GPCR43, utilized as energy source, have anti-carcinogenic properties and also activate innate immunity-related processes (Nicholson, Holmes et al., 2012; Tremaroli and Backhed 2012). In 2006, Samuel et al., demonstrated that fermentation of dietary fructans increases when gnotobiotic mice colonized with *Bacteroides thetaiotaomicron*, and *Methanobrevibacter smithii*. *B. thetaiotaomicron* produces acetate and formate, which is then used for production of methane by *M. smithii*, thereby promoting an efficient carbohydrate fermentation and increased energy production and absorption (Samuel and Gordon 2006). The various species and their role in carbohydrate metabolism in the gut are summarized in Table 62.2.

TABLE 62.2 Role of various microbial species in carbohydrate metabolism

Function	Species
Polysaccahride degradation	*Bacteroides, Faecalibacterium, Prevotella, Roseburia, Ruminococcus*
Monosaccharide breakdown	*Dorea*
Monosaccharides to lactate conversion	*Bifidiobacterium, Lactobacillus*
Monosaccharaide to acetate/butyrate	*Bacteroides, Faecalibacterium, Prevotella, Roseburia, Ruminococcus*
Lactate to butyrate conversion	*Anaerostipes, Eubacterium halii*
Lactate to propionate conversion	*Veillonella, Megamonas, Megasphaera*
Methanogenesis	*Methanobrevibacter*

Processing of proteins

Protein degradation and assimilation is primarily performed by the enzymes of produced in stomach, pancreas, duodenum and intestines. Protein degradation mainly occurs in the distal colon, and it yields both beneficial SCFAs and also potentially toxic metabolites such as ammonia, amines, N-nitroso compounds, phenolic compounds and sulphides (Smith and Macfarlane, 1997). During the late half of the 1980s, Macfarlane et al., teased out the mechanism by which the gut microbiota degrades proteins (Macfarlane, Cummings et al., 1986; Macfarlane, Allison et al., 1988). They demonstrated that when casein or bovine serum albumin were passed over slurries of human feces for a period of 96 hours, TCA soluble peptides, ammonia and volatile fatty acids were produced. The proteolytic bacteria isolated from the fecal samples belonged to *Bacteroides*, *Propionibacterium*, *Streptococcus*, *Clostridium*, *Bacillus* and *Staphylococcus* species. It has now been proven beyond doubt that in irritable bowel syndrome, the gut microbiota is hyperactive and produces extracellular proteolytic enzymes. Some of the species that play a key role in protein digestion are tabulated in Table 62.3.

TABLE 62.3   Some representative microbial species and their function in the protein degradation

Function	Species
Tyrosine digestion	*Clostridium* sp
Cysteine degradation	*Atopobium*, *Megasphaera*, *Prevotella*, *Veillonella*
H_2S production from Cysteine	*Fusobacterium*, *Neisseria*
Leucine breakdown	*Atopobium*, *Clostridium*, *Peptostreptococcus*
Proline modification	*Peptococcus*, *Gamma-proteobacteria*

Lipid modifications

Lipids are another important component of the human diet. Mobilization of lipids, including cholesterol and fat soluble vitamins, from the intestine is carried out by bile acids. Bile acids are synthesized in the liver from cholesterol. Primary bile acids (cholic and chenodeoxycholic acids) are conjugated to glycine in humans (and to taurine in mice). This conjugate is deconjugated in the distal part of the ileum by bacteria and further processed into secondary bile acids. *Clostridium* cluster XI and XIVa have been shown to be responsible for this effect (Ridlon, Kang et al., 2006). Using gnotobiotic and wild type mice, Swann et al. (2010) demonstrated that the amount of conjugated bile acids was higher in germ free animals in comparison to wild type (Swann, Want et al., 2011). Interestingly, the conjugated bile acid concentration was found to be higher in heart and kidney in these animals, whereas in plasma the levels between the germ free and wild type animals were similar. This goes on to suggest that the gut microbiota not only deconjugate bile acids but also modify the chemicals in such a way that they accumulate in heart and kidney. A detailed analysis of proteins and factors responsible for this is required. It is also not known if the modification in bacterial

composition of the gut would have an effect on the tissue specific accumulation of bile acids. *Bilophila wadsworthia* is known to generate H_2S from taurine that is released from the bile acid conjugate. Choline is an important component of cell membranes (Vance, 2008). Choline is an important component of the lipid metabolism and synthesis of very-low-density lipoprotein (VLDL) in the liver. Insufficient levels of choline are associated with altered gut microbiota and hepatosteatosis in both mice (Henao-Mejia, Elinav et al., 2012). Low quantities of Gamma-proteobacteria and high levels of *Erysipelotrichi* in human faeces have been demonstrated to be associate with hepatic steatosis (Spencer, Hamp et al., 2011). Enzymatic activity of host and microbes transforms choline into toxic methylamines. Trimethylamine that is produced by intestinal microbes can be further metabolized to trimethylamine-N-oxide in the liver. These transformations may eventually decrease the levels of choline and have been suggested to trigger non-alcoholic fatty liver disease (NAFLD) in mice. An altered gut microbial composition and its capacity to metabolize choline may have an important role in modulating NAFLD as well as glucose homeostasis (Dumas, Barton et al., 2006).

Vitamin mobilizers

Humans lack the capacity to synthesize vitamins. Most of the vitamins are available to man via the food. In addition, the gut microbial species like Bifidobacteria have the potential to *de novo* synthesize and supply vitamins. Lactic acid bacteria have been shown to *produce folate, riboflavin and cobalamine*. Genetic evidence of cobalamin biosynthesis by *L. reuteri* (strain CRL 1098) has shown involvement of at least 30 genes in *de novo* synthesis of the vitamin. The genetic organization of cob and cbi genes (involved in cobalamine biosynthesis) is similar to that in *Salmonella enterica* and *Listeria innocua* (Santos, Vera et al., 2007). Recently, the genetic pathway responsible for the *de novo* synthesis of vitamin B12 by *L. reuteri* has also been described for more strains of *L. reuteri* (Saulnier, Santos et al., 2011).

D. Microbiome Homeostasis

The concept of resilience in ecological systems was first proposed by the Canadian ecologist C.S. Holling, and he defined it as the time required for an ecosystem to return to an equilibrium or steady-state following a perturbation, and the capacity of a system to absorb disturbance and reorganize while undergoing change so as to retain essentially the same function, structure, identity, and feedbacks (Holling, 1973). As in other ecosystems, the human microbiota, and the gut microbiota specifically, face continuous and potentially disruptive perturbations. Multiple studies have investigated these variations and have concluded that the bacterial community structure remains relatively stable over time in the absence of perturbations (Franks, Harmsen et al., 1998; Zoetendal, Akkermans et al., 1998; Costello, Lauber et al., 2009; Huttenhower, 2012).

Perturbations

The microbial communities are often subjected to perturbations by various factors depending on their localization. Depending on their community composition in reponse to the disturbance inducing agent, the community is able to recover back to its original state, as far as possible. Persistant disturbances lead to major community changes and immune reponses in the body.

However, when disturbances are introduced (in terms of diet or exposure to antibiotics) the gut microbiota composition modifies significantly. The long-term effects of such external influences on the health of the gut microbiota are still to be determined. Spor et al., have discussed in detail the factors that promote stability of healthy microbiota (Spor, Koren et al., 2011). Many studies have demonstrated ecological disturbances in the microbiota communities post antibiotic administration (Jernberg, Lofmark et al., 2007; Dethlefsen and Relman, 2011). Jernberg et al., determined that usage of broad spectrum antibiotic like clindamycin results in rearrangement of species composition in the gut community could be restored to original within 21 days of drug treatment, after which the altered community structure tends to stabilize. The community composition and species representation varies over time with continued usage of clindamycin and after a two year period the species representation is stable and very distinct from the original. Dethlefsen et al., found that use of antibiotic (ciprofloxacin) leads to alteration of the microbial community. These alterations occur on a daily basis and the community structure changes; although the hub-species did not vary over 2 to 5 month periods. In the same study the long term use of antibiotics significantly varied the community composition over a one year period. Members of the bacterial community that are susceptible or resistant to antibiotics contribute significantly to the antibiotic-associated gut microbome perturbations. Antibiotic resistant strains, more often than not, persist in the host gut even in the absence of selective pressure (Karami, Martner et al., 2007; Sommer, Dantas et al., 2009). While the effect of antibiotics on the gut microbiota is short term, there have been reports in which it has been documented to persist for extended periods of time. These long-term imbalances to the microbiome may increase the individual's susceptibility to infections and disease. Excessive use of antibiotics could be fuelling the dramatic increase in conditions such as obesity, type 1 diabetes, IBD, allergies, and asthma, which have increased dramatically in many populations over the past few decades (Blaser and Falkow, 2009). Another aspect of antibiotic exposure is the long-term persistence of antibiotic resistance genes in the human gut. Disrupting microbial colonization by antibiotics is frequently associated with diarrhea, altered gastrointestinal physiology, and abnormal carbohydrate metabolism. Antibiotic-mediated disruption of the microbiome may also result in the proliferation of *C. difficile* and can lead to aggressive bacterial toxin-induced colitis (Reeves, Theriot et al., 2011). Antibiotic use in low-birth-weight neonates can contribute to the onset of necrotizing enterocolitis (NEC), a disease in which microbial dysbiosis is considered to directly contribute to disease onset.

Epidemiologic studies link multiple courses of antibiotics in early childhood with increased risk of Crohn's disease (Hviid, Svanstrom et al., 2011). Tools for measuring perturbations in gut microbiota exist. Yet, the microbial community stability as a measure of the health of the microbiome remains to be elucidated at a larger scale. Table 62.4 provides some causal factors for perturbations in the gut microbiota.

TABLE 62.4　Factors that may lead to gut microbiota perturbations

Factor	*Model*	*Selected References*
Mode of fetus delivery	Humans	Dominguez-Bello et al. (2010), Decker et al. (2011)
Geographic provenance	Humans	De Filippo et al. (2010), Lagier et al. (2012)
Host genotype	Humans	Spor et al. (2011), Li et al. (2012)
	Mice	Kovacs et al. (2011), Ley et al. (2005), Benson et al. (2010)
Nutrition	Humans	Walker et al. (2011), Wu et al. (2011), Smith et al.(2013)
	Animals	Hildebrandt et al. (2009), Turnbaugh et al. (2009), Le Chatelier et al. (2013), Tilg & Moschen (2013)
Antibiotics	Humans	Willing et al. (2011), Jernberg et al. (2007), Dethlefsen and Relman (2011), Ostaff et al. (2013)
	Mice	Yap et al. (2008), Carvalho et al. (2013)
Probiotics	Humans	Rauch and Lynch (2012), Tanabe (2013), Serban (2013),
	Animals	Garcia-Mazcorro et al. (2011)
Age	Humans	Tiihonen et al. (2010), Biagi et al. (2010)
	Mice	Zeng et al. (2013), El Aidy et al. (2013)
Stress	Humans	Konturek et al. (2011)
Fecal therapy	Humans	Vrieze et al. (2013), van Nood (2013), Hamilton et al. (2013), Brandt (2013), Borody et al. (2011, 2012)
	Animals	Damman et al. (2012),

III.　CLINICAL IMPLICATIONS

A.　Probiotics in Use

The primary aim of using probiotics is to promote the establishment of body's naturally occurring gut microbiota. Some probiotics have been demonstrated to prevent diarrhea caused by antibiotics, or antibiotic-related dysbiosis. Investigations on disorders like inflammatory bowel disease (IBD), irritable bowel syndrome (IBS), vaginal infections, and immune enhancement have been shown to subside by application of probiotics in the treatment regimen. Some probiotics have also been investigated in relation to atopic eczema, rheumatoid arthritis, and even liver cirrhosis. Although there is some clinical evidence for the role of probiotics in lowering cholesterol, the results are conflicting. In general, the strongest clinical evidence for probiotics is related to their use in improving gut health and stimulating immune function.

The guidelines for examining the scientific evidence on the functional and safety aspects of probiotics in food have been clearly laid down by FAO [FAO/WHO 2001, 2002, 2011], are to be used when a new probiotic is to be introduced for human use. It is suggested that manufacturers label the genus, species, and strain for each probiotic in a product, along with the number of viable cells of each probiotic strain that will remain up to the end of shelf-life. Some of the more commonly available probiotics, their representative species and manufacturer are listed in Table 62.5.

TABLE 62.5 **Microbial species and strains used in commonly available probiotics across the world.**

Strain	Brand name	Producer
Bifidobacterium animalis DN 173 010	Activia	Danone/Dannon
Bifidobacterium breve Yakult	Bifiene	Yakult
Bifidobacterium infantis 35624	Align	Procter & Gamble
Lactobacillus casei DN-114 001	Actimel, DanActive	Dannon
Lactobacillus casei F19	Cultura	Arla Foods
Lactobacillus casei Shirota	Yakult	Yakult
Lactobacillus johnsonii La1	LC1	Nestlé
Lactobacillus plantarum 299V	GoodBelly, ProViva	NextFoods Probi
Lactobacillus reuteri ATTC 55730	Reuteri	BioGaia Biologics
Lactobacillus rhamnosus ATCC 53013	Vifit	Valio
Lactobacillus rhamnosus LB21	Verum	Norrmejerier
Saccharomyces cerevisiae (boulardii) lyo	DiarSafe, Ultralevure	Wren Laboratories, Biocodex, and others
Combination of Lactobacillus acidophilus CL1285 & Lactobacillus casei Lbc80r	Bio K+	Bio K+ International
Combination of strains of Streptococcus thermophilus, Lactobacillus spp., & Bifidobacterium spp.	VSL#3	Sigma-Tau Pharmaceuticals, Inc.
Combination of Bacillus clausii strains	Enterogermina	Sanofi-Aventis
Lactobacillus fermentum TRF#36		Microbax Hyderabad
Lactobacillus gasserei TRF#8		Microbax Hyderabad
Lactobacillus salivarius TRF#30		Microbax Hyderabad

Besides being mixed with dairy products, probiotics are supplied in the market under various packages – in form of tablets, capsules, and sachets containing the bacteria in freeze-dried form are also available. There is a general concern about the dose and concentration needed for a probiotics to be used for treatment of a particular disorder, when the strain and CFUs vary dramatically between preparations. Generally, over the counter products carry bacteria in the range of 1–10 billion cfu/dose. In contrast, species like *Bifidobacterium infantis* have been demonstrated to be effective in alleviating the symptoms of IBS even at 100 million cfu/day. Another popular product VSL#3 is marketed highlighting its high (and demonstratedly effective, though controversial) 300–450 billion cfu t.i.d. Detailed studies and large scale clinical trials would only be helpful in delimiting the dose and usage of "supportive" bacterial species in designing future probiotics (May, 2008).

Besides the gut, the oral cavity is a site of active colonization and species variability in the human body. The bacterial species residing in the buccal cavity vary in their role from commensal colonizers to pathogens, depending on the kind of environment that is maintained by the host. Periodontitis is a disease that includes both microbial and host responses including microbial colonization, inflammatory and subsequent adaptive immune responses. While the treatment of this disease is essentially focussed on the antibacterial therapy, probiotics provide a promising option. Directed modification of the oral cavity using probiotics has been shown to reduce the pH (Chopra and Mathur, 2013), thereby reducing damage to hard and soft dental tissues. Additionally, the antioxidant produced by the probiotic bacterial species prevent mineral and plaque formation by neutralizing the free electrons in the affected area. A major effect of probiotics is reduction of foul odors by fixing volatile sulfur compounds. Teughels, et al., reported that the subgingival application of a bacterial mixture including *Streptococcus sanguis*, *Streptococcus salivarius*, and *Streptococcus mitis* after scaling and root planning significantly suppressed the re-colonization of *Porphyromona gulae* and *P. intermedia* in a dog model (Teughels, Newman et al., 2007). Burton, et al., reported significant reductions in volatile sulfur compounds for the probiotic group compared to the placebo group when probiotic *Streptococcus* was used (Burton, Chilcott et al., 2006).

Increase in acidogenic and acid resistant species such as mutans streptococci and lactobacilli lead to dental caries, although other speceis like *Actinomyces* spp., *Bifidobacteria*, *Propionibacterium* spp., have similar properties but are not ivolved in caries development. Initial studies tested bacterocin producing species as a treatment of caries. One approach has been to identify food grade and probiotic bacteria which have ability to colonize teeth and influence the supra-gingival plaque. Another approach utilized recombinant strain of *S. mutans* expressing urease, which was shown to reduce the cariogenicity of plaque in an animal model (Caglar, Kargul et al., 2005).

Candida species are generally non-infective and are part of the commensal oral flora in about 50% of healthy subjects. The same species becomes pathogenic under immunosuppressed conditions of the host. In one study, subjects consuming *L. rhammnosus GG* containing cheese exhibited reduction in the prevalence of oral candidosis. In contrast, various lactobacilli could not produce the same effect. Of significance is the finding that *L. acidophilus* treatment was able to clear *Candida* in an animal model (Chopra and Mathur, 2013).

In larynx related prosthetic surgeries, biofilm development is a matter of concern. Probiotics have been demonstrated to reduce this biofilm formation. In Netherlands, anecdotally, consumption of buttermilk, which contains *Lactococcus cremoris, Lactococcus lactis* spp. (Busscher, Bruinsma et al., 1998).

B. Status of Probiotics in Diseases

Cardiovascular disease

Agerholm-Larsen et al., investigated the effect of a probiotics milk product containing a predesigned combination of microbiots (CAUSIDO®) on risk factors involved with cardiovascular diseases in obese patients (Agerholm-Larsen, Raben et al., 2000). The study concluded that the probiotics was effective in reducing the LDL content leading to reduction in cardiovascular risk factors. Another example is of dietary lecithin or phosphatidylcholine (PC). PC is metabolized into choline, betaine and trimethylamine N-oxide (TMAO). Of these lack of choline may lead to non-alcoholic fatty liver disease and/or muscle damage. Choline is metabolized by gut microbiota in to trimethylamine (TMA). TMA is metabolized to oxidized form TMAO. In experimental animals (Wang, Klipfell et al., 2011) discovered that increased dietary choline leads not only to increased plasma levels of TMAO, but also increased plaque development in the arteries of mice prone to atherosclerosis.

Colon cancer

The SYNCAN study (2005) demonstrated that oligofructose and two probiotic strains were able to reduce the expression of colorectal cancer biomarkers. In mice experiments, *Bacillus polyfermenticus*, a common ingredient of probiotics in Asian countries, was shown to suppress the ErbB2 and 3 expression. Additionally, administration of this species in the animals lead to suppression of cyclinD1 expression. Both these effect led to reduction in the growth of colon cancers both *in vitro* as well as in xenografts (Ma, Choi et al., 2010).

Diarrhea

In a randomized clinical trial, Species like *L. reuteri* DCM 17938, *L. rhamnosus* GG, *L. casei* DN-114 001, and *Saccharomyces cerevisiae (boulardii)* have been proven to be effective in reducing the severity and duration of acute infectious diarrhea in children (Francavilla, Lionetti et al., 2012). In the prevention of diarrhea in adults, there is only suggestive evidence that *Lactobacillus* GG, *L. casei* DN-114 001, and *S. boulardii* are effective in some specific settings. In a randomized, double-blind, placebo-controlled pilot study, *L.*

reuteri DCM 55730 was found to be safe and well tolerated as well as more effective (almost 9 folds) than placebo (Cimperman, Bayless et al., 2011). In antibiotic-associated diarrhea, there is strong evidence of efficacy for *S. boulardii* or *L. rhamnosus* GG in adults or children who are receiving antibiotic therapy. Recent research has indicated that L. casei DN-114 001 is effective in hospitalized adult patients for preventing antibiotic-associated diarrhea and C. difficile diarrhea. In a meta-analysis performed by (Goldenberg, Ma et al., 2013) on the use of probiotics for the prevention of *Clostridium difficile*-associated diarrhea (CDAD) in adults and children, the results were a mixed bag. The analysis found that robiotics were able to reduce CDAD by about 64%. In in a double-blind, randomized, placebo-controlled clinical trial, the probiotics VSL#3 was found to prevent antibiotic-associated diarrhea. It is not known if the same probiotics would be as effective with other causes of diarrhea.

Inflammatory bowel disease (IBD)

In recent years, application of probiotics in treating Crohn's disease has gained significant attention. The efficacy of *Lactobacillus GG* has been compared to placebo in patients. In 2005, Bousvaros et al., conducted a randomized double-blind placebo controlled trial to establish the efficacy of *Lactobacillus GG* as adjuvant to standard therapy in the maintenance of remission in patients of Crohn's disease. No difference between the treated and the placebo group was observed (Bousvaros, Guandalini et al., 2005). In contrast another species, *Saccharomyces boulardii*, showed a positive effect and maintained remission for a longer duration (Guslandi, Mezzi et al., 2000). In another study employing the same species demonstrated that the intestinal barrier permeability was also improved after *S. Boulardii* treatment (Garcia Vilela, De Lourdes De Abreu Ferrari et al., 2008). In rpresence of these conflicting reports, a Cochrane review (including seven studies) (Rolfe, Fortun et al., 2006), and a meta-analysis (Rahimi, Nikfar et al., 2008) (including eight randomized placebo-controlled clinical trials) confirmed that probiotics were ineffective in maintaining remission and prevention of recurrence of Crohn's disease.

VSL#3 has been demonstrated in multiple studies to be effective against ulcerative colitis in several studies. VSL#3 efficacy has also been tested on children diagnosed with ulcerative colitis (Miele, Pascarella et al., 2009). In this randomized placebo-controlled trial it was found that VSL#3 treatment was abe to achieve remission levels in 90% of patients in comparison to 4% in placebo. In another study Kruis et al., showed that there was no difference in the use of the probiotic *E. coli Nissle 1917* and mesalazine in the maintenance of remission in ulcerative colitis patients (Kruis, Fric et al., 2004). This only highlights the fact that the constituents of a probiotics need to be carefully determined int he formulation and well-tested. *Lactobacillus GG* has also been demonstrated to be effective in maintenance of remission in patients (Zocco, dal Verme et al., 2006). In a Cochrane review, Mallon et al., (Mallon, McKay et al.,

2007) concluded that probiotics could provide efficacy in the maintenance of remission in patients with mild-moderate ulcerative colitis, whereas only limited efficacy could be predicted for moderate-severe disease.

Lactose malabsorption

Multiple studies have tested the use of probiotics (particularly in dairy products) in improving digestion and reduction of symptoms due to lactose intolerance (Newcomer, Park et al., 1983; Lin, Yen et al., 1998; Jernberg, Lofmark et al., 2010; Shaukat, Levitt et al., 2010; Vermeiren, Van den Abbeele et al., 2012). Species like *Streptococcus thermophilus* and *Lactobacillus delbrueckii* subsp. *bulgaricus* have proven to be effective in bringing about this effect (He et al., 2008). In a recently conducted study on children from five government schools in Indonesia, it was determined that the efficacy of both live or killed microbiota when administered as probiotics reduced the amount of hydrogen as assessed by the breath hydrogen test (Rampengan, Manoppo et al., 2010).

Nonalcoholic fatty liver disease

The gut-liver axis has gained newfound attention in the treatment of liver diseases. Investigations on non-alcoholic fatty liver disease and metabolic syndrome have been conducted to confirm the efficacy of probiotics in these diseases. Recent studies have reported a possible impact of gut micro flora overgrowth on the development of NAFLD/NASH (Nair, Cope et al., 2001; Solga and Diehl, 2003). The species from the gut microbiota are known to release endotoxins like lipopolysaccharides (LPS), which are recognized by the pattern recognition toll-like receptors on liver cells leading to cytokines secretion from liver. It has been proposed that liver injury and fibrosis could be a result of exposure to LPS (Lafyatis and Farina, 2012). Increase in *Escherichia coli* and *Staphylococcus* counts in stool samples have been correlated with encephalopathy and cirrhosis in patients incomparison to healthy controls (Liu, Lazenby et al., 2007). Li et al., showed that the administration probiotic VSL#3 to the ob/ob mice fed with HFD not only altered the animal's obesity but also showed improvements in liver histology and decrease in total fatty acid in liver and serum alanine transaminase (ALT) in serum (Li, Yang et al., 2003). These effects were attributed TNFα, expressed in response to the probiotic (Thurman, Bradford et al., 1997; Yang, Lin et al., 1997; Esposito, Iacono et al., 2009). There was also a reduction in β-oxidation of fatty acids and decrease in insulin resistance. The reduction in TNF-α has also been observed in NAFLD patients, was seen as the reason for the observed (Tilg and Diehl, 2000; Larter and Farrell, 2006). Another probiotic, MIYAIRI 588 has been tested to improve NAFLD related lesions in liver. The major component of this probiotic is a butyric acid producing gram positive anaerobic *Clostridium butyricum* strain. The probiotic has been shown to reduce the expression of tight junction proteins, ZO1 and occluding-1, in the intestine. Also, there was a marked enhancement in the expression of nuclear factor erythroid 2-related factor 2 (Nrf2) resulting in an increase in its targeted antioxidative enzymes leading to suppression of hepatic oxidative stress. The butyrate generated by *C. butyricum* was demonstrated to activate AMP kinase. Interestingly, the report demonstrates that AMPK activation leads to Sirtuin-1 phosphorylation, which eventually activates a cascade from PI3K activation to Akt leading to Nrf2 activity in the liver (Endo, Niioka et al., 2013).

C. Fecal Therapy

The most recent advancement in the field of probiotics has been the development of trnaplantation of healthy set of gut microbiota into the diseased patient's gut by fecal therapy. Though still not popular and difficult to conceive as a treatment option, the transplnation of microbes present in the feces of a healthy individual have proven to improve health of diseased patients. Fecal microbiota transplantation (FMT), intestinal microbiota transfer, fecal transplantation or fecal bacteriotherapy gained popularity with its success for treating *Clostridium difficile* infection (Mattila, Uusitalo-Seppala et al., 2012). The first record of clinical fecal transplant dates back to middle of last century, when in 1958 Eiseman et al., performed a fecal enema in patient of Pseumembranous enterocolitis (Eiseman, Silen et al., 1958). Beneficial effects of FMT have been reported in IBD patients for up to 4 years, when these patients. As this chapter is being written, a randomized controlled double blind trial is currently underway at the Amsterdam Medical Center, investigating the treatment of dysbiosis in ulcerative colitis patients with multiple fecal transplantations.

Differential effects of FMT have also been observed in patients with autoimmune diseases (Baumgart and Sandborn 2012). Bacterial species like *Alcaligens spp.* have been found in the intestinal lymphoid tissue. Systemic immune response has been determined against these bacteria in patients with Crohn's disease (Sonnenberg, Monticelli et al., 2012). The therapeutic efficacy of fecal transplantation has been mostly investigated in Clostridium difficile infections. Approximately 5 million cases of C. difficile infections are attended to in USA each year (O'Brien, Lahue et al., 2007). Decreased and or altered microbial diversity is known to be associated with this infection (Chang, Antonopoulos et al., 2008). A recently concluded open label, randomized controlled trial of fecal transplantation demonstrated that FMT was superior to vancomycin treatment in patients with recurrent *Clostridium difficile* infection (van Nood, Vrieze et al., 2013). The trial also determined that there was a decrease in *Proteobacteria* species and in increase in *Bacteroidetes* species in the patients who received FMT.

D. Probiotic Safety

The risks of probiotics causing infections or modification of metabolic activities, immune system, or even gene transfer into the host cell, are real life concerns (Boyle, Robins-Browne et al., 2006; Williams, 2010). Multiple studies have demonstrated that strains of *Lactobacillus* used in probiotics cause bacteremia in patients with short-bowel syndrome (Kligler and Cohrssen, 2008). Opening and administration of *Saccharomyces* capsules at the bedside of patients with central venous catheter has been reported to cause fungemia (Munoz, Bouza et al., 2005). *Lactobacillus* preparations are contraindicated in individuals with a hypersensitivity to lactose or milk. *S. boulardii* is contraindicated in those with yeast allergy. The review on bacterimia by Boyle et al. (2006) adresses the issue in great detail. In a recent study by Hütt et al., have delineated the methods for assessing evaluation of *in vivo* safety of *Lactobacillus* species in the GI tract (Hutt, Koll et al., 2011). While species like *Lactococcus* and *Lactobacillus* are generally considered safe for use, the other species belonging to the same group, e.g., *Streptococcus* and *Enterococcus*, that are utilized as probiotics can function as opportunistic pathogens.

The risk of *Lactobacillus* infections is estimated at about 1:10 million over a century of probiotic consumption in France (Bernardeau, Guguen et al., 2006). Feeding of premature infants with a commercial *Lactobacillus rhamnosus* GG (named after discoverers Goldin and Gorbach) did not reveal any pathogenic potential of the strain. In contrast, a recent review (Cannon, Lee et al., 2005) reported that of 241 cases of lactobacillemia identified through a Medline search of articles published between 1950 and 2003 Lactobacilli were implicated in endocarditis, bacteremia, and localized infections. *Lactobacillus casei* and *L. rhamnosus* were the commonly associated species with the onset of the disease (Rautio, Jousimies-Somer et al., 1999). In recent years, some cases of liver and spleen abscesses caused by Lactobacilli have been described (Cukovic-Cavka, Likic et al., 2006; Doi, Nakajo et al., 2011), though most of these lactobacillemia cases are associated within the immunosuppressed population. In a recent review on infections due to probiotics it appeared that about 60% of adverse events involved infections due to *L. rhamnosus GG* or *Saccharomyces boulardii* (Whelan and Myers, 2010). In another report the case of a person suffering from a *Lactobacillus*-associated endocarditis after having teeth extracted due to caries appears more logical and natural way of bacterial translocation (Mackay, Taylor et al., 1999). Since the frequency of *Lactobacillus* infected endocarditis is very low (between 0.05 and 0.4%) the results should be handled with care.

In contrast to *Lactobacillus* spp., the genus *Enterococcus* has a higher potential risk. Enterococci are emerging as a major cause of nosocomial infection causing endocarditis, bacteremia, central nervous system infections, neonatal infections, urinary tract infections despite the fact that they are used as starter cultures in the food industry as well as probiotics (Foulquie Moreno, Sarantinopoulos et al., 2006; Franz, Huch et al., 2011). The origin of enterococcal pathogenicity is linked to factors involved in adhesion, translocation, and immune evasion (Johnson, 1994). A multiplex PCR can be used for the detection of the presence of specific enterococcal virulence genes such as *asa*1, *gel*E, *cyl*A, *esp*, and *hly* (Vankerckhoven, Huys et al., 2008). *Enterococcus faecalis* carries a pathogenicity

island of 153 kb containing several virulence factors and is, therefore, believed to be more virulent than *E. faecium*, although horizontal transfer of the entire pathogenicity island into the chromosome of *E. faecium* has been demonstrated (Laverde Gomez, Hendrickx et al., 2011). *Enterococcus faecium* is mostly used as probiotic like in four commercial tablet products in Japan (Hammad and Shimamoto, 2010).

Probiotic *Bacillus* strains are being employed more due to their spore-forming capacity, which provides them an advantage in relation to stability and viability of the probiotic product when this contains spores instead of vegetative cells. Human probiotics produced in Europe, Vietnam and Brazil were shown to contain *B. cereus* spores. Strains belonging to the *B. cereus* group are known to be able to cause two kinds of food-borne illness—one, there is vomiting illness due to the ingestion of food containing the heat-stable toxin cereulide; and the second is a diarrheal infection due to the ingestion of *B. cereus* strains producing heat-labile enterotoxins in the small intestine. There are several methods published to screen *B. cereus* group strains (Wehrle, Moravek et al., 2009). Detection of toxin genes in *B. cereus* allows assessment of the enterotoxic potential of an isolate. Probiotic products can also contain strains of *B. licheniformis*, *B. clausii*, *B. subtilis*, and *B. pumilus*. Although much less important in food-poisoning incidents, strains of several of these *Bacillus* species are also known to be potential producers of heat-stable or heat-labile toxins (De Jonghe, Coorevits et al., 2010).

Another example of probiotic-gone-bad is *Eschericha coli* Nissle strain 1917. It is a well known probiotic prescribed for treatment of ulcerative colitis (Kruis, Fric et al., 2004). On the other hand, this strain has also been associated with severe sepsis in a preterm infant (Guenther, Straube et al., 2010). Using a mice model, the safety of this strain was assessed under defective immunity and varied intestinal microbiota (Gronbach, Eberle et al., 2010). It was found that if both the microbiota and adaptive immunity are defective, the strain may have potentially severe adverse effects. For other *E. coli* strains that would be considered as potential probiotics, it is important to ascertain that they do not belong to one of the pathogenic *E. coli* groups, such as Shiga-toxin producing, enterohaemorrhagic, enterotoxigenic, enteroinvasive, enteropathogenic *E. coli*, etc. (Nataro and Kaper, 1998). Both *Propionibacteria* and *Bifidobacteria* are used in probiotic dairy products. In a recent review it has been shown that in contrast to the cutaneous *Propionibacterium* spp. (the so-called "acnes group"), the dairy *Propionibacterium* spp. are regarded as safe and do not carry any known virulence factor, although *P. Thoenii* and *P. jensenii* strains show hemolytic activity. *Bifidobacterium* is considered one of the safest genera for use as probiotic. Nevertheless, some commensal bifidobacteria have been connected with certain dental infections, pulmonary infections, bacteremia, abscesses, and bloodstream infections. The nonprobiotic *B. dentium* is the only bifidobacterial species classified as a pathogen. *Saccharomyces cerevisiae* var. boulardii has been associated with cases of fungemia (Enache-Angoulvant and Hennequin 2005).

A recently concluded clinical trial suggested that prophylactic probiotics in predicted severe acute pancreatitis patients should not be administered since the probiotic used (Ecologic 641 consisting strains of *Lactobacillus acidophilus*, *L. casei*, *L. salivarius*, *L. lactis*, *Bifidobacterium bifidum*, and *B. lactis* at a total daily dose of 10^{10} bacteria) did not reduce the risk of infectious complications and even increased the risk of mortality (Besselink, van Santvoort et al., 2008). It has been argued after these findings that probiotics should only be contemplated if the integrity of the gastrointestinal tract is not severely compromised (Bjarnason, Adler et al., 2008).

With the advent of next-generation sequencing techniques, the number of full bacterial genomes being sequenced has rapidly increased. Comparative genomics are now being used to find core genes, niche-specific genes, and genes linked to specific probiotic traits. In addition such high throughput techniques should be employed for determination of virulence or antibiotic resistance genes or to find indications of chromosomal integration of horizontally acquired DNA, which would help identify the potential of horizontal transfer of virulence or resistance genes. For example, from a comparative genomic analysis between a probiotic and a clinical *E. faecalis* strain, it seems that several of the enterococcal virulence factors are absent in the probiotic strain *E. faecalis* Symbioflor I (Domann, Hain et al., 2007). Likewise, a genomic comparison between the probiotic strain *E. coli* Nissle 1917 and other *E. coli* strains has indicated the lack of hemolysin and P-fimbrial adhesions in the probiotics strain (Grozdanov, Raasch et al., 2004). Another study indicates that genetic variations and gene expression differences, rather than genomic content of virulence genes per se, contribute to the divergence in the pathogenic traits between *E. coli* strains (Vejborg, Friis et al., 2010).

Another threat from the probiotic bacterial species is generation of secondary biologicals that may be harmful to the host. As described above, some intestinal bacteria are known to convert proteins and their digested products into ammonia, indol, phenols, and biogenic amines (histamine, tyramine, putrescine, etc.). Species belonging to *Lactobacillus* and *Bifidobacterium* are not known to be harmful in this respect. Araya-Kojima et al., measured the enzyme activities related to the generation of ammonia by *Bifidobacterium* species. In comparison to other bacterial species of the gut *Bifidobacterium* sp. have a lower deaminase activity, but a higher ammonia assimilation activity (Araya-Kojima, 1996). Secondary bile acids are important harmful substances that are produced by intestinal bacterial actions on body secretions. They may exhibit carcinogenicity by acting on the mucous-secreting cells and promoting their proliferation, or they may act as promoters of carcinogenesis. Many intestinal bacteria, including *Bifidobacterium* and *Lactobacillus* species, can deconjugate conjugated bile acids. However, *Bifidobacterium* spp., *Lactobacillus* spp., *Leuconostoc lactis subsp. lactis* and *S. thermophilus* have been reported to lack the 7-dehydroxylase activity that is related to the production of secondary bile acids (Ferrari, Pacini et al., 1980; Takahashi and Morotomi, 1994). For *Enterococcus sp*, cytolytic substance and other virulence factors have been reported (Jett, Huycke et al., 1994). It is not yet known if other members of the human microbiota (particularly those harboring the gut) have the toxin generating potential or not. Detailed investigations are required to come to a conclusion and before the probiotics are formulated.

The probiotic, MIYAIRI 588, explained above as being beneficial in NAFLD also sets an example for being cautious while implementing these drugs in treatment regimen. While the authors have deomnstrated succesfully that the probiotic is useful in treating NAFLD (at least in animals), it carries with a risk of increasing the intestinal permeability. This increased permeability will not lead to tranlocation of endotoxins from other bacterial species of the gut remains to be elucidated. Also, if the reduction in expression of ZO-1 and occludin-1 is restricted to intestinal wall only requires further investigations (Endo, Niioka et al., 2013).

While it is interesting to note that a vast variety of microbial species found on the human body have been identified, there are still a significant number of species that cannot be cultured in laboratories and their caharacteristics have not been worked out to full extent. The contribution these species would be making in the health and disease remains elusive. Considering this fact, it is perplexing to come to terms with the new therapeutic modaliies like the fecal therapy. The major issue associated with fecal therapy is that of acceptance as a respectable course of treatment. In most societies, fecal matter is considerd waste and accepting it as medicine is beyond comprehension. While in some middle-eastern tribes, it is a standard practise (e.g., use of camel droppings). The major difficulty that fecal therapy faces is identification of a healthy individual from whose faeces the microbes would be extracted. Use of FMT should be approached with caution, as in a small group of patients with Crohn's disease adverse effects like transient fever, abdominal pains, bloating or even no clinical improvement have been reported (Vermeiren, Van den Abbeele et al., 2012).

IV. FUTURE PERSPECTIVES

With the kind of diversity and distribution the microbes have in the human body, it is difficult to imagine the human body as an individual organism, but rather appears to be a "mobile ecosystem". The microbiota has a systematic influence on human health, some of which was reviewed here. For the progress in this research area, it is essential that the scientific community recognizes the requirement to investigate microbial representatives in human organs in greater details. Since only about 20% of the intestinal microbes are known cultured organisms with known physiological functions, cultivation of the remaining majority, will certainly discover a wealth of novel functions of the ecosystem. Sequencing technologies and bioinformatics tools are expanding their limits at a rapid pace, which would only promote the characterization of microbes at the pace required to formulate disease specific and effective probiotic.

REFERENCES

1. Probiotics and Prebiotics (2008), *World Gastroenterology Organisation Practice Guideline.*

2. Agerholm-Larsen L., Raben A., et al. (2000), Effect of 8 Week Intake of Probiotic Milk Products on Risk Factors for Cardiovascular Diseases, *Eur J. Clin Nutr.*, 54(4):288–297.

3. Araya-Kojima T., Yaeshima T., Ishibashi N. and Shimamura S. (1996), Inhibitory Effects of Human-derived Bifidobacterium on Pathogenic *Escherichia coli Serotype* O-111, *Bioscience Microflora*, 15:17–22.

4. Baumgart D.C. and Sandborn W.J. (2012), Crohn's Disease, 380(9853):1590–1605.

5. Bernardeau M.M., Guguen et al. (2006), Beneficial Lactobacilli in Food and Feed: Long-term Use, Biodiversity and Proposals for Specific and Realistic Safety Assessments, *FEMS Microbiology Reviews*, 30(4):487–513.

6. Besselink M.G., Van Santvoort H.C., et al. (2008), Probiotic Prophylaxis in Predicted Severe Acute Pancreatitis: A Randomised, Double-blind, Placebo-controlled Trial, *Lancet*, 371(9613):651–659.

7. Bjarnason A.S.N. Adler, et al. (2008), Probiotic Prophylaxis in Predicted Severe Acute Pancreatitis, *Lancet*, 372(9633):114–115.

8. Blaser M.J. and Falkow S. (2009), What Are the Consequences of the Disappearing Human Microbiota? *Nat Rev Microbiol.*, 7(12): 887–894.

9. Bousvaros A. Guandalini S., et al. (2005), A Randomized, Double-blind Trial of Lactobacillus GG Versus Placebo in Addition to Standard Maintenance Therapy for Children with Crohn's Disease, *Inflamm Bowel Dis.*, 11(9):833–839.

10. Boyle R.J., Robins-Browne R.M., et al. (2006), Probiotic Use in Clinical Practice: What are the Risks? *Am J. Clin Nutr.*, 83(6):1256–1264; Quiz 1446–1257.

11. Burton J.P., Chilcott C.N., et al. (2006), A Preliminary Study of the Effect of Probiotic *Streptococcus Salivarius* K12 on Oral Malodour Parameters, *J. Appl Microbiol.*, 100(4):754–764.

12. Busscher H.J., Bruinsma G., et al. (1998), The Effect of Buttermilk Consumption on Biofilm Formation on Silicone Rubber Voice Prostheses in an Artificial Throat, *Eur Arch Otorhinolaryngol*, 255(8):410–413.

13. Caglar E., Kargul B., et al. (2005), Bacteriotherapy and Probiotics' Role on Oral Health, *Oral Dis.*, 11(3):131–137.

14. Cannon J.P., Lee T.A., et al. (2005), Pathogenic Relevance of Lactobacillus: A Retrospective Review of Over 200 Cases, European Journal of Clinical Microbiology and Infectious Diseases: Official Publication of the European Society of Clinical Microbiology, 24(1):31–40.

15. Chang J.Y., Antonopoulos D.A., et al. (2008), Decreased Diversity of the Fecal Microbiome in Recurrent Clostridium difficile-associated Diarrhea, *The Journal of Infectious Diseases*, 197(3):435–438.

16. Chopra R. and Mathur S. (2013), Probiotics in Dentistry: A Boon or Sham, *Dent Res J.* (Isfahan), 10(3):302–306.

17. Cimperman L., Bayless G., et al. (2011), A Randomized, Double-blind, Placebo-controlled Pilot Study of *Lactobacillus reuteri* ATCC 55730 for the Prevention of Antibiotic-associated Diarrhea in Hospitalized Adults, *J. Clin Gastroenterol*, 45(9):785–789.

18. Costello E.K., Lauber C.L., et al. (2009), Bacterial Community Variation in Human Body Habitats Across Space and Time, *Science*, 326(5960):1694–1697.

19. Cukovic-Cavka S., Likic R., et al. (2006), *Lactobacillus Acidophilus* as a Cause of Liver Abscess in a NOD2/CARD15–Positive Patient with Crohn's Disease, *Digestion*, 73(2–3):107–110.

20. De Jonghe V., Coorevits A., et al. (2010), Toxinogenic and Spoilage Potential of Aerobic Spore-formers Isolated From Raw Milk, *International Journal of Food Microbiology*, 136(3):318–325.

21. Dethlefsen L. and Relman D.A. (2011), Incomplete Recovery and Individualized Responses of the Human Distal Gut Microbiota to Repeated Antibiotic Perturbation, *Proc Natl Acad Sci.,* USA, 108 Suppl 1:4554–4561.

22. Doi A., Nakajo K., et al. (2011) Splenic Abscess Caused by *Lactobacillus Paracasei, Journal of Infection and Chemotherapy*: Official Journal of the Japan Society of Chemotherapy, 17(1):122–125.

23. Domann E., Hain T., et al. (2007), Comparative Genomic Analysis for the Presence of Potential Enterococcal Virulence Factors in the Probiotic *Enterococcus Faecalis* Strain Symbioflor 1, *International Journal of Medical Microbiology*, IJMM 297(7–8):533–539.

24. Dumas M.E., Barton R.H., et al. (2006), Metabolic Profiling Reveals A Contribution of Gut Microbiota to Fatty Liver Phenotype in Insulin-resistant Mice, *Proc Natl Acad Sci* USA, 103(33):12511–12516.

25. Eiseman B., Silen W., et al. (1958), Fecal Enema as an Adjunct in the Treatment of *Pseudomembranous Enterocolitis, Surgery,* 44(5): 854–859.

26. Enache-Angoulvant A. and Hennequin C. (2005), Invasive Saccharomyces Infection: A Comprehensive Review, Clinical Infectious Diseases: An Official Publication of the *Infectious Diseases Society of America* 41(11):1559–1568.

27. Endo H., Niioka M., et al. (2013), Butyrate-producing Probiotics Reduce Non-alcoholic Fatty Liver Disease Progression in Rats: New Insight into the Probiotics for the Gut-liver Axis, PLoS One 8(5):E63388.

28. Esposito E., Iacono A., et al. (2009), Probiotics Reduce the Inflammatory Response Induced by a High-fat Diet in the Liver of young Rats, *The Journal of Nutrition*, 139(5):905–911.

29. Faust K., Sathirapongsasuti J.F., et al. (2012), Microbial Co-occurrence Relationships in the Human Microbiome, *PLoS Comput Biol* 8(7):E1002606.

30. Ferrari A., Pacini N., et al. (1980), A Note on Bile Acids Transformations by Strains of Bifidobacterium, *The Journal of Applied Bacteriology*, 49(2):193–197.

31. Fierer N., Hamady M., et al. (2008), The Influence of Sex, Handedness, and Washing on the Diversity of Hand Surface Bacteria, *Proc Natl Acad Sci.*, USA., 105(46):17994–17999.

32. Foulquie Moreno M.R., Sarantinopoulos P., et al. (2006), The Role and Application of Enterococci in Food and Health, *International Journal of Food Microbiology*, 106(1):1–24.

33. Francavilla R., Lionetti E., et al. (2012), Randomised Clinical Trial: *Lactobacillus Reuteri* DSM 17938 vs. Placebo in Children with Acute Diarrhoea—a Double-blind Study, *Aliment Pharmacol Ther.*, 36(4):363–369.

34. Franks A.H., Harmsen H.J., et al. (1998), Variations of Bacterial Populations in Human Feces Measured by Fluorescent *in situ* Hybridization with Group-specific 16S RRNA-targeted Oligonucleotide Probes, *Appl Environ Microbiol.*, 64(9):3336–3345.

35. Franz C.M., Huch M., et al. (2011), Enterococci as Probiotics and Their Implications in Food Safety, *International Journal of Food Microbiology*, 151(2):125–140.

36. Garcia Vilela E., De Lourdes De Abreu Ferrari M., et al. (2008), Influence of *Saccharomyces Boulardii* on the Intestinal Permeability of Patients with Crohn's Disease in Remission, *Scand J. Gastroenterol.*, 43(7):842–848.

37. Goldenberg J.Z., Ma S.S., et al. (2013), Probiotics for the Prevention of Clostridium Difficile-associated Diarrhea in Adults and Children, *Cochrane Database Syst Rev.*, 5:CD006095.

38. Gronbach K., Eberle U., et al. (2010), Safety of Probiotic *Escherichia coli* Strain Nissle 1917 Depends on Intestinal Microbiota and Adaptive Immunity of the Host, *Infection and Immunity*, 78(7):3036–3046.

39. Grozdanov L., Raasch C., et al. (2004), Analysis of the Genome Structure of the Nonpathogenic Probiotic *Escherichia coli* Strain Nissle 1917, *Journal of Bacteriology*, 186(16):5432–5441.

40. Guenther K., Straube E., et al. (2010), Sever Sepsis After Probiotic Treatment with *Escherichia coli* NISSLE 1917, *The Pediatric Infectious Disease Journal*, 29(2):188–189.

41. Guslandi M., Mezzi G., et al. (2000), *Saccharomyces Boulardii* in Maintenance Treatment of Crohn's Disease, *Dig Dis Sci.*, 45(7):1462–1464.

42. Hammad A.M. and Shimamoto T. (2010), Towards a Compatible Probiotic-antibiotic Combination Therapy: Assessment of Antimicrobial Resistance in the Japanese Probiotics, *Journal of Applied Microbiology*, 109(4):1349–1360.

43. Henao-Mejia J., Elinav E., et al. (2012), Inflammasome-mediated Dysbiosis Regulates Progression of NAFLD and Obesity, *Nature*, 482(7384):179–185.

44. Holling C.S. (1973), Resilience and Stability of Ecological Systems, *Annual Review of Ecology and Systematics*, 4:1–23.

45. Hutt P., Koll P., et al. (2011), Safety and Persistence of Orally Administered Human Lactobacillus sp. strains in Healthy Adults, *Beneficial Microbes*, 2(1):79–90.

46. Huttenhower C., et al. (2012), Structure, Function and Diversity of the Healthy Human Microbiome, *Nature* 486(7402):207–214.

47. Hviid A., Svanstrom H., et al. (2011), Antibiotic Use and Inflammatory Bowel Diseases in Childhood, *Gut* 60(1):49–54.

48. Jernberg C., Lofmark S., et al. (2007), Long-term Ecological Impacts of Antibiotic Administration on the Human Intestinal Microbiota, *ISME J.*, 1(1):56–66.

49. Jernberg C., Lofmark S., et al. (2010), Long-term Impacts of Antibiotic Exposure on the Human Intestinal Microbiota, *Microbiology*, 156(Pt 11):3216–3223.

50. Jett B.D., Huycke M.M., et al. (1994), Virulence of Enterococci, *Clinical Microbiology Reviews*, 7(4):462–478.

51. Johnson A.P. (1994), The Pathogenicity of Enterococci, *The Journal of Antimicrobial Chemotherapy*, 33(6):1083–1089.

52. Karami N., Martner A., et al. (2007), Transfer of an Ampicillin Resistance Gene Between Two *Escherichia coli* Strains in the Bowel Microbiota of an Infant Treated with Antibiotics, *J. Antimicrob Chemother*, 60(5):1142–1145.

53. Kligler B. and Cohrssen A. (2008), Probiotics, *American Family Physician*, 78(9):1073–1078.

54. Kruis W., Fric P., et al. (2004), Maintaining Remission of Ulcerative Colitis with the Probiotic *Escherichia coli* Nissle 1917 Is as Effective as with Standard Mesalazine, *Gut* 53(11):1617–1623.

55. Lafyatis R. and Farina A. (2012), New Insights into the Mechanisms of Innate Immune Receptor Signalling in Fibrosis, *The Open Rheumatology Journal*, 6:72–79.

56. Larter C.Z. and Farrell G.C. (2006), Insulin Resistance, Adiponectin, Cytokines in NASH: Which Is the Best Target to Treat? *Journal of Hepatology*, 44(2):253–261.

57. Laverde Gomez J.A., Hendrickx A.P., et al. (2011), Intra- and Interspecies Genomic Transfer of the Enterococcus Faecalis Pathogenicity Island, *PLoS* One 6(4):E16720.

58. Li Z., Yang S., et al. (2003), Probiotics and Antibodies to TNF Inhibit Inflammatory Activity and Improve Non-alcoholic Fatty Liver Disease *Hepatology*, 37(2):343–350.

59. Lilly D.M. and Stillwell R.H. (1965), Probiotics: Growth-promoting Factors Produced by Microorganisms, *Science*, 147(3659):747–748.

60. Lin M.Y., Yen C.L., et al. (1998), Management of Lactose Maldigestion by Consuming Milk Containing Lactobacilli, *Digestive Diseases and Sciences*, 43(1):133–137.

61. Liu X., Lazenby A.J., et al. (2007), Resolution of Nonalcoholic Steatohepatits After Gastric Bypass Surgery, *Obesity Surgery*, 17(4):486–492.

62. Ma E.L., Choi Y.J., et al. (2010), The Anticancer Effect of Probiotic *Bacillus Polyfermenticus* on Human Colon Cancer Cells is Mediated Through ErbB2 and ErbB3 Inhibition, *Int J. Cancer*, 127(4):780–790.

63. Macfarlane G.T., Allison C., et al. (1988), Contribution of the Microflora to Proteolysis in the Human Large Intestine, *J. Appl Bacteriol*, 64(1):37–46.

64. Macfarlane G.T., Cummings J.H., et al. (1986), Protein Degradation by Human Intestinal Bacteria, *J. Gen Microbiol*, 132(6):1647–1656.

65. Mackay A.D., Taylor M.B., et al. (1999), Lactobacillus Endocarditis Caused by A Probiotic Organism, *Clinical Microbiology and Infection*: The Official Publication of the European Society of Clinical Microbiology and Infectious Diseases 5(5):290–292.

66. Mallon P., McKay D., et al. (2007), Probiotics for Induction of Remission in Ulcerative Colitis, *Cochrane Database Syst Rev.*, (4):CD005573.

67. Mattila E., Uusitalo-Seppala R., et al. (2012), Fecal Transplantation, Through Colonoscopy, is Effective Therapy for Recurrent Clostridium difficile infection, *Gastroenterology*, 142(3):490–496.

68. Metchnikoff E. (1907), *Essais Optimistes* (Paris), *The Prolongation of Life*, Optimistic Studies, Translated and Edited by P. Chalmers Mitchell, London: Heinemann.

69. Miele E., Pascarella F., et al. (2009), Effect of a Probiotic Preparation (VSL#3), on Induction and Maintenance of Remission in Children with Ulcerative Colitis, *Am J. Gastroenterol,* 104(2):437–443.

70. Munoz P., Bouza E., et al., (2005), *Saccharomyces Cerevisiae Fungemia*: An Emerging Infectious Disease, *Clinical Infectious Diseases,* An Official Publication of the Infectious Diseases Society of America, 40(11):1625–1634.

71. Nair S., Cope K., et al. (2001), Obesity and Female Gender Increase Breath Ethanol Concentration: Potential Implications for the Pathogenesis of Nonalcoholic Steatohepatitis, *The American Journal of Gastroenterology*, 96(4):1200–1204.

72. Nataro J.P. and Kaper J.B. (1998), Diarrheagenic *Escherichia coli*, *Clinical Microbiology Reviews*, 11(1):142–201.

73. Newcomer A.D., Park H.S., et al. (1983), Response of Patients with Irritable Bowel Syndrome and Lactase Deficiency using Unfermented Acidophilus Milk, *The American Journal of Clinical Nutrition*, 38(2):257–263.

74. Nicholson J.K., Holmes E., et al. (2012), Host-gut Microbiota Metabolic Interactions, *Science*, 336(6086):1262–1267.

75. Nissle A. (1918), Die Antagonistische Behandlung Chronischer Darmstörungen Mit Colibakterien, *Med Klin*, 2:29–30.

76. O'Brien J.A., Lahue B.J., et al. (2007), The Emerging Infectious Challenge of Clostridium Difficile-associated Disease in Massachusetts Hospitals: Clinical and Economic Consequences, *Infection Control and Hospital Epidemiology*: The Official Journal of the Society of Hospital Epidemiologists of America, 28(11):1219–1227.

77. Raes J. and Bork P. (2008), Molecular Ecosystems Biology: Towards an Understanding of Community Function, *Nat Rev Microbiol.*, 6(9):693–699.

78. Rahimi R., Nikfar S., et al. (2008), A Meta-analysis on the Efficacy of Probiotics for Maintenance of Remission and Prevention of Clinical and Endoscopic Relapse in Crohn's Disease, *Dig Dis Sci.*, 53(9):2524–2531.

79. Rampengan N.H., Manoppo J., et al. (2010), Comparison of Efficacies between Live and Killed Probiotics in Children with Lactose Malabsorption, *The Southeast Asian Journal of Tropical Medicine and Public Health*, 41(2):474–481.

80. Rautio M., Jousimies-Somer H., et al. (1999), Liver Abscess due to a *Lactobacillus Rhamnosus* Strain Indistinguishable from *L. Rhamnosus* Strain GG, *Clinical Infectious Diseases*: An Official Publication of the Infectious Diseases Society of America, 28(5):1159–1160.

81. Reeves A.E., Theriot C.M., et al. (2011), The Interplay Between Microbiome Dynamics and Pathogen Dynamics in a Murine Model of *Clostridium difficile* Infection, *Gut Microbes*, 2(3):145–158.

82. Ridlon J.M., Kang D.J., et al. (2006), Bile Salt Biotransformations by Human Intestinal Bacteria, *J. Lipid Res.*, 47(2):241–259.

83. Roberfroid M. (2007), Prebiotics: The Concept Revisited, *J. Nutr*, 137(3 Suppl 2):830S-837S.

84. Rolfe V.E., Fortun P.J., et al. (2006), Probiotics for Maintenance of Remission in Crohn's Disease, *Cochrane Database Syst Rev.*, (4):CD004826.

85. Samuel B.S. and Gordon J.I. (2006), A Humanized Gnotobiotic Mouse Model of Host-Archaeal-bacterial Mutualism, *Proc Natl Acad Sci.*, USA, 103(26):10011–10016.

86. Santos F., Vera J.L., et al. (2007), Pseudovitamin B(12), is the Corrinoid Produced by *Lactobacillus Reuteri* CRL1098 Under Anaerobic Conditions, *FEBS Lett.,* 581(25):4865–4870.

87. Saulnier D.M., Santos F., et al. (2011), Exploring Metabolic Pathway Reconstruction and Genome-wide Expression Profiling in Lactobacillus Reuteri to Define Functional Probiotic Features *PLoS One,* 6(4):E18783.

88. Shaukat A., Levitt M.D., et al. (2010), Systematic Review: Effective Management Strategies for Lactose Intolerance, *Annals of Internal Medicine,* 152(12):797–803.

89. Smith E.A. and Macfarlane G.T. (1997), Dissimilatory Amino Acid Metabolism in Human Colonic Bacteria, *Anaerobe*, 3(5):327–337.

90. Solga S.F. and Diehl A.M. (2003), Non-alcoholic Fatty Liver Disease: Lumen-liver Interactions and Possible Role for Probiotics, *Journal of Hepatology*, 38(5):681–687.

91. Sommer M.O., Dantas G., et al. (2009), Functional Characterization of the Antibiotic Resistance Reservoir in the Human Microflora, *Science*, 325(5944):1128–1131.

92. Sonnenberg G.F., Monticelli L.A., et al. (2012), Innate Lymphoid Cells Promote Anatomical Containment of Lymphoid-resident Commensal Bacteria, *Science*, 336(6086):1321–1325.

93. Spencer M.D., Hamp T. J., et al. (2011), Association between Composition of the Human Gastrointestinal Microbiome and Development of Fatty Liver with Choline Deficiency, *Gastroenterology,* 140(3):976–986.

94. Spor A., Koren O., et al. (2011), Unravelling the Effects of the Environment and Host Genotype on the Gut Microbiome, *Nat Rev Microbiol.*, 9(4):279–290.

95. Swann J.R., Want E.J., et al. (2011), Systemic Gut Microbial Modulation of Bile Acid Metabolism in Host Tissue Compartments, *Proc Natl Acad Sci.*, USA, 108 Suppl 1:4523–4530.

96. Takahashi T. and Morotomi M. (1994), Absence of Cholic Acid 7 Alpha-dehydroxylase Activity in the Strains of Lactobacillus and Bifidobacterium, *Journal of Dairy Science*, 77(11):3275–3286.

97. Teughels W., Newman M.G., et al. (2007), Guiding Periodontal Pocket Recolonization: A Proof of Concept, *J. Dent Res.*, 86(11):1078–1082.

98. Thurman R.G., Bradford B.U., et al. (1997), Role of Kupffer Cells, Endotoxin and Free Radicals in Hepatotoxicity Due to Prolonged Alcohol Consumption: Studies in Female and Male Rats, *The Journal of Nutrition*, 127(5 Suppl):903S-906S.

99. Tilg H. and Diehl A.M. (2000), Cytokines in Alcoholic and Nonalcoholic Steatohepatitis, *The New England Journal of Medicine*, 343(20):1467–1476.

100. Tissier H. (1900), Recherchers sur la Flora Intestinale Normale et Pathologique du Nourisson *Thesis*, University of Paris, Paris, France."

101. Tremaroli V. and Backhed F. (2012), Functional Interactions Between the Gut Microbiota and Host Metabolism, *Nature*, 489(7415):242–249.

102. van Nood E., Vrieze A., et al. (2013), Duodenal Infusion of Donor Feces for Recurrent *Clostridium difficile*, *N Engl J. Med.*, 368(5):407–415.

103. Vance D.E. (2008), Role of Phosphatidylcholine Biosynthesis in the Regulation of Lipoprotein Homeostasis, *Curr Opin Lipidol*, 19(3):229–234.

104. Vankerckhoven V., Huys G., et al. (2008), Genotypic Diversity, Antimicrobial Resistance, and Virulence Factors of Human Isolates and Probiotic Cultures Constituting Two Intraspecific Groups of *Enterococcus Faecium* Isolates, Applied and Environmental Microbiology, 74(14):4247–4255.

105. Vejborg R.M., Friis C., et al. (2010), A Virulent Parent with Probiotic Progeny: Comparative Genomics of *Escherichia coli* Strains CFT073, Nissle 1917 and ABU 83972, *Molecular Genetics and Genomics*: MGG 283(5):469–484.

106. Vermeiren J., Van Den Abbeele P., et al. (2012), Decreased Colonization of Fecal *Clostridium Coccoides/Eubacterium rectale* Species From Ulcerative Colitis Patients in an *In Vitro* Dynamic Gut Model with Mucin Environment, *FEMS Microbiol Ecol.*, 79(3):685–696.

107. Wang Z., Klipfell E., et al. (2011), Gut Flora Metabolism of Phosphatidylcholine Promotes Cardiovascular Disease, *Nature*, 472(7341):57–63.

108. Wehrle E., Moravek M., et al. (2009), Comparison of Multiplex PCR, Enzyme Immunoassay and Cell Culture Methods for the Detection of Enterotoxinogenic *Bacillus Cereus*, *Journal of Microbiological Methods*, 78(3):265–270.

109. Whelan K. and Myers C.E. (2010), Safety of Probiotics in Patients Receiving Nutritional Support: A Systematic Review of Case Reports, Randomized Controlled Trials, and Nonrandomized Trials, *The American Journal of Clinical Nutrition*, 91(3):687–703.

110. Williams N.T. (2010), *Probiotics, American Journal of Health-system Pharmacy: AJHP, Official Journal of the American Society of Health-System Pharmacists*, 67(6):449–458.

111. Wirth T., Wang X., et al. (2004), Distinguishing Human Ethnic Groups by Means of Sequences From Helicobacter Pylori, Lessons From Ladakh, *Proc Natl Acad Sci.*, USA, 101(14):4746–4751.

112. Yang S.Q., Lin H.Z., et al. (1997), Obesity Increases Sensitivity to Endotoxin Liver Injury: Implications for the Pathogenesis of Steatohepatitis, *Proceedings of the National Academy of Sciences of the United States of America*, 94(6):2557–2562.

113. Zocco M.A., Dal Verme L.Z., et al. (2006), Efficacy of Lactobacillus GG in Maintaining Remission of Ulcerative Colitis, *Aliment Pharmacol Ther*, 23(11):1567–1574.

114. Zoetendal E.G., Akkermans A.D., et al. (1998), Temperature Gradient Gel Electrophoresis Analysis of 16S RRNA From Human Fecal Samples Reveals Stable and Host-specific Communities of Active Bacteria, *Appl Environ Microbiol.*, 64(10):3854–3859.

SECTION VIII

HUMAN GENETICS, BIOCHEMICAL BASIS OF INHERITANCE, EXPRESSION OF GENETIC INFORMATION, AND GENETIC ENGINEERING

This section provides a glimpse into some of the most exciting developments in Life Sciences during the past several decades. Particularly those in the beginning of this century which have not only impacted the way we understand life but also resulted in a paradigm shift in the practice of medicine. The structure of the genetic material, its organization as a chromatin in the form of a complex of DNA and protein, its replication, transcription and translation of the information from DNA are included. The advent of recombinant DNA technology and the large number of eukaryotic expression systems to express foreign genes are detailed in this section. The chapter on gene expression and its regulation by small interfering RNAs (siRNA) and microRNA (miRNA) throws new insights in mammalian development and also explains the basis of a new paradigm on designing vaccines and controlling infections, clearly with implications in agriculture and human health. The growing realization that epigenetics has an equally important role provides readers a platform to understand the basics of epigenetics and the impact it has made on human health and disease.

This section also addresses certain issues of applied aspects of biology which derive their strength from the principles of human genetics and population genetics. Of special interest to the readers would be the chapters on 'DNA Fingerprinting' and 'Reproductive and Therapeutic Cloning'. The chapter on 'Reproductive and Therapeutic Cloning' will greatly benefit the readers (Refer to Section XIII, Chapter 100). Proteomics and Metabolomics are powerful diagnostic tools in diseases and provide clues to the disease process. The ability to create life at will, the so-called designer life, by combining principles of chemistry and biology has given birth to the area of 'Synthetic Biology' which has profound implications and thus covered in the section.

63

Chromatin Structure and Function
Physiological and Pathophysiological Perspectives

Tapas Kumar Kundu, Rahul Modak, Snehajyoti Chatterjee, Deepthi Sudarshan and Parijat Senapati

CONTENTS

Eukaryotic cells pack a large amount of genetic material inside a relatively small space of nucleus in a very efficient manner. This packaging poses seemingly opposing problems; the highly negatively charged DNA has to be packed in close confined space, on the other hand, the packaging has to be dynamic enough so that the underlying information in the DNA is available to the cellular machinery. Eukaryotic cells have efficiently addressed this issue through formation of 'nucleosome', the primary unit of higher order genome organization. At the primary level, 147 base pair (bp) double stranded DNA is tightly wrapped 1.65 times around an octameric protein core, consisting of two copies of histone H2A, H2B, H3 and H4. This unit, the nucleosome core particle (NCP), is the basic repeating unit of beads on a string structure, the chromatin, which is physiologically visible in interphase nuclei. Hundreds and thousands of such repeating units are connected by 80–200 base pairs of linker DNA. This linear array of nucleosomes bind linker histone H1, high mobility group (HMG) proteins and other non-histone proteins, which help in condensing the chromatin into further higher order structures. Initially the nucleosome arrays are organized in a solenoid like structure of approximately 30 nm diameter, called the '30 nm structure'. Recently, at least 4 different forms of '30 nm fiber' have been identified *in vitro*, which could be the building blocks of higher order chromatin *in vitro*. New *in situ* imaging studies coupled with chromatin conformation capture (3C) techniques have revealed more complex higher order structures, which could not be explained by the simple solenoid basic model. These 30 nm structures or similar basic folds are further condensed and folded back to give rise to more compacted chromatin based on cellular signals.

I. COMPONENTS OF CHROMATIN

A. Histones and Histone Variants

There are several proteins associated with nucleosomes. The nucleosomal core particle is made of a family of highly basic proteins called histones. Histones are highly conserved across different species and coded by different histone gene clusters. There are four core histones, namely H2A, H2B, H3 and H4. All four eukaryotic histones harbor globular domains flanked by extended N and C terminal tails. Globular domains are involved in nucleosome assembly, whereas tails are extended beyond

nucleosomes to interact with regulatory proteins. Histone tails are also the site for protein post-translational modifications, which are essential for regulation of gene expression. Histone H3 and H2B have larger helix α1 and loop L1 compared to H4 and H2A, which results in asymmetric H2A-H2B and H3–H4 heterodimers. Two H3–H4 dimers form a tetramer, which in turn binds to two H2A-H2B dimers, resulting in octameric nucleosome core. The interface of H3–H3 in the central H3–H4 core is designated dyad axis of symmetry (Figure 63.1). Linker histone H1 binds near the entry-exit point of the DNA in nucleosome, at the dyad axis and further organizes another 20 bp DNA, resulting in a complex structure called chromatosome. Linker histone H1 does not harbor histone fold of canonical core histones and its exact location in chromatosome has not been identified yet. In most of the organisms, the multiple copies of histone genes, highly similar in primary sequence, are present as a gene cluster in the genome. Canonical histone genes lack introns and polyA tail in their transcripts. They are coexpressed in S-phase of cell cycle and codes for

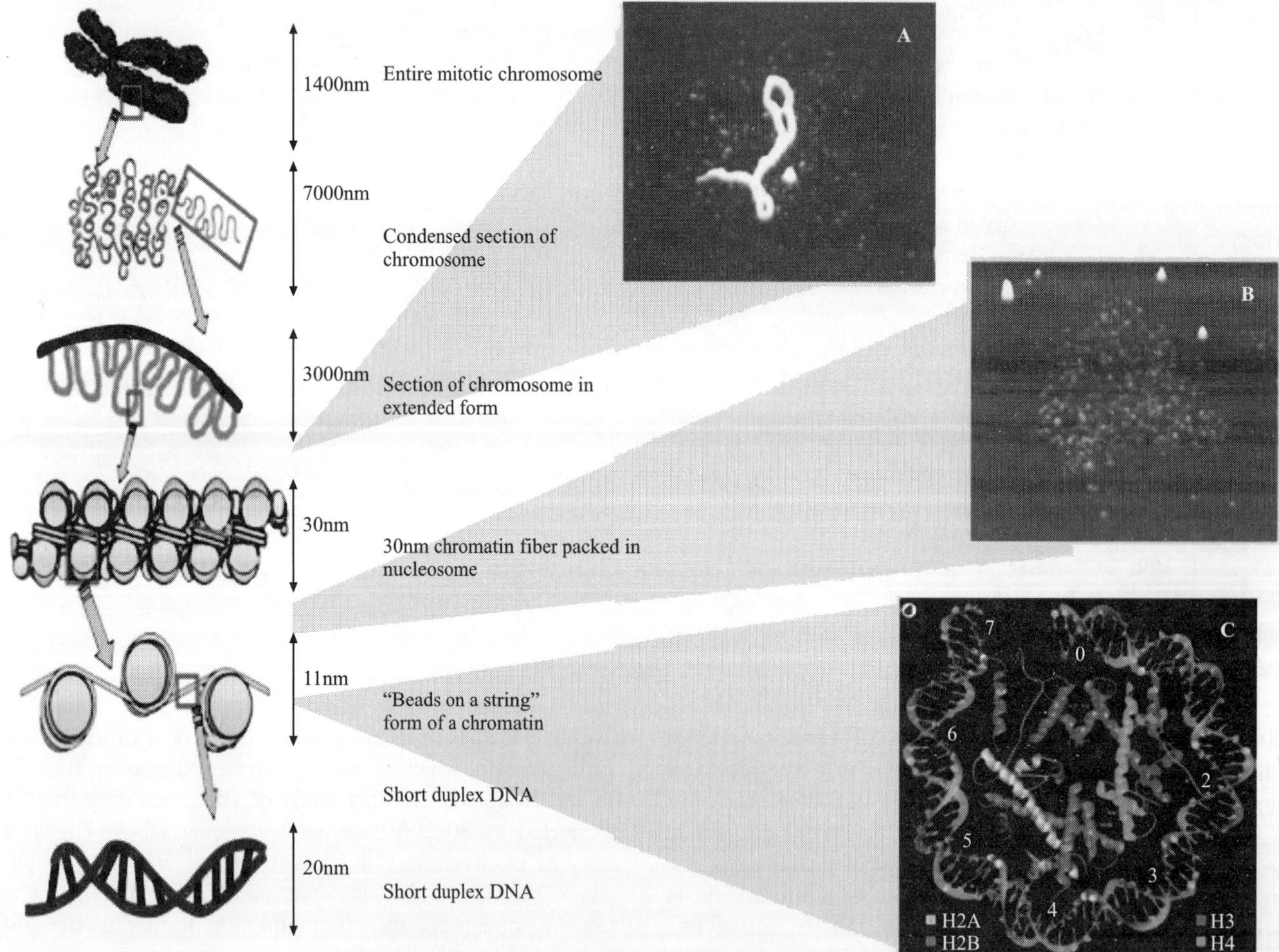

Figure 63.1 **Nucleosome structure and higher order chromatin organization:** The packaging of DNA into higher order chromatin is shown here. In inset is (A) the nucleosome structure showing the histone octamer wrapped by DNA (*Luger K et al., Nature, 1997; 389(6648):251–60)*, (B) Atomic microscopy image of the "bead on string" structure of chromatin (*Kohji Hizume et.al., Biochemistry* **2005***, 44, 12978–12989*), (C) Atomic microscopy image of *in vitro* condensed 30 nm chromatin fibre on addition of linker Histone H1 (*Kohji Hizume et.al., Biochemistry* **2005***, 44, 12978–12989*). (*see Plate 25 for colour figure*)

bulk of core histones. Core histones are deposited during the DNA replication to fill the nucleosome gaps created during the progress of replication fork. Cells also harbor nonallelic variants of the major histones with significant difference in primary sequence. Most variants are single copy genes, which are expressed throughout the cell cycle stages. Histone variants genes have introns and their transcripts are polyadenylated, which renders better posttranscriptional regulations. Histone variants replace the canonical histones at specific locations like centromere and also under different cellular and physiological conditions like DNA damage, differentiation and development which has been further elaborated in Table 63.1. Further it has been observed that only one partner of the histone heterodimer (H3–H4 and H2A-H2B) could be replaced by its variant at a time, which renders least perturbation in the nucleosome structure but at the same time signals for the cascading events of the downstream processes.

B. Nonhistone Chromatin Associated Proteins

Apart from histones, several other nonhistone proteins are also associated with the chromatin. These proteins are involved in chromatin organization, remodeling, DNA replication and gene expression (Figure 63.2). These include, the **high mobility group (HMG)** proteins, which get their name from their increased electrophoretic mobility. HMGs are architectural proteins, which bind to primary chromatin structure (beads on a string) and regulate gene expression. They are also involved in other nuclear and cytoplasmic functions. They play crucial role in the extracellular milieu and their mutation, misexpression and mislocalization leads to severe pathophysiological consequences (Table 63.2). Canonical HMGs share certain basic physical similarities and they all bind to chromatin. Based on their structural features and functional motifs, HMGs are classified

in to 3 families. HMGA groups of proteins have two to three 'AT-hook', HMGB proteins contain two HMG boxes 'Box-A' an 'Box-B' and HMGN proteins contain 'nucleosome binding domain.

Poly(ADP-ribosyl)ation: (PARylation) plays diverse roles in many molecular and cellular processes, like DNA damage detection and repair, chromatin modification, transcription, cell death pathways and mitosis. These processes are critical for many physiological and pathophysiological outcomes, including genome maintenance, carcinogenesis, aging, inflammation, and neuronal function. Linear and branched PARylation of the acceptor protein is catalyzed by PAR polymerases (PARPs). PARPs catalyze polymerization of ADP-ribose units received from nicotinamide adenine dinucleotide (NAD^+) molecules. Among the 18 members of PARP family PARP-1 is the best characterized. PARP enzymes from vertebrates contain highly conserved 'PARP signature' motif. They also have one or more additional motifs including zinc fingers, "BRCA1 C-terminus-like" (BRCT) motifs, ankyrin repeats, macro domains, and WWE domains, each conferring unique properties on the particular PARP protein that contains them. PARP-1 has a highly conserved structural and functional organization including an N-terminal double zinc finger DNA-binding domain (DBD), a nuclear localization signal, a central auto modification domain, and a C-terminal catalytic domain. PARP-1 PARylates itself, apart from core histones, H1 and many other proteins that interacts with PARP-1. PARP-1 binds to a variety of DNA structures, including single- and double-strand breaks, crossovers, cruciforms and supercoils, as well as some specific double stranded sequences. A wide variety of extrinsic and intrinsic stress signals, including those initiated by oxidative, nitrosative, genotoxic, oncogenic, thermal, inflammatory, and metabolic stresses induces PARylation. These responses underlie

TABLE 63.1   Histone variants and their function: disease connection

Histone Variant	Biological Function	Disease
H3.3	Transcriptional activation and DNA repair	
H3.4	Testis specific variant	Male infertility
CenH3 (CENP-A)	Centromere specific variant involved in kinetochore assembly Inherited epigenetic mark through mature spermatozoa	Genomic instability during cancer progression
H2AX	Recruitment of PIKK family of kinase	Male infertility
Phospho H2AX (g-H2AX)	DNA double strand break repair	Tumorigenesis
H2A.Z	Mitosis, Transcription regulation,	
macroH2A	Assembly and maintenance of heterochromatin, ADP ribose binding module, Accumulation on inactive X-chromosome, Repression of chromatin remodeling, transcription initiation and histone acetylation	
H2A-Bbd	Nucleosome destabilization and gene activation	
H2B variants	Gemetogenesis	
H2BFWT	Absence of recruitment of chromosome assembly factor and hence condensed chromatin formation	
Testis specific H2B	Telomere membrane attachment	Male infertility
H1 variants	Transcription repression	

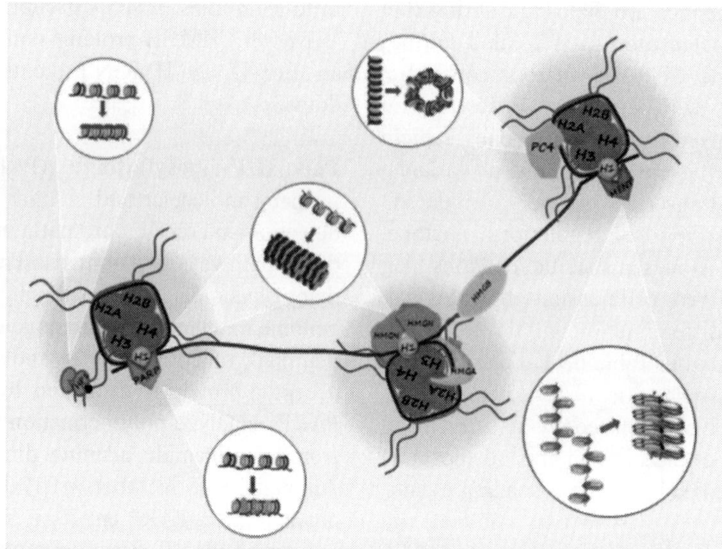

Figure 63.2 Chromatin Associated Proteins and their Functions. The figure illustrates the binding positions of the various Chromatin associated proteins on the nucleosome as discussed in the text. In inset are the mechanisms by which they compact chromatin. (*see Plate 25 for colour figure*)

TABLE 63.2   Chromatin associated proteins and disease

Name	Disease	Mechanism	Reference
HMGA	Pituitary adenomas	HMGA2 overexpression	Biochimica
	Thyroid follicular adenoma and carcinoma	HMGA1, HMGA2 overexpression	et Biophysica Acta (2010) **1799**, 48–54
	Colon carcinomas		
	Pancreatic carcinomas		
	Breast, gastric and lung Cancer		
HMGB1	Most human cancers—breast, colon, lung, prostate, cervical, gastric, hepatocellular carcinoma, leukemia	Overexpression Functional association with hallmarks of cancer	Biochimica et Biophysica Acta (2010) **1799**, 131–140
HMGN	Breast, lung, prostate and bladder cancer, uterine adenocarcinoma, Glioma	Alteration of chromatin fiber structure DNA damage repair	Biochimica et Biophysica Acta (2010) **1799**, 80–85
MENT	Lymphoma	Overexpression	Int J Hematol. (2004) **80**(4), 354–60.
DEK	Subset of acute myeloid leukemia (AML)	chromosomal translocation (6;9) (p23;q34) resulting in fusion protein with nuclear pore complex	Leukemia. 2010 Nov; 24(11):1910–9.
	Juvenile rheumatoid arthritis	Autoantigen	
	Systemic lupus erythematosus	Autoantigen	
	Sarcoidosis	Autoantigen	
PC4	Non small cell Lung Cancer	Overexpression	Cancer Gene Ther. (2012) **19**(10), 690–6
NPM1	Oral cancer	Overexpression and hyperacetylation, altered localization	Mol Cell Biol (2009) **29**(18), 5115–27
	Renal cell carcinoma	altered localization	Cancer Res (2012) **72**(22), 5867–77.
	Lymph node metastasis in colon cancer	Overexpression	J Biomed Sci. (2012) **19**, 53.
	Acute myeloid leukaemia	Mutation and cytoplasmic localization	Leukemia (2009) **23**, 1370–1371
PARP	Breast and ovarian cancer	DNA damage repair	
	Glioblastoma		
	Leukemia		
	Solid tumours		
HP1	Breast, colorectal and prostate cancer	DNA damage repair	Cancer Res. 2000 Jul 1;60(13):3359–63. (Breast cancer) Am J Pathol. 2011 Feb;178(2):672–8. (colorectal cancer) J Mol Endocrinol. 2013 Apr 23;50(3):401–9. (Prostate Cancer)

pathological conditions: cancer, inflammation-related diseases, and metabolic dysregulation, which make PARPs as potent therapeutic targets. PARP-1 is an integral part of chromatin structure, binding to the dyad axis of the nucleosome, which promotes the compaction of nucleosomal arrays into higher order structures in the absence of NAD+. Recent studies have shown that PARP-1 can induce the exclusion of histone H1 and induce the level of HMGB-1 that enhances transcription. The PARylation of DEK, another component of chromatin, promotes the release of DEK from chromatin which can also cause enhancement of transcription. PARP-1 plays significant role in chromatin dynamics also through its interactions with the histones and histone variants. For example, binding of PARP-1 to "non-histone domain" of Macro H2A.1 which leads to inhibition of the PARP-1 enzymatic activity. Thus, this abundant enzyme not only regulates chromatin dynamics but also serves as an architectural base of chromatin.

Heterochromatin protein-1 (HP1) is one of the most extensively studied chromatin associated protein. Phylogenetically HP1 is conserved across many eukaryotes including fission yeast, insects and mammals. HP1 is consistently associated with the heterochromatin regions and telomeres and suppress gene expression. Recent studies have shown that HP1 is associated in different regions on euchromatin and positively regulates gene expression. HP1 is a 206 amino acid protein with two prominent structural motifs, the chromo domain and the chromoshadow domain connected by a hinge region. Chromo domain is involved in chromatin binding and chromoshadow domain is the site for protein-protein interactions. Among the three homologs of HP1 (HP1–α,β,γ), HP1α is most well studied. Apart from chromatin organization, HP1 is involved in metaphase chromatid cohesion, centromere organization and chromatin-nuclear membrane interactions. Depletion of HP1 leads to telomere and chromosomal instability, abnormal chromosome/chromatin segregation, aberrant gene expression that results in various human diseases such as cancer.

Myeloid and erythroid nuclear termination stage-specific protein (MENT) is a nonhistone chromatin associated protein present in lymphocytes and granulocytes. It is a basic 410 amino acid protein belonging to Serpin family of proteins due to the presence of serpin-related conserved reactive center loop (RCL) domain. It also possesses additional M-loop domain (involved in interaction with DNA) and a nuclear localization signal (NLS) domain. Approximately two molecules of MENT bind to 200 bp of DNA, forming condensed and repressed chromatin. M-loop and RCL segments of MENT are involved in the initiation and maintenance of compacted chromatin at terminal stages of granulocyte differentiation. MENT-dependent chromatin condensation requires both a cooperative binding and folding of DNA associated with the M-loop domain and the RCL-dependent MENT oligomerization which in cooperation with linker histones is needed to form self-associated chromatin fibers.

DEK is an abundantly expressed nuclear protein found in all multicellular organisms (animal, plant, fungi except yeast) and in a few unicellular organisms. Although DEK protein vary considerably in length across the species, they share a unique central region called the SAP or SAF box. SAP box is a conserved 35-amino acid DNA binding motif that contains two amphipathic α helices. DEK is a chromatin associated nuclear protein, abundantly expressed in proliferating cells. DEK is an architectural protein, which can bind to both DNA and chromatin and induce structural changes in the chromatin organization. In a subset of advanced myeloid leukemia (AML) patients DEK coding gene has been found to be translocated and fused to the fragment of CAN, coding for the nuclear pore protein NUP 214 in the bone marrow. This chimeric protein is not a proto-oncogene, as the bone marrow cells express high amount of normal DEK and very small percentage of DEK-CAN fusion protein. DEK is also identified as an auto antigen in a relatively high percentage of patients with autoimmune diseases such as juvenile rheumatoid arthritis.

Positive coactivator 4 (PC4): Human positive coactivator 4 (PC4) is a small (14 kDa) highly abundant nuclear protein. It is highly conserved across species. Human PC4 is a 127 amino acid protein with an unstructured N-terminal domain and highly structured C-terminal domain. Though PC4 is a multifunctional protein involved in transcription regulation and DNA repair, primarily it is a non-histone chromatin associated protein. It interacts with both core histone and linker histone H1 and promotes the formation of compact heterochromatin foci. In absence of PC4, the entire genome loses the architectural organization leading to highly hyperacetylated, unstructured chromatin and overexpression of several genes which cause severe cell cycle defects. PC4 interacts with other non-histone chromatin components very specifically, such as HP1α (but not HP1β or HP1γ). The functional significance of such a specific interaction is yet to be found. The unique chromatin organization ability of PC4 could be linked to the maintenance of genome stability and thus may act as a tumor suppressor for several cancers, especially breast cancer.

II. POSTTRANSLATIONAL MODIFICATIONS OF HISTONES AND NONHISTONE PROTEINS

Fluidity of the chromatin is essential for the accessibility of underlying information stored in DNA. Although the chromatin remodelers and chaperones catalyse the chromatin remodelling process, the accessibility of the DNA is regulated by the different protein posttranslational modifications (PTMs) in a context dependent manner. The N-terminal tails of core histones protrude out of the nucleosome and are the major sites for various PTMs. The activity of modifying enzymes and other chromatin associated proteins are also modulated by PTMs. thus, these events orchestrated by the PTMs regulate gene expression in cells. Several PTMs have been identified over a period of time; some of them have been studied in greater detail due to their important physiological outcomes. Deregulation of

PTMs lead to various pathophysiological conditions like defect in development and differentiation, cancer, metabolic disorders, altered immune response, viral infection to name a few.

A. Reversible Methylation

DNA Methylation was the first epigenetic mark identified and perhaps most extensively examined for its various biological roles. DNA methylation generally is associated with gene repression and altered levels of DNA methylation have been associated with various diseases like cancer, diabetes, asthma, HIV infection etc. DNA methylation is catalysed by a family of enzymes called DNA methyltransferases which converts cytosine to 5–methylcytosine (5–meC). In animals DNA methylation predominantly occurs at CpG dinucleotides present in the CpG islands of the genome. DNA demethylation can either be an oxidative process or follow a repair based mechanism. The methylated DNA is a signature of compact chromatin. There are a few methyl CpG binding proteins, such as MeCP2 and MeCP1, which bind to the methylated chromatin and induce the formation of transcriptionally silent, compact chromatin structure. In addition, chromatin methylation and histone deacetylation may occur together and have an additive effect to form the heterochromatin since the enzymes responsible for methylation and deacetylation of histones are present in a single protein complex. Methylated DNA acts as a signal for the methylation of histones present in the nucleosomes.

Protein methylation is another set of very well-known PTM which is catalysed by protein arginine (Arg, R) methyltransferases (PRMTs) and lysine (Lys, K) methyltransferases (KMTs). Histones are methylated at specific R and K residues by a set of specific enzymes. Mono, di- and tri-methylation can happen on lysine residues of H3 and H4. Arginine residues are methylated asymmetrically by type I PRMTs, whereas PRMT5 (type II) catalyzes symmetric methylation. Genereally arginine methylation facilitates acetylation and hence gene activation and it also blocks Lys methylation. The role of Lys methylation is highly context dependent; they can be part of either transcriptionally active or repressive chromatin. Histones could be methylated on different lysine residues by different classes of enzymes. Effect of lysine methylation on the chromatin structure and function depend upon the residue as well as the enzyme involved in the process of methylation. For, example, Histone H3K9me3 by the enzyme G9a and others leads to chromatin compaction. Whereas Histone H3K4me2 is a signature of open chromatin essential for the initiation of transcription process. Protein lysine residues can be demethylated by lysine specific demethylase (LSD1/KDM1) and Jumonji histone demethylases (JHDM). KDM1 contain an amine oxidase (AO) catalytic domain that mediates oxidative demethylation of methylated lysine residues by reducing FAD molecules. JHDMs belong to Fe (II)/2–oxoglutarate-dependent dioxygenase (or hydroxylase) superfamily that catalyse demethylation via multiple steps using the Fe(IV)-oxo species. Methylated arginine residue could also be demethylated by the members of the amine oxidases via the formation of citrulline.

B. Acetylation and Deacetylation

Protein lysine acetylation is one of the key regulators of biological functions of histones and several nonhistone proteins. Transfer of acetyl group from acetyl coenzyme A (acetyl CoA) to specific lysine residues is catalyzed by a group of proteins called lysine/histone acetyltransferases (KATs/ HATs). Based on their cellular localization KATs are classified into nuclear or type A and cytoplasmic or type B HATs. There are only two cytosolic HATs-HAT1 and HAT2 reported till date and they acetylate nascent histones. Nuclear HATs are further classified into 5 families based on their structural and functional differences. There are 3 members of GNAT family—Gcn5, p300/CBP associated factor (PCAF) and ELP3. Gcn5 and PCAF acetylate both histone and nonhistone substrates and they are part of various complexes involved in diverse functions. There are two homologs in p300/CBP family—p300 and CREB binding protein (CBP). Both of them are transcription coactivators and they are involved in similar as well as distinct processes. Tip60, MOZ, MOF, MORF and HBO1 are the major members of MYST family of HATs and they play crucial role in DNA damage repair, development and differentiation. There are few transcription factors—TFIIIC90, ATF2 and TAF1 which have inherent HAT activity and they affect transcription directly. There are also a few nuclear hormone related HATs like SRC4 and ACTR, which act as coactivators. Though they possess histone acetyltransferase activity, they are often part of p300/CBP mediated coactivator complexes. Apart from these families there are a few other HATs like CIITA, CYDL and HAT1, which do not fall under any family. The reversible acetylation mark on histones are removed by histone deacetylases (HDACs), which is essential for gene silencing. There are 3 major classes of HDACs. Class I HDACs (HDAC 1, 2, 3 and 8) are nuclear located and involved in epigenetic regulations. They are always part of multi-enzyme complexes. Class-II HDACs (HDAC 4, 5, 6, 7) are characterized by nucleo-cytoplasmic shuttling. They can function both independently as well as part of a complex and are commonly involved in differentiation. NAD dependent HDACs are grouped under class III and they are generally called sirtuins. Sirtuins are involved in both transcription and metabolism. Acetylated histones are the signature of open and active chromatin whereas deacetylated histones mark the compact chromatin.

C. Phosphorylation

Protein phosphorylation at serine/threonine/tyrosine residues are the most widely occurring post-translational marks involved in a wide variety of biological processes including signal transduction, transcription activation, DNA damage repair, chromatin and chromosome condensation, chromosome segregation, etc. Histone serine phosphorylation in conjunction with other modifications play a crucial role in seemingly opposing processes of transcription activation and chromatin condensation in a context dependent manner.

D. Ubiquitylation and Sumoylation

Lysosomal degradation and ubiquitin proteosome pathways (UPP) are the two major pathways for programmed recycling of the proteins in the eukaryotic cells. Polyubiquitylated proteins are targeted for degradation via 26S proteosome, whereas monoubiqutylation marks the substrate protein for further modifications in a signal dependent manner. Monoubiquitylation of histone H2B at Lys-120 and H2A at Lys-119 plays critical role is transcription initiation and elongation. Recent studies have suggested a positive correlation between histone ubiquitylation and transcription; this may be through interference to chromatin compaction, leading to a more accessible chromatin structure. The nonproteolytic role of UPP system in transcription has also been reported.

Small ubiquitin-like modifier (SUMO) post-translationally modifies many proteins and controls their cellular localization and functions. Sumoylation has been reported to regulate transcription, chromatin structure and DNA repair. It has been reported that sumoylation of histones regulates transcriptional repression.

III. CROSSTALK BETWEEN ACETYLATION AND OTHER EPIGENETIC MODIFICATIONS

One of the most well characterized crosstalk is the histone tail acetylation, which results in charge neutralization of the tail, thus facilitates the weakening of the DNA nucleosome interactions. Acetylation of histones further augments the recruitment of the bromo-domain containing transcription factors including ATP dependent chromatin remodelers, involved in the mobility of nucleosomes over the DNA based on the cellular signals. The next levels of crosstalks are between multiple PTMs, where they either facilitate or negate mutual activities. The first example of this is the phosphorylation of histone H3 Ser10 (H3S10P) which facilitates acetylation of H3 Lys14 (H3K14Ac) and thus activates transcription, which is further augmented by inhibition of HP1 recruitment at methylated H3K9. H2B monoubiquitination being a prerequisite for H3K4 di and tri-methylation by COMPASS is a classic example of inter-histone crosstalk. On the other hand, inhibition of H3K4 methylation by COMPASS by methylated H3 Arg2 is an example of negative interference by PTMs. Bimodal function of methylated H3K4 is typified by recruitment of NuA3 histone acetyltransferase complex and also a histone deacetylase complex, which shows context dependence of the crosstalks.

Since histone acetylation and sumoylation sites overlap considerably, they reciprocally regulate each other's function. Transient loss of sumoylation leads to increase of acetylation and concurrent switch from a repressed to activated state in mammalian cells in a carbon source dependent manner. Similar reciprocal role of acetylation and sumoylation has been observed during the regulation of ETS-domain transcription factor PEA3 and oncoprotein Krüppel-like factor-8 (KLF8).

IV. COMPLEXITY OF EPIGENETIC LANGUAGE

Recent studies have further added another level of complexity by showing the spatio-temporal regulation of the PTMs. Recruitment of oncogenic kinase PIM1 phosphorylates H3S10 at the enhancer region of serum responsive gene FOSL1, which in turn recruits 14-3-3 protein. 14-3-3 complex recruits the HAT-MOF and brings about acetylation of H4K16 at the enhancer. Acetylated H4K16 recruits Brd4 and associated kinase P-TEFb, which phosphorylates RNA pol II and facilitates transcription elongation of FOSL1 gene. This order of recruitment of phospho-H3S10 and acetyl-H4K16 are opposite to that reported earlier and their functions also differ. These observations along with many others have shown that the order of recruitment of PTMs and their readers and writers decide the outcome. Recent evidences have shown that noncoding RNAs play crucial role in recruitment of histone modifiers and thus plays crucial role in transcription regulation. The most well worked out example is the inactivation of one copy of X-chromosome in females by Xist RNA and associated recruitment of repressive chromatin marks.

V. CHROMATIN REMODELLING

A. Chromatin Remodeler

The tightly packed DNA within nucleosome needs to be accessed by the cellular machinery to perform various cellular functions. This process is catalysed by a set of proteins, collectively known as chromatin remodelers. Remodelers alter the chromatin state by moving, altering or evicting the nucleosomes in an ATP dependent manner. In collaboration with other chromatin factors, remodelers regulate packing and unpacking of DNA in a highly controlled manner, which is essential for replication, transcription, repair and recombination. Five basic properties characterize all the remodelers. They have (i) DNA dependent ATPase domain required for DNA translocation motor; (ii) ATPase regulatory domain, (iii) domain for recognition of histone modifications, (iv) domains for protein-protein interaction and (v) high affinity for nucleosome. However, distinctive targeting of remodeler and specialized tasks require remodeler specialization. There are four families of remodelers- SWI/SNF, ISWI, CHD and INO80 family, which differ by the presence of unique flanking domains besides the conserved ATPase domain.

B. Chaperones

Histone chaperones are proteins that stimulate histone transfer to either DNA (deposition) or to another factor or to modifying enzymes. They can evict histones from DNA and also sequester them at different cellular compartments. Histone chaperones can be classified into three categories—(i) independent chaperones that can bind to histones without any partner, e.g., anti-silencing function-1 (Asf1); (ii) multichaperone complexes, e.g., CAF-1 complex and (iii) chaperones that are part of large enzyme complexes, e.g., actin related protein-4 (Arp4) as part of INO80 complex. Asf-1 is H3–H4 donor to CAF1 and HIRA, which in turn deposit the histones onto DNA. Histone chaperones show selectivity for H3–H4 vs. H2A-H2B complexes and also for histone variants, which is essential for their functional outcome. Histone chaperones play critical role in DNA replication, repair, transcription and other chromatin dependent processes.

C. Functional Interaction between Remodeler, Chaperone and PTM in Transcription, Replication and Repair

Chromatin related processes are tightly regulated by interplay between remodelers, chaperones, histone modifying enzymes and chromatin-associated proteins. In many cases they are part of large multifunctional complexes. In response to external or internal cue, histone modifiers induce posttranslation modifications at the core histone tails. These modifications in turn signal for recruitment of remodeler complexes and/ or chaperones. Remodelers and chaperones make the DNA underlying the nucleosome more accessible to the respective cellular machinery.

VI. CHROMATIN STRUCTURE, DYNAMICS AND DISEASE

A. Cancer

Post translational modifications of proteins are one of the key phenomena that regulate several important biological processes like replication, transcription, cell cycle and DNA damage repair. Histone N terminal tails are subjected to various post-translational modifications like acetylation, methylation, phosphorylation, ubiquitination and several others. Lysine acetylation of histone tails generally marks the signature for active chromatin. Lysine acetyltransferases like p300 is also a transcriptional coactivator and thus when bound to the chromatin template recruits the transcriptional machinery and activates chromatin transcription. Therefore most of the lysine acetyltransferases are associated with transcription activation complexes. Acetylation of the histone tails neutralises the charge on it and thus relaxes the compact chromatin structure. Alteration of acetylation of histones and non histone proteins are the components of the key factors that are associated with the hallmarks of cancer. The hallmarks of cancers are the eight essential characteristics or traits acquired by normal cell that governs their transformation to cancerous cells, which includes

(a) Inducing angiogenesis, (b) Resisting cell death, (c) Enabling replicative immortality, (d) Activating invasion and metastasis, (e) Evading growth suppressors, and (f) Sustaining proliferative signalling (g) Reprogramming metabolism and (h) evading the immune destruction. In all the six above mentioned hallmarks, acetylation, phosphorylation and methylation of histones or nonhistone proteins are closely associated (Figure 63.3). Histone hyperacetylation is observed in various cancer cell lines as well as in patient tissues especially in hepatocellular carcinoma and oral cancers. In oral cancer, interferon gamma (IFN-γ), nitric oxide signalling mediates hyper acetylation of p300 (p300 autoacetylation) which in turn activates histone hyperacetylation. Hyperacetylation of histone chaperone nucleophosmin (NPM1) is also observed, which triggers p300 autoacetylation in oral cancer. Autoacetylated p300 is enzymatically hyperactive that marks activation of several genes that triggers angiogenesis and metastasis. Interestingly, over expression of p300 protein levels has been implicated in several cancers including HCC and ESCC. p300 and CBP also forms fusion protein with MOZ in some forms of leukaemia (AML). Also Rubinstein-Taybi syndrome is characterised with mutation in CBP locus and altered CBP acetyltransferase activity that results in malignancy. Tip60, a MYST family lysine acetyltransferase plays crucial roles in DNA damage response. Mutation of Tip60 results in defective DNA double strand break repair and resistance towards apoptosis which is one of the hallmarks of cancers. Tip60 acetylates tumour suppressor protein p53 at specific lysine residue (K120) which triggers the p53 mediated proapoptotic gene expression.

The most common and predominant cause of cancer is either silencing of tumour suppressor gene or constitutively active oncogenes. Lysine and arginine methylation of histones and nonhistone chromatin associated proteins are involved with transcription regulation of various genes associated with cancer. More importantly, the cellular levels of methyltransferase enzymes are perturbed in different diseases including cancers.

B. Pathogenic Infections

Human body constantly encounters multiple external and internal pathogens like viruses, bacteria, fungi etc. Pathogenic interactions induces cascade of signaling events culminating in chromatin remodeling and gene expression in the host tissue. Bacterial infection induces p300–mediated histone H3 and H4 acetylation at specific residues and also phosphorylation. This in-turn induces inflammatory gene expression and host response to infection. Viruses being obligatory internal parasites have to hijack the cellular machinery for its own replication and propagation. Persistent viruses and viruses having latency periods like human immunodeficiency virus (HIV), g-herpesviruses, Kaposi's sarcoma associated virus (KSHV or HHV8) and Epstein–Barr virus (EBV) integrate themselves in the host genome and down-regulate the expression of initiating effector molecules. This is supported by heritable epigenetic changes that regulate expression of cell cycle progression, senescence, survival, inflammation and immune system related

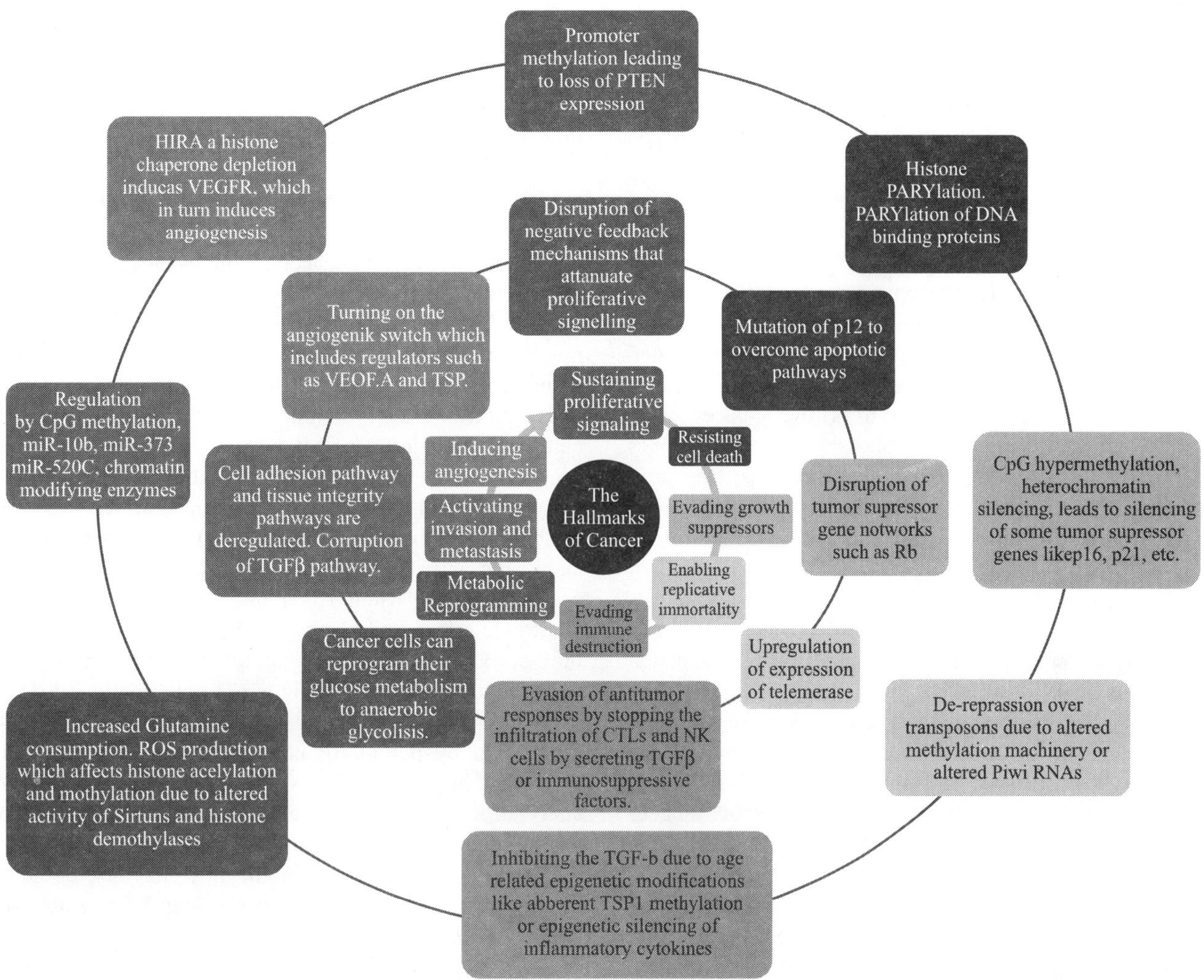

Figure 63.3 Epigenetics and Cancer: The epigenetic Regulation of cancer hallmarks: The innermost circle names the eight hallmarks of cancer, the next circle explains some of the general mechanisms by which these hallmarks are acquired and the outermost circle explains the epigenetic mechanisms that may also contribute to these phenomena. [Modified from *Hanahan and Weinberg; Hallmarks of Cancer: the next generation; Cell 144,2011; 646–673* and *Karthigeyan D et.al, (2012) Cancer: An epigenetic landscape; Subcell Biochem. Epigenetics Development and Disease (Springer) 61, (In Press)].*

genes. Breaking of the latency period is also tightly regulated by epigenetic modifications. Rapidly proliferating viruses also alters the epigenome of the infected cells through different mechanisms. Pathogen infection induced alteration of DNA and protein methylation also significantly contribute to the regulation of gene expression.

Acquired immuno deficiency syndrome

Progression of virus associated diseases is also closely connected with lysine acetylation. Integration of HIV into human genome is finely regulated by p300 and GCN5 mediated acetylation of the viral protein integrase and promotes viral infectivity. Interestingly, the integration of HIV virus is not random in the human genome; rather it favours chromatin sites with transcriptionally active marks like H3K4 trimethylation whereas disfavours transcriptionally silent marks like H3K27

trimethyltion. Acetylation of the viral protein Tat by p300/CBP is essential for the viral gene activation. Nuclear localisation of Tat protein is governed by acetylation of histone chaperone nucleophosmin NPM1 by p300. Viral Tat protein also undergoes methylation by arginine methyltransferase PRMT6 which also methylates viral nucleocapsid. Methylation of Tat imparts a contrasting effect where it reduces the binding affinity of Tat towards TAR protein and inhibits Tat mediated transactivation.

C. Neurodegenerative Diseases

The epigenetic phenomenon governs regulation of gene expression in neuronal cells and shares very close connection with neuronal development, synaptic plasticity and memory formation. A strict balance is maintained between various post translational modifications to fine tune this phenomenon.

Perturbance of activities of chromatin modifying enzymes results in different neurodegenerative disease conditions. Alteration in the balance between acetylation and deacetylation has been observed in neurodegenerative states like Alzheimer's, Parkinson's, Huntington disease and depression. CBP stabilises short term memory into long term memory whereas p300 and PCAF establishes long term memory. Consolidation of spatial memory colocalizes with hyperacetylation of histone H2B and H4 in the dorsal hippocampus and over expression of CBP, p300 and PCAF. Hypoacetylation caused due to down regulation or premature degradation of lysine acetyltransferases like CBP results in apoptosis of neuronal cells. CBP mediated acetylation is essential for spatial memory formation whereas in Alzheimer's disease decreased levels of histone H4 acetylation has been observed. Methylation and phosphorylation of histones are also important for different neuronal functions. Depletion of specific histone methylation marks results in complex behavioural abnormalities.

D. Inflammatory Diseases

Inflammatory response manifestation shares close connection with histone acetylation. Pro inflammatory signals like interleukins induce expression of NF-κB (nuclear factor kappa-light-chain-enhancer of activated B cells) which further results in hyperacetylation of specific lysine residues in histone H4. Hyperacetylation of histones in promoters of genes thus induces expression of inflammatory genes like granulocyte macrophage colony stimulating factor (GM-CSF). Interestingly, regulation of glucose dependent expression of insulin is also controlled by histone acetylation. In diabetes mellitus, histone H4 acetylation and H3K4 methylation which are transcriptional activation marks are increased at the promoters in response to high glucose levels in monocytes for insulin expression. Conversely, at low glucose levels KDAC complexes are recruited which inhibits the gene expression. Thus, in accordance to other disease conditions, inflammatory disorders are also associated with altered balance between acetylation and deacetylation. Protein phosphorylation triggers stimulation of cellular signal for expression of various genes in a context dependent manner. Insulin increases the tyrosine phosphorylation of various insulin receptors and other insulin receptor substrates and proteins important for insulin responsive trafficking machinery like GLUT4.

VII. CHROMATIN MODIFYING ENZYMES AS A TARGET FOR THERAPEUTICS

Altered histone acetylation, methylation and phosphorylation are established candidates as therapeutic targets. Among all the histone posttranslational modifications, reversible lysine acetylation has been extensively studied to target various disease conditions. Modulation of lysine acetylation could be achieved using three different approaches; a) lysine deacetyltransferase inhibitors (KDACi), b) lysine acetyltransferase inhibitors (KATi) and c) lysine acetyltransferase activators (KATa). KDACi inhibits the deacetyltransferase activities of KDAC but they have mostly broad spectrum targets, they indirectly induce histone hyperacetylation. In numerous cancers, including breast cancers compromised histone acetylation levels are reported where KDACi (suberoyl anilide hydroxamic acid, Vorinostat) could potently rescue the acetylation levels and inhibit tumour cell proliferation.

As discussed earlier, oral squamous carcinoma, hepatocellular carcinoma and several other cancers are associated with increased lysine acetyltransferase activities. Small molecule KAT inhibitors could enter inside cancerous cells and efficiently inhibit histone and nonhistone protein acetylation and induces apoptosis. The possible mechanism of action for some of the p300/CBP specific KATi is through inhibition of autoacetylation of the enzyme and thereby reduced enzymatic activity. Therapeutic approach by targeting lysine acetyltransferase using KATi has immense potential as the pleotropic effect could be avoided by using specific inhibitors that could target specific acetyltransferase activity. The therapeutic potential of KATi is not only confined to cancers as it has been explored in other diseases as well. Histone acetylation by p300/CBP is essential for the establishment of HIV pathogenesis. p300 specific KATi reduces the histone acetylation in HIV infected cell lines and inhibits viral proliferation. Targeting HIV infection using small molecule modulators of chromatin modifying enzymes could open the door for possible therapeutics in our fight against AIDS. Other PTMs like histone methylation and phosphorylation are also perturbed in various disease conditions. In some cancers, over expression and hyperactivity of several arginine methyltransferases has been observed. Small molecule inhibitors targeting lysine acetylation and arginine methylation could act as a candidate for combinatorial therapeutics. Expression of kinases which are crucial for cell cycle progression like Aurora kinases are altered in several cancers and causes defects in cell cycle checkpoints, ultimately leading to proliferation. Small molecules targeting these kinases provide an excellent target towards anti cancer therapy. Small molecules that can activate lysine acetylation by activating KAT activity have also been reported. Neurodegenerative states especially Alzheimer's and Parkinson's provides a rather challenging environment, where the acetyltransferase activity of p300/CBP gets compromised and the pool of acetylated histone gets depleted from certain regions of brain. In Alzheimer's disease mice, reduced levels of histone H4 acetylation have been observed which contributes towards the loss of memory. Small molecules that activate the KAT activity by enhancing its autoacetylation could rescue the acetylation on the histones. Activation of histone acetylation in brain of Alzheimer's disease patients is a promising target that is being explored.

64

The Structure of the Genetic Material and its Replication

H.K. Das

CONTENTS

I. What are Genes?
II. Nature of the Genetic Material
III. What is the Structure of DNA?
IV. Special Features of Eukaryotic DNA
V. DNA Replication
VI. Genetic Mutations
VII. Summary

I. WHAT ARE GENES?

The concept of a 'gene' has been undergoing continuous refinement since its very birth. Mendel (1865) has been credited for developing the idea of genes—hereditary units that carry properties (traits) and functions of an organism from parent to the progeny, generation after generation. His experiments on crossing of varieties of pea plants having seeds differing in well-defined characters, suggested the existence of separate hereditary units (genes) for each character (trait) which could segregate and get reassorted independently after crossing.

In 1909, Garrod observed that the human disease, Alcaptonuria, was hereditary and inferred its origin to be in a defective gene. Several other hereditary diseases have been demonstrated to be linked to chromosomal abnormalities. Chromosomes are involved in heredity, because genes are located in the chromosomes. However, it must be understood that genetic defects leading to a disease may not always be traced to visible chromosomal aberrations. This is because a very small change in a gene may also inactivate it, but only large changes show up as an abnormal chromosome. Diseases like Sickle Cell Anaemia or Alcaptonuria are caused by really minute changes.

It became clear soon that a particular character of an organism need not necessarily be controlled by a single gene. By 1915 as many as 13 genes were identified which controlled the eye colour of the fruit fly, Drosophila. As early as 1905, Garrod had postulated from his observations on Alcaptonuria,

that genes do not directly control a character or trait, but control only enzymatic reactions. The conversion of phenylalanine to tyrosine is blocked in persons afflicted with Alcaptonuria and as a result, the accumulated phenylalanine gives rise to phenylpyruvic acid. This concept was established firmly, when in 1941, Beadle demonstrated by genetic crosses and nutritional studies of X-ray generated mutants of the fungus Neurospora, that a gene really controls the synthesis of an enzyme. This concept of "one gene one enzyme" has now been further refined. It became obvious that one gene only codes for one polypeptide. An enzyme composed of two different subunits would thus be controlled by two different genes. (The situation might be little more complex in some cases, however. In the bacterial virus ϕX 174, two different proteins are coded for, in two different frames, by the same segment of a gene.)

II. NATURE OF THE GENETIC MATERIAL

For many years, the nature of the genetic material had remained illusory to biologists. That it is nothing mysterious, but just a chemical substance, became obvious for the first time from the experiments of Griffith (1928). The bacterium, Pneumococcus exists in two forms. The S form (smooth colonies) is virulent, while the R form (rough colonies) is not. Griffith found that heat killed virulent bacteria, when mixed with live avirulent ones, would transform the latter into virulence (tested on mice) and also impart to it the character of smooth colonies, because of the reappearance of the polysaccharide capsule. The transformed bacteria continued to exhibit these newly acquired properties, generation after generation. Obviously, the transformation has been caused by the entry of a chemical substance (which is heat stable and presumably is the genetic material) into the live avirulent cells from the heat killed virulent ones.

In 1944, Avery, MacLeod and McCarty isolated and identified the Pneumococcus transforming principle to be "nucleic acid of the deoxyribose type". The transforming activity was lost completely, if the isolated material was digested with deoxyribonuclease, but was immune to the action of trypsin and chymotrypsin or ribonuclease. It also remained unaffected on extraction of lipid.

In 1951, elegant experiments by Alfred Hershey and Martha Chase lent further support to the idea that the genetic material is deoxyribonucleic acid (DNA). It was clear by then that viruses multiply in their host, because their genetic material gets injected into the host. It was known that the bacterial virus T2 was made up of DNA core and a protein coat. Hershey and Chase demonstrated that it was the DNA that got injected into the host and not the protein. This was done by using viruses whose DNA was labelled with ^{32}P and proteins with ^{35}S. Today we know that except for a few RNA viruses, all living cells and viruses have DNA as their genetic material.

III. WHAT IS THE STRUCTURE OF DNA?

DNA is a linear polynucleotide. A mononucleotide consists of a purine or pyrimidine base—adenine, guanine, thymine or cytosine, linked through one of its ring-nitrogen atom to the 1′ carbon of a 2′-deoxyribose 5′-phosphate (Figure 64.1). Two mononucleotides are linked by a 5′ → 3′ phospho-diester linkage. The backbone of DNA is therefore an alternating deoxyribose-phosphate-deoxyribose-phosphate. Figure 64.2 shows a tetradeoxyribonucleotide. The purine and pyrimidine bases are present as side chains to the backbone. The two ends of a DNA molecule, therefore, are not similar. At one end, there would be a free 5′-OH or phosphate group, while at the other end it would be a free 3′-OH group. These two ends are referred to as 5′ end and 3′ end (Figures 64.3, 64.4 and 64.5).

DNA molecules are very long. The DNA in the enteric bacteria, *Escherichia coli* is about 4 million mononucleotides long and exists in the form of a closed circle. DNA in cells of higher eukaryotes may be much longer. The DNA of the largest chromosome of the fruit fly *Drosophila melanogaster* is about 60 million mononucleotides long and is a linear open ended molecule.

Erwin Chargaff and coworkers found out during the period 1950–1953 that purine and pyrimidine content of the DNA in a large number of different organisms followed an interesting pattern. The molar content of adenine would always be the same as that of thymine and similarly the molar content of guanine would be the same as that of cytosine. The significance of this data was not clear to them.

In 1953, James Watson and Francis Crick made use of this information along with the X-ray diffraction photographs of DNA taken by Rosalind Franklin and Maurice Wilkins, to derive the three dimensional structure of DNA. The important features of this model are the following:

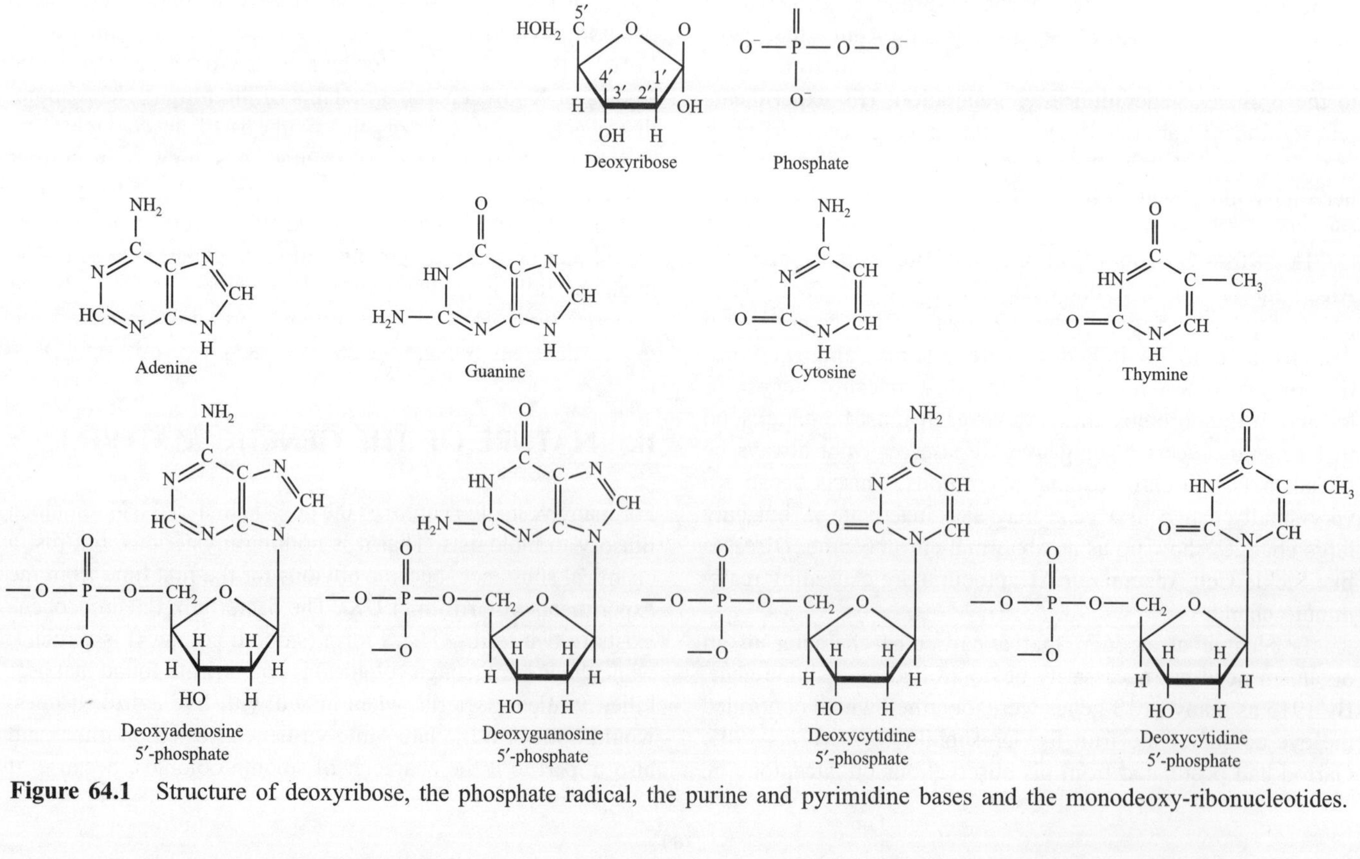

Figure 64.1 Structure of deoxyribose, the phosphate radical, the purine and pyrimidine bases and the monodeoxy-ribonucleotides.

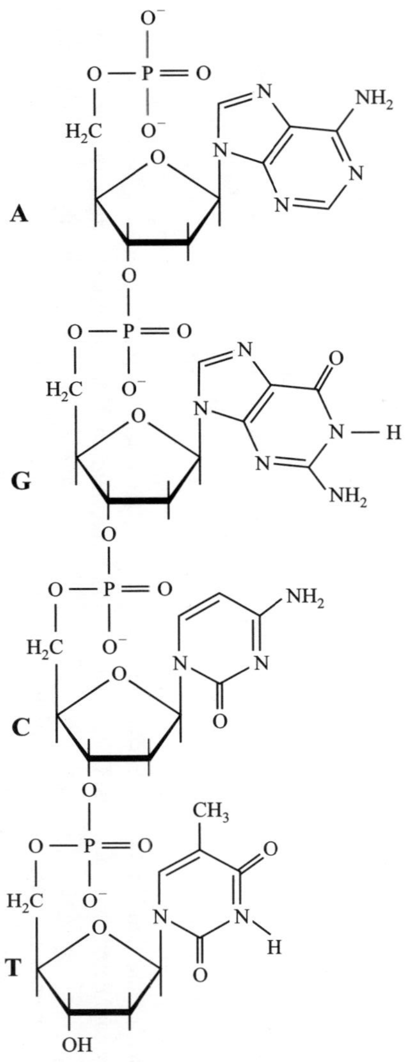

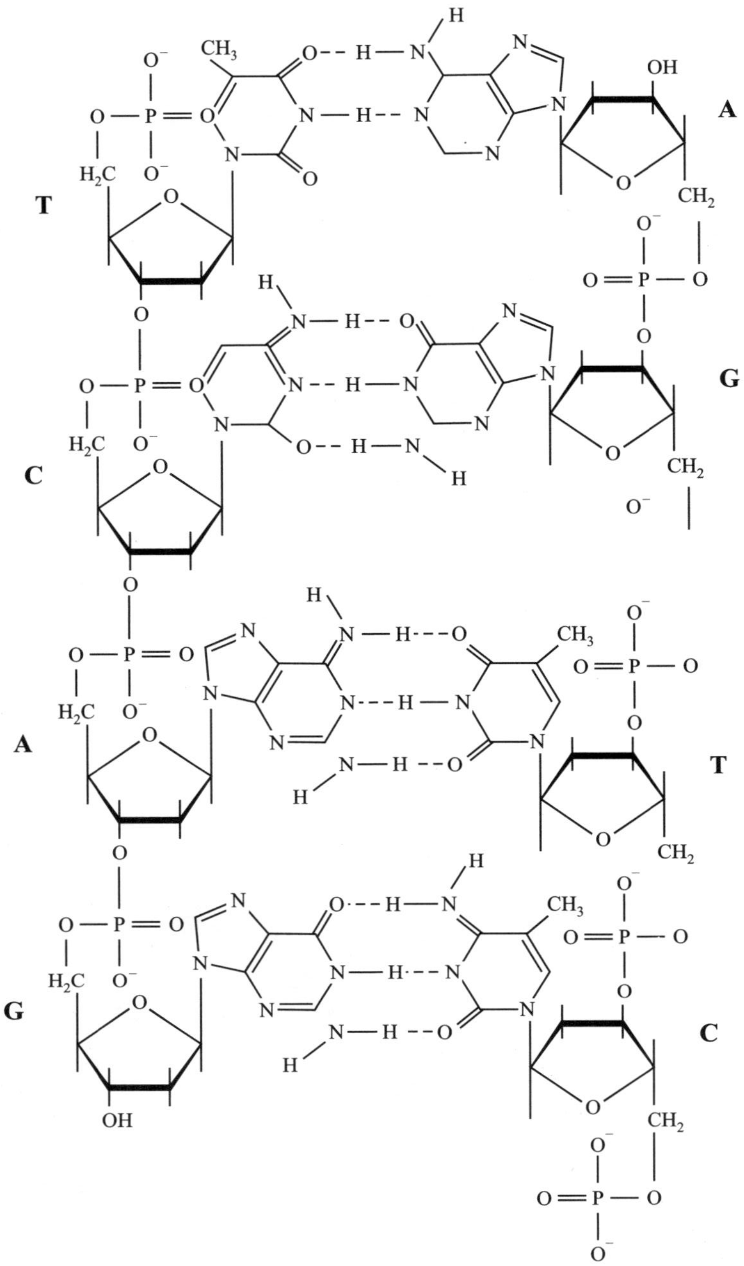

Figure 64.4 Two tetra-deoxyribonucleotides arranged side by side, but in opposite polarity, with hydrogen bonds between complementary bases.

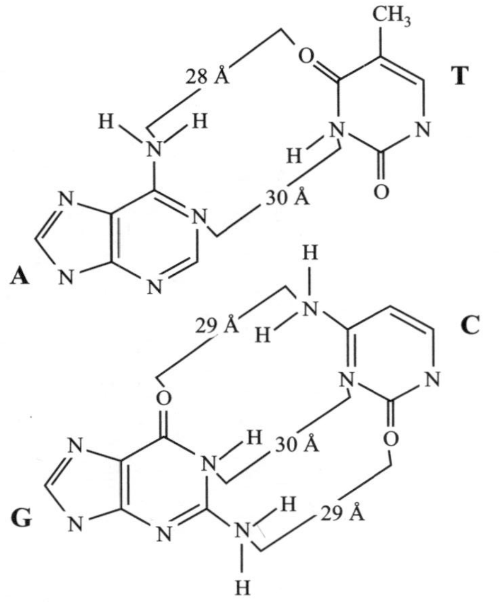

Figure 64.3 Two hydrogen bonds between adenine and thymine, three hydrogen bonds between guanine and cytosine.

Figure 64.2 A tetra-deoxyribonucleotide, in which the mononucleotides are joined by 5′ → 3′ phosphodiester linkages. The 5′-end is at the top and the 3′-end at the bottom.

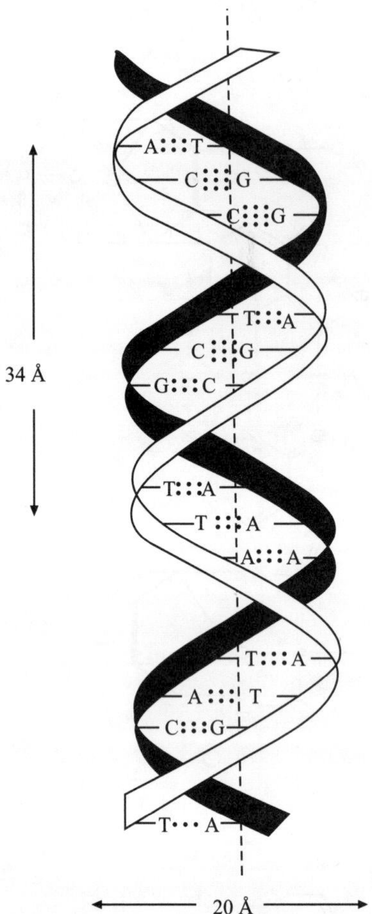

Figure 64.5 The double helix.

(a) The DNA molecule is a duplex of two deoxyribo-polynucleotides arranged side by side in opposite polarity. The 5′-end of one strand and the 3′-end of the other are together at the end of the duplex. These two strands are coiled around a common axis, thus forming a right handed double helix.

(b) The purine and pyrimidine bases project inside from the backbone towards the common axis. The planes of the bases are perpendicular to the axis and at about right angles to the plane of the sugars.

(c) The positions of bases in the two strands are such that an adenine in one strand is always opposite to a thymine in the other and similarly a guanine is opposite to a cytosine. There are two hydrogen bonds between adenine and thymine and three between guanine and cytosine. The sequence of bases in one strand is not restricted in any way, but that of the second would be, so as to maintain complementarity.

(d) The diameter of the double helix is 20 Å. The distance between two adjacent bases along the axis is 3.4 Å and the pitch of the helix is 34 Å. Thus one turn of the helix is completed every ten bases.

Most DNAs encountered in nature are double stranded, but there are some viruses like ϕX174, which have in the virion a single stranded DNA as genetic material. This DNA does not obviously have equal amount of adenine and thymine or equal amounts of guanine and cytosine. Once inside the host, the single stranded viral DNA becomes double stranded as a first step towards replication.

Though adenine, guanine, thymine and cytosine are the bases that are normally found in DNA, some modified bases have also been encountered. DNA of higher plants like wheat, might occasionally contain 5-methyl cytosine. Similarly some bacteria and their viruses may contain in their DNA, small amounts of 6-methyl aminopurine, uracil, 5-hydroxymethyl uracil or 5-hydroxymethyl cytosine.

An alternative structure of DNA has been proposed by V. Sasisekharan and coworkers. This is not a double helix, but consists of two strands of polynucleotides lying side-by-side and held together by hydrogen bonds between complementary bases (Figure 64.6). The individual strands have alternating right and left-handed helical segments approximately five base pairs in length. This model is claimed to be energetically more favourable to the double helix. Under this model, strand separation during replication could be achieved simply by breaking the hydrogen bonds, and no uncoiling motion would be necessary.

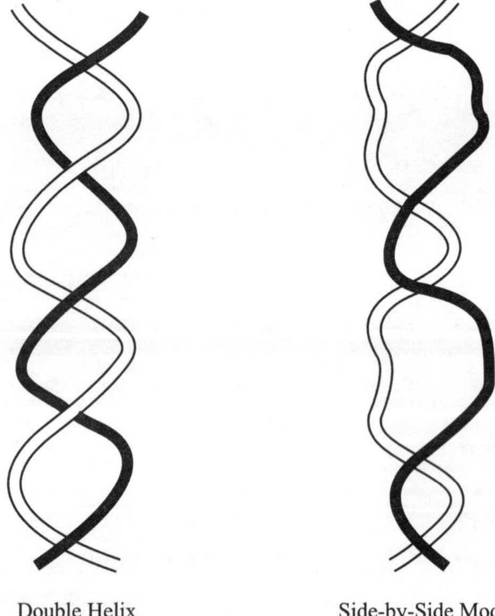

Figure 64.6 The "side-by-side" model of DNA versus the double helix.

Alexander Rich and coworkers have inferred by atomic resolution X-ray crystallographic analysis that the synthetic oligodeoxynucleotide duplex of CpGpCpGpCpG assumes a left-handed conformation (known as the Z form of DNA) when exposed to high salt or ethanol concentration. This group have later demonstrated that antibodies specific to the Z-DNA conformation bind specifically to polytene chromosomes of *Drosophila melanogaster* (fruit fly) in the inter-band regions, suggesting the occurrence of Z form of DNA in material of biological origin.

IV. SPECIAL FEATURES OF EUKARYOTIC DNA

The amount of DNA present in an eukaryotic cell is very much larger than that in any prokaryotic cell. The total DNA in a human cell is equivalent to more than 6 billion nucleotides. This huge amount of DNA is, however, not present as a single duplex chain, as is the case in most prokaryotic cells. The main bulk of the genetic material in an eukaryotic cell is distributed among the various chromosomes found inside the nucleus. A diploid human cell nucleus has 46 chromosomes, which can be visualized by specific staining during mitotic cell division. The chromosomes are of different sizes. Each haploid chromosome, however, has a single molecule of DNA duplex. This DNA is associated with several kinds of small basic proteins which are lysine and/or arginine rich. These proteins are known as histones. Some non-histone proteins are also present in chromosomes. The extranuclear organelles, mitochondria and chloroplasts (in case of plants), also contain some DNA.

The histone-DNA complex in eukaryotic chromosomes is known to occur in the form of tightly folded spheres called "nucleosomes", each flanked between two elongated stretches known as spacer or linker. The core consists of 146 base pairs of DNA wrapped around a globular histone bead containing two each of the histones H2A, H2B, H3 and H4. Such beaded structures are visible in electron micrographs.

A very distinctive feature of eukaryotic DNA is the presence of 'intervening sequences' in most genes. During any gene expression, the sequence of bases in one of the strands of DNA is first 'transcribed' into a complementary sequence of ribopolynucleotide (RNA). The RNA is then 'translated' into a polypeptide, consecutive three bases of RNA coding for one amino acid. The RNA formed on expression of a prokaryotic gene is an exact complementary copy of the relevant DNA strand, this RNA therefore, is colinear with the DNA. The RNA found in the cytoplasm as a result of expression of an eukaryotic gene, however, in most cases, might lack several internal sequences (sometimes quite long) corresponding to the DNA (Figure 64.7). It is now known that inside the nucleus,

the complete colinear transcript is first formed. Subsequently, certain specific segments (introns) are removed and, the remaining portions (exons) are then spliced together. The adenovirus-2 DNA has three introns 100, 2350 and 8050 bases long. The chick ovalbumin gene has as many as seven intron sequences (Figure 64.8).

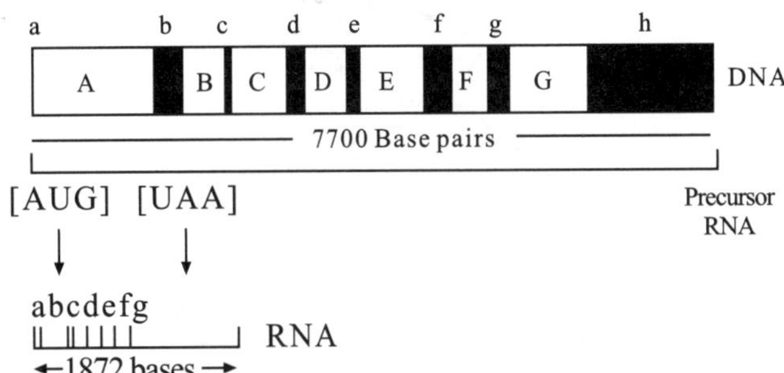

Figure 64.8 Representation of the chick ovalbumin gene and its transcription products. The blank spaces denoted by the capital letters are the intervening sequences (introns) and the black spaces marked by the small letters are the exons. The precursor RNA is full length, but the finished messenger RNA contains sequences only corresponding to the exons. The protein synthesis gets initiated at the AUG in the exon 'b' and is terminated at the UAA in the exon 'h'.

V. DNA REPLICATION

The beauty of the Watson and Crick model of DNA is that it gives an immediate insight into the mechanism of replication of the genetic material. It was predicted by them that specific pairing of the bases would play an important role in replication. Each strand of the duplex could serve as a template for the synthesis of a DNA strand following the rule that adenine would always be opposite thymine and guanine opposite cytosine. Thus two identical daughter duplexes can be obtained from one parent duplex (Figure 64.9).

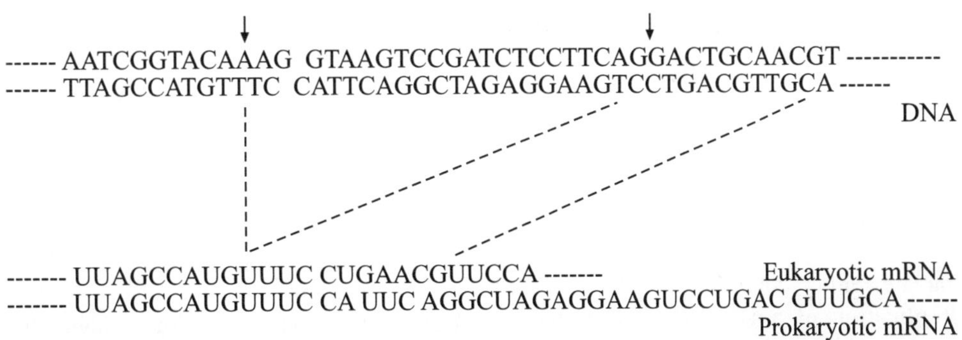

Figure 64.7   Non-colinearity of eukaryotic messenger RNA with DNA. The sequence in DNA between the two arrows is absent in the finished eukaryotic mRNA. Prokaryotic mRNA is exactly complementary to the template strand.

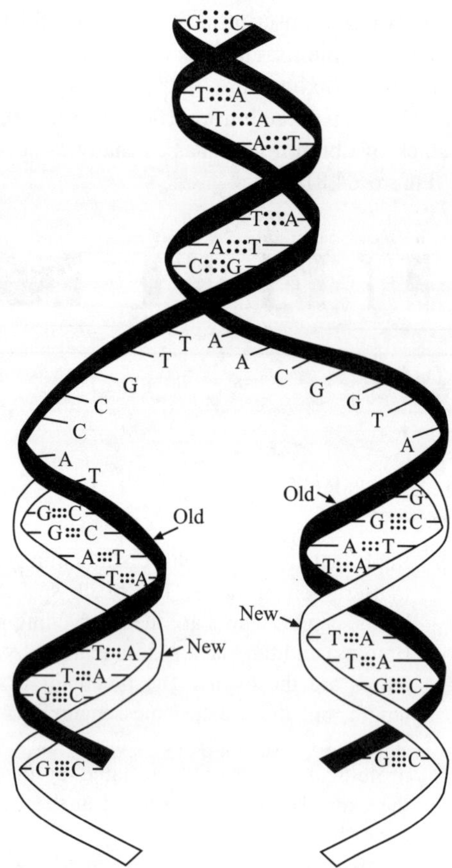

Figure 64.9   DNA replication. Each strand of a duplex serves as a template for the synthesis of a complementary strand, thus producing two identical daughter duplexes.

Mathew Meselson and Franklin Stahl confirmed in 1958 that this was indeed how DNA was replicated. Since one parental strand of the duplex was found conserved in the progeny, this mode was termed "semiconservative". The experiment involved growing of *Escherichia coli* for several generations in a medium containing $^{15}NH_4Cl$ and then transferred to a medium containing $^{14}NH_4Cl$. DNA was isolated periodically from the dividing cells and its buoyant density was determined by equilibrium density gradient sedimentation. After one generation in $^{14}NH_4Cl$, the DNA was of a density expected of a hybrid containing equal amounts of ^{14}N and ^{15}N. After two generations, almost equal amounts of two DNA bands were obtained. One had a density the same as that obtained after one generation, while the other was of a density of pure ^{14}N DNA (Figure 64.10).

The credit for elucidating the enzymatic mechanism of DNA synthesis must go to a large extent to Arthur Kornberg and his coworkers. The following components seem essential for *in vitro* replication of DNA.

1. A DNA template, either double stranded or single stranded. Replication of a DNA duplex must be preceded by a nick (single strand break of the phosphodiester linkage) at the replication origin of the DNA duplex (Figure 64.11).

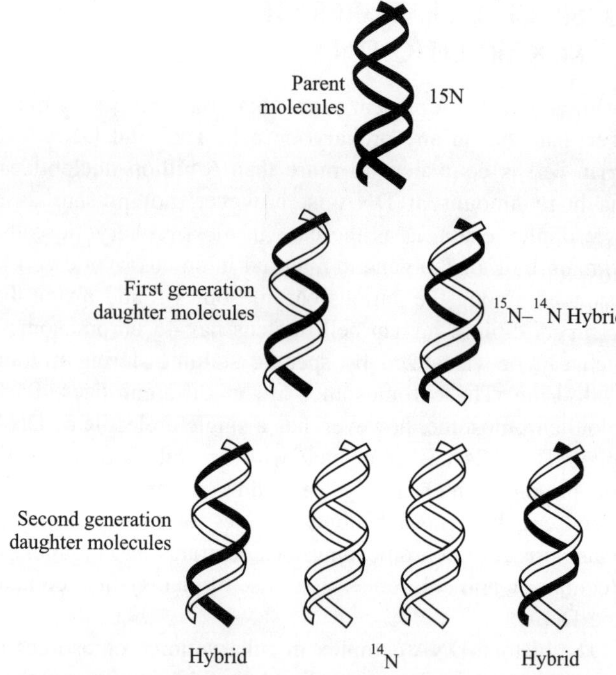

Figure 64.10   Semiconservative replication of DNA.

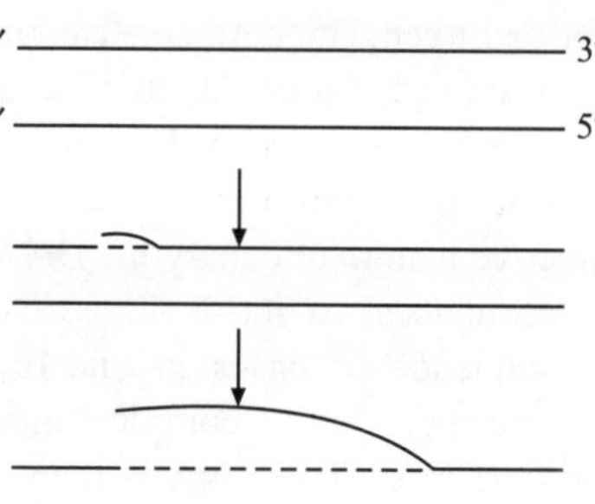

Figure 64.11   DNA synthesis at the nick of a duplex. The broken line represents the newly synthesized DNA.

2. A primer chain with a free 3′-OH group in case of replication of a single stranded DNA. The primer is complementary to the template and it is at the 3′-OH terminus of the primer, where mononucleotides are added one by one by DNA polymerase (Figure 64.12).

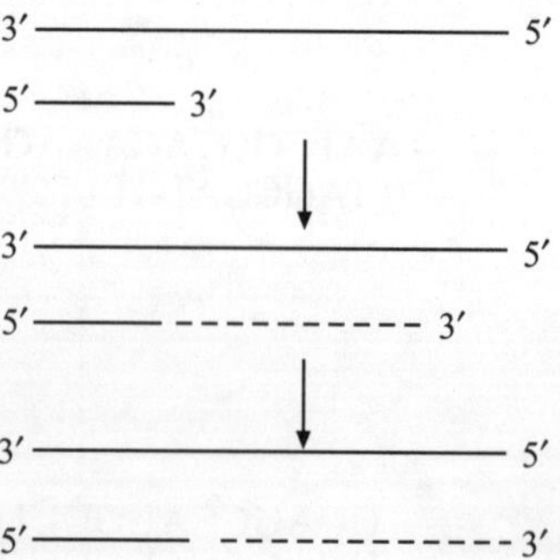

Figure 64.12   Replication of a single stranded DNA. Mononucleotides are added to the 3′-end of the primer. The broken line represents the newly synthesized DNA. So direction of synthesis is 5′ → 3′.

3. All the four deoxyribonucleoside 5′-tiphosphates—deoxyadenosine 5′-triphosphate (dATP), deoxyguanosine 5′-triphosphate (dGTP), deoxythymidine 5′-triphosphate (dTTP) and deoxycytidine 5′-triphosphate (dCTP).

The synthesis of DNA takes place by a "nucleophilic attack of the 3′-OH terminus of the primer on the innermost phosphorus atom of the incoming deoxyribonucleoside triphosphate". A phosphodiester linkage is formed and pyrophosphate is liberated (Figure 64.13). This step is repeated and the DNA strand is elongated in the 5′ → 3′ direction.

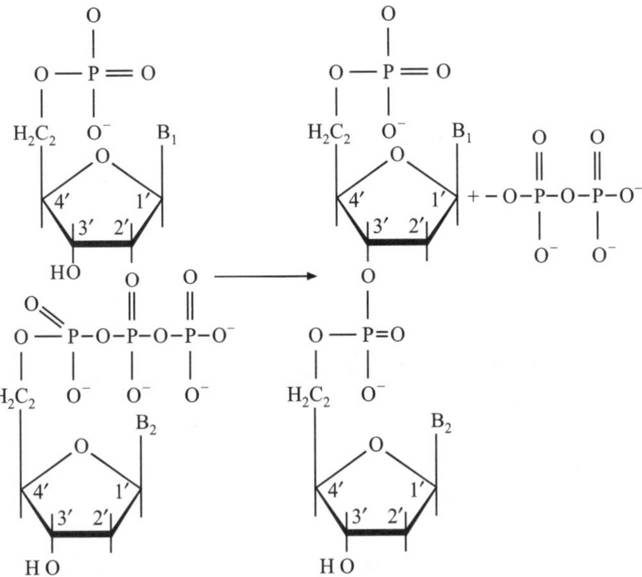

Figure 64.13 Dinucleotide synthesis by the formation of a phosphodiester linkage. Pyro-phosphate is released.

Watson and Crick's complementarity rule is obeyed and the synthesis is thus directed by the template. Both the template strands of a duplex are copied simultaneously.

Altogether five DNA polymerases are now known in *Escherichia coli,* but only DNA polymerase III is definitely thought to be essential for DNA replication *in vivo*. The role of DNA polymerase II is uncertain, while DNA polymerase I has definite role in DNA repair. DNA polymerase I is thought have some role in replication too. DNA polymerase IV is involved in untargeted mutagenesis, while DNA polymerase V carries out trans lesion DNA synthesis or targeted mutagenesis. It has, however, been demonstrated that DNA polymerase V is highly mutagenic even when it carries out gap filling DNA replication.

Apart from the DNA polymerases, many other proteins are also involved in the process of DNA synthesis *in vivo*. Mutations in as many as twenty or so different genes in *Escherichia coli* affect DNA replication.

The DNA duplex in *E. coli* exists as a closed super helical circle. Replication must be preceded or accompanied by localized unwinding, a prerequisite for which, would be the hydrolysis of at least one phosphodiester linkage. It is logical, therefore, that one endonuclease and some of the 'unwinding

proteins' that have been characterized in recent years, would also be involved in DNA replication.

Autoradiographs of replicating DNA from *E. coli* have revealed that replication starts at a unique origin and proceeds simultaneously on both the strands in the same direction, the site of replication appearing as a 'fork'. The important point to remember here is that all the known DNA polymerases, synthesize DNA only in the 5′ → 3′ direction. However, if both the strands have to be replicated simultaneously and in the same direction, one strand must be synthesized in the 3′ → 5′ direction. Reiji Okazaki suggested that the apparen 3′ → 5′ synthesis of one strand could in reality be a discountinuous 5′ → 3′ synthesis in the reverse direction (Figure 64.14). He demonstrated, that indeed small fragments of about 1000 nucleotides length are first formed, which are joined together later by the enzyme DNA ligase.

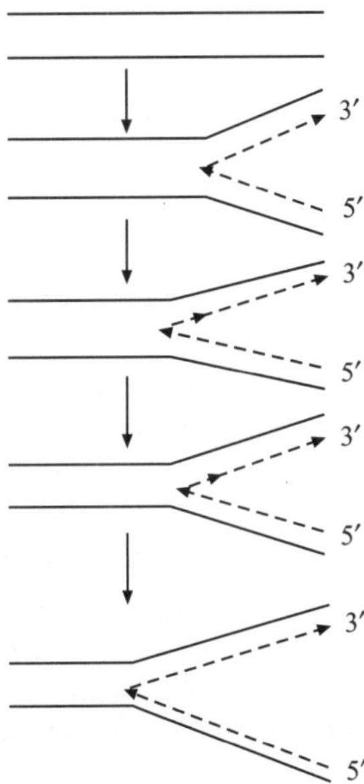

Figure 64.14 The apparent 3′ → 5′ synthesis of one strand is in reality a discontinuous 5′ → 3′ synthesis in the reverse direction. The dotted line is the newly synthesized DNA. The arrows represent the direction of synthesis and also the point where mononucleotides are being added.

Since DNA synthesis starts at a unique site (or sites), it becomes obvious that there has to be some kind of a recognition phenomenon of the specific base sequence signalling replication initiation. None of the DNA polymerases is known to have such recognition capacity, but these enzymes can only add on nucleotide after nucleotide to the 3′-OH of a preexisting primer, and of course, under the direction of a template. How is the primer synthesized? It is now known that the primer is not DNA. RNA polymerases are known to have the capacity to recognise specific nucleotide sequences in the template and initiate synthesis of RNA. A small length of RNA serves as primer for the DNA polymerase. The primer is degraded after it has served its purpose and the gap is filled up, probably by DNA polymerase I (Figure 64.15).

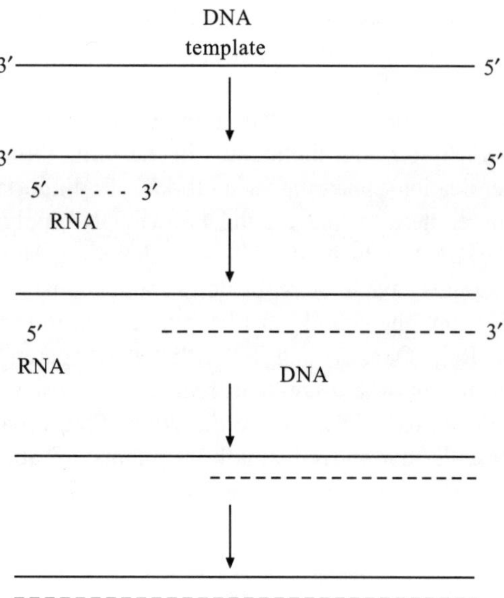

Figure 64.15 RNA as primer for DNA synthesis. The dotted line represents RNA and the broken line newly synthesized DNA strand.

The replication of DNA is thus a very complex phenomenon, in which several other proteins besides DNA polymerases, have to take part.

The autoradiographs of replicating DNA mentioned earlier, have revealed another interesting feature, that simultaneously two replication forks appear at the origin of replication and move in opposite directions. Thus replication is bidirectional (in most cases). In *E. coli* the two forks meet at the replication terminus which is diametrically opposite the origin. This phenomenon has been observed in Chinese Hamster cells also. However, in these eukaryotic cells, replication starts in tandem at several fixed origins along the DNA duplex (Figure 64.16).

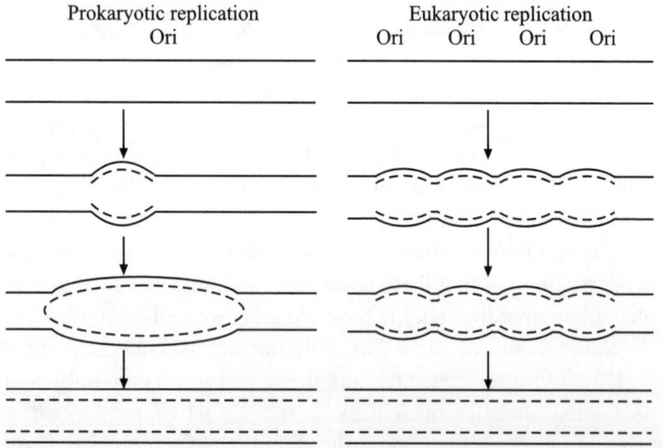

Figure 64.16 Tandem replication of eukaryotic DNA at multiple origins. 'Ori' denotes replication origin(s) and the dotted lines represent newly synthesized DNA. The replication is bidirectional.

A large number of eukaryotic DNA polymerases are known. Table 64.1 lists these.

TABLE 64.1 Eukaryotic DNA polymerases

Name	Function
α (alpha)	DNA replication
β (beta)	Base excision repair
γ (gamma)	Mitochondrial DNA replication
δ (delta)	DNA replication
ε (epsilon)	DNA replication
ζ (zeta)	Bypass DNA synthesis
η (eta)	Bypass DNA synthesis
θ (theta)	DNA repair
ι (iota)	Bypass DNA synthesis
κ (kappa)	Bypass DNA synthesis
λ (lambda)	Base excision repair
μ (mu)	Non-homologous end joining
σ (sigma)	Sister chromatid cohesion
REV1L	Bypass DNA synthesis
TDT	Antigen receptor diversity

VI. GENETIC MUTATIONS

An alteration in a phenotypic character which is perpetuated over generations, is inferred to be due to genetic mutation. A genetic mutation, therefore, must be because of a change in the genetic material. The genetic disease Sickle Cell Anaemia has been traced to a change of the amino acid, glutamic acid to valine in the 6 position from the amino terminal end of the beta chain of haemoglobin. Since the genetic codes for glutamic acid and valine are GAG and GTG respectively, a single base change of A to T in the specific position in DNA, would result in the disease. This is a case of base substitution.

Substitution of one purine by another (e.g., A $\leftrightarrow$ G) or of one pyrimidine by another (e.g., C $\leftrightarrow$ T) (base transition) could be obtained by the action of chemical mutagens like HNO_2 which modifies bases by directly acting on them (Figure 64.17). On the other hand, compounds like ethyl methane sulfonate (EMS) alkylates the 7-position of guanine residues and also that of adenine which results in the cleavage of the base-sugar bond. During replication, any of the four bases can thus be incorporated across this position. Hence, in this case there is no restriction on the nature of the substituent base and both base transition (purine $\rightarrow$ purine) and base transversion (purine $\rightarrow$ pyrimidine) can occur.

It has been postulated that spontaneous mutations may arise during replication, as a consequence of tautomerism or ionisation of bases in the template strand. The base in its rare tautomeric or ionized form would be capable of forming hydrogen bonds with a base not permitted by the Watson-Crick base pairing rule. Thus, C,T, G and A could pair with A, G, T and C respectively. Hence G, A, C or T would be replaced by A, G, T or C respectively, in one strand of the progeny

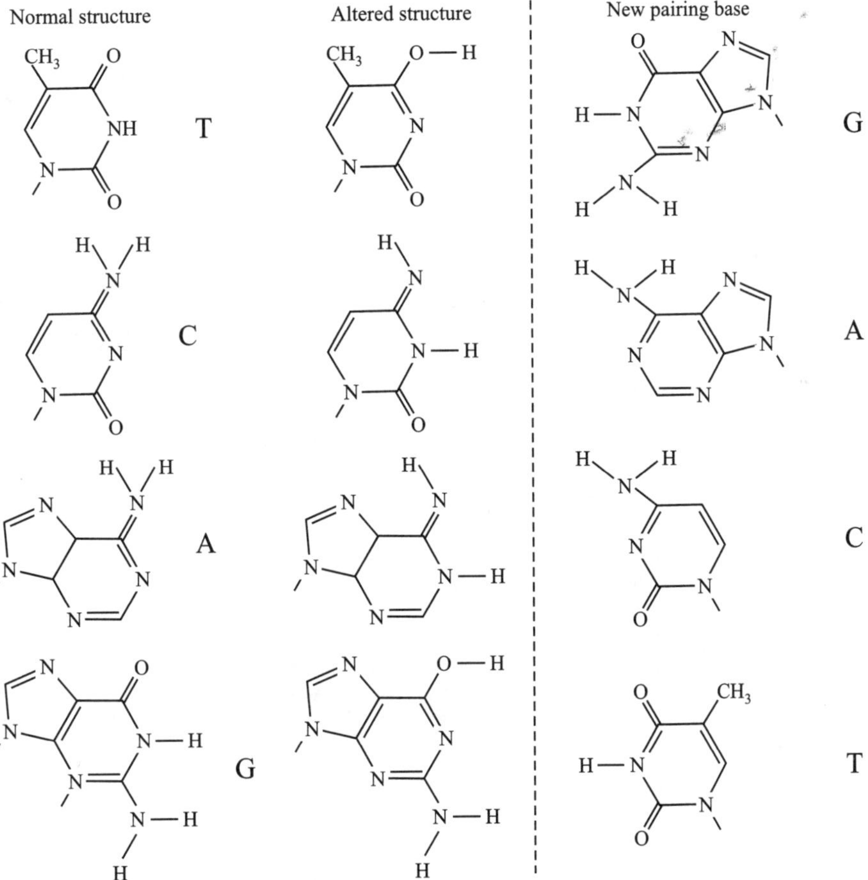

Figure 64.17　Mutagenesis by HNO₂. Cytosine is converted to uracil and adenine to hypoxanthine by the action of HNO₂. Uracil would pair with adenine, while hypoxanthine would pair with cytosine. Thus, in the newly synthesized DNA, adenine would be there where guanine should have been and cytosine would occupy the position where thymine was supposed to be.

Figure 64.18　Tautomerism of bases in the template strand and resultant alteration of pairing specificity. Thus, spontaneous mutation would take place after replication.

DNA (Figure 64.18). During the next replication, both the strands would be mutated.

Mutations are also produced by ultraviolet rays and ionizing radiations. Two adjacent thymines (cytidine too, but less frequently) in the same DNA strand would condense forming a cyclobutane ring on UV irradiation (Figure 64.19).

Ionizing radiations would cause strand breaks. The cells have enzyme systems that would repair these lesions. Mistakes occur during the repair process, particularly during post-irradiation replication repair and as a result, mutations occur.

Mutations can also arise due to insertion or deletion of a single base in the DNA. These are known as "frame shift" mutations, because the whole frame of triplet reading gets altered, resulting in the formation of an altogether changed protein (Figure 64.20).

Deletion or insertion of long sequences of nucleotides are also known. Mutations due to base substitution or frame shift can be reversed. In case of base substitution, a back mutation of the changed base can restore the original one. A frame shift due to addition of a single base might be corrected by deletion of that base or one nearby, thus reverting to the original reading frame. However, mutation caused by deletion of long sequences cannot be reversed for obvious reasons. Insertion of long sequences of nucleotides is mediated by "transposons", which are also known as "jumping genes". Several of these "transposons" have now been isolated and the complete base sequence of these has been worked out. All of them have specific small repeated sequences, either direct or inverted at the two ends.

SUMMARY

Genes control a character or trait by coding for specific enzymes. Chemically, the genetic material of all living cells and most viruses is deoxyribonucleic acid (DNA).

DNA molecules are very long polydeoxyribo-nucleotides and generally exist as a duplex of two complementary strands coiled around a common axis, forming a right handed double helix. The RNA found in the cytoplasm of eukaryotic cells most often is not colinear with the DNA, as is the case with prokaryotic cells. The eukaryotic DNA has "intervening sequences", that are absent in the cytoplasmic RNA transcript.

DNA replication *in vivo* takes place in the "semi-conservative" mode by a complex process. A large number of proteins, besides the DNA polymerases are involved in this.

Genetic mutations are the result of base substitution, nucleotide addition or nucleotide deletion in the DNA.

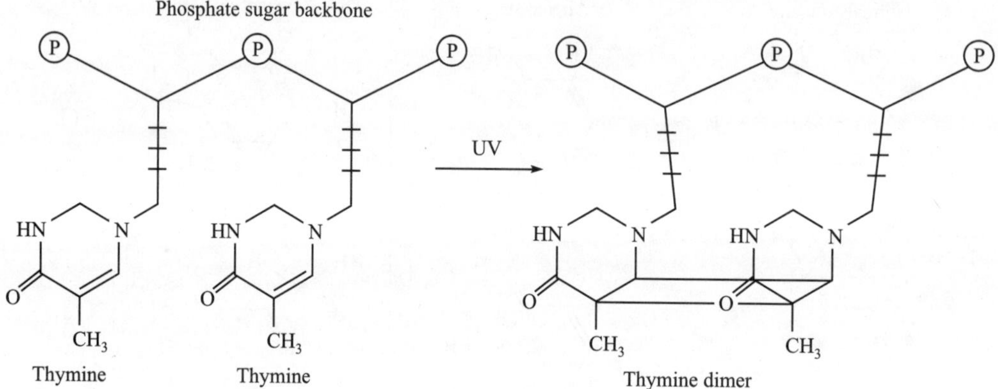

Figure 64.19 Condensation of two adjacent thymines in the same DNA strand forming a cyclobutane ring on UV irradiation.

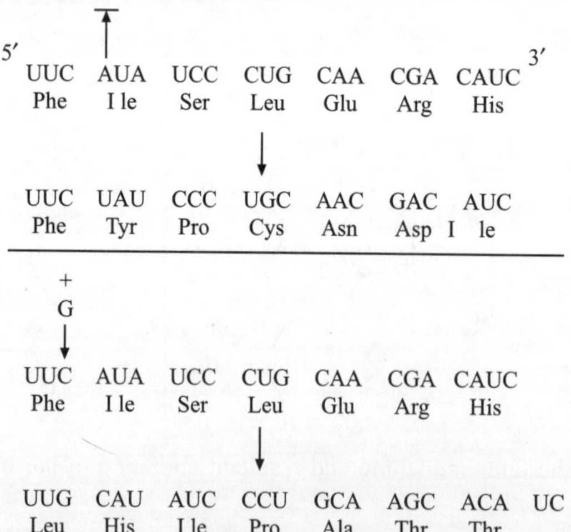

Figure 64.20 Frame shift mutation due to deletion or addition of a single base.

SUGGESTIONS FOR FURTHER READING

Freifelder D. (1987), *Molecular Biology,* 2nd ed., Jones and Bartlett Publishers, Inc., USA.

Hickey R.J. and Malkas L.H. (1997), Mammalian Cell DNA Replication, *Crit. Rev. Eukaryotic Gene Expression,* 7, 125–157.

Kornberg A. (1980), *DNA Replication,* W.H. Freeman and Company, USA.

Lewin B. (1997), *Genes VI,* Oxford University Press, Oxford, UK.

Morin G.B. (1997), Telomere Control of Replicative Life Span, *Expl. Gerontol,* 32, 375–382.

Stent G.S. and Calender, R. (1978), *Molecular Genetics: An Introductory Narrative,* 2nd ed., W.H. Freeman and Company, USA.

Tichan R. (1995), Molecular Machines that Control Genes, *Sci. Am.,* 272, 54–61.

Watson J.D., Hopkins N.H., Roberts J.W., Steitz J.A. and Weiner A.M. (1987), *Molecular Biology of the Gene,* 4th ed., The Benjamin/Cummings Publishing Company, USA.

65

Gene Transcription

Asis Datta and K. Natarajan

CONTENTS

I. INTRODUCTION

Regulation of gene expression is fundamental to how cells respond to environmental and developmental signals. Gene transcription is by far the most prevalent and predominant step for control of gene expression. Indeed it is estimated that roughly a quarter of the human genome is transcribed into RNA. We will come back later in the chapter to the fact that only about 1.5–2% transcripts are translated into proteins. DNA is a chemical polymer known as Deoxyribonucleic acid (DNA) made up of four bases adenine, guanine, cytosine and thymine. DNA is a store-house of genetic information and passed from generation to generation. As you would learn in this chapter, the genetic information in DNA is decoded into RNA that ultimately is translated into functional proteins. The manifestation of this activity, i.e., decoding the genetic information is the central dogma of molecular biology (Figure 65.1) as proposed by Francis Crick in 1956. The first step in this sequence is transcription, by which a new polymer called Ribonucleic acid (RNA) is generated from the DNA molecule.

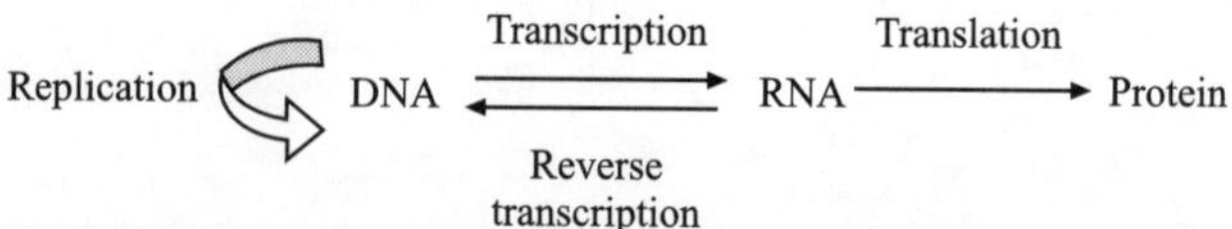

Figure 65.1 Flow of genetic information, called the Central Dogma, involves, DNA replication, synthesis of mRNA from DNA by transcription and synthesis of proteins by translation.

In 1961, Francois Jacob and Jacques Monod extended Crick's hypothesis that the RNA intermediate, which they dubbed messenger RNA or mRNA, would have the following properties, viz., the base composition of RNA would reflect the base composition of the DNA, would have very heterogeneous molecular mass, would be associated with ribosomes and would

"

have a high rate of turnover. These were profound views that have been proven correct by rigorous experiments.

II. RNA—THE PRODUCT OF TRANSCRIPTION

The living cell, whether a prokaryote or eukaryote contains three major types of cellular RNA: ribosomal RNA (rRNA), transfer RNA (tRNA) and messenger RNA (mRNA). In addition, there are other RNAs, involved in DNA synthesis and processing of eukaryotic RNA precursors.

A. Ribosomal RNA

Ribosomal RNA constitutes the bulk of cellular RNA (≥ 80 per cent) and is metabolically stable. The ribosomal RNA gene called rDNA exists in tandem multiple copies in all genomes studied so far. Not every copy of the rDNA gene is transcribed; rather some copies are silenced by the chromatin structure of the individual rDNA gene. The RNAP I transcribes a long precursor rRNA, which is then processed through complicated cleavage steps to yield mature 5.8S, 18S and 28S rRNA. It is an integral component of ribosomes, the cellular organelles responsible for protein synthesis. Prokaryotic ribosome (70S) has two subunits, 50S and 30S, whereas in eukaryotes, ribosome (80S) contains 60S and 40S subunits. The 50S subunit contains 23S and 5S rRNAs and 35 different ribosomal proteins. The 30S subunit contains 16S rRNA and 21 different ribosomal proteins. Eukaryotic ribosome contains larger RNAs (28S, 18S, 5.8S and an additional 5S). Ribosomes are also present in mitochondria and chloroplast. But rRNAs from these organelles have high sequence similarity to those of the prokaryotes.

B. Transfer RNA

tRNA accounts for ~15–20% of cellular RNA and they function as adaptors for amino acids required for protein synthesis. The tRNA genes constitute a large family, of about 274 genes in yeast, having one or more tRNA gene for each of the tRNA codon specificity. They are transcribed by RNAP III as a large precursor tRNA that undergoes both 5' and 3' processing by specific ribonucleases to generate mature tRNA molecules.

C. Messenger RNA

Cellular mRNA constitutes only ~1–5% of total RNA. However, given the critical requirement of the quality and quantity of mRNA, most studies have been carried out on understanding mechanism and regulation of mRNA synthesis. In prokaryotes, protein synthesis occurs cotranscriptionally, even while mRNA synthesis is being completed (Figure 65.2). In contrast, in eukaryotes, where the processes of transcription and translation occur in two different compartments, newly synthesized RNA undergoes processing, usually at both ends (capping at 5'-end and polyadenylation at 3'-end) and frequently internally,

before it is transported to the cytoplasm as mature mRNA. A comparative account of the features of prokaryotic and eukaryotic mRNA is summarized in Table 65.1.

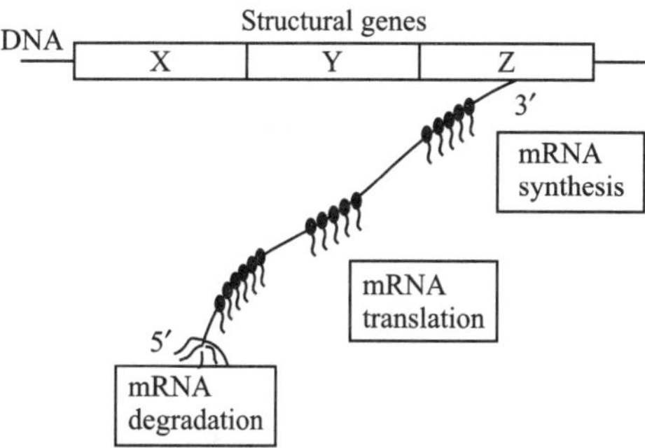

Figure 65.2 Coupled transcription and translation in prokaryotes and the mRNA degration.

TABLE 65.1 **Some Properties of Prokaryotic and Eukaryotic mRNAs**

Prokaryotes	Eukaryotes
mRNA is polycistronic, i.e. it represents a group of adjacent genes	mRNA is monocistronic, i.e. it represents a single gene
mRNA is very unstable. Degradation of mRNA occurs before one round of translation completes (No reinitiation)	mRNA is stable and can be reused for translation (reinitiation)
No poly (A) tail is present	mRNA has a sequence of poly adenylic acid at the 3'-end (60–200 A-residues)
No cap structure is present at the 5'-end	mRNA has a methylated cap at the 5'-end (m^7GpppG)
No mRNP formation occurs	mRNA is associated with proteins in the form of a ribonucleoprotein particle (mRNP)

D. Other RNAs

(i) Small RNAs synthesized by RNA polymerase (the main transcribing enzyme) or DNA primase (a special enzyme) are involved as primer in DNA synthesis. (ii) RNAse P, a processing ribonuclease contains a small RNA (350 nucleotides long) which is essential for the processing of precursor tRNA. (iii) Small nuclear RNAs (snRNAs) complexed with proteins are present in nuclei of eukaryotic cells. These RNAs function in a variety of nuclear processes including pre-mRNA splicing. (iv) Small nucleolar RNAs (snoRNAs) function in processing and chemically modifying rRNAs. (iv) MicroRNAs (miRNAs) are transcribed by RNAP II, and regulate gene expression generally by base-pairing with cognate mRNAs and blocking translation. (v) Small interfering RNAs (siRNAs) transcribed by RNAP II, down regulates gene expression by base-pairing with selective mRNAs and target them for degradation by a pathway involving a complex protein machinery called the Dicer. (vi) Other noncoding RNAs function in diverse cell processes such as telomere synthesis and the long Xist RNA (~17 kb in human) involved in X-chromosome inactivation.

III. CORE PROMOTER, UAS AND ENHANCER ELEMENTS

Gene structure comprises the coding sequence, flanked by non-coding regulatory sequences, whose sizes could vary quite dramatically from one gene to another. The upstream promoter DNA is required for transcription initiation as well as for regulation of transcription. A critical subset is the core promoter DNA, which is essential for accurate transcription, encompasses the transcript initiation site, the conserved DNA sequence motifs (regions of sequence-specific DNA-binding) surrounding the transcription initiation site. In addition, prokaryotic and eukaryotic promoters also contain one or more conserved sequence motifs called Transcription Factor Binding Sites (TFBS) that are sites for binding of transcriptional activator and repressor proteins, collectively known as transcriptional regulators. Because DNA motifs are highly conserved, the location of a promoter in a genomic sequence can be predicted by identifying such regulatory DNA motifs. It is important to note that presence or absence of a regulatory site is in itself not an indicator of a functional significance of the site or the cognate transcriptional regulator. The regulatory sequence elements (TFBS) in the prokaryotic and the yeast (lower eukaryotic) promoters largely reside within the upstream 500-600bp. The TFBS in eukaryotic promoters are either located close to the core promoter called the Proximal promoter or several kilobases away called Enhancers located upstream or downstream of the Transcription Start Site (TSS).

A. Prokaryotes

Three major types of bacterial promoters have been identified based on the arrangement of the conserved sequence motifs. The core promoter contains the −10 element (also called the Pribnow box; 5'TATAAT3'), the −35 element (5'TTGACA3') and the UP element, located further upstream in the promoter. Although these motifs are highly conserved, it is important to note that the exact sequence of the −10 and −35 elements could differ depending on the promoter. The −10 and −35 elements are recognized and bound specifically by the σ subunit of the RNAP, and the UP element is bound by the C-terminal domain of the α subunit of RNAP. Whereas most promoters contain both the −10 and −35 elements, some promoters such as the rRNA gene promoters also contain the UP element for more stable RNAP binding. The *gal* gene promoter in *E. coli* does not contain the −35 element; instead has an additional short sequence element upstream of the −10 element called the extended −10. As you would read later in this chapter, the transcripton of the *E. coli lac* genes in response to carbon source is controlled by two sequence-specific DNA binding proteins the CAP activator and the Lac repressor, which bind to the CAP site and the *lac* operator sites respectively.

B. Eukaryotes

The core promoter in eukaryotes (Figure 65.3) is much more complex than that of the bacterial promoters. The eukaryotic promoters have been classified as those having TATA elements or those with TATA-like sequences. Yeast core promoters have 5'TATAWAWR3', with W and R reprenting A or T and A or G respectively as the TATA element. Interestingly, only 15–20% of the yeast promoters contain this TATA sequence, and rest of the core promoters have 1 or 2 mismatches in this consensus sequence. The core promoters of mammalian cells also have this distribution of TATA and TATA-like elements. Metazoans core promoters contain a TSS denoted as +1, amidst a consensus Initiator (Inr) sequence Py-Py-A-N-T/A-Py-Py, with the conserved A being the +1 site, and Py represents any pyrimidine and N stands for any nucleotide. It is important to note that other eukaryotes also have a designated TSS but the Inr sequence is vastly different. Moreover, the metazoan core promoters are characterized by additional features—those lacking CpG in the vicinity of the Inr and others bearing multiple CpG residues. Interestingly, the promoters lacking CpG residues have precise TSS, and those bearing CpG have multiple TSS. An associated feature is that the CpG-lacking core promoters, but not the CpG-bearing promoters, generally contain canonical TATA element. Studies on core promoters in Drosophila led to the identification of an additional conserved motif RGWCGTG, called Downstream Promoter Element (DPE) located 25–35bp 3' to the Inr, and the Motif Ten Element (MTE), just upstream to the DPE. The DPE functions in conjunction with the Inr sequence. Metazoans also contain metazoan-specific Downstream Core Element. In summary, it is important to note that the eukaryotic core promoter elements are diverse sequence elements and provide diversity to gene promoter function.

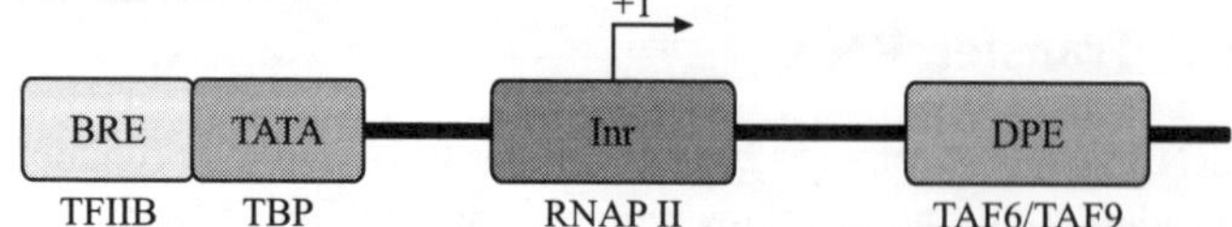

Figure 65.3 Core Promoter structure in eukaryotes.

IV. ENZYMOLOGY OF TRANSCRIPTION

RNA polymerase is the enzyme that is capable of joining ribonucleoside triphosphates by 3'–5' phosphodiester bonds under the direction of a DNA template. It is important to note that, unlike during DNA replication, transcription is initiated by RNAP without a primer.

$$NTP + (NMP)_n \xrightarrow{} (NMP)_{n+1} + PPi$$
$$\text{RNA} \qquad \qquad \text{elongated RNA}$$

The choice of the nucleotides that are to be added to the growing RNA chain is determined by base pairing with the DNA template. One of the strands of DNA would be copied to yield a single-stranded RNA molecule whose nucleotide sequence is complementary and antiparallel to that of the copied DNA strand (Figure 65.4). This DNA strand may be called the 'sense' strand or 'non-coding' strand and its complement the 'non-sense' strand or 'coding strand'. Elongation proceeds in a 5′ to 3′ direction by adding ribonucleotide units to the 3′-OH end of the RNA chain. The overall reaction rate is ≥ 40 nucleotides/second at 37°C (for bacterial RNAP).

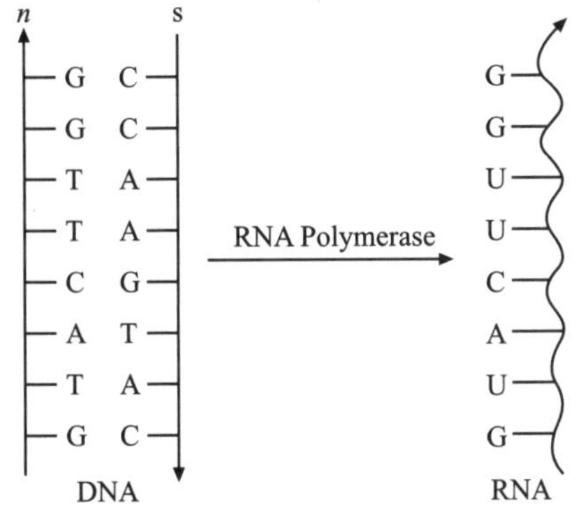

Figure 65.4 Transcription of 'Sense' strand of DNA by RNA polymerase. Note that in RNA, Uracil is incorporated in place of Thymine found in DNA.

As each nucleotide is added to the chain, a pyrophosphate is split off from the incoming nucleoside triphosphate precursor. Only the nucleotide at the 5′-end of the RNA molecule retains its triphosphate group (Figure 65.5).

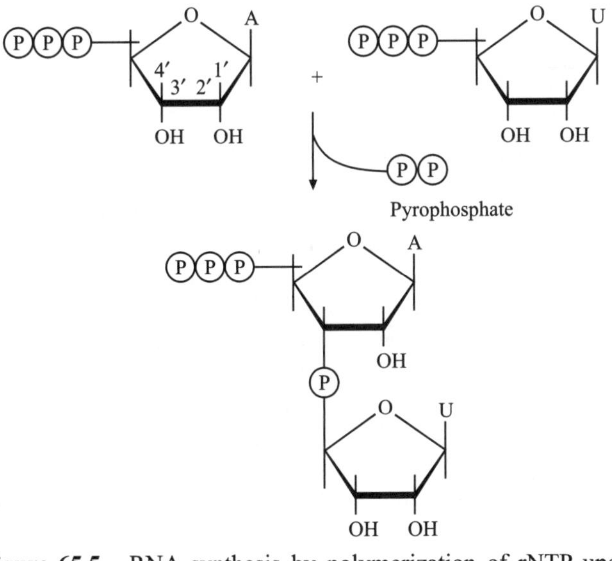

Figure 65.5 RNA synthesis by polymerization of rNTP upon pyrophosphate release and 3′ 5′phosphodiester bond formation.

A. Prokaryotic RNA Polymerase

A single DNA-dependent RNA polymerase, first discovered by Michael Chamberlain, synthesizes all RNAs in prokaryotes, with the exception of the short RNA primers formed during DNA replication by the DNA Primase enzyme. RNAP is composed of α, β, β' and σ subunits in the 2:1:1:1 ratio called the holoenzyme. The purified holoenzyme is capable of correct and efficient synthesis of RNA from a DNA template (Figure 65.6). The holoenzyme can be reversibly separated into two components by phosphocellulose chromatography; the σ polypeptide is released and fails to bind to the phosphocellulose, whereas the remaining core polymerase binds tightly.

Figure 65.6 The composition of core and holo RNA polymerase from *E. coli*. The σ subunit is the specificity factor.

The core polymerase has RNA synthetic activity but cannot initiate transcription at the correct initiation sites (promoters) on the DNA. The other component, the σ subunit confers specificity for RNAP, as without the σ subunit, the core polymerase produces non-specific transcripts from both strands of the DNA molecule. The *E. coli* RNAP has been purified and high resolution crystal structure has been obtained. The molecular details of the RNAP, including that of the σ subunit, interaction with DNA template revealed hitherto unseen details of how the RNAP subunit binds to the UP element, −35 and −10 regions of the promoter. The multiple α-helical regions of the σ subunit makes base-specific contacts to confer specificity. The σ, therefore, is necessary for efficient initiation at the natural promoter sites on the DNA. In addition to the major form of Sigma, σ^{70}, *E. coli* has several variants that can replace σ^{70} under special growth conditions to direct the RNAP to alternative promoters. Moreover, substitutions of sigma factors during the morphological shift of certain bacteria from vegetative phase into a dormant or spore (sporulation) and lytic infection by a bacteriophage, alter the transcriptional specificity and consequently, the expression of new genes occurs (Table 65.2).

B. Eukaryotic RNA Polymerases

Eukaryotic cells contain at least three different RNA polymerases, each responsible for synthesizing a different class of RNA. This defining discovery was made in the laboratories of Robert Roeder and Pierre Chambon. A single RNAP, however, is found in mitochondria and chloroplast. The location and other properties of these polymerases are listed in Table 65.3. All these polymerases have a complex subunit structure of about 10–12 subunits, compared to the five subunit *E. coli* RNA polymease. However, there is sequence conservation between the *E. coli* RNA polymease subunits β and α with subunits in each of the three eukaryotic polymerases. Because each of the polymerases carries out the basic RNA polymerization activity, five of the subunits are shared in RNAPI, RNAPII and RNAPIII (Figure 65.7). X-ray crystal structure showed that the yeast RNAP II and bacterial RNAP have large molecular

TABLE 65.2 Different *E. coli* Sigma Factors that Recognize Promoters with Different Promoter Elements

Factor	Gene	Mass	Use	−35 region	Separation	−10 region
σ^{70}	*rpoD*	70,000	General	TTGACA	16–18 bp	TATAAT
σ^{H}/σ^{32}	*rpoH*	32,000	Heat shock	CCCTTGAA	13–15 bp	CCCGATNT
σ^{E}/σ^{24}	*rpoE*	24,000	Heat shock	?	?	?
σ^{N}/σ^{54}	*rpoN*	54,000	Nitrogen starvation	CTGGNA	6 bp	TTGCA
σ^{F}/σ^{28}	*fliA*	28,000	Flagellar	CTAAA	15 bp	GCCGATAA

TABLE 65.3 Properties of Eukaryotic RNA Polymerases

Properties	Pol I	Pol II	Pol III
Location	Nucleolus	Nucleoplasm	Nucleoplasm
RNAs synthesized	5.8S, 18S and 28S rRNA	mRNA, snoRNA, miRNA, siRNA and most snRNA	tRNA, 5S RNA, and some snRNA
Sensitivity to α-amanitin	Resistant	very sensitive	moderately sensitive
Proportion of cellular activity	50–70%	20–40%	10%

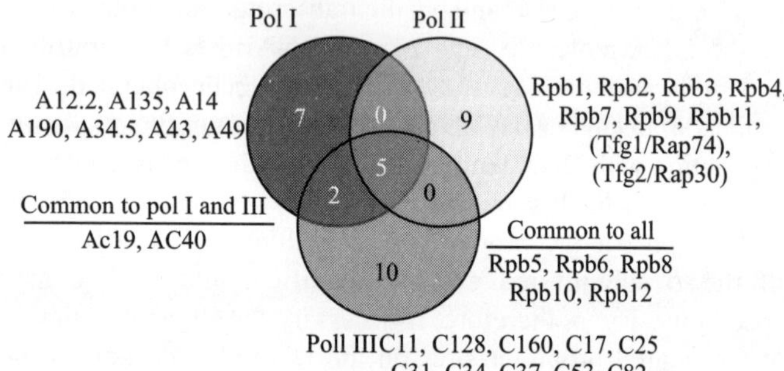

Figure 65.7 Unique and Shared Subunits of the Eukaryotic RNA Polymerase I, II and III. As per the accepted naming convention, the RNAP I specific subunit names have A prefix, that of RNAP III have C prefix, and that of RNAP II with Rpb prefix. The shared RNAP I and III subunits have AC prefix. (*see Plate 25 for colour figure*)

overlap. Although it is not known how each of the 12 subunits contribute to RNA synthesis, the largest subunit of the RNAP II of the yeast *Saccharomyces cerevisiae* has been well-studied.

The carboxyl-terminal domain (CTD) of the largest RNAP II subunit Rpb1 in yeast contains 26 copies of the heptad YSPTSPS amino acid sequence. The Rpb1 in human contains 52 copies of the YSPTSPS heptad amino acid sequence. The heptad repeat is essential for the growth of yeast cells. Indeed the Ser residues in the heptad sequence are phosphorylated differentially during the initiation and elongation phases of transcription. A number of inhibitors of RNA synthesis have been reported and some of them are listed in Table 65.4. Recent genome sequence analysis as well as biochemical experiments have shown that, in addition to RNA polymerases I, II and III, plants have multi-subunit RNA polymerases IV and V involved in transcription of small regulatory RNA molecules.

V. STEPS IN mRNA SYNTHESIS

A. Prokaryotic Transcription

The synthesis of RNA involves (i) binding of RNA polymerase at promoter sites and RNA chain initiation, (ii) chain elongation, and (iii) chain termination. The first step in transcription is the binding of holoenzyme to a specific site in the DNA, called the promoter site. Some short stretches within the promoter appear to be conserved. These are candidates for the signals recognized by RNAP. As discussed above, the six-base pair −10 element (TATAAT) is first recognized by RNAP. This initial binding of RNAP to the double-stranded (ds) promoter DNA is called the closed complex. The next step is the isomerization

TABLE 65.4 Inhibitors of RNA Synthesis

Inhibitors	Nature and Source	Mode of Action
Actinomycin D	Antibiotic produced by *Streptomyces antibioticus*	Insertion (intercalation) of the phenoxazone ring between two G-C base pairs of DNA template.
Rifampicin	Synthetic derivative of a naturally occurring antibiotic rifamycin (Rifamycin is derived from *Streptomyces*)	Binds tightly to the β-subunit of RNA polymerase
α-Amanitin	Toxic substance of poisonous mushroom *Amanita phalloides*	Binds to RNA polymerase II at a low concentration.
Cordycepin (3′-deoxyadenosine)	Substrate analog	Chain termination

leading to the melting of the dsDNA between −11 and +3 with respect to TSS so that the template and non-template strands are locally separated right within the active site of the RNAP. This step leading to the open complex formation involves structural changes in the enzyme.

For mRNA synthesis to begin, the ribonucleotides enter the active site of RNAP through a channel in the enzyme molecule and engage by base-pairing with the template strand at the TSS. The first, or initiating nucleoside-5′-triphosphate, which is usually ATP or GTP, binds to the enzyme. Thus, a new RNA chain has a triphosphate group at its 5′-terminus and a free hydroxyl group at its 3′-terminus.

The enzyme synthesizes several rounds of short RNA strands of upto 10nt called abortive transcripts that are released without the enzyme disengaging from the TSS. However, upon synthesis of RNA longer than 10nt, the RNAP forms a stable ternary complex with the template and the nascent RNA chain. This process commits the enzyme to the elongation step of transcription. After initiation has occurred, chain elongation proceeds in the 5′ → 3′ direction by the successive binding of nucleoside triphosphates complementary to the base in the template strand. As the next section of DNA unwinds, the region of DNA that has been transcribed, regains its double helical conformation (Figure 65.8). The RNAP also carries out two proofreading functions—pyrophosphorolytic editing and hydrolytic editing. Both these processes lead to the removal of the error-containing nucleotide(s) in the nascent RNA chain. The σ subunit dissociates from the rest of the enzyme soon after several polymerization steps and is then reused by another core polymerase for initiation. The core polymerase continues to transcribe the DNA template (Figure 65.9).

The termination step requires the breakage of all hydrogen bonds holding the DNA and RNA hybrid. Defined sequences called terminators stimulate the elongating polymerase to cease transcription by disengaging with the template strand and release

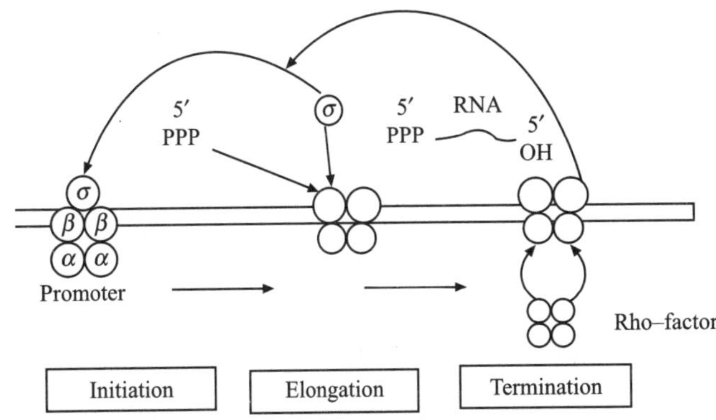

Figure 65.9   Transcription cycle involving recycling of σ factor.

of the RNA chain. There are two kinds of terminators, called Rho-independent and Rho-dependent. The Rho-independent DNA terminator is a stretch of G-C rich inverted repeat sequences with dyad symmetry, flanked by eight A:T base pairs, can be recognized by RNAP itself and can terminate transcription without any additional factors. In this case, the terminator is transcribed, and the ensuing hairpin structure in the RNA is believed to disrupt the elongation complex. The Rho-dependent terminator has less defined RNA elements and requires the Rho factor to terminate transcription. As the RNA exits the polymerase, the oligohexameric Rho protein binds to the single-stranded RNA and using the energy of ATP hydrolysis releases the RNA chain from the DNA-RNAP complex. The specificity of termination is determined by RNA sequence itself, called *rut* sites, of about 40nt that is rich in C residues and do not adopt a strong secondary structure. Termination factors such as NusA, NusB, S10, NusG have also been reported to be involved in rho-dependent termination. Another type of terminator signal is called the attenuator, which precedes a structural gene or between two structural genes. These signals

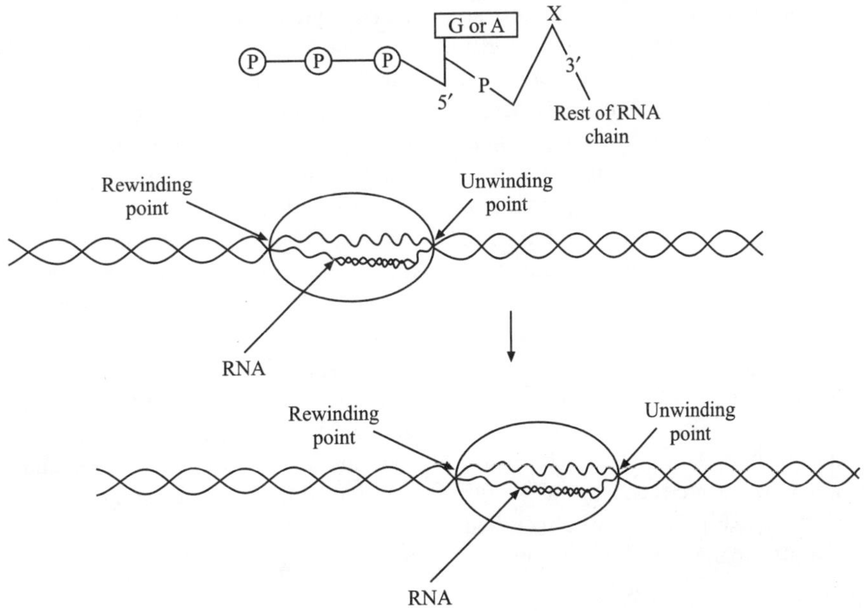

Figure 65.8   Progress of RNA polymerase during transcription.

control gene expression by selective termination of the distal portions of an operon. This phenomenon (attenuation) is widespread among the bacterial operons that are involved in the biosynthesis of amino acids. One of the best studied examples is that of the tryptophan operon. The efficiency of termination depends on the growth conditions of the cell. At a typical terminator with a stretch of U bases, the polymerase pauses, dissociates from the template and releases the RNA chain.

B. Eukaryotic Transcription

Transcription initiation in eukaryotes involves several different initiation factors to assemble the RNA polymerase II to the TSS. Elegant biochemical studies led to the understanding of the steps involved in the assembly of a large multisubunit preinitiation complex (PIC) containing RNAP II and the general transcription factors or GTFs. The assembly process begins with the binding of the TFIID complex, a GTF, composed of TBP and 14 TBP-associated factors (TAFs), to the TATA sequence within the core promoter. The TFIID subunits TAF6 and TAF9 also interacts with the core promoter DNA (Figure 65.3). The TFIID binding leads to bending of the DNA by an 80° angle and this distortion is believed to mark active promoter and to enable subsequent protein assembly both upstream and downstream of the TATA. High resolution X-ray crystal structure of TBP-TATA-TFIIB showed the atomic level details of the interaction of TBP with the TATA element as well as the interaction of the GTF TFIIB with the promoter as well as with TBP. TFIID binding is followed by TFIIA and TFIIB association, both of which interact with TBP as well as with the core promoter DNA. The RNAP II complex along with TFIIH, followed by other GTFs TFIIE and TFIIH assemble to complete the assembly of the preinitiation complex. The PIC is now poised for initiating transcription.

The key trigger for initiation is provided by the TFIIH kinase subunit, which phosphorylates the serine-5 in the CTD (see Section IV.B.). This initiates a series of conformational changes leading to RNAP II activation. Interestingly, shortly after the polymerase transcribes a short RNA (~12nt) and close to the beginning of the coding sequence, the polymerase movement is paused, which is known as promoter proximal pausing, mediated by a set of regulatory proteins that halt or pause elongation. Interestingly, bulk of the RNAP II bound in a eukaryotic genome is in a poised state indicating a general mechanism for priming the genome to be ready to transcribe. This step is reversed by the elongation factors such as P-TEFb. It is important to recognize that these studies were conducted *in vitro* using purified DNA template and purified RNAP II and GTFs. The PIC assembly scheme outlined above is very simplistic and has been seriously challenged in recent times by the discovery of the critical importance of the chromatin environment of the DNA *in vivo*. Moreover, several of the GTFs are also associated with RNAP II and may be recruited as a partial GTF-RNAP complex, called the holoenzyme.

The RNAP II transcription elongation does not proceed smoothly along the coding sequence. During elongation the RNAP, in both bacteria and eukaryotes, is associated with elongation factors to assist the movement of RNAP though the coding sequences. In eukaryotes, the elongation factors also facilitate the RNAP II movement through the nucleosomes in the coding sequence. In addition, RNAP movement creates superhelical region in the coding sequence leading to positive DNA supercoiling and negative supercoiling behind the polymerase. The supercoiling is released by DNA topoisomerase and DNA gyrase in eukaryotes and prokaryotes respectively.

Protein coding regions (exons) of genes in eukaryotes are typically interrupted by noncoding intervening sequences called introns. This was discovered in 1977 independently in the laboratory of Phil Sharp and Richard Roberts, wherein DNA:RNA hybrid of the adenovirus DNA genome and its mRNA were found to be non-colinear as observed in an electron microscope. Indeed subsequent studies established the prevalence of split genes in all eukaryotes with only variability between genomes only in the number of intron per gene and the length of introns.

VI. POST-TRANSCRIPTIONAL PROCESSING OF RNA

RNA molecules synthesized by RNAP are chemically modified or split in a highly selective manner. These post-transcriptional events are called maturation or processing. For example, the product of transcription of rRNA gene in mammalian cells is a 45S precursors, which is then methylated and cleaved into 28S and 18S rRNA molecules. It is now clear that a long rRNA precursor is cut up to yield shorter finished molecules in several eukaryotic cells, like those of yeast, plants, frogs and mice. In bacteria, rRNA precursors are cleaved and methylated (Figure 65.10). Some striking post-transcriptional modifications occur in the biosynthesis of tRNA. For example, *E. coli* tyrosyl tRNA is formed by a cleavage of 41 residues from the 5′-end and 2 bases from the 3′-end of precursor tRNA. Furthermore, tRNA molecules contain a monophosphate rather than a triphosphate at the 5′-end terminus. Some unusual bases (*'unusual'* means a base other than A, G, C or U) also arise by the enzymatic modification of a pre-existing polynucleotide (Figure 65.11).

Thus, it can be seen in many types of cells that rRNA and tRNA are processed into a functional molecule from a precursor RNA. It was interesting to know if messenger RNA was also processed similarly. There was less likelihood of the prokaryotic mRNA being processed, since the ribosomes translate the mRNA while it is still being transcribed. In eukaryotic cells, on the other hand, where transcription and translation are separated in space and time, mRNA processing was a distinct and interesting possibility.

Eukaryotic mRNA have a 5′ Cap (m^{7}GpppG) and a 3′ poly(A) tail (60–200 A-residues). mRNA is first transcribed as a precursor of a very long molecule called heterogenous nuclear RNA (hnRNA). As soon as about 20 bases are transcribed in

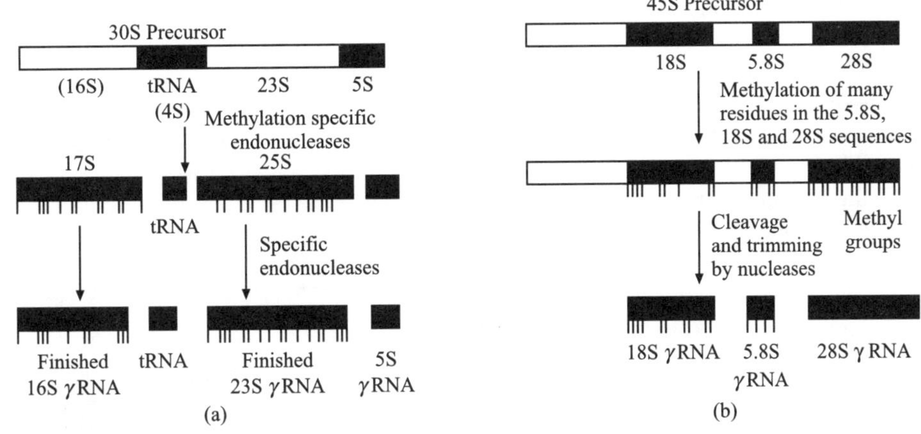

Figure 65.10   Processing of pre-rRNA molecule into mature rRNA: (A) Prokaryotes and (B) Eukaryotes.

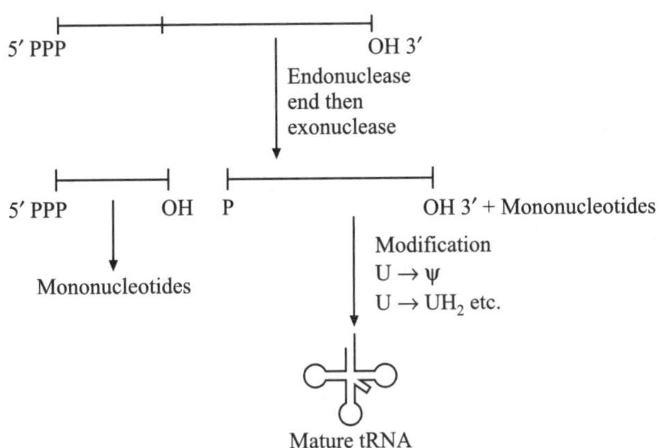

Figure 65.11   Processing of pre-tRNA into a mature rRNA.

the hnRNA, a methylated guanosine block, called a Cap, is added at the 5′-end. The RNA transcript is cleaved at about 20 bases past the signal sequence (AAUAAA) and to this 3′-end, the poly(A) residues are added by an enzyme, called poly(A) polymerase (Figure 65.12). In further processing, the end structure is retained while the interior of the transcript is modified.

Evidence for the shortening of the hnRNA was first made available from the study of Adeno-virus which has a genome of double stranded DNA, that causes upper-respiratory tract infections in human beings. It was conclusively shown that several 'spacer' sequences that are not part of mature mRNA are cleaved off and spliced to give a functional mRNA (Figure 65.10). So, it is evident that genes in eukaryotic cells can be regulated by means of RNA processing. However, the physiological significance of such processing is not clear.

VII.   CHROMATIN AND TRANSCRIPTION

Chromatin organization makes a huge impact on the efficiency of transcription. *In vitro* experiments showed that TBP binding to TATA is largely inhibited when the TATA is within a nucleosome. Indeed, nucleosomal configuration is inhibitory for transcription factor binding, PIC assembly and transcriptional elongation steps *in vivo*. Thus, chromatin reorganization or reconfiguration is an important step in transcriptional regulation. Indeed nucleosomes have a very dynamic structure modulated by two processes—chromatin remodeling and covalent histone modifications. Chromatin remodeling refers to any one of the

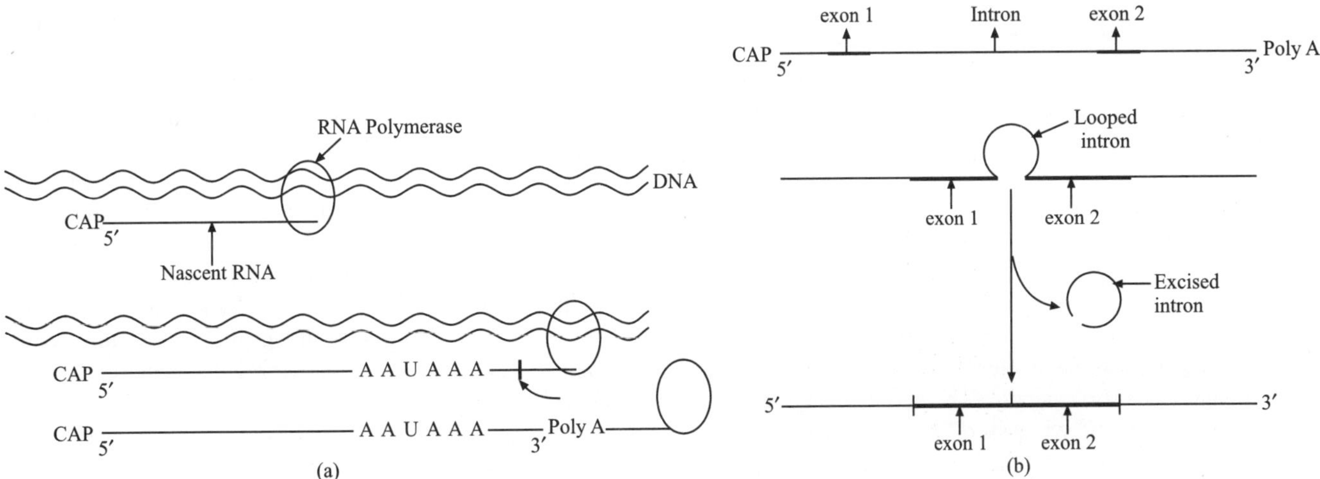

Figure 65.12   Processing of mRNA. (A) Capping and polyadenylation of mRNA. (B) Splicing of mRNA.

steps leading to displacement, repositioning or transplacement of histones in part or nucleosomes as a whole.

Eukaryotes have evolved a very elaborate chromatin remodeling machinery involving ATP-dependent helicases represented by the prototypical yeast SWI/SNF chromatin remodeling complex. The core histones H2A, H2B, H3 and H4 have a highly conserved, unstructured amino-terminal region that undergoes post-translational modifications. For instance, histone H3 has conserved lysines at position 4, 9, 14, 18, 23, 27 and 36. The lysine-4 is subjected to methylation by lysine methyl transferase, KMT, and lysine-9 is acetylated by lysine acetyl transferase or histone acetyl transferases (KATs or HATs). Several different KATs and KMTs have been identified. GCN5 is the first identified transcription-linked HAT/KAT enzyme. Similarly, histone deacetylase and histone demethylases have been identified that participate in the reversal process. Each of the enzymes is recruited to specific sites on the chromatin at distinct phases/times during transcription. The modifications of the histones are carefully regulated, and serve to regulate chromatin structure in defined ways. The acetylation would remove the histone tail from interactions with the positive charge on the DNA backbone, thereby opening up the histone tails for interactions with other chromatin modifying/remodeling machinery. In this context, acetylation of lysine-9 by a HAT allows the recruitment of the SWI/SNF remodeling complex. In fact, the acetylated histones are recognized by histone "readers" such as the bromodomain of the Swi2/Snf2 protein. Thus the chromatin remodeling is intimately linked to histone modifications and regulates transcription. Another fundamental issue here is the mechanism of recognition of the promoters bye these chromatin factors. A large body of work from yeast to humans demonstrated that the job of recruitment of the specific histone modification enzyme complex is carried out by sequence-specific DNA binding transcription factors. An example of such a regulatory mechanism is discussed below.

VIII. TRANSCRIPTIONAL CONTROL

A. Transcriptional Regulation in *E. coli*

In bacteria, gene activity is regulated primarily at the level of transcription. This transcription can be inhibited (negative control) or stimulated (positive control) by protein factors. Negative control is mediated by a repressor molecule binding to a specific site (operator) on a DNA molecule. A mild degree of overlap exists between the promoter and operator sites. The presence of a repressor at operator sites therefore prevents the binding of RNAP to promoter site. Repressor molecules exist in two forms, active and inactive (Figure 65.13). The binding of inducer (small molecules, e.g., β-galactoside that cause the production of enzymes) inactivates the repressor and therefore, repressor cannot bind to operator. The classic example is the mechanism regulating the synthesis of β-galactosidase. When *E. coli* is transferred from glucose medium to a medium containing β-galactoside, the bacterium synthesizes β-galactosidase, an

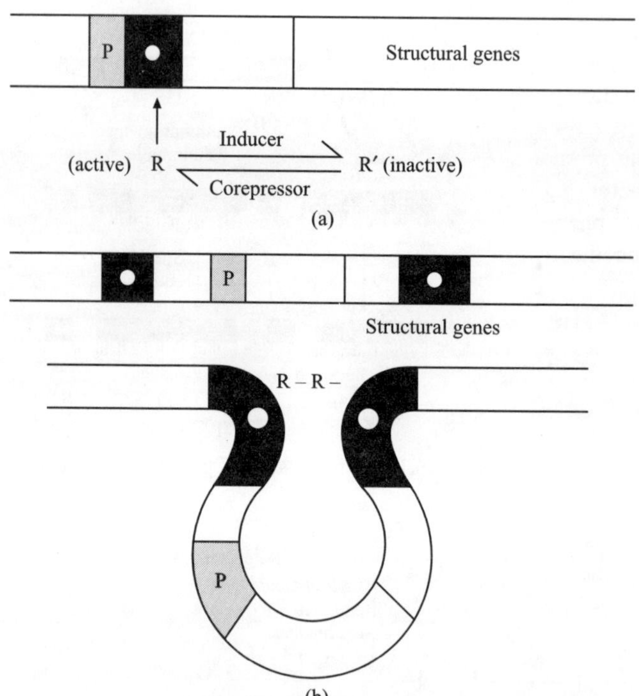

Figure 65.13 Transcriptional regulation in prokaryotes. (a) Activation or inactivation of the repressor (R) by interaction with corepressor or inducer. (b) Repressor binds to operators rendering the promoter nonfunctional.

enzyme which is responsible for the metabolism of lactose. Two other proteins, namely galactoside permease and transacetylase are also synthesized. This is achieved by switching on the synthesis of lac mRNA by inducer. In contrast, the corepressor (small molecules, e.g., tryptophan that prevent the production of enzymes) changes an inactive repressor into an active form. For example, the synthesis of five enzymes of the tryptophan (*trp*) operon occurs only when the tryptophan is limiting. Interestingly, the enzymes of arabinose (*ara*) operon are controlled by a protein (*ara C*) which acts either positively or negatively. In association with inducer (arabinose), this protein binds to promoter to switch on the genes; on the other hand, without inducer, it binds to promoter, thereby preventing transcription. Adhya and co-workers reported a novel type of transcriptional regulation. The *gal* operon in *E. coli* which codes for the proteins that are involved in galactose metabolism harbours an operator upstream of the promoter. Also, it has another operator sequence within the first structural gene. The two operators that flank the promoter region, are brought together by repressor binding and this causes the promoter to loop out, thereby physically excluding the RNAP from binding to the promoter (Figure 65.13).

Sigma factor acts as a positive regulator and binds to the promoter sequence for selective initiation of transcription. Besides sigma factor, positive control is also exerted by the binding of the catabolite activator protein (CAP) in the presence of cyclic AMP (cAMP) to a site in the promoter region. At each promoter, the CAP binding is upstream from the RNAP site. In *in vitro* transcription reactions, CAP-cAMP enhances the binding of RNAP to promoters. It is believed that the binding of CAP-cAMP to the CAP site facilitates RNAP attachment to the promoter. Another positive element, i.e., anti-termination factor (e.g., 'N'-gene product of lambda phage) which controls the phage gene expression has been reported. This protein specifically allows RNAP to read through the terminator.

B. The *β*-interferon Gene—an Example from Metazoans

We will next examine the mechanism of transcriptional regulation in a eukaryotic gene, the human β-interferon (IFN-β) gene. The activation of IFN-β gene is carried out in a series of recruitment steps. The early events initiate the binding of the transcription factor NFκB to the enhancer followed by the ATF-2/c-Jun and IRF3/7 factors in a highly ordered cooperative manner. This leads to the assembly of a complex of transcription regulatory proteins at the enhancers called enhanceosome. This complex then recruits the PCAF histone acetyl transferase, which acetylates the histone tails of the nucleosomes nearby. This leads to the recruitment of the CBP coactivator and consequent release of the PCAF. The CBP then recruits the pre-initiation complex containing RNAP II and GTFs. The SWI/SNF complex is also assembled by the interaction with the acetylated nucleosome, and remodels the nucleosomes in the vicinity of the TSS. This step then enables the TFIID complex to be assembled at the TATA sequence leading to the formation of the complete initiation complex and transcriptional activation of the promoter. Although the exact details of this mechanism is not found in other human or eukaryotic promoters, but the basic principles underlying transcriptional activation is well represented in this model of orchestrated transcriptional activation in response to particular stimulus in eukaryotes from yeast to humans.

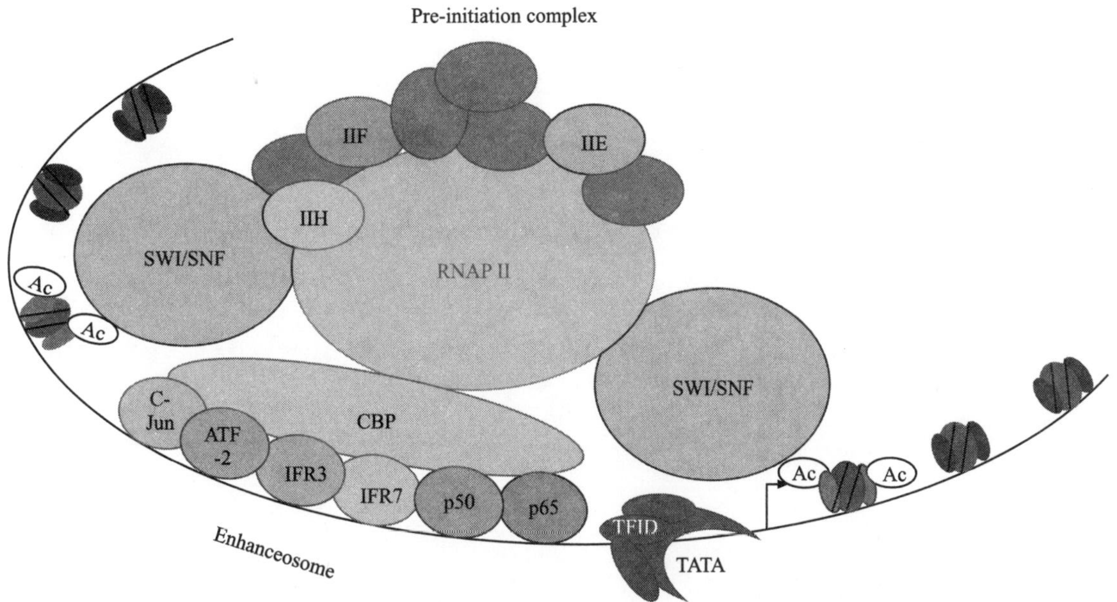

Figure 65.14 Transcriptional regulation in eukaryotes. Schematic view of the composition of the enhanceosome-mediated PIC assembly at the human IFN-b promoter. The transcriptional regulators c-Jun, ATF-2, IRF-3/7, NFkB (p60/p65) constitute a transcriptional regulator complex at the enhancer to form an enhanceosome. The PIC assembly at this b-IFN promoter takes place in a step-wise manner leading to the assembly of the initiation complex. (*see Plate 26 for colour figure*)

SUGGESTIONS FOR FURTHER READING

Alberts B., Johnson A., Lewis J., et al. (200), Molecular Biology of the Cell, 5th ed., New York: Garland Science.

Boyer P.D. (Ed.) (1982), *The Enzymes,* 3rd ed., Vol. 15B, Academic Press.

Breathnach R. and Chambon, P. (1981), *Ann. Rev. Biochem.,* 50, 349–383.

Chambon P. (1981), *Sci. Am.,* 244, 60.

Darnell J.E. Jr. (1982), *Nature,* 297, 365.

Darnell J.E. Jr. (1983), *Sci. Am.,* 249, 72.

Ford E. and Thanos D. (2010), Biochem, *Biochimica et Biophysica Acta* 1799, 328–336.

Latchman D.S. (1996), *N. Engl. J. Med.,* 334, 28–33.

Lewin B. (2004), *Genes,* 8th ed., Pearson Prentice-Hall, New Jersey.

Losick R. and Chamberlin M. (Eds.) (1976), *RNA Polymerase,* Cold Spring Harbor Laboratory, New York.

Losick R. and Pero J. (1980), *Cell,* 25, 582.

Maniatis T., Goodbourn, S. and Fischer J.A. (1987), *Science,* 236, 1237–1245.

Nevins J.R. (1983), *Ann. Rev. Biochem.,* 52, 441.

Workman J.L. and Kingston R.E. (1998), *Annu Rev Biochem.,* 67, 545–79.

Yanofsky C. (1981), *Nature,* 289, 751–758.

Young R.A. (1991), *Ann. Rev. Biochem.,* 60, 689–715.

66

Translation of Genetic Message

Asis Datta and Rakesh K. Tyagi

CONTENTS

The capability of an organism to continuously adjust its metabolic processes in response to the changing environment and time is essential for its existence. To understand the basic processes of life, such as growth and differentiation in living organisms, it thus becomes necessary to have a detailed knowledge of expression of genetic message contained in the genome or in a word, the 'gene expression'. In principle, protein synthesis is the last step in expression of genetic information. Here, messenger RNA (mRNA) is translated to yield a polypeptide chain whose amino acid sequence is determined by the nucleotide sequence of the mRNA.

I. GENETIC CODE

In protein synthesis, the information of the genes has to be translated from the language of the nucleic acids into the language of the proteins, a sequence of amino acids. There must, therefore, exist specific rules of translation and these are called the *genetic code*.

Proteins contain 20 different amino acids, whereas only four different nucleotides are present in DNA. It was, therefore interesting to know how the four base pairs of DNA controlled the sequence of 20 different amino acids in proteins. With the discovery of mRNA as intermediary, the question became how the four bases in mRNA could specify amino acid sequence of a polypeptide. What then is the nature of the genetic code relating mRNA base sequence (or DNA base-pair sequences) to amino acid sequences?

The idea of a triplet code is a natural consequence of the fact that 4 nucleotides either one-by-one or two-by-two ($4^2 = 16$) do not provide enough combinations to code for 20 amino acids. The minimum codon size is therefore three nucleotides, which would provide ($4^3 = 64$) codons that are sufficient to code for 20 amino acids. Genetic experiments later proved that the genetic code is a *triplet code*: three adjacent nucleotides specify an amino acid. For example, pUpUpU (or UUU for short) means phenylalanine. By genetic studies it is now known that genetic code is non-overlapping, commaless and degenerate. Non-overlapping means that the base of one codon is not used in another codon.

Commaless codon means no bases (Punctuations) are present between two codons:

Out of the 64 possible triplets of the four bases, 61 are used to code for amino acids and the remaining three are the nonsense codons, used as punctuation marks to terminate protein synthesis. Since 61 triplets have to specify only 20 amino acids, each amino acid except methionine and tryptophan is

"

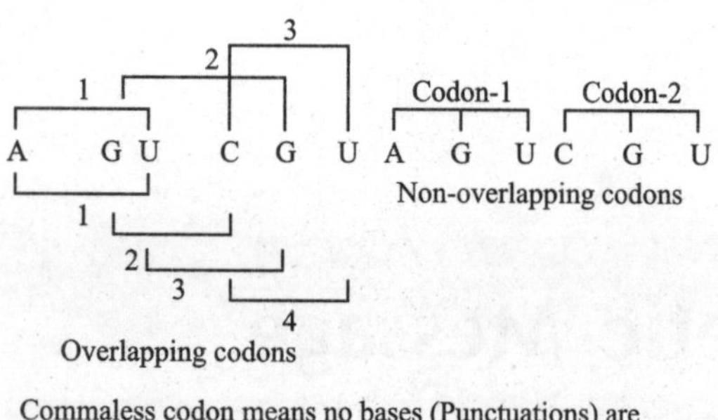

Commaless codon means no bases (Punctuations) are
present between two codons.

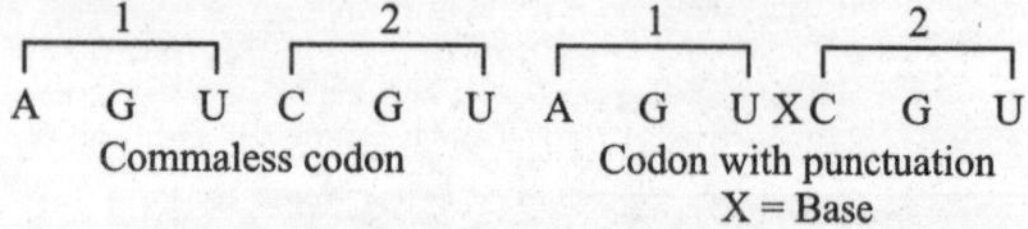

specified by more than one triplet, giving rise to the so called
degeneracy of the code. To explain such phenomenon, Crick
proposed a hypothesis called 'Wobble concept' (Table 66.1).

TABLE 66.1   The 'Wobble' Rules of Codon-Anticodon Pairing

5′ base of anticodon	3′ base of codon
C	G
A	U
U	A or G
G	C or U
I	U, C or A

According to this concept, third base of the anticodon (that
is base at the 5′-end of anticodon of tRNA) is less discriminating
than the other two which means that the pairing of the third base
of the codon has some steric freedom. However, a particular
codon always codes for a particular amino acid, hence there
is no ambiguity of codon.

Deciphering of genetic code

The key to success was discovered by Nirenberg in 1961. Most
cell extracts incorporate amino acids into protein for only a
few minutes. Nirenberg found that, in an extract from *E. coli*
that had apparently died, protein synthesis would restart if he
added some mRNA to it. This was an important discovery. It
made the study of protein synthesis and its regulation amenable
to the methods of enzymology.

Artificial synthesis of mRNA and their translation in
in vitro systems has greatly helped to understand codon
assignments. The discovery of polynucleotide phosphorylase
by Severo Ochoa, and Marianne Grumberg-Manago which
catalyzes the conversion of NDP (nucleotide diphosphates)
to polynucleotides paved the way to prepare synthetic
polynucleotides. Nirenberg and Matthaei reported that addition
of synthetic Poly(U) (polyuridilic acid) to a cell-free system
resulted in the synthesis of polyphenylalanine which means
that UUU codes for phenylalanine. Similar experiments using
homopolymers (e.g., Poly(A) synthesized polylysine and
Poly(C) synthesized polyproline etc.) and heteropolymers
(containing two different nucleotides) were performed. These

and other binding studies of trinucleotides to aminoacyl tRNA
have elucidated the complete genetic code. It is now known
that AUG (sometime GUG) which codes for methionine serves
as *initiation codon*. Hence, all the proteins have methionine as
first amino acid. The codons UAA, UAG and UGA are called
'terminator' codons as they are not read by tRNA, but instead
recognized by protein factors called release factors. Initially,
it was thought that genetic code is universal, i.e., a particular
codon specifies a particular amino acid in all the organisms.
However, some exceptions to the concept of universality are
reported. Such exceptions are found in mitochondria which
contains semi-independent genetic system. In mitochondria of
many species, UGA (which is a terminator codon) codes for
tryptophan. This stop codon also codes for selenocysteine, a
21st amino acid present in some proteins of eukaryotic and
prokaryotic origin. Similarly, in yeast mitochondria CUA
(codon of leucine) codes for threonine. Such changes have
been found in some other cases also. All the 64 triplets and
their codon assignments are shown in Table 66.2.

TABLE 66.2   The Genetic Code

First position (5′ end)	Second position				Third position (3′ end)
	U	*C*	*A*	*G*	
U	Phe	Ser	Tyr	Cys	U
	Phe	Ser	Tyr	Cys	C
	Leu	Ser	Term[a]	Term	A
	Leu	Ser	Term	Trp	G
C	Leu	Pro	His	Arg	U
	Leu	Pro	His	Arg	C
	Leu	Pro	GluN	Arg	A
	Leu	Pro	GluN	Arg	G
A	Ileu	Thr	AspN	Ser	U
	Ileu	Thr	AspN	Ser	C
	Ileu	Thr	Lys	Arg	A
	Met	Thr	Lys	Arg	G
G	Val	Ala	Asp	Gly	U
	Val	Ala	Asp	Gly	C
	Val	Ala	Glu	Gly	A
	Val	Ala	Glu	Gly	G

[a] Chain terminating

Yanofsky and colleagues, in 1966, demonstrated the
colinearity of a gene and its polypeptide product. They found
that the nucleotide sequences in a gene and amino acid
sequences in the corresponding proteins (tryptophan synthetase)
were colinear. In other words, the first three base-pairs (the first
codon) of a gene specify the first amino acid of the protein
polypeptide, the next three base pairs (next codon) specify the
second amino acid and so on in such a fashion that any change
in a base pair of a codon in DNA is reflected in a change in
the corresponding amino acid in the protein.

It has been generally believed that eukaryotic mRNA
typically contains a single start codon and encodes for a single
protein product. However, in recent years alternative translation
start sites and hidden coding potential of eukaryotic mRNAs

have been suggested. Both, experimental and bioinformatic data have revealed instances of one or more start codons that can be used alternatively. It has become evident that ribosomes can initiate translation not only from the annotated start site but also from the nearest downstream AUG codon(s) through leaky scanning. This results in amino-terminal truncated protein isoforms. Thus, alternative translation from downstream AUG codon may produce a protein product either with an isofunctional variant or a variant with distinct properties.

II. MACROMOLECULES INVOLVED IN TRANSLATION

Translation is a very complex and sophisticated process where ribosomes and a number of macromolecules like tRNA, aminoacyl synthetase and many other protein factors are involved.

A. tRNAs and Aminoacyl tRNA Synthetases

Since amino acids cannot recognize the codons on mRNA, transfer RNA molecules (earlier called soluble RNA) function as adaptors with dual function, i.e., recognition of amino acid and codon. Robert Holley, for the first time, sequenced yeast alanyl tRNA which contains 77 nucleotides. After this report, the sequences of many other tRNAs were reported. The unique feature of different tRNAs is their striking similarities among different tRNAs; tRNAs have sedimentation value of 4S and contain 73–93 nucleotides. The sequence bases in tRNA are usually written in the form of clover leaf with single stranded loops and double stranded arms which is shown in Figure 66.1.

tRNAs contain many unusual bases ('unusual' means a base other than A, G, C or U) and some of the bases are methylated. The 3'-end of tRNA has a base-sequence of ACC and the 5'-end of it usually contained 'G', which is phosphorylated. All the tRNAs have TψC, DHU and anticodon loops. However, three-dimensional structure of tRNA is L-shaped (Figure 66.2). The important features of tRNA are summarized as follows:

(i) The 3'-end where amino acid attaches always consists of ACC and a variable fourth nucleotide.

(ii) The first loop which binds to the ribosomal surface contains seven unpaired bases with a sequence of 5'–TψCG–3'.

(iii) The second loop contains variable nucleotides. The function is not known.

(iv) The third loop has seven unpaired bases and contains anticodon to recognized codon (mRNA) by base pairing.

(v) The fourth loop which is rich in dihydro U (D-loop) contains 8–12 unpaired bases.

tRNAs are very specific which means that a specific tRNA always carries a specific amino acid. Each tRNA can be charged with one amino acid. Amino acids are linked

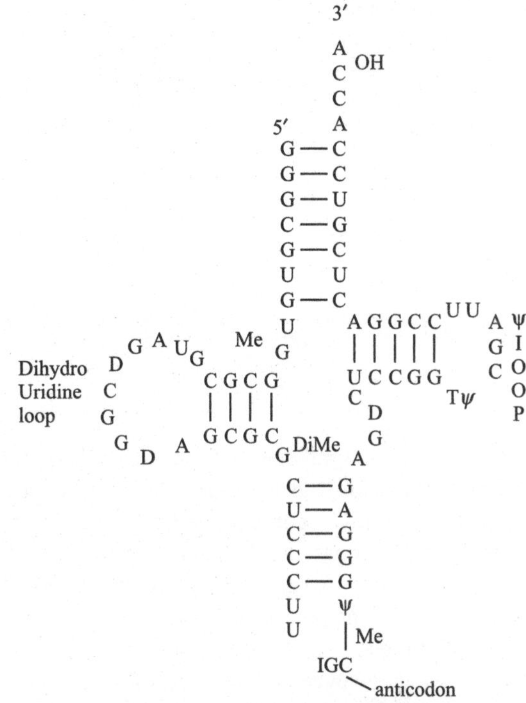

Figure 66.1 The nucleotide sequence of yeast alanine tRNA. Unusual bases are abbreviated as follows: D, dihydrouridine; U, pseudouridine; I, inosine, IMe, methylinosine, GDiMe, dimethylguanosine, GMe, methylguanosine; T, ribothymidine.

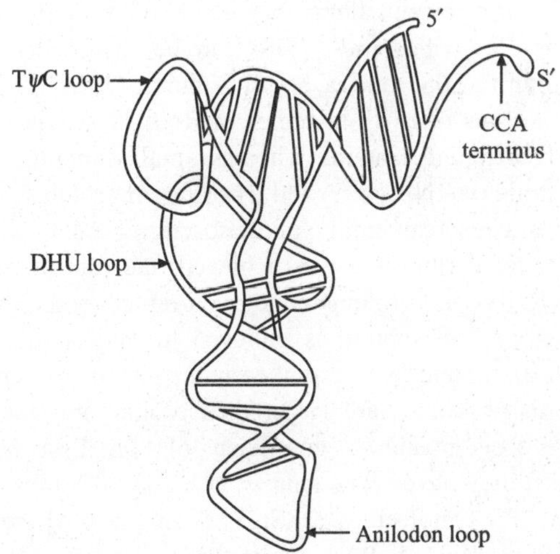

Figure 66.2 A schematic diagram of the three-dimensional structure of yeast phenylalanine tRNA.

to 3'-end of tRNA by means of ester bond which involves carboxyl group of amino acid and hydroxyl group of ribose of the last base of tRNA. The charging of tRNA is carried out by an enzyme called aminoacyl synthetase which shows double specificity like tRNA. The aminoacyl tRNA synthetases recognize specific amino acid on the one hand and specific tRNA on the other hand. Hence, there are different aminoacyl tRNA synthetases for different amino acids. Some aminoacyl tRNA synthetase append an amino acid to the 2'OH group of their cognate tRNA and others do so at the 3′OH group. The charging of tRNA is brought about in two steps:

(a) Activation of amino acid by synthetase

$$AA + ATP \rightleftharpoons AA.AMP + PPi$$

(b) Formation of AA~tRNA

$$AA.AMP + tRNA \rightleftharpoons AAtRNA + AMP$$

i.e.,

$$AA + tRNA \xrightarrow{\text{Aminoacyl tRNA synthetase}} AA.tRNA$$
$$ATP \qquad AMP + PPi$$

Aminoacyl tRNA synthetase accurately selects between very similar aminoacids and therefore, AA~tRNA formation is very accurate. As for example, isoleucyl-tRNA synthetase can precisely discriminate isoleucine from valine though these amino acids differ by only single methyl groups. When a mixture of isoleucine and valine was used as substrate for isoleucyl synthetase, some molecules of valine ~ AMP formation occurs. However, in presence of isoleucyl tRNA, all these valine AMP dissociate into valine and AMP.

B. Ribosomes

The complex process of translation takes place on ribosomes, which can be regarded as the site of protein synthesis. Several ribosomes are present along the length of a single mRNA. In prokaryotes, ribosomes (70S) can be dissociated into a large subunit (50S) and a small subunit (30S). The large subunit consists of two molecules of RNA (23S and 5S) and about 31 different proteins whereas small subunit contains one molecule of RNA (16S) and 21 different proteins. Though the spontaneous reassembly of these components to form a fully functional ribosome was achieved, the role of different kinds of RNAs and proteins is not well understood. However, 16S RNA of 30S subunit is believed to play a role in the initiation of protein synthesis (see latter part of the chapter). Eukaryotic ribosomes are larger than prokaryotic ribosomes and have a sedimentation coefficient of 80S. Like bacterial ribosomes, they have two subunits, 60S and 40S. 60S subunit has three RNA molecules (28S, 5.8S and 5S) whereas 40S contains a single 18S RNA molecule. Eukaryotic ribosomes contain more number of proteins than their prokaryotic counterparts.

Polyribosome (or *polysome*) is a cluster or aggregate of several ribosomes (about 8–10) which are inter-connected with a single mRNA and, are actively and simultaneously translating the same mRNA into protein. Polyribosomes were first discovered and characterized in 1963 by Jonathan Warner, Paul Knopf and Alex Rich. They can be found in three forms: free, cytoskeletal bound and membrane bound. In bacteria they are found free in the cytoplasm while in eukaryotes they are attached to the surface of the rough endoplasmic reticulum and outer membrane of the nucleus.

C. Protein Factors

Several protein factors (non-ribosomal proteins) are involved in protein synthesis. Depending on their function, these factors are called initiation factors (IFs), elongation factors (EFs) and release factors (RFs). Different factors of prokaryotes and eukaryotes are listed in Table 66.3.

III. MECHANISM OF PROTEIN SYNTHESIS

Polypeptide chains always grow in a stepwise manner, one amino acid at a time. The pulse-labelling experiments performed by Howard Dintzis clearly showed that the direction of chain growth is from the amino terminal to the carboxyl terminal. Intact reticulocytes that produce hemoglobin almost exclusively were labeled with (^{3}H)—amino acid for a period shorter than required to synthesize a complete globin chain. Newly synthesized hemoglobin chains were separated and then treated with trypsin. The peptides were analyzed for the distribution of the radioactivity. Radioactivity was found preferentially in peptides close to carboxyl terminal. This suggests that C-terminal residues are the last added, hence the synthesis starts at the N-terminal in a sequential manner.

There are three steps in translation: *initiation*, *elongation*, and *termination*. The steps and mechanisms of protein synthesis are basically identical in prokaryotes and eukaryotes, however, considerable evolutionary changes have occurred in both the ribosomal and the soluble proteins. Moreover, eukaryotic mRNA shows many unique structural features that are absent in prokaryotes.

A. Initiation

Initiation of protein synthesis is the process by which ribosome binds to mRNA and initiator tRNA to form an initiation complex (70S–f–met–tRNAf–mRNA in prokaryotes and 80S–met–tRNAi–mRNA in eukaryotes). Among the three steps, initiation is a very complex process and has been recognized as the rate-limiting step since the selection of specific mRNAs takes place at this stage. As many as three protein factors in prokaryotes and ten in eukaryotes are known to catalyze the formation of initiation complex involved in protein synthesis. In prokaryotes, protein synthesis always starts with formylmethionine whereas in eukaryotes, initiation starts with methionine. The charging of met-tRNA with methionine takes place as discussed above. Later, formylation of met-tRNA is catalyzed by an enzyme called transformylase (Figure 66.3). Formylation of methionine on the initiator tRNA is not absolutely essential for the initiation of protein synthesis. However, it improves the efficiency with which the met-tRNAf is bound by IF–2. Finished polypeptides, however, do not contain any formylated end groups, since an enzyme (deformylase) removes the formyl group from the growing chain. The terminal methionine is also cleaved off from some proteins (not from all) by the enzyme, aminopeptidase.

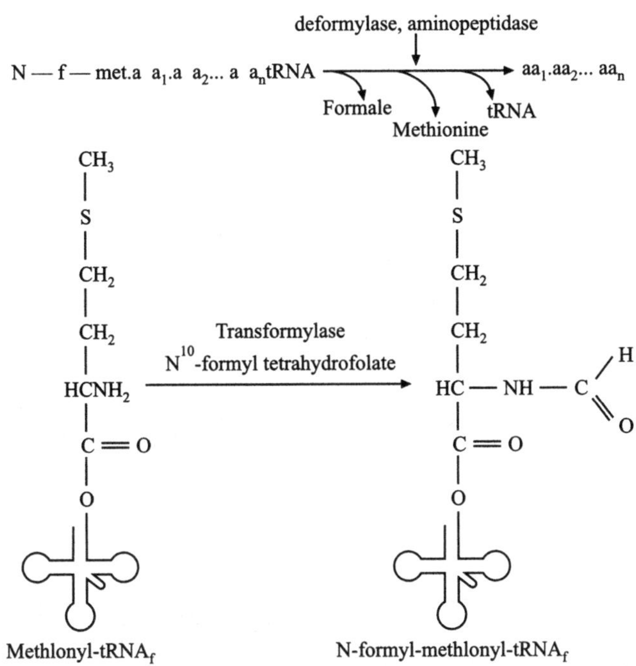

Figure 66.3 Formylation reaction.

Dissociation of subunits of ribosome is a prerequisite for the initiation since initial binding of f-met-tRNAf (as in prokaryotes) or met-tRNAi (as in eukaryotes) occurs on small subunit of ribosome. Under physiological conditions,

the ribosome are in dynamic equilibrium with the subunits (70S $\rightleftharpoons$ 50S + 30S; 80S $\rightleftharpoons$ 60S + 40S). The generation of the ribosomal subunit is the result of the prevention of the association reaction. This is achieved by a protein factor, IF–3 in prokaryotes which binds to the 30S subunit and prevents its association with the large subunit. Similarly, in eukaryotes protein factors eIF–3 and eIF–6 bind to the 40S and 60S subunits respectively and prevent their association.

In prokaryotes, mRNA is first bound to the small ribosomal subunit. Sequence analysis of 16S rRNA and mRNAs suggests that a pyrimidine-rich sequence near the 3′-terminus of the 16S rRNA, base pairs with a complementary 3–10 base purine-rich sequence of mRNA (known as Shine-Dalgarno sequence) centred approximately 10 bases on the 5′ side of the initiation codon, AUG (Figure 66.4). It is believed that this pairing brings AUG into position so that it can bind initiator tRNA.

Figure 66.4 Base-pairing of 3'-end of 16S rRNA and 5'-end of β-galactosidase mRNA.

Three protein factors, identified as IF–1, IF–2 and IF–3 are involved in the formation of 30S initiation complex in prokaryotes (Table 66.3). IF–3 is essential for the binding of mRNA to the 30S subunit, whereas other two factors promote

TABLE 66.3   Characteristics of Protein Factors Involved in Protein Synthesis of Prokaryotes and Eukaryotes

Step	System	Nomenclature	Native mol. weight × 10^{-3}	No. of subunits	Function
Initiation	Prokaryotes	IF–1	9	1	Binding of Ternary complex to 30S
		IF–2	80–90	1	f-Met-tRNAf binding to 30S
		IF–3	23	1	Anti-association of ribosomal subunits and mRNA binding to 30S
	Eukaryotes	eIF–1	15	1	mRNA binding to 40S
		eIF–2	150	3	met-tRNAi binding to 40S
		SP	250	5	Promotes ternary complex formation, rescues Mg^{2+} inhibition, recycling of elF–2
		eIF–3	700	9	Anti-association of ribosomal subunits and stimulation of met-tRNA; binding to 40S
		eIF–4A	50	1	mRNA binding
		eIF–4B	80	1	mRNA binding
		eIF–4C	17	1	40S formation and met-tRNAi binding
		eIF–4D	16	1	Methionyl puromycin formation
		eIF–4E	24	1	Cap recognition during mRNA binding
		eIF–5	150	1	Joining 40S initiation complex with 60S ribosomal subunit and release of initiation factors from initiation complex
		eIF–6			Anti-association activity binds to large ribosomal subunit 60S eIF–2 and 3
Elongation	Prokaryotes	EF–T	43(EFTU)	1	Binding of aminoacyl tRNA to 70S ribosomal A site
			35(EFTS)	1	
	Eukaryotes	EF–G	72	1	Elongation (Translocation)
		eEF–1	55	1	Binding of aminoacyl tRNA to 80S ribosomal A site
			30	1	
		eEF–2	100	1	Same as EF-G
Termination	Prokaryotes	RF–1	47	1	Releases polypeptide chain by recognizing UAA, UAG codons
		RF–2	48	1	Releases polypeptide chain by recognizing UAA and UGA codons
	Eukaryotes	eRF	105	2	Releases polypeptide chain by recognizing all termination codons

the binding of f–met–tRNAf to the mRNA.30S complex. This step requires Mg^{2+} and GTP. Next, the larger subunit joins this complex to form 70S *initiation complex*. Hydrolysis of one GTP is known to occur in the formation of initiation complex. However, in eukaryotes initiation is more complicated than prokaryotes, as more factors are involved. In addition, initiation complex formation shows two striking differences. Firstly, methionine is not formylated and secondly, initiator tRNA binds to 40S subunit prior to binding with mRNA.

In eukaryotes, met-tRNAᵢ in presence of GTP and eIF–2 forms a stable complex called ternary complex which binds to 40S subunit. Next, mRNA binds to form the initiation complex. Four different protein factors—eIF–1, eIF–4A, eIF–4B, eIF–4E seem to be required. However, the precise role of these factors is obscure. Finally, with the help of another protein factor, eIF–5, 60S subunit joins to 40S pre-initiation complex to form 80S initiation complex. At this step, all the initiation factors are released. Initiator factor-2 (eIF–2) is released as eIF–2.GDP complex. An ancillary protein factor (SP) which stimulates the ternary complex formation displaces GDP from eIF–2.GDP complex to give eIF–2-SP complex which in turn can interact with GTP to form eIF–2.GTP; GDP and SP are released and recycled.

Schematic diagram of initiation complex formation in prokaryotes and eukaryotes is shown in Figures 66.5 and 66.6, respectively.

B. Elongation

Formation of initiation complex sets the stage for polypeptide chain elongation. This process can be defined as the addition

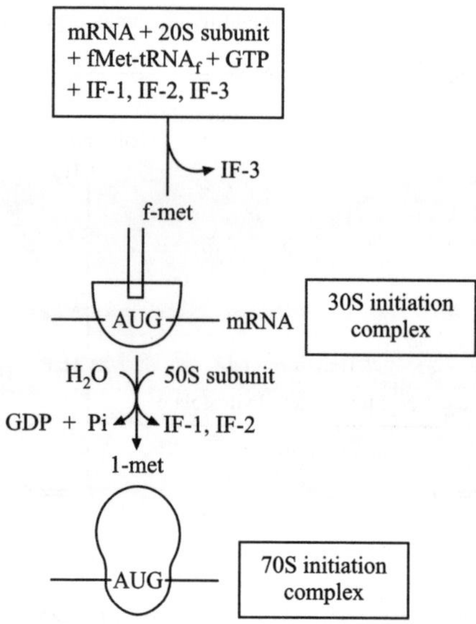

Figure 66.5   Initiation process of protein synthesis in prokaryotes.

of amino acids one at a time to the growing polypeptide in a sequence dictated by mRNA. Elongation involves the relative movement of mRNA and ribosome. The process of elongation consists of three steps (Figure 66.7). First step is codon directed binding of aminoacyl-tRNA to a vacant ribosomal aminoacyl(A)—site adjacent to an occupied peptidyl(P)—site (the ribosome has two adjacent sites, A and P, each of which accommodates a tRNA). This step requires GTP and elongation factor (EF–T) whereby GTP is hydrolyzed to GDP and Pi. EF–T is composed of two components—EF–Tu and EF–Ts. EF–Tu is required to bring aminoacyl tRNAs to the A-site of the ribosome whereas EF–Ts is needed to recycle EF–Tu (Figure 66.8). EF–Tu binds neither formylated nor unformylated tRNAf^met, which is why the initiator tRNA never reads internal AUG or GUG codons. In the presence of GTP, EF–Ts is released and EF–Tu.GTP then interacts with aminoacyl–tRNA to form a

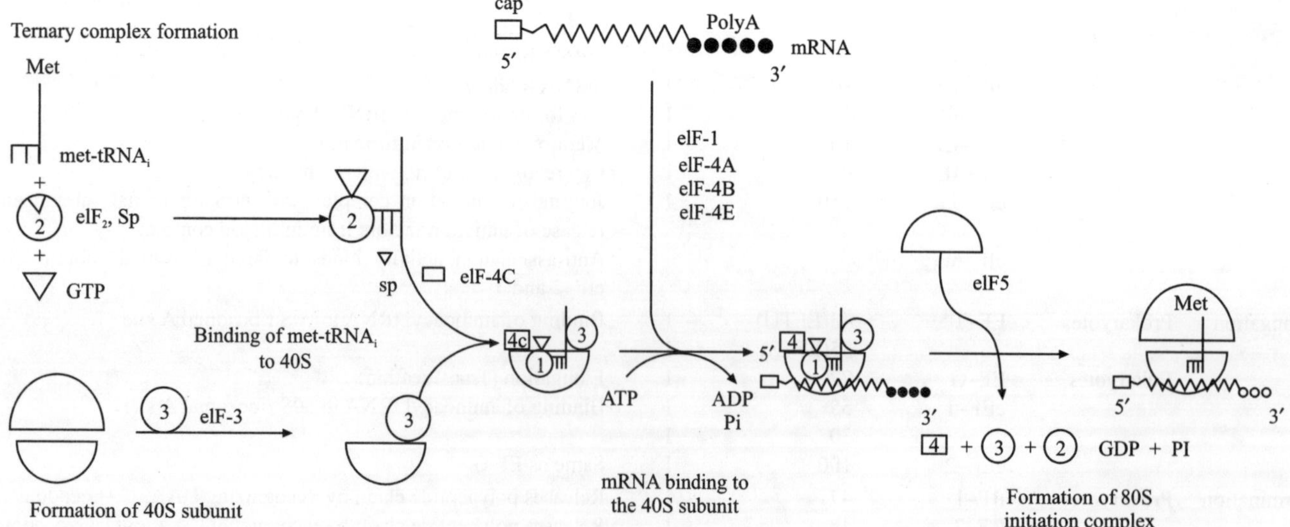

Figure 66.6   Initiation process of protein synthesis in eukaryotes.

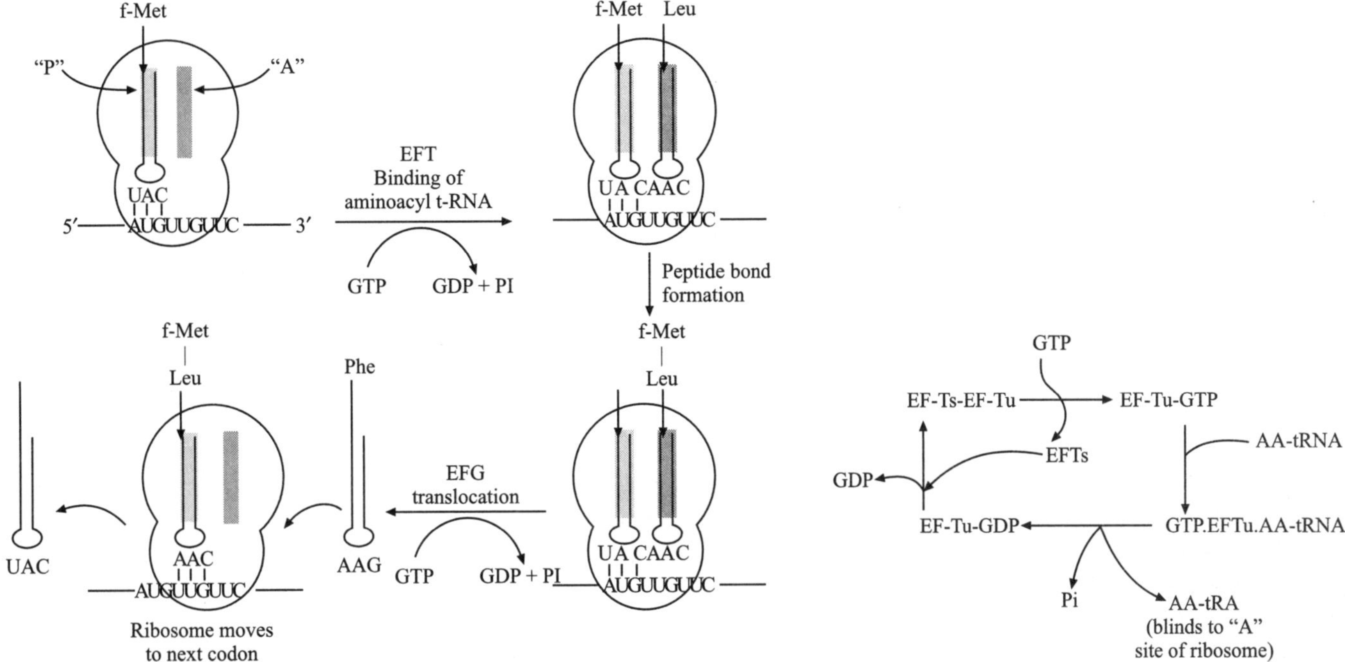

Figure 66.7   Elongation process of protein synthesis.

Figure 66.8   EF–Tu — EF–Ts cycle.

ternary complex which binds to ribosome. After this interaction, EF–Tu.GDP complex is released which in turn cannot bind to GTP unless GDP is released by EF–Ts. Now, the first (initiator) and second aminoacyl-tRNA occupy P and A sites, respectively. In the next step, peptide bond formation takes place; the A-site now bears dipeptidyl-tRNA while the P-site bears the deacylated initiator tRNA. This trans-peptidation reaction is catalyzed by an enzyme called peptidyl transferase which is an integral part of the 50S ribosomal subunit. Puromycin, which bears a structural resemblance to aminoacyl-tRNA inhibits this step (Figure 66.9). It competes with aminoacyl-tRNA as a

substrate for peptidyl transferase and thereby nascent proteins with puromycin attached to C-terminal residue are released from the ribosome. The third step is translocation process by which peptidyl-tRNA is translocated to P-site to allow incoming aminoacyl-tRNA to enter the A-site with concomitant release of discharged tRNA from the P-site. This sets the stage for binding of the third aminoacyl-tRNA and these processes are repeated until the polypeptide chain is completed. Translocation requires a second elongation factor (EF–G) and one more GTP which is hydrolyzed to GDP and Pi. EF–G release is a prerequisite for beginning the next elongation cycle because

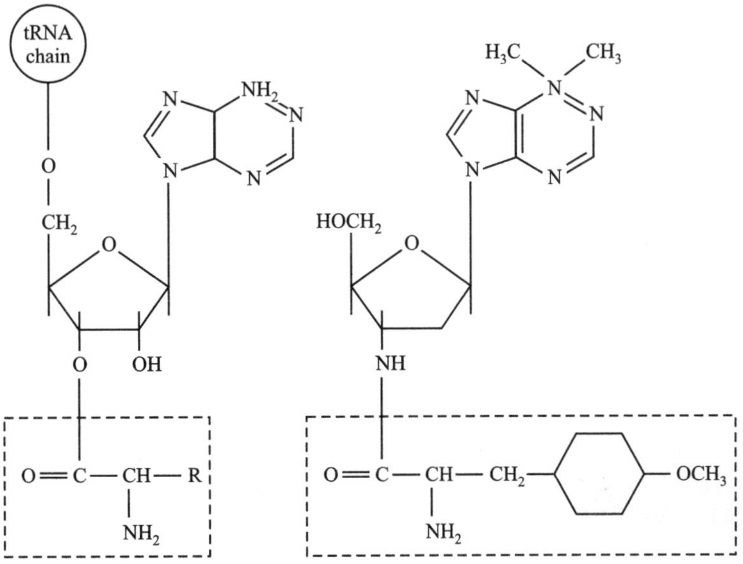

Figure 66.9   Structural resemblance of puromycin to aminoacyl-tRNA.

the ribosomal binding sites for EF–Tu and EF–G are partially overlapping and hence their ribosomal binding is mutually exclusive. EF–T and EF–G correspond to eEF–1 and eEF–2 of eukaryotes.

C. Termination

This process releases the completed polypeptide chain from the ribosome.mRNA complex. This is directed by one of the three terminator codons (UAA, UAG, UGA). When a termination codon is reached by ribosome, termination factor called 'release factors (RF)' catalyze the hydrolysis of the bond between the completed polypeptide chain and tRNA. In prokaryotes, two different release factors (RF-1 and RF-2) with two different codon specificities have been reported. RF-1 recognizes UAA and UAG whereas RF-2 identifies UAA and UGA codons. In addition to RF-1 and RF-2, another factor called RF-3 is known. RF-3 itself does not have any release activity but stimulates the activities of RF-1 and RF-2. In contrast, in eukaryotes there is only one RF which recognizes all the termination codons (Table 66.3).

IV. POST-TRANSLATIONAL MODIFICATION OF PROTEINS

Proteins are known to undergo many post-translational covalent modifications. Phosphorylation and dephosphorylation of protein is one such process which plays a vital role in the regulation of cell metabolism. Phosphorylation of proteins is catalyzed by a group of enzymes called '*protein kinases*'. These enzymes transfer γ-phosphate of ATP to amino acid residues (generally serine or threonine and in some cases tyrosine) of the protein. Many enzymes are known to get phosphorylated and as a result their activities are altered. In addition many structural proteins (ribosomal and membrane proteins) are also phosphorylated:

$$\text{Protein} + \text{ATP} \xrightarrow{\text{Protein kinase}} \text{Protein} - \text{P} + \text{ADP}$$
$$\downarrow \text{Phosphatase}$$
$$\text{Protein} + \text{Pi}$$

The activity of these protein kinases is often regulated by specific effector compounds such as cAMP, cGMP, Ca^{2+}, dsRNA, etc. Depending on the specific effector compounds they are classified into various categories as follows:

Category	Designation	Effector compound
1	cAMP dependent protein kinase	cAMP
2	cGMP dependent protein kinase	cGMP
3	Ca^{2+} dependent protein kinase	Ca^{2+} and calmodulin
4	Double stranded RNA dependent protein kinase	dsRNA
5	cAMP independent protein kinases	No effector compound is known

cAMP dependent protein kinases show their activity only in the presence of cAMP. These enzymes consist of two regulatory and two catalytic subunits (R_2C_2). cAMP activates these enzymes by binding to 'R' following release of active 'C':

$$R_2C_2 + 4cAMP \rightleftharpoons R_2(cAMP)_4 + 2C$$

(Inactive holoenzyme)   (Active catalytic subunit)

These enzymes use only ATP as phosphoryl donor and phosphorylate basic proteins, e.g., histone. cGMP dependent protein kinases are activated in the presence of cGMP. Calcium dependent activities of protein kinases are mediated by calcium binding protein called calmodulin. It is proposed that Ca^{2+} first binds to calmodulin which in turn binds to kinase and activates it. No specific effector compound is required for the activities of cAMP independent kinases. There are two types of cAMP independent kinases which are designated as Type A and B. Type A cAMP independent kinases use either ATP or GTP as phosphoryl donor and phosphorylate acidic proteins, e.g., casein, whereas type B cAMP independent kinases use only ATP as phosphoryl donor and phosphorylate basic proteins (eIF–12 kinases, i.e., active HCI and DAI fall under this category).

Besides *phosphorylation*, protein functions are regulated by *methylation*, *nucleotidylation*, and *ADP-ribosylation*. In these cases also, specific enzymes are involved to modify key enzymes. Moreover, activation of proteins due to proteolytic processing, i.e., removal of a peptide fragment or fragments (e.g., trypsinogen → trypsin; prothrombin → thrombin; proalbumin → albumin; proinsulin → insulin) is known for many years. This type of proteolytic cleavage or cleavages involved in proprotein → protein conversion leads to the production of the final three-dimensional structure.

V. TRANSLATIONAL CONTROL

Most of the control of protein synthesis in both pro- and eukaryotes appears to act at the level of initiation. There can be two kinds of translational control namely, non-selective and *selective translational control*. In case of non-selective control, translation of all the messengers is affected to the same extent. Selective translational control means only a set of messages from a variety of different mRNAs are selected for translation with same or differential efficiency (Figure 66.10). The best examples of selective control are known in bacteria, particularly in the regulation of viral mRNA. As for example, although three genes of RNA viruses (e.g., Qβ) are present in equimolar concentration in infected *E. coli*, the viral proteins are produced in very different amounts.

A. Prokaryotes

In bacteria, the elongation factors and ribosomal proteins (with exception of EF–Tu, present in large amounts and L7/L12, present in four copies per ribosome) are produced in equimolar quantities. Although the genes for these proteins are spread throughout the chromosome and organized into at least

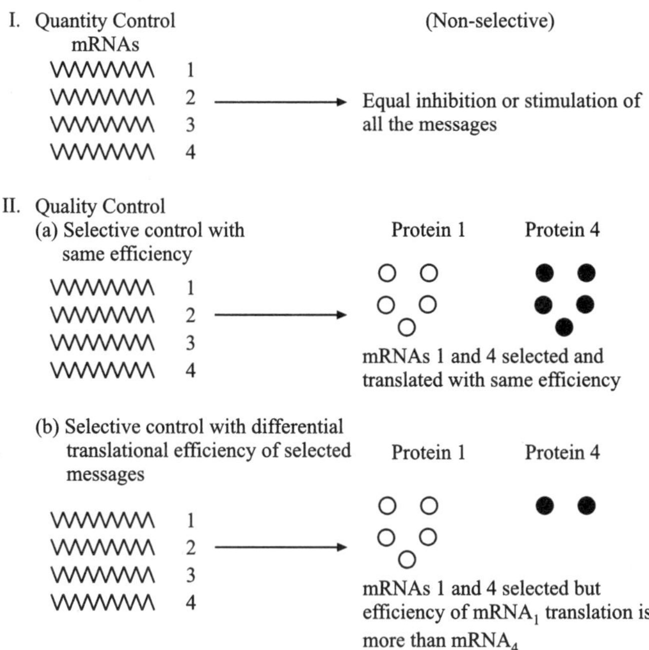

Figure 66.10 Translational control.

10 different operons, the synthesis of these proteins is closely coordinated. In order to explain this phenomenon, Nomura and colleagues have proposed a feedback control mechanism in which a single ribosomal protein, when over produced acts as a *translational repressor*. The repressor binds preferentially to rRNA, but if rRNAs are not available, excess repressors bind to its message to control its own production.

Another type of regulation is known where 'magic spot' nucleotides (ppGpp or pppGpp) control the initiation of rRNA and tRNA chains. Very low levels of these molecules are normally present in bacteria, however, accumulation of large amounts occurs during amino acid starvation. This phenomenon is known as 'stringent response'. In this condition, an uncharged tRNA molecule occupies an 'A' site, and therefore, an idling reaction occurs on the affected ribosome:

$$GDP \text{ (or GTP)} + ATP \rightarrow ppGpp \text{ (or pppGpp)} + AMP$$

The enzyme responsible for the synthesis of 'magic spot' is known as stringent factor.

B. Eukaryotes

In prokaryotes with short lived messenger RNAs, gene expression is probably controlled mainly at the level of transcription. In eukaryotes on the other hand, mRNAs have in general a much longer life span, and gene expression may be controlled not only at the transcriptional, but also at the translational level. Several well characterized regulatory systems have been shown to modulate the overall rate of protein synthesis at the level of translation. Most of the information concerning this comes from reticulocyte and their lysates (Figure 66.11). In reticulocyte, globin synthesis is controlled by the level of heme, the prosthetic group of hemoglobin. It has been known for several years that protein synthesis in reticulocyte lysates is but briefly maintained in the absence of added hemin. Hemin prevent the formation of an inhibitor of chain initiation from a pro-inhibitor (inactive form) of similar molecular weight (pro-inhibitor → inhibitor). This inhibitor (HCI, hemin controlled inhibitor) is a cyclic AMP-independent protein kinase that catalyzes the phosphorylation of the small (38,000 daltons) subunit of the initiation factor eIF–2. This phosphorylation blocks the interaction of eIF–2 with eIF–2 stimulatory protein (SP) without which eIF–2 is unable to form an initiation complex, a pre-requisite for translation.

The addition of double-stranded RNA (dsRNA) activates another protein kinase (DAI), which also phosphorylates the

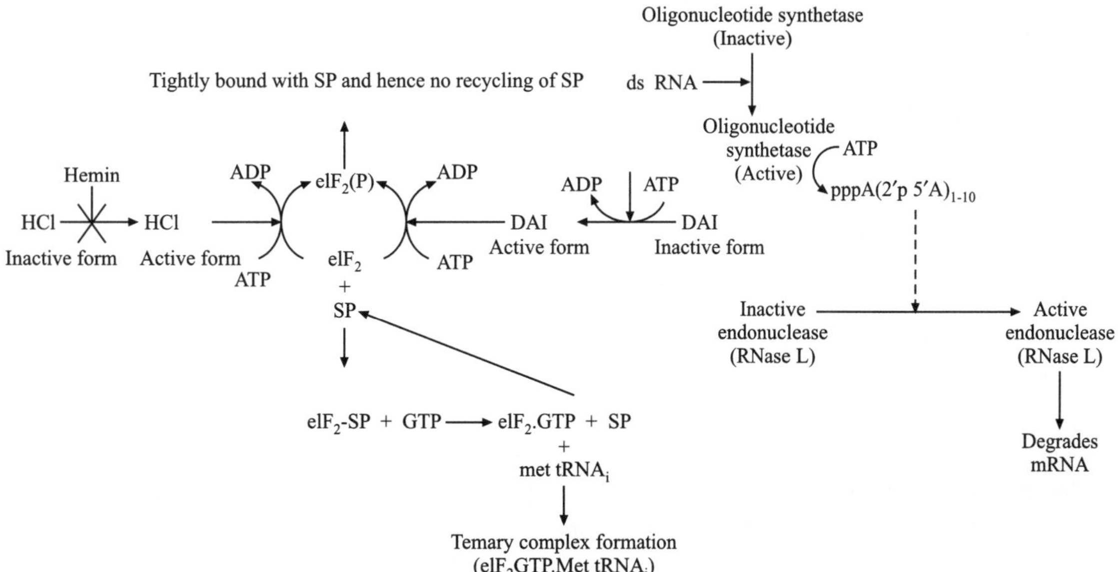

Figure 66.11 Regulation of polypeptide chain initiation by cAMP-independent protein kinases (HCI and DAI).

α-subunit of eIF–2. Synthesis of this cyclic AMP-independent protein kinase may be induced by interferon (a glycoprotein which itself is induced very potently by dsRNA), which specifically inhibits the viral protein synthesis, without affecting host-cell protein synthesis. Besides the activation of this protein kinase, incubation of reticulocyte lysates with dsRNA and ATP also gives rise to the formation of a potent inhibitor, a novel oligonucleotide pppA(2'p 5'A) 1–10 that inhibits protein synthesis at exceedingly low concentrations possibly by activating an endonuclease that degrades both cell and viral mRNAs.

In eukaryotes, nothing much is known about selective translational control of protein synthesis since principles that govern mRNA selection and exact role of factors at this step are not clearly understood. In extracts of polio-virus infected cells, capped messengers are poorly translated whereas translation of polio virus mRNA (which are not capped) is not affected. It is inferred that polio virus infection inactivates 'cap binding protein' (eIF–4E) since translation of capped messages is restored by adding cap binding protein. However, how polio virus infection inactivates cap binding protein is yet to be discovered.

VI. ANTIBIOTICS AND PROTEIN SYNTHESIS

Antibiotics and other inhibitors have been found which block protein synthesis at various steps (Table 66.4). Some of these drugs are very useful in medicine since they inhibit protein synthesis by bacterial 70S ribosomes but do not affect eukaryotic ribosomes. In general, prokaryotic inhibitors block protein synthesis in bacteria, blue green algae, mitochondria and chloroplasts, whereas inhibitors of eukaryotic protein synthesis are active in most cells having 80S ribosomes. Without having functional ribosomes, bacteria cannot survive or cause an infection. It is known that antibiotics cure various infectious diseases by blocking the function of bacterial ribosomes. In recent years, using X-ray diffraction, 3D models have been generated to reveal how different antibiotics bind and act on ribosomes. These models are proving to be of immense help to develop new antibiotics thereby helping save numerous lives and decreasing human suffering. For these groundbreaking structural studies on ribosomes Venkatraman Ramakrishnan, Thomas A. Steitz and Ada E. Yonath were awarded the Nobel Prize in Chemistry during the year 2009.

SUMMARY

The translation of genetic message from the nucleotide sequence of an mRNA into protein product is essential to execute and sustain the basic processes of life like growth and differentiation. Translation is a sophisticated process which is orchestrated by stepwise involvement of a number of macromolecules that includes key players such as ribosomes, tRNA, aminoacyl synthetase and different protein factors termed as initiation, elongation and release factors. There are three major steps in translation: initiation, elongation and termination. Post-translationally, proteins may undergo several covalent modifications including phosphorylation, methylation, nucleotidylation, ADP-ribosylation etc. which influence the function(s) of translated protein products. Though the steps and mechanisms of protein synthesis are basically identical in prokaryotes and eukaryotes, the eukaryotic mRNA shows

TABLE 66.4   Inhibitors of Protein Synthesis

Inhibitor	Site of inhibition	Step inhibited*	Mode of action	Inhibition of pro- or eukaryotes
Tetracycline Aminoglycosides	30S	E	AA~tRNA binding	Pro
e.g., streptomycin	30S	E (I)	AA~tRNA binding	Pro
kanamycin	30S, 50S	E	AA~tRNA binding	Pro
Edeine	30S, 40S	I, E	f-Met-tRNAf	Pro
			AA~tRNA binding	Eu
Aurine tricarboxylic acid	30S, 40S	I (E)	mRNA binding to ribosomes	Pro, Eu
Pactamycin	30S, 40S	I (E)	f-Met and Met-tRNAf binding	Pro, Eu
Chloramphenicol	50S	E	Transpeptidation	Pro
Cycloheximide	60S	E	Transpeptidation (translocation)	Eu
Sparsomycin	50S, 60S	E	Transpeptidation	Eu, Pro
Erythromycin	50S	E	Transpeptidation (translocation)	Pro
Puromycin	50S, 60S	E	Premature chain termination by acting as analog of aminoacyl –tRNA	Pro, Eu
Fusidic acid	EFG, EF–2	E	Translocation	Eu, Pro
Diphtheriatoxin (intact)	EF–2	E	Translocation	Eu
Hygromycin B	EF2, EFG	E	Translocation	Eu, Pro

*E, elongation; I, initiation

many unique structural features that are absent in prokaryotes. Most of the control of protein synthesis in both prokaryotes and eukaryotes appears to act at the translational initiation level via dual mode of controls. In case of 'non-selective control', translation of all messengers is affected to the same extent unlike in case of 'selective control' where only a set of messengers are selected for translation with similar or differential efficiency. With the improved understanding of translational processes, a number of antibiotics and inhibitors of protein synthesis are being designed that are proving useful in medicine and helping save numerous lives from infections and other human sufferings.

SUGGESTIONS FOR FURTHER READING

Herman T. and Westhof E. (1998), *Curr. Opin. Biotechnol.,* 9, 66–73.

Hershey J.W.B., Sonenberg N. and Mathews M.B. (2012), Principle of Translational Control: An Overview, In: Protein Synthesis, *Cold Spring Harbor Laboratory Press*, USA.

Kochetov A.V. (2008), *BioEssays*, 30:683–691.

Krebs J.E., Goldstein E.S. and Kilpatrick S.T. (2009), Lewin's Genes, 10th ed., Jones and Bartlett Publishers.

Lodish H. (1976), *Ann. Rev. Biochem.,* 45, 39–72.

Maitra U., Stringer B.A. and Chaudhari A. (1982), *Ann. Rev. Biochem.,* 51, 869–900.

Merrick W.C. (1992), *Microbiol. Rev.,* 56, 291–315.

Nelson D.L. and Cox M.M. (2012), *Lehninger Principles of Biochemistry*, 6th ed., W.H. Freeman and Company.

Ochoa S. and De Haro (1979), *Ann. Rev. Biochem.,* 48, 549–580.

Puglisi J.D. (2009), *Molecular Cell*, 36, 720–723.

Quagliarello V.J. and Scheld W.M. (1997), *N. Engl. J. Med.,* 336, 708–716.

Voet D. and Voet J.G. (2010), *Biochemistry,* 6th ed., John Wiley and Sons.

Watson J.D., Baker T.A., Bell S.P., Gann A., Levine M. and Losick R. (2007), *Molecular Biology of the Genes,* 6th ed., W.A. Benzamin Inc.

Weissbach H. and Ochoa S. (1976), *Ann. Rev. Biochem.,* 45, 191–216.

Wold F. (1981), *Ann. Rev. Biochem.,* 50, 783.

Zheng N. and Gierasch L. (1996), *Cell,* 86, 849–852.

67

Recombinant DNA Technology in Eukaryotic Gene Expression System

Seyed E. Hasnain, Krishnaveni Mohareer and Sharmistha Banerjee

CONTENTS

I. INTRODUCTION

Recombinant DNA technology and genome sequencing projects have revolutionized the expression and purification of foreign genes in heterologous systems, which have a wide range of applications in research, biotechnology and human medicine. The basic principles of gene transcription, translation and recombinant DNA technology have been dealt in detail in previous chapters. This chapter attempts to give an overview of the application of recombinant DNA technology in foreign gene expression and is focused on eukaryotic genes as foreign genes. Prokaryotic system of expression is very simple, fast and economical as compared to eukaryotic system. However, prokaryotic expression systems have limitations when an eukaryotic gene is attempted to express for large-scale production. The limitations include their inability to overcome the codon bias, lack of machinery for post-transcriptional modification (PTM)s, poor folding of large proteins, membrane proteins or complex proteins. This necessitated the development of technology and understanding to exploit eukaryotic systems for large scale and accurate synthesis of such proteins. This chapter discusses various eukaryotic systems that are used in gene expression for recombinant protein purification.

II. COMPARISON OF PROKARYOTIC VS EUKARYOTIC EXPRESSION SYSTEMS

Prokaryotic system, especially *E. coli* expression system is quite often the universal host of choice for expression of heterologous proteins, be it a prokaryotic or eukaryotic protein. Though there is little doubt that a prokaryotic expression system presents an economical solution to large-scale production of recombinant proteins, it has several shortcomings, especially with reference to expression of eukaryotic proteins or proteins that use an eukaryotic system for expression, like viral proteins. To begin with, while cloning of prokaryotic genes is easier because of absence of exons and introns, eukaryotic genes that are to be over-expressed in prokaryotic systems are required to be synthesized using reverse transcription of completely spliced mRNA into cDNA, followed by amplification of cDNA into a double stranded DNA fragment. The procedure is technically more challenging as it begins with extraction of good quality intact RNA from the eukaryotic source. In addition to that, prokaryotes have limited capacity of post-transcriptional modifications, stable disulphide formation and protein folding. Very frequently, inappropriately folded proteins get accumulated as inclusion bodies that cannot be purified further under native or non-denaturing conditions. This defeats the purpose of making a functionally active recombinant eukaryotic protein. Besides these, bias for codon usage, synthesis of endotoxin free proteins, expression of large proteins etc. have prompted extensive development of eukaryotic expression systems for recombinant proteins.

A comparative chart which points out the advantages and limitations in each system has been shown in Table 67.1.

TABLE 67.1 Comparison of Prokaryotic vs Eukaryotic expression systems

Feature	Prokaryotic expression system	Eukaryotic expression system
Inducible	Yes	Temporal/constitutive
Time of expression	3-12h	1–5 days
Protein folding	Limited	Properly folded (refolding may be necessary in yeast)
PTMs	No	Yes
Protein limit	Limited	No limit
Yield	High	Low
Ease of handling	Yes	To be handled with care
Scale up	Yes	Yes but expensive
Safety	Yes	To be handled with care and safety

III. EUKARYOTIC GENE EXPRESSION SYSTEMS FOR RECOMBINANT PROTEINS: AN OVERVIEW

As discussed above, Eukaryotic gene expression offers several advantages over bacterial expression systems. A typical eukaryotic expression system differs from a prokaryotic system in terms of the promoter, terminator, use of enhancer elements (if necessary), ribosomal binding site and selection marker, each of which is different for each expression system.

Depending on the downstream application and complexity of the protein in question, one could choose an appropriate expression system. As eukaryotic proteins are post-translational modified, each eukaryotic gene system offers different levels and complexities of PTMs. The process of gene expression in eukaryotic systems for recombinant proteins is explained schematically in Figure 67.1.

Typically all eukaryotic vectors have a specific promoter followed by Multiple Cloning Site and a terminator. They also have a bacterial origin of replication and selection markers for both bacterial and host (yeast/insect/mammalian/plant) specific resistance marker to enable selection of non-transformed cells. Some vectors have '*att*' sites wherein gene of interest can be integrated by transposition into these specific sites like in insect vector and plant expression vectors.

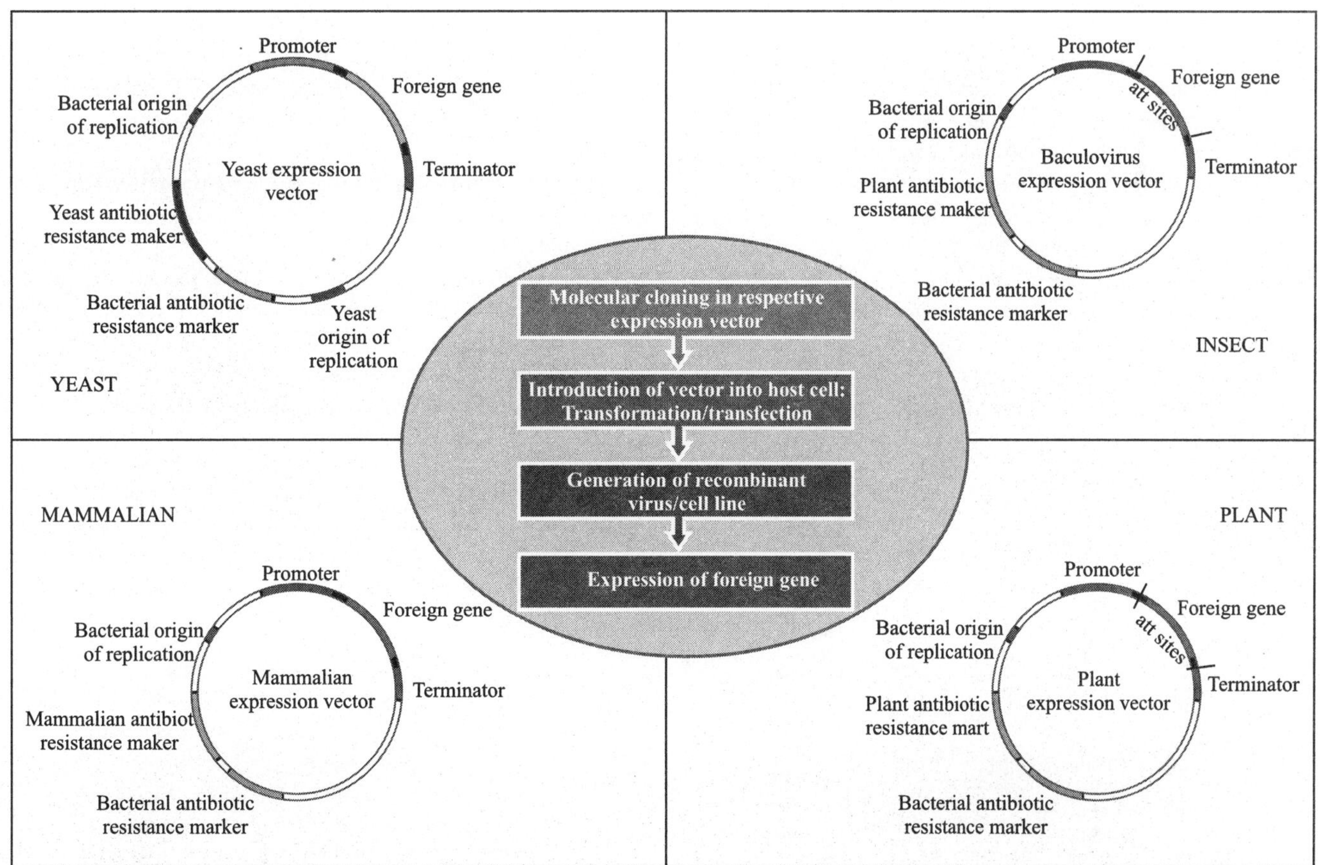

Figure 67.1 Schematic representation of eukaryotic gene expression systems. (*see Plate 26 for colour figure*)

A. Molecular Cloning of Foreign Genes in Eukaryotic Systems

Choice of vector

The foreign gene to be expressed is initially cloned in a suitable expression vector using appropriate restriction sites provided in the multiple cloning site (MCS) of the destination vector. The vectors used for expression can be either plasmid vector or viral vectors. Mostly, the prokaryotic systems and Yeast expression systems use plasmid vectors whereas higher eukaryotic expression systems including insect, mammalian and plant systems often use viral vectors.

Plasmid vectors

The choice of vector is made depending on the downstream application (see Table 67.2 for some typical commercial vectors that are available for each expression system). Some vectors provide *kozak* sequence (the sequence recognized by ribosome for initiation of protein synthesis) whereas in others we need to incorporate the same into the forward primer during PCR amplification of the gene for ensuring expression.

The foreign gene is cloned under host specific promoter, usually a *constitutive* promoter (expression is continuous and sustained) such as *CMV* promoter. For instance in case of over expression of a gene, the gene could be cloned in a vector under host-specific promoter and *transiently* transfected to look out for the effect of the over expression on a particular function of the cell. Besides, certain vectors allow integration of the plasmid genome into host genome by use of selection with markers such as neomycin resistance. Such a cell line in which the expression construct is integrated into its genome is termed as *stable* cell line. They allow indefinite propagation of cell line and avoiding transfection every time.

The foreign gene is introduced into host eukaryotic cell by transformation (yeast), transfection (plant) or electroporation (mammalian/plant) or by recombinant virus (insect/mammalian) or recombinant bacteria (Plant) using suitable vector as described. The vector carries all the suitable features for the expression of the gene of interest. The protein may be expressed transiently (transient transfection) or stably (stable cell lines/virus mediated). The protein may be expressed intracellular or secreted, by using signal sequences, to ease the process of purification.

Viral vectors

In the higher eukaryotic expression systems like insects, plants and mammalian expression systems, virus based approach could be used for protein expression which involves an additional step of preparation of recombinant virus harboring foreign gene and eases the process of introduction of the expression construct (Figure 67.2). The cell line is infected with recombinant virus resulting in the expression of the recombinant protein. The use of recombinant virus in a cell line depends on its *host specificity* and *host range*. For instance Adeno based recombinant viruses can infect HEK293 cells. Similarly, baculovirus is used for recombinant protein production in insect cell lines such as *Sf*9, *Tni*, etc. However, there is considerable variation in infection rates within different insect cell lines.

Choice of promoter

Essentially, the promoters in different expression systems are chosen from the same system, which drive transcription of the associated and have all the genetic elements for recognition of transcription machinery. The promoters are chosen based on their strength from low to high and selection of promoter depends upon the level of expression of the gene of interest. Enhancer elements multiply the fold of expression by several folds. So it can be used when a high level of expression of the gene of interest is desirable. Table 67.2 enlists some of these promoters with their systems of expression.

Constitutive promoters: Constitutive promoters are those promoters that continuously drive the expression of the associated RNA and are not regulated. These promoters are desirable when low and sustained levels of recombinant protein are required. Usually the promoters of housekeeping genes are chosen which fulfill both the requirements. Examples of

TABLE 67.2   Commercial vectors available for expression of Eukaryotic expression systems and their features

System	Commercial vector	Promoter	Terminator	Selection marker	Other features	Host strains/ cell lines
Yeast	pGADT7	ADH1 (Alcohol dehydrogenase)	ADH1	Leu2	HA tag	Saccharomyces cerevisiae, Pichia pastoris
Insect	pFastBac HT a/b/c	polh (polyhedrin)	SV40 Polyadenylation	Gentamycin	His tag	Sf9, Tni, Hifi
Mammalian	pCDNA3.1	CMV	SV40 Polyadenylation	Neomycin	Myc/His tag	HEK 293, CHOK, HeLa, Cos7
Plant	pEarley Gate	(CaMV) Cauliflower mosaic virus 35S promoter and enhancer	3' octopine synthase gene polyadenylation and termination sequence	Chloramphenicol	HA/TAP/ myc,/FLAG tag 'att' sites flank the foreign gene cloned allowing Agrobacterium tumefaciens mediated transformation	Arabidopsis, maize, rice, tobacco, cotton soyabean etc

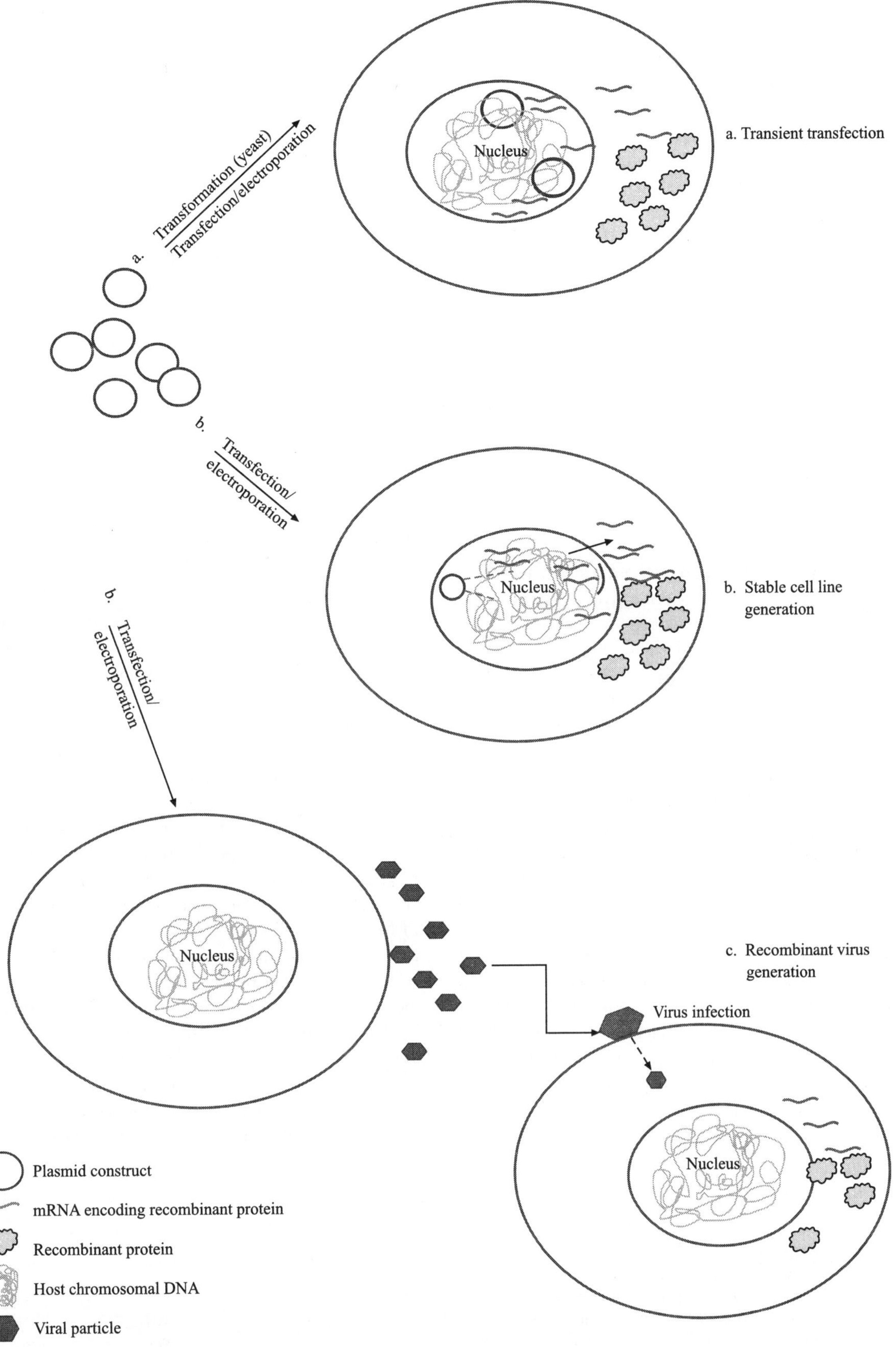

Figure 67.2 Schematic representation of virus dependent and independent eukaryotic gene expression. (*see Plate 27 for colour figure*)

constitutive promoters in yeast include phosphoglycerate kinase, enolase, pyruvate kinase etc; of baculovirus/insect immediate early gene (ie-1), actin etc; of mammalian actin, CMV etc.; and of plants CaMV35s RNA (dicots); actin (monocots) etc.

Regulated promoters: Regulated promoters are activated in a specific condition or stimulus. The regulated promoters can be classified as inducible or repressible. Inducible promoters can be activated in the presence of a specific stimulus whereas the repressible promoters are repressed in the presence of the stimulus. Examples of regulated promoters in yeast include alcohol dehydrogenase, nmt etc; of baculovirus/insect polh, pol10; of mammalian tetR, etc; of plant tissue specific promoters like RBCS (leaf), TA29 (anther); Glutelin (seed); light regulated RBCS; chemically induced tetR, tTA, AlcR, etc.

Heterologous promoters: Heterologous promoters constitute those promoters that are not natively present in the host, belong to other species, but recognized by the host cells. Ex: pathogen promoters CMV, SV40 (Mammalian), CaMV35s (Plant); polh, p10 (baculovirus/insect).

B. Scope of Post-translational Modifications in Eukaryotic Gene Expression System

Post-translational modifications (PTM) in eukaryotic proteins refer to the addition of different moieties or formation of disulfide bridges or proteolytic processing after the polypeptide chain of the protein is formed. It includes acetylation, acylation, ADP-ribosylation, amidation, γ-carboxylation, β-hydroxylation, di-sulfide bond formation, glycosylation, phosphorylation, proteolytic processing and sulfation. The advent of protein therapeutics is mainly dependent on the precise PTMs as it either determines the functional activity of the therapeutic protein or stability in the host; failing which the therapeutic protein cannot be put to use. Depending on the PTM required for a protein, essential for either activity or stability of the protein, the expression system is chosen. Glycosylation is the most common PTM in therapeutic proteins. More than 50% of the human proteins are glycosylated. Absence of glycosylation in congenital disorders leads to severe medical conditions pointing to the importance of glycosylation of proteins. The glycosylation patterns of certain proteins have also been used as disease markers. The nature of the glycoform can affect folding, stability, trafficking, immunogenicity along with the primary functional activity. Glycoform profiles can change from tissue to tissue of the same organism. Some of the therapeutic proteins that are dependent on glycosylation include erythropoietin, antibodies, blood factors, interferons, hormones such as gonadotropins.

Cell culture conditions can be manipulated to optimize PTMs. Temperature, growth rate, media composition optimization might influence productivity and quality of PTMs. Phosphorylation, acetylation, acylation, O-linked glycosylations could be performed even using the Yeast expression system. However, when N-linked glycosylations are required, one has to make a choice depending on the complexity of the glycosylation. Yeast system adds high mannose residues, whereas insect system (*Sf9*) cannot add sialic acid residues. In case of plants, they heavily glycosylate the protein resulting in unwanted glycosylations and it is highly immunogenic when used as a therapeutic protein.

To overcome the limitations of PTMs in different expression systems, there have been numerous attempts to engineer either the protein or the host successfully resulting in a stable or functional protein that can very well serve as therapeutic protein. Protein stability could be engineered by addition of moieties like *in vitro* substituting the stability conferred by glycosyl group. In case of yeast where certain glycosylations are not possible, exogenous introduction of the specific glycosyl transferase can help overcome this problem. In certain cases where the therapeutic protein is to be used as vaccine, the extra glycosylations that might be introduced by plant systems can actually enhance the immunogenic property than native protein. Certain therapeutic proteins such as glucocerobroside (used in treatment of Gauchers disease) have been engineered to have exposed mannose residues enabling efficient uptake by macrophages through the mannose receptors on their surface which otherwise is quickly cleared by hepatocytes. Synthetic PTM such as addition of PEG group (PEGylation, achieved by either enzymatic or direct chemical approach) or acylation can confer stability to IFNs. PEGylation confers protection from proteolytic cleavage and potentially mask immunogenic epitopes as well as reducing renal clearance owing to its increased molecular size. Some of the approved PEGylated therapeutics include IFN (Pegintron), human growth hormone analog (Somavert) and granulocyte-macrophage colony stimulating factor (Neulasta). An example of synthetically acylated therapeutic protein is levemir, insulin, produced from *Saccharomyces cerevisiae*, which increases the shelf life by 3–4 times as that of native insulins.

TABLE 67.3   Comparison of PTMs in different expression system

PTMs	Yeast	Sf9	Mammalian	Adeno	Plant
N-linked glycosylation	High mannose	Simple, no sialic acid	Yes	Yes	Yes
O-linked glycosylation	Yes	Yes	Yes	Yes	Yes
Phosphorylation	Yes	Yes	Yes	Yes	Yes
Acetylation	Yes	Yes	Yes	Yes	Yes
Acylation	Yes	Yes	Yes	Yes	Yes
Gamma-carboxylation	No	No	Yes	Yes	Yes

IV. DIFFERENT EUKARYOTIC EXPRESSION SYSTEMS

A. Yeast Expression System

Yeast expression system offers the simplest eukaryotic environment for expression of heterologous proteins, especially eukaryotic proteins. It can be easily scaled up to high cell mass densities and are simple to handle with availability of well defined media for growth. Additionally, the recombinant protein can be cloned for over-expression and secretion in the fermentation media which renders recovery of the protein less demanding and more cost effective. The yeast system, similar to higher eukaryotic systems, allows post transcriptional modifications like glycosylation, however the glycosylation may be different or heavy in yeast. Therefore, yeast system of expression is most frequently used for those eukaryotic proteins of commercial importance, where glycosylation patterns do no not affect the intrinsic activities of the proteins. Several proteins of pharmacological importance have been produced in yeast including subunit vaccines like HBsAg etc.

Some of the common yeast systems used for recombinant protein expression includes *Saccharomyces cerevisiae* (sugar mold), *Schizosaccharomyces pombe* (fission yeast), *Pichia pastoris* (methylotrophic yeast), *Pichia methanolica* (methylotrophic yeast) etc. Like a typical expression vector, yeast expression vectors have an origin of replication, a selection marker, a promoter sequence and a multiple cloning site (Figure 67.1). Some yeast expression vectors can also have integration sites for integration into yeast genome through homologous recombination.

The yeast expression vectors are mostly 'shuttle' vectors that can be propagated in *E. coli* for selection before transforming into yeast. For this they carry the *E. coli* origin of replication and a selectable marker for the bacterial host, like beta-lactamase for ampicillin resistance. The yeast expression vectors are discussed below.

Integration vector (YIp): These vectors cannot replicate autonomously. These vectors have two yeast fragments for integration into yeast genome through homologous recombination. The integration is usually as a single copy, however vectors containing repetitive DNA sequence like Ty element can integrate at multiple sites in the genome.

Episomal Plasmids (YEp): As these vectors carry yeast origin of replication, these are capable of autonomous replication. Though they have high frequency of transformation and copy-number, they are often not retained despite of selection pressure.

Autonomously replicating plasmids (YRp): These carry a yeast origin of replication (ARS sequence) and can allow several 100-folds propagation of transformed plasmid and are most preferred for over-expression of heterologous proteins.

Centromeric plasmids (YCp): These are low copy-number, unstable plasmids with centromere sequences and yeast origin of replication and hence are capable of autonomous replication. More than an expression system, these are used for regular cloning.

The selection markers in these plasmids are of two types namely, antibiotic markers which are dominant selection markers, and complementation markers Complementation markers are auxotrophic markers. These markers complement auxotrophic mutation in the genome like URA3, LEU2, TRPI and HIS3. It helps in selection of the recombinant yeast in a specific media. For example; URA3 codes for orotidine 5-phosphate decarboxylase (ODCase), an enzyme involved in the pyrimidine ribonucleotide synthesis in yeast. Yeast cells with mutation in URA3 can grow only in presence of either uracil or uridine exogenously provided in media. If such yeast cells are used for expression of hetrologous gene with URA3 marker, the transformed yeast expressing URA3 and hence the linked gene of desire can be selected on the basis of their

TABLE 67.4 Selectable genetic markers in yeast expression systems

Selection marker type	Gene	Remarks
Dominant markers (antibiotic resistance)	KANR	Kanamycin resistance
	TUNR	Tunacamycin resistance
	HYGR	Hygromycin resistance
	CHLRMR	Chloremphenicol resistance
	CUP1	Copper resistance
	G418R	Aminoglycoside resistance
Complement markers	URA3	Used in URA3- or ADE2- yeast strains for complementation in nucleotides biosynthesis pathway
	ADE2	
	HIS3	Complementing auxotrophic mutations in amino acid biosynthesis pathway
	LEU2	
	LYS2	
	TRP1	

growth in uracil or uridine deficient media. Table 67.4 gives a gist of some of the selection markers used in yeast expression vectors. The promoter for yeast expression system can be both constitutive (e.g., GAPDH) or inducible (e.g., ADH2, SUC2). Table 67.5 tabulates some of the promoter elements in the yeast expression system with respect to their properties.

TABLE 67.5   Promoter elements in yeast expression systems

Constitutive promoters	Regulated promoters	Heterologous promoters
Pyk1, Pgk1, Adh1, Eno	Gal1, Gal7, Gal10, Cup1, Met25, Adh2, Pho5	Androgen response element (ARE) Glucocorticoid response element (GRE) Cauliflower Mosaic Virus promoter (CaMV)

Yeast expression also comes with a variety of secretory signals like HSp150, Killer toxin type 1, PHO1 etc. These secretory signals help secretion of the recombinant proteins in the fermentation media for easy recovery.

Yeast cells are most frequently transformed by permeablising the cells with Lithium-acetate or by electroporation. Yet another method is by particle bombardment using DNA coated tungsten microprojectile particles. The 'Biolistics' technology is a physical and most flexible method of DNA delivery in yeasts.

Yeast expression systems heavily glycosyslate the recombinant protein with high mannose chains to N-glycosylated intermediates, the pattern and sequence of which may differ from the naturally occurring proteins. This may be the only major shortcoming of the system which otherwise is commercially more viable system than mammalian expression system for therapeutic mammalian biomolecules. To overcome this, a novel yeast expression system has been engineered that can glycosylate secreted mammalian proteins with similar glycosylation sequence as that found in mammals. The glycosylation pathway of *Pichia pastoris*, the methyltrophic yeast, has been modified to resemble that of mammals. This involved removal of endogenous enzymes that add high mannose chains to N-glycosylated intermediates (Och1p alpha-1-6-mannosyltransferase) and addition of five enzymes that synthesize human oligosaccharides (mannosidases I and II, N-acetylglucosaminyltransferases I and II, and UDP-GlcNAc transporter). This system, can now be used for production of large quantities of uniformly glycosylated humanized proteins. The yeast expression system has the additional advantage of avoiding cross contamination with mammalian viruses while commercial production of mammalian protein, especially therapeutic antibodies.

B. Baculovirus Expression System

Baculovirus became familiar to mankind through the silkworm disease which later became popular as a bio-pesticide since it specifically affects insect cells. In the recent times, baculovirology has become important for other reasons as well including their use in eukaryotic gene expression. Polyhedrin (polh) is an abundant protein in the viral coat structure; the promoter driving the expression is very strong which is second only to the strongest SV40 (from Simian Virus) promoter and is known as workhorse promoter of this system. This promoter was hence forth exploited for expression of foreign proteins in insect host by infection of the recombinant virus encoding a foreign gene instead of the polyhedrin gene. Since this expression system overcomes many of the problems of prokaryotic expression system and presents an edge over yeast expression system and is more easy and economical as compared to higher eukaryotic systems like mammalian systems; it has become a popular system of expression of eukaryotic proteins both for academic and pharmaceutical purposes. Table 67.6 gives a list of proteins expressed using baculovirus expression system for human health care.

Baculoviruses, as other viruses, depend on their host for their propagation and protein synthesis. They infect insects of the phylum arthropoda. When a baculovirus infects insect cell it takes over the host cell machinery for its replication, protein synthesis and finally resulting in lysis of the host cell in about 3–4 days. The virus gene expression occurs in a cascade, wherein the early gene products are involved in late gene expression and late genes in very late gene expression. Polyhedrin is expressed in the late and very late stages of virus infection. So, the polh promoter is active from late phase (48h) and hence the recombinant protein is expressed from late phase onwards. Polyhedrin protein amounts to 50% of total cell protein.

The expression of foreign proteins is initiated by preparation of recombinant virus carrying the gene of interest. The gene of interest is amplified and cloned into a transfer vector (for example, pFastBac series) and this construct is transposed into '*att*' sites present in the bacmid (baculovirus genome contained in a plasmid) of DH10bac strain of *E. coli*. After successful transposition, the recombinant bacmid, containing the entire baculovirus genome with the gene of interest replacing the polyhedrin gene, is transfected into host cells (*Sf*9, *Sf*21, *Tni* etc.) to make viral particles. The titer of these viral particles, is amplified by successive transfection using fresh host cells. Each successive infection increases the viral titer by 10 fold. The host cells are infected with high titer recombinant virus at a Multiplicity Of Infection (MOI) equal to 10. The cells are monitored for expression of foreign protein from late phase onwards (48h post infection).

'polh' promoter is the most commonly used and well characterized promoter for foreign gene expression using baculovirus expression system. However, some proteins do not show optimal biological activity due to aggregation of proteins at higher concentration. This problem could be overcome by use of other relatively strong promoters in baculovirus like p10 promoter. Besides this, sometimes glycosylation of proteins is not optimal in very late phase. This problem has also been addressed by use of early promoters such as 'ie-1' and enhancing their expression by use of polh enhancer elements like hr3 (homologous region 3, an enhancer element).

TABLE 67.6 List recombinant therapeutic proteins produced from baculovirus system

S.No	Recombinant protein	Remark
1.	Vaccines	
	Influenza	Baculovirus expressing trivalent influenza virus hemagglutinin (rHA0 trimeric recombinant
	H5N1	HA (rHA) proteins from two HPAI H5N1 viruses were purified in baculovirus system
	Western equine encephalitis virus (WEEV)	Full-length E1, the E1 ectodomain, an E26KE1 polyprotein precursor, and an artificial, secretable E2E1 chimera under immediate early (ie1), late (p6.9), and very late (polh) promoters respectively were expressed and purified in baculovirus system
2.	Interleukins/interferons	
	• Interleukin-2	Produced in stable Drosophila S2 cells as secreted protein
	• Interleukin 3	Highly active, soluble and stable protein was purified in single step.
	• Interleukin 4	Swine IL-4 was expressed in insect cells in a biologically active form hIL-7 is a
	• Interleukin-6	valuable supplementary agent for immunotherapeutical treatments in patients with
	• Interleukin-7 (Human)	immunodeficiency purified recombinant protein induced similar levels of IFN-gamma
	• Interleukin-18 (IL-18)	as native IL-18 protein in human PBMC and can also specifically bind to EGFR which
	• Interferon-γ (Human, bovine, canine, equine, chicken etc)	can be further developed as novel tumor therapeutic rhIFN-γ protein was expressed in T. ni that had potent antiviral activity recombinant viruses AcNPV-hp40 and AcNPV-hp35
	• Interlekin12 (Human, porcine, bovine, mouse etc)	were co-infected to generate active IL-12
3.	Hormones	
	• CTB -insulin (Cholera toxin B conjugated insulin)	The conjugate protein induces tolerance and protects against diabetes type1 (auto-immune) involving CD4$^+$ CD25$^+$ Foxp3$^+$ Treg cells.
	• Adiponectin	Recombinant adiponectin can be used to treat deficiencies of this hormone which results in type II diabetes and cardiovascular disease.
	• Cecropin B-binding site of luteinizing hormone releasing hormone (CBLHRH')	The anticancer effects of the CB-LHRH' protein was shown through the LHRH receptor, and it is a potential candidate as peptide drug for targeted cancer therapy shows hypercalcemic effect in rats
	• human calcitonin	
	• hTSH glycosylated	Absorbs antibodies to TSH and block hyper activation of thyroid (Graves disease)
	• h prolactin	substitute for pituitary-derived prolactin high levels of protein that binds LHR with high
	• hchorionic gonadotropin-luteinizing hormone receptor ectodomain complex	affinity have been generated.
4.	Growth factors	
	• Human growth hormone	Investigated the activity of hGH in rats which showed a significant gain in body weight.
	• hFSH	EGFL7 helps in the recruitment of endothelial and smooth muscle cells in the process of
	• EGFL7	angiogenesis.

One of the limitations of BEVS is that complex glycosylation patterns including addition of residues such as sialic acid is not possible in insect system. Some of the limitations have been addressed by modification of the host strain such as those stably expressing enzymes associated with mammalian-specific glycosylation. Also, certain cell lines perform better in certain glycosylation patterns than others. Other promoters of choice in baculovirus include p10, ie-1, vp-39, sericin (of *Bombyx mori*, silkworm).

BEVS has been successfully employed in expression and purification of multi-protein complexes. *Bombyx mori* has been utilized as biofactory wherein the recombinant protein is expressed in the whole organism. Usually, in silkworm system of expression, the recombinant protein is tagged with a 'ser' signal which makes the protein water soluble and also eases the downstream processing. This expression system is easy to carry out and economical and therefore of biotechnological importance.

BEVS has been exploited for human health care systems and has produced several proteins (Table 67.6) some of which have already entered the drug market. Apart from production of recombinant proteins, baculovirus has been exploited for various purposes in human health care management. The different applications are briefly discussed.

Display technology

Baculovirus display technology has been used to display peptides/proteins for screening proteins of a particular biological activity. The foreign protein (passenger protein) is tagged to a membrane protein (anchor protein, of viral origin) of a library of proteins expressed in the virus allowing a stable display of foreign protein at the surface. The specific protein expressing clone is identified on the basis of its specific biological activity. Its applications in receptor mediated endocytosis, as vaccine, as biocatalyst, in bioremediation process holds immense potential in human health care systems.

Gene therapy and vaccine

Exploiting the property of baculovirus to infect but not replicate in mammalian cells, its potential as a gene therapy vehicle has been explored in several human diseases in experimental models. They also induce anti-viral effects. Bacmam virus is baculovirus in which the gene therapy candidate is placed under a mammalian promoter. The gene is specifically expressed upon the entry of the virus into mammalian cells. The advantage of Bacmam virus over other methods of gene therapy is that there is no restriction on the length of gene transferred and has high transduction efficiency (70%). Also, multiple genes can be introduced simultaneously. This method has also been used to create stable cell lines by either random integration or site specific recombination by use of 'rep' gene of adenovirus. Tissue specific promoter can be employed to specifically express the foreign gene using Bacmam.

As a vaccine candidate, however, BEVS has certain limitations such as immunogenic viral peptides and consequential inactivation by complement system. Baculovirus mediated gene delivery has been demonstrated to be very efficient in the central nervous system.

Virus-like-particles

Virus like particles refers to virus structural protein assembly excluding the genetic material. They are mostly used as vaccines for various viral diseases like SARS, influenza, HPV, hepatitis, papilloma etc. It can also be used to study assemble of viral particles and it may be possible to design inhibitors.

C. Mammalian Expression Systems

Mammalian expression systems are not preferred for large scale commercial production of biomolecules, mainly for the cost involved in maintaining cell cultures and required infrastructure. Conventionally, mammalian systems cannot produce very large amount of recombinant proteins, maximum going to a few grams per litre for commercial production of some recombinant antibodies. However, these expression systems are most appropriate for synthesis of those recombinant proteins for which stable and accurate post translational modifications like glycosylation, phosphorylation, disulphide bond formation etc. are essential for intrinsic activities.

TABLE 67.7 Therapeutic proteins produced by mammalian expression systems

Category	Protein	Disease
Cytokines	• IFN-a	Certain viral infections
	• Interleukin-2	Skin melanomas, chronic viral infections, cancer
	• Interleukin-11	Prevention of Thrombocytopenia
Hematological agents	• Factor VIII	Treatment of hemophilia
	• Factor IX	
	• Erythropoietin	
	• Anti-hemophilic factor	
Hormones	• Insulin	Diabetes
	• Human growth hormone	Children's growth disorders, adult growth hormone deficiency
	• Follicle stimulating hormone	
Growth factors	• Insulin-like growth factor	Dwarfism
	• Granulocyte-macrophage colony-stimulating factor	Production of white blood cells following chemotherapy, vaccine adjuvant in HIV-infected patients
	• Novel erythropoiesis-stimulating protein	Treating anemia arising from kidney diseases/ cancer therapy agent for repair of osteochondral defects
	• Osteogenic protein	
	• Platelet derived growth factor	Acute wound healing
Antibodies	• Bevacizumab	Colorectal cancer
	• Trastuzumab	Breast cancer
	• Adalimumab	Immune disorders
	• Infliximab	Immune disorders
	• rituximab	Non-Hodgkin lymphoma
Enzymes	• Dnase I	Cystic fibrosis therapy
	• a-galactosidase	Treatment of Fabry's disease
	• Glucagon	Treatment of hypoglycemia
	• Hyaluronidase	Used in ophthalmic surgery in combination with local anesthetics,
	• Laronidase	Treatment of mucopolysaccharidosis I

The proteins in mammalian cells can be expressed either transiently or by making stable cell lines. Mammalian systems are often used to generate secreted proteins. The secretion of the recombinant protein outside the cell simplifies the purification of the recombinant proteins from the culture media.

The most common cell lines used for mammalian expression are Chinese hamster ovary (CHO) cells and Human Embryonic Kidney 293(HEK 293) cells. Both these cell lines are easy to grow and transfect with foreign DNA. The transfection efficiency of HEK 293 cells are as high as >90%. Yet another mammalian cell line which is not only one of the oldest, but also most commonly used is HeLa cell line. This cell line is an immortal cell line that was derived from cervical cancer cells. It has been extensively used in cancer and viral infection studies. However, HeLa cells are not the preferred mammalian system for expression and purification for recombinant proteins.

The gene of interest that requires to be expressed inside a mammalian system is packaged into mammalian expression vectors discussed in the previous sections. Most commonly used expression vectors have heterologous promoters like CMV or SV40, a marker and have adequate multiple cloning site for insertion of gene of desire. As mammalian system has robust splicing machinery, the entire cassette of the gene, with exons and introns can also be cloned. Most of the mammalian expression vectors also carry a termination sequence. This consists of polyadenylation signal sequence AAUAAA present 11-30 bases upstream from polyadenylation site in all eukaryotic mRNAs, except that of yeast. SV40 termination site resembles a typical eukaryotic mRNA transcription termination site. Many vectors, therefore, carry the SV40 termination site after the multiple cloning site.

The vector DNA is then transfected into a mammalian system. Compared to other eukaryotic expression systems, mammalian cells are more resistant to uptake of foreign nucleic acids and require efficient transfection methods and reagents. Some of the methods that are commonly used for mammalian transfections are briefly discussed below.

Chemical methods include liposome and non-liposome based methods. Liposomes are artificially prepared analogues similar to phospholipid bilayer of plasma membrane. Under aqueous conditions, these form spheroid structure, that when mixed with nucleic acids, DNA or RNA, encapsulate them to efficiently deliver across the membranes of mammalian cells. Non-liposomal transfection agents include cationic lipids and polymers that can form micelles. When mixed with DNA in an aqueous condition, the lipophilic portions form the core of micelle while DNA remains exogenously bound for delivery. Other chemical methods include use of polyethyleneimine, calcium phosphate or Dendrimers. Dendrimers are branched, globular macromolecules that sometimes can be potentially toxic for the mammalian cells. For cost effective transfection, polyethyleneimine or calcium phosphate are favoured over others.

Physical methods include electroporation and microinjection of nucleic acids. Electroporation, which involves application of electric current to create transient pores on membrane for transfection, is generally used for suspension cells. Both HEK293 and CHO cells, which are most frequently used for recombinant proteins, are adherent cells and therefore are not electroporated for transfection.

Virus-mediated gene delivery into mammalian cells uses Adenovirus (ssDNA virus), Retrovirus (ssRNA virus) and Lentivirus (RNA virus)-based vectors. These transfection systems have been modified for efficient gene insertions, broaden cell tropism and to make them safer for users. For example, two genes E1 and E3 are deleted in some adenovirus vectors. Deletion of E3 enables larger sequence insertions, while deletion of E1 makes it non-replicative. This makes the adenovirus vector safer for use. Since E1 is essential for virus replication, it is either supplied through a helper plasmid, or the packaging cell line (CHO or HEK 293) should be engineered to produce it. Similarly, the 3′ long terminal repeat elements of Lentiviral vectors are removed for insertion of larger fragments of DNA. This has biosafety implications as the removal of 3′ LTR does not affect packaging, but ensures inactivation of the vector following gene insertion so that live virus cannot be produced.

Commercial and industrial level of production of recombinant proteins, like most therapeutic antibodies in the market, use stable expression of the desired gene in dihydrofolate reductase (DHFR)-deficient CHO cells. These cells are transfected with expression cassette of the gene of interest linked with the selection cassette for DHFR. DHFR converts dihydrofolate to trihydrofolate. Trihydrofolate is essential for synthesis of purines, certain amino acids and thymidylic acid *de novo* by the cells. DHFR deficient CHO cells transfected with DHFR cassette are then selected in presence of an inhibitor, such as, methotexate. Use of this inhibitor in the media kills all the cells other than those that have integrated DHFR selection cassette in their genomes. Increasing the selection pressure by increasing the concentration of inhibitor in the media results into amplification of DHFR gene and with that the linked gene of desire also gets amplified. The stably transfected CHO cells can then be screened for expressing clones. Earlier, the screening procedure used to take at the least two to three months, which has been tremendously reduced by the use of flow cytometer. The stably transfected CHO cells expressing the desired protein can be sorted through flow cytometry and cultured.

D. Plant Expression Systems

Plant cell cultures are only of late being explored as potential systems for commercial production of recombinant proteins. The most successful use of a plant expression system is production of two metabolites Shikonin and Taxol that have potential uses in molecular cancer therapeutics.

Like mammalian systems, plant systems are also appropriate for PTMs. The glycosylation of mammalian proteins is closer to natural pattern and type when expressed in plant systems than in yeast expression systems.

Plant expression systems include derivations from hairy roots, shooty teratomas, immobilized cells and suspension cells. The suspension cells are most preferred system and have been derived from several plant species, including *Arabdiopsis thaliana, Nicotiana tabacum, Oryza sativa, Glycine max, Lycopersicum*, alfalfa, *Catharanthus* etc. Plant cell suspensions are made by agitation of the undifferentiated Callus tissue in shaker flasks or fermentors. The Callus from the source plant is cultivated on soild media with appropriate mixture of hormones to prevent differentiation. The suspension cells are also maintained in the same hormone rich media to avoid further differentiation. Tobacco suspension cells are most regularly used as host cell line as these are easy to propagate and transform.

Like all other eukaryotic expression systems discussed above, the expression vectors used for plant systems are also shuttle vectors that can be selected in bacteria (refer section for Yeast expression system). The promoter elements in these vectors can be heterologus (Cauliflower Mosaic Virus promoter (*CaMV*) 35S promoter), constitutive (octopine synthase and mannopine synthase (ocs)$_3$ mas promoter; ubiquitin promoter from maize) and inducible (rice alpha-amylase RAmy3D promoter which is induced by sugar deprivation). Mostly, the vectors have a leader sequence upstream to MCS that directs the recombinant protein to the secretory pathway.

Transformation of plant cell suspension is brought by recombinant bacterial infection such as, *Agrobacterium*-mediated transformation (ATMT). Yet another new bacterium identified for this purpose is OV14. This is one of the easiest yet most effective methods of transformation of plant cells, where some of the plant species can be transformed by just dipping the callus in suspension of recombinant Agrobacterium. Particle bombardment of plant embryos or callus by exogenously coated DNA to gold or tungsten particles is yet another method. Though not as efficient as ATMT, the plant cells that resist infection by *Agrobacterium* can be transformed using this method. Electroporation and viral transformation (transduction) are additional methods for the purpose.

Though the field of generating genetically modified plants (GMP) is very well developed, but lot requires to be done to make plant expression systems commercially viable for biomolecule synthesis as the major challenges that persist are low yield and easy recovery.

Some of the mammalian proteins of medicinal value that have been successfully produced in plant cell culture is tabulated below in Table 67.8. However, it is to be remembered that these systems are yet not exploited for commercial purposes. Molecular Pharming is a term used to denote the production of biopharmaceuticals in living organisms such as plants. Plant expression systems have also been exploited for generation of antibodies which are often termed as *Plantibodies*. Plant expression system offers great promise; however currently they pose certain limitations which need to be overcome such as- poor yield of the recombinant protein (probably due to poor protein stability); difficulties in downstream processing; presence of non-authentic glycan structures; evidences of horizontal gene transfer; biosafety issues with virus-based systems etc.

V. PROSPECTS AND FUTURE DIRECTIONS

Recombinant proteins are gaining importance with time as they currently contribute 10–15% sales of the total drug market. Therapeutic proteins are those whose intake helps in combating the disease wherein the protein is inadequate, absent or offers novel function. They may also be used to target pharmacological agents to specific cell types/tissues. With the present advancements in molecular biology, recombinant protein production and purification has become very simple. Proteins for therapeutic purposes are produced either by bacterial (*E. coli*) or mammalian expression systems (Lentiviral technology, Lentigen). Table 67.8 lists the different therapeutic proteins available for human use.

Mammalian systems for protein production are dominating the therapeutic protein market for clinical applications. Several features of mammalian expression systems favor their use such as post translational modifications, protein assembly and folding. The first recombinant human protein produced in mammalian expression system approved for use in humans was tissue plasminogen activator. About 60–70%

TABLE 67.8 **Therapeutic proteins produced in plants**

Protein	Host plant	Remarks
Human erythropoietin	*Nicotiana tabacum* (suspension culture)	Used in anemia arising from kidney diseases/cancer treatment etc.
Human anti-trypsin	*Oryza sativa*	Pulmonary emphysema
Human interleukin-2,4	*N. tabacum*	Renal cell carcinoma
Human serum albumin	*N. tabacum*	Restore blood volume in trauma, burns and surgery patients; Hypoalbuminemia
Human granulocyte-macrophage colony-stimulating factor	*O. sativa/N. tabacum*	Stimulates production of white blood cells following chemotherapy
Hepatitis-B surface antigen	*Glycine max, N. tabacum*	As vaccine against hepatitis B
mAB against HBsAg	*N. tabacum*	Treatment of hepatitis B
Bryodin 1	*N. Tabacum*	Cancer therapy
Recombinant ricin	*N. tabacum*	Cancer therapy

of recombinant therapeutic proteins are contributed by mammalian expression systems such as CHO, BHK, HEK 293, human retinal cells which have gained approval by regulatory authorities. Schematic representation of recombinant protein production in mammalian systems is shown in Fig B. Most of the biopharmaceutics produced using mammalian systems are secreted which are either naturally secreted or mediated (through gene construct) to secrete. The main limitation in mammalian expression system is that the yield is very low and hence production costs are high. Improved fermentation tanks have been developed in the past two decades enabling high productivity (from 50 mg/liter to 4.7 g/liter).

Eukaryotic proteins are increasing becoming important for both therapeutic and academic research. This figure captures the several applications of production of eukaryotic gene expression system.

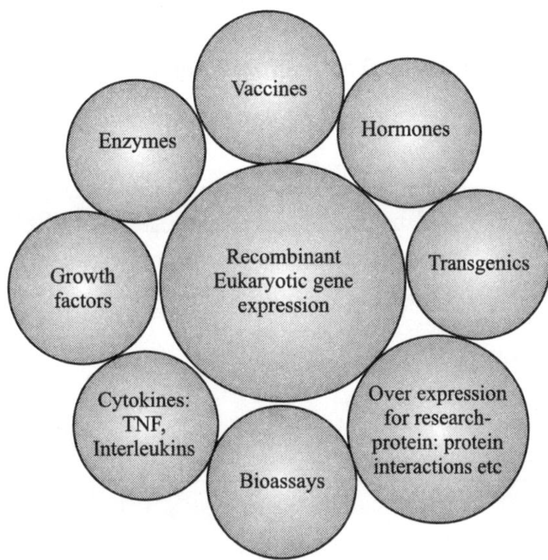

Figure 67.3 Schematic representation of different applications of Recombinant Eukaryotic gene expression.

SUGGESTIONS FOR FURTHER READING

1. Colosimo A., Goncz K.K., Holmes A.R., Kunzelmann K., Novelli G., Malone R.W., Bennett M.J. and Gruenert D.C. (2000), *Bio Techniques*, Vol. 29, 314–331.

2. Michael A.R., Carol A.S. and Jeffrey J.C. (1992), Foreign Gene Expression in Yeast: A Review, *Yeast*, Vol. 8, 423–488.

3. Miller L.K. (1993), Baculoviruses: High-level Expression in Insect Cells, *Curr Opin Genet Dev.*, Vol. 3(1), 97–101.

4. Miller L.K. (1997), *The Baculoviruses,* New York: Plenum Press.

5. Ramchandran A., Bashyam M.D., Viswanathan P., Ghosh S., Kumar M.S. and Hasnain S.E. (2001), The Bountiful and Baffling Baculovirus: The Story of Polyhedrin Transcription, *Curr Sci.*, Vol. 81, 998–1010.

6. Richard G.B., Martin A. and Gleeson G. (1991), Yeast Systems for the Commercial Production of Heterologous Proteins, *Nature Biotech*, Vol. 9, 1067–1072.

7. Geisse S., Gram H., Kleuser B. and Kocher H.P. (1996), Eukaryotic Expression Systems: A Comparison, *Prot. Exp. and Purif.*, Vol. 8, 271–282.

8. Hellwig S., Drossard J., Twyman R.M. and Fischer R. (2004), Plant Cell Cultures for the Production of Recombinant Proteins, *Nature Biotech.*, 2004, Vol. 22(11), 1415–1422.

68

DNA Fingerprinting
Principles and Applications

Seema Mishra, D.S. Negi and Seyed E. Hasnain

CONTENTS

I. DNA FINGERPRINTING: ALEC JEFFREY'S PHENOMENAL DISCOVERY

Nature has established a code or a way of identifying individuals in a species through a remarkable property known as DNA variation. This code was kept secret and hidden from our views until the 1980s when it was cracked open by Professor Sir Alec John Jeffreys at UK in the form of DNA fingerprinting. Little did he realize then that this code could evolve into a widespread phenomenon, with diverse applications across the entire spectrum of life, impacting society. At the University of Leicester, where he was studying DNA variation using mini-satellites (short, repetitive sequences of 10–60 basepair (bp) length distributed throughout the genome), he stumbled upon an X-ray film of the DNA bands, where as expected, he found many mini-satellites present and in addition, also observed that mini-satellites from one individual varied greatly from another's. This discovery of DNA fingerprinting, also known as DNA Typing or Genetic Typing or DNA Profiling, was sheer serendipity and was due to the inability of a student in his laboratory to reproduce Southern blot data, generated earlier by other co-workers in the lab, despite performing identical experiments with just one 'inconsequential' difference, i.e., the target DNA being probed was isolated from a person different from the previous experiments. He was intelligent enough to at once understand the implications, and concluded on the basis of these observations that except for homozygotic identical twins, every person on the earth can have a unique mini-satellite fingerprint, thereby establishing the identity of

that person. In his own words, "It was only after we got that first DNA fingerprint, or set of fingerprints, that the penny dropped that we'd accidentally stumbled upon a method for individual identification, and also for looking at family relationships. I think the penny dropped within about a minute of developing that first X-ray film. So this was a very exciting moment where, literally, my entire life changed in the space of about 60 seconds".

DNA fingerprint refers to the DNA pattern associated with unique genetic makeup of a particular species or individual, which in sum total, is totally different from individuals within that species or between species. From this moment of accidental discovery, DNA fingerprinting has evolved with widespread ramifications ranging from individual identification, paternity testing, identifying biological parents who have had a child through surrogacy and in child swapping cases, pure plant variety identification, in forensics, investigating genetic diversity of wildlife, understanding molecular epidemiology of infectious pathogens, among others.

II. DNA FINGERPRINTING: BASIC PRINCIPLES

The basic premise of the DNA fingerprinting process is simple, although the technological methods have been constantly evolving over the years. *Variations* in DNA sequences in terms of *repeats* or *base pair* (bp) changes are scanned and compared between two individuals. In the initial version of the fingerprinting technology, the variations were visually inspected as *unique* pattern of bands on a gel, separated from each other due to *size* (or length of DNA molecule in question) differences. This unique overall pattern is known as a DNA fingerprint for that particular individual. No two individuals, except homozygotic twins, are genetically identical. If the two individuals are related, there will be many bands in common between these two. Unrelated individuals will have different banding patterns. One of the several applications of DNA fingerprinting is in paternity testing (Figure 68.1). DNA, inherited from father and mother and passed on to their biological progeny, will display common pattern of variable regions or unique bands. Upon a visual comparison of the bands of progeny (D1, D2, S1, S2) with parents ("mom" and "dad"), all the bands in the case of D1 and S1 are also present in mom and dad's lane with some bands corresponding to mom's DNA and some to dad's, leading us to the conclusion that D1 and S1 are their biological daughter and son, respectively. In the case of D2, the banding patterns match with some from mom's lane but not with bands in dad's lane and, hence she has a different biological father. S2 has a different biological mother and father, and hence is a totally unrelated child.

III. DNA FINGERPRINTING: THE PROCESS

Practically speaking, the DNA fingerprints are produced, detected and analysed using the following three basic steps:

1. The DNA from an individual/organism is isolated. While in earlier times, the DNA was fragmented using specific restriction enzymes, in today's technology, the target sequences are amplified using polymerase chain reaction (PCR) corresponding to a *specific* DNA locus amenable to easier amplification and detection.

2. The variation in DNA sequence or locus from one individual/organism to another is scanned. This variation may be in forms of tandem repeats such as mini-satellites (also known as variable number of tandem repeats (VNTRs)), micro-satellites (as compared to mini-satellites, micro-satellites are of 2-6 bp in length, and are also known as short tandem repeats (STRs)), or using still smaller regions with 1 bp difference called single nucleotide polymorphisms (SNPs). Modern era is the era of next-generation sequencing, which is a high-throughput method where hundreds of DNA samples can be sequenced and monitored for variation.

The reason for focussing on the tandem repeats in a DNA sequence is that while most parts of a DNA are highly conservative between individuals, and it is this conservativeness which gives rise to a particular species; some parts of a DNA, such as repeats, are variable and these variable regions give rise to the uniqueness and diversity of individuals. The Human Genome Project estimates that 99.9% of human genome is conserved and just about 0.1% is variable. This variability arises as a result of insertion or deletion of repeats on chromosomes during errors in the crossing-over process in meiosis, and may also arise from other forms of recombination. In fact, it is the *length, location* and *number* of tandem repeats within a mini- or micro-satellite that is highly variable, rather than the *sequence per se*. Single nucleotide polymorphisms, as the name itself suggests, have a difference of *one bp* between individual DNA molecules, and are deciphered on the basis of *sequence* variation.

Mini-satellites usually share a *core* sequence, which is present within each repeating unit in a minisatellite (Figure 68.2).

Microsatellites, as noted above, have a repeat unit of 2–6 bp in length, and one example of a microsatellite at a loci composed of repeating GATA units is shown is Figure 68.3.

SNPs are thought to be present only when one bp change occurs in about 1% of the population, and not when such a change occurs in only two or three individuals (Figure 68.4).

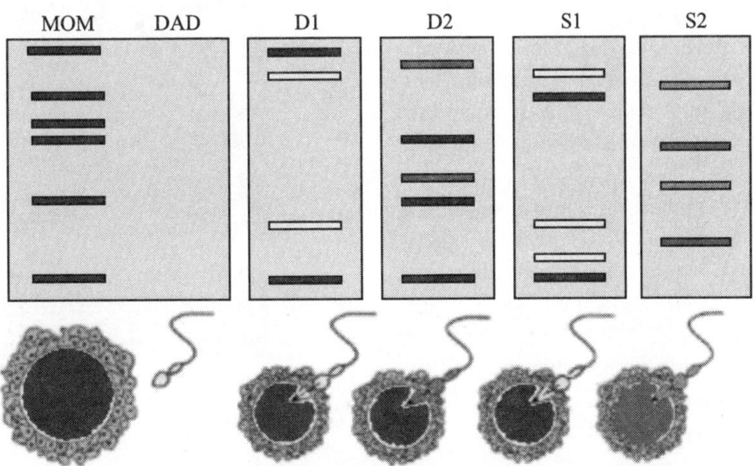

Figure 68.1 Basic premise of establishing individual's uniqueness through DNA fingerprinting. Note that all bands seen in D1 are also present in either mom or dad. *(From URL: http://www.udel.edu/chem/C465/senior/fall97/blood_II/vntr02.gif)* *(see Plate 28 for colour figure)*

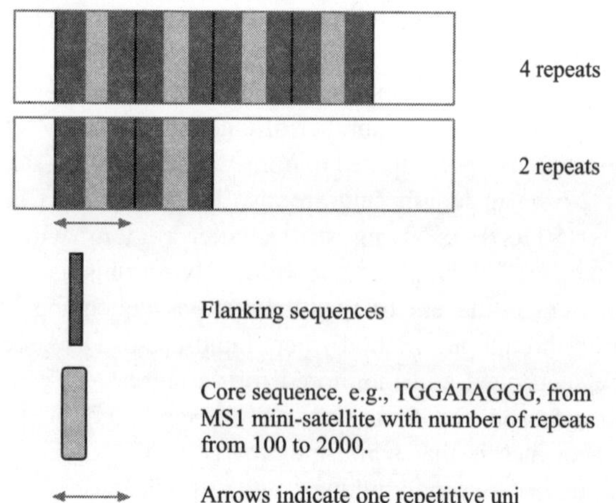

Figure 68.2 Diagrammatic representation of minisatellite at a locus of an individual heterozygous for this locus; the number of minisatellite repeats, and accordingly, the locus genotype is 4, 2. (*see Plate 28 for colour figure*)

AAGG TTAATATATA TAAAGGGTAT GATAGAACAC TTGTCATAGT TTAGAACGAACTAACGATAGATAGATAGATAGATAGATAGATAGATAGAT AGATAGATAG ATAGACAGATTGATAGTTTT TTTTTATCTC ACTAAATAGT CTATAGTAAA

Figure 68.3 Example of a microsatellite repeat unit. (*see Plate 28 for colour figure*)

> Individual 1: CCTTACTTGCA
> Individual 2: CCTTTCTTGCA
> Individual 3: CCTTGCTTGCA

Figure 68.4 SNPs in a general population (SNPs are colored red). (*see Plate 28 for colour figure*)

3. Detection of DNA variation is usually performed by radioactively or fluorescently labelled hybridization *probes* in gel electrophoresis (Figure 68.5) or by capillary electrophoresis (Figure 68.6). Probes are those DNA molecules that are complementary to the locus in question, and to which they can bind and produce a radioactive or fluorescent signal, which can be visually inspected. An example of a probe for the core sequence present in Figure 68.2 would be: ACCTATCCC. Patterns of variability in the form of DNA bands between known and unknown samples are compared with each other as well as with DNA ladders, to deduce DNA size, and/or control fragments. Identical samples with same repeats will show same patterns.

Among several probes designed, the core sequence analysed by Prof. Jeffreys (GGGCAGGAXG, where X is any nucleotide, or GGAGGAGG) is similar to Chi-recombination activator present in *Escherichia coli* and a multi-locus probe was derived based on Chi-like sequences on the premise that recombination is, in part, responsible for the hypervariability of sequences, just as crossing-over error during meiosis is, and so will easily hybridize to hypervariable sequences and detect them.

It is the correct choice of probes that determines an almost accurate outcome. A *combination* of probes selected enhances the accuracy of the results, and one probe should

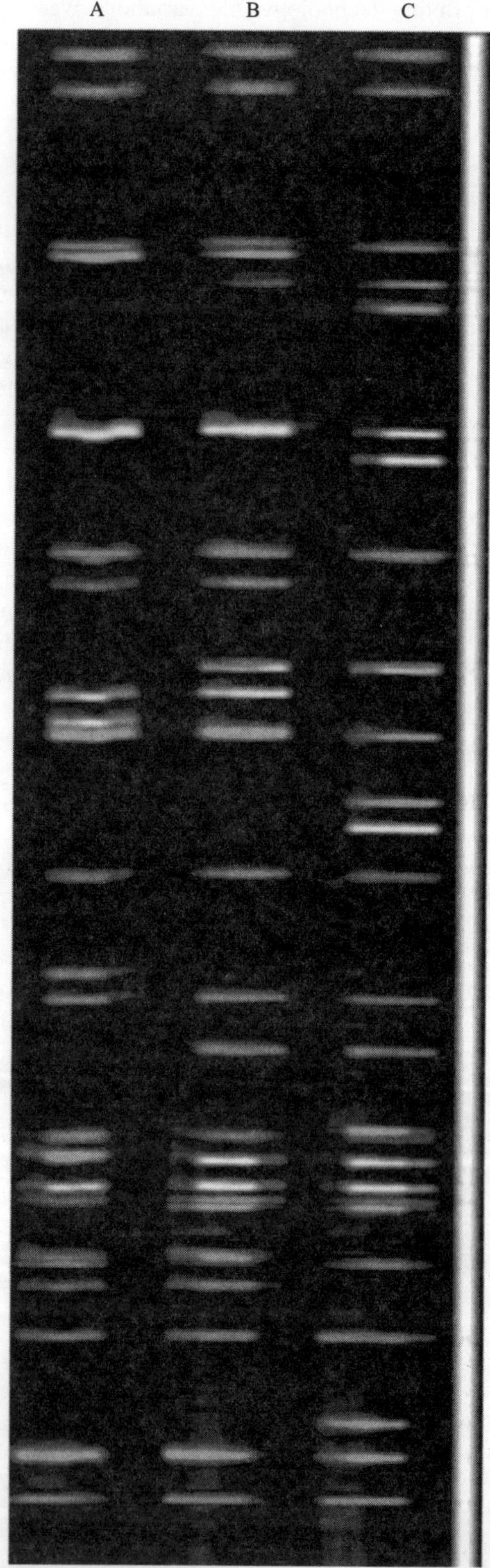

Figure 68.5 A representation of DNA fingerprinting pattern as seen after gel electrophoresis. A, B and C lanes correspond to DNA derived from the mother, her daughter and father. The blue, green, yellow and red bands correspond to different loci analysed. Note that every single band in B is also present in A or C. (*see Plate 28 for colour figure*)

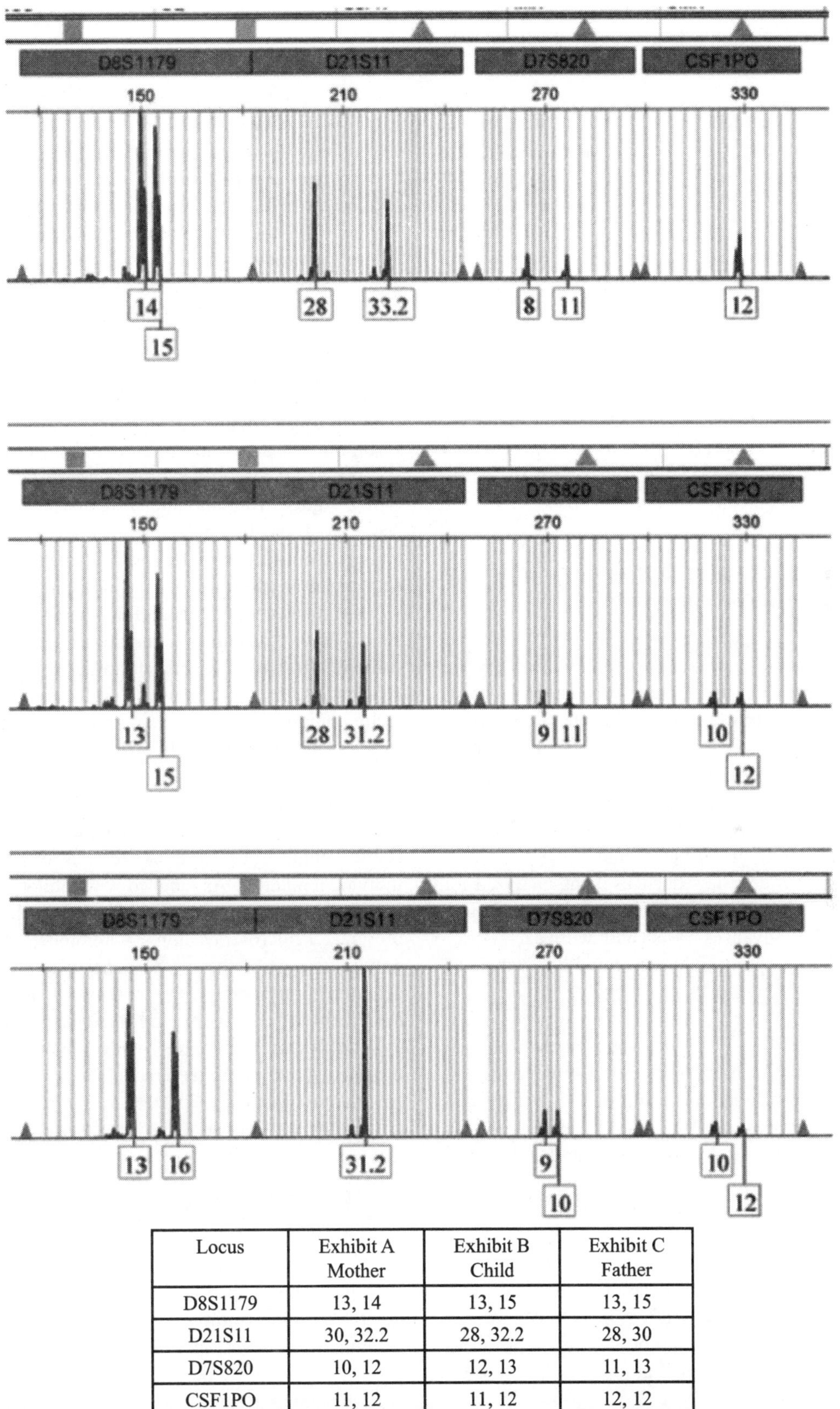

Locus	Exhibit A Mother	Exhibit B Child	Exhibit C Father
D8S1179	13, 14	13, 15	13, 15
D21S11	30, 32.2	28, 32.2	28, 30
D7S820	10, 12	12, 13	11, 13
CSF1PO	11, 12	11, 12	12, 12

Figure 68.6 An example of DNA Fingerprinting carried out by capillary electrophoresis in an automated machine. The x-axis represents length of amplified DNA in bp, and y-axis represents peak height. The electrophoretic gel images were converted into a tabular form where the numbers shown are repeat numbers for a particular loci. It can be seen that of the 4 loci checked, the child has one locus each drawn from the mother and the father. (*see Plate 29 for colour figure*)

be specific to a given locus, for single-locus probes. To rule out the risk of linkage disequilibrium, caution should also be taken that probes should not hybridize to loci that are very close to each other.

Loci to be analysed are chosen based on their allelic variability, and the probes used to detect them can be either a multi-locus (hybridizing to two or more loci) or a single-locus probe (hybridizing to one locus).

Population frequency of alleles is also taken into consideration and is important, as this must be known to determine the uniqueness. An allele is one of the alternative forms of the same genetic locus. For example, if an allele is present at a frequency of 1 in 20 in a population, there will be a chance that $1/20 = 5\%$ of this population would have this allele. Including another probe in this study for an allele that has the frequency of occurrence of 1 in 8, the new frequency of occurrence would be 1 in $(20 \times 8) = 1$ in 160, i.e., only 1 in 160 individuals will show a specific pattern of bands, thereby maximizing specificity.

IV. EVOLUTION OF THE TECHNOLOGY OF DNA FINGERPRINTING

The technological variations have largely evolved due to the ongoing efforts to simplify the tedious procedures so as to reduce the time and most importantly, due to different genome sizes in different species and the need for obtaining hybridization probes specific to each species or a genome, which can help detect the variations. Mini-satellites, micro-satellites, SNPs, RFLPs, AFLPs, RAPDs, SRAP, TRAP, RGAP are actually several variations of one theme.

Based on the size of genetic sequences being analyzed for variation evolving from mini- to micro- to one-bp variation, and the technological advances associated with each, three phases of the DNA fingerprinting history could be established as follows:

Phase 1: Starting from early 1980 to late 1980s. Practical steps involved isolating large quantities of DNA, restriction enzymes digestion, electrophoresis, Southern blotting with radioactively labelled probes, autoradiography and detection for the presence of mini-satellite repeats. The main disadvantage was that these steps were slow, tedious and unfit for obtaining large amounts of DNA from old, fragile samples.

Phase 2: Starting from late 1980s to 2000. The advent of polymerase chain reaction (PCR) changed the technology, with the main advantage of ease of use and requirement of minimal sample amounts. Microsatellites could be detected easily. Old samples and samples present in small amounts could be used for identification.

Phase 3: Starting from 2000s and continuing. It is the era of Next Generation Sequencing (NGS) and detection of variation through SNPs.

V. 21ST CENTURY NOVEL APPROACHES

A global picture of an individual's genome is even better. Intuitive enough to understand, it will provide us with a whole plethora of sequence variations. Minisatellites, microsatellites, SNPs, insertion/deletion (indel) mutations all can come into focus *simultaneously* and *within context*. A *whole genome* sequence can let us achieve exactly that. Sequencing of J. Craig Venter's diploid human genome and comparison with human reference genome from National Center for Biotechnology Information (NCBI) revealed several types of genetic variants, more than 4.1 million, including SNPs, block substitutions and indels, and many novel ones. Advanced and next generation sequencing technologies (IlluminaHiSeq 2000 and 2500, PacBio, Life Sciences' Qdot technology, Oxford Nanopore) have evolved to provide longer sequence readouts, higher resolution (clear sequence data) and lower background noise. As the sequencer's capacity increases, the data generated will be huge, running into 1 terabases (10^{12} base pairs of sequences), and 10 whole human genomes (one human genome is equivalent to over 3 billion (10^9) base pairs) can easily be sequenced in just a single machine run.

For plants with a polyploid genome, with genome size much larger than human, finding sequence polymorphisms is difficult. To overcome this and to get a whole genome profile, Diversity Arrays Technology (DArT) has been developed. This technology uses microarrays without any need for sequencing. A set of genomic DNA library from all plant cultivars with selected genotypes or gene pool is created (after DNA extraction and PCR amplification, fragments are cloned into a cloning plasmid) and arranged in a 384-well plate format and then spotted onto a glass slide to create a microarray. Samples of individual plant DNAs labelled with a fluorescent probe (green in colour) are then used to hybridize to the genomic library present on microarray chips. A reference DNA sample (e.g., polylinker part of the plasmid that was used to generate the genomic library, and which is common to all the array elements) labelled with a dye fluorescing at another wavelength generating a red color is also mixed with sample DNAs. A scanner measures the ratio of green:red signal intensity generated based on relative abundance of DNA in the two cases. Scoring for polymorphism is done using a software present.

VI. BIOINFORMATICS—DATABASES AND TOOLS

High-throughput capabilities and the ever increasing applications of DNA fingerprinting are posing challenges to store the voluminous data sets and extracting key information so as to make sense of the biological data. Without the effective use of such databases and tools in practice, experiments, inference and analyses would be a lot harder and time-consuming. To store, manage, retrieve and analyse such data, many databases and software tools have been developed. These databases and software tools are important in drawing accurate conclusions.

The usage is relatively simple and straightforward, albeit with the requirement of a sound and solid biology background for accurate inference.

VII. EXAMPLES OF SOME PROMINENT APPLICATIONS

DNA Fingerprinting has now become a part of forensic medicine and crime investigations all over the world including India. DNA fingerprinting was started in India by Dr Lalji Singh at the Centre for Cellular and Molecular Biology in Hyderabad. Realizing the potential of this modern technology, the Government of India decided to establish a dedicated Institute, the Centre for DNA Fingerprinting and Diagnostics. Over the years, there have been a number of interesting cases where justice was delivered based on crucial evidence provided by DNA fingerprinting even when the high and the mighty were involved. The almost impeachment of the President of a powerful country for sexual misconduct, the identification of the victims of 9–11 terrorist attack in New York, the identification of the persons involved in the assassination of Rajiv Gandhi, the black buck poaching by a famous Bollywood superstar in Rajasthan, identification of bodies found in a mass grave of victims of the communal riots in Gujarat, the gunning down of innocent villagers by security forces in Chhatisinghpora in Kashmir, etc. are some classic examples of applications of DNA fingerprinting. DNA fingerprinting evidence has also been successfully used in exonerating persons who were serving life term based on circumstantial evidence, thereby forcing the US Supreme Court to order retrial of all cases where DNA evidences were available.

Two very classic examples of DNA fingerprinting from UK and India are described below.

Example 1: The very first use of DNA fingerprinting—Immigration Case

The very first DNA fingerprinting application by Prof. Jeffreys was in an immigration case. A boy in United Kingdom was threatened with deportation by the authorities on the grounds that he was not the biological son of his mother, and that he had a forged passport. Prof. Jeffreys was consulted to establish that this boy was the son and not the nephew of this lady claiming to be the biological mother. This situation was further complicated because of the absence of the boy's father. Prof. Jeffreys took into account the feasibility of using this mother's other sons and his own sample for an unrelated

case to establish a conclusive evidence. DNA fingerprinting conclusively established for the first time in the world that the boy was indeed the biological son of the alleged mother. This path breaking moment was aptly described by Prof. Jeffreys in an interview: *"That was the first DNA case tackled anywhere in the world, and it is still my favorite case because I was there at the tribunal where they dropped the case against the boy, when the mother was told—and just the look in that mother's eyes! She had been fighting the case for two years.*

That was my golden moment. Without DNA, he could have been deported."

Example 2: Basmati rice exports

Basmati rice is a great export earner for India, and pure variety of Basmati rice is favoured by client countries. To account for purity, distinctness and quality of Indian Basmati unambiguously, DNA fingerprinting markers to identify differences between traditional Basmati, evolved Basmati and non-Basmati rice varieties have been derived. Uniqueness of Indian Basmati as compared to Pakistani Basmati was also established using these markers and these have helped authenticate Indian Basmati as a pure product. The markers and methods used in these studies conducted at the Centre for DNA Fingerprinting and Diagnostics (CDFD), have been of immense importance to India's economy, in terms of providing authenticity of the purity of export quality Basmati rice. Consequently, a Centre for Basmati DNA Analysis has been developed by CDFD and Agricultural and Processed Food Products Export Development Authority (APEDA) at CDFD, Hyderabad.

ABBREVIATIONS

PCR	Polymerase Chain Reaction
VNTR	Variable Number of Tandem Repeats
STRs	Short Tandem Repeats
SNP	Single Nucleotide Polymorphism
RFLP	Restriction Fragment Length Polymorphism
AFLP	Amplified Fragment Length Polymorphism
RAPD	Random Amplified Polymorphic DNA
SRAP	Sequence-related Amplified Polymorphism
TRAP	Target region Amplification Polymorphism
RGAP	Resistance Gene-Analog Polymorphism
DArT	Diversity Arrays Technology
SSR	Simple Sequence Repeat

SUGGESTIONS FOR FURTHER READING

Ehtesham N.Z., Das A. and Hasnain S.E. (1992), A Novel Probe for Human DNA Fingerprinting Based on *chi*-like Sequences, *Gene*, 111:261–263.

Ehtesham N.Z. and Hasnain S.E. (1992), A Multi-locus Probe for Human DNA Fingerprinting based on *chi*-like sequences, *Advances in Forensic Haemogenetics*, 4:137–139.

Gitschier J. (2009), The Eureka Moment: An Interview with Sir Alec Jeffreys, *PLoS Genet* 5(12):E1000765.

Jeffreys A.J., Wilson V. and Thein S.L. (1985), Hypervariable 'Minisatellite' Regions in Human DNA, *Nature* 314:67–73.

Jeffreys A.J. (2013), The Man Behind the DNA Fingerprints: An Interview with Professor Sir Alec Jeffreys, *Investigative Genetics*, 4:21.

Jeffreys A.J., Brookfield J.F.Y. and Semeonoff R. (1985), Positive Identification of an Immigration Test-case Using Human DNA Fingerprints, *Nature*, 317:818–819.

Jeffreys A.J. (2005), Genetic Fingerprinting, *Nature Medicine*, 11:1035–1039.

Levy S., Sutton G., Ng P.C., Feuk L., Halpern A.L., et al. (2007), The Diploid Genome Sequence of an Individual Human, *PLoS Biol.*, 5(10):E254.

Nagaraju J., Kathirvel M., Kumar R.R., Siddiq E.A. and Hasnain S.E. (2002), Genetic Analysis of Traditional and Evolved Basmati and Non-Basmati Rice Varieties by Using Fluorescence-based ISSR-PCR and SSR Markers, *Proceedings of National Academy* of *Sciences,* 99:5836–5841.

Pena S.D. (Ed.) (1993), DNA Fingerprinting: State of the Science, Birkhauser, ISBN-13:978–0817627812.

Weising K., Nybom H., Pfenninger M., Wolff K. and Kahl G. (2005), DNA Fingerprinting in Plants: Principles, Methods, and Applications, 2nd ed., CRC Press, ISBN-13:978–0849314889.

69

Essentials of Human Genetics during Post Genomic Era with Emphasis on the Y Chromosome

Deepali Pathak and Sher Ali

CONTENTS

The sequencing of the human genome has brought human genetics into a new era of study resulting in the generation of an explosive amount of information. Application of genomics, proteomics, bioinformatics and related technologies has made it possible to study human genetic diseases on an unprecedented scale, both *in silico* and in the wet lab. However, after more than a decade of intense exploration of the human genome, the burden of human diseases and suffering have only increased across the globe. Heart disease, cancer, and diabetes as well as allergic and autoimmune disorders have all continued to persist. Dynamic interplay of the environment and lifestyle both has been major components in causing hazardous effects on the human health. Therefore, the new era demands not only intensification of more focused research but perhaps also a change in the strategy to understand genetic basis of diseases in the context of environment augmenting more accurate genotype phenotype correlation. Finally, there is an ardent need to undertake translational research based on the epidemiological data to augment health care system in the country.

I. INTRODUCTION

Genetics deals with the molecular structure and function of the genes, its behavior in the context of a cell or organism (e.g., dominance and epigenetics), patterns of inheritance from parents to the offsprings, gene distribution, variation and changes in the populations and Genome-Wide Association Studies. Given that genes are universal to the living organisms, genetics can be applied to the study of all living systems, from viruses and bacteria, through plants and domestic animals, to humans (in the context of medical genetics).

Human genetics is the study of inheritance as it occurs in the human beings. It encompasses a variety of overlapping fields including: classical genetics, cytogenetics, molecular genetics, biochemical genetics, genomics, proteomics, bioinformatics population genetics, developmental genetics, clinical genetics, behavioral genetics and genetic counseling. Yet another aspect of human genetics is cancer biology and autoimmune diseases. Thus, the scope of human genetics has now become much wider and with the development of newer, more powerful analytical tools, more advanced levels of information is generated. This in turn has broadened our understanding of the diagnosis, prognosis and possible measure for disease management, if not for its complete cure.

Human traits are based upon Gregor Mendel's model of inheritance with exceptions in certain instances. Mendel deduced that inheritance depends upon discrete factors, which are now called genes. Genes correspond to regions within the DNA, a molecule composed of a chain of four different types of nucleotides. The sequence of these nucleotides contains genetic information that all organisms inherit. DNA naturally occurs in a double stranded form, with nucleotides on each strand complementary to each other. The sequence of nucleotides of a gene is translated into proteins having chain of amino acids.

This relationship between nucleotide sequence and amino acids is maintained in the form of genetic code. The amino acids in a protein determine how it folds into a three-dimensional shape. This structure is in turn becomes responsible for the protein's function. A change to the DNA can have a dramatic effect on the cell or on the organism causing at times, the genetic anomalies.

Decades before the availability of the sequencing technology, the first differences observed in our genetic composition were the changes in the quantity and structure of chromosomes. These included aneuploidies, rearrangements (which were often associated with diseases), heteromorphisms and fragile sites, all of which were large enough to be identified using a microscope. Subsequently, with the advent of molecular biology, and DNA sequencing in particular, smaller and more abundant alterations were observed. The human genome contains significant numbers of structural rearrangements, such as insertions, deletions, inversions, and large number of tandem repeats. Significant efforts have been made in characterizing human genetic variation at the karyotype and nucleotide level, but knowledge about variation in between these two extremes remains inadequate.

In the human genome, the linkage of a gene to a particular chromosome rarely hints upon its functions and different genes are randomly distributed on different chromosomes. One exception to this is the jumbled up gene distribution on the human Y chromosome. The Y chromosome was considered to be rich in junk DNA, and poor in useful attributes compared to any other chromosomes. It was reported to be continuously degenerating probably due to lack of recombination and therefore, some even believed in its eventual death. Y is grossly heterochromatic and gene poor chromosome that recombines with its stipulated homolog, the X chromosome only at the terminal pseudo-autosomal regions 1 and 2 (PAR1 and PAR2) on the Yp and q regions, respectively.

Functionally, the Y chromosome plays a pivotal role in sex determination and maintenance of the male gender monitoring the testis formation and spermatogenesis. The analysis of the XY females and XX males, Turners and Klinefelters uncovered a particular Y chromosomal fragment which was later named as the sex determining region (*SRY*) located on the short arm of the Y chromosome.

Recent studies have shown a startling level of structural polymorphism in the human genome. About 5% of the human genome is composed of the duplicated sequences. Moreover, segmental duplications have been reported to encompass the mutational hotspots of the chromosomal rearrangements enriching the genome with large scale variations. It is reasonable to expect the differences in the gene copy numbers as another source of genetic variations among humans.

Considering its predominant genetical status, the Y chromosome becomes an important candidate to be undertaken for its in-depth analysis. In this chapter, we present an overview of the Essentials of Human Genetics during Post Genomic Era.

II. HUMAN GENOME

The human *Homo sapiens* genome is stored in 23 pairs of chromosome and in the small mitochondrial DNA. Of these, twenty-two pairs are autosomes while the remaining one pair is sex chromosome. The haploid human genome contains three and a half billion DNA base pairs. The Human Genome Project (HGP) produced a reference sequence of the euchromatic region used worldwide for biomedical research. There are about 20,000-25,000 human protein-coding genes. About 1.5% of the genome codes for proteins, while the rest consists of non-coding RNA genes, regulatory sequences, introns, and noncoding DNA (once known as "junk DNA") (Figure 69.1).

Human genes are distributed unevenly across the chromosomes constituting gene-rich and gene-poor regions. This is correlated with chromosome bands corresponding to GC-content. The significance of these nonrandom patterns of gene density is not well understood. In addition to protein coding genes, the human genome contains thousands of RNA genes, including tRNA, ribosomal RNA, microRNA, and other non-coding RNA genes. Aside from genes and known regulatory sequences, the human genome contains a sizable quantum (by some estimates 97%) of repeat elements the function of which largely remains unknown.

The Human Genome Project (HGP) consortium first published its data on February 15, 2001, in the journal *Nature*. The project had its ideological origin during the mid-1980s, but its intellectual roots stretch back further. Alfred Sturtevant created the first *Drosophila* gene map in 1911. The crucial first step in molecular genome analysis, and in much of the molecular biological research of the last half-century, was the discovery of the double helical structure of the DNA molecule proposed in 1953 by Francis Crick and James Watson for which they received the 1962 Nobel Prize (along with Maurice Wilkins).

The Human Genome Project and a parallel project by Celera Genomics each produced and published the refined version of the human genome sequence in the year 2003. Both of which were a composite of the DNA sequence of several individuals. The human genome has different regulatory sequences controlling gene expression. These are typically short sequences that appear near or within the genes. A systematic understanding of these regulatory sequences and how they act together as a gene regulatory network is only beginning to emerge from computational, high-throughput expression and comparative genomics studies. Some types of non-coding DNA are genetic "switches" that do not encode proteins, but do regulate as to when and where genes are expressed. Identification of these regulatory sequences relies in part on their evolutionary conservation.

All humans have unique genome sequences. In the IHGSC international public-sector Human Genome Project, researchers collected blood (female) or sperm (male) samples from a large number of donors. Only a few of many collected samples were processed as DNA resources. Thus, it is the combined "reference genome" of a small number of anonymous donors. HGP scientists used white blood cells from the blood of two male and two female donors (randomly selected from 20 of each)—each donor yielding a separate DNA library. One of these libraries (RP11) was used considerably more than others, due to quality considerations. The HGP genome is a scaffold for future works in identifying differences among individuals. Although the main sequencing phase of the HGP has been completed, studies of DNA variation continued in the International HapMap Project, whose goal is to identify patterns of single-nucleotide polymorphism (SNP). Key findings

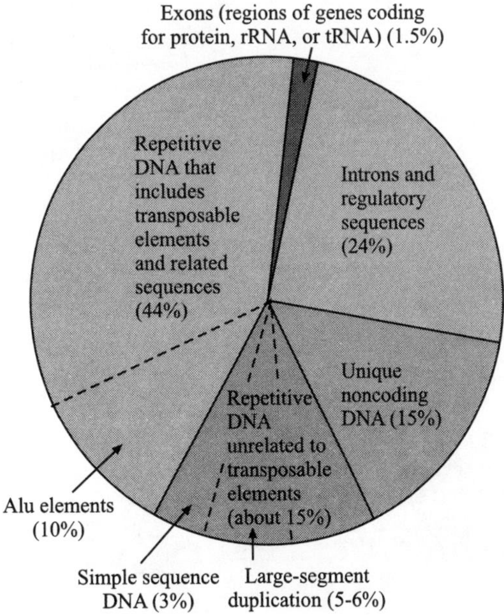

Figure 69.1 Diagrammatic illustration of the human genome showing distribution of different categories of sequences (http://www.carolguze.com/text/442-1-humangenome.shtml).

of the draft (2001) and complete (2004) genome sequences may be accessed through http://www.genome.gov/12011238/.

III. DNA AND CHROMOISOMES

DNA (deoxyribonucleic acid) carries the genetic information in the body's cells. In living organisms, DNA does not usually exist as a single molecule, instead as a pair of molecules that are held together. These two long strands entwine like vines, in the shape of a double helix. The nucleotide repeats contain both the segment of the backbone of the molecule, which holds the chain together, and a nucleobase, which interacts with the other DNA strand in the helix.

Within the cells, the DNA is organized into long structures called chromosomes. It is a single piece of coiled DNA containing many genes, regulatory elements and other nucleotide sequences. Chromosomes also contain DNA-bound proteins, which serve to package the DNA and control its functions. Chromosomes vary widely between different organisms. The DNA molecule may be circular or linear, and can be composed of 100,000 to 10,000,000,000 nucleotides in a long chain. Typically, eukaryotic cells have large linear chromosomes and prokaryotic cells have smaller circular chromosomes, although there are many exceptions to this rule. Cells may contain more than one type of chromosome; for example, mitochondria in most eukaryotes and chloroplasts in plants have their own small chromosomes.

In eukaryotes, nuclear chromosomes are packaged by proteins into a condensed structure called chromatin. Chromosomes may exist either in duplicated or unduplicated forms. Unduplicated chromosomes are single linear strands, whereas duplicated chromosomes contain two identical copies (called chromatids) joined by a common centromere. In the nuclear chromosomes of eukaryotes, the uncondensed DNA exists in a semi-ordered structure, where it is wrapped around histones (structural proteins), forming a composite material to form chromatin. Thus, the chromatin is the complex of DNA and protein found in the eukaryotic nucleus, which packages chromosomes. The structure of chromatin varies significantly during different stages of the cell cycle, according to the requirements of the DNA (Figure 69.2).

IV. GENOMICS OF THE INDIVIDUAL CHROMOSOMES; FUNCTIONS AND DYSFUNCTIONS

Human chromosomes can be divided into autosomes and sex chromosomes. Certain genetic traits are linked to a person's sex and are passed on through the sex chromosomes. The autosomes contain the rest of the hereditary information. Sequencing of the human genome has provided a great deal of information about each of the chromosomes. A chromosome is divided by its centromere into short arm (p) and long arm (q) and can be classified as follows:

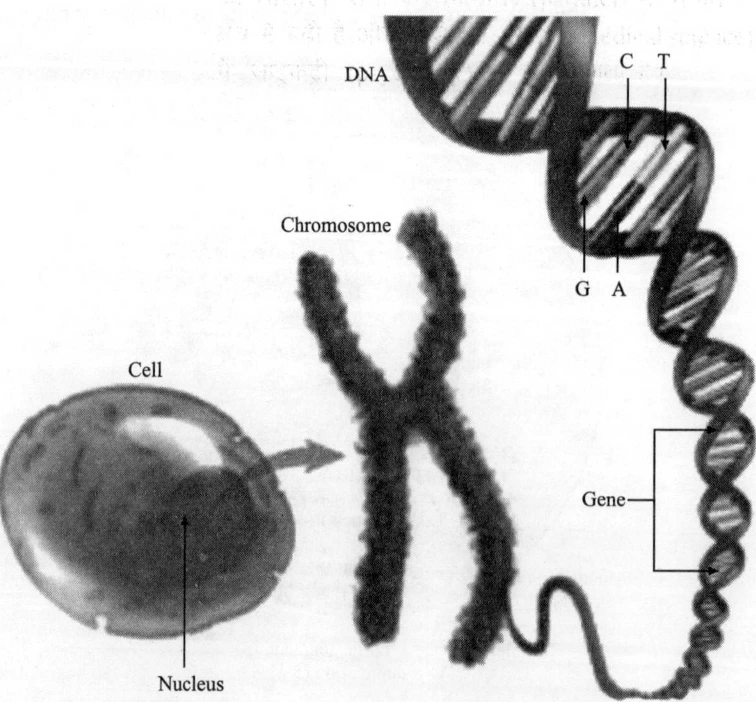

Figure 69.2　A diagrammatic illustration showing relationship of chromosome with DNA (http://2010g09r3bdnawiki.wikispaces.com/%28a%29+Chromosomes)

- Metacentric: If its two arms are equal in length.
- Submetacentric: If arms' lengths are unequal.
- Acrocentric: If the p arm is so short that it is hard to observe, but still present (Figure 69.3).

In the humans, chromosomes are arranged into seven groups (Figure 69.4) based on size and positions of the centromeres as mentioned earlier (Figure 69.3).

1. Group A: chromosomes 1-3 are largest with median centromere.
2. Group B: chromosomes 4-5 are large with sub-median centromere
3. Group C: chromosomes 6-12 and X are medium sized with sub-median centromere
4. Group D: chromosomes 13-15 are medium sized with acrocentric centromere
5. Group E: chromosomes 16-18 are short with median or sub-median centromere
6. Group F: chromosomes 19-20 are short with median centromere
7. Group G: chromosomes 21-22 and Y are very short with acrocentric centromere

In the humans and many other animals, the Y-chromosome contains the gene that triggers the development of the male characteristics. In evolution, this chromosome has lost most of its content and genes, while the X chromosome is similar to the other chromosomes and contains many genes. The X and Y chromosomes form a heterogeneous pair.

V. THE HUMAN Y CHROMOSOME

The Y chromosome is very small, representing only 2-3% of the haploid human genome. Cytogenetic analyses showed three major categories of the Y sequences, the terminal PAR1 and 2, on Yp and Yq, respectively, and the middle male specific portion including euchromatin and heterochromatin. The PAR1 and PAR2 cover approximately 2600 and 300 kb DNA, respectively. The PARs are the regions where the Y chromosome recombines and exchanges genetic materials with the PAR of X chromosome, and hence the genes present in the PARs behave more like that of the autosomal ones. These are also called pseudoautosomal regions. The euchromatic region is distal to the PAR1 and consists of Yp paracentromeric region, centromere, and the Yq paracentromeric region. Lastly, the Y heterochromatin, which is transcriptionally inert, is present in the Yq21 region and consists mainly of the DYZ1 and DYZ2 repeat units. Most important region of the Y chromosome is MSY (male specific region of the Y) or NRY (Non-recombining region of the Y) which includes everything except the PARs. The MSY euchromatin sequence is 23 Mb in total, with 8 Mb on the Yp and 14.5 Mb on the Yq regions of the Y chromosome. In addition to the centromeric heterochromatin and the distal Yq 40 Mb heterochromatic block, the Y chromosome was found to carry a third heterochromatic sequence, the DYZ3. This block contained approximately 3000 tandem arrays of a 125 bp repeat.

A. Origin and Evolution of the Human Y Chromosome

Over the past 240–320 million years of the mammalian evolution, the X and Y chromosomes have evolved from what was originally a pair of ordinary autosomes. During that evolution, just as most of the ancestral X genes were decaying on the Y because of the lack of meiotic recombination, genes which control spermatogenesis arrived on the Y from autosomes. Once on the Y, these formerly autosomal genes amplified into multiple copies, and achieved greater prominence through a process called "gene conversion." Spermatogenesis genes that arrived on the Y, but originated from the autosomes, include the *DAZ* (from autosome 3) and *CDY* genes that are among the seven gene families located in the *AZFc* region. Other spermatogenesis genes on the Y, such as RBMY, have persisted in their original position as that on the X. The

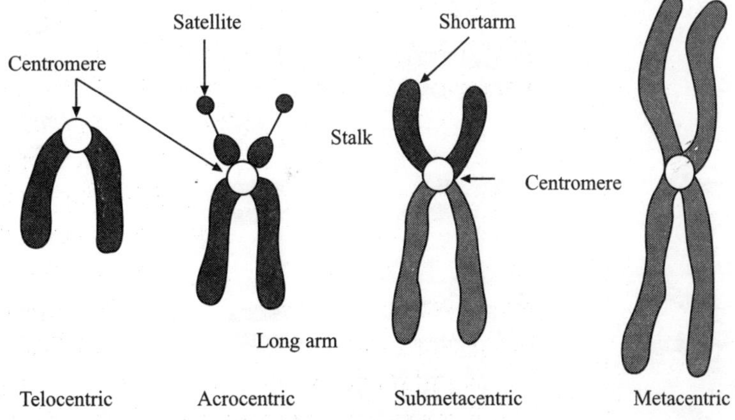

Figure 69.3 Classification of chromosomes based on centromere position (http://www.transtutors.com/homework-help/biology/chromosomes-genetic-disorder/types-of-chromosomes.aspx).

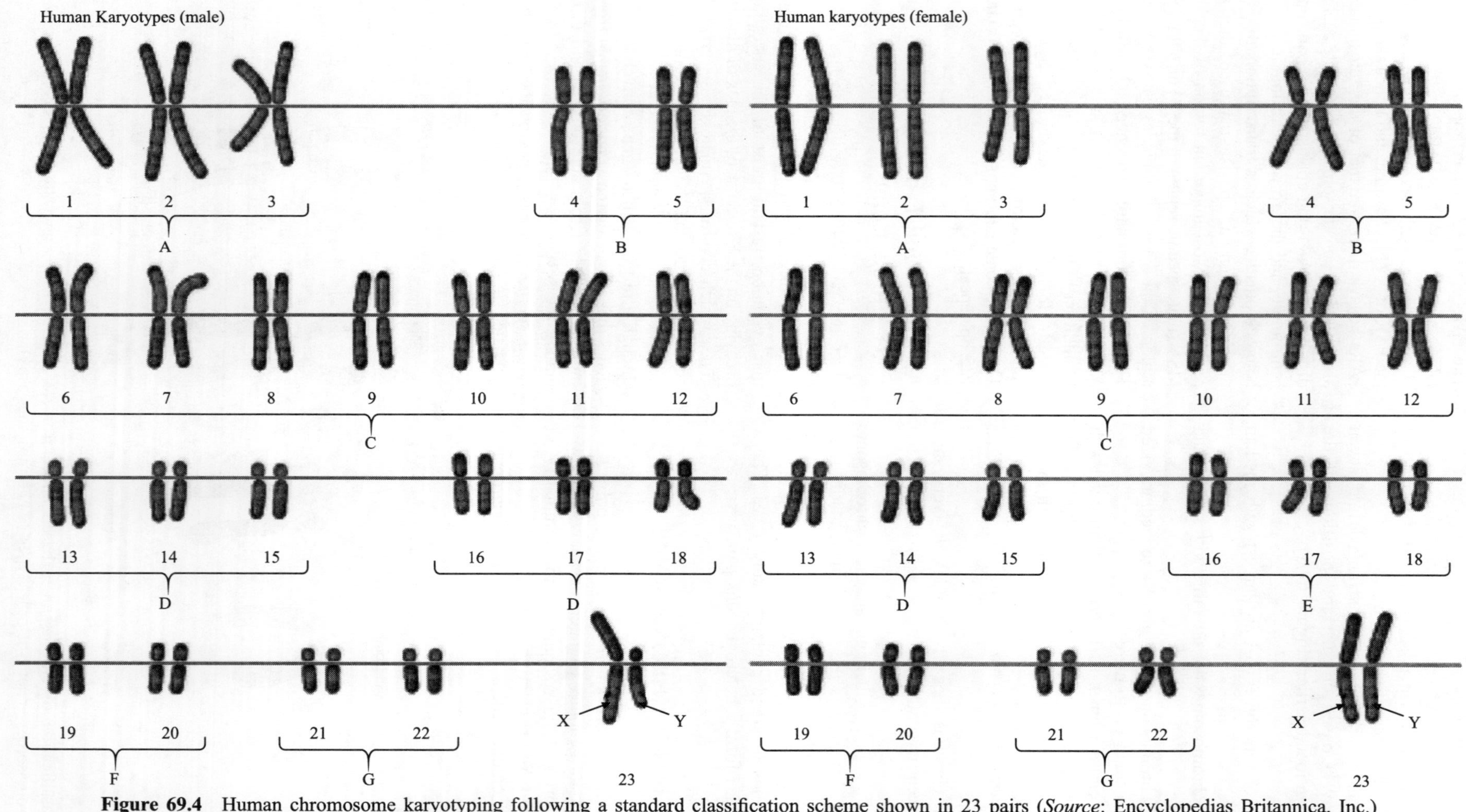

Figure 69.4   Human chromosome karyotyping following a standard classification scheme shown in 23 pairs (*Source*: Encyclopedias Britannica, Inc.) (*see Plate 30 for colour figure*)

ancestral gene that remained on the X chromosome (RBMX) retained its widespread cellular functions, whereas RBMY, on the Y chromosome, evolved into a male-specific function (spermatogenesis).

Recombination between the X and Y chromosome proved harmful—it resulted in males without necessary genes formerly found on the Y chromosome, and females with unnecessary or even harmful genes previously found only on the Y chromosome. As a result, genes beneficial to males accumulated near the sex-determining genes and recombination in this region was suppressed in order to preserve this male specific region. Over time, the Y chromosome changed so as to inhibit the areas around the sex determining genes from recombining at all with the X chromosome. As a result of this process, 95% of the human Y chromosome is unable to recombine.

Three hundred million years ago, the Y chromosome had about 1,400 genes, and now it has about 70-100 genes or so compared with about 1000 on the X chromosome. Comparative genomic analysis, however, reveals that many mammalian species are experiencing a similar loss of function in their heterozygous sex chromosome. Degeneration may simply be the fate of all non-recombining sex chromosomes due to three common evolutionary forces: high mutation rate, inefficient selection and genetic drift. However, nearly all the remaining Y chromosome's genes are found in the X-degenerate regions, which were once identical to sequence on the X chromosome but have since diverged substantially.

Unlike the campliconic sequence, the X-degenerate sequence does not routinely undergo recombination of any sort; so rapid, ongoing gene loss might be expected there. It was later shown that genetic decay has in recent history been minimal, with the human chromosome having lost no further genes during the last six million years, and only one in the last 25 million years. Thus Y chromosome has developed powerful strategy of self-preservation.

The mammalian *SRY* on the short arm of the Y chromosome encodes a nuclear factor-like protein harboring a DNA-binding domain known as the HMG box. The *SOX* genes encode similar protein factor. However, the sequence of the HMG box of *SRY* is at least 60% different with that of *SOX* gene. The functional relationship between *SOX* and *SRY* genes with respect to sex determination is unclear. Thus, it is significant to know more about the evolutionary status of the Y chromosome in addition to the functional relationship between *SRY* and *SOX* genes.

B. Important Genes on Y Chromosome

Excluding pseudoautosomal genes, the human Y includes:

- NRY, with corresponding gene on X chromosome
  - AMELY/AMELX (amelogenin)
  - RPS4Y1/RPS4Y2/RPS4X (Ribosomal protein S4)
- NRY, other
  - AZF1 (azoospermia factor 1)
  - BPY2 (basic protein on the Y chromosome)
  - DAZ1 (deleted in azoospermia)

- DAZ2
- PRKY (protein kinase, Y-linked)
- RBMY1A1 (RNA binding motif protein, Y-linked, family 1, member A1)
- SRY (sex-determining region)
- TSPY (testis-specific protein)
- USP9Y(Ubiquitin specific peptidase 9, Y-linked)
- UTY (ubiquitously transcribed TPR gene on Y chromosome)
- CDY(Testis-specific chromodomain protein Y)
- XKRY (X Kell blood group complex subunit-related, Y-linked)
- ZFY (zinc finger protein)

It is estimated that Y chromosome has about 70–100 genes. Similarly, its size is reported to be about 59–70 million base pairs. However actual size and exact number of genes on the Y are still not clear. What is clear is the fact that together with different repeat elements, every single Y chromosome is unique with respect to its organization and thus calls for in-depth analysis from across the different ethnic races and groups.

C. The Gene *SRY*

The *SRY* mutations have been found to be responsible for the XY gonadal dysgenesis. Functionally, the *SRY* protein binds to DNA in a sequence specific manner which is generally abolished or reduced in the XY gonadal dysgenesis cases. The fact that the XX mice transgenic for the *SRY* develop testis and normal testicular cords indicates that the *SRY* is testis determining factor (TDF) responsible for male phenotype. It may be noted that *SRY* is younger than the Y chromosome and thus was not the original mammalian sex determining gene.

In the humans, the onset of the *SRY* expression defines testis determination, similar to that in mouse. The human genital ridge is formed around 33 d gestations and *SRY* is detected at 41 d in XY embryos. *SRY* peaks at about 44 d gestations when the testis cords become visible. But unlike mouse, the *SRY* expression continues into the adulthood suggesting its probable role in the spermatogenesis.

The *SRY* is located on the Yp11.3, contains no introns in most mammalian species, except for some marsupial ones. The ORF of *SRY* codes for a protein of 204 amino acids (aa), which also includes a 79 aa DNA binding HMG box having two nuclear localization signals. The HMG box is conserved motif for minor groove DNA and has ability to recognize the DNA binding motif A/TAACAAT/A with highest affinity which is enhanced further by phosphorylation. After binding, the *SRY* protein has been reported to bend the DNA by insertion of a non-polar cantilever side chain of Ile68 between the base pairs.

Apart from the HMG box, overall conservation of *SRY* protein is poor amongst mammals. No distinct functional domain motif, which is sometimes found in other transcription factors, has been identified outside the HMG box. *SRY* may be divided into three different regions; the N-terminal domain

(N-TD), HMG box and C-terminal domain (C-TD). The N-TD is some 30–60 amino acids in length, and the C-TD varies amongst mammalian species. However, *mSRY* has an unusual structure in comparison with other *SRY* proteins. Since its N-TD contains only 2 amino acids, the HMG box is essentially N-terminal. It also has a glutamine (Q)-rich domain in the C-TD, which is derived from degenerate CAG repeats. Moreover, there is a "bridge domain" situated between the HMG box and the Q-rich domain.

Although the *SRY* nonsense mutations are scattered throughout the ORF, missense mutations tend to cluster in the central region of this gene which encodes the HMG box. This strongly suggests that the DNA binding domain of the *SRY* is of prime importance in sex determination. The biochemical activities of the *SRY* mutants provide insights into the normal functions of the *SRY* gene. Several mutations lead to the disrupted DNA binding or bending properties. Some mutants show normal bending but abolished DNA binding (V60L, R62G, R75N, L94P, G95R); others exhibit moderate binding (M64I, I90M, S91G, P125L, A113T) and still others show normal wild type binding and bending (S18N, M64R, M64T, F67V, F109S, R133W).

D. Sex Determination and *"SOX"* Gene

SOX genes encode a family of transcription factors that bind to the minor groove of DNA and belong to the HMG (high mobility group) box. This HMG box is a DNA binding domain conserved throughout eukaryotic species. Homologues have been identified in insects, nematodes, amphibians, reptiles, and birds and in a range of mammals. However, HMG boxes can be very diverse in nature, with only a few amino acids being conserved across the species.

There are 20 *SOX* genes present in the humans and mice, and 8 in *Drosophila*. Almost all *SOX* genes show at least 50% amino acid similarity with the HMG box of the *SRY* gene. The family is divided into subgroups according to homology within the HMG domain and other structural motifs, based on the functional assays.

The developmentally important *SOX* family has no singular function. While many *SOX* genes are involved in sex determination, some are also important in the processes such as neuronal development. For example, *SOX*2 and *SOX*3 are involved in the transition of epithelial granule cells in the cerebellum to their migratory state. Granule cells then differentiate to granule neurons, with *SOX*11 being involved in this process. It is thought that some *SOX* genes may be useful in the early diagnosis of childhood brain tumors due to this sequential expression in the cerebellum, making them a target for significant research. One, *SOX9*, has a conserved role in testis determination, as well as in chondrogenesis.

SOX proteins bind to the sequence WWCAAW (W=A/T). They have weak binding and unusually low affinity for DNA. *SOX* are related to the Tcf/Lef1 group of genes which also contain a sequence-specific high mobility group having similar sequence specificity (TWWCAAAG).

SOX genes are classified into the following groups:

* SOXA: SRY
* SOXB1: SOX1, SOX2, SOX3
* SOXB2: SOX14, SOX21
* SOXC: SOX4, SOX11, SOX12
* SOXD: SOX5, SOX6, SOX13
* SOXE: SOX8, SOX9, SOX10
* SOXF: SOX7, SOX17, SOX18
* SOXG: SOX15
* SOXH: SOX30

Genetic Conditions related to *SOX* gene family are:

* campomelic dysplasia
* combined pituitary hormone deficiency
* microphthalmia
* septo-optic dysplasia
* SOX2 anophthalmia syndrome
* Waardenburg syndrome

E. The Y Chromosome Heterochromatin

So far, the analyses of the human chromosomes have largely by-passed the heterochromatic regions (Human Genome Sequencing Consortium, 2001), including a large block of heterochromatic sequences found in the centromeric region of every nuclear chromosome. The human Y chromosome, in addition to the centromeric heterochromatin, harbors a second, much longer heterochromatic block (roughly 40 Mb) that comprises the bulk of the distal long arm. This block is composed by multiple repeats of DYZ1. In the course of the Y chromosome sequencing project, a third heterochromatic block (mentioned above) of approximately 400 kb and comprising 3000 tandem repeats of 125 base pairs (bp) was uncovered. This heterochromatic block interrupts the euchromatin sequences of proximal Yq. The other two heterochromatic blocks also consist of highly amplified tandem repeats of low sequence complexity.

VI. THE HUMAN Y CHROMOSOME AND MALE INFERTILITY

The Y chromosome contains several genes that are involved in spermatogenesis, and deletions in these genes are often found in infertile males. The infertility related to Y chromosome is usually caused by deletions of the azoospermia factor (*AZF*) A, B, or C region. In addition, there are other genes that seem to play equally important roles regulating the fertility in the human males.

A. The Azoospermia Factor (*AZF*) Regions: Structure

The detailed complexity of the *AZF* locus can only be revealed by STS and YAC mapping. The STS-PCR based mapping

led to the identification of submicroscopic deletions, called microdeletions, not detected at the cytogenetic level. It was further postulated that the *AZF* locus consists of three non-overlapping regions termed *AZFa*, *AZFb* and *AZF*c from the proximal to distal long arm of the human Y chromosome.

B. *AZFa* Region

The *AZFa* region is absent in the infertile males. This was referred as *DFFRY* (Drosophila fat facets related), which has been re-named as *USP9Y* (Ubiquitin specific protease 9, Y chromosome). It is a single copy gene, covering almost half of the *AZFa* sequences. Further studies detected two more *AZFa* genes, *DBY* (Dead or death box on the Y chromosome) and *UTY* (ubiquitous TPR motif on the Y). It was suggested that deficiency of the *USP9Y* cause male infertility and the additional loss of *DBY* makes the phenotype worse.

Recent studies have demonstrated the crucial role of *USP9Y* in the spermatogenesis where a 4 bp deletion in this gene leads to a truncated protein in the azoospermic males. Contrastingly, the refinement of *AZFa* map has represented the *DBY* as a major player in the spermatogenesis. It is deleted more frequently than the *USP9Y* and shows testis specific transcript in addition to the ubiquitous transcripts. However, the specific function of the *DBY* in male germ cell development and spermatogenesis is still unknown.

C. *AZFb* Region

In case of *AZFb*, only YAC and BAC clones have been described and a detailed sequence map is still not available. Recently, 38 BAC clones were assembled into a 4.3 Mb *AZFb* and surrounding sequences. The proximal *AZFb* region consists of large repetitive sequences in the form of palindromes but most of it is single copy sequence. A number of known and novel gene families have been mapped into this interval and most of them are testis specific. The analysis of the *AZFb* in several severely infertile males has identified common deletion breakpoints suggesting a specific and crucial role of *AZFb* in spermatogenesis. The microdeletions can remove the *AZFb* alone, parts of *AZFb* or the sequences from the flanking *AZFc* region. Till date, only two genes *EIF1AY* (translation initiation factor IA-isoform Y) and *RBMY* (RNA binding motif on the Y chromosome) have been mapped to the *AZFb* region. A recent addition is the *HSFY* (heat shock transcription factor) on the Y chromosome. The *EIF1AY* codes for testis specific transcript in addition to the ubiquitous transcripts encoding translation initiation factor, eIF-1A but its role in spermatogenesis is completely unknown. The *RBMY* is a family of 20-50 genes and pseudogenes spread over both the arms of Y chromosome with a locus in *AZFb* as well. Consistent with the role in spermatogenesis, the *RBMY* genes are expressed only in the male germline. The actual role of the *RBMY* protein is not yet known but the predicted function is considered to be in pre-mRNA splicing. The *HSFY* characterized by an HSF-type DNA binding domain, related to the HSF2 gene on chromosome 6.

The HSF2 shows higher expression and is active only during, embryogenesis and spermatogenesis.

D. *AZFc* Region

*AZF*c deletions in the distal Yq11 are the most frequently known genetic cause of the human male infertility leading to azoospermia or severe oligozoospermia. The *AZF*c sequence is mainly composed of the large repeated sequence blocks called amplicons that are organized into palindromic structures with >99% arm to arm identity. Such structures frequently undergo recombination and gene conversion, duplications and deletions, discussed in the later sections.

The *AZF*c region contains eight gene families expressing only in the testis, three of which code for the distinct proteins. These gene families are *BPY2*, *CDY1*, *CSPG4LY*, *DAZ*, *TTY3*, *TTY4* and *TTY17*. The *DAZ* (Deleted in Azoospermia) gene exists in four copies, coding for a single RNA binding domain at the N-terminus and the *DAZ* repeats at the C-terminus. Although *DAZ* is not the only gene present in the distal Yq, its prevalent deletion in the infertile males makes it a major *AZF*c candidate. The functions of other *AZF*c candidates are not yet clear except that they share the same characteristics. These are all multicopy, expressed in testis only and thus are male specific. Three *PRY* and *TTY2* genes are identified in the proximal part of the *AZF*c and are possibly not involved in the spermatogenic disruption. Of the two *CDY1* genes of the *AZF*c region, one is present within the *DAZ* locus and the other at the distal end. This suggests that at least one *CDY1* gene is absent in the infertile males with *DAZ* deletion and therefore it may be implicated in spermatogenesis. Since no report is available on the exclusive deletion of the *CDY1* in the infertile males, this hypothesis is not yet confirmed. In the absence of data on it's mapping, it is not considered to be as the major *AZF*c candidate gene so far.

VII. MALE INFERTILITY AND Y CHROMOSOME MICRODELETIONS

Several Y chromosome specific intrinsic deletions have been detected in azoospermic or oligozoospermic males employing PCR based Sequence tagged site (STS) mapping. After the mapping of various PCR based markers (STS's) on the Y chromosome, immediately came the first reports on the interstitial microdeletions in the infertile males. The STS markers fall into three categories:

1. Single copy STSs for the genes such as *SRY*, *DBY*, and *USP9Y* etc. these are most informative since their absence indicates absence of a gene.
2. Markers those are multicopy, but clustered in a specific region such as those for the *DAZ* genes. Their absence is more informative than presence since the positive PCR merely tells about the presence of atleast one *DAZ* gene out of four.

3. Markers that are multicopy and spread all across the Y chromosome such as those for various palindromes, *RBMY, TSPY, CDY* or various inverted repeats. These markers are informative only when their absence uncovers recombination leading to major deletions. Some STSs represent a normal Y chromosome polymorphism as they are frequently absent both in normal and infertile males. The STS mapping pose a problem due to repetitive nature of the Y chromosome and several intrinsic polymorphisms among various ethnic groups. However, all the ethnic group specific STS deletions have not been characterized with a large number of populations.

A. Structural Polymorphisms of the Human Y Chromosome

Studies have shown a remarkable structural polymorphism of the human Y chromosome constituted by segmental duplications and Copy Number Variation (CNV). Among the human chromosomes, the Y bears a large proportion of these duplications and exhibits cytogenetically detectable polymorphisms including length variation of the euchromatin and heterochromatin. The Y chromosome also undergoes inversions and neutral translocations with the autosomes. Unlike the other chromosomes, Y does not require a full-length homologue for recombination and hence the structural polymorphisms seen are attributable to its intrinsic repetitive landscape, as discussed earlier. The polymorphisms can occur among the members of a particular population group or may be detected amongst various ethnic groups with divergent outcomes.

B. Structural Variations on the Short Arm of the Human Y Chromosome

The amelogenin on Yp (*AMELY*) is routinely employed in prenatal sex determination, archaeological analysis, forensic typing and paternity testing. The sequencing of the Y chromosome has shown the deletion/absence of the Yp specific PCR fragments. It was known that a 3.6 Mb segment on the Yp exist in two different orientations due to an inversion polymorphism assisted by IR3. Deletion mapping and haplotype analysis have confirmed that the ICHR or non-allelic homologous recombination (NAHR) between *TSPY* copies on the Yp is responsible for deletion in the males. Haplotyping shows that *TSPY* mediated deletion have arisen seven times independently. The persistence and expansion of deletion lineages suggested that absence of the *PRKY* and *TBL1Y* in addition to *TSPY*, mediated by NAHR, has no major deleterious effects.

C. Structural Variations on the Long Arm of the Y Chromosome

The palindromic structure of the long arm, especially the *AZF* locus mediates several internal deletions, as has been explained

earlier. The unique structure of the *AZF*c region also makes it prone to inversions and gene conversion events, analyzed using STSs and SNVs mapping. This study coupled with the Y based haplotype analysis based on the binary markers showed that the haplogroup N completely lacks the *DAZ1/DAZ2* and the *BPY2.2/BPY2.3* doublets. The normal fertility status of the males suggested that these genes cannot be required for the male fertility. This also confirms that the gene complement of the Y chromosome may differ substantially amongst the individuals and additional variations can be expected from other haplotypes.

In support of this hypothesis, another study was conducted taking one Y chromosome from each of the 47 branches of world's genealogical tree. It was observed that the Yq heterochromatin and *TSPY* arrays changed their lengths >12 times and >23 times, respectively. The inversion of 3.6 Mb IR3/IR3 took place >12 times changing its orientation and the reshuffling of the *AZF*c region took place >20 times. In spite of all these polymorphisms, the copy number of the Y linked genes varies less frequently, raising the possibility of favorable natural selection. It may be noted that the high mutation of the linked loci/amplicon/repeats owing to the absence of a full length homologue and frequent events of the ICHR/NAHR have driven the extensive structural polymorphism of the Y chromosome.

VIII. Y CHROMOSOME ANALYSIS IN PATIENTS WITH SEX CHROMOSOME RELATED ANOMALIES (SCRA)

Patients with XY gonadal dysgenesis carry point mutations in the *SRY* gene which affect it's binding or bending properties. In humans, the *SRY* gene is reported to be single copy and haploid, whereas in some rodents it is present in one or more, monomorphic or polymorphic forms. Screening of *SRY* gene in 16 DNA samples from patients with SCRA showed 2–16 copies of the *SRY* gene of which, one, Oxen (49, XYYYY) had eight copies with sequences different from one another. Of all the DNA samples analyzed from the patients with SCRA, *SRY* gene showed CNP in the range of 2–16. This gene was absent in a Turner patient (p6697) but present in two others, p2C and p4 that showed one and four copies, respectively. An unusually tall Turner patient p65971 was found to have 16 copies, whereas the father of this patient (F-p65971) had two copies of the *SRY* gene. A Swyer syndrome patient, p20 and Oxen showed four and eight copies, respectively. DNA analysis of hypogonadism/Cryptorchidism patients' p65974 and p65973 showed four copies in each. Copy number polymorphism (CNP) might be the possible mechanism for such duplication. It would be of interest to establish if 16 or more copies of the *SRY* gene are always correlated with the long stature of the Turner patients or otherwise.

A classical model suggests that after gene duplication, one copy preserves the ancestral function while the other one is free to conform to a new function. In an alternate duplication, divergence and complementation model, duplicated genes are preserved because each copy loses some but not all of its functions through degenerating mutations. The second model seems to be a more likely explanation for eight copies of the *SRY* gene in Oxen.

The genomics of the human Y chromosome has facilitated analysis of microdeletions in the infertile males for routine diagnosis. Taken together, the data suggest that the Y chromosome microdeletions are one of the most common causes of male infertility.

Analyses of the Y chromosome in 15 clinically diagnosed Turner Syndrome (TS) patients detected high levels of mosaicisms ranging from 45,XO:46,XY = 100:0% in 4; 45,XO:46,XY:46XX = 4:94:2 in 8; and 45,XO:46,XY:46XX = 50:30:20 cells in 3 TS patients, unlike previous reports showing 5–8% cells with Y-material. Also, no ring, marker or di-centric Y was observed in any of the cases. This work demonstrated new types of polymorphisms indicating that no two TS patients have identical genotype-phenotype. Thus, a comprehensive analysis of more number of samples is warranted to uncover consensus on the loci affected, to be able to use them as potential diagnostic markers.

Earlier, no treatment was available for infertile couples when the male had impaired spermatogenesis, which is true even now. However, recent development has shown that infertile couples may avail IVF facility even with apparently 100% abnormal spermatozoa in their ejaculate.

In each category, the morphology and/or motility of the sperm can be normal or abnormal (asthenoteratozoospermia), see Table 69.1.

TABLE 69.1   Classification of Sperm Count

Classification of Sperm Count 1	*Sperm Count in Millions/mL*
Azoospermia	0
Severe oligozoospermia	< 1
Moderate oligozoospermia	1–5
Mild oligozoospermia	5–20
Normal	> 20

Abnormalities
- Aspermia: absence of semen
- Azoospermia: absence of sperm
- Hypospermia: low semen volume
- Oligozoospermia: low sperm count
- Asthenozoospermia: poor sperm motility
- Teratozoospermia: sperm carry more morphological defects than usual

In 1993, microsurgical epididymal sperm aspiration (MESA) in conjunction with ICSI was developed for the treatment of obstructive azoospermia. A few months later, TESE (testicular sperm extraction) was also found to be effective for the majority of cases of non-obstructive azoospermia as well. This is because approximately 60% of azoospermic men with presumably no sperm production actually do have a minute

amount of sperm production in the testis that is just not quantitatively sufficient to spill over into the ejaculate, but is adequate for ICSI. Thus, even men with spermatogenesis so deficient in quantity that no sperm at all can reach the ejaculate, could now have children with the use of TESE-ICSI. There are some common but still unknown factors underlying infertility in patients with Y chromosome related infertility irrespective of their spermatogenic status.

IX.   GENETIC DISORDERS

Chromosome aberrations include numerical or structural ones. Numerical changes are of the two types: polyploidy with changes in the number of sets of chromosomes and aneuploidy with changes in the number of individual chromosomes (e.g., trisomies and monosomies). Structural changes involve the loss or gain of portions of chromosomes. The resulting patient may be said to have "partial monosomy" or "partial trisomy." The normal separation of chromosomes in meiosis I or sister chromatids in meiosis II is termed disjunction. When the separation is not normal, it is called non-disjunction. This results in the production of gametes which have either too many or too few of a particular chromosome, and is a common mechanism for trisomy or monosomy. Nondisjunction can occur during meiosis I or meiosis II, and is responsible for several medical conditions in humans.

The frequency of nondisjunction is high in humans and the results are usually so devastating to the growing zygote that miscarriage occurs very early in the pregnancy. If the individual survives, he or she usually has a set of symptoms—a syndrome—caused by the abnormal dose of each gene product from that chromosome.

A.   Chromosomal Aberrations

Chromosomal anomaly reflects an atypical number of chromosomes or a structural abnormality in one or more chromosomes. This usually occurs when there is an error in cell division during meiosis or mitosis. There are several types of chromosome anomalies. They can be organized into two groups of numerical and structural anomalies (Table 69.2).

B.   Structural Abnormalities

When the chromosome's structure is altered, this can take several forms:

- *Deletions:* A portion of the chromosome is missing or deleted. Known disorders in humans include Wolf-Hirschhorn syndrome, which is caused by partial deletion of the short arm of chromosome 4; and Jacobsen syndrome, also called the terminal 11q deletion disorder.
- *Duplications:* A portion of the chromosome is duplicated, resulting in extra genetic material. Known human disorders include Charcot-Marie-Tooth disease

TABLE 69.2   **Chromosomal Aberrations commonly found in patients with Sex chromosome related anomalies**

S. No.	Chromosomal aberrations	Description	Cause	Characteristics	Diagnosis	Treatment
1.	Down syndrome	Also called Trisomy 21, is a genetic disorder that occurs in approximately 1 of 800 live births.	Nondisjunction, mosaicism and translocations	Associated with mild to moderate learning disabilities, developmental delays, characteristic facial features, and low muscle tone in early infancy. Individuals with Down syndrome also have heart defects, leukemia, early-onset Alzheimer's disease, gastro-intestinal problems, and other health issues. The symptoms of Down syndrome range from mild to severe ones.	Prenatal screening tests includes alpha-fetoprotein (AFP) screening test, the nuchal translucency test, and additional ultrasound screens.	No cure, however, there is treatments and therapies for the physical, medical and cognitive problems associated with Down syndrome. Treatment includes regular checkups and screening Medications, Surgery, Counseling and support.
2.	Patau Syndrome	Trisomy 13 (47, XX, +13), is rare for fetuses with this condition to go to term, so it occurs in only 1 in 15,000 live births.	Nondisjunction	Extra fingers or toes (polydactyly), Deformed feet, known as rocker-bottom feet, Neurological problems such as small head (microcephaly), failure of the brain to divide into halves during gestation (holoprosencephaly), severe mental deficiency, Facial defects such as small eyes (microphthalmia), absent or malformed nose, cleft lip and/or cleft palate, Heart and Kidney defects.	Genetic testing done to confirm the diagnosis. Imaging studies such as computed tomography (CT) or magnetic resonance imaging (MRI) done to look for brain, heart, and kidney defects. An ultrasound of the heart (echocardiogram) is done to detect high frequency of heart defects associated with Patau syndrome.	No cure. Surgery may be necessary to repair heart defects or cleft lip and cleft palate. Physical, occupational, and speech therapy will help individuals with Patau syndrome reach their full developmental potential.
3.	Edward Syndrome	Edwards syndrome is a chromosomal abnormality characterized by the presence of an extra copy of genetic material on the 18th chromosome, either in whole (trisomy 18) or in part (due to translocation).	Meiotic nondisjunction, translocation and mosaicism	Kidney malformations, structural heart defects at birth , intestines protruding outside the body, esophageal atresia, mental retardation, developmental delays, growth deficiency, feeding difficulties, breathing difficulties, and arthrogryposis (a muscle disorder that causes multiple joint contractures at birth).	Physical examination of the infant for abnormalities. X-rays and karyotyping. Potential testing before birth includes maternal serum alpha-fetal protein analysis or screening, amniocentesis, ultrasonography, and chorionic villus sampling. A woman who is pregnant with a child who has Edward's syndrome might have an uncommonly large uterus during the pregnancy because of the presence of extra amniotic fluid. An unusually small placenta might be noted during the birth of the child.	No cure. Treatment today consists of palliative care.
4.	Klinefelter Syndrome	Klinefelter syndrome, also known as the XXY condition. XXY, XXXY, XXXXY is a term used to describe	Nondisjunction	Not all males with the condition have the same symptoms or to the same degree. The XXY condition can affect three main areas of development (1) physical, (2) language, and (3) social.	Physical examination. Study of chromosomal abnormalities through blood and prenatal testing.	The genetic variation is irreversible. Males with Klinefelter syndrome can be given testosterone, a hormone needed for sexual

(Contd.)

TABLE 69.2 Chromosomal Aberrations commonly found in patients with Sex chromosome related anomalies (*Contd.*)

S. No.	Chromosomal aberrations	Description	Cause	Characteristics	Diagnosis	Treatment
		males who have an extra X chromosome in most of their cells.		Basic symptoms include: sparse body hair, enlarged breasts, and wide hips. Small testicles. In some men the penis does not reach adult size. Apart from language and learning problems voices may not be as deep in some individuals.		development. Speech therapy and educational support can help boys who have language or learning problems. Pregnancies possible with IVF technology with surgically removed sperm material from men.
5.	Turner Syndrome	Turner syndrome is a chromosomal condition related to the XO condition where entire X chromosome is absent. However, some cells in the body do have Y chromosome. Thus, there is a mosaicism as pure XO's do not survive. Unfortunately, it is difficult to uncover exact percentage of the cells carrying the Y chromosome in the body. Thus, all turners (female) are not identical with respect to their genotypes and phenotypes.	Nondisjunction and mosaicism	Physical abnormalities, such as short stature, swelling, broad chest, low hairline, low-set ears, and webbed necks. Girls with Turner syndrome typically experience gonadal dysfunction (non-working ovaries), which results in amenorrhea (absence of menstrual cycle) and sterility. Concurrent health concerns are also frequently present, including congenital heart disease, hypothyroidism, diabetes, vision problems, hearing concerns, and many autoimmune diseases. Specific pattern of cognitive deficits is often observed, with particular difficulties in visuospatial, mathematical, and memory areas.	Ultrasound test during pregnancy. Amniocentesis or Chorionic villus samplings, both antenatal tests that detect chromosomal abnormalities, are possible ways to confirm diagnosis. Heart or kidney problems or swelling of the hands and feet present at birth may facilitate diagnosis.	No cure. Several treatments may help with short stature, process of sexual development and learning difficulties. Hormone therapy includes estrogen, progesterone and growth hormones may be required. Pregnancy obtained through IVF technology. Some patients may develop psychological problems and will benefit from psychological therapy.
6.	XXX syndrome	Triple X syndrome is a form of chromosomal variation characterized by the presence of an extra X chromosome in each cell of a human female. It is also referred as 47, XXX.	Nondisjunction	The symptoms vary from person to person, with some women being more affected than others. Tall stature; small head (microcephaly); vertical skinfolds that may cover the inner corners of the eyes (epicanthal folds); delayed development of certain motor skills, speech and language; learning disabilities, such as dyslexia; or weak muscle tone. Delayed language development, EEG abnormalities, motor-coordination problems and auditory-processing disorders, and scoliosis. Premature ovarian failure seems to be more prevalent in these women.	Diagnosed before the baby is born, through CVS (chorionic villus sampling) or amniocentesis, and blood for chromosomal abnormalities.	No cure. Examination by physical, developmental, occupational or speech therapists if developmental or speech problems have occurred. For social problems she may have to be seen by a pediatric psychologist.

(*Contd.*)

TABLE 69.2 Chromosomal Aberrations commonly found in patients with Sex chromosome related anomalies (*Contd.*)

S. No.	Chromosomal aberrations	Description	Cause	Characteristics	Diagnosis	Treatment
7.	Jacobs Syndrome	Jacobs syndrome is a chromosomal genetic syndrome where a human male receives an extra Y-chromosome, giving a total of 47 chromosomes (47, XYY).	Nondisjunction	Learning problems at school delayed emotional maturity. Males with Jacob's syndrome are tall, thin, have acne, speech problems, and reading problems.	47, XYY may be prenatally diagnosed through cytogenetic analysis of cells obtained through such procedures as amniocentesis and chorionic villus sampling.	Effective management relieves pain with analgesics and immobilization using crutches, splints, braces, and restriction of weight bearing to the affected joint. In severe disease, surgery may include arthrodesis or, in severe diabetic neuropathy, amputation. However, surgery risks further damage through nonunion and infection.
8.	Cryptorchidism	Cryptorchidism is the absence of one or both testes from the scrotum. It is the most common birth defect regarding male genitalia.	Unexplained (idiopathic) birth defect. A combination of genetics, maternal health and other environmental factors may disrupt the hormones and physical changes that influence the development of the testicles.	Misplaced testes hidden in the abdomen Missing testicle in male newborn.	An ultrasound may enable the pediatric urologist to locate a non-palpable testicle, particularly if it's located within the groin. Magnetic resonance imaging (MRI) with a contrast agent to locate a testicle in the groin or abdomen. Laparoscopy, direct exploration of the abdomen through a larger incision may be necessary in a small number of more complicated cases.	The primary management of cryptorchidism is surgery, called orchiopexy. Surgical removal of testes - if they cannot be corrected; to avoid risk of later testicular cancer, gonadoblastoma.
9.	Swyer syndrome	Swyer syndrome, or XY gonadal dysgenesis, is a type of hypogonadism in a person whose karyotype is 46, XY.	Mutations in few genes involved in testis development.	The person is externally female with streak gonads, and if left untreated, will not experience puberty. Infertility, Tall stature, Eunuchoid proportion, Absence of menstruation, Ambiguous genitals, Dysgenetic testes, Small uterus, Small fallopian tubes.	Females with delayed puberty, a pelvic ultrasound can be performed to confirm the presence or absence of ovaries. Additionally, blood can be taken for a karyogram, which identifies the full set of chromosomes of the female. Females with Swyer syndrome will have both X and Y chromosomes rather than a set of two X chromosomes, which is normal for females.	Estrogen and progesterone therapy is typically commenced, prompting the development of female characteristics. In some cases, the fibrous streak gonads are surgically removed as a precautionary measure against the development of gonadoblastoma, a type of cancer of the gonads.

type 1A due to duplication of the gene encoding peripheral myelin protein 22 (PMP22) located on chromosome 17.

- *Translocations:* A portion of one chromosome is transferred to another chromosome. There are two main types of translocations:
    - Reciprocal translocation: Segments from two different chromosomes have been exchanged.
    - Robertsonian translocation: An entire chromosome has attached to another at the centromere—in humans these only occur with chromosomes 13, 14, 15, 21 and 22.
- *Inversions:* A portion of the chromosome has broken off, turned upside down and reattached inverting the genetic materials.
- *Insertions:* A portion of one chromosome has been deleted from its normal place and inserted into another chromosome.
- *Rings:* A portion of a chromosome has broken off and formed a circle or ring. This can happen with or without the loss of genetic material.
- *Isochromosome:* Formed by the mirror image copy of a chromosome segment including the centromere.

Chromosome instability syndromes are a group of disorders characterized by chromosomal instability and breakage. They often lead to an increased tendency to develop certain types of malignancies. Few examples of instability and breakage are shown in Table 69.3.

X. PROSPECTS OF HUMAN GENETICS

Genetic testing can be conducted using blood samples and in certain cases even biopsied tissues to isolate DNA for its subsequent analysis. From the blood, chromosome preparation may be made to ascertain the status of the karyotype which is the first level of diagnosis. Other genetic tests include biochemical and enzyme analyses and use of fluorochrome for specific staining. In addition, DNA can be isolated also from the semen samples for focused analysis. Following are the examples where genetic testing may be conducted.

- Carrier screening, which involves identifying the unaffected individuals who carry one copy of a gene for a disease that requires two copies for the disease to appear.
- Preimplantation genetic diagnosis.
- Prenatal diagnostic testing.
- Newborn screening.
- Presymptomatic testing for predicting adult-onset disorders such as Huntington's disease
- Presymptomatic testing for estimating the risk of developing adult-onset cancers and Alzheimer's disease.
- Conformational diagnosis of a symptomatic individual.
- Forensic/identity testing.
- Expressional studies of the genes with cDNA prepared from the semen samples (germline genetics).
- Expressional studies with cDNA prepared from the blood samples.

Currently, more than 1000 genetic tests are available from the testing laboratories of which some are given in the Table 69.4.

XI. GENETIC DISEASES AND EPIGENETICS

Epigenetics is a genetic phenomenon that depends on the chemical modification of DNA or histones, but not on the DNA sequence. Therefore, this form of inheritance is independent of classical Mendelian inheritance. A typical example of epigenetics is DNA methylation, which is important for parent-of-origin-specific gene expression, also known as genome imprinting. DNA methylation adds a methyl group to DNA. It is highly specific and always happens in a region in which a cytosine nucleotide is located next to a guanine nucleotide that is linked by a phosphate called a CpG site. CpG sites are methylated by one of the three enzymes called DNA methyltransferases (DNMTs). Inserting methyl groups changes the structure of DNA, modifying a gene's interactions with the machinery within a cell's nucleus that is needed for transcription. DNA methylation is used in some genes to differentiate its copy inherited from the father and mother, e.g., H19 or CDKN1C non-imprinted allele inherited from the mother or IGF-2 from the father).

Histone modifications are another important form of epigenetics. Histones proteins are the parts of chromatin that makes up chromosomes. When histones are modified after they are translated into protein (i.e., post-translation modification), they can influence how chromatin is arranged, which, in turn, can determine whether the associated chromosomal DNA will be transcribed. If chromatin is not in a compact form, it is active, and the associated DNA can be transcribed. Conversely, if chromatin is condensed, then it is inactive, and DNA transcription does not occur. There are two main ways histones can be modified: acetylation and methylation. These are the chemical processes that add either an acetyl or methyl group to the amino acid located in the histone. Acetylation is usually associated with active chromatin, while deacetylation is generally associated with heterochromatin. On the other hand, histone methylation can be a marker for both active and inactive regions of chromatin.

Environmental stress may modulate the epigenome, which might affect the evolution of the organism in question. For example, ultraviolet (UV) light stress leads to increased homologous recombination in *Arabidopsis thaliana*, even in unstressed progeny, for up to four generations. This seems to be epigenetic phenomenon because the whole population showed changes in each generation. Recently, it was shown that the ATF-2 family of transcription factors is involved in stress-induced epigenome changes. While epigenetic changes are required for normal development and health, they can also be responsible for some diseased conditions (Table 69.4).

TABLE 69.3   Examples of few structural abnormalities seen in patients with Sex chromosome related anomalies

S.No.	Chromosomal instability	Description	Cause	Characteristics	Diagnosis	Treatment
1.	Distal 18q-	Genetic condition caused by a deletion of genetic material within one of the two copies of chromosome 18.	Deletions (distal section of 18q and extending to the tip of the long arm of chromosome 18).	Distal 18q- causes a wide range of medical and developmental disorders. Heart and kidney abnormalities, genital anomalies, the most frequent being cryptorchidism and hypospadias. Hypotonia, Dysmyelination, Strabismus and nystagmus. Myopia has been reported in some individuals. Poor drainage from the middle ears, conductive and/or sensorineural hearing loss. Microcephaly and Hypothyroidism, low IgA levels, psychiatric disorder. Increased incidence of autism.	A routine chromosome analysis, or karyotype, is usually used to make the initial diagnosis. Microarray analysis is also being used to clarify breakpoints. Prenatal diagnosis is possible via amniocentesis of chorionic villus sampling.	Symptomatic routine screenings for thyroid, hearing, and vision problems.
2.	Robertsonian translocation	Chromosomal rearrangement that in humans occurs in the five acrocentric chromosome pairs, nAMELY 13, 14, 15, 21, and 22.	Translocation resulting in chromosomal deletions or addition and result in syndromes of multiple malformations.	Multiple malformations and mental retardation. Robertsonian translocations between chromosomes 13 and 14 lead to the trisomy 13 (Patau) syndrome. The Robertsonian translocations between 14 and 21 and between 21 and 22 results in (trisomy 21 (Down) syndrome. Several forms of cancer are caused by acquired translocations (as opposed to those present from conception). This has been described mainly in leukemia (acute myelogenous leukemia and chronic myelogenous leukemia). Translocations have also been described in solid malignancies such as Ewing's sarcoma. Infertility is also seen in these patients.	Study of chromosomal abnormalities through blood and prenatal testing. Fluorescence in-situ hybridization (FISH) is used for sex determination for X-linked conditions and to investigate the status of embryos from couples where one partner carries a chromosome rearrangement.	Preimplantation genetic diagnosis (PGD) in conjunction with assisted conception using IVF or intracytoplasmic sperm injection (ICSI).
3.	Rreciprocal translocation	Reciprocal translocations are usually an exchange of material between non-homologous chromosomes.	Translocations. Translocation occurring in gametogenesis due to errors in meiosis results in a chromosomal abnormality featured in all the cells of the offspring. On the other hand, translocations in somatic cells, due to errors in mitosis result in abnormalities featured only in the affected cell line.	Most balanced translocation carriers are healthy and do not have any symptoms. But about 6% of them have a range of symptoms which may include autism, intellectual disability, or congenital anomalies. Carriers of balanced reciprocal translocations have increased risks of creating gametes with unbalanced chromosome translocations leading to miscarriages or children with abnormalities. A gene disrupted or disregulated at the breakpoint of the translocation carrier is likely the cause of these symptoms.	Study of chromosomal abnormalities through blood and prenatal testing.	Genetic counseling and genetic testing.

(Contd.)

TABLE 69.3 Examples of few structural abnormalities seen in patients with Sex chromosome related anomalies (*Contd.*)

S.No.	Chromosomal instability	Description	Cause	Characteristics	Diagnosis	Treatment
4.	Chromosome inversion	Rearrangement in which a segment of a chromosome is reversed end to end. Inversions may be paracentric (not involving the centromere) or pericentric containing the centromere).	Inversion	Inversions usually do not cause any abnormalities in the carriers as long as the rearrangement is balanced with no extra or missing genetic information. Individuals which are heterozygous for an inversion, there is an increased production of abnormal chromatids. This leads to lowered fertility due to production of unbalanced gametes. The most common inversion seen in humans is on chromosome 9, at inv (9)(p12q13).	Cytogenetic techniques may be able to detect inversions, or inversions may be inferred from genetic analysis.	Genetic counseling and genetic testing.
5.	Cat-eye syndrome	Rare condition caused by the short arm (p) and a small section of the long arm (q) of human Chromosome 22 being present three (trisomic) or four times (tetrasomic) instead of the usual two times.	Duplication (22pter→q11)	Anal atresia (abnormal obstruction of the anus), Unilateral or bilateral iris coloboma (absence of tissue from the colored part of the eyes), Downward-slanting Palpebral fissures (openings between the upper and lower eyelids), Preauricular pits/tags (small depressions/growths of skin on the outer ears). Cardiac defects, Kidney problems (missing, extra, or under developed kidneys. Short stature, Scoliosis/Skeletal problems. Mental retardation—although most are borderline normal to mildly retarded, and a few even have normal intelligence, CES patients occasionally exhibit moderate to severe retardation. Micrognathia (smaller jaw). Hernias, Cleft palate, Rarer malformations can affect almost any organ.	Prenatal diagnosis of the derivative chromosome 22 associated with cat eye syndrome by fluorescence in situ hybridization.	Treatment is given according to the symptoms the individual is diagnosed with. Surgery is recommended to repair inborn defects in the heart or anus. Patients with short stature are given growth hormone therapy.

TABLE 69.4   List of some currently available DNA-Based Gene Tests

S.No.	Disease	Pathology
1.	Alpha-1-antitrypsin deficiency	Emphysema and liver disease.
2.	Amyotrophic lateral sclerosis	Lou Gehrig's Disease; Progressive motor function loss leading to paralysis and death.
3.	Alzheimer's disease	Late-onset variety of senile dementia.
4.	Ataxia telangiectasia	Progressive brain disorder resulting in loss of muscle control and cancers.
5.	Gaucher disease	Enlarged liver and spleen, bone degeneration.
6.	Inherited breast and ovarian cancer	BRCA1 and 2; early-onset tumors of breasts and ovaries.
7.	Hereditary nonpolyposis colon cancer	Early-onset tumors of colon and sometimes other organs.
8.	Central Core Disease	Mild to severe muscle weakness.
9.	Charcot-Marie-Tooth	Loss of feeling in ends of limbs.
10.	Congenital adrenal hyperplasia	Hormone deficiency; ambiguous genitalia and male pseudohermaphroditism.
11.	Cystic fibrosis	Disease of lung and pancreas resulting in thick mucous accumulations and chronic infections.
12.	Duchenne muscular dystrophy/Becker muscular dystrophy	Severe to mild muscle wasting, deterioration, weakness.
13.	Dystonia (DYT; muscle rigidity, repetitive twisting movements)	Muscle rigidity, repetitive twisting movements.
14.	Emanuel Syndrome (severe mental retardation, abnormal development of the head, heart and kidney problems)	Severe mental retardation, abnormal development of the head, heart and kidney problems.
15.	Fanconi anemia, group	Anemia, leukemia, skeletal deformities.
16.	Factor V-Leiden	Blood-clotting disorder.
17.	Fragile X syndrome	Leading cause of inherited mental retardation.
18.	Galactosemia	Metabolic disorder affects ability to metabolize galactose.
19.	Hemophilia A and B	Bleeding disorders.
20.	Hereditary hemochromatosis	Excess iron storage disorder.
21.	Marfan Syndrome	Connective tissue disorder; tissues of ligaments, blood vessel walls, cartilage, heart valves and other structures abnormally weak.
22.	Mucopolysaccharidosis	Deficiency of enzymes needed to break down long chain sugars called glycosaminoglycans; corneal clouding, joint stiffness, heart disease, mental retardation.
23.	Myotonic dystrophy	Progressive muscle weakness; most common form of adult muscular dystrophy.
24.	Neurofibromatosis type 1	Multiple benign nervous system tumors that can be disfiguring; cancers.
25.	Phenylketonuria	Progressive mental retardation due to missing enzyme; correctable by diet.
26.	Polycystic Kidney Disease	Cysts in the kidneys and other organs.
27.	Adult Polycystic Kidney Disease	Kidney failure and liver disease.
28.	Prader Willi/Angelman syndromes	Decreased motor skills, cognitive impairment, early death.
29.	Sickle cell disease	Blood cell disorder; chronic pain and infections.
30.	Spinocerebellar ataxia, type 1	Involuntary muscle movements, reflex disorders, explosive speech.
31.	Spinal muscular atrophy	Severe, usually lethal progressive muscle-wasting disorder in children.
32.	Tay-Sachs disease	Fatal neurological disease of early childhood; seizures, paralysis.
33.	Thalassemias	Anemia's—reduced red blood cell levels.
34.	Timothy syndrome	Characterized by severe cardiac arrhythmia, webbing of the fingers and toes called syndactyly, autism.

Disrupting any of the three systems that contribute to epigenetic alterations can cause abnormal activation or silencing of the genes. Such disruptions have been associated with cancer, syndromes involving chromosomal instabilities, and mental retardation.

XII. EPIGENETICS AND CANCER

The first human disease to be linked in 1983 with epigenetics was cancer. It was found that diseased tissue from patients with colorectal cancer had less DNA methylation than that from normal tissue of the same patients. Because methylated genes are typically turned off, loss of DNA methylation can cause abnormally high gene activation by altering the arrangement of chromatin. On the other hand, too much methylation can undo the work of protective tumor suppressor genes.

As previously mentioned, DNA methylation occurs at CpG sites, and a majority of CpG cytosines are methylated in mammals. However, there are stretches of DNA near promoter regions that have higher concentrations of CpG sites (known as CpG islands) that are free of methylation in normal cells. These CpG islands become excessively methylated in cancer cells, thereby causing genes that should not be silenced to turn off. This abnormality is the trademark epigenetic change that occurs in tumors and happens early in the development of cancer. Hypermethylation of CpG islands can cause tumors by shutting off tumor-suppressor genes. In fact, these types of changes may be more common in human cancer than DNA sequence mutations.

Furthermore, although epigenetic changes do not alter the sequence of DNA, they can cause mutations. About half of the genes that cause familial or inherited forms of cancer are turned off by methylation. Most of these genes normally suppress tumor formation and help repair DNA, including O^6-methylguanine-DNA methyltransferase (*MGMT*), MLH1 cyclin-dependent kinase inhibitor 2B (*CDKN2B*), and *RASSF1A*.

Hypermethylation can also lead to instability of microsatellites, which represent repetitive DNA. Microsatellites are common in normal individuals, and they usually consist of repeats of the dinucleotide CA. Too much methylation of the promoter of the DNA repair gene *MLH1* can make a microsatellite unstable and lengthen or shorten it. Microsatellite instability has been linked to several cancers such as colorectal, endometrial, ovarian and gastric ones.

A. Epigenetics and Mental Retardation

Fragile X syndrome is the most frequently inherited mental disability, particularly in males. Both sexes can be affected by this condition, but because males have only one X chromosome, one fragile X will impact them more severely. Indeed, fragile X syndrome occurs in approximately 1 in 4,000 males and 1 in 8,000 females. People with this syndrome have severe intellectual disabilities, delayed verbal development, and "autistic-like" behavior.

Fragile X syndrome gets its name from the way the part of the X chromosome appears as if it is hanging by a thread and easily breakable. The syndrome is caused by an abnormality in the *FMR1* (fragile X mental retardation 1) gene. People who do not have fragile X syndrome have 6 to 50 repeats of the trinucleotide CGG in their *FMR1* gene. However, individuals with over 200 repeats have a full mutation, and they usually show symptoms of the syndrome. Too many CGGs cause the CpG islands at the promoter region of the *FMR1* gene to become methylated; normally, they are not. This methylation turns the gene off, stopping the *FMR1* gene from producing an important protein called fragile X mental retardation protein. Loss of this specific protein causes fragile X syndrome. Although a lot of attention has been given to the CGG expansion mutation as the cause of fragile X, the epigenetic change associated with *FMR1* methylation is the real syndrome culprit. Fragile X syndrome is not the only disorder associated with mental retardation that involves epigenetic changes. Other such conditions include Rubenstein-Taybi, Coffin-Lowry, Prader-Willi, Angelman, Beckwith-Wiedemann, ATR-X, and Rett syndromes Combating Diseases with Epigenetic Therapy.

Because several diseases including cancer involve epigenetic changes, it seems reasonable to try to counteract these modifications with epigenetic treatments. These changes seem an ideal target because they are by nature reversible, unlike DNA sequence mutations. The most popular of these treatments aim to alter either DNA methylation or histone acetylation. Inhibitors of DNA methylation can reactivate genes that have been silenced. Two examples of these types of drugs are 5-azacytidine and 5-aza-2′-deoxycytidine. These medications work by acting like the nucleotide cytosine and incorporating themselves into DNA while it is replicating. After they are incorporated into DNA, the drugs block DNMT enzymes from acting, which inhibits DNA methylation. Drugs aimed at histone modifications are called histone deacetylase (HDAC) inhibitors. HDACs are enzymes that remove the acetyl groups from DNA, which condenses chromatin and stops transcription. Blocking this process with HDAC inhibitors turns on gene expression. The most common HDAC inhibitors include phenylb*UTY*ric acid, SAHA, depsipeptide, and valproic acid.

Caution in using epigenetic therapy is necessary because epigenetic processes and changes are so widespread. To be successful, epigenetic treatments must be selective to irregular cells; otherwise, activating gene transcription in normal cells could make them cancerous, so the treatments could cause the very disorders they are trying to counteract. Despite this possible drawback, researchers are finding ways to specifically target abnormal cells with minimal damage to normal cells, and epigenetic therapy is beginning to look increasingly promising.

XIII. HUMAN HEALTH AND ENVIRONMENT

A clean environment is essential for all living being. However, the interactions between the environment and human health are highly complex and difficult to assess. This makes the use of the precautionary principle particularly useful. The best-known health impacts are related to ambient air pollution, poor water quality and insufficient sanitation.

India is one of the most environmentally degraded countries in the world and it is paying heavy health and economic prices for it. The lack of services such as water supply, sanitation, drainage of storm water, treatment and disposal of waste and water and management of solid and hazardous wastes, supply of safe food, water and housing are all unable to keep pace with fast pace of urbanization. Also the unplanned location of industries in urban and sub-urban areas followed by traffic congestion, poor housing, poor drainage and garbage accumulation causes serious pollution problems. These factors not only lead to deteriorating environmental conditions but also have adverse effects on the human health.

Another aspect that is related to human health care system is the Natural background radiation (NBR). Semi-permanent exposure to ionizing radiation leaves a lasting imprint on the genome. Although many of the lesions induced by ionizing radiation are isolated, radiation also induces clustered DNA damage, which represents two or more lesions formed within one or two helical turns of DNA. From biophysical modeling, it was hypothesized that clustered DNA damage induced by radiation is less readily repaired than individual lesions. The dense ionizing radiations can cause highly localized DNA damage and lead to intra-chromosomal breaks. The biologically effective dose received by the coastal population in Kerala is 10000-12000 μSv per year, ~10 times higher than the worldwide average.

It has been shown that there was Tandem duplication and copy number polymorphism of the *SRY* gene in males exposed to NBR. Gene duplication increases the complexity of a genome, particularly when it is not clear if all the copies follow normal levels of expression. When copy number for *DAZ* gene in NBR patients was assessed, NBR males showed 4–16 copies of *DAZ* gene. Since the *DAZ* genes are implicated with the germ cell development, the cells with *DAZ* deletion/duplication are unlikely to survive. Alternatively, an innate mechanism may be operative to protect the germline from the effects of NBR. Accordingly, random microdeletions in the Azoospermia factor (*AZF*) a, b and c regions in 90%, and tandem duplication and copy number polymorphism (CNP) of 11 different Y-linked genes was observed in about 80% of males exposed to NBR. However, these changes remained confined to blood DNA and not in that of semen samples. This indicated that some very powerful innate mechanism is operative to protect germline samples in the males exposed to NBR. In a significant study organizational variation of DYZ1 array in males exposed to NBR was shown. Irrespective of the mechanisms involved, environment seems to contribute to genomic instability of the Y chromosome viz. DYZ1 array, *SRY* and *DAZ* genes, respectively, thereby, supporting the concept of rapid rate of evolution of the Y chromosome.

Several heavy metals present in the environment all over the world are alarmingly unsafe for the human population of which chromium and arsenic are the examples. These metals affect human systems in various ways but their possible genetic consequences remain unknown. In relation to the environmental hazards, the Y chromosome was analyzed in the males exposed to chromium through drinking water (not published). Similar to that of the males exposed to NBR, the chromium exposed (CE) males harbored random Y chromosomal microdeletions but at much lower frequency. Studying the genetic integrity of the human Y chromosome exposed to groundwater arsenic revealed deletion in AZF region of Y chromosome.

XIV. GENETIC VULNERABILITY

Large segments of our population use tobacco, alcohol, and other drugs. Cigarette smoking is common in both industrialized and developing countries. In the United States, over 43 million people use tobacco, and worldwide, over one billion people are tobacco users. Alcohol is the most commonly used and abused substance in the population, and 12.5% of adults in the U.S. develop alcohol dependence during their lifetime. The development of addiction requires the use of a substance and a subsequent chain of behavioral events that leads to addiction. Each step is influenced by environmental and genetic factors, some of which are common to all steps, and others are specific. Genome wide association studies have permitted the discovery of hundreds of genetic variants that alter the risk of developing multiple complex diseases, including type 2 diabetes, Crohn's disease, and Parkinson's disease.

XV. UNDERSTANDING HUMAN GENOME THROUGH BIOINFORMATICS

Bioinformatics is the branch of biology concerned with the acquisition, storage, and analysis of the information found in nucleic acid and protein sequence data. Computers and bioinformatics software are the tools of the trade. Genetic data represent a treasure trove for researchers and companies interested in how genes contribute to our health and well being. Almost half of the genes identified by the Human Genome Project have no known functions even today. Researchers are using bioinformatics to identify genes, establish their functions, and develop gene-based strategies for preventing, diagnosing, and treating the diseases.

A DNA sequencing reaction produces several hundred bases long sequence. Gene sequences typically run for thousands of bases. The largest known gene is Duchenne muscular dystrophy having 2.5 million base pairs. In order to study genes, scientists first assemble long DNA sequences from series of shorter overlapping sequences. Scientists enter their assembled sequences into genetic databases so that other workers may use the data. Since the sequences of the two DNA strands are complementary, it is only necessary to enter the sequence of one DNA strand into a database. By selecting an appropriate computer program, scientists can use sequence data to look for genes, get clues to gene functions, examine genetic variation, and explore evolutionary relationships. Bioinformatics is a young and dynamic science fuelled more towards the path of progress with the development of better computer hardware and softwares. New bioinformatic software are being developed while existing software are continually updated. One of the most important aspects of bioinformatics is identifying genes within a long DNA sequence. Until the development of bioinformatics, the only way to locate genes along the chromosome was to study their behavior in the organism (*in vivo*) or isolate the DNA and study it in a test tube (*in vitro*).

A. Finding Gene Functions

Once a nucleic acid or amino acid sequence has been assembled, bioinformatic analysis can be used to determine if the sequence is similar to that of a known gene. This is where sequences from model organisms are helpful. For example, if we have an unknown human DNA sequence that is associated with the disease cystic fibrosis. A bioinformatic analysis finds a similar sequence from mouse that is associated with a gene that codes for a membrane protein that regulates salt balance. It becomes a good bet that the human sequence also is part of a gene that codes for a membrane protein that regulates salt balance. Determining the similarity of two sequences is not easy but possible. It was recently reported that the genomes of humans and chimpanzees are 96 per cent similar.

Computer programs are available that can be used to see if a particular DNA sequence is similar to any other species that are stored in a sequence database. NCBI (http://www.ncbi.nlm.nih.gov) is one of the most frequently used databases. The database features includes:

- Nucleic acid sequences
- genomic and mRNA, including ESTs
- Protein sequences
- Protein structures
- Genetic and physical maps
- Organism-specific databases
- MedLine (PubMed)
- Online Mendelian Inheritance in Man (OMIM)

Programs for sequence analysis:

- BLAST to search rapidly through sequence databases
- PipMaker (to align 2 genomic DNA sequences)
- Gene finding by ab initio methods (GenScan, GRAIL, etc.)

One of the most popular programs used is called **BLAST** (**B**asic **L**ocal **A**lignment **S**earch **T**ool). Using this program is somewhat like using a search engine on the Internet. The user provides the program with a biological sequence (when using BLAST) or a subject (when using a search engine). In each case, the program compares the input information to the information found in the database. The results are given with the most closely matching items (or sequences) listed first, followed by items (or sequences) that match less well.

B. Whole Genome Sequencing, Functional and Comparative Genomics

Understanding the function of genes and other parts of the genome is known as functional genomics. Efficient interpretation of the functions of human genes and other DNA sequences requires that resources and strategies to be developed to handle large-scale data across the whole genomes. A technically challenging first priority is to generate complete sets of full-length cDNA clones and sequences for human and model-organism genes. Other functional-genomics goals include studies into gene expression and control, creation of mutations that cause loss or alteration of function in nonhuman organisms, and development of experimental and computational methods for protein analyses.

Comparative genomics is the analysis and comparison of genomes from different species. Comparative genomics gives a better understanding of species evolution and to determine the function of genes and noncoding regions of the genome. Most of the functions of human genes have been examined and their counterparts have been studied in simpler model organisms such as the mouse and *Drosophila*. Basic features while comparing genomes includes: sequence similarity, gene location, the length and number of coding regions (called exons) within genes, the amount of noncoding DNA in each genome, and highly conserved regions maintained in organisms as simple as bacteria and as complex as humans. Comparative genomics involves the use of high data computer programs that can assemble multiple genomes and look for regions of sequence similarity. Some of these sequence-similarity tools are accessible over the Internet.

Comparative genomics studies of mammalian genomes suggest that approximately 5% of the human genome has been conserved by evolution since the divergence of extant lineages approximately 200 million years ago. Comparative genomics revealed mice and humans have roughly the same number of nucleotides in their genomes—approximately 3 billion base pairs. This comparable DNA content implies that all mammals contain more or less the same number of genes. There is more than 95% to 98% similarity between related genes in humans and apes. Similarities between mouse and human genes range from about 70% to 90%, with an average of 85% similarity with exceptions to gene to gene variations. Some nucleotide changes are "neutral" and do not yield a significantly altered protein. Others, but probably only a relatively small percentage, introduce changes that could substantially alter what the protein does.

XVI. PERSPECTIVE PREDICTION AND FUTURE STUDIES

Human Genome Project has enabled identification of several genetic abnormalities and will identify many more in due course of time. With this, better and safer treatments are likely to follow. Future perspective of human genome sequencing is to offer a new era of molecular medicine intended not only to treat patients with symptoms, but also to explore the deepest causes of the diseases. Rapid and more accurate diagnostic tests for earlier detection will make it possible to fight against countless maladies.

For every good that a technology can bring to society, there is also a potential for its abuse. Although the Human Genome Project is a great achievement in Science and especially in Genetics, there is a need to address Ethical, social and moral issues. Just as dynamite discovered with good intention to break the rocks was eventually misused to kill people. This may be true for the Human Genome Project, which could be misused in many ways.

CONCLUSION

Over the last decade, researchers have discovered many nuanced features of the human genome, and now they are in the process of fine-tuning the same to expand the understanding of the genomic landscape. But by and large, the dream of applying the knowledge to benefit human health is still a distant dream. Finding usable medical information amid the huge quantum of genomic data is an immense challenge. Knowledge about the DNA variations and its correlation with the occurrence of the diseases affecting mankind would continue to draw the attentions of both, the researchers and policy makers.

SUGGESTIONS FOR FURTHER READINGS

1. Aitken R.J. and Marshall Graves J.A. (2002), The Future of Sex, *Nature*, 415(6875):963

2. Ali S. (2010), Genetic Integrity of the Human Y Chromosome Exposed to Groundwater Arsenic, *BMC Med Genomics.*, 3:35.

3. Centers for Disease Control (CDC), Current Cigarette Smoking Among Adults Aged >18 years–United States, 2010. *Morb. Mortal. Wkly. Rep.*, 59:1135–1140.

4. Egger G., Liang G., Aparicio A. and Jones P.A. (2004), Epigenetics in Human Disease and Prospects for Epigenetic Therapy, *Nature*, 429(6990):457–463.

5. Faerman M., Kahila G., Smith P., Greenblatt C., Stager L., Filon D. and Oppenheim A. (1997), DNA Analysis Reveals the Sex of Infanticide Victims, *Nature*, 385:212–213.

6. Feuk L., Carson A.R. and Scherer S.W. (2006), Structural Variation in the Human Genome, *Nature Rev. Genet.* 7:85–97.

7. Foresta C., Ferlin A. and Moro E. (2000), Deletion and Expression Analysis of AZFa Genes on the Human Y Chromosome Revealed a Major Role for DBY in Male Infertility, *Hum. Mol. Genet.*, 9(8):1161–1169.

8. Goldberg A.D., Allis C.D. and Bernstein E. (2007), Epigenetics: A Landscape Takes Shape, *Cell*, 128(4):635–638.

9. He X. and Zhang J. (2006), Transcriptional Reprogramming and Backup between Duplicate Genes: Is It A Genomewide Phenomenon?, *Genetics*, 172(2):1363–1367.

10. Hindorff L.A., Junkins H.A., Hall P.N., Mehta J.P. and Manolio T.A. (2010), A Catalog of Published Genome-wide Association Studies, www.genome.gov/gwastudies.

11. International Human Genome Sequencing Consortium (2001), Initial Sequencing and Analysis of the Human Genome, *Nature*, 409:860–921.

12. Jobling M.A. and Tyler-Smith C. (2003), The Human Y Chromosome: An Evolutionary Marker Comes of Age, *Nature*, 4:598–612.

13. Penagarikano O., Mulle J.G. and Warren S.T. (2007), The Pathophysiology of Fragile X Syndrome, *Annu Rev Genomics Hum. Genet.*, 8:109–129

14. Santos F.R., Pandya A. and Tyler-Smith C. (1998), Reliability of DNA-based Sex Tests, *Nat. Genet.*, 18:103.

15. Schueler M.G., Higgins A.W., Rudd M.K., Gustashaw K. and Willard H.F. (2001), Genomic and Genetic Definition of a Functional Human Centromere, *Science*, 294:109–115.

16. Skaletsky H., Kuroda-Kawaguchi T., Minx P.J., Cordum H.S., Hillier L., Brown L.G., Repping S., Pyntikova T., Ali J., Bieri T., Chinwalla A., Delehaunty A., Delehaunty K., Du H., Fewell G., Fulton L., Fulton R., Graves T., Hou S.F., Latrielle P., Leonard S., Mardis E., Maupin R., McPherson J., Miner T., Nash W., Nguyen C., Ozersky P., Pepin K., Rock S., Rohlfing T., Scott K., Schultz B., Strong C., Tin-Wollam A., Yang S.P., Waterston R.H., Wilson R.K., Rozen S. and Page D.C. (2003), the Male-specific Region of the Human Y Chromosome is a Mosaic of Discrete Sequence Classes, *Nature*, 423(6942):825–837.

17. World Health Organization (WHO) (2004), Global Status Report on Alcohol (Geneva: World Health Organization).

18. World Health Organization (WHO) (2010), Tobacco Key Facts, Http://www. Who.int/topics/tobacco/facts/en/index.html

19. Graves J.A. (2002), Evolution of the Testis-determining Gene–the Riseand Fall of SRY, Novartis Found, *Symp.*, 244:86–97.

20. Hasin D.S., Stinson F.S., Ogburn E. and Grant B.F. (2007), Prevalence, Correlates, Disability, and Comorbidity of DSM-IV Alcohol Abuse and Dependence in the United States: Results from the National Epidemiologic Survey on Alcohol and Related Conditions, *Archives of General Psychiatry*, 64, 830–842.

70

Epigenetics

Deepti Deobagkar

CONTENTS

I. GENETICS AND EPIGENETICS

Genetics is a branch of science which deals with heredity. It analyses how traits are passed on from generation to generation. Gregor Mendel, a Greek monk, described the principles of independent segregation and assortment of alleles through an elegant series of breeding experiments with pea plants. We inherit our traits or characters from our parents. Parents pass on sets of chromosomes to the progeny in each diploid organism. Meiotic recombination generates a new combination of sets of genes in each gamete. We inherit one set of chromosomes from our mother (maternal) and one from our father (paternal). Genetic information is encoded in the nucleotide sequence of DNA and mutation and recombination are very important in generating alleles, i.e., alternate forms of genes. This genetic blueprint, through a process of genetic information flow, (transcription and translation) results in establishing traits and phenotypes. Organisms differ from other organisms of the same species by small genetic changes caused by mutations or variations. It is becoming increasingly clear that environment and all other external influences also interact in this process of establishing phenotypes. This has been highlighted in the description of "Nature and Nurture" playing a role in shaping our phenotypes. Such processes which play an important role in modulating or modifying the gene expression patterns and thereby phenotypes work without any change in DNA sequence. Such mechanisms which influence and alter gene expression patterns and change the form, structure and function of an organism without any change in DNA nucleotide sequence are called epigenetic processes. *Epigenetics* is a subject area which deals with the study of processes and mechanisms which bring about changes in phenotypes without any change in DNA sequence. Interestingly, evidence has accumulated to describe the role of memory and heritability in such epigenetic mechanisms with possibilities of trans-generational transfer.

Epigenetics (Figure 70.1) coordinates and influences form and function and provides an important component of adaptability to living systems. The development of an organism is governed by networks and cascades of gene regulatory processes. This regulation of gene expression is brought about by altering transcription states. Evidence has accumulated to suggest that epigenetics plays an important role right from the imprinting of alleles to the predisposition to cancer and life style diseases. Epigenetics is also important in the formation of shape, structure and the ability to sustain unfavourable conditions by rapidly and appropriately changing the functioning of a system, thus enabling it to adapt. Along with the phenomena governed by epigenetics, the molecular genetic processes which are involved in the setting up of epigenetic imprint to facilitate adaptability are also being unraveled. It is clear that chromatin remodeling—a process which regulates the active and inactive states of chromatin domains and sets in processes to change chromatin activity states, is involved in establishing

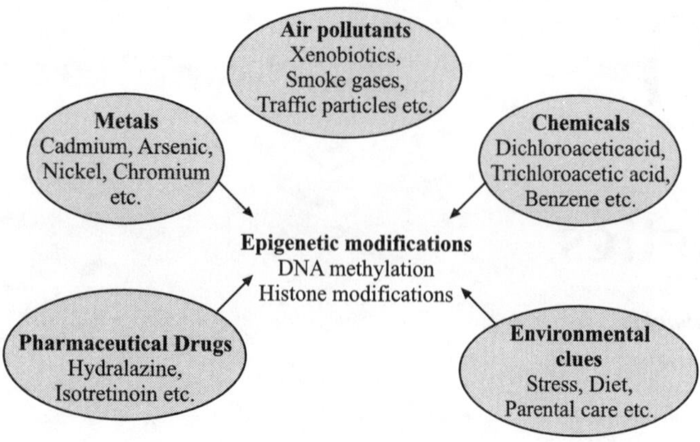

Figure 70.1 Impact of environment on epigenetics.

epigenetic hallmarks. These include DNA modifications, histone modifications, miRNAs, etc. When information about activity states is to be transferred to daughter cells, memory and heritability take on importance. Although much of the epigenetic information is reprogrammed at gamete formation stage or during embryogenesis and throughout life, imprinting effects namely influence of parental origins are known to be trans-generationally transmitted in a few cases.

II. INTRODUCTION TO EPIGENETICS

The genetic information present in the nucleus of an organism carries a blueprint which is believed to be responsible for transmitting hereditary information and governing the development through a very intricately and exquisitely regulated program. This information is encoded in the chromosomes as a sequence of DNA. Organisms differ in the DNA sequence and with the advent of DNA sequencing and genomics the interrelations between gene-transcription-protein and networks which regulate the differentiation processes are being deciphered. Each individual is born with a unique genetic constitution and, except in the case of identical twins, each individual is unique. The alterations in DNA sequence may become apparent in SNPs (small nucleotide polymorphisms or alterations scattered throughout the genome), variations in repeats or rearrangements of genes and these are largely responsible for individual variations and polymorphisms. This information through hereditary processes is transmitted to the next generation. Children inherit a new combination of genes and therefore are not identical to their parents or their siblings. During development and cell differentiation, as a part of normal growth and metabolism, there is an orchestrated regulation of gene expression. This involves signaling cascades and various regulatory networks, including transcription regulatory machinery. This is a very important process since it needs to respond to cell—cell communication signals and external signals in addition to unfolding the inherent program. Eukaryotic DNA is organised as chromatin where nucleosome

is the basic building unit which comprises of 4 core histones (H2A, H2B, H3, H4) and a linker histone (H1) wrapped around 146 bp of DNA. These nucleosomes are organised as beads on a string to form a 30 nm solenoid structure which is further compacted to form chromatin. Chromatin domains are organised as active and inactive domains (based on their transcriptional status). Elaborate and intricate machinery regulates and governs chromatin functions. The remodeling complexes can undergo changes in composition and have the ability to influence and alter activity states of chromosomal domains. When a gene is in active chromatin domain, it is poised for transcription and can form mRNA. This plays a crucial role in the complex cascade of regulation of gene expressions and in establishing networks for appropriate and timely regulation of important processes in cells. These regulations are crucial in proper functioning of an organism. In addition to these genetic factors many epigenetic factors, namely temperature, nutrition, stress, exposures to chemicals and insults, and environmental factors, can also alter states of expressions. The signals are sensed by signaling systems and signal transduction cascades, through coordinated mechanisms leading to responses to stress and adaptations. The epigenetic processes work at the level of chromatin, namely DNA, RNA and proteins and several proteins/ players are involved in modulating these in a developmental stage and cell and tissue specific manner. These processes also show temporal and spatial regulation. One important player in this is a post replicative modification where a methyl group is added onto the nucleotides after their incorporation in the DNA. The enzyme DNA methyl-transferase transfers a methyl group from S adenosyl methionine (SAM) to specific nucleotides in DNA. Methylation is a cis modification and it is a reversible event. Introduction of methyl groups can often influence DNA – protein interactions and DNA conformation. It has been well documented that sequence recognition of many restriction enzymes can be sensitive to presence or absence of methylation on DNA strand. Although activity of some restriction enzymes is unaffected by the presence of methylated bases, in several cases, the presence of methylation hinders digestion by a particular enzyme. There are a few cases where methylation is necessary for recognition of the site for cleavage by restriction enzyme. This provides a good case for alterations of DNA–protein interactions by methylation. In prokaryotes, such as bacteria, DNA methylation plays an important role in marking and distinguishing foreign incoming DNA. It protects the organism from external or parasitic DNA which invades the organism. It has been documented that 6 methyl adenine (6mA) is involved in mismatch repair and *5 methylcytosine* (5mC) in short patch repair. Both 5mC and 6mA seem to regulate gene expression and are extremely important in governing pathogenicity and virulence of pathogenic bacteria. In eukaryotes, 5mC is present in many organisms and a myriad of roles have been proposed to this base modification. These include development and differentiation, regulation of gene expression, X chromosome inactivation, ageing, oncogenesis, etc. DNA methylation forms an important regulatory event in the complex cascade of regulation of gene expression in a eukaryotic cell. In eukaryotes,

5 methyl cytosine has been shown to be present and its level, pattern and occurrence varies from organism to organism.

III. DNA MODIFICATIONS

DNA methylation is one of the most important regulatory processes. Addition of a methyl group to cytosine is a post-replicative modification which can influence DNA protein interactions and DNA conformation. This minor modification has been termed as the 5th base and it adds an additional level of information without changing the DNA sequence. In prokaryotes, methylation is involved in restriction modification, strand discrimination in mismatch repair, regulation of transposition, virulence and pathogenicity, and it has also been implicated in replication. In 1975, Razin and Riggs, Holliday and Pugh, and Sager and Kitchin, independently proposed that cytosine methylation can have multiple roles in regulatory processes such as regulation of gene expression in eukaryotes. The enzyme DNA methyl-transferase has the ability to transfer methyl group from S adenosyl methionine to its substrate DNA. This chemical reaction is catalyzed by enzymes belonging to the DNA methyltransferase family (DNMTs). It is believed that DNMT1 is predominantly responsible for maintaining the preexistent methylation pattern during DNA replication, whereas DNMT3a and DNMT3b are required for de novo DNA methylation. Methylation patterns show memory and heritability and hence are ideal for establishing and maintaining epigenetic imprints. DNA methylation is known to change in a tissue and developmental stage specific manner and it undergoes temporal and spatial changes in eukaryotes.

The molecular genetic mechanisms which govern these processes have been elucidated only in recent years. It has been suggested that DNMT activity regulates global gene expression by hypermethylation of promoter CpG-islands and of regulatory sequences near the promoter (CpG island shores), thus causing gene silencing. The methylation in mammals is proposed to be in CpG dinucleotide predominantly and CpG islands and a set of methylated DNA binding proteins (MBDs such as MeCP1, MECP2) together are believed to regulate gene expression patterns. DNA methylation has been suggested to have a role in development, differentiation, pattern formation, ageing, cancer and metastasis, life style and other susceptibility diseases, governing phenotypes, establishing plasticity, etc. It is generally seen that DNA methylation is more often found in heterochromatic and transcriptionally inactive or silent regions and methylation of promoter upstream sequences modulate regulatory interaction and lead to gene silencing. For example, globin gene is hypermethylated in brain and other tissues but is undermethylated in reticulocytes. In particular, CpG islands (high occurrence of CpG in a short stretch of mammalian genomes where CpG dinucleotides are otherwise relatively less) have been shown to be present upstream of several genes which modulate via methylated DNA binding proteins (MBDs) the binding of RNA polymerase and influence transcription. It has been shown that methylated cytosine is often found in transcriptionaly inactive domains of chromatin. It has therefore been suggested that methylation is thus a necessary and sufficient condition to modulate gene expression. The reticulocyte globin gene family is thus found in active (DNAse 1 sensitive) configuration which upon binding of appropriate trans-acting factors can initiate globin gene expression.

The presence of CpG islands in the 5' regions of some gene and as the name suggests they show presence of increased density of CpG dinucleotide. The CpG doublet occurs at 20% less frequency (than expected) in vertebrate genome. In mammalian genomes CpG islands are often methylated. It has been shown that methylated cytosine can often lead to spontaneous deamination of cytosine leading to conversion of CpG to CpT. In certain positions in the genome the frequency of CpG is 10% higher than the rest of the genome thus reaching the expected frequency. The CpG islands thus have a CpG content of 60% compared to that of the average DNA. They are 1 to 2 Kb long and there are presumably 45,000 such islands in human genome and 15,000 in mouse genome. In many cases CpG islands are upstream of promoter and a specific class of methylated DNA binding proteins (MeCP1 and 2) have been shown to play important role. MeCP1 binds methylated DNA while MeCP2 inhibits chromatin modifications. Mutations in this genes lead to syndromes such as Retts syndrome. Thus there is a correlation and association between DNA methylation, chromatin structure, DNase 1 hypersensitive sites, binding of several transcription factors, and gene expression.

DNA methylation is involved in epigenetic processes. One of the hallmarks of DNA methylation is that it shows memory and heritability. There are several examples of this. It has been shown that coat colour of syngenic mice can be influenced (altered) by diet. Whether the mice are fed methylation proficient or deficient diets determines if they are dark or light in colour. This clearly establishes the ability of DNA methylation to govern genotype-phenotype interactions.

Another modified nucleotide which has gained prominence is the 5 hydroxymethyl cytosine (5HMC). Products of Tet gene deaminate 5 methylcytosine to convert it to 5HMC. On one hand, it has been well documented that this process is a part of demethylation, but on the other hand, it also establishes an additional imprint in the genome. It has been suggested that 5HMC replaces 5mC in stem cells and in mouse embryogenesis, thereby making a way for altering epigenetic imprints and reprogramming. Thus DNA modifications are an important event in establishing information about activity states. It is a cis modification in DNA (a permanent mark along DNA strand) and hence, it can specify epimutation, i.e., change in functionality without any change in DNA sequence. Although these patterns are faithfully transmitted to the next generation, the ability to remove these marks during replication or through protein interactions make them attractive candidates for epigenetic inheritance.

IV. CHROMATIN AND EPIGENETICS

It is well established that eukaryotic DNA is not free in the nucleus but is present as a DNA protein complex, as beads on a string. The basic unit of eukaryotic chromatin is a nucleosome which has two components—the core and the linker. Histone octamers, i.e., two molecules each of H2A, H2B, H3, and H4, are wound around 146 bp of DNA to form a nucleosome core. Nucleosome cores are linked via linker DNA where histone H1 binds at the sites where the DNA enters and leaves nucleosomes. Nucleosomes are then organised to form 30 nm and then solenoid structure. Chromatin contains many regulatory proteins and RNAs in addition to DNA and the basic histone proteins. Our body has a trillion cells and these cells are involved in development and differentiation in generating form, structures and conferring function. In fact, differential and programmed regulation of gene expression is one of the most intricate and important processes in establishing phenotypes and functionality. Whether a gene will be functional or not (transcriptionally active or inactive) will depend upon the structure of chromatin. Traditionally it is well known that eukaryotic chromatin can be described as euchromatin and heterochromatin. The heterochromatin is darkly stained, condensed and late replicating while euchromatin is less densely stained, has open configuration and is accessible to nuclease digestion. Euchromatin harbours all active genes, while heterochromatin carries inactive, i.e., silenced regions. These include centromeres, telomeres non transcribed repeat sequences and inactive genes/domains. Thus whether a gene will be active, i.e., form mRNA (get transcribed) will depend upon the local chromatin configuration. During developmental processes then and subsequent differentiation events orchestration of a coordinated and dynamic modulation of chromatin structure and gene expression is an important key event. Thus, a complex cascade of events can be said to govern these processes which involve multiple players and partners. The brain cell or gastrula needs to have a temporal regulation of gene expression in order to establish, preserve and optimize the function. It is, therefore, extremely important to have the appropriate genes expressed or silenced. The blueprint of genetic program carries this information and also the plan for unfolding this process. Epigenetic processes participate in establishing and maintaining these functional states and can also integrate environmental and other influences along with genetic factors.

Chromatin thus can alternate between different states and this process of establishing and modulating chromatin functionality is called as *chromatin remodeling*. Once established these epigenetic states influence chromatin structure and thereby function can persist through cell divisions creating epigenetic states in which the states of gene function are propagated by a self-perpetuating structure of chromatin. Thus, the states are propagated in a sequence independent manner epigenetically. Regulation of gene expression in tissue and developmental stage specific manner lies at the heart of eukaryotic differentiation. This plays pivotal role in cell differentiation and control of metabolic, catalytic and cellular pathways.

V. CHROMATIN REMODELING

Drosophila has been long known to demonstrate a phenomenon called position effect variegation, where location of a gene on the chromosome can influence its activity state resulting in partial or complete activation or inactivation of a gene. Hairy wing is one such phenotype which shows position effects. When a genetic screen for Su-var (suppressor of hairy wing phenotype) was carried out, a set of genes were identified which were found to be involved in the regulation of this variegation. It became apparent that these genes have homologues in all eukaryotic organisms and are actively involved in chromatin remodeling, influencing activity states of chromatin. Components of many of these processes were then discovered in yeast and subsequently in other animals. Chromatin remodeling is carried out by ATP dependant chromatin remodeling complex. The ATPases subunits of all remodeling assemblies belong to a large superfamily and provide energy from hydrolysis of ATP (Table 70.1). Both transacting factors (activators) and specific *cis* sequences in DNA are involved in influencing and directing remodeling complexes to specific sites in DNA. It involves modulating chromatin accessibility and nucleosome positioning. There are at least five families of complexes involved in remodeling chromatin in eukaryotes: SWI/SNF, ISWI, NURD/Mi-2/CHD, INO80 and SWR1 (Table 70.2). Chromatin remodeling thus involves energy dependant displacement of nucleosomes to alter or influence activity states.

VI. DNA METHYL-TRANSFERASES

DNA methylation involves two processes namely *de novo* methylation, i.e., methylation of hitherto unmethylated DNA to establish new methylation patterns and maintenance methylation where the DNA methylation is faithfully transmitted to the two daughter strands. The DNA methyl-transferase family (DNMT1, DNMT3a, DNMT 3b) catalyse the methylation of cytosine in mammalian DNA. The individual deletion of these three methyltransferases leads to embryonic lethality underscoring the important role of meythyltransferases. DNMT1 has been shown to carry out maintenance methylation where it catalyses the addition of methyl group to the two daughter strands during replication of parental DNA methylation patterns (Figure 70.2). In contrast, DNMT 2 catalyse *de novo* methylation and thus has a role during DNA replication. It has been shown that DNMT3L the catalytically inactive homolog plays a regulatory role. The process of methylation removal from DNA is catalytically difficult and two processes have been described. One is a passive process where methylation of new strand is inhibited at the time of replication or an active process which involves removal of methylated base or methyl group. Recently the existences of TET enzymes which deaminate 5mC to generate hydroxymethyl cytosine have been described. This suggests that conversion of 5mC to hydroxymethyl cytosine in a sequence dependant manner may be another catalytic process which plays a role.

TABLE 70.1   Chromation remodeling complexes and their features

Sr. No	Remodeling complex	Description
1.	SWI/SNF	SWI/SNF remodelers disorder nucleosomes. SWI/SNF complex with the chromatin to destabilise the interaction between DNA and histones and disrupt chromatin and provide access to the transcription-binding domains
2.	ISWI	ISWI isolated from *Drosophila* embryos are involved in transcription and DNA replication through heterochromatin, proper chromatin assembly after DNA replication, replication initiation, and maintenance of higher-order chromatin structures. It can interact with several proteins giving three different chromatin-remodeling complexes in *Drosophila melanogaster*: NURF(nucleosome remodeling factor), CHRAC (chromatin remodeling and assembly complex) and ACF (ATP-utilising chromatin remodeling and assembly Factor).
3.	NURD/Mi-2/CHD	These complexes primarily mediate transcriptional repression in the nucleus as well as are required for the maintenance of pluripotency of embryonic stem cells. NuRD repress transcription via its chromatin-remodeling and histone deacetylase activities. Mi-2/CHD deacetylates chromatin, represses transcription, and regulates development. CHD participates in the remodeling of chromatin by deacetylating histones and is known to promote transcription elongation.
4.	SWR1	SWR1 has many crucial functions including the control of gene regulation and expression, checkpoint regulation, DNA replication and repair, telomer maintenance and chromosomal segregation and as such a component of pathway that are involved in genomic integrity.
5.	INO80	INO80 regulates transcription and is involved in DNA repair and cell cycle checkpoint adaptation. It participates in DNA double-strand break (DSB) repair, nucleotide-excision repair (NER), histone exchange, nucleosome mobilization and positive regulation of transcription from RNA polymerase II promoter.

TABLE 70.2   Histone Modifications

Histone type	Modification residue	Modification type	Enzymes	Function
H1	Ser 27	Phosphorylation	Kinases	Transcriptional activation, chromatin decondensation
	Lys26	Methylation	Histone Methyltransferase	transcriptional silencing
H2A	Lys5 (Mammals)	Acetylation	Histone Acetyltransferase	Transcriptional activation
	Ser139 (Mammalian H2AX) Modified histone	Phosphorylation	Kinases	DNA repair
	Ser 1	Phosphorylation	Kinases	Mitosis, Chromatin assembly, Transcriptional repression
	Lys 119 (Mammals)	Ubiquitination	Ubiquitinating Proteins	Spermatogenesis
	Lys126 (S. cerevisiae)	Sumoylation	SUMO proteins	Transcriptional repression
H2B	Lys12 and Lys15 (Mammals)	Acetylation	Histone Acetyltransferase	Transcriptional activation
	Ser14 (Vertebrates)	Phosphorylation	Kinases	Apoptosis, DNA repair
	Ser33 (D. melanogaster)	Phosphorylation	Kinases	Transcriptional activation
	Lys120 (Mammals)	Ubiquitination		Meiosis
	Lys6 or Lys7 (S. cerevisiae)	Sumoylation	SUMO proteins	Transcriptional Repression
H3	Lys9	Acetylation	Histone Acetyltransferase	Histone deposition, Transcriptional activation
	Lys9	Methylation	Histone Methyltransferase	Transcriptional silencing (tri-Me), Transcriptional repression, Genomic imprinting
	Ser10	Phosphorylation	Kinases	Mitosis, Meiosis, gene activation
	Lys27	Methylation	Histone Methyltransferase	X inactivation (tri-Me)
	Lys4	Biotinylation	Biotinidase	Gene Expression
H4	Lys12	Biotinylation	Biotinidase	DNA damage response
	Lys12	Acetylation	Histone Acetyltransferase	Histone localization, Telomeric silencing, DNA repair
	Lys20	Methylation	Histone Methyltransferase	Transcriptional silencing (mono-Me), Heterochromatin (tri-Me), Checkpoint response
	N-terminal tail (S. cerevisiae)	Ubiquitination	Ubiquitinating Proteins	Transcriptional repression
	Ser1	Phosphorylation	Kinases	Mitosis, Chromatin assembly, DNA repair

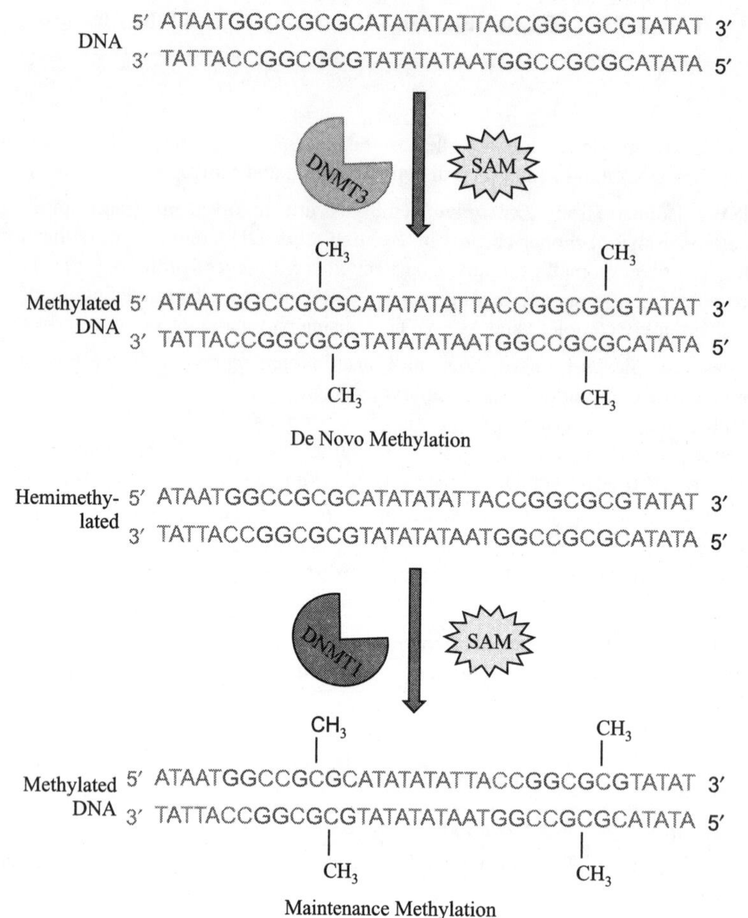

Figure 70.2   Illustration of the processes of de novo methylation and maintenance methylation. (*see Plate 31 for colour figure*)

VII.   POST TRANSLATIONAL MODIFICATIONS OF HISTONES

A key epigenetic process involves the "*histone code*" where histone modifications (…) are the key events in the chromatin remodeling processes in addition to replacement of certain histone with different histone subtypes. These modifications are reversible in response to cellular or extracellular signals. The cascade of signaling processes and networks in cell appear to integrate and cross-talk with remodeling machinery. The histone modifications can be briefly summarised as

1. acetylation of histones
2. methylation of histones
3. phosphorylation of histones
4. sumoylation of histones
5. ubiquitination of histones
6. ADP ribosylation of histones
7. replacement by histone subtypes

Post translational modifications of chromatin are carried out by specific families of enzymes which integrate with remodeling complexes, transcription factors and DNA methylation. Hallmarks of activity states can often be found out by association of a specifically modified or altered histone. For example, acetylation of lysine 9 on histone H3 typically clusters at transcriptionally active loci while methylation of the same lysine is found at transcriptionally silent loci.

A.   Histone Modifications

Chromatin can be configured in transcriptionally active and inactive domains. The hallmarks of activity states are well characterised. Since 146 base pairs of DNA are wound around 2 molecules each of histones H2A, H2B, H3, H4 to form the nucleosome core and along with the linker DNA and histone H1. The beads on the string model of chromatin is the basic unit of chromatin organisation in eukaryotes. This then gets further packaged into 30 nm solenoid and creates the various chromatin domains conducive to the various activity states. Several players are involved in setting up this histone code which is very crucial to establishing and modulating epigenetic states. Modifications such as acetylation, phosphorylation, methylation, sumoylation, ADP-ribosylation, etc. gets established through an intricate and multistep chromatin remodeling processes by employing of specific enzymes. These enzymes add post translational modifications to histones. Some of these modifying enzymes are described here. Histones are highly basic proteins and addition of charged modifications alter the DNA-histone interactions.

Histone acetylation involves addition of acetyl group from acetyl CoA to an amino acid namely a lysine (K) residue in histone protein. Histone acetyl transferases (HATs) were first identified as tumor suppressors and they have also been implicated in several diseases including cancer progression, viral infection and certain breathing disorders. They catalyse the transfer of acetyl group from acetyl coA to specific histone molecules. Introduction of an acetyl group at lysines and other amino acids modify the charge and interactions of histones. Histone deacetylases (HDACs) are enzymes which catalytically remove the acetyl group from specific site in the histone. The modulation of acetylation and deacetylation plays an important role in epigenetic modulation. In fact presence of such modifications at certain precise locations is used to identify activity states of chromatin.

Histone methylation is carried out by histone methyltransferases (HMTs). These enzymes catalyse the transfer of one, two or three methyl groups on histones. H3K9me3 (on histone H3 lysine at 9th position carried 3 methyl groups) and H3K27me3 (histone H3 lysine at 27th position carries three methyl groups) are required for maintenance of facultative and constitutive heterochromatin. Histone methylation is carried out by SET domain proteins (suppressor of variegation, enhancer of Zeste and trithorax) which have 130 amino acids of Set domain which methylates lysines in histone tails.

The current evidence suggests that small noncoding RNA are major components in this modulation of gene expression. It has been suggested that the histone code could be mediated by the effect of small RNAs. The recent discovery and characterization of a vast array of small (21- to 26-nt), non-coding RNAs suggests that there is an RNA component which is possibly involved in epigenetic gene regulation. Small interfering RNAs (miRNAs) can modulate transcriptional gene expression via epigenetic modulation of targeted promoters.

The histone code along with DNA methylation thus is involved in packaging of nucleosomes in higher order structures. The chromatin remodeling machinery along with its partners such as trans activating factors, modifying enzymes takes care of establishing and maintaining activity states of genes and chromosomal domains. Several other protein partners like chromo domain and bromo domain proteins recognize this modified histone code and lead to further compaction. Chromatin structure plays a crucial role during development, differentiation and in integrating environmental and other cues and signals result in altering chromatin states and gene expressions. These altered chromatin states or domains then influence gene expressions. Thus phenotype gets influenced by epigenetic processes. Unless otherwise interfered with these states are stably transmitted during cell division to the daughter cells. The epigenome is tissue-specific, developmentally regulated and highly dynamic. This variability of epigenetic states offers a potential explanation for individual differences in phenotype. Thus the effects of environment, interactions,

stress, exposure to toxic chemicals get reflected in chromatin remodeling through epigenetic processes. Aberrant epigenetic marks are associated with a range of complex pathologies including cancer. It has been proposed that epigenetic processes may explain some complex trait characteristics, such as variable age of onset, fluctuating disease course, parent-of-origin effects and identical twin discordance.

X. CHROMOSOME INACTIVATION

In mammalian females, one of the two X chromosomes is inactivated, i.e., it becomes heterochromatic, late replicating and transcriptionally silenced. Since the males have one X and one Y chromosome while the females have two X chromosomes, this has been thought of as an example of "dosage compensation". The inactive X chromosome (Figure 70.3) carries the hallmarks of heterochromatin and once inactivated early in embryogenesis, the same X continues to be inactive. X chromosome inactivation is thus an example of differential regulation of homologous chromosomes and displays memory and heritability. In marsupial mammals, it is always the paternal X chromosome which gets inactivated. In eutherian mammals early in embryogenesis, in yolk sac and extra-embryonic tissues, paternal X is subjected to inactivation while this becomes random later in embryogenesis. This is an example of differential regulation of homologous chromosomes and this phenomenon shows *imprinting* effects (In imprinting, the behavior (functionality) of genes, chromosomes or chromosomal segments is influenced by parental origin). X chromosome inactivation shows memory and heritability and in all progeny of the cell, same X chromosome continues to be inactivated. If a cell were to have more than one X chromosome, all X chromosomes "in excess of one" get inactivated. The cis acting Xce or Xic locus, which codes for a small noncoding RNA namely XiST (X inactivation specific transcript) has been shown to have a role in establishing and maintaining inactive states. XiST has been shown to localise across inactive X chromosome and recruits polycomb and other groups of proteins to establish the functional states. In

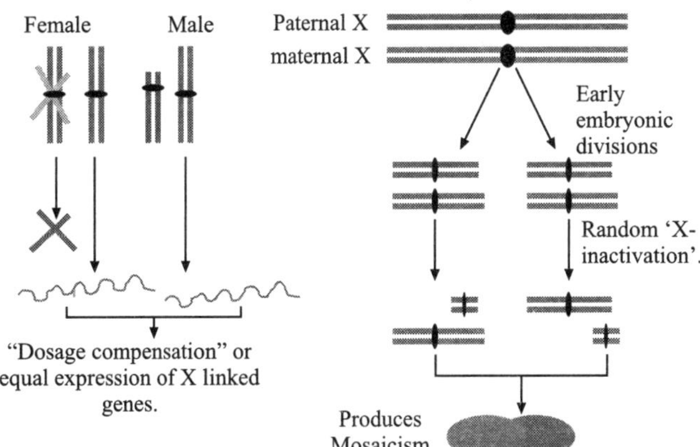

Figure 70.3 Chromosome inactivation. (*see Plate 31 for colour figure*)

addition to DNA methylation, macroH2A, histone methylation and acetylation are involved in establishing the activity states of X chromosomes. This makes the mammalian female a mosaic, since if she is a heterozygous individual; her cells express either the maternal or paternal genes based on their inactivation. In some of her cells, genes from paternal X chromosome are expressed while in some maternal genes are expressed. This actually results in females being less susceptible to several X linked traits such as baldness, haemophilia, etc.

IX. CHROMOSOME IMPRINTNG

Chromosome imprinting (Figures 70.4 and 70.5) represents a phenomenon where the parental origin of a gene influences the functionality of the gene, chromosomal segment or chromosome. The allele therefore may actually be a functional

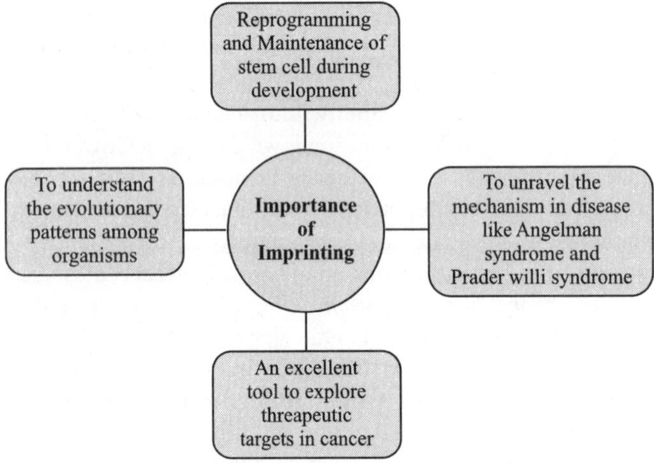

Figure 70.4 Impact and utility of chromosome imprinting.

allele, but can get silenced if it is inherited from mother or father. This behaviour is in stark contrast to Mendelian inheritance where in Punnet square and expression patterns, equal expression (functionality) from both alleles is assumed and expected. It has been shown that in mammals such as mouse and man a few genes (about 80) are subjected to imprinting. The individual is therefore hemizygous for imprinted genes. Although the maternal and paternal alleles have identical sequences, due to imprinting effects these will display different properties or phenotypes. This has very important effects on physiology and can lead to diseases including cancer. These were the first evidences to clearly demonstrate differential role for maternal and paternal alleles in development. Imprinting is thus a phenomenon where it is believed that transmission of a gene from maternal or paternal lineage adds an imprint which then changes the chromatin configuration and structure thereby influencing activity states. A functional gene may be rendered non-functional due to this phenomenon. The marks that are established have been suggested to be differential methylation of DNA and variations in chromatin modifications which seem to be inherited across meiosis and then mitosis. These patterns influence epigenetic memory and are involved in setting up imprinting. Along with DNA methylation differences, polycomb and trithorax group proteins help in setting up the "code" to establish stable transcriptional patterns. One of the classical locus known to be imprinted is responsible for the neurodevelopmental disorders, namely, Prader-Willi syndrome and Angelman's syndrome. Most of the cases are due to deletion or modifications but result in very different presentation of diseases based on which chromosome carries the mutation (namely paternal or maternal). Some human disorders are associated with genomic imprinting, a phenomenon in mammals where the father and mother contribute different epigenetic patterns for specific genomic loci in their germ cells. The

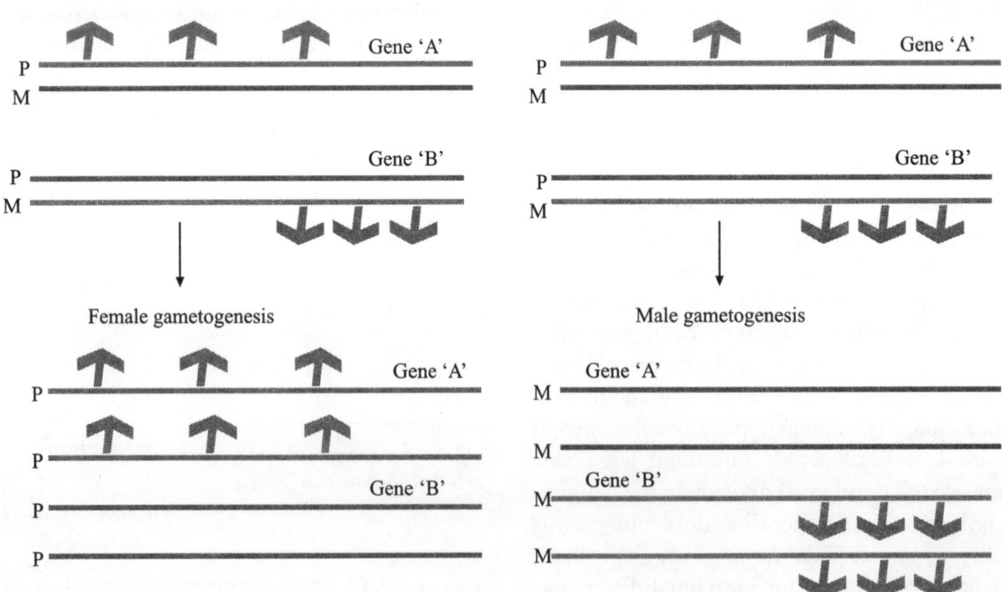

Figure 70.5 Chromosome imprinting and gene silencing. (*see Plate 31 for colour figure*)

best-known case of imprinting in human disorders is that of Angelman syndrome and Prader-Willi syndrome—both can be produced by the same genetic mutation, chromosome 15q partial deletion, and the particular syndrome that will develop depends on whether the mutation is inherited from the child's mother or from their father. This is due to the presence of genomic imprinting in the region. Beckwith-Wiedemann syndrome is also associated with genomic imprinting, often caused by abnormalities in maternal genomic imprinting of a region on chromosome 11.

Second example of imprinting is the IGF locus. The IGF-II gene (insulin like growth factor gene-II) which is inherited from the father gets expressed, but it is silenced when inherited from maternal side. However this dependence is modified for other genes and cases in this locus. There are a few examples of cancer causing genes displaying imprinting effects. More than 50 percent of cancers are believed to be due to epigenetic effects which are termed as epi-mutations. Imprinting thus assumes importance in a number of diseases and also in conferring predisposition.

X. IMPORTANCE AND APPLICATIONS OF EPIGENETICS

Epigenetics can be implicated in several processes and has important implications in medicine, cancer diagnosis, modulation of life style diseases and phenotypes, increasing yields and influencing quantitative characters, etc.

Development

Epiegentic and imprinting processes coordinate environmental influences and external signals in designing form and function. The embryonic development proceeds in a programmed manner with different roles for the maternal and paternal genomes. Although all cells have the same genome, the cells differentiate in a coordinated fashion in different cell types, perform different functions, and respond differently to the environmental signals. As an organism develops, morphogens activate and silence genes giving them different memories. Some cells stay totipotent while some undergo terminal differentiation. Resetting of these memories may be extremely important in the totipotency of plants and formation of stem cells or cloning from differentiated cells in animals. Stem cell retains its pluripotency, continues to divide and can help in organogenesis. Stem cells thus play an important role in rejuvenation and regenerative capabilities. The recuperating and reprogramming capability has been important in the science of induced pluripotent stem cells (iPS) and stem cell therapy. A clear understanding of epigenetic machinery and its regulation will ensure design of a full proof process for reengineering through stem cell treatment.

Medicine

It is becoming apparent that misregulation of epigenetic processes can lead to many diseases. This aspect appears to be important in imprinted diseases and also in multiple situations which arise due to epimutations (alteration in epigenetic state). The therapeutic potential of epigenetic modulators is thus enormous but it is at a very early stage. These may find application in predisposition; life style and exposure related diseases as well as cancer.

Cancer

There is a global effort to characterize cancer genome via next generation sequencing and high throughput methodologies. These allow accurate map generation and association studies of mutations, SNPs, translocations and other chromosomal rearrangements with the cancer phenotype. This has improved accurate diagnosis and identification of stages of cancer and help in better treatment regimes as well as in follow up studies. Biomarkers and candidate genes and changes are thus being marked. In many cancers, in addition to genetic changes, epigenetic processes have been shown to play pivotal role. Exposure to carcinogens (cancer causing agents) appears to modulate epigenetic machinery thereby changing the epigenetic state. It is believed now that a genetic map of cancer gives a partial picture and integration of DNA methylation and epigenetic analysis of cancer tissues may be equally if not more important in cancer diagnosis and analysis.

Evolution

There is clear evidence that epigenetic marks can be transmitted across generations (for 6 generations). More than 100 cases of trans-generational inheritance have been reported in plants and animals. Epigenetics is a means by which environmental and external factors can work on and modulate gene expression in order to alter activity states or phenotypes. This is an example of very carefully and elegantly orchestrated machinery. These changes can show transgenerational inheritance and offer survival advantages and influence selection. The variability and penetrance of certain trait may be explained by altered epigenome. Epigenetic processes seem to give short term advantages by increasing adaptation and modulation of phenotype and behavior. Integration of these mechanisms in evolutionary studies in future will enrich our current models of evolutionary theories.

SUMMARY

Epigenetics is thus a study of a heritable change in gene expression or cellular phenotype caused by processes other than a change in DNA sequence hence the name epi, meaning "over, above, outer". The word "epigenetic code" has been used to describe all such changes ie, the set of epigenetic features that create different phenotype in different cells. One example of epigenetics in evolutionary biology is differentiation. Non genetic factors appear to thus influence the phenotype of an organism. This supports the hypothesis that the genome codes for a range of possibilities and establishment of actual phenotype is governed by many factors (environment, diet, stress, interactions, maternal effects, etc). These factors are termed as epigenetic modulators. This raises an interesting possibility of modulating phenotypes without any genetic change. The range and scope of these modifications will be determined by the genetic component (by setting limits). Instead of gene therapy one may talk of epigenetic therapy which may be easier to administer and less harmful. It is therefore important that future research in this area will unravel the intricacies of this phenomenon and its regulation and provide a clear understanding of molecular genetic and other regulatory mechanisms.

Acknowledgements

I would like to thank all my students Sania, Shriram, Rajesh, Varada, Naina, Chitra for their help in preparing the figures.

SUGGESTIONS FOR FURTHER READING

Lewin's genes X (2010) Krebs J E, Goldstein E S,and Kilpatrik S T pub. Jones and Bartlet

Toxicology and Epigenetics (2012) eds Saura Sahu publishers Wiley

'Epigenetics and chromatin' in Progress in Molecular and Subcellular Biology, Vol. 38 (2005) Springer

71

Micro-RNA and siRNA – Control Gene Expression
Implications in Therapeutics, Vaccines and Pathogenesis

Akhil C. Banerjea, Richa Kapoor and Animesh Banerjea

CONTENTS

I. INTRODUCTION

Sequencing of human genome and knowledge of molecular causes of diseases have opened new areas for generating vaccines against varied health disorders. Knocking down pathogenic genes using RNA interference mechanisms appears to be a new promising approach. RNA interference (RNAi) is referred to as a mechanism involving post transcriptional gene silencing using dsRNA. Andrew Fire and Craig C. Mello in 1998 observed that introduction of dsRNA into *Caenorhabditis elegans* resulted in specific gene silencing.[1] Both shared the Nobel Prize in Physiology or Medicine in 2006 for this revolutionary finding in the field of molecular biology. As dsRNA was found to interfere with the function of its target gene hence the mechanism was named 'RNA interference' (RNAi).

Small RNA molecules like siRNA (small interfering RNA) and miRNA (microRNA) use the RNAi mechanism to either degrade or translationally repress the target messenger RNA.[2]

II. SHORT INTERFERING RNA

The siRNA pathway initiates with the cleavage of long dsRNA by the cytoplasmic enzyme Dicer into siRNA (~ 22 nucleotides). siRNA is incorporated into the RNA induced silencing complex (RISC). The guide strand of siRNA has perfect sequence complimentarity with target messenger RNA. It recognizes the target mRNA and directs it for cleavage (Figure 71.1).

Exogenous introduction of synthetic siRNA in the cell to knockdown mRNAs of disease related genes forms the basis of siRNA therapeutics. It is a very promising approach especially for those diseases for which treatment options are limited.

III. siRNA THERAPEUTICS

Synthetic siRNA was used for the first time in case of Hepatitis B virus infection.[3] To protect infected mice from liver fibrosis, siRNA was administered for Fas mRNA. siRNA can be synthetically designed for any known gene. This makes RNAi a very valuable tool. Antiviral and chemotherapeutic siRNAs have been designed. For instance, disease effector genes being targeted by siRNA are K-RAS in Pancreatic carcinoma,[4] c-raf and bcl-2 in Leukemia,[5] Vif, Nef, Tat, Rev and CCR5 in HIV infection[6,7,8,9] and NP, PA, PB1 in Influenza.[10] Many others are being examined. In a landmark experiment, Banerjea and coworkers have introduced a potent siRNA against HIV-1 Tat gene in hematopoietic stem cells and showed that macrophages and the T-cells that were derived from the these cells were resistant to HIV-1 challenge.[11] Lately, the same group has designed a novel siRNA-Ribozyme chimeric construct to potently knock down hepatitis gene expression[12] and also influenza virus replication.[13]

SiRNA are externally administered in cells and their mode of delivery can be either:

1. Non-viral approach: Using PEG,[14] liposomes (cholesterol)[15]
2. Viral approach: Using Retrovirus,[16] Adenovirus[17]

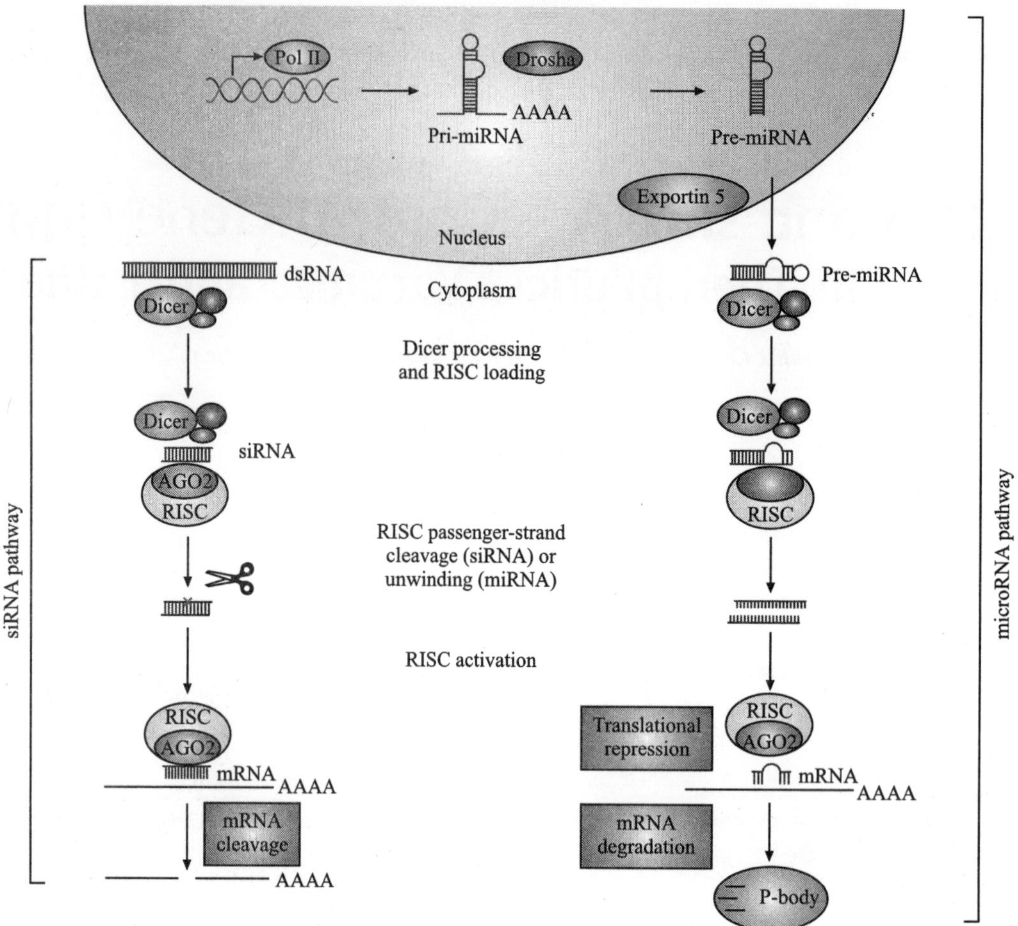

Figure 71.1 Biogenesis of siRNA and miRNA pathway (Nature Review, Drug Discovery; Volume 6, June 2007,443). (*see Plate 32 for colour figure*)

Both the approaches have their limitations, but currently research is going on to overcome them.

SiRNA is not stable for long inside the cell, so while designing of siRNA, various chemical modifications are done. For protection against exonuclease degradation, Phosphorothioate (P = S) backbone is linked at the 3'-end and for endonuclease resistance, M 2'-sugar modification (such as 2'-*O*-methyl or 2'-fluoro) is done.[18–21]

Vaccination based on siRNA has reached the first clinical trials. These are now being referred to as the **'fourth generation vaccines'**.

IV. MICRORNA

MicroRNAs regulate critical biological pathways and processes. Their biogenesis is a multistep process (Figure 71.1) which starts in the nucleus and after many post transcriptional modifications ends in the cytoplasm. Primary microRNA (pri-miRNA) is transcribed from a miRNA gene and is about 70–100nt long hairpin structure. The pri-miRNA is processed in the nucleus by the ribonuclease Drosha to precursor microRNA (pre-miRNA). Then, the pre-miRNA is transported into the cytoplasm by exportin 5. There, a second ribonuclease, Dicer

digests the pre-miRNA resulting in a 21–25nt miRNA. At this stage, the miRNA aligns with target mRNA and forms a RNA Induced Silencing complex. Depending on the level of complementarity between the microRNA and the target sequence, the mRNA can either be translationally repressed (partial) or cleaved (identical).[22–23] MicroRNAs were found to downregulate gene expression by base-pairing with the 3' untranslated regions (3'UTRs) of target messenger RNAs.

A. MicroRNA Therapeutics

MicroRNAs are involved in many signalling pathways such as cell differentiation, proliferation and survival. A single microRNA can target multiple genes and a single gene may contain recognition sites for multiple microRNAs. So, a change in the expression level of a microRNA will in turn affect the expression level of its targets.

In diseased state, the microRNA expression profile of cell gets deregulated. Some microRNA are overexpressed such as in case of Hepatitis C infection, a liver specific miRNA-122 is upregulated. This miRNA-122 is involved in Hepatitis C virus replication and cholesterol metabolism. MiRNA therapeutics is developing these days, which involves the use of anti-microRNA technology to knockdown upregulated microRNAs. This will lead to balanced gene regulation and hence a normal physiological state. Santaris Pharmaceuticals has prepared such kind of microRNA therapeutic for the treatment of Hepatitis C infection. Currently, the clinical trials are going on for this anti miR drug. Many other biotech and pharmaceutical companies have started preparing drugs based on miRNA therapeutics. miRagen Therapeutics in Boulder, USA is focusing on cardiovascular and muscle diseases and Regulus Therapeutics (Carlsbad, USA) is working on liver cancer, immunology and inflammation. Designing of antimiR is a major challenge for these companies. Various aspects such as antimicroRNA stability, mode of administration and pharmacokinetic properties need to be considered.

MicroRNA profile studies in chronic lymphocitic leukemia indicate that miR-15a and miR-16–1 were knocked-out in approximately 69% cases.[24] miR34a has been reported as a biomarker indicating brain ageing.[25] High levels of miR-125b in breast cancer predict poor response to taxol-based treatments *in vitro*.[26] Thus, miRNAs can also be used as diagnostic, prognostic and predictive biomarkers for several diseases and some biological phenomena.

CONCLUSION

SiRNA and miRNA play important roles in gene regulation and can be used as therapeutic tools for treating many diseases. RNAi technology needs to be developed further with regards to its fewer side effects, effective delivery to cells *in vivo* and greater stability. It is a very promising approach and might be of great help to be mankind.

Acknowledgements

The authors acknowledge the help received from Department of Biotechnology, Indian Council of Medical Research, Government of India.

SUGGESTIONS FOR FURTHER READING

1. Fire A., Xu S., Montgomery M.K., Kostas S.A., Driver S.E. and Mello C.C. (1998), Potent and Specific Genetic Interference by double-stranded RNA in *Caenorhabditis Elegans. Nature*, 391 (6669):806–811.

2. Martinez J., Patkaniowska A., Urlaub H., Luhrmann R. and Tuschl T. (2002), *Cell,* 110, 563–574.

3. Song E., Lee S.K., Wang J., Ince N., Ouyang N., Min J., Chen J., Shankar P. and Lieberman J., (2003), RNA Interference Targeting Fas Protects Mice from Fulminant Hepatitis, *Nat Med.,* 9(3):347–351.

4. Brummelkamp T., Bernards R., Agami R. (2002), *Cancer Cell,* 2:243–247.

5. Cioca D., Aoki Y. and Kiyosawa K. (2003), *Cancer Gene Ther.,* 10:125–133.

6. Jacque J.M., Triques K. and Stevenson M. (2002), *Nature,* 418:435–438.

7. Lee N.S., et al. (2002), *Nat., Biotechnol.,*19:500–505.

8. Capodici J., Kariko K. and Weissman D. (2002), *J. Immunol.,* 169:5196–5201.

9. Qin X.F., An D.S., Chen I.S.Y. and Baltimore D. (2003), *Proc Natl Acad Sci.,* USA. 100:183–188.

10. Ge Q., et al. (2003), *Proc Natl Acad Sci.,* USA, 100:2718–2723.

11. Banerjea A., Li M.J., Bauer G., Remling L., Lee N.S., Rossi J. and Akkina R. (2003), Inhibition of HIV-1 by Lentiviral Vector-Transduced siRNAs in T Lymphocytes Differentiated in SCID-hu Mice and CD34[+] Progenitor Cell-Derived Macrophages, *Molecular Therapy,* 8:62–71.

12. Gupta et al. (2008), *Oligonucleotides,* 18, 225–234.

13. Prashant et al. (2010), *Antiviral Res.,* 87, 204–212.

14. Li W. and Szoka F.C. (2007), *Pharm. Res. Adv. Drug Deliv. Rev.,* 24, 438–449.

15. Lorenz C., Hadwiger P., John M., Vornlocher H.P. and Unverzagt C. (2004), *Bioorg. Med. Chem. Lett.,* 14, 4975–4977.

16. Brummelkamp T.R., Bernards R. and Agami R. (2002), *Science,* 296, 550–553.

17. Shen C., Buck A.K., Liu X., Winkler M. and Reske S.N. (2003), *FEBS Lett.,* 539, 111–114.

18. Layzer J.M., et al. (2004), *RNA* 10, 766–771.

19. Choung S., et al. (2006), *Biochem. Biophys. Res. Commun,* 342, 919–927.

20. Allerson C.R., et al. (2005), *J. Med. Chem.,* 48, 901–904.

21. de Fougerolles, A.R., et al. (2005), *Methods Enzymol.,* 392, 278–296.

22. He L. and Hannon G. (2004), *Nat. Rev. Genet.,* 5(7):522–31.

23. Bartel D. (2004), *Cell,* 116(2):281–97.

24. Calin et al. (2002), *Proc Natl Acad Sci.,* USA, 99(24):15524–9.

25. S. Liu S., Liu X. Wang (2011), The P13K-Akt Pathway Inhibits Senescence and Promotes Self-renewal of Human Skin-derived Precursors *in vitro, Aging Cell,* 10–4, 661–674.

26. Zhou et al. (2010), *J. Biol. Chem.,* 285:21496–21507.

72

Agricultural Biotechnology for Food Sufficiency and Benefit to Human Health

Asis Datta and Sumit Ghosh

CONTENTS

I. Introduction
II. What are the Major Limiting Factors on Food Security?
    A. Water
    B. Extreme temperature
    C. Fertile soil
    D. Pests, diseases and weeds
    E. Post-harvest loss
    F. Genetic erosion
    G. Global climate change
    H. Ozone
III. Malnutrition
IV. How does Plant Biotechnology Fit into Crop Improvement Program?
    A. Genetic marker
    B. Genome sequencing
    C. GM technology
V. Global Status of Commercially Grown GM Crops
VI. Biosafety and Environmental Considerations for the Use of Marker Genes in GM Crops
VII. How do Classically Bred Crops Differ from GM Crops?
    Summary

I. INTRODUCTION

Proper human health is only achievable when sufficient quantities of nutritious and safe foods are available throughout the life. Food insecurity and malnutrition lead to serious concerns for human health and even loss of lives in developing countries. The continuous rise in world population along with decline in cultivable lands, not only challenges the supply of global food demand in future but also with the availability of nutritionally balanced foods. In order to ensure food security for future generations, the world must produce 50–100% more food than at present under the adverse environmental conditions. Worldwide 850 million people suffer from malnutrition. It is estimated that over 200 million of the world's hungry are children and at least 5 million die each year from malnutrition. Iron deficiency alone affects about 2 billion people. Vitamin A deficiency affects at least 100 million children each year.

Among them 1.3 to 2.5 million die and over 250,000 others suffer permanent damage such as blindness. Billions of people worldwide are at risk for zinc, iodine and folic acid deficiencies. Thus, upgrading the nutritional quality of food crops is a prime consideration for global food security and the nutritional well being of the world population.

Till date plant breeding practices and use of agricultural chemicals have played a major role in enhancing crop yields. Now time has come for adopting sustainable agricultural practices for increasing crop productivity while maintaining maximum conservation of land, water and other resources. Because of the limited access to gene pool and lack of control over gene transfer, conventional plant breeding can no longer sustain the ever-rising global food demand. Therefore, new technologies must be developed and exploited for crop improvement. Plant biotechnology has the potential to be an integral part of crop improvement efforts as it shows enormous prospects for improving the agronomic performance of crops, for enhancing nutrient content of foods and also for accelerating the breeding process over what can be attained only by means of conventional methods. Several genetically modified (GM) crops are raised by introducing genes for improved agronomic performance and/or enhanced nutrition. It is believed that integration of conventional agricultural practices with modern biotechnology, in a sustainable manner, can fulfil the goal of attaining food security for present as well as future generations. Since their first commercial production in 1996, GM crops have been increasingly cultivated, reaching a global cultivating area of 160 million hectares in 2011. We have noticed several important developments in the area of agricultural biotechnology which are expected to contribute significantly in alleviating some of the major challenges of global society such as food security, sustainability and health. With the growing popularity of genetically modified (GM) crops, concerns have also been expressed regarding their unintended hazardous effects on humans and the environment. Here, we highlighted the major factors limiting food security and human health globally. We described how recent developments in the area of agricultural biotechnology could help meet the current and future challenges in agriculture and health.

II. WHAT ARE THE MAJOR LIMITING FACTORS ON FOOD SECURITY?

The major constrains that limit global food crop production are the availability of water, fertile soil and the incidence of pests, diseases and weeds. The contribution of these factors on limiting agricultural productivity may vary between developing and developed regions of the world. The situation is further aggravated due the global climate change. The major limiting factors on crop production are discussed in the following sections.

A. Water

Adequate amount of water must be available throughout the life cycle of plants for proper growth and tissue expansion. Crop yield is severely affected when water availability is limited to plants during critical stages of growth and reproduction. Among all the biotic and abiotic stresses limiting crop yield, drought has probably the maximum impact. Thus, ensuring sufficient water availability to plants during important developmental stages is a key challenge to increasing crop yield.

Agriculture accounts for around 70% of annual use of global water resources and is usually considered as the main cause behind the increasing global scarcity of freshwater. Maximum amounts of water consumed in agricultural practices are for irrigation. During hot and dry conditions, much higher numbers of irrigation are required compared to less stressed conditions, to produce the same grain yield. Although irrigated areas account for less than 20% of the global area under crop cultivation, it is expected to rise continuously but at slowing rate. Most importantly, nearly 50% of the global food is produced from irrigated lands. Thus, global food supply will be affected seriously if there is decline in irrigated areas or the frequency of irrigation to crops.

The ground water levels in many regions have dropped to alarming levels over the past few years. Among the more noticeable signs of water crisis are the rivers running dry and lakes disappearing. In order to fulfil the aim of reducing agricultural water use combined agronomic, physiological, plant breeding and biotechnological approaches need to be followed. A high priority for the future is to develop elite genotypes that yield significantly while consuming less amounts of water.

Moreover, many parts of world experience heavy crop loss due to river and coastal flooding. Most sensitive areas in this context are the South and Southeast Asia. In India and Bangladesh, 4 million tons of rice, enough to feed 30 million people, are lost each year due to flooding. Rising sea levels leading to exacerbated coastal flooding are predicted to severely affect crop production in many countries.

B. Extreme Temperature

Temperature is an important factor in controlling plant growth, development and reproduction. With the prevalence of extreme climatic conditions, the risk of frequent crop failure will be high. The world is experiencing a rapid rise in temperature at a faster rate than previously predicted. If plants experience high temperature, particularly during flowering, anthesis and fruiting stages, severe yield loss may takes place. Moreover, an increase in temperature caused by climate change is predicted to have various indirect effects including higher water requirements, plant defence response and resistance to diseases. Elevated temperatures may extend the range and severity of diseases. A short episode of combined stresses, such as drought and heat stress, can have very severe effects on crop production. In addition, low temperature stress can also have a catastrophic effect on susceptible plants. Crops belongs to tropical regions are more prone to chilling injury and their cultivation in high latitudes is highly limited. Complete yield failure may be witnessed, in case of fruit crops, if they are exposed to frost at the time of flowering. There is a need for improved crop varieties that can withstand extreme temperatures without having yield penalty.

C. Fertile Soil

Availability of fertile soil is the basic requirement for farming. High quality soil promotes crop productivity when water and nutrients are not limited. Soil quality is mainly determined by the water permeability and drainage capacity, nutrients, organic matter content and microbial population. Soil microbes which are involved in nitrification, nutrient cycling and help plants in nutrient uptake are regarded as beneficial to crop cultivation, whereas denitrifying, disease causing microbes and pests are harmful. Loss of high quality agricultural soil due to erosion and rapid urbanization has serious consequences for global crop production. Soil is considered as a non-renewable resource, at least over non-geological timescales and maintaining soil quality is the fundamental to sustainable agriculture.

Soil salinity is affecting crop production world wide. Globally, more than 800 million ha of land (6% of the world's total land area) are salt affected. This problem has become more prevalent in agricultural land, due to frequent irrigation and extending cultivation into new areas, which resulted in rise of water table and concentrating the salts in the plant root zone. Of the irrigated land, 20% is salt affected compared to 2% in case of dry-land agriculture.

A significant reduction in crop yield may takes place on acidic soil due to aluminium (Al) toxicity. When pH values of soil fall below pH5.5, Al^{3+} is solubilised from aluminosilicate clay minerals and its concentration in soil reaches to the toxic level. Al toxicity is considered as a major constrain to global food security because up to 50% of the world's potentially arable soils are acidic. Moreover, boron toxicity is a worldwide problem affecting crop quality and yield, particularly in arid regions. Up to 17% loss in crop yield has been estimated to due to boron toxicity.

Nitrogen (nitrate or ammonium), phosphorus (phosphate) and potassium greatly influence crop yields. Nitrogen and phosphate fertilizers are becoming either unaffordable or unavailable to many farmers, particularly in the developing countries. This implies that the full potential of many crop genotypes is not attained. An appropriate balance between nitrogen and potassium in soil is essential for performance of crops, as insufficient levels of accessible potassium affect the ability of the plant to take advantage of available nitrogen. Global food production largely depends on the chemical synthesis of nitrogen fertilisers and the mining of rock phosphate. These are non-renewable resources. It is estimated that more than 50% of the nitrogen in the global nitrogen cycle was synthesised industrially in the last 100 years. It is desirable to attain high yield with less reliance on synthetic fertilizers. Biological nitrogen fixation, mainly by *Rhizobium* species and recycling of soil nutrients through the use of green manures, composts and animal manure represent different ways in which dependence on synthetic fertilizers can be minimised.

D. Pests, Diseases and Weeds

Pests, diseases and weeds pose serious threat to global food crop production. Crop losses due to these biotic stresses have been estimated to be 26–40%, for major crops like, wheat, barley, rice, maize, soy, cotton, sugar beet and potato. However, in the absence of effective control measures (e.g. resistant varieties, crop protection chemicals and crop rotations) losses might be 50–80%.

Herbivorous chewing pests such as lepidopteran insects, sucking insects such as aphids and nematodes cause huge crop losses. Many insects are also important vectors of diseases caused by viruses and phytoplasma. Considerable yield reduction to a majority of crops is caused from a range of fungi and oomycetes (fungus-like organisms), bacteria and

viruses. A significant fraction of agricultural resources (water and nutrients) is also diverted to weeds which probably cause highest loss among the biotic constrains on crops. There is an increasing incidence of herbicide resistance among weeds. Thus, herbicides with a range of modes of action and herbicide tolerant crops must be available for effective weed management. Following fungi, oomycete, weed, and virus are considered to be among the serious biological threats to food security.

Puccinia graminis causes stem rust disease in wheat. Ug99, a dangerous new strain appeared in Uganda in 1998 is a major threat to wheat production in Middle East and West Asia. More than 40% yield loss may occur in case of heavy infections. It is estimated that if it reaches Punjab (India) losses could be $3 billion per year and if it spread to United States annual losses may be as high as $10 billion. The big fear is that this rampant fungus may cause famine in India and Pakistan, where small farmers can not afford costly control measures.

Phytopthora infestans which infects solanaceous crops (mostly potato) was responsible for the Irish potato famine during 1850s. Late blight is very frequent in potato fields and still ranks as the world's most devastating potato disease. According to the international Potato Center (Peru), the annual losses, in developing countries, due to this disease is $2.75 billion and fungicide application may cost 10% of total inputs.

Rice blast fungus *Magnaporthe oryzae* may cause up to 100% loss in some rice fields. It is estimated that the amount of rice destroyed each year due to this is sufficient to feed 60 million people. This may cost around $66 billion. Controlling this fungus is an uphill task as it tends to overcome resistance in two to three growing seasons.

Black sigatoka disease of banana and plantation caused by *Mycosphaerella fijiensis* is prevalent in around 100 countries. This fungus may cause up to 50% of yield loss. Crop management practices to contain this fungus may consume 15-50% of fruits final retail price.

Cassava brown streak virus is emerging as a significant threat to cassava production. This virus and cassava mosaic virus together considered as Africa's biggest threat to food security. Yield losses due to brown streak virus reach up to 100%. It is estimated to cause more than $100 billion loss annually.

Striga hermonthica, a parasitic plant infects 20-40 million hectares land in sub-Saharan Africa and may cause 20-100% yield reduction in crops like corn, sorghum, sugarcane, millet. These losses are equivalent to $1 billion per year, affecting around 100 million people.

E. Post-harvest Loss

One of the problems associated with the production and distribution of fleshy fruits and vegetables are their quickly perishable nature. The post-harvest decay process of these crops influences shelf-life, and limits transportation and storage.

These lead to huge losses in developing countries which may reach as high as 50% due to post-harvest decay that farmer can not control. Post-harvest crop loss is one of the sources of food insecurity. Thus, minimizing the postharvest losses of perishable crops is of strategic importance to sustainable agricultural development. The damage to the harvested crop is often caused by pests or fungi. Due to excessive softening during ripening process, fruits and vegetables become more susceptible to pests and pathogens. Thus, enhancement of fruit shelf life by slowing down the softening process and providing resistance against pests and pathogens is among the targets of crop genetic improvement efforts.

F. Genetic Erosion

Genetic variation within a crop species and their relatives is crucial for the agricultural development. Many traits, e.g., disease resistance, drought tolerance, are incorporated into modern crop varieties from other cultivars, landraces and relatives through conventional breeding. These lead to genetic uniformity and a narrowing genetic base of crop species. Future crop improvement efforts completely depend on the availability of diverse genetic resources. Crop genetic diversity has declined sharply in recent times. In India, around 30,000 rice varieties were once grown. However, most land is occupied presently by few high yielding varieties. The preservation of genetic diversity is necessary if crop genetic improvement is to continue.

G. Global Climate Change

Due to global warming, temperature is rising at a faster rate than previously anticipated. The world has experienced a significant increase in atmospheric CO_2 concentrations in the past two centuries. The concentration of atmospheric CO_2 was 270 $\mu mol.mol^{-1}$ in 1750; however, current concentration reached to more than 385 $\mu mol.mol^{-1}$. In addition, the atmospheric level of other green house gases like, methane, ozone and nitrous oxide also increased. It is estimated that the total concentrations of all greenhouse gasses will exceed the value of 550 $\mu mol.mol^{-1}$ by 2050. Due to this greenhouse effect, it has been projected that an average annual mean temperature increase of 3-5°C will occur in the next 50-100 years. The increase in atmospheric CO_2 concentrations will enhance the photosynthesis rate and may lead to higher plant productivity and crop yields. The elevated level of atmospheric CO_2 is predicted to decrease evapotranspiration rate in plants. Although this will improve water use efficiency of most crops, a decrease in evapotranspiration will lead to increase in leaf temperatures and possibly decrease in photosynthesis rate. Besides, drought and increased temperatures are expected to hinder any gain in crop yield that may be brought about by the increased level CO_2 in the atmosphere. Moreover, elevated CO_2 concentrations may also lead to a reduction in grain protein content, due to the nitrogen acquisition gap at elevated CO_2.

The melting of glaciers, owing to global warming, is contributing to the rise in sea levels. Rivers are experiencing shorter and more intense seasonal flows, as well as more flooding. Thus, crop cultivation in low-lands will be affected because of submergence.

H. Ozone

Among the air pollutants, ozone is regarded as the most harmful to plants. Because of human activities, tropospheric O_3 concentrations are increasing at alarming rates. High level of tropospheric O_3 may affect growth and functioning of plant leaf and root, and may also cause a decrease in photosynthesis rate and stomatal conductance. Among all crops, soyabeans and wheat are particularly sensitive. High O_3 concentrations may increase transpiration rate and reduce drought tolerance in crops. It was estimated that high tropospheric ozone concentrations resulted in global crop losses of $14–26 billion in the year 2000. The maximum yield losses for wheat were found to be in India (28%) and China (19%). It has been reported that O_3 will have negative affects on crop quality and on protein contents for a range of crops. Plants tolerance to O_3 is determined by the control of the flux of O_3 into the leaf and the ability of the leaf to deal with oxidative stress through detoxification and repair mechanisms.

III. MALNUTRITION

Inadequate dietary intake results in many forms of macro- and micronutrient malnutrition. Protein energy malnutrition is the most lethal form of malnutrition and affects every fourth child worldwide. In India, one third to one half child death is associated with malnutrition. According to a national survey, 20% of Indian children (below five years of age) suffer from acute malnourishment and 48% suffer from chronic malnourishment. Moreover, 43% of children are underweight. This is twice more than the average figure in sub-Saharan Africa (National Family Health, 2005–06, NFHS-3).

Plant based foods have a poor balance of essential amino acids relative to the needs of animals and humans. Cereals such as maize, wheat, rice, etc. are deficient in Lysine. However, legumes such as soybean, pea, etc. are deficient in sulfur-rich amino acids Methionine and Cystine. Micronutrient malnutrition affects more than half of the world's population, especially women and preschool children in developing countries. Iron, vitamin A, iodine, zinc, and folic acid deficiencies affect large numbers of people, especially children, resulting in significant morbidity and mortality. Many food crops also contain high level of antinutrients such as phytate, oxalic acid, trypsin inhibitors, lectins etc. There is no single solution to the complex problem of malnutrition. The preferred strategy to fight malnutrition will always be to adopt diversity of diet with increased access to fruits and vegetables. However, it is not possible in many regions due to the lack of infrastructure. Another approach has been to upgrade the macro- and micronutrient content of staple crops. Both conventional breeding and modern biotechnology tools are being employed to develop biofortified crops having high nutrition value.

IV. HOW DOES PLANT BIOTECHNOLOGY FIT INTO CROP IMPROVEMENT PROGRAM?

The constraints that limit the food production, globally, include water availability, soil fertility, pests, diseases and weeds. These will be accompanied by adverse effects of climate change. Thus, the aim for crop genetic improvement will be to develop crops that adapt under extreme temperatures, decreased water availability, flooding conditions and saline soil, and evolving threats from the pests and pathogens. Upgrading nutritional value of the food crops is also highly demanding. The following section describes a range of biological science based technologies that could help address these various challenges. Figure 72.1 illustrates the steps involving isolation and functional characterization of useful genes and their application for the crop improvement.

A. Genetic Marker

Most of the existing crop varieties are the product of conventional plant breeding practices involving crosses between genetically distinct parents followed by screening of progeny for desired trait combinations and selection of the best individual. The most challenging part is the screening of individuals for desired traits. Plants have to be tested for phenotypes (e.g. disease/pest resistance, draught tolerance, crop yield, etc.) that may be very difficult to assay when a large number of sample to be analysed. It will be quite easier if instead of assaying the phenotype, segregation of a genetic marker (morphological/biochemical/DNA) linked to the trait

can be monitored in progeny. This method is faster, more reliable and particularly helpful when a tightly linked deleterious trait needs to be removed from the population which requires very rare recombination event.

Morphological markers are in general visually recognised phenotypic traits such as plant height, flower, seed coat colour etc. Biochemical/isozyme markers represent differences among plants in terms of enzyme profile which is determined by gel electrophoresis. The major disadvantages of these markers are that they are affected by environmental and developmental factors and their numbers may be limited. In spite of these constrains, morphological and biochemical markers have been very much useful to plant breeders. In contrast to morphological and biochemical markers, DNA markers are practically unlimited in number and are not influenced by environmental and developmental factors. DNA markers are most widely used in plant breeding programs because of their high abundance. Some of the uses of genetic markers in agriculture and human health are listed below:

- Genome mapping and map-based cloning for gene identification.
- Marker-assisted selection in plant breeding program.
- Protection of crop varieties (assessing purity, stability and varietal identity) and characterization of germplasm.
- Assessing genetic diversity among parents for exploitation of hybrid vigour.
- Association markers with human diseases such as sickel cell anemia, huntingtons disease, tay sachs disease, cystic fibrosis, etc.
- Forensic studies.

DNA marker

Mutations on genetic material lead to variation in DNA sequences that may serve as DNA markers. These mutations may be substitution mutations (point mutations), rearrangements

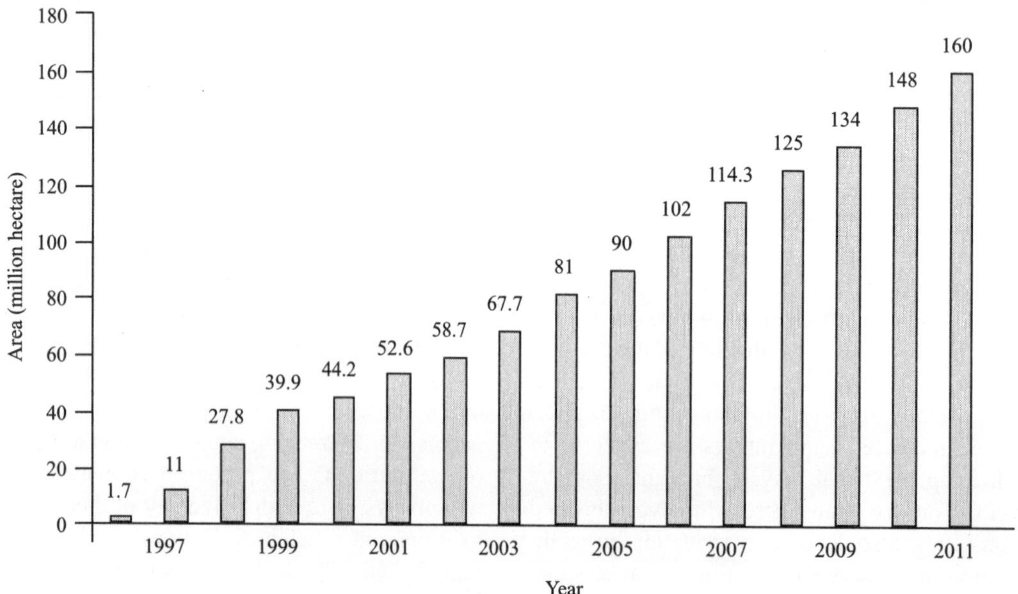

Figure 72.1   Global area under GM crops during 1996 to 2011. (*Source:* James, 2011, ISAAA Briefs No. 43)

(insertions or deletions) or errors in replication of tandemly repeated DNA. Most common DNA markers can be categorised into random amplified polymorphic DNA (RAPD), restriction fragment length polymorphism (RFLP), amplified fragment length polymorphism (AFLP), simple sequence repeat (SSR) or microsatellites and single nucleotide polymorphism (SNP). Except RFLP all these DNA markers are based on polymerase chain reaction (PCR) approach. Table 72.1 summarizes features of some of the DNA markers commonly used in plant breeding and genomic studies. Through DNA markers, high-density genetic map can be constructed for a range of economically important crops that will be helpful in marker-assisted selection (MAS). It is beyond the scope of this chapter to describe on technicalities of how these DNA markers are generated. However, the most desirable characteristics of an ideal DNA marker are mentioned below:

- Marker should be polymorphic in nature so that it is able to differentiate between members of same and different species.
- Codominant markers are highly desirable as they discriminate homozygote and heterozygote organisms.
- Markers should be evenly and frequently distributed throughout the genome of the organism.
- Markers should not be influenced by environmental factors and developmental stages of the organism.

In rice, MAS has been employed to develop flooding tolerant varieties. Other selected examples of application of MAS includes bacterial blight resistance in rice, leaf rust resistance in wheat and soybean, drought tolerance in maize, erect panicle in rice and root-knot nematode resistance in cotton. All these suggest that, with the availability of genome sequences for important crop species, this approach will be increasingly important in plant breeding.

B. Genome Sequencing

To obtain information about all the genes of an organism and all the possible proteins that can be produced, complete DNA sequences of the organism need to be determined. In this regard, modern genome sequencing tools are particularly important. When combined with modern biotechnological tools for assaying gene function and high throughput technologies for the analysis of RNA, proteins and metabolites, genome sequence provides a powerful base for the elucidation of complex cellular processes of an organism.

In addition to model plants Arabidopsis thaliana and Medicago, the complete genome sequences are now available for some important crop species (e.g., rice, maize, sorghum, soya bean, grapevine, potato, tomato and papaya). Moreover, genomes of phytopathogens will be helpful in dissecting the complex plant-pathogen interactions. Genome sequences

TABLE 72.1 DNA markers common in plant breeding and genomic studies

Random amplified polymorphic DNA (RAPD)	Restriction fragment length polymorphism (RFLP)	Amplified fragment length polymorphism (AFLPs)	Simple sequence repeat (SSR)/microsatelites	Single nucleotide polymorphism (SNP)
RAPDs are based on PCR amplification of random DNA sequences with arbitrary oligonucleotide of usually 10 bp in size.	RFLPs differentiate individuals on the basis of the distance between restriction enzyme cleavage sites on genomes.	Basic principle of AFLP is same as RFLP, however; it involves digestion of genomic DNA with restriction enzymes, ligation to specific adapters, and PCR amplification of a subset of DNA using primers containing defined adapter sequences/adapter sequence with one to three additional arbitrary nucleotides.	SSRs differentiate individuals on the basis of how many times a simple sequence (2-3 nucleotides) is repeated.	SNPs discriminate single nucleotide changes (point mutation) within a DNA sequence.
RAPD is a fast, inexpensive and less time consuming technique. It requires small quantity of DNA. Through RAPD, identification of multiple loci using a single primer is possible.	These codominent marker are robust, reliable and transferable accross populations.	AFLPs are highly polymorphic and multiple alleles are generated. It can be used across species.	SSRs are codominant markers. It is technically simple, robust, reliable and transferable between populations. SSRs are highly polymorphic and multiple alleles can be generated.	SNPs are highly polymorphic in nature, frequent in the genome and becoming popular in genomic studies. SNPs can be useful for cultivar discrimination in crops where it is difficult to find polymorphisms.
RAPDs are dominent marker and generally is not transferable across species. It has prolem with reproducibility.	RFLPs are time consuming, expensive, require large amount of high quality of DNA and radioactive labelling.	AFLP are dominant marker involving complicated methodology. It requires large amount of DNA.	DNA sequence information is needed before developing SSRs. Large amount of time and effort are required to design primers for PCR amplification.	DNA sequence information is required for the design of allele specific PCR primers or oligonucleotide probes. This is time consuming and expensive.

are now available for phytopathogens such as Magnaporthe oryzae, Botrytis cinerea, Sclerotinia sclerotiorum, Pseudomonas syringae, Xanthomonas campestris etc. Assigning the biological function to each of the 10,000 to 30,000 genes which have been identified in each species is an important ongoing task.

With the availability of new genome sequencing methodologies which are much more efficient than the older methods, generating the genome sequence data is now straightforward and less time consuming. Currently, it is possible to sequence genomes of several cultivars of a crop in relatively short times. This is helping in determining the genetic basis of differences among the cultivars for the tolerance to environmental stresses, variation in yield and nutritional quality of produce etc. Time-consuming part of a genome project is the computational analysis and annotation of the DNA sequences. Once an annotated genome is available, it is important to identify genes of agricultural importance by applying functional genomic tools. These genes can be targeted in breeding programmes for genetic improvement or they can be inserted into the genome of important crops through GM technology.

C. GM Technology

GM technology aims at addition of new beneficial traits or enhancement of already existing beneficial traits or removal of unfavourable traits from crops by expressing or suppressing the desired gene(s) through genetic engineering approach. Recent advances in the area of molecular biology, plant regeneration and transformation, and other related areas made it possible to isolate the desired genes from any source and introduce into a target plant species, overcoming the reproductive or phylogenetic barrier. Genetic engineering has the ability to contribute in crop improvement programs by enhancing agricultural productivity, reducing crop losses and developing high quality foods in following ways:

- Increase in crop yield through genetic modification of plant growth and development, photosynthetic and nutrient use efficiency, and response to environmental stresses.
- Development of nutritionally balanced and safe foods by removing anti-nutritional factors, increasing the level of proteins having balanced amino acid composition, biofortification with vitamins and essential minerals, and value addition with pharmaceutical molecules to combat diseases.
- Enhancing shelf life of quickly perishable crops by slowing down the natural deterioration process and providing resistance against post-harvest pathogens.

A few examples of how genetic engineering can accelerate crop improvement programs are presented in Table 72.4. Now-a-days, GM technology is followed routinely in research, which has greatly facilitated in advancing research in plant biology for more than two decades. In this section potential benefits of GM in crop improvement will be discussed.

D. GM Crops that Perform better Under Stress

Global food production is severely affected due to biotic and abiotic stresses imposed to crops by insects, fungi, bacteria, viruses, nematodes, weeds, draught, salinity and cold. Together, these stress factors may cause up to 30-60% yield loss; however, in developing countries this loss may be much higher. Insects and weeds had been the primary targets for GM technology. The majority of commercially grown GM crops are either modified for insect resistance and/or herbicide tolerance. The control of insects in crop field is heavily dependent on the use of toxic chemicals, which is expensive, harmful to the environment and also leads to development of resistance. Several genes have been exploited to develop insect resistant transgenic crops. These include *Bacillus thuringiensis* (*Bt*) toxins, protease inhibitors and lectins (Table 72.2). Among these, *Bt* crops received maximum attention worldwide. *Bt* toxins (also known as crystal proteins, Cry) show high insecticidal properties at very low concentration and cause little or no harm to non-target organisms. The *Bt* strategy has been employed to protect crops such as maize, cotton, potato, brassica, brinjal, etc. against various pests and was also found to be effective against nematodes. Use of *Bt* crops resulted in better insect management, higher yield, and reduced insecticide application. The next generation of *Bt* crops are also designed to express a number of different *Bt* genes together to provide resistance against a range of insects. Moreover, herbicide tolerant GM crops have been developed that allow the effective use of broad-spectrum herbicides such as glyphosate and basta to kill weeds in crop fields. These crops have reduced the use of toxic herbicides such as metolachlor and diuron.

Pathogenic fungi not only pose serious threat to crop yield but also produce mycotoxins that contaminate food and may cause concerns to human health. Genetic engineering with pathogenesis-related (*PR*) or resistance (*R*) genes is an important strategy to develop fungal resistance in crops (Table 72.2). Enhanced tolerance to fungal diseases has been conferred in transgenic plants by expressing *chitinase, defensin, thaumatin, Oxalate decarboxylase, Oxalate oxidase, Rpg1, NPR1*, etc. (Table 72.2).

Bacterial and viral diseases which cause significant losses in crop yield are also targeted to develop resistant crop varieties through genetic engineering. Several genes like, *CaPIF1* (Cys-2/His-2 zinc finger), *Pto* (serine/threonine protein kinase), *PFLP* (Ferredoxin-1 protein) and *Bs2* were found to be effective to provide tolerance to bacterial pathogens in transgenic plants (Table 72.2). The expression of transgenes encoding viral coat protein, replication associated protein, ribosome-inactivating protein, ribonuclease, 2,5 oligoadenylate synthetase, ribozyme, and recombinant antibody resulted in development of virus resistance in some crops (Table 72.2). A well-known example is the development of GM papaya resistant to viral infection. Currently, about 90% of papaya grown in the island of Hawaii is genetically modified with a coat protein of papaya ringspot virus (PRSV). Commercial cultivation of GM papaya resulted in a substantial boost in papaya production in that

TABLE 72.2   Some examples of GM crops under research and development for biotic stress tolerance

Target trait	Target gene*	Target crop
Resistance to insect-pest	Snowdrop *lectin* ↑	Potato, rice
	Expression of dsRNA specific to insect vacuolar *ATPase* ↓	Maize
	NaPI and *StPin1A* (PinI and PinII proteinase inhibitors) ↑	Cotton
	Bt toxins (*Cry1Ab, Cry1Ac, Cry3A, Cry1B, Cry1C, Cry1H, Cry9C*)↑	Cotton, corn, potato, eggplant, rice, tomato, *B. napus*, brinjal
	Cowpea *serine protease inhibitor* ↑	Rice, potato
	Oryzacystatin ↑	Potato
	Bean a-AI ↑	Pea
Resistance to bacterial pathogen	*CaPIF1* Cys-2/His-2 zinc finger protein↑	Tomato
	Pto (Serine/threonine protein kinase) ↑	Tomato
	PFLP (Ferredoxin-1 protein) ↑	Tomato
	The *Bs2* resistance gene of pepper↑	Tomato
Resistance to fungal pathogen	*Chitinase*↑	Rice, carrot, *B. napus*, grape, potato, tomato, wheat, strawberry
	Defensin↑	*B. napus*, potato, tomato
	Osmotin↑	Potato
	Rir1b defense gene ↑	Rice
	Thaumatin ↑	Wheat, rice
	Antifungal protein ↑	Wheat
	Oxalate decarboxylase ↑	Tomato, soybean
	Oxalate oxidase↑	Soybean, peanut
	Glucose oxidase↑	Potato
	Rpg1 resistance gene ↑	Barley
	NPR1 gene ↑	Wheat
Resistant to viral pathogen	*T-Rep* replication associated protein gene of the mild strain of Tomato yellow leaf curl virus-Israel) ↑	Tomato
	Ribozyme and antisense genes targeting citrus exocortis viroid ↓	Tomato
	Expression of truncated viral Rep protein ↑	Tomato
	AC1 (replication-associated protein) ↑	Cassava
	A synthetic gene encoding a single chain Fv fragment of an antibody directed against the nuclear inclusion a (NIa) protein of potato virus Y ↑	Potato
	Yeast-derived dsRNA-specific ribonuclease *pac*1 ↑	Potato
	Mammalian *2′–5′ Oligoadenylate synthetase* ↑	Potato
	Ribozyme [R(-)] targeting the minus strand RNA of potato spindle tuber viroid ↑	Potato
	Transcription factors *RF2a* and *RF2b* ↑	Rice
	Viral *replicase* (*Rep*) gene ↑	Rice
	Viral coat protein gene ↑	Rice

*Expression and suppression of the target gene(s) are indicated by upwards (↑) and downwards (↓) arrows, respectively.

region. However, so far no conventional or organic method exists that can effectively contain this devastating virus. GM strategies to produce crops resistant to nematodes which represent a significant threat to crop production in many parts of world especially Africa, Asia and South America include the expression of resistant gene *Mi* and *oryzacystatin-I* gene.

In addition to biotic factors, crop productivity is also affected by several abiotic stress factors. Drought, heat, salinity, cold, and metal stresses are the abiotic factors significantly affecting worldwide crop production. Together, these factors may cause average yield losses of more than 50% for major crops. Several GM crops with an inbuilt capacity to withstand these adverse environmental conditions have been developed which are expected to contribute considerably to reduce yield losses due to such stresses. There are examples where abiotic stress tolerance in transgenic crops has been accomplished by overexpression of osmoprotectant dehydrins, transcription factors and transporters, and also by regulated flux through secondary metabolism (Table 72.3).

Among the various abiotic stresses, drought is considered as the major factor that limits crop production worldwide. GM approaches to develop drought tolerant crops include transgenic expression of bacterial RNA chaperones, NF-Y class of transcriptional regulators, cyanobacterial flavodoxin and silencing to downregulate poly ADP ribose polymerase and farnesyl transferase. Drought tolerance has also been conferred by expression of isopentenyl transferase involved in cytokinin biosynthesis, *LOS5* involved in abscisic acid (ABA) biosynthesis and *ZAT10,* a Cys2/His2-type zinc-finger protein.

Genetic engineering approaches have also been followed to overcoming metal ion toxicity in plants. These involve

TABLE 72.3   Some examples of GM crops under research and development for abiotic stress tolerance

Target trait	*Target gene[*]*	*Target crop*
Drought	*Los5* (ABA biosynthesis, molybdenum cofactor sulfurase) and *ZAT10* (Cys2/His2-type zinc-finger protein) ↑	Rice
	ABF3 (transcription factor) ↑	Rice
	NPK1 (mitogen activated protein kinase) ↑	Maize
	DREB1A (transcription factor) ↑	Wheat
	NF-Y B subunit (transcription factor) ↑	Maize
	FTA (protein farnesyl transferase) ↓	*B. napus*
	HARDY (AP2/ERF-like transcription factor) ↑	Rice
	OCPI1 (proteinase inhibitor) ↑	Rice
	Adc (polyamine synthesis) ↑	Rice
	betA (choline dehydrogenase, glycinebetaine synthesis) ↑	Maize
Drought, salinity, cold	*OsCIPK01* (calcineurin B-like protein)↑	Rice
	HvCBF4 (transcription factor) ↑	Rice
	OsDREB1 (transcription factor) ↑	Rice
	OsCDPK7 (calcium dependent protein kinase) ↑	Rice
	OsCOIN (ring finger protein) ↑	Rice
Drought, salinity	*CBF3/DREB1A* (transcription factor)↑	Rice
	mt1D (mannitol-1-phosphate dehydrogenase, mannitol synthesis)↑	Wheat
	P5CS (pyrroline carboxylate synthase, proline synthesis)↑	Potato, rice, citrus, sugarcane
	SNAC (transcription factor)↑	Rice
	TPS1, TPSP (trehalose synthesis)↑	Tomato, rice
	HVA1 (LEA protein gene)↑	Rice
Drought, heat	*PARP* (Poly ADP-ribose polymerase)↓	*B. napus*
Drought, freezing	*AtDREB1A* (transcription factor)↑	Peanut, potato
	DREB2/DREB3 (transcription factor)↑	Wheat, barley
Salinity, cold	*Glutathione S-transferase*↑	Rice
	SNAC2 (transcription factor)↑	rice
	COX, codA (choline oxidase, glycine betaine synthesis)↑	Rice
	GS2 (chloroplastic glutamine synthetase)↑	Rice
Salinity	*AtNHX1* (vacuolar Na^+/H^+ antiporter)↑	*B. napus*, wheat
	OsNHX1 (vacuolar Na^+/H^+ antiporter)↑	Rice
	OsSOS1 (plasma membrane Na^+/H^+ exchanger)↑	Rice
Heat	*Hsp101* (heat shock protein)↑	Rice
	LeHSP100/ClpB (chloroplast HSP)↑	Tomato
Freezing	*AtCBF1-3* (transcription factor)↑	Potato
	BNCBF5- & 17(CBF/DREB1-like transcription factors) ↑	*B. napus*
Cold	*CaPIF1* (Cys-2/His-2 zinc finger protein) ↑	Tomato
As and Cd tolerance	*AtPCS1* (phytochelatin synthesis) ↑	*B. napus*

[*]Expression and suppression of the target gene(s) are indicated by upwards (↑) and downwards (↓) arrows, respectively.

overcoming aluminium toxicity in plants by expressing genes encoding transporters of organic acids (OAs) such as citrate or malate, which form non-toxic complexes with Al^{3+} ions. Plants tolerant to high level of boron have also been developed by the expression of a boron transporter, *BOR4*.

GM crops with enhanced nutritional value

Plant scientists have been working for years to improve the macro- and micronutrient content of staple crops. Many of the plant proteins, mainly those present in staple crops do not have balanced amino acid composition. Thus, several crops are targeted for the expression of proteins with more desirable amino acid composition (Table 72.3). In this context, it was important to identify genes encoding proteins with balanced amino acid composition. One such example is AmA1, a non-allergic protein having balanced amino acid composition, isolated from Amaranth seeds. With help of GM technology, this gene has been expressed into potato, a protein less starchy crop deficient in many essential amino acids. Tuber-specific expression of AmA1 in potato resulted in a significant increase in the concentration of most essential amino acids. The total protein content in transgenic tubers also increased in comparison

to control plants. To translate this GM approach to farmer's field, seven potato varieties that are adapted to different agro-climatic regions were selected and transformed with *AmA1* gene to improve protein quantity and quality of the tubers. The field performance and biosafety assessment suggest that protein-rich GM potatoes are safe for human consumption and suitable for commercial cultivation.

GM crops with higher levels of bioavailable vitamins, iron and zinc are also in the advanced stage of development (Table 72.4). Vitamin A deficiency is a public health problem in more than 100 countries particularly in Africa and Asia. Vitamin A is essential for good eyesight, and deficiency leads to blindness. Annually, 250,000 preschool children go blind due to vitamin A deficiency. Studies indicate that widespread consumption of β-carotene (provitamin A)-rich GM rice, commonly known as Golden Rice will reduce the incidences of vitamin A deficiency. Transgenic expression of *ferritin*, *phytase* and *nicotianamine synthase* has been among the strategies to develop GM rice, maize and wheat with elevated level of iron (Table 72.4). A major challenge in human health will also be in the area of chronic, noncommunicable diseases, including heart disease, many cancers, type 2 diabetes, and obesity.

TABLE 72.4 **Some examples of GM crops under research and development for increased nutritional value and intended to promote human health**

Target trait	Target gene[*]	Target crop
Quantity and quality of protein		
High protein content and/or balanced amino acid composition	Amaranthus seed storage protein gene *Ama1*↑	Potato
	Rice *10 kDa sulfur-rich prolamin*↑	Potato
	Maize storage protein gene *zein*↑	Potato
	Soybean *glycinin*↑	Rice
	Porcine *alpha-lactalbumin*↑	maize
	Maize *15kDa zain*↑	Soybean
High essential amino acid content	Bacterial *aspartokinase* and *dihydrodipicolinic acid synthase* (lysine biosynthesis) ↑	Canola, soybean (Lys)
	Dzs10 (seed-specific high-methionine storage protein)↑	Maize (Met)
	Threonine synthase↓	Potato (Met)
	OASA1D (feedback-insensitive α subunit of rice *anthranilate synthaseI*)↑	Rice, bean (Trp)
	19 and 22 KDa α-*zeins*↓	Maize (Lys, Trp)
	Brazil nut *albumin* ↑	*B. napus* (Met)
	Lysine-ketoglutarate reductase/saccharopine dehydrogenase (ZLKR/SDH) ↓	Maize (Lys)
	Modified *tRNA[lys]* gene↑	Rice (Lys)
	Sesame *2S albumin*↑	Rice (Met, Cys)
Micronutrients		
Provitamin A	Daffodil *phytoene synthase*, *Erwinia uredovora phytoene desaturase*, *Narcissus pseudonarcissus lycopene b-cyclase*, maize *phytoene synthase*	Rice
	Phytoene synthase↑	Cassava
	Phytoene synthase (CrtB), *phytoene desaturase (CrtI)* and *lycopene beta-cyclase (CrtY)* from Erwinia *uredovora*↑	Potato
	Bacteria *phytoene synthase (CrtB)* ↑	*B. napus*
	Lycopene beta-cyclase↑	Tomato

(Contd.)

TABLE 72.4 **Some examples of GM crops under research and development for increased nutritional value and intended to promote human health** (*Contd.*)

Target trait	Target gene[*]	Target crop
	Erwinia herbicola phytoene synthase (*CrtB*), *phytoene desaturase* (*CrtI*) ↑	Maize
Vitamin B9	Arabidopsis *GTP–cyclohydrolase I and aminodeoxychorismate synthase*↑	Rice
	Synthetic gene based on *GTP cyclohydrolase I, Arabidopsis aminodeoxychorismate synthase*↑	Tomato
Vitamin C	Wheat *dehydroascorbate reductase* ↑	Maize
Provitamin A, vitamin C and B9	Maize *phytoene synthase, Erwinia uredovora carotene desaturase,* rice *dehydroascorbate reductase, E. coli GTP cyclohydrolase* ↑	Maize
Vitamin E	Barley *homogentisic acid geranylgeranyl transferase*↑	Maize
	Arabidopsis *2-methyl-6-phytylbenzoquinol methyltransferase* and *gamma-tocopherol methyltransferase*↑	Soybean
Phyto sterols, carotenoids	Arabidopsis *3-hydroxymethylglutaryl CoA, bacterial 1-deoxy-d-xylulose-5-phosphate synthase*↑	Tomato
Carotenoids	*LeCOP1LIKE* (negative regulator of light signal transduction) ↓	Tomato
	Yeast *S-adenosylmethionine decarboxylase* ↑	Tomato
Carotenoids, flavonoids	*DET1* (regulator of photomorphogenesis) ↓	Tomato
Iron	Soybean *ferritin gene*↑	Rice, lettuce
	Phaseolus vulgaris ferritin, Aspergillus fumigatus phytase, rice *metallothionein*-like protein↑	Rice
	Arabidopsis nicotianamine synthase, Phaseolus vulgaris ferritin, Aspergillus fumigatus phytase↑	Rice
	Soybean *ferritin, Aspergillus fumigatus phytase* ↑	Maize
	Aspergillus fumigatus phytase ↑	Wheat
Zinc	*Arabidopsis nicotianamine synthase, Phaseolus vulgaris ferritin, Aspergillus fumigatus phytase*↑	Rice
	Mouse *metallothionein-I*↑	Tomato
Functional foods		
Anthocyanin	Transcription factors from snapdragon↑	Tomato
Resveratrol	*Stilbene synthase* (*Vst1*) from *Vitis vinifera*↑	Kiwifruits, apple, Rice, tomato
Flavonoids	Petunia *chalcone isomerase*↑	Tomato
	Maize *C1* and *R* transcription factors↑, *flavanone 3-hydroxylase*↓	Soybean
Phenolics	*Hydroxycinnamoyl-CoA quinate: hydroxycinnamoyl transferase*↑	Tomato
Insulin	*1-SST* (*sucrose:sucrose 1-fructosyltransferase*) and *1-FFT* (*fructan: fructan 1-fructosyltransferase*) genes of globe artichoke (*Cynara scolymus*) ↑	Potato

[*]Expression and suppression of the target gene(s) are indicated by upwards (↑) and downwards (↓) arrows, respectively.

Phytochemicals including carotenoids (e.g., provitimin-A, lycopene, lutein), flavonoids (e.g., anthocyanins, flavonones, flavonols), phenolics (e.g., stilbenes, caffeic acid, ferulic acid) and inulin are known for their health promoting effects and thus, are the target for metabolic engineering in food crops for their higher accumulation (Table 72.4). GM tomatoes, commonly known as purple tomato, have been developed with high levels of cancer protecting compounds, such as flavonoids.

Nutritional quality of crops has been enhanced by the removal of anti-nutritional components. An example is the low oxalic acid containing GM tomato developed by transgenic expression of oxalate decarboxylase (*OXDC*) from *Flammulina velutipes*. Consumption of high oxalic acid containing fruits and vegetables is detrimental to human health as oxalate chelates calcium. This leads to precipitation of calcium oxalate in the kidney which causes hyperoxaluria and destruction of renal tissues. Transgenic expression of *OXDC* is considered as an important crop improvement strategy as crops will have two advantages; tolerance to fungi which exploit oxalic acid for pathogenesis and enhanced nutritional quality.

GM crops to reduce post-harvest loss

Postharvest decay process is being slowed down mainly by using post-harvest fungicides and/or keeping fruits and vegetables at low temperatures, which greatly reduce pathogen growth. However, the use of fungicides has been limited by the development of pathogen resistance, the lack of alternative fungicides and their detrimental effects to human health and the environment. Transgenic approaches to enhance shelf life of fruits and vegetables by slowing down the natural softening during ripening and developing pathogen resistant crops are the effective strategies to combat post-harvest decays of fruits and vegetables. In order to delay tomato fruit softening genes involved in cell wall degradation, e.g., polygalacturonase, expansions have been targeted.

Two N-glycan processing enzymes α-mannosidase (α-Man) and β-D-N-acetylhexosaminidase (β-Hex) were targeted for the genetic manipulation of fruits and vegetables through biotechnology. With the help of RNAi technology the expression of α-Man and β-Hex genes was silenced in transgenic tomato plants. Fruits of these GM tomato plants exhibited prolonged shelf life because of the reduced rate of fruit softening during ripening. This novel biotechnological approach was also extended to the non-climacteric fruits of capsicum. Thus, genetic Genetic manipulation of N-glycan processing provided an important strategy to prolong fruit shelf life in both climacteric and non-climacteric fruits. Climacteric fruits require ethylene to complete ripening process; however, non-climacteric fruits do not require ethylene.

GM crops as edible vaccine

In recent years, plants have emerged as alternative system for the large scale cost effective production of foreign proteins or secondary metabolites for industrial or pharmaceutical uses. Moreover, plant based pharmaceutical products provide opportunities for the direct delivery along with food. There are several examples of proteins expressed in plant for their use as vaccines, therapeutics and nutraceuticals. Some of them are in market or undergoing clinical trials (Table 72.5). Vaccines against hepatitis B, cholera, diarrhea, rabises, seasonal and pandemic influenza are undergoing clinical trails. Transgenic maize seeds expressing a Newcastle disease virus fusion protein were found to be effective in conferring protection against the viral challenge in chickens. GM crops that produce pharmaceuticals can be effective for human and veterinary uses to fight against various infectious diseases and metabolic disorders.

TABLE 72.5 **Some of the examples of plant-derived pharmaceuticals in clinical stages of development or on market (*Source:* Nagels et al., 2012, Critical Reviews in Plant Sciences, 31:2, 148-180)**

Product description	Target (Disease, metabolic disorders, etc.)	Host plant	Status of clinical trial	Manufactures
Vaccines				
Hepatitis B antigen (HBsAg)	Hepatitis	Lettuce	Phase I	Thomas Jefferson University, USA
		Potato	Phase II	Arizona State University, USA
Fusion proteins, including epitopes from rabies	Rabies	Spinach	Phase I completed	Thomas Jefferson University, USA
Vibrio cholerae	Cholera	Potato	Phase I	Arizona State University, USA
Heat-labile toxin B subunit of *Escherichia coli*	Diarrhea	Potato	Phase I	Arizona State University, USA
Capsid protein of norwalk virus	Diarrhea	Potato, Tomato	Phase I	Arizona State University, USA
H1N1 vaccine	Seasonal influenza	Tobacco	Phase I	Medicago, USA
H5N1 vaccine	Pandemic influenza	Tobacco	Phase II	Medicago, USA
HN protein of Newcastle disease virus	Newcastle disease (Poultry)	Tobacco suspension cells	USDA approved	Dow Agro Sciences, USA
Viral vaccine mixture	Diseases of horses, dogs, and birds	Tobacco suspension cells	Phase I	Dow Agro Sciences, USA
Poultry vaccine	Coccidiosis infection	Canola	Phase II	Guardian Biosciences, Canada
Antibodies				
CaroRX	Dental caries	Tobacco	EU approved as medical device	Planet Biotechnology, USA
Antibody against hepatitis B	Vaccine purification	Tobacco	On market	CIGB, Cuba
BLX-301 anti-CD20 antibody	Non-Hodgkin's lymphoma	Duckweed	Preclinical phase	Biolex, USA
DoxoRX	Side-effects of cancer therapy	Tobacco	Phase I completed	Planet Biotechnology, USA
RhinoRX	Common cold	Tobacco	Phase I completed	Planet Biotechnology, USA

(Contd.)

TABLE 72.5 Some of the examples of plant-derived pharmaceuticals in clinical stages of development or on market (*Source:* Nagels et al., 2012, Critical Reviews in Plant Sciences, 31:2, 148-180) (*Contd.*)

Product description	Target (Disease, metabolic disorders, etc.)	Host plant	Status of clinical trial	Manufactures
IgG (ICAM1)	Common cold	Tobacco	Phase I	Planet Biotechnology, USA
Therapeutic human proteins				
Gastric lipase, Merispase®	Cystic fibrosis	Maize	On market	Meristem Therapeutics, France
α-Galactosidase	Fabry disease	Tobacco	Preclinical phase	Protalix Biotherapeutics, Israel
Locteron™ (interferon α)	Hepatitis B and C	Duckweed	Phase IIb	Biolex, USA
BLX-155 Thrombolytic drug (plasmin)	Thrombosis (blood clotting)	Duckweed	Phase I	Biolex, USA
Human glucocerebrosidase (prGCD)	Gaucher's disease	Carrot suspension cells	On market in the U.S. and other countries under Expanded Access Programs	Protalix Biotherapeutics, Israel
Insulin	Diabetes	Safflower	Phase III	SemBioSys, Canada
Apo AIMilano (Apolipoprotein)	Cardiovascular	Safflower	Phase I	SemBioSys, Canada
Nutraceuticals				
ISOkine™, DERMOkine™	Human growth factors and cytokines	Barley	On market	ORF Genetics, Iceland
Human intrinsic factor, Coban	Vitamin B12 deficiency	Arabidopsis	On market	Cobento Biotech AS, Denmark
Human lactoferrin	Anti-infection, antiinflammatory	Rice	Advanced, on market as fine chemical	Ventria, USA
Human lysozyme	Anti-infection, antiinflammatory	Rice	Advanced, on market as fine chemical	Ventria, USA

V. GLOBAL STATUS OF COMMERCIALLY GROWN GM CROPS

Since their first commercial cultivation in 1996, GM crops have been increasingly grown in different parts of the world (Figure 72.2). USA is the leading country in commercialization of GM crops with 43% of global area. During the year 2011, about 16.7 million farmers, 90% of these were small resource-poor farmers in developing countries, grew GM crops involving a total of 29 countries and 160 million hectares land. An overall 8% increase of global cultivated area under GM crop was recorded in 2011 (Table 72.6). Among these 29 GM crop growing countries, 19 are developing countries and more developing countries are expected to adopt GM crops in future. These developing countries, together represent ~40% of the global population, collectively grew GM crops on 71.4 million hectares area (44% of global area under GM crops). Major developing countries that grew GM crops are China and India in Asia, Brazil and Argentina in Latin America and South Africa in the continent of Africa. Three new countries, Pakistan, Myanmar and Sweden for the first time in 2010 grew GM crops. In Mexico which is the centre of biodiversity for maize, the first experimental field trials were successfully conducted in 2010 for Bt and herbicide tolerant GM maize. Although 29 countries grew GM crops on commercial basis, a total of 60 countries have granted regulatory approvals since

1996 for import of GM crops for food and feed purpose, and for release into the environment. This included major food importing countries such as Japan, which do not grow GM crops. According to ISAAA report 2011, regulatory approvals for transgenic events were maximum in maize (65) followed by cotton (39), canola (15), potato, and soybean (14 each). Worldwide, herbicide tolerant GM soybean event has received maximum regulatory approval, followed by herbicide tolerant GM maize, insect resistant GM maize and cotton.

Soybean was the principal GM crop in 2011 followed by maize, cotton and canola. Other cultivated GM crops were sugarbeet, alfalfa, papaya, squash, poplar, tomato, potato, and sweet pepper (Table 72.6). Herbicide tolerance crops occupied maximum area (59%) under GM crops. More preference was given to the stacked multiple traits GM crops than single trait crops like insect resistant with 26% and 15% of global area under GM crops in 2010, respectively. The stacked multiple traits GM crops showed fastest growth rate between 2010 and 2011 at 31% growth, compared with 5% and –10% for herbicide tolerance and insect resistance, respectively. This reflected farmer preference for stacked traits. Triple stacked maize conferred resistance to two insect pests along with herbicide tolerance and double stacked maize having insect pest resistance and herbicide tolerance were cultivated in USA and Philippines. GM maize with eight transgenes coding for several insect pest resistant and herbicide tolerant traits was

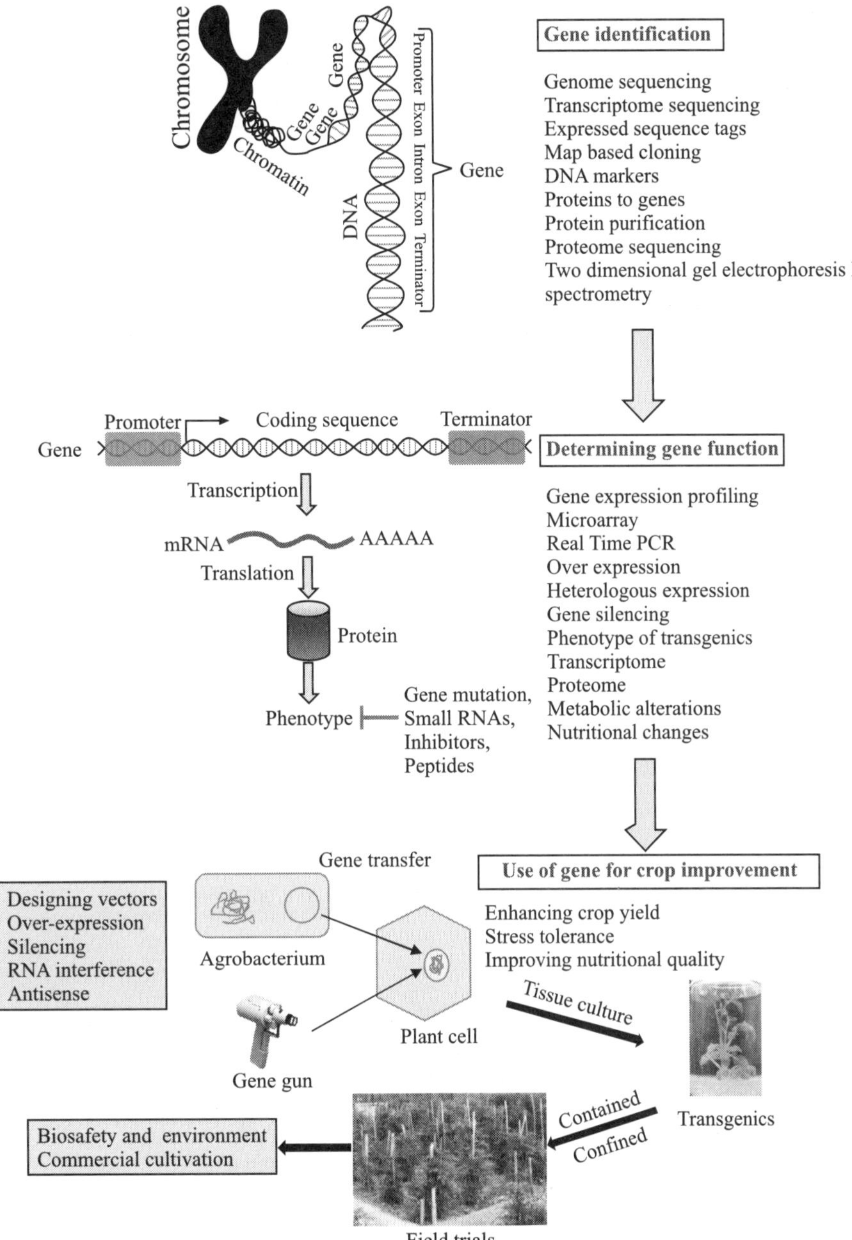

Figure 72.2 Steps involving identification of genes to their use for crop improvement. (*see Plate 32 for colour figure*)

released in the USA and Canada in 2010 with the name of SmartstaxTM. In future, stacked GM crops are expected to offer agronomic input traits for tolerance to biotic and abiotic stresses along with output traits such as high omega-3 oil in soybean or enhanced provitamin A in rice.

In 2011, India grew Bt cotton on 10.6 million hectares of land, representing 88% of the total cotton planted area. Cultivation of Bt cotton has resulted in increased yield, reduction (~50%) in insecticide applications. Bt cotton contributed to the alleviation of poverty of 7 million small resource-poor farmers and their families, in India. It is estimated that farm income of Indian farmers was enhanced from Bt cotton by US$9.4 billion in the period 2002 to 2010 and US$2.5 billion in 2010 alone.

VI. BIOSAFETY AND ENVIRONMENTAL CONSIDERATIONS FOR THE USE OF MARKER GENES IN GM CROPS

Selectable and scorable marker genes are extremely useful in selecting very rare transformation events during the development of GM crops. However, for the commercial purpose, marker gene must be evaluated extensively for biosafety. The environmental and biosafety concerns associated with marker integrated GM plants are the potential to affect non-target organisms, increased weediness or invasiveness, toxicity and allergenicity, pleiotropic effects, and horizontal gene transfer (HGT).

TABLE 72.6   Global area under GM crops (*Source*: James, 2010 and 2011, ISAAA Briefs No. 42 and 43)

Country	Area (Million Hectares)		Cultivated GM crops
	2010	2011	
USA	66.8	69.0	Maize, soybean, cotton, canola, sugarbeet, alfalfa, papaya, squash
Brazil	25.4	30.3	Soybean, maize, cotton
Argentina	22.9	23.7	Soybean, maize, cotton
India	9.4	10.6	Cotton
Canada	8.8	10.4	Canola, maize, soybean, sugarbeet
China	3.5	3.9	Cotton, papaya, poplar, tomato, sweet pepper
Paraguay	2.6	2.8	Soybean
Pakistan	2.4	2.6	Cotton
South Africa	2.2	2.3	Maize, soybean, cotton
Uruguay	1.1	1.3	Soybean, maize
Bolivia	0.9	0.9	Soybean
Australia	0.7	0.7	Cotton, canola
Philippines	0.5	0.6	Maize
Myanmar	0.3	0.3	Cotton
Burkina Faso	0.3	0.3	Cotton
Mexico	0.1	0.2	Cotton, soybean
Spain	0.1	0.1	Maize
Colombia	< 0.1	< 0.1	Cotton
Chile	< 0.1	< 0.1	Maize, soybean, canola
Honduras	< 0.1	< 0.1	Maize
Portugal	< 0.1	< 0.1	Maize
Czech Republic	< 0.1	< 0.1	Maize
Poland	< 0.1	< 0.1	Maize
Egypt	< 0.1	< 0.1	Maize
Slovakia	< 0.1	< 0.1	Maize
Romania	< 0.1	< 0.1	Maize
Sweden	< 0.1	< 0.1	Potato
Costa Rica	< 0.1	< 0.1	Cotton, soybean
Germany	<0.1	<0.1	Potato
Total area	148	160	

As the first and most commonly used selectable marker, the biosafety of neomycin phosphotransferase II (*nptII*) was most extensively evaluated. NPTII confers resistance to aminoglycosides such as kanamycin, neomycin, paromycin, butirosin, gentamycin B and geneticin. According to some reports, NPTII is non-toxic and it is not expected to result in increased weediness or invasiveness and to significantly affect the nontarget organisms. Several studies concluded that transgenic plants expressing the marker gene *nptII* were phenotypically normal and global gene expression patterns were similar to that of non-transgenic controls, thus excluding any pleiotropic effect. NPTII protein was not found to be an acute toxin in mice even at very high level (5 g/kg body weight, a million times more than expected exposure levels. Moreover, NPTII did not show the food allergen like properties and rapidly degraded in simulated gastric fluids and simulated intestinal fluids. The protein had been approved by the FDA in 1994.

Several studies and reports suggested that NPTII in transgenic plants does not pose risk to the health of humans and animals. Moreover, it is also expected that the consequences of HGT of the nptII gene from a GM plant to the gastrointestinal or soil bacteria would be quite insignificant. Although the gene transfer from plants to bacteria has been demonstrated under forced laboratory conditions in the presence of homologous regions in the recipient bacterium, HGT under natural conditions has not been reported so far. The risk that antibiotic resistant genes can be transferred from GM plants to commensal and clinical bacteria should be regarded as almost nil. This is not because the frequency of such gene transfer event is very rare but due to the fact that plethora of these genes are already there in soil bacteria and they are continuously being subjected to selection pressure which limits the possibilities that a gene from a plant can be functionally and stably transferred to bacteria.

In 2004, the scientific panel on GM organisms (GMO Panel) of the European Food Safety Authority (EFSA) concluded that 'the use of *npt*II as a selectable marker in GM plants does not pose a risk to the environment or to human and animal health'. The GMO Panel mentioned the low likelihood of HGT from plants to bacteria, the prevalence of the *npt*II gene in natural bacterial populations and the limited use of kanamycin and neomycin in human and veterinary medicine. However, in 2007 European Medicines Agency (EMEA) indicated that 'aminoglycoside class of antibiotics have become increasingly important in the prevention and treatment of serious invasive bacterial infections in humans, since Gram-negative bacteria (and tuberculosis bacteria) are becoming resistant to other classes of antibiotics'. The EMEA also stressed that, 'although kanamycin and neomycin are used relatively infrequently, the potential development of new chemical entities similar to kanamycin and neomycin should also be taken into account'. In addition, EMEA also mentioned that 'although the veterinary use of kanamycin and neomycin is limited, aminoglycosides as a group are a class of antibiotics critically important for veterinary medicine'. Further, in 2007 the GMO Panel of EFSA agreed with the EMEA that 'the preservation of the therapeutic potential of the aminoglycoside group of antibiotics is important'. The Panel was also of the opinion that 'the therapeutic effect of these antibiotics will not be compromised by the presence of the *npt*II gene in GM plants, given the extremely low probability of gene transfer from plants to bacteria and its subsequent expression'. Furthermore, the GMO Panel considered it very unlikely that 'the presence of the *npt*II gene in GM plants will change the existing widespread prevalence of this antibiotic resistance gene in bacterial sources in the environment'. The GMO Panel also pointed to evidences which indicated that 'integration of the *npt*II gene would only be one of many mechanisms by which bacteria could become resistant to aminoglycosides such as kanamycin'. Therefore, the GMO Panel reiterated earlier conclusions that 'the use of the *npt*II gene as selectable marker in GM plants (and derived food or feed) does not pose a risk to human or animal health or to the environment'. The GMO Panel also confirmed earlier safety assessments of GM plants and derived food/feed comprising the *npt*II gene. Again in 2009, the GMO Panel along with the Panel on Biological Hazards (BIOHAZ) provided a joint scientific opinion on the use of antibiotic resistance genes as marker genes in genetically modified plants. The joint Panel concluded that 'the adverse effects on human health and the environment resulting from the transfer of the antibiotic resistance genes (*npt*II and *aad*A) from GM plants to bacteria, associated with use of GM plants, are unlikely'. However, two members of the BIOHAZ Panel expressed minority opinions on this conclusion.

Safety assessments on the selectable marker genes *nptII*, *hpt*, *bar/pat*, and *manA* and the scorable marker gene *uidA*, have indicated that all have little risk in terms of biosafety. For these well characterized markers, removing of the marker DNA from the GM crops may not be an absolute requirement based on the biosafety reasons. However, marker-free GM plants make the regulatory process before commercialization simple and also improve the consumer acceptance. Thus, it is generally recommended to remove undesired DNA sequences from transgenic plants before commercialization. Moreover, it would permit the recycling of useful marker gene for recurrent transformation of transgenic plants for staking of multiple genes. A suite of strategies (co-transformation, transposition, homologous recombination and site-specific recombination) developed to eliminate the marker gene from transgenics.

VII. HOW DO CLASSICALLY BRED CROPS DIFFER FROM GM CROPS?

Both classically bred and GM crops are the outcomes of genetic modifications to have desirable traits into a new genetic background; however, through different means of gene transfer technology. The whole process of the development of a classically bred variety may involve breeding within species or among closely/distantly related species followed by artificial selection. Plant breeders look for natural genetic variations within a group of crossable species or they create genetic variations through random mutagenesis and hybridization. The classical breeding was in fact initiated long before the advent of Gregor Johann Mendel's principles of genetics. Crop improvement practices over the centuries through conventional plant breeding lead to high yielding crop cultivars that sustain humankind today. The work of Norman Ernest Borlaug during the mid 20th century on the development of high yielding varieties of cereals such as wheat and rice originated green revolution in several parts of Asia. Further, genetically engineered crops developed using modern biotechnological tools have added new dimensions to crop improvement programs. Through these modern technologies, transfer of novel agronomic traits into crops is possible that otherwise would have been very difficult or impossible to introduce following conventional breeding approaches. Through biotechnology, we have access to the massive gene pool which can be exploited for genetic enrichment of crops without the constraint of sexual compatibility. These can result in novel gene combinations that were not even seen before. It is widely accepted that biotechnology may revolutionize crop improvement programs.

Both the conventional breeding and genetic engineering approaches may involve changes in the genetic makeup of an organism with respect to DNA sequence and the order of genes. However, the amount of genetic information modified through the genetic engineering is small and defined as compared to the classical breeding where thousands of uncharacterized genes of an organism may be involved. Moreover, GM crops are the outcome of very precise and targeted modification in the genome and the end product, which may be a protein or metabolite or phenotype, is well characterized. However, crop varieties developed through conventional breeding have high degree of uncertainty about the type of changes introduced in genome.

Furthermore, genetic engineering permits us precise control of when and where a gene product is to be made by linking the desired gene sequence with a regulatory sequence that results in tissue/organ/development/stress-specific expression—an outcome not easy to accomplish with classical breeding.

SUMMARY

Owing to ever-growing world population, the global demand for food is increasing rapidly. There are needs for quality foods that comprise of essential nutrients and functional foods that provide health benefits to humans beyond basic nutrition. As the availability of arable land is decreasing continuously, increase in food production can be achieved by cultivating crop varieties that adapt to extreme soil and climatic conditions and also produce more yields per unit area. In recent times, the use of agrochemicals and high yielding crop varieties developed through conventional plant breeding practices lead to significant boost in crop productivity. Now time has arrived to promote sustainable agricultural practices for increasing crop productivity with utmost conservation of all natural resources. Agriculture biotechnology is proving to be a powerful complement to conventional methods to meet worldwide demand for food with respect to both quality and quantity. Through modern plant biotechnology, we have access to massive gene pool which can be exploited to transfer broad range of traits of interest to crop species, ignoring the sexual barrier. GM crops developed through biotechnology can help meet the demand for high yielding, nutritionally balanced, biotic and abiotic stress tolerant crop varieties. Figure 72.3 summarizes how plant biotechnology can help us to attain food security and promote human health.

Recent advances in the area of plant biotechnology allow introduction of novel traits and their precise and regulated expression into plants. Plant biotechnology has the potential to address various problems in agriculture and society. GM crops that are tolerant to biotic and abiotic stresses can minimize yield losses due to such stresses. Further, the development of GM crops that are tolerant to extreme soil environments can extend crop cultivation under marginal lands. GM strategies can also be employed to enhance the nutritional quality of crops for addressing the problem of malnutrition and providing benefits to human health. Some of the exciting developments include value addition in food crops by enrichment with quality proteins, vitamins, iron, zinc, carotenoids, anthocyanins etc. Other ongoing efforts include the enhancement of shelf life of fruits and vegetables which will allow to significantly reducing the post-harvest losses in perishable crops. Fruit crops are also targeted for the purpose of edible vaccines to combat major diseases.

Although worldwide cultivation of GM crops is escalating each year, concerns are also raised about the unintended and unpredictable pleiotropic effects of GM crops on human health and to the environment. Novel foods developed either by conventional or genetic engineering approach are no different in terms of untended harmful effects they may impart to human health and the environment. A well known example is the conventionally breed insect resistant high psoralens variety of celery that caused skin rashes to the farm workers. So far no harmful effects of GM crops have been documented since their first release to the environment for commercial cultivation and from consumption of GM foods by more than a billion humans and a larger number of animals. However, before being released for the commercial cultivation, the performance of GM crops should be closely examined under field conditions for several generations and these crops must undergo rigorous bio-safety assessments on a case-by-case basis. The unremitting progress made in the last fifteen years of GM crop commercialization in many parts of the world signifies that GM crops are here

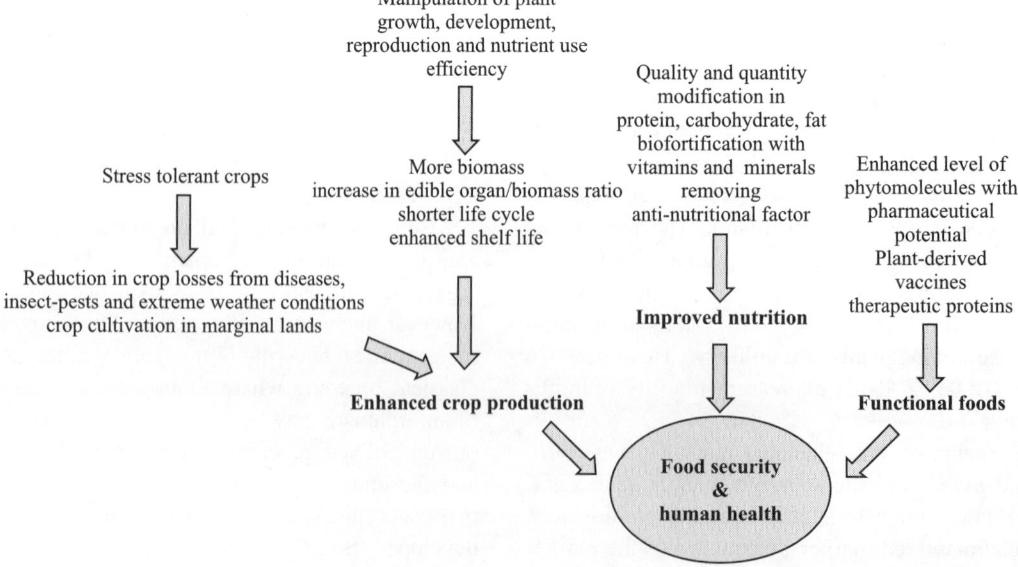

Figure 72.3 Role of plant biotechnology towards achieving food security and promoting human health.

to stay. GM crops which are currently under commercial cultivation are mostly insect resistant Bt crops and/or herbicide tolerant crops. These crops have benefited the farmers through better insect and weed management, higher yields and also reduced chemical pesticide uses. Novel traits once moved into a crop through biotechnology can be transferred to different cultivars suitable for cultivation in different parts of the world and to closely related species through conventional breeding.

It is believed that integration of conventional practices with modern plant biotechnology could make it possible to meet the present as well as future demand for quality foods. Based on the evidences presented here, GM crops may well be regarded as one of the possible means that can help meet aspirations for global agricultural sustainability. Hopefully, the enormous potential of biotechnology will be exploited to the utmost benefit of the human being.

SUGGESTIONS FOR FURTHER READING

Baulcombe D. (2010), Reaping Benefits of Crop Research, *Science,* 327, 761.

Baum J.A., Bogaert T., Clinton W., Heck G.R., Feldmann P., Ilagan O., Johnson S., Plaetinck G., Munyikwa T., Pleau M., Vaughn T. and Roberts J. (2007), Control of Coleopteran Insect Pests Through RNA Interference, *Nature Biotechnol.,* 25, 1322–1326.

Broglie K., Chet I., Holliday M., Cressman R., Biddle P., Knowlton S., Mauvais C.J. and Broglie R. (1991), Transgenic Plants with Enhanced Resistance to the Fungal Pathogen *Rhizoctonia solani, Science,* 254, 1194–1197.

Butelli E., Titta L., Giorgio M., Mock H.P., Matros A., Peterek S., Schijlen E.G., Hall R.D., Bovy A.G., Luo J. and Martin C. (2008), Enrichment of Tomato Fruit with Health-promoting Anthocyanins by Expression of Select Transcription Factors, *Nature Biotechnol.,* 26, 1301–1308.

Cahoon E.B., Hall S.E., Ripp K.G., Ganzke T.S., Hitz W.D. and Coughlan S.J. (2003), Metabolic Redesign of Vitamin E Biosynthesis in Plants for Tocotrienol Production and Increased Antioxidant Content, *Nature Biotechnol.,* 21, 1082–1087.

Capell T., Bassie L. and Christou P. (2004), Modulation of the Polyamine Biosynthetic Pathway in Transgenic Rice Confers Tolerance to Drought Stress, *Proc. Natl. Acad. Sci.,* USA, 101, 9909–9914.

Carpenter J.E. (2010), Peer-reviewed Surveys Indicate Positive Impact of Commercialized GM Crops., *Nature Biotechnol.,* 28, 319–321.

Chakraborty S., Chakraborty N. and Datta A. (2000), Increased Nutritive Value of Transgenic Potato by Expressing a Nonallergenic Seed Albumin Gene from *Amaranthus hypochondriacus, Proc. Natl. Acad. Sci.,* USA. 97, 3724–3729.

Chakraborty S., Chakraborty N., Agrawal L., Ghosh S., Narula K., Shekhar S., Naik P.S., Pande P.C., Chakrborti S.K. and Datta A. (2010), Next Generation Protein Rich Potato by Expressing a Seed Protein Gene AmA1 as a Result of Proteome Rebalancing in Transgenic Tuber, *Proc. Nati. Acad. Sci.,* USA, 107, 17533–17538.

Chaudhary B. and Gaur K. (2009), The Development and Regulation of Bt-brinjal in India, ISAAA Brief No. 38. Ithaca, NY.

Chen Z., Young T.E., Ling J., Chang S.C. and Gallie D.R. (2003), Increasing Vitamin C Content of Plants Through Enhanced Ascorbate Recycling, *Proc. Natl. Acad. Sci.,* USA, 100, 3525–3530.

Dai S., Wei X., Alfonso A.A., Pei L., Duque U.G., Zhang Z., Babb G.M. and Beachy R.N. (2008), Transgenic Rice Plants That Overexpress Transcription Factors RF2a and RF2b are Tolerant to Rice Tungro Virus Replication and Disease, *Proc. Natl. Acad. Sci.,* USA, 105, 21012–21016.

Davison G.M., Davuluri G.R., Van Tuinen A., Fraser P.D., Manfredonia A., Newman R., Burgess D., Brummell D.A., King S.R., Palys J., Uhlig J., Bramley P.M., Pennings H.M. and Bowler C. (2005), Fruit-specific RNAi-mediated Suppression of DET1 Enhances Carotenoid and Flavonoid Content in Tomatoes, *Nature Biotechnol.,* 23, 890–895.

Diaz De La Garza R.I., Gregory J.F. IIIrd. and Hanson A.D. (2007), Folate Biofortification of Tomato Fruit, *Proc. Natl. Acad. Sci.,* USA, 104, 4218–4222.

Duan X., Li X., Xue Q., Abo-El-Saad M., Xu D. and Wu R. (1996), Transgenic Rice Plants Harboring an Introduced Potato Proteinase Inhibitor II Gene are Insect Resistant, *Nature Biotechnol.,* 14, 494–498.

Dunse K.M., Stevens J.A., Laya F.T., Gaspar Y.M., Heath R.L. and Anderson M.A. (2010), Coexpression of Potato Type I and II Proteinase Inhibitors Gives Cotton Plants Protection Against Insect Damage in the Field, *Proc. Natl. Acad. Sci.,* USA, 107, 15011–15015.

Eckardt N.A. (2009), The Future of Science:Food and Water for Life, *Plant Cell,* 21, 368–372.

Falco S.C., Guida T., Locke M., Mauvais J., Sanders C., Ward R.T. and Webber P. (1995), Transgenic Canola and Soybean Seeds with Increased Lysine, *Nature Biotechnol.,* 13, 577–582.

Gao A.G., Hakimi S.M., Mittanck C.A., Wu Y., Woerner B.M., Stark D.M., Shah D.M., Liang J. and Rommens C.M. (2000), Fungal Pathogen Protection in Potato by Expression of a Plant Defensin Peptide, *Nature Biotechnol.,* 18, 1307–1310.

Giovinazzo G., D'Amico L., Paradiso A., Bollini R., Sparvoli F. and DeGara L. (2005), Antioxidant Metabolite Profiles in Tomato Fruit Constitutively Expressing the Grapevine Stilbene Synthase Gene, *Plant Biotechnol. J.* 3, 57–69.

Goto F., Yoshihara T., Shigemoto N., Toki, S. and Takaiwa F. (1999), Iron Fortification of Rice Seed by the Soybean Ferritin Gene, *Nature Biotechnol.,* 17, 282–286.

Hellwege E.M., Czapla S., Jahnke A., Willmitzer L. and Heyer A.G. (2000), Transgenic Potato (*Solanum tuberosum*) Tubers Synthesize the Full Spectrum of Inulin Molecules Naturally Occurring in Globe Artichoke (*Cynara scolymus*) Roots, *Proc. Natl. Acad. Sci.,* USA, 97, 8699–8704.

Horvath H., Rostoks N., Brueggeman R., Steffenson B., Von Wettstein D. and Kleinhofs A. (2003), Genetically Engineered Stem Rust Resistance in Barley using the *Rpg*1 Gene. *Proc. Natl. Acad. Sci.,* USA, 100, 364–369.

Hu H., Dai M., Yao J., Xiao B., Li X., Zhang Q. and Xiong L. (2006), Overexpressing A NAM, ATAF, and CUC (NAC) Transcription Factor Enhances Drought Resistance and Salt Tolerance in Rice, *Proc. Natl. Acad. Sci.,* USA, 103, 12987–12992.

James C. (2011), Global Status of Commercialized Biotech/GM Crops: 2011. ISAAA Briefs No. 43. ISAAA, Ithaca.

Karaba A., Dixit S., Greco R., Aharoni A., Trijatmiko K.R., Marsch-Martinez N., Krishnan A., Nataraja K.N., Udayakumar M. and Pereira A. (2007), Improvement of Water use Efficiency in Rice by Expression of *HARDY*, an Arabidopsis Drought and Salt Tolerance Gene. *Proc. Natl. Acad. Sci.,* USA, 104, 15270–15275.

Keer R.A. (2007), Global Warming is Changing the World, *Science,* 316:188–90.

Kesarwani M., Azam M., Natarajan K., Mehta A. and Datta A. (2000), Oxalate Decarboxylase from *Collybia velutips*: Molecular Cloning and Its Over Expression to Confer Resistance to Fungal Infection in Transgenic Tobacco and Tomato, *J. Biol. Chem.,* 275, 7230–7238.

Lin W., Anuratha C.S., Datta K., Potrykus I., Muthukrishnan S. and Datta S.K. (1995), Genetic Engineering of Rice for Resistance to Sheath Blight, *Nature Biotechnol.,* 13, 686–691.

Lorito M., Woo S.L., Fernandez I.G., Colucci G., Harman G.E., Pintor-Toro J.A., Filippone E., Muccifora S., Lawrence C.B., Zoina A., Tuzun S. and Scala, F. (1998), Genes from Mycoparasitic Fungi as a Source for Improving Plant Resistance to Fungal Pathogens, *Proc. Natl. Acad. Sci.,* USA, 95, 7860–7865.

Martin C., Butelli E., Petroni K. and Tonelli C. (2011), How Can Research on Plants Contribute to Promoting Human Health? *Plant cell,* 23, 1685–1699.

Mehta A. and Datta A. (1991), Oxalate Decarboxylase from *Collybia Velutipes*: Purification, Characterization cDNA Cloning, *J. Biol. Chem.,* 266, 23548–23553.

Mehta R.A., Cassol T., Li N., Ali N., Handa A.K. and Mattoo A.K. (2002), Engineered Polyamine Accumulation in Tomato Enhances Phytonutrient Content, Juice Quality, and Vine Life. *Nature Biotechnol.,* 20, 613–618.

Meli V.S., Ghosh S., Prabha T.N., Chakraborty N., Chakraborty S. and Datta A. (2010), Enhancement of Fruit Shelf Life by Suppressing *N*-glycan Processing Enzymes. *Proc. Natl. Acad. Sci.,*USA, 107, 2413–2418.

Mendelsohn M., Kough J., Vaituzis Z. and Matthews K. (2003), Are Bt Crops Safe? *Nature Biotechnol.,* 21, 1003–1009.

Mittler R. and Blumwald E. (2010), Genetic Engineering for Modern Agriculture: Challenges and Perspectives, *Annu. Rev. Plant Biol.,* 2010. 61:443–62.

Muir S.R., Collins G.J., Robinson S., Hughes S., Bovy A., Ric De Vos C.H., Van Tunen A.J. and Verhoeyen M.E. (2001), Overexpression of Petunia Chalcone Isomerase in Tomato Results in Fruit Containing Increased Levels of Flavonols, *Nature Biotechnol.,* 19, 470–474.

Naqvi S., Zhu C.F., Farre G., Ramessar K., Bassie L., Breitenbach J., Conesa D.P., Ros G., Sandmann G., Capell T. and Christou P. (2009), Transgenic Multivitamin Corn Through Biofortification of Endosperm with Three Vitamins Representing Three Distinct Metabolic Pathways, *Proc. Natl. Acad. Sci.,* USA, 106, 7762–7767.

Nelson D.E., Repetti P.P., Adams T.R., Creelman R.A., Wu J., Warner D.C., Anstrom D.C., Bensen R.J., Castiglioni P.P., Donnarummo M.G., Hinchey B.S., Kumimoto R.W., Maszle D.R., Canales R.D., Krolikowski K.A., Dotson S.B., Gutterson N., Ratcliffe O.J. and Heard J.E. (2007), Plant Nuclear Factor Y (NF-Y) B Subunits Confer Drought Tolerance and Lead to Improved Corn Yields on Water-limited Acres, *Proc. Natl. Acad. Sci.,* USA, 104, 16450–16455.

Newell C.A., Niggeweg R., Michael A.J. and Martin C. (2004), Engineering Plants with Increased Levels of the Antioxidant Chlorogenic Acid, *Nature Biotechnol.,* 22, 746–754.

Paine J.A., Shipton C.A., Chaggar S., Howells R.M., Kennedy M.J., Vernon G., Wright S.Y., Hinchliffe E., Adams J.L., Silverstone A.L. and Drake R. (2005), Improving the Nutritional Value of Golden Rice Through Increased Pro-vitamin A Content, *Nature Biotechnol.,* 23, 482–487.

Pennisi E. (2010), Armed and Dangerous, *Science,* 327, 804–805.

Pusztai A. and Bardocz S. (2006), GMO in Animal Nutrition: Potential Benefits and Risks, In: *Biology of Nutrition in Growing Animals* (Mosenthin R., Zentek J. and Zebrowska T. Eds.), 513–540, Elsevier Limited.

Quist D. and Chapela I.H. (2001), Transgenic DNA Introgressed into Traditional Maize Landraces in Oaxaca, Mexico. *Nature,* 414, 541–543.

Raina A. and Datta A. (1992), Molecular Cloning of a Gene Encoding a Seed Specific Protein with Nutritionally Balanced Amino Acid Composition from *Amaranthus, Proc. Nati. Acad. Sci.,* USA, 89, 11774–11778.

Royal Society (2009), Reaping the Benefits: Science and the Sustainable Intensification of Global Agriculture, Http://royalsociety.org/policy/publications/2009/reaping-benefits (21 Oct 2009).

Sano T., Nagayama A., Ogawa T., Ishida I. and Okada Y. (1997), Transgenic Potato Expressing a Double-stranded RNA-specific Ribonuclease is Resistant to Potato Spindle Tuber Viroid, *Nature Biotechnol.,* 15, 1290–1294.

Shade R.E., Schroeder H.E., Pueyo J.J., Tabe L.M., Murdock L.L., Higgins T.J.V. and Chrispeels M.J. (1994) Transgenic Pea Seeds Expressing the α-amylase Inhibitor of the Common Bean are Resistant to Bruchid Beetles, *Nature Biotechnol.*, 12, 793–796.

Sitch S., Cox P.M., Collins W.J. and Huntingford C. (2007), *Indirect Radiative Forcing of Climate Change through Ozone Effects on the Land-carbon Sink, Nature,* 448, 791–795.

Stein A.J., Sachdev H.P.S. and Qaim M. (2006), Potential Impact and Cost-effectiveness of Golden Rice, *Nature Biotechnol.*, 24, 1200–1201.

Storozhenko S., De Brouwer V., Volckaert M., Navarrete O., Blancquaert D., Zhang G.F., Lambert W. and Van Der Straeten D. (2007), Folate Fortification of Rice by Metabolic Engineering, *Nature Biotechnol.*, 25, 1277–1279.

Tai, T.H., Dahlbeck D., Clark E.T., Gajiwala P., Pasion R., Whalen M.C., Stall R.E. and Staskawicz B.J. (1999), Expression of the *Bs*2 Pepper Gene Confers Resistance to Bacterial Spot Disease in Tomato, *Proc. Natl. Acad. Sci.,* USA, 96, 14153–14158.

Tang X., Xie M., Kim Y.J., Zhou J. and Klessig D.F. (1999), Overexpression of *Pto* Activates Defense Responses and Confers Broad Resistance, *Plant Cell,* 11, 15–29.

The International Service for the Acquisition of Agri-biotech Applications (2010), Agricultural Biotechnology (A Lot More Than Just GM Crops), ISAAA, Ithaca, NY.

Truve E., Aaspollu A., Honkanen J., Puska R., Mehto M., Hassi A., Teeri T.H., Kelve M., Seppanen P. and Saarma M. (1993), Transgenic Potato Plants Expressing Mammalian 2′,5′-oligoadenylate Synthetase are Protected from Potato Virus X Infection under Field Conditions, *Nature Biotechnol.*, 11, 1048–1052.

van Der Meer I.M., Van Eenennaam A.L., Lincoln K., Durrett T.P., Valentin H.E., Shewmaker C.K., Thorne G.M., Jiang J., Baszis S.R., Levering C.K., Aasen E.D., Hao M., Stein J.C., Norris S.R. and Last R.L. (2003), Engineering Vitamin E Content: From Arabidopsis Mutant to Soy Oil, *Plant Cell,* 15, 3007–3019.

Vanderauwera S., De Block M., Van De Steene N., Van De Cotte B., Metzlaff M. and Van Breusegem F. (2007), Silencing of poly (ADP-ribose) Polymerase in Plants Alters Abiotic Stress Signal Transduction, *Proc. Natl. Acad. Sci.,* USA, 104, 15150–15155.

Voesenek L. and Bailey-Serres J. (2009), *Genetics of High-rise Rice, Nature,* 460, 959–960.

Wei J.Z., Hale K., Carta L., Platzer E., Wong C., Fang S.C. and Aroian R.V. (2003), *Bacillus Thuringiensis* Crystal Proteins that Target Nematodes, *Proc. Natl. Acad. Sci.,* USA, 100, 2760–2765.

Welsch R., Arango J., Bar C., Salazar B., Al-Babili S., Beltran J., Chavarriaga P., Ceballos H., Tohme J. and Beyer P. (2010), Provitamin A Accumulation in Cassava (*Manihot esculenta*) Roots Driven by a Single Nucleotide Polymorphism in a Phytoene Synthase Gene, *Plant Cell,* 22, 3348–3356.

Wu X.R., Chen Z.H. and Folk M.R. (2003), Enrichment of Cereal Protein Lysine Content by Altered tRNA(lys) Coding During Protein Synthesis, *Plant Biotechnol. J.*, 1, 187–194.

Yang X., Yie Y., Zhu F., Liu Y., Kang L., Wang X. and Tien P. (1997), Ribozyme-mediated High Resistance Against Potato Spindle Tuber Viroid in Transgenic Potatoes, *Proc. Natl. Acad. Sci.,* USA, 94, 4861–4865.

Ye X., Al-Babili S., Kloti A., Zhang J., Lucca P., Beyer P. and Potrykus I. (2000), Engineering the Provitamin A (beta-carotene) Biosynthetic Pathway into (carotenoid-free) Rice Endosperm, *Science,* 287, 303–305.

Zhang H.X., Hodson J.N., Williams J.P. and Blumwald E. (2001), Engineering Salt-tolerant *Brassica* plants: Characterization of Yield and Seed Oil Quality in Transgenic Plants with Increased Vacuolar Sodium Accumulation, *Proc. Natl. Acad. Sci.,* USA, 98, 12832–12836.

73

Synthetic Biology

Prachee Prakash

CONTENTS

I. WHAT IS SYNTHETIC BIOLOGY

Synthetic biology is an emerging field of research that can be described as the design and construction of novel artificial biological pathways, organisms or devices, or the redesign of existing natural biological systems. More specifically, synthetic biology is the engineering of biology resulting in the synthesis of complex, biological systems with functions that do not exist in nature. The engineering approach of synthetic biology can be applied at all levels of hierarchy of biological structures – from individual molecules to whole cells, tissues and organisms. The ultimate objective of synthetic biology is to produce engineered cells, microbes, and biological systems to perform new, useful functions. Some characterize synthetic biology as "Nano biotechnology." Others claim that synthetic biology is nothing more than genetic engineering. However, the bottom up approach of designing genomes and pathways from the bases/base pairs to higher levels of biological organization is quite different in its implications from changing one or two genes. What seems to make synthetic biology different from other current lines of biological research is the rigorous application of engineering principles (standardization, abstraction and decoupling) to biological research, which indeed offers a new way of doing research in life sciences. Synthetic biology includes the broad redefinition and expansion of biotechnology, with the ultimate goals of being able to design and build engineered biological systems that process information, manipulate chemicals, fabricate materials and structures, produce energy, provide food, and maintain and enhance human health and our environment. Synthetic biology also aims to develop technologies, methods, and biological components that will make the engineering of biology safer, more reliable, more predictable and, ultimately, standardized.

The concept of synthetic biology evolved with the idea that it is possible to extend genetic engineering to be similar to engineering of any hardware. As all life forms are based on roughly the same genetic code, synthetic biology can provide a toolbox of reusable genetic components, equivalent to transistors and switches that could be plugged into circuits at will. Mechanistically, this involves characterization of genetic sequences or the 'parts' that perform needed functions, combine the parts into devices to attain more complex functions, and subsequently insert the devices into cells. The ultimate vision of synthetic biology is towards programming cells to execute functions like producing vast quantities of biofuel from renewable sources, materials, pharmaceuticals, chemicals, food products, security, and other industries that affect our daily lives. The field of synthetic biology is also shedding light on mechanisms of natural biological functions and the organization of biological systems, leading to better engineering of new biological forms and systems.

In literature, the title 'synthetic biology' first appeared in 1980 in an article by Barbara Hobom to describe bacteria that had been genetically engineered using recombinant DNA technology. These bacteria were altered by human intervention (that is, synthetically), and were living systems (therefore biological). In this respect, synthetic biology was largely synonymous with 'bioengineering'. The term 'synthetic biology' was again introduced in 2000 at the annual meeting of the American Chemical Society in San Francisco. Here, the term was used to describe the synthesis of unnatural organic molecules that function in living systems.

A. Synthetic Biology, Inspired by Synthetic Chemistry

The development of synthetic biology is in many ways similar to the development of synthetic chemistry as part of organic chemistry. Organic chemistry started out as a largely analytical science involving purification and characterization of products from natural sources. In that sense, it was a discovery-based science with the objective of elucidating the properties of natural entities. With the advancement of knowledge of structures and chemical properties of natural products, synthetic chemistry started to take shape. In the beginning, the main purpose of synthetic chemistry was to confirm the molecular structures deduced by analysis and to produce more materials for research. Later however, synthetic chemistry developed further to create molecular structures that are not found in nature, some of which mimic or improve on properties found in natural molecules.

Biology has now reached the stage where a sufficient amount of genetic and biochemical data on biological systems has been acquired to enter the synthetic stage. In analogy to synthetic chemistry, synthetic biology is not content with "explaining" or simply reproducing the behaviour of natural systems. Rather, synthetic biology aims to go one step further by building, i.e., synthesizing, novel biological systems from scratch using the design principles observed in nature but with expanded, enhanced and controllable properties. The complexity of such a 'design' goal makes an engineering approach imperative.

B. Engineering of Biology

Synthetic biology with its engineering vision aims to overcome the existing fundamental inabilities in system design and system fabrication, by developing foundational principles and technologies to ultimately enable a systematic forward-engineering of biological systems for improved and novel applications (Figure 73.1)

The biological equivalent to fabrication is genomic scale *de novo* DNA synthesis. The wide spread implementation of standards in biological design has resulted in a large diversity in bioengineering tools (strains, plasmids, expression systems, etc).

There is however a fundamental difference between engineering biology and engineering in other natural sciences such as chemistry or physics: biological systems have the capacity to replicate and to evolve. This fundamental characteristic will interfere at least with the long-term stability of a number of designed systems and will require constant monitoring of the integrity (of crucial parts) of the systems. As the ultimate goal will be to build complex systems into specific hosts, it is likely that interference from mutations becomes a serious issue. However, it is important to point out that the problem as such is not new to biotechnology—rather, it has accompanied every major strain development effort in industrial fermentation and has contributed to the development of appropriate selection programs for sufficiently stable strains and suitable strain storage routines.

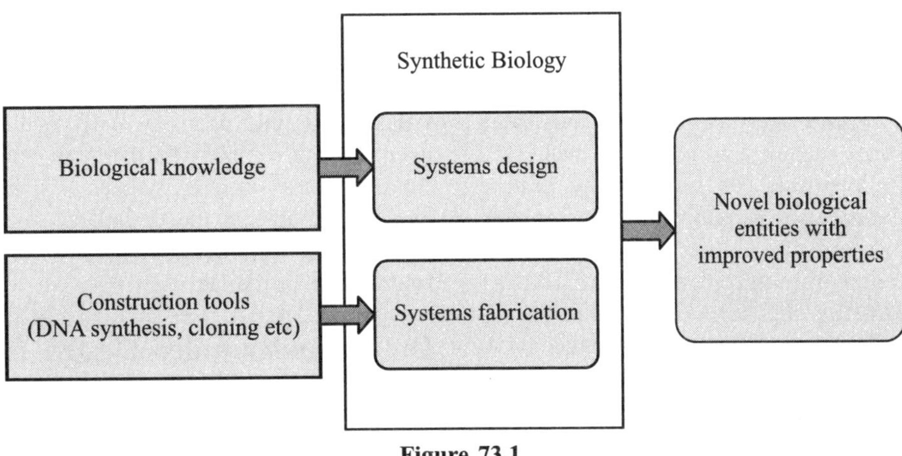

Figure 73.1

Synthetic biology is indeed a novel, true engineering approach to conducting biotechnology. It requires endeavours that exploit at least some of the important elements of an engineering approach, e.g. *de novo* DNA synthesis of an entire pathway, which is composed of genes from various species and includes optimized codons, adapted secondary mRNA structure and tailored regulatory elements. This can be considered a synthetic biology approach to a metabolic engineering problem. Alternatively, the forward-engineering design of genetic modules that can be freely combined to program gene regulation can be considered a synthetic biology approach to genetic engineering.

This chapter would focus on conceptual and methodological approaches to synthetic biology. Later sections would describe the potentials of synthetic biology and how it could contribute to important application areas such as health care, industrial biotechnology and environmental issues. Finally we will discuss the risks, rewards and ethical issues associated with synthetic biology.

II. TOOLS AND TECHNIQUES ENABLING SYNTHETIC BIOLOGY

Synthetic biology aims to follow engineering protocols of design and fabrication, for which it relies heavily on tools for *de novo* DNA synthesis, standardized cloning and others discussed in the following sections.

A. *de novo* DNA Synthesis

DNA *de novo* synthesis is currently performed by assembling overlapping short (25–70 bp long) and chemically synthesized oligonucleotides into longer DNA fragments in a PCR-based assembly process. This technology has already led to the complete reconstruction of some smaller phage genomes such as the polio virus. However, it suffers from two major cost entries: the costs of synthesizing the DNA oligonucleotides by standard phosphoamidite chemistry and the limitations in the accuracy of the chemical synthesis. For example, the *de novo* reconstruction of the phage *ΦX174* genome relied on a 2-fold selection process: (1) oligonucleotides were first gel-purified to ensure the correct length of the set and (2) correctly assembled DNA was recovered from plaques after transformation, effectively providing a positive selection strategy for biological function. Cost and accuracy associated with the current synthesis method will have to be addressed before *de novo* DNA synthesis can become the routine technology that will be required to fabricate more complex biological systems.

Research to optimize the current procedure follows three paths: (1) miniaturizing oligonucleotide production; (2) optimizing sequence verification costs; and (3) eliminating 'false' oligonucleotides by enzymatic and/or hybridization methods.

Over the last five years, synthetic biology has seen a 100-fold increase in the length of genetic material wholly constructed from raw chemicals. Plunging costs of synthesis has allowed a leap past the 1 million base-pair mark, from code to assembly. Comparing this to doubling the diameter of a silicon wafer that can be manufactured that much, going from 1 cm to 1 meter (fabrications) in just five years would have been an incredible achievement.

B. Standardized Cloning

Cloning is another very important fabrication tool in synthetic biology. PCR-based technologies are mostly used to copy an existing DNA fragment for a novel purpose, which is then inserted into different cloning vectors. Recent advances in development of a standardized vector format, allows interoperability of assembled sequences, such as biobricks (http://biobricks.org/). BioBrick parts represent an effort to introduce the engineering principles of abstraction and standardization into synthetic biology. The trademarked words BioBrick and BioBricks are used to refer to a specific "brand" of open source genetic parts as defined via an open technical standards setting process that is led by the BioBricks Foundation.

Through the implementation of such standards, rapid exchange of parts is made possible. Of course, the ideas for standardization need to remain open for optimization. To be successful as an engineering discipline, synthetic biology will need to repeat the corresponding developments of its sister engineering disciplines that have led to highly organized fields such as mechanical and electrical engineering.

C. Microfluidics

The application of microfluidic devices—structures that direct complex flow patterns and chemical operations on the micrometre scale—is rapidly developing into a fundamental experimental technique, driven by systems biology induced requirement for highly automated, parallel experiments. In two areas, the main advantages become very clear: in the area of analysis, where several thousand analyses can be conducted at the same time, and in microTAS (micrototal analysis system, lab-on-a-chip) applications, which allow the miniaturization of entire sequences of processing steps onto chips (e.g., on-chip PCR, on-chip protein purification). Looking towards emerging application areas, synthetic biology will benefit tremendously from this development. For example in the area of biosensors, miniaturization of sample preparation and actual measurement will allow highly precise, robust, and long-term analysis of even difficult samples. Furthermore, in view of the expected future increases in DNA synthesis capacity, microfluidics will be indispensable in the design of economic DNA synthesis protocols that are required to make the synthesis a genuinely routine tool in the lab.

D. Computer Aided Design Tools: Simulation of Gene Circuits

Simulation of gene circuits belongs to the set of basic

technologies that are common to systems biology and synthetic biology. The simulation of biological/biochemical networks was first attempted in the 1960s. Today a large collection of general modeling and simulation tools for biological circuits, both commercial and open source products, is available. In a community effort, development of a standardized model exchange language (Systems Biology Markup Language, SBML), a software broker for SBML-based communication between tools (Systems Biology Workbench, SBW), and software libraries and model repositories has ensured compatibility of currently ~80 modeling and simulation tools.

E. Measurement: Measurement of Noise and Variation

Most of the measurement techniques relevant to synthetic biology are single cell measurements. These techniques include fluorescent labeling of proteins (using GFP and other xFP's), indirect labeling of RNA using RNA-binding fluorescent proteins, and indirect binding of DNA using fluorescent proteins. Such measurements are performed using various microscopy techniques. In some cases, single molecules may be detected and counted. These techniques provide a much more detailed understanding of the cellular processes, in particular gene expression. For example, standard deviations and fluctuations may be measured. These data may be very useful for designing artificial networks at the cellular level.

III. THE FIRST SYNTHETIC LIVING THING

Synthetic biology achieved a major breakthrough in May 2010 with the announcement of creation of the first entirely synthetic life form by the J. Craig Venter Institute. This was the first report of an organism whose genes were made entirely from inanimate chemicals. It was a major technical feat requiring the accurate synthesis of a slightly modified genome of the bacterium *Mycoplasma mycoides*—all 1.08 million base pairs of it—followed by its successful insertion into a related species, *Mycoplasma capricolum*, and then the demonstration of replication of descendant cells exhibiting the characteristics of *M. mycoides*. The entire process of synthesis and assembly of the genome of Mycoplasma is summarized in Figure 73.2.

This was the first report where man-made DNA had booted up a synthetic cell. This initial breakthrough was achieved in bacteria because they are the simplest of all living organisms. For practical applications however, advances in synthetic biology are currently being made in organisms like single cell algae, whose cell structure and metabolism are more suitable for the production of large quantities of food and fuel.

In early 2008, the Venter Institute had announced the assembly of a synthetic *Mycoplasma genitalium* genome. After the publication of this report, the assumption was that the synthetic genome would be running cells in no time. It however took additional two and a half years of work for the researchers at the J. Craig Venter Institute to insert the artificial genetic material—chemically printed, synthesized and assembled—into cells that were then able to grow naturally.

IV. PRACTICAL APPLICATIONS OF SYNTHETIC BIOLOGY

Synthetic biology is a field with enormous scope and potential. In many ways its current status can be compared with the very early days in the development of the computer industry. It has the capacity to change quite fundamentally the way we approach certain key technologies, such as medicine and manufacturing. Potential applications of synthetic biology range very widely across scientific and engineering disciplines, from medicine to energy generation.

A. Synthetic Biology for Biofuels, Biomaterials and Pharmaceuticals

Constructing biosynthetic pathways

Selection of the right biosynthetic pathway is the first step towards engineering of living systems for the production of biofuels, biomaterials and pharmaceuticals. Selection of the ideal host organism for production of these compounds is the next step. Ideally the selected host should have innate ability of these biosynthetic activities. For biofuel production, for instance, certain microorganisms have evolved to be proficient in converting lignocellulosic material to ethanol, biobutanol and other biofuels. These native isolates possess unique catabolic activity, heightened tolerances for toxic materials and a host of enzymes designed to break down the lignocellulosic components. Unfortunately, these highly desired properties exist in pathways that are tightly regulated according to the host's evolved needs therefore may not be suitable in their native state for production scale. A longstanding challenge in metabolic and genetic engineering is determining whether to improve the host's production capacity or whether to transplant the desired genes or pathways into an industrial model host, such as *E. coli* or *S. cerevisiae*; these important considerations and trade-offs are reviewed elsewhere.

The example of the microbial production of biobutanol, a higher energy density alternative to ethanol, provides a useful glimpse into these design trade-offs. Butanol is converted naturally from acetyl-CoA by *Clostridium acetobutylicum*. However, it is produced in low yields and as a mixture with acetone and ethanol, so substantial cellular engineering of a microorganism for which standard molecular biology techniques do not apply is needed to produce usable amounts of butanol. Furthermore, importing the biosynthetic genes into an industrial microbial host can lead to metabolic imbalances. In an altogether different approach, Liao and

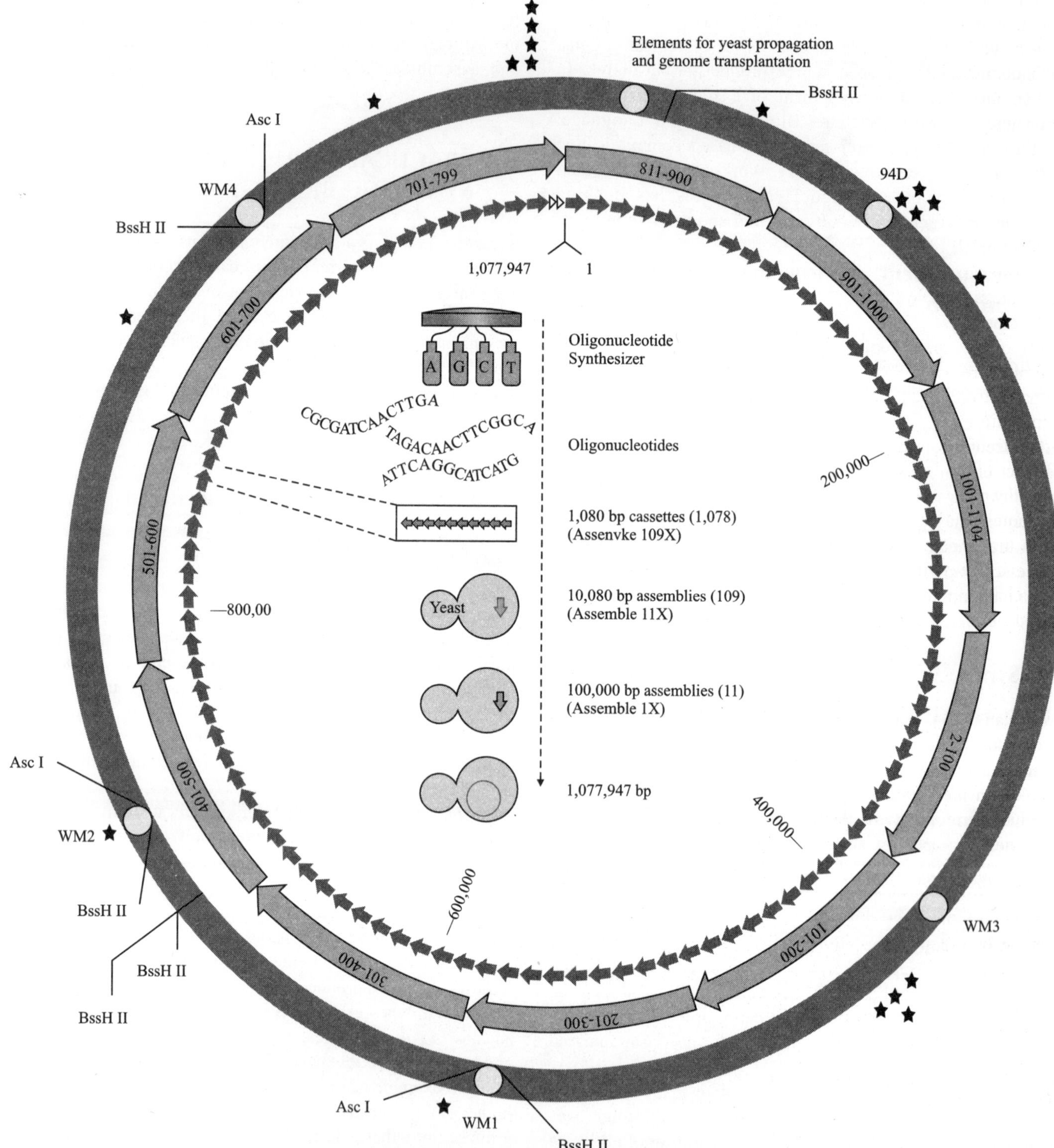

Figure 73.2 Schematic demonstrating the assembly of a synthetic M. mycoides genome in yeast. From Gibson, Venter et al., Science: 329 (2010): 52–56.

colleagues bypassed standard fermentation pathways and found that a broad set of the 2-keto acid intermediates of *E. coli* amino acid biosynthesis could be synthetically shunted to achieve high-yield production of butanol and other higher alcohols in two enzymatic steps.

In an exemplary illustration of this, Keasling and colleagues engineered the microbial production of precursors to the antimalarial drug artemisinin to industrial levels. There are now many such examples of the successful application of synthetic approaches to biosynthetic pathway construction—such approaches have been used in the microbial production of fatty-acid-derived fuels and chemicals (such as fatty esters, fatty alcohols and waxes), and polyketides made from mega enzymes that are encoded by very large synthetic gene clusters.

Optimizing pathway flux

After biosynthetic pathways have been constructed, the expression levels of all of the components need to be orchestrated to optimize metabolic flux and achieve high product titers. A standard approach is to drive the expression of pathway components with strong and exogenously tunable promoters, such as the Pltet, Pllac, and PBAD promoters from the *tet*, *lac* and *ara* operons of *E. coli*, respectively. To this end, there are ongoing synthetic biology efforts to create and characterize more reusable, biological control elements based on promoters for predictably tuning expression levels.

B. Synthetic Biology in Therapeutics

In a relatively short amount of time, synthetic biology has made promising strides in reshaping and streamlining the therapeutic spectrum. Indeed, the rational and model-guided construction of biological parts is enabling new therapeutic platforms, from the identification of disease mechanisms and drug targets to the production and delivery of small molecules.

Synthetic biology in Drug discovery

After a target is identified, whole-cell screening assays can be designed using synthetic biology strategies for drug discovery. As a demonstration of this approach, Fussenegger and colleagues developed a synthetic platform for screening small molecules that could potentiate a *Mycobacterium tuberculosis* antibiotic ethionamide, currently the last line of defence in the treatment of multidrug-resistant tuberculosis, depends on activation by the *M. tuberculosis* enzyme EthA for efficacy. However, due to transcriptional repression of *EthA* by the protein EthR, ethionamide-based therapy is often rendered ineffective. To address this problem, the researchers designed a synthetic mammalian gene circuit that featured an ethR-based trans activator of a reporter gene and used it to screen for and identify inhibitors that could abrogate resistance to ethionamide. Importantly, because the system is a cell basedassay, it intrinsically enriches for inhibitors that

are non-toxic and membrane-permeable to mammalian cells, which are key drug criteria as *M. tuberculosis* is an intracellular pathogen. This framework, in which drug discovery is applied to whole cells that have been engineered with circuits that highlight a pathogenic mechanism, could be extended to other diseases and phenotypes.

Synthetic biology devices have additionally been developed to serve as therapies themselves. Entire engineered viruses and organisms can be programmed to target specific pathogenic agents and pathological mechanisms. For instance, in two separate studies, researchers used engineered bacteriophages to combat antibiotic-resistant bacteria by endowing them with genetic mechanisms that target and thwart bacterial mechanisms for evading antibiotic action.

Synthetically engineered viruses and organisms that are able to sense and link their therapeutic activity to pathological cues may be useful in the treatment of cancer, in which current therapies often indiscriminately attack tumours and normal tissues. For instance, adenoviruses were programmed to couple their replication to the state of the p53 pathway in human cells. Normal p53production would result in inhibition of a crucial viral replication component, whereas a defunct p53 pathway, which is characteristic of tumour cells, would allow viral replication and cell killing.

Synthetic biology is also allowing rare and costly drugs to be manufactured more cost-effectively antibiotics are industrially produced from microbes and fungi, and are therefore widespread and cheap. Conversely, many other drugs are isolated from hosts that are not as amenable to large-scale production and are therefore costly and in short supply. Such drugs include the anti-malaria drug artemisinin and the anticancer drug taxol. Fortunately, global access to drugs is being enabled by hybrid synthetic biology and metabolic engineering strategies for the microbial production of rare natural products. In the case of artemisinin, there exist two biosynthetic pathways for the synthesis of the universal precursors to allisoprenoids, the large and diverse family of natural products of which *artemisinin* is a member. The native isoprenoid pathway found in *Escherichia coli* (the deoxyxylulose 5-phosphate (DXP) pathway) has been difficult to optimize, so instead researchers have synthetically constructed and tested the entire *Saccharomyces cerevisiae* mevalonate-dependent (MEV) pathway in *E. coli* in a piece-wise fashion (for example, by separating the 'top' and 'bottom' operons). The researchers initially used *E. coli* as a simple, orthogonal host platform to construct, debug and optimize the large metabolic pathway. They then linked the optimized heterologous pathway to a codon-optimized form of the plant terpene synthase *ADS* to funnel metabolic production to the specific terpene precursor to artemisinin. This work allowed them to build a full, optimized solution that could be ultimately and seamlessly deployed back into *S. cerevisiae* for cost-effective synthesis and purification of industrial quantities of the immediate drug precursor of artemisinin, FPP, farnesyl pyrophosphate.

CONCLUSION

For a long time, molecular biology has attempted to unravel the molecular mechanisms that are important in cellular function. This knowledge has been exploited by biotechnology to produce enzymes, chemicals and biopharmaceuticals. Now, synthetic biology has moved a step further by building novel biological entities on an ever more complex level for novel applications.

A. Risks and Rewards of Synthetic Biology

It is clear from the sections described above that synthetic biology is an emerging discipline with a huge potential and scope. It opens up the possibility of manipulating living systems and their component parts in a rational way, similar to the way in which engineers design machines, cars or planes. Although we are far away from this point, this is clearly where the field ultimately will lead us. We can expect huge benefits as a result, but also—as with any other potent advance in science—there are risks. It is obvious that genetic manipulation of organisms can be used, or can result by chance, in potentially dangerous modifications for human health or the environment. The possibility of designing a new virus or bacterium could be used by bioterrorists to create new resistant pathogenic strains or organisms, perhaps even engineered to attack genetically specific sub-populations. Thus, the combination of engineering with the possibility of synthesizing whole genomes could be problematic.

It will not be possible to eliminate the possibility of misuse of synthetic biology, any more than for other technologies. However, there are steps that can be taken to minimize risks. For example, controls and regulations can be imposed on 'parts suppliers'. In particular, companies that provide synthetic DNA sequences to order could check what they are making, and to whom they are supplying it. This will require a genomic databank of potential pathogenic microorganisms and viruses, toxic genes and gene circuits. It will probably be useful to set up an international committee that will explore the possible misuse of the technology and develop guidelines of how to prevent it. These guidelines should then be reflected in new laws that will regulate the exchange and access to materials and suppliers.

The basic techniques necessary for conducting some form of synthetic biology already exist and are publicly accessible; so indeed are the genome sequences of many pathogens. Thus, many of the risks of potential abuses exist already; and the tools of synthetic biology themselves offer the most powerful means of counteracting such threats. Insofar as they offer new and effective means of detecting and eliminating harmful biological agents, these tools will also be effective in developing defenses against existing, more conventional means of bioterrorism and bio warfare.

Quite aside from such issues, one can argue an ethical case for developing synthetic biology as a 'biotechnology that works' for the development of new drugs, particularly ones that might provide effective and affordable treatments for diseases such as malaria that present major health hazards and causes of fatalities in developing countries. In this respect, it is encouraging to note that the ethical and safety aspects of synthetic biology have already become an integral part of the discussions in the scientific community.

B. What is the Future of Synthetic Biology?

Synthetic biology strives to make molecular biology more like engineering—with predictable materials and parts that can be put together in predictable ways. As the synthetic cell demonstrates, scientists now have the tools to edit an existing genetic sequence on a computer, use DNA-synthesizing machines to create it in fragments, and stitch these together in the lab. The creation of the synthetic cell, at the J. Craig Venter Institute, hints at a future in which synthetic biologists can redesign living cells to perform whatever tasks they dream of. The future of synthetic biology hinges on the development of reliable means for connecting smaller functional circuits to realize higher-order networks with predictable behaviours.

SUMMARY

Synthetic Biology includes the broad redefinition and expansion of biotechnology, with the ultimate goals of being able to design and build engineered biological systems that process information, manipulate chemicals, fabricate materials and structures, produce energy, provide food, and maintain and enhance human health and our environment.

Synthetic biology will drive industry, research, education and employment in the life sciences in a way that might rival the computer industry's development during the 1970s to the 1990s. Due to the fundamental change in methodology that it entails for the modification of living organisms, synthetic biology may be able to fulfill many of the promises that traditional biotech is still struggling to fulfill, in such important areas as biomedicine, synthesis of biopharmaceuticals, sustainable chemical industry, environment and energy, production of smart materials and biomaterials, security: counter-bioterrorism

While traditional biotechnology has had some notable achievements in several of these areas, they have generally been slow and expensive to develop. A typical approach in bioengineering is to develop cells or molecular components with new functions using empirical, evolutionary processes that may involve screening of vast libraries of candidate systems and are hard to optimize. In essence, today's biotechnologist needs to master a very broad array of complex technologies in order to achieve a goal. By potentially re-organizing biotechnological development in line with the principles of synthetic biology, research & development are likely to proceed much faster and in a much more organized way. Introduction of design rules, separation of design and fabrication, adherence to standardized biological parts, and so on, are likely to aggressively tackle the problems encountered by the traditional empirical approach. Because of its rational, knowledge-based approach to biological design, synthetic biology will allow such goals to be attained more efficiently and economically.

SUGGESTIONS FOR FURTHER READING

Andrianantoandro E., Basu S., Karig D.K. and Weiss R. (2006), Synthetic Biology New Engineering Rules for an Emerging Discipline, *Mol. Syst. Biol.*, 2, 0028.

Benner S.A. and Sismour A.M. (2005), Synthetic Biology, *Nat. Rev. Genet.*, 6, 533–543.

Endy D., Foundations for Engineering Biology, *Nature*, 438(2005), 449–453.

Heinemann M. and Panke S. (2006), Synthetic Biology—Putting Engineering into Biology, *Bioinformatics*, 22, 2790–2799.

Khalil A.S. and Collins J. (2010), Synthetic Biology: Applications Come to Age, In: Nature Reviews (genetics), 11, 367–379.

74

Population Genetics

Partha P. Majumdaar

CONTENTS

I. POPULATION GENETICS: ITS HISTORY AND PERSPECTIVE

The birth of population genetics was concurrent with the rediscovery of Mendel at the turn of the twentieth century. The birth was not smooth. Francis Galton (half-cousin of Charles Darwin) initiated a controversy by emphasizing that Mendelism implied complete "neglect of ancestry". According to Mendel's laws, the genotypes of offspring are completely determined by the genotypes of the parents; individuals of previous generations (such as, grandparental genotypes) had no role to play in the determination of the genotypes of individuals of the present generation. Galton propounded his own "Law of Ancestral Heredity". Further, Galton stated that variation introduced by environmental factors should not be heritable. And, finally, Mendelian inheritance was particulate (i.e., discontinuous) and hence could not form the basis of inheritance of continuous characters, such as, height. Galton's views resulted in polarization of scientists who were thinking about inheritance and genetics, and two schools – with nearly opposing views – were formed; one was the "Biometrical School" led by Karl Pearson and W.F.R. Weldon, and the other was the "Mendelian School" led by William Bateson and R.C. Punnett. A superb description of the history of population genetics has been provided by Provine (1971).

The genius of Ronald A. Fisher brought an end to the war that raged between the two schools. Fisher, in 1918, showed that inheritance of continuous characters could be explained by Mendel's particulate theory. Thus, a synthesis of the arguments that were being made by members of the two schools was obtained, and this marks a landmark in the history of population genetics. Prior to this paper published in 1918, the only other paper that continues to be widely referred to is a letter to the editor that was published in the *Science* magazine in 1908 by the number theorist G.H. Hardy (the Hardy-Weinberg law, see section [2]; Weinberg was a German physician who also simultaneously discovered the law in 1908).

Fisher, Sewall Wright and J.B.S. Haldane laid firm foundations of population genetics in a series of papers that the trio published from about 1925 to about 1940. This foundation was primarily theoretical, but captured the essence of many evolutionary observations. The theoretical foundation was expanded and grounded better by Gustave Malecot and Motoo Kimura in a series of papers published between 1948 and 1955. The first direct genetic observations, especially and the biochemical/molecular genetic level, were obtained by Harry Harris, Richard Lewontin and Jack Hubby in the mid-1960s, which in some ways laid the foundations of empirical population genetics, especially of understanding the genetic structures of contemporary populations and drawing inferences on their evolutionary histories.

Population genetics essentially provides a theoretical and mathematical framework of understanding the genetic basis of population structuring, whether at the macro-level of species or at the micro-level of extant populations, and inferring possible evolutionary mechanisms by which such structuring may have occurred.

II. CONSEQUENCES OF RANDOM MATING IN A POPULATION AND DEVIATIONS

Unless stated otherwise, we with deal will alleles at autosomal loci in diploid organisms, practicing sexual reproduction. In sexually reproducing organisms, two gametes carrying alleles from each of the two parents unite to form the genotype of a child. *Random mating* refers to the situation when mating partners are drawn randomly from a population without regard to their genetic backgrounds. In other words, any two individuals of opposite sexes are equally likely to mate as any other pair of individuals.

Let us consider an autosomal locus with two alleles (*A* and *a*). At this locus, therefore, an individual can be of one of three possible genotypes, namely, *AA* (A allele homozygote), Aa (Heterozygote) and aa (a allele homozygote). We assume that there are a very large number of individuals in the population; technically, we assume that the population size is infinite. In order to draw inferences on the characteristics of this population, we rely on a representative sample of individuals drawn from this population, which we shall denote as N. We denote the number of individuals of genotype AA as N(AA), and similarly N(Aa) and N(aa).

Obviously,

$$N(AA) + N(Aa) + N(aa) = N$$

Although there are three different genotypes, there are only two different alleles (A and a). The first step in being able to derive a quantitative framework for population genetics, is to be able to estimate *allele frequency* from this data. (Note: What is incorrectly referred to as allele *frequency* in population genetics is really the *proportion* of an allele in the population).

The estimation of allele frequency (proportion) is intuitive; we count the number of *A* (or *a*) alleles in the population, and divide it by the total number of alleles present in the population. (Note: This method of allele frequency estimation is called the 'gene counting method', we can arrive at the same estimates by using more refined mathematical tools on a variety of models.)

Let us denote the '*A*' allele frequency as p_A. In order to estimate p_A, we need to count the number of *A* alleles in the population. Each of the $N(Aa)$ individual carries one copy of the A allele, whereas each of the $N(AA)$ individuals carry two copies of the A allele. Hence, the total number of A alleles in the group is:

$$[2 \times N(AA) + N(Aa)]$$

Since there are *N* individuals in the population, and each individual has two alleles, the total number of alleles in the population is 2*N*. Hence, the proportion of allele *A*, allele frequency of *A*, is given by:

$$p_A = \frac{[2 \times N(AA) + N(Aa)]}{2 \times N}$$

The genotype proportions can similarly be estimated as:

$$p_{AA} = \frac{N(AA)}{N}, \quad p_{Aa} = \frac{N(Aa)}{N} \text{ and } p_{aa} = \frac{N(aa)}{N}$$

Let us denote p_{AA}, p_{Aa} and p_{aa} as u, v and $w(u + v + w = 1)$, respectively.

The formal definition of random mating is that, any individual of a randomly mating population is equally likely to mate and produce offspring with any individual (of the other sex) from the same population. Hence, in a large random-mating population, the probability that an individual of the AA genotype mates with an individual of the AA genotype is $(p_{AA} \times p_{AA})$. This specific mating combination, in turn will always yield an AA offspring. Thus, given (or conditional on) that both parents are of genotype AA, the probability of the genotype of an offspring of this mating is equal to 1. Table 74.1 provides the possible mating types, their probabilities and the probabilities of possible genotypes arising out of each mating type.

TABLE 74.1

Mating type	Probability of mating	Probability (Offspring genotype/ Parental Mating type)		
		AA	Aa	Aa
$AA \times AA$	u^2	1	0	0
$AA \times Aa^{(1)}$	$2uv$	1/2	1/2	0
$AA \times aa^{(2)}$	$2uw$	0	1	0
$Aa \times Aa$	v^2	1/4	1/2	1/4
$Aa \times aa^{(2)}$	$2vw$	0	1/2	1/2
$aa \times aa$	w^2	0	0	1

[1] The mating AA × Aa and its reciprocal mating type *Aa* × *AA* have equal probabilities *uv*. Hence, these mating types have together been denoted as *AA* × *Aa* and the two probabilities have been added to yield the total mating probability as 2*uv*.
[2] Same argument and notation as in [1].

Hence, in the next generation, the probability that an offspring will be of genotype AA is

$$(1) \times (u^2) + (1/2) \times (2uv) + (1/4) \times v^2 = (u + v/2)^2 = (p_A)^2$$

Similarly, in the next generation, the probability that an offspring will be of genotype aa is

$$(w + v/2)^2 = (p_a)^2$$

The probability that an offspring will be of genotype Aa is

$$(1/2) \times (2uv) + 1 \times (2uw) + (1/2) \times v^2 + (1/2) \times (2vw)$$
$$= 2(u + v/2)(w + v/2) = 2p_A p_a$$

Note here that we have not specified the initial values of *u*, *v* and *w*. But in a single generation of random mating the frequencies of the genotypes (AA, Aa and aa) will attain the proportions $(p_A^2, 2p_A p_a, p_a^2)$ in the population. If this population continues to practice random mating, the equilibrium will be maintained, or in other words, the genotype frequencies will continue to be in the proportion $(p_A^2, 2p_A p_a, p_a^2)$. This equilibrium is also known as the Hardy-Weinberg Equilibrium (HWE) and the proportions are called the HW proportions. The HW proportions can easily be generalized when there are multiple alleles at a locus.

We emphasize that the genotypic proportions at a locus in an equilibrium population are solely determined by the allele frequencies at that locus. Thus, in populations with initial genotype frequencies of (*AA, Aa, aa*) = (0.1, 0.2, 0.7) or (0.2, 0, 0.8) or (0, 0.4, 0.6), become (0.4, 0.32, 0.64) after a single generation of random mating. This is because, in each of these populations the initial frequencies are 0.2 and 0.8 of alleles A and a, respectively.

It may be noted that, for ease of calculations, we have assumed that the generations are non-overlapping, that is, no mating takes place between two individuals belonging to two different generations. Similar results can be derived for more complex mating scenarios, such as, populations with overlapping generations, but these derivations are beyond the scope of this chapter.

Most organisms, both plants and animals, mate at random. Many assumptions were implicitly made in establishing the principle of HWE and in deriving the HW proportions. These are:

1. The population is assumed to be very large ('infinite'). [In a small ('finite') population, allele and genotype frequencies do not attain equilibrium values and stochastically vary due to a phenomenon called random genetic drift.]
2. No selection or mutation is operating at the locus.
3. The population is not affected by migration (either inward or outward).
4. The allele frequency is the same among both males and females.

If any of these assumptions is violated, then HWE may not hold. We also note that HWE holds for an autosomal locus, but does not hold with respect to a locus that is located on a sex chromosome. Genotype proportions across generations at a locus on the Y-chromosome are trivially determined. Genotype proportions at a locus on the X-chromosome attain equilibrium after an infinite number of generations of random mating; in other words, equilibrium is attained asymptotically. C.C. Li's (1976) textbook may be consulted for details.

We note that there is preferential mate choice with respect to many social and other characteristics. This is called assortative mating. However, when these characteristics are not associated with genotypes at a locus, from genetic considerations mating may still be considered to be random. When in fact assortative mating takes place with respect to physical characteristics, such as skin color, only a small number of loci that determine or are associated with skin color deviate from random mating; with respect to other loci in the human genome, HW proportions hold.

We also note that detection of deviations of genotype frequencies from those expected under HWE requires a large sample size. In other words, even if a population is not practicing random mating, it is not easy to identify this from genotype data unless a large number of individuals are genotyped for that locus. However, systematic genotyping errors result in large deviations of genotype frequencies from HW proportions; therefore, a test of HWE is currently done even when sample sizes are moderate to small to detect genotyping errors as opposed to non-random mating.

III. ASSEMBLY OF GENOMIC VARIANTS

A. Types of Variation

The DNA sequences of any two individuals are not exactly the same. The most abundant are differences at single nucleotide positions, called *single nucleotide variants* (SNVs). At a SNV site, if the frequency of the rarer allele reaches a frequency of 5% in a population, it is said to be a *single nucleotide polymorphism* (SNP). Variations involving consecutive nucleotides are also frequent. Genomes of organisms contain stretches of dinucleotides or trinucleotides that are tandemly repeated. The number of repeats is often variable across individuals. These are called *short tandem repeat* (STR) *variants or polymorphisms*. Sometimes, one or more consecutive nucleotides are found to be inserted or deleted from the genomes of some individuals. These are called *Insertion/Deletion* (InDel) *variants or polymorphisms*. Finally, there are *structural variations* observed across individuals, that involve large chunks of DNA. Structural variants, such as translocations (a portion of DNA from one chromosome moves to another chromosome) or large deletions, usually result in diseases in humans.

B. How these Variants Arise and How They are Maintained

Most SNVs, including small insertions and deletions, arise by random mutation. On the other hand, short tandem repeat variants arise mainly by recombination. Larger variants arise by various mechanisms, which are beyond the scope of this discourse. Variations can be introduced into a population of individuals by migration from another population, if the organism is structured as discrete populations, most matings taking place among individuals who belong to the same population than between two different populations.

If a new variant does not confer either a selective advantage or a selective disadvantage to an individual compared to all pre-existing variants at that locus, then the new variant is said to be selectively neutral. The fate of the variant over generations is largely dictated by chance. The smaller the population size, the stronger is the effect of the chance factor. Consider a locus with two alleles A and a. Consider a population of 1000 individuals in which 250 individuals are of genotype AA, 500 are Aa and 250 are aa. The frequency of allele A in this population is 0.5. Suppose, in the next generation the population size remains the same, but the number of individuals of genotype AA is 260, Aa is 500 and aa is 240. Then, the frequency of allele A in this generation is 0.52; a change of 0.02 from the previous generation just by chance. In the next generation, the frequency of A can change to 0.48, in the next generation to 0.51, etc. In other words, the frequency can drift from one generation to another. If in some generation, the frequency becomes 1 or 0, then no further change is possible. Most newly arising mutations are selectively neutral; their fate is dictated simply by chance.

On the other hand, if the newly arising mutant confers a selective advantage to individuals, then its frequency in the population will quickly rise and will eventually become 1. If it is disadvantageous, it will be quickly lost from the population; that is, its frequency will become 0. The nature of advantage or disadvantage needs to be modeled in order to gain a clear understanding of the rate of change over time. Under natural selection, one simple way in which a mutant can co-exist for a long time is when the mutant confers a selective advantage in the heterozygote. A common example is the persistence of the sickle cell allele in many populations, even though sickle cell homozygotes die early as a result of severe anemia and resultant complications. The sickle cell allele persists in populations in which there is malaria. In a malarial environment, individuals who have only one copy of the sickle cell allele (heterozygotes) are protected against malaria, and hence the heterozygotes enjoy a selective advantage. This situation may be modeled as: fitnesses of Genotypes AA, Aa and aa are, respectively, 1–s, 1 and 1–t, where s and t ($0 < s, t < 1$) are selection coefficients. The change of allele frequency of A in a single generation can then be shown to equal: pq[qt – ps]/w, where w is the average fitness of the genotypes and p is the frequency of the allele a in the present generation. The above equation shows that if the numerator is zero, then there will be no change in allele frequencies over generations. The non-trivial condition when this may be achieved is qt = ps, or when p = t/(s + t) or equivalently when q = s/(s + t). Therefore, if in any generation, the frequency of the A allele attains t/(s + t), an equilibrium is achieved, and the allele frequencies will not change in subsequent generations.

C. Geographical Structuring

When individuals in a population do not form a single random-mating population, but instead comprise several isolated random-mating subpopulations, then the population is said to be sub-structured. Geographical distance between habitats of subpopulations is a major reason for subpopulations to be isolated. A new subpopulation is usually formed from an existing subpopulation, often by emigration of a small number of individuals from the existing subpopulation to a new habitat. The small number of individuals can carry only a small number of genes with them, and the genetic composition of the small subset of individuals is often quite different from the genetic composition of the remaining individuals. These individuals who form a new subpopulation are called 'founders'. Since this subset of individuals is numerically small, their allele frequencies will drift rather strongly and will quickly show large differences compared to those in the parental population. This is called 'founder effect'. Because of founder effect, often frequencies of many single-gene diseases are found to be high in recently founded human populations.

Geographically sub-structured populations show genetic sub-structuring. The extent of sub-structuring is measured as follows:

Consider a biallelic locus, with alleles A and a. Suppose we consider k subpopulations. Suppose the frequencies of alleles A and a in the ith subpopulation are, respectively, p_i and $q_i (i = 1, 2, …, k)$. Let p and q denote the frequencies of these two alleles, averaged over the k subpopulations. Then, the extent of sub-structuring is measured by:

$$F_{ST} = \frac{s_p^2}{(pq)}$$

where s_p^2 = variance in the frequency of allele A among the k subpopulations.

D. Genomic Structuring

Under random mating, the alleles at any locus are combined into genotypes by random union of gametes, as HWE dictates. If we mechanically extend this concept to two bi-allelic loci – locus-1 with two alleles A and a and locus-2 with two alleles B and b – we get the following two-locus genotype proportions:

			Locus-1	
		Allele	*A*	*a*
	Allele	*Frequency*	p_A	p_a
Locus-2	B	p_B	Gamete *AB* $p_A p_B$	Gamete *aB* $p_a p_B$
	b	p_b	Gamete *Ab* $p_A p_b$	Gamete *Ab* $p_a p_b$

The above frequencies—using the product rule of allele frequencies—are valid when the alleles at the two loci associate randomly, in which case the two loci are said to be in linkage equilibrium. Under linkage equilibrium, the gametic frequencies are:

AB : $p_A \times p_B$
Ab : $p_A \times p_b$
aB : $p_a \times p_B$
ab : $p_a \times p_b$

However, there is no reason why the alleles at the two loci cannot associate non-randomly (linkage disequilibrium [LD]). To understand how non-random association between alleles at two loci can come about, we need to appreciate that transmission of one allele at a loci may not be independent of the transmission of an allele at the other loci, especially if they are in close proximity. This is because whole chromosomes are transmitted from parents to offspring, which means that alleles on the same chromosome at different loci will be transmitted together. Recombination breaks chromosomes and thus can cause alleles at two loci not to be associated with one another. The closer two loci are on a chromosome the stronger is the non-random association. It can be mathematically proven (*Li*, 1976) that even if two loci are physically very close, alleles at these loci will eventually show random association, although this can take a long time to attain.

If we denote by D the extent of deviation of gametic frequencies from that expected under linkage equilibrium, then the gametic frequencies can be expressed as:

			Locus-1	
		Allele	*A*	*a*
	Allele	*Frequency*	p_A	P_a
Locus-2	B	p_B	$h(AB) =$ $p_A p_B + D$	$h(aB) =$ $p_a p_B - D$
	b	p_b	$h(Ab) =$ $p_A p_b - D$	$h(ab) =$ $p_a p_b + D$

Linkage disequilibrium (LD) is often quantified using statistics of association between the allelic states at pairs of loci. Consider two biallelic loci, with alleles A/a and B/b, respectively. Let p_A and p_B denote the frequencies of alleles A *and B*, respectively, and h_{AB} the frequency of the AB haplotype. The simplest measure of *LD* is:

$$D = h_{AB} - p_A \times p_B.$$

Recombination that takes place in each generation erodes *LD*. If Dn denotes the *LD* between two loci in generation n, then it can be shown that

$$D_n = (1 - r)D_{n-1} = (1 - r)^2 D_{n-2} = \ldots = (1 - r)^t D_0,$$

where r denotes the recombination fraction between the two loci and D_0 denotes the initial value of *LD* between the two loci.

It may be noted that D is dependent on the frequencies of the alleles, which is undesirable. A normalized measure is:

$$D' = D/D_{max}$$

where $D_{max} = \min (p_A p_B, p_a p_b)$, if $D < 0$; or, $\min (p_A p_b, p_a p_B)$, if $D > 0$, which is a normalized measure of D, normalized by the theoretical maximum for the observed allele frequencies. $D' = 0$ indicates that the two loci are independent, while $D' = 1$ indicates that the two loci are completely dependent.

Another alternative measure, which is also adjusted for allele frequencies, is:

$$r^2 = D^2/(p_A p_a p_B p_b),$$

the squared correlation coefficient between the two loci.

Both measures, D' and r^2, are symmetric, in the sense that the identities of the associated alleles at the two loci are irrelevant, or, in the context of disease gene mapping, which locus is the disease locus and which is the marker locus. All these measures are closely related to each other and to the standard χ^2–statistic for the 2×2 table shown above. However, these measures enjoy different properties; detailed discussion on LD can be found in Devlin and Risch (1995).

IV. RECONSTRUCTION OF GENETIC RELATIONSHIPS AMONG POPULATIONS

Individuals belonging to the same population differ genetically, just as individuals belonging to two different populations also genetically differ. Therefore, there are genetic differences among individuals both within and between populations. It is intuitively expected that two populations that are genetically more similar in the present time than other pairs of populations are also evolutionarily closer; in other words, these two populations are expected to have evolved from a common ancestor more recently. Therefore, conceptually, two measures are required to study genetic relationships of populations: (a) genetic diversity within a population, and (b) genetic distance between two populations.

If p_i denotes the frequency of the ith allele at a locus in a population X, then the probability that two randomly chosen alleles from this population will be identical is $J_X = \sum_i p_i^2$, the summation being taken over all alleles at the locus. The probability of non-identity is, therefore, $H_X = 1 - J_X$, which is a measure of gene diversity among individuals in the population. When multiple loci are considered, the gene diversity is usually expressed as the average of the gene diversities at individual loci. This measure is due to Nei (1973).

In a similar vein, if the frequency of the ith allele at a locus in two populations X and Y are denoted by p_i and q_i, respectively, then the probability that two randomly chosen alleles, one from each population, will be identical is $J_{XY} = \Sigma p_i q_i$, the summation being taken over all alleles at the locus. Nei (1972) defined the standard genetic distance as

$$D_a = -\ln\left(\frac{J_{XY}}{\sqrt{J_X J_Y}} \right)$$

We provide an example based on a study by Chakrabarti that attempted to estimate genomic relationships among three morphologically Mongoloid, Sino-Indian speaking tribal groups (Toto, Mizo, Tharu), and a morphologically proto-Australoid, Austro-Asiatic speaking tribal group (Ho). The Ho were included to serve as an anthropological and linguistic, and possibly genetic, 'outgroup' relative to the three other groups. Based on data of 25 polymorphic DNA markers, gene diversities and Nei's genetic distances were estimated (Table 74.2).

The genetic distances were then used to estimate genomic relationships (Figure 74.1) among the tribal groups using a clustering algorithm (neighbor-joining; description of which is beyond the scope of this chapter). The *a priori* expectation, as stated by the authors, was that the Sino-Indian speaking, Mongoloid groups would bequite similar and distinct from the Austro-Asiatic speaking, proto-Australoid group. The results, however, yielded results contrary to these expectations. The results are in conformity with some earlier observations of anthropologists that the Tharu may have been derived from admixture between a proto-Australoid tribal population of India and a Mongoloid tribal population of Nepal.

TABLE 74.2   Gene diversities (on diagonal) and genetic distances among five tribal populations of India

	Katharia Tharu	*Rana Tharu*	*Toto*	*Mizo*	*Ho*
Katharia Tharu	0.467 ± 0.041				
Rana Tharu	0.0164	0.462 ± 0.038			
Toto	0.0773	0.0517	0.368 ± 0.046		
Mizo	0.0213	0.0172	0.0536	0.448 ± 0.032	
Ho	0.0736	0.0634	0.082	0.0576	0.379 ± 0.040

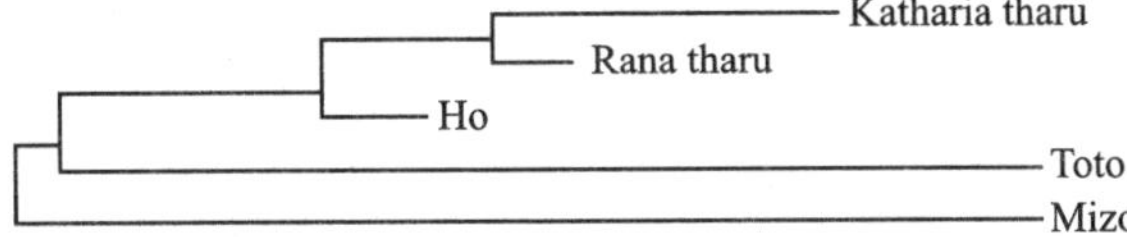

Figure 74.1   Genomic relationships among five tribal populations of India based on data of 25 genomic markers, using Nei's standard genetic distance and neighbor-joining algorithm.

V.   POPULATION GENETICS OF DNA SEQUENCES

With the explosion of DNA sequencing technologies and data, the turn of this century witnessed a major development in population genetics (introduction of the coalescent) that permitted a framework and a series of methodologies for the study of DNA sequences.

The coalescent is a stochastic process that describes the history of recombination and coalescence in a sample of n homologous sequences. The coalescent is best understood if we consider a set of n homologous haploid DNA sequences. The theory of coalescent is derived from the simple fact that all individuals have parents, and some of them have the same parents. So all sequences (or haploid individuals) are derived from or are offspring of haploid individuals of the past. If we take a set of haploid individuals, and assume that generations are discrete, and trace them back in time; after a few generations two of them would have the same parent or a common ancestor. This event of finding or coalescing to a common ancestor is called a common ancestor event or a coalescence event. Thus, if we start with a sample of n individuals (haploid) from a population of size N, then after the first coalescent event, we would be left with $(n-1)$ individuals. Denote by T_n the time/number of generations (*a random variable*) when the first coalescent event had happened. This framework, if extended recursively, would mean that the set of $(n-1)$ ancestors T_n generations back in time would have the second coalescence event after $T_{(n-1)}$ generations. We tacitly assume here that if N is much larger than n, only two individuals can coalesce in a particular generation. Eventually we will be left with 2 individuals to coalesce, and according to our nomenclature they would coalesce after T_2 generations (Figure 74.2). The individual to whom all these n homologous chromosomes coalesce is called the Most Recent Common Ancestor (MRCA) and the time taken for that to happen is called the Time to the Most Recent Common Ancestor (TMRCA). These recursive coalescence events would generate a genealogy, much like a phylogenetic tree.

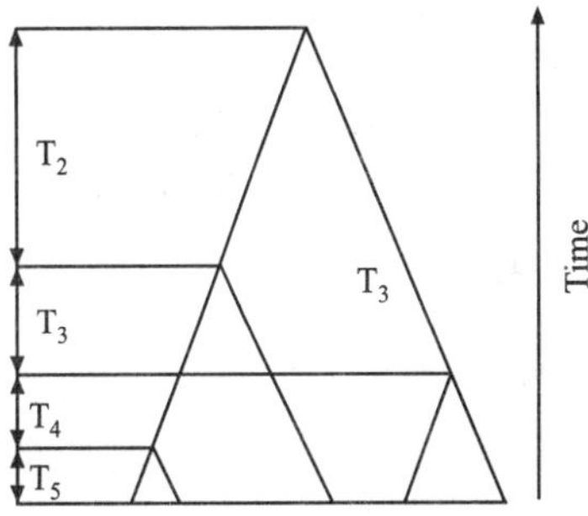

Figure 74.2   A framework for the population-genetic study of 5 DNA sequences using coalescent methods.

As is evident from Figure 1,

$$\text{TMRCA} = T_n + T_{(n-1)} + \ldots + T_2$$

And also the expected time for the MRCA event to happen is

$$E(\text{TMRCA}) = E(T_n) + E(T_{(n-1)}) + \ldots + E(T_2).$$

It can be shown that under assumptions of neutrality and constant population size,

$$E(T_n) = \frac{N}{{}^nC_2}$$

$$E(T_{(n-1)}) = \frac{N}{{}^{(n-1)}C_2}$$

Similarly,

$$E(T_2) = N$$

Thus,

$$E(\text{TMRCA}) = E(T_n) + E(T_{(n-1)}) + \ldots + E(T_2)$$

$$= N\left[\frac{1}{{}^nC_2} + \frac{1}{{}^{(n-1)}C_2} + \cdots + 1\right]$$

For relatively large n, the sum can be shown to be nearly equal to 2. Hence the expected time that a set of n homologous chromosomes would take to coalesce to a single MRCA is approximately 2N. This result has enormous implication in our ability to date past historical events from sequence data. Random neutral mutations, at the rate of μ per generation, are superimposed on the genealogy.

Note 1: E(T$_2$) is approximately E(TMRCA)/2, or half of the expected time to the MRCA is the time when there were only 2 individuals left to coalesce. This would mean that a small number of individuals long ago has left their genetic signatures on most individuals today.

Note 2: To translate these results from haploid to small diploid genetic segments without recombination we can replace N by 2N. For coalescent with recombination, however, we have to resort to a much complex graph called the ancestral recombination graph.

Estimating the time to the most recent common ancestor (TMRCA) of modern humans is of great interest in the study of human evolution and disease. The availability of human mitochondrial DNA sequences allowed the estimation of the age of the common ancestor of maternal lineages, while the recent DNA samples from human Y chromosome provide information about the evolutionary history of paternal lineages. Essential to the inference of the common ancestry of humans are proper estimators of the age of the MRCA of a sample. Templeton (1993) proposed an estimator based on the maximum number (k$_{max}$) of nucleotide differences between two sequences in the sample. Templeton's method has been used to estimate the age of the MRCA of mitochondria and that of the Y chromosome.

Suppose we denote the age (in generations) of the MRCA of a sample by T. Tajima (1983) showed that when the number (n) of DNA sequences in a sample is 2, the mean and variance of T conditional on the number (π) of nucleotide differences between the two sequences are:

$$E\left(\frac{T}{\pi}\right) = \frac{\theta(1+\pi)}{2\mu(1+\theta)} \quad \text{and} \quad V\left(\frac{T}{\pi}\right) \approx \frac{\theta^2(1+\pi)}{4\mu^2(1+\theta)^2}$$

where $\theta = 2N\mu$, N is the effective population size, and μ is the mutation rate per sequence per generation. Templeton (1993) proposed to estimate T by replacing π in equation with k_{max}, the maximum number of nucleotide differences between two sequences in the sample, i.e., the number of nucleotide

differences between the two most divergent sequences in the sample. Templeton also suggested using a gamma distribution to obtain the confidence interval of T.

Hammer (1995) examined a sample of 16 Y-chromosomal DNA sequences each of length 2,654 bp and found $k_{max} = 3$. Hammer assumed the mutation rate to be 1.9×10^{-9} per site per year. Taking 20 years as one human generation yields

$$k = 1.9 \times 10^{-9} \times 2,654 \times 20 = 1.0 \times 10^{-4}.$$

Using Templeton's method and taking N = 4,900, Hammer obtained 188,000 years as the estimate of the age of the MRCA of the sample, and the 95% confidence interval of the age was from 51,000 to 411,000 years.

The above example is a simple method of estimating TMRCA from haploid sequence data. Methods for estimation of TMRCA from whole-genome sequence data have recently been developed by Li and Durbin (2011) that take into account recombination and local ancestry differences.

VI. SOME CONCLUDING REMARKS

With technological improvements and the increasing availability of DNA sequencers, large datasets comprising millions of genomic variants is becoming the norm in population genetics. Population genetic analyses of such large datasets require efficient computational methods and widely applicable models. While such models and methods are being devised, much more work is required. These large datasets are revealing genomic structures and phenomena that need to be taken into account in population genetic modeling. Historically, after the founding of population genetics by Fisher, Haldane and Wright, the neutral theory propounded by Motoo Kimura and James Crow was the major milestone. The coalescent of Kingman provided the next major quantum jump in population genetics. The milestone that is now awaited is a generalized model, with a minimal set of assumptions, of population genetics at the whole-genome level.

SUGGESTIONS FOR FURTHER READING

Crow J.F. and Kimura M. (1970), *An Introduction to Population Genetics Theory*, Harper and Row, USA.

Fisher R.A. (1918), The Correlation between Relatives on the Supposition of Mendelian Inheritance, *Transactions of the Royal Society of Edinburgh*, 52:399–433.

Fisher R.A. (1930), The Genetical Theory of Natural Selection, Oxford: Clarendon.

Haldane J.B.S. (1924), A Mathematical Theory of Natural and Artificial Selection. *Transactions of the Cambridge Philosophical Society*, 23:19–41. (*This was followed by nine more papers with the same title and various subtitles that were published in the Proceedings of the Cambridge Philosophical Society between 1924 and 1934.*)

Hammer M.E. (1995), A Recent Common Ancestry for Human Y Chromosomes, *Nature*, 378:376–378.

Harris H. (1966), Enzyme Polymorphism in Man, *Proceedings of The Royal Society, London*, Series B., 164:298–310.

Hartl D.L. and Clark A.G. (2007), *Principles of Population Genetics*, Sinauer: USA.

Hedrick P. (2011), *Genetics of Populations*, Jones and Bartlett: Sudbury, USA.

Hudson R.R. (1990), Gene Genealogies and the Coalescent Process. In: *Oxford Surveys in Evolutionary Biology*, Vol. 7. (Futuyma D.J. and Antonovics J., Eds.). Oxford: Oxford University Press; 1–44.

Kimura K. (1983), The Neutral Theory of Molecular Evolution, Columbia University Press: USA.

Li C.C. (1976), *First Course in Population Genetics*. California: Boxwood Press.

Li H. and Durbin R. (2011), Inference of Human Population History from Individual Whole-genome Sequences, *Nature,* 475:493–496.

Male´cot G. (1948), *Les Mathe´matiques de l'He´re´dite´*. Paris: Masson. (Extended Translation in The Mathematics of Heredity, San Francisco: WH Freeman, 1969.)

Nei M. (1972), Genetic Distance between Populations, *American Naturalist,* 106:283–92.

Nei M. (1987), *Molecular Evolutionary Genetics*, Columbia University Press: USA.

Provine W.B. (1971), *The Origins of Theoretical Population Genetics*, Chicago: University of Chicago Press.

Templeton A.R. (1993), The Eve Hypotheses: A Genetic Critique and Reanalysis, *American Anthropologist,* 95:51–72.

Tajima F. (1983), Evolutionary Relationship of DNA Sequences in Finite Populations, *Genetics,* 105:437–460.

Wright S. (1931), Evolution in Mendelian Populations, *Genetics,*16:97–159.

Wright S. (1943) Isolation by Distance, *Genetics,* 28:114–138.

Wright S. (1951), The Genetical Structure of Populations, *Annals of Eugenics,* 15:323–354.

SECTION IX

CELL REPLICATION AND CANCER

Elegant mechanisms regulate the cell division, its stasis, and its demise. The disruption of this highly regulated process leads to cancer, which today has become a major cause of death all over the world.

A pertinent question is 'what differentiates stem cells from cancer cells'. Both multiply, the former (stem cells) is useful to regenerate the 'wanted cells'. They differentiate and stop multiplying, whereas cancer cells lose such 'control' signals.

75

Regulation of Cell Cycle and Cell Death

Chandrima Shaha and Ashish Kumar

CONTENTS

I. INTRODUCTION

A. Importance of Cell Death

The body of a multicellular organism consists of trillions of cells. Therefore, a perfect balance between cell proliferation and cell death is necessary for optimum tissue functioning. An imbalance can result in either unwanted cell death or growth that can lead to a variety of diseases. While unlimited cell growth can result in cancer, excessive cell death in the brain can cause neurological diseases. More than normal cell death in the reproductive system can cause infertility. Cell death during development is also essential to provide shape to the body and abnormal cell death during development of an embryo can cause abnormalities. Cells infected with virus or bacteria also needs to be eliminated before they can cause further infection. From the above, it is obvious that cell death is a vital component of our existence and knowledge about regulation of cell death would aid in formulating therapeutics for diseases.

B. Types of Cell Death

Approximately one billion cells die in one normal day in our body to maintain a balance with the rapidly proliferating cells. Cell death can occur via several processes, the primary ones are necrosis, apoptosis, autophagy and entosis. Necrosis is caused by very severe stress from external factors such as infection, toxins or trauma. This form of death can elicit an immune response but death by apoptosis effected by a cellular program called programmed cell death (PCD) is a type of death that does not cause cell lysis and therefore do not initiate an inflammatory reaction. PCD expresses a phenotype that is characterized by a bunch of very typical characters termed as apoptotic phenotype. Apoptosis is a highly conserved and regulated process that is responsible for eliminating unwanted cells or damaged cells from multicellular organisms during embryonic development and cells damaged by stress. Autophagy involves degradation of cell's own components through the machinery involving lysosomes and is a tightly-regulated process that plays a normal part in cell growth, development and homeostasis. Entosis is a death process which occurs due to a neighbouring cell eating up a particular cell when the cell needs to be eliminated. Cell cycle control maintains fidelity of cells by coordinated activity of its components like cyclin dependent kinases, checkpoint controls and repair pathways. Since apoptosis is the most prominent death process and is most widely studied, we will primarily address this process in this chapter.

II.   APOPTOSIS: AN ENERGY CONSUMING SUICIDAL PROCESS

Apoptosis is a form of programmed cell death mediated by intrinsic and extrinsic signals. The term 'apoptosis', coined by Kerr and colleagues refers to a system of cell death marked by cytoplasmic blebbing, chromatin condensation, nuclear fragmentation, and cell shrinkage. All these characteristics makes the cell more amenable to engulfment and clearance by phagocytic cells in the body. These phenotypic characters are preceded by biochemical changes like high molecular weight DNA fragmentation and formation of an oligonucleosomal ladder, externalization of phosphatidylserine from the inner surface of the plasma membrane and proteolytic cleavage of a variety of intracellular substrates. This form of death plays a major role during the normal development and homeostasis in multicellular organisms. It is known that genes that control apoptosis have a profound effect on the malignant phenotype of a cell. The oncogenic mutations that disrupt apoptosis resulting in lesser cell death leads to cancer. Most cytotoxic anticancer agents induce apoptosis and agents that cause highest apoptotic death in dividing cells without much side effects are preferred as best therapeutic agents.

A.   Morphological Features of Apoptosis

Apoptotic features expressed in a cell include cell shrinkage, loss of contact with its neighbouring cells, condensation of chromatin followed by nuclear and DNA fragmentation, plasma membrane blebbing and cell fragmentation into compact membrane-enclosed structures called 'apoptotic bodies'. The apoptotic bodies are engulfed by phagocytes and are removed by specialized mechanisms of corpse disposal without causing an inflammatory response. The above morphological changes are consequences of characteristic molecular and biochemical events occurring in an apoptotic cell, most notably the activation of proteolytic enzymes called caspases.

B.   Caspases

Caspases, a group of cysteine proteases are essential for programmed cell death in a variety of species. At least 14 caspases have been identified in mammals (caspases 1–14). These enzymes recognize tetrapeptide motifs and cleave their substrates on the carboxyl side of an aspartate residue. Individual caspases have distinct substrate specificities that are determined by the pattern of amino acids upstream of the cleavage site. So-called initiator caspases (e.g., caspase-8 and caspase-9) or long prodomain caspases start an avalanche of increasing caspase activity by processing and activating effector or short prodomain caspases (caspase-3,7). The long prodomain caspases undergo proximity induced cleavage and gets activated to cleave the short prodomain caspases. Effector caspases are responsible for the cleavage of cellular substrates like the cytoskeletal structures and other proteins to induce the formation of apoptotic bodies.

C.   Inducers of Apoptosis

There are a variety of apoptosis inducers. The initiation of apoptosis can be generated within the the cell or the stimulus can come from outside the cell.

- *Physiological activators*, TNF family (Fas ligand), transforming growth factor Beta, neurotransmitters (glutamate, dopamine)
- *Damage-related inducers*, heat shock, viral infection, bacterial toxins, oncogenes (myc, rel, E1A), tumor suppressors (p53), cytolytic T cells, oxidants, free-radicals, nutrient deprivation.
- *Therapy-associated agents*, Chemotherapeutic drugs (e.g., cisplatin, nitrogen mustard), Antracyclines (doxorubicin), gamma radiation, UV radiation
- *Toxins*, Ethanol, Beta-amyloid peptide

D.   Pathways of Apoptosis

As mentioned above, there are two distinct primary pathways of apoptosis. Death receptor pathway is mediated by cell surface receptors that are engaged by soluble ligands leading to their activation through adaptor proteins. The mitochondrial pathway on the other hand is activated by intrinsic mechanisms and do not involve cell death receptors. The pathways are summarized schematically in Figure 75.1.

Death receptor pathway

The major players of the death receptor pathway are members of the TNF superfamily of receptors called death receptors that induce a variety of cellular responses including apoptosis, cellular differentiation, and proliferation. Fas is the best characterized member of this family which is a cell surface protein activated by its ligand, FasL. FasL is a 281 amino acid long type II transmembrane molecule belonging to the large TNF family of proteins which bind to and activate members of the TNF-R family. The activated receptors signal the apoptotic response by recruiting Fas associated death domain protein (FADD) to the cytoplasmic death domain (DD) of the receptor to form the death-inducing signaling complex (DISC). FADD consists of two distinct domains, a DD, which binds to the DD of Fas, and a death effector domain (DED), which binds to DEDs on caspase-8 and caspase-10 and regulates binding of the DD to the receptor. Thus binding of ligand to

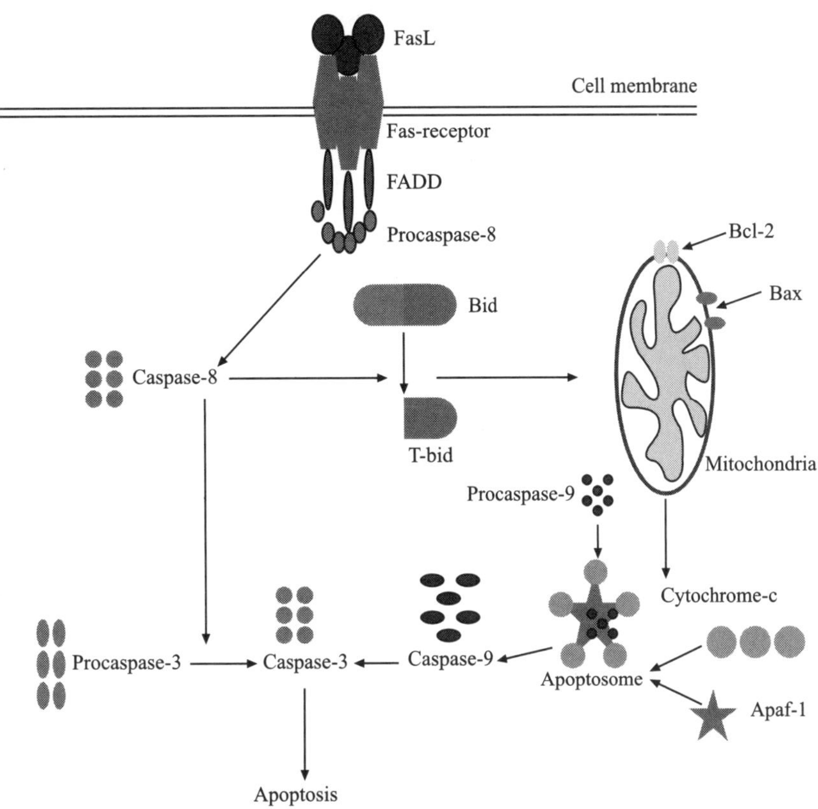

Figure 75.1   Figure shows the components of the death receptor and the mitochondrial pathway of death. (*see Plate 33 for colour figure*)

Fas results in the recruitment of FADD followed by caspases; the induced proximity the initiator caspase results in their dimerization and activation. This leads to cleavage to a fully processed, active form of the caspase that can dissociate from the receptor complex (Figure 75.1). Activation of TNF-R 1 requires an additional adaptor protein, TNF-R associated death domain (TRADD) protein. Binding of TNFα to TNF receptor 1 results in the recruitment of TRADD followed by FADD, again through DD interactions. Similar to the Fas/CD95 receptor, recruitment of FADD to the complex leads to the recruitment and activation of caspase-8. FADD is essential for TNF-induced caspase activation because FADD-deficient cells fail to undergo apoptosis when treated with TNFα and a dominant negative form of FADD that consists of only the DD, can block TNF-induced death. TNF-related apoptosis inducing ligand (TRAIL) can induce apoptosis in tumor cells without affecting normal cells, suggesting that it may be useful for treating cancer. There are two "signaling" receptors DR4 and DR5, which, similar to Fas, contain death domains. In addition, there are two "decoy" receptors, DcR1 and DcR2, that have truncated or completely absent DDs and act as mimics of the Fas.

The mitochondrial death pathway

The basic mitochondrial pathway of apoptosis in vertebrates begins with the permeabilization of the mitochondrial outer membrane. This is brought about by proapoptotic members of the Bcl-2 family. These members of the Bcl-2 family induce the release of proteins from the mitochondrial intermembrane space into the cytosol. Among these released proteins is cytochrome c, which interacts with monomeric APAF-1 to facilitate a conformational change in the latter, leading to its oligomerization and recruitment of caspase-9 to form the apoptosome. Caspase-9, the initiator caspase is activated in this apoptosome and the activated form of the enzyme cleaves and activates caspases-3, the executioner caspase. Caspase-3 then digest key substrates in the cell to produce the phenotypic effects that we see as apoptosis. Other released proteins facilitate caspase activation through inactivation of endogenous inhibitors of caspases, the inhibitor of apoptosis proteins. There are multiple pro-apoptotic members of the Bcl-2 family that include, Bax, Bak, Bad, Bid and Bok and anti-apoptotic members include Bcl-X$_L$, Bcl-w, Bfl-1, and Mcl-1.

III.   GRANZYME MEDIATED APOPTOSIS

Cytotoxic T cells can kill tumor cells and virus-infected cells through the release of perforin and granzyme proteins. The mechanism is that Perforin proteins form pores in the membranes of the target cell, allowing the entry of Granzyme A and Granzyme B in the target cell. Granzyme B induces caspase activation through truncation of Bid protein. Granzyme A can also activate caspase independent pathways. Once in a cell, Granzyme A activates DNA nicking by activating DNAse NM23-H1, a tumor suppressor gene product. The Granzyme pathway of apoptosis is shown in Figure 75.2.

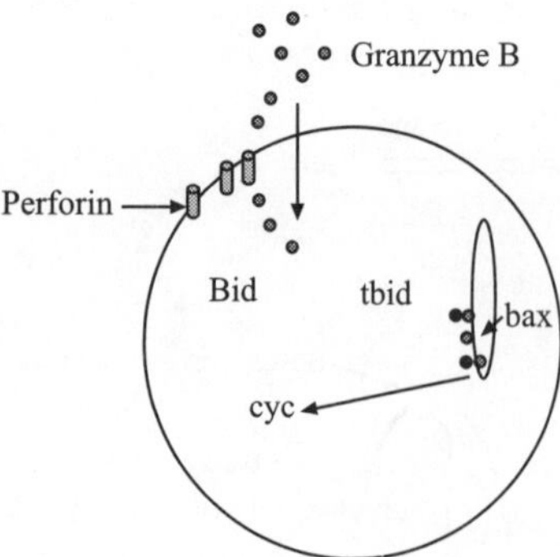

Figure 75.2 Schematic representation of granzyme mediated apoptosis. (*see Plate 33 for colour figure*)

IV. CASPASE INDEPENDENT APOPTOSIS

Apoptosis can also occur through a process that does not require caspases. Mitochondria plays a prominent role in caspase independent apoptosis. Like cytochrome c, apoptosis inducing factor (AIF) is a mitochondrial protein that is normally retained in the intermembrane mitochondrial space. When released from the mitochondria, AIF becomes an active cell killer when it is released to the cytosol. It translocates to the nucleus and triggers DNA fragmentation leading to cell death.

V. AUTOPHAGY

Autophagy is a housekeeping process for the cell meant for degradation of intracellular constituents like damaged, unfolded or long lived proteins, deteriorated organelles and cytoplasm by formation of autophagosome that fuses with lysosome and after degradation recirculates the macromolecules (Figure 75.3). Degradation of cellular components by autophagy is essential for maintaining cellular homeostasis by serving as an energy source in stress conditions. Autophagy is classified into 3 types: macroautophagy, microautophagy and chaperone mediated autophagy. Macroautophagy involves formation of a double membrane structure known as autophagosome in which cellular constituents gets sequestered and undergo degradation on fusion with lysosome. Autophagy generally gets activated in response to stress like nutritional, oxidative or genotoxic stress through inhibition of the protein mTOR (mammalian target of rapamycin) by a complex series of signalling events. Inhibition of mTOR results in formation of double membrane vesicles

called autophagosomes. Further expansion and completion of autophagosome formation is accomplished by recruitment of a protein called LC3 to the autophagosome. In tumor cells, is induced for survival of cells and autophagy avoids cell death by supressing apoptosis. In some situations, autophagy itself acts as a death mechanism, so there are common points in these processes that indicate crosstalk between apoptosis and autophagy. The balance between these two processes is important for determination of survival of a cell.

VI. CELL CYCLE AND ITS REGULATION

Unlike unicellular organisms such as bacteria and yeasts where each cell division gives rise to entire organism, multicellular organisms require a complex sequence of cell division to form a functional organism. The basic function of cell cycle is to duplicate the DNA in the chromosomes and distribute accurately into two genetically identical daughter cells. Further these daughter cells will often continue to divide by going through additional cell cycles.

A. Cell Cycle Phases

Cell cycle is divided into four main stages: G1 phase, S phase, G2 phase (G for Gap) and M phase (M corresponds to Mitosis). G1, G2, S phases are followed by interphase. Mostly, cells consume more time to grow and double their mass of proteins and organelles then duplicate and divide their chromosomes. Numerous preparations for mitosis occur during interphase including replication of DNA. DNA replication and chromosome duplication occurs in S phase. G1 and G2 phase allows cells to monitor the internal and external environment and ensure that preparations are completed and so that cells can proceed into S and M phase. M phase is the period when the contents of cells are actually divided; duplicated chromosomes are separated into two nuclei followed by cytokinesis during which entire cell divides into two daughter cells. M phase lasts only for hours but interphase can extend for days, weeks, or longer and may even enter into a resting phase called G0 phase that varies with cell types and environments.

B. Control of Cell Cycle

Eukaryotes have evolved a complex network of regulators known as cell cycle control system that takes care of proper progression through cell cycle. The progression of cells

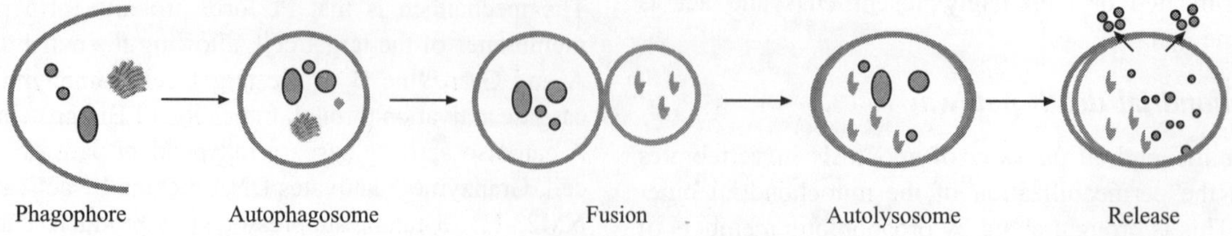

Figure 75.3 Figure shows the steps of autophagic death pathway. (*see Plate 33 for colour figure*)

through various phases of cell cycle is under the control of a number of cyclins and cyclin-dependent kinases (CDKs). Different cyclin/CDK complexes are expressed in specific stages of cell cycle that control various cell cycle events. CDKs are constitutively expressed during cell cycle but in an inactive form. Phosphorylation of CDKs is regulated by their respective binding proteins known as the cyclins. Normally, cells undergo cell cycle without any interruption but once any kind of signal like DNA damage or cell stress occur, it causes arrest in proliferation and resume proliferation once the problem is fixed. Cells may undergo apoptosis and G0 resting rather than going into arrest. In mammalian system cyclin D and and cyclin E forms an active complex that is required for G1 to S phase progression. Overexpression and dysregulation of cyclin D is associated with various cancers. In G1 phase, Cyclin D binds to CDK4 and this complex is critical to phosphorylate retinoblastoma susceptibility protein

(RB). In hypophosphorylated form RB binds to EF2/DP1 and this complex recruits histone deacetylase that results in inhibition of E2F responsive genes. In hyperphosphorylated form (phosphorylation by CDK4/cyclin D) RB dissociates from the RB/EF2/DP1 complex thus activates transcription of E2F responsive genes includes cyclin E, DNA polymerases and several others. In M phase hyperphosphorylated form of RB is again converted into hypo phosphorylated form. Further progession through S phase involves formation of CDK2 and cyclin E complex. The next check point in cycle is G2/M transition. This transition is mediated by E2F responsive genes cyclin A and cyclin B. Cyclin A forms complex with CDK2 and this complex is responsible for regulation of mitotic prophase events. The other mediator is the cyclin B-CDK1 complex which is activated by cdc25 phosphatase that propels the cells beyond prophase and initiates mitosis. Cell cycle regulation is represented schematically in Figure 75.4.

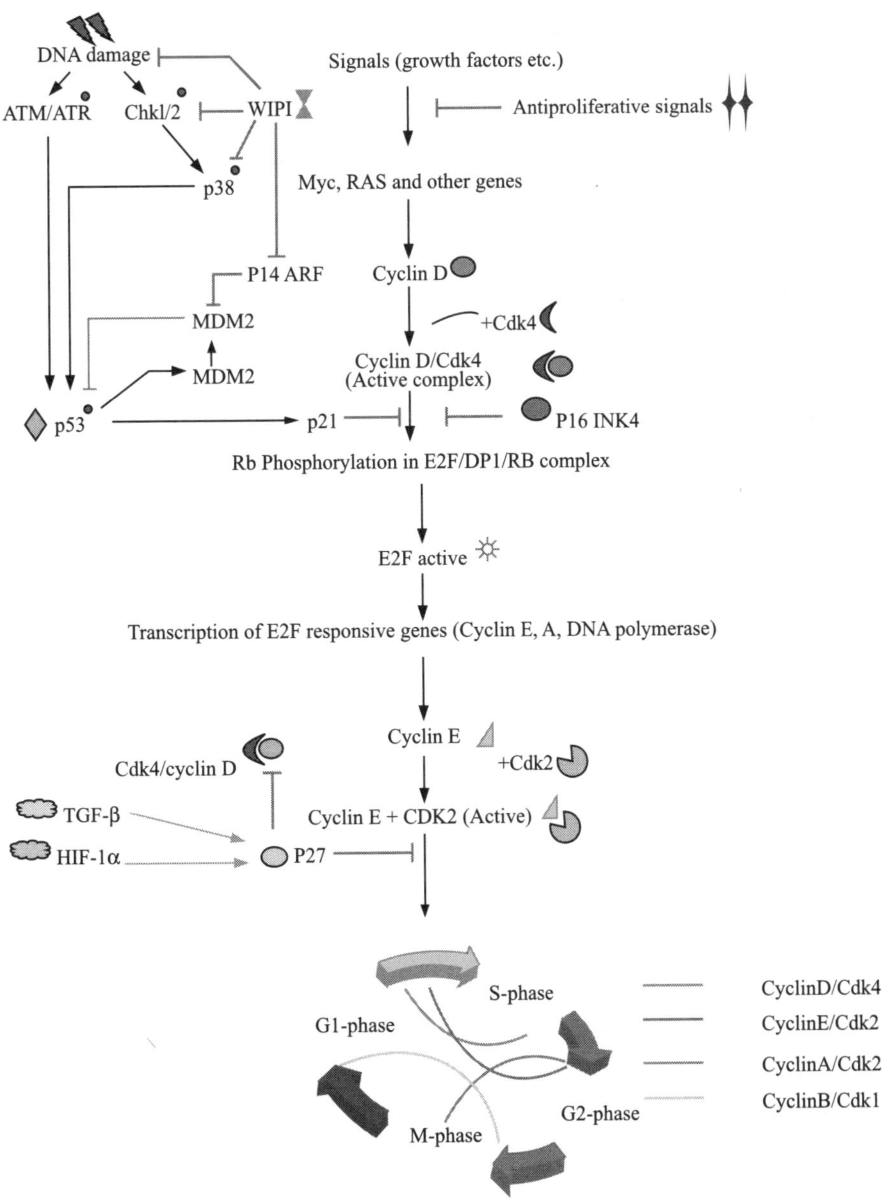

Figure 75.4   A representation of cell cycle and its regulation. (*see Plate 33 for colour figure*)

VII. CELL CYCLE ARREST, APOPTOSIS AND CANCER

Normally, cells have an inbuilt mechanism to sense stress and in response to this cell cycle arrest can occur with repair mechanisms activated. In cancer cells, this regulation is deactivated causing excessive proliferation and uncontrolled growth. p53 is a nodal tumor suppressor gene that have widespread role in cellular regulation and its mutated form is associated with many cancers. Under normal conditions, p53 is rapidly degraded by 26S proteasome that is mediated by two other proteins MDM2 and Jun kinase. In response to genotoxic stress or DNA damage ATM and ATR proteins undergo phosphorylation that causes phosphorylation of p53. This phosphorylation is responsible for dissociation of p53-MDM2 complex and thus stabilization of p53. Activation of p21 by stabilized p53 results in suppression of Rb phosphorylation and therefore removal of E2F block. Normally, when phosphoryalted Rb is bound to E2F, the complex acts as a growth suppressor and prevents progression through the cell cycle and cells remain stalled at G1 phase. Removal of E2F block results in pushing of cells into S phase.

Other crucial nuclear protein is c-myc that acts in early phase of cell cycle and is responsible for pushing cells into G1 phase of cell cycle. C-myc is a regulator for other cell cycle associated genes like cdc25A, cyclinD, E, A, E2F etc. Activation of p53 dependent and p53 independent pathways of apoptosis are also mediated by c-myc. C-myc mediates the activation of proapoptotic genes Bax and also suppression of antiapoptotic genes Bcl 2, Mcl-1. As in S-phase, cells commit themselves towards replication and after S-phase cells are not able to revert back in cell cycle, thus in cell cycle S-phase is a point of no return. In response to stress or DNA damage, p53 either blocks cell cycle so that genomic error could be repaired. If repair is not possible, apoptosis is induced before final commitment to replication. Similar to c-myc, RAS protein regulates normal growth and proliferation of cells. Approximately 15–20% of tumors are linked with mutated form of RAS proteins. RAS is a membrane localised G-protein. Under normal conditions in response to growth factors, GDP bound form of RAS gets converted to GTP-RAS. Activation of RAS activates MAPK and PI3K mediated pathways that in turn promotes translocation of growth promoting factors to nucleus thus promote normal growth and proliferation of cells. On the other hand mutated form of RAS promotes extensive cell proliferation that results in the formation of malignant growth. Normally, expression of RAS increases in response to hyperproliferative signals that result into proapoptotic gene induction to supress excessive proliferation and cancer formation. In cancerous cells, RAS promotes survival mechanisms like autophagy and suppress apoptosis. RAS is responsible for G1 to S-phase transition by phosphorylating and activating Rb. In response to growth signals RAS repress p27 expression by phosphorylation at Thr-187 residue mediated by CyclinA-Cdk2 or CyclinE-Cdk2. Independent of CDK2 mediated repression of p27, RAS mediated activation of ERK/MAPK also degrade p27. Cancer cells supress tumor suppressor genes and proteins and activate the proteins that are responsible for excessive proliferation or survival of cells.

SUMMARY

Cell death is an integral process of our survival. It is required to maintain tissue homeostasis. Cell death can occur by multiple processes, apoptosis being the most prominent. Apoptosis occurs through various pathways of which the most prominent is the intrinsic and extrinsic pathways of death. The primary players in extrinsic pathway are the cell surface receptors while in the intrinsic pathway mitochondria is the main player. Cell cycle is divided in four phases of G1, G2, M, S phase and entry of one phase to the other is regulated by various cyclin D and cyclin dependent kinases (Cdks). Deregulation of apoptosis or cell cycle can cause a variety of diseases, therefore it is important to understand how they operate so as to be able to design therapeutics for treatment.

SUGGESTION FOR FURTHER READINGS

Apoptosis

1. Susan Elmore, Apoptosis: A Review of Programmed Cell Death, *Toxicol Pathol.* (2007), 35(4):495–516. Http://www.ncbi.nlm.nih.gov/pmc/articles/PMC2117903/.

2. Life and Death Partners: Apoptosis, Autophagy and the Cross-talk between them, *Cell Death and Differentiation,* 16, 966–975.

3. Noboru Mizushima, Autophagy: Process and Function, *Genes and Dev.* (2007), 21:2861–2873. http://genesdev.cshlp.org/content/21/22/2861.full

Autophagy

Levine Beth and Kroemer Guido (2008), Autophagy in the Pathogenesis of Disease, *Cell*, 132, 27–42. http://download.cell.com/pdf/PIIS0092867407016856.pdf?intermediate=true.

Cell cycle

Schafer K.A. (1998), Cell Cycle: A Review, Veterinary Pathology, 35,461–478. http://vet.sagepub.com/content/35/6/461.short?rss=1&ssource=mfc.

76

Cancer

Neeta Singh

CONTENTS

I. INTRODUCTION

Cancer is the leading cause of death worldwide, a very common disease. Based on the GLOBOCAN 2012 estimates, about 14.1 million new cancer cases and 8.2 million cancer deaths occur annually, with 57% of the cases and 65% of the deaths in the economically developing world. Of these cancer cases, an estimated 22% are preventable. Cancer occurs in humans of all age groups, and affects a wide variety of tissues and organs. The global burden of cancer continues to increase in economically developing countries largely as a result of the aging and growth of the world population alongside an increasing adoption of cancer-causing behaviors, particularly smoking, lack of physical activity, "Westernized" diets and obesity. According to the World Cancer Research Fund (WCRF), the number of global cancers has increased by 20% in less than a decade and are largely linked to diet. The number is expected to

rise dramatically over the next 10 years. Breast cancer is the most frequently diagnosed cancer and the leading cause of cancer death among females, accounting for 23% of the total cancer cases and 14% of the cancer deaths. Lung cancer is the leading cancer site in males, comprising 17% of the total new cancer cases and 23% of the total cancer deaths. Breast cancer is now also the leading cause of cancer death among females in economically developing countries, a shift from the previous decade during which the most common cause of cancer death was cervical cancer. Further, the mortality burden for lung cancer among females in developing countries is as high as the burden for cervical cancer, with each accounting for 11% of the total female cancer deaths. Although overall cancer incidence rates in the developing world are half of those seen in the developed world in both sexes, the overall cancer mortality rates are generally similar. Cancer survival tends to be poorer in developing countries, most likely because of a combination of diagnosis at the late stage and limited access to timely and standard treatment. A substantial proportion of the worldwide burden of cancer could be prevented through the application of existing cancer control knowledge and by implementing programs for tobacco control, vaccination (for liver and cervical cancers), and early detection and treatment, as well as public health campaigns promoting physical activity and a healthier dietary intake. Although the cancer mortality has declined during the last decade in developed countries, because of the progress in this area in terms of screening, diagnosis and therapeutics, however, the problem still exists and is overwhelming. The current focus is on the biological understanding of the molecular basis of carcinogenesis and biological properties of the tumors.

A.   Definition—What is Cancer?

Cancer is a complex disease where normal cells of the body lose their growth regulation resulting in uncontrolled growth/ proliferation, invasion of local tissues and spread/metastasis. The term cancer is derived from the Greek word "Karkinoma" which means crab. It is named after the crab because of swollen veins around the area of the tumor which resembles a crabs limbs. Neoplasm means "New growth". Oncology is the study of the neoplasms or tumors. Cancer is the term used for all malignant tumors. A neoplasm is an abnormal mass of tissue, the growth of which exceeds and is uncoordinated with that of normal tissue. The entire population of cells within a tumor arises from a single cell that has incurred genetic changes allowing excessive and unregulated proliferation, and hence, tumors are said to be clonal. All tumors, benign and malignant, have two basic components, one is proliferating neoplastic cells that constitute their parenchyma and second is the supporting stroma made up of connective tissue and blood vessels. The parenchymal cells determine the behavior of the tumor and pathological consequences, but the latter is responsible for the tumor growth. There is cross-talk between these cells that influences the growth of the tumor. Some genes controlling growth and interactions with other normal cells are either abnormal in structure or in regulation in cancer cells. In the case of benign tumors there is proliferation, but no invasion and metastasis.

B.   Naming of Different Types of Cancer

Cancer can originate almost anywhere in the body. Nomenclature of tumor is based on the parenchyma. Scientists use a variety of technical names to distinguish many different types of cancer. In general, benign tumors are designated by attaching the suffix—Oma to the cells of origin. Tumors of mesenchymal origin generally follow this rule. For example, a tumor arising from fibroblastic cells is called fibroma, that from cartilage a chondroma, from osteoblasts an osteoma. Classification of epithelial tumor is more complex. Epithelial neoplasms that form glandular patterns as well as tumors derived from glands are termed adenoma. Epithelial neoplasm producing finger like projections are referred to as papillomas, those that form cystic masses, as in the ovary are referred to as cyst adenoma. A neoplasm producing visible projections above a mucosal surface and projects into gastric or colonic lumen is termed a polyp.

Carcinomas, the most common types of cancer, arise from the cells that cover external and internal body surfaces. Lung, breast, and colon are the most frequent cancers of this type.

Sarcomas are cancers arising from cells found in the supporting tissues/ mesenchymal tissue of the body such as bone, cartilage, fat, connective tissue, and muscle.

Lymphomas are cancers that arise in the lymph nodes and tissues of the body's immune system.

Leukemias are cancers of the immature blood cells that grow in the bone marrow and tend to accumulate in large numbers in the bloodstream.

In general, names are created by using different Latin prefixes that stand for the location where the cancer began its unchecked growth. For example, the prefix "osteo" means bone, so a cancer arising in bone is called an osteosarcoma. Similarly, the prefix "adeno" means gland, so a cancer of gland cells is called adenocarcinoma, for example, a breast adenocarcinoma. Examples of sarcomas are fibrosarcoma, lymphosarcoma, leiomosarcoma.

II.   MUTATIONS AND CANCER

If one considers cancer as a disease that results from alterations in the DNA of somatic cells, then it follows that any activity that increases the frequency of genetic mutations is likely to increase the risk of developing cancer. Cancer may begin because of the accumulation of mutations involving oncogenes, tumor suppressor genes, and DNA repair genes. For example, colon cancer can begin with a defect in a tumor suppressor gene that allows excessive cell proliferation. The proliferating cells then tend to acquire additional mutations involving DNA repair genes, other tumor suppressor genes, and many other growth-related genes. Over time, the accumulated damage can yield a highly malignant, metastatic tumor. In addition to all the molecular changes that occur within a cancer cell, the environment around the tumor changes dramatically as well. The cancer cell loses receptors that would normally respond to neighboring cells that call for growth to stop. Instead, tumors amplify their own supply of growth signals. They also flood their neighbors with other signals called cytokines and enzymes called proteases. This action destroys both the basement membrane and surrounding matrix, which lies between the tumor and its path to metastasis—a blood vessel or duct of the lymphatic system. Antimutagens—these are substances which interfere with tumor protein, e.g., Vitamin A reverses pre-cancer conditions, vitamin C and E act as antioxidants, curcumin prevents metastasis, fibre in diet protects against colon cancer.

III. LOSS OF NORMAL GROWTH CONTROL

Cancer arises from a loss of normal growth control. In normal tissues, the rates of new cell growth and old cell death are kept in balance. In cancer, this balance is disrupted. This disruption can result from uncontrolled cell growth or loss of a cell's ability to undergo cell suicide by a process called "apoptosis." Apoptosis, or "cell suicide," is the mechanism by which old or damaged cells normally self-destruct. All cancers arise as a result of genetic alterations. The genes involved in tumorigenesis constitute a specific subset of the genome whose products are involved in cell cycle progression, adhesion of a cell to its neighbour, apoptosis, repair of DNA damage. Approximately 350 different genes have been identified as 'cancer genes' that play a role in development of one type of malignancy. An individual tumor can carry approx. 60–80 different mutations, but a large number of these genes encode proteins that participate as components of a relatively small number of pathways. For example, in gliomas, majority of tumors exhibit mutations that affect three major pathways, i.e., P53, pRB and PI3K. Thus, cancer can be thought of not as a disease of aberrant genes but one of aberrant cellular pathways. Hence, cancer should be looked as a 'pathway disease' rather than a genetic disease. DNA microarray chips are used for gene expression analysis. Tumors are known to be poorly immunogenic and are not eliminated because either tumor response generated is incomplete or is inadequate. Tumor cells can also modulate the immune response generated against them by down-regulating the tumor antigens on their cell surface. Therefore, immune modulation offers a great hope in eradicating the tumor cells.

IV. INVASION AND METASTASIS

Cancers are capable of spreading throughout the body by two mechanisms—invasion and metastasis. Invasion refers to the direct migration and penetration by cancer cells into neighboring tissues. Metastasis refers to the ability of cancer cells to break away from the parent tumor, penetrate into lymphatic and blood vessels, circulate through the bloodstream, and migrate to distant sites in the body where they invade normal tissues and establish secondary tumors. Depending on whether or not they can spread by invasion and metastasis, tumors are classified as being either benign or malignant. *Benign tumors are tumors that cannot spread by invasion or metastasis*; hence, they only grow locally. Malignant tumors are tumors that are capable of spreading by invasion and metastasis. By definition, the term "cancer" applies only to malignant tumors. To diagnose the presence of cancer a sample of the affected tissue is seen under the microscope. Hence, when preliminary symptoms, *Pap test, mammogram, PSA test, FOBT, or colonoscopy* indicate the possible existence of cancer, a biopsy is performed, which is the surgical removal of a small piece of tissue for microscopic examination. For leukemias, a small blood sample serves the same purpose. This microscopic examination confirms whether a tumor is actually present and, if so, whether it is malignant or benign. In addition, microarrays may be used to determine which genes are turned on or off in the sample, or proteomic profiles may be collected for an analysis of protein activity. This information helps in making a more accurate diagnosis and for treatment planning. Metastasis is the most dangerous property of tumor cells. Metastasis is the spread of cancer cells from the primary site of origin to other tissues and organs where they grow as secondary tumors and is a major problem. Studies have uncovered the role of many proteases, certain glycoproteins and glycosphingolipids of the cell surface in metastasis. Changes in oligosaccharide chains of cell glycoproteins may be critical for this phenomenon. Alterations in proteins involved in cell-cell interactions, e.g., integrins, cadherins, and other cell adhesion molecules such as neural cell adhesion molecule (N-CAMs) have been shown to be involved. Another area concerns the blood supply to the tumors. Folkman showed that *tumor growth is angiogenesis-dependent*. Tumor cells can secrete angiogenic growth factors such as basic or acidic fibroblast growth factor (b/a FGF) which promote the proliferation of endothelial cells and the formation of new capillaries. Thus, drugs are being designed to target angiogenesis. Angiostatin and endostatin are some such drugs. They act by blocking the active sites of enzymes that digest plasminogen and collagen, thereby preventing endothelial cells from forming new blood vessels.

V. TUMOR GRADING

Microscopic examination also provides information regarding the likely behavior of a tumor and its responsiveness to treatment. Cancers with highly abnormal cell appearance and large numbers of dividing cells tend to grow more quickly, spread to other organs more frequently, and be less responsive to therapy than cancers whose cells have a more normal appearance. Based on these differences in microscopic appearance, a numerical "grade" is assigned to most cancers. In this grading system, a low number grade (grade I or II) refers to cancers with fewer cell abnormalities than those with higher numbers (grade III, IV).

VI. TUMOR STAGING

After cancer has been diagnosed, the following questions determine how far the disease has progressed:

1. How large is the tumor, and how deeply has it invaded surrounding tissues?
2. Have cancer cells spread to regional lymph nodes?
3. Has the cancer spread (metastasized) to other regions of the body?

Based on the answers to these questions, the cancer is assigned a "stage." A patient's chances for survival are better when cancer is detected at a lower stage.

VII. BASIC PROPERTIES OF A CANCER CELL

Normal cells can be converted to cancer cells by treatment with carcinogenic chemicals, radiation or tumor viruses and the process is called malignant transformation. Cells transformed *in vitro* can cause tumors when introduced into an host animal. The most important morphological and biochemical changes occurring upon transformation are (1) loss of growth control—normal cells respond to inhibitory influences from their environment such as growth factor depletion, but malignant cells under the same conditions continue to grow, piling one on top of another to form clumps or foci, (2) altered morphology—the cells become more rounded in appearance. Oncogenes produce tyrosine kinase, cause abnormal phosphorylation of vinculin, so there is diminished adhesion to substratum, leading to rounded appearance of transformed cells, (3) loss of contact inhibition of growth/movement leading to increased cell density—it is a characteristic of normal cells. Their multiplication is stopped when the cells come into contact with other cells. This is called *contact inhibition*. In cancer cells tight junctions are rare, contact inhibition is lost and adjacent cells continue to grow one on top of the other, to form multilayers, (4) *loss of anchorage dependence*—Normal cells *adhere* to the surface of plastic flask or glass bottle, but cancer cells do not. Vinculin is the protein found in the focal adhesion plates, that is the structure involved in adhesion between cells as well as the basement membrane in case of normal cells. Transformed cells grow without attachment to the surface of the culture dish and can grow in agar, (5) biochemical changes—increased rate of glycolysis, alterations of the cell surface, i.e., changes in the composition of glycoproteins or glycosphingolipids and secretion of certain proteases, (6) alterations in cytoskeleton structures such as actin filaments (7) diminished requirements of growth factors and often increased secretion of growth factors

in the surrounding medium, (8) metastasis—cancer cells have tendency to disintegrate from the main mass and to disseminate to nearby or distant organs. The collagenase and stromolysin released by most cancer cells helps in penetration of cancer cells into surrounding areas (Table 76.1 and Figure 76.1).

TABLE 76.1 Characteristics of Transformed cells

1. Increased saturation density
2. Decreased GF requirements
3. Loss of capacity for growth arrest
4. Loss of dependence on anchorage for growth
5. Changed cell morphology
6. Altered growth habits
7. Loss of contact inhibition
8. Cell surface alterations
9. Ganglioside content of all lipids reduced or altered
10. Fibronectin decreased
11. Altered gene transcription
12. Loss of actin microfilaments
13. Release of proteins (TGF)
14. Protease secretion
15. Immortalization

Normal cells growing in culture depend on growth factors such as EGF, insulin that are present in sera which is normally added to cell culture media. Normal cells in culture exhibit a limited capacity for cell division but cancer cells are seemingly immortal as they continue to divide indefinitely. This is attributed to the presence of *telomerase in cancer cells* and absence in normal cells. Telomerase is a reverse transcriptase enzyme that maintains the telomeres at the ends of the chromosomes. Normal cells maintain their diploid chromosomes complement as they grow and divide, both *in vivo* and *in vitro*. In contrast cancer cells are genetically unstable and have highly aberrant chromosome complement,

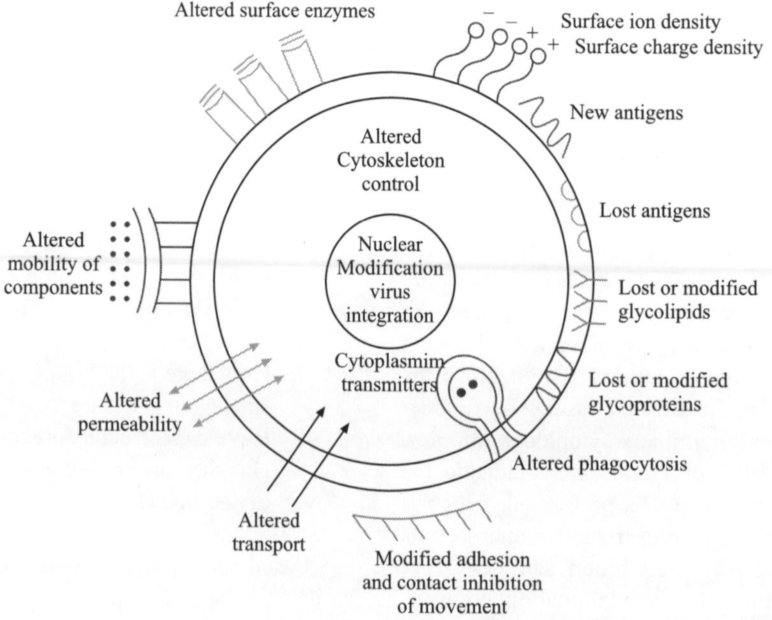

Figure 76.1 Morphological and biochemical changes occurring in a transformed cell.

a condition called aneuploidy. When chromosomal content of a normal cell is disturbed, apoptosis is activated leading to self destruction of the cell. However, cancer cells fail to elicit apoptosis even when their chromosome number becomes deranged. Cancer cells often depend on glycolysis, and an anaerobic metabolic pathway. Under conditions of hypoxia (reduced oxygen), cancer cells activate a transcription factor called HIF (Hypoxia induced factor) that induces the formation of new blood vessels and promotes the migratory properties of the cells, which may contribute to their spread/metastasis of tumor. It is this property of cancer cells that is a major threat to the well being of the entire organism. Most cancer cells delete different enzymes or even whole metabolic pathways such as aerobic glycolysis. Even when oxygen is plentiful, tumor cells continue to generate much of their ATP by glycolysis and the end product of glycolysis is lactic acid, which is secreted into the tumor's micro environment, where it may promote tumor growth. Sialic acid and sialylation—most cancer cells carry more negative charge on their cell surface than normal cells. This abnormality is due to N-acetyl neuraminic acid (NANA) content on the cell membrane. Due to higher content of negative charge cancer cells repel each other leading to less adhesion. Altered sialylation of cell surface glycoproteins and glycolipids is closely related to malignant phenotype of cancer cells including their invasiveness and metastasis potential. Thus, sialidases may be potential targets for cancer diagnosis and therapy. Cell fusion plays an essential role. It may contribute to initiation and progression of cancer and in different stages of tumor progression, including aneuploidy and tumor initiation, origin of cancer stem cells, multi drug resistance and the acquisition of metastatic potential.

VIII. ETIOLOGY OR CAUSES OF CANCER

Cancer is often perceived as a disease that strikes for no apparent reason. Many of the causes of cancer have been identified. Besides intrinsic factors such as heredity, diet, and hormones, studies point to key extrinsic factors that contribute to the cancer's development. All cancers are multifactorial and are generally spontaneous (Figure 76.2). Factors responsible for cancer formation are environmental factors, chemical compounds, both organic and inorganic; physical agents such as radiations—UV, X-rays, Gamma rays; genetic factors such as mutations in specific genes; hormones; viruses and bacteria; oxidative damage to the DNA; etc. The chemical origin of cancer was noted by observations of high incidence of cancer in certain occupational group of people. The environmental and lifestyle exposures were major risk determinants of cancer risk. In 1775, Percivall Pott, a British surgeon made the first known correlation between an environmental agent and cancer. He showed that the high incidence of cancer of the nasal cavity and of the skin of scrotum in chimney sweepers was due to their chronic exposure to soot. Thereafter, the carcinogenic chemicals in soot, were isolated along with several other diverse array of compounds/chemicals that cause cancer. The experimental induction of tumors in animals and neoplastic transformation of cultured cells by chemicals and the analysis of environmentally induced tumors has revealed important concepts and pathways involved in carcinogenesis and has a major impact on human health. Chemical carcinogens are organ specific and target epithelial cells and are genotoxic, i.e., they cause DNA damage and consequently somatic mutations. The mutations can occur by direct exposure to the environmental

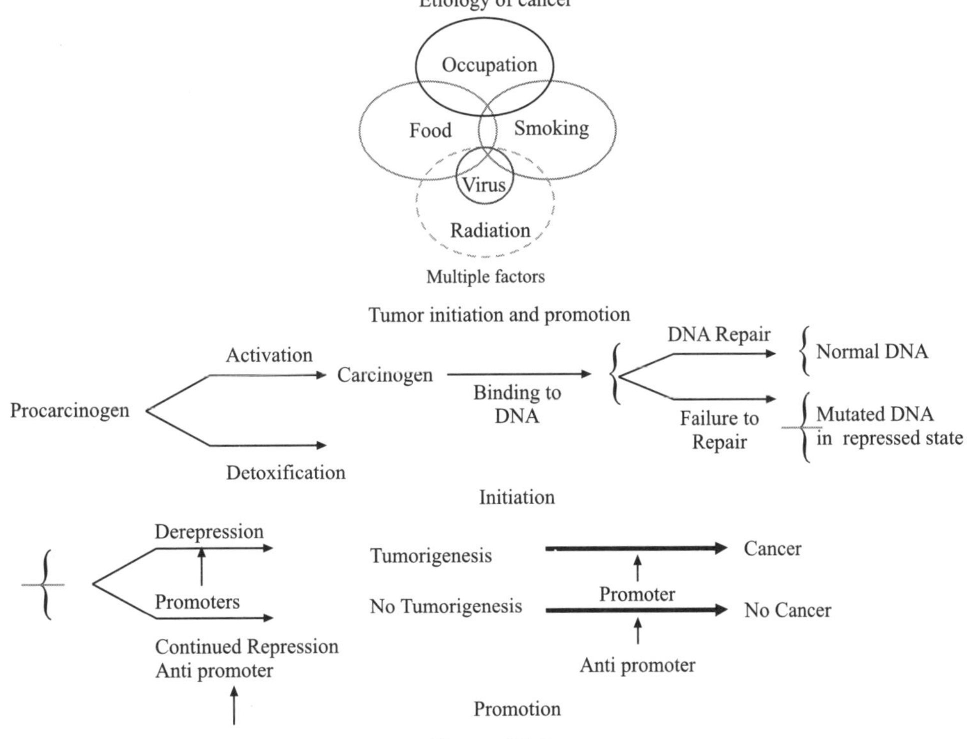

Figure 76.2

chemicals or indirectly by activation of endogenous pathways. The chemicals present in soot, cigarettes, smoke, etc., are shown to act as mutagens or be converted to mutagenic compounds by cellular enzymes. Some of the chemical carcinogens are chemicals found in cigarette smoke, occupational carcinogens such as vinyl choride, benzene, aromatic amines, asbestos, arsenic, aflatoxin, obesity. Genotoxic carcinogens have high chemical reactivity, such as alkylating agents or they can be metabolized to reactive intermediates by the host. The alkylating agents include N-nitroso compounds, aliphatic epoxides, polycyclic aromatic hydrocarbons (PAHs), aflatoxins, mustard, aryl aromatic amines, aminoazodyes, and heterocyclic aromatic amines produced by overheating and charring (Table 76.2a).

TABLE 76.2a   Chemical carcinogens

Occupation	Chemical	Organ affected
Chimney sweepers	Soot	Skin, scrotum
Machine tool cutters	Mineral oil	Skin, scrotum, lung
Dye	Aniline	Bladder
Rubber	$\cong$ Napthylamine	Bladder
Radioisotopes	Ionizing radiation	Skin, Bone
Plastic	Vinyl chloride	Liver
Glass workers (cake oven)	Benz(a) pyrene (coal tar). PAH Cr, Ni, As	Liver, lung, skin
Asbestos	Chrosotile, Amosite	Lung
Radiations	UV, X-ray	Numerous locations

They selectively target purine and pyrimidine bases of DNA leading to base mispairing, deletions, missense and nonsense mutations, and chromosomal breaks. A number of endogenous enzymes activate or detoxify carcinogens and procarcinogens, and in the process transform them into active carcinogens. Carcinogenic agents can directly activate oncogenes, inactivate tumor suppressor genes and cause genomic changes that are associated with autonomous growth, enhanced survival and modified gene expression. In addition to chemicals and radiation, a few viruses can also trigger the development of cancer (Table 76.2b). In general, viruses are small infectious agents that cannot reproduce on their own, but instead enter into living cells and cause the infected cell to produce more copies of the virus. Like cells, viruses store their genetic instructions in large molecules called nucleic acids. In the case of cancer

TABLE 76.2b   Human Cancer Viruses

Viruses	Type of Human Cancer
Epstein-Barr virus	Burkitt's lymphoma
Human papillomavirus	Cervical cancer
Hepatitis B virus	Liver cancer
Human T-cell lypmphotrophic virus	Adult T-cell leukemia
Kaposi's sarcoma-associated herpesvirus	Kaposi's sarcoma

viruses, some of the viral genetic information carried in these nucleic acids is inserted into the chromosomes of the infected cell, and this causes the cell to become malignant. Only a few viruses that infect human cells actually cause cancer. Included in this category are viruses implicated in cervical cancer, liver cancer, and certain lymphomas, leukemias, and sarcomas. Susceptibility to these cancers can sometimes be spread from person to person by infectious viruses, although such events account for only a very small fraction of human cancers. For example, the risk of cervical cancer is increased in women with multiple sexual partners and is especially high in women who marry men whose previous wives had this disease. Transmission of the human papillomavirus (HPV) during sexual relations appears to be involved. Besides chemicals the other types of carcinogenic agents are ionizing radiations and a variety of DNA and RNA containing viruses, depending on the type of nucleic acid found within the mature virus particle, i.e., DNA viruses and RNA viruses.

They all alter the genome. UV radiations are strong mutagens and cause formation of pyrimidine dimers, apurinic or apyrimidinic sites, single and double strand breaks or cross linking of DNA strands. X-rays and gamma rays cause free radical formation in tissues which in turn can interact with DNA and other macromolecules leading to molecular damage. All cancers originate from one aberrant cell, i.e., they are monoclonal in nature. This multiplies to form a tumor mass. A single mutation occurs and as age advances, the number of mutations accumulate thereby increasing the probability of incidence of cancer. That is why cancer it is more a disease of the elderly. Mutagen is any substance that increases the rate of mutation, and thus, can increase the rate of cancer. All carcinogens are mutagens, e.g., X-rays, UV, gamma rays. A wide variety of chemical compounds are carcinogenic. Nearly 80% of human cancers are caused by environmental factors, mainly chemicals (Figure 76.2). Carcinogens can enter the human body by virtue of one's occupation, e.g., benzene, asbestos fibres; diet, e.g., aflatoxin which is produced as mold and given the right temperature and humidity affects several foodstuffs including peanuts in which it was originally discovered as *Aspergillus flavus*; or lifestyle, e.g., cigarette smoking, consumption of alcohol, etc. by virtue of one's occupation—aniline, asbestos fibres, diet-aflatoxin, lifestyle-smoking. Others are PAHs, e.g., benzopyrene, DMBA, aromatic amines, nitroso compounds-dimethynitrosoamine, methylcholanthrene. Tobacco smoke contains more than 3500 chemicals, of which 20 are carcinogens. Specific chemicals in tobacco smoke include PAHs, N–nitrosamines, aromatic amines, ethylene oxide, benzopyrene, nicotine, carbon monoxide, NO_2, carbon soot, and agents that cause oxyradical damage (Table 76.2c). Most constituents cause carcinogen—DNA adduct formation and cause lung cancer. Some carcinogens called direct carcinogens directly interact with the target molecules, whereas others called pro-carcinogens require prior metabolism/activation to become carcinogens and this process of conversion of procarcinogen to carcinogen is called metabolic activation. The enzymes responsible for this metabolic activation are mono-oxygenases and transferases,

TABLE 76.2c   Life Style and cancer

Tobacco	bronchus, bladder, mouth, oesophagus
Betel Nut	mouth
Alcohol	mouth, oesophagus, liver
UV	skin
Aflatoxin	liver

TABLE 76.2d   Heredity and Cancer

Condition	Type of cancer
Hereditary retinoblastoma	Retinoblastoma
Xeroderma pigmentosum	Skin
Wilms' tumor	Kidney
Li-Fraumeni syndrome	Sarcomas, brain, breast, leukemia
Familial adenomatous polyposis	Colon, rectum
Fanconi's anaplastic anemia	Leukemia, liver, skin

and belong to cytochrome P 450 species located in the endoplasmic reticulum. In fact these very enzymes are involved in the metabolism of other xenobiotics, drugs, environmental pollutants. The activity of these enzymes are affected by age, sex, and genetic factors and this helps explain the differences seen in different individuals in their response to exposure to the same chemicals. Many chemicals bind covalently to DNA, RNA and proteins leading to adduct formation. They have been found to interact with the purine, pyrimidine and phosphodiester groups of DNA. The most common sites of attack being guanine and the addition of various carcinogens to the N_2, N_3, N_7, O_6 and O_8 atoms of this base. The covalent interaction of the carcinogens with DNA can result in several types of damage despite existence of DNA repair systems. These unrepaired lesions if they persist for long durations, ultimately lead to mutations critical to carcinogenesis. Assays for screening various chemical compounds for potential carcinogenicity have been developed, which are essentially based on detection of their mutagenecity. One of these is Ame's test. This assay uses a specially constructed strain of Salmonella typhimurium that has a mutation (His-) in a gene that encodes for one of the enzyme involved in the synthesis of histidine. Thus, these particular strains of Salmonella cannot synthesize histidine, which must be present in the medium for growth to occur. When a mutation caused by the carcinogen occurs at the site of histidine, His- mutation occurs, this mutation can restore its reading sequence, converting it to His+. The progeny from bacteria containing such a reverse mutation can now synthesize histidine, and thus, grow in a medium lacking it. Such bacteria can be detected easily, as readily observable and quantifiable colonies growing on agar plates. One problem with this test is that the bacteria does not contain the monooxygenases found in higher animals. Hence, if a compound requires activation to become carcinogenic, this may not occur when bacteria are used. Ames circumvented this problem by incubating the

compounds to be tested in a post mitochondrial supernatant of rat liver, a fraction that contains fragments of endoplasmic reticulum, and thus, most of the enzymes required to activate potential carcinogens. Ames test identifies nearly 90% of known carcinogens and has become a routine test for identification of carcinogenic potential of newly synthesized compounds. However, these compounds should undergo further testing including assessment of carcinogenicity in animals. In certain organs such as skin and liver, carcinogenesis has at least two stages, i.e., initiation and promotion. The classical example is skin. If benzopyrene is applied on the skin of mice no tumors develop. However, if its application is followed by several applications of croton oil (a tumor promoter) many tumors develop subsequently. Application of croton oil by itself does not lead to development of tumors. The stage of carcinogenesis caused by application of benzopyrene is called initiation, this stage is rapid and irreversible and involves irreversible modification of DNA due to one or more mutations (Table 76.3 and Figure 76.2). The second stage resulting from application of croton oil is much slower, can take months or years and is called promotion, with croton oil acting as tumor promoter (Table 76.3 and Figure 76.2). Most carcinogens are capable of acting as both initiating agents as well as promoting agents. The malignancy of tumor cells tends to progress and after getting transformed and becoming a malignant cell, its composition and behavior keeps on changing with a tendency for malignancy to increase. This is manifested by increasingly abnormal karyotype, increasing growth rate, increasing invasion and metastasis, because of fundamental instability of genome of tumor cells. Mutations in DNA repair genes, activation of additional oncogenes may be involved. A number of compounds, e.g., saccharin, phenobarbital, etc can act as tumor promoters. The active agent in croton oil is the phorbol ester 12–o-tetradecanoyl-13–acetate (TPA). It is thought that many promoters increase the proliferation of stem cells. TPA binds to a membrane located protein kinase C and increases the activity of this enzyme, resulting in phosphorylation of a number of membrane proteins, thereby affecting transport and other functions, such as cell division. This is a classical example of transmembrane signaling. Passive smoking, alcohol consumption, diet containing high amount of fat also increase the risk of colon, breast and prostate cancer. Why is it that only some smokers suffer from cancer? This is because of the enzyme glutathione-S-transferase (GST) involved in detoxification of various carcinogens. Five percent of the

TABLE 76.3   Differences between tumor initiation and promoter

Initiation	Promotion
(Azodyes, aromatic amines, nitrosamines, alkylating agents)	(TPA, H_2O_2, Mezerein, Saccharin, Phenobarbital, Keto bile acids).
1. Applied once (single)	1. Multiple applications
2. Irreversible	2. Reversible
3. Rapid	3. Not metabolized
4. Metabolized to derivatives	4. Slow acting
	5. Alters phenotype (epigenetic)

population lacks this enzyme/or its isoenzymes, making such individuals more prone to cancer.

Carcinogenesis is a multistage process in which at each stage several genetic events are involved (Figure 76.3). Chemicals cause genetic damage in different ways, namely the formation carcinogen – DNA adducts, leading to base mutations or gross chromosomal changes. Adducts are formed when a mutagen irreversibly binds to DNA so that it can cause a base substitution, insertion or deletion during DNA replication. Gross chromosomal mutations are chromosomal breaks, gaps or translocations. The level of DNA damage is the biologically effective dose in a target organ and reflects the net result of carcinogen exposure, activation, lack of detoxification, lack of

DNA repair and apoptosis (Figure 76.4). No single biomarker is sufficient for use as a cancer risk marker. Examples of mutagens from food are N-nitrosoamines, occur as contaminants or from cooking practices, heterocyclic amines are formed from overheating of food with creatinine such as meat, chicken, fish and is associated with breast and colon cancer, aflatoxins to liver cancer. PAHs are associated with an increased risk of lung and skin cancer, aryl aromatic amines with bladder cancer.

Viruses for the first time shown that cancer could be transmissible. Oncogenic viruses contain either DNA or RNA as their genome. DNA viruses capable of transforming cells are Papova virus-polyoma virus, simian virus 40 (SV-40) and human papilloma virus (HPV) associated with cervical cancer and head and neck cancer, Adenovirus – adenoviruses -12, -18, -31, Herpes virus – Epstein-Barr virus, herpes virus which causes Kaposi's Sarcoma, Hepadana virus – hepatitis B virus (Table 76.2b). RNA virus or retroviruses are tumor viruses which can transform cells because they carry genes whose products interfere with the cells normal growth regulating activities. Retrovirus type C – murine sarcoma and leukemia viruses, Retrovirus type B – mouse mammary tumor virus. Polyomaviruses and SV 40 viruses have played an important role in development of viral carcinogenesis understanding though they are not responsible for causing any human tumors. They are both small, with a genome size of 5 kb, with circular genomes coding for five to six proteins. These proteins bind to DNA and cause alteration in gene expression, they show cooperative effects, suggesting that one or more process is required for transformation. Some adenovirus cause transformation of certain animal cells. For example, Hepatitis B virus is associated with liver cancer, Epstein-Barr virus is

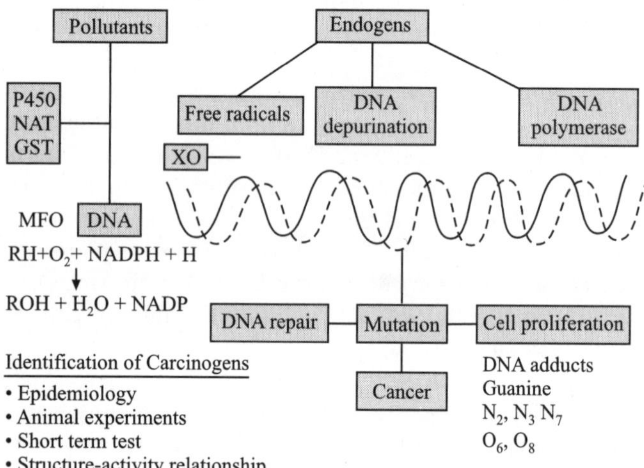

Figure 76.3 Environment-gene interactions.

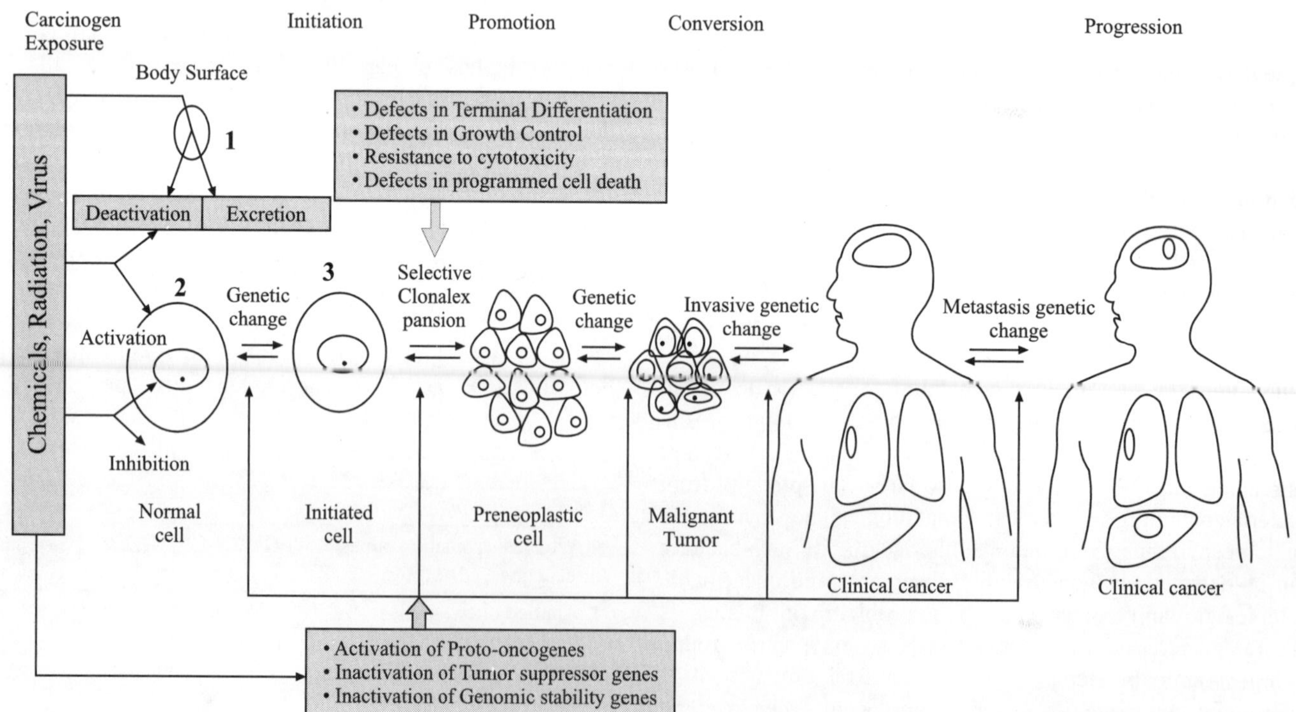

Figure 76.4 Molecular pathogenesis of lung cancer.

associated with Burkitts lymphoma. Viruses are associated with 20% of cancers worldwide. In most cases, these viruses increase a person's risk of developing cancer, rather than being the sole determinant responsible for the disease. Certain gastric lymphoma are associated with the chronic infection by the dwelling of helicobacter pylori (*H. Pylori*), which is also responsible for ulcers. Recent evidence suggests that many of these cancers linked to infections are actually caused by the chronic inflammation triggered by the presence of the pathogen. Causes of some cancers are smoking which causes lung cancer, UV causes skin cancer, asbestos fibers cause mesothelioma. The importance of environmental factors such as diet is seen in studies of the children of parents who moved from Asia to America/Europe. These individuals no longer exhibit a high rate of gastric cancer, as occurs in Asia, but instead are subject to elevated risk of colon and breast cancer, which is characteristic of western countries. Carcinogens act through a multistep process that initiates a series of genetic alterations ("mutations") and stimulates cells to proliferate. A prolonged period of time is usually required for these multiple steps. There can be a delay of several decades between exposure to a carcinogen and the onset of cancer. For example, young people exposed to carcinogens from smoking cigarettes generally do not develop cancer for 20 to 30 years. This period between exposure and onset of disease is the lag time. There is consensus among epidemiologists that some ingredients in the diet, such as fat and alcohol can increase the risk of development of cancer, whereas certain compounds found in fruits and vegetables, tea, can reduce the risk. The cancer suppressing action of nonsteroid anti-inflammatory drugs supports the idea that inflammation plays a major role in the development of various cancers. Cancer is not considered an inherited illness because most cases of cancer, perhaps 80 to 90 percent, occur in people with no family history of the disease. However, a person's chances of developing cancer can be influenced by the inheritance of certain kinds of genetic alterations. These alterations tend to increase an individual's susceptibility to developing cancer in the future. For example, about 5 percent of breast cancers are thought to be due to inheritance of particular form(s) of a "breast cancer susceptibility gene." Because a number of mutations usually must occur for cancer to arise, the chances of developing cancer increase as a person gets older because more time has been available for mutations to accumulate. For example, a 75–year-old person is hundred times more likely to develop colon cancer than a 25–year-old. Because people are living longer today than they did 50 or 100 years ago, they have a longer exposure time to factors that may promote gene changes linked to cancer.

IX. CELLULAR AND MOLECULAR BASIS OF CANCER/GENETICS OF CANCER

Chemicals, radiation, viruses, and heredity all contribute to the development of cancer by triggering changes in a cell's genes (Figure 76.4). Chemicals and radiation act by damaging genes, viruses introduce their own genes into cells, and heredity passes on alterations in genes that make a person more susceptible to cancer. Genes are inherited instructions that reside within a person's chromosomes. Each gene instructs a cell how to build a specific product in most cases, a particular kind of protein. The alterations of genes involved in regulation of cell growth and differentiation is the prime cause for transformation of a normal cell into a cancer cell. Cancer arises from a single cell (monoclonal) by uncontrolled proliferation of this single wayward cell. Genes are altered, or "mutated," in various ways as part of the mechanism by which cancer arises. The simplest type of mutation involves a change in a single base along the base sequence of a particular gene—much like a typographical error in a word that has been misspelled. In other cases, one or more bases may be added or deleted. And sometimes, large segments of a DNA molecule are accidentally repeated, deleted, or moved. Malignant transformation requires more than a single genetic alteration. Two types of genetic alterations make us more likely to develop a particular type of cancer—those that we inherit from our parents (germ-line mutations) and those which occur during a life time (somatic mutations). There are a few types of mutations that we can inherit that make us much more likely to develop cancer. The development of a malignant tumor is a multi-step process characterized by a progression of permanent genetic alterations in a single line of cells, which may occur over the course of many successive cell divisions and take years to complete. Each genetic change may elicit a particular feature of the malignant state such as protection from apoptosis. As these genetic changes gradually occur, the cells in the line become increasingly less responsive to the body's normal regulatory machinery and better able to invade normal tissues. Thus, tumorigenesis requires that the cell responsible for initiating the cancer be capable of a large number of cell division. For example, the common solid tumors such as breast, prostate, colon and lung arise in the epithelial tissues that are engaged in a high level of cell division. Same is true of leukemia, which develop into rapidly dividing blood forming tissues. The cells of these tissues can be divided into three groups (1) Stem cells, which posses unlimited proliferation potential, can self renew and can give rise to all of the tissue (2) Progenitor cells, which are derived from stem cells and posses a limited ability to proliferate (3) The differentiated end product of the tissue, which lack the capability to divide. A number of studies suggest that stem cells are the source of a variety of different types of tumors, given their long life and unlimited division potential, stem cells have the opportunity to accumulate the mutations required for malignant transformation. Alternatively the committed progenitor cells give rise to malignant tumors by acquiring certain properties such as the capacity for unlimited proliferation, as part of the process of tumor progression. As a cancer grows, the cells in the tumor mass are subjected to a type of natural selection that drives the accumulation of cells with properties that are more favourable for tumor growth. The activation of tumor can be considered an epigenetic change, one that results from the activation of gene that is normally repressed. This activation process likely involves a change in the structure of chromatin in and around the gene and or a change in the state of DNA methylation. Once the epigenetic change has occurred, it is transmitted to all of the progeny of that cell and consequently represents a permanent inheritable alteration. Even after they have become malignant, cancer cells continue to accumulate mutations and epigenetic changes that make them increasingly abnormal. This genetic instability makes the disease difficult to treat by conventional chemotherapy because cells often arise within the tumor mass that are resistant to the drug. These genetic changes are often accompanied by histological changes, i.e., changes in the appearance of the cells. The initial changes often produce cells termed 'precancerous' indicating that they have gained some of the properties of a cancer cell, such as loss of certain growth controls but lack the capability of invasion and metastasis. Some tissues often generate benign tumors, which contain cells that have proliferated to form a mass/tumor that posses a threat of becoming malignant, for example moles.

One group of genes implicated in the development of cancer are damaged genes, called "oncogenes." Oncogenes are genes whose presence in certain forms and/or overactivity can stimulate the development of cancer. When oncogenes arise in normal cells, they can contribute to the development of cancer by instructing cells to make proteins that stimulate excessive cell growth and division. Oncogenes are related to normal genes called proto-oncogenes that encode components of the cell's normal growth-control pathways and are derived from gain of functional mutations in proto-oncogenes. Some of these components are growth factors, receptors, signaling enzymes, and transcription factors (Figure 76.5). Growth factors bind to receptors on the cell surface, which activate signaling enzymes inside the cell that, in turn, activate special proteins called transcription factors inside the cell's nucleus. The activated transcription factors "turn on" the genes required for cell growth and proliferation (Figure 76.6). By producing abnormal versions or quantities of cellular growth-control proteins, oncogenes cause a cell's growth-signaling pathway to become hyperactive. To use a simple metaphor, the growth-control pathway is like the gas pedal of an automobile. The more active the pathway, the faster cells grow and divide. The presence of an oncogene is like having a gas pedal that is stuck to the floorboard, causing the cell to continually grow

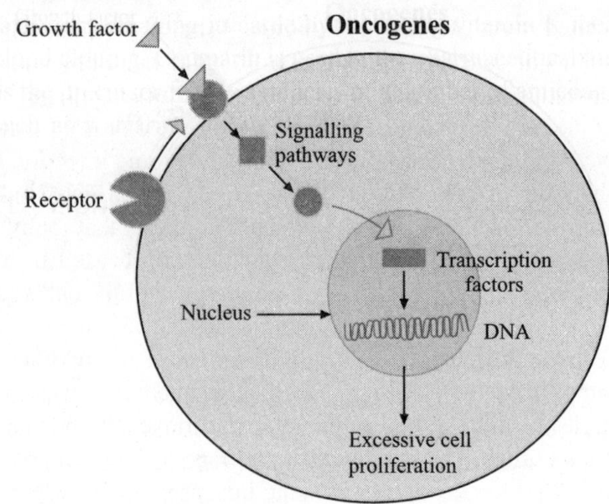

Figure 76.5   Oncogenes. (*see Plate 34 for colour figure*)

and divide. A cancer cell may contain one or more oncogenes, which means that one or more components in this pathway will be abnormal.

A second group of genes implicated in cancer are the "tumor suppressor genes." Tumor suppressor genes are normal genes whose absence can lead to cancer. In other words, if a pair of tumor suppressor genes are either lost from a cell or inactivated by mutation, their functional absence might allow cancer to develop. Individuals who inherit an increased risk of developing cancer often are born with one defective copy of a tumor suppressor gene. Because genes come in pairs (one inherited from each parent), an inherited defect in one copy will not lead to cancer because the other normal copy is still functional. But if the second copy undergoes mutation, the person then may develop cancer because there no longer is any functional copy of the gene. Tumor suppressor genes are a family of normal genes that instruct cells to produce proteins that restrain cell growth and division. Since tumor suppressor genes code for proteins that slow down cell growth and division, the loss of such proteins allows a cell to grow and divide in an uncontrolled fashion (Figure 76.7). Tumor suppressor genes are like the brake pedal of an automobile. The loss of a tumor suppressor gene function is like having a brake pedal that does not function properly, thereby allowing the cell to grow and divide continually. One particular tumor suppressor gene codes

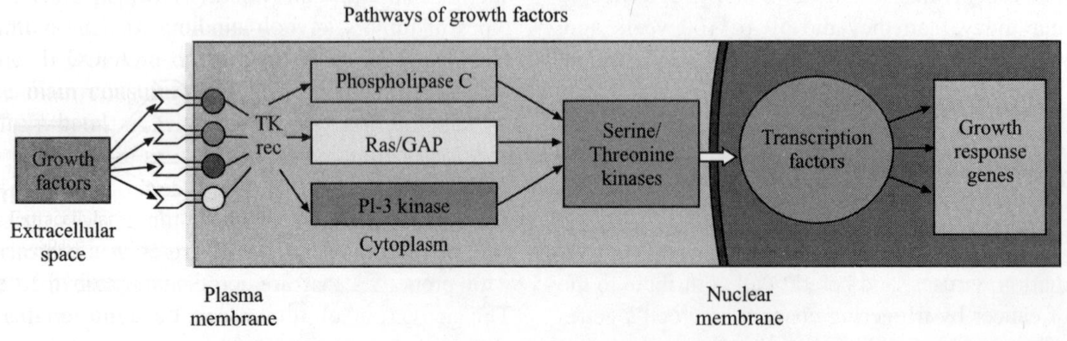

Figure 76.6   Growth factors. (*see Plate 34 for colour figure*)

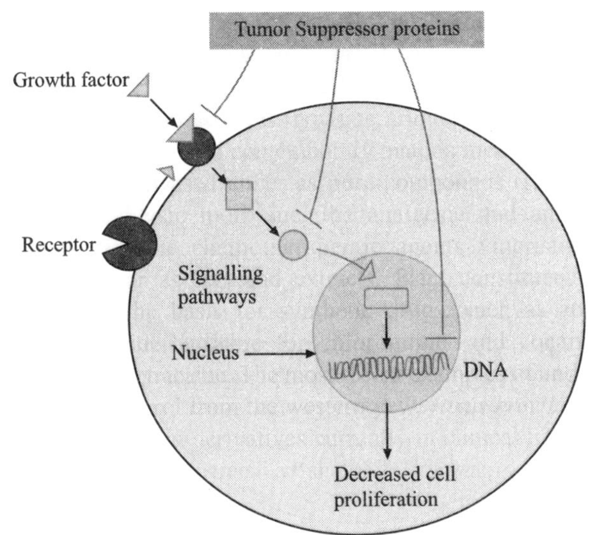

Figure 76.7 Tumor suppressor genes. (*see Plate 34 for colour figure*)

for a protein called "p53" that can trigger cell suicide (apoptosis). In cells that have undergone DNA damage, the p53 protein acts like a brake pedal to halt cell growth and division. If the damage cannot be repaired, the p53 protein eventually initiates cell suicide, thereby preventing the genetically damaged cell from growing out of control. Thus, mutations or polymorphisms in tumor suppressor genes lead to their loss of function or deletion. *One of the most common tumor suppressor genes found to be frequently mutated in a lot of cancers is P53.* This gene is involved in growth, differentiation, cell cycle regulation and apoptosis. Gene replacement strategies involve replacement of the mutated gene with its normal counterpart which would suppress the cancerous phenotype.

X. ONCOGENES

They play a crucial role in carcinogenesis. They are genes capable of causing cancer. They were first recognized as unique genes of tumor-causing viruses that are responsible for the process of transformation. They encode proteins that promote cell proliferation and help in the conversion of a normal cell to a malignant cell. They can lead to genetic instability, prevent apoptosis or promote metastasis. Analysis of the oncogene of the Rous sarcoma virus has been particularly revealing. This retro virus contains four genes namely gag, pol, env, and src. The gag gene codes for group specific antigens of the virus, pol for the reverse transcriptase, env for certain glycoproteins of the viral envelope and *src for tyrosine kinase that is responsible for transformation.* This finding revealed that abnormal phosphorylation of a number of proteins can lead to the various affects seen on transformation of the cell. *Abnormal phosphorylation of vinculin*, a protein found in focal adhesion plaques is one candidate and is *responsible for the rounding of the cells on transformation and their decreased adhesion.* Certain glycolytic enzymes are target proteins for

the src tyrosine kinase leading to increased glycolysis. The product of src may affect a large number of cellular processes by virtue of its ability to phosphorylate various target proteins and enzymes and by stimulating the pathway of synthesis of the poly phosphoionositides. Most, if not all, normal cells contain protein tyrosine kinase activity. Certain receptors such as epidermal growth factor, insulin, platelet derived growth factor are found in both normal and transformed cells, have tyrosine kinase activities that are stimulated upon interaction with their ligands. Thus, tyrosine kinases play an important role in both normal and transformed cells but their amount is raised in transformed cells. In 1976 it was discovered that an oncogene called Src, carried by an RNA tumor virus called avian sarcoma virus, was actually present in the genome of uninfected cells. The oncogene in fact was not a viral gene, but a cellular gene that had become incorporated into the viral genome during a previous infection. Thus, it become evident that cells possess a variety of genes, now referred to as proto-oncogenes, that have the potential to subvert the cells own activities and push the cell towards the malignant form. Viruses apparently incorporated cellular genes into their genomes during their passages through cells. The retention of such genes in their genomes indicated that they must confer a selective advantage on the affected viruses, presumably related to the altered growth properties of transformed cells. The cellular sequences were found to be conserved in a wide range of eukaryotic cells suggesting that they are important components of normal cells. The genes present in normal cells thus have been designated proto-oncogenes and their products are believed to play an important role in normal differentiation and other cellular processes. The abbreviation c-onc (cellular oncogene), e.g., c-ras is used to designate an oncogene present in tumor cells whereas a viral oncogene is designated v-onc, e.g., v-ras.

Mechanism of conversion of proto-oncogene into oncogene

Proto-oncogenes can be activated/converted into oncogenes by several mechanisms which either alter the structure or expression of proto-oncogenes (Table 76.4). The process by which transcription of a gene is increased is called activation. Michael Bishop and Harald Varmus, pioneers in oncogene research were awarded nobel prize in 1989. Today more than 100 human proto-oncogenes are known. They are located on specific chromosomes.

1. Point mutations: The gene can be mutated in such a way that a mutation in the gene alters the structure and function of encoded protein, i.e., it alters the properties of the gene product so that it no longer functions normally. For example, the product of v-ras oncogene is a polypeptide p21 which is related to the G proteins that modulate adenyl cyclase activity, and thus, play a major role in cellular responses to many hormones and drugs. C-ras proto-oncogene from normal human cells and that from various tumors show a difference in only one base, resulting in amino acid substitution at position 12

TABLE 76.4

1. Promoter Insertion – Upstream
   Virus infect cells → DNA copy (cDNA) of their RNA genome is synthesized by reverse transcriptase and cDNA integrated into host genome. Integrated ds DNA is called pro virus

Provirus

A ——— myc — Normal chromosome with inactive myc gene

LTR — LTR — myc — Upstream Insertion 5'–3'
↓ Transcription
≈ Myc mRNA

2. Enhancer Insertion – Downstream

3. Chromosomal Translocation – Places proto oncogene (i.e., myc) under influence of enhancer.

4. Gene Amplification – *Methotrexate is an inhibitor of enzyme dihydrofolate reductase.* Tumor cells become resistant to the action of this drug because the gene for the enzyme is amplified

5. Single Point Mutation – Difference in one amino acid only and affects conformation.

---

of p21 and the position of the mutation varies from one tumor type to the other. This mutation in p21 affects its conformation leading to its diminished GTPase activity which in turn results in chronic stimulation of the activity of adenyl cyclase which normally is dimished when GDP is formed from GTP. This increased adenyl cyclase activity results in several effects on cellular metabolism exerted by increased c-AMP levels affecting various c-AMP dependent kinases, tipping the balance of cellular metabolism to a state favouring transformation or its maintenance. This mechanism involves a change in the structure of the product of the oncogene and not necessarily in its amount/quantity. Thus, the presence of a structurally abnormal key regulator protein in a cell is sufficient to tip the scale towards malignancy. The other four mechanisms involve an increase in amount of the product of an oncogene due to increased transcription but no alteration in the structure. Thus, increased amounts of the product of an oncogene is sufficient to push a cell towards malignancy.

2. Gene amplification: This occurs in a number of tumors. The gene can become duplicated one or more times, resulting in gene amplification and excess production of the encoded protein. Tumors become resistant to the anticancer drug methotrexate which is an inhibitor of the enzyme dihydrofolate reductase. The basis for this phenomenon is that the gene for this enzyme becomes amplified resulting in an increase of several folds in the activity of the enzyme. Increased amounts of products of oncogenes such as c-ras may play a role in the progression of tumor cells to a more malignant state.

3. Chromosomal translocation: Many tumors exhibit chromosomal abnormalities and one type of chromosomal change seen in tumors is translocation. In this a piece of one chromosome is split off and joined to another chromosome. If the second chromosome donates material to the first it is called reciprocal translocation. In Philadelphia chromosome translocation occurs between chromosomes 9 and 22 as in chronic myelogenous leukemias (CML). This produces abnormal juxtaposition of the BCR gene (break point cluster gene) on chromosome 22 with part of c-ABL gene (which normally codes for tyrosine kinase) on chromosome 9 resulting in chimeric BCR-ABL mRNA which encodes bcr-abl protein and displays increased tyrosine kinase activity which in turn transforms the normal cell to a leukemic cell. Another example of reciprocal translocation is Burkitts lymphoma, i.e., cancer of B lymphocytes involving chromosomes 8 and 14 leading to transposition of c-MYC resulting in its activation. A chromosome rearrangement can occur that brings a new DNA sequence/segment from a distant site in the genome into close proximity of the gene, which can either alter the expression of the gene or the nature of the gene product or the structure of the encoded protein. Any of the above alterations can cause a cell to become less responsive to normal growth controls, causing it to behave as a malignant cell. Oncogenes act dominantly, i.e., a single copy of an oncogene can cause the cell to express the altered phenotype.

4. Promoter insertion: When retroviruses infect cells, a DNA copy (cDNA) of their RNA genome is synthesized by reverse transcriptase and the cDNA is integrated into the host genome. This integrated double stranded cDNA is called provirus. The cDNA copies of retrovirus are flanked at both ends by sequences called long term repeats and these sequences are important in the mechanism of proviral integration and can act as promoters of transcription, e.g., infection of chicken B lymphocytes by certain avian leukemia viruses, the provirus becomes integrated near the myc gene. The myc gene is activated by an upstream, adjacent viral long term repeat acting as a promoter resulting in transcription of both the corresponding myc mRNA and translation of its products in such cells and a B cell tumor ensues.

5. Enhancer insertion: In some cases the provirus is inserted downstream from the myc gene or upstream from it but oriented in the reverse direction, nevertheless the myc gene becomes activated. Such activation is not due to promoter insertion, since promoter sequence must be upstream of the gene whose transcription it increases and the sequence must be in the 5′ to 3′ direction. Instead enhancer sequences present in the long term repeat sequences of the retroviruses under consideration appear to be involved. The promoter and enhancer insertion commonly operate in viral carcinogenesis.

Mechanism of action of oncogenes

They may act on key intracellular pathways involved in growth control uncoupling them from a need of an exogenous stimulus, e.g., the product of src acting as tyrosine kinase, the product of ras acting to stimulate adenyl cyclase and the product of myc acting as DNA binding protein. There is considerable evidence that cyclins, cdks and cdk inhibitors are affected in cancer, either by mutation or secondarily. Cdk inhibitors include members of the kinase inhibitory protein (KIP) family, e.g., p21, p27 and p57 and members of the inhibitors of cdk4 (INK4) family, e.g., p16. Increase of various cyclins, e.g., cyclin D, cyclin E and mutations in genes encoding cyclins and members of INK4 family of proteins has been reported. The protein products of tumor suppressor genes such as pRB, p53 and p16 which are frequently mutated in cancer cells play an important role in regulating cell cycle. Growth factors such as PDGF,EGF,TGF beta, etc. are known to affect different types of cells ranging from blood, nervous system, mesenchymal tissues, and epithelial tissues. They exert a mitogenic response on their target cells, play a role in regulating differentiation of stem cells to form various types of mature hematopoietic cells. Growth inhibitory factors exert inhibitory effects on growth of certain cells. Thus, chronic exposure to increased amounts of growth factor or to decreased amount of inhibitory factor could alter the balance of cellular growth. The growth factors act in a paracrine, endocrine or autocrine manner. Growth factors act on the cell cycle and mitosis via transmembrane signaling binding to protein receptors on the plasma membrane of target cells just like peptide hormones. A number of receptors have been found to exhibit protein tyrosine kinase activity reminiscent of the product of the v-src gene. This kinase activity, located in the cytoplasmic domain causes autophosphorylation of the receptor protein and also phosphorylates other target proteins. The receptor ligand complexes are subjected to endocytosis in coated vesicles. For example, PDGF stimulates phospholipase C resulting in hydrolysis of PI 4,5 bisphosphate to form IP3 and DAG. These two second messengers can affect intracellular release of calcium and stimulation of protein kinase C, thus affecting a large number of cellular reactions. The subsequent hydrolysis of DAG by phospholipase A2 liberates arachidonic acid which in turn results in the production of prostaglandins and leukotrienes, which in turn exert many biological activities. Thus, exposure to PDGF can result in rapid (1–2 hours) activation of proto-oncogenes myc and fos. The product of several oncogenes are either growth factors or parts of the receptors for growth factors.

The types of proteins encoded by proto-oncogenes include (1) growth factors (2) growth factor receptors (3) protein kinases and the proteins that activate them (4) proteins that regulate the cell cycle (5) proteins that inhibit apoptosis (6) transcription factors (7) proteins involved in mitosis, tissue invasion and metastasis (Table 76.5).

1. Oncogenes that encode growth factors or their receptors:
The first connection between oncogenes and growth factors was seen in 1983, when it was discovered that the cancer causing simian sarcoma virus contained an oncogene(sis) derived from the cellular gene for platelet derived growth factor (PDGF), a protein present in human blood. Cells transformed with this virus secrete large amounts of PDGF into the medium, which causes the cells to proliferate uncontrollably. Overexpression of PDGF has been implicated in the development of brain tumors (gliomas). Another oncogenic virus, avian erythroblastosis virus was found to carry an oncogene (erb B) that encodes an EGF

TABLE 76.5 Oncogenes as signal transducers

Extracellular	Growth Factors sis, int-1, int-2, hst
Cytoplasm	Growth Factors Receptors erb-B, fms, kit
	Signal Transducers ras, src, raf, abl, mos, crk Bcl-2, Bcl-XL (Cell survival)
Nucleus	Transcription Factors ets, myc, myb, rel, ski, erb-A

receptor that is missing part of the extracellular domain of the protein that binds the growth factor. This altered receptor stimulates the cell constitutively, i.e., regardless of whether or not the growth factor is present in the medium. Thus, cultured cells carrying the altered gene proliferate in an uncontrolled manner. A number of spontaneous human cancers have been found to contain cells with genetic alterations that affect growth factor receptors including EGFR. Mostly the malignant cells contain a much larger number of the receptors on their plasma membrane as compared to normal cells which makes them sensitive to much lower concentrations of the growth factors. Mutations in EGFR occur commonly in lung cancers from non smokers but is not found in smokers who show mutations in the KRAS gene. Thus, lung cancer in non-smokers and smokers have a different course of genetic progression, although the same EGFR-Ras signalling pathway is disturbed. Growth factor receptors have become an important target for therapeutic antibodies which bind to the extracellular domain and small molecule inhibitors which bind to receptors intracellular tyrosine kinase domain.

2. Oncogenes that encode cytoplasmic protein kinases:
Overactive protein kinases function as oncogenes by generating signals that lead to inappropriate cell proliferation or survival. Raf is a serine-threonine protein kinase that heads the MAP kinase cascade, the primary growth-controlling signalling pathway in cells. Mutations that turn 'on' Raf are most likely to convert the proto-oncogene into an oncogene and lead to loss of growth control in the cell. Raf is closely linked to melanoma where BRAF mutations play a causative role in the development of nearly 70% of these cancers. SRC is also a protein kinase that phosphorylates tyrosine residues on protein substrates rather than serine and threonine residues. The substrates of Src are proteins involved in signal transduction, control of the cytoskeleton and cell adhesion. However, SRC mutations are rare in human tumor cells.

3. Oncogenes that encode transcription factors:
A number of oncogenes encode proteins that act as transcription factors. Alterations in the proteins that control cell cycle, and thus, cell division and cell growth can disturb the normal growth pattern of a cell. The oncogene MYC acts as a transcription factor and is one of the first proteins to appear when a cell in quiescent stage (G0) has been stimulated by growth factors to

re-enter the cell cycle and divide. Myc regulates the expression of a large number of proteins and miRNAs involved in cell growth and proliferation. Blocking MYC expression blocks the progression of the cell through G1 phase. MYC gene is one of the proto-oncogenes most commonly altered in human cancers, often being amplified within the genome or rearranged as the result of a chromosomal translocation which removes the MYC gene from its normal regulatory influences and increases its expression level in the cell, producing an excess of the Myc protein. Burkitt's lymphoma seen in African population results from the translocation of a MYC gene to a position adjacent to an antibody gene. The disease occurs primarily in persons infected with Epstein-Barr virus.

4. Oncogenes that encode products that affect apoptosis: The oncogene most linked to apoptosis is Bcl-2, which encodes a membrane-bound protein that inhibits apoptosis. Like MYC, the product of the Bcl-2 gene becomes oncogenic when it is expressed at a higher than normal level or when the gene is translocated to an abnormal site on the chromosome. Certain human lymphoid cancers called follicular B-cell lymphomas are correlated with translocation of the Bcl-2 gene next to a gene that codes for the heavy chain of antibody molecules. Overexpression of Bcl-2 leads to the suppression of apoptosis in lymphoid tissues, allowing abnormal cells to proliferate to form lymphoid tumors. Bcl-2 gene may also play a role by reducing the apoptosis induced by chemotherapy.

XI. TUMOR-SUPPRESSOR GENES

Genes other than oncogenes have been found to play a major role in causation of some cancers. These are tumor suppressor genes, also called recessive genes or anti-oncogenes. They operate differently from oncogenes. Their inactivation as opposed to activation removes certain constraints on growth control and causes cancer. Besides activation of oncogenes, inactivation of tumor suppressor genes are involved in cancer formation. In some cases their activation may be a secondary occurrence associated with transformation rather than a causal event. The transformation of a normal cell to a cancer cell is accompanied by the loss of function of one or more tumor suppressor genes. Nearly two dozen genes have been implicated as tumor suppressor genes in humans. For example, TP53, WT1 which encode transcription factors, RB and p16 which regulate cell cycle, NF-1 components that regulate G proteins, PTEN, a phosphoinositide phosphatase and VHL, a protein that regulates protein degradation. Most of the proteins encoded by tumor suppressor genes act as negative regulators of cell proliferation. The products of tumor suppressor genes also helps to maintain genetic stability, which is why tumors contain such an aberrant karyotype. Some tumor suppressor genes are involved in the development of a wide variety of different cancers.

RB—Tumor suppressor genes also contribute to the development of both inherited and sporadic forms of cancers.

For example, RB is associated with a rare childhood cancer of the retina of the eye called retinoblastoma. This is a malignant tumor of retinal neuroblasts which are precursor cells in the retina. It occurs at high frequency and at young age in members of certain families and it occurs sporadically at an older age among members of the population at large. The fact that retinoblastoma runs in certain families suggested that cancer can be inherited. Retinoblastoma is inherited as a dominant genetic trait because members of high-risk families that develop the disease inherit one normal allele and one abnormal allele. However, 10% of individuals who inherit a chromosome with an RB deletion never develop the retinal cancer. In some cases this tumor is inherited, whereas in others it is not. The genetic basis of retinoblastoma was explained by Alfred Knudson in 1971 who proposed that the development of retinoblastoma requires that both copies of the RB gene of a retinal cell be either eliminated or mutated before the cell can give rise to a retinoblastoma, i.e., the cancer arises as a result of two independent 'hits' in a single cell. In case of hereditary retinoblastoma, the first mutation was present in the germ line cell and the second occurred in retinoblasts. In case of sporadic retinoblastoma or nonhereditary cases, the tumor was thought to develop from a retinal cell in which both copies of the RB gene had undergone successive spontaneous mutation. People who suffer from the inherited form of retinoblastoma are also at high risk of developing other types of tumors later in life, particularly soft tissue sarcomas. Mutations in RB alleles are a common occurrence in sporadic breast, prostate and lung cancers among individuals who have inherited two normal RB alleles.

The cell cycle plays an important role in cell growth and proliferation, and thus, in the development of cancer. The gene involved in formation of retinoblastoma is RB1, located on chromosome 13q14, and its protein product is a 110kDalton nuclear phosphoprotein pRB, whose phosphorylation oscillates during the cell cycle. By using RFLP and Southern blotting it is shown that those individuals suffering with hereditary retinoblastomas were heterozygous in the region of RB1 gene, reflecting one normal and one abnormal allele whereas all tumor specimens examined were homozygous in this region. This phenomenon is called loss of heterozygosity, reflecting that both alleles are mutated in the tumor. The protein encoded by the RB gene, i.e., pRB, helps regulate the passage of cells from the G1 phase of the cell cycle into the S phase, during which DNA synthesis occurs. During the Go or G1 phase, the protein is hypophosphorylated, and its phosphorylation increases in late G1 and early S phase. The hypophosphorylated form of pRB forms complexes with a number of viral proteins such as SV40 large T antigen. pRB is inactive in such complexes, abolishing its negative regulatory effect on cell division. In conditions where oncogenic viruses are not involved, it appears to function by binding in Go/G1 to a set of proteins, some of which are transcription factors that are active in the S phase, and thus, slow cell cycling. Once a cell enters S phase, it invariably proceeds through the remainder of the cell cycle and enters mitosis. The transition from G1 to S phase is accompanied by the activation of many different genes that encode proteins ranging from DNA polymerases to cyclins and histones. E2F family of transcription factors are key targets of pRB and are involved in activating genes required for S phase activities. During G1 phase, E2F proteins are normally bound to pRB, which prevents them from activating a number of genes encoding proteins required for S-phase activities, e.g., cyclin E and DNA polymerase α. E2F-pRB complex is associated with the DNA but acts as a gene repressor. As the end of G1 approaches, the pRB subunit of pRB-E2F complex is phosphorylated by the cyclin-dependent kinases that regulate the G1–S transition. Once phosphorylated, pRB releases its bound E2F, allowing the transcription factor to activate gene expression, which marks the cell's irreversible commitment to enter S phase. A cell that loses pRB activity as a result of RB mutation would be expected to lose its ability to inactivate E2F, thereby removing certain restraints over the entry to S phase. E2F is only one of the many proteins that can bind to pRB, as RB contains at least 16 different sites that can be phosphorylated by cyclin-dependent kinases, suggesting that RB has numerous other functions as well, allowing it to interact with different downstream targets. A number of DNA tumor viruses such as adenovirus, HPV, SV40, encode a protein that binds to pRB, blocking its ability to bind to E2F. By using these pRB blocking proteins, these viruses accomplish the same result as when RB gene is deleted leading to cancer formation. Mutations in this gene are also involved in osteosarcoma and certain other human tumors. pRB is an important factor in small cell lung cancer, adenocarcinoma of prostate, tumors originating from retina, bone and connective tissues.

P53—It is also known as guardian of the genome. In 1990, p53 was recognized as a tumor suppressor gene. The gene gets its name from the product it encodes, p53, which is a polypeptide, a nuclear phosphoprotein with a molecular mass of 53 KDaltons. It is located on short arm of chromosome 17. It binds specific sequences in DNA, acts as G1 checkpoint control for DNA damage. Excessive DNA damage increases its levels, causing G1 specific cell cycle arrest allowing time for DNA repair to take place. It also triggers apoptosis so that damaged cells die, because if the cell cycle were to proceed unchecked the DNA damage would be replicated, introducing permanent mutations into the genome and also in the case if p53 is inactivated as seen in large number of tumors. P53 can activate key genes involved in apoptosis, repress genes involved in cell survival. It binds to various viral proteins, e.g., SV 40 large T antigen forming inactive oligomeric complexes. It is not required for normal cell development but its absence increases the occurrence of tumors and is associated with genome instability. When absent it is responsible for a rare inherited disorder called Li-Fraumeni syndrome. Persons affected with the above syndrome have a very high incidence of various cancers, including breast and brain cancer and leukemia. P53 mutations are the most common mutations/genetic alterations seen in human cancers especially in colon, breast and lung cancers, with nearly half of them containing point mutations or deletions in both alleles of p53 gene. Moreover, tumors composed of cells bearing p53 mutations are correlated with

a poorer survival rate as compared to those containing wild type p53 gene. P53 is a transcription factor that activates the expression of a large number of genes involved in cell cycle regulation and apoptosis. The importance of the transcription regulating role of p53 is evident in the fact that the six mutations most commonly found in p53 in human cancers map in the region of the protein that interacts with DNA thereby impacting the binding of protein to DNA or alters its conformation. P53 is particularly sensitive to mutations in its DNA-binding domain. P53 functions as a tetramer, each subunit of which consists of several domains with different functions. At least 100 different mutations have been detected in this gene which are generally found at highly conserved codons. The mutation spectrum differs among different cancers. For example, transitions occur at CpG dinucleotides frequently in colon, brain and lymphoid tissue and transversions are more common in lung and liver cancers.

P53 encodes a protein called p21 that inhibits the cyclin dependent kinase that normally drives a cell through the G1 check point. As p53 increases in the damaged G1 cell, expression of the p21 gene is activated and progression through the cell cycle is arrested. This gives the cell time to repair the genetic damage before it replicates. When both copies of p53 gene in a cell are mutated so that their product is no longer functional, the cell can no longer produce the p21 inhibitor. Failure to repair DNA damage leads to the production of abnormal cells that have the potential to become malignant. Besides cell cycle arrest, p53 can also direct a genetically damaged cell to commit suicide or apoptosis. The p53 protein leads to apoptosis as a result of several events, including the activation of expression of the BAX gene, whose encoded product Bax initiates apoptosis. p53 is also capable of binding directly to several members of the Bcl-2 family of proteins in a manner that stimulates apoptosis. For example, p53 can bind to Bax proteins at the outer mitochondrial membrane, directly triggering membrane permeabilization and release of apoptotic factors. Thus, the development of drugs, e.g., PRIMA-1 that restore p53 function to mutant p53 proteins has become an active area of research. The level of p53 in a healthy G1 cell is very low but if such a cell sustains genetic damage, the level of p53 rises rapidly. This increase is not due to increased expression of the gene but to an increase in the stability of the protein. In unstressed cells, p53 has a half-life of a few minutes. P53 degradation is facilitated by a protein called MDM2, which binds to p53 and escorts it out of the nucleus and into the cytosol. Once in the cytosol, MDM2 adds ubiquitin molecules to the p53 molecule, leading to its destruction by a proteasome. How does DNA damage lead to stabilization of p53? A protein kinase called ATM is normally activated following DNA damage and p53 is one of the proteins ATM phosphorylates. The phosphorylated p53 is no longer able to interact with MDM2, which stabilizes existing p53 molecules in the nucleus and allows them to activate the expression of genes such as p21 and BAX.

Some tumor cells have been found to contain a wild type p53 gene but extra copies of MDM2. Such cells produce excessive amounts of MDM2, which prevents p53 build up to levels to stop cell cycle or induce apoptosis following DNA damage. Thus, drugs are being developed that block the interaction between MDM2 and p53 in an attempt to restore p53 activity in cancer cells that retain p53. Gene knock out studies have shown that mice lacking the gene encoding MDM2 die at an early age of development because their cells undergo p53 dependent apoptosis, whereas mice lacking genes that encode both MDM2 and p53 (double knockouts) survive to adulthood but are highly prone to cancer. This observation illustrates an important principle in cancer genetics: even if a 'crucial' gene such as RB or p53 is not mutated or deleted, the function of that gene can be affected as a result of alterations in other genes whose products are part of the same pathway as the 'crucial' gene. In this case, overexpression of MDM2 can have the same effect as the absence of p53. As long as the tumor suppressor pathway is blocked, the tumor suppressor gene itself need not be mutated. Because of its ability to trigger apoptosis, p53 plays a pivotal role in treatment of cancer by radiation and chemotherapy. Cancer cells that have sustained DNA damage are more likely to become apoptotic as long as they have a functioning p53 gene. If cancer cells lose p53 function, they often cannot be directed into apoptosis and become highly resistant to further treatment. This is the reason why tumors that lack a functional p53 gene such as colon cancer, prostate and pancreatic cancer respond much more poorly to radiation and chemotherapy than tumors that possess a wild type copy of the gene such as testicular cancer and childhood ALL. Besides directing a potential cancer cell into either growth arrest or apoptosis, recent studies have indicated that p53 also controls signaling pathways that lead to cellular senescence. Senescent cells remain alive and metabolically active but are permanently arrested in a non dividing state. Studies suggest that oncogene activation triggers a period of accelerated division after which the senescence program takes effect as is apparent during the formation of benign moles. One of the pathways leading to senescence requires expression of a tumor suppressor gene called INK4a which is often disabled in human cancers.

Other tumor suppressor genes

There are other tumor suppressor genes besides RB and P53. Familial adenomatous polyposis colic (FAP) is an inherited disorder in which individuals develop several premalignant polyps (adenomas) from epithelial cells that line the colon wall. If not removed they are likely to progress to a malignant state. Patients with this condition were found to contain a deletion of a small portion of chromosome 5, which is the site of tumor suppressor gene APC. The conversion of cells from polyp to malignant state is presumably gained by the accumulation of additional mutations, including those in p53. Mutated APC genes are found in both inherited as well as sporadic colon tumors suggesting that the gene plays a major role in the development of this disease. The protein encoded by the APC gene binds to a number of different proteins. APC suppresses the Wnt pathway, which activates the transcription of MYC and CCND1 genes that promote cell proliferation.

The presence of mutated APC DNA has been found in blood of persons with early-stage colon cancer. Similarly, 5–10% of breast cancer are due to inheritance of a gene that predisposes the individual to develop the disease and BRCA1 and BRCA2 are identified for this. BRCA mutations also predispose a women to develop ovarian cancer. Whether a cell lives or dies following a particular insult depends on the balance between pro-apoptotic and anti-apoptotic signals. Mutations that affect this balance, such as those that contribute to overexpression of kinase PKB (AKT) or lipid kinase PI3K, can shift this balance in favour of cell survival, providing a potential cancer cell with a tremendous advantage. Another such protein is the lipid phosphatase, PTEN, which removes the phosphate group from the 3–position of PIP3, converting the molecule into PI (4, 5)P2 which cannot activate PKB. Mutations in PTEN cause a rare hereditary disease characterized by an increased risk of cancer and such mutations are also found in a variety of sporadic cancers. PTEN genes are often rendered non functional by epigenetic mechanisms, such as DNA methylation, which silences transcription of the gene.

XII. ALTERATIONS IN DNA REPAIR GENES

A third type of genes implicated in cancer are called "DNA repair genes." DNA repair genes code for proteins whose normal function is to correct errors that arise when cells duplicate their DNA prior to cell division. Mutations in DNA repair genes can lead to a failure in repair, which in turn allows subsequent mutations to accumulate (Figure 76.8). People with a condition called xeroderma pigmentosum have an inherited defect in a DNA repair gene. As a result, they cannot effectively repair the

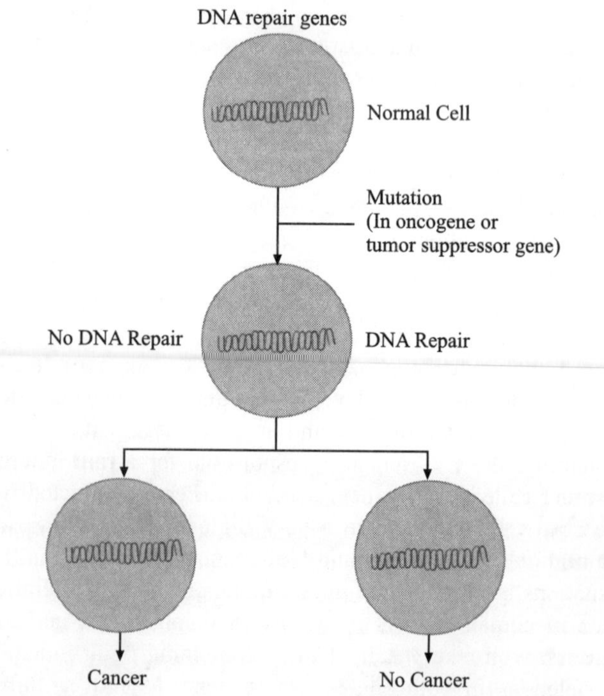

Figure 76.8 DNA repair genes. (*see Plate 34 for colour figure*)

DNA damage that normally occurs when skin cells are exposed to sunlight, and so they exhibit an abnormally high incidence of skin cancer. DNA repair defects have been identified in a number of cancer prone individuals, making them susceptible to transformation by chemical and physical carcinogens. DNA repair process requires the cooperative efforts of a substantial number of proteins, including proteins that recognize the lesion, remove it and replace it with complementary nucleotides. If any of these proteins is defective, the affected cell displays a high mutation rate and is called a 'mutator' phenotype. Cells with a mutator phenotype are likely to incur secondary mutations in both tumor suppressor genes and oncogenes which increase their risk of malignancy. BRCA proteins are part of one or more large multiprotein complexes that respond to DNA damage and activate DNA repair. Cells with mutant BRCA proteins contain unrepaired DNA and exhibit a highly aneuploid karyotpe. Unlike other tumor suppressor genes, the BRCA genes are not mutated in sporadic cancer although BRCA1 is often epigenetically silenced. The human cancer susceptibility genes BRCA 1 and -2, and ATM participate in repair of carcinogen-induced DNA ds breaks in pathways linked to homologous recombination. Certain forms of hereditary colon cancer also involve defects in DNA repair. Defective repair manifested by microsatellite instability appears to be directly linked to development of mutations in gene encoding growth factor receptor, which if inactivated permits escape from normal growth control. Inability to repair damage may lead to activation or inactivation of various genes in tumors and allow them to diverge from their normal parental genes. Mismatch repair genes are involved in the development of hereditary nonpolyposis colon cancer (HNPCC). The chain of events that lead to loss of growth control in HNPCC include mutations in genes encoding proteins of the mismatch repair system, leading to defective mismatch repair. These unrepaired mutations in turn lead to microsatellite instability, which in turn affects the growth factor receptor II leading to defective TGF beta receptor and thereby cells lose inhibitory growth response to TGF beta. The mismatch repair system corrects errors which occur while the DNA is being copied, e.g., insertion of an incorrect nucleotide. It involves six proteins which recognize the abnormal site, cut it out, and replace it with a correct sequence. The other DNA repair system, i.e., nucleotide excision repair system commonly removes carcinogen – DNA adducts. It removes larger areas of DNA damaged by chemicals or irradiation and involves 12 proteins. HNPCC is a common syndrome characterized by the early onset of cancer of the colon and other organs. Cells from this type of cancer show microsatellite instability. Microsatellites are short sequences of 2,3,4 bases of DNA repeated many times in normal cells but vary in length in cancer cells indicating that tumors had either gained or lost sequences, suggesting that the mismatch repair was not working. Mutations in the mismatch genes such as hMSH2, hMLH1 and hPMS2 account for most of the cases of HNPCC.

Determining of genetic susceptibility can be done by phenotyping or genotyping. A genetic polymorphism, e.g., single nucleotide polymorphism (SNP) is defined as a genetic variant present in at least 1% of the population. Lung cancer has been extensively studied for gene-environment interactions and, e.g., of frequently studied genetic polymorphisms in tobacco related cancers are glutathione–S–transferase M1 (GSTM1), cytochrome P 4501A1 (CYP1A1) genes, and DNA repair genes. For breast cancer there are variations in estrogen metabolism by CYP1A1, CYP1B1, CYP1A2, CYP17 and other genes can affect hormonal levels, resulting in altered cell proliferation or DNA damage and subsequent cancer risk. Determining of haplotypes that combine multiple SNP data are more predictive for cancer. Studying of mutations in P53 is suited for the study of cancer etiology, exposure, susceptibility because P53 is involved in many cellular processes. P53 mutations frequently vary by organ site and histological subtype. DNA methylation in gene promoter region is an important regulator of gene expression. Hypermethylation is a common finding in breast, colon and lung cancer. Mutations in six or seven genes may be necessary for tumor development - Familial adenomatous polyposis is an autosomal dominant inherited disorder which predisposes to colorectal cancer. These patients develop hundreds of colorectal adenomas, some of which undergo malignant change. The major features of development of colorectal cancer include the following.

1. It is a multistep process.
2. Mutations in APC initiate the process and tumor progression results from mutations in other genes.
3. Mutations in K-ras oncogene are involved.
4. Mutations in p53 and other tumor suppressor genes are also involved.
5. Mutations in genes affecting DNA repair accelerate the overall process.
6. Overall 6 to 7 genes are initially involved.
7. Additional mutations are necessary to permit invasion and metastasis.

Mutations in APC gene initiate the neoplastic process. The APC protein has been found to interact with certain other proteins including the cytoplasmic protein beta catenin, which in turn interacts with a cytoplasmic domain of E-cadherin, a cell surface protein that is a component of the adheren junctions involved in cell-cell interactions. This interaction promotes degradation of beta catenin. This requires interaction of glycogen synthase kinase 3 which phosphorylates APC protein. GSK 3 is a component of the WNT signaling pathway. When the pathway is active, GSK3 is inhibited and beta catenin is not phosphorylated by it and not degraded. When APC protein is inactivated by say mutations in the APC gene beta catenin accumulates and complexes with Tcf-4 a transcription factor leading to activation of certain colonic cell genes. If beta catenin itself is mutated as in some cases of colorectal cancer, it is no longer subjected to regulation by APC and GSK3 and accumulates and complexes with Tcf-4. One of the genes that is activated is c-MYC, the product of which is a DNA binding protein which in turn brings about cell proliferation.

Genetic susceptibility genes

The identification and characterization of genes that modify risk for cancer development has shown variation in susceptibility to chemically induced carcinogens at specific tissue sites. Genetically determined differences in the affinity for aryl hydrocarbon hydrolase receptor or other differences in metabolic rocessing of carcinogens has a major impact on experimental cancer risk. Genes such as phospholipase A2 gene, APC gene, a receptor type tyrosine phosphatase and STK6/15 – an aurora kinase have been shown to influence susceptibility to organ specific cancer induction in humans. A number of genes that increase the susceptibility to cancer have been isolated. Some cancers such as breast cancer show an increased familial incidence. Genes have been isolated which increase the susceptibility to cancer. They include RB1 responsible for retinoblastoma, P53 responsible for Li Fraumeni syndrome, APC for adenopolyposis of the colon, BRCA1 and -2 for breast and ovarian cancer, WT1 for Wilms tumor (Table 76.2d). Different methods and models have been used for assessment of chemical carcinogens for humans and for population based risk assessment including mutagenesis assays, mammalian cell cultures, animal studies, classic epidemiology, namely gene environment interactions. These genes are being used for screening of cancer.

XIII. TUMOR MARKERS

The biochemical profile of a highly malignant cell may be very different from that of a newly transformed or a normal cell. The biochemical changes include increased RNA and DNA synthesis and increased activity of ribonucleotide reductase, increased rates of aerobic and anaerobic glycolysis, decreased catabolism of pyrimidines, alterations of isoenzyme profiles, synthesis of fetal proteins, loss of differentiated biochemical functions and inappropriate synthesis of certain growth factors and hormones. Some of these changes are secondary to rapid growth rate and others due to chromosomal instability. These fast growing cells try to maximize the anabolic process involved in growth such as DNA and RNA synthesis and decrease catabolic functions such as catabolism of pyrimidines. Analysis of biochemical profile of such cells is very important in choosing chemotherapy.

Tumor markers are enzymes, proteins, hormones released from the tumor cells and can be detected in the plasma or serum (Table 76.6). They play an important role in management of cancer. They can be used for screening, monitoring the response to therapy, detection of early recurrence, for diagnosis to differentiate between benign and malignant tumors, for prognosis to see the tumor load to see whether the disease is curable or not, for localization, for staging, i.e., defining extent of disease, and for therapy. No single tumor marker is useful for all types of cancer or for all patients with a given type of cancer. Hence it is recommended to use a battery of tumor markers. Tumor markers are often detected in advanced

TABLE 76.6   Tumor markers

1. Oncofetal antigens	e.g. $\cong$ fetoprotein (liver, germ cell) carcino – embryonic antigen (colon, lung, breast, pancreas)
2. Pregnancy-associated protein	$\cong$ hCG, placental proteins (trophoblast, germ cell)
3. Enzymes and isoenzymes as tumor markers	(a) Enzymes synthesized by tumor cell, e.g., glycosyltransferese II, LDH isoenzymes (b) Enzymes derived from organ tissue with metastatic carcinoma from liver, bone, prostate – isoenzyme of alkaline PO_4ase – Acid phosphatase $\cong$ Glutamyl transferase, 5′ nucleotidase
4. Hormones as tumor markers	calcitonin (Thyroid), serotonin, insulin, glucagon, gastrin, PTH, ACTH, GH
5. Monoclonal immunoglobulins	Ig G, A, M, D, E
6. Steroid receptors	Estrogen and progesterone receptors
7. Polyamines and nucleoside	
8. Cellular markers	T & B lymphocytes

stages of cancer rather than early stage when they would have been more helpful.

Some of the tumor markers are:

Alpha Fetoprotein (AFP): It is a 70K Dalton fetal albumin with similarities to adult albumin. It is increased in blood in hepatocellular carcinoma, germ cell tumors, teratocarcinoma of ovary. Normal value is less than 15ng/L. A value of 300ng/L is often associated with cancer, but high levels are also seen in nonmalignant liver disease. The gene for AFP is located in chromosome 4.

Carcinoembryonic antigen (CEA): It is a carbohydrate antigen. Its level is increased in colorectal, breast, colon, lung, gastric, ovarian, pancreatic and uterine cancer. It can also be increased in inflammatory bowel disease, pancreatitis and liver disease.

Beta oncofetal antigen: It is an oncofetal product. Its levels are increased in serum in pancreatic cancer.

Cancer antigen 125 (CA-125): It is a glycoprotein with a mol wt of 10 million. It is so named as it reacted with a monoclonal antibody, originally termed OC-125. It is increased in serum in ovarian, pancreatic and digestive tract cancers. Normal blood levels are less than 35U/ml.

CA-19.9: It is a carbohydrate antigen. Its level is increased in serum in gastric carcinoma.

Human chorionic gonadotropin (hCG), Beta-HCG: It is a 45K Dalton glycoprotein having alpha and beta subunits. It is synthesized by normal syncytiotrophoblasts. It is increased in hydatidiform mole, choriocarcinoma, and germ cell tumors and in about 60% of testicular cancers, trophoblastic tumors.

Calcitonin (CA): It is a hormonal tumor marker. It is associated with thyroid cancer.

Prostatic acid phosphatase (PAP): It is associated with prostate cancer. An increase in its level indicates metastasis.

Alkaline phosphatase: This enzyme helps in detecting bone secondaries.

Tissue polypeptide antigen (TPA): It is a common human carcinoma antigen produced during G2 phase and released into surrounding fluids during mitosis. It is useful to assess the activity of the tumor and is seen in blood as long as the tumor cells proliferate. It helps in detecting general cancer load.

Prostate specific antigen (PSA): It is produced by the secretory epithelium of prostate gland into the seminal fluid where it is necessary for the liquefaction of seminal coagulum. It is a 32kb glycoprotein, a protease complexed in serum with α-1 antitrypsin. It is increased in blood in prostate cancer. Normal levels are less than 4ng/L. It is useful in monitoring recurrence or persistence of prostate cancer after surgery.

Other tumor markers are estrogen receptor, progesterone receptor, nuclear matrix protein, beta-2 microglobulin, prostatic acid phosphatase, immunoglobulins, hydroxyproline, calcitonin, alkaline phosphatase, etc. HER2/neu or erbB-2 or EGER2, neuron specific enolase, thyroglobulin, etc. Immunoglobulins are increased in serum in multiple myeloma. Bence Jones proteins are increased in multiple myeloma.

XIV. CANCER TREATMENT—ANTICANCER DRUGS

The prime modalities of treatment are surgery, chemotherapy/radiotherapy but they lack the specificity to kill cancer cells as they also simultaneously damage normal cells, as evidenced by the side effects that accompany chemotherapy and radiotherapy (Table 76.7). Surgery and radiotherapy are most effective in reducing the initial tumor load. Gene therapy is the alteration, insertion or deletion of genes from a cell with an aim to treat disease. It involves gene transfer in order to correct genetic defects or to express therapeutics within or near target cells. Drugs used in cancer chemotherapy act at various biochemical sites. The problem is to make available drugs that kill tumor cells effectively and spare normal cells. The effectiveness of cytotoxic drugs is directly proportional to the doubling time of the tumors and is inversely proportional to the number of cancer cells. Since a typical feature of malignant cells is unrestrained growth, hence these drugs are used because they inhibit DNA synthesis. These drugs affect all the cells in the body which are in the proliferating phase. Thus, besides the proliferating cancerous cells these drugs also affect the dividing normal cells in the body, i.e., of the GIT, hematopoietic system/bone marrow, hair follicles, gonads, and hence lead to their toxic side effects. Tumors that contain a high growth fraction are usually more responsive than cells that are dormant and in the G1 phase of the cell cycle. During therapy it is important to eliminate all tumor stem/clonogenic cells, since they have the potential for unlimited replication. High dose intermittent therapy is more likely to accomplish this rather than low dose continuous therapy, because it exposes tumor cells to higher level of the agent used. Combination therapy, i.e., combining 3 to 4 agents has proved successful in many cancers because the drugs may act synergistically/or by different signaling pathways involved in carcinogenesis. This also delays the onset of resistance and drug toxicity is also less.

Common anticancer drugs are

Alkylating agents: Nitrogen mustard gas was the first non hormonal agent that showed clinical antitumor activity. Its derivatives mustargen, vincristine, procarbazine, prednisone are used currently for treatment of Hodgkin's disease as well as other cancers. Other alkylating agents are cyclophosphamide, ifosfamide, melphalan and chlorambucil, mitomycin an antibiotic, busulfan. All compounds produce cytotoxicity by forming covalent interstrand crosslinks in DNA such as

TABLE 76.7 Cancer Treatment

1. Surgery			
2. Irradiation			
3. Chemotherapy	Alkylating agents	Antimetabolites	Antibiotics
	↓	↓	↓
	Busulphan	Methotrexate	Bleomycin
	Melphelan	5–Fluorouracil	Adriamycin,
	Cyclophosphamide		Mitomycin
			Actinomycin D
4. Immunotherapy (interferon)			
5. Hyperthermia			
6. Dietary factors			

G-X-C/G-Y-C as opposed to G-C/C-G. Some alkylating agents are monofunctional and react with only one strand of DNA, others are bifunctional and react with an atom on each of the two strands of DNA to produce a cross link that covalently links the two strands of the DNA double helix. There is greater frequency of these cross links at N7 atom of guanylates of DNA. Example is *Cyclophosphamide*. It is a cell cycle non specific drug, a derivative of nitrogen mustard. It is an alkylating agent, causing cross linkage between adjacent bases of the double helix. Thus, *DNA strands cannot separate and new DNA synthesis is blocked*. It is used for treatment of breast cancer along with doxorubicin or with methotrexate and 5–fluorouracil. It is given orally or intravenously. It is rapidly absorbed and has high bioavailability. It is inactive *in vitro* and is metabolized by cytochrome P-450 enzyme in liver to active species. Nitrosourea—it is member of alkyl nitrosoamines. Its hydrazine derivatives are procarbazine and dacarbazine. They inhibit mono amine oxidases. Used with other chemotherapeutic agents for melanomas and Hodgkins disease. *Common toxicities* of alkylating agents are *hematopoietic, GIT, gonadal, pulmonary, alopecia and immune suppression*.

It is a cell cycle specific anticancer drug. It has structural similarity to folic acid and inhibits dihydrofolate reductase thereby preventing the formation of tetrahydrofolate which is necessary for incorporation of C2 and C8 of purines and C5 methyl group in thymidine (Figure 76.9). Thus, there is inhibition of DNA synthesis and consequently cell division.

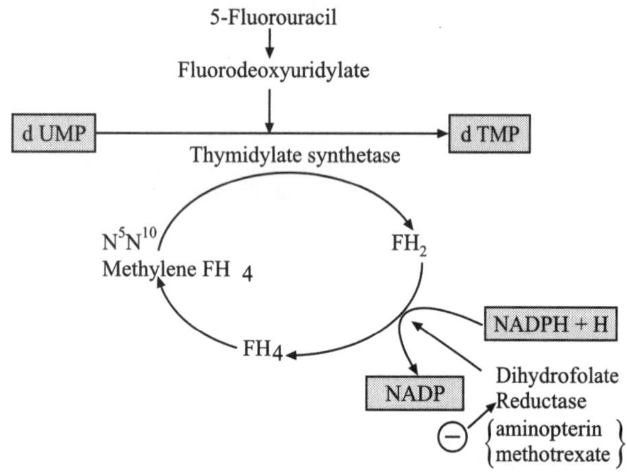

Figure 76.9 Mechanism of action of some chemotherapeutic agents.

5–Fluorouracil—It is a cell cycle specific drug, an analogue of uracil or thymine. It inhibits thymidylate synthase, thereby reducing the conversion of dUMP to dTMP (Figure 9). Thus, thymine incorporation in the DNA and consequently DNA synthesis are prevented.

Other agents

Mitomycin: It is an antibiotic derived from Streptomyces. It causes cross links between DNA strands, preventing separation of the strands and inhibits new DNA synthesis.

Vincristine and Vinblastine: They are alkaloids isolated from the leaves of vinca rosea. It has spindle movement inhibitory effect on cells undergoing cell division. Thus, the cells are arrested in metaphase and cell division is inhibited.

Azaserine: It is a derivative of serine and is a glutamine analog. It blocks the reactions where glutamine takes part in. It inhibits the first step of purine synthesis by inhibiting phosphoribosylamine synthesis, step 4 of purine synthesis, i.e., formyl glycinamide ribonucleotide and guanosine mannophosphate synthesis. Thus, purine synthesis and thereby DNA synthesis are inhibited.

6–Mercaptopurine: It is a purine analog. It is activated into 6–mercaptopurine ribotide, which prevents the amination of IMP to AMP, thus reducing the availability of AMP. This leads to inhibition of synthesis of DNA and cell division.

Thioguanine: It is also a purine analog and its mechanism of action is similar to that of 6–mercaptopurine.

L-asparaginase: It is an enzyme that converts asparagine to aspartic acid by removing one molecule of ammonia when asparaginase is injected. Blood asparagine is lowered leading to asparagine starvation and destruction of cancer cells which are deficient in asparagine synthetase. Normal cells are spared as they contain asparagine synthetase.

Cisplatin: It is a platinum compound, a cell cycle non-specific drug. It forms intrastrand DNA adducts.

Bleomycin: It is a cell cycle specific drug, an antibiotic which acts on the G2–M phase.

Etoposide: It is a podophyllotoxin. It stabilizes topoisomerase-II DNA cleavage complexes.

Adriamycin: It is an anthracyclin. It causes topoisomerase mediated breaks in DNA.

Drug resistance

A major problem in cancer chemotherapy is development of drug resistance to the drug used over a period of several months despite showing a good initial response. Resistance may also be intrinsic, i.e., the particular drug is never effective against the target tumor cells, e.g., bacteria are sensitive to penicillin as it inhibits their cell wall synthesis but since eukaryotic cells lack a cell wall, hence it is not effective in them. Intrinsic resistance can also be high metabolic inactivation of the drug by cytochrome P450. The resistance of cancer cells to chemotherapeutic agents also reflects their high mutation rates. And its development is frequently the cause of failure of cancer chemotherapy. The mechanisms of drug resistance include (1) the target enzyme or pathway is deleted (2) alternate minor pathways for drug catabolism are opened (3) decreased drug uptake (4) increased synthesis of enzymes degrading the drug (5) increased efflux (6) insufficient activation of drug (7) increased activation of drug (8) drug sequestration (9) mutation in target enzyme (10) increase of target enzyme (11) rapid repair mechanism.

Sometimes resistance is found for several several drugs simultaneously and this is called multidrug resistance. Multidrug resistance in cancer chemotherapy is due to over expression of membrane protein called *P-glycoprotein* (P is pleiotropic) which acts as an energy dependent efflux pump of drugs, and thus, diminishing their efficacy. It pumps hydrophobic drugs out of the cell by an ATP dependent process (ABC protein) thereby reducing their cellular concentration. This leads to drug resistance. Efforts are being made to develop inhibitors of P-glycoprotein. Drugs such as verapamil and cyclosporine have shown promise. P-glycoprotein is present in normal organs such as kidneys and gut and plays a role in the excretion of potentially toxic compounds by these organs. Another mechanism of drug resistance is gene amplification. When methotrexate is given, tumor cells increase 400 times the synthesis of the enzyme.

New strategies for combating cancer

Cancer is currently treated by surgery, chemotherapy and radiotherapy. However, these conventional approaches are not usually successful in curing a patient of metastatic cancer. Because the above therapies kill large numbers of normal cells, along with cancer cells, chemotherapy and radiation tend to have serious side effects, in addition to having limited curative value for most advanced cancers. Based on new insights into the molecular basis of malignancy there is hope that these would be replaced by 'targeted therapies'. A therapy can be 'targeted' by (1) targeting to attack only cancer cells and not normal cells (2) targeting a particular protein whose inactivation leaves the cancer cells unable to grow or survive (3) targeting the cancer cells of a particular patient based on unique pattern of somatic mutations. Although the cure rate for most cancers has not improved significantly over the past 50 years, there is hope that targeted therapies may be better. Some of the anticancer strategies are:

Immunotherapy: The immune system has evolved to recognize and destroy foreign materials but since cancer is derived from one's own cells, the tumor cells do contain proteins that are not expressed by their normal counterparts, or mutated proteins (e.g., *Ras*) that are different from those present in normal cells, they are still basically host proteins present in host cells. As a result the immune system fails to recognize these proteins as inappropriate. Tumor cells evolve a number of mechanisms that allow them to escape immune destruction. Many different strategies have been formulated to overcome these hurdles and artificially stimulate the immune system to mount a more vigorous attack against tumor cells. For example, immune cells, typically *dendritic cells* are isolated from the patient, stimulated *in vitro*, allowed to proliferate in culture, and then reintroduced into the patient. Sometimes the isolated immune cells are genetically modified before proliferation to enhance their tumor—attacking potential thus serving as cancer vaccines.

Immunotherapy depends on antibodies or immune cells to attack tumor cells. In recent years, two broad treatment strategies involving the immune system have been pursued, i.e., passive immunotherapy, which involves administering antibodies as therapeutic agents. These antibodies recognize and bind to specific proteins on the surface of the tumor cells being targeted. Once bound, the antibody orchestrates an attack on the cell carried out by other elements of the immune system. At present 20 or *more monoclonal antibodies have been approved to treat cancer*. Herceptin is a humanized antibody directed against a cell surface receptor HER2 that binds a growth factor that stimulates the proliferation of breast cancer cells by inhibiting the activation of the growth factor receptor and stimulates receptor internalization. Nearly 25% of breast cancers show overexpression of HER2 gene, thus making them especially sensitive to growth factor stimulation. Generally patients with HER2 have a poor prognosis. But herceptin has reduced their chance of recurrence by 50% in a four year period. Similarly, the humanized antibody Rituxan has been very effective for the treatment of non-Hodgkin's B-cell lymphoma. It binds to cell surface protein *CD 20* that is present on the *malignant B* cells, and thus, inhibits cell growth and induces the cells to undergo apoptosis. Another fully humanized antibody is Vectibix directed against EGF receptor which has shown promise for colon cancer. A new generation of antibodies are being developed that contain a radioactive atom or a toxic compound conjugated to the antibody molecule. This antibody targets the complex to the cancer cell and the associated atom or compound kills the targeted cell. Active or adoptive immunotherapy is an approach that tries to get a person's own immune system. Micro RNAs (miRNAs) are tiny regulatory RNAs that negatively regulate the expression of targets mRNAs. Since cancer arises as a result of abnormal gene expression, hence it was thought that miRNAs would be involved in tumorigenesis. In 2002, it was reported that the locus that encodes two microRNAs, miR-15a and miR-16, was either deleted or underexpressed in most cases of chronic lymphocytic leukemia. It was later found that these two miRNAs act to inhibit the expression of the mRNA that encodes the antiapoptotic protein Bcl-2. In the absence of the miRNAs, the oncogenic Bcl-2 protein is overexpressed, which promotes development of leukemia. Because these miRNAs act to inhibit tumorigenesis, they can be thought of as tumor suppressors. The locus that encodes miR-15a and miR-16 is also deleted in other types of cancer suggesting their widespread importance. Similarly, the expression of oncogenes RAS and MYC has been shown to be inhibited by miRNA, i.e., let-7. Some miRNAs act like oncogenes rather than tumor suppressors. One specific cluster of miRNA genes is overexpressed during the formation of certain lymphomas. Their overexpression could be because the gene cluster encoding them is present in increased numbers (amplified) in the tumor cells, or it can be because the gene cluster is over transcribed as a result of an overactive transcription factor. The abnormal expression of miRNAs has also been implicated as a causal factor in tumor invasion and metastasis. A number of microarray studies suggest that human tumors have a characteristic mRNA expression profile which may serve as a sensitive biomarker to identify the tumor type and the best treatment option/anticancer therapeutics.

The vaccine that has reached the most advanced stage of testing is 'Provenge', which is designed to treat patients with advanced prostate cancer that is no longer responsive to hormone treatment. Immune cells are isolated from the blood, exposed to prostate—specific protein, i.e., *prostatic acid phosphatase* that is expressed by this cancer, and then reinfused into the body. Another personalized vaccine is DC Vax which has shown remarkable promise in the treatment of brain cancer. The ultimate goal of cancer immunotherapy is to vaccinate people with antigen that would prevent them from ever developing life threatening cancers.

Targeted inhibitors/therapies: Cancer cells contain proteins present in abnormal concentrations or show abnormal activity. The growth and or survival of the tumor cells is dependent on continued activity of one or more of these proteins. A number of compounds have been synthesized which inhibit the activity of these cancer promoting proteins. These compounds are tested in vitro for their effectiveness using 60 different human cancer cell lines. This is followed by testing them against mice carrying human tumor transplants (xenografts). Some of these small molecule targeted therapies approved by FDA include *Gleevec a tyrosine kinase inhibitor which targets BCR-ABL, PDGFR; Iressa and Tareva both targeting EGFR; Sutant targeting VEGFR, PDGFR; Tykerb targeting EGFR, HER2.* All these compounds are tyrosine kinase inhibitors. Tamoxifen, Raloxifene target estrogen receptor by blocking estrogen action. Aromasin, Avimidex target aromatase by inhibiting synthesis of estrogen. Genasense targets Bcl-2 and ABT-737 targets Bcl-XL by inhibiting synthesis of these anti and pro-apoptotic proteins. *Nutlins, RITA and PRIMA-1 target p53.* Gleevec initially showed remarkable success in patients with chronic myelogenous leukemia (CML) which is due to a translocation that brings the proto-oncogenes ABL in contact with another gene BCL to form a chimeric BCR-ABL gene. Blood cells carrying this translocation express high level of Ab tyrosine kinase activity, thus causing the cells to proliferate uncontrollably and initiate the process of tumorigenesis. The drug has been in use for several years. Patients have to continue to take the drug to remain in remission but patients who begin treatment at an advanced stage, develop drug resistance due to mutations in the ABL portion of the fusion gene. The second generation of targeted inhibitors remain active against most of the mutated forms of the ABL kinase thereby suggesting that an ideal drug regimen should consist of a cocktail of several different inhibitors that target different parts of the same protein, ensuring that drug—resistant mutants do not emerge. Unfortunately none of the small molecule inhibitors studied and tested has been known to fully stop the growth of any of the common solid cancers. One reason may be that these tumors are more complex genetically and the cells involve a multitude of aberrant signaling pathways or only a fraction of the patients are sensitive to a particular drug. For example, with Iressa which is used for lung cancer, it was found that responders and nonresponders had mutations that affected different regions of the EGFR protein. This finding verifies that targeted cancer therapies will ultimately have to be personalized/tailor made. Another reason for failure

to develop more effective targeted therapies may be that the agents are not targeting the appropriate cells within the tumor. The tumor is a heterogenous mass of cells rather than a homogenous mass as was initially thought, as a result of ongoing genetic changes. While most of the cells of tumor may be proliferating rapidly, they have a limited long-term potential to sustain the primary tumor or initiate a secondary tumor. Instead a relatively small number of cells scattered throughout the tumor are responsible for maintaining the tumor and promoting its metastasis. These special cells are known as *cancer stem cells*. Efforts are underway to identify cancer stem cells in various types of tumors and learn about these properties in order to have an impact on drug development.

Inhibition of angiogenesis: As a tumor grows in size, it stimulates the formation of new blood vessels or angiogenesis to deliver oxygen and nutrients to the rapidly growing tumor. However, blood vessels also provide a means for the cancer cells to escape and spread to other sites leading to formation of secondary tumors. In 1971, Nobel Laureate Judah Folkman suggested anti angiogenesis as an anticancer therapy. Cancer cells promote angiogenesis by secreting growth factors such as VEGF, that act on the endothelial cells of surrounding blood vessels, stimulating them to proliferate and develop into new vessels. The naturally occurring inhibitors of angiogenesis are endostatin and thrombospondin. Others drugs developed include antibodies, and synthetic compounds directed against integrins, growth factors, growth factor receptors. Tumors treated with these inhibitors did not become resistant as inhibitors of angiogenesis target normal genetically stable endothelial cells which continue to respond to the presence of these agents and do not interfere with normal physiologic activities because angiogenesis is not a required activity in a mature adult. They act on cells lining the blood stream, which are directly accessible to blood borne drugs and are effective against many different types of tumors. The most promising results in inhibiting angiogenesis in human tumors have been obtained with *Avastin*, a humanized antibody that is directed against *VEGF*, the endothelial cell growth factor overexpressed in most solid tumors. It blocks VEGF from binding to and activating its receptor VEGFR, and has been shown to prolong the life of patients with metastatic colorectal cancer by several months when given with standard chemotherapy. According to one hypothesis, choking off the growth of blood vessels in a tumor might even have a negative impact by creating a more oxygen deficient environment for the tumor cells and driving them to seek out other sites in the body.

New and developing cancer therapies: Antisense therapy directed against key oncogenes. Cell cycle inhibitors which selectively inhibit key steps of cell cycle of cancer cells. Differentiation stimulators such as trans retinoic acid. Apoptosis promoters—develop agents that stimulate apoptosis selectively in tumor cells. Farnesyl transferase inhibitors by inhibiting the function of Ras proteins. Metastasis inhibitors inhibit various proteases such as metalloproteases. Protein kinase inhibitors inhibit various protein kinases involved in signaling. Telomerase inhibitors act on tumor cells in which telomerase is active. Ribozymes can degrade RNA, e.g., specific mRNAs that are unique to tumor cells. Vaccines—peptides are injected from relatively tumor specific proteins to boost the T cell immune response. Gene therapy replaces key tumor suppressor proteins, e.g., p53, pRB or mutant repair enzymes.

Screening: The best anticancer strategy is early detection. The earlier a cancer is discovered, the greater the chance of survival. Hence screening tests could significantly impact in lowering cancer related deaths. Cancer screening involves detection of unsuspected cancers in an asymptomatic population. Screening tests suitable for large numbers of healthy people must be relatively affordable, safe, noninvasive procedures. Screening for cancer can lead to earlier diagnosis in specific cases. A number of screening procedures are currently in use, e.g., breast self examination, mammography for detecting breast cancer; pap smears for cervical cancer, Prostate specific antigen (PSA) for prostate cancer, fecal occult blood testing, colonoscopy for colorectal cancer. Other biological blood biomarkers include mutant DNA, circulating tumor cells, abnormal metabolites, etc. Advances in genomics and proteomics will help in development of new screening tests based on the relative levels of various proteins present in the blood. Genomic screening tests apply for persons whose family history suggests that they might carry mutations in the BRCA1 gene and consequently be at risk of development of breast cancer.

XV. CANCER PREVENTION

Cancer prevention is defined as the reduction of cancer mortality via reduction in the incidence of cancer. It is the action taken to lower the chances of getting cancer. This can be achieved by avoiding exposure to carcinogens, pursuing a healthy life style and dietary practices, chemoprevention or early detection strategies that can result in reversal of precancerous lesions example colorectal polyps. Tobacco addiction is a major cause of cancer, with tobacco smoking and use of smokeless tobacco being a major health hazard. Smoking is the main cause of many kinds of cancer, especially oral and lung cancer. Hence, once should avoid use of tobacco products. The balance between detoxification and metabolic activation determine the susceptibility of smokers to cancer. Quitting smoking reduces the risk of lung cancer. Prolonged or repeated exposure to certain types of radiations such as UV from sunlight should be avoided, to prevent skin cancer, by avoiding going in strong direct sunlight and by wearing protective clothing. A number of mutations must occur for cancer to arise, hence the chances of developing cancer increases as a person gets older, as mutations accumulate with increasing age. Evidence suggest that high intake of animal fat increases risk of breast cancer and colon cancer. Low fat diets decrease the risk of cancer. Lack of physical exercise and adult weight gain increase risk of cancer. There is a strong link between obesity and endometrial cancer.

Thus, balancing calorie intake with adequate physical exercise is important. Dietary fiber constitutes plant polysaccharides which resist hydrolysis by the digestive enzymes. The property that makes dietary fiber a cancer preventing agent are its bulking effect, which reduces the transit time in colon and the binding of potentially carcinogenic chemicals. A diet low in fat has been associated with reduced serum estrogen levels, and thus, may play a role in reducing risk of breast cancer through reduction in intestinal absorption of estrogens excreted via the biliary system. Fruits and vegetables may be major dietary contributors to cancer prevention as they are rich in potential anticarcinogenic substances. They contain antioxidant vitamins and minerals, vitamin C, carotenoids, potassium, folate, other vitamins and are a god source of fiber. Thus, diets high in them decrease risk of many cancers. Retinoids are also shown to prevent carcinogenesis. Vitamin E and Vitamin C act as antioxidants and help in preventing the damage caused by free radicals especially in lipid peroxidation. Curcumin is also known to prevent mutations. Prevention can be taken to avoid exposure to viruses that have been implicated in human cancer, by adopting healthy lifestyles.

A. Apoptosis

Tumorigenesis occurs due to increased cell proliferation and decreased cell death or both. Apoptosis counters cell proliferation. It is a highly organized and tightly regulated mechanism of cell suicide/cell death, important to maintain cellular homeostasis in multicellular organisms. It plays an important role in physiological processes such as embryogenesis, tissue remodeling, normal development and morphogenesis. It is also known as programmed cell death and the program is intrinsic to the cell. In contrast necrosis which is due to external factors such as hypoxia and toxins is more a pathological kind of cell death. Morphological features of apoptosis include cell shrinkage, cell membrane blebbing, chromatin condensation, endonucleolytic cleavage of DNA and formation of apoptotic bodies. There is no inflammation. Necrosis includes cell swelling, bursting of the cell, release of lysosomal enzymes, accompanied by inflammation. One or more endonucleases degrade the DNA leading to ladder pattern formation on running this DNA on agarose gel for apoptosis and a smear appears for the random breakdown of DNA in case of necrosis. Two signaling mechanisms are involved in *apoptosis*. One is the extrinsic pathway, i.e., the receptor mediated pathway by *Fas/Fas ligand, TNF alpha*, and the other is *the intrinsic pathway*, i.e., the *mitochondrial mediated pathway*. Many signals are capable of modulating cell death. Apoptosis can be initiating by extrinsic pathway via death receptor which triggers a caspase cascade. Intracellular damage by intrinsic pathway involves Bcl-2 family of pro and anti apoptotic proteins as well as release of cytochrome C from the mitochondria to further activate caspases. Genetics of cell death has been studied in the nematode/worm Caenorhabditis elegans. Two sets of genes have been identified, i.e., pro-apoptotic and anti-apoptotic genes. The pro-apoptotic genes include P53,

Bax, BclXs, Bad (members of the Bcl 2 family) and the *anti-apoptotic ones* are Bcl 2, BclXl, *survivin*, etc. Bcl-2′ is anti apoptotic and extends cell survival by blocking apoptosis. It regulates cell death resulting in resistance to apoptosis, that enables accumulation of addition genetic alterations. Bcl-2 family of proteins is upstream of irreversible cellular damage in the apoptotic pathway. Many of the pro-apoptotic molecules are located in the cytoplasm or cytoskeleton and undergo post translational modification, after a death signal which allows them to target and integrate into the mitochondrial membrane.

Caspases are cysteine proteases expressed as inactive proenzymes and are activated by proteolytic cleavage after a death stimulus. Caspases function in initiation of apoptosis in response to pro-apoptotic signals and in the subsequent effector pathways to disassemble the cell. Hence, they are separated into initiator, effector and execution caspases that carry out a co-ordinated programme of proteolysis. Death effector caspase 8 and 10 mediate recruitment of initiator caspase to death receptor. Caspase 9 is involved in the mitochondrial pathway and casaspe 3 and 7 are the downstream executioner caspase, where both these caspases merge. TNFα and Fas/Fas ligand initiate cell death via receptor pathway. The intrinsic cell death pathway triggers release of cytochrome C from the mitochondria resulting in the formation of caspase 3 activating complex the apoptosome. Active effector caspases carry out their role in cell death through proteolytic cleavage of anti-apoptotic proteins, as well as the degradation of cell structures such as nuclear lamina. Inhibitor apoptosis proteininclude BIR. The mitochondrial dysfunction that occurs in cell death is manifest as alterations in the transmembrane potential, the release of proteins from the inner membrane and production of reactive oxygen species. One link is P53 which is mutated in a number of tumor types. P53 expression is induced in response to a variety of cellular stress, including DNA damage, hypoxia and oncogene activation, resulting in cell cycle arrest or apoptosis. Apoptotic pathways provide molecular targets for design of new therapeutics especially to promote apoptosis in cancer cells as cancer cells frequently possess apoptotic defects.

B. Telomeres and Telomerase

Each round of DNA replication leaves 50–200bp DNA unreplicated at the 3′ end. Telomeres are structures composed of DNA and proteins that are located at the ends of chromosomes of eukaryotes and confer stability to them. They function to cap chromosome ends and prevent end to end recombination and thereby maintain chromosomal integrity. In humans they consist of long arrays of double stranded TTAGGG repeats, a G rich 3′ single stranded overhang and are associated with repeat binding proteins. The length and sequence of telomere varies from species to species. It is 5 to 15 Kb in humans and 20–80 Kb in mice. Double strand break repair proteins assume a protective role at the telomeres. Human telomeres form telomere loop (T loop) and displacement loop (D loop) structure. In humans telomeric repeats are bound to TRF1 and TRF2 (telomeric repeat binding factors). Another protein Pot1

(protective factor 1) binds to single stranded human telomeric 3′ overhangs. Telomere maintenance and cellular response to telomere dysfunction play a crucial role in tumorigenesis, genomic instability and aging. DNA polymerase operating in cell cycle requires RNA primer for reverse strand synthesis, resulting in incomplete DNA replication of telomeres during each cell division. The solution to the end replication problem is the telomere synthesizing enzyme telomerase, a specialized ribonucleoprotein complex with reverse transcriptase activity. The telomerase holoenzyme is a large multi sub-unit complex that includes essential telomerase RNA (hTERC) molecule that has one segment that is complementary to the TTAGGG repeat, serving as a template for the addition of telomere repeats, and a telomerase reverse transcriptase (hTERT) catalytic sub-unit. It has the ability to translocate along the DNA. In normal cells telomerase levels are insufficient to maintain telomere length, resulting in progressive shortening of telomeres with each cell division. Telomerase is expressed by germ cells, stem cells, epidermal skin cells, follicular hair cells and early embryonic cells but is not expressed by most somatic cells in humans. The loss of telomeres serves as a mitotic clock that limits the number of cell divisions and cellular life span. Many tumors have shorter length of telomeres than normal cells and telomerase activity is present in most tumors but is lacking in normal cells. Thus, telomeres and telomerase activity are explored as useful targets for anticancer therapy. Measurement of telomerase activity also helps in staging of some tumors. However, normal cells with high proliferation rate have been found to have high telomerase activity and telomerase null mice have been shown to form tumors.

Critically short telomeres manifest as cellular senescence response, termed Hayflick limit, or mortality stage I, M1. Loss of tumor suppressor P53, RB pathway leads to additional cell division beyond Hayflick limit, leading to activation of senescence programme, and to telomere erosion, loss of capping function of telomeres, resulting in increased chromosomal instability, loss of cell viability, resulting in cellular crisis or mortality stage II, M2, leading to cell death and growth arrest. Activation of telomerase or activation of alternate lengthening of telomere (ALT) mechanism. Thus, there is telomere based crisis especially during early stages of neoplastic development. Increased telomerase activity is seen in >80% of all human cancers. A preceeding and transient period of telomere shortening and dysfunction contribute to carcinogenesis. Telomeres allow cells to distinguish chromosomal ends from breaks in DNA. Short telomeres induce a DNA damage response. Progressive loss of telomeres has been seen in aging. Once the telomere shrinks to a certain level, the cells can no longer divide. Its metabolism slows down, it ages and dies. Both apoptosis and senescence block tumor growth. Telomerase inhibitors are being used to selectively kill cancer cells.

SUMMARY

Cancer is a major cause of mortality worldwide. Carcinogenesis is the pathway that leads to cancer. Cancer is a disease of damaged DNA, comprising of a series of genetic mutations that can transform normal cells to cancerous cells. Cancer is a complex disease where normal cells of the body lose their growth regulation resulting in uncontrolled growth/proliferation, invasion of local tissues and spread/metastasis. They are transported by lymphatic and blood vessels and invade normal tissues elsewhere in the body and grow at new locations. Humans of all ages develop cancer, affecting a wide variety of organs. Tumors are classified as benign or malignant. Benign tumor grows locally and show no metastasis unlike the malignant tumors which show invasion and metastasis. In early stages cancer may not show any symptoms. That is why screening is important. Tumor grading provides information on the likely behavior of a tumor and its responsiveness to treatment. A low grade tumor refers to cancer cell with fewer abnormalities. Tumor staging is based on size, invasion and spread of cancer. A patient's chances of survival are better if the cancer is detected at an early stage. Cancer is multifactorial and is caused by intrinsic factors such as hereditary/genetic, diet and hormones, and by extrinsic environmental factors, viruses and bacteria. Physical agents such as UV rays, X-rays, gamma rays; chemical carcinogens such as polycyclic aromatic hydrocarbons, nitrosamines, aromatic amines, alkylating agents; and DNA and RNA viruses such as retroviruses, papovavirus, Epstein barr virus, herpes virus being the causative factors. Chemicals and radiations act by damaging genes, virus introduce their own genes into cells and hereditary passes on alterations in genes that make a person more susceptible to cancer. Genes can be mutated in different ways, such as point mutation, insertion, deletion, translocation and transversion. Cancer linked virus insert and change genes responsible for cell growth. Oxidative stress plays an important role in increasing the mutation rate. Carcinogenesis involves initiation, promotion and progression. Cancer cells show altered morphology such as large number of irregular shaped dividing cells and loss of normal specialized cell features. Some of the characteristics of a transformed cell include altered cell morphology, increased cell density, loss of anchorage dependence, loss of contact inhibition, alteration in cytoskeletal structure, and diminished requirement of growth factors. Cancer tends to involve multiple mutations. One group of genes implicated in development of cancer are mutated oncogenes which then accelerate cell growth and division. Oncogenes are related to normal genes called proto-oncogenes that encode normal cell growth factors, their receptors, signaling enzymes such as protein kinases, transcription factors and products that affect apoptosis. Oncogenes arise from mutations of proto-oncogenes. The conversion of proto-oncogenes to oncogenes involves promoter/enhancer insertion, chromosomal translocation, gene amplification and point mutation. Tumor suppressor genes are normal genes whose absence leads to cancer. They affect transcription of genes involved in cell cycle regulation. Mutation or inactivation of tumor suppressor genes can lead to cancer since they code for proteins that slow down cell growth and cell division. Other type of genes implicated in cancer are DNA repair genes, and mutations in them can lead to a failure to repair, allowing mutations to accumulate. Other genes involved in growth regulation which are abnormal in cancer cells include apoptosis related genes, and growth factors and their receptors. Thus, cancer progression involves multiple hits/mutations in several of these genes before the development of the malignant tumor. Metastasis involves the spread of cancer cells from the primary site of origin to other tissues and organs where they grow as secondary tumors, and is a major problem.

Many proteases, glycoproteins, glycosphingolipids of the cell surface, changes in oligosaccharide chains of cell glycoproteins play an important role in this phenomenon. Alterations in proteins involved in cell-cell interactions, e.g., integrins, cadherins, cell adhesion molecules such as neural cell adhesion molecule (N-CAMs) have also been shown to be involved. The blood supply to the tumors, i.e., angiogenesis is important for their growth and spread. Tumor cells can secrete angiogenic growth factors which promote the proliferation of endothelial cells and the formation of new capillaries. Certain enzymes, hormones, proteins are released into the blood and act as tumor markers for screening, diagnosis and monitoring response to therapy. Cancer treatment essentially involves surgery, chemotherapy, immunotherapy and radiotherapy. Chemotherapeutics include antimetabolites, alkylating agents, antibiotics, cell cycle inhibitors, apoptosis stimulators, differentiation stimulators, metastasis inhibitors, i.e., anti-angiogenic agents. Monoclonal antibodies, gene therapy, etc. are also being used. Multidrug resistance is an important phenomenon, with P-glycoprotein playing an important role in this.

Cancer prevention is important. Since carcinogens trigger cancer people can reduce their cancer risk by avoiding such agents, having a healthy diet and life style. One should avoid exposure to carcinogens at work and one should limit intake of fat and calories as they have been shown to be associated with increased cancer risk, especially of colon cancer. Consumption of fruits and vegetables has been strongly correlated with a reduction in cancer risk, as they contain protective antioxidants and other compounds. Intake of dietary fiber has shown an inverse correlation with cancer. It increases bowel motility and the binding properties of fiber reduce the exposure of gut to carcinogens. An inverse correlation between cancer and intake of foodstuffs rich in antioxidants has also been shown. Tumorigenesis occurs due to increased cell proliferation and decreased cell death or both. Apoptosis counters cell proliferation. Apoptotic pathways provide molecular targets for design of new therapeutics especially to promote apoptosis in cancer cells. Telomere maintenance and cellular response to telomere dysfunction play a crucial role in tumorigenesis. Many tumors have shorter length of telomeres than normal cells and telomerase activity is present in most tumors. Thus, telomeres and telomerase activity are explored as useful targets for anticancer therapy.

SUGGESTIONS FOR FURTHER READING

Abbruzzere J.L. and Lippmann S.H. (2004), The Convergence of Cancer Prevention and Therapy in Early Phase Clinical Drug Development, *Cancer Cell*, 6, 321.

Alderton G.K. and Bordon Y. (2012), Tumor Immunotherapy–Leukocytes Take up Fight, *Nat. Rev. Immunol.*, 12, 237.

Conradt B. (2002), With a Little Help from Your Friends: Cells Don't Die Alone, *Nat. Cell Biol.*, 4, E139.

Chang E.H., Wei M.C., et al. (2001), Bcl-2, Bcl-XL, Sequesters BH3 Domain only Molecule Preventing Bax and BAK Mediated Mitochondrial Apoptosis, *Mol. Cell*, 8, 705.

Daniel N.N. (2007), Bcl-2 Family Protein: Critical Checkpoints of Apoptotic Cell Death, *Clin. Cancer Res.*,, 13, 7254.

Hanash S.M., Baik C.S. and Kallioniemi O. (2011), Emerging Molecular Biomarkers—Blood Based Strategies to Detect and Monitor Cancer, *Nat. Rev. Clin. Oncol.*, 8, 142.

Hengartner M.O. (2000), The Biochemistry of Apoptosis, *Nature*, 407, 770.

Ki Won Lee, Ann M. Bode and Zingang Dong (2011), Nat. Rev. *Cancer,* 11, 211.

Levine A.J. (2011), Introduction: The Changing Directions of p53 Research, *Genes Cancer*, 2, 382.

Liston P., Fong W.G. and Korneluk R.G. (2003), The Inhibition of Apoptosis; There is More to Life than Bcl-2. *Oncogene.*, 22, 8568.

Miller U. and Marx J. (1998), Apoptosis, *Science*, 281, 1301.

Podlevsky J.D. and Chen J.J. (2012), It all Comes Together at the Ends, Telomerase Structure Function and Biogenesis, *Mut. Res.*, 730, 3.

Prescott J., Wentzensen I.M., Savage S.A. and De Vivo I. (2012), Epidemiological Evidence for a Role or Telomere Dysfunction in Cancer Etiology, *Mut. Res.*, 730, 75.

Proteni M.M., Nagle C.M. and Webb P.M. (2012), Obesity and Ovarian Cancer Survival: A Systemic Review and Meta-analysis, *Cancer Prev. Res.*, 1–10, Doi-10.1158/1940–6207.CAPR-12–0048.

Shay J.W. and Wright W.E. (2002), Telomerase a Target for Cancer Therapeutics, *Cancer Cell*, 2, 257.

Shay J.N. and Wright W.E. (2011), Role of Telomere and Telomerase in Cancer, *Semin. Cancer Biol.*, 6, 349.

Shi Y. (2002), Mechanism of Caspase Activation and Inhibition During Apoptosis, *Mol. Cell.*, 9, 459.

Singh A., Sharma H., Salhan S., Gupta S.D., Bhatla N., Jain S.K., et al. (2004), Evaluation of Expression of Apoptosis Related Protein and their Correlation with HPV, Telomerase Activity and Apoptotic Index in Cervical Cancer, *Pathobiology*, 71, 314.

Tait S.W.G. and Green D.R. (2010), Mitochondria and Cell Death; Outer Membrane Permeabilization and Beyond. *Nat. Rev. Mol. Cell Bio.*, 11, 621.

Wong K.K. and De Pinho R.A. (2003), Walking the Telomere Plank into Cancer, *J Natl. Cancer Inst.*, 95, 1184.

Wong J.M. and Collins K. (2003), Telomere Maintenance and Disease, Lancet, 362, 983.

SECTION X

STEM CELLS AND REGENERATIVE THERAPY

An important discovery in recent years has been on stem cells, or root cells, which divide and differentiate to give rise to various organs and tissues. Depending on the stage, these are pluripotent, multivalent and univalent. Their role in the formation of various organs is obvious. They also constitute a valuable resource for regenerative therapy. It has also been realized that the adult tissues have a reserve of stem cells with capacity to multiply to fill in the gap or renew the lost cells. Chapter 38 (in Section III) covers the use of stem cells in restoring vision, which is now a clinical reality and is available to patients.

77

Pluripotent Stem Cells from Human Embryos and Adult Gonads

Deepa Bhartiya, Punam Nagvenkar, Kalpana Sriraman and Sandhya Anand

CONTENTS

I. INTRODUCTION

Stem cells are today one of the most fascinating areas for both biologists and clinicians and have become a new hope for the development of novel cell therapies for regenerative medicine. Compared to adult stem cells, pluripotent stem cells (PSCs) have much more plasticity and they have the ability to differentiate into all three germ layers viz. ectoderm, endoderm and mesoderm. Besides human embryonic stem (hES) cells from spare human embryos, PSCs are also present in adult body tissues. It is believed that pluripotent stem cells from epiblast stage embryo while migrating to the gonadal ridge during early embryonic development as primordial germ cells also migrate to other body organs and persist throughout life and serve as a backup pool of relatively quiescent stem cells to maintain homeostasis by giving rise to tissue committed stem cells or 'progenitors' which divide rapidly and differentiate into various cell types. The present chapter is a brief description of pluripotent stem cells, hES cells isolated from inner cell mass of spare human embryos and very small embryonic-like stem cells (VSELs) isolated from adult ovary and testis.

II. HUMAN EMBRYONIC STEM CELLS

Embryonic stem cells are versatile by nature. As their name suggests, ES cells are derived from embryos, more specifically from the inner cell mass of the blastocyst. During embryonic development, on day 5–6 post-fertilization, the fertilized human egg grows into a ball of cells called the blastocyst. The blastocyst comprises of two structures viz., an outer layer of cells, the trophectoderm that surrounds a fluid-filled cavity known as the blastocoel and a group of approximately 30–40 cells on the interior that forms the inner cell mass (ICM). Human ES cells are defined by distinct properties (i) self-renewal—ability to proliferate indefinitely (ii) pluripotency—give rise to cells of all the three embryonic germ lineages viz. ectoderm, mesoderm and endoderm and (iii) genetic stability over long time in culture. The hES cells possess a high nucleus-to-cytoplasm ratio, exclude Hoechst dye and have long telomeres indicating their ability to replicate for long periods. ES cells exhibit high activity of endogenous alkaline phosphatase and express nuclear (Oct-4, Nanog) and cell-surface (SSEA-3/4, TRA-1-60/81) markers of pluripotency. When cultured in suspension they tend to clump together to form embryoid bodies—an aggregate of cells containing all three embryonic germ layers. Besides, they can form teratomas when injected in immune deficient (SCID) mice, are clonogenic and are capable of producing chimeras when injected into blastocysts in the mouse model. ES cells were first derived from mouse embryos in 1981 by Evans and Kaufman. A breakthrough occurred in 1998 with the derivation of hES cells by Thomson and colleagues. Although similar in their characteristics, some differences exist between mouse ES (mES) cells and hES cells (Table 77.1).

TABLE 77.1   Comparison of mouse and human ES cells

Properties and cell markers	Mouse ES cells	Human ES cells
Oct-4	+	+
Nanog	+	+
SSEA-1	+	-
SSEA-3/4	-	+
TRA-1-60/81	-	+
Alkaline phosphatase	+	+
Telomerase activity	+	+
Formation of embryoid bodies	+	+
Teratoma formation	+	+
Chimera formation	+	+

A. ES Cell Derivation, Culture and Differentiation

Since the first report on derivation of hES cell lines in 1998, at least 1071 hES cell lines have been derived world-wide. Besides spare human blastocysts, hES cell lines have also been derived from morula stage embryos, abnormally developing and arrested embryos, single blastomeres of 8-cell stage embryos and 4-cell stage embryos. Seven well-characterized hES cell lines derived using spare human embryos have been developed in the Indian subcontinent by Mandal et al, Inamdar et al and Kumar et al of which two, KIND-1 and KIND-2 were developed by our group, and propagated on human feeders by manual cut and paste method.

The derivation of ES cells is a multi-stage process (Figure 77.1) that first involves the isolation of ICM from the surrounding trophectoderm of the blastocyst. This is an extremely crucial and technically challenging step for the successful generation of ES cell lines. The ICM can be separated by various approaches depending upon the quality and grade of the blastocyst. The methods include mechanical isolation, laser microsurgery, use of acid Tyrode's or recombinant protease to digest the zona pellucida, immunosurgery—an antibody mediated dissolution of the trophectoderm and direct culture of spontaneously hatched blastocyst.

The resulting ICM cells are then plated in culture dishes containing growth medium and coated with a layer of cells known as "feeder" that are inactivated so that they are not dividing and expanding. The feeder layer provides a surface for attachment of the ICM and secretes growth factors into the

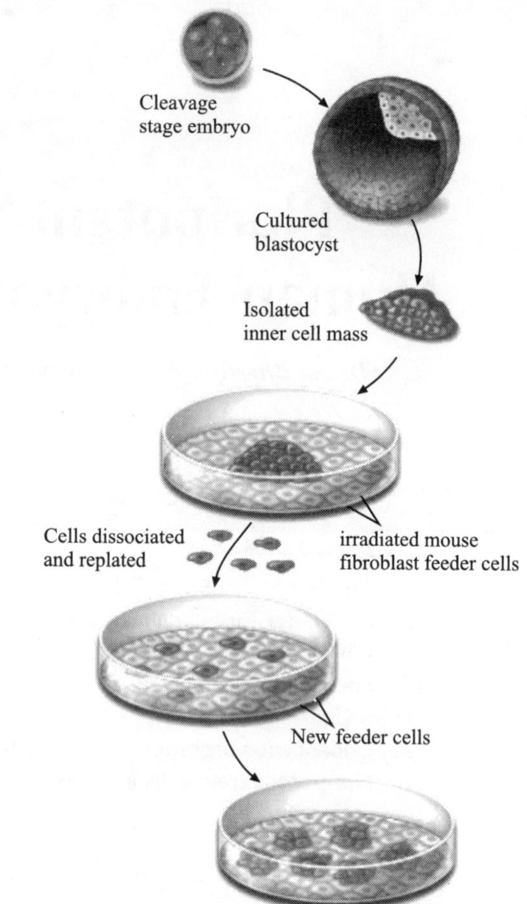

Figure 77.1   Schematic representation for derivation of embryonic stem cells. (Adapted from http://stemcells.nih.gov/) (*see Plate 35 for colour figure*)

culture medium that sustains ES cells in an undifferentiated state. Both mouse embryonic fibroblast (MEF) and human feeders (fetal muscle, skin, foreskin, adult fallopian tube epithelial cells) have been used. Conventionally, MEFs are best used between three and six passages whereas human feeders can be maintained for >60 population doublings. Mouse ES cells can be grown *in vitro* without feeder layers if the cytokine leukemia inhibitory factor (LIF) is added to the culture medium, however human ES cells do not respond to LIF.

After 7-8 days in culture, the attached ICM cells proliferate gradually to form colonies. The colonies are dissociated mechanically and replated in the same culture conditions. The ES cells are further expanded by mechanical or enzymatic passaging of the cells every 4-7 days. The cell population doubling time for hES cells is higher (30-35 hr) compared to mES cells (12–15 hr). Once the cell line is established it can be propagated and cryopreserved for future use. Furthermore, culture conditions have been developed for feeder-free culture of ES cells using a reduced growth factor basement membrane matrix and a more defined medium which supports feeder-free growth of ES cells.

ES cells have the ability to form different cell types in our body. Various studies have shown that ES cells can be differentiated into neuronal, hematopoietic, endothelial, muscle, cardiac, pancreatic, hepatic lineages. Moreover ES cells have also been shown to differentiate *in vitro* into female and male gametes. Essentially, ES cells can be differentiated spontaneously by embryoid body (EB) formation or by directed differentiation using a cocktail of growth factors. On removal of the feeder layer, ES cells tend to cluster together in suspension culture on a non-adherent surface to form 3D aggregates known as embryoid bodies that may be simple or cystic. After plating on gelatin-coated plates EBs differentiate spontaneously into many cell types. Alternatively, EBs can be dissociated, replated and the monolayer cultures can be treated with growth factors that can activate different signalling pathways for further cell differentiation. Several growth factors have been shown to direct differentiation of ES cells. (i) Activin-A and transforming growth factor (TGF-β1) mainly induce mesodermal cells (ii) Retinoic acid (RA), epidermal growth factor (EGF), bone morphogenic protein 4 (BMP-4), and basic fibroblast growth factor (bFGF) activate ectodermal and mesodermal cells (iii) β nerve growth factor (NGF) and hepatocyte growth factor (HGF) differentiate all three embryonic germ layers. Directed differentiation is a more controlled process involving stage specific sequential addition of growth inducers and inhibitors which are known to effect key pathways. For example Activin A and BMP4 are two such growth factors which have been used widely for cardiogenic differentiation. Although hES stem cell lines are similar with respect to self-renewal and expression of pluripotency markers, published literature however suggests that they exhibit differences in their differentiation ability under identical culture conditions.

B. Ethical Guidelines for Human ES Cell Research in India

Embryonic stem cell research holds great promise in the arena of regenerative therapy. However, it is associated with several ethical and social issues such as destruction of human embryos to create hES cell lines. Any research on human beings as subjects of medical or scientific experimentation need to follow the principles outlined in the "Ethical Guidelines for Biomedical Research on Human Participants" issued by the Indian Council of Medical Research (www.icmr.nic.in/bioethics). Moreover, the Indian Council of Medical Research and the Department of Biotechnology have jointly drafted the "Guidelines for Stem Cell Research and Therapy" in India (http://icmr.nic.in/stem_cell/stem_cell_guidelines.pdf). All institutions and investigators, both public and private, involved in stem cell research are expected to constitute a functional Institutional Committee for Stem Cell Research and Therapy (IC-SCRT) with appropriate expertise as indicated in the guidelines and must be registered with the National Apex Committee for Stem Cell Research (NAC-SCR). Human ES cell research is a permissible area of research. The establishment of hES cell lines from spare, supernumerary, donated embryos is permitted with prior approval of the IC-SCR and Institutional Ethics Committee and after obtaining written informed consent from the donor as per the guidelines. Once the cell line is established, it has to be registered with the IC-SCR and NAC-SCRT.

C. Potential Use of ES Cells

The remarkable features of hES cells have served as an important breakthrough for basic research and has great potential for regenerative medicine. ES cells may act as key research tools for understanding the complex events that occur during embryonic development which may explain the causes of birth defects. They are ideal candidates for studying apoptosis in early stage of embryo, mechanism of differentiation, mutagenesis, immune rejection and aging. Human ES cells and their derivatives may be used for testing therapeutic drug efficacy and toxicity. They also have wide applications in tissue engineering. Following their culture on polymer scaffold, it has been reported to coax stem cells to form tissues with characteristics of developing human cartilage, liver, neurons and blood vessels.

Despite being associated with the risk of inducing teratomas and immune rejection, the vital potential application of hES cells is the generation of cells and tissues that could be used for cell-based therapies. Human ES cells directed to differentiate into specific cell types offer the possibility of a renewable source of replacement cells and tissues to treat a myriad of diseases and disabilities including Parkinson's and Alzheimer's diseases, spinal cord injury, burns, heart failure and diabetes etc. The first FDA-approved phase-1 clinical trial for safety began with Geron's (Menlo Park, CA, USA) GRNOPC1 derived oligodendrocyte progenitor cells to treat complete thoracic-level spinal cord injury (Mayor, 2010). The trial was initially stalled for occurrence of microscopic cysts in animal transplants but was later approved (Alper, 2009; De Francesco, 2009). However, in November 2011 Geron dropped out of stem cell research for financial reasons and said that they would continue to monitor existing patients, and were attempting to find a partner that could continue their research. The recent success of a prospective clinical study of Advanced Cell Technology (CA and MA, USA) to establish the safety and tolerability of subretinal transplantation of hES cell-derived retinal pigment epithelium (RPE) in patients with Stargardt's Macular Dystrophy (SMD) and Dry age-related Macular Degeneration (Dry AMD) represents an important step towards therapeutic use of hES cells (Schwartz et al., 2012). Although long-term follow up is essential and eye is an immune-privileged site; it is still encouraging to note that there are no associated signs of hyperproliferation, tumorigenicity, ectopic tissue formation, or immune-rejection after 4 months of transplantation.

III. STEM CELLS IN OVARY

The long-held dogma in female biology is that women and other mammalian females are born with fixed and non-renewing pool of germ cells, which are enclosed in structures called follicles. The number of follicles decreases with age due to ovulation or atresia and their exhaustion leads to menopause. This theory that is believed to occur in only female mammals is considerably different from germ cell development in invertebrate females in which germ-line stem cells renew the pool of oocytes in adult ovaries. This evolutionary disparity seems to be mammalian female specific as all metazoan males are known to have spermatogonial stem cells to maintain spermatogenesis throughout life. The theory of non renewing pool of oocytes was put-forth by Sir Solomon Zuckerman in 1951 solely based on his personal interpretation of the data available at that time, that none of the available data is inconsistent with non renewing pool of oocytes endowed in females at birth. This doctrine was challenged in recent years by several researchers across globe and has provided strong evidence for post-natal oogenesis and new follicle formation.

A. Evidence for Presence of Stem Cells in Ovary and Post-natal Oogenesis

The first breakthrough that challenged the 50 years old theory put-forth by Sir Zuckerman came from a paper published in 2004 by Dr Jonathan Tilly group (Johnson et al., 2004). Through simple and elegant experiments conducted using mouse ovaries, they put-forth evidence that were inconsistent with the existing non-renewing pool of oocyte theory. Through assessing the number of healthy and atretic follicles in postnatal mouse ovaries, they demonstrated that there is high discordance between changes in non-atretic follicle numbers and the corresponding incidence of atresia. Through histological studies, they showed the presence of mitotically active germ cells and cells expressing meiotic entry markers. They also showed wild-type ovaries grafted into transgenic female mice with ubiquitous expression of green fluorescent protein (GFP) become infiltrated with GFP-positive germ cells that form follicles.

They proposed an extra-ovarian, i.e., bone marrow origin for these stem cells based on studies of parabiotic exchange of blood between different mice. Since then, many have corroborated presence of possible ovarian stem cells and post-natal oogenesis using different techniques, although various researchers have questioned the theory for origin of ovarian stem cells by Tilly group. Another study by an independent group quantified all healthy follicles in C57BL/6 mouse ovaries from day 1 to 200 using unbiased stereological methods and showed that primordial follicle numbers remain relatively constant and suggested that active folliculogenesis persist in postnatal ovaries. Some studies have shown increased oocyte numbers in adult ovaries by either pharmacological enhancement of basal oogenesis or oocyte regeneration after pathological insult that initially depletes the resting follicle pool. Tilly's group further reported that aged mouse ovaries possess pre-meiotic germ cells that differentiate into oocytes on transfer into a young ovarian environment. Existing controversies and

recent advances in the field were recently reviewed (Notarriani 2011; Bhartiya et al., 2013).

A key finding supporting claims that adult mouse ovaries retain the capacity for oogenesis was provided by Zou and co-workers, who published isolation and propagation of cells similar to male spermatogonial stem cells from neonatal and adult mouse ovaries (Zou et al., 2009). These cells, termed female germ-line stem cells (FGSC) were isolated based on presence of very conserved germ cell marker (MVH-Mouse Vasa Homolog) and were cultured on mouse embryonic fetal fibroblasts using a medium which was used for culturing male germ stem cells. They successfully cultured them for more than 15 months and showed that these cells can reconstitute ovarian function in adult female mice rendered sterile by treatment with busulphan and cyclophosphamide. Furthermore, mating trials performed in females transplanted with cultured stem cells modified to express GFP, yielded offspring containing the GFP transgene in their genomic DNA. The presence of FGSCs in postnatal mouse ovary was also confirmed using transgenic mice that express GFP under the control of Oct-4 promoter, an important stem cell marker.

Studies on human ovarian stem cells are relatively few in number because of scarcity of the ovarian tissue for research. First study came in 2005 which showed that scraped surface epithelium of postmenopausal human ovary develops into oocyte-like structures. Another study identified putative stem cells in ovarian sections and also in scraped ovarian surface epithelium (OSE) of postmenopausal women and those with premature ovarian failure. These stem cells express pluripotent transcripts Oct-4, Sox2 and Nanog, expressed cell surface antigen SSEA-4 and differentiated into oocyte-like structures and parthenotes *in vitro*. Our study showed the presence of pluripotent VSELs in ovaries (OSE), which can be easily isolated by gentle scraping of ovarian surface epithelium in adult rabbit, sheep, monkey and peri-menopausal women (Bhartiya et al., 2013 and 2014). These stem cells spontaneously differentiate into oocyte-like structures and parthenotes *in vitro* in agreement with studies mentioned above. Very recently, Tilly group have isolated and cultured ovarian germ stem cells from ovaries of reproductive age women through FACS sorting and have shown that injection of the human germline cells, engineered to stably express GFP, into human ovarian cortical biopsies leads to formation of follicles containing GFP-positive oocytes 1–2 weeks after xenotransplantation into immunodeficient female mice.

The interesting aspect of the studies mentioned above is that most of them have localized presence of stem cells to OSE. During embryonic development, germ cells are formed from primordial germ cells (PGCs) that originate from proximal epiblast. During development, PGCs migrate from epiblast to developing gonads and gets associated with the somatic cells to form follicles. There is evidence for OSE forming the partial source for granulosa cells in the developing follicles and hence it is understandable to find stem cells associated with OSE. This is further supported by studies that have shown that OSE have multipotent characteristics. Several studies including our study have also shown that OSE cells on culture show epithelial-mesenchymal transition to form putative granulosa cells.

B. Pluripotent Stem Cells in Ovary

Presence of pluripotent stem cells in adult human and mice ovary has been demonstrated by many groups. Our group has identified two distinct types of stem cells in OSE of human and other mammalian species. The two stem cells are very small embryonic-like stem cells (VSELs) that are pluripotent and slightly larger progenitor committed cells termed ovarian germ stem cells (OGSCs). The pluripotent VSELs are probably the PGCs or their precursors persisting into adulthood. Based on this study in ovarian stem cells and other literature, a probable model for oogenesis and follicular assembly in adult mammalian ovaries was proposed (Figure 77.2).

IV. STEM CELLS IN TESTIS

Spermatogenesis is a unique biological phenomenon in which successive mitotic and meiotic divisions give rise to the only cell type in the body capable of germline genetic transmission, i.e., gamete spermatozoa. Testis comprises of germ cells and somatic cells which are organized in a well defined manner in small units termed seminiferous tubules.

The germ cell population comprises of spermatogonial stem cells (SSCs, undifferentiated and differentiated), spermatocytes (primary and secondary), spermatids (round and elongating) and spermatozoa. SSCs undergo proliferation, meiotic divisions and differentiation to result in the formation of sperm during a process termed spermatogenesis. Spermatogenesis takes place within the seminiferous tubules of the testis and the process is divided into 3 main stages (i) spermatogenic stage, (ii) meiotic stage and (iii) spermiogenesis stage.

Spermatogenic stage: It is the mitotic stage of spermatogenesis wherein the stem cells in the testis, i.e., spermatogonial stem cells (SSCs) undergo series of mitotic divisions to form differentiated spermatogonia.

In rodents, A single (A_s) spermatogonia are the SSCs. They are capable of self-renewing and also give rise to A paired (A_{pr}) spermatogonia that further differentiate into A aligned (A_{al}) spermatogonia. Similarly in humans, A dark (A_d) spermatogonia are the SSCs that form A pale (A_p) spermatogonia which further forms type B spermatogonial (Figure 77.3).

Meiotic stage: The differentiated spermatogonia formed in the spermatogonial stage enter meiosis and forms tetraploid primary spermatocytes and further secondary spermatocytes. These secondary spermatocytes form haploid round spermatids after second meiotic division.

Spermiogenesis stage: The spermatids formed after meiosis enter spermiogenesis in which they undergo dramatic nuclear and cytoplasmic transformations to produce spermatozoa.

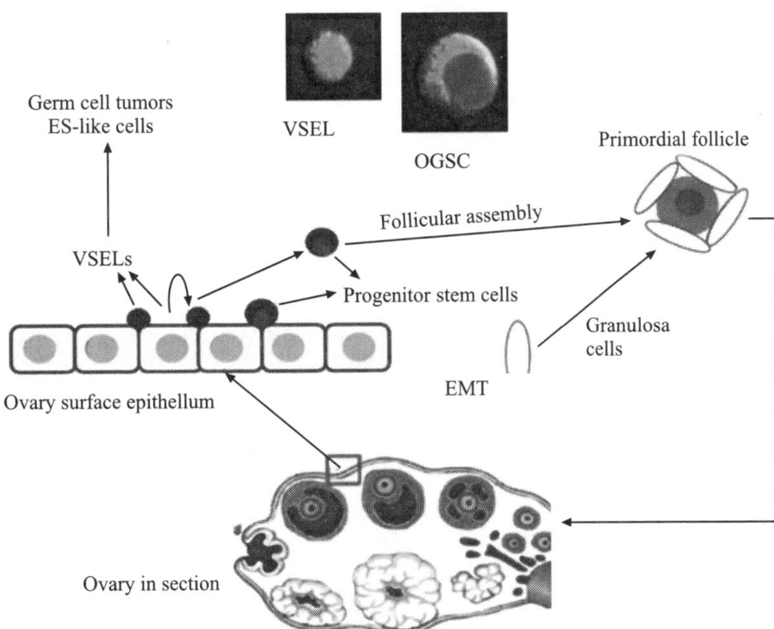

Figure 77.2 Proposed model for postnatal oogenesis in adult mammalian ovary—Pluripotent stem cells with nuclear Oct-4 (VSELs) are located in the ovary surface epithelium (OSE). These cells undergo asymmetric cell division and give rise to cells with cytoplasmic Oct-4 (OGSCs, which intensely stain with Haematoxylin). The OGSCs undergo further proliferation, meiosis and differentiation to assemble into primordial follicles in the OSE. The granulosa cells are formed by the epithelial cells that undergo epithelial mesenchymal transition. As the follicles grow and further mature they shift into the ovarian medulla. Confocal images represent VSEL and OGSC isolated by scraping the surface epithelium of peri-menopausal human ovary and immuno-stained for OCT-4. VSELs have nuclear OCT-4 whereas the OGSCs have cytoplasmic OCT-4. (*see Plate 35 for colour figure*)

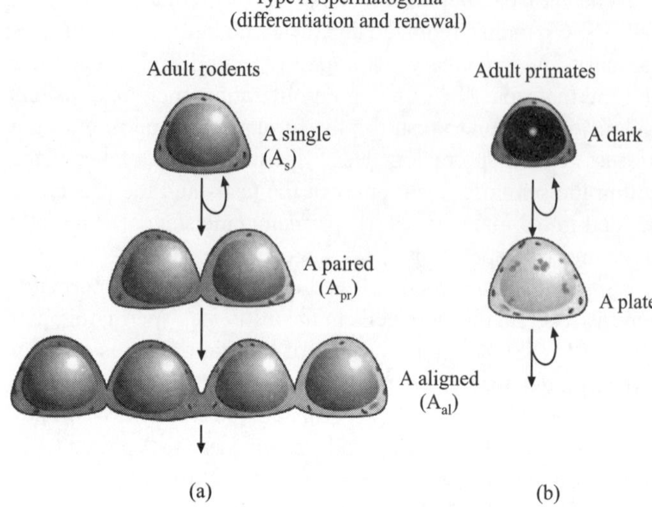

Figure 77.3 Schematic representation of self renewal and differentiation in spermatogonia of rodents and primates. (*see Plate 35 for colour figure*)

The process of spermatogenesis is tightly regulated and involves complex interplay of the somatic cells and the germ cells in testis. Major cell types in the testis are depicted in Figure 77.4.

A. Somatic Cells

They form the microenvironment for the SSCs and are a source of growth factors, hormones and other cytokines which act in paracrine/autocrine manner to maintain homeostasis. Major somatic cells in testis are (i) Sertoli cells also named nurse cells lightly so because of its undisputed role in nurturing the germ cells by being the site of hormonal action (e.g., FSH), secreting numerous regulatory molecules (e.g., Inhibin, SCF, GDNF, lactate, TGF, bFGF, IGF, etc.) and by providing physical support to growing germ cells. (ii) Leydig cells are present in the interstitium between the seminiferous tubules and are involved in androgen production. (iii) Somatic myoid cells together with extracellular matrix (ECM) constitute the basement membrane of the seminiferous tubules and (iv) Vascular network/blood vessels—the interstitium between the tubules also comprises of blood vessels. Though the role of vascular network as somatic niche is not largely known, there are recent reports that attribute this vascular network with localization of undifferentiated spermatogonia along specific portions of the basement membrane.

B. Pluripotent ES-like Stem Cells in Adult Mammalian Testis

Various studies have reported that the primitive testicular spermatogonial stem cells represent the only population of stem cells in adult body that can be reprogrammed to pluripotent state by a process termed dedifferentiation in mice as well as in humans and give rise to ES-like colonies in culture. However, the efficiency of this process remains very poor. There is also an emerging argument about the authenticity of these ES-like colonies derived from testicular tissue. Besides the possibility of dedifferentiation of SSCs to pluripotent state, it has also been reported that pluripotent stem cells may exist in adult testis which could give rise to ES-like colonies in both mice and men. Pluripotent VSELs were first reported in mice testis by flow cytometry (Zuba-Surma et al., 2008) and have been demonstrated in both humans and mice by our group (Bhartiya et al., 2013) and extensively characterized.

The basis for isolation was sorting VSELs using positive and negative selection of markers specific for VSELs (CD45-, Lin -, Sca-1+) and were reported to be 9.55 ±10.04 × 10^3 cells/gm in testicular tissue. In human testis, the presence of VSELs was reported using various techniques like immunolocalization, in situ hybridization and qPCR. VSELs were localized next

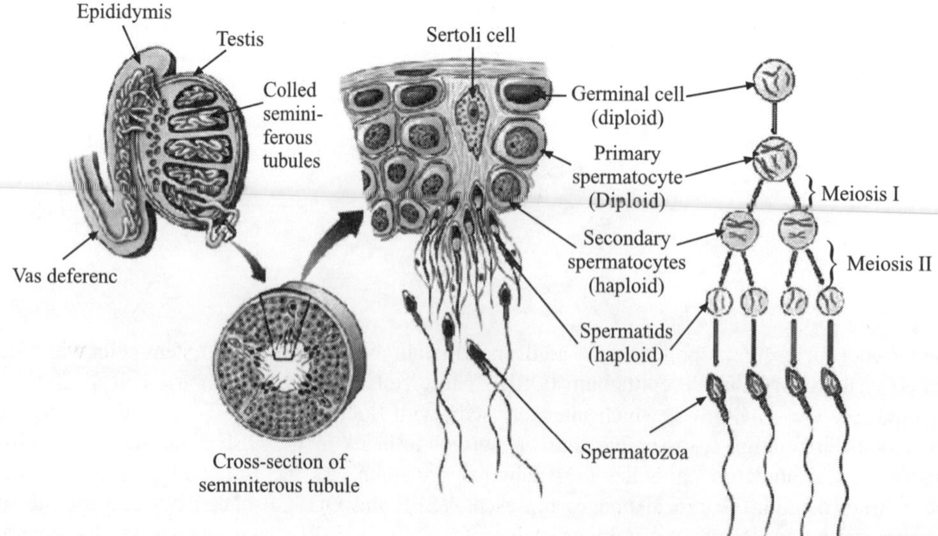

Figure 77.4 Schematic representation of spermatogenesis in humans showing interaction between different cell types.

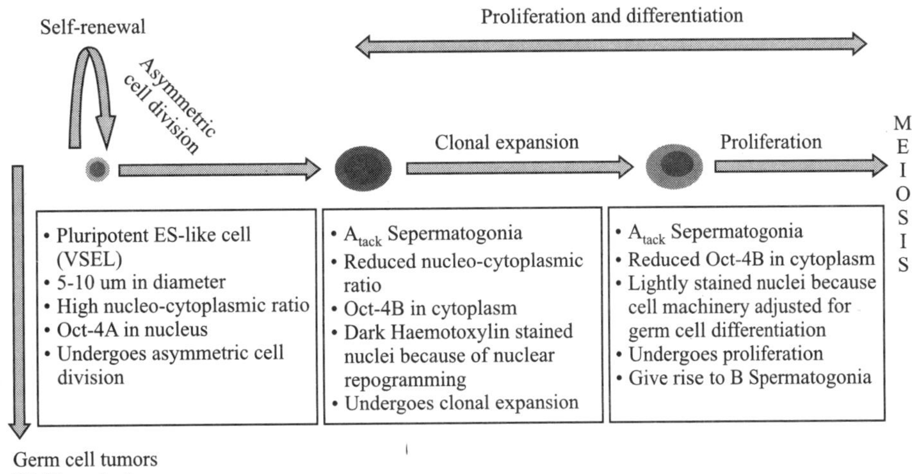

Figure 77.5 Revised scheme for premeiotic development of germ cells in adult human testis. It is hypothesized that small and rare ES-like cells undergo asymmetric cell division to give rise to Adark spermatogonial cells with cytoplasmic Oct-4B. These cells undergo extensive nuclear reprogramming by epigenetic changes resulting in the dark appearance of their nuclei and later transform into Apale spermatogonial stem cells (SSCs) with normal pale nucleus and decreased expression of Oct-4B. Apale SSCs proliferate and undergo further differentiation and meiosis to complete the spermatogenesis cycle.

to basement membrane of seminiferous tubules. We studied differential localization and expression of two major transcripts of Oct-4, viz. Oct-4A and Oct-4B in adult human testis. A novel population of 5–10 μm pluripotent stem cells with nuclear Oct-4A was identified by in situ hybridization and immunolocalization studies. Besides Oct-4, other pluripotent markers like Nanog and TERT were also detected by RT-PCR. VSELs are relatively quiescent stem cells which undergo asymmetric cell division to renew themselves and give rise to A$_{dark}$ SSCs which proliferate extensively and undergo lineage-specific differentiation. Besides maintaining normal testicular homeostasis, the VSELs may also be implicated in germ cell tumors and ES-like colonies that have recently been derived from adult human testicular tissue (Bhartiya et al., 2012).

The possibility of existing sub-population of SSCs with pluripotent/primitive characteristics is not new. Various reports in literature which are suggestive of their presence are listed below

- A side population of stem cells can be isolated from adult testis (Lassalle et al., 2004)
- A subpopulation of SSCs exists in human testis with pluripotent characteristics which can repopulate the testis (Izadyar et al., 2011)
- In SSCs culture, SSEA-4 and GFR alpha do not co-localize rather co-exist as two different cell populations (Lim et al., 2010)

Thus our work has provided novel information on basic spermatogenesis by identifying a stem cell more basic to SSCs in both mouse and human testis as depicted below.

CONCLUSIONS

Pluripotent stem cells can be isolated from the inner cell mass of embryos to derive ES cell lines. Besides stem cells with similar pluripotent characteristics exist in adult body tissues including gonads. One may refer to extensive work by Ratacjzak and group on VSELs in adult somatic tissues (Ratajczak et al, 2014). Our work on VSELs in bone marrow, cord blood, and cord tissue is also interesting (Bhartiya et al., 2012b, c). To conclude, pluripotent stem cells isolated from spare human embryos or adult body tissues including gonads need to be exploited for their clinical potential. Since VSELs can be isolated from an autologus source easily, they will not have the problem of immune rejection nor do they form teratomas. Thus both the concerns on using ES cells (immune rejection and teratoma formation) for regenerative medicine are taken care by the VSELs. But at the same time compared to ES cells, VSELs are relatively quiescent. Thus it will be interesting to follow research progress in the area of regenerative medicine using ES cells and VSELs.

ABBREVIATIONS

EB	Embryoid body
ES cells	Embryonic stem cells
hES cells	Human embryonic stem cells
ICM	Inner cell mass
IC-SCRT	Institutional Committee for Stem Cell Research and Therapy
MEF	Mouse embryonic fibroblast
mES	Mouse embryonic stem cells

NAC-SCRT	National Apex Committee for Stem Cell Research and Therapy	PGCs	Primordial germ cells
		PSCs	Pluripotent stem cells
OGSCs	Ovarian germ stem cells	SSCs	Spermatogonial stem cells
OSE	Ovarian surface epithelium	VSELs	Very small embryonic-like stem cells

SUGGESTIONS FOR FURTHER READING

Alper J. (2009), Geron Gets Green Light for Human Trial of ES-cell Derived Product, *Nat Biotechnol.*, 27:213–214.

Bhartiya D. and Singh J. (2014), FSH-FSHR3–Stem Cells in Ovary Surface Epithelium: Basis for Adult Ovarian Biology, Failure, Aging and Cancer, *Reproduction*, Pii: REP-14–0220.

Bhartiya D., Sriraman K. and Parte S. (2012), Stem Cell Interaction with Somatic Niche may Hold the Key to Fertility Restoration in Cancer Patients, *Obstet Gynecol Int.*, 921082.

Bhartiya D., Shaikh A., Nagvenkar P., et al., Very Small Embryonic-like Stem Cells with Maximum Regenerative Potential Get Discarded During Cord Blood Banking and Bone Marrow Processing for Autologous Stem Cell Therapy, Stem Cells Dev 2012b; 21:1–6.

Bhartiya D., Unni S., Parte S. and Anand S. (2013), HYPERLINK "http://www.ncbi.nlm.nih.gov/pubmed/23509758" Very Small Embryonic-like Stem Cells: Implications in Reproductive Biology, *Biomed Res Int.*, 2013:682326.

De Francesco L. (2009), Fits and Start for Geron (2003), *Nat Biotechnol,* 27:877. Dettin L., Ravindranath N., Hofmann M.C., et al., Morphological Characterization of the Spermatogonial Subtypes in the Neonatal Mouse Testis, *Biol Reprod.*, 69:1565–1571.

ICMR. http://icmr.nic.in/stem_cell/stem_cell_guidelines.pdf. Indian Council of Medical Research, India.

ICMR. http://www.icmr.nic.in/bioethics. Indian Council of Medical Research, India.

Inamdar M.S., Venu P., Srinivas M.S., et al., Derivation and Characterization of Two Sibling Human Embryonic Stem Cell Lines from Discarded Grade III Embryos, *Stem Cells Dev* 2009; 18:423–433.

Izadyar F., Wong J., Maki C., et al. (2011), Identification and Characterization of Repopulating Spermatogonial Stem Cells from the Adult Human Testis. *Hum Reprod,* 26:1296–1306.

Johnson J., Canning J., Kaneko T., et al. (2004), Germline Stem Cells and Follicular Renewal in the Postnatal Mammalian Ovary. *Nature,* 428:145–150.

Kumar N., Hinduja I., Nagvenkar P., et al. (2009), Derivation and Characterization of Two Genetically Unique Human Embryonic Stem Cell Lines on In-house-derived Human Feeders, *Stem Cells Dev,* 18:67–77.

Lassalle B., Bastos H., Louis J.P., et al. (2004), 'Side Population' Cells in Adult Mouse Testis Express Bcrp1 Gene and are Enriched in Spermatogonia and Germinal Stem Cells, *Development*, 131:479–87.

Lim J.J., Sung S.Y., Kim H.J., et al. (2010), Long Term Proliferation and Characterization of Human Spermatogonial Stem Cells Obtained from Obstructive and Non-obstructive Azoospermia under Exogenous Feeder-free Culture Conditions, *Cell Prolif,* 43:405–417.

Mandal A., Tipnis S., Pal R., et al. (2006), Characterization and *in vitro* Differentiation Potential of a New Human Embryonic Stem Cell Line, ReliCellhES1, *Differentiation*, 74:81–90.

Mandal A., Bhowmik S., Patki A., et al. (2010), Derivation, Characterization and Gene Expression Profile of Two New Human ES Cell Lines from India, *Stem Cell Res.*, 5:173–187.

Mayor S. (2010), First Patient Enters Trial to Test Stem Cells in Spinal Injury, *BMJ.*, 341:c5724.

NIH Stem Cell Information Http://stemcells.nih.gov/

Notarianni E. (2011), Reinterpretation of Evidence Advanced for Neo-oogenesis in Mammals, in Terms of a Finite Oocyte Reserve, *J. Ovarian Res.*, 4:1.

Ratajczak M.Z., Marycz K., Poniewierska-Baran A., et al. (2014), Very Small Embryonic-like Stem Cells as a Novel Developmental Concept and the Hierarchy of the Stem Cell Compartment, *Adv Med Sci.*, 59(2):273–280.

Schwartz S.D., Hubschman J.P., Heilwell G., et al. (2012), Embryonic Stem Cell Trials for Macular Degeneration: A Preliminary Report, *Lancet*, 379:713–720.

Zou K., Yuan Z., Yang Z., et al. (2009), Production of Offspring from a Germline Stem Cell Line Derived from Neonatal Ovaries, *Nat Cell Biol*, 5:631–636.

78

Stem Cells and Hepatic Regeneration

Anupam Kumar

CONTENTS

Liver is an extremely important vital organ. The "strategic" location of liver in relation to the food supply via the portal vein, allow it to function as a biochemical defence against toxic chemicals entering through food as well as a re-processor of absorbed food ingredients. It acts as; a complex metabolic bio-reactor of the body, performs a myriad of functions like metabolism and re-processing of absorbed food, Synthesis of various plasma proteins, detoxification and metabolism of exogenous drugs and endogenous toxic metabolites, removal of metabolic wastes and toxins, immune and hormonal modulation etc. These wide arrays of functions performed by liver towards the rest of the body have been safeguarded by evolutionary conserved unique phenomenal capacity to regenerate. Unlike other organs and tissues in response to tissue damage normal liver recovers lost mass without any scars. After partial hepactomy as little as 25% of a liver can regenerate back into a whole liver. This unique property of liver allows it to rescue the structural and functional integrity without jeopardizing the original architecture and function. During the course of development of advanced liver disease this safeguard is either impaired or lost that leads to collapse of the liver which subsequently jeopardize the viability of the entire organism. Till now Liver transplantation is the only life-saving procedure for these patients. Many of its associated disadvantages, such as relative shortage of donors, operative risks, GVHR (Graft-Versus-Host Reaction) associated death and a high cost limits its potential benefits. Restoring the normal regenerative capacity of the liver holds great promises in the management of liver disease patients. The advances made over the past few years in the field of stem cell and developmental biology have provided substantial understanding of the molecular and cellular biology of hepatic regeneration. In the following chapter we will first briefly describe the embryonal origin of different hepatic cells and their maturation then we will discuss the various aspect of hepatic regeneration in normal and diseased liver.

I. CELLULAR ORGANIZATION OF LIVER

An appreciation of liver architecture is essential for the understanding of hepatic stem cell biology and regeneration. Histological section of liver shows homogeneous landscape of hepatic cells periodically infiltrated with vascular tissue and bile ducts. As described in Figure 78.1, the basic architectural unit of liver is liver lobule. Located along the lobule perimeter, the portal triad consists of a small portal vein, hepatic artery, and bile duct. Blood enters the liver from the portal vein and hepatic artery, and it flows through liver sinusoids toward the central vein. The lobule consists of plates of polarized epithelial cells (usually one cell thick in mammals) called Hepatocytes. Hepatocytes are the major parenchymal cell type of the liver and account for more than 70% of liver volume. They support the vide array of liver functions such as protein secretion, bile secretion, cholesterol metabolism, detoxification,

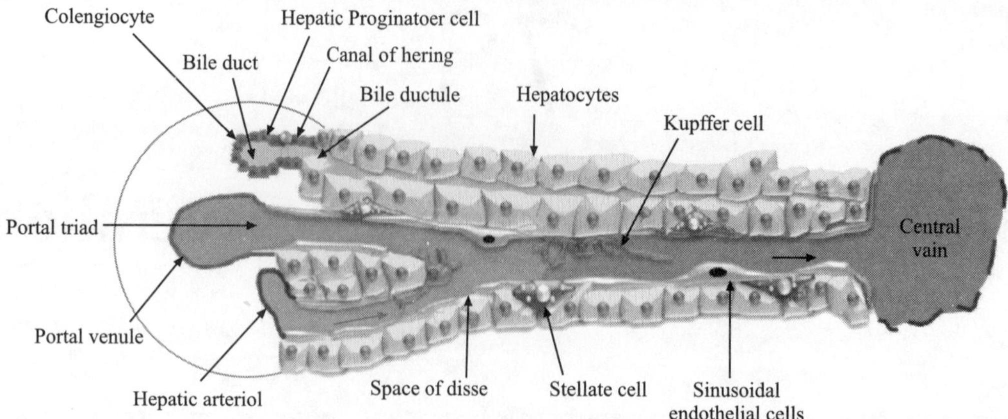

Figure 78.1 Diagrammatic representation of Structural and cellular organisation of hepatic lobule. The portal triad consists of bile ducts, hepatic artery, and portal vein. Mixed blood from the hepatic artery and portal vein flows past hepatocytes through the sinusoids, covered with fenestrated endothelial cells to the central vein. Bile produced by the hepatocytes is collected in the bile canaliculus and flows towards the bile duct. The Canal of Hering is the junction between the hepatic plate and the bile ducts. This is the region where oval cell precursors reside. (*see Plate 35 for colour figure*)

urea metabolism, glucose/glycogen metabolism etc. Basolateral surface of hepatocytes are lined with fenestrated endothelium known as Liver Sinusoidal Endothelial Cells (LSEC). They are highly specialized cells which allow transfer of molecules between serum and hepatocytes, acts as scavenger of macromolecular waste, associated with production of various cytokines, antigen presentation and blood clotting. Apart from LSEC liver also consist of endothelial cells which support the liver vasculature and contribute to parenchymal zonation. Tight junctions formed between neighboring hepatocytes generate a canaliculus that surrounds each hepatocyte and is responsible for collection of bile acids and bile salts that are transported across the hepatocyte's apical surface. Bile collected by the canaliculi is carried to the bile ducts within the portal triad and subsequently transported for storage in the gall bladder. Bile ducts are lined by epithelial cells called cholangiocyte/bile duct cell. Cholangiocytes constitutes approximately 3% of the total liver. They form bile ducts to transport bile, control rate of bile flow and secrete water and bicarbonate to control pH of bile. The connecting channel between the biliary duct and intralobular bile canalicular system is partly lined by both hepatocytes and cholangiocytes called Canal of Hering. The canal of Hering also consist of a small population of cells with oval nucleus and a small rim of cytoplasm known as Hepatic Progenitor Cells (HPC) (commonly not observed in normal adult liver and seen mostly in acute or chronic injury to liver). HPCs show intermediate morphology and immunophenotype of both hepatocytes and cholangiocytes. They are bipotential and differentiate to both hepatocyte and cholangiocyte lineage (details will be discussed in subsequent section of this chapter). The space between the LSEC and Hepatocytes are called *Space of Disse*. Space of Disse or perisinusoidal space is lined with cells of mesenchymal origin called *Hepatic stellate cells* (HSC). HSCs consist of ~1.4% of total liver cell. They are involved in maintenance of extracellular matrix, Vitamin A and retinoid storage and control of microvascular tone. In response to injury they get activated to become myofibroblasts

and contribute towards regenerative response to injury by their secreted cytokines and growth factors. Liver sinusoid also consists of resident macrophages called *Kupffer cells*. They consist of ~2% of total liver cell volume and act as scavengers of foreign material. In response to injury they produce various cytokines and proteases for regeneration and repair. Hepatic lobule also consists a very small percentage of resident liver natural killer cells called *Pit cell* that is mainly associated with cytotoxic activity.

II. DEVELOPMENT AND ORIGIN OF HEPATIC CELLS

As discussed in the previous section liver cells consist of both epithelial origin (hepatocytes and cholangiocytes) that form the parenchymal mass of the liver and mesenchymal origin (HSC, LSEC, endothelial cells, Kupffer cells and Pit cells) that constitute the nonparenchymal mass of the liver. Parenchymal cells are derived from endoderm while nonparenchymal cells are mesodermal in origin except Kupffer cells which originate from extraembryonic yolk sac (summarized in Figure 78.2). Development of both parenchymal and nonparenchymal cells occur in coordinated fashion. Inductive signals from nonparenchymal cells regulate the growth and maturation of parenchymal cells and vice versa. Unlike parenchymal cells cellular and molecular basis of nonparenchymal cell development is not clearly understood. Hence in the following section we will mainly discuss the developmental origin of parenchymal cells.

Development of hepatoblast from endoderm: At around embryonic day (E) 7.0 definitive endoderm emerges from the primitive streak to displace extraembryonic endoderm of the yolk sac. Shortly after this, endoderm invaginates to form a portal at the anterior region of the developing embryo that ultimately defines the foregut of the mouse. There are three distinct domains of hepatic progenitor cells that are

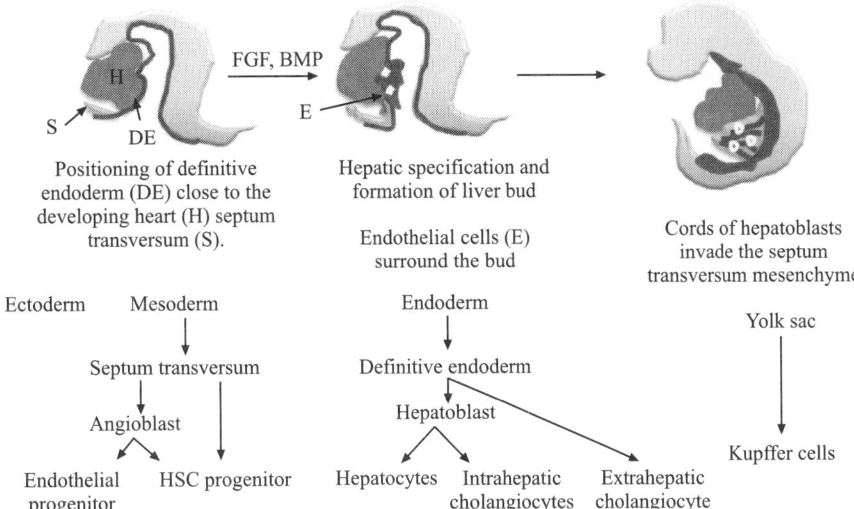

Figure 78.2 Diagrammatic representation of developmental origin of different liver cells. Details given in the text. (*see Plate 35 for colour figure*)

located in the medial and bilateral regions of the foregut. As the foregut closes, the progenitor cells within these regions converge to lie adjacent to the developing heart and in close apposition to regions of lateral plate mesoderm. Lateral plate mesoderm ultimately generates the mesothelial cells of the proepicardium and septum transversum. In human, first sign of liver development appeared during the 4th gestational week (Carnegie embryonal stage 11), where local thickening of the endoderm lining the ventral wall of the very distal part of the foregut (point, where the fetal intestine meets the yolk sac) forms the hepatic primordium. At the end of the 4th gestational week (Carnegie embryonal stages 12 and 13), the proliferative activity of the cells of the hepatic primordium and septum transversum results in the development of hepatic diverticulum, that buds from the ventral wall of the foregut. The cranial part of the hepatic diverticulum developed into the liver and caudal part of the hepatic diverticulum forms the extrahepatic bile ducts, the cystic duct and the gallbladder. By the 6th gestational week (Carnegie embryonal stage 16), epithelial cells of endoblastic origin (forming the early liver bud) rapidly invade the caudal part of the septum transversum, and form anastomosed cords separated by vascular channels. The mesenchyme of the septum transversum forms the connective tissue associated with the intrahepatic vessels, including sinusoids and portal vessels, and to the connective capsule of the liver, the Glisson's capsule. Developing cardiac mesoderm and septum transversum plays a crucial instructive role during the induction of hepatic cell fate in both human and mice liver development. They express various growth factors like fibroblast growth factor (FGF) Bone morphogenic factor (BMP) and transforming growth factor (TGF-β). FGF and BMP positively regulate the induction of hepatic fate while TGF-β signaling acts to restrict endoderm specification. FGF-mediated specification of hepatic cell fate is concentration dependent and this appears to be controlled by position of the endoderm relative to the heart. In response to the inductive cues from the heart and mesenchyme, the cells forming the hepatic endoderm (lie proximal to the sinus

venosus) acquire a columnar morphology and express several hepatic genes including Albumin, Afp, Ttr (transthyretin), Rbp (retinol binding protein), and the transcription factor Hnf4a, all of which are reliable indicators of early hepatic cell fate. This transition in cellular morphology results in a thickening of epithelium, which bulges into the surrounding stroma. The basal face of the diverticulum is surrounded by a matrix which contains laminin, nidogen, type IV collagen, fibronectin, and heparin sulfate proteoglycans. At around 21 somites in the mouse, and the 6th gestational week in human nuclear migration within the epithelial cells results in further change in cellular morphology into pseudostratified epithelial. The matrix surrounding the basal surface of the epithelium is then degraded and E-cadherin expression is downregulated in the hepatic cells as they delaminate and invade the surrounding stroma as migrating cords of hepatoblasts. Migration of hepatoblasts, the hepatic progenitor cells into the stroma is facilitated by degradation of surrounding matrix by metalloproteinases MMP-14 (secreted by liver bud) and MMP-2 (secreted by surrounding mesenchyme) (summarized in Figure 78.2).

A. Hepatoblast Expansion and Maturation

As soon as primary liver bud emerges from endoderm, angioblasts or endothelial cells surround it and separate it from the septum transversum mesenchyme. These endothelial cells and their precursors plays an integral role in controlling growth and fate of hepatoblast together with the development of hepatic vasculature. Inductive signals from the surrounding mesenchymal cells guide the development of hepatic parenchyma. As discussed above at the time of hepatic induction, septum transversum mesenchymal cells provide inductive signals including fibroblast growth factors and bone morphogenetic proteins to initiate hepatic specification of endodermal cells. Later on after formation of hepatic bud angioblast, endothelial cells and mesothelial fibroblast guide the growth, maturation and fate of hepatic stem cells and hepatoblsts. Angioblast cells provide matrix and paracrine support for the growth of hepatic stem cells. Coordinated maturation of parenchymal and mesenchymal cells is critical for the proper development of liver. Angioblast cells further give rise to endothelial and hepatic stellate progenitor cells (origin of hepatic stellate progenitor cells is not clearly understood. Scientists with one line of thought believe that they originate from angioblast and others believe that they originate from mesothelial mesenchymal cells) which provide the matrix and paracrine support for the growth of hepatoblast cells. The exact mechanisms through which bipotential hepatoblasts decide to become hepatocytes or biliary epithelial cells are still unclear. Based on recent studies it is believed that association of hepatoblasts with either hepatic stellate progenitor or endothelial progenitor cells decide fate of hepatoblast. Endothelial progenitor cells secrete matrix and paracrine factors, which guide the growth and differentiation of hepatoblasts into hepatocyte lineages. While the matrix and paracrine factors secreted by hepatic stellate progenitor cells guide the growth and differentiation of hepatoblasts into cholangiocyte lineages. Cells that follow a hepatocyte cell fate progressively mature and, during the remainder of both embryonic and postnatal development, accumulate the gene expression and physiological profile of mature hepatic parenchymal cells. Cellular origins of extrahepatic and intrahepatic cholangiocytes are different. While the extrahepatic Cholangiocytes derive directly from the endoderm, cholangiocytes that line intrahepatic bile ducts arise from hepatoblasts.

B. Molecular Signalling Associated with Development of Hepatic Cells

Maturation of hepatocytes is facilitated through an expanding and complex network of transcription factors that regulate hepatocyte gene expression. Six transcription factors, (HNF1a, HNF1b, FoxA2, HNF4a1, HNF6, and LRH-1 [Nr5a2]), were found to form the core of this regulatory circuit. They occupy each other's promoters as well as the promoters of peripheral hepatic transcription factors and guide the maturation of hepatocyte from hepatic progenitor cells. HGF regulate the expression of most of these transcription factors. Endothelial progenitor cells are the main source of HGF in developing liver. In mouse embryos, differentiation towards a biliary epithelial cell phenotype appears to be promoted by Notch signaling pathways and antagonized by hepatocyte growth factor, which may in turn promote hepatocyte differentiation. Hepatic stellate progenitor cells are the main source of Notch in a developing liver. Wnt, beta-catenin signaling is important signaling cascade that is critical for successful development of liver. In the prestreak embryo at E5.5, a gradient of Wnt activity controlled by regional expression of Wnts and Wnt antagonists (sFRP and Dkk) promote gastrulation. During gastrultion the region expressing Wnt ultimately gives rise to the anterior visceral endoderm from which the gut tube derives. Later on at the time of gut tube patterning around E8, suppression of Wnt activity through enhanced sFRP5 expression promotes hepatic competence in the ventral foregut endoderm. Immediately after foregut patterning

by E9, again Wnt signaling resumes activity, and regulates the morphological transition of hepatic endoderm cells from columnar to pseudostratified. In conjunction with HGF/Met and FGF/FGFR signaling, Wnt/beta-catenin signaling regulates the expansion of the bipotential hepatoblasts comprising the liver bud. In embryonal stage Wnt/beta-catenin activity is essential for both biliary epithelial cell (BEC) differentiation and hepatocyte differentiation from hepatoblasts. While Wnt activity promotes BEC differentiation, the combination of Wnt and HGF appears to promote to hepatocyte differentiation.

III. STEM CELLS IN HEPATIC REGENERATION

Stem cells are biological cells found in all multicellular organisms, that have the capacity to self renew (divide to generate identical stem cells) and differentiate into diverse specialized cell types. In mammals, there are two broad types of stem cells: *embryonic stem cells*, which are isolated from the inner cell mass of blastocysts, and *adult stem cells*, which are found in various tissues and are responsible for the normal turnover of differentiated cells as well as repair and regeneration of damage tissues. Adult mammalian tissues also contain the progeny of stem cells that are committed to differentiate into particular cell type and lack the capacity of self renewal known as *Progenitor cells*. Repair and regeneration of an adult mammalian liver is mediated by stem and progenitor cells of both hepatic and non-hepatic origin.

A. Hepatic Stem and Progenitor Cells

Adult liver consist of both parenchymal and nonparenchymal stem cells. Two different kind of parenchymal progenitor cells Hepatic Stem cells (hHpSCs) and hepatoblasts (hHBs) are present in the both fetal and adult liver (ref 15). *Hepatic Stem cells (hHpSCs)* are multipotent stem cells located within the liver's stem cell compartments (i.e. ductal plates in fetal and neonatal livers, and canals of Hering in pediatric and adult livers). They constitute 0.5%–2% of the parenchyma of livers of all age donors. The hHpSCs cells range in size from 7–9 μm in diameter and have a high nucleus-to-cytoplasm ratio. They are tolerant of ischemia and can remain viable in cadaveric livers for up to 6 days after asystolic death. They express epithelial cell adhesion molecule (EpCAM), neural cell adhesion molecule (NCAM), CD133, CXCR4, SOX9, SOX17, FOXA2, cytokeratins (CK) 8/18/19, Hedgehog proteins (Sonic and Indian), intranuclear telomerase protein, claudin 3, and MDR1. They show weak or negligible expression of albumin and major histocompatibility complex (MHC) antigens and are negative for alpha-fetoprotein (AFP), intercellular, adhesion molecule (ICAM-1), P450s, hemopoietic markers (e.g., CD34/38/45/90, glycophorin), endothelial (e.g., vascular endothelial growth factor receptor [VEGFr], CD31, von Willebrand factor), mesenchymal cell markers (e.g., CD146, desmin, vitamin A, CD105). hHpSCs differentiate to Hepatoblast. *Hepatoblasts (hHBs)* are diploid bipotent cells giving rise to hepatocytic and cholangiocytic lineages, associated with precursors of both endothelia and hepatic stellate cells, and the liver's probable transit amplifying cells. They reside throughout parenchyma of fetal and neonatal livers and as single or small cell aggregates tethered to the ends of canals of Hering in adult livers. With age, hHBs decline to <0.01% of the parenchymal cells in postnatal livers. They are larger in size (10–12 µm) with higher amounts of cytoplasm than hHpSCs. hHBs have an antigenic profile that overlaps with hHpSCs. Similar to hHpSCs they show positive expression of CXCR4, CD133, SOX17, MDR1, cytokeratins (CK) 8/18 and 19, Hedgehog proteins (Sonic and Indian), and null expression of late P450s (e.g., P450-3A) hemopoietic, endothelial or mesenchymal cell markers. Unlike hHpSCs they shows reduction in EpCAM levels with primary localization to plasma membrane surfaces; filamentous CK14 and CK19, elevated albumin levels with discrete cytoplasmic packaging; switch from NCAM to ICAM-1, Positive for CK7; and strong positive expression of hepatic-specific AFP. hHBs further differentiate to both hepatocyte or cholangiocyte lineages. Nonparenchymal stem/progenitor cells mainly includes hepatic stellate cell progenitors, endothelial cell progenitors and sinusoidal endothelial cell progenitors. *Hepatic stellate cell progenitors* are bipolar cells, <10 µm in size with their nucleus at one end. They express CD146 and have very low levels of desmin, α-smooth muscle actin (ASMA), vitamin A, and lipids. They are negative for glial fibrillar acidic protein. Liver consists two different populations of *endothelial progenitor cells (EPCs)*. EPCs with antigenic phenotype of CD133+/Cd45-/CD31+ give rise to endothelial cell lining of most of the hepatic vasculature while EPCs with CD133+/CD45+/Cd31+ phenotype differentiate to form sinusoidal endothelial cells. Endothelial progenitor cells are present in very less amount in adult liver and mainly participate in the normal turnover of endothelial cells. After liver injury repopulation of liver endothelial cells is mainly driven by the bone marrow derived EPSc and Bone marrow derived sinusoidal endothelial progenitor cells. Progenitor cells for Kupfer's cells are present in the adult liver or not is not known-most of the studies suggest that bone marrow derived monocytes act as source of Kupffer cells in adult liver but recent finding in animal model suggest that during normal tissue turn over renewal of Kupffer cells are mainly through the yolk sac originated innate Kupffer cells rather than bone marrow monocytes.

B. Non-hepatic Stem and Progenitor Cells

Normal turnover of hepatic cells are mainly mediated by in house stem and progenitor cells (discussed in previous section). In case of tissue damage bone marrow derived stem and progenitor cells migrate to liver and play a significant role in liver repair and regeneration.

In mammals soon after liver progenitors invade the surrounding mesenchyme, the fetal liver is colonized by hematopoietic progenitors and transiently becomes the principal hematopoietic organ. In the fetal liver immature parenchymal and non parenchymal progenitor cells generate an environment that supports hematopoiesis. Conversely, hematopoietic cells within the fetal liver provide cytokines and growth factors (e.g. Oncostatin M) which support the growth and maturation of hepatic progenitor cells. With the maturation of liver this support is lost and hematopoietic cells migrate to bone marrow. This symbiotic relationship between liver and hematopoietic stem and progenitor cells remain intact even in adults. *Hematopoietic stem cells* (HSC) and progenitor cells migrate to liver in response to injury and provide tropical support for activation and maturation of hepatic stem and progenitor cells. Apart from HSC, bone marrow also contains various stem and proginators cells such as *Mesenchymal Stem Cells (MSC), Endothelial Progenitor cells* (EPC) and *Liver sinusoidal endothelial Progenitor cells* etc. In course of hepatic injury these bone marrow derived stem and progenitor cells also migrate and engraft to liver. Very little is known about the physiological relevance of bone marrow-derived cells engrafting in liver with injury. Based on recent studies it is believed that bone marrow derived stem and progenitor cells help in repair and regeneration of injured liver by (a) providing various cytokines and growth factor for activation and differentiation of hepatic stem cells, (b) matrix remodelling, (c) source of various non parenchymal cells such as monocytes for Kupffer cells, EPC for endothelial cells, sinusoidal endothelial Progenitor cells for sinusoidal endothelial and MSCs for myofibroblast, (d) transdifferentiation or fusion to give hepatocyte like cells.

IV. REGENERATION OF ADULT LIVER

A. Normal Liver Tissue Turnover

Hepatocyte replacement occurs relatively slowly; the average life span of adult hepatocytes ranges from 200 to 300 days. Two opposing hypothesis have been proposed to explain liver tissue turnover. According to "streaming liver," hypothesis young hepatocytes or cholangiocytes originate from hepatic stem/progenitor cells present in the portal zone. Young hepatocytes then subsequently migrate and mature towards the central vein. Based on some evidence against the streaming liver hypothesis "self replicating model" was proposed. This model suggests that liver maintenance is mostly achieved directly by cell division of hepatocytes and bile duct epithelial cells. Although the contribution of stem cells to normal liver turnover in adult animals is debated and requires further studies. Some recent line of evidence further support the "streaming liver," hypothesis. During the normal turnover of liver dying hepatocytes in pericentral zone gives the positive signal to hepatic stem cell compartment of zone 1 that leads to the activation and differentiation of hepatic stem cells. Differentiating stem cells migrate from stem cell compartment towards Zone 3. Recently a group of studies suggest that hepatic parenchymal cells are highly plastic in nature, depending on the nature of damage and need both mature hepatocytes and cholangiocytes and dedifferentiate to hepatic progenitor cells. Changing composition of matrix components and paracrine signaling of associated

962 *Textbook of Biochemistry, Biotechnology, Allied and Molecular Medicine*

mesenchymal partner form portal to central zone modulate growth and differentiation of hepatic parenchymal cells.

Adult liver consist of different maturational lineage of hepatic cells from the stem cells in zone 1 (periportal), through the midacinar region (zone 2), to the most mature cells and apoptotic cells found pericentrally in zone 3. Cells found in the biliary tree are still needed to be defined. Parenchymal cells are closely associated with lineages of mesenchymal cells, and their maturation is coordinated. Each lineage stage of hepatic cells consists of distinct mesenchymal cell partners. They are distinguishable by their morphology, ploidy, antigens, biochemical traits, gene expression, and ability to divide (as summarized in Table 78.1). Lineage hierarchy and structural and functional integrity of parenchymal cells are mainly governed by gradients of paracrine signals (soluble factors and insoluble extracellular matrix components) between epithelial-mesenchymal partners, mechanical forces, and feedback loop signals derived from late lineage cells.

Paracrine Signalling between Epithelial-Mesenchymal Partners: Similar to hepatic ontogeny paracrine signaling between parenchymal and associated nonparnechymal cells is the primary regulator of the hepatic cell maturation and function in adult liver repair and regeneration. Coordinated maturation of [parenchymal]:[mesenchymal] cell associations in liver starts with [hHpSCs]:[angioblasts] and splits into lineages of [hepatocyte]:[endothelia] and [cholangiocyte]:[stellate cells], giving rise to lineage-dependent gradients of paracrine signals governing biological responses of cells at each lineage stage. Adult hepatic stem and progenitor cells are usually quiescent. Late lineage stage cells plays an important role in maintaining the quiescent stage or activation of hepatic stem and progenitor cells by producing various positive and negative signaling regulators, including bile salts, various soluble factors and components of the matrix. Identity and specific function of most of these factors are yet to be established.

Acetylcholine stimulates proliferation of stem/progenitor cells and cholangiocytes expressing M3 acetylcholine receptors. In normal liver and even after partial hepatectomy, late lineage stage hepatocytes lacking M3 receptors release acetyl cholinesterase into the bile that delivers it to zone 1 where it destroys acetylcholine in the stem cell niche, thus blocking proliferation of stem/progenitor cells and cholangiocytes. During the normal tissue turnover dying zone 3 hepatocytes produce positive regulators such as hepatopoietin, that stimulate stem/progenitors expansion. Once activated these progenitor cells matured along the line of portal to central zone.

B. Regeneration after Tissue Injury

Adult liver regeneration has three basic arms (A) self replication of parenchymal and nonparenchymal cells (B) activation, migration and differentiation of hepatic stem/progenitor cells and (C) activation, migration and engraftment of bone marrow derived stem and progenitor cells. Self replication of parenchymal cells is the primary mode of rescue of parenchymal cell loss after tissue injury. When this mode of hepatic regeneration is overwhelmed or impaired hepatic stem or progenitor cell mediated hepatic regeneration get activated and rescue the parenchymal loss. Role of non hepatic, BM- derived stem and progenitor cell is minimal in normal tissue turnover, in response to tissue injury these cells migrate and get engrafted to the damaged liver facilitating the regeneration process by i) providing the tropical and matrix support for regenerating parenchymal cells and act as cellular source for different non parenchymal hepatic cells such as monocyte—Kupffer cells, EPC—endothelial and sinusoidal endothelial cells, MSC—HSC/myofibroblast etc. Homeostasis between all these three arms of hepatic regeneration is required for successful hepatic regeneration. Type and severity of liver tissue damage are the major rate limiting factors which decide extant of contribution of different arm of hepatic regeneration. To simplify our

TABLE 78.1   Different lineage of hepatic parenchymal cells and their associated non parenchymal cells

	Zone 1				Zone 2		Zone 3
Lineage state	1	2	3	4	5	6	7
Parenchymal cells	Hepatic stem cell	Hepatoblast	Committed progenitors	Deploid adult cells			Tetrapolid pericentral parenchymal cells
				Periportal Parenchymal cell	Midacinar hepatocytes	Pericentral diploid hepatocytes	
Size	7-10 mm	10–12 mm	Small Cholangiocyte (6–7 mm) hepatocytes (12–15 mm)	Large cholangiocyte (14 mm) hepatocytes (18–20 mm)	22–25 mm	25–30 mm	>30 mm
Mesenchyma I partner	Angioblast	Endothelial and stellate cell precursor		Stellate cell-cholangiocyte Endothelial cells Hepatocytes	Endothelia	Endothelia	Endothelia
Matrix	Type III, IV & V collagens, laminins, hyaluronans, CS-PGs, HS, PGs			<<<<<Gradient>>>>		Type III, IV and VI collagens, HS, PGs	

understanding of complex cellular and molecular process of hepatic regeneration, following section is further divided into three parts (i) **Regeneration in Partial hepatectomy**, will discuss the various cellular and molecular events of self replication of hepatocytes in context with partial hepatectomy (ii) **Regeneration after selective loss of Parenchymal cells** will discuss the cellular and molecular events of hepatic stem and progenitor cells activation, migration and differentiation (iii) **BMCs and hepatic regeneration,** cellular and molecular events of bone marrow derived stem and progenitor cells activation, migration and liver engraftment and their potential contribution in hepatic regeneration.

(i) Regeneration in partial hepatectomy

The restoration of liver volume after partial hepatectomy depends primarily on the proliferation of hepatocytes. After partial hepatectomy, zonal and lineage hierarchy of hepatic cells are maintained hence feedback loop signaling is essentially intact. In this case stem cell compartment is usually quiescent, DNA synthesis occurs in cells across the liver plates but only a portion of the cells undergo cytokinesis, yielding increased numbers of polyploid cells, higher numbers of apoptotic cells, and more rapid turnover of the liver with restoration of the normal ploidy profiles within weeks. Under nonpathological and static conditions, hepatocytes are in a quiescent state (G0 phase). Activation and proliferation of hepatocytes is mainly regulated by the primary and auxiliary mitogens. Primary mitogens includes HGF and ligands of EGFR (EGF, TGFα, Heparin Binding-EGF, Amphiregulin) that positively regulate the hepatocyte proliferation and TGF-β that negatively regulate the hepatocyte proliferation. Auxiliary mitogens include norepinephrine, TNF, IL6, bile acids, insulin and serotonin etc. In static condition anti mitotic effect of matrix bound TGF-β, matrix bound inactive HGF and basal level of auxiliary mitogens keep hepatocytes in quiescent stage. After partial hepatectomy, increased sinusoidal blood flow in the remnant liver increases shear stress. Increased shear stress and effective concentration of various gut derived factors, which act as the main stimulus to regeneration (Figure 78.3). It regulates liver size, growth and organ atrophy. Increased shear stress on liver cells leads to the increased production of NO by endothelial cells followed by increase in sinusoidal endothelial permeability to circulating hepatotrophic substances and gut derived factors (ref-1). It also causes matrix remodelling and increases hypoxia. These inductive signals leads to the proliferation of mature hepatic parenchymal and nonparenchymal cells in coordinated manner. Kinetic studies from animal model of liver regeneration suggest that in response to hepactomy hepatocytes first inter into the cell cycle followed by Kupffer and Hepatic stellate cells, Biliary ductular cells and lastly sinusoidal endothelial cells. Cellular proliferation of hepatocytes begins in the periportal region and proceeds toward the central vein. Though the molecular and cellular basis of hepatocyte induction and proliferation is well studied still further studies are required to understand the induction and proliferation of cholangiocytes and other non parenchymal cells after PHx. Hence in the following section we will mainly discuss hepatic regeneration in terms of hepatocyte proliferation.

Hepatocyte induction and proliferation: Increase in gut-derived factors (such as lipopolysaccharide LPS, inflammatory mediators of the innate immune response complement factors C3a and C5a, and immunoglobulin superfamily proteins intercellular adhesion molecules ICAMs) in portal blood and subsequently in liver sinusoids, activate Küpffer cells and sinusoidal endothelial cells. Activated Küpffer cells come in close contact with parenchymal hepatocytes and hepatic stellate cells. It causes the activation of hepatic stellate cells and prime the hepatocytes to become responsive to growth factors through their release of tumor necrosis factor (TNFα) and interleukin 6 (Il6) cytokines in a paracrine manner. This paracrine and endocrine signals from Küpffer cells initiate the first cascade of transcriptional activity in hepatocytes (priming phase). During this phase, pre-existing transcription factors such as NFκB, Stat3, and AP-1 get activated by post-translational modifications without de novo protein synthesis. This allows rapid expression of genes that promote hepatocytes to leave their quiescent G0 state and enter the cell cycle (G0-G1 transition). Activated NFκB by TNFα, LPS, C3a, C5a, and ICAMs in Küpffer cells leads to the production of cytokines IL-6. IL6 produced by Kupffer's cells bind to Gp130 receptor present on hepatocytes and activate STAT3. Activated Stat3 leads to de novo synthesis of various early response proteins like MYC, GADD45, FOS, JUNB, and EGR1 required for G0-G1 and G1-S transitions. After triggering the G0 to G1 phase transition by cytokines further progression of Cell cycle requires synergistic signals from growth factors. c-Met-HGF and EGFR-EGFR ligands, are the two main growth factor signaling, which further regulate the growth and division of hepatocytes. In the abscnce of these signaling cells inter into the cell cycle but fail to divide, induce apoptosis and die.

Activated endothelial cells first decrease the production of Ang2 that leads to the decrease in the production of TGF-beta by these cells and hence release the TGF-beta growth inhibitory brack from hepatocytes. Activated Stellate cells secrete growth factor SDF1 which increase the expression of of its receptor CXCR7 on endothelial cells. Increased expression of CXCR7 together with CXCR4 leads to the production of hepatocyte growth factor HGF and wnt2a. Hemodynamic changes in liver also initiate matrix remodelling. It causes the activation of matrix degrading enzyme urokinase that leads to remodeling and degradation of the pericellular matrix. HGF is therefore released and activated by urokinase, where it can have a local mitogenic effect on hepatocytes. Altogether this increase in local hepatocyte mitogens HGF and decrease in TGF-beta leads to rapid proliferation of hepatocytes. The ligand for c-Met, HGF is also secreted by macrophages and activated HSC in liver. Apart from liver HGF is also secreted by other organs such as lung, spleen and kidney. EGFR ligands in the liver is secreted by macrophage and liver endothelial cells and outside liver by salivary gland and duodenum which secrete EGF. Activation of both c-met and EGFR receptor leads to the activation of various

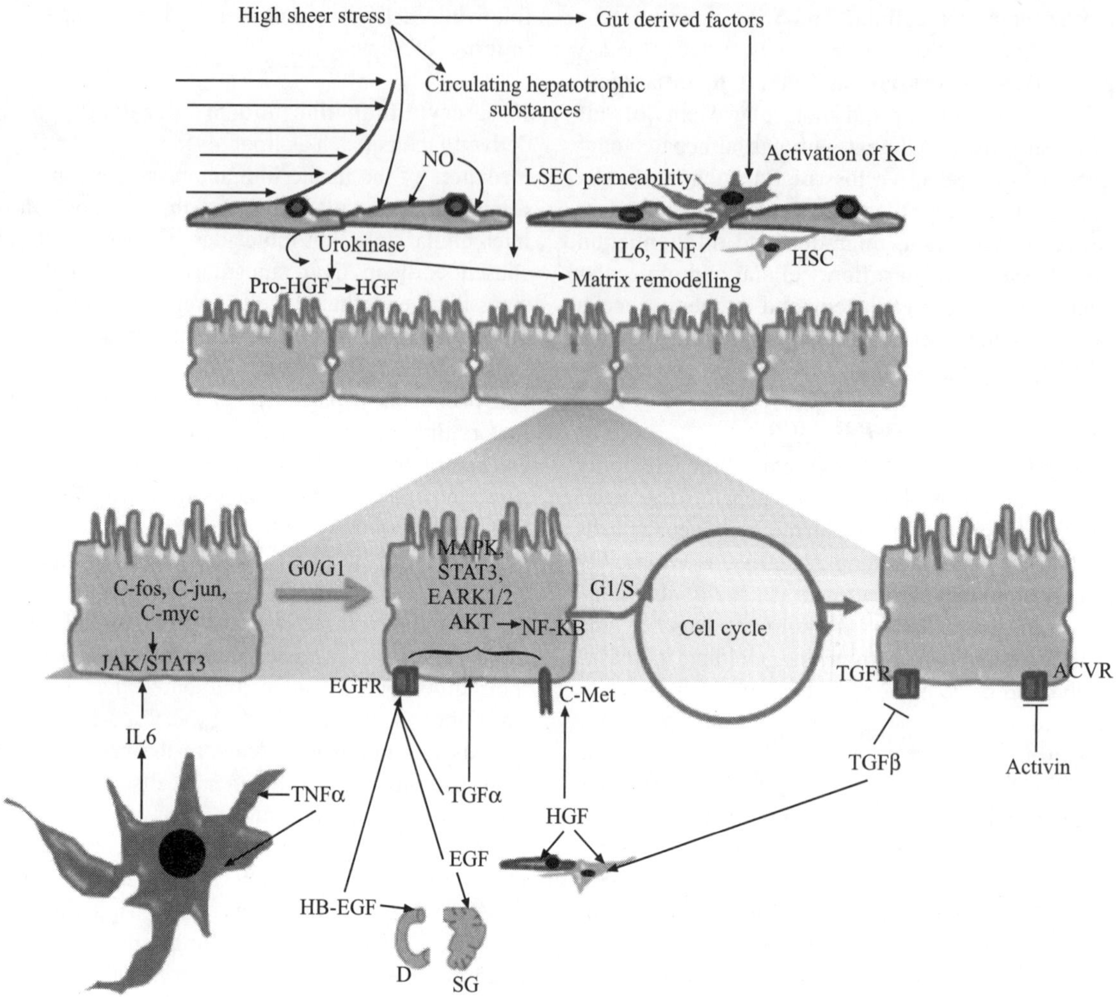

Figure 78.3   Diagrammatic representation of hepatocyte proliferation after PHx. (details given in the text). In response to PHx increased portal blood flow increase the concentration of gut derived factor and circulating hepatotrophic substances as well as increased shear stress on endothelial surface in hepatocytes. Increased gut derived factors active the Kupffer cells which subsequently prime the hepatocyte for further mitotic division and make them responsive for mitogens. Increase shear stress on endothelial cells increase the NO production and sinusoidal endothelial permeability to circulating hepatotrophic substances. It also causes the matrix remodelling and increase the local bioavailability of various hepatic mitogens required for further growth and division of hepatocytes. H—hepatocyte; KC—Kupffer Cell; SEC—sinusoidal Endothelial Cells; HSC—Hepatic stellate cells; D—duodenum; SG—salivary gland.

downstream signaling cascade such as RAS-MAPkinase, PI3K-AKT, PLCγ, and Stat signaling pathways which further control the proliferation, differentiation, migration, and survival of the cell. Among the cytokines TNF produced by Kupffer cells also has both pro and anti mitotogenic effect on hepatocytes. It has a promitogenic effect if it can activate NFκB. If this activation does not occur, TNF can induce apoptosis in hepatocytes. NFκB activation is dependent on activation of Akt. Both MET and EGFR are strong inducers of Akt activation. In the event when MET and EGFR fail to act, absence of Akt activation leads to failure of NFκB activation and TNF induces apoptosis of hepatocytes. TNF also causes induction of TACE, a plasma membrane associated protease which controls activation of TGFα (EGFR ligands) by hepatocytes during regeneration which act as mitogens for hepatocytes in autocrine fashion. Other blood born growth factors such as insulin, serotonin, norepinephrine etc also act as co mitogens for various hepatic cells. Norepinephrine is normally cleared by liver; reduction in liver mass after partial hepatectomy diminished clearance and may cause a rise in catecholamines. Increased catecholamines in blood increase the production of EGFR ligands in burner's gland. It also causes increased production of HGF by other organs such as lung, spleen and kidney. Hence there is increase concentration of HGF and EGFR ligands in the circulation. It also causes increase activation of HGF and EGF receptor via alpha-1 adrenergic receptor. Activated stellate cells also secrete ligands for Hh(hedgehog) and Notch which help in proliferation of cholangiocytes and ductular cells. In the letter phage of liver regeneration sinusoidal endothelial cells increase the production of Ang2 leading to VEGFR2 and Tia2 dependent replication of endothelial cells and revascularization of newly produced hepatocytes.

Non parenchymal cell induction and proliferation: The mechanisms responsible for priming of non-parenchymal liver cells during regeneration are not clearly understood. Hepatocytes are the first to undergo proliferation, and regeneration after partial hepatectomy. Regenerating hepatocytes produce growth factors such as PDGF,VEGF, FGF1, FGF2, SCF, Angiopoietins 1 and 2, and TGFα which act as mitogens for hepatic stellate and endothelial cells. Proliferation of endothelial cells restores sinusoidal network. VEGF produced by regenerating hepatocytes attract endothelial cells which penetrate the clusters of newly proliferated yet not fully vascularised hepatocytes and re-establish a sinusoidal network. Kupffer cells have not been clearly proven to proliferate during regeneration and their numbers may be affected by migration from precursor cells in the bone marrow. Regeneration of Hepatic stellate cells after partial hepatectomy is not well understood. Some line of evidence suggest that regeneration of stellate cells is dependent on expression of neurotrophin receptors.

Termination of hepatocyte proliferation: Liver index (ratio of liver weight/body weight) is tightly regulated and depends on metabolic demands of the organism. Liver regeneration after PHx stops precisely when a preoperative liver index is restored. Precise cellular and molecular mechanism regulating this process is not clearly understood. A significant research effort made in this area, highlighted several molecules as potential terminators of hepatocyte proliferation, among them suppressors of cytokine signaling Socs proteins, plasminogen activator inhibitor PAI-1 protein, TGFβ- and activin-ligands as well as ECM integrin-linked kinase (ILK) are the most prominent terminators of hepatocyte proliferation. Hepatocyte proliferation starts with the remodelling of extercellular matrix followed by the replication of non parenchymal cells and final reconstitution of liver histology. Towards the end of liver regeneration TGFβ- and activin-ligands induced signaling plays an important role in assembly of hepatic tissue. It is produced predominantly by stellate cells. It stimulates synthesis of multiple extracellular matrix proteins from mesenchymal cells. Regenerating hepatocytes escape from the mito-inhibitory effects of TGFβ by reducing the expression of TGFb1 receptors I, II. It also stimulates tubulogenesis and formation of neovascular structures. Reassembly of the extracellular matrix and the sinusoidal capillary network provides matrix driven signaling that terminates the regenerative process either by direct signaling through integrins or signaling induced by TGFβ (bound to the newly synthesized decorin and again exerting a "tonic" mito-inhibitory effect). Newly synthesized matrix is also be capable of binding HGF (a protein with high affinity to glycosaminoglycans and heparin) and preventing it from being activated by urokinase. TGFβ also inhibit the expression of urokinase. These set of events would bring hepatocytes back into a state of quiescence, surrounded by HGF (bound to glycosaminoglycans) and TGFb1 (bound to decorin). Reconstitution of ECM also activates the integrin-linked kinase (ILK) resulting in increased production of C/EBPα and decreased production of C/EBPβ that suppresses the proliferation of hepatocytes.

(ii) Regeneration after selective loss of Parenchymal cells

Tremendous mitotic capacity of mature hepatocytes provide an efficient means by which the normal liver can regain its liver mass, as discussed in previous section. However, this regenerative capacity is overwhelmed during acute or chronic injury to the liver as occurs with repeated drug exposures, radiation, or certain viral infections like hepatitis B or C. This leads to the selective loss of late lineage cells. Selective loss of matured hepatocytes either due to acute or chronic liver injury results in secretion of cytokines and growth factor by dying hepatocytes and muting of the feedback loop signaling that activates rapid cell division of early lineage stage cells in stem cell compartment. In response, hepatic progenitor cells HPCs (hepatic stem and their early lineage cells) undergo rapid hyperplastic growth (mitosis) followed by their migration and differentiation from periportal to pericentral zone. These phenomena, is commonly known as the *"oval cell response"* in rodents and the *"ductular reactions"* in humans. Unlike hepatocyte mediated liver regeneration, molecular and cellular basis of HPC mediated liver regeneration is not well understood. HPC mediated liver regeneration consists of four steps: activation, proliferation, migration, and differentiation altogether known as HPC response. The mechanisms controlling the HPC response are under intense investigation. In general, although many of the signals that control liver regeneration in the normal liver (i.e., via hepatocyte replication) and hepatic differentiation in developmental stages are involved in HPC-mediated regeneration (summarised in Figure 78.4). Multiple parenchymal and nonparenchymal liver cells, inflammatory cells and bone marrow derived stem cells are present during hepatic progenitor activation, which provide the supporting signals for HPC response.

Acute and chronic damage to liver leads to the production of pro-inflammatory tumour necrosis factor (TNF) such as TNFα and TWEAK (TNF-like weak induction of apoptosis), by Kupffer cells and monocytes respectively, both of which appear to play pivotal role in HPC activation. TWEAK mediates pro-proliferative effects directly on HPCs via the Fn14 receptor and activates pro-proliferative NFκB signaling cascade. It also activates the proliferation of non parenchymal mesenchymal cells of liver. It modulates the cellular activity of HPC response by its receptor TNF R1 and TNF R2. GP130 mediated signaling cascade is another important regulator of HPC response. A variety of cytokines, including interleukin 6 (IL6), oncostatin M (OSM) and leukaemia inhibitor factor (LIF), act through this signaling pathway. Binding of these cytokines to gp130 receptor leads to the homodimerisation, and activation of downstream JAK (Janus kinase)/STAT (signal transductor and activator of transcription) and ERK (extracellular signal-regulated kinase) pathways that regulate the various aspects of HPC response. IL-6 is the best characterised gp130 activators; it is produced by a variety of cell types including macrophages, fibroblasts and endothelia. IL-6 signals via the type I cytokine receptor CD126 (IL-6Rα) together with the signal transducing

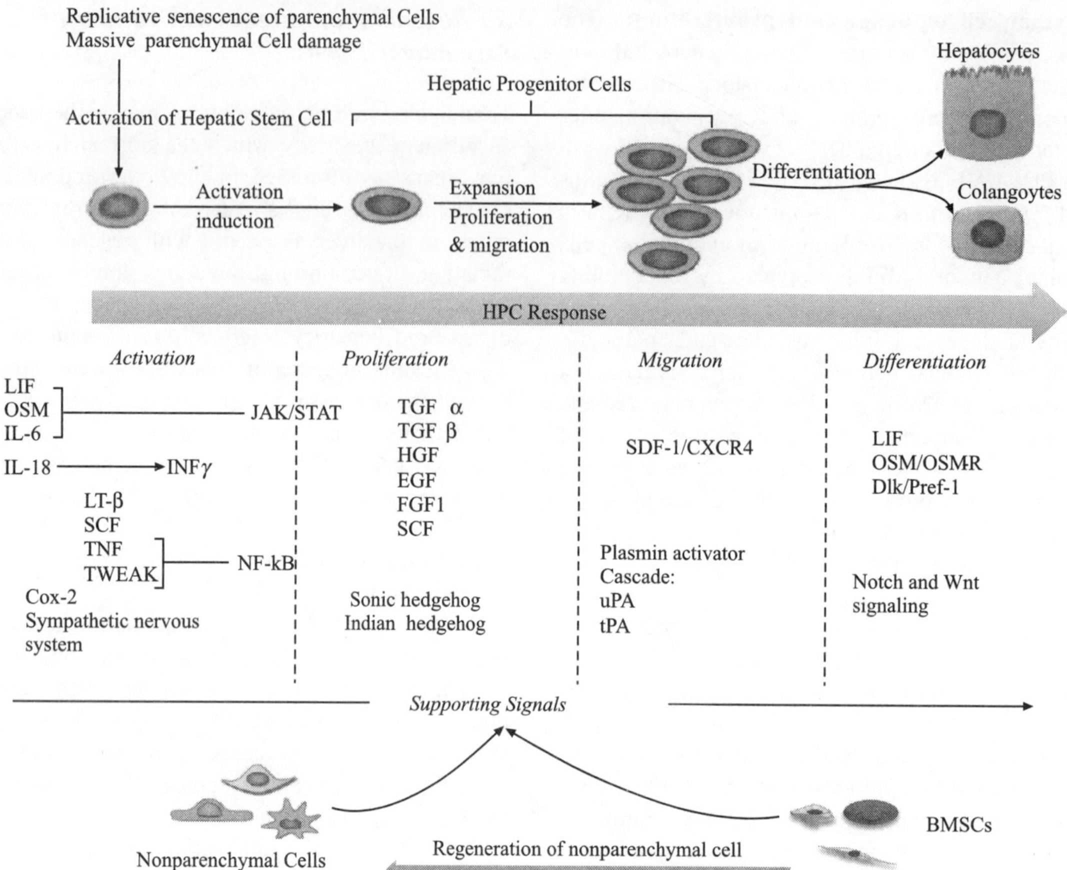

Figure 78.4 Diagrammatic representation of HPC mediated hepatic regeneration: in response to replicative senescence and/or massive loss of hepatic parenchyma, Hepatic stem and progenitor cells get activated. They migrate from their compartment and subsequently differentiate into the hepatocytes or cholangiocytes. Various cytokines and growth factors produced by dying parenchymal cells and surrounding non parenchymal and bone marrow derived cells regulate the activation, proliferation, migration and differentiation of HPC. Bone marrow derived stem cells also act as a source for nonparenchymal regeneration. Original figure modified from Ref 4. (*see Plate 36 for colour figure*)

gp130 homodimer principally activating STAT3 and regulate the proliferation and migration of hepatic stem and progenitor cells. Another cytokines like LIF and OSM both participate in a variety of processes including the regulation of growth and differentiation of HPCs. The action of LIF is mediated via the LIF receptor (LIFR), which is composed of LIFRβ and gp130. Its downstream effect in HPCs occurs predominantly via STAT1. OSM also activates gp130 either via its own OSM receptor (OSMRβ) subunit, or via LIFR. OSM influences extrahepatic progenitor cell activity and extracellular matrix (ECM) deposition. It is produced by hepatic macrophages and promotes the proliferation and differentiation of HPCs. Inflammatory cytokines such as INFγ and INFα also play an important role in HPC response. IFNα promotes differentiation of HPCs into hepatocyte lineage where as INFγ indirectly affects the activation of HPC by inhibiting the proliferation of mature hepatocytes. Primary mitogens for hepatic regeneration HGF, ligands of EGFR and TGFβ also modulate the HPC response. HGF regulates the growth and differentiation of HPCs but unlike hepatocytes, HGF alone is not sufficient to activate HPCs. Ligands of EGFR such as TGFα and EGF act as growth

factor for the activation of HPCs. The most prominent mitogen for HPC activation is Stem cell factor (SCF) acts via the c-kit receptor present on the HPCs and regulates the growth and expansion of HPCs. The chemokine stromal-cell-derived factor 1 (SDF-1) uniquely binds to the CXCR4 receptor and plays a variety of roles including cell trafficking, proliferation and organogenesis of hematopoitic stem cells, it also regulates the proliferation and migration of hepatic stem and progenitor cells. Similar to the development of liver Notch and wnt signaling plays an important role in HPC mediated liver regeneration. While, Notch is required for the expansion of HPC and their commitment to cholangiocyte lineage, Wnt signaling regulate the commitment of HPCs to hepatocyte lineage.

(iii) BMCs and hepatic regeneration

Bone marrow cells and liver has evolutionary conserved developmental relationship. In Nonhepatic stem cell section we have briefly discussed the developmental relationship between liver and hematopoietic cells, and the type of different bone marrow derived stem and progenitor cells involved in hepatic regeneration. In normal hepatic tissue turnover bone marrow cells do not have any significant role. However, in response to liver tissue injury bone marrow cells get activated, migrate and get engrafted into the liver and support the hepatic repair and regeneration. Very little is known about the physiological relevance of bone marrow-derived cells engrafting in the liver with injury. What role do the cell type(s) involved play in the development and/or prevention of disease? What signals from the liver regulate the migration and engraftment of different bone marrow cells? Based on *in vitro* and *in vivo* studies it is believed that the balance of the intrahepatic microenvironment, combined with the intrinsic intrahepatic regenerative capacity, determines both the extent and nature of the liver-specific bone marrow stem cell response. Bone marrow cells consist of three major type of stem and progenitor cells viz. Hematopoietic stem and progenitor cells, mesenchymal stem cells and endothelial progenitor cells (as discussed previously in non hepatic stem cell section). As described in the Figure 78.5, in response to hepatic injury regenerating liver increased the expression of CXC chemokine, Stromal derived factor-1 (SDF-1) and granulocyte colony stimulating factor G-CSF. G-CSF helps in the mobilization of hematopoietic stem and progenitor cells from the bone marrow and SDF-1 helps in their homing and engraftment to the liver. Similarly VEGF (vescular endothelial growth factor) produced by regenerating hepatocytes signals the activation, migration and engraftment of endothelial progenitor cells from bone marrow to the liver. Unlike HSCs and EPCs molecular signals responsible for the activation, migration and engraftment of bone marrow derived MSCs are not known.

Once engrafted to the liver what role these different cells do is not clearly understood. Initially because of in vitro hepatic differentiation potential of HSCs and MSCs it was believed that these cells might directly transdifferentiate to hepatocytes or cholangiocytes. However, recent line of evidence suggests that bone marrow derived stem and progenitor cells help in repair and regeneration of injured liver by (a) providing various cytokines and growth factor for activation and differentiation of hepatic stem cells, (b) matrix remodelling, (c) source of various non parenchymal cells such as monocytes for Kupffer's cells, EPC for endothelial cells, sinusoidal endothelial progenitor cells for sinusoidal endothelial and MSCs for myofibroblast. Hence contribute significantly to non parenchymal regeneration (d) Transdifferentiation or fusion to give hepatocyte like cells in case of extensive damage to hepatic parenchyma.

V. REGENERATION IN DISEASED LIVER

As discussed in the previous sections that liver is a highly regenerative organ. It can effectively rescue the lost mass even after acute or repetitive damage and recover from even substantial necrosis caused by toxins or a viral infection. This efficient biological response of hepatic regeneration is orchestrated by synergistic interaction between different parenchymal and nonparenchymal hepatic, cells as well as infiltrating non hepatic cells of bone marrow origin via secreted growth factors, cytokines and matrix components. Epithelial

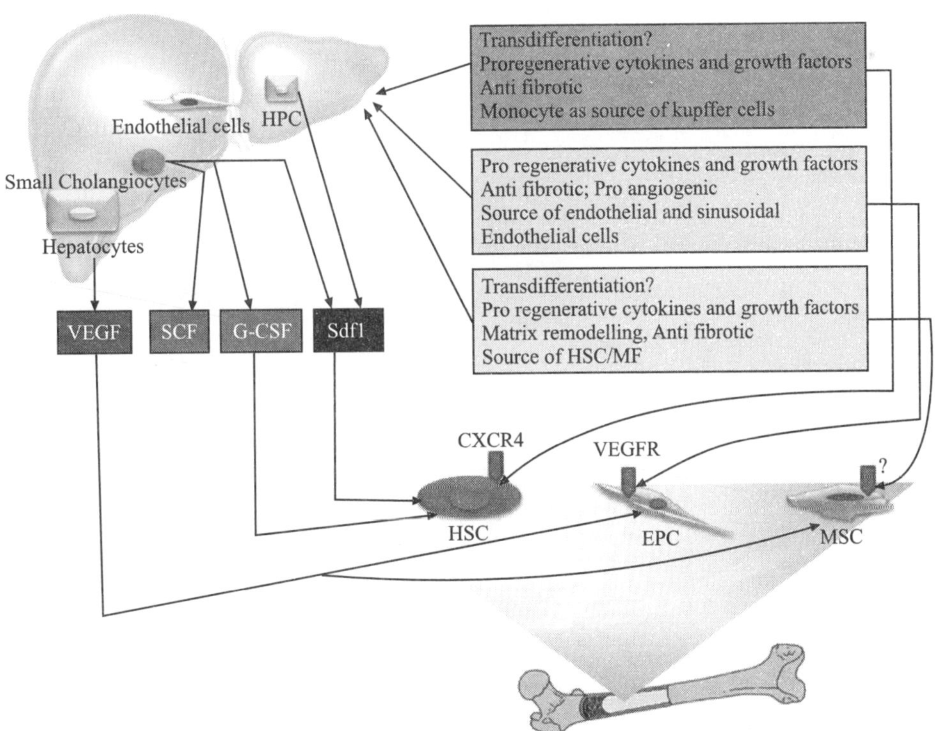

Figure 78.5 Crosstalk of Liver and bone marrow in hepatic regeneration and repair: In response to liver damage regenerating liver produce different cytokines and chemokines that leads to the activation, migration and engraftment of bone marrow derived cells to the liver where they directly or indirectly participate in hepatic regeneration. (*see Plate 36 for colour figure*)

plasticity of hepatic parenchyma in response to damage helps in efficient recovery of lost parenchymal cells and associated nonparenchymal resident liver cells and recruited bone marrow cells helps in preparing ground for efficient epithelial regeneration as well as rebuilding of lost or damaged native tissue architecture. Hence inductive regeneration of both parenchymal and nonparenchymal hepatic cells is critical for complete regeneration of hepatic mass. Depending upon the type and severity of damage, rescue of hepatic mass takes place either by change in size and/or ploidy, self replication of parenchymal cells or by activation and differentiation of hepatic stem and progenitor cells or both (Summarized in Figure 78.6). Though our understanding form the animal model of liver disease suggest that during prolonged liver injury, such as that seen in chronic hepatitis due to viral infection or alcohol

excess, there is excessive deposition of ECM, and hepatocyte and biliary epithelia cell proliferation is impaired, which compromises regeneration. Stellate cell–derived factors, such as HGF and hedgehog, fail to stimulate liver growth, and stellate cells deposit scar tissue, which inhibits hepatocyte and hepatic progenitor cell–mediated regeneration resulting in reduced liver mass and function. Furthermore, the vasculature structures within the liver failed to support the parenchymal regeneration and aggravate pathological angiogenesis and fibrosis of tissue eventually lead to organ failure and liver cancer. Precise mechanism of impaired hepatic regeneration in advance liver disease is still under investigation. In acute liver failure (ALF) both viral and non viral acute insult to the liver leads to massive necrosis of matured hepatocytes. In early phage of injury, some remaining hepatocytes have self replicating potential

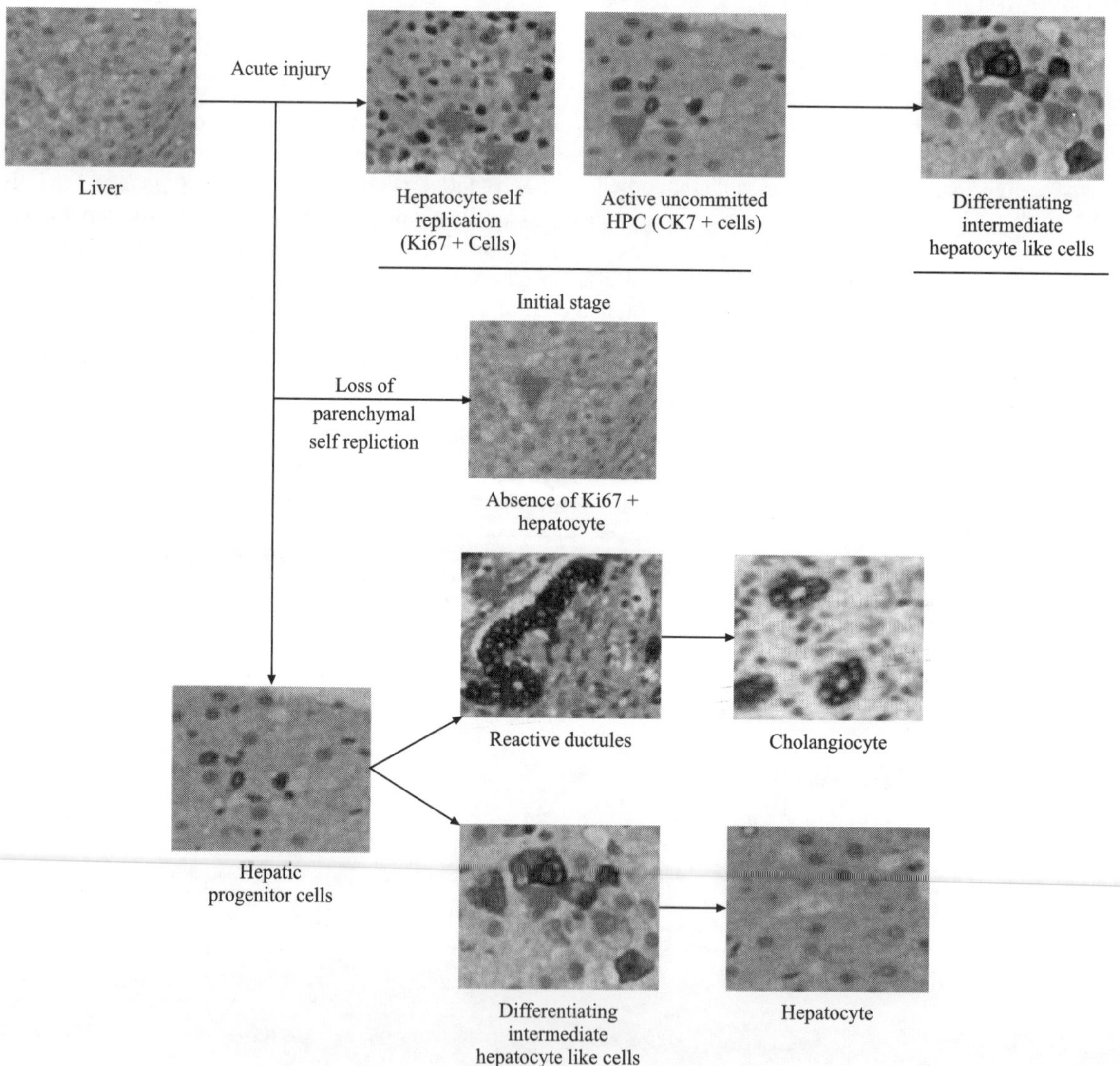

Figure 78.6   Regeneration in Diseased Liver: (details given in the text) presence of Ki67 positive hepatocyte is the indication of hepatocyte Self replication; CK7 positivity and morphology of cells shows the HPC (small CK7+ cells with high N/C ratio), reactive ductules (small CK7+ Cholangiocytes) in portal area and small intermediate hepatocyte (small hepatocytes with faint cytoplasmic CK7 positivity) like cells in periportal and midacinar area (IHC photomicrograph is taken from our unpublished data). (*see Plate 37 for colour figure*)

and try to compensate the loss. Dying hepatocytes also gives selection presser on HPCs, in response HPCs get activated and try to compensate the loss. In early phase of acute insult, HPCs are activated but they are not committed to any lineage, and self replication of remaining hepatocytes tries to compensate the loss. Due to ongoing toxicity in later stage of injury remaining hepatocytes show ballooning and anisokaryosis, and loses its self replication power, while HPCs shows differentiation into later lineage cells, as evident by presence of increased intermediate hepatocytes and reactive ductules. If acute insult is managed in these patients, most of the liver can regenerate to normal volume, while failure to manage the insults increases the ongoing damage and regeneration fails to meet the demand, subsequently leading to liver failure. Unlike acute injury in chronic liver damage, persistent inflammation and oxidative stress leads to telomere shortening and replicative senescence of hepatocytes. Hence, first line of hepatic regeneration is not functional and liver tries to compensate the loss through HPC mediated regeneration extant of HPC activation is similar in both chronic (CLD) and acute on chronic (ACLF) liver disease but degree of differentiation is comparatively higher in ACLF. In chronic liver injury even though there is a profound ductular response, it seems that differentiation signals are either insufficient for—or inhibitory to—hepatic ductular/progenitor mediated regeneration. Liver failure can ensue, implying that this form of regeneration is often inadequate. Further understanding of these processes in human liver injury is needed for both understanding the underlying mechanism of regeneration failure in diseased liver as well as therapy development.

REFERENCES

1. Abshagen K., Eipel C. and Vollmar B. (2012), A Critical Appraisal of the Hemodynamic Signal Driving Liver Regeneration, *Langenbecks Arch Surg.*, 397(4):579–90.

2. Asahina K., Hepatic Stellate Cell Progenitor Cells, *J. Gastroenterol Hepatol.*, 27 Suppl 2:80–4.

3. Dalakas E., Newsome P.N., Harrison D.J. and Plevris J.N. (2005), Hematopoietic Stem Cell Trafficking in Liver Injury, *FASEB J.*, 19(10):1225–31.

4. Duncan A.W., Dorrell C. and Grompe M. (2009), Stem Cells and Liver Regeneration, *Gastroenterology*, 137(2):466–81.

5. Eckersley-Maslin M.A., Warner F.J., Grzelak C.A., McCaughan G.W. and Shackel N.A. (2009), Bone Marrow Stem Cells and the Liver: Are They Relevant? *J. Gastroenterol Epatol.*, 24(10):1608–16. Sancho-Bru P., Najimi M., Caruso M., Pauwelyn K., Cantz T., Forbes S. and Roskams T.

6. Fujiyoshi M. and Ozaki M. (2011), Molecular Mechanisms of Liver Regeneration and Protection for Treatment of Liver Dysfunction and Diseases, *J. Hepatobiliary Pancreat Sci.*, 18(1):13–22.

7. Gilchrist E.S. and Plevris J.N. (2010), Bone Marrow-derived Stem Cells in Liver Repair: 10 Years Down the Line, *Liver Transpl.*, 16(2):118–29.

8. i-Tayeb K., Lemaigre F.P. and Duncan S.A. (2010), Organogenesis and Development of the Liver, *Dev Cell.*, 16;18(2):175–89.

9. Kurinna S. and Barton M.C. (2011), Cascades of Transcription Regulation During Liver Regeneration, *Int J. Biochem Cell Biol.*, 43(2):189–97.

10. Michalopoulos G.K. (2010), Liver Regeneration After Partial Hepatectomy: Critical Analysis of Mechanistic Dilemmas, *Am J. Pathol.*, 176(1):2–13. Epub 2009 Dec 17.

11. Michalopoulos G.K. and Liver Regeneration, *J. Cell Physiol.*, 213(2):286–300.

12. Ott M., Gehling U., Sokal E., Verfaillie C.M. and Muraca M. (2009), Stem and Progenitor Cells for Liver Repopulation: Can We Standardise the Process from Bench to Bedside? *Gut.*, 58(4):594–603.

13. Stutchfield B.M. and Forbes S.J. (2012), Liver Sinusoidal Endothelial Cells in Disease –And for Therapy? *J. Hepatol.*, Doi:pii: S0168–8278(12)00627–7.

14. Tanaka M., Itoh T., Tanimizu N. and Miyajima A. (2011), Liver Stem/progenitor Cells: Their Characteristics and Regulatory Mechanisms, *J. Biochem.*, 149(3):231–9.

15. Turner R., Lozoya O., Wang Y., Cardinale V., Gaudio E., Alpini G., Mendel G., Wauthier E., Barbier C., Alvaro D. and Reid L.M. (2011), Human Hepatic Stem Cell and Maturational Liver Lineage Biology, *Hepatology*, 53(3):1035–45.

16. Wandzioch E. and Zaret K.S. (2009), Dynamic Signaling Network for the Specification of Embryonic Pancreas and Liver Progenitors, *Science.*, 26;324(5935):1707–10.

17. Zhao R. and Duncan S.A. (2005), Embryonic Development of the Liver, *Hepatology.*, 41(5):956–67. Review.

79

Bone Marrow Cells and Tissue Regeneration

Asok Mukhopadhyay

CONTENTS

I. INTRODUCTION

Bone marrow (BM) is usually seen as a reservoir of blood and immune cells, which provides support to an adult organism through out the life time. Hematopoietic stem cells (HSCs) of the BM self-renew, commit to multipotent progenitors (MPPs) and lineage restricted common lymphoid/myeloid progenitors (CLPs/CMPs) cells, which finally give rise to terminally differentiated cells for physiological functions. Osteoblastic niche in the marrow maintains HSCs in the quiescent state, which then activated and undergoes self-renewal division (asymmetric), in response to stress or stimuli. The BM is most prolific tissue in mammals, in adult humans the daily output of blood cells is nearly 400 billion [1]. The BM tissue is composed of a highly heterogeneous population of both hematopoietic and non-hematopoietic cells, originated from common mesoderm germ layer. The detail of the cellular composition and respective functions will be discussed in other Chapters of the book. Studies in the past have shown that besides HSCs, human and mouse BM tissue contains *multipotent* adult stem cells (ASCs), these are known as *mesenchymal* stem cells (MSCs) and *multipotent adult progenitor* cells (MAPCs). These three classes of stem cells not only involve in the regeneration of hematopoietic system, but also found to play crucial roles in the repair and regeneration of non-hematopoitic tissues.

II. AUTONOMOUS REGENERATION OF TISSUE

Each organism regenerates tissues/organs, though the extent of regeneration may vary from species to species. Amphibians (e.g., frog), reptiles (e.g., lizard) regenerate complex body structures, like tails and limbs; newts generate lens. In cases of vertebra the regeneration process has limitation, which is termed as *maintenance regeneration*. In mammals, including humans, the maintenance regeneration is essential to preserve the integrity of the tissue and to perform normal functions.

In mammals the regeneration could occur in three levels: molecular, single cell, and tissue. The nature of regeneration is governed by the type of tissue and the extent of injury. In molecular level of regeneration, the original cells do not die but the turnover of protein synthesis is increased to compensate their degradation due to stress (example, synthesis of certain proteins in cardiomyocytes in case of elevated blood pressure). Single cell level regeneration is found to occur in the regeneration of axons in transected sensory and motor neurons. Tissue level regeneration is important in the regenerative medicine. It is based on three main criteria: (i) tissue specific stem/precursor cells or competent circulating progenitor cells are present, (ii) damaged tissue has regeneration potential by eliciting microenvironmental cues, and (iii) inhibitors that

970

suppress normal regeneration process are absent locally and systemically.

In mammals, the regeneration of damage tissues is taken place following three mechanisms: *compensatory hyperplasia, activation of adult tissue specific progenitor/stem cells,* and *dedifferentiation followed by redifferentiation.* In compensatory hyperplasia, mature tissue specific cells perform normal functions and at the same time undergo proliferation to maintain tissue homeostasis. Liver (hepatocytes), pancreas, and blood vessel are amongst few important tissues regenerate by compensatory hyperplasia (Figure 79.1A). Activation of ASCs is the alternate way by which most of the mammalian tissues regenerate. It is believed that after the *embryonic life,* tissue specific ASCs specific to all three germ layers reside in the specialized niches in the respective tissues. In response to the regenerating cues these ASCs proliferate and differentiate into mature cells, thus support the organisms throughout the life-time [2]. Including liver and blood vessels, many other tissues of humans are regenerated from ASCs, these are bone marrow, epithelia, skeletal muscle, bone, etc. (Figures 79.1B). A list of tissue specific ASCs is shown in the Table 79.1. The third alternate process by which tissue may regenerate is dedifferentiation followed by redifferentiation. Dedifferentiation is the loss of phenotype and functional specialization of mature cells of one lineage by which they converted into progenitor cells of the replacement tissue (Figure 79.1C). Dedifferentiation followed by redifferentiation is rare in humans except in case of few malignancies.

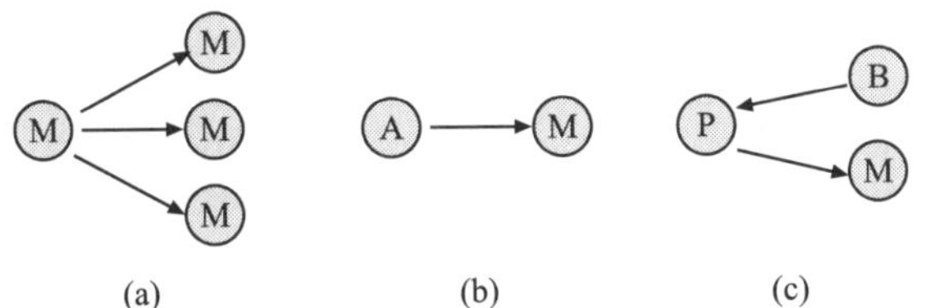

Figure 79.1 Mechanisms for the regeneration of damage tissues in mammals. (a) Compensatory hyperplasia, (b) Activation of adult tissue specific progenitor/stem cells, and (c) Dedifferentiation followed by redifferentiation.

III. LIMITATION IN AUTONOMOUS REGENERATION

In humans and other mammals, maintenance regeneration is slow, but a tightly regulated process. Through this, tissue homeostasis is naturally achieved, such as regeneration of bone marrow following irradiation, regeneration of skin following an injury due to burn or incision, etc. As mentioned above, the autonomous or natural regeneration of tissue or organ occurs either by compensatory hyperplasia and/or by activation of tissue specific ASCs. The autonomous tissue regeneration has been found to be compromised in many pathological conditions. The failure of autonomous tissue regeneration could be due to three main reasons: (i) lack of regeneration-competent cells, (ii) absence of inductive microenvironment, and (iii) synthesis of inhibitory factors. Few of the pathophysiological states in which natural tissue regeneration is inhibited:

Aging: The natural regeneration process is gradually compromised with increase in age of the subject. These may be due to overall aging of the subject and/or aging of stem/progenitor cells that involve in the regeneration of the tissue/organ. Besides this chronological aging, environmental factors (e.g., accumulation of reactive oxygen species, etc.) may influence the aging process. Epidermal stem cells are found to be inherently aging resistant and that local environmental and systemic factors modulate aging of skin. Aging of tissue may slowdown autonomous regeneration process. For example, in case of younger subject ischemia, vascular trauma and pro-angiogenic chemokines stimulate mobilization of endothelial progenitor cells (EPCs) from bone marrow into the peripheral blood, which later home into the diseased organ for vasculogenesis. Aging may reduce mobilization of EPCs due to decrease in the synthesis of angiogenesis factor (VEGF) or lowering EPC number in the bone marrow, inefficient homing into traumatized tissue/organ, etc.

Degeneration of tissues/cells: In case of certain neurological diseases, like Parkinson's, the dopaminergic neurons are irreversibly degenerated. These degenerated cells are not autonomously regenerated.

TABLE 79.1 **Adult stem cells and their tissue specific locations**

Cell type	Tissue-specific location	Cells or tissue produced
Hematopoietic stem cells	Bone marrow and peripheral blood	Bone marrow and all blood cells
Mesenchymal stem cells	Bone marrow	Stroma cells, bone, cartilage, tendon, muscle, adipose tissue.
Neural stem cells	Central nervous system (CNS)	Neurons, astrocytes, oligodendrocytes
Oval cells	Canals of Hering (near bile ductules in liver	Hepatocytes, bile ductular cells
Pancreatic stem cells	Intraislet	Beta cells
Skeletal muscle stem cells or satellite cells	Muscle fibers	Skeletal muscle cells
Keratinocyte (skin stem cells)	Basal layer of epidermis	Epidermis, hair follicles
Epithelial stem cells of lung	Tracheal basal and mucus-secreting cells, bronchiolar clara cells	Mucus and ciliated cells, type I and II pneumocytes
Stem cells of intestinal epithelium	Epithelial cells located around the base of each crypt	Paneth's cells, brush-border enterocytes, mucus-secreting goblet cells, etc.

Absence of inductive cues and the presence of inhibitory factors: Despite the presence of competent cells and microenvironmental cues, the damaged tissue may not regenerate. This could be due to the presence of local and/or systemic inhibitory factors; some of them belonging to the TGFβ super family of proteins. In case the tissue injury is repetitive in nature, instead of spontaneous natural regeneration, the tissue is repaired by *fibrosis* or *scar* tissue formation. The scar tissue adversely modulates autonomous regeneration of damaged organ/tissue and its normal physiological functions.

IV. CIRCULATION OF BONE MARROW CELLS IS ASSOCIATED WITH TISSUE INJURY

Studies in animals have shown that in case of stress, like tissue injury or infection, the number of circulating bone marrow cells are increased several folds than the baseline values of healthy animals. In case of human subjects, similar trend of circulating bone marrow cells has been noticed under various pathological conditions. There are many physiological functions that are performed by the circulating BM cells: (a) control inflammatory responses (e.g., neutrophils to kill the invading microbes, macrophages to clear up the necrotic tissue), (b) initiation of angiogenesis (EPCs form new vasculature in ischemic tissue), (c) repair of tissue (MNCs and/or MSCs provide growth and anti-apoptotic factors to the injured tissue) and regeneration of tissue (BM-derived stem cells differentiate into target tissue). It is believed that after tissue injury, ASCs replace necrotic cells as first line of action. If tissue specific stem cells are inadequate, BM-derived stem cells are homed to the damaged tissue to replenish the pool for tissue regeneration. Under normal physiological conditions tissue specific ASCs maintain homeostasis and BM-derived stem cells have little or no role in replacing apoptotic tissue [3]. In case of acute and extensive injury, the demand for tissue repair and/or regeneration can not met by the ASCs alone. BM-derived circulating stem cells act as a backup system and are engrafted to the damaged tissue. Later, these cells undergo differentiation into organ-specific cells and help to recover from the compromised state.

The BM is the largest reservoir of multipotent stem cells, which have been demonstrated to give rise to most known adult cell lineages [4]. As mentioned earlier, human and mouse bone marrow is reported to contain many distinct classes of stem cells, different from HSCs and MSCs, these are know as multipotent adult progenitor cells (MAPCs), BM-derived multipotent stem cells (BMSCs), and very small embryonic-like (VSEL) cells. In vivo and in vitro studies have shown that these cells exhibit the capacity for differentiation into cells of all 3 germ layers. Despite the presence of large pool of multiple adult stem cells in the BM tissue, except HSCs, none of them does mobilize in adequate number in response to the signals from the injured tissues. As a result, the number of stress-signal-induced differentiated BM-derived stem cells is considered inadequate, compared with tissue-derived stem cells, for the fulfillment of the demand by the regenerating tissue.

V. ROLE OF BONE MARROW CELLS IN TISSUE REPAIR: THE PARACRINE EFFECT

It is apparent from the earlier section that stress-induced circulatory BM-derived stem cells may not be enough either to repair or regenerate the damaged tissue/organ. This is due to the homing of cells to the damaged tissue to the extent one-tenth than actually needed. The therapeutic effect of BM cells has been noticed in cases when dense population of cells is transplanted in the damaged organ. In case of cell therapy the damaged tissue has been restored to normalcy by the process of repair and/or regeneration. Repair refers to the restoration of tissue/organ from existing damage one by deriving the benefits of the paracrine effect of the engrafted donor cells, here the bone marrow cells. It is important to remember that in this case the donor cells eventually leave the damaged tissue once the repair process is completed. In contrast, regeneration involves physical incorporation of the engrafted bone marrow cells and their differentiation into target tissue. This section will deal the paracrine effect of the BM cells in repair of the tissue.

Since the results of the preclinical studies published until now have shown that stem cell differentiation (or fusion) alone cannot account for enhanced cardiac function, endogenous pathway of cardiac repair has become the acceptable mechanism in case of human subjects. It has been considered than the therapeutic benefit of BM cells in case of myocardial infarction (MI) is primarily due to the paracrine effect of the BM-derived stem cells. This conclusion has been drawn based on the following evidences: (a) low levels of engraftment of donor cells, (b) low degree of cell differentiation, and (c) no correlation between the type of cells transplanted and the functional effect observed. The paracrine effects of MSCs and bone marrow mononuclear (BM-MNCs) cells have been demonstrated in treatment of ischemic heart. It has been found that MSCs alone or BM-MNCs secrete a broad range of proangeogenic, proaeteriogenic, and antiapoptotic factors, such as vascular endothelial growth factor (VEGF), fibroblast growth factor-2 (FGF-2), Angiopoietin-1, interleukin-1/6/11 (IL-1/6/11), insulin-like growth factor-1 (IGF-1), hepatocyte growth factor (HGF), etc. These secreted cytokines are expected to be involved in promoting angiogenesis, activation of existing cardiac progenitor cells, their proliferation and survival. Further, local injection of MSC-conditioned medium enhances collateral perfusion and remodeling in a murine model of hind-limb ischemia, reduces tissue atrophy and limb damage, and improves limb function. The proposed mechanisms of action of MSCs in cardiovascular repair are shown in Figure 2. The soluble factors, secreted by the cells, promote vascular and cardiomyocyte regeneration, support myocardial protection, and prevent/reverse remodeling in the ischemically injured ventricle. It is felt that by combined actions of above, the damaged cardiac tissue repairs and physiological function is partly restored.

VI. ROLE OF BONE MARROW CELLS IN TISSUE REGENERATION: THE CELLULAR DIFFERENTIATION OR FUSION

As bone marrow contains endothelial progenitor cells (EPCs), it is thought that besides paracrine effect BM-MNCs therapy could potentially improve cardiac regeneration (Figure 79.2). G-CSF mobilized human CD34$^+$ cells, injected into MI athymic rats, are homed into the infarct zone and incorporate into newly formed chimeric coronary vascular system. Later, many reports ascertain the capacity of other BM-derived cells to participate in mascular regeneration in the infracted heart and in the ischemic hind limb, these include EPC, HSCs, BM-MNCs, umbilical cord blood stem cells and MSCs. In human bone marrow-derived stem cells (hBMSCs) can differentiate into ECs, smooth muscle cells (SMCs), and cardiomyocytes (CMCs) [5]. The transplantation of these cells into infracted myocardium attenuates cardiac dysfunction both by paracrine effects and *de novo* differentiation of hBMSCs into myocardial tissues. Preclinical studies in the late 90s' and early 2000 have demonstrated that BM-derived stem cells exhibit a remarkable plasticity than previously thought. Successive animal experiments showed BM-derived stem cells not only transdifferentiate into cardiomyocytes, they can also form hepatocytes, epithelial cells of the gastrointestinal tract and lung, muscle, neurons and pancreatic β-cells. ASC plasticity or transdifferentiation or direct differentiation refers to the conversion of developmentally committed cells into cells of another lineage (within or across the germ layer) without fusion with the target cells or through dedifferentiating into the primitive stage (Figure 79.3).

Adult stem cells plasticity study has been criticized by many investigating groups, mainly due to lack of reproducibility of some of the earlier findings and *spontaneous fusion* between

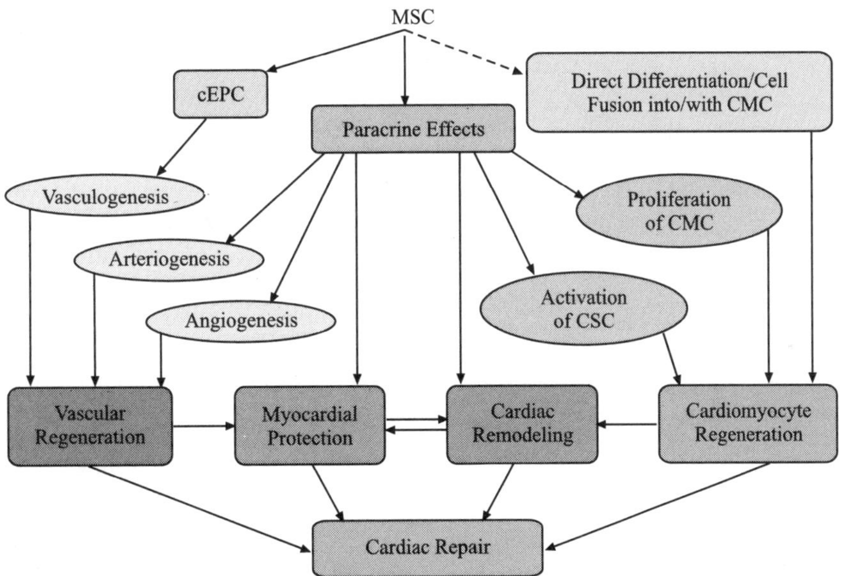

Figure 79.2 Proposed mechanism for the repair and regeneration of myocardial tissue. MSCs are shown to have paracrine role for repair of damaged cadiomyocytes. Cardiac regeneration may also occur by direct differentiation/cell fusion between MSCs and endogenous cells. cEPCs, circulating endothelial progenitor cells; CSCs, cardiac stem cells; CMCs, cardiomyocytes.

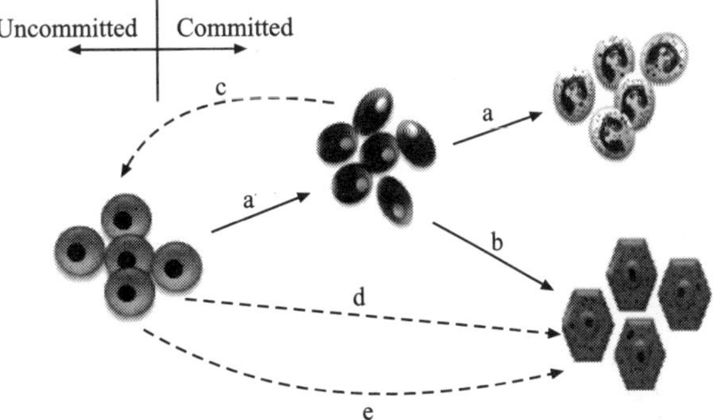

Figure 79.3 Mechanisms of direct differentiation/trans-differentiation. A normal pathway of differentiation is shown by path 'a'. In the case of direct differentiation, the target cells (polygonal) will originate from the intermediate committed cells following path 'b'. In this case the target cells will neither originate following path 'c-d' (combined effect of dedifferentiation followed by redifferentiation) nor from the uncommitted cells following path 'e'. (*see Plate 37 for colour figure*)

the donor and the target cells. However, few well-controlled studies have clearly demonstrated that bone marrow-derived cells can home in the wide variety of tissues of the recipient without cell fusion, though the number of cells is believed to be low to perform any substantial regeneration of the damaged tissue. Therefore, for future therapeutic applications, there is a need to understand the control of tissue-specific differentiation of BM-derived stem cells. There are three important issues, connected with the plasticity of BM cells, which need to be elaborately investigated at the stage of preclinical investigation. These studies are important due to the fact that clinical trials are rarely permitted by the regulatory agencies and there are limitations in execution level of a clinical trial. In this chapter, some of the important issues in the context of liver regeneration by BM-derived stem cells are elaborated using mouse as an experimental model. However, almost similar analyses will be necessary in each non-hematological application of BM-derived stem cells (Figure 79.4).

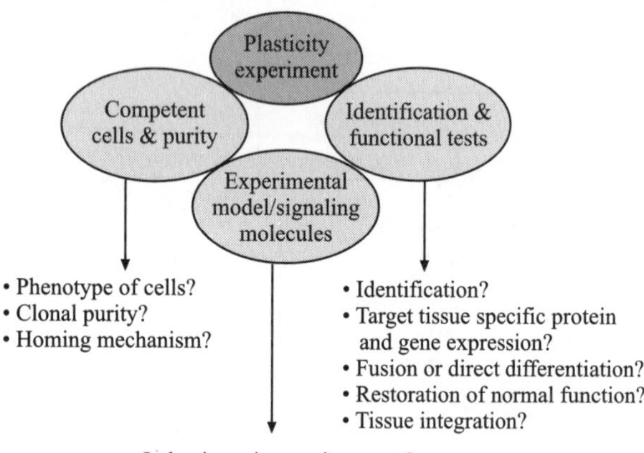

Figure 79.4 Experimental proof of principal for direct differentiation. There are three main factors that should be given importance for any successful plasticity experiments: the presence of competent cells and their purity, the experimental model and signaling molecules involved, and the identification of the donor-derived cells and functional recovery. Proposed tests under each factor are crucial to confirm the mechanism of cellular plasticity.

(a) Presence of competent cells and purity

BM cells are highly heterogeneous, and only a small fraction of them with a distinct phenotype can trans-differentiate into hepatocytes. Identification of these competent cells is essential before establishing cellular plasticity. The competent cells should exhibit the unique properties of being able to repopulate the tissue of origin (in this case bone marrow), and at the same time, to differentiate into hepatocytes. Rat and human BM-derived hepatic stem cells, phenotypically characterized as β_2m^- Thy-1^+, are found to express liver-specific genes. These cells undergo differentiation into mature hepatocytes and integrate into the rat hepatic plates. Highly purified human umbilical cord blood cells, characterized as Lin$^-$CD45$^+$CD38$^-$CD34$^{+/-}$C1qR$_p^+$,

are capable of repopulating BM tissue and can differentiate into human hepatocytes in NOD/SCID mouse [6].

(b) Experimental model and signaling molecules

In *vivo* studies have demonstrated that the extent of direct differentiation may vary with the experimental models. Lethally irradiated mouse model seems to exhibit discouraging results in liver regeneration. Fetal liver development and liver regeneration following partial hepatectomy (PH) or acetaminophen/CCl$_4$ administration are considered to be the appropriate experimental models. The reason for this has been assumed to be microenvironmental cues. The combination of toxic insult to the liver and HGH treatment may improve the engraftment of BM-derived cells, their differentiation, and the proliferation of donor-specific hepatic cells.

Other important aspects pertaining to the experimental model are the time and the route of transplantation after perturbing the liver. Tissue damage is generally followed by pro-inflammatory responses, which attract immune cells to the site of injury to clear necrotic tissue. Liver injury by administration of CCl$_4$ or acetaminophen is not an exception of this process. To avoid immune reactions, many investigators delay transplantation of cells. This has been particularly observed in case of clinical trials in MI. In the case of liver regeneration by BM-derived cells, it is recommended that the graft should be transplanted between 36 to 48h post-injury by CCl$_4$ or acetaminophen. MSCs are known to be hypo-immunogeneic and immunomodulatory in nature, so pro-inflammatory responses are expected to be minimum. The route of transplantation of cells is another variable. In systemic infusion, the donor cells is dispersed and home to different organs, especially in the lungs, which reduces liver-specific engraftment. To improve homing of cells in the liver, the best route of delivery is intra-portal. In the case of MI, the best routes of cell delivery would be pericardium or endocardium.

(c) Identification of donor cells and functional recovery

Identification: Donor cells can be identified by many techniques, some of which generate quantitative information on engraftment. The popular techniques followed for identification of donor cells in experimental models are described in the Table 79.2A. Each of this technique has its own advantages and limitations. The most convenient method to identify donor cells is using green fluorescence protein (GFP)-expressing BM stem cells. Both qualitative and quantitative analyses on engraftment would be possible on the basis of three-color flow cytometric analysis (GFP and hepatic markers) and immunohistochemistry (IHC) of the liver tissue sections.

Target tissue specific gene and protein expression: In case of direct differentiation, donor cells-specific gene expressions are expected to be completely shutdown, instead hepatic genes and proteins are expressed. In the case of fusion between donor and recipient cells, it may be possible that both donor and recipient antigens are co-expressed in the resultant cells. By analyzing only hepatic genes and proteins expression in donor-derived cells, one cannot distinguish direct differentiation (plasticity) from cell fusion. This event will become more critical if BM cells are fused with polyploid hepatocytes or cardiomyocytes. In the case of diploid-diploid fusion, it is expected that the resultant cells will possess comparable amounts of transcription factors to both parental cells and will, thus, manifest both of their characteristics. Liver is composed of more than 50% polyploid hepatocytes, so hepatic transcription factors may dominate in the fused cells. Therefore, it is imperative that the newly formed hepatocytes are critically analyzed with respect to genes and proteins of hepatocytes as well as of BM cells.

Analysis of cell fusion: To claim direct differentiation of BM cells into hepatocytes (or cardiomyocytes), it is essential that donor-recipient fusion occurs minimally or nil. In somatic tetraploid fused cells, there is apprehension of neoplastic transformation and induction of cell-cycle arrest in G1 phase. The most important issue is that the resultant fused hepatocytes tend to show chromosomal abnormalities, same as in the tumor cells. Thus, thorough analyses of donor-derived hepatocytes are warranted to prove or disprove fusion, if any. Fusion of recipient hepatocytes with BM-derived stem cells can be conveniently examined by three independent assays (Table 79.2B).

Restoration of normal liver function: For genetically deficient mouse strains, beside normal liver function and histology, the most important test is determining the expression of wild type protein in the host liver. In case of tyrosinemia type I, the functionally restored mouse after BM stem cell therapy become independent of 2–(2–nitro-4–trifluoro-methylbenzyol)-1,3–cyclohexanedione (NTBC) [7]. This is possible through direct differentiation of BM-derived cells or fusion with the host hepatocytes.

Tissue integration: Finally, it is essential to show that BM-derived hepatocytes are integrated in the native hepatic plate. The recipient liver will recognize donor-derived hepatic cells if they are integration with the host tissue.

VII. PRESENT THERAPEUTIC APPLICATIONS OF BM CELLS AND LIMITATIONS

Advancing stem cell biology into clinic requires success in all three key elements of translational research: basic research, preclinical studies and clinical trials. Many clinical trials, conducted in the pretext of '*medical innovation*', have sidestepped preclinical studies. Furthermore, the experimental conditions, nature of control experiments and final analyses in the preclinical setting are more exhaustive and sophisticated than that followed in clinical setting. Therefore, it is always crucial to interpret the results of the clinical studies of BM cell therapy with more caution.

In case of human subjects, the post therapy investigations are primarily based on imaging techniques, such as echocardiography (ECHO), magnetic resonance imaging (MRI), positron emission tomography (PET), and single photon emission computer tomography (SPECT). None of this non-invasive technique can precisely identify/predict the donor cells in the regenerating tissue. This makes difficult for direct correlation of functional recovery with the engrafted cells. However, overall functions of the organs can be well predicted.

TABLE 79.2   Methods for identification of donor cells in recipient liver and fusion analysis

A. Techniques	*Remarks*
Fluorescent dye • PKH-26 • CFSE	Cells can be identified by flowcytometry or IHC. Dyes are toxic to the cells and have a tendency to leach out with time; intensity reduces with proliferation of cells.
Sex chromosome • FISH • *Sry* gene by qPCR	Female (XX) donor and male recipient (XY) mouse are generally used. FISH is performed in tissue section and single cell preparation. *Sry*-qPCR is a simpler method than FISH, it requires >98% pure target cells for analysis.
Transgene marker • lacZ (Rosa26 mouse) • eGFP [C57BL/6–Tg(UBCGFP)30Scha/J]	Require transgenic mouse. Bright field and/or fluorescent microscope can be used for detection. Transgene may be immuno-geneic to the host. lacZ system often gives false positive value due to leakage of β-galactosidase. In GFP-expressing donor cells, the engraftability can be determined using flowcytometry.
B. Techniques	*Remarks*
FISH	Female (XX) donor and male recipient (XY) mouse are generally used. FISH is performed in tissue section and single cell preparation. The assay is preceded with IHC or IC of the target cells. Karyotypes: fused cells- XXXY/XXY; unfused-XX/XXXX.
Sry gene qPCR	Female (XX) donor and male recipient (XY) mouse are generally used. In case of fusion the chromosome of donor-derived hepatocytes will contain *Sry* gene, which can be quantitatively determined. Ploidy can not be determined.
Cre/lox recombination system	The recipient mouse is the conditional *Cre* reporter Rosa26R. In case of fusion loxP-flanked stop cassette is excised and LacZ is expressed in fused cells and identified by X-gal staining of β-galactosidase.

CFSE: Carboxyfluorescein succinimidyl ester; FISH: Fluorescence *in situ* hybridization;
IC: Immnocytochemistry; IHC: Immunohistochemistry.

The use of super-paramagnetic nanoparticles in combination with MRI can provide increase sensitivity and high-resolution molecular imaging. In order to track down donor cells using MRI, attempt has been made to load cells with these particles before transplantation in large animals. The method has not been tested in human subjects.

Autologous bone marrow cell therapy for nonhematopoietic tissues has become popular due to involvement of minimum risk and simple procedure. As a result, patients suffering from MI, stroke, critical limb ischemia (CLI), neurodegenerative disorders, metabolic diseases and liver cirrhosis, osteogenesis imperfecta, etc. are turn out to be the subjects for clinical and experimental therapy. Most popular clinical trials of autologous BM cells are in the cases of MI, stroke and CLI. Recently, two concise reviews with meta-analyses of the clinical trials in MI and CLI have been published [8,9]. Analyses of above clinical trials propose that autologous BM cells therapy is safe for patients of MI, CLI and other diseases. Further, results suggest that in some patients the recovery or functional improvement has been significantly high compare to the control arms. However, many reports show no appreciable beneficial effect of the transplantation.

In comparison with clinical studies, preclinical results are found to be more encouraging. This is primarily due to following reasons:

(i) Preclinical experiments are conducted in more control manner, like in each case a consistent time gap is maintained between an injury is established and the cells are transplanted. Further, number and type of cells transplanted, the route of transplantation are maintained similar. In clinical trails, above parameters are usually changed according to the convenience of the patients that impact on the overall results.

(ii) In most of the clinical trials BM-MNCs are used, only a few of them are competent for differentiation.

(iii) Whole body distribution of donor cells, transplanted through intravenous route, may reduce the availability of the cells at the target site below optimum number.

(iv) Cellular and molecular level analyses would be possible in experimental small animals, thus optimum perturbation of the organ is normally achieved to improve the therapeutic efficiency.

(v) Identification of donor cells and the cellular mechanism of differentiation can be conveniently studied in preclinical models.

SUMMARY

- Human bone marrow tissue contains multipotent adult stem cells, which have non-hematopoietic tissue repair/regeneration potential.
- Beneficial effect of BM cells may be due to trans-differentiation into target cells and/or paracrine effect, depending on the type of tissue and experimental parameters.
- Damaged tissue-induced signals may not mobilize enough competent cells for regeneration of the tissue, which can be ensured by locally transplanting cells at the site of injury.
- To achieve maximum benefit of the basic research, results of the preclinical studies should be the starting point of any clinical trails.
- Cell fusion and long-term safety study are essential in the studies of BM cell therapy for nonhematopoietic organs.

SUGGESTION FOR FURTHER READING

1. Blau H.M., Pavlath G.K., Hardeman E.C., Chiu C.P., Silberstein L. and Webster S.G., et al. (1985), Plasticity of the Differentiated State, *Science*, 230,758–766.

2. Jang Y.Y., Collector M.I., Baylin S.B., Diehl A.M. and Sharkis S.J. (2004), Hematopoietic Stem Cells Convert into Liver Cells Within Days Without Fusion, *Nat. Cell Biol.*, 6, 532–539.

3. Korf-Klingebiel M., Kempf T., Sauer T., Brinkmann E., Fischer P. and Meyer G.P., et al. (2008), Bone Marrow Cells are a Rich Source of Growth Factors and Cytokines: Implications for Cell Therapy Trials After Myocardial Infarction, *Euro. Heart J.*, 29, 2851–2858.

4. Michael A.L., Zbinden S., Epstein S.E. and Murry C.E. (2007), Cell-based Therapy for Myocardial Ischemia and Infarction: Pathophysiological Mechanisms, *Annu. Rev. Pathol. Mech. Dis.*, 2, 307–339.

5. Raff M. (2003), Adult Stem Cell Plasticity: Fact or Artifact? *Annu. Rev. Cell Dev. Biol.*, 19,1–22.

6. Scadden D. and Srivastava A. (2012), Advancing Stem Cell Biology Toward Stem Cell Therapeutics, *Cell Stem Cell*, 10,149–150.

7. Theise N.D. (2010), Stem Cell Plasticity: Recapping the Decade, Mapping the Future, *Expt. Hematol.*, 38, 529–539.

8. Wang X., Willenbring H., Akkari Y., Torimaru Y., Foster M. and Al-Dhalimy M., et al. (2003), Cell Fusion is the Principal Source of Bone-marrow-derived Hepatocytes, *Nature*, 422, 897–901.

9. Hofmann-Amtenbrink M., Von Rechenberg B. and Hofmann H. (2009), Superparamagnetic Nanoparticles for Biomedical Applications, In: *Nanostructured Materials for Biomedical Applications* M.C. Tan (Ed.), Transworld Research Network, Trivandrum, India. 119–149 (ISBN: 978–81–7895–397–7).

REFERENCES

1. Koller M.R. and Palsson B.O. Tissue Engineering: Reconstitution of Human Hematopoiesis *ex vivo*. *Biotechnol. Bioengg.*, 42, 909–930.

2. Weissman I.L. (2000), Stem Cells: Units of Development, Units of Regeneration, and Units in Evolution, *Cell*, 100,157–168.

3. Anderson D.J., Gage F.H. and Weissman I.L. (2001), Can Stem Cells Cross Lineage Boundaries? *Nat. Med.* 7, 393–395.

4. Vieyra D.S., Jackson K.A. and Goodell M.A. (2005), Plasticity and Tissue Regenerative Potential of Bone Marrow-derived Cells, *Stem Cell Rev.*, 1, 65–69.

5. Yoon Y.S., Wecker A., Heyd L., Park J.S., Tkebuchava T. and Kusano K., et al. (2005), Clonally Expanded Novel Multipotent Stem Cells from Human Bone Marrow Regenerate Myocardium After Myocardial Infarction, *J. Clin. Invest.*, 115, 326–338.

6. Danet G.H., Luongo J.L., Butler G., Lu M.M., Tenner A.J. and Simon M.C., et al. (2002), C1qRp Defines a New Human Stem Cell Population with Hematopoietic and Hepatic Potential, *Proc. Natl. Acad. Sci.*, USA, 99,10441–10445.

7. Lagasse E., Connors H., Al-Dhalimy M., Reitsma M., Dohse M. and Osborne L., et al. (2000), Purified Hematopoietic Stem Cells can Differentiate into Hepatocytes *in vivo*, *Nat. Med.*, 6,1229–1234.

8. Jeevanantham V., Butler M., Saad A., Abdel-Latif A., Zuba-Surma E.K. and Dawn B. Adult Bone Marrow Cell Therapy Improves Survival and Induces Long-term Improvement in Cardiac Parameters: A Systematic Review and Meta-analysis, Circulation 126, 551–568.

9. Botti C., Maione C., Coppola A., Sica V. and Cobellis G. (2012), Autologous Bone Marrow Cell Therapy for Peripheral Arterial Disease, Stem Cells and Cloning: *Adv. and App.* 5, 5–14.

SECTION XI

HORMONES

In multicellular organisms it is necessary to regulate and coordinate the activities of various organs. This is achieved through the agency of hormones and the nervous system. In fact, an intrinsic relationship exists between the two. The secretion of many hormones is regulated by factors elaborated by the nerve cells. The central nervous system (CNS) is itself subject to regulation by hormones and in recent years receptors for many of the sex steroids and corticoids have been mapped in the brain. Hormones influence the synthesis of proteins and enzyme activities in the brain. They also programme the circuitry within the central nervous system when given at discrete stages of development. It is for this reason that it has been considered appropriate to include chapters on the relationship of the CNS and endocrines as also on hormones and behaviour states. A number of behaviour states have been found to have an organic basis.

Two other chapters are on subjects which are not fully covered in the usual textbooks. These pertain to the biological rhythms and the role played by the pineal gland. Leaving aside birds and lower forms of life with biological rhythms, the higher developed mammalian organisms also manifest rhythms, such as in secretion of hormones. Many hormones are secreted during the sleep period and are high early in the morning and low late in the evening. Similarly, enzyme activities in many cells are not constant throughout the day, but increase or decrease at various hours. These aspects open up new possibilities: optimal therapeutic efficacy may be attained by administration at some, but not other, hours of the day.

Chapters have been devoted to the pituitary hormones, the hormones of the thyroid, the adrenals and the pancreas. Each chapter covers not only the chemistry, biology and mode to action of these hormones but attempts have been made as well to describe the clinical disorders associated with the dysfunction of an endocrine gland.

80

Hormones—General Introduction

G.P. Talwar

<table>
<tr><td colspan="2" align="center">CONTENTS</td></tr>
<tr><td>I.</td><td>Intercellular Communication through Chemical Mediators
  A. Hormones
  B. Parahormones
  C. Pheromones
  D. Phytohormones</td></tr>
<tr><td>II.</td><td>Functions of the Hormones</td></tr>
<tr><td>III.</td><td>Mechanism of Action of Hormones
  A. Quick acting hormones
  B. Growth promoting and developmental hormones</td></tr>
<tr><td>IV.</td><td>Assay of Hormones
Summary</td></tr>
</table>

I. INTERCELLULAR COMMUNICATION THROUGH CHEMICAL MEDIATORS

A. Hormones

In multicellular organisms, there is a need for *coordination* of the activities of various organs. This task is achieved by an *integrated interplay of the central nervous system and the hormones*. The word hormone is derived from the Greek *hormos* meaning to excite, to arouse, to set in motion. Hormones are products of specialized tissues, of endocrine or ductless glands. *They are poured into circulation on discrete demand, and are effective at minute concentrations in influencing the activities of other organs.*

Figure 80.1 illustrates the location of the endocrine glands and some other tissues producing hormone-like substances (*para hormones*) in the human body. Their interrelationship is shown in Figure 80.2. Their principal products and functions are summarized in Table 80.1. As will be evident from this table, substances of diverse chemical nature can have hormonal properties. They can be:

(i) amino acid derivatives such as the thyroid hormones, epinephrine, norepinephrine, serotonin, etc.

(ii) short length polypeptides such as the posterior pituitary hormones oxytocin, ADH; release factors elaborated by the hypothalamus, MSH, ACTH, etc.

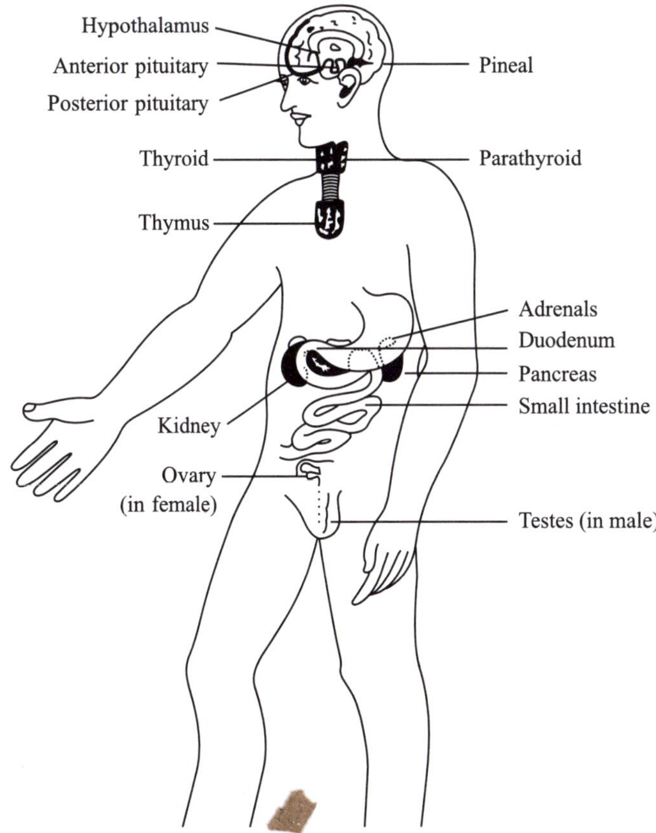

Figure 80.1 Organs producing hormones and parahormones.

981

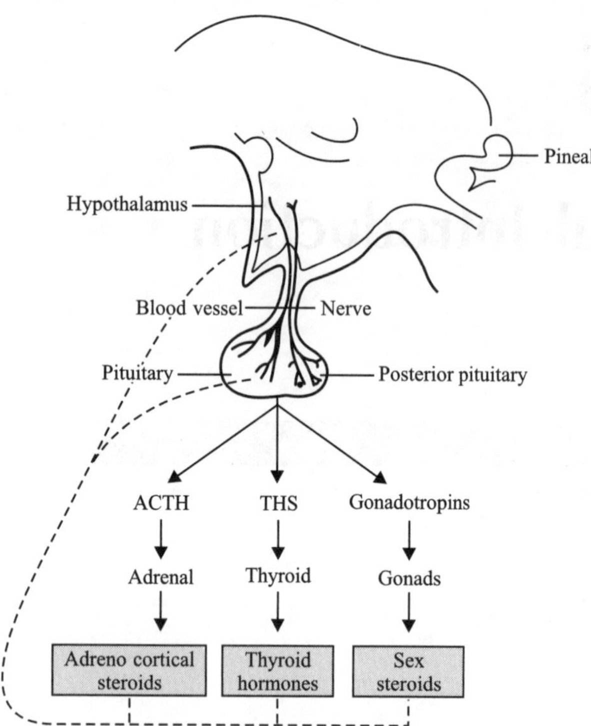

Figure 80.2   Feedback control of secretion of hormones. Solid arrows indicate stimulatory effects and dotted lines inhibitory influences.

(iii) Proteins and glycoproteins, e.g., insulin, growth hormone, TSH, LH, FSH, parahormone, prolactin, etc.

(iv) steroids, e.g., adrenal cortical hormones; sex hormones, estrogens, progesterone, androgens.

B. Parahormones

Broadly speaking, hormones can be considered as a part of the general system in which *chemical messengers bring about cell to cell interaction*. The term hormone has been traditionally, though inappropriately, restricted to substances produced by the endocrine glands. Chemical mediators for a variety of other systems are usually grouped under parahormones. Notable examples are the following:

1. **Gastrointestinal tract hormones:** Act as triggers for secretion of pancreatic, gastric and intestinal juices, and for flow of bile (Chapter 51).

2. **Neurotransmitters:** Products of neurons, released at the synaptic junctions making possible the communication of the nerve impulse to a neighbouring cell (Chapter 26).

3. **Kidney as a source of hormones:** Kidney elaborates a factor, erythropoietin, that stimulates the formation of erythrocytes from stem cells of the bone marrow. The signal for the formation of erythropoietin is hypoxia or diminished oxygen tension in the blood. Kidney also releases a proteolytic enzyme renin under certain conditions of emergency, which gives rise to the formation of a peptide angiotensin that causes vasoconstriction, increased cardiac output and elevation of blood pressure.

4. **Lymphoid cell products:** Sensitized lymphocytes when cultured in the presence of antigens, secrete a number of biologically active peptides that influence the activities of other lymphoid cells and macrophages (Section XIV). Thymus is believed to elaborate humoral factors that may have a role in the differentiation of bone marrow stem cells into a class of immunocompetent lymphocytes (T cells).

5. **Chalones:** In recent years, a number of glycoproteins and polypeptides have been extracted from various tissues that inhibit the synthesis of DNA and multiplication of cells. These substances may have an important regulatory role in generation of cells and in maintaining the size of organs.

C. Pheromones

Chemical substances have also been observed to mediate social interaction within some species. In bees and in termites, the queen bee and the reproductive castes secrete substances that are ingested by the 'workers' that prevent their differentiation to those castes. The pheromone excreted by the mandibular glands of the queen bee has been identified as 9-ketodecanoic acid. It inhibits the development of the ovaries in worker bees and also restricts them from constructing royal cells for the rearing of new queens. In insects many of the disciplined social activities are believed to be regulated by substances such as the 'trail' factor, alarm substances, etc. Several substances may be acting as sex attractants. The base of most perfumes is constituted by musk, a secretion from the preputial glands of a deer and musk-rat, the secretion from the perineal glands of the civet cat. The reproductive functions in mice are affected by smells. The odour of a strange male mouse causes resorption of foetus in a newly pregnant female mouse.

D. Phytohormones

Regulation by chemical mediators is not restricted to animal systems. Lacking a nervous system, plants are predominantly dependent on chemical messengers. A number of regulatory substances such as auxins, gibberellins, leaf growth substances, root growth regulators, kinins and florigens have been discovered and described. Their main action is the promotion of growth of a certain part of the plant. Their judicious use can be of economic advantage.

II. FUNCTIONS OF THE HORMONES

The functions of various hormones are summarized in Table 80.1. Details will be given in the succeeding chapters. The following generalizations can be made on the type of functions performed by hormones.

TABLE 80.1

Organ	Product secreted	Sites of action	Biological effects
Brain			
(a) Neurosecretory cells of hypothalamus	Oxytocin	Uterus	Contraction of the uterus (a) to facilitate the ascent of spermatozoa (b) delivery of the fetus
		Myoepithelial cells of mammary gland	Ejection of milk
	Vasopressin or antidiuretic hormone (ADH)	Kidney	Increases absorption of water
	Release factors of anterior pituitary hormones	Anterior pituitary	*Stimulate* or *inhibit* the secretion
(b) Neurons	Neurotransmitters Acetyl choline Serotonin	Neurons Synapses	Transmission of nerve impulse
	GABA Adrenaline Noradrenaline Dopamine etc.	Neuromuscular junctions	
(c) Pineal	Melatonin (synthesis and secretion promoted by darkness and inhibited by light)	Gonads (and other sites?)	Photoperiodic influences on reproductive functions in rats Biorhythms
		Melanophores	Blanching of frogs' skin
Pituitary			
(a) Anterior lobe	Growth hormone or somatotropic hormone (STH)	Almost all organs of the body	Growth of bones, cartilage, muscles and viscera; promotes synthesis of proteins in tissues; influences carbohydrate and fat metabolism; mobilizes body mechanisms in situations of nutritional scarcity
	ACTH	Adrenals	Synthesis and secretion of adrenal hormones
		Adipose tissue	Mobilization of fats
	TSH	Thyroid	Synthesis and secretion of thyroid hormones
	Gonadotropins FSH, LH	Ovaries in females	Development of follicles, ovulation; secretion of estrogens and progesterone
		Testes in males	Spermatogenesis, development and function of testes, secretion of testosterone
	Prolactin	Mammary glands	Development of mammary glands in combination with other hormones, estrogens, progesterone, STH, adrenal corticoids, lactation
(b) Pars intermedia	Melanocyte stimulating hormone (MSH)	Melanocytes	Pigmentation of skin
Thyroid	Thyroxine (T_4)	Several organs	Increase in basal metabolic rate and in consumption of oxygen by cells; growth promotion and development, metamorphosis of tadpoles
	Calcitonin	–	Lowers the level of calcium in blood
Parathyroid	Parathyroid hormone	Bones Kidney	Mobilizes calcium from bones to raise serum levels of calcium; increases excretion of phosphate
Thymus	Partially characterized humoral factors	Stem cells from bone marrow	Development of precursor cells to lymphocytes eventually competent for cell-mediated immunity and having a cooperative role also in production of antibodies
Pancreas	Insulin	Muscles, liver, adipose tissue and many other organs	Reduces blood glucose levels by promoting its uptake by peripheral cells; increases utilization of carbohydrates; increases deposition of glycogen, stimulates lipogenesis; anabolic effects on protein synthesis
	Glucagon	Liver, adipose tissue	Hyperglycemic factor mobilizes liver glycogen stores to replenish blood glucose level Enhances conversion of amino acids to carbohydrates

(Contd.)

TABLE 80.1 (*Contd.*)

Organ	Product secreted	Sites of action	Biological effects
Adrenal cortex	Gluco-corticoids principal member (hydrocortisone) mineral corticoids (aldosterone) androgens	Several organs	Retention of sodium and excretion of potassium; Maintenance of extracellular volume; cardiovascular functions; muscular and kidney functions, adaptation to stress (trauma, cold, heat, toxins, infections, fasting, forced exercise, etc.), immunosuppressive action; promotion of gluconeogenesis anti-inflammatory action
Adrenal medulla	Epinephrine Norepinephrine Dopamine	Heart, muscles, liver and other organs	Increase the heart rate; raise systolic blood pressure; mobilize glucose from liver and muscles; anxiety
Ovaries	Estrogens	Female reproductive organs and secondary sex organs, uterus, vagina, mammary glands, pituitary, brain	Promotes growth of uterus, mammary glands and vaginal epithelium; feedback effect on secretion of gonadotropins; influences behaviour and psychic patterns
	Progesterone	-do-	Acts synergistically and in some cases antagonistically to estrogens; preparation of the uterus for implantation of the blastocyst; maintenance of pregnancy; regulation of the accessory organs
Testes	Androgens	Male reproductive and accessory organs, muscles	Spermatogenesis, seminal plasma secretions, hair patterns, skeletal configuration, voice changes; regulation of sebaceous gland activity; general anabolic hormone stimulates protein synthesis
Stomach	Gastrin	Gastric glands	Stimulation of gastric secretions
Duodenum	Secretin	Pancreas	Stimulates the secretion of pancreatic fluids
	Cholecystokinin	Gall bladder	Release of bile
Intestines	Pancreozymin	Pancreas	Stimulates the secretion of pancreatic fluids
	Enterogastrone	Stomach	Inhibition of secretion of hydrochloric acid; diminishes motility of the stomach
Kidney	Erythropoietin	Hematopoietic organs	Stimulation of the formation of blood cells
	Renin-Angiotensin	Cardiovascular system, adrenals, kidney	Vasoconstriction; Elevation of blood pressure; Increased cardiac output; Autonomous system

A. Homeostasis

Hormones help maintain the constancy of the 'milieu interieur' (internal environment), vital for sustenance of life.

B. Regulation of Metabolism

Hormones influence the metabolism of carbohydrates, fats, proteins, and minerals directing their synthesis, storage, mobilization and utilization according to needs.

C. Growth

The growth of bones, viscera and various types of tissues is under the control of one or more than one hormone acting synergistically.

D. Differentiation and Development

Several hormones (e.g., thyroid hormones and sex steroids) have a critical role in morphogenesis and differentiation of tissues and structures at appropriate stages of development. The presence of a Nerve Growth Factor in the submaxillary gland of the mouse has been reported. It has a growth promoting and mitotic effect on ganglia.

E. Behaviour

An important role of hormones in behaviour is being increasingly realized.

F. Adaptation to Stresses and Strains

The human body has remarkable capacity to adjust and adapt to needs and environmental demands, a trait mediated in no small way by the agency of hormones.

G. Hormones and Resistance

Hormones contribute to resistance by (i) effects on lymphoid organs and immune responses; (ii) by confinement of infective zone through suppression of inflammatory response by steroids like gluco-corticoids; (iii) by activation of the elimination mechanisms through hormonal activation of lysosomal enzymes.

H. Reproduction

Reproductive organs are highly sensitive to hormones. The development and maturation of gametes are under hormonal influence in both males and females. Ovulation, preparation of the uterus for implantation of the fertilized ovum are again dependent on hormones. Hormones are also required for the maintenance of pregnancy (Section XIII).

III. MECHANISM OF ACTION OF HORMONES

Hormones can be classified into two categories:
 A. Quick acting hormones: implicit in homeostasis
 B. Growth promoting and developmental hormones, where the action of the hormone is expressed after a lag period.

A. Quick Acting Hormones

The first category of hormones include biogenic amines (adrenaline, noradrenaline), anterior pituitary tropic hormones (ACTH, TSH, LH), parahormones, vasopressin, glucagon, etc. These hormones act initially with membrane bound "receptors". Hormone-Receptor interaction results in the activation of adenyl cyclase, an enzyme that acts on ATP and generates a cyclic nucleotide 3′, 5′, AMP. Cyclic AMP in turn activates protein kinase(s) that are competent to phosphorylate serine moiety in several enzymes and membrane structures. With phosphorylation (limited to 2 to 4 serines per molecule), the conformation of the enzyme protein changes and it passes from an inactive state to an 'active' form capable of catalyzing a reaction (Figure 80.3).

The reverse can also happen, where the phosphorylation process can convert a biologically active molecule into an inactive enzyme, e.g., example (2) in Figure 80.3. Thus a common signal (3′, 5′ AMP) generated by, say, glucagon or epinephrine in liver cells, causes by a common mode of action (phosphorylation of proteinic enzymes), an activation of phosphorylase and an inhibition of glycogen synthetase. The net result is the mobilization of stored glycogen to replenish blood glucose levels. Simultaneously the conversion of glucose phosphate to glycogen reserves is inhibited. Cyclic AMP has been discussed more fully elsewhere in this book (Chapter 81).

B. Growth Promoting and Developmental Hormones

Among the hormones falling in this category are growth hormone, thyroid hormones and sex steroids (estrogens, progestrogens and androgens). The growth effect is observed after a lag period of several hours, sometimes days, while the hormone reaches the target tissue within minutes of administration, implying thereby that this biological effect is distant from the initial site of interaction of the hormone. In the target tissue, it is recognized and bound by specific macromolecules as exemplified by "receptors" for estrogens (Chapter 94). The hormone-receptor protein complex migrates rapidly to the nuclear compartment, where it activates the synthesis of early species of RNA that code for the synthesis of early proteins(s). The process builds up with more prominent synthesis of new ribosomes and messenger RNA. Hypertrophy of the cell takes place leading sometimes to hyperplasia.

The hormones may not act uniquely through an activation of the transcription and translation processes. There may be additional effects on the membrane permeability facilitating the uptake of amino acids and other metabolites. Some hormones may also require cell division and differentiation of stem cells to those competent to respond to the hormone (Chapters 77 and 78).

IV. ASSAY OF HORMONES

Hormones can be estimated in blood, urine and other fluids by a variety of methods, viz., chemical, physical, biological and immunological. Radioimmuno assays are now available for not only the proteinic hormones but also for steroid hormones. These enable the determination of most hormones at nanogram

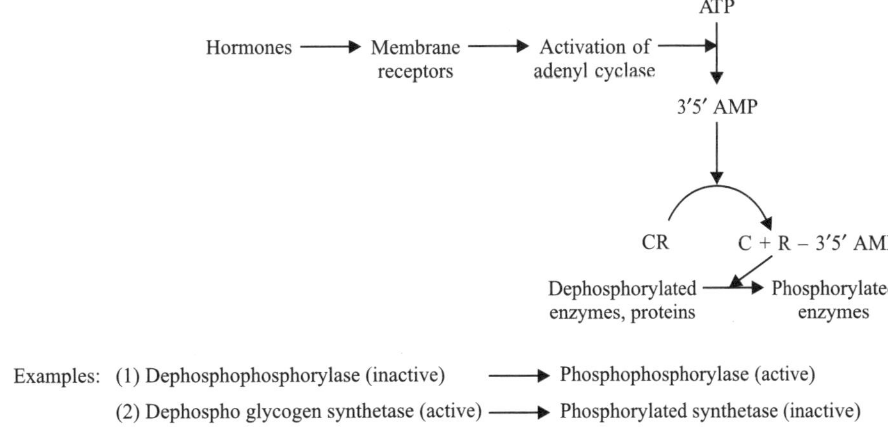

Figure 80.3

(10^{-9}, i.e., one thousand millionth of a gram) and picogram (10^{-12}) levels. The assay requires the use of antibodies to the hormone. The hormone is made radioactive by ^3H or ^{14}C or by tagging ^{125}I to the molecule, and reacted with the antibody. The presence of unknown concentrations of the non-radioactive hormone in plasma or biological fluid causes a proportionate dilution of the radioactivity precipitable by the antiserum under standard conditions permitting an evaluation of its content.

Enzyme immuno assays EIAs do not require radioisotopes nor radioactivity counting equipment. They can be performed in the laboratory or in the field. The principle is similar to radioimmunoassays except that an enzyme is the marker instead of an isotope, which is estimated by its catalytic activity of converting a substrate into a product. EIAs are today available for a variety of hormones, drugs, bacterial and parasitic antigens.

The detection of HCG by immunological method in the urine is a simple and early test for the diagnosis of pregnancy. Besides tests based on agglutination of latex or erythrocytes, enzyme linked immunoassays (ELISA) are highly sensitive methods.

SUMMARY

In multicellular organisms, cell to cell communication takes place through the agency of the central nervous system and chemical mediators (hormones and parahormones). These substances (listed in Table 80.1) could be amino acid derivatives, polypeptides of short or long chain, glycoproteins, or steroids. They are effective in minute concentrations. They have a variety of important functions in the body such as maintenance of homeostasis; regulation of the metabolism of carbohydrates, fats, proteins and minerals; in promoting growth of bones, viscera and other tissues; in differentiation and development; in behaviour; in adaptation of the body to stresses and strains; in building up of resistance to foreign substances, and in reproductive function.

Hormones act on their target organs by specific binding to "receptors" located either on the membranes (for most of proteinic hormones and biogenic amines) or to proteinic macromolecules within the cells (steroid hormones). The interaction leads to any one or more of the following consequences: (i) activation of adenyl cyclase, generation of 3′, 5′ cyclic AMP, the second messenger in the action of many hormones, (ii) alteration of membrane permeability to ions, and metabolites, and (iii) activation of transcription and translation processes. Hormonal effects accrue by a chain of interconnected reactions.

Radioimmune and enzyme immunoassays permit the determination of hormones at nanogram levels in biological fluids. Immunological methods are also simple and quick approaches for early diagnosis of pregnancy.

Chemical mediators are not confined to animal systems; plants have a variety of hormonal regulators (phytohormones) influencing general growth of the plant or of its components, leaves, roots, flowers, etc. Pheromones are chemical substances excreted by animals that have an important role in social interactions within the same species (e.g. sex attractants; trail, alarm substances, inhibitors of sexual development in bees, termites, pheromones in mice, etc.).

SUGGESTIONS FOR FURTHER READING

Becker K.L. (Ed.) (1995), *Principles and Practice of Endocrinology and Metabolism,* 2nd ed., J.B. Lippincott Co., Philadelphia.

Clegg A.G. and P.C. (1969), *Hormones, Cells and Organisms,* Meinemann, London.

Hardman J.G., Robison G.A. and Sutherland, E.W. (1971), Cyclic Nucleotides, *Annual Review of Physiology,* Vol. 33, 311.

McKerns F. (Ed.), *Sex Steroids—Molecular Mechanisms, Biochemical Endocrinology Monographs,* Vol. III, Appleton Century-Crofts, New York.

Talwar G.P. (1972), Growth Promoting and Developmental Hormones—Some Thoughts on Their Mode of Action, *Int. J. Biochem.,* 3, 39.

Talwar G.P. (Ed.) (1983), *Non Isotopic Immunoassays and their Applications,* Vikas Publishers.

Turner C.D., *General Endocrinology,* W.B. Saunders Company, Philadelphia.

Williams-Ashman H.G. and Reddi A.H. (1971), Actions of Vertebrate Sex Hormones, *Annual Review of Physiology,* Vol. 33, 31, Victor E. Hall (Ed.), (Palo Alto, California, USA).

Wilson J.D. and Foster D.W., Kronenberg H.M. and Larsen P.R. (Eds.) (1998), Williams *Textbook of Endocrinology,* 9th ed., W.B. Saunders Company, Philadelphia.

81

Regulation of Cell Function
Receptors for Hormones and Signal Transduction Mechanisms

Shail K. Sharma

I. REGULATION OF CELL FUNCTION

One of the fundamental features of living organisms is their adaptability to constantly changing external environment. The complex mammalian system—a multicellular organism—is essentially an ensemble of a large variety of cells, having highly specialized functions. The development of such an organism is an extremely complex process involving differentiation, multiplication and organization of cells in an orderly pattern. Survival of such an organism calls for a coordination of metabolic activities of various types of cells. In higher organisms, intercellular communication takes place by way of nerves *(neurotransmitters)* or by way of *chemical messengers (hormones)* circulating in the blood. An obvious barrier between two cells and between blood and cells is the plasma membrane—the outer covering of the cell. The selective permeability of the plasma membrane, and the interaction of its components with a variety of effector agents keep the cell in a highly organized and differentiated form. Communication between two cell types, presumably at the level of plasma membrane, must influence the metabolism of the cells through the regulation of enzymatic activities. Regulation of metabolic pathways may be excercised at three basic levels.

1. Modification of the biological activity of allosteric enzymes by stimulatory or inhibitory effector molecules through feedback, polymerization or depolymerization of enzymes, and by influencing the energy-linked processes.
2. Hormonal regulation, whereby, hormones specifically bind to the target tissue and stimulate or inhibit the formation of second messengers which, in turn, regulate metabolic activity.
3. Induction or repression of enzymes exerted at the level of cell nucleus whereby the concentration of an enzyme protein is increased or decreased.

Hormone action begins with the binding of the hormone molecule to its *receptor* on (or in) a target cell, which induces a conformational change in the receptor, that is transmitted to its active site (if receptor is an enzyme) or to other macromolecules. In either case a chain of events is elicited that ultimately affects a vast array of metabolic processes ranging from alteration in enzyme activities to changes in gene expression. These changes lead to profound alterations in cell growth, morphology and function.

II. KINETICS OF HORMONE (LIGAND BINDING TO THE RECEPTOR

The key property of the binding of hormone to its receptor is its association with a biological response (function). The kinetics of receptor binding studies are essentially equivalent to those of enzyme kinetics.

Receptors, as do other proteins, bind their ligands according to the law of mass action.

$$R + L = R \cdot L$$

Here R and L represent receptor and ligand (hormone) and the reactions dissociation constant is expressed as:

$$KL = \frac{[R][L]}{[R \cdot L]} = \frac{[R]_T - [R \cdot L][L]}{[R \cdot L]} \tag{1}$$

where the total receptor concentration,

$[R]_T = [R] + [R.L]$. Equation (1) may be rearranged to a form analogous to the Michaelis-Menten equation of enzyme kinetics.

$$Y = \frac{[R \cdot L]}{[R]_T} = \frac{[L]}{K_L + [L]} \tag{2}$$

where Y is the fractional occupation of the ligand binding sites Equation 2 represents a hyperbolic curve (Figure 81.1a) in which KL may be operationally defined as the ligand concentration at which the receptor is half maximally occupied by ligand. In (Figure 81.1b) a linear form of the equation is plotted. Equation (1) may be rearranged

$$\frac{[R \cdot L]}{[L]} = \frac{([R]_T - [R \cdot L])}{K_L} \tag{3}$$

In keeping with the customary receptor binding nomenclature, Eq. (3) is redefined -[RL] as B for bound ligand), L as F (for free ligand) and [R]_T as Bmax. Then Eq. (3) becomes:

$$\frac{B}{F} = \frac{(B_{max} - B)}{K_L} \tag{4}$$

A plot of B/F verses B (Figure 81.1b), which is known as the Scatchard plot (after George Scatchard—its originator), therefore yields straight line of slope $- 1/K_L$ whose intercept on the B axis is B_{max}. Using this method, the ligand bound to the receptor and free may be determined. Once the receptor-binding parameters for one ligand have been determined, the dissociation constant of other ligands for the same ligand binding site may be determined through competition binding assays. This is analogous to the competitive inhibition of a Michaelis-Menten enzyme

$$R + L \longleftrightarrow R. L$$
$$+$$
$$I$$
$$\updownarrow$$
$$R. I + L \rightarrow \text{no reaction}$$

where I is the competing ligand whose dissociation constant with the receptor is expressed as

$$K_i = \frac{[R][1]}{[R \cdot I]} \tag{5}$$

$$[R \cdot L] = \frac{[R]_T [L]}{K_L \left(1 + \dfrac{[I]}{K_i}\right) + [L]} \tag{6}$$

The relative affinities of a ligand and an inhibitor may be determined by dividing Eq. 6 in the presence of inhibitor with that in the absence of inhibitor.

$$\frac{[R \cdot L]_I}{[R \cdot L]} = \frac{K_L + [L]}{K_L \left(1 + \dfrac{[I]}{K_i}\right) + [L]} \tag{7}$$

When this ratio is 0.5 (50%), the competitor concentration is referred to as [I_{50}] thus solving Eq (7) for K_i at 50 % inhibition.

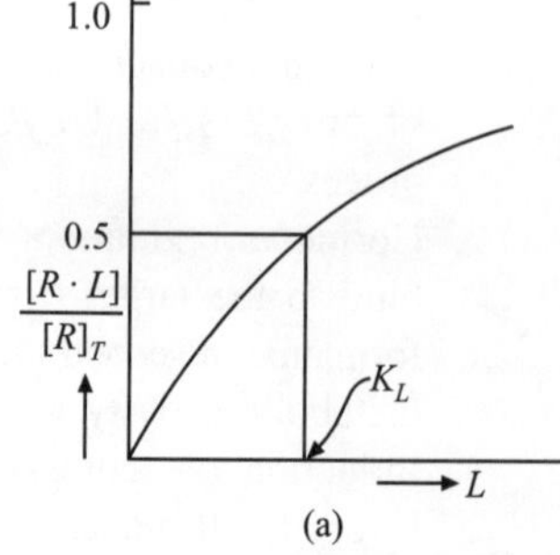

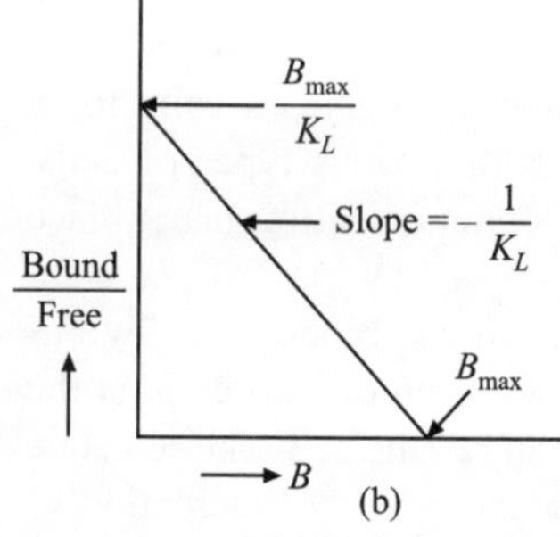

Figure 81.1 The binding of ligand to receptor: (a) Hyperbolic plot (b) Scatchard plot, $B = [R \cdot L]$ $F = [L]$ and $B_{max} = [R]_T$.

$$K_i = \frac{[I_{50}]}{I + \frac{[L]}{[K_L]}} \qquad (8)$$

Through these procedures it has been shown that the hormone epinephrine tightly binds to β adrenoreceptors (KL = 5×10^{-6} M) and the antagonist propranolol binds even more tightly ($K_L = 0.0034 \times 10^{-6}$ M).

III. RECEPTOR TYPES

Receptors could be divided into four classes;

1. Receptors that are *transmembrane proteins* are coupled through G proteins to an enzyme adenylyl cyclase or to a plasma membrane phospholipase C specific for plasma membrane lipids. In these cases, the binding of the hormone to the receptor generates *second messengers,* such as *cyclic AMP, diacylglycerol, inositol-1,4,5-trisphosphate* and *calcium* that mediate the hormonal signal in the cell.
2. Receptors for growth factors such as *insulin* and *epidermal growth factor* (EGF)—are membrane spanning proteins whose cytoplasmic domains are themselves activated as tyrosine specific protein kinases, when the hormone binds to the receptor.
3. Receptors that are *transmembrane ion channels,* open or close when the *neurotransmitter (e.g.acetylcholine)* binds to the receptor, thereby initiating or stopping the flow of specific ions such as Na^+ or K^+ into/out of a cell. Such ligand-gated channels are most conspicuously associated with neurotransmission.
4. Receptors for steroid hormones are cytoplasmic and nuclear proteins that on binding the respective hormones are activated as transcription factors.

In this chapter the membrane associated hormone receptors (mentioned in 1 above) which generate diffusible intracellular signals called *second messengers* will be discussed.

IV. SIGNAL TRANSDUCTION PATHWAY— ADENYLYL CYCLASE SYSTEM

E.W. Sutherland during an investigation of factors controlling the breakdown of glycogen in liver discovered that in liver homogenates, the breakdown of glycogen was enhanced

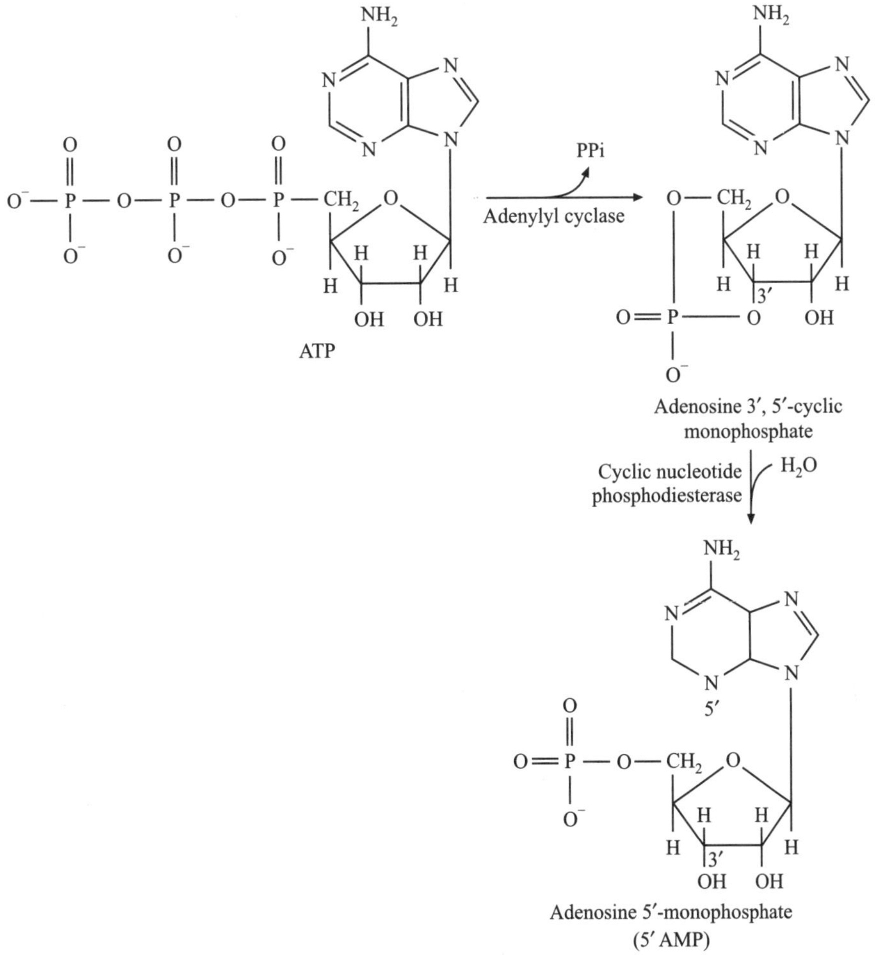

Figure 81.2 Synthesis of cAMP from ATP by adenylyl cyclase and its degradation by cyclic AMP phosphodiesterase.

by a heat-stable factor which they later identified as cyclic AMP (cyclic adenosine 3′, 5′ monophosphate). They further demonstrated that the binding of the hormone to receptors led to an accumulation of cyclic AMP by accelerating the conversion of adenosine triphosphate (ATP) to cyclic AMP (Figure 81.2).

Sutherland proposed cyclic AMP as the *second messenger* in hormone action, and for this discovery he was awarded Nobel prize in 1971.

Four components of the cyclic AMP system are:

1. Specific receptor for the hormone
2. G-proteins couple the receptor to the enzyme adenylyl cyclase that catalyzes cyclic AMP formation.
3. CyclicAMP dependent protein kinases which phosphorylate the target enzymes within the cells, altering their activities.
4. Cyclic nucleotide phosphodiesterase which degrades cyclic AMP to 5′ AMP and terminates the intracellular signal.

The action of the hormone begins with its binding to specific receptor protein. This binding is tight (but non-covalent), stereospecific and will bind only to the hormone ligand or molecules with a closely similar three–dimensional structure. Structural analogues (synthetic) which bind to the receptor and produce a functional response in the cell are called *agonists,* and the analogues that bind to the same receptor (sometimes with higher affinity than the agonist) but do not produce the physiological response are called *antagonists.* Hormone receptors that are coupled through G proteins (GTP binding proteins) to the adenylyl cyclase are integral membrane proteins with 7 hydrophobic regions, thus traversing the lipid layer 7 times. The binding of hormone to the receptor results in conformational change in the receptor which allows its interaction with G proteins in the signal transduction pathway (Figure 81.3).

The next element in the signal transducing pathway is a protein called stimulatory G-proteins (Gs). It consists of three subunits, α, β, γ, of which Gα binds GTP or GDP. When GDP is bound to the α subunit Gs is inactive. Binding of the hormone to the receptor catalyzes the displacement of the bound GDP by GTP and simultaneously dissociating the α subunit from the $\beta\gamma$ subunits. Gsα with GTP bound moves in the plane of the membrane from the receptor to a nearby protein—adenylyl cyclase and activates this enzyme, which in turn catalyzes the synthesis of cyclic AMP from ATP. *Activation of adenylyl cyclase by Gsα is self limiting;* Gsα has a weak GTPase activity and turns itself off by hydrolyzing the bound GTP to GDP. This now inactive Gsα dissociates from adenylyl cyclase and reassociates with the $\beta\gamma$ subunits in the membrane. Once Cyclic AMP is formed the hormone is no longer required and it acts as the second messenger—hormone being the first messenger, which mimics the hormone action in the cell. Cyclic AMP—the intracellular second messenger is short-lived; it is quickly inactivated by cyclic nucleotide phosphodiesterase to 5′ AMP. Methylxanthines such as *theophylline and caffeine* (components of tea and coffee) inhibit the phosphodiesterase thereby potentiating the action of agents that act through adenylyl cyclase (Figure 81.4).

A unique feature of hormonal regulation through second messengers is the *amplification cascade effect.* In the absence of external stimulus the intracellular concentration of cyclic AMP has been shown to be very low in the order of 10^{-7} M to 10^{-8} M. Binding of a few hormone molecules to the receptors catalytically activate several Gs molecules which activate adenylyl cyclase and produce manifold increase in the intracellular concentration of cyclicAMP which, in turn, activates many more molecules of the next enzyme in the cascade. *In this way, there is a large and rapid amplification of the incoming signal, and within minutes the physiological response is observed.*

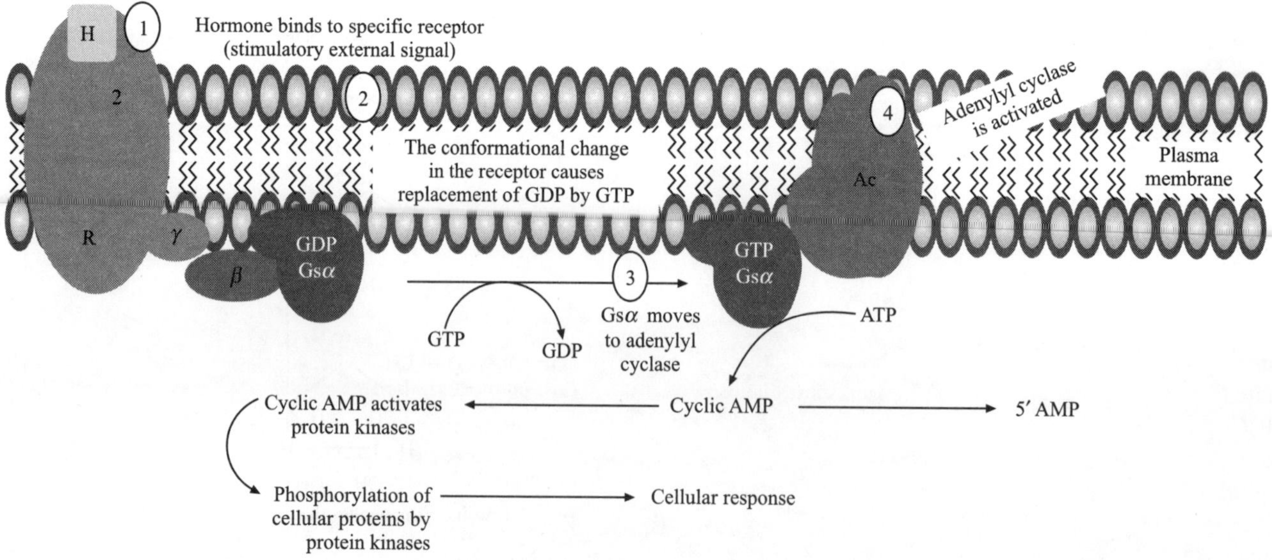

Figure 81.3 Mechanism of receptor-activated adenylyl cyclase. (*see Plate 37 for colour figure*)

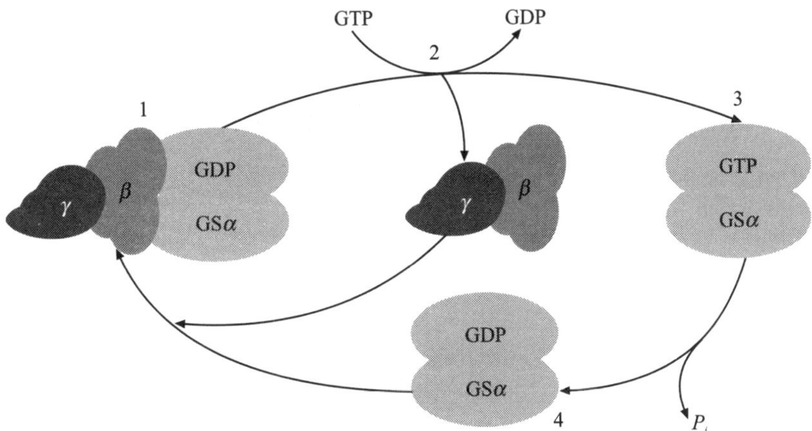

Figure 81.4 Regulation of Gs activity. (1) Gs with GDP is inactive, i.e., it can not activate adenylyl cyclase; (2) Binding of hormone to the adenylyl cyclase causes displacement of bound GDP by GTP; (3) Gs with GTP bound is active and activates adenylyl cyclase. The β and γ subunits dissociate from α; (4) GTP bound to Gs is hydrolysed by the intrinsic GTPase at Gs, turning it off. The inactivee α subunit reassociates with the $\beta\gamma$ subunits. (*see Plate 38 for colour figure*)

V. SOME HORMONE RECEPTORS INHIBIT ADENYLYL CYCLASE

The inhibition of adenylyl cyclase is mediated by "inhibitory G protein Gi" which probably has the same β and γ subunits but has different Gα subunit. When the hormone binds to the receptor, Giα subunit exchanges (similar to Gs) bound GDP for GTP and dissociates from the $\beta\gamma$ subunits. Giα however, inhibits adenylyl cyclase and decreases cyclic AMP concentration.

Gs and Gi are members of a large family of related signal transducing—G-proteins. This family also includes :

1. Gq, which forms a link in the phosphoinositide signaling system (Section XI).
2. Gt (transducin) which transduces visual stimuli.
3. Golf, which is involved in odorant signal transduction in olfactory sensory neurons.

VI. CYCLIC AMP DEPENDENT PROTEIN KINASES

Cyclic AMP dependent protein kinases called protein kinase A are allosterically activated by cAMP. The inactive form of cAMP dependent protein kinase contains two catalytic and two regulatory subunits. The tetrameric R_2–C_2 complex is catalytically inactive, because an autoinhibitory domain of each R subunit occupies the substrate binding site of each C subunit. When cAMP binds to two sites on each of the two R subunits the R subunits undergo a conformational change and the R_2C_2 complex dissociates to yield two free catalytically active C subunits. The cAMP dependent protein kinases can phosphorylate the hydroxyl groups of threonine and serine residues in a number of important enzymes in different kinds of target cells. Other protein kinases phosphorylate tyrosine residues in many proteins.

$$R_2C_2 \text{ (tetramer-inactive)} + 4cAMP \rightarrow$$
$$2C \text{ (catalytic subunits – active)} + R2(cAMP)4$$

TABLE 81.1 Some enzymes regulated by cAMP dependent phosphorylation

Enzyme	Pathway
Glycogen synthase	Glycogen synthesis
Pb kinase	Glycogen breakdown
Acetyl CoA-carboxylase	Fatty acid syntheses
Pyruvate dehydrogenase complex	Pyruvate oxidation to acetyl CoA
Triacylglycerol lipase	TAG mobilization/fatty acid oxidation
Phosphofructokinase fructose-2,6 – bisphos-phatase	Glycolysis/gluconeogenesis

VII. CYCLIC AMP ACTS AS A SECOND MESSENGER FOR A NUMBER OF REGULATORY MOLECULES

In addition to hormones listed in Table 81.2, growth factors and other regulatory molecules also act by changing the intracellular levels of cyclic AMP.

In the adenylyl cyclase system an extracellular signal (hormone) such as epinephrine or PGE_1 can have quite different effects on different tissues and cell types. For example, epinephrine by binding to specific receptors in hepatocytes, stimulates the intracellular cyclic AMP concentration and the physiological effect is increased blood glucose.

The same epinephrine when bound to specific receptors on adipocytes stimulates cyclic AMP formation which, inturn, causes lipolysis increasing free fatty acid concentration in blood. In the same adipocytes (adipose tissue) prostaglandin $E_1(PGE_1)$ binds to specific receptors, inhibits adenylyl cyclase and lowers the cyclic AMP concentration, thus slowing lipolysis triggered by epinephrine.

In short, an extracellular signal epinephrine or PGE_1 can have different effects on different tissues or cell types, depending on (1) the type of receptors, (2) the presence of Gs or Gi, and (3) the enzymatic machinery of the cell, i.e., the set of enzymes susceptible to phosphorylation by cyclic AMP.

TABLE 81.2   Some Hormones which act by increasing cyclic AMP concentration

Hormone	Tissue	Physiological responses
Epinephrine	Liver	Glycogenolysis
	Muscle	Glycogenolysis
	Adipose	Lipolysis
Glucagon	Liver	Glycogenolysis
	Adipose	Lipolysis
ACTH	Adrenal cortex	Steroidogenesis
PTH	Renal cortex	Phosphaturia
	Bone	Calcium resorption
TSH	Thyroid	Thyroglobulin hydrolysis, Iodination
LH	Ovary	Steroidogenesis
FSH	Corpusluteum	Steroidogenisis
	Adipose	Lipolysis

VIII. SIGNALLING PATHWAYS REGULATING CREB-DEPENDENT TRANSCRIPTION

CREB (cAMP response element binding) is cellular transcription factor. It binds to certain DNA sequences called CRE *(cAMP response elements)*, thereby increasing or decreasing the transcription of downstream genes. When a signal arrives at the cell surface, it activates the corresponding receptor, which leads to the production of second messengers such as cAMP, calcium and diacylglycerol, which in turn activate the specific protein kinases. Protein kinases are translocated to the cell nucleus, where they phosphorylate CREB protein which binds to the CRE region on DNA. The binding is further coactivated by CBP (CREB-binding protein) allowing to switch certain genes on or off. CREB has many functions in many different organs (Figure 81.5).

IX. CYCLIC GMP AS A SECOND MESSENGER

Guanosine 3′ 5′ cyclic monophosphate (cyclic GMP) functions as a second messenger in the cells of the *intestinal lining, heart,*

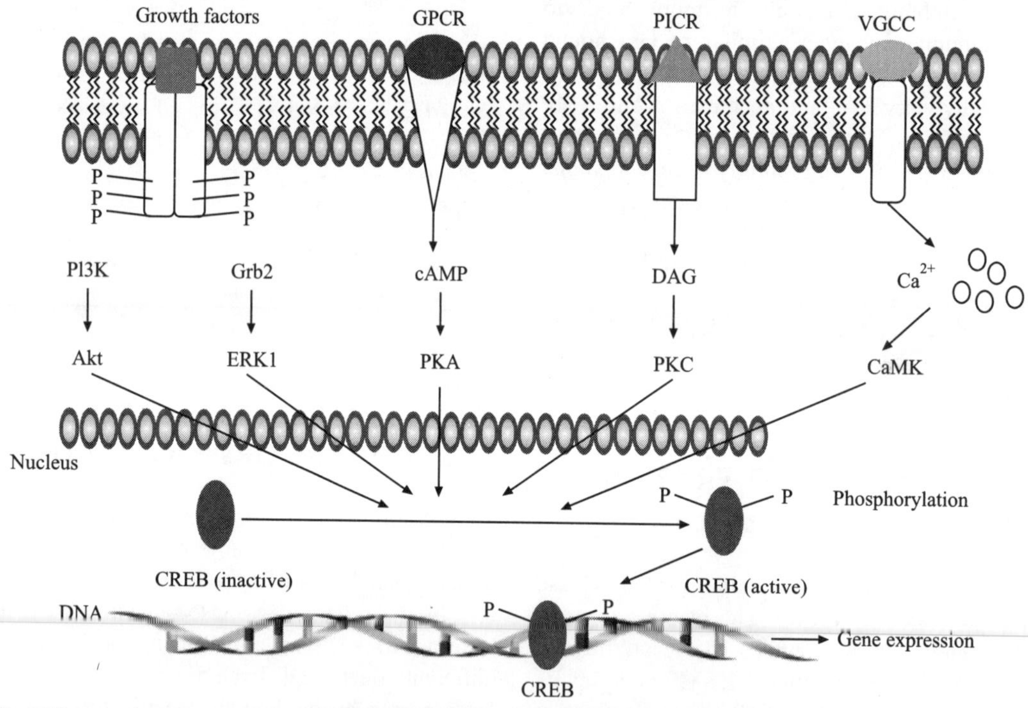

Figure 81.5   Signalling pathways regulating CREB dependent transcription. A signal arrives at the cell surface and activates the corresponding receptor: Growth factors such as Insulin, IGF (insulin like growth factor); EGF (epidermal growth factor); PDGF (platelet derived growth factor), GPCR (G-protein coupled receptor), PICR (phosphoinositide coupled receptor), VGCC (voltage gated calcium channels-in neurons), lead to formation of second messengers such as cAMP, diacylglycerol (DG) and calcium. These, in turn, activate protein kinases-P13K (phosphatidylinositol 3 kinase); AKT (protein kinase B); PKA (protein kinase A); PKC (protein kinase C); DG (diacylglycerol); Grb2 (mitogen activated protein kinase); ERK (extracellular signal-regulated protein kinase); CaMK (calmodulin dependent kinase. Active protein kinases move to the cell nucleus where they activate CREB (cAMP response element binding) protein by phosphorylation. Activated CREB binds to the CRE (cAMP response element) region on DNA. This binding is further coactivated by another protein CBP (CREB binding protein). This allows to switch on or off certain genes. (*see Plate 38 for colour figure*)

blood vessels, brain and the collecting ducts of the kidneys. The enzyme guanylate cyclase which catalyzes the formation of cyclic GMP from GTP exists as two isozymes—one is the integral membrane protein of the plasma membrane with the hormone binding domain on the outerface and the other is soluble. The hormone *atrial natriuretic factor (ANF)* released by cells in the *atrium of heart,* activates guanylate cyclase in the cells of the collecting ducts of the kidney resulting in increased concentration of cyclic GMP which, in turn, triggers increased renal excretion of sodium and consequently water. In the intestinal epithelial cells, membrane bound guanylate cyclase is also activated by a heat labile *bacterial endotoxin* (a small peptide) produced by *E.Coli* (common bacteria present in the gut). The resulting elevation in cyclic GMP causes decreased absorption of water by the intestinal epithelium, producing *diarrhea* characteristic of this toxin. A second guanylate cyclase is a cytosolic protein with a tightly associated heme group. This enzyme is activated by its natural ligand *Nitric Oxide (NO).* In many human tissues NO is produced from arginine by a Ca^{2+} calcium dependent mixed function oxidase—NO synthase, NO diffuses from the cells into nearby cells, where it binds to the heme group of guanylate cyclase and activates this enzyme producing cyclic GMP. Several *nitrovasodialators*-such as *nitroglycerine tablets* are used in the treatment of *angina (angina pectoris is a condition caused by insufficient blood flow to the heart muscle leading to chest pain).* Once inside the cell nitroglycerine tablets are spontaneously broken to NO which stimulates the soluble guanylate cyclase, cyclic GMP is formed which causes cardiac smooth muscle relaxation. Cyclic GMP acts directly on target molecules like ion channels, can activate cGMP protein kinase G that can phosphorylate many proteins. *Protein kinase G differs from the cAMP kinases in that the regulatory and catalytic domains are part of the same molecule rather than being encoded by separate genes.*

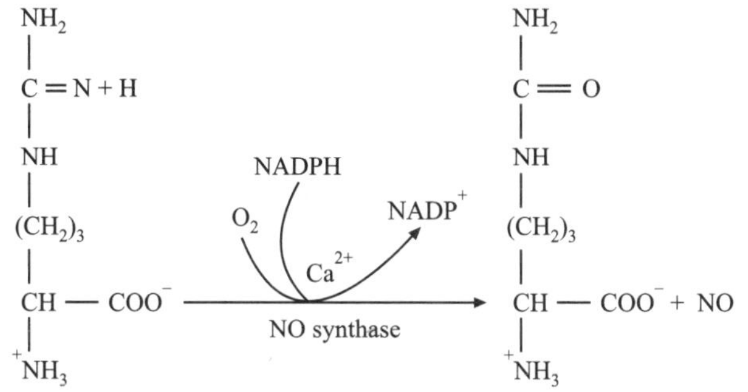

Arginine → (NO synthase, with O_2, NADPH, Ca^{2+} → $NADP^+$) → Citrulline + NO

X. NITRIC OXIDE REGULATORY FUNCTION IN ERECTION OF PENIS

The penis physiological states of flaccidity or erection, result from the contraction or relaxation of smooth muscle cells in the corpora cavernosa (CSMCs). Tonic sympathetic activity, releases norepinephrine and other agonists that generate contractile signals in the CSMCs, with the cooperation of endothelium derived messengers. Through activation of membrane receptors in the CSMCs they raise the concentration of IP3 and diacylglycerol. This results in a transient increase in cytosolic calcium concentration that starts the contractile response which is further sustained by the parallel agonist induced activation of a calcium sensitizing mechanism involving the *RhoA/ Rho-kinase* pathway. Overexpression of the latter might contribute to several vascular disorders as *hypertension, vasospasm or erectile dysfunction.* On stimulation the cavernosa nerves release *nitric oxide* that starts the erectile response. They also release acetylcholine that further stimulates the endothelium to generate a more sustained release of NO which diffuses into CSMCs and increases the intracellular levels of cyclic GMP which dcreases and deactivates the calcium sensitizing mechanism, thus relaxing CSMCs. This main physiological pathway for CSMCs relaxation is helped by the cAMP pathway activated by various intracellular messengers from *neural or paracrine sources.*

Different phosphodiestearse enzymes (PDEs) inactivate the cyclic nucleotides thereby limiting the erectogenic action. Indeed the pharmacological *inhibition of PDEs especially the cGMP specific PDE-5,* greatly enhances the erectile response. There are crosstalk mechanisms between cAMP and cGMP pathway that offer additional possibilities for the pharmaco therapy of erectile dysfunction.

XI. PHOSPHOINOSITIDE PATHWAY

Several hormones *(e.g., catecholamines, vasopressin and angiotensin)* as well as some neurotransmitters (e.g., acetylcholine acting on pancreatic acinar cells to stimulate secretion of digestive enzymes or on β cells of pancreas to stimulate insulin secretion) bind the specific receptors on the outside of the cell and transmit the signal by a G-protein (Go or Gq) which then activates a phosphodiesterase (phospholipase C) that cleaves the polar inositol trisphosphate (IP$_3$) from phosphatidyl inositol-4,5 bis-phosphate (P1P$_2$). IP$_3$ enters the cytoplasm whereas diacylglycerol (DG) moiety remains in the membrane. *Both IP$_3$ and diacyglycerol are second messengers. IP$_3$ binds to receptors on the endoplasmic reticulum and stimulates the release of sequestered calcium into the cytoplasm.* Calcium gets bound to calmodulin which then activates the calcium-calmodulin dependent protein kinases. Diacylglycerol alongwith phosphatidylserine (PS) activates a membranc associated protein kinase C which in the presence of calcium phosphorylates threonine and serine residues of proteins. Thus, the second messengers generated in the phosphoinositide pathway activate several kinases (similar to the cAMP dependent protein kinases) and thus modify the activity of rate-limiting enzymes in metabolic pathways. In addition to generating the second messengers the breakdown of P1P$_2$ also stimulates the production of *arachidonic acid* (one of the main fatty acid found in the diacylglycerol moiety of P1P$_2$). Arachidonic acid is the precursor of *eicosanoids,* which function as local hormones IP$_3$ is rapidly degraded to IP$_2$ (which is inactive) and then to inositol. Diacylglycerol is phosphorylated and converted to CDP-diacylglycerol, which combines with inositol to form phosphatidylinositol which is subsequently phosphorylated to P1P$_2$. *The degradation and resynthesis of P1P$_2$ completes the phosphatidyl inositol cycle* (Figure 81.6).

XII. SOME RECEPTORS HAVE ENDOGENOUS KINASE ACTIVITY

The receptor for insulin is itself a protein kinase. The insulin receptor has two identical α chains with insulin binding domain towards outer face of the plasma membrane, and two transmembrane β subunits, with their carboxy termini on the cytosolic face, having tyrosine kinase domain. Insulin binding to the α chains activates the tyrosine kinase activity of the β chains which transfers a phosphate group from ATP to the hydroxyl group of tyrosine (not serine or threonine). Individuals with insulin-resistant diabetes secrete insulin normally, but their tissues do not respond to their own or injected insulin, due to a mutation in the tyrosine kinase domain of the insulin receptor. Variety of other growth factor receptors such as EGF and PDGF have structural homology with insulin receptor.

XIII. INSULIN RESISTENCE, OBESITY AND METABOLIC SYNDROME

Metabolic syndrome consists of a group of metabolic changes, being the most important problem in insulin resistence. Other important components of this syndrome are *abdominal obesity, hypertension and hyperlipedemia/hypercholesterolemia.* Psychiatric patients have a greater risk oto develop metabolic syndrome.

In contrast to the systemic metabolism of glucose, the liver is the primary metabolic clearing house for four specific foodstuffs that have been associated with the development of Metabolic syndrome: Trans-fats, Branched chain amino acids, Ethanol and Fructose. These four substrates (1) are not insulin regulated and (2) deliver metabolic intermediates to hepatic mitichondria without an appropriate pop-off mechanism. Excessive fatty acid derivatives interfere with hepatic insulin signal transduction. Reactive oxygen species (ROS) accumulate which can not be quenched by adjacent peroxisomes. These ROS reach the endoplasmic reticulum, leading to a compensatory process termed the *"unfolded protein response"* driving further insulin resistence and eventually insulin deficiency. No obvious drug target exists in this pathway, thus the only rational therapeutic approaches remain

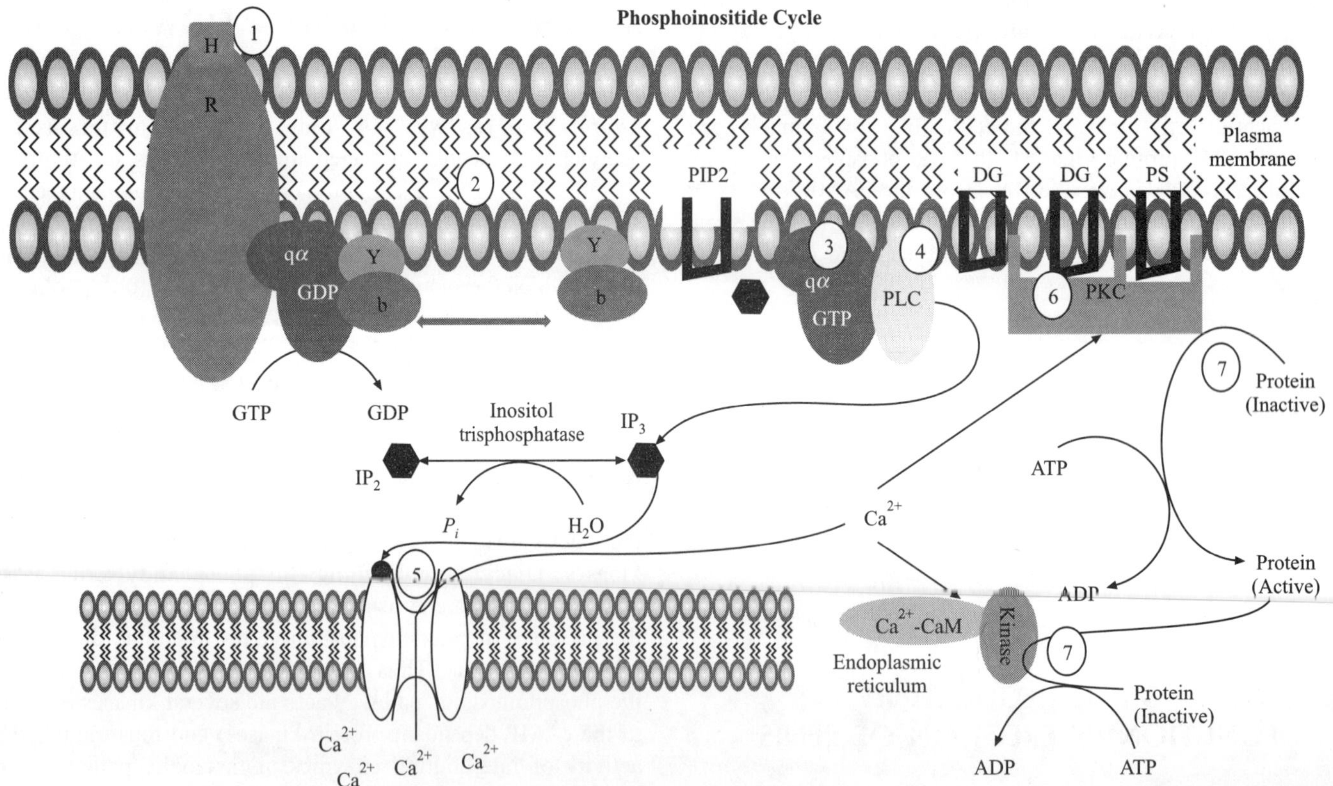

Figure 81.6 Phosphoinositide cycle. (1) Hormone binds to the specific receptor; (2) The occupied receptor causes GDP-GTP exchange on Gqα; (3) Gqα moves to phospholipase C (PLC); (4) Active PLC cleaves phosphatidylinositol-4,5 bisphosphate (PIP$_2$) to inositol trisphosphate (IP$_3$) and diacylglycerol (DG); IP3 binds to endoplasmic reticulum, causing release of stored Ca^{2+}; (6) DG and Ca^{2+} activate protein kinase C (PKC) at the inner surface of the plasma membrane. This activation also requires phosphatidyl serine (PS) another membrane phospholipid; (7) active protein kinase C and Ca^{2+}-calmodulin dependent protein kinase phosphorylate proteins and produce the cellular response. (*see Plate 38 for colour figure*)

(1) altering hepatic substrate availability (dietary modification), (2) reducing hepatic substrate flux (high fiber diet), or (3) increasing mitochondrial efficiency (exercise).

XIV. DEFECTS IN SIGNAL TRANSDUCTION PATHWAYS ARE THE CAUSES OF CERTAIN DISEASES

A. Microbial toxins

Vibriocholerae (the bacteria that causes cholera) secretes the *cholera toxin* (an 87-KD protein with AB5 subunits). The toxin (CT) binds to its cell surface receptor ganglioside GM1. CT is taken into the cell via endocytosis.The process results in CT activation through the proteolytic cleavage and disulfide bond reduction of the A subunits to A1 and A2. A1 is released into the cell cytosol where it catalyzes the transfer of ADP-ribose from NAD^+ to the α subunit of Gs, blocking the GTPase activity of $GS\alpha$, and thereby rendering it permanently activated. The net result is continuous activation of adenylyl cyclase of intestinal epithelial cells and high concentration of cyclic AMP triggers continuous secretion of Cl^-, HCO_3^- and water into the intestinal lumen. The results is massive diarrhoea which, if untreated, often results in the death of victims due to dehydration.

The pertussis toxin produced by Bordetella pertussis catalyzes ADP–ribosylation of Gi, GDP displacement by GTP and blocking inhibition of adenylyl cyclase by Gi. This defect produces symptoms of whooping cough, including hypersensitivity to histamine (Figure 81.7).

B. Hyper Parathyroidism and Pseudo-hypoparathyridoism

The first and still dominant use of cyclic AMP determinations is in *hyperparathyroidism*. An increase in the circulating parathyroid hormone levels leads to increased *tubular secretion* of cyclic AMP. Measurement of urinary cyclic AMP, when corrected for glomerular filtration rate is a useful and routine diagnostic aid for diagnosis of primary hyperparathyroidism.

In *pseudohypoparathyroidism* in humans there is a genetic mutation in which parathyroid hormone receptor adenylyl cyclase system is defective.

C. Proliferative Disorders

Tumors and cancer are the result of uncontrolled cell division. Some well-studied growth-factors are EGF (Epidermal growth factor), NGF (Nerve growth factor), FGF (Fibroblast growth factor), PDGF (Platelet derived growth factor), erythropoietin and a family of proteins called lymphokines, which include

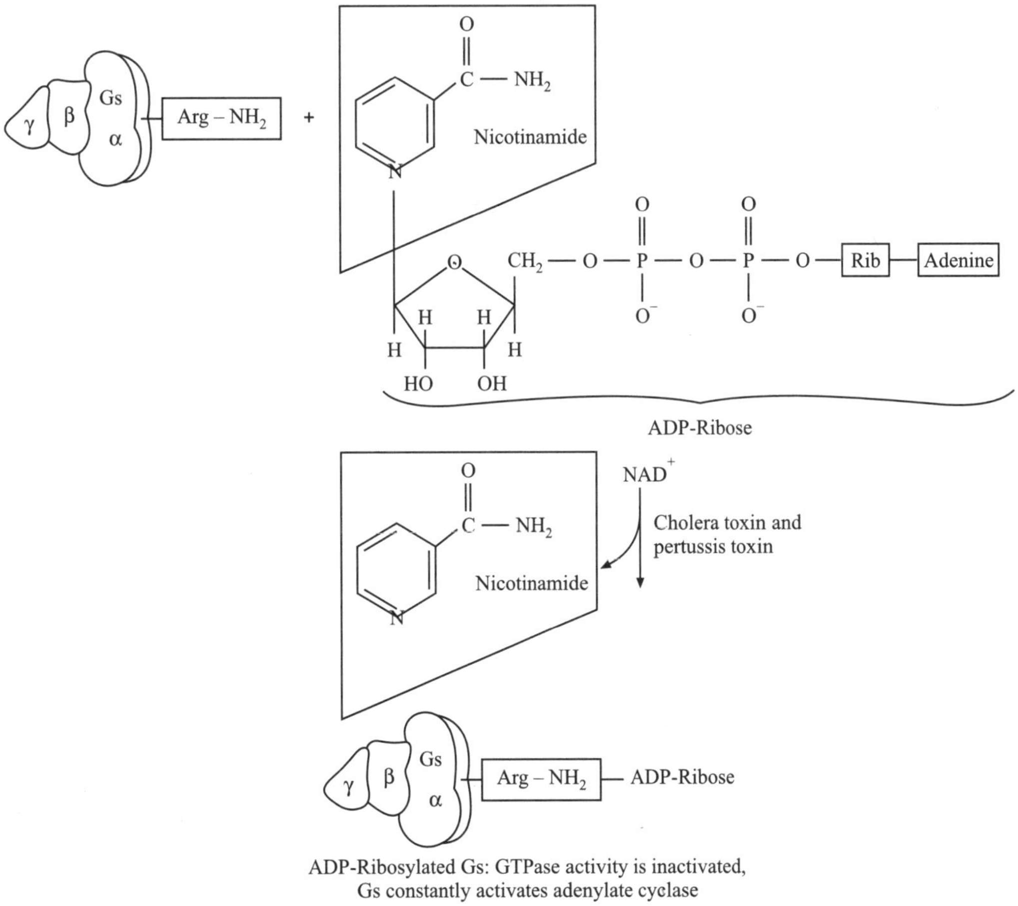

Figure 81.7 Mechanism of action of cholera toxin and pertussis toxin

interleukins and interferon γ. In addition some extracellular factors antagonize the effects of growth factors (mentioned above) slowing or inhibiting cell proliferation, e.g., transforming growth factor β (TGF β), and tumor necrosis factor (TNF). Similar to hormones growth factors also act through cell surface receptors and by similar mechanisms: the production of intracellular second messengers, protein phosphorylation and ultimately gene expression.

Many types of cancer are the result of abnormal signal-transducing proteins, which lead to continual production of the signal for cell division. The mutated genes that encode these defective signaling proteins are called *oncogenes* Many *viral oncogenes* encode unregulated tyrosine kinase activities, and in some cases, the oncogene product is nearly identical to the normal cell-receptor, but the signal binding site is defective or missing, e.g., the *erbB oncogene product,* a protein called erb B, is essentially identical to receptor for EGF, except that erbB protein lacks the domain that normally binds EGF. *The erbB-2 oncogene is commonly associated with adenocarcinomas of the breast, stomach and ovary.*

Other signal transducing proteins with oncogene analogs are the GTP-binding (G) proteins. One well characterized oncogene, ras, encodes a protein with normal GTP binding but no GTPase activity. When a ras protein is produced in the cell, it remains always in the activated form, regardless of the signals coming through the normal receptors. Again, the result is unregulated growth-cancer. Mutation in ras is associated with lung and colon and pancreatic carcinomas.

The action of a group of compounds known as *tumor promoters* can also be understood in the light of signal transductions. The best understood are phorbol esters (synthetic) potent activators of *protein kinase C.* They apparantely mimic cellular diacyglycerol as second messenger but unlike naturally formed diacylglycerol they are not rapidly metabolized.

XV. PROTEIN PHOSPHORYLATION AND DEPHOSPHORYLATION ARE CENTRAL TO CELLULAR CONTROL

Whether the signal transductions involve adenylyl cyclase, a transmembrane receptor tyrosine-kinase, PLPase C, or an ion channel, a common factor is the eventual regulation of the activity of a protein kinase. The number of protein kinases has grown remarkably since their discovery by E.G. Krebs and E.H. Fisher in 1959. Although many other types of covalent modifications are known to occur on proteins, it is clear that phosphorylations make up the vast majority of known regulatory modifications of proteins. The addition of phosphate group to a Ser, Thr, or Tyrosine residue introduces a bulky, highly charged group into a region that was only moderately polar. When the modified side chain is located in a region of the protein critical to its three-dimensional structure, phosphorylation can be extended to have dramatic effects on protein conformations and thus on catalytic activity of the protein. To serve as an effective regulator mechanism these phosphorylations must be reversible, allowing the regulated enzyme to return to their pre-stimulus levels, when the hormone signal stops. Cells contain a family of phospho-protein phosphatases that hydrolyze specific phosphoserine, phosphothreonine and phosphotyrosine esters releasing Pi. Phosphatases are as important as the protein kinases in regulating cellular processes and metabolism.

SUGGESTIONS FOR FURTHER READING

Lemmon M.A. and Schlissinger J. (2010), Signalling by Receptor Tyrosine Kinases, *Cell*, 141:1117–34.

Pierce K.L., Premont R.T. and Lefkowitz R.J. (2002), Seven Transmembrane Receptors, *Nature Rev, Mol Cell Biol.*, 3:639–650.

82

Relationship of
Central Nervous System and Endocrines

G.P. Talwar

CONTENTS

I. INTRODUCTION

Endocrine glands which secrete a number of hormones, fulfil a very important physiological role in the body, as they provide important mechanisms of chemical regulation. With the evolution of the unicellular living organism into a multicellular organism, need arose for the development of fluid internal environment, named by Claude Bernard as the *"milieu interieur"*. As all the cells of the body 'exchange' with the internal environment for fulfilling their requirements, for an optimum functioning of these cells it is essential that the composition of the internal environment remains somewhat constant and does not change much. This was described by Claude Bernard as the *'constancy of milieu interieur'*, also termed by Cannon as *"homeostasis"*. As the activities of the various cells, tissues, organs and systems in any living organism have to be coordinated and integrated with each other to achieve homeostasis, efficient control systems have been evolved in the body. Chemical control, slowly acting and long lasting, is provided by the secretions of glands, both exocrines as well as endocrines. Nervous control, quick acting and short-lived, is provided by the nervous system. Even the nervous system ultimately brings about its effects by the liberation of chemical transmitters at the nerve endings. Nervous system is also involved in the chemical control, by its direct or indirect regulation of the secretory activities of both exocrine and endocrine glands.

There are various endocrine glands in the body. Possibly the most important among these is the *pituitary gland (hypophysis)*, whose secretory activity is under the direct influence of the central nervous system; and this in turn influences the secretory activity of other endocrine glands through its trophic hormones. The pituitary gland consists of two lobes: a posterior lobe which is also called *neurohypophysis,* and an anterior lobe also called *adenohypophysis.*

Secretory activity of both these lobes of hypophysis is influenced through a small nervous region called hypothalamus which is a part of the diencephalon and is lying ventral to (below) the thalamus at the base of the brain. Hypothalamus constitutes the walls and the floor of the III ventricle, and extends from a position slightly rostral to the optic chiasma caudally to the mammillary bodies. Rostral to hypothalamus is the septal region and caudal to it is the midbrain tegmentum. On the ventral surface of the hypothalamus lies the optic chiasma in front; behind this there are three protuberances, the median eminence (*tuber cinereum*) on the middle hypothalamus and the two mammillary bodies on the posterior hypothalamus. From the median eminence arises the *infundibulum* (pituitary stalk) which connects it to the posterior lobe of hypophysis; through this passes the hypothalamo-hypophysial tract (Figure 82.1).

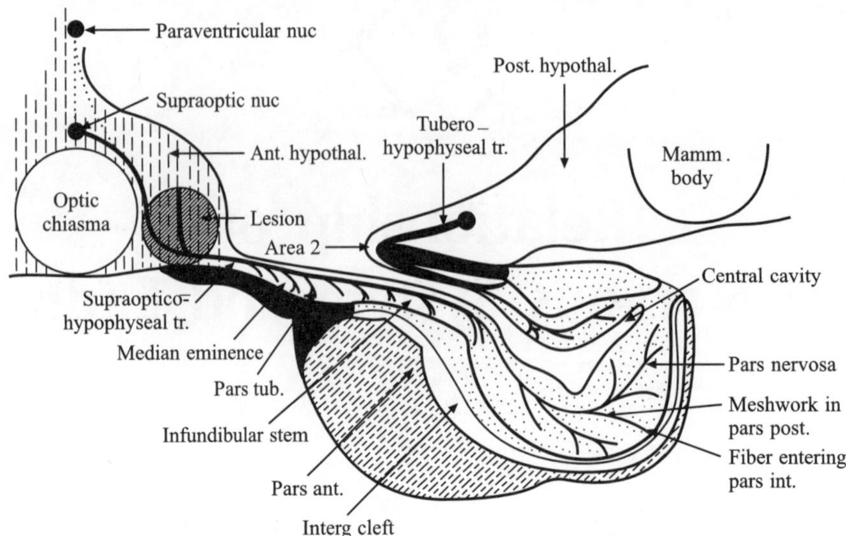

Figure 82.1   Longitudinal section of cat's hypothalamus, showing supraoptico-hypophysial tract. (From Fisher, Ingram and Ranson, *Diabetes Insipidus and the Neuro-hormonal Control of Water Balance*, Edwards Bros., 1938.)

Along the infundibular stalk run a number of portal blood vessels which connect the capillary system of the median eminence to the blood capillaries (*sinuses*) in the anterior lobe of hypophysis (Figure 82.2). The blood supply to the hypothalamic region is more profuse than to many other brain structures.

Thus the hypothalamus is connected to the posterior lobe of hypophysis (neurohypophysis) by nervous connections of the hypothalamohypophysial tract which originates in the supraoptic and paraventricular nuclei and proceeds through the median eminence and pituitary stalk to the posterior lobe. On the other hand, between hypothalamus and the anterior lobe of hypophysis there are no neuronal connections: the median eminence of the hypothalamus being connected with the anterior lobe of hypophysis only through vascular channels provided by the portal vessels. Hypothalamus, therefore, influences the neurohypophysial activity through its nervous connections (Figure 82.1) by which neurohumors produced in the hypothalamus are transmitted and delivered in the posterior lobe of hypophysis; while it influences the adeno-hypophysial activity by releasing some chemical factors in the median eminence which are then transported through portal vessels to anterior lobe of hypophysis. As these hypothalamic chemical factors influence the secretion of trophic hormones from the anterior hypophysis, these are termed *"release factors"*. The trophic hormones secreted by the anterior hypophysis then influence the secretory activity of other endocrine glands (Figure 82.2).

There are only two hormones which are released in the posterior lobe of hypophysis, from where they enter the general circulation and thus produce their effects in the body. These are: (i) Antidiuretic hormone (ADH—also called *'Vasopressin'*) and (ii) Oxytocin.

Both these hormones are synthesized in the hypothalamus, in the supraoptic and paraventricular nuclei respectively, and they are then transmitted and released in the posterior hypophysis at the nerve terminations.

The anterior lobe of hypophysis on the other hand secretes the following six hormones—FSH, LH, LTH, TSH, ACTH and growth hormone (Chapter 61).

As the secretion of these six trophic hormones is influenced by the 'release factors', the hypothalamus is expected to produce at least six release factors, each one helping in the release of one of the above mentioned trophic hormones from the anterior hypophysis. Although recent work has provided experimental evidence for the presence of all these six release factors, it has been possible to purify and identify the chemical nature of only a few of them. All these release factors are synthesized in the hypothalamus, are released in the median eminence (near the capillary system from which arise the portal blood vessels), and from there are transported in the portal blood and delivered into anterior hypophysis, where they help to synthesize as well as release the trophic hormone.

It is thus apparent that the hypothalamic region of the central nervous system plays a very important role in regulating and adjusting the secretory activity of both the posterior lobe as well as the anterior lobe of the hypophysis (Figure 82.3). The trophic hormones of the anterior hypophysis in their turn influence the secretory activity of the other endocrine glands. Further, the hypothalamus through its connections with the sympathetic motor nerves influences the secretory activity of the adrenal medulla. It is thus evident that the secretory activity of the anterior hypophysis and through it of ovaries, testes, thyroid and adrenal cortex is influenced from the hypothalamus through the agency of neurosecretions, while the activity of posterior hypophysis and adrenal medulla is directly regulated by the hypothalamus through nervous connections. The relationship between hypothalamus and the remaining two endocrine glands is not very clear. The insulin secretion from the islets of the pancreas is possibly influenced more through the

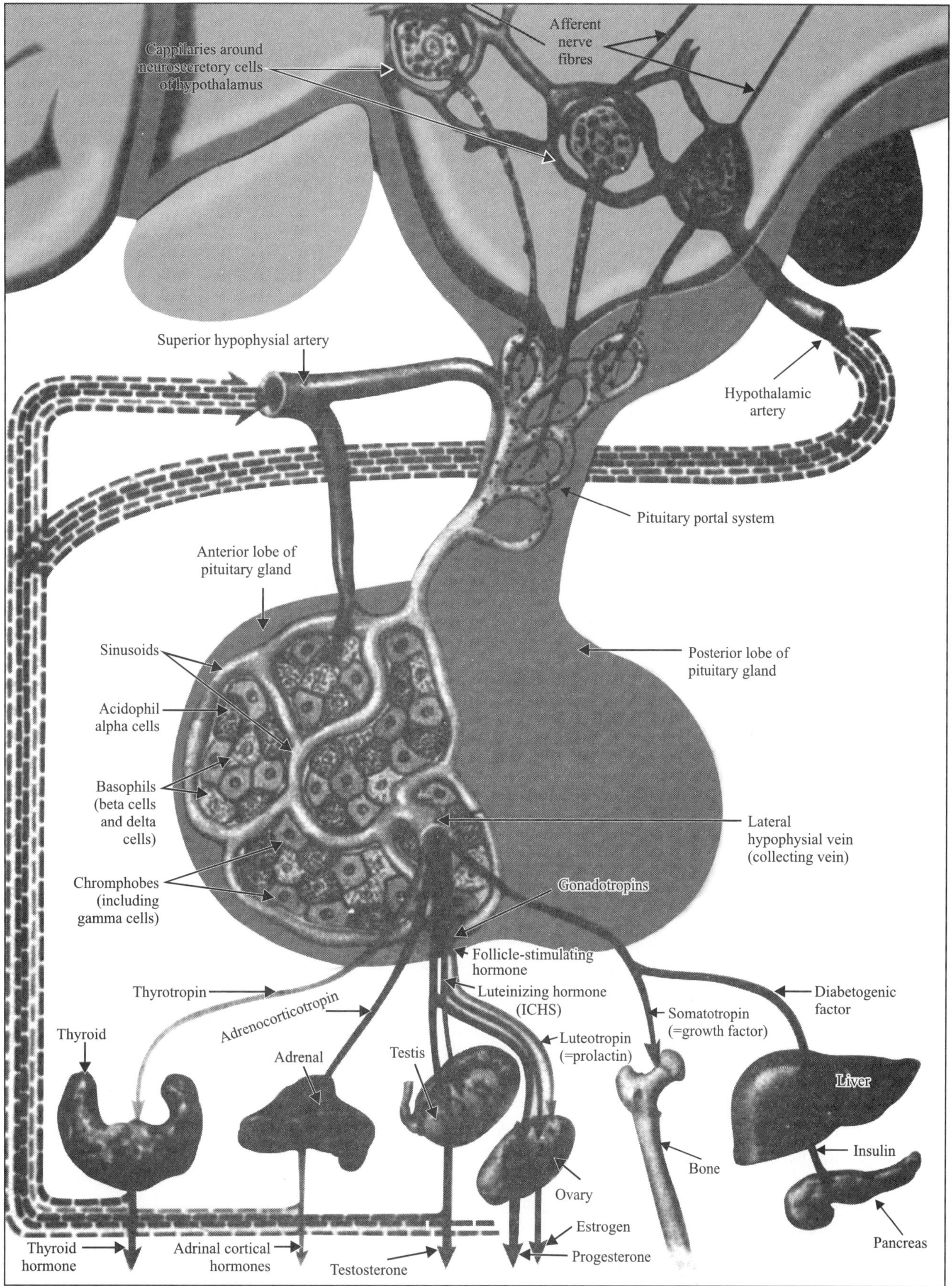

Figure 82.2   Diagrammatic representation of hypothalamic regulation of anterior pituitary tropic hormones through portal vessels. (From Netter, *Ciba Collection of Medical Illustrations*, 1, 1958.)

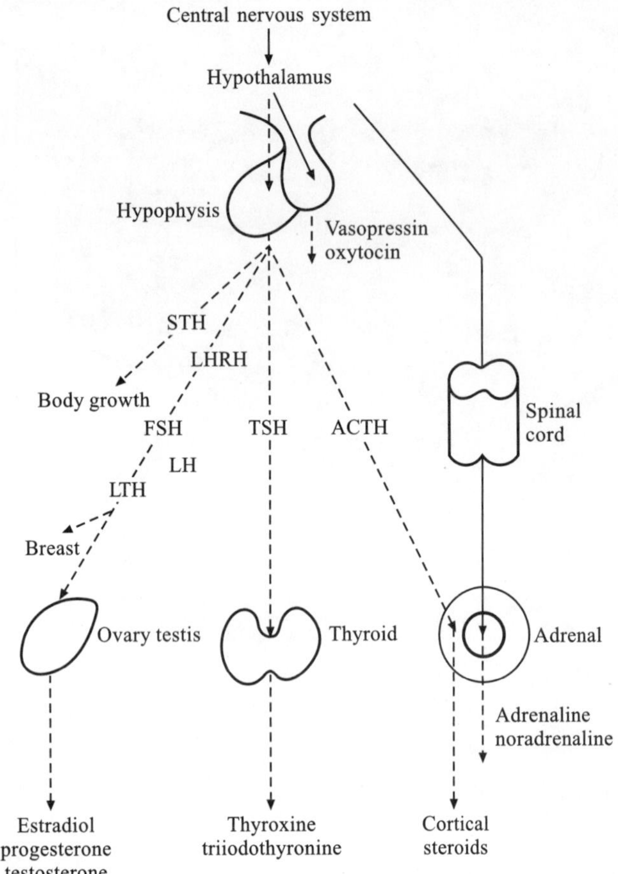

Figure 82.3   The relationship between the nervous system and endocrine system, nerve pathway, vascular pathway. Note that two glands, the adrenal medulla and neurohypophysis are regulated by a direct nerve supply whereas the activity of other glands is controlled by humoral or hormonal agents carried in the blood stream. (From Best and Taylor, *The Physiological Basis of Medical Practice,* 8th edition, William and Wilkins Company, 1967.)

circulating levels of blood glucose; however it is also influenced by autonomic nerves which are under the regulation of the hypothalamus. The regulation of parathyroid function appears to be directly affected by the blood concentrations of calcium, and the involvement of hypothalamus in this, if any, is not clear.

The hypothalamus thus is the key central nervous region through which nervous controls over endocrines are mediated. Other parts of the central nervous system, such as the reticular formation of the brain stem, the temporal lobes and their nuclei like the amygdala and the hippocampus, the frontal lobes, and certain other central nervous regions produce their effects, if any, on the endocrines acting through the hypothalamus.

In spite of the hypothalamus exerting a very important regulating and controlling influence over the secretory activity of the endocrines, these glands themselves also have a certain amount of automatic activity of their own. This is revealed by the fact that in the hypophysectomized animals, which removes the effects of both hypothalamus and pituitary gland, the thyroid and adrenal cortex still show some secretory activity but at a very low rate, while the gonads show no signs of endocrine function. On the other hand, if the pituitary gland is left intact but is disconnected from the hypo-thalamus by section of the pituitary stalk, still there is no gonadal endocrine function, demonstrating that independent anterior pituitary is not capable of secretion of FSH and LH. However, in such a preparation a certain amount of TSH, ACTH and LTH are still released by hypophysis, whereby the thyroid and adrenal cortex show functional activity which is greater than that seen in the hypo-physectomized animal, although it is less than in the intact animal. It thus appears that the secretion of gonodal trophic hormones from the hypophysis and the endocrine activity of gonads is entirely regulated through the central nervous mechanisms operating through the hypothalamus, while a certain level of thyroid and adrenal cortical activity can be generated and further influenced from the hypophysis even without any operating nervous controls. However, without the nervous control it is not possible for even these two glands to show full scale activity, and further their activities cannot be increased under the influence of environmental stresses for which a stimulating effect operating through hypothalamus is essentially required.

Just as the trophic hormone of the hypophysis stimulates the activity of the target endocrine gland, the hormone secreted by the target gland in its turn regulates the secretion of the particular trophic hormone. More the level of circulating hormone in the blood, less will be the secretion of the trophic hormone which is specific for this. This phenomenon is called the *"negative feedback"*. Experimental evidence suggests that this negative 'feedback' by the thyroid and the adrenal cortical hormones takes place in the anterior hypophysis, while the negative feedback of gonadal hormones occurs primarily in the hypothalamic region. The level of thyroxin circulating in the blood acts on the anterior hypophysis and decreases the secretion of TSH. Similarly, circulating levels of adrenal corticoids decrease the secretion of ACTH from hypophysis. On the other hand, circulating levels of estrogens, progestrone, and androgens affect the activity of the hypothalamus and change the gonadotrophic secretions by influencing the secretion of the release factors. Some feedback of gonadal hormones directly into the hypophysis has also been experimentally shown.

II. REGULATION OF NEUROHYPOPHYSIAL FUNCTION

As stated earlier, two hormones, viz. *antidiuretic hormone* (ADH), which produces increased absorption of water through renal tubules (which also has a vasoconstrictive effect hence called vasopressin), and *oxytocin,* which contracts the gravid uterus and produces other similar effects on plain muscle, are released in the neurohypophysis. It has now been established that these hormones are synthesized by the hypothalamic neurones and are then transported in the hypothalamo-hypophysial tract to be released in the posterior pituitary.

A. Osmoreceptors

The release of antidiuretic hormone (ADH) depends upon the presence of *"osmoreceptors"* in the supraoptic and paraventricular nuclei of the hypothalamus. Whenever there is an increase of osmotic pressure of the circulating blood, more ADH is secreted, which by conserving body water tends to correct the osmotic imbalance. Dehydration produced by any means by decreasing plasma volume and increasing osmolarity releases ADH through these osmoreceptors. It has also been shown that haemorrhage and other mechanisms which change blood volume, also influence the secretion of ADH through the agency of baroreceptors (volume receptors) situated at the periphery, possibly in the chambers of the heart. Even when a preparation is made in which the hypothalamus is disconnected from other areas of the central nervous system, but its connections with hypophysis are left intact, it will respond to osmotic changes produced by infusion of hypertonic saline by secreting ADH but will not respond to changes in blood volume as produced by haemorrhage.

The hypothalamo-hypophysial mechanism is thus an important mechanism for the regulation of water balance in the body. This it does by conservation of water through ADH secretion. Lesions or injuries to this mechanism, either to the supraoptic nuclei, or in the hypothalamo-hypophysial tract result in *"diabetes insipidus"*, through loss of ADH secretion.

N.B.:—It may further be noted that another 'thirst' regulating mechanism is also situated in the lateral hypothalamus, which directly regulates water intake. It is thus by the dual mechanisms of hypothalamo-neurohypophysial (ADH) and the hypothalamic "thirst centres" that the regulation of body water is brought about.

B. Oxytocin

Oxytocin release is induced by certain stimuli, especially sucking of the teats of the mammary glands. This way suckling of the milk by the infants brings about further contraction of the uterus.

It has been further suggested that the hypothalamic neural mechanisms regulating the release of ADH and oxytocin are separate; possibly supraoptic nuclei play a more important role in release of ADH, while paraventricular nuclei are more important for oxytocin.

C. Melanocyte Stimulating Hormone (MSH)

The melanophore expanding effect, shown to be present in the posterior lobe extracts, and considered to be a secretion of the pars intermedials of the pituitary in certain species, possibly plays an important role in certain amphibia in regulating the expansion of melanophores in response to photic stimulation. Although the evidence is conflicting, it has been suggested that there may be both melanocyte stimulating hormone (MSH) releasing and release inhibiting factors in the hypothalamus. It

is further suggested by some that ACTH and MSH secretion may be controlled by a common mechanism.

III. REGULATION OF ADENOHYPOPHYSIAL FUNCTIONS

It has already been stressed that the hypothalamus influences secretion of the trophic hormones of the adenohypophysis ACTH, TSH, STH and gonadotropins (FHS, LH and LTH) by *"release factors"* which are formed in the hypothalamus and then pass on to the interior pituitary through the portal vessels. Also, the secretion of these trophic hormones from the hypophysis is ordinarily adjusted through negative feedback into the hypophysis and the hypothalamus by the circulating level of hormones of the stimulated glands. Further, although the regulation of gonadotrophins and gonadal activity entirely depend on hypothalamic influences, the thyroid and the adrenal glands have a certain intrinsic or automatic activity which is further influenced through the hypothalamus by environmental changes of temperature and stress, etc.

A. Adrenocorticotropic Hormone (ACTH)

Its secretion is increased by the activation of hypothalamus in conditions of stress. As a result of this a neurohumor called CRF (*corticotrophic releasing factor*) is liberated in the median eminence, passes to hypophysis in portal vessels, and there increases secretion of ACTH. In spite of many years of efforts the CRF has not still been isolated and identified. Crude acid extracts of hypothalamus show ACTH releasing activity. While CRF has been purified to a great degree, further purification has not been possible due to its instability and loss of activity. Hypothalamic CRF content is shown to be increased under stress, as well as following adrenalectomy (possibly there is some negative feedback of adrenal corticoids into hypothalamus as well). There have been suggestions that ADH may, to some extent be able to produce release of ACTH from anterior hypophysis, but this has not been well established. The CRF and ADH (vasopressin) are possibly not identical.

Hypothalamic activation (during stress) also stimulates sympathetic supply to adrenal medulla, and the released adrenaline going to hypophysis further increases ACTH secretion. ACTH in its turn stimulates the adrenal cortical activity.

Some experimental studies have also suggested that certain hypothalamic regions on stimulation inhibit the release of ACTH. This may be the result of inhibition of CRF (or even the production of some inhibitory releasing factor).

B. Thyroid Stimulating Hormone (TSH)

Its secretion is increased by the activation of hypothalamus in those conditions in which body metabolism requires to be increased (cold exposure). A neurohumor called TRF (*thyrotrophin releasing factor*) is liberated in the median eminence, passes to hypophysis, and leads to increased TSH secretion, which in its turn increases thyroxin secretion. This mechanism also explains why in nervous (emotional) individuals TSH secretion is raised and may sometimes lead to hyperthyroidism (*Grave's disease*). The identification, chemical characterization and synthesis of TRF has already been achieved.

The hypothalamic regions involved in the regulation of TSH secretion appear to be localized in the median eminence and anterior hypothalamus. Stimulation of median eminence increases TSH activity. On the other hand, some experimental studies have suggested that anterior hypothalamic stimulation inhibits releases of TSH. This hypothalamic region also contains thermosensitive neurones responding to rise in temperature, and so integration through this hypothalamic region can decrease TSH secretion when body temperature tends to rise. This may be resulting from inhibition of TRF (or even from the production of some inhibitory releasing factor).

C. Growth Hormone (GH), or Somatotrophic Hormone (STH)

Evidence for hypothalamic regulation of growth hormone is now generally accepted. It appears that growth hormone is controlled from the hypothalamus through a growth hormone releasing factors (GRF) and a *growth hormone inhibiting factor* (GIF). Crude hypothalamic extracts mostly show GRF activity. However, after further fractionation of hypothalamic extracts, some fractions are found to contain GRF and others GIF. There are many factors involved in the regulation of GH secretion, probably the most important being hypoglycemia or falling blood glucose, moderate exercise, and other stressful stimuli also increase GH secretion.

D. Gonadotrophic Hormones

Hypothalamus regulates the secretion of gonadotrophic hormones (FSH, LH and prolactin) from the hypophysis. These in their turn determine the secretory activity of gonads, and also produce ovulation and spermatogenesis.

The gonadotrophin secretion from the hypophysis is mostly under the control of hypothalamus. Two neurohumors, called *follicle stimulating hormone releasing factor* (FRF), and *luteinizing hormone releasing factor* (LRF) are liberated in the median eminence, pass to hypophysis in portal vessels, and there increase secretion of FSH and LH respectively. No definitive work has been reported on the chemical nature of FRF; while LRF has been purified, although some consider that the preparation so identified is not totally purified. The action of purified LRF is very short and its intracarotid injection induces release of LH even in a few minutes. Some studies have been done to identify the sites for production and storage of these releasing factors which appear to be concentrated in the arcuate nucleus and the preoptic area of the hypothalamus. Further a group of fine non-myelinated fibres arising in the floor of the third ventricle and terminating in the median eminence have been shown to be associated with the secretion of hypophysial trophic releasing factors.

In the male of the species, the FSH and LH are being continuously secreted, and so the hypothalamus is continuously producing both the releasing factors. In the female of the species there is an inverse relationship between the secretion of FSH and LH occurring cyclically. This cyclic activity in the female is a hypothalamic function, adjusted by cyclic variations in the liberation of FRF and LRF. This can be shown by removing pituitary gland from a male animal and transplanting it under the hypothalamus of a female animal, after which the secretion of gonadotrophins from this hypophysis follows the female pattern.

Another hypothalamic neurohumor *prolactin inhibiting factor* (PIF) inhibits the secretion of prolactin from hypophysis. So prolactin secretion only occurs when secretion of PIF is inhibited by some peripheral sensory inputs. Presence of a *prolactin releasing factor* (PRF) has also been shown in some species.

Spermatogenesis in the sexually mature male occurs continuously, while ovulation in sexually mature female occurs only periodically. Ovulation results from sudden release of a relatively large quantity of LH from hypophysis under the influence of LRF. In certain species, like ferrets and birds, hypothalamic mechanisms leading to ovulation are stimulated by visual sensations. In still others, like rabbits, these hypothalamic ovulating mechanisms are stimulated by afferents coming from vagina and cervix during the act of copulation. In human females and females of some other mammals, hypothalamic mechanisms function in a rhythmic cyclic manner, accompanied by the periodic secretion of FSH and LH. These periods determine the menstrual cycle in women (and primates), and oestrus cycles in the other mammals.

It has already been stated that the gonadal hormones (estrogens and progestrone in female and androgens in male) circulating in the blood provide negative feedback into the hypothalamic region. Estrogens inhibit the release of FRF, while progestrone inhibits the release of LRF. This fact is made use of for blocking ovulation with the use of oral contraceptives which mainly contain progestins and to some extent estrogens. These steroids, feeding back into the hypothalamus, depress the release of LRF, thus blocking the surge of LH essential for ovulation. It has further been demonstrated experimentally that estrogens can act as a positive stimulus for triggering LH release during the estrous cycle of the adult female rat.

Even in the species, like rat, which have spontaneous cyclic mechanism for the release of LH to produce ovulation, this can be disturbed by the injection of male hormone into the newly born female. Such an injection results in a permanently anoestrus female, which is like a male. The pituitary of such an androgenised anovulatory rat contains large stores of LH as well as total gonadotrophins, as compared with their amounts present in the pituitary of a normal female. Thus, the gonado-trophin characterizations of the pituitary gland from female and male are similar. It is concluded from these studies that the male hormones affect the neural mechanisms regulating gonadotrophin storage and release, thus converting the cyclic release of gonadotrophins into a tonic release of these hormones.

ADDENDUM

IV. CHEMICAL STRUCTURE OF SOME OF THE HYPOTHALAMIC RELEASE HORMONES

Painstaking research in many laboratories but notably of Schally, Guillemin and McCann has resulted in the isolation purification of a number of biologically active polypeptides from the hypo-thalamus. To achieve this, several kilogram batches of bovine, porcine and ovine hypothalmi had to be processed. The primary structure of some of the release factors with either a stimulatory or inhibitory action on anterior pituitary hormones has been elucidated. This information is summarized below:

The determination of primary structure has been followed up by chemical synthesis of these polypeptides. Thus, a new era has emerged. Hormones whose availability from tissue extracts was scarce and expensive can now be made available in large quantities at reasonable rates through the industrial channels, with the result that these can be utilized for clinical purposes. TRF and LRF are already finding therapeutic uses.

A. TSH Release Hormone (TRF or TRH)

It is a simple tripeptide (Figure 82.4). It stimulates not only the synthesis and release of TSH, but also of prolactin from the anterior pituitary.

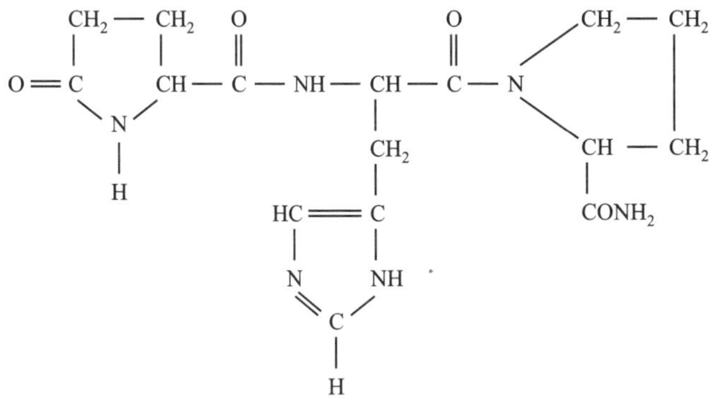

L — pyroglutamyl — L — histidyl — L — prolineamide

Figure 82.4 Primary structure of TRH (TRF).

B. Gonadotrophin Release Hormone

(Pyro) Glu–His–Trp–Ser–Tyr–Gly–Leu–Arg–Pro–Gly–NH$_2$

Figure 81.5 Primary structure of LHRH.

(Ala^4)–LH–RH: pGlu–His–Trp–Ala–Tyr–Gly–Leu–Arg–Pro–Gly–NH$_2$
(Phe^5)–LH–RH: pGlu–His–Trp–Ser–Phe–Gly–Leu–Arg–Pro–Gly–NH$_2$
(Ala^4) (Phe^5)–LH–RH: pGlu–His–Trp–Ala–Phe–Gly–Leu–Arg–Pro–Gly–NH$_2$

Figure 82.6 Primary structures of three synthetic analogs of LHRH modified in position 4 & 5 and found to possess LHRH activities in the order of 8.6%, 44.3% and 1.01% respectively.

C. Somatostatin (Growth Hormone—Release Inhibitory Hormone)

H–Ala–Gly–Cys–Lys–Asn–Phe–Phe–Trp–Lys–Thr–Phe–Thr–Ser–Cys–OH
(Release inhibitory hormone)

Figure 82.7 Primary structure of somatostatin, a hypothalamic peptide that inhibits the secretion of pituitary growth hormone.

83

Pituitary Hormones

G.P. Talwar

CONTENTS

I. GENERAL INTRODUCTION TO PITUITARY HORMONES

The pituitary is a unique endocrine gland. It is the source of nine hormones, six secreted by anterior pituitary (namely, ACTH, TSH, LH, FSH, prolactin and growth hormone), two by posterior pituitary (oxytocin and vasopressin) and one by pars intermedia (MSH). The majority of the anterior pituitary hormones (at least four out of six) are tropic hormones for other endocrine glands. TSH acts on the thyroid, ACTH on adrenals and LH, FSH on the gonads. In each case these tropic hormones stimulate the synthesis and secretion of hormones characteristic of the target endocrine glands. Pituitary is thus a 'master' endocrine gland secreting not only some valuable hormones (like the growth hormone, prolactin, oxytocin and ADH), but also controlling the activity of many other endocrine glands through the agency of the tropic hormones. A well regulated feedback system is operative by which the products of the target endocrine glands regulate the level of the tropic hormones. The central nervous system is intimately associated with pituitary function (Chapter 81).

The pituitary, a tiny gland in the base of the brain, has a reddish grey appearance and measures, in the human, about 12 mm transversely, 8 mm in its anteroposterior diameter and

6 mm in its vertical dimension. It weighs about half a gram in the adult male. The pituitary of the adult female is slightly heavier about 600 mg, whereas in multiparous women it may attain a size of about 700 mg. The hormones secreted by the male and female pituitaries are chemically identical. Extracts from pituitaries of either sex can be used for therapeutic purposes. The main difference in the male and female pituitary function is the pattern of the secretion of hormones. The female has a cyclic pattern with distinct secretory profiles of gonadotropins, whereas in males there is a continuous tonic secretion of these hormones without a cyclic pattern.

The anterior pituitary has three types of cells distinguishable by their staining characteristics. The acidophils secrete the somatotropin (or growth hormone) and prolactin. It is not certain whether these hormones are secreted by the same cell or by two different sub-classes of cells with acidophilic staining characteristics. The basophils can be subdivided into four sub-classes by finer staining techniques: the β^1 cells are the source of ACTH and MSH, the β^2 cells of TSH, the Δ^1 cells of LH and Δ^2 cells of FSH. The chromophobes secrete ACTH.

II. POSTERIOR PITUITARY HORMONES

A. Chemical Nature

The posterior pituitary is the source of two hormones: (i) oxytocin and (ii) vasopressin or ADH (antidiuretic hormone). Both are octapeptides. They differ from each other in only two amino acids at positions 3 and 7, the rest of the molecule is identical. Their structures (Figure 83.1) were elucidated by Du Vigneaud and collaborators who accomplished also the chemical synthesis in the laboratory, a work that received a Nobel prize.

While most mammals have the arginine vaso-pressin as depicted in Figure 83.1, the pig and the hippopotamus make lysine vasopressin instead of arginine vasopressin, i.e., the amino acid at position 7 is lysine instead of arginine. In birds, amphibia and fishes, the hormone molecules have essentially this backbone but differ in terms of minor changes at one or more amino acids. The substitutions are at positions 3, 4, or 7. The study of the molecular structure of these hormones provides yet another fascinating example of the inherent relationship between various living organisms and of evolution as it has taken place at the molecular level. It is also evident that mutations at one point, e.g., amino acid 7 as in the case of lysine arginine in vasopressin do not lead to gross changes in biological activity. But mutations at both positions 3 and 7 shift the biological activity of the molecule considerably as is the case with oxytocin and vasopressin.

B. Origin, Regulation and Secretion

The posterior pituitary or the neurohypophysis is constituted of nerve fibres emanating from the paraventricular and supraoptic nuclei in the hypothalamus. The two hormones secreted from

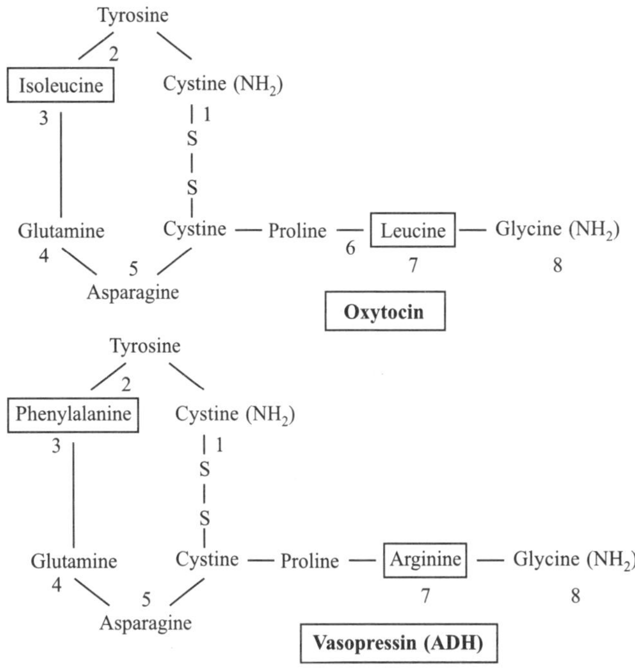

Figure 83.1

the gland are, properly speaking, neuronal products. They are synthesized in the body of the neurons and are stored in the nerve endings from which they are released into the blood. The stimulus for their release is perceived by appropriate neuronal receptors and is communicated by neuronal channels. For example, if the body loses water, the rise in blood osmotic properties is sensed by 'osmoreceptors' located mostly in supraoptic nuclei. Impulses are generated to release ADH, which acts on the distal renal tubules to promote enhanced reabsorption of water. The net result is the gearing in of the body mechanisms for saving the loss of water.

Neurogenic stimuli are also implicit in the release of oxytocin such as those elicited by the suckling of the breasts.

C. Synthesis

Both hormones secreted by the neurohypophysis are composed of 8 amino acid residues. The mode of their synthesis is not clear. One of the view points is that they are synthesized as a part of a larger precursor proteinic molecule, which is later cleaved to give rise to the octapeptide. If this mechanism be true, then the synthesis of these hormones would be similar to the general pattern for synthesis of proteins. There would be the requirement of messenger RNA, ribosomes, tRNA and other factors for initiation, elongation, termination and release of the polypeptide chain. This hypothesis would also require cleavage to operate at a discrete site in the precursor molecule. This can be brought about by the presence of either a protease with peculiar specificity or by the conformation of the precursor protein, to be such that the protease gets access to the protein molecule only at a point where an octapeptide can be split off. While no unequivocal evidence is available for the validity of this hypothesis the presence of a precursor protein has been suggested by the work of Sachs. A carrier protein "neurophysin" with strong binding affinity for ADH has also been described.

An alternate hypothesis for synthesis of posterior pituitary hormone and for that matter other Release hormones elaborated in the hypothalamus would conceive of a serial participation of a group of enzymes with specificities to join amino acids in a definite order. Such mechanisms do operate for synthesis of carbohydrate moieties of glycoproteins, where the sugars are added in a definite sequence.

D.   Biological Actions, Functions

ADH: Posterior pituitary hormones do not appear to be vital for survival of the organism. The main effect noticed on ablation of the gland is diuresis and increased thirst. As much as 30 litres of water may be lost in the urine in 24 h, a state described as *diabetes insipidus*. The condition is remedied by posterior pituitary extracts given subcutaneously or applied on the nasal mucosa. ADH or vasopressin is the active principle for this action. The anti-diuretic effect is exercised by enhancement of the "facultative" reabsorption of water from the distal tubules of the kidney (Chapter 32). The hormone exerts its effect by activation of the adenyl cyclase of target cells; 3′, 5′ cyclic AMP so generated, alters in some way the permeability of the cells to water. *Alcohol* causes diuresis by suppression of the action of ADH. Other *drugs* such as morphine, nicotine and barbiturates *decrease* the urine output by *increasing the secretion of ADH.*

The vasopressor activity of ADH is only manifest at pharmacological doses. At physiological levels, this hormone has probably no important vasoconstrictor action.

Oxytocin: This hormone induces contractility of (a) uterine musculature and (b) myoepithelial cells of the mammary glands. It is secreted in response to neurogenic stimuli originating from the reproductive tract or by suckling of breasts. Its function is to help in the ejection of milk. In the reproductive tract it is believed to promote the ascent of spermatozoa following coitus. It is also implicit in the contraction of the uterus at the time of parturition (see also the role of prostaglandins, Chapter 9).

E.   Summary

Oxytocin and vasopressin (ADH) are two octa-peptide hormones synthesized by cells of hypothalamus and stored in the neurohypophysis. Their molecular composition is known.

They differ from each other in terms of 2 amino acids, while the remaining six are identical. The secretion of ADH is prompted by increase in osmotic properties of blood as sensed by 'osmoreceptors'. It acts on the distal tubules of the kidney promoting the facultative absorption of water. In its absence 'diabetes insipidus' results with marked polyuria and large intake of H_2O. Ethyl alcohol causes diuresis by suppression of the action of ADH. Many drugs such as morphine, nicotine, barbiturates decrease urine output by increasing the secretion of ADH. Oxytocin is released in response to neurogenic stimuli of suckling, coitus, etc. It helps in ejection of milk by its action on myoepithelial cells of the mammary glands. It is also believed to promote the ascent of spermatozoa in female reproductive tract. It contributes to the contraction of uterine musculature during parturition.

III.   HORMONES OF THE INTERMEDIARY LOBE OF THE PITUITARY

Melanocyte stimulating hormone (MSH) is a short length polypeptide originating from pars intermedia of the pituitary. Primary structures of the hormones purified from pig pituitary are shown in Figure 83.2. It will be observed that a full molecule of alpha MSH and a part of the molecule of beta MSH have amino acid sequences identical to that of ACTH. This is probably the basis of MSH-like action of high doses of ACTH on pigmentation. The interrelation in biological activity of these hormones is complicated by the fact that while ACTH itself has this effect, glucocorticoids, secreted in response to ACTH by the adrenals, block the secretion of MSH by the pars intermedia.

MSH acts on melanocytes or chromophore cells distributed mostly in the dermis and epidermis of the skin. These cells develop from the neural ectoderm. They have the enzymes to synthesize melanin from an amino acid tyrosine by a pathway given in Figure 83.3. The pigment is stored in form of dense granules in areas adjacent to the nuclear membrane. MSH promotes in the first instance the dispersion of this pigment to the entire cellular processes, that makes these cells dark in appearance. Consequently the colour of the skin is darkened. The dispersion of the pigment also has a chain reaction on synthesis of more of melanin. MSH thus influences both the dispersion and the synthesis or total cell content of the pigment, even though the primary action may only be on the dispersion step. In case the stored perinuclear pigment granules are not dispersed over prolonged periods,

Alpha MSH   *Ac–Ser–Tyr–Ser–Met–Glu–His–Phe–Arg–Trp–Gly–Lys–Pro–Val–NH$_2$
(13 amino acids)

Beta MSH   Asp–Glu–Gly–Pro–Tyr–Lys (Met–Glu–His–Phe–Arg–Trp–Gly)–Ser–Pro–Pro–Lys–Asp
(18 amino acids)   *Acetyl

Figure 83.2   Primary structure of alpha and beta MSH from pig pituitary αMSH—the amino acid sequence is that of first 13 amino acids of ACTH except that the N-terminal amino acid serine is acetylated and the C-terminal valine is in the carboxyamide form. βMSH—amino acid sequence between the brackets is identical with amino acids 4 to 10 of ACTH.

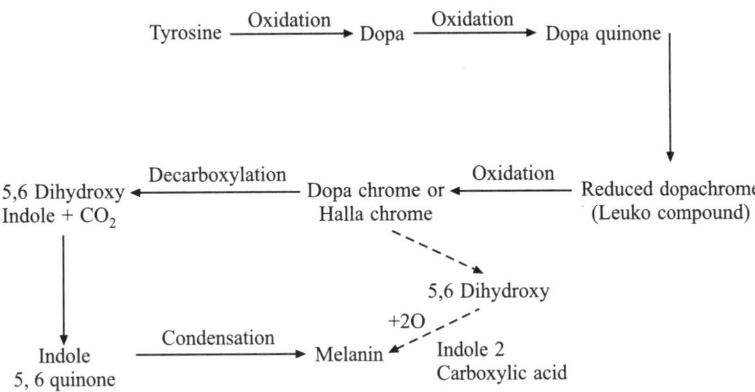

Figure 83.3 Outline of the pathways leading to the formation of melanin from the precursor amino acid tyrosine. The three oxidation steps and the decarboxylation step are all enzyme mediated reactions. The first enzyme in the series oxidising tyrosine is tyrosinase. In albinism, a genetically determined defect, this enzyme is absent.

the pigment is progressively lost and reduced in content in the cells. It is surmised that there is not much difference in the number of melanocytes per unit skin area among various races. The difference primarily resides in the content of the pigment (partially dependent on MSH action for its dispersion) but also in genetic factors controlling the elaboration of the enzymes, necessary for synthesis of this pigment. Epinephrine and even more strongly norepinephrine have a blanching action by interfering in the action of MSH.

A more complete discussion on melanin and other skin pigment is given elsewhere in this book (Chapters 23).

IV. PITUITARY SOMATOTROPIN— GROWTH HORMONE

A. Chemical Nature, Primary Structure

Growth hormone (GH) has been purified from a number of sources, e.g., bovine, ovine, porcine, human, simian, rat, fish, etc. In all cases it has been found to be a simple protein. The human growth hormone (HGH) has a molecular weight of about 21,500 and is composed of 191 amino acids. The order in which these amino acids are linked to each other to generate the biologically active GH molecule, is known and is shown in Figure 83.4. There is a close molecular resemblance between GH and prolactin, so much so that, in some species the same molecule seems to fulfil the biological activity of both hormones. There is however good evidence to show that in most species, GH and prolactin are two related but distinct hormones.

B. Species Specificity, Source of HGH

Some animals like the rat can utilize and respond to GH from widely different sources (bovine, ovine, human, in fact almost all those tried except fish GH), but the human being has narrow species specificity. The effective hormone for human use is either human or that from the monkey. Consequently the hormone used for radioimmuno-assays in human biological fluids, or for therapy in subjects with hypopituitary function is extracted from human pituitaries collected from postmortem cases. The supply is fairly limited and has imposed restrictions on the widespread use of the hormone for needy cases. Two alternate sources are being tried and hold promise. Several laboratories are engaged in developing cell lines that can synthesize the hormone. The cell lines are developed from pituitary tumours removed from acromegaly patients. Some cell lines with properties of synthesizing and secreting hormones have already been successfully developed from rat pituitary tumours. This approach may offer a novel source material for this hormone in the years to come. Another breakthrough to resolve the limitation in supply of adequate amounts of human growth hormone is the reported success by C.H. Li et al. of the chemical synthesis of this hormone, although the polypeptide synthesized so far has only 10% of HGH potency and 5% lactogenic activity of the natural hormone. A synthetic hormone with better biological activity should be available in course of time in large amounts and, hopefully, at reasonable cost.

The gene for human growth hormone has been cloned by the DNA recombinant technology. It has been expressed successfully in a bacteria, cell and the cloned protein is testing is available for therapeutic purposes.

C. Factors Influencing the Release of GH

The secretion of growth hormone is not limited to periods of active somatic growth, c.g., foetal and adolescent phase but continues over the entire span of life (Table 83.1) suggesting that this hormone has several functions besides that of promoting somatic growth. The fact that somatic growth does not continue all through life in spite of adequate levels of GH in circulation, points to the importance of "end organ responsiveness" as a factor that determines the biological effects of the hormone on various organs. The cartilage and bones stop growing at a certain age even though the hormone is present in the circulation and is also producing somatomedins.

The secretion of the hormone from the pituitary is controlled by a stimulating release hormone and also an inhibitory factor elaborated in the hypothalamus

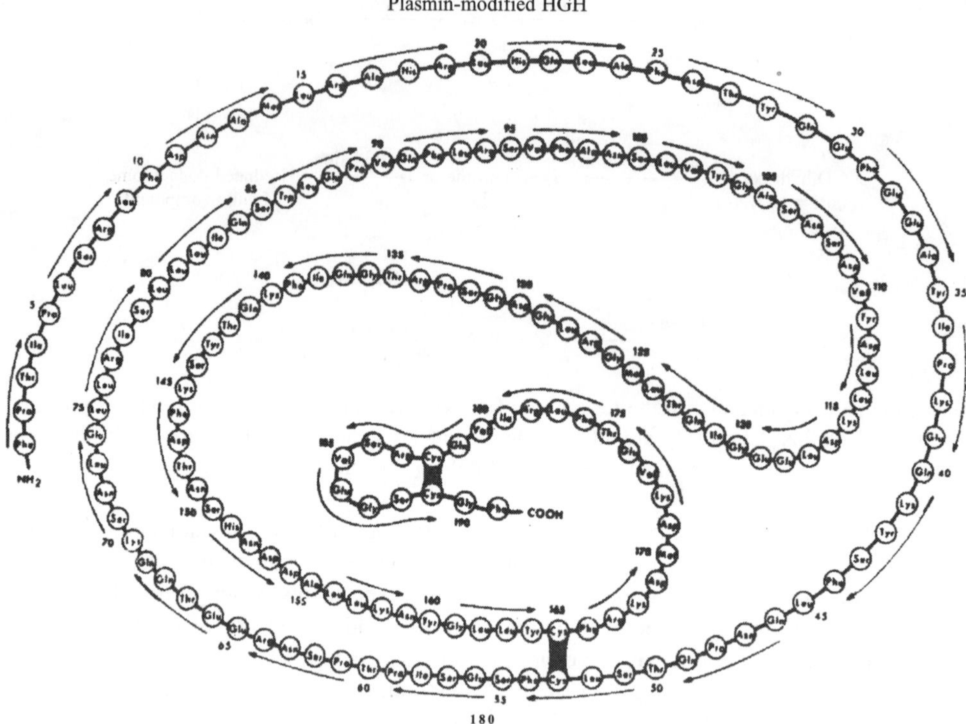

Figure 83.4   Primary structure of human growth hormone. (Courtesy Prof. C.H. Li)

TABLE 83.1   Fasting levels of HGH in normal children and adults

	Mean ± S.D. HGH µg/ml
Birth to 3 days (Cord Blood*)	40.2 ± 10.6*
Less than 1 yr.	1.0 ± 0.8
1–5 yr.	5.2 ± 3.4
6–10 yr.	4.4 ± 3.9
11–15 yr.	4.9 ± 3.3
Adults	1.2 ± 1.0

*Ref. Otto Westphal, Human Growth Hormone—A methodological and clinical study, *Acta Paed. Scand., Suppl.*, 182, 1968.

(Chapter 60). The release of growth hormone is subject to a large number of influences. Hypoglycemia, either resulting from starvation or by administration of insulin, is a potent stimulus for release of GH. Amino acids like arginine induce release of the hormone. Stress also causes release of GH. The hormone is released during sleep. The biological half-life of endogenous plasma GH is around 20 to 30 minutes in humans.

D.   Functions and Metabolic Actions of GH

(a) *Promotion of general growth*

GH has a generalized, growth promoting effect on almost all organs of the body. Its deficiency in infancy leads to stunted growth, dwarfism, whereas hyperactivity causes gigantism. It has a supporting role for the development of viscera and glands. One of the earliest assays used for its biological activity was the increase in weight of (i) plateaued rats whose body weight had attained a maximum, or (ii) hypophysectomized animals. It also possesses the biological activity of lactogenic hormone. The hormone has a wide spectrum of action both in terms of the tissues affected, and in the range of the effects elicited. To some extent this ubiquity of action may be due to the widespread presence of "receptors" for binding of the hormone on various types of cells. It is also possible that some of the effects associated with the hormone are not a direct property of the hormone molecule, but are elicited by its partial degradation products or by hormones and mediators released from other tissues by its action.

1. Somatomedins and action on cartilage and bones: A well established assay for the biological activity of the hormone is the growth of tibial epiphysis in rodents. Before the closure of epiphysis, it promotes the lengthening of bones. Hyperactivity of its secretion leads to extraordinary height, characteristic of gigantism [cases have been recorded up to 8 ft. 3¼″ weighing 395 pounds (179 kg) at age 18, shoe size 34]. Hypopituitary function is associated with stunted growth. The active principle for bone growth is not GH directly but a plasma factor called somatomedin or sulfation factor elaborated, probably, in the liver (and perhaps in other tissues) by the action of growth hormone on these tissues. Somatomedin (SM) is insulin-like in many of its traits but is different from insulin. SM is a polypeptide or a group of closely related peptides of mol wt. 5000–7000 daltons. SM mimics insulin in its action on adipose tissue, but does not cross-react immunologically with insulin. It acts as a growth factor for fibroblasts and human glia-like cells. Hypophysectomy of animals results in the removal of SM from circulation. The serum from GH treated animals has an action, *in vitro,* on cartilage tissue, in which the incorporation of ^{35}S-sulphate into mucopolysaccharides; or proline into collagen; and of uridine and thymidine into RNA and DNA is markedly stimulated.

2. Effect on lymphoid tissue and immune response: Of late, a relationship between thymus and pituitary is being recognized. In Snell-Bagg strain of dwarf mice which are deficient in pituitary function, there is an inadequate development of the thymus. The animals manifest runting syndrome or wasting disease caused by the absence or deficiency of thymus derived lymphocytes and consequent impairment of the immune response. The condition is alleviated by administration of the pituitary extracts or growth hormone. Systematic studies have shown that GH can stimulate by about 30%, *in vitro,* the uptake of radioactive precursors into RNA and other metabolic activities in isolated thymocytes. There is an obligatory requirement for Ca^{++} in the medium for the stimulatory effect of the hormone.

(b) Metabolic actions

1. Anabolic action—effect on RNA and protein metabolism: GH is an anabolic hormone. It induces a positive nitrogen balance by promoting the synthesis of proteins rather than by decreasing catabolism. The positive effect of the hormone on protein synthesis is a result of the following actions.

    (a) It facilitates the transport of amino acids into cells as gauged by the uptake of radioactive amino acids by isolated diaphragms incubated with the hormone, *in vitro.*

    (b) The ribosomes prepared from liver of hypophysectomized rats are observed to be less efficient for protein synthesis, a defect that is rectified by administration of the hormone to the animals.

    (c) It stimulates the synthesis of RNA in liver, muscles and, perhaps, other tissues. The effect is on the synthesis of both the messenger and ribosomal RNA. The total number of protein synthesizing units are increased in the cells. The hormone activates the rate of synthesis of RNA.

2. Effect on carbohydrate and fat metabolism: The hormone has, paradoxically, both an insulin-like and insulin-antagonistic type of effect on carbohydrate and fat metabolism. In hypophysectomized rats, the initial, though transient, effect is insulin-like, namely, the enhanced uptake of glucose and lipogenesis. This is soon followed by insulin-antagonistic type of action, namely resistance to the uptake of glucose,

resulting in hyperglycemia, increased lipolysis and mobilization of lipids with rise in serum NEFA (nonesterified fatty acids) levels. The latter is the dominant action of the hormone observed in normal individuals. The hormone is thus hyperglycemic and mobilizes lipids from adipose tissue. GH can, in a way, be conceived to be a hormone with conservation properties. Its actions on the metabolic activities are such as to make the best in the face of scarcity. It is secreted during starvation and in conditions of falling blood glucose level. It contributes to the restoration of blood glucose to normal levels essential for proper brain function, by limiting the transfer of glucose to cells. For energy needs, the fat reserves in adipose tissue are mobilized. The protein synthesis is rendered more efficient. In contrast, insulin can be considered to be a hormone regulating the direction of metabolic pathways in situations of abundance. It is secreted in response to high blood glucose levels. Its function is to transfer the glucose from blood to peripheral cells and to promote the storage of excess carbohydrates and calories in the form of glycogen and fats.

E. Mechanism of Action of Growth Hormone

The manner in which this hormone acts on a variety of tissues is not fully known. There is growing evidence to suggest that one of the sites of its action are the membranes. The hormone causes alterations in the helical content of the membrane-bound proteins of some cells. The permeability of the diaphragm cells to take up amino acids and glucose is changed. Direct evidence for the location of the hormone along, or in close proximity to cell surface has been obtained by experimental studies on thymocytes in our laboratory, employing radio-autographic and high resolution immunoenzymatic methods for detection of the hormone. Another observation supporting the location of the hormone on membranes is the change that it causes in the cell surface charge of thymocytes. The interaction of the hormone with "receptors" on membranes leads to a membrane flux with modulation of the relationship of the cell to the extracellular environment. Some of the biological actions of the hormone would, thus, be a direct consequence of the uptake of metabolites and displacement of cations.

The interaction of the hormone may not be restricted to membranes. The hormone is also implicit directly, or more likely indirectly, in stabilization of key regulatory enzymes such as phosphofructokinase.

It also has an influence on the transcriptional and translational processes. The rate of synthesis of RNA is enhanced. While all species of RNA are increased quantitatively, the predominant effect is on the synthesis of more ribosomes.

The hormone promotes the synthesis of proteins by a combination of two actions: (i) by a facilitated uptake of amino acids, (ii) by improving the efficiency of ribosomes for the synthesis of proteins. The latter could well be a consequence of the enhanced synthesis of ribosomes and mRNA.

Figure 83.5 resumes the current information on the mode of action of this hormone.

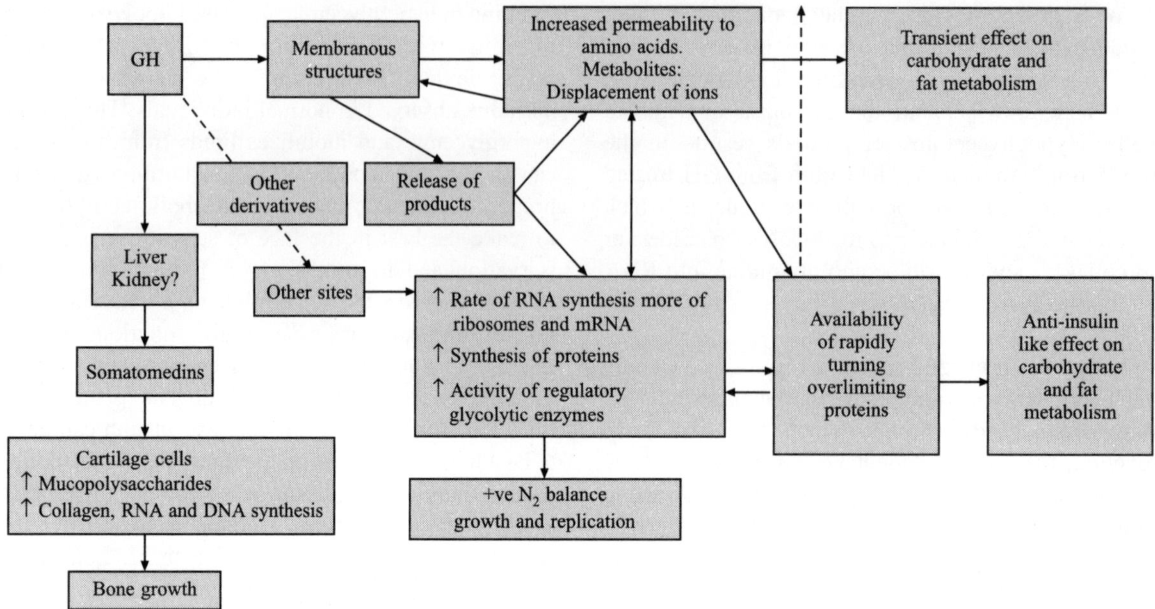

Figure 83.5   Schematic representation of possible mechanism of action of pituitary growth hormone.

V.   GROWTH FACTORS

A.   Growth Factors in Blood

(a) Insulin

Insulin can stimulate the growth of a variety of cells, *in vitro*. The concentration of insulin required for this purpose is however very large as compared to its normal physiological concentration in blood. Any direct role of insulin as a growth factor, *in vivo,* therefore appears to be unlikely. Another insulin-like growth promoting activity is present in normal serum which is different from insulin since it is not suppressed by anti-insulin antibodies. This activity was therefore named as *non-suppressible insulin-like activity* or *NSILA* in short. NSILA has been purified from human serum and has been shown to be a polypeptide, containing 70 amino acid residues. NSILA is 50 to 100 times more efficient as compared to insulin in stimulating DNA synthesis in embryonic chick fibroblasts.

(b) Somatomedins

Somatomedins are a set of peptides isolated from serum which are thought to mediate the action of growth hormone on cartilage and bones. These peptides are generated in the body (probably in the liver and kidneys) in response to growth hormone. Absence of growth hormone in circulation as a consequence of hypophysectomy results in the depletion of somatomedins in serum. Acromegaly patients on the other hand have elevated serum somatomedin levels. It has been proposed that stunted growth in spite of normal serum levels of growth hormone, as seen in African pygmies and in *Laron syndrome,* may be due to a defect at the level of production of somatomedins in response to growth hormone.

Though all somatomedin peptides have a molecular weight of approximately 7000, they have been named as somatomedins A, B and C depending upon whether the peptides are neutral, acidic or basic in nature respectively. Somatomedins have profound effect on the metabolic activities of cartilage cells, *in vitro*. Paradoxically, however, *in vivo* administration of these peptides does not elicit significant effect on cartilage and bone growth.

(c) Erythropoietin

Erythropoietin is a growth factor in serum which stimulates the generation of erythrocytes. Presence of such an activity in serum was established by experiments utilizing parabiotic rats (two rats having common blood circulation obtained by experimentally joining their blood vessels by thin tubes). If one rat was subjected to hypoxia, elevated erythropoietic activity was observed in the bone marrow of the other rat. Since hypoxia or low oxygen tension is known to stimulate erythropoietic activity, it was apparent that the animal under hypoxic stimulus secreted some erythropoiesis stimulating substance which influenced the other rat through common circulation. The substance was named erythropoietin.

Erythropoietin from human serum appears to be a glycoprotein with a molecular weight of 23,000. It acts on the proerythroblast cells and induces their division and differentiation to erythrocytes.

B.   Growth Factors from Organs

(a) Nerve Growth Factor (NGF)

NGF has been isolated from the submaxillary glands of adult rats. Another rich source of NGF is cobra venom. This factor exercises a trophic influence on the neuronal cells both *in vitro* and *in vivo*. Administration of anti NGF antibodies in chick embryos results in a virtual derangement of the sympathetic and sensory nervous system. NGF is a polypeptide having 118 amino acids residues. Structurally, NGF shows marked resemblance to proinsulin, the precursor molecule of insulin.

(b) Epidermal Growth Factor (EGF)

EGF is another polypeptide (mol. wt. 6045) isolated from mouse submaxillary glands. Administration of EGF to newborn mice results in early eye opening in these animals. EGF has a potent mitogenic effect not only on the cultured epithelial cells but also on various other types of cells like fibroblasts. The significance of the presence of high concentrations of NGF and EGF in submaxillary glands is not understood. Removal of submaxillary glands does not lead to a deficiency of these growth factors in circulation.

(c) Fibroblast Growth Factor (FGF)

FGF is a basic polypeptide (mol. wt. 13,400) and has been isolated from bovine pituitaries. The brain has a higher concentration of this factor. FGF may be responsible for the neurotrophic effect of nerves which results in the regeneration of a limb after amputation in chicken embryos. FGF stimulates markedly the proliferation of fibroblast cells, *in vitro*. An *Ovarian Growth Factor,* different from gonadotropic hormones, which has also been isolated from bovine pituitaries, has a mitogenic effect on the cultured ovarian cells.

C. Mechanism of Action of Growth Factors

Receptors for various growth factors have been shown to be present in target cells. These receptors may be linked to the adenyl cyclase and phosphodiesterase enzyme system. Most growth factors rapidly cause a depression of cyclic AMP levels inside the cells, which may act as a primary signal for cell division. Depression in cyclic AMP levels is followed by an increased transport of nutrients, increased RNA synthesis, aggregation of ribosomes to polysomes, increase in protein synthesis, DNA synthesis, leading finally to cell division. This response of a target cell to a growth factor has been termed as a *pleiotypic response.* It should be noted that a lowering of intra-cellular levels of cyclic AMP results in a pleiotypic response. An increase in intracellular levels of cyclic AMP generally leads to an inhibition of proliferative activity which, in many cases, is conducive to cell differentiation and maturation.

D. Physiological Role of Growth Factors

Growth factors have a relevance in the development of multicellular organisms. A human being consisting of about 100 million million cells develops from a single fertilized ovum. Cells have, therefore, to proliferate and differentiate in order to perform the specialized functions essential for life. An array of growth factors is involved in the regulation of the process of growth, and a complex interaction of growth promoting and differentiation inducing signals results, ultimately, in the development of different types of tissues. In adult life, growth factors may be involved in repair processes like wound healing, and also in maintaining the normal turnover of various cell types.

Growth hormone is of course the single most important growth factor, *in vivo.* Administration of growth hormone to hypophysectomized animals completely restores their growth. *In vitro,* however, growth hormone has only meagre, if any, effect on the proliferation of cells. It is not unlikely that the growth promoting action of growth hormone, *in vivo,* is mediated through various growth factors like somatomedins, etc.

VI. ACTH (ADRENOCORTICOTROPHIC HORMONE)

A. (a) Chemical structure

39 amino acids comprise this polypeptide hormone. Molecular weight is 4500. It is now possible to synthesise this hormone chemically. The first 24 amino acids with complete N terminal are necessary for full biological activity, sequence 25–30 of amino acids are determinants of immuno-reactivity and are responsible for the species differences.

(b) Factors influencing release of ACTH

Corticotropin releasing factor (CRF) is present in the ventral nuclei of hypothalamus. CRF regulates the secretion of ACTH. It is sensitive to stress stimuli as pyrogens and hypoglycaemia. Vasopressin has a marked CRF action; however, it is *per se* not the physiological mediator of ACTH secretion.

(c) Physiological functions

ACTH stimulates synthesis and release of adrenocortical steroids except aldosterone which is affected only to a minor degree. The process of steroidogenesis is dependent on numerous enzymatic steps influenced by cyclic AMP and phosphorylase. Glycogenolysis, synthesis of proteins and depletion of cholesterol and ascorbic acid are involved in the production of the hormone. ACTH has also extra-adrenal effects in animals, i.e., lipolytic action, both *in vitro* and *in vivo*—this effect is not so far clearly proven in man.

B. Methods of Assay

Earlier bioassay techniques using hypophysectomized rats were employed for determination of increase in steroid output in the adrenal vein for quantation of ACTH values. Adrenal ascorbic acid depletion test was also considered to be reliable and specific bioassay method for ACTH.

Of late immunoassays employing double antibody for separation or charcoal or talc have increased sensitivity and precision in the results obtained. ACTH values in normal fasting healthy individual are 10–60 pg/ml (Feller-Aubert, 1971). Biological half-life is about 5–15 min. A circadian rhythm has been observed; values are maximal in the early morning and lowest in late evening, spectrum is similar to that observed in case of cortisol.

C. ACTH and Related Clinical Disorders

In differential diagnosis of adrenal hyperfunctioning or hypofunctioning states, determination of ACTH can be of considerable help. Table 83.2 summarises the present knowledge on this subject.

Clinical abnormalities of ACTH secretion: Basophilic adenoma can result in clinical picture of Cushing's syndrome. This is typically akin to state seen with elevated plasma cortisol, skin pigmentation is seen in some of these instances and related to high level of beta MSH.

Deficiency of ACTH may be the result of panhypopituitarism or rarely be due to its isolated deficiency. Patients present with weakness, hypoglycaemia or weight loss. Changes in relation to lack of minor corticoid deficiency or androgen lack are less predominant than usually met with when similar clinical condition is due to primary adrenal gland deficiency (Addison's disease).

VII. LUTEINIZING HORMONE (LH), FOLLICLE STIMULATING HORMONE (FSH), AND THYROID STIMULATING HORMONE (TSH)

(a) Chemistry

All the three hormones have been isolated from the anterior pituitary of a variety of species, in relatively pure form, free of contamination from one another. The physicochemical properties of these glycoproteins are listed in Table 83.3, while FSH of all species so far tested has sialic acid, not all LH and TSH preparations appear to have significant amounts of sialic acid. All the three glycoproteins contain other carbohydrate residues like hexosamines and free hexoses. On removal of sialic acid from FSH, it loses its biological activity as tested, *in vivo*. *In vitro* incubation of asialo FSH, however, shows that the biological activity is still retained, suggesting that sialic acid may be needed for effective transportation of the hormone in blood to the target tissue. Glycosylated hormones have longer half life in circulation. While both LH and TSH are easily denatured by treatment with urea and guanidine hydrochloride, FSH appears not to lose activity when incubated with 6M urea or 4M guanidine HCl. Further, LH appears to be much more susceptible to the action of proteolytic enzymes like chymotrypsin and trypsin than FSH. The above two methods have been used by some investigators for cleansing the FSH molecule of contaminating LH activity.

Obtaining FSH and TSH preparations free of LH contamination has for a considerable period been a major problem. The available LH preparations, however, have been relatively free of FSH and TSH contamination. Recent evidence has shown that *each* of the three glycoprotein hormones are made up of two dissimilar subunits—α and β, each of approximately 15,000 in molecular weight. The α *subunit* within a species appears *to be common to all the three glycoprotein hormones*, the β *subunit apparently being responsible for expression of biological activity*. The β subunit in itself, however, is not biologically active; recombination of α of anyone of the three and β subunits to regain the original activity is possible, the β subunit used, determining the kind of activity regained. In other words, while the β sub-unit of LH can combine with the α subunit of FSH or TSH to give LH activity, the subunit of FSH or TSH has to combine with the α sub-unit of any of the three hormones to reproduce FSH

TABLE 83.2   Endocrine Profile in Adrenal Functions

			Plasma ACTH	*Plasma cortisol*	*Remarks*
1.	A.	Adrenal hyperfunction			
		(1) Bilateral adrenal hyperplasia	Normal	Raised	Plasma cortisol responds to dexamethasone suppression
		(2) Adrenal tumors	Undetectable, less than 10 pg/ml	Raised	
		(3) Ectopic ACTH (as from non-endocrine tumors)	Raised more than 200 pg/ml	Normal	
	B.	Pituitary (ACTH) hyperfunction 'Nelson' syndrome	Very high, more than 900 pg/ml	Normal or low	
2.		Adrenal hypofunction			
	A.	Primary			
		(1) Addison's disease	Raised more than 350 pg/ml	Low	After hydrocortisone, immunoreactive ACTH disappears, circadian rhythm stays intact.
	B.	Secondary			
		(1) Hypothalamic or pituitary	Normal or low	Low	
		(2) Long-term corticoid therapy	Undetectable	Normal	

TABLE 83.3   Physicochemical Properties of Glycoprotein Hormones

Hormone	*Mol. wt.*	*pI*	*No. of sub-units*	*Carbohydrate content** (%)*				
				Total %	*Hexose**	*Hexosamine**	*Fucose**	*Sialic* acid*
Luteinizing *Hormone* (LH)								
Human	28000	5.4	2	12	10	11.1	1.7	–
Ovine	30000	7.3	2	15	12	10.7	2.9	–
Follicle Stimulating Hormone (FSH)								
Human	35000–41000	3.0–5.6	2	26	39	30.0	–	8
Ovine	25000–30000	4.5	2	14	5.7	4.5	1.1	2.8
Human Menopausal Gonadotropin								
LH	51000	–	–	17	8.9	6.5	–	1.8
FSH	28000	–	–	25	14–18	8–17	–	2–9
Human Chorionic Gonadotropin (HCG)	27000–48000	2.9	2	31	18	14	1	8
Pregnant Mare Serum Gonadotropin (PMSG)	28000	1.8	–	45	18.6	14.1	1.4	10.4
Thyroid Stimulating Hormone (TSH) Human	25000–28000	–	2	15	9.4	13	1.3	1.7

*Number of residues per molecule.

**Values are approximate. Variation is observed.

and TSH activities respectively. The dissociation to subunits can be brought about at an acid pH (e.g., in propionic acid solution), recombination being facilitated by incubating in an alkaline medium. The reader is referred here to a few excellent reviews on the subject.

(b) Biology and immunology

In keeping with the names suggested, the three glycoprotein hormones stimulate their respective target tissues in the following manner:

(i) FSH—stimulates the growth of the ovarian follicle in the female and apparently the Sertoli cells of the testis in the male.

(ii) LH or ICSH—contributes to the growth of the ovarian follicle responsible for ovulation and differentiation of granulosa to luteal cells; regulates corpus luteum function; and stimulates interstitial cell growth and function in both the ovary and testis.

(iii) TSH—stimulates the growth of the thyroid follicle, and regulates the synthesis and secretion of thyroglobulin into the intrafollicular colloid.

Removal of the pituitary leads to the involution of the gonads and thyroid (in addition to the adrenal cortex and others), supplementation with purified LH, FSH or TSH leading to regeneration of the specific target tissue. LH and FSH in many instances have been shown to act synergestically, for instance the right proportion of the two being needed for the full development of the ovarian follicle.

The circulating level of these hormones is normally low, the plasma levels of the target tissue hormones regulating, via the hypothalamus and also the pituitary, the secretion of the tropic hormones from the pituitary. Following castration or after menopause has set in women, the secretion of the gonadotropin hormones from the pituitary is vastly increased and the excess finds its way into the urine. The urine of menopausal women has therefore been used as a rich source of gonadotropins, and fractions (human menopausal gonadotropin—HMG) rich in FSH and LH activity have been isolated and used for therapeutic purposes. The urinary gonadotropin while retaining the original biological activity appears to be smaller in size when compared to the native hormone, perhaps due to metabolic changes the native hormone undergoes in the kidney and the liver.

The secretion of TSH is also influenced by the status of the thyroid, hypothyroidism or thyroidectomy leading to increased secretion of pituitary TSH.

High levels of blood thyroxine would lead to reduction in TSH output from the pituitary; excess production of TSH from the pituitary contributes to the establishment of hyperthyroidism and eventually goiter.

Considerable information is available on the immunology of gonadotropins. These make excellent antigens and the antibodies produced against gonadotropin of a species cross-react with the same hormone of other species, though the degree of cross-reactivity may vary from species to species. Well characterized LH and FSH antisera have been used in addition to radioimmunoassay, to specifically neutralize endogenous gonadotropin activity. The latter has been extremely useful in establishing the essential role of LH in induction of ovulation and regulation of corpus luteum function.

Immunochemical techniques have also been used in detecting the homology that exists between the α-subunit of LH, FSH and TSH of a species and the β-subunit of a single hormone of different species.

(c) Biochemical action

All the three hormones increase cyclic AMP output in their target tissues. This is apparently done by enhancing the activity of the membrane-bound enzyme, adenyl cyclase. Evidence available hitherto strongly suggests that the hormones bind to specific receptors located in the membranes of the cells of the target tissue. An increase in nucleic acid and protein synthesis in the target tissue is noticed following stimulation with the specific hormone. These increases are apparently fore-runners of expression of hormonal activity, as the use of inhibitors of nucleic acid and protein synthesis have been shown to abolish the physiological activity associated with the hormone.

While no unequivocal proof is available on the ability of FSH to stimulate steroidogenesis, LH has been clearly shown to stimulate the synthesis of both male and female sex steroids. LH is known to accelerate cholesterol turnover, probably by activating the enzyme cholesterol esterase; it also enhances the conversion of cholesterol to pregnenolone and progesterone. LH appears to make available enough reducing equivalents (NADPH) essential for the latter conversion. Whether LH achieves this (enhancement in steroidogenesis) by acting at a single locus or has multiple points of attack is not clearly known. In addition to cyclic AMP, prostaglandins have been suggested as possible mediators of LH action.

While the role of FSH in mammals appears to be confined to the stimulation of maturation and growth of ovarian follicles, Sertoli cells of the testis and spermatogenesis (from spermatogonium to spermatid stage), in the lower order of animals (e.g., reptiles) it appears also to be capable of stimulating steroidogenesis. Recently, FSH has been shown under specific conditions to promote estrogen synthesis by stimulating conversion of androgen to estrogen.

TSH which is known to stimulate cyclic AMP production in the thyroid follicular cell has been shown to enhance all the processes connected with production of thyroid hormone, namely iodide trapping, organification of iodide, synthesis of iodothyronines and the release of metabolically active thyroid hormone (T_4 and T_3) from thyroglobulin into the blood stream, etc. Since all these individual processes have been shown to be capable of stimulation by cyclic AMP, a strong suggestion is made that the principal action of TSH is one of enhancing adenyl cyclase activity which, in turn, increases intracellular levels of cyclic AMP.

(d) Gonadotropins of the placenta

Some of the female mammals (e.g., the primate and the equine) during the early phase of pregnancy secrete, from the trophoblastic cells of the maternal placenta (chorionic tissue in the primate and endometrial cups in the equine) gonadotropic hormones which have a high biological activity. HCG or MCG (human or monkey chorionic gonadotropin) and PMSG (pregnant mare serum gonadotropin) are glycoproteins of an approximate molecular weight of 30,000. Their carbohydrate content, in particular their sialic acid content, is very much higher than that of the pituitary gonadotropins. While HCG and MCG have pre-dominantly LH-like activity, PMSG has exhibited a high degree of FSH activity. HCG has been shown to have a longer biological half-life than pituitary LH or FSH. HCG has also been shown to consist of two subunits; considerable amount of homology has been demonstrated between the β-subunit of HCG and the β-subunit of human LH.

Both HCG and PMSG cease to exhibit their *in vivo* activity if they are denuded of their sialic acid. The asialo HCG has been tested in *in vitro* systems and shown to retain its biological activities. The principal function of these placental gonadotropins appears to be one of stimulating the corpus luteum of pregnancy to secrete the much needed progesterone in relatively large quantities.

Abbreviations

TSH	Thyroid Stimulating Hormone
ACTH	Adrenocorticotropic Hormone
LH	Luteinizing Hormone also called Interstitial Cell Stimulating Hormone (ICSH)
FSH	Follicle Stimulating Hormone.

SUGGESTIONS FOR FURTHER READING

Apter D. (1997), Development of the Hypothalamic-pituitary-ovarian Axis, *Ann. N.Y. Acad. Sci.,* 816, 9–21.

Daughaday W.H. (1971), Regulation of Skeletal Growth by Sulfation Factor, *Adv. Int. Medicine,* 17, 237–263.

Gospodarowicz D. and Moran J.S. (1976), Growth Factors in Mammalian Cell Culture, *Annual Review of Biochemistry,* Vol. 45, 531–558.

Hall Kerstin and Luft Rolf (1974), Growth Hormone and Somatomedin, *Adv. Metabolic Disorders,* 7, 1–36.

Knobil E. and Hotchkiss J. (1964), Growth Hormone: A Review, *Annual Review of Physiology,* 26, 47–74.

Korner A. (1965), Growth Hormone Control of Biosynthesis of Protein and Ribonucleic Acid, *Recent Progr. Horm. Res.,* 21, 205–240.

Li C.H. (1973), Human Growth Hormone: Perspectives on Its Chemistry and Biology, *Adv. Human Growth Horm. Res.,* 321–348.

Li C.H. (1982), *Hormone, Protein and Peptides,* Academic Press, New York.

Luft R. and Hall K. (Eds.) (1975), Somatomedins and Some Other Growth Factors, *Advances in Metabolic Disorders,* Vol. 8.

Moudgal N.R. (Ed.) (1974), *Gonadotropins and Gonadal Function,* Academic Press, New York.

Mukherjee M., Kochu Pillai N. and Talwar G.P. (1971), Growth Hormone: A Review, *J. Sci. & Industr. Res.,* 30, 473–479.

O'Malley B.W. and Birnbaumer L. (1978), *Hormone Receptors, Receptors and Hormone Action,* Academic Press, New York.

Talwar G.P., Pandian M.R., Kumar Nirbhay, Hanjan S.N.S., Saxena R.K., Krishnaraj R. and Gupta, S.L. (1975), Mechanism of Action of Pituitary Growth Hormone, *Recent Prog. Horm. Res.,* 31, 141–174.

Vaitukitis J.L., Ross G.T., Braunstein G.D. and Rayford P.L. (1976), Gonadotropins and Their Subunits—Basic and Chemical Studies, *Recent Progress in Hormone Research,* 32, 289.

Von Werder K. (1996), *Pituitary Adenomas: From Basic Research to Diagnosis and Therapy,* Elsevier Science, New York.

Wolstenholme G.E.W., Knight J.J. and Churchill A. (Eds.) Ltd. London (1970), *Hormones and Immune Response,* Ciba Foundation, Study Group No. 36.

Williams R.H. and Wilson J.D. (1998), Williams *Textbook of Endocrinology,* 9th ed., W.B. Saunders Company, Philadelphia.

84

The Pineal Gland

Ajay Kumar

CONTENTS

I. INTRODUCTION

The pineal complex or epiphysis cerebri of vertebrates has been an enigma since the time of Galen. It is even mentioned as the 'seat of the soul' in the philosophical discourses of Rene Descartes. In the first half of this century, speculation was rife with regard to the function of the pineal, which ranged from a vestigial organ to an endocrine gland. In the last two decades, scientists in diverse disciplines have made noteworthy discoveries regarding the function of the pineal. Some of the important landmarks which have contributed appreciably towards the understanding of the pineal physiology are the chemical isolation of a unique compound, melatonin, from the bovine pineal glands, by Lerner and his co-workers in 1958, the demonstration in the pineal of an exclusive enzymatic system involved in melatonin synthesis by Axelrod and his co-workers in 1961, and the discovery of the regulation of pineal activity by environmental lighting, independently by Quay, and Wurtman and co-workers in 1963.

II. ANATOMY OF THE PINEAL

A. Embryology, Morphology and Organization of Pineal Components

The pineal complex lies dorsally in the diencephalic region of the vertebrate brain. It arises along the midline between the developing habenular commissure anteriorly, and the posterior commissure and subcommissural organ posteriorly. Two primordia constitute the embryonic beginnings of the pineal, each derived from the lateral edge of the embryonic anterior neural plate. The two primordia fuse to form a single pineal after the formation of the neural tube. Experimentally, either of the primordia can be induced to develop into a normal pineal organ without the other, suggesting that the early progenitors of vertebrates had paired pineal organs.

The fully developed pineal complex is composed of the vesicular parietal organ (parapineal) and the stalked, sometimes glandular epiphysis also known as the pineal. The paraphysis, often erroneously included as a pineal component, is in fact a telencephalic circumventricular organ not related to pineal structure or function. The dorsal sac present near the pineal complex becomes associated with the choroid plexus in higher vertebrates.

In the most primitive living vertebrate, the lamprey, the pineal system is very extensive. Both the epiphysis and the parapineal organs are situated within the cranium. The epiphysis is located at the end of a long connecting stalk and it overlies the parapineal organ. The innervation to the epiphysis and parapineal originates mainly from the posterior commissure—subcommissural organ and habenular commissure and nuclei, respectively. The elaborate pineal system of reptiles has innervation similar to that of the lamprey. But in the frog, the nerve from the frontal organ penetrates the epiphysis and intermingles with the fibres of the organ. The habenular fibres reach the anterior part of the epiphysis and thus that region of epiphysis is the more likely homologue of the lizard parietal eye than the frontal organ. The pineal organ of birds and mammals is a single solid mass without any vestige of a lumen in the adult. In humans, the epiphysis is slightly larger than a pea and weighs 100–180 mg. It is wedged deeply between the cerebral hemispheres and lies between the superior colliculi and below the posterior border of the corpus callosum. The pineal is very small and rudimentary in owls, procavians, elephants, opossums, and whales. It is absent in dugong, crocodiles, alligators, armadillos, sloths, and anteaters.

1. Vascularization: The pineal in lower vertebrates is supplied by capillary networks and loops arising from the pial membrane and surrounding the periphery of the organ. The pineal is richly vascularized as it is closely associated with the system that drains the choroid plexus, the dorsal sac, the paraphysis, and the subcommissural organ. The pineal of the pigeon is supplied by the posterior meningeal artery which arises from the posterior cerebral artery. In mammals, the capillary network goes deeply through the parenchyma of the pineal organs. Neither in the lower vertebrates nor in the higher vertebrates is there any evidence for an epithalamo-epiphyseal portal circulation.

2. Innervation: In lower vertebrates, the pineal innervation is associated with the habenular commissure and nuclei, the posterior commissure and, perhaps, the subcommissural organ. However, to date, the autonomic neural component has not been demonstrated in the pineal organs of lower vertebrates. In contrast to this, the innervation of the pineal in mammals is exclusively from the postganglionic sympathetic fibres from the superior cervical ganglia which enter the pineal via paired nervi conarii from the tentorium cerebelli and intermingle with the parenchyma. In birds, sympathetic innervation to the pineal is provided by projections from the superior cervical ganglia.

B. Cytology of Saccular Pineal System of Lower Vertebrates

A photoreceptive function was attributed to the pineal by anatomists of the nineteenth century on the basis of the description of the pineal eye of the reptile, *Sphenodon*, commonly known as Tuatara or 'the living fossil'. The arrangement of the pineal structure in this animal is reminiscent of a typical vertebrate eye. A well developed parietal eye is also present in lizards and in some fishes. Three types of cells have been recognized in the pineal organs of fish, amphibians, and reptiles. They are: (a) photoreceptor or sensory cells, (b) supporting cells, and (c) ganglion cells. The light and electron microscopic studies indicate that the sensory cells display many cytological features similar to those of retinal rods and cones in the arrangement of their inner and outer segments. The outer segment may undergo a continuous cycle of degeneration and replacement. The strongest argument for light perception is furnished by electrophysiological evidence that the pineal organs in fish and frog are capable of producing neural responses to darkness and to photic stimuli of various wavelengths. Thus, in lower vertebrates the basic functional unit of the pineal system is the photoreceptor concerned with the transmission of photic sensory information.

C. Cytology of Parenchymal Pineal System of Birds and Mammals

The pineal of mammals is composed of the characteristic pinealocytes, pineocytes, or the pineal parenchymal cells, which are arranged in circular clusters or lobules. Each pineocyte (Figure 84.1) gives off numerous processes from its cell body which permeate the pineal stroma and end in relation with small blood vessels. In addition, glial cells are also present. Birds have, besides pineocytes and glial cells, large cells associated with lamellar bodies, which are generally considered to be vestigial photoreceptors. As mentioned earlier, the main neural components are the efferent autonomic (adrenergic) fibres which originate from the superior cervical ganglia and invade the pineal tissue via two tracts and terminate directly among the parenchymal cells. The innervation of the pineal is thus analogous to that of the adrenal medulla.

III. BIOCHEMISTRY AND PHARMACOLOGY OF THE PINEAL

The biochemically active mammalian pineal gland is rich in biogenic amines such as serotonin, noradrenaline, dopamine and histamine. The most significant biochemical discovery made on the pineal gland was the isolation of a unique methoxylated indole compound, melatonin, by Lerner and his collaborators in 1958. This compound isolated from bovine pineal glands blanches the skin of frogs by bringing about concentration of

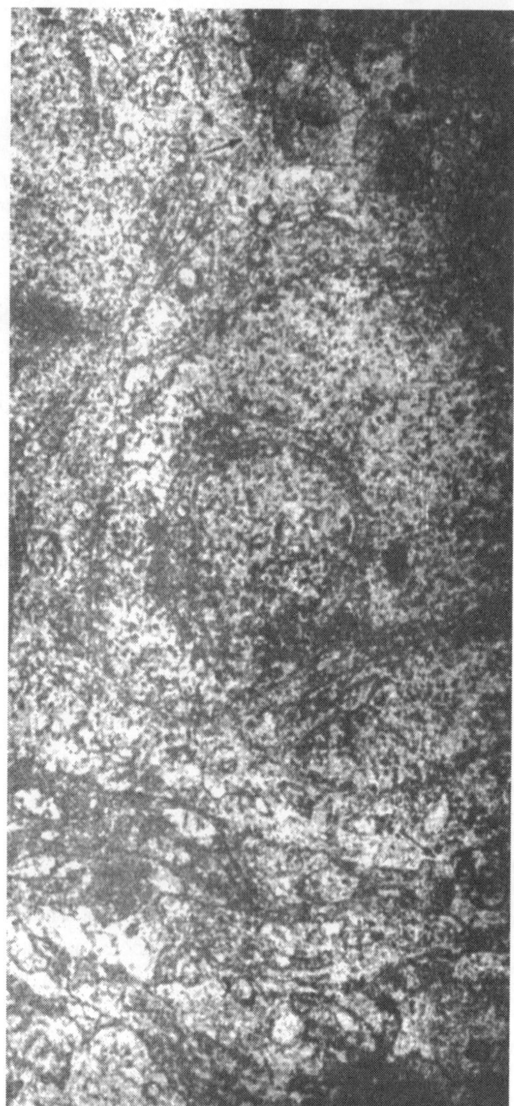

Figure 84.1 An electron micrograph of the pineal parenchyma of the epiphysis of a squirrel monkey showing the main body of a pineocyte in the centre, its spatial relationship with other pineocytes (P), and the cell processes within the parenchyma. The nucleus of the central pineocyte is indented, and large and small processes (arrows) extend from the main cell body. Top arrow shows microtubules. Astrocyte processes (A) are packed with filaments. Unidentified dense inclusions (D) are also seen occasionally. Glutaraldehyde fixation, lead and uranyl acetate stained. × 8,150 (from Wurtman et al. (1968), *The Pineal,* Academic Press).

pigment granules in the melanophores. It is effective even at a concentration of a trillionth of a gram per ml of medium.

Melatonin has a wide distribution and is present in the pineal of mammals, birds, reptiles, amphibians and fish. It is also present in the eyes of lower vertebrates and in the peripheral nerves of mammals. But the peripheral nerves lack the enzyme system required to synthesize melatonin.

Melatonin (5-methoxy-*N*-acetyltryptamine) is produced in the mammalian pineal by *O*-methylation of *N*-acetylserotonin and this reaction is catalyzed by the enzyme, hydroxy-indole-*O*-methyl transferase (HIOMT). This enzyme is present in the pineals of all vertebrate species. It is also found in the brain of amphibians, as well as in the eyes of fish, amphibians, reptiles (except snakes) and birds. *N*-acetylserotonin is formed by *N*-acetylation of serotonin (5-hydroxy-tryptamine) which is present in the mammalian pineal gland (Figure 84.2). The circulating melatonin is rapidly metabolized in the liver to 6-hydroxymelatonin.

Noradrenaline is present in the pineal sympathetic nerve endings and is either synthesized in the nerves or taken up from circulation. The sympathetic nerves of the pineal also contain serotonin and this is also synthesized in the pineal parenchymal cells from the amino acid, tryptophan. The concentrations of both noradrenaline and serotonin vary in the rat pineal depending upon the time of the day or night. Many drugs modify the fate of noradrenaline, serotonin and melatonin in the body.

Melatonin is proposed as a pineal hormone because (1) HIOMT enzyme is uniquely localized in the pineal of mammals, (2) melatonin is selectively taken up by a few tissues, and (3) melatonin produces a number of physiological effects when administered to animals.

IV. PHOTIC REGULATION OF PINEAL FUNCTION

Day-length (photoperiod) influences the endocrine glands through the photoreceptors and the nervous system. The principal photoreceptor is the eye from where the impulses are conveyed to special areas of the brain which act as neuroendocrine transducers. The hypothalamus-median eminence area is one such transducer. The hypothalamic hormones secreted by neurosecretory neurons are directly released into the hypothalamo-hypophyseal portal circulation and eventually reach the anterior pituitary where they activate or inhibit the pituitary cells to produce the tropic hormones (Chapters 59 and 60). The pineal gland is also considered as a similar neuroendocrine transducer which possibly converts the environmental lighting input to an endocrine output in the form of melatonin hormone.

A. Effect of Light on the Pineal

It is generally agreed that environmental lighting is a primary regulator of pineal activity. The weight of the pineal gland is influenced by environmental lighting. The pineal is inhibited by light and stimulated or released from inhibition in darkness, inasmuch as exposure of rats to continuous light results in a significant decrease in pineal weight. Also, the pineal weight varies with a daily rhythm; it is lowest at the end of the daily photoperiod. Work on pineal cytology has revealed that the pineal parenchymal cells are considerably more active during the dark period than during the lighted period.

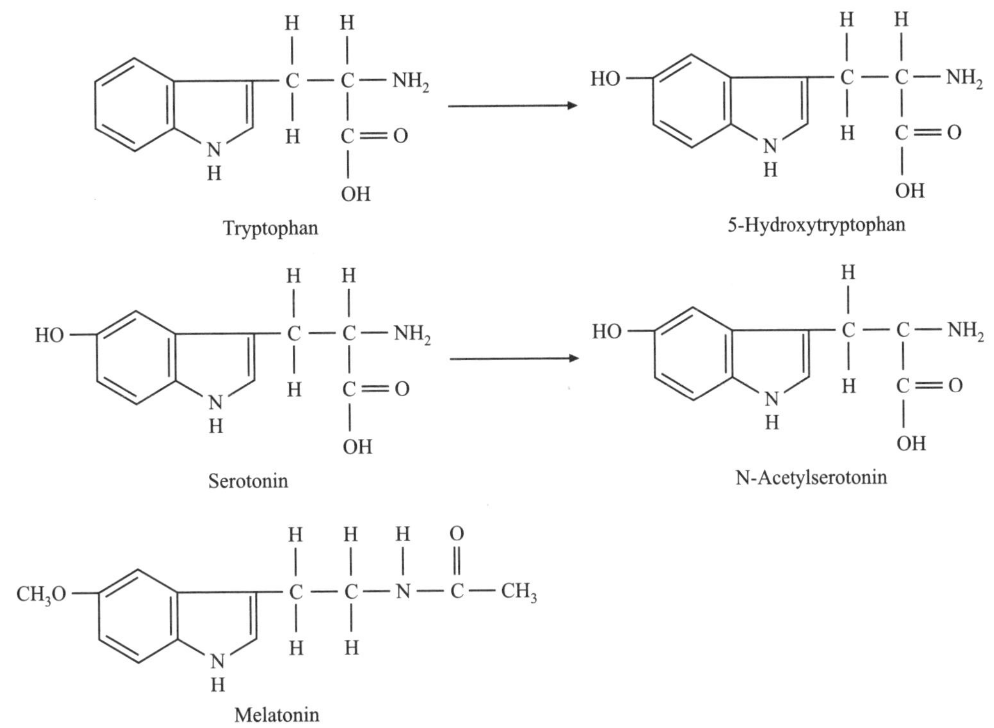

Tryptophan

5-Hydroxytryptophan

Serotonin

N-Acetylserotonin

Melatonin

Figure 84.2 Biosynthesis of melatonin from tryptophan in the pineal gland. Step 1 is catalyzed by tryptophan hydroxylase; Step 2 by L-aromatic acid decarboxylase; Step 3 by N-acetylating enzyme; and Step 4 by hydroxyindole-*O*-methyl transferase. (From Wurtman et al. (1968), *The Pineal*, Academic Press.)

B.   Pineal Mediated Effects of Light

Polyestrus species such as the rat show constant vaginal estrus on exposure to continuous illumination. This is significantly reduced by treating rats with bovine pineal extract. Injection of pineal extracts into rats brings about depression of gonadal weights and function. Melatonin reduces the incidence of light-induced estrus in rats, whereas other pineal substances such as serotonin and histamine have no such effect. Thus, it seems clear that the pineal gland contains a factor (melatonin) which inhibits gonadal function in the rat.

C.   Light and Melatonin Synthesis

Melatonin synthesis involves the rate-limiting step of *O*-methylation of *N*-acetylserotonin which is catalyzed by the enzyme HIOMT. Continuous light induces a marked decline in the activity of HIOMT and thereby a significant reduction in the formation of melatonin. There is general agreement that the pineal gland gets the information on environmental lighting through its sympathetic nerves originating from the superior cervical ganglia. Transection of the pineal nerve supply makes HIOMT activity independent of environmental lighting. Very little is known about the location of the central nervous tracts which transmit photic information to the brain centres regulating the sympathetic nervous system. The proposed neural connections between the eyes and pineal are diagrammatically shown in Figure 84.3. Thus, the pineal is considered as a neuroendocrine transducer where the sympathetic nervous input regulated by light, is converted into a secretory (melatonin) output.

D.   Pineal and Biological Rhythms

Most animals are exposed to 24 h light-dark cycles of the day and night as well as to the annual day-length (photoperiod) cycle. These cycles influence and regulate many rhythmic changes in diverse systems of organisms (Chapter 64).

The daily (circadian, or about a day) changes in the pineal HIOMT activity and noradrenaline content are totally dependent upon environmental lighting, whereas pineal serotonin rhythm persists even in the absence of photoperiodic information. The rate of melatonin synthesis, as revealed by HIOMT activity, in rats exposed to 12 h light and 12 h dark cycles shows a circadian rhythm. HIOMT activity is maximum at midnight and minimum at 6 p.m. (Figure 84.4). Rats kept in constant darkness have considerably more HIOMT activity at all times than those exposed to constant light although the expected rhythmic variations are absent in both groups. Further, no rhythm is observed in HIOMT activity in rats subjected to bilateral superior cervical ganglionectomy. Thus, the circadian rhythm in pineal HIOMT activity is dependent on environmental lighting (Figure 84.5).

In birds, a cycle of melatonin content of the pineal similar to that seen in mammals has been reported. Even in a blinded chicken, diurnal periodicity of pineal melatonin content, indistinguishable from that of intact ones, has been demonstrated. Further, after bilateral superior cervical ganglionectomy, which

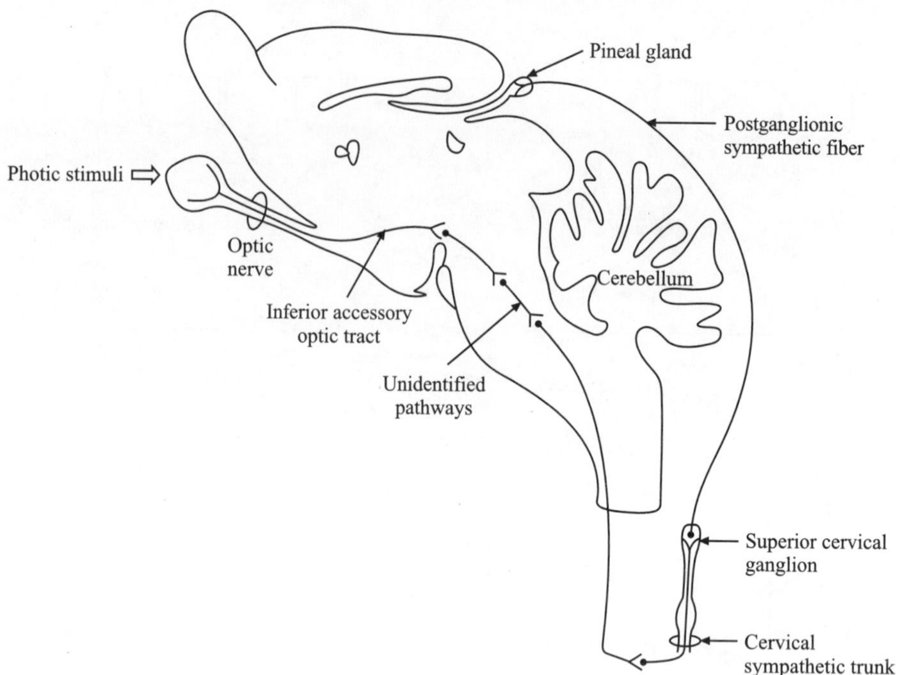

Figure 84.3   Schematic diagram of neural pathways between the eyes and the pineal gland (From Reiter and Fraschini (1969), *Neuroendocrinology, 5,* 219).

denervates the pineal, the pineal melatonin rhythm does not change significantly.

Pineal serotonin content also varies in a circadian rhythmic fashion in rats subjected to 14 h light and 10 h darkness cycles. It is highest around noon and lowest around midnight and this rhythm persists even in continuous darkness or after blinding. But the pineal serotonin rhythm is abolished when rats are placed in continuous light or when the sympathetic nerves to the pineal are cut (Figure 84.6). More information will be necessary to understand the factors producing the serotonin rhythm.

The noradrenaline content of the pineal also shows a circadian rhythm: it is lowest at the end of the daylight period (7 p.m.) and highest at 7 a.m. This rhythm is abolished in blinded rats. Thus, the pineal noradrenaline rhythm is almost similar to that of HIOMT activity and indicates that melatonin synthesis is influenced by the sympathetic neuro-transmitter, the norepinephrine.

V. PHYSIOLOGY OF THE PINEAL GLAND

A. Role of the Pineal Complex in Lower Vertebrates

Mammalian pineal extract or melatonin causes concentration of pigment granules in melanophores in the skin of anuran tadpoles—an effect opposite to that of the hypophyseal melanocyte-stimulating hormone. Both intact and eyeless larval amphibians blanch when placed in darkness and this response is abolished by pinealectomy. Experimental evidence suggests that the pineal may be involved in background adaptation and

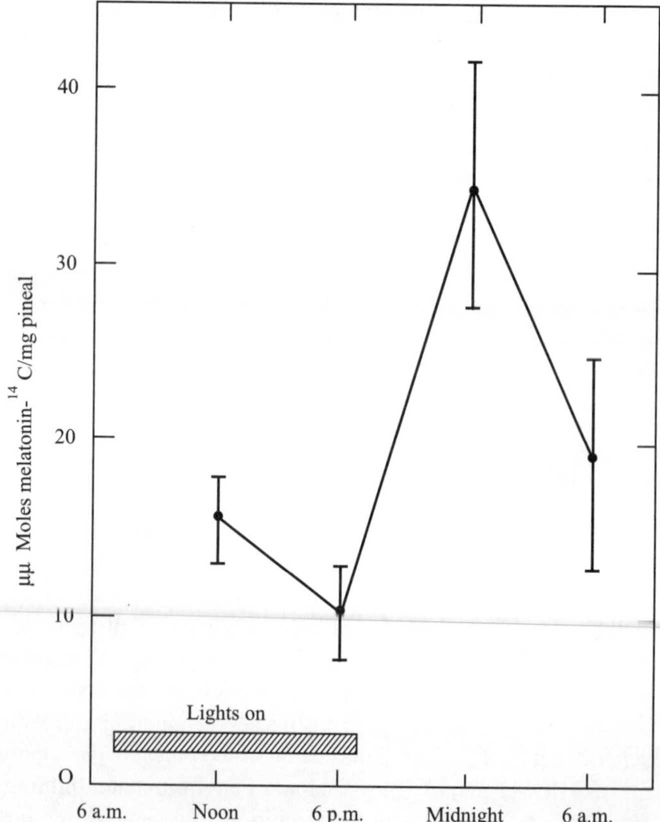

Figure 84.4   Circadian rhythmic changes in the pineal HIOMT activity expressed as micromicromoles of melatonin formed per mg of pineal per hour. Vertical lines represent standard errors of the mean. (From Axelrod et al. (1965), *J. Biol. Chem., 240,* 949.)

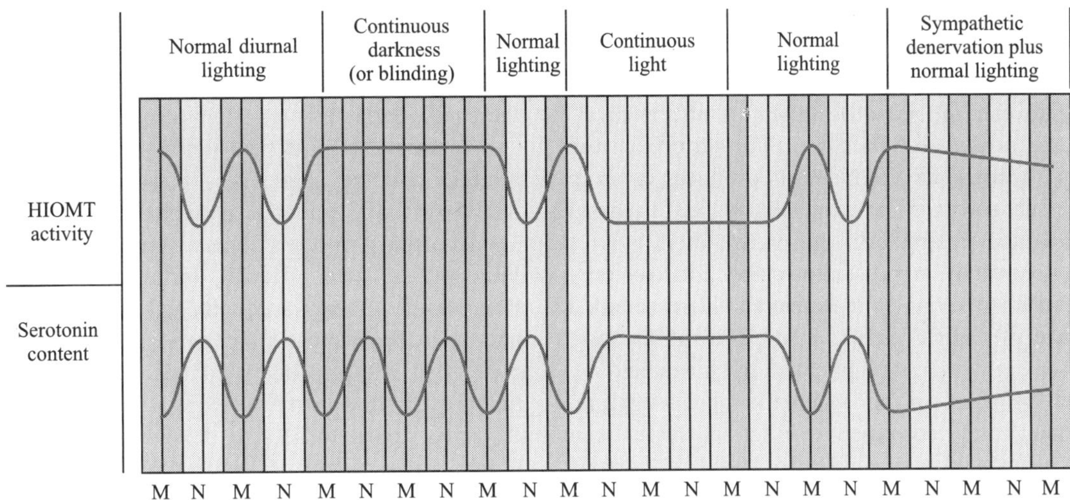

Figure 84.5 Circadian rhythmic changes in the activity of HIOMT and serotonin content in the rat pineal gland. Serotonin content is highest at noon (N), whereas HIOMT activity is highest at midnight (M). The HIOMT rhythm is dependent on environmental lighting and it is abolished when rats are kept in continuous light or darkness or after blinding. The serotonin rhythm persists in continuous darkness or after blinding but it abolished after rats are subjected to continuous illumination. Both rhythms are depressed when the sympathetic nerves to the pineal are cut. Dark areas denote darkness. (From Wurtman and Axelrod (1965), *Sci. Amer., 213,* 50.)

in responses to light and darkness. Nevertheless, it is yet to be conclusively shown that the pineal in lower vertebrates is the exclusive source of melatonin and that melatonin is the pigment-concentrating principal. The enzyme system (HIOMT) necessary for melatonin synthesis is present in the pineal as well as in other neural tissues, including the retina of lower vertebrates in contrast to its exclusive localization in the mammalian pineal. On the other hand, the concept that the pineal organs in lower vertebrates are photoreceptors is not only strengthened by the morphological resemblance to photoreceptors in the vertebrate retina, but also by the neurophysiological evidence that the pineal organs produce neural responses on exposure to darkness and light. Thus, pineal organs in lower vertebrates are photoreceptive, besides being one of the sites for melatonin synthesis.

B. Role of the Pineal Gland in Higher Vertebrates

1. Effects on skin and hair: Mammalian melanocytes, unlike those of lower vertebrates, do not have the capacity for melanin aggregation or dispersion. Melatonin does not elicit a direct melanocytic response in mammals. In weasels, melatonin has been claimed by some workers to initiate changes in the central nervous system and endocrines which result in moulting and growth of white pelage. Melatonin inhibits growth of hair in intact and pinealectomized male mice and pinealectomy leads to increased hair growth.

2. Effects on the thyroid: There is little information on the relationship between the pineal and the thyroid. Pinealectomy in rats results in a marked hypertrophy of the thyroid cells with a concomitant increase in I^{131} uptake. The effects of pinealectomy on the thyroid are counteracted by pineal extract

administration. Melatonin reduces the height of follicular cells as also thyroid hyperplasia induced by propyl thiouracil treatment. The pineal may influence the thyroid indirectly by acting on the neural centres that regulate thyrotropin or on the pituitary itself or directly at the level of the thyroid.

3. Effects on the adrenal cortex: Farrel and his co-workers have shown a relationship between the pineal and the adrenal cortex. They have isolated a factor called 'adrenoglomerulotropin' from pineal extracts which specifically stimulates aldosterone production by the adrenal cortex. Later, the work has revealed that the pineal is not specifically responsible for the regulation of aldosterone synthesis. At present, the data are too few and controversial to define a specific role for the pineal in the regulation of adrenal mineralocorticoid activity.

4. Role of the pineal gland in reproduction: Recent studies in pineal physiology clearly indicate that the pineal affects reproductive physiology in many vertebrates, especially in mammals. Three experimental procedures such as pinealectomy, administration of pineal extracts or purified pineal principles, and the activation of the pineal by exposure of animals to short photoperiod have been employed to clarify pineal-gonadal interrelations.

(a) *Effects of pinealectomy:* In both birds and mammals pinealectomy modifies gonadal growth and function. Pinealectomy in adult cocks induces a significant enlargement of testes and combs. In immature rats, pinealectomy results in premature opening of the vagina and enlargement of the ovaries and uteri in females, and enlargement of testes, seminal vesicles and prostate in males. In adult female rats, removal of the pineal results in an increase in the incidence of estrus vaginal smears. Pinealectomy in adult male rats produces hypertrophy of the seminal vesicles and prostate gland. In the male hamsters, removal of the pineal prevents the seasonal involution of gonads.

(b) *Light, the pineal and gonads:*   Short-term exposure of rats to constant illumination produces ovarian enlargement and an increase in the incidence of vaginal estrus. If the treatment is prolonged, their ovaries eventually become significantly smaller. Pinealectomy also results in ovarian enlargement and increase in the incidence of vaginal estrus in rats. But the biphasic effect of light is not duplicated by pinealectomy. The interrelationship among environmental lighting, the pineal, and reproductive functions have been clearly elucidated in the hamster. In the adult male hamster, reproductive regression follows exposure to short photoperiod only when the pineal is present. Similarly, blinding also induces testicular regression in the hamster with intact pineal. Thus, the pineal gland in the male hamster may be an intermediary between day-length and seasonal breeding. In the female hamster, the uterine weights are depressed after the restriction of light and this effect is reversed by pinealectomy.

(c) *Effects of pineal extracts and melatonin:*   Prior to the isolation of melatonin from the pineal glands, crude pineal extracts were used to understand the nature of the gonad-inhibiting factor. Administration of pineal extracts depresses gonadal growth and also inhibits ovulation in rats. The ovarian enlargement that follows removal of the pineal, as well as the constant light-induced persistent vaginal estrus in rats, are inhibited by injection of pineal extracts. The consensus is that administration of pineal extracts is inhibitory to general reproductive physiology.

Administration of melatonin generally mimics the effects of pineal extract and also reverses the changes that follow pinealectomy. The sites of action of the pineal substance melatonin are still under investigation. Administered melatonin is concentrated in the gonads, brain, and pituitary of rats. It has been suggested that melatonin may inhibit the synthesis and/or release of the luteinizing hormone, but has no effect on the follicle-stimulating hormone. Thus, melatonin may exert its effects through a direct action on the neuroendocrine centres in the brain. This, however, does not preclude a direct action of melatonin at the level of the target organ.

VI. PINEAL TUMOURS

Precocious puberty is noticed in most boys afflicted with pineal neoplasms. Many workers have postulated that pineal tumours induce precocious puberty by inhibiting the synthesis and secretion of the pineal hormone which delays the maturation process. The pineal neoplasms are localized and even encapsulated, but they also invade the brain tissue

and sometimes metastasize in spinal cord, lung and bone. Most pineal tumours are called parenchymal pinealomas and contain two cell types, the large spherical epithelial cells and the small, dark-staining, lymphocyte-like cells.

Teratomas also account for 10–15% of all reported pineal tumours and they often contain adenocarcinoma tissue and mucus-secreting columnar epithelial cells.

According to Kitay, pineal tumours result in precocious puberty, delayed puberty, or no gonadal involvement. Precocious puberty frequently develops in patients with teratomas. Pinealomas are commonly associated with delayed puberty or with secondary gonadal failure. The consensus is that precocious puberty develops because of the failure of the damaged pineal to secrete the hormone (melatonin) which normally inhibts gonadal functioning. The human pineal gland in the adult undergoes partial calcification; its effect on pineal function is not known.

SUMMARY

Investigations on the physiology of the pineal complex of lower vertebrates highlight the variability in the responses of fish, amphibians and reptiles which, in part, is due to the wide variety of species used, various protocols, and differences in the methods of evaluation. The consensus is that the epiphyseal complex, which is directly or indirectly responsive to light or darkness, helps the organism adapt to its environment. In amphibians, there is strong evidence to show that melatonin is a hormone which normally regulates body blanching. Melatonin is not formed exclusively in the pineal of lower vertebrates inasmuch as it is also formed in the eyes and brains of fish and amphibians. The pineal complex may also contain other substances whose physiological role has been barely studied.

In birds the available evidence indicates that the pineal gland may regulate gonadal functions through the mediation of indole amines, but the supporting evidence is inconclusive.

Investigations on the mammalian pineal gland conclusively prove that it is capable of modifying the functions of some endocrine organs, especially the gonads. Pinealectomy induces precocious gonadal growth in immature animals and also brings about a transient hypertrophy of the reproductive organs in adults. Administration of pineal extracts counteracts the effects of pinealectomy. The pineal is anti-gonadotropic and is activated by darkness. Presumably, the pineal plays an important role in modulating the reproductive activity of animals in relation to seasonal changes in the environment.

The mammalian pineal gland has the unique capacity to form melatonin (*N*-acetyl-5-metho-xytryptamine) which is synthesized as follows: tryptophan → 5-hydroxy-tryptophan → serotonin → *N*-acetylserotonin → melatonin. The last step is catalyzed by the enzyme hydroxy-indole-*O*-methyl transferase (HIOMT) which is localized only in the mammalian pineal gland. The HIOMT activity shows a circadian periodicity; its activity is highest around midnight and lowest around 6 p.m.

Environmental lighting information reaches the pineal via the eye, the inferior accessory optic tract, the sympathetic nervous system, the pinealocyte. Noradrenaline liberated by the sympathetic nerves stimulates the synthesis of melatonin.

Although a great deal is known about the pineal, many questions still remain unanswered. Melatonin accounts for most of the activity of the pineal gland. Nonetheless, the role of other methoxylated indoles is not known. The route of secretion of melatonin is not yet clarified. It is assumed that melatonin is released into the peripheral circulation because melatonin is present in the peripheral nerves where it cannot be synthesized. The site of action of melatonin is still controversial. The evidence is in favour of a neural site of action.

In conclusion it can be stated that the pineal gland is no more an enigma because its potential is now fully established. Nevertheless, the statement of Kelly (1962) is still appropriate: 'What has been termed "*the* pineal problem" still remains open—and open from many angles'.

SUGGESTIONS FOR FURTHER READING

Axelrod J. and Wurtman R.J. (1966), the Formation, Metabolism and Some Actions of Melatonin, A Pineal Gland Substance, in *Endocrines and the Central Nervous System* (Ed. R. Levine), Vol. XLIII, 200, Williams and Wilkins, Baltimore. (Account on Synthesis, Action and Metabolism of Melatonin.)

Bagnara J.T. and Hadley M.E. (1970), Endocrinology of the Amphibian Pineal, *Amer. Zool.,* 10, 201–216. (Up-to-date Review of the Role of the Pineal in Amphibians.)

Dickson R.B. and Salomon D.S. (1998), *Hormones and Growth Factors in Development and Neoplasia,* New York, John Wiley and Sons.

Kelly D.E. (1962), Pineal Organs: Photoreception, Secretion, and Development, *Amer. Scientist,* 50, 597–625. (Review of Pineal Morphology, Development, and Photoreception.)

Kitay J.I. and Altschcule M.D. (1954), *The Pineal Gland,* Harvard Univ. Press, Cambridge. (Gives Account of Pineal Including Clinical Aspects Up to 1954.)

Quay W.B. (1970), Endocrine Effect of the Mammalian Pineal, *Amer. Zool.,* 10, 237–246. (Current Status of Pineal Endocrine Effects and the Possible Mechanisms for Their Mediation Are Reviewed.)

Reiter R.J. (1972), the Role of the Pineal in Reproduction, in *Reproductive Biology* (Eds. H. Balin and S. Glasser), Excerpta Medica, Amsterdam, 71–114. (Up-to-date Account of Reproductive Functions of the Pineal in Birds and Mammals.)

Reiter R.J. and Fraschini F. (1969), Endocrine Aspects of the Mammalian Pineal Gland: A Review, *Neuroendocrinology,* 5, 219–255. (Review of Articles on the Endocrine Functional Aspects of the Mammalian Pineal Gland.)

Thakker R.V. (1997), *Molecular Genetics of Endocrine Disorders,* New York, Chapman and Hall.

Williams R.H. and Wilson J.D. (1998), Williams *Textbook of Endocrinology,* 9th ed., Philadelphia, W.B. Saunders Co.

Wurtman R.J., Axelrod J. and Kelly D.E. (1968), *The Pineal,* Academic Press, New York. (Gives Comprehensive Account of the Pineal Anatomy, Biochemistry and Physiology in All Vertebrates.)

Wurtman R.J. and Axelrod J. (1965), the Pineal Gland, *Sci. Amer.,* 213, 50. (Gives Popular Account of the Pineal.)

85

Melatonin

Uma Sinha

CONTENTS

I. INTRODUCTION

Melatonin is a chronobiotic hormone with various pleomorphic actions. It is synthesized and secreted from the pineal gland. Phylogenetically pineal gland is originated from photoreceptor cells but the human gland does not possess this property. Pinealocytes act as neuroendocrine transducers to secrete melatonin during the dark phase of the day/night cycle and therefore, melatonin is also called the 'hormone of darkness'. Its peak secretion during night acts as a hormonal signal to indicate the 'time of day' and 'time of year' to all tissues and thus, melatonin exclusively regulates the body's chronological pacemaker or 'Zeitgeber'. Melatonin is found in wide range of animal species and also in the seeds and leaves of a number of plants. In the photoperiodic animals like sheep and hamsters melatonin synchronizes the onset of seasonal reproductive cycle, whereas in humans, melatonin primarily regulates the circadian clock to maintain sleep pattern and core body temperature.

Melatonin is N-acetyl-5-methoxy tryptamine, an indolamine derivative of the essential amino acid tryptophan. The word melatonin is derived from Greek words 'melas' and 'tonein' due to the property of melatonin to lighten tadpole skin. The central pacemaker of suprachiasmatic nucleus (SCN) of anterior hypothalamus receives input about the light/dark cycle via retino-hypothalamic tract and drives the nighttime surge of endogenous melatonin. This nocturnal melatonin persists even in absence of alternating dark/light cycle indicating that this internal rhythm is free running. Moreover, visual blindness does not necessarily cause circadian impairment so far melanopsin-containing retinal ganglion cells remain connected to SCN. SCN ablation in primates appears to disrupt but fails to abolish the biological rhythm completely. Similarly, pinealectomy in mammals has not been seen to disturb the circadian cycle to any large extent.

II. DISCOVERY OF MELATONIN

It was the beginning of twentieth century when mammalian pineal tissue was identified to contain an active substance that had blanching effect on tadpole skin. Ultimately in 1958, Lerner et al., isolated the pineal hormone from bovine gland and identified it as melatonin (N-acetyl-5-methoxy tryptamine). Subsequently it was demonstrated that melatonin is synthesized in the pineal gland from its circulating amino acid precursor tryptophan which is then converted to serotonin and undergoes acetylation and methylation. In 1960s several animal studies showed that melatonin synthesis decreases in presence of ambient light and reaches its peak in darkness. Since then, considerable progress has been made towards understanding the molecular biology, chemical nature and pharmacology of melatonergic system in view of the potential application of melatonin in several therapeutic corners.

III. BIOLOGY OF MELATONIN

SCN controls the circadian pattern of melatonin release. Photosensory information from neuroretina is relayed through retinohypothalamic tract to SCN of hypothalmus. Fibers from SCN descend to the spinal cord with projections to superior cervical ganglia from which postganglionic adrenergic neurons

1024

containing norepinephrine and neuropeptide Y come to innervate the pineal gland. Norepinephrine is the final neurotransmitter in the process of melatonin secretion from pinealocytes. During light phase, SCN activity remains high and norepinephrine levels are low. Under reduced adrenergic activity, tryptophan is converted to serotonin in a two-step process but at this stage, serotonin is not readily converted to melatonin. With the beginning of dark phase, SCN becomes silent and superior cervical ganglion releases norepinephrine activity increases to activate β1 adrenergic receptors on pinealocytes. This results in increased intracellular cAMP/protein kinase A signaling leading to melatonin synthesis. Neurotransmitters and the rate-limiting enzyme (N-acetyl transferase) of this synthetic pathway are also expressed in a circadian fashion to regulate biological time-bound melatonin surge. Longer the dark phase more is the duration of melatonin secretion.

Melatonin biosynthesis

AANAT = arylalkylamine N-acetyltransferase (the rate limiting enzyme with strong diurnal rhythm); NE = Norepinephrine; HIOMT = hydroxy indole-o-methyl transferase.

Melatonin, once synthesized, is not stored; rather diffuses into circulation and cerebrospinal fluid of third ventricle. Human studies by Lynch et al. in 1975, demonstrated that nocturnal circulatory melatonin levels in young adults are 10-40 times higher than those in daytime. Plasma melatonin concentration increases from 2–10 pg/ml in light phase to 100–200 pg/ml during darkness. Though melatonin rhythm varies widely among person to person; normally it reaches its maximum level in the middle of night and starts decreasing before sunrise. A number of extrapineal sources of melatonin have been identified such as gastrointestinal tract, bone marrow, platelets, lymphocytes and skin where melatonin may have its local action. Retina can also generate melatonin in a minor quantity. Intracellular concentration of melatonin is higher than its circulatory level indicating preferential cellular uptake or storage. Gastrointestinal tract, gall bladder and bile also contain melatonin in higher concentration. It is also present in other body fluids like urine, saliva, semen and breast milk maintaining its night specific surge. Melatonin is lipid soluble; circulates in plasma being 70% bound to albumin. Its transport is not restricted by the blood-brain barrier. It has a bi-exponential half-life, the first distribution half-life being 2 minutes and the second one is 20 minutes.

Nocturnal exposure to high ambient light acutely suppresses melatonin secretion. Exercise and postural changes can reduce plasma melatonin levels. Melatonin synthesis is also modified by agents like α2 agonists, β antagonists and benzodiazepines. Mammalian pineal gland receives both central and peripheral innervations and contains receptors for monoamines, amino acids and peptides. Hence, α and β adrenergics, D1-dopaminergic, μ and δ opioids, VIP/PHI/PACAP (vasoactive intestinal peptide/peptide histidine isoleucin/pituitary adenylate cyclase activating peptide) receptors have been found to stimulate pineal melatonin synthesis and D2-dopaminergic, α2 adrenergic, M1-cholinergic,

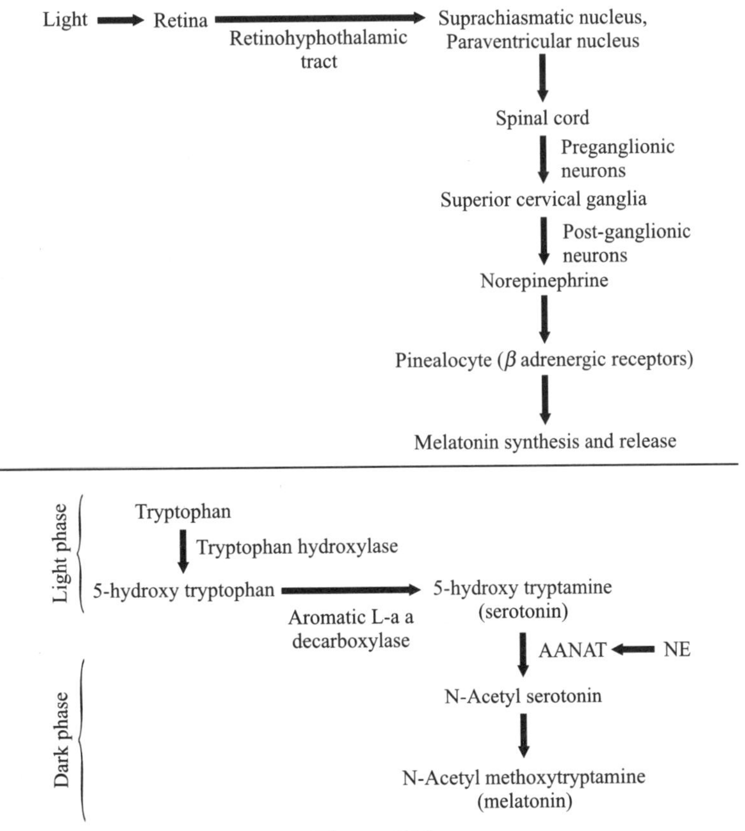

Figure 85.1

GABA, glutamate and tachykinin receptors are involved in inhibition of melatonin synthesis.

Melatonin is metabolized in liver by cytochrome P450-specific monooxygenase enzymes with hydroxylation and conjugation. In non-hepatic tissues melatonin is metabolized through deacetylation. It is also partially metabolized non-enzymatically in all cells, specially in the brain. It is excreted in the urine and feces primarily in the form of 6-sulfatoxymelatonin and only 2% of melatonin is excreted unchanged into urine and saliva. Human infants and fetuses do not synthesize melatonin but get the supply from maternal milk and placenta respectively. Rhythmic melatonin production increases rapidly after 9–12 weeks postnatal life with highest nocturnal levels achieved within 5 years of age. Thereafter melatonin concentrations tend to decline with aging.

IV. MELATONIN RECEPTORS

Melatonin acts through activation of at least two types of high affinity G-protein coupled membrane receptors- MT1 and MT2, consisting of 350 and 362 amino acids respectively. A third type of melatonin receptor (Mel$_{1c}$) first cloned from *Xenopus laevis*, is found only in non-mammals. MT3 is another low affinity melatonin binding site which is basically a melatonin-sensitive form of human quinone reductase 2 enzyme. As melatonin is a very small lipid soluble compound it can easily cross cell membrane and bind to specific nuclear receptors of RAR (retinoic acid receptor) family and also to calmodulin. Melatonin is thought to modulate calcium/calmodulin signaling through either G-protein coupled receptors or direct interaction with calmodulin. Melatonin is one of the endogenous calmodulin antagonists.

Effector arms of melatonin membrane-receptor signaling pathway involve adenylyl cyclase, phospholipase C, phospholipase A2, guanylyl cyclase, and calcium and potassium channels. These receptors are found in wide array of tissues including retina, brain SCN, pars tuberalis of anterior pituitary, aorta, cerebral, coronary and other peripheral arteries, kidney (fetal), pancreas, duodenum, caecum, colon, appendix, ovaries, placenta, cardiac ventricles, adipocytes and immune precursor cells. In the human SCN, MT2 receptors seem to be either absent or expressed at very low levels. Endogenous pineal melatonin mainly through MT1 receptors inhibits neuronal firing of the internal biological clock as a hormonal feedback and produces phase shift. Melatonin receptor mRNA has recently been amplified from various parts of human brain outside SCN such as thalamus, hippocampus, cerebellum and most parts of cerebral cortex. Nocturnal melatonin surge activates MT1 receptors and inhibit prolactin release from pars tuberalis to modulate circadian and seasonal rhythms.

In vascular bed, melatonin has been found to exert opposite responses as MT1 activation produces vasoconstriction and MT2 induces vasodilation. Prostate tumors, breast carcinoma and colon carcinoma cells are also found to express melatonin receptors (mainly MT1 receptors). Progressive reduction in receptor expression as a consequence of degeneration has been observed in Parkinson's disease (involvement of substantia nigra and amygdala) and in Alzheimer's disease (involvement of pineal gland and cortex). Receptor polymorphism with altered melatonin signaling has been related to several disease processes like diabetes mellitus, rheumatoid arthritis, polycystic ovarian syndrome and schizophrenia.

V. FUNCTIONS OF MELATONIN

Melatonin represents a biological timing signal to all the cell lines and tissues of the body. It acts as a stable peripheral biomarker of timing of the central biological clock. Melatonin is involved in diverse physiological processes like consolidation of sleep and circadian rhythm, cellular growth and neoplasia, immune activity, antioxidation and mitochondrial respiration. Thus melatonin is more than a hormone due to its versatility. Melatonin has multiple receptor-mediated functions acting through widespread membrane receptors and nuclear receptors but unlike a hormone, it has receptor-independent actions too.

1. *Biological clock and melatonin*

Circadian rhythms are biological processes that have a 24-hour periodicity even in absence of external cues. The master biological clock located in hypothalamus is composed of a group of specialized neurons in SCN. This central circadian pacemaker controls a number of complex physiological functions like nocturnal melatonin surge, regulation of sleep/wake cycle, maintaining core body temperature, hormonal secretion, food intake and metabolism. This nighttime rhythmic output of melatonin plays key role in the regulation of biological timing. During daytime photic signals fall on retina to stimulate photoreceptors and generate neurochemical signals to the primary visual cortex as well as to the non-visual centers of brain like SCN. Input of daylight resets the circadian pacemaker of SCN to synchronize internal biological clock to the environmental 24 hour day/night cycle. Whenever there is a change in the timing of ambient light/dark cycle, the internal clock is self adjusted by advancing/delaying its time-phase, a phenomenon known as phase shift. Hence, phase shift can change the phasing of all physiological rhythms. These changes are experienced during night-shift works, transcontinental long-haul travels and space flights. Exposure to bright light during night hours disrupts the biological clock as it instantly stops melatonin production. During 24-hour cycle, rising melatonin levels are associated with decreasing core body temperature, cortisol levels and alertness and as melatonin levels fall, core temperature, REM (rapid eye movement) sleep propensity and cortisol concentrations rise.

2. *Sleep and melatonin*

In humans, the sleep/wake cycle is internally synchronized with 24 hour circulatory melatonin cycle and the body temperature rhythm. Maximal sleepiness coincides with the peak melatonin levels and minimal core temperature.

Endogenous melatonin levels reach their peak 3-5 hours after the darkness falls while the core body temperature reaches a nadir between 3:00 hr and 6:00 hr. Onset of sleep occurs usually 5-6 hours before the lowest core body temperature. Melatonin is known to reduce sleep onset latency, determined by polysomnography as LPS (latency to persistent sleep), increase sleep efficiency and the total duration of sleep in primary insomnia though the evidence of direct sleep inducing property of endogenous melatonin is less clear-cut. Sleep promoting effect of melatonin has been linked to the inhibition of neuronal firing by activation of MT1 receptors in SCN. Exogenous melatonin produces a phase advance when administered in the late afternoon whereas early morning dose of melatonin causes a phase delay by antagonizing the effect of daylight. Sleep disorders in children specially with concomitant developmental disorders has been successfully treated with melatonin showing sleep restoration in 70-90% cases. Smith-Magenis Syndrome, a severe form of altered circadian rhythm, manifests as excessive daytime sleepiness and nocturnal insomnia. Treatment with beta-blockers reduces daytime melatonin levels and sleepiness. Melatonin also seems to be useful in the management of insomnia in elderly.

3. Endocrine functions

Melatonin affects the secretions from other endocrine glands in a paracrine fashion. It modulates pituitary hormone release, testosterone production by testes and cortisol secretion from adrenal cortex.

4. Reproductive functions

Melatonin regulates seasonal reproductive cycle in seasonal breeders. It acts as antigonadotropic; inhibits LH, FSH and GnRH. In mating seasons daylight duration increases, melatonin levels fall and gonads get activated. In human reproductive system also, melatonin may mediate seasonal fluctuations. Melatonin levels are elevated in male hypogonadotropic hypogonadism and infertility. In females, stress and exercise induced amenorrhoea is associated with high melatonin levels. Melatonin has also been implicated in sexual maturation as it is having inhibitory action at the onset of puberty.

5. Effect on blood pressure

During night blood pressure, heart rate and cardiac activity are low when plasma melatonin levels remain high. Similarly, all these vital parameters increase in the early morning inceasing the risk of stroke and myocardial infarction when melatonin starts falling. Nocturnal melatonin surge seems to be blunted in patients with nocturnal hypertension. Though the exact association between melatonin and cardiovascular events is yet to be determined, patients with cardiac disorders are found to have lower plasma melatonin levels in comparison with their normal counterparts.

6. Anticancer effect

Endogenous melatonin suppresses tumorigenesis via several mechanisms.

- Cell cycle arrest
- Dedifferentiation of cancer cells
- Antiangiogenesis
- Induction of apoptosis
- Antioxidation and free radical scavenging action
- Immunomodulation
- Aromatase inhibition
- Telomerase inhibition
- Inhibition of growth factor uptake
- Increased quinone reductase activity

7. Immunomodulation

Melatonin has an immunoenhancing effect. It activates T-lymphocytes, monocytes, natural killer cells and induces cell dependent cytotoxicity and antibody dependence responses. At physiological nocturnal circulatory concentrations melatonin can reduce type 2 anti inflammatory cytokine IL-10 and stimulate IL-1, IL-2, IL-6, Il-12 and TNF-α production. Melatonin has been shown to reverse immunosuppressive effects of acute stress, chemotherapy and viral infections due to its stimulatory effects on the immune system.

8. Antioxidation

Melatonin has antioxidant property in supraphysiologic concentrations. This is one important non-receptor mediated non hormonal activity of melatonin. Free radical scavenging action of melatonin requires no specific receptors. Proposed mechanisms are as follows:

- Direct detoxification of free radicals is done by melatonin which can donate one or more electrons to free radicals. It potentially binds with hydroxyl and hyperoxide radicals.
- Stimulation of activities of antioxidative enzymes like superoxide dismutase, glutathione peroxidase, glutathione reductase.
- Inhibition of activities of pro-oxidative enzymes including lipoxygenase.
- Melatonin promotes synthesis of glutathione.
- Melatonin also influences mitochondrial respiration and generation of free radicals.

Melatonin is around 4 times and 14 times stronger scavenger than glutathione and mannitol respectively.

VI. MELATONIN ANALOGUES

Melatonin analogues are melatonin receptor (MT1/MT2) agonists having longer half-life, prolonged effects and greater potency as compared to exogenous melatonin. Much attention has been focused on the development of more potent melatonin analogues because of the following reasons – (1) very short half-life of melatonin in circulation; (2) plasma melatonin levels vary widely among individuals with a lack of consistency on its therapeutic value, (3) its sleep promoting efficacy is not uniform; (4) its rapid metabolism after oral administration of fast release preparations. Here, melatonin agonists provide the clinician with new tools to address insomnia. The melatonin analogues ramelteon, agomelatine and tasimelteon are examples of this strategy. The main advantage is that these drugs are very safe and tolerable. However, their use in circadian rhythm sleep disorders has not yet been fully explored.

Ramelteon: It is a synthetic tricyclic (indenofuran) derivative of melatonin. It acts as a selective agonist for MT1/MT2 receptors without significant affinity to other receptors. Compared to melatonin, it has 3-16 times higher affinity for melatonergic receptors, more selectivity for MT1 type. It has a longer half-life of 1-2 hours. In 2005, ramelteon was approved by FDA for the treatment of insomnia. In several clinical studies ramelteon has been used in the dose range of 4–32 mg/day to evaluate safety and efficacy in chronic insomnia. Oral preparation of 8 mg at 30 minutes before desired sleep is the usual recommendation to promote sleep.

Agomelatine: It is a naphthalenic compound with both melatonergic and antidepressant property. It is the agonist of MT1/MT2 receptors and antagonist to 5-HT$_{2c}$ serotonin receptors (found in frontal cortex, hippocampus, amygdale etc involved in mood and cognitive function). Agomelatine has been licensed for major depressive disorders and bipolar affective disorders. Depending on the circadian phase (day/night) of administration, drug activity differs. Agomelatine given before sleep seems to have immediate sleep-promoting melatonergic effect but when given during daytime it is potentially anti-hypnotic and maintains alertness. As a hypnotic drug 25 mg evening dose of agomelatine effectively increases the duration of non-REM sleep and reduces circadian rhythm disturbance. As an antidepressant, agomelatine has unique non-aminergic mechanism with low relapse rate, early onset of action and excellent safety profile.

Tasimelteon: It is another MT1/MT2 agonist which has recently completed phase III clinical trials in treating insomnia. FDA in 2010 has approved this orphan drug for blind persons without any light perception with non-24-hour sleep-wake disorders.

P1roperties of melatonin and melatonergic agonists

	Melatonin	Ramelteon	Agomelatine
Half-life	45 min	1–2 hr	1–2 hr
Relative potency	MT1: 1 MT2: 1	MT1: 8 MT2: 3	MT1: 1 MT2: 1
Protein binding	70%	82%	95%
Recommended daily dose	2–3mg	8–16 mg	25–50 mg

VII. THERAPEUTIC APPLICATIONS OF MELATONIN

Functions of melatonin in humans are still matter of investigations. Different pharmacological applications have been proposed to melatonin and its analogues for the treatment of sleep disorders, mood disorders, neurodegenerative pathologies and cancer therapy. Long term safety data though awaited, exogenous melatonin has got very scanty reports of possible side effects like nightmare, fatigue, nausea, hypotension and abdominal pain. Melatonin may interfere with calcium channel blockers. Its use may be contraindicated in chronic fatigue syndrome. But, more rigorous clinical studies are needed to prove potential benefits of melatonin and also to rule out possible toxicities.

Sleep disorders: Melatonin exhibits both hypnotic and chronobiotic properties. It has been used for the treatment of age-related insomnia as well as of other primary and secondary insomnia. A recent consensus of the British Association for Psychopharmacology on evidence-based treatment of insomnia, parasomnia, and circadian rhythm sleep disorders concluded that melatonin is the drug of first-choice when a hypnotic is indicated in patients over 55 year. Both melatonin and its synthetic analogues exhibit mild but acute hypnotic effects with reduced sleep onset latency. In contrast to recommended doses of Ramelteon (4 or 8 mg/day) and Agomelatine (25 mg/day), melatonin induces sleep only at a daily dose of 0.1 to 0.3 mg of immediate release preparation.

Jet lag: Jet lag is a universally common problem of the frequent airline passengers who use to fly across a number of time zones. It occurs due to desynchronization between body's circadian rhythm and the new dark/light cycle at the traveler's destination. Depending on the number of time zone crossed and the direction of travel, jet lag causes sleep disturbance, daytime fatigue, irritability and loss of mental efficiency. Melatonin has been found to be remarkably effective in preventing or reducing the complaints related to jet lag, specifically when taken at the beginning of darkness. Controlled studies have evidenced that melatonin at a daily dose from 0.5 to 5 mg is equally effective when taken between 10 pm and 12 pm, before the sleep time in destination area. The benefit seems to be greater, the more are the time zones crossed and for eastward flights. Melatonin is now indicated in adult traveller crossing more than five time zones, specially eastwards who had previous attacks of jet lag.

Circadian rhythm sleep disorders (CRSDs): Circadian rhythm sleep disorders (CRSDs) are characterized by lack of synchronization between a person's biological clock and environmental 24-hour schedule of light/dark cycle. Melatonin has been used therapeutically to re-entrain the disrupted circadian rhythms and to induce sleep in delayed sleep phase syndromes and in shift-workers. Exogenous melatonin when administered in the late afternoon produces a phase advance. Melatonin receptor agonists may also be tried in this purpose. Melatonin can be helpful in the treatment of seasonal affective disorders. Oral melatonin preparation of 10 mg in the evening may be suitable for the prophylaxis of cluster headache in some patients.

Oxidative stress: Melatonin has been used to combat the exposure of intense oxidative stress; as for example, Japan has recently used melatonin to prevent disaster induced by ionizing radiation.

Cancer therapy: Many studies have shown that melatonin has a preventive or protective role in the field of clinical oncology. Melatonin has been shown to exhibit cytotoxic effects on cancer cells and has thus been used as a chemotherapeutic agent. Both at physiological and pharmacological concentrations melatonin acts as a differentiating agent in some cancers lowering their invasion and metastasis whereas, in others melatonin induces cell death. Melatonin may also be used as adjuvant therapy in progressive cancer and for prevention of chemotherapy and radiotherapy induced toxicities.

Increased incidence of cancer has been reported in persons chronically exposed to higher-than-usual artificial magnetic fields associated with reduced nocturnal melatonin secretion. But the scope of melatonin in terms of the optimal dose, route, bioavailability and timing of administration of melatonin depending on the pathological type of cancer is yet to be established.

Immunotherapy: As an immunoenhancing agent, melatonin has been tried in viral and bacterial infections and in septic shock. Melatonin maintains mitochondrial homeostasis counterbalancing the effects of reactive nitrogen compounds and its administration is thought to be beneficial in the management of sepsis. In severe sepsis, circadian melatonin secretion is impaired as cytokines like IL-1β and TNF-α drastically reduce pineal hormone secretion. Melatonin inhibits nitric oxide production in sepsis by inhibiting nitric oxide sythase enzyme. Melatonin is also protective against vascular injury during endotoxic shock. RBC deformation may also be prevented by melatonin administration. Melatonin inhibits endothelin, the potent vasoconstrictor and protects against septic shock. In animal models of experimental sepsis, melatonin has been found to prevent multi-organ dysfunction, circulatory failure and mitochondrial damage.

Neurodegenerative disorders: Melatonin has got a potential therapeutic value as a neuroprotective agent. Further clinical trials are needed to explore this molecule in preventing disease processes like Alzheimer's disease, Parkinson's disease, amyotropic lateral sclerosis, Huntington's disease and traumatic brain injury.

SUMMARY

1. Melatonin is a hormone (methoxyindole) synthesized within the pineal gland and secreted during the night.
2. The hormone contributes to the regulation of biological clock. Melatonin is involved in facilitation of sleep, maintaining core body temperature, inhibition of cancer development and growth, and the enhancement of immune function.
3. Melatonin is used therapeutically for the management of sleep disorders and jet lag, for the resynchronization of circadian rhythms in blindness and shift work, for the prevention as well as adjuvant therapy in cancer and also for preventing the progression of Alzheimer's disease and other neurodegenerative disorders.
4. The development of selective melatonin agonists has enhanced the prospects of manipulating the melatonergic system to the treatment of a wide range of sleep disorders.

SUGGESTIONS FOR FURTHER READING

1. Bukowska A. (2011), Anticarcinogenic Role of Melatonin—Potential Mechanisms., *Med Pr* 62:425–434.
2. Reiter R., Tan D. and Fuentes-Broto L. (2010), Melatonin: A Multitasking Molecule. *Prog Brain Res.,* 181:127–151.
3. Rothman S. and Mattson M. (2012), Sleep Disturbances in Alzheimer's and Parkinson's Diseases. *Neuromolecular Med.,* 14:194–204.
4. Srinivasan V., Spence D., Pandi-Perumal S., Trakht I. and Cardinali, D. (2008), Jet Lag: Therapeutic use of Melatonin and Possible Application of Melatonin Analogs, *Travel Med Infect Dis.,* 6:17–28.
5. Zawilska J., Skene, D. and Arendt J. (2009), Physiology and Pharmacology of Melatonin in Relation to Biological Rhythms, *Pharmacol Rep.,* 61:383–410.
6. Zee P. and Goldstein C. (2010), Treatment of Shift Work Disorder and Jet Lag, *Curr Treat Options Neurol* 12:396–411.

86

Clinical Disorders of Pituitary Function

Nikhil Tandon and Sachin Chiittawar

CONTENTS

I. PITUITARY ANATOMY, PHYSIOLOGY

The pituitary gland originates from nasopharynx and is situated within the sella turcica. Pituitary gland together with the hypothalamus, controls the structural integrity and function of endocrine glands including the thyroid, adrenal, gonads, bones and mammary tissue. The pituitary stalk is a meeting point of hypothalamus and pituitary and plays crucial role in enabling hypothalamic control of pituitary and its functions. Control of sexual function, fertility, linear and organ growth, lactation, stress responses, energy, appetite, and temperature regulation requires an intact hypothalamic pituitary axis.

A. Pituitary Anatomy

The anterior lobe, the posterior pituitary and a vestigial intermediate lobe forms the pituitary which is situated within the bony sella turcica under the dural diaphragma sella. The pituitary stalk connects to the median eminence of the hypothalamus. The adult pituitary weighs approximately 600 mg (range, 400 to 900 mg) and measures approximately 13 mm in the longest transverse diameter, 6 to 9 mm in vertical height, and about 9 mm anteroposteriorly. The sella turcica forms the bony roof of the sphenoid sinus. The lateral walls are made up of bony and dural tissue and surround the cavernous sinuses, which are traversed by the third, fourth and sixth cranial nerves and the internal carotid arteries. The dural roof protects the gland from compression by fluctuant cerebrospinal fluid pressure. The optic chiasm, located anterior to the pituitary stalk, is directly above the diaphragma sella. The posterior pituitary gland is directly innervated by supraopticohypophyseal and tuberohypophyseal nerve tracts of the posterior stalk. The anterior pituitary consists of corticotrophs, somatotrophs, lactotrophs, thyrotrophs and gonadotrophs, which secrete the adrenocorticotropic hormone (ACTH), growth hormone (GH), prolactin (PRL), thyrotropin or thyroid stimulating hormone (TSH), leutinizing hormone (LH) and follicular stimulating hormone (FSH) respectively. Each cell type is under highly specific signal controls that regulate its differentiated gene expression.

B. Pituitary Blood Supply

The pituitary enjoys an abundant blood supply derived from several sources. The superior hypophyseal arteries branch from the internal carotid arteries to supply the hypothalamus, where they form a capillary network in the median eminence. Long and short hypophyseal portal vessels originate from the infundibular plexuses and the stalk respectively. These vessels form the hypothalamic portal circulation. There are 10 portal vessels originating from median eminence, which run over ventral surface of anterior pituitary stalk and drain to adeno pituitary, anastomosing with capillaries from the neurohypophysis. The circulation is predominantly through hypothalamus to pituitary, making pituitary the most irrigated region of body [0.8 ml/g/min]. 70–90 % of adeno-pituitary irrigation comes from major portal vessels and the remainder from lesser ones. Arterial supply by a local artery is still not established. Inferior hypophyseal arteries supplies post pituitary but also provide capsular arteries to anterior pituitary.

C. Pituitary Development

The pituitary has a dual origin. The posterior lobe of the gland is derived from the neuroectoderm, while the anterior and intermediate lobes originate from the hypophysial placode. These two parts of the gland closely interact both physiologically and developmentally. Pituitary stem cells give rise to acidophilic (somatotroph and lactotroph) and basophilic (corticotroph, thyrotroph and gonadotroph) cells. At 6 weeks, corticotroph cells are morphologically identifiable, while immunoreactive ACTH is detectable by 7 weeks. At 8 weeks, somatotroph cells are evident with abundant immunoreactive cytoplasmic GH expression. Glycoprotein hormone-secreting cells express at 12 weeks, with differentiated thyrotrophs and gonadotrophs expressing immunoreactive subunits for TSH and for LH and FSH, respectively. Fully differentiated PRL-expressing lactotrophs become evident late in gestation (after 24 weeks).

D. Pituitary Transcription Factors

Pituitary development from Rathke's pouch involves complex interplay of lineage specific transcription factors expressed in pluripotent precursor cells and gradients of locally produced growth factors. The transcription factor Prop-1 induces pituitary development of Pit-1 specific lineages as well as gonadotrophs. The transcription factor Pit-1 determines cell specific expression of GH, PRL, and TSH in somatotropes, lactotropes and thyrotropes. DAX-1 and T-Pit-1 results in a series of rare selective or combined pituitary hormone deficits.

II. DISORDERS OF ANTERIOR PITUITARY

The disorders of anterior pituitary are consequent to a complex interplay of hypothalamic pituitary axis, including pituitary hypersecretion and hyposecretion, sellar enlargement and visual loss. The most common manifestation in adults is a pituitary adenoma which could either be non-secreting or hyper-secreting; though large tumours can result in hyposecretion of piututiary hormones due to compression of normal pituitary tissue. The most common manifestation of these disorders include headache, hypogonadism, and in case of large tumours with suprasellar extension, even visual loss. In children pituitary adenomas are a rarity. The common neoplams in the pediatric age group are craniopharyngiomas and hypothalamic masses which present as GH deficiency, delayed puberty and diabetes insipidus, prior to development of headache, visual loss and other CNS manifestations.

A. Pituitary Hypersecretion

The three clinically common presentations of pituitary hypersecretion are excess of PRL (usually presenting as galactorrhoea-amenorrhoea in women and hypogonadism in men), GH (acromegaly) and ACTH (Cushing's syndrome) levels.

Prolatinomas

Prolactin hypersecretion is the most common endocrine abnormality. It usually manifests as galactorrhoea (milk secretion from the breasts), and hypogonadotropic hypogonadism. Generally prolactinomas are microadenomas, less commonly are macroadenomas, and tend to grow slowly. High prolactin levels can be a consequence of intake of drugs like ranitidine, phenytoin, antipsychotics, anti-depressants, oral contraceptives, and alpha methyl dopa. Other conditions associated with elevated serum prolactin levels include, pregnancy, hypothalamic compression of stalk, hypothyroidism, liver and renal disorder or rarely macroprolactinemia (prolactin is usually single polypeptide chain of 22 kd protein, but sometimes is secreted as, a usually inactive, larger molecule varying in size from 50–150 kD). Detailed clinical examination, history and assessment of medication are required to identify the cause of hyperprolactinemia. In the absence of other identifiable causes of prolactin excess, and if PRL is >200 ng/ml, there is a strong possibility that a pituitary adenoma (prolactinoma) is responsible for elevated prolactin levels. Satisfactory control of prolactin levels can be achieved in microprolactinomas and most macroprolactinomas by using dopamine agonists (bromocriptine, cabergoline). Surgical intervention is rarely required if the tumor causes visual impairment (by compressing the optic chiasma), is resistant to therapy with dopamine agonists, or undergoes apoplexy. Drug resistant tumors can also be subjected to chemotherapy, i.e, temozolamide, and rarely radiotherapy.

Acromegaly

Growth hormone hypersecretion is the second most common cause of pituitary hypersecretion. The commonest cause for GH excess is a GH secreting pituitary adenoma, though rarely eutopic or ectopic secretion of GHRH can also lead to GH excess. Familial and syndromic association constitutes a small percentage of cases. GH excess manifesting in childhood (before fusion of skeletal epiphyses) presents as gigantism, while in adults acral (bony and soft tissue) enlargement are prominent clinical features. Acral enlargement, hyperhidrosis, skin tags, seborrhoea, greasy nose, joint pains, paraesthesias, headache, hypertension, cardiomegaly, diabetes, and hypogonadism form the main features of syndrome. Excess GH secretion leads to increase in levels of IGF-1 and increased GH and IGF-1 leads to the pathological presentation of this disorder. Clinical presentation is confirmed by high IGF-1 levels (age matched) and GH non-suppressibility, i.e, serum GH levels > 1 ng/ml after a 75 gram oral glucose load. These biochemical abnormalities are supported by pituitary imaging revealing a pituitary adenoma. The treatment of acromegaly rests on the resectability of the tumor, with the preferred option being surgical removal of the pituitary adenoma. If the tumor is not completely resectable, e.g., there is parasellar spread involving the cavernous sinus or there is residual disease after surgery, medical therapy in the form of somatostatin receptor analogues like octreotide, octreotide LAR, Lanreotide LAR/autogel can be tried. Dopamine agonist like bromcroptine and cabergoline have also been used, albeit with limited success, in the medical management of GH excess. GH receptor antagonists like Pegvisomant can be a choice in some cases, either as monotherapy or in combination with somatostatin receptor analogues. Radiotherapy, both conventional and stereotactic, can also be used in surgically non-resectable lesions.

Cushing's syndrome (CS) is a complex endocrine disorder with potential for serious health consequences if not promptly and adequately treated. Incidence of CS varies in different studies from 0.7 to 2.4 per million populations per year. Nearly 80% of cases of endogenous hypercortisolism are ACTH dependent, ACTH source being the pituitary in about 70 to 80% and ectopic in 10 to 15%. The ACTH source may remain occult in a few cases for many years in spite of extensive investigations. Primary adrenal disease (adrenal adenoma, adrenal carcinoma) accounts for about 15 to 20% of cases. CS is more common among certain high risk groups, e.g., 2% to 5% among patients with type 2 diabetes mellitus with poor blood glucose control, and 0.5% to 1% among hypertensive patients. CS patients experience significant clinical burden as a result of the physical illness, associated comorbidities like hypertension, diabetes mellitus, vulnerability to infections and associated psychopathology. Disease onset is insidious, developing over months and years. Obesity (especially centripetal fat distribution), hypertension, glucose intolerance and gonadal dysfunction like amenorrhoea and impotence are common manifestations. Other common clinical features include, mooning of face, plethora, hyperpigmentation, easy bruisability, thinning of skin, osteopenia, psychological disturbances, fungal infections, striae, proximal muscle weakness, acne and poor wound healing. After ruling out exogenous steroid intake, the diagnosis of CS requires identification of hypercortisolemia by 2 or more screening tests (overnight dexamethasone suppression test [ONDST], low dose dexamethasone suppression test [LDDST], late night sallvary cortisol and 24 hour urinary free cortisol). Once the diagnosis of CS is made, ACTH levels are measured to make a diagnosis of ACTH dependent CS (plasma ACTH > 20 pg/ml) or ACTH independent CS (plasma ACTH < 10 pg/ml). A high dose dexamethasone suppression test may help to suggest the diagnosis of pituitary ACTH excess if post dexamethasone levels are less than 50% of those prior to administering dexamethasone (2 mg every 6 hours for 48 hours). Dynamic MRI with gadolinium enhancement of the pituitary-hypothalamus is recommended to localize the pituitary adenoma in those with ACTH dependent CS. ACTH dependent CS patients with normal or inconclusive (lesion < 5 mm) findings on MRI sella are further investigated to localize ectopic source of ACTH with contrast-enhanced CT scan of neck, chest, abdomen and pelvis and inferior petrosal sinus sampling (IPSS). Management of Cushing's disease depends on identification and resection of pituitary adenoma by transsphenoidal resection of tumor. Patients with non resectable tumor, unidentifiable source of ACTH and very severe hypercortisolemia, or those medically unfit for surgery are either subjected to bilateral adrenalectomy or medical treatment with drugs like ketoconazole, metyrapone, octreotide analogues, dopamine agonists or adrenolytics like mitotane. Nonresectable pituitary tumors are subjected to gamma knife surgery as a modality of treatment. Temozolamide therapy is coming up for resistant tumors and in rare cases of pituitary carcinomas.

Nelson's syndrome – patient undergoing bilateral adrenalectomy for hypercortisolemia may present later with logarithmic increment in ACTH levels and aggressive enlargement of pituitary adenomas threatening vision and cutaneous/mucosal hyperpigmentation. With improved understanding of CS, newer imaging modalities enabling better tumour localiarction and stringent follow up, this has now become a rare presentation.

Thyrotropin secreting pituitary adenomas – TSH secreting pituitary adenomas are a rare cause of hyperthyroidism, goitre and high TSH levels. The biochemical findings are similar to those seen in thyroid hormone resistance. Identification of tumor in imaging helps in making the diagnosis. Surgical removal of tumor forms the main line of treatment.

Gonadotropin secreting pituitary tumor are very rare tumors. While most pituitary tumors secrete alpha subunits of FSH, these tumors secrete both LH and FSH and present as macroadenomas causing visual disturbances and hypogonadism. Surgical removal is the main treatment, but because of large size and non-resectability some of them may require radiotherapy.

Nonfunctional pituitary adenomas form 10% of cases identified incidentally during neuroimaging, most of which do not require treatment. However, if the tumour is large or compromising vision, it needs to be removed surgically.

Pituitary carcinomas are extremely rare tumors with one or more metastasis in brain.

B.   Pituitary Hypofunctioning

Hypopituitarism manifests as loss of secretion of one or more anterior pituitary hormones and if there is loss of two or more hormones it is known as panhypopituitarism. It is a slow insidious process depending on the amount of destruction of pituitary or deficiency of the hypothalamic secretory factors and respectively classified as primary and secondary hypopituitarism. Disorder of diverse etiologies can be remembered by mnemonic "nine Is", i.e., invasive, infarction, infiltrative, injury, immunologic, iatrogenic, infectious, idiopathic and isolated.

Pituitary macroadenomas, craniopharyngiomas, sellar and parasellar masses are known to cause hypopituitarism because of mass effect. Postpartum pituitary necrosis, i.e., Sheehan's syndrome presents as loss of lactation and hypopituitarism. Pituitary apoplexy in macro- or micro-adenomas leads to hemodynamic instability, loss of vision, headache, ophthalmoplegia, and meningismus, is an endocrine emergency which later presents as hypopituitarism. Infiltrative disorders like sarcoidosis, hemochromatosis and Langerhan's histiocytosis causes loss of one or more pituitary function. Head injury leading to avulsion of the stalk can manifest as loss of anterior and posterior pituitary functions. Surgery and radiotherapy for pituitary adenoma result in hypopituitarism. Tuberculosis, syphilis and mycotic infections are implicated for the same but are presently extremely rare because of improvements in anti-microbial therapy.

Isolated GH deficiency can be because of sporadic, autosomal dominant, recessive or X linked cause. Isolated ACTH deficiency is a rare illness most often observed after lymphocytic hypophysitis. The combination of anosmia, hyposmia, hypogonadism, eunuchoidal body proportions, cleft lip, and cleft palate associated with X linked inheritance has been identified as Kallman syndrome. KAL-1, FGFR-1, FGF-8, PROKR2, and PROK2 mutations are responsible for X linked forms. Isolated TSH and PRL deficiencies are rare. Multiple pituitary hormone deficiencies result from abnormal pituitary development related to mutations in genes encoding pituitary function: PIT-1 (TSH, GH and PRL) and PROP1 (TSH, GH, PRL, LH, FSH and ACTH).

Clinical features indicating GHD in the neonate are: (1) hypoglycemia, prolonged jaundice, microphallus, or traumatic delivery; (2) history of cranial irradiation; (3) head trauma or central nervous system infection; (4) consanguinity and/or an affected family member; and (5) craniofacial midline abnormalities. It is recognized that short stature is often the only feature present. Adult GHD presents as loss of lean body mass, increase in fat mass, bone loss, fatigue, lethargy, and cardiovascular comorbidity.

Hypogonadism in women presents as amenorrhoea, while in men the main manifestations are erectile dysfunction, loss of libido and reduced shaving frequency. PRL deficiency when isolated usually presents as lactation failure after delivery. Deficiency of ACTH presents as weakness, nausea, vomiting, anorexia, weight loss and hypotension. Since the mineralocorticoid axis is intact, dehydration and hyponatremia is less common.

C. Multiple Pituitary Hormone Deficiency

Patients of pan-hypopituitarism are slightly overweight with fine wrinkling of skin over face, loss of body and pubic hairs, atrophied breasts (women) and reduced testicular volume (men). Postural hyopotension, bradycardia, decreased muscle strength, and delayed tendon reflexes occur in some cases. Other features related with loss of vision and other focal neurological deficits depends on presence of large intrasellar and parasellar mass. Clinical features include anaemia, hypoglycaemia, hyponatremia, hypoadrenalism, water retention, and low voltage complex in electrocardiogram. There is no hyperpigmentation and less hyperkalemia unlike primary adrenal insufficiency.

The diagnostic tests to establish various pituitary hormone deficiencies are outlined in Table 86.1.

Treatment of cortisol deficiency is done by replacement of steroids (hydrocortisone 15-25mg/d, or prednisolone 5–7.5 mg/dL. For secondary hypothyroidism, thyroxine replacement titrated to achieve a high normal serum T4 level is advocated. Females with hypogonadism who are premenopausal are given estrogen and progesterone replacement in cyclical form. Women requiring to conceive need replacement with HCG and FSH as per protocol to achieve ovulation induction. Men with hypogonadism are given testosterone replacement till 55 years of age and those requiring children are subjected to hCG alone or in combination with recombinant FSH for fertility. Recombinant human GH is available for treatment in GHD children and adults and is given s.c as per weight

III. DISORDERS OF POSTERIOR PITUITARY

The posterior pituitary is not a gland but is an extension of distal axons of the magnocellular neurons of the hypothalamus located in supraoptic and paraventricular nucleus. Vasopressin and oxytocin are two hormones released by the posterior pituitary which regulates water metabolism, parturition and lactation.

A. Vasopressin Deficiency

Diabetes insipidus (DI) belongs to the group of polyuric and polydipsic diseases, hereditary or acquired disorders mainly associated with either an inadequate arginine vasopressin (AVP) secretion or renal response to AVP, which clinically results in hypotonic polyuria and a compensatory or underlying polydipsia. Central or neurogenic DI (CDI) occurs mainly due to lesions of the neurohypophysis or the hypothalamic median eminence resulting in deficient synthesis and/or release of AVP. A number of acquired and congenital disorders may cause CDI with variable clinical manifestation, depending on the extent of neuronal destruction. Usually 80 to 90% of the magnocellular neurons in the hypothalamus need to be damaged before symptoms of DI arise. Nephrogenic DI (NDI) results from renal resistance to the antidiuretic action of AVP, which may also be due to acquired or inherited conditions. The diagnosis of DI is made on clinical grounds by strict intake, output monitoring and if urine output is more than 3 to 3.5 litres per day, the patient is subjected to water deprivation test to establish the diagnosis of a polyuric state and its etiology. Desmopressin is available in nasal sprays, tablets and injections for treating CDI.

Inappropriate antidiuretic hormone secretion (SIADH) implies states wherein there is excessive secretion of vasopressin

TABLE 86.1   Pituitary hormone tests

Hormone	Test	sampling	Remarks(normal response)
Growth hormone	Insulin tolerance test (ITT)	Samples at every 30 min till 120 minutes	At blood sugar < 40 mg/dl, GH should be > >3 mcg/l
	L-Arginine test: 30 grams iv over 30 min GHRH test L- Dopa test		GH > 3 mcg/l is a normal response
Prolactin	TRH 200-500 mcg i.v	Every 20 min, 3 samples	Increment in PRL > 200% of baseline or PRL > 2 mcg/l
ACTH	ITT	Every 30 min till 90 min	Increment in cortisol > 7 mcg/dl, or cortisol > 20 mcg/dl
	CRH test 1 mcg/kg ovine CRH	Every 30 min till 120 min	Increase in ACTH levels 2 to 3 times of basal values or cortisol> 20 mcg/dl
	Metyrapone test	Baseline 30 and 60 minutes	Cortisol < 4 mcg/dl, 11 DOC > 7.5mcg/dl,and ACTH > 75 pg/ml
	Cosyntropin 250 mcg i.v or i.m		Cortisol > 21 mcg/dl, aldosterone>4 ng/dl
	Low dose cosyntropin test, 1mcg i.v		Cortisol > 21 mcg/dl
	3 day cosyntropin stimulation test 250 mcg i.v 8 hourly		Cortisol > 21 mcg/dl
TSH	Baseline T4,TSH		Low free T4 levels not correlating with TSH
	TRH test 200-500 mcg i.v		Increment of TSH > 5mU/L
LH,FSH	Basal LH, FSH, testosterone, estradiol levels GnRH 100 mcg i.v	Every 30 min till 60 min	All at low levels than normal. LH > 10 IU/L

secondary to lesion in central nervous system or in thorax, resulting in dilutional hyponatraemia. We need to rule out adrenal insufficiency and hypothyroidism before treating SIADH. Treatment involves fluid restriction, i.e., 500 ml lesser than urine output, and salt intake. If the patient is hemodynamically stable, loop diuretics can be administered to increase free water clearance. Nonpeptide AVP antagonists are now available which can block the effect of AVP.

SUGGESTIONS FOR FURTHER READING

1. Gardner D.G., Shoback D., *Greenspan's Basic and Clinical Endocrinology,* 9th ed., McGraw Hill Lange.

2. Harrison's *Principles of Internal Medicine*, 18th ed., McGraw Hill.

3. Melmed S., Polonsky K.S., Larsen R., et al., William's *Textbook of Endocrinology*, 12th ed., Saunders Elsevier.

4. Jameson J.L., De Groot L.J., et al., *Endocrinology Adult and Pediatric*, 6th ed., Saunders Elsevier.

5. Wilson T.A., Rose S.R., Cohen P., Rogol A.D., Backeljauw P., et al. (2013), Update of Guidelines for the use of Growth Hormone in Children: The Lawson Wilkins Pediatric Endocrinology Society Drug and Therapeutics Committee, *J. Pediatr*, 143:415–21.

6. Fenske W. and Allolio B. (2012), Current State and Future Perspectives in the Diagnosis of Diabetes Insipidus: A Clinical Review. *J Clin Endocrinol Metab* 97:3426–3437.

7. Schlechte J.A. (2007), Long-term Management of Prolactinomas, *J. Clin Endocrinol Metab.*, 92:2861–2865; Doi:10.1210/jc.2007–0836.

8. Biller B.M.K., Grossman A.B., Stewart P.M., Melmed S., Bertagna X., Bertherat J., et al. (2008), Treatment of Adrenocorticotropin-Dependent Cushing's Syndrome: A Consensus Statement, *J. Clin Endocrinol Metab.*, 93:2454–2462; Doi:10.1210/jc.2007–2734.

87

The Adrenals

Shail K. Sharma

CONTENTS

I. HORMONES OF THE ADRENAL CORTEX

In 1855, Thomas Addison described a clinical condition due to a disease of the suprarenal gland. About half a century later it was established that the outer portion of the gland is essential for life, but how it exercises its function was not known till 1924. The curative effect of the adrenal extract was shown when its administration revived a patient suffering from Addison's disease (see subsection L of this Section).

There are two adrenal glands which are located just above the kidneys. Each of them consists of two distinct parts, a central part, the *medulla,* derived from the ectodermal cells of the neural crest, and an outer part, the *cortex* derived from mesodermal cells. In sections the cortex is seen to be composed of two distinct zones. The outermost zone is the zona glomerulosa, which secretes the mineral-corticoid aldosterone. The inner zones, zona fasiculata and reticularis, secrete the glucocorticoid cortisol as well as the adrenal androgens. Adreno-corticotropic hormone (ACTH) secreted by the anterior pituitary, stimulates the cells of the inner zone to secrete glucocorticoids. Aldosterone secretion is not under the influence of ACTH. The adrenals during foetal life have a structure and function different from the adult gland. The foetus of 8 weeks has adrenal glands of the same size as kidney, but they are reduced to one-third of their size at birth.

A. Adrenocortical Steroids

More than 30 corticosteroids have been isolated from the human adrenal cortex but only a few are secreted in large enough quantities, which are physiologically active hormones. The adrenal cortex synthesizes two classes of steroids: the corticosteroids (glucocorticoids and mineral corticoids) which have 21 carbon atoms, and the androgens which have 19 carbon atoms. In human beings, cortisol is the main glucocorticoid and aldosterone is the main mineralcorticoid. The normal human adrenal glands (Figure 87.1) secrete each day about 10 mg of cortisol and 0.125 mg of aldosterone. Trace amounts of progesterone and dehydroepi-androsterone are also secreted.

B. Chemistry

All the steroids from the adrenal gland have a common basic structure. They contain the cyclopentanoperhydro-phenanthrene nuclues. The corticosteroids are made from cholesterol which in turn is synthesized from Acetyl CoA. Cortisol, corticosterone and aldosterone have a side chain attached to C_{17}. The side chain consists of a keto group (C = O) at position 20 and a primary alcohol (CH_2OH) group at position 21. Cortisol has alpha hydroxyl group at C_{17} (alpha groups lie below the plane of the ring and are shown by a dotted line). The hydroxyl group at position 11 of the natural corticosteroids is in the beta position (beta groups lie above the plane of the ring and are shown by a solid line). The hydroxyl group at position 11 is essential for glucocorticoid activity, and 11-deoxycorticosterone which lacks this group has very weak effects on carbohydrate metabolism. All active steroids except DHEA have a double bond (designated by the sign Δ) between carbon atoms 4 and 5 and an oxygen atom at position 3. The mineralcorticoid aldosterone has an aldehyde (–CHO) group attached to carbon 18, and can exist in the aldehyde or hemiacetal form. The androgens have a keto group on carbon atom 17 and are therefore called 17-keto or 17-oxosteroids (Figure 87.1).

C. Biosynthesis of Corticosteroids

The adrenal cortex has the highest concentration of ascorbic acid in the body and its cholesterol content is second only to nervous tissue. Adrenocorticotropic hormone (ACTH) of the anterior pituitary gland stimulates the adrenal cortex to produce corticosteroids. Simultaneously there is a decrease in cholesterol and ascorbic acid content of the gland. The precursor of corticosteroids is cholesterol, whereas ascorbic acid may donate electrons.

D. Biosynthesis of Cortisol

As shown in Figure 87.1, the cholesterol molecule contains 27 carbon atoms and one hydroxyl group at position 3. Its conversion to cortisol involves the removal of six carbon atoms from the side chain and four oxygen atoms are introduced. The biosynthetic pathways are summarized in Figure 87.1. In the mitochondria the side chain of cholesterol is hydroxylated at C_{20} and C_{22} positions with the formation of 20α, 22 dihydroxycholesterol. Its conversion to pregnenolone is catalyzed by desmolase and involves the cleavage of the bond between C_{20} and C_{22}. NADPH and oxygen are the cofactors in all the three reactions (pregnenolone is also an intermediate

in the formation of testosterone in the testis and estrogens in the ovary). The conversion of pregnenolone to progesterone involves a 3-β dehydrogenase which catalyzes the oxidation of 3 hydroxyl to a 3 keto group and a Δ5-3 keto-isomerase, catalyzes the isomerization of Δ5 double bond to a Δ4 double bond. Progesterone is in turn hydroxylated at C_{17} and C_{21} with the formation of 17-hydroxyprogesterone and 11-deoxycortisol. In the final step 11-deoxycortisol is hydroxylated at C_{11} with the formation of cortisol. In an alternate pathway (not shown in Figure 87.1) pregnenolone is first hydroxylated at C_{17} and the resulting 17-hydroxy pregnenolone undergoes the dehydrogenation at C_3 and isomerization of the double bond from Δ5 to Δ4 with the formation of 17-hydroxy-progesterone. 17α hydroxylase is absent in zona glomerulosa and C_{17} must be hydroxylated before C_{21}, whereas hydroxylation at C_{11} may occur at any stage. The most effective donor of hydrogen and electrons for the hydroxylases is NADPH, generated in the mitochondria by the NADP-specific isocitrate dehydrogenase. The oxygen atom of the incorporated hydroxyl group comes from O_2. NADPH transfers its electrons to a flavoprotein which are then transferred to adrenodoxin (a non-heme iron protein), which in turn transfers an electron to the oxidized form of cytochrome P_{450}. The reduced cytochrome P_{450} is then oxidized, and in the process one oxygen atom of the molecule goes into the substrate whereas the other is reduced to water.

The rate limiting step of cortisol synthesis is the conversion of cholesterol to pregnenolone catalyzed by the enzyme system desmolase and is stimulated by cyclic AMP. ACTH binds to specific receptor protein and stimulates cyclic AMP formation by activating adenylate cyclase in the cells of adrenal cortex which synthesize costicosteroids.

E. Biosynthesis of Aldosterone

The synthesis of aldosterone (Figure 87.1) in the adrenal cortex is restricted to the zona glomerulosa which functions independently from the other zones. Progesterone is hydroxylated at C_{21} to form 11-deoxycorticosterone. The latter is again hydroxylated at C_{11} to corticosterone, followed by hydroxylation at C_{18} to 18-hydroxycorticosterone. The 18-hydroxycorticosterone is then dehydrogenated to aldosterone by an enzyme found only in the mitochondria of zona glomerulosa.

F. Biosynthesis of Androgens and Estrogens

17α-hydroxypregnenolone and 17α-hydroxyprogesterone (in a reaction similar to that catalyzed by desmolase) lead to the formation of DHEA, the major precursor of 17-ketosteroids. DHEA is readily converted to Δ^4-androstenedione, which is hydroxylated at C_{11} to form 11β-hydroxyandrostenedione and later into 11-ketoandrostenedione. While DHEA has an androgen potency of 40–60 per cent that of testosterone, Δ^4-androstenedione is only 20 per cent potent. These steroids account for most of the androgens produced by the adrenal cortex. The fact that certain adrenal tumours are associated with feminization, and castrated women secrete estrogens which

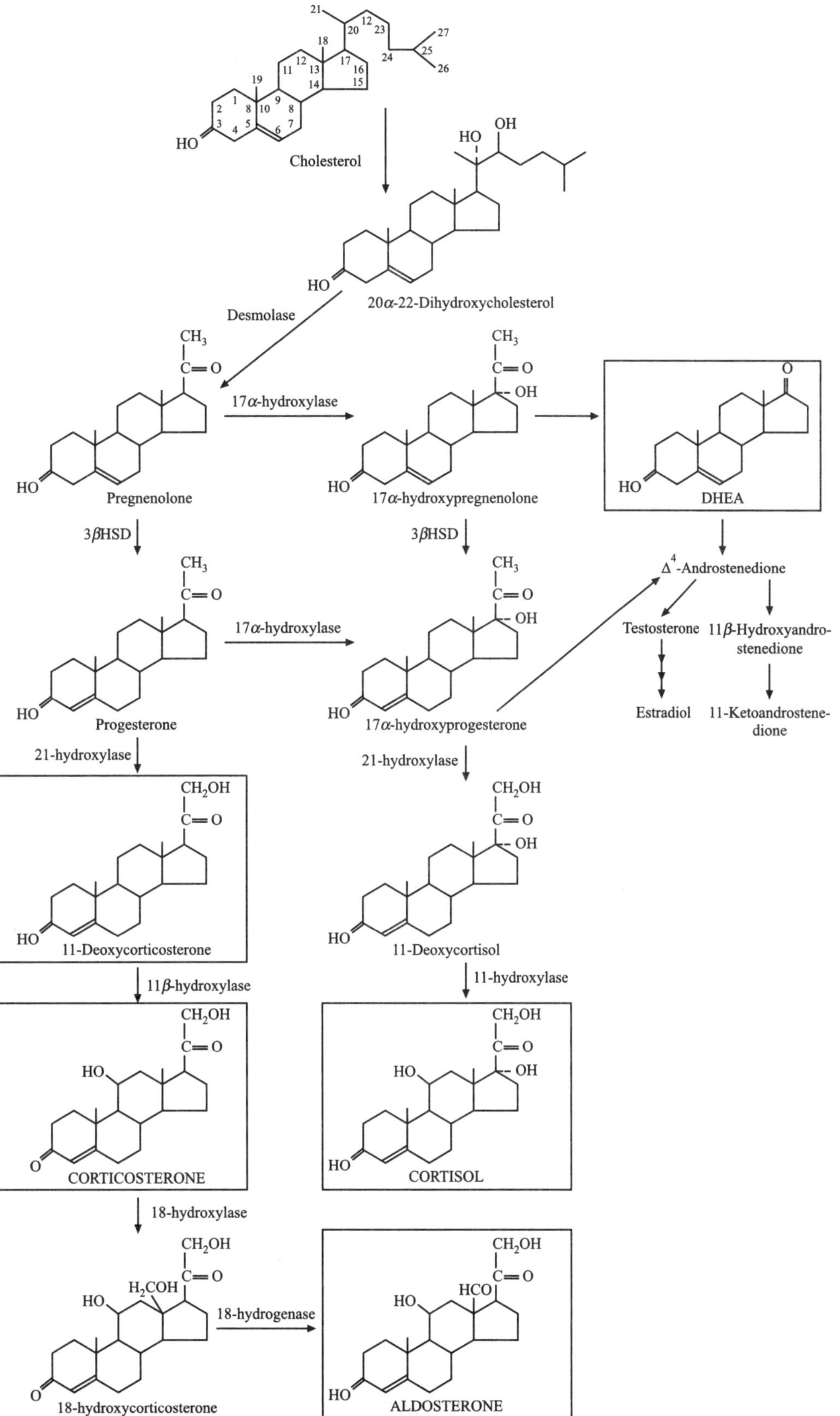

Figure 87.1 Synthesis of Corticosteroids.

disappear when adrenals are removed, shows that adrenal glands can produce estrogens. Estrogens may be formed by oxidation of testosterone at position 19, followed by demethylation and aromatization (Figure 87.1).

It is important to note that 17, 21 and 11 hydroxylases are highly specific in the biosynthesis of corticosteroids and the absence of one or the other results in genetic disorders which are discussed in Section I of this chapter.

G. Secretion of Corticosteroids, Transport, Metabolism and Excretion

The adrenal cortex shows a regular rhythmic (circadian) variation in activity. In a normal person the highest plasma values of corticosteroids are likely to be found around 8.00 a.m., with a slow steady fall to a minimum between 9.00 p.m. and midnight. The new cycle then begins afresh with a rise in plasma level that starts at about 4 a.m. while the subject is still asleep. These fluctuations are largely due to a rhythmic and corresponding rise and fall in plasma ACTH levels. Diurnal rhythm of these steroid levels in plasma is liable to be almost abolished in hypothyroidism, whereas in spontaneous hyperthyroidism or that due to tri-iodothyronine administration it is exaggerated.

In acute illness, after major surgical operations, and under emotional stresses the 17-hydroxy-corticosteroid levels rise in plasma.

Under physiological conditions about 75% of circulating cortisol is reversibly bound to a α-globulin termed *transcortin* which is produced in the liver, 15% is bound to albumin and 10% is free. It is the unbound fraction of cortisol (10% of the total), that enters into cells for exerting its metabolic effects or to be transformed to inactive products. Since *transcortin* is produced in liver, when transcortin concentration is lowered in liver disease, free cortisol level rises in plasma. Similarly, during pregnancy or administration of estrogens, when transcortin level increases, bound cortisol also rises. Aldosterone is transported chiefly, in combination with albumin. The half-life of cortisol at normal plasma concentration is about 60–80 minutes and that of aldosterone about 30 minutes.

The main site of cortisol degradation is the liver and involves reduction and conjugation reactions (Figure 87.2). In the presence of reductases with NADPH as cofactor, the

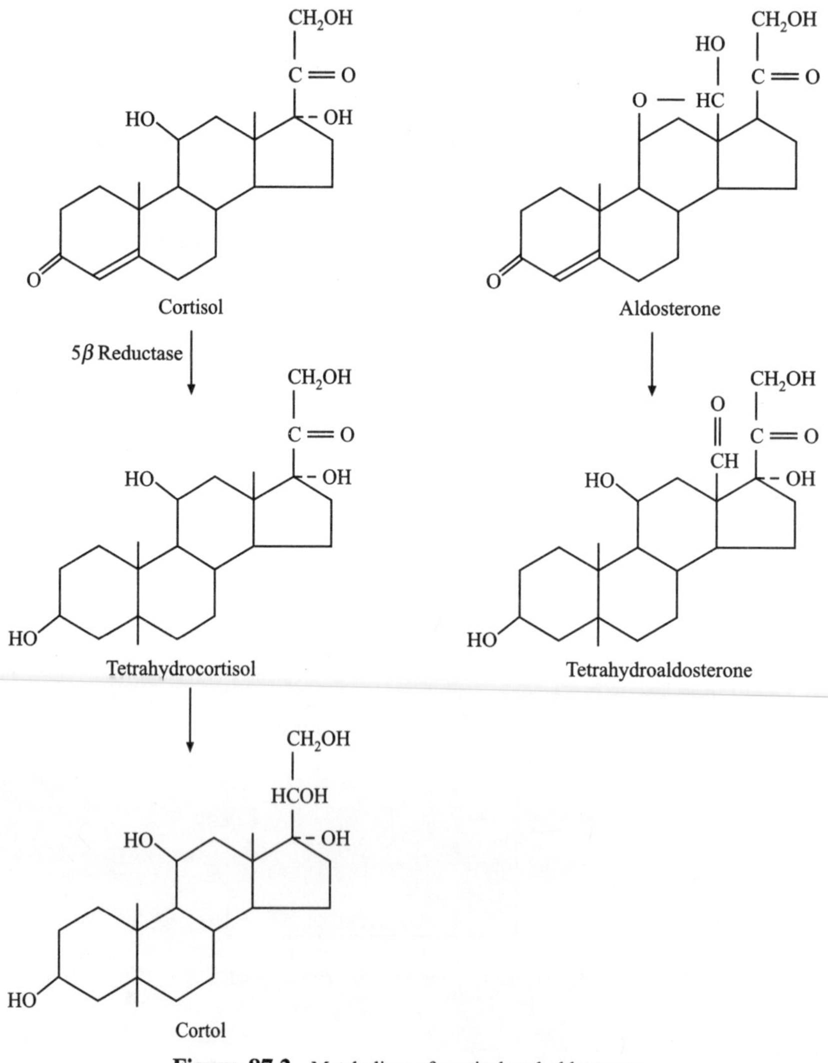

Figure 87.2 Metabolism of cortisol and aldosterone.

double bond at position 4–5 and the keto group at position 3 are reduced. Most of these reduced steroids are conjugated at the 3-hydroxyl group by enzymatic reactions that take place in the liver and, to a lesser degree in the kidney. The resultant sulphate esters and glucuronides form water-soluble derivatives and are excreted in the urine. The degradation of cortisol is increased in hyperthyroidism and decreased in hypothyroidism, liver disease and old age (Figure 87.2).

H. Physiological Functions and Pharmacological Effects

Corticosteroids have diverse effects on carbohydrate, protein and lipid metabolism. They maintain the fluid and electrolyte balance and are important in the preservation of normal functions of the cardiovascular system, the immune system, the kidney, the skeletal muscle, the endocrine and the nervous system. Another important function of corticosteroids is to endow the body with the capacity to resist stressful circumstances such as noxious stimuli and environmental changes.

Physiological functions

1. *Carbohydrate and protein metabolism:* The effect of glucocorticoids on intermediary metabolism in liver is to stimulate gluconeogenesis from amino acids and glycerol, and cause increased glycogen synthesis and glucose release. In the peripheral tissues, cortisol diminishes glucose utilization, increases protein breakdown and activates lipolysis, thus, providing amino acids and glycerol to the liver for gluconeogenesis.

In adipose tissue, skin, fibroblasts thymocytes and polymorphonuclear leukocytes, cortisol inhibits glucose transport by translocation of the glucose transport protein from plasma membrane to an intracellular location. These effects of cortisol in peripheral tissues are associated with a number of catabolic actions, which include atrophy of lymphoid tissue, decreased muscle mass and negative nitrogen balance. Amino acids mobilized from a number of tissues in response to glucocorticoids reach the liver where they are utilized for glucose synthesis. In the liver glucocorticoids induce the transcription of a number of enzymes involved in gluconeogenesis and amino acid metabolism: Phosphoenol pyruvate carboxykinase (PEPCK), glucose-6-phosphatase and fructose-2, 6-bis phosphatase. Studies on the molecular basis of PEPCK gene expression have indicated an interplay among glucocorticoids, insulin, glucagon, and catecholamines.

2. *Lipid metabolism:* Glucocorticoids have a dramatic effect on redistribution of the body fat as seen in Cushings Syndrome. Other effects of glucocorticoids are the permissive facilitation of the effect of growth hormone, and β adrenergic receptor agonists which result in lipolysis in the adipocytes, with increase in free fatty acids. With respect to fat distribution, there is increased fat in the back of the neck ("buffalo hump"), face (moonface), and supraclavicular area coupled with loss of fat in the extremities.

Electrolyte and water balance

Mineralcorticoid aldosterone is the most potent naturally occurring corticosteroid which regulates the fluid and electrolyte balance. Aldosterone acts on the distal tubules and collecting ducts of the kidney to enhance reabsorption of Na^+ from the tubular fluid; it also increases the urinary excretion of both K^+ and H^+. Similar effects of aldosterone are also seen in colon, salivary glands, and sweat glands. Thus the primary features of hyper-aldosteronism are positive Na^+ balance with consequent expansion of the extracellular fluid volume, normal or slight increases in plasma Na^+ concentration, hypokalemia and alkalosis. Mineralcorticoid deficiency leads to Na^+ wasting and contraction of the extracellular fluid volume, hyponatremia, hyperkalemia and acidosis. Aldosterone exerts its effects on Na^+ and K^+ homeostasis primarily via its action on the principal cells of the distal renal tubules and collecting ducts, while the effect on H^+ excretion are largely exerted in the intercalated cells. Aldosterone acts by increasing the number of open Na^+ and K^+ channels in the luminal membrane leading to an increase in Na^+ uptake into the tubular cells. Aldosterone also directly increases the activity of Na^+, K^+-ATPase in the basolateral membrane, returning Na^+ to the systemic circulation in exchange for K^+.

General effects

Cortisol augments the secretion of HCl and pepsinogen by the gastric mucosa and trypsinogen by the pancreas. This may be the basis of the ulcerative lesions of gastrointestinal tract on prolonged steroid therapy. Corticosteroids exert a number of indirect effects on the CNS, through maintenance of blood pressure, plasma glucose concentration and electrolyte concentrations. The presence of steroid receptors in the brain has lead to increasing recognition of direct effects of corticosteroids on the CNS, including effects on mood, behaviour and brain excitability. The most striking effects of corticosteroids on the cardiovascular system result from mineralcorticoid induced changes in renal Na^+ excretion. Permissive concentrations of corticosteroids are required for normal function of the skeletal muscle.

Anti-inflammatory and immunosuppressive actions

Cortisol is commonly used to *suppress inflammatory responses* of any kind. It is given to patients with rheumatoid arthritis. In fact, whenever some of the symptoms of a disease can be attributed to the inflammatory reaction produced in that disease, steroid therapy is expected to relieve such symptoms often completely. Cortisol is also effective when applied locally at the site of inflammation. It is probably due to its effect on capillary dilatation, which results in reduction in the permeability of the vessels and in the extravasation (infiltration or effusion from a blood vessel) of cells and fluid. Thus, there is reduction in the number of both polymorphs and macrophages which leave the bloodstream for the inflamed area. Multiple mechanisms are involved in the suppression of inflammation by glucocorticoids. It is now clear that glucocorticoids inhibit the production by multiple cells of factors that are critical in generating the inflammatory response. As a result, there is decreased release of vasoactive and chemoattractive factors, diminished secretion of lipolytic and proteolytic enzymes, decreased extravasation of leukocytes to areas of injury, and ultimately decreased fibrosis. Glucocorticoids also exert profound effects on specific host immune responses, at least partly through their effects on cytokine production. These factors include interferon gamma, granulocyte/monocyte colony-stimulating factors (GM-CSF), interleukins (IL-1 IL-2, IL-3, IL-6), and tumour necrosis factor α (TNF-α). Cytokine network plays an essential role in the integrated actions of macrophages/monocytes, T-lymphocytes and B-lymphocytes towards mounting immune responses to various pathogens, and therefore inhibition of cytokine synthesis and action by glucocorticoids in the immune response is suppressed. However, steroid therapy to control inflammation is not all that good. Since cortisol also suppresses the normal tissue immune response, the continuous repair required in mild infections in its presence is not possible. In fact some of the most important effects of cortisol are exerted on lymphocytes. Even a single dose of any active glucocorticoid to a normal person may lead to oedema of the lymphoid tissue. With continued steroid administration the number of lymphocytes diminish and there is suppression of mitosis resulting in lymphopenia. Cortisol exerts a *pronounced anti-allergic action*. It is commonly used to treat patients receiving organ transplants such as heterologous kidney, since it interferes with the immune response for rejections.

I.   Hypothalamic–Pituitary–Adrenal Axis

These three organs collectively are referred to as hypothalamic–pituitary–adrenal (HPA) axis, which maintains appropriate levels of glucocorticoids. HPA axis is regulated at three levels: diurnal rhythm in basal steroidogenesis, negative feedback by adrenal corticosteroids and marked increases in steroidogenesis in response to stress. Stressful stimuli can override these normal negative feedback control mechanisms leading to marked increases in plasma concentration of adrenocortical steroids. The diurnal rhythm is entrained by higher neuronal centres in response to sleep-wake cycles, such that levels of ACTH peak in the early morning hours, causing the circulating glucocorticoid levels to peak at about 8.00 a.m.

A fall (from normal) in the $(Na^+)/(K^+)$ ratio of blood circulating through the adrenal cortex causes the cells to secrete more aldosterone. The mechanism is as follows. The kidney glomeruli contain the juxtaglomerular apparatus that secretes an enzyme, renin. When the Na^+ concentration and blood volume falls, renin is secreted and acts on a particular α_1 globulin in the blood plasma to hydrolyze it, removing a peptide containing 10 amino acid residues. This peptide is known as angiotensin I and has the sequence: Asp-Arg-Val-Tyr-Ile-His-Pro-Phe-His-Leu. As the venous blood carrying this peptide passes through the lungs, another enzyme in the lungs causes the hydrolysis of angiotensin I liberating the C-terminal His-Leu fragment. The remaining peptide is physiologically active. It has 8 amino acids and is called *angiotensin II*. It causes an increase in the arterial

blood pressure and stimulates the secretion of aldosterone by the adrenal cortex zone of the glomerulosa.

Angiotensin converting enzyme inhibitors (ACE) are the *first-line therapeutic agents for treating hypertension in patients*. ACE inhibitors inhibit angiotensin converting enzyme (a compotent of the blood pressure-regulating renin-angiotensin system) thereby decreasing the tension of blood vessels, and blood volume thus lowering blood pressure. ACE inhibitors block the conversion of angiotensin I to angiotensin II.

J.   Mechanism of Action of Corticosteroids

Like other hormones, corticosteroids interact with specific receptor proteins in target tissues to regulate gene expression, thereby changing the levels of proteins synthesized by the target tissues. As shown in Figure 87.3, the glucocorticoid receptor (GR) resides predominantly in the cytoplasm in an inactive

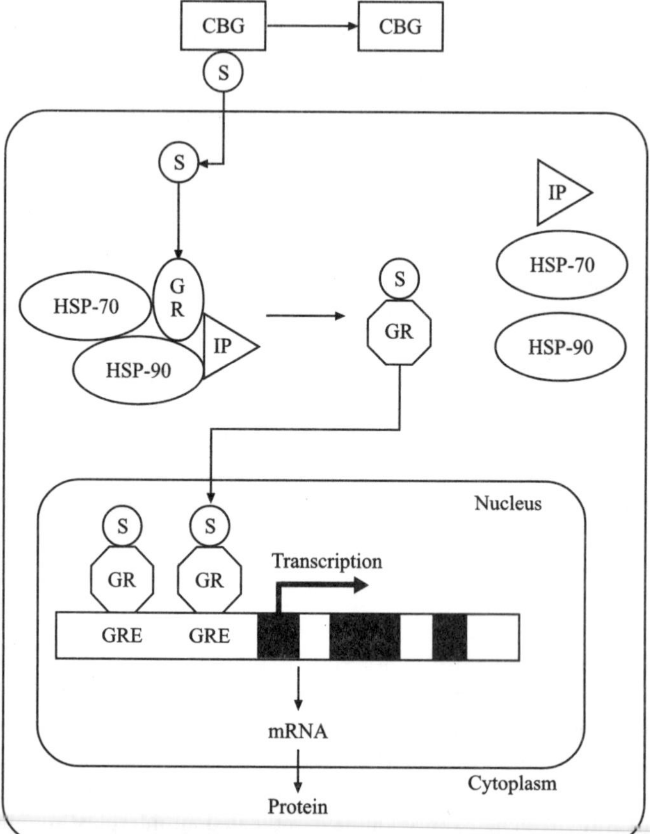

Figure 87.3   Mechanism of action of the glucocorticoid receptor. Glucocorticoid steroids (S) enter cells, bind to glucocorticoid receptor (GR) to change the GR conformation (indicated by change in GR shape) S-GR complex enters the nucleus, and activates expression of certain target cell genes. HSP-70 and HSP-90 are 70-KDa and 90-KDa heat shock proteins; IP, is the 56-KDa immunophilin GRE glucocorticoid-response elements in the DNA that recognize and bind the GR. Within the gene introns (unshaded) and exons (shaded).

form until it binds the glucocorticoid steroid ligand {S}. Steroid binding results in receptor activation and its translocation to the nucleus. The inactive GR is found as a complex with other proteins, including the heat shock protein (HSP) 90—a member of the heat shock family of stress-induced proteins—HSP70, and a 56-KDa immunophilin, which binds the immuno suppressive agent cyclosporine. Following steroid ligand binding, HSP90 and several of the other associated proteins dissociate and (S-GR) is directed into the nucleus where it interacts with specific DNA sequences within the regulatory regions of affected genes termed glucocorticoid-responsive elements (GREs) and provide specificity to the induction of gene transcription by glucocorticoids. The currently known GRE DNA sequence is GGTACAnnn TGTTCT (n is any nucleotide). Genes whose expression is negatively regulated by glucocorticoids have been identified. One such example is the negative regulation by glucocorticoids of POMC gene expression in corticotropes. Other genes that are negatively regulated by glucocorticoids include those that code for a number of cytokines, regulatory molecules that play key roles in the immune and inflammatory networks, and for collagenase and stromelysin, enzymes which play key roles in the destruction of joints as seen in inflammatory arthritis.

Mineralcorticoid receptor

Like the glucocorticoid receptor, the mineralcorticoid receptor also is a ligand-activated transcription factor and binds to a very similar (not identical), hormone-responsive element; unlike the glucocorticoid receptor, the mineralcorticoid receptor has a restricted expression, the principal sites being the kidney, colon, salivary glands, sweat glands and hippocampus.

After binding to mineralcorticoid receptors in responsive cells, aldosterone activates the expression of several genes, the most studied of which is that encoding Na^+, K^+-ATPase. This protein is found in the basolateral membrane of tubular cells and generates the electrochemical gradient that drives monovalent cations (i.e., Na^+ and K^+) through their respective membrane channels.

Receptor-independent mechanism for corticosteroid specific action

Recent studies have identified a steroid metabolizing enzyme, 11β-hydroxysteroid dehydrogenase, particularly in the kidney, colon and salivary glands (the mineralcorticoid responsive tissues) which metabolizes the glucocorticoids such as cortisol to receptor-inactive 11-keto derivative that does not bind to the receptor. Aldosterone resists metabolism by this enzyme, because its predominant form under physiological conditions is the hemiacetal derivative which is resistant to 11β-hydroxysteroid dehydrogenase action.

K. Hypofunction of the Adrenal Cortex

Failure of the adrenal cortex by a destructive primary lesion of the adrenal glands such as tuberculosis or autoimmune adrenalitis causes *Addison's disease.* Sometimes, lack of

ACTH stimulation causes atrophy of the adrenal cortex with depression of function.

1. **Primary hypoadrenalism:** This is due to destruction of the adrenal glands by the so-called idiopathic fibrosis, by tuberculosis or by invasion by malignant tissue. This nearly always results in a deficiency of aldosterone, cortisol and corticosterone formation. The characteristic feature of this group is that ACTH does not stimulate the adrenal cortex to produce corticosteroids. The feedback mechanism for ACTH release does not operate and over production of ACTH is often associated with an increased secretion of melanophore stimulating hormone, and ACTH itself having MSH like activity, is the main cause of the well known *bronzy pigmentation* of the skin in the Addison's disease.

2. **Anterior pituitary disease:** In this case the adrenal cortex is not itself diseased, but the lack of ACTH availability results in a degree of atrophy of the gland which is secondary and usually severe. However, complete recovery is possible if adequate supply of ACTH is given from exogenous sources. In this type of hypoadrenalism there is little or no disturbance in electrolyte metabolism as aldosterone secretion is not affected.

3. **Failure of ACTH releasing mechanisms:** The feedback mechanism does not operate and ACTH release does not occur. This results in atrophy of the adrenal gland.

Other biochemical features of *Addison's* disease are described in the section on functions of cortisol. The general symptoms are *loss of appetite, rapid loss of weight, a general feeling of weakness and easy fatigability and a low degree of resistance to infection.* In adrenal cortex insufficiency the patient cannot readily cope with challenges to homeostasis by noxious stimuli. Thus haemorrhage, physical trauma, infectious agents, sensitizing antigens or certain chemicals may produce fatal consequences. There is marked sensitivity to insulin and thyroxine (*see* Physiology of Stress discussed later in this chapter).

L. Hyperfunction of the Adrenal Cortex

This is a condition in which effects of excessive cortisol dominate the picture. It may result from adrenocortical cell tumours in the adrenal gland or of extra adrenal locus, or excessive cortisol treatment.

1. **Cushing's syndrome:** In Cushing's syndrome (Plate 87.1) there is a general depletion of proteins with wastage in muscle, skin, blood vessels and bone, resulting in general weakness, haemorrhage, etc. The excessive release of glucose from the liver causes hyperglycaemia and the overload on the pancreas to secrete insulin may result in diabetes. The hyperglycaemia leads to the

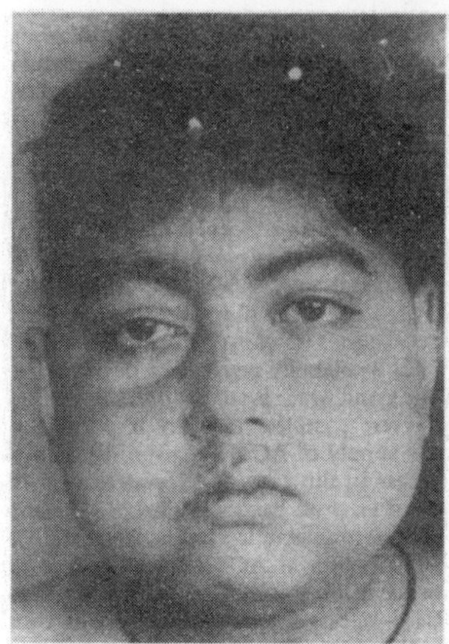

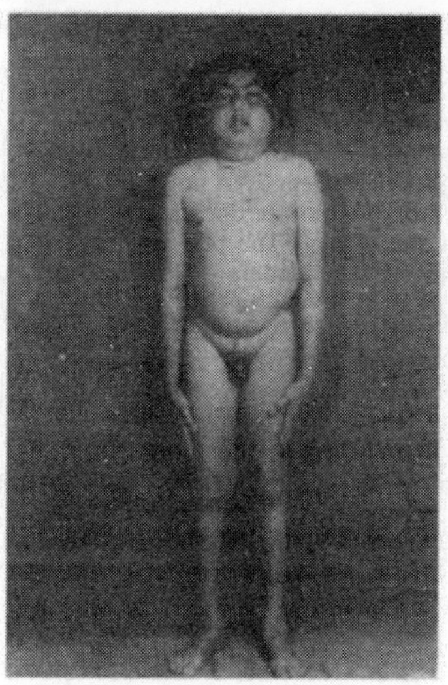

Plate 87.1   Cushing's syndrome. Male, age 18 years. Note the moon face, fish mouth and truncal obesity. The patient was operated upon and adrenal tumour was removed. (Photographs by kind courtesy of Prof. M.M.S. Ahuja, Professor of Medicine, A.I.I.M.S.)

deposition of fat with a peculiar distribution causing a characteristic buffalo hump and moon face. The earlier sign of Cushing's syndrome may be loss of the normal circadian rhythm and a raised midnight cortisol level. Estimation of 17-ketosteroids is also useful in diagnosis of Cushing's syndrome.

2. **Hyperaldosteronism:**   *Primary aldosteronism:* The symptom of primary aldosterone excess may be divided into three groups: (a) those due to severe potassium loss, (b) those associated with the hypertension that is almost invariably present, and (c) those associated with the disturbances of renal function. Due to potassium lack and rise in sodium content, there is muscle weakness. The hypertension is frequently severe and headaches which may be both persistent and severe, are a usual feature. The renal changes may arise due to sustained hypertension or the prolonged and severe hypokalaemia. These changes render the kidney particularly vulnerable to recurrent infection.

M. Genetic Abnormalities in Adrenal Cortical Function

Adrenogenital syndrome: This may occur as a result of congenital abnormalities or acquired adrenal disease. There is hypersecretion of intermediate steroids with sex hormone activity. The manifestations vary according to the age and sex of the patient. Adult women become masculine in appearance and the condition is called *adrenal virilism*. The voice *deepens, menstruation ceases, the breasts atrophy and hair may grow on the face, chest and limbs. In the adult male there is an overgrowth of hair, enlargement of the penis and increased sexual desire. In children puberty appears prematurely and unusal muscular development and rapid growth is seen* (Plate 87.2).

C_{17}, C_{21} and C_{11} hydroxylase occupy key positions in the synthesis of corticosteroids. The congenital adrenal hyperplasia are due to deficiencies in one or more enzymes essential for normal synthesis of steroid hormones.

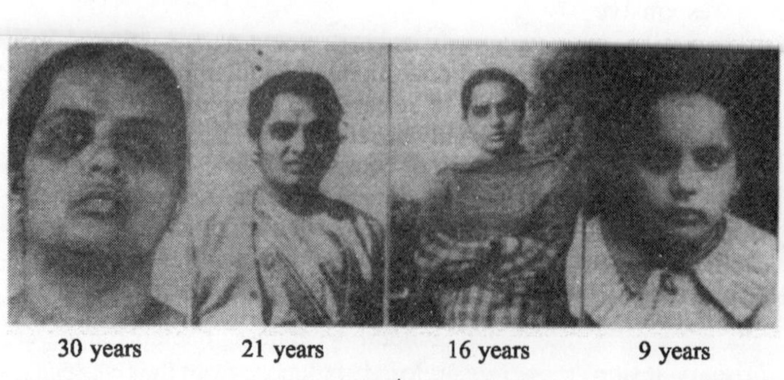

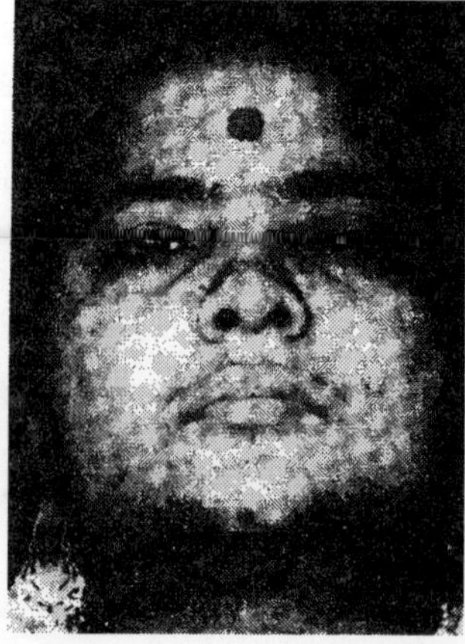

Plate 87.2   A. Adrenogenital syndrome. See features of hirsutism and virilization. B. Marked hirsutism in a female age 28 years. (Photographs by kind courtesy of Prof. M.M.S. Ahuja, Professor of Medicine, A.I.I.M.S.)

The absence of 21 hydroxylase results in inadequate production of cortisol, corticosterone and aldosterone. Since cortisol controls ACTH release by a feedback mechanism, the net result is an increase in ACTH secretion which cannot be stopped and excessive amounts of pregnenolone, progesterone and 17 OH progesterone accumulate and are diverted to androgen synthesis.

Inadequacy of 11-hydroxylation reaction also interferes with normal production of cortisol, corticosterone and aldosterone resulting in adrenocortical hyperplasia due to excessive ACTH production. In this case, 11-deoxycortisol and 11-deoxycorticosterone accumulate.

The clinical abnormality in cases of deficiency of 17-hydroxylase is the inability of adrenal gland to form cortisol, while corticosterone and deoxy-corticosterone formation is normal. Deficiency of 17-hydroxylase appears to affect the ovary too. Formation of androgens and estrogens is not possible. As a result the patient is phenotypically female but is unable to mature, and primary amenorrhea occurs.

In 3β-hydroxysteroid dehydrogenase deficiency, masculinization of females occurs but is also more extreme than that seen in the 21-hydroxylase defect and the major secretory product is DHEA (Figure 87.1).

N. Synthetic Adrenal Steroids

The natural steroids have a large number of pharmacological actions, some of which are desirable, such as the anti-inflammatory effect, and some less desirable or definitely disadvantageous. The goal is to seek substances so that less desirable effects of the drug are suppressed and the more desirable ones enhanced (Table 87.1).

TABLE 87.1 Activity of Some Natural and Synthetic Steroids Relative to that of Cortisol

Compound	Anti-inflammatory potency	Na^+-retaining potency	Duration of action
Cortisol	1	1	S
Fludrocortisone	10	125	S
Prednisone and Prednisolone	4	0.8	I
Betamethasone and Dexamethasone	25	0	L

S, short (i.e., 8–12 hour biological half-life);
I, intermediate (i.e., 12–36 hour biological half-life);
L, long (i.e., 36–72 hour biological half-life).

II. HORMONES OF THE ADRENAL MEDULLA

Catecholamines are released by the adrenal medulla and the sympathetic nervous system. The nerve supply of adrenal medullary cells is analogous to that of the postganglionic sympathetic neurons. Norepinephrine is the major neurotransmitter in the peripheral sympathetic nervous system, epinephrine is the primary hormone secreted by the adrenal medulla. Both the amines are water soluble and are derived from the essential amino acid tyrosine via 3, 4-dihydroxyphenylalanine (DOPA), which is decarboxylated to the third catecholamine, 3, 4-dihydroxyphenylethylamine (Dopamine). Dopamine is the third naturally occurring catecholamine and is present in the CNS and periphery. Catecholamines are involved in regulating a host of physiological functions particularly in integrating responses to variety of stresses that would otherwise threaten homeostatic mechanisms.

A. Biosynthesis of Catecholamines

The initial step in the biosynthesis of catecholamines is the hydroxylation of tyrosine to 3, 4-dihydroxyphenylalanine (DOPA). The enzyme tyrosine hydroxylase catalyzes this reaction. It utilizes molecular oxygen which is activated by tetrahydrobiopterin so that one atom of oxygen goes onto the aromatic ring of the substrate and the other oxidizes the coenzyme.

Tetrahydrobiopterin + Tyrosine + O_2

Dihydrobiopterin + DOPA + H_2O

The formation of DOPA is the rate-limiting step in the biosynthesis of epinephrine. The next step is the decarboxylation of DOPA to dopamine (3,4-dihydroxyphenylethylamine) and is catalyzed by L-amino acid decarboxylase, which uses pyridoxal phosphate as the cofactor. This enzyme is not specific for DOPA and is widely distributed in the kidney, liver, and adrenal medullary tissue. Some of the nerves in the central nervous system have specially high concentrations of dopamine. In fact the decarboxylation steps in the formation of histamine, tyramine and serotonin are all catalyzed by this enzyme. Inhibitors of the L-amino acid decarboxylase have been widely used for therapeutic modification of catecholamine biosynthesis.

Parkinsons disease is loss of dopaminergic neurons in a region of the midbrain-substantia nigra. L-DOPA (L-DOPA) is converted into dopamine in the dopaminergic neurons) and dopanine agonists are used to relieve the symptoms.

In the nerves and adrenal medulla dopamine is oxidized, on the β-carbon of the side chain producing norepinephrine. The reaction is catalyzed by dopamine β-hydroxylase. The enzyme is also not specific and can convert a variety of phenyl-ethylamines to their β hydroxylated derivatives. The enzyme with a M.W. of 290 Kd is a mixed function oxidase containing copper, and uses ascorbate (vitamin C) as a second reducing agent in reaction with molecular oxygen. Dopamine β-hydroxylase is localized in the catecholamine storage particles of adrenergic tissues and dopamine must be taken up by these granules before hydroxylation can occur. Norepinephrine is the transmitter in sympathetic nerves and at some synapses in the central nervous system, and is the final product in these tissues. In the adrenal medulla phenylethanolamine-N-methyl transferase catalyzes the methylation of norepinephrine to epinephrine. This enzyme is inhibited by its product epinephrine. In mammals, the transferase is absent during the early period of fetal development and its concentration is increased by the administration of glucocorticoids. Therefore, the cortex and medulla, although derived from two separate embryonic regions, are together for a biological function. Venous drainage, with blood containing the highest concentration of steroids from the cortex, bathes the medullary cells.

Epinephrine comprises 80% of the total amines in the adrenal medulla, norepinephrine 10%, and the remaining 10% are other amines. In nervous tissue, phenylethanolamine-N-methyltransferase is absent. Therefore in this tissue biosynthesis of catecholamines stops at the level of norepinephrine (Figure 87.4).

B. Storage and Release of Epinephrine

Epinephrine synthesized in the adrenal medulla is stored in the form of granules (0.1–0.5 μm in diameter) in the chromaffin vesicles. These membrane bound vesicles contain about 20 per cent epinephrine and about 4 per cent ATP. Two pools of epinephrine can be detected, one that is kept from active metabolism by cytoplasmic enzymes, which has a half-life of about 24 hours and an active pool (with a half-life of 2 hours) that can be released by sympathetic nerve activity. Nerve impulses reaching the medulla cause the exocytosis of epinephrine from these granules into the surrounding extra-

cellular fluid and then into blood. Under normal conditions epinephrine level in blood is about 0.06 micro gram per litre (about 10^{-10} M) but during an alarm reaction the concentration of epinephrine in blood may increase in seconds by thousand fold. Epinephrine prepares the body for emergency in several ways which are discussed in Section D below.

C. Metabolism of Epinephrine

The amines from the blood are mainly catabolized in the liver. The major route of catabolism is by methylation, and oxidative deamination. Both norepinephrine and epinephrine are first oxidatively deaminated by monoamine oxidase (MAO) to 3, 4-dihydroxyphenylglycolaldehyde (DOPGAL) and then either reduced to 3, 4-dihydroxy-phenylethylene glycol (DOPEG) or oxidized to 3, 4-dihydroxy-mandelic acid (DOMA). Alternatively, they can be first methylated by catechol O-methyl transferase (COMT) to metanephrine or normetanephrine. Products of either type of reaction are then metabolized by COMT or MAO to form the major excretory products, 3 methoxy-4-hydroxyphenyl-ethylene glycol (MOPEG) or 3 methoxy-4-hydroxyphenyl glycol aldehyde (MOPGAL) and 3 methoxy-4-hydroxy mandelic acid (VMA). VMA is the major metabolite of catecholamines excreted in the urine. Measurement of the concentrations of catecholamines and their metabolites in blood and urine is useful in the diagnosis of pheo-chromocytoma, a catecholamine secreting tumour of the adrenal medulla.

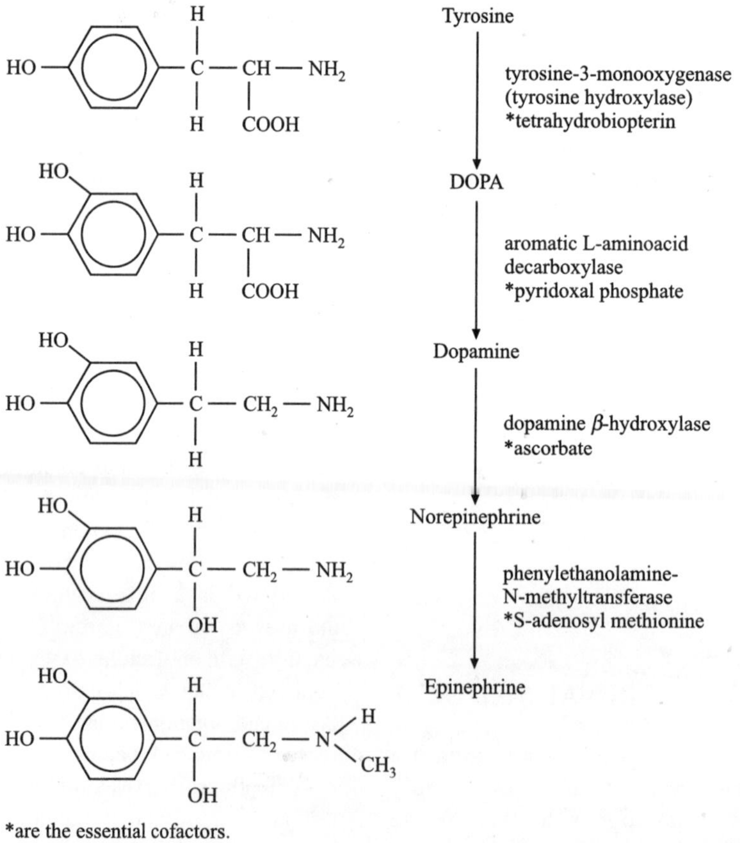

*are the essential cofactors.

Figure 87.4 Biosynthesis of dopamine norepinephrine and epinephrine.

The glycol and to some extent the o-methylated amines may be conjugated to the corresponding sulphates or glucoronides. The enzyme COMT is found in cytoplasm of most tissues and exists as a mixture of isoenzymes. The highest activity is found in the liver and kidney. The enzyme is specific for the catechol grouping and requires magnesium.

Monoamine oxidase (MAO)—a copper containing flavoprotein (M.W. = 290 Kd) catalyzes the oxidative deamination of compounds having the amino group attached to the terminal carbon atom as in epinephrine, norepinephrine and dopamine. It is particularly concentrated in the mitochondria of the liver, gastrointestinal mucosa and sympathetic nerves.

After an intravenous injection of labelled epinephrine to a human subject more than 80% was recovered in the urine in the first 48 hours as methylated derivates, 6% as epinephrine, 7% as 3-methoxy-4-hydroxyphenylglycol sulphate and 2% as 3, 4-dihydroxymandelic acid. The metabolic pathway is shown in Figure 87.5.

Monoamine oxidase (MAO) inhibitors are a class of anidepresent drugs prescribed for the treatment of depression. MAO inhibitors prevent the breakdown of monoamine neurotransmitters and thereby increase their availability.

D. Functions of Epinephrine

Epinephrine acts on the liver and skeletal muscle as well as the heart and the vascular system. On the membranes of the target tissues there are two kinds of adrenergic receptors which were designated alpha and beta receptors. More recently the development of selective agonists and antagonists that act at

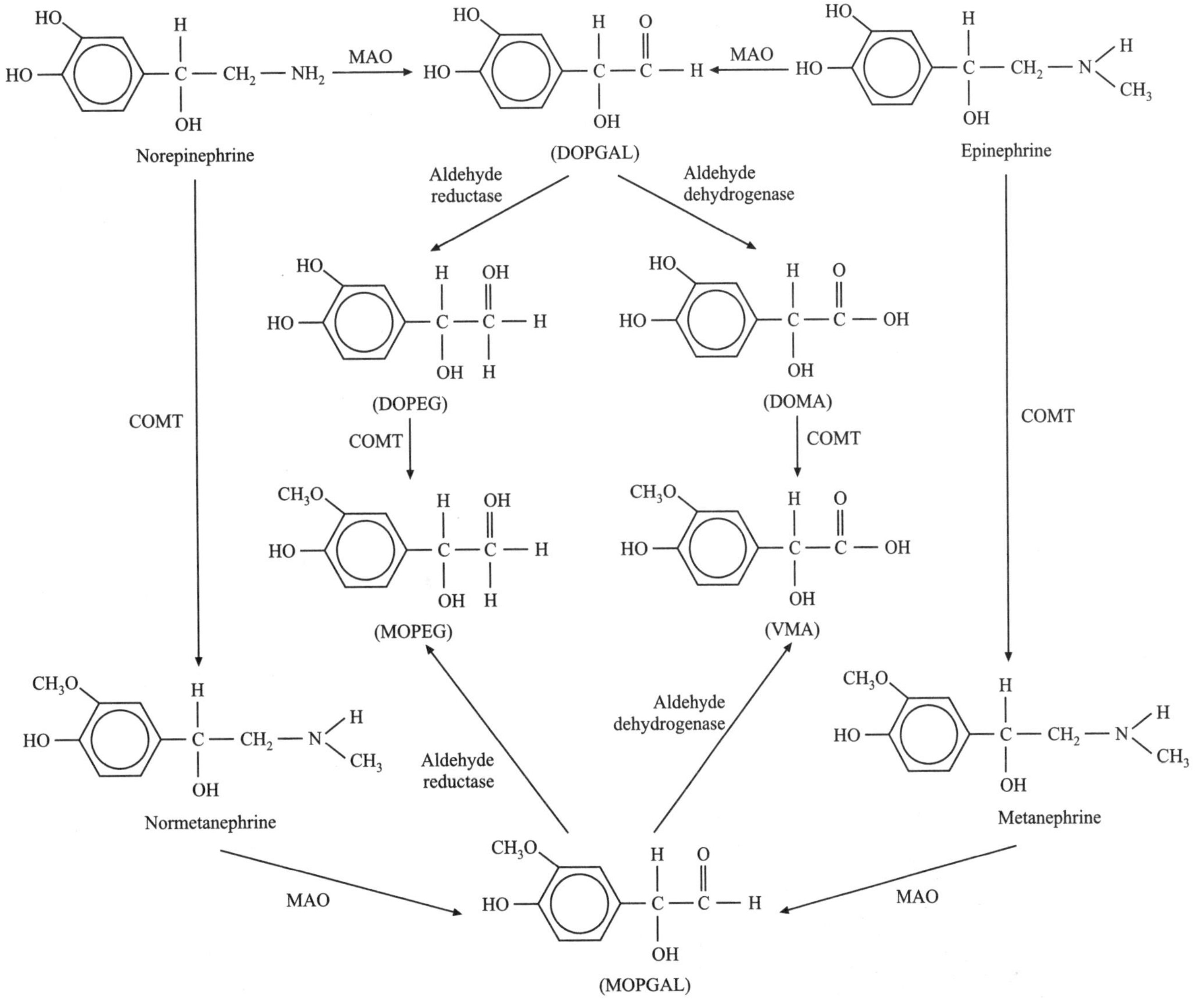

Figure 87.5 Metabolism of catecholamines. MAO, Monoamineoxidase; COMT, Catechol-o Methyltransferase; DOPGAL, 3,4-dihydroxyphenyl-glycolaldehyde; DOPEG, 3,4-dihydroxyphenylethyleneglycol; DOMA, 3,4-dihydroxymandelic acid; MOPEG, 3-methoxy-4-hydroxyphenylethyleneglycol; VMA, 3-methoxy-4-hydroxymandelic acid; MOPGAL, 3-methoxy-4-hydroxyphenylglycol aldehyde.

adrenergic receptors has allowed their further classification into subtypes.

Many responses mediated by beta receptors are associated with activation of adenylate cyclase and intracellular accumulation of cyclic AMP. The epinephrine induced glycogenolysis in liver and skeletal muscle, lipolysis in adipose tissue, positive inotropic effect on the heart and increased amylase secretion from salivary glands all appear to be mediated by cyclic AMP. There is strong evidence that alpha receptors are sites at which the formation of cyclic AMP is inhibited.

The most striking effects of epinephrine are those elicited by severe alterations in external or internal environment and are adjusted to the *preservation of circulation and supply of metabolic energy*. Epinephrine increases the heart rate, the heart output and the blood pressure. It stimulates glycogen breakdown in liver, with release of glucose in blood. It promotes the anaerobic breakdown of the glycogen of skeletal muscle into lactate via glycolysis, and ATP produced during this process is used for muscular contraction. In life-threatening situations when the cardiovascular system is in acute collapse, epinephrine is used as one of the valuable drugs. Epinephrine also relieves the symptoms of acute asthma by relaxing the smooth muscle surrounding the bronchioles of the lung.

Epinephrine is one of the *most potent vasopressor drugs* known. The mechanism attributable to epinephrine is three-fold:

1. A direct myocardial stimulation that increases the strength of ventricular contraction (Positive inotropic action).
2. Increased heart rate (Positive chronotropic action).
3. Vasoconstriction in many vascular beds, especially in the skin mucosa and kidney along with marked constriction of the veins. Positive inotropic and positive chronotropic responses are due to stimulation of β-receptors with increase in cyclic AMP resulting in enhanced phosphorylation of proteins and metabolism of the heart muscle. Direct activation of voltage sensitive Ca^{2+} channels by Gs may also contribute to the inotropic response. Vasoconstriction is mediated by binding of epinephrine to alpha receptors. Renal blood vessel constriction increases the release of renal pressor substances which in turn constrict the blood vessels throughout the body and thus increase the peripheral resistance against which the heart pumps. It is this pumping against great resistance which increases blood pressure.

Effects on smooth muscle

Gastrointestinal smooth muscle is, in general, relaxed by epinephrine. This effect is due to activation of both alpha and beta receptors. Intestinal tone and the frequency and amplitude of spontaneous contractions are reduced ($alpha_1$, $alpha_2$, $beta_2$ receptors are involved). Stomach is usually relaxed and the pyloric and ileocecal sphincters are contracted ($alpha_1$-receptors).

Effects on respiration

Epinephrine affects respiration primarily by relaxing the bronchial muscle. It *has a powerful bronchodilator action and is used in bronchial asthma* (β-receptors). By its vasoconstricting action, it causes shrinkage of the mucosa and reduces the secretion of the mucus (α-receptors). Epinephrine also inhibits the antigen induced release of inflammatory mediators from most cells (β_2-receptors). Epinephrine has variety of therapeutic uses based on its actions on blood vessels, heart and bronchial muscle.

The most common uses of *epinephrine are to relieve respiratory distress due to bronchospasm*.

It provides rapid relief from hypersensitivity reactions to drugs and other allergens.

Effects on metabolism of liver, heart and skeletal muscle

The mechanism is summarized as follows:

(a) Epinephrine interacts with specific beta receptors in the membranes of liver and muscle cell and activates the enzyme adenylyl cyclase. This results in an increase in intracellular cyclic AMP levels.
(b) Cyclic AMP activates protein kinase which catalyzes the phosphorylation of phosphorylase kinase.
(c) Active phosphorylase kinase catalyzes the conversion of inactive phosphorylase (phosphorylase b) to active phosphorylase (phosphorylase a).
(d) Active phosphorylase catalyzes the break-down of glycogen to glucose-1-phosphate, from which glucose-6-phosphate and then free glucose are formed and rule in blood.

Each step in this sequence is a magnification of the previous response. Therefore, the stimulation of glycogen breakdown by epinephrine occurs by amplification cascade in the range of about 25-million fold. The cascade of epinephrine effect (Figure 87.6) up to the point of glucose-6-phosphate is identical in liver, heart and skeletal muscle. Since muscle lacks glucose-6-phosphatase, the end product of glycogenolysis in this tissue is not glucose but lactate. Therefore, epinephrine induced glycogen breakdown in muscle results in formation of glucose-6-phosphate, which via glycolysis is metabolized to lactate, ATP derived from glycolysis is used for muscular contraction. The cascade effect of epinephrine in liver (Figure 87.6) is also set in operation by the pancreatic hormone glucagon.

While epinephrine via cyclic AMP *stimulates glycogen breakdown in liver, it inhibits glycogen synthesis*. Glycogen synthase also exists in two forms. The dephospho form of glycogen synthase is active. Cyclic AMP promotes the phosphorylation of the dephospho (active) form of glycogen synthase to its phosphorylated inactive form. Therefore, the same stimulus (epinephrine) that causes increased glycogen breakdown, also inhibits glycogen synthesis. In this way all the liver glycogen is converted to glucose, which is released in the blood thus providing the muscles with maximal supply of fuel.

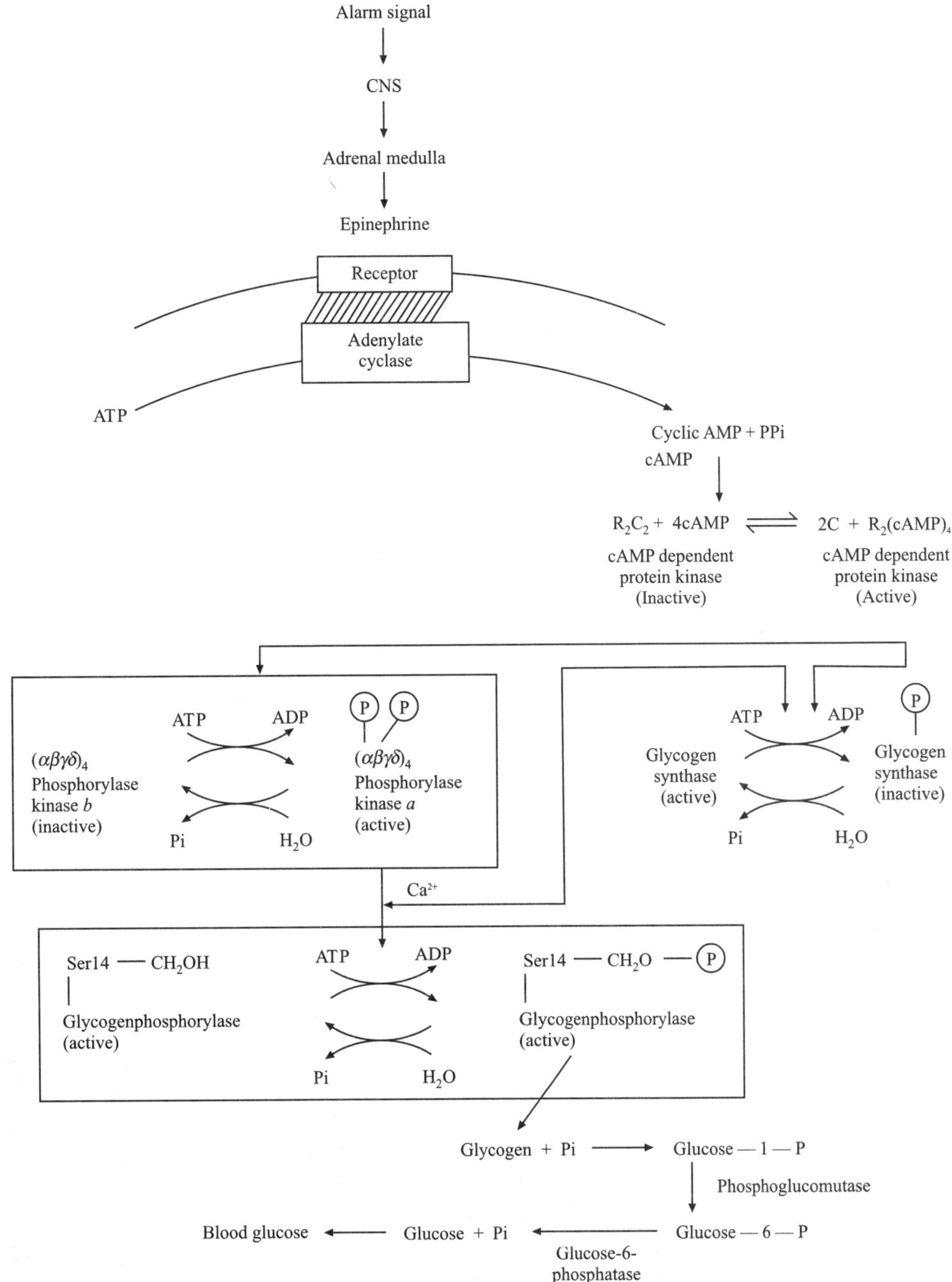

Figure 87.6 Mechanism of action of Epinephrime.

When the stimulus is removed and the skeletal muscle can relax, and heart slows down, another sequence of events occur. Epinephrine is removed by a sequence of reactions described above (half life in blood is about 10 seconds) and cyclic AMP is no longer formed, at the elevated rate. After hydrolysis of cyclic AMP by cyclic AMP phosphodiesterase, protein kinases loose their activity and all the enzymes return to their original resting state.

In *adipose tissue, epinephrine* via cyclic AMP *stimulates lipolysis*, resulting in increased mobilization of triglycerides to free fatty acids. Therefore, the alarm signal that leads to secretion of epinephrine causes the release of free fatty acids from the tryglyceride store in adipose tissue, for use by the skeletal muscle. When insulin is bound to the cell surface receptors of adipocytes it cancels the effects of epinephrine by decreasing the effect on lipase. Some of the effects of epinephrine on release of free fatty acids are counteracted by prostaglandins.

Major biochemical responses to adrenergic impulses are summarized in Table 87.2.

TABLE 87.2   Biochemical Responses to Adrenergic Impulses

Effector organ	Receptor type	Biochemical response
Skeletal muscle	β_2	Increased contractility glycogenolysis, K^+ uptake
Liver	β_2	Glycogenolysis
Pancreas		
Acini	$\alpha*$	Decreased secretion[+]
Islets (β cells)	α_2	Decreased secretion[+++]
	β_2	Increased secretion[+]
Fat cells	$\alpha_2, \beta_1 (\beta_3)$	Lipolysis[+++] (thermogenesis)
Salivary glands	α_1	K^+ and water secretion[+]
	$\beta*$	Amylase secretion[+]
Lacrymal glands	$\alpha*$	Secretion[+]
Pineal gland	$\beta*$	Melatonin synthesis
Posterior pituitary	β_1	Anti-diuretic hormone secretion

*receptor subtypes have not been identified in humans.

The responses of epinephrine mediated by β-receptors are associated with activation of adenylate cyclase and intracellular accumulation of cyclic AMP. Binding of epinephrine to the α-receptors activates the phospholipase c β isoform through the Gq family of G protein. This results in the formation of diacylglycerol and inositoltriphosphate and the latter stimulates release of calcium from endoplasmic reticulum and finallly causes activation of protein kinase C, which in turn causes phosphorylation of channel proteins, ion pumps and ion exchange proteins (Ca^{2+} transferase ATPase). Alpha adrenergic stimulation of phospholipase A_2 results in formation of Arachidonate, a precursor of cyclooxygenase and lipooxygenase pathways, thus generating prostaglandins and leukotrienes. In most smooth muscles the mobilization of intracellular Ca^{2+} causes contraction as a result of activation of Ca^{2+} sensitive protein kinases such as phosphorylation of light chain of myosin which is associated with the development of tension. In contrast, the increased concentration of intracellular Ca^{2+} that results from alpha$_1$ adrenergic receptor stimulation in gastrointestinal smooth muscle causes hyperpolarization and relaxation by activation of Ca^{2+} dependent K^+ channel.

E.   Hyperfunction of the Adrenal Medulla

Hyperfunction is due to chromaffin tissue tumours, termed pheochromocytomas. These tumours are characterized by excessive production, storage and release of catecholamines. Plasma levels of norepinephrine and epinephrine may rise to more than 500 times normal. An increase in free fatty acids and increased urinary excretion of these amines and 3-methoxy-4-hydroxy-mandelic (VMA) acid is generally seen. The dominating

features are signs of excessive β-receptor stimulation such as tachycardia, hypertension, elevation of BMR and glucosuria.

III.   PHYSIOLOGY OF STRESS

Any threat to body homeostasis or equilibrium is what we call *stress*. Selye's experiments showed that continuous exposure to cold, exercise, low oxygen tension, burns, wounds, acute and chronic infections, toxic agents, and shock producing drugs, etc., produced definite changes in the structure and chemical composition of the body. These diverse type of changes were manifested as enlargement of adrenal cortex, thymolymphatic involution and intestinal ulcers. Besides these, there was a sharp fall in the number of circulating eosinophils. To the totality of these changes the name *general adaptation syndrome* was given. It develops in three stages: alarm reaction, stage of resistance, and the stage of exhaustion. During the alarm reaction the cells of the adrenal cortex discharge stored lipid granules, cholesterol and ascorbic acid, and an increase in the level of active cortical steroids is seen in the blood. Conversely, in the stage of resistance, the cortex accumulates an abundant reserve of secretory granules. After prolonged exposure to any of the noxious agents, the acquired adaptation is eventually lost, and the phase of exhaustion sets in. (Adrenal exhaustion is a rare cause of death in organic disease but is an important factor when organic disease has affected the adrenal glands or when prolonged steroid therapy has rendered the glands either atrophic or non-responsive to the stimulus). Many nervous and emotional disturbances, high blood pressure, gastric ulcers, certain types of rheumatic, allergic, cardio-vascular and renal diseases appear to be diseases of adaptation. *Adaptability is the most distinctive characteristic of life and the penalty for its failure is disease.*

Stress therefore can also be described as the state manifested by a specific syndrome which consists of all the non-specifically induced changes within a biological system. Thus stress has its own characteristic form and composition and it shows itself as a specific syndrome, yet it is non-specifically induced. (A non-specifically found change is one that affects all or most parts of a system without selectivity and can be produced by many agents.)

The general belief that only serious disease or intensive physical and mental injury can cause stress is not true. In fact any emotional activity can cause stress; stress is not necessarily bad, the only prerequisition is that your system should be prepared to take it. The same stress which may make one person sick, may be an invigorating experience for another. It is indeed the constant interplay between our mental and bodily reaction which influence the strains of everyday existence.

Response of human adrenal gland to stress

Exposure to anoxia (equivalent to 30 minutes at 27,000 ft) or cold (4 hours at 4°C), *provoked a 40–60% fall in circulating eosinophils*, which is the most sensitive indicator of stress. Plasma levels of *cortisol are higher* in *anxious subjects.* In general surgery, plasma 17-OH cortisol shows a moderate rise during induction of anaesthesia, the rise persists during and after surgery, (the magnitude is proportional to the duration and severity of the operative procedure) and lasts for 6 hours after completion of the operation. Increased production of cortisol with a resultant rise in the blood level, is the usual response to almost any major threat to the body equilibrium on homeostasis. The reaction of the adrenal gland results from nervous stimuli reaching the region of the hypothalamus and then stimulating the pituitary to produce ACTH. If cortisol is not produced in sufficient quantity, sudden collapse may result from stress and may even prove fatal.

SUGGESTIONS FOR FURTHER READING

Martin E. Feder (1999), Heat Shock Proteins, Molecular Chaperones and the Stress Response, *Ann. Rev. Physiol*, 61:243–282.

Miller W.L. and Auchus R.J. (2011), The Molecular Biology, Biochemistry and Physiology of Human Steroidogenesis and its Disorders, Endocr Rev., 32:81–151.

Tsai L.C. and Beavo J.A. (2011), The Roles of Cyclic Nucleotide Phosphodiesterases in Steroidogenesis, *Curr. Opin Pharmacol.*, 6:670–675.

88

The Thyroid Gland

Kalpana Luthra

I. INTRODUCTION

The thyroid, one of the largest endocrine gland in the body, regulates major metabolic and endocrine activities of the body such as *fuel combustion in the body, protein synthesis and response to other hormones*. It participates in these processes by producing thyroid hormones, principally thyroxine (T_4) and triiodothyronine (T_3). The thyroid hormones are amino acid derived hormones, 'tyrosine' being the amino acid precursor which is iodinated to form monoiodotrosine (MIT) and diiodotyrosine (DIT) that condense to form T_4 and T_3. These hormones regulate the rate of metabolism and affect the growth and rate of function of many other systems in the body. Thyroid hormones influence most functions in the human body and exert their action on every cell (Table 88.1). *Thermogenesis* (calorigenic effect) is the major effect of thyroid hormone and it is mediated by *uncoupling of oxidative phosphorylation*. Thyroxine (T_4) increases the basal metabolic rate (BMR) and metabolic activities of the cell by promoting RNA and protein synthesis. Along with growth hormone, thyroid hormones are essential for early growth and development. Thyroid hormones stimulate mitochondrial oxygen consumption and production

of ATP. T_3 induces an increase in Na^+-K^+-ATPase activity. Inhibition of the Na^+-K^+-ATPase activity by ouabain, and other sodium pump antagonists markedly reduces the action of thyroid hormones on heat production and oxygen consumption. Elevated T_3 levels leads to protein catabolism and negative nitrogen balance. Both *excess and deficiency of thyroxine can cause disorders*. Thyrotoxicosis or hyperthyroidism, as observed in Grave's Disease, is caused by an excess of circulating free T_4, free T_3, or both. Hypothyroidism is a condition where there is a deficiency of T_4, T_3, or both.

Hyperthyroidism: There is a marked loss of body weight; increased rate of gluconeogenesis and breakdown of carbohydrates. Fatty acid metabolism is increased with increased degradation of cholesterol, leading to a decrease in blood cholesterol levels.

Hypothyroidism: The most common cause for hypothyroidism is primary thyroid disease such as in autoimmune thyroiditis, e.g., Hashimoto's thyroiditis, leading to myxedema in adults, females are more affected than males. Other primary causes include thyroidectomy and radiation therapy. Anti-thyroid

1050

TABLE 88.1   Major roles of Thyroid Hormones in Mammals

Feedback Inhibition of hypothalamic TRH and pituitary TSH secretion
Promotes the action of many other hormones:
    Enhances lipolytic response of adipose tissue to hormones
    Required for the growth-promoting activity of GH
Regulates basal metabolic rate:
    Increases mitochondrial oxidative phosphorylation
Required for hepatic conversion of carotenes to vitamin A
Required for bone growth and maturation
Required for nervous system differentiation in early development
Required for pituitary prolactin and growth hormone synthesis
Increases the rate of intestinal glucose absorption
Increases human red blood cell Ca^{2+} ATPase activity
Induces synthesis of a number of enzymes:
    Na^+/K^+ - ATPase (or a protein component or activator of the sodium pump)
    Carbarmoyl phosphate synthetase (a phosphotransferase)
    α-Lactalbumin (lactose synthetase system proteins)
    Hepatic pyruvate carboxylase (converts pyruvate to oxaloacetate)
    Chromatin protein kinase
    Mitochondrial α-glycerophosphate dehydrogenase
    Malic dehydrogenase (converts malic acid to oxaloacetate)
    Hyaluronidase (dissolves intercellular ground substance)
Induction of cellular proteins (other than enzymes):
    Prolactin, growth hormone, lung surfactants, brain nerve growth factor (NGF)

drugs, lithium and para-aminosalicylic acid can lead to hypothyroidism. Congenital hypothyroidism is seen in iodine deficiency, absent or ectopic thyroid gland, dysfunctions in hormone production and due to mutations in the TSH receptor. Iodine deficiency may lead to euthyroid goiter (Figure 88.1) wherein TSH levels are elevated causing continued stimulation of the thyroid gland leading to hyperplasia and goiter. Secondary hypothyroidism is due to diseases of the pituitary or hypothalamus. Tumour of the pituitary, pituitary surgery, irradiation infiltration, Sheehan's syndrome and TSH deficiency are the factors that lead to hypothyroidism.

Symptoms of hypothyroidism include lethargy, intolerance to cold, tolerance to heat, slow heart rate, weight gain, dry coarse skin, slow responses and sluggishness. Hypothyroidism in children leads to mental and physical retardation (cretinism).

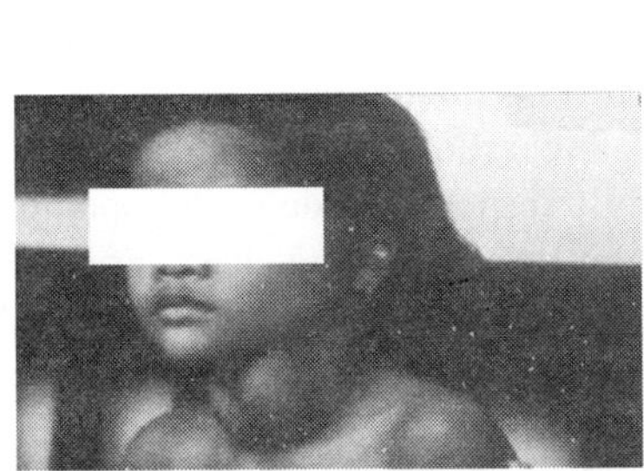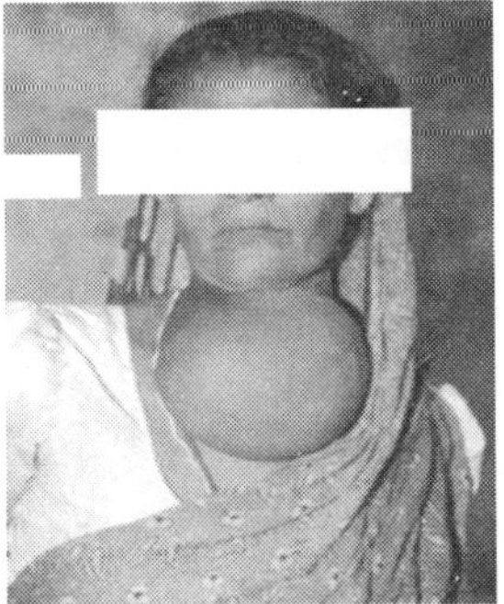

Figure 88.1   Images of individuals suffering from euthyroid goitre.

Prompt diagnosis and treatment in cretinism can prevent irreversible retardation in growth and development.

Some of the major effects of hyperthyroidism and hypothyroidism are summarized in Table 88.2.

TABLE 88.2   Major effects of Hyperthyroidism and Hypothyroidism in humans

Hyperthyroidism	*Hypothyroidism*
Elevated T_4-T_3 levels	Decreased (or absent) T_4-T_3 levels
Elevated basal metabolic rate (BMR)	Low basal metabolic rate (BMR)
Increased perspiration	Decreased perspiration
Rapid pulse (increased cardiac output, hypertension)	Slow pulse (decreased cardiac output, hypotension)
Increased body temperature with intolerance to heat	Lowered body temperature with intolerance to cold
Warm, moist palms, nervousness, anxiety, excitability, restlessness, insomnia	Coarse, dry skin, Lethargy, depression, paranoia, sleepiness, tiredness
Weight loss	Weight gain
Muscle wasting	Loss of hair, dry and brittle texture
Increased appetite	Edema of face and eyelids
Menstrual irregularities	Menstrual irregularities
Goiter (primary or secondary origin)	Goiter (may or may not be present)

II. IODINE METABOLISM IN THE THYROID GLAND

The thyroid gland in humans is located in the neck, anterior to the trachea and consists of two lobes. The normal thyroid is 12 to 20 grams in size and is highly vascular. It is composed of numerous spherical follicles in which thyroid follicular cells surround secreted colloidal fluid containing large amounts of thyroglobulin (TG), the protein precursor of thyroid hormones. The thyroid follicular cells are polarized, with the basolateral surface facing the bloodstream and the apical surface facing the follicular lumen. Thyroid hormone secretion is regulated by thyroid-stimulating hormone (TSH). Binding of TSH to its receptors on the basolateral surface of the follicular cells leads to reabsorption of TG from the follicular lumen followed by proteolysis within the cytoplasm of the cells to release thyroid hormones for secretion into the bloodstream.

The thyroid gland is unique in its ability to abstract from circulation, and store in its acini, large quantities of the essential micronutrient, iodine. Iodine is a member of the halogen family of elements, and is scarce in the terrestrial environment. Iodine scarcity is particularly severe in lands far away from the sea or high above the sea level, because rainfall leaches away the water soluble iodine containing minerals from the soil. Therefore iodine storage ability was an important survival factor during the course of evolution of life from aquatic to terrestrial forms.

The thyroid gland abstracts iodine from the extracellular fluid. The normal iodine concentration of extracellular fluid in humans living in an iodine sufficient environment is approximately 0.2 micrograms % (μg%). However, in many parts of the world with an environmental iodine deficiency, it may be 0.05 μg% or less, while in countries like Japan where iodine intake is high, it may be 2.0 μg%.

The iodine in the extracellular fluid is primarily cleared by the thyroid gland and the kidneys. Only a small proportion of iodine in the extracellular compartment is cleared by sweat, saliva and gastric glands. The renal clearance of iodine in humans is relatively constant, however, the thyroidal clearance of iodide varies from environment to environment depending upon iodine availability, and the stage of iodine repletion of the gland.

The thyroid abstracts iodide from the extracellular fluid against a gradient by an energy dependent concentrating mechanism. Under normal conditions of iodine availability, a thyroid serum (T:S) concentration ratio of 20:40 is reached in humans by virtue of this 'active transport'. However, in iodine deficient environments, under the influence of TSH, this ratio may increase over a hundredfold.

The main source of iodide in the extracellular fluid is dietary iodine absorbed from the gut. Daily requirement of iodine is 150–200 μg/day. There are wide variations in the daily iodine intake among humans from one geographic area to the other. Thus, while the average daily intake in the U.S.A. is estimated to be about 500 μg, in many parts of the world with incidence of endemic goitre, it is less than 20 μg/day. The iodine contained in the diet is rapidly and efficiently absorbed from the gut, and normally there is no significant loss of dietary iodine in the stool. Dietary ingredients that prevent utilization of iodine are called goitrogens. Goitrogens are present in maize, millet, sweet potatoes, bamboo shoots and beans.

The thyroid gland itself is the second major source of iodide to reach the extracellular fluid. The estimated daily discharge of iodide by the thyroid is approximately 60 μg in iodine rich environments. Iodide released by the non-thyroidal tissues of the body as a result of deiodination of the metabolizing thyroid hormones, constitutes the third largest contribution to the iodide in the extracellular compartment.

In iodine sufficient environments, the estimated total iodide content of the extracellular compartment is 200–300 μg, while the intrathyroidal iodine stores can be as high as 8–10 mg. The iodine equivalent of thyronines secreted by the gland daily is estimated to be around 60 μg. The major bulk of this iodine is released by deiodination in the peripheral tissues, and either recycled back to the thyroid or excreted through the kidneys. Figure 88.2 depicts the quantitative relationship between the different iodine compartments of the body.

The metabolic handling of iodine by the thyroid gland can be depicted as a cyclical process which starts with concentration of iodine by active transport and ends in the

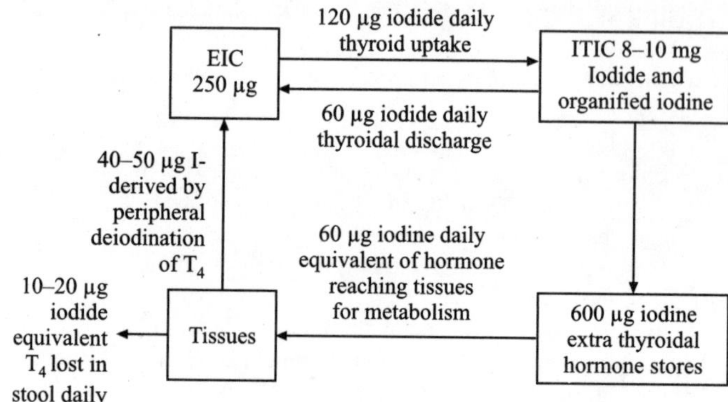

Figure 88.2   The quantitative relationship between the different iodide compartments of the body. EIC –Extracellular iodide compartment. ITIC: Intrathyroidal iodide compartment. Compartment **A** stores iodine mainly as protein-bound T₄ and T₃ in circulation.

generation of iodide by dehalogenation of iodinated tyrosyl residues (Figure 88.3). The iodide thus generated is partly discharged into the extrathyroidal compartment and partly reutilized. There are five distinct steps in this cyclic process:

1. Active transport of iodide into the thyroid from the extracellular fluid;
2. Oxidation of iodine
3. Organification of iodide (iodination of tyrosyl residues of thyroglobulin (TG));
4. Coupling (intramolecular thyronine synthesis);
5. Proteolysis of thyroglobulin (TG) and release of thyroid hormones and iodotyrosines;
6. Dehalogenation and release of non-hormonal iodine for recycling.

Each one of the above steps is important in the biosynthesis of thyroid hormones, and genetically determined or environmentally induced, or both, defects in one or more of these steps lead to functional abnormalities of the thyroid.

Active transport of iodide

The iodine concentrating ability of the thyroid gland is traditionally evaluated by determining the thyroid: serum concentration (T:S) ratio of radioactive iodine achieved after equilibrating with the isotope. The T:S ratio at equilibrium conditions reflects the balance between influx and efflux of iodide across the thyroid cell membrane. Though other tissues such as the salivary gland, mammary gland, choroid plexus, skin and placenta can concentrate iodine, the thyroidal membrane seems to have the highest efficiency in this respect. The mechanism that actively transports iodide (iodide pump) across the thyroid membrane has the following characteristics:

1. It is an energy dependent process requiring continued generation of ATP.
2. It is inhibited by inhibitors of oxidative phosphorylation
3. Transport is competitively inhibited by anions such as

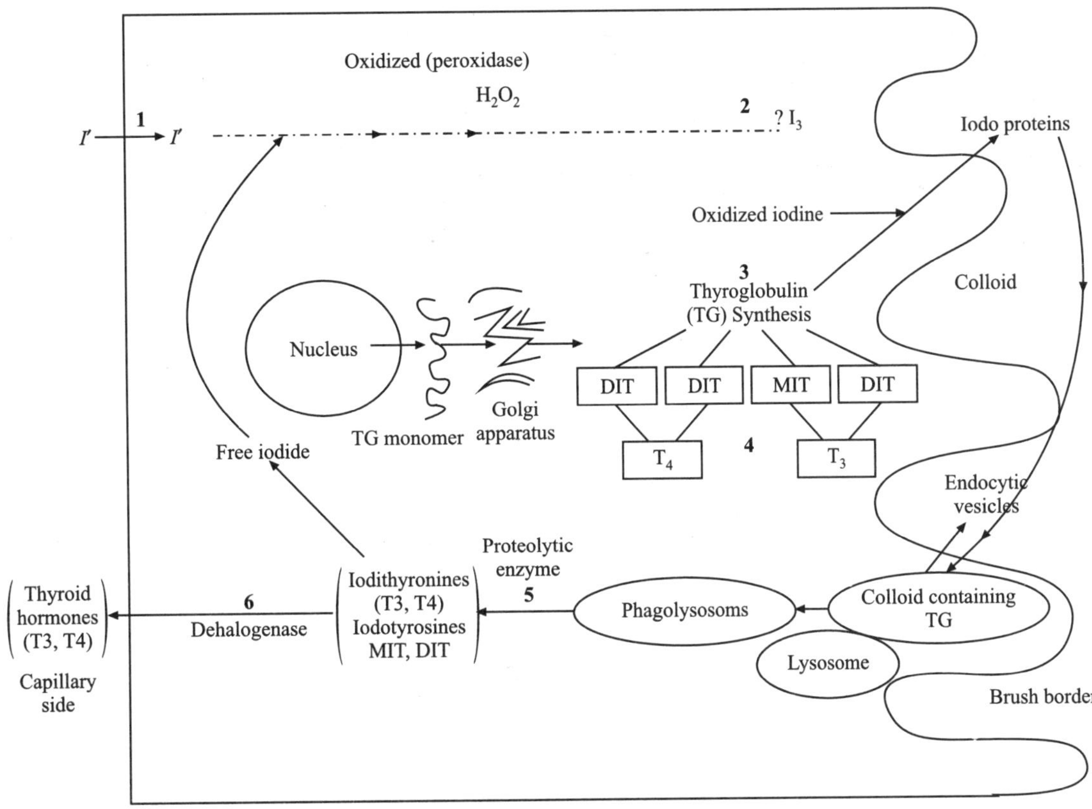

Figure 88.3 The iodine cycle

thiocyanate, perchlorate, tetrafluoroborate, pertechnitate and nitrate.

4. The iodide transport mechanism is ouabain sensitive.

A large number of factors are known to influence iodide transport activity of the thyroid. TSH, and thyroid stimulating immunoglobulins such as LATS (Long Acting Thyroid Stimulating antibodies) and HTS (Human Thyroid Stimulator) are among the biologically important stimulators of the thyroidal iodide pump. Anions such as perchlorate, thiocyanate and pertechnitate competitively inhibit iodide transport activity. Interestingly, iodide itself in large concentrations inhibits iodide transport activity.

Genetically determined enzymatic defects of thyroidal iodide transport have been described in man. Such individuals develop enlargement and ultimately functional failure of the thyroid.

III. STEPS INVOLVED IN THYROID HORMONE SYNTHESIS AND REGULATION

A. Oxidation of iodine

The iodide taken up by the thyroid cell is oxidized to active iodine by the enzyme thyroperoxidase (TPO). The hydrogen peroxide required for this step is produced by an NADPH dependent reaction in the Hexose Monophosphate (HMP) shunt pathway. Oxidation of iodine is stimulated by TSH and inhibited by antithyroid drugs such as thiourea, thiouracil and methimazole.

B. Organification of Iodide

The iodine is next incorporated into the tyrosyl residues in peptide linkage in the thyroglobulin (TG) molecule. Thyroglobulin, synthesized by the thyroid follicular cells, is a 660 kDa protein containing 115 tyrosine residues, of which 35 residues can be iodinated. Iodination of tyrosine takes place on the TG molecule in the follicular space leading to the production of mono-iodotyrosine (MIT) and di-iodotyrosine (DIT) (Figure 88.4).

Iodide itself, when present beyond an optimum concentration in the gland, has been shown to inhibit its organification. This phenomenon, known as the Wolf-Chaikoff effect, is responsible for the long recognized anti-thyroid action of pharmacological doses of iodide.

C. Coupling Reaction

Coupling of the tyrosine residues occurs on the surface of the follicular cells. When two di-iodotyrosine (DIT) molecules are coupled, one molecule of tetra-iodothyronine (T$_4$) or thyroxine is formed (Figure 88.3). The T$_4$ formed are attached to the thyroglobulin molecule. Tri-iodothyronine (T$_3$) may be formed by deiodination of T$_4$. Under physiological conditions, 99% of the hormone produced in the thyroid gland is T$_4$.

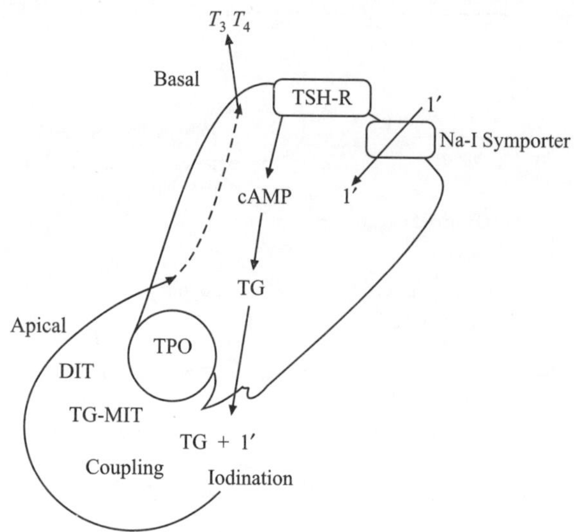

Figure 88.4 Thyroid hormone synthesis and regulation. Thyroid follicles are formed by thyroid epithelial cells surrounding proteinaceous colloid, which contains thyroglobulin. Follicular cells, which are polarized, synthesize thyroglobulin and carry out thyroid hormone biosynthesis. TSH-R, thyroid-stimulating hormone receptor; TG, thyroglobulin; TPO, thyroid peroxidase; DIT, diiodotyrosine; MIT, monoiodotyrosine.

D. Proteolysis of Thyroglobulin (TG) and Release of Thyroid Hormones

As detailed above, the iodide trapped in the thyroid is organified and stored as iodotyrosines and iodothyronines in the matrix of the TG molecules. TG, which forms the bulk of the iodoproteins and is stored in the thyroid acini, has to be resorbed and proteolysed for the release of the glandular stores of hormones and iodine. The cytological and cytochemical events that lead to the above process are initiated and promoted largely by the action of thyroid-stimulating hormone (TSH) of the anterior pituitary. TSH stimulation leads to endocytosis of the colloid by the apical membrane of the acinar epithelium. The colloid droplets thus resorbed fuse with cytoplasmic lysosomal particles to form phagolysosomes. Inside the phagolysosomes, proteolysis of TG occurs as a result of the presence of acid proteases and peptidases. The iodotyrosines and iodothyronines thus liberated are released into the cytoplasm. Microtubules and microfilaments present in the thyroid epithelium play an important role in the coordinated migration and fusion of colloid droplets and lysosomal particles intracellularly, and in the secretory activity of the acinar epithelium.

The iodothyronines (mainly T_4 and T_3) released intracellularly ultimately reach the circulation. However, the iodotyrosines (MIT and DIT) are deiodinated intra-cytoplasmically by a NADH-dependent deiodinase. The iodide thus released is partly reutilized in the hormone synthetic process while a small fraction of it leaks out into the extrathyroidal iodide pool.

IV. REGULATION OF SECRETION AND PERIPHERAL METABOLISM

A. Auto-regulation

The biosynthetic and secretory activities outlined above are regulated to some degree by an auto-regulatory mechanism of the thyroid (Figure 88.4). Processes such as iodide transport, organification as well as proteolysis which govern the secretion of thyroid hormones all come under this auto-regulatory mechanism. Iodine plays an important role in this process. Thus the iodine concentrating ability of the thyroidal membrane shows a sharp decline when intrathyroidal concentration of organified iodine goes beyond a certain optimum. The mechanism of this iodine-induced reduction of iodide transport maxima of the thyroidal membrane is not known. However, it is postulated that there is an iodinated inhibitor of iodide transport whose concentration and action vary with the total organic iodine content of the gland. Organification of iodine in the thyroid is also subjected to a similar auto-regulatory process. With the increasing concentration of iodine, the synthesis of organic iodine progresses up to a point beyond which there is a progressive decline in glandular organic iodine. This iodide-induced decline of organic iodination is associated with a progressive decrease in thyronine (T_4 and T_3) content as reflected in low T_4:DIT ratios. Subsequently, with greater iodide concentrations, DIT content also decreases, as evidenced by progressively increasing MIT:DIT ratios. The inhibition of organification of iodine brought about by the higher concentrations of iodide is known as the Wolf-Chaikoff effect as mentioned earlier. The effect is transient and the gland escapes from it quickly, presumably, as a result of the auto-regulatory inhibition of thyroidal iodine transport as described above.

Iodide, in pharmacological doses, also effectively inhibits the proteolysis of thyroglobulin and hence the release of T_4 and T_3 from the gland. This action of iodine is believed to be due to its inhibitory effect on thyroidal adenylyl cyclase activity, which is responsible for the endocytosis and hydrolysis of thyroglobulin.

Thus, the important steps in the biosynthesis and secretion of thyroid hormones are regulated by the degree of iodine availability, through autonomous intrathyroidal mechanisms.

B. TSH-mediated Regulation

Superimposed on this auto-regulation is the neuroendocrine regulation of thyroid function through the hypothalamo-hypophyseal axis. The glycoprotein TSH, secreted by the thyrotrophs of the anterior pituitary, plays a major role in the regulation of the metabolic and secretory processes of the thyroid gland (Figure 88.5).

TSH regulates thyroid gland function through the TSH receptor (TSH-R), a seven transmembrane G-protein coupled receptor (GPCR). The TSH-R is coupled to Gsα, the α subunit of stimulatory G protein, which activates adenylyl cyclase,

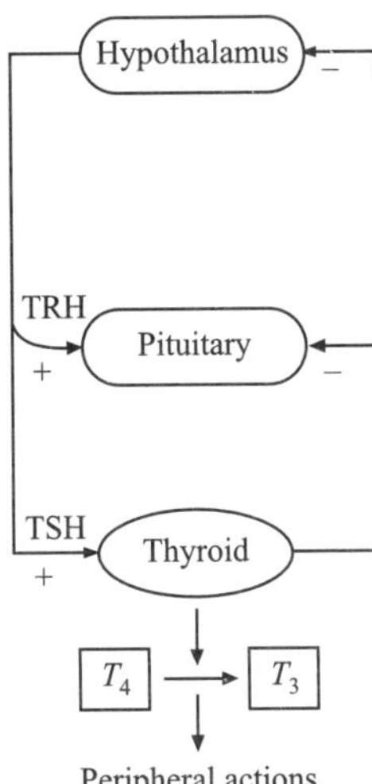

Figure 88.5 Regulation of thyroid hormone synthesis and secretion—Thyroid hormones T_4 and T_3 feed back to inhibit hypothalamic production of thyrotropin releasing hormone (TRH) and pituitary production of thyroid stimulating hormone (TSH). TSH stimulates thyroid gland production of T_4 and T_3.

resulting in increased production of cAMP. TSH also stimulates phosphatidylinositol turnover by activating phospholipase C.

In addition to TSH which is the major hormone that regulates thyroid gland growth and function, some of the locally produced growth factors such as insulin-like growth factor I (IGF-I), epidermal growth factor (EGF), transforming growth factor β (TGF-β), endothelins and certain cytokines also influence thyroid hormone synthesis. For example, in acromegaly, increased levels of growth hormone and IGF-I are associated with goitre.

The biochemical events that follow the TSH-induced generation of cyclic AMP are not known. Cyclic AMP-dependent thyroidal protein kinases are believed to be involved in the triggering of a large number of intracellular events which ultimately result in the metabolic and morphological changes that are brought about by TSH.

TSH brings about dramatic changes in the metabolism and morphology of the thyroid gland. There is overall stimulation of intrathyroidal iodine metabolism. Accelerated iodine transport as reflected in an increase in T:S ratio of radioactive iodine, increase in iodide organification and coupling as seen in the higher yield of T_3 and T_4, for a given degree of iodide availability, increase in the rate of synthesis of thyroglobulin as well as increase in the rate of proteolytic dissolution of the thyroglobulin stored in the colloid, and consequent increase in the rate of secretion of thyroid hormones, all are events that follow the action of TSH on the thyroidal epithelium. Apart from its specific stimulatory effect on iodine metabolism, TSH also brings about stimulation of many other metabolic pathways in the thyroid. Thus, oxidation of glucose through the hexose monophosphate shunt pathway is enhanced by TSH. TSH also has been shown to increase thyroidal phospholipid and nucleic acid synthesis.

It has now been clearly established that TSH release from the pituitary is regulated by two mutually opposing influences: one of feedback inhibition brought about by circulating T_3 or T_4 and the other of stimulation of release of TSH from the pituitary. The latter influence has now been shown to be due to the action of a tripeptide neurohormone thyrotropin releasing hormone (TRH) secreted by the hypothalamus. TRH has been identified as pyroglutamyl-histidyl-prolineamide (Figure 88.5). It is released into the hypophyseal portal circulation, through which it reaches in high concentrations to act on specific receptors on the plasma membrane of thyrotrophs. TRH action on thyrotrophs has now been shown to be mediated through the adenylyl cyclase-cAMP second messenger system which, ultimately through the mediation of protein kinases, brings about the release of TSH stored in the form of secretory granules.

The negative feedback influences of T_3 and T_4, on the secretory activity of the pituitary thyrotrophs has been shown to be through the inhibition of TRH action (Figure. 88.5). This T_3 or T_4 mediated inhibition can be suppressed by inhibitors of protein synthesis such as cyclophosphamide. Also, it can be overcome by excess TRH. These observations seem to suggest that a protein elaborated in the thyrotrophs under the influence of T_3 or T_4 competitively inhibits TRH action on the thyrotrophs.

Thyroid function is thus regulated by an intrinsic auto-regulatory mechanism based on iodine availability as well as by the action of the trophic hormone, TSH, whose secretion is controlled by feedback inhibition by thyroid hormones as well as by the secretory stimulus of TRH elaborated by the hypothalamus.

C. Peripheral Metabolism of Thyroid Hormones T_4 and T_3

Tetra- and tri-iodothyronines (T_4 and T_3) constitute the predominant forms of thyroid hormones in circulation. There are at least two other thyronines which have been recently demonstrated in the human circulation, namely r-T_3 and 3, 3′ di-iodo-thyronine (3, 3′ T_2). The reported normal levels of circulating thyronines, their daily production rate and metabolic turnover are given in Table 88.3.

The estimated extrathyroidal pool size of thyroxine or T_4 in euthyroid individuals is approximately 900 µg, of which roughly 50% is extracellular and bound to the circulating carrier proteins, thyroxine binding globulin (TBG) and thyroxine binding prealbumin (TBPA). The remaining 50% is present intracellularly, mainly in tissues like liver, kidney, muscle, and skin. A normal adult thyroid secretes 90–100 µg of thyroxine (T_4) daily. Thyroid is the only known source of thyroxine in the body. The daily thyroxine secretion is kept fairly constant at this level, in the face of marked fluctuations in the daily iodide intake. The daily metabolic turnover of T_4 is estimated to be about 90 µg. 80% of this is accounted for by deiodination of T_4 in the peripheral tissues. The remaining 20% is excreted through the bile, as sulphates and glucuronides.

TABLE 88.3   Iodothyronines produced in the throid gland

Thyronine	Structural formula	Normal concentration in serum per 100 ml	Daily production rate	Thyroidal secretion per day
Thyroxine	(structural formula)	7–10 µg	90–100 µg	90–100 µg
Tri-iodothyronine	(structural formula)	100–140 µg	5–8 µg	28 µg
Reverse tri-iodothyronine	(structural formula)	40 µg	37 µg	1–2 µg

The daily production rate of T_3 is reported to be approximately 28 µg/day, whereas the estimated daily thyroidal secretion of T_3 is 5–8 µg. Thus, less than 25% of the T_3 produced daily in the body is secreted by the thyroid. The remaining T_3 is generated by the mono-deiodination of the phenolic ring of T_4 molecule in the peripheral tissues. This mono-deiodination occurs primarily in the liver, kidney and muscle. As T_3 is calorigenically 5 times more potent than T_4, and as more than 50% of T_4 is converted in the peripheral tissues to T_3, it is proposed that T_4 is a prohormone of the biologically active 3,5,3′ tri-iodothyronine (T_3). The total extrathyroidal pool of T_3 is estimated to be around 40 µg, of which more than 80% is intracellular, mainly in muscle and skin.

An interesting fact about the peripheral metabolism of T_4 is its conversion to 3.3′, 5′, T_3 or r-T_3, a calorigenically inactive isomer of T_3. Its daily production rate is estimated to be about 37 µg. Only about 3% of this is accounted for by thyroidal secretion. The remaining is produced in peripheral tissues by mono-deiodination of T_4 (Figure 88.6).

Both r-T_3 and T_3 are further deiodinated in the peripheral tissues. The product of deiodination of tri-iodinated thyronines includes 3,3′ T_2 which has recently been shown to be present in measurable quantities in human circulation. The factors that determine the rate of degradation of T_4, T_3 and r-T_3 in the peripheral tissues are multiple. The binding by carrier proteins such as THG and TBPA play an important role in controlling the rate of degradation of thyronines in peripheral tissues. Also, higher the concentrations of free T_4 and T_3, greater the activity of the degrading enzymes in the tissues. Certain drugs also influence the degradation of T_4 in the periphery. Thus propylthiouracil (PTU) inhibits mono deiodination of T_4 to generate T_3. Drugs like diphenylhydantoin and phenobarbitone are known to induce degradative enzymes and augment the disposal rate of T_4.

A significant observation made recently about the peripheral metabolism of T_4 is its dual conversion to the calorigenically active (T_3) and inactive (r-T_3) tri-iodinated thyronines by two different mono-deiodinating processes (Table 88.3). It is now recognized that many abnormal states which affect energy metabolism in the body such as starvation, chronic illnesses, and old age, influence the peripheral conversion of T_4, such that r-T_3 is preferentially produced in place of T_3. Teleologically, this can be interpreted as an adaptive mechanism of the body to conserve energy. The precise enzymatic mechanisms which respond selectively to generate either T_3 or r-T_3, based on the dictates of energy metabolism in the body are not known. Nevertheless, it appears that the regulation of thyroid hormone activity resides as much in its peripheral metabolism as its glandular production.

V.   MECHANISM OF ACTION OF THYROID HORMONES

A.   Thyroid Hormone Transport

Thyroid hormones from the blood circulation enter cells by passive diffusion and via specific transporters like the monocarboxylate 8 (MCT8) transporter. Mutations in the MCT8 gene have been identified in patients with thyroid function abnormalities such as low T_4, high T_3 and high TSH. On entering cells, thyroid hormones act primarily by binding to nuclear receptors. Besides, they also mediate their actions by stimulating plasma membrane and mitochondrial enzymatic responses.

B.   Nuclear Thyroid Hormone Receptors

Thyroid hormones bind nuclear thyroid hormone receptors (TRs) α and β with high affinity. Both TRα and TRβ are

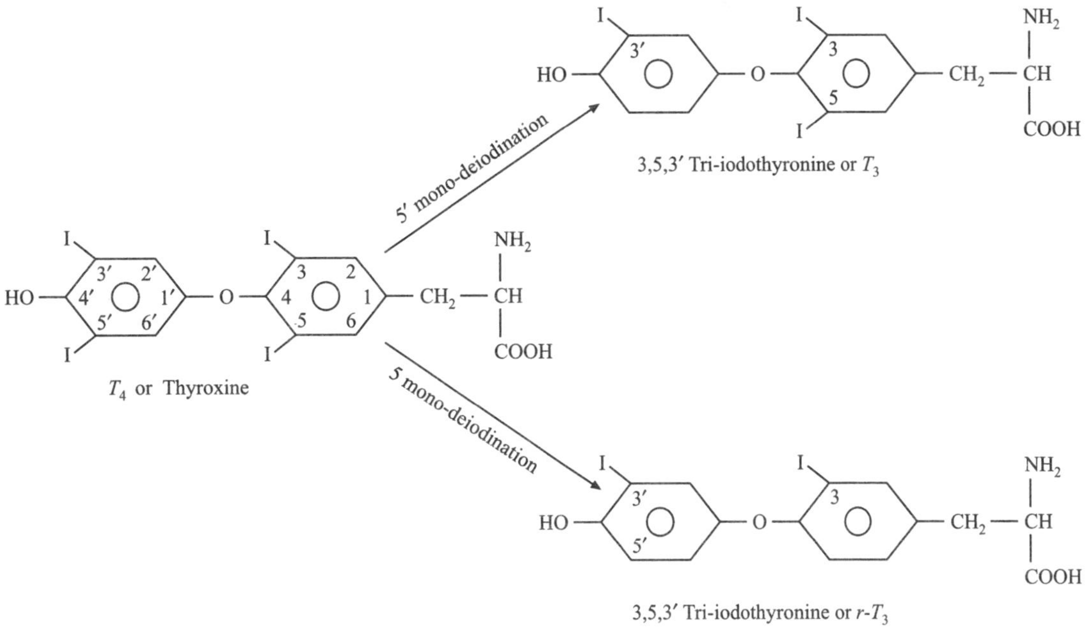

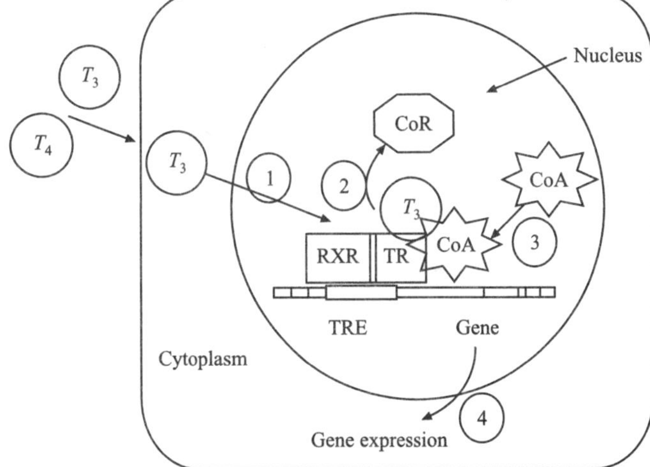

Figure 88.6 Mechanism for dual metabolic routes of T_4.

expressed in most tissues with varying expression levels among different organs. Both receptors are variably spliced to form distinct isoforms.

The TRs contain a central DNA binding domain and a C-terminal ligand binding domain. They bind to specific DNA sequences, termed thyroid response elements (TREs), in the promoter regions of target genes (Figure 88.7). The receptors bind as homodimers, or more commonly, as heterodimers ith retinoic acid X receptors (RXRs). The activated receptor can either stimulate gene transcription (e.g., Myosin heavy chain α) or inhibit transcription (e.g., TSH β-subunit gene), depending on the nature of the regulatory elements in the target genes.

Thyroid hormone T_3 is bound with 10–15 times higher affinity than T_4, hence it has higher hormonal potency than T_4. Though T_4 is produced in excess of T_3, receptors are mainly occupied by T_3, reflecting T_4 to T_3 conversion by peripheral tissues, greater T_3 bioavailability in plasma and greater affinity of TR for T_3. Binding of the thyroid hormone to its receptors (TRs) induces conformational changes in the receptors that modify its interactions with accessory transcription factors. In the absence of thyroid hormone binding, the receptors bind to co-repressor proteins that inhibit gene transcription. Binding of thyroid hormone dissociates the co-repressors and allows the recruitment of coactivators that enhance transcription.

VII. MEASUREMENTS OF THYROID HORMONES AND THYROID FUNCTION

Altered levels of T_4 and T_3 lead to corresponding changes in the levels of TSH. Hence, thyroid functioning is first assessed by determining whether TSH is normal, suppressed or elevated. A normal TSH level majorly excludes a primary abnormality of

Figure 88.7 Thyroid hormone signaling mediated via thyroid hormone receptors. In the nucleus of a cell, the thyroid hormone receptor (TR) and retinoid X receptor (RXR) form heterodimers that bind specifically to thyroid hormone response elements (TRE) in the promoter regions of target genes. In the absence of hormone, TR binds co-repressor (CoR) proteins that silence gene expression. The numbers refer to a series of ordered reactions that occur in response to thyroid hormone: (1) T_4 or T_3 enters the nucleus; (2) T_3 binding dissociates CoR from TR; (3) Coactivators (CoA) are recruited to the T_3-bound receptor; (4) gene expression is altered.

thyroid function. Immunochemiluminometric assays (ICMAs) for TSH are sensitive enough to discriminate between the lower limit of the reference range and the reduced values that are observed due to suppression of TSH in thyrotoxicosis (overproduction of thyroid hormones). The findings of abnormal TSH levels must be followed by measurements of circulating thyroid hormone levels to confirm the diagnosis of hyperthyroidism (suppressed TSH) or hypothyroidism (elevated TSH). Estimation of serum total T_4 and total T_3 are made by radioimmuunoassays or ELISA. T_4 and T_3 are mostly protein bound, and several factors such as illness, medications and genetic factors can alter protein binding. It is therefore desirable to measure the free or unbound hormone levels, which correspond to the biologically available hormone pool.

Total thyroid hormone levels are raised when TBG is increased due to estrogens (pregnancy, oral contraceptives, hormone therapy, tamoxifen), and lowered when TBG binding is reduced (androgens, nephrotic syndrome). Genetic disorders, illness and various drugs such as phenytoin, salicylates and non-steroidal anti-inflammatory drugs (NSAIDs) can interfere with thyroid hormone binding. However, in such cases, unbound thyroid hormone levels are normal and hence, assays that measure unbound hormone are preferred.

In hyperthyroidism, thyroid hormone levels T_3 and T_4 are increased while TSH level is reduced due to feedback inhibition. In hypothyroidism, T_3 and T_4 levels are reduced with elevated levels of TSH due to lack of feedback inhibition. However, when hypothyroidism is due to a primary defect in the hypothalamus or pituitary, levels of TSH, T_3 and T_4 are decreased. Hyperthyroidism is observed in Grave's disease, toxic goitre, excess intake of thyroid hormones or rarely due to TSH secreting tumours of the pituitary.

SUGGESTIONS FOR FURTHER READING

1. *Endocrinology*, 5th Ed., Mac E. Hadley (Ed.), 312–335.
2. Harrison's *Principles of Internal Medicine,* 18th ed., Vol. II, 2911–2939.

89

The Pancreas
Insulin and Glucagon

Nikhil Tandon and Sameer Aggarwal

CONTENTS

I. INTRODUCTION

Pancreatic beta cells are found in the islets of Langerhans, which range in size from few hundred to a few thousand endocrine cells. Islets are anatomically and functionally distinct from pancreatic exocrine tissue (source of pancreatic enzymes). Normal subjects have about one million islets that in total weigh 1 to 2 grams and constitute 1–2% of the pancreatic mass.

Islets vary in size from 50 to 300 micrometers in diameter. They are composed of several types of cells. The constituent cells of the islets include centrally located beta cells (70%), which are surrounded by alpha cells that secrete glucagon, smaller numbers of delta cells that secrete somatostatin, and PP cells that secrete pancreatic polypeptide. These cells communicate with each other through extracellular spaces and gap junctions. A neurovascular bundle containing arterioles and sympathetic and parasympathetic nerves enters each islet through the central core of beta cells. The arterioles branch to form capillaries that pass between the cells to the periphery of the islet and then enter the portal venous circulation.

II. INSULIN

Insulin is a 51 amino acid anabolic peptide-hormone that is secreted by the beta cells. It consists of two chains (A and B) connected by disulfide bonds. One of its primary functions is the stimulation of glucose uptake from the systemic circulation, as well as the suppression of hepatic gluconeogenesis, thereby serving a primary role in glucose homeostasis. The work of Banting, Best, Collip and MacCleod in the 1920's resulted in the identification of a substance in extracts of pancreas that had the remarkable ability to reduce blood glucose levels in diabetic animals. By 1923, these pancreas extracts were being used to successfully treat patients with diabetes.

A. Insulin Biogenesis and Mechanism of Release

Insulin was the first peptide hormone discovered. With the illustration of the primary sequence of insulin by Sanger in the mid 1950's it became known that insulin was a two chain heterodimer consisting of a 21 amino acid A-chain linked to a

30 residue B chain by two disulfide bonds derived from cysteine residues (A7-B7; and A20-B19). An intrachain disulfide bond also exists in the A-chain (A6-A11) (Figure 89.1a). Proinsulin, a 9 kd protein, contains both the A- and B-chain of insulin in a continuous single chain joined through an intervening region called the C-peptide. The C-peptide is a variable length peptide segment, consisting of 26-31 residues depending on the species, which links the carboxy terminus of the B-chain to the amino terminus of the A-chain via two dibasic residue links (Arg-Arg and Lys-Arg) as shown in Figure 1b. Proinsulin is cleaved at those dibasic links to release two chain insulin and free C-peptide.

Chan et al., subsequently discovered that there was an additional precursor of insulin, preproinsulin. Preproinsulin is a 12 kd single chain polypeptide which consists of proinsulin extended at the amino terminus by a 24 amino acid signal peptide region of hydrophobic residues. The initial mRNA transcript is modified via excision of the two intervening sequences, capping of the 5′ terminus by 7-methyl guanosine, and polyadenylation of the 3′ terminus to produce a mature mRNA product. This mRNA product codes for preproinsulin, which is translated on the rough endoplasmic reticulum (RER) and subsequently translocated into the RER lumen via a series of interactions of the signal peptide with the signal recognition

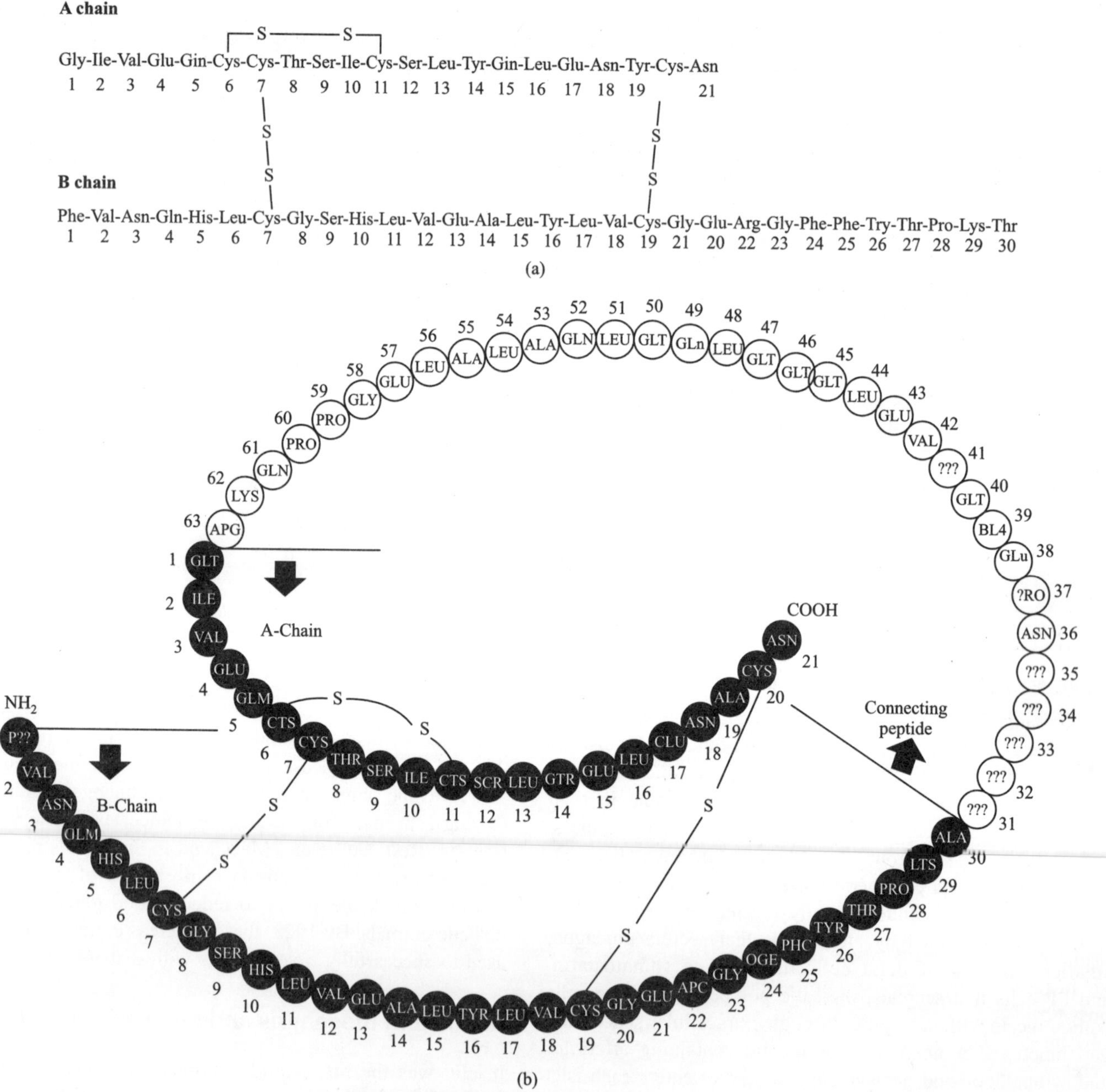

Figure 89.1   Primary structures of porcine insulin and porcine proinsulin. The primary sequence of porcine insulin (a) as determined by Sanger and co-workers; and proinsulin (b). The sequence of human insulin is identical to that of porcine insulin except for the change of AlaB30 to ThrB30 in human insulin.

particle (SRP) and the SRP-receptor in the RER membrane. The signal peptide is then cleaved off in the lumen of the RER by a signal peptidase. Proinsulin undergoes rapid folding and disulfide bond formation to generate the native tertiary structure of proinsulin. Proinsulin is then transported to the Golgi apparatus where it is packaged into secretory granules and then converted to native insulin and C-peptide. The signal peptide is rapidly degraded in the RER and is therefore not a normal secretory product of the beta cells. The conversion process may begin in the Trans Golgi network but continues in the condensing vacuoles (early secretory granules). The products are stored in mature secretory vesicles, and secreted in equimolar amounts along with small amounts (3%) of proinsulin and intermediate cleavage products. Glucose, in addition to stimulating insulin secretion, is also a direct stimulator of insulin gene transcription, and insulin mRNA translation and stability.

Proinsulin has many of the physical properties of insulin despite the larger size of this molecule. Proinsulin has been shown to aggregate, forming dimers and Zn^{2+} coordinated hexamers in a manner similar to insulin, have a comparable isoelectric point and solubility, as well as, react with insulin antisera. It was also found that proinsulin is a full agonist of insulin and displays 3–5% biological activity. In 1969, it was demonstrated that proteolytic processing of proinsulin occurred in the Golgi apparatus. The subsequent conversion of proinsulin to insulin is initiated in the Trans Golgi and proceeds for several hours within secretory granules as they mature in the cytosol in preparation for secretion. In rat islets the conversion of proinsulin to insulin begins about 30 min after synthesis and resembles first order reaction kinetics having a $t/2$ of approximately 30–60 minutes.

The conversion of proinsulin to insulin occurs by combined action of two types of proteases: one with trypsin-like endoprotease activity to cleave at the dibasic residues at each end of the C-peptide, and another with exopeptidase activity to remove the basic residues left after tryptic cleavage. Previous studies have also demonstrated that mixtures of pancreatic trypsin and carboxypeptidase B could convert proinsulin to insulin in vitro.

Secretory granules in the beta cells undergo maturation in the cytosol. Studies suggest that newly synthesized insulin probably forms crystals with zinc that is transported into secretory granules; and these reside in the dense core of the beta cell granules with the soluble C-peptide located in the less dense or clear periphery of the granule. Proinsulin is also known to crystallize with insulin in small amounts, probably as mixed hexamers. The side chain of histidine B10 of each insulin molecule is known to coordinate with zinc, stabilizing the hexamer. Glucose is the primary regulator of insulin biosynthesis and secretion but other hormones and chemical substances also play an important role. Glucose causes an increase in cAMP levels by a adenylate cyclase independent mechanism. cAMP then exerts its effects via a mechanism involving protein kinase A, leading to the phosphorylation and activation of certain key proteins. Glucose and cAMP

rapidly increase translation and transcription of insulin mRNA. Insulin mRNA normally turns over slowly, with a half life of approximately 30 hours at normal or below normal levels. However, elevated glucose increases the half life of insulin mRNA as much as three fold. Calcium dependent exocytosis of secretory granules is the main mechanism of secretion in both glucose-stimulated and basal states. Other stimulators of insulin secretion include glucagon, glucagon-like peptide; cholecystokinin; and gastric inhibitory peptide, all acting via specific receptors on the beta cell. Inhibitors of insulin secretion include catecholamines (adrenaline and nonadrenaline) which interact with adrenergic receptors on the beta cell membrane and somatostatin.

B. Biochemistry and Electrophysiology of Insulin Secretion

Insulin secretion from the beta cell is inadequate not only in the diabetic state, but even in the prediabetic state.

Insulin is stored in large dense core vesicles and is released via exocytosis. This is a multistep process that consists of the transport of the secretor vesicles to the plasma membrane, then docking, priming, and finally fusion of the vesicle with the plasma membrane. It is well known that this process is regulated by nutrients, other hormones, and neurotransmitters to cause the electrical depolarization of the beta cell with the consequent release of insulin. However, only a small portion of the insulin stored in vesicles in the beta cell is released even under maximum stimulation. This suggests that the systemic insulin levels are therefore regulated by secretion instead of synthesis or storage pools.

The mechanism of coupling glucose metabolism to insulin secretion resides in the electrical excitability of the beta cell. A large number of ion channels, pumps, and transporters contribute to intracellular calcium concentration, as well as other ions, to form the membrane potential (Vm) of the beta cell of around 70 mV when extracellular glucose is ~3 mM. Beta cells are electrically excitable and that glucose and sulfonylureas (known insulin secretagogues) control this excitability. ATP-sensitive potassium (KATP) channels maintain the resting membrane potential of the beta cell.

KATP channel conductance predominates in the resting beta cells. These potassium channels belong to the inward rectifier (Kir) subfamily. These channels conduct K^+ current into the cell more readily than to the outside of the cell.

KATP channels are unique in the inward rectifier family because they require an auxiliary subunit, the sulfonylurea receptor (SUR1), to function. The SUR1 is a member of the ATP binding cassette (ABC) family of membrane proteins, which includes the cystic fibrosis transmembrane conductance regulator (CFTR) chloride channel among others. KATP channels in the beta cells consist of Kir6.2 subunits surrounded by their accompanying SUR1 subunit.

Elevation of glucose concentration to >8–10 mM results in the depolarization of the beta cells. Glucose is taken up into the beta cells by the GLUT2 transporter and is metabolized via glycolysis and in the mitochondria to generate ATP. This alters the ATP/ADP ratio, which causes closure of the KATP channel and depolarization of the cell via the decreased K^+ permeability. ATP inhibits KATP channels and ADP opens them. Mutations in either the Kir, or SUR1, can result in persistent activation with hyperinsulinemia and hypoglycemia in infancy.

Intracellular Ca^{2+} concentrations play as the key coupling factor between membrane depolarization and vesicular exocytosis in beta cells. These voltage dependent Ca^{2+} channels open upon membrane depolarization caused by Kir channel closure; and the resulting Ca^{2+} influx leads to insulin secretion.

Several hormones and neurotransmitters regulate insulin secretion in addition to the voltage sensitive pathways. Molecules such as epinephrine, galanin, somatostatin, acetylcholine, and glucagon-like peptide (GLP) help regulate insulin secretion by binding to their cognate receptors. Cholecystokinin and acetylcholine potentiate insulin secretion via phosphoinositide catabolism with the subsequent mobilization of intracellular calcium. Other potentiators of insulin secretion, such as GLP-1, and glucose-dependent insulinotropic polypeptide (GIP), bind to their respective heterotrimeric G-protein coupled receptors to activate adenyl cyclase thereby increasing intracellular cAMP, and subsequently cause the activation of protein kinase A (PKA). Stimulation of either PKC or PKA alters second messenger systems in the beta cell and can chemically modify ion channels to cause a direct influence on insulin secretion.

The process of insulin secretion is a complex set of events; and research continues to provide new insights on beta cell molecular biology and electrophysiology for the purpose of obtaining a better understanding of insulin action, and its dysregulation that results in diabetes.

C. Insulin Structure

There are several crystal forms of insulin that have been solved, all of which display a general similarity to the 2-Zn insulin hexamer initially described by Hodgkin and her coworkers.

The 2-Zn insulin hexamer (MW~36000) consists of six molecules of insulin (MW~6000) arranged as three dimeric units which possesses a three-fold symmetry axis. The dimers (MW~12000) possess a pseudo two-fold symmetry axis which is perpendicular to the three fold axis of rotation. Although each monomer of the dimers has the same peptide backbone structure, they are not identical in the arrangement of certain side chains, breaking the perfect two fold symmetry.

The 2-Zn insulin structure has provided a great deal of information about the hydrophobic, solvent exposed and potential binding surfaces of insulin. Several 2D NMR solution structures of the insulin hexamer, dimer, and monomer have recently been described.

Insulin exists primarily as a monomer at low concentrations (~10–6 M) and forms dimers at higher concentrations at neutral pH. At high concentrations and in the presence of zinc ions, insulin aggregates further to form hexameric complexes.

Control of insulin and glucagon secretion

The control of insulin secretion is complex. The most potent metabolic stimuli to insulin secretion are glucose and amino acids that act synergistically. Triglycerides and fatty acids have only a small stimulatory effect on insulin release. In response to an oral glucose load, insulin secretion occurs in two phases. The first phase represents the release of insulin stored in secretory granules. Approximately 10 minutes later when pre-formed granules have been depleted, there is a more gradual and sustained increase in insulin release that can last for several hours in normal individuals and is dependent on *de novo* synthesis of the hormone.

The secretion of glucagon is stimulated by a reduction in blood glucose concentration. It is also stimulated by a protein meal, particularly by the amino acids, alanine and arginine. These secretory responses are suppressed by the presence of insulin and glucose respectively (reducing transcription of the glucagon gene). Thus, ingestion of an ordinary meal produces much less variation in glucagon secretion than in that of insulin because of insulin's suppressive effect on glucagon release.

The secretion of both insulin and glucagon is potentiated by gastrointestinal hormones (termed incretins) that are released in response to orally ingested nutrients. Thus, an oral glucose load stimulates a greater insulin response than intravenous administration because one or more gastrointestinal hormones are released.

In addition to metabolic stimuli and hormones from the gut, insulin and glucagon secretion are also controlled by neural and paracrine mechanisms. Sympathetic, parasympathetic and peptidergic nerves innervate the islets. Sympathetic nerves stimulate insulin release via δ-adrenergic receptors (and inhibit via α-adrenergic receptors) whilst parasympathetic vagal nerves stimulate both insulin and glucagon release. This innervation accounts for the increase in insulin secretion that may occur before the entry of food into the gastrointestinal tract. Finally, somatostatin from the δ-cells inhibits the release of both hormones through a paracrine action.

Two gastrointestinal hormones which have significant effects on insulin secretion and glucose regulation are the glucagon-like peptides (principally glucagon-like peptide-1, GLP-1) and glucose-dependent insulinotropic peptide (GIP). Both of these gut hormones constitute the class of molecules referred to as the incretins. Incretins are molecules associated with food intake-stimulation of insulin secretion from the pancreas.

Briefly, GLP-1 is derived from the product of the proglucagon gene. This gene encodes a preprotein that is differentially cleaved dependent upon the tissue in which it is synthesized. For example, in pancreatic α-cells prohormone convertase 2 action leads to the release of glucagon. In the gut prohormone convertase 1/3 action leads to release of several peptides including GLP-1. Upon oral intake of food GLP-1 is secreted from intestinal enteroendocrine L-cells that are found predominantly in the ileum and colon but also to some extent in the duodenum and jejunum. Bioactive GLP-1 consists of 2 forms; GLP-1(7–37) and GLP-1(7–36) amide, with the latter form constituting the majority (80%) of the circulating hormone.

The primary physiological responses to GLP-1 are glucose-dependent insulin secretion, inhibition of glucagon secretion and inhibition of gastric acid secretion and gastric emptying. The latter effect will lead to increased satiety with reduced food intake along. The action of GLP-1 at the level of insulin and glucagon secretion results in significant reduction in circulating levels of glucose following nutrient intake. This activity has obvious significance in the context of diabetes, in particular the hyperglycemia associated with poorly controlled type 2 diabetes. The glucose lowering activity of GLP-1 is highly transient as the half-life of this hormone in the circulation is less than 2 minutes. Removal of bioactive GLP-1 is a consequence of N-terminal proteolysis catalyzed by dipeptidylpeptidase IV (DPP IV).

D. Measurements of Insulin in Blood

The techniques for measuring insulin generally fall into the following two groups: biological and immunological assays.

Biological assays

These are used to measure the potency of insulin solutions or the amount of insulin in biological fluids like plasma. Biological assays can be further divided into:

1. *In vivo* assays: These are rarely used currently for measuring insulin
2. *In vitro* assays: Electrochemiluminescence Assay

The Electrochemiluminescence assay is a two-site non-competitive assay, using monoclonal antibodies, with stated negligible cross-reactivity with intact proinsulin. The electrochemiluminescence assay employs two monoclonal antibodies which together are specific for human insulin.

The test is based on Sandwich principle. In a first 9-min incubation, plasma insulin, a biotinylated anti-insulin antibody, and a second monoclonal anti-insulin antibody labeled with an electrochemiluminescent ruthenium complex react to form a sandwich complex. After the addition of streptavidin-coated microparticles, during a 9-min incubation phase, the complex binds to the solid phase. Both assays are calibrated against the WHO 66/304 reference preparation.

The reaction mixture is aspirated into the measuring cell where the microparticles are magnetically captured onto the surface of the electrode and unbound substances removed. Application of a voltage to the electrode then induces chemiluminescent emission which is measured by a photomultiplier. The amount of light produced is directly proportional to the amount of insulin in the sample.

Results are determined via a calibration curve which is instrument-specifically generated by 2-point calibration and a master curve provided via the reagent barcode.

Proinsulin

Proinsulin is the prohormone precursor to insulin which is encoded by the *INS* gene in humans.

Synthesis and post-translational modification: Proinsulin is synthesized in the endoplasmic reticulum, where it is folded and its disulfide bonds are oxidized. It is then transported to the Golgi apparatus where it is packaged into secretory vesicles, and where it is processed by a series of proteases to form mature insulin. Mature insulin has 35 fewer amino acids; 4 are removed altogether, and the remaining 31 form the C-peptide. The C-peptide is abstracted from the center of the proinsulin sequence while the two other ends (the B chain and A chain) remain connected by disulfide bonds.

Immunogenicity

When insulin was originally purified from bovine or porcine pancreata, all the proinsulin was not fully removed. With the introduction of highly purified porcine insulin (99% pure), this ceased to be a significant clinical issue.

Clinical Application: Alterations in proinsulin release have been documented in subjects with type 2 diabetes. Using insulin immunoassays, approximately 15 percent of circulating immunoreactive insulin is comprised of proinsulin in healthy subjects. In patients with type 2 diabetes, this proportion is approximately doubled. Milder alterations have also been demonstrated in subjects with prediabetes or at risk of developing type 2 diabetes. The explanation for this disproportionate proinsulinemia is debated. It is believed to be either the result of a primary abnormality in beta-cell processing or an increase in secretory demand on the beta-cell, resulting in the premature release of secretory granules that have not had sufficient time to fully process the proinsulin within them.

Lipid metabolism and diabetes

Insulin is an anabolic hormone and promotes lipid synthesis and suppresses lipid degradation. Recent studies indicate that the transcription factor steroid regulatory element-binding protein (SREBP)-1c is a major mediator of insulin action on the expression of glucokinase and lipogenesis-related genes in the liver. In addition to promoting lipogenesis in the liver, insulin also stimulates lipid synthesis enzymes (fatty acid synthase, acetyl-CoA carboxylase) and inhibits lipolysis in adipose tissue. The anti-lipolysis effect of insulin is primarily mediated by inhibition of hormone sensitive lipase through a mechanism that involves activation of a cAMP-specific phosphodiesterase.

The increase in the intracellular deposition of triglycerides (TG) in muscles, liver and pancreas in subjects prone to diabetes is well documented and demonstrated to affect glucose metabolism by interfering with insulin signaling and insulin secretion. The obesity often associated with type 2 diabetes is mainly central, resulting in the overload of abdominal adipocytes with TG and reducing fat depot capacity to protect other tissues from utilizing a large proportion of dietary fat.

In contrast to subcutaneous adipocytes, the central adipocytes exhibit a high rate of basal lipolysis and are highly sensitive to fat mobilizing hormones, but respond poorly to lipolysis restraining insulin.

Protein metabolism and diabetes

Insulin deficiency is a protein catabolic state. *In vivo* studies have shown that insulin enhances short-side-chain amino acid intracellular uptake, stimulates transcription and translation of RNA, increases the gene expression of albumin and other proteins and inhibits liver protein breakdown enzymes. In patients with Type 1 Diabetes insulin deficiency increases protein breakdown and increases amino acid oxidation and that these effects are reversed by insulin treatment. Protein synthesis in the insulin-deprived state is also increased, although to a lesser extent than protein breakdown. This increased whole-body protein synthesis is reduced with an insulin infusion; thus the effects of insulin are largely mediated through its effects on protein breakdown. The observed effects of insulin deficiency in diabetic patients vary in different body compartments; most of the effects of insulin on protein synthesis appear to occur in non-muscular tissues especially in the splanchnic area. In addition, insulin has a differential effect on hepatic protein synthesis, i.e., inhibits fibrinogen synthesis and promotes albumin synthesis. The net protein anabolism due to insulin occurs largely in skeletal muscle. In patients with type 2 diabetes these effects are not noted, presumably because of residual endogenous insulin secretion.

Carbohydrate metabolism and diabetes

In diabetes there is an increased gluconeogenesis. Increased glucagon levels are present in diabetes and hyperglucagonemia stimulates gluconeogenesis. In absence of insulin, less protein synthesis occurs in muscle and therefore blood amino acid levels rise. Increased amino acids are a source for gluconeogenesis.

The enzyme activity that catalyzes the conversion of pyruvate and other two carbon metabolic fragments to glucose is increased. These include phosphoenolpyruvate carboxykinase, which help the conversion of oxaloacetate to phosphenolpyruvate. They also include fructose 1, 6-diphosphatase, which catalyzes the conversion of fructose diphosphate to fructose 6-phosphate, and glucose 6-phosphatase. Increased acetyl-CoA increases pyruvate carboxylase activity, and insulin deficiency increases the supply of acetyl-CoA since lipogenesis is decreased. Pyruvate carboxylase catalyzes the conversion of pyruvate to oxaloacetate

II. GLUCAGON

Glucagon is a 29 amino acid polypeptide (molecular weight of 3485 daltons) which was discovered as a "contaminant" hyperglycemic factor in pancreatic extracts by Kimball and Murlin in 1923 and finally sequenced by Bromer and Behrens in the late 1950s. There has recently been renewed interest in glucagon because of the findings suggestive that up to 1/3 of

improved glycemic control using a new class of medications was due to suppression of the hormone in the postprandial state.

A. Biosynthesis

Glucagon is synthesized in and secreted from A-cells of pancreatic islets. Normally, these cells constitute approximately 15% to 20% of the total islet cell mass. In most species, A-cells are located at the periphery of islets juxtaposed to both B-cells, which secrete insulin, and D-cells, which secrete somatostatin which can inhibit both insulin and glucagon secretion. In humans, however, A-cells are scattered throughout the islet. Granules in A-cells containing glucagon differ in ultrastructure from those in B and D-cells, containing respectively insulin and somatostatin, in having an electron-dense core and no halo.

Glucagon is synthesized initially as a 160 amino acid prohormone (proglucagon) of approximately 12 kd whose gene is encoded on chromosome 2. Proglucagon ultimately undergoes cleavage into four peptides. The whole process takes about 90 minutes. All of these peptides are immunoreactive, but only the 3485-d molecule is biologically active. L-cells of the small intestine synthesize an identical proglucagon molecule but different processing results in the formation of different polypeptides, of which glucagon-like peptide 1 and 2 are probably of most physiologic importance.

B. Plasma Glucagon

Plasma immunoreactive glucagon concentration varies considerably from individual to individual. The main factors responsible for this variation are the specificity of the antiserum used in the immunoassay and the relative proportion of the total immunoreactivity accounted for by the 3485-d molecule.

Normally arterial and peripheral venous plasma immunoreactive glucagon concentrations range between 25 and 150 pg/ml (1.0–5.0×10–8 M) after a 12- to 16-hour fast. Portal venous levels can average 1.5 to 3.0 times those present in arterial blood because of extraction of glucagon by the liver. As with other peptide hormones, circulating glucagon immunoreactivity is heterogeneous. By using chromatography, four immunoreactive species with apparent molecular weights of >40,000, 9000, 3500, and 2000 have been found. There is considerable individual and species variation in the proportions of each component found in plasma. In early studies, the 3500-d species usually constituted only about 25% of total plasma glucagon immunoreactivity due to nonspecificity of assay. With subsequent improvements and extraction procedures now available, it probably accounts for 90–95%. The 9000-d molecule, which in some assays has similar immunoreactivity but substantially less bioactivity than the 3500-d molecule. Increased amounts of the 9000-d molecule are found in the plasma of patients with the glucagonoma syndrome, with renal failure, and hepatocellular damage or carcinoma of the pancreas. The 2000-d molecule probably represents an inactive degradation product of glucagon.

Glucagon, stored within A-cells in distinctive granules, is secreted by a process called emiocytosis, which involves migration of secretory granules to the periphery of cells, fusion of granules with the plasma membrane, and extrusion of granule contents into the extracellular space.

Like insulin, secretion of glucagon involves A-cell substrate metabolism and consequent signals which affect cellular potassium and calcium channels and cAMP levels as well as protein kinase A and C.

Most substrates (glucose, free fatty acids and ketone bodies) except for amino acids suppress glucagon secretion. Inhibition of the metabolism of these substrates prevents the inhibition suggesting, in contrast to insulin secretion by beta cells, that generation of ATP inhibits secretion. While this is consistent with the reciprocal roles of insulin and glucagon in glucose homeostasis, a definitive explanation for this difference remains to be elucidated.

A-cells contain ATP-sensitive potassium channels as well as sulfonylurea, adrenergic, insulin and somatostatin receptors. Sulfonylurea receptors and ATP-sensitive potassium channels are associated with both plasma membranes and secretory granule membranes. Sulfonylureas stimulate glucagon release under appropriate conditions. This effect is dependent on protein kinase C, mimicked by inhibitors of mitochondrial ATP-sensitive potassium channels and inhibited by K+ channel openers (diazoxide).

It has been proposed that the effect of sulfonylureas on A-cell granules involves alterations in granule pH which renders them more competent for emiocytosis. Since metabolism of substrates would be expected to generate ATP and thus inhibit ATP-sensitive potassium channels like sulfonylureas, it is difficult to reconcile these observations into a consistent molecular mechanism for acute regulation of glucagon secretion. However, it has been proposed that a decrease in the intra-alpha cell ATP/ADP ratio activates adenylate cyclase and the resultant increase in cyclic AMP stimulates protein kinase A, which causes opening of calcium channels and an increase in intra-A cell calcium, which triggers glucagon release. These observations, however, are limited to the experimental animal data and full effect of sulfonylurea on glucagon secretion in humans is not known.

In vivo secretion of glucagon is the net result of the influence of substrate, neural, ionic, hormonal, and local factors on islet A-cell function. The plasma concentration of glucagon depends on the balance between rates of secretion and degradation and also on the sampling site (e.g., peripheral venous versus portal venous). Basal (non-stimulated) secretion rates of glucagon can be estimated from data on portal venous-arterial differences and portal venous plasma flow rates. It should be pointed out, however, that these values underestimate secretion of glucagon and merely represent post-hepatic delivery of glucagon.

C. Glucagon Catabolism

In normal humans, the metabolic clearance rate of glucagon which ranges from 7 to 14 ml/kg/min is independent of the prevailing plasma glucagon level. Normal rates occur in patients with diabetes or liver disease, whereas reduction has been found in renal failure and starvation. Thus, the liver and kidney seem to be the major sites of glucagon catabolism, but the relative contribution of each remains unclear.

Early reports suggested that the liver was not a major site of glucagon degradation. These observations may, however, be explained if the heterogeneity of circulating glucagon immunoreactivity is taken into account. When portal venous and peripheral venous plasma is subjected to gel filtration, it seems that the liver does not appreciably extract the biologically inactive 9000- and > 40,000-d plasma glucagon immunoreactivity. Thus, the portal-peripheral gradient of glucagon immunoreactivity is almost totally accounted for by extraction of the biologically active 3500-d molecule; this averages approximately 60% and results in a portal-peripheral gradient of 2.5 to 3 for the biologically active molecule.

It has long been known that the kidney is capable of degrading glucagon. Arteriovenous gradients across the kidney in normal animals infused with glucagon indicate extraction of 23% to 39% of the presented glucagon. Because less than 2% of the extracted hormone appears in urine and because non-filtering kidneys continue to extract appreciable amounts of glucagon, it seems that both tubular reabsorption and postglomerular capillary tubular uptake precede renal parenchymal degradation of glucagon. The hyperglucagonemia found in patients with chronic renal failure is due primarily to decreased clearance of the 9000-d molecule and cannot be accounted for by increased secretion of glucagon (3500-d molecule) or its decreased catabolism. Bilateral nephrectomy decreases the glucagon metabolic clearance rate of 3500-d glucagon approximately 30%. Consequently, liver and kidney can account for 80% to 90% of the metabolic clearance of the biologically active glucagon fraction of plasma glucagon immunoreactivity.

D. Regulation of Glucagon Secretion

Glucose is the most important physiologic regulator of glucagon secretion. Hyperglycemia decreases and hypoglycemia increases glucagon secretion. In vitro studies, such as those using the isolated perfused pancreas in which most variables operative in vivo can be controlled, indicate that the Alpha cell is as exquisitely sensitive to changes in the ambient extracellular glucose concentration as is the beta cell; thus, glucose suppresses basal and stimulated glucagon release at concentrations as low as 5 mM glucose (90 mg/dl). To some extent the inhibition of glucagon secretion is dependent on concomitant stimulation of insulin release. *In vivo*, a decrease in plasma glucose of 1 to 2 mM increases plasma glucagon.

Other substrates also influence glucagon secretion. Various amino acids stimulate Alpha cell release of glucagon, while free fatty acids and ketone bodies suppress glucagon secretion. Amino acid stimulation of glucagon release may be important in preventing hypoglycemia, which might otherwise occur because of insulin release accompanying ingestion of a noncarbohydrate meal. Suppression of glucagon secretion by free fatty acids and ketone bodies may be part of a negative feedback system regulating ketogenesis.

The islets of Langerhans are richly innervated. Like insulin release, glucagon secretion is influenced by both sympathetic and parasympathetic nervous sytems; epinephrine, norepinephrine, and acetylcholine and electrical stimulation of mixed pancreatic, splanchnic, and vagus nerves augment glucagon release. Both alpha cells and beta cell secretion are influenced in the same direction by parasympathetic (i.e., increase), β-adrenergic (i.e., increase), and α-adrenergic (i.e., decrease) mechanisms. The observation that glucagon secretion is increased by epinephrine, while insulin release is simultaneously decreased, can best be explained by postulating that the alpha cell contains a preponderance of β-adrenergic receptors, while the B-cell contains a preponderance of a-adrenergic receptors. Neural input to the alpha cell is probably important in modulating the increases in plasma glucagon observed during stress and perhaps also after mixed meals.

A variety of hormones have variable effects on alpha cell function depending on type of experiment and ambient glucose concentration. Epinephrine, gastrin, pancreozymin, vasoactive intestinal peptide, and gastric inhibitory polypeptide increase glucagon release while secretin apparently suppresses glucagon secretion. Whether these represent true physiologic interactions or merely pharmacologic effects is unclear. Hyperglucagonemia, relative or absolute, has been found in states of growth hormone, cortisol, and thyroid hormone excess. Conceivably, this might play a role in the associated abnormalities of carbohydrate and lipid metabolism.

Alterations in nutrition also influence alpha cell function. Acute ingestion of pure or high carbohydrate meals suppresses glucagon release, whereas pure or high protein-containing meals stimulate glucagon release. Concomitant changes in plasma glucose and amino acid levels are probably responsible for these changes. Prolonged (i.e., weeks or days) alterations in diet also alter alpha cell function. During total starvation, there is an acute increase in plasma glucagon lasting 1 to 2 days, probably as a result of increased secretion. Prolonged ingestion of high-carbohydrate or isocaloric high-fat diet decreases basal and meal-stimulated plasma glucagon levels. Conversely, low-carbohydrate diets or high-protein diets increase basal and stimulated glucagon secretion. In obesity, increased plasma glucagon responses have been reported.

Hypoglycemia stimulates glucagon secretion through both intra-islet and central nervous system mediated autonomic signals. Within the islets low glucose concentrations increase Alpha cell glucagon secretion directly and, by reducing Beta cell secretion, decrease tonic Alpha cell inhibition by insulin. Autonomic adrenergic (i.e., norepinephrine), cholinergic, and peptidergic neural and adrenomedullary hormonal (epinephrine) signals, triggered by hypoglycemia, may also contribute.

Glucagon secretion is also controlled by incretins, glucagon like-peptide (GLP)-1 and gastric inhibitory polypeptide (GIP. They are produced by L-cell and K-cells in the small intestine, respectively. GLP-1 lowers glucagon concentration in humans while preserving glucagon release in response to hypoglycemia. It is not known whether GLP-1 directly affects alpha cell or has indirect effects via stimulation of insulin production by beta cells. The latter hypothesis is favored because no GLP-1 receptors have been identified in the membrane of alpha cells and direct application of GLP-1 to A cells did not affect glucagon secretion. In contrast, GIP stimulates glucagon production in the isolated perfused rat pancreas and healthy individuals. Both GLP-1 and GIP concentrations increase after a meal in healthy and Type 2 DM patients suggesting that our understanding of the physiology of incretin effects on glucagon production is presently incomplete. To make the picture even more complicated, a potential for reciprocal control of incretins by glucagon was recently demonstrated by Meier et al. In the experiments on healthy and type 2 DM volunteers, it was demonstrated that post-prandial glucagon may exert tonic suppression on GLP-1 secretion.

Another dimension of glucagon regulation may involve dysfunction in glucagon receptors, which are encoded by the GCGR (glucagon receptor) gene and expressed abundantly in the liver and kidney and to a lesser extent in heart, adipocytes, lymphoblasts, spleen, endocrine pancreas, brain, retina, adrenal gland and the gastrointestinal tract. In the liver, glucagon receptors are located mainly in hepatocytes, but can also be found on the surface of Kupffer cells. Interestingly, glucagon receptors in the pancreas are predominately located on beta cells, which suggests a bidirectional feedback mechanism. Further evidence supporting this hypothesis is the fact that glucagon, at physiological concentrations, was found to stimulate insulin release via these receptors. The genetic deletion of glucagon receptors leads to alpha cell hyperplasia, hyperglucagonemia, elevated GLP-1, resistance to diet-induced obesity, and increased lean body mass. Whether any activating functional or genetic changes in glucagon receptors are present in humans with diabetes mellitus or insulin resistance remains to be determined.

In conclusion, control of glucagon secretion is complex and dependent on substrate as well as hormone concentration and interaction. Its production and secretion are affected by nutrients, such plasma glucose level, amino acids, free fatty acids, autonomic nervous system, gut-derived incretins, and paracrine effect of B-cell products such as insulin, Zn, and GABA.

E. Role of Glucagon in Fuel Homeostasis

Although adipocytes have glucagon receptors and there is evidence that alterations in plasma glucagon can affect lipolysis, the main target organ of glucagon is the liver.

Carbohydrate homeostasis

Glucagon is a potent stimulator of hepatic glycogenolysis, gluconeogenesis, and ketogenesis *in vitro*. These actions of glucagon and the increases in plasma glucagon observed during hypoglycemia, exercise, trauma, infection, and other stress proves that glucagon is important in the maintenance of euglycemia in the post-absorptive state and at times when there are increased demands for fuels. Under these conditions, when B-cell function is normal, the major action of glucagon would be to counteract the actions of insulin on storage of glucose and other fuels. Conversely, when B-cell function is deficient, glucagon could increase the metabolic consequences of insulin deficiency and be an important determinant in pathogenesis of hyperglycemia and hyperketonemia found in diabetes.

Evidence for the role of glucagon in glucose homeostasis has been provided from studies employing somatostatin; a potent inhibitor of glucagon and insulin secretion which does not itself directly affect substrate metabolism at doses used *in vivo*. Infusion of somatostatin in normal man results in an acute decrease in the glucose production rate, which is accompanied by a decrease in plasma glucose; this occurs despite a concomitant decrease in plasma insulin and can be prevented by replacement infusion of glucagon. These observations suggest that in the post-absorptive state, glucagon action on the liver balances insulin action on the liver to maintain an appropriate output of glucose to match glucose utilization and, therefore, maintain stable euglycemia. With prolongation of the glucagon deficiency during infusion of somatostatin, glucose production does not exceed normal rates. These changes reflect the effects of the concomitant insulin deficiency and the unopposed actions of other counter-regulatory factors. When insulin deficiency is avoided by infusion of replacement amounts of insulin along with somatostatin, which results in an isolated deficiency of glucagon, plasma glucose decreases more than that observed during infusion of somatostatin alone, and both it and the glucose production rate remain suppressed below normal.

In addition to a role for glucagon in the maintenance of euglycemia by antagonizing the effects of post-absorptive (i.e., low) plasma insulin concentrations, there is evidence that glucagon acts in the defense against hypoglycemia by antagonizing the effects of excess plasma insulin. When hypoglycemia is produced in humans by injection of insulin, release of glucagon is stimulated along with that of other counter-regulatory hormones when the plasma glucose decreases below 3.8 mM (~68 mg/dl). Restoration of euglycemia is due to a compensatory increase in hepatic glucose production. Although secretion of catecholamines, growth hormone, and cortisol are stimulated along with that of glucagon, only the increases in plasma glucagon and catecholamines coincide with or precede the compensatory increase in the glucose production rate. The effects of glucagon during restoration of euglycemia involve both glycogenolysis and gluconeogenesis, predominantly the former.

There is also evidence for the role of glucagon in disposal of ingested carbohydrate. The liver is the main organ responsible for clearance of glucose appearing in the portal vein after ingestion of carbohydrate and presumably also that derived from a meal. The increase in the portal venous insulin and glucose concentrations act to promote formation of glycogen from the ingested glucose. Suppression of glucagon secretion

is probably also important in the decrease in endogenous glucose output and in the formation of glycogen from the ingested glucose. In people with Type 1 diabetes, incapable of insulin secretion, suppression of increase in plasma glucagon following ingestion of a mixed meal or glucose load improves postprandial glucose tolerance. Moreover, the effectiveness of exogenous insulin in preventing postprandial hyperglycemia and improving diabetic control is markedly augmented when glucagon secretion is suppressed by somatostatin.

Homeostasis

Circulating levels of ketone bodies (e.g., acetone, acetoacetic acid, and β-hydroxybutyrate) are determined by the net balance between rates of ketone body production and removal. The plasma ketone body concentration and insulin seem to be the major factors affecting removal of ketone bodies by tissue. There is considerable data, mainly from animal studies, that indicate that glucagon may play a key role in the formation of ketone bodies.

Ketone body formation results from β-oxidation of free fatty acids derived from intra- and extrahepatic sources. Two key factors are necessary for ketone body formation: sufficient substrate in the form of free fatty acids and a shift in the hepatic handling of free fatty acids from triglyceride synthesis (i.e., esterification) to oxidation. Glucagon directly acts on the liver *in vitro* to augment ketogenesis. This is thought to involve the promotion of transport of free fatty acids across the mitochondrial membrane by acylcarnitine transferase, an important rate-limiting step for free fatty acid oxidation. Glucagon apparently does not directly affect this enzyme but indirectly causes its activation by lowering intra-hepatic levels of malonylcoenzyme A (CoA), an inhibitor of acylcarnitine transferase. It has been postulated that glucagon is essential for switching the liver to a ketogenic mode (i.e. from an organ primarily esterifying free fatty acids to one oxidizing them) to permit maximal rates of ketogenesis to occur.

Pharmacologic doses of glucagon given as a bolus have been reported to increase both plasma free fatty acid and ketone body concentrations in normal subjects despite concomitant increases in plasma insulin. Following acute withdrawal from insulin in patients with Type 1 diabetes, the expected hyperketonemia can be markedly attenuated by suppression of glucagon secretion with somatostatin. Under such conditions (e.g., insulin withdrawal and somatostatin administration), infusion of physiologic amounts of glucagon, producing circulating glucagon concentrations less than those reported in ketoacidosis, results in a marked degree of hyperketonemia. There are two prerequisites for glucagon to stimulate ketogenesis: adequate substrate (e.g., free fatty acids) and insulin deficiency or the inability to increase plasma insulin concentrations.

F. Mechanism of Action

It is well established that the actions of glucagon on glycogenolysis, gluconeogenesis, and ketogenesis are mediated mainly by cAMP. Binding of glucagon with its receptor activates the catalytic subunit of the membrane bound enzyme adenylate cyclase, which catalyzes the conversion of adenosine triphosphate (ATP) to cAMP, which in turn leads to activation of intracellular kinase. For glycogenolysis, these results in phosphorylation of phosphorylase, which activates the enzyme and desphosphorylation of glycogen synthase which inactivates the enzyme. Thus, glycogen formation is inhibited and glycogen breakdown stimulated.

The action of glucagon on gluconeogenesis is more complex and involves several steps. Glucagon increases hepatic uptake of amino acids, but its main effect is intrahepatic. Glucagon stimulates gluconeogenesis mainly by increasing the rate of phosphoenolpyruvate production and decreasing the rate of its disposal by pyruvate kinase.

Stimulation of ketogenesis by glucagon is linked to some of the biochemical steps involved in its stimulation of gluconeogenesis, namely, an inhibition of glycolysis. The rate-limiting step of ketogenesis is the transport of fatty acid CoA esters across the mitochondrial membrane where they undergo β-oxidation. The enzyme catalyzing this transfer is fatty acid carnitine acyl transferase II. This enzyme is inhibited by malonyl-CoA. The inhibition of glycolysis by glucagon lowers intracellular levels of malonyl-CoA and results in activation of the fatty acid CoA acyl transferase.

G.   Glucagon Secretion in Diabetes Mellitus

In diabetes, plasma glucagon concentrations are either increased in an absolute sense or are "normal" but inappropriate for the prevailing plasma glucose concentration; they are markedly increased in diabetic ketoacidosis. In contrast to the normal situation, carbohydrate ingestion does not appropriately suppress plasma glucagon in people with impaired glucose tolerance and diabetes. This failure to suppress glucagon secretion leads to excessive appearance in plasma of glucose released from the liver which has been correlated with plasma insulin: glucagon molar ratios. Excessive increases in plasma glucagon are observed with protein meals, mixed meals, and infusion of amino acids. Some of these abnormalities, such as the fasting hyperglucagonemia and excessive responses to infusion of arginine, protein, or mixed-meal ingestion, can be improved or corrected by administration of physiologic quantities of insulin, suggesting that they were, in part, the result of insulin deficiency. Studies have demonstrated that infusion of GLP-1 receptor agonist exenatide has resulted in up to 50% reduction of endogenous glucose production and most of the effect was achieved by suppression of postprandial glucagon secretion.

In contrast to the above, acute administration of physiologic or even pharmacologic amounts of insulin have not been able to correct abnormal alpha cell responses to glucose in human diabetes. These observations suggest that abnormal alpha cell responses to glucose may not be solely due to insulin deficiency. Evidence for a selective defect in A-cell glucose recognition independent of insulin deficiency is provided by the findings that plasma glucagon can be suppressed normally in human diabetes by elevation of circulating free fatty acid levels but not by hyperglycemia, and that the diabetic Alpha cell fails to respond appropriately to hypoglycemia or to hyperglycemia. As noted above, recently appreciated regulators of glucagon production GLP-1 and GIP with former suppressing and latter inhibiting the hormone secretion, may play important role in glucose dysregulation in pre-diabetes and diabetes.

H. Metabolic Consequences of Alpha Cell Dysfunction in Diabetes Mellitus Patients

At present, the evidence suggests that the full-blown manifestations of diabetes cannot be explained solely on the basis of insulin deficiency, and that abnormal alpha cell function is an important determinant of the magnitude of hyperglycemia and hyperketonemia found in diabetes. The evidence for this can be summarized as follows. Fasting hyperglycemia and insulin requirements are lower in pancreatectomized patients lacking glucagon. Moreover, in such individuals and in insulin-dependent diabetics whose glucagon secretion is suppressed with somatostatin, hyperglycemia and hyperketonemia following acute withdrawal of insulin are markedly diminished. The failure to suppress glucagon secretion appropriately after meal ingestion increases postprandial hyperglycemia in people with impaired glucose tolerance and diabetes.

The failure of hypoglycemia to stimulate glucagon secretion in people with type 1 diabetes and in those with type 2 diabetes and marked B-cell dysfunction increases the risk for severe hypoglycemia in these individuals. The role of glucagon in maintenance of fasting euglycemia has been recently studied. In the setting of optimal insulin replacement in type 1 DM patients, the infusion of somatostatin caused hypoglycemia which was rescued by exogenous glucagon administration. The concentration of plasma insulin is critical for glucagon production during hypoglycemia of type 1 DM. Alpha cells retain their ability to produce glucagon in hypoinsulinemic hypoglycemia but not during hyperinsulinemic hypoglycemia. These two mechanistic studies provided us explanations how excessive insulin replacement leads to hypoglycemia in type 1 DM patients. Thus abnormalities in alpha cell function and regulation play an important role not only in the pathophysiology of metabolic abnormalities in diabetes mellitus but also in its management.

CONCLUSION

Insulin, which is secreted from pancreatic δ cells, is the key hormone in regulating glucose metabolism. Insulin secretion is a highly dynamic process regulated by complex mechanisms. It is regulated by nutrient status, hormonal factors such as gastrointestinal hormone incretins (i.e., glucagon-like peptide 1 [GLP-1] and glucose-dependent insulinotropic and neural factors. In addition, the regulation of insulin secretion is a multi-tiered process, occurring at the level of the single δ cell, the pancreatic islet, the whole pancreas, and the intact organism. Thus, *in vivo*, the dynamics of insulin secretion is the consequence of an integration of all of these systems.

In the last few years, several new anti-hyperglycemic agents for treatment of Type 2 DM representing the class of incretins were introduced. In addition to insulinotrophic effects, they were shown to contribute to glucose-dependent glucagon suppression which was seen as an additional benefit of the new medication class. The latter effect has been reflected in dose-dependent reduction of fasting and postprandial plasma glucose levels.

SUGGESTIONS FOR FURTHER READING

Braun M., Ramracheya R. and Rorsman P. (2012), Autocrine Regulation of Insulin Secretion, *Diabetes Obes. Metab.*, 14 Suppl., 3:143–151.

Buse J.B., Polonsky K.S. and Burant C.F., Disorders of Carbohydrate and Metabolism, In: Melmed S., Polonsky K.S., Larsen P.R. and Kronenberg H.M. (Eds.), *Williams Textbook of Endocrinology*, 12th ed., Saunders Elsevier, 1371–1435.

Chevenne D., Trivin F. and Porquet D. (1999), Insulin Assays and Reference Values, *Diabetes Metab.*, 25:459–476.

DeFronzo R.A., Bonadonna R.C. and Ferrannini E. (1992), Pathogenesis of NIDDM. A Balanced Overview, *Diabetes Care*, 15(3):318–68.

Dunning B.E. and Gerich J.E. (2007), The Role of Alpha-cell Dysregulation in Fasting and Postprandial Hyperglycemia in Type 2 Diabetes and Therapeutic Implications, *Endocr. Rev.*, 28:253–283.

Endocrine Function of the Pancreas and Regulation of Carbohydrate Metabolism, In: Barrett K., Brooks H., Boitano S. Barman S. (Eds.), *Ganong's Review of Medical Physiology*, 23rd ed., Lange McGraw Hill; 2010, 316–8.

Rubenstein A.H., Melani F., Pilkis S., and Steiner D.F. (1969), Proinsulin. Secretion, Metabolism, Immunological and Biological Properties. *Postgrad Med J.*, 45: Suppl:476–481.

90

Diabetes Mellitus

Nikhil Tandon and Yashdeep Gupta

CONTENTS

Diabetes mellitus (DM) refers to a group of metabolic diseases characterized by hyperglycemia resulting from defects in insulin secretion, insulin action, or both. The metabolic dysregulation associated with DM is associated with long-term damage, dysfunction, and failure of different organs, especially the eyes, kidneys, nerves, heart, and blood vessels.

According to the International Diabetes Federation (IDF) Diabetes Atlas 6th edition there are already 382 million people with diabetes. In addition, there are about 280 million with impaired glucose tolerance (IGT). The number is likely to grow to over 590 million people with diabetes and almost 400 million with IGT by 2030. Almost one-fifth of the world's people with diabetes live in just seven countries of IDF South-East Asia region. Current estimates indicate that 8.3% of the adult population in South-East Asia region, or 72 million people, have diabetes in 2013, 65.1 million of whom are in India. The number of people with diabetes in India has been projected to increase to 101.2 million by 2030. Diabetes is the most common cause of blindness in those of working age, the most common single cause of end-stage renal failure worldwide, and the consequences of neuropathy make it the most common cause of non-traumatic lower limb amputation. Mortality from ischemic heart disease and stroke is 2–4 fold higher than in the age- and sex-matched non-diabetic population.

I. CLASSIFICATION

DM is classified on the basis of the pathogenic process that

leads to hyperglycemia, as opposed to earlier criteria such as age of onset or type of therapy. Table 90.1 provides a current classification of diabetes mellitus. The two broad categories of DM are designated type 1 and type 2. Other specific types of diabetes include genetic defects in insulin secretion or action, mitochondrial abnormalities, and a host of conditions that impair glucose tolerance. GDM is defined as diabetes diagnosed during pregnancy that is not clearly overt diabetes. It does not include women with type 1 or type 2 diabetes mellitus. The traditional definition of GDM "glucose tolerance with onset or first recognition in pregnancy" included women with unknown pre-existing diabetes; particularly type 2 diabetes mellitus (T2DM). The terms insulin-dependent diabetes mellitus (IDDM) and non-insulin-dependent diabetes mellitus (NIDDM) are obsolete now. Firstly, since many individuals with type 2 DM eventually require insulin treatment for control of glycemia, the use of the term NIDDM generated considerable confusion. Secondly, the age is not a criterion in the classification system. While type 1 DM most commonly develops before the age of 30 years, in some individuals it develop after the age 30 years. Although type 2 DM more typically develops with increasing age, it is now being diagnosed more frequently in children and young adults, particularly in obese adolescents.

A. Type 1 Diabetes Mellitus

This form of diabetes accounts for 5–10% of those with diabetes. Type 1 DM results from autoimmune beta cell destruction, and most, but not all, individuals have evidence

TABLE 90.1 Etiologic classification of diabetes mellitus

I. Type 1 Diabetes
    A. Immune mediated
    B. Idiopathic

II. Type 2 Diabetes

III. Other specific types of diabetes

   A. Genetic defects of beta cell function characterised by mutation in MODY 3 (HNF-1a); MODY 1 (HNF-4a); MODY 2 (glucokinase); Other very rare forms of MODY (e.g. MODY 4, insulin promoter factoer-1; MODY 6: NeuroD1; MODY 7: carbodxyl ester lipase); Transient neonatal diabetes (most commonly SAC/HYAMi imprinting defect on 6q24); Permanent neonatal diabetes (most commonly KCNJ11 gene encoding Kir6.2 subunit of b-cell KATP channel); Mitochondrial DNA; Other

   B. Genetic defects in insulin action: Type A insulin resistance; Leprechaunism; Rabson-Mendenhall syndrome; Lipoatrophic diabetes; Others

   C. Diseases of the exocrine pancreas: Pancreatitis; Trauma/pancreatectomy; Neoplasia; Cystic fibrosis; Hemochromatosis; Fibrocalculous pancreatopathy; Others

   D. Endocrinopathies: Acromegaly; Cushing's syndrome; Glucagonoma; Phechromocytoma; Hyperthyrodism; Somatostatinoma; Aldosteronoma; Others

   E. Drug or chemical induced: Vacor; Pentamidine; Nicotinic acid; Glucocorticoids; Thyroid hormone; Diazoxide; β-Adrenergic agonists; Thiazides; Dilantin; α-Interferon; Others

   F. Infections: Congenital rubella; Cytomegalovirus; Others

   G. Uncommon forms of immune-mediated diabetes: "Stiff-man" syndrome; Anti-insulin receptor antibodies; Others

   H. Other genetic syndromes sometimes associated with diabetes: Down syndrome; Klinefelter syndrome; Tumer syndrome; Wolfram syndrome; Friedreich ataxia; Huntington chorea; Laurence-Moon-Biedl syndrome; Myotonic dystrophy; Porphyria; Prader-Willi syndrome; Others

IV. Gestational diabetes mellitus

of pancreatic islet-directed autoimmunity. American Diabetes Association (ADA) classification categorizes this further into two entities: type 1A (immune mediated) and type 1B (idiopathic) diabetes mellitus. Although only a minority of patients with type 1 diabetes fall into category of type 1B diabetes, most of them are of African or Asian ancestry. Autoimmune destruction of β-cells has multiple genetic predispositions and is also related to environmental factors that are still poorly defined. The major susceptibility gene for type 1 DM is located in the HLA region on chromosome 6. Polymorphisms in the HLA complex account for 40–50% of the genetic risk of developing type 1 DM. Most individuals with type 1 DM have the HLA DR3 and/or DR4 haplotype. Type 1 or autoimmune diabetes is strongly associated with

HLA-DR (major histocompatibility complex, class II, DR) and HLA-DQ (major histocompatibility complex, class II, DQ) genes. The HLA DQA1*0301–DQB1*0302 and DQA1*0501–DQB1*0201 haplotypes, alone or in combination, may account for up to 90% of children and young adults with type 1 diabetes. These two haplotypes may be present in 30%–40% of a Caucasian population, and HLA is therefore necessary, but not sufficient for disease. The haplotype DQA1*0102, DQB1*0602 is extremely rare in individuals with type 1 DM (<1%) and appears to provide protection from this disease. In addition to MHC class II associations, genome-wide association studies have confirmed that the following non-HLA genetic factors increase the risk for type 1 diabetes, both in first-degree relatives of type 1 diabetic patients and in the general population: INS VNTR, CTLA4, PTPN22, and others. The concordance of type 1 DM among monozygotic twins ranges between 40 and 60%, while that among dizygotic twins ranges between 10–19%. Although more than 85% of T1DM occurs in individuals with no previous first-degree family history, the risk among first-degree relatives is about 15 times higher than the general population. An affected father confers a 6–9% risk of T1DM to his offspring compared to 2–4% if the mother is affected and up to 30% risk if both parents are affected.

Various autoantibodies have been used as markers of the autoimmune destruction of pancreatic beta cells. These include, autoantibodies to islet cell cytoplasm (ICA), to native insulin [referred to as "insulin autoantibodies" (IAA)], to the 65-kDa isoform of glutamic acid decarboxylase (GAD65A), to two insulinoma antigen 2 proteins [IA-2A and IA-2bA (also known as phogrin)], and to three variants of zinc transporter 8 (ZnT8A). Autoantibody markers of immune destruction are usually present in 85% to 90% of individuals with type 1 diabetes when fasting hyperglycemia is initially detected. After years of type 1 diabetes, some antibodies fall below detection limits, but antibodies to GAD65A usually remain detectable. Patients with type 1A diabetes have a significantly increased risk of other autoimmune disorders, including celiac disease, Graves' disease, thyroiditis, Addison's disease, and pernicious anemia. As many as 1 in 4 females with type 1 diabetes have autoimmune thyroid disease, whereas 1 in 280 patients develop adrenal autoantibodies and adrenal insufficiency. Type 1B diabetes does not appear to be immune mediated and has not been associated with any HLA haplotypes. However, individuals with this form of diabetes do suffer from episodic ketoacidosis and exhibit varying degrees of insulin deficiency between episodes. The degree to which environmental events are important in the pathogenetic sequence is unresolved. There is compelling evidence that such events can initiate the pathogenetic processes in genetically predisposed animals, but the evidence for their role in type 1 diabetes in human beings is less secure. Putative environmental triggers that in genetically susceptible individuals might play a role in initiating the disease process, include viral infections, chemical toxins, or exposure to cows' milk proteins in early infancy. The interactions of genetic, environmental, and immunologic factors ultimately lead to the destruction of the pancreatic beta cells and insulin deficiency. The probable temporal development of type 1 diabetes is described below in Figure 90.1.

B. Type 2 Diabetes Mellitus

Type 2 diabetes mellitus is a complex endocrine and metabolic disorder. Although the primary defect is controversial, most studies support that insulin resistance precedes an insulin secretory defect, but that diabetes develops only when insulin secretion becomes inadequate. Insulin resistance begins many years before the onset of type 2 diabetes as a result of the interaction of genetic and several environmental factors. Key genes, including TCF7L2, PPARG, KCNJ11, SLC30A8, HHEX, CDKN2A, IGF2BP2, CDKAL1 and FTO act in conjunction with environmental factors, including pregnancy, physical inactivity, quality and quantity of nutrients, puberty and ageing, to promote adiposity, impair β-cell function, and impair insulin action. Most prominent is a variant of the transcription factor 7-like 2 gene that has been associated with type 2 diabetes in several populations. Positive family history confers a 2.4 fold increased risk for type 2 diabetes. The concordance of type 2 diabetes in identical twins is between 70 and 90%. 15–25% of first-degree relatives of patients with type 2 diabetes develop impaired glucose tolerance or diabetes. The lifetime risk (at age 80 years) for type 2 diabetes has been calculated to be 38% if one parent had type 2 diabetes. If both parents are affected, the prevalence of type 2 diabetes in the offspring is estimated to approach 60% by the age of 60 years.

Overweight and obesity are major contributors to the development of insulin resistance and impaired glucose tolerance. Obesity, particularly visceral or central, is very common in type 2 diabetes (80% or more are obese in T2DM, in India around 40% of T2DM are either normal weight or lean). In the early stages of the disorder, glucose tolerance remains near-normal, despite insulin resistance, because the pancreatic beta cells compensate by increasing insulin output. As insulin resistance and compensatory hyperinsulinemia progresses, the pancreatic islets in certain individuals are unable to sustain the hyperinsulinemic state required to maintain euglycemia. A further decline in insulin secretion and an increase in hepatic glucose production lead to overt diabetes. Insulin resistance in muscle and liver and β-cell failure represent the core pathophysiologic defects in type 2 diabetes. In addition to the muscle, liver and β-cell (triumvariate), the fat cell (accelerated lipolysis), gastrointestinal tract (incretin deficiency/resistance), α-cell (hyperglucagonemia), kidney (increased glucose

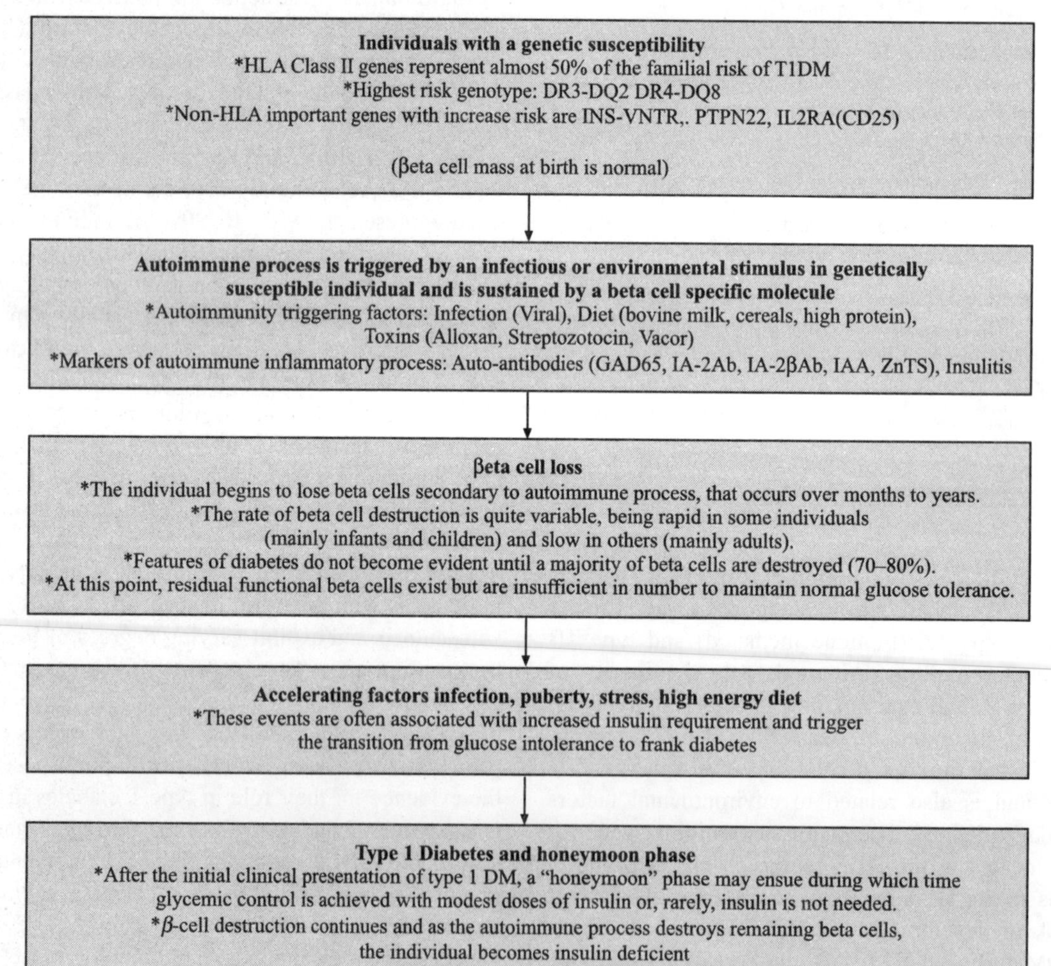

Figure 90.1 Temporal development of type 1 diabetes.

absorption) and brain (insulin resistance) all play important roles in the development of glucose intolerance in type 2 diabetic individuals. Collectively, these eight players comprise the ominous octet. The factors implicated in pathophysiology of T2DM are depicted in Figure 90.2.

C. Other Specific Types of Diabetes

Genetic defects of the β-cell: Monogenic diabetes, caused by single gene mutations, accounts for approximately 2% of all diabetes cases. The most prevalent phenotype is maturity-onset diabetes of the young (MODY). MODY is characterized by dominant inheritance of early-onset (typically before 25 years of age), usually not requiring insulin. The term MODY describes a heterogeneous group of disorders caused by mutations in genes important to beta cell development, function and regulation, glucose sensing, and in the insulin gene itself. Mutations in at least eight different genes can cause MODY (Table 90.2).

Heterozygous GCK mutations cause impaired glucokinase activity resulting in stable, mild hyperglycaemia that rarely requires treatment. HNF1A mutations cause a progressive insulin secretory defect that is sensitive to sulphonylureas, most often resulting in improved glycaemic control compared with other diabetes treatment. MODY owing to mutations in the HNF4A gene results in a similar phenotype, including sensitivity to sulphonylurea treatment. HNF1B mutations most frequently cause developmental renal disease (particularly renal cysts) but may also cause MODY in isolation or may cause the renal cysts and diabetes syndrome (RCAD syndrome). Mutations in NEUROD1, PDX1 (IPF1), CEL and INS are rare causes of MODY. The expected clinical course, complications and associated extra-pancreatic features vary based upon the underlying molecular genetic defect. Correctly identifying MODY has important implications for treatment, surveillance of complications and associated extra-pancreatic disorders, and identification of affected and at-risk family members.

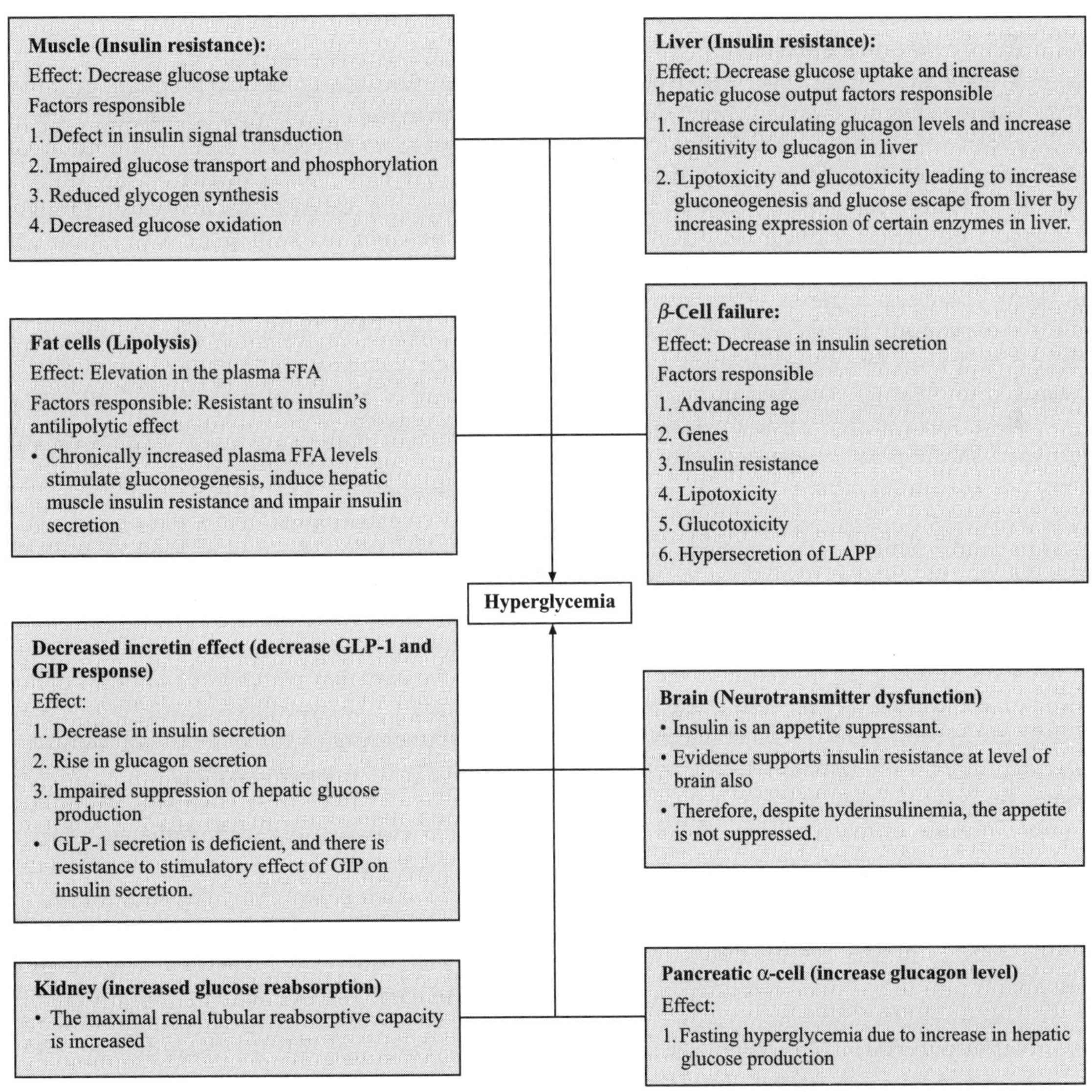

Figure 90.2 Pathophysiology of type 2 diabetes mellitus.

TABLE 90.2   Classification of Maturity onset diabetes of the young

Phenotype	Affected gene	Clinical features
MODY 1	HNF 4-α	• Presentation in adolescence or early adulthood • Possible history of neonatal macrosomia and/or neonatal hypoglycemia
MODY 2	Glucokinase	• Mild hyperglycemia (in early childhood)
MODY 3	HNF 1-α	• Progressive insulin secretory defect with presentation in adolescence or early aulthood • Low renal glucose threshold
MODY 4	IPF-1	• Rare • Pancreas agenesis and neonaal diabetes mellitus in homozygote
MODY 5	HNF 1-β	• Progressive diabetes, often with renal disease • Renal and genitourinary abnormalities • Pancreatic atrophy, excrine pancreatic dysfunction • Hyperuricaemia, gout, and abnormal liver function tests may occur
MODY 6	NEURODI	• Very rare • Neonatal diabetes in homozygote
MODY 7	Carbodyl-ester lipase; CEL	• Pancreatic atrophy, exocrine pancreatic dysfunction
INS-MODY	Insulin	• Very rare • Mild to severe diabetes

Maturity-onset diabetes of the young should be considered in any individual carrying a diagnosis of either type 1 or type 2 diabetes with atypical features for these polygenic disorders. This includes the absence of islet cell autoantibodies (type 1B) at the time of diagnosis or evidence of continued endogenous insulin outside of the honeymoon period in type 1 diabetes. A diagnosis of type 2 diabetes in a young individual who is not significantly overweight or who lacks hallmarks of insulin resistance, including acanthosis nigricans or elevated fasting insulin, should be questioned. The presentation of MODY-related diabetes can be delayed for decades beyond the original definition of onset before age 25. Diabetes in two or more consecutive generations in a pattern consistent with autosomal dominant inheritance should prompt consideration of MODY (but spontaneous, *de novo* cases occur).

Genetic defects in insulin action: Genetic abnormalities in the insulin receptor can give rise to rare but well-described syndromes characterised by severe insulin resistance. Clinically these patients have acanthosis nigricans and hyperandrogenism. Mutations of the gene encoding the α subunit of the insulin receptor can lead to leprechaunism or the less severe Rabson-Mendenhall syndrome. Type A insulin resistance affects mainly adolescent girls and shares many features with the polycystic ovary syndrome. In 25% of cases there is a mutation of the tyrosine kinase domain of the β-subunit of the insulin receptor. Rare mutations of the human preproinsulin gene can lead to abnormal levels of insulin precursors. Such patients are heterozygous (homozygosity would be incompatible with life), and develop diabetes in later life in response to other factors such as obesity.

Diseases of the exocrine pancreas: Many pancreatic diseases can cause diabetes, but in total they account for <1% of all cases. Any process that diffusely injures the pancreas can cause diabetes. Acquired processes include pancreatitis, trauma, infection, pancreatectomy, and pancreatic carcinoma. With the exception of that caused by cancer, damage to the pancreas must be extensive for diabetes to occur. However, adenocarcinomas that involve only a small portion of the pancreas have been associated with diabetes. This implies a mechanism other than simple reduction in β-cell mass. If extensive enough, cystic fibrosis and hemochromatosis will also damage β-cells and impair insulin secretion. Fibrocalculous pancreatopathy may be accompanied by abdominal pain radiating to the back and pancreatic calcifications identified on X-ray examination. There may be fibrosis of pancreas and calcium stones in the exocrine ducts.

Endocrinopathies: Cortisol, growth hormone, glucagon, and the catecholamines (epinephrine and norepinephrine) can antagonize insulin action. Tumors that produce these hormones in excess lead to Cushing's syndrome, acromegaly, glucagonoma, and pheochromocytoma, respectively. The importance of recognizing these secondary forms of diabetes lies in the fact that resection of the underlying tumor can cure diabetes. The hyperglycemia that is seen in the setting of aldosterone-producing adenomas and somatostatinomas results from alteration of insulin secretion.

Drug- or chemical-induced diabetes: Many drugs can impair insulin secretion. These drugs may not cause diabetes by themselves, but they may precipitate diabetes in individuals with insulin resistance. Certain toxins such as Vacor (a rat poison) and intravenous pentamidine can permanently destroy pancreatic β-cells. There are also many drugs and hormones that can impair insulin action. Examples include nicotinic acid and glucocorticoids. The list shown in Table 90.1 reflects the more commonly recognized drug-, hormone-, or toxin-induced forms of diabetes.

Infections: Certain viral infections, including rubella and coxsackie B virus, have been associated with diabetes. Some studies suggest that a viral infection can trigger autoimmune destruction of β cells in genetically predisposed individuals, leading to autoimmune type 1 diabetes.

Uncommon forms of immune-mediated diabetes: The stiff-man syndrome is a rare neurologic syndrome characterized by spasticity of the axial muscles. It is associated with very high titers of anti-GAD antibodies, and up to one third of patients develops diabetes. Autoantibodies directed against the insulin receptor represent another rare cause of diabetes (referred to as type B insulin resistance). The syndrome is typically seen in African American females in association with other autoimmune diseases (most commonly systemic lupus erythematosus).

Other genetic syndromes sometimes associated with diabetes: Many genetic syndromes are associated by an increased incidence of diabetes. These include the chromosomal abnormalities of Down syndrome, Klinefelter syndrome and Turner syndrome. Wolfram's syndrome is an autosomal recessive disorder characterized by insulin-deficient diabetes and the absence of β-cells at autopsy. Additional manifestations include diabetes inspidus, hypogonadism, optic atrophy, and neural deafness. Other syndromes are listed in Table 90.1.

D. Gestational Diabetes Mellitus (GDM)

GDM is defined as diabetes diagnosed during pregnancy that is not clearly overt diabetes. It does not include women with type 1 or type 2 diabetes mellitus. GDM carries risks for the mother and neonate. Women should be screened for undiagnosed type 2 diabetes at the first prenatal visit in those with risk factors, using standard diagnostic criteria. As per new ADA criteria, a woman is assigned a diagnosis of pre-existing diabetes if any of the following are present at first visit: fasting plasma glucose (FPG) $\geq$ 126 mg/dl; HbA1c $\geq$ 6.5%; random plasma glucose $\geq$ 200 mg/dl (confirmed by FPG or HbA1c). If FPG at first visit is <126 mg/dl but $\geq$ 92 mg/dl, the woman is diagnosed to have GDM. The woman without pre-existing diabetes or GDM at first visit should undergo 75 g, 2 h OGTT at 24–28 weeks gestation. The diagnosis of GDM is made if any of the following plasma glucose values is met; FPG $\geq$ 92 mg/dl; 1 h $\geq$ 180 mg/dl or 2 h $\geq$ 153 mg/dl. ADA recommends screening for diabetes at 6–12 weeks postpartum, every three years thereafter, but annually in women with prediabetes, using non-pregnant (OGTT) criteria.

II. INTERMEDIARY METABOLISM OF DIABETES

For further details the reader is referred to Chapter 89 (The Pancreas: Insulin and Glucagon).

III. DIAGNOSIS OF DIABETES

A. Clinical Presentation

The classic manifestations of diabetes mellitus include hyperglycaemia, secondary to insulin insufficiency. As high glucose in the bloodstream is filtered through the kidneys, osmotic balance causes excess urination resulting in polyuria. As more water is excreted, the body requires more water intake and, because thirst mechanisms are intact, thirst increases with resulting polydipsia. Loss of water as well as well as loss of muscle mass and fat mass all contribute to acute or subacute weight loss at the time of diagnosis of type 1 diabetes mellitus. Fatigue and weakness probably occur as a result of decreased glucose utilization and subtle electrolyte and/or mineral abnormalities as well as clinical and subclinical dehydration.

The initial presentation may be with acute complications of diabetes, diabetic ketoacidosis (DKA) and hyperglycaemic hyperosmolar state (HHS). DKA was formerly considered a hallmark of type 1 diabetes, but also occurs in individuals who lack immunologic features of type 1 DM and who can sometimes subsequently be treated with oral glucose-lowering agents. Classic type 2 diabetes in children and adolescents sometimes presents with ketoacidosis, but almost always with moderate to severe insulin resistance because of obesity. HHS is primarily seen in individuals with type 2 diabetes. Both disorders are associated with absolute or relative insulin deficiency, volume depletion, and acid-base abnormalities.

Since type 2 DM often has a long asymptomatic period of hyperglycaemia, many individuals with type 2 DM have chronic complications at the time of diagnosis. Chronic complications can be vascular or nonvascular. The vascular complications of DM are further subdivided into microvascular (retinopathy, neuropathy, nephropathy) and macrovascular complications [coronary heart disease (CHD), peripheral arterial disease (PAD), cerebrovascular disease]. Nonvascular complications include problems such as gastroparesis, infections, and skin changes. Long standing diabetes may be associated with hearing loss.

B. Laboratory Diagnosis of Diabetes Mellitus

For decades, the diagnosis of diabetes was based on plasma glucose criteria, either the fasting plasma glucose (FPG) or the 2-h value in the 75-g oral glucose tolerance test (OGTT). In 2009, an International Expert Committee that included representatives of the ADA, the International Diabetes Federation (IDF), and the European Association for the Study of Diabetes (EASD) recommended the use of the A1C test to diagnose diabetes, with a threshold of $\geq$6.5%, and ADA adopted this criterion in 2010 (refer to subsection on glycated hemoglobin).

Glucose levels

Before 1997, the diagnosis of diabetes was defined by the ADA and the World Health Organization (WHO) as a fasting plasma glucose level of 140 mg per deciliter (7.8 mmol per liter) or more or a 2-hour plasma glucose level of 200 mg per deciliter (11.1 mmol per liter) or more during an oral glucose-tolerance test (OGTT) conducted with a standard loading dose of 75 g. This definition was based on earlier recommendations from the National Diabetes Data Group. These values were originally chosen on the basis of the risk of future symptoms of uncontrolled hyperglycemia. In 1997, with recommendations from the Expert Committee on the Diagnosis and Classification of Diabetes Mellitus, the ADA and the WHO lowered the diagnostic threshold to a fasting plasma glucose level of 126 mg per deciliter (7.0 mmol per liter)—the level at which a unique microvascular complication of diabetes, retinopathy, becomes detectable. The OGTT identifies more patients as having diabetes than the fasting plasma glucose test, but the former test has drawbacks, including greater expense and complexity and lower reproducibility. Thus, the fasting plasma glucose test has been the preferred test in the United States. The diagnosis is confirmed by repeat testing on a separate day. In symptomatic patients, a random plasma glucose level of 200 mg per deciliter or more also establishes the diagnosis and does not require confirmation.

An excellent review on Guidelines and Recommendations for Laboratory Analysis in the Diagnosis and Management of Diabetes Mellitus has been published by ADA in 2011. Some salient points have been captured below. Blood for FPG should be drawn in the morning after an overnight fast (no caloric intake for at least 8 h), during which time the individual may consume water *ad libitum*. Evidence reveals diurnal variation in FPG, with the mean FPG being higher in the morning than in the afternoon, indicating that many diabetes cases would be missed in patients seen in the afternoon. There is decrease in glucose concentration in whole blood *ex vivo* due to glycolysis. The rate of glycolysis—reported to average 5%–7%/h [approximately 0.6 mmol/L (10 mg/dL)]—varies with the glucose concentration, temperature, leukocyte count, and other factors. Such decreases in glucose concentration will lead to missed diabetes diagnoses in the large proportion of the population who have glucose concentrations near the cut points for diagnosis of diabetes. The commonly used glycolysis inhibitors are unable to prevent short-term glycolysis. Glycolysis can be attenuated by inhibiting enolase with sodium fluoride (2.5 mg/mL of blood) or, less commonly, lithium iodoacetate (0.5 mg/mL of blood). Although fluoride helps to maintain long term glucose stability, the rates of decline in the glucose concentration in the first hour after sample collection are virtually identical for tubes with and without fluoride, and glycolysis continues for up to 4 h in samples containing fluoride. After 4 h, the concentration of glucose in whole blood in the presence of fluoride remains stable for 72 h at room temperature (leukocytosis will increase glycolysis even in the presence of fluoride if the leukocyte count is very high). Loss of glucose

can be minimized in two classic ways: (1) immediate separation of plasma from blood cells after blood collection [the glucose concentration is stable for 8 h at 25°C and 72 h at 4°C in separated, nonhemolyzed, sterile serum without fluoride]; and (2) placing the blood tube in an ice–water slurry immediately after blood collection and separating the plasma from the cells within 30 min. These methods are not always practical and are not widely used. Glucose can be measured in whole blood, serum, or plasma, but plasma is recommended for diagnosis. Glucose concentrations are approximately 11% higher in plasma than in whole blood if the hematocrit is normal. In contrast, some more recent studies found that glucose concentrations are slightly higher in plasma than in serum. The observed differences were approximately 0.2 mmol/L (3.6 mg/dL), or approximately 0.9–2%. Other studies have found that glucose values measured in serum and plasma are essentially the same. Given these findings, it is unlikely that values for plasma and serum glucose will be substantially different when glucose is assayed with current instruments, and any differences will be small compared with the day-to-day biological variation of glucose. Clinical organizations do not recommend the measurement of glucose in serum (rather than plasma) for the diagnosis of diabetes. Use of plasma allows samples to be centrifuged promptly to prevent glycolysis without waiting for the blood to clot. The glucose concentrations in capillary blood obtained during an OGTT are significantly higher than those in venous blood [mean, 1.7 mmol/L (30 mg/dL), which is equivalent to 20%–25% higher], probably owing to glucose consumption in the tissues. In contrast, the mean difference in fasting samples is only 0.1 mmol/L (2 mg/dL).

Knowledge of intra-individual (within-person) variation in FPG concentrations is essential for meaningful interpretation of patient values. The intra-individual coefficient of variation in one study was 6.4% for the FPG and 16.7% for the 2-hour PG value, compared with less than 2% for HbA1C. Similar findings were obtained from an analysis of 685 adults from NHANES III, in which the mean within-person variation in FPG measured 2–4 weeks apart was 5.7%. If a within person biological CV of 5.7% is applied to a true glucose concentration of (126 mg/dL), the 95% CI would encompass glucose concentrations of (112–140 mg/dL). If the analytical CV of the glucose assay (approximately 3%) is included, the 95% CI is approximately ±6.4%. Thus, the 95% CI for a fasting glucose concentration of 7.0 mmol/L (126 mg/dL) would be 110–142 mg/dL. Use of an assay CV of 3% only (excluding biological variation) would yield a 95% 118–134 mg/dL among laboratories, for a true glucose concentration of 7.0 mmol/L (126 mg/dL). Performing the same calculations at the cut-off for impaired fasting glucose yields a 95% CI of 100 mg/dl (6.4%), i.e., (87–113 mg/dL). One should bear in mind that these intervals include 95% of the results and that the remaining 5% will be outside this interval. Thus, the biological variation is substantially greater than the analytical variation.

Urine glucose testing is not recommended for routine care of patients with diabetes mellitus. Semi-quantitative urine glucose testing, once the hallmark of diabetes care in the home setting, has now been replaced by self monitoring of blood glucose (SMBG). Semi quantitative urine glucose monitoring should be considered only for patients who are unable or refuse to perform SMBG. The rationale behind, that although urine glucose is detectable in patients with grossly increased blood glucose concentrations, it provides no information about blood glucose concentrations below the variable renal glucose threshold [approximately 10 mmol/L (180 mg/dL)]. Semi-quantitative urine glucose tests also cannot distinguish between euglycemia and hypoglycemia. Furthermore, the extent to which the kidney concentrates the urine will affect urine glucose concentrations, and only mean glucose values between voidings are reflected.

Glucose in plasma, serum or whole blood is measured exclusively by enzymatic methods. Hexokinase or glucose oxidase is used in virtually all analyses. Very few laboratories use glucose dehydrogenase method. Commercially available strips for urine glucose testing use the glucose oxidase reaction. Test methods that detect reducing substances are not recommended because they are subject to numerous interferences, including numerous drugs and nonglucose sugars. When used, single voided urine samples are recommended. Second-voided specimens do not appear to offer any significant advantage over first-voided specimens.

Glycated hemoglobin

Glycated hemoglobin has long been used in the management of established diabetes as a biomarker of long-term glycemic control. Levels of this end product of non-enzymatic glycation of the most prevalent protein in blood correlate well (though not perfectly) with average ambient blood glucose levels during the previous 2 to 3 months. Epidemiologic datasets show a similar relationship between A1C and risk of retinopathy as has been shown for the corresponding FPG and 2-h plasma glucose thresholds. The diagnostic test should be performed using a method that is certified by the National Glycohemoglobin Standardization Program (NGSP) and standardized or traceable to the Diabetes Control and Complications Trial (DCCT) reference assay. The AIC has several advantages to the FPG and OGTT, including greater convenience, since fasting is not required. Low biologic variability, marker of long-term glycemia, stablity during acute illness, sample stability in vial, global standardization, and close association of results with complications adds to its advantages. Despite some advantages, the use of glycated hemoglobin testing has its limitations. Depending on the assay, spuriously low values may occur in patients with certain hemoglobinopathies (e.g., sickle cell disease and thalassemia) or who have increased red-cell turnover (e.g., hemolytic anemia and spherocytosis) or stage 4 or 5 chronic kidney disease, especially if the patient is receiving erythropoietin. In contrast, falsely high glycated haemoglobin levels have been reported in association with iron deficiency and other states of decreased red-cell turnover. Some investigators have reported a "glycation gap," or different glycated haemoglobin levels in patients with the same mean ambient blood glucose levels. This phenomenon may result from genetically determined altered access of glucose to the intracellular compartment (where hemoglobin resides), although this hypothesis is controversial. Inconsistencies in the correlations between glycated hemoglobin and other measures of ambient glycemia have also been reported in different ethnic and racial groups, findings that suggest genetic influences on haemoglobin glycation. For example, blacks appear to have slightly higher glycated hemoglobin levels (an absolute increase of 0.2 to 0.3 percentage points) than whites. It is unclear whether this observation reflects differences in rates of postprandial hyperglycemia or in glycation rates. These potential pitfalls must be recognized when glycated hemoglobin testing is used for diagnosis, especially for prediabetes, since the cut-off points for this state are already somewhat arbitrary.

Additional laboratory studies

Ketone testing: The ketone bodies acetoacetate (AcAc), acetone, and b-hydroxybutyric acid (bHBA) are catabolic products of free fatty acids. Ketone bodies are usually present in urine and blood, but in very low concentrations (e.g., total serum ketones, 0.5 mmol/L). The two major mechanisms for high ketone concentrations in patients with diabetes are increased production from triglycerides and decreased utilization in the liver—both of which are due to an absolute or relative insulin deficiency and increased level of counter-regulatory hormones, including cortisol, epinephrine, glucagon, and growth hormone. Increased ketone concentrations detected in patients with known diabetes or in previously undiagnosed patients presenting with hyperglycemia suggest impending or established DKA, a medical emergency. The ADA recommends that ketosis-prone patients with diabetes check urine or blood ketones in situations characterized by deterioration in glycemic control in order to detect and pre-empt the development of DKA. Ketones measured in urine or blood in the home setting by patients with diabetes and in the clinic/hospital setting should be considered only an adjunct to the diagnosis of DKA, as per recommendation.

Autoimmune markers: Islet cell autoantibodies are not recommended for routine diagnosis of diabetes, but standardized islet cell autoantibody tests may be used for classification of diabetes in adults and in prospective studies of children at genetic risk for type 1 diabetes after HLA typing at birth. Screening patients with type 2 diabetes for islet cell autoantibodies is not recommended at present. Screening for islet cell autoantibodies in relatives of patients with type 1 diabetes or in persons from the general population is also not recommended at present.

Albuminuria: Annual testing for albuminuria in patients without clinical proteinuria should begin in pubertal or postpubertal individuals 5 years after diagnosis of type 1 diabetes and at the time of diagnosis of type 2 diabetes, regardless of treatment. Acceptable samples to test for increased urinary albumin excretion are timed collections (e.g., 12 or 24 h) for measurement of the albumin concentration and timed or untimed samples for measurement of the albumin–creatinine ratio. The optimal time for spot urine collection is the early morning. All collections should be at the same time of day to minimize variation. The patient should not have ingested food within the preceding 2 h but should be well hydrated (i.e., not volume Depleted). The analytical CV of methods to measure low levels of albuminuria should be <15%.

Genetic markers: Routine measurement of genetic markers is not of value at this time for the diagnosis or management of patients with type 1 diabetes. For selected diabetic syndromes, including neonatal diabetes, valuable information can be obtained with definition of diabetes-associated mutations. There is no role for routine genetic testing in patients with type 2 diabetes as per recommendation. These studies should be confined to the research setting and evaluation of specific syndromes.

Management: The goals of therapy for type 1 or type 2 DM are to:

(i) Eliminate symptoms related to hyperglycemia
(ii) Reduce or eliminate the long term microvascular and macrovascular complications of DM.
(iii) Allow the patient to achieve as normal as lifestyle as possible.

To reach these goals, each patient should be provided with the educational and pharmacologic resources necessary to reach this level, and monitor DM-related complications. As per ADA, A1C is primary goal and in general should be less than 7%. Pre-prandial capillary plasma glucose between 70-130 mg% and peak postprandial capillary plasma glucose less than 180 mg% are recommended goals for a diabetic patient as per ADA, but with the statement of individualization of goals for each patient. Because the complications of DM are related to glycaemic control, normoglycemia or near normoglycemia is desirable. The patients with diabetes should receive education about nutrition, exercise, care of diabetes during illness, and medications to lower the plasma glucose. The general aspects for treatment for type 1 and type 2 diabetes are described below. The care of the individuals must also include treatment of conditions associated with diabetes like hypertension, dyslipidemia.

Type 1 Diabetes Mellitus: For type 1 diabetes, the goal is to design and implement insulin regimens that mimic physiologic insulin secretion. As individuals with type 1 DM partially or completely lack endogenous insulin production, administration of basal insulin is essential for regulating glycogen breakdown, gluconeogenesis, lipolysis, and ketogenesis. Likewise, insulin replacement for meals should be appropriate for the carbohydrate intake and promote normal glucose utilization and storage.

Type 2 Diabetes Mellitus: In addition to glycaemic control, the care of individuals with type 2 DM must also include treatment of conditions associated with type 2 DM (obesity, hypertension, dyslipidemia, cardiovascular disease) as well as management of diabetes related complications. Management of type 2 DM should begin with medical nutrition therapy (MNT) along with institution of an exercise regimen to promote weight loss and increase insulin sensitivity. Pharmacologic approaches to the management of type 2 DM include oral glucose lowering agents, insulin and other agents that improve glucose control. Oral glucose lowering agents target different pathophysiologic processes in type 2 DM. Based on their mechanisms of action, glucose lowering agents are subdivided into agents that stimulate insulin secretion, reduce hepatic glucose output improve peripheral glucose utilization by increase insulin sensitivity and action, and enhance GLP-1 action.

Metformin, which is the only bigaunide in clinical use currently, reduces hepatic glucose production and improves peripheral glucose utilization slightly. Insulin secretagogues, sulfonylurea and non-sulfonylurea, stimulate insulin secretion by interacting with the ATP-sensitive potassium channel on the beta cell. Glimepiride and glipizide can be given in a single daily dose and are preferred over glibenclamide. Repaglinide and nateglinide are non-sulfonylurea secretagogues, but also interact with the ATP-sensitive potassium channel. GLP-1 agonists (exenatide, liraglutide) and dipeptidyl peptidase-4 inhibitors (Linagliptin, Saxagliptin, Sitagliptin, Vildagliptin) act as GLP-1 agonist or enhance endogenous GLP-1 activity respectively and are approved for the treatment of type 2 diabetes. Agents in this class do not cause hypoglycaemia if given as mono-therapy, because of the glucose-dependent nature of incretin-stimulated insulin secretion. α-Glucosidase inhibitors (acarbose, miglitol, voglibose) reduce postprandial hyperglycaemia by delaying glucose absorption; they do not affect glucose utilization or insulin secretion. Thiazolidinediones reduce insulin resistance by binding to the PPAR-γ (peroxisome proliferator-activated receptor γ) nuclear receptor (which forms a heterodimer with the retinoid X receptor). Thiazolidinediones promote a redistribution of fat from central to peripheral locations. Circulating insulin levels decrease with the use of thiazolidinediones, indicating a reduction in insulin resistance. Pioglitazone, used as 15–45 mg/day in a single daily dose belongs to thiazolidinediones class of drugs. The bile acid-binding resin colesevelam has been approved for the treatment of type 2 DM. Evidence indicates that bile acids, by signalling through nuclear receptors, may have a role in metabolism. Since bile acid-binding resins are minimally absorbed into the systemic circulation, how bile acid-binding resins lower blood glucose is not known. Bromocriptine, a dopamine receptor agonist has been approved by the FDA for the treatment of type 2 DM. Insulin should be considered as the initial therapy in type 2 DM, particularly in lean individuals or those with severe weight loss, in individuals with underlying renal or hepatic disease that precludes oral glucose lowering agents, or in individuals who are hospitalized or acutely ill. The level of hyperglycaemia should influence the initial choice of therapy. Assuming maximal benefit of MNT and increased physical activity has been realized, patients with mild to moderate hyperglycaemia (FPG, 200–250 mg%) often respond well to a single, oral glucose lowering agent. Patients with more severe hyperglycaemia may respond partially but are unlikely to achieve normoglycemia with oral monotherapy. A stepwise approach that starts with a single agent and adds a second agent to achieve the glycaemic agent can be used. Insulin therapy, as discussed above can be used as an initial therapy in certain situations.

Assessment of long-term glycaemic control: Measurement of glycated haemoglobin is the standard method of assessing long term glycaemic control when plasma glucose is consistently elevated, there is an increase in nonenzymatic glycation of haemoglobin; this alteration reflects the glycaemic history over the previous 2–3 months, since erythrocytes have an average life span of 120 days (glycaemic level in the preceding month contributes about 50% to the A1C value).

A1C should be measured in all individuals with DM during their initial evaluation and as a part of their comprehensive diabetes care. As the primary predictor of long term complications of DM, A1C should mirror to a certain extent the short term measurements of SMBG. These two measurements are complementary in that recent intercurrent illnesses may impact the SMBG measurements but not the A1C. Likewise, postprandial and nocturnal hyperglycemia may not be detected by the SMBG of fasting and pre-prandial capillary plasma glucose but will be reflected in the A1C.

The degree of glycation of other proteins, such as albumin, can be used as an alternative indicator of glycaemic control when the A1C is inaccurate (haemolytic anaemia, hemoglobinopathies). The fructosamine assay (measuring glycated albumin) reflects the glycaemic status over the prior 2 weeks. Alternative assays of glycaemic control should not be routinely used since studies demonstrating that it accurately predicts the complications of DM are lacking.

CONCLUSION

Diabetes mellitus is diagnosed by identifying chronic hyperglycemia. DM is classified on the basis of the pathogenic process that leads to hyperglycemia rather than on basis of age of onset or type of therapy. Type 1 DM results from autoimmune beta cell destruction and is responsible for 5-10% of those with diabetes. Type 2 diabetes mellitus is a complex endocrine and metabolic disorder. Insulin resistance in muscle and liver and β-cell failure represent the core pathophysiologic defects in type 2 diabetes, but in addition to these, other factors contribute in pathophysiology. The diagnosis of diabetes is based on plasma glucose criteria, either the fasting plasma glucose (FPG) or the 2-h value in the 75-g oral glucose tolerance test (OGTT). HbA1c was incorporated as diagnostic criterion in 2009. In certain situations random value can be used for diagnosis. The management aims at eliminating symptoms related to hyperglycemia, reduce or eliminate the long term microvascular and macrovascular complications of DM and allow the patient to achieve as normal lifestyle as possible. The care of the individuals must also include treatment of conditions associated with diabetes like hypertension, dyslipidemia. Lifestyle modification along with appropriate drug therapy should be used to achieve the goals.

SUGGESTIONS FOR FURTHER READINGS

1. American Diabetes Association (1969), Standards of Medical Care in Diabetes, *Diabetes Care*, 36:S11–S66.

2. Sacks D.B., et al. (2011), Guidelines and Recommendations for Laboratory Analysis in the Diagnosis and Management of Diabetes Mellitus, *Diabetes Care,* 34:e61–99.

3. Longo D.L., Fauci A.S., Kasper D.L., et al. (Eds.), (2012), Harrison's *Principles of Internal Medicine*, 18th ed., McGraw Hill.

4. Jameson J.L. and Degroot L.J. (Eds), (2010), *Endocrinology*, 6th ed., *Philadelphia*, Elsevier.

5. DeFronzo R.A. (2009), Banting Lecture. from the Triumvirate to the Ominous Octet: A New Paradigm for the Treatment of Type 2 Diabetes Mellitus, *Diabetes*, 58:773–95.

SECTION XII

HOMEOSTASIS

Cells in multicellular organisms such as the human body function in a special environment and cannot tolerate changes beyond narrow limits in the composition of this medium. In most terminal cases, the imminent cause of death is the disturbance in the water and electrolyte balance. The fundamentals and practical approaches towards regulation of water and electrolyte balance have been discussed.

91

Water and Electrolyte Balance

Archana Singh

CONTENTS

I. INTRODUCTION

Water constitutes some 60% of the body weight in adults (males 55–70%, females 45–60%); in infants water may account for 65–78% of the body weight. Thus, an average male weighing 70 kg has 42 litres of water in his body. The fractional water content decreases with age and is highest in infants and children.

The total body water is determined by intravenous injection of Deuterium Oxide (D_2O) or Tritium Oxide. Since these substances diffuse evenly through the entire body like water does, the degree of dilution of the injected substance is used to calculate the amount of total body water.

The correlation of lean body mass[1], e.g. the body weight free of the stored fat to total body water is of practical importance. It has been shown that the total body water (TBW) bears a constant relationship to the lean body mass.

$$TBW = 73.2\% \text{ of Lean Body Mass.}$$

Stored fat does not contain water. Thus the total body water of 42 litres in 70 kg man represents 60% of his body weight. To illustrate, if there is an increase in his body weight to 80 kg due to fat accumulation, his total body water would still remain constant at 42 litres but would now constitute only 52.5% of his new body weight of 80 kg. It is for this reason that in females and obese individuals water constitutes a lower

---

1 The total fat-free portion of body weight consists of protein, water, glycogen and minerals.

percentage of body weight (TBW = 45–60%) than in thin individuals (TBW = 55–70%). Also, Asian adult populations have higher body fat percentage than Europeans for the same Body Mass Index (BMI).

While calculating the fluid requirements of acutely ill patients there is a real risk of infusing larger than required volumes of fluids into obese patients. This can be avoided by calculating the requirements in relation to the Lean Body Mass.

A. Distribution of Body Water

For convenience of discussion, body fluid may be divided into two main compartments:

1. Extracellular fluid (ECF), that is constituted by plasma, interstitial fluid, and transcellular fluid.
2. Intracellular Fluid (ICF).

1. Extracellular fluid: The chief function of ECF is to act as a conduit between cells and organs and also to facilitate maintenance of the ionic composition and volume of the intracellular fluid.

The ECF constitutes about 30% of the total body water. In a 70 kg male with a TBW of 42 litres, 12.5 litres lie in the extracellular fluid.

The extracellular fluid volume is determined by the degree of dilution attained by a substance which on injection remains outside the cells. Measurement of extracellular volume is less reliable than that of total body water and no ideal marker has been found. The most suitable marker is Sodium thiosulphate and the volume of ECF as calculated by this method (termed as Thiosulphate Space) is approximately 17% of the body weight.

Other diffusible substances like insulin, mannitol or Na^{24} give misleading figures, because of their variable permeability in certain parts of the extracellular fluid.

(a) *Plasma:* This is the fluid which lies within the heart and blood vessels and approximates 45ml/kg body weight or 4.5% of body weight in a healthy adult and about 6% of total body water. The plasma is route of rapid transit.

In a 70 kg man the plasma volume is about 3 litres. It is best determined by the intravenous injection of a known amount of RIHSA (Radio–iodinated human serum albumin) and working out the degree of dilution in the circulating plasma. Due to its combination with albumin the radio-iodine remains within the vascular bed and a fairly accurate determination can be made.

(b) *Interstitial fluid:* This includes lymph and the fluid which constitutes the immediate surrounding environment of all cells in the body. It is obtained by subtracting the plasma volume from the volume of extracellular fluid. In a healthy adult it approximates 8.5–10 litres (21–23%) of the total body water of 42 litres. In comparison to the plasma, the interstitial fluid is a slow supply zone, and flows around the cell

thus permitting the whole cell surface to be used as an area of exchange.

(c) *Transcellular fluid:* This includes fluid within the lumen of the secretory organs and anatomical spaces, viz. the gastrointestinal and biliary secretions, urine, fluid in joints and chambers of the eye, cerebrospinal fluid and fluid in the thyroid, gonads and other secretory glands. This heterogeneous but important group comprises about 2.5% of the total body water.

2. Intracellular fluid: The intracellular fluid volume is derived by subtracting the volume of extracellular fluid from the total body water. The ICF constitutes approximately 70% of the TBW. Thus, in a 70 kg adult with a total body water content of 42 litres, the intracellular fluid is about 30 litres. Of this 24–26 litres lies within the soft tissues (excluding fat), whereas about 4 litres lies in bone. Whereas the intracellular water of soft tissue readily participates in all fluid exchanges in the body; the water contained within bones does not do so.

The volumes of total body water, extracellular fluid and intracellular fluid are shown in Table 91.1.

TABLE 91.1   Water content in different body compartment (percentage of total body water)

Intracellular volume	~ 30 L (65–70%)
Extracellular volume	~12.5 L (30–33 %)
Interstitial volume	8.5–10 L (21–23%)
Plasma volume	3.0 L (6–7%)
Transcellular volume	1.0 L (2.5–3.0%)

B. Daily Water Turnover and Water Balance

There is a large daily internal turnover of water within the body. About 180 litres of fluid is filtered by the glomeruli into the renal tubules with only 1 to 1.5 litres excreted as urine. Salivary, gastrointestinal, biliary and pancreatic juices amounting to 8–12 litres of are being discharged into the gut lumen. Most of the water in these secretions is reabsorbed by the intestines, and only 150–250 ml is lost in the faeces. In the same manner the cerebrospinal fluid is being constantly secreted and reabsorbed 3 to 5 times every day. Thus, with a total water content of only 42 litres, the body maintains a daily turnover of approximately 190–195 litres. This highly efficient system of control is akin to a commercial organization conducting business worth 195 crores with a capital outlay of only 42 crores!

It is of the utmost importance for the body to regulate the loss of water from the body in accordance with the physiological needs of the body. This regulation is achieved primarily by the antidiuretic hormone (ADH or Arginine Vasopressin, AVP) secreted by the posterior pituitary in tandem with other hormones. In a healthy individual the water intake and water output are finely regulated to maintain the total water content of the body at a constant level.

1. Water Intake: This includes water being supplied to the body by endogenous or exogenous sources:

(a) Endogenous water represents the water derived from oxidative processes within the body and amounts to 300–500 ml daily. It may be increased in states of increased metabolic activity (hyperthyroidism, high fever, and leukaemia) and decreased during starvation.

(b) Exogenous intake is derived from ingestion of food and water. It is important to remember that fresh fruit and vegetables have high water content.

The amount of water and other liquids drunk by a person varies widely with climate and personal habits of the individual. Some persons may drink 2–3 litres of water daily and others may ingest only a litre in the same environment. Generally people living in a hot climate drink much more water than those living in temperate regions.

Water intake is modified by a feeling of 'thirst'. This sensation is produced when the total water content of the body is reduced. There is some evidence that a thirst centre is located in the hypothalamus.

2. Water Output: Water is lost from the body through four routes – skin, lungs, faeces, and urine.

(a) *Loss through skin:* An invisible water loss takes place by persistent diffusion through the skin which is chiefly a mechanism of eliminating heat. This loss is not under the control of any of the body system but the extent of loss depends mainly upon the atmospheric temperature and humidity. The average daily loss through invisible perspiration is 600 ml of water though this may be much greater in hot and dry weather.

Water loss through visible perspiration is brought about by the activity of the sweat glands and is subject to extreme variation. In hot tropical weather more than a litre of water may be lost within a few hours by profuse sweating, whereas in cold weather, the water loss by sweating is negligible.

(b) *Loss through lungs:* Expired air is saturated with water vapour (expired air has more water than inspired air), causing a daily loss of about 400 ml of water. In fever and hot, dry weather, this avenue may account for loss of larger quantities of water.

(c) *Faecal loss:* Water loss through this route varies from 150–250 ml daily. An excessive amount of water may be lost during diarrhea.

(d) *Urine:* In health, the urinary output ranges between 1–2 litres per day, whereas the loss of water through skin, respiration and faeces does not vary much. The adjustments in urine volume are mediated through the action of AVP secreted from posterior pituitary in response to changes in volume and osmolality of ECF. Excessive water loss and consequent increased osmotic pressure of the plasma causes a release of AVP into the circulation resulting in increased absorption of water from the collecting ducts of the kidney while decreased plasma osmolality (e.g., drinking of 1–2 litres of water) suppresses AVP secretion resulting in decreased tubular reabsorption of water and increase in the urinary output (Water Diuresis). The adaptations in permeability and mediation of hormonal action are effected by the differential expression of *aquaporins* in the nephron; these are a family of proteins that act as water channels.

The urinary output is suitably modified to maintain the water balance after adjusting for the amount of water loss via the skin, lungs, and faeces. For this reason, the specific gravity of the urine bears a direct relationship to the total urinary output—the specific gravity being high when urinary output falls due to increased tubular reabsorption of water and low when the reabsorption is reduced to excrete more water.

Besides the action of the anti-diuretic hormone, renal excretion of water is also governed by the renal handling of sodium. Water is reabsorbed and excreted along with sodium by the kidney.

3. Daily water requirements: These fluctuate widely depending upon the climate and the state of the body. In temperate climates, in the absence of marked sweating an adult requires 1.5 litres fluid in addition to the volume of the urine passed. This is used mainly to compensate for the losses through skin, lungs, and faeces.

The water requirements of infants and children are relatively higher because of greater metabolic activity, poor concentrating power of the kidneys, and larger surface area per unit of body weight causing a higher water loss from the skin.

II. PRINCIPAL ELECTROLYTES

The handling of some of the principal electrolytes by the body may now be discussed.

The existing composition of ECF and ICF and the underlying homeostatic processes for maintenance of the same stem from an evolutionary adaptation that was a result of dietary intake patterns in primitive times. Diets in pre-historic times were very rich in potassium (due to high intake of fruits, vegetables and meats) and low in sodium. Therefore physiologic mechanisms were adapted to secreting potassium and conserving sodium. These mechanisms also evolved for delivery of large amounts of sodium to the renal tubules and recycling of potassium within the kidney to enable potassium excretion.

A. Sodium

1. Location: The body contains 3800–4000 mEq of sodium. Of this, 1700–1750 mEq (44%) is contained in the ECF (145 mEq × 12 = 1740 mEq), 1800–1850 mEq (47%) is contained in bone. Barring the 800–1000 mEq contained in bone, the entire sodium content of the body amounting to about 3000 mEq is osmotically active, water soluble and exchangeable (Table 91.2).

TABLE 91.2   Sodium distribution in body compartment

Compartment	Content
Extracellular fluid	1750 mEq
Soft tissues	400 mEq
Bone	
(a) Exchangeable	850 mEq
(b) Non- exchangeable	1000 mEq
Exchangeable:	1000 mEq
Non-exchangeable:	3000 mEq
Total:	1000 mEq
	4000 mEq

2. Control of sodium excretion: In spite of a large reserve of sodium, the body exercises a meticulous control over its excretion. Serum Na^+ is the primary determinant of extracellular volume and thereby osmolality and is maintained within a narrow range by fine-tuned regulatory mechanisms[2].

The daily sodium excretion varies from 80–100 mEq (5–6 grams of sodium chloride). Small quantities are excreted in the sweat and faeces; most of the sodium is lost through urine.

Some 24,000 mEq (560 grams) of sodium is present in the glomerular filtrate every day. About 80% is reabsorbed in the proximal convoluted tubule along with water through isoosmotic reabsorption. This is not under any hormonal control.

The reabsorption of sodium by the proximal convoluted tubule is an active process involving work on the part of tubular cells. The concomitant reabsorption of water with sodium is a secondary phenomenon. If mannitol is administered, it is filtered by the glomeruli and exerts an osmotic pressure in the proximal tubule as it is not reabsorbed. This does not permit the reabsorption of water resulting in diuresis (osmotic diuresis). Sodium is still reabsorbed, showing that the reabsorption of sodium is primary and unaffected by changes in the tonicity of the filtrate.

The remaining filtered sodium traverses the loop of Henle where a small amount of it is reabsorbed. Further, reabsorption occurs in the distal convoluted tubule under the influence of mineralocorticoids (aldosterone and deoxycorticosterone) secreted by the adrenal cortex (facultative reabsorption). The source of energy for reabsorption of sodium at various nephron segments is Na^+-K^+-ATPase located on the basolateral membrane.

As a result of sodium reabsorption in the proximal and distal convoluted tubules, only a very small fraction (80–100 mEq) of the sodium filtered by glomeruli is excreted every day in the urine.

A large unchecked loss of sodium occurs in patients with Addison's disease. Conversely, in primary aldosteronism due to an adrenal tumor or patients receiving corticosteroids, there is sodium retention and edema due to simultaneous retention of sodium and water by the body. Abnormal losses of sodium in cases of Addison's disease can be checked by injection of aldosterone or oral administration of corticosteroids.

Following an injury or operation, the daily sodium excretion may fall to 20 mEq or less due to renal conservation of sodium. The degree and duration of enhanced sodium retention depends upon the severity of injury or operation and may last for 2–5 days. It is now believed that the kidney exercises specific control over the reabsorption of sodium and the chloride follows sodium movement as a passive anion. Like sodium, most of the filtered chloride is reabsorbed in the proximal tubule. About 1/5th of the filtered chloride is reabsorbed in the distal tubule along with sodium. Variations in urinary excretion of chloride closely parallel those of sodium.

B.   Potassium

1. Location: Potassium is the principal intracellular cation. 98% of the total potassium content of the body (3500 mEq) is intracellular; three-fourth of it is contained in the skeletal muscles. Twelve litres of ECF contain a mere 60 mEq of potassium at a concentration of 4–5.2 mEq per litre. Experiments with radioactive potassium have shown that about 95% of the total body potassium is exchangeable. The distribution of potassium is influenced by various factors including membrane bound sodium pump, acid base status, hormones (insulin, catecholamines) osmolality, etc. The daily potassium requirement is 2–3 grams (80–100 mEq). It is widely distributed in foods; foods with high potassium content include meats and fishes and fruits and vegetables like potatoes, soybean, tomatoes, apricots and bananas. Except for very small quantities, excreted in the sweat and faeces (10%), the bulk of potassium is excreted in the urine (90%) to maintain constant levels in the cells and ECF.

2. Control of potassium excretion: Besides their differential predominance in the ECF and ICF, there is a striking difference in the manner in which the kidneys handle sodium and potassium. While only 80% of the filtered sodium is reabsorbed in the proximal convoluted tubules, there is almost total reabsorption of potassium in the proximal tubules (both by active and passive mechanisms). There is then an active excretion of potassium by the cells of the distal convoluted tubule[3]. Due to this some 80–100 mEq of potassium is excreted in the urine. It is, therefore necessary to supplement the oral intake of potassium in patients receiving mercurial diuretics.

---

[2] For a conceptual understanding, this means that unlike other electrolytes (e.g. potassium, magnesium) a change in total body Na^+ is not accompanied by a concomitant change in serum Na^+ since proportionate changes in water content also occurs via regulation through ADH.

[3] A Na^+–K^+ ATPase located on the basolateral side of the cortical collecting duct transfers K^+ into and Na^+ out of the cell. The luminal Na^+ entering the cell leads to a negative luminal potential which causes reabsorption of chloride. This reabsorption does not have a 1:1 ratio, as a result of which the charge imbalance is corrected by secretion of K^+ into the lumen.

The administration of ACTH, corticosteroids and diuretics also increase the urinary excretion of potassium.

There is now evidence that K^+ and H^+ ions compete for a common transport mechanism in the distal tubule. K+ excretion is favoured in alkalosis, while K^+ depletion favours the excretion of H^+ ions. The reader is referred to Section V of this chapter for a discussion of acid-base balance.

III. COMPOSITION OF BODY FLUIDS

A. Total Body Water

Though the total amount of water in the body remains fairly constant, its distribution between various compartments is largely guided by osmotic forces. This in turn is governed by the concentration of various solutes dissolved in the body water. The body solutes may be described as belonging to three main categories. Compounds/molecules that cannot move freely between cells and extracellular fluid are called "effective osmols" since their restricted mobility between compartments contributes to osmolality. This increase in their concentration in the ECF causes a shift of fluid into the ECF from cells.

1. Organic compounds of small molecular size: These are mainly the breakdown products of metabolites like glucose, amino acids, urea, fatty acids (obtained through the diet) and lactate ions .

All these substances diffuse freely across cell membranes and are not of great importance in the movement of water across the various compartments except in large concentrations when they help to retain water in the body.

2. Organic compounds of large molecular size: These are mainly the proteins. The proteins of extracellular fluid are mainly "locked" in the plasma and influence exchanges between the plasma and interstitial fluid by exerting their osmotic effect.

Intracellular proteins usually do not leave the cells. Important exceptions to this are considered along with the composition of intracellular fluid.

3. Inorganic electrolytes: These occur in large quantities in both extracellular and intracellular fluids and dissociate readily into their constituent ions or radicals. These comprise the single most important factor in the transfer and movement of water and electrolytes between the three divisions of the extracellular fluid and also between the extracellular and intracellular compartments.

The maintenance of relatively constant electrolyte composition in the various body fluids is an important part of homeostasis.

For an intelligent understanding of electrolyte balance of the body a few commonly used terms may first be defined.

Compounds, which dissociate readily in solution into their constituents ion or radicals are defined as electrolytes. To cite two common examples, NaCl (sodium chloride) does not exist in the body fluids as a compound but as a mixture of positively charged Na^+ ions (cations) and negatively charged Cl^- ions (anions). Similarly, sodium bicarbonate ($NaHCO_3$) does not exist as $NaHCO_3$ but as a mixture of cations (Na^+) and anions (HCO_3^-).

In discussing electrolyte changes within the body it is both convenient and desirable to discuss the behavior of individual ions rather than compounds.

The gram equivalent weight (generally termed as equivalent weight) of any molecule, radical or ion may be defined as its weight in grams which would combine with or displace 1 gram of hydrogen. By implication the gram equivalent weight of any molecule or ion will combine with the gram equivalent weight of another reactive unit. The equivalent weight of a monovalent ion or radical (Na^+, K^+, Cl^-, HCO_3^-) is the same as its atomic weight, whereas the equivalent weight of a divalent ion or radical (Ca^{++}, Mg^{++}, SO_4^-) is half its atomic weight. Thus in any chemical reaction in the body, 23 grams of Na, 39 grams of K, 18 grams of NH_4 mutually replace each other or combine with 35.5 grams Cl^-, 48 grams SO_4^{--} or 61 grams of HCO_3^-. All these figures represent the equivalent weight of the various ions or radicals.

When considering the chemical reactions and physiologic concentrations in the body, a gram equivalent is an inconvenient entity to deal with. Hence the term milliequivalent (mEq) has been introduced.

Thus 1 gram equivalent = 1000 milliequivalents. To revert to a previous comparison, one mEq of any ion or radical will combine with or replace one mEq of any other ion or radical. This concept of the use of milliequivalent has greatly simplified the study of electrolyte balance in the body.

To convert the concentration of any substance from milligrams/100 ml to mEq/litre, the following simple formula can be applied:

$$\frac{\text{Concentration in mg per litre} \times \text{Valency}}{\text{Atomic weight}} = \text{Concentration (mEq/litre)}$$

The following example illustrates the utility of this formula:

For plasma concentrations of sodium and calcium in mEq/litre.

Plasma sodium = 322 mg per 100 ml; Valency = 1

$$\frac{322 \times 10 \times 1}{23} = 140 \text{ mEq/litre}$$

Serum calcium = 10 mg per 100 ml; Valency = 2

$$\frac{10 \times 10 \times 2}{40} = 5 \text{ mEq/litre}$$

With this background knowledge of electrolytes, let us examine the chemical composition of extracellular and intracellular fluid.

B. Composition of ECF and ICF

The composition of ECF and ICF is shown in Table 91.3.

TABLE 91.3   Concentration of Electrolytes in Body Fluids

Extracellular fluid				Intracellular fluid			
Cations		Anions		Cations		Anions	
Na^+	142	Cl^-	105	K^+	155	PO^-	90
K^+	5	HCO_3^-	27	Na^+	12	$Protein^{--}$	60
Ca^{++}	5	HPO_4^-	2	Mg^{++}	15	HCO_3^-	8
Mg^{++}	2	Org.	4	Ca^{++}	2	Cl^-	8
		Acids Proteins	16			Others	18
Total	154		154		184		184

All values are in milliequivalents per litre.

The cationic and anionic equivalences must balance in both extracellular and intracellular fluids. This is necessary to maintain electrical equilibrium. In all electrolyte exchanges in the body this must be maintained. Thus, if there is loss of a cation from the ECF, it must be made up by corresponding changes in other cations and anions to maintain a balance. No loss of cations or anions can be left 'uncovered' by the body.

The concentrations of electrolytes in plasma are easily measured and thus direct estimates of their values are possible. The intracellular concentrations, however, of the electrolytes cannot be measured accurately and are estimated by subtracting the extracellular value from the total tissue value. Also, intracellular electrolyte compositions vary between tissues. For example, muscle contains about 3 mEq/L of chloride anion while its RBC level is 75 mEq/L .

There is a marked difference between the concentration of principal cations and anions in the ECF and ICF. Thus, Na^+ is the principal extracellular cation (140–145 mEq/litre), the concentration of K^+ concentration of 155 mEq/litre, Na^+ being only 12 mEq/litre. There are also remarkable differences in the anion concentration of ECF and ICF. Secondly, the intracellular fluid is hypertonic as compared to the extracellular fluid.

The reasons for this remarkable difference in the composition of the ECF and ICF are:

(a) The presence of ion specific, energy dependent, transport pumps, e.g., membrane bound sodium pump (Na^+, K^+ – ATPase);

(b) The effect of Gibbs-Donnan equilibrium[4] across the membrane resulting in an electrochemical gradient.

(c) The selective permeability of the cell membrane to water, organic solutes of small mol. wt., e.g., glucose, urea, and amino acids, etc.

It is thus not governed simply by the concentration gradient of these ions across the cell membrane. A living cell at 37°C provided with a suitable energy substrate will continue to maintain its intracellular concentration of Na^+ and K^+ ions. However, if cellular activity is interrupted by cooling the cell to 0°C or by addition of a metabolic poison, the electrolytes in the ECF and ICF soon equilibrate according to their concentration gradients. Sodium enters the cell, while potassium leaks out of it.

C. Permeability of Cell Membrane to proteins

By and large, proteins lie locked within cells and constitute a substantial part of intracellular anions. Being non diffusible and polyvalent, they contribute to the Gibbs-Donnan effect. The proteins are responsible for the colloidal osmotic pressure (oncotic pressure) and regulate the distribution of water between plasma HSF as illustrated in Starlings hypothesis. A few notable exceptions of proteins traversing the normal cell membrane are as follows.

(a) Proteins may enter liver cells. The liver cells also synthesize albumin and transfer it across the cell membrane into the circulation.

(b) Maternal antibodies may cross the placental membrane to enter the foetal circulation.

(c) Injected antigens may enter antibody producing cells in the body.

As already stated, water diffuses freely across all cell membranes, its movement being determined chiefly by the concentration of osmotically active electrolyte ions.

Regulation of body fluid and electrolytes: Water, Na^+ and K^+ metabolism is interrelated and regulated by osmotic and hormonal mechanisms in a complex manner to achieve homeostasis. The chief hormones involved in this are antidiuretic hormone (ADH or Arginine Vasopressin, AVP), the renin-angiotensin-aldosterone axis, along with some others like the natriuretic peptides (Atrial, Brain and C-type natriuretic peptide). The volume and osmolality of ECF, acting through the water intake/output areas of hypothalamus, are regulated

---

[4] Two solutions of differing ionic concentrations, separated by a semi-permeable membrane will establish equilibrium such that the ionic concentration is equal on both sides of the membrane (if the solutes can move freely; if not, the even though diffusible ion concentration will vary, the ionic concentrations shall be equal).

by the anti-diuretic hormone secreted from posterior pituitary. To elaborate, increased effective osmolality leading to the secretion of ADH by the posterior pituitary is caused by the stimulation of osmoreceptors in the hypothalamus. This leads to stimulation of the thirst center in the cerebral cortex in turn sending impulses into the supraoptic and paraventricular nuclei that produce ADH. ADH secretion is extremely sensitive to changes in effective osmolality and a rise of only 2–3% can lead to ADH secretion for production of concentrated urine. The other non-osmotic stimuli regulating ADH secretion are beyond the scope of this chapter and are not being discussed.

The renin-angiotensin-aldosterone system plays a crucial role in maintaining the sodium, potassium and water balance. Aldosterone increases potassium excretion by the renal tubules and conserves sodium. In fact, alterations in body salt content are primarily achieved by changes in renal salt output. The increase in activity of plasma renin attenuates the effects of volume depletion caused by either salt loss, or regulates increased serum K^+ through feedback loops operating with aldosterone.[5] Abnormalities in plasma renin and aldosterone activity are either caused by or lead to aberrations in K^+ potassium metabolism. In a primary deficiency of renin, aldosterone activity is low while an increase in renin activity will always lead to increase in aldosterone secretion (unless increase in plasma renin activity is due to a primary defect in aldosterone secretion).

The Atrial Natriuretic peptide, a 28-amino acid peptide secreted by the heart, is the most well studied for its action on increasing urinary sodium excretion in response to high extracellular fluid and blood volume.

IV. DISEASE STATES

A. Changes in Osmolality and Intracellular Volume

Changes in osmolality (increase/decrease) and the corresponding changes in intracellular volume cause signs and symptoms that are related partly to changes in volume as well as to those in electrolyte concentration. Most of these changes are discussed in the context of hypernatremia/hyponatremia and the accompanying changes in osmolality. Symptoms like easy fatigability and muscle cramps in hyponatremia are due to reduced effective vascular volume. When hyponatremia is caused by hyperglycemia[6], the signs are due to hyperosmolality and cell dehydration. While hyperosmolality may be caused by accumulation of effective solutes other than sodium

(i.e., hyperosmolality can occur without hypernatremia), hypernatremia is always associated with hyperosmolality. The signs and symptoms range from depression of the mental state manifesting as lethargy, muscular symptoms include muscular rigidity, tremor, hyperreflexia, etc.

For intracellular changes, different cells of the body have differing capacities to regulate intracellular volume (volume regulatory increase or decrease) in response to changes in effective osmolality. Brain cells and RBCs, for example, can adapt by volume regulation to changes in osmolality while muscle cells cannot and the latter depend on restoration of normal physiological osmolality for regaining their intracellular volume to normalcy.

B. Water Depletion (Dehydration)

Simple water depletion is uncommon. It is generally accompanied by a concomitant loss of electrolytes. Simple water depletion may occur in unconscious patients or in those who are unable to swallow due to painful oropharyngeal conditions. There may also be an increased loss from lungs in high fever or through tracheostomy.

Dehydration may be isotonic, hypotonic or hypertonic depending on the quantity of salt loss in proportion to water that has been lost. When salt is lost along with water from the GI tract or by removal of ascitic fluid, the osmolality of body fluids is adjusted to **isotonicity** by intake of water or excretion resulting in decrease in intracellular volume. In **hypertonic dehydration**, there is water deficit due to insufficient water intake or loss due to osmotic diuresis/osmotic diarrhea usually alongwith a deficit in drinking leading to a hypernatremic state. **Hypotonic dehydration** occurs usually when loss of salt containing body fluids is compensated by replacement with plain water or a solution with lower electrolyte concentration than that which has been lost.

The chief clinical features of dehydration are weakness, thirst and oliguria. Thirst is due to stimulation of the drinking centre in the hypothalamus caused by increase in the osmolality[7] of the ECF. There is also cellular dehydration and loss of potassium from the cells during periods of intense thirst. The kidney conserves water by producing smaller quantity of urine with a high specific gravity under the influence of ADH.

The Goals of therapy in a patient requiring salt and water replacement is divided into 3 chief components:

- Identification and compensation of existing deficits.
- Ensuring supply of daily basal water and electrolyte requirements.

---

[5] Renal salt output is regulated by both humoral mechanisms and by physical factors such as changes in GFR and capillary oncotic and hydrostatic pressures. The different sites of Na^+ absorption in the nephron use different mechanisms of Na^+ entry.

[6] Hyperglycemia leads to a reduction in serum sodium. The change in serum sodium and change in glucose concentration in a physiologically normal adult is 1.5 mEq/L of Na^+ for 100 mg/dL of glucose.

[7] Osmolality is the number of milliosmoles/kg (mOsm/kg) of solvent. This is measured in the clinical laboratory using an osmometer. Osmolarity is the number of milliosmoles/liter (mOsm/L) of solution.

- Quantification and provision for replacement of ongoing losses.

Treatment: Oral intake should be resumed or increased whenever possible. In unconscious patients, administration of fluids through an intragastric tube is safe and reliable.

If oral intake is impossible, 5% dextrose is given intravenously. The dextrose is metabolized and water is restored to the circulation. Restoration of normal hydration is heralded by diuresis. Care should be taken not to overload the circulation, especially in the elderly. Any concomitant loss of sodium or other electrolytes should be rectified.

C. Disorders of Sodium Balance

Water and salt loss in sodium excess are discussed below:

1. Sodium deficiency (hyponatremia—Decrease in plasma Na^+ <130–135 mmol/L): Pure sodium deficiency is uncommon in clinical practice. This may occur if, in a patient with loss of water and salt, the water is replaced by drinking or by intravenous infusion of glucose, resulting in a sodium deficiency.

Several clinical states can lead an abnormal or excessive loss of sodium from the body. These include:

(i) Gastroenteritis with diarrhea and vomiting.
(ii) Small gut obstruction resulting in a loss of large volume of digestive fluids into the intestinal lumen or to the exterior through vomiting.
(iii) Loss of gastrointestinal secretion by gastric aspiration or through biliary, pancreatic or intestinal fistulae.
(iv) Burns
(v) Use of mercurial diuretics
(vi) Addison's disease

Due to loss of sodium, there is fall in the tonicity of ECF. Water diffuses from ECF into the cells causing a swelling of the cells. The hypotonicity of the ECF thus produced, may lead to a temporary diuresis may be induced due to suppression of ADH secretion. As a result of loss of water from ECF due to these two factors the plasma volume may fall. This may be severe enough to produce circulatory failure.

The most important response of the body to the loss of sodium is renal. There is an almost complete reabsorption of sodium by the kidney and the daily urinary excretion of sodium may fall to 20 mEq or less. Ultimately sodium and chloride may disappear from the urine. If the sodium loss is persistent and replacement is not carried out, serious disturbances may follow. It is important to realize that a normal sodium concentration in ECF is necessary for the proper functioning of several tissues of the body including nerves, skeletal muscle, cardiac muscles and above all the cells of the renal tubule. In severe sodium depletion, renal function may be impaired with a blood urea

rising to 50–100 mg per 100 ml. Urea clearance is decreased and drinking of 1–2 litres of water does not produce expected diuresis.

A severe sodium deficiency thus produces a fall in the tonicity as well as total volume of the plasma. In severe cases with a fall of over 20% in plasma sodium level, tonicity may be sacrificed to preserve the plasma volume.

In the presence of severe sodium deficit, the kidney may not be able to perform its function of maintaining the acid base balance. This is discussed later.

2. Water and salt loss: In most of the clinical states listed above there is a concomitant loss of water and sodium from the body in the concentration as in ECF (isotonic loss). The kidney responds readily by effecting a more efficient reabsorption of both water and sodium to maintain the volume and tonicity as near to normal as possible.

Clinical features begin to appear when three litres of ECF and 300 mEq of sodium have been lost. The clinical features are—a dry skin, reduced sweating and a small volume of high specific gravity urine. With the continued loss of water and sodium, maintenance of plasma volume may become impossible and a state of circulatory failure may supervene.

In states of water and sodium deficiency, the regulatory mechanisms of the body, chiefly the kidney, strive to maintain a normal ECF at a normal electrolyte concentration as far as possible as also to maintain the acid-base balance. However, in severe sodium deficiency there may be conflict between these three regulatory functions. Under these circumstances the homeostatic mechanisms adopt the following order of priority:

(i) The volume of ECF.
(ii) The electrolyte concentration
(iii) Hydrogen ion concentration

Correction of sodium deficiency is best done by intravenous infusion of normal saline (0.9% NaCl solution). The amount infused depends upon the requirements of each case.

3. Sodium excess (hypernatremia): This is much less common as compared to sodium deficiency. It is very difficult to raise serum sodium and cause hypernatremia when thirst and water drinking mechanisms are intact. Thus when hypernatremia occurs, it's almost always due to inability/unavailability of water to drink. A rise in sodium concentration is always associated with increased effective plasma osmolality[8]. The commonest cause is an overzealous intravenous administration of saline. The risk is greater in the post-operative period or in patients receiving corticosteroids, causing renal conservation of administered sodium.

Water is retained along with the sodium and the patient may show puffiness of the face or oedema over the sacrum. Due to increase in the plasma volume, a dangerous pulmonary oedema may occur. The infusion of saline should

---

[8] cf. hyponatremia where hyponatremia may not always be associated with hypoosmolality.

be discontinued forthwith. In acute cases, diuretics are used to increase urinary output.

D. Potassium Balance Disorders

Potassium is the chief intracellular cation. Potassium levels are maintained in the ECF within a very narrow range and are critical to survival. Following injury, operative stress etc. involving cellular breakdown or abnormal loss of K^+ ions (see hypokalaemia), and acidosis, potassium moves from the cells into the ECF. The kidneys excrete the mobilized potassium readily to maintain normal potassium concentration in the ECF. The degree and duration of the potassium loss varies with the extent of the trauma or duration of starvation, but since the body has large exchangeable reserve of potassium, symptoms of hypokalemia may not appear unless the loss of potassium amounts to 500–700 mEq, which represents 15–20% of the total potassium content of the body.

1. Potassium deficiency (hypokalemia): It is to be appreciated that the potassium content of the most of the gastrointestinal secretions is higher than that of plasma (Table 91.4).

TABLE 91.4 Electrolyte constituents of Intestinal Secretions

Secretion	Daily volume	Concentration in mEq/litre			
		Na^+	K^+	Cl^-	HCO_3^-
Gastric juice	2000 ml	50	10	50	–
Intestinal juice	3000 ml	140	10	100	25
Bile	600 ml	140	5	100	30
Pancreatic juice	700 ml	140	5	70	70
Plasma	–	140	5	100	25

All values are expressed in mEq/litre

As a result of the adaptive mechanisms stated above, it is important to appreciate and understand that hypokalemia may be either due to redistribution or due to a true K^+ deficit in the body.

Potassium depletion can readily follow the loss of large volumes of gastrointestinal secretions. Patients with prolonged or recurrent diarrhea, intestinal obstruction or those with external duodenal or biliary fistulae are the usual subjects.

Patients with pyloric stenosis or those requiring prolonged gastric aspiration are particularly prone to developing potassium deficiency. Following extensive sodium depletion, the sodium content of gastric secretion is reduced to minimize the sodium losses but the loss of potassium continues unabated as the potassium content of the gastric mucus is high. Over a period of few weeks a patient may develop potassium deficit of upto 1000 mEq. Since 98% of the body potassium is intracellular and readily exchangeable, the intracellular potassium leaks into the ECF and is lost through the kidney thereby exacerbating the potassium depletion.

Hypokalemia due to redistribution occurs in alkalosis or by β_2–agonist drugs that stimulate Na^+–K^+ ATPase. Chronic metabolic acidosis may also cause hypokalemia due to decreased renal reabsorption.

Clinical features

Apathy, muscle weakness and lack of interest in the surroundings are the predominant features of hypokalemia. There is marked weakness, the muscle tone is poor, and the tendon reflexes may be weak or abolished. There is usually considerable abdominal distension with feeble bowel sounds or even paralytic ileus.

Characteristic electrocardiographic changes of prolonged Q-T interval, depression of S-T segment and flattening or inversion of T-wave help in confirming the diagnosis, but their absence does not exclude hypokalemia.

It should be stressed that the serum potassium level may be near normal in a patient who has lost 200–300 mEq of the potassium from the cells. The extracellular concentration tends to be maintained as potassium is being constantly mobilized from the cells. The presence of clinical states stated above should suggest the need for potassium supplemented therapy. It is imperative to anticipate potassium deficiency in a likely subject, rather than await the development of serious clinical situation.

Whenever oral feeding is possible, 2–3 grams of potassium chloride is given as a mixture or in capsules. Milk, meat and fruit juice are rich in potassium and should be added to the diet. In urgent and seriously ill patients potassium chloride may be added to the intravenous infusion.

Care should be taken to ensure that the patient has an adequate urinary output as an intravenous infusion of potassium may raise the plasma level to a dangerous limit and induce diastolic arrest of the heart if there is suppression of renal function.

2. Potassium excess (hyperkalemia): This is much less common as compared to hypokalemia. It may occur in rhabdomyolysis or hemolysis, in digitalis toxicity and also in hyporeninemichypoaldosteronism. With a normally functioning kidney, additional potassium intake can be taken care of by increasing the renal secretion of potassium through the distal convoluted tubules. The chief danger of the development of hyperkalemia is in patients where, due to reasons already explained above, intracellular potassium is being mobilized into the ECF and the renal capacity to excrete it is severely impaired. In the event of renal failure the excessive amounts of potassium in ECF cannot be excreted consequently causing an increase in the level of plasma potassium. The same situation can be created by the rapid intravenous infusion of a solution of potassium chloride to correct potassium deficiency. A rise in plasma potassium concentration to 7 mEq/litre threatens life due to the risk of diastolic arrest of the heart.

Determination in clinical laboratories

Serum Na^+ and K^+ are commonly determined by (1) flame photometry, (2) Ion specific electrodes, and recently (3) Chromogenic ionophore based methods.

V. ACID-BASE EQUILIBRIUM

One of the most important features of homeostasis is the maintenance of a suitable chemical environment for cellular function. Besides keeping a constant electrolyte composition of the ECF, the body also regulates the pH of this fluid within very narrow limits. Serious consequences ensue if the pH is allowed to fall below 7.35 or rise above 7.45.

A. Acids and Bases

The pH is defined in relation to the hydrogen ion concentration (remember that the concentrations are expressed as logarithmic values). An acid may be defined as a substance which provides hydrogen ions or protons, increases the (H^+) ion concentration of the solution and thus lowers the pH. A base may be defined as a substance which accepts hydrogen ions or protons, decreases the (H^+) ion concentration of the solution and thus raises the pH. Further acids and bases may be strong or weak depending on their degree of affinity for H^+. Various weak acids are formed as a result of normal or abnormal metabolism in the body. These are H_2CO_3, lactic, ketoacids, sulphate (HSO_4^-), and phosphate (HPO_4^{--}) etc.

B. Buffer Systems in the Blood

If a solution contains a weak acid and its salt with a strong base, the combination is termed a buffer pair; this can absorb or adjust a limited number of hydrogen or hydroxyl ions, which may be introduced into the solution. The pH of this solution thus remains practically unchanged.

Three such buffer pairs exist in the blood.

1. Carbonic acid–Sodium bicarbonate $(H_2CO_3–NaHCO_3)$[9]
2. Plasma proteins and hemoglobin
3. Sodium dihydrogen phosphate–disodium hydrogen phosphate $(NaH_2PO_4–Na_2HPO_4)$

The reader should recollect the Henderson-Hasselbach equation in order to understand the weak acid, base and buffer relationships as well as the biochemical evaluation of acid-base disorders.

$$pH = pKa + \log \frac{[Salt]}{[Acid]}$$

The capacity of the buffer or the amount of acid or base which can be added to this system without a significant change in the pH is determined by:

(i) Absolute concentration of the components of the buffer system.

(ii) The total amount of the buffer solution.

Using this criteria, it is obvious that the mono and dihydrogen phosphate buffer pair can contribute little towards maintaining the pH since the blood phosphate content is very low. Also, while the chief value of the haemoglobin lies in the efficient transport of CO_2 from tissues to the lungs and not so much on its buffering capacity, it must be appreciated that proteins especially albumin account for the greatest portion of the nonbicarbonate buffer value of plasma while hemoglobin accounts for the major part of the buffer value of erythrocyte fluid.

The buffer pair $H_2CO_3 – HCO_3$ is the most important of the three. In health the ratio HCO_3/H_2CO_3 remains constant at 20;

$$\frac{\text{Plasma concentration of HCO}}{\text{Plasma concentration of H}_2\text{CO}} = \frac{22 - 27 \text{ mEq/litre}}{1.25 - 1.35 \text{ mEq/litre}} = 20$$

Despite the fact that the pK for the bicarbonate/carbonic acid pair is 6.1, whereas the normal plasma pH is 7.4, it is an effective buffer because the lungs can readily dispose of, or retain CO_2. Also, the renal tubules can adjust the rate of reclamation of bicarbonate from the glomerular filtrate.[10]

By several mechanisms (discussed below) the ratio HCO_3/H_2CO_3 is maintained at a constant level. Any alterations in one are immediately offset or compensated by proportionate alterations in the other to maintain the ratio at 20:1. Ordinarily the plasma bicarbonate and carbonic acid concentrations determine the pH of blood.

It is seldom realized that 1 to 1.5 kg of carbon dioxide are being produced daily in the tissues. In addition, the body manufactures and handles a large load of non-volatile acid anions. These include phosphates as end products of phospholipid metabolism, sulphates derived from oxidation of sulphur containing amino acids (e.g., methionine and cystine)[11] and lactate in muscular exercise. In terms of acid disposal, a 70 kg person disposes about 20 mol of carbon dioxide and 70–100 mmol of nonvolatile acids. Besides these normal acid metabolites, the body has to excrete keto acids in abnormal states like starvation and diabetes. Thus, the body has to handle a sizeable load of CO_2 and other metabolic products which constantly challenge the acid-base equilibrium. The two chief regulators of this balance are the lungs and kidneys. Only the essential features of these regulatory mechanisms are mentioned here. Students are advised to familiarize themselves thoroughly with the principle of gaseous exchange in the alveoli and the functions of the kidney for a complete understanding of the role of lungs and kidneys in regulating acid-base equilibrium.

---

[9] Bicarbonate is the second largest fraction of plasma anions (after chloride)

[10] Generally buffers work best at resisting pH changes when the pH lies ± 1 pH unit of its pK i.e. when ratio of acid/base is within the range of 10:1 or 1:10.

[11] The source of sulphuric acid is protein and in general proteins of animal sources have higher amounts of sulphur than those of plant origin for any given amount of protein. The total amount of acid or alkali content from diet depends not only on the sulphur content but also on the alkali content, which is present as salts of organic acids. For example, milk has a net alkali value while meat has a net acid value.

C.   Role of the Lungs

A small fraction of the CO_2 produced in the tissues is dissolved in the plasma and carried as H_2CO_3 and about 25% combines with haemoglobin to form carbamino-haemoglobin. The rest diffuses into the red cells where it is converted under the influence of carbonic anhydrase, into H_2CO_3. This dissociates into H+ and HCO_3^- ions. The HCO_3^- in turn diffuses out of the red cells and its place is taken by chloride (chloride shift).

In the lungs these changes are reversed, resulting in the release of carbon dioxide from the blood, from the pulmonary capillaries into the alveoli and then into the expired air.

It is thus easy to appreciate that the changes in the pulmonary ventilation or disturbances in gaseous exchanges in the alveoli can interfere with the plasma HCO_3/H_2CO_3 ratio and thus disturb the acid base equilibrium. The CO_2 of the blood (normal 40 mmHg) reflects the concentration of H_2CO_3 in the ECF (normal 1.25–1.35 mEq/litre). In a patient with disturbances of acid-base balance, changes in pCO_2 indicate that the disturbance is due to respiratory factors.

The isohydric shift: is the mechanism whereby the carbon dioxide and pH are maintained at essentially constant levels between the arterial and venous blood. The mechanisms that ensure that the additional load of H^+ ions generated (by the formation of carbonic acid) does not influence pH include binding of H^+ ions to plasma buffers, formation of *carbamino* compounds, binding on H^+ ions to hemoglobin (this is mediated via the shift of the oxygen binding curve to the right and subsequent release of oxygen; this is accompanied by binding of H^+ ions to the hemoglobin molecule). Some remaining H^+ ions are buffered by the non-bicarbonate buffers of the erythrocytic fluid. It is also important to appreciate that there is a concomitant shift of chloride ions *into the RBC* to compensate for the change in erythrocyte membrane potential (it becomes less negative) that results from the isohydric shift. This is the reason for the venous plasma chloride concentrations being somewhat lower than its arterial concentration.

D.   Role of Kidneys

The kidneys play a major role in the regulation of acid-base balance. The three renal mechanisms namely sodium conservation and H^+ ion secretion, bicarbonate reabsorption and ammonia formation have been discussed in detail elsewhere.

E.   Disorders of Acid-Base Balance

1. Normal values: For an understanding and management of disorders of acid-base balance certain concepts and normal values have to be remembered:

   (i)   **pH of blood:**   Normal range 7.35 to 7.45.

   (ii)   **pCO_2:**   This denotes the partial pressure of CO_2 in blood (normal 40 mmHg). Disturbances in the acid-base balance due to ventilatory disorders bring about change in the pCO_2.

   (iii)   **Standard bicarbonate (or alkali reserve):**   This is the bicarbonate concentration of fully oxygenated blood which has been equilibrated with CO_2 at 40 mm Hg at 37° C to exclude possible respiratory compensation of an altered bicarbonate level (Range 22–27 mEq/litre).

   (iv)   **Base deficit or excess:**   Normally the cations and anions may be considered in two categories, e.g., fixed or buffer ions, Normally the ECF cations (chiefly Na^+, K^+ and Ca^{++}) are all 'fixed' in the sense that their concentrations are unaffected by moderate variations in pH. The anions, however, may be fixed (Cl^-, SO_4^{--}, PO_4^{---}) or buffer anions (HCO_3^-).

Base excess or deficit may be expressed as an amount by which the sum of all cations exceeds or fall short of the sum of all fixed anions (Cl^-, PO_4^{--}) inmEq. In practice, this difference equals the plasma HCO_3^- concentration. The amount of this base will increase or decrease according to the amount of HCO_3^- (buffer anion) available to combine with it.

The base deficit is useful for calculating the amount of HCO_3^- or ammonium chloride required to correct acidosis or alkalosis. For this purpose:

Base excess/Base deficit $\times$ 0.3 $\times$ Body weight in kg = Base excess/Base deficit in mEq in ECF.

2. Acidosis and Alkalosis: In clinical practice, disturbances of acid-base balance are described as acidosis or alkalosis. These actually represent the pathologic process leading to acidic or alkaline pH while acidemia and alkalemia refer to acidic or alkaline pH respectively. Conditions causing acidosis lead to accumulation of H^+ ions or loss of base, chiefly bicarbonate, and a fall in the pH of blood. Alkalosis reflects an accumulation of base or excessive loss of H^+ ions with a consequent rise in pH of the blood.

It has already been pointed out that alterations in the HCO_3^-/H_2CO_3 ratio are the chief cause of a disturbance in the acid-base balance. Since, H_2CO_3 level is governed mainly by pulmonary ventilation and alterations in HCO_3^- level by metabolic changes, acidosis and alkalosis may be termed as 'Respiratory or Metabolic' depending upon the cause of this disturbance.

(a) *Types of acidosis:*   This may be metabolic or respiratory depending upon the respiratory factors.

   (i)   *Metabolic acidosis* occurs when there is reduction in HCO_3^- (usually with Na^+) levels in the body. This reduction may be a result of a primary increase in acid production (extrarenal) or a reduction in net acid excretion (renal acidosis). Conditions under which this occurs include loss of upper intestinal secretions by vomiting, due to intestinal obstruction or diarrhea. It may also follow the accumulation of fixed acid anions like PO_4^{---}, sulphate and lactate in starvation or severe muscular activity. An important cause is the accumulation of lactic and pyruvic acids in an ischaemic liver (causing tissue hypoxia) during a cardiac by-pass. On restoration of circulation, the accumulated acids cause a steep fall in the pH. Retention of acid metabolites is a common feature of renal insufficiency because of the failure of the kidney to reabsorb Na^+ and HCO_3^- as also to excrete H^+ ions. Lowering of pH stimulates the respiratory centre with increased rate of pulmonary ventilation and excretion of CO_2 from the alveoli in an effort to restore HCO_3^-/H_2CO_3 ratio to normal.

Treatment of metabolic acidosis requires restoration of fluid and electrolytes, particularly of sodium. An intravenous infusion of 50–100 ml of 8.4% sodium bicarbonate may be given (1 ml of 8.4% sodium bicarbonate = 1 mEq HCO_3^-). Acidosis due to renal disease poses a very serious problem due to a persistent loss of Na^+ and HCO_3^-. This requires repeated corrective treatment.

(ii) *Respiratory acidosis* results from impaired pulmonary ventilation or gaseous exchange in the alveoli. This situation is produced in closed circuit anaesthesia with improper CO_2 absorption, chronic emphysema interfering with alveolar gaseous exchange or depression of the respiratory centre by barbiturate, morphine, or head injuries. There is a rise in pCO_2 with an increase in the H_2CO_3 level. The kidneys compensate by excreting an excess of H^+ ions and converse HCO_3^-. Treatment consists of improving ventilation and removal of underlying cause if possible.

(b) *Types of Alkalosis:* Alkalosis results from accumulation of base or loss of H^+ ions. Like acidosis, it may result from metabolic or respiratory factors.

(i) *Metabolic alkalosis.* Implies the presence of increased HCO_3^- . This precludes a mechanism to both increase plasma HCO_3^- and maintain an increased HCO_3^- level which would otherwise be adjusted for by increased renal excretion of HCO_3^-, i.e., bicarbonaturia. Persistent vomiting or prolonged gastric aspiration is the chief cause. The loss of gastric secretion produces a selective loss of H^+ and Cl^- ions. The loss of Cl^- is compensated by an increase in the HCO_3^- concentration of the plasma and the volume depletion and K^+ loss increase the renal threshold of HCO_3^- thereby maintaining a high HCO_3^-. Occasionally the prolonged and excessive use of alkalies by patients with peptic ulcer may also raise the plasma HCO_3^- levels.

The association of K^+ loss with alkalosis in cases of pyloric stenosis is now well established. In these cases, K^+ loss and loss of H^+ ions co-exist because gastric mucus is rich in K^+. The hypokalaemia is compensated by a shift of intracellular potassium to the ECF. This is replaced by an intracellular shift of Na^+ and H^+ ions. This sometimes produces a paradox of intracellular acidosis of the renal tubules which continue to secrete K^+ and H^+ ions. There may thus be acidic urine even though the plasma HCO_3^- is high.

Compensation of metabolic alkalosis is attempted by the lungs. The respiration is of the Cheyne-Stokes type with brief periods of apnoea during which CO_2 is allowed to accumulate and initiate another respiratory cycle. Latent or overt tetany may be seen in metabolic alkalosis. With a rising pH, the ionized calcium of the plasma combines with phosphate to lower the serum calcium. An associated deficiency of magnesium has also been considered responsible for the tetany in alkalosis. In a typical case, the blood pH and standard bicarbonate are both raised and the urine is highly alkaline. Since most cases show dehydration with a deficit of both Na^+ and K^+, the first step is to correct this with an adequate infusion of saline and potassium chloride. As the Na^+ concentration of ECF is restored, sodium is excreted by the kidney along with HCO_3^- and the alkalosis is readily corrected. The cause of vomiting may also require surgical correction.

The administration of ammonium chloride in only rarely required to correct metabolic alkalosis.

(ii) *Respiratory alkalosis.* It is uncommon and results from a decreased p CO_2. This may be due to excretion of alveolar CO_2 by hyperventilation as in fever or hot, dry weather. It may also occur during anaesthesia with manual control of the breathing causing expulsion of CO_2. The fall in the alveolar CO_2 may induce periods of apnoea due to lack of stimulation to the respiratory centre and respiratory arrest may take place. The kidneys attempt compensation by increased excretion of HCO_3^-. The chief diagnostic guide to respiratory alkalosis is a decreased pCO_2. Treatment should be aimed at correcting hyperventilation. In the event of severe depression of breathing due to fall in pCO_2, inhalation of a mixture containing 10% CO_2 may help.

Laboratory evaluation of acid-base status

The acid base status can be precisely measured using pH, pCO_2 and pO_2 electrodes. The modern blood gas analysers are fully automated and apart from measuring pH, pCO_2, they calculate a number of parameters like bicarbonate, base excess, O_2 saturation, etc. Hence a full picture of acid-base balance of the patient can be obtained which helps in planning an appropriate therapy.

REFERENCES AND SOURCES

1. Moshe Shike, A. Catharine Ross, Benjamin Caballero and Robert J. Cousins (Eds.) (2006), *Modern Nutrition in Health and Disease,* 10th ed., Chapter 8, Lippincott Williams and Wilkins, Baltimore, US.

2. Carl A. Burtis, Edward R. Ashwood and David E. Bruns (2012), *Tietz Textbook of Clinical Chemistry and Molecular Diagnostics*, 5th ed., Chapter 49, Elsevier Publications.

SECTION XIII

CHEMISTRY AND BIOLOGY OF REPRODUCTION

An essential definition of living organisms is their ability to reproduce. In mammals, reproduction takes place by the fusion of a male gamete (sperm) and the female gamete (egg). Along with the gametes male and female sex hormones are produced which play a variety of functions. Implantation of the embryo into the uterus results in the onset of pregnancy. The newborn should be fed on mother's milk, hence lactation is covered also in a dedicated chapter. It is realized that infertility is on an increase in recent years. However, the desire to have a child is fulfilled by assisted reproductive technology offered by a growing number of clinics. On the other hand, many couples with active sexual intercourse like to restrict the number of children, hence there is a need for contraceptives and family planning methods. Techniques enabling reproductive cloning of animals are also covered in a separate chapter in this section.

92

Male Reproductive System and Sexual Dysfunction

Subeer Majumdar and Indrashis Bhattacharya

CONTENTS

I. INTRODUCTION

Reproduction is the process by which living organisms create descendants. In this chapter a brief outline of the *male reproductive system*, the *biology of mammalian spermatogenesis* and the *male sexual disorders* including *primary testicular failure, testicular dysgenesis syndrome, erectile dysfunction* and *male infertility* are discussed. Since detailed stepwise studies and experimentations including those during embryonic development are easily done using rat or mice as a model of mammalian spermatogenesis, references have been made about them. In addition to that, the current status of the probable remedies available for certain disorders are also highlighted.

II. MALE REPRODUCTIVE SYSTEM

The male reproductive system consists of testes and a series of ducts and glands (Figure 92.1). Sperm are produced within the testes and are transported through the reproductive ducts. These ducts include the *epididymis, ductus (vas) deferens, ejaculatory duct* and *urethra*. The accessory reproductive glands

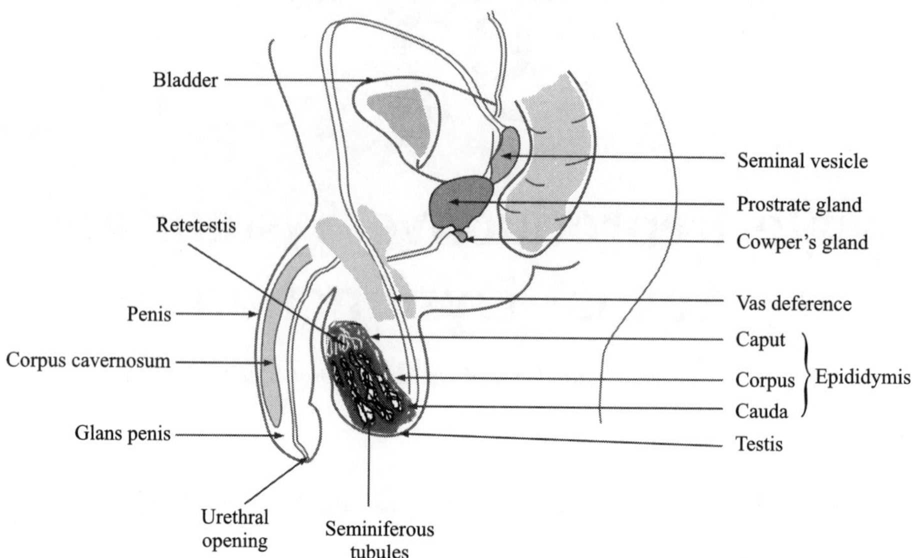

Figure 92.1   Male reproductive system of human.

produce secretions that become part of semen, the fluid that is ejaculated from the *urethra*. These glands include the *seminal vesicles*, *prostate gland*, and *bulbourethral glands*.

A. Scrotum

The two testicles are each held in a fleshy sac called the scrotum. The major function of the scrotal sac is to keep the testes cooler than 37°C, which is important *for spermatogenesis* and survival of male germ cells (Gc).

B. Testes

The testes (singular, testis) are mainly responsible for production of spermatozoa (sperm) and hormones. Testosterone (T), the male sex steroid, is produced in the testes and regulates the production of sperm as well as development of secondary sex characteristics in males during the onset of puberty.

(a) **Seminiferous Tubules:**   Each testis contains several seminiferous tubules where spermatogenesis takes place. Once the sperm are produced, they move from the seminiferous tubules into the rete testis, situated on apical side of the testis (Figure 92.1).

(b) **Interstitial Cells:**   In between the seminiferous tubules here are interstitial spaces called as interstitium. The *Leydig cells* (Lc) are found embedded in these spaces (Figure 92.2A, B) and are responsible for secreting T. Macrophages are also present in this interstitial region to regulate the local inflammatory responses.

(c) **Sertoli Cells (Sc):**   Sertoli cells are specialized columnar epithelial cells hanging on the inner side of the seminiferous tubule originating from the basement of the tubule (Figure 92.2A). These cells are the 'nursing' cells of the testes feeding the germ cells.

They express both follicle-stimulating hormone (FSH)-receptor (FSH-R) and androgen receptor (AR) which are known targets of FSH and T, respectively. The main function of these cells are to hold and nurture various developing germ cells (Gc) under the influence of FSH and T through the stages of spermatogenesis. Moreover, they provide the structural support and establish the blood testes barrier (BTB).

(d) **Peritubularmyoid cells (PTc):**   Peritubularmyoid cells (PTc) are the another somatic cell type of the testes which surrounds seminiferous tubules making its outer most layer (Figure 92.2A). These cells also express AR (but not FSH-R) and presumably crosstalk with Sc and Lc to regulate spermatogenesis.

(e) **Efferent ductules:**   Sperm are transported out of the testis into the epididymis through a series of ductules at apical side of the testis known as efferent ductules.

(f) **Blood Supply:**   The testes receive blood through the testicular arteries (male gonadal or internal spermatic artery). Venous blood is drained by the testicular veins. The right testicular vein drains directly into the inferior vena cava. The left testicular vein drains into the left renal vein.

C. Epididymis

Vasa efferentia from the rete testis open into the epididymis which is a highly coiled tubule (Figure 92.1). Within the epididymis, sperm complete their maturation and their flagella become functional. The epididymis has three parts:

(a) **Head or caput epididymis:**   It is the proximal part of the epididymis. It receives the sperm from the testis.

(b) **Body or corpus epididymis:**   It is the highly convoluted middle part of the epididymis.

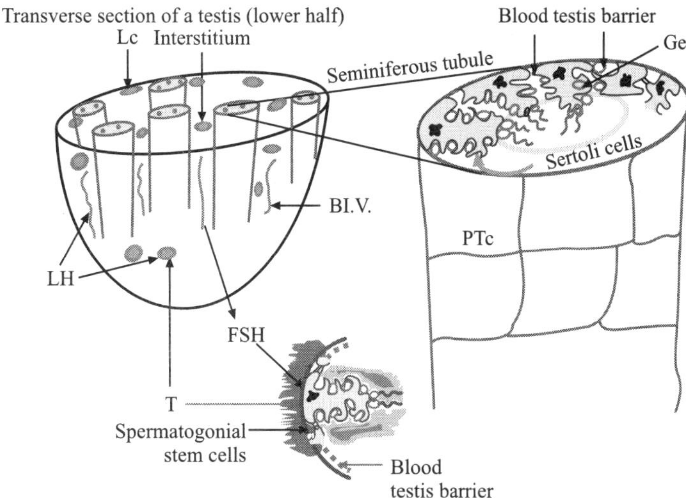

Figure 92.2A Overview of Testes depicting location of various testicular cells. Lc = Leydig cell, BI.V.= blood vessel, PTc = Peitubular cell, Gc = Germ cell, T = Testosterone. (*see Plate 39 for colour figure*)

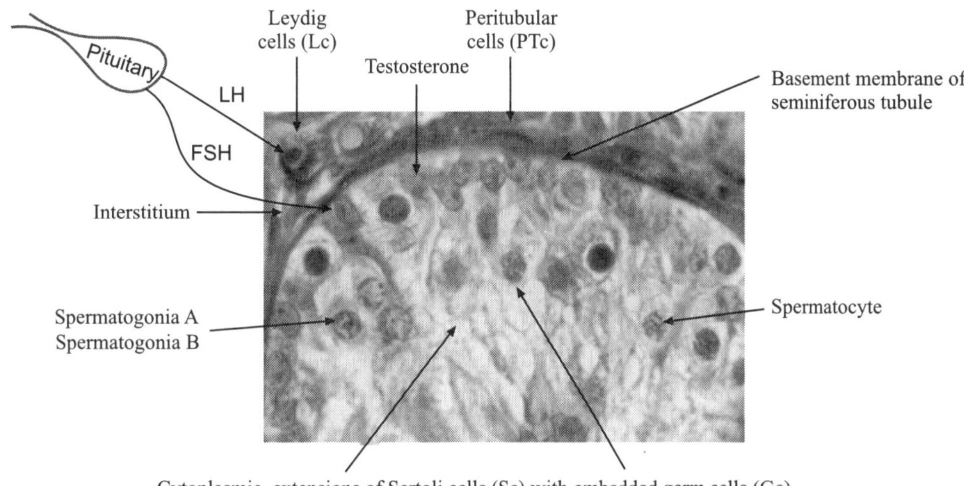

Figure 92.2B A portion of pubertal primate seminiferous tubule and interstitium (cross section) stained with eosin and hematoxylene. Site of action of androgen and of gonadotropins are also shown. (*see Plate 39 for colour figure*)

(c) **Tail or cauda epididymis:** It is the last part of the epididymis that takes part in storing and delivering sperm to the vas deferens. Epididymis retains sperm for some time, gives nourishment to it. The cauda epididymis continues to form less convoluted vas deferens.

D. Ductus (Vas) Deferens

The ductus (vas) deferens, also called sperm duct, or, spermatic deferens, extends from the epididymis in the scrotum on its own side into the abdominal cavity through the inguinal canal (Figure 92.1). The inguinal canal is an opening in the abdominal wall for the spermatic cord (a connective tissue sheath that contains the ductus deferens, testicular blood vessels, and nerves). The smooth muscle layer of the ductus deferens contracts in waves of peristalsis during ejaculation for smooth passage of epididymal sperms.

E. Seminal Vesicles

A pair of seminal vesicles is present posterior to the urinary bladder (Figure 92.1). The duct of each seminal vesicle joins the ductus (vas) deferens on that side to form the ejaculatory duct. The development and function of this organ is totally dependent upon androgen, mainly DHT (5α-dihydrotestosterone), a metabolite of T. The seminal vesicles secrete a significant proportion of the fluid that ultimately becomes semen. *Lipofuscin* granules from dead epithelial cells give the secretion its yellowish color. About 50–70% of the seminal fluid in humans originates from the seminal vesicles, but is not expelled in the first ejaculate fractions which are dominated by spermatozoa and *zinc-rich prostatic fluid*. Seminal vesicle fluid is alkaline, resulting in human semen having a mildly alkaline pH. The alkalinity of semen helps in neutralizing the acidity of the vaginal tract prolonging the lifespan of sperm. Moreover, the vesicle produces a substance that causes the semen to become sticky/jelly-like after ejaculation, which is thought to be useful in keeping the semen near the womb. The thick secretions from the seminal vesicles contain proteins, enzymes, *fructose, mucus, vitamin C, flavins, phosphorylcholine* and *prostaglandins*. The high *fructose* concentrations provide source of nutrient energy for the spermatozoa when stored in semen.

F.   Ejaculatory Ducts

There are two ejaculatory ducts, each receives sperm from the ductus deferens and the secretions of the seminal vesicle on its own side. Both ejaculatory ducts empty into the single urethra.

G.   Prostate Gland

The prostate gland is a muscular gland that surrounds the first inch of the urethra as it emerges from the bladder (Figure 92.1). Like that of the seminal vesicle, the development and function of this organ is totally dependent upon androgen (mainly DHT). The smooth muscle of the prostate gland contracts during ejaculation to contribute to the expulsion of semen from the urethra. The function of the prostate is to secrete a slightly acidic fluid, milky or white in appearance, that usually constitutes 20–30% of the volume of the semen along with spermatozoa and seminal vesicle fluid. Prostatic secretions vary among species. They are generally composed of simple sugars and are often slightly acidic. In human prostatic secretions, the protein content is less than 1% and includes *proteolytic enzymes, prostatic acid phosphatase, beta-microseminoprotein* and *prostate-specific antigen*.

H.   Bulbourethral or Cowper's Glands

The bulbourethral or Cowper's glands are located below the prostate gland and empty into the urethra (Figure 92.1). During sexual arousal each of the gland produces a clear, salty, viscous secretion known as pre-ejaculate. This fluid helps to lubricate the urethra for spermatozoa to pass through, neutralizing traces of acidic urine in the urethra, and helps flush out any residual urine or foreign matter. It is possible for this fluid to pick up sperm, remaining in the urethral bulb from previous ejaculations, and carry them out prior to the next ejaculation. The Cowper's gland also produces some amount of *prostate-specific antigen (PSA)*, and Cowper's tumors may increase *PSA* to a higher level generating doubt about prostate cancer.

I.   Penis

The penis is an external genital organ (Figure 92.1). The distal end of the penis is called the glans penis and is covered with a fold of skin called the prepuce or foreskin. Within the penis are masses of erectile tissue. Each consists of a framework of smooth muscle and connective tissue that contains blood sinuses, which are large, irregular vascular channels.

J.   Urethra

The urethra, which is the last part of the urinary tract, traverses the corpus spongiosum and its opening, known as the meatus which lies on the tip of the glans penis. It is a passage for urine as well as for the ejaculation of semen (Figure 92.1).

A brief detail of above mentioned aspects is summarized in Table 92.1.

III.   BIOLOGY AND REGULATION OF SPERMATOGENESIS

Testis plays a major role in the development of male reproductive system and provides the site of spermatogenesis. Spermatogenesis is a complicated process facilitating transmission of the genetic patrimony and, thus, perpetuation of the species. In true sense, robust spermatogenesis is initiated at puberty in the seminiferous tubule heralding differentiation of spermatogonial cells into advanced germ cells (Figure 92.2B). Mammalian spermatogenesis is classically divided into three 3 phases. In the first—the proliferative or mitotic phase—spermatogonia undergo a series of mitotic divisions. In the second—the meiotic phase—the spermatocytes undergo two consecutive divisions to produce the haploid spermatids. In the third—spermiogenesis—spermatids differentiate into spermatozoa. The entire process is tightly regulated by paracrine, autocrine and endocrine pathways, an array of structural elements and chemical factors modulating the activity of somatic and germinal component.

A.   Phases of Spermatogenesis

Mitosis

The germ line in the mammalian gonads originates from primordial germ cells (PGCs), which are specified at embryonic day (E) 6.5 (in mouse embryo). These bipotential PGCs migrate, proliferate, and colonize to the gonadal ridge at E11 irrespective of the sex of the embryo. Between E12.5 and E14.5, female PGCs in the ovary enter meiosis, which is signaled by retinoic acid (RA). By contrast, male PGCs in the developing testis do not enter meiosis because the somatic fetal Sertoli cells produce the enzyme CYP26B1, which degrades RA. Instead, male PGCs, now called gonocytes, undergo G1/G0 mitotic cell cycle arrest. By postnatal day 3, these gonocytes re-enter the cell cycle, begin to migrate toward the basement membrane of the testis cords and differentiate into spermatogonial stem cells (Sscs), which reside on the basement membrane (Figure 92.3A). and are the foundation of spermatogenesis to ensure production of sperm throughout the life. In laboratory rodents, such as rats and mice, four classes of spermatogonia are present: undifferentiated type A spermatogonia [A single (A_s), A paired (A_{pr}), A aligned (A_{al})]; differentiated type A spermatogonia (A_1, A_2, A_3, A_4); intermediate spermatogonia

TABLE 92.1   Overview of male reproductive system

Name	Location	Role
Bulbourethral glands	Pea sized organs posterior to the prostate on either side of the urethra.	Secretion of gelatinous seminal fluid called pre-ejaculate that helps to lubricate the urethra for spermatozoa to pass through, and flush out any residual urine or foreign matter.
Leydig cells	Adjacent to the seminiferous tubules in the interstitium.	Production of T.
Cremaster muscle	Covers the testes.	Raises and lowers scrotum to help regulate temperature and promote spermatogenesis. Voluntary and involuntary contraction.
Dartos muscle	Layer of smooth muscular fiber outside the external spermatic fascia but below the skin.	Contraction by wrinkling to decrease surface area available for heat loss to testicles, or expansion to increase surface area available to promote heat loss; also helps raise and lower scrotum to help regulate temperature.
Efferent ductules	Part of the testes and connect the rete testis with the epididymis.	Pathway for sperm to be in the epididymis.
Epididymis	Tightly coiled duct lying just outside each testis connecting efferent ducts to vas deferens.	Storage and maturation of sperm.
Penis	Three columns of erectile tissue: two corpora cavernosa and one corpus spongiosum. Urethra passes through penis.	Male reproductive organ and also male organ for urination.
Prostate gland	Surrounds the urethra below the urinary bladder.	Stores and secretes a clear, slightly alkaline fluid constituting up to one-third of the volume of semen. Raise vaginal pH.(25–30% of semen).
Scrotum	Pouch of skin and muscle that holds testicles.	Regulates temperature at slightly below body temperature.
Semen	Usually white but can be yellow, gray or pink (blood stained). After ejaculation, semen first goes through a clotting process and then becomes more liquid.	Components are sperm and "seminal plasma". Seminal plasma is produced by contributions from the seminal vesicle, prostate, and bulbourethral glands.
Seminal vesicles	Convoluted structure attached to vas deferens near the base of the urinary bladder.	About 65–75% of the seminal fluid in humans originates from the seminal vesicles. Contain proteins, enzymes, fructose, mucus, vitamin C, flavins, phosphorylcholine and prostaglandins. High fructose concentrations provide nutrient energy for the spermatozoa as they travel through the female reproductive system.
Seminiferous tubules	Long coiled structure contained in the chambers of the testis; joins with vas deferens.	Site of meiosis and generation of sperm.
Sertoli cells	Junctions of the Sc form the blood-testis barrier, a structure that partitions the interstitial blood compartment of the testis from the adluminal compartment of the seminiferous tubules.	Responsible for nurturing and development of Gc, provides both secretory and structural support; under the signaling of FSH and T.
Urethra	Connects bladder to outside body, about 8 inches long.	Tubular structure that receives urine from bladder and carries it to outside of the body. Also passage for sperm.
Vas deferens	Muscular tubes connecting the left and right epididymis to the ejaculatory ducts to move sperm. Each tube is about 30 cm long.	During ejaculation the smooth muscle in the vas deferens wall contracts, propelling sperm forward. Sperm are transferred from the vas deferens into the urethra, collecting fluids from accessory sex glands en route.

(In); and type B spermatogonia (Figure 92.3.B–C). However, in primates, including humans un-differentiated spermatogonia is further classified into A_{dark} and A_{pale} spermatogonia. Irrespective of the species, different spermatogonial classes can be characterized by light microscopy and transmission electron microscopy according to the presence and distribution of heterochromatin. The continuity of the spermatogenesis is dependent upon the presence of *spermatogonial stem cells*

(Ssc, considered to be A_s spermatogonia in rodents and mainly A_{dark} in primates) capable of self-renewal and differentiation, located in the specific "niches" associated with highly vascular beds of the seminiferous epithelium.

Meiosis

B-spermatogonia divide by mitosis forming two preleptotene spermatocytes, cells representing the beginning of meiotic

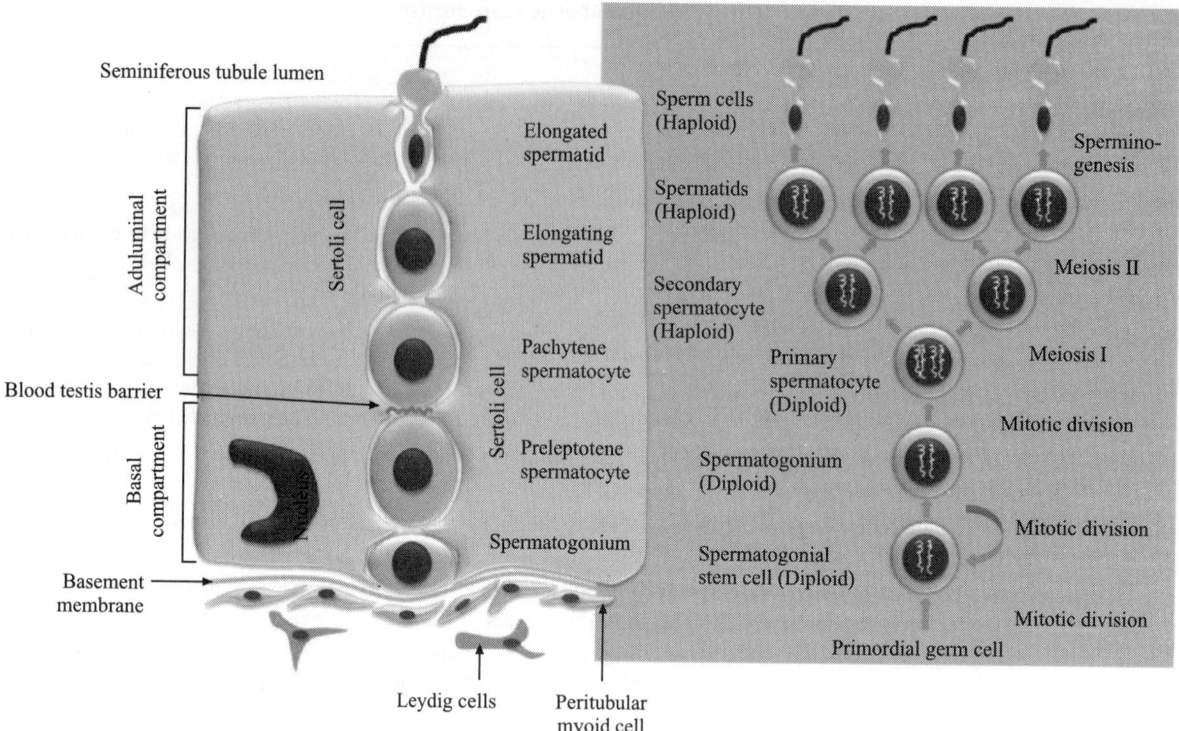

Figure 92.3A Process of spermatogenesis, showing location and stages of male Gc differentiation.

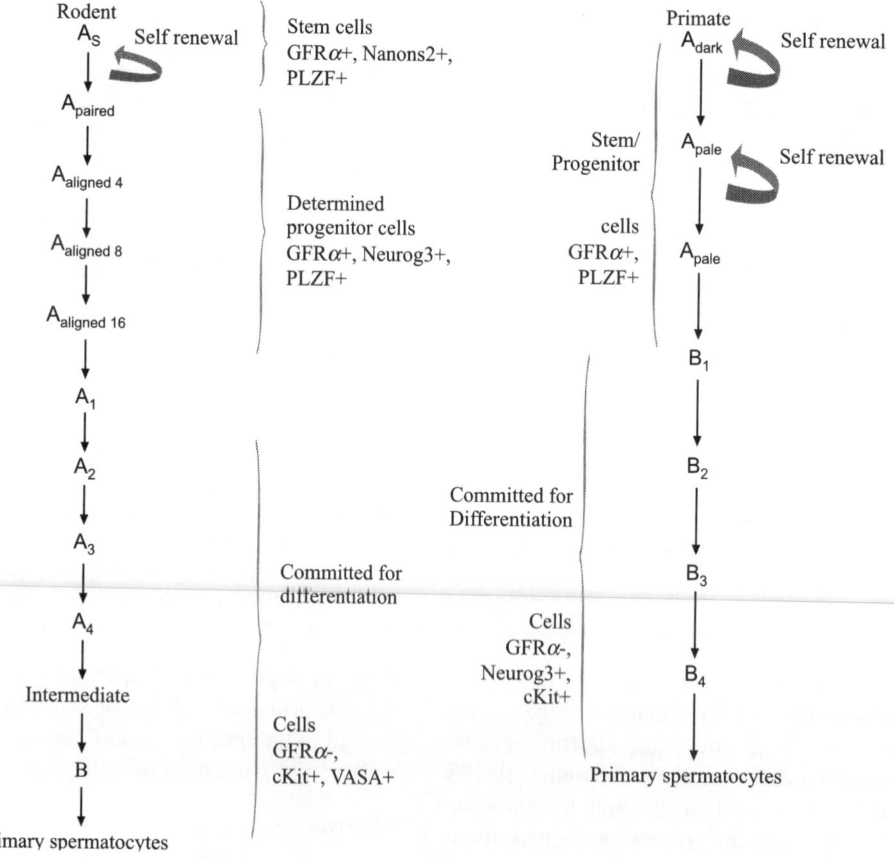

Figure 92.3B Steps of male Gc differentiation.

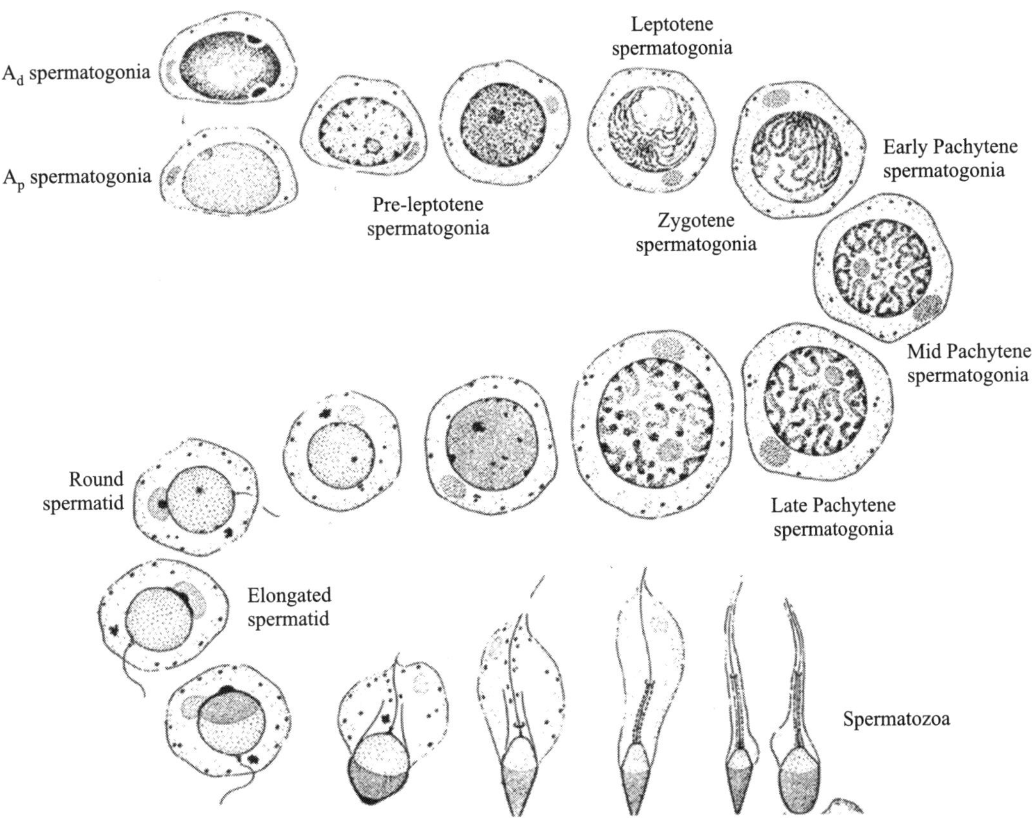

Figure 92.3C   Phases of primate spermatogenesis.

prophase (Figure 92.3A). These small cells rest on the basement membrane, but leptotene and zygotene spermatocytes become transitional and move across the blood-testis-barrier (BTB) made by Sc-Sc junctions. Preleptotene, leptotene and zygotene spermatocytes are located in specific stages of seminiferous epithelial cycle and are identifiable by routine microscopy, although fixation artifact results in the leptotene and zygotene cells appearing to be attached to the basement membrane. Based on association of various Gc, a portion of seminiferous tubule is staged. Spermatocytes are found in all stages, because meiosis is a prolonged period of spermatogenesis that extends over approximately 14 days in the mouse. This cellular division goes through three categories, all occurring in stage XII: (a) meiosis I, the division of 4n cells; (b) formation of secondary spermatocytes (2n), which are larger than step 1 spermatids, but rarely are found as the only spermatocytes in a tubular cross section; and (c) meiosis II, the division of 2n secondary spermatocytes to form haploid (1n) round spermatids. A stage specific appearance of developing Gc in the rodent testes is summarized in Table 92.2.

Spermiogenesis

The transformation of spherical, haploid spermatids (1n) into elongated, highly condensed and mature spermatozoa that are released into the seminiferous tubule lumen is called spermiogenesis (Figure 92.3C). The differentiation of spermatids proceeds through at least 4 prolonged steps (or phases): Golgi, capping, acrosomal, and maturation.

TABLE 92.2   **Postnatal developmental schedule of rat and mouse testis**

Stage	Description of developmental stage	Age in rat	Age in mouse
I	Spermatogonia and somatic cells only	Days 6–7	Days 6–7
II	Initiation of meiosis I—leptotene cells	Days 13–14	Day 10
III	Appearance of zygotene cells	Days 17–18	Day 12
IV	Appearance of early pachytene cells	Days 19–20	Day 14
V	Appearance of late pachytene cells	Days 22–23	Days 17–18
VI	Appearance of round spermatids	Days 24–25	Days 20–21
VII	Appearance of elongating spermatids	Days 30–31	Days 24–25
VIII	Appearance of elongated spermatozoa	Days 36–37	Days 27–28
VIII	Mature sperm in the testes and spermiation	Days 43–45	Days 33–35

These steps are useful for the identification of specific stages in the cycle of the seminiferous epithelium in cross section of testis.

(a) **Golgi phase:**   Golgi apparatus is very important during the early steps of spermiogenesis, as the formation of the acrosome is dependent upon this organelle's ability to produce vesicles and granules containing the enzymatic components of the acrosomic system that will cover the developing sperm nucleus. Differentiation of the first three steps of round spermatid formation involves appearance of prominent Golgi apparatus. Step 1 spermatids have a small, perinuclear Golgi region without an acrosomic vesicle or granule. Subsequent steps 2–3 show proacrosomal vesicles and granules within the Golgi apparatus, with the formation of a single, large acrosomal granule within a larger vesicle that will indent the nucleus (Figure 92.3C).

(b) **Capping phase:**   Capping involves steps 4–5 round spermatids, where the acrosomic granule touches the nuclear envelope and the vesicle begins to flatten into a small cap over the nuclear surface. In steps 6–7, the acrosomic vesicle becomes very thin and the granule flattens. Step 8 is the last step of round spermatid when the acrosome flattens over approximately 1/3 of the nuclear surface. In late stage VIII of seminiferous tubule, step 8 nuclei begin to change shape.

(c) **Acrosomal phase:**   Steps 9–14 involve migration of the acrosomal system over the ventral surface of the elongating spermatid nucleus (Figure 92.3C). This migration of the acrosome is completed approximately by step 14 spermatid and is difficult to identify in typical histological sections due to its presence in different planes of sections and angles or orientation. Thus, recognition of specific stages of spermatogenesis will typically rely on the acrosomal system observed in the round spermatids, rather than in the elongated cells. These spermatid steps also involve condensation of the chromatin, as the chromosomes are packed more tightly and stain more intensely with hematoxylin.

(d) **Maturation phase:**   Steps 15–16 appear across stages III-VIII and show fewer changes in nuclear shape and acrosomal migration. The nucleus continues to condense and the acrosome matures into a thin structure that protrudes at the apex but covers nearly all the nucleus, except for that portion connected to the tail. Protamines are small, arginine-rich, nuclearproteins that replace histones at this stage and this is essential for sperm head condensation and DNA stabilization. Excess cytoplasm is removed in Stages VII-VIII, resulting in the formation of prominent cytoplasmic lobes and residual bodies, which contain unused mitochondria, ribosomes, lipids, vesicles and other cytoplasmic components.

B.   The Cycle and Wave of Spermatogenesis

Gc within each layer of the seminiferous epithelium change in synchrony with the other layers over time, producing the sequence of stages described above. The cells do not migrate laterally along the length of the seminiferous tubule; however, an unusual successive order of the stages is observed, whereby sequential stages occur with repetition along the length of the tubules, in a 'wave' of the seminiferous epithelium. That is, at least in the rodent, stage I is followed by II, followed by III, etc. through stage XIV, which is then repeated by stage I. The stages are found in ascending order from the rete testis to the center of the seminiferous tubule, where the stages are reversed. The wave is produced by synchronous development of clonal units of Gc through a mechanism of biochemical signaling that remains a subject of inquiry.

C.   Sperm Production

The most important role of the testes is to produce sperm. In most mammals, each spermatogenic cycle lasts around 9 to 12 days, whereas the total duration of spermatogenesis lasts nearly 40 to 54 days [35 days and 43 days in case of mice and rats respectively]. Particularly in humans, the entire spermatogenic process is very long and lasts not more than 64 days. As a general pattern for mammals, and probably related to the synchronized development of different Gc types per seminiferous tubule cross-sections (stages), each phase of spermatogenesis (spermatogonia, spermatocyte, and spermatid) lasts approximately one third of the duration of the entire process.

D.   Sperm Maturation

Mammalian spermatozoa acquire the fertilizing potential in the epididymis. Prominent among the hallmarks of epididymal sperm maturation is the proximal-distal migration of the cytoplasmic droplet (CD), the last remnant of the spermatogenic cell cytoplasm, down the sperm flagellum. The definition was first applied to *in vivo* fertilization where spermatozoa were removed from the *caput, corpus* or *cauda* epididymidis and inseminated into the vagina, uterus or oviduct in different species. As transport through the epididymis takes approximately a week, these osmotic encounters are extremely gradual so that isovolumetric regulation (IVR), the slow movement of osmolytes and water that do not impinge on cell volume, might occur. In this scenario, as spermatozoa move through the epididymis, they would sequentially encounter *glycerophosphocholine (GPC,* providing a driving force for water efflux and IVR) and then solutes (*myo-inositol, L-carnitine, taurine, glutamate*) that would be taken up into the cells as a result of their high concentration. The result would be osmolyte loads for spermatozoa of the order: *cauda> corpus > caput.* This speculation of osmotically-driven solute uptake needs to be tested experimentally.

E.   Endocrine Regulation of Spermatogenesis

The hypothalamo–hypophyseal–testicular axis

The hypothalamo–hypophyseal–testicular axis is the classic example of an endocrine regulatory circuit, with hierarchical cascades of regulatory feedback loops (Figure 92.4). Specific hypothalamic nuclei (like mediobasal and arcuate) synthesize the decapeptide *gonadotrophin-releasing hormone* (GnRH), which stimulates the gonadotrophin secretion from gonadotroph cells present in the anterior pituitary gland. In the median eminence, the axon terminals of GnRH neurons make contact with the hypophyseal portal vessels, which transport the releasing hormone, secreted in pulses of 60–90 min intervals, to the anterior pituitary. However, recent studies have elucidated the involvement of several other hypothalamic hormones in the maturation and fine-tuning of GnRH neurons. These include classical neurotransmitters such as *noradrenaline, glutamate* and *γ-aminobutyric acid (GABA), neuropeptide Y, galanin-like peptide, opioid peptides* and *orexins*. Very recently, two novel regulators of this neuronal network, *kisspeptin*, a product of the *KiSS1* gene acting via the *G protein-coupled receptor 54 (GPR-54)* and Neurokinin B has been demonstrated as the essential gatekeeper of puberty and fertility as humans and mice lacking functional *GPR54* and/or *KiSS1* genes display severe hypogonadotrophic hypogonadism. Majority of *GnRH* synthesizing neurons express *GPR54* and *kisspeptin* is able to evoke extraordinarily potent depolarization in these cells, coupled to robust *GnRH* and gonadotrophin secretory responses. However, Neurokinin B probably augments the GnRH release from these cells somewhere upstream of the GPR-54. Interestingly, it is important to note here that there is a fundamental difference between most primates including humans and laboratory rodents in terms of the phases and duration of testis development. Two features in humans that particularly differ from laboratory rodents are 1) the occurrence of an extended postnatal period of hypothalamic-pituitary-testicular hormone activation, for about 4–6 months after birth,

during which period T levels remain within the low adult range (in laboratory rodents any such activation lasts for a matter of hours at around the time of birth) and 2) after this period, in the human and other primates, there is an extended phase of childhood testicular quiescence, unlike rodents, when activity of hypothalamus is low and relatively little happens in terms of testicular development.

Gonadotrophins

The two gonadotrophins, LH and FSH act on testis. They are synthesized differentially by the single pituitary cell type depending upon pulse frequency and amplitude of GnRH (produced by hypothalamic neurons) stimuli (Figure 92.4). They are heterodimers of a common α-subunit to which hormone specific β-subunit is attached conferring the hormonal specificity. Two N-linked carbohydrate side chains are coupled to the α-subunit, one to LH-β and two to FSH-β. The termini of the carbohydrates in LH are heavily sulphated (50%), in FSH they are mainly sialylated. This difference underlies the longer half-life of FSH (3–4 hrs) than LH (20 min) in the circulation. Both the gonadotrophins exist in a number of isoforms, due to the microheterogeneity of their carbohydrate moieties. They vary in bioactivity, and their relative proportions in circulation are regulated by androgen. However, the physiological significance of this variability remains unknown.

The gonadotropin receptors

The LH receptor (LH-R) and FSH receptor (FSH-R) are structurally related glycoproteins with a molecular mass of about 80 kD, and belong to class A rhodopsin-like G-protein coupled receptor (GPCR). They have a transmembrane domain that traverses the plasma membrane as 7 α-helices connected

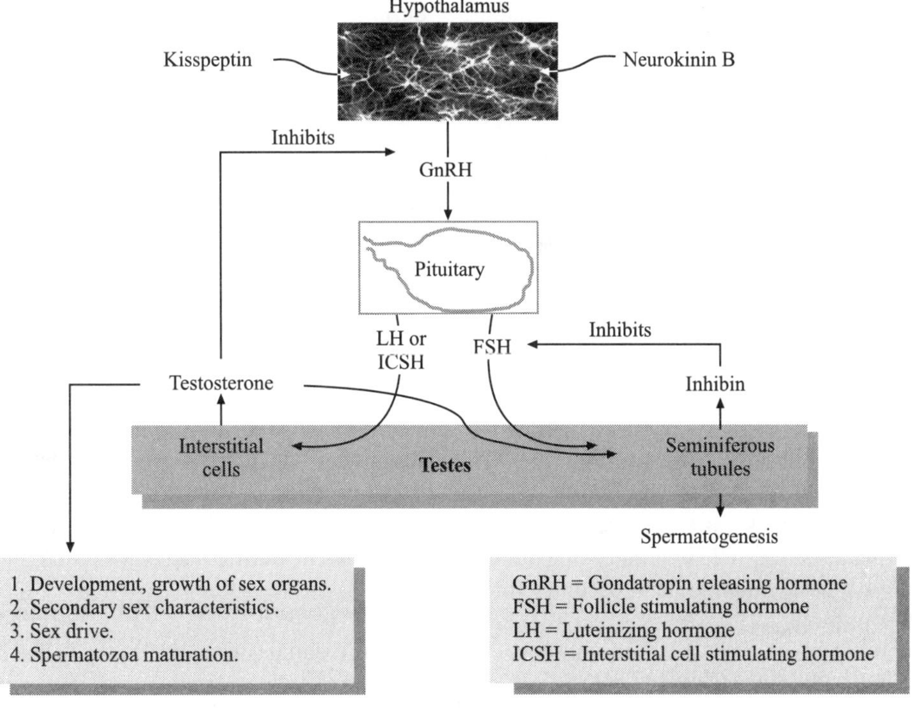

Figure 92.4 Hypothalamo hypophyseal testicular axis

by 3 extracellular and 3 intracellular loops. The transmembrane part is the integral part in signal transduction of gonadotropins across the plasma membrane.

Mechanisms of action and physiological effects of gonadotrophins

The molecular events mediating the actions of LH and FSH on testicular Leydig cells (Lc) and Sertoli cells (Sc), respectively, are principally similar. The molecular mechanisms of gonadotrophin action in the primate testis have not been studied in great detail, and much of the information is based on animal experiments and *in vitro* studies. LH induces the synthesis and secretion of androgens mainly, T by testicular Lc. Both FSH-R and Androgen Receptor (AR) are located on testicular Sc. A synergistic effect of FSH and T regulates transcriptional activity modulating secretion of various growth factors and metabolites by Sc to support the initiation and maintenance of Gc differentiation (Figure 92.2B).

FSH-R mediated signaling pathway

FSH binding to its receptor, i.e., FSH-R is known to activate at least 5 different signaling pathways (including cyclic-AMP dependent protein kinase A or PKA in particular) in Sc and these signaling pathways crosstalk to organize the final desired cellular consequences. It is important to note that FSH down regulates and up regulates FSH-R expression in Sc in a time dependent manner. Some of the putative downstream target genes of FSH in Sc which has a potential role in spermatogenesis are: *P450 Cyp19 aromatase, FSH-R, Stem cell factor, Glial cell line derived nurotrophic factor, transferrin, androgen binding protein, Inhibin α and Inhibin β, etc.*

AR mediated signaling pathway

AR is a type of nuclear receptor which is activated by binding of either of the androgenic hormones, T or dihydrotestosterone (DHT). In the testis, Lc, PTc and Sc express AR. No AR is expressed in Gc of the mature testis. The differential role of AR present in different somatic testicular cells are controversial. However, data generated from cell specific knock outs of AR have revealed that AR present in the Sc is the key regulator of spermatogenesis that directs the formation of BTB and also governs the completion of the meiotic differentiation of the developring Gc in the testes. Unlike rodents, adult AR levels increase and decrease in a cyclical fashion. Specifically, analyses by *in situ* hybridization and immuno-cytochemistry revealed that AR levels progressively increase during cell association stages II through VII of the spermatogenic cycle and then decline sharply during or immediately after stage VII to become barely detectable in stages IX–XIII. The levels of AR are highest in stage VII and thus this stage is thought to be the most regulated by and sensitive to T. In primates including human, AR is less active during childhood and becomes fully active during and after puberty.

Feedback regulation of gonadotrophin secretion

The regulatory feedback of testicular origin exert their effect at the level of hypothalamus and pituitary via gonadal steroid, i.e., T and peptide hormones like *inhibin* and *activin* to maintain the functional balance of gonadotrophin secretion (Figure 92.4). T (or at least partly after its conversion to estradiol) suppresses GnRH secretion at the hypothalamic level and gonadotrophin synthesis in the pituitary gland in rats and as well as in primates. However, aromatization (conversion of T to estradiol) is not mandatory, since the nonaromatizable T metabolite, DHT, is also effective in this feedback loop. intestresigly, this T mediated feedback also directs the inhibition of synthesis of gonadotrophin (LH) at the pituitary level as evident from *in vivo* and *in vitro* studies in rats but not in rhesus monkeys. Although testicular steroids regulate FSH production, at the pituitary level, FSH is mainly under negative control of a Sc product called inhibin.

Role of FSH in spermatogenesis

Since Gc do not express FSH-R, the action of FSH on spermatogenesis is indirect and occurs via testicular Sc. Although FSH is not indispensible for murine spermatogenesis since FSH β and FSH-R knock out male mice are fertile, the size of the Sc population is set during the onset of puberty under the influence of FSH which is the major determinant of the maximal sperm output during adulthood. Moreover, FSH is known to up regulate the survival and differentiation rate of Gc.

Role of T in spermatogenesis

The main targets of T action are meiosis and spermiogenesis. All the somatic cells present in the testes express AR. Generation of Gc-specific AR knockout (KO) mouse (G-AR$^{-/y}$) showed normal spermatogenesis. Together, these reports provide *sufficient* evidences suggesting the essential roles of AR during spermatogenesis might come from T induced cell–cell communication in a paracrine fashion. Somatic cell specific AR knock out mice models have been generated. PTc- specific AR-KO mice (PM-AR$^{-/y}$), showed oligospermia with normal fertility or sterility. On the other hand, all Sc specific AR-KO males generated by three independent groups are sterile. Hence, it is obvious that AR present in Sc is very crucial to determine male fertility.

One study suggests that the elevated levels of AR normally present in stage VIII may be responsible for increased T-dependent transport of BTB proteins from the apical to the basal side of the BTB to reseal the barrier after preleptotene spermatocytes pass through. Three tight junction components (*occludin, claudin 11, claudin 3*) are down-regulated in the absence of AR indicating that T activates these genes via AR. Another action of T is to remodel and strengthen connections with post meiotic Gc at the elongated spermatid stage of development during stages VII–VIII.

Role of gonadotropins from human clinical case studies

The differential action of LH (via T) and FSH which is essential for spermatogenesis remains a controversial issue in human. It is noteworthy that the hormonal (FSH and T) requirements for initiation of the first spermatogenic wave during puberty, for its maintenance in adulthood, and for the re-initiation after transient gonadotropic suppression are totally different phenomena of the entire successive process of spermatogenesis. In infertile men, prolonged hCG treatment alone, via augmentation of T production, is reported to initiate spermatogenesis. However, these individuals may not have been totally deprived of FSH. Treatment of healthy men with T suppresses both gonadotrophins and intra-testicular T while maintaining peripheral T action. As a result, spermatogenesis is severely suppressed. If these men receive injections of LH or hCG, their spermatogenesis is restored due to the restoration of high intra-testicular T, although their FSH level remains suppressed. In such individuals, the sperm counts remain about 50% of the normal individuals. This suggests that the re-initiation of spermatogenesis is possible with T alone, but its quantitative recovery also needs FSH. When FSH was added during the treatment regimen, spermatogenesis was fully restored. In T-suppressed men, treatment with purified FSH was able to marginally stimulate spermatogenesis, though not quantitatively, suggesting that quantitatively normal spermatogenesis requires support of both FSH and T. Interestingly, men with inactivating FSH-R mutations are normally masculinized and not azoopermic (no sperm in the semen) but subfertile with reduced testicular size and poor sperm quality. However, the sporadic men with inactivating FSH-β mutation (affecting bioactivity of FSH hormone) have been found to be azoospermic. The reason for the discrepancy between phenotypes of the hormone and receptor inactivation remains unanswered. Observations from both the FSH-β and FSH-R knock-out mice are in agreement with phenotype of the FSH-R deficient men, where spermatogenesis, though quantitatively and qualitatively suppressed, may be possible without FSH action.

IV. MALE SEXUAL DYSFUNCTION

Research in the last decade has revealed various aspects of male reproductive health of human such as the rising incidence of *primary testicular failure, testicular cancer; declining semen quality*; increasing frequencies of *, cryptorchidism, hypospadias, precocious male puberty with testotoxicosis, erectile disorders and male infertility.*

A. Primary Testicular Failure

Testis is the major endocrine gland of the male reproductive system. Endocrine testicular failure results in T deficiency. In primary endocrine testicular failure, a decline in T secretion resulting in a condition termed *hypoandrogenism* is caused by a deficiency or absence of Lc function. However, the phenotype of primary exocrine testicular failure is *male infertility*. Clinically relevant diseases described here are *anorchia, Lc hypoplasia, Gc aplasia, spermatogenetic arrest, hypospermatogenesis* and *numerical chromosomal abnormalities* as well as *Y chromosome microdeletions.*

(a) Anorchia: Anorchia is defined as the complete absence of testicular tissue in genetically and phenotypically male patients. In unilateral anorchia testicular tissue is still present on the contralateral side. Pure anorchia has to be differentiated from conditions with ambiguous and intersex genitalia. However, a clinically important differential diagnosis is cryptorchidism and testicular atrophy, where testicular tissue is detectable.

(b) Lc hypoplasia: Lc hypoplasia is a rare disease with an autosomal recessive pattern of inheritance and estimated incidence of 1:1,000,000. Lc are unable to develop because of inactivating mutations of the LH-R that fails to provide the necessary stimulation of intracellular pathways. Men with Lc hypoplasia have very low serum T and high LH levels.

(c) Spermatogenic failure: While endocrine testicular failure causes hypogonadism; spermatogenic failure—defined as exocrine testicular failure—leads to *male infertility*. Spermatogenic failure might be caused by hypothalamic, pituitary, testicular or post-testicular disorders. Various testicular etiologies of spermatogenic failure may lead to the same histopathological pattern. In this sense, *Gc aplasia, spermatogenic arrest,* and *hypospermatogenesis* as described below have to be considered as a description of certain histopathologic phenotypes of spermatogenic failure, and not as manifestation of single disease entity.

(d) Gc Aplasia (or Sc only Syndrome): Gc aplasia or Sertoli cell only (SCO) syndrome is a histopathologic phenotype where the seminiferous tubules are reduced in diameter, and contain only Sc but no other cells involved in spermatogenesis.

Gc aplasia or SCO syndrome is one common cause of non-obstructive azoospermia. In congenital germ cell aplasia, the primordial germ cells do not migrate from the yolk sac into the future gonads or do not survive in the epithelium of the seminiferous tubule. Chromosomal abnormalities, especially microdeletions of the Y chromosome, are predominant genetic causes for complete germ cell aplasia. Moreover, anti-neoplastic therapy with radiation or chemotherapy may cause complete loss of Gc. Other reasons include viral infections of the testes such as mumps orchitis. Gc aplasia can also occur in maldescended testes.

(e) Spermatogenic Arrest: Spermatogenic arrest is also not a specific primary testicular failure, but a histopathological description of the interruption of normal Gc maturation at the level of a specific cell type including that of spermatogonial arrest; spermatocyte arrest or spermatid arrest. A definite diagnosis can only be made by multiple testicular biopsies. Testicular volume, FSH and inhibin B may lie in their normal ranges, but may also be elevated or decreased, respectively. The arrest may be caused by genetic or by secondary influences. Genetic etiologies include *trisomy, balanced-autosomal anomalies (translocations, inversions) or deletions in the Y chromosome (Yq11).*

(f) Hypospermatogenesis: This is the third major histological grouping in primary exocrine testicular failure in which all Gc types, including mature spermatids, is present in some or all tubules but are mildly, moderately or severely reduced in number. Some patients have the appearance of complete Gc aplasia in some tubules but with complete spermatogenesis in adjacent tubules (sometimes called 'focal' germinal cell aplasia) while others have the appearance of an excess number of precursor Gc in relation to the number of mature spermatids in the epithelium. In most patients with hypospermatogenesis, testicular volume is reduced. FSH is elevated in most, but not all patients, with serum levels correlating positively with the proportion of tubules with Gc aplasia.

(g) Numerical chromosomal abnormalities: *Klinefelter Syndrome:* This syndrome was first described by Harry Klinefelter in 1942 as a clinical condition with small testes, azoospermia, gynecomastia and an elevated serum FSH. In 1959, the chromosomal basis of the disorder was described. Subsequently the diagnosis of Klinefelter syndrome has required the demonstration of the 47, XXY karyotype or one of its rarer variants. The prevalence of Klinefelter syndrome appears to be approximately 1 in 660 males. It appears that at least half of the cases remain undiagnosed and untreated throughout life.

XX-Male Syndrome: The XX-Male Syndrome is characterized by the combination of male external genitalia, testicular differentiation of the gonads and a 46,XX karyotype by conventional cytogenetic analysis. This disorder shows a prevalence of 1:9,000 to 1:20,000. Applying molecular methods, it has been demonstrated that about 3 of 4 XX-males have Y chromosomal material translocated onto the tip of one X chromosome. Translocation of a DNA-segment which contains the testis-determining gene (SRY = Sex Determining Region Y) from the Y to the X chromosome takes place during paternal meiosis. The presence of the gene SRY is sufficient to cause the initially indifferent (bipotential) gonad to develop into a testis.

XYY-Syndrome: Most 47,XYY males have no health problems distinct from those of 46,XY males. The diagnosis relies entirely on the cytogenetic demonstration of two Y chromosomes with an otherwise normal karyotype. The non-mosaic chromosomal aneuploidy is caused by nondysjunction in paternal meiosis. Usually the finding is incidental, occurring when karyotyping is done for unrelated issues. The prevalence among unselected newborns appears to be 1:1000. Men with 47,XYY-syndrome have serum levels of T and gonadotropins, as well as testicular volumes, comparable to those of normal healthy men. Most men with 47,XYY-syndrome have normal fertility.

(h) Structural chromosome abnormalities: Structural chromosome abnormalities encompass pathological alterations of chromosome structure that are detectable through light-microscopic examination of banded metaphase preparations. Smaller lesions that are only detectable with molecular probes are not termed structural chromosomal abnormalities.

Structural abnormalities of the autosomes

Balanced autosomal anomalies may interfere with the meiotic pairing of the chromosomes and thus adversely affect spermatogenesis. These abnormalities often do not display a typical clinical phenotype. The presence and extent of disturbed fertility cannot be foreseen in individual cases. The same balanced autosomal aberration can have a severe effect on spermatogenesis in one patient and none at all in another patient. Even brothers with the same pathological karyotype can have widely differing sperm densities. So far, no clinical or laboratory parameter in an infertile male is known which can reliably indicate the presence of an autosomal structural anomaly. Therefore, in cases of unclear azoospermia or severe oligozoospermia, karyotyping is generally advised.

Structural Abnormalities of Sex Chromosomes

X chromosome—The X chromosome contains numerous genes essential to survival. Every major deletion of this chromosome has a lethal effect in the male sex. Translocations between the X chromosome and an autosome usually result in disturbed spermatogenesis, whereas inversions of the X chromosome do not substantially affect male fertility.

Y chromosome—The intact Y chromosome is essential for the male reproductive system. The male-specific region of the Y chromosome (MSY) differentiates the sexes and comprises 95% of the chromosome length. The SRY gene is localized on the short arm of the Y chromosome and it influences differentiation of the bipotential embryonic gonad towards the genesis of testis and accessory reproductive organs. The long arm of the Y chromosome contains areas responsible for the regular spermatogenesis. Short arm deletions of the Y chromosome that encompass the sex determining SRY gene result in sex reversal. Clinically, affected men appear as phenotypically female individuals with somatic signs of **Turner's syndrome**. If the deletion affects the long arm, the phenotype will be male. Loss of the heterochromatic part of the Y chromosome's long arm (Yq12) leaves general and reproductive health unaffected. Deletions of the euchromatic part of the Y chromosome's long arm (Yq11) may affect spermatogenesis, because Yq11 harbors loci essential for spermatogenesis. In addition to deletions, a series of structural anomalies of the Y chromosome are known. Pericentric inversions are without consequence. An isodicentric Y chromosome is a more complex aberration nearly always occurring as a mosaic with a 45, X-cell line. The phenotype may be male, female or intermediate. Patients with a male phenotype are usually infertile. These patients also have an increased risk of developing testicular tumors. Reciprocal translocations between the Y chromosome and one of the autosomes are rare. In most cases spermatogenesis is severely disturbed, however, several men with this karyotype are fertile. Translocations between the X and Y chromosomes occur in several variations; often the karyotype is unbalanced. The correlation between karyotype and clinical presentation is complex. The phenotype may be male or female, fertility may be normal or disturbed.

Y chromosome microdeletions—The long arm of the Y chromosome contains four partially overlapping but discrete regions that are essential for normal spermatogenesis. The loss of one of these regions, designated as *AZF (azoospermia factor)a, P5/proximal P1 (AZFb), P5/distal P1 (AZFb)* and *AZFc (or b2/b4),* can lead to infertility. The deleted regions are usually of submicroscopic dimensions and are known as Y chromosome microdeletions. Their prevalence in azoospermic men lies between 5–10% and in oligospermic men between 2–5%. Clinically, the patients present with severely disturbed spermatogenesis; endocrine testicular function may or may not be affected by the microdeletion as in other cases of spermatogenetic failure. Testicular histopathology varies from *complete or focal Sertoli-cell-only* pattern to *spermatogenic arrest* or *hypospermatogenesis* with qualitatively normal but quantitatively severely reduced spermatogenesis. In azoospermic men, the presence of a complete deletion of AZFa or AZFb seems to be associated with uniform Gc aplasia and negative prognostic value for testicular sperm retrieval. No clinical parameter can help distinguishing patients with microdeletions of the Y chromosome from infertile men without microdeletion. It should be noted that Y chromosome microdeletions have also been described in men with proven fertility.

B. The Testicular Dysgenesis Syndrome (TDS)

Several lifestyle (obesity, smoking) and environmental-related (exposure to traffic exhaust fumes, dioxins, combustion products, pesticides, food additives, persistent pollutants such as DDT, polychlorinated biphenyls) factors have appeared to be negatively associated with both the perinatal and adult testes, emphasizing the importance of environmental/lifestyle impacts throughout the life course. Such observations have recently given rise to the *testicular dysgenesis syndrome (TDS) hypothesis* which proposes that each of these disorders may have a common fetal origin triggered by maldevelopment of the testis, leading to malfunctional testes. These new aspects of poor semen quality, testis cancer, undescended testis and hypospadias are hypothesized to be symptoms of one underlying entity, *TDS,* which is increasingly becoming common due to variety of adverse environmental influences.

(a) Testicular germ cell (Gc) cancer: Gc proliferate and differentiate normally during fetal life and in the postnatal period to facilitate normal spermatogenesis in adulthood. However, Gc differentiation in perinatal life and the regulatory processes involved are poorly understood. Based on rodent studies, there are three important steps of this process (as discussed earlier in the text see Biology of Spermatogenesis section). At first, the fetal Gc lose their pluripotency, by expressing a number of *differentiating markers* shared with embryonic stem cells (e.g., *OCT-4).* Failure to execute this differentiation step results in persistence of pluripotency characteristics, which is dangerous since such cells can potentially differentiate into various tissues. Such a failure is thought to be the underlying mechanism via which *carcinoma-in situ (CIS)* cells are formed in boys, and it is from these CIS cells that *testicular germ cell cancer (TGCC)* subsequently develops in young adulthood. Incidence of TGCC has increased progressively in many western nations, predominantly North European countries over the past 60 years. Among Caucasians it is the commonest cancer in young men.

(b) Cryptorchidism and Hypospadias: Testicular descent into the scrotum normally occurs by birth in boys and failure of testicular descent (*cryptorchidism*), especially when this extends into puberty and adulthood, results in absence of spermatogenesis. The testes descend into the scrotum in order to keep their temperature about 3–4°C below core body temperature, a requirement for efficient spermatogenesis. Heat exposure causes *hypoxia* and *oxidative stress* responses in the Gc, manifest as increased expression of *hypoxia inducible factor 1α, haemoxygenase 1, glutathione peroxidase 1* and *glutathione-S-transferase-α,* which induce Gc apoptosis. Studies with lorry and taxi drivers, who remain seated for a long time limiting temperature regulation of the scrotum, have also produced evidence of poor semen quality. Cryptorchidism (undescended testis) and *hypospadias* (birth defect of urethra), are also associated with increased risk of TGCC.

(c) Effects of maternal lifestyle/environment on son's testes: The incidence of female obesity has increased considerably, and one preliminary study has suggested that high maternal Body mass index (BMI) negatively affects semen quality of her son(s) when they grow to adulthood, with indirect evidence pointing towards reduced Sc number. Exposure to a wide range of *environmental chemicals (ECs)* has increased, several of which are clearly documented to have intrinsic *endocrine-disrupting activity,* in particular *anti-androgenic activity.* Environmental anti-androgens acting as endocrine disrupters also have potential adverse effect on male reproductive health. For example, exposure of a range of persistent ECs (e.g., PCBs, polybrominated compounds, DDT or certain phthalates) present in cosmetics/toiletries/medications during pregnancy are known to increase the risk of TDS disorders human male offspring in adulthood.

C. Precocious Male Puberty and Testotoxicosis

Testotoxicosis is a form of gonadotropin-independent precocious puberty in which boys experience early onset and progression of puberty. Patients have accelerated growth, early development of secondary sexual characteristics and usually reduced adult height. Testotoxicosis is caused by an *activating mutation of the LH-R,* leading to increased levels of sex steroids in the face of low LH. Therapy for this has, therefore, traditionally targeted steroidogenesis. However, the drugs used have been associated with side effects. More recently, a combination of an oral *anti-androgen spironolactone* and an *aromatase inhibitor testolactone* decreased height velocity and improved predicted height. A phase II study in testotoxicosis is currently underway, exploring the combination of a highly selective *anti-androgen, bicalutamide,* and the potent *aromatase inhibitor, anastrozole.* These agents are well tolerated in the populations in which they have been studied and effectively inhibit testosterone activity and estrogen production, in adult patients.

D.   Male Erectile Dysfunction

Erectile dysfunction is defined as "the inability to achieve or maintain an erection sufficient for sexual intercourse". This is one of the most common psychosocially taxing sexual dysfunctions in men. Physiology, aetiology and therapy of this disorder is mentioned below:

(a) Prevalence*:*   Erectile dysfunction is common between 40–70 years of age in human males. Fifty two per cent of men experience some degree of impotency—mild in 17.1%, moderate in 25.2%, and complete in 9.6%. Complete impotency was reported by 5% of men at 40 years of age and 15% at 70 years of age. However, a higher prevalence of complete impotence was seen in men with concomitant illnesses. Erectile dysfunction is more common with advancing age, and since the aged population will increase, its prevalence will continue to rise.

(b) Aetiology: Normal erectile function requires the coordination of psychological, hormonal, neurological, vascular, and cavernosal factors. Alteration in any one of these factors is sufficient to cause erectile dysfunction. Not uncommonly, a combination of factors is involved.

Hormonal factors: The role of T in erectile dysfunction is not clear. Some men continue to achieve erection even after castration. The fall in free serum testosterone and increases in concentrations of sex hormone binding globulin with aging may be associated with loss of libido and reduced frequency of erection, but restoration of normal testosterone concentrations does not usually improve sexual function. Patients with *hyperprolactinaemia,* frequently associated with low T values, can develop low libido and erectile dysfunction by unknown mechanisms. T replacement treatment, without correction of concurrent hyperprolactinaemia, does not resolve erectile dysfunction associated with hyperprolactinaemia.

Local conditions: Poor blood supply as a result of congenital malformations or trauma is a less common cause of erectile dysfunction that can affect the young male. Peyronie's disease is a specific condition of the penis in which the development of fibrous plaques in the tunica albuginea, sometimes extending into the erectile tissue, may cause pain (in the early inflammatory stage) and penile deviation, making coitus impossible (Figure 92.5). Inability to retain pressurized blood in the corpus cavernosum follows disruption of the veno-occlusive mechanism, which can be caused by Peyronie's disease or appear due to the result of trauma or surgery.

Chronic systemic illness: Diabetes mellitus, heart disease, and hypertension are all commonly associated with erectile dysfunction.

Drug induced erectile dysfunction: Around 25% of erectile failure seen in clinic is caused by medication. Erectile dysfunction may affect 10–20% of patients taking *thiazide diuretics,* and to a lesser extent to patients who are using β *blocking drugs.* This may be a result of reduced perfusion pressure, as blood pressure falls in response to the medication, or probably a direct (but unknown) effect on smooth muscle. Further support for this mechanism comes from the observation that treatment of hypertension with the α adrenergic receptor blockers is not associated with erectile failure. Erectile dysfunction commonly complicates antidepressant treatment with both *monoamine oxidase inhibitors* and *tricyclic antidepressants. Benzodiazepines* and selective serotonin reuptake inhibitors have been reported to cause erectile failure, decreased libido, or ejaculatory problems. *Cimetidine, digoxin,* and *metoclopramide* cause erectile dysfunction, as do anabolic

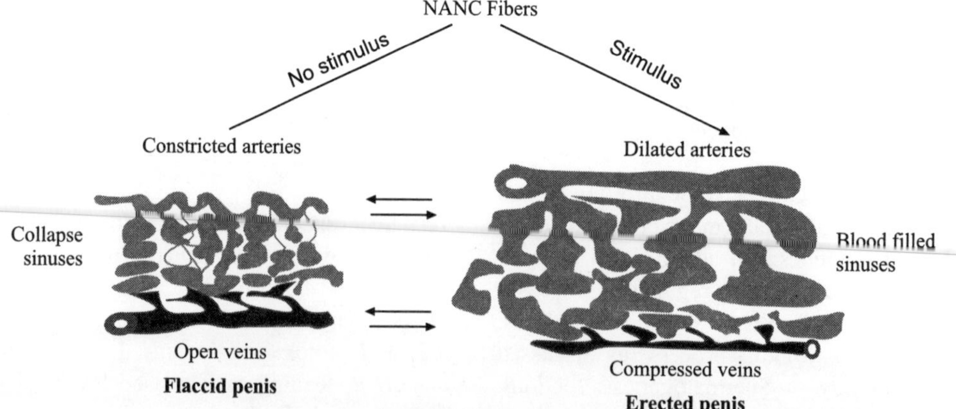

Fgirue 92.5   Major events involved in mediating penile erection. Following sexual arousal, non-adrenergic non-cholinergic (NANC) fibres release Nitrous Oxide (NO) in both pudendal arteries and cavernosal smooth muscle. NO is also released from the vascular endothelium of these tissues. NO elicits relaxation of the arteries and cavernosum, which results in a greater inflow of blood relative to outflow. As a consequence, the subtunical venules are compressed against the tunica albuginea resulting in veno-occlusion and a rise in intracavernosal pressure and erection ensues. Adapted from, Jeremy et al. 2007. *International J. of Impotence Research* 19, 265–280.

steroids, either through a direct effect on penile tissues or through suppression of normal androgen production. Up to 75% of alcoholic patients have erectile dysfunction.

Psychogenic causes: In young men, psychogenic influences are the most likely causes of intermittent erectile failure. Anxiety about "performance" may result in inhibitory sympathetic nervous system activity, and anticipatory anxiety can make the condition self-perpetuating.

(c) Treatments: *Psychosexual counseling*—Patients who have a sizeable psychogenic component may be helped by psychosexual counseling. Since the recognition that an organic element is present in most patients, this approach is increasingly being used in conjunction with pharmacological treatment.

Mechanical devices: The simplest and least expensive treatment is a vacuum constriction device. Air is pumped out of the cylinder with the hand held pump to create a vacuum and cause an erection.

Hormonal therapy: Testosterone may improve erectile dysfunction in some patients with diagnosed hypogonadism. T should not be used in eugonadal men with erectile dysfunction as it may enhance prostatic hyperplasia or promote the growth of occult prostate cancer.

Vascular surgery: Arterial reconstructive surgery is sometimes indicated in men with arterial occlusive disease, but careful selection of patients is required. The best results are obtained in young patients with isolated arterial lesions following trauma. Venous surgery, with extensive ligation of the veins that drain the corpora cavernosa, is sometimes used as the last resort before the implantation of a penile prosthesis in young men with veno-occlusive disease. The results are generally poor as only 30% of patients report long term improvement.

Drug treatment: Drugs that are currently available have limited effectiveness. *Trazodone*, given as a single agent, has been effective in some studies, but not others. Side effects, such as sleepiness and gastrointestinal discomfort are common and limit its use. Trazodone combined with Yohimbine has been recommended, although little scientific proof of a synergistic effect exists. Yohimbine is the main alkaloid from the bark of the tree called *Coryanthe johimbe* K. Schum (yohimbehe tree), particularly growing in Central Africa.

Other therapies: Other therapies include agents acting locally in the corpora cavernosa and oral drugs. Future oral therapies include (i) phosphodiesterase (PDE) inhibitors, (ii) dopaminergic agents, (iii) *α-Adrenergic* blockers and (iv) neuropeptides.

(i) *PDE inhibitors:* PDE inhibitors are a group of new therapeutic agents that block cGMP degradation via PDE inhibition and by this mechanism enhance the effect of nitric oxide (NO) in inducing smooth muscle relaxation and erection. Five different PDE isoforms (PDE1–PDE5) have been described and found to be present at various concentrations in human tissues. The main PDE activity in human corpus cavernosum is due to PDE5, with PDE2 and PDE3 also identified. The observation of specific PDE isoform distribution inside human corpora cavernosa led to the study of a possible use of a potent and selective inhibitor of the cGMP–PDE5 in the treatment of male erectile dysfunction. The most common adverse events reported were headache, flushing, visual disturbances, dyspepsia and muscle aches. Further studies on the pharmacokinetics and safety of the drug are in progress. Commonly known PDE inhibitors are Sildenafil (Vaigra^R), tadalafil, and Vardenafil. Until the launch of sildenafil and the other PDE5 inhibitors, Yohimbine was the most prescribed substance worldwide for the treatment of ED.

(ii) *Dopaminergic agents:* The aphrodisiac effects of apomorphine (APO) alkaloids, central dopaminergic-receptor agonists derived from water lily tubers, have been known for several centuries. Clinical trials are in progress with sublingual tablets containing 2, 4 and 6 mg APO. Preliminary results of these studies performed in men with non-organic impotence showed satisfactory erections in up to 70% of patients in home use. However, adverse events were reported by >50% of patients,, i.e., nausea, yawning and drowsiness.

(iii) *α-Adrenergic antagonists:* The *α*-adrenergic receptor antagonist *phentolamine* has been widely used for intracavernous vasoactive injection therapy and it has been suggested that its systemic administration may exert both central and peripheral erectogenic effects. Phentolamine at a dose of 40–60 mg administered buccally before sexual intercourse resulted in satisfactory erections in ~50% of subjects with organic impotence without producing any important side-effects. Clinical trials with sublingual phentolamine are in progress.

E. Male Infertility

About 61 countries are facing major fertility related problems. About 15% of couples worldwide are childless due to infertility. Approximately 50% of such situation is because of the male factor and of which half is idiopathic (causes unknown) in nature. The cornerstone of diagnosis for male infertility is the routine semen analysis in which sperm concentration, motility, morphology and the presence of other cells (spermatogenic and white blood cells) are assessed, along with indicators of the patency and function of the genital tract (volume, liquefaction time, pH and concentration of fructose which is product of seminal vesicles). This analysis, however, is not a test of fertility potential but indicates infertility only in the case of azoospermia (i.e., absence of sperm in the ejaculate); analysis of accessory sex gland products (like fructose) in semen is done to check whether there is any obstruction in the duct(s). The majority of endocrine defects are relatively simple to diagnose and treat because of the availability of tests to diagnose and hormones/drugs to treat such defects. However, it is noteworthy that in certain patients suffering from male infertility, hormonal therapy fails to initiate spermatogenesis.

In past decade, studies of functional genomics using several transgenic mouse models have identified some proven candidate male infertility associated genes. Due to lack of a rigorous clinical correlation and evaluation, these etiologies are not established properly in clinical situation. Hence, it is crucial to evaluate molecular mechanism related to intracellular (within the cell) and intercellular (between 2 different cells) events within testis influencing spermatogenesis, specifically within seminiferous epithelium, for divulging hormone-independent causes of infertility residing within the testis.

CONCLUDING REMARKS

The field of male gamete biology has significantly progressed in the last two decades starting from I) transplantation of male germinal stem cells from infertile males to donors of different age or species (rodents, large animals and primates) to restore fertility, II) standardization and characterization of *in vitro* Ssc cultures and III) *in vitro* generation of mammalian spermatozoa. Although basic knowledge about governance of spermatogenesis is very well understood with reference to endocrine regulation of this system, there are several local cell-cell interactions in the testes important for regulation of functional sperm production by testes using paracrine and autocrine actions. It is now possible to explore greater details of these phenomena with the help of genetic information obtained from purified cell cultures of each kind from the testes. Use of recently developed techniques of high throughput DNA microarray and functional genomics, involving genetically modified transgenic animal models, may substantially help in achieving this goal. Identification of important genes using these techniques may explain basis of hormone independent regulation of normal sperm production and will provide strong basis for better diagnosis and treatment of male infertility in near future.

REFERENCES AND SUGGESTED READINGS

1. Sharpe R.M. (1994), in Knobil E. and Neil J.D., *The Physiology of Reproduction,* New York, NY: Raven Press: 1363–1434.
2. Plant T.M. (2008), *J. Neuroendocrinol*, 20(6):719–26.
3. De Gendt and Verhoeven G. (2012), *Mol. Cel. Endocrinology*, 352:13–25.
4. Allan C.M. and Handelsman D.J. (2005), *Endocrine*, 26:235–239.
5. Kumar T.R. (2005), *Reproduction*, 130:293–302.
6. Huhtaniemi I.T. and Themmen A.P. (2005), *Endocrine.*, 26:207–217.
7. Behre H.M. et al., Endotext. Com, Endocrinology of Male Reproduction, Chapter 6. http://www.endotext.org/male/male6/maleframe6.htm
8. Reiter E.O. and E. Norjavaara (2005), *PediatrEndocrinol Rev.*, 3(2):77–86.
9. Wagner G. and De Tejada (1998), *BMJ*, 36:678–685.
10. Matzuk M.M. and Lamb D.J. (2008), *Nat. Med.* 14:1197–1213.
11. Sharpe R.M. (2010), *Philos Trans R SocLond B Biol Sci.*, 27;365(1546):1697–712.
12. Porst H., et al. (2013), *J. Sex.* Med. 10:130–171.

93

Androgens
Biosynthesis and its Regulation, Source and Blood Levels Metabolism, Biological Actions and Clinical Aspects

Vijayakumar Govindaraj and A.J. Rao

CONTENTS

I. INTRODUCTION

Androgens are a group of steroid hormones which have been found in all vertebrates studied so far and are necessary for spermatogenesis and maintenance of secondary sexual characters in the male. The term testosterone was coined in 1935 when Ernest Liqueur isolated it from bull testes. The testicular transplantation experiments by John Hunter in 1786 were the beginning of these studies and testosterone was chemically synthesized independently in 1935 by Leoplold Ruzicka and Adolf Butenandt. It is synthesized mainly in the testes in the males and in the ovaries in the females. It is also synthesized to some extent by the adrenal and small quantities are synthesized in the brain also.

II. BIOSYNTHESIS OF TESTOSTERONE

In the males, the site of synthesis of testosterone is testes and the Leydig cells which are located in the interstitium of the testes are the main site of synthesis of testosterone. The precursor for the steroid hormone synthesis is cholesterol that is synthesized from acetate from which several intermediates are synthesized, catalyzed by specific enzymes (Figure 93.1). The important intermediate in the synthesis of cholesterol is isoprenoid unit. A series of enzymes convert the side chain at C-17 of the cholesterol precursor to a hydroxyl group, transfer double bond from C-6 to C-4 and oxidize the hydroxyl group at C-3 to a carbonyl group. The general pathways for synthesis of testosterone both in the male and female and in the adrenal are more or less same although they differ significantly in quantitative aspects. The plasma testosterone concentration in the adult human male plasma is about 7 to 8 times more than the concentration in the adult human female plasma. By the action of cytochrome P-450 and dehydorgenase dependent enzymes, the various androgens are synthesized. The first intermediate in the pathway is 20, 22 hydroxy cholesterol. This is acted upon by desmolase which is the side chain cleavage enzyme resulting in the formation of pregnenolone. Thus, from cholesterol which is C-27 compound, pregnenolone, a C-21 compound is formed. From this intermediate, testosterone is synthesized by two pathways which are Δ^5, 3β hydroxyl pathway and Δ^4, ketosteroid pathway. In the rodents the Δ^4 pathway predominates and the C-21 pregnenolone is converted to progesterone by 3β-hydroxysteroid dehydrogenase/Δ^5-Δ^4-isomerase (3β-HSD). 3β-HSD belongs to a group of short chain dehydrogenases of which type II human enzyme is specific for the gonad. The intermediates in the Δ^5 pathway are 17-hydroxy pregnenolone, dehydroepiandrosterone (DHEA) and androstenediol. And the Δ^5 pathway is the predominant pathway in the humans. In the Δ^4 pathway the intermediates are 17-hydroxy progesterone and androstenedione and testosterone. All the intermediates in the Δ^5 pathway can be converted to testosterone by 3β-OHSD. One of the intermediates in the Δ^5 pathway namely dehydro epiandrosterone is produced in large quantities by adrenal and this can be converted in to androstenediol by 17β-hydroxysterioid dehydrogenase (17β-OHSD) or into androstenedione by 3β-Hydroxy steroid dehydrogenase (3β-OHSD) which again can be converted into testosterone by 17β-OHSD. This reaction is reversible under *in-vitro* conditions However, the reaction is generally unidirectional under *in-vivo* conditions. 3β-HSD can convert androstenediol into testosterone. Thus, in both the pathways the C-21 precursors are converted to C-19 testosterone by the appropriate enzymes.

III. REGULATION OF ANDROGEN SYNTHESIS AND SECRETION

The regulation of synthesis of testosterone by the Leydig cells in the testis is primarily under the control of pituitary Luteinizing hormone (LH) which in turn is regulated by Gonadotropin-releasing hormone (GnRH) produced in the hypothalamus and secreted into the hyphothalamo-hypophyseal portal system. The synthesis and secretion of GnRH and LH in turn are under the feed back control of testosterone. High levels of circulating testosterone inhibit the secretion of GnRH and LH. And low levels of testosterone are associated with a compensatory rise in the secretion of LH. Administration of estrogen and progesterone also cause a fall in plasma levels of LH and consequently testosterone. Although these effects of estrogen are mediated via the hypothalamo-pituitary-axis (HPA), there is evidence to indicate that estrogen has a direct inhibitory effect on testosterone production by Leydig cells under *in-vitro* condition. The receptors for the neuroendocrine feedback are in the hypothalamus and these are susceptible to influences that may stimulate or inhibit the synthesis and secretion of GnRH. The secretion of GnRH is pulsaltile and every increase in LH is preceded by a pulse of GnRH, although every pulse of GnRH need not necessarily result in an increase in LH. Understandably these sequences of events regulate the synthesis and secretion of testosterone. Administration of antiestrogen (Clomiphene Citrate) causes a rise in plasma levels of LH and testosterone, presumably through stimulation of GnRH. This is a good test for hypothalamo-pituitary-gonadal function. Androgen secretion from the ovary is also under the control of pituitary LH.

LH acts by binding to a specific G-protein coupled seven trans-membrane receptor. This binding results in activation of signaling cascade which in turn results in stimulation of adenyl cyclase which results in increase of cAMP (3'-5'-cyclic adenosine monophosphate) levels and activate cAMP dependent protein kinase (PKA). Activation of PKA results in two pathways in stimulation of testosterone synthesis by LH. One occurs within minutes of addition LH and is the acute response which is a result of increased cholesterol transport into the mitochondria. The source of cholesterol needed for synthesis of testosterone in the adrenal and gonad varies depending on the species in question. While both low density lipoprotein (LDL) and high density lipoprotein (HDL) can serve as source of cholesterol as precursor for testosterone synthesis, in humans LDL is the preferred source of cholesterol and in rodents HDL is the preferred source. Recent studies have clearly established that the rate limiting step in steroid production is the delivery of cholesterol to the P450scc complex into the mitochondria. This transport of cholesterol into mitochondria is regulated by the Steroidogenic Acute Regulatory Protein (StAR). The synthesis of StAR is stimulated by LH in Leydig cells which in turn results in an acute increase in testosterone production within minutes. The importance of StAR in steroidogenesis became evident by the clinical disorder congenital adrenal hyperplasia in which mutations in StAR gene lead to the failure of the resulting defective protein to transport and deliver cholesterol to P450Scc, resulting in decreased steroidogenesis in the adrenal and gonads.

In contrast to the acute stimulation of testosterone production in Leydig cells by LH, the chronic response needs

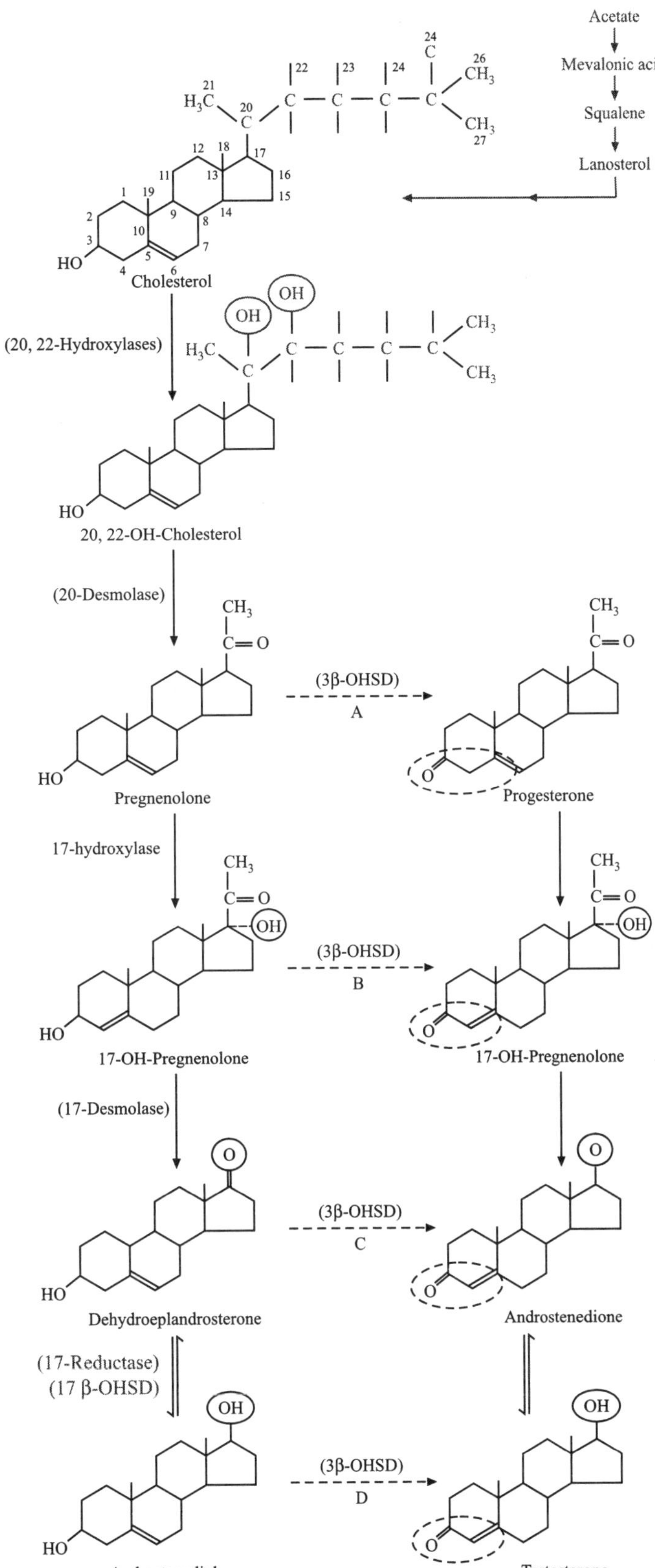

Figure 93.1   The major pathways of androgen biosynthesis are summarized. The D5, 3β-hydroxyl pathway is shown on the left and the D4, ketosteroid pathway on the right. Synthesis of androstenedione and testosterone from C21 steroid precursors along routes A and B, and from C19 steroid precursors along routes C and D. The enzymes taking part at different steps are shown in parentheses, and the sites of changes in the steroid molecule are indicated by encircling—dotted line for horizontal steps and the other vertical steps. (3β-OHSD indicates 3beta-hydroxy steroid dehydrogenase).

several hours and involves transcriptional activation of genes which code for the genes for the enzymes involved the synthesis of testosterone. These are P450scc and P450c17, 3β-HSD and 17β-HSD and the increase in these enzymes involves both transcriptional and post transcriptional mechanisms. It is to be noted that most of the information on the role and mechanism of LH action in stimulation of testosterone production in Leydig cells has been obtained from *in-vitro* studies using purified Leydig cells from mice or rats as well as MA-10 mouse Leydig tumor cells. While these studies clearly established the importance of LH in testosterone production by Leydig cells, several other reports are available which also suggest a role for Follicle-stimulating hormone (FSH), Prolactin, Growth Hormones, Arginine Vasopressin and Interleukins, in regulation of testosterone synthesis by Leydig cells. Although initial studies with purified preparations of FSH indicated that FSH addition resulted in an increase in testosterone production by Leydig cells, the use of recombinant preparation of FSH and FSH preparations from which LH contamination was removed by passing through LH antibody column have clearly established that FSH has no direct role in stimulation of testosterone synthesis. While prolactin receptors have been detected on Leydig cells, its role remains unclear. PRL (Prolactin) receptor knockout mice have normal circulating levels of testosterone and fertility is not affected in the animals. However, clinical studies with both male and female hyperprolactinemic subjects clearly suggest a role for prolactin in regulation of reproduction although the exact mechanism by which PRL affects fertility is still not clear. Both clinical and rodent studies suggest minor role for Growth Hormone (GH) in testosterone production by Leydig cells and in Leydig cells *in-vitro* addition of LH results in an increase in IGF which is known to be increased in response GH addition.

Administration of human chorionic gonadotropin (hCG) which has LH like activity to normal hyphophysectomized and prepubertal males produces a prompt rise in plasma testosterone. hCG administration to normal females also increases the urinary excretion of DHEA. There is evidence to suggest that the androgen secretion from human fetal testis is stimulated mainly by maternal hCG. Clinical data which report inactivating mutations of LH receptor which produce XY infants with Leydig cell agenesis and female external genetalia support such suggestion.

IV. SOURCE AND BLOOD LEVELS OF ANDROGENS

A. In Fetal Life, Neonatal Period, Infancy and Childhood

Fetal testis starts secreting potent androgen probably by the ninth week of fetal life. The Leydig cells begin to proliferate at about the eighth fetal week, and the peak level of growth of these cells as well as their capacity to convert steroid precursors to testosterone have been demonstrated in a fetus of 12 to 13 weeks. During fetal life there is a relative deficiency of 3β-hydroxy-dehydrogenase activity in the adrenal cortex. As a result, a large amount of dehydroepiandrosterone, either free or sulphate-conjugated, is secreted by the fetal adrenal cortex and much of it is converted in the fetal liver to 16 alpha-hydroxy-dehydroepiandrosterone sulphate (16α-OH-DHEAS). The latter compound is transported to the placenta and serves as the major precursor for estriol synthesis in the placenta. In view of the enzyme deficiency, it is unlikely that the fetal adrenal cortex normally secretes either androstenedione or testosterone in significant amounts, although under abnormal circumstances it may do so.

In the immediate neonatal period, the plasma level of testosterone is elevated in both sexes and there is no significant sex difference. After birth, the plasma level falls quickly to values that are at or below the normal female range, and remain at this level until puberty. Interestingly, in childhood the plasma androstenedione levels are same as that in the adult male, and the ratio of androstenedione: testosterone is about two. In a male infant, testosterone level is less 30ng/dL, while in boys aged 14-15 yrs testosterone levels are 8-53 ng/dl.

B. In Adolescence and Puberty

As puberty approaches in the male there is a gradual increase in the plasma levels of LH and testosterone. Measurable changes occur many months prior to physical changes. Demonstration of high levels of testosterone in a boy with delayed pubescence is an assurance that puberty will eventually set in.

C. In Adult Life

The testis secretes mainly testosterone and small amounts of androstenedione and DHEA. The major androgen secreted by the ovary is androstenedione, and small amounts of DHEA and testosterone are also produced. The major androgen secreted by the adrenal cortex is DHEA, and small amounts of androstenedione and 11β-OH-andro-stenedione are also produced by this gland.

In adult males, more than 95% of the testosterone in the blood is secreted as such by the testis, and the remaining 5% is derived from peripheral conversion of androstenedione. The consensus is that the Leydig cells are the major source of androgens in the testis. The peripheral plasma level of testosterone in adult males is between 3.5 to 10.5 μg/L.

In normal females, about 40% of testosterone is secreted directly from the ovary, and about 60% of circulating testosterone is derived from peripheral conversion of dehydro-epiandrosterone and androstenedione contributed mainly by the adrenal cortex and to a small extent by the ovary. The plasma level of testosterone in adult females is about ten times lower than that in adult males, while in female infants testosterone is less than 10 ng/dL. In blood, testosterone is bound to a specific plasma protein namely; Sex hormone binding globulin (SHBG) and only approximately 2% of testosterone is unbound and available as free testosterone for biological action. The level of SHBG in plasma is a strong indicator of the total testosterone concentration among normal men. Consequently the factors that influence the levels of SHBG levels which, for example, are low in obese men and increased in hyperthyroid subjects. Testosterone can also be measured in saliva and salivary concentrations of testosterone are correlated with free testosterone in serum.

D. In Old Age

There is no abrupt fall in the plasma levels of testosterone in old age, although low values may be encountered in some men in their seventies and above. At the age of 80, the levels of testosterone in men may account for only 20% of what it was in the age of 20. This decline in testosterone takes place slowly, begin as early as his mid-30s, and can lead to higher risk for developing obesity, diabetes, and heart disease.

The deficiency of testosterone can also develop number of severe symptoms, including decrease in stamina and muscle mass, low libido, depression, anxiety, and reduced cognitive functions. The changes due to testosterone deficiency syndrome (Andropause) in males are comparable of female menopause. Contrary to menopasue, symptoms of andropause appear over a longer period of time and are frequently unnoticed for a short time or are simply blamed on "getting older."

Several symptoms and signs of testosterone deficiency can be reversed by restoring the testosterone levels by a number of methods. The use of testosterone supplements and even testosterone itself cause men to feel more competent, virile and less depressed.

V. METABOLISM

Androgens circulate in blood, in bound as well as unbound forms; the unbound form is physiologically active. Although testosterone is bound to albumin, this binding is weak and non-specific compared to binding with β-globulin. The latter type of binding is more in the female and it is increased with estrogen therapy and in pregnancy.

The androgens are interconvertible to varying extents. The metabolism of testosterone is tissue specific and as such, the many of the products formed in different tissues may have a specific biological role. Tissues like the liver, skin and skeletal muscles are the sites of androgen inactivation and transformations. Although the liver is undoubtedly the principal organ responsible for metabolic removal of testosterone from the plasma, extra-hepatic metabolism of testosterone certainly occurs. However, the precise contribution by the skin, muscle and other tissues to testosterone clearance has not been determined. Although the blood levels of testosterone do not fall in old age, the rate of clearance of testosterone from plasma is low in older men, probably due to lower metabolism in extra-hepatic tissues which may utilize less testosterone with advancing age. Testosterone is metabolized into two biologically

active products dihydrotestosterone (DHT) and estradiol. Testosterone is converted into DHT by the enzyme steroid 5-alpha-reductase by the reduction of the double bond in ring A of testosterone. Other related steroids which have Δ^5, 3 keto structure such as progesterone, 17-alpha-hydroxyl-progesterone, androstendione and cortisol are also substrates for this enzyme. The presence of the enzyme has been demonstrated in testis, epididymis, ovary, adrenal, brain, kidney, prostate, liver and non genital skin although the distribution of the type I and type II 5-α reductase in these tissues varies. In prostate, the biologically active form is DHT and Fenestride which is an inhibitor of 5α-reductase is employed as therapeutic agent in benign prostatic hyperplasia (BPH) and testosterone dependent prostate cancer. About 5% of free dehydroepiandrosterone is covered into androstenedione and testosterone. Androstenedione and testosterone are interconvertible. About 60% of the daily production of testosterone in the human female is derived from peripheral conversion of androstenedione. On the other hand, about 30% of daily production of androstenedione arises from circulating testosterone. A small portion of circulating androstenedione arises from side chain cleavage of circulating C21 steroids. The microsomal enzyme aromatase P450 converts testosterone into estradiol. It has been purified and cloned from human placenta initially and subsequently from variety of sources. It is also expressed in Leydig cells of the adult testis as well as in brain. In testis, it is up-regulated by LH and in Sertoli cells from immature rat and non human primate and in granulose cells, aromatase is stimulated by FSH. Interestingly, the promoter sequences of the aromatase gene are tissue specific although the translated protein is same in all tissues. Recent studies have established that testosterone is converted to estradiol by brain aromatse. This conversion in the brain has been shown to mediate masculinization in the rodent and neonatal administration of estradiol to female rats on day 2 and 3 after birth results in permanent irreversible changes as adults such as loss of cyclicity and infertility.

Degradation of androgens occurs mainly in the liver, where, in the presence of 5α- and 5β-steroid reductases, testosterone and androstenedione are converted into androsterone and etiocholanolone either along '17-keto' or '17-hydroxyl' pathway (Figure 93.2). In the second pathway, these are converted first into 5α- and 5β-androstanediols, which in turn may undergo oxidation to androsterone and etiocholanolone. In the third pathway, androstenedione and testosterone are converted into estrone and estradiol respectively. Dehydroepiandrosterone sulphate may be converted into free dehydroepiandrosterone, and thence to andro-stenedione, which may be metabolized as above.

Conjugation of androgen metabolites with glucuronic or sulphuric acid occurs mainly in the liver and, to a certain extent, in the kidney. The rate of excretion of androgen glucuronides through the kidney is rapid, but that of sulphate esters is much slower. A portion of esterified androgen metabolites is excreted into the bile; a major part of these is, however, reabsorbed to undergo entero-hepatic circulation.

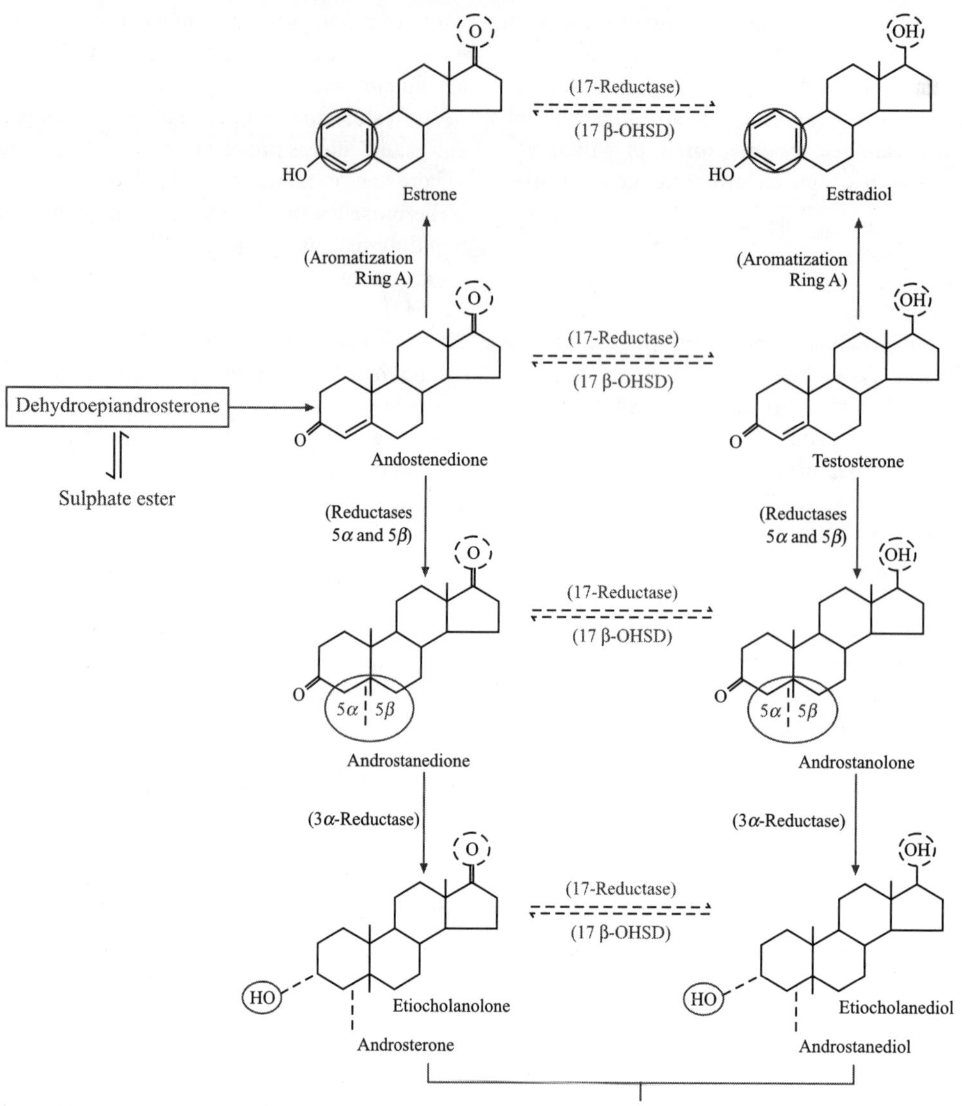

Figure 93.2   The major peripheral interconversions and metabolism of androgens are shown. The 17-keto pathway is shown on the left and the 17-hydroxyl pathway on the right. The enzymes taking part at different steps are shown in parentheses, and the sites of changes in the steroid molecule are indicated by encircling—dotted line for horizontal steps and the other for vertical steps. (17-OHSD indicates 17-hydroxy steroid dehydrogenase).

VI.   ASSAY OF BIOLOGICAL ACTIVITY OF ANDROGENS

The biological activity of androgens can be assayed by monitoring a variety of parameters using mice, rats or birds. Based on these studies the following methods are generally employed to assay the biological action of androgens, (1) Chick comb method; (2) Increase in the weight of seminal vesicles and prostate gland in immature or castrate rats; (3) Increase in the weight of the Levator ani muscle in castrated immature rats; (4) Increase in the beta glucuronidase activity in the kidney of mouse.

As judged by the findings of different bioassays employed, of the naturally occurring androgens, testosterone is the most potent one, followed by dihydrotestosterone (90%), androstanediol (60%), androstenedione (20%), dehydroepiandrosterone (10%) and androsterone in the order of potency. In contrast to assessing activity by injecting in to the experimental animal, oral administration can also be employed to assess the activity (Table 93.1).

It is evident that the androgens tested are more potent when the increase in weight of prostate is considered.

VII.   PHYSIOLOGICAL ACTIONS OF ANDROGENS

The actions of androgens are diverse and depending on the stage of development and sex, androgens exert different actions. Testosterone effects can also be classified by the age of usual occurrence. For postnatal effects in both males and females, these are mostly dependent on the levels and duration of circulating free testosterone.

TABLE 93.1 Comparison of the Oral Activity of Various Androgens (mg daily) in Castrated Rats administered 1 mg daily for 10 days

Substance*	Weight of seminal vesicles (mg)	Weight of prostate (mg)
Methyltestosterone	14	40
Testosterone	35	90
Testosterone propionate	20	57
Androsterone	15	50
Controls	12	42

* Compounds were dissolved in 50% ethyl alcohol.

The following is a brief summary of the physiological actions of testosterone in the male, although depending on the biological end point the potency of the particular androgen may vary. Androgens promote spermatogenensis directly, prolong the epidydimal sperm life; inhibit spermatogenensis by blocking LH and GnRH via pituitary and hypothalamus. Stimulates growth and secretory activity of seminal vesicle, prostate, coagulating bulbourethral and preputial glands, growth of penis and scrotum; stimulates secondary sexual characteristics such as distribution of body hair (beard and auxiliary hair), configuration of body, comb wattles spurs, feathers of birds, clasping pads of amphibian, dorsal spine of certain fishes, pitch of voice, behavior, sexual and aggressiveness and finally metabolism; promotes nitrogen retention, protein anabolic action, synthesis of certain enzymes; increased storage of creatine.

As mentioned above, the physiological action varies depending on the stage of development. The prenatal androgen effects such as genital virilization (midline fusion, phallic urethra, scrotal thinning and rugation, phallic enlargement), development of prostate and seminal vesicles occur during two different stages, between 4 and 6 weeks of the gestation. In the second trimester the androgen level are associated with gender identity; this period changes the femininization or masculinization of the fetus and can be a good indicator of feminine or mascular behaviors such as sex typed behavior than an adult's own levels. During pregnancy, the testosterone levels of a mother affects the behavior greater than the testosterone levels of own daughter.

A. Differentiation of Reproductive System in Embryonic Life

At the seventh week of intrauterine life the fetus is equipped with primordia of both male and female genital ducts. In the presence of functional testes the mullerian or female structures involute, while the Wolffian or male ducts, from which the epididymis, vas deferens and seminal vesicles are derived, complete their development. It is believed that all these changes are brought about by a non-steroidal "duct-organizing substance" (or male duct evocator), which is secreted by fetal testes but is distinct from ordinary androgen. Fetal testicular androgen may help in stabilizing the male duct development.

At the eighth week of fetal life the external genitalia of both sexes are identical and have the capacity to differentiate in either direction. As in the case of genital ducts there is an inherent tendency for the external genitalia to feminize and this requires no hormonal stimulant. Normal differentiation of these structures along male lines will occur only if androgenic stimulation is received during early fetal life, i.e., between 8th and 12th weeks, and this is normally provided by fetal testicular androgen. Androgenic hormones stimulate growth of the genital tubercle, and induce fusion of the urethral folds and labioscrotal swellings. Deficient androgen secretion from the testis or failure of target tissues to respond to androgenic stimulation would result in varying degrees of failure of masculinization producing ambiguity of external genitalia (Figures 93.3 and 93.4a). Conversely, if the female fetus is subjected *in utero,* to androgenic stimulation from any extragonadal source her external genitalia will show varying degrees of masculinization, ranging from simple clitoral hypertrophy to the formation of a normal looking penis (Figure 93.4b). Thus deficiency of androgen in the male and its excess in the female in early fetal life may lead to similar types of ambiguous external genitalia (Figure 93.4).

Early infancy effects

The androgen effects of early infancy are poorly understood. In male infants, the increase in testosterone levels during first weeks of life, and the levels are maintained in a pubertal range for a few months, but usually reach the barely detectable levels of childhood by 4–6 months of age. The purpose of this increase in humans is unexplained. It has been assumed that masculinization of brain is taking place as no significant changes have been noticed in other body tissues. Interestingly, the male brain is masculinized by testosterone, which move across the blood-brain barrier and enters into the male brain, converted into estrogen by the aromatase, whereas female fetuses have alpha-fetoprotein which binds the estrogen so that female brains are not affected.

During first six months after birth, the male primates including human exhibit very active testes and the levels of testosterone can reach the adult levels. This period is called 'the neonatal testosterone surge'. This type of surge does not occur in females. It has been demonstrated that milk made of soya beans, contain very high levels of chemicals called isoflavones, which are phytoestrogens (literally, plant-made chemicals that mimic the female sex hormone, oestradiol). It has been observed that in a study carried out with marmosets (which always delivers twins one of which served as unfed control and the other as soya milk fed group) twins fed with soy formula milk exhibited considerable inhibition of their 'neonatal testosterone surge' in contrast to their control twin brothers. This alteration was associated with a great increase in the numbers of Leydig cells in the testicles of the soy formula milk neonatal marmosets, which was surprising as these are the cells that make the testosterone. Indeed, there is every reason to believe that human male infants supplemented with soy formula milk will exhibit same suppression of their neonatal testosterone surge.

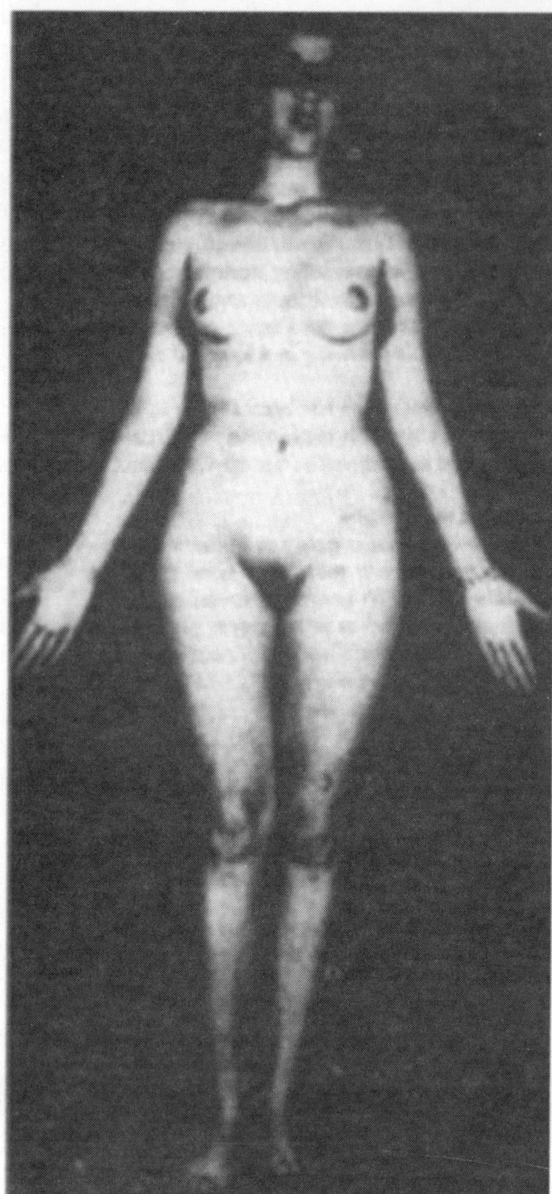

Figure 93.3 A case of testicular feminization syndrome. Because of the non-responsiveness of the end-organs, the external genitalia have not masculinized in spite of the normal levels of androgen secreted by the testis. The unopposed action of testicular estrogen has caused breast development.

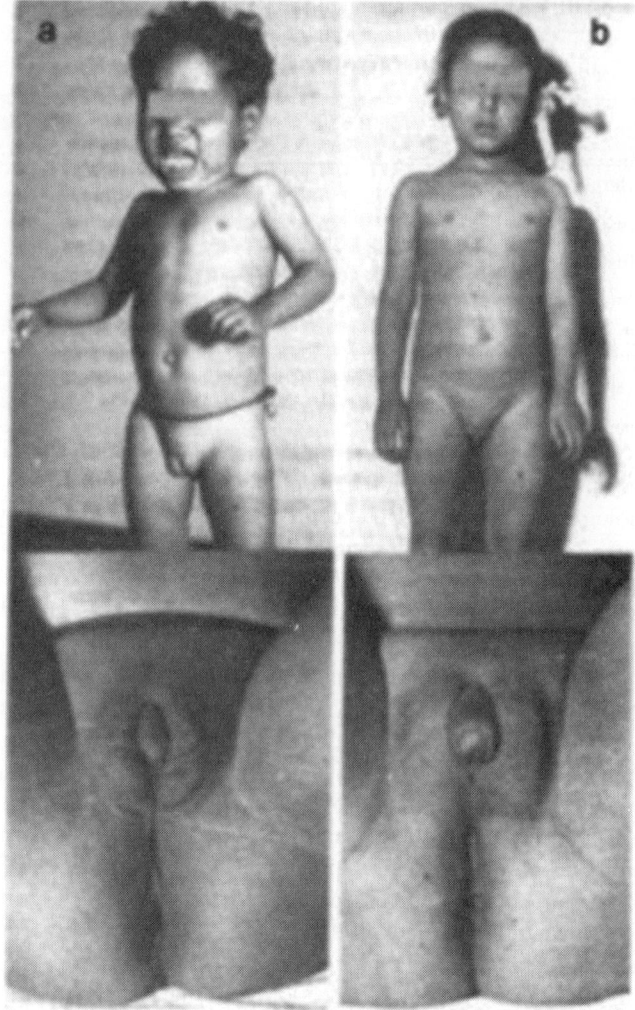

Figure 93.4 (a) A case of male pseudo-hermaphroditism with 46 XY chromosome complement. The ambiguous external genitalia are presumably due to lack of androgen action in early fetal life. (b) A case of female pseudo-hermaphroditism due to congenital adrenal hyperplasia, with 46 XX chromosome complement. The ambiguous external genitalia are due to the action of excess androgen secreted from the adrenal cortex due to enzyme defect in early fetal life. Note the similarity of genital abnormalities in two patients of opposite chromosomal and gonadal sex.

It has not been possible to give clear answer to the query whether this observed effect following feeding of soybean milk is of concern. The reason is that we do not know yet the function(s) of the neonatal surge is in boys, though effects on growth of the penis and prostate gland and effects on the immune system are suspected.

General effects during pre-peripubertal, pubertal and adult stages.

Pre-peripubertal effects are the first noticeable change of increasing androgen levels at the final stage of puberty, taking place in both boys and girls. It includes adult-type body odour, rise in oiliness of skin and hair, appearance of pubic, axillary hair, growth spurt, etc. During pubertal period, the testosterone levels increase sharply in the males, while in females the levels are low. Increased testosterone levels results in increased muscle strength and mass, deepening of voice, growth of the Adam's apple, growth of spermatogenic tissue in testicles, fertility, and completion of bone maturation and termination of growth. In males, these are usual late pubertal effects, and occur in women after extended periods of heightened levels of free testosterone in the blood.

During adult stages, testosterone plays a key role in sperm development and activates certain genes in Sertoli cells which help in differentiation of spermatogonia. Testosterone regulates acute hypothalamic–pituitary–adrenal axis activity, cognitive function and physical energy and maintenance of muscle mass. Testosterone regulates the population of thromboxane A2 receptors (TXA2) on megakaryocytes and platelets and hence platelet aggregation in humans. TXA2 is an important pathophysiologic mediator for cardiovascular diseases. The high androgen levels in both clinical populations and healthy women are associated with menstrual cycle irregularities. The effects of testosterone are more evident in adult males than in females, but are likely to be essential to both sexes and some of these effects may decline as testosterone levels decrease with of age.

B. Growth and Maintenance of Sexual Organs

The increase in size of the penis, scrotum and testis at the time of puberty is promoted by androgens (Figure 93.5). Accompanying the testicular growth the sack of the scrotum becomes elongated and pendulous, and rugal folds appear in the scrotal skin. The penis and scrotum become pigmented.

Androgen stimulates the growth of seminiferous tubules in the testis. It has been shown that testosterone alone can stimulate and maintain spermatogenesis in hypophysectomized animals, but under normal circumstances it probably assists the pituitary follicle stimulating hormone (FSH) in its action on the germinal epithelium.

Androgen stimulates the growth and secretory activities of the secondary sex organs such as the epididymis, vas deferens, prostate and seminal vesicles. The maturation of spermatozoa during their passage through the epididymis, where they acquire motility and fertilizing ability, is androgen-dependent. Varying degrees of breast development which is noted in a majority of boys during pubertal development may be partly due to androgenic stimulation in addition to the action of estrogen elaborated by the testis (Figure 93.5).

C. Changes in Non-sexual Tissues

1. **General growth:** There is a generalized effect on the growth process of many tissues as a result of androgenic stimulation. Due to somatic and general anabolic effects of androgen rapid growth of bone and muscle tissue occurs and there is a spurt in linear growth.

2. **Hair growth:** At puberty under the influence of androgen pubic hair appears and gradually grows upwards to form typical diamond shaped pattern of male escutcheon. Hairs develop on the lip and face, and the scalp hair line in the frontotemporal region

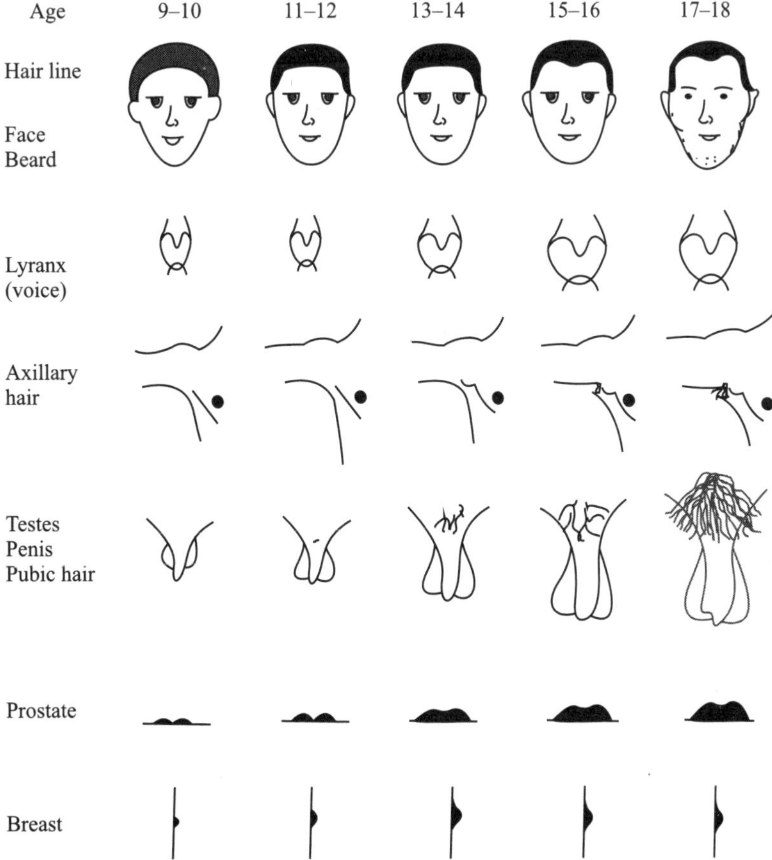

Figure 93.5 (Taken from Dorfman and Shipley, 1956). Note the androgen dependent pubertal changes.

undergoes regression (Figure 93.5). Hairs also appear in the axilla, on the body, extremity and anal region.

3. **Changes in skin:**   Secretion from axillary glands and sebaceous glands of the skin is increased. Oiliness increases and some degree of acne are noticeable in most boys. Skin pigmentation, especially manifestation of tanning effect of ultraviolet light, is intensified by androgen.

4. **Voice change:**   Androgen causes enlargement of the larynx, and thickening and enlargement of vocal cords resulting in a lowered pitch of voice.

5. **Others:**   Androgen has been shown to cause enlargement of kidney. It may cause an increase in the red cell, and haemoglobin, content of blood.

D.   Effects on Metabolism

The growth changes noted at puberty and the maintenance of muscle mass and bone tissue of adult males are due to potent anabolic action of the androgens. Retention of nitrogen, potassium, phosphorus, and calcium is a manifestation of this action. Increased protein synthesis and decreased amino acid catabolism are responsible for the nitrogen-retaining effect of testosterone. Androgen therapy decreases creatinuria probably indicating that more creatine is made available to ATP to form creatine phosphate, which plays an important role in the carbohydrate metabolism of muscle.

E.   Effects on Psychosexual Behaviour

Androgen produces a more aggressive attitude and stimulates libido and sexual potency. Psychosexual development is a very gradual process. Although it becomes accelerated at puberty it begins well before sexual maturity. The rate at which the physiological, emotional and genital manifestations of sex drive and behavior becomes fused into usual adult patterns depends largely upon social conditioning of the child. Androgen probably plays a very important role, but is not the sole determinant of sex drive and expression.

F.   Effects of Androgen in the Female

Androgen has specific effects on the ovary, uterus, vagina, oviduct, clitoris, and mammary gland. In pharmacological doses androgens can cause uterine stimulation, potentiate the action of gonadotrophin on the ovary, have a progesterone like action on the vagina, and produce clitoral enlargement. It can exert a stimulating action on the nipple, mammary duct and acini. In relatively larger doses it inhibits lactation. Normally, adrenal androgens cause pubic and axillary hair growth in women.

Although the influence of androgen on psycho-sexual behavior in women has not been studied extensively, it is believed that adrenal androgens normally maintains libido in human females. Virilizing tumors and androgen administration to women has been report to cause an increase in the libido, sensitivity of external genitalia and responsiveness to sexual stimulation, and to produce a more intense sexual gratification accompanying intercourse.

VIII.   CLINICAL ASPECTS

A.   Deficient Androgen Secretion

Clinically, androgen deficiency is encountered more often than the excess. Deficiency may arise from primary defect in the testis and Leydig cells, secondary defect in the testis or as a result of more generalized endocrine disorder. Primary defects are encountered following castration, and in cases of anorchia and bilateral cryptorchidism, in some cases of Klinefelter's syndrome, oligo- and azoospermia, and leprosy. Secondary defects may be associated with hypogonadotrophic hypogonadism, pan-hypopituitarism, cirrhosis of the liver, and haemochromatosis.

A clinical manifestation of androgen deficiency varies depending upon the degree of failure and the time of its onset. Following pre-pubertal castration in man, the secondary sex organs including the penis, scrotum, prostate, seminal vesicles and epididymis fail to develop. The pitch of the voice remains high, pubic and axillary hairs are sparse, no regression in frontotemporal hairline occurs, muscle strength may be significantly diminished, and the body growth may be eunuchoid. Varying degrees of these manifestations are observed in other forms of hypogonadism (Figure 93.6). Following post-pubertal castration varying degrees of regression in secondary sex characters may be noted. The masculine voice persists, there may be only slight decrease in penile dimension, and the facial hair diminishes, if at all, only very slowly over a period of years. On the other hand, the prostate glands regress very promptly and ejaculate decreases or disappear. Body weight is lost, and physical vigor and emotional stability may be impaired. Hot flushes may be experienced.

The influence of castration on psychosexual behavior in man, particularly in the post-pubertal period, does not provide any ready answer to the importance of androgens. The effects of pre-pubertal castration are perhaps more definite and extensive deviations from the normal man including failure of ejaculation are noted. In post-pubertal castrates the sexual behavior pattern varies extensively, some showing impotence and diminished libido, while others may show sexual responsiveness.

B.   Failure of End-organ Responsiveness to Androgen

In the testicular feminization syndrome, because of failure of end-organ responsiveness to androgen, the external genitalia fail to masculinize in spite of the presence of normal levels of testicular androgen. As a result, a genotypic male with testis is born as a phenotypic female (Figure 93.3). Since the testicular estrogen gets a chance to act unopposed, the breasts develop as in a female. In cases with incomplete failure of end-organ

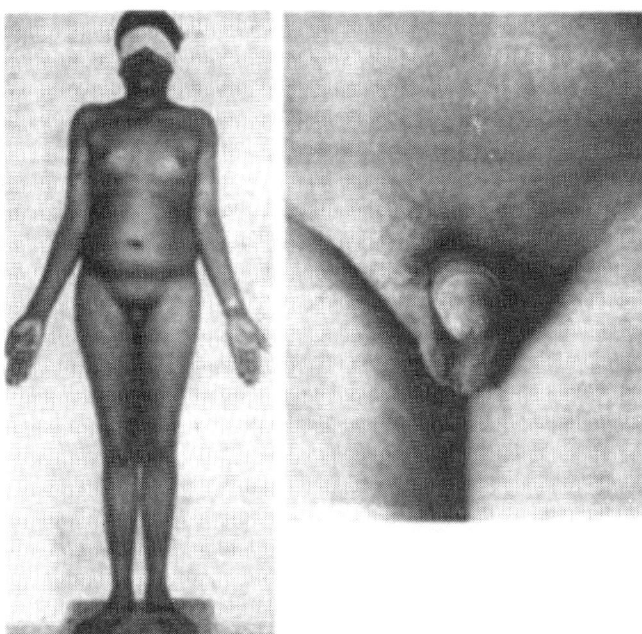

Figure 93.6 A case of Klinefelter's syndrome. Note the effects of androgen deficiency—eunochoid features, small penis and scrotum and sparse pubic hair.

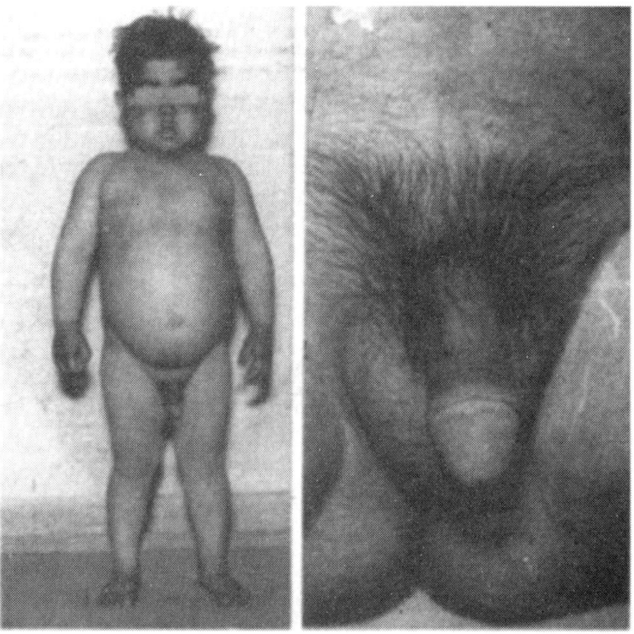

Figure 93.7 A male child aged about 5 years with Cushing's syndrome. Note the enlargement of phallas and growth of pubic hair.

responsiveness varying degrees of failure of masculinization occur giving rise to ambiguous external genitalia.

C. Excess Androgen Secretion

1. Pre-pubertal: Excess of androgen before puberty will cause isosexual precocity in males, and heterosexual precocity and virilization in females. In the male, this condition may result from increased secretion of LH arising from an idiopathic cause or due to hypothalamic disorder or a pineal tumor. In other cases excess androgen may be secreted by the interstitial cell tumor of the testis. In both sexes excess androgen may be secreted from the adrenal cortical tumor or hyperplasia, as is observed in virilizing tumor, Cushing's syndrome, and in congenital adrenal hyperplasia resulting from enzyme defect (Figures 93.7 and 93.8). In pre-pubertal girls excess androgen causes clitoral enlargement, development of pubic and axillary hair, deepening of the voice and premature advancement of epiphyseal development. Acne may appear and muscle may become prominent. Facial hair does not appear except in older children. In both sexes androgen excess causes a sudden spurt of linear growth, but the ultimate stature may be stunted due to premature closure of epiphyses resulting from continued exposure to androgen excess.

2. Post-pubertal: Pure androgen excess in a normal man may not produce any clinically recognizable manifestation. An increase in 17-ketosteroids and a decrease in spermatogenesis may be noted. Such androgen excess may arise from Leydig cell tumor of the testis or adrenal cortical disorders such as Cushing's syndrome, virilizing adrenal tumor and congenital adrenal hyperplasia.

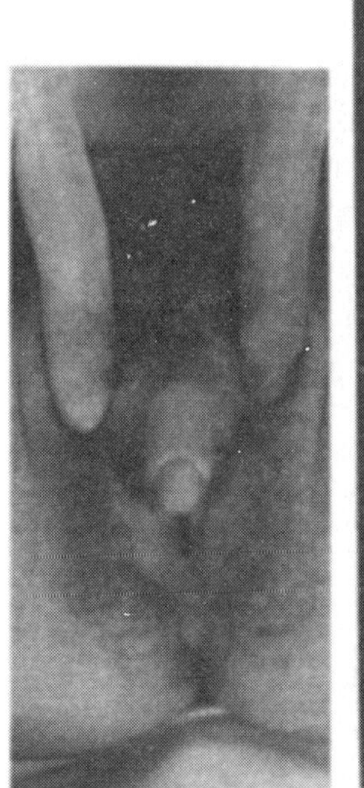

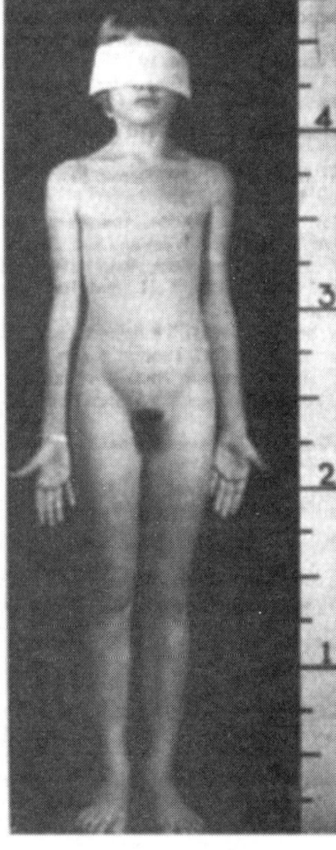

Figure 93.8 A female child aged 7 years with congenital adrenal hyperplasia. Note the spurt of growth in height, clitoral enlargement and pubic hair growth caused by excess androgen secreted by the adrenal cortex due to enzymatic defect.

In women with androgen excess, hirsutism is the earliest and most prominent symptom. Pubic hair becomes masculine, coarse hair develops on the face, extremities and the chest. With further androgenization signs of defeminization appear, the breasts shrink in size, body and face become angular because of the decrease in subcutaneous fat, menses decrease, ovulation may cease and uterus may regress. With further progression, signs of virilization appear—clitoris enlarges, acne may appear, voice deepens, body musculature may become prominent, and regression of frontal hair lines and even baldness may appear. Such androgen excess in women may result from ovarian disorders such as polycystic ovary syndrome, hyperthecosis of the ovary and virilizing tumors of the ovary, or due to adrenal cortical dysfunctions such as Cushing's syndrome, virilizing tumor of the adrenal cortex and variant of congenital adrenal hyperplasia.

D. Androgen Therapy

1. Hypogonadism: In conditions of androgen deficiency arising from either primary or secondary defects in the testis, androgen therapy is beneficial. As mentioned earlier, such conditions are encountered following castration, and in anorchia, eunuchoidism, hypogonadotrophic hypogonadism and panhypopituitarism.

2. Impotence: Both sex drive and erectile function of man are sustained by a combination of hormonal and psychic forces. Most often it is difficult to ascertain whether a hormonal deficiency exists in cases of impotency. A course of trial therapy with androgen is perhaps the best procedure in questionable cases. Failure to respond to an adequate dose of androgen given for 2 to 3 weeks rules out fairly well, androgen deficiency as the cause of the disorder. A positive response to such therapy, however, does not establish the diagnosis unless one excludes the effect of suggestion by appropriate placebo therapy.

3. Sterility: Although it has been demonstrated in animals that androgen has a direct stimulatory action on spermatogenesis, administration of androgen in therapeutic doses causes inhibition of FSH secretion thus causing further suppression of sperm production. Administration of relatively high doses of androgen for a rebound phenomenon has been found beneficial in certain oligospermic individuals. Another form of effective therapy is administration of androgen along with gonadotrophin.

4. Breast cancer: Since breast cancer in the female often retains the estrogen dependency of normal breast tissues, androgen has been used effectively to counteract estrogen action in some inoperable cases.

5. Menopause: Androgen has been used effectively along with estrogen for the relief of menopausal symptoms as well as for the treatment of post-menopausal osteoporosis.

6. Debility and tissue healing: Androgens and other anabolic steroids may prove beneficial during the period of recovery from debilitating conditions and during tissue healing, including fractures.

7. Osteoporosis: This condition, which is characterized by the reduction in bone mass, is found in Cushing's syndrome hypogonadism, rheumatoid arthritis, malnutrition, during corticoid therapy, and many other instances. Use of anabolic steroids along with high protein and high calcium diet may be beneficial.

8. Contraindications: Androgen therapy is contraindicated in cancer of the prostate and male breast.

IX. MEDICAL USES

The main use of testosterone is for the treatment of hypogonadal males who have insufficient or no natural endogenous testosterone production. The rationale for using hormone replacement therapy [Testosterone Replacement Therapy (TRT)], is to keep blood serum testosterone levels within the normal range. Apart from hormone replacement, like every hormone, the testosterone or other anabolic steroids has also been given for a variety of other conditions and purposes, with anecdotal success but with adverse side effects examples include reducing infertility, correcting lack of libido or erectile dysfunction, correcting osteoporosis, encouraging penile enlargement, encouraging height growth, encouraging bone marrow stimulation and reversing the effects of anemia, and even appetite stimulation. By the late 1940's testosterone was being sold as an anti-aging wonder drug. The decrease in testosterone production with age has led to many studies of androgen replacement therapy. Testosterone therapy has potential to cause virilizing effects. Testosterone is given to transsexual men as part of the HRT, to have normal male testosterone level. Similarly, transsexual women are commonly recommended anti-androgens to reduce testosterone levels in the body and allow for the effects of estrogen to develop.

Testosterone patches, a medicated patches contain testosterone are efficient in treating low libido or hypoactive sexual desire disorder *(HSDD)* in post-menopausal women. Low libido may appear as a symptom or consequence of hormonal contraceptive use. Testosterone therapies in postmenopausal women may be used for prevention of loss of bone density, maintenance of muscle mass, for alleviating certain depression, energy levels and to protect healthy body fat levels. Adverse effects of testosterone therapy in women may be protected by prevention of hair-reduction, acne vulgaris, etc. There is a theoretical concern about the risks of testosterone therapy like breast or gynecological cancers, and additional studies are needed to ensure any such risks more clearly.

A. Hormone Replacement Therapy (HRT)

The testosterone levels decrease with advancing age in human beings and this result in 'Andropause'. There is discrepancy about when to treat aging men with testosterone replacement therapy. The American Society of Andrology suggested that the testosterone replacement therapy in aging men is signified when both clinical symptoms and signs suggestive of androgen deficiency and reduced testosterone levels are present. The American Association of Clinical Endocrinologists defined hypogonadism as a free testosterone level that is below the lower limit of normal for young adult control subjects. Earlier, age related decrease in free testosterone was accepted as normal. But now, they are not considering as normal. Many men have normal sexual function even if their testosterone levels decrease into the age-adjusted lower normal range. Patients with average testosterone levels deserve a clinical trial of testosterone.

In spite of age or disease etiology, men with a total testosterone level less than approximately 300 ng/dL frequently develop signs and symptoms associated with classic hypogonadism, which can have consequences for their long term health. Notably, there is some variation in what is considered the threshold total testosterone level for indicating hypogonadism, with the Endocrine Society and the American Association of Clinical Endocrinologists (AACE) recognizing 200 ng/dL as the threshold and the US Food and Drug Administration (FDA) recognizing 300 ng/dL. There is not total agreement on the threshold of testosterone value below which a man would be considered hypogonadal. Currently there are no standards as to when to treat women. Identification of inadequate testosterone in an aging male by symptoms alone can be difficult.

The available options for the testosterone replacement are intramuscular injection, transdermal patches and gels, subcutaneous pellets, and oral therapy. Each option has side effects, and advantages and disadvantage. The effects of testosterone supplementation comprise small side effects such as acne and oily skin, and more important complications such as increased hematocrit which can require venipuncture in order to treat, worsening of sleep apnea and prostate cancer growth in individuals who have subjected to androgen deprivation. Additional adverse effect may be significant hair loss and/or thinning of the hair. This may be avoided with Finasteride, a synthetic type-2 5α-reductase inhibitor prevents conversion of testosterone to dihydrotestosterone (DHT). Finasteride is approved for the treatment of benign prostatic hyperplasia (BPH) and male pattern baldness (MPB).

B. Benefits and Adverse Effects

Observational and experimental data show that physiological replacement of testosterone results improvement in insulin resistance, obesity, dyslipidaemia and sexual dysfunction as well as improved quality of life. Testosterone therapy can improve quality of life in aging men because aging is accompanied by declining testosterone levels that may contribute to decreases in muscle mass, bone density, libido, stamina, and cognition. However, there are no long-term interventional studies to assess the effect of testosterone replacement on mortality in men with low testosterone levels. The exogenous testosterone

supplementation can have various side effects. Fluoxymesterone and methyltestosterone are synthetic derivatives of testosterone. Methyltestosterone and Fluoxymesterone are no longer prescribed by physicians given their poor safety record, and testosterone replacement in men does have a very good safety record as evidenced by over sixty years of medical use in hypogonadal men. A 2006 article in *'The Journal of Urology'* pointed out that Prostate cancer may evolve into clinically apparent within months to a few years after the commencement of testosterone treatment. Physicians suggesting testosterone supplementation and patients who receiving it should be aware of this risk, and prostate-specific antigen test and DRE (Digital Rectal Examination) should be carried out on a regular basis during the treatment.

C. Androgen Use and Abuse in Sports

The first documented use of testosterone by an athlete as a performance-enhancing compound in sport started in the 1950s by Russian weightlifters. Testosterone is considered to be a form of doping (Use of banned performance-enhancing drugs) in most competitive sports. The medical community debated that there was no evidence that testosterone improved athletic performance for many years. Bhasin and co-workers, in 1996, showed that supraphysiological doses of testosterone certainly increase the performance enhancing benefits, what the athletes had known for four decades. Anabolic steroids including testosterone have also been used to increase muscle development and stamina. Following continuous scandals and publicity in the 1980s, the use of drugs to enhance performance is considered unethical by most international sports organizations and especially the International Olympic Committee. Androgen abuse in elite sports is protected by the WADA (World Anti-Doping Agency) which standardizes sports doping tests internationally. The reasons for the ban are mainly the alleged health risks of performance-enhancing drugs, the equality of opportunity for athletes, and the exemplary effect of "clean" ("doping-free") sports for the public. Testosterone and other anabolic steroids were classified as "controlled substance" under United States Controlled Substance Act.

Various methods are used for identifying testosterone abuse; most of them are based on a urine test. The ratio of the concentration of testosterone to the concentration of epitestosterone (T/E ratio) as determined in urine by gas chromatography/mass spectrometry is the most commonly used method to confirm testosterone abuse by athletes. A T/E ratio higher than 6:1 has been considered as proof of abuse. The methods for detecting administration of exogenous testosterone depend on falsification of the normal hormone profile in the patient's urine. Several markers for detecting administration of exogenous testosterone, it has been observed that high-dose testosterone administration resulted in dose-dependent suppression of both serum and urinary LH since the secretion of testosterone is under the control of luteinizing hormone (LH). It has been suggested that the urinary T/LH ration more that 30 is considered for testosterone abuse. The technique widely used for the measurement of T/E ratios, is gas chromatography coupled to isotope ratio mass spectrometry (GC-IRMS). This technique measures the ratio of carbon-12 to carbon-13 isotope, because of the different pathways used in the preparation of the natural and synthetic forms of substance like Testosterone. Both increased T/E values and suppressed urinary LH are being used for detecting testosterone abuse.

SUMMARY

Androgens are a group of steroid hormones produced mainly by testis in the male. Of the several androgens produced in the male, testosterone is the most important and potent one, indispensable for maintenance of spermatogenesis and libido and other secondary sexual characters. A variety of assays are available to assess the biological activity of androgens. Testosterone is biosynthesized from the precursor cholesterol and LH is the main hormone involved in the regulation of testosterone production by Leydig cells. This chapter also describes the metabolism of testosterone and the main metabolites produced. The physiological actions of androgen vary depending on the stage of development and a variety of developmental defects and abnormalities are described following the deficiency or excess of androgens. In addition the medical uses of androgen are also described.

SUGGESTIONS FOR FURTHER READING

Abraham Morgentaler (2006), Testosterone and Prostate Cancer: An Historical Perspective on A Modern Myth, *European Urology,* 50, 935–939.

Ahmed A. Alomary, Monique Vallée, Laura E. O'Dell, George F. Koob, Robert H. Purdy and Robert L. Fitzgerald (2006), Acutely Administered Ethanol Participates in Testosterone Synthesis and Increases Testosterone in Rat Brain. *Alcoholism: Clinical and Experimental Research,* Vol. 27, Issue 1, 38–43.

Briggs M.H. and Brotherton J. (Eds.) (1970), *Steroid Biochemistry and Pharmacology,* Academic Press, London, 121.

Degroot Leslie J. and Jameson J. Larry (Eds.) (2006), The Male Reproduction in *Endocrinology,* 5th ed., Saunders, Elsevier, 3087–3325.

Dorfman R.I. and Shipley R.A. (Eds.) (1956), *Androgens: Biochemistry, Physiology and Clinical Significance,* John Wiley and Sons, Inc., New York.

Eberlein W.R., Winter J. and Rosenfield R.L. (1967), The Androgens, In: *Hormones in Blood,* Gray, C.H. and Bacharach, A.L. (Eds.), Vol. 2, Academic Press, London, 187.

Freeman E.R., Bloom D.A. and McGuire E.J. (2001), *A Brief History of Testosterone, Journal of Urology,* 165, 371–373.

Gaylis F.D., Lin D.W., Ignatoff J.M., Amling C.L., Tutrone R.F. and Cosgrove D.J. (August 2005), Prostate Cancer in Men using Testosterone Supplementation, *J. Urol.,* 174(2):534–8.

Greenblatt R.B. and Roy Somnath (1963), Virilizing Adrenal Tumor, In: *The Hirsute Female,* Greenblatt R.B., C.C. Thomas (Eds.), Springfield, Illinois.

Hamilton, D.W. (1972), The Mammalian Epididymis, In: *Reproductive Biology,* Balin, H. and Glasser, S. (Eds.), Excerpta Medica Foundation, Amsterdam, 268.

Joel, C.A. (Ed.) (1971), *Fertility Disturbances in Men and Women,* S. Karger, 207 and 226.

Larry D. Bowers (2008), Testosterone Doping: Dealing with Genetic Differences in Metabolism and Excretion, *The Journal of Clinical Endocrinology and Metabolism,* Vol. 93, No.7, 2469–2471.

Margaret M. McCarthy (2008), Estradiol and the Developing Brain, Physiol Rev., 88:91–134.

Nieschlag E., Behre H.M., Bouchard P., Corrales J.J., Jones T.H., Stalla G.K., Webb S.M. and Wu F.C. (2004), *Testosterone replacement therapy: current trends and future directions,* Human Reproduction Update 10, 409–419.

Paulsen C.A. (1968), The Testes, in *Textbook of Endocrinology,* 4th ed., Williams, R.H. (Ed.), W.B. Saunders Co., Philadelphia, 405.

Peters R.J.B., Rijk J.C.W., Bovee T.F.H., Nijrolder A.W.J.M., Lommen A. and Nielen M.W.F. (2010), Identification of Anabolic Steroids and Derivatives using Bioassay-guided Fractionation, UHPLC/TOFMS Analysis and Accurate Mass Database Searching, Analytica Chimica Acta, Vol. 664, Issue 1, 77–88.

Rosenburg E. and Paulsen C.A. (Eds.) (1970), *The Human Testis,* Plenum Press, New York. Roy, Somnath (1963), An Outline for Clinical and Laboratory Investigations of Hirsute Female, In: *The Hirsute Female,* Greenblatt, R.B. (Ed.), C.C. Thomas, Springfield, Illinois.

Sharpe R.M., Martin B., Morris K., Greig I., McKinnell C., McNeilly A.S. and Walker M. (2002), Infant Feeding with Soy Formula Milk: Effects on the Testis and on Blood Testosterone Levels in Marmoset Monkeys During the Period of Neonatal Testicular Activity, *Hum Reprod.,* (7):1692–703.

Sriraman V. and Rao A.J. (2005), FSH, The Neglected Sibling: Evidence for Its Role in Regulation of Spermatogenesis and Leydig Cell Function, *Indian Journal of Experimental Biology,* 43:993–1000.

Sriraman V., Sairam M.R. and Jagannadha Rao A. (2003), Evaluation of Relative Roles of LH and FSH in Regulation of Differentiation of Leydig Cells using an Ethane 1, 2–Dimethylsulfonate-treated Adult Rat Model, *J. of Endocrinology,* 176:151–161.

Sriraman V., Sairam M.R. and Jagannadha Rao A. (2003), Evaluation of Relative Role of LH and FSH in Restoration of Spermatogenesis using EDS Treated Adult Rat. *Reproductive Biomedicine Online*, 8:167–174.

Sriraman Venkataraman and Jagannadha Rao A. (2004), Evaluation of the Role of FSH in Regulation of Leydig Cell Function During Different Stages of its Differentiation, *Molecular and Cellular Endocrinology*, 224:73–82.

Steinberger E. and Steinberger A. (1972), Testis: Basic and Clinical Aspects, In: *Reproductive Biology,* Balin, H. and Glasser, S. (Eds.), Excerpta Medica, Amsterdam, 144.

Stephen J. Winters and Barbara J. Clark (2003), Testosterone Synthesis, Transport, and Metabolism in *Androgens in Health and Disease* (*Contemporary Endocrinology*), Carrie Bagatell, William J. Bremner (Eds.) Humana Press, 3.

Van Wyk J.J. and Grumback M.M. (1968), Disorders of Sex Differentiation, In: *Textbook of Endocrinology,* 4th ed., Williams R.H. (Ed.), W.B. Saunders Co., Philadelphia, 537.

Vida J.A. Ed. (1969), *Androgens and Anabolic Agents*, Academic Press.

Wood R.I. and Stanton S.J. (2012), Testosterone and Sport: Current Perspectives. *Horm Behav,* 61(1):147–55, Epub 2011 Oct 1.

Zarrow M.X., Yochim J.M. and McCarthy J.L. (1964), Experimental Endocrinology, A Sourcebook of Basic Techniques, Academic Press, New York, 109–123.

94

Ovulation and Corpus Luteum Function, Menstrual Cycles and Ovarian Steroids

Ruby Dhar

CONTENTS

The processes described in this chapter form an integral part of female reproductive system and are required for maintenance of normal physiology. The female reproductive system includes the two **ovaries** and the female reproductive tract, two **fallopian tubes**, a **uterus** and a **vagina.** These structures together are also termed as **the female internal genitilia**.

The major function of the ovary is to produce mature and fertilizable ova, besides synthesizing predominant female sex hormones. The dynamic relationships between the different components of the reproductive axis in the adult female are such that this reproductive process occurs in a cyclic fashion, in an orderly sequence of events. All these sequences are very well coordinated between hormonal secretion and morphologic changes in various organs.

The reproductive cycle of the female primate can be divided into three states (1) the **follicular phase**, the time for follicular growth; (2) the **ovulatory period**, when final maturation of the oocyte and its release into the reproductive tract occurs; and (3) **the luteal phase**, when a newly formed corpus luteum secretes hormones in preparation for implantation. If the egg is not fertilized and implantation does not occur, a new cycle is initiated (a cyclic phenomenon) as the activity of corpus luteum wanes. If the fertilized egg implants into the uterus, the luteal phase is prolonged and becomes the progestational phase of the pregnancy that follows.

I. FEMALE REPRODUCTIVE SYSTEM: STRUCTURE AND FUNCTION

Broadly speaking female reproductive system can be divided into two major components: internal and external organs.

A. The Internal Reproductive

The main function of the internal structure involves the production of mature ova, leading to its ovulation, maintenance of the menstrual cycle and provides a safe home equipped with necessary requirements for the developing fetus (Figure 94.1a).

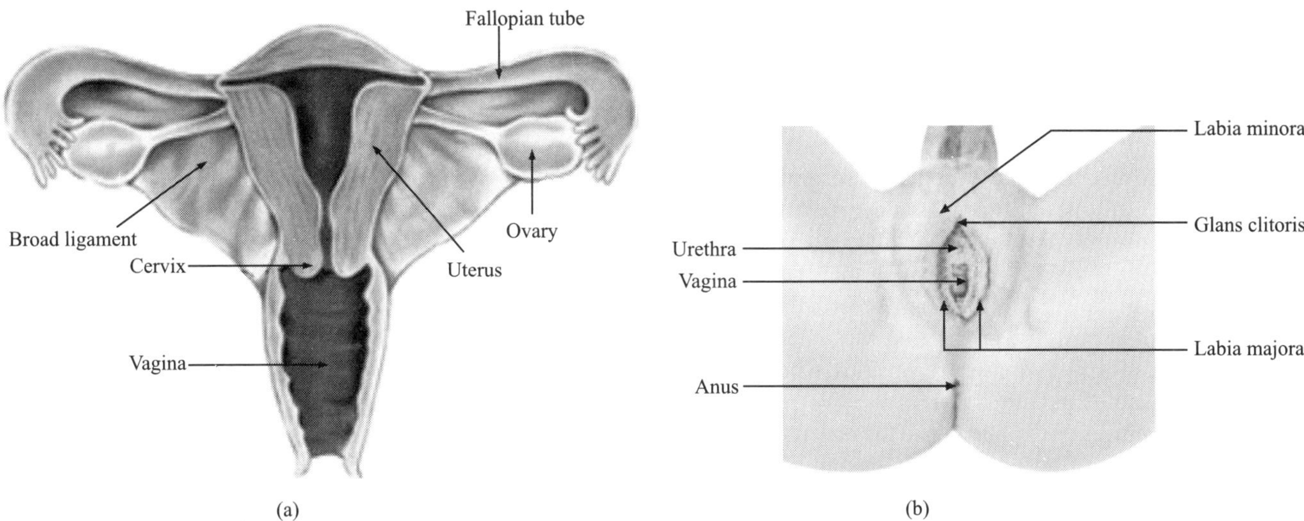

Figure 94.1   (a) Diagram showing the female internal reproductive organs, (b) Diagram showing the female external reproductive organs.

- **Uterus (womb):**   The uterus is a hollow thick-walled muscular organ lying between the urinary bladder and rectum. This pear-shaped organ is the home to the developing fetus and also the source of bleeding during menstruation. The uterus is divided into two parts: the cervix, which is the lower part that opens into the vagina, and the main body of the uterus, called the corpus. The corpus can easily expand to hold a developing baby. A channel through the cervix allows sperm to enter and menstrual blood to exit.
- **Ovaries:**   The ovaries are small, oval-shaped glands that are located on either side of the uterus in the upper pelvic cavity. The ovaries produce eggs and hormones.
- **Fallopian tubes:**   Also known as **uterine tubes** are the narrow tubes that are attached to the upper part of the uterus and serve as tunnels for the ova (egg cells) to travel from the ovaries to the uterus. Conception, the fertilization of an egg by a sperm, normally occurs in the fallopian tubes. The fertilized egg then moves to the uterus, where it implants into the lining of the uterine wall.
- **Vagina:**   The vagina is a canal that joins the cervix (the lower part of uterus) to the outside of the body. It is also known as the birth canal.

B.   The External Structure

The function of the external female reproductive structures (the genitals) is two fold: To enable sperm to enter the body and to protect the internal genital organs from infectious organisms. The main external structures of the female reproductive system include, (Figure 94.1b):

- **Labia majora:**   The labia majora enclose and protect the other external reproductive organs. Literally translated as "large lips, " the labia majora are relatively large and fleshy. The labia majora contain sweat and oil-secreting glands. After puberty, the labia majora are covered with hair.

- **Labia minora:**   Literally translated as "small lips, " the labia minora can be very small or up to 2 inches wide. They lie just inside the labia majora, and surround the openings to the vagina (the canal that joins the lower part of the uterus to the outside of the body) and urethra (the tube that carries urine from the bladder to the outside of the body).
- **Bartholin's glands:**   These glands are located besides the vaginal opening and produce a fluid (mucus) secretion.
- **Clitoris:**   The two labia minora meet at the clitoris, a small, sensitive protrusion that is comparable to the penis in males. The clitoris is covered by a fold of skin, called the prepuce, which is similar to the foreskin at the end of the penis. Like the penis, the clitoris is very sensitive to stimulation and can become erect.

C.   Ovarian Function

The ovary, like the testis, serves a dual purpose : (1) oogenesis, the production of gametes—the ova; and (2) secretion of the female sex hormones, estrogens and progesterone. Prior to ovulation, the gametogenic and endocrine functions take place in a specific structure-the follicle-and after ovulation the follicle differentiates into a corpus luteum, which has only an endocrine function.

Oogenesis

At birth, a female's ovaries contain an estimated total of 1 million ova, and no new ones appear after birth. Thus, in marked contrast to the male, the newborn female already has all the germ cells she will ever have. Only a few, perhaps 400, are destined to reach full maturity during her active reproductive life. All others degenerate at some point in their development so that few remain by the time a woman reaches approximately 50 years of age. As a result of this developmental pattern, the ova that are released near age 50 are 30 to 35 years older than those ovulated just after puberty. It has been suggested that certain defects, more common among children of older women are the result of aging changes in the ovum.

During early fetal development, the primitive germ cells, or **oogonia** (*singular oogonium*), a term analogous to spermatogonia in the male, undergo numerous mitotic divisions. At some point in fetal life, the oogonia cease dividing, and from this point on no new germ cells are generated. In the fetus, all the oogonia then develop into **primary oocytes**, which begin a first meiotic division by replicating their DNA. They do not, however, complete the division. Accordingly, all the germ cells present at birth are primary oocytes containing 46 chromosomes, each with two sister chromatids. The cells are said to be in a state of meiotic arrest.

This state continues until puberty and the onset of renewed activity in the ovaries. Indeed, the first meiotic division is completed only just before an oocyte is about to be released from the ovary. This division is analogous to the division of the primary spermatocyte, and each daughter cell receives 23 chromosomes, each with two daughter cells, the **secondary oocyte**, retains virtually all the cytoplasm. The other, termed the first **polar body**, is very small and adheres to the secondary oocyte. Thus, the primary oocyte, which is already as large as the mature ovum will be, passes on to the secondary oocyte half of it's chromosomes but almost all its nutrient-rich cytoplasm.

The second meiotic division occurs, in a uterine tube, after ovulation but only if the secondary oocyte is fertilized, that is, penetrated by the sperm. As a result of this division the daughter cells each receive 23 chromosomes, each with a single chromatid. Once again, one daughter cell, now termed a **mature ovum,** also known as "fertilized ovum", retains nearly all the cytoplasm, whereas the other, termed the second polar body. is very small and nonfunctional.

Interestingly, the first polar body also undergoes a second meiotic division at the same time the secondary oocyte does? However, as far as is known, neither of these resulting cells nor the second polar body fertilized, and they all eventually disintegrate. The net result of oogenesis is each primary oocyte produces only one fertilizable ovum. In contrast each primary spermatocyte produces four viable spermatogonia (Figure 94.2).

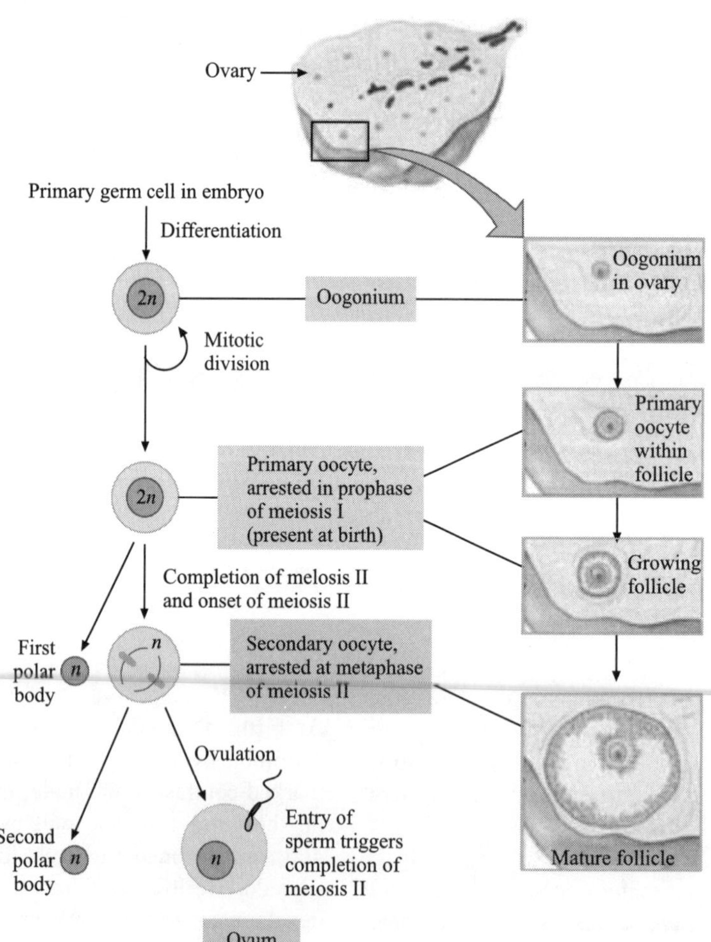

Figure 94.2 Meiotic and mitotic divisions during oogenesis in females. The primary oocyte undergoes its first meiotic division, just before ovulation. The secondary oocyte is ovulated and undergoes the second meiotic division, after being fertilized by a sperm. The polar bodies, devoid of their cytoplasm eventually disintegrate and undergo atresia.

II. FOLLICLE GROWTH

Throughout their life in the ovaries, the oocytes exist in structures known as **follicles**.

In the adult ovary, most of these follicles are either **primordial follicles** or, the next stage of development, **primary follicles.** Both of these follicle types consist of one primary oocyte surrounded by a single layer of cells called **granulosa cells**, the primary follicles differing from the primordial follicles only in the shape of the granulosa cells.

Further development from the primary follicle stage is characterized by an increase in size of the ooycte and a proliferation of the granulosa cells. The oocyte becomes separated from the granulosa cells by a thick layer of material, the **zona pellucida**, secreted by the granulosa cells and oocyte. Despite the presence of the zona pellucida, however the inner layer of granulosa cells remains intimately associated with the oocyte by means of cytoplasmic processes that traverse the zona pellucida and form gap junctions with the oocyte.

Through these gap junctions, nutrients and chemical messengers are passed to the oocyte. The follicle grows as new cell layers are formed, not only from mitosis of the original granulose cells but as a result of the differentiation of specialized ovarian connective tissue cells. Thus, the follicle now consists of the oocyte, layers of granulosa cells surrounding the oocyte, and the outer layers of cells known as **theca**.

When the follicle reaches a certain size, a fuild filled space, the **antrum,** begins to form in the midst of the granulosa cells as a result of the fluid they secrete. By the time the antrum begins to form, the oocyte has reached full size. From this point on, most of the follicle's enlargement is due to the expanding antrum (Figure 94.3a and b).

The progression from primordial follicle to primary follicle to the preantral and small-antral stages occurs continuously in the adult ovary, so that any moment there are always present a sizable number of prenatal follicles and a few small-antral follicles. Then, at the beginning of each menstrual cycle, several of these follicles begin to develop into larger antral follicles. Into one week into the cycle, a selection process occurs: Only the largest follicle, the **dominant follicle**, continues to develop, and the others that had begun to enlarge during that week degenerate, a process termed **atresia**. Thus once a follicle has matured beyond a certain point, it is committed to go on and ovulate its oocyte or more commonly, to become atretic. However, on occasion, two or more follicles reach maturity, and more than one oocyte may be ovulated. This is the common most cause of multiple births. In such cases the siblings are fraternal, not identical.

Ultimately, the oocyte is surrounded by granulosa cells that project from the inner follicle wall into the antrum and are termed the cumulus oophorous. The mature follicle (also termed a Graafian follicle) becomes so large that it balloons out on the surface of the ovary. Ovulation occurs when the walls of the follicle and ovary at this site rupture and the oocyte, surrounded by its adhering zona pellucida and several layers of granulose cells is carried out of the ovary and onto the ovarian surface by the antral fluid.

All this happens on approximately day 14 of the cycle.

III. OVULATION

Ovulation is induced by the sudden release of large amounts of gonadotropins (LH) from the pituitary gland (gonadotropin surge). Species can be differentiated as to whether the trigger to the surge of gonadotropins is external or internal. Species in which the trigger to the surge is external, such as rabbit, cat, mink and ferret are termed *reflex ovulators*. The ovulatory surge is induced by mating and genital stimuli, transmitted via neural pathways connecting the copuloreceptive area to

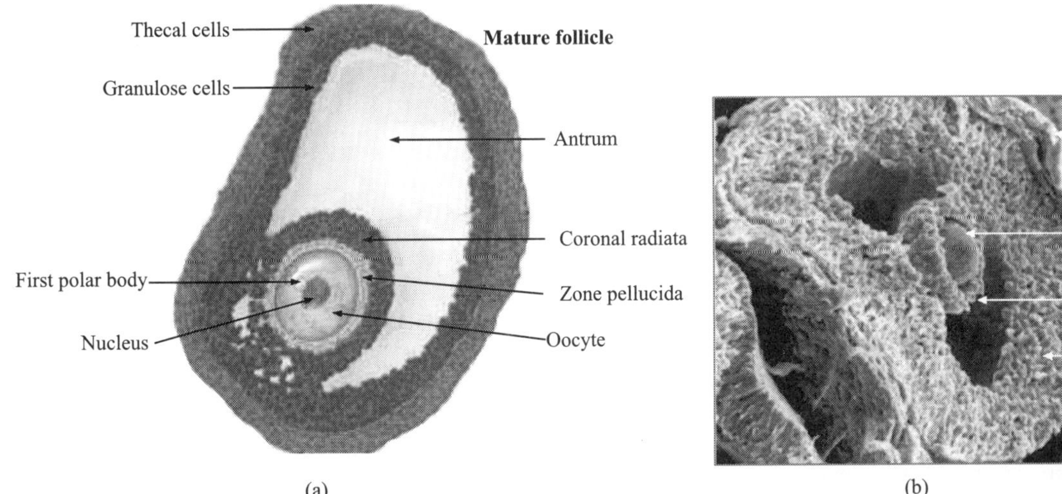

Figure 94.3 (a) Illustration of the Human female follicle. A follicle is an anatomical structure in which the primary oocyte develops. Granulosa cells within the follicle surround the oocyte, which in turn are enclosed within a layer of extracellular matrix called the thecal cells, (b) Scanning electron micrograph of a developing human secondary follicle.

the hypothalamus, stimulate **luteinizing hormone** (LH) **and Follicle stimulating hormone** (FSH) release. In these species, the ovarian steroids merely act to promote sexual receptivity. Species in which the trigger to the surge is internal, such as rat, guinea, pig, sheep, cow and primate including human are termed *spontaneous ovulators*. The ovulatory surge is triggered by changing endogenous ovarian steroids titers that accompany follicular maturation.

A. The Ovulatory Process

Ovulation is the end process of a series of events initiated by the godotropin surge and resulting in the release of a mature fertilizable egg from Graafian follicle. Unfortunately, although there is wealth of literature on the subject, the precise sequence of local events within the follicle that lead to rupture of the follicular wall and expulsion of the egg is not known. These mechanisms most probably relate to the interaction of a number of factors, each of which by itself may not be the primary mover, but which may complement each other's action.

Studies over all these years show that ovulation is initiated by the gonnadotropin surge, which occurs in response to the long loop estradiol positive feedback, the signal to the brain and pituitary that the dominant follicle has attained maturity. The gonadotropin surge terminates estradiol synthesis; the theca cell now changes from an androgen to a progesterone-secreting tissue (the preovulatory progesterone rise). Studies in rodents have shown that blocking progesterone can reduce ovulatory rate. Whether this indicates a direct action of progesterone on the ovulatory process or whether progesterone acts through the regulation of other products, such as prostaglandins is unknown (Figure 94.4).

Vascular changes in the preovulatory follicle occur within minutes of the LH surge.

The multilayered capillary plexus within the theca dilates causing hyperemia, a prelude to the ovulatory process. This phenomenon is perhaps related to the rapid release of histamine and/or other kinins by local mast cells. About 6 hours into the LH surge, there is increasing ovarian blood flow due to decreased vascular resistance; increase in capillary and venule permeability leading to an increase in interstitial fluid volume.

Protein synthesis is largely initiated by the gonadotropin surge and remains an ongoing process throughout the ovulatory period. This increased synthesis may, in part, reflect the production of enzymes involved in follicle wall degradation. The release of proteolytic enzymes has been described,

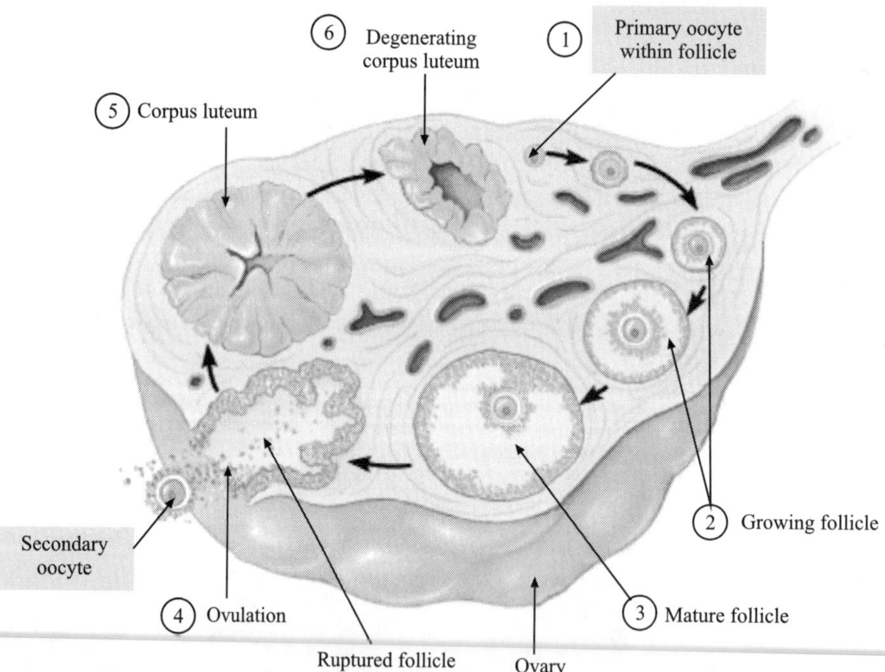

Figure 94.4   Illustration of Human Ovarian cycle, showing various stages. All of these stages occur sequentially inside the ovary in a very controlled manner under the influence of various hormones. (1) In a primary follicle, a primary oocyte is surrounded by a single layer of granulosa cells, (2) Under the influence of local paracrines, granulosa cells proliferate and form the zona pellucida around the oocyte, (3) Mature follicle forms under the influence of FSH after about two weeks of rapid growth. At this stage the follicle has expanded antrum; the oocyte, which by now has developed into a secondary oocyte, is displaced to one side, (4) At midcycle, in response to a burst in LH secretion, the mature follicle, bulging on the ovarian surface, ruptures and releases the oocyte, resulting in ovulation and ending the follicular phase. (5) Entering into the luteal phase, the ruptured follicle develop into a corpus luteum under the influence of LH. The corpus luteum continues to grow and secrete progesterone and estrogens that prepare the uterus for implantation of a fertilzed ovum. (6) After 14 days, if the ferlilized ovum does not implant in the uterus, the corpus luteum degenerates, the luteal phase ends and a marks the beginning of the menstrual cycle under the influence of a changing hormonal milieu.

enzymatic degradation of the wall of the preovulatory follicle is a primary hypothesis to explain rupture. The plasmin may be suitable candidate in this process. This protease is generated following secretion by the cell of plasminogen activator in response to FSH. This occurs within 2 hours of exposure to the gonadotropin surge. Plasmin inturn activates collagenase, which stimulates proteolysis and disintegration of collagen, the primary component responsible for follicular wall tensile strength. This then probably results in a reduced tensile strength of the follicular wall to the point at which it ruptures.

The involvement of ovarian innervation consists of adrenergic fibers, which are distributed to the theca externa and theca interna. Adrenergic neurons within follicular walls are activated by LH and secrete norepinephrine in response. The contractility of whole ovaries is influenced by adrenergic compounds and adrenergic agonists can enhance the ovulatory response to the gonadotropin stimulus; yet, ovulation can occur even after denervation, thus casting doubt on a primary adrenergic role in the ovulatory process.

Prostaglandins are another group of substances released by LH which have been implicated in the phenomenon of ovulation. The two main prostaglandins within the ovary are prostaglandin E2 (PGE2) and prostaglandin F2-a (PGF2-a). Local ovarian prostaglandin levels rise as the time of ovulation approaches. Prostaglandins probably contribute to the process of ovulation through various pathways, such as affecting smooth muscle contractility. Contraction of smooth muscle in postcapillary venules may increase intracapillary pressure and increase transsudation, thus contributing to an increase in free fluid pressure within the follicle and by activating proteolytic enzymes especially those associated with collagen degradation.

IV. CORPUS LUTEUM: BEGINNING OF THE LUTEAL PHASE

The ruptured follicle left behind in the ovary after the release of the ovum changes rapidly as the granulosa and theca cells remaining in the remnant follicle undergo a dramatic structural and functional transformation.

A. Formation of Corpus Luteum

These old follicular cells form the **corpus luteum (CL)**, a process called **luteinization**. The follicular-turned-luteal-cells enlarge and are converted into very active steroid hormone-producing tissue. Abundant storage of cholesterol, the steroid precursor molecule, in lipid droplets within corpus luteum gives this tissue yellowish appearance, hence it's name (*corpus* means" body"; *luteum* means "yellow")

The corpus luteum secretes into the blood abundant quantities of progesterone, along with smaller amounts of estrogens. Estrogens secretion in the follicular phase followed by progesterone secretion in the luteal phase is essential for preparing the uterus for implantation of a fertilized ovum. The

CL becomes fully functional within 4 days after ovulation, but it continues to increase in size for another 4 or 5 days.

B. Degeneration of the Corpus Luteum

If the released ovum is not fertilized and does not implant, the corpus luteum degenerates within about 14 days after its formation. The luteual cells degenerate and are phagocytized, and connective tissue rapidly fills in to form a fibrous tissue mass known as the corpus albicans("white body"). The luteul phase is now over, and one ovarian cycle complete. A new wave of follicular development, which begins when degeneration of the old CL is completed, signals the onset of a new follicular phase (Figure 94.4).

C. Corpus Luteum of Pregnancy

If fertilization and implantation do occur, the corpus luteum continues to grow and produce increasing quantities of progesterone and estrogens instead of degenerating. Now called the corpus luteum of pregnancy, this ovarian structure persists until pregnancy ends. It provides the hormones essential for maintaining pregnancy until the developing placenta can take over this crucial function.

D. Control of the Corpus Luteum

LH "maintains" the corpus luteum-that is, after triggering development of the CL, LH stimulates ongoing steroid hormone secretion by this ovarian structure. Under the influence of LH, the CL secretes both progesterone and estrogens, with progesterone being its most abundant hormonal product. The plasma progesterone level increases for the first time during the luteal phase. No progesterone is secreted during the follicular phase. Therefore, the follicular phase is dominated by estrogens and the luteal phase by progesterone.

A transitory drop in the level of circulating estrogens occurs at midcycle as the estrogens-secreting follicle meets its demise at ovulation. The estrogens level climbs again during the luteal phase because of the CL's activity, although it does not reach the same peak as during the follicular phase. Even though a high level of estrogens stimulates LH secretion, progesterone, which dominates the luteal phase, powerfully inhibits LH secretion as well as FSH secretion by acting at both the hypothalamic ARC nucleus (arcuate nucleus) and the anterior pituitary. Inhibition of FSH and LH by progesterone prevents new follicular maturation and ovulation during the luteal phase. Under progesterone's influence, the reproductive system is gearing up to support the just-released ovum, should it be fertilized, instead of preparing other ova for release.

The corpus luteum functions for an average of 2 weeks and then degenerates if fertilization does not occur. The mechanisms that govern degeneration of the CL are not fully understood. The declining level of circulating LH, driven down by inhibitory actions of progesterone, undoubtedly contributes to the CL's downfall. Prostaglandins and estrogens released by the luteal cells themselves may play a role. Demise of the CL terminates the luteal phase and sets the stage for a new follicular phase. As the CL degenerates, plasma progesterone and estrogens levels fall rapidly because these hormones are no longer being produced. Withdrawal of the inhibitory effects of these hormones on the hypothalamus allows FSH and tonic LH secretion to modestly increase again. Under the influence of these gonadotropic hormones, another batch of primary follicles is induced to mature as a new follicular phase begins.

V. MENSTRUAL PHASE

The menstrual phase is the most overt phase, characterized by discharge of blood and endometrial debris from vagina. By convention, the first day of menstruation is considered the start of a new cycle. It coincides with the end of the ovarian luteal phase and onset of a new follicular phase (Figure 94.5). As the corpus luteum degenerates because fertilization and implantation of the ovum released during the preceding cycle did not take place circulating levels of progesterone and estrogens drop precipitously. Because the net effect of progesterone and estrogens is to prepare the endometrium for implantation of a fertilized ovum, withdrawal of these steroids deprives the high vascular, nutrient-rich uterine lining of its hormonal support.

The reduction in ovarian hormone levels also stimulates release of a uterine prostaglandin that causes vasoconstriction of the endometrial vessels, disrupting the blood supply to the endometrium. The subsequent reduction in O2 delivery causes death of the endometrium, including its blood vessels. The resulting bleeding through the disintegrating vessels flushes the dying endometrial tissue into the uterine lumen. Most of the uterine sloughs during each menstrual period except for a deep, thin layer of epithelial cells and glands, from which the endometrium will regenerate. The same local prostaglandin also stimulates mild rhythmic contractions of the uterine myometrium. These contractions help expel the blood and endometrial debris from the uterine cavity out through vagina as **menstrual flow**. Excessive uterine contractions caused by prostaglandin overproduction produce the **dysmenorrhea** (*menstrual cramps*) some women experience.

Ovarian cycle is divided into three phases: Follicular, ovulation and luteal phase. The first day of bleeding is marked as the day 1 of the menstrual cycle. The first half of the ovarian cycle, the follicular phase, the ovarian follicle secretes estrogens, under the influence of FSH, LH and estrogens itself. The rising moderate levels of estrogens inhibit FSH secretion, which declines during the last part of follicular phase, with less suppression of LH secretion, which continues to rise throughout the follicle phase.

With the follicular output of estrogens at its peak, triggers also a LH secretion. This LH secretion around midcycle, leads to ovulation of the mature follicle. Estrogens levels go down after ovulation and the old ruptured follicle is transformed into corpus

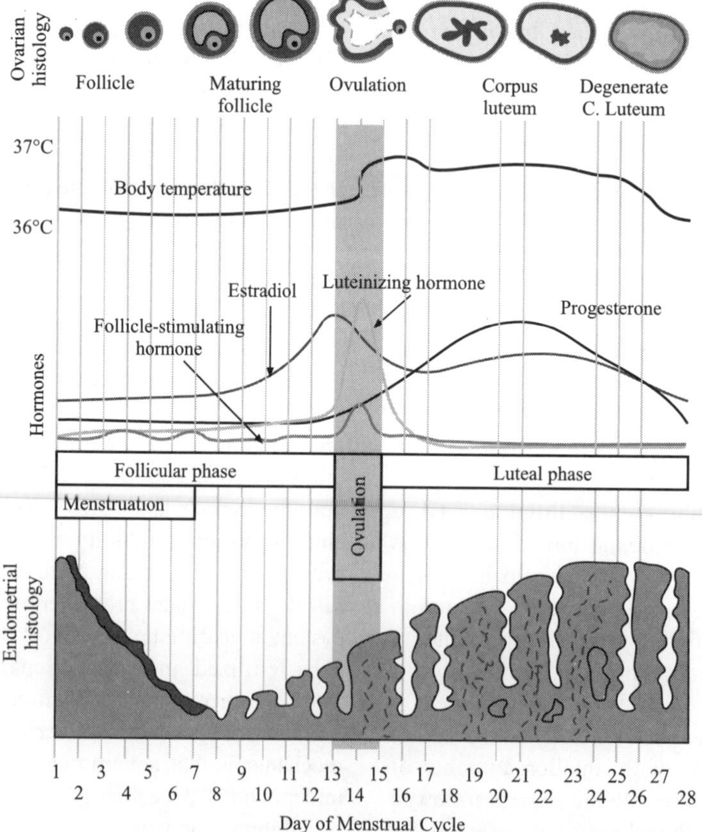

Figure 94.5   Diagram showing the correlation between hormonal levels and cyclic uterine changes. (*see Plate 39 for colour figure*)

luteum, which secretes progesterone and estrogens, during the last half of the cycle, called the luteal phase. Progesterone strongly inhibits both FSH and LH, which continue to decrease throughout the luteal phase. The corpus luteum degenerates in about two weeks if the released ovum has not been fertilized and implanted in the uterus. Progesterone and estrogens levels sharply decrease when the corpus luteum degenerates, removing the inhibitor influences on FSH and LH. As these anterior pituitary hormone levels start to rise again, they stimulate the batch of new follicle as a new follicular phase begins.

The uterus also undergoes various changes under the influence of ovarian hormones. Early in the follicular phase, the highly vascularized, nutrient-rich endometrial lining is slogged off (the uterine menstrual phase). This sloughing results from the withdrawal of estrogens and progesterone, when the old corpus luteum degenerates. Late in the follicular phase, the rising levels of estrogens cause the endometrium to thicken the uterine proliferative phase). After ovulation, progesterone from the corpus luteum brings about vascular and secretory changes in the estrogens-primed endometrium to produce a suitable environment for implantation (uterine secretory or progestational phase).

When the corpus luteum degenerates, a new ovarian follicular phase and uterine menstrual phase begin.

Under the influence of progesterone there is a also some slight fluctuation in body temperature.

The temperatures before ovulation fluctuate in a low range, and the temperatures after ovulation fluctuate towards a slightly high range

The average blood loss during a single menstrual period is 50 to 150 ml. Blood that sleeps slowly through the degenerating endometrium clots within the uterine cavity, and then is acted on by fibrinolysin, a fibrin dissolver that breaks down the fibrin forming the meshwork of the clot. Therefore, blood in the menstrual flow usually does not clot, however at times when blood is not exposed to sufficient fibrinolysin, blood clots may appear during profuse menstrual flow. In addition to the blood and endometrial debris, large numbers of leukocytes are found in menstrual flow. These white blood cells play an important defense role in helping the raw endometrium resist infection.

Menstruation typically lasts for about 5 to 7 days after degeneration of the CL, coinciding in time with early part of the ovarian follicular phase. Withdrawal of progesterone and estrogens on degeneration of the CL leads simultaneously to sloughing of the endometrium (menstruation) and development of new follicles in the ovary under the influence of gonadotropic hormone levels. The drop in gonadal hormone secretion removes inhibitory influences from the hypothalamus and anterior pituitary, so FSH and LH secretion increases and a new follicular phase begins.

VI. MENOPAUSE

The cessation of a woman's menstrual cycles at menopause sometime between the ages of 45 and 55 has traditionally been attributed to the limited supply of ovarian follicles present at birth. The termination of reproductive potential in a middle-aged woman is "preprogrammed" at her own birth. Recent evidence suggests, however, that a midlife hypothalamic change instead of aging ovaries may trigger the onset of menopause. Evolutionarily, menopause may have developed as a mechanism that prevented pregnancy in women beyond the time that they could rear a child before their own death.

Menopause is preceded by a period of progressive ovarian failure characterized by increasingly number of irregular cycles and dwindling estrogens levels. This entire period of transition from sexual maturity to cessation of reproductive capability is known as the climacteric, or perimenopause. Ovarian estrogens production declines from as much as 300 mg per day to essentially nothing. Postmenopausal women are not completely devoid of estrogens, however, because adipose tissue, the liver, and the adrenal cortex continue to produce up to 20 mg of estrogens per day. In addition to the ending of ovarian and menstrual cycles, the loss of ovarian estrogens following menopause brings about many physical and emotional changes. These changes include vaginal dryness, which can cause discomfort during sex, and gradual atrophy of the genital organs. However, postmenopausal women still have a sex drive because of their adrenal androgens.

Estrogens has widespread physiological actions beyond the reproductive system, the dramatic loss of ovarian estrogens in menopause affects other body systems, most notably the skeleton and the cardiovascular system. Estrogens helps build strong bones, shielding premenopausal women from the bone-dissolving osteoclasts and diminishing activity of the bone-building osteocblasts. The result is decreased bone density and a greater incidence of bone fractures.

Estrogens also help modulate the actions of epinephrine and norepinephrine on the anterior walls by promoting local release of the vasodilator nitric oxide. The menopausal downfall of estrogens leads to unstable control of blood flow, especially in the skin vessels. Transient increases in the flow of warm blood through these superficial vessels are responsible for the "hot flashes" that frequently accompany menopause. Vasomotor stability is gradually restored in postmenopausal women so that hot flashes eventually cease.

VII. OVARIAN HORMONES

The ovulatory process also produces the two ovarian hormones, estrogens and progesterone.

Estrogens denotes not a single specific hormone but rather a group of steroid hormones which have similar effects on the female reproductive tract. These include estradiol, estrone, and estriol, of which the first is the major estrogens secreted by the ovaries. Estrogens is produced by the developing follicle before ovulation; these stimulates the glands of the cervix to secrete a particular type of mucus ("mucus with fertile characteristics") which is essential for the sperm to pass through the cervix and reach the ovum. Estrogens also stimulates growth of the endometrium lining the uterus (womb), helps in maintaining uterine tube contractions and ciliary activity. Estrogenic stimulation maintains the uterus, uterine tubes, vagina, the extended genetalia, and the breast. Estrogens are also responsible for the female body configuration; narrow shoulders, broad hips, and the characteristic female fat deposition in the hips, abdomen and breasts. Estrogens, have a nuclear receptor ligand (for its receptor estrogens receptor-alpha; *ESR1*) acts on the cell nucleus and can regulate gene transcription.

Progesterone, the second ovarian steroid hormone, is secreted in very small amounts by the granulosa cells, just prior to ovulation, but its major source is the corpus luteum. After ovulation, progesterone and estradiol are produced by the corpus luteum which is formed from the ruptured follicle. The effects of progesterone are less widespread than those estrogens, the endometrial changes being the most prominent. Progesterone causes the abrupt change in the mucus which occurs immediately after ovulation and defines the Peak symptom. Progesterone also prepares the estrogens-primed endometrium for implantation of the fertilized ovum. Progesterone exercises important effects on the breasts, the uterine tubes, and the uterine smooth muscle. Progesterone also causes "cornification"-an increase in layering of the cells lining the vagina, and the microscopic examination of some of these cells provides an indicator of ovulation.

In the absence of pregnancy, production of estrogens and progesterone begins to decline approximately 7 days after ovulation and this result in shedding of the endometrium as menstrual bleeding 11–16 days after ovulation.

Two peptide hormones are also produced by the ovaries, **inhibin** by the granulosa cells and **relaxin** by the corpus luteum, which play important role in regulating various stages of this process.

Androgens in women

Testosterone is not a uniquely male hormone but is found in a very low concentration in the blood of normal women as a result of production by the ovaries and adrenal glands as well as peripheral conversion of adrenal androgens. Androgens other than testosterone are found in some concentrations in the blood of women.

Adrenal androgens do play several important roles in the female, notably maintenance of sexual drive. In several disease states, the female adrenals may secrete abnormally large quantities of androgens, which produce virilism: The female fat distribution disappears, a beard appears along with the male body-hair distribution, the voice lowers in pitch, the skeletal muscle mass enlarges, the clitoris enlarges and the breasts diminish in size.

SUMMARY

The menstrual cycle is the scientific term for the physiological changes that occur in fertile women for the purposes of sexual reproduction and fertilization. This chapter highlights the various steps and regulatory mechanisms involved during this process. The cyclic repetition of this process every month, under the control of the endocrine system, is crucial for maintenance of female fertility and is also crucial to be able to give birth to a new life on earth. The entire cycle is commonly divided into three phases: the follicular phase, ovulation, and the luteal phase. It is also occasionally misclassified using the uterine cycle: menstruation, proliferative phase, and secretory phase. Menstrual cycles are counted from the first day of menstrual bleeding. Hormonal contraception interferes with the normal hormonal changes with the aim of preventing reproduction. Stimulated by gradually increasing amounts of estrogens in the follicular phase, discharges of blood (menses) slow then stop, and the lining of the uterus thickens. Follicles in the ovary begin developing under the influence of a complex interplay of hormones, and after several days one or occasionally two become dominant. Non-dominant follicles atrophy and die. Approximately mid-cycle, 24–36 hours after the Luteinizing Hormone (LH) surges, the dominant follicle releases an ovum, or egg, an event called ovulation. After ovulation, the egg only lives for 24 hours or less without fertilization while the remains of the dominant follicle in the ovary become a corpus luteum; which has a primary function of producing large amounts of progesterone. Under the influence of progesterone, the endometrium (uterine lining) changes to prepare for potential implantation of an embryo to establish a pregnancy. If implantation does not occur within approximately two weeks, the corpus luteum will involute, causing sharp drops in levels of both progesterone and estrogens. A drop of these hormones causes the uterus to shed its lining and egg in a process termed menstruation. This is marked by the beginning of a new cycle in the female.

SUGGESTIONS FOR FURTHER READING

1. Ovarian Follicular Growth and Development in Mammals Biology of Reproduction 50(1994), 225–232.

2. Sayoko Makabe and Blerkom Van Jonathan (2006), Atlas of Human Female Reproductive Function: Ovarian Development to Early Embryogenesis After *In vitro* Fertilization Taylor and Francis Group.

3. Physiological Changes Associated with the Menstrual Cycle: A Review. *Obstet Gynecol Surv.*, 64(1):58–72.

4. An Integrated View on the Luteal Phase: Diagnosis and Treatment in Subfertility, *Clin Endocrinol* (Oxf)., 77(4):500–7.

5. Estrogens Signaling Multiple Pathways to Impact Gene Transcription, Curr Genomics (2006), 7(8):497–508.

6. Eli Y. Adashi (2001), Ovulation: Evolving Scientific and Clinical Concepts, Springer.

95

Fertilization, Early Mammalian Development and Implantation

Jayshree Sengupta and Debabrata Ghosh

CONTENTS

I. INTRODUCTION

Fertilization is the union of male and female gametes leading to the development of a new individual. It occurs after a complex and integrated series of preliminary events. The release of oocyte is initiated by the concerted actions of endocrine and neuroendocrine systems and the ovulated oocyte is carried to the ampulla where generally fertilization occurs. Spermatozoa undergo a series of biochemical and morphological changes in the female genital tract before they fertilize the oocyte.

II. STRUCTURE OF SPERM

A spermatozoon consists of a head and a tail. The tail may be subdivided into a neck, mid-piece, principal piece and end piece. The shape of the sperm head varies from species to species and is composed of 2 parts, the nucleus and the acrosome which covers the anterior two thirds of the nucleus. Figure 95.1 gives a diagrammatical representation of a spermatozoon.

The acrosome consists of an electron dense matrix and is surrounded by the acrosomal membrane; posterior to the acrosome the nucleus is surrounded by a laminar structure called the post nuclear cap. The main components of the tail are the axial filament complex which is made up of several

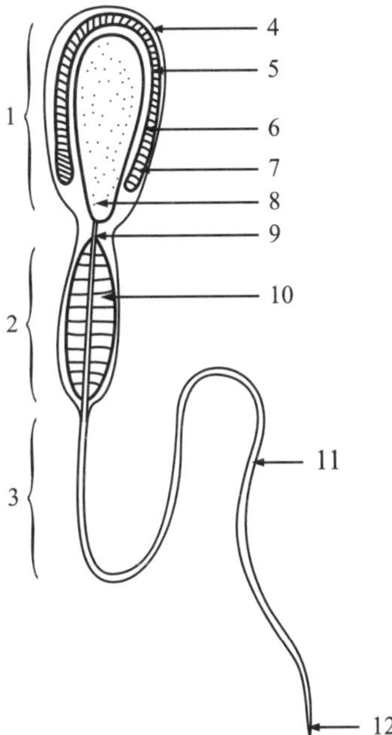

Figure 95.1 Diagrammatic sketch (longitudinal section) of mammalian sperm. 1. Head; 2. Midpiece; 3. Tail; 4. Plasma membrane; 5. Outer acrosomal membrane; 6. Inner acrosomal membrane; 7. Equatorial region; 8. Nuclear chromatin; 9. Neck; 10. Mitochondrial sheath; 11. Tail piece; 12. End piece.

proteins. The flagellum also contains ATPase which can cleave ATP for the production of energy required for sperm motility. The sperm flagellum contains spermosin, a myosin like protein, and flactin, an actin like protein, actomyosin and even tropomyosin like proteins. The mitochondrial sheath in the midpiece is particulary rich in phospholipid bound to proteins and it contains the complete cytochrome system and a number of oxidizing enzymes. A peculiar feature of the germ cell enzyme content is the presence of lactic dehyrogenase isoenzyme LDH-X or LDH-IV.

III. SPERM TRANSPORT IN THE FEMALE GENITAL TRACT

Spermatozoa and seminal plasma are deposited in the posterior vaginal fornix at ejaculation. They penetrate into the cervical mucus very rapidly and the coagulum is liquefied by fibrinolytic enzymes. The physical characteristic of cervical mucus varies in different phases of the menstrual cycle and is able to restrict fertility by limiting passage to spermatozoa. It is scant thick and viscous during most of the cycle and becomes thinner and copious during mid cycle favourable for sperm entry. Changes in the property of the mucus are determined by the changing levels of oestrogen and progesterone. Sperm motility has been shown to be maximal in cervical mucus at mid-cycle and sperm movement occurs in straighted lines in the 'maximal' grade of cervical mucus which coincides with the oestrogen peak. Spermatozoa pass rapidly into the uterus and increased uterine activity may assist sperm transport. The passage of spermatozoa into the oviduct is restricted by various barriers especially at the utero-tubal junction. The passage of spermatozoa into the ampullary region of the oviduct is aided by muscle contractions in the ciliary sperms towards the ampulla (Figure 95.2). Spermatozoa are found in the human oviduct a few minutes after coitus and these may represent a highly selective sample of highly fertilizable and acceptable gametes capable of achieving fertilization. Human spermatozoa may remain mobile in the oviduct for several days and retain their fertility for 48 h.

Excess numbers of spermatozoa in uterus and oviducts are removed by phagocytosis.

IV. STRUCTURE OF OOCYTE

The oocyte is shed from the Graafian follicle by the process of ovulation which occurs at a precise time after onset of LH surge. The timing of ovulation in relation to the reproductive cycle varies considerably among species and may be dependent upon diurnal rhythms or be induced by coitus as in the rabbit. The ovulated egg is released from the ovary and it passes into the oviduct through its fimbriated end. In several mammalian species, the ovary together with oviduct remains encapsulated within the bursa, a fine membranous capsule.

The presence of cilia in fimbriated as well as the fluid currents generated by movement of oviductal infundibulum lead the cummulus enclosed oocyte into the oviduct. Once the egg reaches the oviduct muscular contractions lead the egg into the ampullary region where fertilization will occur. The mammalian oocyte is around 0.15 to 0.20 mm in diameter and is surrounded by a plasma membrane.

The egg is surrounded by a refractile acellular membrane, the zona pellucida which is about 0.02 mm in thickness and it is separated from the egg plasma membrane by a space, the perivitelline space. The freshly ovulated egg is surrounded by cummulus oophorus which consists of layers of granulosa cells enclosed in a jelly like matrix substance. The cells closest to the zona are arranged radially, the corona radiata. The intracellular matrix contains hyaluronic acid, mucopolysaccharides and various proteins (Figure 95.3). The cummulus oophorus disperse within 2 hours of insemination, although corona radiata remain attached to the dividing egg up to 48 h *in vitro* and *in vivo*, and it persist for 2–3 days after ovulation has occurred. In other mammalian species like, cow, sheep, goat, rabbit and rodents eggs are usually denuded of corona cells generally by 24 h, if fertilization has occurred. The function of these cells is not known for certain, however, studies of human preimplantation embryos showed secretion of progesterone, estradiol and prostaglandins by the corona cells *in vitro*.

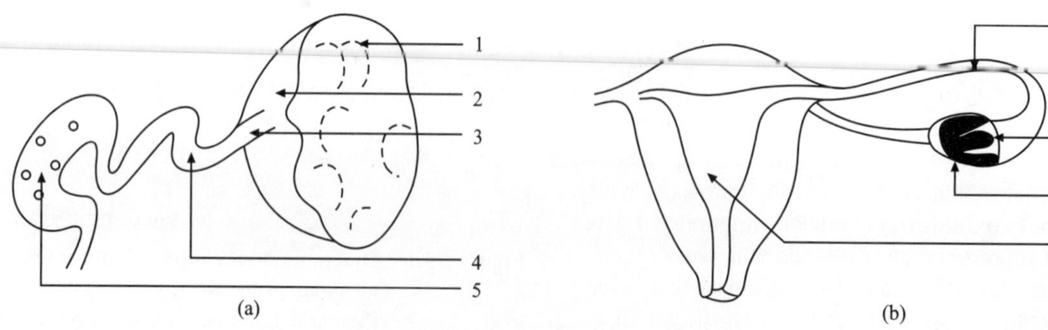

(a) (b)

Figure 95.2 (a) Relationship of fimbria to the ovarian surface in the rat. 1. Ovary; 2. Periovarial sac (tissue capsule); 3. Fimbrial end of the oviduct; 4. Oviduct; 5. Ampulla containing freshly ovulated eggs. (b) Relationship of fimbria to the ovarian surface in the human. 1. Fallopian tube; 2. Fimbria; 3. Ovary; 4. Uterine cavity.

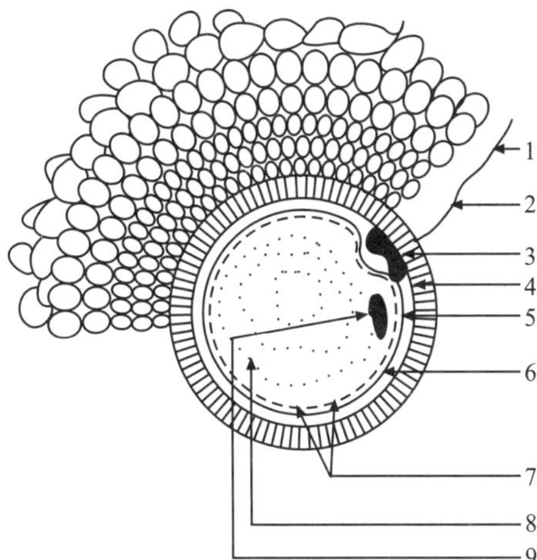

Figure 95.3 Structure of freshly ovulated eggs: 1. Cumulus oophorus; 2 Corona radiata; 3. First polar body; 4. Zona pellucida; 5. Perivitelline space; 6. Plasma membrane; 7. Cortical granules; 8. Vitellus; 9. Second meiotic spindle.

V. CAPACITATION

Spermatozoa undergo various physiological changes in the female reproductive tract before they are able to penetrate into oocytes. The initial change is known as capacitation and this permits the gametes to undergo their next, second change known as the acrosome reaction. Capacitation involves subtle physicochemical changes in the spermatozoal membrane and it requires a period of several hours in most mammalian species. Capacitation normally occurs in the oviduct and it can also be induced by exposing spermatozoa to relatively simple and defined culture solutions. The changes that are occurring in the membrane of the spermatozoa during capacitation are not known for certain. However, the process is believed to involve the removal of surface coats from the membranes. Inhibitory substances in the male reproductive tract, and in the lower portions of the female reproductive tract may bind to the plasma membrane to prevent capacitation and thus maintain the integrity of the acrosome. Removal of these deposits can induce vesiculation or disruption of plasma membranes in the region overlying the acrosome, which will facilitate permeability of the sperm membrane to ions and initiate the second process, that is the acrosomal reaction. Capicitation is associated with increased respiratory activity and motility of sperms. Compounds may be present in the plasma of cauda epididymis and in seminal plasma which prevent capacitation. A compound has been isolated in the epididymal plasma of hamsters which was shown to prevent capacitation and acrosome reaction *in vitro*. A glycoprotein with a molecular weight of 37kDa is deposited on rat spermatozoa during its passage through the male reproductive tract. These compounds termed as 'decapacitation factors' are present in seminal plasma. Sialoproteins or glycoproteins which are inhibitors of specific enzymes could also be involved in decapacitation. Proteinase inhibitors present in cervical mucus could also play a specific role in inhibiting capacitation. A detailed search has been made for ligands which can bind to the plasma membrane and induce capacitation. Enzymes such as α-amylase, and β-gluronidase may expose receptive sites on membrane. *In vitro* studies have revealed that a diverse number of compounds bind to the acrosome or modify its properties and they include ligands involved in capacitation. These compounds include enzymes such as lactic dehydrogenase (LDH), cAMP and hCG-β present on human spermatozoa. Proteins in serum or follicular fluid could be involved in capacitation because they induce acrosome reaction. One of them has been highly purified and identified as albumin. Albumin may also potentiate the binding of catecholamines to the sperm membrane as adrenergic receptors on spermatozoa are involved in capacitation, by enhancing sperm motility and inducing the acrosome reaction. The catecholamines by acting via cAMP may mediate Ca^{++} fluxe across the membrane and thereby induce activation of acrosomal enzymes.

VI. ACROSOME REACTION

Capacitation allows for the acrosomal reaction which is a prerequisite for fertilization in mammals. It presumably exposes enzymes which permit spermatozoa to penetrate between the commulus cells and attach to and/or penetrate through zona pellucida of the oocyte. This reaction occurs when a spermatozoon is in the vicinity of, or in contact with the cummulus mass. It involves multiple points of fusion between the plasma membrane and the outer acrosomal membrane followed by vesicle formation and shedding of these membranes from the sperm head. This process results in the release of acrosomal enzymes like hyaluronidase which permits the sperm to penetrate between cummulus mass and coronal cells. There is also movement of calcium ions in the acrosome reaction which has been related to enzyme activation, microfilament contractions and membrane fusion processes.

The acrosome represents a specialized lysosome on account of its content of hydrolytic enzymes like, hyaluronidase, proteinase, β-N-acetyl glucosaminidase, neuraminidase, acid phosphatase, aryl sulphatase, phospholipase A. Acrosin plays a key role in zona pellucida penetration. It is a typical serine proteinase cleaving carboxyl groups of arginine and lysine. Sperm heads contain an acrosin inhibitor—acrostatin, it remains to be seen whether acrosin and acrostatin are somehow complexed together in the sperm. Acrosin exists in the inactive proacrosin form. Activation of proacrosin probably occurs as an immediate consequence of the acrosome reaction. Acrosin participates in egg fertilization by enabling the spermatozoon to cross the zona pellucida. The other lysosomal hydrolytic enzymes may also play active role in the sperm-egg interaction.

The binding of spermatozoa to the zona pellucida and their movement through it, is highly species specific. Receptors are thought to be present on sperm plasma membrane which binds to the zona pellucida. In most mammalian species including the human, zona pellucida consists of three families of acidic glycoproteins known as ZP1, ZP2 and ZP3. Primary sperm binding is mediated by ZP3 and subsequently consolidated by sperm interaction with ZP2. Thus, ZP3 plays a potent role in the acrosome reaction.

The process of fertilization has been divided into five distinct stages as shown in Figure 95.4:

Stage 0: Capacitated spermatozoon passes through the corona radiata and zona pellucida. The penetration occurs in a curved oblique manner with a slit left in the zona. Acrosin is widely believed to be the zona lysin. On entering the perivitelline space the sperm head lies flat on the vitellus and its plasma membranes fuses with that of the vitellus.

Stage 1: The head then sinks into the egg cytoplasm, the cortical granules of the egg disappear and the second maturation division of the oocyte is resumed. As fusion occurs, the egg surface contracts and this may help in incorporating the spermatozoon into egg cytoplasm. Cortical granules are present in egg cytoplasm just beneath the plasma membrane. Fusion initiates changes in membrane surface and leads to exocytosis of cortical granules and with their discharge on to the surface of the plasma membrane the major black to polyspermy occurs. The enzymes present in cortical granules are though to inactivate the sperm-egg binding reaction and causes hardening of the membrane. After gamete fusion, exocytosis of the oocyte cortical granule (cortical reaction) leads to partial hydrolysis of ZP3 and ZP2 by a cortical granule protease. The resulting

ZP3f and ZP2f forms have impaired sperm binding ability, leading to a block in polyspermy.

Stage 2: The sperm head swells and is engulfed by the cytoplasm. The second polar body of the oocyte is extruded into the perivitelline space.

Stage 3: The sperm head forms the male pronucleus, the mid piece though separated from the head remains near the male pronucleus. The egg chromatin forms the female pronucleus.

Stage 4: Pronuclei enlarge, in some species the female pronucleus is distinguished by its smaller size and asymmetrical distribution of DNA, nucleoli appear and nuclear envelope formed.

Stage 5: It is the stage of syngamy. The male and female pronuclei come into contact in a subcentral position. The nuclear membrane break, the nucleoli fade out and the chromosome groups complete their condensation and assemble together, uniting the maternal and paternal hereditary contributions. The zygote stage in mammalian development is formed and it is in the prophase of the first cleavage division.

VII.   *IN VITRO* FERTILIZATION

Fertilization of mammalian egg *in vitro* can be achieved in several species including man and the conditions required in

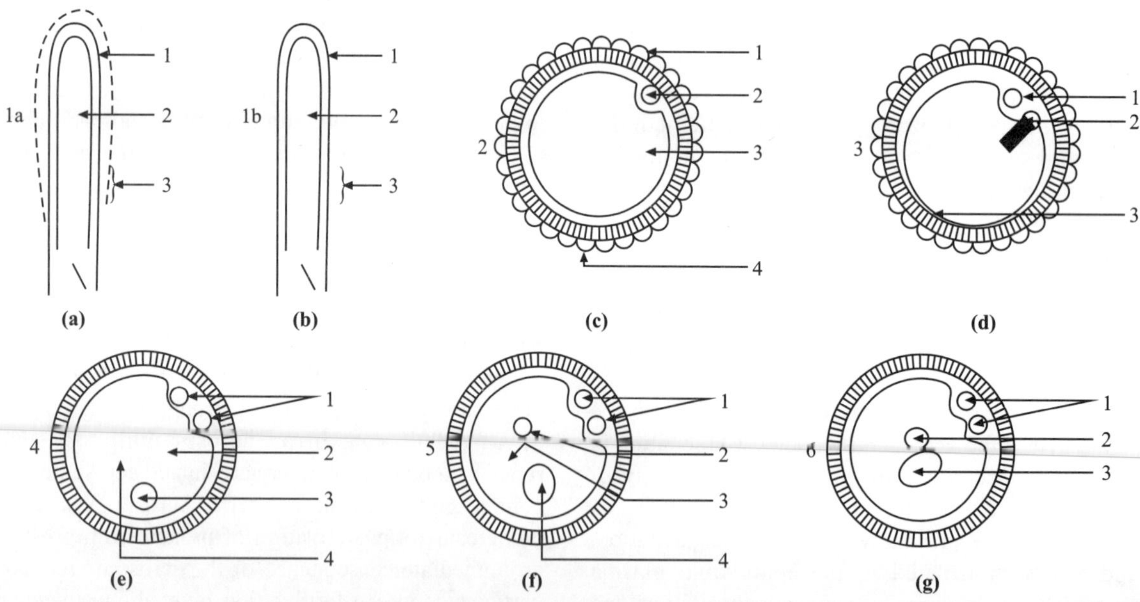

Figure 95.4   Sequence of structural changes in the sperm and the egg during fertilization. (a) and (b) Sperm undergoing acrosome reaction. (a) 1. Vesicles; 2. Inner acrosomal membrane; 3. Nucleus; 4. Equatorial segment. (b) 1. Inner acrosomal membrane; 2. Nucleus; 3. Equatorial segment. (c) Penetration of the zona pellucida (stages 0–1). 1. Zona pellucida; 2. 1st polar body; 3. 2nd meiotic spindle; 4. Sperm. (d)–(g) Sperm entry into the vitellus (stages 2–5). (d) 1. 1st polar body; 2. 2nd polar body; 3. Sperm. (e) 1. Polar bodies; 2. Female chromatin; 3. Sperm head; 4. Mid-piece. (f) 1. Polar bodies; 2. Female pronucleus; 3. Male pronucleus; 4. Mid-piece. (g) 1. Polar bodies; 2. Female pronucleus; 3. Male pronucleus.

this process have been well defined. The techniques used for fertilization *in vitro* are well standardized, and simple media such as Tyrode's or Earle's saline are reinforced with serum, follicular fluid or bovine serum albumin. The pH is kept at 7.2 or 7.3 and the osmotic pressure close to that in plasma. The gas phases used are 5% CO_2 in air, or 5% O_2 and 90% nitrogen. A reduced O_2 tension has been recommended to avoid inhibitory effects on oocyte maturation, and embryo cleavage after fertilization has taken place. Temperatures ranging from 33–34° C to 37° C have been used. Ejaculated spermatozoa have been used for fertilization of human eggs *in vitro* and the oocytes obtained by aspiration from ovaries of patients by laparoscopy just before ovulation was expected. Studies have shown that critical attention must be paid to the timing of ova collection following the LH surge. Successful fertilization and formation of cleavage stages of human preimplantation embryos have been reported as also delivery of healthy babies after the reimplantation of embryos into the mother. This method has proved to be very useful in treatment of infertility.

Recently assisted fertilization programmes have been developed to circumvent the barriers that prevent sperm access to ooplasm, namely the zona pellucida and the ooplasmic membrane. Successful fertilization, embryo development, pregnancy, and birth have been achieved after partial zonadissection (PZD), subzonal insemination (SUZI) and intra-cytoplasmic sperm injection (ICSI). ICSI can also be achieved with epididymal or testicular spermatozoa in cases of obstructive azoospermia or non-obstructive azoospermia due to Sertoli cell only syndrome or maturation arrest.

VIII. ABNORMAL FERTILIZATION

Polyploid conditions arise through errors in fertilization of the first cleavage. Triploidy is due to the presence of an additional set of chromosomes, either maternal or paternal. Tetraploidy arise due to suppression of the first cleavage division. Trisomy and monosomy may arise through errors in cell division, during meiosis or mitosis. The majority of foetuses die during gestation. The use of molecular biology tools to analyse specific genes in blastomeres or trophectoderm excised from human embryo fertilized and grown *in vitro* allows for preimplantation diagnoses of monosomy and trisomy associated with chromosomes 13, 18, 21 and X/Y. These monosomies and trisomies are the ones that develop to birth. Replacement of embryo free of these conditions will ensure that such births are avoided

IX. PREIMPLANTATION DEVELOPMENT AND IMPLANTATION OF THE MAMMALIAN EMBRYO

The embryo after fertilization goes through a number of cell cleavages. Embryo cell cleavage occurs with mitotic divisions and the cytoplasm of the oocyte is parcelled out into blastomeres with any increase in the size of the embryo (Figure 95.5).

The nucleus of cytoplasm ratio declines becoming more similar to that found in somatic ells. Cell division leads to the formation of a stage called morula. The cells of morula stage are in the shape of a ball of cells, and sometimes their cell membranes are not clearly discernible as they remain in a compacted form. This is followed by the appearance of cavity (blastocoels) in it, forming the blastocyst (Figure 95.6). The cells of the blastocyst are arranged in an outer single layer of cells, trophoblast and the inner layer of cells called the inner cell mass. The cells of inner cell mass form the embryo proper while the trophoblast cells lead to formation of extraembryonic (chorionic) membranes. The fluid in the blastocoels is formed by the active transport of ions and an inward movement of water. The zona pellucid is retained till the blastocyst stage. It is then shed by the action of embryo itself through hatching or it undergoes lysis by the action of uterine proteases. The zona free form of blastocyst under goes implantation.

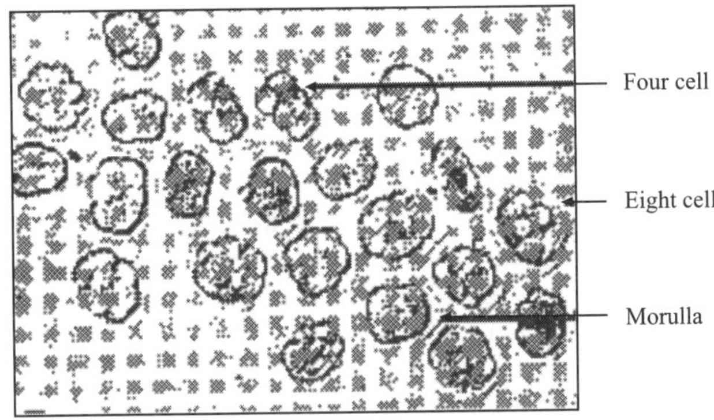

Figure 95.5   Cleavage stage mouse embryos developing *in vitro*.

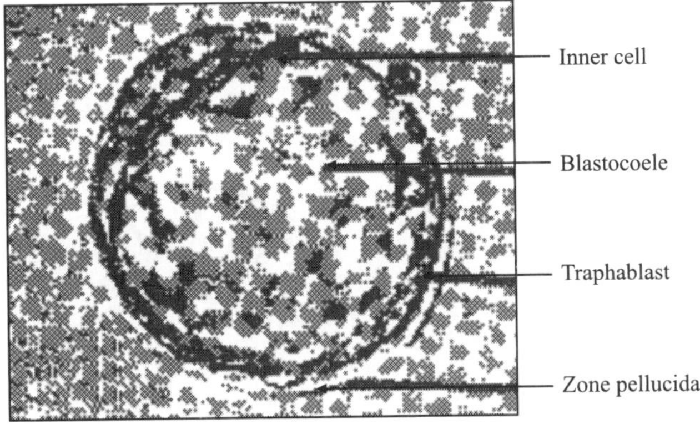

Figure 95.6   A mouse blastocyst developed *in vitro*.

The embryo after its fertilization in the ampullary region of the oviduct is slowly transported down the oviduct to reach the uterus. Embryo transport processes are probably under the regulation of steroid hormones as well as neurotransmitter substances which will regulate the flow of cilia and fluid in the tube as well as the contractile machinery of oviduct musculature. A proper coordination must exist in embryo development and transport mechanisms so that the fertilized egg reaches the uterine cavity in age and stage synchronized manner. If the arrival of the fertilized egg into the uterus, it leads to its expulsion from the genital tract or it gets degenerated.

In primates, including human transfer of early stage embryo (even one cell stage) into uterine cavity may not hinder its development to blastocyst and subsequent implantation and birth, but the probability of successful implantation is higher when the embryo transportation in the uterine cavity is age and stage synchronized. In lower mammals, however, the uterine environment is hostile for pre-morula stage egg for its development and differentiation. The uterus undergoes a series of morphological and biochemical changes which render the uterus receptive to the blastocyst. The period of uterine receptivity is for a limited time only. Once this period is over, the uterus again becomes refractory to the blastocyst. In lower mammalian species the window for implantation is very narrow and extends for 24–48 hours. However, in primates the extent of the period of uterine receptivity for blastocyst implantation is much wider and may extend for 4–7 days in the mid-luteal phase of the cycle. Steroid hormones are necessary for conditioning the uterus for ovo-implantation. The hormonal requirements for the induction of uterine receptivity are well established for various mammalian species, though we have not much understanding of these events and their control mechanisms in primates. It has been documented that in the primates, the luteal phase ovarian estrogen is not essential and that progesterone alone is sufficient for inducing secretory phase differentiation. However, a permissive role of estrogen from local endometrial and embryonic sources for supporting the implantation reaction cannot be ruled out. The factors which control the preimplantation development and differentiation of mammalian embryo remain as yet unknown. The need for a close synchrony between embryo and endometrial development and differentiation is recognized. The presence of several growth factors and cytokines as well as their receptors in embryonic and endometrial cells has been documented. It has been suggested that autocrine and paracrine actions of such cytokines and growth factors play a role in modulating embryonic development and endometrial receptivity for implantation. Implantation involves the attachment of the blastocyst to the wall of the uterus with, in some cases the embedding of the entire ovum in the uterine wall. Implantation is essentially an interactive process between two cell types: the trophectodermal cells with the uterine luminal epithelium and it is associated with cellular phagocytosis/autolysis of endometrial cells causing a remodelling or reprocessing of its cellular matrix. There is extensive involvement of lysosomes and their hydrolytic enzymes at sites of implantation. With blastocyst adhesion and attachment to the uterus, the stromal cells start to differentiate into large polyploid decidual cells and form the decidual zone surrounding implanting nidus. The timing of the implantation process relative to the day of fertilization varies from species to species. In the laboratory rodent species like in the rat it occurs on the evening of day 5 of pregnancy and in the mouse on the evening of day 4, while in the human as well as in the rhesus monkey it occurs around days 7–10 of gestation (Figure 95.7).

Understanding of the process of human implantation is highly limited. Based on studies done in rhesus monkey and baboon it has been shown that blastocysts adhere to uterine epithelium along its embryonal pole. Penetration of luminal epithelium occurs around day 9 by syncytiotrophoblast cells which penetrate and breach the basal lamina. This is followed by a planar expansion of trophoblast cells laterally forming a trophoblast plate of cells. At this time there is some degree of fusion of cyto and syncytiotrophoblast cells with cells of endometrium forming large, multinucleated cells known as

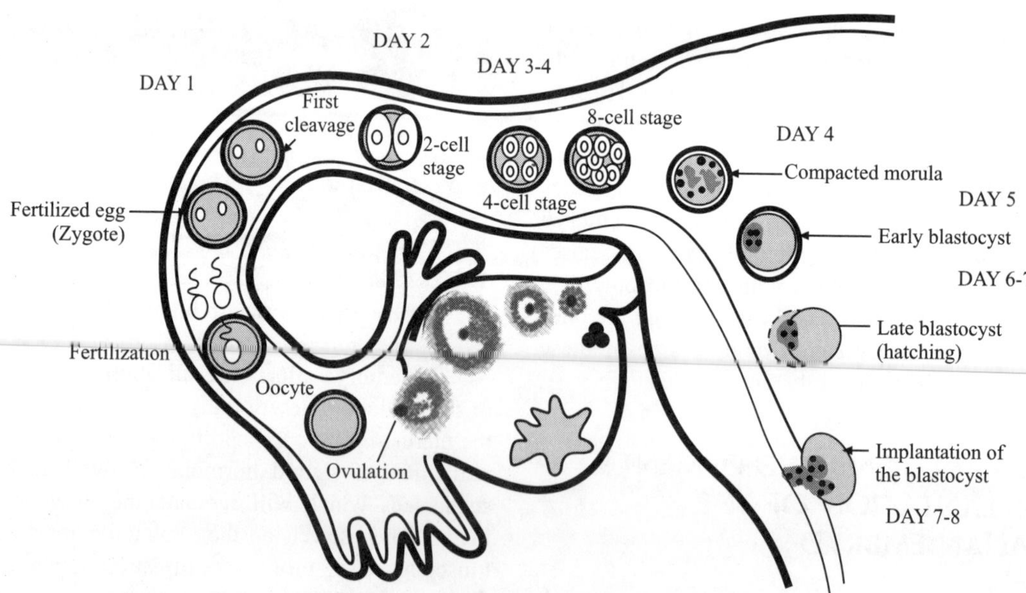

Figure 95.7   Typical synchrony between developmental age and stage of preimplantation embryo starting with fertilization on day 0 followed by blastocyst hatching on days 6 to 7 and blastocyst implantation on days 7 to 8 after ovulation in the human. Concomittantly uterine endometrium undergoes a series of physiological adaptive changes making it receptive to the blastocyst.

teterokaryons. Proliferation of cyto and syncytiotrophoblast cells lead to differentiation in syncytiotrophoblasts from proliferative to absorptive cell type. Lacunae are formed between these cells and as trophoblasts tap maternal blood vessels and glands, the lacunae get infiltrated with maternal blood and primary villi are formed. Invasive cytotrophoblast cells are found to invade glands and arterioles. The area beneath and surrounding the implanted blastocyst becomes highly edematous with dilated venules. The formation of secondary villi occurs with infiltration of extraembryonic mesodermal cells between cytotrophoblast cells forming embryonic compartment. The biochemical events associated with trophoblast adhesion and invasion to maternal endometrium as well as cellular remodelling of endometrial matrix during implantation remain largely unknown. It has been suggested that a number of growth factors, cytokines, metalloproteases and tissue inhibitors of proteases play a role in these biochemical events.

X. CONTROL OF IMPLANTATION

Potential anti-implantation agents fall under three categories: hormones, cell adhesion molecules, cytokines and growth factors. The likely targets for these agents are: impairment of endometrial receptivity, impairment of embryo's ability to implant, and impairment of embryo-uterine interaction. Mifepristone, a potent antiprogestin, administered during early post-ovulatory stage has been found to be effective in providing pregnancy protection by inducing endometrial desynchrony through altered glandular and vascular functions.

SUGGESTIONS FOR FURTHER READING

Austin C.R. and Short R.V. (1972), *Germ Cells and Fertilization,* Cambridge University Press.

Edwards R.G. (1980), *Conception in the Human Female,* Academic Press.

Johnson M.H. (2013), *Essential Reproduction,* Wiley-Blackwell.

Neill J.D. (2005), *Knobil and Neill's Physiology of Reproduction*, Academic Press.

96

Biochemistry of Pregnancy

Jayshree Sengupta and Debabrata Ghosh

<table>
<tr><td colspan="2" align="center">CONTENTS</td></tr>
<tr><td>I.</td><td>Introduction</td></tr>
<tr><td>II.</td><td>Fertilization and Implantation of Ovum</td></tr>
<tr><td>III.</td><td>Hormones of Pregnancy</td></tr>
<tr><td></td><td>A. Progesterone</td></tr>
<tr><td></td><td>B. Estrogens</td></tr>
<tr><td></td><td>C. Human placental lactogen (HPL)</td></tr>
<tr><td></td><td>D. Human chorionic gonadotrophin (HCG)</td></tr>
<tr><td></td><td>E. Human chorionic thyrotrophic hormone (HCT)</td></tr>
<tr><td>IV.</td><td>Concept of Fetoplacental Unit</td></tr>
<tr><td></td><td>A. Fetoplacental function tests</td></tr>
<tr><td>V.</td><td>Metabolic Changes during Pregnancy</td></tr>
<tr><td></td><td>A. Weight gain</td></tr>
<tr><td></td><td>B. Water metabolism</td></tr>
<tr><td></td><td>C. Fat metabolism</td></tr>
<tr><td></td><td>D. Protein metabolism</td></tr>
<tr><td></td><td>D. Mineral exchange</td></tr>
</table>

I. INTRODUCTION

The human female from the onset of sexual maturity prepares, approximately each lunar month, for the possible occurrence of pregnancy. Ovulation occurs at about mid-cycle and the released ovum is swept through the open end of the fallopian tube into the oviduct where it meets with a sperm and fertilization takes place. With fertilization the ovum completes its maturation with extrusion of second polar body into the periviteltive space. Once the ovum gets fertilized a series of events takes place leading to pregnancy which culminates in the birth of a new offspring and helps in the propagation of the species. This chapter deals with the fertilization and implantation of the ovum, hormones of pregnancy, pregnancy diagnosis, concept of fetoplacental unit, fetoplacental function tests and metabolic changes during pregnancy.

II. FERTILIZATION, GROWTH AND DIFFERENTIATION OF OVUM

It takes less than 15 minutes after coitus for the sperm to be transported through the uterus and fallopian tube, to the site of fertilization. This represents quite a rapid movement of the sperm and the female genital tract helps in this movement. After ovulation an ovum is fertilizable for about 24 hours. At the same time, the lifespan of the sperm in the female genital tract is between 24–72 hours. If coitus occurs at the appropriate time when the ovum and sperm meet, fertilization can take place. The fertilization takes place either before the ovum enters the fallopian tube or just on its entry into the tubes. The fertilized ovum is transported through the fallopian tube and reaches the uterus. It stays in the uterus for 4–5 days before the implantation takes place in the endometrium (see Chapter 95 for details).

The growth and differentiation of embryo is a critically timed event depending upon a precise hormonal milieu. Differentiation is accompanied by changes in the ultra structure with elaboration of the endoplasmic reticulum, numerous ribosomes and modifications of the mitochondria. These indicate active protein synthesis in morulae and blastocyst. RNA and DNA synthesis occurs with sharp increase in the activities of several enzymes during the stage of transition of morula and blastocyst. RNA and DNA synthesis occurs with sharp increase in the activities of several enzymes during the stage of transition of morula to blastocyst. Glucose metabolism in the developing embryo is crucial in sustaining growth and differentiation by providing the energy source stage several such specific proteins are secreted into the uterine lumen. There is a considerable debate whether the preimplantation stage embryo can synthesize and secrete hormones. Most workers agree that by day 10–12 of gestation, hCG is is secreted by the trophoblast cells and is detectable in maternal plasma.

III. BLASTOCYST IMPLANTATION

Implantation is associated with an increasing degree of interaction between blastocyst and the uterine endometrium. The process of implantation is a continuum of events which begin with the apposition of a zona free blastocyst on the endometrial lining and is completed with the formation of the placenta. Implantation leads to a more intimate physical and morphological relationship between the blastocyst and endometrium. In humans there is hemochorial type of placentation. The sequence of morphological events at implantation are apposition, adhesion of the blastocyst to the uterine epithelium, penetration of endometrial epithelium and tapping of maternal blood Vessels by trophoblast cells as the embryo gets lodged within the endometrial stroma.

Implantation occurs on days 6–8 of gestation. There is increase capillary permeability at the site of embryo attachment and this is one of the first sign of implantation. The first stage in the implantation process is apposition of trphoblast to endometrial cells. Microvilli present on both syncytiotrophoblast and endometrium cell surfaces become considerably to occur. Surface glycoproteins and charge particles are considered to play role in apposition and adhesion of blastocyst onto the uterine epithelial cells. The trophoblast cells covering the inner cell mass first penetrate through the endometrial cells. At this stage the trophoblast contiguous with, and invading the endometrium, coalesce to become an amorphous multinucleated syncytium. These syncytiotrophoblast cells are derived from cytotrophoblast hyperplasia cells. The uterus responds to the presence of the blastocyst with oedema hyperplasia and hypertrophy of the endothelial cells. Finally, decidualization of stromal fibroblasts occurs. During the menstrual cycle, stromal cells undergo predecidual changes associated with very high levels of progesterone but true decidualization is an integral response of the uterus to blastocyst implantation. Decidual cells are large polyploid cells, rich in glycogen and lipid. Distinct gap junctions are found between adjacent cells. The decidual cells clearly have secretory functions as they possess large Golgi complexes and well developed granular endoplasmic reticulum. They may serve to provide nutrition to the developing conceptus, protect the embryo, protect maternal tissue from inasive trophoblast cells and also have endocrine function such as synthesis of prolactin and other peptide and protein factors.

Two days after implantation the conceptus is already buried in the endometrium and the trophoblast cells constitute masses of syncytial cells bearing large polyploidy nuclei. At

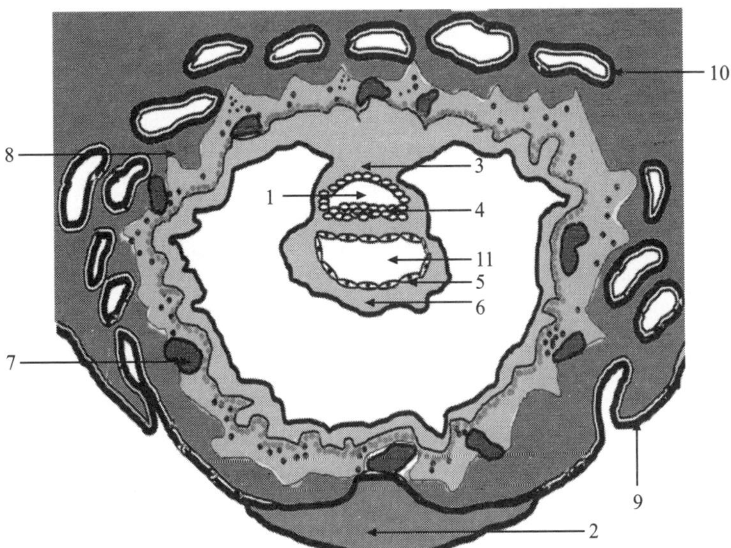

Figure 96.1   Typical implantation stage embryo on day 8-9 after fertilization. 1, amniotic cavity; 2, blood clot at the site of initial implantation; 3, body stalk, or connective stalk later forming the placental cord region with placental blood vessels; 4, embryonic ectoderm that contributes to embryonic and placental membrane development; 5, endoderm that contributes to embryo development; 6, mesoderm, consisting of both embryonic mesoderm (in the trilaminar embryonic disc) and extraembryonic mesoderm (outside the trilaminar embryonic disc); 7, maternal spiral arteries; 8, trophoblast that contributes to placental development; 9, uterine epithelium the epithelial layer that lines the unerus; 10, uterine glands that secrete nutrients to support the initial growth both before and after implantation; 11, yolk sac which is the endoderm-lined and extraembryonic mesoderm-covered cavity and it eventually contribute to the fetal gastrointestinal tract, blood and blood vessels. After Ramsey E (The Placenta, Praeger Publishers, 1982).

the blastocyst stage, the blastocoels or cavity is lined by a single layer of trophoblast stage, the blastocoel or cavity is lined by a single layer of trophoblast cells and the inner cell mass lies at one pole. Cells of the inner cell mass differentiate into endoderm and ectoderm to form the germ disc. As the embryo gets implanted there is oedema in the endometrium with the vessels becoming hyperaemic and sinusoidal at the site of attachment. Implantation usually occurs in the posterior wall of the uterus near the fundus.

Cytotrophoblast cells are large and polyhedral and overlie and embryonic disc. The syncytiotrophoblast cells surround small capillaries and thus begins the formation of trophoblastic lacunae become continuous with maternal vessels thus beginning the formation of hemochorial placenta. Hemochorial placenta means that the chorionic villi are floating in the maternal pool of blood. The surrounding stromal cells become decidualized initially showing predecidual characteristics. The cytotrophoblast forms clumps of tissue which project into syncytiotrophoblast. These represent the primary chorionic villi. Maternal arteries enlarge and form sinusoids at the implantation site. The uterine endometrium is now decidualized and the epithelium at the site of blastocyst invasion is repaired. The decidua is identified into three prominent regions: deciduas basalis at the embryonic pole, deciduas capsularis over the rest of chorion covering the sac and the deciduas parietalis over the remainder of the uterus (Figure 96.1). The syncytiotrophoblast becomes folded and the lacunae coalesce to form intervillous spaces.

IV. MATERNAL-FETAL-PLACENTAL UNIT

Neither the placenta nor the fetus alone can lead to normal growth and development of the fetus. It is the maternal-foeto-placental unit which is responsible for the the maintenance of pregnancy and parturition. The different sterroidogenic pathways by which steroid production occurs with the help of maternal, placental and fetal compartments functioning as a well coordinated unit are now well established. The placenta consists of apposition of fetal and maternal tissues for purposes of physiologic exchange. Placenta formation begins by about day 13 of gestation and is completed by 10 to 11 weeks. The maternal blood vessels are tapped by syncytiotrophoblast cells.

The mature placenta is a flat disc-like structure weighing about 500 g. The umbilical cord is attached near its centre and the fused amnion and chorion leave (membranes) attached to it, all round the periphery. The placenta has a dual origin, fetal and maternal. The fetal component of the placenta consists of the chorion and the villi which arise from the chorionic plate. The maternal of the chorion and the villi which arise from the chorionic plate. The maternal part consists of the decidua basalis. The placenta has a dual circulation. (i) The foetoplacental circulation which delivers deoxygenated blood from the fetus to the chorionic villi by means of two umbilical arteries and returns oxygenated blood to the fetus through a single umbilical vein. (ii) The uteroplacental circulation which delivers oxygenated maternal blood to the intervillous space through multiple spiral arteries that penetrate the basal plate. Deoxygenated blood is drained back through the veins leaving the chorionic plate. The maternal and fetal circulations are separated by a placental barrier. The barrier allows for simple diffusional exchange between the maternal and fetal circulations only for low molecular weight substances such a Na^+, water, urea or non-polar molecules such as cholesterol and non-conjugated steroids. Hexose sugars, conjugated steroids, amino acids, nucleotides, water soluble vitamins and plasma proteins gain an entry into the fetal circulation only through special transport mechanisms. The fetus is unable to synthesize glucose since the enzymes required for gluconeogenesis are not active till birth. Glucose is transferred to the fetus through facilitated diffusion in which concentration gradients help and also through special carrier systems. Similarly, active transport mechanisms help in providing the fetus with the other essential nutrients.

V. HORMONES OF PREGNANCY

In pregnancy the placenta secretes large quantities of progesterone, estrogens, human chorionic gonadotropin and human placental lactogen. The pattern of secretion and functions of these hormones are discussed below:

A. Progesterone

Progesterone is essential for the maintenance of pregnancy. In the initial stages of gestation progesterone is secreted by the corpus luteum of the ovary. At a later stage, at about 56 days of gestation, the placenta takes over the function of corpus luteum and starts secreting progesterone and estrogens. At this stage, even if the corpus luteum is removed, pregnancy will continue normally. The rate of secretion of progesterone starts increasing after implantation and increases at a much higher rate during the latter half of pregnancy. It is estimated that the placenta may be producing as much as 1 g of progesterone per day towards the end of pregnancy.

Progesterone has a number of functions in the maintenance of pregnancy. It causes decidual cells to develop in the endometrium. These play an important role in the nutrition of the embryo. Progesterone decreases the contractility of the uterus thus preventing spontaneous abortion. It also plays a role in the development of the fertilized ovum prior to implantation by producing secretions in the fallopian tubes and in the uterus for the nutrition of the developing morula and blastocyst. Progesterone also prepares the breast for lactation.

B. Estrogens

The levels of estrogens continue to increase after implantation. As seen in the case of progesterone, the levels of estrogens also increase rapidly during the latter half of pregnancy. All the three estrogens, viz., estrone, estradiol and estriol are produced during pregnancy. However, estriol constitutes to the extent of 90 per cent of the total estrogen secreted during the latter part of pregnancy. It is estimated that the levels of estriol increase to the extent of a thousand fold in the latter stages of the pregnancy compared with the mid-luteal phase levels of the estrogen.

The estrogens cause an enlargement of the uterus and breast, and growth of the breast glandular tissue. These hormones also cause an enlargement of the female external genitalia and relax various pelvic ligaments, enabling an easy delivery of the fetus.

C. Human Placental Lactogen (HPL)

Human placental lactogen (HPL) starts being secreted at about the 5th week of pregnancy and its rate of secretion increases with the progress of pregnancy. HPL has a prolactin-like effect on the breasts. It also has some similarities to growth hormone in inhibiting the action of insulin in glucose transport through cell membrane. HPL thus promotes cellular growth.

D. Human Chorionic Gonadotrophin (HCG)

HCG is secreted by the trophoblastic tissues. The secretion of HCG rapidly increases after implantation, reaching a maximum at about 7 weeks after ovulation. There is now evidence that HCG is secreted by embryo even before implantation. The levels of HCG then decline to a low value at 16 weeks after ovulation. The low levels then persist till about the time of parturition.

The main function of HCG is to support implantation, and the corpus luteum at the end of the menstrual cycle and cause the corpus luteum to secrete even larger quantities of progesterone and estrogens. The sex hormones then produce the desired effects for growth and maintenance of the endometrium and pregnancy. HCG also causes an increase in the size of the corpus luteum. HCG produces an interstitial cell-stimulating effect on the testis, thus resulting in the production of testosterone in the male fetus. The small quantities of testosterone greatly influence the formation of male organs. The testosterone secretion also causes a descent of the fetal testis into the scrotum.

E. Human Chorionic Thyrotrophic Hormone (HCT)

HCT is secreted from the placenta during early stages of pregnancy. It is a glycoprotein hormone distinct from TSH of pituitary origin. HCT may play a role in the development of the fetal thyroid which differentiates by 10 weeks of pregnancy.

Pregnancy diagnosis test: The pregnancy diagnosis tests are based upon the detection of human chorionic gonadotrophin (HCG) in the urine. The HCG is synthesized in the placenta throughout the pregnancy but production per g of placenta is highest during the first trimester. HCG can be detected in the serum and urine 8–10 days after fertilization, the peak values are found at about 60 days of gestation and by 80 days the values drop to a lower level and remain low till delivery.

The various assay methods for the diagnosis of pregnancy attempt to detect HCG. The accuracy of these tests is of the order of 99 per cent. The bioassay procedures like Aschheim-Zondek test (hemorrhagic follicles or corpora lutea in the ovary), Friedman test (ovulation in the adult rabbit), Xenopus test (production of ova in female frogs or discharge of spermatozoa from male frogs), immature rat ovarian hyperaemia test (hyperaemia of the ovary), immunologic test of pregnancy, estimation of micro amounts of steroid hormones, constitute the various pregnancy diagnosis test. The bioassay procedures are lengthy and time-consuming. The immunologic test, because of its higher sensitivity, accuracy, specificity and ease of performance as well as low cost, finds widespread use. The test is based upon the fact that the HCG can inhibit the haemagglutination reaction of HCG treated red blood cells with antiserum. Enzyme linked immunoassay (ELISA) using colour end product are also available and packaged as detection kits for individual use.

Endocrine function of foetoplacental unit: According to this concept, the fetus and placenta together form a functional unit to carry out synthesis of steroidal hormones. These steroidal biosynthetic reactions cannot be carried out singularly by the placenta or fetus independently. The fetus is endowed with a great capacity to sulphurylate steroids but it has no appreciable sulphatase activities and, therefore, cannot hydrolyse steroid sulphates. On the other hand, the placenta has large amounts of certain sulphatase activities but little sulphurylation ability. This selective enzyme distribution seems to regulate the amounts of steroids made available for fetal, placental or maternal metabolism. The steroid sulphates has little biological activity compared with the free steroids. Therefore, the mechanism of sulphurylation present in many tissues of the fetus protects it from the biological effects of placental hormones.

The biosynthesis of progesterone takes place in the placenta when cholesterol is converted into pregnenolone and then into progesterone. Therefore, practically all progesterone formed in the fetoplacental unit is elaborated by the placental compartment. On the other hand, the reverse is true concerning estrogen synthesis. If the fetoplacental circulation is interrupted by ligation of the cord only a slight decrease in urinary pregnanediol results, but the estrogen excretion shows a very marked and immediate drop. The placental pregnenolone is converted into dehydroepiandrosterone and its sulphate in the fetal adrenal (Figure 96.2). The dehydroepiandrosterone liberated is converted to androstenedione and testosterone and then into estrone and estradiol but not into estriol. The formation of estriol requires fetal contribution. The fetal liver contains large quantities of 16 α-hydroxylating enzyme which is absent in the placenta. Both dehydroepiandrosterone sulphate and estrone sulphate are first 16 α-hydroxylated in the fetal liver. These 16 α-hydroxylated compounds are then hydrolysed and converted into estriol by the placenta.

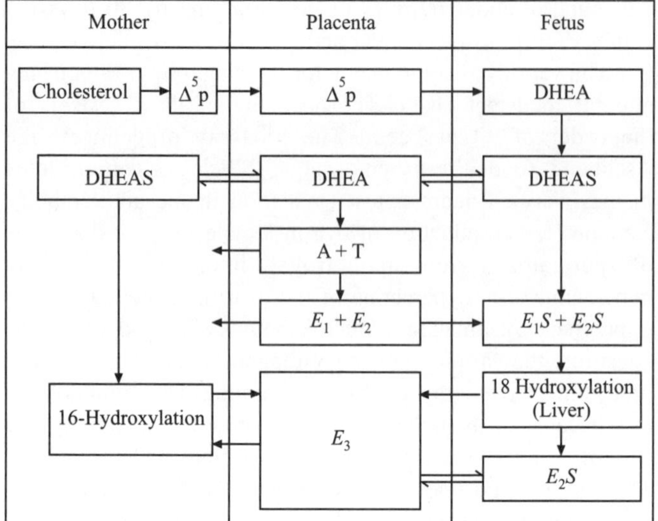

Figure 96.2 General concept of the principal pathways of estrogen synthesis in the fetoplacental unit. Δ^2p = Pregnanediol; DHEA = Dehydroepiandrosterone; DHEAS = Dehydroepiandrosterone sulphate; A = Androstenedione; T = Testosterone; E_1 = Estrone; E_1S = Estrone sulphate; E_2 = Estradiol; E_2S = Estradiol sulphate; E_3 = Estriol; E_3S = Estriol sulphate.

Fetoplacental function tests: A number of fetoplacental function tests have been developed to assess the health of the fetus during the course of a normal or abnormal pregnancy. Human pregnancy is accompanied by a spectacular rise in both steroidal and protein hormones (Figure 96.3). The fetoplacental function tests are based upon the assessment of the levels of these hormones. Among the hormones which have been used for the fetoplacental tests are human placental lactogen (HPL), human chorionic gonadotrophin (HCG), progesterone and its metabolite pregnanediol, serum estradiol, urinary estriol and a number of enzymes produced

by the placenta, such as alkaline phosphatase. However, hormones having a considerably faster disappearance rate than the enzymes, have been generally preferred for fetoplacental tests because these assays take the minimum time between fetoplacental compromise and its earliest possible detection.

HPL, HCG and progesterone are produced by the placenta and a measurement of these hormones and pregnanediol in normal and pathological pregnancy will largely tell us about the health of the placenta and its contribution to the health of the fetus. However, the other hormones like serum estradiol and urinary estriol are produced by the fetoplacental unit and, therefore, their measurement will tell us the total picture of the well-being of the fetus and the placenta. The different hormonal indices may provide useful guidelines for proper management and control of pregnancies associated with various pathological conditions. An accurate assessment of the fetoplacental function is essential to determine the optimal chances of fetal survival if pregnancy has to be terminated. In cases of pathological pregnancies based upon the fetoplacental indices the obstetrician can terminate the pregnancy at an earlier stage and save the fetus.

All other tests, except urinary estriol, require radio-immunoassay procedures involving expensive instrumentation and reagents. Based on the fact that a rapid rise in serum HPL starts about 5–6 weeks earlier than a similar rise is noted in both urinary estriol and serum estradiol, the estimation of HPL appears to be more useful in cases of spontaneous abortions in the early stages of pregnancy.

The levels of serum estradiol or estriol closely follow the pattern of urinary estriol in normal pregnancy and pathological pregnancies. However, urinary estriol being a simple and rapid method, has been extensively used for the assessment of the fetoplacental function (Figure 96.4). Serial measurements of the hormone would be necessary to determine whether fetal jeopardy is present and timely intervention may be made to save the fetus. The measurement of serum hormones, in addition to

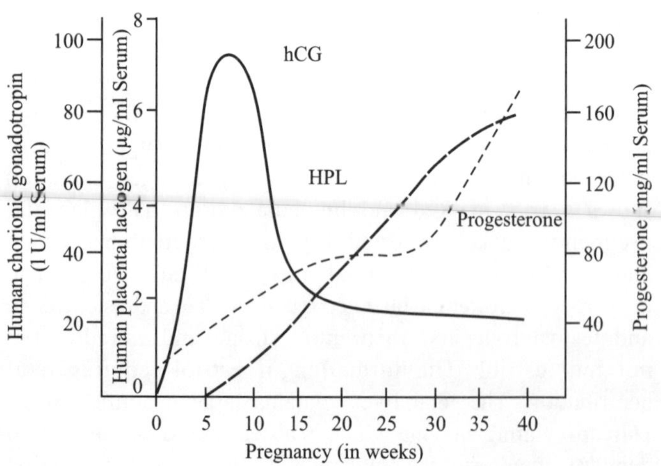

Figure 96.3 Schematic presentation of changes in human placental lactogen (HPL), human chorionic gonadotrophin (HCG) and progesterone levels throughout pregnancy.

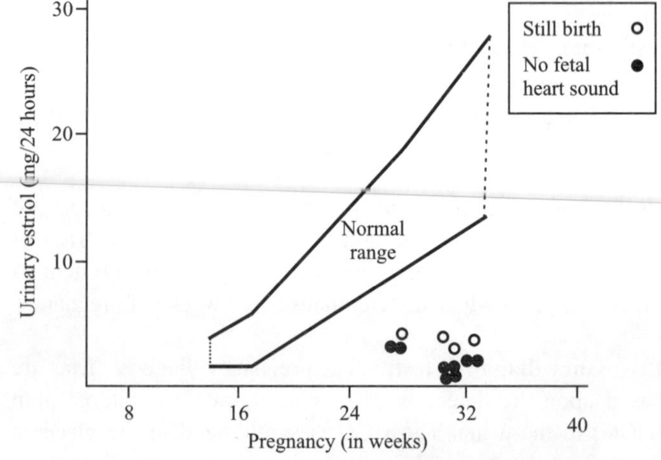

Figure 96.4 Urinary estriol level in pathological pregnancies as compared with range in normal pregnancies. Joint points indicate values in the same patient.

urinary estriol, would provide an additional help in high risk pregnancies as well as in instances when the urinary estriol is suspected of providing misleading information.

VI. METABOLIC CHANGES DURING PREGNANCY

A number of metabolic changes take place in the mother during pregnancy. The presence of a growing fetus, formation of placenta and the secretion of hormones cause a number of metabolic changes in the mother. Some of the important metabolic changes are briefly discussed below:

A. Weight Gain

During pregnancy the average weight gain is reported be 7–10.5 kg, most of it occurs during the second and third trimester. Fetus, amniotic fluid, placenta and fetal membranes contribute about 5 kg and the uterus, breasts, fluid retention and lipid accumulation contribute the rest.

B. Water Metabolism

A marked biochemical change in pregnancy is the increased water retention. Water is retained in the fetus, placenta and amniotic fluid to the extent of 3.5 litres. Retention in the breast, uterus and other tissues makes for another 1 litre. The increase in blood volume accounts for about 1.3 litres. Sodium retention takes place in the mother to the extent of about 12 g during the second and third trimesters of pregnancy. Increased glomerular filtration rate (GFR) and increased progesterone during pregnancy cause fluid and sodium loss. However, the sodium and fluid balance is maintained by significant increase deoxycorticosterone which prevent sodium loss.

Marked water retention with the lowering of plasma osmolarity in pregnancy may result in edema in the legs during late pregnancy because of a rise in venous pressure due to compression by the gravid uterus. Rise in blood volume and lowered hematocrit in pregnancy result in decreased osmotic pressure enhancing the escape of fluid from the capillaries into the tissue bed. The reduction in serum concentration of sodium and associated anions is responsible for further lowering of plasma osmolarity.

C. Lipid Metabolism

During pregnancy there is a rise in plasma concentration of triglyceride and cholesterol and possibly of phospholipid concentration. There is enhanced mobilization of lipids from adipose tissue to raise the plasma free fatty acid (FFA) level. This most marked in the last month of gestation. Human placenta lactogen (hPL), also called human chorionic somatomammotrophin has pronounced growth promoting and lactogenic properties. It also plays a major role in the glucose 'sparing' effect during pregnancy by helping to mobilize maternal FFA for the mother's skeletal and cardiac muscles and diverting the glucose to the placenta and fetus. There is also pronounced fat storage during pregnancy. This may be due to a combination of factors such as increased calorie intake and high levels of oestrogen which tends to promote fat deposition.

D. Carbohydrate and Protein Metabolism

During pregnancy metabolism is essentially anabolic. The demands of the growing fetus are met by the mother. The fetus derives its energy almost totally from glucose, which is passed through the placenta by facilitated diffusion. Blood insulin level is raised in pregnancy, however, insulin during pregnancy shows a reduced half life possibly due to anti insulin factor(s) secreted by the placenta. During pregnancy adequate diet intake leads to substantial nitrogen retention.

E. Mineral Exchange

If the mother's diet is not adequate it may lead to a number of mineral deficiencies. Such deficiencies often occur for calcium, iron, phosphates, and vitamins. Balance studies show that during pregnancy the mother stores about 50 g of calcium and 35–40 g of phosphorus. Only half the calcium goes to the fetus, the rest being stored in the mother's tissues. It is presumably in preparation for lactation. The need of iron by the fetus and the mother during pregnancy far exceeds the normal store of non-hemoglobin iron in the mother at the time of pregnancy. If sufficient iron is not present in the food, the mother may become anaemic. The obstetrician generally prescribes multivitamins containing iron to make up for the deficiency of iron and vitamins.

SUGGESTIONS FOR FURTHER READING

Johnson M.H. (2013), *Essential Reproduction,* Wiley-Blackwell.

Neill J.D. (2005), *Knobil and Neill's Physiology of Reproduction*, Academic Press.

97

Lactation

Jayshree Sengupta and Debabrata Ghosh

CONTENTS

I. INTRODUCTION

Lactation involves all the processes underlying the capacity and ability of the breast to secret milk. This complex process allows for the synthesis of milk, its transport through the alveolar cell membrane and finally its release from the mammary gland. This ability is essentially a characteristic of mammals and perhaps with the exception of humans it is as essential for survival of the mammalian species as reproduction.

II. MAMMARY GROWTH

Although mammary glands are present in both sexes they are usually functional in the female only. During reproductive life (from puberty to menopause) the female mammary glands in human form a roughly hemispherical organ situated on the anterior thoracic wall. The main anatomical characteristics of the gland are as follows. Behind the gland lie the pectoral muscles above and the serratus anterior below, and laterally and behind is the rib cage. The mammary gland extends from the second rib above to the sixth costal cartilage below. Medially it reaches the edge of the sternum and laterally the mid-axillary line. In general the gland is composed of a mixture of fatty and glandular tissue interspersed with fibrous tissue septa. It can be roughly divided into fifteen to twenty lobes each of which can be further subdivided into a number of lobules. Scattered in the lobules and mixed in the fatty tissue are groups of alveoli or milk-secreting glands. Small ducts lead from the collections of alveoli and each duct expands to form a lactiferous sinus which in turn opens directly upon the surface of the nipple.

In the female the development of the mammary gland proceeds during various stages of the sexual life of the animal, namely birth to puberty, during recurrent sexual cycles, pregnancy, lactation and finally involution.

At birth in the female the duct development has progressed very little beyond that apparent in the male at this stage. The prepubertal development is characterized by a gradual increase in the extent of the mammary duct system (Figure 97.1). In general no alveolar development occurs in the prepubertal period. In well nourished girls the breast may increase in size before puberty owing to increase in the fatty connective tissue. At puberty, under the influence of the hormonal changes which take place at this time, the breast increases in size until adult dimensions are reached. Microscopically apart from the fat and connective tissue, ducts form the major part of the actual breast parenchyma seen in the resting state. The terminal ducts have end-buds consisting of alveolar cells and it is the hormonal state of the female that influences the development of mammary gland.

The hormones which play a role in the development of mammary glands are broadly classified as mammogenic hormones. The ovarian hormones—estrogen and progesterone—play a significant role in the gland development but require the support of the hormones of the anterior pituitary and adrenal steroids as well. Among the anterior pituitary hormones, growth hormone and prolactin are required along with the

1150

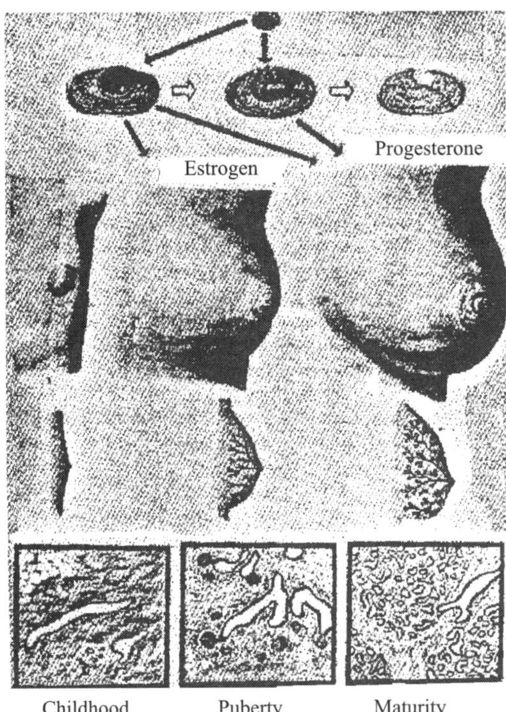

Figure 97.1 The developmental stages of the mammary gland. After Oppenheimer, E. (*The Ciba Collection of Medical Illustrations, Volume 2, Reproductive System*).

ovarian steroids, and the particular combination of hormones or hormone ratios necessary for optimal development vary with different species. In the human, clinical studies indicate that patients with Sheehan's disease do not lactate and their breasts atrophy. Similarly, patients with ovarian agenesis do not show breast development even with intact pituitary. It therefore appears that in, general, estrogens affect the ductal development while progesterone influences the growth of the alveoli and for these changes to occur, prolactin and growth hormone are necessary. Other hormones such as thyroid and insulin appear to have an indirect role in breast development primarily by their effects on metabolic processes yielding energy. Figure 97.2 shows the hormonal involvement in the development of the mammary gland.

Experimental studies on animals indicate that the extent of alveolar development under the hormonal treatment may have some bearing in neoplasia as well. In mice it has been shown that a particular strain which possessed neither the genetic susceptibility to mammary tumour formation, nor the mammary tumour virus showed the least alveolar response to the hormonal combination used. It was concluded that the mammary tumour virus influenced the extent of alveolar development under the hormonal treatment but had no relation to overall gland growth. In a sexually mature virgin female the type of mammary development is related to the type of cycle characteristic of the species in question. The typical hormonal pattern of the menstrual cycle consists of a rise in estrogen level during the proliferative phase, which reaches a peak at

the time of ovulation (day 14 in a 28 day cycle). There is then a slight drop followed by a rise to a higher peak just before menstruation starts. During this latter half of the cycle there is also a rise in the progesterone level to a peak about a week before menstruation. These hormonal changes are dependent upon and in turn influence, the pituitary gonadotropic hormones. Thus, in women the ovarian cycle is accompanied by cyclic changes in the mammary parenchyma, lobule-alveolar growth occurring during luteal phase followed by complete regression in menstruation. Clinically many women notice fullness, tightness, heaviness and occasionally pain in the breasts just before menstruation. In the normal woman the significance of subjective symptoms is questionable, although in pathological conditions increase of symptoms may be important.

An obvious point, but one sometimes forgotten is that breast development for lactation takes place in pregnancy. During human pregnancy, the breasts enlarge and there is a considerable increase in alveolar structures. Two main phases of mammary development can be distinguished during pregnancy, namely the growth phase and the secretory phase. The essential difference between hyperplasia of the mammary parenchyma which is characteristic of the first half or two-thirds of pregnancy and the hypertrophy of the mammary gland due to distension of the epithelial cells and alveolar lumina with secretion which occurs during the later stages of this condition, is thus emphasized. It should be understood however that there is no sharp division between these two phases of mammary gland development, inasmuch as there is evidence that glandular growth may proceed side by side with the gradual development of the secretory process. In the case of mice, formation of buds occur along the walls of the ducts and particularly at their ends. The end-buds elongate to form intralobular ducts, secondary buds later forming along the wall and at the ends of these. Gradually these elongate to interlobular ducts at the end of which more buds form which develop into alveoli.

Endocrinologically, pregnancy can be divided into two phases: an initial ovarian one, when the ovary forms the steroids that maintain pregnancy, followed by the placental phase when the steroid hormones originate in the fetoplacental unit. Initially the excretion of chorionic gonadotropin increases rapidly reaching a peak at the 80th to 90th day after the last period. The secretion of estrogen and progesterone by the corpus luteum is stimulated by chorionic gonadotropin. The urinary values of pregnancy, pregnenediol and of estrogen rise relatively slowly. After 60–90 days of pregnancy, the steroid hormones produced by the placenta can maintain pregnancy. Thus the increased circulating levels of ovarian steroids are predominantly responsible for the mammary gland development in pregnancy. The prolactin- and growth hormone-like factor produced by the placenta may also have considerable influence on the mammary gland development. In this connection it is to be noted that these maternal hormones are also responsible for the mammary hyperplasia that frequently occurs in the newborn infant. In these breasts a number of ducts with epithelial lining, the future milk ducts and glandular lobules,

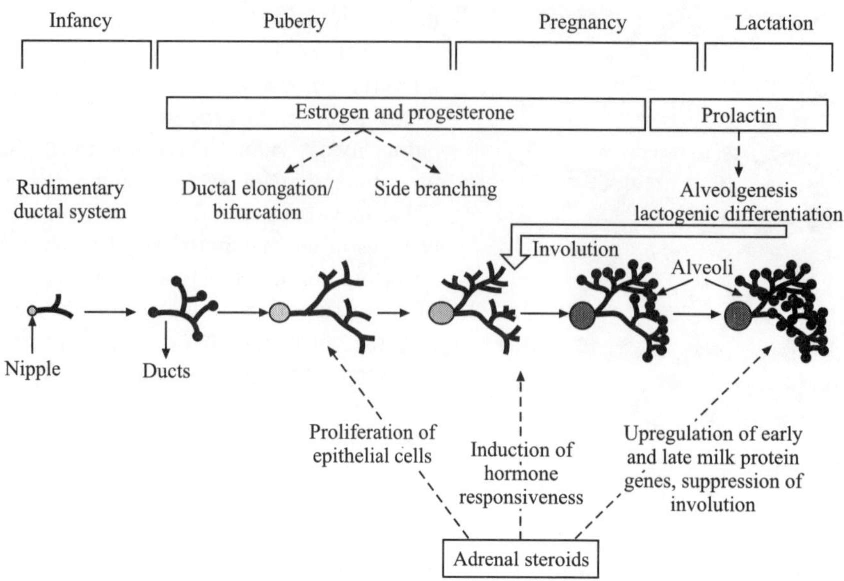

Figure 97.2   The sequences of hormonal regulation of the development of the mammary glands. At birth, the mammary glands in male and female infants are identical. The prominence of the neonatal breast and production of nipple secretion are due to maternal estrtogens and progesterone and their withdrawal at birth results in atrophied ducts. An integrated mechanism involving a number of hormones are required for initiating mammary alveolar ductal development at puberty. Estrogen, progesterone, insulin, cortiocosteroids and thyroid hormones are also involved in promoting mamogenesis. About 10–15 galactophores open at the nipple. At the base of the nipple, each glactophore dilates into the galactophore sinus where the milk collects. Approximately 10–15 of the lobes are potentially fuctional. Each lobe is separated from the other by dense fibrous tissue. Further subdivision of each lobe into lobules is accomplished by interlobular septa. In the presence of progesterone and estrogen, budding of the terminal ducts and formation of alveolar epithelium occurs. Eestrogen surge at puberty causes deposition of fat in the breast. During pregnancy high levels of estrogen, progesterone and prolactin cause further growth and development of the mammary glands. Increased estrogen levels cause ductal arborisation and differentiation of epithelial cells into ductal, alveolar and myoepithelial cells. It also stimulates the synthesis and release of prolactin. In the presence of estrogen and prolactin, progesterone stimulates alveolar proliferation and inhibits lactose synthesis. High levels of estrogen and progesterone inhibit the secretory effect of prolactin on mammary alveolar epithelium. The initial hormonal stimulation of the mammary gland appears during the first few weeks of pregnancy. By the second trimester, colostrums appears in the alveoli. By the third trimester, alveoli contain a significant amount of colostrums with marked expansion of vascular supply to the breast.

can easily be seen. The stimulation is sometimes sufficient to cause the infants to secrete milk for a short time after birth.

III. LACTATION

Although there is some synthesis and storage of milk before parturition, the copious flow of milk occurs at or soon after parturition. Milk is secreted whether the child is born alive or dead, and whether or not put to the breast. Complete failure of milk secretion is a rare event. Milk secretion is seen even after premature delivery even as early as the 16th week of gestation. In contrast, the ability to establish a successful lactation varies greatly. For a short time, following parturition, the mammary alveoli and lobules increase in size due to increase in secretory activity and this period varies with the species. However, it is not clear whether there is actually an increase in glandular tissue and if so whether this is due to cellular hypertrophy or to increase in the number of cells. However, mitosis is rarely seen during full lactation which is

usually seven days post-partum. Thus there appears to be no continued growth of the mammary parenchyma during lactation. The idea that the onset of lactation was due to the abolition of an inhibitory influence operating during pregnancy is an old one. The origin of these inhibitors was believed to be the corpus luteum of the placenta and these ideas are essentially in agreement with the current concepts and the inhibitors are believed to be hormones.

The hormones responsible for the initiation of milk secretion are termed lactogenic hormones. The hormonal changes responsible for the production of milk have been postulated from work in animals. One of the widely accepted theories for the initiation of lactation is the double threshold theory of Folley. According to this theory, the low levels of estrogen which can stimulate lactation can be rendered inhibitory by appropriate levels of progesterone and this is considered to be the normal inhibitory influence of pregnancy. The falling level of progesterone during parturition removes the inhibition and is replaced by the positive lactogenic effect of estrogen alone which is to enhance the secretion of prolactin resulting in lactation. Another theory of the initiation of lactation is proposed by Meites who suggests that milk secretion after parturition is due to both an increase in the circulating levels of prolactin and the combined inhibitory action of estrogen and progesterone during pregnancy, on the mammary gland, making it refractory to prolactin.

Once lactation has been initiated the maintenance of successful lactation depends on a variety of factors which are termed galactopoietic factors. The term galactopoiesis, however, is used to denote the experimental stimulation of established lactation. The galactopoietic factors can be broadly classified into two classes namely those that cause secretion and those that cause ejection. However, the two processes are closely related since milk production does not continue unless milk is removed from the breast. The milk ejection reflex is an active though an unconscious process on the part of the lactating animals. The reflex results in a sudden rise in the milk pressure in the gland after stimulation of the sensory nerve endings in the nipple by suckling or milking stimulus.

When the nipple is stimulated the effector tissues around the alveoli called the myoepithelium contract and milk is transported from the alveoli to the duct. The suckling reflex involves a neuro-endocrine arc. Sensory nerves leading from the nipple to higher nervous centres are initially stimulated and these stimuli cause the release of posterior pituitary oxytocic hormone. This appears to be the specific humoral agent that is involved in the transport of milk from the alveoli to the duct system. In lactating women the occurrence of reflex is accompanied by a sensation of tightness and fullness of the breast. During nursing many women have uterine contractions, a phenomenon suggesting the release of oxytocin. Similarly, administration of oxytocin to lactating animals including humans results in the transport of milk from the alveolus to the duct. Also, during diuresis, suckling causes marked decrease in the urine flow because the neurohumoral reflex from the nipple stimulation brings about release of posterior pituitary vasopressin simultaneously with oxytocin. Electrical stimulation of supra-opticohypophyseal tract also causes transport of milk.

In addition to the involvement of oxytocin, the pituitary-adrenal axis play a crucial role in the maintenance of lactation, and once again the particular hormones involved depend on the animal species. Growth hormone, ACTH, prolactin and thyroid hormone have all been found to be galactopoietic in nature. Prolactin seems to be the most crucial hormone concerned with the maintenance of lactation. Whether prolactin acts in the secreting gland by maintaining the structural integrity of the mammary glands or by maintaining their metabolic activity is not clear. Experimental evidence shows that the hypothalamus produces a prolactin inhibiting factor which prevents the secretion of prolactin by the anterior pituitary. Clinically, to inhibit milk secretion in mothers not wishing to nurse their child, estrogens and testosterone are often used. The assumption is that large doses of either will inhibit prolactin secretion by their action on prolactin inhibiting factor. Conversely, clinical use of chloropromazine to stimulate lactation is based on the assumption that chloropromazine can lower the concentration of prolactin inhibiting factor in the hypothalamus.

Growth hormone alone does not seem to be crucial but milk secretion is usually enhanced when growth hormone is administered along with prolactin and ACTH. The effect of ACTH in maintenance of milk secretion is through its role on the adrenals to secrete the steroids which in turn play a role in lactation. Evidence for this is again provided by experimental work where adrenalectomised rats when injected with cortisone or deoxycorticosterone or both are able to maintain lactation as opposed to the control animals. The thyroid hormone acts as a galactopoietic factor again by affecting metabolic activity of the gland and increasing milk secretion.

IV. INVOLUTION

The regressive changes in the mammary structure and function which follow either the post-lactational weaning or menopause are collectively called involution (Figure 97.2). During weaning involution changes quickly follow failure to withdraw the secretion. During involution the alveoli collapse and disintegrate and the lobules gradually disappear. Frequently vacuolization, pyknosis, and karyolysis are observed. The resting post-lactational mammary gland often resembles that of the adult virgin with the exception perhaps of the ducts which can be more ramified.

The absence of suckling which follows weaning seems to be the principal reason for the mammary involution. However, it is the hormonal control mediated by suckling rather than the lack of removal of milk that leads to mammary involution. Failure to evoke the release of oxytocin leads to an inhibition in the release of prolactin and the absence of prolactin leads to changes in the structural integrity of the mammary gland. This hypothesis seems to be valid since in experimental systems, ligation of the galactophores to prevent the escape of milk followed by prolactin administration alone can retard involution. Similarly oxytocin can retard mammary regression in the rat after weaning at the fourth day. In this connection it is interesting to note that oxytocin shows no retarding effect on the mammary involution in the adrenalectomized rats, and the defect can be restored with 9α-fluorocortisol. In this regard it should also be mentioned that reserpine is strikingly active in retarding mammary involution in the lactating rat after weaning at fourth day. Thus, once again, the pituitary-adrenal axis plays an important role in involution.

The post-menopausal involution of the mammary gland is due to ovarian aging when responsiveness to gonadotropins decreases to such low levels that ovarian secretion of estrogen becomes too slight to exert any effect on the mammary gland. The mammary gland, therefore, undergoes regressive changes once again. The post-menopausal mammary gland may not be markedly different from that of the pre-pubertal gland except in pathological cases.

V. ABNORMALITIES OF THE MAMMARY GLAND

A. Congenital Anomalies

The occurrence of aberrant breast tissue is known as polymastia and the presence of supernumerary nipples is called polythelia. These abnormalities are very rare and so is the absence of one or both breasts (amastia). If the patient has normal ovarian function, the assumption must be made that the target lacks sensitivity to hormones.

Excessive development of the breasts may also occur either during adolescence or occasionally during mature life. Pregnancy can enhance the condition and often the patient will require plastic surgery since hormonal therapy using androgens can also lead to virilization.

B. Gynecomastia

This is the condition that occurs physiologically in normal males resulting in breast enlargement. Marked degrees of breast development in adolescent males or the onset of gynecomastia in later life may indicate the presence of an estrogen-secreting tumour of the adrenal. Choriogenic tumours and more rarely, interstitial cell and granulosa cell tumours of the testes may produce gynecomastia.

V. COMPLICATIONS OF LACTATION

Galactorrhea

The production of milk spontaneously without a preceding pregnancy or the continuation of lactation after weaning is termed galactorrhea. Persistent lactation probably represents an effect of continued secretion of prolactin. In humans persistence of lactation and failure to re-establish a normal menstrual function after pregnancy, is called the Chiari-Frommel syndrome. This is best treated by cyclic administration of estrogen, or estrogen and gestagens, in amounts sufficient to cause withdrawal bleeding. This usually stops lactation. Clomiphene in small doses also sometimes causes lactation to stop.

Specific complications intrinsic to lactation processes include engorgement of breasts, sore nipples, mastitis and breast abscess. Engorgement is usually due to accumulation of milk due to the failure in the ejection reflex and insufficient suckling. The pain and swelling due to engorgement can inhibit further ejection and continued statis of the milk in the breast will finally lead to failure of lactation. Breast infections in the form of mastitis and breast abscess represent very serious complications of both early and later lactation. If an abscess develops it should be drained surgically.

VII. MAMMARY TUMOURS

Breast cancer is one of the most commonly occurring malignant disease in women. Cancer of the breast occurs at all ages but it is most common in women between the ages of 40–60 years. It appears that the supporting connective tissue of the breast is not more liable to pathological changes than the fibrous connective tissue in other parts of the body. Therefore, tumour of the breast consisting of only connective tissue is very rare.

Most of the histologic classification of mammary cancer can be divided into three types based on the site of origin of the epithelial cells: (1) cancer arising in the epithelium of the ducts on the surface of the nipple, (2) cancer arising in the epithelium of the ducts—duct carcinoma, and (3) cancer arising in terminal ducts and acini-terminal duct and acinae carcinoma. Once the tumour has begun to spread within the breast numerous events alter its morphologic structure and microscopic appearances. Furthermore, these conditions vary in various parts of the same breast. Those malignant tumours that resemble their parent tissue, in that their continued growth depends to a greater or lesser extent on the presence of specific circulating hormones, are called hormone dependent tumours; these usually comprise 30–40% of all human breast cancers. The failure of a tumour to demonstrate a clinical response to a specific endocrine therapy is regarded as evidence of autonomy. There is some evidence to suggest that pre-menopausal and post-menopausal breast cancer respond differently to endocrine manipulation as a result of their having originated in different hormonal environments.

SUGGESTIONS FOR FURTHER READING

Cunningham F.G., et al. (2009), *William's Obstetrics*, McGraw-Hill.

Johnson M.H. (2013), *Essential Reproduction*, Wiley-Blackwell.

98

Assisted Reproductive Technologies

Rajesh Naz

<table>
<tr><td colspan="2" style="text-align:center">CONTENTS</td></tr>
<tr><td>I.</td><td>Introduction</td></tr>
<tr><td>II.</td><td>Infertility
A. Female infertility
B. Male infertility</td></tr>
<tr><td>III.</td><td>Assisted Reproductive Technologies
A. Natural rhythm method
B. Intrauterine insemination technique (IUI)
C. In Vitro fertilization (IVF) procedure
D. Intracytoplasmic sperm injection (ICSI) procedure
Summary</td></tr>
</table>

I. INTRODUCTION

Infertility is defined as the inability to conceive a child after unprotected sexual intercourse. World Health Organization considers infertility if the couple is unable to conceive after two years, but most of the Infertility Clinics define infertility/subfertility after one year of regular sexual intercourse. Primary infertility is infertility where the couple has never conceived a child. Secondary infertility is defined as problems with fertility after at least one conception. Infertility is on the rise worldwide. In the US, one out of every five to six couples has some fertility problem. Several reasons have been described for this rise in infertility, including delayed marriages, environmental factors, and delayed childbirth. The cause of infertility can lay either in the woman, man, or both. Generally in the clinics, the couples are examined together to find out the cause of infertility.

II. INFERTILITY

A. Female Infertility

Infertility solely due to the female constitutes approximately 40% of infertile cases. For conception to occur, the woman has to cycle normally and ovulate. The egg has to be fertilized with viable, normal sperm at the ampulla-isthmus junction, divide, and the developing embryo has to implant at the uterine wall. The embryo subsequently develops into the fetus, which grows inside the amniotic sac during the gestation period. Failure in any of these steps can cause infertility. There are five major causes of female infertility. These are ovulatory dysfunction, including polycystic ovarian syndrome (PCOS), tubal blockage, endometriosis, immune infertility, and sexually transmitted diseases (STDs).

Ovulatory dysfunction constitutes approximately 20–25% of female infertility. A normal regular cycle (22–35 days, commonly 28 days) indicates, though does not confirm, that the woman is ovulating normally. Abnormality in the menstrual period (time and/or quantity of blood discharge) is suggestive of some problem in ovulation. Menses that is heavy, prolonged, or irregular (menorrhagia/polymenorrhea/metrorrhagia/oligomenorrhea/amenorrhea), suggests some problem in ovulation. The most common cause of abnormal cycles is polycystic ovarian syndrome (PCOS). PCOS causes an imbalance of hormones, leading to ovarian dysfunction and infertility.

About 25–35% of infertility cases are a result of a tubal problem in the female. Fallopian tubes may have a complete blockage at a distal end of the tube (hydrosalpinx) or a partial obstruction.

Endometriosis accounts for 5–10% of infertility. Endometriosis is a disorder in which cells from the lining of the uterus grow in the peritoneum, fallopian tubes, and several other places outside the uterus. It is generally attributed to retrograde menses. Endometriosis can cause infertility by mechanical obstruction and/or by upregulation of cytotoxic immune cells and cytokines in the endometrial cells/tissue, which are secreted and can be spermatoxic and embryotoxic.

Other causes of female infertility include the presence of antisperm antibodies (ASA), which can affect sperm function, and infections of the genital tract, especially Chlamydia and *Neisseria gonorrhoeae*. The STDs, if left untreated, can cause pelvic inflammatory disease (PID) that can damage the fallopian tubes, uterus, and surrounding tissue, leading to infertility.

B. Male Infertility

Male infertility is the sole cause in 20% and a contributing factor in 30–40% of infertility in infertile couples. Male infertility can be caused by low sperm production, a genital tract obstruction, and/or defective sperm function.

Varicocele is the most common cause of male infertility. Varicocele can occur with or without infertility. Other causes of male infertility include infections, cryptorchidism, obstruction of vas deferens, and immune infertility. Defective endocrine milieu and genetic factors can affect male fertility. Male fertility can easily be diagnosed by semen analysis following the WHO criteria. Low sperm count, low motility (percent and progressive), and a high percentage of abnormal sperm and leukocytes in the semen are an indicator of male infertility. The presence of ASA can also cause infertility. These various factors, including ASA, that can affect spermatogenesis, sperm motility, and sperm capacitation/ acrosome reaction/zona binding can lead to infertility. Male infertility is relatively easier to diagnose and treat compared to female infertility.

In 15% of cases, the reason for infertility cannot be diagnosed in either partner, leaving a mystery to why pregnancy has not occurred. This is called idiopathic infertility. As better tools for diagnosis are emerging, many cases of idiopathic infertility are now better diagnosed and their etiology is better defined.

III. ASSISTED REPRODUCTIVE TECHNOLOGIES

A. Natural Rhythm Method

Depending upon the pathophysiology of infertility, an appropriate treatment modality is formulated. The proper diagnosis is an important part of formulating a treatment modality. The first line of treatment is to educate the infertile couple about timing intercourse with the woman's ovulation. There are various methods to determine the ovulatory phase, which include using a basal body temperature (BBT) chart, identifying a fern pattern in the cervical mucus, and determining an LH surge.

B. Intrauterine Insemination Technique (IUI)

If coinciding the timing of intercourse with ovulation does not work in 3–5 menstrual cycles, the next step is to try intrauterine insemination (IUI). IUI can be performed using partner or donor sperm A sufficient number of motile sperm isolated from the semen is delivered directly in the uterus, coinciding with the time of ovulation. IUI takes care of infertility caused by cervical mucus hostility in the female partner and also obliterates the factors in the semen that can affect sperm motility and function. The IUI can be coinciding with the natural ovulatory cycle or the woman can be superovulated with various hormone treatments to increase the chances of fertilization. With superovulation, there are chances of multiple births. If the husband's sperm do not work and are defective, the couple is recommended to use the fertile donor sperm. An IUI using donor sperm is referred to as artificial insemination. The success rates of IUI range from 15–20% per cycle, for up to three to five cycles. If a woman cannot conceive after five cycles of IUI, then she is recommended to undergo assisted reproductive technology treatments described below.

The Center for Disease Control and Prevention defines ART to include "all fertility treatments in which both eggs and sperm are handled". In general, ART procedures involve surgically removing eggs from a woman's ovaries, combining them with sperm in the laboratory, and returning them to the woman's body. According to the CDC, they do not include treatments in which only sperm are handled, for example IUI or artificial insemination. They also do not include procedures in which a woman takes medicine only to stimulate egg production without the intention of having eggs retrieved.

According to the CDC's 2010 preliminary ART success rates report, during 2010, 154,417 ART cycles were performed at 443 reporting clinics in the United States, resulting in 47,102 deliveries of 61,561 infants. Although the use of ART is still relatively less common as compared to the potential demand, its use has doubled over the past decade. Today, over 1% of all infants born in the U.S. every year are conceived using some form of ART.

C. *In Vitro* Fertilization (IVF) Procedure

In November 1977, Dr. Robert Edwards and Dr. Patrick Steptoe performed a procedure, that was later named "in vitro fertilization (IVF)". Louise Brown was the world's first "test tube" baby of this procedure. Her parents had been trying for nine years to conceive, but faced challenges with a fallopian tube blockage. In 2010, Dr. Edwards received the Nobel Prize in Physiology and Medicine for discovering the IVF procedure.

IVF has become the most effective and most popular form of ART. In this procedure, fertilization takes place outside the body. It is helpful in treating various causes of infertility, both in females and males. Recent data indicates that the live birth rates approaching natural fecundity can be achieved using ART, depending upon the couples age and embryo quality.

1. Ovarian stimulation

The first step in an IVF procedure is ovarian stimulation, used to develop multiple eggs in the ovaries. This is done by administering a regimen of fertility drugs, usually oral or injectable gonadotropins, which induce ovulation. Medication used for ovarian stimulation involves gonadotropin formulations, such as clomiphene/serophene, gonal-F/follistim, and repronex/pergonal/menopur. Spontaneous ovulation is blocked using GnRH (Gonadotrophin-releasing hormone) antagonists that prevent the natural surge of LH. The follicular growth is monitored in the ovaries using ultrasound and serum estradiol levels. Once a sufficient number of follicles have reached maturation, it is time to retrieve the eggs. Human chrionic gonadotrophin (hCG) is injected to induce further maturation. Timing is critical, and transvaginal oocyte retrieval (TVOR) must be performed 34–36 hours after the hCG injections to ensure that the eggs are fully mature but have not yet ruptured. An ultrasound-guided needle is used to carefully retrieve mature eggs from the follicles. Next, the eggs and sperm need to be prepared. Eggs are removed from surrounding cells, and eggs with maximal chances of pregnancy are selected. Semen is prepared using a sperm washing procedure, where the sperm are separated from inactive cells and seminal fluid. Then, the sperm and egg are incubated together in a growth medium for about 18 hours. Once fertilization is confirmed and cell division is taking place, the egg is now considered an embryo.

On the second or third day of a woman's cycle, she starts taking medications which will simulate follicle growth by synthesizing follicle-stimulating hormone (FSH), usually clomiphene (oral pills), follitropin (subcutaneous injections), or a combination of the two. A GnRH agonist (luprolide or buserelin) and GnRH antagonists (cetrorelix) is also given to prevent spontaneous follicle rupture by preventing the pituitary gland from producing LH. A woman's response to her medications is closely monitored by vaginal ultrasounds or blood estrogen levels. The medication may be increased or decreased if the estrogen level is too high or low. The patient will continue this daily medication for the next 6–7 days under frequent monitoring to make sure she is not producing too many eggs and the eggs are maturing at an appropriate speed. Once the desirable number of eggs has reached the suitable maturation, an hCG injection is given to trigger ovulation. Approximately 34–36 hours after hCG administration the ovulation occurs, so the egg retrieval is timed accordingly. The hCG is essential to stimulate the natural LH surge in a woman, as well as initiate the final growth spurt of the developing oocytes. After the hCG injection, the oocyte becomes loosely bound to the surrounding cells and can be more easily aspirated during the egg retrieval procedure. Sometimes, women can face complication due to medication, which can lead to ovarian hyperstimulation syndrome (OHSS). In this situation, the ovaries become enlarged, and nausea, diarrhea, and temporary weight gain can occur. This is generally resolved in one to two weeks. In these cases, the stimulation protocol is stopped, and the retrieval procedure is aborted.

2. Egg retrieval

Timing is critical in the egg retrieval. It must be performed within 34–36 hours after the hCG injection. A delay in the procedure can make the eggs from the follicles ovulate into the peritoneum, where it is difficult to retrieve them. The most commonly used technique to collect eggs from unruptured follicles in the ovaries is a vaginal ultrasound-guided egg retrieval procedure. When IVF started in the late 1970's, the egg retrieval was performed under general anesthesia using laparoscopy. The vaginal ultrasound-guided egg retrieval is easier and performed with the woman given mild sedatives, sometimes with paracervical local anesthesia. In this procedure, a probe attached to a small, hollow needle is inserted into the vagina using ultrasound guidance. The needle is advanced further into the reproductive tract into the ovary to gently suck fluid from the follicles containing eggs. After suction, the follicular fluid is promptly analyzed for the presence of oocytes. Sometimes there may be follicles without eggs. This is called empty follicle syndrome, and occurs in approximately 2–5% of IVF cases. The procedure is performed on both ovaries, when possible, and the maximum number of eggs is retrieved. The procedure lasts for an hour or so, and the woman can go home after the retrieval. Afterwards, the woman can experience some pain or discomfort, which is easily treatable. Minimal vaginal bleeding may also occur for a day or two following egg retrieval. Antibiotics/steroids may be prescribed to prevent infection/inflammation.

Due to guidelines which regulate that only 1–3 embryos can be transferred in one cycle, several times not all eggs retrieved are used for fertilization in the same cycle. These eggs or subsequent embryos are cryopreserved for use in subsequent cycles if the woman does not become pregnant. Cryopreservation is the process in which eggs/embryos/sperm are preserved by slow freezing in different media and cryopreservants to maintain their viability.

3. Fertilization

After retrieval of the eggs and separation from the follicular fluids, they are added to a dish containing a culture medium to be incubated at 37° C with 5% CO_2 for several hours. The male partner is then asked to provide semen. The semen sample is passed through the percoll gradient to separate the motile sperm from the other cells and seminal fluid. Approximately 20,000 to 100,000 motile sperm are added to each egg and kept in the incubator for fertilization. After eighteen hours of incubation, the eggs are examined for fertilization, and noted by the presence of two pronuclei. The zygote is transferred to another dish containing an embryo culture medium. After twelve hours, the zygote is observed for cytokinesis, which will subsequently develop into the morula and hatching blastocyst. Embryo quality is judged based on several morphological criteria and growth pattern. Usually, a good quality embryo is defined as having an expected number of cells without fragmentation (Figure 98.1). Once the embryos are three to five days old, they are ready to be transferred into the uterus.

4. Embryo transfer

The embryo transfer is a quick procedure generally done without any analgesic or sedative. The patient usually lies in the knee-chest position for easy delivery. First, the cervix is rinsed of any mucus if present. One to three embryos are loaded in the tip of a thin catheter, which is then passed through the cervical canal into the uterine cavity. Then, the embryos, along with a small amount of medium, are pushed from the catheter into the cavity. The catheter is carefully removed and checked that all embryos have indeed been transferred. After the transfer, a woman is generally recommended to refrain from any intense physical activity for at least a 24 hour period.

The stage at which the embryos are transferred varies from clinic to clinic. Some clinics prefer to transfer at an earlier stage (~8–16 cell), while others prefer to transfer at the morula to blastocyst stage. The transfer at each of these stages has its own advantages and disadvantages. Later stages of development are preferred if chromosomal analysis of the

blastomeres by preimplantation genetic diagnosis (PGD) is required (described below).

When the IVF procedure started during the 1980's, laparoscopic retrieval of the oocytes was performed instead of ultrasound-guided. During that period, two procedures, namely gamete intrafallopian transfer (GIFT) and zygote intrafallopian transfer (ZIFT), were employed, which provided better results at the time. Now, these procedures are not used as frequently due to the development of new, easier methods of egg retrieval and better methods/media for embryo growth.

In a GIFT procedure, the gametes (oocytes and sperm) are retrieved and mixed, and then transferred into the fallopian tube by using a laparoscope. Here, the fertilization takes place *in vivo*, rather than *in vitro*. In a ZIFT procedure, the eggs are fertilized outside and transferred into the fallopian tube at the zygote stage to further develop.

5. Implantation

A three-day old embryo still has two days of development before it can implant in the uterus, while a five-day old embryo is ready to implant when transferred. A woman is generally given

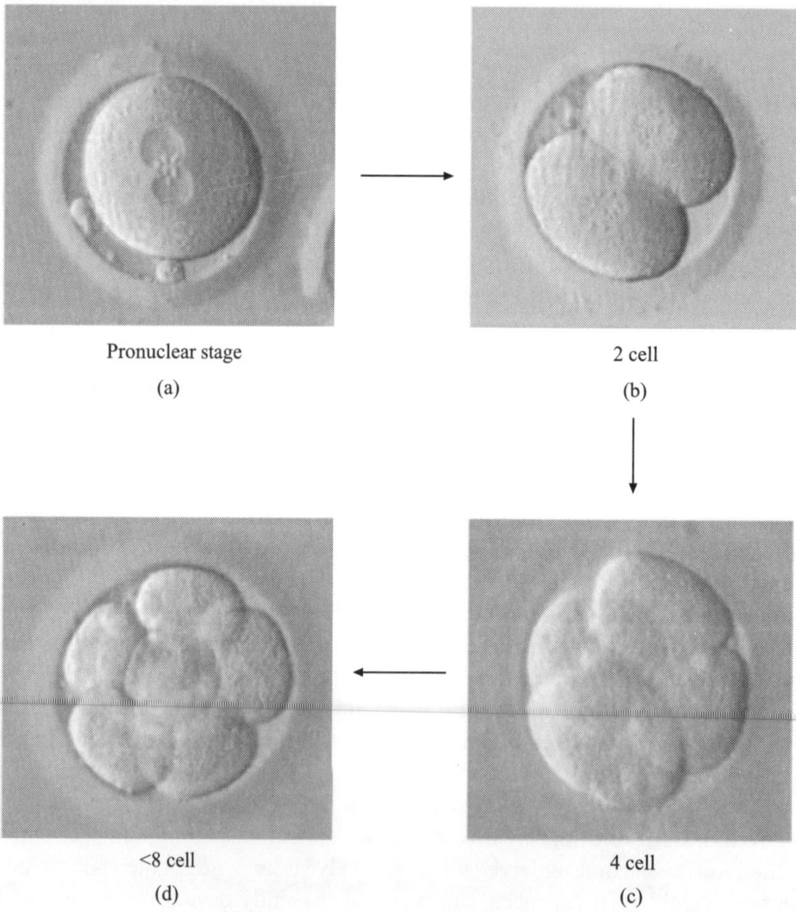

Figure 98.1 Various developmental stages of human embryos after fertilization of human egg with human sperm after IVF procedure. Approximately 16–18 hours after sperm and egg co-incubation, one can see fertilization by the presence of two pronuclei in the egg (a). The pronuclear stage zygote divides into two cell (b), then four cell (c), and then into multi-cellular morula (d) which subsequently develops into hatching blastocyst. *Courtesy of Dr. Roger Toffle, M.D., and Dr. Melanie Clemmer, Ph.D.*

a progesterone vaginal suppository to enhance endometrial growth for implantation. Approximately one-two weeks after implantation, the fetal-placental unit will start producing an increasing amount of detectable hCG, confirming that implantation has indeed occurred. The implantation is further confirmed by the presence of a heartbeat using ultrasonography.

6. Success rates

Determining a success rate in IVF involves taking into consideration the age of the woman, the number of embryos transferred, the use of patient or donor eggs, and the condition of the embryos (fresh vs. cryopreserved). There is also a difference between the success of clinical pregnancy detected by hCG, versus the success of a live birth. The following rates are from the CDC's 2010 National Summary of ART treatments and include the percentage of IVF cycles resulting in a live birth using fresh, patient eggs:

- Under age 35: 30–35%
- Ages 35–37: 25%
- Ages 38–40: 15–20%
- Ages 41 and up: 6–10%

7. Limitations of IVF

IVF is usually the first ART treatment modality used for infertile couples, however it has its own limitations. Like any other procedure, IVF is not a panacea for all kinds of infertility and is not suitable for everyone. Men with low sperm count, poor sperm motility, or sperm with the inability to bind and penetrate to the egg may not have high success using IVF. Men with azoospermia, or lack of any sperm in the semen, the IVF procedure has limited application. Men with these limitations may need to consider a more complex version of IVF, called the intracytoplasmic sperm injection (ICSI). Also, it has limited application in women who develop ovarian hyperstimulation syndrome (OHSS) and have severe polycystic ovarian syndrome (PCOS). Also, the age of the partners play an important role in the success of IVF outcome.

8. Future research directions

Currently, there are three major areas being researched to enhance success rates of IVF. These include: (1) the development of a procedure to enhance in vitro maturation of the immature eggs, (2) better criteria for the selection of good quality embryos to transfer, and (3) better cryopreservation technology to enhance quality and longevity of storage of eggs/embryos.

In cases where immature eggs are retrieved from the ovaries, a method to stimulate in vitro maturation (IVM) of these eggs would be desirable. IVM procedures involve culturing the immature eggs in media containing FSH and LH (menotropin medication). Currently, IVM procedures are performed with low success rates (44.3% IVF live birth rate with mature eggs versus 16.5% IVF live birth rate using immature eggs matured in vitro).

Selecting better embryos to be transferred to the uterus would enhance rates of IVF. Preimplantation genetic diagnosis (PGD) is the process in which an embryo is screened for defects in DNA or the presence of genes that cause serious, heritable diseases. Screening is performed following successful fertilization and embryo growth, and prior to transfer to the mother. Using polymerase chain reaction (PCR) or fluorescence in situ hybridization (FISH), chromosomal irregularities, most commonly Down syndrome, are recognized. Older mothers, especially >37 years old, have a drastically increased risk of having a child with chromosomal abnormalities. Couples with specific, inherited disorders, as well as high-risk couples (spontaneous abortion, failed implantation, etc.), are most likely to pursue these diagnostic procedures.

Cryopreservation technology has recently become a common practice for storing excess eggs and embryos that cannot be transferred due to regulatory limitations imposed on the number of embryos one can transfer. It has become increasingly popular as more couples elect to only transfer one or two embryos into the uterus. It has also become an alternative for patients with cancer and other diseases with a demand to preserve oocytes for use in the future. Better methods are needed that can improve cryopreservation technology.

D. Intracytoplasmic Sperm Injection (ICSI) Procedure

Intracytoplasmic sperm injection (ICSI) is a modified version of IVF in which a single sperm cell is directly injected into the egg, instead of co-incubating sperm and egg to fertilize. In this procedure, stimulation of the ovaries and egg retrieval are performed as in IVF described above. This procedure was first discovered by Gianpiero Palmero and André van Steirteghem in 1991. Since then, it has become a highly successful and routinely used procedure in infertility clinics.

Once the sperm and eggs have been extracted and prepared, the fertilization procedure can be performed. This surgery is done under a microscope along with the use of micromanipulation tools (Figure 98.2). The egg is held in place using a small pipette with gentle suction being applied. A single sperm cell in a small amount of medium is selected using a thin micropipette with the sperm head going towards the tip of the pipette. Once loaded, the micropipette containing the sperm cell is penetrated into the immobilized oocyte, through the zona pellucida and sperm membrane, and the sperm cell is released under the visualization of the microscope. After the sperm injection, the micropipette is slowly withdrawn without damaging the egg, which is then released from the suction pipette. The egg is placed in a special culture medium and checked for fertilization after 12–16 hours. If fertilization has taken place, the zygote is placed in a culture medium for embryonic growth, as described above for IVF.

The selection of the injected sperm cell is based on morphological criteria and motility. A highly motile sperm with good morphology is desired. If the patient has no sperm in the semen, a procedure called testicular sperm extraction (TESE) is performed. This is a desired treatment for men who have azoospermia, obstruction/absence of the vas deferens,

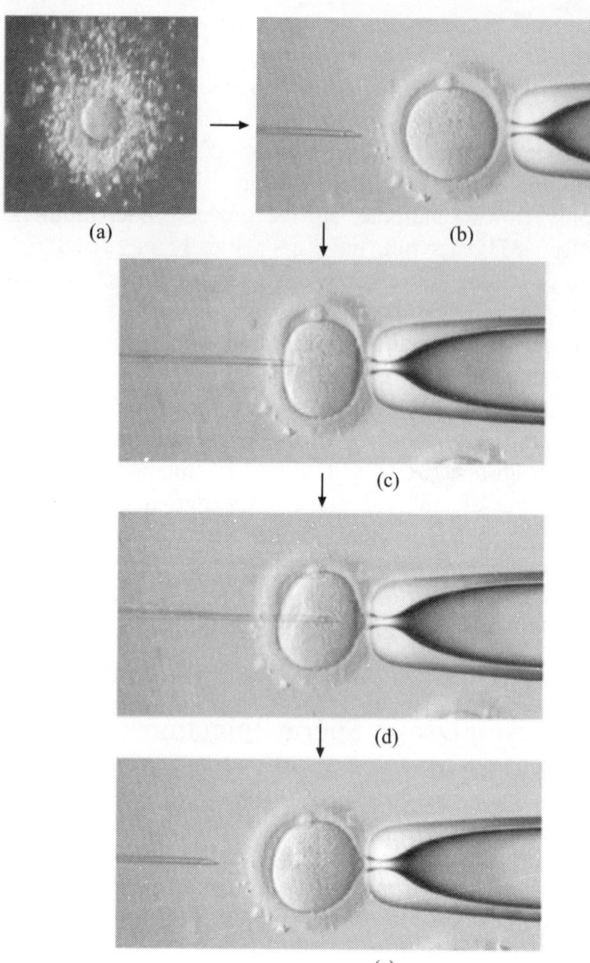

Figure 98.2   The human egg retrieved from the ovary is surrounded by cumulus mass containing cumulus and granulosa cells (a). Incubation of the egg with hyaluronidase is used to digest the surrounding somatic cells and the oocyte is free from the cumulus and granulosa cells. The egg is immobilized by light suction and held with the micropipette, and the sperm is injected into the cytoplasm of the oocyte without damaging the egg (b-e). *Courtesy of Dr. Roger Toffle, M.D., and Dr. Melanie Clemmer, Ph.D.*

and for men whose sperm are unable to fertilize the eggs in a previous IVF procedure.

1. Testicular sperm extraction (TESE) procedure

A testicular sperm extraction procedure involves the surgical removal of a small piece of testicular tissue from which sperm can be obtained. A small incision is made in the scrotum under general or local anesthesia and a sample of the testicular tissue (50–100 mg) is extracted in the culture medium. Using a small microscope, small tubules containing sperm are identified and a small sample (50–100 mg) of tissue is extracted and culture medium.

As reported by the CDC, from 1999 to 2008, the ICSI success rate (as for other ART procedures) has increased. In 1999, the average success rate, for all ages, of an ICSI procedure using fresh eggs resulting in a live birth was 30.1% (with an IVF average of 32.5%). In 2008, the success rate was reported to be 36.2% (with an IVF average of 38.0%). Success rates of ICSI resulting in a live birth using donor eggs was significantly higher, reported at 54.0% in 2008 (with an IVF average using donor eggs of 57.7%).

2. Limitations of ICSI

ICSI is a desired procedure with couples with male factor infertility. Limitations for men could involve the retrieval or selection of a sperm cell which has a viable, intact membrane, without aneuploidy and DNA fragmentation. Currently, selection of the best quality sperm depends entirely on motility and other morphological features. These characteristics may have no correlation to the quality of the DNA held inside the sperm. In a natural conception, only the most resilient sperm cells make it to the egg for fertilization. ICSI removes this process of natural selection.

3. Future research directions

The most critical research needed for ICSI is the development of a non-invasive, non-damaging, and reliable method for selecting a good sperm cell. Current research is focused on finding such a method for sperm selection which has membrane integrity with no aneuploidy or DNA fragmentation. At this time, such a method is not available.

SUMMARY

In conclusion, assisted reproductive technologies (ART) are defined as the procedures which handle both eggs and sperm outside the body to be used for the treatment of infertility. The use of ART has tremendously increased in infertiltiy clinics due to increased infertility rates. In the US, one out of every five to six couples who want to have children is infertile. Infertility can be caused by female, male, or both partners of an infertile couple. There are various etiologies of infertility, leading to several treatment modalities. The first line of treatment is based upon planning the time of intercourse to the ovulatory phase of the female partner cycle. If unsuccessful, then more involved and invasive procedures are sorted out. The next line of treatment is IUI with husband or donor sperm, with or without the ovarian stimulation of the female partner. IUI is not technically defined as an ART procedure because only the sperm are handled, instead of both gametes. The next line of treatment is the use of ART procedures. The most commonly used ART procedure is in vitro fertilization (IVF). IVF can be used to treat both male and female infertility, and it can treat various forms of infertility caused by different etiologies. If there is a difficulty in obtaining the sperm, due to its absence/

low quantity in the semen and/or defective spermato-genesis, a modified form of IVF is employed, called ICSI. In ICSI, a single sperm retrieved from semen and/or the testes is injected directly into the egg. ICSI bypasses the various parameters required for successful fertilization, such as motility, capacitation/acrosome reaction, hyperactivation, sperm binding, and penetration. ICSI, along with IVF, has become a very popular procedure for infertility treatment. This was recently recognized by awarding a Nobel Prize in Physiology and Medicine in 2010 to Dr. Robert Edwards that discovered IVF. To further increase the success rates in ART procedures, various methods of in vitro maturation (IVM) of oocytes, preimplantation genetic diagnostics (PGD), and cryopreservation are being researched. At this time (2012), it is estimated that over one million babies have been born using ART procedures, and some of them have become parents and becoming grandparents.

Acknowledgements

The author would like to thank Erin Barthelmess for helping him in typing the article.

REFERENCES AND SUGGESTIONS FOR FURTHER READING

1. Steptoe P.C. and Edwards R.G. (1978), Birth After the Reimplantation of a Human Embryo, *Lancet* 2:366.

2. Palmero G., Joris H., Devroey M.T. and Van Steirteghem A.C. (1992), Pregnancies After Intracytoplasmic Injection of Single Spermatozoon into an Oocyte, *Lancet* 340: 17–18.

3. Van Voorhis B. (2007), *In vitro* Fertilization, *N. Engl. J. Med.,* 356: 379–386.

4. Jain T. and Gupta R. (2007), Trends in the use of Intracytoplasmic Sperm Injection in the United States, *N. Engl. J. Med.,* 357:251–257.

5. CDC (2008), Assisted Reproductive Technology Success Rates. National Summary and Fertility Clinic Reports, National Center for Chronic Disease and Prevention and Health Promotion, Atlanta, Georgia: Center for Disease Control and Prevention.

5. Rebar R. and DeCherney A. (2004), Assisted Reproductive Technology in the United States, *N. Engl. J. Med.* 350:1603–1604.

6. Davies M., Moore V., Willson K., Essen P., Priest K., Scott H., Haan E. and Chan A. (2012), Reproductive Technologies and the Risk of Birth Defects, *N. Engl. J. Med.,* 366: 1803–1813.

8. Practice Committee of Society for Assisted Reproductive Technology and Practice Committee of American Society for Reproductive Medicine (2012), Elective Single-embryo Transfer, *Fertil. Steril.* 97: 429–444.

9. Raman J.D. and Schlegel P.N. (2003), Testicular Sperm Extraction with Intracytoplasmic Sperm Injection is Successful for the Treatment of Nonobstructive Azoospermia Associated with Cryptorchidism, *J. Urol.,* 170: 1287–1290.

10. Palmero G.D., Neri Q.V., Monahan D., Kocent J. and Rosenwaks Z. (2012), Development and Current Application of Assisted Fertilization, *Fertil. Steril.* 97: 248–259.

99

Control of Fertility
Present Practice and Future Perspectives

G.P. Talwar

CONTENTS

I. THE GLOBAL DEMOGRAPHIC TRENDS

The world population is increasing at an alarming pace. In 1980, it was a little above 4 billion, it is expected to cross the 6 billion mark by the year 2000. In other words, in these two decades half as many more people will be added to the figure accumulated on the surface of this earth since the origin of man till the year 1980. The 1981 census showed that the rate of growth of population in India during the previous 10 years was about 2.4%. In 1991 census, a perceptible fall has taken place. However, even now, around 18 million people are added to our population every year, which is equal to the *total* population of a country like Australia. This increment imposes a heavy burden on the community. To maintain status quo 400 million jobs are required every year. Family planning has become a pressing necessity.

II. CURRENTLY AVAILABLE METHODS OF CONTRACEPTION

A number of methods are available. They have enabled successfully the control of family size in developed countries and more recently in China. These can be classified into five categories.

A. Barrier Methods

The most widely used barrier device is *condom*. They are particularly popular in Japan and Thailand. Technological developments have ushered in extra thin membrane sheaths, marketed in a variety of colours with or without extra lubricants. The public sector product Nirodh is now available as a better grade superior unit. When utilized properly, the condoms have a satisfactory efficacy. The device can be employed as and when required. No medication of either the male or the female partner is involved. Consequently there are no side effects except those of misuse (possible pregnancy). The main demands are discipline, motivation and of course availability of the material at the time and place of need. An extra advantage in the use of condoms is the lack of overt transmission of venereal diseases.

Diaphragm, microbicides cum spermicides

Usable in the female seeks to cover the passage of the sperm to the cervix. It has a fairly high failure rate, presumably because the sizes vary and optimal fit is not achieved. Its use is on the wane.

Along with the diaphragm, the use of foams, jellies and creams containing *spermicides* is invariably recommended to increase the overall efficacy. Commercial preparations at present use Nonoxyl-9 (N-9)—a powerful detergent—as a base. The frequent use of N-9 containing spermicides causes inflammation and mucosal damage, which may enhance the transmission of some sexually transmitted infections such as HIV. Research is in progress around the world to develop better spermicides with effective antimicrobial and antiviral action as well. A polyherbal formulation containing purified Neem (*Azadirachta indica*) extracts, saponins from Sappindus mukerosi and an essential oil of plant origin has shown anti *N. gonorrhea*, anti-candida, anti-herpes and anti-HIV properties. It has cured abnormal vaginal discharge due to a variety of pathogens during phase I/II clinical trials conducted with Praneem polyherbal cream in India, Egypt and Dominican Republic. The cream has also potent spermicidal properties.

An improved Polyhedral formulaion named BASANT has been made by Talwar et al., . It contains 95% purified curcumin, Aloe Vera, purified extracts of Amla (*Emblica officinalis*), purified Neem (*Azadirachta indica*) leaves extract with pharmacopially approved excipients. BASANT is dispensed as cream and as powder in veg. (cellulose) capsules, which can easily be inserted in the vagina with washed fingers. BASANT has inhibitory action on a wide spectrum of genital pathogens: *Chlamydia trachomatis* (both pre infection and post-infection), *N. gonorrhea* including strains resistant to penicillin, tetracycline nalidixic acid and ciproflaxin. It has inhibitory action on *Candida glabrata*, *Candida albicans* and *C. tropicals*. BASANT exercises a high virucidal action against Human Immunodeficiency Virus (HIV), inhibiting both the T cell tropic and M cell tropic strains of HIV. It also prevents the entry of Human Papilloma Virus (HPV) type 16 in Hela cells. In an ongoing clinical trial in women at entry stage of cervical dysplasia, molecularly positive for HPV, administration intravaginally of one capsule of BASANT every night for 30 days resulted in elimination of the pathogenic strain of HPV from infected cervical cells and return of cylological changes from abnormal to normal type. Thus, besides the inhibitory actions manifested by BASANT on a variety of pathogens, BASANT has manifested an important therapeutic action at early stage of cervical cancer. It has also been observed to regress/cure vaginosis, a frequent ailment in women suffering from abnormal vaginal discharge.

B.   Intra-Uterine Devices (IUD)

The antifertility action of a foreign body in the uterus was known since ancient times in Egypt and India. A small stone was inserted in the uterus of caravan camels to ward off pregnancy during long journeys. In medieval period, intra-uterine pessaries were made from metals.

IUDs were introduced in the family planning programme about 25 years back. A variety of IUDs differing in shape have been proposed. Most are made of silastic polymer with or without additional metallic components. Examples are the Lippe's loop, the Dalken shield, Soonawala Device, Antigon etc.

The first generation IUDs had limited acceptors, inspite of the advantage of one time insertion for a long period of protection. The probable reasons for expulsion were bleeding disorders and pain. These side effects have been reduced in the second generation of IUDs currently offered, which are the copper IUDs and the copper silver IUDs. The protective efficacy of these devices is also better than the first generation IUDs.

The mechanism of action of IUD is partially understood. It prevents primarily the implantation of the blastocyst. The device induces an inflammatory reaction in the endometrium. The copper containing IUDs have additional action by the slow release of copper, influencing metabolic reactions of endometrial cells. Silver may contribute Galvanic currents.

C.   Contraceptive Steroids

Feedback control exists between the steroid hormones produced by the ovaries and the pituitary gonadotrophins which stimulate their synthesis. This property was the basis of utilizing progestogens and estrogens to inhibit the secretion of gonadotrophins and thereby block the ovulation. Natural hormones, progesterone and estradiol were not employed as they have a short biological half-life. Instead of these, synthetic steroids with progestogenic and estrogenic activity were used. The earliest mode in which these were administered was the oral route, the familiar *oral pills* consisting of a combination of a pro-gestogen with small quantities of estrogens. These are taken daily for 21 days in the month. Oral pills have found widespread use in western countries. They have high efficacy. The incidence of side effects is low. The risk of cardiovascular disorders increases several fold in women who are smokers.

In recent years some other modalities of using contraceptive steroids have evolved. These include the *injectibles*. These are depot preparations dispensed in an oily base. The two steroids commonly employed are (i) medroxyprogesterone acetate or depo-provera and (ii) norethindrone acetate. The former is in use in some parts of Thailand. It has a slightly longer span (3 months) than the injectibles based on the second steroid whose efficacy extends over 1 to 2 months. Depo-provera went into disrepute by the observation on causation of breast cancer in beagle dogs. No such effect has been observed in primates and other animal species.

D. Termination of Pregnancy

In India, and several other countries of the world, medical termination of unwanted pregnancy is permitted. It should be done in the first trimester, the earlier the better. Early pregnancy can be aborted by evacuation, a procedure simplified enough to make it applicable in outpatient department without need of overnight hospitalization. It is estimated that 50 million abortions are done in the world every year. The contribution of this service to the control of population growth is not negligible.

E. Sterilization, Male and Female

It is a one time procedure and is accepted by nearly a quarter of the total opting for family planning. In the male, vasectomy and in the female tubal ligation or tubectomy is surgically performed to block the passage of sperm or egg respectively. Surgical refinements have made the operation simple and applicable even in rural camps. Adequate aseptic conditions should however be assured to prevent infections and unnecessary complications. With proper surgery, efficacy is assured with no observable side effects. The main limitation of the method is that it is by and large a terminal procedure and although reversal surgery is possible, the restoration of fertility has only been achieved in a small percentage of cases.

To sum up, the commonly available methods for family planning are: (1) condoms, (2) oral pills or injectible depot preparations of contraceptive steroids, (3) Intra-Uterine Devices, in particular the copper IUD, (4) Termination, and (5) Sterilization. Each has its merits and limitations. Choice is available to suit different needs. In the section that follows, mention will be made of the new developments and improvements of the existing methods.

III. NEW DEVELOPMENTS AND IMPROVEMENTS OF THE PRESENTLY AVAILABLE METHODS

A. The Subdermal Implant-Norplant

The Population Council has developed this implant which consists of a set of six silastic capsules filled with the progestogen, levonorgestrel. These are introduced beneath the skin under local anesthesia. The release of the steroid is at a steady rate and ensures effective contraception for five years. The annual pregnancy rate is below 0.5 per 100 woman-years. The implants can be removed at will even before 5 years by a minor surgery. The side effect of this implant is the irregular bleeding which is however not heavy. It was well tolerated in countries where it was tested.

The method seems to be useful for young women who have completed their family and who require contraception for long periods but want to retain the option for another pregnancy. A simpler version of this implant consisting of only two rods, instead of the present six capsules, will soon be available which would be easier to introduce under the skin.

The Norplant is on large scale phase IV use in Egypt, Indonesia and several other countries. Evaluation of the device on a couple of thousands of cases is being done in India by the Indian Council of Medical Research.

B. Progestin Releasing IUD

Scientists in Finland have developed a levonorgestral releasing IUD, which is effective for more than 5 years. This IUD diminishes markedly the amount and duration of bleeding. However during the first three months, the number of days of spotting is increased compared to the copper releasing IUD of the same shape. The blood loss is scanty.

The local release of the progestogen suppresses the endometrium and renders it insensitive to estrogens. Oligomenorrhea and amenorrhea develop in spite of completely normal ovarian endocrine function. The incidence of amenorrhea is high during the first 2 years of the insertion of the device, but thereafter there is regular scanty bleeding over 5 years of use, possibly due to a small part of the endometrium coming under normal control of ovarian steroids.

The efficacy of the device is high. Its apparent disadvantage of inducing prolonged amenorrhea as perceived in some cultures may be an advantage in the sense that hemoglobin levels are improved.

C. Anti LHRH Vaccines, Multiple Applications

LHRH is a decapeptide made by nerve cells in the hypothalamus. It is a master molecule regulating in turn the secretion of FSH and LH which act on the gonads to make sperm and testosterone (in males) and eggs and estrogens, progesterone in females. LHRH is a largely conserved molecule in mammals and is common to mouse, rat, dog, pig or humans. Furthermore, it is a unisex molecule, identical in males and females.

Immunization against LHRH can be employed for a variety of purposes, such as:

1. Control of fertility of wild animals and stray dogs.
2. Improvement of quality of meat from male pigs. After they have attained maturity and optimum weight, males produce odorous derivatives of androgens. The production of these can be prevented by anti-LHRH immunization. Thus the meat from such male animals can be made free of such odors.
3. Prostate growth is promoted by androgens. LHRH antagonists are in use for therapeutic purposes in prostate cancers, so long as these are androgen dependent. The vaccine against LHRH would be a cheaper option requiring only periodical intake.

Talwar et al. made a fairly immunogenic vaccine by employing a decapeptide in which glycine at position 6 was replaced by D-lysine, which increased the biological half life of LHRH, resisting catabolic degradation by the native enzyme cutting the molecule at glycine. Furthermore, the extra NH2 group of lysine was available for linkage to the carrier, Tetanus or Diptheria Toxoid through a spacer. This vaccine adsorbed on alum provoked good and effective anti-LHRH antibody response bringing down testosterone and Prostate Specific Antigen PSA in patients suffering from carcinoma of prostate. The vaccine was used clinically in India at the All India Institute of Medical Sciences, New Delhi, the Postgraduate Institute of Medical Education and Research, Chandigarh and in Austria in Urology Clinic, Salzburg.

Typical results obtained in patients of carcinoma of prostate on immunization with the Anti-LHRH vaccine are given below:

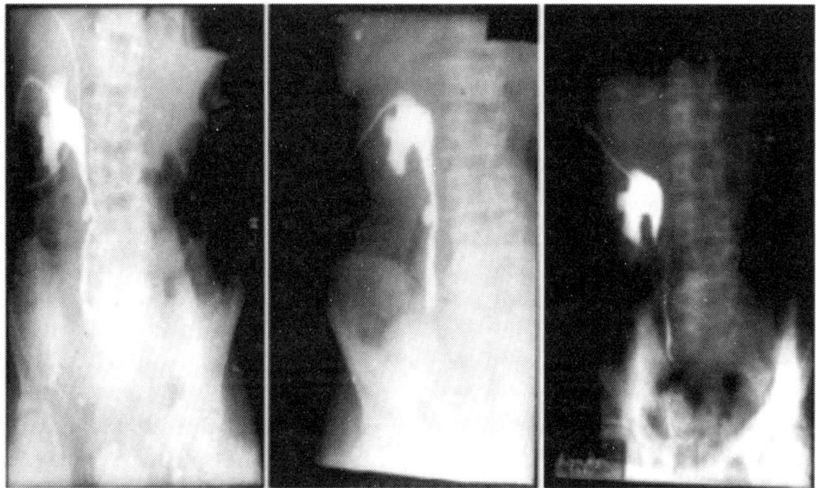

Figure 99.1 Nephrostograms showing the noticeable reduction of prostatic tissue mass at various stages of immunization with anti-LHRH vaccine of a patient in Chandigarh.

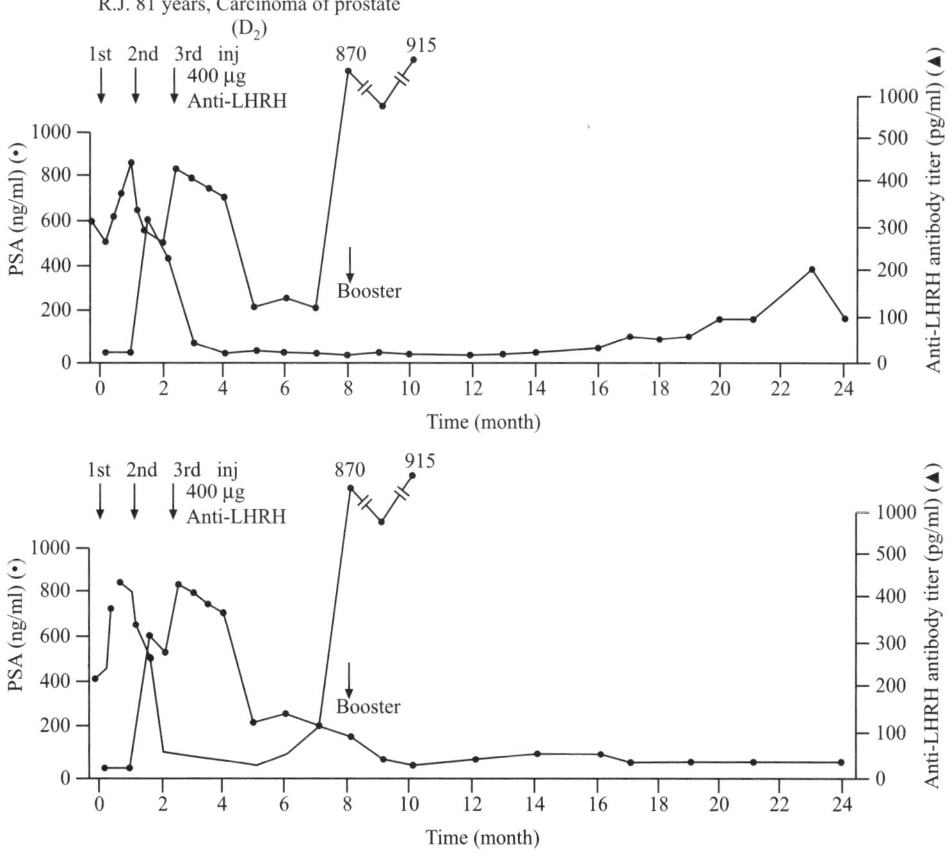

Figure 99.2 Effect of anti-LHRH vaccine on a patient with advanced carcinoma of prostate in Salzburg. With the generation of antibodies, testosterone and prostatic specific antigen (PSA) levels fall and stayed low for several months.

In addition to the above mentioned application of the Anti-LHRH vaccine in patients of carcinoma of prostate, immunization against LHRH could be employed as a non-surgical alternate to orchiectomy with advantage of reversibility. Immunization reduces the testosterone production, spermatogenesis, and atrophies the testes. These are reversible changes and return to normal values on decline of anti-LHRH antibodies.

D. Prostaglandins and Anti-progesterone Compound(s) for Menstrual Regulation and Termination of Pregnancy

The stimulatory action of prostaglandins on smooth muscle contraction is known. Natural prostaglandins however get metabolized fast and inactivated by enzymes during circulation. A series of derivatives and analogues have been synthesized which prolong the biological half-life of the prostaglandins. Investigations have been carried out on their use in the form of intravaginal suppositories. A success rate of above 80% has been recorded for termination of pregnancy with the best derivatives given at optimum concentration.

Roussel UCLAF in France have come out with a compound Ru 486 which blocks the uterine receptors for progesterone. As gestation demands progesterone for sustenance, the antiprogestational action of this compound leads to termination of pregnancy. The compound could also be used for menstrual regulation as a monthly medication.

Another idea lanced out is the combined therapeutic use of prostaglandins and the anti- progesterone compound. Their mode of action is at two different points, and their combination has resulted in dosage regimes causing abortion in 91–96% of cases during early gestation.

E. Emergency Contraception

An increasing need has arisen to ward off pregnancy following an unplanned intercourse, rape or incest. Several companies are marketing the "day after" pills containing the synthetic steroids in high dose. The knowledge about such intervention has to be diffused to a wider public, in particular the adolescents.

A new, more physiological and longer time window method employing preformed anti hCG antibodies may become available in the near future. The antibodies would have a half-life of about 20 days. In case these are taken within 7 days of the intercourse, these would prevent the implantation of the embryo on the uterus and thus the on-set of pregnancy. These would also be effective in termination of early gestation if their use is made in the post implantation period.

IV. NEW METHODS AT PHASE II/III CLINICAL TRIALS STAGE

Methods mentioned in this section may become operative in the family planning programmes in about 3–7 years if no serious difficulties are encountered.

A. Gossypol

This compound extracted from cotton seeds was used by accident in China by several thousand males causing azospermia. The compound binds with sperm proteins. Its toxicity is high in rodents. However in humans only hypokalemia was observed in China. Due to its low efficacy versus toxicity ratio in conventional experimental animals, it is unlikely that it will easily be approved for human usage in other countries. Several derivatives of gossypol have been made and research is in progress to develop compounds preserving efficacy but devoid of toxicity. Gossypol has a pronounced anti-viral action on herpes simplex virus without affecting the host cell. One of the propositions made is on its use in spermicide creams. However the strong yellow colour and staining that it produces may render it unacceptable in vaginal preparations. More recently it has been used in HIV patients with indications that its virucidal action and diminution of sperms in ejaculate may diminish the chances of HIV transmission by males.

B. Immunological Approaches to Contraception

Vaccines have helped admirably in the past for the control and containment of communicable diseases. Antibodies capable of inactivating hormones or proteins, critical to the success of reproduction, could in principle intercept fertility. This idea has been tested successfully in experimental animals including the subhuman primates, giving rise to the concept of feasibility of the *birth control vaccines*. In recent years after due toxicology and safety studies, phase I clinical trials have been conducted on three birth control vaccines. The HSD-HCG vaccine has also passed through phase II efficacy trials demonstrating safety, reversibility and efficacy of this vaccine in preventing pregnancy in women.

Mammalian reproduction results from the union of gametes contributed by male and female partners. Both the sperm and egg have immunologically specific proteins on the surface. The early embryo has a number of stage specific antigens, some of which are cross-reactive with tumour antigens. The early embryo produces and secretes substances such as the human chorionic gonadotrophin (hCG), which act as signal(s) of conception to the mother and which help prepare the uterus for establishment and sustenance of pregnancy. Other substances made are the EPF and LIF. The secretion of HCG occurs before the implantation. *In vitro* fertilized human eggs make hCG which appears to have a role in nidation of the embryo to the uterus. Anti hCG antibodies prevent implantation and thereby the initiation of pregnancy. Any one of these can be intercepted by antibodies to regulate fertility. The genesis of gametes is under the control of pituitary gonadotrophins, which are in turn regulated by the hypothalamic gonadotrophin release hormone (GnRH) or LHRH. Antibodies neutralizing these can also interfere in fertility. There are thus numerous possibilities of immuno-interception and more than one birth control vaccine is theoretically feasible. Research is currently being done to develop the following vaccines:

1. Counter hCG for control of female fertility
2. Counter GnRH or LHRH for control of "both male" and female fertility
3. Counter FSH for control of male fertility
4. Anti Zona Pellucida (Female)
5. Anti Sperm (usable preferably in females)

Antibodies against vitamin riboflavin carrier proteins abort pregnancy. The observations to-date however indicate the abortion to be at a fairly advanced stage of pregnancy. The vitamin carrier protein has been observed by P.R. Adiga to be present on sperm with the result that antibodies agglutinate sperm. Future studies may indicate the contraceptive advantage of immunizing both males and females against this protein.

The most advanced birth control vaccine at this stage is the one directed against hCG. Two parallel approaches have been made. One utilizes the beta subunit of hCG and the other a carboxyterminus peptide (CTP) of 37 amino acids of the beta sub-unit. The former is more immunogenic and produces antibodies with better neutralizing capacity of hCG bioactivity. The antibodies produced however have a degree of cross-reaction with LH. Long term studies discount the possible hazards of LH cross-section; the cross-reaction is found beneficial and contributory to control of fertility.

The first prototype vaccine developed by G.P. Talwar et al. on which phase I clinical trials were carried out in India, Finland, Sweden, Chile and Brazil consisted of the beta subunit of hCG linked to tetanus toxoid. The vaccine generated anti-tetanus and anti hCG antibodies in 61 out of 63 subjects tested. The antibody response was reversible. No side effects were noted in any clinic. The limitation of this vaccine was the large variability of antibody titres from individual to individual; those with low titres were prone to pregnancy. Further research has led to improved vaccine formulations.

The improved hCG vaccine employs a hetero species dimer (HSD) instead of β-hCG as immunogen. HSD is composed of β-hCG associated noncovalently with alpha subunit of ovine luteinizing hormone. The ability of alpha subunits to combine noncovalently with β subunits is conserved across species. HSD has higher immunogenicity and the antibodies generated by it have better bioneutralization capacity than those induced by βhCG-TT vaccine. The improved vaccine is HSD linked to either tetanus toxoid (TT) or diphtheria toxoid (DT) as carrier. HSD-TT/DT vaccine has undergone toxicology studies and has received the clearance of the Drugs Controller of India and Institutional Ethics Committees for phase I/phase II clinical trials in women. The phase I trials conducted in 5 centres in India showed that the vaccine is safe, devoid of side effects and is reversible, e.g. the antibodies decline to near zero levels in course of time in the absence of boosters. This vaccine has also completed phase II efficacy trials. Women of proven fertility, sexually active were protected from becoming pregnant, so long as the antibody titres were above 50 ng/ml. The contraceptive effect was exercised without block of ovulation or disturbance of menstrual regularity. Women regained fertility, conceived and carried pregnancy to term normally when the antibody titres declined below 35 ng/ml. This is the first birth control vaccine for which firm evidence is available on safety, reversibility and efficacy. The scientific foundations of fertility control vaccine have been laid. Further R&D is, however, required to make the vaccine at low cost by the DNA recombinant technology and to deliver the multiple doses at a single contact point through biodegradable microspheres. Experimental work has shown the feasibility of accomplishing these objectives.

Advances in the Anti-hCG Vaccine; a Recombinant Vaccine against hCG

hCG-β was fused at the C-terminal with β subunit of heat labile enterotoxin of *Escherichia coli*. Base sequence of the construct as shown in Figure 99.3 was expressed in *Pischia pastoris* to obtain hCG-LTB nucleotide sequence.

Besides the protein version of hCG β – LTB, the vaccine was also prepared as DNA after cloning the hCG-β gene in a plasmid which is approved by US FDA for human use. The recombinant hCG-β adsorbed on alum and given along with *Mycobacterium indicus* pranii (MiP) as adjuvant evoked antibody response in every mouse immunized (100% positivity of response). Furthermore the antibodies were effective in neutralizing the bioactivity of hCG. The titers attained were well over 50 ng/ml, infact over 100 times higher than the protective threshold against pregnancy in sexually active women, as established by previous Phase II efficacy trials. Figure 99.4 shows the antibody response in Balb c mice. Three injections of primary immunization, were required when the immunization is done for the first time, followed by a booster which generated above protective threshold response in all mice lasting for over 240 days.

As the immune response varies with the genetic makeup of the individual, studies were also carried out in mice of 5 different genetic strains. Although the complete quantum of response varied with genetic makeup, as expected, all mice made antibodies of titers beyond 50 ng/ml bioneutralization capacity. These results point to the high immunogenicity of the recombinant hCG-β – LTB vaccine.

Of interest is the potential of immunization with the DNA version of the recombinant vaccine, as 2 primary injections, followed by the 3rd injection of the recombinant hCG-β protein. This combination produced equally good, if not much higher titers. DNA vaccines are attractive option. They are heat stable and do not require cold chain. They are also cheaper to make.

The recombinant vaccine hCG-LTB has received approval of the RCGM, the National Committee for Recombinant genetically made products. It is scheduled to undergo appropriate Toxicology studies and then go back to clinical trials. There is every hope that it will prove to be effective in preventing unwanted pregnancy in women without impairment of ovulation and without derangement of menstrual regularity and bleeding profiles.

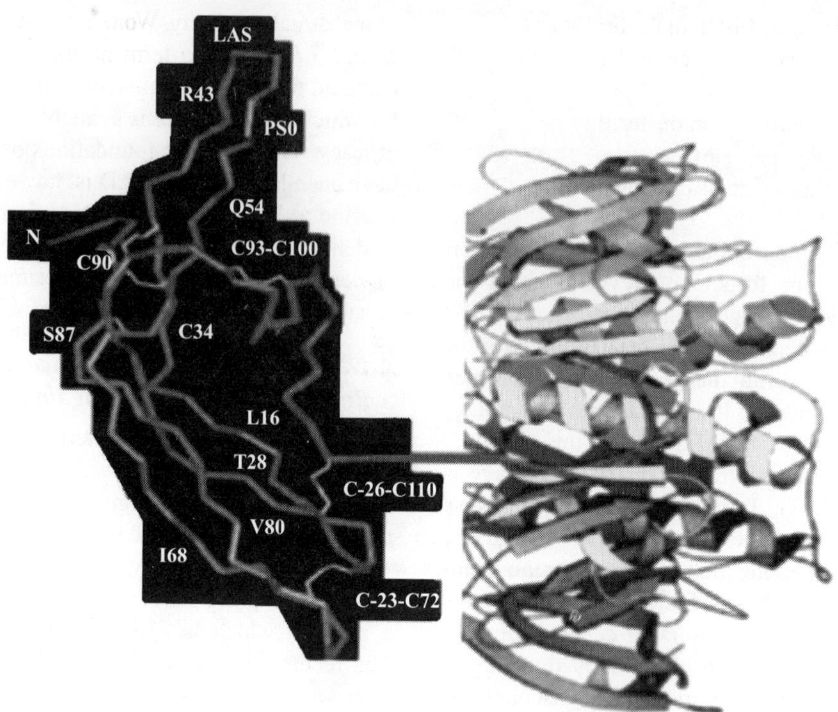

Figure 99.3   Base sequence and conceptualized structure of LTB fused at C-terminal end of hCGβ. (*see Plate 40 for colour figure*)

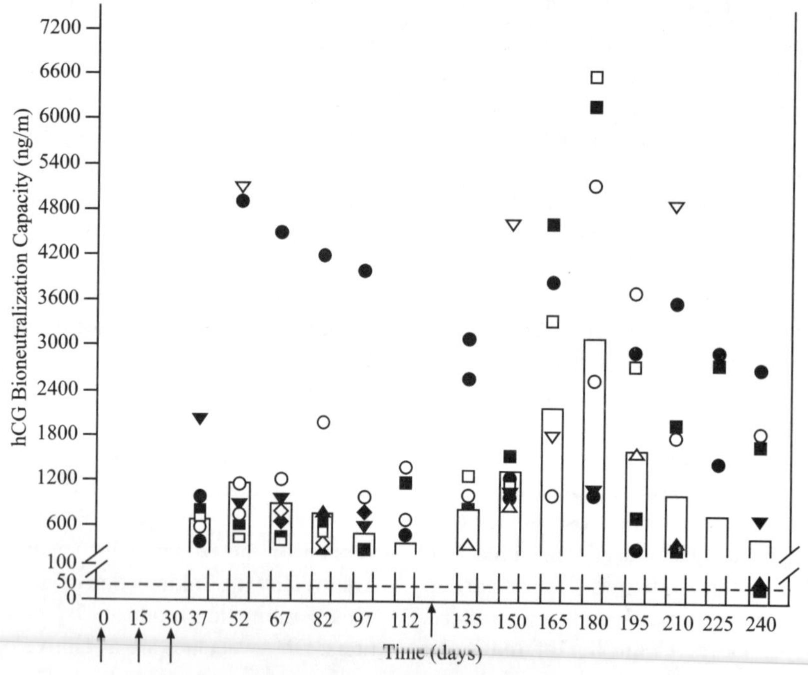

Figure 99.4   BALB/C mice were immunized with hCG-LTB + MiP. The symbols give the titers in each mouse at various time points. Bars give the geometric means. 50ng/ml is the titers required to present pregnancy in women as per previous Phase II efficacy trials. (Reproduced from Puruswani & Talwar Vaccine, 2011, 29: 2341–48). (*see Plate 40 for colour figure*)

Ectopic (unexpected) expression of hCG by Advanced Stage Terminal Cancers

During the last 20 years, a number of publications have appeared, which report the expression of hCG by a variety of Cancers of different origin (Review by Talwar et al., 2011). Invariably cancers at this stage are refractory to the currently available drugs. A question arises as to whether such cancers can be controlled/treated with the Anti-hCG antibodies.

Experiments have been carried out both in vitro and in vivo. Cells from a cell line developed by ATCC of a patient dying from such terminal stage cancer, in culture, are inhibited in a dose dependent manner by Anti-hCG antibodies (Figure 99.5).

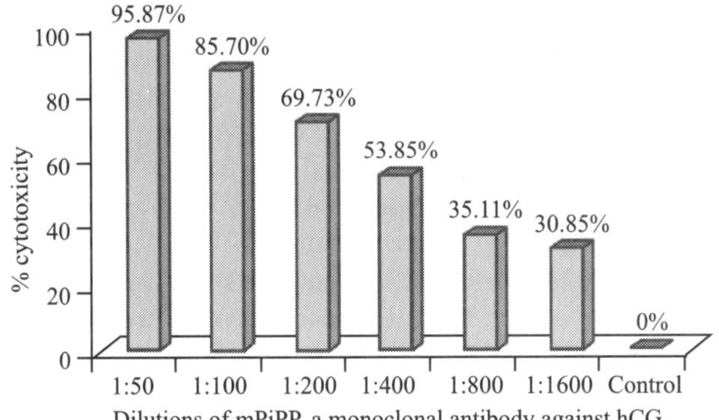

Figure 99.5 Dose dependent cytotoxicity exercised by a monoclonal anti-hCG antibody cPiPP on A-549 Lung Cancer Cells. (Talwar et al., *J. Cell Sci. Ther.*, 2014, 5:159)

Also, the antibodies block the growth of Chago lung cancer in nude mice by Anti-hCG antibodies (Figure 99.6).

While therapeutic antibodies are approved in many countries for clinical use, these are fairly expensive and demand repeated use. A vaccine would be a more affordable option.

Passive immunization approaches

The concept of immuno-interception by antibodies received unambiguous support by the demonstration that antibodies could indeed interfere with fertility. Hybrid cell clones producing in unlimited amounts antibodies of selected characteristics have been developed against several reproductive tract hormones and proteins. The anti-hCG monoclonal P_3 W80 developed in India terminated pregnancy in chimpanzees. The chimpanzee chorionic gonadotrophin is strongly cross-reactive with hCG, if not identical. The endocrinology of pregnancy in chimps is similar to the human. There is every reason to believe that the results in chimpanzees will be applicable to humans.

In summary, active immunization (vaccines) and passive use of antibodies have been demonstrated to control fertility in a number of animal species including monkeys, baboons, marmosets and chimpanzees. The anti-hCG vaccines have already entered the phase of clinical testing. Other vaccines based on sperm antigens are likely to follow. Polyvalent vaccines engendering antibodies intercepting fertility at more than one point may be the preferred second generation birth control vaccines. The design of these vaccines, e.g., linkage of a reproductive tract hormonal subunit or protein with a carrier such as tetanus toxoid or cholera alligns fertility control with immuno-prophylaxis as the antibodies are produced against both components of the vaccine. In future vaccines, this concept is likely to find extension. Diversified immunoprophylactic benefit will be achieved by use of additional carriers. Cholera toxin chain B has already been evaluated and is a valuable supplement to tetanus. Diptheria toxin, hepatitis B-surface

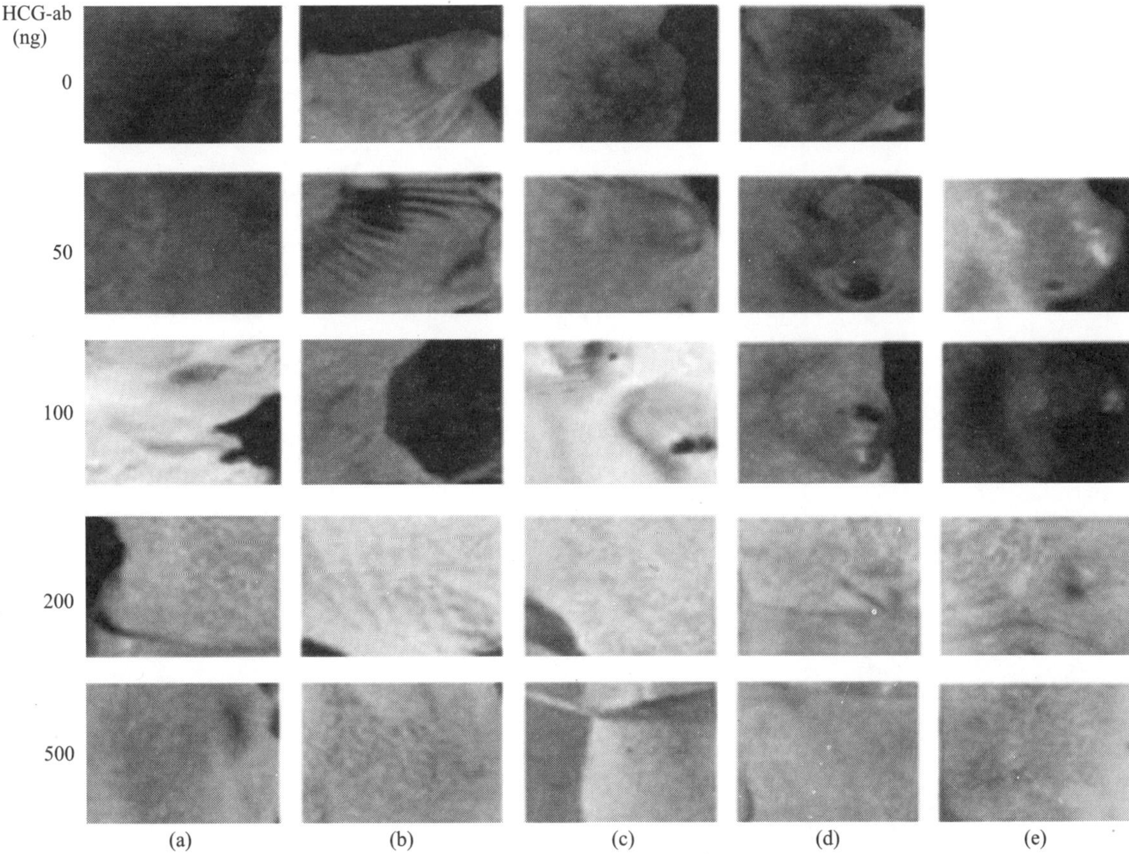

Figure 99.6 Dose dependent Prevention of Human lung Chago cell tumor by antibodies in nude mice. (Talwar et al., *J. Clin. Cell. Immunol.*, 2014, 5:247) (*see Plate 40 for colour figure*) (*see Plate 40 for colour figure*)

protein and possibly sporozoite coat protein of Plasmodium falciparum are other potential carriers that may be employed for making the polyvalent double purpose vaccines for control of fertility and prevention of diseases. Short peptides corresponding to T non B lymphocyte determinants can replace with advantage the full tetanus and other carriers for enlisting T cell help to induce antibody response to "self" antigens such as hCG and LHRH. A cocktail of such peptides would generate immune response from people of different genetic backgrounds.

An important technological advance is the possibility of "humanizing" mouse monoclonal antibodies. These can be further engineered to enhance affinity for the target epitope. The genes of desired antibodies can be cloned and expressed in either *bacteria* or *plants* for production of large quantities of antibodies at low price. In near future, the passive use of such antibodies is likely to find practical applications for control of fertility. Preformed anti-hCG antibodies will be usable for emergency contraception with wider time window after an unprotected sexual intercourse. These would also be usable for short term "vacation" contraception, each injection providing efficacy and protection for 2 to 3 months. A distinct advantage from active immunization is the delivery of adequate antibodies to act as sure efficacy in every recipient.

SUGGESTIONS FOR FURTHER READING

Christine K. Mauck, M. Cordero, Henry Gabelnick, Jeff Spieler and Roberto Rivera, (Eds.) (1994), *Barrier Contraceptives, Current Status and Future Prospects,* Wiley-Liss, New York.

Talwar G.P. (1997), Vaccines for Control of Fertility and Hormone-dependent Cancers, *Immunology and Cell Biology,* 75, 184.

Talwar G.P., Singh Om, Pal Rahul, Chatterjee N., Sahai P., Dhall Kamala, Kaur Jasvinder, Das S.K., Suri Sushma, Buckshee Kamal, Saraya L., and Saxena Badri N. (1994), A Vaccine that Prevents Pregnancy in Women, *Proc. Natl. Acam. Sci.,* USA, 91, 8532.

Talwar G.P. and Raghupathy Raj (Eds.) (1995), *Birth Control Vaccines,* R.G. Landes Co. Medical Intelligence Unit Monograph.

Regine Sitruk-Ware and Wayne Bardin C. (Eds.) (1992), *Contraception, Newer Pharmacological Agents, Devices and Delivery Systems,* Marcel Dekker Inc., New York.

100

Reproductive and Therapeutic Cloning

Subeer Majumdar, Neerja Wadhwa and Indrashis Bhattacharya

CONTENTS

I. HISTORY: A JOURNEY FROM FROGS TO MICE

We believe now that the cell fates are neither restrictive nor irreversible during development, but cell biologists for long believed that the terminally differentiated cells lost their potential of producing other cell types. The first study about nuclear transplantation was proposed at the end of 19th century, to test whether nuclei from differentiated cells remain genetically equivalent to zygotic nuclei. In the beginning, most nuclear transfer experiments were conducted on fish or amphibian eggs due to their large size and abundant resources. In 1952, Briggs and King pioneered the technique of nuclear transplantation by injecting nuclei from *Rana pipiens* frog blastomeres into previously enucleated eggs, generating early cleavage embryos. Only a few years later, Gurdon conducted similar experiment with a different species of frog and succeeded in transferring nuclei from highly specialized tadpole intestinal cells into ultraviolet-light-irradiated (for damaging nuclear DNA) oocytes, obtaining not only tadpoles but also fully grown adult frogs. The ultimate goal of this approach was to test whether nuclei taken from fully differentiated cells were equivalent to the zygotic nucleus ('nuclear equivalence') or whether differentiation was associated with irreversible loss of genetic information ('nuclear differentiation'). This has remained a long-standing question in developmental biology. Several years later, McGrath and Solter announced the first successful cloning of mice.

II. METHOD: SOMATIC CELL NUCLEAR TRANSFER

Clearly, 1997 was the year of the clone when Ian Wilmut and coworkers announced the birth of the first ever clone of an adult mammal reprogramming somatic cell nucleus. The cover illustration of the journal *Nature* (27 February 1997) announced the birth of Dolly, the ewe cloned from an adult sheep in Scotland, and the journal *Science* (19 December 1997) proclaimed Dolly to be the "breakthrough" of the year. Newspapers, news magazines, radio, and television were even more fervent in their reports of Dolly. The achievement was significant because the ability to clone mammals by simple nuclear transfer was thought to be biologically impossible and Dolly's birth dismissed the dogma because Dolly was not created naturally using haploid germ cells but was derived from the nuclear genome (diploid) of a quiescent-induced mammary cell with the support of cytoplasm originating from an enucleated oocyte. Even now, a number of puppies, kittens and calves, all cloned from adult cells, continue to excite people. To date, adult sheep, cattle, mice, pigs, goats, cats, pigs and rabbits have been cloned using Somatic Cell Nuclear Transfer (SCNT). Furthermore, nuclei from frozen tissues have also been transplanted successfully into enucleated oocytes a decade after tissue freezing. Such experiments were carried out with the aim of eventually using cryopreserved cells for cell-based therapies in future and also for cloning of animals that are endangered or nearing extinction.

The conventional method of SCNT is relatively the same for all animals. The first step is to collect the somatic cells from the animal that will be cloned. The hardest part of SCNT is to remove maternal DNA from an oocyte at metaphase II. Once this is done, the diploid somatic nucleus can be inserted into an egg cytoplasm. The somatic cells could be used immediately after isolation or stored in the laboratory for later use. The grouped somatic cell nucleus and egg cytoplasm are then chemically activated to allow the cloned embryo to begin development. The successfully developed embryos are then transferred to surrogate recipients. This conventional method has been successfully used to generate several species including cattle (Figure 100.1).

In SCNT, somatic cells used as donor can originate from a wide variety of cell types ranging from embryonic blastomeres all the way up to adult cells. Mammary epithelial cells, ovarian cumulus cells, fibroblast cells from skin and internal organs, Sertoli cells, macrophage and blood leukocytes have been successfully utilized for nuclear transfer.

Various strategies have been employed to modify donor cells prior to transfer in ooplast which includes synchrony of the cell cycle stage of donor cells, as well as synchrony between donor cells and recipient oocytes, using somatic cells from donors of various ages, various tissue origins, various passages and various culture conditions.

Reporter genes, like enhanced green fluorescent protein (EGFP) had been introduced into cultured cells which were used as source of donor nuclei for SCNT leading to the generation of transgenic green embryos (Figure 100.2). This suggested that SCNT could be used as a method to create transgenic animals. SCNT based transgenic animal production has significant advantages over the previously employed microinjection of foreign DNA into pronuclei of zygotes. This cell-based transgenesis is compatible with gene targeting and allows both, the addition of a specific gene and the deletion of an endogenous gene. Therefore, cloning can also be used as one of the ways

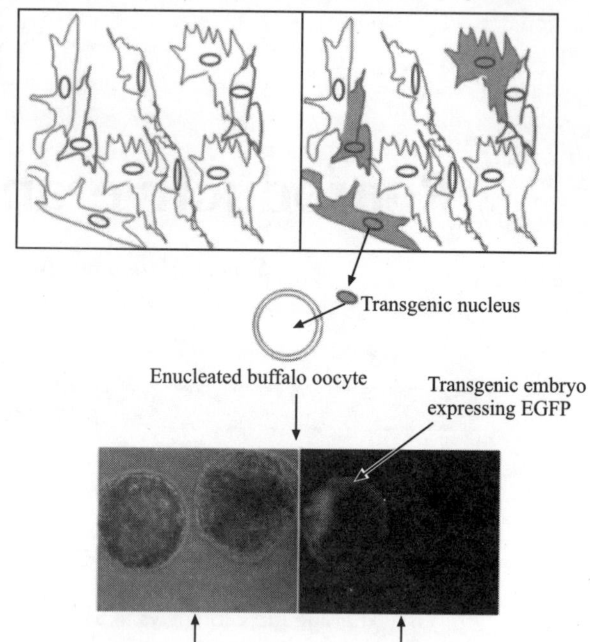

Figure 100.2   Transgenesis via cloning. (*see Plate 41 for colour figure*)

to generate transgenic cattle or buffaloes avoiding cumbersome male pronuclear DNA microinjection method.

III.   NUCLEAR REPROGRAMMING

Nuclear reprogramming is a set of epigenetic changes (those not involving a change in DNA sequence) required for a nucleus to change developmental fates. Upon nuclear transfer, the components within enucleated oocyte change the fate of the donor nucleus from its original status (e.g., skin, granulosa

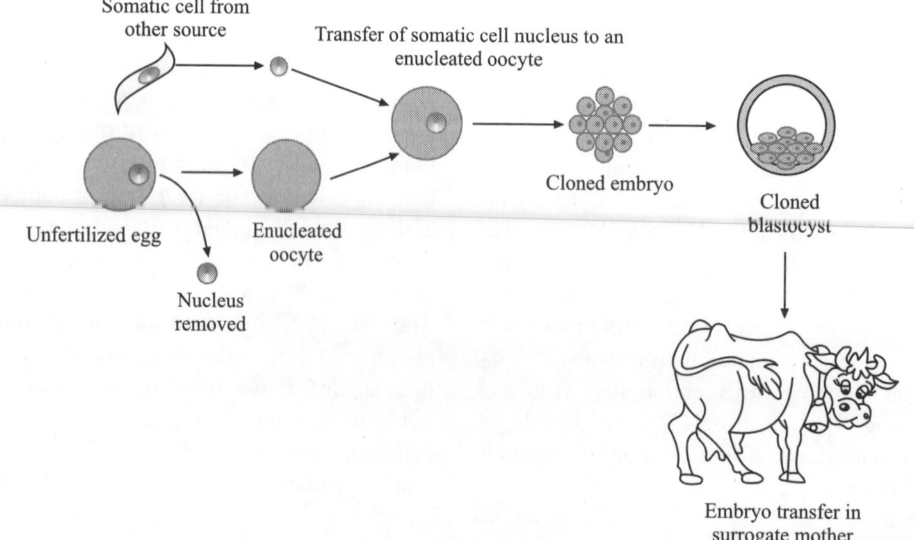

Figure 100.1   A cartoon of SCNT showing reproductive cloning. (*see Plate 41 for colour figure*)

etc.) to that of a zygotic nucleus. Following nuclear transfer, the donor nucleus often fails to express early embryonic genes and to establish a normal embryonic pattern of chromatin modifications. These defects correlate with the low number of cloned embryos being able to produce blastocysts and/or adult animals. Improper nuclear reprogramming of the donor nucleus in the oocyte is thought to be a major hurdle in the cloning process. In cattle, studies have shown that abnormal methylation as well as gene expression is responsible for inappropriate reprogramming of nucleus upon nuclear transfer. Apart from these factors, several unidentified reprogramming factors present in oocytes are capable of initiating a cascade of events that can reset the mature, specialized nucleus of the cell back to an undifferentiated, embryonic state.

Therefore, for successful cloning, efficient reprogramming of gene expression of the donor nucleus is necessary. This requires silencing of native gene expression of the transferred donor nucleus and activation of an embryonic pattern of gene expression. Recent observations indicate that reprogramming of the recipient nucleus by ooplasm may be initiated by early events that occur soon after nuclear transfer, and it continues as the development progresses through cleavage and probably to gastrulation. Since reprogramming is slow and progressive, cloned embryos have dramatically altered characteristics in comparison to fertilized embryos.

IV. POTENTIAL APPLICATIONS OF SOMATIC CLONING

One of the most important practical utilities of cloning is in producing clones of high milk yielding cattle and buffalo. Over the years, several calves (cows) had been cloned for this purpose, worldwide. Recently, in 2008, Deshun Shi of China, using fetal fibroblasts and adult ovarian granulosa cells, cloned first water buffalo. Later, in India, embryos of water buffaloes were cloned using another technique, i.e., Hand-made cloning (HMC) technique, which is a modification of the "Conventional Cloning Technique". This technique, described in the year 2001, by Gabor Vajta of Australia is cheaper and simpler because it does not require micromanipulation of oocyte. Handmade cloning is done with micro blades and free-hand cell surgery under the dissecting microscope. The recipient egg is split in half and the nucleus containing half is discarded. The other half lacking nucleus is fused with a donor somatic cell and the fused egg half is fused again with a second enucleate half egg before it is activated (with a mild electric shock) to produce an embryo that is then implanted in a surrogate mother. This technique has gained popularity and, a cloned buffalo named "Garima" was born in India following this technique.

Somatic cloning is still at early stage of development although significant improvements have been made in the past years that in turn generated potential for its practical applications. At present, the main reason to clone farm animals is to preserve the breeding capacity of genetically elite animals (proven through progeny testing), particularly males, to prevent loss of valuable genetic characteristics. It may have usefulness in conservation of endangered species or for producing transgenic animals for pharmaceutical protein production in milk or xeno-transplantation, by *in vitro* gene manipulation in the nuclei of cultured cells prior to their use for cloning.

During SCNT, genetic modification can be performed in cultured cells and the nucleus of the modified cell selected prior to animal production. The production of an animal from a single nucleus removes the problems associated with mosaicism; all of the cells within the resultant animal will contain the modification that will be transmitted through the germ line. All the animals produced will be transgenic and producing multiple copies from the cultured cells can accelerate flock or herd generation. The experiments that led to 'Dolly' involved the use of mammary epithelial cells. The use of this cell line can be related to the potential screening of transgenic cells for qualitatively and quantitatively better milk production *in vitro* prior to production of animal. Thus, it may be possible to predict expression level and select the best cell prior to use of its nucleus for animal production. In order to carry out these modifications the cultured cell populations must be amenable to transfection and selection to maintain their ability to be used as successful nuclear donor. 'Polly', a nuclear transfer lamb transgenic for human factor IX was derived from a transfected, selected fetal fibroblast cell population. Later, several cloned animals have been produced carrying an addition or deletion of a single allele of the gene.

Genetically modified domestic animals may play an important role in agriculture and biomedical research. In the former, field animals with both qualitatively and quantitatively improved traits (for example, cattle with altered casein ratio in milk or increased beef production) can be introduced, but genetic modification may also be used to achieve more ambitious goals such as decreasing the impact on the environment, increasing disease resistance, and creating males that produce offspring with a desired gender. In biomedical applications, important goals are to establish living bioreactors for mass production of various proteins for medical applications; organ donors for xeno-transplantation in humans and models of human diseases.

Clones of cells with most desired traits can be used for making large number of cloned transgenic animals simultaneously carrying multiple genetic modifications. This may play a major role in a range of therapies given below:

Biopharmaceuticals

Human or animal proteins may be produced in a range of tissues and bodily fluids including blood, urine and milk. A range of therapeutic proteins is being produced in the milk of transgenic animals including alpha-1-antitrypsin (for treatment of cystic fibrosis) factor IX (for treatment of hemophilia B). Nuclear transfer may facilitate the removal of unwanted endogenous genes to aid purification, for example, the replacement of gene for bovine serum albumin in cattle with gene of human serum albumin (HSA), in order to produce and purify large amounts of HSA from cattle for the treatment of men with burn injury.

Nutraceuticals

Cloning can be exploited for generating Nutraceuticals which are usually meant for the medical benefits but which may include the modification of animal milk to enhance nutritional value and the removal of specific proteins which may not be suitable or induce allergy, e.g., β-lactoglobulin in cattle milk.

Xenotransplantation

This includes use of animal organs and other tissues for human transplantation. Physiologically, pig organs are similar to those of human and are considered suitable for transplantation. However, there is a major problem with organ rejection. Although all of the mechanisms that are involved in this rejection are not completely understood, it is known that a major antigen involved is α-1,3-galactose. This is present on pig cells but is not found in humans who therefore mount an immune response. Nuclear transfer from cultured porcine cells in which the pig gene coding for α-1,3-galactosyl transferase is knocked out may be potentially suitable for acceptable organ transplantation. Such animals have now been reported by a number of groups. Potential organs and tissues for transplantation include the heart, lungs, kidneys, liver and pancreatic islets.

Disease models

At present, many of the animal models available for the study of human genetic disorders are only available in mice. This species may not manifest the same clinical symptoms as in humans. Precise genetic modification of somatic cells *in vitro*, coupled to nuclear transfer will allow the production of disease models in clones of species which are physiologically more similar to the human in order to follow disease progression or to assess the benefits of any potential new therapies (including gene therapy). An example of this is cystic fibrosis, in mice this disease manifests in the gut and not the chest, however, subhuman primate may provide a better model. Since gestation in large animals is longer, cloning is the only way to produce more animals in less time, concurrently.

Cloning in stem cell therapies

The use of embryonic stem (ES) cells as a source for the production of specific differentiated cell lineages provides numerous opportunities for human therapies. At the present time, stem cell populations have been isolated from embryonic fetal and adult tissues and studies are underway in numerous laboratories on the controlling lineage-specific differentiation. The practical aspects of cell transplantation with regard to function, longevity and rejection may differ depending upon the *cell origin*. Cloning followed by ES cell isolation and differentiation may prove to be a valuable tool as it provides a possible route for autologous transplantation and since ES cells are not differentiated, they are destined to differentiate into a kind of cell (Figure 100.3). SCNT may be used as a tool to produce ES cells for therapeutic purposes called as therapeutic cloning. The goal is not to create cloned human

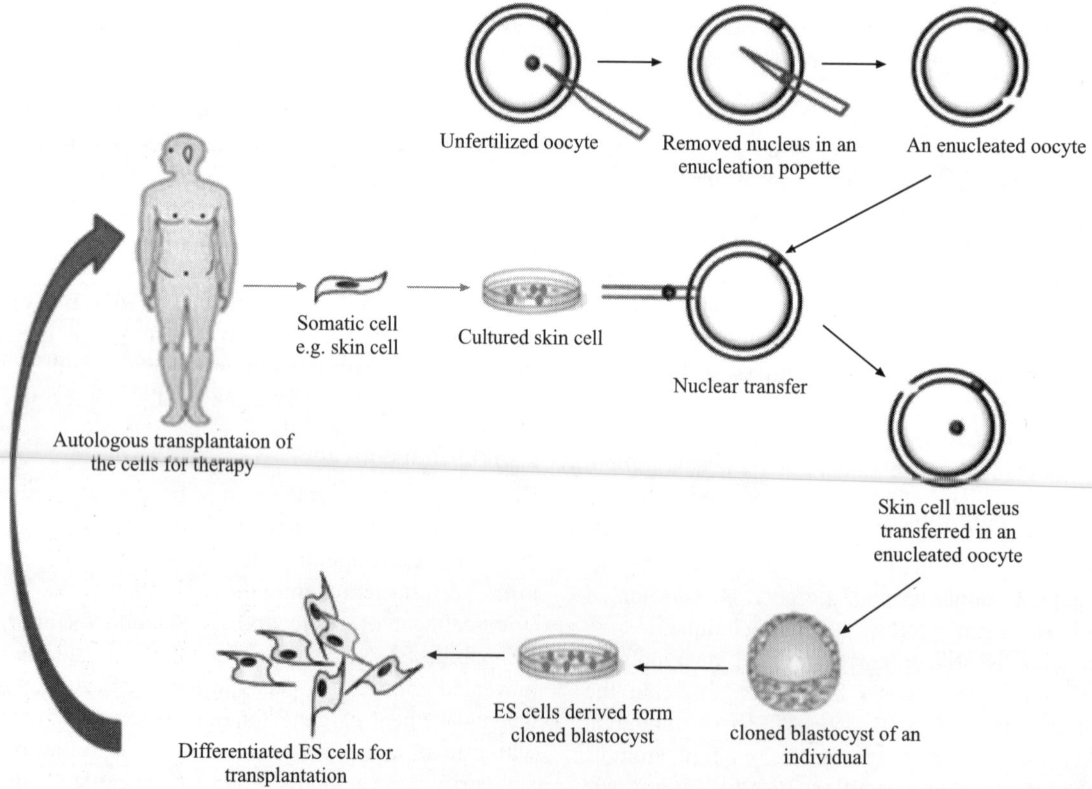

Figure 100.3 A flow diagram of SCNT for therapeutic cloning.

beings (called "reproductive cloning") but rather to harvest stem cells that can be used to study human development and to potentially treat diseases. Therapeutic cloning is achieved by creating ES cells in the hopes of treating ailments such as diabetes and Alzheimer's in a specific person. It is being used to generate tissues and organs for transplants. To do this, cells of patient can be cultured and gene manipulations *in vitro* may be done, if required, before using the nucleus for insertion into an enucleated egg. After the egg containing the patient's DNA starts to divide, ES cells that can be transformed into any type of tissue would be harvested. The stem cells would be used to generate an organ or tissue that is genetically matched to the recipient. In theory, the cloned tissue/organ could then be transplanted into the patient without the risk of tissue rejection. If organs could be generated from cloned embryo of human, the need for organ donation could be significantly curtailed.

Phenomenal progress is being made in generating tissue-engineered skin, cartilage, bone, and blood vessels that are being tested in patients. However, limited knowledge of the molecular pathways of tissue growth and differentiation has brought in hiatus in understanding the process well enough to channel the *in vitro* differentiation of human ES cells into all of the various cell replacement types.

V. ECONOMIC VIABILITY OF SCNT IN ITS PRESENT FORM

Although SCNT opened up a wealth of opportunities, but success of cloning is still limited. Published literature suggests that though SCNT may produce cloned embryos for *in vitro* manipulation and studies, more than 100 nuclear transfer procedures or attempts could be required to produce one viable clone. Currently, the efficiency of live births upon nuclear transfer is between 0–10%, i.e., 0–10 live births after transfer of 100 cloned embryos into recipient surrogate mothers. Even Dolly was just one cloned offspring that resulted after 277 attempts (0.3% efficiency), similar inefficiencies are still being described for cloning adult animals. Incomplete or incorrect reprogramming of the transferred nucleus may underlie low success rate of cloning.

In addition to low success rates, cloned animals tend to have more compromised immune function and higher rates of infection, tumor growth and other disorders. Studies have shown that cloned mice live in poor health and die early. About a third of the cloned calves born alive have died young, and many of them were abnormally large. For example, in 2003, Dolly developed severe arthritis and lung disease and died at the age of 6 years instead of expected life expectancy of 12 years. It was said that a contributing factor to Dolly's death may be that she could have been born with a genetic age of six years, the same age as the sheep from which she was cloned. One basis for this hypothesis was the finding that Dolly's telomeres were found to be short, which typically is a result of the ageing process.

Developmental defects, including abnormalities in cloned fetuses and placentas, in addition to high rates of pregnancy loss and neonatal death have been encountered by most of the research teams studying somatic cloning. Although reprogramming of a somatic cell nucleus can successfully generate an embryo, long-term stability of reprogrammed nucleus that is differentiated to various cell types, is yet to be determined. For example, in case of a person whose skin or blood cell is used as a donor to generate cloned embryo to obtain ES cells for induced differentiation into cardiomyocytes for transplantation to restore heart function, the question arises whether the transplanted cardiomyocytes will ever express the characters of original somatic cell (skin/blood cell) or will never revert back?

Any animal created using SCNT is not truly an identical clone of the donor animal. Only the clone's nuclear DNA is the same as that of the donor, however, some of the clone's genetic material is also contributed by mitochondrial DNA that had been left over in the cytoplasm of the enucleated egg. Since, mitochondria have a broad range of critical functions in cellular energy supply, cell signaling and programmed cell death, inadequate or perturbed mitochondrial functions may adversely affect SCNT success. Animals cloned by SCNT bear a mixture of mitochondrial genes from the donor somatic cell carried over along with nucleus and from the egg cytoplasm, a condition referred to as heteroplasmy. In nature, only the maternal parent provides mitochondrial genes. When the cloned female is mated to a cloned male the offspring are all heteroplasmic and likewise, when the cloned female is mated with a normal male the offspring will always be heteroplasmic. However, when a cloned male is mated with a normal female the offspring will all be normal homoplasmic, carrying the mitochondria of the female parent. In the case of transgenic clones, or humanized animal clones the mother may continue to carry transgenic or human cells from the embryo. During pregnancy there is transplacental traffic of fetal cells into the maternal circulation. Remarkably, cells of fetal origin can then persist for decades in the mother which could be detectable in the circulation and in a wide range of tissues.

Somatic cloning is also considered to be one of the expensive, laborious and time consuming technique that requires sophisticated equipments like micropipette puller, micromanipulators, piezo drillers etc. along with skilled manpower. Studies of cloning had been performed in various countries including India. Extremely low efficiency of cloned animal and compromised health reported so far suggests that present technique of somatic cloning may not be economically viable. However retrieval of ES cells from cloned embryo *in vitro* may be the most fruitful outcome of this technique in its present form for the purpose of cell based therapies via autologous transplantation.

Cloning and Stem cells earned Nobel Prize in Physiology or Medicine in 2012

The research by John Gurdon of the Gurdon Institute, Cambridge, UK was recognized for his lifelong endeavor to understand the process by which an adult cell nucleus can be reprogrammed to start development all over again when transplanted into an egg. Research by Shinya Yamanaka of Kyoto University, Japan led to successful reprogramming of adult cells to the pluripotent state *in vitro* which they called as iPS. Their observations have changed the way we thought about the stability of the differentiated cell state. The discoveries of both scientists fetched them Nobel Prize in Physiology or Medicine in 2012. Yamanaka's discovery built on work done by Gurdon, who more than 40 years earlier showed in frogs that the genetic program of a mature cell could be "reset" to its embryonic state.

VI. THERAPEUTIC CLONING vs. iPS

In 2006, Takahashi and Yamanaka made a groundbreaking discovery, when by introducing four specific genes associated with embryonic development into adult mouse cells, the cells were reprogrammed to resemble ES cells and named as iPS. The four transcription factors (*Oct4, Sox2, Klf4 and Myc*) were introduced in both mouse and human somatic cells. These iPS cells closely resemble ES cells in terms of global gene expression, epigenetic profile, long-term self-renewal capacity and developmental potential. This approach does not require generation or destruction of embryos, avoiding ethical issues regarding research on ES cells. Therefore, the use of 'custom-made' adult cells derived from human iPS cells might ultimately allow the treatment of patients with debilitating degenerative disorders. Since these autologous cells also have DNA identical to the recipient, it is possible that the necessity of cloning for cell-based therapy will fade out.

However, Zhao and colleagues have shown that some iPS cells were immunogenic, raising concerns about their therapeutic use. They found that ES cells derived from blastocysts from a given genetic background grow into teratomas on transplantation into mice of the same genetic background. The immune system therefore 'tolerates' autologous ES cells. In contrast, autologous iPS cells, reprogrammed *in vitro* from fetal fibroblasts by viral or non-viral genetic approaches, elicit an unexpected immune reaction in genetically identical mice upon transplantation, resulting in their rejection.

Despite the robustness in iPS cell technology, human SCNT research remains an important approach for *regenerative medicine*. Firstly, SCNT is advantageous over iPS cell in therapeutic potential for certain disorders, for example, mitochondrially transmitted diseases. The human mitochondrial genome is transmitted cytoplasmically from the oocyte and carries its own DNA that encodes for 13 polypeptides. Mitochondrial DNA (mtDNA) genes have *a much higher mutation rate* than nuclear DNA genes. Mutations in mitochondria genes lead to many mitochondrially transmitted diseases. Since patient's somatic cells might contain mutated mitochondria, proof-of-concept experiments implied that SCNT might be a prior option to generate therapeutic pluripotent cells without native mtDNA because in cloning, mitochondrial DNA comes from an another individual donating recipient enucleated oocyte.

In case of cloning, the time required for the reprogramming process by oocyte is very short, and usually occurs within hours, while the factor-induced process for iPS takes days or weeks. Longer induction might allow accumulation of harmful genetic or epigenetic changes on the genome. Indeed, iPS cells displayed certain differences to ES cells in global gene expression profiles and epigenetic state such as DNA methylation. Also, certain deficiencies in differentiation have been reported for human iPS cells. In addition, during the process of induced reprogramming, the maintenance of genome integrity has been challenged with reports showing that *de novo* mutations and copy number variations might be introduced in iPS cells. Thus, it is essential to carefully compare the properties of induced reprogramming to those mediated by oocyte to fully understand the reprogramming mechanisms and their roles in regenerative medicine. In mice, iPS cells were found to harbor a residual epigenetic signature characteristic of their donor cells. These so-called epigenetic memories restrict their fate choice following differentiation. In contrast, mouse pluripotent cells generated through cloning did not appear to show such memories. There are several proteins in oocyte that might have functions in chromatin decondensation, DNA demethylation and transcriptional activation. It is felt that, research to generate normal pluripotent cells *via* SCNT will help to unravel role of these important proteins in the egg, use of which may further help in fine tuning the speed and safety of iPS cell generation and acceptability of its usage in cell based therapy.

At present, Somatic cloning is advantageous over iPS. The ability to create autologous embryos represents the first step towards generating immune-compatible stem cells that could be used to overcome the problem of immune rejection in regenerative medicine. Also, in case of cloned embryos, mtDNA is from the oocyte that is relatively young and fresh while nuclear material is from somatic cell which may be from any individual aged or young. While in case of iPS, cytoplast including mtDNA is as old as nuclear material of the individual. Therefore, vigor of ES cells generated by cloning may be more in comparison to iPS.

Other than these, the conventional iPS cell technique suffers from some drawbacks that may hinder many practical applications of the technology. First, permanent genome modification of reprogrammed somatic cells by introduction of the exogenous genes themselves (i.e., any of the four reprogramming factors) or even fragments from delivering vectors could be very harmful. However, progress has been made in recent time overcoming these safety concerns by using non-integrating gene delivery approaches or reprogramming proteins. In addition, the reprogramming process induced by the four factors generally entails a step-wise progression of non-specific events that give rise to pluripotent cells as one of many possible cell fate destinations, resulting in low efficiency and slow kinetics. This intrinsically random nature of the reprogramming process of iPS may also pose risks such as residual epigenetic memory, inappropriately induced epigenetic changes and accumulation and selection of other subtle genetic and epigenetic abnormalities during the reprogramming process. However, careful examination of these issues can help us determine whether such differences pose a hurdle in the potential therapeutic use of iPS cells.

SUMMARY

Somatic cell cloning of mammals by nuclear transfer is relatively new with many potential applications. However, at the current stage of development, the reprogramming by nuclear transfer is still inefficient. Further efforts are needed to perfect this technology and extend it to its fullest potential for therapeutic purposes in man and for cloning in economically important farm animals.

SUGGESTION FOR FURTHER READINGS

1. Briggs and King (1952), Transplantation of Living Nuclei from Blastula Cells into Enucleated Frogs' Eggs, *Proc. Natl. Acad. Sci.*,USA, 38(5):455–463.

2. Gurdon J.B. (1960), Factors Responsible for the Abnormal Development of Embryos Obtained by Nuclear Transplantation in *Xenopus Laevis*, *J.Embryol. Exp. Morphol.*, 8:327–340.

3. Wilmut I., Schnieke A.E., McWhir J., Kind A.J. and Campbell K.H.S. (1997), Viable Offspring Derived from Fetal and Adult Mammalian Cells, *Nature,* 385:810–813.

4. Campbell K.H.S. (2002), A Background to Nuclear Transfer and its Applications in Agriculture and Human Therapeutic Medicine, *J. Anat.*, 200:267–275.

5. Hodges C.A. and Stice S.L. (2003), Generation of Bovine Transgenics using Somatic Cell Nuclear Transfer, *Reprod. Biol. Endocrinol.* 1:81–87.

6. Takahashi K. and Yamanaka S. (2006), Induction of Pluripotent Stem Cells from Mouse Embryonic and Adult Fibroblast Cultures by Defined Factors, *Cell*, 126:663–676.

7. Wadhwa N., Kunj N., Tiwari S., Saraiya M. and Majumdar S.S. (2009), Optimization of Embryo Culture Conditions for Increasing Efficiency of Cloning in Buffalo (*Bubalus bubalis*) and Generation of Transgenic Embryos via Cloning, *Clon. Stem Cells*, 11(3):387–395.

8. Zhao T., Zhang Z.N., Rong Z. and Xu Y. (2011), Immunogenicity of Induced Pluripotent Stem Cells, *Nature,* 474:212–216.

9. Pan G., Wang T., Yao H. and Pei D. (2012), Somatic Cell Reprogramming for Regenerative Medicine: SCNT vs iPS Cells, *Bioessays*, 34:472–476.

SECTION XIV

IMMUNOLOGY

The ability to generate an immunological response to a "nonself" entity is a fascinating trait of the multicellular organism. This ability does not exist in the same form in unicellular organisms. The system has a memory and operates through the collaborative interaction of more than one type of cell. The immune system is an exciting area of current biological research. It is also one of those disciplines which have important applications in medicine—both in preventive as well as in treatment and handling of various diseases with an immunological basis. The importance of this system cannot, thus, to overemphasized.

101

Innate and Adaptive Immunity

Raj Raghupati

CONTENTS

I. INTRODUCTION

All organisms on this planet from bacteria to invertebrates to humans face some potential threats to their individuality, to their integrity and to their very survival. Bacteria face the potential threat of attack by bacteriophages, while aquatic animals and amphibians have to survive bacterial attacks, and birds and reptiles are vulnerable to viral infections. Mammals are constantly threatened by potential attacks by *external* agents such as viruses, bacteria, fungi and protozoal parasites and moreover are susceptible to *internal* attack by tumours that arise from time to time. Every organism is host to innumerable parasites. Parasites like bacteria and viruses can cause infectious diseases that can debilitate and even kill, parasites can cause epidemics that can potentially wreak havoc on an entire species. Given the amazing range of deleterious effects that an incredible variety of pathogens can cause, ranging from mild illness and morbidity to death, and given the fact that organisms have survived and flourished over millions of years, it stands to reason that organisms would have evolved special mechanisms and processes that *protect* them from pathogens, that *defend* their integrity and their organs, tissues and cells from pathogenic attack. Indeed, to protect our individuality and integrity, we have developed a large variety of defense reactions, an immense range of molecules and cells that orchestrate exquisite defense responses, in fact a whole *system* consisting of molecules, cells, tissues and organs, referred to as the *immune* system, a fascinating network of specialized cells and molecules.

Immunology is the study of our defense system, reactions that protect us by mounting cellular and molecular actions against pathogens; the word Immunology derives from the Latin "immunis" which means "free of burden". While the word *immunis* centuries ago meant to be "free from the burden of taxes', in its modern derivation the word "immunity" means "freedom from the burden of infection". Thus a person resistant to a certain disease or protected from a disease is considered "immune" to it, and the state of being specifically resistant to a certain disease is described by the word "immunity".

The immune system is beautiful, logical, incredible and fascinating; it is also intricate, complex and sophisticated. it is extremely adaptable, has exquisite specificity, is extremely diverse and has memory. *Adaptability, specificity, diversity and memory* are terms that have special connotations in Immunology and these terms will be explained later in this chapter. The immune system is all about recognizing and responding; recognizing specific molecules and parts of molecules on pathogens and responding to these pathogens, mounting reactions against them.

This chapter describes the process of immune responses against infections from the time a pathogen enters the body until it is eliminated; however, we need to first set the stage by discussing the *structure* of the immune system, in other words the organs and cells of the immune system.

II. ORGANS AND CELLS OF THE IMMUNE SYSTEM

The immune system consists of a large number and variety of cells, tissues and organs distributed throughout the body. Phagocytes and lymphocytes are the two main cell types that mediate immune responses. These cell types are distributed throughout the body in lymphoid organs that can be broadly divided, on the basis of their function, into two groups - primary and secondary lymphoid organs. Primary lymphoid organs include the thymus and the bone marrow. It is here in these primary lymphoid organs that immature lymphocytes undergo hematopoietic maturation into mature lymphocytes with specificity to particular antigens. Specificity implies the ability of molecules of the immune system to recognize and interact with pathogenic molecules, or parts of these molecules, in a unique and exclusive manner without binding to other molecules. In immunology, we use the term "antigens" to refer to such molecules; an antigen is defined as any substance that binds specifically to antibodies and T cell receptors, which are immune recognition molecules described later in this chapter. Thus, the thymus and bone marrow are primary lymphoid organs wherein lymphocytes develop and mature and become committed to recognize particular antigens.

A. Lymphocytes

Three kinds of lymphocytes operate in the immune system—T lymphocytes, B lymphocytes and Natural Killer (NK) cells. These cells all originate in the bone marrow, immature T cells migrate to the thymus in order to mature, while B cells mature in the bone marrow. During the process of maturation in the thymus, T lymphocytes end up expressing on their surface unique antigen-recognizing receptors called T cell receptors (TCR). TCR are extremely diverse, as diverse as are the antigens that can be encountered in this universe! As there exist an immense number of different pathogens with immense numbers of different molecules, there also exists an incredibly large number of different T cell receptors that are specific to the large number of different antigens. T cells undergo maturation and differentiation in the thymus and gain specificity, i.e., the ability to recognize particular unique antigens.

T lymphocytes are of three main kinds; T helper (T_H) cells, T cytotoxic (T_C) cells and T regulatory (T_{REG}) cells. T_H and T_C cells are functionally different, but can also be distinguished by identifying molecules on their surfaces that act as *markers*; T_H cells express CD4 glycoproteins on their cell membranes, while T_C cells express CD8 glycoproteins. T_H cells mature in the thymus and later differentiate into effector T_H cells which then go on to "*help*" other cells, hence their name "*helper*" T cells. What do they help? T_H cells help activate B cells, T_C cells, macrophages and other cells of the immune system. Mature T_C cells on the other hand differentiate into effector cells called cytotoxic T lymphocytes (CTL) which recognize and kill virus-infected cells, tumour cells and cells of foreign tissue grafts. T_{REG} cells have the interesting ability to *suppress* immune responses, thus these cells *regulate* or control immune responses.

B cells originate in the bone marrow and undergo maturation in the bone marrow. During the process of maturation, B cells end up expressing, on their surfaces, special antigen-recognition receptors called immunoglobulins or antibodies. A given B cell expresses a particular antibody specific to a given antigen; as there are millions of different antigens in the universe, we have millions of *different* B cells with specificities for different antigens. Immature B cells develop into mature, but *naïve* B cells; these are called naïve because they have not encountered their specific antigen yet. Upon recognizing and interacting with the corresponding antigen, these naïve B cells undergo differentiation into effector B cells—plasma cells and memory B cells. Plasma cells secrete large quantities of specific antibodies that play key roles in eliminating the offending antigen. As for memory B cells, more on that later.

B. Primary Lymphoid Organs

1. The thymus: The thymus is a bi-lobed organ located behind the breastbone extending up to the upper edge of the heart. Each of these lobes consists of an outer layer called the cortex and an inner compartment termed medulla. In addition to T cells at different stages of development, the thymus contains several cell types including thymic stromal cells, dendritic cells and macrophages. We should recall that the role of the thymus is to nurture the development of T cells of different specificities; immature T cells arrive in the thymus from the bone marrow and then undergo maturation and development into mature T cells with a huge array of specificities. A critically important series of events, collectively called Thymic Education, occurs in the thymus. T cells with an awesome range of T cell receptor specificities are initially generated in the thymus by gene rearrangement, in a seemingly random manner. This includes, for example, T cells that are specific for antigens of the human immunodeficiency virus, of Mycobacterium tuberculosis, of tetanus toxin, of plants, of cereals, of all kinds of foreign proteins, and interestingly also T cells that are specific to "self" antigens. In Immunology, "self" refers to one's body, i.e., to a given person. So, T cells are randomly generated even against antigens of the "self", e.g antigens of the heart, the brain, red blood cells and white blood cells. Is this not dangerous, is this not potentially lethal? Yes, indeed immune reactions against self antigens can potentially lead to *autoimmune* reactions (immune reactivity against self antigens) and autoimmune disease and these need to be prevented. This is the process of thymic education comes in to play. During T cell development, those T cells which have T cell receptors specific for self antigens are killed. Thus, the thymus, a primary lymphoid organ plays crucial roles in helping develop T cells with the immense range of receptors capable of recognizing an immense range of antigens while at the same time eliminating those T cells that can recognize self antigens.

2. The bone marrow: The bone marrow is basically the cellular content of bone cavities which are a labyrinth of compartments and pathways consisting of lymphocytes, macrophages, granulocytes and fat cells. It is actually the site of *hematopoiesis*, which is the process of formation of all blood cells, red and white, in the body. Lymphocytes, macrophages, granulocytes and red blood cells are all generated in the bone marrow. In the context of B cell development, we should understand that B cells originate and develop in the bone marrow, a primary lymphoid organ. From hematopoietic stem cells arise immature B cells which then proceed to develop through a series of stages to become full-fledged mature B cells with an amazing range of specificities. For example, we develop some B cells specific for an antigen from the influenza virus, some B cells to another antigen from a flower, some specific to antigens in the malarial parasite, in fact we develop B cells specific for millions and millions of *different* antigens. It is estimated that an average human adult has B cells specific to 10^{10} *different* antigens! In other words, B cells with an astounding variety of antibodies with different specificities develop in the bone marrow. As in the case of T cell development in the thymus, the generation of B cell specificities also occurs by gene rearrangement in a random manner. And just as in the case of T cell development, some B cells may develop with the potential to produce

antibodies against self antigens and this is likely to be quite dangerous; fortunately for us, a process of selection within the bone marrow gets rid of B cells that make self-reactive antibodies.

Thus, the primary lymphoid organs, the thymus and bone marrow, are lymphoid organs where T cells and B cells mature. What then are *secondary* lymphoid organs and what is their function? Secondary lymphoid organs are tissues and organs where mature lymphocytes encounter, recognize and interact with antigens. Secondary lymphoid organs help trap antigens and provide an environment for lymphocytes to interact with these antigens. Thus, primary lymphoid organs are organs where lymphocytes develop and mature while secondary lymphoid organs are the sites where mature lymphocytes are lodged for immune responses to be initiated. Secondary lymphoid organs include the spleen, lymph nodes and diffuse lymphoid tissues in different parts of the body.

C. Secondary Lymphoid Organs

1. The spleen: The spleen, located in the upper left abdominal cavity, is a major site of immune responses. Antigens that are carried by blood are handled in the spleen; as blood circulates, antigens in the blood are filtered out in the spleen and thus this is where blood-borne pathogens and antigens are encountered by lymphocytes that are lodged here. The outer layer of the spleen, the capsule, sends out several projections that compartmentalize the spleen into the red pulp where aged red blood cells are destroyed and the white pulp which consists mainly of T cells, B cells and macrophages. Antigens are brought into the spleen by blood and are trapped by dendritic cells within the spleen; and this is where T cells and B cells interact with, and get activated by, antigens. In other words, immune responses are initiated here. We will soon see how immune responses are initiated in secondary lymphoid organs when various cells interact with antigens.

2. Lymph nodes: If the spleen is where blood-borne antigens, lymph nodes are where antigens in *lymph,* stimulate lymphocytes in order to initiate immune responses. The plasma of blood flowing through blood vessels leaks through vessel walls to collect in tissue spaces as interstitial fluid; this interstitial fluid consists of blood molecules and lymphocytes and circulates through a network of lymphatic vessels. This lymphatic system comprises of numerous lymph nodes all over the body connected by a vast network of lymphatic vessels that carry lymph back to blood circulation. Lymph nodes are secondary lymphoid organs containing lymphocytes, macrophages and dendritic cells; they filter out and trap antigens that are carried by lymph circulating via the lymphatic system. Lymph nodes occur at the junctions of lymphatic vessels, and are thus strategically located to trap antigens present in tissue spaces, interstitial fluids and lymph. Lymph nodes contain all the cells necessary to activate T and B lymphocytes and thus initiate different kinds of immune responses mediated by T lymphocytes and B lymphocytes.

In addition to the spleen and lymph nodes, which are discrete secondary lymphoid organs, there are diffuse collections of lymphoid tissues in the mucous membranes lining the respiratory, digestive and urogenital tracts. These *mucosa-associated lymphoid tissues* contain T cells, B cells and macrophages, i.e., the basic cellular components required to mount immune responses. These tissues handle antigens that are encountered in the respiratory tract, the digestive tract and the urogenital tract. In the gastrointestinal tract, for instance, are present specialized structures called Peyer's patches which contain cells that trap and transport antigens to antigen-presenting cells, T cells and B cells. Similarly, the cutaneous-associated lymphoid tissue present in the skin play useful functions in responding to antigens that enter through the skin.

D. Monocytes, Macrophages, Granulocytes and Dendritic Cells

In addition to lymphocytes described above, monocytes, macrophages, granulocytes and dendritic cells also play critical roles in immunity. Like lymphocytes, all these cells originate in the bone marrow from hematopoietic stem cells which differentiate along different pathways to yield this impressive variety of cells. Monocytes and macrophages are both mononuclear phagocytes; monocytes are found in blood circulation while macrophages abound in tissues. Monocytes circulate in the blood for a while, before settling down in the different tissues of the body and differentiating into tissue macrophages. Now, macrophages can be found everywhere in the body, in all organs! They exist as Kupffer cells in the liver, alveolar cells in the lungs, mesangial cells in the kidney, as histiocytes in the bone marrow, as macrophages in the spleen, thymus and lymph nodes, as peritoneal cells in peritoneal fluid and microglial cells in the central nervous system. Considering their critical roles in the immune system it is but logical that macrophages be present in all tissues and organs. Monocytes and macrophages, as we shall soon see, have the crucial function of initiating immune responses, by *presenting* antigens to T and B cells.

Neutrophils, eosinophils and basophils are all granulocytes that differ on the basis on morphology and cytoplasmic granules that they contain. Neutrophils and eosinophils are phagocytic, while basophils are non-phagocytic; all three types of granulocytes secrete anti-microbial substances, while in addition to this, neutrophils use phagocytosis as one of their weapons against microbes.

Dendritic cells are immune cells with long membranous processes and are present in interstitial spaces, i.e., spaces between cells of organs and tissues. They are adept at capturing antigens and transporting them to where the action is, i.e., where immune responses are initiated. They are very efficient *antigen-presenting* cells, a term that will soon be clear.

The immune system thus consists of a vast network of organs, tissues and cells; (i) primary lymphoid organs where lymphocytes develop and mature, (ii) secondary lymphoid organs where lymphocytes meet antigens and where immune responses are stimulated, and (iii) diffuse tissues containing lymphocytes and macrophages in the different tracts of the body. Cells of the immune system range from (i) the different kinds of lymphocytes that respond to antigenic challenge, (ii) monocytes, macrophages and dendritic cells that present antigen and thus provoke immune responses and (iii) granulocytes that have effector functions in the immune system. It should also be noted that the cells of the immune system are not stationary, just stuck in one tissue or organ. Like policemen who patrol the streets and lanes of cities, lymphocytes continuously *recirculate* – they transit through blood circulation, stop off in the spleen for a while before travelling through blood again to enter the different lymph nodes and back into circulation again; lymphocytes traverse through blood and lymphatic circulation all the time, thus providing ample opportunity for them to encounter lurking antigens.

Having glimpsed the structure of the immune system, let's proceed to describe immunity itself. Immunity as the state of being protected from infectious disease. Immunity can be classified into two components or two systems; one of them is referred to as *Innate Immunity*. The word innate implies something that is inborn, inherent, not established by conditioning or learning. The other system of immunity is called *Adaptive* or *Acquired*, and this type is established by conditioning, by adapting, by learning; but more on this type later. Innate immunity is also called *Natural* immunity, as it does not have to be developed by experience, i.e., by encountering antigens. It is also described as *Non-specific* immunity, because it is far less specific than Adaptive immunity. Cells, molecules and mechanisms of innate immunity are already present and ready to act even before an infection sets in and are thus not as specific as Adaptive immunity which develops specifically in response to every infection or every exposure to an antigen.

II. INNATE IMMUNITY

Innate immune barriers are the first defense barriers that pathogens encounter and have to overcome if they are to be successful invaders. These mechanisms include physical, biochemical and cellular ones.

A. Physical Innate Immune Barriers

Skin and mucous membranes are the most obvious and effective physical barriers. An intact skin is a rather efficient barrier against microbes, containing as it does an epidermal layer of dead cells supported underneath by the dermal layer. The gastrointestinal, respiratory and urinary-genital tracts are lined with mucous membranes which serve as an effective anatomical barrier to microbial invaders. These membranes are sticky in nature and thus help trap microbes; they also contain cilia the beating motion of which retards microbial movement further inward.

B. Biochemical Innate Immune Barriers

Clearly simple physical action will not be sufficient to block the entry of highly capable and invasive pathogens; the layers of skin and mucous membrane are also endowed with biochemical protectors. The *fatty acids present in sebum*, the oily secretions of sebaceous glands, have antimicrobial action, as do some peptides secreted by the skin and epithelial cells. Similarly anti-microbial peptides perform a policing action along mucous membranes lining the various tracts, and enzymes present in the saliva, stomach and small intestines perform innate biochemical protection. The enzyme *lysozyme* that is present in mucous membranes and in tears contributes to protection against bacteria by *degrading their cell walls*. We should note that these are all *innate* protective mechanisms as they are already present before any encounter with pathogen, and we should also note that these mechanisms are not specific to any antigen, i.e., they are not specific to any pathogens as they are reactive against a wide range of pathogens.

Other molecules also deserve our attention; these include interferon-α (IFNα) and proteins of the complement system. IFNa is a group of several molecules called cytokines which can be generally defined as molecules produced by white blood cells and other cells of the body that influence the duration and intensity of immune responses and play key roles as signaling molecules in the immune system. IFNα is secreted by macrophages when they are infected by viruses, and in a rather altruistic manner this secretion of IFNα helps neighbouring cells become resistant to viral infection.

Complement is actually a group of about 30 serum proteins found in blood circulation normally in an inactivated state; complement proteins can get activated directly upon binding to microorganisms and this results in a series of sequential activation steps culminating in the destruction of the membranes of microbes. Thus, complement proteins play key roles in killing pathogens; not content with this, when complement proteins are activated, some of them fragment and serve as very powerful attractors of immune cells to sites of infection (Figure 101.1). In other words, complement fragments are potent chemotactic factors and this helps in immune defense by marshalling more immune cells to areas where infection has occurred. The importance of complement molecules is underlined by the fact that children with inherited deficiencies in complement suffer from increased susceptibility to infections.

C. Pattern Recognition Receptors

Molecules like lysozyme and some of the complement proteins are now understood to recognize *patterns* on microorganisms; these patterns are such that they are found only on microorganisms and not in the host, e.g., humans. We should realize that this recognition of "patterns" or a class of molecules is not really a very specific recognition. In immunology, when we refer to specificity we often use phrases such as "high"

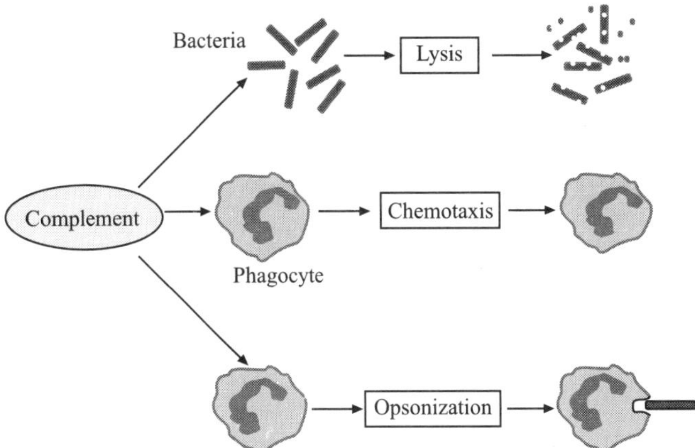

Figure 101.1   Roles of complement in the immune system.

and "low" specificity, high specificity implying the ability of a given receptor to bind uniquely and exclusively to a given ligand and not to others, low specificity implying the property of binding to several related molecules or structures. Adaptive immunity is highly specific as we shall soon see, while innate immunity is either non-specific (e.g., skin, mucous membranes, fatty acids etc) or is of low specificity. Pattern Recognition Receptors (PRRs) (e.g., some complement proteins) recognize some common motifs or molecular patterns on microorganisms and are thus of low specificity; these patterns on microbes are termed Pathogen-associated Molecular Patterns (PAMPs). PRRs are, by definition, part of innate immunity as they exist before microbes invade and are not *specifically* activated; when microbes carrying Pathogen-associated Membrane Patterns (PAMPs) are encountered by corresponding Pattern Recognition Receptors (PRRs) the microbes are recognized and attacked (Figure 101.2).

PRRs may be *soluble (e.g. lysozyme, complement)* or may be expressed on the cell-surface of immune cells, such as the Toll-like Receptors (TLRs). "Toll" is a receptor protein expressed on cell surfaces initially identified in flies; and similar or related molecules are called Toll-like receptors or TLRs which are basically receptors expressed on cell surfaces,primarily of macrophages and neutrophils, and which recognize some common patterns or motifs on pathogens. Eleven TLRs, TLR1 to TLR11, have been identified so far; each TLR recognizes a pattern or motif on a class of microbes, rather than one particular specific molecule. For example, TLR1 recognizes lipopeptides on Mycobacteria, TLR2 recognizes protozoa in addition to bacteria, TLR3 recognizes a common pattern in several viruses, TLR4 recognizes lipopolysaccharide in Gram-negative bacteria and TLR6 recognizes fungi. In fact, TLR4 confers resistance to high lethal doses of lipopolysaccharide (LPS) indicating the importance of this receptor in immunity. The success of the TLR system can be attributed to the fact that they recognize Pathogen-associated Membrane Patterns on or in microbes.

Having discussed physical and biochemical innate immune barriers, let us turn our attention to effector cells of the innate immune system.

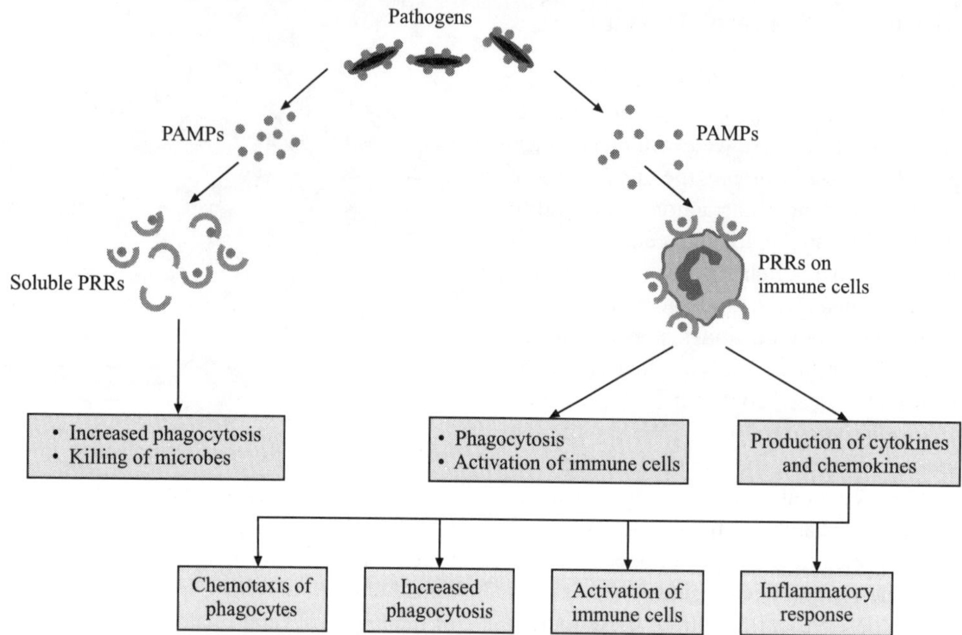

Figure 101.2   Roles of Pathogen-associated Molecular Patterns (PAMPs) and Pathogen Recognition Receptors (PRRs) in innate immunity.

D. Cellular Innate Immune Barriers

Macrophages, monocytes, natural killer (NK) cells and neutrophils are the principal cellular effectors of innate immunity. These cells were mentioned earlier when we discussed cells of the immune system, but in this section we focus on their functions in innate immunity. Most prominent among the diverse functions of macrophages, is its phagocytic ability. Phagocytes (i.e., "cells that eat") are cells such as macrophages, monocytes and neutrophils that are endowed with the property of phagocytosis, i.e., the ingestion and digestion of particulate substances (such as bacteria and viruses). Phagocytosis consists of several steps – recognition of the bacterium, attachment to it, engulfment, exposure of bacteria to degradative enzymes, killing and digestion of bacterium and release of degraded material from the cell. The immunological benefit of this property is obvious and indeed phagocytes like macrophages are considered the first line of defense against infectious pathogens. Macrophages can recognize a range of microbes by virtue of the TLRs on their surface and this can result in the activation of macrophages, which leads to an increase in several activities—phagocytosis, killing of internalized microbes, production of mediators that are toxic to microorganisms and activation of T cells. One of the many toxic mediators that macrophages contain is nitric oxide, a potent killer of bacteria, fungi, protozoa and worms.

Natural killer (NK) cells are also effectors in the innate immune system. NK cells are large, granular lymphocytes capable of killing virus-infected cells. Just as macrophages and neutrophils are an important barrier against bacteria, *NK cells constitute an important barrier against viruses.* Viruses necessarily need to infect cells in order to propagate themselves by replication and thus virus-infected cells are major sources of viruses. NK cells target these infected cells and thus destroy the very "factories" that harbor and synthesize more and more viruses. We should bear in mind that NK cells do not have the specificity against antigens that T cells do; NK cells do not express T cell receptors on their surfaces. Adaptive immunity against viruses takes several days to develop; during this period, NK cells play important roles in killing virus-infected cells thus restricting viral replication. NK cells are present in blood circulation and in the spleen, lymph nodes and peritoneum. *In addition to its well-documented anti-viral activity, NK cells kill tumour cells and thus participate in anti-tumour immunity as well.*

Neutrophils are highly effective members of innate immunity, endowed as they are with robust phagocytic ability as well as the ability to secrete numerous toxic anti-microbial substances. When microbes infect a tissue site, several cells rush to this site of infection and earliest among them are neutrophils. The production of neutrophils is enhanced during infections and their importance to immunity is emphasized by the fact that people with a deficiency of neutrophils suffer from increased incidence of bacterial and fungal infections. Neutrophils bind to microbes using surface TLRs. In addition to killing microbes by phagocytosis, neutrophils are also bestowed with several anti-microbial peptides and toxic substances such as hypochlorous acid, hydrogen peroxide and nitric oxide; ingested microbes are killed by these toxic mediators.

E. Inflammation

The last of the innate immune reactions that we will highlight is *inflammation*. The phenomenon of inflammation has been known for over two thousand years; it's cardinal features of redness, swelling, heat, pain and loss of function are well known. Generally viewed as a nuisance, inflammation was first described as a defense reaction involving blood vessels and a local cellular reaction in the late 1700s. When microbes manage to sneak through physical barriers such as the skin and mucous membranes, they may initiate local infection and tissue damage that sets off a series of cellular events collectively referred to as an inflammatory response which helps fight the microbe and limit the infection. How does an inflammatory response develop? The first step is the *production*, by local macrophages and endothelial cells, of *vasodilatory factors* that cause the dilation of local blood vessels which allows higher flow of blood to the area (Figure 101.3). As more blood flows into the area, naturally more white blood cells reach the area. This is aided significantly by chemotactic factors; chemotaxis is the movement of cells from one area to another under the influence of factors such as chemokines which are basically cytokines with chemotactic ability. This local production of chemokines attracts large numbers of white blood cells, particularly neutrophils, to the inflamed region. Leukocytes so attracted to the site are stopped from flowing further inside blood vessels by adhering to special molecules called *adhesion molecules* on the endothelial cells of blood vessels; *leukocytes are then encouraged to squeeze between endothelial cells to squeeze out of blood vessels and enter the tissue spaces in the site of infection* (Figure 101.3). This process in which leukocytes move out of intact blood vessels is called *extravasation* which is encouraged by local mediators including chemokines.

Once neutrophils, monocytes and lymphocytes reach the relevant tissue site they proceed to attack the microbes; neutrophils and monocytes phagocytose the microbes and lymphocytes get activated. TLRs on the surfaces of monocytes, macrophages and neutrophils help recognize pathogen-associated membrane patterns. An important activity in this cascade of events is the production of *pro-inflammatory cytokines* which as the name implies enhance inflammation by activating a variety of cells such as macrophages and lymphocytes (Figure 101.3). In short, inflammation is a beneficial defense reaction that in effect accomplishes the attraction of leukocytes to the area of infection and activating them to make them efficient effector cells to combat the infection. This is much like signaling police forces to rush to a neighbourhood where violence has just occurred and encouraging the policemen to combat the violence!

Now that we have discussed innate immune mechanisms, let's turn our attention to Adaptive Immunity.

III. ADAPTIVE (ACQUIRED) IMMUNITY

A. Features of Adaptive Immunity

Innate immunity is that system of cells and molecules that are pre-positioned ahead of infections and that jump into the fray to combat microbes with no time lost but without much specificity. The other form of immunity, Adaptive or Acquired Immunity relies predominantly on lymphocytes and develops *specifically* in response to an infection, after the infection, to *specifically* attack the invading pathogen. Generally, innate immune barriers are the first lines of defense against pathogens

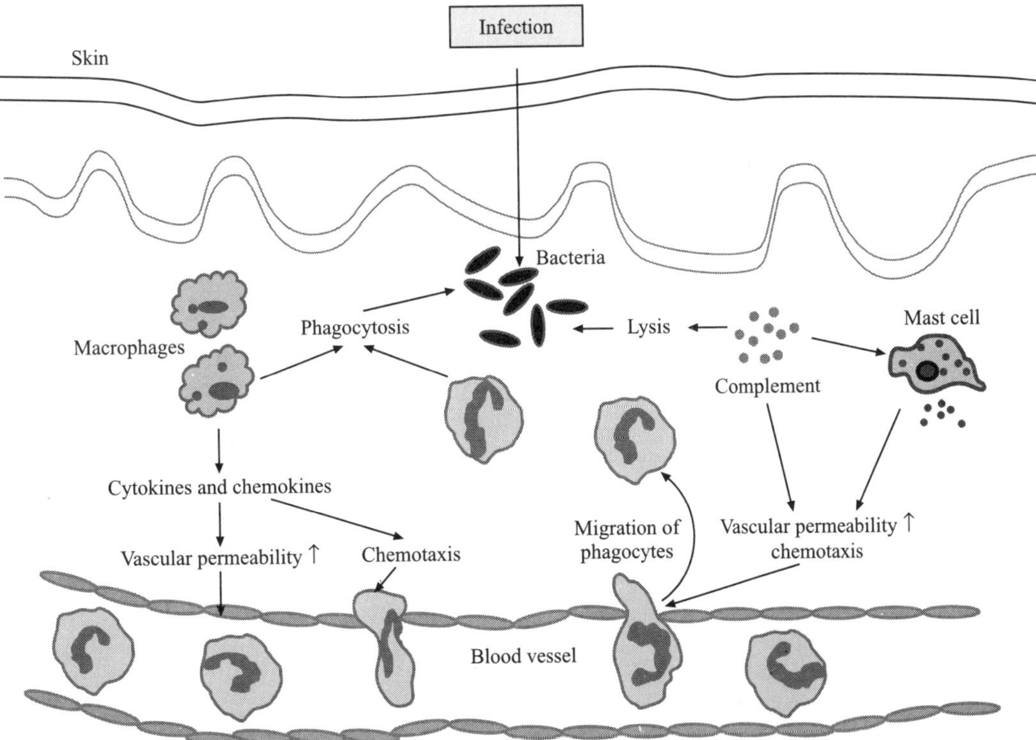

Figure 101.3 The inflammatory response as an immune defense mechanism.

and they manage to protect us quite effectively. Adaptive immunity comes into play if the pathogens breach the barriers of innate immunity or if innate immunity is unable to adequately contain the infection. What is the significance of the phrases "adaptive" and "acquired"? This system of immunity adjusts according to each infection, to each microbe, to each antigen; it adapts according to the pathogen and its antigens, which is why this form of immunity is called *adaptive*. It is also referred to as acquired immunity, because it does not pre-exist, it has to be developed, obtained, acquired. Another term for adaptive immunity is "specific" immunity, because unlike innate immunity which has no or low specificity, adaptive immunity is highly specific as we shall see.

One of the main characteristics that distinguishes adaptive immunity from innate immunity has already been mentioned—specificity. *Adaptive immunity also has the features of (i) high diversity, (ii) the ability to discriminate between self and non-self antigens and (iii) memory*. Adaptive immunity is extremely diverse; millions upon millions of *different*, unique antigens can be recognized by this system. This form of immunity has the crucial ability to distinguish self from non-selfor foreign; self molecules are *tolerated*, while foreign molecules are recognized and responded against. This ability has been described earlier when we discussed thymic education. Acquired immunity also has the remarkable ability to "remember" infections, to remember encounters with antigens; this ability ensures that when we are exposed to an infectious agent for the second time, our immune system responds specifically, faster and with a stronger response. As we discuss immune responses, we shall elaborate on these features of adaptive immunity—antigen specificity, diversity of recognition, immunologic memory and the ability to react to non-self without reacting to self.

Adaptive immunity can be classified as (i) Humoral Immunity, primarily mediated by antibodies, and (ii) Cell-mediated Immunity that is mediated by T lymphocytes and macrophages (Figure 101.4).

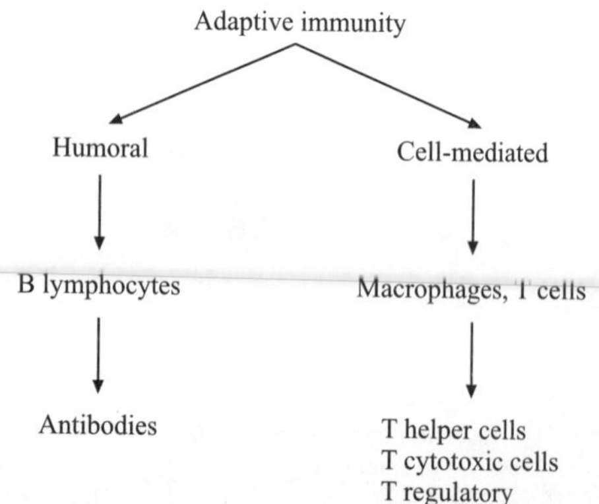

Figure 101.4   Arms of adaptive immunity.

Now let's examine the entire process of an immune response, from the time a pathogen enters the body, to the initial activation of T and B lymphocytes to the development of full-fledged immune responses that eliminate the infection.

B.   Internalization and Processing of Antigens

Adaptive immune responses are stimulated when cells of the immune system encounter antigens. We defined antigens as substances that bind specifically to immune recognition molecules, i.e., antibodies or T cell receptors. Bacteria, viruses, protozoa, cells, tissues and vaccines all contain antigens. The *internalization of antigen* by relevant cells is an essential first step in the activation of adaptive immune responses. A short while after internalization of antigens, say bacteria or a foreign protein, or after foreign proteins are generated within the cell such as after viral infection of a cell and replication of the virus, the surfaces of these cells display parts or segments of these antigens. What happens in the interim? What happens is the degradation of antigens by enzymes into small peptides, a step that is referred to as "*antigen-processing*".

Protein antigens are processed along two different pathways within APC; one pathway, the *endocytic pathway*, handles exogenous or extracellular antigens that are taken into the cell by endocytosis or phagocytosis, while the other pathway, the *cytosolic pathway* deals with intracellular antigens that are produced within the cell.

1. Processing of exogenous antigens: The cells of the immune system that first encounter and interact with exogenous or extracellular antigens (e.g. bacteria, viral particles, foreign proteins, inhaled molecules, ingested antigens, vaccines), are collectively called "*antigen-presenting cells*" (APC). Macrophages, dendritic cells and B lymphocytes all serve as specialized, professional APCs; their function is to "process" and "present" antigens; the terms "processing" and "presenting" have special meanings in immunology. Macrophages and dendritic cells internalize antigens such as bacteria and viruses by phagocytosis and/or endocytosis. B lymphocytes also serve as antigen-presenting cells but they are not phagocytic; they internalize antigens by receptor-mediated endocytosis. The antibody molecules on B cell surfaces act as receptors for specific antigens which are internalized after binding.

Once antigens are internalized by phagocytosis or endocytosis, they are taken through the endocytic pathway; they are digested by enzymes into smaller peptides ranging from *10 to 15 amino acids* in length. These fragments of antigen proteins are called *epitopes*, defined as segments or portions of antigens that are recognized by T cells. And it is these epitopes that are then transported to the surface of APCs to be displayed on the cell membrane, to be "*presented*" as it were. So, antigen-processing is the breaking down of large protein antigens into smaller epitopes and transporting them to the cell surface; and antigen-presentation is the displaying of these peptides on cell surfaces, to be "seen" or recognized by T lymphocytes.

What is particularly fascinating is that these "processed" peptides are not presented on the cell surface of APC as native peptides by themselves; they are actually *displayed attached to, or associated with, special molecules* called the *Major Histocompatibility Complex (MHC)* molecules. Epitopes are bound to these MHC molecules inside the APC during antigen-processing so that when they are transported to the cell surface they are actually presented as epitope-MHC complexes. These peptide-MHC complexes are recognized by specific T helper (T_H) cells which then get activated; thus the whole purpose behind the processing and presentation of antigens is to convert antigens to a form that can be recognized by T_H cells. This form consists of MHC molecules presenting or displaying processed epitopes; more about what these MHC molecules after we discuss the processing of endogenous antigens.

2. Processing of endogenous antigens: Endogenous antigens are those that are not internalized by phagocytosis or endocytosis, but are generated within the cell itself; endogenous antigens include viral proteins that are synthesized within infected cells and tumour antigens that arise within tumour cells. Endogenous proteins are degraded by cytosolic proteases after which fragmented peptides are transported to the rough endoplasmic reticulum where these peptides are bound by MHC molecules. Peptide-bound MHC molecules are transported to the Golgi complex and from there to the cell surface.

Several important points should be noted. One of the first cells that gets activated by antigens, regardless of whether they are bacteria or viruses or other foreign proteins, is the T_H cell; but these T_H cells do not recognize the native antigens, they *only recognize antigens that are processed and then presented on cell surfaces*. Exogenous antigens are internalized and then fragmented into peptides, associated with MHC molecules before being presented on surfaces of antigen-presenting cells. Endogenous antigens are degraded into peptides while they are being generated inside cells; the peptides then get attached to MHC molecules before being presented on the surfaces of virus-infected cells or tumour cells. It should be noted that the MHC molecules that present endogenous antigens are of a different *class* or group from those that present exogenous antigens; *class I MHC molecules present endogenous antigenic peptides, while class II MHC molecules present exogenous peptides*. But what are MHC molecules? What is their structure and how do they function? While a whole chapter in this book focuses on MHC, we will present a brief summary here.

C. The Major Histocompatibility Complex (MHC)

MHC molecules are now known to have extremely important roles in presenting antigens to T lymphocytes, but interestingly enough they were first identified not as "antigen-presenters" but as molecules that are involved in the rejection of transplanted tissue grafts. When cells, tissues or organs from one individual are transferred into another unrelated individual, these are recognized as "foreign" and are destroyed by the immune system; they are *rejected*. The molecules that are recognized as foreign are proteins of the major histocompatibility complex; the primary function of these molecules is to present epitopes. Thus, the MHC is a complex of genes that code for cell surface proteins that present antigenic peptides to T lymphocytes. MHC molecules are extremely polymorphic, i.e., exist as an immense number of alleles in the population (which is why foreign grafts are recognized as "different" and are rejected). MHC molecules are present in mammals, birds, amphibians and fish; in humans the MHC is called HLA or human leukocyte antigens. The phrases MHC and HLA are used synonymously.

There are two main types of MHC molecules and these are called class I and class II. Class I molecules consist of a large polymorphic glycoprotein that is linked to β2-glycoprotein, while class II molecules are made up of two glycoproteins encoded by two different genes in the MHC. In humans, three class I MHC molecules exist; HLA-A, HLA-B and HLA-C, and three class II MHC molecules are present, HLA-DP, HLA-DQ and HLA-DR (Figure 101.5). A third class of molecules, class III, have also been identified but these are not relevant to histocompatibility.

In addition to differences in their structures, MHC molecules also differ in terms of the cells on which they are expressed. Class I molecules are expressed on almost all nucleated cells while class II molecules are expressed only on APC, i.e., on macrophages, dendritic cells and B cells. This should remind us that exogenous antigens are presented by APC; thus, it follows logically that exogenous antigenic peptides are presented on APC surfaces by class II MHC molecules. APC are "professional" antigen-presenters; these cells internalize extracellular antigens, process them into small peptides and present them on their plasma membranes. Let's contrast this to presentation of endogenous antigens; any cell can be potentially infected by viruses and any cell can potentially become neoplastic and become a tumour cell—antigens generated within these cells are presented by class I MHC molecules which, as mentioned above, are expressed on just about all nucleated cells in the body.

D. Antigen-presentation and Activation of T_H Cells

T cells and B cells are both lymphocytes with the ability to recognize antigens; however, B cells by virtue of antibody

Complex	Major Histocompatibility Complex							
Class	II			III		I		
Gene	DP	DQ	DR	C	TNF	B	C	A
Molecule	HLA-DP	HLA-DQ	HLA-DR	C	TNF	HLA-B	HLA-C	HLA-A

Figure 101.5 MHC genes and molecules in humans. (*see Plate 41 for colour figure*)

molecules on their surfaces recognize native antigens directly, while T cells are "blind" to antigen. They can recognize antigenic peptides only in the context of MHC molecules, i.e., only when these peptides are presented by MHC molecules (Figure 101.6). As explained earlier, exogenous antigens are presented by class II MHC molecules, while endogenous antigens are presented by class I MHC molecules. Both kinds of epitope-MHC complexes are recognized by specific T lymphocytes; however, epitopes presented by class II MHC molecules are recognized by specific T helper (T$_H$) lymphocytes while epitopes presented by class I MHC molecules are recognized by specific cytotoxic T (T$_C$) cells.

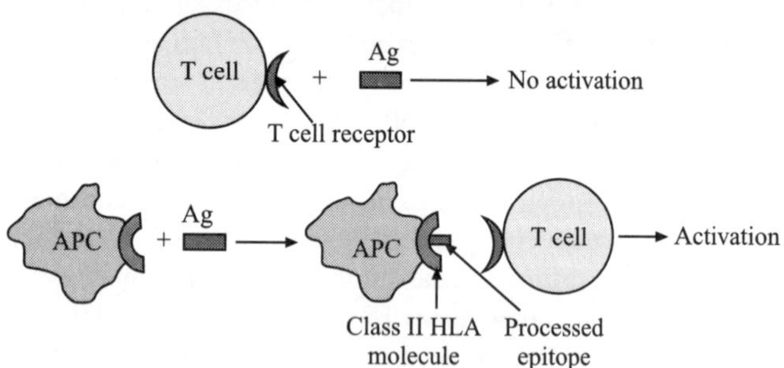

Figure 101.6   Activation of T cells by epitopes complexed with MHC molecules. (*see Plate 41 for colour figure*)

In addition to recognizing epitope-MHC complexes via T cell receptors, T cells also use certain *accessory molecules* to either strengthen the binding between T cells and antigen-presenting cells or to transmit signals or both. Some of these molecules are classified as CD (cluster of differentiation) molecules, which are a group of molecules that are useful as markers that help us identify different cell types. For example, T helper cells express on their surface a membrane glycoprotein called CD4 that binds to MHC Class II molecules on the surfaces of antigen-presenting cells; the binding of CD4 to MHC Class II molecules enhances the binding of T cells to APC, and CD4 also transmits signals to T cells which contributes to T cells getting activated. Cytotoxic T cells express another CD molecule, CD8, which has an affinity for Class I MHC molecules. CD8 contributes to increased binding between T$_C$ cells and target cells expressing epitope-class I MHC molecules.

It should be noted that the recognition of epitope-MHC complexes is extremely *specific* and extremely *diverse*; the recognition is done by specific T cell receptors (TCR) on T cells surfaces and there are millions and millions of *different* TCRs with specificity for different epitope-MHC complexes. Interestingly, while the TCR performs antigen recognition, it does not transmit activation signals to the T cell. Transduction of activation signals is done by a complex of five proteins, collectively referred to as CD3, which are complexed with TCR molecules; these polypeptides have long cytoplasmic tails that activate nuclear transcription factors. In addition to stimulation of the TCR-CD3 complex, T cells also need to be stimulated by the cytokine IL-1 secreted by APC. Finally, once specific T$_H$ cells recognize their corresponding epitope-MHC complexes aided in binding by CD4 molecules

and stimulation by the cytokine IL-1, these specific T$_H$ cells get *activated*. If the processing of antigens into epitopes and the presentation of these epitopes with MHC molecules is an important first step in immune responses, then activation of specific T$_H$ cells is an essential next step. Within a few minutes of signals transduced by CD3 molecules, a series of activation events take place. Several transcription factors, enzymes and cytokines are produced in sequence; specific T$_H$ cells thus activated proliferate and differentiate into two main kinds of cells, effector T cells that "help" immune responses and memory T cells that contribute to immunologic memory described earlier. In other words, T$_H$ cells that are normally in a "resting" state are turned on by specific epitope-MHC complexes and are stimulated to develop into effector T$_H$ cells that *help* turn on effector immune responses.

E.   T$_H$1 and T$_H$2 cells

Several subsets of T$_H$ cells exist, of which the two important ones in this context are T$_H$1 and T$_H$2 cells; these two T$_H$ subsets are functionally different and they secrete different cytokines. T$_H$1 cells secrete the cytokines interleukin (IL)-2, interferon (IFN)-γ and tumor necrosis factor (TNF)-β. These helper T cells help stimulate *"cell-mediated immunity"*. On the other hand, T$_H$2 cells secrete the cytokines IL-4, IL-5, IL-6, IL-10 and IL-13 and help activate *"humoral immunity"*. T$_H$ cells differentiate into either T$_H$1 or T$_H$2 cells depending on which cytokines are present in the environment during differentiation. Which T$_H$ cell predominates after differentiation influences the stimulation of immune responses along humoral versus cell-mediated pathways (Figure 101.7), and this stimulation is attributable to cytokines produced by the T$_H$ cells.

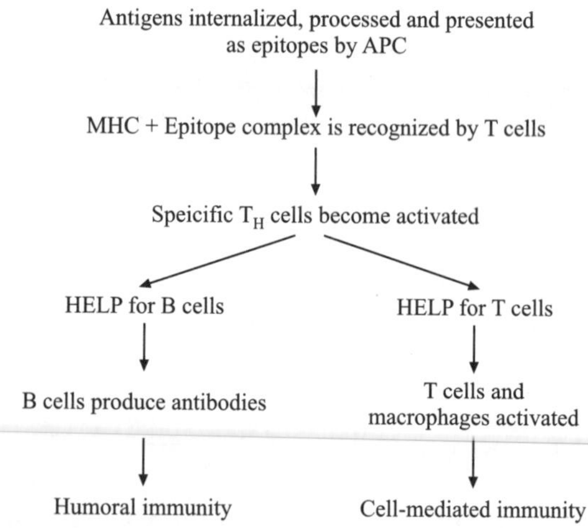

Figure 101.7   Activation of humoral and cellular immunity by T$_H$1 and T$_H$2 cells.

IV.   ACTIVATION OF HUMORAL IMMUNITY

T$_H$2 cells provide help to B cells and thus activate what is called Humoral Immunity. What is humoral immunity? Immune responses are broadly divided into two classes or groups of responses, humoral and cell-mediated. Humoral responses are those that are mediated by antibodies in fluids such as blood plasma, lymph and tissue fluids (in fact the word "humor" means fluid in Latin). Thus, B cells, which secrete antibodies, are responsible for humoral immune responses. Cell-mediated immune responses are those that are mediated by specific T cells and other cells such as macrophages, NK cells and granulocytes.

Humoral immunity is helped, i.e., activated, by T_H2 cells. What does this process of "help" entail? How do T_H2 cells help B cells? Most antigens are "T-dependent", i.e., they require help from T cells in order for them to stimulate the production of specific antibodies, which is why T_H2 cells are first activated. These activated specific T_H2 cells then go on to "help" activate humoral immunity, i.e., B cell responses; T_H2 cells cause the activation, division (proliferation) and differentiation of resting B cells into antibody-secreting cells (Figure 101.8).

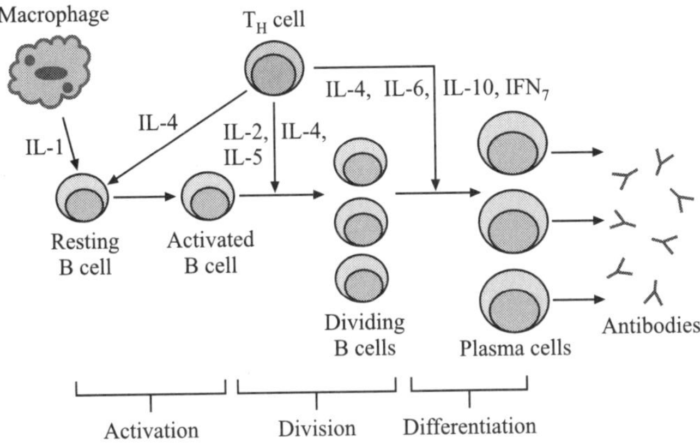

Figure 101.8 Roles of T helper cells in inducing humoral immunity. (*see Plate 41 for colour figure*)

Resting B cells require two signals in order to become activated; one signal is provided when antigen interacts with specific immunoglobulins on B cell surfaces, the second signal is provided by T_H2 cells when CD40 molecules on B cells bind to the CD40 ligand expressed on the surface of T_H2 cells. Once specific B cells are activated, they start expressing receptors for three key cytokines, IL-2, IL-4 and IL-5 produced by T_H2 cells; these cytokines "help" by inducing the activated B cells to start proliferating (Figure 101.8). We should bear in mind that before any immune response occurs, there exist only small numbers of B cells specific for each antigen and these small numbers have to be induced to proliferate into large numbers of specific B cells in order that sufficient amounts of antibodies are produced to combat infections. Thus, the second stage of induction of humoral immunity is the proliferation of specific B cells, influenced by help mediated by cytokines produced by T_H2 cells. These proliferating B cells then differentiate into two kinds of cells – (i) plasma cells that secrete antibodies and (ii) memory B cells that serve as 'reserve forces' for immune responses against future attacks by the same pathogen. This stimulus for the differentiation of B cells is also provided by T_H2 cells. Thus T_H2 cells play very crucial roles in the

activation, proliferation and differentiation of B cells, in other words in the activation of humoral immunity, primarily via the production of T_H2 cytokines. These cytokines in essence "help" stimulate humoral immunity.

The process of T_H2 help for B cells results in the development of plasma cells that produce large quantities of specific antibodies which then participate in a variety of defense reactions.

A. Antibodies

Antibodies or immunoglobulins are basically glycoproteins produced by B cells that specifically recognize epitopes on antigens and help in humoral immune defenses against these antigens. The secreted antibodies are identical to the immunoglobulin expressed on the surface of the B cell. Antibodies consist of two identical light chain polypeptides approximately 22,000 daltons in weight and two identical heavy chain polypeptides about 50,000 daltons, linked by disulphide bridges and non-covalent bonds. A basic immunoglobulin unit, consisting of two identical light chains and two identical heavy chains, is about 150,000 daltons in weight (Figure 101.9).

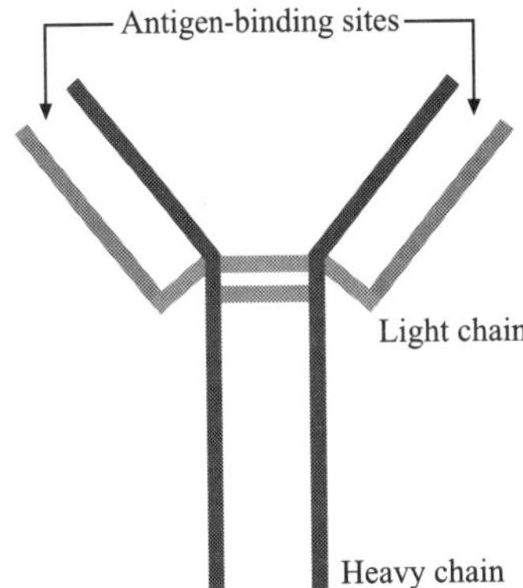

Figure 101.9 Structure of an immunoglobulin molecule. (*see Plate 42 for colour figure*)

Antibodies are extremely diverse, in other words the polypeptides of antibody molecules are amazingly diverse and are specific for an amazingly diverse range of different antigens. This diversity in the structure is actually due to diversity in what is called the *variable region* of the light and heavy chains of immunoglobulin molecules. Each of the light and heavy chains has an amino-terminal variable region, which accounts for the incredible diversity or variability between antibodies, and a relatively constant (i.e., non-variable) carboxy-terminal portion called the *constant region*, or the Fc region. Antibodies bind to antigen via their amino-terminal ends which are called Fab (for Fragment antigen binding) units; a single immunoglobulin unit has two identical antigen-binding sites. Amino acid sequences within the variable region contribute to the antigen binding sites of antibodies, and these sequences are responsible for antigen-specificity.

There are five major *classes* of immunoglubulins, based on the type of heavy chain that they contain. These antibody classes are designated as IgG, IgM, IgA, IgE and IgD. These five classes of immunoglobulins all have the basic two-heavy-chain two-light-chain structure, but they differ from one another in the amino acid sequence of their heavy chains. The main differences between these five subclasses is therefore in their Fc regions and because of this, they have different effector functions.

IgG antibodies consist of the classical single immunoglobulin unit consisting of two heavy chains and two light chains and are the most abundant antibody in blood circulation as well as non-mucosal tissues. These antibodies can also enter into intercellular tissue spaces and thus provide immune protection both in blood circulation as well as in tissue spaces. The IgG class of antibody is the only class that is transferred from the pregnant mother to her fetus thus playing important roles in the immune defense of the fetus.

IgM antibodies are pentamers, i.e., they consist of five basic units of two heavy chains and two light chains each, and thus they have 10 antigen-binding sites.

IgM antibodies are found only in blood circulation and are very efficient in binding pathogens with repeating identical epitopes; because of this feature IgM antibodies are very efficient at agglutination, i.e., the clumping or aggregation of particles such as bacteria and viruses. They are also very efficient at binding and activating complement molecules, which then proceed to carry out several anti-microbial actions such as lysing bacteria. IgM is also expressed on B cells where they play a role in antigen recognition by B cells.

IgA antibodies exist either as monomers (consisting of the basic two heavy and two light chains) or as dimers; IgA antibodies are also called *secretory antibodies* because they are the principal antibody class found in secretions such as the saliva, tears and breast milk. Just as IgG antibodies afford protection to the developing fetus, IgA antibodies provide protection to the developing newborn via mother's milk. IgA antibodies are also secreted into mucosal surfaces of the respiratory and gastro-intestinal tracts, and play important roles in providing immune protection along these tracts.

IgE antibodies also exist as monomers and are found in extremely minute levels in the blood circulation of normal individuals but are present in high levels in people with allergies. IgE antibodies mediate allergic reactions; these are type I hypersensitivity reactions and are discussed in greater detail in another chapter in this section. IgE is bound to mast cells via Fc receptors for IgE on mast cell surfaces, and contribute to inflammatory reactions that aid immune defense reactions.

IgD is also present at extremely low levels in secreted form, but is present on B cell surfaces; it is believed to play roles in the activation of B cells.

B. Protective Roles of Antibodies in Defense Against Infections

Pathogens are generally first challenged by innate immune barriers which endeavor to prevent them from establishing an infection. If however these pathogens succeed in breaching innate barriers, then adaptive or acquired immunity steps in the form of humoral and/or cell-mediated immunity. How do humoral immune responses defend us from infectious microbes? Antibodies have several direct and indirect effects, as shown in Figure 101.10.

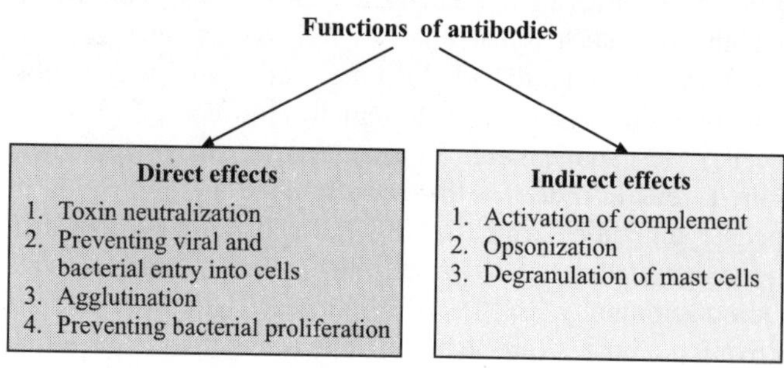

Figure 101.10   Functions of antibodies.

Viruses need to bind to host cells in order to establish infection; antibodies are quite effective at blocking the binding of viruses to host cells and also at inhibiting the fusion of the viral envelope with the cell membrane of host cells. IgG and IgM antibodies in the blood and IgA antibodies in mucous secretions block viral binding very efficiently. Antibodies also perform an interesting function called "opsonization" or enhanced phagocytosis. As mentioned earlier, antibodies have antigen-binding sites on their N-terminal ends, these are called Fab (fragment, antigen binding); the C-terminal end, which is the constant segment, is called Fc (fragment, crystallizable – as this segment crystallizes when chilled). Many cells such as macrophages, granulocytes, NK cells and lymphocytes, have receptors for the Fc portion of immunoglobulins; these receptors mediate the binding of immunoglobulins (via their Fc segments) by cells. Thus, when specific antibodies bind to viruses, their Fc portions stick out and can be bound by cells such as macrophages and neutrophils; as mentioned earlier these cells are phagocytic, and upon binding to viruses via the Fc portions of specific anti-viral antibodies, phagocytosis is enhanced significantly. This process is called opsonization which is also mediated by complement molecules that bind to viruses either directly or via antibodies that are bound to viruses and are then recognized by complement receptors on phagocytes (Figure 101.1). Complement molecules (considered part of humoral immunity) also contribute to the neutralization of viruses by killing enveloped viruses. Further details on complement and their mechanisms of action are covered in greater detail in another chapter in this section. Finally, antibodies also serve to "agglutinate" viral particles; agglutination or aggregation or clumping of viruses leads to viruses being arrested and prevented from spreading.

How does humoral immunity defend us against bacterial infections? Antibodies block the attachment of bacteria to host cells (Figure 101.11); IgA antibodies are particularly useful in this regard given that considerable exposure of bacteria to the host occurs in sites such as the respiratory and gastrointestinal tracts. Bacteria are eliminated effectively

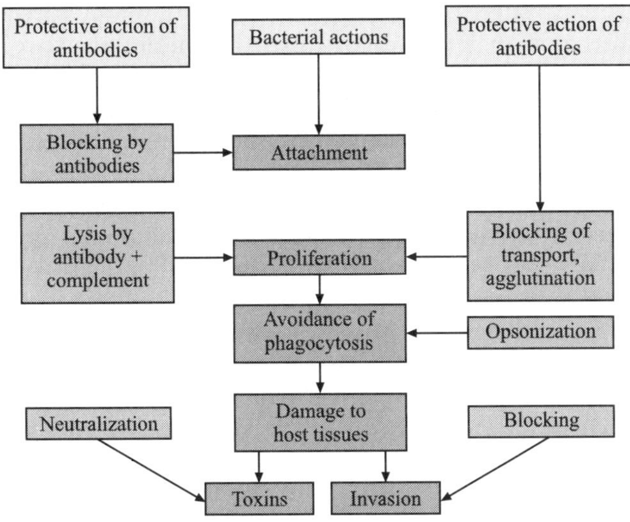

Figure 101.11 Antibodies versus bacteria.

by opsonization mediated by antibodies and by complement molecules; antibodies basically prepare bacteria for attachment and internalization by phagocytes such as neutrophils and macrophages, by effectively opsonizing them. In fact, the binding of bacteria by antibodies and/or complement, makes them far more vulnerable to adherence by phagocytes and subsequent killing. Complement molecules can cause the lysis of bacteria both by binding to bacterial cells either directly or indirectly by binding to specific anti-bacterial antibodies (Figure 101.10). Agglutination of bacteria by antibodies prevents bacteria from spreading and from invading tissues. Some bacteria mediate pathogenesis by toxins that they secrete; anti-toxin antibodies are efficient at neutralizing these toxins and rendering them ineffective. Also, once toxins are bound by antibodies they are prevented from diffusing away and are phagocytosed and thus eliminated. Just as IgG and IgM antibodies patrol blood circulation, mucosal surfaces are well protected by IgA antibodies. These antibodies provide protection in fluids such as saliva, nasal secretions and tears, and also along mucosal tracts of the respiratory and intestinal systems. Here, IgA antibodies bind to bacteria and viruses and prevent them from adhering to host cells, consequently blocking bacterial colonization.

V. ACTIVATION OF CELL-MEDIATED IMMUNITY

As mentioned earlier, adaptive immunity consists of two types of responses, (i) humoral immunity described above and (ii) cell-mediated immunity (CMI) which consists of immune responses against intracellular bacteria and viruses mediated by antigen-specific T cells and other cells such as macrophages, natural killer (NK) cells and granulocytes (Figure 101.12). Antibodies and complement, the effectors of humoral immunity, certainly play very important roles in defending us; however, they are limited to acting against microbes that are *outside* the

cell and cannot attack or neutralize viruses and bacteria that have entered the cell. This is where cell-mediated immunity (CMI) comes in. Cell-mediated responses are geared towards destroying cells that harbour intracellular pathogens.

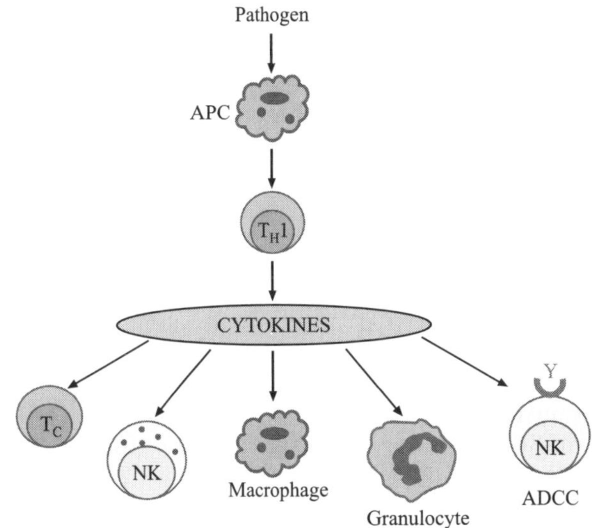

Figure 101.12 Effector cells of cell-mediated immune responses. (*see Plate 42 for colour figure*)

Just as T_H2 cells activate humoral immunity by "helping" activate B cells, T_H1 cells activate cell-mediated immunity. T_H1 cells have a pattern of cytokine secretion that is different from that of T_H2 cells; T_H1 cells secrete IFN-γ and TNF-β in addition to IL-2. T_H1 activate cytotoxic T (T_C) cells to become effector cytotoxic T lymphocytes (CTL) that kill virus-infected cells and they also activate macrophages increasing their microbicidal capability. In addition, T_H1 cells activate NK cells. The activation of these three cell types – cytotoxic T lymphocytes, macrophages and NK cells – results in cell-mediated immunity (Figure 101.12).

As T_H1 and T_H2 cells activate different immune responses, which T_H subset gets activated first and more strongly influences the outcome of immune responses, i.e., whether humoral or cell-mediated immunity gets activated more strongly. It should also be noted that the different biological functions of T_H1 and T_H2 cells are attributable to the different patterns of cytokines that they secrete, with T_H1 cells secreting cytokines that activate cytotoxic T lymphocytes and macrophages and T_H2 cells activating B cells and thus humoral immunity. The "help" rendered by T helper cells is thus due to the kinds of cytokines they secrete.

In addition to the classification of T cells described earlier, T cells can also be classified on the basis of the *type* of T cell receptor they express on their surfaces. Most of the T cells found in blood circulation express T cell receptors made up of α and β polypeptide chains; these T cells are called $\alpha\beta$ T cells, $\alpha\beta$ T cell receptors have an extremely high degree of specificity for antigens and are therefore major participants in adaptive immunity. The other kind of T cell receptor is the $\gamma\delta$ T cell receptor; $\gamma\delta$ T cells are not normally found in high numbers in blood circulation unlike $\alpha\beta$ T cells. Instead they are found in higher numbers in the skin and along the intestinal mucosa. Unlike $\alpha\beta$ T cells that have exquisite specificity for individual antigens, $\gamma\delta$ T cells appear to recognize classes or kinds of microbial antigens, suggesting that these cells function in a more "innate-like" manner.

A. Protective Roles of Cell-mediated Immunity in Defense Against Infections

It would be fair to make a general statement that humoral immunity eliminates infections that occur *outside* our cells; antibodies and complement are geared to destroy and neutralize pathogens such as extracellular bacteria and free viruses. Similarly, cell-mediated immune responses combat bacteria and viruses that have invaded and entered cells, i.e., intracellular pathogens. How does CMI manage to do this? Through the actions of cytotoxic T lymphocyte, macrophages and NK cells.

Cytotoxic T cells (T_C) are activated by cytokines produced by T_H1 cells and they differentiate into effector cells called cytotoxic T lymphocytes (CTL). These cells specifically recognize target cells and kill them; target cells in this context are virus-infected cells that have on their surfaces HLA Class I molecules presenting viral epitopes and also tumour cells that similarly express tumour antigen epitope-MHC class I complexes on their surfaces. By the way, it should be pointed out that cells like dendritic cells and macrophages that present epitopes in the context of HLA class II molecules are called antigen-presenting cells, and cells that present antigenic epitopes in the context of HLA class I, in effect, are also antigen-presenting cells; however, the convention in immunology is to refer to the former as antigen-presenting cells and to the latter as "target" cells. It is because of the fact that virus-infected cells become *targets* of cytotoxic T cells, cells that present epitope-MHC class I complexes are called target cells. Target cells (e.g., virus-infected cells) present epitope-MHC class I complexes to cytotoxic T lymphocytes.

CTLs, like T_C, express CD8 molecules and the affinity between CD8 molecules on CTL for MHC class I molecules on target cells serves to strengthen the attachment of CTL to target cells. Specificity of recognition of the right target cells is ensured by the need for specific recognition first by T_H1 cells and then by T_C cells. The process of killing of specifically recognized target cells is quite fascinating. Within minutes of binding of epitope-MHC complexes on target cells by specific TCR-CD3 complexes on CTL, granules are released from the CTL aimed in the direction of the target cell. These granules contain *perforin* and *granzymes*; perforin molecules form pores, i..e. drill holes, on cell membranes of target cells after which granzymes enter the target cell through these pores. Upon entering the cytoplasm of target cells, granzyme molecules initiate a process known as *apoptosis*; this process, also known as programmed cell death, causes the death of target cells by nuclear fragmentation and blebbing. Granzymes activate apoptotic pathways in target cells which commit a sort of "cell suicide". In addition to the perforin-granzyme pathway,

CTLs kill target cells by apoptotic signals given by special CTL surface molecules called FasL to Fas molecules on target cells.In addition to these two mechanisms, CTLs secrete the cytokine TNF-β which gives apoptotic signals to target cells. Thus, CTLs give signals to target cells to undergo apoptosis and die; this interaction takes a few minutes after which the CTL detaches itself and moves on to the next target giving it the same "kiss of death". A single CTL can kill several target cells in series. DNA fragmentation in target cells occurs within a few minutes of receiving signals from CTLs and during cell-mediated immune responses to viral infections a large number of virus-infected cells are thus killed.

We should recall that class I MHC molecules are expressed on almost all nucleated cells in the body, which implies that almost any nucleated cell that is infected by viruses can be recognized and killed by CTL. What does this killing of infected cells achieve? Viruses use infected cells as "factories" in which they replicate themselves generating larger and larger numbers of viruses, and CTL-mediated killing accomplishes the task of killing these virus factories, thus preventing the synthesis of more viruses. The process of fragmentation of viral DNA during apoptosis also ensures that viral replication is halted.

Natural killer (NK) cells also kill virus-infected target cells and tumour cells. These killer cells use mechanisms similar to those used by CTL, i.e., perforin and granzymes, and thus NK cells also kill virus-infected cells by inducing apoptosis. However, CTLs recognize target cells with great specificity using T cell receptors with specificity for epitope-MHC complexes, while NK cells several activating and inhibiting receptors that mediate killing of target cells in a *non-antigen-specific* manner. NK cells play useful roles in immunity, by participating in killing virus-infected cells during early infection when CTLs have not yet been activated and also by killing virus-infected cells that do not express MHC molecules which would be ignored by CTLs.

A third mechanism for killing virus-infected cells is by the process of antibody-dependent cell-mediated cytotoxicity (ADCC). In this process, cells such as macrophages, monocytes, neutrophlls and NK cells, which have Fc receptors on their surfaces, bind to Fc portions of immunoglobulins that are attached to virus-infected cells followed by the release of cytokines like TNF-α by macrophages, lytic enzymes by macrophages and neutrophils and perforin + granzyme by NK cells. These molecules then bring about the killing of target cells which have specific antibody molecules attached to their surfaces, with the antibody molecule serving as a sort of bridge between these killer cells and target cells.

Macrophages have been mentioned several times in this chapter as very efficient participants in immunity against pathogens. They play important roles in phagocytosis and ADCC. T_H1 cells secrete cytokines such as IFN-γ that activate macrophages to enhance their phagocytosis and killing of bacteria. In turn activated macrophages secrete a cytokine called IL-12 that serves to stimulate further development of T_H1 cells. Thus, T_H1 cells and macrophages act in concert to combat intracellular infections. Some bacteria upon intracellular infection of macrophages shut down enzymic machinery that would otherwise digest them; T_H1 cells activate these macrophages, induce the action of enzymes and thus enhance killing of intracellular pathogens. At sites of infection by intracellular pathogens such as *Mycobacterium tuberculosis*, a chronic inflammatory response ensues; this reaction consists of T_H1 cells that activate macrophages which develop into multinucleated giant cells; these cells form a sort of a barrier around the site of infection. This tissue is called granuloma and it serves to limit the spread of the bacteria; the activation of macrophages by T_H1 cells also helps suppress the proliferation of bacteria. This reaction is termed delayed-type hypersensitivity which is basically a cell-mediated immune response that is aimed at killing ingested microbes in a more efficient manner.

Thus, cell-mediated immune reactions against viruses are mediated by cytotoxic T lymphocytes, natural killer cells and macrophages while CMI against intracellular bacteria are mediated by T_H1 cells and macrophages.

B. Cytokines

The critical importance of cytokines to the immune system would have become apparent by the numerous times that they have been alluded to in this chapter. Cytokines are low molecular weight soluble messenger proteins secreted by cells of the immune system and by other cells that influence immune reactions. More than 40 cytokines have been identified so far. Cytokines bind to receptors on cells and either stimulate or inhibit differentiation, proliferation or functioning of cells. We saw earlier that the cytokine IFN-α produced by virus-infected cells, inhibits the replication of viruses in neighbouring cells. When we discussed the activation of T lymphocytes, we noted that the cytokine IL-1, secreted by APC, activates T_H cells thus playing an important role in the initiation of immune responses. Similarly, cytokines produced by T_H2 cells are indispensable to the stimulation of humoral immunity, as these cytokines cause the activation, division and differentiation of B cells. In cell-mediated immune reactions, cytokines produced by T_H1 cells cause the activation of cytotoxic T lymphocytes, macrophages and NK cells. The cytokine TNF-β produced by cytotoxic T lymphocytes contributes to the apoptosis of target cells. Which cytokines are present during the time that T cells differentiate into T_H1 and T_H2 is a decisive factor in which of these two cells predominate; and, the cytokines produced by T_H1 and T_H2 cells influence the development of cell-mediated and humoral immunity, respectively. Thus, cytokines influence the type of immunity developed and the outcome. They are involved in hematopoieisis, initiation of immune responses, cell proliferation and differentiation, inhibition of responses and the killing of cells. This is a very small list of the myriad functions that cytokines perform. In short, cytokines play major roles in influencing the duration and intensity of immune responses.

In addition to T_H1 and T_H2 subsets, a recently discovered subset is the T_H17 subset; these cells predominantly secrete the cytokine IL-17 which affects several cells such as endothelial cells and epithelial cells. T_H17 cells induce the production of

chemokines and other pro-inflammatory cytokines which attract and activate neutrophils, thus amplifying inflammatory reactions (Figure 101.13). These cytokines also induce the secretion of antimicrobial peptides, that along with inflammatory reactions help in defense against microbes. However, these cells are also implicated in potent immunopathologic reactions and autoimmune tissue damage.

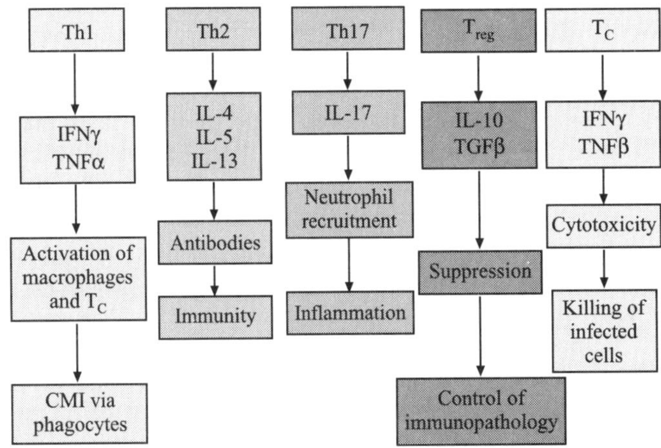

Figure 101.13 Effector cells in cell-mediated immune responses.

VI. PROBLEMS RELATED TO THE IMMUNE SYSTEM

We will highlight three scenarios; one in which the immune system is deficient, another in which the immune system may attack self antigens and third in which the immune system may react to antigens in such an abnormal manner as to cause damage to host tissues.

The immune system does indeed perform the challenging task of defending us from a huge array of pathogens that assail us via the air we breathe, the food and water we ingest and from the things we touch. It has at its disposal an impressive arsenal of molecules, cells and mechanisms to mediate defenses against pathogens. So essential is the immune system to our survival that if we lack a crucial molecule like a member of the complement family, or a critical lymphocyte type that may be present in low numbers or may be absent, a condition called *immunodeficiency* results. The absence or paucity of key elements of the immune system, i.e., critical molecules or cells, resulting in poor or no immune responses of certain types is immunodeficiency. Immunodeficiency diseases may result from deficiencies in innate and adaptive immunity. Phagocyte function could be impaired or certain key components of the complement system could be absent or in low levels, leading to defects in innate immunity. B cell deficiencies would of course lead to poor or no production of certain classes of antibodies; similarly, defects in T cell function or a lack of T cells would result in major repercussions on immune function.

The problem mentioned above relates to the inability of the immune system to mount immune responses against foreign invaders, i.e., pathogens. Imagine another scenario in which the immune system turns against the host and starts attacking host molecules, cells or tissues. This was anticipated almost two centuries ago when it was suggested that the immune system, in addition to reacting against "foreign" or "non-self", could also attack "self". This condition called "autoimmunity", in simple terms, is immunity against self antigens. We noted earlier in this chapter that T and B lymphocytes undergo a process of education and tolerance-induction whereby self-tolerance is established, i.e., our immune systems are trained to recognize non-self and to not recognize, i.e. to tolerate self. T cells and B cells that recognize self, are eliminated during this process of tolerance induction; however, some of these cells may not have been eliminated and can, in later life, be activated to mount responses against self. Or, regulatory T cells that maintain tolerance, could be bypassed and anti-self reactions could ensue. When autoimmune reactions result in damage to cells and tissues, *autoimmune diseases* may result. This subject is dealt with in another chapter in this book.

Immune responses against antigens involve local inflammatory reactions mediated by antibodies, complement, T lymphocytes, macrophages etc that remove the offending antigen; this is generally the case. However, sometimes the immune system may mediate inflammatory reactions that could be harmful to such an extent that tissues may be injured. Such responses may actually be exaggerated, as in mounting a hyperactive reaction, resulting in tissue damage; in other cases the responses may be inappropriate, as in stimulating the production of IgE antibodies instead of IgG. Such immune reactions are called *hypersensitivity* reactions and are discussed in detail in another chapter in this section.

SUMMARY

Here is a simple analogy that provides a framework for understanding the immune system. Imagine a country that has two unfriendly neighbours on two sides and an ocean on the third. The neighbouring countries pose a threat to our country which is vulnerable to attack led by armies that operate over land, by air forces that attack via the skies and naval forces that approach our country via the sea. This can be likened to our bodies which are vulnerable to pathogens that can enter through different routes – via the food and water we ingest, the air that we breathe and things that we touch.

Fortunately this country is well supported with defense forces; an army consisting of tanks, soldiers and guns that repel attacks over land, an air force consisting of jet fighters, drones and missiles that defend us in the air and a naval force comprised of battle ships and aircraft carriers that protect us from invaders that approach over the sea. Similarly, we are protected by a whole host of protective mechanisms consisting of cells and molecules; we have innate and adaptive immune systems, humoral and cell-mediated immunity, antibodies, complement, cytotoxic T cells and natural killer cells and macrophages and granulocytes, to name but a few of our immune effectors. For different kinds of pathogens, we have different kinds of effective immune responses; for pathogens that enter through different routes we have appropriate immune defense mechanisms.

Sometimes countries do not have effective defense mechanisms and are thus easy to attack; we have a similar situation, immunodeficiency, which can render afflicted people incapable of defending themselves from pathogenic invasions. Sometimes defense personnel can mistake a harmless intruder for a dangerous invader and can engage in an inappropriate or exaggerated defensive response that ends up damaging the infrastructure of the country. In the immune system, we have a similar response – hypersensitivity reactions that are directed at innocuous molecules and which damage host tissues. Finally, defense forces might inadvertently react against their own citizens; the immune system is no different in that it sometimes attacks "self" tissues and organs causing autoimmune diseases.

Despite these deficiencies and problems, the immune system is a marvel of evolution, protecting us from an incredible range of pathogens using a fascinating array of cells, molecules and tissues to mount defense reactions 24/7.

SUGGESTIONS FOR FURTHER READING

Basset C., et al. (2003), Innate Immunity and Pathogen-host Interaction, *Vaccine* 21:s2/12.

Bromburg J.S. (2004), The Beginnings of T-B Collaboration, *Journal of Immunology*, 173:7.

Carroll M.C. (2004), The Complement System in Regulation of Adaptive Immunity, *Nature Immunology*, 5:981.

Delves P.J., Martin S.J., Burton D.R. and Roitt I.M. (2011), *Roitt's Essential Immunology*, 12th ed., Wiley-Blackwell, Sussex, UK.

Fitzgerald K.A., et al. (2001), The Cytokine Facts Book, 2nd ed., Academic Press, New York.

Kaufmann S.H., Sher A. and Ahmad R. (2002), *Immunology of Infectious Diseases*, ASM Press, Washington DC, USA.

Kindt T.J., Goldsby R.A. and Osborne B.A. (2007), *Kuby Immunology*, 6th ed.,W.H. Freeman and Co., New York, USA.

Kunkel E.J. and Butcher E.C. (2002), Chemokines and the Tissue-specific Migration of Lymphocytes, Immunity, 16:1.

Lemaitre B. (2004), The Road to Toll, *Nature Reviews Immunology*, 4:521.

Murphy K. (2012), Janeway's *Immunobiology*, Garland Science Publishers, New York, USA.

Russell J.H. and Ley T.J. (2002), Lymphocyte-mediated Cytotoxicity, *Annual Review of Immunology*, 20:370.

102

Immunoglobulins

G.P. Talwar

CONTENTS

I. GENERAL INTRODUCTION

Anti (foreign) bodies are generated in a multicellular organism when a substance "foreign" to the organism is injected. The antibodies are globular proteins. It was believed for a long time that antibodies were all gamma globulins. With more information on the different classes of antibodies, it is noticed that all antibodies need not necessarily co-precipitate with gamma globulins nor manifest electrophoretic mobility characteristic of gamma globulins. Some of them migrate faster than gamma globulins on electrophoresis. A large number of antibodies, however, have the physico-chemical characteristics of gamma globulins.

There are five classes of immunoglobulins. These are IgG, IgA, IgM, IgD and IgE. The most abundant immunoglobulin in serum is IgG (about 80 per cent of the total). The normal levels of the various classes of immunoglobulins in human blood are given in Table 102.1. There are variations in their contents in some physiological and pathological conditions. The properties and biological functions of different classes of immunoglobulins are summarized in Tables 102.2, 102.3, and 102.4.

TABLE 102.1 Serum Levels of Three Major Classes of Immunoglobulins in Health and Disease

	IgG i.u./ml*	IgA i.u./ml**	IgM i.u./ml***
Normal healthy (Indian subjects)	172 (120–246)[a]	121' (55–268)	175 (108–284)
Typhoid	187 (53–305)	94 (72–156)	368 (248–791)
Rheumatoid arthritis	237 ± 14.99[b]	156 ± 16.38	238 ± 26.92
Systemic Lupus erythematosus (SLE)	230	236	154
Amoebiasis	232 (103–774)	89 (30–363)	112 (52–180)
Lepromatous[c] Leprosy	14.35 ± 0.05	2.77 ± 0.05	1.23 ± 0.03
Tuberculoid[c] Leprosy	9.7 ± 0.04	1.07 ± 0.10	1.2 ± 0.10

* 1 i.u. of IgG = 80.4 μg/ml of serum.
** 1 i.u. of IgA = 14.2 μg/ml of serum.
*** 1 i.u. of IgM = 8.47 μg/ml of serum.

a Figures in parentheses denote range.
b Standard error of the mean.
c mg 100 ml plasma.

TABLE 102.2   Physio-chemical Properties of Various Classes of Human Immunoglobulins

	W.H.O. nomenclature	IgG	IgA	IgM	IgD	IgE
1.	Concentration in serum (mg/ml)	8–16	1.4–4	0.5–2	0–0.4	17–450 ng $(10^{-9}$ g)/ml
2.	Sedimentation coefficient (S)	7	Serum: 7, 9, 11, 13 Secretions: 11 (with J and secretory piece)	19	7	8
3.	Molecular weight	150,000	Serum: 160,000 and Polymers Secretions: 385,000	900,000	185,000	200,000
4.	Heavy chains	Gamma (γ)	Alpha (α)	Mu (μ)	Delta (δ)	Epsilon (ε)
5.	Light chains	Kappa (κ) Lambda (λ)	κ, λ	κ, λ	κ, λ	κ, λ
6.	Other chains	–	J chain SC	J chain	–	–
7.	Usual molecular form	Monomer	Monomer Dimer	Pentamer Hexamer	Monomer	Monomer
8.	S–S loops in heavy chain	4	4	5	5	5
9.	Valency for antigen binding	2	2, 4	10, 12	2	2
10.	Serum half-life (days)	23	6	5	3	2.5

TABLE 102.3   Biological Properties of Different Classes of Human Immunoglobulins

		IgG	IgA	IgM	IgD	IgE
1.	Complement fixation	+	–	+	?	–
2.	Placental transfer	+	–	–	–	–
3.	Receptors for binding to macrophages	+	–	–	–	–
4.	Fixation to Mast cells	–	–	–	–	+

II. BASIC STRUCTURE AND PROPERTIES OF IMMUNOGLOBULINS

A. IgG

IgG is a 4-peptide unit composed of two light chains and two heavy chains. The words light and heavy denote the relative size of the polypeptide chains. The light chains have approximately 214 amino acids whereas the heavy chain is made up of 440 amino acids. In a given immunoglobulin molecule, the two light and the two heavy chains are identical. The light and heavy chains are linked to each other by S–S bonds, as depicted in Figure 102.1.

Purified IgG is a Y-shaped molecule when visualized in electron microscope by negative staining. This has been concluded from the shape of the antigen antibody complexes against a divalent hapten, bis-N-dinitrophenyl (DNP)–octa-methylene–diamine. In the electron-micrographs reported by Valentine and Green, a series of geometric forms are seen representing the different structures expected from a Y-shaped molecule with combining site at the end of each of the arms of the Y complexed with this divalent hapten (Figure 102.2). There is flexibility around the hinge region.

Light chains: The light chains are of two types in IgG as well as in other classes, e.g., these do not vary with the class of

TABLE 102.4   Biological Functions of Immunoglobulins

Early, low avidity antibody	IgM
Late, high avidity antibody	IgG
Opsonization via Fc cell receptors	IgG
Opsonization via complement activation — classical pathway — alternative pathway	IgM, IgG IgG, IgA, IgD
Viral neutralization	IgG, IgA
Histamine release from mast cells/basophils	IgE
Function as receptors for antigens on B-cells	IgM. IgD

immunoglobulins, it is the heavy chain of immunoglobulins that determines the class. The two types of light chains are called kappa (κ) and lambda (λ). These chains are distinguishable from each other on the basis of their primary amino acid sequence. Antisera raised against κ or λ type of chain give serological reactions with the homologous corresponding light chains but not with the other. A given antibody would contain either kappa or lambda type of light chains.

The amino acid sequence analysis of Bence Jones proteins, which are light chains, obtained from patients suffering from multiple myeloma (a tumour of antibody producing cell) has shown that the light chains are divisible in two parts. The first half (1–108 amino acids from the N-terminal end) is variable from one antibody to the other, whereas the second

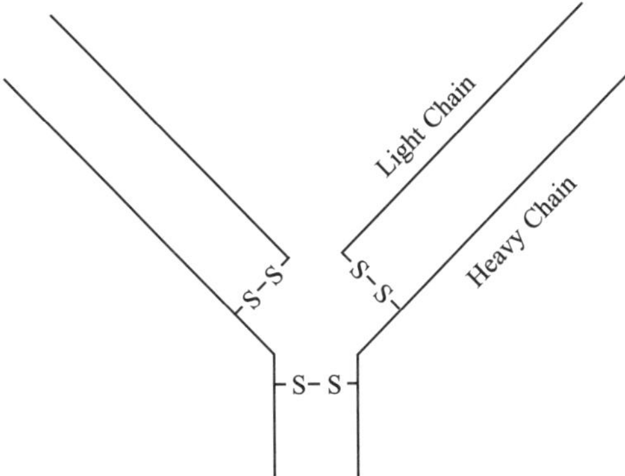

Figure 102.1 IgG has four polypeptide chains, two light and two heavy. The light chains are joined to the heavy chains through –S–S– bonds. Similarly, the two heavy chains are joined to each other through –S–S– linkages.

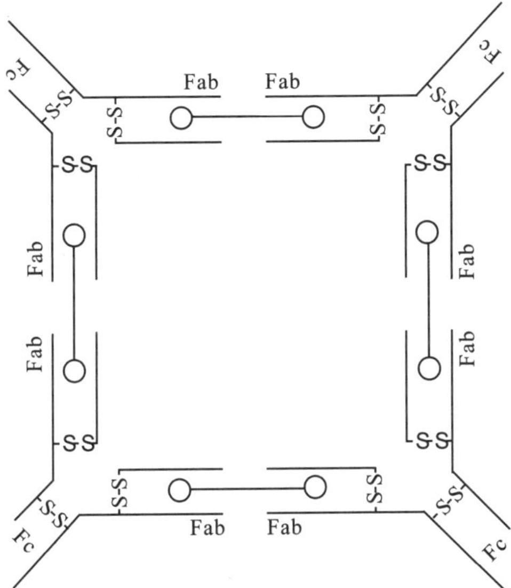

Figure 102.2 IgG is a Y shaped molecule. Figure shows diagrammtically a tetramer type of disposition consequent upon the binding by four IgG antibody molecules with a dimer antigen (O—O).

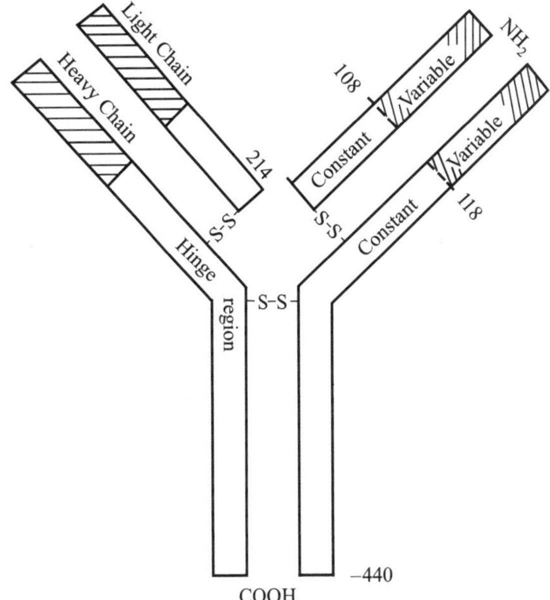

Figure 102.3 *Variable and constant regions in IgG.* In IgG molecule, the amino terminal half (1–108) of light chain and amino terminal quarter (1–118) of the heavy chain is the variable part, and contains the antibody combining (Fab) fragment. The antibody molecule has a hinge region rich in proline residues which prevent the peptide chain assuming α-helical conformation. The hinge region makes possible the moving over of the combining sites and assume different angles.

half (109–214 amino acids) is more or less constant in its amino acid sequence in the myeloma proteins belonging to a given class of light chains. These regions are designated as "V" (variable) and "C" (constant) parts of the polypeptide chain (Figure 102.3). It is obvious that the difference between one antibody molecule and the other of the same class would reside in the variable part of this chain, the constant part being characteristic of the type.

Within the light chain there are two interchain S–S loops, one in the variable region of the chain and the other in the constant part (Figure 102.4).

Heavy chains: The analysis of myeloma proteins from man and mouse has given valuable information on the primary structure of the heavy chains. The heavy chains also have a variable region extending from 1 to 118 amino acids from the N-terminal end. The sequence of amino acids in this region

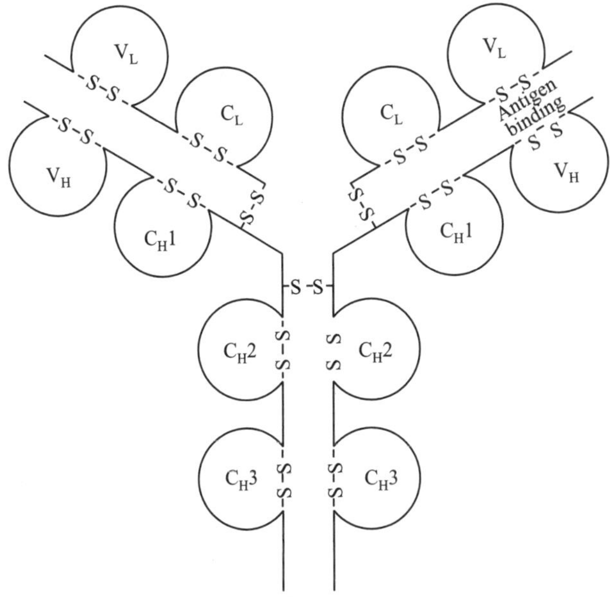

Figure 102.4 *Interchain –S–S– loops in IgG.* The light chains have two –S–S– linkages formed between cystine residues present at position 23–88 and 134–194. The heavy chain has four of these linkages between residues at 22–96, 144–200, 261–321 and 397–425. The –S–S– linkages form two disulphide loops in each light chain and four in the heavy chain.

varies from one myeloma protein to the other. The remaining 119–440 amino acids constitute the "constant" part of the heavy chain (Figure 102.3). It will be observed that this length is approximately three times as much as the constant part of the light chain. A hypothesis has been advanced that the constant part of the heavy chain may have resulted from first a duplication of the genes corresponding to the 'C' region of the light chain followed by a recombination of the two chains.

Like the light chain, there are three intrachain –S–S loops in the constant region of the heavy chain (Figure 102.4). These divide the constant region of the antibody molecule into three segments. The segment (C_H2) has the receptor for the binding of the complement, and the segment (C_H3) is the portion which binds to macrophages.

The *class specificity* pertains essentially to the *constant part of the heavy chain* of the immunoglobulin molecule. The difference between the five classes of immunoglobulins stems from the variation in the heavy chains, e.g., IgG class of immunoglobulins have the gamma heavy chains, IgA have alpha, IgM μ, IgD δ and IgE ε chains. These heavy chains differ from each other serologically and in their primary amino acid sequence.

The antibody: As mentioned above, IgG is composed of two light and two heavy chains joined to each other through S–S bonds. The antigen binding site is situated in the variable part of the light and the heavy chains. Both light and heavy chains contribute to the optimum binding of the antigen. From the basic structure, outlined in Figure 102.3, it can be seen that there are two antigen binding sites in each IgG molecule.

If IgG is subjected to enzymatic digestion, it is noted that papain cleaves the immunoglobulin molecule, producing two monovalent antigen binding fragments called (Fab) and one crystallizable fragment termed as (Fc). Pepsin digestion gives one F (ab')$_2$ fragment (fragment with two antigen binding sites) and the Fc, as shown in Figure 102.5. The attack by both enzymes is primarily around the 'hinge' region which has the maximum elasticity in the antibody molecule.

Antigen binding site: The antigen binding property is present in the variable portion of the antibody molecule. Amino acid sequence analysis of a number of myeloma proteins has revealed that within the variable portion of the light and heavy chains, there are regions where there are frequent variations in amino acid residues. These "hyper-variable" spots are involved in antigen binding, a postulate that is supported by X-ray analysis. The hypervariable regions in light chains are situated between (i) amino acids 24 to 34, (ii) 50 to 56, and (iii) 89 to 97 (Figure 102.6). There should be corresponding three hypervariable regions in the variable part of the heavy chain namely V_H (variable) D (Diversity) and J_H (Joining) and light chain has two hypervariable regions V_L and J_L. These regions come fairly close to each other spatially by the folding of the chains. Each globular domain consists of a band shaped assembly of 7–9 anti-parallel β strands. The antigen binding site is constituted by the apposition of the hypervariable

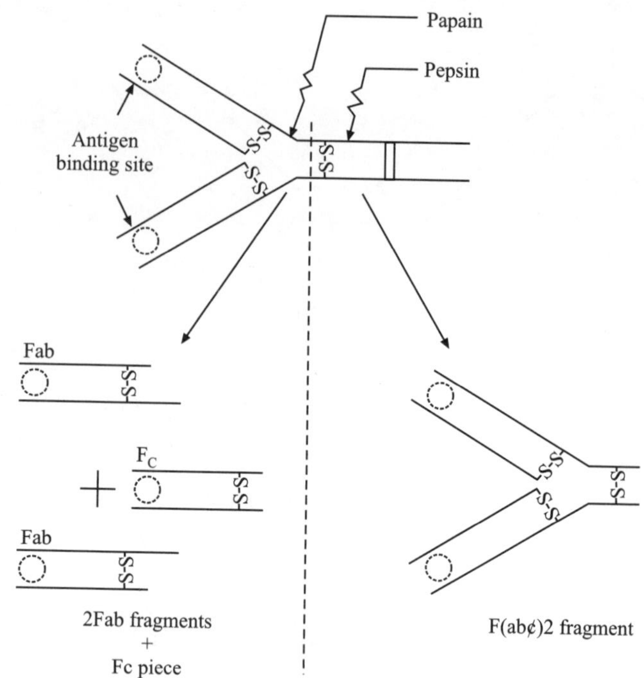

Figure 102.5   IgG can be split into three fragments on treatment with papain at the position indicated in the figure. Two of these are identical and are able to combine univalently with antigen. They are therefore called Fab Fragments. The third does not bind with antigen and is called the Fc (Crystallizable) fragment. Fc is believed to direct the biological activity of the molecule. Pepsin cleaves at a point below the –S–S– link of the heavy chain giving rise to a bivalent F (ab')$_2$ fragment in which both the antigen binding sites of the immunoglobulin remain associated.

regions in the light and heavy chains (Figure 102.7). Glycine residues are invariably present around the hypervariable sites in all immunoglobulins and Kabat and Wu have brilliantly suggested that the flexibility of the bond angle in this amino acid provides the plasticity for the stereospecific folding of the antibody binding sites with epitopes on the antigen.

Fc: The Fc part of the antibody molecule is endowed with traits that impart direction and capabilities to the biological activity of the molecule. It is this part of the antibody molecule that determines the "homing" properties. IgG is transferred through the placenta but not other classes of immunoglobulins. The Fc part of the molecule has "receptors" for the fixation of complement—IgG and IgM bind the complement but not other classes of immunoglobulins. The complement fixation property is located in the C_H2 domain of the heavy chain (Figure 102.6). The cytophilic properties reside in the C_H3 part of the molecule. Thus, although the antigen binding and specificity of the antibody for reaction with an antigen is in the Fab part of the antibody molecule, the heavy chain tail end is important for the effector functions.

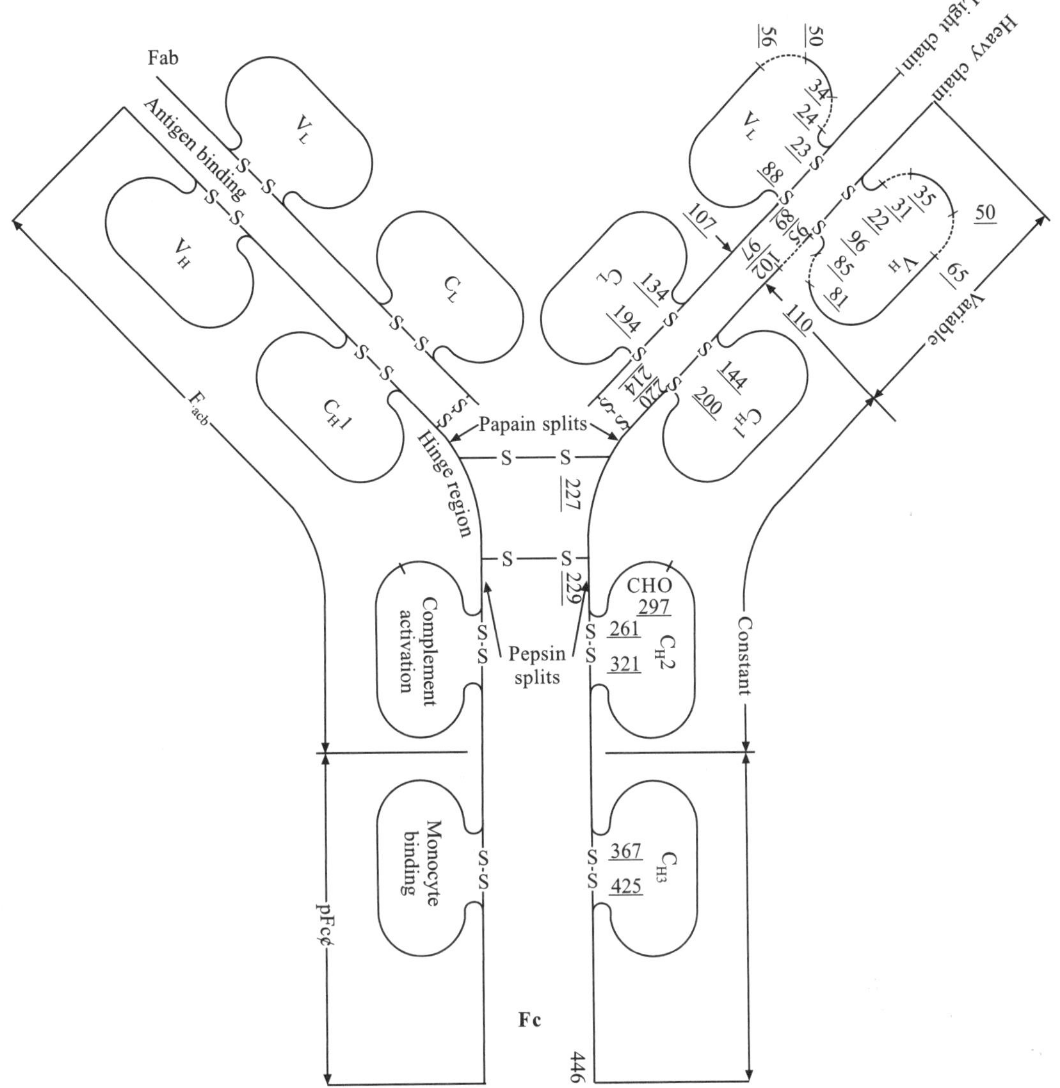

Figure 102.6 Schematic diagram of human IgG heavy chain subclass G_1 light chain kappa as conceived by Kabat. The two light and heavy polypeptide chains are held by interchain disulphide bonds. The amino acid residues are numbered starting from the N-terminal end. Fab portion has the antigen binding site. The constant region of the light chains and the three constant regions of the heavy chains, each containing disulphide loops exist as separate domains or homology regions. Each domain is specialized for a specific function. Thus C_H2 region in IgG binds complement while adherence to the monocyte surface is at the terminal C_H3 region.

Genetic variability, allotypes: Besides the structural heterogeneity in heavy chains due to class and subclass specificity, allotypic variants have also been recognised. The allotypes are genetically inherited and differ in a couple of amino acids at a given point, e.g., Gm^+ individuals would show in each IgG_1 molecule the sequence: Asp–Glu–Leu–Thr–Lys. Another individual with Gm-negative inheritance would have in IgG_1 the sequence: Met–Glu–Glu–Thr–Lys, i.e., two amino acids are different in the two cases. These differences are not large enough to alter the class and sub-class characteristics. About 25 genetic markers (Gm) groups have been identified on the heavy chain and three (Inv groups) on the light chain.

B. Other Classes of Immunoglobulins

1. IgA: IgA is again basically made up of two light chains (λ or κ) and two heavy chains (alpha). IgA is the second largest class of immunoglobulin in the serum. It accounts for about 10–15% of the total in contrast to IgG which is predominant (about 80%). In the serum, IgA is present as a 7 S monomer, but tends to form polymers by associating with another polypeptide J-chain.

IgA is the important immunoglobulin in seromucous secretions, such as the saliva, secretions of respiratory and gastrointestinal tract, tears, colostrum, etc. It is synthesized locally by the plasma cells and secreted as a dimer which combines

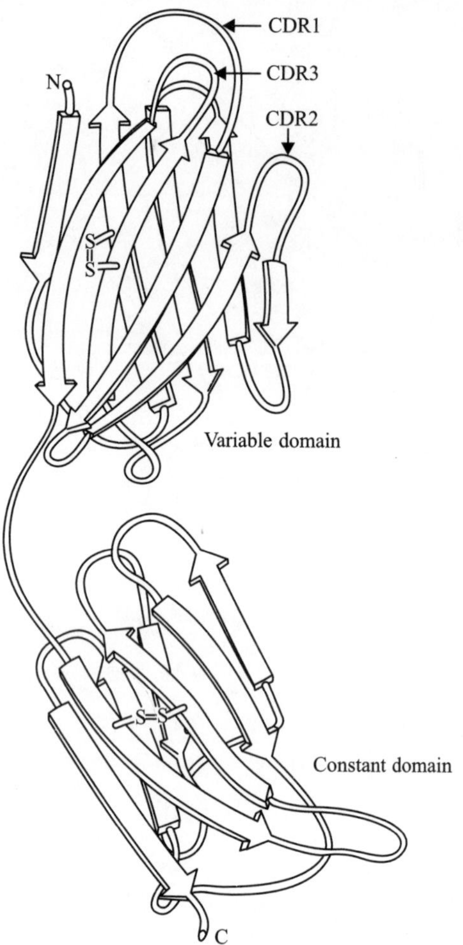

Figure 102.7 Three-dimensional structure of a light chain. Each globular domain consists of a barrel shaped assembly of 7–9 antiparallel β strands. The hypervariable regions project outward from the amino terminal end.

with another protein—the secretory piece (mol. weight 60,000) which is synthesized by adjacent epithelial cells (Figure 102.8). IgA_2 dimer + secretory piece complex is resistant to proteolytic enzymes.

This immunoglobulin has apparently a role in local defence. By adherence, it prevents the entry of pathogens in mucous linings. Orally given polio vaccine presumably elicits local immunity in the gastro-intestinal tract, the tract through which viral infection is received. Attempts are being made to develop an aerosol vaccine against flu with the idea of inducing immunity against the virus at the site at which the infective agent passes into the body, e.g., respiratory tract. Oral vaccines capable of eliciting IgA formation and cell-mediated immunity in G.I. tract locally against cholera and other enteric infections should have merits over the systemic vaccines.

2. IgM: IgM is sometimes referred to as the macroglobulin. It has a molecular weight of 900,000 and sedimentation constant of 19 S. Its heavy chain is a unit longer than the size of the heavy chains in IgG and other immunoglobulins; it has a C_H4 domain, besides C_H2 and C_H3. Furthermore, IgM is a polymer of five immunoglobulin molecules, each consisting of two light and two heavy chains. The polymerization is achieved by association with J-chain (Figure 102.9). Theoretically it has capacity to bind ten antigen molecules, but practically it binds five, presumably due to steric restrictions.

IgM is the type of antibody formed in the early stages of immunization. Later on the response shifts to IgG type of antibodies. By virtue of the fact that IgM has multivalent sites, it is highly effective as an agglutinating antibody. It has also the property of binding the complement. IgM antibodies are thus cytolytic. IgM is primarily present in blood. It is the major immunoglobulin involved in defence in the initial stages of infections.

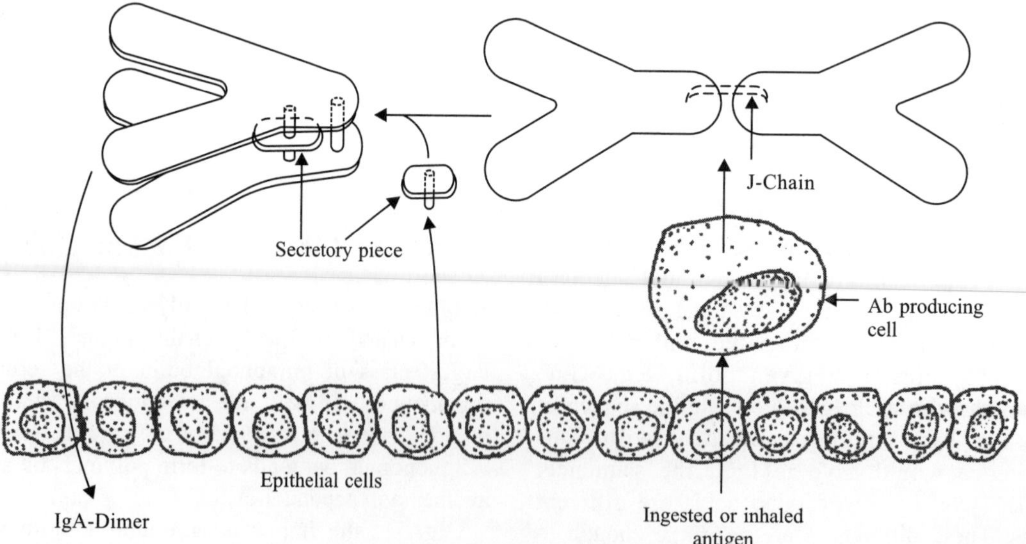

Figure 102.8 Schematic diagram of IgA immunoglobulin in seromucous secretion. IgA is present as a dimer formed in association with another polypeptide—J-chain. A secretory piece contributed by the epithelial cells is further associated with this immunoglobulin rendering it resistant to the proteolytic enzymes.

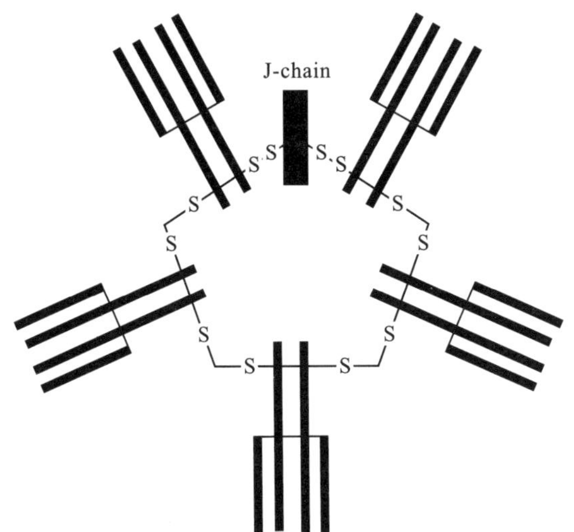

Figure 102.9 Schematic model of IgM. This macro-globulin is composed of five 4-peptide units linked to each other through –S–S– bond and J-chain.

3. IgD: The discovery of this class of immunoglobulin took place from a myeloma protein, which was found to be devoid of reactivity with anti IgG, IgM and IgA heavy chain specific sera. It cross-reacted, however, with anti light chain sera. It was a four-peptide unit as other immunoglobulins—two light and two heavy chains.

The heavy chain, characteristic of this class of immunoglobulin, was named as delta and the class, IgD.

IgD is coexpressed on B-lymphocytes that also bear surface IgM. It is rarely secreted in significant amounts and only traces are normally found in blood. On these B-cells both classes of heavy chains are produced by alternative splicing of a single m-RNA and have identical antigen specificity. The surface IgD binds antigens and is involved in signal transduction similar to IgM. The presence of IgD on virgin B-cells suggests a role in B-cell differentiation.

4. IgE: IgE is present in extremely low amounts in the blood (17–450 ng/ml; ng = 10^{-9}g). The proportion of IgE synthesizing lymphocytes is extremely small out of the total immunocompetent cells in normal healthy humans. This immunoglobulin by virtue of its cytophilic properties is *mostly* bound to cells and is present in only trace quantities in circulation. The cytophilic trait is a function of the class specific heavy chain—the epsilon chain which has receptors for binding to mast cells in the skin and elsewhere. The antigen-antibody reaction for this class of immunoglobulin takes place on the cell surface which results in the degranulation of the mast cells and release of vasoactive amines. These compounds elicit symptoms characteristic of allergy, e.g., vasodilation, oedema, rash, hay fever, asthma, etc.

IgE levels are raised in several tropical parasitic diseases. It has a definite defensive function in these circumstances, the histamine released by antigen-antibody reaction helps in ejection of the parasites. IgE and eosinophils have also been shown in recent studies of Houba et al., to contribute to defence process in schistosomiasis.

III. SUBCLASSES OF IgG AND IgA

With the accumulation of data on serology and sequence analysis of IgG myeloma proteins, it has become evident that there are subclasses of this immunoglobulin. Four subclasses have been recognized: IgG_1, IgG_2, IgG_3 and IgG_4. The corresponding heavy chains would be γ_1, γ_2, γ_3, and γ_4. These have considerable homologies and react with anti IgG sera, but each has some structural features characteristic of the subclass. The differences are mainly in amino acid sequence here and there and in disulphide bridging. These differences give rise to some variations in their biological behaviour. Table 102.5 summarizes the properties and characteristics of the IgG subclasses.

Two subclasses of IgA, viz., IgA_1, IgA_2 have been identified.

TABLE 102.5   Comparison of Human IgG Subclasses

	IgG_1	IgG_2	IgG_3	IgG_4
Per cent total Ig in normal serum	65	23	8	4
–S–S– between heavy chains	2	4	5	2
Complement fixation	+ +	+	+ + +	0
PCA reactivity	+ + +	0	+ + +	+ + +
Macrophage receptors	+ + +	+	+ + +	
Placental transfer	+ + +	+	+ + +	+ + +
Blocking IgE binding	–	–	–	+
Binding to heterologous skin	+ +	–	+ +	+ +
Antibody dominance	Anti Rh	Anti Dextran	Anti Rh	Anti factor VIII
Half life in serum (days)	23	23	7	23

IV. ANTIBODY DIVERSITY

The immune system has a capacity to recognize a hundred million different determinants. Each of their epitopes is read by a unique sequence in the variable part of the antibody. This diversity exists prior to contact with the antigen.

The mechanism of this variability was one of the most intriguing question in immunology. The discovery of distinct variable and constant regions raised the question of their association to form a complete Ig gene. In the 1980s, Susama Tonegawa discovered that V and C genes are far apart in germ line DNA but are closely associated in the DNA of plasma cells. How is this diversity generated?

The human immunoglobulin gene loci are found on three chromosomes, the H chain locus is on chromosome 14, the K chain on chromosome 2 and ↗ chain locus on chromosome 22. The V regions of Ig heavy chain molecules are assembled from three multigene families of DNA segments: the variable (V_H) segment, diversity (D) segment and joining (J_H) segments. These segments are brought together by recombination of germ line DNA (Figure 102.10). The heavy chain locus on chromosome 14 has fifty functional V_H segments, thirty D segments and six J_H segments. The D and J_H segments join first followed by V_H to form the V_H D J_F region. The C_H region genes are brought adjacent to the V region genes through RNA splicing of a primary transcript.

Expression of four genes is tissue specific. Only lymphoid cells contain the transactivating factors necessary for H-chain gene activation.

The variable regions of light chains are also assembled in a similar manner from their loci in chromosome 2(K) and 22(λ). Translation of H and L chain mRNA occur in the rough endoplasmic reticulum. Covalent attachment of H and L chains occur in progressive stages H + H $\rightarrow$ H$_2$ $\rightarrow$ H$_2$L $\rightarrow$ H$_2$L$_2$. Protein Disulphide Isomerase (PDI) and Immunoglobulin heavy chain binding protein (BiP) are the chaperones believed to be involved in Ig assembly. The many different combinations of V_H D J_H and $V_L J_L$ segments and heavy and light chain associations account for VDJ combinational diversity.

The V, D and J segments are separated in the germ line by thousands of nucleotides but are able to unite due to a precise site-specific recombination process (Figure 102.11). Specific sequences known as Recombination Signal Sequences (RSS) are present downstream of V_H, on either side of D and upstream of J_H as shown in Figure 102.12. There are two types of RSS—a consensus heptamer (CACT(A)GTG and a consensus nonamer GGTTTTTGT separated by a 12 bp or 23 bp spacer. The heptamers line is adjacent to the gene segments.

The heptameric and nonameric sequences on the 3′ side of V_H are complementary to the RSS on the 5′ side of D segment. Recombination always occurs between segments with a 23 bp spacer and a 12 bp spacer—called the 12–23 bp spacer rule

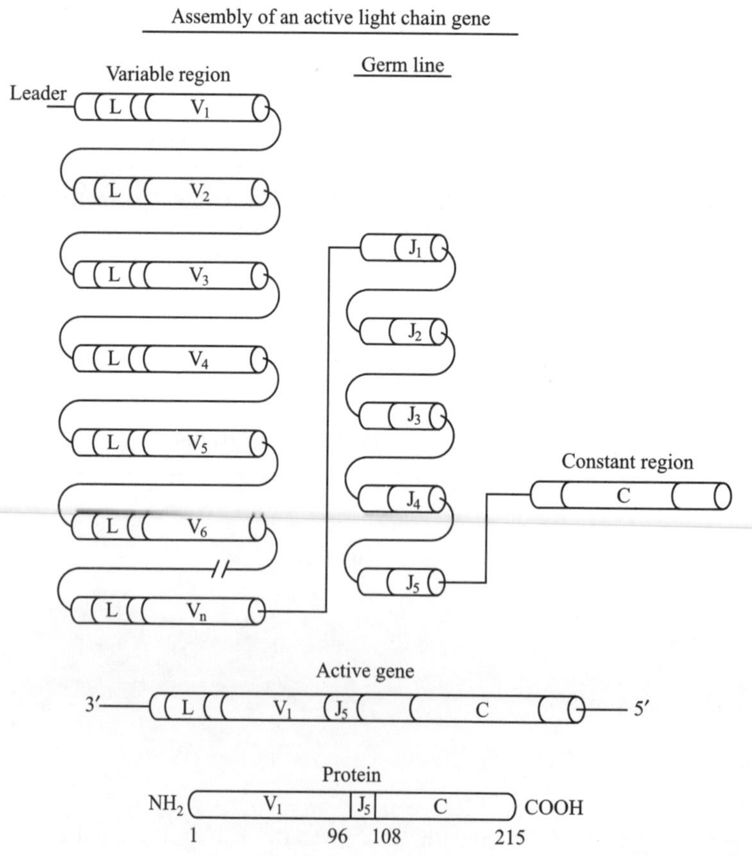

Figure 102.10 Assembly of an active light chain gene. [Biology of Immunological diseases (ed.) Dixon.]

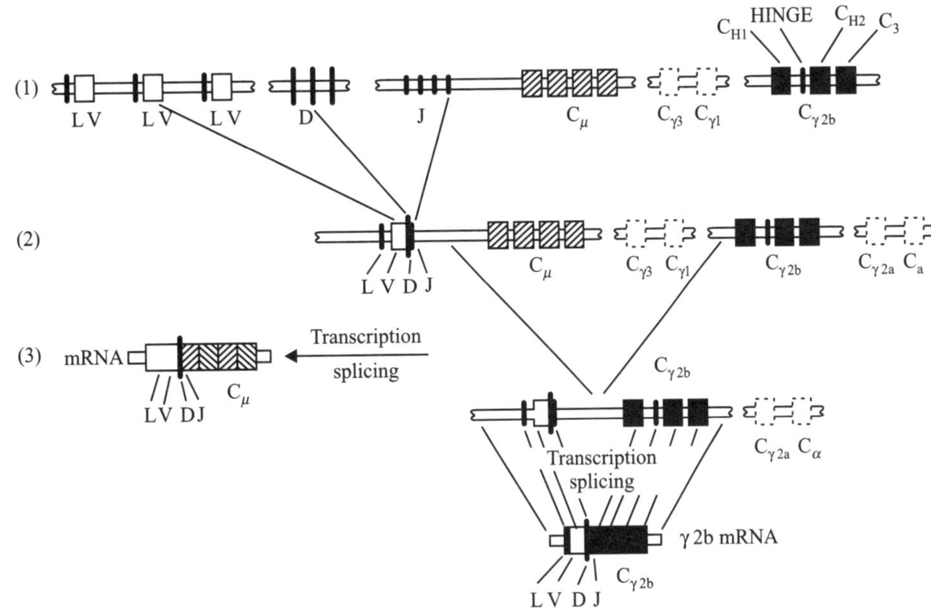

Figure 102.11 Recombination events leading to complete Cr2b gene which encodes for IgG$_{2b}$ Immunoglobulin molecule. (1) Gene before rearrangement; (2) After V, D, J joining; (3) Heavy chain switch deleting in between germ line genome, e.g., C$_\upsilon$, C$_{r3}$, C$_{rl}$ to give rise to VDJ-C$_{r2b}$ processed segment.

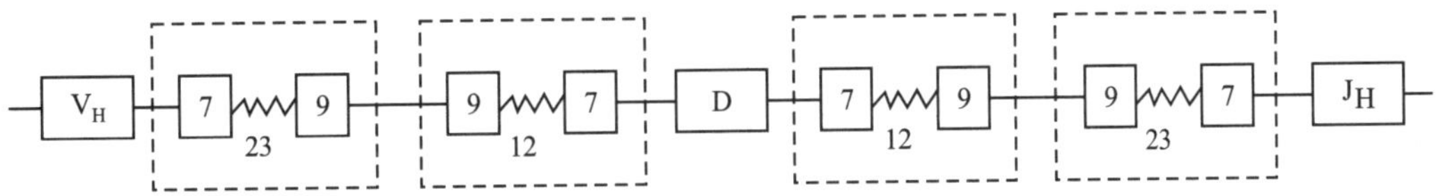

Figure 102.12 Positions of RSS in the heavy chain variable region locus.

(Figure 102.13). The RSS separated by spacers of the same length do not recombine. Two recombination activation genes RAG–1 and RAG–2 have been identified.

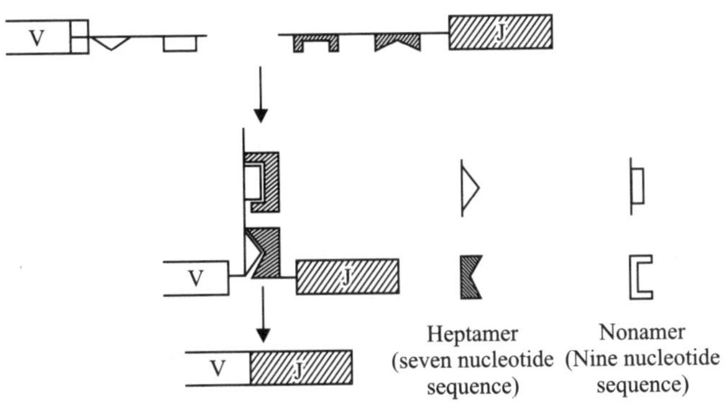

Figure 102.13 Diagrammatic representation of somatic recombinations in case of "V" light chain. (Details in the text.)

Other mechanisms which cause further antibody diversity are N segment and P nucleotide additions and somatic mutations. Insertion of N-nucleotides (guanine rich) by action of Terminal Deoxyribonucleotide Transferase (TdT), addition of P-nucleotides (palindromic), deletion of nucleotides by exonuclease at the splice joint are some of the mechanisms involved.

Antigen induced B-cell proliferation activates a hypermutation mechanism by which point mutations are introduced into the V regions and flanking sequences. This phenomenon, unique to B-cells, is responsible for affinity maturation or progressive production of higher affinity antibodies as the immune response continues.

A *single* B-cell expresses *only one pair of immunoglobulin* H and L chain genes. The remaining alleles are inactivated in a process known as allelic exclusion.

V. IMMUNOGLOBULIN CLASS SWITCHING

After antigenic stimulations, IgM bearing B-cells differentiate into plasma cells that secrete antibody of the IgG, IgA or IgE class. This is mediated by a switch sequence, tandem repeats of GAGCT 2–10 Kb in length present upstream of C genes except C$_\delta$. Recombination between switch sequences of S regions 5′ of C$_\mu$ and the S regions 5′ of each of C$_\gamma$, C$_\alpha$ and C$_\varepsilon$ result in deletion of DNA between the two S regions. The original specificity is retained while the effector function associate with the C$_H$ region is altered (Figure 102.11).

VI. HYBRIDOMAS

An important technological advance took place a few years ago, which enables obtaining unlimited quantities of an antibody with *defined* characteristics. The monoclonal antibodies are produced by a clone or family of cells derived from a single progenitor that is why these are homogeneous in composition. Monoclonal antibodies are products of "hybrid" cells created by fusion of the two cells, one of them is a myeloma cell with genetic potential of multiplying perpetually, the other has a gene which produces code for the desired antibody. The hybrid cell thus has properties of both parents, e.g., of unlimited divisions and capability of producing a given antibody. (For details see Chapter 106).

Monoclonal antibodies are of extremely high titre, e.g., 1 to 10 million, which was rarely possible by conventional immunization. The hybrid cells can be propagated as ascites in mouse peritoneum or *in vitro*. These antibodies have found application in:

1. Immuno-diagnostics
2. For imaging of tumours
3. As a vehicle for delivery of drugs
4. Immunotherapy

5. For purification of antigen by affinity chromatography.

Monoclonal antibodies have been instrumental in providing following basic information in the field of immunology:

1. Since MoAb binds to a single antigenic determinant, it is possible to know the fine structure of an antigen.
2. MoAb's against cells of haemopoietic system have enabled identification and separation of different cell types, which is going to help in better understanding of immune system.
3. MoAb is being utilized as a tool in understanding the process of cell differentiation.

VII. ANTIBODY ENGINEERING

In recent years, recombinant DNA techniques have revolutionized the isolation and production of antibodies. Phage Display techniques are used to prepare antibody libraries. "CDR grafting" allows "humanisation" of monoclonal mouse antibodies by grafting the CDR regions on to a human Ig framework. These have found important applications in immunotherapy of cancers.

SUGGESTIONS FOR FURTHER READING

Baltimore D. and Davies D.R. (1986), Immunoglobulins, *Immunology,* 6, 121.

Essar C. and Radbruch A. (1990), Immunoglobulin Class Switching, *Annual Rev. Immunol.,* 8, 717.

Fongerean M. and Dausset J. (Eds.) (1984), *Progress in Immunology,* Vol. V., Academic Press, London.

Janeway C., Traver P., Hunt S. and Walport M. (1997), *Immunobiology,* 3rd ed., London, Garland.

Jerne N.K. (1973), The Immune System, *Scientific American,* Vol. 229, No. 1, 52–60.

Kabat E.A. (1973), General Features of Antibody Molecule, In: *Specific Receptors of Antibodies, Antigens and Cells,* 3rd Int. Convoc, Immunol., Buffalo, N.Y., 4–30, Karger, Basel.

Natvig J.B. and Kunkel H.G. (1973), Human Immunoglobulins, Classes, Subclasses, Genetic Variants and Idiotypes, *Advances in Immunology,* Vol. 16.

Papers from 5th International Congress of Immunology (1982), *Scientific American,* 246, 72.

Roitt I.M. (1997), *Essential Immunology,* 9th ed., Blackwell Science, Oxford.

Talwar, G.P., J.C. Gupta, Yogesh Kumar et al. (2014), immunological approaches for treatment of advanced stage cancers invariably refractory to drugs, *J. Clin. Cell. Immunol.* 5: 247, doi: 10, 4172/2155–9899, 100247.

Tonegawa S. (1983), Somatic Generation of Antibody Diversity, *Nature,* 302, 575.

103

Genetics of Antibody Diversity

Alpana Sharma

CONTENTS

I. INTRODUCTION

The immune system of a vertebrate is capable of responding to an essentially infinite array of foreign antigens by stimulating the production of molecules called immunoglobulins (Ig). The organisation and expression of these Ig genes is the basis of variability which explains the constancy and variation displayed by the protein molecules.

Two theoretical gene models for Ig were proposed by Dreyer and Bennett in 1965, which Tonegawa and Hazumi proved experimentally later in 1975. Tonegawa won Nobel Prize for this work in 1987. They proposed the following facts:

(i) Heavy and light chain of a single Ig is encoded by two separate genes. One for the variable region (V) and other for the constant region (C).
(ii) The shuffling of these two genes allow them to come in close proximity to form a single Ig heavy (H) and light (L) chain.
(iii) Plethora of genes exists for variable region but only a single gene exists for constant region for each class or subclass of the antibody.

II. IMMUNOGLOBULN MULTIGENE FAMILIES

H-chain and L-chain genes of immunoglobulin component (kappa, lambda and heavy) are organised into clusters called *gene families*. These genes for H-chain and L-chain are located on different chromosomes.

A. Light Chain Genes

Two types of light chains are present, kappa (κ) or lambda (λ). Each consists of three gene segments Variable (V), Joining (J) and Constant (C) (Figure 103.1).

- Two coding sequences in the variable region (V and J).
- One coding sequence in the constant region (C).

The number of functional genes coding Ig light chains (κ or λ) for the variable and constant regions varies. In addition to these gene segments, human λ-light chains contain many non-functional *pseudogenes* for the variable, joining and constant gene segments. Intervening non-coding DNA sequences exist between different gene segments of both κ and λ chains. The genes for κ and λ chains in humans are located on chromosome 2 and 22, respectively.

B. Heavy Chain Genes

As proposed by Leroy Hood, various gene segments encode for Ig heavy chains. Each H chain is encoded by four gene segments located on chromosome 14.

The variable region is encoded by three genes: V_H (Variable), D_H (diversity) and J_H (joining). The diversity gene segment (D) provides a major contribution to antibody diversity. H-chain germ-line DNA contains many genes. Series of constant gene segments encode for the constant region of Ig heavy chain isotypes. In humans, the C_H gene segments are arranged sequentially in the fixed order: C_μ, C_δ, C_γ, C_ε and C_α (Figure 103.1). This order is related to the sequential expression of the Ig classes in the course of B-cell development and the initial IgM response of a B cell to its first encounter with an antigen.

C. Immunoglobulin Gene Rearrangement

In germ-line DNA, Ig genes are not transcribed and translated into H or L chains until they are shuffled. During B-cell maturation in the bone marrow, some of these gene segments are randomly shuffled and reorganised by a dynamic genetic system and generate thousands of combinations.

(i) The H-chain genes are rearranged first, followed by the L-chain genes. The variable region of L chains (κ or λ) has two gene segments: V and J. For the expression of κ or λ light chains, any of the Vκ genes can combine with any of the Jκ genes (Figure 103.2).

(ii) H-chain gene rearrangement starts in the variable region. Here, a diversity gene (D) segment is also present between the V and J gene segments. Rearrangement in the variable region begins with the joining of the diversity gene segment with the joining (J) gene segment. After D_H and J_H join, the V_H gene segment joins with it, finally generating a V_H D_H J_H gene segment in the variable region (Figure 103.3).

(iii) After rearrangement of the VDJ segment in the variable region, RNA polymerase binds to the promoter sequence and transcribes the entire H-chain variable region genes, including introns.

(iv) Finally, introns are spliced off to produce a complete functional chain.

III. ANTIBODY DIVERSITY

One of the important structure–function relationships to antigen recognition is antibody diversity. An individual is capable of making an enormous number of structurally distinct antibodies, each with a distinct specificity. The presence of a large number of antibodies that bind different antigens/pathogens is called **antibody diversity**, and the total collection of antibodies with different specificities is called the **antibody repertoire**. This diversity is mainly of genetic origin, where the recombination of different gene segments plays an important role. The genetic mechanisms that can generate such large antibody repertoire occur exclusively in lymphocytes.

A. Mechanisms of Antibody Diversity

The mechanisms/factors responsible for total antibody diversity are as follows:

(a) Multigene organisation of Ig chains: As discussed earlier, in germ-line DNA sequences, three separate multigene families encode all Ig genes. The presence of enormous multiple germ-line V, D and J gene segments contribute to the diversity of the antigen binding sites in antibodies.

(b) Combinatorial joining of different gene segments: The random rearrangement of multiple copies of the V, D and J gene segments further adds to antibody diversity. Many different variable regions generates by selecting different combinations of these segments.

(c) Junctional flexibility: Multiple copies of different gene segments join randomly and are rearranged into many unique combinations. These combinations are guided by some specific, two-turn (23 base pair) and one-turn (12 base pair) recombination signal sequences (RSS) that helps in

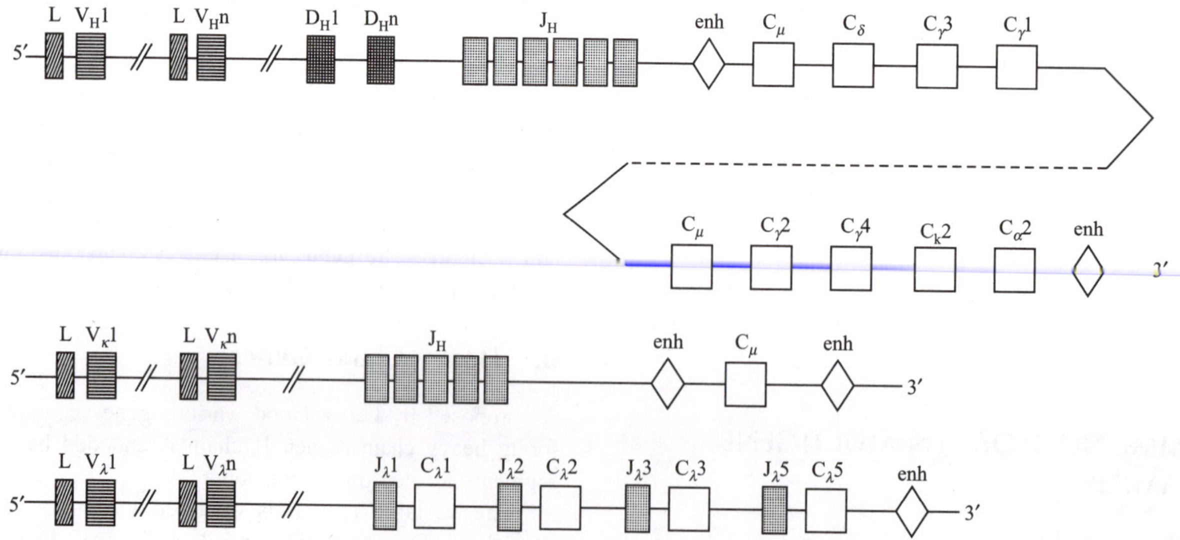

Figure 103.1   Germ line organization of human immunoglobulin genes. Schematic representation of gene segments which are indicated as follows: L, Leader; V, Variable; D, Diversity; J, Joining; C, Constant; enh, Enhancer. Pseudogenes are not shown for simplicity.

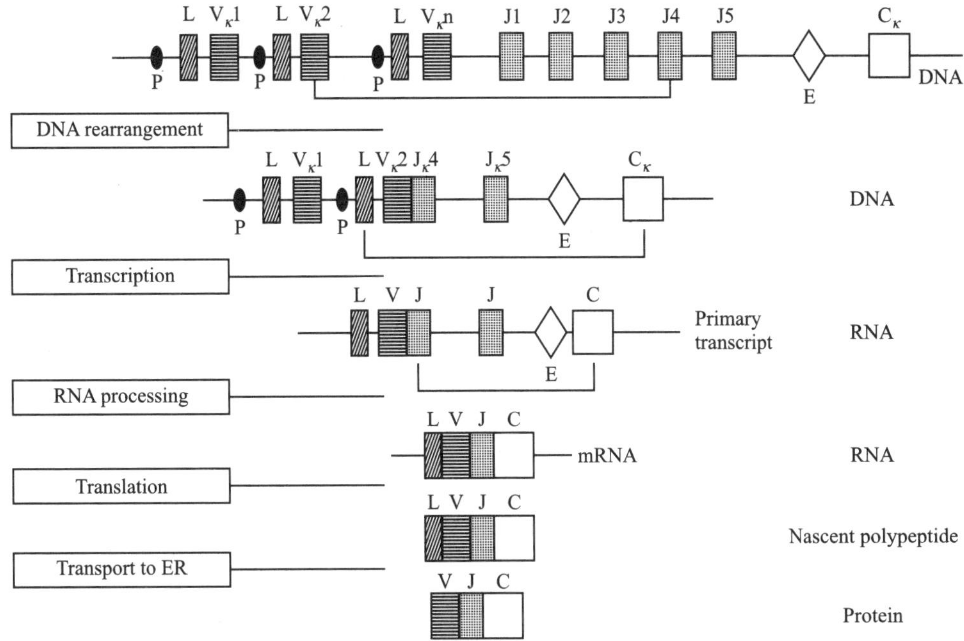

Figure 103.2   Rearrangement of κ light chain gene segment and RNA processing events required to generate a κ chain protein.

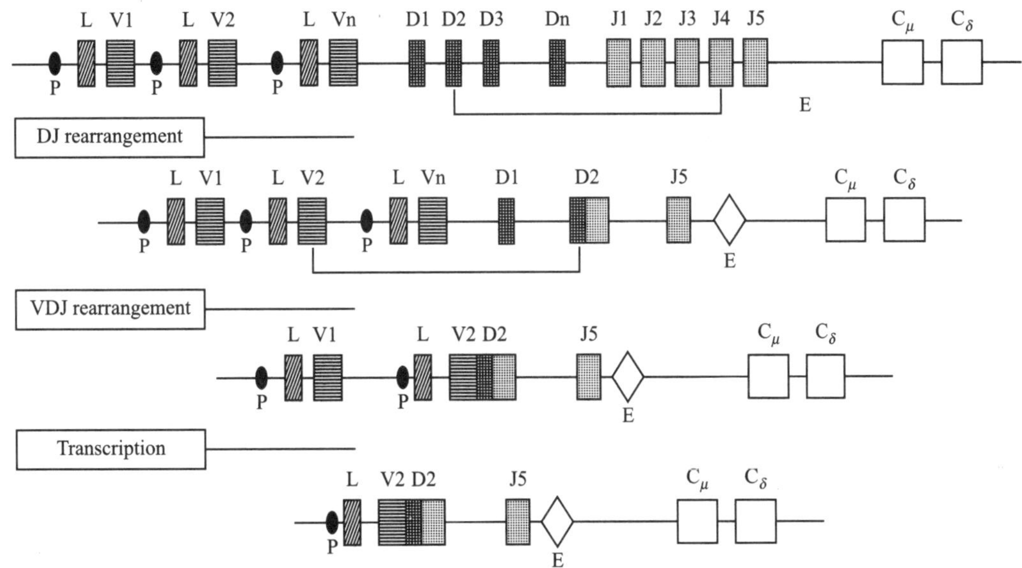

Figure 103.3   Rearrangement of heavy chain gene segment and generation of heavy chain primary transcript.

rearrangement of genes (Figure 103.4). The recombination of different gene segments between RSS and coding sequences is catalyzed by enzymes called recombinases, which are the products of recombination activating genes 1 and 2 (RAG1 and RAG2) (Figure 103.5).

Non-productive arrangement: The joining of coding sequences may lead to deletion or frame shift mutation. If such a mutation produces a stop codon, then immunoglobulin chain synthesis stops hence non-productive arrangement.

Productive rearrangement: Imprecise joining leads to change in amino acid sequence, but synthesis of the Ig chain continues. It encodes alternative amino acids at each coding joint that increases antibody diversity.

(d) P-nucleotide addition: In both heavy and light chains, diversity of complementarity determining region 3 (CDR3) is significantly increased by the addition and deletion of nucleotides in the formation of junctions between gene segments. The added nucleotides are called P-nucleotides because they make up palindromic sequences added to the ends of the gene segments (Figure 103.5).

(e) N-nucleotides addition: Short stretches of amino acid sequences not encoded by the V, D or J gene segments are called N-nucleotides because they are non-template encoded. N-nucleotides are found especially in the V–D and D–J junctions of the assembled H-chain gene. These are added by the action of

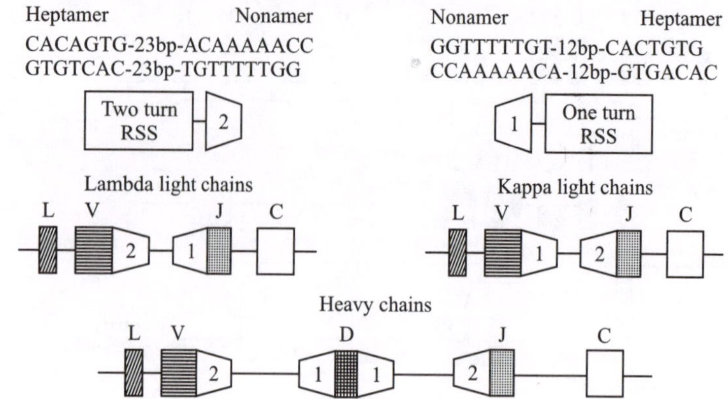

Figure 103.4   Nucleotide sequence of recombination signal sequences and their location in immunoglobulin DNA.

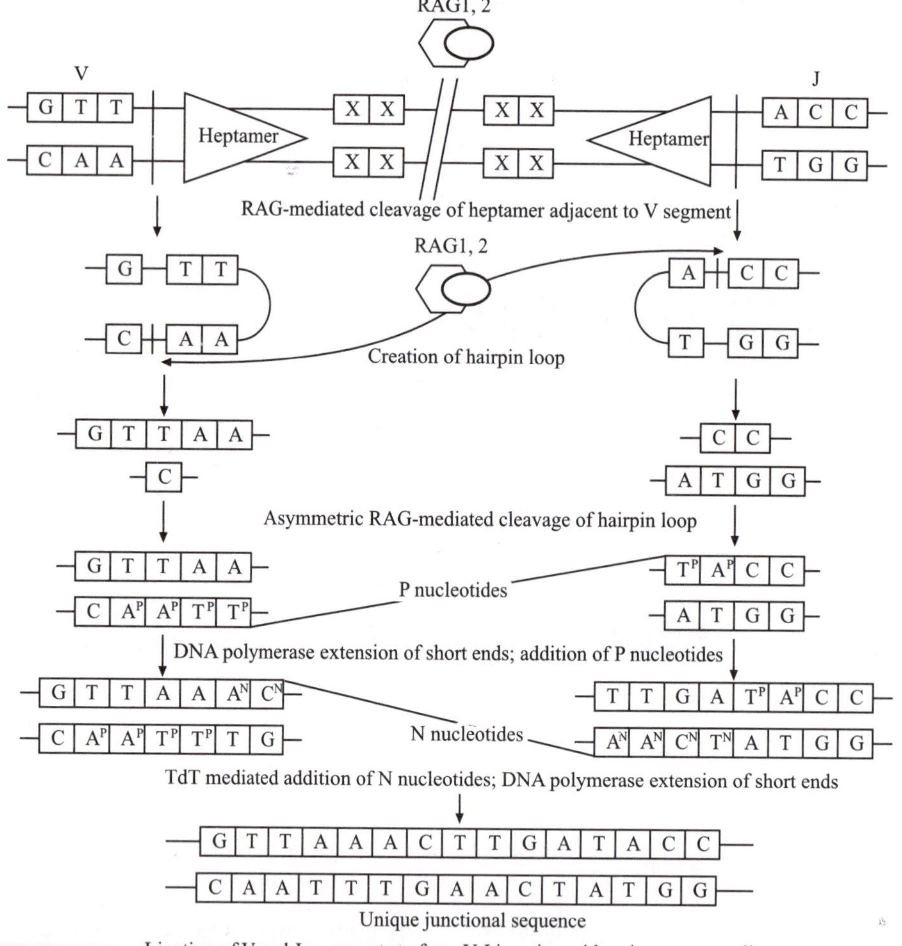

Figure 103.5   Depiction of the process of recombination of immunoglobulin gene segments and generation of novel sequences previously not present in germ-line by addition of P- and N-nucleotides.

Terminal deoxyribonucleotide transferases (TdT). N-nucleotide additions are less common in light chains (Figure 103.5).

(f) Somatic hypermutation: This is an additional mechanism that promotes antibody diversity throughout the V region after assembly of functional immunoglobulin genes. Somatic hypermutation introduces bases by point mutation (substitution) into the V region of the rearranged heavy- and light-chain genes. The mutation rate is a million times higher than the spontaneous mutation rate in other genes and hence known as hypermutation. Somatic hypermutation influence the overall affinity for antigens.

G. Association of heavy and light chains: Combinatorial associations of Heavy and any one of Light chains (κ or

TABLE 103.1   Overall antibody diversity in humans

Joining of germ-line gene segments	Possible number of combinations
Heavy chain V-D-J joining	V_H D_H J_H $51 \times 27 \times 6 = 8262$
Light chain	
1. Kappa chain V_κ-J_κ joining	V_κ J_κ $40 \times 5 = 200$
2. Lambda chain V_λ-J_λ joining	V_λ J_λ $30 \times 4 = 120$
Association of H- and L-chain combinations	$8262\,(200 + 120) = 2.64 \times 10^6$

λ) variable regions to form the antigen binding site of an immunoglobulin contribute further to antibody diversity. An approximate calculation is give below:

1.6×10^4 H-chain genes, 1.2×10^3 L-chain genes

Minimum number of combinations of H and L chains: 1.9×10^7.

The combination of all these mechanisms of diversity results in the generation of a vast repertoire of antibody specificity from a relatively limited number of genes. Some studies have estimated that antibody diversity might be as high as 10^{11}, but some of their mechanisms are not yet known.

IV. CLASS SWITCHING

On antigen stimulation, the heavy chain variable unit (VDJ) can join any constant gene segment of H-chain and express the particular antibody class. This process is called class switching. The exact mechanism is unclear but evidence suggests that the DNA flanking regions/sequences are located 2–3 kb upstream of each constant H-chain gene segment (C_H), except C_δ. These are called switch sites (S) and are composed of multiple copies of short repeated sequences (GAGCT and TGGGG). Class-specific recombinases bind to switch sites and facilitate recombination. Therefore, the expression of a particular antibody class depends on the specificity of the recombinase proteins expressed (Figure 103.6).

Class switching depends on the interplay of three factors:

- Switch regions/sites
- Switch recombinase
- Cytokine signals (IL-4)

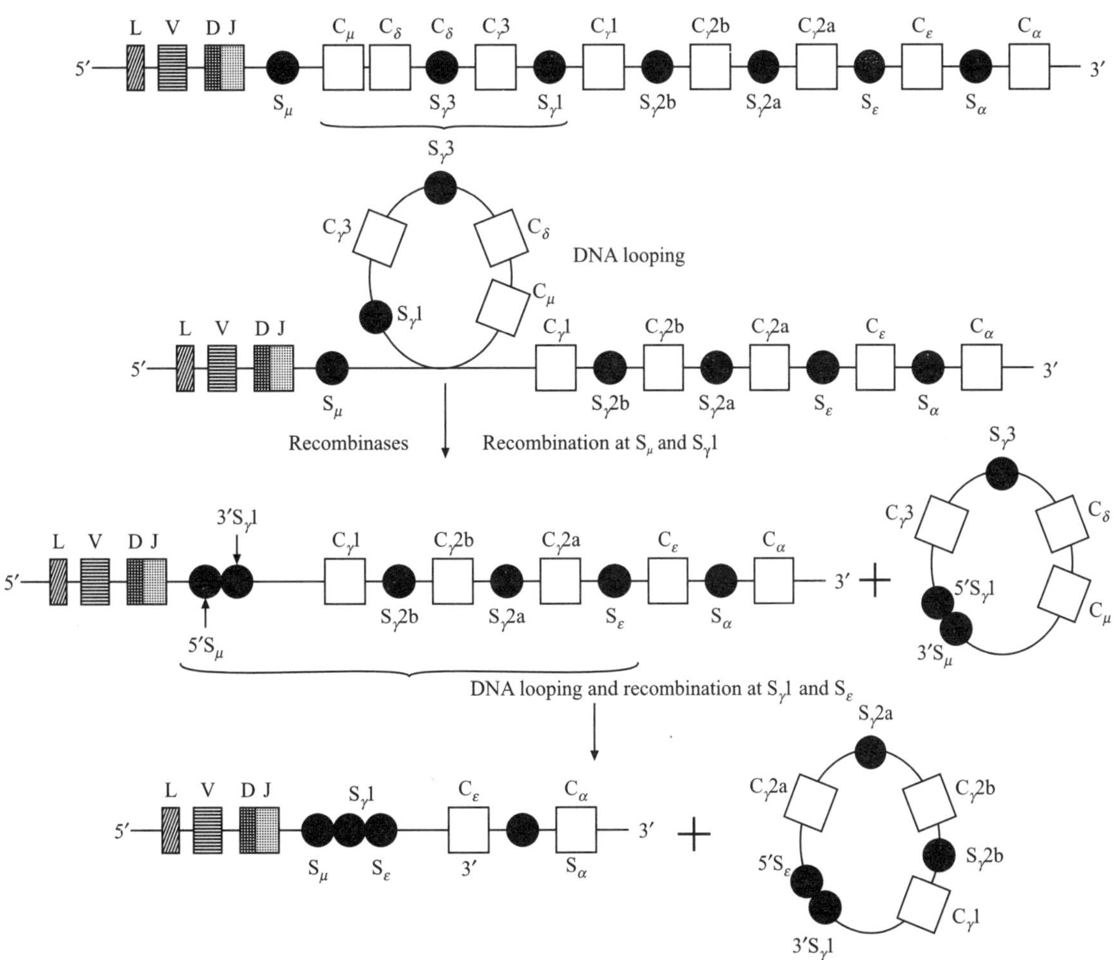

Figure 103.6   Proposed mechanism for class switching. A switch site (•) is located upstream from each C_H segment except C_δ. Circular excision products containing portions of switch sites are generated when sequential class switching occurs.

V.   CLINICAL RELEVANCE

Immunoglobulin genes and their organization has made it possible to develop engineered antibodies for therapeutic purposes to treat various lymphomas and autoimmune diseases. The genetic variation and mutability of the human immunodeficiency virus (HIV) has created a plethora of constantly changing antigens which generate diverse immune responses. This poses a major challenges in developing an HIV vaccine.

Immunoglobulin class switch recombination deficiencies (Ig CSR deficiencies) or hyper IgM syndromes (HIGM) are a group of primary immunodeficiency diseases characterised by defective CD40 signalling of B cells.

SUGGESTIONS FOR FURTHER READING

Abbas, A.K. et al., 2012, *Cellular and Molecular Immunology*, 8th edition, Elsevier Saunders, Philadelphia.

Delves, P.J. et al., 2011, *Roitt's Essential Immunology* (Desktop edition), 12th edition, Wiley-Blackwell.

Gupta, B.K. and Sharma A. 2012, *Immunology: Basic Concepts*, Peepee Publishers, India.

Kindt, T.J., et al., 2006, *Kuby Immunology*, 6th edition, WH Freeman and Company, New York.

Susumu Tonegawa, 1983, Somatic generation of antibody diversity, *Nature*, 302, 575–58.

104

MHC and Transplantation

Narinder Kumar Mehra and Gurvinder Kaur

CONTENTS

I. INTRODUCTION

Transplantation is the eventual treatment for functional failure of many organs. The success of transplantation depends on the degree of histocompatibility between donor and recipient, among other factors. The genetic difference between recipient and the donor is determined by a set of immune response genes clustered together as human leukocyte antigen (HLA) system. These antigens play an important role in immune discrimination between self and non-self (foreign) molecules.

Recent advances in histocompatibility testing including HLA typing technologies, detection of anti donor specific antibodies with high specificities and sensitivities and discovery of novel immunosuppressant drugs have drastically improved success rates of hematopoietic stem cell and organ transplantation. Transplantation immunology has evolved extensively both from bench to bed and from bed to bench side and this translation has been obtained mainly through elaborate collaborations across borders.

This chapter describes topics relating to very basics of histocompatibility and how these are tested and utilized for patients undergoing organ or hematopoietic stem cell transplantation. Various researchable facets of transplantation and associated challenges have also been described briefly and readers are encouraged to refer to further reading material mentioned in the end and elsewhere.

II. MAJOR HISTOCOMPATIBILITY COMPLEX

The blood group system (ABO), the major histocompatibility complex (MHC) and the minor histocompatibility antigens (mHA) that include the male specific H-Y antigen system constitute the overall histocompatibility system in humans. The MHC of man, referred to as the HLA system, is located on the short arm of chromosome 6 and spans over 4 megabases of DNA. The main biological function of the MHC is to bind peptide fragments derived from protein antigens from pathogens or mismatched transplants and display them on the surface of antigen presenting cells (APCs) evoking recognition by T cell receptors and induction of downstream immune responses.

Some of the striking features of the MHC are its extraordinary high gene polymorphism that varies in different populations, strong linkage disequilibrium across multiple loci and its multi-peptide binding ability rendering it an important immune response control system. The main clinical applications of HLA testing include selection of donors for stem cell and organ transplantation, paternity determination, identification of predisposition to various diseases and drug hypersensitivity reactions. It also serves as an excellent minigenome model for anthropological studies.

The human MHC consists of three distinct clusters of genes, a centromeric '*class II region*' which has HLA-DP, DQ and DR loci, a telomeric '*class I region*' that contains the classical HLA-A,B,C and a central 'non-HLA' or '*class III region*' which contains genes whose products are involved in the complement cascade (C2, C4 and properdin factor Bf) and various immunological/ non immunological functions (Figure 104.1).

Though only about one-thousandth of the total human genome, the system contains genes that have an important role to perform in biology and medicine. Major advances in the HLA system have taken place during the past two decades whereby the understanding has moved from that of two sets of genes whose products represent the major barrier to transplantation to a relatively greater understanding of the genetic complexity of the system and its applications in biology and medicine.

A. Historical Perspective

The history of MHC dates back to the early twentieth century when Karl Landsteiner in 1900 discovered the ABO blood groups, laying the basis for blood transfusions. Matching of these antigens had a major effect on the survival of transplanted tumors. Following this, an extensive search for antigens on white blood cells was undertaken. In 1937 Peter Gorer, a British pathologist, conducted experiments involving tissue transplantation in inbred strains of mice and first introduced the term 'tissue and transplantation antigens'. The fact that

rejection of allografts was caused by an immune response was first demonstrated in rabbits by the British biologist, Peter B Medawar in 1944 for which he was awarded the Nobel prize in 1960 shared with Frank McFarlane Burnet. Later, George Snell in 1948 coined the term 'Major Histocompatibility Complex' because the MHC products acted as strong transplantation antigens involved in graft rejection.

The first evidence for the existence of HLA in man was obtained in 1954 by Jean Dausset in France who observed that patients whose sera contained 'leucoagglutinins' had received a large number of blood transfusions than those lacking such antibodies. He termed the leucoagglutinin thus identified as 'MAC', but later became HLA-A2+A28. Another important landmark in the history of MHC is 1958 when Rose Payne in the USA and Jon vanRood in Holland showed simultaneously and in independent studies that pregnancy per se provided an effective stimulus for the induction of leukocyte isoantibodies in mothers against paternal antigens expressed by the fetus. Based on studies on sera obtained from pregnant women, John vonRood discovered the diallelic leucocyte antigen system 4a, 4b (now Bw4 and Bw6). Subsequently, Terasaki in 1964 introduced the lymphocytotoxicity test for serological testing of HLA antigens using a 60-well microtest tray. This test recognized as microlymphocytotoxicity test is valid even today and the trays are referred to as 'Terasaki trays'.

Later, Kissmeyer-Nielsen and coworkers in 1969 described crossing over between two loci thus establishing the existence of two independent loci, LA (now HLA-A) and 4 (later HLA–B). The third locus, AJ (now HLA-C) was discovered in 1970 by Sondberg and coworkers. In 1975, Thorsby and Piazza described an alternative approach of mixed lymphocyte culture (MLC) for studying the HLA complex. Using this, the lymphocyte defined (LD) HLA-D locus was gradually designated for differences in surface antigens leading to a strong mixed lymphocyte reaction. The first gene product detected by all antibodies that mapped to the HLA-D region was called HLA-D related or HLA-DR.

An important contribution was made by Benacerraf, McDevitt and others leading to basic understanding of the role of MHC gene products in immune regulation. They first reported in 1963 that the antibody response in guinea pigs to

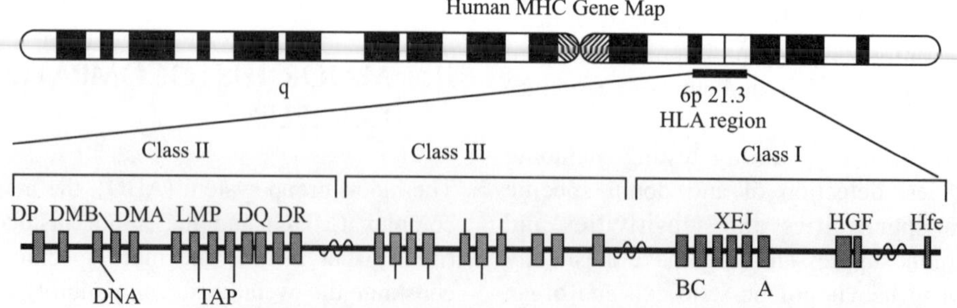

Figure 104.1   Human major histocompatibility complex (MHC) with chromosomal location and gene map showing multiple genes arranged as class I, III and II groups on the short arm of chromosome 6 (6p21.3). The class I region is located towards telomeric end, class II is located towards centromeric side while class III genes are interposed in between. For further details, please refer to the text. (*see Plate 42 for colour figure*)

a particular synthetic polypeptide antigen was controlled by a single gene. Genes controlling specific immune responses were later called as immune response genes. The pioneering efforts of George Snell, Jean Dausset and Baruj Benaceraff for discovery of the MHC and understanding of its biological importance were recognized when they shared the 1980 Nobel Prize in medicine and physiology. The concept that the MHC antigens were directly involved in T cell recognition of antigens was first shown in 1974 by Rolf Zinkernagel and Peter Doherty in Australia. This phenomena of T cell recognition of antigens was called MHC restriction and they received the Noble prize in 1996 for such elegant studies.

Over the years, sufficient evidence had been gathered by Dausset and vonRood and coworkers that the HLA antigens are strong histocompatibility antigens. This was further extended and confirmed by Ceppellini and Amos and their groups. They showed that skin grafts between HLA identical siblings had a longer survival than between siblings mismatched for 1 or more loci. The first data reporting correlation between HLA matching and kidney survival was presented by P Terasaki. Eventually, there have been several remarkable studies on the role of MHC matching in both organ and stem cell transplantation.

Another important discovery was the association of HLA antigens with disease that was first elucidated between HLA-B27 and ankylosing spondylitis by Brewerton et al in 1973. Since then, several diseases have been reported to be associated with HLA antigens and more recently also with drug hypersensitivities.

A lot of studies on the structure of MHC antigens were carried out in 1960s and 1970s. After this period, in 1980s, Strominger et al, Bjorkman et al and others used x ray crystallography and described detailed molecular structures of HLA class I and class II molecules.

The year 1964 is also considered a turning point in the history of the HLA system when productive international collaboration was initiated in the form of the first International Histocompatibility Workshop (IHWC) held at Durham. Since then sixteen IHWCs have been conducted. United efforts and long-term exchange of biological reagents, materials, techniques and ideas during such workshops have been largely responsible for many of the discoveries in the MHC and their applications in biology and medicine.

III. GENOMIC ORGANIZATION OF HLA MOLECULES

Both class I and II gene products show important differences in structure, function and tissue distribution. Matching donors and recipients for HLA-A,B and C loci of class I and of HLA-DR/DQ loci in the class II region during organ and bone marrow transplantation is critical to the graft outcome.

A. HLA Class I Genes

HLA class I gene products are expressed on all nucleated cells. Of the 36 loci that have been described so far, HLA-A, B and C are the most important. As per the November 2012 records of IMGT/HLA statistics database (http://www.ebi.ac.uk/ipd/imgt/hla/stats.html), 2132 HLA-A, 2798 HLA-B and 1672 HLA-C alleles have been identified. These molecules are heterodimeric glycoproteins consisting of an MHC encoded alpha or heavy chain (44 kDa) and non MHC encoded light chain (beta-2 microglobulin b2m; 12Da) (Figure 104.2).

The extracellular portion of the heavy chain is folded into three globular domains (α1, α2, α3), each of which are ~90 amino acids long and encoded by separate exons (numbers 2 to 4). The transmembrane region consists of ~23-25 amino acids and is encoded by exon 5 while the cytoplasmic domain is ~30 aa long and transcribed from exon 6. The

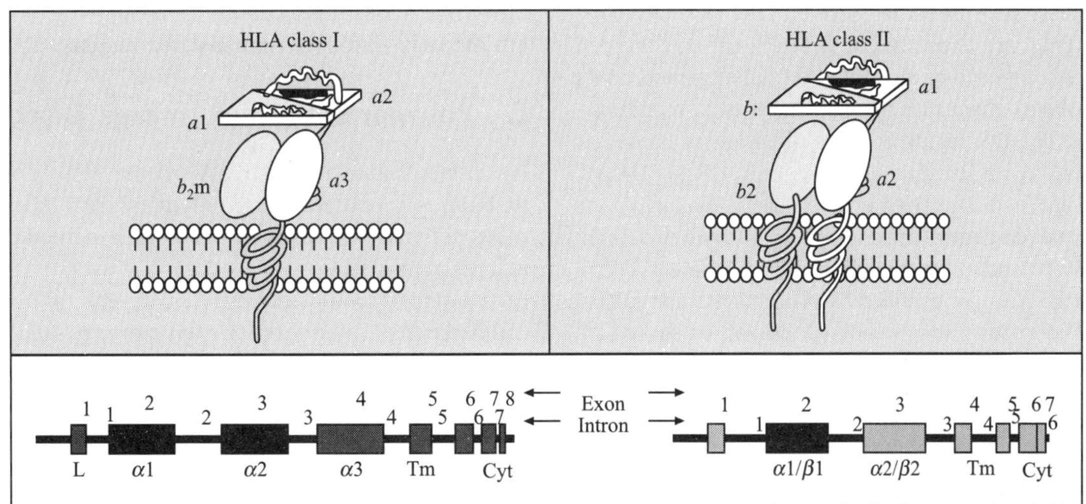

Figure 104.2   Schematic view of HLA class II and class I molecular structures showing the peptide binding cleft formed between a1 and b1 domains in case of class II and a1 and a2 in case of class I molecules. The membrane proximal domains (a2, b2 of class II and b2m and a3 of class I) are conserved and non-polymorphic. The molecular gene structure showing exons and introns is shown in the lower portion of the figure. (*see Plate 42 for colour figure*)

alpha chain associates non covalently with the β2-m. Cells deficient in β2m also lack HLA class I expression. X-ray crystallography of class I molecules has revealed that the region distal from the membrane is formed by α1 and α2 domains which take part in peptide binding. Functionally, MHC class I molecules bind to CD8 co-receptor molecules on the T cells, and therefore, present non self antigens to CD8+ T cells. MHC class II molecules, on the other hand, bind to CD4 co-receptor molecules on T cells, and hence, they present antigens to CD4+ T-cells.

The nonclassical class I genes include HLA-E, F and G as well as pseudogenes HLA-H,J,K,L that are not expressed. The nonclassical class I genes have relatively limited polymorphism, differential tissue distribution and have evolved specialized functions in both innate and adaptive immunity. The HLA-E binds nonamer signal peptides derived from leader sequences of classical class I and HLA-G molecules. The complex formed between HLA-E and signal peptide is known to bind to the inhibitory CD94/ NKG2 heterodimer found on NK cells. It therefore represents as an indicator of classical class I expression and ensures cell integrity against NK cell killing. HLA-G exists in various isoforms and is expressed differentially at immunepriviledged sites like the cornea, thymus and trophoblast. It plays a significant role in pregnancy. It is analogous to HLA-E as it binds leader sequence derived signal peptides and binds with leukocyte Ig-like inhibitory receptors, LIR1 and LIR2. Another important nonclassical class I gene is the HFE gene whose mutations have been strongly associated with hemochromatosis. Other than classical and nonclassical class I molecules there are MHC class I chain related molecules (MIC) that are however, located in class III region of the MHC.

B.   HLA Class II Genes

The HLA class II region of the human MHC extends over 1000 kb and has atleast six subregions, termed DR, DQ, DO, DN, DM and DP. They are expressed as heterodimers on the cell surface and are composed of an 34kDa α (alpha) gene product and 29kDa β (beta) chain that traverse the plasma membrane (Figure 104.2). The DR region consists of ten genes named as DRA and DRB1 to DRB9. Among these, only DRB1, DRB3, DRB4, DRB5 and invariant DRA encode for proteins while the remaining are pseudo genes. The conventional serology defined DR1 to DR18 are coded by the DRB1 locus, whereas DR52 and DR53 specificities are encoded by HLA-DRB3 and DRB4 respectively. The exons 2 and 3 code for α1, α2 or β1 and β2 domains of class II molecules respectively. The DQ subregion contains 5 genes, DQA1, DQA2, DQB1, DQB2 and DQB3, of which DQA2, DQB2 and DQB3 are not expressed. The DQA1 and DQB1 molecules can be expressed in cis or trans configuration and exist as heterodimers. The DP subregion is composed of two a and two b genes, where DPA2 and DPB2 are pseudgenes and not expressed. In addition to above, there are important genes in the HLA class II region, for example, ABC transporter genes (TAP1 and TAP2) that are associated with antigen processing or transport of antigen peptide. Others include Tapasin, PSMB8 (proteasome subunit beta type 8 also called LMP7) and PSMB9 (or LMP2), concerned with antigen processing machinery.

The HLA class II tissue distribution is generally restricted to specific cells of the immune system involved in antigen presentation (APCs), namely B lymphocytes, macrophages, monocytes, epithelial cells, dendritic cells, Langerhan's cells and Kupffer cells. Like the DR, DQ molecules also present peptides to CD4 T cells. Currently, 1638 DRB1, 734 DQB1 and 498 DPB1 alleles have been identified using molecular technologies.

C.   Class III Genes (Central Genes)

The MHC central region is most gene rich and interposed in between class I and class II. Atleast 39 functional genes have been described in a 1000 kb stretch of DNA within the class III region and include the complement genes C2, C4 and Bf, the MICA, MICB, TNF-b and others with either immunological or non immunological functions. A tandem repeat of CYP21 and complement 4 genes also coexists in this region and their copy number variations have been studied in relation to autoimmune and other diseases.

D.   Nomenclature of HLA

The HLA nomenclature was recently revised in April 2010 (http://hla.alleles.org/). As per this new nomenclature, a colon (:) has been introduced into the allele names and acts as a delimiter of separate fields. Thus, the allele HLA-A*02010101 is designated as HLA-A*02:01:01:01, where the first 2 digits following the star sign represent the HLA allele family, which often corresponds to broad serological HLA antigen (in this case HLA-A2). The 3rd and 4th digits correspond to the order in which the sequences were determined, 5th and 6th digits the synonymous substitutions in coding region and 7th and 8th digits in introns or untranslated regions.

E.   Polymorphism and Linkage Disequilibrium

A unique feature of the HLA system closely related to its function is its extraordinary polymorphism. Indeed this is the most polymorphic segment of human genome. Since each individual inherits two alleles of each of the HLA loci (one from each parent), theoretically there are several million phenotypic combinations (150 billion or even more) possible and reflects nature's gift for survival to human species. However, in transplantation context, this represents a major barrier in the pursuit of finding HLA matched donors.

An anthropological analyses of the MHC in different populations has revealed that the degree of polymorphism in the HLA system in a particular population may not actually be as extensive as indicated theoretically. This is because certain alleles are associated with each other or inherited more frequently than would be expected on the basis of their individual gene frequencies. This non random association of allele of multiple HLA loci found together on the same HLA haplotype is termed 'linkage disequilibrium' (LD). Multiple factors or mechanisms are involved in the establishment of LD that interplay during evolution and include chance breeding, natural selection, microbial pressure, genetic drift and others.

A complete set of alleles on the same chromosome is usually inherited (a *haplotype*) and the two HLA haplotypes in an individual derived after family testing constitute the *genotype*; the total HLA antigen profile is the *phenotype*. Siblings in a family have a 25% chance of being either totally HLA-identical or totally unidentical and 50% chance of being haploidentical (sharing one parental haplotype only) (Figure 104.3).

Most people are heterozygous for each of the HLA loci due to extensive polymorphism known for each of them. However, sometimes both parents may have one and the same allele combination on one of their HLA haplotypes. This is more commonly observed in ethnicities that practice consanguinity. Biologically, a heterozygous combination is more advantageous to an individual as it offers a capacity to present a wider range of pathogens.

F. Laboratory Testing of HLA

Histocompatibility testing has undergone tremendous evolution over the years. From earlier days of comparatively crude serology, HLA has come a long way to the present sophistication of cellular, biochemical and molecular procedures.

The earlier commonly used serological tissue typing techniques have almost phased out in the recent past and have been swiped by molecular technologies since these can identify HLA alleles with better resolution and precision (Figure 104. 4). Molecular matching of HLA is therefore a prioritized approach for selection of donor and recipient pairs. Using these procedures, an appreciable number of novel alleles and unique haplotypes have been discovered in the Indian and other populations.

HLA studies at the DNA level started in 1986 initially using the Southern blot technique using specific restriction enzymes and full length cDNA probes for class II genes. The restriction fragment length polymorphisms (RFLPs) thus obtained were found to be helpful in defining further polymorphisms in the HLA-DR region genes. With the introduction of PCR by K Mulllis in 1985, it became possible to amplify specific HLA gene sequences and further study their polymorphisms. In the early 1990's, use of sequence specific oligonucleotide probes (SSOPs) and sequence specific primers (SSP) for specific HLA class I and class II alleles allowed more direct analysis of relevant gene sequences encoding the polymorphic determinants.

Serological methods

The conventional method for testing HLA antigens used to be the serology based Microlymphocytotoxicity technique first developed by Terasaki in 1964 (Terasaki and McClleland 1964). The technique is based on antigen-antibody reaction on the immune cell surface, followed by complement-mediated lysis. The discrimination of HLA specificities in this technique

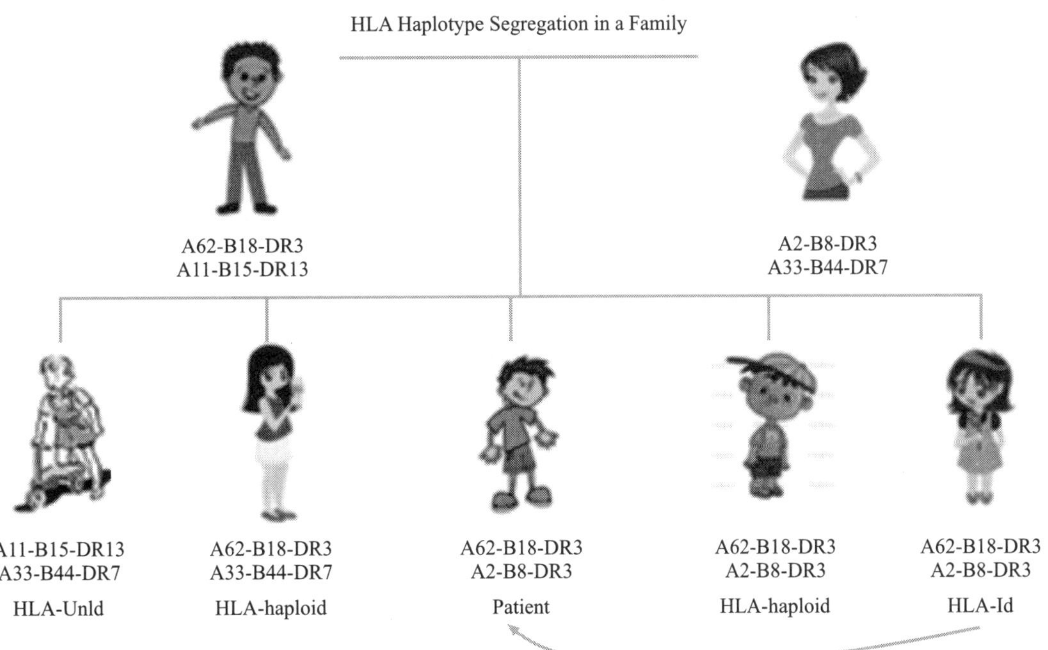

Figure 104.3   Segregation of HLA haplotypes in a family. The sibs inherit one haplotype from each parent and accordingly four different haplotypic combinations are possible. In this case, the patient is HLA identical with his eldest sister; HLA unidentical with his youngest sister and shares only one haplotype (haplo-identical) with his younger sister and elder brother.

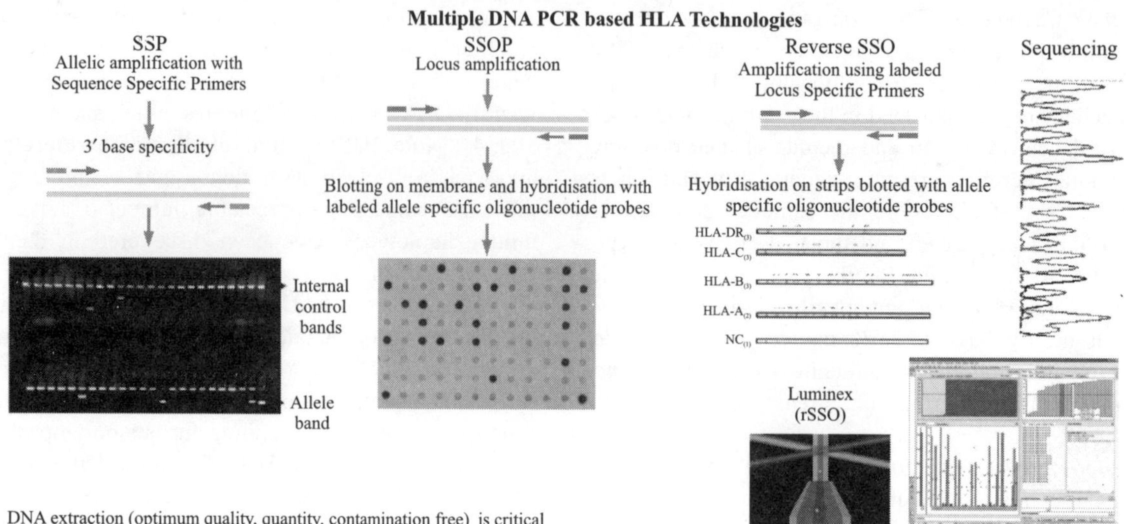

Figure 104.4 Multiple molecular technologies that are used for HLA typing. The PCR-SSP technique involves amplification of HLA alleles using allele specific primers while PCR-SSOP is similar to dot blotting and allele specific probes are used for hybridization and detected. Reverse SSO, as the name suggests, utilizes DNA sample in solution and the allele specific probes are immobilized in reverse. Sequencing and luminex are relatively more recent techniques and can read alleles at high resolution. Most of the histocompatibility testing laboratories worldwide use more than one technique(s) for selection of donors for transplantation.

is based on the recognition of different conformational structures of each molecule expressed on the cell surface and therefore requires a large number of monospecific or multispecific antisera for clear delineation of absence or presence of particular antigens. Sources for such antibodies are limited, except for monoclonal antibodies and are not generally renewable. Moreover several antibodies are generally required to assign each antigenic specificity due to the complex pattern of serological cross reactivity between individual HLA determinants. Often two antigens with the same amino acids at positions critical for protein folding tend to have similar three-dimensional structure and thus would be bound by the same antibody, a phenomenon called 'crossreactivity'. Alleles with different nucleotide sequences but with same conformational structures of the expressed protein cannot be distinguished by serology. Therefore identification of new alleles is difficult by serological techniques unless a very large number of mono and multispecific sera are used to different epitopes of the three dimensional antigenic structure.

DNA based technologies

DNA based methodologies for HLA typing were first employed during the 10th IHW held in New York during the end of 1987. Since then these technologies have been further refined as more suitable and reliable methods of HLA typing and routine clinical testing. In general, these methods are based on discrimination of specific nucleotide sequence differences between any two alleles and allow better resolution among them. In contrast to the rapid development of HLA class II typing strategies, DNA typing for HLA class I alleles has been slow due to the inherent complexity of polymorphism in HLA class I loci, multitude of polymorphic HLA class I alleles

with at least six genes and 12 pseudogenes within the region and high degree of homology and sharing of sequence motifs between loci. A variety of PCR based techniques are currently available depending on resolution of the test required and the number of samples to be tested. As a result of application of DNA based techniques, a large number of new HLA alleles have been discovered and characterized.

PCR-SSOP: Polymerase Chain reaction based-Sequence Specific Oligonucleotide Probes. The technique is well established as a low to high resolution typing tool for HLA class II antigens according to the sequence and number of probes used for testing and is analogous to dot blotting. The PCR-SSOP technique is specially suited to testing of large number of samples at the same time. The technique, however, poses problems in generation of ideal, monospecific probe sequences for identification of single alleles at high resolution, standardization of temperatures for hybridization and critical washing of probes and false negative/positive reactions. Another recent advancement of this technique is the use of reverse SSO and is described below.

Reverse SSO: This testing template is the forerunner to the DNA chips wherein a set of oligonucleotides probes (to specific polymorphic motifs or regions of variation in MHC) are immobilized on nylon membranes. The membranes are then hybridized with the amplified target DNA and detected for positive hybridization. The technique is suitable for routine testing of small sample numbers, especially as in forensic cases.

PCR-SSP: Polymerase Chain reaction based-Sequence Specific Primers is a reliable technique for both class I and class II alleles and analogous to ARMS-PCR. Complementary oligonucleotides flanking polymorphic sequences in different regions of the target DNA are used as a set of primer pairs to amplify regions of differences between two alleles. Only allele specific complementary oligonucleotides would prime the chain reaction, which is then visualized as an amplified band on an agarose gel. Relatively intermediate to high resolution typing can be carried out using this technology. The SSP technique is more suitable for small number of samples.

Luminex typing: This is a reverse SSO based technique where probes are bound to florescent/ magnetic beads and are detected using a sensitive flow cytometry based luminex system. It is relatively intermediate to high resolution technique and can type several samples simultaneously in a short period of time. It has become a common and more acceptable technology during the recent years.

SBT or Sequence based typing has gained immense importance during the last few years as the ultimate technique for high resolution typing. It is based on the identification of complete exonic sequence of the allele in question and thus allows no discrepancies. A PCR reaction is performed that amplifies both alleles in a heterozygous individual. The two alleles are then sequenced and positions of heterozygosity are analyzed using computer programs that compare patterns obtained with combination of all known alleles and assign possible HLA types. The technique is highly suitable for Class II typing but class I alleles due to their extensive polymorphism place great demands on the quality of the sequencing analysis. High degree of sequence homology among different HLA class I loci including the less polymorphic HLA-E, F, G, genes and several pseudogenes cause multiple problems in investigation of a single specific locus. Considering the increasing number of newly detected alleles sequencing is fast becoming the most reasonable and suitable typing approach for both HLA class I and II alleles.

The more recent advanced technologies of reverse SSOs and sequencing, real time PCRs, luminex platform, mass spectroscopy and microarrays have now opened up a whole new dimension of studying polymorphisms at single nucleotide level and resolved several ambiguities. Because of the availability of such superior technologies, the thrust has now switched from antigen matching to epitope matching to nucleotide matching. This has enhanced our ability to identify and interpret differences at the high resolution level and to match the recipient-patient pairs leading to high grade histocompatibility.

IV. BIOLOGICAL SIGNIFICANCE OF MHC

Although the HLA system was originally discovered because of its prime need in organ transplantation, its overriding influence in governing immune responses has revealed that it plays an important role in susceptibility to diseases. Hence, HLA is an excellent clinical aid in the confirmatory diagnosis of several diseases (e.g spondyloarthropathies, celiac disease, narcolepsy etc), estimation of 'risk' for disease development in families and possible prognosis of certain diseases.

A. Peptide-MHC Interactions

The initiation of cellular and humoral immune responses requires successful interaction between T lymphocytes and antigen presenting cells. The major biological function of the MHC is to bind and present peptides to T cells in a MHC restricted manner. While class I MHC molecules are important for the presentation of foreign peptides to cytotoxic T lymphocytes (CTLs), the helper T cells recognize antigens only in association with class II molecules. MHC class I and II molecules also differ in terms of their antigen processing pathways, peptide binding and is discussed briefly in the following section.

The peptide binding anchor residues in class I molecules are placed at the terminal ends of the peptide binding groove while in class II molecules, the specificity of the interaction is determined by evenly distributed binding forces. The MHC class I molecules bind peptides of 9 or 10 amino acids but class II can accommodate peptides that are much longer, average of 14–18 amino acids. Unlike class I, the class II groove is open at both ends with the flanking portions of the peptide protruding out of the groove's ends. The MHC class I cleft is has rather closed ends and the peptide binding pockets particularly B and F are deeply buried inside the cleft. In relation to the MHC class II motifs, relative positions of five peptide binding pockets designated P1, P4, P6, P7 and P9 have been defined, each of which accommodates specific peptide side chains.

The peptide processing and loading pathway in class I involves LMP proteasome mediated digestion of endogenous proteins into 8-10 amino acid long peptides. The peptides are transported by TAP molecules into the lumen of endoplasmic reticulum (ER) where these bind class I molecules, induce a conformational change and facilitate their export to cell surface. In case of class II pathway, endocytosed exogenous antigens are digested by proteolytic enzymes in endosomes into small peptide fragments. These peptides subsequently bind to class II molecules that have undergone Invariant chain cleavage thus opening up cleft for peptide occupancy in the post Golgi compartment in the cytoplasm.

Cell surface expression of MHC is important for antigen presentation to T cells but also important for NK cell functions. Among the non immune functions of the MHC, the noteworthy ones are its interactions with other receptors on cell surface, in particular with transferrin receptor, epidermal growth factor and various hormone receptors and signal transduction.

B. Role of HLA Matching in Transplantation

HLA is one of the main etiological determinants of tissue compatibility. Several studies have illustrated that more precise HLA matching between donor and patient significantly facilitates engraftment, improves overall transplant survival and reduces the incidence of transplant related mortality. Recent advances in HLA typing have undergone tremendous revolution and led to better grade of matching between donors and patients. Their wide-spread use along with better understanding of HLA immunobiology has revolutionized the consequences of transplantation and improved transplant outcomes substantially. Data generated by most transplant centers have indicated an HLA matching effect with excellent graft outcome based on the degree of match grade, while poorly matched recipients are more likely to raise antibodies against the graft. The next section of this chapter provides details on immunological significance of MHC system in the context of organ transplantation followed by hematopoietic stem cell transplantation.

Allorecognition

Allorecognition refers to T cell recognition of genetically encoded polymorphism between members of the same species. The recipient's CD4 T cells can recognize tissue alloantigens by direct, indirect, and semidirect pathways of allorecognition. Direct allorecognition is the interaction of recipient T cells through the T-cell receptor (TCR) with intact allogeneic major histocompatibility complex (MHC)-peptide complexes presented by donor-derived APCs, including DCs. Indirect allorecognition occurs when peptides derived from donor MHC or minor histocompatibility (miH) antigen are degraded by antigen processing pathways and presented by recipient APCs. Semi-direct allorecognition is the capture of donor MHC peptide complexes by recipient APCs (Figure 104.5).

The direct pathway is of particular relevance during acute rejection. Donor derived APCs expressing donor alloantigens rapidly migrate from the graft and enter the secondary lymphoid tissues, where they can encounter and prime the allospecific T cells. Cytotoxic T cells, B cells macrophages and NK cells are recruited in a graft undergoing rejection. The main mechanisms implicated for graft destruction are cell mediated cytotoxicity, delayed type hypersensitivity and antibody dependent cellular cytotoxicity.

Transplants carried out between HLA mismatched donor-recipient pairs have a greater chance of rejection due to development of donor specific antibodies as compared to those between better matched grafts. Data collated at the CTS registry has indicated that over a period of time, the impact of HLA matching becomes distinctively evident, due to deteriorating graft survival rates in poorly matched grafts. The data provide evidence to suggest that better the match grade between the recipient and the donor the lesser are the chances of antibody development and subsequent rejection.

Antibody mediated graft rejection

Efforts are always made to ensure that patient and donor are matched for HLA gene loci as completely as possible. However, this is difficult to achieve and immunosuppressive drugs are given to control the host immunity caused by HLA disparity. The aim is to prevent or control rejection and allow the recipient to develop long term acceptance (or tolerance) of the graft. Rejection can be mediated by antibodies, lymphocytes or both and can manifest as hyperacute (occurs in the early post-transplant period, within hours), acute (occurs at any

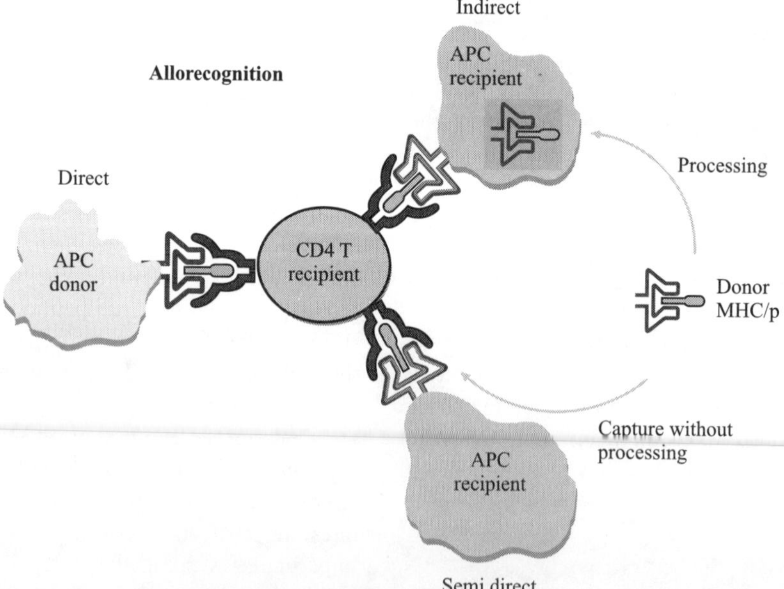

Figure 104.5 The pathways of allorecognition that cause problems in organ transplantation as a result of HLA mismatching. In direct pathway, the alloreactive responses of recipient CD4 T cells to donor APCs expressing incompatible antigens. In indirect pathway, allogeneic HLA antigens are taken up and processed by the recipient APCs and presented in the context of autologous (self) HLA molecules to recipient T cells. APCs trigger CD4 and CD8 T cells, and as a result, local and systemic immune response develops. Cytokine recruitment and activation of specific T cells, NK cells and macrophage mediated mechanisms lead to cytotoxicity and finally the allograft destruction. In semi direct allorecognition, the donor MHC peptide complex is captured and displayed to recipient T cells.

time) and chronic rejection (slowly developing process with progressive decline in graft function).

The past decade has witnessed three important developments that have influenced our understanding of the antibody mediated rejection (AMR). These are i) characterization of detrimental effect of donor specific anti-HLA antibodies on allograft survival and their association with AMR, ii) advent of newer and more sensitive techniques for detection of donor specific antibodies (DSA) and iii) revision in Banff criteria for assessment of graft pathology. Several studies have reported that the pre and the post transplant development of IgG donor directed antibodies predict the risk of both acute and chronic allograft rejection.

Hyperacute rejection is caused by preformed donor specific alloantibodies in a presensitised recipient. Humoral presensitization may be caused by a previous transplant, blood transfusions or pregnancy. The alloantibody mediated rejection is initiated by activation of the complement cascade leading to release of various inflammatory mediators and the initiation of the coagulation and fibrinolytic systems. Hyperacute rejection is manifested by rapid vascular constriction, oedema and thrombotic occlusion.

The injury in acute rejection (AR) is caused by T cells (T cell mediated rejection) and antibodies (humoral rejection), either alone or together. It typically appears during the first 1-6 weeks after transplantation and declines sharply after the first 6 months. The allograft vascular endothelium is the primary target of the initial stage of cellular rejection and lymphocytes exert a direct cytotoxic effect on graft parenchymal cells, and mediate a delayed type hypersensitivity (DTH) response.

Chronic rejection, affects most long term transplant survivors and is characterized by vasculopathy, fibrosis and a progressive loss of the organ function. Chronic rejection is mediated by low grade, persistent DTH response and activated macrophages secrete mesenchymal cell growth factors leading to fibrosis. Persistent viral infections also induce cellular immune responses. The donor specific alloreactive T cells and the chronic ischemia secondary to injury of the blood vessels by antibody or cell mediated mechanisms also contribute. Vascular occlusion may occur as a result of smooth muscle cell proliferation in the intima of the arterial walls.

Alloantibodies preferentially attack peritubular and glomerular capillaries of the transplanted kidney. The target antigens in AMR are most often situated on the endothelium resulting in the histopathological findings of acute (glomerulitis, peritubular capillaritis) and chronic (transplant glomerulopathy) vascular injury. Endothelial damage also results in platelet activation and microthrombi formation. Of critical importance is the C4d deposition which is regarded as an important surrogate marker for the diagnosis of hyperacute and acute humoral rejection. Recent reports have provided a convincing proof that C4d deposition is strongly associated with the development of HLA class I and II antibodies directed against the donor antigens. Presently this is the most widely accepted marker of complement fixing circulating antibodies to the endothelium.

Antibody detection techniques

The two main goals of organ transplantation are i) increasing access to transplantation for better quality of life and ii) improving graft outcome so that the grafted organ has the ability to stay for significantly longer duration of time in the new host. The most important factor in organ transplantation is the meticulous evaluation of the possible occurrence of antidonor antibodies before transplantation. Since antibodies cause graft rejection, technologies for their detection must be sufficiently sensitive to predict hyperacute or humoral rejection and adequately specific to determine immunological failure. An optimal combination of the two elements is necessary to determine the predictive value of individual techniques. Recently, methods for antibody detection have evolved remarkably from the conventional cell based assays to the more advanced solid phase systems. Currently, anti-HLA antibodies can be detected by a number of techniques that include i) target donor cell-based crossmatch assays, such as complement dependent cytotoxicity (CDC) or flow cytometry cross-match (FCXM) ii) HLA protein-based (solid phase) assays, such as an enzyme linked immunoabsorbent assay (ELISA) or HLA antigen-coated fluorescence bead assay systems. Serum treatment with dithiothreitol (DTT) is used to distinguish clinically less relevant IgM type antibodies.

Complement-Dependent Cytotoxicity crossmatch (CDCXM) Assay: Introduced in 1960s, the CDC crossmatch assay still remains an essential requirement before transplantation. In this assay, target lymphocytes are killed by antibody-activated complement when the recipient has pre-existing antibodies or those developed de novo following transplantation. The main advantage of this assay is that it specifically detects antibodies that are capable of activating the complement. It is the only assay with functional readout which may best mimic the in vivo reality. The greatest drawback of the CDC assay is, however, that it often misses low titer antibodies resulting in false negatives or may give rise to false positive results due to detection of non-HLA antibodies and autoantibodies or high background reading due to spontaneous cell death (especially B cells). In addition, it is relatively more time-consuming, requires the isolation of T and/or B lymphocyte subpopulations from donor peripheral blood, and detects only complement fixing antibodies (IgG and IgM). Further, the test results are dependent largely on the level of cell surface antigen expression and the assay is less sensitive in detecting antidonor antibodies than the flow cytometry based crossmatch.

Flow Cytometry Crossmatch (FCXM):   Flow Crossmatch technique came into being primarily to address some of the problems inherent with the CDC assay. Since the test could detect both complement fixing (IgG and IgM) as well as non fixing (IgA etc) antibodies, it readily gained greater acceptance. It is a less subjective and quantitative method with a 10 to 250 fold greater sensitivity than the CDC test and is able to detect both low titre circulating alloantibodies as well as those directed against HLA Class I (Tcell FCXM) and/or Class II (B-cell FCXM) determinants. Further, there is no requirement of physically isolating lymphocyte subpopulations. In contrast to the traditional CDC test, FCXM is not a functional test because it provides information only on the extent of binding of the donor specific antibody to its potential donor target (expressed as channel shift) and not killing of the target. Accordingly, a positive FCXM (channel shift >50) does not necessarily mean that the bound antibodies have any pathological effect on the target cells. Antibody detection by flowcytometry bears no relation to the ability of that antibody to activate complement and this is the major disadvantage of this technique.

Purified HLA Antigen-Based Enzyme-Linked Immuno-absorbent Assay (HLA-Ag- ELISA):   An ELISA based HLA antibody detection method was commercially developed in the mid 1990s as a more sensitive and less time consuming technique for detecting primarily anti HLA antibodies. A panel of mixture containing class I and/or Class II HLA antigens, purified from culture supernatants or cell lysates from Epstein–Barr virus transformed lymphoblastoid cell lines are coated on to the ELISA plates. After incubation of the coated plates with the test serum, the bound antibodies are detected using a peroxidase conjugated antihuman immunoglobulin antibody. Absorbance is read using an ELISA reader and the assay results analyzed using software. More recently, single HLA Class I and Class II antigens, instead of purified antigen mixture have been tried as coating antigens. This purified HLA antigen-based ELISA test is highly sensitive, comparable to that of flowcytometry or the recently adopted Luminex based assay system. It has been argued that the high sensitivity of this assay might be due to the greater amount of HLA antigens used to coat the ELISA plate. Whether this is acceptable as a clear indicator of the amount of antibody detected is questionable because the test utilizes much higher quantity of HLA antigens as compared to the normal biologic expression on the immune cell.

Luminex Technology: This latest luminex acquired technology has completely transformed the histocompatibility laboratory's approach and ability to detect HLA antibodies. Current data indicate that the Luminex platform is one of the most sensitive of the solid phase antibody detection techniques that enables identification of antibody specificities in highly reactive sera. The technique has been successfully applied to monitor patient antibody profiles in relation to particular specificities such as following antibody removal strategies at pre-transplant stage and for identification of donor specific antibodies posttransplantation. The biggest advantage of this technique is the use of single antigen bead system that permits detection of epitope specific antibodies to individual and specific HLA alleles and thus provides information on the virtual cross-match. The Mean Fluorescence Intensity (MFI) of the beads is a measure of the amount and strength of the antibody present. Most centres concur that peak MFIs >6000 is associated with poor graft outcomes, although, controversy exists on the optimum MFI cut-offs for

classifying the antibody as positive or negative particularly if it is HLA-DSA. Most centres agree that HLA-DSA with MFI >2000 should be considered clinically significant. In our centre, MFI>1000 is not ignored because it indicates doubtful positivity of donor specific antibodies, leading to accelerated acute rejection in a proportion of patients. Others have also suggested that the cut-off value of 1000 for DSA-MFI provides a more valuable result in the context of clinical correlation. Further, peak MFI in the recipient serum is considered a better indicator of graft survival rather than the MFI in the current sera, particularly if it is HLA-DSA. Being highly sensitive, the luminex technology often provides false positive results leading to unnecessary delays on the waiting list. The biggest limitation of the technology is that it does not discriminate between complement fixing and noncomplement fixing antibodies. This is important in the light of the recent observation suggesting that non-complement fixing antibodies (IgG2 and IgG4) could be as deleterious to the graft as the complement fixing ones (IgG1 and IgG3). Further the current panel of single HLA flow beads do not cover the whole diversity of HLA alleles.

Hence populations like the Asian Indian that are characterised by a large number of novel alleles and unique HLA haplotypes will be at a disadvantage because antibodies raised against such HLA alleles are likely to be missed. Another limitation of this technology relates to the question that not all HLA antibodies detected by single flow beads are in fact clinically relevant. The assay may also provide false positive results and the detected HLA antibodies may only be of limited clinical relevance even in the presence of a true positive result. Luminex beads accommodate both intact as well as denatured HLA molecules. The 'natural' HLA antibodies present in non-immunised individuals react preferentially with these denatured molecules. A recent study has shown that antibodies reactive to these denatured HLA molecules are clinically irrelevant. Further, binding of the solubilized class I molecules on beads may alter the tertiary structure of the antigen leading to false negative and/or positive reactions when human monoclonal antibodies directed against class I molecules are used as detecting reagents. Similar reactions may occur with sera from potential transplant recipients.

C. Pre And Post Transplant Antibody Screening

A meticulous *'crossmatch test'* for possible occurrence of anti donor antibodies is always performed before any transplantation. This is similar to the direct cross-match done prior to blood transfusions; donor lymphocytes instead of red cells are tested with patient serum.

Anti HLA antibodies: Pre transplantation level

There is a universal agreement that pre-transplant HLA class I and/or class II antibodies alone or in combination are associated with poor allograft outcome following renal transplantation. Detection of these antibodies has become more precise and specific with the use of Luminex single antigen bead assay system. However the exact clinical significance of antibodies defined by this assay in the absence of a positive CDC cross-match needs further evaluation. Based on the current experience, an assessment of the gradient of risk prediction can be of immense value in helping the transplant team to plan a pre-transplant work up. Studies have shown that patients who develop both anti HLA class I and II antibodies experienced more frequent rejection episodes than those having either of these antibodies. The latter study has also highlighted the importance of A, B and DR mismatches in the presence of HLA class I and II antibodies. It has also been reported that the presence of class I and class II IgG DSA as detected by single antigen bead in the pre-transplant sera of renal recipients is indicative of an increased risk for graft failure. Donor-specific HLA-DQ and DP antibodies are detectable only by the Luminex and flow cytometry assays and they generally represent a low to intermediate level of immunological risk. However, if such antibodies give rise to an IgG +ve B-cell CDC cross-match test, the level of immunological risk rises to intermediate or even high. HLA-DP specific alloantibodies have traditionally been thought to be of minimal significance because of the relatively low expression of DP antigens on renal endothelial cells. However recent studies have suggested that donor specific HLA-DP antibodies can mediate both acute and chronic allograft rejection just the same way as the anti DR antibodies can. A recent study has demonstrated the influence of pretransplant donor specific HLA-C and DP antibodies on the development of acute humoral rejection episodes and subsequently leading to graft failure in most of the cases. The study also revealed that the presence of anti-DP antibodies exerted more deleterious effect on graft outcome as compared to anti-C antibodies. Since presence of antibodies is a major deterrent for the selection of a potential donor, an assessment of risk for graft rejection could provide a guide to donor selection. It is important to know the level of antibodies detected by more sensitive assays in the patient serum as the low levels of pretransplant donor specific antibodies are frequently clinically irrelevant. Similarly, detection by highly sensitive methods should be critically analyzed before taking a clinical decision.

Anti HLA and donor specific antibodies (DSA): de novo development

HLA antibodies that develop de novo post transplantation and not necessarily donor specific, are associated with poor graft survival. It is, therefore, advisable to monitor routinely the post transplant development of such antibodies, as a predictive marker for allograft function. The impact of de novo development of class I and/or class II anti HLA antibodies on acute or chronic allograft rejection is a subject of intense investigation. The development of such antibodies varies between and among individual patients. There is lack of consensus on the time of onset of these antibodies, their intensity, duration and associated graft dysfunction. It has been shown that de novo HLA class II DSA have greater influence on the development of antibody mediated injury and allograft failure than those against HLA class I alleles. Alloantibodies formed within the first year of transplantation are much more damaging than those formed after completion of one year of transplantation. Recently, the role of anti DQ donor specific antibodies has been highlighted with inferior renal allograft outcome. This is an important observation since HLA-DQ expression is known to be upregulated on endothelial cells following renal transplantation and may thus be an efficient target for HLA-DQ antibodies. These results have highlighted the clinical importance of HLA-DQ matching (in addition to HLA-DR) in renal transplantation in defining the risk associated with the development of DQ donor specific antibodies. It has been demonstrated that de novo development of non donor specific HLA antibodies are also associated with poor graft survival. The presence of circulating anti HLA antibodies along with positive C4d stating on allograft biopsies presents a strong correlation with the development of acute and chronic rejection in kidney transplant recipient. Hence, the very strong association between de novo production of donor specific HLA antibodies and allograft failure highlights the importance of monitoring for post-transplant HLA antibodies as a predictive biomarker for allograft function. The clinical significance of IgM type anti-HLA antibodies in relation to allograft rejection is not clear. Studies using CDC and Flowcytometry methods have suggested that IgM antibodies are generally not detrimental to the graft. Others have however shown that a positive T cell IgM flowcytometry crossmatch at the time of transplant could be a risk factor for allograft rejection.

C1q fixing antibodies

C1q is the first component of classical complement pathway. Activation of the pathway begins with binding of the antibody with C1q. Once activated, the classical pathway leads to the formation of membrane attack complex (MAC) resulting in cell lysis and cell death. It is now recognized that all DSAs do not lead to the development of AMR. In this respect, recent data have provided evidence to suggest that only those DSAs that are able to activate C1q are more likely to be associated with a higher risk of AMR as compared to the C1q negative group. In fact, only 47% of HLA-Ab specificities identified by Luminex–IgG could bind C1q, suggesting that about half of the IgG antibodies may not have an immediate or even late adverse effect on the graft. In this complex and ever evolving field, one can conclude that complement mediated injury initiated by DSAs is responsible for most early AMR episodes. These studies have opened up novel therapeutic approaches towards preventing AMR, by reducing deleterious donor specific antibodies through inhibition of complement activation.

Non-HLA antibodies

In recent years, the focus has also shifted to the role of antibodies other than those against HLA molecules in renal transplantation. Such "non-HLA" antibodies may also be clinically important in the development of allograft rejection although the extent of their harmful effect is not yet fully clear.

Anti-endothelial cell antibodies (AECA): Endothelial cells are critical in that they provide the initial contact point between the recipient's immune system and the transplanted allograft. Over the years greater recognition of the role of AECAs in renal transplantation has provided the researchers with an additional parameter to find an answer to all those unexplained allograft losses. Several investigators have reported a significant correlation between the development of AECAs with hyperacute, accelerated and acute renal rejections even in situations involving HLA-identical sibling transplants. The potential antigenic targets of AECAs that have been implicated with renal allograft rejection include angiotensin II type 1 receptor, vimentin, agrin (glomerular basement protein) and MICA. The AECAs are mainly of IgG2 and IgG4 subclasses that do not activate complement and exert their action through endothelial cell activation, proliferation or apoptosis.

Major Histocompatibility Complex Class I Related Chain A (MICA): MICA is a highly polymorphic gene complex located within the MHC class I region of chromosome 6 and is known to have at least 100 alleles. It is expressed constitutively on endothelial cells, gut epithelial cells, skin-derived fibroblasts, monocytes and keratinocytes but not on immunocompetent cells. An association of MICA antibodies with poor graft outcome in renal transplant recipients has been recognized in several studies. Pretransplant MICA antibodies in kidney recipients were found to have decreased graft survival, even when well matched for HLA antigens and in the absence of HLA specific antibodies. We had earlier reported that presence of HLA-DSA and anti-donor MICA antibodies together in the recipient is a potentially disastrous combination with very poor graft outcomes even leading to hyperacute rejection in some cases. The exact mechanism by which individuals develop antibodies to MICA is not clear. Both MICA and MICB are highly polymorphic antigenic molecules present on endothelial cells and are therefore an easy target of humoral immunity associated with irreversible rejection of kidney allografts. It has been suggested that although pregnancy can induce such antibodies, the role of blood transfusions is debatable.

D. Selection of Optimal Donors for Organ Transplantation

Matching strategies for renal transplantation consider only HLA-A, B and DR loci. In renal and other solid organ transplantation, a direct relationship exists between graft survival and the level of matching. Two groups of donors, live and deceased (brain dead), have been used in kidney transplantation. More than 80% of transplants done in Europe, North America and Australia use deceased donors. In other countries, lack of facilities, cultural or religious considerations prohibit the use of cadaver donors and hence living related or unrelated donors are used. The deceased donors and recipients do not generally show HLA haplotype identity with each other.

Four types of live donors are available for renal transplantation. Family donors include parents, siblings and offsprings and their match status is determined by the parental haplotype inheritance and whether they share one, both or no HLA haplotypes (Table 104.1).

TABLE 104.1 A comparison of probability of HLA matching in different donors for organ transplantation

Donor category	Match probability	Match designation
Family donors		
Parents	50% match or one haplotype match	Haploidentical
Siblings	25% chance of full match,	HLA identical
	25% chance of no haplotype match,	HLA unidentical
	50% chance of one haplotype match	Haploidentical
Offsprings	50% match or one haplotype match	Haploidentical
Deceased	Generally poor	Variable number of mismatches
Spousal	Generally poor	Variable number of mismatches
Unrelated	Generally poor	Variable number of mismatches

Some HLA mismatches are more immunopotent than others. It has been reported that matching for HLA antigen splits or the specific epitopes leads to superior graft survival.

Most grafts from unrelated/ deceased donors are mismatched at several loci. The 20 year graft survival of 60% seen in grafts obtained from zero mismatch HLA identical sibling donor is significantly reduced to 41% for the HLA haplo-identical related grafts and 33 % for those involving poorly matched deceased donors as is seen in the long-term renal graft survival compiled by the CTS registry (Figure 104.6). The 'half-life' (period at which at least half the grafts are still functioning well) is 28.7 years for well matched HLA-identical sibling grafts, 16.4 years for haploidentical donors and only 13.8 years for deceased donor grafts which are generally mismatched. There has been no significant improvement in the half-life of latter grafts even following introduction of highly effective immunosuppressive agents including tacrolimus and sirolimus.

E. HLA Matching for HSCT

In HSCT, the level of HLA matching is much more critical than solid organ transplants and is limited to HLA-identical sibling donors. This is because of the risk of graft versus host disease (GVHD) where T cells in the stem cell graft react against allogeneic host HLA molecules expressed on all tissues.

Four main donor categories are considered for HSCT: (1) HLA-identical sibling, (2) family donors other than siblings

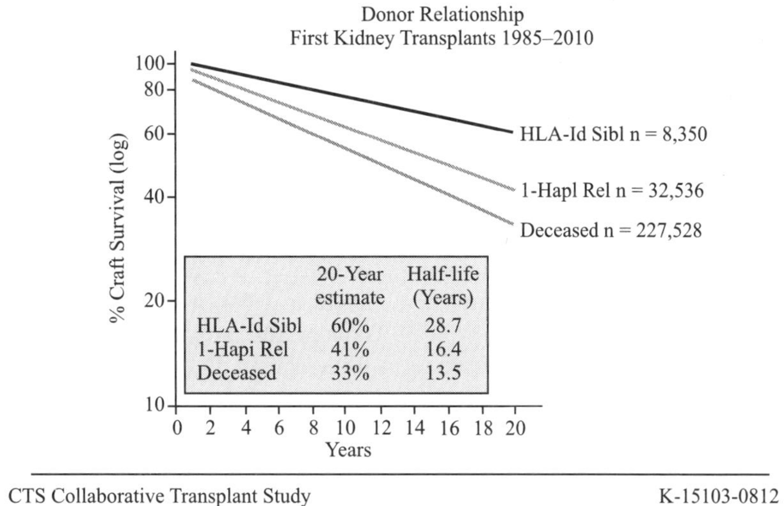

Figure 104.6 CTS report of relative % graft survival following kidney transplantation from donors who are HLA identical siblings or haploidentical relatives or deceased. Notice the difference in survival among three different donor sources.

that have been defined by extended family testing and who share HLA-A,B,C,DR and DQ alleles, (3) HLA-identical cord blood following antenatal testing, and (4) unrelated HLA matched donors obtained through voluntary marrow registries. A comparison of different HSC sources for allogeneic transplant is shown in Table 104.2

Although HLA matching remains a critical factor for successful transplantation, additional factors including disease status, patient age, conditioning regimen and source of cells (Bone marrow or peripheral blood stem cells or cord blood) are also important and can be custom-tailored for patients to a large extent.

F. Alternate Sources of Stem Cell Donors

While searching for donors other than HLA matched siblings, it is imperative to make a distinction between HLA genotypically identical sibling and family donors who are HLA phenotypically identical but haploidentical. Such donors share one inherited haplotype with the patient and the second noninherited haplotype may contain mismatches for non tested HLA loci. In haploidentical transplantation, the mismatched haplotype of the donor can originate from either of the parents. Such mismatched haplotypes are referred to as noninherited maternal

antigens (NIMA) or noninherited paternal antigens (NIPA). It has been reported that non T cell depleted bone marrow grafts donated by haploidentical siblings to recipients mismatched for NIPA and transplants donated by parents caused more acute and chronic GVHD and mortality than those donated by haploidentical siblings mismatched for NIMA. Similarly, the mother to child haploidentical transplants involved less GVHD than father to child transplantations.

Cord blood can overcome HLA barrier

The main advantages of using cord blood as an alternate source for HSCT are better tolerance despite HLA disparity, immediate availability of cells and reduced incidence cum severity of acute GVHD. A limiting factor, however, is the lower number of HSCs in a unit of cord blood leading to delayed neutrophil recovery post transplantation.

Unrelated donor marrow registries

The best HSCT results are seen in patients with a genotypically HLA-identical sibling. The probability of finding such a donor is approximately 30% to 35% whereas most of the remaining patients have to approach global unrelated voluntary donor marrow registries in search of a matched unrelated voluntary donor.

TABLE 104.2 **Different possible sources of HSCs for allogeneic transplantation**

Feature	BM	PBSC	UCB
HLA matching	Crucial. At least 6/8 HLA match	Crucial. At least 6/8 HLA match	Permissive. At least 4/6 HLA match
Time to neutrophil engraftment	Slower	Faster	Slowest
CD34+ cell dose	Usually sufficient	Sufficient	Limiting (Second unit can be incorporated)
Risk of GVHD	Low	High	Lower
Next donation (DLI)	Available	Available	Not available (Second unit may be given)

BM = Bone marrow, PBSC = Peripheral blood stem cells, UCB = Umbilical cord blood

The Bone Marrow Donors Worldwide (BMDW) is a large global networking constellation of 67 stem cell donor registries from 49 countries, and 47 cord blood banks from 31 countries. The current number of donors and cord blood units in the BMDW database is: 20,710,862 (20,150,517 donors and 560,345 CBU's). The largest of the unrelated registries is the US based National Marrow Donor Program (NMDP) with > 4 million donors and the Anthony Nolan Research Centre in UK that enlists several hundred thousand voluntary donors. Presently, such registries occur in almost all active transplant centres around the world.

A significant challenge in the identification of HLA matched donors from the random population relates to the extreme degree of diversity of the HLA system. A simple statistical analysis reveals several thousand million possible phenotypic combinations of HLA genes in a population thereby making it virtually impossible to find a perfect match donor for transplantation. Further, the distribution of HLA genes and specific HLA haplotypes varies between different ethnic and racial groups. These variations may reflect different selection pressures on populations subject to geographically distinct environments. Variations may occur at three levels:

(i) Some HLA genes are ubiquitous and although they occur in all populations, their frequencies vary significantly among populations.

(ii) Certain HLA genes may be restricted to a specific population, e.g., HLAA*02:11 is the most prevalent HLA-A2 subtype among Indians but is least prevalent in Caucasians.

(iii) Certain haplotypes may be population specific. For example, the autoimmune associated haplotype HLA-A1B8DR3, commonly found in type 1 diabetes in Whites is rarely observed in Indian patients. The latter on the other hand, have a unique family of DR3 haplotypes that differ at several intervening loci.

Asian Indian Donor Marrow Registry (AIDMR): A large global donor pool has been established in various worldwide registries; however, Asian Indians are poorly represented in them. Although NMDP has made an enormous effort to recruit minority donors, the chances for African Americans, Asian Americans or Asians of finding a fully matched unrelated donor are vastly lower than for Caucasian patients. Population studies based on genetic diversity of HLA indicate that although the gene pool of Asian Indians is comparable to those of Caucasoids and the Orientals, they appear to constitute a 'transition zone' between major populations of the west and the east. This is obvious from the occurrence of several 'novel alleles' and 'unique haplotypes' that occur in the Indian population. With this perspective an Asian Indian Donor Marrow Registry (AIDMR) was established at the All India Institute of Medical Sciences, New Delhi in 1994. It has now close to 3000 registered voluntary donors and provides searches to nation wide requests and from abroad. A few sister registries with limited unrelated voluntary donor pool have also been established at Mumbai and other regions in India and are connected to global online registry search portals.

SUMMARY AND CONCLUSIONS

A lucid understanding of histocompatibility per se, its requirements, major limitations and development of modern tools for its testing are major footsteps in the path leading to successful transplantation, both organ as well as stem cell. Recent advances in immunological techniques promise improved management of transplant patients by predicting rejection episodes before the onset of irreversible and terminal damage. HLA matching is beneficial in both short as well as long term survival of stem cell transplants, renal allografts both in live related as well as deceased donors. The poor match status between donor and recipient is frequently associated with more vigorous antibody responses and warrant careful examination. Some of the key areas of futuristic research in the area of clinical transplantation include upgradation of functional assays for determining the extent of alloreactivity in conjunction with high resolution HLA matching and discover 'permissible mismatches' distinct from 'taboo mismatches'. Continuing progress in understanding molecular mechanisms of graft rejection may lead to the ultimate goal of long term acceptance of organ allograft through tolerance induction.

Acknowledgements: The authors thank the Department of Biotechnology (DBT), Ministry of Science and Technology, Government of India and the Indian Council of Medical Research (ICMR) for financial assistance.

SUGGESTIONS FOR FURTHER READING

1. Mehra N.K., Kaur G., McCluskey J., Christiansen F. and Claas F. (2010), *The HLA Complex in Biology and Medicine: A Resource Book*. Jaypee Brothers Medical Publishers (P) Ltd., New Delhi.

2. Marsh S.G.E., Albert E.D., Bodmer W.E., et al. (2010), Nomenclature for Factors of the HLA System, *Tissue Antigens,* 75:291–455.

3. Abbas A.K., Lichtman A.K. and Pillai S., *Cellular and Molecular Immunology*, Philadelphia: WB Saunders.

4. Terasaki P. (1990), History of H.L.A., Ten Collections, Los Angeles, CA, UCLA Tissue Typing Laboratory, USA.

5. Mehra N.K., Siddiqui J., Baranwal A., Goswami S. and Kaur G. (2013), Clinical Relevance of Antibody Development in Renal Transplantation, Annals of NYAS.

6. Tait B.D., Susal C., Gebel H.M., et al. (2013), Consensus Guidelines on the Testing and Clinical Management Issues Associated with HLA and Non HLA Antibodies in Transplantation, *Transplantation*, 95:19–45.

105

Complement System

L.M. Srivastava and Tapasya Srivastava

CONTENTS

One of the major components of the immune system is complement, a group of self-reacting proteins in blood serum and other body fluids that plays an important role as a mediator of immune and allergic reactions. Complement represents humoral arm of natural host defense mechanism and is essential for survival. The name derives from the ability of the complement system to complement certain reactivities initiated by the antibodies. At least thirty plasma proteins constitute the complement system. All of these proteins are glycoprotein in nature.

Molecular weight, and serum concentrations of these proteins are presented in Table 105.1. These proteins are divided into four functionally separate groups. Two groups—the classical and the alternative pathway proteins, are concerned with generation of enzymes which cleave C3 and C5. The third group forms a multimolecular protein complex which is responsible for complement mediated cytolysis and the fourth group constitutes a group of proteins which modulate complement activation.

Complement system once activated kills certain bacteria, protozoa, fungi, viruses as well as cells of higher organism. In addition, it exerts its varied functions by particle bound C3b serving as ligand for corresponding receptors on phagocytic cells, fluid phase humoral mediators such as C3a, C4b, C5a, C3b and C3e which have various effects particularly in inflammatory reactions and membrane attack complex comprising C5–C9. Thus, complement activation forms a major part of natural host defense affording a range of mediators possessing immunoinflammatory potency (Table 105.2). However,

complement may also damage host tissue as in immune complex diseases. Some proteins of the complement system exist in circulation as precursor molecule (non-activated form) and may be activated through three well-defined pathways, namely:

(a) Classical pathway
(b) Alternative (Properdin) pathway
(c) Mannose binding ligand (MBL) pathway

Activations via the classical and alternative pathway is mutually dependent, because as would be evident through the latter part of the chapter, alternative pathway may be activated in the course of specific classical activation, since the latter produces (C3b) which can then trigger the feedback cycle under favourable conditions. Mannose binding ligand pathway (MBL) can be activated by mannose, a carbohydrate that is found in the cell wall of fungi, bacteria and viruses and produces C3b and then follows the sequence similar to classical and alternative pathways.

I. CLASSICAL PATHWAY

Classical pathway (Figure 105.1), a calcium dependent pathway, comprises the reaction steps of four plasma proteins, i.e., first (C1), second (C2), third (C3) and fourth (C4) complement proteins. IgM, IgG3 and IgG1 are the best complement fixing

TABLE 105.1   **Complement Proteins characterized by biochemical techniques important in the complement pathway**

Component	Approximated serum concentrations (u-g/ml)	Molecular weight
Classical pathway		
Clq	70	410,000
Clr	34	170,000
CIs	31	85,000
C4	600	206,000
C2	25	1 17,000
Alternative pathway		
D	1	24,000
C3	1,300	195,000
B	200	95,000
Membrane attack complex (MAC)		
C5	80	180,000
C6	60	128,000
C7	55	120,000
C8	55	150,000
C9	60	79,000
Soluble control proteins **Positive regulation**		
Properdin	25	220,000
Negative regulation		
Cl-INH	200	105,000
C4BP	250	550,000
Factor H	500	150,000
Factor I	34	90,000
Anaphylatoxin inactivator (Carboxypeptidase B)	35	280,000
S.protein (vitronectin)	500	80,000
SP-40, 40 (clusterin)	60	80,000
Receptors		
Complement receptor 1 (CR1)	—	190,000-250,000
Complement receptor 2 (CR2)	—	145,000
Complement receptor 3 (CR3)	—	170,000 (a chain) 95,000 (p chain)
Complement receptor 4 (CR4)	—	150,000 (a chain) 95,000 (P chain)
C5a receptor (C5aR)	—	42,000
Membrane regulatory proteins		
Decay accelerating factor (DAF)	—	70
Membrane cofactor protein (MCP)	—	45-70
CD59	—	18-20
HRF/C8-binding protein	—	65

human antibody isotypes with a general order of activity: IgM > IgG3 > IgG1 > IgG2 >> IgG4. Apart from antibodies, many other proteins can activate this pathway either by providing binding sites for C1 or by acting as a C1q surrogate protein. These proteins include mannose binding protein, some viruses, tissue damage products (DNA, mitochondrial membranes, mitochondrial cardiolipin), G reactive protein and serum amyloid P. Activation by these substances is initiated by high affinity binding of Cl (Clq) to the activator.

The first reacting component of the classical pathway, i.e., C1 is the calcium dependent trimolecular complex of three distinct proteins termed C1q, C1r and C1s. In native form, in the circulation C1 contains one molecule of C1q and two polypeptide chains each of C1r and C1s (Clq C1r2 C1s2).

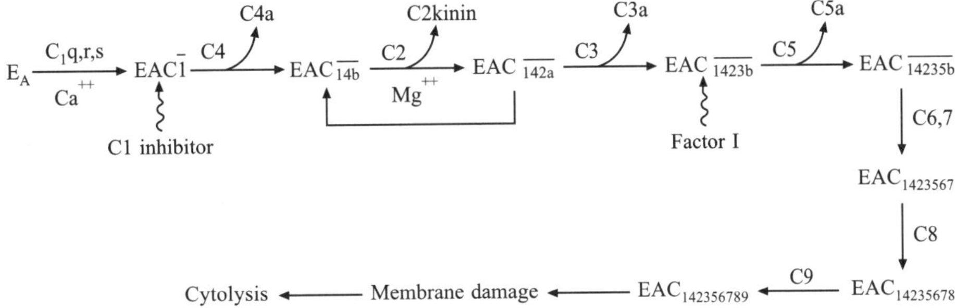

Figure 105.1 Sequence of activation of complement components via the classical pathway. E-erythrocyte or any antigen, A-antibody.

TABLE 105.2 Complement Activities in Host Defense against Infection

Complement components or fragments	Functional activity
C14, C1423	Virus neutralization
C3a, C5a	"Anaphylatoxin" (histamine release, capillary dialation)
C3 and C5 fragments	Chemotaxis of PMNs, monocytes
C567	Eosionophils
C3b	Opsonization
C3b, C3d	Enhanced induction of anti-body formation
C3b	Stimulation of B-cell lymphokine production
C3b	Enhancement of antibody dependent cellular cytotoxicity
C3 cleavage product	Induction of granulocytosis
C5 cleavage product	Neutropenia (C5a)
C1–9	Lysis of bacteria, viruses, virus-infected cells, tumour cells, mycoplasma, protozoa and spirochetes

For activation of C1, C1q must bind the heavy chain of two antibody molecules (Figure 105.2). Thus, a single IgM molecule can activate C1. C1q binding site on IgG is in $C\gamma2$ domain and is strongly influenced by Glu 318, Lys 320, Lys 322 and Ser 331 and to a lesser extent by other positions between 316 and 338. Activation of complement is probably equally efficient whether the antigen is a soluble protein precipitated by antibody or is an antibody coated erythrocyte.

C1q the recognition molecule of the classical pathway is a globulin. It consists of collagen like triple helix distributed amongst six identical subunits. Each subunit contains three different polypeptide chains termed A, B and C. Amino acid sequencing of these three chains has shown that they are very similar, each having about eighty residues of typical collagen like sequence beginning close to the N-terminus. All three chains contain repeating triplet Gly-X-Y, where Y is often hydroxyproline or hydroxylysine. The carboxy terminal regions are very similar and these form a globular unit at the end of each triple helix section. The globular protein structure at the

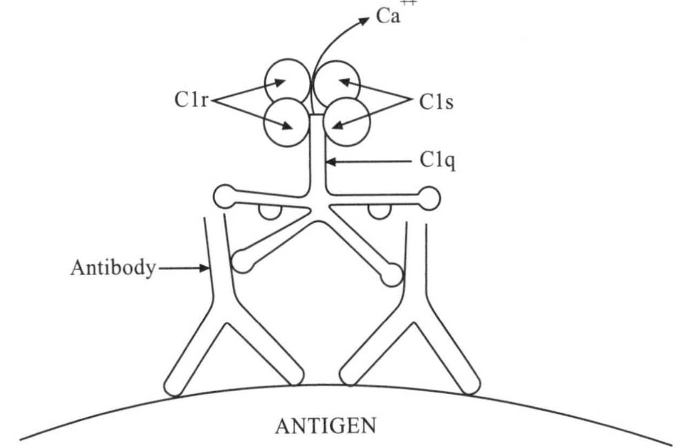

Figure 105.2 Binding of C1 to the antigen-antibody complex. C1q binds to the CH_2 domain of the Fc portion of the antibody.

carboxy terminus recognizes the changes in the IgG or IgM molecules in immune complexes and the collagen like helix binds the C1r and C1s subcomponents.

Binding of C1q subunit to the activator triggers intramolecular events in the C1 molecule that facilitates cleavages and thus activation of two C1r Zymogen polypeptide chains in C1. The activated Clr enzymes then cleave a peptide bond in C1s and thereby renders both C1s polypeptide chains in the C1 molecule enzymatically active. Activated C1s is a protease and is also called as C1 esterase. Both C1r and CIs can be inactivated by C1 esterase inhibitor (C1-INH), hence preventing generalized activation. Natural substrates for C1 esterase are C4 and C2.

C4 consists of three disulfide linked polypeptide chains α, β and γ. Activated C1s cleaves a single peptide bond in the α chain of C4 to release a fragment C4a from the amino terminal end. The larger fragment C4b has a labile binding site which allows it to bind covalently to the cell membranes via an ester or amide linkage or to the immunoglobulin in the immediate neighbourhood of the C1 complex. Thus, C4b is the first of the complement components to be directly attached to the target cell membrane. This dual type of bond formation can modify to a certain extent, the site, amount and type (C4A vs. C4B allotype) of C4b binding and hence subsequent complement activation. Only 5% of C4 gets attached to the target while the rest is inactivated in surrounding fluid phase by reaction of thioester site with H_2O. After C4, C2 is cleaved by the adjacent activated C1S into two fragments the smaller C2b and larger C2a. The complex C4bC2a is held together with magnesium ions and has protease activity. It is able to cleave C3 and is known as C3 convertase. Recent studies have shown that some C3b molecules can covalently bind to a specific site on C4b during activation. This facilitates binding of CR1 which can bind both C3b and C4b at three separate ligand binding sites. Various breakdown products of C3 and their biological activity are listed in Table 105.3.

C3 is the bulk protein of the complement system. Upon cleavage of C3 by C3 convertase, again a smaller peptide C3a is released. The point of cleavage in C3 is an arginyl serine bond in position 77. The major fragment C3b can be fixed to the cell surface by an extremely short lived binding site. This site is based on an internal thioester which in the native molecule is protected from access of nucleophilic material but becomes exposed upon fission of the C3a, a biologically active peptide chain.

The thioester can be hydrolysed by reaction with water or react with a hydroxyl group of a surface constituent. In the former case the C3b fragment loses its binding capacity and remains in the solution, haemolytically inactive. In the latter case it is covalently fixed to the cell surface by an ester or amide group. Once C3b is fixed the C423 complex attains enzymatic activity and is called as C5-convertase. The limited proteolysis of C5 by C5 convertase leads to the separation of C5a, a biologically active peptide and C5b which forms a noncovalent rather a stable complex with C6.

The interaction of early acting components (C14235) is enzymic in nature, whereas interaction between C5b, C6, C7, C8 and C9 is non-enzymatic and through noncovalent probably, hydrophobic bonds.

Complexing of C5b with C6 initiates the formation of membrane attack complex. C5b6 attaches reversibly to cell membrane. Then C7 combines to C5b6 and the resulting C5b67 complex now becomes firmly inserted in the lipid bilayer of cell membrane. The interaction of C8 with the complex results in deeper incorporation into the lipid bilayer but incomplete disruption of the cell membrane. To this complex, C9 binds initially as a monomer and then polymerizes which disturbs the integrity of the lipid surface, probably punching holes and hence leading to osmotic cell rupture.

II. REGULATION OF CLASSICAL PATHWAY

Complement pathway can be tightly controlled by certain proteins which can act at specific sites of the pathway using different strategies to block activation. The proteins may be in fluid phase, on cell membranes and in matrix. The general function of these proteins is to inactivate a specific component irreversibly or maintain it in an inactive form or to dissociate components from multi-protein complexes like C3/C5 convertases. The requirement for mechanism to control complement activation is obvious, because variety of biologically active products are generated during activation. During the early stages of complement activation, cleavage of C4, C3 and C5 exposes their labile binding sites which for less than 100 milli-seconds can bind to cell membrane or other structure.

The natural spontaneous decay of C3 and C5 convertases of classical pathway offers a further level of control. C2a decays from C4b2a, and C4b2a3b.

A third level of control is provided by the presence of certain plasma proteins which modulate the activity of certain complement components.

Cl-Inhibitor

C1-inhibitor is an α-2 neuraminoglycoprotein. It provides the main control mechanism of activated C1. C1-inhibitor binds rapidly, firmly and irreversibly to the activated forms of C1r and C1s, thereby inhibiting and blocking the action of activated C1. The protease trapping 'bait site" of C1-INH is the bond between Arg 444 and Thr 445.

C3b Inactivator (Factor I)

C3b Inactivator cleaves C3b into C3c and α-2D and results in inactivation of C3b. Factor I was purified as a protein that can cleave and inactivate C3b and C4b with the help of cofactors. The cleavage is accomplished in a highly regulated form to molecules that cannot act as part of C5 convertase.

C4 binding protein

C4 binding protein is a glycoprotein which binds to C4b. Following the formation of classical C3 convertase, (C4b, 2a) C4 binding protein binds with C4b and accelerates the

TABLE 105.3 Properties and Biological Activities of C3 Fragments

Fragment	Molecular weight	Polypeptide chains	Biological activity
C3a	9,000	1	Anaphylatoxin: Histamine release, cellular enzyme release, smooth muscle contraction
C3b	170,000	2	Subunit of C3/C5 convertases of classical and alternative pathways, ligand for cellular C3b receptor, binding to biological particles
iC3b	170,000	3	Ligand for cellular iC3b receptor (CR3)
C3c	140,000	3	Precursor of C3e
C3d	25,000	1	Ligand for cellular C3d receptor (CR2)
C3e	12,000	1	Induction of leukocytosis

decay of the enzyme, probably by displacing C2a from the complex.

Anaphylotoxin Inactivator

Carboxypeptidases—The action of these plasma proteins is to catalyze C-terminal cleavage of Arg to *des* Arg on C5a and C3a, hence resulting in proteins with 2–3 orders less cell stimulating activity.

Membrane regulatory proteins

A series of membrane proteins which can block complement activation at specific sites has been described over the last few years. These proteins can inhibit assembly of functional C3 and C5 convertases and MAC, hence stopping the cascade.

MCP (CD46)

It is a widely distributed surface membrane human C3 binding protein and acts as a cofactor for cleavage of C3/C4 into their inactive forms C3bi and C4bi. The erythrocytes are an exception as they show no expression of this protein.

CD59 (Protectin, Membrane inhibitor of reactive lysis)

This protein is widely distributed on erythrocytes, monocytes, granulocytes, platelets, endothelial cells as well as on cells of reproductive and nervous tissues. CD59 functions by binding to C8 in C5b-8 complex and hence blocks effective incorporation of C9. It may also bind C9 in MAC and block subsequent polymerization of C9 and formation of transmembrane pores.

HRF/C8 binding protein

This protein blocks the deposition of C8 and C9 (but not C7) into membranes.

C3 membrane proteinases

A protein designated p57 has been observed on complement resistant melanoma cell lines with the ability to cleave and inactivate C3b and hence mediate resistance associated with tumours.

Matrix proteins

Decorin: It can bind C1q with a high affinity at physiologic ionic strength and alter its activity.

III. ALTERNATIVE (PROPERDIN) PATHWAY (Figure 105.3)

Alternative pathway is a well adapted system of natural resistance to infection and is not a strictly immunological process as specific antibodies are not essentially involved. Early reacting complement components, i.e., C1, C4 and C2 do not participate.

Alternative pathway can be activated by lipopolysaccharide as is found in Gram-negative bacterial walls, Zymozan, inulin, agarose, complex polysaccharide, bacterial endotoxins and aggregated IgA and IgE. Many different particles and cells

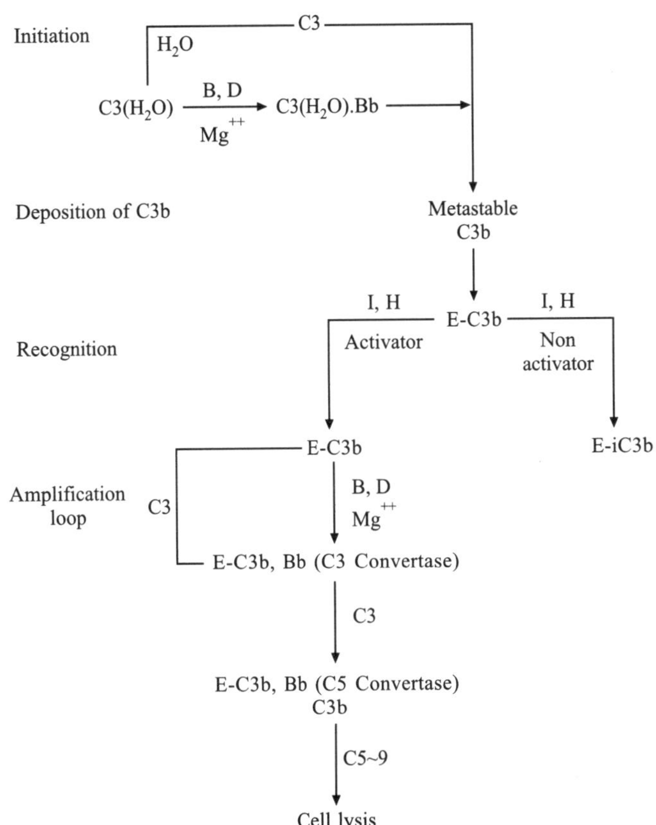

Figure 105.3 Activation of the alternative pathway of the complement system. E = antigen.

activate the alternative pathway, these include bacteria, yeast, viruses, virus infected cells, tumour cells and parasites, C3 nephritic factor (C3Nef). Although a great deal is known about the molecular events comprising activation, the common denominator among the various activators has yet to be described in chemical terms. Probably it is carbohydrate in nature.

The pathway involves six serum proteins. C3, factor B and Factor D form the positive feedback amplification system and factor H, factor I and properdin function as regulators.

The exact mechanism of the initiation of alternative pathway has been controversial since its discovery. C3 plays a central role in alternative pathway. Factor B binds to C3b in the presence of magnesium to form a C3b, B complex. When bound to C3b, only then factor B is susceptible to proteolytic activation by factor D, a serine protease. Cleavage of factor B releases a fragment Ba, leaving the C3b, Bb complex, C3b, Bb complex is the C3 convertase of the alternative pathway and is the central enzyme of this pathway. Bb portion of this enzyme is a serine protease which is capable of cleaving C3 into C3a and metastable C3b. The C3b,Bb enzyme in the presence of a second molecule of C3b, also cleaves C5. This releases anaphylatoxin C5a and activates the membrane attack pathway of the complement.

Although alternative pathway was recognised through the discovery of properdin, it has recently been shown, that it is not an essential component of the alternative pathway. Properdin binds to the C3 convertase, thus increasing the stability of this complex and greatly enhances the activity of this pathway. It also partially counteracts the destablizing activity of factor H.

IV. REGULATION OF ALTERNATIVE PATHWAY

Factor H: Factor H is defined as AP regulator with two distinct enzymatic activities. These include an acceleration of slow and spontaneous decay of Bb from C3b, hence dissociating both C3 and C5 convertase. The other is a cofactor activity for Factor I to mediate cleavage of C3b into a hemolytically inactive form C3bi on targets like bacteria and immune complexes as well as on free soluble C3b-Bb containing complexes.

The ability of the Factor H to distinguish C3b deposition between self and non-self targets has been supported by the presence of polyanionic binding site.

Membrane regulatory protein

DAF: DAF acts by binding C3b or C4b on the cell (Decay Accelerating Factor/CD55) membrane and markedly increases the spontaneous decay of both classical pathway C4b2a and alternative pathway C3bBb complexes.

C3 and C5 convertases of the alternative pathway also undergo spontaneous decay where Bb decays from C3bBb and $Bb(C_3b)_nP$.

V. AMPLIFICATION OF ALTERNATIVE PATHWAY (Figure 105.4)

Amplification refers to the C3b dependent feedback process. C3b and factor B combine to form a stable bimolecular complex C3bB which has no enzymatic activity until acted on by factor D. The enzyme C3bBb is capable of generating many molecules of C3b by cleaving C3. In the presence of excess factor B, more C3 convertase is formed. In the absence of control, the process resembles a chain reaction. Thus the attachment of one C3b to a particle could result in deposition of many C3b molecules in a very short time.

C3b formed by classical or alternative pathway activation or by the action of enzymes such as trypsin, and plasmin can trigger the alternative pathway, in a positive feed back fashion Therefore C3b, once formed, should perpetually stimulate the

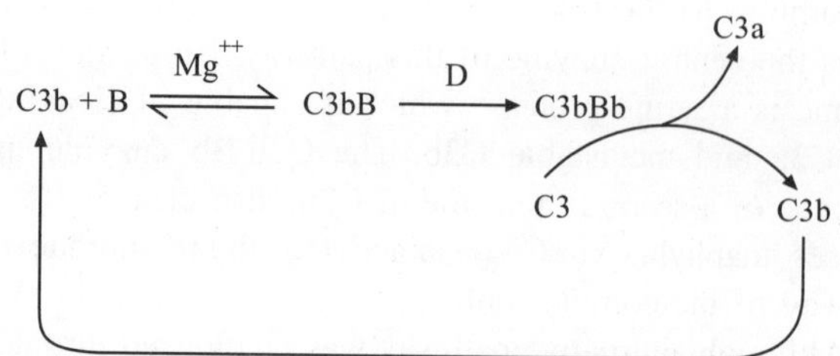

Figure 105.4   Amplification loop of the alternative pathway.

alternative pathway until C3b and/or factor B are completely consumed. However, this does not occur because of the presence of factor H and factor I in the plasma, which act in concert to limit the alternative pathway activation.

The Lectin pathway

The Lectin pathway originates with host proteins binding microbial surfaces. Lectins are proteins that recognize and bind to specific carbohydrate targets (since the lectin that activates complement binds to mannose residues. Some scientists refer this as MB-Lectin pathway or mannan binding lectin pathway). The lectin pathway, similar to alternative pathway, does not depend on antibody for its activation. However, the mechanism is more or less like that of the classical pathway, since after initiation, it proceeds through the action of C4 and C2 to produce a C5 convertase.

The lectin pathway is activated by the binding of Mannose Binding Lectin (MBL) to mannose residues on glycoproteins or carbohydrates on the surface of microorganisms. MBL is an acute phase protein produced in inflammatory responses and function is similar to Clq which it resembles in structure.

After MBL binds to the surface of a cell or pathogen, MBL associated serine proteases MASP -1 and MASP -2 bind to MBL. The active complex so formed by this association causes cleavage and activation of C4 and C2. The MASP -1 and -2 proteins due to structural similarities to Clr and Cls mimic their activities / function thereby activating the C2-C4 components to generate a C5 convertase without the requirement for specific antibody binding. Such a phenomenon of activation represent a significant innate defense mechanism similar to alternative pathway but utilizing the components of classical pathway except for CI proteins.

An overview of the above said three pathways activating the complement system including sites of control or negative regulation by specific inhibitory proteins is depicted in Figure 105.5.

VI. BIOLOGICAL ACTIVITIES OF COMPLEMENT

(i) Opsonization and phagocytosis

A fundamental role of complement is the ingestion of C3 bound microbial targets by macrophages and neutrophils. The ingestion is facilitated by a cooperative binding mediated by CR1, CR3, CR4 and by binding to ligands like IgG on matrix proteins. Specific signalling mechanisms including Protein kinase C and phospholipase like PLC and PLD may be involved. Further, complement receptor mediated phagocytosis can be augmented by certain cytokines.

(ii) MAC formation and cytolysis

MAC mediates the lysis of target membranes by osmotic gradient disruption in non-nucleated target cells. The effect on nucleated cells is very complex and involves cell activation,

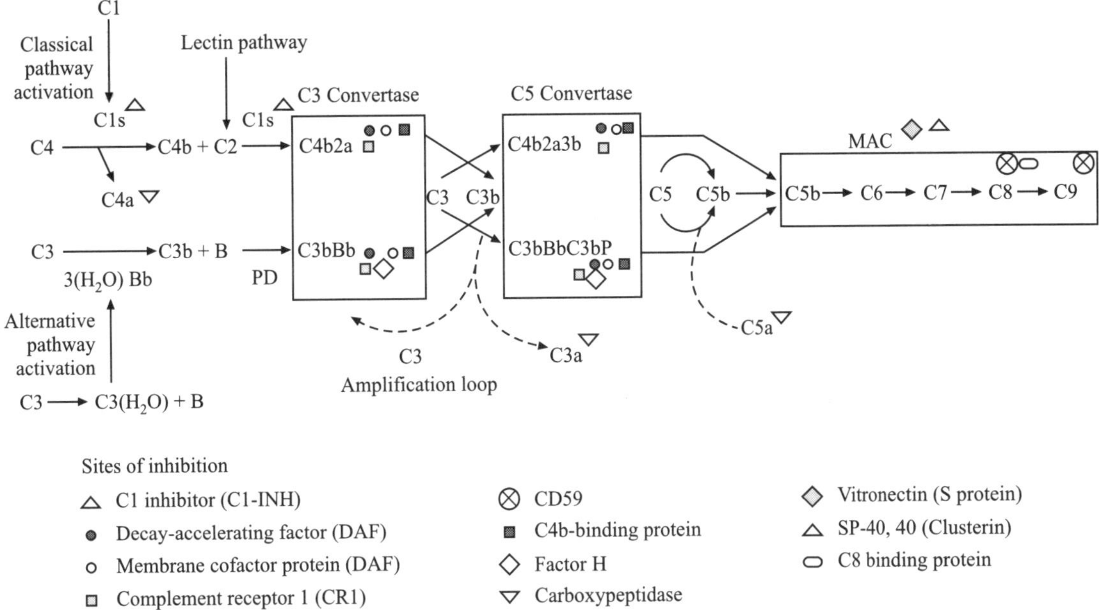

Figure 105.5 Schematic outline of complement pathway including sites of control or negative regulation by specific inhibitory proteins.

recovery from MAC effects, and secondary phenotypic changes. Before the lysis, an increase in Ca^{2+} intracellular level (both from extracellular and intracellular sources) occurs followed by PKC activation and generation of DAG and ceramide. Other byproducts generated are arachidonic acid and different eicosanoid species, reactive oxygen metabolites and IL-1β.

The cell under attack effectively tries to remove MAC from its surface either through vesiculation (as in neutrophils or erythrocytes) or through endocytosis. The secondary phenotypic changes are specific for certain cell types. The human glomerular epithelial cells respond to C5b-8 and C5b-9 by increasing collagen synthesis which in turn induces glomerular sclerosis associated to complement deposition. There are other reports on activation of neutral proteases with the C5b-9 insertion into membranes. Overall, this removal of MAC from cell surfaces has a complement system involvement in human diseases.

(iii) Anaphylatoxins and Chemotaxis

Both C3a and C5a, the cleavage products of C3 and C5 by their respective convertases or proteolytic enzymes possess anaphylatoxin activity. Anaphylotaxins bind to receptors on the membrane of mast cells and basophils with resulting degranulation. These granules contain vasoactive amines such as histamine. Histamine release is associated with increased vascular permeability. Smooth muscle contraction is induced by anaphylatoxins which act on specific receptors on smooth muscle cells.

C3a and C5a are chemotactic for polymorphonuclear leukocytes and macrophages. The trimolecular complex C5b67 has also been said to possess chemotactic properties.

(iv) Modulation of immune complex mediated effects

The interaction of immune complexes with the proteins of the complement system completely alters their biological activities.

The presence of C3b or C4b on immune complexes permits their interaction with their specific receptors on the surfaces of certain cells such as human erythrocytes, mononuclear phagocytes, polymorphonuclear leukocytes and B lymphocytes. This interaction is termed as immune adherence. The biological consequences of this phenomenon vary depending upon the types of cell involved, for instance phagocytic cells show increased phagocytosis and increased intracellular killing of bacteria.

(v) Role of complement system in the clearance of immune complex from circulation.

The significance of the complement system in clearing immune complexes is visualized in patients with auto immune disease systemic lupus erythematosus (SLE). These individuals produce large quantities of immune complexes and suffer tissue damage as a result of complement mediated lysis and generation of Type II or Type III hypersensitivity reactions. Although complement plays a significant role in the development of tissue damage in SLE, paradoxically the deficiencies of C1, C2, C4 and CR1 predispose an individual to SLE. It seems the complement deficiencies interfere with effective solubilization and clearance of immune complexes and as a result these complexes persist, leading to tissue damage in effect by the same system whose deficiency was to be blamed for such a happening.

The coating/attachment of soluble immune complexes with C3b is considered to facilitate their binding to CR1 on erythrocytes and erythrocytes in an individual accounts for about 90% of CR1 in the blood. This being the prime reason for erythrocytes to play a significant role in binding C3b-coated immune complexes and carrying these complexes to the liver and spleen. In these organs, the immune complexes are detached from the blood cells and phagocytosed thereby preventing their deposition in tissues (Figure 105.6). In SLE patients, deficiencies in Cl, C2 and C4 each contribute to reduced levels of C3b on immune complex thereby inhibiting their clearance. The lower levels of CR1 expressed on the erythrocytes of SLE patients may also interfere with the proper binding and clearance of immune complexes.

(vi) Antiviral activity of complement

Neutralization of herpes-type viruses by antibody is enhanced by Cl, C4 and C2. RNA tumour viruses possess a Clq receptor, which binds Clq, with activation of Cl. Virolysis occurs in human but not guinea pig serum.

Lysis of antibody coated measles virus infected cells occurs by the alternative pathway. Antibody dependent virolysis by leukocytes is enhanced by the addition of complement.

VII.   COMPLEMENT SYSTEM IN HUMAN DISEASES

(i) Clinical measurement of complement

Many assays of complement components like measurements of total C3, C4, Factor B and functional measurements of classical pathway and MAC (CH50), AP50-measure of AP and MAC and C1 esterase inhibitor (C1-INH) levels in patients with hereditary angioedema are of utility. Complement deficient states are best diagnosed using these tests. In addition, SLE patients where the complement components levels tend to follow disease activity are also monitored. The pattern of complement deficiency in diseases is outlined in Table 105.4.

TABLE 105.4   Genetic Deficiency of Complement in Man and Association with Disease States

Component	Clinical association
C1s	SLE like syndrome
C1r	Recurrent bacterial infection, SLE like disease, chronic glomerulonephritis
C4	SLE like syndrome, DLE
C2	Glomerulonephritis, SLE, Discoid LE, chronic vasculitis, Anaphylactoid purpura
C3	Recurrent infections, skin eruptions, fever arthrolagias
C5	SLE and recurrent infections, gonococcal infection
C6	Repeated gonococcemia, meningococcal meningitis
C7	Raynaud's phenomenon
C8	Repeated gonococcemia, SLE-like syndrome
C1 Inhibitor	Hereditary angioedema, SLE, DLE, progressive glomerulonephritis
C3b Inactivator	Recurrent infection

Over the past two decades, it has become increasingly apparent that complement activation may play a pathogenetic role. The conclusions are based upon:

1. Finding of hypocomplementaemia in patients, the degree of which parallels disease activity.
2. Deposition of complement components at the site of tissue injury.
3. Hypersynthesis and hypercatabolism of complement components in disease states.
4. Observation that the lesions of complement mediated tissue injury in experimental animals resemble the lesions in human diseases.

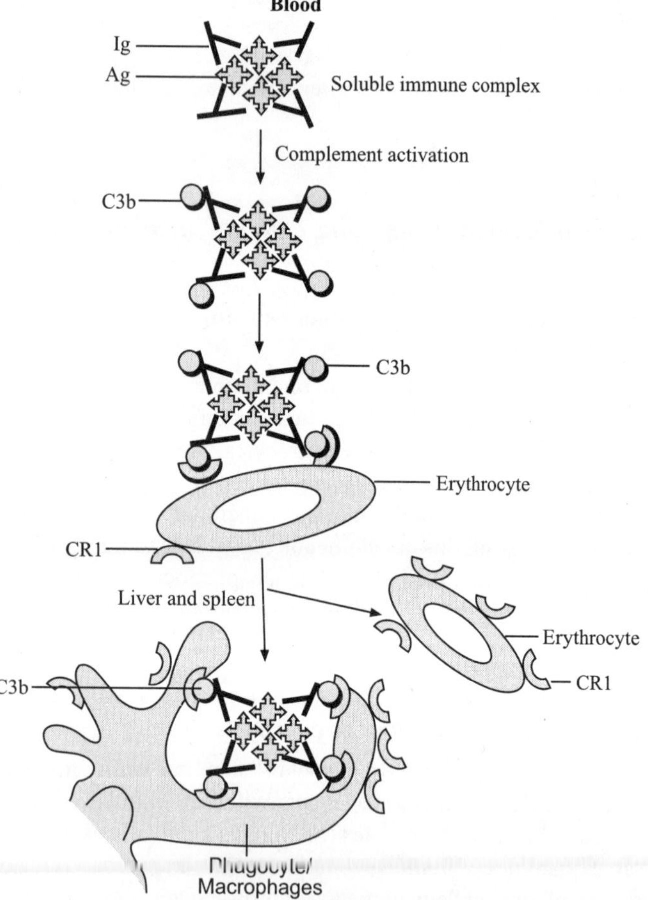

Figure 105.6   Diagrammatic representation of the clearance of circulating immune complexes by reaction with CR1 on erythrocytes and removal of these complexes by receptors on macrophages in the liver and spleen. Because erythrocytes have fewer receptors than macrophages, the latter can detach the complexes from the erythrocytes as they pass through the liver or spleen. Deficiency in this process can lead to renal damage due to accumulation of immune complexes.

Complement system has been found to play a pathogenic role in diseases like systemic lupus erythematosus (SLE), Rheumatoid arthritis, Bronchial asthma, Infectious diseases and skin diseases. Documentation of genetically determined deficiencies of complement components has greatly enhanced our knowledge of the biological significance of complement system in terms of resistance to bacterial infection. Deficiencies of all complement components are inherited in autosomal recessive or codominant fashion. Homozygous deficiency of Clq, C3, C5, C6, C7 and C8 are associated with recurrent bacterial infections. Table 105.4 shows some reported examples of genetic deficiency of complement component in man.

(ii) Complement deficiencies

These deficiencies are classified either as inherited or as acquired. Inherited deficiencies due to specific genetic defects (like a gene deletion, loss of splicing sites or generation of stop codons). Acquired deficiencies, on the contrary, are due to consumption of complement components by excessive activates (which may be ICs in SLE) or deficiencies of regulatory proteins.

(iii) Complement as an effector pathway in tissue damage

Complement is an essential component of pathophysiology of many diseases like some forms of vasculitis, rheumatoid arthritis, adult respiratory distress syndrome, SLE, many classes of substantial advantage for cells.

(iv) Evasion of complement activities by pathogens

Many pathogens utilize complement receptors and regulatory proteins to gain access to target cells. Bacteria, viruses and parasitic organisms can evade immune responses by expressing complement regulators on their surface. Some malignancies also have been demonstrated to express complement regulatory proteins like MCP and DAF, hence escaping the cytolysis by anti-tumour antibody and complement.

Complement system has also been implicated in both host resistance to HIV-1 and infection of cells by CD4 independent mechanism. A high complement activation is observed for HIV positive patients. The viral particles were able to activate both the alternative and classical complement pathways. The HIV-1 gp41 but not gp120 can activate the classical complement pathway *in vitro,* independently of the antibody, due to its ability to interact with C portion of C1. The consequent C3 deposition and receptor binding is, however, evaded due to an efficient means of resistance manifested by the virus which proves protective to the virus.

(v) Ab-dependent cellular cytotoxicity and natural killer cell activity

Complement binding to target cells or organisms can augment both antibody dependent cellular cytotoxicity (ADCC) and NK cell activities through the expression of CR1 and CR3. The presence of DAF on either target cell or effector cell *in vitro* can block NK activity, and whether this effect is complement dependent needs to be investigated to ascertain that the presence of DAF can really help target cells resist the cytotoxic effects of NK cells.

(vi) Reproduction and pregnancy

The maternal immune system, during pregnancy, is unavoidably exposed to allogeneic and semiallogeneic tissues in the form of sperm, seminal plasma components and developing foetus and placenta. Complement fixing antibodies with a specificity to sperm antigens have been proposed to mediate infertility. Also, complement activation pathway proteins have been observed in cervical mucus, vagina and ovarian follicular fluid. However, the evidence of complement mediated damage is negligible. The reason is the presence of DAF, MCP and CD59 at high levels on surface of cells of the reproductive system. Sperms also express regulatory proteins on such cells which help prevent: (i) their lysis in the female reproductive tract, (ii) renal diseases, (iii) hemolytic anaemias, and (iv) myocardial infarction.

VIII.   BIOSYNTHESIS OF COMPLEMENT PROTEINS

Four cell types namely hepatocytes, mononuclear phagocytes, fibroblasts and the epithelial cells of gastrointestinal and genitourinary tracts have been shown to be capable of synthesizing complement components.

IX.   GENETIC POLYMORPHISM OF COMPLEMENT COMPONENTS

C4, C2, C3, C6 and factor B exhibit genetically determined polymorphism. The polymorphic variability of complement components is determined by autosomal co-dominant genes. The polymorphic variants of C2, C4 and factor B are controlled by genes on the short arm of chromosome 6, and the genes controlling the major histocompatibility complex.

X.   RECEPTORS FOR COMPLEMENT COMPONENTS

Many biologic effects of complement are mediated by high affinity specific receptors for fragments of complement proteins (Table 105.5). The effects include both clearance of complement coated antigens and activation of cells of the immune system. Complement activation fragments for which the receptors are well characterized are C1q, C3, C4, C5a, C3a, C4a, Factor B (Ba and Bb) and Factor H.

(i)   The C1q receptor has been observed on a variety of cells like leukocytes, endothelial cells, fibroblasts, platelets and epithelial cells. The major effect is opsonization of immune complex targets and phagocyte

TABLE 105.5   Complement binding receptors

Receptor	Major Ligands	Functional Role	Cellular Localization
CR1 (CD35)	C3b, C4b	Blocks formation of C3 convertase; binds immune complexes to cells.	Erythrocytes, neutrophils, monocytes, macrophages, eosinophils, follicular dentritic cells, B cells, certain T cells
CR2 (CD21)	C3d,C3dg, C3bi	Part of B-cell co-receptors, binds Epstein – Barr virus	B cells, follicular dendritic cells, certain T cells
CR3 (CD 1 lb/18)	C3bi	Bind cell-adhesion molecules on neutrophils, facilitating their extravasation; bind immune complexes, thereby enhancing their phagocytosis	Monocytes, B cells, macrophages, neutrophils, natural killer cells, certain T cells.
CR4 (CDllc/18)	C3bi	Not well understood	Monocytes, B cells, macrophages, neutrophils, natural killer cells, certain T cells.
C3a/C4a Receptor	C3a, C4a	Induces degranulation of mast cells and basophils	Mast cells, basophils, granulocytes
C5a Receptor	C5a	Induces degranulation of mast cells and basophils	Mast cells, basophils, granulocytes, monocytes, macrophages, platelets, endothelial cells
Cl Receptor	Clq	Opsonization of Immune Complex targets	Leucocytes, endothelial cells, fibroblasts, platelets, smooth muscle cells and epithelial cells.

activation with a subsequent promotion of oxidative burst in polymorphonuclear cells; an enhanced ADCC and cytokine release by macrophages.

The four major complement receptors are CR1, CR2, CR3 and CR4.

(ii) CR1/CD35 is one of the well described proteins which functions as a high affinity receptor for C3 and C4 fragments. CR1 is the major receptor for C3b fragment of C3 and C4b fragment of C4. The major form of CR1 is a 190 kDa protein. CR1 has also a regulatory activity and can accelerate the decay of both classical and alternative pathways C3 convertases in addition to providing cofactor activity for C3 and C4 cleavages. It is present on primate erythrocytes, on B and K cells, monocytes, macrophages, neutrophils, eosinophils, kidney glomerular cells and mast cells.

(iii) CR2 promotes antigen processing and presentation of C3d around targets. Also, it has a potential to interact with antioncogene proteins P53 and P68. CR2 is expressed exclusively on B lymphocytes and binds specifically to C3d.

(iv) CR3: The general binding affinity of C3 fragments to CR3 is C3bi > C3b > C3d. CR3 exhibits lectin like activities and is proposed as an organizer of cytoskeletal events necessary for adhesive interactions and membrane reorganization during phagocytosis. CR3 receptors have been identified more recently and bind to iC3b. CR3 exist on B lymphocytes, K cells, granulocytes, monocytes and macrophages.

(v) CR4: It also binds the C3bi fragment of complement but its biological activities are not well understood. C5a, C3a and C4a receptors bind to the corresponding anaphylatoxins and exert many biologic effects which include leucocyte chemotaxis, aggregation of neutrophils and platelets, increase in capillary permeability and increased relase of IL-1 and IL-6.

CR1 promotes the binding of immune complexes to neutrophils and mononuclear phagocytes, thereby enhancing their ingestion. Recently CR1 has been demonstrated, to be a powerful inhibitor of both classical and alternative pathway, C3 and C5 convertases. Therefore CR1 might function as a membrane associated inhibitor of complement cascade and might provide cells that interact with immune complexes with a mechanism to circumvent the damaging effect of complement activation in their vicinity.

The idea that CR1 may function as a complement inhibitor provides an explanation for the puzzling finding of CR1 on the epithelial cells of kidney glomeruli. The glomerular capillary wall behaves like a sieve that restricts the passage of molecules with increasing molecular radius. All components of the alternative pathway have molecular weight smaller than 220,000 while the control proteins H and I have much higher molecular weight. It is conceivable that under pathologic circumstances complement component of the alternative pathway, but lesser amounts of control proteins can traverse the basement membrane and reach the urinary space. If low concentration of control proteins are present, the cascade could be activated, leading to damage of the epithelial cells. The presence of CR1 on the membranes of epithelial cells might inhibit this activation.

Complement Regulatory molecules: applications in therapy

Due to its important role in many types of inflammatory and autoimmune diseases, many potentially efficacious inhibitors have been developed.

Non-recombinant inhibitors

Heparin, an anticoagulant has been observed as an inhibitor of complement activation at various points. Another class is substituted isocoumarins—a class of serine protease inhibitors. Other inhibitors like leupeptin and DFP can also exert the same effect. Intravenous immunoglobulin (IVIG) has been used for many autoimmune diseases as it can block C3 binding to opsonized erythrocytes, hence blocking inflammation *in vivo*.

Recombinant soluble complement regulatory proteins

The soluble forms of biologically active membrane complement receptors CR1 and CR2 as well as regulatory proteins DAF, MCP and CD59 have proved effective in inhibition of complement activation at specific sites.

SCR1—recombinant soluble human CR1—can block activation of both alternative and classical pathways by its decay accelerating and cofactor activities and hence has proved to be an ideal candidate for use as a drug. SCR1 has proved efficient in animal models of diseases like myocardial infarction, limb and gut ischaemia, thermal injury and transplantation. Other membrane complement regulators have also been utilized in their recombinant soluble forms like MCP, DAF and CD59. The advantage of these over the SCR1 is that they are more selective in their inhibitory activities.

SUGESSTIONS OF FURTHER READINGS

1. Carrol M.C. (2000), The Role of Complement in B-cell Activation and Tolerance, *Adv. Immunol.*, 74:61.

2. Colten H.R. and Rosen, F.S. (1992), Complement Deficiencies, *Ann. Rev. Immunol.*, 10, 809–834.

3. Favored H.W., Vander Walle G.R., Nauwynck H.J. and Pensaert M.B. (2002), Virus Complement Evasion Strategies, *J. Gen. Virol.* M., 1–5.

4. Fujita T. (2002), Evolution of the Lectin-complement Pathway and its Role in Innate Immunity, Nature Reviews, *Immunol.*, 2:346–348.

5. Kirschfink M. (1997), Controlling the Complement System in Inflammation, *Immunopharamacology*, 38, 51–62.

6. Lambris J.D., Reid K.M.B. and Volanosis (1999), The Evolution, Structure Biology and Pathophysiology of Complement, *Immunol., Today*, 20, 207–11.

7. Lindahl G., Sisbring U. and Johnson E. (2000), Human Complement Regulators: A Major Target for Pathogenic Microorganism, *Current Opin. Immunol.*, 1–2, 44.

8. Makrides S.C. (1998), Therapeutic Inhibition of Complement System, *Pharmacol.* Rev., 50, 59.

9. Morgan B.P. and Walpart M.J. (1991), Complement Deficiency and Disease, *Immunol. Today J.*, 2, 301–6.

10. Muller-Eberhard H.J. (1998), Molecular Organization and Function of the Complement System, *Ann. Rev. Biochem.*, 57, 321.

11. Nielsen G.H., Fischer E.M. and Leslie R.G.G. (2000), The Role of Complement in the Acquired Immune Response, *Immunology*, 100, 4–12

12. Nonaka M. (2000), Origin and Evolution of the Complement System. *Curr. Top Microbol. Immunol.* 248, 37.

13. Play Fair, John and Bancroft, Gregory (2008), In: *Infection and Immunity,* 3rd ed., Oxford University Press, Oxford, U.K.

14. Volankis J.E. (1995), Transcriptional Regulation of Complement Genes; *Ann. Rev. Immunol.*, 13, 277–305.

106

Monoclonal Antibodies

S.K. Gupta and Nachiket Shembekar

I. INTRODUCTION

Köhler and Milstein (1975) opened up new avenues in immunology by demonstrating that somatic cell hybridization could be used to generate hybrid cells capable of growing in culture and secreting antibody molecules of predefined specificities. The hybridoma technology has already fulfilled the long lasting desire of immunologists, i.e., the routine production of large amounts of homogeneous antibody against a wide variety of antigens. Some of the important characteristics of antibodies obtained by this technique in contrast to those obtained from conventionally immunized animals are summarized in Table 106.1.

II. PRINCIPLE OF GENERATING HYBRID CELLS

Hybridoma technique involves the fusion of cells (capable of making antibody of predefined specificity) obtained either from spleen or from lymph node of animals, previously immunized with the desired antigen, with a myeloma cell line, which is capable of growing in tissue culture. The broad outline of the procedure is shown in Figure 106.1. Polyethylene glycol (PEG-1500) is used as a fusion agent. Spleen cells or cells obtained from lymph node die-off in due course of time in tissue culture medium. The myeloma cells used for fusion are defective in the enzyme hypoxanthine guanine phosphoribosyl transferase (HGPRT), the key enzyme for *Salvage* Pathway for DNA synthesis. The HGPRT defective myeloma cells are selected for their resistance to grow in purine analog, 8-azaguanine. In a selective medium consisting of hypoxanthine, aminopterin

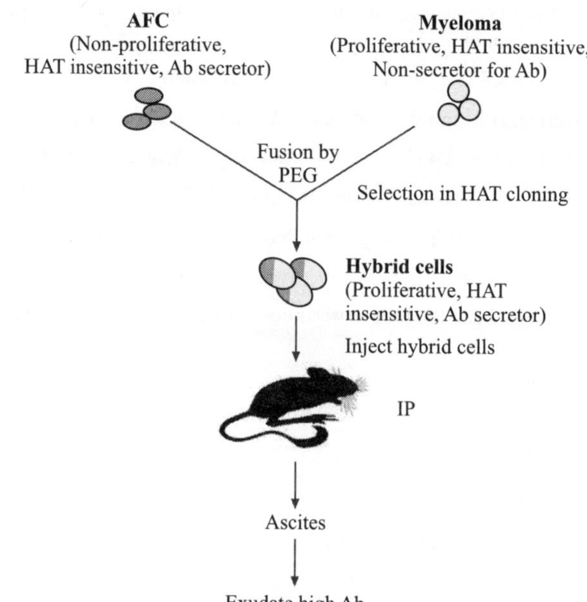

Figure 106.1 Schematic diagram to illustrate the principle of generating hybrid cell clones. AFC, Antibody forming cell; PEG, Polyethylene glycol; HAT, Medium containing hypoxanthine, aminopterin and thymidine; Ab-Antibody; IP-Intraperitoneal. (*see Plate 43 for colour figure*)

and thymidine (HAT), the myeloma cells (HGPRT⁻) die-off as aminopterin (a folic acid analog) blocks the main pathway of DNA synthesis. Only the hybrid cells survive in HAT selective medium, since the myeloma cells provide the ability to grow in tissue culture and the spleen cells contribute the functional

TABLE 106.1 Properties of Monoclonal and Conventional Antibodies

Property	Conventional antibody	Monoclonal antibody
Useful antibody content	Low (0.1–1.0 mg/ml)	High (0.5—5 mg/ml)
Composition	Heterogeneous	Homogeneous
Specificity	Variable from animal to animal	Highly consistent
Cross-reaction with other antigens	Partial with antigens bearing common antigenic determinants	Usually absent but complete if antibody binds to a common determinant
Class and sub-class of immunoglobulins	Typical mixture of all	Only one, may be any
Supply	Limited	Abundant

HGPRT enzyme necessary to overcome the aminopterin block.

Myeloma cells defective in the enzyme, thymidine kinase (TK), can also be successfully used. Such variants of myeloma cells are selected by their resistance for growing in pyrimidine analog, 5-bromodeoxyuridine.

III. INTRASPECIES HYBRIDOMA FOR PRODUCTION OF MURINE MONOCLONAL ANTIBODIES

Hybridomas between mouse myeloma and spleen or lymph node cells obtained from immunized mice have been frequently developed. A number of mouse myeloma cell lines are available for hybridoma production (Table 106.2). Earlier fusion with P3-X63-Ag8 myeloma, secretor of kappa light chain and γ heavy chain had the disadvantage that the hybrid cells secreted immunoglobulins made up of heavy and/or light chains contributed from both parent cells. Such antibodies will have lower avidity as well as lower titre. To circumvent this problem, variants of this cell line such as X63-Ag8.653 or SP2/0-Agl4, which do not synthesize or secrete immunoglobulins, are now widely used. To immunize animals there is no set rule for doing this. Trials with different immunization schedules will only give an idea about the best method for a given antigen. It has been found that boosting by an intravenous injection, 3 to 4 days prior to the removal of spleen for fusion, greatly enhances the frequency of specific hybridoma production. Unlike the conventional methods of antibody production, the antigen of interest need not be pure for immunization, as selection of specific hybridomas at the later stage will avoid the

necessity for purification of the antigen. Choice of the proper strain of mouse is also very important. As mouse myeloma cell lines are derived from BALB/c, this strain of mouse is preferred for immunization.

Fusion of immune cells with mouse myeloma is brought about by polyethylene glycol (PEG). However, the main problem has been the relatively low frequency of fusion. Using PEG and unfractionated spleen cells, it has generally been possible to obtain 1 hybrid for every 2×10^5 spleen cells. This results in approximately 1000 hybrids/spleen, which is adequate if good immunogens are being used, but is a serious problem for weak immunogens. However, fusion produces an apparent enrichment with respect to specific hybrids, which may be due to selective fusion of myeloma cells to replicating B cells. To increase further the fusion frequency, various investigators have explored the use of different molecular weight PEG, the importance of pH during fusion, the ratio of spleen to myeloma cells (between 10:1 and 1:1 have been used successfully), different batches of fetal calf serum/fetal bovine serum, which are summarized in Current Topics in Microbiology and Immunology 81, 1978.

Selection of specific hybrid cell clone involves assaying for the presence of specific antibodies in the culture supernatants. Antibodies can be detected by any of the following methods:

(i) Radioimmunoassay using ^{125}I-antigen under investigation
(ii) Enzyme linked immunoadsorbant assay (ELISA)
(iii) Direct or indirect binding assay using ^{125}I-protein A
(iv) Indirect immunofluorescent assay
(v) Antibody mediated ^{51}Cr release cytotoxicity assay
(vi) Haemagglutination test

After antibody forming hybrid cells are identified, quite often on mass culture, the production of antibody is lost. This is mostly due to the overgrowth of non-specific hybrids or loss of chromosomes or segregation of light and/or heavy chain genes of immunoglobulin molecules. The problem due to overgrowth of non-specific hybrids can usually be avoided by cloning of positive hybrids by either of the following methods: (i) limiting dilution technique, or (ii) soft agar cloning. Positive stable hybrids can be stored by freezing them in liquid nitrogen and can be revived, whenever required.

Growth of hybrid cells in tissue culture can produce antibody levels from 10 to 100 μg/ml of culture medium. By

TABLE 106.2 Mouse Myeloma Cell Lines

Cell Line	Chromosome no.	Derived from	IgChain
P3-X63-Ag8	65	MOPC 21, BALB/c	κ, γl
P3/NSI/I-Ag4-1	65	X-65-Ag8	κ*
X63-Ag8. 653	58	X-63-Ag8	None
SP2/O-Ag14	72	X-63-Ag8 x BALB/c hybridoma	None

*Only synthesize, do not secrete

this way unlimited quantities of monoclonal antibody (MAb) can be generated although the antibody molecules will be contaminated with the serum used for growing of the hybrid cells. Most of the hybrid cells obtained can, however, be grown as ascites in syngenic or nude mice. By this way, very high titres of specific antibody ranging from 0.5 to 5 mg/ml can be obtained.

Besides mouse myeloma, rat myeloma cell line (210.RCY3. Ag1.2.3) has also been developed. This cell line has been used successfully to generate rat-rat hybridomas.

IV. INTERSPECIES HYBRIDS

A number of investigators have fused mouse myeloma with immune cells obtained from rat, rabbit and even human. Two main advantages of mouse-mouse hybrids are their compatibility to grow in conventional murine host and greater knowledge about the mouse immunogenetics. Interspecies hybridomas are invariably rejected by conventional rats and mice. However, these can be propagated in athymic (nude) animals or conventional animals which have been immuno-suppressed using a regimen of anti-lymphocyte serum and/or total body irradiation.

Interspecies rat-mouse hybridomas do offer a number of distinct advantages over mouse hybridomas, such as: (i) Lewis rat κ light chain allotype is non-cross-reactive with BALB/c κ light chain, which therefore can serve as one of the useful parameters for quantifying the amount of rat immunoglobulin (Ig) in a given culture supernatant; (ii) Karyotype of rat (40 chromosomes, 14 metacentric) is different from mouse (40 chromosomes, none metacentric) can be a useful parameter for assessing the stability of interspecies hybridomas; (iii) rats are bigger in size than mice and one can therefore, repeatedly sample portions of the spleen at different time points from an individual rat after antigen challenge.

A number of mouse X human hybrids have been developed, which secrete human antibody against Forssman antigen, human mammary carcinoma cells, keyhole limpet hemocyanin (KLH) and tetanus toxoid. Mouse-human interspecies hybridomas preferentially segregate human chromosomes. However, loss of human chromosomes from mouse X human hybridomas is not random. Human chromosome 14 (heavy chain) and 22 (light chain-λ) are preferentially retained, whereas chromosome 2 (light chain-κ) is preferentially lost.

V. HUMAN MONOCLONAL ANTIBODIES

Human MAbs are desirable for immunoglobulin therapy and are preferred to conventional murine fusion products as the former will be less immunogenic in humans than the xenogeneic mouse immunoglobulins. Human MAbs have been produced mainly by the following ways:

(i) *By fusing human lymphocytes with murine myeloma:* See interspecies hybrids

(ii) *By fusing human lymphocytes with human plasmacytoma/ lymphoblastoid cells:* Since the chromosomal constitution of intraspecies human hybrids is much more stable, human X human hybridomas are more likely to be a useful source of specific human MAbs. The limitation in this approach has been the availability of suitable human plasmacytomas or lymphoblastoid cell lines as one of the fusion partners. They secrete light and/or heavy chain and divide slowly in the culture. Several plasmacytoma cell lines have been developed (Table 106.3). SKO-007 human myeloma that originated from the IgE-producing myeloma (U-266) has been rendered deficient in the purine salvage enzyme, HGPRT. The first reported human hybridoma producing MAbs against 2, 4 dinitrophenyl (DNP) has employed this cell line.

TABLE 106.3 Human Myeloma Cell Lines Used for Human Hybridoma Production

Cell Line	Type	Origin
SKO-007	Human myeloma	Derived from fusion of U-266 human myeloma cell line with lymphoid cells from patient with Hodgkin's disease
KR-12	Human myeloma	Generated by fusion of human lymphoblastoid cell line KR-4 and human plasmacytoma cell line RPMI8226
GM1500 6TG-A12	Human myeloma	6-thioguanine resistant mutant of EBV-transformed GM1500 lymphoma line
K6H6/B5	Human-mouse heterohybridoma	Generated by fusion of NS-1-Ag4 myeloma cells and malignant lymphoid cells from patient with nodular lymphoma
Karpas 707H	Human myeloma	Generated by fusion of 707 cell line from multiple myeloma patient and EBV-transformed 164 B cells
SHM-D33	Human-mouse heterohybridoma	Generated by fusion of human myeloma cell line FU-266 with the mouse myeloma cell line P3X63Ag8
HMMA 2.5	Human-mouse heterohybridoma	Fusion efficient clone derived by limiting dilution of clone from the parental line HMMA2.11TG/O; generated by fusion of the mouse myeloma cell line P3X63Ag8.653 with bone marrow mononuclear cells from a patient with IgA myeloma
MFP-2	Human-mouse heterotrioma	Generated by fusion of murine myeloma cell line with a human myeloma cell line, yielding the intermediate heteromyeloma B6B11, followed by fusion with a human lymphocytes

The paucity of human myeloma lines has prompted other investigators to construct human X human hybrids with lymphoblastoid cell lines (LCLs). The first LCL fusion partner GM1500 6TG-A12 secreting IgG2, κ was established from a patient with multiple myeloma. The cell line has been used to produce IgM antibody specific for measles virus nucleocapsids as well as islet cells in the pancreas. Table 106.3 lists some of the important human myeloma cell lines that have been used to generate human hybrid cell clones secreting MAbs.

Lymphocytes isolated from peripheral blood, bone marrow, spleen, tonsil or lymph node have been used for fusion. The number of B cells secreting specific antibody are very few in peripheral blood. Their number can be increased by stimulating these cells *in vitro* with pockweed mitogen (PWM) or antigen or a combination of both.

(iii) *By Epstein-Barr virus (EBV) transformation of human lymphocytes:* EBV isolated from B95-8 marmoset cell line is a lymphotropic herpes virus which transforms normal B lymphocytes, making them capable of culture as established lines. Using EBV technique, a number of hybrid cell clones secreting human MAbs against streptococcal carbohydrate A, tetanus toxoid, Rhesus antigen D, etc have been developed. The technique involves two steps: (a) the enrichment of specific antibody producing cells; (b) immortalization of these cells by EBV infection (Figure 106.2). The main limitations of the EBV technique are the low quantities of antibody produced (1 µg/ml) and the relative instability (<8 months) of these lines. The relative instability and low amount of antibody produced by EBV transformed cell lines and low frequency of specific hybrids obtained by fusion of human lymphocytes with murine or human plasmacytomas/lymphoblastoid cells can be overcome by fusion of specific EBV transformed B cell lines with appropriate myeloma partner.

(iv) *Generation of humanized antibodies by genetic engineering:* Using genetic engineering (recombinant DNA technology), it has been possible to "humanize" defined mouse MAbs. The first generation of "humanized" MAbs were that of the "chimeric MAbs" consisting of mouse variable regions linked to human constant domains of light and heavy chains (Morrison et al., 1984; Figure 106.3). It took a decade for the first chimeric MAb, abciximab for homeostasis, to be approved by FDA in 1994 (Faulds and Sorkin, 1994). Abciximab is made from the Fab fragments of an immunoglobulin that targets the glycoprotein IIb/IIIa receptor on the platelet membrane. It is a platelet aggregation inhibitor and mainly used during and after coronary artery procedures like angioplasty to prevent platelets from sticking together and causing thrombus (blood clot) formation within the coronary artery. Rituximab, another chimeric MAb was developed to treat lymphoma, leukemia, transplant rejection and some autoimmune disorders. The MAb against the protein CD20, destroys both normal and malignant B cells that have CD20 on their surfaces. Cetuximab is an epidermal growth factor receptor (EGFR) inhibitor, given by intravenous infusion for treatment of metastatic colorectal, and head and neck cancer.

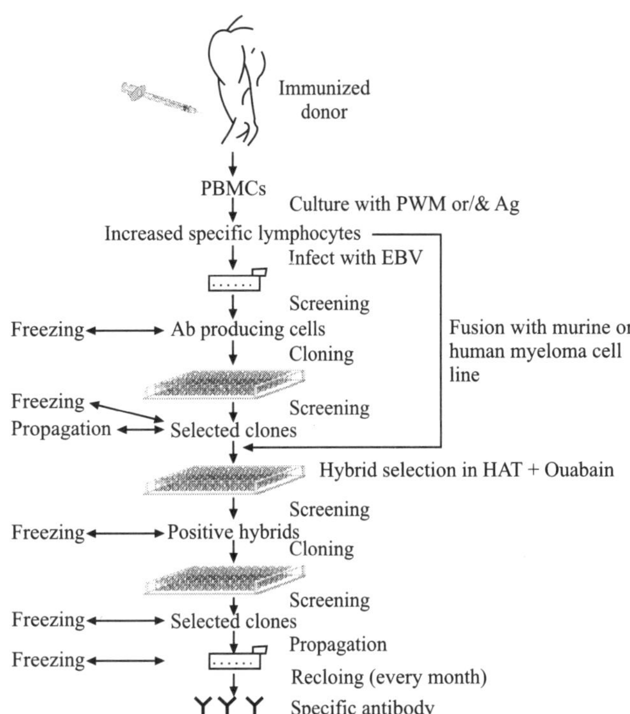

Figure 106.2 Generation of human monoclonal antibodies using Epstein-Barr virus. PBMCs, peripheral blood mononuclear cells; PWM, pockweed mitogen; Ag, antigen; EBV, Epstein-Barr virus; Ab, antibody; HAT, selection medium containing hypoxanthine, aminopterin and thymidine.

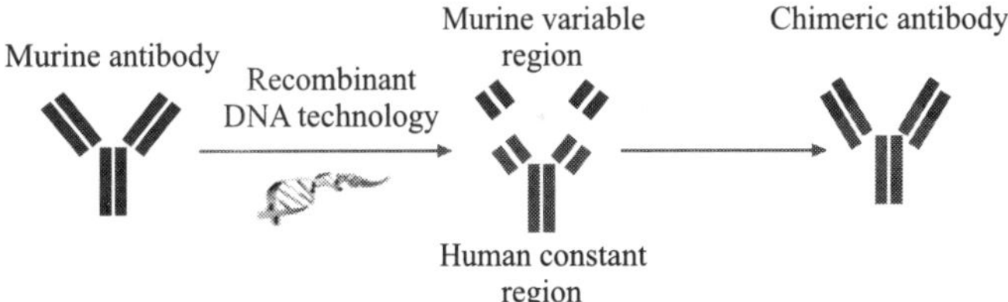

Figure 106.3 Schematic representation of the methodology to generate chimeric antibody. (*see Plate 43 for colour figure*)

To further reduce the murine content of these antibodies, only complementarity determining regions (CDR) were grafted onto the human framework regions giving rise to "humanized" antibodies (Kettleborough et al., 1991; Figure 106.4). The first humanized MAb, Zenapax for kidney transplant rejection, was approved for clinical use by FDA in 1997 (Vincenti et al., 1998). After this, till 2010 several humanized antibodies got license for their application as a potential therapeutic agent. These versions (chimeric and humanized) constituted majority of candidates in clinical study during 1990s and even today half of the MAbs currently in the market are either chimeric or humanized products. Humanization alleviated the induced human-anti-murine antibody (HAMA) response to certain degree, but there remained some other critical issues as the humanization process is technically demanding and might result in reduced antigen binding affinity and decreased efficacy.

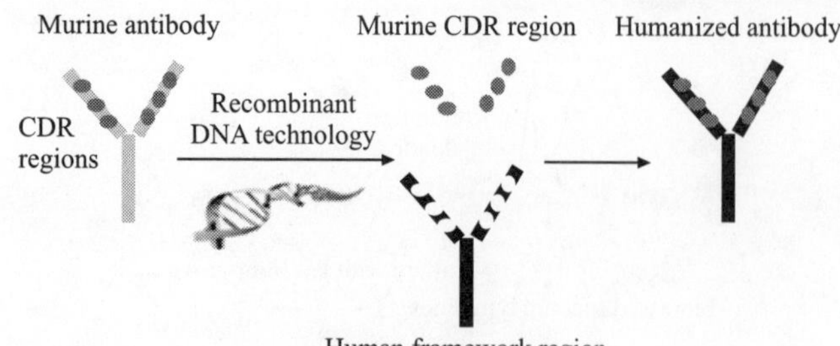

Figure 106.4 Schematic representation of the methodology to generate humanized antibody. CDR, complementarity determining region. (*see Plate 43 for colour figure*)

(v) *Using phage-display antibody library:* Phage display technology is an efficient tool to generate high affinity recombinant antibodies (Figure 106.5). Developments in the field of molecular biology have made it possible to amplify variable regions of immunoglobulin genes. A phage-display antibody library consists of expression of variable region of antibodies (single chain variable fragment (scFv) or Fab fragments) on the surface protein of bacteriophage (Vaughan et al., 1996). Such phage-displayed antibodies are readily panned against the desired immunogen and subsequent to elution step are re-infected on bacterial hosts. The selected antibodies can be either solubely expressed in the bacterial expression system or the specific antibody variable region can be excised and cloned into whole human antibody expression vector to produce fully human MAbs. Humira™ (Adalimumab), the first fully human MAb derived from a bacteriophage displayed antibody library, was approved by the FDA in 2003 for the treatment of rheumatoid arthritis (Weinblatt et al., 2003; Table 106.4).

(vi) *Using transgenic humanized mice:* Kohler and Milstein's classical hybridoma technique showed rather disappointing results in the human hybridoma context which inspired the development of humanized transgenic mice technology. The knockout mice wherein mouse immunoglobulin genes are replaced by human immunoglobulin genes produce fully human MAbs upon vaccination. The spleens of such mice are used to generate MAbs using conventional hybridoma technique. Recently, technology has also been developed to introduce human chromosomes 14 and 2 into heavy and light chain gene deficient mice (Russell et al., 2000; Lonberg, 2005). Vectibix, an anti-EGFR antibody approved for colorectal cancer therapy in 2006, was the first fully human therapeutic antibody derived from the transgenic mice (Chua and Cunningham, 2006; Table 106.4). The drawback of this technology is that only certain biological industries are the proprietors of these transgenic mice and hence this technology is not available to scientists.

(vii) *Novel approach for generating human MAbs:* Recently, an innovative method has been developed for successful generation and cloning of antigen specific human MAbs from the B cells of vaccinated individuals (Wrammert et al., 2008, Smith et al., 2009). Figure 106.5 schematically represents the general scheme of this procedure. The method describes the generation of influenza specific antibodies by vaccinating

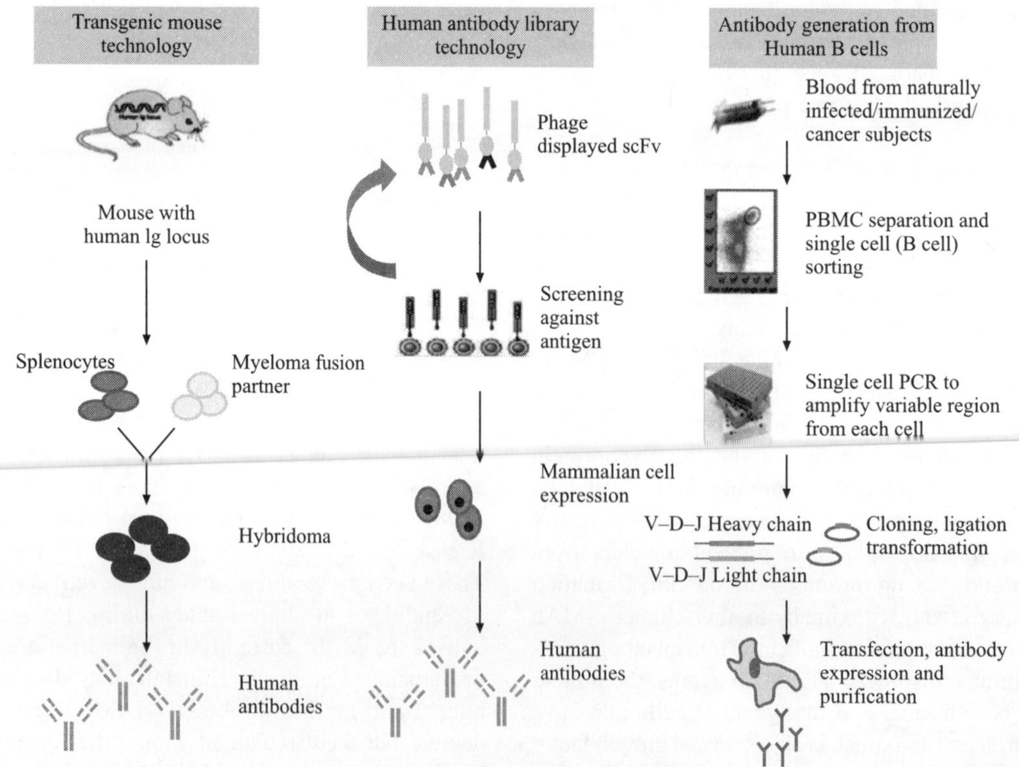

Figure 106.5 Schematic representation of various methodologies to generate human monoclonal antibodies. scFv, single chain variable fragment; PBMC, peripheral blood mononuclear cells; PCR, polymerase chain reaction. (*see Plate 43 for colour figure*)

TABLE 106.4   Therapeutically Approved Human MAbs and Their Applications

Product name (Trade name)	Indication	Company (FDA approval)	Source
Adalimumab (HUMIRA[TM])	Autoimmune disorders like rheumatoid arthritis, Psoriatic arthritis, Morbus Chron	**Abbott** (2002)	Phage display technology
Panitumumab (Vectibix[TM])	Metastatic colorectal carcinoma	**Amgen** (2006)	Transgenic mice
Golimumab (Simponi[TM])	Rheumatoid and Psoriatic arthritis, Active ankylosing spondylitis	**Centocor** (2009)	Transgenic mice
Canakinumab (Ilaris[TM])	Cryopyrin associated periodic syndromes	**Novartis** (2009)	Transgenic mice
Ustekinumab (Stelara[TM])	Plaque psoriasis	**Centocor** (2009)	Transgenic mice
Ofatumumab (Arzerra[TM])	Chronic lymphocytic leukaemia	**GSK** (2009)	Transgenic mice
Denosumab (Prolia[TM])	Treatment of postmenopausal osteoporosis	**Amgen** (2010)	Transgenic mice

individuals and collecting their blood after 7 days. Antibody secreting B cells are then sorted using described markers and single cell PCR is carried out to amplify variable regions from each of the sorted cells. Subsequently, the variable regions are cloned and antibodies are expressed in mammalian system. Vaccination offers an advantage that the majority of the isolated antibodies are antigen specific. However, this technique can also be applied in cases of natural infections or certain autoimmune disorders since during or soon after natural infections there is significantly high repertoire of antibody secreting cells (Smith et al., 2009).

VI.   CURRENT APPLICATIONS AND FURTHER POTENTIALS OF THE MONOCLONAL ANTIBODIES

With the development of hybridoma technique, MAbs can now be prepared against any substance capable of eliciting an antibody response. To produce enormous amounts of MAbs, one requires small amount of antigen and this technique can be usefully exploited in raising antibodies against antigens available in limited supply. The unique ability of antibody molecules to bind specifically to the target molecules can serve to diagnose and treat various medical conditions. An important feature of the use of MAb is the possibility to obtain perfectly standardized reagents which can be maintained stable and reproduce indefinitely. Such reagents may be used as international standards, for quantitative and diagnostic immunoassays for *inter alia,* hormones, enzymes, immunoglobulins, drugs and for various bacterial, viral and parasitic diseases of man. Monoclonal antibodies (MAbs) have been used in the study of the hormone receptors. Because monoclonality does not necessarily exclude cross-reactivity, anti-hormone MAbs could be used to identify the basis of antigenic cross-reactivity between related hormones. One consequence of this might be development of more precise radioimmunoassay.

Monoclonal antibodies (MAbs) offer excellent tools for the purification of antigens using usual immunochemical methods, such as solid phase immunoadsorbent columns. Further, the use of MAbs should considerably help in the identification of antigenic determinants on parasite membranes or virus which are involved in effector mechanisms related to the development of immunity to the parasite or virus. Many classification schemes of microorganisms like viruses are based on their antigenic analysis and for this purpose MAbs-based reagents capable of defining unambiguously, for example, antigenic variants of a given type of virus isolated from various geographical locations are of immense utility. These reagents, since they could be made in large amounts, could potentially serve as ideal reference typing reagents for parasites or viruses from different parts of the world. MAbs have shown good potential to isolate critical immunogenic determinants for subunit vaccine production.

Use of MAbs in human tissues typing has great potentials of clinical application and use in transplantation surgery. Monoclonal antibodies directed against human, mouse and rat major histocompatibility antigens are available. Monoclonal antibodies are ideal tools for the identification and characterization of tumour specific antigens and may become valuable in the classification of tumours. Epitopes diversity and distribution, the frequency of that epitope in a particular tumour type and the stage of the tumour's life in which the epitopes appear can all be explored using MAbs. Monoclonal antibodies have already been produced against many tumours including neuroblatoma, teratocarcinoma and acute lymphocytic leukaemia. Tumour specific MAbs including those directed against secreted tumour products, could be used for early diagnosis, monitoring of the response to cancer drugs as well as cancer immunotherapy. Monoclonal antibodies can also be used as delivery agents for drugs, e.g. daunamycin or ricin-A chain at tumour specific sites. Monoclonal antibodies can also be tagged with radioisotopes and thus can be used for radio-immuno-detection/imaging of cancer.

As each MAb is individual antibody species, it may be used to study the diversity of antibody responses to a single antigenic determinant. Amino acid sequencing of two MAbs that differ only in their idiotypes would be helpful in analyzing the structural basis of idiotype.

Of far greater potential is the therapeutic use of MAbs in protection studies. The striking finding that MAbs derived against rabies virus protect experimental animals from subsequent lethal challenge by the virus provides great hope for the future clinical applications of these as well as other anti-viral and anti-parasite reagents. The traditional therapy areas in the MAb market have been oncology, autoimmune and inflammatory disorders, allergic disorders and transplantation. By 2010, approximately 30 FDA approved therapeutic antibodies are available in the market with overall revenue of 48 billion US dollars. Over 40 MAbs are entering in clinical studies per year since 2007. Currently, 20% of all biopharmaceuticals in clinical trials are MAbs. More than 75% of the marketed therapeutic antibodies currently available are meant for treatment of cancers and autoimmune disorders. Approximately, 10% of the therapeutic antibodies are used for treatment of allergic disorders and infectious diseases. Introduction of newer therapeutic antibodies against chronic conditions like Asthma will greatly expand the market. More than 18 MAbs are under development for HIV, viral hepatitis, rabies, anthrax etc. Attempts are also being made to make therapeutic antibodies against non-traditional area such as macular degeneration, bone loss, Alzheimer, sepsis etc. One of the key advantages of MAb based therapeutics includes reduction in the debilitating side effects currently experienced with radiation and chemotherapy, offering patients an alternative and more effective therapy. New structure antibodies, such as bispecific and fusion as well as multivalent antibodies are also being developed and will go to clinical studies in the near future.

Acknowledgements

S.K. Gupta would like to acknowledge TATA Innovation Fellowship by Department of Biotechnology, Government of India that facilitated writing this review article.

SUGGESTIONS FOR FURTHER READING INCLUDING CITED REFERENCES

Chua Y.J. and Cunningham D. (2006), Panitumumab, *Drugs Today (Barc)*, 42, 711–779.

Faulds D. and Sorkin E.M. (1994), Abciximab (c7E3 Fab): A Review of its Pharmacology and Therapeutic Potential in Ischaemic Heart Disease, *Drugs*, 48, 583–598.

Kennett R.H., McKearn T.J. and Bechtol K.B. (Eds.) (1980), *Monoclonal Antibodies Hybridomas: A New Dimension in Biological Analysis*, Plenum Press, New York.

Kettleborough C.A., Saldanha J., Heath V.J., Morrison C.J. and Bendig M.M. (1991), Humanization of A Mouse Monoclonal Antibody by CDR-grafting: The Importance of Framework Residues on Loop Conformation, *Protein Eng*, 4, 773–783.

Köhler G. and Milstein C. (1975), Continuous Cultures of Fused Cells Secreting Antibody of Predefined Specificity, *Nature*, 256, 495–497.

Kozbor D. and Roder J.C. (1983), The Production of Monoclonal Antibodies from Human Lymphocytes, *Immunology Today*, 4, 72–79.

Langone J.L. and Vanakis H.V. (Eds.) (1983), *Methods in Enzymology*, Vol. 92, Immunochemical Techniques (Part E: Monoclonal Antibodies and General Immunoassay Methods), Academic Press, New York.

Lonberg N. (2005), Human Antibodies from Transgenic Animals, *Nat. Biotechnol*, 23, 1117–1125.

McMichael A. and Fabre J. (Eds.) (1982), *Monoclonal Antibodies in Clinical Medicine*, Academic Press, New York.

Meichers F., Potter M. and Warner N. (Eds.) (1978), *Current Topics in Microbiology and Immunology*, Vol. 81, Lymphocyte Hybridomas, Springer Verlag, Berlin, Heidelberg, New York.

Morrison S.L., Johnson M.J., Herzenberg L.A. and Oi V.T. (1984), Chimeric Human Antibody Molecules: Mouse Antigen-binding Domains with Human Constant Region Domains, *Proc Natl Acad Sci.*, USA, 81, 6851–6855.

Russell N.D., Corvalan J.R., Gallo M.L., Davis C.G. and Pirofski L. (2000), Production of Protective Human Antipneumococcal Antibodies by Transgenic Mice with Human Immunoglobulin Loci. *Infect Immun*, 68, 1820–1826.

Smith K., Garman L., Wrammert J., Zheng N.Y., Capra J.D., Ahmed R. and Wilson P.C. (2009), Rapid Generation of Fully Human Monoclonal Antibodies Specific to a Vaccinating Antigen, *Nat Protoc*, 3, 372–384.

Staines N.A. and Lew A.M. (1980), Whither Monoclonal Antibodies? *Immunology*, 40, 287.

Thompson K.M. (1988), Human Monoclonal Antibodies, *Immunol. Today*, 9, 113.

Vaughan T.J., Williams A.J., Pritchard K., Osbourn J.K., Pope A.R., Earnshaw J.C., McCafferty J., Hodits R.A., Wilton J. and Johnson K.S. (1996), Human Antibodies with Sub-nanomolar Affinities Isolated from a Large Non-immunized Phage Display Library, *Nat Biotechnol*, 14, 309–314.

Vincenti F., Kirkman R., Light S., Bumgardner G., Pescovitz M., Halloran P., Neylan J., Wilkinson A., Ekberg H., Gaston R., Backman L. and Burdick J. (1998), Interleukin-2–receptor Blockade with Daclizumab to Prevent Acute Rejection in Renal Transplantation, *N Engl J. Med.*, 338, 161–165.

Weinblatt M.E., Keystone E.C., Furst D.E., Moreland L.W., Weisman M.H., Birbara C.A., Teoh, L.A., Fischkoff S.A. and Chartash E.K. (2003), Adalimumab, A Fully Human Anti-tumor Necrosis Factor Alpha Monoclonal Antibody, for the Treatment of Rheumatoid Arthritis in Patients Taking Concomitant Methotrexate: The ARMADA Trial, *Arthritis Rheum*, 48, 35–45.

Winter G. and Milstein C. (1991), Man-made Antibodies, *Nature*, 349, 293–299.

Wrammert J., Smith K., Miller J., Langley W.A., Kokko K., Larsen C., Zheng N.Y., Mays I., Garman L., Helms C., James J., Air G.M., Capra J.D., Ahmed R. and Wilson P.C. (2008), Rapid Cloning of High-affinity Human Monoclonal Antibodies Against Influenza Virus, *Nature*, 7195:667–71.

107

Auto-immune Disorders
Pathogenesis, Diagnosis and Treatment

Amita Aggarwal

CONTENTS

Autoimmune diseases are chronic immune-inflammatory diseases characterized by diverse clinical manifestations and an array of autoantibodies or presence of autoreactive T cells. The autoimmune diseases can be broadly classified, as organ specific in which one organ is predominantly affected like Insulin dependent diabetes mellitus (IDDM), or as non-organ specific where multiple organ systems of the body are affected like systemic lupus erythematosus (SLE). These diseases develop due to breakage of tolerance to self-antigens.

I. HOW IS TOLERANCE TO SELF-ANTIGEN GENERATED?

During development of immune system, the body tries to eliminate all T and B cells that react against self-antigens and select those that react against foreign antigens. In T cells this happens in the thymus, when a developing thymocyte recognizes MHC loaded with self-antigen it either undergoes apoptosis, the so called negative selection or is converted into a regulatory T cell. Similarly, in developing B cell when the B cell receptor recognizes self-antigen in the bone marrow, it is either deleted or it changes its B cell receptor by receptor editing or it is prevented from coming out of the bone marrow.

However these mechanisms of central tolerance (tolerance in primary lymphoid organs) are not perfect and a small number of autoreactive T and B cells reach the periphery in all healthy individuals. In the periphery there are multiple mechanisms to keep these autoreactive cells in check.

1. **Lack of exposure to self-antigens:** Certain tissues like brain and eye are protected from circulating cells due to presence of blood-brain or blood-eye barrier. Thus, autoreactive cells do not encounter the self-antigen and thus do not get activated

2. **Lack of co-stimulation:** T cell activation requires 2 signals for optimum activation, one from MHC-antigen-TCR interaction and the other from B7-CD28 or other co-stimulatory molecules. In the periphery, during homeostasis, self-antigens are presented by APCs that lack co-stimulatory molecules leading to anergy rather than activation of T cells.

3. **Suppression by regulatory T cells:** Regulatory T cells either generated in the thymus or developed in the periphery actively suppress autoreactive T cells by secretion of inhibitory cytokines like IL-10, TGF beta or by contact dependent mechanisms.

4. **Expression of Fas by tissues:** Certain tissues in the periphery have high expression of Fas on their surface and when it interacts with Fas ligand on autoreactive T cell it leads to apoptosis of T cells.

5. **Suppression by regulatory B cells:** Autoreactive B cells are also kept under check by regulatory B cells, similar to T cells.

II. HOW IS TOLERANCE TO SELF-ANTIGENS BROKEN IN AUTOIMMUNE DISEASES?

Though all of us have autoreactive T and B cells in our circulation, majority of us never develop an autoimmune disease suggesting that the balance between autoreactive cells and regulatory mechanisms needs to be tilted to develop disease.

Failure of central tolerance mechanisms

1. **Increased generation of autoreactive cells:** In certain genetic diseases like APECED where there is a defect in expression of self-antigens in the thymus due to defect in autoimmune regulator (AIRE) gene, the developing cells are not able to undergo negative selection leading to egress of large number of autoreactive T cells to periphery. This leads to autoimmunity affecting multiple organs like parathyroids, islet cells, intestine etc.

2. **Decreased generation of Regulatory T cells:** Defects in Foxp3 transcription factor gene which is crucial for development of regulatory T cells leads to decreased frequency or absence of Tregs. This leads to a syndrome called IPEX (immunodeficiency, polyendocrinopathy, ectodermal dysplasia, X-linked).

Failure of peripheral tolerance mechanisms

1. **Break in the barrier:** In trauma/disease in one eye, the blood-eye barrier is broken and autoreactive T cells reactive to eye antigens can now enter the eye as well as eye antigens can enter the circulation leading to activation of auto-reactive T cells. This leads to immuno-inflammation leading to sympathetic opthalmitis in the opposite eye. Similarly, post MI pericarditis can occur due to release of self-antigens during myocardial damage.

2. **Presence of co-stimulation along with self antigens:** Whenever there is any infection at any site, the APCs and DCs get activated and start expressing high levels of co-stimulatory molecules like CD80 and CD86. In addition, at site of infection there is release of auto-antigens due to tissue damage. These two factors make the mileu suitable for expansion of autoimmune T cells. In most situations this is a transitory phenomenon but if it is not regulated well, this can result in autoimmune disease.

   Microbial products like lipopolysaccharide (LPS), proteoglycan (PG) can activate pattern recognition receptors called Toll like receptors (TLR) on APCs again leading to production of pro-inflammatory cytokines and up-regulation of co-stimulatory molecules.

3. **Increased Generation of self-antigens:** Self antigens may be generated during tissue injury but the most common pathway of self-antigen generation is apoptosis. Apoptosis, or programmed cell death leads to formation of apoptotic bodies containing self DNA and other cellular antigens. In physiological condition they are recognized by scavenger receptors on APCs and cleared at the local site without activating APCs. However, if the amount of apoptotic bodies generation is increased or their uptake is defective, presence of apoptotic bodies can lead to autoimmunity as seen in SLE. In SLE there is increased apoptosis of lymphocytes, keratinocytes etc as well as there is defect in uptake of this material by macrophages leading to generation of autoantibodies to antigens like DNA, Ro and La which are present in apoptotic bodies.

4. **Environmental factors:** In addition to this environmental factors play an important role these include microbial agents, chemicals, vaccines and trauma. These agents can work by multiple mechanisms like

   1. Molecular mimicry
   2. Alteration of host proteins
   3. Exposure of self-antigen
   4. Provision of pro-inflammatory mileu

Molecular mimicry: This is a phenomenon whereby the immune response generated against a pathogen also reacts against host antigen as the self-antigen is similar to the pathogen antigen. The best example is acute rheumatic fever where immune response directed against M protein of Group B streptococci reacts against cardiac myosin and brain tissue leading to development of acute rheumatic fever. Similar mechanisms are postulated for association of *Campylobactor jejuni* with Gullian-Barre syndrome

Alteration of Host antigen: Treatment with high dose penicillin leads to binding of penicillin to host antigens and generation of autoimmune hemolytic anemia, hydralazine can bind to histones and lead to generation of drug induced lupus. *Pyrophormanas ginigivalis* leads to citrullination of host proteins thus providing autoantigen for generation of antibodies to cyclic citrullinated peptides in rheumatoid arthritis.

III. PATHOGENESIS OF AUTOIMMUNE DISEASE

Pathogenesis of autoimmune disease is a multistep process where host genetics interacts with environmental factors and results in dysregulated immune response leading to persistent immune-inflammation. This inflammation results in tissue injury and clinical disease. With availability of multiple genome wide association studies (GWAS) it has become evident that there are multiple genes that predispose to autoimmunity in general and there are some disease specific susceptibility genes. Most of the genes implicated have an immunological function and result in persistent inflammation.

The effector pathway consists of Th1 and Th17 cells that produce pro-inflammatory cytokines like IFN-gamma, TNF-alpha, IL-17 and IL23 etc. In addition, autoantibodies along with antigen form immune complexes (ICs) and these ICs activate complement leading to generation of anaphylotoxins like C5a and C3a. They attract neutrophils, increase capillary permeability leading to tissue damage. In SLE the IC mediated damage play a major role in nephritis whereas Th1 and Th17 cells play a major role in synovitis of rheumatoid arthritis or insulitis in IDDM.

A schematic picture of inflammation in autoimmune disease is given below (Figure 107.1).

Spectrum of autoimmune diseases: Autoimmune diseases affect almost 4–5% of world population the most common being autoimmune thyroid diseases, rheumatoid arthritis, Vitiligo and IDDM. However, any organ can be affected by autoimmune disease from brain to skin. Some autoimmune diseases affect predominantly only one organ whereas some other have a more wide spread manifestations the so called non-organ specific or systemic autoimmune diseases. Systemic lupus erythematosus is the best example of non-organ specific disease.

Diagnosis of autoimmune diseases

The diagnosis of an autoimmune disease is based on constellation of signs and symptoms. In organ specific autoimmune diseases there is either hypofunction of that organ like less thyroxine production in autoimmune hypothyroidism or less insulin production in IDDM or there is dysfunction like presence of diarrhea in ulcerative colitis or Crohn's disease where intestine is affected or joint pains in rheumatoid arthritis.

In SLE which is a non-organ specific disease there can be involvement of any organ but it mainly presents with fever, joint pains, skin rash, oral ulcers, nephritis and brain involvement.

The clinical diagnosis is further supported by laboratory data. Routine laboratory tests show evidence of organ dysfunction like cytopenias in SLE, elevated creatinine or abnormal urine examination in renal disease, liver enzyme abnormalities in hepatic involvement etc.

Since presence of autoantibodies or autoreactive T cells is a hallmark of autoimmune diseases, detection of autoantibodies in serum of patient or presence of antibodies in situ in tissue biopsy helps in diagnosis. Detection of autoreactive T cells is a tedious process thus is not used in clinic but is used as a research tool.

Autoantibodies can be detected by agglutination, indirect immunofluorescence (IIF), ELISA, radio-immunoassay or western blotting. Rheumatoid factor, an autoantibody seen in Rheumatoid arthritis, SLE, Sjogren's syndrome etc was traditionally detected by latex agglutination but is now done by nephlometry as that is a quantitative and totally automated precipitation reaction.

Antinuclear antibodies are antibodies directed against various nuclear antigens that are seen in systemic autoimmune disease and are usually detected by IIF using cultured Hep2 cells. On IIF various patterns like homogeneous, speckeled, nucleloar and centromere can be seen which give a rough idea about the antigenic specificity.

Similarly, antithyroid peroxidase antibodies for autoimmune hypothyroidism, anti-acetylcholine antibodies for myasthenia gravis, anti-dsDNA antibodies are now a days detected by solid phase assays which provide quantitative results. However, *presence of autoantibodies does not mean there is autoimmune disease and nor does absence of autoantibodies exclude a disease*. The sensitivity of different autoantibodies varies from 70–90% and the specificity varies from 75–90%. (Table 107.1)

Skin and kidney biopsy can be used to look for immuno-globulins or complement deposits on direct immunofluorescence assay Deposits at dermo-epidermal junction are seen in SLE and Pemphigoid whereas intra-epidermal deposits are seen in pemphigus. In SLE, kidney biopsy shows presence of immunoglobulin, complement and fibrin in glomeruli.

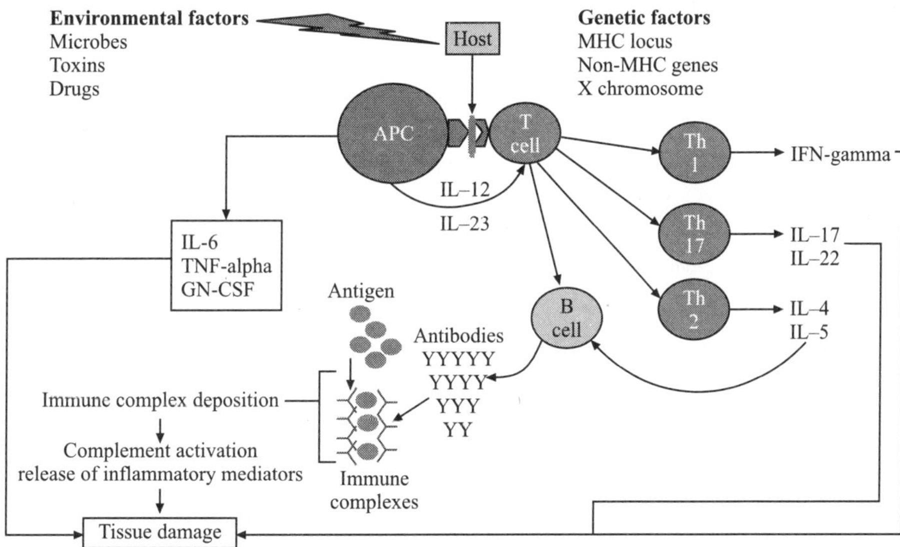

Figure 107.1   Complex interaction of host genetic factors and environmental trigger initiates an immune response against self-antigen (●) leading to release of pro-inflammatory mediators from antigen presenting cell (APC) and activation of naïve T cell (T cell). The T cell differentiates into Th1, Th2 or Th17 cells. The cytokines produced from these cells either directly cause tissue damage or activate other cells to release pro-inflammatory mediators. In addition Th2 cells activate autoreactive B cells to produce autoantibodies that form immune complexes with self-antigen. Immune complexes get deposited in tissues like kidney, joint, skin and brain and activate complement mediated tissue damage. Resultant tissue damage releases more self-antigens and the cycle is perpetuated resulting in significant tissue damage and organ dysfunction. (*see Plate 44 for colour figure*)

TABLE 107.1   Commonly used autoantibodies and the diseases in which they are present

Antibody	Method of detection	Diseases in which present
Rheumatoid factor	Agglutination, Nephlometry	Rheumatoid arthritis (70%), Connective tissue diseases, chronic infections (Leprosy, TB, Malaria)
Anti-nuclear antibody	IIF, ELISA	SLE (98%), Sjogren syndrome (75%), Myositis, Rheumatoid arthritis, juvenile idiopathic arthritis, chronic active hepatitis
Anti-parietal cell antibodies	IIF	Pernicious anemia, atrophic gastritis
Anti-glomerular basement membrane antibodies	ELISA	Good pasture's syndrome
Anti-mitochondrial antibodies	IIF, Immunoblotting	Primary biliary cirrhosis
Antibodies to Glutamic acid decarboxylase (GAD)	Immunoprecipitation assay	Insulin dependent diabetes mellitus
Islet cell antibodies	IIF	Insulin dependent diabetes mellitus
Antibodies to thyroid peroxidase (TPO)	ELISA	Autoimmune hypothyroidism
Antibodies to TSH	Radio-immunoassay	Throtoxicosis
Anti-neutrophil cytoplasmic antibodies	IIF, ELISA	Wegener's Granulomatosis, Microscopic PAN, idiopathic crescentic glomerulonephritis
Antibodies to phospholipids	ELISA	Recurrent venous/arterial thrombosis, recurrent fetal loss

IIF: Indirect immunofluorescence assay; ELISA: enzyme linked immunosorbent assay; PAN polyarteritisnodosa

Treatment

Treatment in a patient with autoimmune disease is dependent on clinical severity and organ involved. In some autoimmune disease replacement of deficient hormone is sufficient to correct the clinical features like Eltroxine in hypothyroidism, Insulin in IDDM. In others, immuno-suppressive agents are used to tame the exaggerated immune response. These immuno-suppressive agents can be non-specific that is they cause general immune suppression like corticosteroids, anti-metabolites etc. or could be targeted to T cells, B cells, co-stimulatory molecules or effector molecules (Table 107.2).

The availability of biological agents has changed outcome of some autoimmune diseases remarkably, especially rheumatoid arthritis, juvenile arthritis, Crohn's disease and spondyloarthropathies. However, in others like SLE the search is still on for better targeted therapy. With better understanding of the underlying immune mechanisms in future it will be possible to design targeted therapies with minimal side effects for better control of autoimmune diseases.

TABLE 107.2   Immunosuppressive drugs, main mechanism and clinical utility

Drug	Main targets	Use
Non-specific immunosuppressive agents		
Corticosteroids	Lymphocytes, phagocytes, cytokine production	Most autoimmune diseases
Azahtioprine	Inhibits purine synthesis	SLE, Crohn's disease, ulcerative colitis, Autoimmune cytopenias
Cyclophosphamide	Inhibits DNA synthesis	SLE, Autoimmune cytopenias, Vasculitis
Mycophenolatemofetil	Inhibits Pyrimidine synthesis	SLE, Autoimmune cytopenias, Vasculitis,
Methotrexate	Interferes with folate metabolism, increases adenosine levels	Rheumatoid arthritis, Psoriasis, SLE, Myositis, Ulcerative colitis
Specific immunosuppressive agents		
Cyclosporine	Decreases IL-2 production	SLE, Psoriasis, Uveitis
Rapamycin	Inhibits IL-2 function	SLE
Anti-TNF agents	Inhibit TNF	Rheumatoid arthritis, Crohn's disease, Psoriasis, Ankylosing spondylitis
Anti-CD20 antibody	Inhibits B cells	Rheumatoid arthritis, SLE, Immune cytopenias
CTLA4 Ig	Inhibits B7-CD28 interaction Prevents T cell co-stimulation	Rheumatoid arthritis, SLE
Anti-Blys antibody	Blocks B cell stimulator (BLyS)	SLE
Anti-IL12/IL23	Blocks Th 17 cells	Psoriasis, Ankylosing spondylitis

SLE: systemic lupus erythematosus

SUGGESTIONS FOR FURTHER READING

1. Kamradt T. and Mitchison N.A. (2001), Tolerance and Autoimmunity, *N Engl J. Med.*, 344:655–64.

2. Goodnow C.C. (2001), *Pathways for Self-tolerance and the Treatment of Autoimmune Diseases*, Lancet, 357:2115–21.

3. Cho J.H. and Gregersen P.K. Genomics and the Multifactorial Nature of Human Autoimmune Disease, *N Engl J. Med.* 365:1612–23.

4. McInnes I.B. and Schett G. (2011),The Pathogenesis of Rheumatoid Arthritis, *N Engl J. Med.*, 365:2205–19.

5. Tsokos G.C., Systemic Lupus Erythematosus, *N Engl J. Med.*, 365:2110–21.

6. Galli M. Clinical Utility of Laboratory Tests Used to Identify-antiphospholipid Antibodies and to Diagnose the Antiphospholipid Syndrome, *SeminThrombHemost*, 34:329–34.

7. Isermann B., Ritzel R., Zorn M., Schilling T. and Nawroth P.P. (2007), Auto-antibodies in Diabetes Mellitus: Current Utility and Perspectives, *Exp Clin Endocrinol Diabetes*, 115:483–90.

8. Wiik A.S. (2005), Anti-nuclear Autoantibodies: Clinical Utility for Diagnosis, Prognosis, Monitoring, and Planning of Treatment Strategy in Systemic Immune-inflammatory Diseases, *Sc and J Rheumatol.*, 34:260–8.

SECTION XV

EMERGING INFECTIONS AND VACCINES UNDER DEVELOPMENT

Dengue and Chikungunya are recent infections making people suffer in Delhi and other parts of the country. Also, many deaths from Japanese encephalitis have been reported from some parts of India for which an improved vaccine is being developed in a national laboratory, besides what is available at present for the epidemics.

108

Dengue

Navin Khanna

CONTENTS

I. INTRODUCTION

Dengue, a mosquito-borne viral disease, is caused by any of four dengue virus (DENV) serotypes, DENV-1, –2, –3 and –4. These belong to the genus *Flavivirus* of the *Flaviviridae* family. Other members include pathogenic flaviviruses such as West Nile virus (WNV), Japanese encephalitis virus (JEV), Tick-borne encephalitis virus (TBEV), yellow fever virus (YFV), St. Louis encephalitis virus (SLEV), Murray Valley encephalitis virus (MVLV), Omsk hemorrhagic fever and Kyasanur forest disease viruses. Female *Aedes* mosquito serves as the vector for transmission of DENVs. Infection with DENV results in varying degrees of pathological conditions, ranging from a mild febrile illness, dengue fever (DF), to severe and potentially fatal dengue hemorrhagic fever (DHF) and dengue shock syndrome (DSS). Currently, there is no vaccine or effective antiviral against the DENVs. The spread of DENVs has been attributed to massive unplanned urbanization, overpopulation, increasing global travel and the inability to eradicate mosquito vectors.

Recently, a dengue virus was discovered from infected human sample during dengue outbreak in Malaysia's Sarawak State in Borneo. Not only was its genomic sequence different from the known four DENV serotypes, but also the antibodies elicited by it in monkeys and humans. Hence, this Malaysian strain is being considered as the fifth serotype. Though this serotype is suspected to be circulating among non-human

1253

primates (macaques) in Borneo, it does not have a sustained human transmission cycle and is believed to have caused only one outbreak till now. Thus, it is considered as sylvatic, circulating primarily in non-human primates.

II. HISTORY

Though dengue may have been reported as early as the last century of the first millennium in China, the first dengue epidemic is believed to have occurred in 1779–1780. The critical event which led to the spread of dengue during the last century has been World War II. The movement of troops and the destruction caused to the environment led to the spread of the mosquito vector and the virus in Western Pacific and, especially, Southeast Asia, which has been a hyperendemic area since then. The causative agent of Dengue was identified in the early part of the twentieth century. Albert Sabin was the first person to isolate DENV-1 (Hawaii) and DENV-2 (New Guinea C) in 1944. Later in 1956, DENV-3 and -4 were isolated.

III. EPIDEMIOLOGY

Dengue has become one of the most widespread re-emerging diseases. There has been a rise in the number of dengue cases and also in the countries reporting such cases (Figure 108.1).

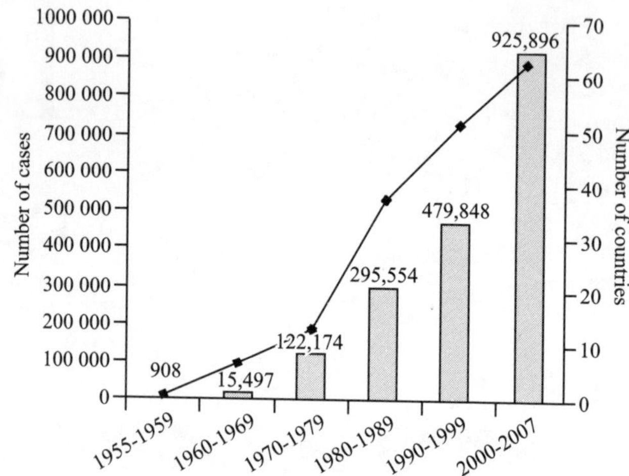

Figure 108.1 Rise in the number of dengue cases and countries reporting dengue. Plot of average annual number of dengue cases (DF and DHF) reported to WHO and the number of countries reporting dengue cases in recent decades.

It is highly endemic in the tropical countries of the world (Figure 108.2).

This disease is rapidly spreading to other parts of the world, and is no longer restricted to the tropics because of the change in climate and the increase in human movement. It is said to have become a global burden causing ~390 million infections per year.

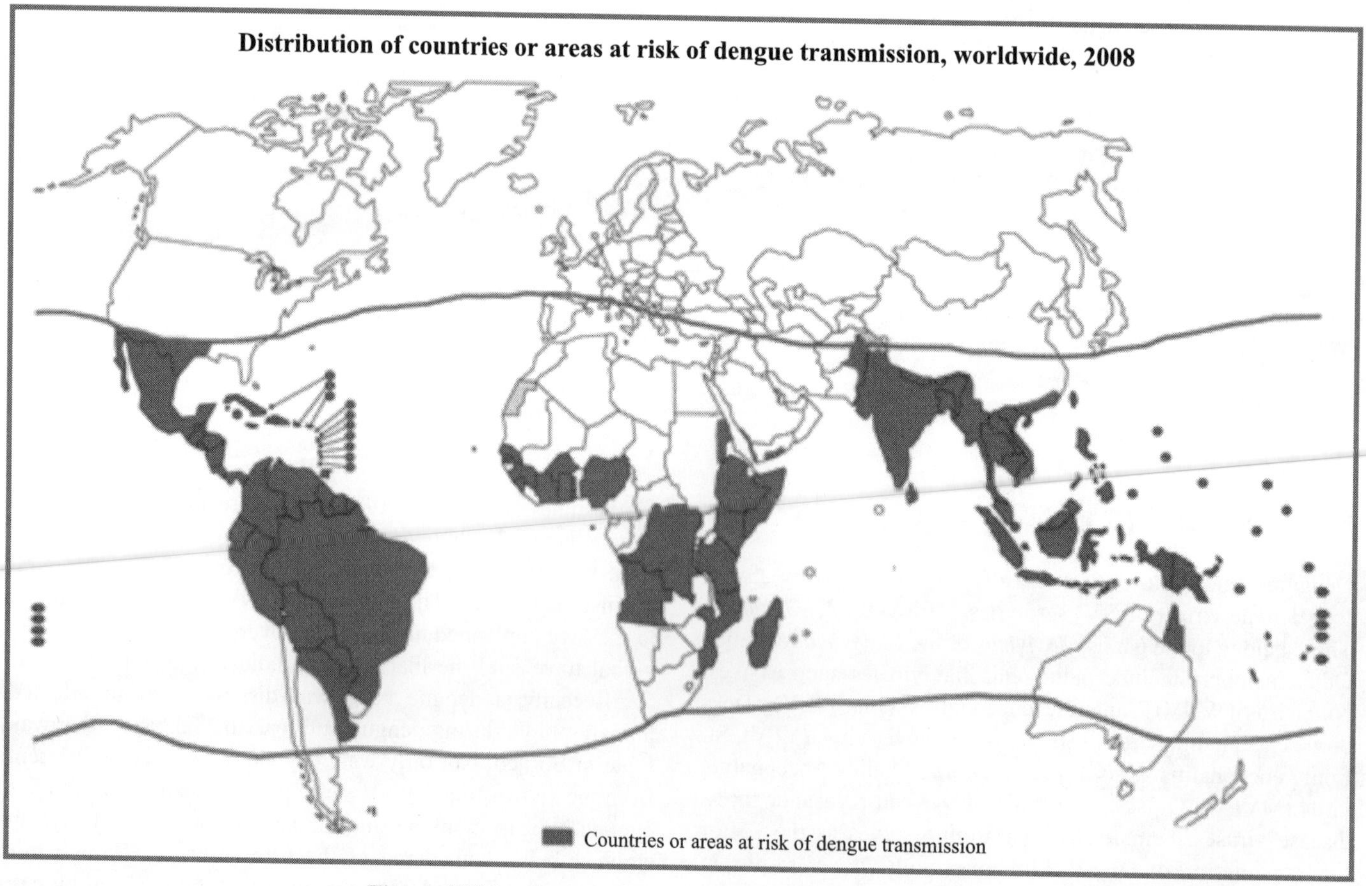

Figure 108.2 Countries/areas affected with dengue in 2008.

IV. ENTOMOLOGY

The female *Aedes* mosquito which spreads dengue in the human population, unlike its male counterpart, feeds on blood (hematophagous). Since, it can also feed on other biological materials, like pollen and fruit juice, it is an optional hematophage. Figure 108.3 gives a pictorial representation of the male and female *Aedes aegypti*, its taxonomical classification and features.

The life cycle of *Aedes* mosquito (Figure 108.4), depending upon the extent of feeding, lasts for 8–10 days at room temperature. It consists of two phases: aquatic (larvae, pupae) phase and terrestrial (eggs, adults) phase.

V. DENGUE TRANSMISSION

A. Dengue Transmission Cycles

Dengue virus serotypes are maintained by two ecologically and evolutionarily distinct transmission cycles: Sylvatic cycle and Human cycle (Table 108.1, Figure 108.5). Humans are the major amplifying reservoir for the virus, and are the only host to manifest symptoms of dengue infection. Certain species of lower primates have also been shown to be susceptible to dengue virus infection, although the duration and magnitude

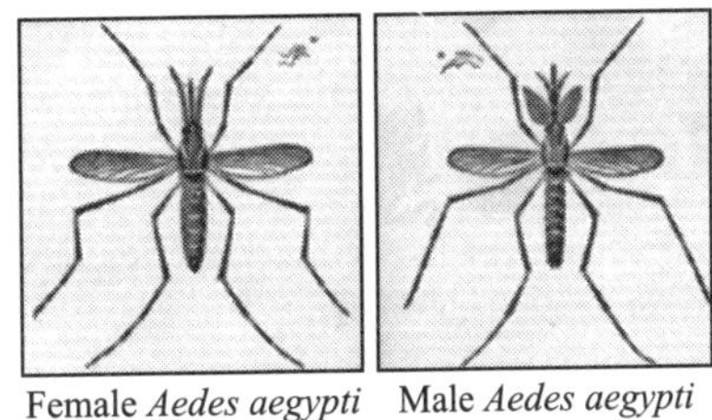

Kingdom	Animalia
Phyllum	Arthropoda
Class	Insecta
Order	Diptera
Family	Culicidae
Subfamily	Culicinae
Genus	Aedes
Subgenera	Stegomyia
Species	aegypti

(b)

Features of Female Aedes aegypti

- Colonizes domestic breeding sites
- Takes several blood meals before laying eggs
- Does not lay all eggs at one site
- Has a short flight range
- Feeds almost entirely on humans during daylight hours
- Endophilic, endophagous

(c)

Figure 108.3 The mosquito vector. (A) Female and male *Aedes aegypti*, (B) Taxonomical classification of *Aedes aegypti* and (C) General features of female *Aedes aegypti*.

of viremia tends to be much lower than that in humans. They constitute the sylvatic cycle of dengue transmission. Transfer of dengue virus from an infected mosquito to its eggs through

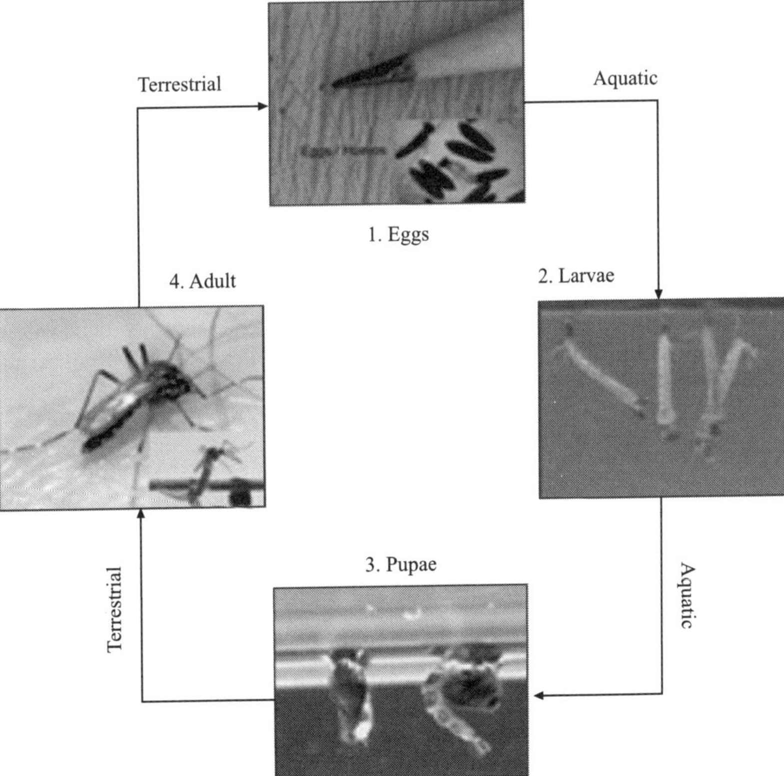

Figure 108.4 General life cycle of *Aedes* mosquito, characterized by aquatic and terrestrial phases. Female mosquitoes lay eggs on the inner wet walls of water-filled container. Eggs are inundated with water to result in hatching of larvae (picture 1, inset). Larvae (picture 2) feed on microorganisms and particulate organic matter to undergo developmental changes. Larvae metamorphose into pupae (picture 3). The pupae, which do not feed, undergo changes in their form until adult mosquito is formed and released into its terrestrial phase (picture 4). (*see Plate 44 for colour figure*)

TABLE 108.1   Host, Vector and location details of the two dengue transmission cycles

Transmission cycle	Host	Vector	Area of prevalence
Sylvatic	Non-human primates like African green monkey, Guinea baboon, patas monkey	Arboreal *Aedes* mosquito	Western and central Africa, peninsular Malaysia
Human	Humans	Domestic *Aedes aegypti* subsp. *aegypti* and peridomestic *Aedes albopictus* mosquitoes	Tropics and subtropics

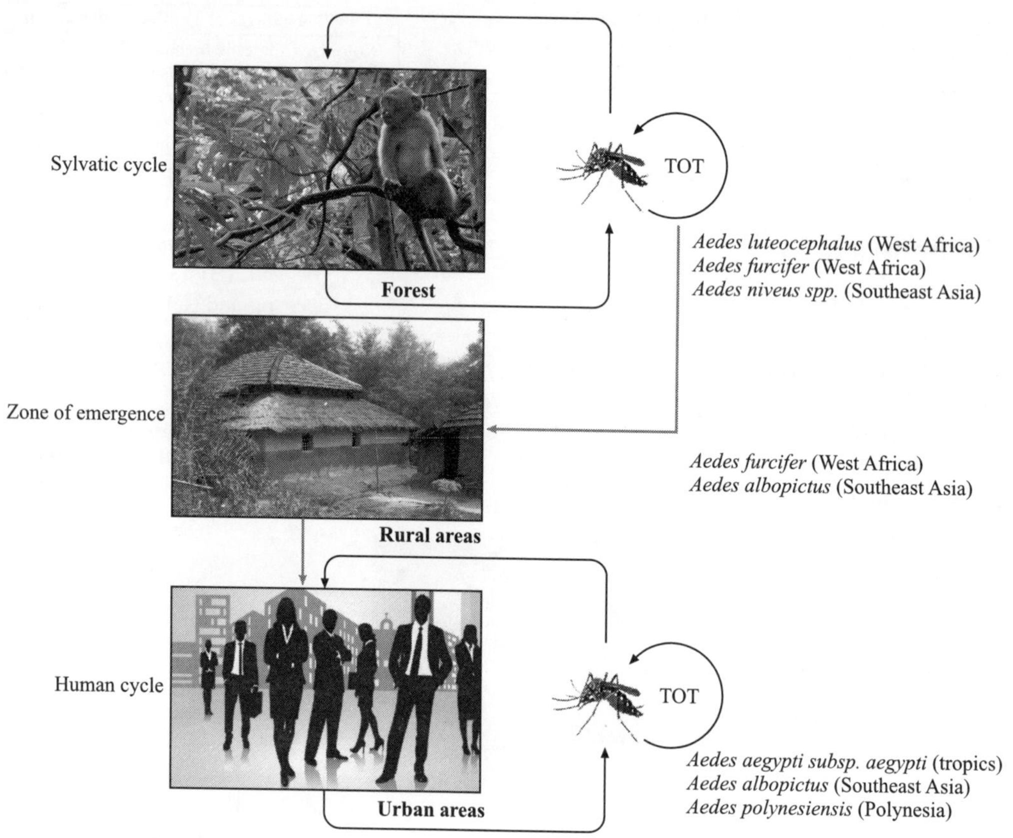

Figure 108.5   Human and sylvatic cycles of dengue virus transmission. Zone of emergence denotes the area where sylvatic cycles contact human populations in rural areas. The transmission of dengue virus from an infected mosquito to its eggs takes place through TOT. The commonly prevalent vectors (*Aedes* sp.), in each of the two cycles and in zone of emergence, have been indicated (Vasilakis et al., *Nat Rev Microbiol*, 9: 532–541, 2011).

transovarial transmission (TOT) is not known in all species. *In vitro* experiments, designed to study the capability of dengue sylvatic strains to infect human cells, have revealed that they are equally capable of infecting human cells. Hence, it has been inferred that substantial adaptive barriers do not exist between sylvatic dengue strains and the human host. Thus, because of absence of adaptive barriers and presence of ecological disturbances, it is believed that sylvatic dengue strains have the potential to shift from non-human to human hosts.

B.   Human Cycle of Dengue Virus Transmission

The human cycle of dengue virus transmission is illustrated in Figure 108.6. For the transmission of DENV to mosquito,

the female *A. aegypti* must bite an infected human during the viremic phase of illness, which generally lasts 2–7 days. Once infective, a mosquito is capable of transmitting the virus to susceptible individuals for the rest of its life.

After ingestion of a virus-containing blood meal, the epithelial cells lining the midgut of the mosquito get infected. This is followed by the escape of virus from midgut epithelium into the haemocele and infection of salivary glands. DENVs secreted into the saliva are transmitted to another human during probing and feeding. Other tissues, like foregut epithelium, nervous tissue, epidermal cells, ovary, fat body and hemocytes, also get infected. DENVs also replicate in the female mosquito's genital tract and may infect the progeny through TOT.

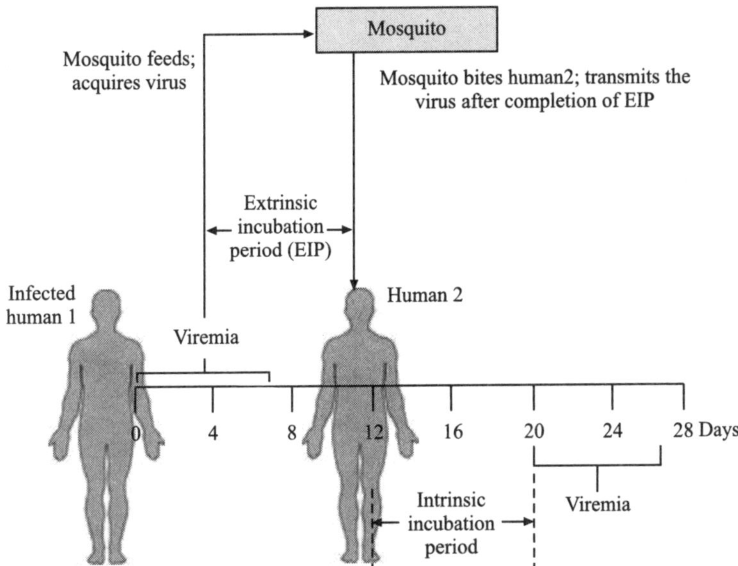

Figure 108.6 DENV transmission cycle in humans. The mosquito bites an infected human (human 1), in viremic phase and acquires the virus. In the mosquito, the virus goes through an extrinsic incubation period, which lasts for 8–12 days, during which it replicates, disseminates and travels back to the salivary gland. Mosquito transmits the virus to another human (human 2), after completion of this period. In the human host, the virus undergoes intrinsic incubation period during which it replicates in the target organs, infects white blood cells and lymphatic tissues. The virus circulates in the blood of infected humans for 2–7 days (viremic phase), approximately at the same time as they have fever. Mosquitoes may now acquire the virus when they bite human 2 during viremic phase, resulting in continuation of the cycle.

VI. SYNDROME

A. Dengue Disease Manifestations

The spectrum of clinical illness ranges from an inapparent, mild, non-specific viral fever (dengue fever, DF), to severe and fatal hemorrhagic disease, known as Dengue Hemorrhagic Fever (DHF)/Dengue Shock Syndrome (DSS). Figure 108.7 illustrates the classification of dengue disease.

It has been found that 50–90% of the dengue cases are asymptomatic. The severe cases of DHF/DSS constitute a very small part of total number of dengue infections and, hence, have been positioned at the tip of the dengue disease pyramid in Figure 108.8A. The major clinical signs and symptoms of DF and DHF/DSS are indicated in Figure 108.8B.

B. Dengue Fever (DF)

It is a self-limiting fever usually lasting for 2–7 days, accompanied by severe headache, sore throat, muscle and joint pain, ocular pain, anorexia, nausea, vomiting, mucosal membrane bleeding, rash and enlargement of liver.

C. Dengue Hemorrhagic Fever/Dengue Shock Syndrome (DHF/DSS)

DHF has been divided into four grades, depending on the severity of the symptoms (Figure 108.9), with grade IV being the most severe.

At onset, the symptoms of DHF appear virtually identical to those of DF, but after a few days the patient's condition

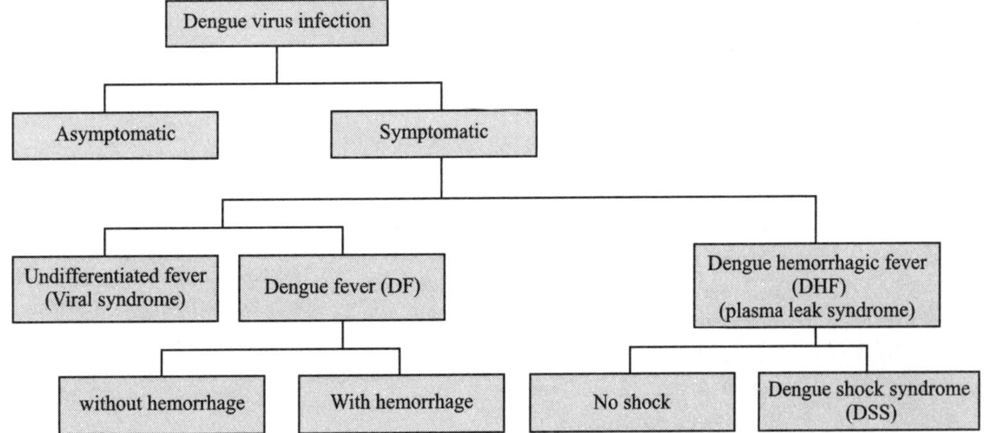

Figure 108.7 Classification of dengue disease.

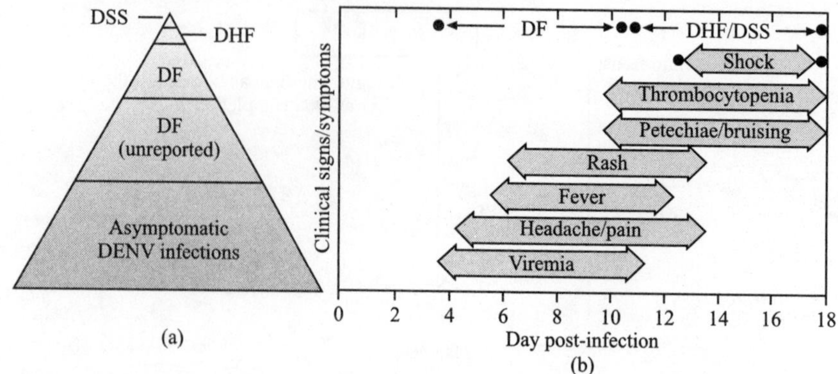

Figure 108.8   (A) Pyramid of dengue disease (Kyle and Harris, Annu Rev Microbiol 62: 71–92, 2008). (B) Time course of clinical signs and symptoms of dengue disease (Murphy and Whitehead, *Annu. Rev Immunol*, 29: 587–619, 2011).

Grade	Symptoms			
I	Fever (2–7 days)	Petechiae, bruising or (*) tourniquest test	Increased vascular permeability	Thrombocytopenia
II		↓	↓	
III		Other hemorrhagic manifestations	Rising hematocrit Hypoproteinemia Serous effusion	Coagulopathy
IV			↓ Hypovolemia Weak pulse Hypotension	
		Severe bleeding ←	→ Profound shock ←	→ Disseminated intravascular coagulopathy

Figure 108.9   The four grades of DHF and their symptoms (Murphy and Whitehead, *Annu Rev Immunol*, 29: 587–619, 2011).

deteriorates rapidly, with signs of fever, spontaneous hemorrhage, overt bleeding, thrombocytopenia and pleural effusion, due to increased vascular permeability leading to plasma loss.

DSS is the severe form of DHF, which includes rapid fall in pulse pressure (< 20 mmHg), clammy skin and restlessness. It lasts for 1–2 days, during which an appropriate volume replacement therapy becomes vital for the recovery of patient. Lack of timely and appropriate medical care of such patients may lead to death.

VII.   PRIMARY DENGUE INFECTION

The first exposure of an individual to any of the four DENV serotypes is known as primary dengue infection. It may result in symptomatic infection. In primary infection, high titers of Immunoglobulin M (IgM) and Immunoglobulin G (IgG) antibodies appear in 3–5 and 6–10 days, respectively, after the onset of infection. The presence of IgM antibody is transient, disappearing in 2–3 months, whereas IgG antibody persists for life. Theoretically, an individual living in a dengue-endemic area can have up to four dengue infections during his/her lifetime. It is believed that sequential or secondary infection,

especially with a different serotype, increases the risk of severe manifestation of the disease (DHF/DSS).

VIII.   SECONDARY DENGUE INFECTION

A secondary infection, with a previously un-encountered DENV serotype, usually results in classical DF. However, 2–3% of secondary infection cases develop into DHF, which may progress to DSS and death.

The primary dengue infection leads to development of moderate viremia and serotype specific antibodies. This provides long lasting immunity to the homologous virus, but the protective immunity towards heterologous serotype is short-lived. During a second infection with a different serotype, the presence of low amounts of heterotypic antibodies (which form complexes with DENVs) promote the access of the virus to monocytes, *via* Fc receptors, leading to an increase in viral load and severity of the disease (Figure 108.10). This

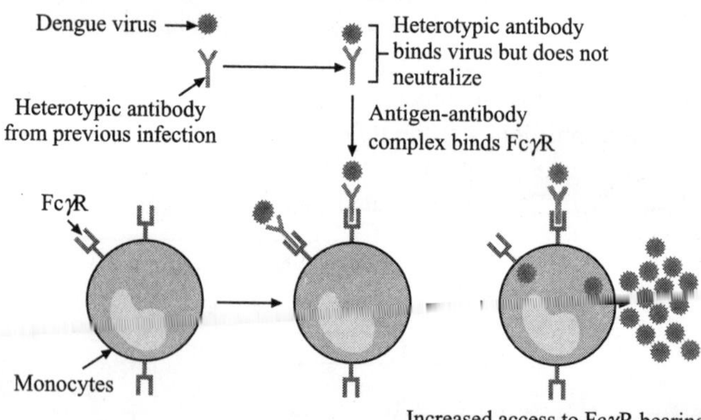

Figure 108.10   Schematic illustration of the ADE phenomenon. DENV binds to heterotypic antibodies present at sub-neutralizing concentrations from previous infection. The complex is internalized by Fcγ receptor bearing cells and leads to an increase in viral load and severity of infection (Murphy and Whitehead, *Annu. Rev Immunol*, 29: 587–619, 2011). (*see Plate 44 for colour figure*)

phenomenon is known as Antibody Dependent Enhancement (ADE). Furthermore, the immune clearance of infected monocytes by T-cells may itself have pathological consequences through the release of cytokines, complement activation and release of γ-interferon, which, in turn, up-regulates expression of Fc receptors, further increasing the enhancement activity. Experimental evidence for ADE comes from studies on non-human primates and AG129 mice, where DENV titers increased on passive transfer of DENV envelope-specific monoclonal antibodies at sub-neutralizing concentrations. Although ADE has been found to result in disease severity, neither all the severe cases are associated with secondary infection nor do all cases of secondary infection progress to DHF/DSS.

IX. DENGUE VIRUS BIOLOGY

A. Virus Structure

The structure of DENV-2 (PR159–S1 strain) was determined by fitting the known structure of the envelope (E) glycoprotein of a related flavivirus, TBEV, into the cryo-electron density map of DENV-2. It was revealed that DENV is an enveloped virus of ~50nm diameter with a characteristically smooth surface. Cryo-electron microscopic studies of the mature DENV-2 virion particle have revealed that the E glycoprotein head-to-tail homodimers cover the entire virion giving it a smooth surface (Figure 108.11).

The E protein, which constitutes the glycoprotein shell of the virus, plays a critical role in several aspects of DENV biology like host cell receptor recognition, membrane fusion and virion morphogenesis. It is also the major target of neutralizing immune response, and is the most highly conserved flavivirus structural protein, with an amino acid similarity of 40%. The amino acid similarity of E protein within serotypes and between serotypes ranges from 90–96% and 60–70%, respectively. Located beneath the E glycoprotein is the membrane protein, followed by the lipid bilayer covered nucleocapsid containing the viral genome. The viral genome is single stranded RNA of positive polarity. The ~11kb genomic RNA has a 5′ methylated cap, but lacks 3′ poly-A tail. It contains a single open reading frame (ORF) flanked by non-coding regions (NCRs) at the 5′ and 3′ ends. The 5′ and 3′ NCRs contain secondary structures and conserved sequences which are involved in regulation of viral replication. The single ORF encodes a single polyprotein precursor, which is co-/post-translationally cleaved by the action of both cellular and virus-derived proteases to generate three structural and seven non-structural proteins. The structure and genomic organization of DENV are presented in Figure 108.11. The dengue viral proteins and their major functions are listed in Table 108.2.

Structural analysis of the E protein has shown that it consists of three domains: Envelope domain (ED) I, II and III. The eight-stranded β-barrel domain, located at the center, is EDI. Envelope Domain II (EDII) is the dimerization domain and contains an internal fusion loop at the tip, which is vital in the membrane fusion process and, hence, in the release of viral genome into the cytoplasm of the target cell. The domain with immunoglobulin-like fold is Envelope Domain III (EDIII), which is involved in binding the receptor on the target cell. EDI and EDIII have been shown to contain, predominantly,

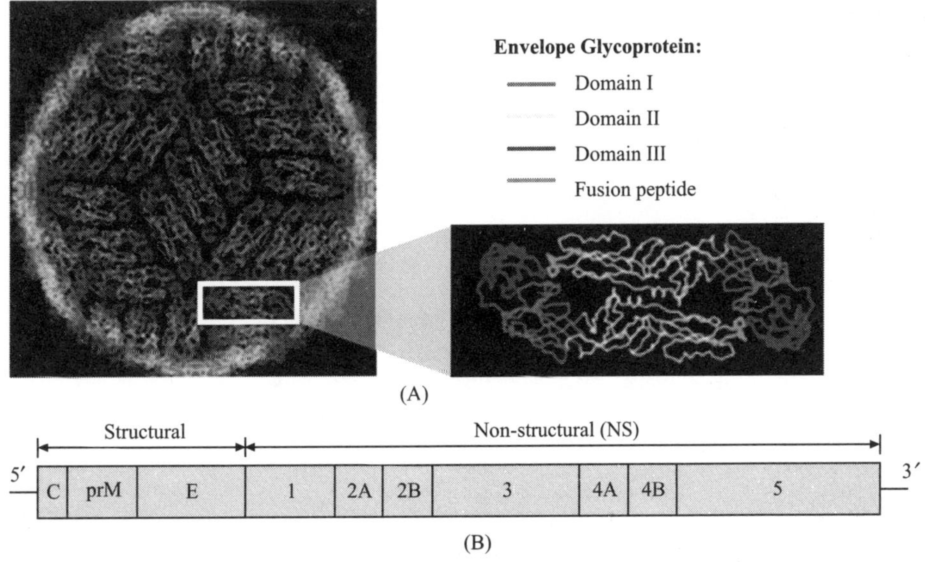

Figure 108.11 DENV structure and genomic organization. (A) Computer-generated structure of DENV based on cryo-electron microscopy. The white rectangle indicates an E homodimer. Shown alongside is an enlarged view of the homodimer, where domains I, II, III and fusion peptide have been represented in red, yellow, blue and green colour, respectively (Kuhn et al, Cell 108: 717–725, 2002). (B) A schematic representation of the DENV genome which encodes a polyprotein. Viral and host cell proteases co- and post-translationally cleave this polyprotein into structural (C, prM and E) and non-structural (NS1, 2A, 2B, 3, 4A, 4B and 5) protein components. Symbols C, prM, E and NS denote capsid protein, precursor membrane protein, envelope protein and non-structural proteins, respectively. (*see Plate 44 for colour figure*)

TABLE 108.2   Dengue virus proteins and their predicted functions

	Protein	Function
Structural	Capsid (C)	Binds and stabilizes viral RNA
	Precursor Membrane/Membrane (prM/M)	'Pr' peptide of prM prevents pre-mature fusion, Membrane protein (M) forms ion channels
	Envelope (E)	Recognition and binding to the host cell, involved in uncoating of virus by enabling fusion of viral and endosomal membranes
Non-Structural	NS1	Viral replication
	NS2A	Probably associated with replicase and plays a role in RNA accumulation
	NS2B	NS3 protease cofactor
	NS3	Protease, helicase, nucleoside triphosphatase activity
	NS4A	Role in membrane modulation
	NS4B	Enables the virus to escape host's innate immune response
	NS5	RNA dependent-RNA polymerase, methyltransferase activity

subcomplex- and type-specific epitopes. EDII contains the major flavivirus group and sub-group cross-reactive epitopes. It has been found that virus-specific neutralizing antibodies are largely raised against the epitopes of EDIII.

The ~100aa long EDIII which is an independently folding immunoglobulin (Ig)-like domain is stabilized by a single disulfide bond. It is exposed and accessible on the virion surface and has been identified to be involved in the binding of virion to the host cell surface receptor. This domain contains multiple-type- and sub-type-specific neutralizing epitopes. It has been observed that the infectivity of DENV can be blocked by recombinant EDIII protein and EDIII-specific mAbs. Thus, these attributes of EDIII have led to its emergence as a promising diagnostic and vaccine candidate.

B.   Cellular Targets of DENV

DENV is known to target, primarily, three organ systems: immune system, liver and endothelial cells. Table 108.3 lists the mammalian target cells and their receptors of DENV. It is believed that many of these act as accessory receptors acting to concentrate DENV at the site of the primary receptor whose identity is not yet known.

The other known mechanism of uptake of DENV occurs in the presence of sub-neutralizing concentration of anti-DENV antibodies in cells bearing Fc receptors (monocytes, macrophages). The uptake is through antibody-dependent and Fc receptor-mediated endocytosis resulting in ADE, referred to earlier.

C.   Life Cycle

The salient steps in the life-cycle of DENV are summarized in Figure 108.12. DENV enters the target cell through receptor-mediated endocytosis. The low pH within the endosomal vesicle causes conformational change in the E protein transforming the dimers to trimers. This leads to exposure of the fusion peptide of E protein, leading to fusion of the viral membrane with the endosomal membrane and concomitant release of viral RNA into the cytoplasm. Due to its positive polarity, the genomic RNA is directly translated into a single polyprotein precursor at the rough endoplasmic reticulum (ER). It is co- and post-translationally processed, by host- and virus-encoded proteases into the structural proteins C, prM and E, and seven non-structural proteins, NS1, 2A, 2B, 3, 4A, 4B and 5. Translation at rough ER also assists localization of the proteins to its luminal, membrane or cytoplasmic side. The viral replicase components, NS3 and NS5, bind to 3′ NCR of the genomic RNA to initiate replication, resulting in the generation of the complementary (-) strand, which, in turn, serves as a template for additional (+) strand synthesis. In the later stages, the (+) sense RNA genome associates with capsid

TABLE 108.3   Mammalian cell receptors of dengue virus*

Receptor	Cell/Tissue expression	Serotype
Heparansulphate (Sulfated glycosaminoglycan)	Vero, BHK-21, SW-13	DENV-1–4
nL$_{c4}$Cer (Glycosphingolipid)	Vero, BHK-21, K562	DENV-1–4
DC-SIGN/L-SIGN (Dendritic cell-specific lectin, CD209)	Dendritic cell, macrophage	DENV-1–4
Mannose receptor (possesses lectin activity)	Macrophage	DENV-1–4
HSP70/HSP90 (Heat shock proteins; Expression on plasma membrane)	HepG2, Macrophage, SK-SY-5Y	DENV-2
GRP78 (Chaperon; Expression on plasma membrane)	HepG2	DENV-2
Laminin receptor	HepG2, PS cone D cells	DENV-1–3
CD-14 associated protein (associated with LPS receptor)	Monocyte, macrophage	DENV-2

* Hidari and Suzuki, *Trop Med Health*, 39 (4): 37–43, 2011

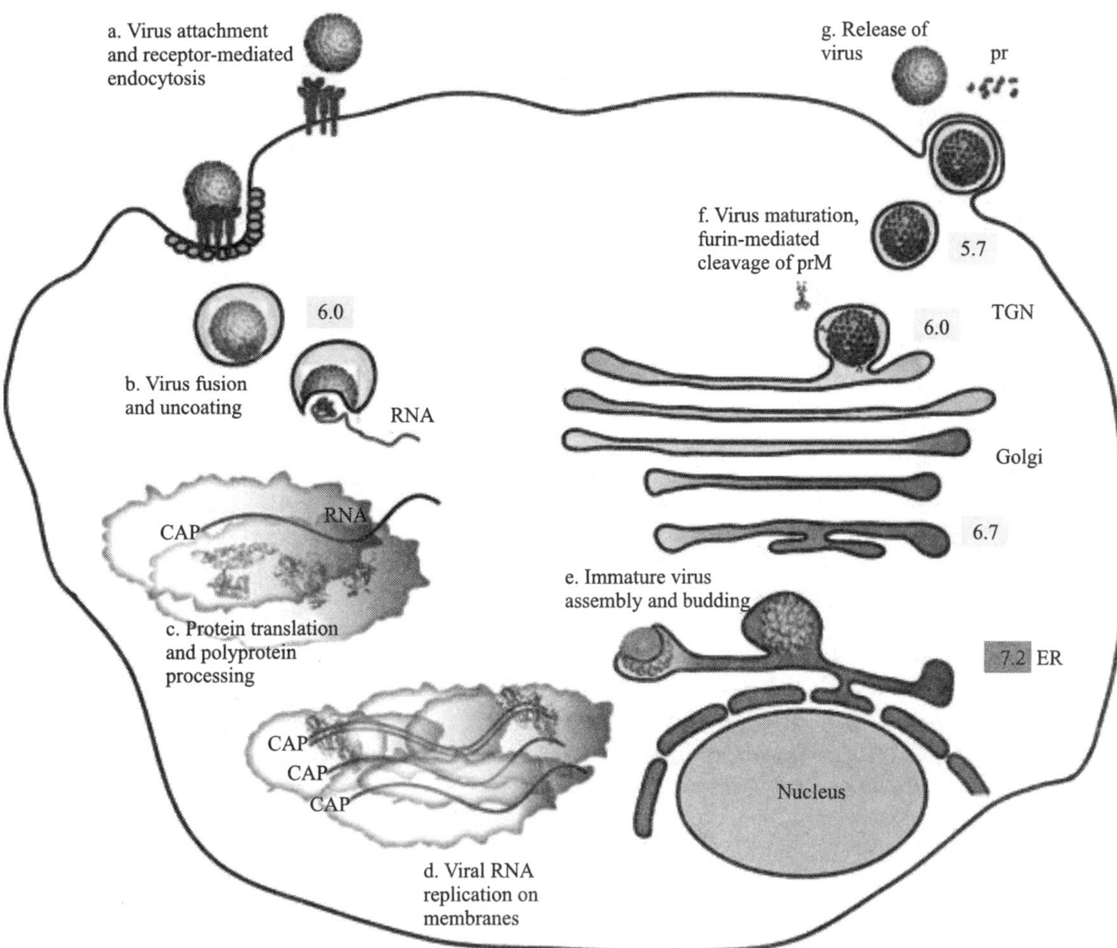

Figure 108.12 DENV life cycle. The virus enters the target cell through receptor-mediated endocytosis (a). Endosomal membrane fuses with the viral membrane to release viral RNA ('+' strand) into the cell cytoplasm (b). The viral RNA is translated into polyprotein, which is cleaved by host cell and virus-encoded proteases into structural and non-structural proteins (c). Non-structural proteins replicate viral RNA (d), which gets associated with the structural proteins to generate immature virions (e). Immature virions get transported through cell secretory pathway and get converted into matured, infectious particles in TGN by cleavage of 'pr' peptide (f). The 'pr' peptide gets released with the release of virus (g) (Perera et al, *Antivir Res*, 80: 11–22, 2008). (*see Plate 45 for colour figure*)

proteins to form RNA-capsid complex. This nucleocapsid, buds into ER lumen acquiring the host lipid membrane, which is decorated by prM-E heterodimers, to form immature virions. Heterodimerization with prM prevents premature exposure of the E protein fusion loop within the vesicles. Immature virions are transported through cell secretory pathway, and virus maturation takes place in trans-golgi network (TGN). Exposure to low pH in the vesicle triggers host furin-assisted cleavage of 'pr' peptide from prM, resulting in the formation of E-E homodimer and maturation of particles. However, the 'pr' peptide remains associated with E protein in order to prevent pre-mature fusion. It is liberated only when the virion is exocytosed and released into the extracellular milieu.

The process of cleavage of the 'pr' peptide is inefficient, leading to the release of a mixture containing both mature and immature particles. In an immature particle, the 'pr' peptide prevents the exposure of fusogenic domain, thus inhibiting the fusion of endosomal and viral membrane. Such immature particles should, ideally, be non-infectious; however, these particles are believed to acquire infectivity towards immune cells bearing Fc receptors *via* complexation with prM-specific antibodies. Such prM-specific antibodies which are present only during secondary infection contribute to ADE.

D. Pathogenesis

The mechanisms leading to the severe manifestations of DENV infections are still not completely understood, but there are multiple factors contributing to it. Host determinants, like age and gender, are associated with severe dengue and death. On the basis of epidemiological data, children especially females, aged 1–5 years, have been found to be more prone to severe disease. Few individuals have been found to have greater genetic susceptibility to DHF/DSS, perhaps mediated by differences in human leukocyte antigen (HLA) genotype. For example, in Thailand, severe disease has been found to be often associated with genotypes HLA-A*0207 for DENV-1and HLA *B52 for DENV-2. Some associations are protective as well, for example HLA DR alleles in Mexican and Vietnamese populations. Various single nucleotide polymorphisms (SNPs) have been found to be associated with protection and severity of disease.

Both viral factors and host immune response determine the severity of infection. Upon inoculation of DENV after mosquito bite, Langerhan's cells are infected first, which migrate to the lymph nodes, and the virus subsequently infects cells of macrophageal-monocytic lineage. Infection gets amplified and the virus spreads through lymphatic and vascular system. The replication efficiency of DENV in its target cell (like dendritic cells, monocytes, macrophages, endothelial cell linings of blood vessels, bone marrow stromal cells, and liver cells) as well as its tropism, collectively determines viral load measured in blood. In the cases of secondary infection, ADE (mediated, for example, by prM-specific antibodies) has been implicated in the increase in viral load and consequent disease severity. It has been reported that high viral load over-stimulates cross-reactive T-cells, which leads to production of high levels of pro-inflammatory cytokines and other mediators. High levels of immunomediators result in coagulopathy and plasma leakage. The levels of immunomediators like IL-6, IL-8, IL-10, and IL-18 have been reported to result in suppression of hemopoiesis, and hence, decreased blood thrombogenicity. Eventually, it results in thrombocytopenia and platelet dysfunction, leading to increased capillary fragility, which is clinically manifested as petechiae, easy bruising and gastrointestinal mucosal bleeding. Increased vascular permeability is further assisted by the generation of antibodies with the ability to cross-react with endothelial cell linings of blood vessels, platelets and plasmin, leading to a further increase in vascular permeability and coagulopathy.

Apart from these, a host of other factors are implicated in enhancement of viral load and level of immunomediators: (i) DENV affects host's Ubiquitin Proteasome System (UPS), which is involved in maintenance of protein homeostasis by regulated clearance of misfolded or obsolete protein. It has been reported that upon DENV infection, expression levels of key components of UPS, which act as viral replication promoting factors, are enhanced, leading to an increase in viral load. (ii) DENV capsid protein is believed to be accumulated on the surface of lipid droplets (LDs) in DENV-infected cells. It has been observed that on DENV infection, LDs are degraded *via* autophagy (induced by DENV NS4A), to release lipids, which are used for generation of energy by β-oxidation. Apart from inducing lipid catabolism, DENV (primarily by NS3), simultaneously, induces lipid anabolic pathway, presumably, to facilitate expansion of membranous replication complexes, leading to increase in viral load. (iii) Type 1 interferon (IFN) response, triggered by viral RNA acts as the primary antiviral response. DENV also induces type 1 IFN response in many cells. The viral RNA is recognized by pattern recognition receptors like RIG-I, Mda-5 and TLR-3. Antiviral effect of IFN is brought about by IFN-mediated stimulation of various genes in a time-dependent manner. It has been reported that

DENV replication is inhibited in cells with pre-activated IFN-stimulated genes, unlike cells where these genes were activated few hours post-infection. Such cells, where IFN-stimulated genes were activated post-infection, were found to be resistant to IFN-1 and -2. Virus gains resistance to host IFN response by many ways. DENV NS2A, NS4A and NS4B proteins inhibit STAT-1 phosphorylation. Blocking of Tyk-2 phosphorylation has also been reported. NS2B-3 protease inhibits production of IFN; NS5 binds STAT-2 and stimulates its degradation *via* proteasomal pathway. It has been reported that STAT-2 is not degraded in mice and hence it plays a critical role in restricting DENV replication. Thus, DENV load is increased by overcoming host's innate immune response. (iv) During polyprotein processing, viral protease NS2B-3 cleaves NS4AB into NS4A and NS4B precursor known as 2KNS4B, which carries a signal peptide '2K'. In the lumen of ER, the signal peptide is cleaved by host signalase to release mature NS4B, which is integrated in the ER membrane. Recently, the effect of maturation process of NS4B was studied on THP-1 monocytic cell line. It was observed that maturation process of NS4B, primarily cleavage of signal peptide from 2KNS4B, resulted in greater transcript levels of IL-6, IL-8 and IP10, as compared to NS4B and NS5 (known to increase the levels of immunomediators) alone. Hence, NS4B maturation has been implicated for enhanced levels of severe dengue disease associated chemokines and cytokines.

X. DENGUE DIAGNOSTICS

Dengue disease may be asymptomatic or manifest symptoms ranging from mild fever to severe hemorrhage and shock syndrome, making clinical diagnosis difficult. Hence, a lot of emphasis is being made on the development of laboratory diagnostic methods which could enable quick and accurate detection of dengue. Timely diagnosis of this disease may prevent its progression to severe hemorrhagic conditions and death. Laboratory diagnosis of dengue may be based on isolation of virus, detection of viral genome, viral antigens or dengue-specific antibodies (Table 108.4). Currently, a number of kits are commercially available for the diagnosis of dengue on the basis of detection of IgG, IgM and NS1 antigen, either alone or in combination (Table 108.5), using rapid tests (based on immune-chromatography) or ELISAs. Panbio Limited, Standard Diagnostics and BioRad have been the major players involved in the development of these kits. An Indian company J. Mitra & Co Pvt. Ltd. in collaboration with International Centre for Genetic Engineering and Biotechnology (ICGEB), New Delhi, has also developed some of these kits which are already in the market. PCR-based diagnostic kits for the detection of dengue viral genome have not yet been commercialized.

The accuracy of each of these tests depends on the stage of disease progression in the patient (Figure 108.13). For example, if the patient sera sample has been collected post-viremia, diagnostic test based on isolation of virus or based on detection of viral nucleic acid would be inappropriate. Hence,

TABLE 108.4   The various dengue diagnostic tests and their features

Diagnostic tests	Patient dample	Merits	Demerits
Virus isolation: Virus present in the sample is cultured in mosquito or mammalian cell lines	Sera, whole blood, plasma leukocytes, tissues	Most specific test result, can distinguish between serotypes	Time consuming, allows detection only in viremic phase, requires extensive laboratory facilities and expertise
Nucleic acid test: Detection of presence of viral genome using PCR techniques	Tissues, whole blood, serum, plasma	High sensitivity and specificity, can distinguish between serotypes, allows early detection	Can detect patients in viremic phase only, expensive, requires expertise, cannot distinguish between primary and secondary infection
Detection of viral antigens: Detection of NS1 in sera by ELISA/rapid tests or by immuno-histological study of tissues from patient	Sera or tissues	Inexpensive, easy to perform, allows early detection	Shorter window period of detection, cannot differentiate between serotypes, sensitivity lower than virus isolation and nucleic acid detection
Detection of dengue-specifc antibodies: Detection of IgM or IgG seroconversion by ELISA or rapid test	Sera, whole blood, plasma	Least expensive, easy to perform, can distinguish between primary and secondary infection	False-positive results due to cross-reactivity with other co-circulating flaviviruses, false negative when IgM is low, cannot differentiate between serotypes, requires two samples to detect seroconversion delaying dengue disease confirmation

TABLE 108.5   Commercially available dengue diagnostic kits

Kits	Detects			Format	Major companies
	NS1	IgM	IgG		
Dengue combo	✓	✓	✓	POCT	Standard Diagnostics, J. Mitra, Millennium Biotechnology
Dengue Duo	✗	✓	✓	ELISA	Panbio, Standard Diagnostics, Diagnostic Automation/Cortez Diagnostics, Novatec Immundiagnostica
				POCT	Panbio, J. Mitra, Standard Diagnostics, Millennium Biotechnology, Diagnostic Automation/Cortez Diagnostics, Orgenics, Omega, Focus Diagnostics
				Immuno-fluorescence	Euroimmun
Dengue IgM	✗	✓	✗	ELISA	Panbio, J. Mitra, Standard Diagnostics, Diagnostic Automation/Cortez Diagnostics, Euroimmun, Orgenics, Omega, Focus Diagnostics, GenWay biotech
				POCT	Millennium Biotechnology, Orgenics
Dengue IgG	✗	✗	✓	ELISA	Panbio, J. Mitra, Standard Diagnostics, Diagnostic Automation/Cortez Diagnostics, Euroimmun, Orgenics, Omega, Focus Diagnostics, GenWay biotech
				POCT	Millennium Biotechnology, Orgenics
Dengue NS1	✓	✗	✗	ELISA	Panbio, Biorad, J. Mitra, Standard Diagnostics, Diagnostic Automation/Cortez Diagnostics
				POCT	Panbio, J. Mitra, Biorad, Standard, Millennium Biotechnology Diagnostics, Diagnostic Automation/Cortez Diagnostics, Focus Diagnostics

POCT stands for Point of Care Testing or Rapid Tests

selection of the most suitable diagnostic assay would only be possible by taking clinical parameters of the patient into consideration. In case of secondary infection (Figure 108.14), presence of NS1, IgG (elevated titers) and IgM allows confirmed early diagnosis of dengue in more than 80% of the cases.

These diagnostic assays, depending upon their specificity, may either confirm dengue illness or suggest probable dengue infection (Figure 108.15). In most of the cases, combinations of these tests have to be used for diagnosis. Currently, efforts are being made to develop diagnostic assays which could confirm dengue infection during its initial phase of progression. This would allow necessary actions to be taken at an early stage, preventing its progression to severe form.

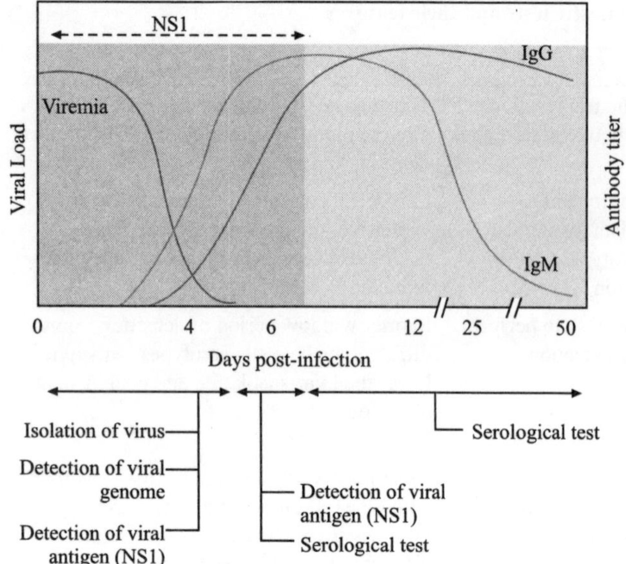

Figure 108.13 A schematic representation of viral load, viral antigen (NS1) and immune response during the course of primary dengue infection. The diagnostic tests available for the different stage of illness (marked in blue, pink and green background) have also been listed (Guzman et al., *Nat Rev Microbiol*, 8(12): S7–16, 2010). (*see Plate 45 for colour figure*)

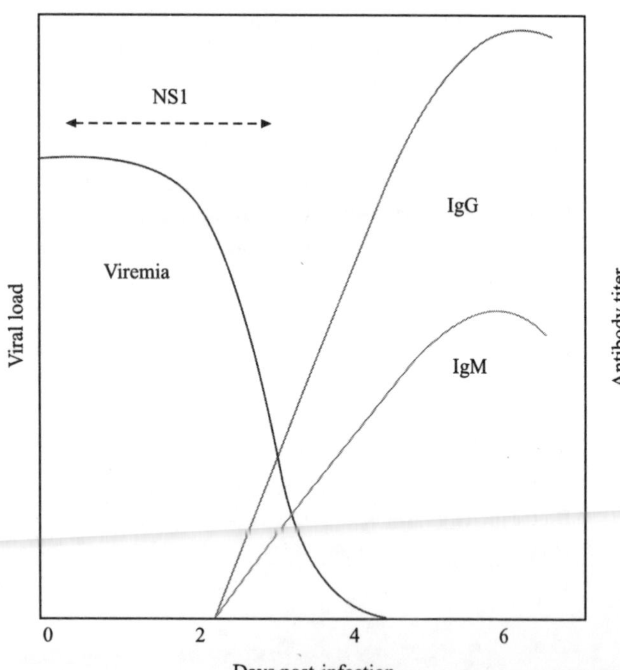

Figure 108.14 A schematic representation of viral load, viral antigen (NS1) and immune response during the course of secondary dengue infection. It is characterized by a narrower window period for detection of NS1 antigen, elevated IgG and low IgM titers. (*see Plate 45 for colour figure*)

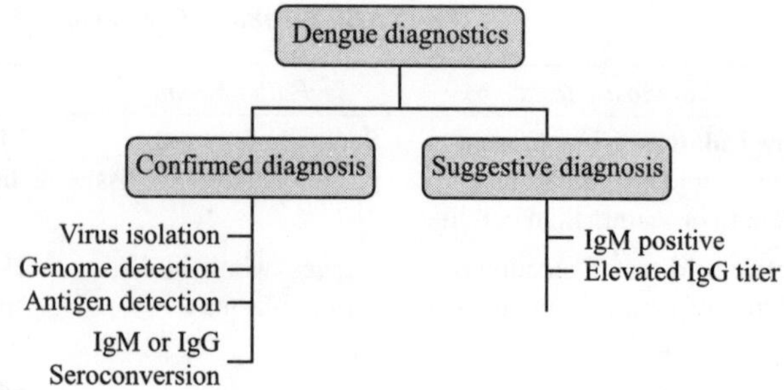

Figure 108.15 List of laboratory diagnostic tests used for dengue. The tests which confirm dengue infection and those which suggest a probable dengue infection have been separately listed (Guzman et al., Nat Rev Microbiol 8(12): S7–16, 2010).

XI. DENGUE VACCINE CANDIDATES

Till now there is no vaccine approved for human use against dengue. There are significant hurdles in the development of dengue vaccine, the most important being the requirement to generate a balanced immune response against all four serotypes to avoid ADE and the lack of an appropriate animal model.

A. Dengue Vaccine Strategies

There are different types of vaccines, like live-attenuated virus-, inactivated virus-, subunit-, virus vector- and plasmid-based vaccines, as listed in Table 108.6.

B. Dengue Vaccine Candidates in Clinical Trials

Table 108.7 gives a list of major potential dengue vaccine candidates in the clinical phase of development.

ChimeriVax™-DENV1–4 (CVD) is a tetravalent dengue chimeric live, attenuated virus vaccine, based on licensed yellow fever vaccine 17D. It is being developed by Sanofi Pasteur and is under Clinical Phase III. It was constructed by replacement of structural genes of live, attenuated yellow fever virus vaccine 17D with structural genes from each DENV serotype (Figure 108.16), resulting in the generation of four monovalent vaccines. Currently, a mix of all four monovalent chimeric viruses (tetravalent formulation) is being used for clinical studies, after successful immunization with monovalent CVD-2 in clinical phase I study. There are a few issues regarding this chimeric virus vaccine. First, the genetic basis of its attenuation is not known. Secondly, on immunization of monkeys with a mixture of the monovalent chimeric viruses, it was found that the level of replication of each serotype was different (viral interference). This phenomenon has been observed in human trials as well. Thirdly, full seroconversion requires 2–3 multiple doses administered over a 1 year period. Partial seroconversion of vaccine recipients in hyperendemic areas where multiple DENV serotypes co-circulate may pose safety concerns from the viewpoint of ADE.

TABLE 108.6 An overview of the advantages and disadvantages of various vaccine strategies

Vaccine Type	Advantages	Disadvantages
Live attenuated virus (LAV)	Robust, lasting and broad immunity; lower production cost	Difficulty in attenuation; genetic instability; possibility of reversion; interference in the case of multi-component LAV vaccines
Purified Inactivated Virus (PIV)	Reduced reactogenicity; better suitability for immune-compromised individuals; no scope for viral intereference	Less broad, potent and durable immune response; requires the use of adjuvants
Recombinant subunit vaccines	Lower reactogenicity; suitable for immune-compromised individuals; safe; can be produced in large amounts	Less broad, potent and durable immune response; requires the use of adjuvants
DNA vaccines	*In vivo* expression of antigens; thermostable	Difficulty in cellular uptake and antigen expression
Virus-vectored	Allows *in vivo* expression of antigens; thermostable; allows efficient antigen delivery	Pre-existing immunity to viral vaccine vectors could reduce its immunogenicity

TABLE 108.7 Dengue vaccine candidates in clinical trials

Developer	Strategy	Clinical phase			
		I	II	III	IV
Sanofi Pasteur	YF17D backbone carrying DENV prM and E genes				
GSK/WRAIR	Empirical attenuation of DENVs				
Inviragen, CDC, Shantha Biotechnics	DENV-2 attenuated virus backbone carrying prM and E genes of remaining 3 DENV serotypes, separately				
NIH, NIAID, LID, Biological E, Panacea Biotech, Butantan, Vabiotech	DENVs with a 30 nucleotide attenuating deletion in the 3'NCR				
Merck/Hawaii Biotech	Recombinant E ectodomain expressed in insect cells				
GSK/WRAIR	Inactivated DENV				
Vical/WRAIR/NMRC	Plasmid DNA encoding DENV prM and E genes				

Blocks in green, blue and red denote phase completed, under study and on hold, respectively.
Abbreviations used: GSK: GlaxoSmithKline; WRAIR: Walter Reed Army Institute of Research, USA; CDC: Centers for Disease Control and Prevention, USA; NIH: National Institute of Health, USA; NIAID: National Institute of Allergy and Infectious Diseases, USA; LID: Laboratory of Infectious Diseases; NMRC: Naval Medical Research Center, USA. (*see Plate 46 for colour figure*)

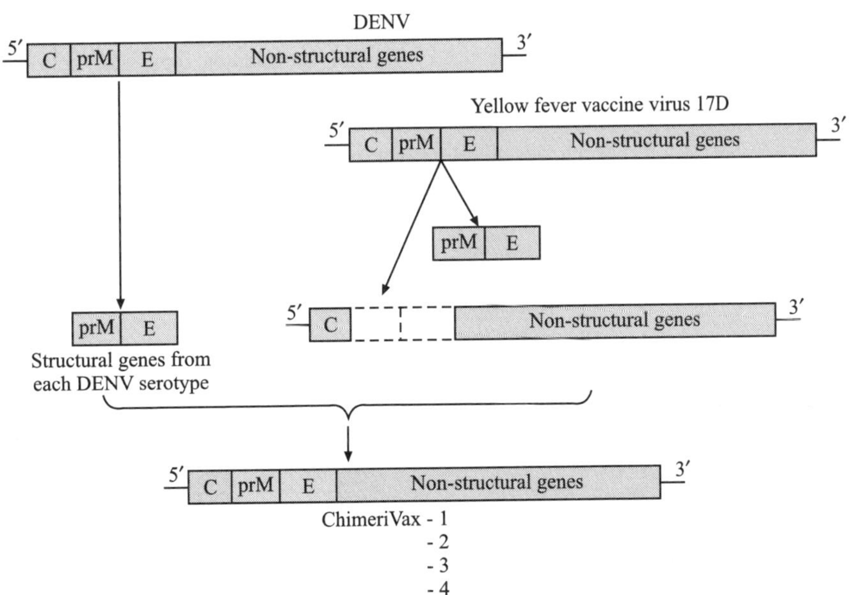

Figure 108.16 Illustration of strategy used for the development of ChimeriVax-DENV (CVD) vaccine, where structural genes of live, attenuated yellow fever virus vaccine, 17D, were replaced with structural genes from each DENV serotype, separately (Murphy and Whitehead, *Annu. Rev. Immunol.*, 29: 587–619, 2011). (*see Plate 46 for colour figure*)

Another LAV-based vaccine candidate, developed by Walter Reed Army Institute of Research (WRAIR) and GlaxoSmithKline (GSK), reached phase II of clinical development. DENVs were attenuated by a serial passage in Primary Dog Kidney (PDK) cells. The last passage was made in rhesus lung cells, in order to adapt the attenuated DENVs for animal test in rhesus monkeys. The four monovalent vaccine strains were also tested in clinical trials. Unlike DENV-2, -3 and -4 vaccine strains, DENV-1 strain was found to be under-attenuated, and, hence, led to the development of fever and rashes in 40% of the human subjects under study. DENV-4 strain was found to be over-attenuated. The passage numbers for all the strains were optimized to attain a balance and tetravalent formulation of the improved strains was tested in phase II trial in various countries. The initial results indicated that the formulation was well tolerated and immunogenic, but manifested varying degrees of viral interference. This vaccine candidate is not being pursued further.

The NIH (National Institute of Health) developed potential vaccine candidate, currently in clinical phase II, is also based on live, attenuated virus vaccine strategy. It has been developed by creating genetically defined attenuating mutations in 3' NCR (Figure 108.17).

DENV-1 and -4 were attenuated by 3' 172–143 deletion mutation (Δ30). Attenuated DENV-2 and -3 could not be obtained by this approach. They were, instead, generated using DENV-4Δ30. The prM and E genes of DENV-4Δ30 were replaced with corresponding genes from DENV-2 or DENV-3 to generate attenuated DENV-2/4Δ30 and DENV-3/4Δ30 chimeric viruses, respectively. Tetravalent formulation of these viruses was studied in rhesus monkey and they were found to be immunogenic and attenuated. Phase I study of these attenuated viruses revealed DENV-3/4Δ30 chimeric virus to be less infective in humans. Thus, a new live, attenuated, chimeric DENV-3 has been generated by replacing 3' NCR of DENV-3 with that of DENV-4Δ30 to obtain DENV-3–3'DENV4Δ30. This chimeric virus has appeared to be safe and immunogenic, in the initial Phase I study. It has also been reported that administration of monovalent DENV vaccine in an individual with pre-existing heterotypic antibody, did not lead to disease enhancement. A tetravalent formulation of these attenuated viruses is being currently studied. Phase I results indicated that administration of specific dose of these viruses resulted in low viremia but high infectivity and immunogenicity, and thus, is a promising vaccine candidate.

Recombinant subunit-based vaccine, being developed by Merck/Hawaii Biotech, uses the DENV E subunit ectodomain (called 80%E), purified from Drosophila S2 expression system. It is the first subunit-based dengue vaccine candidate to have reached clinical phase. DENV-1E subunit is being tested in the first Phase I evaluation. Pre-clinical studies involved testing of DENV-2E and -4E subunits in rhesus monkeys with various adjuvants.

The LAV vaccine, DENVax, being developed by Center for Disease Control and Prevention (CDC), Inviragen and ShanthaBiotechnics, is based on DENV2–PDK-53 vaccine strain of Mahidol University, Thailand. It is an attenuated strain of DENV-2 generated by serial passage in PDK cells. The genetic basis of its attenuation has been established. The infectious cDNA of DENV2–PDK-53 was used as the genetic backbone to construct chimeric dengue viruses by replacing its prM and E regions with the corresponding regions from DENV-1, -3 and -4. After transfection of Vero cells with the recombinant chimeric viruses, viruses were prepared by three sequential plaque purifications to reduce variation and ensure safety of master seed culture. Tetravalent formulation of these viruses, tested in AG129 mice and cynomolgus monkeys, was found to be safe and immunogenic. This vaccine candidate has entered phase II study.

C. Next Generation Dengue Vaccine Candidates

A number of vaccine candidates are in pre-clinical phase of development (Table 108.8). These vaccine candidates ensure the availability of alternate options in case of failure of vaccine candidates under clinical trials. These, potential, next generation vaccine candidates may also serve to be better vaccines in terms of efficacy, stability, dose and immunization schedule. They can also be used along with advanced vaccine candidates, for example, by prime-boost strategy, to further improve the immunogenicity.

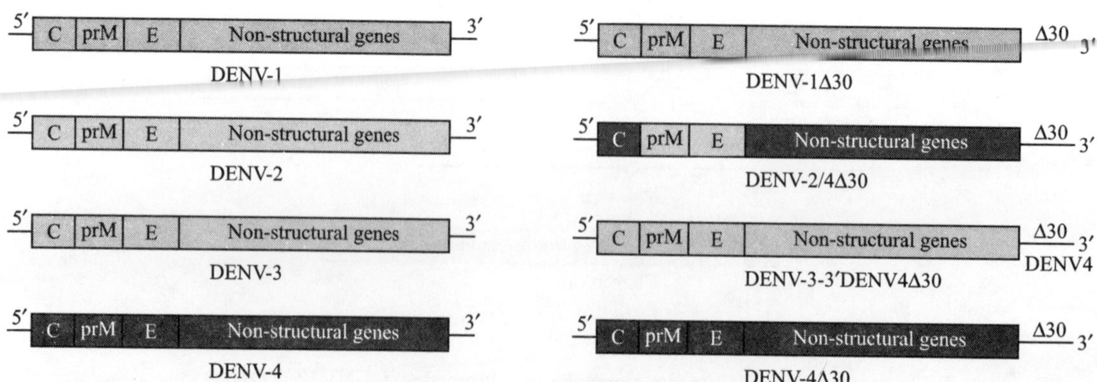

Figure 108.17   Illustration of the differences between the native and attenuated DENVs developed by NIH (Murphy and Whitehead, *Annu. Rev. Immunol.*, 29: 587–619, 2011). (*see Plate 46 for colour figure*)

TABLE 108.8 Dengue vaccine candidates in pre-clinical phase of development*

Strategy	Developer	Vaccine Features
Live attenuated virus vaccine (LAV)	• FIOCRUZ	• DENV prM/E expressed from live attenuated chimeric YF17D/DENV
	• Chiang Mai University/ Mahidol University/ NSTDA/ BioNet-Asia	• DENV/DENV chimeric viruses
Purified Inactivated Virus (PIV)	• NMRC	• Psoralen-inactivated DENV
Recombinant subunits	• IPK and CIGB	• Chimera of DENV EDIII and p64k of *Neisseria meningitides*; Monovalent EDIIIs; DENV EDIII- capsid fusion mixed with oligonucleotides
	• VaxInnate	• EDIII-flagellin fusion protein
	• ICGEB	• Tetravalent chimeric EDIII
	• NHRI	• Consensus EDIII; chimeric consensus EDIII-Ag473 lipoprotein of *Neisseria meningitides*
	• Cytos Biotechnology	• EDIII chemically coupled to Qß VLPs
	• ICGEB	• EDIII-HBsAg VLPs and ectoE-based VLPs
	• Kobe University	• prM/E VLPs
DNA vaccines	• Inovio Pharmaceuticals	• Tetravalent chimeric EDIII with inbuilt cleavage sites
	• Kobe University	• prM/E expressed from plasmid vector
	• CDC	• prM/E expressed from plasmid vector
Virus-vectored vaccines	• ICGEB	• Tetravalent chimeric EDIII expressed from replication-deficient adenovirus vector
	• GenPhar/NMRC	• Bivalent prM/E expressed from replication-deficient adenovirus vector
	• UNC	• EDIII, E85 or prM/E expressed from single-cycle VEE virus vector
	• UTMB	• prM/E expressed from single-cycle West Nile virus vector
	• Themis Bioscience/Institut Pasteur	• Tetravalent EDIII and DENV-1 ectoM expressed from live-attenuated measles virus vector

Abbreviations used: CDC: Centers for Disease Control and Prevention, USA; CIGB: Center for Genetic Engineering and Biotechnology, Cuba; FIOCRUZ: Oswaldo Cruz Foundation, Brazil; ICGEB: International Center for Genetic Engineering and Biotechnology, India; IPK: Pedro Kourí Tropical Medicine Institute, Cuba; NHRI: National Health Research Institutes, Taiwan; NMRC: Naval Medical Research Center, USA; NSTDA: National Science and Technology Development Agency, Thailand; UNC: University of North Carolina at Chapel Hill, USA; UTMB: University of Texas Medical Branch, USA; WRAIR: Walter Reed Army Institute of Research, USA

*Schmitz et al, Vaccine 29(42): 7276–7284, 2011

D. Envelope Domain III-based Subunit Vaccine Candidates

Subunit vaccines have come into focus because of the problem of unbalanced immune response generated by LAV due to viral interference. In this regard, EDIII has shown a lot of promise because it is one of the most critical domains against which neutralizing antibodies are elicited. It contains multiple-type and sub-type specific neutralizing epitopes, but lacks cross-reactive and enhancing epitopes. In order to avoid the problem of ADE, many groups have focused on utilization of EDIII in generation of a potential subunit-based dengue vaccine candidate. EDIII has been fused with other proteins to further enhance its immunogenicity. Table 108.9 lists the various EDIII-based candidates being developed as potential dengue subunit vaccine.

SUMMARY

Dengue is an *Aedes* mosquito borne-viral disease caused by any of the four serotypes DENV-1, -2, -3, -4 belonging to the *Flaviviridae* family. Though this disease is highly endemic in tropical and sub-tropical region, it has become a global burden with nearly 40% of the world's population at the risk of contracting this disease because of increased urbanization, travel and destruction to the environment. DENV infects humans primarily, except in parts of Africa and Malaysia where sylvatic cycle of DENV infection has been reported. It often leads to the development of self-limiting fever, but some cases may progress to severe DHF and DSS or even death. The first exposure of an individual to a particular DENV serotype leads to development of primary infection, which is often characterized by mild fever. Subsequent infection of that individual with DENV of a

TABLE 108.9   EDIII-based potential dengue subunit vaccine candidates

Dengue envelope domain III	Fusion partner	Expression host
EDIII-2 (alongwith its transmembrane region and N-terminal sequence of NS1)	Staphylococcal protein A	*E. coli*
EDIIIs	Maltose Binding Protein	*E. coli*
EDIIIs	P64k of *Neisseria mennigitidis*	*E. coli*
EDIII-2 epitopes	Parvovirus B19 capsid	Baculovirus
EDIII-2	EDIII-4	*E. coli*
EDIII-4	-	*E. coli*
EDIIIs of all four serotypes linked by flexible glycine linker	-	*P. pastoris*
EDIII-2	DENV-2 capsid	*E. coli*
EDIII-2	-	*P. pastoris*
Consensus EDIII lipidated	-	*E. coli*
EDIII multi-epitope peptide	Universal T helper cell epitope, PADRE	Synthetic

different serotype, known as secondary infection, might lead to severe disease due to ADE. As a result, accurate detection of infection at an early stage becomes important. Dengue diagnosis is done using a combination of tests, which are decided on the basis of clinical symptoms of the patient. A number of kits are commercially available in order to facilitate and allow early diagnosis, which would enable restriction of disease from progressing to severe form. The treatment of dengue is largely supportive in nature and no antiviral therapy is available yet. A lot of efforts are also being made in development of vaccines against this disease because of its global burden and impact. Efforts in the field of dengue vaccine development have brought few vaccine candidates to the clinical phase of development,

with an attenuated virus-based vaccine in clinical phase III. In case of dengue, the risk of ADE makes it mandatory for a successful vaccine to induce a balanced immune response against all four DENV serotypes. It has been observed that tetravalent formulation of attenuated DENVs may not lead to generation of balanced immune response because of the difference in replicative potential of each serotype. In this context, subunit-based vaccines are now drawing attention because of their obvious safety advantages. The research being done on the study of dengue disease pathogenesis, diagnostics, DENV inhibitor and vaccine development would together assist in controlling the spread and severity of this disease.

SUGGESTIONS FOR FURTHER READING

Guzman M.G., Halstead S.B., Artsob H., Buchy P., Farrar J., Gubler D.J., Hunsperger E., Kroeger A., Margolis H.S., Martinez E., Nathan M.B., Pelegrino J.L., Simmons C., Yoksan S. and Peeling R.W. (2010), Dengue: A Continuing Global Threat, *Nat Rev Microbiol*, 8(12 Suppl):S7–16.

Mukhopadhyay S., Kuhn R.J. and Rossmann M.G. (2005), A Structural Perspective of the Flavivirus Life Cycle. *Nat Rev Microbiol.*, 3(1):13–22.

Murphy B.R. and Whitehead S.S. (2011), Immune Response to Dengue Virus and Prospects for A Vaccine, *Annu Rev Immunol.*, 29:587–619.

Schmitz, J., Roehrig J., Barrett A. and Hombach J. (2011), Next Generation Dengue Vaccines: A Review of Candidates in Preclinical Development, Vaccine.

Swaminathan S., Batra G. and Khanna N. (2010), Dengue Vaccines: State of the Art. *Expert Opin Ther Pat.*, 20(6):819–835.

109

Chikungunya

Navin Khanna, S.M. Talha and S. Swaminathan

CONTENTS

I. INTRODUCTION

Chikungunya fever (CHIKF) is a viral disease caused by chikungunya virus (CHIKV), which is transmitted by *Aedes* mosquitoes. CHIKV is an arthropod-borne virus (arbovirus) of the genus *Alphavirus,* belonging to the Togaviridae family. CHIKV, an enzootic virus, is found mainly in tropical and subtropical regions of Africa, in the Indian Ocean Islands, and in south and Southeast Asia. The word chikungunya, which is used to describe both the virus and the disease, is derived from African dialect Swahili or Makonde that means "to walk bent over". The characteristic symptoms of the disease are fever, skin rash, arthralgia (joint pain) and possibly myalgia (muscle pain). Most of the symptoms resolve with time but the joint pain in some patients can last for years and can be so rigorous that they adopt a bent or stooping posture. Although CHIKV infections are rarely fatal, the polyarthralgia, typical of human disease, is severely debilitating and often persists for several months or even years, resulting in a massive burden of disease and disability that extensively affects quality of life and economy.

II. HISTORY

It is believed that the CHIKV originated in Africa. CHIKF is specifically a tropical disease. Phylogenetic studies have revealed that there are three distinct genotypes of CHIKV based on the gene sequences of the Envelope protein (E1), namely West African, East/Central/South African and Asian. It was estimated that Asian genotype emerged between 50 and 310 years ago, and the West and East African genotypes diverged between 100 and 840 years ago. The detection of CHIKV was first reported in 1952 in the Makonde Plateau in Africa. It was later detected in Thailand in 1958. In India, CHIKV was first detected in 1963 in West Bengal. It is reported that the CHIKV outbreaks in India during 1963–73 were caused by Asian genotype, whereas outbreaks since 2005 have been caused by Central/East African genotype. CHIKV remained essentially neglected until the epidemic outbreak in 2005 in the islands of the Indian Ocean, which attracted the attention of the whole world. This outbreak made CHIKV and CHIKF important subjects of study (Figure 109.1).

III. EPIDEMIOLOGY

CHIKF epidemics are characterized by long periods of more than 10 years between the epidemics. Before the year 2000, large outbreaks of CHIKV were rare, but since then outbreaks have become more frequent. Since 2001, major outbreaks have been documented in the islands of Indian Ocean including

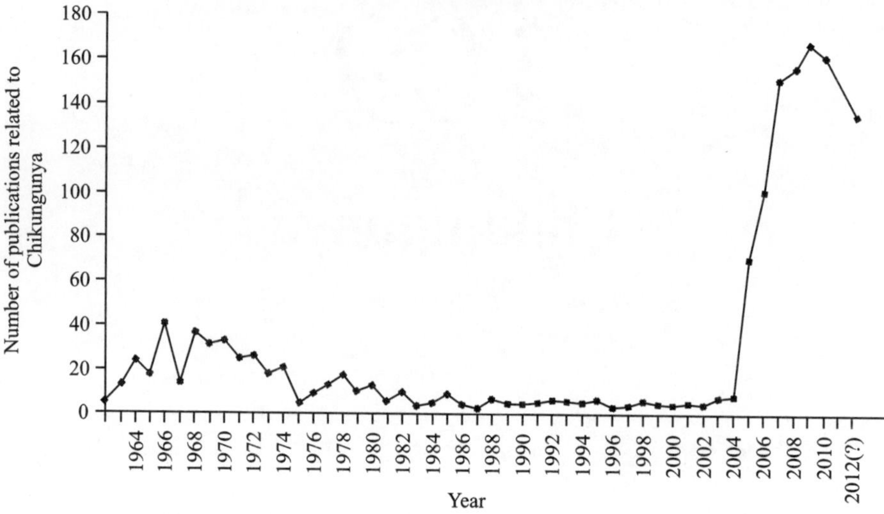

Figure 109.1 Overview of CHIKV-related publications from 1963 to 2012.

Mauritius, Mayotte, Madagascar and Réunion Island. The outbreak in Réunion Island infected one-third of its population (population 770,000), with 237 deaths ascribed directly or indirectly to CHIKV. In 2006 CHIKV outbreak was noted in India with an immense magnitude affecting ~1.4 million people. Many countries have reported the outbreak of CHIKV (Table 109.1). With the increase in global travel for the purpose of tourism, education and employment etc., the risk of CHIKV spreading to non-endemic areas has also increased. Several cases have been reported where CHIKV was brought by travelers to European countries, the United States and Australia. The virus, once thought to be associated with the tropical regions only, is entrenching itself in the non-tropical regions also, thus becoming a global burden.

IV. ENTOMOLOGY

CHIKV is a mosquito-borne virus, transmitted mainly by the bites of *Aedes* mosquitoes. Several *Aedes* species, e.g., *Ae. furcifer, Ae. vittatus, Ae. fulgens, Ae. luteocephalus, Ae. dalzieli, Ae. vigilax,* and *Ae. camptorhynchites,* transmit the virus in Africa . In the regions of Asia and the Indian Ocean, the main CHIKV vectors are anthropophilic hematophagous (feeding on blood) female *Ae. aegypti* and *Ae. albopictus* mosquitoes. *Ae. aegypti,* the urban mosquito, maintains a close association with humans and is likely to be responsible for large regional and global outbreaks. In India, the dominant carrier of chikungunya virus is *Ae. aegypti,* which breeds mainly in stored fresh water in urban and semi-urban environments. An adaptive mutation in the CHIKV has led the *Ae. albopictus* to become a major vector for the virus. *Ae. albopictus* has shown a remarkable capacity to adapt to human beings and to urbanization, allowing it to supersede *Ae. aegypti* in many places (including China, the Seychelles, and Hawaii). It is particularly resilient, and can survive in both rural and urban environments. *Ae. albopictus* mosquito's eggs are highly resistant and can remain viable

TABLE 109.1   Outbreaks of CHIKV in chronological order

Year	Location	Approximate no. of reported cases
1952	Tanzania	?
1966	Vietnam	?
1968	Uganda	?
1998, 2006	Malaysia	514271
2001–08	Indonesia	19238
2005–06	Réunion Island	266076
2005–06	Mayotte	6346
2006	India	1.39 million
2006	Mauritius	15760
2006–07	Gabon	48000
2006–07	Maldives	11879
2007	Sri Lanka	40000
2007	Italy	248
2008–09	Singapore	1033
2009	Thailand	11669
2010	Gabon	1208
2011	Republic of Congo	8000
2011–12	Cambodia	1300

throughout the dry season, giving rise to larvae and adults during the following rainy season, and to become a secondary but important vector of dengue and other arboviruses. It is very difficult to avoid *Ae. albopictus* bites and to control this mosquito species in tropical, human-modified ecosystems with modern infrastructures, irrigation, and massive solid waste production. *Ae. albopictus* has been introduced into several European countries (Belgium, Bosnia, Croatia, France, Greece, the Netherlands, Serbia, Spain, and Switzerland) and also to Central America, Brazil, and the United States. Returning travelers with high viremia could be a source of virus and could cause a CHIKV outbreak.

V. CHIKUNGUNYA TRANSMISSION CYCLES

CHIKV is typically circulated in two distinct transmission cycles: (1) sylvatic and (2) human-mosquito-human (Figure 109.2, Table 109.2). In Africa, the virus is maintained in a sylvatic transmission cycle between non-human primates and forest-dwelling *Aedes* species mosquitoes. The scale of epidemics is smaller in sylvatic cycle and is confined within Africa. The outbreaks have tended to affect small populations and appear to be heavily dependent upon the sylvatic mosquito densities that increase with periods of heavy rainfall. During epidemics, CHIKV can circulate between human beings and mosquitoes without the need for animal reservoirs. Prior to 2005, *Ae. aegypti* was considered the primary vector in Asia, but the spread of *Ae. albopictus* into Africa and Europe, combined with adaptive mutations in the CHIKV, has dramatically increased the role of this species in transmission among humans.

VI. CLINICAL PRESENTATION OF CHIKUNGUNYA INFECTION

After the initial infection CHIKV spreads rapidly in the body. During this time, there is a silent incubation period that ranges from 1 to 12 days (average 2–4 days). After the incubation period the clinical onset is abrupt, with high fever (>102° F), rash, headache, back pain, myalgia, and arthralgia. Arthralgia can be intense, affecting mainly the extremities (ankles, wrists and phalanges), but also the large joints. Few infected people

TABLE 109.2   CHIKV transmission cycles

Transmission cycle	Host	Vector	Area of prevalence
Sylvatic	Non-human primates such as monkeys	forest-dwelling *Aedes* mosquitoes (*Ae. furcifer, Ae. taylori, Ae. luteocephalus, Ae. africanus*, and *Ae. neoafricanus*)	Africa (Senegal, Cote d'Ivoire, Central African Republic and South Africa)
Human-mosquito-human	human	*Ae. aegyti Ae. albopictus*	Tropics and sub-tropics

exhibit asymptomatic infection (5%–18%), representing mainly the age group of <25 years. However, accompanying fever and arthralgia are commonly occurring symptoms. Rash is the least reliable symptom present in as few as 19% of patients.

Debilitating polyarthralgia is a prominent feature in most of the CHIKF cases. Almost all patients with CHIKV infection have arthralgia that usually affects more than one joint. Joints that are already affected by pre-existing disorders such as osteoarthritis are particularly more susceptible. The acute signs and symptoms of CHIKV infection usually resolve within 1–2 weeks. In this period most of the patients have partial improvement in arthralgia. At this stage, some patients recover completely, but in many patients the arthralgia can persist for months or even years. The chronic arthralgia normally varies in intensity and relapses, but usually affects the same joints that were affected during the acute phase. Additional symptoms that have been reported in CHIKV-infected patients include fatigue, nausea, vomiting, diarrhea and conjunctivitis, that were present with varying regularity and severity.

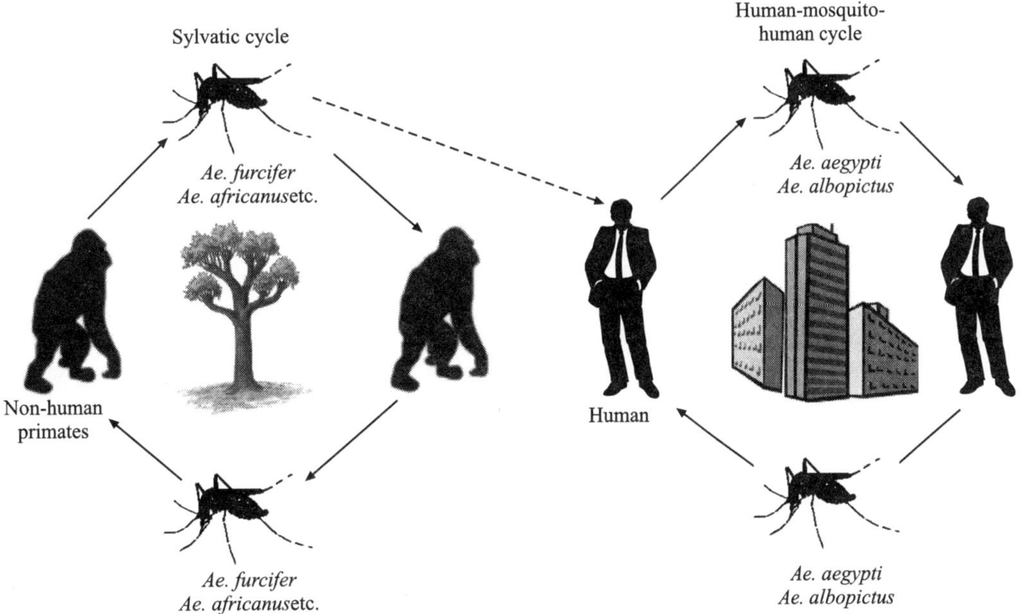

Figure 109.2   Sylvatic and human-mosquito-human cycles of CHIKV transmission illustrating the specific vectors and hosts.

CHIKV is generally not considered to be a neurotropic virus. However, neurological involvement in CHIKV infection has also been reported and neurological complications include meningoencephalitis, myeloneuropathy, Guillan-Barré syndrome, acute flaccid paralysis, and palsies. Mortality has also been reported in association with CHIKV infection, but its relationship with CHIKV infection has not been established. Pre-existing conditions apparently pose a common risk factor with CHIKV-associated mortality. Vertical (mother to child) transmission of CHIKV has also been observed in some of the cases. CHIKV dissemination and target organs are shown in Figure 109.3.

CHIKV infections are typically underreported because the disease symptoms overlap widely with dengue, malaria and many other acute infectious tropical diseases. CHIKV and DENV infections are among the most difficult infections to distinguish. Following CHIKV infection, symptom onset is usually more abrupt, fever is shorter lived, and rash, conjunctival infection (redness of the white sclera of the eye) and arthralgia are more common than in dengue. A comparison of clinical characteristics in CHIKV and DENV infections is presented in Table 109.3.

TABLE 109.3 Comparison of clinical characteristics in CHIKV and DENV infections

Clinical characteristics	CHIKV	DENV
Fever, debility	Common	Common
Polyarthritis	Very common, edematous	✗
Myalgia	Possible	Very common
Rash	Days 1-4, important skin edema	Days 3-7
Tenosynovitis	✓	✗
Leukopenia	✗	✓
Thrombocytopaenia	✗	✓
Hypotension	Possible	Common, days 5-7
Retro-orbital pain	Rare	Common
Minor bleeding	Rare	Common
Chronic polyarthritis	Up to 1 year or more	✗
Second stage	Possible; tenosynovitis at month 2-3	Fatigue up to 3 months

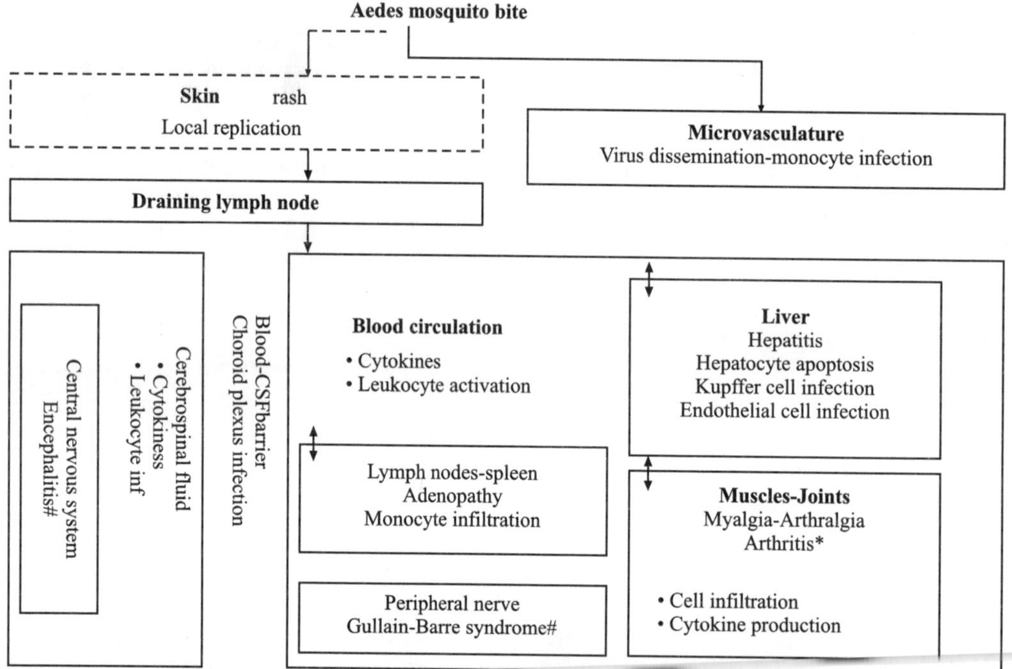

Figure 109.3 Virus dissemination and target organs. Following inoculation with CHIKV through a mosquito bite, the virus directly enters the subcutaneous capillaries, with some viruses infecting susceptible cells in the skin, such as macrophages or fibroblasts and endothelial cells. Local viral replication seems to be minor and limited in time, with the locally produced virus probably being transported to secondary lymphoid organs close to the site of inoculation. The blood carries most viruses, as free virions or in the form of infected monocytes, to the target organs, the liver, muscle, joints, and remote lymphoid organs. In these tissues, infection is associated with a marked infiltration of mononuclear cells, including macrophages, particularly when viral replication occurs. The pathological events associated with tissue infection are mostly subclinical in the liver (hepatocyte apoptosis) and lymphoid organs (adenopathy), whereas mononuclear cell infiltration and viral replication in the muscles and joints are associated with very strong pain, with some of the patients presenting arthritis. #Guillain-Barré syndrome and encephalitis are very rare events. *True arthritis remains a rare event (from 2% to 10%). (Redrawn from Dupuis-Maguiraga et al. (2012). *PLoS Negl Trop Dis*, 6(3): e1446. doi:10.1371/journal.pntd.0001446)

VII. CHIKUNGUNYA VIRUS

A. Structure

CHIKV belongs to the *Alphavirus* genus of the *Togaviridae* family. CHIKV is a small, spherical enveloped virus of about 60-70 nm in diameter. A model for the whole glycoprotein layer of the alphavirus particle was built by fitting the structure of the CHIKV E2–E1 heterodimer into the available cryo-EM map of the Sindbis virus (SINV) (Figure 109.4). The CHIKV genome is 11.8 kb single-stranded positive sense RNA that consists of a 5′cap non-translatable region (NTR) followed by non-structural proteins (nsP1-4) and structural proteins (C-E3-E2-6K-E1) encoding sequence, and a 3′ terminal poly-A tail. The 5′ NTR of CHIKV is composed of 76 nucleotides (nt) and 3′ NTR is composed of 526 nt. The 3′ NTR of alphaviruses are believed to contain several cis-acting conserved motifs, named as conserved repeated sequence elements (RSEs), which regulate viral RNA synthesis. The genome of the CHIKV, similar to the other alphaviruses, contains two open reading frames (ORFs) of 7424 nt and 3732 nt that encode non-structural and structural proteins of CHIKV, respectively. The junction (J) region of CHIKV which is also non-translatable is composed of 68 nt. Using SINV as a model, it has been

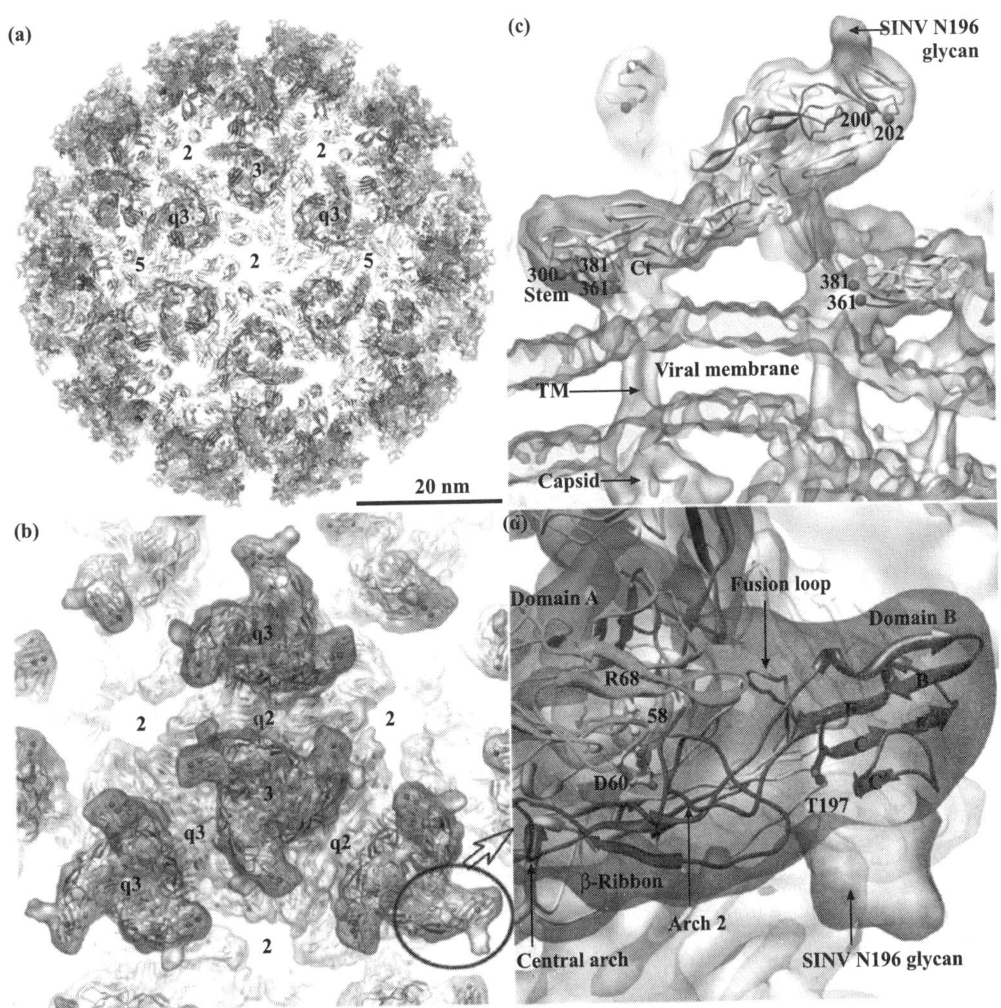

Figure 109.4 CHIKV structure. (a) The alphavirus T = 4 icosahedral surface glycoprotein shell. Atomic model of the 240 chikungunya E2–E1 heterodimers arranged as 80 spikes. E2 domain A is coloured cyan, B dark green, C pink and the β-ribbon dark purple, and E1 sandy brown. (b) Top view of the spikes. The 9 Å resolution cryo-EM map of the SINV is represented as a grey transparent surface. The T = 4 icosahedral symmetry and quasi-symmetry axes are indicated. The black circle marks the close-up view of d. (c) Side view of a spike. Red spheres mark the transitional epitopes on E2 domain B and on E1 domain III, in a region buried at interspike contacts. Regions of the map not fitted in this work (stem region, transmembrane (TM) domains, capsid and viral membrane) are indicated. (d) Close-up of the spike top. The fusion loop (labelled) is capped by E2 domain B. The extra density extending away of domain B corresponds to an N-linked glycan in SINV (attached to N196, corresponding to Thr 197 in CHIKV, labelled). In domain A, the side chains of Asp 60 and the buried Arg 68, shown in yellow sticks and labelled, make a stabilizing salt bridge. Position 58, a virulence determinant is marked in red. (Adapted from Voss et al. (2010). *Nature*, **468**: 709-12. doi:10.1038/nature0955). (*see Plate 47 for colour figure*)

shown that the J region contains an internal promoter for transcription of the subgenomic mRNA (26S mRNA), and the start site and 5′-NTR leader sequence of the 26S mRNA. The genomic organization and viral proteins of CHIKV are illustrated in Figure 109.5. The functions of the major CHIKV proteins are summarized in Table 109.4.

TABLE 109.4   Major CHIKV proteins and their predicted functions

Proteins		Functions
Non-Structural	nsP1	(-) strand synthesis, RNA capping
	nsP2	Helicase, proteinase
	nsP3	RNA synthesis
	nsP4	RNA-dependent RNA polymerase
Structural	C	Capsid
	E2	Envelope glycoprotein
	E1	Envelope glycoprotein

B.   Life Cycle

The transmission cycle of the virus requires infection of female mosquitoes via a viremic blood meal and, after a suitable extrinsic incubation period, transmission to another vertebrate host during subsequent feeding. Enveloped viruses utilize membrane-bound receptor(s) for entry into specific target host cells. During infection, alphaviruses enter vertebrate host cells via receptor-mediated endocytosis. CHIKV is able to replicate both in mosquitoes and higher vertebrate cells, however the mosquito and human cell surface receptor(s) have not been identified so far. In the invertebrate host (mosquito), virus replication takes place in the midgut and salivary gland. In addition, both sites also represent two physical barriers which limit CHIKV dissemination in mosquitoes.

The spread of CHIKV to the susceptible cells is mediated by two viral glycoproteins, E1 and E2, which carry the main antigenic determinants and form an icosahedral shell at the virion surface. Glycoprotein E2 is responsible for receptor binding and E1 for membrane fusion. The ability of CHIKV to bind human cells and to replicate in cell cultures has been documented. CHIKV infection of epithelial HEK293T mammalian cells is dependent on functional Eps15 and independent of clathrin heavy chain. Eps15 is a component of clathrin-coated pits where it interacts with adaptor protein (AP)-2, which is the major clathrin adaptor complex. Upon attachment to cellular receptors, CHIKV is rapidly internalized and delivered to endosomes. The steps of the CHIKV replicative cycle, after CHIKV has been routed to the endosomes, have not been reported so far. General information regarding these steps of CHIKV replicative cycle can be deduced from the information available for other alphaviruses. Upon entry, alphavirus particles undergo disassembly, delivering genomic RNA into the cytoplasm of the host cells. The viral genome is then translated from the two ORFs to generate the nonstructural and structural polyproteins (Figure 109.5). Nonstructural proteins (nsPs) of CHIKV are encoded by an open reading frame (ORF) of 7424 nucleotide. This ORF encodes a polyprotein precursor of 2474 amino acids termed nsP123 that produces the different nsPs after proteolytic cleavage. According to information available for related *Alphaviruses*, after synthesis and maturation, the nsP123 precursor is expected to complex with the free nsP4 protein forming a replication complex that catalyzes the synthesis of the negative strand RNA. In *Alphaviruses*, cleavage of the structural polyprotein occurs co-translationally, beginning with the auto-proteolytic cleavage of the capsid (C) protein from the polyprotein. C protein is then available to bind with newly synthesized RNA, recognizing specific packaging signals in the 5′ half of the genome, so that only full-length genomic RNA is packaged

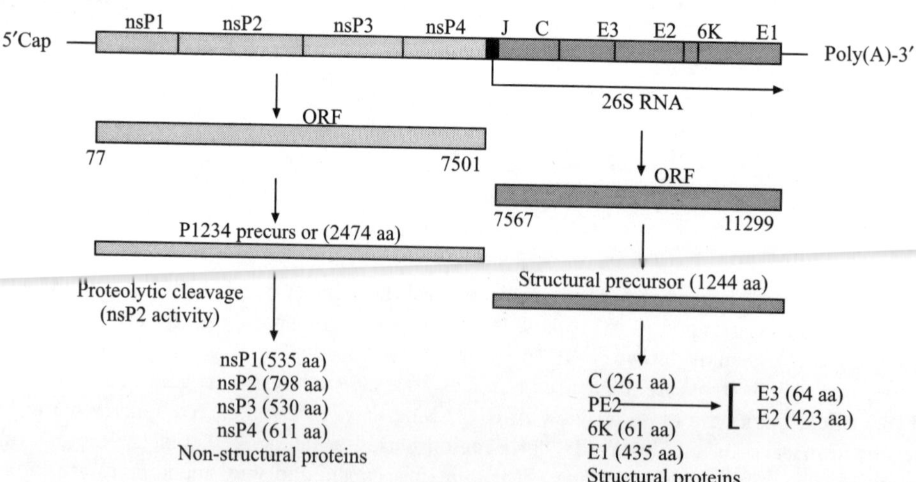

Figure 109.5   Genome organization of CHIKV and its gene products. The CHIKV genome is 11.8 kb RNA with 5′ cap structures and 3′ poly(A) tail. Non-translatable regions (NTR) are present at the 5′ and 3′ proximal sequences of CHIKV genome. The junction region (J) is also non-coding. A subgenomic positive-strand mRNA termed as 26S RNA, serves as the mRNA for the synthesis of the viral structural proteins. The different non-structural proteins (nsP1–nsP4) and structural proteins (C, Capsid; E1, E2, E3, envelope; 6K) are generated after proteolytic cleavage of polyprotein precursors.

into nucleocapsid-like particles. The E3 protein acts as a signal sequence for internalization of the remaining polyprotein into the endoplasmic reticulum, where it is processed by host signal peptidase. Similarly, the 6K protein works as a signal sequence for the downstream processing of the E1 protein. Upon synthesis, the E2 glycoprotein precursor, PE2 and E1 glycoproteins interact with each other to form heterodimers. These heterodimer complexes are then transported from the endoplasmic reticulum to the cell surface via the Golgi complex. During this step, PE2 is cleaved by a cellular furin or furin-like proteinase to generate E2 and E3. Interactions between the C protein and the cytoplasmic domain of the E2 protein start the budding process through the cell membrane. The E1-E2 heterodimers form an envelope around nucleocapsid like particles. Upon release from cells, the virions acquire a membrane bilayer derived from the host cell plasma membrane. According to the available information for the *Alphaviruses*, a possible CHIKV replicative cycle is presented in Figure 109.6.

C. Pathogenesis

Little information is available regarding the specific mechanisms of CHIKV pathogenesis. Molecular pathogenesis of CHIKV may be similar to the observations made on the molecular and cellular aspects of Ross River virus (RRV) that also causes epidemics of fever, polyarthritis (including rheumatic symptoms), and mucocutaneous manifestations similar to that seen with CHIKF. During RRV infection, the T-cell–mediated immune response is assumed to play a role in persistence of symptoms. Patients who quickly recover from RRV have a predominance of CD8+ T cells in contrast to patients with chronic disease where CD4+ cells are more prevalent in the synovial fluid. Thus, a defective cell-mediated immune response (CMI), where CD8+ T cells are absent or inactive, has been thought to be the cause of chronic disease and viral persistence. Furthermore, treatment of *in vitro* infected macrophages with CD8+ T cells, produced by vaccination of mice with RRV capsid protein, results in complete clearance of the infection. Another possible pathway of pathogenesis due to RRV infection involves the activity of the signaling receptor, complement receptor 3 (CR3). CR3 deficiency in recombinant mice reduced the expression of certain pro-inflammatory and cytotoxic effectors within infected tissues, resulting in decreased tissue destruction. Finally, viral antigens in the synovial fluid of RRV infected patients with chronic joint pain draw inflammatory infiltrates to the affected joints. It is assumed that persistently infected macrophages cause pathogenesis via tissue destruction, because clearance of macrophage from infected tissue has been

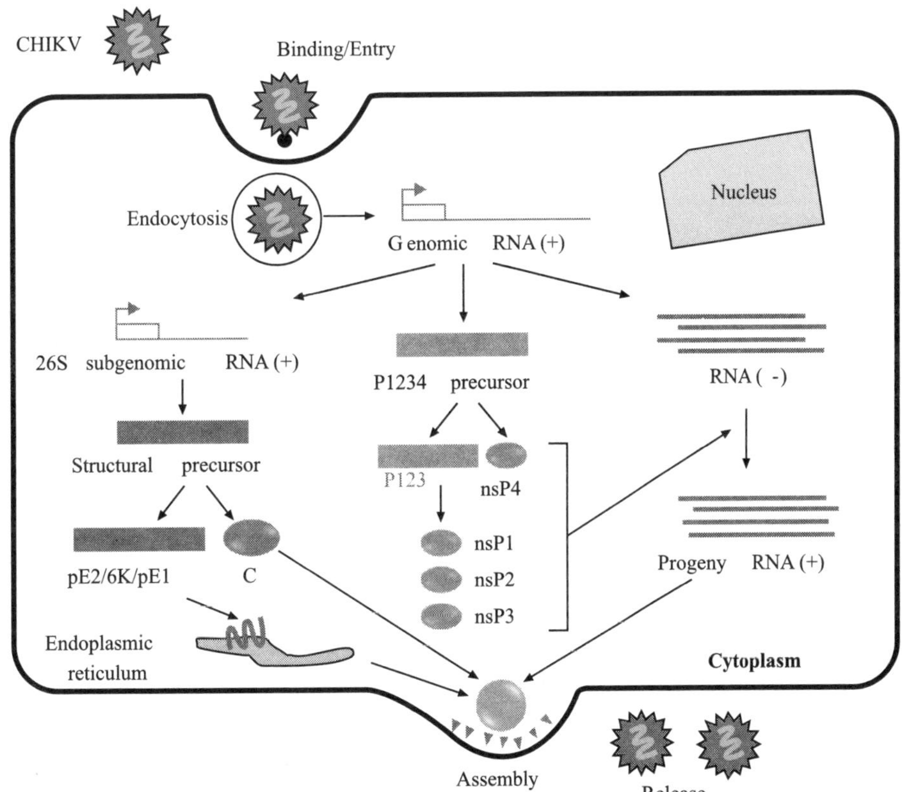

Figure 109.6 Replication cycle of Chikungunya virus inside infected cells. Replication occurs in the cytoplasm. Positive-strand genomic RNA acts directly as mRNA and is partially translated to produce non-structural proteins (nsP1-4). These proteins are responsible for replication and formation of positive-strand progeny RNA. 26 S subgenomic RNA is translated to produce the structural proteins such as Capsid and protein E2-6E-E1 through structural precursor. Assembly of the CHIKV occurs at the cell surface, and the envelope is acquired as the virus buds from the cell and, release and maturation take place almost simultaneously. (*see Plate 47 for colour figure*)

directly associated with a decrease in symptoms. Since RRV infects macrophages, it is possible that chemokines secreted by these infected cells attract additional macrophages that are also infected subsequently, hence propagating the cycle. Although virus could not be cultured from synovial fluid in CHIKV patients with prolonged joint pain, the presence of high-titered specific antibodies indicates the persistent presence of the antigen, and the similarities to RRV infection requires further research in this area for the better understanding of common pathogenic pathways.

VIII. CHIKUNGUNYA DIAGNOSTICS

CHIKV infection is diagnosed on the basis of clinical, epidemiological, and laboratory-based tests. An acute onset of high fever with severe arthralgia that is not explained by any other medical condition is likely to be a case of CHIKV infection. If the patient has lived in or visited epidemic areas, the chances of it being a CHIKV infection are high. Confirmation through laboratory testing is critical, in order to distinguish it from various diseases with similar clinical symptoms, such as dengue fever, other alphaviruses infections and arthritic diseases.

When selecting a CHIKV diagnostic test, the most important parameter is the amount of elapsed time since the onset of symptoms. Immune response against CHIKV is illustrated in Figure 109.7. CHIKV can be detected and isolated by culturing with mosquito cells (C6/36), mammalian cells (Vero), or by intracerebral inoculation of 1-day-old mice. The presence of early antibody apparently prevents isolation of the virus, hence virus isolation has been shown to be most successful in antibody-negative samples collected on or before the second day of illness. Virus isolation method is time consuming and can take several days. Moreover, high sensitivity can be attained

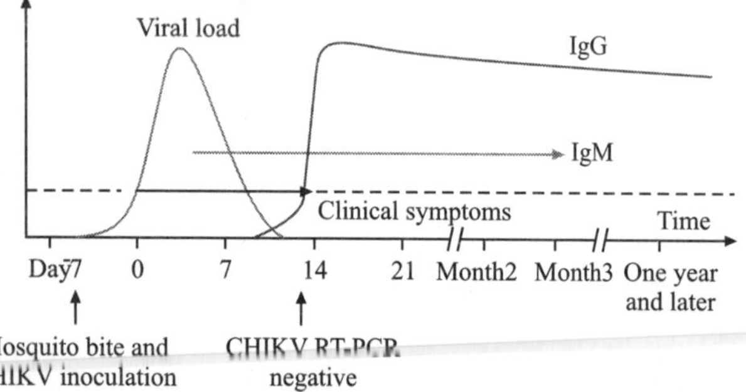

Figure 109.7 Illustration of immune response against CHIKV infection. Following inoculation by mosquito bite, CHIKV viremia in blood lasts between 2 and 10 days (viral load, red line). infected individuals experience an acute onset of disease and clinical symptoms 2–4 days after infection. Clinical symptoms (fever) disappear when CHIKV viremia drops below the level of detection. Presence of IgG and IgM are shown in blue and green lines, respectively. (*see Plate 48 for colour figure*)

in the viremic phase only. This method is still the preferred method for detecting the CHIKV strain. Confirmation of the diagnosis in the acute phase (within the first week of disease onset) of illness can be done by detection of viral nucleic acid in serum samples by reverse-transcriptase PCR (RT-PCR). RT-PCR is faster and comparatively more sensitive method and is also useful in quantifying viral mRNA. RT-PCR can detect viral nucleic acid in samples obtained from 1 day before onset of symptoms up to day 7, even from the CHIKV antibody-positive patients. Although PCR-based methods are very useful for early and accurate diagnosis of CHIKV, the high cost of the assay and requirement of skilled technicians and thermal cycler limit its applicability only to well-equipped laboratories.

Classic serological detection of CHIKV can be carried out by assays such as ELISA, immunofluorescence and haemagglutination inhibition. Antigen capture ELISA, developed by using the antibodies raised by immunizing animals with purified virus preparation, have been described for detection of antigen in serum samples and cerebrospinal fluid collected as early as day 2 after onset of symptoms. Indirect immunofluorescence and ELISA are rapid and sensitive methods for detection of an immune response to chikungunya and can differentiate between IgG and IgM antibodies. A specific IgM antibody response is usually detectable between 2 and 7 days after onset of fever with ELISA and immunofluorescence. An IgG antibody response is usually detectable from days 5 to 6. IgG antibody persists for years, whereas IgM antibody activity usually decreases to undetectable levels by month 3–4 post-infection. Various in-house ELISA techniques that use whole antigen or recombinant capsid or envelope antigens have been described for the detection of IgM and IgG. Some of the commercially available CHIKV immunoassay kits are listed in Table 109.5.

IX. CHIKUNGUNYA VACCINES

The wide dissemination of CHIKV is linked with genetic mutations that facilitate its adaptation to a new vector, *Ae. albopictus,* which survives in temperate climates and is widely distributed. CHIKV continues to cause considerable morbidity resulting in significant economic burden. There are no approved antiviral treatments currently available for CHIKV. The only recommended treatments for CHIKV-induced arthralgia are non-steroidal anti-inflammatory drugs. The explosive widespread outbreaks of CHIKV, combined with the limitation of drugs for the treatment of CHIKV, necessitates a safe and effective vaccine which would be the best weapon for protecting at-risk populations against CHIKV-induced disease. Epidemiological evidence suggests that CHIKV infection results in lifelong protective immunity against the virus.

A live-attenuated CHIKV vaccine was developed by the US Army, through the serial passage of the CHIKV in MRC-5 cells resulting in a virus (TSI-GSD-218), which carried several genetic changes. This vaccine was assessed in phase 2 clinical

TABLE 109.5   Commercially available CHIKV diagnostic immunoassay kits

Kits	Detection		Format	Manufacturers
	IgM	IgG		
CHIKV IgM	✓	×	ELISA	IBL International (Hamburg, Germany), Standard Diagnostics (Korea), Novatech Immundiagnostica (Dietzenbach, Germany)
	P	O	Rapid	CTK Biotech Inc (San Diego, USA), Standard Diagnostics (Korea), J. Mitra & Co Pvt. Ltd. (New Delhi, India), Bhat Bio-Tech India (P) Ltd. (Bangalore, India), Tulip Diagnostics (P) Ltd. (Goa, India)
	P	O	Immuno- fluorescence	EUROIMMUN AG (Luebeck, Germany)
CHIKV IgG	×	✓	ELISA	Novatech Immundiagnostica (Dietzenbach, Germany)
	O	P	Immuno-fluorescence	EUROIMMUN AG (Luebeck, Germany)

trials encompassing 59 healthy volunteers. Eight percent of the volunteers experienced transient arthralgia, while 98% of the volunteers developed CHIKV neutralizing antibodies. Further assessment of this vaccine was discontinued by the US Army in 2000 because of the change in research priorities. However, live CHIKV vaccines raise serious concerns. A live vaccine could possibly induce chronic rheumatism. Moreover, the widespread use of such vaccine in the human population could trigger mosquito transmission or lead to chronic infection or viral reversion.

There are always certain risk factors associated with live attenuated or inactivated viral vaccines. In order to avoid these potential problems, alternative strategies like DNA vaccine, chimeric virus vaccine, VLP vaccine, recombinant vaccine and subunit vaccine have been followed. Summary of different CHIKV vaccine studies have been presented in Table 109.6. To date, many experimental CHIKV vaccines have been developed and found to be effective in mice and/or monkeys, but none of them have been licensed. So far, there has been limited success in developing a safe and efficacious CHIKV vaccine.

TABLE 109.6   CHIKV vaccines

Year	Country	Vaccine	Formulation	Production methods	Animal models
1967	USA	African CHIKV 168 virus, original isolate from Tanganyika Territory	Formalin-inactivated	Harvested from green monkey kidney cells, chick-embryo or suckling-mouse-brain	Mice, rhesus monkeys
1970	USA	African CHIKV strain 168	1. Formalin-inactivated 2. Tween–ether treated	Harvested from green monkey kidney cells	Male Swiss Bagg strain mice
1971	USA	Thailand (designated 15561)	Formalin-inactivated	Harvested from green monkey kidney cells	Mice, human volunteers
1972	USA	African CHIKV strain 168	Formalin-inactivated	Harvested from green monkey kidney cells or concentrated chick-embryo suspension cultures	Male Swiss Bagg strain mice
1986	USA	CHIKV 181/clone 25 (origin from CHIKV strain 15561)	Attenuated viruses	Serial plaque-to-plaque passages in MRC-5 cultures	Mice (ICR strain and CD-1 strain), Monkey (male and female rhesus macaques)
2000	USA	TSI-GSD-218 (origin from CHIKV 181/clone 25)	Attenuated viruses	Lyophilized supernatant from CHIKV 181/clone 25 infected MRC-5 cells	Adult volunteers (47 men, 26 women)
2008	USA	Chimeric alphavirus/ CHIKV vaccine viruses	Attenuated chimeric viruses	Réunion Island CHIKV strain for cDNA production, SINV, VEEV and EEEV as alphavirus backbone. Chimeric viruses were produced from BHK-21 cells	Mice (NIH swiss and female C57BL/6)
2008	India & USA	CHIKV-capsid and envelope genes cloned in expression vector	DNA vaccine	NCBI database of all CHIKV and predict consensus sequence, codon-optimized and inserted into pVax1 expression vector	Female C57BL/6 mice

(Contd.)

TABLE 109.6   CHIKV vaccines (*Contd.*)

Year	Country	Vaccine	Formulation	Production methods	Animal models
2010	USA	Virus-like particles resembling replication-competent alphaviruses	Virus like particle (VLP) vaccine	Co-transfection of 293T cells with CHIKV structural proteins encoding plasmids (strains 37997 and LR2006 OPY-1), transducing vector and a packaging plasmid expressing all HIV-1 structural proteins except envelope	BALB/c mice, monkeys (rhesus macaques)
2011	USA & India	CHIKV-envelope construct (E3+E2+E1) cloned in expression vector	DNA vaccine	consensus sequence predicted from NCBI database for CHIKV, codon-optimized and inserted into pVax1 expression vector	BALB/c mice, monkeys (rhesus macaques)
2011	USA	CHIKV strain LR 2006-OPY1	Attenuated viruses	Attenuation by altering the expression of the CHIKV structural proteins with an encephalomyocarditis (EMCV) IRES	Mice (CD-1, and A129)
2011	USA, Australia	Structural polypeptide of CHIKV (strain isolate LR2006 OPY1) expressed in adenovirus vector	Recombinant CHIKV vaccine	non-replicating complex adenovirus vector encoding the structural polyprotein of CHIKV in HEK293 packaging cell line	Female mice (CD-1 and C57BL/6)
2011	USA	Chimeric alphaviruses encoding CHIKV-specific structural proteins	Chimeric alphaviruses	Chimeric alphaviruses with VEEV-derived nonstructural and, CHIKV-specific structural proteins (expression mediated by the EMCV IRES)	Mice (CD-1, and A129)
2012	India	Recombinant envelope proteins (E1 and E2) of CHIKV	Subunit vaccine	*E. coli* expressed recombinant CHIKV envelope proteins along with various adjuvants (FCA, montanide ISA720 and alum)	BALB/c mice

SUMMARY

CHIKF is caused by CHIKV, which is an arbovirus transmitted by *Aedes* mosquitoes. CHIKV is typically circulated in two distinct transmission cycles, sylvatic and human-mosquito-human. The latter cycle transmits CHIKV among humans vectored by *Ae. aegypti* and *Ae. albopictus*. The disease is typically characterized by an acute illness with fever, skin rash, and debilitating arthralgia. CHIKV infections are rarely fatal and severe, but in some cases deaths have been reported attributed directly or indirectly to CHIKV. CHIKV epidemics are characterized by long inter-epidemic periods of typically more than 10 years. Since its first report in 1952, CHIKV has affected millions of people in different outbreaks. The worst outbreaks were observed in Reunion Island in 2005-06 affecting one-third of its population and in India in 2006 affecting 1.39 million people. Although, CHIKV was initially thought to be associated with tropical regions only, cases from non-tropical regions have also been reported, where the virus was imported through international travelers. Thus CHIKV infection is becoming a global burden. CHIKV infections are typically underreported because the disease symptoms have similarities with dengue and many other acute infectious tropical diseases. CHIKV diagnosis can be done using several techniques, depending upon the clinical manifestation of the patient. Many diagnostic kits are available in the market facilitating early diagnosis of the infection. Since there is a lack of established antiviral treatment against CHIKV, CHIKF is treated symptomatically. The most commonly recommended treatment for CHIKF is non-steroidal anti-inflammatory drug combined with bed rest and fluids. Life cycle of CHIKV and its molecular pathogenesis are not clearly understood so far. The best way to prevent CHIKV infection, is the eradication of vector mosquitoes, but this strategy is not feasible because of increased urbanization, global warming and, evolvement of the virus and the vector. In such circumstances, a vaccine becomes the last resort to prevent CHIKV infections in at-risk populations. Efforts are being made towards the development of a safe and efficacious vaccine against CHIKV. Many experimental CHIKV vaccines, such as inactivated virus vaccine, attenuated virus vaccine, chimeric virus vaccine, DNA vaccine, subunit vaccine and VLP-based vaccines, have been developed so far, and found to be effective in mice and/or monkeys, but none of them have been licensed. In future, research providing new insight into CHIKV receptor, life cycle and pathogenesis will aid in developing efficient CHIKV inhibitor, antiviral therapy and vaccine.

SUGGESTIONS FOR FURTHER READING

1. Dupuis-Maguiraga L., Noret M., Brun S., Le Grand R., Gras G. and Roques P. (2012), Chikungunya Disease: Infection-associated Markers from the Acute to the Chronic Phase of Arbovirus-induced Arthralgia, *PLoS Negl Trop Dis.*, 6:e1446.

2. Kam Y.W., Ong E.K., Renia L., Tong J.C. and Ng L.F. (2009), Immuno-biology of Chikungunya and Implications for Disease Intervention, *Microbes Infect*, 11:1186–96.

3. Schwartz O. and Albert M.L. (2010), Biology and Pathogenesis of Chikungunya Virus, *Nat Rev Microbiol.*, 8:491–500.

4. Solignat M., Gay B., Higgs S., Briant L. and Devaux C. (2009), Replication Cycle of Chikungunya: A Re-emerging Arbovirus, *Virology*, 393:183–97.

5. Thiboutot M.M., Kannan S., Kawalekar O.U., Shedlock D.J., Khan A.S., Sarangan G., Srikanth P., Weiner D.B. and Muthumani K. (2010), Chikungunya: A Potentially Emerging Epidemic? *PLoS Negl Trop Dis.*, 4:E623.

6. Voss J.E., Vaney M.C., Duquerroy S., Vonrhein C., Girard-Blanc C., Crublet E., Thompson A., Bricogne G. and Rey F.A. (2010), Glycoprotein Organization of Chikungunya Virus Particles Revealed by X-ray Crystallography, *Nature*, 468:709–12.

110

Japanese Encephalitis

Sudhanshu Vrati and Kaushik Bharati

CONTENTS

I. JAPANESE ENCEPHALITIS: AN INTRODUCTION

Japanese encephalitis (JE) is so called simply because of the fact that it was first reported from Japan, and the major complication is encephalitis (inflammation of the brain). JE is the most important form of viral encephalitis commonly known as brain fever that mostly affects children and young adolescents in Asia. JE is caused by a virus called Japanese encephalitis virus (JEV). Though outbreaks of encephalitis attributed to JEV were reported in Japan as early as 1871, it was not until 1924 that JEV was isolated from a clinical case in the first recorded JE epidemic from Japan. The celebrated "Nakayama strain" was isolated in 1935 from the brain of a dying JE patient. Moreover, the mode of transmission, by Culicine mosquitoes, was not elucidated till 1950 (Table 110.1).

The disease is endemic in most parts of Southeast Asia, China, India and Oceania. The disease is continuously expanding its geographical territory. Over the last decade or so, it has spread to hitherto unaffected regions, such as the western parts of India, Karachi (Pakistan), Western provinces of Papua New Guinea and the Torres Strait islands of Northern Australia. There is a realistic possibility that JE could spread further. The changing geographical distribution pattern of JE is presented in Figure 110.1.

A. Japanese Encephalitis Virus: The Etiologic Agent of JE

Japanese encephalitis virus (JEV), which causes JE is an arthropod-borne virus (arbovirus) that belongs to the family *Flaviviridae* and genus Flavivirus. The prototype virus of this genus is the yellow fever virus (Latin flavus: yellow). The other Flaviviruses that cause human disease are dengue virus (DEN), St. Louis encephalitis virus (SLE), Murray Valley encephalitis virus (MVE), West Nile virus (WNV) and tick-borne encephalitis virus (TBE). Another flavivirus of importance in Karnataka (India) is the Kyasanur Forest disease virus (KFD), which causes a febrile illness, popularly called "monkey fever" (Table 110.2). JEV is spread by the bite of infected Culicine mosquitoes, predominantly *Culex tritaeniorhynchus*. The major amplifying vertebrate hosts are domestic pigs and ardeid wading birds, such as herons and egrets. A number of other animals become naturally infected with JEV, including donkeys, chicken, ducks, water buffalos, cattle, sheep, mice, snakes and frogs, but their role in JEV transmission is questionable. However, there is evidence that bats could play a role in overwintering of the virus, as well as viral persistence, since transplacental transmission has been demonstrated in bats. It has been suggested that bats, migratory birds as well as wind-blown mosquitoes that have been infected with JEV, may be responsible for introducing the virus to hitherto unaffected geographical regions.

TABLE 110.1   Japanese Encephalitis – Major Incidents

Year	Incident
1871	First recorded clinical case of JE, reported from Japan
1924	Large outbreak of JE in Japan with >6,000 cases and a fatality rate of 60%; Isolation of JEV from human brain
1933	First cases of JE reported from the Korean peninsula
1935	Isolation of Nakayama strain of JEV
1938	Isolation of JEV from *Culex tritaeniorhynchus* mosquitoes
1940	First cases of JE reported from the Chinese Mainland
1950	First cases of JE reported from the Philippines
1950s	Elucidation of transmission cycle of JEV, with pigs and ardeid birds identified as amplifying hosts and *Culex tritaeniorhynchus* as primary vector species
1955	First cases of JE reported from Vellore, India
1965	Major epidemic in northern Vietnam
1969; 1970	Major epidemic in Chiang Mai Valley, Thailand
1973	First epidemic in India, in the state of West Bengal
1978	Major epidemic in Terai region of Nepal
1983	JE reaches Pakistan, the furthest geographical extension to the West
1985-86; 1987	Major epidemics in Sri Lanka
1995	JE reaches Papua New Guinea and Torres Strait islands (Australia), the furthest geographical extension to the South
2005	Major epidemic in Gorakhpur, Uttar Pradesh state of India. 5,737 cases, with 1,344 deaths; India imports live-attenuated SA 14-14-2 vaccine from China

Figure 110.1   Changing geographical distribution pattern of Japanese encephalitis.

TABLE 110.2 Major Human Flaviviruses and their Endemic Areas

Virus	Endemic areas
Japanese encephalitis	Asia and Oceania
Yellow fever	South America and Africa
Dengue	Tropics, worldwide
West Nile	Europe, Africa, Asia and North America
St Louis encephalitis	North and South America
Murray Valley encephalitis	Australia
Tick-borne encephalitis	Europe and Asia

B. The JEV Life Cycle

The JEV life cycle normally involve pigs that act as the amplifying hosts, mosquitoes that act as the vectors and humans that act as dead-end hosts. Usually after two 4-day amplification cycles in pigs, approximately 20% of the pigs become seroconverted. Mosquitoes become infected by feeding on the viremic pigs. After this, the virus undergoes a 7–14 day extrinsic incubation period (EIP) in mosquitoes, where it multiplies in various organs and body compartments. After the EIP, mosquitoes infect other pigs, as a result of which almost 100% of the swine population becomes seroconverted. Blood meal from these highly viremic pigs is followed by another round of EIP, following which, the virus spills over to the human population and clinical cases start to appear. Humans become infected coincidentally when they encroach this enzootic cycle between mosquitoes, birds and pigs. Viral titers in humans are not high enough to cause further transmission, which is why humans are regarded as dead-end hosts. The life-cycle of JEV in nature is depicted in Figure 110.2.

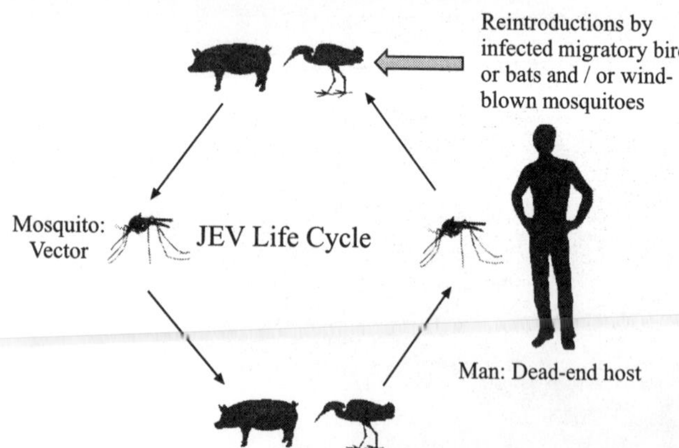

Figure 110.2 JEV Life Cycle.

C. Structural Organization of the JEV Genome

JEV is spherical in shape and is ~50 nm in diameter. Its nucleocapsid core is surrounded by an envelope. Its single-stranded, positive sense RNA genome contains a single long open reading frame (ORF) flanked by 5′- and 3′-untranslated regions (UTRs), which have secondary structures that are essential for the initiation of translation and for replication. The 5′ end of the genome has a type 1 cap, while the 3′ end lacks a poly-A tail. The viral genome is ~11 kb, and codes for a single polyprotein of ~3400 amino acids that is cleaved into 3 structural and 7 non-structural proteins. The structural proteins are capsid (C), envelope (E), and membrane (M). The non-structural proteins are NS1, NS2A, NS2B, NS3, NS4A, NS4B, and NS5.

JEV structural proteins

As the name suggests, the structural proteins constitute the structural component of the virion. Importantly, the nucleocapsid is formed by the C protein. The M protein is formed by the cleavage and removal of the N-terminal segment from its precursor, the pre-membrane protein (prM). The prM helps in the proper folding of the E protein, which is a typical membrane glycoprotein, consisting of a C-terminal membrane-anchorage domain, and the remaining portion forms the outer structural protein component of the virus. The E protein is the major virion antigen responsible for a number of important processes that include virion assembly, receptor binding, and membrane fusion. The E protein is the most important viral protein from a vaccine development standpoint, as most of the neutralizing epitopes are located on its domain III.

JEV non-structural proteins

The non-structural proteins generally perform the "house-keeping functions" of the virus. The NS1 protein is a glycosylated protein that is believed to be involved in the assembly and release of virions. NS2A and NS2B are low molecular-weight proteins that are thought to be involved in the processing of other viral proteins. NS3 protein is conserved among flaviviruses and has protease and nucleotide triphosphatase / helicase activities. NS4A and NS4B are small proteins whose functions are not clear, although they may be involved in the membrane localization of NS3 and NS5 through protein-protein interactions, or in the formation of the genomic RNA replication complex. NS5 protein is the largest and most conserved protein and is the viral RNA-dependent RNA-polymerase (RdRp). The genome organization of JEV is depicted in Figure 110.3.

D. JEV Genotypes

Four distinct genetic subtypes or genotypes of JEV have been identified on the basis of nucleotide sequence data of C/PrM and E genes and phylogenetic analysis. Genotype I includes isolates from Northern Thailand, Cambodia and Korea. Genotype II includes isolates from Southern Thailand, Malaysia, Indonesia and Northern Australia. Genotype III includes isolates from mostly temperate regions of Asia, including Japan, China, Taiwan, the Philippines and the Asian subcontinent. Genotype IV includes some isolates from Indonesia. In addition, based

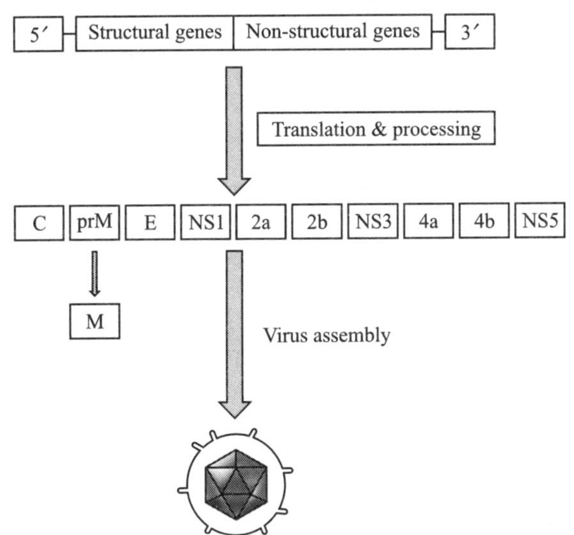

Figure 110.3   JEV genome organization. The genome is a single-stranded, plus-sense RNA molecule containing a long open reading frame (ORF) encoding the viral polyprotein with 5′ and 3′ untranslated regions (UTRs). The encoded proteins subsequently self-assemble into complete JEV particles.

on phylogenetic evidence, the Muar strain of JEV isolated in Singapore in 1952 from a patient who originated in Muar (Malaysia), may represent a fifth genotype.

All the genotypes differ from each other by about 10–20% at the nucleotide level and 2–6% at the amino acid level, and belong to the same serotype and are similar in terms of virulence and host preference. As a consequence, any JEV strain used for the development of a vaccine antigen may be expected to confer protection against all other genotypes. Current JEV vaccines are based on genotype III strains and it has been demonstrated that these vaccines confer protection against heterologous JEV strains, although the neutralizing antibody titers in these instances are lower than those against homologous strains.

II. EPIDEMIOLOGY OF JE

JEV is distributed throughout the temperate and tropical regions of Southern and Eastern Asia. Two major epidemiological patterns of disease are observed—endemic or epidemic. In the Northern temperate areas, such as Northern Vietnam, Northern Thailand, Korea, Japan, Taiwan, China, Nepal, and Northern India, JE occurs in the form of epidemics during the summer/monsoon months. In Southern tropical areas, such as Southern Vietnam, Southern Thailand, Indonesia, Malaysia, Philippines, Sri Lanka and Southern India JE is endemic. In regions where the tropical and temperate climes intermingle, the disease pattern also becomes superimposed. The disease burden is estimated to be ~175,000 cases annually, which is probably a gross under-estimate, since under-reporting often occurs, as reflected in the number of reported JE cases, which is only 50,000 annually (Table 110.3).

TABLE 110.3   Japanese Encephalitis – Facts & Figures

Causative agent	Japanese encephalitis virus (JEV): Single-stranded, positive-sense, RNA virus. Family: *Flaviviridae*. Genus: *Flavivirus*.
Geographic range	Indian subcontinent (including Pakistan), South Asia, Southeast Asia, China, Oceania, Northern Australia
Major vectors	Paddy-breeding mosquitoes of the *Culex vishnui* subgroup, particularly *Cx. tritaeniorhynchus*. Other important secondary or regional vectors include *Cx. gelidus, Cx. fuscocephala, Cx. pipiens, Cx. annulirostris*
Major vertebrate hosts	Domestic pigs (amplifying hosts); ardeid wading birds such as the black-crowned night heron (*Nycticorax nycticorax*), plumed egret (*Egretta intermedia*), and little egret (*Egretta garzetta*) (primary enzootic hosts); domestic fowls; migratory birds
High risk groups	Infants and children below 10 years in endemic areas; the elderly in endemic areas; non-immune adults (e.g. travelers from non-endemic countries visiting JE endemic areas for more than a month); immunocompromised individuals
Annual estimated cases	~175,000
Annual reported cases	~ 50,000
Annual reported morbidity	~ 15,000
Annual reported mortality	~ 10,000

JE is predominantly a disease of children and serological tests have shown that when the children reach adulthood, all have been exposed to the virus and therefore exhibit neutralizing antibody (NtAb) titers. However, non-immune adults coming from non-endemic regions of the world and staying in endemic areas for more than a month are likely to get infected. Another instance when adults become infected is when the virus spreads to new geographical locations, as has been the case in Nepal and Northern Australia. Immunocompromised adults are also susceptible to JEV infection, as are the elderly, possibly due to waning immunity.

JE is essentially a disease of rural areas, although populations residing in peri-urban areas are now-a-days also afflicted, possibly due to unplanned constructions that encroach agricultural areas, thereby disrupting the vector population dynamics, leading to human JE cases. Moreover, climate change will definitely have a drastic effect on the epidemiology of JE in the coming decades, as it is likely to alter the mosquito breeding patterns.

Of the 50,000 JE cases reported annually, 10,000 prove to be fatal. Those who survive the disease are very often left with neurological and psychiatric problems, which can sometimes be life-long in the absence of proper medical care. JE, therefore, is a serious public health problem, even though its severity is often under-stated and under-appreciated.

A. Pathogenesis

As has been highlighted above, JEV transmission occurs by the bite of infected Culex mosquitoes. Following the mosquito bite, the virus initially replicates locally, after which it spreads to the blood stream, causing a transient viremic phase. The virus also multiplies in the regional lymph nodes. Studies indicate that in normal circumstances, the virus enters the CNS by passage across the cerebrovascular endothelium, rather than across the olfactory membrane, where the blood-brain barrier (BBB) is scanty or lacking. Moreover, a recent study indicates that the permeability of the BBB is differentially altered in response to JEV infection, leading to entry of the virus particles into the cerebrum, as the initial site of virus entry into the CNS. The olfactory route of transmission is important in case of laboratory workers, especially during administration of injections into mice with the live virus, where aerosols can easily enter the nasal cavity if appropriate protective measures are not taken. JEV has also been reported to infect the developing fetus transplacentally and cause abortions.

B. Mechanism of Neuronal Damage

Within the CNS, JEV replicates in the neurons, more specifically within their secretory system, involving the rough endoplasmic reticulum (RER) and Golgi apparatus, eventually leading to their destruction as the virions mature and the infection spreads to other neighboring neurons. Although there is no direct evidence from humans, neuronal apoptosis has been demonstrated in flavivirus disease models, both *in vitro*, as well as *in vivo*. It has been demonstrated that activation of microglial cells leads to production of proinflammatory cytokines that may elicit neuronal death. It has been suggested that an elevation in proinflammatory cytokine levels in the cerebrospinal fluid (CSF) indicates a poor prognosis. Other mechanisms of neuronal damage that have been suggested include astrocyte activation and nitric oxide (NO)-mediated damage. The latter has received particular attention due to the fact the JEV has been found to induce the expression of inducible NO synthase (iNOS), the major enzyme in NO synthesis, thought to be a key component of the host innate immune response. Importantly, the elucidation of the mechanism of neuronal damage may lead to identification of targets for drug intervention.

III. CLINICAL FEATURES OF JE

JE infections are often asymptomatic, but might follow a course resembling a mild non-specific febrile illness. Approximately 1 in 300 cases of JE result in symptomatic disease, which depends on 4 major factors; (i) route of entry, (ii) titer of virus, (iii) neurovirulence of virus, and (iv) host factors, such as age, health, genetic make-up, and pre-existing immunity. The first signs of infection appear after an incubation period of 1–6 days, but may take as long as 15 days to manifest. The onset of the disease may be acute or gradual. The disease may be crudely divided into three stages: (i) a prodromal stage, before central nervous system (CNS) involvement occurs, (ii) CNS stage, where the virus infects the CNS, and (iii) a convalescent stage, where either marked improvement occurs or the CNS symptoms may persist. The prodromal stage is marked by fever above 38°C, chills, myalgia, malaise, headaches accompanied by nausea, vomiting and abdominal pain. Besides these common symptoms, in some instances, gastric hemorrhage, thrombocytopenia and liver dysfunction have also been observed. These non-specific signs may continue for 2–4 days or the patient's condition may deteriorate rapidly. CNS involvement (day 3–5) is marked by a progressive decline in alertness, often leading to coma. CNS infection can result in encephalitis, meningitis or myelitis, or a combination of all the three, in the form of meningoencephalomyelitis.

A. Movement Disorders

A Parkinson's like syndrome, characterized by dull mask-like facies, tremors and cogwheel rigidity have been observed in JE cases. Convulsions may be experienced in up to 87% of patients. The meningeal syndrome predominates with painful neck stiffness. Motor paralysis including hemiplegia and tetraplegia may also be present. A poliomyelitis-like acute flaccid paralysis can also occur. Other movement disorders include opisthotonus (Figure 110.4), dystonia (Figure 110.5), oro-facial dyskinesias (e.g., lip-smacking), hemiparesis, choreoathetosis and gaze palsy. Abulia is another striking feature observed in JE patients.

It is important to detect early clinical signs of CNS involvement, such as abnormal oculocephalic reflexes, acute onset hemiparesis with hypertonia and decorticate

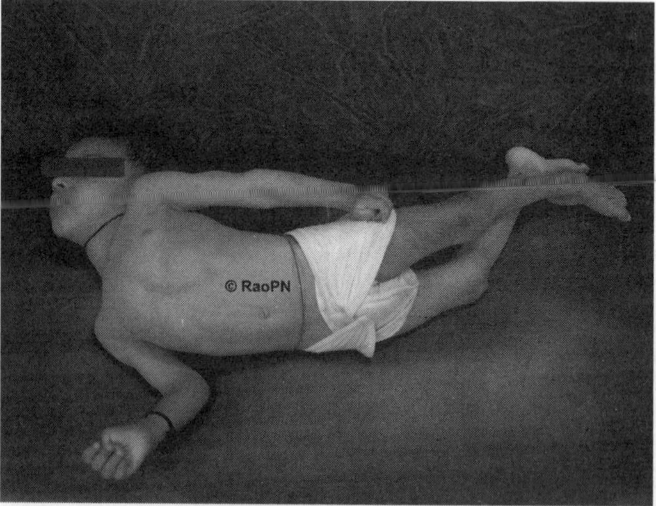

Figure 110.4   Opisthotonus in a boy with JE (Photo: Courtesy Dr. P. Nagabhushana Rao).

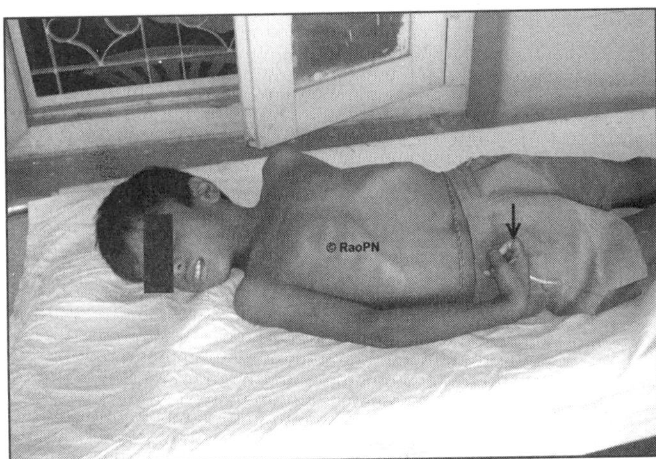

Figure 110.5 Dystonia in the right hand of a boy with JE (Photo: Courtesy Dr. P. Nagabhushana Rao).

and decerebrate posturing, which help in the early clinical identification of intracranial hypertension. Effective management of intracranial hypertension will dictate whether the patient survives or dies. Clinical diagnosis can be buttressed by various neuroimaging techniques.

B. Neuroimaging

Neuroimaging techniques such as Computed Tomography (CT Scan) and Magnetic Resonance Imaging (MRI) carried out in JE patients have revealed extensive bilateral thalamic lesions (Figure 110.6). In a JE-endemic region, bilateral lesions of the thalamus are indicative of JE. It has been suggested that the movement disorders are the clinical correlates of damage to the thalamus and other parts of the brain. Lesions to sites such as the former, as well as that to the lentiform nucleus and

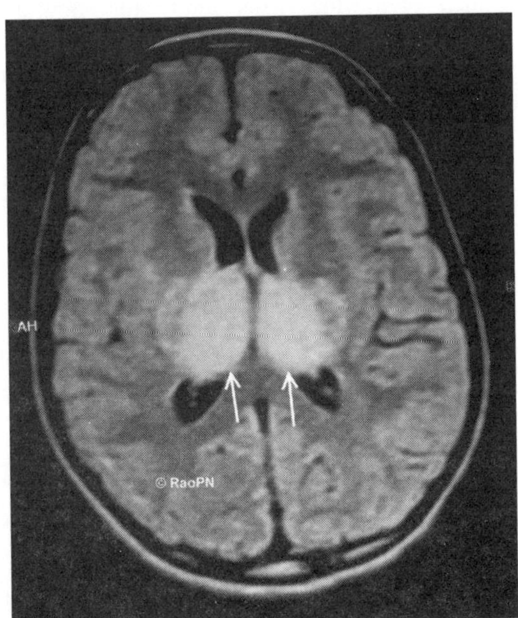

Figure 110.6 MRI scan showing hyperintense thalami (Photo: Courtesy Dr. P. Nagabhushana Rao).

basal ganglia could result in Parkinsonism. Other complications such as abulia could be due to thalamic lesions disrupting the prefrontal cortex-caudate-pallido-thalamocortical circuit. Dystonia results from putaminal and possibly thalamic lesions. Another powerful technology is the Single-Photon Emission Computed Tomography (SPECT). SPECT may be useful as a diagnostic tool in the early stages of JE.

C. Prognosis

Fatality is seen in 20–30% of the cases, with signs of acute cerebral edema or severe respiratory distress from pulmonary edema. Children who survive the disease usually regain neurological function over the following weeks to months. However, almost half of the survivors are left with serious neuropsychiatric sequelae. These include a persistently altered sensorium, epileptic seizures and severe mental retardation. These neurological complications are very often understated in the medical literature. One should understand that the patients with neurological deficits are very often in their early childhood and hence, have to bear with these complications for the rest of their lives, given the fact that most of them cannot afford treatment. Moreover, even those who have a good recovery, often have minor sequelae like learning and behavioral problems, giving rise to stigmatization. Hence, the human face of the disease burden is enormously more than what the mere statistics suggest.

IV. DIAGNOSIS

A. Differential Diagnosis

Diagnosis of JEV infection should be made within an epidemiological context. Because of clinical, biological and epidemiological similarities, three other viral diseases should be considered in the differential diagnosis (DD). These are (i) Herpes Simplex Virus (HSV) Encephalitis, (ii) Dengue, and (iii) West Nile Encephalitis. Moreover, other CNS infections should also be kept in the DD. These include bacterial and fungal meningitis, tuberculosis, cerebral malaria, leptospirosis, tetanus and typhoid encephalopathy. Moreover, enteroviruses, paramyxoviruses, rabies virus, Chikungunya virus and Nipah virus should also be kept in mind. Some non-infectious diseases that exhibit CNS manifestations include tumors, cerebrovascular accidents, Reye's syndrome, toxic and alcoholic encephalopathies, and epilepsy. Hence, it is of paramount importance to spend some time in taking down the case history. However, in case of epidemics, occurring in JE endemic areas, it is often easier to rule out the other causes from the DD.

B. Physical Examination

A thorough physical examination is crucial. The level of consciousness should be established with a quantitative scale such as the Glasgow Coma Scale (GCS) and any seizures arising out of the infection should be immediately treated. Any other mental and behavioural abnormalities should be documented. The examination should search for any other causes of altered levels of consciousness.

C. Lumbar Puncture

If the condition of the patient permits, a lumbar puncture should be performed, since the initial CSF findings, such as opening pressure, cell count, glucose and protein levels are crucial for initially establishing a CNS infection. The CSF findings will demonstrate whether the infection is viral or bacterial in origin. Moreover, downstream laboratory testing, by techniques such as MAC ELISA or PCR will pin-point the incriminating organism, and thus provide a guide for management. JE should be suspected if biochemical investigations reveal: (i) high CSF opening pressure (>250 mm), (ii) moderate CSF pleocytosis (10–100 cells/mm^3), (iii) mildly increased CSF protein (50–200 mg/dL), but (iv) normal levels of CSF glucose.

D. Rapid Diagnostic Tests

Confirmation of a suspected case of JE requires laboratory diagnosis, which relies on virus isolation or demonstration of virus specific antigen or antibody in the CSF or serum. The humoral immune response to JEV infection involves early production of IgM antibodies in both serum and CSF, followed by IgG production. Hence, the IgM-capture ELISA (MAC ELISA), which detects specific IgM in CSF or serum of almost all patients within 7 days of onset of disease has become the practical standard for the diagnosis of JE. Three MAC ELISA kits are being manufactured commercially. These are the (i) JE-Dengue IgM Combo ELISA Test (Panbio Limited), (ii) JE IgM ELISA (InBios International, Inc.), and (iii) JEV CheX Kit (XCyton Diagnostics Ltd.). These tests use cell culture-derived inactivated virus or recombinant E protein as the antigen; the Panbio test also uses recombinant Dengue 1-4 antigens and can detect both JE and Dengue antibodies. The Panbio kit has shown 90% agreement to the USAMC-AFRIMS (United States Army Medical Component – Armed Forces Research Institute of Medical Sciences) standard, the internationally accepted standard diagnostic test for JE, when both JE and dengue IgM-positive samples were considered in the analysis. In this situation, sensitivity was good for the InBios and XCyton kits, but lower for the Panbio kit. However, specificity was low for both the InBios and XCyton kits as a result of cross-reactivity with dengue antibodies. When dengue cross-reactivity was eliminated, all the three kits had specificities of 96% or above. If pricing is competitive, then the Panbio kit would have a distinct advantage over the other two.

E. Molecular Diagnostics

A viral genome amplification using RT-PCR technique allows rapid detection of viral RNA in the CSF of JE patients. Another molecular technique, the reverse transcription-loop-mediated isothermal amplification (RT-LAMP) assay is a rapid real-time detection system for JEV, and the results can be obtained within 30 minutes under isothermal conditions at 63°C. This test could become useful in low resource settings as it does not require a thermocycler. However, these molecular tests are not currently used for routine diagnostic purposes.

V. TREATMENT

There is currently no specific treatment for JE. The antiviral, ribavirin has been evaluated for therapeutic applicability in JE patients in a clinical trial in children in the state of Uttar Pradesh (India).This antiviral agent did not have any effect in reducing early mortality associated with JE. Although isolated case studies have indicated that IFN-α could be an effective and promising agent for the treatment of JE, a randomized, double-blind, placebo-controlled trial of IFN-α2a did not improve the outcome of JE patients. Likewise, the efficacy of intravenous immunoglobulin for treatment of JE cases is also inconclusive.

Although specific antivirals against JEV are still lacking, a number of approaches are under laboratory scale development against this virus. It has been demonstrated that JEV, as well as other flaviviruses that share a similar genome organization are susceptible to a broad spectrum of agents, including chemical compounds, natural products, siRNAs and DNAzymes. Other efforts have included targeted drug discovery approaches that have been aimed to find specific inhibitors against JEV. These have been briefly highlighted below.

A. Targeted Drug Discovery Efforts to find Specific Inhibitors of JEV

The members of the Flavivirus family all follow a similar replication strategy. So, there is a high possibility that viral inhibitors can be discovered that have a broad-spectrum of antiviral activity. The targeted approach towards drug discovery involves, first identifying the targets, and then designing molecules that could act as inhibitors of these selected molecular targets.

Polymerase inhibitors

The polymerases are the most promising targets for development of antiviral agents. The RdRp is a multimeric complex and is essential for viral replication. Importantly, since this molecular complex is encoded by the viral genome, and has no cellular counterpart, there is no chance of cellular toxicity.

Protease inhibitors

Another attractive target for drug development is the viral protease. Viral protease inhibitors are already being used in clinical practice. The HIV-protease inhibitors are examples of this class of drugs that are components of the highly active anti-retroviral therapy (HAART). The flavivirus proteases could also be targeted in a similar fashion. However, it needs to be investigated whether these viral serine proteases are sufficiently different from the cellular serine proteases, so that toxicity issues do not arise.

Virus entry inhibitors

Virus entry inhibitors are ideal antiviral agents. These agents are in many ways superior to the polymerase or protease inhibitors as issues of toxicity and emergence of resistant strains does not occur in this class of antiviral agents. In fact, virus entry inhibitors have been approved for use against HIV-1.

Capping inhibitors

The flaviviruses, like other RNA viruses, replicate in the cytoplasm, and have evolved their own capping enzymes that are independent of the host. Since 5′ capping is an essential component of viral replication, the methyltransferase enzyme, which is an essential component of the capping machinery, is an attractive target for antiviral drug development.

Helicase inhibitors

The NS3 is the helicase of flaviviruses. It is an essential component for viral replication. It is involved in unwinding the double-stranded RNA intermediate during genome replication and acts in a multimeric complex with NS5. This enzyme is also an attractive target for antiviral drug development.

VI. MANAGEMENT

Management essentially involves measures to control both the immediate complications of infection, including seizures and raised intracranial pressure, and the longer-term consequences of neurologic impairment, such as limb contractures and bed sores. Corticosteroids have been investigated in the treatment of JE, but they failed to show any beneficial effects in acute JE cases. Hence, management is essentially symptomatic and supportive. The major management strategies during the acute phase and convalescent or recovery phase are briefly discussed below.

A. Management during the Acute Phase

Management of patients during the acute phase of the disease is of the utmost importance because the treatment given in this period decides whether the patients live or die. Here, the knowledge, experience, skills and dexterity of the clinical staff is called into play. This period is a real testing time for the health infrastructure of the hospital where the patients are managed. Everything, from laboratory diagnostics to the actual treatment administered, has to be well coordinated for the patients' survival.

Management of fever

One of the major clinical signs of children being brought to hospital is fever. Management of fever is a clinical priority, mainly because of the fact that high fever can lead to increase in intracranial pressure by increasing cerebral metabolism and cerebral blood flow, leading to cerebral edema. Antipyretics such as paracetamol may be used along with sponging in order to lower the temperature.

Management of seizures

Seizures are very common in JE patients, particularly in children. Seizures may be generalized tonic-clonic seizures, or more subtle, such as twitching, which may occur in a digit or around the lips or eyes. Uncontrolled seizures can lead to raised intracranial pressure, leading to increased metabolism and other physiological changes that can further aggravate the intracranial hypertension, operating in a positive-feedback vicious cycle, which can lead to brain herniation. Seizures can often be managed with phenytoin and benzodiazepines. If not, intubation and artificial ventilation should be initiated so that higher doses of sedatives and anticonvulsants can be administered. The patient should be monitored by carrying out an electro-encephalogram (EEG) from time to time. EEG should be carried out after administration of anticonvulsants and should be continued until seizures subside. Coma should be monitored by the GCS score. Other parameters such as blood pressure, urine output, serum osmolality, oxygen saturation and central venous pressure should be monitored until the condition of the patient improves.

Management of raised intracranial pressure

From the foregoing discussion, it is evident that raised intracranial pressure is one the most life-threatening complications of JE infection. Management of raised intracranial pressure is crucial and it is important to keep the patient's head in a midline position, tilted at 30. Keeping the head tilted improves the cerebral perfusion pressure, a key determinant in cerebral circulation, influenced by the intracranial pressure. Moreover, keeping the head elevated increases CSF drainage and maximizes cerebral venous return. Importantly, the head should lie on the midline in order to prevent any obstruction in the venous drainage through the jugular vein. Hyperventilation can reduce the intracranial pressure by bringing about a decrease in cerebral blood flow, which in its turn is brought about by a decrease in the CO_2 tension (pCO_2) caused by the hyperventilation itself. Intracranial pressure begins to fall within seconds to minutes, stabilizes in about 30 minutes, and returns back to the original value in about an hour. Although hyperventilation is useful for bringing down the intracranial pressure, prolonged hyperventilation can, in fact, be harmful, and may worsen the outcome. Hyperventilation should always be withdrawn gradually. Mannitol is commonly used to control raised intracranial pressure. The action of mannitol is dependent on the rate of administration – higher infusion rates lead to a rapid fall in intracranial pressure that is short-lived, while slower infusion rates bring about a sustained decrease in the intracranial pressure. However, overdosing with mannitol should be avoided at all costs, as this can aggravate cerebral edema. Loop diuretics such as frusemide can be used alone, or in combination with mannitol to lower the elevated intracranial pressure. Frusemide alone causes slow reduction in intracranial pressure, but when combined with mannitol, the fall in pressure is rapid and much more sustained than when either drug is used alone.

A. Management during the Convalescent Phase

Management during the convalescent or recovery phase is no less important than that during the acute phase. This phase is often neglected, because the danger of death has passed. But it is in this phase that the real struggle for the surviving patients begins. It may be recalled that ~50% of the survivors, usually small children, have severe neuropsychiatric sequelae, which are likely to be life-long. A thorough neurological and neuropsychiatric examination of the patients is required, soon after discharge from hospital. This will help assess the condition of the patient for leading a normal life. Intellectual and behavioural problems are common in children recovering from JE. These conditions should be treated properly and promptly as these can lead to absenteeism from school and thus hamper normal social functioning. Hence, management of neurological deficits is a challenging task, and requires a lot of patience, encouragement and supportive care, both on the part of the healthcare provider, as well as the family of the affected child. Moreover, treatment of neurological problems is a costly affair and is particularly difficult in the resource-poor developing countries of Asia, where JE is most prevalent.

VI. JE VACCINES

Currently, there are three JE vaccines, one of which is licensed for use in China and four other Asian countries, while the other two are in late stages of development. Two of the vaccines are cell culture-based, while the third is genetically-engineered.

A. SA14-14-2

This vaccine is based on the genetically stable, neuro-attenuated SA14-14-2 strain of JEV that is derived from serial passage of the SA14 strain of the virus in primary hamster kidney (PHK) cells, and has been developed by the Chengdu Institute of Biological Products, China. As the vaccine is produced on primary cells, the manufacturing process includes detailed screening for endogenous and adventitious viruses and is currently being manufactured as per WHO Guidelines for production of live JE vaccines for human use. The vaccine elicits broad-spectrum immunity against heterologous JEV strains. Reversion to neurovirulence is considered highly unlikely. This single-dose vaccine has been licensed for use in China since 1988, where over 200 million children have been successfully immunized so far, with a brilliant safety record. Recently the SA14-14-2 vaccine has been licensed for use in India, Nepal, Sri Lanka and South Korea. In India,

the imported SA 14-14-2 vaccine has been used to immunize over 9.3 million children (aged between 1 and 15 years) in the summer of 2006 in 4 states. A major advantage of this vaccine is that it is inexpensive and hence would be affordable by the economically weaker countries of Asia where JE is endemic.

B. IXIARO®

IXIARO® is a purified, formalin-inactivated, whole-virus JE vaccine, manufactured by Intercell, an Austrian company. It is based on the SA14-14-2 strain of JEV adapted to grow on Vero cells. After extensive clinical trials of IXIARO®, using a two-dose regimen, the vaccine was approved by the US Food and Drug Administration (USFDA) on March 30, 2009. It has also been approved by the European Commission and the Australian Therapeutic Goods Administration (TGA). The vaccine is currently licensed for use in USA, Australia and Europe. The vaccine is currently undergoing regulatory clearance in India, and will be available soon and will be marketed by Biological E.

C. IMOJEV®

Besides the above two JE vaccines that are commercially available, a very innovative genetically engineered third generation JE vaccine is in late stages of clinical development. This uses the proprietary Chimeri-Vax™ platform, developed at the St. Louis University Health Sciences Center and by Acambis. This recombinant DNA vaccine is based on the Yellow Fever (YF) virus 17D vaccine strain as a backbone, but with the envelope and pre-membrane protein genes of YF virus replaced by those of JEV. Importantly, YF 17D has been extensively used as a live-attenuated yellow fever vaccine for over 60 years, with excellent record of safety and efficacy. Clinical trials using a single dose of IMOJEV® have shown the vaccine to be well tolerated as well as effective. Importantly, pre-existing immunity to yellow fever virus did not dampen the immune response. An update on the current vaccines against JE is given in Table 110.4.

VII. OTHER PREVENTIVE MEASURES

JE control measures may be three-pronged, namely (i) changes in pig rearing techniques, (ii) vector control, and (iii) prophylactic vaccination of susceptible human populations. Since the former two approaches have their limitations, it is the third that has to be relied upon to keep this disease at bay. However, it must be stressed that the other approaches should not be neglected altogether. In the Southeast Asian countries, where pig-rearing is widely practiced, the pigsties should be located far away from human dwellings, although this is not always practicable. Vaccination of swine has also been suggested and practiced in countries like Japan. But a universal and sustained swine vaccination effort is likely to be

TABLE 110.4   Vaccines against Japanese Encephalitis

Vaccine name	Type	Virus strain	Manufacturer	Dosing	Status
SA14-14-2	PHK cell-cultured; Live-attenuated	SA14-14-2	Chengdu Institute of Biological Products, China	Single dose	Licensed by respective National Regulatory Authorities for use in China, India, Nepal, Sri Lanka and South Korea; WHO approval awaited.
IXIARO®	Vero cell-cultured; Purified-inactivated	SA14-14-2	Intercell, Austria	Two doses	USFDA approved. Licensed in USA, Australia and Europe in 2009.
IMOJEV®	Infectious clone; Live-attenuated	JEV SA14-14-2 and YF 17D chimeric infectious clone	Acambis, UK; Sanofi Pasteur	Single dose	Phase III clinical trials in USA, Australia and Asia. Marketing authorization sought by Sanofi Pasteur from Australian and Thai authorities.

a costly affair, and thus, out of reach of most resource-poor JE-endemic countries. If pig-rearing is practiced in large pig-farms, located far away from human dwellings, and managed by co-operatives instead of following family-centric pig-rearing practices, then this problem could be overcome to a large extent. Mosquito control by insecticides has largely been found to be ineffective, impractical and costly. However, mosquito larva control by biological means such as keeping larvivorous fish in the paddy fields is an alternative and eco-friendly approach that may be adopted. Avoiding mosquito bites by wearing full-sleeved shirts, restricting outdoor activities at dawn and dusk, and using mosquito repellents, could dramatically reduce the incidence of JE. Hence, what is truly required for JE control is the adoption of an integrated approach, including improved agricultural practices, improved living standards, greater health awareness, as well as a sustained mass childhood vaccination program. It should, however, be noted that unlike smallpox and polio, for which humans are the only host, JE is a zoonotic disease with large animal reservoirs and hence cannot be totally eradicated.

CONCLUDING REMARKS

The JE problem, unlike many other Public Health problems, is multifaceted and complex. This complexity stems from the fact that we are still not totally aware of the true magnitude of the problem. This, in turn, stems from the lack of good quality disease burden data. In many countries, such as Bangladesh, Cambodia, Indonesia, North Korea, Laos, Myanmar, Papua New Guinea, and Pakistan, there is definite lack of data regarding the distribution and Public Health importance of JE. Until and unless solid disease burden data is available, National Regulatory Authorities in the respective JE-endemic countries cannot be convinced about the importance of implementation of a childhood immunization program for JE. Hence, much more aggressive and thorough epidemiological studies are required. This is especially true in areas like Pakistan and possibly, beyond. We know that JE cases have been documented from Karachi. But in countries like Afghanistan, where JE surveillance systems are non-existent, the virus could well be circulating since the necessary vectors are readily available. This is a challenge for the epidemiologists.

Two main prerequisites for controlling JE are (i) greater political will, and (ii) sound financial resources. With a good combination of these two factors, initiation and sustaining of childhood vaccination programs would become much easier. There is ample evidence from the wealthier JE-endemic Asian countries like Japan, South Korea and Taiwan, what a good vaccination program can achieve. This type of vaccination program is urgently required in countries like India, which has a high disease burden, and thereby contributes a large share of the JE burden to the Asian region. Vaccination programs have been initiated in two states, but a national immunization program is required. For implementation of any vaccination program, adequate trained staff, a functional cold-chain, good roads for transportation of vaccines and a good campaigning mechanism, for educating the masses, all contribute to acceptability of the vaccine.

In the near future, greater challenges are likely to occur. These may stem from newer issues of this day and age. Climate change could well be one such issue. Changing weather patterns, especially rainfall patterns, could lead to altering agricultural practices, which would lead to unpredictable changes in mosquito breeding patterns. Moreover, changing bird migration patterns brought about by climate change, could lead to introduction of JE to new geographical locations. A sustained and integrated approach would definitely lead to proper management of JE in the near future, given the fact that good quality and effective vaccines are on the horizon.

SUMMARY

The present chapter on Japanese encephalitis provides an up-to-date review of the current status of disease in the area. The chapter begins with a discussion on JEV, the etiologic agent of JE, including its life cycle, genome organization, as well as the various genotypes circulating in the region. This is followed by the epidemiological aspects, including pathogenesis as well as the mechanism of neuronal damage, which is followed by a section on clinical aspects of JE. The subsequent sections address aspects pertaining to diagnosis, treatment, as well as the current status on JE vaccines, which are the cornerstone for JE control. The chapter has been written in the light of current international knowledge on the subject.

SUGGESTIONS FOR FURTHER READING

1. Bharati K. and Vrati S. (2010), Japanese Encephalitis Vaccines: Current Status and Future Prospects. *Proc Natl Acad Sci.,* India, 80(Section-B; Part III):179–189.

2. Bharati K. and Vrati (2006), Japanese Encephalitis: Development of New Candidate Vaccines, *Expert Rev Anti Infect Ther.,* 4:313–324.

3. Gould E.A. and Solomon T. (2008), Pathogenic Flaviviruses, *Lancet,* 371:500–509.

4. Halstead S.B. and Thomas S.J. (2010), New Vaccines for Japanese Encephalitis, *Curr Infect Dis Rep.,* 12:174–180.

5. Mackenzie J.S., Williams D.T. and Smith D.W. (2007), Japanese Encephalitis Virus: The Geographic Distribution Incidence and Spread of a Virus with a Propensity to Emerge in New Areas, *Perspect Med Virol.,* 16:201–268.

6. World Health Organization, Safety of Japanese Encephalitis Vaccination in India. *Wkly Epidemiol Rec.,* 82:17–24.

SECTION XVI

VACCINES

Vaccines are the most cost-effective tools for prevention of a variety of infections. Mere use of a few vaccines for immunization of children has reduced their mortality rate drastically. Vaccines enabled the eradication of smallpox and elimination of poliomyelitis. The role of these vaccines have been covered as also new vaccines on influenza. On horizon are vaccines for malaria, and perhaps cancer.

111

Universal Immunization Program
Vaccines for Preventing Diptheria, Pertusis, Tetanus, Tetravalent and Pentavalent Vaccine

Aditi Sinha and Manoj Kumar

CONTENTS

I. INTRODUCTION

The word 'immunity' is derived from the Latin word *'immunis'* that refers to the state of protection from infectious disease. The immune system has evolved as defense system to protect mammals from invading microorganisms and malignant disorders. The innate immune system is primitive, non-specific, has no memory and provides the first line of defense against infections. The adaptive immune system is highly evolved, specific and has memory characterized by a rise in immune response that serves to eliminate the microorganisms.

II. UNIVERSAL IMMUNIZATION PROGRAM

Immunization, considered one of the most cost effective health interventions of all times, has helped in the eradication of small pox, elimination of poliomyelitis from several countries, and a significant decline in incidence of measles, tetanus and diphtheria. Rapid advances in research have enabled the introduction of more potent vaccines against a wide spectrum of infections. However, thousands of children in India continue to die from vaccine preventable diseases each year.

The Expanded Program of Immunization (EPI), organized under by the World Health Organization (WHO) in 1974, was the first global initiative at immunization, and focused on immunization of children below 5 years and pregnant women. Since the launch of the EPI, millions of deaths have been prevented every year by immunization of infant through national immunization programmes. The first diseases targeted by the EPI were diphtheria, whooping cough, tetanus, measles, poliomyelitis and tuberculosis. When adopted by India in 1978, the EPI included only Bacillus Calmette–Guérin (BCG), diphtheria toxoid, pertussis and tetatus toxoid (DPT), oral poliomyelitis (OPV) and typhoid vaccines, and its coverage was limited to urban areas.

Global policies for immunization and establishment of the goal of providing universal immunization for all children by 1990 were established in 1977, and this goal was considered an essential element of the WHO strategy to achieve health for all by 2000. In India, The Universal Immunization Program (UIP) was introduced in 1985 in order to improve immunization coverage within India and to extend the focus to beyond infancy. Subsequent years saw the addition of measles vaccine to the Program and the exclusion of typhoid vaccine. Further, vitamin A supplementation was introduced in 1990 and Polio National Immunization Days were initiated in 1995. The UIP remains the largest immunization program in the world and caters to 27 million infants and 30 million pregnant women annually. The program has been among the Technology Missions since 1986 and was monitored under the 20 point program by the Prime Minister's Office. Support to the UIP is available internationally through the Child Vaccine Initiative (CVI) and Global Alliance for Vaccines and Immunization (GAVI). The UIP has remained an essential part of the Child Survival and Safe Motherhood (CSSM) program in 1992, the Reproductive and Child Health (RCH) program I in 1997, and the RCH II and National Rural Health Mission (NRHM) since 2005. Under the UIP, some states have initiated universal immunization against hepatitis B virus since 2002.

While the UIP achieved a significant decline in the incidence of vaccine preventable diseases, the findings of the National Family Health Survey 2005–06 suggested that only 43.5% of children in India received all the primary vaccines by 12 months of age, and this coverage was below 30% in Uttar Pradesh, Rajasthan and Arunachal Pradesh. The Immunization Strengthening Project of the Government of India was launched with the intent to strengthen routine immunization rates to increase the percentage of fully immunized children to above 80%; eliminate and eradicate polio; review and develop a new immunization program based on the availability of newer vaccines, recent advances in vaccine production and cold chain technologies and the changing epidemiology of diseases; and improve vaccination surveillance and monitoring. This program is being implemented through the National Institute of Health and Family Welfare and regional training centers, and has improved immunization coverage in 50 poorly performing districts in 8 priority states. Additional vaccines have been added to the UIP, and the current schedule is shown in Table 111.1. Strengthening of immunization coverage remains an important component of the NRHM. Measures proposed to improve vaccination rates include mobilization of children and pregnant women by Accredited Social Health Activist (ASHA) and link workers to increase coverage and decrease dropouts, provision of auto-disposable syringes and other disposables, and strengthening of cold chain maintenance and immunization monitoring systems.

The proportion of 1–yr-old children immunized against measles is an important indicator for immunization coverage within the country. According to the Coverage Evaluation Survey 2009, the national coverage of measles immunization of 1–yr-olds has reached 74.1%, and 61% children are completely immunized with one dose each of BCG and measles and three doses of DPT and OPV. However, these proportions compare poorly with the statistics from elsewhere, and the vaccination coverage remains far from satisfactory. Globally, an estimated 85% of children under one year of age had received at least three doses of DTP vaccine (DTP3) in 2010. One of the factors responsible for this is the highly focused attention on polio eradication program for last 13 years at the cost of other health programs including immunization against other vaccine preventable diseases. An urgent need at present is to strengthen routine immunization coverage in the country with EPI vaccines.

An equally pressing need is to introduce more vaccines in EPI. Most countries, including the majority of low-income countries have added hepatitis B and Haemophilus influenzae type b (Hib) to their routine infant immunization schedules, and an increasing number are in the process of making immunization with pneumococcal conjugate vaccine and rotavirus vaccines mandatory. Recognizing the relevance of disease prevention by a variety of these vaccines, the Indian Academy of Pediatrics has widened its recommendations to include a variety of vaccines in the immunization schedule (Table 111.1). While many new vaccines have been introduced in the Indian market in the last couple of decades, their exorbitant pricing ensures that most vaccines are accessible only to the select few that can afford them. The Government has introduced some additional vaccines such as MMR, Hep B and Hib into the EPI in certain districts. It is essential that coverage with these vaccines is improved and more relevant and cost effective vaccines are introduced. While self-reliance in vaccine technology and self-sufficiency in vaccine production are policy objectives, these require implementation to avoid the need to import expensive vaccines. Vaccine production by indigenous manufacturers should be encouraged to reduce costs, decrease dependence on imports and ensure availability of vaccines specifically needed by India (e.g., typhoid) and custom made to Indian requirements (rotavirus and pneumococcal vaccines). An adverse event monitoring system should be set up to improve immunization safety and accountability. Finally, a system to monitor the incidence of vaccine preventable diseases and conduct appropriate epidemiological studies is necessary to make evidence based decisions on incorporation of new vaccines in the EPI and to study the impact of vaccine adminstration on disease incidence, serotype replacement and epidemiologic shift.

III. VACCINATION TO PREVENT DIPTHERIA, PERTUSIS AND TETANUS

Vaccination against diphtheria, tetanus and pertussis (DTP) have constituted the cornerstone of childhood immunization programs for decades. The global burden of these three diseases has declined rapidly since the introduction of the combined

TABLE 111.1 Comparison of the National Immunization Schedule and the 2012 recommendations of the Indian Academy of Pediatrics

Age	National Immunization Program	IAP recommendation
0 (at birth)	BCG, OPV_0, Hep B_0[*]	BCG, OPV_0, Hep B_1
6 weeks	$DTwP_1$, OPV_1, Hep B_1[*]	$DTwP_1$/$DTaP_1$, IPV_1[$], Hep B_2, Hib_1, $Rotavirus_1$, PCV_1
10 weeks	$DTwP_2$, OPV_2, Hep B_2[*]	$DTwP_2$/$DTaP_2$, IPV_2[$], Hib_2, $Rotavirus_2$, PCV_2
14 weeks	$DTwP_3$, OPV_3, Hep B_3[*]	$DTwP_3$/$DTaP_3$, IPV_3[$], Hib_3, $Rotavirus_3$, PCV_3
6 months	-	OPV B_1, Hep B_3
9 months	Measles, Vitamin A_1	OPV_2, Measles
12 months	-	Hep A_1
15 months	-	MMR_1, $Varicella_1$, PCV B_1
16–24 months	DTwP B_1, OPV B_1, Vitamin A_2**, Japanese Encephalitis[^]	(16–18 months) DTwP B_1/DTaP B_1, IPV B_1[$], Hib B_1 (18 months) Hep A_2
2 years	–	$Typhoid_1$
5 years	DT B_2	DTwP B_2/DTaP B_2, OPV_3, MMR_2, $Varicella_2$, $Typhoid_2$[$$]
10 years	TT	Tdap/ Td[#], HPV
16 years	TT	-
Pregnancy	2 doses of TT 1 month apart	-

B_1 first booster dose; B_2 second booster dose; BCG Bacillus Calmette Guérin vaccine; DT diphtheria toxoid with tetanus toxoid; DTwP diphtheria toxoid, tetanus toxoid, whole cell killed pertussis vaccine; DTaP diphtheria toxoid, tetanus toxoid, acellular pertussis vaccine; Hep B hepatitis B vaccine; Hib hemophilus b vaccine; HPV human papilloma virus vaccine; IPV Inactivated polio vaccine; MMR measles, mumps and rubella vaccine; OPV oral poliovirus vaccine; PCV7 pneumococcal heptavalent conjugate vaccine; Td tetanus toxoid with reduced dose diphtheria; Tdap tetanus toxoid with reduced dose diphtheria and pertussis vaccine; TT tetanus toxoid;

[*]Implemented in selected states, districts and cities

[**]Subsequently, one dose every 6 months up to the age of 5 years

[^]SA 14–14–2 vaccine, in select endemic districts

[$] Given with OPV where affordable

[$$] Revaccination required every 3 years

[#]Td preferred over TT

vaccine in immunization programs worldwide, with substantial decline in rates of infant mortality. However, the incidence of diphtheria and pertussis cases has increased recently, particularly in developed countries, and neonatal and maternal tetanus continue to be an important cause of preventable mortality in many developing regions.

A. Burden of Disease

Diphtheria

Diphtheria is an acute respiratory or cutaneous illness caused by exotoxin-producing *Corynebacterium diphtheriae*. Morbidity and mortality are due to the bacterial toxin that may cause obstructive pseudomembranes in the upper respiratory tract (croup) or damage to myocardium and other tissues. While most infections with *C. diphtheria*, are asymptomatic or run a relatively mild clinical course, the disease is severe in non-immune individuals, particularly young children. During epidemics, respiratory diphtheria has a case fatality rate of 5–10%, with mortality in up to 20% of children below 5 years. While the incidence has declined substantially in most

countries due to implementation of routine immunizations, the disease remains endemic in several countries in Africa, Asia and eastern Europe with poor EPI coverage, occurring mostly as sporadic cases or in small outbreaks. Due to varying immunization coverage, India continues to account for 50–85% of the global burden of disease and more than two-thirds of about 5000 deaths reported annually. Diphtheria is also an emerging concern in countries with high EPI coverage and low rates of natural infections, with adolescents and adults rendered susceptible to diphtheria as a result of waning immunity.

Tetanus

Tetanus is caused by *Clostridium tetani*, a toxin-producing anaerobe that mulitplies in dirty necrotic wounds to produce tetanospasmin, an extremely potent neurotoxin. This toxin blocks inhibitory neurotransmitters in the central nervous system and causes muscle stiffness and spasms typical of generalized tetanus.

The disease may affect any age group, but the majority of cases occur in developing countries among newborn babies or in mothers following unclean deliveries and poor postnatal hygiene. Case-fatality rates are high even with the best supportive care. Tetanus may also affect children and adults following injuries. While routine immunization has resulted in elimination of neonatal tetanus (less than 1 case per 1000 live births in every district) from many regions, maternal and neonatal tetanus continue to claim 180,000 lives annually in 48 countries, chiefly in Asia and Africa.

Pertussis

Pertussis or 'whooping cough' is an acute respiratory illness caused by *Bordetella pertussis*. The disease is highly contagious with a secondary attack rate of up to 90% among non-immune household contacts. Typical presentation includes a prodromal catarrhal phase followed by the onset of the typical paroxysmal cough or 'whoop', particularly at night and associated with vomiting. Young infants below 6 months may show severe illness, including cyanosis, apnea and death. The course of disease is often atypical in older children, adolescents and adults and pertussis may be unrecognized. Case fatality rate for pertussis is approximately 0.2%, mortality being most common in the first three months of life. Despite high rates of global immunization, 16 million cases of pertussis are reported each year with mortality in 195,000 cases, 95% of which occur in children in developing countries. The incidence is increasing even in developed countries, particularly in adolescents and adults due to waning immunity.

B. Available Vaccines

The whole cell vaccine, introduced in 1940s, combines diphtheria and tetanus toxoids with inactivated whole cell pertussis bacteria. The common occurrence of minor local adverse effects and rare systemic reactions with the whole cell pertussis (wP) vaccines prompted the development of less reactogenic vaccines. Acellular DTP vaccines (aP), first licensed in 1981 in Japan, contains the pertussis component as one to five purified antigens. These vaccines have recently been introduced in many countries, including India, and include preparations with an improved reactogenicity profile and/or additional vaccines such as *Haemophilus influenzae* type b, hepatitis B virus and inactivated poliovirus (Table 111.2).

The main aim of vaccination is to reduce the risk of severe pertussis in infancy through >90% coverage worldwide with 3 doses of high quality wP or aP vaccines. Immunizations for infants and young children include diphtheria toxoid, tetanus toxoid, and either acellular or whole-cell pertussis vaccine. Both diphtheria and tetanus toxoid induce satisfactory immune responses in infants aged below 6 weeks, but DTwP or DTaP vaccines are recommended only for infants aged 6 weeks or older in order to improve seroconversion for pertussis. The pertussis antigens that are included in acellular pertussis vaccines and the amount of diphtheria and tetanus toxoids vary depending upon the vaccine product (Table 111.2). Vaccines for children younger than 7 years (DTaP, DTwP) contain more diphtheria toxoid (at least 25–30 IU per dose) and pertussis antigens and may contain more tetanus toxoid than vaccines used to boost immunity in older children and adults (tetanus toxoid, reduced diphtheria toxoid, dT; along with reduced acellular pertussis vaccine, Tdap). This reduction of diphtheria toxoid potency minimizes reactogenicity at the injection site but is still sufficient to provoke an antibody response in older children and in adults. A vaccine that includes diphtheria and tetanus toxoids, but not pertussis (DT) is available for children younger than 7 years in whom it is desirable to avoid pertussis vaccine.

TABLE 111.2   Commonly used acellular vaccines against diphtheria, pertussis and tetanus and their constituents

Type of vaccine	Brand name	Pertussis component	Diphtheria toxoid, Lf	Tetanus toxoid, Lf
DTaP	Tripacel, Daptacel	Detoxified pertussis toxin (PT) 10 μg; filamentous hemagglutinin (FHA) 5 μg; pertactin 3 μg; fimbria (types 2 and 3) 5 μg; aluminium as adjuvant	15	5
	Infanrix	PT 25 μg; FHA 25 μg; pertactin 8 μg; aluminium and polysorbate	25	10
	Tripedia	PT 23.4 μg; FHA 23.4 μg	6.7	5
DTaP-IPV*	Pentaxim	PT 25 μg; FHA 25 μg	>30	>40
DtaP-IPV-HB-Hib**	Infanrix Hexa	PT 25 μg; FHA 25 μg; pertactin 8 μg; aluminium and polysorbate	25	10
Tdap	Adacel	PT 2.5 μg; FHA 5 μg; pertactin 3 μg; fimbria (types 2 and 3) 5 μg; aluminium as adjuvant	2	5
	Boostrix	PT 8 μg; FHA 8 μg; pertactin 2.5 μg; aluminium and polysorbate	2.5	5

* Also contains inactivated poliovirus vaccine (IPV) type 1 40 Dantigen units (DU), type 2 8 DU and type 3 32 DU

** Also contains IPV type 1 40 Dantigen units (DU), type 2 8 DU and type 3 32 DU; hepatitis B (HB) surface antigen 10 μg; purified capsular polysaccharide of Hemophilus influenzae type B (HiB) covalently bound to 20–40 μg of tetanus toxoid (Not available in India)

Lf flocculation unit (1 Lf unit is the amount of toxoid that flocculates 1 unit of an international reference antitoxin)

Conjugate vaccines that contain diphtheria toxoid or CRM197, a nontoxic variant of diphtheria toxin (e.g., pneumococcal conjugate vaccine, meningococcal conjugate vaccine), are not substitutes for diphtheria toxoid immunization. The conjugate *Haemophilus influenzae* vaccines that contain tetanus toxoid (PRP-T, ActHIB, Hiberix) are not a substitute for tetanus toxoid immunization.

C. Choice of Vaccine

Both acellular (aP) and whole cell pertussis (wP) containing vaccines have excellent safety records, and protection against severe pertussis in infancy and early childhood can be obtained with similar efficacy (80–90%) with wP or aP vaccine. Changing among or within the wP and aP vaccine groups is unlikely to interfere with the safety or immunogenicity of these vaccines. The risk of non-serious adverse reactions is lower with acellular than whole-cell pertussis vaccines (Table 111.3). However, the aP-containing vaccines continue to be significantly more expensive than wP-containing vaccines and aP-containing vaccines are unlikely to be currently affordable in most developing countries. In countries where the higher risk of adverse effects with wP is likely to influence rates of vaccination, aP is the vaccine of choice for children 6 weeks through 6 years of age.

Interchanging between or within the wP and aP vaccine groups is unlikely to interfere with safety or immunogenicity. When feasible, the same brand of vaccine should be used for all doses, but an opportunity to administer a dose should not be missed because the brand used for earlier doses is not available; limited data suggest that using different brands to complete the primary series does not adversely affect vaccine safety or immunogenicity.

In view of the adverse effects associated with whole cell pertussis vaccine, only acellular vaccines are recommended for use in children older than 6 years (beyond 7th birthday), and the content of diphtheria toxoid and pertussis antigens is reduced (Tdap). Alternatively, the booster should contain only tetanus toxoid with (dT) or without (TT) reduced content of diphtheria toxoid.

D. Immunization Schedule

Primary immunization

All infants and children aged 6 weeks through 6 years should receive routine immunization with diphtheria toxoid, tetanus toxoid, and acellular pertussis vaccine. However, the recommended schedule varies considerably between countries.

TABLE 111.3 **Comparison of adverse reactions after vaccination with acellular and whole cell DPT vaccines**

Adverse effect	*Acellular DTP (n=1818) %*	*Whole cell DTP (n=371)*
Redness 1–20 mm	31.4	56.3
Redness >20 mm	3.3	16.4
Swelling 1–20 mm	20.1	38.5
Swelling >20 mm	4.2	22.4
Moderate pain*	6.5	25.9
Severe pain*	0.4	14.3
Moderate fussiness**	12.4	29.1
Severe fussiness**	4.7	12.4
Use of antipyretic	55.9	83.3
Drowsy	42.7	62.0
Anorexia	21.7	35.0
Vomiting	12.6	13.7
Long term effects, %	Acellular DTP (n=1517)	Whole cell DTP (n=305)
Hospitalization	7.4	7.2
Developmental delay	0.6	1.0
Seizure(s)	1.1	0.7
Failure to thrive	0.5	0.3
Pertussis	0	0
Cough >2 wk	0.7	1.0
Cough >4 wk	0.7	1.0
Other serious infection	2.6	2.6

* Moderate, cried or protested to touch; severe, cried when leg moved

**Moderate, prolonged crying & refusal to play; severe, persistent drying & could not be comforted Modified from: Decker MD, Edwards KM, Steinhoff MC, Rennels MB, Pichichero ME, Englund JA, Anderson EL, Deloria MA, Reed GF. Comparison of 13 acellular pertussis vaccines: adverse reactions. Pediatrics. 1995; 96: 557–66.

The WHO recommends that the first dose of DTwP or DTaP vaccine should be administered not earlier than 6 weeks of age, and subsequent doses be given 4–8 weeks (at least 4 weeks) apart, at 10–14 and 14–18 weeks of age. The last dose of the primary series should be completed by the age of 6 months. Neonatal immunization against pertussis is not recommended. For optimal immunization against tetanus, a schedule including 5 doses of tetanus toxoid in childhood is recommended. For children who have an absolute contraindication to pertussis vaccine, diphtheria-tetanus toxoid (DT) is an acceptable alternative.

Booster doses

A booster dose (DTaP or DTwP) is recommended for children 1–6 years of age, preferably during the second year of life. Many national immunization programs offer 1–2 booster doses containing pertussis and diphtheria components, one at 15–24 months of age and a second at age 4–7 years. Booster doses are expected to extend the duration of immunological protection against pertussis for at least 6 years and compensate for the loss of natural diphtheria boosting. The minimum recommended interval between the primary schedule and the first booster dose is 6 months. However, the first booster (fourth dose of vaccine) may be given as early as 12 months of age if the child is considered unlikely to return for a visit at the recommended age (15–18 months) and is considered valid if at least 4 months have elapsed since the third dose. The minimum age for the second booster (fifth dose) is 4 years, and the minimum recommended interval between the fourth and fifth doses is 6 months. The second booster may be omitted if the child received the fourth dose (first booster) after 4 years of age. No more than 6 doses of diphtheria and tetanus toxoids should be given before the seventh birthday.

The optimal timing of and number of diphtheria-containing booster doses should be based on epidemiological surveillance as well as on immunological and programmatic considerations. Boosters of tetanus toxoid are ideally administered in childhood (to complete five doses by 4–7 years) and an additional dose during adolescence. Tetanus boosters may use either DTP or DT/Td vaccines depending on the child's age; Td should be used beyond 7 years of age.

Catch up immunization

Children who are unimmunized or under-immunized should undergo catch up immunization. Catch up immunization is particularly important for the following groups of children: (i) Close contact of individual with pertussis disease (should complete the primary series with minimal intervals); (ii) diphtheria disease or close contact with an individual with diphtheria disease; (iii) tetanus disease. Children whose vaccination series has been interrupted should have their series resumed, without repeating previous doses.

For previously unimmunized children aged 4 months to 6 years (before the seventh birthday), both the WHO and Centers for Disease Control (CDC) Advisory Committee on Immunization Practices (ACIP) recommend 4–5 doses of diphtheria, tentatnus and pertussis containing vaccines (DTwP or DTaP). The first two doses are administered 2 months apart, and a third dose is given after 6–12 months. Following completion of the primary immunization schedule, the fourth dose (first booster) is administered at least 6 months after the third dose. While a fifth dose may be given 6 months later to obtain long term protection, it is not necessary if the child received the fourth dose after 4 years of age.

Only aP-containing vaccines should be used for vaccination in individuals 7 years or older. The recommended schedule for primary immunizations of children 7–10 years includes one dose of Tdap, followed by two doses of dT after 1 and 6 months. For previously unimmunized older children and adults or those where the immunization status is unclear one dose of Tdap or dT should be administered instead of the 10–yearly booster; two more doses of dT may be administered to complete the schedule.

Boosters in adolescents and adults

The WHO emphasizes that current evidence is insufficient to support the addition of vaccine boosters in adolescents and adults, or vaccination of pregnant women and close household contacts ('cocooning'), for achieving the primary goal of reducing severe pertussis in infants. However, countries with demonstrable nosocomial transmission of the disease are encouraged to implement pertussis vaccination with emphasis on maternity and pediatric staff, if economically and logistically feasible. The ACIP recommends a single dose of Tdap for healthcare personnel who have not previously received Tdap and who have direct patient contact. The WHO encourages careful epidemiological surveillance of pertussis and surveys comparing age-specific incidences in countries with different vaccine booster policies. Further study of the safety and effectiveness of earlier and/or additional booster doses is necessary before changes to the routine immunization schedule are recommended.

Revaccination of adults against diphtheria (and tetanus) every 10 years may be necessary to sustain immunity in some epidemiological settings. Particular attention should be given to revaccination of health-care workers who may have occupational exposure to *C. diphtheriae*. People living in low-endemic or non-endemic areas should receive booster doses of diphtheria toxoid approximately 10 years after completing the primary series and subsequently every 10 years throughout life. For example, the ACIP recommends a single booster dose of Tdap at age 11 or 12 years in children who have completed a childhood vaccination series, followed by booster doses of Td every 10 years.

Since booster responses can still be elicited after intervals of 25–30 years, primary immunization is not required to be repeated when boosters are delayed. Combined diphtheria toxoid and tetanus toxoid is preferred to TT when tetanus prophylaxis is needed following injuries. For this purpose, DTaP (DT, if pertussis vaccine contraindicated) is used in children <7 years old and Td in older children and adults (Tdap is preferred if the patient is previously not immuized against pertussis).

Variations in the schedule

Based on considerations of cost and availability of acellular vaccines as well the risk of adverse effects, the Indian Academy of Pediatrics recommends providing parents the choice to opt for either DTaP or DTwP for immunization in children 1–6 years of age. On the other hand, the American Academy of Pediatrics (AAP) advocates the use of DTaP vaccine to avoid adverse effects.

National recommendations for the primary series vary considerably, including immunization at the ages of 6 weeks, 10 weeks and 14 weeks; at 2 months, 3 months and 4 months; at 3 months, 4 months and 5 months; and at 2 months, 4 months and 6 months. While the IAP and EPI in India endorse the use of the first schedule, the AAP recommends delaying vaccination to ensure robust immunological responses, with the vaccine administered at 2 months, 4 months and 6 months, 15 through 18 months, and 4 to 6 years of age.

Although most countries continue to use full-dose DTaP vaccines for booster doses in young children, there is some evidence to suggest that reactogenicity and immunogenicity are lower with the use of reduced antigen content in the 4–6 year age group. For this reasons, the immunization programs in the UK and Germany are now using reduced-dose DTaP or DTaP-IPV booster vaccines for the 3–6 year age-group.

Special circumstances

History of pertussis disease: Natural infection with diphtheria and tetanus disease does not induce immunity. Individuals with diphtheria or tetanus disease should receive diphtheria-tetanus toxoid during their convalescence (even if they were completely immunized before their illness). Well-documented pertussis disease (e.g., positive culture or epidemiologic linkage to a culture-proven case) confers short-term immunity. However, the duration of protection is unknown. DTaP or Tdap should be used to complete the primary series.

Neurologic disease: The administration of diphtheria and tetanus toxoids may sometimes be deferred until one year of age in children in whom neurologic conditions may be evolving and the physician is keen to defer pertussis immunization. However, the risk of a tetanus-prone wound increases as the child becomes ambulatory, and the decision to administer DT or DTaP/DTwP must be made. The number of doses of DT vaccine needed to complete the primary series depends upon the child's age at the time of the first dose, with a total of 4 doses advised for vaccination initiated at <12 months, and 3 doses if begun at ≥12 months of age.

Vaccine mix-up: If Tdap has been administered inadvertently (instead of DTaP) to a child younger than 7 years it should be not be counted as a valid dose if it is administered for the first, second, or third dose. DTaP should be administered on the same day or as soon as possible to keep the child on schedule. The interval between the replacement dose of DTaP and the subsequent dose should be at least 4 weeks. If Tdap was administered as the fourth or fifth dose, it should be counted as valid.

E. Dose, Route, Storage and Handling

Diphtheria toxoid, tetanus toxoid, and pertussis vaccines are administered intramuscularly in a dose of 0.5 mL in the anterolateral thigh in children younger than three years and the deltoid in older children. Observational studies show that local reactions requiring medical attention are fewer with administration in the anterolateral thigh than in the deltoid and less common following DTaP than DTwP. The vaccines can be administered at the same visit as other recommended vaccines. DTaP vaccine should not be mixed in the same syringe with another vaccine.

The vaccines should be stored continuously at 2–8°C. During transport, the vaccine may be out of refrigeration for as long as four days, but it should be refrigerated immediately upon arrival. Freezing reduces the potency of the tetanus component; vaccine exposed to freezing temperature should not be administered.

F. Efficacy of Existing Vaccines

The efficacy and effectiveness of diphtheria, tetanus, and pertussis immunization vary depending upon the number of doses received and the definition used for efficacy, i.e., levels of antibody correlated with protection or protection from disease. Vaccine effectiveness is a measure of how well a vaccine works to protect against an infection when the vaccine is used in routine circumstances in the community.

After a series of 4 appropriately spaced doses of diphtheria-tetanus toxoid-containing vaccines, approximately 95% of infants and children achieve levels of diphtheria antitoxin correlated with protection (>0.1 IU of antitoxin/mL) and virtually all infants and children achieve levels of tetanus antitoxin correlated with protection (>0.1 international unit of antitoxin/mL). The estimated efficacy of diphtheria toxoid in the prevention of diphtheria disease is high. Case control studies show that 3 or more doses of the toxoid induces 95.5% (92.1–97.4%) protective efficacy among children aged <15 years, and the protection increases to 98.4% (96.5–99.3%) after 5 or more doses of the vaccine. The clinical efficacy of TT for the prevention of neonatal tetanus was initially shown by two clinical trials undertaken in the early 1960s, in places with an annual neonatal tetanus incidence of about 80 cases per 1000 live births. In one trial, three doses of TT without adjuvant (equivalent to about two doses of aluminium-adsorbed tetanus toxoid) had an efficacy of 94% for the prevention of deaths from neonatal tetanus; in the second, no deaths from neonatal tetanus were seen in neonates born to mothers who had received 2–3 doses of aluminium-adsorbed TT within the previous 5 years. Subsequent observational and case-control studies assessing the effect of tetanus toxoid on neonatal tetanus incidence or mortality have consistently noted vaccine effectiveness to be 80% or better. One hospital-based study showed a 90% reduction in maternal tetanus in women aged 20–40 years.

In a systematic review, the efficacy of acellular pertussis vaccines containing ≥3 pertussis antigens in preventing typical pertussis was demonstrated to be about 85%, while that of vaccines containing <3 pertussis antigens ranged from 59% to 75%. When compared with DTwP, DTaP was as effective or more effective, with fewer adverse effects. However, the protection provided by acellular pertussis vaccine decreases with time, with effectiveness decreasing from 98% during 12 months since the last dose of DTaP, to 71% by ≥5 years.

G. Adverse Reactions

Adverse reactions to diphtheria-tetanus-pertussis vaccines may include, (from most to least frequent): (i) mild local and systemic reactions; (ii) entire limb swelling; (iii) persistent, inconsolable crying (≥3 hours); (iv) hypotonic-hyporesponsive episode (collapse or shock-like state); (v) seizures; (vi) fever ≥105°F (40.5°C); (vii) anaphylaxis. Acellular pertussis-containing vaccines have fewer local reactions, fever, and systemic symptoms than whole-cell pertussis-containing vaccines (Table 111.3).

Mild local and systemic reactions are the most common adverse reactions to DTaP and include fever, redness and tenderness at the injection site, drowsiness or poor appetite, fussiness (seen in 20–33% cases each) and vomiting (2%). These reactions occur within several hours of vaccination and resolve spontaneously without sequelae. The rate of fever is higher among infants receiving the pentavalent combination vaccine (DTaP-HepB-IPV) than among those who receive the vaccines separately.

Extensive local erythema or swelling are more common among children who receive 4 or 5 consecutive doses of DTaP than those who receive a mixture of DTaP and DTwP. The occurrence of an extensive local reaction after the fourth dose is not predictive of a similar reaction after the fifth dose. In a retrospective review, such reactions to the fifth dose of DTaP were uncommon, seen in 43.5 per 10,000 injections), were more common if the vaccine was administered in the arm than in the thigh (47.4 versus 32.1 per 10,000 injections), and were not prevented by prophylactic administration of acetaminophen or ibuprofen. While swelling of the entire limb is less common than with the whole cell vaccine, it has been reported in 2 to 3% infants receiving the fourth or fifth dose of DTaP. Limb swelling may be accompanied by erythema, pain and fever and may occasionally interfere with ambulation.

Prolonged crying and hypotonic-hyporesponsive episodes are severe events associated with pertussis containing vaccines. Neither reaction is known to be associated with long-term sequelae.

Seizures following pertussis-containing vaccines are usually febrile seizures, usually occurring within 24 hours of vaccination. The risk of seizures depends upon the vaccine preparation, being highest with DTwP vaccine (adjusted relative risk 5.7), increased with the DTaP-IPV-Hib combination vaccine (hazard ratio [HR] 6.02, 95% CI 2.86–12.65 with first dose, and HR 3.94, 95% CI 2.18–7.10 with the second dose) and not increased compared to baseline in infants receiving DTaP (incidence rate ratio 0.87, 95% CI 0.72–1.05). However,

the absolute risk of seizures is small and is not associated with an increased risk of epilepsy or neurodevelopmental disability. Rarely, vaccine associated seizures may be the initial presentation of infantile myoclonic epilepsy, which should be considered before attributing the seizures to vaccination

H. Contraindications and Precautions

Contraindications

Absolute contraindications to diphtheria-, tetanus-, and/or pertussis-containing immunizations include: (i) anaphylactic reaction to the diphtheria-, tetanus-, and/or pertussis-containing vaccine or vaccine constituent (contraindication to subsequent doses of all components); (ii) Encephalopathy within 7 days of the administration of a previous dose of the vaccine without another identifiable cause (contraindication to subsequent doses of pertussis vaccine); (iii) Progressive neurologic disorder, including infantile spasms, uncontrolled epilepsy, progressive encephalopathy (administration of pertussis-containing vaccines should be deferred until the neurologic status is clarified and stabilized); (iv) history of anaphylactic reaction to latex (contraindication to DTaP vaccines that contain latex). Children younger than 7 years who have a contraindication to pertussis immunization should not receive pertussis-containing vaccine but may receive DT.

Precautions

These are conditions that might increase the risk for a serious reaction to immunization and carry an unknown risk of a subsequent similar reaction. Previous adverse reactions to DT/dT and/or TT that constitute precautions to subsequent doses of include: (i) Guillain-Barré syndrome (GBS) within six weeks of TTor DT/dT containing vaccine; (ii) history of Arthus-type reaction (a specific type III, immune complex mediated hypersensitivity reaction characterized by severe pain, swelling, induration, edema, hemorrhage and/or necrosis at injection site) following previous dose of DT/dT and/or TT containing vaccine.

Precautions to immunization with pertussis-containing vaccines include: (i) Moderate or severe illness with or without fever (immunization should be administered upon recovery); (ii) The occurrence of any of the following conditions (without another identifiable cause) after administration of a pertussis-containing vaccine (DTwP or DTaP): temperature ≥105°F (40.5°C) within 48 hours of vaccine; hypotonic-hyporesponsive episode (collapse or shock-like state) within 48 hours of vaccine; seizure (with or without fever) within three days of vaccine; and persistent, inconsolable crying for ≥3 hours within 48 hours of vaccine. While the risk of recurrence of such a reaction may justify discontinuing administration of pertussis vaccine, in circumstances such as an outbreak of pertussis, the benefits of immunization may outweigh the risks. While several reports describe uneventful vaccination with aP vaccine in children with previous severe conditions (e.g., hypotonic-hyporesponsive episode) after immunization with whole-cell pertussis vaccine, such events have been described with the former vaccines as well.

Temporary or indefinite deferral of pertussis immunization may be warranted for infants and children with certain neurologic conditions or children who are undergoing evaluation for these conditions. Decisions regarding pertussis immunization in such children should be made on a case-by-case basis, after weighing the risks and benefits of immunization, reassessed at each subsequent immunization visits. The following neurologic conditions that are considered precautions to administration of pertussis-containing vaccines:

(i) *Conditions that predispose to seizures or neurologic deterioration:* To avoid confusion about causation, children with unstable or evolving neurologic conditions that predispose to seizures or neurologic deterioration (e.g., tuberous sclerosis, inherited metabolic or degenerative conditions) should not receive pertussis vaccine until the diagnosis and prognosis of the disorder are ascertained, with administration of pertussis-containing vaccine should be reconsidered at each visit, particularly if the condition is resolved, corrected, or controlled.

(ii) *History of seizures:* While children with epilepsy have an increased risk of seizures after pertussis vaccine, there is no evidence that vaccine-associated seizures induce permanent brain damage, cause epilepsy, or aggravate or affect the prognosis of the underlying neurologic disorder. Pertussis immunization should be deferred in children with a history of seizures until a progressive neurologic condition is excluded; may be administered to infants and children with well-controlled seizures under cover of acetaminophen or other antipyretic (started before immunization every 6 hours during the ensuing 24 hours).

Not contraindications or precautions

Conditions that are not considered contraindications or precautions to diphtheria, tetanus and/or pertussis containing immunization include: (i) mild acute illness with or without fever; (ii) family history of seizure or sudden infant death syndrome; (iii) stable or resolved neurologic condition (e.g., controlled idiopathic epilepsy, cerebral palsy, developmental delay); (iii) immunosuppression; and (iv) extensive local reaction after the fourth dose of DTaP.

V. COMBINATION VACCINES BASED ON THE TRIPLE VACCINE

A. Vaccine Preparations

The number of antigens available as vaccines for providing protection against childhood infections has increased to almost 25, leading to a dramatic increase in the number of immunizations recommended for children in the first two years of life. In the United States children are expected to receive vaccines against as many as 14 diseases in the first 2 years, leading to as many as 6–8 injections in a single visit to the pediatrician. Even within India, the immunization schedule recommended by the IAP in 2012 includes as many as 5 vaccines at certain ages. A complex immunization program makes it increasingly more challenging to incorporate new vaccines into the schedule. Parents may worry that the child's immature immune system will not be able to 'handle' multiple vaccines simultaneously or be distressed at the pain accompanying vaccines, particularly when more vaccines are given concurrently. Concerns about the safety of vaccines children receive are among the factors that influence parents' decision to vaccinate their children.

Simplifying immunization schedules by combining multiple vaccines into a single syringe has been reported to have numerous positive effects, as listed in Table 111.4. The concept of combining vaccines for protection against multiple diseases is not new, and began with the combination of individual diphtheria, tetanus, and pertussis (DTP) vaccines into a single vaccine that was first used to vaccinate infants and children in 1940s. The DPT vaccine has been the cornerstone of pediatric immunization programs, and has been modified over the years in an attempt to add other vaccines to the combination and to replace reactogenic components to improve its adverse effect profile. As discussed above, an important improvement was the introduction of the less reactogenic acellular pertussis vaccine in the early 1990s which lead to the combination of DTaP with other routine vaccines such as inactivated polio vaccine (IPV), *Haemophilus influenzae* vaccine (Hib), and hepatitis B vaccine (HepB). Development is currently focused on maximizing the number of vaccines that can be combined in a single formulation and strategies to provide protection against pertussis before the commencement of routine infant immunization.

Table 111.4 Advantages of combination vaccines

Fewer injections
Reduced trauma to the infant
Higher rates of compliance with complete vaccination schedules
Better vaccination coverage
Timely vaccination as the vaccination schedule gets completed on time
Reduced administration costs
Smaller storage space requirements, and
Allows incorporation of new vaccines into immunization schedules

B. Composition of Combination Vaccines

DTaP vaccines from different manufactures are very similar in their composition, with the main differences being related to the number, amount, and detoxification method of the pertussis components (Table 111.2). Furthermore, some combinations contain additional vaccines such as HepB, Hib, and IPV in addition to the DTaP base.

Table 111.5 provides a list of combination vaccines based on DTaP or DtwP that are licensed for use internationally and within the country. As shown in Table 111.2, the available combinations vary in the content of individual components, and in the number and type of antigens combined, from trivalent to hexavalent preparations. Other licensed combinations in use are bivalent (dT) to tetravalent booster (e.g., DTaP-IPV) vaccines for use in adults and adolescents. Infanrix™, the leading DTaP vaccine used worldwide, provides the backbone of the hexavalent vaccine Infanrix™-hexa which is licensed in Europe and protects against six different diseases is adjuvanted with both aluminum hydroxide (for DTaP) and aluminum phosphate (for HepB vaccine) adjuvants.

C.   Challenges Faced when Combining Vaccines

Although combination vaccines are associated with clear benefits, the main challenge in their development is the risk that the efficacy or safety of the combination would be less than that seen with the administration of the vaccines separately. Immunological, physical, and/or chemical interactions between the combined components have the potential to alter the immune response to specific components. Furthermore, if the vaccines to be combined have differing immunization schedules, consolidation of these should also not negatively affect immunogenicity, efficacy, or safety. Finally, the combined product should not have reduced stability. Uncommon transport and storage conditions and complicated bedside mixing could hamper the development of a combination vaccine. Some examples given below illustrate the difficulties encountered in combining vaccines and the options used to circumvent the concerns.

Reduced Hib response with DTaP-Hib combination

Many DTaP based combination vaccines are reported to demonstrate interference with the immune response to the Hib component, with resulting reduction in antibody titers to the polyribosylribitol phosphate antigen of Hib confirming reports of reduced immunogenicity in animal models. The extent of immune interference is reported to be lower in DTwP-based combination vaccines, attributed possibly to the adjuvant effect of the whole-cell pertussis component. However, the precise mechanism by which the Hib response is reduced after combination with DTaP is incompletely understood and may result from complex interactions between Hib and TT, FHA, and aluminum hydroxide adjuvant. Possible mechanisms of immune interference include: (i) competition between TT-specific and PRP-specific B cells for the Hib PRP-TT conjugate antigen, (ii) suppression of PRP response by clonal expansion of TT-specific B cells; (iii) physical prevention of binding of Hib PRP-TT conjugate antigen to PRP-specific B cells by the TT carrier protein; (iv) incompatibility of Hib with the alum adjuvant of DTaP. While some authors have attributed the rise in incidence of cases of Hib in the UK and Ireland in late 1990s to the reduced immunogenicity of Hib in DTaP-based combination vaccines, other factors may explain the increase in incidence, including decreasing herd immunity, the short UK immunization schedule, and the lack of a booster immunization

TABLE 111.5   Combination vaccines based on DPT vaccines licensed for use internationally and in India

Vaccine	Available internationally	Available in India
dT		BE-Td, SII Td-Vac
DT		Dual antigen
DTaP	Tripedia, Daptacel, Infanrix, Certiva, Acel-imune	Tripacel, Infanrix
dTaP	Adacel, Boostrix	Boostrix, Adacel
DTaP–HB	Tritanrix	Not available
DTaP–HB–IPV	Pediatrix	Not available
DTaP–HB–IPV/Hib	Infanrix hexa	Not available
DTaP–Hib	TriHIBit; Infanrix+Hib; Tetramune	Tripacel+ActHib, Infanrix+Hiberix
DTaP–Hib-IPV	Pentacel; Infanrix+IPV+Hib	Pentaxim
DTwP		Tripvac, Triple antigen, Comvac3
DTwP–Hib		Easy-4, Quadrovax, Shan-4, TetractHib; Triple antigen+HibPro
DTwP-HB		Ecovac-4, Shantetra, Q-VAC, Tritanrix-HB, Tripvac-HB, Comvac-4-4-
DTwP–HB-Hib	Tritanrix-Hib	Easy-5, Pentavac, Comvac-5, Shan-5; Qvac+HibPro, Hiberix+Tratanrix
dT-IPV	dT polio adsorbed	Not available
DTaP–IPV	Quadracel; Infanrix+IPV	Not available
Hib–DT	Vaxem HIB	Not available
Hib-TT conjugate	ActHIB	Not available

in the second year of life. It has also been suggested that the current serological surrogate of efficacy for Hib vaccines, an anti-PRP levels of 1.0 μg/ml, is arbitrary and that antibody responses below this threshold are not associated with impaired antibody function or loss of immune memory. With current evidence favoring the use of DTaP based Hib vaccines, many such combinations have been licensed.

Combining immunization schedules for DTaP-HepB vaccine

Consolidation of the different immunization schedules of the vaccines proposed to be combined is a significant impediment in implementing vaccine combination. In order to prevent perinatal transmission of hepatitis B infection in countries with high rates of hepatitis B surface antigenemia, the WHO recommends that the first HepB vaccine dose be administered <24 hours after birth. For the same reason, the ACIP and IAP recommend a birth dose of hepatitis B vaccine for all infants in the USA and India, respectively, which should be followed by two more doses of the vaccine at 1 and 6 months. In contrast, the primary immunization series of DTaP requires doses at 2, 4, and 6 months. Therefore, combining the vaccines requires that the monovalent HepB be given at birth followed by combination of HepB and DTaP at 2, 4, and 6 months, resulting in an unnecessary fourth dose of HepB at 6 months of life.

Another consideration with the combination of DTaP with HepB is concern over the immunological response to the HepB component. The serological response for HepB was inferior when the combination of DTaP-HepB administered at 2, 4, and 6 months as compared to separate administration of HepB at birth, 1 and 6 months and DTaP at 2, 4, and 6 months; however, antibody levels were above the level considered seroprotective in 99% of the subjects. Administration of the hexavalent DTaP-HBV-IPV/Hib vaccine at 2, 4, and 6 months after a dose of HepB vaccine at birth did not adversely impact protective anti-HBs titers and was not more reactogenic than the same combination given without the birth dose of HepB. Nevertheless, concerns over its reduced efficacy for hepatitis B, particularly on the long term, led to the suspension of a hexavalent combination vaccine (Hexavac) in 2005.

Improving the safety and tolerability profile of DTP-based vaccines

Safety and tolerability of vaccine combinations have to be ensured before vaccines are licensed for use. The substitution of whole cell pertussis with acellular pertussis antigens in DTP-based vaccines was a major advance in improving its reactogenicity profile. Multiple studies have corroborated the increased risk of systemic as well as local reactions with whole cell when compared to acellular vaccines, whether as DTaP alone or in combinations with IPV, Hib and HBV vaccines. However, reactions are still seen with all DTaP vaccines, notably an increased frequency and severity of erythema and swelling, particularly with booster immunizations given as the fourth or fifth dose.

D. Future Directions

The chief concern with the currently available DTaP-based combination vaccines, whether pentavalent or hexavalent, is the reported decreases in antibody titers to Hib when combined with other components in a single syringe. While further improvements in vaccine immunogenicity is desirable, clinical studies using these vaccines have demonstrated satisfactory efficacy and safety profiles, comparable to that with separate administration of these antigens. The ACIP recommends that combination vaccines be used whenever possible, and IAP endorses the use of these vaccines where afforded. As new vaccines targeting previously unaddressed diseases are added to the vaccination calendar, it is mandatory that any concerns with the safety and efficacy of currently available combination vaccines are addressed.

Addition of pneumococcal or meningococcal components

Addition of pneumococcal or meningococcal vaccines to DTaP-based vaccines has been attempted. While there is no clinical data for such combination vaccines, there is considerable information on the safety and efficacy of co-administration of DTaP with pneumococcal or meningococcal vaccine vaccines. Multiple clinical studies evaluating concurrent administration of DTaP-based combination vaccines with a sevenvalent pneumococcal conjugate vaccine found no significant immunological interference between the vaccines. However, a large study found that the serological responses to several antigens were reduced when DTaP-HepB-Hib/IPV was adminsitered along with a sevenvalent pneumococcal conjugate vaccine, as compared to administration of DTaP-HepB-Hib/IPV or pneumococcal vaccine, although the rates of seroprotection and vaccine response were comparable at 96.8–100%. Other studies show that co-administration of pneumococcal or meningococcal conjugate vaccines with hexavalent DTaP-HepB-Hib/IPV has reasonable efficacy, with immunogenicity retained for meningococcal serogroup C vaccines, both CRM197–conjugated vaccines (Meningitec™ and Menjugate™) and the tetanus toxoid conjugate vaccine (NeisVac-C™).

Use of novel carrier proteins for conjugate vaccines

Changing either the antigens or the carrier proteins used for conjugates may result in improved immunogenciity of vaccine components. New approaches to carrier protein technology, with novel carrier proteins replacing TT, have been devised. In one approach a candidate 11–valent pneumococcal conjugate vaccine contains two carrier proteins, TT and DT, with the amount of TT reduced further by conjugating to DT those polysaccharides that require the largest carrier amounts. In another approach, full-length protein carriers such as DT and TT are replaced with peptides containing T helper cell epitopes and lacking B cell epitopes; the former are expected to enable a strong helper effect to the conjugated antigen without inducing an antibody response to the carrier itself.

E. Novel Adjuvants

The use of adjuvants more potent than alum is expected to be useful by extending the period of protection from vaccine combinations administered in infancy into adolescence and adulthood. The reduced antibody responses to Hib when delivered in combination with DTaP are attributed to an incompatibility with alum adjuvant, and experiments show that anti-PRP antibody levels are several fold lower when Hib alone is adsorbed to aluminum hydroxide adjuvant. Furthermore, reduction in Hib Use of alternative adjuvants, such as oil-in-water emulsion MF59 and polylactide co-glycolide (PLG) microparticles, with established vaccine antigens such as DT, TT, HepB (HBsAg), *N meningitides* serotype C conjugate (MenC) and a *N meningitides* serotype B recombinant antigen (MenB), may improve immunogenicity and overcome the reduced responses observed with Hib and HepB in DTaP combinations. This strategy may additionally enable antigen dose reduction and reduction in the number of vaccine doses required.

Ensuring consistency in vaccine quality

An emerging need for all vaccines, whether single agent or as a combination, is the development of improved *in vitro* assays for measuring batch-to-batch consistency of vaccine production. This is particularly important for combination vaccines where one or more components may affect immunogenicity. With a decline in *in vivo* potency testing on animals, *in vitro* approaches as a correlate to *in vivo* protection are now being applied for some components of combination vaccines, e.g., HepB. An enzyme-linked immunosorbent assay (ELISA) has been developed to quantify DT antigen in DTP-based combination vaccines. Similar assays for TT have also been described. Antibody-based assays have the potential to measure antigen that is adsorbed to alum, avoiding potentially destructive methods otherwise required for the elution of antigens prior to quantification. In addition, antibody-based assays are specific to the target antigen and are more sensitive than other biochemical and biophysical tests applied to these toxins (such as gel electrophoresis, size-exclusion chromatography, and circular dichroism) which are crude and characterize only unformulated bulk antigens due to interference from other components in the final vaccine product.

SUMMARY

Disease prevention and eradication are the ultimate and immediate goals of immunization. Recommendations on vaccines against diphtheria, pertussis and tetanus are reviewed and summarized. Combination vaccines are expected to overcome key challenges to maintaining coverage through simplification of vaccine schedules and reduction of injections required. However, multiple technical challenges require to be overcome in maintaining immunogenicity and safety when combining vaccines. Additional studies are needed to further explore safety and the ideal timing of vaccination of novel combination vaccines Based on DTaP vaccine.

SUGGESTIONS FOR FURTHER READING

Indian Academy of Pediatrics Committee on Immunization (IAPCOI), Consensus Recommendations on Immunization and IAP Immunization Timetable (2012), *Indian Pediatr.*, 49:549–64.

[No Authors Listed] Recommended Immunization Schedules for Persons Aged 0 Through 18 years: United States (2012), *MMWR Morb Mortal Wkly Rep.*, 61:1–4.

Skibinski D.A.G., Baudner B.C., Singh M. and O'Hagan D.T., Combination Vaccines, *J. Glob Infect Dis.*, 3:63–72

http://www.who.int/immunization

http://www.iapindia.org/immunisation

http://www.aap.org/immunization/

http://www.cdc.gov/vaccines/

www.vaccinesafety.edu

112

Eradication of Poliomyelitis
Triumph of Biotechnology

T. Jacob John

CONTENTS

I. Introduction
II. History
III. Classification
IV. Morphology of virus
    A. Structure of viral genome
    B. Replication of virus
    C. Immunity
    D. Pathophysiology of poliomyelitis
    E. Paralytic poliomyelitis
IV. Laboratory Diagnosis of Poliomyelitis
V. Vaccines
    A. Inactivated polio vaccine (IPV)
    B. Live attenuated vaccine (oral polio vaccine)
    C. Vaccine derived poliomyelitis
VI. Eradication in India

I. INTRODUCTION

Human mastery over microbial pathogens began with their discovery as cause of disease in the late 19th century (by Louis Pasteur and Robert Koch). Techniques were developed for detecting them and for their cultivation (or culturing) in media in laboratory ware under specified conditions. Since then Medicine gradually demanded that every infectious disease has to be diagnosed ideally by its causative pathogen or 'aetiology'.

A second major advancement was the creation of vaccines against some microbial pathogens. In fact, the very first vaccine was made even before pathogens were discovered. Vaccine against smallpox was designed by the *in vivo* human adaptation of a zoonotic pathogen of low virulence in humans, namely cowpox or vaccinia (by Edward Jenner). Although microbes had not been discovered yet in the late 18th century, the principle of transmission of infectious 'something' was obvious from careful observations. Cowpox vaccination protected against smallpox. After the discovery of an increasing list of pathogens, laboratory manipulations were developed to develop vaccines against the more important ones. Today over 26 diseases can be prevented by immunization with vaccines.

Human mastery continued as anti-microbial therapeutics were discovered or invented. The more potent or successful ones were against bacteria and fungi, but we also have a number of effective anti-viral drugs in clinical use today. Against polioviruses we do not have an anti-viral drug, but we have two vaccines, the inactivated poliovirus vaccine (IPV) and the live attenuated oral poliovirus vaccine (OPV).

Parallel with the advancements in clinical medicine, Public Health also developed with supporting scientific tools, the most important being epidemiology that brought together medicine, biology, behavior dynamics and mathematics. Epidemiologically, human mastery over pathogens could be achieved at various levels, namely, control, elimination and eradication. Control means the reduction of incidence of disease in a specific geographic community, using intervention tools, to a pre-desired level, and confirming that reduction was due to intervention. Elimination means the control level is zero. In other words, zero incidence of a disease in a defined geographic community, achieved through interventions using biotechnological tools, and so proved by epidemiology, is elimination. Usually the defined population for elimination is at least one whole country, if not a continent. Eradication is elimination at global level.

Smallpox was eradicated by 1976, using smallpox vaccination in carefully crafted tactics. The next disease targeted for eradication using vaccine as the major tool of intervention is poliomyelitis. Diseases like yaws and Guinea Worm disease (Dracunculiasis) are also under eradication mode, using antibiotics (yaws) and interruption of the complex life cycle of pathogen (Dracunculiasis). Diseases that could be eradicated have some specific criteria. There must be suitable diagnostics available, to prove that the pathogen no longer infects anyone. Excellent diagnostic tools including molecular virology techniques and reagents are available against polio. There must be effective intervention tool: against polio there are two vaccines, each with some advantages and disadvantages. There must not be any extra-human reservoir of the pathogen; for polio there is none as humans are the only natural hosts. In summary polio fulfills all criteria of global level eradicability.

Polioviruses are the cause of poliomyelitis, a systemic infectious disease that affects the central nervous system (CNS) causing paralysis. The name of the disease (*polios,* "gray"; *myelos,* "marrow" or "spinal cord"), widely known as polio, is descriptive of the pathologic lesions that involve neurons in the gray matter, predominantly in the anterior horn of the spinal cord. Older, less commonly used names for the disease include *infantile paralysis* and *Heine-Medin disease.* Paralytic poliomyelitis has been completely controlled in developed countries and almost all the developing nations except a few, through immunization programs. Inactivated (IPV) or live-attenuated vaccine (OPV), or both are being administered, and global eradication is anticipated to occur in next few years.

II. HISTORY

The history of Enteroviruses (EV) is very much the history of poliovirus (PV). Many of the PV milestones are landmarks in the study of EV and, in fact, all of Virology. The evidence of polio has been since antiquity. An Egyptian stele dating back to 1580–1350 BC shows a priest with polio like leg deformity. The First reported case was by German orthopedist Jakob von Heine in 1840. In 1890 Karl Osker Medin of Uppsala University of Sweden published an article on the natural history and complications of the disease. Though there have been several outbreaks the first recorded outbreak was in 1894 in Vermont of USA where 132 cases were recorded. In 1907 Ivar Wickman a Swedish pediatrician called it Heine-Medin disease based on the report of two earlier clinicians. In 1908 Landsteiner and Popper were able to confirm the viral etiology of disease. They ground the spinal cord of the infected person and after filtering the suspension they inoculated the filtrate into monkeys which produced symptoms. This proved that the disease causing agent was a filterable particle. In 1916 there was

a devastating epidemic in New York with more than 9000 cases & over 2,000 deaths. In 1948 Enders-Weller-Robbins (Nobel Prize 1954) cultured poliovirus in *vitro* in embryonic tissue. This work lead to the introduction of viruses in tissue cultures. 1955 was one of the land mark years in polio history. Jonas Salk in USA produced killed polio vaccine (IPV) with which immunization for polioviruses started. In 1961 Albert Sabin pioneered and made the oral polio vaccine (OPV). As vaccines became available several organizations came up with the polio immunization programs. Expanded Program for Immunization (EPI) in 1974, Global Polio Eradication Initiative (GPEI) by World Health Organisation (WHO) in 1988 are noteworthy. Polio vaccination programme started in India just few years after discovery of the vaccine but national level immunization started only since 1995. India reported its last polio case due to wild virus in 2011 though vaccine derived poliomyelitis is still being reported.

III. CLASSIFICATION

The family *Picornaviridae* contains six genera namely Aphthovirus, Cardiovirus, Enterovirus, Hepatovirus and Rhinovirus. It derives its name from the small size of the viruses (*pico* = small) and their RNA genome. They are acid-resistant, and thus are able to survive the low pH in the stomach. They replicate in the small intestine and can be isolated from faeces and are spread by the faeco–oral route. Poliovirus was the first genus—indeed, of the whole family—to be isolated. This was done by Enders, Weller, and Robins in 1948 when they cultured it in monkey kidney cells. They received a Nobel Prize for the same. Soon after the isolation of the polioviruses, similar viruses were discovered.

TABLE 112.1   Genus and main syndromes caused by them

Genus	*Syndromes*
Enterovirus	Infections of the central nervous system, heart, skeletal muscles, skin, and mucous membranes
Hepatovirus	Hepatitis A
Rhinovirus	Common colds
Aphthovirus	Foot and mouth disease of cattle (rarely in humans)
Cardiovirus	Encephalitis and myocarditis
Parechovirus	Diarrhea

Polio virus is classified under the genus enterovirus and have three serotypes: Poliovirus-1, Poliovirus-2 and Poliovirus-3

IV. MORPHOLOGY OF VIRUS

Poliovirus is 25–30 nm spherical particle. The capsid (protein coat that encloses viral genome) has 60 protomers arranged in icosahedral symmetry and each of them have one molecule of four viral protein (VP1-VP4) .

An icosahedron is an object with

20 faces, each an equilateral triangle

12 vertices, each formed where the vertices of five triangles meet

30 edges, at each of which the sides of two triangles meet.

In poliovirus the icosahedral particle have following protein arrangement

- 20 faces with 3 protomers each
- 12 apices occupied by 5 copies of VP1, and the center of each face occupied by VP2 and VP3 alternatively.

Some remarkable studies have been performed with polio and rhinoviruses using X-ray crystallography. These analyses have helped to pinpoint functional areas in the virion, the most interesting being the 'canyon' or cleft where the cell receptor binding site is located. All four viral proteins (VP1-VP4) are anti parallel β array forming a barrel known as beta barrel. Narrow end of β barrel of VP1 is at the apex, the corresponding sides VP2 and VP3 alternate around the axis, VP4 forms lattice within the pentamer.

A. Structure of Viral Genome

Poliovirus is a RNA virus with a positive sense single strand linear RNA where the RNA acts as the mRNA. The size is 7.4 kb with polyadenylated 3′ end and a viral protein VPg at 5′ terminus. Genome is translated as a large polyprotein which is then cleaved into individual four viral proteins. Capsid or structural proteins are encoded towards the 5′ terminal half and nonstructural proteins coded from middle and 3′ terminal half of the genome.

B. Replication of Virus

Polio virus multiplies in cytoplasm of enteric cells or neurons. Total duration of replication is about 6 to 8 hours. Replication process involves:

1. Entry
2. Uncoating
3. Protein synthesis
4. Genome replication
5. Assembly
6. Release

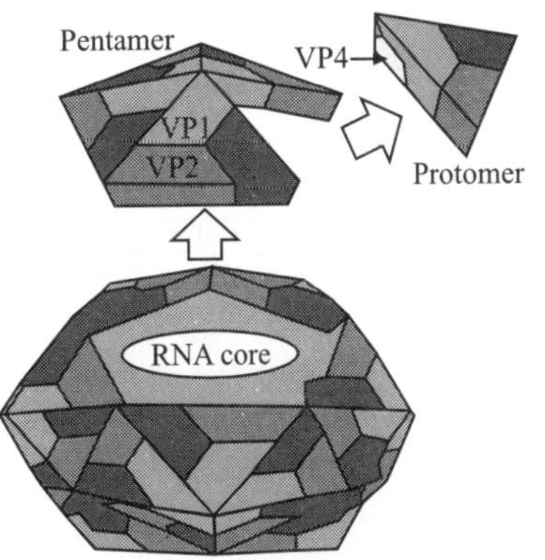

Figure 112.2 Replication of poliovirus in cell cytoplasm of cells. Note that cells nucleus is not effected in infected cell.

Details of the processes

1. Entry: The entry of virus into the enterocyte or neurons is receptor mediated. Receptors are found on the cells surface. The receptor is CD155 protein, member of immunoglobulin gene super family. This receptor binds to the canyon on the virus and the virus thus enters the cell.

2. Uncoating: Uncoating is the process where nucleic acid of virus comes out of the capsid. This happens once the virus enters the cell. Binding to receptor brings conformational changes in viral capsid which allows the interaction with plasma membrane forming a pore through which VP3 plug opens, releasing viral RNA into host cell cytoplasm.

3. Translation: Translation in poliovirus differs from eukaryotes in that it is 5′cap independent because viral RNA

Figure 112.1 Icosahedral symmetry of poliovirus. Protein arrangements with RNA core is seen in the picture in the left side. (*see Plate 48 for colour figure*)

lacks 5′ cap. Instead, VPg is covalently bound to 5′ cap. The following steps ensue:

- VPg protein dissociates from 5′ end.
- 40S ribosome binds to internal sequence called *internal ribosome entry site* (IRES).
- At this point host cell translation is shut-off by cleaving of eIF4G (initiation factor 4) by 2A (a viral protein that act as protease).
- Template reading occurs in 3′→5′ direction while translation occurs in the opposite direction.
- Structural and non-structural proteins (replication enzymes) are produced in the form of large polyprotein which is then cleaved to form individual proteins by viral coded 2A and 3C proteins.

Following proteins are produced when the polyprotein is cleaved

TABLE 112.2   Different viral proteins and their functions are enumerated

Product	Function	
VP1-VP4	Structural Proteins	
2A	Proteases	Initial Clevage—Cleaves Host Eif4g
3C		Later Cleavages
3D	Polymerase	
2C	Nucleotide Tri-Phosphatase	
VPg (3B)	Uridylylated VPg Primes Translation	

4. Viral genome (RNA) replication: As stated earlier the genome of the poliovirus is a positive sense RNA which is used as template for the production of negative sense RNA. Negative sense RNA synthesis is required to produce positive sense genome. To produce this negative sense RNA, circularization of viral genome (positive sense) is necessary. This happens when ribonucleoprotein (RNP) at 5′ end connects 5′ end with 3′end. Viral protein 3D a polymerase mediates the replication.

5. Assembly: Once the polyprotein molecule is produced it has to be cleaved for virus maturation to occur. The initial cleavage products are VP0, VP1 and VP3 but VP4 is not yet obtained. This initial cleaving is mediated by a viral protein 3CD.

$$\text{Polyprotein} \xrightarrow{\text{3CD}^{\text{pro}}} \text{VP0} + \text{VP1} + \text{VP3}$$

Once these protein molecules are produced, protomers (5S) self-assemble into 14S pentamers, and pentamers assemble into 80S empty capsids. This empty capsid is virion molecule without the RNA genome in it. These empty particles are noninfectious. Now RNA is threaded into the empty capsid producing a 150S provirion in which VP0 is uncleaved. Provirion may also be formed alternatively by assembling pentamers around the RNA genome. Virus is fully matured only when VP0 is cleaved to produce VP2 and VP4. Cleaving of VP0 is the final stage of the viral assembly and maturation. This produces 160S infectious virion.

6. Release: While virus is still multiplying within the host cells (enterocytes and neurons), the host cell normal function is disrupted. One prominent ill effect that virus causes to the cell is that it shuts off host cell protein synthesis. These cells are slowly degraded and with disintegration of the cell, newly produced viruses are released which then infect other cells.

C.   Immunity

Not every individual that is infected by poliovirus proceed to paralytic manifestation. It is seen only in very small proportion of infected individuals. Infected individuals may show varying degrees of clinical manifestations, many of the infected individuals being asymptomatic.

All individuals infected by poliovirus including asymptomatic ones eventually develop immunity against the virus. Antibody appears early in infection and lasts for life. These antibodies protect the individual from subsequent viral attack and paralytic poliomyelitis. Antibodies are also transferred to fetus from mother or even to child during breast feeding

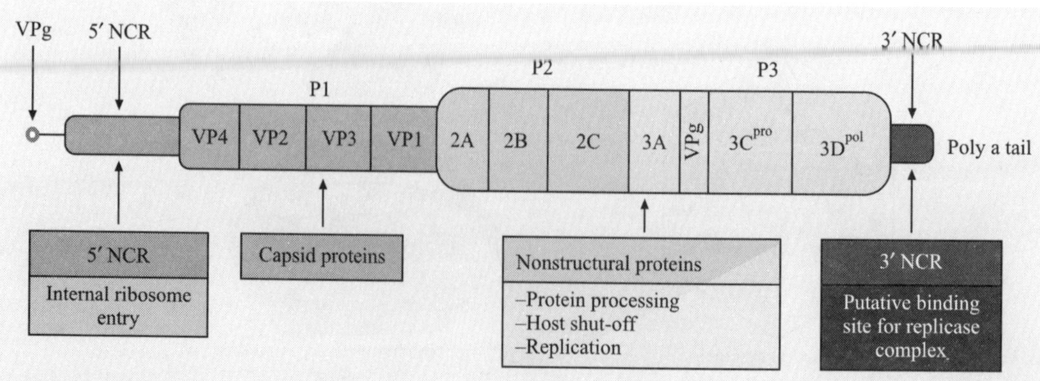

Figure 112.3   Schematic representation of poliovirus genome and the products of genes.

providing passive immunity that last for less than six 6 months. Children are susceptible to disease after 6 months of age if they are not vaccinated.

Following types of immunity play role:

A. Innate – this is the earliest activated immune response. Interferons and Natural Killer cells play protective role against paralysis in most cases.

B. Humoral – neutralizing antibodies are produced against the different antigenic sites of VP1, VP2, VP3 proteins. These however do not provide protection against gut infection. Thus a person may develop gut associated symptoms but will not progress to paralysis

C. Cell mediated – This type of immunity plays a role in viral clearance. Cell lysis is mediated through CD8+ and CD4+ cell response. These cells recognize the antigenic epitope on VP1.

D. Pathophysiology of Poliomyelitis

Humans are the only natural host and reservoir of polioviruses. It is possible to produce experimental infection in other primates, and can be made to replicate in sub primate mammals.

Pathogenesis

Virus enter the body by ingestion of contaminated materials or by inhalation of the virus in aerosols. Incubation time 7-14 days (2-35). Following entry, they replicate in the gut, naso pharyngeal mucosa and adjacent lymphoid tissues. There is primary viremia which seeds virus in the reticuloendothelial system. Further viral spread is not seen in asymptomatic infections. The virus at this point elicits the formation of type-specific antibodies. In symptomatic cases virus replicates in the RE system which results in a secondary viremia where there is high viral load. At this point there is minor nonspecific illness related to abortive poliomyelitis. Now virus reaches the gray matter of spinal cord and brain to infect the neurons specially the motor and aoutonomic neurons. Route taken by the virus to reach these areas is uncertain. Some studies propose that the seeding from muscle to CNS occur via peripheral nerve fibres rather than haematogenous route. Nerve cell damage with inflammation is the hallmark of the infection specially affecting the anterior horns of spinal cord, and the motor nuclei in pons and medulla. In severe cases spinal cord and cranial motor nuclei are involved. There is no sensory defect.

Polioviruses can be isolated from the spinal cord for only the first few days after the onset of paralysis but inflammation last for months.

BOX 112.1   Over view of pathogenesis

- Entry – ingestion or inhalation
- Primary multiplication in Peyer's patches, tonsil, neck lymph nodes
- Cell degenerate – viremia
- Virus reach the axons of peripheral nerve to reach CNS-multiply
- Attacks specially anterior horn cells of spinal cord, Severe cases-brain stem, reticular formation
- Damage or destruction of neurons accompanied by inflammation

Clinical features

When an individual susceptible to infection is exposed to the virus, clinical features manifest. They range from asymptomatic infection to devastating paralytic poliomyelitis. Most infections are subclinical; only about 1% of infections result in clinical illness. Following table summarizes the type of infections and their prevalence

TABLE 112.3   Summery of clinical outcomes in Poliovirus infected individuals

Type	Prevalence	Features	Virus yield
Asymptomatic	90–95%		Oropharynx Stool
Abortive	4-8%	Flu like symptoms Sore throat GI manifestation Recovery in few days	Throat Stool
Nonparalytic poliomyelitis (aseptic mengitis)	~ 1%	Meningeal irritation Stiff neck, myalgia Rapid and complete recovery	Stool Throat
Paralytic poliomyelitis	<0.5%	See the text	Stool CSF Throat

E. Paralytic Poliomyelitis

This is seen only in about 0.5% infections. A biphasic course of minor and major illnesses is observed. The minor illnesss lasts for 1-3 days coinciding with viremia, showing symptoms of abortive diesease. 2–5 days following it, the the patient appears to be recovering and remains symptom-free for 2 to 5 days. Then there is abrupt onset of major symptoms heralded by severe mylgia, fever, chills, paresthesia, localized cutaneous hyperesthesia, fasciculation, CSF pleocytosis, flaccid paralysis which is predominantly asymmetrical paralysis affecting some muscles and sparing others. Lower limbs are affected more than the upper. Sensory loss is extremely rare.

In bulbar poliomyelitis which involves brain stem, specially the 9th and 10th cranial nerve are affected leading to dysphagia, nasal speech, dyspnoea and other life threatening symptoms. These lead to anxiety because of inability to swallow and breathe. If the circulatory and respiratory centres are involved then the outcome might be extremely fatal.

Polio encephalitis is a rare entity which is mostly seen in infants. Spastic paralysis sometime may be observed.

Post polio syndrome is an entity that is observed in some patients with previous poliovirus attack. It is seen many years (25–30) after recovery from acute illness. Fatigue and paralysis typically affect the same muscles involved in prior infection and in addition other muscles also can be involved. This may be due to slow progressive degeneration of infected nerve over a period of time. Diagnosis of such cases is established based on history, EMG, CT & MRI of the spinal cord. There are no drugs for treating such cases. Physiotherapy to assist movement and steroid to reduce pain might prove helpful.

IV. LABORATORY DIAGNOSIS OF POLIOMYELITIS

Once paralysis occurs, it is irreversible. Diagnosis of poliomyelitis and the earlier milder forms are important because of the risk of the spread of infection in the community. Laboratory diagnosis of poliomyelitis can be divided in two categories

1. Direct methods: These methods provide the direct evidences of presence or absence of virus. Since viruses are not visible even under ordinary microscopes alternate techniques are employed. These methods are more sensitive and thus reliable. Most commonly used technique being the viral culture in various cell lines. Newer techniques which are based on viral nucleic acid detection and amplification (molecular methods) are now in wide use. They are more sensitive, specific and less laborious.

(a) Viral culture: Virus may be cultured on selected cell lines like BMK cells, Vero cell, MRC5 cells and A549 cells. Most commonly acquired specimen is stool sample but rectal swab and nasopharyngeal swabs are also acceptable specimens. After incubating for 2 to 7 days these cells can be observed for characteristic cytopathic effects of a infected cell. The typical cytopathic effect described for Poliovirus is – round and refractile, teardrop like shrinking and pyknosis of nucleus, and cell detachment from each other.

Once characteristic CPE is observed the culture should be confirmed by specific methods.

Neutralization assay is one of the tests for confirmation. Here poliovirus specific antiserum is used to neutralize the virus. Antibody treated culture fluids are re-cultured and inhibition of growth is observed i.e. there is no CPE. Alternatively a PCR can be performed on culture fluid.

(b) Molecular methods: Molecular methods including polymerase chain reaction (PCR—conventional, RT PCR or real time PCR), Nucleic Acid Sequence Based Assay (NASBA) are being increasingly used due to their increased reliability and availability. They also have less turnaround time thus results are available quicker. These tests are more expensive and require technical expertise.

These methods are useful not only in detection of virus but also in serotyping in the same run. There are several primers available e.g. Pan Poliovirus oligonucleotide primer which is useful primer in serotype specific detection including vaccine strain. This primer targets the VP1 coding region. Multiplex PCR that can detect all 6 strains (3 wild and 3 vaccine strains) is also available.

Samples used for these assays are the same as for culture. CSF can also be used. CSF sample is found to be more sensitive than blood samples in some cases of polio.

2. Indirect methods: These methods provide indirect evidences of infection. Individuals infected with poliovirus will produce specific antibodies against the virus. Thus detection of specific antibodies against poliovirus suggests that the individual is infected. Serological tests to detect antibodies are available though they are not widely used due to their limited sensitivity and specificity. A person will always show positive results irrespective of current or past infection. Interpretation of these tests should be done carefully.

V.   VACCINES

Both IPV(killed) and live-attenuated OPV are available for over 50 years now. Newer molecular techniques paved way for several recombinant vaccines. There was a dramatic reduction in polio as soon as the killed vaccines became available by 1955. But a few percentage of vaccinised individuals who took three or more doses of vaccine developed paralytic poliomyelitis. The introduction in 1960s of live-attenuated OPV by Albert Sabin saw wide acceptance due to simplicity

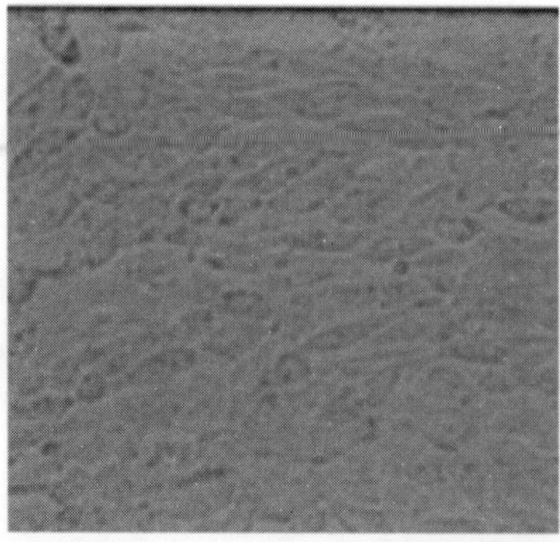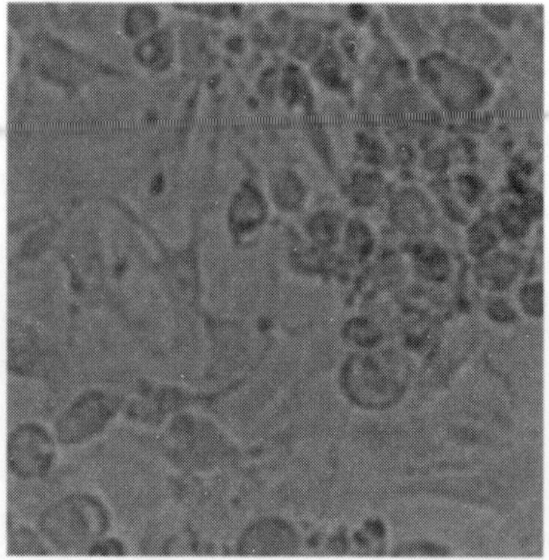

Figure 112.4   Normal Vero cell lines (right) and poliovirus infected Vero cell lines with CPE (left).

in administritaion, better immunogenicity and low cost with induction of gastrointestinal immunity. However OPV carries the risk of causing vaccine-associated paralytic poliomyelitis.

A. Inactivated Polio Vaccine (IPV)

Jonas E Salk in 1955 introduced the first inactivated polio vaccine. Though early vaccines had low potency, the burden of the disease in comparison was so high that the vaccine turned out to be an extremely important breakthrough. Inactivation was achieved by treating the virus with formalin (1:1000) at 37°C for 12–14 days. Over a period of time newer improved IPV have become available with increased potency, efficacy (100% efficacy after 2nd dose) and additional advantage of clubbing with other childhood vaccines such as DPT. This vaccine was licensed in India in 2006. IPV is administered intramuscular.

Dose: 0.5 ml/dose. 3 doses at 1–2 month interval

1st dose at 6 weeks of life booster every 3–5 years till 18 years of age.

TABLE 112.4   Advantages and disadvantages of IPV

Advantage	Disadvantage
Safe in immunocompromised people	GI manifestation can occur
Safer than OPV	Slow immunity—not suitable in epidemics
Can be used in old age as a first time vaccine	Injection should be avoided when paralysis noted
Safe in pregnancy	Repeated booster doses essential

B. Live Attenuated Vaccine (Oral Polio Vaccine)

It was developed by Albert Sabin 1961 by serially passaging the wild type virus in primary monkey kidney cell or human diploid cells. It's a trivalent vaccine with 3:1:3 ratio of type 1, 2 and 3 viruses respectively. Since it contains live viruses, vaccines have to be stored frozen (–30°C). If thawed it should be kept at 10°C and used within 30 days.

Dosage:

- 2 drops using WHO recommended dropper
- First dose at birth then at 6, 10, 14 weeks
- All pulse polio immunization days till 5 years of age.

C. Vaccine Derived Poliomyelitis

This entity also known as vaccine associated paralytic poliomyelitis was first reported in Haiti & Dominican Republic in the year of 2000. Here the vaccine strain of virus reverts back to the virulent one. This is exclusively associated with OPV specially in immunodeficient people below 18 years of age and poliovirus type 2 virus is most predominantly responsible. Rapid sequencing methods to differentiate between wild and vaccine type viruses are available.

TABLE 112.5   Advantage and disadvantages of OPV

Advantage	Disadvantage
Systemic and intestinal immunity	Cannot be used in immuno-deficient conditions
Relatively inexpensive	Monkeys needed for safety testing
Herd immunity – good for epidemics	Outbreaks – Vaccine derived paralytic poliomyelitis
Quick immunity	
Easy to administer	
Stabilized vaccine – easy handling, transporting and storage	

Eradication

As mentioned earlier in this chapter, large scale immunization started as soon as the vaccine became available. IPV was introduced in 1955 followed by OPV in 1961. World health Assembly passed a resolution in 1974 towards immunizing children against several childhood diseases under the banner of Expanded Programme on Immunization which included vaccination against polio also. This resulted in eradication of Polio in most of the European countries. In 1988 Global Polio Eradication Initiative (GPEI) with eradication of Polio by the year of 2000 as its goal was taken up by the WHO.

The goal and strategies of GPEI can be summarized in Box 112.2.

BOX 112.2   The goal and strategies of GPEI

Goals
- No case of clinical poliomyelitis associated with wild poliovirus by the year 2000.
- No wild poliovirus found worldwide despite intensive efforts.

Strategies
- High routine vaccine coverage in infants who are younger than 1 year with at least 3 doses of OPV.
- Administration of supplementary dose of OPV to all children below 5 years during National Immunization days (NIDs) to rapidly interrupt the transmission.
- Surveillance of acute flaccid paralysis and investigate if they were actually due to polio.
- Mop-up vaccination campaign in a particular polio defined area by house to house and child to child immunization.

As a result of this extensive steps and strategies polio cases have fallen drastically. In 1994 Americas (36 nations) were certified polio free and in 2000 western pacific regions including China achieved polio free status.

VI. ERADICATION IN INDIA

The disease had always been endemic in India and was grossly ignored before the 1980s. A study in 1979 showed that about

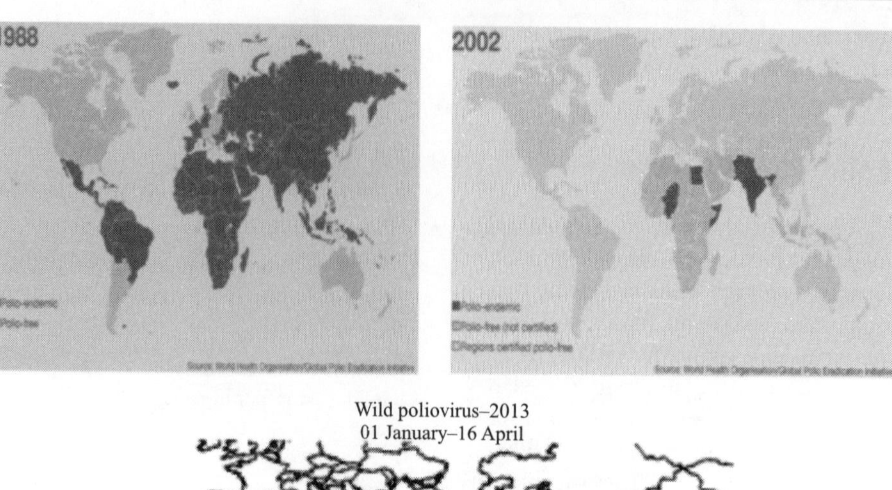

Figure 112.5   These figures show drastic decline in polio cases from 1988 when GPEI was started till the year 2002. In figure C taken from WHO, the three endemic countries namely Pakistan, Afghanistan and Nigeria are indicated in yellow.

70000 were affected annually and on the other hand official reported data showed cases which were 5–8 times low in number. In India although formal immunization with OPV started in 1994, polio eradication on a full fledged scale started only in 1995. Mass OPV administration under the banner of Pulse Polio became one of the landmarks in polio eradication programs in India. All children below the age of 5 years were actively immunized irrespective of their prior immunization history. In 1997 National Polio Survey Project (NPSP) was started to keep a close watch on acute flaccid paralysis cases and their investigation to rule out polio. The last case of Poliovirus type two was reported from the state of Uttar Pradesh in 1999 followed by type 3 in 2010. A greater mile stone was achieved when the last case of Polio in India was reported in January 2011 from the state of West Bengal which was the last type 1 virus in India.

SUGGESTIONS FOR FURTHER READING

Arie J. Zuckerman, Jangu E. Banatvala, John R. Pattison, Paul D. Griffiths and Barry D. Schoub, *Principles and Practice of Clinical Virology*, 5th ed., Chapter 14.

Gerald L. Mandell, John E. Bennett and Raphael Dolin, *Principles and Practice of Infectious,* 7th ed., Vol. 2, Chapter 171.

James Versalovic: *Manual of Clinical Microbiology*, 10th ed., Vol. 2, Chapter 85.

Knipe David M. and Howley Peter M., *Fields Virology,* 5th ed., Vol. 2, Chapter 1.

Topley and Wilson, *Microbiology and Microbial Infection,* 10th ed., Virology, Chapter 40.

Internet Readings

http://npspindia.org/

http://www.cdc.gov/

http://www.who.int/

http://www.polioeradication.org/

113

Influenza Viruses and Vaccines

Rafi Ahmed

CONTENTS

I. THE VIRUS

A. Ecology, Structure and Replication

Influenza viruses are members of the *Orthomyxoviridae* family. They are antigenically classified into three different types A, B and C (1). These three types also differ in many other aspects such as host range, pathogenicity, and transmission. Influenza A viruses are the main culprits of seasonal influenza epidemics and occasional pandemics. Whereas aquatic birds are considered the primary reservoir for influenza A viruses (2). The virus can also be isolated from a variety of other hosts including humans, pigs, horses, whales, seals and mink (2). In contrast, influenza B and C viruses have a more restricted host range. Unlike the case in humans where the virus replicates primarily in the respiratory tract, the site of viral replication in aquatic birds is in the cells lining the gastro-intestinal tract (2). It is not uncommon for influenza A viruses to cross the interspecies barrier causing significant morbidity and mortality in humans.

Influenza A viruses are enveloped, negative sense, single stranded RNA viruses with a segmented genome (1). The genome of influenza consists of eight segments in case of influenza A and B viruses, and seven segments in case of influenza C viruses.

The eight segments of influenza A viral genome code for up to 11 proteins. Three of the proteins are expressed on virus surface. First, the hemagglutinin (HA), which

is a homotrimeric glycoprotein that mediates initial virus attachment to host epithelial cells, and the subsequent virus entry (Figure 113.1) (3). HA also mediates virus fusion with the endosomal membrane, which in turn allows for the virus replication machinery to be transported into the cytosol (3). Second, the neuraminidase (NA), which is a homotetramer that cleaves budding viral particles from the host infected cells (4). The third surface glycoprotein is the matrix protein 2 (M2), produced as a splice variant of the matrix gene, and acts as an ion channel (5) that is responsible for acidifying the viral core, an essential step during viral fusion with the endosomal

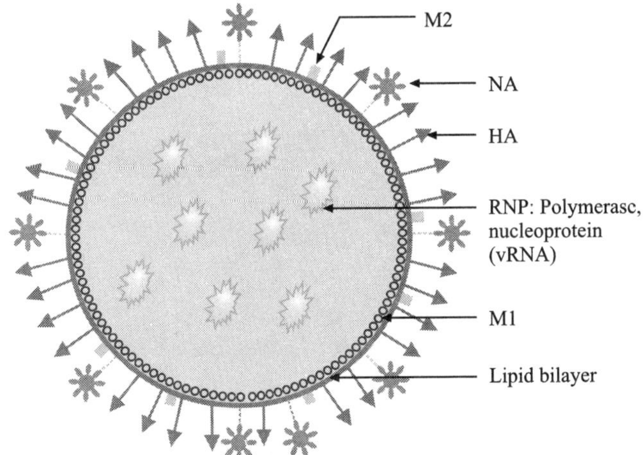

Figure 113.1 Structure of Influenza A viruses. (*see Plate 48 for colour figure*)

membrane (1). Influenza A viruses are further classified based on the antigenicity of their surface HA and NA molecules. While HA is the major target of virus neutralizing antibodies, NA and M2 are the major targets of the most commonly used anti-influenza antiviral drugs. Matrix protein 1 or M1 binds to the cytosolic domains of HA, NA and M2 forming a septum between the viral membrane and viral internal proteins (1) (Figure 113.1). Four internal viral proteins, the nucleoprotein (NP), the RNA polymerase basic 1 and 2 (PB1 and PB2) and the RNA polymerase acidic (PA), form the ribonucleoprotein (RNP) complex which represents the viral replication apparatus (1). NS1 or non-structural protein 1 inhibits host innate antiviral immune response by interfering with virus recognition through binding to double stranded RNA (6). The nuclear export protein or NEP (also known as NS2) is a splice variant of the NS gene and it is essential in exporting newly formed virus RNPs from the infected cell nucleus into cytosol (1). Finally, PB1-F2, which is produced by an alternate reading frame of the PB1 gene in some strains and recent studies have suggested that it is involved in viral pathogenesis (7).

Influenza A virus replication is initiated by binding of HA to the sialic acid receptor on the host cell surface (Figure 113.2). Viral tropism is determined by the type of the terminal sialic acid linkage to galactose on host cells. In general, human influenza viruses bind to sialic acids with α2-6 linkage while avian influenza viruses bind to sialic acid with α2-3 linkage (1). Next, the virus enters the host cell by clathrin-coated receptor mediated endocytosis (8). The acidic environment within the endosome induces a conformational change within the HA leading to its fusion with the vesicular membrane and the subsequent release of the viral RNP complex into the cytosol (Figure 113.2) (8). The proton influx mediated by the M2 ion channel induces the dissociation of the M1 from the

viral RNP, resulting in uncoating of the virus. Viral RNPs are then translocated into the host cell nucleus where transcription and replication of the viral genome occur. With the help of NEP or NS2, the progeny viral RNPs are translocated into the cytosol and then to the cell membrane where the viral surface protein are assembled. NA cleaves the budding viral particle to release it from infected cells (Figure 113.2) (8). Finally, for the budding virus particles to be infective, proteolytic cleavage of the HA by host proteases must occur (8).

B.   Seasonal Epidemics and Pandemic Potential

In the northern hemisphere, seasonal influenza viruses cause an annual epidemic that usually lasts from the month of November to late April or May with the peak number of infections occurring in February (9). The most affected age populations in terms of infection rate and severity of infection are young children (< 2 years old), and the elderly (> 65 years old) (9). This is mostly due to the general weakness of the immune system associated with these age groups. Other risk groups include individuals with chronic medical conditions and immunocompromised individuals. Annually, seasonal influenza viruses are responsible for approximately 36,000 deaths in the US alone (9).

In the last four decades, three different subtypes of seasonal influenza viruses have circulated in the human population; two influenza A viruses with H1N1 and H3N2 subtypes and influenza B viruses (10). Annual vaccination programs have been developed and are effective in reducing the incidence of seasonal influenza-associated complications (9).

The incubation period after seasonal influenza virus infection ranges from one to four days (9). It is typically characterized by mild respiratory symptoms including fever,

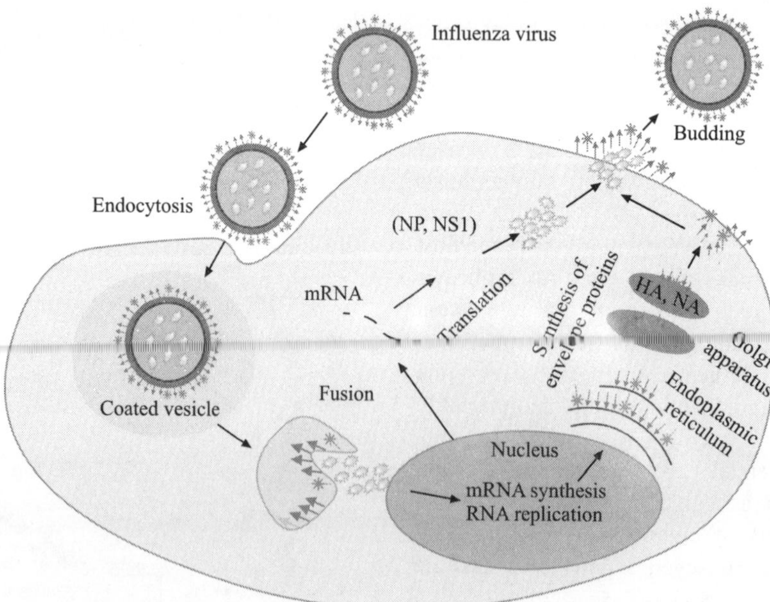

Figure 113.2   Influenza virus replication cycle. Binding and entry of the virus, fusion with endosomal membrane and release of viral RNA, replication within the nucleus, synthesis of structural and envelope proteins, budding and release of virions capable of infecting neighboring epithelial cells. (*see Plate 48 for colour figure*)

myalgia, headache, malaise, nonproductive cough, sore throat, and rhinitis (9). The majority of influenza infections resolve within a week (9). However, influenza virus infections can cause primary influenza viral pneumonia; exacerbate underlying medical conditions (e.g., pulmonary or cardiac disease); lead to secondary bacterial pneumonia, sinusitis, or otitis media; or contribute to coinfections with other viral or bacterial pathogens (9).

In order for any influenza virus to cause a pandemic, two conditions must be met; (1) a large segment of the human population has to be immunologically naïve to that virus; (2) the virus must acquire the ability to transmit efficiently among humans. The emergence of such viruses can occur by genetic reassortment of a human influenza virus with a non-human virus leading to a reassortant virus with a novel HA and internal genes that allow it to replicate efficiently in the human host (11). A pandemic influenza virus can also be directly introduced to the human population via interspecies transmission (11). These two mechanisms are not mutually exclusive and a combination of the two processes usually leads to the genesis of a pandemic virus.

During the twentieth century, three influenza pandemics occurred (12). The first one was the 1918 Spanish influenza. The causative virus was an avian H1N virus, and the number of deaths as a consequence of this pandemic is estimated to be 50 million (13, 14). The virus infection was not systemic and secondary bacterial infection was the leading cause of death (13, 14). Recently, the 1918 H1N1 virus was recreated using reverse genetics and studies on rodents and non-human primates revealed an aberrantly intense innate immune response associated with virus infection (13, 14). This fact might explain the fact that young adults were the most affected age group.

In 1957, an H2N2 influenza virus caused the second influenza pandemic that is also known as the Asian influenza (15). The virus emerged in Sothern China, and the number of deaths within the US is estimated to be 70,000 (15). The virus resulted from a genetic reassortment event with the HA, NA and PB1 genes coming from an avian influenza virus while the rest of genes being contributed by a human virus (15). Unlike the 1918 pandemic, secondary bacterial infection was not the major cause of death.

The last pandemic of the twentieth century was the Hong Kong influenza. The causative agent was an H3N2 influenza virus that emerged in 1968 and has been circulating in humans ever since (15). The pandemic caused 34,000 deaths in the US (15). It is speculated that pre-existing immunity against the N2 contributed to the decreased mortality (15).

In April of 2009, the US Centers for Disease Control and Prevention (CDC) reported two cases of a novel H1N1 influenza virus infection (16). It was retrospectively shown that these cases represented the continued spread of this virus, subsequently labeled pandemic A (H1N1) 2009 (H1N1pdm), from an ongoing outbreak in Mexico (17). The virus genome was quickly sequenced and shown to result from a novel reassortant between lineages of influenza viruses known to circulate widely in swine (18). Within a matter of weeks the virus had spread to all other continents, signaling the beginning of the first influenza pandemic in the 21st century.

C. Immune Response

Influenza A virus infection in humans is acute and the clearance of the virus is mediated by several arms of the immune system. Initially, influenza infection triggers a robust innate immune response in the form of a massive secretion of inflammatory cytokines and influx of inflammatory immune cells into the respiratory tract (6). The early clinical symptoms associated with influenza infection are caused by these innate immune responses. These early responses are essential not only for limiting virus replication, but also for priming the virus specific adaptive immune responses. On the other hand, influenza viruses have developed several evasion strategies such as encoding for the NS1 protein, which blocks the triggering of the host interferon (IFN α/β) response through sequestering viral RNA (6).

The innate immune response

Our knowledge of the contribution of the innate immune system to the overall immune response to pathogens and vaccination has expanded over the last decade (19-21). Recognition of conserved motifs that are associated with microbial infection is the initial signal for the generation of an appropriate innate and adaptive immune response (19-21). These motifs can be either derived from the invading microbes, and thus called pathogen associated molecular patterns (PAMPs) or they can be endogenous danger signals caused by the infection (danger associated molecular patterns or DAMPs). The host cell sensors that evolved to recognize these PAMPs and DAMPs are known as the pattern recognition receptors or PRRs. Infection by influenza viruses can be detected by three major families of PRRs; the toll-like receptors (TLRs), the NOD-like receptors (NLRs) and the RIG-like receptors (RLRs) (19).

The humoral immune response

The hemagglutinin (HA): As mentioned earlier, HA is a viral glycoprotein that mediates virus binding to lung epithelial cells leading to virus entry via receptor-mediated endocytosis. In addition, HA also mediates virus fusion with endosomal membrane, releasing viral genomic materials into the cytosol (22). This "invasion" role makes HA the primary target of protective antibodies generated by the host after infection or induced by influenza vaccines. Such antibodies, if induced to sufficiently high titers, can neutralize the virus, i.e. completely prevent viral binding and invasion of airway epithelial cells (Figure 113.3). There are several defined neutralizing antibody recognition sites on the HA globular head, and those antibodies can efficiently neutralize virus infectivity. However, antigenic variation at these sites thwarts the development of long-term, protective influenza vaccines (23, 24). For influenza H1 HA, five immunodominant epitopes (Sa, Sb, Ca1, Ca2, and Cb) have been characterized (Figure 113.3) (25, 26). Expectedly, these epitopes are located around the receptor-binding domain

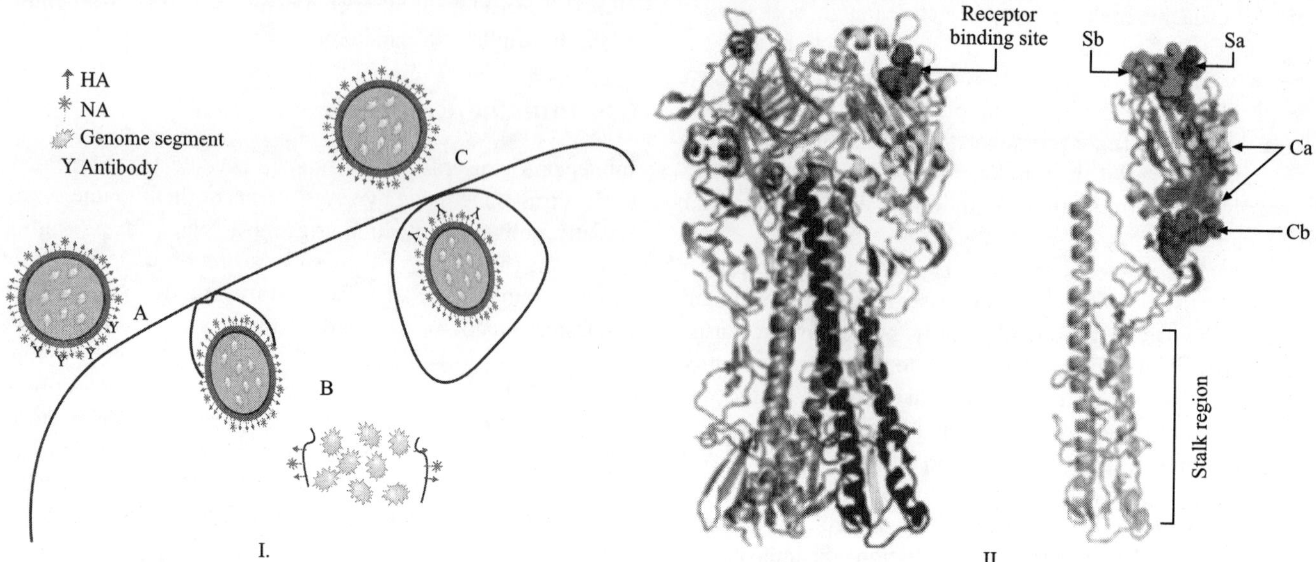

Figure 113.3   Antibody responses to influenza viruses. I. Mechanisms of influenza virus neutralization. A. Blocking of virus binding (Anti-HA head Abs) B. Blocking of virus fusion (Anti-HA stem Abs). C. Blocking of virus release (Anti-NA Abs). II. Crystal structure of influenza HA. (Left) A structural overview of the trimeric HA molecule. One monomer is colored in yellow (HA1 subunit) and blue (HA2 subunit). The residues forming the receptor-binding site colored in red. (Right) Ribbon diagram of a monomeric form of the H1 HA showing prominent HA epitopes. (*see Plate 48 for colour figure*)

within the HA globular head. While antibodies targeting such epitopes are potentially neutralizing and protective, they are strain-specific, and thus lack the much-desired broad cross-neutralizing activity to different HA subtypes.

Thus, the challenge of eliciting broadly protective anti-HA antibodies is two-fold; first the ability of such antibodies to cross-react with different HA subtypes and second, the ability to cross-react with escape variants of the original HA that the antibodies were raised against. *Drifted strains are often generated within a single influenza season as a result of the persistent immune pressure driving frequent genetic and, consequently, antigenic variation allowing the drifted virus to escape the existing immunity.* This process is known as "antigenic drift". Therefore, and despite annual flu vaccination programs, seasonal influenza infections cause a considerable amount of morbidity and mortality (9). A more serious situation occurs *when two influenza A viruses co-infect the same host* cell, producing progeny virions that acquire HA molecules that might have not been introduced widely into the human population (e.g., avian or swine HA), a phenomenon termed "antigenic shift" (11)

A third potential viral immune evasion mechanism is acquisition of an HA molecule that had once circulated in the human population but ceased to circulate for decades. The human population has thus become mostly naïve to the reintroduced HA molecule. In either of the last two scenarios, the potential of emergence of a pandemic influenza virus increases especially if the newly reassortant virus acquires the ability to transmit among humans.

It was suggested that such broad specificities are unlikely to exist in principle, or repeated infection would have selected for them. However, several humans studies have now established

the existence of antibodies that target epitopes within the highly conserved HA stem region (Figure 113.3). These antibodies showed a broad neutralization potential across many HA subtypes (27). Additionally, the conservation of this region among different HA subtypes indicates certain structural restrictions that could limit antigenic escape (27). Nevertheless, the immunogenicity and protective potential of antibodies targeting the HA stem region remain to be fully elucidated.

The neuraminidase (NA):   NA-specific antibodies are believed to interfere with the virus replication cycle, causing an overall "yield reduction" of virus infection (23, 24). *Unlike anti-HA antibodies, NA-specific antibodies do not block infection initiation, but interfere with virus budding, and thus can significantly reduce morbidity and mortality* (Figure 113.3) (23, 24). Similar to HA, NA exhibits a high degree of variability in terms of antigenicity in response to immune pressure rendering anti-NA-mediated protection rather inconsistent (23, 24). However, it was shown that antibodies elicited against human NA1 could partially protect against H5N1 infection (28).

The Matrix protein 2 ectodomain (M2e):   A third viral transmembrane protein antibody target is the matrix protein 2 (M2). Early studies in mice showed that anti-M2 antibodies provided partial protection (24). These results fuelled speculation that M2 might be the long sought target of a universal influenza vaccine, since, in comparison to HA and NA molecules, M2 proteins are highly conserved among different influenza subtypes (24). However, disappointingly, the anti-M2- mediated protection shown in mice has not been as dramatic in other, more relevant animal models (29).

Cytotoxic T-lymphocytes (CTLs)

CTLs are essential for effective clearance of influenza virus infection (30). There are two key mechanisms via which CTLs help in controlling influenza virus infection. The first one is by direct killing of virus-infected cells. The second mechanism is through the production of various proinflammatory cytokines. In addition, the generation of virus specific memory CTLs significantly enhances virus clearance upon reexposure (30). The majority of CTL epitopes are derived from internal virus proteins such as NP and PB1 (30). These proteins are relatively conserved in comparison to HA, the main target of B-cell responses. Therefore, CTLs are important for generating a cross protective anti-influenza immune responses (31).

Immune correlates of protection

The immunogenicity of influenza virus vaccines is typically tested by the hemagglutination inhibition (HI) assay. In general, the sensitivity and specificity of this assay is poor. Virus microneutralization (MN) assay is another assay that quantitatively measures the levels of neutralizing anti-HA antibodies and it is believed to be more sensitive and more functional than the HI assay (32). HI titers of 40 or higher are generally considered protective assuming that the higher the HI or MN titers the higher the probability of protection (32). The enzyme-linked immunosorbent assay or ELISA has also been used to determine the levels of anti-HA antibodies (32). In ELISA, however, all (neutralizing and non-neutralizing) antibodies that bind to HA are measured. Although ELISA is rapid and more sensitive than HI and MN, there is no information regarding the correlation between antibody levels and protection. Moreover, the use of whole virus preparations instead of purified HA as a coating antigens adds antibodies that target the internal protein such as anti-nucleoprotein antibodies to the equation. The clinical implications of non-neutralizing antibodies are not fully understood (32). Although these antibodies are more likely to be cross-reactive with different HA subtypes, the mechanism(s) through which they may be contributing in viral clearance is not well studied.

II. THE VACCINE

A. Historical Perspective

The earliest trials to develop an approved influenza vaccine dates back to the 1940s. It was developed by the US military to be used in the Second World War, and it was an inactivated whole virus vaccine. These vaccines were prepared by high-speed centrifugation of virus preparations propagated in embryonated chicken eggs or mouse lungs. This form of influenza vaccines was predominantly used till the 1960s, when the development of continuous-flow zonal ultra centrifugation allowed the separation of virus-rich fraction out of the allantoic fluid of embryonated chicken eggs. In 1968, "split" influenza vaccines, in which the major surface glycoproteins (HA & NA) are enriched, were licensed. The same manufacturing process is still being used to produce the current inactivated influenza vaccines.

B. Annual Production Cycle

Due to the rapid evolution of influenza viruses, the antigenic composition of influenza vaccines is reviewed annually. Influenza vaccines are produced from viruses grown in hens' eggs. The vaccine production process starts with the selection of two influenza A strains (H1N1 and H3N2) and one or two influenza B strain to be included in the vaccine. The selection process starts by collection and characterization of circulating influenza viruses by national influenza surveillance centers and WHO collaboration centers around the globe. After studying the dominant strains and associated mutations, a panel of WHO-affiliated influenza experts decides on which strain is to be used as the vaccine seed strain. This decision is made each year in February for the following northern hemisphere winter season, and in September for the following southern hemisphere winter season. Next, the vaccine seed strains are incubated 2–3 days in embryonated hens' eggs. Next, allantoic fluid is harvested and virus particles are inactivated and disrupted and the HA and NA are then purified. The process from strain selection to the final vaccine administration may take up to 6-8 months humans.

C. Types of Influenza Vaccines

Inactivated virus vaccines

Most of the currently licensed influenza vaccines are in the form of inactivated antigen preparations. As stated above, *all vaccines work primarily through the generation of antibodies to HA*. As such, they are potent, but prone to antigenic changes such as would be expected with the emergence of a new strain in a population. Clinical studies have shown that inactivated vaccines have an efficacy, measured by reducing serologically confirmed influenza illnesses, of 70% in the age group between 14-60 years (33). The efficacy is reduced with both infants and the elderly (34). The first of the inactivated vaccine formulation is whole virion, the experimental use of which dates back to 1940s (Figure 113.5). The second is split virion, which is derived by disrupting whole virus particles with disinfectants and finally, the subunit form, which is prepared by enriching for the viral surface glycoproteins HA and neuraminidase (NA) following disruption of viral particles. Although arguably more immunogenic, the reactogenicity associated with inactivated whole virus vaccine preparations, particularly in infants and children drove the development of the split vaccine technology back in the 1960s (35). Split and subunit vaccines have subsequently proven to be safe and have been delivered to millions of people (Figure 113.5) (36). Unfortunately the relatively poor immunogenicity of split vaccines means that at least two doses of vaccine must be provided to generate protective immune responses in naïve

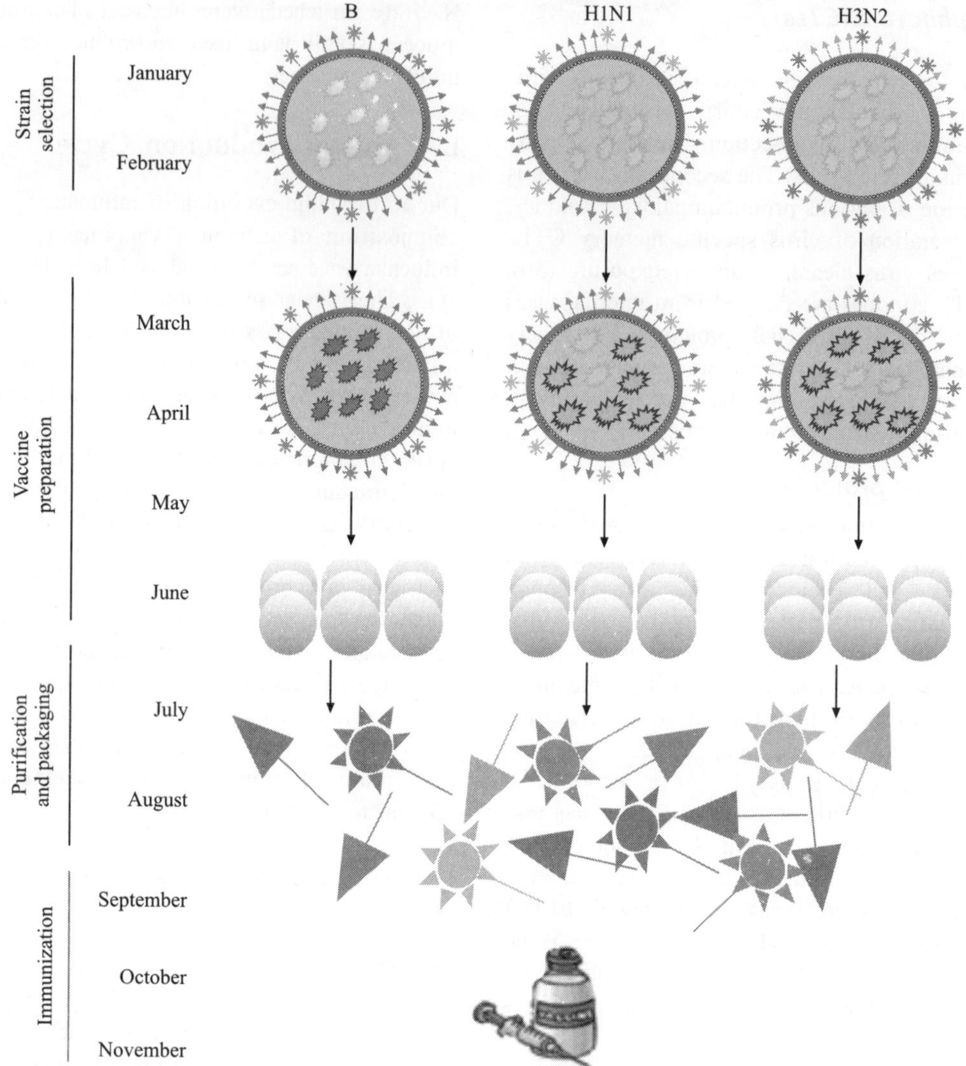

Figure 113.4   The annual cycle of influenza vaccine production. (*see Plate 49 for colour figure*)

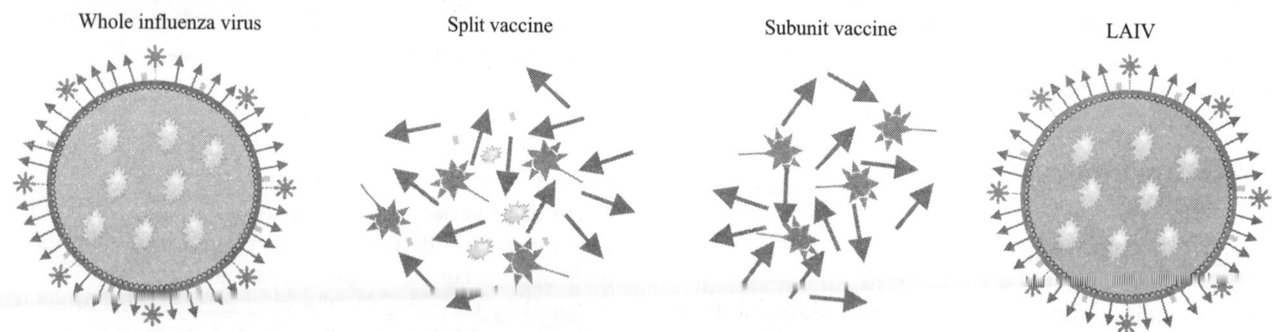

Figure 113.5   Types of influenza vaccines. (*see Plate 49 for colour figure*)

individuals (37). In the event of a release of a novel and virulent strain of influenza, a two-dose schedule for vaccination poses logistical and temporal issues.

Live attenuated virus vaccines

Unlike inactivated vaccines, live attenuated influenza virus vaccines (LAIVs) are administered by intranasal inoculation of replication competent virus. In these vaccine viruses, the HA and NA of the target strain is introduced into a backbone of an attenuated, cold-adapted virus (38, 39). Cold-adapted influenza viruses are attenuated by adaptation to grow at 25° C (38, 39). The first adapted influenza A virus (commonly referred to as the master strain) is A/Ann Arbor/6/60-H2N2 (38, 39). The resulting vaccine virus has the antigenic phenotype of the target strain but the attenuated phenotype of the master strain.

The perceived advantages of the LAIV approach is that both a local neutralizing antibody and a cell-mediated response can theoretically be generated. Cell mediated immunity is an attractive goal for influenza vaccines due to the fact that such immunity targets the more conserved viral proteins (31). In an ideal scenario, such immunity would protect across a range of different virus strains. In practice, in the few side-by-side comparisons of LAIV and inactivated vaccines in a seasonal influenza setting, the theoretical advantage of a broadened immune response is only seen in the younger cohorts (40). This is hypothesized to be due to the over attenuation of the viruses in an immunologically primed population. Within the context of a release of a novel strain, however, the population will essentially be naïve and LAIV could have an important role to play in terms of corresponding vaccine strategies.

Despite some advantages, LAIV do suffer from some of the same problems, as do their inactivated counterparts. LAIV will also likely require two doses to elicit optimal immune responses, and their generation time is not substantially different from inactivated vaccines. LAIV also have some unique drawbacks in that it is possible not all HA and NA combinations will form viable viruses on the attenuated backbone, the vaccine virus must be able to infect the human upper respiratory tract (potentially an issue for avian strains), and potential safety risks of administering a live virus into a population before the target strain is widespread.

D. Adjuvants

Why adjuvants are needed for influenza vaccines?

The poor immunogenicity of split or subunit influenza vaccines is a major challenge for the rapid response to an emerging virus. The impact is two-fold, first higher quantities of antigen are needed per dose to elicit a protective immune response. Second, two doses will likely be needed for vaccine recipients to attain these protective responses. The use of adjuvants to overcome poor immunogenicity and for antigen sparing is not new to vaccinology, indeed, adjuvants such as alum salts for several human vaccines started several decades ago (41, 42). In addition to their ability to enhance vaccine immunogenicity, adjuvants such as the oil-in-water emulsion adjuvant MF59 have been shown to expand the breadth of an immune response to influenza vaccines (43, 44). MF59 induced the generation of higher serum antibody titers as well as more cross reactive responses when administered with split or subunit H5N1 vaccines (43, 44). Although these vaccines are unlikely to protect across subtypes, they do appear to be more cross-reactive across the variations within a subtype.

General mechanisms of action of adjuvants

Antigen presenting cells (APCs) such as dendritic cells (DCs) and macrophages are the main target immune cell population for most vaccine adjuvants (45). Once activated, APCs play a vital role in priming of antigen specific CD4 T-cells, the maestro of the adaptive immune response (45). Activation of APCs typically leads to upregulation of surface costimulatory molecules such as CD40 and CD80/86 mediate their interaction with CD4 T-cells (45). Naïve B-cells have the capacity to present antigens to CD4 T-cells as well. There are two broad categories of vaccine adjuvants based on their perceived mechanism of action. In both categories, proper activation of APCs and the subsequent adaptive immune response is the target. These two categories are delivery systems and immune potentiators (46).

The most common adjuvants used as delivery systems are mineral salts and emulsions. As their name implies they are designed to act as a carrier for the vaccine antigen. The precise way through which they boost the immune response is not entirely clear. Forming a depot of the vaccine antigen at the injection site and delaying antigen clearance are the leading hypotheses. In addition, the impartation of a particulate nature to vaccine antigens by these adjuvants is believed to increase antigen uptake by APCs. Regardless of the precise mechanism, improved antigen presentation by APCs seems to be the general outcome of using adjuvants as vaccine vehicles.

Immune potentiators are usually ligands for the pattern recognition receptors (PRRs) we previously discussed (see the innate immune section). The most characterized as vaccine adjuvants are the ligands for TLRs (46). Upon binding of such adjuvants to their specific PRRs, they initiate a signaling cascade that result in secretion of various proinflammatory cytokines and chemokines. These proinflammatory "messages" do not only activate APCs but also they dictate the type of

TABLE 113.1   Adjuvants that were shown to significantly boost influenza vaccines in humans (adapted from (46))

Adjuvant	*Type*	*Mechanism of action*	*Licensed*
Aluminum salts or alum	Mineral salts	Depot Activation of the NLRP3 inflammasome	Yes
MF59 and AS03	Oil-in-Water emulsion	Unknown Tissue inflammation	Yes
Cholera toxin (CT) and *E.coli* heat labile enterotoxin	Microbial derived	PRRs-independent immune potentiation	No
CpG	Microbial derived	Stimulation of TLR9	No
Muramyl tripeptide	Microbial derived	Stimulation of TLR2	No
IL-2	Cytokine	Direct stimulation of the immune response	No

the adaptive immune response to be activated as well (45). For example signaling through most TLRs leads to secretion of IFN-gamma and IL-12 which in turn favors the activation of specific subset of CD4 T-cells, the Th1 cells (45). The latter cells have been shown to provide the essential cytokines needed for proliferation of cytotoxic T-lymphocytes (CTLs) (45). Another subset of immune potentiators would be the direct administration of cytokines such as IL-2 in combination with the vaccine antigens (46).

Mineral salts

Mineral salts are the most commonly used vaccine adjuvants in humans (47). Insoluble aluminum salts or alum are the only adjuvant that is approved for use in human influenza vaccines in the US. Chemically alum composition is variable with aluminum oxyhydroxide and aluminum hydroxyphosphate being the most commonly used derivatives (48). The use of calcium derivatives has also been reported (48).

There are several mechanisms that have been proposed to explain alum adjuvanticity. Firstly, as a classic delivery system, alum is believed to form a depot of the vaccine antigen at the site of injection and to improve antigen uptake by APCs (45). Recently, several reports have shown that alum is capable of inducing a local inflammation at the injection site and this inflammation is correlated with its immune enhancing effects (45). These reports have revealed that alum mediates NLRP3 inflammasome activation and the consequential release of the proinflammatory cytokines IL-1 beta and IL-18 (49). Whether the deletion of this pathway significantly impacts the adjuvant effects of alum is still controversial (50-52).

Alum has been used with various forms of influenza vaccines for decades (46). When used with subunit H5N1 influenza vaccines, the results were inconsistent. While Bresson et al observed a greater response to the adjuvanted vaccine formulation; this was only at the highest dose (30 μg) of the vaccine (37). Keitel et al. observed more frequent seroconversion responses with the adjuvanted vaccine at the 7.5 μg dose group with no significant differences among other groups (53). Additional studies found no evidence for a significant adjuvant effect of alum (54). Overall evaluation of alum-adjuvanted subunit H5N1 influenza vaccines in humans indicates that alum effect is modest. This is equally true for whole virus influenza vaccines as well (55).

Emulsions

Emulsions are the second most commonly used adjuvants in human vaccines. Emulsions are prepared by mixing two immiscible liquid phases (aqueous and oily phases) in the presence of surfactants. There are two types of emulsions based on the nature of the dispersed (or inner) vs. the dispersing (or outer) liquid phase; water-in-oil (w/o) and oil-in-water (o/w) emulsions. Despite being efficient adjuvants, w/o emulsion adjuvants such as complete Freund's adjuvant (CFA) are no longer used in human vaccines due to their high reactogenicity

(47). On the other hand, o/w emulsions such as MF59 and AS03 have been shown to be effective and safer adjuvants (46).

The mechanism of action of emulsion adjuvants is unknown. Initially it was thought that similar to other vaccine vehicles, emulsions form a depot of the vaccine antigen allowing its slow release from the injection site. However, studies with labeled gD2 from type 2 herpes simplex virus (HSV) revealed similar release kinetics between the unadjuvanted and the MF59 adjuvanted antigen (56). Similarly to alum, it is proposed that MF59 improve antigen uptake by APCs (57). More recently studies with MF59 adjuvanted vaccines have shown that MF59 stimulates a potent localized inflammatory response at the site of injection (58). Unlike alum the innate immune pathway(s) involved in emulsion-based adjuvants induced inflammation is not clear.

III. CONCLUSIONS

One of the most striking features of influenza virus infection is the speed with which the virus replicates once infection is established in the respiratory epithelium. Preformed HA-specific antibodies in the serum or, preferably, on the airway mucosal surface can block viral entry and the subsequent establishment of infection (24). A single infection with any strain of influenza viruses elicits a lifelong antibody-mediated protection against homologous challenge. This immunity is primarily mediated by the antibodies generated against the HA glycoprotein. While influenza vaccination establishes immune memory against the particular influenza strains included in the vaccine, it does not guarantee protection against infection with antigenically different strains or even slightly drifted variants of the same strain (23), which necessitates the annual revision of the influenza strains included in the vaccine. Therefore, an ideal influenza vaccine would have the following characteristics compared to current influenza vaccines:

1. *Broader protection:* This can be achieved by designing new HA immunogens that can refocus the immune responses towards conserved epitopes such as those within the HA stalk and the receptor-binding site.
2. *Increased immunogenicity:* This can be achieved by adding adjuvants to the vaccine formulation. Conventional adjuvants such as alum and MF59 are already being used with some influenza vaccines.
3. *Longer-lasting immunity:* This can be achieved by developing new adjuvants that can selectively prolong the immune response to influenza vaccines.
4. *Increased yield:* This can be achieved by either improving the backbone of the vaccine seed strains.
5. *Faster production time:* This can be achieved by adopting other ways of vaccine production such as cell-culture based vaccines.

REFERENCES

1. Cheung T.K. and Poon L.L. (2007), Biology of Influenza a Virus, *Ann N Y Acad Sci.,*1102, 1.

2. Webster R.G., Bean W.J., Gorman O.T., Chambers T.M. and Kawaoka Y. (1992), Evolution and Ecology of Influenza A Viruses, *Microbiol Rev.* 56, 152.

3. Skehel J.J. and Wiley D.C. (2000), Receptor Binding and Membrane Fusion in Virus Entry: The Influenza Hemagglutinin, *Annu Rev Biochem.,* 69, 531.

4. Gamblin S.J. and Skehel J.J. (2010), Influenza Hemagglutinin and Neuraminidase Membrane Glycoproteins, *J. Biol Chem.,* 285, 28403.

5. Pinto L.H., Holsinger L.J. and Lamb R.A. (1992), Influenza Virus M2 Protein has Ion Channel Activity, *Cell,* 69, 517.

6. Fernandez-Sesma A. (2007), The Influenza Virus NS1 Protein: Inhibitor of Innate and Adaptive Immunity, *Infect Disord Drug Targets,* 7, 336.

7. Chen W., et al. (2001), A Novel Influenza A Virus Mitochondrial Protein That Induces Cell Death, *Nat Med.,* 7, 1306.

8. Mikulasova A., Vareckova E. and Fodor E. (2000), Transcription and Replication of the Influenza a Virus Genome, *Acta Virol.* 44, 273.

9. Fiore A.E., et al. (2008), Prevention and Control of Influenza: Recommendations of the Advisory Committee on Immunization Practices (ACIP), 2008, *MMWR Recomm Rep.,* 57.

10. Hay A.J., Gregory V., Douglas A.R. and Lin Y.P. (2001), The Evolution of Human Influenza Viruses, *Philos Trans R Soc Lond B Biol Sci.,* 356, 1861.

11. Webby R.J. and Webster R.G. (2003), Are We Ready for Pandemic Influenza? *Science,* 302, 1519.

12. Neumann G., Noda T. and Kawaoka Y. (2009), Emergence and Pandemic Potential of Swine-origin H1N1 Influenza Virus, *Nature,* 459, 931.

13. Tumpey T.M. and Belser J.A. (2009), Resurrected Pandemic Influenza Viruses, *Annu Rev Microbiol.,* 63, 79.

14. Taubenberger J.K., Reid T.A., Janczewski T.A. and Fanning T.G. (2001), Integrating Historical, Clinical and Molecular Genetic Data in Order to Explain the Origin and Virulence of the 1918 Spanish Influenza Virus, *Philos Trans R Soc Lond B Biol Sci.,* 356, 1829.

15. Kilbourne E.D. (1997), Perspectives on Pandemics: A Research Agenda, *J. Infect Dis.,* 176 Suppl 1, S29.

16. Ellebedy A.H. and Webby R.J. (2009), Influenza Vaccines, *Vaccine,* 27 Suppl 4, D65.

17. Dawood F.S., et al. (2009), Emergence of A Novel Swine-origin Influenza A (H1N1), Virus in Humans, *N Engl J. Med.,* 360, 2605.

18. Smith G.J., et al. (2009), Origins and Evolutionary Genomics of the 2009 Swine-origin H1N1 Influenza A Epidemic, *Nature,* 459, 1122.

19. Ichinohe T., Iwasaki A. and Hasegawa H. (2008), Innate Sensors of Influenza Virus: Clues to Developing Better Intranasal Vaccines, *Expert Rev Vaccines.,* 7, 1435.

20. Hargreaves D.C. and Medzhitov R. (2005), Innate Sensors of Microbial Infection, *J. Clin Immunol.,* 25, 503.

21. Kanneganti T.D. (2010), Central Roles of NLRs and Inflammasomes in Viral Infection, *Nat Rev Immunol.,* 10, 688.

22. Dutch R.E., Jardetzky T.S. and Lamb R.A. (2000), Virus Membrane Fusion Proteins: Biological Machines that Undergo a Metamorphosis, *Biosci Rep.,* 20, 597.

23. Ellebedy A.H., et al. (2010), Impact of Prior Seasonal Influenza Vaccination and Infection on Pandemic A (H1N1), Influenza Virus Replication in Ferrets, *Vaccine.*

24. Gerhard W. (2001), The Role of the Antibody Response in Influenza Virus Infection, *Curr Top Microbiol Immunol.,* 260, 171.

25. Caton A.J., Brownlee G.G., Yewdell J.W. and Gerhard W. (1982), The Antigenic Structure of the Influenza Virus A/PR/8/34 Hemagglutinin (H1 Subtype), *Cell,* 31, 417.

26. Wilson I.A. and Cox N.J. (1990), Structural Basis of Immune Recognition of Influenza Virus Hemagglutinin, *Annual Review of Immunology,* 8, 737.

27. Wei C.J., et al. (2010), Induction of Broadly Neutralizing H1N1 Influenza Antibodies by Vaccination, *Science,* 329, 1060.

28. Sandbulte M.R., et al. (2007), Cross-reactive Neuraminidase Antibodies Afford Partial Protection Against H5N1 in Mice and are Present in Unexposed Humans, *PLoS Med.,* 4, E59.

29. Fan J., et al. (2004), Preclinical Study of Influenza Virus A M2 Peptide Conjugate Vaccines in Mice, Ferrets, and Rhesus Monkeys, *Vaccine,* 22, 2993.

30. Valkenburg S.A., et al. (2008), Immunity to Seasonal and Pandemic Influenza A Viruses, *Microbes Infect,.*

31. Stambas J., et al. (2008), Killer T Cells in Influenza, *Pharmacol Ther.,* 120, 186.

32. Prabakaran M., et al. (2009), Development of Epitope-blocking ELISA for Universal Detection of Antibodies to Human H5N1 Influenza Viruses, *PLoS One* 4, E4566.

33. Jefferson T., et al. (2010), Vaccines for Preventing Influenza in Healthy Adults, *Cochrane Database Syst Rev.,* CD001269.

34. Jefferson T., et al. (2010), Vaccines for Preventing Influenza in the Elderly, *Cochrane Database Syst Rev.,* CD004876.

35. Mostow S.R., Schoenbaum S.C., Dowdle W.R., Coleman M.T. and Kaye H.S. (1969), Studies with Inactivated Influenza Vaccines Purified by Zonal Centrifugation, 1. Adverse Reactions and Serological Responses, *Bull World Health Organ.,* 41, 525.

36. Peck F.B. Jr. (1968), Purified Influenza Virus Vaccine, A Study of Viral Reactivity and Antigenicity, *JAMA* 206, 2277.

37. Bresson J.L., et al. (2006), Safety and Immunogenicity of an Inactivated Split-virion Influenza A/Vietnam/1194/2004 (H5N1), Vaccine: Phase I Randomised Trial, *Lancet,* 367, 1657.

38. Wareing M.D. and Tannock G.A. (2001), Live Attenuated Vaccines Against Influenza: An Historical Review, *Vaccine* 19, 3320.

39. Maassab H.F., Heilman C.A. and Herlocher M.L. (1990), Cold-adapted Influenza Viruses for Use as Live Vaccines for Man, *Adv Biotechnol Processes,* 14, 203.

40. Monto A.S., et al. (2009), Comparative Efficacy of Inactivated and Live Attenuated Influenza Vaccines, *N Engl J. Med.,* 361, 1260.

41. Aimanianda V., Haensler J., Lacroix-Desmazes S., Kaveri S.V. and Bayry J. (2009), Novel Cellular and Molecular Mechanisms of Induction of Immune Responses by Aluminum Adjuvants, *Trends Pharmacol Sci.,* 30, 287.

42. Bodewes R., Rimmelzwaan G.F. and Osterhaus A.D. (2010), Animal Models for the Preclinical Evaluation of Candidate Influenza Vaccines, *Expert Rev Vaccines,* 9, 59.

43. Dormitzer P.R., et al. (2011), Influenza Vaccine Immunology, *Immunol Rev.,* 239, 167.

44. Khurana S., et al. (2010), Vaccines with MF59 Adjuvant Expand the Antibody Repertoire to Target Protective Sites of Pandemic Avian H5N1 Influenza Virus, *Sci Transl Med.,* 2, 15ra5.

45. McKee A.S., MacLeod M.K., Kappler J.W. and Marrack P. (2010), Immune Mechanisms of Protection: Can Adjuvants Rise to the Challenge? *BMC Biol.,* 8, 37.

46. Atmar R.L. and Keitel W.A. (2009), Adjuvants for Pandemic Influenza Vaccines, *Curr Top Microbiol Immunol.,* 333, 323.

47. Vogel F.R. (2000), Improving Vaccine Performance with Adjuvants, *Clin Infect Dis.,* 30 Suppl 3, S266.

48. Hem S.L. and Hogenesch H. (2007), Relationship between Physical and Chemical Properties of Aluminum-containing Adjuvants and Immunopotentiation, *Expert Rev Vaccines,* 6, 685.

49. Eisenbarth S.C., Colegio O.R., Connor W.O., Sutterwala F.S. and Flavell R.A. (2008), Crucial Role for the Nalp3 Inflammasome in the Immunostimulatory Properties of Aluminium Adjuvants, *Nature,* 453, 1122.

50. Spreafico R., Ricciardi-Castagnoli P. and Mortellaro A. (2010), The Controversial Relationship between NLRP3, Alum, Danger Signals and the Next-generation Adjuvants, *Eur J. Immunol.,* 40, 638.

51. McKee A.S., et al. (2009), Alum Induces Innate Immune Responses through Macrophage and Mast Cell Sensors, but these Sensors are not Required for Alum to Act as an Adjuvant for Specific Immunity, *J Immunol.,* 183, 4403.

52. Franchi L., Nunez G. (2008), The Nlrp3 Inflammasome is Critical for Aluminium Hydroxide-mediated IL-1beta Secretion but Dispensable for Adjuvant Activity, *Eur J. Immunol.,* 38, 2085.

53. Keitel W.A., et al. (2008), Safety and Immunogenicity of an Inactivated Influenza A/H5N1 Vaccine Given with or without Aluminum Hydroxide to Healthy Adults: Results of a Phase I-II Randomized Clinical Trial., *J. Infect Dis.,* 198, 1309.

54. Bernstein D.I., et al. (2008), Effects of Adjuvants on the Safety and Immunogenicity of an Avian Influenza H5N1 Vaccine in Adults, *J. Infect Dis.,* 197, 667.

55. Ehrlich H.J., et al. (2008), A Clinical Trial of a Whole-virus H5N1 Vaccine Derived from Cell Culture, *N Engl J. Med.,* 358, 2573.

56. Dupuis M., McDonald D.M. and Ott G. (1999), Distribution of Adjuvant MF59 and Antigen GD2 after Intramuscular Injection in Mice. *Vaccine,* 18, 434.

57. Dupuis M., et al. (1998), Dendritic Cells Internalize Vaccine Adjuvant after Intramuscular Injection, *Cell Immunol* 186, 18.

58. Mosca F., et al. (2008), Molecular and Cellular Signatures of Human Vaccine Adjuvants, *Proc Natl Acad Sci.,* USA, 105, 10501.

114

DNA Vaccines

Syamlal Roy, Shantanabha Das and Rajatava Basu

CONTENTS

I. VACCINE: AN INTRODUCTION

Although the demonstration in 1776 by Edward Jenner that vaccinia virus could protect against small pox was epochal, he was following the path opened by the ancients who had used the small poxvirus itself in the practice of variolation. The work of Loius Pasteur on chicken cholera opened the way to vaccine development in laboratory. In April 1880, Pasteur reported to the French Academy of Science that "…. Chicken cholera is produced by a microscopic parasite (now known as *Pasteurella multocida*), that there exists an attenuated virus (Pasteur was using "virus" in the ancient sense of the word) of that disease, and that one or more inoculations of these attenuated virus can preserve the animals from the mortal effects of a latter inoculation…let me be permitted to use the word 'vaccinate' to express the act of inoculating a chicken with the attenuated virus". Thus the word 'vaccinate' was extended beyond vaccinia and came to have its modern meaning. Today epidemic infectious diseases of children for which there are vaccines have virtually disappeared from industrialized countries such as the United States. These are remarkable successes, which together with clean water and antibiotics have profoundly affected human society. In addition, a new field of microbiology and immunology has evolved, called "vaccinology" that comprises not only the vaccine development but also use of vaccines and their effect on public health.

II. THE DIFFERENT TYPES OF VACCINES

The first generation vaccines were made up of whole organisms, either killed or live attenuated. *The live attenuated organisms had the advantage of inducing both humoral and cellular immunity*. The killed parasites when used for vaccination seldom induced adequate cellular immune response. But the live attenuated vaccines had the problem of chance to revert back to virulent form inside the host and often had adverse side effects in immune compromised individuals. To overcome the problems of whole organism vaccines, the second generation vaccines adopted the used of isolated antigens in protein or peptide form.

They can be of specific antigens that gives high immune stimulation, inactivated toxins to induce immune response or protein subunits vaccines either alone or combined with carrier proteins. Although the second generation vaccines were highly immunogenic, giving rise to high antibody titers, they were very poor inducer of cellular immunity. This problem was addressed with the advent of third generation vaccines, which are mainly DNA encoding the antigen either administered in naked DNA form or through a viral delivery system.

III. DISCOVERY OF DNA VACCINES

In the 1980s, the observation that direct *in vitro* and *in vivo* gene transfer of recombinant DNA by a variety of techniques (e.g., complexed with liposomes, viral gene transfer etc.) resulted in expression of protein paved the way for DNA vaccine research. These approaches included using formulations of DNA with liposomes or proteoliposomes, polylysine-glycoprotein carrier complex.

In 1990 Wolff et al., published their seminal study (*Science* 247:1465–68, 1990) of "plasmid or naked" DNA vaccination in vivo, it was shown that *direct intramuscular inoculation of plasmid DNA encoding several different reporter genes could induce protein expression within the muscle cells*. This study provided a strong basis for the notion that purified/recombinant nucleic acids ("naked DNA") can be delivered in vivo and can direct protein expression.

These observations were further extended in a study by Tang et al. (*Nature* 356:152–54, 1992), who demonstrated that mice injected with plasmid DNA encoding hGH could elicit antigen-specific antibody responses. Subsequently, demonstrations by Ulmer et al., (*Science* 259:1745–49, 1993) and Robinson et al. (*Vaccine* 11:957–60, 1993) that DNA vaccines could protect mice or chickens, respectively, from influenza infection provided a remarkable example of how DNA vaccination could mediate protective immunity.

IV. AN INTRODUCTION TO DNA VACCINES

Immunization with naked DNA is a new approach, which promises to revolutionize the prevention and treatment of infectious diseases. The gene encoding the vaccine candidate is cloned in a mammalian expression vector and the DNA is injected directly into muscle or skin. Surprisingly, the DNA containing the expression cassette (e.g., an expression plasmid containing the desired gene) is taken up by cells and translocated to the nucleus, where it is transcribed into RNA and then translated in the cytoplasm. The efficiency of uptake and the expression of DNA must be extremely low, but there is abundant evidence that it is sufficient to provoke immune responses in both T and B cells. There are convincing data that antigen presentation is not done directly via skin or muscle cells but, rather, is done via either class I or class II MHC molecules on professional antigen-presenting cells. A

large body of literature indicates that both CD4+ and CD8+ mediated responses are induced, making a DNA vaccine attractive for a vaccine development against intracellular pathogens like *Leishmania*. In addition to being able to induce the appropriate immune responses, *DNA vaccines are attractive because they ensure appropriate folding of the polypeptide, produce the antigen over long periods, and do not require adjuvants*. Another advantage is that the technology for production is very simple. *DNA is stable, has a long shelf life, and does not require a strict cold chain for distribution*. Concerns rose in relation to safety, such as integration of the DNA into the mammalian genome and induction of autoimmune disease or cancer, have not been substantiated to date. Several DNA vaccines are in advanced clinical trials; these include a malaria vaccine, a mycobacterial vaccine, and an HIV vaccine.

Remarkably, in the early 1990s it was discovered that bacterial DNA plasmids containing foreign genes can induce antigen production in muscle cells; this antigen is carried by antigen presenting cells to the bone marrow, where immune responses are generated. However, the passage from mice to humans has been difficult: DNA vaccine has given only transient and low level antibody responses in primates and humans. Nevertheless such vaccine has the potential to be protective. Despite low antibody induction in humans, DNA vaccination is of interest for several reasons: (a) it often produces good cellular immune responses; (b) DNA coding for vaccine antigens appears to induce excellent immunologic memory; (c) priming by DNA might conceivable be used to orient the immune system in a Th1 direction to induce allergic disease, particularly if the gene for cytokine is incorporated.

The efficacy of DNA immunization may be enhanced by the administration through the skin or by adjuvantation. Codon optimization for expression in mammalian cells is also critical. Another powerful strategy to induce cellular responses uses vectors, microbes that are naturally or artificially infective for humans, in which foreign genetic information has been inserted. Experimental vectors include a wide range of viruses and bacteria, but the ones most explored are pox, alpha, flavi, adeno viruses and BCG, *Shigella* and *Salmonella* species. The enteric bacterial vectors are given orally and depend for their action on the injection into intestinal cells of DNA plasmids carrying foreign genes. Most currently available vaccines induce antibody responses capable of mediating long-term protection. By contrast, vaccines for diseases requiring cell-mediated immunity (e.g., HIV, malaria, leishmaniasis and tuberculosis) are either not available or not uniformly effective. Thus, understanding the mechanisms by which long-lived cellular immune responses are generated following vaccination will have importance in the rational design of vaccines for the aforementioned diseases. Over the past 5–10 years, DNA vaccination has been shown to stimulate long-lived humoral and cellular immune responses *in vivo* in a variety of animal models. More recently, there has been a substantial effort in understanding the cellular mechanisms by which DNA vaccines elicit T cell responses.

V. KEY ATTRIBUTES OF DNA VACCINES

- Most importantly DNA vaccines induce all three arms of adaptive immunity [antibodies, helper T cells, cytotoxic T lymphocytes (CTLs)], and even innate immune responses.
- DNA vaccines can be manipulated easily and hence are rapid to construct, a critical attribute for making vaccines against an emerging pandemic threat.
- Use of conserved, highly immunogenic proteins only, while excluding other undesirable proteins (e.g., hypermutable proteins, proteins that serve as decoys for effective immune response or proteins that mask critical epitopes).
- It can facilitate the modification of immune response by addition of cytokine/chemokine, etc in the coding sequence.
- They are also relatively easy to manufacture, and the process is generic, in contrast to the complicated processes needed for vaccines such as attenuated viruses.
- *Stability of DNA vaccine can potentially eliminate the need of cold chain from vaccine distribution.*

VI. BASIC REQUIREMENTS FOR A PLASMID DNA VACCINE VECTOR

The standard DNA vaccine consists of the specific gene(s) of interest cloned into a bacterial plasmid engineered for optimal expression in eukaryotic cells. Essential features include a strong promoter for optimal expression in mammalian cells (i.e., derived from cytomegalovirus), an origin of replication allowing for growth in bacteria, a bacterial antibiotic resistance gene and incorporation of polyadenylation sequences to stabilize mRNA transcripts. Notably, DNA vaccines also contain specific nucleotide sequences that play a critical role in the immunogenicity of these vaccines. *This specific motif consists of an unmethylated cytosine–phosphate–guanosine (CpG) dinucleotide with optimal flanking regions composed of two 5′ purines and two 3′ pyrimidines.* A diagram of a hypothetical plasmid DNA vaccine is illustrated in Figure 114.1.

VII. CpG MOTIFS AND IMMUNOSTIMULATORY SEQUENCES

In 1984, the immunostimulatory properties of bacterial DNA were reported after an anti-tumor activity of a DNA fraction derived from mycobacteria was discovered (Tokunaga et al., 1984). A few years later, in a seminal report it was shown that the immunostimulatory activity of bacterial DNA was due to the presence of unmethylated CpG dinucleotides. Importantly, the activity was specific for bacterial DNA, as mammalian genomic DNA was found to be completely inactive. In addition, short synthetic oligonucleotides were also potent inducers of B cell proliferation so long as they contained an unmethylated CpG in a particular stimulatory sequence context (CpG motif).

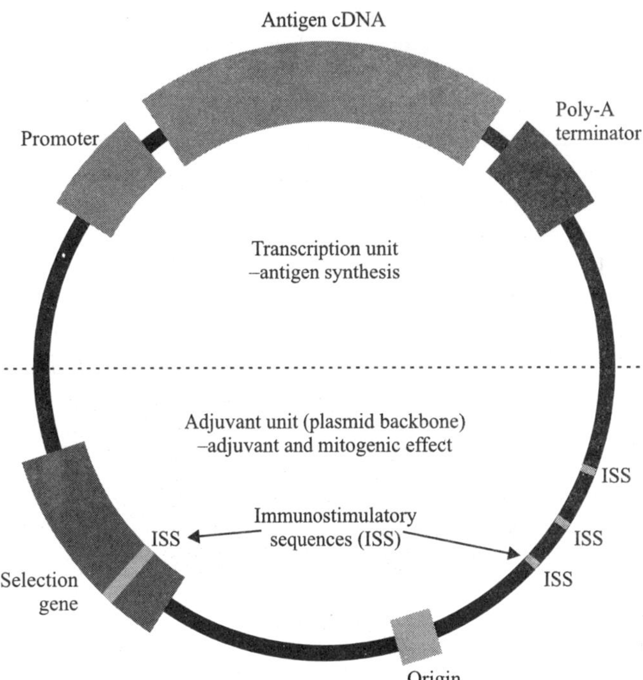

Figure 114.1 Diagram of a hypothetical plasmid DNA vaccine: The transcription unit consists of the promoter, antigen coding gene and poly-A terminator sequence. The expression is controlled by a strong promoter (usually a viral promoter) active in mammalian cells. Genes encoding cytokines and co-stimulatory molecules can be added to the antigen encoding gene to improve potency. Expression terminates at poly-A terminator. The bacterial selection gene is used for the production of the plasmid and usually contains many CpG motifs. Newer approaches are steadily eliminating the use of antibiotics during plasmid manufacture for safety concerns. CpG motifs are also present in the plasmid backbone and increase the innate immune response further enhancing the adaptive immunity conferred by the encoded gene.

This aspect of genetic immunization that has recently received significant attention is the *immunostimulatory activity of DNA itself.* It has been well demonstrated that DNA from bacteria, but not vertebrates, can induce a nonspecific immune response. It has been proposed that this is due to differences in the frequency of unmethylated cytosine-phosphate-guanine (CpG) dinucleotides found in the two genomes. While this dinucleotide appears in bacterial DNA at the expected frequency of 1 in 16, it is 10- to 20-fold less frequent in vertebrate DNA, a phenomenon termed 'CpG suppression'. In vertebrates the majority of these dinucleotides contain a methylated cytosine, whereas in bacteria they are unmethylated.

Krieg and colleagues showed that oligonucleotides containing one or more CpG dinucleotides could trigger B cell proliferation and immunoglobulin secretion. Immune activation was optimized when the dinucleotide was flanked by two 5′ purines and two 3′ pyrimidines, forming a 'CpG motif' or 'immunostimulatory sequence'. This immune response was not seen with otherwise identical oligonucleotides that did not contain CpG motifs or whose CpG dinucleotides had been methylated by CpG methylase. Bacterial DNA can also stimulate natural killer (NK) cell activity and cellular immunity through the induction of inflammatory cytokines such as interferons (IFNs) and interleukin 12 (IL-12). The relevance of these findings for genetic immunization became clear when Sato et al. reported that a DNA vaccine whose plasmid backbone contained CpG immune-stimulatory sequence (ISS) induced a more vigorous antibody and CTL response than an otherwise identical vaccine that did not contain the ISS, despite a higher level of gene expression produced by the latter plasmid. The immune response to the vaccine lacking CpG motif could be restored to normal levels by co-administering it with non-coding plasmid containing immune-stimolatorysequence. This study demonstrated that the plasmid backbone itself could have a significant booster effect on DNA vaccine-induced immunity, and indicates the role of CpG-containing ISS in initiating the response.

Subsequently, studies have confirmed that CpG motifs can enhance immunity to DNA vaccination as they preferentially induce a type 1 helper T lymphocyte (Th1) response. CpG-containing plasmids or synthetic oligonucleotides (ODN) induce production of several Th1-associated cytokines like IFN-γ, and IL-12 (but not IL-4), increase the IgG2a:IgG1 ratio, and decrease immunoglobulin E antibody production. The inherent Th1-inducing activity of bacterial DNA is a reason behind the ability of most DNA vaccines to initiate a predominant Th1 response when injected intramuscularly or intradermally. There does not appear to be any preferential cellular uptake of plasmids or oligonucleotides with and without immunostimulatory motifs. The first step to initiate cellular responses to CpG DNA is internalization of the DNA followed by endosomal acidification. Then CpG motif is recognized and signalling is initiated via Toll like receptor-9 (TLR9). DNA vaccines with immune-stimulatory motifs administered by conventional routes (i.e., skin or muscle) have not shown any significant adverse effects. Thus, the discovery that DNA vaccine can have potent immune-enhancing and immune-regulatory activities has provided another variable that can be used and manipulated in designing constructs for genetic immunization.

A. Types of CpG DNA and CpG-ODN

In general, nuclease-resistant phosphorothioate (PS)-ODNs are much more potent at activating B cells compared to the same sequence with a phosphodiester (PO) backbone but there are some differences in the immune stimulatory effects of CpG motifs in PS-ODN compared to PO-ODN. One of the non-bridging oxygens in PO-ODN is replaced by sulfur in to derive the PS-ODN. PS-ODN bearing suboptimal CpG motifs are less likely than PO-ODN to drive B cell proliferation, especially if the CpG is followed by a G. PS-ODN without CG motifs are also frequently observed to drive the proliferation of murine

and human B cells, although to a more limited degree than that which occurs with CpG ODN. In contrast, PS-ODN are generally less active at activating macrophages or NK cells, compared to ODN in which at least part of the backbone is PO. ODN containing PO backbones are particularly effective at activating NK cells and inducing IFN-γ production from plasmacytoid DC precursors, for which reason they have been termed as CpG-A ODN. The highest degrees of NK cell activation and IFN-γ production occur with ODN in which the 5' and 3' ends are PS-modified, and the center portion is PO. These chimeric ODN have a high degree of nuclease resistance provided by the PS-modified ends, yet they also retain potent NK-stimulating effects because the CpG motif has a PO backbone. CpG-A ODN are generally similar to bacterial DNA in their strong activation of NK cells, but the IFN-γ expression induced by an optimal CpG-A ODN is far higher, and the level of B cell activation generally lower, than that induced by the bacterial DNA. Even in comparing DNA from different bacteria, there are striking differences in the levels of immune activation, with *E. coli* DNA tending to be much more stimulatory than that from *C. perfringens* or *Streptococcal* or *Staphylococcal sp.* In contrast to CpG-A ODN, CpG motifs in nuclease-resistant PS backbones have dramatically enhanced B cell stimulatory properties but reduced NK stimulation, despite being much more stable than PO-ODN. CpG-B ODN as a class has a fully PS-modified backbone with one or more CpG motifs and no poly G motif. Even though PS-ODN as a family have similar properties, there are still qualitative differences in their effects, and some are substantially more effective at inducing the expression of TNF-α than others. To some degree then, every DNA molecule containing CpG motifs must be considered as a separate agent

B. CpG DNA and its Activation of Diverse Immune Cells

In general, CpG DNA stimulates T cells, B cells, NK cells, and antigen-presenting cells, both in form of plasmid DNA or in the form of synthetic ODN. The extent of immune stimulatory effects of an ODN depends on the nucleotides flanking the CpG dinucleotide. Together with the one or two bases on its 5' and 3' sides, the CpG dinucleotide comprises a CpG motif. Among all possible base combinations, optimal CpG motifs for activating mouse cells have the general formula: purine-purine-CG-pyrimidine-pyrimidine. However, the best CpG motif was found to be GACGTT for mouse and. GTCGTT for activating human cells. The human GTCGTT motif also appears to be optimal for many other vertebrate species, including cow, sheep, cat, dog, goat, horse, pig, and chicken. The immune stimulatory effects of the ODN are enhanced if the ODN has a TpC dinucleotide on the 5' end and are pyrimidine rich on the 3' end. The immune regulatory properties of DNA are contingent on the numbers of the CpG motifs as well as the spacing of them, Presence of flanking sequences in the ODN, and the structural attributes of the ODN backbone also affect immune stimulatory properties. Long stretches of CpG motifs sequences are not necessary for immune stimulation as sequences as short as six bases exhibit activity.

Antigen-presenting cells

Distinct subsets of murine and human dendritic cells (DC) and DC precursors express different subsets of TLRs that enable them to induce different patterns of immune responses to different pathogens. Among human DC precursors, the plasmacytoid pre-DC (pDC, sometimes called DC2) strongly express TLR9, while myeloid CD11cC pre-DC do not, instead expressing different TLRs such as TLR4 that cause them to be activated by a different set of pathogen molecules including LPS. Among human DC subsets, so far only the pDC are clearly demonstrated to be directly activated by CpG DNA. pDC are also notable for their role as the primary source of the IFN-α that is rapidly produced in response to viral infections. It is surprising that the two classes of CpG ODN have quite different activities upon plasmacytoid pre-DC IFN-α production: optimal CpG-A ODN induce extraordinarily high levels of IFN-α production, but CpG-B ODN induce relatively little, despite similar abilities to induce DC maturation and TNF-α production.

Similar to human pDC, murine bone marrow–derived DCs and Langerhans cells are activated by CpG-ODN to express costimulatory molecules, and secrete IL-12 and IL-6. CpG efficiently primes the DC for high-level IL-12 p70 production in response to CD40 crosslinking. In combination, CD40 ligation and CpG DNA induce synergistic IL-12 p70 production in murine splenic DC, including both the CD8α+ and CD8α- subsets. These stimulatory effects of CpG DNA enhance the ability of the DC to activate allogeneic T cells. Subcutaneous administration of CpG DNA also leads to in vivo activation of skin Langerhans cells to upregulate costimulatory molecules and produce IL-12. These CpG-activated Langerhans cells change morphology and migrate out of the skin within 2 hours. In summary, CpG DNA induces a pattern of Th1-like immune activation in human and murine DC and DC precursors and activates their migration.

Purified human monocytes do not express TLR-9 and are not activated by CpG ODN. Nevertheless, CpG DNA treatment of human PBMC or whole blood secondarily activates the monocytes to express increased levels of CD40 and CD69 and to produce IL-6 and TNF-α, though with delayed kinetics compared to LPS. In murine macrophage/monocytes CpG DNA causes the direct activation of NF·κB and initiation of cytokine expression including TNF-α. At low concentrations, CpG DNA and LPS show synergy for inducing macrophage nitric oxide production and monocyte cytokine production. This synergy is reportedly regulated in part at the level of NF·κB and in part at a posttranscriptional level. ODN can have backbone-dependent effect on macrophages—PS but not PO ODN has chemoattractant effects on primary macrophages.

B & T lymphocytes

CpG-ODNs are strong stimulators of B cells. CpG-ODN induces B cells to secrete IL-6 and IL-10. Induction of IL-6 by B cells on CpG-ODN stimulation IL-6 expression enables them to secrete IgM. On the other hand, CpG-induced IL-10 production functions counter-regulate pro-inflammatory activities of the immune system. CpG DNA can indirectly influence B cells by inducing NK cells to secrete IFN-γ, which in turn stimulates B cells. Mice genetically deficient in IFN-γ produce strikingly reduced IL-6 and IgM response to CpG DNA. CpG-activated B cells also express increased levels of the Fcγ receptors, Class II MHC, CD80, and CD86.

CpG DNA generally has not been reported to have direct stimulatory effects on resting T cells. In murine mixed cell populations, the type I interferons produced by CpG-stimulated adherent cells stimulate T cells to produce some activation and costimulatory molecules but also inhibit their proliferative response to TCR ligation. However, if T cells are highly purified away from adherent cells, then CpG DNA synergistically enhances the proliferative response of murine T cells to TCR ligation.

Natural killer cells

Since the initial observation that mycobacterial DNA activates murine NK cells to make IFN-γ and to have increased lytic activity, many additional studies have confirmed and extended this finding for mouse and human cells. It's the palindromic sequences in mycobacterial DNA that were responsible for the NK activation. The presence of a palindrome in an ODN probably leads to the formation of a double-stranded duplex region that stabilizes the structure against degradation compared to an ODN without a duplex. Palindromes are not required for NK stimulation but simply CpG motifs in appropriate base contexts are required, and the stimulatory effects are dramatically enhanced if the ODN is stabilized against nuclease degradation by modifying the ends with PS linkages. However, if the CpG motifs themselves are modified with PS linkages, rather than just the ends, then the magnitude of activation is much less than that with the same sequence in a PO backbone. The major source of the initial IFN-γ production in CpG-stimulated murine spleen cells appears to be NK cells. The stimulatory effects of CpG DNA on murine NK cells are not direct but require either the presence of adherent cells or their CpG-conditioned supernatants, which contain IL-12, TNF-α, and type I interferons. Thus, CpG DNA appears to act as a costimulatory signal for murine NK cells.

Role of CpG-DNA in T-helper differentiation

Primarily CpG DNA elicits Th1-type of immune response both in vitro and in vivo. Systemic administration of CpG DNA initiates a systemic Th1-like immune response. Similarly, subcutaneous injection of CpG into the footpad of a mouse induces IL-12 and IFN-γ production in the draining lymph nodes. As a result of CpG administration, DC become a more prominent population in the lymph nodes and exhibits an activated phenotype with increased expression of costimulatory molecules. CpG-induced Th1 differentiation is sustained as mice resist *Leishmania* challenge by eliciting an antigen-specific Th1 response. Intradermal or intranasal deliveries of the CpG are also known to initiate Th1 type of immune response.

C. Role of CpG as a Vaccine Adjuvant

Several recent reviews have covered the potent adjuvant activity of CpG DNA in Vaccination with various antigens and with DNA vaccines, which has become well established. CpG DNA is a stronger Th1-like adjuvant for inducing B cell and T cell responses than the gold standard, complete Freund's adjuvant, as measured by its ability to drive the differentiation of CTL and IFN-γ secreting T cells. Moreover, CpG DNA can be used for vaccination not only through parental routes, but also by means of oral or mucosal vaccination. The ODN activate APC to present antigen more effectively and therefore have to be injected in the same location, but not necessarily at the same time; the antigen can be given a week or more after the CpG injection and outstanding immune response to the antigen are still seen.

VIII.   THE MECHANISM OF ANTIGEN PRESENTATION BY DNA VACCINES

One of the most exciting developments in DNA vaccination over the past few years has been elucidating the mechanism by which the antigen encoded by the bacterial plasmid is processed and presented to the immune system *in vivo*. First it is important to note that the amount of antigen produced *in vivo* after DNA inoculation is usually in the picogram to nanogram range. Given the relatively small amounts of protein synthesized by DNA vaccination, the efficient induction of immune responses must relate to the type of APC transfected and/or the immune-enhancing properties of the DNA itself (i.e., CpG motifs). This section will focus on the cellular mechanisms by which DNA vaccines elicit responses *in vivo*. There are at least three mechanisms by which DNA vaccines are processed and presented *in vivo* to elicit immune responses. We have listed and discussed these mechanisms in the chronological order in which they were studied: firstly direct priming by somatic cells (e.g., myocytes and keratinocytes); secondly direct transfection of professional APCs (i.e., DCs); and thirdly 'cross-priming' in which plasmid DNA transfects a somatic cell and/or professional APC and the secreted protein is processed by untransfected DCs and presented to T cells. A diagrammatic representation of mechanism of DNA vaccine action is displayed in Figure 114.2.

A. Direct Priming by Somatic Cells

Studies showing that DNA vaccination could lead to protein expression and induction of CTL responses were done by injecting the DNA directly into the muscle. It was subsequently shown that adoptive transfer of stably transfected myoblasts expressing an influenza nucleoprotein protected mice from infectious challenge. These latter data provided definitive evidence that the expression of viral protein by muscle cells *in vivo* is sufficient for CTL-mediated protection. The question remained, however, as to whether CTL responses were directly

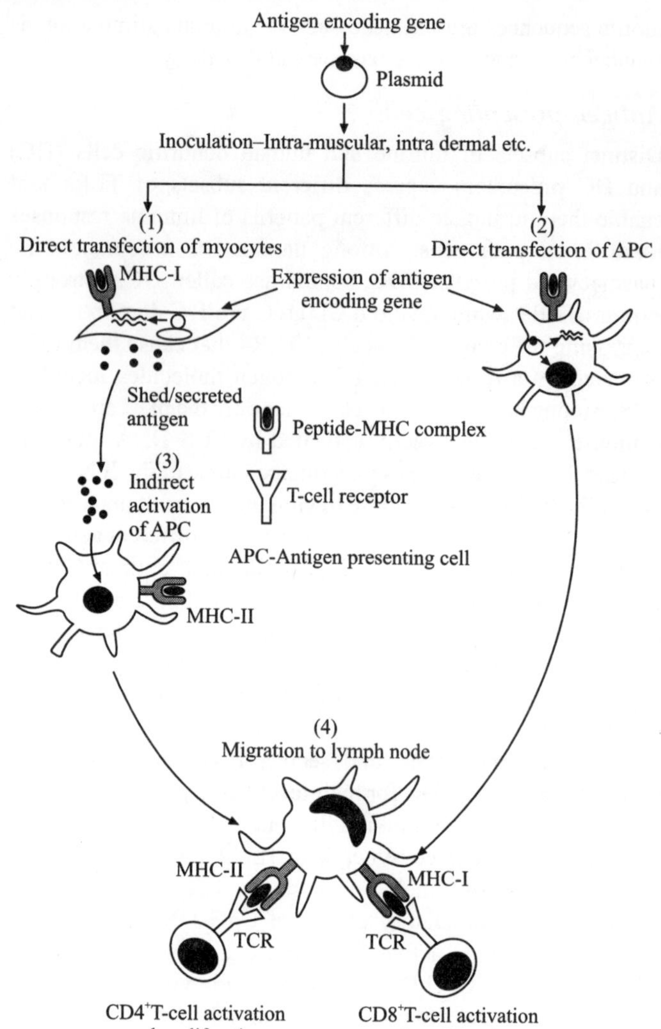

Figure 114.2   Mechanism of immune system activation by DNA vaccine: The DNA vaccine is introduced to the either the skin (subcutaneous or intradermal) or in the muscle. The DNA is taken up by the host cellular machinery. The plasmid can enter the host cell nucleus of either (1) the myocytes or (2) the antigen presenting cells (APCs). Here the plasmid components initiate transcription of the antigen gene and the product is folded into the native conformation by the host cellular machinery in cytoplasm. These proteins can be processed by the endogenous pathway and antigenic peptides can be presented in association with MHC-I molecule. Some of the shed or secreted antigens can be taken up by APCs and processed by the exogenous pathway to present antigen peptides in association with MHC-II molecules (3). These antigen loaded APCs travel through the afferent lymphatic vessels to the draining lymph nodes (4). In the draining lymph node the engagement of MHC bound antigen peptides with TCR and activation of co-stimulatory molecules activate antigen specific CD4+ and CD8+ T-cell activation and proliferation.

induced by myocytes expressing protein or indirectly by transfer of protein from myocytes to professional APCs. To assess the direct role that muscle cells play in the induction of cellular immune responses following DNA vaccination, bone marrow chimera experiments were used. In these studies, it was shown that antigen-specific CTL responses could not be induced by non-bone-marrow-derived (muscle) cells following intramuscular vaccination with DNA. Moreover, in a separate study, removing the muscle immediately (within 10 minutes) after intramuscular immunization with DNA did not alter the subsequent immune response. Taken together, these results suggest that although muscle cells can have a role in regulating immune responses following DNA vaccination, it is likely that this effect is indirect and occurs via cross-priming. In addition to intramuscular inoculation with DNA, the other major route by which DNA vaccines are administered is via the skin. The skin is a rich source of somatic cells such as keratinocytes and fibroblasts as well as potent bone-marrow-derived APCs such as Langerhans cells. In this regard, there have been several studies that have evaluated the various cells in the skin for their role in regulating immune responses following DNA vaccination. First, in contrast to the data discussed above for muscle cells, several studies have shown that optimal priming of immune responses following DNA immunization via the skin occurred when the vaccination site was left intact. Moreover, these studies also showed that immunity was induced from transfected APCs migrating from the skin and/or from non-lymphoid cells that transfer antigen for presentation by bone-marrow-derived APCs within the lymph node. In addition, transfected somatic cells such as keratinocytes in the skin may represent a reservoir of antigen to influence the magnitude and maintenance of the response.

B. Direct Transfection of APC

In addition to the role of somatic cells in priming immune responses following DNA vaccination, investigators also evaluated the role of bone-marrow-derived cells in this process. For these bone marrow chimera experiments, parent cells injected into F1-bone-marrow-reconstituted mice created a mismatch between the haplotypes of somatic cells and bone-marrow-derived cells. The immune response following DNA immunization was found to be restricted to the haplotype of reconstituted bone marrow. These data provided clear evidence that bone-marrow-derived cells were critical in priming immune responses after DNA vaccination. This work served as a prelude to understanding the specific type of bone-marrow-derived APC that was regulating the immune response after DNA vaccination. It was later shown that small numbers of DCs (0.4%) were transfected following DNA vaccination. Moreover, isolated DCs but not B cells were able to efficiently present antigen to T cell lines or hybridomas *in vitro*. Furthermore, direct *in vivo* visualization of draining lymph nodes after gene-gun vaccination demonstrated co-localization of protein expression from a reporter gene within a cell that had morphologic indices consistent with a DC. Taken together,

these data suggest that small numbers of directly transfected DCs are critical for priming immune responses following DNA vaccination. Finally, one interesting and exciting new area of research has been the delineation of DCs into subsets based on function and lineage. In the mouse, CD11c+CD8+ cells are denoted 'lymphoid' DCs whereas CD11c+CD8– cells are denoted 'myeloid' DCs. As it relates to the induction of cellular immune responses, CD11c+CD8+ DCs are enriched in their IL-12-producing capacity. Thus, because DNA vaccines readily transfect DCs and preferentially elicit Th1 responses (depending on the route of immunization), it is interesting to speculate whether these effects are mediated by direct transfection of lymphoid DCs by DNA.

C. Cross-priming

A final mechanism that has generated a substantial amount of recent interest is cross-priming. The genesis for this phenomenon was based on the observation that peptides derived from exogenous sources rather than from the classical endogenous pathway can present on MHC class I molecules *in vivo*. This induction of CD8+ T cell responses without *de novo* antigen synthesis within the APCs has been termed 'cross-priming'. For DNA vaccination, a confluence of data suggests that immune responses following DNA vaccination can be induced from somatic cells such as myocytes, keratinocytes or nonlymphoid cells by transfer of antigen to professional bone-marrow-derived APCs. In addition, recent reports in other systems have elegantly shown that cross-priming can occur when professional APCs process secreted peptides or proteins by phagocytosis of apoptotic bodies from other cells. In summary, it is clear that bone-marrow-derived APCs and, in particular, DCs are important in the initiation of immune responses following DNA vaccination. This could occur by direct priming of T cells by DCs and/or by antigen transfer (cross-priming) from bone-marrow-derived cells and/or somatic cells such as myocytes or keratinocytes. In addition, somatic cells may be important in augmenting and/or maintaining the immune response following DNA vaccination.

IX. DNA VACCINES IN DISEASE MODELS

The ability of DNA vaccines and CpG ODNs to preferentially generate Th1 responses in mouse models has been shown to be useful for preventing or treating intracellular infections requiring cellular immune responses. Although DNA vaccination has been used in a variety of infectious disease models, this section of the review will focus on examples of diseases in which Th1 cells are essential (i.e., leishmaniasis, malaria and tuberculosis).

A. Leishmaniasis

L. major has served as a premier model for understanding the regulation of Th1 and Th2 cells *in vivo*. It was the original model showing that IL-12 treatment at the time of infection induced a Th1 response, allowing susceptible BALB/c mice to control infection. More recently, treatment with CpG ODNs was found to enhance production of IFN-γ (a Th1 cytokine) and inhibit IL-4 (a Th2 cytokine) in BALB/c mice that were already infected with *L. major*. These striking data suggested that Th1-type responses could be induced in the course of an ongoing Th2 response. Thus, CpG ODNs offered a potential advantage over IL-12 in enhancing immunity for an ongoing disease. The mechanisms responsible for this are a longer duration of action for CpG ODNs and the induction of a wider spectrum of cytokines than that resulting from IL-12. As an extension to these data, CpG ODNs could be considered as a vaccine by providing enhancement of the immune response (innate immunity) in a non-antigen-specific manner that could be sufficient for short-term protection for diseases in which IFN-γ is important. In this regard, this might be an interesting option for travelers going to areas endemic for *leishmania* or malaria in which short-term immunity would be protective. With regard to a prophylactic vaccine against *L. major*, vaccination with plasmid DNA encoding a specific leishmanial antigen has been shown to be more effective than vaccination with leishmanial protein plus IL-12 protein in sustaining Th1 responses and controlling infection long-term. These data provided evidence that induction of Th1 responses *in vivo* following vaccination with protein plus IL-12 may not be sustained and raise the important issue as to how long-term Th1 responses are maintained by a vaccine. Reasons for the enhanced efficacy of DNA vaccination over protein plus adjuvant may include low levels of persistent antigen, and/or IL-12 induced by the CpG in the plasmid DNA. In addition, DNA vaccination — in contrast to protein plus IL-12 vaccination— induces CD8+ IFN-γ-producing cells that are likely to have an important role in mediating protection. Finally, it was also shown in the previous study that vaccination with leishmanial protein could induce long term immunity and sustain Th1 responses if given with IL-12 DNA. These data strongly suggest that persistent IL-12 may be required to sustain a sufficient number of effector Th1 cells able to control infection long-term.

Prophylactic vaccination with a plasmid encoding Kinetoplastid Membrane Protein-11 (KMP-11) is able to protect hamsters from experimental infection with the viscerotropic *L. donovani* parasites. This protection is correlated with a strong cellular immune response in vaccinated animals.

B. Malaria

Although both humoral and cellular immunity have a role in protection against certain stages of malaria infection, it is clear that the cellular immune response — through production of IFN-γ is critical in mediating protective immunity against the liver stage of the disease. There have been many mouse and primate studies evaluating DNA vaccination for protection against malaria. In a seminal study, a group of normal human volunteers were vaccinated intramuscularly three times with varying doses of plasmid DNA encoding a malaria circumsporozoite antigen. After the last boost, antigen-specific CTL activity could be detected following a period of *in vitro* culture. This study was important in providing the first demonstration that a cellular immune response could be elicited in humans by a DNA vaccine; however it remains to be determined whether the precursor CTLs generated would have been sufficient for protection against challenge. Moreover, it is clear from other mouse and non-human primate models that, although DNA vaccination may be effective for priming immune responses, further immunizations with either antigenic protein or replication-deficient vaccinia virus encoding the antigen(s) may be required to optimally boost these responses to a sufficient level for protection.

C. Tuberculosis

M. tuberculosis infection represents one of the most significant causes of mortality worldwide. With regard to prophylactic vaccination, bacille Calmette–Guérin (BCG) is the only currently approved vaccine; however, the protective efficacy of BCG may range from 0% to 80% worldwide. Thus, first and foremost, there is an urgent need for more-effective vaccines to prevent infection. Furthermore, there have recently been two additional issues that relate to established disease. The first is the evolution of multidrug-resistant strains. The second is based on a recent report in a small group of patients showing that relapse of pulmonary tuberculosis after curative therapy was due to exogenous re-infection rather than re-activation. Taken together, these data suggest that novel treatment regimens may allow for a shorter duration of therapy leading to greater compliance and a reduction in the development of drug-resistant strains as well as provide protection against re-activation or re-infection. Over the past few years, there have been several studies that have examined all of these areas.

SUGGESTIONS FOR FURTHER READING

1. Jon A. Wolff., et al. (1992), Expression of Naked Plasmids by Cultured Myotubes and Entry of Plasmids into T Tubules and Caveolae of Mammalian Skeletal Muscle, *Journal of Cell Science*, 103:1249–1259.
2. Margaret A. Liu., (2011), DNA Vaccines: An Historical Perspective and View to the Future, Immunological Reviews; 239:62–84.
3. Gurunathan Sanjay, et al. (2000) DNA Vaccines: Immunology, Application, and Optimization, *Annual Reviews of Immunology*, 18:927–974.
4. Tokunaga T., et al. (1984), Antitumor Activity of Deoxyribonucleic Acid Fraction from *Mycobacterium bovis* BCG. I. Isolation, Physicochemicalcharacterization, and Antitumor Activity, *Journal of National Cancer Institute*,72(4):955–62.
5. Montgomery D.L., et al. (1997), DNA Vaccines, Pharmacology and Therapeutics, 74:195–205.
6. John J. Donnelly (1997), DNA Vaccines, *Annual Reviews of Immunology*, 15:617–48.
7. Sanjay Gurunathan, et al. (2000), DNA Vaccines: A Key for Inducing Long-term Cellular Immunity, *Current Opinion in Immunology*,12:442–447.

115

Vaccines for Cholera

Amit Ghosh

CONTENTS

I. INTRODUCTION

Disease cholera caused by the gram negative protobacterium *Vibrio cholerae*, is a potentially lethal diarrheal disease marked by rice-water stool and vomiting. Symptoms of cholera are primarily due to a toxin, cholera toxin (CT), encoded by the genes present on a lysogenic bacteriophage CTXφ, harbored by all cholera toxin producing strains of *Vibrio cholerae* (Figure 115.1). Cholera affects both adults and children, but the incidence rate is highest among the children under 5 years of age. It is an ancient disease, which is thought to have originated in India from where it spread to the rest of the world. Seven pandemics of cholera, which is simultaneous or near-simultaneous occurrence of epidemics in many parts of the world over an extended period of time, have been recorded. Although the disease has largely been controlled in the developed countries through improved sanitation, supply of safe drinking water and good food hygiene, it continues to remain a scourge in the developing countries in many of which it is endemic. Alarmmingly also, the burden of cholera is increasing in the recent years. Not only cholera epidemics are breaking out in newer areas but also the disease is re-emerging in countries from where it disappeared long time ago. In the past decade, catastrophic epidemics of cholera occurred in many parts of the world. Many countries like Afgantstan, Bangladesh, Guinea-Bissau, Haiti, Indonesia, Iraq, Kenya, Liberia, Malwai, Mozambique, Nigeria, South Africa, Sudan, Vietnam, Zimbabwe etc. have witnessed major outbreaks.

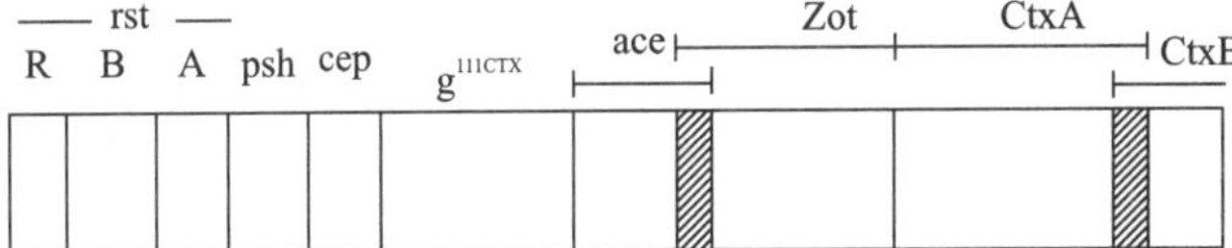

Figure 115.1   Schematic structure of CTXφ. (not to scale)

In 2010, an epidemic of cholera broke out in Haiti, a country which was free from the disease for the last hundred years, and affected more than 500000 people, killing around eight thousand. Although the case fatality ratio, which is defined as the percentage of persons dying in a given period due to a specific illness, was not as high as it was in the Zimbabawean epidemic of 2008 where it was 5%, Haitian outbreak more than amply demonstrated what devastation cholera can cause even in this century. The disease is now endemic in more than 50 countries. According to the World Health Organization (WHO) every year cholera affects around 3–3.5 million people globally, killing between 100 to 120 thousand. In the recent times, most major epidemics of cholera have occurred in the coastal regions in which climatological factors are thought to play a significant role. It is envisaged that rising environmental temperature due to global warming could lead to an increase in the globally incidence of cholera in the coming years.

Cholera can be treated successfully by the replacement of the fluid lost due to diarrhea through oral rehydration. Antibiotic treatment help shorten the duration of the disease and reduce stool volume. However, these are wasteful and expensive options. Moreover, apart from the fact that, as seen many times in the past, such interventions are not very easy to implement in times of war or natural disasters. It is also being observed that of late *V. cholerae* strains are developing resistance to the drugs commonly used in the treatment of cholera prompting WHO to recommend that antibiotics should be given only to those patients who are suffering from serious dehydration. Furthermore, it is now well established that *V. cholerae* can exist both in free-living state or in association with marine planktons in a culturable form or in a viable but non-culturable form making its eradication an unachievable goal. These observations considered together, underscore the need for the development of effective prophylactic measures to control the disease. Thus, besides improved sanitation and supply of safe drinking water, development of an efficacious vaccine could be a very powerful tool for controlling cholera.

II. BACKGROUND

A. Classification of *Vibrio Cholerae* and Epidemiology

V. cholerae strains can be differentiated on the basis of the o-antigen portion of their lypopolysacharide (LPS), which consists of three regions—Lipid A, core-polysaccharide (Core-PS) and O-lypopolysacharide (O-PS or O-Ag) (Figure 115.2). Serogrouping of the *V. cholerae* strains is done on the basis of their O-PS. So far more than 200 serogroups of *V. cholerae* have been described, out of which only two, namely, O1 and O139, differing in composition of their core-PS and O-PS, have been associated with epidemics (Figure 115.3). Most strains belonging to these serogroups produce cholera toxin, while majority of the non-O1, non-O139 strains do not. *V.*

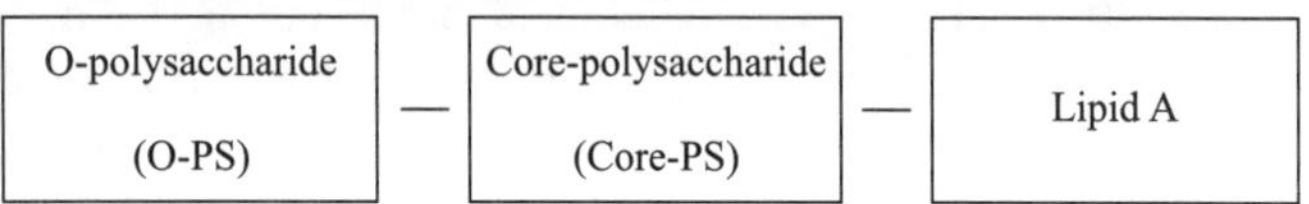

Figure 115.2 Schematic diagram of LPS. (not to scale)

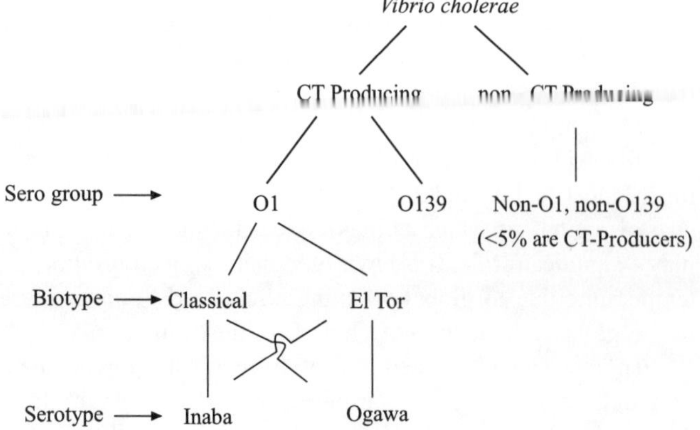

Figure 115.3 Classification of *V. cholerae* strains.

cholerae strains belonging to O1 serogroup are further sub-divided into three distinct serotypes namely, Inaba, Ogawa and Hikojima. While the anti-serum raised against the Inaba strain does not cross-react with the Ogawa strains and vice-versa, Hikojima strains are agglutinated by both anti-Inaba and anti-Ogawa anti-sera. Hikojima strains are not stable and are rarely encountered. Inaba and Ogawa strains, which differ from each other in the composition of their O-PS in contrast, are found widely. Strains of *V. cholerae* O1, whether Inaba or Ogawa, are further sub-classified into two genotypically and phenotypically distinct biotypes—the classical and the El Tor, differentiable from each other on the basis of a few simple tests. The other epidemic strain of *V. cholerae*—*V. cholerae* O139 appeared for the first time in 1992 and caused major outbreaks in parts of India and Bangladesh. This strain originated from a *V. cholerae* O1 El Tor strain by lateral transfer from an unknown source, a genomic island encoding O139 antigens leading to the replacement of some of the genes coding for the O1 antigens. *V. cholerae* O139 is very similar to a typical O1 El Tor strain, also the cholera toxin elaborated by it is almost identical to the one produced by the El Tor strains. However, it does not produce LPS and unlike the O1 strains, is covered by a polysaccharide capsule.

As has been stated earlier, while cholera is endemic in many parts of the world with regular seasonal outbreaks, epidemic cholera occurs when a pathogenic strain of *V. cholerae* enters a cholera–free region for the first time (as had happened in Haiti in 2010) or a new pathogenic strain emerges (e.g., emergence of *V. cholerae* O139 in 1991). People living in endemic areas develop natural immunity against cholera, which increases with age due to repeated exposure to the disease. In such areas, highest incidence and the most severe form of the disease are found among children. In an epidemic setting in contrast, both adults and children face equal risk because of the absence of natural immunity in both due to their lack of prior exposure to the disease. Since 1817, seven pandemics of cholera have been documented. Fifth and Sixth pandemics were determined to be due to *V. cholerae* O1 classical strains and it is presumed that the earlier pandemics were also due to such strains. First six pandemics originated in India. The seventh one, which started in the Sulwesi Island in Indonesia in 1961, was caused unlike the previous pandemics, by a *V. cholerae* O1 El Tor strain. *V. cholerae* El Tor was first isolated in Egypt in 1905 and was not known to be responsible for any major outbreak. Seventh pandemic has been bigger than all the previous ones in terms of the number of people affected, geographical expanse and duration. While the first six pandemics lasted between 5 to 25 years, the seventh pandemic is continuing unabated even after 50 years of its onset. Cholera, as has been pointed out earlier, has now become endemic in many parts of the world including the Indian subcontinent, South East Asia and parts of Africa. When *V. cholerae* O139 emerged for the first time in 1992 and caused havoc in the coastal regions of India and Bangladesh, there was considerable apprehension that it could be ushering in a new pandemic—the eighth one. However, the fear proved to be unfounded and cholera caused by

V. cholerae O139 has remained confined only in parts of Asia. Ninety eight percent of all cholera cases in the world today are due to *V. cholerae* O1 El Tor strains.

B. Pathogenesis and Immune Response to Infection

Cholera is transmitted by faceo-oral route through contaminated food and water. After ingestion, *V. cholerae* has to overcome the acid barrier of the stomach and negotiate several other obstacles to reach small intestine to cause the disease. Once it reaches the small intestine, aided by toxin-coregulated pilus (TCP), essential for *V. cholerae*'s binding to the intestinal mucosa, hemaglutinin protease and its flagellum it successfully colonises the gut, and it begins to elaborate the cholera toxin (CT) and a number of other virulence factors. CT is a pentameric protein made up of one "A" and five "B" subunits of Mr. 27.3 KDa and 11.6 KDa, respectively. Toxicity of the cholera toxin is due to the "A" subunit. "B" subunit, which is responsible for the binding of CT to the ganglioside GM1 receptors on the epithelial cell surface, does not have any toxic property. It is primarily responsible for evoking host immune response against CT. Binding of the "B" subunit pentamer causes a change in the membrane permeability of the cells that allows the "A" subunit to translocate into the cytosol of the target cells where it is proteolitically cleaved into two fragments—enzymatically active A1 (Mr 21.8KDa) and A2 (Mr 5.4 KDa). A1 subunit then activates the host adenylate cyclase through the mediation of GTP-binding proteins (G-proteins) present in the cells, causing an increase in the level of cyclic AMP(cAMP) within the cells. Cyclic AMP thus generated in turn activates cAMP dependent protein kinase to cause protein phosporylation, alteration in ion transport and secretion of the body fluid into the gut lumen leading to diarrhea, which if untreated can be fatal. *Thus, while the symptoms of cholera are mainly due to the action of the "A" subunit, it is the "B" subunit, against which the anti-toxin host immune response is directed.*

Vibrio cholerae infection elicits both systemic and local mucosal response affording protection against a subsequent attack by the strains belonging to the same serogroup. Thus, although both *V. cholerae* O1 and O139 strains produce almost identical CT and elicit similar serum antitoxin response, infection with *V. cholerae* O1 does not offer protection against *V. cholerae* O139 and vice versa, pointing to the relative unimportance of anti-toxin immunity in comparison with the antibacterial immunity, in providing protection against cholera. Although the exact host immunological mechanism that orchestrates protection against *V. cholerae* is not known, it is generally accepted that *V. cholerae* being a non-invasive pathogen that colonizes the gut without destroying brush borders or invading exterocytes, protection against cholera is thought primarily to be due to the synergestic action of anti-LPS and anti-CT secretary antibodies (SIgA) produced locally in the gut lumen. Elevated levels of anti-LPS and anti-CT, SI$_g$A antibodies have been detected in the intestine

and serum of the patients recovering from cholera. However, no correlation between the levels of these and the immunity against cholera has been seen. The only known marker of protective immunity has been found to be the serum vibriocidal antibody, a complement fixing antibody consisting mainly of IgM, which is directed primarily against the O-antigen of *V. cholerae* O1 LPS. Vibriocidal antibodies are so called because they lyse *V. cholerae* cells *in vitro* in the presence of guinea pig complement.

It is now well documented that individuals with blood group O are at greatest risk of contracting cholera followed by those with blood groups A and B. Nevertheless, no matter what the blood group of an individual is, there is an inverse correlation between the level of vibriocidal antibody in serum and the person's susceptibility to cholera caused by *V. cholerae* O1 (This co-relation does not hold for *V. cholerae* O139). In population and household contact studies carried out in Bangladesh, it was seen that every two fold increase in serum vibriocidal antibody titre was associated with a near fifty percent drop in the risk of infection by *V. cholerae* O1. It was further observed that in the endemic areas where people are "exposed" to cholera again and again, level of vibriocidal antibody in the serum increases with age. However, although the complement fixing vibriocidal antibody remains the best characterised serological marker of immunity against cholera, the role played by it in providing protection against cholera is not understood; Even after the decline of the vibriocidal antibody titre to an undetectable level immunity to the disease continues. Furthermore, threshold vibriocidal abtibody level above which complete protection against cholera is attainable is also not known. All these suggest that the complement fixing vibriocidal antibody in serum is actually a surrogate marker of the other factors responsible for the protection. Recent experiments suggest that long-term protection is mediated possibly by memory B cells directed primarily against LPS.

In addition to the vibriocidal antibodies against LPS, antibodies against MSHA, TCP and outer membrane proteins, have also been detected in convalescing patients. Their roles in protection against the disease however, remain an open question.

III. CHOLERA VACCINES

Since 1884, when Jaime Ferran first attempted to vaccinate people against cholera with a naturally attenuated strain of *V. cholerae* in Spain, most vaccines during the following 90 years or so had been injectable (parenteral) ones. Serious attempts to vaccinate people with such vaccines began in earnest in the 1960s and a large variety of such parenteral vaccines—killed whole cell *V. cholerae*, purified LPS, toxoids, killed whole cells in combination with the B–subunit of CT and various other adjuvants, have been used both in volunteer studies, and in field trials. In the end, their use was discontinued because it was found that even the most effective ones among them had a protective efficacy of only 50%, and that too for a maximum of six months. Besides, they all caused adverse reactions in the vaccinees and were not effective in controlling the transmission and spread of cholera during epidemics. However, use of these vaccines and subsequent analyses of the data obtained generated a wealth of valuable information as summarised below:

(i) Both killed or live attenuated whole cells administered through the parenteral route induce similar serological responses and evoke for a short duration a significant increase in gut SI_gA in the already immunologically primed individuals.

(ii) In endemic areas, vaccines were more effective in adults than in the children, indicating that protective antibodies are evoked more readily in those who are already immunologically primed through prior antigenic contact.

(iii) LPS plays a crucial role in protection. In a large scale field trial a parenteral vaccine consisting of LPS alone could provide significant protection to the vaccinees.

(iv) Toxoids by themselves can offer little protection.

IV. ORAL VACCINES

As stated earlier, transmission of cholera takes place when *V. cholerae* enters the human body through mouth by the way of contaminated food for drink. After reaching the intestine and colonizing it, *V. cholerae* and the toxin(s) it elaborate(s) remain confined in the intestine and evoke in the lymphatic tissues beneath the mucosal surface, proliferation of IgA, primarily responsible for immunity against cholera. This knowledge, along with the observation that patients who recover from a bout of the disease develop resistance to further attack for atleast three years, suggested that to maximize the production of the intestinal IgA, the vaccine should perhaps be delivered through the oral route—the route naturally taken by *V. cholerae* for causing infection. It was also realised that such a vaccine would be much easier to administer and unlike the injectable ones would be free from the risk of transmitting blood borne infections. All these considerations eventually shifted the focus of vaccine development from the creation of "newer" parenteral vaccines to the development of the oral ones.

For the development of oral vaccines, two different approaches have been pursued; in one approach. Killed cells of *V. cholerae* (WC) alone or in combination with the "B" subunit of the cholera toxic (WC-CTB), obtained in pure form either by chemical extraction or by recombinant DNA technology, have been used as vaccines. While in the other, use has been made of live attenuated strains of *V. cholerae*.

A. Killed Vaccines

Although a number of non-living vaccines were developed in 1960s, the first killed vaccine to undergo large scale field trials [Box] both in endemic and non-endemic areas, besides undergoing extensive evaluation in volunteers, was a vaccine

developed in 1976 by Swedish Scientists, comprising killed *V. cholerae* O1 of different serotypes alone or in combination with the B subunit of CT. The rationale for the addition of the B subunit in the killed vaccine formulation was the discovery that intestinal mucosa, which is incapable of mounting effective immune response against non-living foreign substances, can be induced to do so if co-administered with CT or CT-B. In volunteer studies with North American volunteers both WC and WC-CTB given in 3 doses at 2 week intervals, apart from not causing any adverse reactions, induced significant rise in vibriocidal antibody titres – 71 % in the recipient of WC and 89 % in those, who were administered WC-CTB. When these volunteers were challenged with a virulent strain of *V. cholerae* El tor, Inaba (N16961) one month after the completion of the immunization schedule, protective efficacies of 56% and 64% were seen in the recipients of WC and WC-CTB, respectively.

Clinical Trials: Any research project that evaluates the effect of a health related intervention (e.g., administration of drugs, vaccines, treatment protocols etc) in human subjects. Clinical Trials are conducted in several phases. Phase 1 trials are designed to examine the safety of a biomedical intervention under consideration (e.g., vaccine) in a small number of people. Phase 2 trials examine the efficacy of a biomedical intervention (e.g., vaccine) in a larger group of persons. It also confirms the safety of the intervention established in phase 1. Phase 3 is similar to phase 2 but involves a much bigger group of people (several hundred to thousands). It also compares the efficacy of the intervention under study (e.g., vaccine) with those of other similar interventions already available.

Double blind Placebo Controlled Randomized Trials
These are trials in which neither the volunteers/patients nor the investigators/researchers know who is being administrated the agent under study and who is being given the placebo.

Challenge Studies: Studies in which healthy human volunteers are intentionally administrated an agent (e.g., a Virulent organism) under carefully controlled conditions to test the efficacy of vaccines, drugs etc.

In a large scale trial in Bangladesh the killed whole cell vaccine (WC), which contained per dose 10^{11} heat and formalin inactivated *V. cholerae* O1 of both sero-and biotypes, evoked a two fold rise in anti Ogawa and anti-Inaba vibriocidal titres when administered in 3 doses 1 weeks apart and during the first 6 months provided protection to 58% of all vaccinees above 5 years of age. It also did not cause any adverse reactions. In the vaccine formulation heat inactivation of *V. cholerae* ensured maximization of the expression of the cell-wall LPS, while the formalin killed cells compensated for the antigens destroyed by the heat treatment.

In a large scale field trial conducted in Bangladesh WC-CTB (with B subunit obtained in pure form from CT by chemical extraction) administered along with bicarbonate buffer to prevent CTB from being damaged by gastric acid, showed greater efficacy compared to WC alone in terms of the protection provided in the first year of surveillance; During the first six months after vaccination, level of protection provided by WC-CTB was 85% in contrast to 56% provided by WC alone. After 12 months, these figures fell to 62% and 52%, respectively. However, at three years the cumulative protective efficacy became almost the same for both the vaccines. These results also showed that WC does not require the adjuvantcity of CTB to elicit, albeit to a lesser degree, mucosal immune response. Mechanism underlying this phenomenon remains unclear. It was also found that WC-B was equally protective against both mild and severe forms of cholera, and protected adults and children equally well. In 1991 a new version of this WC-B vaccine, in which CTB prepared by chemical extraction was replaced with CTB produced by genetic engineering techniques was produced and licensed in Sweden under the trade name Dukoral. This new version was subsequently tested widely in Peru and results similar to Bangladesh trial were obtained. Here also in the first 6 months it demonstrated a protective efficacy of around 85% among all two dose recipients. In spite of these positive features, WC-B (Dukoral), cholera being primarily a disease of the poor, did not find acceptance for routine use because of its high price, caused due to the high manufacturing cost of pure CTB. It however, found use as a vaccine for traveller's diarrhea for some time.

Taking cue from the 1985 Bangladesh trial, where WC alone showed a protective efficacy of around 60%, Government of Vietnam, in order to have an affordable vaccine, for its own public health programme launched in the 1980s, initiated a programme for the development of a variant from of Dukoral without CTB under a technology transfer agreement with the Swedish scientists.

Besides the omission of CTB to bring down the cost, the other change that was made in the variant was the replacement of formalin killed *V. cholerae* O1 classical strain Cairo 50 with the formalin killed *V. cholerae* O1 classical strain 569B, known to be a good producer of the toxin coregulated pillus (TCP), a potentially powerful protective immunogen. Elimination of CTB in the formulation not only reduced the cost of vaccine drastically but also obviated the need for the administration of buffer before administering the vaccine. In an open trial in Bangladesh, where individuals in alternative households, were given 2 doses of the vaccine 2 weeks apart, 66 % protection was seen for both adults and children over one year of age. Higher level of protection (66%), provided by this vaccine compared to that (60%) provided by WC in the 1985 Bangladesh trial, was thought to be due to the higher LPS content of this new vaccine as the result of the different method of standardization adopted in its preparation. Later, after the advent of *V. cholerae* O139 in 1992, this vaccine was further modified and made bivalent by the inclusion of killed *V. cholerae* O139. This new formulation was licensed and commercialized in Vietnam under the trade name ORCVAX. Given in 2 doses, 2 weeks apart, ORCVAX induced a significant rise in anti-O1 vibriocidal titre in 60% of the vaccinees. In contrast, anti-O139 vibriocidal response could be detected only in 40% of the vaccine recipients. It has been conjectured that this low anti-O139 vibriocidal response

was due probably either to the sub-optimal level of *V. cholerae* O139 cells present in the formulation or the failure of the O139 antigens to induce a "booster effect" in the vaccinees because of their lack of prior or inadequate exposure to *V. cholerae* O139. It was seen subsequently that overall vaccine efficacy 3-5 year after dosing was 50%. ORCVAX though licensed in Vietnam, did not find international acceptance as its production and standardization were not in compliance with the World Health Organization norms. Besides, it was found to contain residual level of Cholera toxin (CT) coming from *V. choelrae* 569 B, a high producer of CT. To make it acceptable internationally, ORCVAX was modified further and *V. choelrae* 569 B was replaced by heat killed classical O1 Inaba strain cairo 48 and formalin killed classical O1 Ogawa cairo 50.

This modified vaccine is now produced and licensed in Vietnam as mORCVAX and to make it internationally acceptable, as the Vietnam National Regulatory Authority is not recognized by the WHO, as Shancol in India. Administered in 2 doses 2 Weeks apart, Shancol has been shown to provide 66% protection to both adults and children over 1 year. Quite curiously, the level of protection provided by the vaccine after 1 year of dosing was only 45% which rose the 72 % in the second year. No satisfactory explanation for this unexpected result has been offered.

B. Live Oral Vaccines

Development of killed oral vaccine affording a significant level of protection has been a great achievement. But a primary drawback of these vaccines has been the requirement of multiple doses. Besides, these vaccines could protect children under six months of age for only a very short period – less than six months. So alternative routes of vaccine development have been pursued—the most prominent one being the one where live attenuated strains of *V. cholerae* have been sought to be used as live oral vaccines.

Although the idea of live oral cholera vaccine came from Jaime Ferran himself, no attempt to develop such a vaccine was made till 1969, when some naturally occurring non-toxigenic strains of *V. cholerae* were attempted to be used as vaccines. All of these efforts failed. Nevertheless this line of research continued to be pursued because live vibrios are more immunogenic than the killed ones—they are taken up more readily by the M cells, the major antigen sampling cells in the gut located in the Peyer's patches. Furthermore, it has been argued that antigens which could be important in invoking immunity against cholera but are expressed only *in vivo*, will be made only by the live strains and that such antigens will be absent in a killed vaccine. Based on these considerations several putative live oral vaccines have been created following two different paths. In one, attenuated strains have been derived from toxigenic strains by "mutating" (through chemical mutagenesis) or deleting *ctxA* by genetic engineering techniques. In the second approach non-toxigenic *V. cholerae* O1 strains have been genetically engineered to make them capable of elaborating the immunogenic B subunit of CT.

Texas Star SR

Earliest vaccine to be made following the first approach was Texas Star SR. It was created by disabling the *ctxA* expression in *V. cholerae* O1 El Tor strain 3083 by Nitrosoguanidine (NTG) mutagenesis. This mutated strain named Texas star, could produce only the "B" but not the "A" subunit of CT. Administered at doses ranging from 10^5 to 5×10^{10} organisms in one or two doses 1 week apart, Texas Star SR induced a significant rise in serum vibriocidal antibody titre. In a volunteer trial it showed a protective efficacy of 61% against challenge doses, which caused cholera in 70–80% of the recipients, who did not receive the vaccine. Texas Star, however, was not taken forward because of the unacceptable level of adverse reactions it caused in 24% of the vaccinees and also because it was likely that NTG mutagenesis produced unidentified mutations(s) in the genome besides the one(s) in *the ctxA*. Moreover, as the impairment of *ctxA* expression was due to a point mutation(s) (NTG induces point mutations) the possibility of the vaccine strain reverting to wild type could not be ignored.

Although Texas Star was not successful as a vaccine, it demonstrated convincingly that an attenuated strain of *V. cholerae* used as a live vaccine can elicit significant level of immune response and provide protection. It thus established the feasibility of such an approach for the development of live attenuated oral cholera vaccines.

CVD 103 Hg^R

After the advent of the recombinant DNA technology, to address the issue of possible reversion of vaccine strains to the wild type, attenuated strains were created by deletion of the *ctxA* gene or a part thereof. Most of the putative vaccine strains thus created, although highly immunogenic, were found in volunteer studies to cause unacceptable levels of adverse reactions in the vaccinees. One strain, CVD-103, (developed at the Centre for vaccine Development, University of Maryland, USA), engineered from the classical Inaba strain 569B by deleting 94% of the *ctxA* gene, however, turned out to be highly immunogenic and yet did not cause any adverse reaction in the recipients. To make the strain easily identifiable a gene for mercury resistance was subsequently inserted into its cryptic hemolysin locus. In a challenge study, a single dose ($2 - 8 \times 10^8$ viable cells) of the resulting vaccine strain designated CVD-103 Hg^R administered to North American volunteers along with buffer registered a 4 fold or more rise in vibriocidal antibody titre in 91% of the recipients. This study also showed that CVD-103 Hg^R had a protective efficacy of 82–100% against *V. cholerae* O1 classical and 62–67% against *V. cholerae* O1 Eltor. It was later found, however, that to obtain a similar level of immunity in of the developing countries, a 10 fold increase in the administered dose (5×10^9 viable cells) was required.

Encouraged by these results a field trial involving more than 65000 adults and children were conducted in Indonesia in 1992. After the first year, the vaccine displayed a protective efficacy of only 25%. In spite of this it was decided to extend this trial because very few cholera cases that occurred during the first 6 months of vaccination, made the proper assessment of its short term efficacy difficult. Disappointingly, however, it showed no significant protection in any year of the follow up and the protective efficacy at 4 years of follow up was only 14%. Later, in a trial in Micronesia, where this vaccine (2×10^9 viable cells per dose) was introduced in the middle of an outbreak, an efficacy of 79% was seen. However, as the vaccine was introduced in the middle of an outbreak, that is after the study population had already been exposed to *V. cholerae* O1, and also because the study was neither randomised nor blinded, it is felt that results need confirmation. Nevertheless, as a single dose ($2 - 8 \times 10^8$ viable cells) CVD-103Hg^R provided protection to North American Volunteers for 24 weeks, 1 week after dosing, this vaccine was manufactured and sold in some countries for sometime as a traveller's vaccine under the trade names Orochol and Mutachol. At present, these are no longer manufactured.

Peru 15

Commercially known as CholeraGarde, Peru-15 was made by attenuating an El Tor Inaba strains isolated in Peru in 1991. Attenuation was achieved first by deleting the CT element along with the flanking *att*RS1 regions and then selecting for a non-motile variant (as reactogenecity of live attenuated vaccines was thought to be due their motility). This strain has been tested both in the USA and in Asia. In a randomized, double blind trial in the USA, a single dose (2×10^8 cells) of Peru-15 induced 4 fold or more rise in vibriocidal antibody titre in the vaccinees, and demonstrated a protective efficacy of 93% when the vaccinees were challenged with a virulent strain of *V. cholerae,* 3 months after the administration of the vaccine. In a limited study in Bangladesh, Peru 15 was found to be safe for adults, children and infants. A single dose of 2×10^8 cells induced significant level of sero conversion in 75% of the vaccinees. The overall figure for children less than 5 years of age was 77%. The Vaccine has not yet undergone a field trial.

V. cholerae 638

This vaccine strain made in Cuba is derived from a *V. cholerae* O1 El tor Ogawa strain, isolated in Peru during the 1991 epidemic. Attenuation was done by deleting the CTX prophage. To make the strain easily differentiable from other *V. cholerae* strains, a reporter gene *cel*A encoding endoglucanase A from *Clostridium thermocellum* was inserted into its *hap* locus, which codes for hemagglutinn/protease, a minor virulence factor. In a randomized, double blind, placebo controlled trial with 45 volunteers in Cuba, where there is no environmental exposure to *V. cholerae*, 96% of the Vaccine recipients, after administration of a single dose of 10^9 cells, registered a 4 fold or more rise in vibriocidal antibody titre. When the volunteers were challenged with a virulent O1 El Tor Ogawa strain one month after vaccination, none of the volunteers who had been vaccinated suffered from diarrhea. In contrast 78% of the unvaccinated volunteers had diarrhea. In this limited study, *V. cholerae* 638 displayed a protective efficacy of 100%.

Subsequently, in a randomized double blind trial involving 120 volunteers, in Mozambique, a cholera endemic area, a single dose of 2×10^9 cells, seroconverted 97% of the Vaccinees.

VA 1.3 and VA 1.4

All the live oral vaccines described above were derived from toxigenic clinical isolates of *V. cholerae* O1 and hence had the potential, even after attenuation, to cause mild dirrhoea due to the presence of unknown diarrhoeagenic factors. The vaccine strain VA1.3 developed in India in contrast, has been developed from a non-toxigenic clinical isolate of *V. cholerae* O1 inaba strain naturally devoid of CTX phage element and totally non-reactogenic in biological assays. Cholera toxin B subunit gene ctxB was introduced into the strain at its cryplic *hly* locus, coding for hemolysin, to ensure production of CTB by VA1.3. Also, to make the strain easily identifiable an Amplicillin marker was linked to *ctx*B. In a randomised placebo control human volunteer trial involving 309 men aged between 16 and 15 years, a single dose (5×10^9 cells) of VA1.3 elicited 4 fold or more increase in vibriocidal antibody titre in 57% of the vaccinees. In a subsequent development the Ampicillin marker was removed from VA1.3 to create VA1.4. In a randomized double blind trial conducted in Kolkata, a cholera endemic zone, after the administration of $5–7 \times 10^8$ cells in a single dose VA 1.4 elicited a 4 fold or more increase in the vibriocidal antibody titre in 66% of the recipients.

Composition of the vaccines described above and their dosing schedules are given in Table 115.1.

C. Vaccines against *V. cholerae* O139

Although 98% of the cholera cases in the world today are due to *V. cholerae* O1 El Tor strains. Attempts have been made to develop vaccines, both killed and live attenuated, against *V. cholerae* O139 as well. As has been mentioned elsewhere in this article, killed WC cholera vaccine in its later formulations, mORCVAX and Shancol, were made bivalent by the inclusion of killed *V. cholerae* O139. So far as the live oral vaccines are concerned, two, namely, Bengal 15 derived from a *V. cholerae* O139 strain M10 following essentially the same strategy that was followed for the creation of Peru 15 and CVD112 produced from *V. cholerae* O139 strain A11387 using the same approach as that used for developing CVD 103Hg^R, have been tested in a small number of American volunteers and have been found to be safe with protective efficacies of 84 and 83%, respectively.

CONCLUSION

Cholera continues to be a major scourge in most of the developing countries of the world. In recent times it has re-appeared in places from where it disappeared long ago. Even though good sanitation and supply of safe water can contain and perhaps eliminate the disease, the sheer magnitude of the task makes it, for the most governments, a goal unattainable in near future. Importance of cholera vaccines as an useful tool to

TABLE 115.1   Composition of the vaccines described in the text and their dosing

Vaccine	Composition	Dosing
WC	Three strain of *V. cholerae* O1 : 2.5x1010 each of heat killed *V. cholerae* O1 classical, Cairo48 (Inaba); Cairo 50(Ogawa); Formalin killed *V. cholerae* O1 El Tor, Phil 6973 (Inaba), Cairo 50 (Ogawa)	3 doses 6 weeks apart
WC-CTB	As above plus 1 mg chemically extracted pure CTB	Do
WC-rCTB (Dukoral)	As in WC-CTB above excepting that CTB is produced by recombinant DNA techniques.	Two doses 2 weeks apart
ORCVAX	Four strain of *V. cholerae* O1: Same as WC excepting that 2.5×10^{10} formalin killed *V. cholerae* O1 classical 569B (Inaba) replaced 2.5×10^{10} formalin killed Cairo 50	Do
ORCVAX/Shancol	Three strains of *V. cholerae* O1 and one strain of *V. cholerae* O139: 300 ELISA units each of heat killed *V. cholerae* O1 classical Cairo 50 (Ogawa), Cairo 48(Ogawa) and formalin killed *V. cholerae* O1 Cairo 50 (Ogawa) plus 600 ELISA units each of formalin killed *V. cholerae* O1 El Tor, Phil 6973 (Inaba) and *V. cholerae* O139 4260B	Do
CVD 103 HgR Orochol/Mutachol	Genetically modified live attenuated *V. cholerae* O1 classical strain 569B (Inaba)	Single dose. 5×10^9 lyophilised cells administered with buffer
Peru-15 (Cholera Garde)	Genetically modified live attenuated *V. cholerae* O1 El Tor strain C6709 (Inaba)	Single dose. $5 \times 10^8 – 1 \times 10^9$ organisms administered with buffer
VC 638	Genetically modified live attenuated *V. cholerae* O1 El Tor strain 07258 (Ogawa)	Single dose. 5×10^9 cells administered with buffer
VA1.3	Genetically modified live attenuated non-toxigenic *V. cholerae* O1 El Tor strain VA1 (Inaba)	Single dose. 5×10^9 cells administered with buffer
VA1.4	Same as above without the Amp marker	Single dose. $5 – 7 \times 10^8$ cells administered with buffer

contain and control the disease is being increasingly recognized. In a recent policy paper WHO has endorsed the use of oral cholera vaccines as a part of the cholera control programme. As this chapter shows, over the last 30 year a number of new vaccines have been developed and atleast one of them, Shancol, have reached the market. Although it constitutes a major progress, it still suffers from two drawbacks- requirement of two doses and relatively high price, USD 1.85 per dose. The single dose live oral CVD-103 Hg^R, which showed great promise in limited studies failed in a large scale field trial. The others in development are yet to reach this stage. Work on improving these vaccines or developing newer ones therefore must continue.

SUGGESTION FOR FURTHER READING

1. Ali M., Lopez A.L., You A.Y., Kim Y.E., et al. (2012), Bull. WHO 90:209–218.

2. Kaper J.B., Morris Jr., G.J. and Levine M.M. (1996), *Clin. Microbiol.*, Rev., 8:48–86.

3. Harris J.B., LaRocque R., Quadri F., Ryan E., et al. (2012), *Lancet*, 379:2466–2476.

4. Colwell R.R. (1996), *Science*, 274:2025–2031.

5. Waldor M.K. and Mekalanos, J.J. (1996), *Science*, 272:1910–1914.

6. Safa A., Nair G.B. and Kong R.Y. (2009), *Trends Microbiol.*, 18:46–54.

7. Chatterjee S.N. and Chaudhuri K. (2003), Biochim., *Biophys. Acta.*, 1639:65–79.

8. Ghosh A., Thungapathra M. and Ghosh R.K. (1996), *Indian J. Med. Res.*, 104:60–75.

9. Pasetti M.F. and Levine M.M. (2012), *Clin Vaccine Immunol.*, 19:1707–1711.

10. Lopez A.L., Clemens J.D., Deen J. and Jodar L. (2008), *Human Vaccines*, 4:165–169.

11. Clemens J.D., Sack D.A., Harris J.R., Chakraborty J., et al. (1998), *Lancet*, ii: 124–127

12. Trach D.D., Clemens J.D., Ke M.D., Thuy H.D., et al. (1997), *Lancet*, 349:231–235.

13. Mahalanabis D., Lopez A.L., Sur D., Deen J., et al. (2008), *PLOS One*, 3:E2323

14. Sur D., Lopez A.L., Kanungo S., Paisley A., et al. (2009), *Lancet*, 374:1694–1702

15. Chowdhury M.I., Sheik A. and Qadri F. (2009), *Expert Rev Vaccines* 8:1643–1652.

16. Garcia H.M., Thomson R., Valera R., Fando R., et al. (2011), *Vacci Monitor*, 20:1–8

17. Mahalanabis D., Ramamurthy T., Nair G.B., Ghosh A., et al. *Vaccine*, 27:4850–4856.

18. Shin S., Desai S.N., Sah B.K. and Clemens J.D. (2011), *Clin Infect Diseases*, 52:1343–1349.

116

Development of Vaccines for Malaria
Challenges and Prospects

Chetan Chitnis

CONTENTS

I. Introduction
II. Malaria Parasite Life Cycle and Malaria Pathogenesis
III. Immunity to Malaria
IV. Targeting Different Life Cycle Stages of Malaria Parasites with Vaccines
   A. Pre-erythrocytic stage vaccines
   B. Blood stage vaccines
   C. Transmission blocking vaccines
   Summary

I. INTRODUCTION

Malaria remains a major public health problem in many parts of the tropical world. Human malaria is caused by five *Plasmodium* species, namely, *Plasmodium falciparum*, *P. vivax*, *P. malariae*, *P. ovale* and *P. knowlesi*. Around 500 million cases of malaria, which lead to around 1 million deaths, are reported worldwide every year. *P. falciparum* infections are the most lethal and account for the vast majority of deaths attributed to malaria. *P. vivax*, geographically is the most widespread *Plasmodium* species, is responsible for around 80 million malaria cases every year.

Malaria is transmitted by different species of *Anopheles* mosquitoes. Methods used to control malaria include indoor spraying of insecticides to control the vector population, use of insecticide-treated bednets to limit exposure of humans to infected mosquitos and early diagnosis and treatment to protect against malaria and limit transmission. Intensive application of such methods can significantly reduce malaria transmission. However, these methods are unlikely to be economically sustainable in the resource poor settings where malaria is commonly endemic. The development of a malaria vaccine could provide a valuable cost-effective tool, which in conjunction with the tools described above, may enable the effective control and ultimately elimination of malaria from endemic regions.

II. MALARIA PARASITE LIFE CYCLE AND MALARIA PATHOGENESIS

Infection in humans begins with the injection of *Plasmodium* sporozoites by infected female *Anopheles* mosquitoes during a blood meal. The injected sporozoites cross the vascular endothelium lining blood capillaries to find their way into the bloodstream. The sporozoites ride the bloodstream to reach the liver where they invade hepatocytes through specific receptor-ligand interactions and reside within a parasitophorous vacuole (PV). The parasites multiply within the PV by schizogeny and differentiate into merozoites that are released into the bloodstream to initiate the blood stage of the parasite life cycle.

In the blood stage, *Plasmodium* merozoites invade and multiply within host erythrocytes. Repeated cycles of invasion, multiplication, egress and re-invasion lead to increase in blood stage parasitemia. The blood stage of the life cycle is responsible for all the clinical symptoms of malaria. The rise of blood stage parasitemia beyond the fever threshold results in development of clinical symptoms including high fever accompanied by shivering and chills that are characteristic of malaria. The failure to control blood stage parasitemia, especially in case of *P. falciparum* infections, can lead to the development of severe malaria syndromes such as cerebral malaria, which is characterized by the patient slipping into coma and may be accompanied by other severe symptoms such as respiratory distress and multi-organ failure. Cerebral malaria has a high fatality rate even when good medical treatment is available. One of the key pathogenic mechanisms responsible for severe malaria syndromes such as cerebral malaria is the property of cytoadherence, which refers to the ability of *P. falciparum*-infected erythrocytes to adhere to receptors on vascular endothelium leading to sequestration of the parasite in the capillaries and deep vasculature of different host organs. Sequestration in the capillaries of the brain obstructs blood flow leading to the complications of cerebral malaria. The parasite proteins that are expressed on the surface of infected erythrocytes and mediate adhesion of *P. falciparum*-infected erythrocytes with endothelial receptors belong to the *P. falciparum* erythrocyte membrane protein-1 (PfEMP1) family and are encoded by *var* genes. Expression of PfEMP1 undergoes antigenic variation as a result of switching of *var* gene expression. Antigenic variation enables the parasite to avoid host antibody responses targeted against PfEMP1 on the infected erythrocyte surface.

During blood stage growth some of the parasites differentiate into male and female gametocytes, which are picked up by female *Anopheles* mosquitoes during a blood meal. Following transfer to the mosquito midgut, the male and female gametocytes differentiate into gametes, which fertilize to form a zygote. The zygote transforms into an ookinete, which traverses the mosquito midgut and forms an oocyst on the outer surface of the midgut wall. Thousands of sporozoites develop in each oocyst and are released in the hemocoel when mature. The sporozoites traverse the hemocoel to reach the thorax where they bind and invade salivary glands. The sporozoites are now ready to be injected into the human host again thus completing the parasite life cycle.

III. IMMUNITY TO MALARIA

Studies on the epidemiology of malaria in endemic regions with high transmission reveal that the incidence of *P. falciparum* and *P. vivax* malaria is highest in young children (1 to 5 year age group) and reduces with age. The incidence of severe malaria and deaths due to *P. falciparum* infections are also highest in young children and reduce in older children and adults. These observations suggest that residents of malaria endemic regions develop immunity to malaria following repeated exposure to malaria over a period of time. Parasite rates as well as parasite densities in case of both *P. falciparum* and *P. vivax* are lower in adults compared to children suggesting that naturally acquired anti-parasite immunity enables adults to limit parasite growth.

Immunity to malaria thus takes long to develop and requires multiple infections over a number of years. Moreover, sterilizing immunity is never acquired and even adults in endemic areas are infected although they may not develop clinical symptoms. Residents of malaria endemic regions develop antibodies against diverse parasite proteins over time following repeated exposure to *P. falciparum* and *P. vivax*. However, there are no clear correlates of protective immunity and it has not been possible to define 'neutralizing antibody titres' as in the case of viral infections.

The lack of clear immune correlates has made it difficult to develop a clear rationale for a vaccine that can elicit protection against malaria. However, the following experiments provide evidence to indicate that it should be possible to develop vaccines against malaria. First, it has been shown that transfer of immunoglobulins (IgG) purified from immune adults residing in a *P. falciparum* endemic region to infected individuals successfully cleared blood stage *P. falciparum* parasites leading to full recovery (1). This observation suggests that it should be possible to elicit antibodies against key *P. falciparum* blood stage antigens that will limit parasite growth and provide protection against *P. falciparum* malaria. Secondly, it has been demonstrated that immunization of mice with irradiated *P. berghei* sporozoites and naïve adults with irradiated *P. falciparum* sporozoites elicits protection against challenge with *P. berghei* and *P. falciparum* sporozoites, respectively (2, 3). Finally, it has been possible to achieve protective efficacy of 30–50% in children residing in malaria endemic regions of Africa against natural challenge with *P. falciparum* (4–7). These observations indicate that it is possible to elicit protective immune responses against malaria parasites.

IV. TARGETING DIFFERENT STAGES OF THE MALARIA PARASITE LIFE CYCLE WITH VACCINES

The different stages of the life cycle of malaria parasites offer unique opportunities for intervention (Figure 116.1). *Plasmodium* sporozoites that are injected into the human host during a blood meal by *Anopheles* mosquitoes are susceptible to antibodies that target its surface antigens on sporozoites and prevent invasion of hepatocytes (Figure 116.1). One can attempt to elicit antibody responses against sporozoite surface antigens such as the circumsporozoite protein (CSP) or thrombospondin related adhesive protein (TRAP), which mediate interactions with host receptors during hepatocyte invasion. Cytotoxic T cells can target infected hepatocytes by recognizing peptides derived from parasite antigens presented in the context of major histocompatibility class I (MHC 1) molecules on the hepatocyte surface. CD8+ T cell epitopes from parasite antigens such as liver stage antigen 1 (LSA-1) that are expressed in infected hepatocytes can be used to develop malaria vaccines that elicit cytotoxic T cell responses that clear such infected hepatocytes. Secretion of cytokines such as interferon-γ (IFN-γ) by such cellular immune responses can also inhibit growth of infected hepatocytes. Merozoites released into the bloodstream from mature liver stage schizonts infect erythrocytes to establish the blood stage of the infection. Multiple cycles of invasion, multiplication, egress and re-invasion leads to a rise in blood stage parasitemia. All the clinical symptoms of malaria are attributed to the blood

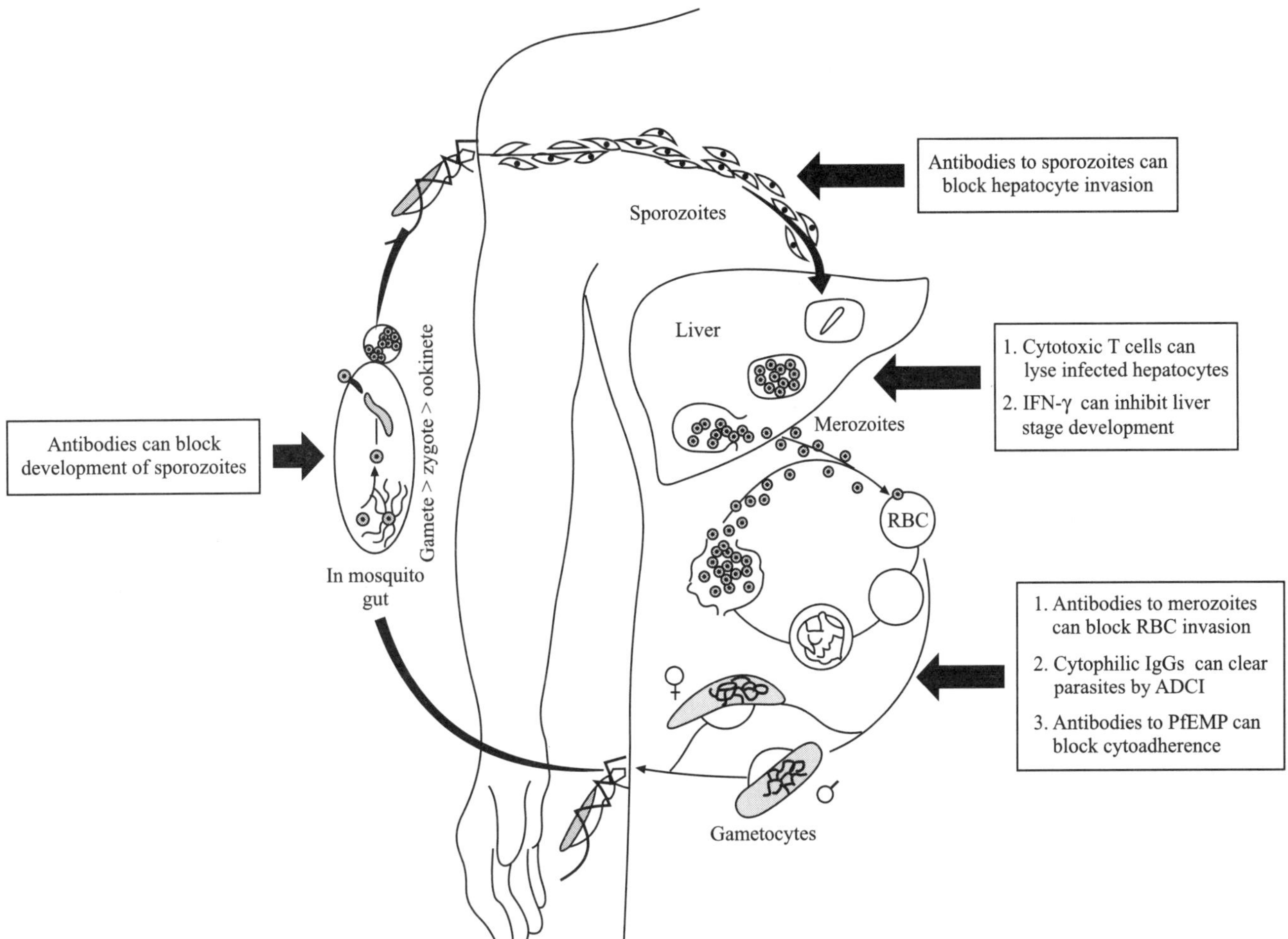

Figure 116.1　Life cycle stages of malaria parasites and points for intervention. Antibodies against sporozoite surface proteins can inhibit hepatocyte invasion. Cellular immune responses directed against infected hepatocytes can inhibit intracellular growth and development of merozoites. Merozoites released in the bloodstream can be targeted with antibodies that block invasion of erythrocytes or lead to clearance of merozoites by antibody dependent cellular inhibition (ADCI). Antibodies against the variant surface antigen, PfEMP1, which also mediates cytoadherence, can block cytoadherence and prevent severe malaria. Antibodies directed against transmission stages can be transferred to the mosquito midgut where they can inhibit development of malaria parasites in the vector.

stage of the parasite life cycle. Targeting blood stage malaria parasites to limit growth is thus critical to elicit protection against malaria. Antibodies may be elicited against merozoite proteins that mediate interactions with erythrocyte receptors during invasion to block key receptor-ligand interactions and inhibit erythrocyte invasion. Following invasion, *P. falciparum* expresses proteins on the surface of infected erythrocytes that bind receptors on endothelial cells lining capillary walls leading to sequestration of infected erythrocytes in the vasculature of host organs. This phenomenon referred to as cytoadherence allows *P. falciparum* to avoid passage through the spleen where infected erythrocytes may be cleared. Sequestration of infected erythrocytes in brain capillaries is implicated in the severe malaria syndrome of cerebral malaria and adhesion in the placenta leads to complications in pregnancy. Antibodies elicited against parasite ligands expressed on the surface of *P. falciparum* – infected erythrocytes can limit cytoadherence

to prevent syndromes such as cerebral malaria or placental sequestration providing protection against severe disease. One of the challenges in raising antibodies against cytoadherence ligands is the extent of sequence diversity found in PfEMP1 family members and the continuous switching of PfEMP1 expression by the process of antigenic variation. Finally, antibodies elicited against either the sexual stages such as gametes or mosquito stages such as ookinetes can block development of sporozoites to inhibit transmission of malaria parasites. Such vaccines will not protect the individual against malaria but will prevent the transmission of malaria in the community. Ultimately, a combination of liver stage, blood stage and transmission blocking vaccine candidates may be required to effectively prevent malaria and inhibit its transmission. We will review below the rationale for and progress made in development of malaria vaccines against the liver, blood and transmission stages of malaria parasites.

A. Pre-erythrocytic Vaccines

Infection in the human host commences with the injection of a small number (<100) of sporozoites through the bite of an infected *Anopheles* mosquito. *Plasmodium* sporozoites specifically invade host hepatocytes. The major surface protein on sporozoites is referred to as the circumsporozoite protein (CSP). The N-terminus of *P. falciparum* CSP (PfCSP) contains a conserved region, referred to as Region I, which contains multiple basic residues that play a role in binding to salivary glands (Figure 116.2). The central region of PfCSP contains multiple tetrapeptide (NANP/NVDP) repeats and is followed by a conserved cysteine-containing region, Region II (RII), that shares homology with the mammalian protein thrombospondin (TSP), which mediates cell-cell interactions (Figure 116.2). The tetrapeptide repeat region is immuno-dominant. Naturally acquired antibodies against PfCSP following natural infection primarily recognize the NANP/NVDP repeat region. The receptor-binding domain of PfCSP, which mediates binding to hepatocyte receptors, has been mapped to the TSP-like region (RII). One of the leading malaria vaccine candidates, RTS, S, being developed by GlaxoSmithKline Biologicals and Walter Reed Army Institute of Research is based on PfCSP. RTS, S contains part of the NANP repeat region (R) and thrombospondin-like RII region of PfCSP (T) fused with hepatitis B surface antigen (S). The fusion protein (RTS) was co-expressed with wild type hepatitis B surface antigen (S) to allow assembly of virus-like-particles (VLPs) that are referred to as RTS, S. RTS, S formulated with strong adjuvants such as AS02 (oil-in-water emulsion of monophosphoryl lipid A (MPL) and saponin QS21) and AS01 (MPL and QS21 in liposomes) yielded 30–50% protection against challenge by *P. falciparum* infected *Anopheles* mosquitoes in various Phase IIa challenge trials (4). Phase IIb field trials conducted in children residing in *P. falciparum* endemic regions of Africa demonstrated that immunization with RTS, S resulted in 30–50% reduction in clinical malaria cases during first 12 months of follow up (4). The Phase IIb trials also suggested that RTS, S may have higher efficacy against severe disease and may thus protect against mortality associated with *P. falciparum* malaria. RTS, S is currently being tested in a Phase III efficacy trial involving ~15,000 children and infants residing in 11 endemic sites in 7 African countries. Interim results in 5–17 month old infants after 1 year follow up demonstrated efficacy of 50.4% against clinical malaria and efficacy of 45% against severe malaria (5). However, interim results in younger 6–12 week old infants

indicated that efficacy of RTS, S for protection against clinical malaria was only 30.1% and efficacy against severe malaria was only 26% (6).

Other sporozoite or liver stage parasite antigens that are under development as subunit vaccines for malaria include the *P. falciparum* thrombospondin repeat adhesive protein (PfTRAP) and *P. falciparum* liver stage antigen-1 (PfLSA-1). A prime-boost strategy using fowlpox virus (FP9) and modified vaccine virus (MVA) to deliver PfTRAP fused to a string of multiple T cell epitopes from different pre-erythrocytic antigens (ME-TRAP) was tested in children residing in a *P. falciparum* endemic region of Africa. PfTRAP, like PfCSP, contains the conserved thrombospondin-like region, and mediates interaction with hepatocytes. Although the prime-boost regimen elicited modest antibody and cellular immune responses against parasite epitopes, there was no protection against *P. falciparum* malaria in a Phase IIb field trial conducted in a malaria endemic region of Kenya (7). LSA-1 is expressed in *P. falciparum* infected hepatocytes. Immunization with recombinant LSA-1 formulated with adjuvants AS01 and AS02 elicited moderate antibody and cellular immune responses that failed to protect naïve adults against *P. falciparum* challenge in a Phase IIa clinical trial (8).

As mentioned above, immunization with irradiated *P. falciparum* sporozoites provides protection against sporozoite challenge. Methods to produce such attenuated sporozoites under good manufacturing practice (GMP) for use as a vaccine against malaria have been established by Sanaria Inc. The attenuated sporozoites are stored in liquid nitrogen and thawed prior to immunization. Initial attempts to immunize volunteers with attenuated sporozoites by intramuscular injection did not yield any protection against subsequent *P. falciparum* challenge (9). Clinical trials to test immunization with such attenuated sporozoites by intravenous delivery are currently underway.

B. Blood Stage Vaccines

As mentioned above, all the clinical symptoms of malaria are attributed to the blood stage of the malaria parasite life cycle during which *Plasmodium* merozoites invade and multiply within host erythrocytes. Blood stage vaccines that elicit antibodies against merozoite antigens to inhibit blood stage growth can help control parasite densities and protect against clinical malaria. The leading blood stage malaria vaccine candidates under development include *P. falciparum* merozoite surface antigen 1 (PfMSP1), apical merozoite antigen-1 (PfAMA1), 175 kd erythrocyte binding antigen (PfEBA175)

Figure 116.2 Structure of *P. falciparum* circumsporozoite protein (PfCSP). PfCSP has a signal sequence at the N-terminus and a consensus sequence for attachment of a glycosylphosphatidylinositol (GPI) anchor at the C-terminus. The extracellular region contains a tetrapeptide repeat region that is flanked by conserved sequences such as region I, which contains conserved basic amino acid residues, and region II, which contains a conserved thrombospondin-like region that mediates interactions with hepatocyte receptors.

and its *P. vivax* homolog, Duffy binding protein (PvDBP). Antibodies against these antigens have been shown to inhibit erythrocyte invasion and reduce blood stage growth *in vitro*. Antibodies against merozoite antigens may also clear blood stage parasites by the method of antibody-dependent cellular inhibition (ADCI). In this mechanism, monocytes recognize cytophilic antibodies bound to merozoites leading to clearance of merozoites by opsonization or killing by secretion of cytocidal factors. A vaccine candidate based on glutamic rich protein (GLURP) and MSP3, both of which elicit such cytophilic antibodies is under development. We will briefly review progress made with each of these malaria vaccine candidates.

PfMSP1

PfMSP1 is a 195 kD protein that covers the surface of *P. falciparum* merozoites and is anchored to the surface by a glycophospatidyl inositol (GPI) anchor. It contains multiple conserved amino acid stretches that are interspersed with variable regions. The C-terminus of PfMSP1 contains two conserved cysteine-rich regions with homology to epidermal growth factor (EGF)-like domains. PfMSP1 is extensively processed by proteases prior to invasion. A 42 kD C-terminal fragment of PfMSP1 (PfMSP1$_{42}$) remains on the merozoite surface after initial processing events. Finally only a 19 kD, C-terminal fragment, which contains the conserved, cysteine-rich EGF-like domains, remains attached to the merozoite as it invades host erythrocytes. Antibodies raised against the C-terminal PfMSP1$_{19}$ fragment inhibit erythrocyte invasion *in vitro* and presence of naturally acquired antibodies against PfMSP1$_{19}$ in sera of individuals residing in endemic regions is associated with protection against *P. falciparum* malaria. Recombinant PfMSP1$_{42}$ formulated with AS02A failed to provide any protection against natural challenge by *P. falciparum* in endemic settings (10). The failure to provide

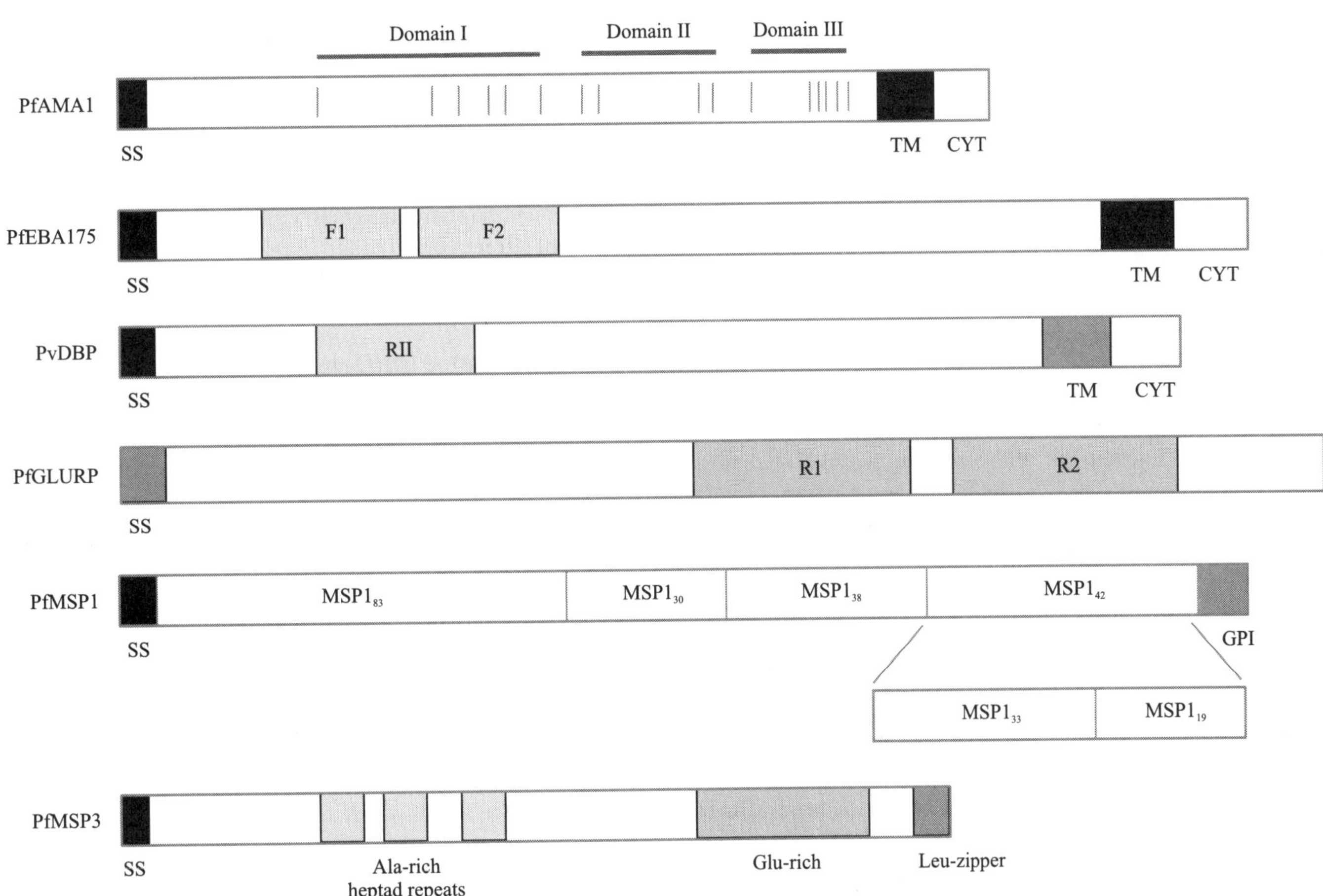

Figure 116.3 Blood stage antigens from *P. falciparum* merozoites. Parasite proteins from the merozoite surface and apical organelles that can be targeted with antibodies include *P. falciparum* apical merozoite antigen-1 (PfAMA1), 175 kD erythrocyte binding antigen (PfEBA175), *P. vivax* Duffy binding protein (PvDBP), *P. falciparum* glutamate rich protein (GLURP), *P. falciparum* merozoite surface protein-1 (PfMSP1) and *P. falciparum* merozoite protein-3 (PfMSP3). PfAMA1 contains three sub-domains with conserved cysteines (vertical blue bars) that form structurally important intra-domain disulfide bonds. The receptor-binding domains of PfEBA175 and PvDBP lie in regions F1-F2 and RII respectively. GLURP contains two regions, R1 and R2, that are rich in glutamates. PfMSP1, which is attached to the merozoite surface by a glycosylphoshatidyl inositol (GPI) anchor, is proteolytically cleaved into multiple fragments during invasion. T C-terminal PfMSP142 fragment is finally cleaved to leave a conserved 19kD fragment on the merozoite surface referred to as PfMSP1$_{19}$. PfMSP3 contains a alanine-rich heptad repeat and a glutamate-rich region in the extracellular domain.

protection is likely due to inability to elicit high titer growth inhibitory antibodies against PfMSP1$_{42}$ in humans and presence of polymorphisms in the target antigen.

PfAMA-1

PfAMA1 is localized in membrane bound organelles referred to as micronemes at the apical end of *P. falciparum* merozoites. PfAMA1 is secreted to the merozoite surface from micronemes just before invasion. It contains three conserved cysteine-rich subdomains that contain intra-domain disulfide linkages. Recombinant PfAMA1 is highly immunogenic and immunization of laboratory animals elicits highly potent growth inhibitory antibodies. However, *P. falciparum* field isolates display extensive polymorphisms in PfAMA1 sequence. Antibodies raised against PfAMA1 display lower growth inhibition titers against heterologous *P. falciparum* strains. Recombinant PfAMA1 formulated with human compatible adjuvants failed to provide protection against *P. falciparum* in Phase IIb field trials conducted in malaria endemic regions of Africa (11). Extensive polymorphism found in PfAMA1 in *P. falciparum* field isolates is thought to be one of the reasons for the lack of efficacy of a vaccine based on PfAMA1. It may be necessary to combine multiple PfAMA1 variants in a vaccine to cover the wide sequence diversity in PfAMA1 and provide protection against diverse *P. falciparum* isolates.

PfEBA175 and *P. vivax* homolog PvDBP

A family of parasite proteins belonging to the erythrocyte binding protein family mediate specific interactions with erythrocyte receptors during invasion. The 175 kD erythrocyte binding antigen (EBA175) binds sialic acid residues on glycophorin A (gly A), its paralogs EBA181, EBA140 and EBL1 mediate interactions with other erythrocyte receptors and its homologue from *P. vivax*, Duffy binding protein (PvDBP) binds the Duffy blood group antigen. The extracellular domains of EBPs contain two conserved cysteine-rich regions referred to as regions II and VI. Of these, the N-terminal cysteine-rich region, region II, of each EBP has been shown to mediate binding to erythrocyte receptors. Malaria parasites have thus used a conserved cysteine-rich motif to generate parasite ligands that can bind diverse receptors with specificity.

P. vivax is completely dependent on interaction with the Duffy antigen on human erythrocytes for invasion. Duffy negative erythrocytes are resistant to invasion by *P. vivax*. Antibodies against the receptor-binding region II of PvDBP (PvDBPII) inhibit erythrocyte invasion by *P. vivax in vitro*. The receptor-binding residues mapped on PvDBPII are highly conserved in diverse *P. vivax* field isolates. Antibodies raised against recombinant PvDBPII have been shown to inhibit binding of diverse PvDBPII variants to Duffy antigen suggesting that polymorphisms will not be a problem for a vaccine based on PvDBP. Importantly, field studies have demonstrated that acquisition of binding inhibitory antibodies against PvDBPII upon natural exposure to *P. vivax* infection is associated with

reduced risk of subsequent *P. vivax* infections (12). Together, these data provide support for development of PvDBPII as a vaccine candidate for *P. vivax* malaria.

Unlike *P. vivax*, *P. falciparum* can use multiple pathways for invasion of human erythrocytes. EBA175 binds sialic acid residues on gly A to mediate invasion by sialic acid dependent invasion pathway. The receptor-binding domain of EBA175 has been mapped to the conserved, cysteine-rich region at the N-terminus referred to as region II. Antibodies raised against region II have been shown to inhibit erythrocyte invasion by diverse *P. falciparum* strains that use alternative invasion pathways (13). These studies validate the inclusion of EBA175 region II in a blood stage vaccine for *P. falciparum* malaria. However, it may be useful to combine EBA175 with other merozoite proteins to target multiple pathways and block erythrocyte invasion efficiently.

GLURP-PfMSP3

Cytophilic antibodies (e.g., IgG2a and IgG3) against merozoite proteins can lead to clearance of parasites by antibody dependent cellular inhibition (ADCI) mechanisms that involve co-operation with monocytes. Two merozoite surface proteins that serve as targets for ADCI include glutamate rich protein (GLURP) and merozoite surface protein-3 (PfMSP3). A fusion peptide, GMZ2, containing B cell epitopes for cytophilic antibodies from GLURP and PfMSP3 was produced as a synthetic peptide and is currently being tested in clinical trials to test efficacy in Africa (14).

Other blood stage antigens and combinations

Erythrocyte invasion is a complex multi-step process that is mediated by a cascade of interactions between parasite ligands and host receptors. A number of novel parasite antigens that mediate these interactions during erythrocyte invasion have been recently identified. For example, a family of *P. falciparum* proteins referred to as Rh proteins that share homology with *P. vivax* reticulocyte binding proteins has been shown to mediate erythrocyte binding. Like other invasion related proteins, PfRH proteins are also localized in apical organelles and are secreted to the merozoite surface during invasion. Other novel apical merozoite proteins that mediate binding to erythrocytes during invasion include the apical asparagine rich protein (AARP) and thrombospondin-related apical merozoite protein (TRAMP). The receptor-binding domains of these invasion related parasite proteins have been mapped to conserved internal regions. Antibodies against individual invasion related parasite proteins do not inhibit invasion with high efficiency. Combination of antibodies against at least three invasion proteins (e.g., antibodies against PfEBA175, PfRH2 and PfAARP) leads to synergistic inhibition of invasion with high efficiency (>80% inhibition at antibody concentrations of 3.3 mg/ml against each antigen) (15). Such combinations, which inhibit invasion of diverse *P. falciparum* strains with high efficiency, should form the basis of next generation blood stage vaccines for *P. falciparum* malaria.

C. Transmission Blocking Vaccines

Male and female gametes picked up by an *Anopheles* mosquito during a blood meal fertilize in the mosquito midgut to form zygotes, which develop into ookinetes that traverse the midgut epithelium and form oocysts on the outer wall of the midgut. Sporozoites that develop in the oocyst are released and invade salivary glands. Development stages such as gametes, zygote and ookinete can be targeted by antibodies in the mosquito midgut to block development of transmission stages. Antibodies directed against gametocyte antigens such as Pfs48/45 inhibit fertilization and formation of zygotes in the mosquito midgut to interrupt transmission. Pfs48/45 contains two conserved 6-cys motifs in its extracellular domain. Sera raised against Pfs48/45 have been shown to block development of transmission stages in standard membrane feeding assays (SMFA). Two other transmission blocking vaccine candidates include Pfs25 and Pfs28, which are localized on the surface of ookinetes. Antibodies raised against these antigens inhibit midgut traversal by ookinetes to block oocyst development. While immunization of laboratory animals with these recombinant vaccine candidates has yielded high titer inhibitory antibodies that block oocyst development in SMFA, clinical trials with recombinant Pfs25 formulated with human compatible adjuvants did not yield such potent inhibitory antibodies (16). There is a need to develop novel delivery platforms that elicit long lasting high titer antibodies against such transmission blocking vaccine candidates to effectively block transmission.

SUMMARY

There is an urgent need to develop vaccines that provide protection against *P. falciparum* and *P. vivax* malaria. Recombinant subunit vaccines that target different stages of the malaria parasite life cycle are under development. Pre-erythrocytic stage vaccines either elicit antibodies against the sporozoite surface to block invasion of hepatocytes or elicit cellular immune responses to clear infected hepatocytes. Blood stage vaccines target parasite antigens to block invasion of erythrocytes by *Plasmodium* merozoites. Targeting a combination of key merozoite proteins involved in invasion is likely to be necessary to elicit potent invasion inhibitory antibodies. Vaccines based on antigens expressed in sexual or mosquito stages elicit antibodies that block transmission of malaria parasites. It may be necessary to combine vaccines targeting different stages to develop a malaria vaccine with high efficacy.

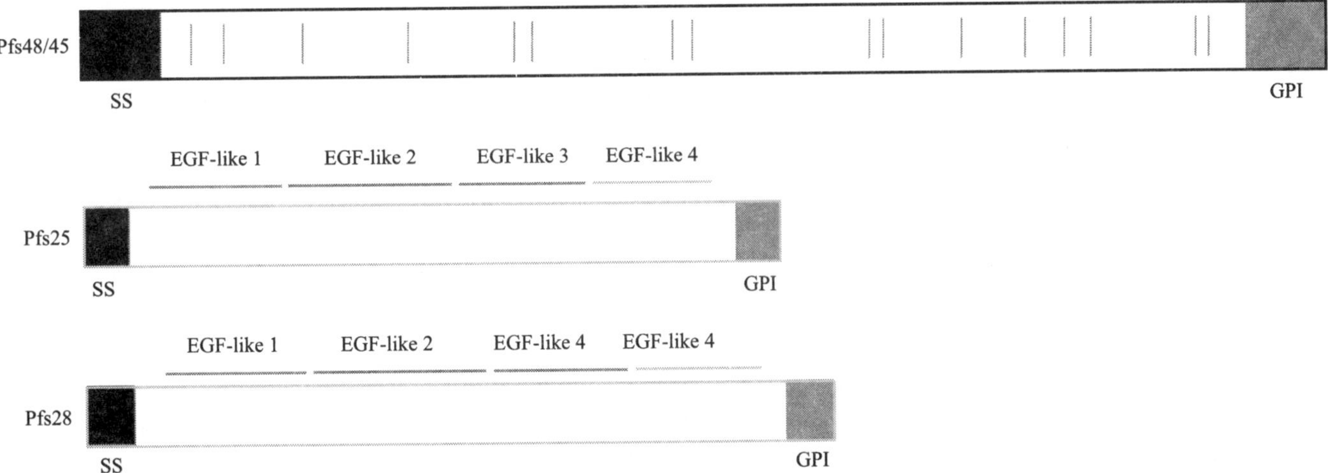

Figure 116.4   Antigens from transmission stages of malaria parasites. Pfs48/45, which is expressed on gametocytes, contains conserved cysteines (vertical blue bars). Pfs25 and Pfs28, which are expressed on surface of ookinetes, contain multiple cysteine-rich domains with homology to epidermal growth factor-like (EGF-like) regions.

REFERENCES

1. Cohen S., McGregor I. and Carrington S. (1961), Gamma-globulin and Acquired Immunity to Human Malaria, *Nature*, 192:733–7.

2. Nussenzweig R.S., Vanderberg J., Most H. and Orton C. (1966), Protective Immunity Produced by the Injection of X-irradiated Sporozoites of *Plasmodium berghei, Nature*, 216:160–2.

3. Clyde D.F., Most H., McCarthy V.C. and Vanderberg J.P. (1973), Immunization of Man Against Sporozite-induced Falciparum Malaria, *Am. J. Med. Sci.,* 266, 169–177.

4. Ballou W.R. (2009), The Development of the RTS,S Malaria Vaccine Candidate: Challenges and Lessons, *Parasite Immunology*, 31:492–500.

5. The RTS, Clinical Trials Partnership (2011), First Results of Phase 3 Trial of RTS, S/AS01 Malaria Vaccine in African Children, *N Engl J. Med.,* 365:1863–75.

6. The RTS, S Clinical Trials Partnership (2012), A Phase 3 Trial of RTS,S/AS01 Malaria Vaccine in African Infants, *N Engl J. Med.,* 367: 2284–2295.

7. Bejon P., Mwacharo J., Kai O., Mwangi T., Milligan T., Todryk S., Keating S., Lang T., Lowe B., Gikonyo C., Molyneux C., Fegan G., Gilbert S.C., Peshu N., Marsh K. and Hill A.V.S. (2006), A Phase 2b Randomised Trial of the Candidate Malaria Vaccines FP9 ME-TRAP and MVA ME-TRAP Among Children in Kenya, *PLoS Clinical Trial,* 1(6):e29.

8. Cummings J.F., Spring M.D., Schwenk R.J., Ockenhouse C.F., Kester K.E., Polhemus M.E., Walsh D.E., Yoon I.K., Prosperi C., Juompan L.Y., Lanar D.E., Krzych U., Hall B.T., Ware L.A., Stewart V.A., Williams J., Dowler M., Nielsen R.K., Collette Hillier J., Giersing B.K., Dubovsky F., Malkin E., Tucker K., Dubois M.C., Cohen J.D., Ballou W.R. and Heppner Jr. D.G. (2010), Recombinant Liver Stage Antigen-1 (LSA-1) Formulated with AS01 or AS02 is Safe, Elicits High Titer Antibody and Induces IFN-⁺/IL-2 CD4+ T Cells But Does Not Protect Against Experimental *Plasmodium falciparum* Infection, *Vaccine,* 28:5135–5144.

9. Epstein J.E., Tewari K., Lyke K.E., Sim B.K., Billingsley P.F., Laurens M.B., Gunasekera A., Chakravarty S., James E.R., Sedegah M., Richman A., Velmurugan S., Reyes S., Li M., Tucker K., Ahumada A., Ruben A.J., Li T., Stafford R., Eappen A.G., Tamminga C., Bennett J.W., Ockenhouse C.F., Murphy J.R., Komisar J., Thomas N., Loyevsky M., Birkett A., Plowe C.V., Loucq C., Edelman R., Richie T.L., Seder R.A. and Hoffman S.L. (2011), Live Attenuated Malaria Vaccine Designed to Protect through Hepatic CD8⁺ T Cell Immunity, *Science,* 334:475–480.

10. Ogutu B.R., Apollo O.J., McKinney D., Okoth W., Siangla J., Dubovsky F., Tucker K., Waitumbi J.N., Diggs C., Wittes J., Malkin E., Leach A., Soisson L.A., Milman J.B.,Otieno L., Holland C.A., Polhemus M., Remich S.A., Ockenhouse C.F., Cohen J., Ballou W.R., Martin S.K., Angov E., Stewart V.A., Lyon J.A., Heppner D.G. and Withers M.R. (2009), Blood Stage Malaria Vaccine Eliciting High Antigen-specific Antibody Concentrations Confers No Protection to Young Children in Western Kenya, *PLoS One,* 4(3):e4708.

11. Thera M.A., Doumbo O.K., Coulibaly D., Laurens M.B., Ouattara A., Kone A.K., Guindo A.B., Traore K., Traore I., Kouriba B., Diallo D.A., Diarra I., Daou M., Dolo A., Tolo Y., Sissoko M.S., Niangaly A., Sissoko M., Takala-Harrison S., Lyke K.E., Wu Y., Blackwelder W.C., Godeaux O., Vekemans J., Dubois M.C., Ballou W.R., Cohen J., Thompson D., Dube T., Soisson L., Diggs C.L., House B., Lanar D.E., Dutta S., Heppner D.G. Jr. and Plowe C.V. (2011), A Field Trial to Assess a Blood-stage Malaria Vaccine, *N Engl J. Med.,* 365: 1004–1013.

12. King C.L., Michon P., Shakri A.R., Marcotty A., Stanisic D., Zimmerman P.A., Cole-Tobian J.L., Mueller I. and Chitnis C.E. (2008), Naturally Acquired Duffy-binding Protein-specific Binding Inhibitory Antibodies Confer Protection from Blood-stage Plasmodium Vivax Infection, *Proc Natl Acad Sci.,* USA, 105:8363–8368.

13. Jiang L., Gaur D., Mu J., Zhou H., Long C.A. and Miller L.H. (2011), Evidence for Erythrocyte-binding Antigen 175 as a Component of a Ligand-blocking Blood-stage Malaria Vaccine, *Proc Natl Acad Sci.,* USA, 108:7553–7558.

14. Sirima S.B., Cousens S. and Druilhe P. (2011), Protection Against Malaria by MSP3 Candidate Vaccine, *N Engl J. Med.,* 365:1062–1064.

15. Pandey A.K., Reddy K.S., Sahar T., Gupta S., Singh H., Reddy E.J., Asad M., Siddiqui F.A., Gupta P., Singh B., More K.R., Mohmmed A., Chitnis C.E., Chauhan V.S. and Gaur D. (2013), Identification of a Potent Combination of Key *Plasmodium falciparum* merozoite Antigens that Elicit Strain-transcending Parasite Neutralizing Antibodies, *Inf. Immun.,* 81:441–451.

SUGGESTIONS FOR FURTHER READING

1. Riley E.M. and Stewart V.A. (2013), Immune Mechanisms in Malaria: New Insights in Vaccine Development, *Nat Med.,* 19:168–178.

2. Crompton P.D., Pierce S.K. and Miller L.H. (2010), Advances and Challenges in Malaria Vaccine Development, *J. Clin Invest.,* 120: 4168–4178.

3. Birkett A.J., Moorthy V.S., Loucq C., Chitnis C.E. and Kaslow D.E. (2013), Malaria Vaccine R&D in the Decade of Vaccines: Breakthroughs, Challenges and Opportunities, *Vaccine,* 31 Suppl 2:B233–43.

117

BCG, Pros & Cons & New/Improved Vaccines for Tuberculosis

Abu Salim Mustafa

CONTENTS

I. HISTORY OF BCG

The currently licensed vaccine to immunize humans against tuberculosis is known as Bacillus Calmette-Guérin (BCG), which has been widely used around the world as a prophylactic measure against tuberculosis for about 90 years. The parent BCG vaccine strain was derived from a virulent strain of *Mycobacterium bovis*, a primary cause of tuberculosis in cattle. *M. bovis* was isolated in 1902 by Nocard at the Pasteur Institute of Paris, France, from the milk of a cow suffering from tuberculous mastitis. This isolated strain of *M. bovis* was transferred to the Pasteur Institute of Lille, France, and used by Albert Calmette and Camille Guérin to study bovine tuberculosis. However, *M. bovis* bacilli used to grow in clumps in vitro, and in order to minimize mycobacterial clumping for optimization of infection in animals, they added ox bile to the glycerol-soaked potato slices and used it as a medium to culture *M. bovis*. Within a few months of growth on this medium, they observed alterations in colony morphology and when injected into guinea pigs, the bacilli were less virulent than the original *M. bovis*. Appreciating the importance of reduced virulence in terms of vaccine development, Calmette and Guérin continued the serial *in vitro* passaging of the Nocard's *M. bovis* strain at 3-weekly intervals for 13 years (1908–1921), leading to a total of 231 passages. When administered at different doses and by different routes, the lab-adapted *M. bovis* BCG strain was well tolerated and did not produce tuberculous lesions in various animal models, including guinea pigs, rabbits, dogs, cattle, horses, chickens and non-human primates. However, these cultures maintained the clumped morphology, and the same physical properties and exhibited continued immunogenicity in animal models, i.e., chimpanzees, guinea pigs, mice, and cattle. Furthermore, BCG vaccination protected cows against challenge with virulent *M. bovis*. These results established the safety and efficacy of BCG vaccine in experimental animals.

In 1921, the BCG strain was used for the first time in humans at the request of a French physician, who wanted to protect an infant born to a mother, who died of tuberculosis a few hours after the childbirth, and who was under the care of a grandmother suffering from tuberculosis. Calmette administered orally the culture of the BCG strain in three doses of 2 mg each (6 mg total; ~2.4 × 10^8 bacilli). On the follow-up, there were no serious side effects and the child did not develop any sign of tuberculosis. Based on this first success, over the next year, additional newborns were vaccinated and no ill effects were reported. Thus, for the first time, a safe and apparently effective vaccine was available for protection against human tuberculosis. The Pasteur Institute in Lille started to mass-produce the BCG vaccine for medical use and human applications.

The protective efficacy of BCG on a relatively large scale was determined in high-risk children (born to mothers with tuberculosis or having contact with tuberculosis patients) between 1921 to 1927. The children were divided into BCG-vaccinated and non-vaccinated groups. Among 969 vaccinated children, 3.9% died of TB or unspecified causes, whereas the death rate in a comparable group of unvaccinated children was relatively much higher, i.e., 32.6% of them died. Based on these interesting results, the League of Nations, in 1928, recommended widespread vaccination with BCG. Hence, the BCG vaccine became internationally accepted and mass vaccinations were first started in Europe, i.e., France and Scandinavian countries. After the Second World War in 1945, the use of BCG vaccine against tuberculosis increased in Europe and the developing world. Since 1966, as part of a World Health Organization (WHO) initiative, the BCG vaccine is being lyophilized and provided in a powder form. Furthermore, since 1974, BCG vaccination has been included in the WHO Expanded Program on Immunization. It is estimated that over 4 billion people have been immunized with BCG and about 100 million doses of BCG are administered annually, which makes it the most widely used vaccine in humans.

II. ROUTE AND DOSE OF BCG IN HUMANS

Between 1921 and 1924, Calmette administered the culture of BCG vaccine orally in about 300 children in a total dose of 6 mg divided in three doses of 2 mg each. After demonstrating the safety of the BCG vaccine at this dose and route, Calmette decided to increase the total dose of BCG to 30 mg divided in three doses of 10 mg delivered orally. The reason Calmette chose the oral route was the fact that the natural route for infection by *M. bovis* is the gastrointestinal tract, and therefore oral route could be optimal for induction of protective immunity. In addition to oral route, parenteral routes (cutaneous and subcutaneous) were also attempted in children, but due to local reactions, these routes were unacceptable to the children's relatives and the oral route was continued for immunization with BCG. Between 1924 and 1928, 114,000 children were immunized orally without serious complications.

Although, the BCG vaccination started in children via the oral route, soon the vaccination program was extended to include adults, but in this population BCG was first given subcutaneously, and in considerably smaller doses. This was primarily based on the observations made in studies comparing oral and subcutaneous administration of BCG in adults between 1924 to 1926. In these studies, the nurses in a hospital of Oslo (Norway, a Scandinavian country) observed that the oral administration of BCG failed to produce consistent allergic response to the purified protein derivative (PPD) of *M. tuberculosis*. In agreement with the hypothesis at the time that the cutaneous tissue is an important source of antibodies required for protection against the disease, the nurses decided to apply the vaccine via the subcutaneous route. To demonstrate the workability of the concept of subcutaneous

administration, BCG vaccine was first injected into two "non-allergic" (naive) individuals, who subsequently became positive to the cutaneous tuberculin test using PPD. Further work showed that the subcutaneous immunization produced positivity to the tuberculin skin test in considerably larger proportion of vaccinated people (about 90% of the cases), whereas oral immunization yielded tuberculin positivity only in 40% of the recipients. Furthermore, it was demonstrated that subcutaneous administration of BCG at the dosage used did not cause unacceptable side effects. Since the parenteral administration of BCG could enable significantly higher degree of "allergic" reactions to PPD in tuberculin skin test, one of the criteria accepted as evidence of immunity against tuberculosis, the oral route was abandoned and the subcutaneous and then intradermal administration was recommended. There were two other reasons to discontinue the oral administration of BCG. One was the reduction (1 to 2 log) in the viability of BCG upon exposure to gastric secretions and low pH, and thus very high concentrations of bacilli were required for sensitization of the vaccine recipients. Second, oral administration of BCG was frequently associated with cervical lymphadenopathy. Consequently, the subcutaneous, followed by intradermal route became popular, especially after 1927, when Wallgreen perfected the intradermal vaccination, inoculating 0.1 mg of BCG in individuals of any age, negative to the tuberculin skin test.

III. WORLD-WIDE DISTRIBUTION AND SUBSTRAINS OF BCG

After the presentation of results to the National Academy of Medicine in Paris in 1924, showing protection against tuberculosis in children immunized with BCG, the Pasteur Institute of Lille was authorized to distribute samples of the bacillus to other laboratories all over the world. This distribution occurred in two phases. In the first phase between 1924 and 1926, at least 34 countries received the culture of BCG and, in the second phase after 1927, another 26 countries received cultures of BCG from the Pasteur Institute. In the recorded literature related to BCG, at least 49 substrains have been produced and used in humans at one time or another in various parts of the world, and could be traced to the original BCG seed lot produced at the Pasteur Institute. The original BCG was not cloned, and distributed strains have been propagated and preserved throughout the world under varying conditions. These differences created a variety of related "BCG" vaccines that today have varying genotypic and phenotypic characteristics. Thus, the original BCG vaccine strain has developed into several different substrains that differ at several levels, including antigenicity, protective efficacy, genetic composition, or protein profile. The comparative analyses of *M. tuberculosis*, *M. bovis* and *M. bovis* BCG genomes have identified region of difference (RD) and tandem duplication (DU) markers in BCG substrains. BCG substrains are classified into four groups (I-IV) based on RD and DU markers.

Furthermore, single-nucleotide polymorphisms (SNPs) that are unique to particular BCG substrains or shared among substrains have also been identified. Some of these SNPs have functional implications for the affected genes, e.g., a SNP in *mma3* (*BCG0692c*) is responsible for the lack of methoxymycolate production in late BCG substrains that were distributed after 1927.

IV. GENOMICS OF BCG

The sequencing of mycobacterial genomes has shown that *M. tuberculosis* has a larger genome size than *M. bovis* and *BCG;* 4.41 Mb for *M. tuberculosis* H37Rv versus 4.34 Mb for *M. bovis* AF2122/97 and 4.37 Mb for BCG substrains Japan and Pasteur. The comparative genomic studies have identified a total of 16RDs, of which 15 RDs are present in all strains of *M. tuberculosis* but absent in some or all substrains of BCG (Table 117.1). Among the RDs absent in all BCG substrains, except RD1, other RDs (RD4, RD5, RD6, RD7, RD8, RD9, RD10, RD11, RD12, RD13) are also absent in *M. bovis*. Thus the primary deletion in attenuation of *M. bovis* to BCG was most likely RD1, a 9.5kb DNA segment (Figure 117.1). RD2, a 10.7kb DNA segment conserved in all virulent laboratory and clinical strains of *M. bovis* and *M. tuberculosis*, was deleted only from BCG substrains derived from the BCG Pasteur strain after 1927 (Figure 117.1). RD14 is deleted in BCG Pasteur, RD15 in BCG Frappier and Connaught and RD16 in BCG Moreau (Table 117.1, Figure 117.1).

TABLE 117.1 *M. tuberculosis* (H37Rv) RDs absent in BCG

RD designation	Absence in BCG substrains
RD1, RD4, RD5, RD6, RD7, RD8, RD9, RD10, RD11, RD12, RD13	All substrains
RD2	Danish, Prague, Glaxo, Meriux, Frappier, Tice, Pasteur, Connaught and Phipps
RD14	Pasteur
RD15	Frappier, Connaught
RD16	Moreau

RD3 is not shown because it is absent in some clinical isolates of *M. tuberculosis* as well.

Based on RD1 and RD2 deletions among BCG vaccine substrains, two main groups of BCG have been suggested (Figure 117.1, Table 117.2). The first group includes BCG Russia, Moreau, Japan, Sweden and Birkhaug, which were distributed between 1921 to 1926 and have only RD1 deletion. The second group, distributed after 1927, includes BCG Prague, Glaxo, Merieux, Danish, Tice, Frappier, Connaught, Phipps and Pasteur and have both RD1 and RD2 deletions (Figure 117.1, Table 117.2). The genes for two major antigenic proteins of *M. bovis*, MPB70 and MPB83 are still present but not expressed in post-1927 sub-strains, which may have implications for protective efficacy. Based on tandem duplication markers, each of the above two groups have further been divided into two DU types each, i.e., Russia, Moreau, Japan as DU2 group I and Sweden and Birkhaug as DU2 group II; and Prague, Glaxo, Merieux and Danishas DU2 group III, and Tice, Frappier, Connaught, Phipps and Pasteur as DU2 group IV (Figure 117.1).

TABLE 117.2 Specific deletions of *M. bovis* **genomic regions in substrains of BCG**

Substrain	Deleted region (important antigens)
Russia	RD1 (ESAT6 and CFP10)
Moreau	RD1 (ESAT6 and CFP10) and RD16
Japan	RD1 (ESAT6 and CFP10)
Sweden	RD1 (ESAT6 and CFP10) and IS6110
Birkhaug	RD1 (ESAT6 and CFP10) and IS6110
Prague	RD1 (ESAT6, CFP10), IS6110 and RD2 (MPT64)
Glaxo	RD1 (ESAT6 and CFP10), IS6110 and RD2 (MPT64)
Meriux	RD1 (ESAT6 and CFP10), IS6110 and RD2 (MPT64)
Danish	RD1 (ESAT6 and CFP10), IS6110 and RD2 (MPT64)
Tice	RD1 (ESAT6 and CFP10), IS6110 and RD2 (MPT64)
Frappier	RD1 (ESAT6 and CFP10), IS6110, RD2 (MPT64) and RD15
Connaught	RD1 (ESAT6 and CFP10), IS6110, RD2 (MPT64) and RD15
Phipps	RD1 (ESAT6 and CFP10), IS6110 and RD2 (MPT64)
Pasteur	RD1 (ESAT6 and CFP10), IS6110, RD2 (MPT64) and RD14

V. PROS AND CONS OF BCG

Although, BCG is the world's most widely used vaccine, and is directed against the world's leading cause of infectious disease problem, it is one of the most controversial vaccines in current use. Meta-analysis of vaccine trials have shown that BCG immunization *provides significant protection against TB in children, and has about 80% efficacy against severe forms of TB, including tuberculous meningitis and milliary TB in all age groups. However, evidence for protection against pulmonary TB in adolescents and adults, which carries the major burden of morbidity and mortality from this disease, remains doubtful.* This is because the *efficacy estimates* from clinical trials, observational case control studies and contact studies range from *0 to 80%*. Furthermore, BCG vaccination has been estimated to prevent only 5% of all potentially vaccine preventable deaths due to TB. As a result of the *variability in efficacy*, the impact of BCG on the global TB epidemic has been negligible. The reasons for the variable protective efficacy of BCG are not fully known but several hypotheses have been proposed, including differences among the vaccine strains used in clinical studies, exposure of trial populations to environmental mycobacteria and parasites, nutritional or

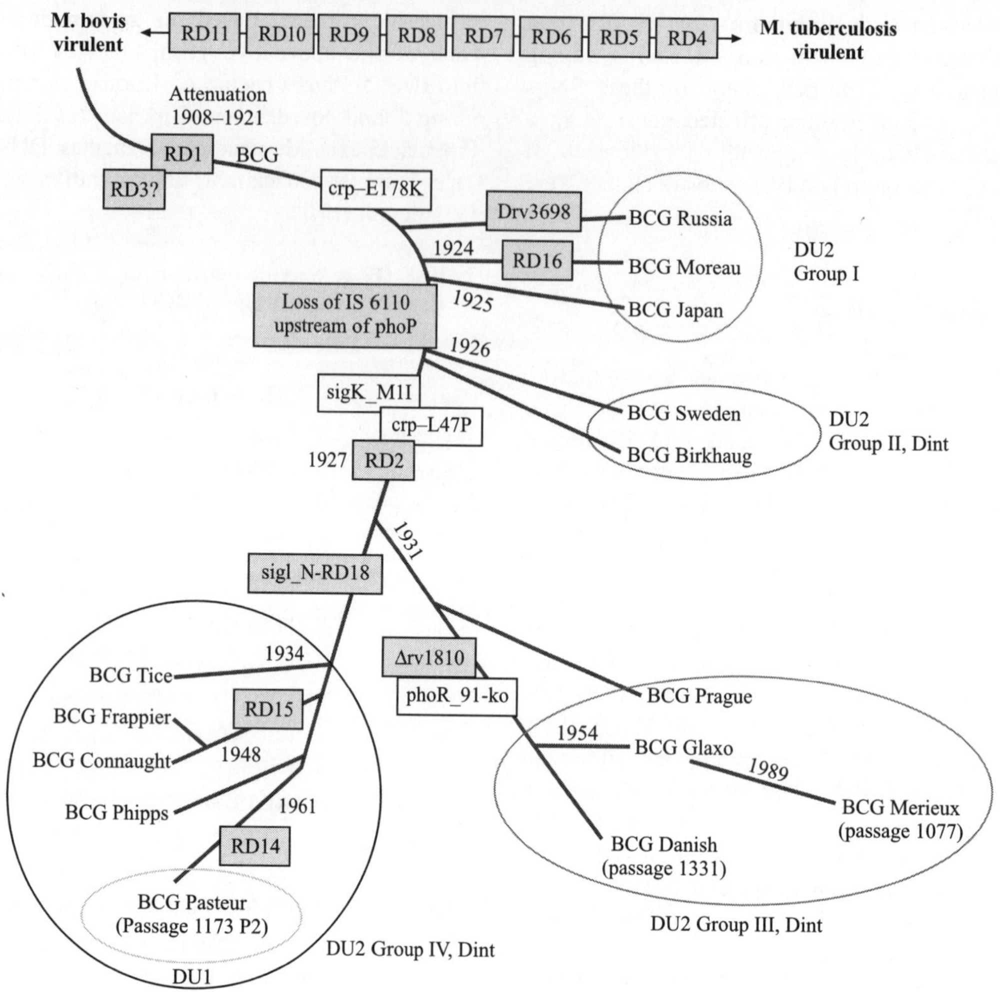

Figure 117.1   Genealogy of BCG vaccines, adapted from Brosch et al. (2007). *Proc. Natl. Acad. Sci.* USA, 104, 5596–5601. The scheme shows the position of genetic markers, including RD markers, some strain-specific deletions, and the distribution of vaccines into four groups based on tandem duplication markers.

genetic differences in human populations, differences in trial methods, variations among clinical *M. tuberculosis* strains and environmental influences such as sunlight exposure, and poor cold-chain maintenance etc. These explanations are not mutually exclusive and all may contribute to the heterogeneity observed in the efficacy of BCG vaccine.

The implementation of BCG vaccination faces two additional problems. Firstly, vaccination with BCG induces a delayed type hypersensitivity (DTH) skin response that cannot be distinguished from infection with *M. tuberculosis*, and therefore it compromises the use of PPD of *M. tuberculosis* in tuberculin skin tests for diagnostic or epidemiological investigations. Secondly, BCG is *a live vaccine with potential to cause disease by itself, particularly in immunocompromised subjects*, and therefore it cannot be used in all groups of people. In this regard, *WHO* recommends that children with symptoms of HIV or AIDS should receive all the vaccines *except* BCG. The shortcomings of BCG vaccines have provided the impetus for research to identify alternative and safer vaccine candidates, which could either replace or supplement BCG. The predictions suggest that an effective TB vaccine can save millions of lives in future and in the absence of an effective vaccine it will not be possible to achieve the WHO goal of eradicating TB from the world by 2050.

VI. NEW/IMPROVED VACCINES FOR TUBERCULOSIS

BCG is clearly insufficient for worldwide TB control. Thus, there is a strong need to develop new vaccines that can either boost BCG's initial priming and protective effects, or replace BCG.

The research during the past 20 years has led to the identification of a large number of new candidate vaccines against tuberculosis, which could be broadly classified into two major groups, whole mycobacterial and subunit vaccines (Table 117.3).

TABLE 117.3 Classification of candidate vaccine against tuberculosis

1. Whole mycobacterial vaccines:
   (a) Live recombinant BCG and other mycobacterial vector-based vaccines
   (b) Live attenuated *M. tuberculosis*
   (c) Killed mycobacteria
2. Subunit vaccines:
   (a) Viral vector-based vaccines
   (b) Fusion proteins along with appropriate adjuvants

VII. WHOLE MYCOBACTERIAL VACCINES

The whole mycobacterial vaccines could be living or killed. The development of new live TB vaccines requires replacing BCG by either improved recombinant (r)BCG and non-pathogenic mycobacteria or by attenuated *M. tuberculosis* through deletion of essential/virulent genes. Among the ways to improve BCG include introducing new immunodominant antigens of *M. tuberculosis* absent in BCG or by overexpressing immunodominant antigens that BCG already expresses by itself but probably not sufficiently high throughout all phases of infection, e.g. Ag85B. Another way to improve BCG is by introducing genetic modifications for superior targeting of essential immune pathways, e.g., by enhancing or facilitating cross-priming; and to inhibit its ability to neutralize maturation of phagosomes. Two improved rBCG vaccines (rBCG30 and VPM1002) have extensively tested in animal models and have entered into clinical studies for safety and efficacy in humans. The results available suggest that both of the replacement vaccines *were more potent and safer than the currently used BCG vaccine.* In addition to BCG, other non-pathogenic mycobacterial hosts have also been attempted to express antigens of *M. tuberculosis* and to be used as live recombinant mycobacterial vaccines. In this context, another new type of live mycobacterial vaccine against TB is the *esx*-3 mutant of *M. smegmatis* complemented with *M. tuberculosis esx-3* locus. This recombinant *M. smegmatis* induced strong bactericidal and protective activity against *M. tuberculosis* in the mouse model, and in some cases even leading to sterile eradication of *M. tuberculosis*, which may eventually solve the problem of reactivation diseases in TB.

Among the live mycobacterial vaccines, another strategy has been to attenuate *M. tuberculosis* by deletion of essential metabolic genes to create auxotrophic mutants; or the deletion of major virulence genes and their regulators. In both strategies, at least two independent genetic loci of *M. tuberculosis* should be deleted to avoid reversion to virulence. The most advanced among the attenuated *M. tuberculosis* vaccine candidates is the MTBVAC01, which was constructed by disruption of two sets of genes associated with *M. tuberculosis* virulence, i.e., the transcriptional regulator gene *phoP* and the *fadD*26 genes required for the production of Phthiocerol and phthiodiolonedimycocerosates. In addition to recombinant and attenuated live mycobacteria as prophylactic agents, several

killed mycobacterial vaccines are under development for therapeutic applications. They act by inducing appropriate protective immune responses leading to faster clearance of bacteria and shortening the duration of chemotherapy against TB. These include the RUTI vaccine, which consists of fragmented, detoxified and heat-inactivated *M. tuberculosis*, formulated in liposomes, and saline suspensions of two species of non-tuberculous whole-cell killed mycobacteria (*M. indicuspranii* and *M. vaccae)* and whole-cell extract of *M. smegmatis* (Table 117.4).

VIII. SUBUNIT VACCINES

A subunit vaccine consists of a few important antigens of *M. tuberculosis* that are capable of inducing protective immunity. It is more likely to be safer because the subunit vaccine consists of only a few selected molecules rather than the thousands of molecules of different types that make up a whole bacterium. Since a subunit vaccine can be constructed so as to eliminate irrelevant or even immunosuppressive components of the whole bacterium, it might induce a stronger protective immune response than a whole-bacterium vaccine. These vaccines use a prime-boost strategy to complement the immune response induced by BCG. Two types of subunit vaccines have been developed. These include recombinant fusion proteins consisting of two or more immunodominant proteins of *M. tuberculosis* and viral vectors that express one or more mycobacterial proteins.

The viral vector-based vaccines contain an immunogenic, replication deficient modified viral vector backbone, such as the modified vaccinia Ankara in the case of the candidate vaccine Oxford MVA85A/Aeras-485, or a modified adenovirus in the case of Crecell Ad35/AERAS-402 (Table 117.4). Adjuvanted vaccines contain one or more recombinant antigens of *M. tuberculosis*, which are co-administered with an appropriate adjuvant to enhance the immune response (Table 117.4). The subunit vaccines can be delivered safely into the human host regardless of immune status of the recipients. Although, subunit vaccines can theoretically be used as priming vaccines, they are considered to be mostly useful as booster vaccines on top of BCG-, recombinant BCG-, or attenuated *M. tuberculosis*-priming vaccines. The subunit vaccines are then expected to boost strong, long-lived immune responses in already primed individuals.

IX. NEW ANTI-TB VACCINES IN CLINICAL TRIALS

The clinical trials for vaccines to be used in humans begin with small scale trials that primarily determine the safety of candidate vaccines usually in healthy adults (phase I, small size, $n = 12–80$). After successfully completing phase I, the vaccines are further tested for safety and immunogenicity in target population (phase 2a, medium size, $n = 50–300$),

TABLE 117.4 New candidate TB vaccines for pre-exposure, prime-boost or immunotherapeutic applications undergoing clinical trials in humans

Type	Candidate vaccine	Description	Status of clinical trial
rBCG (pre-exposure/prime vaccination)			
	rBCG30	rBCG expressing antigen 85B	Phase 1
	VPM1002	rBCG Tice strain expressing listeriolysin and urease deletion	Phase 2a
Whole bacterial vaccines (boost, post infection immunotherapy)			
	RUTI	Detoxified *M. tuberculosis* in liposomes	Phase 2a
	M. indicus pranii	Killed bacteria	Phase 3
	M. vaccae	Killed bacteria	Phase 3
	M. smegmatis	Whole cell extract	Phase I
Viral vector (pre-exposure/booster vaccinations)			
	Oxford MVA85A/Aeras-485	Modified vaccine Ankara-expressing antigen 85A	Phase 2b
	Crucell Ad35/Aeras-402	Replication deficient adenovirus 35-expressing antigens 85A, 85B, TB10.4	Phase 2b
	AdAg85A	Replication deficient adenovirus 5-expressing antigen 85A	Phase I
Recombinant fusion protein in adjuvant (pre-exposure/booster vaccination)			
	Hybrid I + IC31	Fusion of Ag85B and ESAT-6 in adjuvant IC31	Phase 1
	Hybrid 56 + IC31	Fusion of Ag85B, ESAT-6 and Rv2660c in adjuvant IC31	Phase 1
	Hybrid I + CAF01	Fusion of Ag85B and ESAT-6 in adjuvant CAF01	Phase 1
	M72+AS01 or AS02	Fusion of Rv1196 and Rv0125 in adjuvant AS01 or AS02	Phase 2a
	Aeras-404:HyVac4+IC31	Fusion of Ag85B and TB10.4 in adjuvant IC31	Phase I

followed by safety, immunogenicity and efficacy in target population (phase 2b, larger size, $n = 1000$–5000) and efficacy in target population (phase 3, very large size, $n \Rightarrow 5000$). The vaccines are licensed for human use only after successfully completing the phase 3. Post-licensure surveillance of vaccines is conducted under phase 4.

At least fifteen new TB vaccines have been evaluated in clinical trials to date, many in high burden settings, and 14 are actively being tested currently. An overview of the new candidate TB vaccines undergoing clinical trials in humans is given in Table 117.4. Of these, at least 4 have entered or completed Phase 2b or Phase 3 testing.

X. SUMMARY

BCG has been used as a vaccine against TB for about 90 years and more than 4 billion doses of the vaccine have been delivered. Although, BCG provides significant protection against TB in children and extra-pulmonary TB in adults, it has failed to have an impact on the global epidemiology of TB due to failure to protect against pulmonary TB in adults,

which is the major manifestation of the diseases in humans. In the recent past, studies have been carried out to identify new vaccine candidates to improve upon the safety and efficacy of BCG. These candidates include live and subunit vaccines for prophylactic applications and killed mycobacteria for therapeutic applications. Both of these vaccine strategies have shown promise in animals models of TB and clinical trials in humans are ongoing. Among the prophylactic vaccines, three different strains of recombinant BCG, three different viral vector-based and five different fusion protein-based subunit vaccines are at different stages of clinical trials in humans. Similarly four different preparations of non-living mycobacteria have also shown promising results in therapeutic applications in humans. All these results, suggest that the availability of improved vaccines against TB is a possibility in the near future.

Acknowledgment

The work was supported by Kuwait University Research Sector grants MI01/10 and GM01/01.

SUGGESTIONS FOR FURTHER READING

Brosch R., et al. (2007), Genome Plasticity of BCG and Impact on Vaccine Efficacy, *Proceedings of National Academy of Sciences* USA, 104:5596–5601.

Checkley A.M. and McShane H. (2011), Tuberculosis Vaccines: Progress and Challenges, *Trends in Pharmacological Sciences*, 32:601–606.

Corbel M.J., Fruth U., Griffiths E. and Kneizevic I. (2004), Report on A WHO Consultation on the Characterization of BCG Strains, Imperial College, London, *Vaccine*, 22:2675–2680.

Hawkridge T. and Mahomed H. (2011), Prospects for a New, Safer and More Effective TB Vaccine, *Paediatric Respiratory Review*, 12:46–51.

Liu J., Tran V., Leung A.S., Alexander D.C. and Zhu B. (2009), BCG Vaccines: Their Mechanisms of Attenuation and Impact on Safety and Protective Efficacy, *Human Vaccine*, 5:70–78.

Mustafa A.S. (2005), Progress Towards the Development of New Anti-tuberculosis Vaccines, In: *Focus on Tuberculosis Research*, (Ed.), Lucy T. Smithe, 47–76, Nova Science Publishers, Inc., New York.

Mustafa A.S. (2009), Vaccine Potential of *Mycobacterium tuberculosis*-specific Genomic Regions: *In vitro* Studies in Humans, Expert Reviews Vaccines 8:1309–1312.

Mustafa A.S. (2012), What's New in the Development of Tuberculosis Vaccines, *Medical Principles and Practice* 21:195–196.

Ottenhoff T.H.M. and Kaufmann S.H.E. (2012), Vaccines Against Tuberculosis: Where Are We and Where Do We Need To Go? *PLoS Pathogens*, 8:E1002607.

118

An Immunotherapeutic Vaccine for Leprosy

G.P. Talwar

CONTENTS

I. INTRODUCTION

Leprosy is caused by a Mycobacteria, *Mycobacterium leprae* discovered over two centuries back by a Norwegian, Armauer Hansen. His thesis for Doctorate is of only one page, the shortest ever written. He was unable to grow these mycobacteria in any medium.

The majority of humans (99%) exposed to this microorganism do not develop leprosy. Those who develop, manifest a disease ranging from the pauci-bacillary, a usually single limited lesion, termed the *Tuberculoid leprosy* (TT) to the other extreme of the spectrum, the *Lepromatous leprosy* (LL), where the disease is widely disseminated and macrophages are loaded with *M. leprae*. It is the LL patients, who perpetuate the infection and transmit it to others. In case LL macrophages are rendered inhospitable for growth and multiplication of *M. leprae*, no source of transmission of *M. leprae* will remain and leprosy would be amenable to eradication.

II. DEVELOPMENT OF EXPERIMENTAL SYSTEMS

Talwar et al. developed this vaccine against leprosy. The approach adopted was to find an organism, a mycobacteria, which could make LL patients immunologically reactive to *M. leprae*.

Normally LL patients are Lepromin negative. They are devoid of delayed hypersensitivity rection to killed *M. leprae*. To begin with, along with my student A D Krishnan, an *in vitro* system was developed, in which *M. leprae* could grow. Macrophages were developed from healthy human blood monocytes. These could be infected with *M. leprae* derived from lepromas of LL patients. The lepromas contained both live and dead *M. leprae*. How could the live *M. leprae* be differentiated from the dead, is not an easy task. *M. leprae* is a slow grower, taking 13 days to divide. Counting of *M. leprae* in the infected macrophages was unpractical not only for the difficulty of long time culture, but also because the inoculum initially used would contain variable and unknown number of live and dead organisms forming a globus mass. A novel approach was adopted. Luckily macrophages do not multiply in *in vitro* culture system and should not make DNA, whereas the bacteria, if they grow should make DNA, howsoever slowly. Thus, incorporation of ^{3}H-thymidine into DNA could serve as an index of *M. leprae* growth or the lack of it. This was tested and confirmed by pioneering studies. These studies revealed that lymphocytes of tuberculoid leprosy patients react to *M. leprae in vivo* (Delayed Hypersensitivity reactions) and *in vitro* whereas such reactions do not occur with lymphocytes from lepromatous leprosy patients. They were negative to Mitsuda reactions, to subcutaneous or intradermal injection of killed leproma extracts. It was thus argued and proved by experimentation that co-culture of lymphocytes with *M. leprae* generates a conductive signal, for the growth or otherwise of *M. leprae* in host cells, which could be assessed by incorporation of ^{3}H-Thymidine.

III. SEARCH FOR AN IMMUNOLOGICALLY CROSS-REACTIVE MYCOBACTERIA

A panel of known (defined) mycobacteria, as well as atypical isolates in clinical laboratories was employed under code words to screen their reactivity with T cells of polar tuberculoid leprosy (TT) and lepromatous leprosy (LL) patients. *M. leprae* provokes blast transformation and generation of cytokines influencing macrophages from T cells of TT patients, but fails to do so with cells from LL patients. Our search was for an organism which evoked reactivity with cells of both TT and LL patients. By such studies, 5 mycobacterial strains were shortlisted from a panel of 16 cultivable mycobacteria. These were then tested *in vivo* in patients of TT and LL leprosy in Leprosy control centers not only in Delhi but also those located in various parts of the country, so as to take care of the possible environmental sensitivities. Lepromins were prepared from not only *M. leprae*, but also from the shortlisted mycobacteria. Delayed hypersensitivity tests (DHT) were conducted with these preparations in patients, eventually to select an organism which evoked DHT reactions in both TT and LL patients. This turned out to be the Mycobacteria coded as w. Figure 118.1 is a micrograph of this marvelous mycobacteria, now named *Mycobacterium indicus pranii* (MiP).

Mw (now MiP) was a non-pathogenic mycobacteria, but could evoke the desirable immunological reactions in both live and autoclaved form. The latter property was valuable in shortening the endless toxicology studies required for a new, hitherto undefined mycobacteria.

IV. CLINICAL STUDIES

Based on positive data from extensive experimental studies,

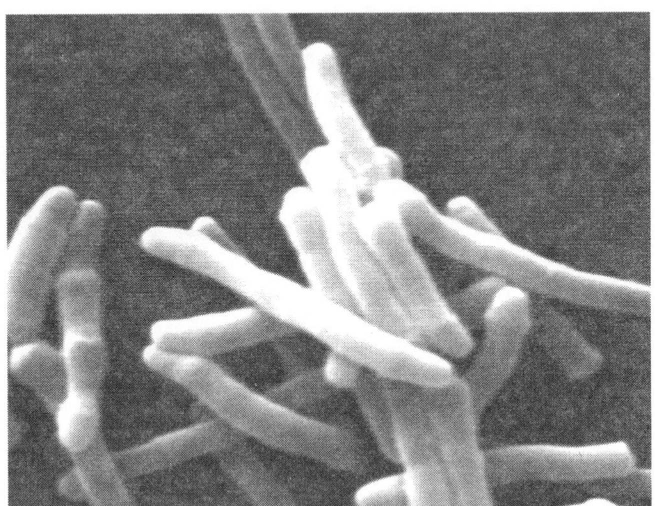

Figure 118.1 Micrograph of Mw (*Mycobacterium indicus pranii*).

permission was granted by the Drugs Controller General of India and Institutional Ethics Committee for its use in killed autoclaved form as adjunct to standard multidrug therapy (MDT) regime in multibacillary LL and BL patients. Patients receiving Mw (now *Mycobacterium indicus pranii*, MiP) *had faster clearance of bacilli, and recovered fully in shorter time periods than those receiving MDT alone.* What was amazing was the *dramatic normalcy returning to disfiguration that the disease brought about in patients receiving Mw,* quite distinguishable from patients becoming bacillary negative in lesions, but retaining a part or more of the apparent disfigurations on treatment with MDT alone. Figures 118.2(a)–(g) represent a few patients who received treatment with MDT + Mw (MiP).

The fact that inclusion of Mw for treatment, improved immune responses of patients against *M. leprae,* is provided by a large number of patients becoming lepromin positive, while those on MDT (Multi Drugs Treatment) alone, even after becoming bacillary negative after prolonged treatment did not become responsive to lepromin (Figure 118.3).

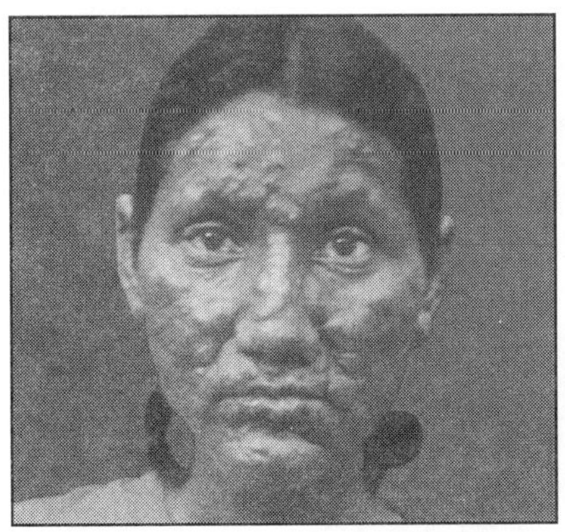

Patient code: MD   Initial

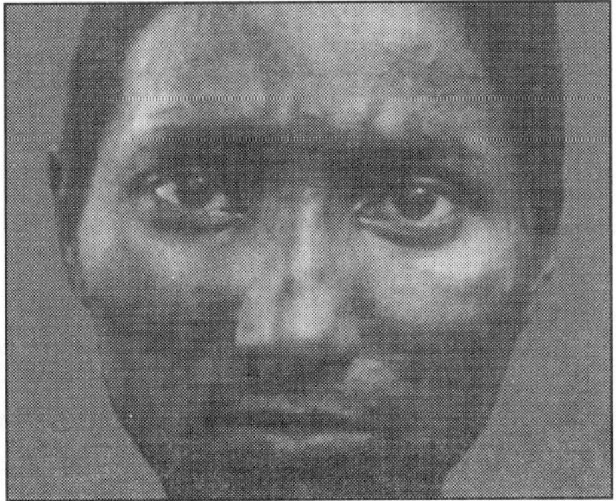

After 4 doses

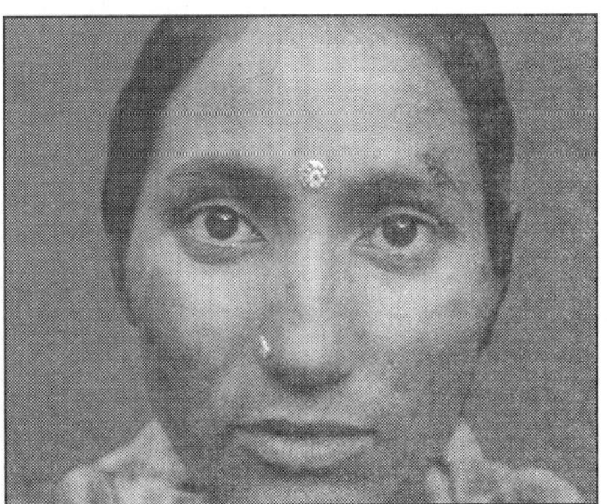

There after 1 year

(a)

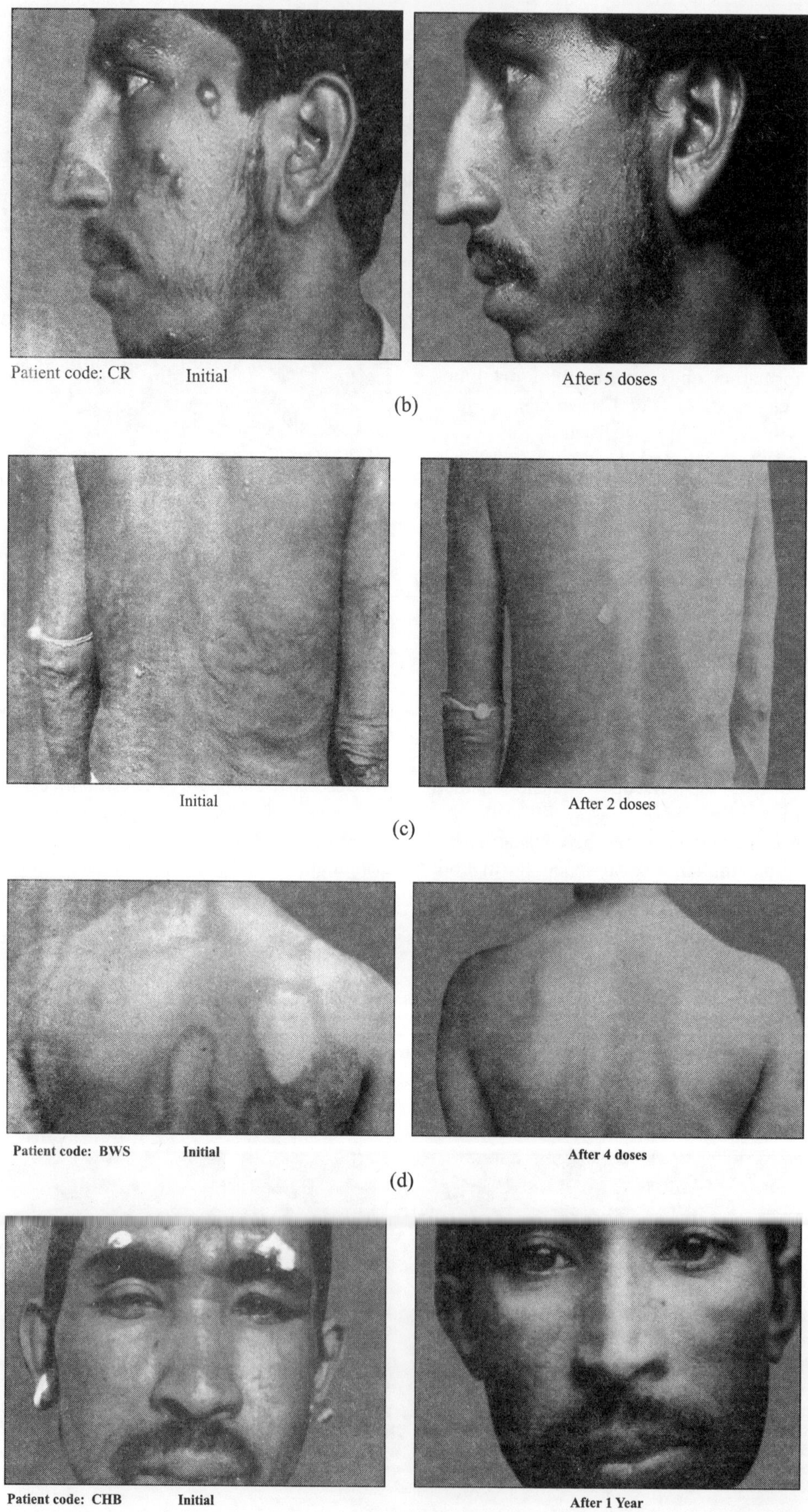

Patient code: CR   Initial
After 5 doses
(b)
Initial
After 2 doses
(c)
Patient code: BWS   Initial
After 4 doses
(d)
Patient code: CHB   Initial
After 1 Year
(e)

Patient code: RAB

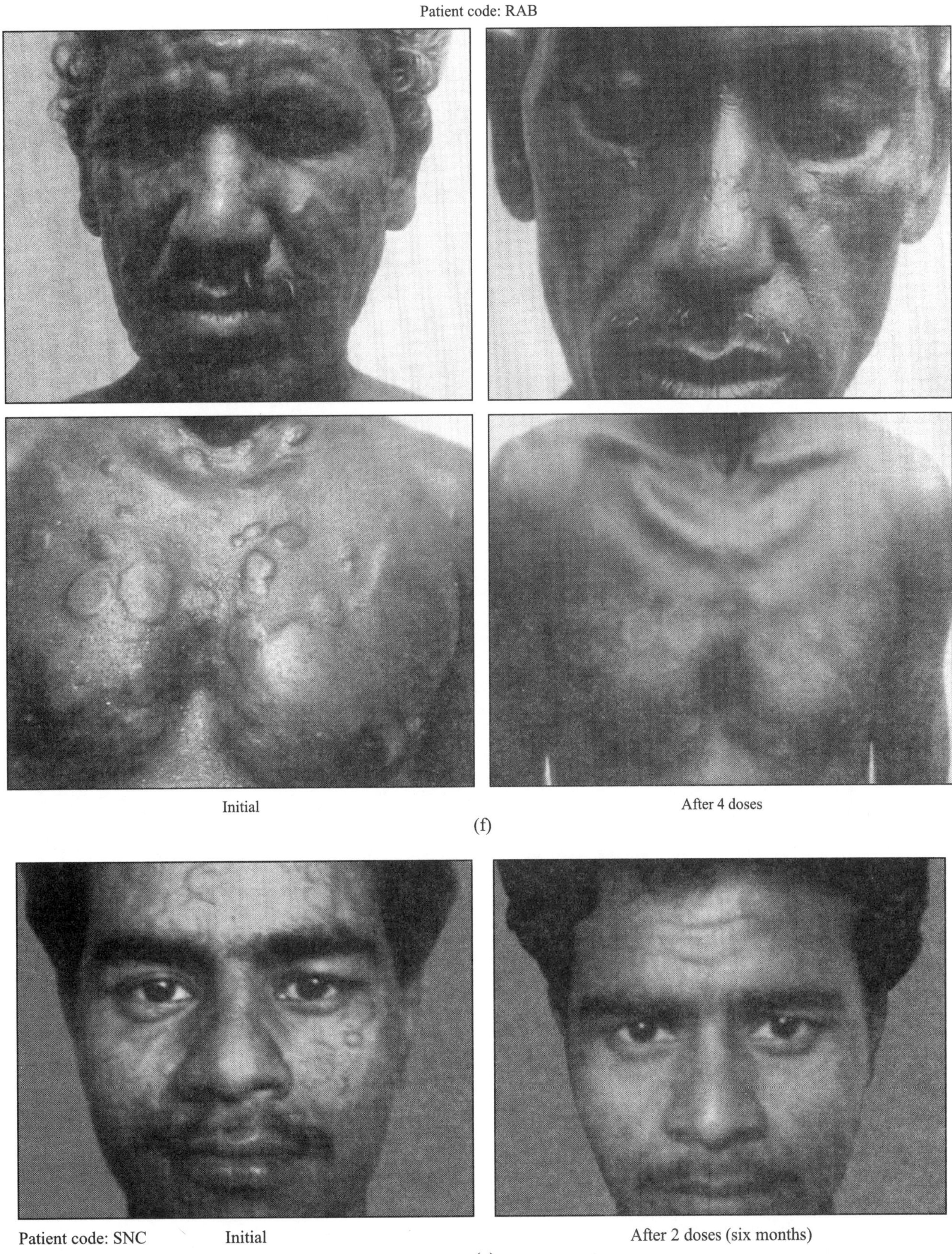

Figure 118.2 Some representative cases of LL/BL multibacillary patients treated with MDT plus Mw (*Mycobacterium indicus pranii*). (*see Plates 50–51 for colour figure*)

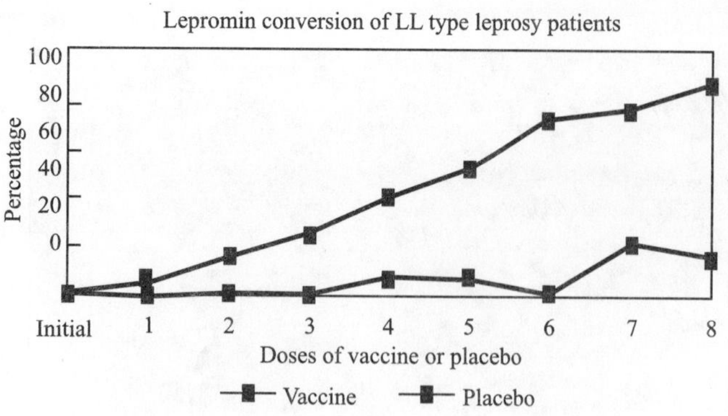

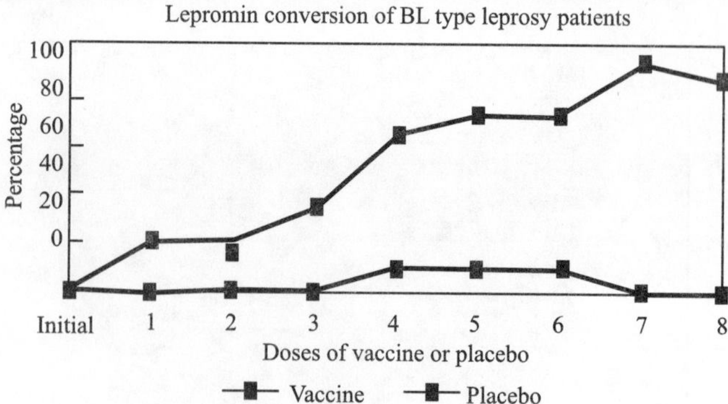

Figure 118.3   DHT response to Lepromin in patients receiving treatment with drugs alone or with Mw (MiP).

V. MIP EVOKED RESPONSE IN SLOW RESPONDERS OR DRUG RESISTANT PATIENTS

Evidence that MiP causes reduction in Bacillary Index (BI) of the multibacillary leprosy patients is provided by the valuable studies that bright and enterprising Dr. Zahir Ahmed (at that time doing post graduation with me) did in 14 "slow" responder leprosy patients. These patients showed hardly any decline of BI (Bacterial Index) in their lesions and or skin biopsies over several months of treatment with MDT. *Inclusion of the Mw (now MiP) vaccine heralded a notable fall in BI and their recovery* (Figure 118.4).

VI. CLEARANCE OF *M. LEPRAE* FROM PERIPHERAL NERVES

One of the main lesions caused by *M. leprae* is a loss of sensation from peripheral parts of the body. The Bacillus invades the peripheral nerves. It is also the site from where clearance of bacilli by MDT is slow, requiring long term persistent medication. Zaheer examined the nerve biopsies from a few patients to indicate that inclusion of Mw (MiP) in treatment caused better clearance of *M. leprae* from peripheral nerves than with MDT alone (Table 118.1).

Field trials

After obtaining positive results from trials in 3 hospitals of Delhi namely the All India Institute of Medical Sciences (AIIMS), Safdarjung and Lal Bahadur Shastri Hospitals under eminent Professors of Dermatology, field trials were undertaken along with the National Leprosy Control Programme in Kanpur Dehat, which at that time was an endemic area of leprosy. It was composed of 272 villages with a population of 420,823. Double blind coded trials (code kept by ICMR) were undertaken in these villages. 24,060 family members and healthy threshold contacts of index leprosy cases were also given, the vaccine or Placebo (Tetanus toxoid at 1/8th dose) to determine the protective efficacy of the Mw (MiP) vaccine against leprosy. Surveys were carried out after 3, 6 and 9 years to determine the fresh cases amongst the family contacts receiving the immunization. On opening of the code, it was observed that the Mw vaccine had a protective efficacy of 68.6% at 3 years, 59% at 6 years and 39.3% after 9 years of immunization with Mw (MiP) against leprosy. *Thus over and above the significant therapeutic action of the Mw (MiP) vaccine as adjunct to standard drugs, the vaccine had also a prophylactic protective effect in preventing household contacts constantly exposed to M. leprae to get Leprosy.*

VII. APPROVAL OF MW (MIP) BY THE DRUGS CONTROLLER GENERAL OF INDIA (DCGI)

On the basis of the overall findings, the Mw (MiP) vaccine received approval of the Drugs Controller General of India and subsequently also of US FDA. It is the only vaccine of its type at present in the whole world. It is licensed to M/s Cadilla Pharma and the product is available for public use.

TABLE 118.1   Effect of inclusion of the MiP vaccine on clearance of *M. leprae* from peripheral nerves of multibacillary patients

Group	Type	Initial Bacterial Index (BI)	Final BI	Initial Histopathology	Final Histopathology	Nerve *BI
Vaccine +	LL	4	0	LL	NSI	0
MDT	BL	3	0	BL	NSI	0
	LL	5	0	LL	BL	0
	BL	2.33	0	BL	NSI	0
MDT alone	LL	4.33	0	LL	LL	2
	LL	3.16	0	BL	BL	1
	BL	2.13	0	LL	LL	0

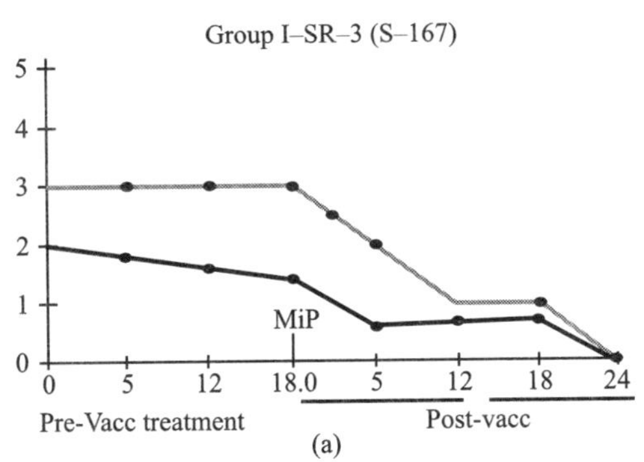

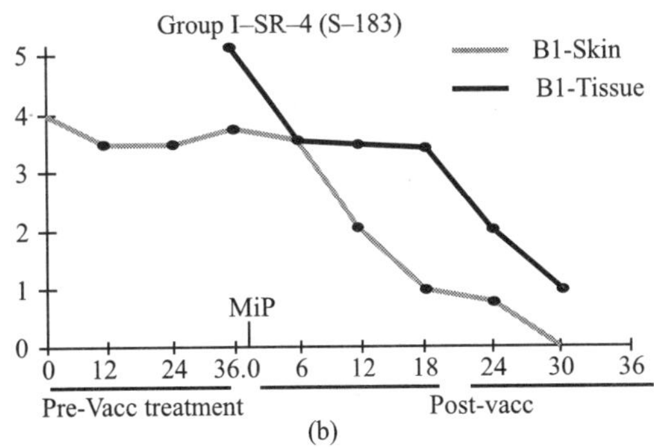

Figure 118.4　Bacillary Index (BI) of two slow responder (SR-3 and SR-4) patients to MDT, on MDT alone and on administration of MiP vaccine, which led to the BI decline much faster.

VIII.　UTILITY IN TUBERCULOSIS

Mycobacterium indicus pranii (Mw) shares antigens with both *M. leprae* and *M. tuberculosis*. Approved by the DCGI and available to public through the licensee M/s Cadilla Pharma Ahmedabad, it has been employed by Practitioners for treatment of category II, difficult to treat tuberculosis patients as adjunct to standard therapy. Patients become AFB (Acid Fast Bacillus) negative much earlier than those receiving the standard treatment with drug alone (Figure 118.5). Furthermore the relapses are distinctly lower (Figure 118.6) and Table 118.2.

Dr. Sangeeta Bhaskar at the National Institute of Immunology, New Delhi has carried out extensive studies

TABLE 118.2　**Outcome of Therapy (CAT II Tuberculosis)**

	Cured	
Mw + MDT	48/49*	97.96%
MDT alone	21/27**	77.77%

* Defaulter for 6 doses, sputum negative after intensive phase
** 2–2+ No effect of therapy
　　4–1+ No effect of therapy

on the protective effect of live and autoclaved MiP (Mw) on tuberculosis.

It is further observed that Mw (MiP) is effective in both killed and live state, whereas BCG at present employed for

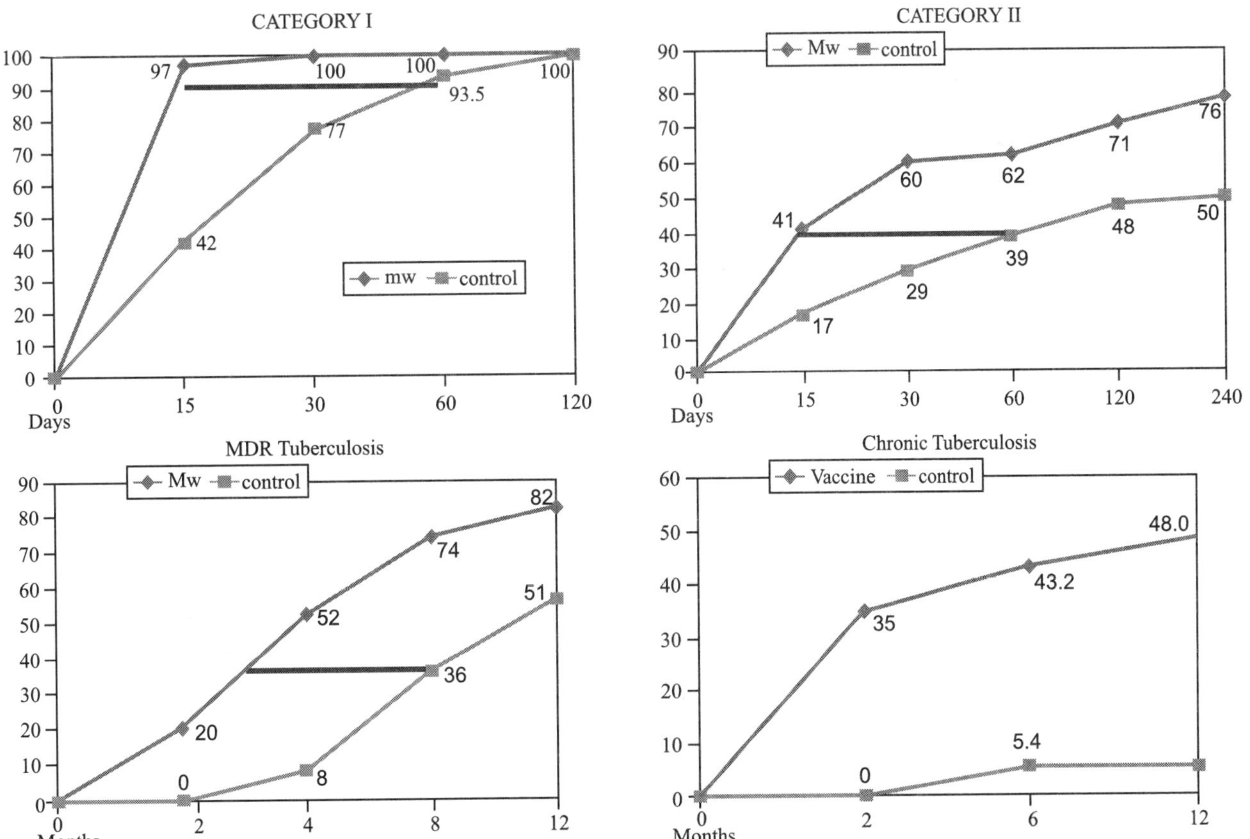

Figure 118.5　Improved therapy of Tuberculosis patients by supplementation of multidrugs regimen with injection of Mw (MiP). (*see Plate 52 for colour figure*)

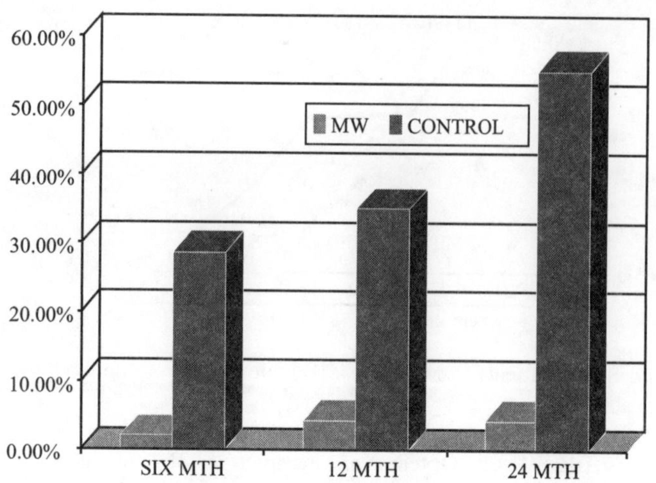

Figure 118.6   Relapse Rate in Category II after completion of therapy. (*see Plate 52 for colour figure*)

immunization in Public Helath Program is only active in live state. BCG as vaccine against tuberculosis has a variable record of protective immunity in various countries. In Uganda, it was effective whereas in India it failed to accord significant immunity against tuberculosis as per ICMR trials.

Being given that MiP (Mw) is a general invigorator of immunity. Dipankar Nandi at the Institute of Sciences, Bangalore has investigated the possible impact of immunization with MiP (Mw) on SP_2O Myelomas in mice. Interestingly he observed both a preventive and therapeutic action of Mw vaccine on mouse Myelomas (Figure 118.7).

Dr. Nandi has carried out extensive studies on immunological parameters influenced by MiP (Mw) vaccination. These are reported in recent papers (see References).

IX. CURE OF ANOGENITAL WARTS

A most unexpected therapeutic action of MiP (Mw) was reported by Prof. Sumesh Gupta at AIIMS. Patients suffering from fairly ugly warts on penis and rectum were given local injections of MiP after initial inoculation in deltoid region. Lo and behold, their warts were cleared as shown in Figure 118.8. The therapy worked in both HIV negative and HIV positive subjects.

X. GENE SEQUENCING OF Mw

Professors Seyed Hasnain, Aklesh Tyagi and Anil Tyagi carried out a genomic sequencing of Mw (Saini et al., PLOS one 2009; 4: e 6263; Talwar et al., Impact. Genet. Evol. 2008; 8: 100–101). No Mycobacteria of such genetic makeup has so

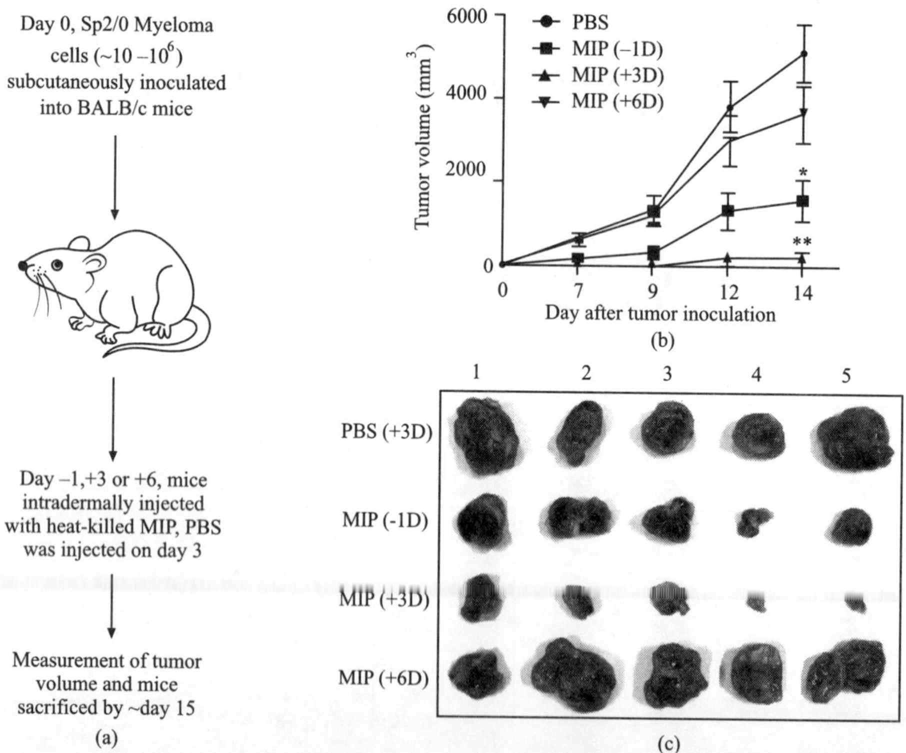

Figure 118.7   MIP treatment suppresses tumor growth and induces a Th1 cytokine response. (a) General outline of the *in vivo* experiment protocol. (b) Comparison of the anti-tumor effects of MIP administered at different time points. Cohorts of ten mice were inoculated s.c. with $\sim 10^7$ Sp2/0 cells. Mice were injected i.d. with a single dose of MIP ($\sim 5 \times 10^8$) either one day (-1D) before or 3 (+3D) or 6 (+6D) days after tumor inoculation. Mice injected i.d. with PBS on day 3 were included as controls. The growth of tumors (mean $\pm$ SD mm^3) at indicated days post implantation. (c) Representative photographs of solid tumors from different treatment groups dissected on day 14. (*see Plate 52 for colour figure*)

far been reported in World Data Bank. It was given the name of *Mycobacterium indicus pranii*, Pran is my first name used by family and friends. NII is the abbreviation of the National Institute of Immunology of which I was the first Founder Director, where much of the work relating to clinical trials of the vaccine was done.

XI. MIP, A POTENT ADJUVANT

A recombinant vaccine against hCG has been developed by us. The immunogenicity of the vaccine is substantially enhanced by MiP as adjuvant, as is evident from Figure 118.9(a), (b).

MiP is an invigorator of both Th_1 and Th_2 immune responses.

SUMMARY

A vaccine based on a non-pathogenic mycobacterium coded as Mw now named *Mycobacterium indicus pranii* (MiP) has been developed for leprosy. It has prophylactic action in protecting household contacts of leprosy patients from contacting the disease. Given as adjunct to standard multidrug regime, in multidisciplinary patients it expedites bacterial clearance and shortens the full recovery period. It renders also a large number of patients from lepromin negative to lepromin positive status. MiP is also effective in treating of category II, difficult to treat tuberculosis patients, reducing also substantially the relapse rate. In contrats to BCG, MiP is effective in both live and killed state. It protects also genetic strains of mice, not protected by BCG. Additional applications of MiP are in treatment of Ano-genital works and as potent adjuvant to enhance antibody response to the birth control vaccine hCGp-LTB.

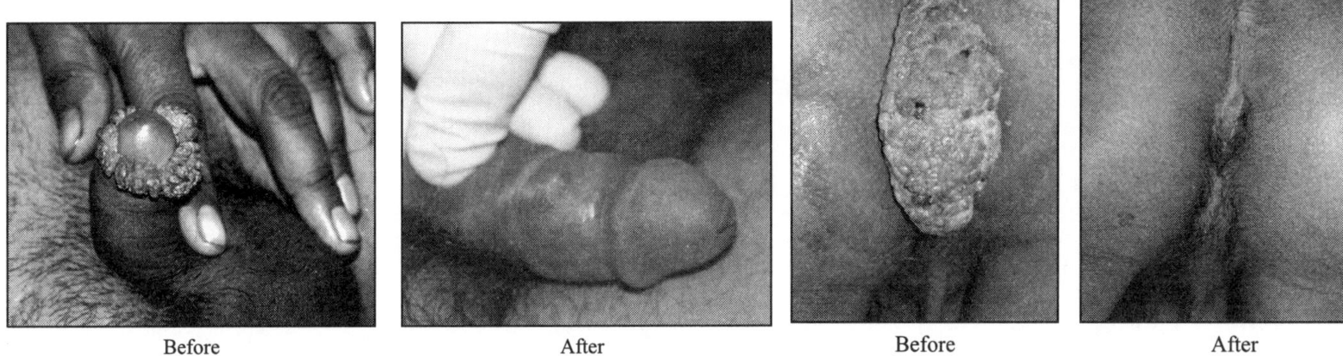

Figure 118.8 Therapeutic action of MiP on Ano-genital warts in patients under Prof. Sumesh Gupta, at in All India Institute of Medical Sciences, New Delhi. (*see Plate 52 for colour figure*)

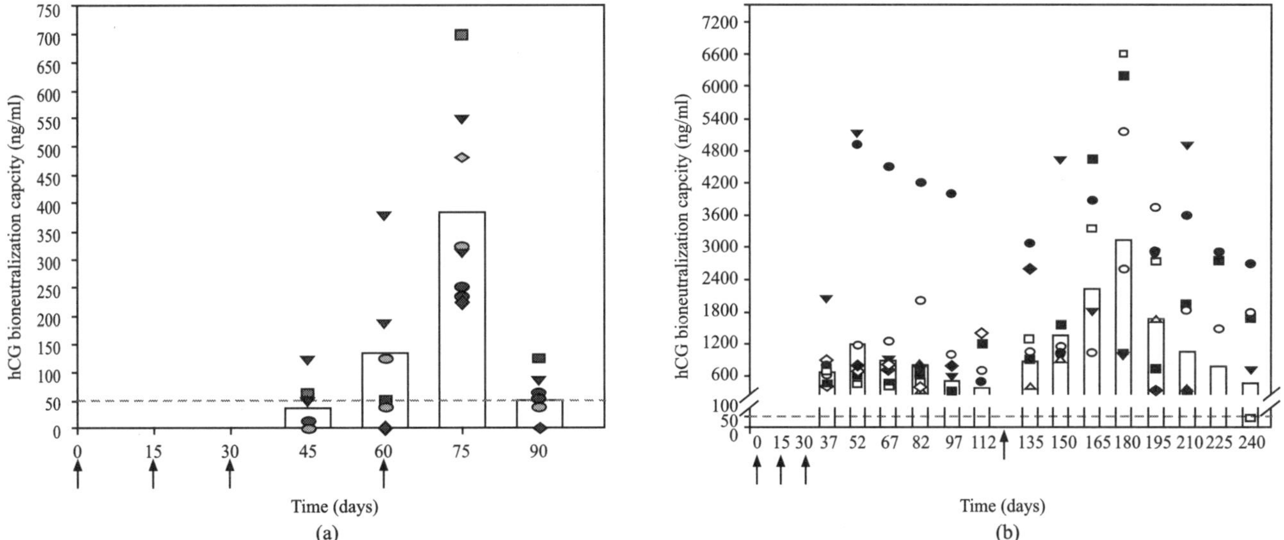

Figure 118.9 High bio-effective antibody response generated by hCGβ-LTB in Balb c mice given along with *Mycobacterium indicus pranii* (MiP) as adjuvant. Bars give the geometric mean of bioneutralisation capacity determined by inhibition of 125 I-hCG binding to rat leydig cell receptors. The symbols represent the titres in individual mice and bar the geometric means.

119

Immunotherapeutic and Immunoprotective Role of *Mycobacterium indicus pranii*

Sangeeta Bhaskar

CONTENTS

I. TUBERCULOSIS

As tuberculosis continues to be a burden, there are concerted efforts to find new vaccines to combat this problem. The current vaccine against tuberculosis (TB), *Mycobacterium bovis* BCG, fails to protect against the most prevalent disease form, pulmonary TB in adults. Other mycobacterial strains which share cross-reactive antigens with *M. tuberculosis* have been considered as alternatives to *M. bovis* for vaccine use. One such strain, *Mycobacterium w*, had been evaluated for its immunomodulatory properties in leprosy. Recently it has been suggested that *M. w* be renamed as *Mycobacterium indicus pranii* (MIP) to avoid confusion with virulent *M. tuberculosis*-W (Beijing strain). A vaccine against leprosy based on killed MIP is approved for human use, where it has resulted in clinical improvement, accelerated bacterial clearance, and increased immune responses to *Mycobacterium leprae* antigens. *Mycobacterium indicus pranii* is nonpathogenic cultivable mycobacterium which shares antigens not only with *M. leprae* but also with *M. tuberculosis*. Further biochemical, molecular and phylogenetic analysis has shown MIP to be distinct species but closely related to *M. avium intracellulare*. Initial studies had shown that vaccination with killed MIP induces protection against tuberculosis in *Mycobacterium bovis* BCG responder, as well as BCG nonresponder, strains of mice.

A. Protective Efficacy of MIP

Protective potential of MIP and the underlying immune responses were further studied in the animal models of tuberculosis. It is generally known that live bacteria impart greater protection than killed bacteria. It may be that persistence of live bacteria in the host for some time results in a robust memory response. In order to accelerate the control of initial *M. tuberculosis* infection, vaccine-induced immunity in the lung must be further enhanced. Inhalation of aerosols provides a noninvasive delivery system that physically targets the lung as the desired site of pharmacological effect. This route of immunization induced both local and systemic immunity. Protective efficacy of MIP immunization in both live and killed forms through parenteral route and by aerosol immunization, was compared with that of BCG. MIP immunization by both the routes parenteral and aerosol gave higher protection than BCG in the animal models of tuberculosis.

MIP induces Th1 type of response whether given in killed or live form. It activates macrophages, as well as lymphocytes. Besides activation of CD4+ T cells and IFN-γ secretion, MIP also induces significant CD8+ T cell cytotoxic activity and macrophage effector functions. Activated macrophages inhibit the intracellular growth and multiplication of *M. tuberculosis*.

Protective efficacy and modulation of post-infection immune response was evaluated in susceptible guinea pig model of TB. Pathogenesis in guinea pig model resembles human tuberculosis. Reduced bacterial loads, improved pathology and organized granulomatous response were observed at different post-infection time points in the MIP-immunized group. MIP-immunization resulted in heightened protective Th1 response as compared to BCG group, after infection with *M.tb* and a balanced Th1 versus immunosuppressive response at late chronic stage of infection. A higher antigen presenting cells activation was observed inside the granuloma of infected lungs in the MIP-immunized group. MIP-immunization provided significantly better long term protection as compared to BCG against TB.

B. MIP Immunotherapy as an Adjunct to Chemotherapy for Treatment of TB

TB is treatable by drugs, and the World Health Organization has promoted the "directly observed therapy" to improve compliance as chemotherapy needs to be taken for 6–9 month long duration. Patients who default on therapy have severe risk of relapsing and acquiring drug resistance. Hence, multidimensional approach is desired to effectively eradicate the actively multiplying as well as persistent bacteria from the infected lungs.

During the course of mycobacterial infection, the protective Th1 immune response is dominated by immunosuppressive Th2 response which could facilitate the survival and progression of infection. Immunotherapy could be a potential tool in antimycobacterial therapy since it may boost the Th1 immune response in infected patients and might act as a force in shifting the immune response from Th2 to Th1 type. This protective immune response can act in synergy to antibacterial chemotherapy. Efficacy of MIP as an adjunct to standard chemotherapy for tuberculosis, when given by aerosol or parenteral route was evaluated. Immune cells accumulation in the lung and in-situ expression of different cytokine genes were analysed to characterize the modulation of immune response with MIP-treatment in *M.tb*-infected guinea pigs. Protection with MIP-treatment was associated with reduced bacterial loads and pulmonary pathology. A balanced inflammatory and suppressive immune response in later part of the drug plus immunotherapy treatment was observed which was perhaps useful in the process of restoration of normal tissue architecture and controlling the initial inflammatory reaction. The results indicated that MIP might not be able to reduce the treatment duration, but could be potentially very useful in eradicating the persistent bacteria when given with chemotherapy. Aerosol route of MIP immunotherapy can play a very important role for inducing immediate local immune response in the lung.

C. Clinical Trial

Protective efficacy of MIP was evaluated in a rural population of about 30,000 people belonging to 272 villages in Kanpur (India). The population was vaccinated with two doses of MIP at six month interval. The vaccine/placebo was given to healthy contacts of leprosy patients who had no evidence of suffering from tuberculosis. Vaccination with MIP was protective against pulmonary tuberculosis and the incidence of tuberculosis was significantly less in the MIP vaccinated group in this randomized placebo controlled study with a follow-up of 13 years. Thus killed vaccine, is safe and well tolerated. Being a killed vaccine there is no danger of precipitating latent infection in some populations as is the case with BCG. It can be used for both leprosy and TB and especially in countries where both the diseases affect the population. Available results indicate, it may be used both immunoprophylactically as well as therapeutically.

Besides immunoprophylaxis MIP also has shown immunotherapeutic properties for the treatment of pulmonary tuberculosis. Sputum conversion has been reported to appear more rapidly in patients who received DOTS +MIP as compared to DOTS alone.

D. Safety of MIP Immunisation

Acute and sub-acute toxicity studies were undertaken with autoclaved saline suspension of MIP in mice and guinea pigs. There was no evidence of toxicity in both these animals. Infection with Live MIP in mice and guinea pigs is self-limiting and short-lived, which is important for the induction of a memory response and also from safety point of view.

Presence of persistent antigens which provide strong T-cell stimuli over extended periods of time activate effector T cells efficiently but exhaust memory. What is required is a vaccine that can persist long enough to generate memory T cells but gets cleared by the host without generating adverse pathology. It is probable that the immune response induced by a viable vaccine which persists for only a restricted time period best fulfills these requirements.

II. LEISHMANIA

Leishmania donovani, a protozoan parasite, causes a strong immunosuppression in a susceptible host and inflicts the fatal disease visceral leishmaniasis. Relatively high toxicity, low therapeutic index, and failure in reinstating host-protective antileishmanial immune responses have made antileishmanial drugs patient non-compliant and an immunomodulatory treatment a necessity. Antileishmanial efficacy of combination of MIP and an antileishmanial drug, Amphotericin B (AmpB) was analysed. MIP alone or with a suboptimal dose of AmpB offered significant protection against *L. donovani* infection by activating the macrophages. MIP induced collateral host-protective effects while the AmpB chemotherapy eliminated the parasites and thereby the parasite-induced immunosuppression.

III. CANCER

Immune system plays a crucial role in protecting the host

against cancer. There is strong evidence for the existence of an effective cancer immunosurveillance process in human and mice. In tumor bearing host, however, immune system is often not able to mount an effective response, primarily because of negative regulatory mechanisms employed by the growing tumor. Th1 branch of the immune system, which employs T cells, NK cells and macrophages, play major role in combating cancer but growing cancers actively suppress immune response and disregulate the activity of these effector cells. Cancer immunotherapy has emerged as a powerful method to treat tumors and numerous agents that enhance the potential of the host immune system are being evaluated in clinical trials. Since long, bacteria and bacterial products have been tried for the treatment of cancer. Starting from the practical observation of tumor regression in individuals with concomitant bacterial infection to systematic efforts of William Coley who used the extracts of pyogenic bacteria to treat sarcomas. BCG has been used for the treatment of superficial bladder cancer for over 25 years. BCG-cell wall components have been administered to patients for postoperative treatment of cancer, producing good prognosis. Indeed, it has been observed that different mycobacterial species differ widely in their antitumor potential and there is need of sincere efforts to look for potent candidate species. *Mycobacterium indicus pranii* (MIP) is non-pathogenic, soil derived, rapidly growing, atypical mycobacterium. Polyphasic taxonomic analysis has established it as a distinct species. There are some key differences between BCG and MIP. BCG is attenuated form of pathogenic organism *M. bovis* which survives in the host cells for months and has been shown to cause toxicity in some patients while MIP is non-pathogenic and cleared from the host body in few weeks.

A. Immunotherapeutic Activity of MIP in Cancer

A couple of small clinical studies, where MIP was given to lung cancer and bladder cancer patients, indicated its beneficial effect in the management of the disease. As induction of Th1 type of immune response is crucial to overcome the immuno suppressive tumor microenvironment, the immunotherapeutic potential of MIP in the mouse model of tumor was analysed. It was observed that tumor growth was delayed and volume of tumors were significantly less in the MIP treated group as compared to control. Higher frequency as well as higher expression of phenotypic activation markers were observed on tumor infiltrating macrophages and T cells of MIP treated mice which also reflected their enhanced functional activity. In addition to induction of IFN-γ, TNF-α and IL-12 secretion, higher NK cell and CD8+ T cell cytotoxic activity were observed.

Peritumoral injection of MIP also resulted in significantly less number of tumor infiltrating regulatory T cells. Treg cells inhibit the development and effector functions of tumor specific T cells. They accumulate in the tumor microenvironment and suppress cytotoxic T cell responses against the tumors. Elimination of these suppressor T cells from tumor uncovers natural antitumor response. Immunotherapy with MIP

modulated these suppressive factors and simultaneously activated the pro-inflammatory Th1 type of response within the tumor microenvironment hence, this could be a potentially effective approach which could be combined with chemotherapy to combat cancer successfully.

To evaluate and delineate the mechanisms by which autoclaved MIP enhances anti-tumor responses, the growth of solid tumors consisting of Sp2/0 (myeloma) and EL4 (thymoma) cells were studied in BALB/c and C57BL/6 mice, respectively. Treatment of tumor bearing mice with MIP greatly suppressed tumor growth, lowered IL6 but increased IL12 and IFN γ and IL2 amounts in sera. Also, increase in CD8 + T cell mediated lysis of specific tumor targets was observed.

There are several notable features of the use of MIP for tumor therapy. First, the efficacy of MIP in reducing tumors in two distinct tumor models suggests that MIP may be effective against a wide range of tumors. This aspect is important as *M. vaccae*, together with chemotherapy, was shown to be ineffective against squamous carcinoma of the lung although efficacy was observed in adenocarcinoma. Second, the use of "dead" MIP in patients reduces the risk of bacterial proliferation and this aspect is relevant in case of immunocompromised patients as observed with "live" BCG vaccination.

Adjuvant effects mediated by mycobacteria and their constituents are known. However, there may be differences in the mechanism of action by different mycobacterial species. Direct contact between "live" BCG and tumor cells is important for the anti-tumor efficacy of BCG. On the other hand, "dead" MIP injected at a site distant from tumor inoculation reduced the tumor growth.

IV. POTENT ADJUVANT ACTIVITY OF MIP WITH RECOMBINANT ANTI-HCG VACCINE

Mycobacterium indicus pranii (MIP) is a potent immuno-modulator for a recombinant vaccine against human chorionic gonadotropin. Human use-permissible adjuvant, MIP significantly enhanced the antibody response to recombinant anti-hCG vaccine. Previous Phase II efficacy trials in sexually active women have demonstrated the prevention of pregnancy at hCG bioneutralization titers of 50 ng/ml or more. MIP injection along with recombinant anti-hCG vaccine increased the antibody titers several fold higher in the FVB, SJL, C3H, and C57Bl/6 strains of mice. In addition, the duration of the antibody response was prolonged. The vaccine given together with MIP, induced both Th1 and Th2 response, which was reflected in the production of IgG1 and also a high proportion of IgG2a and IgG2b antibodies. This is of interest with regards to potential use of the vaccine for treatment of hCG-expressing tumors. There is ectopic expression of hCG by a variety of cancers at advanced stages. Such cancers are invariably refractory to available drugs and have poor prognosis and adverse survival. In animal models of hCG expressing tumors, MiP has shown promising results.

V. MIP IMMUNOTHERAPY IN PATIENTS WITH ANO-GENITAL WARTS

Intralesional immunotherapy with skin test antigens and vaccines has been found to be effective in the management of genital and extragenital warts. In an open label pilot study, intralesional immunotherapy with killed MIP in patients having ano-genital warts proved to be effective. Intralesional injections were repeated weekly until either complete clearance or a maximum of 10 injections were given. Genital warts cleared completely. The treatment was well tolerated by the majority of the patients. No recurrence was seen after a mean follow-up of 5.1 months. Intralesional immunotherapy of ano-genital warts with MIP vaccine seems to be a promising new approach.

A valuable property of MIP immunotherapy is its virtual lack of adverse side effects as this has been administered to thousands of patients in clinical trials of tuberculosis, and leprosy with no reported problems. Another great advantage is the low cost of production. There is a need to further understand the MIP mediated modulation of immune response at molecular level. Also, MIP is a general immunomodulator and enhances immunity to other diseases, e.g., HIV, psoriasis. MIP retains its immunologic potential even after it is killed, unlike BCG.

SUGGESTIONS FOR FURTHER READING

Ahmad F., Mani J., Kumar P., Haridas S., Upadhyay P. and Bhaskar S. (2011), Activation of Anti-tumor Immune Response and Reduction of Regulatory T Cells with *Mycobacterium indicus pranii* (MIP) Therapy in Tumor Bearing Mice, *PLoS ONE*, 6(9):E25424.

Ahmed N., Saini V., Raghuvanshi S., Khurana J.P., Tyagi A.K., et al. (2007), Molecular Analysis of a Leprosy Immunotherapeutic Bacillus Provides Insights into Mycobacterium Evolution, *PLoS ONE*, 2(10):E968.

Gupta A., Ahmad F.J., Ahmad F., Gupta U.D., Natarajan M., Katoch V.M. and Bhaskar S. (2012), Protective Efficacy of *Mycobacterium indicus pranii* Against Tuberculosis and Underlying Local Lung Immune Responses in Guinea Pig Model, *Vaccine*, 30(43):6198–209.

Gupta A., Ahmad F.J., Ahmad F., Gupta U.D., Natarajan M., Katoch V. and Bhaskar S. (2012), Efficacy of *Mycobacterium indicus pranii* Immunotherapy as an Adjunct to Chemotherapy for Tuberculosis and Underlying Immune Responses in the Lung, *PLoS ONE*, 7(7):E39215.

Gupta A., Geetha N., Mani J., Upadhyay P., Katoch V.M., Natrajan M., Gupta U.D. and Bhaskar S. (2009), Immunogenicity and Protective Efficacy of "Mycobacterium W" Against *Mycobacterium Tuberculosis* in Mice Immunized with Live Versus Heat-killed M.w by the Aerosol or Parenteral Route, *Infect Immun.*, 77(1):223–31.

Katoch K., Singh Padam, Adhikari T., Benara S.K., Singh H.B., Chauhan D.S., Sharma V.D., Lavania M., Sachan A.S. and Katoch V.M. (2008), Potential of *Mw* as a Prophylactic Vaccine Against Pulmonary Tuberculosis, *Vaccine*, 26, 1228–1234.

Purswani Shilpi, Talwar G.P., Vohra Richa, Pal Rahul, Panda Amulya K., Lohiya Nirmal K. and Gupta Jagdish C. (2011), *Mycobacterium indicus pranii* is a Potent Immunomodulator for a Recombinant Vaccine Against Human Chorionic Gonadotropin, *Journal of Reproductive Immunology*, (91), 24–30.

Rakshit Srabanti, Ponnusamy Manikandan, Papanna Sumitha, Saha Banishree, Ahmed Asma and Nandi Dipankar (2012), Immunotherapeutic Efficacy of *Mycobacterium Indicus Pranii* in Eliciting Anti-tumor T Cell Responses: Critical Roles of IFNγ, *Int. J. Cancer*, 130, 865–875.

Saini Vikram, Raghuvanshi Saurabh, Talwar Gursaran P., Ahmed Niyaz, Khurana Jitendra P., Hasnain Seyed E., Tyagi Akhilesh K. and Tyagi Anil K. (2009), Polyphasic Taxonomic Analysis Establishes *Mycobacterium indicus pranii* as a Distinct Species, *PLoS ONE*. 4(7):E6263.

120

Cancer Immunotherapy with *Mycobacterium indicus pranii*

Dipankar Nandi and Srabanti Rakshit

CONTENTS

I. Introduction
II. MIP as an Immunoadjuvant in Cancer
   A. Cytokines play key roles in MIP mediated anti-tumor T cell responses
   B. Roles of T cells in MIP mediated anti-tumor response
III. The Combinatory Therapy of MIP and Cyclophosphamide Greatly Reduces Tumor Growth
IV. Highlights

I. INTRODUCTION

Mycobacterium w, rechristened as *Mycobacterium indicus pranii* (MIP), is a soil derived, rapidly growing, atypical saprophytic bacterium which was originally identified to stimulate cell mediated responses in patients suffering from leprosy. Together with standard multidrug treatment, MIP vaccination in leprosy patients led to enhanced clearance of bacilli and reduction in the recovery time. After undergoing initial clinical trials, it was approved as an adjunct to multi drug treatment for leprosy and administered as an autoclaved preparation (Immuvac/Cadi-05). MIP is evolutionarily related to the *M. avium intracellulare* complex according to molecular and biochemical analysis and is strategically placed in a transitory position between slow growing intracellular pathogens and saprophytic non-pathogenic mycobacteria species. Although MIP possesses all the virulence genes, it lacks the critical genes required for its survival and multiplication within macrophages. Taxonomic studies established MIP to be distinct from other mycobacterial members and more recently, complete sequencing of the MIP genome revealed that higher antigenic potential of MIP is responsible for its unique ability for immunomodulation.

Live BCG, the attenuated form of *M. bovis*, displays its adjuvant activity only in contact with tumor cells and pathogens. In contrast, MIP retains its immunologic potential even after it is killed. It has been found to share antigens with *Mycobacterium leprae* and *Mycobacterium tuberculosis*. In an experimental model of tuberculosis in mice and guinea pig, MIP immunizations have been shown to be protective, including mice strains that are less responsive to BCG. Importantly, either as stand alone or as an adjunct to chemotherapy, MIP immunization in humans has shown prophylactic efficacy against pulmonary tuberculosis. Also, MIP enhances immunity to other diseases by displaying characteristics of a general immunomodulator: it enhances immune responses against HIV, psoriasis and leishmaniasis.

II. MIP AS AN IMMUNOADJUVANT IN CANCER

Novel and alternative cancer treatment options are warranted for patients with advanced solid tumors who either show resistance to conventional anticancer strategies or experience frequent relapse after therapy. In addition, new anticancer therapies can only be successful if they selectively kill cancer cells but exhibit restricted toxicity towards normal cells. In 1890, the pioneer of cancer immunotherapy Dr. William Coley injected extracts of *Stretococcicus pyogenes* and *Serratia marcescens* (Coley's toxin; CT) that led to spontaneous tumor regression and considerable remission in patients with inoperable metastatic sarcomas. Currently, CT is undergoing phase 1 clinical trial for NY-ESO-1 expressing cancers. Nowadays, live or heat-killed, attenuated or genetically modified non-pathogenic bacteria are being investigated as potential anti-tumor agents, which either induce direct cytotoxicity or deliver tumoricidal molecules as they are proficient in inhibiting tumor growth by selective multiplication within tumors. Hence, proper and in-depth understanding of mechanisms of action of these bioimmunoadjuvants is necessary as they could emerge as promising candidates for effective cancer management.

Notably, members of the genus *Mycobacterium*, which belong to the actinomycetales family, display widespread adjuvant activity and ameliorate cancer progression by potentiating the host immune response. Immunization with BCG, live attenuated form *Mycobacterium bovis*, minimizes the chances of advent of childhood leukemias, breast cancer and melanoma and is also most effective against non-muscle invasive bladder cancer. In addition, immunotherapy with heat-killed suspension of *M. vaccae* (SRL172) has significantly improved quality of life and enhanced overall survival time in patients with renal cancer, lung adenocarcinoma and breast cancer.

MIP too has been found to be effective against lung cancer, bladder cancer and gonadotropin-sensitive tumors. However, the detailed mechanism by which it exerts its immunomodulatory activity via mounting of a powerful host anti-tumor response was not investigated earlier. Thus, the purpose of our study was to evaluate the underlying mechanisms involved in the immunotherapeutic effects of MIP using two different tumor models: Sp2/0 myeloma in BALB/c and EL4 thymoma in C57BL/6 mice. A single intradermal injection of MIP three days after implantation of tumor cells significantly reduced the size of subcutaneous tumors compared to PBS-treated mice in both the tumor systems.

A. Cytokines Play Key Roles in MIP Mediated Anti-tumor Response

Cytokines, chemokines and growth factors are key molecules that participate in dynamic cross-talk between host immune cells and cancer cells and, thereby, modulate the outcome of the immune response either against or in favor of tumorigenesis. Cancer progression or regression depends upon critical interplay between cytokines released by the tumor cells and by tumor infiltrating immune cells which includes both lymphocytes and myeloid populations. Cytokines in the tumor microenvironment can either be Th1/ Th2 or belong to the regulatory phenotype and critical balance in the cytokine milieu will determine the fate either elimination of tumor or extensive tumor growth followed by invasion and metastasis. Importantly, substantial evidence suggests that disease progression in some cancers may be due to a Th2 bias, whereby, tumors escape host immune responses by preferentially enhancing Th2 and downregulating Th1 immune responses.

A recent study regarding the potential mechanism for the immunotherapeutic effects of BCG, *M. vaccae* and *M. obuense* in cancer clearly states that circulating myeloid dendritic cells secrete IL-12, IL-1β and TNF-α which indirectly activate $\gamma\delta$ T-cell mediated anti-tumor effector responses via induction of IFN-γ and TNF-α. Hence, the role of cytokines in MIP mediated immune response was addressed. Interestingly, pattern of cytokine production in the serum of MIP immunized mice were significantly modulated during tumor progression whereas basal cytokine amounts in non-tumor mice injected with MIP or PBS were comparable to that of control un-injected mice. MIP treatment led to considerable reduction in serum IL-6 amounts which otherwise consequently increased with tumor progression in PBS injected mice. Clinical studies have predicted that deregulated and excess production of IL-6 hampers tumor regression due to inhibition of cancer cell apoptosis, stimulation of angiogenesis and drug resistance. Hence, immunotherapy with MIP is likely to be beneficial in patients suffering from different cancers including multiple myeloma. In addition, IL-4 amounts were less, however, no appreciable difference was observed with MIP treatment. On the contrary, the amounts of IFN-γ and IL-12, the two principal cytokines that play crucial roles in anti-tumor immunity were significantly enhanced in the sera from mice injected with MIP compared with PBS treated mice which clearly suggests that MIP predominantly skews the cytokine profile towards Th1 type. This aspect is of utmost importance as MIP immunotherapy could counteract tumor escape mechanism and restore the Th1 immune responses required to promote anti-cancer immunity.

B. Roles of T Cells in MIP Mediated Anti-tumor Response

Numerous reports have demonstrated that most mycobacterial strains exert a Th1 mediated anti-tumor immune response during cancer therapy. However, detailed investigations on the role of T cells during implementation of tumor immunotherapy is not delineated. Cancer immunosurveillance involves the close coordination between innate and adaptive cells of the immune system in order to detect aberrant cancer cells and eliminate them by mounting a strong anti-tumor immune response. Previously, CD8[+] T cells which are responsible for direct cytolytic activity, were thought to be the main players of this immunoediting process; however importance of CD4[+] T cells in orchestrating of immune responses during cancer immunotherapy is dramatically amplified over the past decade and has been shown to be indispensable for optimal immune function.

It was important to elucidate whether MIP mediated anti-tumor immune response is dependent on both CD8[+] T cell induced tumor cytotoxicity and CD4[+] T cell induced cytotoxicity coupled with antigen specific immune response of either of them. The functional roles of CD4[+] and CD8[+] T cells in the above mentioned processes were studied by depleting these populations using specific antibodies *in vitro*. Notably, antigen specific stimulation of splenocytes from MIP treated mice with tumor lysates for 5 days greatly reduced BrdU incorporation of specific target cells without affecting non-specific tumor targets demonstrating tumor specificity. Simultaneously, depletion of CD8[+], but not CD4[+] T, cells almost completely reversed the drop in Sp2/0 cell proliferation indicating that *CD8[+] T cells are primarily involved in MIP mediated lymphocyte-induced cytotoxicity*. Similarly, stimulated splenocytes belonging to MIP treated mice produced enhanced amounts of IFN-γ and IL-2 but negligible IL-4 in the culture supernatants. Likewise depletion of T cell subsets confirmed that CD4$^+$ Th1 cells are solely responsible for mounting a strong antigen-specific anti-tumor recall response.

To ensure that the efficacy of MIP was not tumor cell specific, the potency of MIP was tested in an *aggressive tumor, EL4 thymoma. MIP treatment successfully reduced tumor size and induced potent T cell mediated anti-tumor immune responses* in wild-type C57BL/6 mice, however, its immunomodulatory activity was significantly diminished in *Ifnγ$^{/-}$* mice, strengthening the fact that *IFN-γ is indispensable for MIP efficacy*. In addition, MIP mediated anti-tumor immune responses were severely downregulated in the immunodeficient NOD-SCID mice stressing the functional relevance of the activation of the complex immune network by MIP. Thus, our study was the first published report to shedls some light on the mechanistic insights of MIP based immunotherapy of cancers and the highlights of which are summarized in Figure 120.1.

III. THE COMBINATORY THERAPY OF MIP AND CYCLOPHOSPHAMIDE GREATLY REDUCES TUMOR GROWTH

Despite significant progress in understanding of tumor biology and introduction of new effective chemotherapeutic drugs, various cancers have unfortunately remained incurable. Moreover, the use of multiple and high dose of curative chemotherapeutic strategies is restricted primarily due to elevated levels of toxicity accompanied by harmful side-effects that may lead to severe and long-term adverse complications and also hamper the quality of life. In addition, resistance to these drugs is the principal cause for relapse in majority of patients. Nevertheless, complete cure is very rarely achieved due to persistence of residual disease and exemplify the need of newer therapeutic regimens to counter tumors.

In recent years, cancer immunotherapy has emerged as a novel treatment option that enhances the power of host

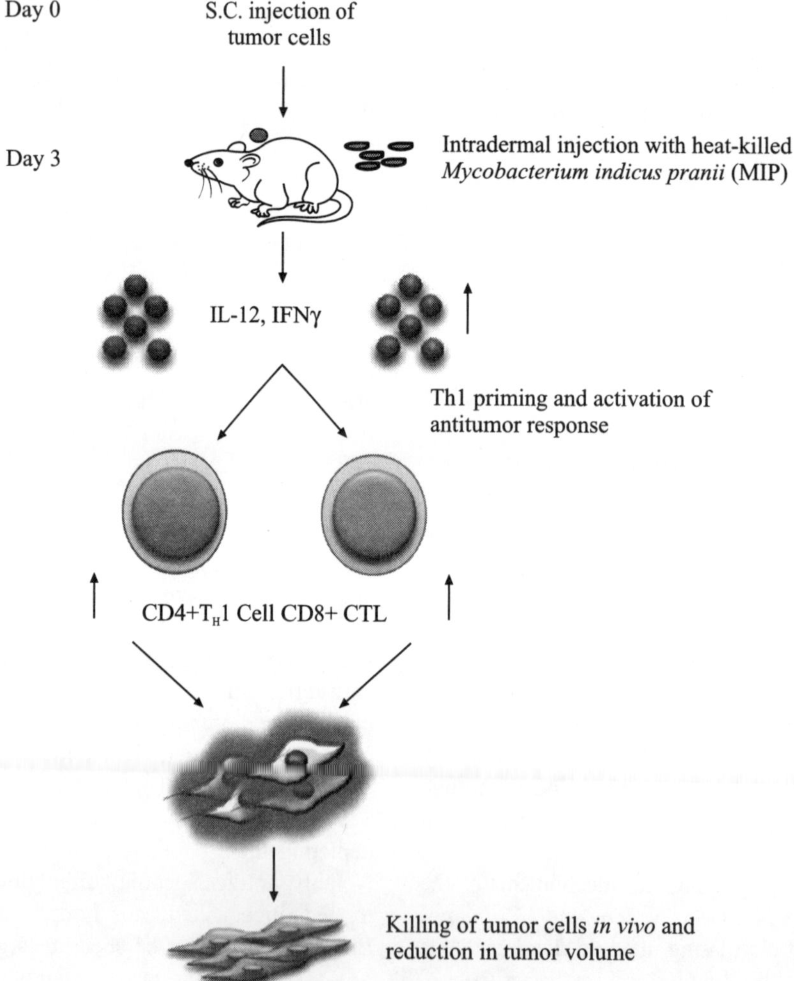

Figure 120.1   Schematic representation for the mechanism of MIP mediated activation of antitumor immune response. Treatment of mice with MIP 3 days after implantation of tumor cells results in *high amounts of IL-12 and IFN-γ* which lead to increased anti-tumor immune responses mediated by CD4$^+$ and CD8$^+$ T cells. Importantly, the MIP-induced IFN-γ plays a key role in enhancing *in vivo* anti-tumor T cell responses leading to reduced tumor burden. (*see Plate 53 for colour figure*)

immune system to induce and sustain the endogenous antitumor responses responsible for selective eradication of tumors. Numerous immunotherapy-based approaches are being currently evaluated in clinical trials that either directly acts on tumors or indirectly as immunoadjuvants when administered alone or in combination with chemotherapeutic drugs to improve the latter's efficacy by reducing their toxic side-effects.

Cyclophosphamide is widely used in chemo-adoptive immunotherapy and is known to suppress regulatory T cells which lead to enhanced cytolytic activity against the tumor by the adoptively transferred, activated effector cells. Cyclophosphamide therapy also leads to cardiac, lung, renal, bladder toxicity and alopecia. MIP was ineffective when administered on day 6 post-inoculation of tumor cells. Therefore, we wanted to investigate whether chemo-immunotherapy comprising of MIP and Cyclophosphamide will lead to synergistic tumor regression even when administered at a later stage when the tumors have already developed.

MIP administered alone on day 7 post-inoculation of Sp2/0 cells did not significantly reduce tumor volume nor did cyclophosphamide alone at low doses, i.e., 1, 5 and 15 mg/kg body weight cyclophosphamide administered alone either on the same day as MIP or one day prior to or one day after MIP injection had no effect on tumor volume. Combination therapy was effective only when cyclophosphamide was administered one day after MIP and greatly retarded tumor incidence and reduced tumor volume. The synergistic efficacy of cyclophosphamide was best noticed even at suboptimal dose of 15 mg/kg body weight when given in combination with MIP, thereby reducing toxic side effects seen at high doses when administered alone. However, cyclophosphamide administered alone at 50 mg/kg body weight was equally effective and no further reduction was observed upon combined therapy with MIP.

IL-12p70 and IFN-γ amounts in combo treated mice was enhanced, whereas, IL-6, IL-1β and TNFα were reduced compared to that with MIP treatment alone. TGFβ was reduced to a greater extent in Cyclophosphamide treated mice compared to MIP treatment but was significantly reduced in the combo group. IL-10 levels were similarly reduced in the combo treated group but interestingly, IL-10 amounts were somewhat higher in the MIP treated group. *Thus, the combinatorial therapy mounts a robust Th1 response and also leads to suppression of regulatory T cell cytokines.* Hence, combination therapy of *Mycobacterium indicus pranii* (MIP) and Cyclophosphamide *induces almost complete tumor eradication in two independent tumor models: Sp2/0 myeloma and EL4 thymoma when administered at early stages of tumor progression.* Interestingly, combination therapy also reduces tumor volume in a synergistic manner even when administered at later stages of tumor development (United States Patent no. 8,367,075 B2; date Feb 5, 2013).

IV. HIGHLIGHTS

Studies have shown that MIP exerts its anti-tumor effects by enhancing the cellular arm of the immune responses, especially T cells. The role of IFN-γ is critical in this response as the MIP-mediated anti-tumor response is abrogated in mice lacking IFN-γ. Importantly, the cotherapy of MIP with Cyclophosphamide has shown great efficacy in lowering tumor growth in vivo. Further studies are required to understand the mechanisms by which MIP exerts its actions and the clinical relevance and implications of these studies.

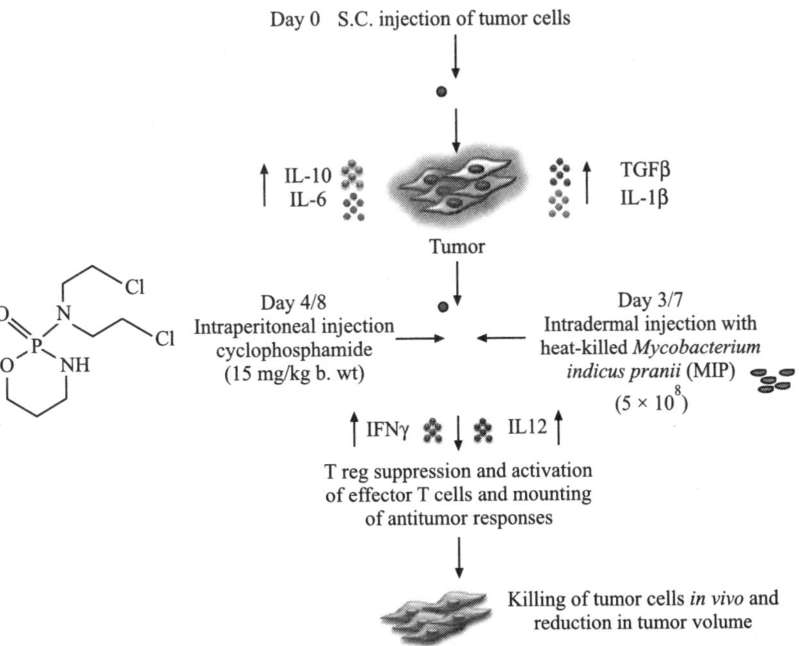

Figure 120.2 Schematic representation of the combined therapy of cancer using MIP and low dose of cyclophosphamide. Cyclophosphamide is administered one day after MIP injection and synergistic reduction in tumor volume is achieved even at later stages of tumor progression by modulation of effector and regulatory T cell cytokines. (*see Plate 53 for colour figure*)

SUGGESTIONS FOR FURTHER READING

1. Talwar G.P. (1978), Towards Development of a Vaccine Against Leprosy, Introduction, *Lepr.* India, 50:492–497.

2. Saini V., Raghuvanshi S., Khurana J.P., Ahmed N., Hasnain S.E., and Tyagi A.K. (2012), Massive Gene Acquisitions in *Mycobacterium indicus pranii* Provide a Perspective on Mycobacterial Evolution, *Nucleic Acids Res.*, 40:10832–10850.

3. Gupta A., Geetha N., Mani J., Upadhyay P., Katoch V.M., Natrajan M., Gupta U.D. and Bhaskar S. (2009), Immunogenicity and Protective Efficacy of "*Mycobacterium w*" against *Mycobacterium tuberculosis* in Mice Immunized with Live Versus Heat-killed Mw by the Aerosol or Parenteral Route, *Infect Immun.*, 77:223–231.

4. Hoption Cann S.A., Van Netten J.P. and Van Netten C. (2003), Dr. William Coley and Tumour Regression: A Place in History or in the Future, *Postgrad Med J.*, 79:672–680.

5. Grange J.M., Bottasso O., Stanford C.A. and Stanford J.L. (2008), The Use of Mycobacterial Adjuvant-based Agents for Immunotherapy of Cancer, *Vaccine*, 26:4984–4990.

6. Rakshit S., Ponnusamy M., Papanna S., Saha B., Ahmed A. and Nandi D. (2012), Immunotherapeutic Efficacy of *Mycobacterium indicus pranii* in Eliciting Anti-tumor T Cell Responses: Critical Roles of IFNγ, *Int J. Cancer*, 130:865–875.

7. Dranoff G. (2004), Cytokines in Cancer Pathogenesis and Cancer Therapy, *Nat Rev Cancer*, 4:11–22.

8. Hanahan D. and Weinberg R.A. (2011), Hallmarks of Cancer: The Next Generation, *Cell*, 144:646–674.

9. Vanneman M. and Dranoff G. (2012), Combining Immunotherapy and Targeted Therapies in Cancer Treatment, *Nat Rev Cancer*, 12: 237–251.

121

Vaccines for Cancer

S. Biswas

CONTENTS

I. INTRODUCTION

The cancer, a major con-communicable disease is a global health threat with substantial huge economic expenses. The cost of cancer in United States alone is proposed to increase from US$125 billion in 2010 to US$207 billion by 2020 (Mariotto et al., 2011). Although there was great improvement in reducing the mortality from cardiovascular diseases and other communicable diseases by preventive efforts, the cancer is still a major challenges at advanced, mostly metastatic stages. The screening methods have increased for some cases, but early detection is not yet possible for most malignancies. In fact the targeted therapies are emerging, but treatments with these costly agents are not economically sustainable. Though the world population is increasing towards 7 billion, the cancer care has become an important urgent goal in the current scenario. The Cancer vaccines offer a unique therapeutic modality where they initiate a dynamic process of turn on host's own immune system. This has the potential to change the initial patient responses. In 2005, there was a high hope for therapeutic vaccines, with more than ten products designed as tumor associated antigens were taken for phase III trial, but nearly 5 years later, several of these vaccines failed completely. None of these was approved by the US FDA or the Europian Medicine agencies (EMEA). Still recent preclinical and clinical experiments have shown that well-designed clinical trials with appropriate administration of vaccines with new paradigms of combination therapies, may ultimately lead to the use of cancer vaccines for the treatment of many types of cancers.

This chapter will review and discuss the major areas of investigations of vaccines (a) cancer types and clinical exploration (b) diverse vaccine platforms including peptides and proteins, recombinant viral vectors (poxviral, adenoviral, and alphavirus) or bacterial vectors (e.g., Listeria), whole tumor cell vaccines, dendritic cell vaccines, and dendritic cell/tumor cell fusions etc (c) appropriate clinical trial design and end points (d) multiple strategies for combining vaccines and other therapies (e) the status of cancer vaccines in ongoing clinical trials (f) newer generation of vaccines and vaccines strategies.

II. CANCER TYPES AND CLINICAL EXPLORATION

Both genetic and environmental factors are responsible for developing cancers by aberrant growth regulation of stem cell population, or by the dedifferentiation of more mature cell types. In normal cases, in response to injury the cells proliferate, to replace the cells that undergo apoptotic cell death. Genetic variations or other environmental manipulation leads to disrupt this orders leading to cancer development. The environmental factors that are responsible for causing the mutation in DNA are called carcinogens and there are many types, e.g., the UV rays, industrial pollutants etc. Less than 10% are hereditary. Breast cancer, colon cancer, ovarian cancer and uterine cancers are known to have inherited tendencies.

Cancers are classified based on their origins. 90% human cancers fall in the category of carcinomas that arise in the epithelium. Adenocarcinomas develop in an organ/gland. For example, prostate cancer, cervical cancer, lung cancer, liver cancer, colon cancer, kidney cancer, thyroid cancer, pancreatic cancers. Squamous cell carcinomas are cancers that originate in the skin. Melanomas originate in the skin, usually in the pigment cells while Sarcomas (like Bone, breast cancer and soft tissue cancer) originate in the supporting tissues of the body, such as bone, muscle and blood vessels. Brain cancers are the Gliomas that develop in the nerve tissue. Leukemias and Lymphomas develop in blood and lymph glands, respectively. For example, Acute lymphoblastic leukemia, Chronic lymphocytic leukemia (CLL), Acute myelogenous leukemia (AML), Chronic myelogenous leukemia (CML), Hairy cell leukemia (HCL), T-cell prolymphocytic leukemia (T-PLL), Hodgkins and Non Hodgkins. The Leukemia and Lymphoma Society (LLS), founded in 1949, is the world's largest voluntary health organization dedicated to funding blood cancer research, education and patient services. Still today mostly doctors have three ways of dealing with cancer: cutting (surgery), burning (radiation), and poisoning (chemotherapy) of cancer. Another recent treatment for cancer is hormone therapy which involves anything that deals with manipulation of the body's hormones to treat the cancer including administration of hormones and drugs or removal of hormone glands to kill cancer or to prevent further cancerous growth. Gene therapy is also one of the important parameters to correct a genetic disorder by inserting a functional gene into an organism to replace a defective one. Numerous *in vitro* and preclinical animal models, testing a wide variety of gene therapy agents, have shown remarkable efficacy. But oncologists have long sought a fourth path in using the body's own immune system to attack tumors. Lots of compounds of biological origin used in the immune response can now be made in the laboratory. Interferons, Interleukin 2 (IL2) and monoclonal antibodies used to boost the immune response to help the body kill cancer cells. Major disadvantages of chemotherapy and radiotherapy are reducing immunity with the drop of white blood cells made in the bone marrow. The major objective of cancer vaccine is to stimulate the immune system to be able to recognise the abnormal cancer cell and finally wipe it out.

III. IMMUNE MODULATORS IN THERAPEUTIC CANCER VACCINES

Current advances in our understanding of immunology allowed the design of vaccines using immune modulators as defined molecular adjuvants, rationally to improve the immune responses. It is already well established from numerous preclinical and clinical trials that the diversity of immune responses that can be elicited in anti-tumor responses comprise the generation of CD8 and CD4 T-cell responses, natural killer cells and macrophages, cytokines, and various types of antibodies. Due to limitations to get sufficient amount of peripheral blood available for assay as well as the difficulty in obtaining biopsy specimens pre- and post-therapy, the evaluation of the proper range of immune responses is tricky. But it is now understandable that the immune system has evolved qualitatively different types of responses. For example, distinct subsets of helper T cells, such as TH1, TH2 and TH17, are effective at protecting against different pathogens (Zhu et al. 2010). Follicular helper T cells (TFH cells) produce interleukin 21 (IL-21) for differentiation of B cells and generation of memory B cells (Crotty 2011). The differentiation memory CD4+ and CD8+ T cells can be subcategorized into central memory and effector memory cell subsets based on their distinct functions. Research during the past decade has identified a fundamental role for the innate immune system in sensing vaccines and adjuvants and in programming protective immune responses. The innate immune system can sense microbes through pattern-recognition receptors (PRRs), such as the Toll-like receptors (TLRs), which are expressed by various cells, including dendritic cells (DCs) (Iwasaki and Medzhitov 2010). There are many subsets of functionally distinct DCs, and it is now clear that the DC subset, as well as the nature of the PRR, have a key role in determining the magnitude and quality of adaptive immune responses.

Based on the immune modulation studies for cancer vaccines, the immune modulators can be classified by two types, one that strengthen the response or change its quality, and ones that take away negative regulatory mechanisms or check point inhibitors, so as to allow the response to reach its full potential. Jay A. Berzofsky's group from Center for Cancer Research, National Institute of Health, USA explained the former the "push" and the latter the "pull" of their "push-pull" strategy (Berzofsky et al. 2004). They proposed the model to increase first the immunogenicity of the T-cell epitopes by substituting amino acids that interact with the MHC molecule to increase affinity, without altering the three-dimensional configuration seen by the T cell receptor. Next, they proposed to push the response with cytokines, co-stimulatory molecules and Toll like receptor (TLR) ligands as molecular adjuvants, not only to increase the magnitude of the response, but also to improve the quality of the response. Still this is not enough because of the negative regulatory cells, surface molecules and cytokines that are often present and inhibit the immune response. So it needs to overcome by blocking the negative regulation of (the "pull" of the approach) Treg, reg NKT, MDSC, CTLA-4, PD-1, TGF-β etc. (Berzofsky et al. 2004) (Figure 121.1).

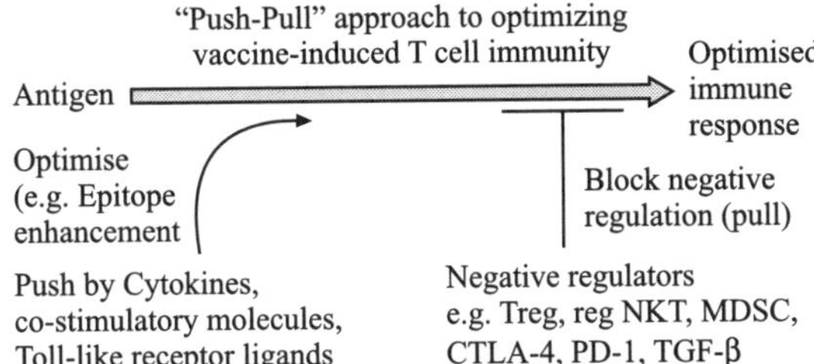

Figure 121.1   The Push-pull approach showing the basic skeleton of a vaccine as an antigen used to induce an immune response (concept from Berzofsky et al., 2004).

IV. MICROBES ASSOCIATED WITH CANCER AND VACCINES

It is already well established that microbes cause or contribute to 15–25% of all cancers diagnosed worldwide each year. The percentage is lower in the developed countries than the developing countries. The U.S. Food and Drug Administration (FDA) has approved two vaccines, Gardasil and Cervarix, that protect against infection by the two types of HPV types 16 and 18, that cause approximately 70% of all cases of cervical cancer worldwide (Doorbar, 2006). The FDA has also approved a preventive cancer vaccine against Hepatitis B (HBV) infection. Chronic HBV infection can lead to liver cancer. The original HBV vaccine was approved in 1981, making it the first cancer preventive vaccine to be successfully developed and marketed. Today, most children in the developed and developing countries are vaccinated against HBV shortly after birth and after hepatitis B virus-related liver transplantation (Takaki et al., 2013). The recently introduced HCV vaccine is going towards the clinical trials (Lauer, 2013). Several other active clinical trials for possible cancer treatment vaccines are stipulated. (http://www.cancer.gov/clinicaltrials/search)

The International Agency for Research on Cancer (IARC) has classified several microbes as carcinogenic. The major infectious agents, viruses, bacteria, and parasites associated with cancers are listed in the Table 121.1.

To develop effective cancer treatment vaccines, it is important to understand how immune system cells and cancer cells interact. The immune system often does not "see" cancer cells as dangerous or foreign, as it generally does microbes. Therefore, the immune system does not mount a strong attack against the cancer cells. All cancer preventive vaccines approved by the FDA to date have been made using antigens from microbes that cause or contribute to the development of cancer.

V. DIVERSE VACCINE PLATFORMS

Broadly we can classify vaccines in two types, specific or universal. Of course a more appealing cancer vaccine would be the universal that could fight cancer cells regardless of the cancer type. Based on treatment category we classified cancer vaccines in broadly two categories:

1. **Prophylactic/Preventive** vaccines—that prevent to develop cancer in healthy people.
2. **Therapeutic/Treatment** vaccines—that treat an existing cancer by strengthening the body's natural defenses against the cancer.

Cancer cells can carry both self antigens and cancer-associated antigens. The cancer-associated antigens mark the cancer cells as abnormal, or foreign, and can cause B cells and killer T cells to mount an attack against them. Cancer cells may also make much larger amounts of certain self antigens than normal cells.

TABLE 121.1   **The major infectious agents associated with cancer**

Infectious agents	Type of organism	Associated Cancers
Hepatitis B virus (HBV)	virus	Hepatocellular carcinoma
Hepatitis C virus (HCV)	virus	Hepatocellular carcinoma
Human papillomavirus (HPV) types 16 and 18, as well as other HPV types	virus	Cervical cancer; Vaginal cancer; Vulvar cancer; Oropharyngeal cancer, Anal cancer; Penile cancer; Squamous cell carcinoma of the skin
Epstein-Barr virus	virus	Burkitt lymphoma; non-Hodgkin lymphoma; Hodgkin lymphoma; Nasopharyngeal carcinoma (cancer of the upper part of the throat behind the nose)
Kaposi sarcoma-associated herpesvirus (KSHV), also known as human herpesvirus 8 (HHV8)	virus	Kaposi sarcoma
Human T-cell lymphotropic virus type 1 (HTLV1)	virus	Adult T-cell leukemia/lymphoma
Helicobacter pylori	bacterium	Stomach cancer; Mucosa-associated lymphoid tissue (MALT) lymphoma
Schistosomes (Schistosomahematobium)	parasite	bladder cancer
Liver flukes (Opisthorchisviverrini)	parasite	Cholangiocarcinoma

Here is the list of approaches for cancer vaccines being developed.

1. Antigen vaccines
2. Whole tumor cell vaccines
3. Combining vaccines with immune modulators
4. Anti-Idiotype antibody-based vaccines
5. Dendritic cell/Tumor fusion vaccines
6. DNA/vector based vaccines
7. Stem cell as cancer vaccines

It is also a of fact that unlike vaccines for other diseases that prevent the occurrence of the diseases, still there is no vaccine under development that can prevent the onset of specific cancer. Cancer preventive vaccines target infectious agents that cause or contribute to the development of cancer (Frazer et al., 2007). They are similar to traditional vaccines, which help prevent infectious diseases, such as measles or polio.

VI. ANTIGEN VACCINES

In this approach the tumor-specific proteins displayed on a tumor cell are used to stimulate the immune system, especially T lymphocytes, to generate killer T cells to attack the cancer cells that carry the specific antigen. These antigens are the proteins that make up the outer surface of the viruses. Scientists have learned how to mass-produce many antigens in the lab. Researchers are also creating synthetic versions of antigens in the laboratory for use in cancer preventive vaccines. In doing this, they often modify the chemical structure of the antigens to stimulate immune responses that are stronger than those caused by the original antigens. This new technology means that large amounts of these very specific antigens can now be given to many patients. Now a cancer vaccine must induce an immune response to 'self' rather than to 'non-self' antigens without provoking the harmful autoimmunity. It happens when vaccine administered after tumor formation is exerting immunosuppressive effects. Promising results have been achieved in animal models by vaccination with whole cells from autologous or allogeneic tumors (Le et al., 2010) or with cDNA libraries (Kottke et al., 2011).

VII. WHOLE TUMOR CELL VACCINES

In this strategy the vaccines are made up of actual cancer cells removed during surgery. The cells are treated in the lab, usually with radiations, to reduce multiplication. In some cases, doctors give the vaccine along with substances called adjuvants that increase the immune response. Using the whole tumour cell may expose the immune system to a large number of cancer antigens.

There are two basic kinds of tumour cell vaccines:

A. Autologous vaccines: Autologous means "coming from the self." These vaccines are made from killed tumor cells taken from the same person in whom they will later be used. Autologous cancer cells may be reinjected shortly after surgery, or they may be grown in the lab or frozen and given later. The major difficulty in this process is that it can be expensive to create a new, unique vaccine for each patient. There may not be enough usable cells from the removed tumor to make a vaccine.

B. Allogeneic vaccines: Allogeneic means "coming from another." These vaccines use cells of a particular cancer type that originally came from someone other than the patient being treated. The cells for the vaccine are grown in the lab from a stock of cancer cells kept for that purpose. Some allogeneic tumor vaccines use a mixture of cells which were removed from several patients. Modified whole cell cancer vaccines represent one form of immunotherapy currently in development and clinical trials. Such vaccines can be made more immunogenic by transforming tumor cells to express co-stimulatory molecules, such as Granulocyte Macrophage-colony stimulating factor (GM-CSF). The transduced cell lines are usually irradiated to prevent further cell division upon vaccination (Gruiji et al., 2008).

The major advantage of this strategy is that the multiple tumor antigens can be targeted at once, generating immune responses to more than one tumor antigen.

Figure 121.2 gives an example of amplified immune system approach by granulocyte-macrophage colony-stimulating factor (GM-CSF). GM-CSF along with irradiated vaccine cells, pull towards dendritic cells (DCs) to the site of antigen for processing and presentation to DC cells. DCs can also be stimulated by monoclonal antibodies (mAb) that bind to specific tumor antigens on the vaccine cell surface via their Fc receptor recognization site. Toll-like receptor (TLR) agonists or immunomodulatory chemotherapic agents such as paclitaxel can also inspire DCs through TLRs to up regulate co-stimulatory molecules by increasing cytokine production, and enhance antigen processing and presentation. Secondly, the DCs process and present tumor antigen derived from the vaccinating cells to CD4 and CD8T cells. Tumor cell killing occurs when the TCR, expressed on effector CD8. Tumor antigen is presented in the form of peptide/major histocompatibility complex (MHC) complexes on antigen-presenting cells (APCs); T cells bind this complex with their T-cell receptors (TCRs). Additional signals are required from anti-CD40, anti1BB, and anti-OX40 etc. The activation and proliferation of tumor antigen-specific T cells can also be increased with the use of blocking antibodies to immune checkpoint molecules such as CTLA-4 and PD-1. On the other site, the APCs and T cells can be suppressed by inhibitory cytokines and molecules such as TGF and IL-10 secreted by suppressive immune cell populations like myeloid-derived suppressor cells (MDSC) and T-regulatory cells (Tregs). Chemotherapy (such as cyclophosphamide and gemcitibine) and radiation, when used at immunomodulatory doses, can be used to inhibit these populations.

So, as per push-pull approach, the T cells can work synergistically with traditional treatments such as chemotherapy,

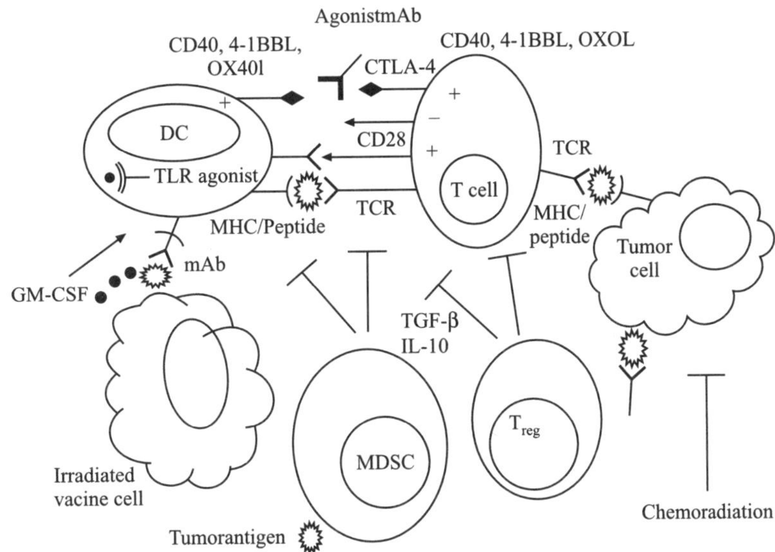

Figure 121.2   The vaccine approach with granulocyte-macrophage colony-stimulating factor (GM-CSF) associated immune modulation [concept adapted from (Gupta and Emens 2010)].

radiation therapy, or monoclonal antibodies to kill or inhibit tumor cells. Whole tumour cell vaccines are currently being developed to treat threatening cancers such as Acute Myeloid Leukemia (AML).

VIII.   COMBINED VACCINES WITH IMMUNE MODULATORS

There are many preclinical studies that have established the influence of the tumor micro environment in reducing the effectiveness of vaccine monotherapy. In addition to issues of vascular permeability, the tumor micro environment will in many cases include regulatory T cells (Tregs), myeloid derived-suppressor cells (MDSCs), and soluble factors such as TGF and interleukin (IL)-10, which are also capable of inhibiting immune-mediated tumor destruction as mentioned in beginning. So the use of certain chemotherapeutic agents, local radiation of tumor, and small molecule targeted therapeutics have all been shown (Figure 121.3) in preclinical studies to be capable of enhancing vaccine efficacy, and evidence is emerging in clinical studies how to best employ these therapeutics in combination with vaccine nowadays. Certain small molecule targeted therapeutics such as a pro-apoptotic protein BCL-2 inhibitor and tyrosine kinase inhibitors (TKIs) have been shown to differentially inhibit T_{regs} and MDSCs compared to effector T cells, resulting in greater immune responses.

IX.   ANTI-IDIOTYPE ANTIBODY BASED VACCINES

Lots of efforts towards developing vaccines against human malignancies have been frustrated by the lack of identification of a tumor specific antigen. Despite advancements in therapy, indolent lymphomas remain incurable, and disease relapse is predictable. In fact after complete remissions are achieved, small volumes of tumor cells remain. B-cell malignancies are composed of a clonal proliferation of mature resting and reactive lymphocytes that synthesize a single immunoglobins (Ig) molecule with unique heavy and light chain variable regions on their cell surface. The idiotypic determinants of the surface Ig molecules of a B-cell lymphoma can thus serve as a tumor specific marker for the malignant clone. Scientists have studied intensively how to make these anti-idiotype antibodies in the laboratory. Idiotype vaccination was first reported to induce tumor resistance in a mouse plasmacytoma model in the 1970s. Later, Kaminski and colleagues showed that by conjugating the carrier protein, Keyhole Limpet Hemocyanin, to the vaccine, its potential immunogenicity was enhanced. Subsequent research finally reproduced anti-idiotype models of multiple myeloma, B-cell lymphoma, and leukemia (Schuster et al., 2011). Further studies also demonstrated that production of CD8 T cells could be induced by attracting antigen-presenting cells, including dendritic cells, to the vaccination site.

X.   DENDTRITIC CELL/TUMOR FUSION CELL VACCINES

Dendritic cells (DCs) are already characterized for a complex network of highly potent antigen presenting cells (APC) that express many costimulatory molecules and are uniquely capable of inducing primary immunity. Scientists utilize the cancer vaccine strategies using native DC populations through the introduction of tumor antigens and immune adjuvants that recruit DCs to the vaccine bed and facilitate their uptake and presentation of the introduced antigen(s). DCs with this

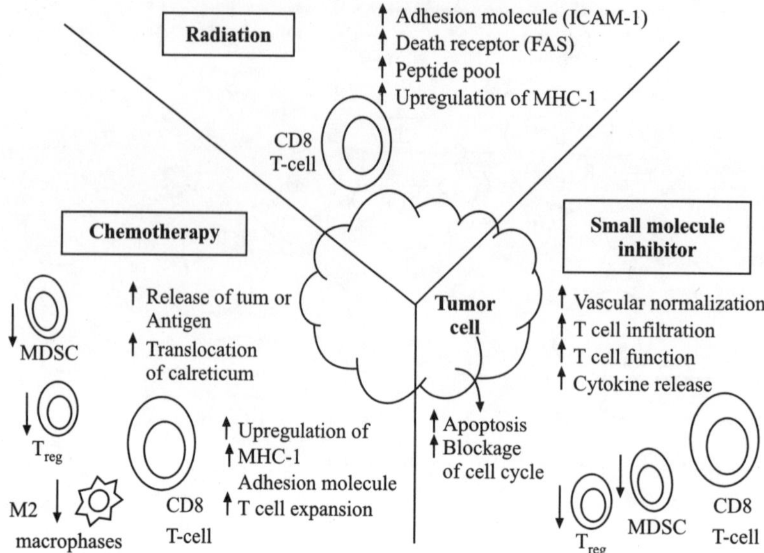

Figure 121.3   Combination approach including the synergy between radiation therapy, chemotherapy, or small molecule inhibitors and immunotherapy [adapted from (Hodge et al., 2012)].

activated phenotype may be generated *ex vivo* through the culture of progenitor populations with appropriate cytokines. However, creating DCs that maintain an activated phenotype *in vivo* and the capacity to migrate to sites of T-cell traffic poses a significant challenge to increase the potential immunogenicity through appropriate Human Leukocyte Antigen (HLA).

A promising strategy is recently developed using a potent cell based tumor vaccine that involves the fusion of patient derived DCs with autologous tumor cells. This strategy optimizes both Class I and Class II HLA pathway through processing newly synthesized and internalized antigens generating a balanced CD4- and CD8 mediated response that is critical for a sustained immunologic response. In an animal model, MC38 cells were transduced to express MUC1 as a tumor marker and fused with syngeneic DCs. DCs were shown to coexpress MUC1 and costimulatory molecules derived from the tumor and DC fusion partner, respectively (Figure 121.4). Animals vaccinated with DC/MC38-MUC1 fusions were protected from an otherwise lethal challenge of tumor cells consisting of MC38-MUC1 or wild-type MC38 cells. Preclinical studies have demonstrated fusion cell–mediated stimulation of tumor-specific helper and cytotoxic T-cell responses.

Previously it was shown that the protection from tumor challenge was most effective in the presence of intact class I and II mediated presentation (Koido et al., 2005). In another recent study, patient-derived breast cancer cells were fused with autologous DCs and found that the fusion cells were highly effective in stimulating the expansion of cytotoxic T cells with the capacity to lyse autologous breast cancer cells *in vitro*. The efficacy of DC/tumor fusions in generating tumor specific immunologic responses has been validated in animal and preclinical human studies across multiple tumor models, including multiple myeloma, osteosarcoma, glioma, ovarian cancer, myeloid leukemia, hepatocellular carcinoma, prostate cancer (Cathelin et al., 2011; Lee, 2011).

XI. DNA/VECTOR BASED VACCINES

Sometime the antigen vaccines become less effective over time. This is because the immune system recognizes them as foreign body and quickly destroys them. Without any further stimulation, the immune system often returns to its normal. So, in contrast to the peptide and protein-based vaccine strategies, recombinant vectors provide a platform for the expression of one or more full-length gene segments. The simplest method is to use a DNA plasmid encoding the gene(s) of interest, although *in vivo* expression is often difficult to achieve. In contrast, genes can be engineered into viral vectors for vaccination or gene therapy approaches.

This offers the advantage of (i) easy manufacturing (ii) easy delivery to patients (iii) easy storage and (iv) may provide additional nonspecific immune adjuvant effects, especially if there is a strong host immune response against the vector. The vaccine strategies recruit both CD8+ and CD4+ T cells enabling presentation of peptide via both MHC I and MHC II. Only disadvantage of this method especially if using oncogenic DNA, is the potential of the DNA to integrate into the genome of the cell that takes it up, thus potentially promoting malignancy.

In the vector based vaccines doctors use special delivery systems (called vectors) to make them more effective. They are not really a separate category of vaccine, e.g., there are vector-based antigen vaccines and vector-based DNA vaccines. Vectors are special viruses, bacteria, yeast cells, or other structures that can be used to get antigens or DNA into the body. The first viral vector used was vaccinia, a poxvirus, over 20 years ago. In contrast to poxviruses that have been used for expression of tumor antigens and immunostimulatory genes to induce anti-tumor immunity, an attenuated herpes simplex virus (HSV), type 1 has been generated for direct lytic destruction of tumor cells. Allovectin-7 (Vical, San Diego, CA) is a bicistronic

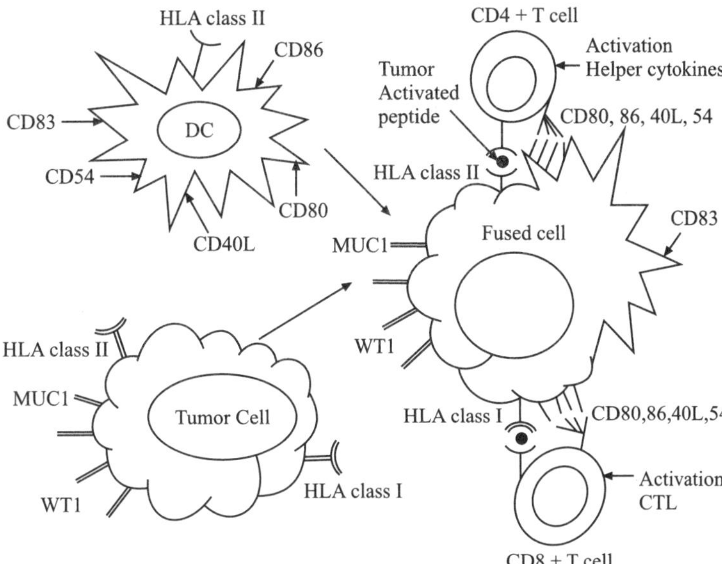

Figure 121.4 The dendritic cell/tumor fusions represent a huge diversity of tumor associated specific and non-specific antigens, in the context of DC-derived costimulatory molecules. The fused cell induce both HLA class I and class II responses, presenting tumor antigens to both cytotoxic and helper T-cell populations (adapted from Avigan et al., 2012).

plasmid made up of a cationic lipid complex containing the DNA sequences encoding HLA-B7 and microglobulin. This DNA vaccine was produced for direct injection into accessible tumors and results in expression of the full MHC class I HLA-B7 complex, which is expected to induce both local and systemic antitumor immune responses resulting in tumor eradication (Slingluff et al., 2009). The use of recombinant poxviruses for vaccination against cancer has been described for a number of cancers. Recombinant MVA-5T4 (also known as TroVax, Oxford Biomedica, Oxford, UK) was tested in several phase II clinical trials (Harrop et al., 2011). 5T4 is a cell surface glycoprotein that is expressed on placenta, at low levels in normal pituitary tissue and at very high levels in more than 90% clear cell cancers and papillary RCCs. OnvoVexGM-CSF-treated patients also exhibited decreased number of regulatory CD4 and CD8T cells and myeloid-derived suppressor cells within the tumor micro environment of injected melanoma lesions. The OncoVex GM-CSF Pivotal Trial in Melanoma (OPTIM) has been initiated (Kaufman and Bines, 2010).

XII. STEM CELLS AS CANCER VACCINES

Recently a group of scientists from University of Michigan have discovered a new paradigm for immunotherapy against cancer by priming antibodies and T cells with cancer stem cells. They examined the vaccination effects produced by cancer stem cell (CSCs) enriched populations from histologically distinct tumors after their inoculation into different syngeneic immunocompetent hosts in mouse model (Ning et al., 2012). These CSCs were immunogenic and more effective as an antigen source than unselected tumor cells in inducing protective antitumor immunity. Another interesting approach also suggested the application of embryonic stem (ES) cell

technology to immunize against cancer (Li et al., 2009). The study demonstrates whether vaccination of mice with human ES or induced pluripotent stem (iPS) cell lines could activate immunological response against shared antigens expressed by the primitive normal cells and the colon carcinoma cell line CT26, the so-called oncofetal antigens. The authors suggest that if the cancer stem cell concept is valid, then immunization with stem cells should prove valuable. However, it is possible to argue that human ES cells as a stem cell entity are defined merely on the basis of functional characteristics observed *in vitro* and that they are not likely to represent authentic stem cells formed during embryogenesis or in adult somatic tissues. If this is the case, it would be difficult to accept at face value claims that the observed protective effect was attributable to specific immunogenicity against tumor stem cells. The researchers also found that cytotoxic T lymphocytes harvested from cancer stem cell vaccinated hosts were capable of killing cancer stem cells *in vitro*. But at this moment it is important to determine the therapeutic efficacy of established tumors with CSC-based vaccines.

XIII. CANCER VACCINES AND CLINICAL TRIALS

Therapeutic anticancer vaccines function by drawing out or enhancing an immune response that specifically targets tumor-associated antigens. After extensive efforts made to develop the clinically useful anticancer vaccines, few Phase III clinical trials boosting this immunotherapeutic strategy have achieved their primary endpoint. Most therapeutic cancer vaccines available today require doctors to first remove the patient's immune cells from the body, then reprogram them and reintroduce them back into the body. The new approach, which was first reported to eliminate tumors in mice in Science Translational Medicine in 2009, instead uses a disk-like sponge about the size of a fingernail that is made from US Food and Drug Administration (FDA)-approved polymers.

Sipuleucel-T (Provenge®) is the first cancer vaccine approved by the FDA in April 2010. The FDA approval of sipuleucel-T was based on a statistically and clinically significant improvement in overall survival in men with asymptomatic or minimally symptomatic metastatic castrate-resistant prostate cancer (CRPC). Sipuleucel-T consists of autologous peripheral blood mononuclear cells, including antigen presenting cells (APCs), which have been incubated with a recombinant fusion protein comprising the prostate protein prostatic acid phosphatase (PAP) fused to the cytokine, granulocyte-macrophage colony-stimulating factor (GM-CSF). Administration of the PAP/GM-CSF antigen directly into rats induced antibody responses but failed to induce autoimmune prostatitis. The sipuleucel-T phase III studies in metastatic CRPC demonstrated 22% reduction in the risk of death, the 4.1-month median survival benefit, and 38% increase in estimated survival at 3 years.

Yervoy, the drug recently in the market using the immune system against melanoma, made by Bristol Myers Squibb was approved by US regulators in 2011. Other important finding was in some carcinomas where distinct tumor-associated antigens (TAAs), such as prostate-specific antigen (PSA) were expressed for prostate cancer. There are numerous TAAs, such as *MUC-1* and carcinoembryonic antigen (*CEA*), that are shared among different carcinoma types. TRICOM Vaccine plate form also have been successfully developed that comprises of a recombinant vaccinia virus (rV-) prime and multiple recombinant avipox (fowlpox, rF-) booster vaccinations (Madan et al., 2012). Each vector contains transgenes for one or more TAAs and transgenes for three costimulatory molecules (B7.1 [CD80], ICAM-1 [CD54], and LFA-3 [CD58], designated TRICOM) to enhance immune system. Prophage Series cancer vaccines are autologous therapies derived from cells extracted from the patient's tumor.

Phase II trial of Prophage Series G-100 (HSPPC-96) for newly diagnosed glioblastoma multiforme (GBM) was first reported at the 81st American Association of Neurological Surgeons (AANS) Annual Scientific Meeting in May 2013. Very recently the company Agenus Inc. announced their Phase II trial of patients with newly diagnosed GBM treated with HSPPC-96 in combination with the current standard radiation and temozolomide therapy. Their study reflects 18 month median progression free survival (PFS), which represents a 160% increase versus current standard care alone. Agenus expects the final trial results of this study to be published in a scientific journal in 2014.

The company OncoSec Medical Inc. is developing its advanced-stage ImmunoPulse DNA-based immunotherapy and NeoPulse therapy with principle of electroporation to deliver plasmid IL-12 to treat solid tumors. Their programs include three Phase II clinical trials for ImmunoPulse targeting lethal skin cancers. The Biopharmaceutical Company ISA Pharmaceuticals B.V. is currently focusing on rationally designed, fully synthetic therapeutic vaccines against cancer and persistent viral infections. Recently they initiated a Phase I/II clinical study of its lead candidate ISA101 in HIV-positive men suffering from anal intraepithelial neoplasia (AIN). The MOLOGEN AG performed their final analysis of the clinical trial phase I/II with cancer vaccine MGN1601.

AstraZeneca is now moving quickly to recover a future for olaparib, the Inhibitors of poly(ADP-ribose) polymerase (PARP), after it prematurely terminated its ovarian cancer programme (Garber). PARP inhibitors could kill cancer cells with pre-existing DNA repair defects, specifically mutations in the breast cancer susceptibility 1 (*BRCA1*) and *BRCA2* genes, which are involved in double-strand DNA repair through homologous recombination.

Based on the current literature atleast 23 Phase II/III clinical trials testing 17 distinct therapeutic anticancer vaccines appeared as completed. 18 of these studies had failed to achieve their total objectives, while 4 had succeeded so far. Within these 4 successful trials, 50% had defined tumor stage as part of the criteria of patient inclusion in the study (Ogi and Aruga). Many of the other anticancer vaccination studies were completed and have been incorporated for immune monitoring to correlate the therapeutic responses. The responses in vaccine clinical trials are being monitored for vaccine-induced CD8 T cells via ELISPOT or by fluorescence-activated cell sorting (FACS)-based assays. The studies measure the amount of cytokine, such as interferon, being induced by CD8 T cells when presented with a single tumor antigen peptide epitope. This, of course, does not measure the lytic activity of these T cells, the avidity of T cells, nor the breadth of CD8 T-cell response to other epitopes of the antigen in the vaccine; this evaluation also does not measure the contribution of CD4 T cells, NK cells, and the inhibitory aspects of regulatory cells that may be present,

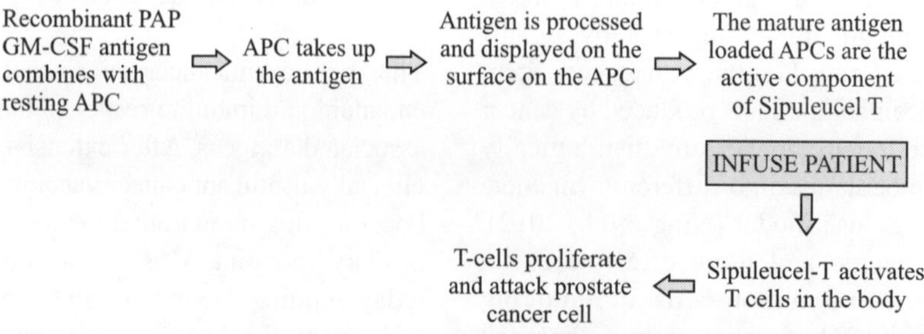

Figure 121.5 The Mechanism of action of Sipuleucel-T [concept adpted from (Di Lorenzo et al., 2011)].

This will be an important area for further investigation as more studies are now reporting good correlations between immune endpoints and clinical outcomes. In addition, there has been considerable interest in the identification of patient specific and tumor specific biomarkers that may predict therapeutic response and/or clinical outcomes. These studies would help select patients more likely to respond to a particular vaccine approach and might identify new strategies for improving the potency of individual vaccines so that more patients might benefit from this immunotherapy.

SUMMARY

Several factors must be considered in developing a vaccine for use in cancer patients. Firstly, it needs to recognize a unique tumor antigen. Secondly, an effective adjuvant must be identified to promote an immunogenic response. Thirdly, it should determine the clinical setting in which the vaccine is likely to be most effective, as well as the patients who are most likely to benefit from such an immunization. Finally, doctors must judge the feasibility of large-scale vaccine production, with regard to both cost and preparation time. The use of various vaccine vehicle as described allogeneic, whole tumor cells, peptide or protein-pulsed antigen-presenting cells (including dendritic cells), recombinant DNA and viral vectors, and recombinant *Saccharomyces* (yeast) are providing evidence of clinical benefits. For last several years several clinical studies have provided us a much deeper understanding of the mechanisms that control the immune response, as well as the mechanisms by which tumors evade the immune system. With a variety of agents now available that can modulate these mechanisms, we can expect an acceleration of progress in the years ahead, with the field delivering new treatments for our patients that are both more effective and better tolerated than traditional cancer therapies.

REFERENCES

Avigan D., Rosenblatt J. and Kufe D., Dendritic/tumor Fusion Cells as Cancer Vaccines, *Semin Oncol.*, 39(3):287–295.

Berzofsky J.A., Terabe M., Oh S., Belyakov I.M., Ahlers J.D., Janik J.E. and Morris J.C. (2004), Progress on New Vaccine Strategies for the Immunotherapy and Prevention of Cancer, *The Journal of Clinical Investigation,* 113(11):1515–1525.

Cathelin D., Nicolas A., Bouchot A., Fraszczak J., Labbe J. and Bonnotte B. (2011), Dendritic Cell-tumor Cell Hybrids and Immunotherapy: What's Next? *Cytotherapy*, 13(7):774–785.

Crotty S. (2011), Follicular Helper CD4 T Cells (TFH), *Annual Review of Immunology*, 29:621–663.

de Gruijl T.D., van den Eertwegh A.J., Pinedo H.M., Scheper R.J. (2008), Whole-cell Cancer Vaccination: From Autologous to Allogenic Tumor and Dendritic Cell-based Vaccines. *Cancer Immunol. Immunother.*, 57(10): 1569–77.

Di Lorenzo G., Buonerba C. and Kantoff P.W., Immunotherapy for the Treatment of Prostate Cancer, *Nat Rev Clin Oncol.*, 8(9):551–561.

Doorbar J. (2006), Molecular Biology of Human Papillomavirus Infection and Cervical Cancer, *Clinical Science*, 110(5):525–541.

Frazer I.H., Lowy D.R. and Schiller J.T. (2007), Prevention of Cancer Through Immunization: Prospects and Challenges for the 21st Century, *European Journal of Immunology*, 37 Suppl 1:S148–155.

Garber K., PARP Inhibitors Bounce Back, *Nat Rev Drug Discov.*, 12(10):725–727.

Gupta R. and Emens L.A. GM-CSF-secreting Vaccines for Solid Tumors: Moving Forward, *Discov Med.*, 10(50):52–60.

Harrop R., Shingler W.H., McDonald M., Treasure P., Amato R.J., Hawkins R.E., Kaufman H.L., De Belin J., Kelleher M., Goonewardena M. and Naylor, S. (2011), MVA-5T4–induced Immune Responses are an Early Marker of Efficacy in Renal Cancer Patients, *Cancer Immunology, Immunotherapy*: CII60(6):829–837.

Hodge J.W., Ardiani A., Farsaci B., Kwilas A.R. and Gameiro S.R., The Tipping Point for Combination Therapy: Cancer Vaccines with Radiation, Chemotherapy, or Targeted Small Molecule Inhibitors, *Semin Oncol.*, 39(3):323–339.

Iwasaki A. and Medzhitov R. (2010), Regulation of Adaptive Immunity by the Innate Immune System, *Science,* 327(5963):291–295.

Kaufman H.L. and Bines S.D. (2010), OPTIM Trial: A Phase III Trial of an Oncolytic Herpes Virus Encoding GM-CSF for Unresectable Stage III or IV Melanoma, *Future Oncology*, 6(6):941–949.

Koido S., Nikrui N., Ohana M., Xia J., Tanaka Y., Liu C., Durfee J.K., Lerner A. and Gong J. (2005), Assessment of Fusion Cells from Patient-derived Ovarian Carcinoma Cells and Dendritic Cells as a Vaccine for Clinical Use, *Gynecologic Oncology*, 99(2):462–471.

Kottke T., Errington F., Pulido J., Galivo F., Thompson J., Wongthida P., Diaz R.M., Chong H., Ilett, E., Chester J., Pandha H., Harrington K., Selby P., Melcher A. and Vile R. (2011), Broad Antigenic Coverage Induced by Vaccination with Virus-based CDNA Libraries Cures Established Tumors, *Nature Medicine*, 17(7):854–859.

Lauer G.M. (2013), Immune Responses to Hepatitis C Virus (HCV) Infection and the Prospects for an Effective HCV Vaccine or Immunotherapies, *The Journal of Infectious Diseases*, 207 Suppl 1:S7–S12.

Le D.T., Pardoll D.M. and Jaffee E.M. (2010), Cellular Vaccine Approaches, *Cancer Journal,* 16(4):304–310.

Lee W.T. (2011), Dendritic Cell-tumor Cell Fusion Vaccines, *Advances in Experimental Medicine and Biology*, 713:177–186.

Li Y., Zeng H., Xu R.H., Liu B. and Li Z. (2009), Vaccination with Human Pluripotent Stem Cells Generates a Broad Spectrum of Immunological and Clinical Responses Against Colon Cancer, *Stem Cells*, 27(12):3103–3111.

Madan R.A., Bilusic M., Heery C., Schlom J. and Gulley J.L. (2012), Clinical Evaluation of TRICOM Vector Therapeutic Cancer Vaccines, *Seminars in Oncology*, 39(3):296–304.

Mariotto A.B., Yabroff K.R., Shao Y., Feuer E.J. and Brown M.L. (2011), Projections of the Cost of Cancer Care in the United States: 2010–2020, *Journal of the National Cancer Institute*, 103(2):117–128.

Ning N., Pan Q., Zheng F., Teitz-Tennenbaum S., Egenti M., Yet J., Li M., Ginestier C., Wicha M.S., Moyer J.S., Prince M.E., Xu Y., Zhang X.L., Huang S., Chang A.E. and Li Q. (2012), Cancer Stem Cell Vaccination Confers Significant Antitumor Immunity, *Cancer Research*, 72(7):1853–1864.

Ogi C. and Aruga A., Immunological Monitoring of Anticancer Vaccines in Clinical Trials. *Oncoimmunology*, 2(8):E26012.

Schuster S.J., Neelapu S.S., Gause B.L., Janik J.E., Muggia F.M., Gockerman J.P., Winter J.N., Flowers C.R., Nikcevich D.A., Sotomayor E.M., McGaughey D.S., Jaffe E.S., Chong E.A., Reynolds C.W., Berry D.A., Santos C.F., Popa M.A., McCord A.M. and Kwak L.W. (2011), Vaccination with Patient-specific Tumor-derived Antigen in First Remission Improves Disease-free Survival in Follicular Lymphoma, *Journal of Clinical Oncology*: *Official Journal of The American Society of Clinical Oncology*, 29(20):2787–2794.

Slingluff C.L. Jr., Petroni G.R., Olson W.C., Smolkin M.E., Ross M.I., Haas N.B., Grosh W.W., Boisvert M.E., Kirkwood J.M. and Chianese-Bullock K.A. (2009), Effect of Granulocyte/macrophage Colony-stimulating Factor on Circulating CD8+ and CD4+ T-cell Responses to a Multipeptide Melanoma Vaccine: Outcome of a Multicenter Randomized Trial, *Clinical Cancer Research*: *An Official Journal of The American Association for Cancer Research*, 15(22):7036–7044.

Takaki A., Yagi T., Yasunaka T., Sadamori H., Shinoura S., Umeda Y., Yoshida R., Sato D., Nobuoka D., Utsumi M., Yasuda Y., Nakayama E., Miyake Y., Ikeda F., Shiraha H., Nouso K., Fujiwara T. and Yamamoto K. (2013), Which Patients Respond Best to Hepatitis B Vaccination After a Hepatitis B Virus-related Liver Transplantation? *Journal of Gastroenterology*.

Zhu J., Yamane H. and Paul W.E. (2010), Differentiation of Effector CD4 T Cell Populations (*). *Annual Review of Immunology*, 28:445–489.

SUGGESTIONS FOR FURHTHER READINGS

Adrian Bot, MihailObrocea Francesco and M. Marincola (Eds.) (2011), *Cancer Vaccines: From Research to Clinical Practice*, CRC Press.

Hilal Arnouk (Ed.) (2012), Advancements in Tumor Immunotherapy and Cancer Vaccines, Publisher: InTech; ISBN 978–953–307–998–1.

Samir N. Khleif (Ed.) (2015), Tumor Immunology and Cancer Vaccines, Cancer Treatment and Research, Vol. 123, ISBN: 978–1–4020–8119–4 (Print) 978–0–387–27545–1 (Online).

SECTION XVII

COMPLEMENTARY, ALLIED AND INTEGRATED MEDICINE

Modern medicine available at present cannot treat all maladies. The patient has to make recourse to alternate systems of medicine, be it Homeopathy, Ayurveda or Unani. Also, traditionally known herbs are made use of. Some of these, such as Curcumin has been the focus of research in various laboratories of the world. Trials have shown it to be highly safe, intake up to 8 gm per day could produce no observable side-effects. Its mode of action at molecular level has been elucidated. It is a highly useful compound with anti-cancer properties. Yoga is today recognized the world over for its invigorating properties for the body, endocrine system and nervous system. This section includes a brief but important information from specialists in these areas.

122

Ayurveda and Natural Product Drug Discovery

Bhushan Patwardhan

CONTENTS

I. AYURVEDA—THE ANCIENT SCIENCE OF LIFE

Ayurveda remains one of the most ancient and yet living traditions practiced widely in India, Sri Lanka and other countries and has a sound philosophical and experiential basis. Atharvaveda (around 1200 BC), Charak Samhita and Sushrut Samhita (1000–500 BC) are main classics that give a detailed description of over 700 herbs. The Brihadatrayee consisting of Charaka, Sushruta and Vagbhata are the main Ayurvedic classics, which describe some of its original and profound concepts. The most contemporary commentary on the Brihadatrayee is that of M.S. Valiathan in his Legacy series[1]. Currently, with over 400,000 registered Ayurveda practitioners, Government of India has formal structures to regulate issues related to quality, safety, efficacy and practice of herbal medicine. With unique holistic approach, Ayurvedic medicines are usually customized to an individual constitution. Ayurvedic philosophy strongly believes that human being is a microcosm of nature and all the five basic elements of nature are present in him. The original philosophical basis of Ayurveda is in Sankhya meaning knowing the truth. Sankhya is the philosophy of creation and is based on the twenty four principles or elements of the universe and Prakriti a female principle and Purusha a male principle, together they evolve Mahad – universal intellect. Its first manifestation is Ahamkar or Ego, which further manifests into five basic principles or Panchamahabhootas. From this concept the Ayurvedic philosophy has been based and further developed. Some of the fundamentals of Ayurvedic philosophy are: Five basic elements of nature are called as Panch (five) Maha (great) Bhootas (elements). These Panchamahaboota include: Earth (Prithvi), Water (Aap), Fire (Tejas), Air (Vayu) and Ether (Akash). Earth element relates to sense of smell, Water relates to taste, Fire relates to vision, Air relates to touch and Ether relates to hearing. Thus these five elements are connected with five sensory organs: ear, tongue, eye, skin and nose, respectively. These five elements in various permutations and combination and proportions form all living bodies including the humans. Therefore every human being has its own character presented in form of its Prakriti or Constitution—something like a genetic structure determined at the time of birth.

The five basic principles or Panchamahabhoota in the human body and their combination, proportions, determine the constitution or Prakriti of that individual and is presented in form of Tri-Doshas or three humors namely Vata (Air and Ether), Pitta (Fire and Water) and Kapha (Water and Earth). Water predominates in the overall constitution of human body as it is part of Kapha and Pitta constitution. There are three main constitutions, seven types of secondary and many subtle combinations of Vata, Pitta, Kapha (VPK) depending on the individual proportions. As per the Ayurvedic principles, an individual is born with a definite Prakriti, which remains unaltered throughout the life span quite similar to the genome. However, the combination of the basic elements that govern patho-physiological changes in the body can alter its response due to variety of causative factors including diet, environment, life-style, infections and like. The disease prevention, health promotion or Ayurvedic therapeutics revolves around all the strategies of obtaining back the original balance of the basic elements by manipulations of materials and procedures so as to remove imbalance of Tri-Dosha and attain the original stable constitution.

Human constitution or Prakriti can be determined and a wide spectrum of indicators is available for this purpose. These indicators are based on past history; observations, physical examination, psychological profile, and social, cultural, ethnic aspects also serve as compounding factors. Since this determination of Prakriti has subjective considerations of individual and physician, it might vary due to differences in their understanding. However, attempts have been made to bring in sufficient mechanisms for objectivity and this exercise can be carried out mechanically by using a suitable computer and software. Determination of correct Prakriti is very vital since the entire further inventions and therapeutics is based on that consideration.

Every material either organic or inorganic has certain attributes. Ayurveda has illustrated twenty such attributes or Gunas. These attributes have a definite effect on TriDoshas and then on the Panchamahabhootas. Thus, one can selectively use variety of combinations of materials to achieve a desired, pre-determined impact to bring in equilibrium of imbalance TriDoshas. The Ayurvedic pharmacology and principles therapeutics are based on the actions and counteractions of selectively used materials with predetermined combination of these attributes for desired balance of TriDoshas. Another important feature of Ayurvedic pharmacology is its depth in understanding effects of these materials after they enter in the human body and the effects thereof on the three humors KPV. This Ayurvedic Pharmacokinetics and Pharmacodynamics is very vital in therapeutics and dietetics. Ayurveda has described six tastes as identified by tongue or Rasa: Sweet, Sour, Salty, Pungent, Bitter and Astringent. These substances will produce two types of effects (Hot or Cold) during the digestive phase, which is called as Virya. The same material will produce post-digestive effects of sweet, sour or pungent type, which is called as Vipaka. Generally it is observed that sweet and salty tastes will have sweet Vipaka; Sour taste has sour and pungent, bitter, astringent have pungent Vipaka. Rasa, Virya and Vipaka directly influence the TriDosha, Nutrition and Tissue development of SaptaDhatu: Rasa (plasma), Rakta (blood), Mansa (Muscle), Meda (Adipose), Asthi (Bones), Majja (Nerves) and Shukra (semen). Exhaustive information is available in Ayurvedic literature that can be converted into a large database giving information of various foods, herbs, medicines and other materials with their taste, actions and utility in different disorders. Again, this information is primarily used to maintain equilibrium of TriDoshas to maintain or regain health. An innovative method to help quantitative representations of various Ayurvedic concepts including, Prakriti, Rasa, Guna and such has been developed by Indian Institute of Chemical Technology. This technology has been registered as Herboprint and is patented. Herboprint essentially gives a three-dimensional HPLC fingerprint with Ayurvedic property profile.

Ayurvedic database available in classic texts has many applications. First, it can be used for bio-prospecting exercise to identify new sources of medicines. Second, it can give information about likely effects ranging from primary taste to its post digestive effects. Third, it gives information about safety, efficacy along with possible indications and contraindications. Fourth, it can give information of therapeutic potential and selective benefits to people with different constitutions. Most importantly, it will greatly facilitate intentional, focused and safe natural product drug discovery and development.

II. NATURAL PRODUCT DRUG DISCOVERY

Natural products including plants, animals and minerals have been the basis of treatment of human diseases. History of medicine goes back practically to the existence of human being. The current accepted modern medicine or allopathy has gradually developed over the years of scientific and observational efforts of scientists—however, the basis of its development remains in the roots of traditional medicine and therapies. The history of medicine includes many ludicrous therapies. Nevertheless, the ancient wisdom has been the basis of modern medicine and will remain as one important source of future medicine and therapeutics. Even during the early part of this century, plants were a vital source of raw material for medicines. The future of natural product drug discovery will be inspired by the traditional knowledge, it will be more holistic, personalized and involve wise use of ancient and modern therapeutic skills in complementary manner so that maximum benefits can be given to patients and community[2].

Greek physician Galen (129–200 AD) devised the first Pharmacopoeia describing the appearance, properties and use of many plants of his time. The foundations of modern pharmaceutical industry were laid when techniques were developed to produce synthetic replacements for many of the medicines that had been derived from the forest. Natural Product Chemistry actually began with the work of Serturner who first isolated morphine from opium. This in-turn was obtained from opium poppy (*Papaver somniferum*) by processes that have been used for over 5000 years. Many such similar developments followed. Quinine from Cinchona tree had its origin in the royal households of South American Incas. Before the first European explorers arrived, the native people of the Americas had developed complex medical systems complete with diagnosis and treatment of physical as well as spiritual illnesses. Indigenous peoples derived medicines and poisons from thousands of plants. Review of some of the plants that originated from Central and South America indicate that most of them either had potentially toxic or poisonous characters or were from food sources. In the early 1500's, Indian fever bark was one of the first medicinal plants to find appreciative consumers in Europe. Taken from the cinchona tree (*Cinchona officinalis*), the bark was used as an infusion by native people of the Andes and Amazon highlands to treat fevers. Jesuit missionaries brought the bark back to Europe. By the early sixteenth century, this medicine was known as "Jesuit fever bark," quite a transformation. The name coca (*Erythroxylum coca*) comes from an Aymara word meaning simply "tree." In Andean cultures, the leaves of the coca tree have been primarily chewed to obtain the benefits. From ancient times, indigenous people have added an alkaline such as crushed seashells or burnt plant ashes to the leaves in order to activate the pharmacologically part of coca. In 1860 a German chemist isolated the chemical responsible for the plant's power, now known as cocaine. Carl Koler found cocaine could act as a local anesthetic in eye surgery. As the years passed, scientists found cocaine paralyzed nerve endings responsible for transmitting pain. As a local anesthetic, it revolutionized several surgical and dental procedures. Pot curare Arrow head poison used in the East Amazon is predominately from the species *Strychnos guianensis*. Tube curare in the West Amazon is from *Chrondrodendron tomentosum* from which the curare in modern medicine is made from and named as tubocurarine. The jaborandi tree (*Pilocarpus jaborandi*) secretes alkaloid-rich oil. Several substances are extracted from this aromatic oil, including the alkaloid pilocarpine, a weapon against the blinding disease glaucoma. American Indians used pineapple (*Ananas comosus*) poultices to reduce inflammation in wounds and other skin injuries, to aid digestion and to cure stomachache on the island of Guadeloupe. In 1891 an enzyme that broke down proteins (bromelain) was isolated from the fresh juice of pineapple was found to break down blood clots. Even now, many pharmaceuticals have their origin in botanicals. Just few more to mention include Atropine, Hyoscine, Digoxin, Cholchicine, Emetine, and like. Reserpine an anti-hypertensive alkaloid (*Rauwolfia serpentina*) became available as a result of work carried out by Ciba in India. It is pertinent to note that most of these early discoveries are based on potentially poisonous sources based on little traditions. A large number of molecules have come out of the Ayurvedic clinical base including Rauwolfia alkaloids for hypertension, Psoralens in Vitiligo, Holarrhena alkaloids in Amoebiasis, Guggulsterones as hypolipidemic agents, Piperidines as bioavailability enhancers, Baccosides in mental retention, Picrosides in hepatic protection, Curcumines in inflammation, Withanolides, and many other steroidal lactones and glycosides as immunomodulators. In the future, modern medicine could well come to be based on such ancient eastern time-tested remedies, developed using advanced technologies from the West. In this process rationale and science will be key attributes. Traditional medicine-inspired drug discovery and development is therefore considered to be an innovative, efficient, faster and affordable strategy.

III. DISCOVERING MEDICINES OR POISONS?

One of the major problems with traditional indigenous medicine is discovering a reliable 'living tradition' rather than relying upon second hand accounts of their value and use. In many parts of the world the indigenous system of medicine have almost completely broken down and disappeared. This includes mostly developed countries and some developing countries where indigenous population is marginalized. In others, the system is fragmented with the use of indigenous materials being limited to small tribal and geographical areas as in many parts of Africa. In anthropological terms these are 'little traditions' while the Ayurvedic Indian and traditional Chinese systems are living 'great traditions'. Although, the little traditions would have excellent knowledge about medicinal and poisonous properties of botanicals, the researchers have mainly exploited poisonous sources. This may be primarily because of many reasons. First, it is relatively easy to present and demonstrate poisonous characteristics of botanicals. Second, there may not be a written documentation and by word of mouth poisonous characters get predominance. Third, for an outsider, poisonous characteristics differentiate in between ordinary and extraordinary material for pharmaceutical development. Fourth, a considerable time period is required to demonstrate true medicinal activities with proven safety profile. As against, great traditions have relatively organized database, and more exhaustive description of botanical material is available that can be tested using modern scientific methods. Ayurveda and Chinese Medical systems thus have an important role in bio-prospecting of new medicines.

IV. GREAT TRADITIONS – KEY TO DISCOVERY

Many of the major pharmaceutical corporations have renewed their strategies in favor of natural product drug discovery. However, the 'Discovery Engines' that are largely used are still based on the knowledge and information obtained from 'little traditions' and is based on comparatively short-duration experiences. An overview of natural product Drug Discovery will be useful to support our hypothesis that most of the natural materials particularly of botanical origin that have been explored for producing or designing newer pharmaceuticals is largely from potentially poisonous sources. The examples can be drawn from Quinine in the last century to Taxol of the present century are mentioned in previous sections. And although, there is a re-generation of natural product drug research, still the real 'great traditions' remain under-explored and they are really the true sources of future medicines.

A major thrust for scientific research on Ayurveda was given by R.A. Mashelkar through his Golden Triangle and New Millennium Indian Technology Leadership Initiative (NMITLI) initiatives, which brought CSIR, ICMR and AYUSH institutions together to generate evidence-based Ayurveda. The Council for Scientific & Industrial Research (CSIR) supported an Ayurveda-based herbal drug development project under the NMITLI program. Three projects were supported for Ayurveda based herbal drug development for hepatitis, diabetes and arthritis. Following several rounds of national level consultation involving Ayurvedic scholars, several drugs were short listed. They entered a parallel track of open-label observational studies by selected Vaidyas and animal pharmacology studies for safety and efficacy.

Another good example of systematic natural product drug development is from GP Talwar's group, where a polyherbal cream (Basant) has been formulated using diferuloylmethane (curcumin), purified extracts of *Emblica officinalis* (Amla), purified saponins from *Sapindus mukorossi*, Aloe vera and rose water along with pharmacopoeially approved excipients and preservatives. Basant inhibits the growth of WHO strains and clinical isolates of *Neisseria gonorrhoeae*, including those resistant to penicillin, tetracycline, nalidixic acid and ciprofloxacin. It has pronounced inhibitory action against Candida glabrata, Candida albicans and Candida tropicalis isolated from women with vulvovaginal candidiasis, including three isolates resistant to azole drugs and amphotericin B[3,4].

Origin of all such knowledge in India is in the Great Tradition of Ayurveda, which is a living tradition of practice even today. Indian health care consists of medical pluralism and Ayurveda still remains dominating even as compared to the modern medicine particularly for treatments of variety of chronic disease conditions. India has about 45,000 plant species; medicinal properties have been assigned to several thousand. About 2000 figure frequently in the literature; indigenous systems commonly employ about 500. Some recent work in drug development relates to species of Commiphora (used as a hypolipidaemic agent), Picrorhiza (which is hepatoprotective), Bacopa (memory enhancer), Curcuma (antiinflammatory) and Asclepias (cardiotonic).

V. REVERSE PHARMACOLOGY

Combining the strengths of knowledge base of traditional systems such as Ayurveda with the dramatic power of combinatorial sciences coupled with HTS will help in the generation of structure-activity libraries. Ayurvedic knowledge and experiential database can provide new functional leads to reduce time, money and toxicity—the three main hurdles in the drug development. These records are particularly valuable since effectively these medicines have been tested for thousands of years on people. Efforts are underway to establish pharmacoepidemiological evidence-base to Ayurvedic medicines, safety and practice[5]. Development of standardized herbal formulations is underway as an initiative of Council for Scientific and Industrial Research (CSIR). Randomized controlled clinical trials for Rheumatoid and Osteoarthritis, Hepatoprotectives, Hypolipidemic agents, Asthma, Parkinson's disease, and many other disorders have reasonably established clinical efficacy and a review of some exemplary evidence-based researches and approaches has now resulted in wider acceptance of Ayurvedic medicines[6].

Thus the Ayurvedic knowledge database allows drug researchers to start from a well-tested and safe botanical material. With Ayurveda, the normal drug discovery course of 'Laboratory to Clinics' actually becomes from 'Clinics to Laboratories'—a true Reverse Pharmacology Approach. In this process 'Safety' remains the most important starting point and the efficacy becomes a matter of validation. A golden triangle consisting of Ayurveda—Modern medicine—Science will converge into a real discovery engine that can make newer, safer, cheaper and effective drugs. Efforts of Talwar Research Foundation to develop standardized, evidence based, polyherbal formulations like Praneem and Basant offer good hope for future[7].

VI. REVIEW OF TOP TEN BOTANICALS

Charak Samhita is one of the most cited Ayurvedic classic books that we have referred to select Top 10 Ayurvedic botanicals. For ready reference we have given one indicative key reference to each of them. To augment this effort we have short-listed some reviews, database and compendium generally covering research on most of the popular Ayurvedic drugs.

Neem (*Azadirachta indica*): Commonly known as Neem tree. 'Praneem', a polyherbal formulation developed by renowned scientist G P Talwar, has successfully completed Phase II efficacy study for treatment of abnormal vaginal discharge due to reproductive tract infections that act as co-factors for HPV persistence. Potential anti-HPV activity of Praneem in women infected with high risk HPV type 16 has been studied and results show that a 30-day intra-vaginal application of the Praneem can result in elimination of HPV infection from the uterine cervix[8,9]. Various extracts were studied for castor oil induced diarrhea; gastrointestinal motility and prostaglandin E2 induced enteropooling. Results of the study suggests the prostaglandin inhibition primarily and mild inhibition of gastrointestinal motility as secondary effect. The ethanolic extract was studied for carbon tetrachloride induced hepatic changes in albino rats. The changes were assessed by serum enzyme profile and hepatic triglycerides levels, histological changes in liver and pentobarbitone sleeping time as a functional parameter. There was significant reversal of biochemical, histological and functional changes induced by carbon tetrachloride in rats by ethanol extract treatment. Early epithelialisation with acceptable scar and maintenance and function of the concerned part are important goals in burn care. A study was carried out to evaluate tolerability and efficacy of a topical formulation of *Azadirachta indica* in patients with burns. *Azadirachta indica* was reported to be safe and effective prohealer for burns, which lead to a better cosmetic outcome.

Shunthi (*Zingiber officinale*): *Zingiber officinale* (Ginger) is commonly used in household medicines since ages. It is also known as Mahaaushadi meaning a great medicine. It is used in coughs, bronchitis and as a stomach tonic. Anticancer drugs like Cisplatin causes nausea, vomitting and inhibition of gastric emptying. The anti emetic effect of the acetone and ethanolic extract of ginger against cisplatin-induced emesis in drugs was evaluated. Significant reversal of cisplatin-induced delay in gastric emptying was seen. Therefore ginger as an antiemetic for cancer chemotherapy may also be useful in improving gastrointestinal side effects associated with cancer chemotherapy. The antiemetic effect of ginger may be due to the scavenging activity of gingerol. Another study showed that the 5 HT 3–receptor-blocking activity could also be the reason for the antiemetic effect. Antioxidant activity of ginger is extensively studied. The free radical scavenging and the antiserotonergic action of ginger could be suggested as possible mechanism to reverse pyrogallol induced delay in gastric emptying. The combination of the ginger acetone extract along with the antioxidants Vitamin C and Vitamin E produced a significant reverse as compared to ginger acetone extract alone. Thus the beneficial effects of the ginger extract in improving symptoms such as abdominal discomfort and bloating are seen.

Guduchi (*Tinospora cordifolia*): *Tinospora cordifolia* a plant belonging to the Rasayana group of Charaka Samhita is given for prevention of diseases and strengthening of both physical and mental health. It has been extensively studied for its immunostimulant activity. Chronic liver damage in humans and experimental animals is characterized by deposition of fibrous tissues in liver. Kupffer cells are considered as major determinant of outcome of liver injury. This activity was studied in a model of chronic liver disease. Study was undertaken to evaluate effects of *Tinospora cordifolia* in a rat model of cholestasis. Results showed reduced endotoxicaemia after tinospora cordifolia therapy. The Phase I clinical study carried out on healthy volunteers to evaluate safety did not reveal any alterations in biochemical, immunological or radiological parameters. Maximum clinical exploration was done in patients with obstructive jaundice. Tinospora was used as an add-on regime to conventional therapy. This was associated with decrease in septicaemia. It was also found to potentiate neutrophil activity. The neutrophils of these patients also showed an increase in phagocytic activity, expressed a feeling of well being, and improved appetite. No side effects were observed with *Tinospora cordifolia*.

Shatavari (*Asparagus racemosus*): The possible antioxidant effects of crude extract and a purified fraction of *Asparagus racemosus* against membrane damage induced by the free radicals generated during gama-radiation were examined in rat liver mitochondria. The extracts have potent antioxidant properties *in vitro* in mitochondrial membranes of rat liver. It also protected superoxide dismutase and protein thiols from inactivation. This contributes to the cellular antioxidant defense system. The hypothesis that macrophages appear to play a pivotal role in the development of intraperitoneal adhesions, modulation of macrophage activity, therefore is likely to provide a tool for prevention of adhesions effect of *Asparagus racemosus* an immuno stimulant drug was evaluated in albino rats with intra peritoneal adhesions induced by caecal rubbing. After treatment with *Asparagus racemosus* a significant decrease was observed in the adhesion scores. There was significant increase in activity of macrophages too. This study can help for prevention and management of post-operative adhesions.

Ashwagandha (*Withania somnifera*): Ashwagandha is a relatively well-studied plant. Chemopreventive activity was shown in experimentally induced fibro sarcoma tumors. Ashwagandha significantly reduced the tumor incidence, tumor volume and enhanced the survival of the mice. Liver biochemical parameters revealed a significant modulated of reduced glutathione, lipid peroxidases, glutathione-s-transferase, catalase, superoxide dismutase in extract treateed mice. The mechanism of chemopreventive activity of *Withania somnifera* extract may be due to its antioxidant and detoxifying properties. Effect of pre- and co-treatment of hydroalcoholic root extract of *Withania somnifera* at different doses was investigated against Isoproterenol induced myocardial infarction in rats. Results were confirmed by histopathological findings. Ashwagandha was compared with Ginseng for adaptogenic and antistress activity and was found to be a better drug of choice with additional anti-inflammatory activity. Immunomodulatory activity has also been demonstrated in experimental immuneinflammation and cyclophosphamide induced myelo-suppression.

Pippali (*Piper longum*): The ethanolic extract of *Piper longum* fruit was tested for cytotoxicity using brine shrimp lethality test. It showed potent cytotoxic activity. Piperine is a constituent of various spices that are used as common food additives. The reproductive toxicity of piperine was studied in albino mice. Piperine increased the period of diestrous phase, which seemed to result in decreased mating performance and fertility. Prostaglandin E-1 induced acute inflammation of rat paw was significantly reduced after piperine treatment. Thus piperine was shown to interfere with several crucial reproductive events in mammalian model. A reverse phase HPLC method to determine pierine in different medicinally used Piperine samples has been developed. The method is precise, sensitive, reproducible and easy to perform. The sensitivity of the method was 0.1 mcg and the linearity was observed in the range of 0.1 mcg to 0.8 mcg. This method can be used for the detection and monitoring of Piperine. A spectrophotometric method has been developed to estimate piperine in Piper species. This method is simple rapid and economical and based on the identification of piperine by TLC and on the UV absorbance maxima in alcohol at 328 nm.

Haridra (*Curcuma longa*): Haridra (Turmeric) has been studied for effects of curcumin, a yellow pigment of the spice turmeric on the mutagenecity of several environmental mutagens in the Salmonella microsome test with or without Aroctor 1254 induced rat liver homogenate (S-9 mix). With Salmonella typhimurium strain TA-98 in the presence of S-9 mix curcumin inhibited the mutagenicity of bidi and cigarette smoke condensates, tobacco and masheri extracts benzopyrine and dimethylbenzoanthracene in a dose dependent manner. This indicates that curcumin may alter the metabolic activation and detoxification of mutagens. Curcumin and its derivatives have significant abilities to protect plasmid pBR 322 DNA against single strand breaks induced by singlet oxygen a reactive oxygen species with potential genotoxic/mutagenic properties. The protective ability of curcumin was higher than that of well-known biological antioxidants—alpha-tocopherol and beta-carotene. This partly explains the anti carcinogenic and antimutagenic properies. High Performance Thin Layer Chromatography method for Quantitative analysis of guggul and turmeric has been devised which is precise, specific, reproducible and cost effective.

Vidanga (Embelia ribes):   Embelin is a plant benzoquinone from berries of *Embelia ribes* with proven antispermatogenic effects. Administration of embelin results in certain metabolic changes in lipid metabolism in primary and accessory reproductive tissues and serum, which however returns to near normal biochemical make up once the drug regimen, is withdrawn. The biodistribution study of embelin was studied by Gupta et al Administration of embelin in rats by subcutaneous injection or oral route caused significant tissue deposition of drug in testis. Epididymis, seminal vesicles, ventral prostrate, liver, kidney, lung, spleen, heart, brain and intestine. A TLC method for identification and spectrophotometric method for estimation of embelin has been devised. It is simple, rapid, economical and can be used for standardization and monitoring of embelin in *Embelia ribes*.

Bramhi (*Bacopa monnieri*):   *Bacopa monnieri* contains Bacoside A as the active constituent. It is reported to be active in walker carcinoma the ethanolic extract is toxic to the Sarcoma-180 cell line. An investigation to confirm the cytotoxic activity was performed in the brine shrimp lethality assay and active constituents were assayed. The ethanol fraction and bacoside A showed potent cytotoxic activity. Bacoside A was responsible for other biological activities also. Anti convulsant profile of different extracts of *Bacopa monnieri* in rats was studied using maximum electro shock seizures in rats after administration and pentylene tetrazole (PTZ) test in mice and rats. Overall Bramhi has faster onset of action and time/dose responses were qualitatively similar to phenytoin. Hence its use as an antiepileptic drug can be supported. Studies on the mast cell stabilising activity of extracts of Bacopa were tested in vitro for mast cell stabilising effects. The methanolic factors exhibited potent activity comparable to disodium cromoglycate a known mast cell stabiliser. The results indicate its potential use in allergic conditions.

Tulasi (*Ocimum sanctum*):   The essential oil of *Ocimum sanctum* and eugenol tested in vitro showed potent anthelmintic activity in the Caenorhabditis elegans model. The essential oil of Ocimum and eugenol showed potent anthelmintic activity. Oxidative stress is one of the major risk factors for age related cataract development. Ocimum has natural antioxidants and has been evaluated against galactose induced experimental cataract development in rats. It possesses anti cataract activity. The therapeutic and prophylactic value of *Ocimum sanctum* in treatment of myocardial infarction was evaluated by studying the cardioprotective effeect. Histopathological findings confirmed the cardioprotective effect.

VII. AYURVEDIC PRODUCTS WITH EVIDENCE-BASE

Zandopa [*Mucuna pruriens (MP)*] (Zandu): MP contains a compound called L-DOPA which is a direct precursor of the neurotransmitter dopamine. Leads from usage of MP in Ayurvedic practice encouraged researchers to explore the plant in detail (at Zandu Pharmaceuticals Mumbai). Experimental findings suggest that MP contains water-soluble ingredients that either have an intrinsic dopa-decarboxylase inhibitor (DDCI)-like activity or mitigate the need for an add-on DDCI to ameliorate parkinsonism. The copper chelating property may be one of the mechanisms by which MP cotyledon powder exerts its protective effects on DNA. Symptomatic and Neuroprotective Efficacy of MP Seed Extract has been studied experimentally. Clinical study reveals contribution of L-DOPA in the recovery of Parkinson' disease. The clinical studies with MP seed powder suggests that this natural source of L-dopa might possess advantages over conventional L-dopa preparations in the long term management of Parkinson's disease. Experimental study suggests that MP endocarp may contain unidentified antiparkinsonian compounds in addition to L-DOPA, or it may have adjuvants that enhance the efficacy of L-DOPA. n-propanol extract of MP, which contain negligible amount of L-DOPA has shown significant neuroprotecctive activity, suggesting that some components other than L-DOPA might also be responsible for anti-Parkinson' property of seeds. The failure of MP endocarp to significantly affect dopamine metabolism in the striatonigral tract along with its ability to improve Parkinsonian symptoms in the 6-hydorxydopamine animal model and humans may suggest that its antiparkinson effect may be due to components other than levodopa or that it has an levodopa enhancing effect. MP was found to be an effective treatment for patients with Parkinson's disease. US patents of MP (patent no 6106839 and 6340474) are filed for its efficacy in the treatment of nervous system, including Parkinson's disease and potentiation of growth hormone. (Khandelwal VM, Koneri R, Balaraman R, Kandhavelu MS. Biological Activities of Some Indian medicinal plants. *Journal of Advanced Pharmacy Education & Research* 2011; 1:12–44).

Immumod [*Tinospora cordifolia* (TC)] (Wockhardt): TC stimulates activation of macrophages that leads to increase in GM-CSF, which leads to leucocytosis and improved neutrophil function. It has been studied for its adaptogenic effect. Protection against induced infections in mice and rats was observed in groups pre-treated with TC. TC extract showed beneficial effect in allergic rhinitis, diabetic foot ulcers and burns. The aqueous extract of TC was found to enhance phagocytosis in vitro. Further aqueous and ethanolic extracts also induced an increase in antibody production in vivo. TC has been shown to have hepatoprotective and immunomodulatory properties in experimental studies, on surgical outcome in patients with malignant obstructive jaundice. Extracts of TC have been studied in various areas such as diabetes, liver damage, free radical mediated injury, infections, stress and cancer. The immunomodulatory, diuretic, anti-inflammatory, analgesic, anticholinesterase and gastrointestinal protective effects have been studied. Effect of TC in HIV patients has been studied. Tinocordin is used as an adjuvant immunomodulator along with antitubercular drugs safely. (Sood P, Ludhadia SK, Puri S, Paul G. Role of Immunomodulator Tinocordin in Sputum-Positive Fresh Cases of Pulmonary Tuberculosis. Chest 2010;138(4):supp749A).

Guglip [*Commiphora mukul* (CM)] (Cipla): Central Drug Research Institute (CDRI) has developed gugulipid, a cholesterol lowering drug, taking the lead from plant *Commiphora mukul* used in Ayurveda. Farnesoid x receptor (FXR) is a key transcriptional regulator for the maintenance of cholesterol and bile acid homeostasis. Guggulsterone has been established is an antagonist at (FXR). The antagonism has been proposed as a mechanism for its hypolipidemic effect. Bile salt export pump (BSEP) is an efflux transporter responsible for removal of cholesterol metabolites, bile acids from the liver. Guggulsterone upregulates the BSEP, that favors cholesterol metabolism into bile acids. This may be an another possible mechanism for its hypolipidemic activity. Experimental and clinical studies demonstrate guggulsterone extracts as an effective treatment for hypercholesterolemia and hypertriglyceridemia. However recent review questions efficacy of Guglip in dyslipidaemia. (Yu BZ, Kaimal R, Bai S, El Sayed KA, Tatulian SA, Apitz RJ, et al. Effect of guggulsterone and cembranoids of Commiphora mukul on pancreatic phospholipase A(2): role in hypocholesterolemia. J Nat Prod. 2009;72(1):24-8).

Hyponidd (Charak): Hyponidd contains *Yashad Bhasma, Shilajit, Momordica charanti, Curcuma longa, Cassia auriculata, Emblica officinalis, Eugenia jambolana, Enicostemma littorale, Gymnema sylvestre, Pterocarpus marsupium, Tinospora cordifolia, Melia azadirachta* and *Swertia chirata.* Hyponidd exhibits antihyperglycaemic and antioxidant activity in streptozotocin-induced diabetic rats. Safety and efficacy of the formulation has been established through a double blind, placebo controlled study in Type 2 Diabetic patients with secondary failure to oral drugs. It content D-chiro–inositol (an insulin-sensitizing agent) is as efficacious as metformin and better tolerated than metformin. (Sridharan K, Mohan R, Ramaratnam S, Panneerselvam D. Ayurvedic Treatments for diabetes mellitus. Cochrane Database of Systematic Reviews 2011, Issue 12. Art. No.: CD008288. DOI: 10.1002/14651858. CD008288.pub2).

Rumalaya (Himalaya): This is one of the popular formulations widely used for various musculoskeletal disorders including arthritis. It *contains Mahayoograj Guggul, Shankh bhasma, Shilajeet, Swarnamakshik Bhasma, Abelmoschus moschatus, Rubia cordifolia, Moringa pterygosperma and Tribulus terrestris.* A double blind placebo-controlled trial conducted at All India Institute of Medical Sciences, efficacy of Rumalaya was comparable with Diclofenac sodium with lesser side effects. Although there are several studies on Rumalaya, the ingredients of contemporary Rumalaya tablet differ from those used in the studies conducted in 80s and 90s. (Rastogi S, Bansal R. Efficacy of Rumalaya Tablets in Arthritis – A Double Blind Placebo Controlled Trial. The Antiseptic 2001;98(5):172-73).

VIII. THE FLIPSIDE

Popularity always comes with a risk of misuse, which is true for every sector and so is for Ayurveda. With growing acceptance for Ayurveda, there is mushrooming growth of so-called Ayurvedic pharmaceutical companies. Many Ayurvedic versions of popular brands, especially soaps, shampoos, toothpastes, hair oils, face wash, are available in market. The companies are mainly targeting the health problems and diseases for which contemporary medicine do not provide a satisfactory solution. It ranges from acne to AIDS and includes hair fall, obesity, infertility, joint pain, brain disorders, cancer, and heart diseases. The aggressive promotional strategies, tele marketing, tall claims, advertisement with big shots from television and film industry influence and lure the desperate health seekers. The products neither are developed through robust and credible research findings nor follow good manufacturing practices. Although the schedule of Drugs and Magic Remedies (Objectionable Advertisement) Act, 1954 prohibits tall claims of complete cure and enlists 54 diseases in the schedule, numerous advertisements are appearing in the print and electronic media with exaggerated claims exploiting gullible people.

Government has recognized Gujarat Ayurveda University, Jamnagar, as a centre for 'National Pharmacovigilance Programme for Ayurvedic Siddha and Unani Drugs' since 2008. Ayurvedic products like Stresscom (Dabur), Ilogen excel (Pankajakasthuri Herbals), Diabecon (Himalaya), Palsinuron (SG Phyto Pharma), ASMON (Herbochem Remedies), Sookshma Triphala (Ayurveda Rasashala), M2 Tone (Charak), Trichup Hair Oil (Vasu), TORCHNIL (Dr Palep's MRF) and may more are extensively used by medical practitioners. However, research publications for many products are not readily available online. In addition, the publication in peer reviewed indexed journals are very less. Therefore, scientific research to establish evidence-based Ayurveda is really the need of the hour[10].

Future perspectives

In true sense, Ayurveda is not just herbal medicine but a holistic system that offers personalized medicine and complete health care. Ayurveda classifies the whole human population into three major constitutions, Kapha, Pitta and Vata, Prakritis. The Ayurveda database of human constitution, disease, detailed symptoms, logic of treatment and drug-activity libraries may provide new leads in individualized, standardized and uniform treatment, making medicine more a science and less an art to practice. Ayurveda is uniquely patient oriented where the Ayurvedic physician diagnoses, treats and dispenses medicine to every individual patient. This important principle forms the basis for a personalized medicine. This specific prescription is not limited to therapeutic medicine but may also include supportive interventions, diet and life-style advice so as to regain physiological balance, finally resulting in removal of the disorder.

This question of genetic basis to Ayurvedic concept of Prakriti is of keen interest to Indian scientists. A number of teams are now investigating the correlation between Ayurvedic phenotypes and individual human genotypes. Our group's pioneering works has had a catalytic influence and several groups in India and abroad have undertaken research projects in areas related to Ayugenomics[11]. Several good papers have been published, and many more will come in the near future. Thus from the first proof of concept paper published on AyuGenomics in 2005 has now reached a higher level of scientific appreciation with papers appearing in PNAS and other high impact journals[12]. The CSIR has recently established a new research initiative named Ayu*r*genomics, which is an integrative approach of Ayurveda and Genomics for the

discovery of predictive markers for preventive and personalized medicine. Such efforts have certainly helped Ayurveda to receive a moderate recognition among the global scientific community and have opened new vistas for development of future medicine.

SUMMARY

Natural products including plants, animals and minerals have been the basis of treatment of human diseases. A considerable research on pharmacognosy, chemistry, pharmacology and clinical therapeutics has been carried out on medicinal plants described in Ayurveda. This chapter gives a primer on Ayurveda, which is widely in use in India and Sri Lanka. Ayurveda inspired drug development can follow a reverse pharmacology path and reduce time and cost of development. Ayurveda is much more than a mere source of herbal drugs. Its unique holistic approach is based on individual Prakriti, which can form basis of future personalized medicine.

SUGGESTIONS FOR FURTHER READING

Aggarwal S., Negi S., Jha P., et al. (2010), EGLN1 Involvement in High-altitude Adaptation Revealed through Genetic Analysis of Extreme Constitution Types Defined in Ayurveda, PNAS, Doi/10.1073/pnas.1006108107.

Bhengraj A.R., Goyal A., Talwar G.P. and Mittal A. (2009), Assessment of Antichlamydial Effects of a Novel Polyherbal Tablet Basant, *Sex Transm Infect.*, 85(7):561.

Garg S., Talwar G.P. and Upadhyay S.N. (1998), Immunocontraceptive Activity Guided Fractionation and Characterization of Active Constituents of Neem (*Azadirachta Indica*) Seed Extracts, *J. Ethnopharmacol.*, 60(3):235–46.

Joglekar N.S., Joshi N.S., Deshpande S.S., et al. (2010), Acceptability and Adherence: Findings from a Phase II Study of a Candidate Vaginal Microbicide, 'Praneem Polyherbal Tablet', in Pune, India, Transactions of the Royal Society of Tropical Medicine and Hygiene,104;6, 412–415.

Patwardhan B. and Bodeker G. (2008), Ayurvedic Genomics: Establishing a Genetic Basis for Mind-body Typologies, *Journal of Alternative and Complementary Medicine,* 14(5):571–6.

Patwardhan B. and Mashelkar R.A. (2009), Traditional Medicine-inspired Approaches to Drug Discovery: Can Ayurveda Show the Way Forward? *Drug Discov Today*, 14(15–16):804–11.

Patwardhan Bhushan (2007), *Drug Discovery and Development*, New India Publishing Agency, New Delhi.

Patwardhan Bhushan (2012), A Quest for Evidence-based Ayurveda: Lessons Learned, *Current Science*, 102(10), 1406–1417.

Salhan S., Tripathi V., Sehgal R., Kumar G., Talwar G.P. and Chatterjee A. (2009), A Phase II Randomized Controlled Trial to Evaluate the Safety and Efficacy of Praneem Polyherbal Vaginal Tablets Compared with Betadine Vaginal Pessary in Women with Symptoms of Abnormal Vaginal Discharge, *Asia Pac J. Public Health*, 21(4):461–8.

Talwar G.P., Dar S.A., Rai M.K., et al. (2008), A Novel Polyherbal Microbicide with Inhibitory Effect on Bacterial, Fungal and Viral Genital Pathogens, *International Journal of Antimicrobial Agents*, 32(2), 180–185.

Vaidya A.D.B., Vaidya R.A. and Nagaral S.I. (2001), Ayurveda and a Different Level of Evidence: From Lord Macaulay to Lord Walton (1835–2001 AD), *Journal of Association of Physicians*, India (JAPI), 49:534–537.

Vaidya Ashok, et al., (2003), Ayurvedaic Pharmacoepidemiology—a New Discipline, *Journal of Association of Physicians*, India (JAPI), 51; 528.

Valiathan M.S., *Legacy of Caraka* (2003), *Legacy of Susruta* (2007) and *Legacy of Vagbhata* (2010), Orient Longman, Chennai.

123

Healing by Homoeopathy

Lal Singh

CONTENTS

I. GENERAL INTRODUCTION

As we all know art and science are inter related and inseparable. Every art has its foundation in science, and every science finds its expression in art.

Homoeopathy is both an art and science. The successful homoeopathic physician must be both an artist and a scientist. Theory and practice are two sides of the same coin for a physician.

Homoeopathy was founded and developed into a logical scientific system by Dr. Samuel Hahnemann (1755-1843) under the principles of Inductive Method of Science. Its practice is governed by the principle of SYMPTOM-SIMILARITY, which is the application in medicine of the universal principle of mutual action and reaction formulated by Sir Issac Newton in his third law of motion "action and reaction are equal and opposite".

Dr. Hahnemann paid attention principally on the spiritual and dynamic aspects of the man. He believed living organisms act and react under the influence of external and internal factors——mental, psychic or physical.

Homoeopathy is a refined natural system of Medicine based on the natural law of 'SIMILIA SIMILIBUS CURANTUR', i.e., 'let like be cured by like' and that's how word homoeopathy came in, "Homoeos" means "similar", "pathos" means "suffering". Dr. Hahnemann concluded that medicines cure diseases only because they can produce similar symptoms in healthy individuals. As a pre requisite to clear understanding of the subject, as well as to attain efficiency in practical application of its principles, it is assumed that homoeopathy is a complete system of therapeutic medication. As a scientific system it is made up of certain facts, laws, rules, and methods, each of which is the integral part of the whole system.

Laws of nature are derived by experience and observations with regards to the course of events and phenomena's. All changes in nature are the result of the reciprocal action (action and reaction) of bodies and forces. Living organisms receive external substances and they force them into life, yet not always they produce the changes. There is LAW OF CAUSATION which states that no internal effect can arise without an external cause, and that the effect itself may in turn become a cause of further changes. Therefore, it is necessary to know the state of body before the change happens. According to LAW OF SPECIFICATION every change of form or function in organism is accompanied by a corresponding changed combination of matter. Hence, when we see any physical phenomena undergoing a change in living organism, it is actually chemico vital changes that are going on. So we can say, Disease is neither an action nor a reaction, but a changed state of an organism caused by the interaction of an external cause with the internal constituents of the organism, resulting in a new form of the whole.

Hahnemann starts with the conception of life as a real or substantial power or principle, having laws of its own, and refers all the phenomena of health and disease to it under two names—"The Dynamis" and "The Life Force". Dynamis, the life power, the substance, the thing itself and Life Force, the action of the power. Later he changed vital force into Vital Principle.

From this conception arises the dynamical theory of disease upon which is based the Hahnemann pathology—*that disease is always primarily a dynamic (or functional) disturbance of the vital principle.* Life is substantial, objective, primary, originating power or principle and not merely a condition or mode of motion. From this conception arises dynamical theory of disease.

There came an era to realize that there is something more important than matter and materialism, than germs and germicides, than serums and vaccines, than mechanics and mechanisms, than pathological processes and products. That 'something' is a fuller knowledge and realization of the spiritual nature of life or mind in organism; of life or mind as a spiritual, entiative power or principle manifesting itself in and through the physical organism of which it is the architect and builder as well as the tenant.

II. RELATION OF BACTERIOLOGY WITH HOMOEOPATHY

Earlier and still many physicians consider the invading micro-organism as the cause of infections, i.e., "will o' the wisp", "specifics for disease", while Dr. Hahnemann pursued cure is always individual never generalized, since, no two cases, even of the same disease, are ever the same. Dr. Kent has said that bacteria are only the results of disease; he calls them scavengers. John Paterson has demonstrated that the very same non-pathogenic bacteria present in normal healthy body may become pathogenic if there is any disturbance in the condition of the host. For disease to happen, the body must participate and hence occurs the inter relation between the host, agent and environment, which can result in disease.

Hahnemann says all chronic diseases show constancy, and if not healed they increase within whole man's lifetime, and cannot be diminished in strength by the most robust constitutions even. They never pass away by themselves, but increase and aggravate until death. They must therefore, have for their origin and foundation constant chronic miasms, whereby their parasitical existence in human organism is enabled to continually rise and grow. Miasm word was used by Dr. Hahnemann to express the morbific emanations from putrescent organic matter, animal or vegetable, and sometimes the effluvia arising from bodies of those affected by certain diseases, some of which were regarded as infectious and others not.

III. HOLISTIC CONCEPT OF DISEASE

What we today relate as psychosomatic diseases, was actually found ages back by Dr. Hahnemann. He showed inter relationship between body, mind and soul. It's not only the specific organ which gets diseased it is the whole organism which suffers and needs to be treated. The human body stands as a single entity with the mind and body as an indivisible continuum and the old concept of mind-body dualism seems more obsolete than ever. Homoeopathy aims at creating, bring about internal harmony of the mind, body and spirit, must operate in such a way that there is complete peace, there is no disturbance in the body as a whole or in any part of the body. Practice of Homoeopathy, tries to cure the man as a whole, it creates balance where there is imbalance.

IV. PREPRATION OF HOMOEOPATHIC MEDICINES BY POTENTIATION

"Homoeopathic potentiation is a mathematico-mechanical process for the reduction, according to scale, of crude, inert or poisonous medical substances to a state of physical solubility, physiological assimilability and therapeutic activity and harmlessness, for use as homoeopathic healing remedies."

The primary object of potentiation is to reduce all substances designed for therapeutic use to "a state of approximately perfect solution or complete ionization, which is fully accomplished only by infinite dilution," (Arrhenius). The greater the dilution, the higher the degree of ionization until, at infinite dilution, ionization is complete and therapeutic activity conditionally greatest. (Dr. S. Close).

Mechanical triturition's of one part of the substance with nine, or ninety nine parts of pure crystalline sugar of milk, according to decimal or centicimal scale of dilution is done of minerals, organic or inorganic substances. This process is continued to any degree, recording and numbering at each step is done to know the degree of dilution and potentiation. The resulting products are known as "potencies" or "dilutions", bearing the name of the medicine and number of the dilution.

V. DRUG POTENTIAL

Any agent or substance which can modify the action of life principle medicinally must do so by virtue of its inherent dynamic energy; and its action is governed fundamentally by the same dynamical laws which govern the operation of the life principle physiologically and pathologically.

Potential is determining the intensity and direction of the force. Drug potential is to find out the variation, a drug is causing from one condition or state to another. Peculiar affinity or attraction exists between the sick organism and the drug which is capable of producing symptoms in a healthy organism similar to those of the sickness.

A dose of medicine is placed on the tongue, in contact with the sentient nerves of the organism, from which it is distributed throughout the entire nervous system.

Every drug acts on organism and every organism reacts differently to each drug, led to the recognition of the specific character of drug action and led to 'Doctrine Of Elective Affinities'. The Doctrine Of Specifics applies to disease as well as to drugs, but is limited to the individual.

Small doses are administered for cure and it can be best explained by Dr. Bier:

(a) All cells of the body are not sick;
(b) The finely subdivided remedy goes past the healthy cells because they have no attraction for it;
(c) The sick cells have less resistance and are more responsive to stimuli. The minimum dose affects these hypersensitive sick cells and stimulates them to reaction. Similar remedy induces normal reaction. If the remedy is dissimilar its action is not curative

There is always cause behind disease, this cause gets active under certain conditions and action of disease producing agent is always modified by the peculiar character and condition of the individual and his environment. Quality of drug action is governed by the quantity of the drug used.

The theory of the drug potential, dynamic theory of life, the law of similars and the law of potentiation, together makes the fundamental principles in the Hahnemann philosophy.

VI. MECHANISM OF ACTION OF HOMOEOPATHIC DRUGS

Mechanism of action is defined as "the mechanism by which a pharmacologically-active substance produces an effect on a living organism or in a biochemical system…"

Samuel Hahnemann explained that their mechanism of action is predicated on what we would today refer to as a form of electro-chemical energy. The current scientific zeal to slice and dice everything in the known world into ever decreasing particles has resulted in high dilution research.

Homoeopathic drugs are either present as matter, these are called tinctures and low potencies (below Avogadro's number). In other they are present in potency form in which very low chemical content is present but the energy packets called photons are present. These photons move through nerves and reach at site of disease at the speed of the nerve impulse (100 m/sec).

Dr. Hahnemann states **therapeutic law of nature** as

"A weaker dynamic affection is permanently extinguished in the living organism by a stronger one, if the latter (whilst different in kind) is very similar to the former in its manifestation"

Modus operandi of homoeopathic cure

Medicine administered on the basis of symptom similarity, affects the morbidly deranged dynamic vital force through medium of sentient faculty of nerves and thus because of primary action produces a similar but stronger disease against the existing one. This artificial disease (medicinally created) is always stronger than natural disease because medicines act unconditionally (at all times, under all circumstances, on every individual), dose regulation is under our hands. As per above law, the natural disease will be removed permanently by the stronger similar medicinal disease which will occupy precisely the same place as the former. Now there is only medicinal disease on the vital force. This vital force employs increased energy to extinct the medicinal disease. As well as, this medicinal disease is gradually becoming weaker due to minuteness of dose and shorter and fixed duration of action of medicinal agent. Now as medicinal disease is growing weaker, vital force can easily empower it and hence, gets free from both natural as well as medicinal disease and thus, health is restored.

Natural disease is removed by the primary action of the administered medicine and the substituted medicinal one is removed by the secondary curative action of the vital force.

VII. DRUG PROVING IN HOMOEOPATHY

Homoeopathy is relying upon the symptoms as the language of nature to express the disease. Totality of symptoms constitutes the nature and quality and all that is to be known of the disease. Dr. Hahnemann built Materia Medica by proving on healthy persons and to start with, he started proving on himself. Drugs are the substances which have the capacity to affect the development of disease and / or its expression through systems. For acquiring the knowledge about drug-effects, experiments have been done on healthy human beings by the process of Drug Proving (Table 123.1).

VIII. CONCEPT OF MIASMS

Homeopaths regard the patient's symptom profile as a systemic manifestation of an underlying chronic disorder called a miasm. Miasms are serious disturbances of what homeopaths call the patient's vital force that are inherited from parents at the time of conception. Hahnemann believed that the parents' basic lifestyle, their emotional condition and habitual diet, and even the atmospheric conditions at the time of conception would affect the number and severity of miasms passed on to the child. Hahnemann himself distinguished three miasms: the psoric, which he considered the most universal source of chronic disease in humans; the syphilitic; and the sycotic, which he attributed to gonorrhea. Later homeopaths identified two additional miasms, the cancernic and the tuberculinic. Constitutional prescribing evaluates the person's current state or miasmic picture, and selects a remedy intended to correct or balance that state.

TABLE 123.1   The pre-requisites for the prover, drug & the physician are as under

Prover	Drugs	Physician
Must be healthy physically and mentally. Should be honest, lover of truth that he can narrate symptoms exactly.	Must be pure. Must be genuine.	Should be unprejudiced. Should give full attention to the prover.
Should have the capacity to express symptoms in exact manner.	Must be of known origin.	Should be a person of sound sense.
Should be keen observer.	Should be full of energy.	Should have fidelity to his prover.
Should be intelligent	Indigenous drug procured fresh juice should be mixed with alcohol then should be given.	Should be tactful.
Should be introspective that can look within himself.	Exotic drug should be procured fresh, made into tincture of dry powder, to be taken with equal volume of water.	Should be Circumspective.
Should be conscientious.	Gum and resin, salts should be dissolved in water then to be given.	Should have patience of an eminent degree.
Must avoid all sorts of dissipation and disturbing passions.	Medicines obtained in dry form are made into powder should be mixed with hot water and taken as infusion.	Must take caution during enquiry.
		Should have knowledge of human nature.
		Should have the capacity of proper judgment.
		Should have clear knowledge about narcotic, alternating action, idiosyncrasy, and surrogate.

IX. CONCEPT OF INDIVIDUALISATION AND CONSTITUTION

As no two remedies are alike, similarlily no two patients are the same for homoeopathic physician. Therefore general and mental case taking is given special emphasis in homoeopathy so as to discriminate between two medicines and thus between two patients. Dr. Hahnemann clearly stated that "each medicine produces particular effects in the body of man, and no other medicinal substance can create any that are precisely similar". In homeopathy each individual is considered as a distinct entity with certain characteristics of his own and based on these characteristics we choose an identical medicinal picture.

Some points necessary to perform Individualisation are:

- Complete symptoms.
- Modalities and Concomitant symptom.
- Signs, appearance, behavior, gait and expressions.
- Mental reactions and symptoms.
- Past History, Family History, and Personal History

Constitution is the physical make-up of the body including the mode of performance of its functions, activity of its metabolic processes, the manner and degree of its reaction to the stimuli and its power of resistance to the attack of pathogenic organism. The sum total of emotional, physical & intellectual characteristics of individual characterize his constitution.

'The constitutional remedy is a picture of the sum total of the strengths and weaknesses of the person, mentally, emotionally and physically. It is in the early undiagnosable stages of illness that we must find the constitutional remedy.'

Constitutional prescribing is not the same as chronic prescribing, since the patient does not have to exhibit any sign of chronic disease pathology to require a constitutional remedy. Constitutional prescribing does not simply address the present disturbance, but also the past and future. It is treating the patient's susceptibility (which may include the propensity to repeated acute and any existing chronic disease), by addressing the inner disturbance which gives rise to outward disease symptoms. It treats the manifest, the previously manifest, and also the potential to manifest.

The idea of constitutional typing arose long before homeopathy: 'The conception of a patient in terms of healing agent is found even in Paracelsus ... stress in our drugs the cure and not the causes because the healing shows us the cause'.

X. CONCEPT OF CURE: HERING'S LAWS OF CURE

The homeopathic laws of cure were outlined by Constantine Hering, a student of Hahnemann who came to the United States in the 1830s. Hering enunciated three laws or principles of the patterns of healing that are used by homeopaths to evaluate the effectiveness of specific remedies and the overall progress of constitutional prescribing:

Healing progresses from the deepest parts of the organism to the external parts. Homeopaths consider the person's mental and emotional dimensions, together with the brain, heart, and other vital organs, as a person's deepest parts. The skin, hands, and feet are considered the external parts.

Symptoms appear or disappear in the reverse of their chronological order of appearance. In terms of constitutional treatment, this law means that miasms acquired later in life will resolve before earlier ones. Healing proceeds from the upper to the lower parts of the body.

XI. ANY SIDE EFFECTS OF THIS SYSTEM

Many conventional medicines just suppress symptoms without allowing the body to heal, sometimes resulting in harmful side effects. But Homoeopathy, instead of suppressing works gently with the body to bring healing and also reduces the percentage of iatrogenic diseases because of minuteness of doses.

Homoeopathic doses especially mother tinctures or the one with high dilutions should be taken under administered care of a homoeopathic specialist, to avoid any adverse effect.

CONCLUSION

We see principles of Homoeopathy discovered and evolved with time, experimentations and increasing enlightenment. Our access to investigations for the functions of the body has increased our knowledge of life processes, which is used to deal with human sufferings. But, in spite of all this knowledge, no guiding principles have been discovered by dominant school of medicine that is sure and give certain indications in the field of therapeutics.

Dr. Hahnemann observed that any drug is poisonous if given in large doses. Therefore, it is for sure homoeopathic physicians recognize the dangers of synthetic drugs. The ability of qualified, knowledgeable homoeopath makes it easy for him to correlate symptoms and easily traces further consequences which could be vital, and, he knows the dangers of suppression and palliation.

Homoeopathy acts equally well in acute cases as in the chronic ones. The results of our remedies in infections such as Staphylococcal, Streptococcal, Pneumonia etc. have been more remarkable than in any other system of medication. Homoeopath knows 'the more acute the case, the more the infection strikes at the life of the patient, the more clearly indicative of the symptoms. Obscuration of symptoms (unless produced by crude drugging) is very rare in a case of acute infections.'

To conclude, I can say that I praise all major modes of treatments (allopathy, homoeopathy, ayurvedic, natural, unnani, yogic, physiotherapy). All have their own pros and cons but are made to benefit patient but amongst all, Homoeopathy has the minimum demerits because it is the only system which never uses anything crude or pathological; neither its medicines produce any side effects; easily acceptable, easy to carry, convenient doses and above all its effectiveness empowers it over all other systems.

Our mission is to set new standards for homoeopathy, making its science, the most popular system of Medicine. Benefiting the suffering humanity by preventing, early detecting and curing diseases; making everybody healthy by using all its natural laws and remedies.

SUMMARY

In the article Healing By Homoeopathy by Dr. Lal Singh it is learned that Homoeopathy founded by Dr. Hahnemann is both an art and science governed by the principle of symptom similarity, SimiliaSimilibusCurantur. Any physical change that occurs in living organism is because of chemico vital changes that are going in organism. Homoeopathy is based on the dynamis and vital principle. For any disease to occur, the body must participate and hence occurs the inter relation between the host, agent and environment, which ultimately results in disease. It's not only the specific organ which gets diseased it is the whole organism which is suffering as there exists a relationship between body, mind and soul. Homoeopathic medicines are prepared by process of potentiation and small doses are administered for cure. These medicines act through sentient faculty of nerves and reach at site of disease. Homoeopathic medicines are proved on healthy individuals and their symptom changes are noted in form of materiamedica. As per homoeopathic system no two patients are alike and similarly there is no same medicine for any specific disease. It works on concept of individualization and constitution. Cure in homoeopathy is based on Hering's law of cure, i.e., cure takes place from inward to outward, in reverse order of their chronological appearance, from upper to lower parts of the body. Hence, Homoeopathy doesn't suppress but gently brings out with healing, and is equally effective in both acute and chronic diseases.

Hence we see that homoeopathy is to Benefit the suffering humanity by preventing, early detecting and curing diseases; making everybody healthy by using all its natural laws and remedies.

SUGGESTIONS FOR FURTHER READING

Lectures on Homoeopathic Philosophy by Dr. J.T. Kent.

Organon of Medicine by Dr. Samuel Hahnemann.

Organon of Medicine by Dr. Shivnarayan Ganguli.

The Chronic Diseases Their Peculiar Nature and Their Homoeopathic Cure by Dr. Samuel Hahnemann.

The Genius of Homoeopathy Lectures and Essays of Homoeopathic Philosophy by Dr. Stuart Close.

The Lesser Writings of Samuel Hahnemann by R.E. Dudgeon.

The Principles and Art of Cure by Homoeopathy by Dr. H.A. Robert.

www.similima.com

124

Essentials of Unani System of Medicine

S.K. Jain and Asim Ali Khan

CONTENTS

I. Introduction
II. The Physiological Concepts
III. Maintenance of Health
IV. Concept of Diseases
V. Methods of Diagnosis
VI. Modes of Treatment
VII. Frequently Used Single and Compound Unani Formulations
VIII. Current Research
Conclusion

I. INTRODUCTION

During the last century medical science has seen major developments that have changed the entire gamut of health management. Human quest for treatment of ailments has its origin from the period when the civilization started. Each civilization developed its own medical system that was based on natural products. With development in chemical sciences, the synthetic drugs evolved. The modern system of medicine (Allopathic system) is based on synthetic molecules as drugs. However, many of these molecules have originally been isolated from natural sources.

Despite popularity of modern medicines, a large proportion of Indian population, especially rural and tribal people still depend on Indigenous systems of Medicine. These systems have their own advantages and disadvantages; though the benefits of these medicines and their scientific basis still remain the topics of debate.

Allopathic system, while very effective in treating the acute conditions, is relatively less beneficial in treating the chronic conditions as well as many of the modern life style related ailments. Besides, most of the chemicals have side effects, some of which are difficult to be acceptable. New multidrug resistant strains (MDR-strains) of a number of pathogens have emerged that have further complicated the management of infectious diseases. Many diseases that were thought to be almost eradicated (or fully controlled) have re-emerged (tuberculosis, malaria etc. to name a few). The natural products on the other hand, often have relatively fewer side effects. Besides, the Indigenous system takes a more holistic approach for health care than the molecular medicine that primarily treats the symptoms of a disease in isolation.

This chapter deals with one of the Indian systems of Medicine, namely, the Unani system of medicine. Efforts have been made to briefly describe its principles and basis of treatment. As will be seen, many of the concepts of Unani system are similar to modern concepts of health and disease, though these may have different names or terminology.

Unani system of Medicine, as the name indicates, owes its origin to Greece. It was Greek philosopher physician Hippocrates (460–377 BC), who put forward the theoretical framework of Unani medicine, freed this practice from the realm of superstition and magic and gave it a shape of science. He provided a health management theory based on humors. According to him, the human body contains four humors viz, Blood (*Dam*), Phlegm (*Balgham*), Yellow bile (*Safra*) and Black bile (*Sauda*). As long as these humors remain in perfect equilibrium, the body remains in health and in the absence of this equilibrium, the disease sets in.

After Hippocrates, a number of Greek scholars further enriched this system. Galen (131–210 AD), probably the most important name in the series, subjected this knowledge to experimentation and devised a number of polyherbal recipes, which are still used by Unani Medicine practitioners. Avicenna defined Unani Medicine as *"The science by which we learn the various states of the body in health and when not in health, and the means by which health is likely to be lost and when lost is likely to be restored."*

Going through various phases and geographical locations such as Italy and Persia, Unani Medicine reached the Arab world. Around 8th century AD, most of the literature of Unani Medicine was translated from Greek to Arabic. At this stage much of the Persian knowledge was also incorporated into it.

During the reign of Caliph Haroon Rashid (786–809 AD) Baghdad became a seat of great learning; the State provided patronage to Unani medical scholars and invited many of the Indian Medicine scholars (Vaidyas). Unani scholars shared their knowledge with them and enriched the System considerably. Important Ayurvedic books like 'Susrata Samhyata' were also translated into Arabic. The eminent scholars of that period include *Rhazes, Hunain and Avicenna*. Hunain's treaties on *ophthalmology*, Rhazes work on *smallpox and measles* and Avicenna's *Canon of Medicine* are some remarkable development in medical field.

Canon of Medicine, an encyclopedia of medical knowledge used to be considered as the '*Bible of Medicine*' has remained the main source of learning all through Arab, Persian and European world for over six centuries. During this period, Arabs made three significant contributions to this system, which include *medicinal chemistry, organization of pharmacy and establishment of hospitals*. Many other pharmaceutical procedures like distillation, sublimation, calcinations and fermentation etc. were also developed and refined by them.

Unani Medicine got enriched by imbibing a synthesis of contemporary systems of traditional medicine in Egypt, Syria, Iraq, Iran, China, India and other Far East countries. This system of medicine is popular in different parts of the world with different names, such as Unani (Greek-o-Arab Medicine), Arabic Medicine, Tibb-e-Sunnati, Traditional Iranian Medicine and Eastern Medicine.

In India, Unani Medicine was introduced by Arabs around 14th century AD when Mongols ravaged Central Asian region like Tabriz and Shiraz; many scholars/physicians of Unani medicine from these places fled to India and received considerable support and State patronage by kings of Delhi. During 13th to 17th century, Unani Medicine flourished in India. However, due to rich tradition of already existing Ayurveda system of medicine and temperament of Indians, Unani Medicine was not easily accepted by the masses. Besides, the Unani scholars were not content with the native herbs. Over the years many of these Indians herbs were subjected to experimentation, which further enriched the Unani system. Thus a hybrid amalgam of Unani (Greek-o-Arab) and some Ayurvedic medical practice was gradually evolved, which is known as '*Unani Tibb*'.

The translations of many Arabic and Persian Unani medicine texts were also done in Urdu and other native languages and this system of medicine gradually got acceptance by Indian population and it became the major health care system during the Mogul period. However, during British rule, it suffered a setback. Hakim Ajmal Khan (1868–1927) took the lead in Unani medicine and established a school of Indian Medicine in Delhi known as '*Ayurvedic and Unani Tibbia College*' and '*Hindustani Dawakhana*'. Later Hakim Abdul Hameed (1908–99) founded Hamdard Tibbi College, Majeedia hospital, Jamia Hamdard (University), New Delhi along with other institutions in different parts of the country.

After independence, along with other Indigenous healthcare systems, Unani medicine also gathered momentum. In 1946, the Conference of Ministers resolved to provide sufficient facilities for research in Ayurveda and Unani. In 1969 Central Council of Research in Indian Medicine and Homoeopathy (CCRIMH) was established, which was later splitted to make four separate research councils, one each for *Unani, Ayurveda and Siddha, Yoga and Naturopathy, and Homeopathy*. Government of India enacted 'Indian Medicine Central Council Act 1970 and established *The Central Council of Indian Medicine (CCIM)*. At present, a separate Ministry of AYUSH (for Ayurveda, Yoga, Unani, Siddha and Homeopathy) has been established by the Government of India. Currently there are a number of organizations for Unani medicine such as the National Institute of Unani Medicine (Bangalore, India), Centre Council of Research in Unani Medicine (CCRUM), over 40 teaching and training institutions, number of hospitals, dispensaries and private clinics. Besides, collaborations have occurred with leading research organizations of the country like CSIR and ICMR etc. Presently, the focus of research in Unani Medicine is on vitiligo, arthritis group of disorders, hepatic disorders, psychosomatic disorders and life style disorders.

The popularity of Unani system of Medicine has attracted increasing global attention and a number of countries are planning to establish and revive their traditional systems.

II. THE PHYSIOLOGICAL CONCEPTS

Hippocrates described that disease is a natural process and symptoms arise due to the reactions of the body to disease process and the duty of a physician is to aid and support the body to overcome the disease process.

Humoural and temperament theories are the backbone of this system. According to Unani Medicine, human body is composed of seven natural components (*Umur al Tabai'yah*). This term is derived from the word '*Tabi*' meaning "*Physis*". The term physiology has also its origin from the Greek word '*Physiologikos*' meaning "knowledge of the nature" or in other terms "*Tabi'at*". The ancient Greek physicians used the term 'Physiology' to denote origin and nature of things.

Tabi'at is a power which all living beings posses to maintain good health. It takes care of the smooth functioning of all bodily functions and fight against the disease. Hippocrates laid down the theory; '*the nature heals*' and the job of physician is only to follow the power of '*Tabi'at*' and not to antagonize it.

According to Unani Medicine, seven natural components are responsible for maintenance of health. Loss or misbalance in any of these states may lead to disease or even death. These seven natural components (*Umur al Tabai'yah*) include:

1. Elements (Arkan)
2. Temperament (Mizaj)
3. Humors (Akhlat)
4. Organs or Members (Aaza)
5. Pneuma or vital breaths (Arwah)
6. Faculties (Quwa)
7. Functions or Operations (Afa'al)

Elements (Arkan)

Among the seven natural principles, first is '*Arkan*'. According to Avicenna 'Arkan' or elements are simple indivisible matters which provide the primary components for the human body and others. These cannot be further resolved into simpler entities and all types of matter found in the nature are formed by the combinations of these substances (chemical combinations). Observations with regard to number and properties of these Arkan have always been changing with times. Some of the scholars in ancient times maintained that there is only one element, some forwarded two element theory and later on few scholars put forth that there are only three elements.

Hippocrates and his followers like Aristotle (384–322 BC), Galen (131–210 AD) and Avicenna (980–1037 AD) accepted the theory of four elements. They maintained that there are four primary elements, *Air, Fire, Water and Earth*. Two of these, i.e., fire and air are light; whereas, other two, i.e., earth and water are heavy. These elements form the basic blocks of all substances in nature including the human body. These four elements also determine the four states of the matter. Air stands for the gaseous state, water stands for the liquid state, earth stands for the solid state and fire stands for the matter which has been transformed into heat.

Earth is a simple substance which should be regarded as being normally situated in the centre of the other elements. In temperament it is *Cold* and *Dry*. In nature it serves the purpose of making the compounds firm, stable and lasting. Water is a simple substance that is *Cold* and *Moist*. Moisture is the quality which makes it capable of ready dispersion and aggregation and which enables it to accept varied shapes of an unstable kind. Air is *Hot* and *Moist* in temperament, whereas, Fire is *Hot* and *Dry* in temperament. It is necessary for the proper maturation of the constituents and for making them light. The two heavy elements (*Earth and Water)* are essential for the formation of organs, while the two light elements (*Air and Fire*) are required for the production and movements of the vital energies (Kulliyate Qanoon Part I Paras 22–28).

The properties of elements have been described in terms of their temperament that is expressed in four quality forms; Heat, Cold, Moistness and Dryness. Each element contains two properties in which one is active and other is passive. For example, in fire heat is an active property, which has the productive capacity, while dryness is the passive property. Two active properties can never be found in an element at any given time, i.e., an element cannot simultaneously be both hot and cold. Every element has a certain temperament or '*Mizaj*', which makes every compound, differ in nature as far as chemical and physical states are concerned.

Temperament (Mizaj)

This is the second component out of seven natural principles. Avicenna has defined *Mizaj* as the new state of a matter having different quality from that having present in the elements or compounds before coming into intermixture, which results from action and reaction from the contrary qualities present in the elements when they are mixed together. The resultant uniform state of equilibrium achieved after the combination of one or more elements is called as temperament.

Temperament is a quality resulting from the interaction of opposite qualities present in elements consisting of minute particles so that most of the particles of each of the elements may touch most of the others. Thus, when these particles act and react on one another with their properties, there emerges from their total properties, a uniform quality which is present in all of them. This is the temperament. As the primary qualities of the elements are *Heat, Cold, Dryness* and *Moisture*, the organization as a whole of either a newly formed or of a disintegrating body is a resultant of these qualities.

Types of Temperament: According to Unani Medicine theory there are two types of intermixtures
- Simple intermixture (*Imtizaj sazij*) and
- Factual intermixture (*Imtizaj haqeequi*)

When two or more elements are simply mixed together, wherein, their individual properties continue to remain the same as before, it is called as a *Simple Intermixture*, e.g., mixture of sugar and water to form a syrup. But when the elements or other compounds are mixed in such a way that their original properties are changed altogether and entirely a different state is formed, it is called a *Real Intermixture*.

The factors that are responsible for factual temperament (*Imtizaj haqeequi*) are:
- Chemical affinity (*Ulfat kimiyawiyah*) and
- Chemical repulsion (*Nafrat kimiyawiyah*)

Ancient Unani scholars had an understanding that certain elements have a natural affinity to combine with other elements or compounds and they labeled this property as *Ulfat kimiyawiyah* or chemical affinity. Such compounds readily and easily combine with each other. Whereas, certain elements do not posses this quality, hence, do not form any compound when put together. This is labeled as *Nafrat kimiyawiyah* or chemical repulsion.

Classification of temperament (*Mizaj*): According to Alladdin Qarshi there are nine kinds of temperament. One is normal/equable (*Mizaj Muatadil*) and eight are imbalanced/inequable (*Mizaj Ghair Muatadil*). Normal or equable temperament is further classified into two types.

Real equable temperament (*Muatadil Haqeequi*) wherein, the contrary properties and quantities of the participating elements are in perfect equilibrium. This is in fact nonexistent in the world.

Tibbi equable temperament (*Muatadil Tibbi*) is the condition where contrary quality and quantity of participating elements in a compound are not equal, but are in perfect balance, i.e., homeostatic according to the required properties of that compound.

The *Muatadil Tibbi,* with regard to species, races, individuals and organs has been further classified into eight kinds.

Imbalanced/Inequable Temperament (*Mizaj Ghair Muatadil*): When the contrary qualities of elements in a compound are imbalanced and in unequal quantities, it is called as imbalanced/inequable temperament (*Mizaj Ghair Muatadil*). Inequable temperaments, whether considered in respect of race or group, an individual or organ are of eight kinds and all of which are counter parts of the equable temperament. Primarily, such imbalanced or inequable temperament has two types; simple inequable temperament (*Mizaj Ghair Muatadil Mufrad*) and Compound inequable temperament (*Mizaj Ghair Muatadil Murakkab*).

In the simple inequable temperament, a single quality predominates more than others. It could be *Hot, Cold, Dry or Moist.* Whereas, in inequable compound temperament, two contrary qualities out of four are more dominating. They could be *Hot and Dry, Hot and Moist, Cold and Dry* and *Cold and Moist.*

Health can be maintained only by the maintenance of normal temperament (*Mizaj Muatadil*) within the body parts and the entire body, which means the maintenance of static or constant conditions (Homeostasis) in the internal environment of the cells or whole body.

The *Tabiyat functions in two ways;* one, under physiological conditions it maintains the homeostasis in the internal environment of the body for proper functioning of the cells, tissues and organs, second, in case body becomes diseased or derangement of temperament takes place, *Tabiyat* fights against the disease and in appropriate conditions homeostasis is regained.

Temperament of organs: Each individual and every organ has a temperament which is the most appropriate and best suited for its functions and condition according to its power of endurance. Man has been bestowed upon with the most equable temperament compatible with his faculties through which he acts and reacts.

As Unani Medicine maintains the four qualities viz. Heat, Cold, Moistness and Dryness to express the temperament of various body organs, these have been classified into four categories.

Hot organs (*Aaza e Ha'rra*): The hot organs are the ones in which blood supply and metabolic activities are comparatively more intense. Of all the organs, heart is the hottest as it is the centre of the forces (energies), then comes liver and then flesh (muscle). These are the organs, where oxygen consumption is comparatively more than other organs.

Cold organs (*Aaza e Barida*): The cold organs are those in which blood supply and metabolic activities are comparatively less intense. Of all the organs, bone is coldest,

then comes cartilage, ligament, nerves followed by spinal cord and then brain. Here oxygen consumption is less than the hot organs.

Moist Organs (Aaza e Ratebah): The moist organs are more humid than the other category. Phlegm is most moist. After this is fat, then comes loose flesh, i.e., glands. Brain, spinal cord, breasts, testicles and lungs are also moist in temperament. These are the organs which contain more water than others.

Dry Organs (Aaza e Yabisa): The dry organs are drier than the other category. Of all the organs of the body, hair is the driest, then come bones, cartilage, tendons and nerves in order. These organs have comparatively less quantity of water.

The bodies of children and youth are moderately hot and the bodies of the middle aged and old are cold. However, bodies of children are moister for meeting the requirements of growth in comparison to youth. Middle aged and the older along with cold temperament are drier too.

If sex is considered with reference to the difference of temperament, it is observed that females are colder than males. People living in northern countries in general are moister than the people in western countries.

Humours (Akhlat)

Humours (*Akhlat*) is the third natural principle. This concept occupies the central place in Unani system of medicine. The humoural theory was initially put forward by Hippocrates, about 2500 years ago, in the book '*Tabiat al Insaan*'. He mentioned that every person has a unique humoural constitution which represents his or her state of health. In perfect balance and right proportion with regard to quality and quantity of these humours, the body continues to remain healthy and wrong proportion and irregular distribution with regard to their quality and quantity constitutes disease.

This system combines physiology and pathology and explains health and disease on humoural basis. According to Unani Medicine, the body is composed of three parts; solid, liquid and gas. The solid parts are known as Organs (*Aaza*), liquid parts as Fluids (*Rutubat or Akhlat*) and gaseous parts as Pneuma (*Arwah*). Avicenna maintains that there are two types of humours; good and bad. Good humour has the capacity to provide a compensation for what is destroyed by the recipient. Whereas, the residual or bad humour is the one which does not have this property and rarely changes into good humour.

Much emphasis has been given to the color of humours and accordingly they have been classified in four categories; red, white, yellow and black.

- All fluids of the body which are red in color are called as blood (*Khilt Ahmar or Dam*)
- All white and colorless fluids of the body irrespective of their locations are called as phlegm (*Khilt Abyadh or Balgham*)
- All fluids of the body which are yellow in color are called as yellow bile (*Khilt Asfar or Safra*)
- All fluids of the body which are black or blue in color are called as black bile (*Khilt Aswad or Sauda*).

Blood is the admixture of all the four humours which remain recessive under the influence of red color of blood. It is '*Hot and Moist*' in temperament. Abu Sahal Masihi in his book 'Miat' mentioned that in normal (*Taba'i*) blood all humours are in right proportion with regard to their quality and quantity. Qarshi mentions that it should be normal in color, free from bad odour, moderate in viscosity and sweet in taste. Abnormal blood differs from normal blood in respect to color, odour, viscosity and taste and is of two kinds:

- That which has deviated from good temperament not because of something mixing with it, but because its temperament has in itself become vitiated.
- That which has changed because a vitiated humour has entered into it. It has two kinds; either the bad humour has found access to it from outside and corrupted it, or the bad humour is generated in blood itself.

The main functions of blood include providing nutrition and replacement, keeping the body warm and maintaining the body temperature, transporting all materials to various locations in the body, promoting growth of the body in adolescence and helping in the generation of innate heat by supplying fuel to the human body. It also provides beauty and shine to the skin.

Phlegm is in fact a genus which has different kinds, each having diverse properties and it performs diverse functions in the body. However, there is a divided opinion about fixing its temperament. It is considered to be blood which has yet not completely metabolized. It provides nutrition to the body by getting converted into blood, provides moisture to the organs and sub serves nutrition to the organs which are phlegmatic in temperament such as the brain.

It can also be normal or abnormal. The normal phlegm is fit to become blood at any time, whereas the abnormal phlegm is of different excessive consistencies.

Yellow Bile is '*Hot and Dry*' in temperament. It helps blood to penetrate into narrow vascularities (*Ibne Nafis*). It gives some nutrition to organs like lungs, helps in washing excreta and other residues from the intestines, helps in digestion of the food (*Tabari*) and increases the peristaltic movements of intestines, thus helping in expulsion of excreta (*Ibn Nafis*).

Black Bile is '*Cold and Dry*' in temperament. It bestows the density and consistency upon the blood, provides nutrition to certain organs like bones and cartilage and increases the appetite.

Based on the constitution of humours that depends on the proportion of blood, phlegm, yellow bile and black bile, respectively in the body, the temperament of a person would be Sanguine, Phlegmatic, Choleric or Melancholic. To maintain the correct humoural balance there is a power of self-preservation or adjustment known as '*Medicatrix Naturae*,' in the body. If this power weakens, imbalance in the humoural composition occurs and disease sets in. The medicines used in this system help the body to regain this power to an optimum level, thereby, restoring humoural balance and thus retaining health.

Organs (Aaza)

This is the fourth component out of seven natural principles. Organs of the body are derived primarily from the coarser and solid particles of the humours, as compared to the pneuma, which is formed out of finer and lighter constituents of the humours. Organs may be divided into two classes:

- Simple organs (Aaza *Mufrad*)
- Compound organs (Aaza *Murakkab*)

Simple organs are those, whose smallest part exactly resembles the whole organ. In other words they are homogenous in structure. For example a tiny piece of cartillage is still a cartilage or tiny part of nerve is still a nerve. Simple organs include bone, cartillage, ligament, nerve, tendon, membranes, fat, muscle, arteries and veins.

Compound organs (Aaza *Murakkab*) are basically heterogeneous in structure and are composed of various single organs. Some ancient scholars have classified these compound organs according to their functions also.

Vital organs are the organs which are important for carrying out the essential functions of the body; hence, these are important for preservation of individuals and their species. There are three vital organs necessary for the maintenance of life of an individual. Along with these, one more vital organ is necessary and that is for the preservation of race. These vital organs are:

(i) Heart (Source of the vital power).
(ii) Brain (Seat of all mental faculties, sensations and movements, it is sub served by nerves).
(iii) Liver (Center of the nutritive faculties. It is sub served by veins).
(iv) Testis (or ovary) (Concerned with the propagation of the race. It is the seat of forming generative elements).

Pneuma (Arwah)

This is the fifth component out of seven natural principles that is absolutely essential for the sustenance of life and without which the maintenance of life is impossible. Air is considered to be composed of a number of substances, the most important and the one on which life depends is the pneuma or vital air. When inspired, this air enters the lungs and is purified, it reaches alveoli and then heart and from there to all the parts of the body through the arteries. This atmospheric air after becoming the part of the body is called *Ruh*. The waste that comes out with this air is exhaled out by the lungs. This pneuma or *Ruh* is important to keep the body alive, to produce energy and to keep the body functioning.

Faculties (Quwa)

This is the sixth component of seven natural principles. The faculties (Quwa) are that property of the body with which the phenomenon of life is manifested. The faculties provide the basis for different bodily functions. The smooth functioning of these faculties depends on the continuous supply of *Ruh* through which certain *Lateef Akhlat* get oxidized to provide energy.

There are three kinds of faculties; Natural (*Taba'i*), Psychic or mental (*Nafsania*) and Vital (*Haiwania*). Accordingly, many scholars have classified three kinds of *Ruh* viz natural pneuma (*Ruh Taba'i),* vital pneuma *(Ruh Haiwani)* and psychic pneuma (*Ruh nafsani*). Many scholars including Galen are of the opinion that there is one principal organ for each faculty, through which the functions of the respective faculty emerge.

Natural faculty is responsible for ingestion, digestion, assimilation, metabolism and excretion of the food materials. It is a stepwise process that starts in the mouth, then in stomach which produces chyle (known as *Hazme Medi*), then in liver which converts it into blood and three humours (*Hazme kabidi*). Finally, the metabolism takes place in tissues, which assimilate the digestive products (*Hazme uzvi*). The energizing agent for these metabolic changes is innate heat acting through appropriate faculties that reside in nutritive organs. Liver is the centre of this faculty for nutrition and growth. The other part of this faculty located in testis/ovaries is for the preservation of race and is also responsible for the sexual functions, formation of germinal fluid and the fertilization and further development of the ovum.

Psychic or mental pneuma (nervous system) consists of the faculty of cognition and the faculty of movement. Brain is the centre of the nervous system and performs its various functions. Vital system conditions and suitably prepares the vital force for the sensory and the motor functions of the brain. It also carries vital force into the organs and tissues for their life and vigour. This system is centered in the heart and functions through it. (Kulliyate Qanoon Part I para 259).

Unani theory also advocates that the natural faculty is further sub served by four other important faculties. These include;

Attractive (*Quwwate Jaziba*)
Retentive (*Quwwate Masika*)
Digestive (*Quwwate Hazima*) and
Expulsive (*Quwwate Dafia*).

Functions (Afa'al)

This is the seventh and last component of natural principles. *Afa'al* is the term in Unani medicine used for the normal or special performance of a part of the body. It is considered that some of the functions are carried out by one single faculty like retention when it is called as simple function (*Afa'al Mufrida*). While certain other functions are carried out by two or more faculties at any given time, like the act of digestion of food in gastrointestinal tract, when it is called as compound function (*Afa'al Murakkaba*).

Maintenance of Health: Unani system of medicine believes in the influence of surroundings, atmospheric and ecological conditions on the health of human beings. It focuses more on preventive measures rather than curative. The Unani Medicine doctrine advocates of six essential factors or causes that are important for maintaining good health. These causes having undeniable influence on every human being are responsible for producing and maintaining humours as well as body temperament and thereby represent the causes of health and disease. Following the good practice of these factors, the body remains healthy and less prone to diseases. These six essential factors or causes called as '*Asbab e Sitta Zarooriyah*' include;

1. Air (*Hawa*)
2. Foods & Drinks *(Makulat w mashroobat)*
3. Body movements & Repose (*Harakat w sukoon e badniyah*)
4. Mental movements & Repose (*Harakat w sukoon e nafsania*)
5. Sleep and wakefulness (*Naum w yaqzah*) and
6. Retention and Evacuation (*Istafragh w Ihtibas*)

In addition there are certain non-essential components also. These do not concern everybody and therefore, do not necessarily influence each and every human body. Such non essential components called as '*Asbabe ghair zarooriyah*' include;

1. Habits
2. Habitat
3. Profession
4. Sex
5. Temperament
6. Social factors
7. Cosmic and terrestrial influences
8. Other influential factors which are antagonistic to nature and body health like micro-organisms, natural and ionizing radiations, electricity and other natural forces etc.

Air (Hawa): By means of respiration process, fresh and clean air provides reinforcement and life support to human body and through exhalation process also does the purification. Avicenna mentioned that pure air, not mixed with any extraneous substance which is contrary to the temperament of lungs, is effective in attaining health and preserving it. However, if altered, it acts contrary to its function. This air is subjected to normal as well as abnormal changes. Normal changes are considered to be the seasonal changes. It is the source of life that provides source for the activation of energies to form body fluids and maintain life. Good and fresh air should be pure and free from dust, smoke and vapors from lakes, ditches and damp lands; it should not come from the areas of dense forests and from near the fig and walnut trees and should be from open and high lands and not from closed boundaries. With variations in seasons, air accordingly produces different actions. Hot air dissolves the body fluids, causes more sweating, increases thirst, reduces urine and weakens digestion. Cold air helps improvement of digestion, increase in appetite and quantity of urine and is not considered good for the patients with nervine disorders.

Moist air help soften the skin, improves the skin turgor and complexion. Whereas, the dry air acts opposite in action and dries the skin.

Food and Drinks *(Makulat w Mashroobat)* are essential for life as they are the source of nutrition to the body. In order to remain healthy, one needs to have fresh and balanced food as well as drinks in appropriate quantities.

Food is used by the body for producing the four humours and the quantity and quality of humours is directly related to the quantity and the quality of the food consumed.

The substances that enter the body act in three ways; some are altered by the body but do not make any alteration in the body, others are those that are altered by the body and also cause changes in body and the third type are those that alter the body but do not get altered by the body. The substances could be homogeneous to the body or heterogeneous. The inputs that are homogeneous become a part of the body and are absolute nutriment *(Ghiza e Mutlaq)* whereas, the one which is not homogeneous to the body is a temperate medicine *(Dawa e Muatadil)*. Any substance that is not altered at all by the body, but alters the body, up to disturbing levels is known as absolute poison *(Samm e Mutlaq)*.

Certain foods have medicinal properties and are known as *Medicinal foods*. Out of these some are more drugs than food, whereas, some are more food than drugs. Some foods are light *(Ghiza e Lateef)*, some heavy *(Ghiza e Kaseef)* and some are moderate *(Ghiza e Muatadil)*. Light food produces thin blood, while heavy food is the one which produces thick blood. Light foods rich in nutrition include wine, extract of meat *(Ma'ul Laham)* and half fried or boiled yolk of eggs. Light food poor in nutrition include honey diluted with rose water, vegetables which are moderate in property and consistency and fruits like apples and pomegranates. Heavy foods poor in nutrition include cheese and dried meat. Heavy foods rich in nutrition include beef. Biochemistry of body depends on the quality of food one consumes. Above the nutritional qualities of food, Unani medicine also recognizes the inherent qualities of food.

Following principles of eating are recommended to remain healthy and avoid illness:

- Too many varieties of foods should not be consumed at one time.
- Meals should be chewed properly.
- On one occasion food should not be eaten for a long duration.
- Foods should be consumed as close to their natural forms as possible.
- Too much water during meals should not be consumed as it may disturb digestion.
- Seasonal foods should be preferred.

Water is equally important to consume. It helps liquefy food and carries it to vessels and channels. The water is of different kinds mainly on the basis of certain contents mixed with it. Water that runs on earth is better than the one which flows on stones, because earth cleanses the water and absorbs the extraneous matter on it, whereas, stones do not do so. Well and canal waters are worse than the spring waters. Water from hail and snow are thick. Water containing iron is considered advantageous, as it strengthens the viscera and stimulates appetite. Moderately cold water is considered to be the best water for the people of equable temperaments. Salt water causes emaciation.

Body movements and repose *(Harakat w Sukoon e Badniyah)*: Movement is essential to life. Even in the condition when body is at rest, activities keep going within the body. A balance between movement and repose is needed for maintaining health. Increase in movements results in increase of heat. When the movements are light and for short duration heat is associated with moisture. But with the increase in intensity and duration of movements, there is a decrease in moistness and increase in dryness *(Hot and Dry)*. One can have the movements according to his/her individual capacity. If movements continue for undesired levels, it results into serious loss of heat and loss of functions.

Mental movements and repose *(Harakat w Sukoon e Nafsania)* is also central in maintaining health. Emotional states like anger, joy, fear, depression, passion and mood swings etc. have their direct impacts on the body. Temperament of an individual is influenced by these states accordingly. Nervous excitements lead to certain diseases such as anxiety and hypertension etc.

Sleep and wakefulness *(Naum w Yaqzah)*: Sleep rejuvenates physical and psychological performance. Requirement of sleep may differ from individual to individual and in different age groups to regain the lost energy and maintaining balance in the temperament. Infants and children need more sleep. Women require more sleep than men. Usually 6-8 hours of sleep is essential for a normal adult. For elderly, sleep and rest is important as it produces more moistness in them. Avicenna in his book Canon of Medicine mentions that balanced sleep and wakefulness indicate the equability of the temperament, especially of the brain. Excessive sleep denotes excess of moisture and cold and excessive wakefulness denotes excess of dryness and heat, especially in the brain. Moderate sleep is moistening and warming.

He further mentioned that the best sleep shall be moderate and properly timed. It should be deep and at a time when the food has passed down from the upper part of the stomach. Sleep on an empty stomach is bad.

Retention and evacuation *(Istafragh w Ihtibas)*: Retention of nutritive substances and excretion of waste is necessary to remain healthy. Excretion is basically the way of detoxifying the body. The waste products are the result of various digestive and metabolic activities. Retention of matter which must be expelled from the body causes certain diseases. Depletion of matter which must be retained causes coldness in temperament. Sometimes depletion makes the temperament hot as may be in the case of excessive depletion of phlegm.

Unani scholars believe that ultimate cure is possible only when the elimination of toxins takes place from the body at the proper time and in a suitable manner. In such cases when waste products are accumulated in the body instead of being eliminated properly, physician induces different evacuation methods like laxation, diuresis and cupping etc.

IV. CONCEPT OF DISEASES

Unani Medicine believes that the correct mix and proportion of blood, phlegm, yellow bile and black bile in the body keeps a person healthy. Whereas, any misbalance in the ratio of their quantities and quality leads to disease. Concept of disease in Unani medicine is based on pathology (*Ilmul amraz*), wherein, the abnormalities of structure and functions are studied. The whole concept revolves around three factors namely; Causes (*Asabab*), Disease (*Marz*) and Symptoms (*Arz*).

Galen is of the view that there are three states of human body; Health (*Sehat*), Disease (*Marz)* and Intermediate state (*Halat e Salisa*). Health is the state of body through which human body maintains its functions keeping a balance of temperaments and composition of humours. Disease is the state of human body which is contrary to the health. Intermediate state *(Halat e salisa)* is the state of human body when there is neither complete health nor established disease for example the bodies of old and the convalescent and children.

Cause of diseases

Causes for the disease precede the clinical manifestation and lead to new state of body. Causes of disease can be divided into two types; external causes (*Asbab e Badia*) and internal causes (*Asbab e Batina*). External causes include excessive heat and excessive cold, extreme weather and injury etc.

Microorganisms are cause of a number of infectious diseases. Unani scholars recognized the presence of invisible fermenting agents as source of disease. Effect of these pathogenic organisms is put under the title of external causes (*Asbab e Badia*). Sepsis is produced by the entry of microbes (*Ajsam e Khabeesa*) in the body leading to disturbance in the proportion of humours in qualitative as well as quantitative terms.

The internal and external causes disturb the humoural constitution and may even lead to putrefaction that weakens the defensive mechanism (immune system) of the body (*Madicatrix Naturae*). It paves the way for microbes to grow and multiply. If the humoural defense gets too weak these microbes set the disease. In case *Madicatrix Naturae* is still powerful enough, these microbes do not find the favorable environment for their survival and are automatically killed.

With this understanding, the physician helps in regaining the balance of humours by rectifying the factors that have caused the imbalance and eventually treats the disease by creating the right humoural atmosphere in the body that is not favorable for the growth of these microbes.

Diseases can be of two kinds—**Simple** and **Compound**

Simple diseases are either the diseases of temperament or diseases of structure and according to the causes could be classified into three groups:

(i) **Due to abnormal temperament (*Sue Mizaj*):** Abnormality in the temperament leading to disease could be *Hot, Wet, Dry Cold, Hot and Dry; Hot and Wet; Cold and Dry; Cold and Wet.*

(ii) **Due to abnormal structure (*Sue Tarkeeb*):** It may be of three kinds; Diseases **of formation, Diseases of size** (it may again be of two kinds; *diseases due to increase in size*, e.g., elephantiasis and *diseases due to decrease in size*, e.g., tongue atrophy) and **diseases of number** (it may again be of two kinds; *diseases due to increase in number*, e.g., normal as supernumerary finger and abnormal as kidney stones and *diseases due to decrease in number* again having two types *congenital and acquired).*

(iii) **Due to loss of continuity/adhesion (*Tafarruq w Ittisal*)**, e.g., wounds and injuries, traumas and abnormal healing, scratches and abrasions of skin and ulcers etc.

Compound diseases: When two or more simple diseases combine together, it is known as compound disease. A simple example of the compound disease could be abscess/swelling, as it is a disease of altered temperament associated with matter. This includes the diseases of temperament, structure and also size.

Avicenna mentions that most of the diseases have four periods; the period of onset, period of advancement, the period of culmination and the period of decline. Diseases may be primary or secondary and also could be acute (*Haa'd*) or chronic (*Muzmin*). Sometimes it may involve other organs also when it is known as *Marz e Shirki*. Diseases could be hereditary, seasonal and infectious also.

V. METHODS OF DIAGNOSIS

In Unani system of medicine diagnosis of a disease or state of health is based on following methods:

- Case history
- Physical examination
- Systemic examination
- Investigations and day to day observations

Through detailed case history, physical and systemic examination, observation and detailed account of symptoms and signs, Unani physicians try to locate the problem. Recording of detailed case history of present and past illness, personal, family, social and occupational history is very important to understand the patient and his problem in holistic manner.

As most of disorders are due to the imbalance in temperament and humoural disturbances, it is important to determine the temperament. A number of criteria (symptoms/signs) are followed for understanding different temperament types which are of great help in reaching to some conclusion. Domination of certain humour necessarily influences the temperament. Based on the domination of the four humours, humans have also been broadly classified into **four types of personalities.** These are:

(i) **Sanguine (*Damvi Mizaj*)** people have *Hot and Moist* temperament. They are active, have good appetite, have strong and full pulse and pass colored and concentrated (reddish) urine.

(ii) **Bilious (*Safravi Mizaj*)** people have *Hot and Dry* temperament. Such individuals have a thin and lean body, have strong and rapid pulse, have prominent vessels, are energetic and intelligent and pass yellow colored urine.

(iii) **Phlegmatic (*Balghami Mizaj*)** people have *Cold* temperament. Such personalities are flaccid and obese, have whitish complexion, are dull in movements and reflexes, pass white color urine, and have lack of thirst.

(iv) **Melancholic (*Saudavi Mizaj*)** people have *Cold and Dry* temperament. They are dark complexioned and thin. They pass black or reddish black urine and their pulse is slow.

Ten parameters have been identified to determine the individual's temperament; Based on the observation of each one of these parameters physician may allocate the temperament of an individual. These criteria include; *Touch Built, Complexion, Hairs of the body, Forms of the body, Quality of passiveness of organs, Sleep and wakefulness, Bodily functions, Excreta of the body and Psychic factors.*

Signs of temporary and acquired qualities in temperament

Excessive Heat: Feelings of uncomfortable heat, such as from fevers, excessive thirst, burning epigastrium, bitter taste in mouth, weak and rapid pulse, feeling of comfort from cold things especially during summers.

Excessive cold: Weak digestion, diminished desire for drinks, laxity of joints, predisposition for phlegmatic types of fevers and catarrhal conditions, cold things upset these, whereas hot things give comfort. Suffer basically during winters.

Excessive moisture: Signs are almost similar to cold conditions. They have puffiness of the face, excessive salivation and nasal secretion, tendency to diarrhea and dyspepsia, excessive sleep and puffiness of eyelids.

Excessive dryness: Signs include dryness of the skin, insomnia, wasting, moist things feel refreshing and suffer greatly during autumn. (Kulliyate Qanoon Part 2 paras 1025–1028).

Symptoms of the disease

Certain signs and symptoms indicate the disease itself as the case of rapid pulse in fever. Some indicate location of the disease, e.g., serrate pulse in case of pain in the chest is suggestive of swelling in pleura and diaphragm. Wavy pulse in chest pain indicates that swelling may be in substance of lungs. Some of these signs/symptoms indicate internal states, e.g., tremors of the lips reveal vomiting.

While making the diagnosis great emphasis is laid on the observation and examination of *Pulse, Urine and Stool.*

Pulse (*Nabz*) is the movement of arteries which comprises expansion and contraction. Each pulse beat has four components, i.e., two movements and two pauses; *movement of expansion, rest between expansion and contraction, movement of contraction and rest between contraction and expansion.*

Unani scholars have mentioned ten features through which state of the pulse and the body could be known; these are known as *Addilla e Nabz.* These ten features include; *Degree of expansion, Impact of the pulse beat on fingers, Duration of each movement, Duration of rest period, Texture of the artery, Emptiness and fullness of the artery, Touch of the artery (heat and cold feel of the pulse), Equality and inequality of the pulse and Rhythm of the pulse.*

Degree of the pulse is noted by three dimensions of the pulse, i.e., *length, breadth and height.* Accordingly the *pulse has nine simple kinds* and *twenty seven compound varieties.* Nine simple Pulse kinds include;

Long, Short and Moderate, Broad, Narrow and Moderate, Low, Elevated and Moderate. Compound Pulse is derived through the combinations of these simple pulse.

Based on Stroke effects on fingers, it has three varieties; *Strong, Weak and Moderate.* Based on the Duration of each movement it has three varieties; *Rapid, Slow and Moderate.* Based on Texture of the artery, it has three varieties; *Soft, Hard and Moderate.* Based on state of the artery, it has three varieties; *Full Pulse, Empty Pulse and Moderate Pulse.* Based on Feel, it has three varieties; *Hot, Cold and Moderate.* Based on the Duration of the rest, it has three varieties; *Quick, Sluggish and Moderate.* Based on equality and inequality, it may be either *Equal* or *Inequal* and so on.

Urine (*Boul*) is an excretory product. Through its quantity and qualities, urine is indicative of various body states in health and disease. Normal quantity of urine passed daily varies from 700 ml to 2400 ml. Output of urine depends on the intake of fluids. Physiologically, in cold weather and after good intake of fluids one passes more urine. Quantity of urine reduces in case of less intake of fluids, or in hot weather. Pathologically, quantity of urine may increase in certain disorders like diabetes mellitus etc. It may also reduce in certain disorders like fevers, diarrhea, dehydration and extensive burns etc.

For observation and examination of urine, seven features (*Dalail e Baul*) have been identified by the Unani scholars. These include; *Color, Viscosity, Odor, Clearness or Turbidity, Sediments, Quantity and Foam.* These features may change due to certain climatic conditions, type of food eaten, intake of water and other fluids, use of drugs and in different age groups as well as gender. These factors may also change during the course of certain diseases. Urine may be of five colors; viz. *Yellow, Green, Red, Black and White* suggesting clues for various disorders of the body. These individual colors may further have various shades. For example yellow colored urine may have five shades like; reddish yellow, orange yellow, saffron yellow, flame yellow etc. Green colored urine may also have five shades like emerald green, pistachios green etc. Similarly, red colored urine could be of four shades. All these colors/shades of urine provide important information about various states of body.

Sediments in urine provide information through their *Substance, Quality, Quantity, Shape, Structure and Position* etc. Sedimentation could be normal or abnormal. Abnormal sediments may also be of different kinds like; *Flaky, Fleshy, Purulent and Mucoid* etc.

Urine also differs in *Hot and Cold* temperaments. Dark urine with a strong odour indicates the *Hot* temperament whereas, a colorless urine with weak or no smell suggests *Cold* temperament.

Presently, besides the above mentioned physical parameters, Unani physicians also get urine tested chemically and microscopically.

Stool (*Baraz*): Great emphasis is laid on the examination of faeces while making the diagnosis. Observations are made through the quantity of faeces, wheather it is equal to the quantity of food intake or more or less than it. In some cases, scantiness of faeces indicates the weak expulsive faculty.

Inferences are drawn based on certain points, which include; *Quantity, Color, Odour, Consistency and Froth.*

Normally stool is flame colored. Pallor or slightly white colored stools may be due to lack of bile mix, as in the case of obstruction of bile duct. Black stools may be due to the intake of Iron. Red may be in case of bleeding. Green and bluish stools are suggestive of extinction of innate heat. In case of dysentery, it may be with mucus. In terms of consistency, it may be normal or dry and hard or watery.

Healthy stools are compact and homogeneous. Their watery and solid content are intimately mixed. They pass easily without causing irritation. They have a yellowish color. They do not have very offensive odour. Such stools are passed without any flatus and are free from foam. They pass regularly at about the same time.

In an effort to get more precise information and to reach correct diagnosis, Unani physicians also take help of the tools and techniques developed through modern research in the areas of microscopic examination, chemistry, microbiology, bacteriology, enzyme studies and radiology etc.

VI. MODES OF TREATMENT

Unani system of medicine does not limit itself to the treatment of only symptoms of diseases, but has a comprehensive approach for curing the diseases and preservation of health through a holistic approach. It uses drugs in their natural sources and forms. This system of medicine has a wide range of therapeutic approaches that include; *Diet, Herbs, Metals, Minerals, Precious Stones & Non Drug Therapies.* A balance should also be maintained between mind and the body so that metabolic processes can take place easily and the body waste gets evacuated.

The principles of treatment are based on temperament theory. Treatment is usually based on the opposite temperament, i.e., '*Ilaj bil Zid*'. Management of diseases in Unani system of medicine is based on four methods of treatment which include:

- Regimenal Therapy (*Ilaj bil Tadbeer*)
- Diet-o-Therapy (*Ilaj bil Ghiza*)
- Pharmacotherapy (*Ilaj bil Dawa*)
- Surgery (*Jarahat*)

Regimenal therapy includes Exercise, Massage, Fomentation, Turkish Bath, Aroma therapy, Emesis, Purging, Enema, Sitz Bath, Cauterization, Leeching, Cupping and Venesection etc.

In diet o therapy management is tried through prescribing various types of diets and by regulation of ingredients, quantity and quality of food.

Pharmacotherapy deals with the use of naturally occurring drugs. In most of the cases these drugs are from natural sources and are herbal in origin. However, certain mineral and animal origin drugs are also prescribed. Certain drugs which are considered toxic in nature are purified (through specific methods) before use.

Though, both single drugs and compound formulations are used to manage disorders, use of single drugs (or their combinations in raw forms) is preferred in Unani system of medicine. Compound formulations are usually used as a second choice or for synergistic actions.

Both Greek and Arab scholars encouraged the use of poly pharmacy in Unani system of medicine and devised many compound formulations which are effectively used in managing many complex disorders. Standardisation and quality control has been given an important place in Unani pharmacy to minimise the side effects and toxicity of the drugs.

Surgery has also been used in managing certain disorders. Many instruments were also developed by the ancient Unani scholars. These days at various centers only minor surgery is in practice.

VII. FREQUENTLY USED SINGLE AND COMPOUND UNANI FORMULATIONS

See Table 124.1.

VIII. CURRENT RESEARCH IN UNANI MEDICINE

Concerted efforts are being made to validate the Unani medicines using modern scientific parameters. The research being carried out at various centers in India and certain other countries include; standardization of single and compound Unani formulations, efficacy based evaluation, revalidation of pharmacological activities (reported in classical Unani literature) based on new modules/guidelines of regulatory

TABLE 124.1   **Commonly used single drugs/compound formulation, their therapeutic uses and recommended dosages**

Single drug/Compound unani formulation	Indications/Action/Uses	Dosage
Roghan Aamla Sada	Graying of hairs; prevents drying of scalp; acts as hair tonic.	Apply daily on scalp (according to need)
Amla (Emblica officinalis)	Provide energy to brain; maintains eye sight; hair Tonic	3–5 gm (powder/decoction)
Kishneez (Coriander)	Headache; Palpitation; provides energy to brain	5–7 gm
Ustukhodoos (Lavendula steochas)	Nervine/brain disorders; sinusitis/rhinitis; headache	5–7 gm (powder/decoction)
Sapistan (Sebstan)	Coryza; rhinitis; throat congestion	5–9 pc (decoction)
Asalussoos (Liqourice)	Bronchial asthma; pharyngitis; laryngitis; cough	3–7 gm (decoction)
Abresham (Silk cocoon)	Palpitation; IHD	3–5 gm (decoction)
Khamira Gauzaban Ambari Jawahar wala	Effective brain tonic	5 gm
Kasni (Chicorium intybus)	Hepatitis; spleenomegaly; jaundice	3–5 gm (Distilled extract)
Itrifal Muqawwi-e-Dimagh	Provides energy to brain; good for eye sight; cold; headache.	7–10 gm with water
Majoon-e-Brahmi	Provides energy to brain; good for memory enhancement.	7–10 gm with milk
Itrifal Ustukhodoos	Nervine/brain disorders; sinusitis/rhinitis; headache	5–7 gm
Itrifal Shahtra	Effective in skin disorders like itching, ringworm, dermatitis, pruritis	5–10 gm
Majoon-e-Ushba	Effective in skin disorders like itching, ringworm, dermatitis, pruritis; gout	5–10 gm
Lauq-e-Katan	Bronchial asthma; COPD; allergic cough	7–10 gm twice a day.
Hab-e-Nazla	Common cold; rhinitis	2 pills twice a day
Laooq-e-Sapistan Kheyarshambri	Coryza; pharyngitis; laryngitis	7–10 gm with lukewarm water twice a day
Sharbat Toot Siyah	Prevents pain of Throat and Pharynx; laryngitis	25 ml with luke warm water twice a day
Sharbat Zoofa Murakkab	Cough; cold; bronchial asthma	20–30 ml with lukewarm water twice a day
Sualeen	Throat irritation; URTI	1–2 tablets twice or thrice a day
Khamira Abresham Hkm Arshadwala	Palpitation; IHD	5–6 gm once or twice a day
Jawarish Jalinoos	Loss of appetite; indigestion; provides strength to stomach	5–8 gm after meals
Hab e Kabid Naushadri	Hepatitis; hepatomegaly; loss of appettite	2 pills after meals.
Majoon Dabid-ul-Ward	Hepatitis; hepatomegaly	5–7 gm with arq–e–makoh & arq e kasni
Arqe Ma'ul leham Makoh Kasni Wala	Hepatitis; good for strengthening stomach	125 ml with Sharbat–e–Kasni
Sharbat-e-Belgiri	Checks diarrhea and dysentery	25–50 ml
Majoon Injeer	Chronic constipation .	7–10 gm with milk
Qurs Mulaiyen	It removes morbid matters from stomach and intestines; useful in softening of stools	1–2 pills at bed time.
Hab e Muqil	Hemorrhoids	2–3 pills
Majoon Hajrul Yahood	Renal / ureteric calculi	5–8 gm
Majoon Sang-e-Sarmahi	Renal / ureteric calculi	5–8 gm
Sharbat Aaloo Baloo	Renal / ureteric calculi; diuretic	25–50 ml
Laboob-e-Saghir	Sexual debility; spermatogenesis	7–10 gm with milk
Majoon Punmbadana	Sexual debility; spermatogenesis	7–10 gm with milk
Majoon Sa'alab	Sexual debility; spermatogenesis	7–10 gm with milk
Masturine	Menstrual Disorders; leucorrhoea; uterine disorders	20–30 ml
Hab e Asgand	Arthritis; Gout; Low backache	2 pills once or twice a day
Habe Suranjan	Arthritis	2 pills once or twice a day
Jawarish Amla Sada	Cardiac and brain tonic; improves digestion and strengthens stomach	6–10 gm twice a day.
Khamira Marwarid	Cardiac tonic; removes general debility	5 gm
Sharbat-e-Faulad	General debility; haematinic	20–30 ml

bodies, reformulation of certain pharmacopeia preparations, modification of existing formulations to make them more patient compliant and application of new formulation techniques to Unani medicines like micronization and nanonization etc.

CONCLUSION

Unani system of medicine is one of the Indian systems of medicine. There are well established principles on which the system works. Many of these principles evolved with time and experience of the practitioners. Other systems of medicines influenced the original concepts and the present Unani System is an amalgamation of different Indigenous systems. Unlike allopathic system, in Unani system of Medicine each patient is considered different from other and a medicine regime of treatment is tailor made on the individual characteristics of the patient.

Like other Indigenous systems, this system also lacks systematic scientific basis and standardization in many areas. However, this does not undermine the importance of the system, as the system has proven itself to be effective over many centuries. Substantial number of people practice, believe in and get benefited by this system. Present day Unani physicians are well trained in *Tibbi* colleges that function on the same lines as the medical colleges. These offer a four and a half year (plus one year internship) well structured program leading to BUMS degree. Postgraduate courses leading to MD (Unani) are also offered by many of the colleges. Institutions like Jamia Hamdard are also conducting PhD in Unani medicine.

Efforts are being made to carry out advanced research, clinical trials and standardization of the Unani medicines using modern tools. The acceptability of these drugs has increased recently and a number of countries (both developing and developed) have started looking to the oriental medical system for the treatment of many chronic and life style related diseases. During the last few decades, interest in natural products and indigenous systems of medicine has been revived. Present day thinking has evolved the concept of integrated medicine where the combination of Allopathic system and (at least one of the) Indian systems of medicine could be used in combination to complement each other. This concept of integrated medicine that is being evolved takes a more holistic view than any one system. It is expected that in coming years there will be more amalgamation of various systems and there will be more inter-system consultations and collaborations for treating a patient.

SUGGESTIONS FOR FURTHER READING

1. *Ifada e Kabir*, Taleef Allama Allauddin Qarshi (Translation by Kabiruddin), Faisal Publications, Deoband, UP, India.
2. *Introduction to Al Umur al Tabi'yah* (*Principles of Human Physiology in Tibb*) by S.I. Ahmad, Published by N. Ishtiyaq, Delhi, India.
3. *Al Qanun Fil Tibb, Book I*, by Avicenna, Translated Under the Supervision of Hkm. Abdul Hameed at Jamia Hamdard, New Delhi, India.
4. *Makhzanal Mufradat*, Taleef by Kabiruddin, Delhi.
5. *Qarabadeen e Majeedi*, Taleef by Kabiruddin, Pub. Ajanta Offset Ltd. Delhi.
6. *Amraz o Ilaj* by Hamdard Dawakhana, Delhi.
7. *National Formulary of Unani Medicine*, Ministry of Health and FW, Govt. of India, New Delhi.
8. *Tibb e Akbar* by Mohammad Akbar Arzani, Fasial Publications, UP.
9. *Kitab al Mansouri* by Abubakr Mohammad Bin Zakriya Razi.
10. *Theories and Philosophies of Medicine* (2nd ed. 1973), Compiled by Department of Philosophy of Medicine and Science, Institute of History of Medicine and Medical Research, New Delhi, India.
11. *Kulliyat e Umoor Tabi'ya*, by Taskheer Ahmad, Nizami Publications, Lucknow, UP, India.
12. *Unani—The Science of Greek o Arab Medicine* by J Ahmad and A. Qadeer, Publication Roli Books, Lustre Press.
13. *Kulliyate Nafeesi* Part II, Translated by Kabiruddin.
14. *Hypertension and Unani Medicine* by Asim Ali Khan, C Publications, Distributed by Global Media Publications, New Delhi, India.
15. *Usool e Tibb by Kamaluddin*, Aligarh, UP, India.

125
Herbal Medicines and Human Health

S.K. Jain and Deepshikha P. Katare

CONTENTS

I. INTRODUCTION

Staying healthy with treasure of nature is the Ayurvedic philosophy of health. Ayurveda literally means "the science of life". Ayurveda is an oriental approach that explains how specific nature products from the herbs and foods keep us healthy.

Hippocrates, the father of Western medicine, advised his followers "let food be your medicine and medicine your food". All cultures have understood the intimate link between food and medicine and most people today intuitively understand the relationship between diet and health, which is also the key for understanding of herbal medicine. A "food" model of herbs is at the core of herbal medicine. It is quite opposed to the mainstream view of botanicals whose "drug" model of herbs is based upon conventional pharmacological understanding. It also constitutes the essential link between nutrition and herbals as inseparable therapeutic modalities and the key to the connection between prevention and treatment of chronic illness. A compound that exists in the diet as food ingredient serves as medicine when intentionally deployed by the physician and transforms prevention into treatment. In other words, for herbalists, the ultimate difference between food and medicine is primarily that of intention. It has been recognized throughout the history that medically useful material can be extracted from nature. Today's physician are concentrating on the traditional system of medicines to avoid the toxicity or the allopathic system of medicines.

Herbal drugs have been used since ancient times as medicine for the treatment of a range of diseases and medicinal plants have played a key role in health management. In spite of great advances observed in modern medicine in recent decades, plants still make important contribution to health care. Medicinal plants are distributed worldwide, but they are most abundant in tropical countries. In India a wide variety of medicinal plants are present, mostly in forests and wildness and only about 10% of the known medicinal plants of India are restricted to non-forest habitats. However it is estimated that 20-40% of all the plants found in India are used for medicinal purpose. (The world average stands at 12.5%).

In recent years the interest in drugs derived from higher plants, especially the phytotherapeutic has increased substantially. Almost 25% of all modern medicines are directly or indirectly derived from higher plants. In some particular cases, such as antitumoral and antimicrobial drugs, about 60% of the medicines currently available in the market and most of those in the late stages of clinical trials are derived from natural products, mainly from higher plants.

WHO estimates that 65–80% of the world's population, primarily from developing countries, depends on plants for primary health care. Herbal medicinal preparations are very popular in developing countries that have long tradition in the use of medicinal plants and also in some developed countries such as Germany, France, Italy and the United States.

The basic use of plants in medicine will continue in the future, as a source of therapeutic agents, and as raw material base for the extraction of semi-synthetic chemical compounds such as cosmetics, perfumes and food industries. Popularity of healthcare plant-derived products has also been in the cosmetic industry. Due to lack of systematic scientific research and standards many of the herbal drugs are used as supplements.

Herbs are defined botanically as a specific subgroup of plants, but herbal supplements are loosely defined as any plant-containing or plant-derived product. Herbal remedies have been commonly used in different cultures since ancient time, and still serve as the main means of therapeutic medical management in many countries. The use of herbal remedies is commonly referred to as "unconventional" or "alternative" medicine.

II. SECONDARY METABOLITES (PHARMACEUTICALLY ACTIVE COMPOUNDS)

The production of secondary metabolites forms part of the plant's defense system, for example, the constitutive production of antifeedants and phytoanticipins, and the inducible production of phytoalexins. Plant secondary metabolism has multiple functions. These included the rate as mediators in the interaction of the plant with its environment, such as

plant–insect, plant–microorganism and plant–plant interactions; facilitator of plant reproduction, for example by attracting the pollinators and in male fertility. Secondary metabolites also determine important aspects of human food quality (taste, colour and smell), and plant pigments are important for the diversity of ornamental plants and flowers. Besides many plant secondary metabolites are used for the production of medicines, dyes, insecticides, flavors and fragrances. Many of the secondary metabolites have been implicated as medicine. The sum of all of the chemical reactions that take place in an organism constitutes metabolism, which is a dynamic process. The molecular composition of a cell at any given time represent the balance between their synthesis and degradation. Much of carbon and energy ends up in synthesis of protein, nucleic acids, lipids and other molecules that are common to all organisms. However, in plants a significant proportion of assimilated carbon and energy is diverted to the synthesis of the secondary product molecules that may have no obvious role in its growth and development.

The secondary products can be divided into four major classes:

   (i) Terpenes
  (ii) Steroids, Saponins, Cyanogenic glycosides etc.
 (iii) Phenolic compounds
 (iv) Alkaloids.

Terpenoids

The terpenoids are functionally and chemically diverse group of molecules. These are lipophilic substances derived from the basic five carbon unit (Figure 125.1). The terpenoids family includes plant hormones (gibberellins and abscisic acid); the carotenoid pigments, sterols such as ergosterol, sitosteral, cholesterol and the sterol derivatives as well as, many of the essential oils that give the plants their distinctive odours and flavors. All terpenes and terpene derivative share a common biosynthetic pathway called Mevalonic Acid Pathway, named after the key intermediate.

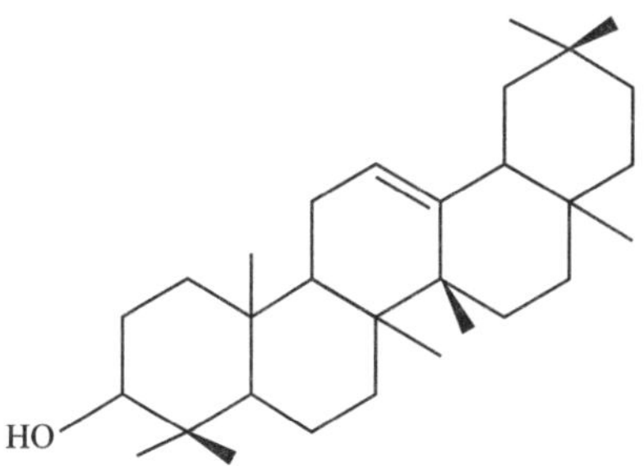

Figure 125.1 Pertacyclic triterpenoid (*β*-Amyrin).

A. Steroids and Sterols

Steroids are triterpenoids and are synthesized from the acylic triterpenoid. Steroids with an alcohol group are known as sterols. The most abundant sterols in higher plants are stigma-sterol and sito-sterol which together constitute more than 70% of the total sterols. Sterols are planner molecule and their packing tend to increase the viscosity and enhance the stability of membranes. Some sterols may have a protective function, such as the phytoecolysones, which have structure similar to the insect molting hormones when ingested by insect herbivores, phytoecdysores disrupt the insects molting cycle.

B.   Saponins

Saponins are terpene glycosides. These may be steroid glycosides, steroid alkaloids and triterpene glycosides. These may occur as aglycones which are known as sapogenins. The combination of hydrophobic triterpene with a hydrophilic sugar gives saponines the properties of a surfactant or detergent. The name saponin is derived from the soapwort or bouncing bet which at one time was employed as a soap substitute. Commercially, saponins from the bark of *Quillaja saponaria* have been used as surfactants in photographic film in shampoo, liquid detergents and beverages.

C.   Cardiac Glycosides

Cardiac glycosides or cardinolides have a wide distribution. The two principal cardinolides present in *Digitalis* spp are digitoxin and its close analog digoxin. *Digitalis* is also the source of a saponin, digitoni. *Digitalis* has been used for its therapeutic value in treating heart condition such as arteriosclerosis.

D.   Phenolics

Phenolics or polyphenols are a large and chemically diverse family of compounds ranging from simple phenolic acids to very large and complex polymers such as tannins and lignin.

Phenolic acids have their potential protective role, against oxidative damage diseases (coronary heart disease, stroke, and cancers). Plant phenolic compounds are diverse in structure but are characterised by hydroxylated aromatic rings (e.g., flavan-3-ols). Phenol, the parent compound is used as an disinfectant and for chemical synthesis. Examples of phenolic compounds are capsaicin, the pungent compound of chilli peppers, tyrosine, an amino acid and the neurotransmitters such as serotonin, dopamine, adrenaline, and noradrenaline; L-DOPA, a drug to treat Parkinson's disease, eugenol, the main constituent of the essential oil of clove, chavibetol from betel.

E.   Coumarins

The coumarins are a widespread group of lactone formed by ring closure of hydroxycinnamic acid. Coumarin itself is not toxic but it can be converted to a toxic product dicoumarol by fungi that is typically found in moldy hay. Dicoumarel causes fatal hemorrhaging in cattle by inhibiting vitamin K mediated blood clotting. Coumarin is used in the pharmaceutical industry as the precursor for the synthesis of a number of anticoagulants such as warfarin.

F.   Lignin

Lignin is a highly branched polymer of three simple phenolic alconals. Gymnosperm lignin is comprised mainly of coniferyl alcohol subunits while angiosperm lignin is a mixture of coniferyl and sinapyl alcohol subunits. The alcohols are oxidized to free radicals by a ubiquitous plant enzyme called peroxide. Lignin is found in cell walls, especially the secondary walls of treachery element, the elements in the xylem, where it contributes mechanical strength and rigidity to woody stems. Lignins and other phenolic derivatives accumulate at the site of fungal penetration presumably slowing the rate of cell degradation.

G.   Tannins

The name tannins is derived from the use of plant extracts to tan animal hides and convert it in leather. Condensed tannins are polymers of flavonoid units linked by bonds that are not hydrolyed but can be oxidized by strong acids to release anthocyanidins.

H.   Alkaloids

Alkaloids-(the term is linguistically derived from the Arabic word *a/-qali*, the plant from which soda was first obtained) are nitrogenous compounds that constitute the pharmacologically active "basic principles" of predominantly, though not exclusively, flowering plants. Since the identification of the first alkaloid, morphine, from the opium poppy, *Papaver somniferum*, by Sertürner in 1806 and more than 10,000 alkaloids have been isolated so far and their structures elucidated. The use of alkaloid-containing plant extracts as potions, medicines, and poisons can be traced back almost to the start of civilization. Famed examples include Socrates' death in 399 B.C. by consumption of coniine-containing hem lock (*Conium maculatum*) and Cleopatra's use during the last century B.C. of atropine-containing extracts of Egyptian henbane (*Hyoscyamus muficus*) to dilate her pupils and thereby appear more alluring. Medieval European women utilized extracts of deadly nightshade, *Atropa belladonna*, for the same purpose, hence the name belladonna. Although coniine is too toxic to find therapeutic use today outside of homeopathy, tropicamide, an anticholinergic that is a synthetic derivative of atropine, is routinely used in eye examinations to dilate the pupil. Tropicamide has also shown promise as an early diagnostic tool for the detection of Alzheimer's disease. A tonic prepared from the bark of *Cinchona officinalis* that contains the antimalarial drug quinine greatly facilitated European exploration and inhabitation of the tropics during the past two centuries.

Alkaloids have been used for variety of pharmacological purposes such as muscle relaxants, tranquilizers pain killers and mind altering drugs. Most alkaloids are synthesized from a few common amino acids (tyrosine, tryptophane and lysine). Approximately 25% of contemporary materia medica is derived from plants and used either as pure compounds (such as the narcotic analgesic morphine, the analgesic and antitussive codeine, and the chemotherapeutic agents vincristine and vinblastine) or as teas and extracts. Plant constituents have also served the basis for synthetic drugs, such as atropine for tropicamide, quinine for chloroquine, and cocaine for procaine and tetracaine. The most recent examples of anticancer alkaloids are taxol from the western yew, *Taxus brevifolia*, and camptothecin (and derivatives currently in clinical trials) from the Chinese "happy tree," *xi* shu (*Camptotheca acuminata*),

III. ACTION OF SOME HERBAL DRUGS AND THEIR ISOLATED COMPOUNDS

A. Hepatoprotective Herbs

Silymarin: *Silybum marianum*, commonly known as 'milk thistle' (other names include Marian Thistle, Mary Thistle, Saint Mary's Thistle, Mediterranean Milk Thistle, Variegated Thistle and Scotch Thistle) belongs to Family: Asteraceae/ Compositae. It is one of the oldest and thoroughly researched plant that are used for the treatment of liver diseases. The active principle, silymarin is a complex mixture of four flavolignans namely, silybin, silydianin, iso-sylibin and silychristin (Figure 125.2). It is known to protect liver against

Figure 125.2 *Silybum marianum* plant, its flower and structures of flavonolignan isomers of silymarin.

the toxic effects of CCl_4, N-nitrodiethyl amine, acetaminophene, ethanol, galactosamine, iron overload and *Amanita phalloidis* toxin etc.

Recent findings suggest that silymarin and garlic have a synergistic effect, and could be used as hepatoprotective agents against hepatotoxicity.

Mechanism of action

The preclinical studies using different hepatotoxic substances showed that silymarin has multiple actions as a hepatoprotective agent. The antioxidant property and cell-regenerating functions as a result of increased protein synthesis are considered as most important.

Antioxidant properties

Free radicals, including the superoxide radical, hydroxyl radical (.OH), hydrogen peroxide (H_2O_2), and lipid peroxide radicals have been implicated in liver diseases. These reactive oxygen species (ROS) are produced as a normal consequence of biochemical processes in the body and as a result of increased exposure to xenobiotics. The mechanism of free radical damage include ROS-induced peroxidation of polyunsaturated fatty acid in the cell membrane bilayer, which causes a chain reaction of lipid peroxidation, thus damaging the cellular membrane and causing further oxidation of membrane lipids and proteins. Subsequently cell contents including DNA, RNA, and other cellular components are damaged. The cytoprotective effects of silymarin are mainly attributable to its antioxidant and free radical scavenging properties. Silymarin can also interact directly with cell membrane components to prevent any abnormalities in the content of lipid fraction responsible for maintaining normal fluidity.

Stimulation of protein synthesis

Silymarin can enter inside the nucleus and act on RNA polymerase enzymes resulting in increased ribosomal formation. This in turn hastens protein and DNA synthesis. This action has important therapeutic implications in the repair of damaged hepatocytes and restoration of normal functions of liver.

Anti-inflammatory actions

The inhibitory effect on 5-lipoxygenase pathway resulting in inhibition of leukotriene synthesis is a pivotal pharmacological property of silymarin. Leukotriene (B4) synthesis was reduced while prostaglandin (E2) synthesis was not affected at higher concentrations of use of silybinin.

Antifibrotic action

Liver fibrosis can result in remodeling of liver architecture leading to hepatic insufficiency, portal hypertension and hepatic encephalopathy. These processes involve complex interplay of cells and mediators. In the initial phase there will be proliferation of hepatic parenchymal cells. The conversion of hepatic stellate cells (HSC) into myofibroblast is considered as the central event in fibrogenesis. Silymarin inhibits NF-κ-B and also retards HSC activation. It also inhibits protein

kinases and other kinases involved in signal transduction and may interact with intracellular signaling pathways.

Drug and toxin related liver damage

Drugs in common use can cause toxic effects on the liver which can mimic almost every natural diseases of the liver. About 2 per cent of all causes of jaundice in hospitalized patients are drug induced. About a quarter of cases of fulminant hepatic failure are thought to be drug related. Early detection is needed to identify a drug related hepatic reaction. Severity is greatly increased if the drug is continued after symptoms develop or if the serum liver enzymes continue to rise.

Hepatocellular injury due to drugs seems to be the primary event. This is rarely due to the drug itself and a toxic metabolite is usually responsible. The drug metabolizing enzymes activate chemically stable drugs to produce electrophilic metabolites. These potent agents bind covalently to liver molecules such as proteins and fatty acids which are essential to the life of the hepatocyte and necrosis ensues. This follows exhaustion of intracellular substances such as glutathione which are capable of preferentially conjugating with toxic metabolites. In addition, metabolites with an unpaired electron are produced by oxidative reactions of cytochrome P450. This free radical can also bind covalently to proteins and to unsaturated fatty acids of cell membranes. This results in lipid peroxidation and membrane damage. The end result is hepatocyte death related to failure to pump calcium from the cytosol and to depressed mitochondrial function. Drugs like paracetamol and halothane produce clear cut zone 3 necrosis of liver.

Silymarin has a regulatory action on cellular and mitochondrial membrane permeability in association with an increase in membrane stability against xenobiotic injury. It can prevent the absorption of toxins into the hepatocytes by occupying the binding sites as well as inhibiting many transport proteins at the membrane. These actions along with antiperoxidative property make silymarin a suitable candidate for the treatment of toxic liver diseases.

Adverse events

Silymarin is reported to have a very good safety profile. Both animal and human studies showed that silymarin is non toxic even when given at high doses (>1500 mg/day). However, a laxative effect is noted at these doses may be due to increased bile secretion and bile flow.

B. HERBS WITH ANTICANCEROUS PROPERTY

Population studies have shown that people in South East Asian countries have far lower risks of developing most cancers as compared to those in North America, and it is considered that the consumption of foods such as garlic, ginger, turmeric, soy and cruciferous vegetables play a key role in this "chemoprevention". These chemopreventive compounds are dietary ingredients, which being food derived, are considered pharmacologically safe. These dietary ingredients contain a wide range of different molecules such as organosulfur compounds from garlic, polyphenols from green tea and curcumin from turmeric.

Most chemopreventive dietary compounds interact with several of these targets simultaneously and often synergistically. For example *green tea polyphenols* can influence signal transduction factors, inhibit cyclooxygenase-2 (COX-2), promote cell cycle arrest, increase apoptosis and disable multidrug resistance pumps. Curcumin from turmeric has been found to influence over 60 such molecular targets in the cancer process.

These chemopreventive agents have diverse mechanism of action. They have effect on the cell cycle, transcription factors (NF-kappa B and AP-1), cell differentiation, enzyme (cyclooxygenase-2), have antimutagenic activity (DNA repair), growth factor analogue and induction of antioxidant enzymes. The chemopreventive agents have been shown to suppress cell proliferation, inhibit growth factor signaling pathway, induce apoptosis, inhibit NF-kappaB, activator protein-1, inhibit JAK-STAT activation pathway, inhibit angiogenesis, inhibit COX-2, inhibition of multiple drug resistance (MDR) and act as cell cycle interfering agents (Figure 125.3)

Taxol, a natural product isolated initially from the Pacific Yew (*Taxus brevifolia*) in the late 1980s destroys the spindle, leading to a loss of chromosome segregation with consequent inhibition of cell division and cell death. Interestingly, it is likely that most of the herbal medicines or mixtures exert their antitumor activities via the apoptotic process.

Herbs that induce apoptosis via tumor suppressor gene p53

Bupleurum falcatum Saikosaponin D, one of the major components of *Bupleurum falcatum* which could be extracted from other species of Bupleurum and from related genus, is used for the treatment of various liver diseases in traditional Chinese medicine. It can inhibit the proliferation of the lung cancer cell line, A549, with the IC50 value at 10.18 + 0.09.

Curcuma longa Curcumin, a phenolic compound of the rhizome of the plant *Curcuma longa* has anti-inflammatory, antioxidant and anticancer activities. A549 and H1299 human lung cancer cell lines were used for the present study and found that the growth inhibitory effect of curcumin was concentration dependent in both cell lines. The IC50s at 24 hours exposure of curcumin were 50 and 40 M in A549 and H1299 cells, respectively.

Acacetin Acacetin (5,7-dihydroxy-4′-methoxyflavone), a flavonoid compound, has been reported to possess anti-peroxidative, anti-inflammatory and antiplasmodial effects. The proliferation inhibitory effect of acacetin was observed to be in a dose dependent manner with A549 human lung cancer cell lines. Its IC50 value was 9.46 mM.

Isoliquiritigenin Isoliquiritigenin (ISL), a flavonoid found in licorice and shallot, is a potent anti-oxidant with anti-inflammatory antiplatelet aggregation and cancer preventing properties The ISL inhibited the proliferation of A549 cells with IC50 value at 27.14 mol/L.

Herbs that promote apoptosis via death signaling molecules

Angelica sinensis The root of *Angelica sinensis*, also known as "Danggui", is a popular herbal medicine and has been widely used in China for gynecological diseases for a long time. Bio-based assay for extracts of *Angelica sinensis* showed that the acetone extract (AE-AS) was dose-dependent in antiproliferative effects on A549, HT29, DBTRG-05MG and J5 human cancer cells.

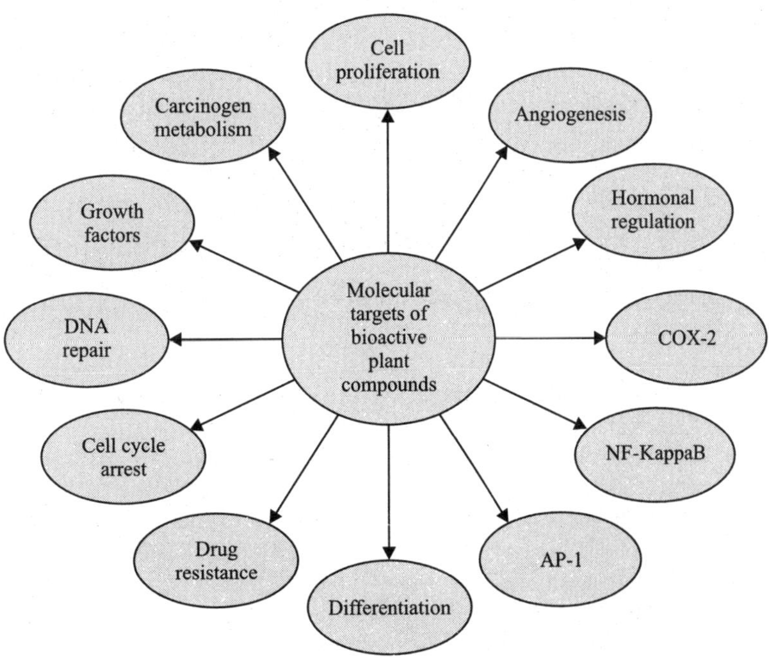

Figure 125.3 Molecular targets for active principles of herbal drugs.

Bupleurum scorzonerifolium The acetone extract of *Bupleurum scorzonerifolium* (AE-BS) showed a dose-dependently antiproliferative effect on the proliferation of A549 human lung cancers.

Solanum incanum A purified compound from the *Solanum incanum* herb, solamargine (SM), has been shown to inhibit the growth of human tumor cells, e.g., colon, prostate, breast and human hepatoma cells. The IC50 of solamargine of A549 cells was 2.9 uM.

Typically, normal p53 function plays a crucial role in inducing apoptosis and cell cycle checkpoints in human cells following DNA damage. This has been further supported by the finding that p53 is the most commonly mutated tumor suppressor gene. Moreover, the chemo-sensitivity of cancer cells to chemotherapy agents is greatly influenced when the function of p53 is abrogated. However, it appeared that the antiproliferative activities of herbs in human lung cancer cells is induced via an apoptotic system with or without p53 pathway. Some herbs were observed to have an interaction with the Fas/FasL system which is a key signaling transduction pathway of apoptosis in cells and tissues. Ligation of Fas by agonistic antibody or its mature ligand induces receptor oligomerization and formation of a death-inducing signaling complex (DISC), followed by activation of caspase-8, then further activating a series caspase cascade resulting in cell apoptotic death. FasL is a tumor necrosis factor related type II membrane protein.

C. Herbs that Relieve Inflammation (Pain, Swelling, Redness, and Fever)

(a) Herbs to relieve pain

   (i) White willow bark (Usual dose is 240 mg/day), meadowsweet, wild lettuce.
   (ii) Omega-3 fatty acids (1.5–5 gm/day with meals).
   (iii) Green tea (3–4 cups decaffeinated tea or 300–400 mg extract/day).
   (iv) Pycnogenol (Maritime pine bark—100–200 mg/day).
   (v) Frankincense (Boswellia—300–500 mg 2 or 3 times daily).
   (vi) Cat's Claw (Uncaria Tomemtosa—20–60 mg/day.
   (vii) Ashwaganda demonstrates anti-inflammatory properties and pain relief for arthritis.
   (viii) Devil's Claw demonstrates anti-inflammatory properties and pain relief similar to cortisone—good for arthritis and other inflammatory pain
   (ix) Ginger root—Ginger may also be used as tea, massaged into the skin, or added to the bath. When used as a bath for Rheumatoid arthritis, boil one cup of chopped ginger root for 20 minutes. Place the liquid in the tub of warm water. Stay in the bath about 30 minutes.
   (x) Capsaicin (from Cayenne and other hot peppers)—may be applied directly to the skin as an infusion (tea), used in ointments for herpes (shingles), neuralgias, diabetic neuritis, etc., or massaged into the tissues around joints for osteoarthritis and other aches and pains Turmeric (curcumin)—(400–600 mg three times daily) anti-inflammatory, arthritis, menstrual pain, topical antiseptic, and wound healing agent—may be used internally as tea, capsule, etc. or applied externally as ointment, paste, etc.
   (xi) Common poppies—Use 8–10 petals as infusion (tea) or syrup. Use 2–3 seedpods (unripe, fresh or dried) as decoction.
   Note: common poppies provide analgesic (pain relieving) qualities, but do not contain addicting drugs.

(b) Herbs with analgesic qualities for topical (on skin or mucus membranes) use

   (i) Cloves—apply clove or a drop of clove oil to an aching tooth. Use tea (decoction) as a mouthwash or rinse.
   (ii) Aloe Vera—Gel from fresh cut leaf brings quick relief when applied directly to all types of burns, abrasions, rash, and other skin conditions.

(c) Herbs to relieve itching

   (i) Aloe vera, witch hazel, jewelweed when applied as juice, infusions (tea), or poultices.
   (ii) Plantain (crushed fresh leaves, or infusion)—bee stings, bites of various kinds, hemorrhoids.

D. Herbs Affecting the Brain and Nervous System

(a) Herbs that relieve driver fatigue (often used as lozenges)

   (i) Cinnamon: May increase alertness and decrease driver fatigue and frustration at the wheel. Cinnamon odors boost motivation and performance.
   (ii) Peppermint: Peppermint odor may increase alertness, boost motivation and performance, and decrease driver fatigue.

(b) Herbs for the nerves

   (i) Lettuce, wild lettuce, lavender, linden, poppy, rose petals, willow, catnip, hops, valerian, chamomile, St. John's Wort, ginko biloba—prepared as infusions of decoctions.
   (ii) Alzheimer's disease: Turmeric reduces the action of a number of genes that promote inflammation as it is linked to heart disease, colon and some other cancers, and Alzheimer's disease (**Note:** India, where turmeric is regularly used in food, has the world's lowest incidence of Alzheimer's disease). Other Cox-2 inhibitors may also be beneficial for preventing and managing patients with Alzheimer's disease.

E. Herbs for Female Symptoms: Breast Pain, Menstrual Disorders, Menopause

(i) Orange tree, basil, wormwood, parsley, and calendula (infusion or decoctions of roots, leaves, flowers or seeds).

(ii) Tea Tree oil, lavender oil (also frequently used in "natural" cosmetics, medications, etc.).

(iii) Pycnogenol (maritime pine bark extract)—effective in treatment of endometriosis at 30 mg. twice daily.

(iv) Goldenseal (infusion of roots), Witch Hazel (infusion of leaves or bark), grapevine (decoction of leaves), useful for heavy periods.

F. Herbs for Prostate

(i) Saw Palmetto: 50–100 grams daily of fresh fruit or as decoction.

(ii) Pumpkin seeds: Eat a handful of seeds 2–3 times daily, raw or cooked.

(iii) Phytosterol flower pollen extract called Cernitin and Prostaphil, sugar cane pulp, soybean products as a regular part of diet.

(iv) Evening Primrose: 2–4 grams daily as capsules.

(v) Other Hormone modulators—licorice, skullcap, ginseng, chase tree, partridge berry, lavender oil, nettle—as infusions (tea), capsules, etc.

G. Herbs that Enhance Immune Functions

(i) Echinacea, astragalus, licorice, skullcap, ginseng, rose hips, garlic, turmeric (curcumin), oregano.

(ii) Strawberry, violet, and many other leaves very high in vitamin C.

(iii) Most berries and fruits are high in antioxidants and other protective phyto-nutrients.

H. Herbs Effective Against Invading Organisms— Parasites, Bacteria, Viruses, etc.

(a) Herbs with antibiotic qualities

(i) Calendula (decoction of flowers), grapefruit seeds or extract, garlic, onion, aloe vera.

(ii) Turmeric (curcumin), Oregano—antibiotic and antiviral (flu, colds).

(iii) Black elderberry, anti-viral, effective for most flu viruses.

(b) Urinary antiseptic herbs

(i) Garlic—raw, extracts, or decoctions.

(ii) Cranberry juice, Thyme (infusion of leaves).

(iii) Bearberry (upland cranberry)—Use decoction with 50–60 grams of dry, ground leaves. Soak 3–4 hours before boiling for 15 minutes.

(c) Herbs for vaginal infections

(i) Douche (irrigation) White willow (decoction of bark), pomegranate (infusion of flowers and bark), goldenseal (infusions of plant), myrtle (infusion of leaves and berries), Turmeric—anti-viral and anti-bacterial, healing.

(d) AIDS

(i) These herbs used in conjunction of optimal lifestyle practices are said to be effective in the management of AIDS—golden seal, garlic, turmeric (curcumin).

(e) Herbs effective against some parasites

(i) Garlic—as garlic oil or as an enema for worms—a garlic clove in the anus is said to be helpful for pin-worms.

(ii) Onion—Use the juice of one raw onion daily. It may be mixed in a blender with honey, lemon, tomato, or carrot juice and taken in small amounts during the day.

(iii) Wormwood—10–20 grams per liter of water infusion (tea) (Too dangerous and sour for children).

(iv) Pumpkin seeds—Crush 200–400 grams of seeds (fresh, dry, or cooked). Eat 1/3 of this at each meal with nothing else but water and carrots for one day. Take a purgative one hour following last "meal."

(v) Myrrh—oleo gum resin from the stem—10 mg/kg weight/day for 3 days—said to be very effective against *Schistosoma hematobium*.

(vi) Turmeric—Parasites, including round worms, malaria, and scabies.

(vii) Papaya and papaya seeds for intestinal worms, and for malaria.

(viii) Ginger root for Giardia and fish worms.

I. Herbs Useful for the Digestive System

(a) Herbs that neutralize acid, coat the lining of the stomach, relieve spasm, etc.

(i) Peppermint—may be used as tea, lozenge (peppermint oil tablets are commercially available—meta-analysis studies suggest that peppermint oil is more effective than many commercial anti-spasmotic medications on the market). **Note:** use with caution in presence of esophageal reflux symptoms!

(ii) Raspberry leaf tea, carrot, or cabbage (cooked, raw, or as juice), cassava, oats (cooked), olive oil, avocado, Slippery Elm (reduces gastric and intestinal inflammation), blueberry (bilberry) may substitute for Zantac.

(iii) Ginger (nausea and motion sickness).

(iv) Turmeric (used for gastrointestinal symptoms, reduces inflammation).

(v) Fennel (gastrointestinal spasms).

(vi) Dill, caraway, basil, marjoram, calendula blossoms or licorice root (infusion of seeds or leaves).

(vii) African red tea (Rooibos bush) said to be very effective for infant colic.

(b) Diarrhea

(i) Garlic: Use as decoction, as sauce diluted with water, or eat raw cloves.
(ii) Thyme flowers, bramble (blackberry or other), grapevine, (juice or a decoction of leaves and buds).
(iii) Papaya, cassava, apples, and other fruit.

(c) Constipation

(i) High complex carbohydrate diet—whole grains, fruits, vegetables, legumes, etc.
(ii) Psyllium, flax seed, senna, many others.
(iii) Molasses—take daily, adjusting dosage to desired effect. (**Note:** It may take several days to realize the optimum benefit.) Unstable diabetics must exercise caution.

(d) Purging, cleansing

(i) Cascara sagrada, powder, or decoction of bark, castor oil.

J.   Herbs Useful for the Respiratory System

(a) Herbs for the throat

(i) Wild cherry bark (tea), Horehound (liquid extract), Garlic in oil, Turmeric in honey.
(ii) Gargles with hot sage tea or many of the anti-inflammatory herbs listed above (Y, 1, a).
(iii) Elderberry (juice of berries or tea of leaves)—colds, flu, sore throats.

(b) Cough

(i) Eucalyptus—use as a steam inhalation with Eucalyptus leaves in boiling water for inhalation—drink tea made from several leaves (20–30 grams per liter water) or 2–3 drops of essence per glass of water, or use as syrup or lozenges containing essence of eucalyptus.
(ii) Thyme: Steam inhalation with 2–3 drops of essence per liter of water; 2–3 drops on handkerchief for inhaling; Drink 2–3 drops in water three times daily.
(iii) Onion: Expectorant—loosens secretions—use raw, ground in blender and mixed with honey and/or lemon. Sip as needed for cough and congestion.
(iv) Poppy: Syrup made with 8–10 petals/cup; infusion with 6–8 petals per cup or decoction of 2–3 green-dry seedpods in 1/2 cup water, use 2–3 tablespoons full before bedtime.
(v) Hyssop: Use infusion of 50–60 grams per liter of water, or 2–3 drops or essence per cup and drink 1 cup 3–4 times daily.
(vi) Linden: infusions of flowers or decoctions of bark 20–40 grams per liter. Drink 3–4 cups per day with honey.

(vii) Ginger, horehound, juniper, plantain, licorice, and raw garlic are other commonly available herbs with good qualities for chest congestion.

K.   Herbs Useful for the Eyes

(i) Infusions and decoctions of cornflower, tea, oak bark, witch hazel, plantain, garden violets, fennel, grapevine, rose and briars.
(ii) Drops of Aloe Vera are a safe, effective agent to reduce inflammation.

L.   Herbs Useful for the Skin

(i) Aloe Vera gel—burns, sunburn, wounds.
(ii) Evening primrose oil—healing agent.
(iii) For poultices—plantain, carrots, cabbage, onion, ivy leaves.
(iv) For infusions and decoctions for compresses—pansy, rose, witch hazel, calendula, lemon, oak, and walnut.
(v) Tea tree oil—candida and other fungus infections.
(vi) Turmeric—antibiotic, antiseptic and healing properties—used in Asia to beautify the skin.
(vii) Pine pitch applied directly to superficial wounds to enhance healing.
(viii) Geranium oil—control bleeding, useful for shingles (herpes zoster pain).
(ix) Plantain—available worldwide—crushed leaves useful as poultice for bites, stings, healing wounds and ulcers.

IV.   COMMON MISCONCEPTIONS ABOUT THE USE OF HERBAL DRUGS

(a) Herbal medicine is safe because it is natural: After researching the literature, one can truly say that at the very least, herbal medicine is safer than conventional drugs. The primary reason for this is that the herbal medicine incorporates the whole plant material with its active ingredients intact, rendering a more holistic, gentler approach. This is not true of drugs. Moreover, most medicinal plants have a very broad range between their effective and toxic doses. However, safety is based on proper usage. Over the last 10 years, most of the reports due to adverse reactions have been associated with the misuse of herbal medicine, especially those for weight loss.

(b) Herbal medicine lacks scientific validation: The scientific validation of herbal medicine has been going on for hundreds of years, just not in this country but most clinical studies on herbal medicine come from countries that still appreciate and embrace this discipline—Japan, Russia, China, Germany and many European nations. However, times and interest have changed in America. Some of the most respected peer-review medical journals such as Journal of the American Medical Association, the New England Journal of Medicine, etc. are now publishing clinical studies on herbal medicine quite frequently. In fact, there are now more than a dozen medical journals whose primary focus is publishing the latest research on plant medicines. Some of these journals (a few available via the internet) are The International Journal of Alternative Medicine, Journal of Naturopathic Medicine, Journal of Natural Products, Journal of Ethnopharmacology, European Journal of Herbal Medicine, Planta Medica, Phytotherapy, Phytotherapy Research, and Phytomedicine.

(c) Never take herbal medicine and conventional drugs together: There is nothing wrong with using herbal medicine and conventional medicine together so long as the two are carefully coordinated for compatibility, dose, etc. Properly applied, one can indeed bring about a complimentary beneficial effect. In fact, traditional medicine (including herbs) is sometimes called complementary medicine. However, case in point: if one is taking drugs to lower cholesterol and decides to add red yeast, garlic, or guggul lipid to enhance the effect, these supplements may result in increasing the potency of the drugs. While this particular interaction may not be life-threatening, it may sometimes be wise for your health-care provider to modify the dose of the drugs. Unfortunately, most patients who use herbal medicine along with their conventional drugs do not disclose this to their health-care provider. Until the fact is generally known among the public that herbs and drugs are both potent agents that can cause harm, the advice not to take them together remains true.

Herbs, when used for healing, are medicines, and must be treated like medicine! They should only be used when specifically needed, and then, only in modest amounts. More is not always better! Herbs too can be toxic in excessive doses. This is a rapidly expanding field of research with new findings published on a regular basis. For most people, what is most important is that one knows a few important herbs that are readily available in time of need.

V. SOME OF THE COMMONLY USED HERBS WITH MEDICINAL PROPERTIES

1. Curcuma longa (Turmeric): It is a rhizomatous herbaceous perennial plant of the ginger family. Curcumin is a polyphenol derived from the (*Curcuma longa*). Among polyphenols, curcumin is currently one of the most studied polyphenol (Figure 125.4). It is a hydrophobic, low molecular weight polyphenol.

It possesses diverse anti-inflammatory and anti-cancer properties following oral or topical administration. Apart from curcumin potent antioxidant capacity at neutral and acidic pH, its mechanisms of action include inhibition of several cell signaling pathways at multiple levels, effects on cellular enzymes such as cyclooxygenase and glutathione S-transferases, immuno-modulation and effects on angiogenesis and cell–cell adhesion. Curcumin ability to affect gene transcription and to induce apoptosis is likely to be of particular relevance to cancer chemoprevention and chemotherapy in patients.

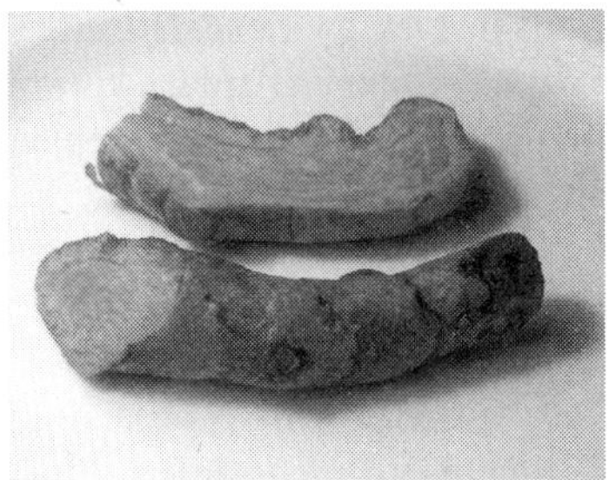

Figure 125.4   The plant and rhizomes of *curcuma longa* and the structure of curcumin.

Although curcumin, its low systemic bioavailability following oral dosing may limit access of sufficient concentrations for pharmacological effect in certain tissues, the attainment of biologically active levels in the gastrointestinal tract has been demonstrated in animals and humans. Further, the bioavailability of curcumin can be enhanced by making the nanopreparations.

A list of studies describing anti-tumor effects of curcumin in animals in shown in Table 125.1.

TABLE 125.1   Antitumor effects of curcumin

Tumor	Route	Dose	Model
Ascites	IP	50 mg/kg	Ascites
Ascites	IP	50 mg/kg	Ascites
Lung Metastasis	Diet	2% w/w	Orthotopic
Breast	Diet	1% w/w	Orthotopic
Colon	IV	40 mg/kg	Xenograft
Gastric Cancer	Oral	50–200 mg/kg	Xenograft
Giloblastoma	IT	10 mg/kg	Orthotopic
HCC		100–200 mg/kg	Orthotopic
Hepatoma	Oral	50–200 mg/kg	Xenograft
Hepatoma		20 µg/kg	Ascites
HNSCC	Sub acute	50–250 µ mol/L	Xenograft
Leukemia	Oral	50–200 mg/kg	Xenograft
Melanoma	IP	25 mg/kg	Xenograft
Melanoma	Oral	200 nmol/Kg	Metastasis
Ovarian	IP	500 mg/kg	Xenograft
Pancreas	IV	40 mg/kg	Orthotopic
Pancreas	Gavage	1 gm/kg	Xenograft
Prostate	Diet	2% w/w	IV
Prostate	Gavage	5 mg/day	Xenograft

IP – intraperitoneal; IT – intratumoral; IV – intravenous

2. Phyllanthus emblica (*Emblica officinalis*): The Indian gooseberry, or aamla from Sanskrit **amalika**, is a deciduous tree of the Phyllanthaceae family. It is known for its edible fruits that are the richest natural source of Vitamin C, alkaloids, Enzymes, Tannin. Amla is known for its adaptogenic properties immune support herbal supplement of the body. Amla also has natural bioflavonoids. It is the most used herb among all herbs in Ayurvedic medicine system. Amla also retains most of *its vitamin content even in dried form. Amla contains high amounts of vitamin C (ascorbic acid), more then 445 mg per 100 g of pulp.* This is around 20 times more then the vitamin C found in Orange juice. Amla contains high concentration of many minerals (also zinc, copper, chromium) and amino acids, more then found in Apple.

Amla herb is also used in many Hair dyes, Shampoo, Detergents and Laxative products apart from its herbal treatment uses.

Figure 125.5 *Emblica officinalis* plant and furits.

It has undergone preliminary research, demonstrating *in vitro* antiviral and antimicrobial properties. It has very strong anti-salmonella activity and protects against typhoid. There are evidences in *in vitro* studies that its extracts induce apoptosis and modify gene expression in osteoclasts involved in rheumatoid arthritis and osteoporosis. It may prove to have potential activity against some cancers. One recent animal study found treatment with *E. officinalis* reduced severity of acute pancreatitis (induced by L-arginine in rats). It also promoted the spontaneous repair and regeneration process of the pancreas.

A human pilot study demonstrated a reduction of blood cholesterol levels in both normal and hypercholesterolemic men with treatment. Another recent study with alloxan-induced diabetic rats given an aqueous amla fruit extract has shown significant decrease of the blood glucose, as well as triglyceridemic levels and an improvement of the liver function caused by a normalization of the liver-specific enzyme alanine transaminase activity.

Although these fruits are reputed to contain high amounts of ascorbic acid (vitamin C), 445 mg/100g, the specific contents are disputed, and the overall antioxidant strength of amla may derive instead from its high density of ellagitannins[12] such as emblicanin A (37%), emblicanin B (33%), punigluconin (12%) and pedunculagin (14%). It also contains punicafolin and phyllanemblinin A, B, C, D, E and F.[14]

The fruit also contains other polyphenols: flavonoids, kaempferol, ellagic acid and gallic acid.

Medicinal use

In traditional Indian medicine, dried and fresh fruits of the plant are used. All parts of the plant are used in various Ayurvedic/ Unani medicine (*chyavanprash* and *Jawarish amla*) herbal preparations, including the fruit, seed, leaves, root, bark and flowers. According to Ayurveda, aamla fruit is sour (*amla*) and astringent (*kashaya*) in taste (*rasa*), with sweet (*madhura*), bitter (*tikta*) and pungent (*katu*) secondary tastes (*anurasas*). Its qualities (*gunas*) are light (*laghu*) and dry (*ruksha*), the postdigestive effect (*vipaka*) is sweet (*madhura*), and its energy (*virya*) is to cooling (*shita*).

According to Ayurveda, aamla balances all three doshas. While aamla is unusual in that it contains five out of the six tastes recognized by Ayurved, it is most important to recognize the effects of the "virya", or potency, and "vipaka", or post-digestive effect. Considered in this light, aamla is particularly helpful in reducing *pitta* due to its cooling energy and balances both Pitta and *vata* by virtue of its sweet taste. The *kapha* is balanced primarily due to its drying action. It may be used as a *rasayana* (rejuvenative) to promote longevity, and traditionally to enhance digestion (*dipanapachana*), treat constipation (*anuloma*), reduce fever (*jvaraghna*), purify the blood (*raktaprasadana*), reduce cough (*kasahara*), alleviate asthma (*svasahara*), strengthen the heart (*hrdaya*), benefit the eyes (*chakshushya*), stimulate hair growth (*romasanjana*), enliven the body (*jivaniya*), and enhance intellect (*medhya*).

In Ayurvedic polyherbal formulations, Indian gooseberry is a common constituent, and most notably is the primary ingredient in an ancient herbal *rasayana* called *Chyawanprash*. This formula, which contains 43 herbal ingredients as well as clarified butter, sesame oil, sugar cane juice, and honey, was first mentioned in the Charaka Samhita as a premier rejuvenative compound.

In Chinese traditional therapy, this fruit is called *yuganzi*, which is used to cure throat inflammation.

Emblica officinalis tea may ameliorate diabetic neuropathy. In rats it significantly reduced blood glucose, food intake, water intake and urine output in diabetic rats compared with the non diabetic control group.

3. *Ocimum tenuiflorum* (Holy Basil also *tulsi, tulasī*): It is an aromatic plant in the family Lamiaceae which is native throughout the Old World tropics and widespread as a cultivated plant and an escaped weed. It is an erect, much branched subshrub, 30–60 cm tall with hairy stems and simple, opposite, green leaves that are strongly scented. Leaves have petioles, and are ovate, up to 5 cm long, usually slightly toothed. The flowers are purplish in elongate racemes in close whorls. The two main morphotypes cultivated in India and Nepal are green leaved (Sri or Lakshmi *tulsi*) and purple-leaved (Krishna *tulsi*).

Tulsi is cultivated for religious and medicinal purposes, and for its essential oil. It is widely known across South Asia as a medicinal plant and an herbal tea, commonly used in

Figure 125.6   *Ocimum sanctum* plant and its active compounds.

Ayurveda, and has an important role within the Vaishnavite tradition of Hinduism, in which devotees perform worship involving *tulsi* plants or leaves.

The variety of *Ocimum tenuiflorum* used in Thai cuisine is referred to as **Thai holy basil**, it is not be confused with Thai basil, which is a variety of *Ocimum basilicum*.

Recent studies suggest *tulsi* may be *a COX-2 inhibitor*, like *many modern painkillers*, due to its high concentration of eugenol. One small study showed it to reduce blood glucose levels in type 2 diabetics when combined with hypoglycemic drugs. The same study showed significant *reduction in total cholesterol levels* with *tulsi*. Another study showed its beneficial effect on blood glucose levels is due to its antioxidant properties. Tulsi also shows some promise for protection from radiation poisoning and cataracts. It has anti-oxidant properties and can repair cells damaged by exposure to radiation. The fixed oil has demonstrated antihyperlipidemic and cardioprotective effects in rats fed a high fat diet. Experimental studies have shown an alcoholic extract of *tulsi* modulates immunity, thus promoting immune system function. Some of the main chemical constituents of *tulsi* are: oleanolic acid, ursolic acid, rosmarinic acid, eugenol, carvacrol, linalool, β-caryophyllene (about 8%), β-elemene (c.11.0%), and germacrene D (about 2%). β-Elemene has been studied for its potential anticancer properties, but human clinical trials have yet to confirm its effectiveness.

O. sanctum extracts are antibacterial (against *E. coli*, *S. aureus* and *P. aeruginosa*).

Tulsi has been used for thousands of years in Ayurveda for its diverse healing properties. It is mentioned in the Charaka Samhita, an ancient Ayurvedic text. *Tulsi* is considered to be an adaptogen, balancing different processes in the body, and helpful for adapting to stress. Marked by its strong aroma and

astringent taste, it is regarded in Ayurveda as a kind of "elixir of life" and believed to promote longevity.

Tulsi extracts are used in ayurvedic remedies for common colds, headaches, stomach disorders, inflammation, heart disease, various forms of poisoning, and malaria. Traditionally, *tulsi* is taken in many forms: as herbal tea, dried powder, fresh leaf, or mixed with *ghee*. Essential oil extracted from Karpoora *tulsi* is mostly used for medicinal purposes and in herbal cosmetics, and is widely used in skin preparations due to its antibacterial activity. For centuries, the dried leaves have been mixed with stored grains to repel insects.

4. Cuminum cyminum Cumin (sometimes spelled cummin; *Cuminum cyminum*): It is a flowering plant in the family Apiaceae, native from the east Mediterranean to India. Its seeds (each one contained within a fruit, which is dried) are used in the cuisines of many different cultures, in both whole and ground form.

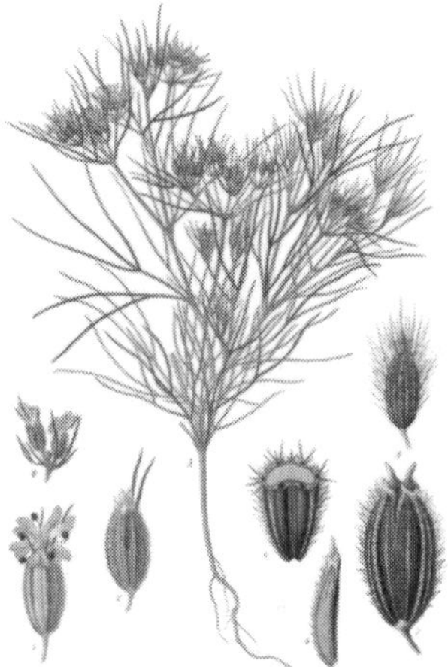

Figure 125.7   *Cuminum cyminum* plant, its flowers and the fruits.

Cumin has been in use since ancient times. Seeds excavated at the Syrian site Tell ed-Der have been dated to the second millennium BC. They have also been reported from several New Kingdom levels of ancient Egyptian archaeological sites. It was originally cultivated in Iran and Mediterranean region. There are several different types of cumin but the most famous ones are black and green cumin which are both used in Persian cuisine.

Today, it is mostly grown in Iran, Uzbekistan, Tajikistan, Turkey, Morocco, Egypt, India, Syria, Mexico, Chile, and China. The plant occurs as a rare casual in the British Isles, mainly in southern England, but the frequency of its occurrence has declined greatly. According to the Botanical Society of the British Isles' most recent Atlas, there has been only one confirmed record since 2000.

5. Psoralea corylifolia (popularly known as babchi): It is a herbaceous weed, which grows throughout the plains of India. It is an erect annual plant of about 0.6–1.2 m height. Stem and Branches grooved, studded with conspicuous glands and with a few appressed and spreading white hairs.

Figure 125.8   *Psoralea corylifolia* plant and structure of its active compound, psoralan.

Its seeds have been described by the ancient Hindu physicians as 'hot and dry' and according to some cold and dry, laxative, fragrant, stimulant, and aphrodisiac. The drug has been considered to be so efficacious in leprosy that it was given the name of 'kushtanasini'. In inflammatory diseases of the skin leucoderma and psoriasis it is given both as local application and by the mouth. The seeds are also used as an anthelmintic, diuretic and diaphoretic in febrile conditions.

Dose: 3-6 gm of the drug in powder form.

It is also used in Unani medicine for skin conditions, particularly leucoderma, anti-souda, balghami, fever anti-helmintic and sedative for internal ulcers. The seeds are bitter and acrid; purgative stomachic, vulnerary, stimulant, aphrodisiac; improve appetite; good for scabies, biliousness and cure blood diseases. In Siddha its seeds are used to treat glandular swellings, skin diseases, leprosy, and as a laxative, carminative, tonic and stimulant.

In Southern India, it is used as stomachic and deobstruent and prescribed in lepra and other cutaneous diseases. In Konkan, the seeds are used in making perfumed oil, which is applied to the skin. In Ceylon, in case of snakebite the seeds are ground with water and the liquid poured into each nostril in stupor and coma. In China and Malaya, the seeds are regarded as tonic and aphrodisiac and they are used in certain cutaneous diseases. In Indo China, the fruit is prescribed in stomachache, spermatorrhoea and certain skin diseases. In North of Annam, the seed are macerated in alcohol and the liqueur is given in rheumatism and in women's diseases. According to Dymock several species of *Psoralea* have been medicinally used in America and have been found to act as gentle, stimulating and toxic nervines.

Other uses: Externally seeds are used for vitiligo, pollakiuria, psoriasis and alopecia. Also used to treat premature ejaculation, enuresis, backache, knee pain and pollakiuria. It is used to treat menstruation disorders. Seed extract of *P.*

corylifolia is also suggested to be useful as a remedy for bone fracture, osteomalacia and osteoporosis. It is used against strigidity of limbs in epilepsy, and also against loss of memory. It cures haemetemesis, diabetes, fever, cough, oedema, anaemia, piles, dysuria, poisonous affections and ulcers. Psoralen is also used to treat uterine hemorrhage.

6. *Catharanthus roseus* (Madagascar Periwinkle): It is a species of *Catharanthus* native and endemic to Madagascar. Synonyms include *Vinca rosea* (the basionym), *Ammocallis rosea*, and *Lochnera rosea*; other English names occasionally used include Cape Periwinkle, Rose Periwinkle, Rosy Periwinkle, and "Old-maid". In the wild, it is an endangered plant; the main cause of decline is habitat destruction by slash and burn agriculture. It is also however widely cultivated and is naturalised in subtropical and tropical areas of the world.

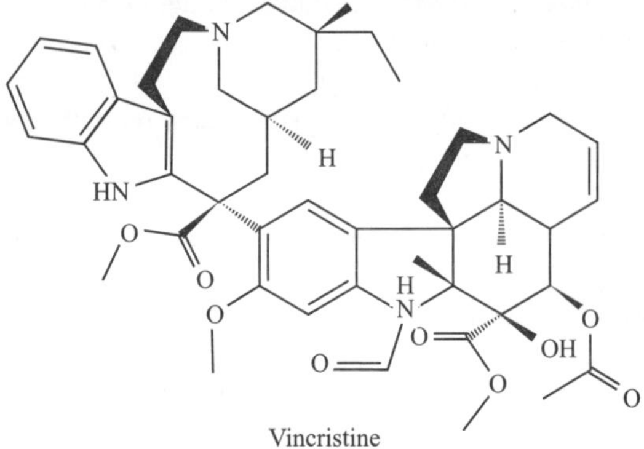

Figure 125.9   *Catharanthus roseus* plant and structure of vincristine.

In Ayurveda the extract of its roots and shoots is used against several diseases. In traditional Chinese medicine, its extracts have been used against many diseases, including diabetes, malaria, and Hodgkin's disease. The substances vinblastine and vincristine extracted from the plant are used in the treatment of leukemia.

7. *Bacopa monnieri* (Waterhyssop, Brahmi, Thyme-leafed gratiola, Water hyssop): It is a perennial, creeping herb whose habitat includes wetlands and muddy shores.

Figure 125.10   *Bacopa monnieri* plant and structure of bacoside A, one of its active compounds.

Bacopa monnieri has many chemical constituents including alkaloids (brahmine and herpestine), saponins (d-mannitol and hersaponin acid A, and monnierin), flavonoids (luteolin and apigenin). It also contains significant amounts of betulic acid, stigmasterol, beta-sitosterol, and bacopasaponins (bacosides A, bacosides B, bacopaside II, bacopaside I, bacopaside X, bacopasaponin C, bacopaside N2). The minor components include bacopasaponin F, bacopasaponin E, bacopaside N1, bacopaside III, bacopaside IV, and bacopaside V).

In rats, bacosides A enhance antioxidant defenses, increasing superoxide dismutase (SOD), catalase (CAT) and glutathione peroxidase (GPX) activity. Laboratory studies on rats indicate that extracts of the plant improve memory capacity. Recent studies suggest Bacopa significantly improves cognitive ability including an accelerated rate of learning and enhanced memory. These effects are optimized after a 12 week period. The sulfhydryl and polyphenol components of *Bacopa monnieri* extract have also been shown to impact the oxidative stress cascade by scavenging reactive oxygen species, inhibiting lipoxygenase activity and reducing divalent metals.[13] This mechanism of action may explain the effect of *Bacopa monnieri* extract in reducing beta amyloid deposits in mice with Alzheimer's disease. *B. monnieri* has a demonstrated ability to reverse diazepam-induced amnesia in the Morris water maze test. The mechanism of action is unknown. In some trials, bacopacide extract did not restore or enhance memory, but improved retention. In others including a randomized clinical trial of 98 healthy older people (over 55 years) *Bacopa* significantly improved memory and retention.

Brahmi, can enhance immune function by increasing immunoglobulin production. It may regulate antibody production by augmenting both Th1 and Th2 cytokine production. *Bacopa* may also increase the effects of calcium channel blocker antihypertensive drugs and their doses may need to be reduced if using *Bacopa* as a supplement. It may also cause a lower heart rate, and increase secretions in the stomach, intestines, and urinary tract. The increase in secretions may irritate ulcers and urinary tract obstructions.

8. Artemisia annua (also known as Sweet Wormwood, Sweet Annie, Sweet Sagewort or Annual Wormwood; Chinese: *qīnghāo*): It is a common type of wormwood that is native to temperate Asia, but naturalized throughout the world.

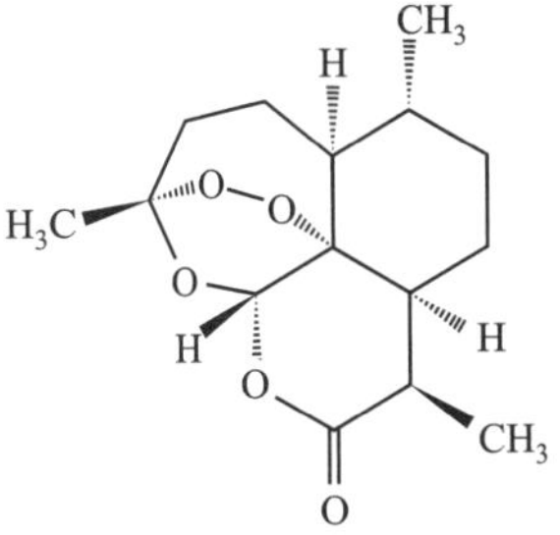

Figure 125.11   *Artemisia annua* plant and Artemisinin, the anti-malarial drug obtained from it.

Artemisinin itself is a sesquiterpene lactone with an endoperoxide bridge and has been produced semi-synthetically as an antimalarial drug. The efficacy of tea made from *A. annua* in the treatment of malaria is contentious. According to some authors, artemesinin is not soluble in water and the concentrations in these infusions are considered insufficient to treat malaria. Other researchers have claimed that *Artemisia annua* contains a cocktail of *anti-malarial substances*, and insist that clinical trials be conducted to demonstrate scientifically that artemisia tea is effective in treating malaria. This simpler use may be a cheaper alternative to commercial pharmaceuticals, and may enable health dispensaries in the tropics to be more self-reliant in their malaria treatment. In 2004, the Ethiopian Ministry of Health changed Ethiopia's first line anti-malarial drug from Fansidar, a sulfadoxine agent which has an average 36% treatment failure rate, to Coartem, a drug therapy containing artemesinin which is 100% effective when used correctly, despite a worldwide shortage at the time of the needed derivative from *A. annua*.

VI. COMMERCIAL PREPARATIONS OF SOME HERBAL DRUGS

Name of herb	Company name	Drug Name	Product type
Curcuma longa (Turmeric)	Thai Herbal Products Co., Ltd.	Curmin Capsule	Capsule
	Himalaya Herbal Health care	Diabecon	Capsule
		Geriforte	Tablet, syrup
		Ophthcare	Eyedrop
		Purim	Tablet
		V-gel	Gel
		Talket	Syrup, capsule
		Foot Care Cream	Cream
		Geriforte Vet	Liquid, powder
		Scavon	Powderspray
		Bresol	Tablet, syrup
	Physician Naturals	Ultra Pure Curcumin Powder C3 Complex	Powder
		Super Bio-Enteric Curcumin w/Bioperine Enteric Coated	Tablet
	AOR CANADA	CurcuVIVA	Capsule
Silybum marianum (blessed milk thistle)	JSTS Pharmaceutical Co., Ltd	Milk Thistle Extract	
	Fouda Pharmacy (Egypt)	alpha hepadox 20c 2st.	Capsule
		amino ts 20c 2st.	Caspsule
		bee s 20c 2st.	Capsule
		c mune 20c 2st.	Capsule
		Cura 20c 2st	Capsule
		Hepanox	Capsule
	WholeHealth	Myristin CMO Kit	Capsule
		Complete Milk Thistle Extract	Capsule
	APP Chem	Milk Thistle Extract	Powder
Phyllanthus emblica (Amla)	Bioprex labs (pune)	Amla Extract	Dried Powder
	Himalaya Herbal Health care	Abana	Tablet
		Bonnisan	Drops, liquid
		Mentat	Syrup
		Pilex	Tablet, ointment
		Septilin	Drop, syrup
		Styplon	Tablet
		Revitalizing hair oil	Oil
		Digyton	Tablet
		Styplon Vet	Bolus
		Amalaki	Capsule
		Chyavanprasha	
	Majoria Drugs	Amalaki	Capsule

CONCLUDING REMARKS

India has a treasure of medicinal plants and herbs. Many of these have strong medicinal properties against various diseases. A large proportion of Indian population especially the rural and tribal peoples depend on herbal medicines for their health needs. Many of these herbs are part of our daily food. Minor adjustment in their quantity, addition or deletion of some food ingredients can metamorphose these herbs into medicine. Traditionally these practices have been prevalent in every household in all the countries for centuries. The traditional Indian system of medicine, the Ayurveda, uses plants and plant products as therapeutic agents. This is an ancient system that has survived for centuries. However, during last two centuries, after the advent of purified chemicals as medicine, large proportion of population started using allopathic system of medicine and slowly and slowly, the herbal drugs have been pushed to back burner. One of the reasons for this has been the lack of systematic scientific research for the effectiveness of these drugs and variation in various formulations as the standards are missing. Further, the concentration of secondary metabolites can vary in plants, sometimes very significantly depending on the place of cultivation, its climate, season, time of harvesting and other growth conditions.

A good proportion of allopathic medicines have their origin in plant products. Once the structure of the compound was elucidated, it was synthesized and synthetic molecule is being used today. A number of molecules have very complex structure (artemisinin, a WHO recommended anti-malarial drug, for example) and are not easy to synthesized chemically in a cost effective manner. Such drugs are produced by and isolated from plants even today. Of course modern biotechnological techniques are being used to enhance their concentration.

Due to emergence of a large number of MDR strains of pathogens, failure of pure chemical based drugs in managing many of the chronic and life style derived diseases and relatively higher side effects of many pure chemicals has revived the interest in traditional systems of medicines. A number of countries (both developing and some developed) are reviving their traditional medicare systems. The growing global trend is to go for an integrated medical system where synergism is made between the modern medicine and traditional medicines (almost all of these are based on natural products, mostly plant derived).

In India a number of traditional systems are existing. These include Ayurveda, Unani, Siddha, Homeopathy, Yoga etc., besides the Chinese medicines and Tibetan medicines. After independence, special attention has been given to revive and standardize these medical systems. A number of Institutes like CDRI, CIMAP, IIIM etc have carried out substantial research on various aspects of herbal medicines. Efforts have been made to screen the plants, identify their active principles and develop standardized formulations after carrying out safety studies. Systematic studies have removed many of the misconceptions about herbal drugs. As a result, the acceptance of herbal medicines amongst the urban population has increased. The scenario of health care may change in this century and the interaction between a physician and a Vaid or Hakim will increase and for many diseases an integrated regimen of drugs may be used. It is therefore, necessary that a modern health specialist should also have the basic knowledge of herbs, plant products and food additives.

SUGGESTIONS FOR FURTHER READING

1. Srividya A.R., Navangul M.V. and Vishnuvarthan V.J., *Anti-Hepatotoxic Potential from Medicinal Plants: Hepatoprotective Activity*, Lambert Academic Publisher.

2. Patel V.A., Joshi D.Y. and Soni G., *Herbal Plant Use for Screening of Hepatotoxicity in Wistar Rats: Anti-Hepatotoxicity,* Lambert Academic Publisher.

3. Sherlock S. and Dooley J., *Diseases of the Liver and Biliary System*, 11th ed., John Wiley & Sons.

4. Ved D.K. and Krishnan R., Tropical Indian Medicinal Plants, Propagation Methods S. Oommen, *Buisness Horizans*, Pharmaceutical Publisher, New Delhi.

5. Premila M.S., *Ayurvedic Herbs: A Clinical Guide to The Healing Plants of Traditional Indian Medicine*, Routledge, Talyor and Francis Group.

6. Warrier P.K., Nambiar V.P.K. and Ramankutty C., *Indian Medicinal Plants: A Compendium of 500 Species*, Vol. 5, Orient Longmann, Pvt. Ltd.

7. Earnst E., *Herbal Medicine: A Concise Overview for Professionals*, Butterworth-Heinemann, Jordan Hill, Oxford.

8. Buhner S.H., *Herbal Antibiotics*, Edwards Brothers Malloy, USA

9. Sivarajan V.V. and Balachandran I., *Ayurvedic Drugs and their Plant Sources*, Oxford and IBH Publishing Company.

10. Arora R., *Medicinal Plant Biotechnology*, CABI, UK.

11. Prajapati N.D., *A Handbook of Medicinal Plants: A Complete Source Book*, Agrobios, India.

12. Daniel M., *Medicinal Plants: Chemistry and Properties*, Science Publisher, New Hampshre, USA.

13. Sanyal D. and Singh Bishen and Singh Mahendrapal, *Vegetable Drugs of India*, Publisher.

126

Antiinflammatory Lifestyle and Spices
How are They Linked?

Bharat B. Aggarwal, Sahdeo Prasad, Bokyung Sung, and Subash C. Gupta

CONTENTS

I. INTRODUCTION

Inflammation is the first response of the immune system to tissue injury, infection, or irritation and is characterized by increased blood flow to the tissue causing increased temperature (calor), redness (rubor), swelling (tumor), pain (dolor), and loss of function (functio laesa) of the organ involved. The first four classical signs were described by the Roman physician Cornelius Celsus (ca. 30 BC–38 AD), whereas loss of function was added by Galen; the origination of the fifth sign, however, has also been ascribed to Thomas Sydenham and Rudolph Virchow (1870). In fact, the word *inflammation*, the root cause of most chronic diseases, itself is derived from the Latin word *inflamacio*, meaning "to set a fire." Pathologically, inflammation is represented by adding suffix "-itis" such as in *bronchitis*, *esophagitis*, *gastritis*, *colitis*, *pancreatitis*, *prostitis*, *cervicitis*, and *hepatitis*. However, some conditions such as asthma and pneumonia do not follow this convention. Inflammation can be divided into acute or short-term and chronic or long-term. Short-term or acute inflammation is a protective attempt by the organism to remove injurious stimuli and to initiate the healing process. Long-term or chronic inflammation, however, leads to several inflammatory disorders and chronic diseases.

II. INFLAMMATION LINKED TO CHRONIC DISEASE

In the 19th century, the German physician Rudolf Virchow first suggested a link between inflammation and chronic disease. Chronic inflammation damages cells of the brain, heart, arterial walls, and other body structures, leading to various inflammatory diseases such as cancer, heart diseases, diabetes, pulmonary disease, and neurological disease (Figure 126.1).

Usually, inflammation arises from everyday lifestyle factors including dietary agents (such as fried foods, meat, and high

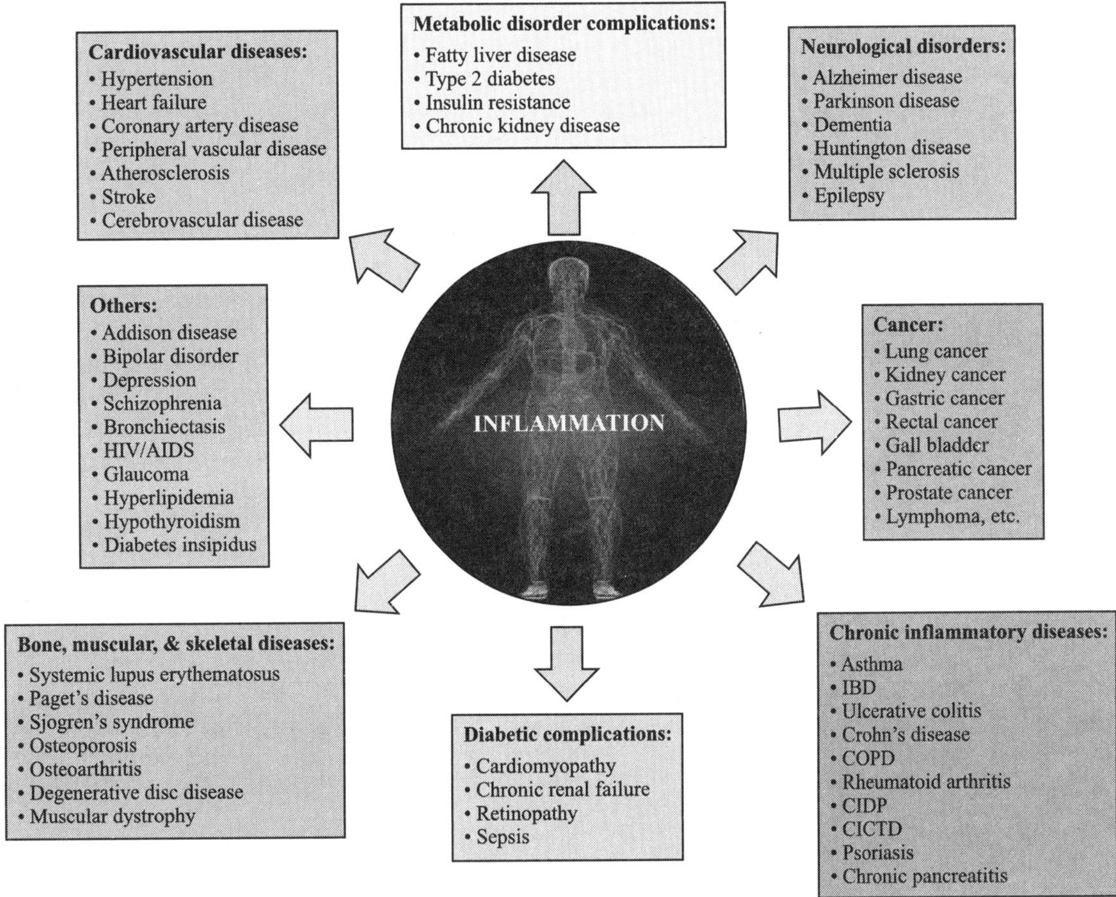

AIDS, acquired immunodeficiency syndrome; CICTD, chronic Inflammatory connective tissue diseases; CIDP, chronic inflammatory demyelinatingpolyneuropathy; COPD, chronic obstructive pulmonary disease; HIV, human immunodeficiency virus; IBD, inflammatory bowel disease

Figure 126.1   Inflammation linked to the several chronic diseases.

fat), tobacco, UV radiation, alcohol, environmental pollutants (such as diesel, CO, and heavy metals), stress (psychological, chemical, physical, and mechanical), obesity, and infectious agents (such as bacteria and viruses). These inflammatory factors at the molecular level induce the expression of inflammatory transcription factors (such as nuclear factor-kappaB [NF-κB] and signal transducer and activator of transcription [STAT] 3), inflammatory enzymes (such as cyclooxygenase [COX]-2, matrix metalloproteinase [MMP]-9, 5-lipoxygenase [LOX], and Phospholipase A2 [PLA2]), inflammatory cytokines (such as tumor necrosis factor [TNF], interleukin [IL] 1, IL-6, IL-8), and chemokines. When these inflammatory factors persist, chronic inflammation and further chronic diseases can develop.

Cancer is one of the major diseases caused by chronic inflammation. Recognition of the functional relationship between inflammation and cancer is not new. In 1863, Virchow hypothesized that the origin of cancer was at sites of chronic inflammation, based in part on his hypothesis that some classes of irritants, together with the tissue injury and ensuing inflammation these irritants cause, enhance cell proliferation. Today, the causal relationship between inflammation and cancer is more widely accepted. Various proinflammatory biomarkers have been found to be elevated in several cancer types. Among them, constitutively active transcription factor NF-κB has been identified in the tissue of patients in most cancers including leukemia and lymphoma and cancers of the prostate, breast, oral cavity, liver, pancreas, colon, and ovaries. Activation of NF-κB has been linked with metastasis of prostate cancer, high recurrence and poor survival in squamous cell carcinoma, oral tumor progression, chemoradiation resistance, progression of lung cancer, reduced survival in ovarian cancer, and aggressiveness of breast cancer. Another transcription factor, STAT3, was found to be activated in 82% of patients with late-stage prostate cancer, and a higher level of STAT3 activation was correlated with more severe disease and shorter patient survival times. Altered levels of cytokines and chemokines may also play a role in cancer- and cancer treatment-related cognitive difficulties. In patients with benign prostate hyperplasia, the levels of cytokines IL-1α, IL-6, and TNF-α were elevated with increased prostate-specific antigen serum levels.

Inflammation is typically present in patients with heart disease and stroke and is believed to be a sign of atherogenic response. Other than low-density lipid, C-reactive protein

(CRP), a hepatic acute phase reactant found in blood, is a marker for inflammation in the body. High levels of this protein are associated with increased risk of heart disease. It has been reported that in human aortic endothelial cells, CRP induces IL-8 synthesis and secretion via upregulation of NF-κB activity. Also in angina patients, through the activation of NF-κB, CRP amplifies and perpetuates the inflammatory component of acute coronary syndromes and influences clinical outcome.

Diabetes is a proinflammatory state with increased levels of circulating cytokines, suggesting a causal role for inflammation in its etiology. Inflammation has been shown to be provoked by immune cells called macrophages, which leads to insulin resistance and then to type 2 diabetes. The levels of inflammatory biomarkers, including CRP and IL-6, have been shown to be significantly higher in patients with diabetes. IL-1 has been shown to impair insulin secretion and to induce β-cell apoptosis. Abnormal expression of NF-κB has also been shown to be involved in the development of diabetic microvascular and neuropathic complications, further indicating an association between inflammation and diabetes.

Another inflammatory chronic disease is multiple sclerosis, an autoimmune disease that affects the brain and spinal cord. In multiple sclerosis, inflammation damages the myelin sheath, the protective covering that surrounds nerve cells, slowing or stopping nerve signals. As more and more nerves are affected, a person experiences progressive interference with functions controlled by the nervous system such as vision, speech, walking, writing, and memory. In patients with multiple sclerosis, levels of COX-2–derived prostaglandins and of activated NF-κB are elevated in the cerebrospinal fluid. Levels of cytokines, such as IL-1α, IL-2, IL-4, IL-6, IL-10, interferon (IFN)-γ, transforming growth factor (TGF)-β1 and 2, and TNF-α, have been also found to be elevated in the frozen sections of central nervous system tissue from multiple sclerosis patients. In addition, increased levels of phosphorylated STAT3 have been reported in T cells of multiple sclerosis patients.

Crohn's disease or inflammatory bowel disease is another chronic inflammatory disease. It causes inflammation of the lining of the digestive tract, which can lead to abdominal pain, severe diarrhea, and even malnutrition. It usually affects the intestines but may occur anywhere from the mouth to the end of the rectum. Crohn's disease appears to be primarily an overactive response of Th1 cells that leads to overproduction of various cytokines, such as IL-2, IL-12, IL-18, interferon-γ, and TNF-α. Levels of activated NF-κB have also been higher in lamina propria biopsy specimens from patients with Crohn's disease than in specimens from patients with ulcerative colitis or from controls. It has been reported that in cells from a smooth muscle cell line isolated from the ileum of a patient with Crohn's disease, IL-8 transcription was associated with remarkable spontaneous activation of NF-κB, indicating that NF-κB is involved in the regulation of the inflammatory response in Crohn's disease. Activated STAT3 was also reported in mucosal biopsy specimens and in isolated lamina propria mononuclear cells from inflamed tissues from patients with Crohn's disease.

Rheumatoid arthritis, an autoimmune disorder, is also associated with chronic inflammation. Excess levels of cytokines such as CRP, TNF-α, IL-6, IL-1β, and IL-8 have been shown to cause or contribute to this inflammatory syndrome. Activated STAT3 has been found in patients with rheumatoid arthritis and is known to be a key mediator of both chronic inflammation and joint destruction. Activated STAT3 either directly or indirectly induces expression of the IL-6 family of cytokines, which has led to further activation of STAT3 in murine osteoblastic and fibroblastic cells. High induction of NF-κB in the synovial membranes of rheumatoid arthritis patients, which may regulate the production of cytokines at the site of synovial inflammation, has also been reported. These studies have shown that inflammatory transcription factors and cytokines are strongly associated with several chronic diseases including rheumatoid arthritis.

III. MODULATION OF INFLAMMATORY PATHWAYS BY SPICES AND SPICE-DERIVED NUTRACEUTICALS

Chronic diseases, such as heart disease, stroke, cancer, chronic respiratory diseases, and diabetes, are by far the leading cause of mortality in the world and are responsible for 60% of all deaths. The risk factors for most of the chronic diseases are common and modifiable, indicating that these diseases can be prevented. An estimated 80% of premature heart disease, stroke, and type 2 diabetes, and 40% of cancer could be avoided through lifestyle changes that include eating healthy foods, getting regular physical activity, and avoiding the use of tobacco. Although numerous factors to avoid in the prevention of chronic diseases are known, diet plays a major role that could delay or prevent the incidence effectively. For example, relatively low intake of fruits and vegetables is a risk factor for many of the most important chronic diseases, whereas greater consumption of natural products—including spices, nuts, whole-grain cereals, legumes, fruits, and vegetables (Figure 126.2)—is associated with a lower risk of many diseases.

Among the dietary agents, spices have their own importance in the prevention and treatment of chronic diseases. Spices are used all over the world to add flavor, taste, and nutritional value to food. A spice could be a dried seed, fruit, root, bark, or flower of a plant. The use of spices has shaped a large part of the world's history. For example, the ancient Egyptians pioneered maritime trade to fetch the incense of Arabia; Greco-Roman navigators found their way to India for pepper and ginger; Columbus sailed west for spices; Vasco de Gama sailed east for them; and Magellan sailed across the Pacific Ocean on the same quest. Despite globalization, persons in Asian countries are still the largest consumers of spices. In ancient times, many spices were used as medicines for treating several diseases such as rheumatism, body ache, intestinal worms, diarrhea, intermittent fevers, hepatic diseases, urinary discharges, dyspepsia, inflammation, constipation, and dental diseases. The U.S. Food and Drug Administration has stated that a spice is an "aromatic

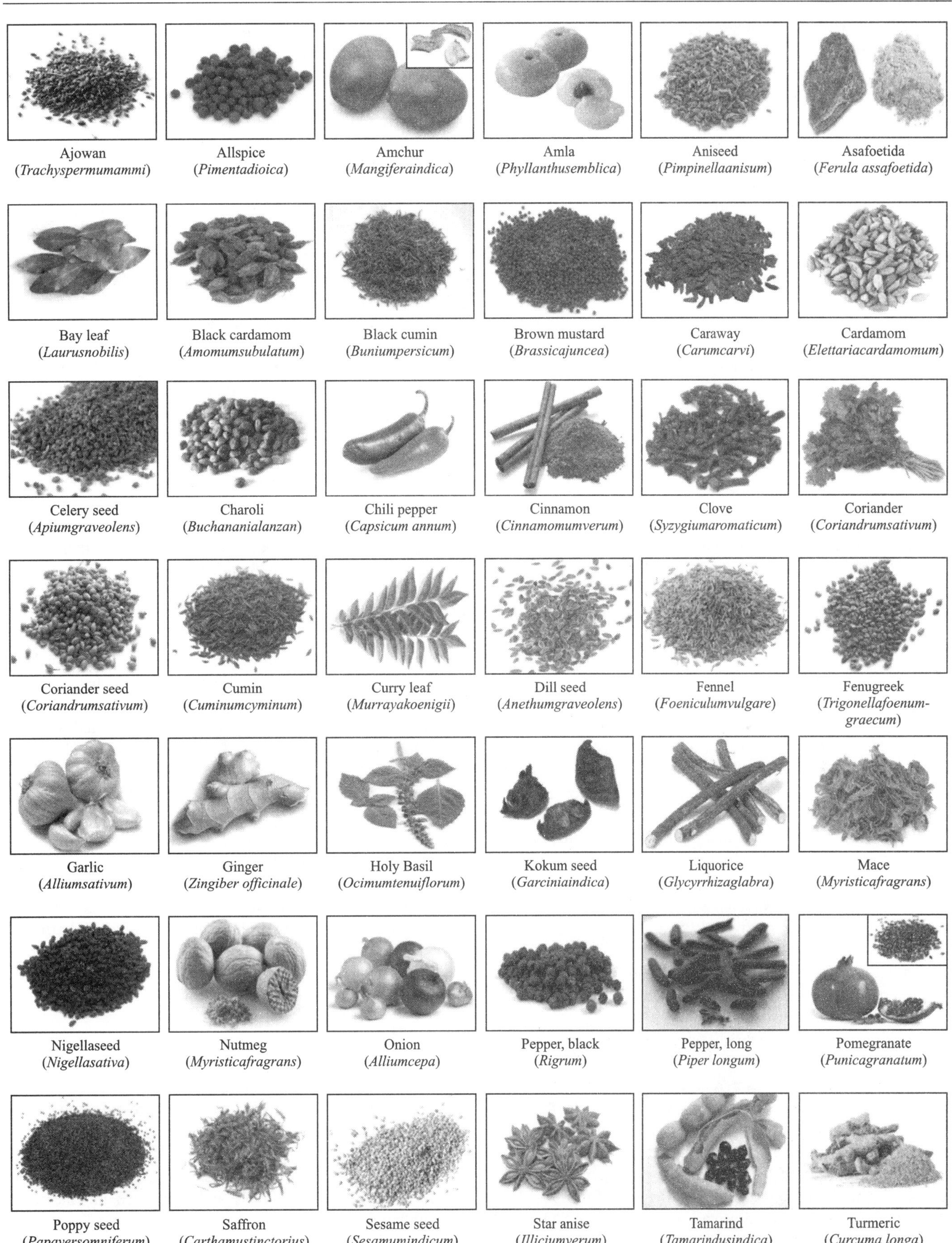

Figure 126.2 Commonly used spices, which exhibit anti-inflammatory properties.

vegetable substance in the whole, broken, or ground form, the significant function of which in food is seasoning rather than nutrition" and from which "no portion of any volatile oil or other flavoring principle has been removed."

A number of spices and spice components (Figure 126.3), such as curcumin (which is found in turmeric), diallyl sulfide (garlic), thymoquinone (black cumin), capsaicin (red chili), gingerol (ginger), anethole (licorice), diosgenin (fenugreek), and eugenol (clove, cinnamon) possess therapeutic and preventive potential against several chronic diseases. Besides these phytochemicals ellagic acid (clove), ferulic acid (fennel, mustard, sesame), apigenin (coriander, parsley), betulinic acid (rosemary), kaempferol (clove, fenugreek), sesamin (sesame), piperine (pepper), limonene (rosemary), and garcinol (kokum), have shown potential to act against chronic diseases, we will focus on some of these major spices and their derivatives. These spice-derived nutraceuticals modulate numerous inflammatory biomarkers, which include inflammatory transcription factors, inflammatory enzymes, kinases, cytokines, and chemokines (Figure 126.4).

A. Curcumin

Curcumin, a bioactive component of turmeric, has been shown to play an important role in the prevention and treatment of various proinflammatory chronic diseases including neurodegenerative, cardiovascular, pulmonary, metabolic, autoimmune, and malignant diseases. The activity of curcumin against chronic diseases was associated with inhibition of inflammatory biomarkers. Numerous in vitro studies have reported that curcumin has the ability to inhibit proinflammatory transcription factors NF-κB and STAT3. It inhibits NF-κB activation in several types of cells including breast cancer cells, gastric cancer cells, pancreatic cancer cells, renal epithelial cells, biliary cancer cells, and other cells. STAT3 has also been shown to be inhibited by curcumin in different cell lines including Hodgkin lymphoma cells, biliary cancer cells, cutaneous T-cell lymphoma cells, ovarian and endometrial cancer cells pancreatic cancer cells, multiple myeloma. Studies have revealed that a low dose of curcumin inhibited the lipopolysaccharide (LPS)-induced production of TNF-α and IL-I by a human monocytic macrophage cell line. In addition, curcumin significantly inhibited the release of nitric oxide (NO), prostaglandin (PGE)2, and proinflammatory cytokines in LPS-stimulated BV2 microglia.

In inflammation-related diabetes, curcumin suppresses the proinflammatory transcription factors NF-κB, STAT3, and Wnt/beta-catenin, and it activates peroxisome proliferator-activated receptor-gamma (PPARγ) and Nrf2 cell-signaling pathways, thus leading to the downregulation of adipokines, including TNF, IL-6, resistin, leptin, and monocyte chemotactic protein-1, and the upregulation of adiponectin and other gene products. Curcumin suppresses proinflammatory cytokines IL-1β, IL-6, monocyte chemotactic protein-1 (MCP-1), and MMP-9 and transcription factors activator protein (AP)-1 and NF-κB in aortic tissue. In a mice model, curcumin also decreased

the myocardial mRNA levels of IL-6, MCP-1, and TNF-α. Curcumin inhibits experimental allergic encephalomyelitis by blocking IL-12 through STAT3 signaling in T cells, suggesting it would be effective in the treatment of various chronic diseases including multiple sclerosis.

B. Diallyl sulfide

Allium vegetables have been shown to have beneficial health effects against several chronic diseases. Diallyl sulfide (DAS), an organosulfur compound present in garlic, is well known for its anti-inflammatory properties in several experimental models. In primary synovial cells and chondrocytes from osteoarthritic patients, DAS inhibited inflammatory molecules PGE2 and COX-2 expression. Activated NF-κB was also blocked by DAS. Garlic oil and its component diallyl disulfide or diallyl trisulfide suppressed endotoxin-induced ulceration and apoptosis in the intestinal mucosa through the inhibition of inducible nitric oxide synthase (iNOS) activity in the rat model. In an animal model of angiogenesis, DAS treatment significantly reduced the production of proinflammatory cytokines such as IL-1β, IL-6, TNF-α, and GM-CSF in the serum. DAS has also been reported to inhibit cytokines such as IL-6, TNF-α and CRP in diabetic mice. These studies indicates that the organosulfur compound DAS has anti-inflammatory properties.

C. Thymoquinone

Thymoquinone (TQ) is an active ingredient of the seed oil extract of *Nigella sativa* Linn (black cumin). TQ has a variety of beneficial properties including anti-inflammatory activities. The anti-inflammatory activity of TQ was associated with the inhibition of both COX-2 and LOX pathways of arachidonate metabolism . The use of TQ has also been shown to have anti-inflammatory effects in several inflammatory diseases, including experimental allergic encephalomyelitis (EAE), colitis, arthritis, and cancer. In pancreatic cancer cells, TQ dose- and time-dependently reduced synthesis of MCP-1, TNF-α, IL-1β, and COX-2. It also inhibited the constitutive and TNF-α–mediated activation of NF-κB in pancreatic cancer cells. TQ exhibited anti-inflammatory activity on both human rheumatoid arthritis fibroblast-like synoviocytes and a rat adjuvant-induced arthritis model of rheumatoid arthritis.

D. Capsaicin

Capsaicin is a spicy component of red chili and has been shown to improve inflammatory disease. One study showed that capsaicin transcriptionally COX-2 expression and PGE2 production. Capsaicin also inhibits LPS-elicited NF-κB and AP-1 activation and IFN-γ–elicited STAT1 activation in macrophages. In septic rats, capsaicin treatment increased anti-inflammatory IL-10 levels and attenuated the increases in proinflammatory cytokines (TNF-α, IL-6), NO$_x$, and tissue malondialdehyde (MDA), indicating that it is a potent anti-inflammatory agent. Capsaicin also inhibited the expression

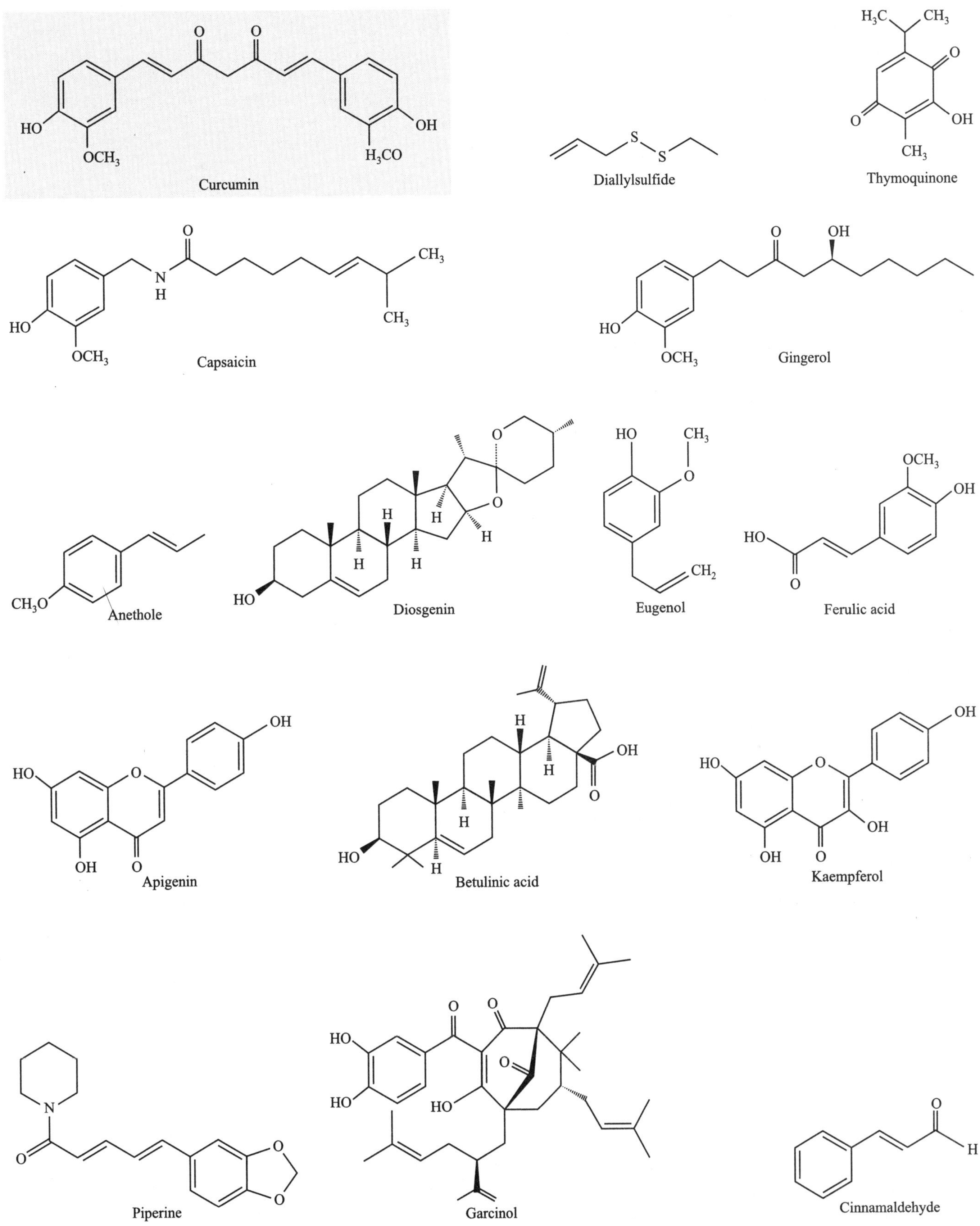

Figure 126.3 Chemical structure of bioactive compounds present in anti-inflammatory spices.

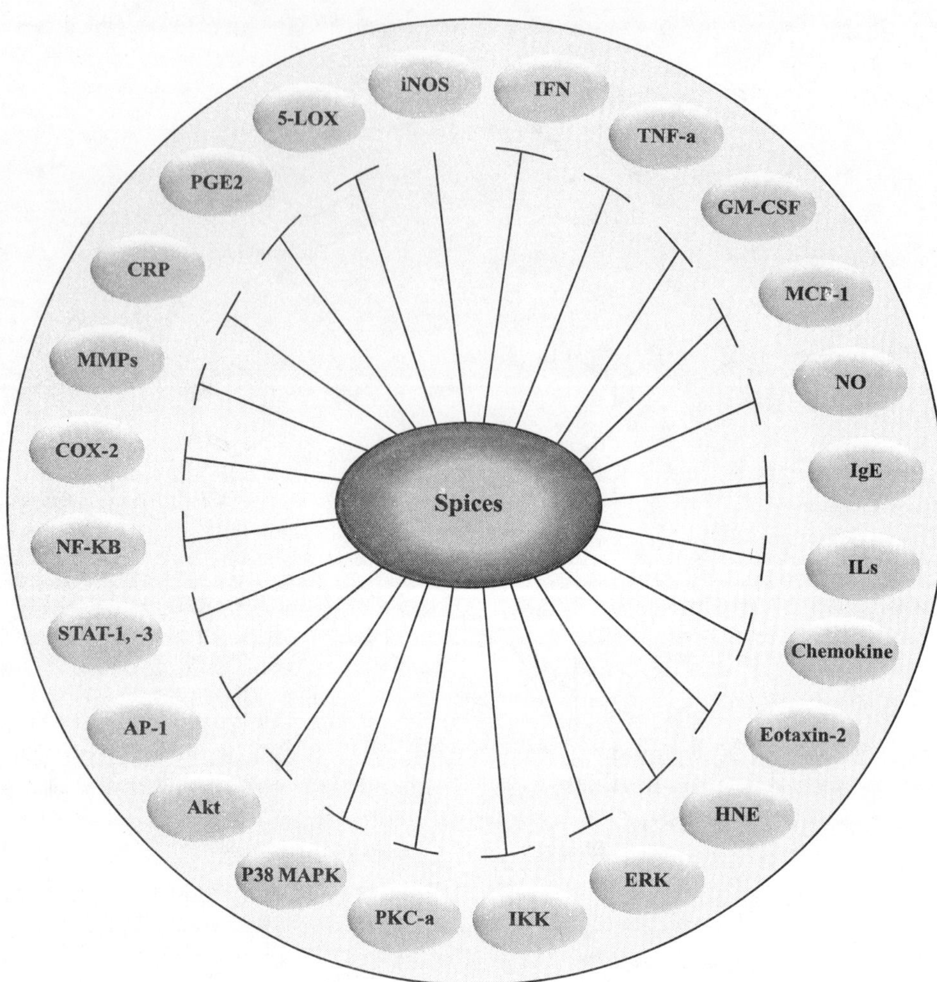

Figure 126.4 Molecular targets of anti-inflammatory spices.

of IL-6 and MCP-1 mRNAs and protein release from the adipose tissues and adipocytes of obese mice, whereas it enhanced the expression of the adiponectin gene and protein. The action of capsaicin was found to be associated with NF-κB inactivation and/or PPARγ activation. This study showed that capsaicin attenuates obesity-induced inflammation and related complications.

E. Gingerol

Gingerol is the pungent phenolic substance of ginger, which contains pronounced antioxidative and anti-inflammatory activities. Ginger itself in aqueous extract exerted anti-inflammatory properties by suppressing eosinophilia in a mouse model of Th2-mediated pulmonary inflammation. It has been also reported that crude ginger extract has the ability to inhibit joint swelling in an animal model of rheumatoid arthritis and streptococcal cell wall–induced arthritis. Its active component, 6-gingerol, has been shown to suppress inflammation in both in vitro and in vivo models. In a skin carcinogenesis model of mice, 6-gingerol inhibited inflammation by inhibiting epidermal ornithine decarboxylase activity. In vitro studies showed

that 6-gingerol inhibits the production of proinflammatory cytokines from LPS-stimulated murine peritoneal macrophages. It has also been reported that 6-gingerol exhibits an anti-inflammatory effect in macrophages by decreasing iNOS and TNF-α expression mediated through suppression of IkBα phosphorylation, NF-κB nuclear activation, and protein kinase C (PKC)-α translocation in LPS-stimulated macrophages.

F. Anethole

Anethole, a chief constituent of anise, camphor, and fennel, has been shown to block both inflammation and carcinogenesis. It has been reported that anethole-induced inhibition of inflammation is through suppression of TNF signaling. It inhibited TNF-induced NF-κB activation in cancer cells. Anethole also blocked the NF-κB activation induced by a variety of other inflammatory agents. Anetholdithiolthione, a derivative of anethole, also inhibited NF-κB activation in response to a variety of stimuli in human Jurkat T cells. These studies indicate that anetholerepresent anti-inflammatory activities by inhibiting inflammatory cytokines and transcription factors.

G. Diosgenin

Diosgenin is a sapogenin that is present in *Trigonella foenum-graecum* (fenugreek) plant. It has shown to exhibit dose-dependent attenuation of subacute intestinal inflammation induced by indomethacin in rats. Diosgenin has also been shown to inhibit intestinal inflammation, including the occurrence of diarrhea, the infiltration and degranulation of mast cells, and the presence of mucin-containing goblet cells in the duodenum. The inflammatory factor COX-2 has found to be inhibited by diosgenin, which leads to suppression of proliferation, cell cycle arrest, and induction of apoptosis in the human osteosarcoma cell line. Diosgenin also suppressed TNF-induced NF-κB activation through inhibition of Akt activation. In addition, diosgenin inhibited another inflammatory transcription factor STAT3 at both constitutive and inducible level with no effect on STAT5. Diosgenin has been reported to inhibit inflammation induced in obesity. A study has shown that diosgenin inhibited inflammatory cytokines TNF-α, monocyte chemoattractant protein-1, and nitric oxide in 3T3-L1 adipocytes when cocultured with RAW 264 macrophages. Furthermore, diosgenin has been shown to inhibit production of ROS, IL-1, IL-6, and TNF-α. Inhibition of these inflammatory mediators appeared to be at the transcriptional level. In an animal model of diabetes, fenugreek ameliorated diabetes by promoting adipocyte differentiation and inhibiting inflammation in adipose tissues. Fenugreek (2%) inhibited macrophage infiltration into adipose tissues and decreased the mRNA expression levels of inflammatory genes. These studies have suggested that diosgenin can ameliorate chronic diseases by improving inflammation.

H. Eugenol

Eugenol is an aromatic molecule found in several plants including clove and cinnamon that is widely used in dentistry for analgesic and antiseptic purposes. Despite the frequent applications of eugenol in the practice of dentistry, its anti-inflammatory activities are being investigated. The methanolic extract of the cortex of *Eugenia caryophyllata* Thunberg (clove) was found to potently inhibit PGE2 and COX-2 production in LPS-activated mouse macrophage RAW264.7 cells. Since NF-κB is a key transcriptional factor in the expression of inflammatory cytokines, bis-eugenol inhibited NF-κB activation in RAW264.7 murine macrophages. Beside these, Eugenol and isoeugenol inhibited LPS-dependent production of NO, which was due to the inhibition of protein synthesis of iNOS, IL-1β, and TNF-α.

Eugenol also inhibited the proliferation of colon cancer HT-29 cells and inflammation by inhibiting the mRNA expression of COX-2, but not COX-1. In animals, eugenol suppressed markers of tumor promotion and inflammation. It inhibited activities of ornithine Decarboxylase (ODC) and iNOS and COX-2 expression, and levels of proinflammatory cytokines. Eugenol also inhibited the upstream signaling molecule NF-κB, which regulates the expression of these genes. Eugenol

has shown to inhibit 5-LOX, which is the key enzyme in the biosynthetic pathway of leukotrienes. Furthermore, eugenol was found to significantly inhibit the formation of leukotrienes C(4) in polymorphonuclear leukocytes. Since leukotrienes are implicated in the pathophysiology of allergic and inflammatory disorders such as asthma, allergic rhinitis, arthritis, inflammatory bowel disease, and psoriasis, their inhibition by eugenol will play important role in the management of these chronic diseases.

I. Ferulic Acid

Ferulic acid is one of the components of asafoetida, the dried latex from the giant fennel (*Ferula communis*) and sesame (*Sesamum indicum*). It has several medicinal benefits including antioxidant and anti-inflammatory activities. The anti-inflammatory activities of ferulic acid were associated with suppression of inflammatory cytokines and transcription factors. It significantly inhibited the release of proinflammatory factors such as TNF-α and NO and suppressed proliferation of spleen cells. In a rat model, ferulic acid exerts a protective effect against nicotine toxicity by modulating inflammation and oxidative stress. It inhibited expression of COX-2 and NF-κB in lung and liver tissues of nicotine-treated rats. Ferulic acid may play an important role in improving complications of chronic diseases. It has been shown that in chondrocytes, ferulic acid decreased hydrogen peroxide–induced IL-1β, TNF-α, MMP-1, and MMP-13 gene expression, which indicates its importance in the treatment of osteoarthritis. Another study also suggested its protective and therapeutic efficacy in diabetic nephropathy by reducing oxidative stress and inflammation. Ferulic acid treatment in diabetic rats decreased oxidative stress markers and MCP-1 levels in the rats' urine and in supernatants of cultured podocytes.

J. Apigenin

Apigenin, a flavonoid abundant in fruits, vegetables, and spices, including coriander and parsley, exhibits antiproliferative and anti-inflammatory activities. Apigenin blocks expression of COX-2 and iNOS in LPS-activated macrophages. The inhibition of TNF-α and IL-1β by apigenin was also observed in LPS-activated macrophages, which again indicates its anti-inflammatory properties. Apigenin also blocked the LPS-induced activation of inflammatory transcription factor NF-κB. Another transcription factor, STAT1, was also inhibited by apigenin. Apigenin suppressed the expression of TNF-α, IL-8, IL-6, GM-CSF, and COX-2 by decreasing the intracellular Ca(2+) level and inhibiting NF-κB activation in the human mast HMC-1 cell line. Apigenin has exhibited anti-inflammatory properties not only in an *in vitro* model but also in animals, specifically, in a murine asthma model. In this model, apigenin attenuated allergen-induced airway inflammation by decreasing the degree of inflammatory cell infiltration, airway hyperresponsiveness, and total immunoglobulin E levels. Apigenin downregulated COX-2 expression in lupus T cells, B cells, and antigen-presenting cells, and its use initiated apoptosis in these cells in animal model of lupus disease. Thus, it is effective in downmodulating several chronic diseases associated with inflammation.

K. Betulinic Acid

Betulinic acid is a naturally occurring pentacyclic triterpenoid found in the bark of several species of plants including rosemary. The anti-inflammatory activities of betulinic acid have been shown in both in vitro and in vivo studies. In carrageenin- and serotonin-induced rat paw edema, betulinic acid showed significant anti-inflammatory activity. The anti-inflammatory activity of betulinic acid was also associated with inhibition of NF-κB and its regulated gene products including COX-2. Further study on human bone marrow cells showed that betulinic acid inhibited LPS-induced COX-2 protein expression and PGE(2) production and attenuated LPS-induced ERK and Akt phosphorylation, which contributes to its anti-inflammatory efficacy. It has also been reported that betulinic acid inhibits inflammation and acute lung damage induced by LPS in male Sprague-Dawley rats. Oral administration of 25 mg/kg of betulinic acid for 7 days suppressed LPS-induced increased expression of TNF-α, TGF-β1, and iNOS and restored the antioxidant enzymes (GSH and SOD) and lipid peroxidation. These results suggest that betulinic acid is endowed with anti-inflammatory and antioxidant properties that protect the lung against the deleterious actions of LPS.

L. Kaempferol

Kaempferol is one of the most commonly found dietary flavonoids in plants such as clove and fenugreek. Flavonoids have been reported to lower inflammation and oxidative stress and to exert other positive effects in chronic inflammatory diseases. The inhibition of inflammation by kaempferol has been associated with inhibition of inflammatory cytokines. Kaempferol inhibited cytokines TNF-α and IL-1β and gene expression in J774.2 macrophages. Kaempferol has also been shown to inhibit iNOS along with NO production. In addition, kaempferol inhibited the activation of NF-κB and STAT1 as well as iNOS in LPS-activated macrophages. In aged Sprague-Dawley rats, kaempferol treatment inhibited NF-κB activation and expression of its target genes COX-2, iNOS, and MCP-1, indicating the anti-inflammatory effect of kaempferol in an aged animal model. Studies have shown that kaempferol also down-regulates key asthmatic features in established allergic lung disease in animals. Administration of kaempferol glycoside to ovalbumin-sensitized Balb/c mice impaired Th2 cytokine production (IL-5 and IL-13) and did not induce a Th1 pattern of inflammation, suggesting its potential clinical use against allergic asthma. Another study revealed that kaempferol inhibits carrageenan-induced inflammation in a rat air pouch model. Administration of kaempferol (50 and 100 mg/kg) significantly inhibited carrageenan-induced production of nitrite, PGE(2) generation, and COX-2 expression. Infiltration of the cells into the rat granuloma air pouch was also significantly inhibited

by kaempferol, which significantly contributes to kaempferol's anti-inflammatory activity. Recently, kaempferol was shown to inhibit ulcerative colitis in a rat model. Kaempferol also inhibited plasma levels of NO and PGE(2) in rats.

M. Piperine

Piperine is a plant alkaloid found mainly in the spices *Piper longum* Linn and *Piper nigrum* Linn. It has a long history of medical benefits and has been used as a common food additive all over the world. Piperine has shown various medicinal activities including anti-inflammatory activities. In ovalbumin-induced asthma in a murine model, it inhibited inflammation and exhibited a therapeutic mechanism of asthma treatment. It reduced Th2 cytokines (IL-4, IL-5), eosinophil infiltration, and chemokine, eotaxin-2, and IL-13 mRNA expression in lung tissue. Piperine also reduced IL-4, IL-5, and eotaxin levels in bronchoalveolar lavage fluid, as well as histamine- and ovalbumin-specific immunoglobulin E production in serum.

It has been reported that piperine inhibits inflammation in arthritis. In rat models of carrageenan-induced acute paw pain and arthritis, piperine inhibited the expression of IL-6 and MMP-13 and reduced the production of PGE2. It also inhibited the migration of AP-1, but not that of NF-κB, into the nucleus in IL-1β–treated synoviocytes. In addition, piperine significantly reduced the inflammatory area in the ankle joints of mice. Piperine also inhibits invasion and migration of tumor cells by suppressing NF-κB and AP-1 activation and PKCα and ERK phosphorylation. In an arthritic model, piperine inhibited the levels of lysosomal enzymes, lipid peroxidation, TNF-α, and paw volume and restored the activities of antioxidant status treatment. Piperine inhibited LPS-induced endotoxin shock, leukocyte accumulation, and the production of TNF-α. In addition, it suppressed LPS-induced expression of type 1 IFN mRNA and activation of STAT-1. In an animal model of cardiovascular disease, supplementation of piperine normalized blood pressure, improved glucose tolerance and reactivity of aortic rings, reduced plasma parameters of oxidative stress and inflammation, attenuated cardiac and hepatic inflammatory cell infiltration and fibrosis, and improved liver function. Piperine administration also had a beneficial effect on cerebral ischemia–induced inflammation in male Wistar and reduced infarct volume and neuronal loss. It reduced levels of proinflammatory cytokines IL-1β, IL-6, and TNF-α and the expression of COX-2, NOS-2, and NF-κB in the ischemic group.

N. Garcinol

Garcinol, harvested from the fruit rind of *Garcinia indica* (kokum), has been used traditionally in tropical regions, where it has been appreciated for centuries. In the past few years, its biological properties, including its anticarcinogenic and anti-inflammatory activities, have started to be elucidated. In cell-free assays, garcinol inhibited the activity of purified 5-LOX and blocked PGE2 synthesis. In LPS-activated macrophages, garcinol inhibited the expression of iNOS and COX-2. It also strongly blocked the activation of eukaryotic transcription factor NF-κB. Suppression of the NF-κB signaling pathway by garcinol led to apoptosis-induced cancer cells, which was consistent with the down-regulation of NF-κB–regulated genes. Thus, garcinol acts as a potential chemopreventive and/or therapeutic agent through the suppression of inflammatory biomarkers.

SUMMARY

Although spices have been used for centuries for many purposes, including flavoring agents, colorants, and preservatives, recently accumulated results from *in vitro*, animal, and clinical studies have shown evidence that spices are associated with chronic inflammatory diseases. Spices exhibit antioxidants and anti-inflammatory properties and modulate aberrant signaling pathways in cells, thereby suppressing the onset of chronic diseases such as cancer, osteoporosis, cardiovascular disease, diabetes, multiple sclerosis, and Crohn's disease. To date, the majority of spices and their active components have been tested *in vitro*, in animal models, or in preclinical conditions; only a few spices or their active components have been tested in clinical trials. More clinical trials and other studies of spices are needed to elucidate their efficacy in the treatment and prevention of human chronic diseases, including cancer.

SUGGESTIONS FOR FURTHER READING

Aggarwal B.B. (2010), Targeting Inflammation-induced Obesity and Metabolic Diseases by Curcumin and Other Nutraceuticals, *Annu Rev Nutr.*, 21;30:173–99.

Aggarwal B.B. and Shishodia S. (2004), Suppression of the Nuclear Factor-kappaB Activation Pathway by Spice-derived Phytochemicals: Reasoning for Seasoning, *Ann N Y Acad Sci.*, Dec;1030:434–41.

Aggarwal B.B. and Sung B. (2009), Pharmacological Basis for the Role of Curcumin in Chronic Diseases: An Age-old Spice with Modern Targets, *Trends Pharmacol Sci.*, 30(2):85–94.

Aggarwal B.B., Gupta S.C. and Sung B. (2013), Curcumin: An Orally Bioavailable Blocker of TNF and other Pro-inflammatory Biomarkers, *Br J. Pharmacol.*, 169(8):1672–92.

Aggarwal B.B., Kunnumakkara A.B., Harikumar K.B., Tharakan S.T., Sung B. and Anand P. (2008), Potential of Spice-derived Phytochemicals for Cancer Prevention, *Planta Med.*, 74(13):1560–9.

Aggarwal B.B., Sundaram C., Malani N. and Ichikawa H. (2007), Curcumin: The Indian Solid Gold, *Adv Exp Med Biol.*, 595:1–75.

Aggarwal B.B., Van Kuiken M.E., Iyer L.H., Harikumar K.B. and Sung B. (2009), Molecular Targets of Nutraceuticals Derived from Dietary Spices: Potential Role in Suppression of Inflammation and Tumorigenesis, *Exp Biol Med.* (Maywood), 234(8):825–49.

Anand P., Sundaram C., Jhurani S., Kunnumakkara A.B. and Aggarwal B.B. (2008), Curcumin and Cancer: An "old-age" Disease with an "age-old" Solution, *Cancer Lett.*, 18;267(1):133–64.

Goel A., Jhurani S. and Aggarwal B.B. (2008), Multi-targeted Therapy by Curcumin: How Spicy is it? *Mol Nutr Food Res.*, 52(9):1010–30.

Gupta S.C., Kim J.H., Prasad S., Aggarwal B.B. (2010), Regulation of Survival, Proliferation, Invasion, Angiogenesis, and Metastasis of Tumor Cells through Modulation of Inflammatory Pathways by Nutraceuticals, *Cancer Metastasis Rev.* 29(3):405–34.

Gupta S.C., Kismali G. and Aggarwal B.B. (2013), Curcumin, a Component of Turmeric: From Farm to Pharmacy, *Biofactors*, 39(1):2–13.

Gupta S.C., Patchva S. and Aggarwal B.B. (2013), Therapeutic Roles of Curcumin: Lessons Learned from Clinical Trials, AAPS J., 15(1):195–218.

Gupta S.C., Sung B., Kim J.H., Prasad S., Li S. and Aggarwal B.B. (2013), Multitargeting by Turmeric, The Golden Spice: From Kitchen to Clinic, *Mol Nutr Food Res.*, 57(9):1510–28.

Jagetia G.C. and Aggarwal B.B. (2007), "Spicing up" of the Immune System by Curcumin, *J. Clin Immunol.*, 27(1):19–35.

Prasad S., Gupta S.C., Tyagi A.K. and Aggarwal B.B. (2014), Curcumin, a Component of Golden Spice: From Bedside to Bench and Back, *Biotechnol Adv.*, 32(6):1053–64.

Prasad S., Phromnoi K., Yadav V.R., Chaturvedi M.M. and Aggarwal B.B. (2010), Targeting Inflammatory Pathways by Flavonoids for Prevention and Treatment of Cancer, *Planta Med.* 76(11):1044–63.

Prasad S., Tyagi A.K. and Aggarwal B.B. (2014), Recent Developments in Delivery, Bioavailability, Absorption and Metabolism of Curcumin: The Golden Pigment from Golden Spice, *Cancer Res Treat.*, 46(1):2–18.

Ralhan R., Pandey M.K. and Aggarwal B.B. (2009), Nuclear Factor-kappa B Links Carcinogenic and Chemopreventive Agents, Front Biosci (School ed.), 1;1:45–60.

Sung B., Prasad S., Yadav V.R. and Aggarwal B.B. (2012), Cancer Cell Signaling Pathways Targeted by Spice-derived Nutraceuticals, *Nutr Cancer*, 64(2):173–97.

Sung B., Prasad S., Yadav V.R., Lavasanifar A. and Aggarwal B.B. (2011), Cancer and Diet: How are they Related? *Free Radic Res.*, 45(8):864–79.

Yadav V.R., Prasad S., Sung B. and Aggarwal B.B. (2011), The Role of Chalcones in Suppression of NF-*k*B-mediated Inflammation and Cancer, *Int Immunopharmacol*, 11(3):295–309.

127

Wonders of Curcumin
Haldi is Healthy for Jaldi Recovery

Bharat B. Aggarwal and Subash C. Gupta

CONTENTS

I. WHAT IS CURCUMIN AND WHAT DOES IT DO?

Curcumin is the active ingredient (nutraceutical) of the dietary spice turmeric (*Curcuma longa*) and has been consumed for medicinal purposes for thousands of years in Asian countries. Since ancient times, curcumin has been preferred as potential therapeutics for most chronic diseases because of its safety, affordability, long-term use, and ability to target multiple cell signaling pathways.

Curcumin was first discovered about two centuries ago by Harvard College laboratory scientists Vogel and Pelletier. Vogel and Pelletier reported the isolation of "yellow coloring-matter" from the rhizomes of *Curcuma longa* (turmeric) and named it curcumin. Later, this substance was found to be a mixture of resin and turmeric oil. In 1842, Vogel Jr. obtained a pure preparation of curcumin but did not report its formula. In the decades that followed, several chemists reported possible structures of curcumin. However, it was not until 1910 that Milobedzka and Lampe identified the chemical structure of curcumin as diferuloylmethane, or 1,6-heptadiene-3,5-dione-1,7-bis (4-hydroxy-3-methoxyphenyl)-(1E, 6E). Further work by the same group in 1913 resulted in the synthesis of the compound. Subsequently, Srinivasan separated and quantified the components of curcumin by chromatography.

Although turmeric, the major source of curcumin, has been consumed as a dietary spice and a cure for human ailments for thousands of years in Asian countries, the biological characteristics of curcumin were not scientifically identified until the mid-twentieth century. In a paper published in Nature in 1949, Schraufstatter and colleagues reported that curcumin is a biologically active compound that has anti-bacterial properties. Despite those findings, only five papers were published on curcumin during the next two decades. In the 1970s, curcumin became the subject of scientific investigation and was shown to possess cholesterol-lowering, anti-diabetic, anti-inflammatory, and anti-oxidant activities. We were the first to demonstrate that curcumin exhibits anti-inflammatory activity by suppressing the pro-inflammatory transcription nuclear factor (NF)-kB in 1995. We also delineated the molecular mechanism by which curcumin inhibits NF-kB activation.

The interest in curcumin research has increased dramatically over the years. As of December 2014, more than 7000 articles on curcumin were listed in the National Institutes of Health PubMed database (www.ncbi.nlm.nih.gov/sites/entrez), all of which appeared mostly in the last decade. We now know that curcumin can modulate multiple signaling pathways in either a direct or indirect manner. The direct molecular targets of curcumin are shown in Figure 127.1. This polyphenol has been demonstrated to possess activities in animal models of many human diseases including cancer, lung diseases, neurological diseases, liver diseases, metabolic diseases, autoimmune diseases, cardiovascular diseases, and various other inflammatory diseases. In human clinical trials, curcumin has been found to be safe and efficacious, and the U.S. Food and Drug Administration has approved curcumin as a "generally regarded as safe" compound. To date, more than 65 clinical trials involving over 1000 patients have been completed while more than 35 clinical trials are further evaluating the efficacy of this polyphenol against numerous human ailments.

Curcumin is now regarded as a "new drug" with great potential and is being used as a supplement in several countries. For example, in India, turmeric containing curcumin has been used in curries; in Japan, it is popularly served in tea; in Thailand, it is used in cosmetics; in China, it is used as a colorant; in Korea, it is served in drinks; in Malaysia, it is used as an antiseptic; in Pakistan, people use it as an anti-inflammatory agent to get relief from gastrointestinal discomfort; and in the United States, it is used in mustard sauce, cheese, butter, and chips, as a preservative and a coloring agent. Curcumin is marketed in several forms including capsules, tablets, ointments, energy drinks, soaps, and cosmetics.

In the following sections, we review the biological activities of curcumin, with a special focus on its anti-oxidant, anti-inflammatory, and anti-microbial activities as well as its efficacy for human diseases including cancer, cardiovascular, neurological, diabetes, skin, depression, arthritis, inflammatory bowel disease and uveitis.

II. BIOLOGICAL ACTIVITIES OF CURCUMIN

A. Curcumin as an Anti-oxidant Agent

Curcumin has been demonstrated to possess anti-oxidant and free-radical scavenging activities. The anti-oxidant activity of curcumin can arise either from the hydroxyl group or the methylene group of the β-diketone (heptadiene-dione) moiety. Curcumin has been found to be atleast 10 times more active as an antioxidant than even vitamin E. Curcumin prevents the oxidation of hemoglobin and inhibits lipid peroxidation. The antioxidant activity of curcumin could be mediated through anti oxidant enzymes such as superoxide dismutase, catalase, and glutathione peroxidase. The suppression of lipid peroxidation by curcumin could lead to suppression of inflammation.

B. Curcumin as an Anti-inflammatory Agent

Inflammation is a part of the host defense system which counteracts insults incurred by internal or external stimuli. The clinical and fundamental signs of inflammation include redness, swelling, heat, pain, and loss of function. Inflammation can be classified as acute or chronic. While *acute inflammation* is an immediate response of the body and is required to ward off harmful pathogens, *chronic inflammation* is a delayed response and has been associated with numerous human chronic diseases.

At the molecular level, inflammation is regulated by numerous molecules and factors, including cytokines [interleukin (IL)-1, IL-2, IL-6, IL-12, tumor necrosis factor (TNF)-α, TNF-β), chemokines (monocyte chemo, attractant protein 1, IL-8), pro-inflammatory transcription factors (NF-κB, STAT3), pro-inflammatory enzymes [cyclooxygenase-2 (COX-2), 5-lipooxygenase (5-LOX), 12-LOX, matrix metaloproteinases (MMPs)], prostate-specific antigen (PSA), C-reactive protein, adhesion molecules (intercellular adhesion molecule [ICAM-1], vascular cell adhesion molecule [VCAM]-1, endothelial-leukocute adhesion molecule [ELAM]-1), vascular endothelial growth factor (VEGF), and TWIST. Among all these mediators, *NF-κB is the central regulator of inflammation.*

Extensive research using a wide range of *in vitro* and *in vivo* models over the past several years has indicated that curcumin can reduce inflammatory response by regulating the production of inflammatory molecules. Our laboratory was the first to demonstrate that *curcumin is a potent inhibitor of NF-κB and STAT3 pathways.* The hydroxyphenyl unit in curcumin has been shown to be crucial to its anti-inflammatory activity. In osteoarthritic-affected dogs, curcumin exhibits anti-inflammatory activities. Curcumin has been shown to exhibit therapeutic potential in various chronic illnesses in which inflammation is known to play a major role including Alzheimer's disease, Parkinson's disease, multiple sclerosis, epilepsy, cerebral injury, cardiovascular diseases, cancer, allergy, asthma, bronchitis, colitis, rheumatoid arthritis, renal ischemia, psoriasis, diabetes, obesity, depression, fatigue, and AIDS.

C. Curcumin as an Anti-microbial Agent

The anti-fungal activities of curcumin are evident from its ability to display *anti-fungal* properties against Candida strains. Curcumin also improves the activity of common azole and polyene anti-fungals. Curcumin also possess *anti-viral* activities.

D. Curcumin as an Anticancer Agent

During the past two decades, our laboratory and others have demonstrated curcumin's potential as both a chemopreventive and a chemotherapeutic agent against cancer in rodent models.

The *chemopreventive efficacy of curcumin for colon cancer* is particularly well established. Other common cancers in which curcumin has shown protective effects include esophageal, lung, kidney, stomach, liver, oral, breast, bladder, leukemia, skin, small intestine, pancreatic, brain, and prostate cancers. Accumulating evidence over the past several years has indicated that curcumin can be used for the treatment of established cancers as well. Most of these studies have used orthotopic or xenotransplant rodent models and have employed curcumin either alone or in combination with existing therapies. Curcumin has shown potential for treatment of the following transplanted human cancers: cholangiocarcinoma, lymphoma, and melanoma, and prostate, pancreatic, colorectal, hepatocellular, breast, ovarian, and bladder cancers. Curcumin has also demonstrated potential to reduce cancer-associated symptoms such as fatigue, neuropathic pain, and cognitive deficit.

E. Curcumin for Cardiovascular Diseases

Numerous reports have indicated that inflammation plays a major role in most cardiovascular diseases (CVDs). Because of its anti-inflammatory activities, curcumin has been demonstrated to mediate its effects against CVDs through diverse mechanisms. Several studies have suggested that curcumin protects the heart from ischemia/reperfusion injury. The proliferation of peripheral blood mononuclear cells (PBMCs) and vascular smooth muscle cells (VSMCs), which are hallmarks of atherosclerosis, is inhibited by curcumin. Curcumin prevents the oxidation of low density lipoproteins (LDLs), inhibits platelet aggregation, and reduces the incidence of myocardial infarction. Curcumin has also been shown to improve the lipid profiles in patients with acute coronary syndrome.

Atherosclerosis is a condition in which fatty materials such as cholesterol accumulate and thickens the artery wall. This is a chronic disease that normally remains asymptomatic for decades. Curcumin has been demonstrated to reduce serum levels of cholesterol and lipid peroxides and to increase HDL cholesterol.

F. Curcumin for Neurological Diseases

Curcumin has been shown to exhibit activity against various neurologic diseases, including Alzheimer's disease, multiple sclerosis, Parkinson's disease, epilepsy, cerebral injury, age-associated neurodegeneration, schizopherenia, Spongiform encephalopathies, neuropathic pain, and depression. Curcumin can suppress oxidative damage, inflammation, cognitive deficits, and amyloid accumulation in Alzheimer's disease. Curcumin inhibits experimental allergic encephalomyelitis by blocking interleukin (IL)-12 signaling in T cells, suggesting its potential for the treatment of multiple sclerosis.

G. Curcumin for Diabetes

Curcumin has been shown to improve the symptoms associated with diabetes. Curcumin act as an anti-diabetic agent and maintains the normal structure of the kidney in diabetic mice model. Curcumin also exhibits anti-hyperglycemic effect and improves insulin sensitivity; these actions were attributed in part to its anti-inflammatory properties and anti-lipolytic effects. Curcumin has been suggested to be used as a beneficial adjuvant therapy in type 2 diabetes mellitus (T2DM). Curcumin appears to be a potent glucose-lowering agent in T2DM mice, but it exhibits no effects in non-diabetic mice. Dietary curcumin also partially reversed the abnormalities in plasma albumin, urea, creatine, and inorganic phosphorus in diabetic animals.

H. Curcumin for Skin Diseases

Curcumin has been shown to be effective against different skin diseases including skin carcinogenesis, psoriasis, scleroderma, vitiligo, and dermatitis. Numerous reports suggest that curcumin accelerates wound healing. In addition, curcumin also prevents the formation of scars and plays a role in muscle regeneration following trauma. Psoriasis is a chronic inflammatory skin disease characterized by thick, red, scaly lesions that may appear on any part of the body. The disease exists in five different forms—plaque, guttate, inverse, pustular, and erythrodermic—of which plaque psoriasis is most common. The disease affects approximately 2% of the population worldwide and is associated with increased cardiovascular risk. Elevations of activity in phosphorylase kinase (PhK), a serine/threonine-specific protein kinase, have been correlated with pathogenesis of psoriasis. Therefore, agents with potential to inhibit PhK activity can be useful for the treatment of psoriasis. One of the early studies from our own laboratory indicated that curcumin is a noncompetitive inhibitor of PhK, with a Ki of 75 μM.

Vitiligo is a skin disorder in which the cells producing pigment (color) in the skin (melanocytes) are destroyed, resulting in white patches that appear on the skin on different parts of the body. Although what causes damage to melanocytes remains unclear, oxidative stress has been implicated in the pathogenesis of the disease. Because of its anti-oxidant property, curcumin seems to be a therapeutic option for the treatment of vitiligo.

I. Curcumin for Depression

Mounting evidence over the past several years has indicated curcumin's efficacy in various animal models of psychiatric disorders including depression. The anti-depressant activity of curcumin has been shown to be potentiated by the concomitant administration of fluoxetine, venlafaxine, or bupropion. In mice, piperine, an agent known to enhance bioavailability has been demonstrated to augment the anti-depressant action and penetration of curcumin to the brain.

J. Curcumin for Arthritis

Arthritis is a chronic disease that results from the inflammation of one or more joints. It usually results from dysregulation of pro-inflammatory cytokines (e.g., TNF, IL-1β) and pro-inflammatory enzymes that mediate the production of prostaglandins (e.g., COX-2) and leukotrienes (e.g., lipoxygenase), together with the expression of adhesion molecules and matrix metalloproteinases. Although more than 100 different kinds of arthritis have been reported, the three most common forms are osteoarthritis, rheumatoid arthritis, and gout. Curcumin has been shown to possess antiarthritic and antirheumatic effects, most likely through the down-regulation of COX-2, TNF, and other inflammatory cytokines.

K. Curcumin for Inflammatory Bowel Disease

Inflammatory bowel disease (IBD) is a condition in which the intestines become inflamed. Although the etiology of IBD is not clearly known, it appears to be driven by inflammatory cytokines such as TNF-α. Two major types of IBD are ulcerative colitis and Crohn disease. Whereas ulcerative colitis is limited to the colon, Crohn disease can involve any part of the gastrointestinal tract from the mouth to the anus. Another mild-to-moderate form of ulcerative colitis is called ulcerative proctitis, which involves inflammation of the rectum. Patients with IBD have a significantly higher risk of developing colon cancer than the general population has. Curcumin has demonstrated potential against ulcerative colitis, Crohn's disease, and ulcerative proctitis.

L. Curcumin for Uveitis

Uveitis is an inflammation of the uvea, the middle layer of the eye. Uveitis is a major cause of visual impairment and has been estimated to account for 10% to 15% of all cases of total blindness in the United States. Depending on the anatomical localization and visible signs of the disease, uveitis can be classified into anterior, posterior, pan, and intermediate. The course of the disease can be acute, chronic (>3-month duration), and recurrent. Corticosteroids are normally used for treatment of uveitis. However, the adverse effects associated with these drugs limit their use. Curcumin has demonstrated potential against chronic anterior uveitis and recurrent anterior uveitis.

M. Other Activities of Curcumin

In addition to the activities indicated above, curcumin possess numerous other properties such as wound healing, anti-nociceptive activity in ganglion neurons, anti-parasitic activity, schistosomicidal activity against *Schistosomamansoni* adult worms, anti-malarial activity, and nematocidal activity. The potential of curcumin to enhance memory, to ameliorate morphine addiction as well as for conditions such as fibrosis, aging, asthma, endometriosis, and muscle wasting has also been reported.

N. Curcumin in the Clinic

The extensive studies from cell-based and animal models have formed a solid basis for evaluating the safety and efficacy of curcumin against a plethora of human diseases. Curcumin's clinical efficacy against human biliary diseases was first studied in 1937. In this study, curcumin produced remarkably good results against cholecystitis. Since this initial discovery, observations from more than 65 human clinical trials of curcumin, which included more than 1000 patients, have been published, and more than 35 other clinical trials are under way to further evaluate the efficacy of this polyphenol against human diseases. Among the most common human diseases against which curcumin has exhibited activities in humans include cardiovascular disease, arthritis, uveitis, cancer, ulcerative proctitis, Crohn disease, ulcerative colitis, peptic ulcer, gastric ulcer, idiopathic orbital inflammatory pseudotumor, oral lichen planus, gastric inflammation, vitiligo, psoriasis, acute coronary syndrome, atherosclerosis, diabetes, Dejerine-Sottas disease, diabetic nephropathy, diabetic microangiopathy, lupus nephritis, renal conditions, acquired immunodeficiency syndrome, irritable bowel disease, tropical pancreatitis, β-thalassemia, cholecystitis, and chronic bacterial prostatitis. The safety, tolerability, and nontoxicity of curcumin at high doses have been well-established by human clinical trials. In these clinical trials, curcumin was used either alone or in combination with other agents such as gemcitabine, soy isoflavones, bioperine, quercetin, mesalamine, acetylcysteine, prednisone, lactoferrin, piperine, docetaxel, sulfasalazine, and pantoprazole. Although the molecular basis for curcumin's efficacy against some of these diseases is still not completely known, this polyphenol has been shown to modulate numerous signaling molecules including proinflammatory cytokines [(TNF)-α, (IL)-1β, IL-6)], apoptotic proteins, NF-κB, COX-2, STAT3, IkappaB kinase beta (IKKβ), endothelin-1, malondialdehyde, C-reactive protein, prostaglandin E2, glutathione-S-transferase (GST), prostate specific antigen (PSA), VCAM1, GSH, pepsinogen, phosphorylase kinase (PhK), transferrin receptor, total cholesterol, transforming growth factor beta (TGF-β), triglyceride, creatinine, hemoxygenase-1, antioxidants, aspartate transaminase, and alanine transaminase in human participants. In most of the clinical trials, either a mixture of curcuminoids or turmeric from which curcumin is derived was used; pure curcumin has been used in only a few studies. Although curcumin has shown efficacy against numerous human ailments, poor bioavailability due to poor absorption, rapid metabolism, and rapid systemic elimination limits its therapeutic efficacy. As a result, numerous approaches including the use of adjuvants, nanoparticles, liposomes, phospholipid complexes, and structural analogues have been used to increase the bioavailability of curcumin in human participants. The bioavailability of curcumin has also been shown to be greatly enhanced by reconstituting curcumin with the non-curcuminoid components of turmeric.

IV. CURCUMIN: A TIP OF THE ICEBERG

Curcumin is one of the most widely studied components of turmeric as indicated by more than 7000 publications, most of which appeared in the last decade. Although initially curcumin was believed to contribute most of the activities of turmeric, recent research has identified numerous other chemical entities with similar activities in turmeric. Studies have indicated that turmeric contains more than 300 different components. Cell-based assays conducted in our laboratory have indicated that curcumin is less potent in inhibiting cancer growth than is turmeric containing an equivalent amount of curcumin. Accumulating evidence over the past decade has suggested that aqueous extracts of turmeric that lack curcumin exhibit antioxidant and corneal wound healing activities, suppress hepatitis B virus replication by enhancing p53 levels, exhibit anti-Helicobacter pylori activity, stimulate insulin release and mimic insulin action, and exhibit anti depressant activity in mice. In addition, these aqueous extracts were found to be effective in preventing cancer. Studies have indicated that curcumin-free aqueous extracts of turmeric is as potent as curcumin-containing turmeric or curcumin alone in suppressing tumor formation in mice. Curcumin-free aqueous extracts of turmeric have been shown to protect against DNA damage induced by fuel smoke condensate in human lymphocytes. Of note, turmeric has been found to be more effective than curcumin in suppressing streptozotocin-induced diabetic cataracts in rats and in reducing blood glucose levels in type 2 diabetic KK-A mice.

Over all, above examples provide sufficient evidence that whole turmeric may exhibit superior activities compared with curcumin alone. Considering these evidences and the fact that turmeric has been used as a spice for at least 2500 years, we believe that by using curcumin alone, we might be limiting ourselves from the various utilities of turmeric.

SUMMARY

Curcumin is an active polyphenol of the turmeric that has been consumed in the Asian countries since ancient time to treat such human ailments as acne, psoriasis, dermatitis, and rash. Curcumin is a highly pleiotropic molecule with the potential to modulate the biological activity of a number of signaling molecules. Accumulating evidence from preclinical and clinical studies have indicated the safety and efficacy of curcumin. Although being used as a supplement in several countries, this polyphenol has not yet been approved for the treatment of any human diseases. It is the hope of the authors that numerous ongoing studies will help to move this fascinating molecule to the forefront of therapeutics for human use.

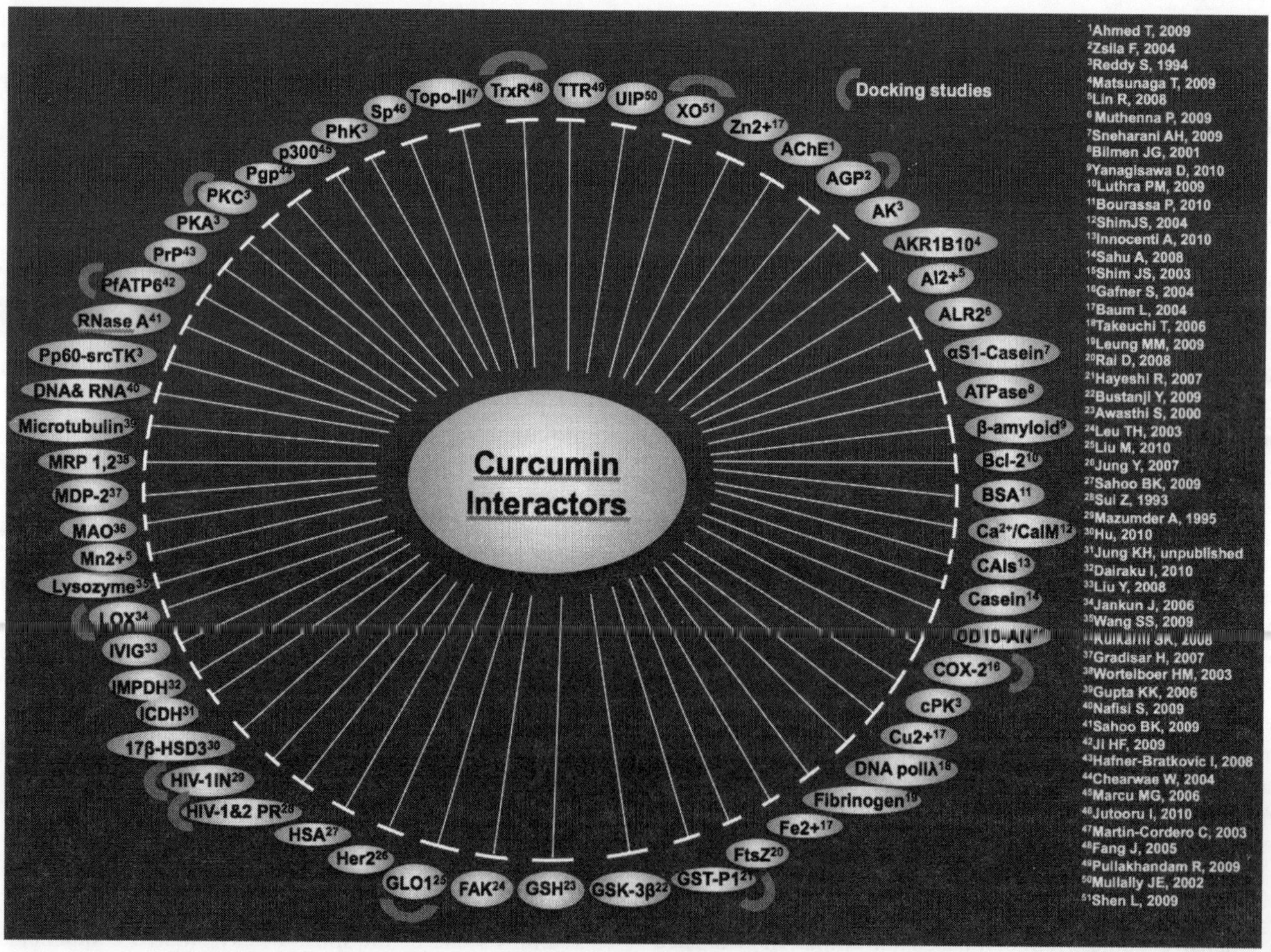

Figure 127.1 The direct molecular targets of curcumin.

SUGGESTIONS FOR FURTHER READING

1. Aggarwal B.B. and Sung B. (2009), Pharmacological Basis for the Role of Curcumin in Chronic Diseases: An Age-old Spice with Modern Targets, *Trends Pharmacol Sci.*, 30(2):85–94.

2. Gupta S.C., Prasad S., Kim J.H., Patchva S., Webb L.J., Priyadarsini I.K., et al. (2011), Multitargeting by Curcumin as Revealed by Molecular Interaction Studies, *Nat Prod Rep.*, 28(12):1937–55.

3. Gupta S.C., Patchva S. and Aggarwal B.B. (2012), Therapeutic Roles of Curcumin: Lessons Learned from Clinical Trials, The AAPS Journal DOI: 10.1208/s12248–12012–19432–12248.

4. Aggarwal B.B., et al. (2003), Anticancer Potential of Curcumin: Preclinical and Clinical Studies, *Anticancer Res.*, 23, 363–398.

5. Singh S. and Aggarwal B.B. (1995), Activation of Transcription Factor NF-kappa B is Suppressed by Curcumin (diferuloylmethane) [corrected], *J. Biol. Chem.*, 270:24995–5000.

128

Yoga as a Therapeutic Tool in Modern Medicine

Ramesh Bijlani

CONTENTS

Lifestyle disorders such as obesity, hypertension, coronary heart disease, diabetes mellitus, stroke, bronchial asthma and cancer have emerged as major public health problems during the last few decades. This has drawn the attention of modern medicine towards the ancient discipline of yoga because yoga embodies the finest lifestyle ever discovered by mankind. Therefore, a doctor practicing medicine in the twenty-first century cannot afford to ignore yoga. Even if he himself cannot teach yoga to his patients, he should at least be able to provide them an informed opinion about the role, possibilities and limitations of yoga in the treatment of various disorders. Further, learning yoga might inspire the doctor to make his life healthier, happier and more meaningful by bringing yoga into his own life.

I. HOW YOGA ENTERED MODERN MEDICINE

After the conquest of infectious diseases, nutritional deficiencies and endocrine disorders within a period of just about 50 years (1900–1950), modern medicine went through a humbling experience. Although life expectancy more than doubled in those parts of the world which benefited the most from the spectacular advances made in the first half of the twentieth century, people now started dying of a different set of diseases: cardiovascular disease, complications of diabetes, and cancer. History has repeated itself in India. In 1947, our life expectancy was about 30 years; today it is more than sixty. The new killers of mankind, moreover, did not lend themselves to simplistic solutions of the type that had worked in case of infectious diseases and nutritional deficiencies. In those cases, giving vaccines, the right antibiotic or the missing vitamin had resulted in miraculous prevention or cure. The new killers were multi-factorial in etiology. Their roots lay in the lifestyle, of which *mental stress* was a major accompaniment. Hence, there began a search for better strategies for managing stress. This is how yoga entered modern medicine—as a way of life that could influence our diet, physical activity, and use of alcohol and tobacco; and as a tool for stress management. Apart from clinical studies, which have shown the efficacy of yoga as a combination of a healthy lifestyle and a tool for mental peace, further support for the approach has come from developments in psychoneuroimmunology (PNI). PNI has established the overwhelming importance of *mental peace for optimal immune function*. Further, immunocompetence has been shown to be closely linked to the self-healing mechanisms of the body. Finally, self-healing has been shown to operate not only in case of acute infections, but also in case of chronic non-communicable diseases such as cardiovascular disease, diabetes and cancer. The popular tendency to equate yoga with a set of postures and breathing practices is neither in keeping with the philosophy of yoga nor in tune with the latest developments in modern medicine. Correcting this distortion will be one of the aims of this chapter.

II. WHAT IS YOGA?

Yoga is basically a change that improves the person as a whole. The person normally functions as a combination of the body and the mind. *Yoga improves both the body and the mind.* The mind further has two aspects: feeling and thinking. Yoga improves both our feelings and our thinking. Thus yoga makes a person physically more fit, emotionally more stable, and mentally more agile. We may spend only an hour a day on the yogic techniques designed to improve the body-mind complex. Within the work that we do, as a doctor, as a teacher or as a scientist, we have choices to make, attitudes to adopt. Yoga guides us on how to make these choices, what attitude to adopt. By making the choices, and by adopting the attitude towards which yoga guides us, we find that life becomes happier, more meaningful, and therefore more fulfilling. Moreover, PNI has shown that when we are happier, we also become healthier. Thus yoga can give us everything that we are generally looking for, but for that we have to bring yoga into our daily lives, into everything we do. That is what Sri Aurobindo meant when he said, "All life is yoga". What this celebrated quote means is that all life gives us an opportunity for the practice of yoga. The way we can use the opportunity provided by life to practice yoga is to organize life around our highest and noblest aspirations. That leads to choices based on love and compassion, and an attitude based on selfless service. In short, *yoga gives us* the techniques that improve the body-mind complex, *and also* the guidance on how to use the complex.

A. Schools of Yoga

While looking upon yoga as a journey towards self-perfection is the underlying spirit common to all schools of yoga, the emphasis and form may differ enormously.

Hatha yoga is the school of yoga that emphasizes physical perfection. The asanas (postures) and pranayamas (breathing practices), with which yoga is commonly identified, have their origin in hatha yoga.

Raja Yoga, also known as *Patanjali's yoga* or *Ashtanga (eight-limbed) yoga*, emphasizes mental perfection. Raja yoga is the source of one of the most comprehensive techniques of meditation. The eight limbs of raja yoga have been listed in Table 128.1.

The yoga of the *Gita*, or the *Triple path*, emphasizes bringing yoga into everyday life. Further, it encourages the seeker to start the journey with action (*karma*), knowledge (*jnana*) or devotion (*bhakti*), depending on his temperament and circumstances. Thus, the triple path has a 'student-centred' approach, which accounts for its enormous universal appeal.

Integral yoga of Sri Aurobindo and the Mother is a powerful synthesis of the traditional schools of yoga. It seizes on the central principles of all the major traditional schools of yoga and harmonizes them into a way of life. It does not let the spirit of yoga get overshadowed by the form. One of its central features is that, while being consistent with Vedanta, it looks upon the world not as an illusion but as a manifestation

TABLE 128.1   The Eight Limbs of Raja Yoga

1. Yama
   (a) Ahimsa (non-violence): do not hurt (through thought, word or deed)
   (b) Satya (truth): do not tell a lie
   (c) Asteya (non-stealing): do not steal
   (d) Brahmcharya (celibacy): do not be attached to sensory pleasures
   (e) Aparigraha (non-receiving): do not be greedy
2. Niyama
   (a) Shaucha (cleanliness): internal and external purification
   (b) Santosh (contentment): being satisfied with one's lot
   (c) Tapas (austerities): discipline
   (d) Swadhyaya (study): studying the Self, studying on one's own
   (e) Iswarpranidhan (surrender to God): accepting God's will
3. Asanas (posture)
4. Pranayama (control of prana/'breath')
5. Pratyahara (sensory withdrawal)
6. Dharana (concentration)
7. Dhyana (meditation)
8. Samadhi (superconsciousness)

of the Divine. Therefore, it does not reject the world, but seeks to transform it to befit the One that it manifests. Transforming the world rather than stopping at individual salvation is a unique feature of integral yoga (Table 128.2).

TABLE 128.2   Central Principles of Different Schools of Yoga

School of yoga	*Central principle*
Hatha yoga	Perfection of the body
Raja yoga	Perfection of the mind
Triple path	Bring yoga into daily life Three alternative approaches Dominant approach left to the seeker
Integral yoga	All above + Do not stop at salvation of the individual

B. Techniques of Yoga

As discussed above, yogic techniques are designed to improve the performance of the body-mind complex. The three principal yogic techniques are asanas, pranayamas and meditation.

Asanas

Asanas are of three types: relaxing postures, stretches, and meditative postures. The value of the *relaxing postures* should not be underestimated. They involve a voluntary effort to relax the body. It has been shown that electromyographic activity ceases only when a person makes a voluntary effort to relax. *Shavasana* (the cadaveric posture) is one of the classical relaxing postures. In *shavasana*, the person not only relaxes the body completely, but also breathes slowly and deeply. As a result, *shavasana* relaxes both the body and the mind.

Further, by adding elements of meditation, autosuggestion and imagery, *shavasana* can become a rejuvenating and healing experience (Figure 128.1).

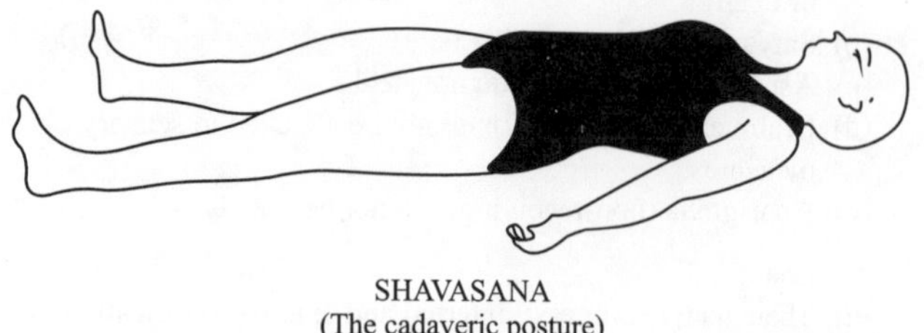

SHAVASANA
(The cadaveric posture)

Figure 128.1 Shavasana, a relaxing posture. The person lies down with the feet about a foot apart, hands a little away from the body, the palms facing upwards, and the eyes closed. Breathing is slow and deep. While breathing in, the abdomen should move out; and while breathing out, the abdomen should move in.

Yogic *stretches* are the most visible part of yoga. A typical stretch starts with an initial position in which the person is relaxed. From there he goes to a final position through a series of slow, gentle and graceful movements. The final position is a position of intense stretch. The stretch is intense, but it should be an enjoyable stretch, and the person should be stable in *his/her* final position. The stretch is maintained for some time (usually for 10-20 seconds), without shaking or getting wobbly. Then the person returns to the initial position or a relaxing posture, once again through a slow, gentle and graceful movement. How we go into the posture, how we maintain it, and how we come out of it, are just as important as the final position itself. Further, the entire sequence is performed with full awareness. Once a person is no longer a beginner, going into the posture, and coming out of it, may be synchronized with the breathing cycle. A relaxing posture may be inserted between two stretches, specially if the stretch is strenuous. The sequence of the stretches is so arranged that every pose is followed by a counter-pose. Thus, if one posture involves flexion of the spine, the next one involves extension. As a result, the stretch of one posture is neutralized by the next posture (Figure 128.2). It is because of the gentle movements, the relaxation built into the session, and the sequence of the asanas that at the end of the session the person is relaxed rather than exhausted. Additional factors such as the sweet voice of the instructor, the pleasant atmosphere, and soothing background music may further add to the relaxing effect. Although yogic practices are not strenuous, the improvement in cardiorespiratory fitness is comparable to that achieved by other far more intense exercises such as soccer or swimming. The discrepancy is explained by the fact that during yogic practices there are remarkable fluctuations in thoracic and abdominal pressures. Hence the cardiorespiratory reflexes are jogged to the same extent as in strenuous exercises.

SETUBANDHASANA
(The bridge posture)

(a)

PAVANMUKTASANA
(Wind releasing posture)

PASHCHIMATANASANA
(The back stretching posture)

(b)

KONASANA
(The inclined plane)

USHTRASANA
(The camel posture)

(c)

PARVATASANA
(The mountain posture)

Figure 128.2 Every pose is followed by its counter-pose. Pavanmuktasana is the counter-pose for setubandhasana (A), konasana for pashchimatanasana (B), and parvatasana for ushtrasana (C).

Pranayamas

Pranayamas are practices involving voluntary control of the pattern of breathing. One of the simplest, and yet very helpful pranayamas is slow and deep (the deepest that the person can) breathing, spending more time on breathing out than on breathing in. There are others that involve hyperventilation (*kapalabhati* and *bhastrika*), alternate nostril breathing (*anulom-vilom*), and deep breathing coupled with chanting 'nnn…' during expiration (*bhramari*). Pranayamas contribute significantly to the improvements in cardiorespiratory fitness and autonomic balance achieved by yogic techniques.

Meditation

The basic principle of meditation is to subdue the ordinary mental activity so that the person can look deeper within. Clearing the clutter created by the ordinary chaotic activity of the mind allows the deeper clarity and peace to emerge. After studying a variety of meditative techniques, Herbert Benson concluded that they all have four basic features: a quiet environment; a comfortable posture; a mental device; and a passive attitude. The mental devices are broadly of two types: involving either concentration or mindfulness. Patanjali's eight-limbed yoga has elements of both—*dharana* involving concentration, and *dhyana* involving mindfulness (Table 128.1). Depending upon individual preference, one may choose one or the other, or mix both. Physiologically, one finds a sharp decline in metabolic rate, and predominance of alpha waves in the EEG during meditation.

C. How Yoga Affects both Lifestyle and Stress

Keeping the body-mind complex fit is an important part of yoga. It is important because the practitioner has identified the mission of his life, and to fulfill that, the basic requirement is that the instrument that he has been provided with should function optimally. In yoga, many of the physical aspects of lifestyle that keep the body-mind complex fit are collectively referred to as physical culture. The principal components of physical culture are the physical practices such as asanas and pranayamas; a balanced diet in just the right quantity; not putting into the body harmful substances such as tobacco and alcohol; and just the right duration of good quality sleep. As important as physical culture is the mental discipline, of which meditation is the most visible part. In yoga, these lifestyle factors are a means to better health, which is a requirement for right thought and action, which in turn is necessary for spiritual growth, which is the purpose of life (Figure 128.3). Thus, health is not an end in itself, but a means to a higher

end. When lifestyle gets woven into a philosophy of life, as in yoga, the individual not only follows a healthy lifestyle, he follows it willingly and happily. That makes all the difference because the same things may be done out of fear of disease and death, or because the individual enjoys them. If the individual does something because he has to, the good that the action might do may be neutralized by the stress created by the compulsion. For example, if the individual is eating fruits and vegetables but thinking of omelets and cutlets, the good done by eating right may be wiped out by the stress due to missing the forbidden foods. Now that the mind-body connection is on a very firm scientific footing, it makes sense that the healthy lifestyle would be more effective if the right things are done not because one has to, but because one wants to.

Yoga is not just a healthy lifestyle, it is also a stress buster. The common perception is that yoga reduces stress through meditation. This is rather simplistic and reflects the tendency to expect too much from too little. Twenty minutes of meditation cannot compensate for the tension and turmoil of twenty-four hours unless one has learnt to look at the inevitable difficulties, problems and traumatic events of life with a different attitude. Yoga enables us to look at events of life in a spiritual light. Meditation might help in the introspection required for adopting a spiritual attitude to events and circumstances of life. Depending on the person's level of spiritual development, the spiritual worldview allows three approaches in every adverse situation. *First*, the situation is seen as an expression of the divine will, and therefore there is no choice but to accept it. *Second*, the situation is seen as an expression of not only the divine will but also divine wisdom, and therefore there must be something good about it, which may not be obvious because of the limitations of the human intellect. *Third*, the situation is an expression of not only the divine will but also divine wisdom, and therefore there must be something good about it, and at least one positive feature can always be seen in it – the situation can serve as an opportunity for spiritual growth, which is the purpose of life.

III. YOGA AS THERAPY

Yoga is not a system of medicine. It is primarily a spiritual discipline, of which one of the by-products is good health. Ayurveda made use of this fact long ago by taking some elements of the discipline and used them primarily for promotion and restoration of health; modern medicine is doing so now. However, modern man has little understanding of this background, and this affects his attitude to and expectations from yoga. First, he wants to use yoga for cure more than for prevention of disease, whereas yoga is much more relevant to prevention. Secondly, he tears the yogic techniques out of context, and wants to use them as a substitute for pills, the side effects of which he does not want. The fact is that it is not an either/or situation. Medication may help control the disease fast, whereas yoga can go to the roots and lead slowly to self-healing. Thus, medication may buy precious time, during

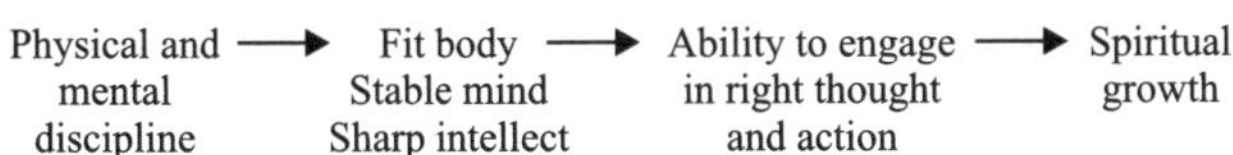

Figure 128.3 In yoga, what looks like an end in itself is the means to a higher end.

which yoga can start reversing the disease process. After yoga has resulted in some self-healing, the person may be able to reduce the dose of medication, or stop it altogether. Finally, there is a tendency to expect the same specificity from yogic techniques as from pills. Thus, patients want to learn just a few asanas that will cure them of diabetes or back pain or whatever ailment they have. In yoga, that degree of specificity neither exists nor is encouraged. It is much better to treat the disease as an opportunity that has brought the person to yoga, which would help his present disease, prevent diseases that he might have otherwise got in future, and, if he has the opening, enable him to live a more meaningful life. Therefore, the attitude of a good yoga therapist to the yogic techniques is to make every patient do the same set of judiciously chosen asanas and pranayamas except those that are contra-indicated in the disease that a given patient has. Further, while he is teaching the techniques, he also introduces his patients to the other components of physical culture, and also makes them aware of the spiritual dimension of yoga.

A. Lifestyle Modification Programs

Modern medicine uses yoga as a major tool in lifestyle modification programs that are in tune with the latest advances in mind-body medicine. These programs involve, *first*, physical activity, improvements in diet, stopping smoking and drug abuse, and sleep hygiene. *Secondly*, there is patient education so that the patient can fit lifestyle changes into an overall philosophy of life. The patient also learns to look at life differently, so that he enjoys lifestyle changes, and is not affected by the ups and downs of life. Knowledge is also imparted to improve the patient's understanding of his disease, because that helps him take care of it better. *Thirdly*, the daily routine includes meditation, soothing music, autosuggestion, imagery and visualization. *Finally*, facilities are also made available for individualized counseling on lifestyle and psychological issues that the patient might have. These programs are particularly useful for those having obesity, hypertension, coronary artery disease, diabetes mellitus, back pain, cervical spondylosis, irritable bowel syndrome, bronchial asthma, tension headache and polycystic ovary syndrome. But the list of lifestyle disorders is growing, and it is being increasingly realized that there is hardly any disease that does not have a psychological component in its genesis, or which does not benefit from a mind-body approach such as yoga. There is also a trend towards a rising incidence of disorders in which the patient has physical symptoms but there is no organic disorder that might explain the symptoms. These cases, in which the explanation of the physical symptoms seems to be the underlying psychological distress, also benefit from these programs, specially if coupled with adequate individualized counseling. That these programs do work has been substantiated by many studies, the pioneering ones being those of Dean Ornish on coronary heart disease in which he showed that a yoga-based lifestyle intervention led to angiographically demonstrable reduction in the stenosis of coronary arteries within one year. Programs along these lines were initiated at the All India Institute of Medical Sciences, New Delhi, in the year 2000.

B. How Do Yoga-based Lifestyle Modification Programs Work?

The human body is a wonderful machine. If abused by an unhealthy lifestyle, specially if it is accompanied by mental stress, the body undergoes slow but cumulative damage. Because of the physiological reserve, however, there may be no symptoms for a few decades. But eventually the slow accumulation of damage reaches a level where one or more lifestyle negligence may manifest; which disease will manifest in a given individual depends upon his weak point, which in turn may depend largely on hereditary factors. Fortunately, it has now been realized that this process is not a one-way street. Just as negligence can lead to slow accumulation of damage, correcting the physical and psychological factors that led to the disease can also lead to slow reversal of damage. Thus, there are potent self-healing mechanisms that can lead to partial or total recovery from even chronic degenerative diseases such as coronary artery disease and diabetes mellitus. This is a major paradigm shift in modern medicine. Till at least the middle of the twentieth century, self-healing mechanisms were thought to exist only in case of acute disorders such as common cold or diarrhea. Chronic degenerative diseases were believed to, at best, remain the same, but were generally thought to get worse with time. That self-healing is possible also in chronic degenerative diseases is a revolutionary discovery that has been consolidated by several studies such as those of Dean Ornish on coronary patients. Thus, there is a healing system in the body. The healing system is a super-system, its principal components being the immune system, the endocrine system and the nervous system. However, for the healing system to work, right conditions are necessary. The right conditions are just the opposite of what leads to the lifestyle disorders. In short, the right conditions are: moderate physical activity; a good balanced diet in just the right quantity—neither too much nor too little; keeping away from smoking, alcohol and other forms of drug abuse; good quality sleep for just the right duration—neither too much nor too little; and mental peace. Several studies have demonstrated the effects of all these factors on health indicators—favourable effects when these factors are present, and adverse effects when absent. The objective indicators studied most frequently in such studies have been those related to immunocompetence, but the breadth of studies is now expanding.

The conditions that promote self-healing act primarily by creating an appropriate biochemical environment for the diseased cells. The environment may be created by alterations in the neurotransmitter profile, which in turn may influence the endocrine profile, leading finally to immunomodulation through paracrine and autocrine mechanisms. In some cases, the neurotransmitters or cytokines may affect the disease process much like the drugs that the doctors use. But the chemicals released from the endogenous pharmacy are far better because they are part of a well-orchestrated response that leads to the release of the right chemical at the right place in just the right amount, and up-regulation of the receptors for the chemical in the organ where its action is required. On the other hand, doctors administer chemicals in large doses. Per chance, a small quantity of the chemical reaches the right place and gives some relief. But quite a bit of the chemical also reaches those places in the body where it is not required; that is what gives rise to 'side effects', a euphemism for unpleasant and undesirable effects. A case in point is morphine, which is a very potent analgesic, but is used in medicine with great caution because of its being strongly addictive. But endorphins, which are very similar to morphine, are a part of the endogenous pain relief system. Endogenously released endorphins provide relief from pain but have no undesirable effect. That is why, a substance that is freely released by everybody's brain whenever required has to be regulated by strict laws if it has to be sold by a chemist!

In short, mind-body approaches such as yoga act through a number of mechanisms. They rectify the flaws of the unhealthy lifestyle that might have contributed to the disease, and thereby arrest further progress of the pathogenic process. Secondly, they give the person the confidence that he will get better, and relieve the stress associated with the illness and also the stress that might be irrespective of the illness. All these factors improve immunocompetence and promote self-healing. Finally, the neuromodulation brought about by positive thinking releases chemicals from the endogenous pharmacy that might have a direct effect on the symptoms or the disease process. By making the best use of the self-healing mechanisms of the body, mind-body approaches such as yoga reduce our dependence on interventions like drugs, surgery and radiotherapy.

CONCLUSION

Mind-body approaches such as yoga have a place at virtually all levels of prevention and management of disease. If used regularly early in life, they promote good health and reduce the likelihood of disease. If in spite of practicing yoga, a person does fall ill, self-healing mechanisms accelerate recovery from the illness. If self-healing is not possible, the endogenous pharmacy reduces the intensity of the symptoms. Further, if the disease is incurable, the person is not miserable – rather he is at peace because he can see even in the disease an opportunity for spiritual growth, which is the purpose of life. That is what has distinguished great yogis from ordinary mortals. The yogis have not been immortal, and many of them have died of incurable painful diseases, but their mental peace was not disturbed by the illness. Finally, when the person is dying, the true practitioner of yoga dies in peace because he looks upon death not as the end of life but as a mere episode in the long journey of the soul. Thus, yoga has a role at all stages of medical practice, right from primary prevention through palliative care. As is obvious, the popular tendency

to view yoga as a close cousin of physiotherapy reflects a poor understanding of what yoga is, and also leads to gross under-use of a valuable tool.

SUMMARY

Towards the middle of the twentieth century, diseases such as heart disease, diabetes and cancer emerged as the major killers of mankind. An unhealthy lifestyle, along with mental stress with which it is invariably associated, contributes significantly to the causation of these diseases. In its search for healthy lifestyles and strategies for stress management, modern medicine rediscovered ancient discipline such as yoga because these disciplines combine superb lifestyles with potent prescriptions for lasting mental peace. Further impetus for the integration of mind-body approaches like yoga into modern medicine came from developments in psychoneuroimmunology.

Yoga makes us *physically fit, emotionally stable, and intellectually agile*. The principal techniques that yoga uses for improving the body-mind complex are asanas, pranayamas and meditation. Yoga not only improves the body-mind complex, it also tells us how to use this equipment. While looking upon yoga as a journey towards self-perfection is the underlying spirit common to all schools of yoga, the emphasis and form differs in different schools of yoga. Thus, hatha yoga emphasizes physical perfection, whereas raja yoga emphasizes mental perfection. The yoga of the Gita encourages us to bring yoga into daily life, and provides the option of beginning with action, devotion or knowledge, depending upon the inclinations of the seeker. Integral yoga of Sri Aurobindo and the Mother synthesizes the central principles of all the major schools of yoga into a harmonious whole, and emphasizes that yoga should go beyond individual salvation to embrace raising the world consciousness.

In yoga, many of the physical aspects of lifestyle that keep the body-mind complex fit are collectively referred to as physical culture. The principal components of physical culture are the practices such as asanas and pranayamas; a balanced diet in just the right quantity; not putting into the body harmful substances such as tobacco and alcohol; and just the right duration of good quality sleep. As important as physical culture is the mental discipline, of which meditation is the most visible part. In yoga, these lifestyle factors are a means to better health, which is a requirement for right thought and action, which in turn is necessary for spiritual growth, which is the purpose of life. Yoga is not just a healthy lifestyle, it is also a stress buster. Yoga relieves stress by helping us look at all conditions of life, including adverse circumstances and traumatic events, as triggers or opportunities for spiritual growth, which is the purpose of life.

Modern medicine uses yoga as a major tool in lifestyle modification programs that are in tune with the latest advances in mind-body medicine. These programs involve, first, physical activity, improvements in diet, stopping smoking and drug abuse, and sleep hygiene. Secondly, there is patient education, which deals with the disease process, lifestyle, and stress management. The daily routine includes meditation, soothing music, autosuggestion, imagery and visualization. Finally, facilities are also made available for individualized counseling on lifestyle and psychological issues that the patient might have. By making the best use of the self-healing mechanisms of the body, mind-body approaches such as yoga reduce our dependence on interventions like drugs, surgery and radiotherapy. These approaches can play a role at all stages of medical practice, right from primary prevention through palliative care.

SUGGESTIONS FOR FURTHER READING

Benson Herbert (1975), *The Relaxation Response*, Avon Books, New York.

Bhatia R., Dureja G.P., Tripathi M., Bhattacharjee M., Bijlani R.L. and Mathur R. (2007), Role of Temporalis Muscle Over-activity in Chronic Tension Type Headache: Effect of Yoga Based Management, Indian *J. Physiol Pharmacol.,* 51:333–344.

Bijlani R. (2011), Indianizing Psychiatry (Letter), *Indian J. Psychiatry*, 53(4):373–374.

Bijlani R.L. (2008), Yoga: An Ancient Tool in Modern Medicine, Editorial. *Natl Med J.* India, 21:215–216.

Bijlani R.L., Vempati R.P., Yadav R.K., Ray R.B., Gupta V., Sharma R., Mehta N. and Mahapatra S.C. (2005), A Brief but Comprehensive Lifestyle Education Program Based on Yoga Reduces Risk Factors for Cardiovascular Disease and Diabetes Mellitus, *J. Altern Complement Med.,* 11:267–274.

Bijlani Ramesh (2008), *Back to Health through Yoga*, Rupa, New Delhi.

Brownstein, Arthur (2005), *Extraordinary Healing: The Amazing Power of Your Body's Secret Healing System*, Gig Harbor: Harbor Press.

Chopra Deepak (1989), *Quantum Healing, Exploring the Frontiers of Mind/Body Medicine*, Bantam, New York.

Cousins Norman (1979), *Anatomy of an Illness as Perceived by the Patient,* W.W. Norton, New York.

Dossey Larry (1999), *Reinventing Medicine: Beyond Mind-Body to a New Era of Healing*, HarperSanFransisco, New York.

Gupta N., Khera S., Vempati R.P., Sharma R. and Bijlani R.L. (2006), Effect of Yoga-based Lifestyle Intervention on State and Trait Anxiety, *Indian J. Physiol Pharmacol.,* 50:41–47.

Manjunatha S., Vempati R.P., Ghosh D. and Bijlani R.L. (2005), An Investigation into the Acute and Long-term Effects of Selected Yogic Postures on Fasting and Postprandial Glycemia and Insulinemia in Healthy Young Subjects, *Indian J. Physiol Pharmacol.,* 49:319–324.

Mohan A., Sharma R. and Bijlani R.L. (2011), Effect of Meditation on Stress-induced Changes in Cognitive Functions. *J. Altern Compliment Med.,* 17(3):207–212.

Ornish D., Brown S.E., Scherwitz L.W., Billings J.H., Armstrong W.T., Ports T.A., McLanahan S.M., Kirkeeide R.L., Brand R.J. and Gould K.L. (1990), Can Lifestyle Changes Reverse Coronary Heart Disease? *Lancet,* 336:129–133.

Ornish Dean (1996), *Dr. Dean Ornish's Program for Reversing Heart Disease*, Ivy Books, New York.

Ornish Dean (1996), *Love and Survival:* 8 *Pathways to Intimacy and Health,* Harper Collins, New York.

Read Nick (2005), *Sick and Tyred: Healing the Illnesses Doctors Cannot Cure*, Phoenix, London.

Siegel Bernie (1990), *Peace, Love and Healing*, Arrow, London.

Siegel Bernie (1998), *Love, Medicine and Miracles*, Arrow Books, London.

Vempati R.P., Bijlani R.L. and Deepak K.K. (2009), The Efficacy of a Comprehensive Lifestyle Modification Programme Based on Yoga in the Management of Bronchial Asthma: A Randomized Controlled Trial, *BMC Pulmonary Medicine,* 9:37.

Weil Andrew (1995), *Health and Healing: Conventional and alternative medicine—The Principles, the Practice*, Warner, London.

129

Mind, Molecules and Medicine

Ramesh Bijlani

CONTENTS

Since the middle of the twentieth century, modern medicine has been going through a double revolution. On one hand, advances in cellular and molecular biology have brought within the realm of possibility techniques such as gene therapy and stem cell therapy. On the other hand, the emergence of disorders rooted in the modern lifestyle, of which mental stress is a major accompaniment, has turned the attention of modern medicine to ancient disciplines such as yoga, which combine superb lifestyles with potent prescriptions for lasting mental peace. Thus, while medicine is targeting cells, genes, receptors and individual molecules for reaching the roots of disease, it is also looking at holistic disciplines that treat the body, mind and soul as one unit. While the two revolutions seem to be poles apart, they are in many ways interrelated. The revolution at the molecular end of the spectrum has helped place the mind-body relationship on a scientific footing. It has been shown that psychological changes have physiological consequences that are reflected in alterations in immunocompetence and neuroendocrine secretions. By showing a link between the mind, which is subjective, and biochemical profile, which is objective, these studies have made the mind-body connection acceptable to modern science, which has an inherent mistrust of subjective phenomena. The biochemical profile can be assessed in a laboratory, the results expressed in figures and analysed statistically. The quantifiable data that techniques in molecular biology have generated has helped place the mind-body relationship on a firm scientific footing. What makes this development relevant to medicine is the fact that the physiological and biochemical consequences of psychological changes are extremely relevant to health and disease. Aided once again by the advances in molecular biology, the biochemical environment that promotes or prevents health and healing at cellular level has reached a new level of understanding. Further, it has been found that stress creates an environment that promotes cellular damage and poor protection against disease. On the other hand, relaxation provides the cells an environment that strengthens our defence mechanisms and promotes self-healing. As a result, modern scientific medicine has now willingly incorporated into its armamentarium mind-body approaches like yoga to promote relaxation and positive thinking. This chapter will focus on the holistic end of the revolution in modern medicine, concentrating particularly on the biochemical correlates of psychological states with special reference to health and healing.

I. HOW DID WE GET THERE?

Let us trace the progression of ideas on the mind-molecule relationship since the beginning of the twentieth century. A convenient point to start is Walter Cannon's demonstration of the effects of anger on the body mediated by the sympatho-adrenal system using adrenaline and noradrenaline as neurotransmitters and hormones (Cannon, 1939). From here we might jump to the classic work of Roger Guillemin and many others starting in the 1950s on the hypothalamo-hypophysial relationship (Guillemin, 2005). Their work and its extensions gave us by the 1960s a scheme that has been summarized in Figure 129.1. This scheme made it easy to understand how thoughts and feelings could affect a wide variety of systems in the body not only through adrenal medullary secretions but also through alterations in the secretions of the adrenal cortex, the thyroid and the gonads. Thus, it became possible to explain rationally how the maternal instinct could stimulate ejection of milk from the breast, and psychological changes could lead to impotence or menstrual irregularities. However, till recently this knowledge remained primarily an interesting aspect of physiology. In the 1950s, modern medicine had not much use for such knowledge because it was going through the euphoria generated by advances such as the discovery of antibiotics, vaccines, vitamins and insulin and other hormones. These advances gave us tools that worked wonders without depending on the mind-body relationship. The result was that modern medicine sent the mind-body relationship into exile. A major milestone in the return of the mind-body relationship to clinical medicine was probably the publication of Norman Counsins' classic, *The Anatomy of an Illness* (Cousins, 1979). Norman Cousins was neither a physician nor a scientist by qualification, but was an original thinker. After getting ankylosing spondylitis, an essentially 'incurable' autoimmune disorder, following a stressful period in his life, he linked his disease to the mental stress. He then asked a simple question that modern medicine had forgotten to ask: 'if stress had led to the illness, could happiness restore health?' Based on this logic, he literally cured himself of the 'incurable' disease virtually through a combination of laughter and vitamin C. Eventually, Norman Cousins was invited to join the faculty of the UCLA School of Medicine, and he went on to stimulate physically, financially and intellectually a vast body of research that culminated in the area of knowledge today known as psychoneuroimmunology (Cousins, 1989). By the year 2000, the mind-body relationship had a firm scientific foundation, and the simple scheme of the 1960s had been transformed into a complex web (Figure 129.2). The immune system was now a part of the neuroendocrine circuit, and even more significantly, the relationship was now shown to work in both directions. On one hand, the brain influences the endocrine and immune systems; conversely, hormones and the cytokines released by the immune system affect the neurochemistry of the brain. Further, although the focus so far has been on the immune

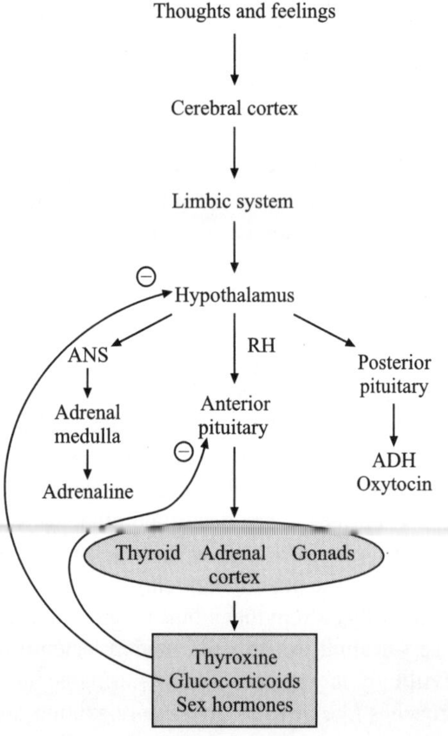

Figure 129.1   Knowledge of the hypothalamo-hypophysial axis and the sympatho-adrenal axis, as shown here, had been worked out by the 1960s, and gave the basic scientific foundation of the mind-body relationship

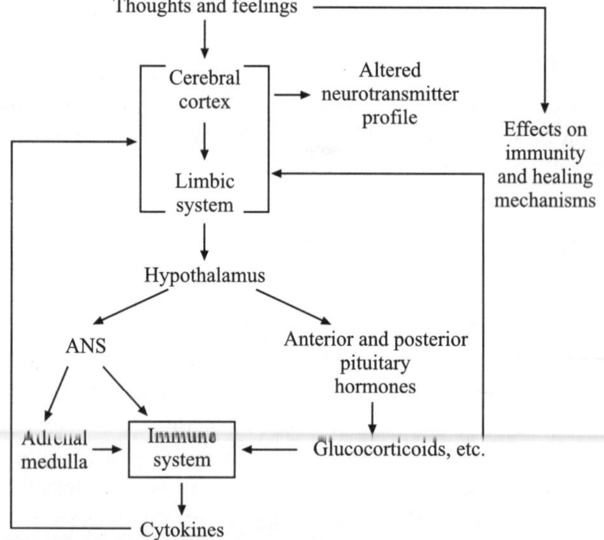

Figure 129.2   The emergence and development of psychoneuro-immunology, and the discovery of multiple bi-directional interactions between the mind, the nervous system, the endocrine system and the immune system put the mind-body relationship on a firm scientific footing by the year 2000. Some of the details shown in Figure 129.1 have been omitted here for the sake of clarity.

system, it is gradually becoming clearer that blood, the immune system, the nervous system and the endocrine system, along with many paracrine and autocrine mechanisms, form part of a ubiquitous system that may be termed the healing system (Brownstein, 2005). A glimpse of the ramifications of the recent developments may be had from Figure 129.3.

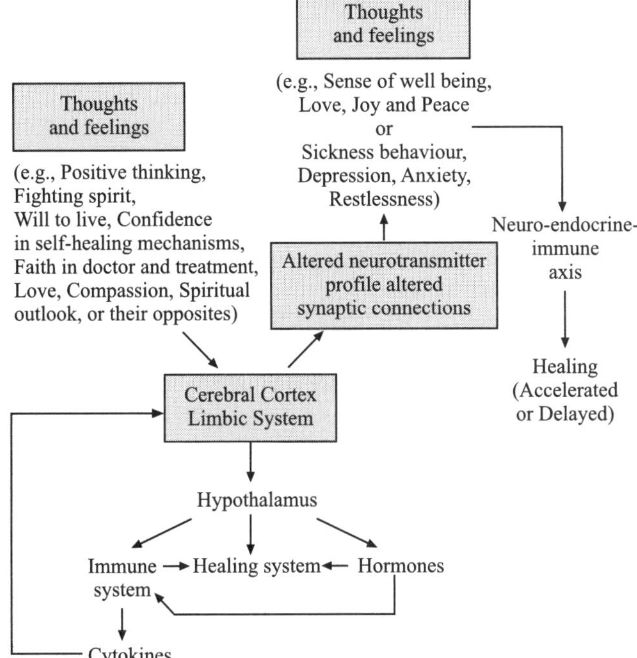

Figure 129.3 A corollary to the establishment of psychoneuro-immunology as a scientific discipline was its application to health and healing. Several experimental, clinical and epidemiological studies have contributed to the scientific basis of mind-body medicine depicted here. Some of the details shown in Figures 129.1 and 129.2 have been omitted here for the sake of clarity.

II. BIOLOGICAL CORRELATES OF PSYCHOLOGICAL STATES

George Bernard Shaw, while condemning animal experiments designed for studying the physiology of pain, which inevitably involve some pain to the animal, had once remarked that he could not understand why the physiology of pleasure had remained unexplored by medical scientists. Today, at least that criticism would be invalid because the biological correlates of positive as well as negative emotional states have been studied in considerable detail.

A. Stress

Physiologically, stress does not have an altogether negative connotation. Any situation that threatens homeostasis is stress. Stress may be acute or chronic.

Acute stress

Acute stress expresses itself through sympatho-adrenal discharge. Various parts of the body show the effects of sympathetic over-activity (such as an increase in the heart rate). In response to acute stress, the blood levels of catecholamines also go up. Physiologically, both a hard blow and a passionate kiss constitute acute stress, and the bodily manifestations of both are similar, although at the emotional level one is unpleasant and the other pleasant. The sympatho-adrenal discharge helps increase the blood flow and delivery of oxygen to the appropriate organs and mobilizes glycogen reserves of the body. Thus, the additional requirements of the body during acute stress are met by the physiological response. Acute stress occurs in both animals and man, and the body's response to it is a useful coping mechanism, a mechanism that helps preserve homeostasis.

Chronic stress

Chronic stress is much more common in human beings than animals. Having limited intelligence, animals tend to live in the here and now. But human beings are full of regrets about the past and anxieties about the future. Therefore they cannot enjoy the present, and suffer the effects of chronic stress. However, the harmful effects of chronic stress became apparent initially through experiments on rats performed by Hans Selye in the 1930s. He found that irrespective of the type of stress—cold, heat, exercise or surgery—the effects were peptic ulcer, enlarged adrenals, and atrophic lymph nodes. These effects were mediated largely by the elevated glucocorticoid secretion from the adrenal cortex. While the overall effects of glucocorticoids shift the metabolism towards mobilization of fuel reserves, and away from deposition of carbohydrate, fat or protein reserves, and are therefore useful to the body in coping with stress, excessive secretion going on for too long has deleterious effects.

B. Pain, Pain Relief and Pleasure

Pain is a stressful situation, and may be acute or chronic. Therefore, much of the above discussion on stress applies also to pain. But there are some molecules relatively specific to pain which need special consideration. Besides, pain may be physical or emotional.

Endorphins

The nervous system has structures not only for perception of pain but also for relief from pain. The endogenous pain relief system is activated by severe pain, a fact experienced by all of us, and expressed in poetic language by the great Urdu poet, Ghalib: *"Ishrat-e-katra hai dariya mein fana ho jana/Dard ka had se guzarna hai dawa ho jana"* (Just as a drop of water gets lost in a river, when pain crosses a limit, it cures itself).

The neurotransmitters employed by the endogenous pain relief system belong to a group collectively called the endorphins (from endogenous morphine). Morphine is an extremely powerful time-honoured pain killer, but it is avoided because of its strong addictive properties. When morphine receptors were detected in the nervous system by Solomon Snyder and his student Candace Pert (Pert and Snyder, 1973), it was a puzzle as to why receptors existed for a substance that was used only as a drug. How had nature anticipated that we would use morphine for pain relief, and furnished the appropriate neurons with receptors for the drug? The mystery was solved when it was discovered by John Hughes at the laboratory of Hans Kosterlitz in Aberdeen, Scotland, that the brain did have a morphine-like substance (Hughes, 1975). The substance was subsequently located as a neurotransmitter in the endogenous pain relief system. Later it was found that endogenous opioid peptides are not just neurotransmitters in the brain and spinal cord that may act post-synaptically or pre-synaptically, but are also released into the blood stream by the pituitary. There is a whole family of endorphins, all of them having the amino acid sequence tyr-gly-gly-phe. The simplest of endorphins are leu-enkephalin and met-enkephalin, which are both pentapeptides. β-endorphin is a thirty-amino acid polypeptide that is expressed primarily in the anterior pituitary, and is released along with ACTH in response to stress. There are three major classes of opioid receptors—μ, δ and κ. All these receptors belong to the G-protein coupled class of receptors. Analgesic activity of morphine and enkephalins depends on their binding to μ receptors. Naloxone also binds to μ receptors but does not activate the receptor. Therefore, naloxone acts as a competitive inhibitor of morphine and enkephalins, displacing them from the μ receptors.

Endorphins are involved not only in the pain relief system, but are also associated with pleasure and euphoria (Relief from pain is also a pleasure, and makes a person euphoric!). Besides pain, endorphins are released also by physical exercise, and possibly also by palatable food, love and sex—in short, by just about anything that is pleasurable. Thus endorphins are the biochemical correlate of pleasure, and of the addictive quality of pleasurable activities. Thus, endorphins, like the hypothalamus, may influence, directly or indirectly, a wide variety of autonomic functions. In fact, hypothalamus is one of the major areas of the brain where endorphins are employed as neurotransmitters.

Cannabinoids

Like morphine, cannabis is also addictive, and has been used by many to achieve the hallucination of being in a state of bliss. Now it has been found that there are similar substances produced in the body, which have been collectively named endocannabinoids (i.e., endogenous cannabinoids). The best known members of the group are anandamide (Sanskrit, *anand*, bliss) and 2-AG. Both anandamide (*N*-arachidonoylethanolamide, AEA) and 2-AG (2-arachidonoylglycerol) are derivatives of arachidonic acid. There are at least two types of cannabinoid receptors (CB1 and CB2, both G-protein coupled receptors), with indications that many more will be discovered soon. CB1 receptors are found in the brain, and CB2 receptors in the immune system. Cannabinoids are 'reverse neurotransmitters',

i.e. they are released by the post-synaptic neuron and act on the pre-synaptic neuron to reduce the release of the conventional neurotransmitter from the pre-synaptic neuron. Thus cannabinoids help the post-syntaptic neuron regulate the traffic of signals impinging on it. This phenomenon is called neuromodulation, and works through retrograde signaling.

Functionally, endocannabinoids seem to be part of a ubiquitous system involved in several regulatory functions. Only bits and pieces of information are available, but indications are that when all pieces of the puzzle have been discovered and put together, the endocannabinoid system will turn out to be a very fascinating component of the wisdom of the body. Cannabinoids have a unique role in the stress response. Glucocorticoids make an important contribution to the stress response, but their prolonged secretion has several deleterious effects. In case of non-threatening chronic stress, prolonged secretion of glucocorticoids is prevented by habituation of the hypothalamo-pituitary-adrenal (HPA) axis via 2-AG. On the other hand, the sensitivity of the HPA axis to novel stimuli is maintained by a decrease in anandamide secretion. A similar moderation of the desirable arousal and the undesirable aggression and anxiety in response to novel stimuli is also mediated by the endocannabinoid system. Cannabinoids are also involved in regulation of autonomic functions, food intake, reproductive system, memory, immunity and relief from pain.

Dopamine

Dopamine is a neurotransmitter in the brain. Major dopaminergic pathways in the brain exist in the prefrontal cortex, basal ganglia, limbic system, hypothalamus, substantia nigra and the ventral tegmental area. Dopamine acts via at least five types of receptors, named D1 through D5. Dopamine is degraded by reuptake and subsequent enzymatic breakdown. The reuptake is via the dopamine transporter (DAT_1), and breakdown by monoamine oxidase (MAO) in parts of the brain other than the prefrontal cortex. In the prefrontal cortex, the reuptake is by the norepinephrine transporter (NET), and the breakdown by catechol-o-methyl-transferase (COMT). The DAT_1 pathway is much faster than the NET pathway. Dopamine is a major component of the reward-driven learning system of the brain. Anything that gives us pleasure may be considered a reward. If an activity leads to a reward, we try to remember it, and then we try to repeat the activity. Thus, we learn to engage in activities that lead to a reward. One of the activities that give pleasure is novelty. When we have a good experience for the first time, it is enthralling and exciting. However, to get the thrills of new experiences, we have to be impulsive, willing to take risks, and have an exploratory temperament. Such behaviour, called novelty-seeking behaviour, has been found to have a strong genetic basis, and depends to a significant extent upon the high level of expression of the gene that encodes the D4 dopamine receptor in the limbic system and hypothalamus. Extroverts tend to have novelty-seeking personalities, and show higher levels of dopamine in the brain than the introverts. Several drugs of addiction such as cocaine and methamphetamine act by releasing dopamine in the brain, which in turn may be responsible for the pleasure that these drugs give.

Serotonin

Within the last few decades, serotonin has acquired the distinction of becoming popular as the happiness molecule. Much of the serotonin in the body is in the gut, where it regulates motility, but serotonin is also a neurotransmitter in the central nervous system. Serotonergic neurons are located principally in the raphe nuclei of the medullary reticular formation. Serotonin has seven known types of receptors, 5-HT_1 through 5-HT_7. They are mostly G-protein coupled receptors, with the exception of the 5-HT_3 receptor, which is a ligand-gated Na^+ and K^+ cation channel receptor.

Serotonin is involved in regulation of food intake. When a person is in presence of food, dopamine is released, which stimulates appetite. But as a person eats and starts feeling full, serotonin is released. Serotonin inhibits the secretion of dopamine, which reduces the motivation to eat. A high carbohydrate, low protein diet, like the traditional Indian diet, releases serotonin, and may thus promote a sense of well-being while at the same time limiting food intake (Fernstrom and Wurtman, 1971).

Serotonin is a key molecule in the induction of slow wave (non-REM) sleep. The serotonergic efferents projecting from the raphe nuclei to substantia gelatinosa of the spinal cord form a part of the endogenous pain relief system. These efferents act by reducing the transmission of pain impulses by the pain afferents in the spinal cord.

Serotonin is degraded by reuptake followed by enzymatic breakdown. The reuptake is mediated by a specific serotonin receptor (SERT). Amphetamine and tricyclic anti-depressants inhibit serotonin reuptake, and thereby allow serotonin levels to build up, leading to a sense of well-being, which may go on to euphoria. The latest compounds in the category are the specific serotonin reuptake inhibitors (SSRI), fluoxetine and sertraline. Being 'specific', they have fewer side effects than the older drugs. Cocaine and Ecstasy (3,4-methylenedioxy-N-methylamphetamine, or MDMA) also inhibit serotonin reuptake, which may contribute to their psychological effects.

C. Love and Compassion

Not just pain and pleasure, but also higher rungs of emotions such as love, compassion and empathy have biochemical correlates. The best-known molecule associated with these emotions is oxytocin.

Oxytocin

Oxytocin is a hypothalamic hormone stored and released by the posterior pituitary. For long it was considered to have only two functions: facilitating parturition by making the uterus contract, and ejection of milk. It was a hormone looking for a function in the male. But now oxytocin is considered to be primarily a neuromodulator in the brain. The neuromodulator activity is not due to the oxytocin released into the blood stream by the posterior pituitary because oxytocin cannot cross the blood-brain barrier. The neuromodulator activity is mediated by the release of oxytocin within the brain by hypothalamic neurons. Oxytocin receptors, which are G-protein coupled receptors, are found particularly in the limbic system, hypothalamus and the brainstem. An interaction between serotonin and oxytocin has been suggested by a study showing that the prosocial behaviour induced by the serotonin agonist MDMA may be at least partly due to release of oxytocin (Thompson et al., 2007).

Oxytocin is particularly active during positive emotional states, and has been nicknamed 'the love hormone'. Blood oxytocin levels increase in both men and women during sexual arousal. Any positive social interaction also releases oxytocin into the blood. Oxytocin has been associated with feelings of love, trust, empathy, contentment, peace and a sense of security; and with suppression of fear, anxiety and excessive caution. Intranasal administration of oxytocin promotes trust and willingness to take risks associated with social interactions (Kosfeld et al., 2005). In a closely related study that used functional magnetic resonance imaging (fMRI) for recording the activity of amygdala, intranasal administration of oxytocin was shown to reduce the activation of amygdala by fear-inducing visual stimuli (Kirsch et al., 2005). Possibly it is the modulation of behaviour by oxytocin that makes a young couple throw caution to the winds and indulge in socially unacceptable activities. Oxytocin seems to contribute to bonding between partners, and between the mother and her children. Interestingly, more oxytocin was released in women while having physical interactions with non-biological children than with biological children (Bick and Dozier, 2010). One of the explanations offered for the observation is that oxytocin is important for initiating the mother-child bond, not for maintaining it. Another explanation offered is that in case of non-biological children, the oxytocin surge possibly played an important role in reducing the anxiety associated with interacting with an unfamiliar child. In a double blind placebo-controlled study, intranasal administration of oxytocin to heterosexual couples increased positive communication behaviours between the partners during couple conflict discussions, and reduced their salivary cortisol levels (Ditzen et al., 2009). Oxytocin reduces inflammation by suppressing the release of certain cytokines. There is a study by Gouin et al. (2010) in which positive social interactions were found to accelerate wound healing, which was attributed to the reduction in inflammation mediated by oxytocin. Oxytocin has, in fact, been shown to reduce several indicators of inflammation such as IL-4, IL-6 and TNF-alpha (Clodi et al., 2008). Since there is an element of sub-clinical inflammation also in atherosclerosis and type 2 diabetes, oxytocin might be one of the biochemical explanations for the positive effect of love and intimacy in these conditions (Ornish, 1998).

Other hormones

Although oxytocin is the best studied 'love hormone', arginine-vasopressin (AVP) and sex hormones are also emerging as important elements in the biochemistry of emotions. The levels of endogenous oxytocin (OT) and AVP levels generally covary, and both together are a good biochemical indicator of the quality of marital relationship and physiological stress. While higher oxytocin levels were associated with more positive communication behaviours, higher vasopressin levels were related to fewer negative communication behaviours. Further, individuals in the upper oxytocin quartile and women in the upper vasopressin quartile healed experimental wounds faster than the remainder of the sample. (Gouin et al., 2010). That is why Semir Zeki has proposed that attachments result from a push-pull mechanism. On one hand, there is activation of the reward neural circuitry that leads to the thrill of companionship; on the other hand, there is suppression of the circuitry associated with critical assessment of people (Zeki, 2007). Although oxytocin and AVP are both involved in forming attachments, AVP seems to be associated primarily with male-typical social behaviours, such as aggression and pair-bonding (Heinrichs and Domes, 2008). It has been reported that there is an association between one of the human AVPR1A repeat polymorphisms (RS3) and traits reflecting pair-bonding behaviour in men, including partner bonding, perceived marital problems, and marital status. The results suggest that the well-established influence of AVP on pair-bonding in voles may be relevant also to human beings (Walum et al., 2008). Induction of an immune response by injecting lipopolysachharide (LPS) led to the release of AVP, and the release was mediated by interleukin-6 (IL-6) (Palin et al., 2009). Besides oxytocin and AVP, prolactin has also been implicated in the positive health outcomes of loving and intimate relationships (Neumann, 2009). The effect of oxytocin on neuronal activity in the medial amygdala seems to be important for social recognition in both sexes. In females, estrogens facilitate this process by stimulating oxytocin production in the hypothalamus, and up-regulating oxytocin receptors in the medial amygdala. Estrogens may also facilitate social recognition through non-genomic actions in the hippocampus. In males, on the other hand, androgen-dependent action of AVP seems to be more important for social recognition. Overall, it appears that modulation of OT and AVP by sex hormones regulates and fine tunes social recognition and other related behaviors such as social bonds and social hierarchies in a sex-specific manner (Gabor et al., 2012).

The broader implications of these very recent studies are clear, although the exact mechanisms require more studies. The broader implications are that loving interactions, and absence of negative interactions, specially between mother and child, and between romantically involved partners, are associated with higher oxytocin and AVP levels, better immunity, and faster wound healing. In short, love and intimacy translate into better health outcomes, and the link is mediated by oxytocin and AVP. But the mechanisms are complex. Either the social interactions may determine the hormonal secretions, or the genetically determined capacity to secrete hormones (or hormone receptors) may determine the quality of social interactions, or to some extent, both may be true.

D. Meditation and Altered States of Consciousness

Although there are several techniques of meditation, the general principle underlying all of them is to reduce the turbulence resulting from random mental activity that normally characterizes the awake state. Most people who try to meditate confess that they are unable to make much of a dent on the chaotic traffic of unwanted stray thoughts. However, some spiritual practitioners, after years of practice, are able to silence the surface activity of the mind for long periods of time. When they do that, their mind does not go completely blank: instead, they get what is often called a spiritual (or mystic) experience, but a neutral scientific description of the experience would be 'an altered state of consciousness'. There are several studies on meditation, but very few during spiritual experiences. Further, compared to the large number of studies on the physiological correlates of meditation employing electroencephalography (EEG) and brain imaging techniques, there are very few on the biochemical correlates. The imaging techniques have revealed a shift in the areas of the brain that are active during meditation. In general, areas of the brain involved in perceiving stimuli that orient us to our surroundings, such as the parietal cortex, become silent, whereas some other areas, such as the anterior cingulate cortex become active. But the exact areas that are active during meditation differ with the style of meditation. For example, in guided meditation, since the meditator is trying to understand the instructions, the prefrontal cortex is active (Cahn and Polich, 2006). Therefore, it may be assumed that during meditation the neurochemistry of the brain would also shift towards the neurotransmitters predominantly employed by areas that are active during the practice. There are a few studies that support this assumption.

A study on the dopaminergic transmission in the striatum using ^{11}C-raclopride, a competitive inhibitor of dopamine D2 receptors, found that ^{11}C-raclopride binding was reduced during yoga nidra meditation, indicating a 65% increase in dopaminergic transmission (Kjaer et al., 2002). The authors have hypothesized that this leads to a suppression of cortico-striatal glutamatergic transmission, which in turn leads to the reduced desire for motor activity seen during yoga nidra. An acute effect of meditation is to accentuate the mid-night increase in plasma melatonin level (Tooley et al., 2000). This might explain partly how meditation can facilitate falling asleep, and possibly some other benefits attributed to meditation such as improved immunity and slowing down of aging. A study on transcendental meditation revealed that after four months of daily meditation, the basal as well as average level of cortisol is lower, but the cortisol response to acute stress higher (MacLean et al., 1997). These findings indicate a lower level of stress, and an enhanced capacity to cope with a challenge. Studies on Sudarshan Kriya Yoga (SKY), which includes meditation, have found lower levels of blood lactate, higher levels of superoxide dismutase, glutathione and catalase (Sharma et al., 2003) and a better antioxidant status both at the enzyme activity and RNA level in the practitioners of SKY. Further, the lifespan of lymphocytes was prolonged in these subjects due to up-regulation of antiapoptotic genes and prosurvival genes, possibly leading to better immunity and better ability to cope with stress (Sharma et al., 2008).

Chemically-induced altered states of consciousness

While there are changes in the neurotransmitter profile seen during meditative states and spiritual experiences, states of consciousness bearing a close resemblance with spiritual experiences can be induced by chemicals that alter the neurochemistry of the brain. These chemicals include nitrous oxide, lysergic acid (LSD) and dimethyltryptamine (DMT). However, the resemblance between spiritual experiences and chemically-induced experiences is superficial. Two important differences exist between these contrasting situations. First, chemically-induced experiences are temporary. They last only as long as the person is under the influence of the chemical. On the other hand, spiritual experiences have a tendency to gradually outlast the meditative practice. Secondly, and more importantly, the core of the spiritual experience is the transforming influence it has on the individual. A person whose consciousness is elevated by meditation shows a progressive loss of vanity, desires, cravings and aversions, and a remarkable growth in compassion (Austin 1998). These behavioral changes are singularly missing in one who has gone through apparently similar drug-induced experiences. This all-important difference holds in spite of similar brain areas being activated or inactivated, and similar neurotransmitters being released in both cases. Thus there is no short-cut to enlightenment. Bliss may have an associated neurochemistry but bliss does not reside in chemical molecules.

III. IMPLICATIONS FOR HEALTH AND DISEASE

The mind-body relationship has three principal participants: the nervous system, which works through neurotransmitters; the endocrine system, which secretes hormones; and the immune system, which employs cytokines for many of its functions. Further, the same chemical is frequently a neurotransmitter in one situation, and a hormone or a cytokine in another. This complex web linking the mind with these versatile molecules has innumerable implications for health and disease, only a few of which are known, and still fewer will be briefly touched upon here.

A. Psychoneuroimmunology

Psychoneuroimmunology (PNI), which as the name partly suggests, is the study of the relationship between the mind and the immune system, as mediated by the nervous and the endocrine systems. In general, chronic stress, depression and negative emotions decrease or derail immunocompetence, whereas relaxation, cheerfulness and positive thinking enhance it. However, in acute stress, immunocompetence might improve, and the teleological explanation that has been given for the observation is that an animal that has sustained an acute injury will recover better if the improvement in immunity is able to prevent the wound from getting infected (Dhabhar & McEwen, 1999). The link between the mind and the immune system is mediated at least partly by the innervation of the lymphoid tissue, and by the presence of receptors for various neurotransmitters and hormones on lymphocytes, monocytes, neutrophils and eosinophils (Sarkar et al., 2009). The derangement of immune system by mental stress may increase the risk not only of infections but also that of allergies, autoimmune disorders, and cancer. The immune system is involved also in the pathophysiology of atherosclerosis and insulin resistance, and hence PNI might explain partly the contribution that mental stress makes to the higher risk of coronary artery disease and diabetes mellitus.

Let us touch very briefly on some of the key mechanisms unraveled by PNI and their clinical implications (Malarkey & Mills 2007). Cytokines, that are important participants in the immune mechanisms are produced not only by immune cells but also by neural and endocrine cells. For example, anterior pituitary cells have co-localization of hormones and interleukins, and adrenal cortex produces IL-6 and TNF-α, which may regulate glucocorticoid secretion by autocrine or paracrine mechanisms. Immune cells produce the hypothalamic hormone, corticotrophin releasing hormone (CRH), and also the anterior pituitary hormones, adrenocorticotropic hormone (ACTH), growth hormone (GH) and prolactin. GH, in turn, increases INF-γ production by immune cells; and prolactin stimulates lymphocyte proliferation. Hormones affect the production of pro-inflammatory cytokines such as TNF-α, which in turn act at the level of glucocorticoid receptors to induce glucocorticoid resistance. The cytokine IL-6 reduces the expression of hepatic GH receptors, and thereby induces GH resistance. Prolactin and GH have significant immunomodulatory effects in the thymus and bone marrow. CRH reduces the production of the gonadotropin releasing hormone (GnRH), and glucocorticoids inhibit the release of the luteinizing hormone (LH), estrogen and progesterone. The cytokines, IL-1 and TNF-α also inhibit the production of GnRH and LH. Immune cells have β-2 adrenergic receptors, and the sensitivity or density of these receptors is affected by stress. Thus, the complexity as well as implications of the hypothalamo-pituitary axis (HPA) and the hypothalamo-sympatho-adrenal medullary axis (SAM) have multiplied enormously within the last few decades as a result of developments in PNI. The interactions, and the direction of the interactions, within the rapidly growing web have yet to be worked out in detail, but some consequences of the interactions may be spelt out. Elevation of catecholamines and glucocorticoids due to acute stress may improve the immunity but pose the risk of cardiac decompensation in a susceptible person. Persistent elevation of glucocorticoids and catecholamines due to chronic stress may lead to subclinical Cushing's disease, with abdominal obesity and its sequelae, such as the insulin resistance syndrome. Persistent elevation of glucocorticoids and catecholamines due to chronic stress also reduces the TH-1/TH-2 ratio, thereby increasing the risk of infections, allergies, and bronchial asthma. The suppression of GnRH secretion by CRH may be behind the frequent association between chronic stress and amenorrhoea and polycystic ovary syndrome. The levels of both ACTH and glucocorticoids in chronic stress may be high, but may be accompanied by glucocorticoid resistance. This combination may contribute to steroid-resistant asthma, ulcerative colitis and Crohn's disease. Glucocorticoid resistance also increases the TH-1/TH-2 ratio, thereby increasing the risk of rheumatoid arthritis, multiple sclerosis, type 1 diabetes mellitus, and autoimmune thyroiditis. Persistent elevation of glucocorticoids in the cerebrospinal fluid (CSF) accompanied by glucocorticoid resistance also has implications for the nervous system, specially in the very young, in whom it may lead to cognitive impairment, as has been shown in studies on rodents deprived of maternal love and care soon after birth (Weaver et al., 2004).

PNI has been a very significant development in terms of bringing about a paradigm shift in basic as well as clinical sciences. In basic sciences, it has blurred the classical distinctions between neurotransmitters, hormones and cytokines. In clinical sciences, it has taken the mind-body relationship out of the realm of anecdotes to that of evidence-based medicine.

B. Healing System of the Body

The healing of a bony fracture or a surgical incision is the result of finely orchestrated events that cannot be clearly localized to a single system. Healing invariably involves at least the blood cells, the immune system, and the nervous and endocrine systems, besides plenty of paracrine and autocrine interactions. Thus the healing system may itself be considered a super-system based on the coordinated activity of several systems (Brownstein, 2005). Since the mind is closely linked to the systems involved in healing, it is not surprising that chronic stress delays wound healing, and the relationship may involve more than PNI (Walburn et al., 2009). Although the focus of research so far has been on the immune system, the relationship of the mind in relation to healing is unlikely to be confined to the immune system.

C. Illness Behaviour

The relationship between the mind and the immune system works both ways. The neurotransmitters released by the brain affect the immune system; on the other hand, the cytokines released by the immune system affect the neurons, and consequently behaviour. For example, interleukin-1 not only contributes to fighting infection directly but also induces sleep, which might help fight infection indirectly by enforcing physical rest. Several cytokines have now been shown to induce a whole complex of behavioural changes such as reduced appetite, tiredness, anhedonia (reduced pleasure from otherwise rewarding stimuli), mild depression, and a generalized feeling of being unwell. These are symptoms of which we all have personal experience. Now we also know that these symptoms are due to the neural effects of the same cytokines that "orchestrate immune defense and homeostasis" (Moon et al., 2011). The complex has been called illness (or sickness) behaviour. Like sleep, illness behaviour also possibly helps in recovery by making the person withdraw from activity and social interactions.

D. Psychiatric Illnesses

Considerable progress has been made in the delineation of the neurochemical correlates of the major psychiatric illnesses. This knowledge came initially from the site of action of the drugs found to be useful in the treatment of these illnesses. Knowledge arrived at by this indirect route has been further examined and supplemented by brain imaging, post-mortem studies, and studies on animal models of mental illnesses. Progress in this field has, in turn, helped in the development of psychopharmacological agents that are more effective than the older drugs and have fewer side effects. By way of illustration, we shall consider here only three mental illnesses.

Depression

Depression is one of the commonest mood disorders, affecting almost everyone at least a few times during life. Depression is not a single illness, but a group of disorders. However, depression has been linked to abnormalities of the subgenual region of the prefrontal cortex. This part of the brain has extensive connections with the noradrenergic, serotonergic and dopaminergic systems of the brain. The most commonly used antidepressants are the selective serotonin reuptake inhibitors such as fluoxetine, paroxetine and sertraline. By inhibiting the reuptake of the serotonin released as a neurotransmitter, these drugs allow serotonin levels to build up at the synapses. But there are indications that depression does not result from a simple reduction in serotonergic transmission. There are other elements such as excessive hypothalamic secretion of CRH, and the consequently elevated secretion of ACTH and glucocorticoids, and it is the neural effects of excess glucocorticoid secretion that might change the neurotransmitter profile, or the sensitivity to neurotransmitters in a direction that induces depression (Bondy, 2005). Depression has also been linked to cytokines, and this might particularly explain the depression associated with illness behaviour (Maes, 2011).

Schizophrenia

As in case of depression, clues about the neurochemistry of schizophrenia have also come primarily from pharmacological studies. Drugs like L-DOPA, which increase the level of dopamine in the brain, can induce psychotic symptoms resembling schizophrenia. There is a strong correlation between the antipsychotic potency of a drug and its ability to block dopamine D2 receptors. Of the several dopaminergic systems in the brain, overactivity of dopaminergic transmission in schizophrenia is thought to be restricted to the mesolimbic system, specially the projections to the nucleus accumbens. Nucleus accumbens receives dopaminergic inputs from the amygdala, hippocampus and the temporal lobe, processes this information, and then sends its output to the hypothalamus and frontal lobes. Excessive activity of this circuitry due to overactive dopaminergic inputs to the nucleus accumbens may account for the positive symptoms of schizophrenia, such as hallucinations and paranoia (Kandel, 2000). Patients of schizophrenia who do not respond to D2 receptor blockers respond to newer drugs that antagonize not only D2 but also D3 and D4 dopaminergic receptors and serotonergic receptors. This gave rise to the dopamine/serotonin hypothesis, i.e. overactivity of these systems is the neurochemical lesion in schizophrenia. However, later studies revealed that this overactivity may be secondary to primary abnormality in the glutamate system. These studies suggest that the primary abnormality may be a loss of NMDA receptors, particularly in interneurons. Loss of inhibition leads to a secondary overstimulation of the glutamergic and monoaminergic systems. Thus schizophrenia seems to be characterized by disinhibition in the cortico-limbic networks (Kehrer et al., 2008). The lack of cortical inhibition seen in schizophrenia has also been attributed to deficiency of GABA-ergic transmission. Studies suggest a possible GABA(A) dysfunction early in the disease course, whereas alterations in GABA(B) function seem to occur later in the disease's progression (Hasan et al., 2012). There are also indications that cytokines might serve as markers of schizophrenia, but this knowledge is yet to be incorporated into a coherent scheme. Thus, some cytokines (IL-1β, IL-6, and TGF-β) may be 'state markers' for acute exacerbations of schizophrenia, while others (IL-12, IFN-γ, TNF-α, and sIL-2R) may be 'trait markers', i.e. personality traits suggestive of schizophrenia (Miller et al., 2011).

Autism

Autism is a heterogeneous group of behavioural disorders affecting very young children, in which genetic, autoimmune, and environmental factors might make a significant contribution to the pathogenesis. The symptoms include socially unacceptable behaviour, attention deficit, repetitive and stereotyped behaviour, and poor communication. As in case of schizophrenia, the disorder has been attributed to impaired excitatory/inhibitory balance. However, when it comes to detail, there is no consensus, the suggestions varying from deficient to excessive glutamtergic transmission (Bernardi et al., 2011). In their own study, Bernardi et al. (2011) found lower glutamate concentrations in the right anterior cingulate cortex, and lower myo-inositol concentrations in the left temporo-parietal junction in autistic children than in healthy controls. There are also indications that in some cases of autism, the pathogenesis may involve local inflammation of the nervous system. In support of this hypothesis are studies such as those of Li et al. (2009) that have found elevated levels of the cytokines IL-6, TNFα, IL-8, GM-CSF and IFNγ in the frontal cortices of autistic subjects. The wide range in findings is in keeping with the fact that autism includes a spectrum of disorders with partial overlap in symptoms.

IV. APPLICATIONS IN MEDICINE

The discovery of mind-molecule correlations has been exploited for prevention and management of disease as discussed briefly below.

A. Creating Happiness

High brain serotonin levels have been correlated with happiness, or at least a sense of well-being. In 1971 was published a study, which has now become a classic, that demonstrated that a high carbohydrate diet led to an insulin-mediated increase in blood and brain tryptophan levels, which in turn elevated brain serotonin levels (Fernstrom and Wurtman, 1971). Following this study, many more have been conducted with the expectation that a high carbohydrate diet would elevate the mood, and the results have not been disappointing. One may speculate whether the capacity of Indians to stay happy in spite of many problems is due to their high carbohydrate diet.

Soon after its development, the second generation antidepressant, fluoxetine (Prozac or Fludac) made waves as the pill that promotes happiness. Fluoxetine acts as a selective serotonin reuptake inhibitor (SSRI) in the nervous system, and thereby raises the serotonin levels in the brain. Like a high carbohydrate diet, serotonin might also improve the subjective sense of well-being. But the deeper question that may be raised is whether genuine happiness can really be packed in a diet or a pill, or are these mere stop-gap arrangements to make a dismal situation tolerable.

B. Placebo

Doctors have known for long, and it agrees with common sense, that if the patient expects a treatment to work, it is likely to work irrespective of what the treatment is. The healing effect of suggestion and expectation has also been used by giving the patient a harmless but pharmacologically inert substance such as sugar, packed in a pill or a capsule, which the patient takes under the impression that he is taking a drug. The inert substance is called a placebo (literally, 'I please') because it has been given simply to please the patient. Not long ago, giving a placebo to a patient was considered a form of cheating, which was acceptable only if specific treatment for the disease was not known. But now, with a better understanding of the healing powers of the mind, the placebo is looked at with a good deal of respect. It is being increasingly acknowledged that quite a significant fraction of the therapeutic effect of even pharmacologically active substances is placebo effect. That is why, the colour and brand of a drug could affect its efficacy (Autret et al., 2012). In a PET study it was found that the placebo induced activation of opioid neurotransmission in a number of brain regions. These included the rostral anterior cingulate, orbitofrontal and dorsolateral prefrontal cortex, anterior and posterior insula, nucleus accumbens, amygdala, thalamus, hypothalamus, and periaqueductal grey (Zubieta and Stohler, 2009). Thus the placebo triggers in the patient the release of healing molecules from the rich endogenous pharmacy. This pharmacy delivers the right chemicals in just the right dose at only the right place for the right duration. Further, release of a chemical from the endogenous pharmacy is part of a series of coordinated events, which may include up-regulation of the receptors for the chemical at the sites where its action is really required. This is the ideal way to deliver a healing chemical, and leads to no side effects. In contrast, if a similar chemical is given in a pill, capsule or injection, it is given in a large dose, it circulates all over the body, and acts at the sites where its action is needed, as well as at sites where its action is not required. The actions at the sites where the drug is not required leads to side effects. Therefore, the current way of looking at a placebo is as an acceptable way of promoting self-healing. What is the harm in using it if it works, and has no side effects? The placebo may be considered a concession to the human consciousness, which is rooted primarily in the physical. Using the powers of the mind to heal the body is much easier if a physical object such as a pill is used as a trigger, than to use these powers without any physical aid.

C. Mind-body Medicine

After much research and reflection, Larry Dossey has reached the conclusion that modern medicine left behind the era of physical medicine to enter the era of mind-body medicine around the year 1950 (Dossey, 1993; Dossey, 1999). That does not mean that the spectacular advances made during the heyday of physical medicine were rejected, but what happened was that a new dimension was added to physical medicine. The new dimension was based on the scientific validation of the mind-body relationship. Now that the mind-body relationship has a firm scientific foundation, our understanding of health and disease has expanded, with significant implications for prevention and management of all diseases. For example, in an infectious disease, the invading microorganism is necessary but not sufficient to produce an illness. Illness will be seen only if the host resistance is low. If a thousand people eat contaminated food at a party, not all get food poisoning. And the ones who get the poisoning are the ones whose immunity is weak. Immunocompetence depends on many factors, of which the mental well-being is one. Thus mental stress weakens the immunity, and weakening of the immunity makes a person susceptible to infectious diseases. Once the disease is there, recovery from the disease also needs good immunity. The antibiotics reduce the bacterial population, and thereby make the task of the immune system easier, but elimination of the very last germ needs good immunity. That is why a person under stress gets repeated infections. This is relevant also to cancer. It is immunological surveillance that eliminates potentially cancerous cells every time there is a dysregulation of the cell cycle somewhere in the body. Further, if a tumour has developed, surgery and other modes of treatment only reduce the population of tumour cells. Total cure can come only if the last cancer cell lurking somewhere in the body is eliminated, and that can be done only by the immune system.

That is why, complete recovery from cancer is invariably seen only in patients who have hope, a fighting spirit, a will to live, and are confident that they will overcome the cancer. Mental stress is a major contributor to chronic degenerative diseases such as coronary artery disease and diabetes. Further, it has been shown through several studies that the mind-body relationship works both ways even in these diseases. Removal of mental stress can help create the environment in the body in which self-healing mechanisms can work, and even chronic degenerative diseases can start reversing when such an environment is created. Thus, mind-body medicine uses the powers of the patient's mind to help heal the patient's body. That does not mean that it rejects drugs or surgery. Mind-body medicine is not an either/or approach. It enriches physical medicine; it does not replace physical medicine. It expands the armamentarium of the physician by giving him one more tool. But this tool depends on the patient harnessing the powers of his own mind. Therefore, it is a tool that requires active participation of the patient in his own treatment. The task of the physician is to promote positive thinking in the patient in order to facilitate self-healing. Therefore, the physician needs a tool to promote positive thinking, and that is where ancient disciplines like yoga become relevant.

Yoga as a tool in mind-body medicine

Yoga is a way of life, and also a view of life. It is a way of life that is designed to make the individual physically fit, emotionally stable, and intellectually agile. It is a view of life that gives meaning and purpose to life. It is a view of life that makes a person look at *all* events and circumstances of life as opportunities for spiritual growth. It is a view of life that provides potent and infallible prescriptions for lasting mental peace. It is easy to understand that if yoga is viewed in this comprehensive manner, it becomes a valuable tool in mind-body medicine. That is how some of the best medical centers in the world, including the All India Institute of Medical Sciences, have used yoga as an integral part of modern scientific medicine (Bijlani et al., 2005).

CLOSING THOUGHTS

It is important to bear in mind that the biological correlates of psychological states are mere associations. Associations do not prove a cause and effect relationship; and even where a cause and effect relationship seems likely, the association does not identify the cause from the effect. In science, there is a tendency to consider consciousness to be an emergent property of the brain. As a corollary, a neuroscientist would instinctively consider the neuronal activity or the neurotransmitter profile to be the cause, and the psychological state to be the effect. But science has no evidence whatsoever to establish that this indeed is the cause-effect relationship. The fact that inducing a neurochemical change by, say, giving a drug, induces a psychological change does not prove that the chemical change is *always* the cause. The fact that neurons are necessary to manifest a psychological state also does not prove that neuronal activity is the cause. For example, the fact that a bulb is necessary to get light does not prove that the bulb is the cause, or the source, of the light. The cause is the invisible electricity, which needs a bulb to manifest as light. Thus the bulb is only a piece of apparatus that allows a certain potential of electricity to manifest; the fan or the computer allow other potentials of electricity to manifest. This is the line of thought that spiritual philosophers adopt in relation to also the brain and consciousness. They consider the brain to be only an apparatus that allows consciousness to manifest in a certain way. In this view, consciousness, or the psychological state precedes the neuronal activity and neurochemistry. As scientists, we have to remember that neither the scientific nor the philosophical view is based on irrefutable evidence. And as far as rational analysis goes, it is possible to argue endlessly but it is impossible to prove, or disprove, either point of view.

Mind-body medicine is the latest, but not the last incarnation of modern scientific medicine. From a broader perspective, the mind and the body influencing each other is not surprising because they are just two different modes of expression of the same consciousness. Larry Dossey has predicted that the next milestone in the development of modern medicine will be its entry into the era of non-local medicine (Dossey, 1993; Dossey, 1999). The concept of non-locality recognizes the fact that the consciousness of an individual is not confined to his body. Further, the consciousness of different individuals is part of a larger field of universal consciousness. Thus the patient's body and mind, and the healer's body and mind, are different modes of expression of the universal consciousness. Hence, in spite of a clear boundary separating their bodies, there is a continuity connecting the consciousness of different individuals. Therefore, not only the patient's mind, even the healer's mind has the power to influence the patient's body. For this influence, it is not necessary that the healer be a doctor. The good intentions, or prayers, of any well-wisher of the patient can also influence the patient's health positively. These are revolutionary ideas, but evidence in their favour is mounting so rapidly that medical scientists have already softened their stand against such heresies. At the other extreme, advances in molecular biology and biotechnology are promising to make stem cell therapy and gene therapy routine interventions. Time is not far when medical practice will be radically affected by these developments. Since these developments have been rather rapid, medical curricula have not kept pace with them. Therefore, medical students of today, who will be in practice for the next half century, will have a lot of self-learning to do, both in terms of the way they have been taught to think, and the way they have been taught to treat.

SUMMARY

The mind-body relationship stands now on a firm scientific footing. Psychological changes have physiological consequences; mental states have biochemical correlates. Adrenaline and glucocorticioids have long been associated with the stress response. Endorphins, earlier known as part of the endogenous pain relief system, are now looked upon as the biochemical correlate of pleasure, and of the addictive quality of pleasurable activities. Endocannabinoids fine tune the stress response by attenuating the glucocorticoid response to non-threatening chronic stress while still retaining the capacity to cope with novel stressful stimuli. Dopamine is a major component of the reward-driven learning system of the brain. Serotonin, the 'happiness molecule', is involved also in pain relief, food intake, and slow wave sleep. Oxytocin, working as a neuromodulator, has been associated with feelings of love, trust, empathy, contentment, peace and a sense of security; and with suppression of fear, anxiety and excessive caution. Besides oxytocin, arginine vasopressin (AVP) also is involved in forming attachments. Further, AVP seems to be associated with male-typical social behaviours, such as aggression and pair-bonding. Neurochemical alterations are seen during altered states of consciousness that may characterize meditative practices. An improvement in health-related indicators such as antioxidant activity and immunocompetence has been reported in regular meditators. Rapid advances have also taken place in our understanding of the neurochemistry of psychiatric illnesses.

The discovery of the physiological and biochemical consequences of mental states has wide-ranging implications for health and disease. Knowledge of the neurochemistry of psychiatric illnesses has facilitated the development of better antidepressants and antipsychotic drugs. The emergence of psychoneuroimmunology (PNI) as a scientific discipline has provided plausible explanations for the clinically observed relationship between mental stress and infections, allergies, autoimmune diseases and cancer. Further, the mind-body relationship, as substantiated by PNI, works both ways. Not only stress impairs immunocompetence, peace and joy enhance it. Peace and joy, love and compassion, and positive thinking and confidence in self-healing, create the biochemical environment that facilitates healing. The positive health outcomes of positive emotions form the basis of mind-body medicine, the latest incarnation of modern medicine. Mind-body medicine uses the power of the patient's mind to facilitate healing. That has made ancient mind-body approaches such as yoga, which incorporate potent prescriptions for positive thinking and lasting mental peace, valuable tools in the practice of medicine.

REFERENCES

Austin J.H. (1998), *Zen and the Brain: Toward an Understanding of Meditation and Consciousness*, The MIT Press, Cambridge, 579–584.

Autret A., Valade D. and Debiais S. (2012), Placebo and Other Psychological Interactions in Headache Treatment, *J. Headache Pain,* 13:191–198.

Bernardi S., Anagnostou E., Shen J., Alexander Kolevzon, Buxbaum J.D., Hollander E., Hof P.R. and Fana J. (2011), *in vivo* 1H-magnetic Resonance Spectroscopy Study of the Attentional Networks in Autism., *Brain Res,* 1380:198–205.

Bick J. and Dozier M. (2010), Mothers' and Children's Concentrations of Oxytocin Following Close, Physical Interactions with Biological and Non-biological Children, *Dev Psychobiol.,* 52:100–107.

Bijlani R.L., Vempati R.P., Yadav R.K., Ray R.B., Gupta V., Sharma R., Mehta N. and Mahapatra S.C. (2005), A Brief but Comprehensive Lifestyle Education Program Based on Yoga Reduces Risk Factors for Cardiovascular Disease and Diabetes Mellitus, *J. Altern Complement Med.,* 11:267–274.

Bondy B. (2005), Pharmacogenomics in Depression and Antidepressants, *Dialogues Clin Neurosci.,* 7:223–230.

Brownstein A. (2005), *Extraordinary Healing, The Amazing Power of Your Body's Secret Healing System,* Gig Harbor WA: Harbor Press.

Cahn B.R. and Polich J. (2006), Meditation States and Traits: EEG, ERP, and Neuroimaging Studies, *Psychological Bulletin,* 132:180–211.

Cannon W.B. *The Wisdom of the Body,* WW Norton, New York.

Clodi M., Vila G., Geyeregger R., Riedl M., Stulnig T.M., Struck J., Luger T.A. and Luger A. (2008), Oxytocin Alleviates the Neuroendocrine and Cytokine Response to Bacterial Endotoxin in Healthy Men, *Am J. Physiol Endocrinol Metab.,* 295:E686–E691.

Cousins N. (1979), *Anatomy of an Illness as Perceived by the Patient—Reflections on Healing and Regeneration,* W.W. Norton/Bantam Books, New York.

Cousins N. (1989), *Head First, The Biology of Hope and the Healing Power of the Human Spirit,* Penguin, New York.

Dhabhar F.S. and McEwen B.S. (1999), Enhancing Versus Suppressive Effects of Stress Hormones on Skin Immune Function, *Proc Natl Acad Sci,* USA, 96:1059–1064.

Ditzen B., Schaer M., Gabriel B., Bodenmann G., Ehlert U. and Heinrichs M. (2009), Intranasal Oxytocin Increases Positive Communication and Reduces Cortisol Levels During Couple Conflict, *Biol Psychiatry,* 65:728–731.

Dossey L. (1993), *Healing Words,* HarperCollins, New York.

Dossey L. (1999), *Reinventing Medicine: Beyond Mind-body to a New Era of Healing,* HarperSanFransisco, New York, (A Division of HarperCollins).

Fernstrom J.D. and Wurtman R.J. (1971), Brain Serotonin Content: Increase Following Ingestion of Carbohydrate Diet, *Science,* 174:1023–1025.

Gabor C.S., Phan A., Clipperton-Allen A.E., Kavaliers M. and Choleris E. (2012), Interplay of Oxytocin, Vasopressin, and Sex Hormones in the Regulation of Social Recognition, *Behav Neurosci.,* 126:97–109.

Gouin J.P., Carter S., Pournajafi-Nazarloo H., Glaser R., Malarkey W.B., Loving T.J., Stowell J. and Kiecolt-Glaser J.K. (2010), Marital Behavior, Oxytocin, Vasopressin, and Wound Healing, *Psychoneuroendocrinology,* 35:1082–1090.

Guillemin R. (2005), Hypothalamic Hormones a.k.a. Hypothalamic Releasing Factors, *J. Endocrinol.,* 184:11–28.

Hasan A., Wobrock T., Grefkes C., Labusga M., Levold K., Schneider-Axmann T., Falkai P., Müller H., Klosterkötter J. and Bechdolf A. (2012), Deficient Inhibitory Cortical Networks in Antipsychotic-naive Subjects at Risk of Developing First-episode Psychosis and First-episode Schizophrenia Patients: A Cross-sectional Study, *Biol Psychiatry.,* [Epub Ahead of Print] PMID: 22502988.

Heinrichs M. and Domes G. (2008), Neuropeptides and Social Behaviour: Effects of Oxytocin and Vasopressin in Humans, *Prog Brain Res.,* 170:337–350.

Hughes J. (1975), Isolation of an Endogenous Compound from the Brain with Pharmacological Properties Similar to Morphine, *Brain Research,* 88:295–308.

Kandel E.R. (2000), Disorders of Thought and Volition: Schizophrenia. In: Kandel E.R., Schwartz J.H., Jessell T.M. (Eds.), *Principles of Neural Science,* McGraw Hill, 4th ed., Chapter 60, New York, 1188–1208.

Kehrer C., Maziashvili N., Dugladze T. and Gloveli T. (2008), Altered Excitatory-inhibitory Balance in the NMDA-hypofunction Model of Schizophrenia, *Front Mol Neurosci.,* 1:6. Epub PMID: 18946539. [PubMed] Free PMC Article, Prepublished online 2008 March 19. Doi: 10.3389/neuro.02.006.2008. PMCID: PMC2525998.

Kirsch P., Esslinger C., Chen Q., Mier D., Lis S., Siddhanti S., Gruppe H., Mattay V.S., Gallhofer B. and Meyer-Lindenberg A. (2005), Oxytocin Modulates Neural Circuitry for Social Cognition and Fear in Humans. *J. Neurosci.,* 25:11489–11493.

Kjaer T.W., Bertelsen C., Piccini P., Brooks D., Alving J. and Lou H.C. (2002), Increased Dopamine Tone during Meditation-induced Change of Consciousness, *Brain Res Cogn Brain Res.,* 13:255–259.

Kosfeld M., Heinrichs M., Zak P.J., Fischbacher U. and Fehr E. (2005), Oxytocin Increases Trust in Humans, *Nature,* 435(7042):673–676.

Li X., Chauhan A., Sheikh A.M., Patil S., Chauhan V., Li X.M., Ji L. and Brown T. (2009), Malik M. Elevated Immune Response in the Brain of Autistic Patients. *J. Neuroimmunol.,* 207:111–116.

MacLean C.R.K., Walton K.G., Wenneberg S.R., Levitsky D.K., Mandarino J.P., Waziri R., Hillis S.L. and Schneider R.H. (1997), Effects of the Transcendental Meditation Program on Adaptive Mechanisms: Changes in Hormone Levels and Responses to Stress After 4 Months of Practice, Psychoneuroendocrinology, 22:277–295.

Maes M. (2011), Depression is an Inflammatory Disease, But Cell-mediated Immune Activation is the Key Component of Depression, *Prog Neuropsychopharmacol Biol Psychiatry.,* 35:664–675.

Malarkey W.B. and Mills P.J. (2007), Endocrinology: The Active Partner in PNI Research, *Brain Behav Immun.,* 21:161–168.

Miller B.J., Buckley P., Seabolt W., Mellor A. and Kirkpatrick B. (2011), Meta-analysis of Cytokine Alterations in Schizophrenia: Clinical Status and Antipsychotic Effects, *Biol Psychiatry,* 70:663–671.

Moon M.L., McNeil L.K. and Freund G.G. (2011), Macrophages Make Me Sick: How Macrophage Activation States Influence Sickness Behavior, *Psychoneuroendocrinology,* 36:1431–1440.

Neumann I.D. (2009), The Advantage of Social Living: Brain Neuropeptides Mediate the Beneficial Consequences of Sex and Motherhood, *Front Neuroendocrinol,* 30:483–496.

Ornish D. (1998), *Love and Survival:* 8 *Pathways to Intimacy and Health,* HarperPerennial, New York.

Palin K., Moreau M.L., Sauvant J., Orcel H., Nadjar A., Duvoid-Guillou A., Dudit J., Rabie A. and Moos F. (2009), Interleukin-6 Activates Arginine Vasopressin Neurons in the Supraoptic Nucleus During Immune Challenge in Rats, *Am J. Physiol Endocrinol Metab.,* 296(6):E1289–E1299.

Pert C.B. and Snyder S.H. (1973), Opiate Receptor: Demonstration in Nervous Tissue, *Science,* 179:1011–1014.

Sarkar C., Basu B., Chakroborty D., Dasgupta P.S. and Basu S. (2010), The Immunoregulatory Role of Dopamine: An Update, *Brain, Behavior, and Immunity.,* 24:525–528.

Sharma H., Datta P., Singh A., Sen S., Bhardwaj N.K., Kochupillai V. and Singh N. (2008), Gene Expression Profiling in Practitioners of Sudarshan Kriya, *J. Psychosom Res.,* 64:213–218.

Sharma H., Sen S., Singh A., Bhardwaj N.K, Kochupillai V. and Singh N. (2003), Sudarshan Kriya Practitioners Exhibit Better Antioxidant Status and Lower Blood Lactate Levels, *Biol Psychol.,* 63:281–291.

Thompson M.R., Callaghan P.D., Hunt G.E., Cornish J.L. and McGregor I.S. (2007), A Role for Oxytocin and 5–HT(1A) Receptors in the Prosocial Effects of 3,4–methylenedioxymethamphetamine (ecstasy), *Neuroscience,* 146:509–514.

Tooley G.A., Armstrong S.M., Norman T.R. and Sali A. (2000), Acute Increases in Night-time Plasma Melatonin Levels Following a Period of Meditation, *Biol Psychol.,* 53:69–78.

Walburn J., Vedhara K., Hankins M., Rixon L. and Weinman J. (2009), Psychological Stress and Wound Healing in Humans: A Systematic Review and Meta-analysis, *J. Psychosom Res.,* 67:253–271.

Walum H., Westberg L., Henningsson S., Neiderhiser J.M., Reiss D., Igl W., Ganiban J.M., Spotts E.L., Pedersen N.L., Eriksson E. and Lichtenstein P. (2008), Genetic Variation in the Vasopressin Receptor 1a Gene (AVPR1A) Associates with Pair-bonding Behavior in Humans, *Proc Natl Acad Sci.,* USA, 105:14153–4156.

Weaver I.C., Cervoni N., Champagne F.A., D'Alessio A.C., Sharma S., Seck J.R., Dymov S., Szyf M. and Meaney M.J. (2004), Epigenetic Programming by Maternal Behavior, *Nat Neurosci.,* 7:847–854.

Zeki S. (2007), The Neurobiology of Love, *FEBS Lett.,* 581:2575–2579.

Zubieta J.K. and Stohler C.S. (2009), Neurobiological Mechanisms of Placebo Responses, *Ann N Y Acad Sci.,* 1156:198–210.

SUGGESTIONS FOR FURTHER READING

Benson Herbert (1997), *Timeless Healing: The Power and Biology of Belief, Fireside,* New York.

Brownstein Arthur (2005), *Extraordinary Healing: The Amazing Power of Your Body's Secret Healing System,* Harbor Press, Gig Harbor.

Chopra Deepak (1989), *Quantum Healing, Exploring the Frontiers of Mind/Body Medicine, Bantam,* New York.

Cousins Norman (1979), *Anatomy of an Illness as Perceived by the Patient,* W.W. Norton, New York.

Cousins Norman (1989), *Head First: The Biology of Hope and the Healing Power of the Human Spirit,* Penguin, New York.

Dossey Larry (1993), *Healing Words: The Power of Prayer and the Practice of Medicine,* Harper Collins, New York.

Dossey Larry (1999), *Reinventing Medicine: Beyond Mind-Body to a New Era of Healing,* Harper San Fransisco, New York.

Ornish Dean (1996), *Dr. Dean Ornish's Program for Reversing Heart Disease,* Ivy Books, New York.

Ornish Dean (1998), *Love and Survival: 8 Pathways to Intimacy and Health,* Harper Collins, New York.

Pert Candace (1997), *Molecules of Emotion: The Science Behind Mind-Body Medicine,* Simon and Schuster, New York.

Read Nick (2005), *Sick and Tired: Healing the Illnesses Doctors Cannot Cure,* Phoenix, London.

Sapolsky Robert (1998), *Why Zebras Don't Get Ulcers: An Updated Guide to Stress, Stress-related Diseases and Coping,* W.H. Freeman, New York.

Siegel Bernie (1990), *Peace, Love and Healing,* Arrow, London.

Siegel Bernie. Love (1998), *Medicine and Miracles,* Arrow Books, London.

Weil Andrew (1995), *Health and Healing: Conventional and Alternative Medicine—The Principles, the Practice,* Warner, London.

SECTION XVIII

PRINCIPLES OF CHEMOTHERAPY

Drugs are widely used for treatment of patients. It is thus relevant to bring home to the reader the general principles of chemotherapy.

130

General Principles of Chemotherapy with Particular Reference to Antimetabolites

(Late) Ranjit Roy Chaudhury and Sangeeta Sharma

CONTENTS

I. HISTORY

Pasteur and Joubert in 1877 observed that anthrax bacilli grew rapidly when inoculated into sterile urine but failed to multiply and soon died if one of the common bacteria of the air was introduced into the urine. This observation that a microbe can produce a chemical which can kill another micro-organism led to the development of the modern antibiotics.

The credit of introducing the modern concepts of chemotherapy goes to Paul Ehrlich (1907) whose brilliant and painstaking work in the beginning of the present century led to the development of arsenicals such as atoxyl, tryparsamide and arsphenamine. The term "chemotherapy" was used by him to define a particular type of study having as its aim the discovery of synthetic chemical substances acting specifically on infective organisms. The most that he expected from these was maximum "parasitotropic" and minimal "organotropic" effect.

The development of chemotherapy during the present century is one of the most important therapeutic advances made in the history of medicine. Ehrlich could hardly have foreseen that, within 50 years of his discovery of arsenicals, penicillin would be discovered which could kill microorganism without harming the host, a term now referred to as selective toxicity.

The discovery in 1932 of prontosil—a sulphonamide azo dye—in the management of streptococcal infections led to the sudden spurt of interest in chemotherapy and antimetabolites. Trefouels and his associates (1935) showed that prontosil was first converted to sulphanilamide which was the active agent and Woods (1940) showed that there was a close resemblance between the structures of sulphanil-amide and para-aminobenzoic acid, an essential metabolite for the bacteria.

The chance observation of Fleming in 1928, while studying staphylococcus variants in the laboratory at St.

"

Mary's Hospital in London that a mould contaminating staphylococci plate caused the bacteria in the vicinity to undergo lysis led to the discovery of penicillin. Florey, Chain and Abraham (1940) demonstrated its efficacy against bacterial infection without much toxicity to the host. Since then there has been a steady stream of new chemotherapeutic agents. Streptomycin, tetracycline, newer beta lactam penicillins, the four generations of cephalosporins, the aminoglycosides, rifampin, the quinolones, amphotericin B and antibiotics which have anticancer properties.

Although significant advances have taken place in the availability of antibiotics, it has also led to the development of resistant strains of bacteria which do not respond to antibiotics to which they were previously sensitive. Notable among these are the multiple drug resistant tubercule bacilli and vancomycin resistant staphylococcus aureus. Treating patients harbouring drug resistant strains of microorganisms seems to be a challenge which the scientists of the twenty-first century are facing.

II. AGENTS AVAILABLE

A. Antibiotics

These agents can be divided into two major subgroups:

(a) Bactericidal drugs—drugs which are capable of killing the bacteria, e.g. penicillin, gentamicin, rifampin, metronidazole, vancomycin.
(b) Bacteriostatic drugs—drugs which inhibit the growth of microorganism. These drugs are effective when the host's immune system is functioning, e.g., sulfonamides, tetracyclines, macrolides, linezolid, chloramphenicol, clindamycin.

Yet another way to classify them is according to their source. These drugs can be obtained from both natural and synthetic sources.

(a) Chemicals, e.g., sulphonamides, isoniazid, para-amino-salicyclic acid and quinolones.
(b) Antibiotics, e.g., penicillin, cephaloridine, rifampin and vancomycin.

An antibiotic is, therefore, defined as a chemical substance obtained from a living organism which is capable of destroying another microorganism. However, this term is more often used loosely and includes even those agents which have a chemical source such as sulphonamides and quinolones.

B. Antimetabolites

These are classified into the following subgroups:

(a) antimicrobials
(b) anticancer drugs

(c) insecticides
(d) miscellaneous.

III. MECHANISM OF ACTION

A. Sulphonamide

The object of all chemotherapy is to produce toxic or lethal effects on the invading parasites with minimal toxicity to the host. This is known as selective toxicity. To accomplish this, a step in the biochemical synthetic pathway of the microorganism which is not present in the host cell has to be attacked. This point can be further elaborated by the example of sulphanilamide. It was observed that when a 0.01% sulphanilamide solution was added to a culture of haemolytic streptococci, its further growth was inhibited. In in vivo experiments it was shown that the polymorphonuclear leukocytes were eventually responsible for the destruction of the bacteria. The antimicrobial activity of sulphonamide is based on the theory of competitive inhibition put forward by Woods and Fildes (1940). The salient features of this hypothesis are:

1. Para aminobenzoic acid (PABA) is essential for the growth and cellular metabolism of certain bacteria. In other words, it is an essential metabolite.
2. There is a close similarity between the chemical structures of sulphanilamide and PABA. (Figure 130.1)
3. A concentrate from haemolytic streptococci antagonized the action of the sulphanilamide.
4. PABA inhibited the effect of sulphanilamide, in vitro.
5. Very low concentrations of PABA (1:5000 to 1:25000) were enough to prevent the bacteriostatic action of sulphanilamide.
6. The degree of susceptibility of an organism depended in part on its ability to synthesize PABA. Resistant strains synthesized sufficient PABA to antagonize all the sulphanilamide present in the medium.
7. PABA is an essential constituent of folic acid. Sulphanilamide-susceptible organisms are unable to form the essential cofactor, i.e. tetrahydrofolic acid. This is due to competitive inhibition. Because of the close resemblance between the chemical structures of PABA and sulphanilamide, the latter gets incorporated and nucleic acid synthesis is hampered. The content of folate in the bacterial cell wall becomes diluted and a stage is reached when the level is insufficient to continue replication. The reason why the host cell is not affected is that it does not depend on the synthesis of folic acid by the body but utilizes the preformed folic acid in the diet. A few other examples where there is a close similarity between the drug and essential metabolites (Figure 130.2) are given below:

Besides these agents, there are various other groups of drugs which act as antimetabolites.

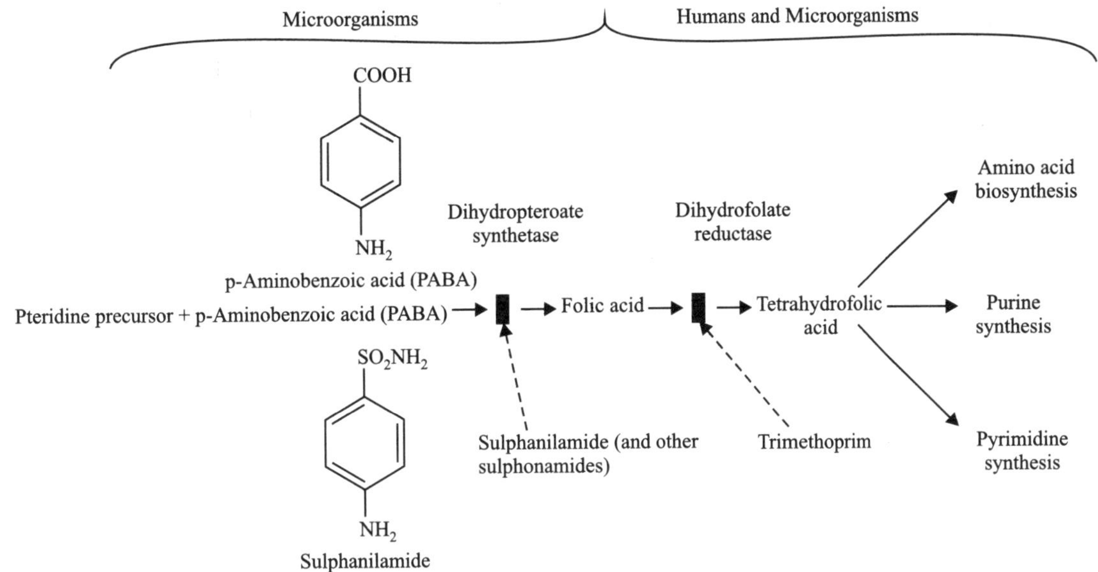

Figure 130.1   Inhibition of tetrahydrofolate synthesis by sulphonamides and trimethoprim.

Drug	*Essential Metabolite*
(a) Penicillin	D-alanyl-D-alanine
(b) Pyrazinamide	Nicotinamide
(c) D-cycloserine	D-alanine
(d) Azaserine	Glutamine.

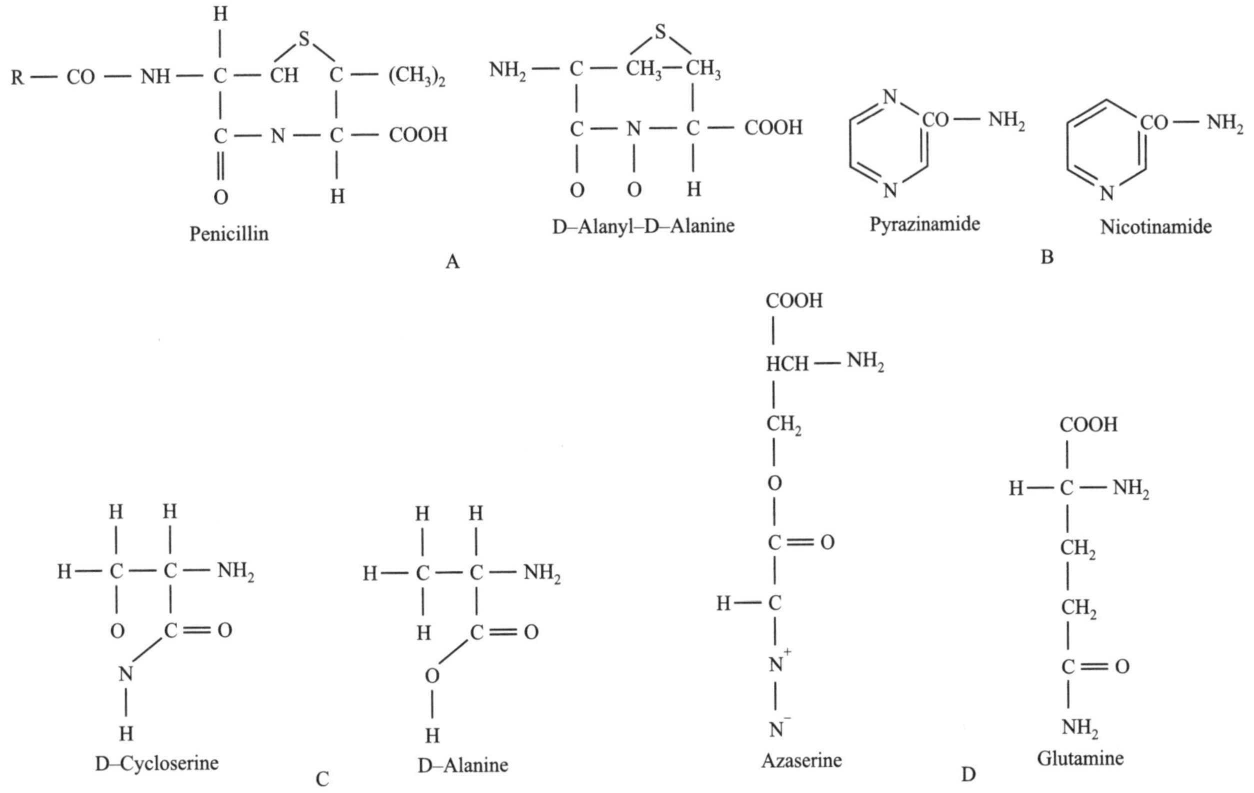

Figure 130.2   Drugs bearing structural similarity to metabolites inhibited.

B.   Anticancer Drugs

On the basis of their effects on the proliferation of tumour cells and haematopoietic stem cells in experimentally induced tumours, the anticancer drugs can be classified into three groups. The first group (nitrogen mustard and X-ray) decimated both the haematopoietic stem cells and the tumour cells to the same degree. The second group which included drugs like methotrexate, azaserines, 6–thioguanine and 6–mercaptopurine exhibited a marked differential cell death (five-hundred fold), but this difference diminished with increased doses. The third group consisted of agents (5–fluorouracil) which at even higher doses produced as much as ten-thousand fold greater death of the tumour cells. The major site of action of methotrexate as shown below is the inhibition of the enzyme dihydrofolate reductase (Fig. 130.3), which catalyzed the reduction of folate and dihydrofolate to tetrahydrofolate.

Folic acid   →   Folic acid   →   Dihydrofolic   →   Tetrahydrofolic
                                         acid   acid

Conjugates   1. Methotrexate   2. Methotrexate
in food

Figure 130.3

1 and 2 are the sites catalyzed by the enzyme dihydrofolate reductase.

A few examples of antimetabolites and essential dietary factors are given below. There is a close structural similarity between these two (Figure 130.4).

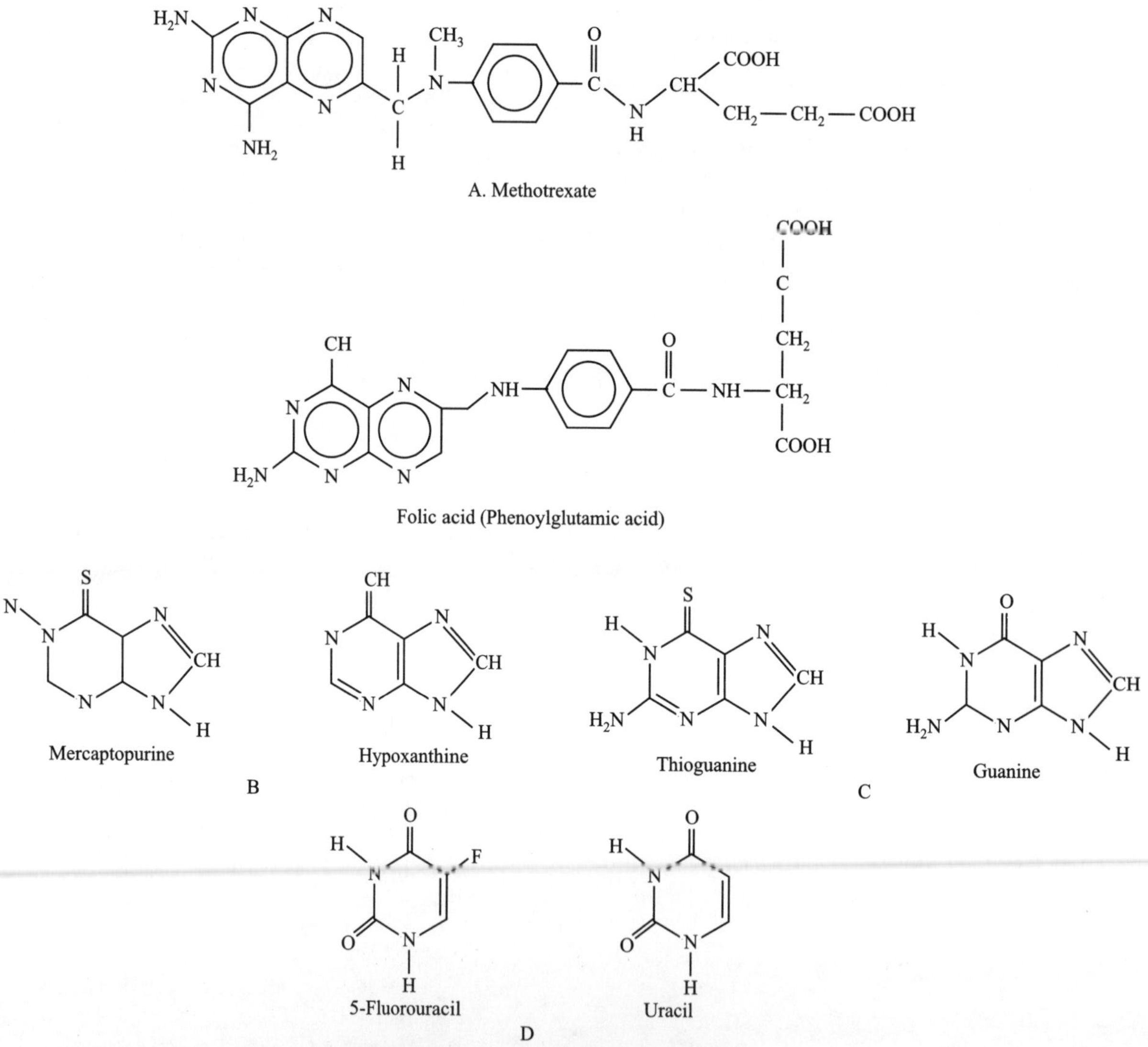

A. Methotrexate

Folic acid (Phenoylglutamic acid)

Mercaptopurine   Hypoxanthine   Thioguanine   Guanine

B   C

5-Fluorouracil   Uracil

D

Figure 130.4   Structural similarity between

Antimetabolite		*Essential dietary factor*
1. Methotrexate	and	Folic acid
2. 6–Mercaptopurine	and	Hypoxanthine
3. 6–Thioguanine	and	Guanine
4. 5–Fluorouracil	and	Uracil.

A folic acid antagonist such as methotrexate is a useful drug in the treatment of acute leukemia in children, and choriocarcinoma in women. Other antimetabolites like 5–fluorouracil, 6–mercaptopurine and azaserine are also used in the treatment of malignant diseases of tissues such as the breast, gastro-intestinal tract, cervix, and in acute leukemia. Newer folate antagonists have been identified that exploit the differences between the folate influx system in certain tumours and those in normal tissue.

The compound 5, 10–dideazatetrahydrofolate is transported into some tumour cells much more efficiently than into normal tissue and is undergoing clinical trials. Yet another compound, trimetrexate, which is an analog with high lipid solubility has been synthesized. This compound, though not proved to be effective as an antitumour drug, has proven beneficial in the treatment of Pneumocystis carinii infection.

C. Insecticides

1,1,1, trichloro-2–2–bis (P-chlorophenyl) ethane (DDT), gammexane and chlordan are highly lethal towards a number of disease-carrying insects and if not used with caution there is a slight danger to humans. DDT penetrates the chitinous exoskeleton of many arthropods (lice) and exerts its lethal action by acting on the nervous tissue. Organophosphorus compounds such as tetraethylpyrophosphate (TEPP) inhibits the enzyme acetylcholinesterase (AChE) so that acetylcholine is not hydrolyzed normally at nerve endings, following passage of nerve impulses at these junctions. A number of organophosphorus compounds are now available that are used as insecticides. These agents exert their action by AChE-irreversible inhibition of the enzyme of the insects. Organophosphorus compounds with selective toxicity to insects have been sought. Malathion is one such compound which is more rapidly metabolized to inactive forms in higher animals than in insects; thus it is less harmful to man and other animals.

D. Miscellaneous Antimetabolites

Disulfiram (Antabuse) is a drug used in the treatment of alcoholism. The compound acts as an inhibitor of an enzyme or enzymes involved in the normal metabolism of the aldehyde resulting from an early step in the oxidation of alcohol (Figure 130.5).

IV. CLASSIFICATION OF ANTIMICROBIAL DRUGS

Based on their mechanism of action, the antimicrobial drugs are classified as follows:

1. Agents disrupting the bacterial cell wall
   — Penicillins, Cephalosporins, carbapenems, Monobactams, Vancomycin, Bacitracin, Cycloserine, Miconazole, Ketoconazole, Clotrimazole.

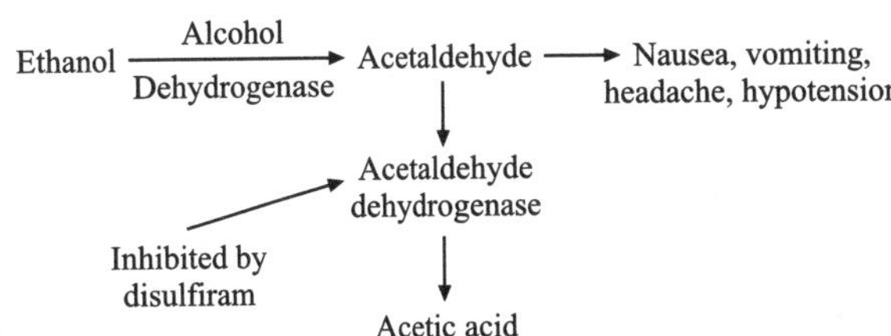

Figure 130.5   Enzyme inhibition by disulfiram resulting in acetaldehyde accumulation leading to nausea, headache, flushing and hypotension.

2. Agents acting directly on the cell membrane and affecting permeability
   — Polymixin, Nystatin, Amphotericin B.
3. Agent acting on the bacterial ribosomes causing reversible inhibition of protein synthesis
   — Chloramphenicol, Tetracycline, Erythromycin and Clindamycin.
4. Agents that bind to 30 S ribosomal subunit and alter protein synthesis
   — Aminoglycoside antibiotics—Gentamicin, Amikacin.
5. Agents affecting nucleic acid metabolism
   (a) Inhibit DNA-dependent RNA polymerase— Rifampin
   (b) Inhibit DNA gyrase—Quinolone, e.g., Ciprofloxacin.
6. Antimetabolites which block specific metabolic steps that are essential to microorganisms
   — Trimethoprim and the Sulphonamides (Cotrimoxazole), Pyrimethamine.
7. Nucleic acid analogues which bind to viral enzymes that are essential for DNA synthesis, thus halting viral replication
   — AZT (Zidovudine), Acyclovir.

V. AN IDEAL CHEMOTHERAPEUTIC AGENT

An ideal chemotherapeutic agent should fulfil the following criteria

1. Selective and effective antimicrobial activity against microorganisms.
2. Should be bactericidal and not bacteriostatic.
3. Should not induce bacterial resistance.
4. Should have a wide margin of safety and be free from side effects.
5. Hypersensitivity and damage to vital organs should not be there.
6. Should be equally effective for infections in body fluids, exudates, plasma proteins and tissues.
7. Should be water-soluble and stable at room temperature.
8. Should be orally effective.
9. Optimal levels should develop rapidly at the site of infection and be maintained for long intervals.
10. The cost should be low.

VI. RATIONAL USE OF ANTIBIOTICS

Antibiotics are the most important weapons in our hands. Each one of them have been invented after spending considerable amount of time, energy and money. Therefore, we cannot afford to lose them. We must exercise considerable restraint in prescribing antibacterials and restrict the use of antibacterials to only certain definite indications.

Indications for antibacterial therapy

1. **Definitive therapy:** This is for proven bacterial infections. Antibiotics (read antibacterials) are drugs to tackle bacteria and hence should be restricted for the treatment of bacterial infections only. This may sound silly, but most doctors seem to forget this simple fact! Attempts should be made to confirm the bacterial infection by means of staining of secretions/fluids/exudates, culture and sensitivity, serological tests and other tests. Based on the reports, a **narrow spectrum, least toxic, easy-to-administer and cheap drug** should be prescribed.

2. **Empirical therapy:** Empirical antibacterial therapy should be restricted to critical cases, when time is inadequate for identification and isolation of the bacteria and reasonably strong doubt of bacterial infection exists: septicemic shock/ sepsis syndrome, immunocompromised patients with severe systemic infection, hectic temperature, neutrophilic leukocytosis, raised ESR etc. In such situations, drugs that cover the most probable infective agent(s) should be used.

3. **Prophylactic therapy:** Antimicrobial prophylaxis is administered to susceptible patients to prevent specific infections that can cause definite detrimental effect. These include antitubercular prophylaxis, anti rheumatic prophylaxis, anti endocarditis prophylaxis and prophylactic use of antimicrobials in invasive medical procedures etc. In all these situations, only narrow spectrum and specific drugs are used. It should be remembered that there is NO single prophylaxis to 'prevent all' possible bacterial infections.

Source of infection: Community acquired infections are less likely to be resistant whereas hospital acquired infections are likely to be resistant and more difficult to treat (e.g. Pseudomonas, MRSA etc.).

Type of infection: Infections can be localised or extensive; mild or severe; superficial or deep seated; acute, sub acute or chronic and extracellular or intracellular. For extensive, severe, deep seated, chronic and intracellular infections, higher and more frequent dose, longer duration of therapy, combinations, lipophilic drugs may have to be used.

Severity of infection: In treating the severe infections, drugs should be administered by only intravenous route to ensure adequate blood levels. Only bactericidal drugs should be used to ensure faster clearance of the infection. If the site of infection is known, narrow spectrum drugs should be used. If the site is unknown, attempt should be made to cover all possible organisms, including drug resistant Staphylococcus, Pseudomonas and anaerobes. A combination of penicillins/3rd generation cephalosporins, aminoglycosides and metronidazole may be used. The dose should be higher and more frequent. Whenever possible, a switch to oral therapy should be made.

Isolation and sensitivity: As far as possible and if the situation permits, antibacterials can be started only after the sensitivity report is available. Narrow spectrum, least toxic, easy to administer and cheapest of the effective drugs should be chosen. If the patient is responding to the drug that has already been started, it should not be changed even if the in vitro report suggests otherwise.

VII. FACTORS INFLUENCING THE SELECTION OF THE CHEMOTHERAPEUTIC AGENT

A large number of agents are available and many a time there are more than one chemotherapeutic agents, which are effective. There are likely to be occasions when the physician may face difficulties in selecting the most suitable drug. The object is to give the patient maximal benefit with least chances of toxicity. The "benefit versus risk" ratio should always be assessed before starting therapy. This can only be achieved if the physician has:

1. A working knowledge of common pathogenic organisms.
2. Knowledge of the spectrum of activity of chemotherapeutic agents.
3. Information on the sensitivity pattern of the micro-organisms and alternatives available in chemotherapy.
4. Information on the pharmacokinetic of antimicrobials.

Table 130.1 gives a list of commonly encountered pathogens and the drugs used for these. Although in most situations the decision to employ a particular chemotherapeutic agent depends on the sensitivity of the microorganisms, there are occasions when other factors besides "the bug and drug" have to be kept in mind. (Figure 130.6). Table 130.2 gives routes of administration, contraindications, common adverse effects and interactions.

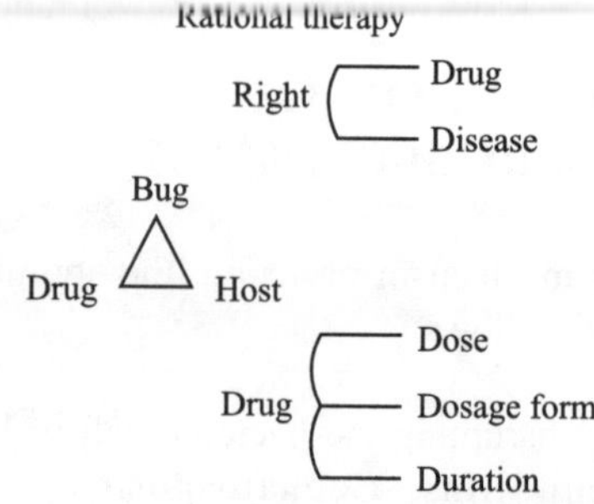

Figure 130.6 Essential background information for the rational prescribing of antimicrobials.

TABLE 130.1 Antimicrobial selection

Site of infection	*Type of infection*	*Drugs of choice*	*Duration of treatment*
Upper respiratory tract infection (URTI)	Throat infections	Erythromycin or amoxycillin or penicillin V	10 days for Group A Streptococcal infections, 10–14 days for mycoplasma.
	Sinusitis	Amoxycillin or doxycycline or erythromycin	2 weeks
	Otitis media	Amoxycillin or erythromycin	7 days
Lower respiratory tract infection (LRTI)	Bronchitis	Amoxycillin or co-trimoxazole or tetracycline	7 days
	Uncomplicated pneumonia	Amoxycillin or benzyl penicillin or macrolide	7 days
	Severe pneumonia of unknown aetiology	Erythromycin plus cefuroxime or cefotaxime	7–10 days
	Suspected atypical pneumonia	Macrolide	10–14 days
	Hospital acquired pneumonia	Cefotaxime or ceftazidime or an antipseudomonal penicillin plus an aminoglycoside	7–10 days
Urinary tract infection (UTI)	Lower urinary tract infection	Fluoroquinolone or Co-trimoxazole or amoxycillin or nitrofurantoin	3–7 days
	Upper urinary tract infection, pyelonephritis or prostatitis	Fluoroquinolones or co-trimoxazole or gentamicin or cephalosporins	10–28 days
Pelvic inflammatory disease (PID)		Doxycycline plus metronidazole	Metronidazole for 7–14 days and doxycycline for 14–21 days
Skin infections	Impetigo	Topical fusidic acid or mupirocin; erythromycin, if wide spread	7–10 days
	Cellulitis	Amoxycillin plus cloxacillin or erythromycin or co-amoxyclav	7–10 days
Peri-surgical prophylaxis	Clean (cardiac, vascular, neurologic or orthopaedic)	Cefazolin	Single dose within 60 minutes before incision
	Clean-contaminated (head and neck, high-risk gastroduodenal or biliary tract surgery; caesarian section; hysterectomy)	Cefazolin	Single dose within 60 minutes before incision
	Clean-contaminated (colorectal surgery or appendicectomy)	Cefoxitin or 3rd generation Cephalosporins plus metronidazole	One to three doses
	Dirty (ruptured viscus)	Cefoxitin plus gentamicin or 3rd gen. Cephalosporins plus metronidazole	One to three doses
	Dirty (traumatic wound)	Cefazolin	Before and for 3–5 days after trauma

Adapted from Srinivas Kakkilaya B. Rational use of antibiotics. http://www.rationalmedicine.org/antibiotics.htm acccessed on August 20, 2013.

Note: Intraoperative redosing at one to two half lives of the antibiotic is recommended in longer duration of the procedure or if there is excessive loss of blood (>1500 ml) during the procedure, extensive burns.

A. Related to the Host

1. Age: Patients at both extremes of age handle drugs differently, primarily due to differences in body size and kidney function. Most paediatric drug dosing is guided by weight. In geriatric patients, the serum creatinine level alone is not completely reflective of kidney function, and the creatinine clearance should be estimated by factoring in age and weight for these patients. Therefore, the dose of those drugs which are excreted as such through the kidneys have to be reduced. Similarly, liver enzymes are not so well developed early in life as in adulthood. High levels of sulphonamides can be seen in the new born as the acetylation process in the liver is deficient. Chloramphenicol can cause "gray baby" syndrome in neonates as the hepatic metabolizing enzymes (glucuronyl transferase) are not well developed.

2. Genetic factors: Inherited differences in man cause individual variations in response to drugs. Persons who have a deficiency of glucose-6–phosphate dehydrogenase (G6PD) may develop haemolysis when administered drugs like primaquine, sulphonamides and nitrofuran derivatives. Potentially fatal hypersensitivity reaction with abacavir is significantly higher in patients with the human leukocyte antigen allele HLA-B*5701.

3. Pregnancy: Those drugs which cross the blood placental barrier may prove hazardous to the growing foetus. Varying degrees of hearing loss have been observed in babies born to mothers who were on streptomycin therapy during pregnancy. Other drugs, such as tetracyclines and chloramphenicol, have well-described foetal or neonatal adverse effects and should be avoided. Penicillins, cephalosporins, and macrolides have historically been the most commonly used antimicrobial agents considered safe in pregnancy. Organogenesis, in utero, takes place from day 18–38 of pregnancy and therefore drugs should be used very cautiously during the first trimester of pregnancy. In general, however, human studies on the safety of many antimicrobial agents in pregnancy and lactation are limited, and antimicrobial agents should be prescribed with caution.

TABLE 130.2 Doses, adverse effects and contraindications of commonly used antimicrobial agents

Drug	Route of administration	Dose and frequency of administration	Contra indications/ caution	Interactions	Adverse effects
Penicillin G	IM, IV	10–20 L Units every 2 to 6 h.	Hypersensitivity	None significant	Hypersensitivity-rashes, anaphylaxis, fever, joint pains, angioedema, serum sickness like reaction, blood disorders, CNS toxicity in high doses; colitis and diarrhoea; ampicillin rashes are more common in infectious mononucleosis, chronic lymphatic leukemia and HIV infection.
Penicillin V	Oral	400–800 mg every 6–8 hours			
Ampicillin	Oral, IV	New born (<1 week):25–50 mg/kg 12 hourly; 1–4 weeks: 100–200 mg/kg/day 8th hourly; Older children: Same dose 6 hourly; Adults: 1–12 g/day, 6 hourly			
Amoxycillin	Oral, IV	Oral: 250 mg-1g every 8 hourly; Severe RTI: 3 g 12 hourly IV: 500 mg -1 g every 6–8 h; Children- 50–100 mg/kg in 3–4 doses.			
Cephalexin (1st gen.)	Oral	250mg-1 g every 6 hours	Hypersensitivity, renal failure, porphyria		Hypersensitivity rashes, joint pains, rarely anaphylaxis; abdominal discomfort; colitis; erythema multiforme; reversible interstitial nephritis; transient hepatitis; blood disorders; dizziness etc.
Cefazolin (1st gen.)	IV	1–1.5 g every 6 hours			
Cefadroxil	Oral	1 g every 12 hours			
Cefuroxime (2nd gen)	IV, oral (axetil)	Up to 3 g every 8 hours IV 250–500 mg 12 hourly oral			
Cefotaxime (3rd gen.)	IV	1–2 g every 4–8 hours			
Ceftriaxone (3rd gen.)	IV	1–2 g every 12–24 hours			
Ceftazidime (3rd gen., anti pseudomonal)	IV	1–2 g every 8 hours			
Cefoperazone (3rd gen., anti pseudomonal)	IV	1.5 to 4 g, 6–8 hourly			
Gentamicin	IM, IV	Adults: 2 mg/kg loading dose, then 3–5 mg/kg per day, every 8 hours Age < 2 y: 2–2.5 mg/kg every 8 h.	Renal failure; myasthenia gravis	Ototoxicity may worsen with potentially ototoxic diuretics like furosemide	Ototoxicity, nephrotoxicity, more common in the elderly and in renal failure; may impair neuromuscular transmission and may cause transient myasthenia
Amikacin	IM, IV	15–22.5 mg/kg/day in 2–3 equally divided doses			

(Contd.)

TABLE 130.2 Doses, adverse effects and contraindications of commonly used antimicrobial agents (*Contd.*)

Drug	Route of administration	Dose and frequency of administration	Contra indications/ caution	Interactions	Adverse effects
Tetracycline	Oral	Adults: 1–2 g/day in 2–4 doses. Children >8 y: 25–50 mg/kg/day in 2–4 doses	Pregnancy, lactation; renal impairment; children below 8 years; SLE	Antacids reduce absorption; anticonvulsants increase metabolism of doxycycline; calcium salts, oral iron and zinc reduce absorption; sucralfate and bismuth salts reduce absorption.	Nausea, vomiting diarrhoea; erythema (discontinue); benign intracranial hypertension; colitis
Demeclocycline	Oral	150 mg 6 hourly or 300 mg 12 hourly in adults			
Doxycycline	Oral	100 mg twice a day on the first day, then 100 mg once or twice a day depending on the severity of infection			
Erythromycin	Oral	>8 y: 250–500 mg every 6 hours or 0.5–1 g every 12 h; Age 2–8 years: 250 mg every 6 hours; Age < 2 y: 125 mg every 6 hours	Hepatic and renal impairment; prolonged QT interval; porphyria; reduce dose of erythromycin estolate in liver disease and in pregnancy	Avoid use with astemizole, terfenadine, pimozide, midazolam, zopiclone, cisapride; increases theophylline levels	Nausea, vomiting, abdominal discomfort; diarrhoea; rashes; arrhythmias; cholestatic jaundice
Azithromycin	Oral, parenteral	5 day regimen: 10 mg/kg PO day 1 (max dose 500 mg/24 h) followed by 5 mg/kg/24 h PO (max dose: 250 mg/24 h) on days 2–5 3 day regimen: 10 mg/kg/24 h PO for 3 days (max dose: 500 mg/24 h); 1 day regimen: 30 mg/kg/24 h PO × 1 (max dose: 1500 mg/24 h)	Documented hypersensitivity, and history of cholestatic jaundice, hepatic impairment with prior azithromycin use; Use with caution in impaired hepatic function, GFR < 10 mL/min	Aluminum- and magnesium-containing antacids decrease absorption. Oral dosage forms may be administered with or without food. Intravenous administration is over 1–3 h; do not give as a bolus or IM injection.	Vomiting, diarrhoea and nausea, epigastric discomfort; rarely hepatotoxicity and hypersensitivity
Norfloxacin	Oral	400 mg twice daily for 3–7 days for UTI	Age <18 years; pregnancy and lactation; epilepsy; hepatic and renal impairment	Increased risk of convulsions with NSAIDs; antacids reduce absorption; anticoagulant effects of coumarins enhanced; iron and zinc reduce absorption; increases levels of theophylline, avoid concomitant use.	Nausea, vomiting, diarrhoea, headache, dizziness, sleep disorders, rash, pruritus, anaphylaxis, increase in urea, creatinine and liver enzymes, arthralgia and myalgia, blood disorders, restlessness, hallucinations, depression, tendon damage
Ciprofloxacin	Oral, IV	Oral: UTI: 250–500 mg twice daily; LRTI: 500–750 mg twice a day IV: 200–400 mg over 30–60 minutes, twice a day			
Co-trimoxazole	Oral	Adult: 960 mg twice daily; 480 mg twice daily if used for >14 days. Children: 120–480 mg 12 hourly	Hepatic and renal impairment; blood disorders; G6PD deficiency; breast feeding	Enhanced effects of warfarin, sulfonylureas, phenytoin	Nausea, vomiting, rash, Steven Johnson syndrome (discontinue), blood disorders (discontinue), glossitis, arthralgia, liver damage, colitis, eosinophilia, tinnitus, interstitial nephritis

(Contd.)

TABLE 130.2   Doses, adverse effects and contraindications of commonly used antimicrobial agents (*Contd.*)

Drug	Route of administration	Dose and frequency of administration	Contra indications/ caution	Interactions	Adverse effects
Metronidazole	Oral, IV	800 mg initially, then 400 mg 8 hourly; IV: 500 mg 8 hourly. Children: 7.5 mg/kg 8 hourly	Hepatic failure, pregnancy and lactation,	Disulfiram like reaction with alcohol; enhances effect of anticoagulants, phenytoin; increases toxicity of lithium	Nausea, unpleasant taste, furred tongue, headache, drowsiness, anaphylaxis, rashes.
Nitrofurantoin	Oral	50–100 mg every 6 hours	Renal impairment; G6PD deficiency; full term pregnancy and lactation; major organ failure	Magnesium trisilicate retards absorption	Anorexia, nausea, vomiting, acute and chronic pulmonary reactions; SLE like syndrome; rashes etc.

Adapted from Srinivas Kakkilaya B. Rational use of antibiotics. http://www.rationalmedicine.org/antibiotics.htm acccessed on August 20, 2013.

4. Concurrent disease: Intramuscularly administered penicillin is absorbed to a lesser degree in patients having diabetes mellitus.

5. History of Allergy or Intolerance: There is a greater incidence of hypersensitivity reactions with antimicrobial agents in those patients who have atopic allergy. A history of antimicrobial allergy or intolerance should be routinely obtained in the evaluation and management of infection.

6. History of Recent Antimicrobial Use: Eliciting a history of exposure to antimicrobial agents in the recent past (approximately 3 months) can also help in selection of antimicrobial therapy. Because the causative microorganism for a current episode of infection emerged under the selective pressure of a recently used antimicrobial agent, it is likely to be resistant to that drug and/or drug class, and an alternative agent should be used. Pharyngitis caused by S. pyogenes responds to penicillin, but if the throat contains penicillinase producing organisms as its normal flora, there can be therapeutic failure.

7. Hepatic and renal function: Because the kidney and the liver are the primary organs responsible for elimination of drugs from the body, it is important to determine how well they are functioning during antimicrobial administration. In most cases, one is concerned with dose reduction to prevent accumulation and toxicity in patients with reduced renal or hepatic function. The dose and frequency of antibiotics such as gentamicin has to be altered in case of renal disease based on serum creatinine values. However, sometimes doses might need to be increased to avoid underdosing in young healthy patients with rapid renal elimination or those with rapid hepatic metabolism due to enzyme induction by concomitant use of drugs such as rifampin or phenytoin.

8. Host defence mechanisms: The anti-microbial drug is either able to kill the micro-organism and such agents are known as bactericidal drugs (e.g., Penicillin, Cephalosporin, Rifampin, Gentamicin) or prevent their growth—bacteriostatic drugs (Tetracycline, Chloramphenicol). Such drugs require the host's immune system to be functionally normal. Subjects whose immune system is compromised such as those on anticancer-chemotherapy or having organ transplant and are on immuno-suppressive drugs, would need a bactericidal drug to effectively combat an infection.

9. Duration of therapy: The antimicrobials should be continued till such periods when microorganisms have been completely eradicated from the body. It is a good practice to continue the antibiotic for 2–3 days after all the signs of infection have been cleared (Figure 130.7).

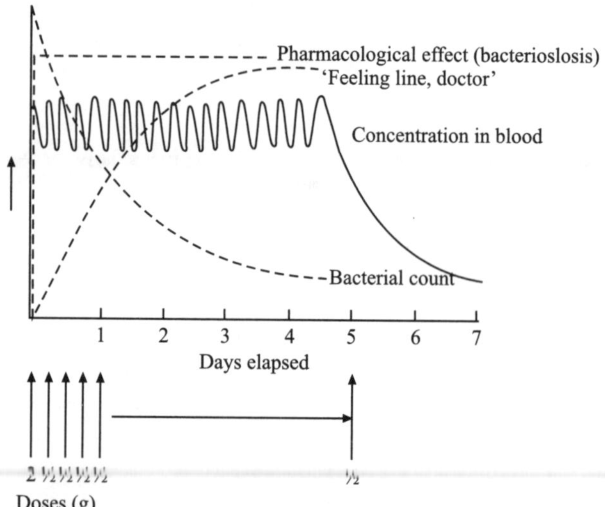

Note. The treatment should not be discontinued on day 5 of the treatment, but should be continued for another 2 or 3 days. This would ensure complete eradication of the bacteria.

Figure 130.7   Time course of bacteriostatic chemotherapy.

B.   Related to the Microorganisms

1. The problem of resistance: With the advances and innovations in chemotherapy, bacteria change in ways to become resistant to the antibiotics which they are previously sensitive to, i.e., the previous effective treatment is no longer capable of controlling the same infection. When the bacteria become resistant to most antibiotics, they are referred to as "superbugs". Bacteria have found their self-defenses against anti-microbials by acquiring genetic adaptations in the form of tolerance, mutations, transfer of R-factor from one bacteria to another through plasmid, acquisition of enzymes resistant to anti-microbials.

The microorganisms capable of developing resistance are *M. tuberculosis*, certain strains of *S. aureus, E. coli*, Salmonella, Shigella, Klebsiella, and Pseudomonas. The list of drugs to which resistance has developed include streptomycin, sulphonamides, tetracyclines, chloramphenicol, kanamycin, neomycin, penicillin, ampicillin, cephalosporins and gentamicin. In most of the situations bacterial sensitivity tests must be performed before giving the antibiotic. This problem of resistance can be sometimes overcome by the introduction of a new drug such as cloxacillin for resistant strains of *S. aureus* to penicillin, but again mutant and resistant strains can develop to the new drug. Another way of delaying the development of resistance is by the concurrent use of two or more agents such as ethambutol, rifampin and isoniazid in the treatment of tuberculosis. Administration of penicillin and an aminoglycoside proves to be much better in enterococcal infections.

2. The anatomic location of infection: The efficacy of antimicrobial agents also depends on their capacity to achieve a concentration equal to or greater than the MIC at the site of infection. Antimicrobial concentrations attained at some sites (e.g., ocular fluid, CSF, abscess cavity, prostate, and bone) are often much lower than serum levels e.g., first- and second-generation cephalosporins and macrolides do not cross the blood-brain barrier and are not recommended for central nervous system infections. Fluoroquinolones achieve high concentrations in the prostate and are preferred oral agents for the treatment of prostatitis. Many antibiotics (e.g., aminoglycosides) are less active in the low-oxygen, low-pH, and high-protein environment of abscesses, and drainage of abscesses to enhance antimicrobial efficacy is recommended when possible. Agents in the same class can differ from one another; for example, moxifloxacin does not achieve significant urinary concentrations because of its low renal excretion and is therefore not suitable for treatment of UTIs; in contrast, both levofloxacin and ciprofloxacin are excellent choices for UTIs caused by susceptible bacteria. The presence of foreign bodies at the site of infection also affects antimicrobial activity.

Too much reliance on chemotherapy without surgical drainage, in situations when there is a collection of pus in anatomical sites such as the pleural cavity, may result in failure. Then there are conditions like osteomyelitis when antibiotic therapy alone is not effective, due to relative avascularity of the site. In order to have effective concentrations in the bone, very high blood levels and thus high doses of antibiotics are needed.

3. Mixed infection: Bronchiectasis, peritonitis, urinary tract infections and otitis media are conditions in which more than one microorganism may be the cause. Sometimes these may respond to a single antimicrobial agent, but there can be situations when two agents have to be used. In such a state, two antimicrobial agents should be used individually and in their full therapeutic doses.

C. Related to the Chemotherapeutic Agent

Along with host factors, the pharmacodynamic properties of antimicrobial agents may also be important in establishing a dosing regimen. Specifically, this relates to the concept of time-dependent vs concentration-dependent killing. Drugs that exhibit time-dependent activity (β-lactams and vancomycin) have relatively slow bactericidal action; therefore, it is important that the serum concentration exceeds the MIC for the duration of the dosing interval, either via continuous infusion or frequent dosing. In contrast, drugs that exhibit concentration-dependent killing (aminoglycosides, fluoroquinolones, metronidazole, and daptomycin) have enhanced bactericidal activity as the serum concentration is increased.

Combined use of anti-microbials

When two different antibiotics are given, one of the following three results may be seen depending on whether the drug is bacteriostatic or bactericidal. The situations can be:

1.	Bactericidal drug (penicillin)	+ Bactericidal drug: (aminoglycoside)	Supra additive effect 'synergism'
2.	Bactericidal drug (penicillin)	+ Bacteriostatic drug: (tetracycline)	Diminished effect 'antagonism'
3.	Bacteriostatic drug	+ Bacteriostatic drug:	Simple addition 'additive'

It is believed that if an agent inhibiting cell wall synthesis is mixed with another cell wall inhibitor, the effect is additive. On the other hand, if in the combination one agent inhibits cell wall synthesis (penicillin) and another suppresses protein synthesis (aminoglycoside) the action is markedly enhanced 'synergism'. The objective of combining more than one anti-microbial drugs is indicated in some conditions e.g., mixed infections, tuberculosis, HIV/AIDs mainly

1. To achieve synergism (Penicillin/ampicillin + streptomycin/gentamycin in Enterococcal sub-acute bacterial endocarditis)
2. To prevent emergence of resistance (Rifampin + isoniazid in tuberculosis, combination antiretroviral agents in HIV/AIDs)
3. To decrease the severity and incidence of adverse drug effects (Penicillin G + streptomycin in Enterococcal

sub-acute bacterial endocarditis, as dose of streptomycin can be reduced in combination).

4. To broaden the spectrum of action (In mixed infections e.g., cephalosporins+ metronidazole/clindamycin or unknown severe infections. However, for the therapy of unknown severe infections. Whenever combination therapy is desired, the two agents should be given separately and in full therapeutic doses and not as fixed dose combinations. The disadvantages of such fixed dose combinations are:

(a) Inadequate dose
(b) False sense of security
(c) Sometimes, reduced therapeutic efficacy
(d) May lead to resistance.

Fixed dose combinations of antibiotics in which the drugs have been mixed on the basis of in vitro determinations have not proved superior to the results obtained by a single agent administration.

Disadvantages of combining anti-microbials: Combining the anti-microbials is not always useful. The following are the disadvantages of combining anti-microbials:

1. Promote irrational prescriptions without weighing benefit: risk ratio.
2. Increase adverse drug effects sometimes.
3. Increase risk of superinfections (new infection secondary to the prolonged use of broad spectrum anti-microbial drug due to suppression of normal flora of intestine) and pseudomembranous enterocolitis.
4. Adds to the cost of treatment.

The following combinations are irrational, not useful or even harmful:

1. Combinations of bactericidal with bacteristatic drugs (e.g. Penicillins with tetracyclines)
2. Combinations of drugs with similar toxicity (e.g., chloramphenicol and sulfonamides)
3. Combining drugs for non-existing 'mixed infections' (e.g., ciprofloxacin + metronidazole/tinidazole).

VII. CHEMOPROPHYLAXIS

Chemotherapeutic agents have been used in attempts to prevent infection on the assumption that such a therapy is effective in discouraging the implantation of microorganisms. Such a therapy, therefore, has not been always wisely employed. Clinical studies have shown that there are only some areas where chemoprophylaxis is highly effective, e.g., the use of penicillin G (Benzathine penicillin) to prevent invasion by a group A streptococci for preventing rheumatic heart disease or giving isoniazid to an infant whose mother has tuberculosis.

A. Drug Interaction

Drug-drug interactions (DDI) can be defined as the modulation of the pharmacologic activity of one drug (i.e., the object drug) by the prior or concomitant administration of another drug (i.e., the precipitant drug). In these reactions, the pharmacologic properties of the object drug and/or the precipitant drug can be either severely enhanced or diminished. The interaction can be potentiated, or synergistic, or antagonistic. Knowledge of drug interactions enables a physician to not only prevent harmful interactions, avoid drug toxicity but also used to draw therapeutic benefits.

Drug interaction can be beneficial allowing lower doses of the drug(s) being administered while still achieving therapeutic serum drug levels. For example probenecid a drug which decreases excretion of penicillin, when given together dose of penicillin can be reduced. Other examples are use of penicillin G + streptomycin for enterococci; use of gentamicin + carbenicillin in *P. aeruginosa* infections). One fixed dose combination—cotrimoxazole—which has so far stood the test of time and acts by sequential blockade as shown below is a clinically useful drug combination.

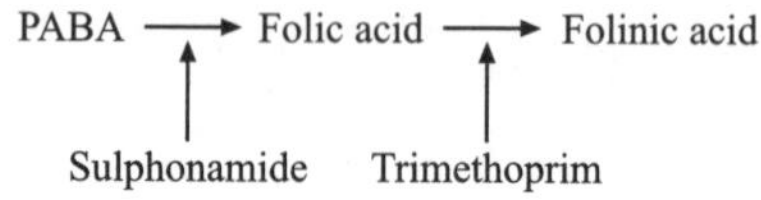

The other examples are beta-lactam antibiotics with beta-lactamase inhibitors, ampicillin or amoxycillin + sulbactam and amoxicillin + clavulanic acid.

Most of the time drug interactions are harmful. These can be in terms of lack of efficacy due to enhanced metabolism of the drug or additive toxicity. All cephalosporines when given with amino glycosides (gentamicin, amikacin) increase renal toxicity, therefore, needing close patient monitoring for renal toxicity. Increased bone marrow toxicity of zidovudine and ganciclovir and peripheral neuropathy, pancreatitis and lactic acidosis with didanosine and Stavudine (DDI/D4T). Zidoivdine (AZT) and Stavudine (d4T) reduces antiviral effect.

Thus DDI carry potential risk both outside and inside the body. Chemical or physical incompatibilities are also well known with antibiotics given by intravenous infusions (e.g., IV. chloramphenicol is incompatible with cephalothin, polymixin B, sulfadiazine and tetracyclines; penicillin G is incompatible with amphotericin B and lincomycin). Gentamicin is physically/chemically incompatible with most beta-lactams resulting in loss of antibiotic effect.

Difficulties in treating TB in HIV patients may arise due to interactions with rifampicin, a potent inducer of liver enzymes. Many ARVs contraindicate the use of rifampicin, while others may require dose modification.

Many drugs are reversibly bound to plasma proteins and when bound they are pharmacologically inert. Metheotrexate may be displaced by sulphonamides causing pancytopenia when used in treating selected cases of psoriasis. Tolbutamide, an oral hypoglycemic agent, is avidly bound to plasma proteins;

if simultaneously sulphaphenazole is added, hypoglycemia results.

The magnitude of the drug interactions problem increases significantly in certain patient populations—the elderly, critical ill patients, and patients undergoing complicated surgical procedures and as the number of medications taken each day increases. Drug interactions that may be of minor clinical significance in patients with less severe forms of a disease but can cause significant exacerbation of the clinical condition in patients with more severe forms of the disease.

Not only do drug interactions present a danger to the patient, but they can also greatly increase healthcare costs. Patients on a combination of drugs require hospitalization or more laboratory tests to monitor resolution of these interactions. Whenever there is an unexpected response to a drug, consider "Drug - Drug Interaction" as one of the possibility for the unexpected response.

B. Adverse Effects and Problems Associated with the use of Anti-microbial Therapy

1. Normal bacterial flora resides in oral cavity, Gastrointestinal tract, respiratory tract, vagina and on skin. Latter provides resistance to the colonization of other pathogenic bacteria. Anti-microbial therapy markedly reduces this normal commensal flora allowing the invasion of opportunistic pathogenic micro-organisms leading to a new infection known as superinfection. Superinfection can be caused by 3rd generation cephalosporins, clindamycin, tetracyclins, chloramphenicol. Superinfection may manifest in the form of bloody diarrhoea, dehydration, abdominal pain also known as pseudomembranous enterocolitis. Causative organisms responsible for this condition are mainly *Clostridium difficile* or Candida. This condition can be treated by use of metronidazole or vancomycin

2. Nutritional deficiency may result with prolonged use of anti-microbial therapy, as latter suppress the normal flora of intestine which is responsible for production of vitamin B.

3. Local irritation and pain at the site of injection.

4. Diarrhoea, loss of appetite, e.g., mainly ampicillin, erythromycin etc.

5. Brown discolouration of teeth (deciduous and permanent teeth), e.g., with tetracycline.

6. Grey baby syndrome (seen in pre-mature infants due to inability to metabolize chloramphenicol adequately)

7. Anaphylaxis and allergic reactions can occur with any antimicrobial. Extra precaution to be taken while giving injectable antibiotics, e.g., penicillins. Emergency kit should be available to tackle any untoward reaction.

8. Ototoxicity (hearing impairment), with aminoglycosides, chloroquine (anti-malarial).

9. Nephrotoxicity (kidney toxicity) with aminoglycosides, demeclocycline and anti-fungal like amphoteracin-B.

10. Phototoxicity (skin rash in the presence of sun light) with tetracycline.

11. Hepatotoxicity (liver toxicity) with chloramphenicol, rifampicin etc.

Masking of infection sometimes occurs in the presence of antimicrobials, e.g., streptomycin taken for upper respiratory tract infections generally masks the underlying tuberculosis in the patient.

No antibiotic, or for that matter any drug, is free from toxic effects. In fact antibiotics are much more toxic than most of the commonly used drugs. Therefore, the most important decision in the use of antibiotics is whether they should be used at all. Before starting antibiotics the physician must ensure that (a) the infection is definitely present, (b) the infection is treatable, and (c) the indications for antibiotic therapy are compelling enough to warrant exposing the patient to their possible hazards.

VIII. MISUSE OF ANTIMICROBIAL AGENTS

One of the most commonly misued group of drugs are the antimicrobial agents. Antibiotics are most often prescribed unnecessarily in fever, sore throat, diarrhoea. Since these diseases are most often due to the viral infections and antibacterials have no role to play in their management. Use of antibacterials in non-bacterial illness results only in the destruction of susceptible bacteria and selective proliferation of resistant bacteria, thus aiding the propagation of bacterial drug resistance. Sometimes antibiotics are given in inadequate doses and for shorter durations and therefore are ineffective and in fact may lead to drug resistance. There is always a fancy for using a new drug. The efficacy and safety of a drug is only known after it has been in clinical use for sometime. It is, therefore, better to use a well tried drug rather than a new one.

Cost considerations in antibiotic selection and antimicrobial stewardship

Due consideration should also be given to the cost of the therapy which is dependent on many factors, in addition, to the purchase price of a particular agent and may include administration costs, prolonged hospitalization as a consequence of adverse effects, the cost of serum concentration monitoring, and clinical efficacy (Table 130.3). Oral therapy is generally less expensive, potentially associated with fewer adverse effects, and can result in considerable cost savings by facilitating earlier dismissal and a shortened hospital stay. All things being equal, the use of an expensive drug against a cheaper one, but equally effective, should be discouraged. Cost assumes significance particularly in India with limited public spending on health. Cheaper drugs like doxycycline or co-trimoxazole would be as effective as the costlier clarithromycin or cephalosporins in the management of LRTI. Antimicrobial stewardship programs are aimed at "optimizing antimicrobial selection, dosing, route, and duration of therapy" to maximize clinical cure or prevention of infection while limiting the unintended consequences, such as the emergence of resistance, adverse drug events, and cost.

TABLE 130.3   Cost of Antimicrobials: Wide apart

Antimicrobial	Cost (Rupees)	
Oral	*Minimum retail price*	*Maximum retail price*
Levoflxacin/5 tab	44	475
Ofloxacin 200 mg/10 tab	32	310
Ampicillin 500 mg/10 cap	24	55
Cefixime 200 mg/10 tab	79	200
Cefuroxine 500 mg/1 tab	21	83
Ceftazidime 250 mg/vial	69.41	109.50
Azithromycin 500 mg/3 tab	60	92
Roxithromycin 150 mg/10 tab	60	73.50
Injectables		
Cefotaime 1 g vial	30	87
Ceftriaxone 1 g inj	55	76
Levofloxacin 5 mg/ml	100	290

The cost is is based on the prices in MIMS, May 2013.

At a meeting of the Berlin Medical Society held on the 13th of February 1907, Paul Ehrlich had, "What we want is a chemotherepia specifica, that is, we are looking for agents, that on the one hand are able to kill certain parasites, without on the other hand causing too much harm to the organism when applied in doses necessary to kill these parasites". We have come a long way from where we started, a number of vulnerable points in the bacterial cell wall and protein synthesis are now known, but the much desired property of the antimicrobial agent—killing the parasite without any effect on the host remains to be achieved. The emergence of resistant strains which is keeping pace with the development of newer antibiotics is another challenge which the scientists of today have to face.

Antimicrobial resistance (AMR)

Antimicrobial resistance (AMR) is resistance of a microorganism to an antimicrobial medicine to which it was originally sensitive. Resistant organisms (they include bacteria, fungi, viruses and some parasites) are able to withstand attack by antimicrobial medicines, such as antibiotics, antifungals, antivirals, and antimalarials, so that standard treatments become ineffective and infections persists and increasing risk of spread to others. The evolution of resistant strains is a natural phenomenon that happens when microorganisms are exposed to antimicrobial drugs, and resistant traits can be exchanged between certain types of bacteria. Resistance to antimicrobial agents is one of the greatest problems faced by the medical community. Antimicrobial resistance, initially a problem in hospitals and developing countries, today affects the world at large. These powerful weapons, have been rendered less effective or totally ineffective only because of the misuse of antimicrobial medicines which accelerates this natural phenomenon.

Resistance to antimicrobial agents can be due to various mechanisms:

1. Inability of the drug to reach the organisms
2. Inactivation of the drug
3. Alteration in the target

Resistance may be acquired by mutation and passed onto the next generations. It may also be acquired by horizontal transfer from a donor cell by transformation, transduction or conjugation.

Antimicrobial resistance a global concern

- Antimicrobial resistance is a growing health issue since infections caused by resistant microorganisms often fail to respond to conventional treatment, resulting in prolonged illness, greater risk of death and higher costs and increasing the risk of spreading resistant microorganisms to others.
- Tuberculosis strains resistant to isoniazid and rifampicin (multidrug-resistance - MDR-TB) require treatment courses that are much longer and less effective. Current problems with antimicrobial resistance are predominantly being detected in the following organisms and diseases: MRSA, VRE, *E. coli*, Salmonella, Campylobacter and in gonorrhea, pneumonia, tuberculosis, influenza, HIV, malaria and others.
- Resistance to earlier generation antimalarial medicines such as chloroquine and sulfadoxine-pyrimethamine is widespread in most malaria-endemic countries.
- A high percentage of hospital-acquired infections are caused by highly resistant bacteria such as methicillin-resistant Staphylococcus aureus (MRSA) or multidrug-resistant Gram-negative bacteria.

New resistance mechanisms have emerged, making the latest generation of antibiotics virtually ineffective. The achievements of modern medicine are put at risk by AMR. Without effective antimicrobials for care and prevention of infections, the success of treatments such as organ transplantation, cancer chemotherapy and major surgery would be compromised. The growth of global trade and travel allows resistant microorganisms to be spread rapidly to distant countries and continents through humans and food.

Causes of antimicrobial drug resistance

There are many causes of antimicrobial drug resistance including selective pressure, mutation, gene transfer, societal pressures, pattern of antibiotic usage inappropriate drug use, inadequate diagnostics, hospital use and agricultural use of drugs. Antibiotic usage greatly affects the number of resistant organisms which develop. Overuse of broad-spectrum antibiotics, such as second- and third-generation cephalosporins, greatly hastens the development of methicillin resistance. Other factors contributing towards resistance include incorrect diagnosis, wrong dosage and wrong duration of therapy with antibiotics, unnecessary prescriptions, improper use of antibiotics by patients, and the use of antibiotics as livestock food additives for growth promotion. Poor infection control practices encourage the spread of AMR. Increase in poverty, over crowded living areas, crowded day care centers all contribute in spreading the resistant bacterial infection. The tremendous increase in the size of the high risk populations because of immunocompromise, the increased frequency of invasive medical interventions and prolonged survival of patients with chronic debilitating disease have amplified the problem.

Preventing emergence of antibiotic resistance

Availability of effective antibiotics has revolutionized public health and has been responsible for enabling countless advancement in medical care. The ongoing explosion of antibiotic-resistant infections continues to plague health care. Prevention of infection is the best policy since AMR cannot be halted. Therefore, containment is a next best option. Improving antimicrobial use must be a key action for containment.

Containment requires a multi-disciplinary approach. The emergence of antimicrobial resistance can be prevented or delayed through restricting non-human use of anti-microbials as growth promoters and judicious prescribing in humans since using anti-microbials causes resistance. Therefore, these drugs should only be used to treat infections. Rational use of anti-microbials is characterized by avoidance of antibiotic treatment for community-acquired, mostly viral, upper respiratory tract infections; use of narrow-spectrum antibiotics when possible; and use of antibiotics for the shortest duration that is effective for the treatment of a particular clinical syndrome. Prevention of antimicrobial drug resistance is aided by reducing infections by a healthy lifestyle, hand washing, and other good hygiene methods.

National and local guidelines for antimicrobial use (antibiotic policy) along with infection control policy are definitely the need of the hour for prevention of emergence and spread of antimicrobial resistance. It is a collective responsibility of governments, organizations and individuals to take action.

SUGGESTIONS FOR FURTHER READING

1. Brunton L.L., Blumenthal D.K., Murri N., Dandan R.H. and Knollmann B.C. (2011), Goodman and Gilman's the Pharmacological Basis of Therapeutics, 12th ed., McGraw-Hill, New York, 1029–1056.

2. Weatherall D.J., Ledingam J.G.G. and Warrel D.A. (Eds.) (1996), Oxford Text Book of Medicine, 3rd ed., Vol. 1, 295–314.

3. Lewis R. the Rise of Antibiotic-Resistant Infections Available At http://dwb4.unl.edu/chem/chem869k/chem869klinks/www.fda.gov/fdac/features/795_antibio.html

4. Antimicrobial Resistance. WHO Fact Sheet N°338 (2010), Available at Http://www.who.int/mediacentre/factsheets/fs338/en/

5. Antibiotic/Antimicrobial Resistance, CDC Website at Http://198.246.98.21/drugresistance/index.htm

SECTION XIX

CLINICAL BIOCHEMISTRY

Cardiovascular, renal, hepatic and other organ functions and dysfunctions are covered in various chapters of the Section, as also hematology, anemia and leukemia. Thus repetition has been avoided. The Section comprises only two chapters, one on 'Biological Fluids' (blood, urine, CSF) and the other on 'Enzymes and Isoenzymes', which are the part of investigations for clinical diagnosis and evaluation of organ functions.

131

Biological Fluids

N.C. Sharma, Parul Singla and L.M. Srivastava

CONTENTS

I. COLLECTION AND PRESERVATION OF BIOLOGICAL FLUIDS

A. Blood

There are a number of ways by which blood samples may be collected: from the capillary, vein and artery. In order to obtain comparable, results, most blood samples are collected in the *postabsorptive state,* that is, 12 to 14 hours after the last meal with the patient in the resting state to minimize effects due to exercise and intestinal absorption. Samples collected in the morning before breakfast generally conform with these requirements.

Serum: To obtain serum, the blood in the syringe is gently transferred to a clean and dry glass centrifuge tube. The tube is allowed to stand at room temperature until blood has clotted. It should then be kept in a warm place until the clot has retracted and the serum separated. Retraction of the clot may be assisted by gentle loosening from the walls of the tube. After centrifugation for 5 minutes at 1000–1500 revolutions per minute, the clear serum is removed and analysed or stored.

Plasma: The blood from the syringe is gently poured into a container containing an anticoagulant, mixed promptly and centrifuged to separate plasma. If the analysis cannot be done immediately, plasma should be stored under suitable conditions.

Anticoagulants: These substances prevent the coagulation (clotting) of blood. These include fluoride, oxalate, citrate, ethylenediamine tetra-acetic acid and heparin. Oxalate, citrate, ethyle-nediamine tetra-acetic acid and fluoride act by combining with Ca^{++} of the blood and thus preventing its participation in the clotting process. Oxalate and fluoride form insoluble calcium salts while citrate and ethylenediamine tetraacetic acid form soluble but unionized complexes. Heparin acts catalytically by mediating complexing of anti-thrombin III with thrombin (or other serine proteases), forming an inactive ternary complex (Murano, 1980). Heparin is the most satisfactory anticoagulant because it does not alter the blood constituents, although it is more expensive than oxalates and other anticoagulants.

For most analytical work only two types of anticoagulant containers are sufficient. One containing oxalate-fluoride (for blood glucose), and the other containing heparin. 2 mg of potassium or sodium oxalate (+2 mg of sodium fluoride for glucose), 5 mg of sodium citrate, 100 I.U. of sodium heparin and 10 I.U. of calcium heparin are adequate for 1 ml of blood.

Choice of the sample: Serum is more suitable than whole blood or plasma. If a capillary blood collection is done, whole blood is more convenient. The following points are important in the selection of a sample for analysis.

1. If the substance to be analysed is equally distributed between erythrocytes and plasma (or serum), it is immaterial which is used.
2. There are many substances in the red blood cells (RBC) that have adverse effects on the accuracy and specificity of the determination, by interfering in the reaction. For instance, glutathione and ergothioneine (both reducing agents present in the erythrocytes) are also estimated as glucose by the old methods (Folin-Wu method), and thus non-specific high values are obtained. In such cases it is better to use plasma or serum. As a rule, one should avoid the use of whole blood in any method in which first the blood is laked.
3. In many diseases the clinically significant variations occur only in plasma or serum. For instance, serum chloride varies much more, in disease, in serum than in whole blood, since the RBC content of chloride is relatively stable. In such conditions the use of serum is more suited.
4. If there are large differences in the concentrations between the RBCs and plasma, variations in haematocrit will greatly affect the results of analysis on whole blood.
5. It is easier to avoid haemolysis in the production of serum but when the separation and analysis must be done quickly, plasma is usually preferred.

B. Urine

Urine is the simplest, safest, cheepest and user friendly body fluid for analysis. Samples of the urine obtained at different times in a day greatly vary in their composition and any abnormal substance may be excreted only at a certain period of the day. It is, therefore, necessary for quantitative analysis to collect urine for 24 hours, and the results are expressed as the amount of the substance in the 24 hour urine sample. A 24 hour specimen is usually collected by discarding the first morning sample and collecting the urine voided up to and including the first specimen of the next morning. For most of the qualitative tests, the first morning fresh sample is suitable as it is likely to be more concentrated than any other voided during the 24 hours and is more uniform from day to day.

Sometimes it is important to obtain urine from women by catheterization in order to avoid contamination by material from the vagina. If the urine cannot be analysed immediately, or a 24 hour sample is to be collected, suitable preservatives are added to avoid decomposition and other changes that may occur on storage.

Preservatives: For determination of nitrogen, ammonia, calcium and phosphorus, 100 ml of 1N HCl is adequate for 24 hour urine sample. For proteins, sodium, potassium, and uric acid estimations 10 ml of toluene per 24 hour specimen is used. 5 ml of a 5% aqueous solution of hibitane (chlorhexidine diacetate) is suitable for glucose, 10 ml of glacial acetic acid for ascorbic acid and 5-hydroxyinolylacetic acid, 1 ml of concentrated sulphuric acid for catecholamines, 3 ml of 25% sulphuric acid for vanilmandelic acid (VMA) and 5 g of sodium bicarbonate for coproporphyrin and urobilinogen are used per 24 hour urine samples. For hormone estimations the urine is refrigerated.

C. Cerebrospinal Fluid (CSF)

CSF is usually obtained by spinal puncture for the purposes of analysis for diagnosis. The puncture is usually carried out in the lumbar region between the third and fourth, or between the fourth and fifth lumbar vertebrae.

Lumbar puncture: It is usually carried out with the subject lying on the side and with the torso flexed to produce the maximum separation of the vertebrae. The lumbar area is disinfected with an antiseptic. A line from the crest of the ileum will indicate the appropriate site for the puncture. The space between the appropriate spinous processes may be identified by a palpable depression in the skin. Under local anaesthesia at the appropriate site the needle is inserted slowly at right angles to both axes. When a give is felt, the tip of the needle is in the subarachnoid space. The stylet is removed, and the clear fluid drips from the needle if it is in proper place. The fluid is collected in two types of containers. One tube may contain potassium oxalate or heparin if the fluid tends to coagulate; the other is a plain container. For glucose estimation the tube should contain fluoride also. The stylet is then replaced, and the needle is removed. The needle is covered with a sterile dressing. The subject is kept in bed overnight to minimize the chance of lumbar puncture headache.

D. Safety and Blood Handling

The first description of acquired immunodeficiency syndrome (AIDS) in the beginning of 1982 and subsequent recognition of the fact that transfusion of contaminated blood is one of the most important routes through which the causative human immunodeficiency virus (HIV) is transmitted have caused worldwide a lot of concern. Not only AIDS but hepatitis, syphilis, malaria and several other infections can also be transmitted by transfusing contaminated blood. The safety of blood is often questioned and handling of blood requires special attention in blood banks and clinical laboratories. Healthcare workers, practitioners, patients, phlebotomist, laboratory personnel and support staff are at maximum risk primarily because of contaminated needles, hands or gloves, skin puncture by needles or broken specimen containers, spill or splash of specimen, contaminated work surface, contact with contaminated waste, puncture wounds and cuts. Since there is no effective treatment and no vaccine for AIDS, HIV infection is on the rise and it has really reached pandemic dimensions. It is, therefore, essential to learn to live with the virus. Universal work precautions and good laboratory procedures must always be followed when handling blood and other body fluids.

II. CARDINAL TESTS FOR DIAGNOSIS

A large variety of physical, chemical, immunological, and microscopic tests are employed for routine investigations for the diagnosis and follow up of disease. An abnormality in the functions of the body (or tissues) in most cases represents an abnormality of the functions of the constituting cells which are reflected in the fluids bathing them and in turn in the plasma and urine. For instance, in diabetes mellitus, the cells do not metabolize or metabolize glucose to a little extent depending on the severity of the disease. As a result of this, glucose accumulation takes place and its level rises in the blood. In many instances, abnormal constituents appear in blood, and finally in urine, such as ketone bodies in ketosis. The analysis of blood and urine thus provides the physician with information of great clinical importance for the diagnosis and follow-up of disease. The analysis is performed on a number of body fluids and juices such as blood, urine, cerebrospinal fluid, bile, gastric juice, pancreatic juice, etc. Among these, because of the ease with which the samples may be obtained, and the changes that occur in disease, blood and urine analysis have continued to play a dominant role in clinical chemistry.

Some of the important tests in blood and urine that are employed in clinical practice have been described below. A number of other tests for organ, endocrine and metabolic functions, electrolytes and vitamins have been described elsewhere in respective chapters. The summary of the normal values and the clinical interpretations of these tests is given in Table 131.1.

A. Analysis of Blood

1. Proteins: Serum or plasma is a very complex mixture of a variety of proteins and their determination is clinically of much value in diagnosis. Proteins in blood and other body fluids can be studied by numerous methods. A detailed description of the diagnostic methods employed in the study of the proteins is outside the scope of this chapter. Advanced texts on the subject may be consulted to appreciate the newer methods and their clinical usefulness. Recent methods for separation, detection, and quantitation of proteins in complex mixtures use one or more of chromatographic, electrophoretic, or immunological approaches routinely. Immunodiffusion, single and two dimensional electrophoresis and immunoelectrophoresis in gels are specific and highly sensitive techniques. These methods can efficiently separate, detect, and quantitate various proteins in one microlitre (or less) of plasma. Radioimmunological (RIA), or enzyme linked immunosorbent assay (ELISA) or Western blotting employ specific or monoclonal antibodies and can measure a few nanograms of a single protein or a non-protein substance (steroid hormones, etc.) in a complex mixture such as plasma or urine. These methods have been over simplified and in most cases test kits are commercially available.

The proteins that are frequently measured have been discussed earlier (see Chapter 6). Here only an account of their variations in diseases and the interpretations of the results will be described.

(a) *Albumin:* Increase in serum albumin levels is seldom encountered clinically. It may be observed in dehydration with diminution in the water content of blood. On the other hand, low values are observed in many disease states. Oedema is the most prominent feature of reduced serum albumin concentration (hypoalbuminaemia). The lowered colloidal osmotic pressure tends to reduce the passage of water from interstitial tissue spaces into the capillaries. In frank oedema (non-cardiac) albumin falls below 2.5 ± 0.2 g/100 ml. It has been experimentally observed that oedema occurs if albumin falls below 0.8 g per 100 ml and disappears when its level rises to 1 g per 100 ml. Hypoalbuminaemia may be due to the following factors:

(i) NEPHRITIS AND NEPHROSIS with excessive loss of albumin in urine. In chronic nephritis as much as 60 g albumin may be lost daily in urine.

TABLE 131.1 Lipoprotein composition of normal human serum and their variation in disease

Class of lipo-protein	Normal values (Mean ± S.D.) mg/100 ml	Interpretations
		Low density lipoproteins
Sf 0–12	352 ± 79	High in xanthoma tendinosum, myxoedema, coronary heart disease, nephrotic syndrome.
Sf 12–20	49 ± 21	High in xanthoma tendinosum, xanthoma tuberosum, glycogen storage disease, myxoedema, coronary heart disease, nephrotic syndrome.
Sf 20–100	92 ± 51	High in xanthoma tendinosum, xanthoma tuberosum, essential hyperlipaemia, diabetic acidosis, glycogen storage disease, coronary heart disease, nephrotic syndrome.
Sf 100–400	53 ± 63	High in hyperlipaemia diabetic acidosis, glycogen storage disease, coronary heart disease, nephrotic syndrome.
		High density lipoproteins
HDL_1	17	High in xanthoma tendinosum, essential hyperlipaemia, diabetic acidosis, glycogen storage disease.
HDL_2	88	Low in xanthoma tendinosum, glycogen storage disease, nephrotic syndrome, chronic biliary obstruction.
HDL_3	193	Low in xanthoma tendinosum, chronic biliary obstruction.

(ii) INADEQUATE SUPPLY OF PROTEIN: Impaired protein digestion and absorption in peptic ulcer, advanced tuberculosis and malignancy, gastrointestinal malignancy, pancreatic diseases cause a fall in the albumin levels in blood.

(iii) IMPAIRED SYNTHESIS in chronic hepatic diseases particularly cirrhosis, infections, severe anaemias, cachectic states.

(iv) EXCESSIVE CATABOLISM in association with either defective utilization or inadequate supply as in uncontrolled diabetes mellitus, thyrotoxicosis, prolonged febrile illness and injury.

(v) Genetic Disorders: Analbuminaemia (absence of albumin) may be observed in genetically determined defects in albumin synthesis.

(vi) PHYSIOLOGICAL: A steady fall in the albumin concentration may occur during normal pregnancy due to diminution in the albumin fraction, in toxaemia of pregnancy. Synthesis of proteins in the foetus constitutes a drain upon the maternal organism.

(b) *Globulins:* A globulin fraction is a complex mixture of proteins. Over 30 components have been identified and characterized. Frequently, in routine clinical analysis, albumin and globulins fractions are separated by salt precipitation and by paper electrophoresis. Sometimes, *globulin reactions* are also studied.

GLOBULIN REACTIONS: The qualitative reactions that produce abnormal results when abnormalities in one or more globulins occur are termed as globulin reactions. Those frequently used are the thymol turbidity, cephalin-cholesterol flocculation and the zinc sulphate turbidity tests. These methods are in general less accurate and depend on the state of balance between certain stabilizing and precipitating factors. In the presence of the availability of the better screening chemical and immunological methods, the use of these tests is becoming less important. Nonetheless, these tests are routinely used for clinical diagnosis.

ELECTROPHORESIS OF SERUM PROTEINS: The proteins of plasma (or serum) and tissues are interrelated. By studying the plasma proteins, therefore, a great deal can be learned concerning the general position of protein metabolism in health and in disease. Electrophoresis provides an important method of their fractionation and study. Electrophoretically, plasma proteins may be fractionated into albumin, α_1-globulin, α_2-globulin, β-globulin, fibrinogen and γ-globulins.

α-GLOBULINS: This is a heterogeneous group of proteins and normal plasma values are 0.5 to 1.3 g per 100 ml. Low levels are observed in infective hepatitis, nephrotic syndrome, portal cirrhosis, multiple myeloma, and hypothyroidism. High values occur in a great variety of conditions associated with destructive or inflammatory lesions, acute and chronic infections, severe burns and other skin lesions, extensive malignancy, myocardial infarction; glomerulonephritis, hepatic cirrhosis, rheumatoid arthritis, etc.

β-GLOBULINS: Normal values range from 0.6–1.2 g per 100 ml plasma. Low values occur rarely but may be sometimes observed in multiple myeloma. High values, mainly due to an increase in the β-lipoproteins, occur in disorders associated with hyperlipaemia such as in nephrotic syndrome, obstructive jaundice, diabetes mellitus, etc. A better correlation of β-lipoproteins levels with that of serum lipid or lipoprotein can be made by separating them by boundary electrophoresis rather than paper electrophoresis.

γ-GLOBULINS: These are the circulating antibodies in blood and are also termed as immunoglobulins. They are constituted of a number of closely related proteins possessing the entire antibody activity of serum (see Chapter 102).

The γ-globulin fraction is often increased in infections and other disorders that evoke an immunological response. The most striking increased values are observed in viral and protozoal diseases, extensive tissue destruction or suppuration, extensive granulomatous proliferation, multiple myeloma, hepatic cirrhosis, and hypothyroidism. Low values can occur in genetically determined impaired and defective synthesis (primary agamma-globulinaemia). In this condition circulating γ-globulins are usually less than 0.1 g/100 ml and the low levels are due to the defect in plasma cell formation which is associated with the synthesis of γ-globulins. In many acquired diseases also there is inadequate synthesis of γ-globulins (secondary hypogamma-globulinaemia). This refers to the involvement of the reticuloendothelial system and may be observed in multiple myeloma, leukaemia, lymphomatous diseases, etc.

(c) *Fibrinogen:* High values occur in multiple myeloma, nephrosis (up to 1000 mg/100 ml), hepatitis, tissue destruction, most acute infections except typhoid fever, pneumococcal pneumonia. Low values are observed in congenital afibrinogenaemia due to genetic defect in synthesis, hepatic insufficiency, and obstetric complications.

(d) *Protein bound iodine (PBI):* Protein bound iodine (PBI) is a measure of the total circulating thyroid hormone. In serum or plasma most of the thyroxine is bound to proteins. These are thyroxine binding globulin (TBG), and the thyroxine binding prealbumin (TBPA). Thyroxine binding proteins act as specific carriers for the hormone. Excess of the hormone is bound to albumin. Normal values for PBI range from 4 to 8 µg/100 ml serum. Some inorganic iodine is also present in serum and therefore, while measuring PBI care should be taken to exclude the inorganic iodine and any iodine compounds used therapeutically or for diagnosis. Inorganic iodine and iodotyrosines may be easily removed by extracting serum with *n*-butanol. PBI, TBG, and butanol extractable iodine (BEI) measurements are clinically very important to assess the thyroid function.

TBG is high in pregnancy and after administration of estrogens to non-pregnant individuals. In nephrosis and after treatment with androgens and steroids, TBG levels are low. When TBG levels are normal, high and low PBI values correspond to *hyper* and *hypo-thyroidism*. Elevated PBI levels are also observed in pregnancy.

In the diagnosis of hypothyroidism determination of PBI is of great value in children and adults. Hypercholesterolaemia, reduced basal metabolic rate, and low voltage with depressed or inverted T waves in electrocardiogram may provide additional evidence to confirm the disease. Also, liberation of fatty acids from the tissues is decreased.

(e) *Specific antibodies, antigens and DNA probes:* In almost all infections, the requirement of diagnosis is to detect the specific microorganism or its marker antigenic molecules, antibodies and specific DNA sequences. Immunoassays for the detection of antibodies and antigens of human immunodeficiency virus (HIV), hepatitis-B virus (HBV), *M. tuberculosis, S. typhi, T. pallidum,* etc. are widely used. These assays are very sensitive and can detect less than 1.0 ng quantities but are not sensitive enough to rule out presence of organism in blood particularly when the target organism cannot be cultured as is often the case with viral infections. This raises serious doubts especially when blood is tested for its suitability for transfusion. Failure to detect organism specific antigens or antibodies does not rule out infectivity of blood. Recent advances in the gene manipulation technology, hybridization of a nucleic acid probe to a target DNA or RNA sequence has made possible detection of single-copy gene sequences (1–10 pg) in filter hybridization. Application of polymerase chain reaction (PCR), that is, repeated cycles of denaturation, priming and extension, results in a rapid exponential accumulation of a specific target sequence. After only 10 cycles, the target sequence is amplified 1000-fold. Ou and his colleagues (1988) were able to detect HIV-1 DNA in peripheral blood mononuclear cells in less than a day whereas virus isolation took 3–4 weeks.

2. Blood glucose: In the postabsorptive state (12 to 14 hours after the last meal) the *true sugar* or *sugar* in the blood is mostly glucose. The values in 2417 healthy normal Indians by the method of Sharma and Sur (1966, 1972) are in the range 60 to 100 mg per 100 ml (79.4 ± 10.1 S.D.) and by the autoanalyzer method in 471 subjects are 76.08 ± 18.43 mg per 100 ml. In adults blood glucose values are not significantly influenced by sex, dietary habits, and socio-economic status. Blood glucose levels, however, change with the methods used for its estimation. Older methods such as that of Folin-Wu are not specific and also estimate non-sugar *reducing substances* (saccharoids) of blood and give higher values (80–120 mg/100 ml). Saccharoids are chiefly present in red blood cells and include glutathione and ergothioneine. The recent methods such as glucose oxidase, autoanalyzer and that of Sharma and Sur achieve better specificity and give identical values differing maximally by less than 5 per cent. Glucose oxidase based estimation is oversimplified and do yourself or bed-side or point-of-care tests using a drop of finger-prick blood are available commercially which provide results in just under 10 seconds.

Blood glucose estimation is of great clinical importance for the diagnosis and treatment of diabetes mellitus where high fasting values (hyperglycaemia) are obtained depending on the severity of the disease. Hyperglycaemia is also observed in vomiting of pregnancy, asphyxia, nephritis, anaesthesia, pneumonia, dehydration, and certain endocrine disturbances such as hyper-adrenalism, hyperthyroidism, and hyperpituitarism.

Hypoglycaemia signifies a decrease in blood glucose level below normal limits and may be observed in certain physiological conditions such as pregnancy, starvation, severe prolonged exercise, and lactation. Low values in diseases occur in insulin excess, deficiency of insulin antagonists, deficiency of available glycogen and renal glycosuria.

Glucose tolerance tests are performed for diagnosis of cases of pre- or latent diabetes mellitus (see Chapter 12).

Measurement of glycated proteins primarily glycated haemoglobin (GHb or HbAlc) is effective in monitoring long term glucose control in people with diabetes mellitus. It provides a retrospective index of the integrated plasma glucose values over a period of 6-8 weeks. Glycation is nonenzymatic addition of sugar residue to amino group of proteins. HbAlc is formed by the condensation of glucose with the N-terminal valine residue of each beta chain of haemoglobin A to form an unstable Schiffs base. This Schiff's base may either dissociate or undergo Amadori rearrangement to form stable ketamine. Formation of glycated haemoglobin is essentially irreversible, depends on the life span of the red blood cell (average 120 days) and directly proportional to the concentration of glucose in the blood. Glycated haemoglobin is free of day to day glucose fluctuations and are unaffected by the recent exercise and food ingestion. For a normal individual with normal blood glucose levels the HbAlc is less than 6%. In diabetes mellitus patients HbAlc value of more than 7% indicates poorly controlled diabetes. Patients with haemolytic disease or recent blood loss may exhibit substantial reduction in HbAlc levels. The most common methods used for its determination is Ion exchange chromatography, HPLC, electrophoresis, isoelectric focusing, affinity chromatography and photometry.

3. Lipids: Even today, the clinical chemistry laboratories offer the clinician only a meagre aid for the study of the status of lipid metabolism in their patients. Total lipids, total and free fatty acids, and esterified and free cholesterol are frequently analysed. The analysis of plasma lipoprotein fractions is of great clinical significance as most of the lipids do not exist individually in plasma but occur in the form of a number of lipoproteins.

Many diseases characterized by the distribution patterns of cholesterol, triglycerides, phospholipids in fact represent changes in the lipoprotein fractions.

(a) *Lipoproteins* (see Chapter 6): Since these are separated by ultracentrifugation by the method of floatation instead of sedimentation, plasma lipoproteins are characterized by floatation constants (*Sf*) and their electrophoretic mobility. The higher the *Sf* value, the lower is the density of the lipoprotein.

Lipoproteins show a characteristic and definite distribution in diseases. *Sf* 20–100 fractions register markedly high values in essential hyperlipaemia, diabetic acidosis, glycogen storage disease, and significant diminution of the HDL lipoproteins. In xanthoma tuberosum (a familial disease with characteristic deposition of lipid in extensor surface of elbows, ankles, buttocks, knees and hands), xanthoma tendinosum (believed to be a familial disease, characterized by external lipid deposition mainly in the tendon regions of the body), nephrotic syndrome, and biliary obstruction, low density lipoproteins are increased in serum with a highly significant reduction in HDL_2 lipoproteins. Table 131.1 shows the normal values of various lipoproteins and their clinical significance.

The measurement of serum lipoproteins provides new biochemical information indicative of an abnormal metabolic situation prior to the appearance of clinically manifested disease and, therefore, requires serious attention for inclusion in clinical diagnosis.

TOTAL LIPIDS: The total plasma lipid concentration in the postabsorptive state of normal subjects ranges from 340–800 mg per 100 ml. Though of limited clinical value the estimation of total lipids gives some information of the overall metabolism of lipids. It is valuable for fat tolerance experiments and in the screening of hyperlipaemia. In essential familial hyperlipaemia, values as high as 5600 mg/100 ml are not uncommon.

FATTY ACIDS: Total fatty acids constitute about 50% of the total lipids (7–20 meq/1). They occur in plasma in two forms, *free fatty acids* (0.3–0.6 meq/1) and *bound fatty acids*. Free fatty acids (FFA) are also termed as *unesterified fatty acids* (UFA) or *nonesterified fatty acids* (NEFA). This represents the most active fraction of all with a turnover sometimes of 3000 calories per day. In diabetic acidosis, FFA may be increased to 1.5 meq/1. FFA level rises in exercise, starvation, emotional stress, hyperthyroidism and on administration of catecholamines. It is due to increased lipolysis of triglycerides in adipose tissue.

(b) *Triglycerides:* The concentration of triglycerides in normal plasma is 60–130 mg/100 ml (3.2 meq of triglyceride fatty acids). They contain about 25% palmitic acid, 4% stearic acid, 10% linoleic acid and 41% oleic acid. Other, fatty acids constitute about 20%. The composition appears to be identical in all lipoprotein fractions, indicating a free exchange of triglycerides between the various lipoprotein fractions.

LIPID PHOSPHORUS: The concentration in normal subjects ranges from 6–15 mg of lipid P per 100 ml. Out of the total, about 70% is lecithin, about 20% is sphingomyelin and the rest includes cephalins, inositol, phospholipids, lysolecithin, and other compounds. Phospholipids are best expressed in terms of meq of phospholipid fatty acids. There are 2 meq of fatty acids for each 31 mg of P.

Individual lipids vary absolutely and selectively in a variety of physiological and disease states. However, free cholesterol/total cholesterol and total cholesterol/phospholipid ratios remain fairly constant. In liver disease the former fails to hold this relationship.

(c) *Cholesterol:* The total serum cholesterol level in normal individuals ranges from, 135 to 260 mg per 100 ml. Values from 250 to 300 mg per 100 ml, once considered normal, are now doubted very much. The values are very low in newborn infants (30–90 mg/100 ml), rise rapidly in early life for the first few months. Cholesterol seems to rise gently with age. Seasonal variations are also observed in serum cholesterol levels. In summer and spring values are low and the highest values are observed at the time of fall of winter. Cholesterol is present in both ester and non-ester forms. About 60–80% of the total is present in the ester form.

HYPERCHOLESTEROLAEMIA: When the level of serum cholesterol is higher than the normal range, the condition is known as hypercholesterolaemia and signifies an abnormality. Hypercholesterolaemia is seen in a number of metabolic and other diseases and some of these are mentioned below.

DIABETES MELLITUS: Serum cholesterol concentration has been regarded by many workers as an index of lipid metabolism and of prognosis in diabetics. Changes in the cholesterol level in diabetic children are not so marked. In diabetic coma, higher values are rather consistently observed. In a large majority of patients with hypercholesterolaemia in diabetes mellitus, both free and ester fractions are raised. The ratio of ester and free cholesterol remains normal even if the total cholesterol values are as high as 400–500 mg/100 ml. In diabetes mellitus the reason of hypercholesterolaemia seems to be related to the metabolism of acetate which is produced in increased amounts from the catabolism of fatty acids, carbohydrates, and amino acids. Because of insulin deficiency and diminished supply of oxaloacetate, acetate is not utilized at a normal rate for the synthesis of fatty acids. Administration of insulin usually lowers the serum cholesterol level. Cholesterol estimation may be a useful measure of the adequacy of the control of diabetes mellitus.

NEPHROTIC SYNDROME: Hypercholesterolaemia is a constant feature of the nephrotic syndrome. Sometimes very high values for the plasma cholesterol are observed (2200 mg/100 ml). The total lipids also often rise beyond 2 g/100 ml. The serum looks milky. This hypercholesterolaemia is accompanied by the presence of increased amounts of cholesterol in the urine and its deposition in the renal tubules.

ATHEROSCLEROSIS: The incidence of atherosclerosis is high in diseases associated with hypercholesterolaemia such as diabetes mellitus, nephrosis, idiopathic hyperlipaemia and hypothyroidism. A considerable number of patients suffering from atherosclerosis do not show abnormal rise in serum cholesterol. This, of course, does not mean that a derangement in lipid metabolism or transport is of no clinical value in this disease. In spite of the extensive studies on atherosclerosis, the definite role of lipids in general and of cholesterol in particular has not yet been established. The measurement of plasma cholesterol and triglyceride levels are still of great value in the management of hyperlipoproteinaemia.

HEPATIC AND BILIARY TRACT DISEASES: High cholesterol levels are usually observed in jaundice due to uncomplicated common duct obstruction. The values rise more than that of bilirubin and return to normal with the release of obstruction. Usually with a rise in serum bilirubin, free cholesterol also rises and therefore esterified cholesterol decreases progressively. The mechanism of these changes is not understood. In obstruction, decline in esterified cholesterol cannot be a reliable parameter indicative of superimposed hepatocellular damage.

Hypercholesterolaemia may also occur in biliary cirrhosis and mild hepatocellular jaundice (hepatitis).

OTHER CONDITIONS: Hypercholesterolaemia is also observed in untreated hypothyroidism, following acute haemorrhage, hypertrophic osteoarthritis, senile cataract, psoriasis and dermatitis.

The relationship of the occurrence of hypercholesterolaemia to a particular disease is not understood in most cases, nevertheless, physicians have been compelled to seek methods to reduce the level of serum cholesterol. One of the most promising types of blood cholesterol-reducing management appears to be the substitution of fats rich in polyunsaturated fatty acids for the highly saturated fats of the normal diet. Other treatments include: (1) a high intake of a plant sterol, sitosterol, which perhaps precipitates a large portion of the intestinal cholesterol, thereby preventing its absorption, (2) use of cholestyramine which combines with bile acids in the intestine and prevents their reabsorption, (3) oral administration of large doses of niacine, (4) use of estrogens and (5) use of HMG-CoA reductase inhibitors (statins).

HYPOCHOLESTEROLAEMIA: Lower than normal values of plasma cholesterol refers to hypocholesterolaemia. The plasma cholesterol is low in pernicious anaemia during relapse, in haemolytic jaundice, hepatocellular damage (particularly ester fraction), acute infectious diseases, hyperthyroidism, cachexia, prostatic and intestinal obstructions, arthritis, sprue, celiac disease, and during epileptic seizures.

TERMINAL STATES: In non-nephrotic glomerulo-nephritis, the total fat, fatty acids, and phospholipids in the plasma are increased but cholesterol is usually normal or sub-normal. The variability of plasma cholesterol in chronic glomerullonephritis is of much prognostic value particularly when associated with increased nitrogen retention.

Low cholesterol values are also observed (50–100 mg/100 ml) in the terminal stages of a number of diseases without nitrogen retention. These include congestive heart failure, carcinoma, acute pancreatitis, pulmonary tuberculosis, coronary artery occlusion, diabetes mellitus, and bacterial endocarditis.

The mechanism underlying the occurrence of hypo-cholesterolaemia in terminal stages of these diseases is not known. Anaemia and cachexia, which are frequently present in these patients, may be the responsible factors. However, these factors are frequently absent in those dying of pneumonia, coronary artery occlusion, congestive and left-side heart failure, and peritonitis. Whatever may be the mechanism, the development of hypocholesterolaemia under these conditions is of great prognostic significance particularly in diabetes

mellitus and chronic glomerulonephritis, in which the plasma cholesterol may previously have been elevated.

LIPOIDOSIS AND XANTHOMATOSES: This group of disorders has been long known but remains ill-defined. The main observation is the infiltration of tissues by cells of the reticuloendothelial system, which are packed with lipids. Because of their microscopic appearance they are also called *foam cells*. Infiltration may be confined to one or more tissues or it may be widespread. The infiltrating cells involve especially flat tissues (skin or tendon) which may take on the appearance of tumours having a yellowish colour due, apparently, to dissolved carotenoids; these have been termed *xanthomas*. The general disease refers to *lipoidosis, and xanthomatosis* to the condition in which cholesterol is the chief lipid ingredient of the cells.

B. Clearance Tests

The constancy of the composition of blood is maintained by what the kidneys keep and not by what the mouth ingests. Kidneys are by and large the principal regulators of the internal environment of the body. Present knowledge of the physiology and biochemistry of the kidneys permits rather precise interpretations of its functions. Often glomerular filtration rate, renal blood flow and maximum tubular reabsorption, and excretion are measured. The methods employed for the study of the functional activity of the kidneys include (1) tests based on the clearances of the substances, (2) excretion of dyes, and (3) the concentrating capacity.

Clearance: Clearance refers to the volume of plasma in millilitres which contains the amount of a particular substance removed per minute by renal excretion. This also represents the minimum volume of blood required to furnish the amount of the substance excreted in the urine in one minute. The concept of kidney clearance has contributed, to a large extent, to our understanding of the renal function in health and disease.

If P represents the concentrations of the substance in plasma (mg/100 ml), U in the urine (mg/100 ml), and V the volume (ml) of urine formed per minute then UV equals the amount of the substance excreted per minute in the urine. UV/P is, therefore, the virtual volume of plasma that contains the amount, of the substance removed by renal excretion per minute (or cleared of the substance per minute) and is the clearance of that substance.

Clearance tests are performed to measure (a) glomerular filtration rate, (b) filtration plus tubular excretion, (c) renal blood plasma flow, (d) filtration and excretion fractions, and (e) urea clearance.

(a) *Urea clearance:* When the urine volumes are fairly large, the rate of elimination of urea is directly proportional to the plasma urea concentration. This relationship holds good only when the urine volume exceeds a certain limit, the *augmentation limit*. For normal adults it is 2 ml per minute. The urea clearance under these conditions is termed *maximum urea clearance* (C_m) and is given by:

$$C_m = \frac{UV}{P}$$

C_m in normal adults ranges from 64 to 99 ml. The average value for an adult of an average size of 1.73 square metres surface area is 75 ml of plasma per minute.

When the urine output is less than 2 ml per minute (below the augmentation limit), a certain volume of plasma will be cleared of urea per minute and the clearance is termed *standard clearance* (C_s). It is given by

$$C_s = \frac{U\sqrt{V}}{P}$$

C_s varies normally from 41 to 65 ml and the average value in adults (of surface area of 1.73 square metres) is 54 ml per minute.

Since the urea clearance, urine volume and augmentation limit vary with changes in the body surface area, a correction should be made in the above formula in the case of subjects clearly above and below the average adult size. The corrected expressions are as follows:

$$C_m = \frac{UV}{P} \times \frac{1.73}{A}$$

$$C_s = \frac{U}{P} \times \sqrt{V} \times \frac{1.73}{A}$$

where A is the body surface area in square metres and 1.73 is the body surface area of average adult. However, no correction is required for persons between 158 and 180 cm in height, since the error involved is not more than 5%.

Urea clearance is increased by ingestion of caffeine, milk, and by small quantities of epinephrine. It is diminished by the administration of vasopressin or large quantities of epinephrine. Low values are also observed during relapse in pernicious anaemia and other severe anaemias.

In interpreting urea clearance in diseases, it must be kept in mind that this is not an accurate measure of the individual phases of renal function. Glomerular filtration rate, renal blood flow, and tubular reabsorption or excretion have to be determined to reveal discreate abnormalities of renal physiology. It is, therefore, of great clinical value to gauge overall renal function.

(b) *Glomerular filtration rate* (GFR): For the measurement of GFR, inulin or creatinine clearance is determined. Inulin (a polysaccharide) is not metabolized by the body and following its intravenous injection, it is removed quantitatively by glomerular filtration and is neither reabsorbed nor excreted by the tubules.

It follows, therefore, that the inulin clearance (UV/P) represents the volume of glomerular filtrate formed per min. Creatinine clearance is also frequently employed as a routine test to determine GFR. It has an advantage over inulin clearance in that it requires no intravenous injection, the analytical procedure involved is relatively simpler and it is less influenced by decrease in the rate of urine flow. However, it is not as accurate and is subject to errors in measurement. Under normal conditions, the endogenous creatinine clearance ranges from 75 to 125 per cent of the inulin clearance. Employing inulin clearance the average rate of glomerular filtration is 125 ml per minute (about 70 ml per square metre of body surface). Similar values are obtained if mannitol, sorbitol, and dulcitol are used in place of inulin. In normal subjects and often in diseases, under average conditions of the rate of urine flow (2 ml per min. or more), urea clearance values usually run parallel to those of inulin clearance and the latter may be calculated by dividing urea clearance by 0.6. This proportionality does not hold at low urine volumes.

Many other useful physiological deductions can be made by determining GFR.

1. Water reabsorbed in the tubules per min.:
   = GFR – Rate of urine flow (bladder).
2. Proportion of substance X reabsorbed in the tubules:

$$= \frac{\text{Inulin clearance} - \text{Clearance of } X}{\text{Inulin clearance}}$$

For example,

$$\frac{\text{Inulin clearance} - \text{Urea clearance}}{\text{Inulin clearance}} = \frac{125 - 75}{125} = 0.4$$

This indicates that 40 per cent of urea of the glomerular filtrate is reabsorbed during its passage through the tubules.

Glomerular filtration is affected by renal blood flow, glomerular pressure and plasma osmotic pressure, capsular pressure, and the filtration surface area. Any abnormality in these factors indicates clinical disorders and altered glomerular filtration rate. Decreased volumes of glomerular filtrate are observed in conditions associated with reduced renal blood flow (marked fall in systemic arterial blood pressure due to left ventricular or peripheral circulatory failure, congestive heart failure, essential hypertension, constriction or sclerosis of the glomerular arterioles, all types of diffuse glomerular damage).

(c) *Renal blood flow:* If a substance is completely removed from the plasma during its single passage through the kidneys, the clearance of that substance (UV/P) is equal to the effective plasma flow, i.e. the flow through active renal excretory tissue and if the haematocrit value is also known, renal blood flow can be easily calculated. Diodrast and p-amino hippurate (PAH) are the substances used for the determination of renal blood flow. More than 90 per cent of the plasma content of these substances are removed during a passage through the kidneys when their concentrations are low (0.5–2.5 mg of Diodrast iodine or PAH per 100 ml). Normal value for the renal plasma flow for men is about 700 ± 135 ml and for women 600 ± 100 ml per 1.73 square metre body surface. Corresponding renal blood flow values are about 1275 ± 245 ml and 1090 ± 180 ml, respectively. In older subjects values are somewhat lower. Renal blood flow shows diurnal variation, decreases during exercise and is influenced by emotional states.

Lower values for renal blood flow are seen in (1) diseases associated with diminished cardiac output or arterial blood pressure. These include congestive heart failure, coronary artery occlusion, shock, etc., (2) organic diseases of the renal vascular system such as renal arteriosclerosis, nephrosclerosis, glomerulonephritis, glomerulosclerosis, and periarteritis, (3) reduced mass of the functioning kidney tissue such as renal hypoplasia, polycystic disease, pyelonephritis, tuberculosis, and malignancy, and (4) elevated local resistance to blood flow such as in early essential hypertension and systemic arterial hypotension.

(d) *Maximum tubular excretory capacity:* It has been pointed out that at low concentrations, Diodrast and PAH are both almost completely removed from the plasma in a single passage by the combination of both glomerular filtration and tubular excretion. As the plasma concentration of these substances is raised, a value is reached at which the absolute amount of the substances removed by the tubules, reaches a maximum. For achieving this maximum, the plasma concentration of the substance must be high enough to put a load of about 1.5 times the maximum that the tubules can remove. If still greater concentrations are employed, no greater amount will be removed by the tubules. However, the amount in the glomerular filtrate will rise with the plasma concentration. This *maximum excretory capacity of the tubules* (Tm) is about 77 mg per min. For PAH ($Tm_{PAH} = 77$) and 50 mg per min. for Diodrast ($Tm_D = 50$). Values are somewhat lower in females.

Tm is determined by simultaneously measuring inulin and Diodrast (or PAH) clearances at high plasma Diodrast (or PAH) levels. The results are expressed as mg of Diodrast iodine, or PAH removed per min. by tubular excretion. Tm is independent of glomerular activity and reflects the amount of functioning renal tubular tissue (i.e., the *tubular excretory mass* of the kidneys). If a Tm value is 50 per cent of normal, it means the absence or destruction of half the normal number of tubules. A value of 5 per cent of normal signifies that only one twentieth of the normal number of tubules is present or capable of function. In the uraemic stage of chronic Bright's disease, not only is the number of tubules greatly reduced, but those present are incapable of normal maximal function.

(e) *Maximum tubular reabsorption capacity:* This is followed in a manner identical to that used for the study of excretory Tm. In this case a substance that is extensively reabsorbed rather than excreted by the renal tubules is employed. Glucose is usually used to study the maximum reabsorptive capacity of the tubules. At an arterial plasma glucose concentration of say 100 mg per 100 ml and GFR of 120 ml per min., 120 mg of glucose are removed per minute in the glomerular filtrate. Normally, the entire amount of glucose is reabsorbed into the blood. If the concentration of glucose is raised, a level will be achieved at which the full capacity of the tubular reabsorption system is reached and no more of the glucose will be reabsorbed. The excess glucose under this condition will pass into the urine

from tubular fluid. This maximum tubular reabsorptive capacity for glucose (Tm_G) is about 350 mg per min.

The clinical significance of this factor is much the same as in the case of excretory Tm for PAH or Diodrast but it applies in this instance to tubular reabsorptive mass.

The plasma flow per unit of functional tubular reabsorptive, mass can be determined by measuring.

$$\frac{\text{PAH or Diodrast clearance}}{Tm_G} = \text{ml per mg}$$

Its normal value in subjects of 1.73 square metre body surface area is 2 ± 0.4 ml per mg.

Glomerular function in terms of functioning nephrons (filtration per unit functioning nephrons) may be calculated by:

$$\frac{\text{GFR (Insulin clearance)}}{Tm_{PAH}} = \text{ml per mg}$$

Normal value in subjects of 1.73 square metre body surface area is 1.7 ± 0.4 ml per mg. Increased values indicate hyperfiltration and lower values hypofiltration in residual nephrons.

C. Analysis of Urine

Urine analysis provides information of great clinical importance chiefly on (1) the state of kidneys, (2) net nutritional balance, and (3) on the formation of pathological constituents if these can be excreted by the kidneys. It includes physical, chemical and microscopic (sediment) analysis.

1. Physical analysis: This involves the interpretation of the volume, colour, odour, specific gravity, and reaction (pH) of urine.

(a) *Volume:* The volume of the 24 hour urine for men ranges from 1000 to 1500 ml. The values are somewhat lower in women. However, it depends on water intake and climate. Out of the total water consumed about one-third to two-thirds is eliminated daily in the urine, the rest by the skin, lungs, and faeces. In hot weather or after violent exercise the loss through the skin may be considerably higher, whereas almost no water may be lost by this route if the outside air is saturated with moisture. Gross variations in the urinary volume occur in a number of disease states.

Oliguria refers to the condition of reduced volume of urine, and *anuria* means a total lack of urine excretion. Oliguria occurs in disorders associated with fever because a large volume of water is lost in sweat and by the lungs. If wrong blood is transfused into a patient, some red cells disintegrate and release free haemoglobin which damages the glomerular membrane and escapes into the renal tubules. This haemoglobin precipitates and blocks the tubules preventing the excretion of urine, resulting therefore, in oliguria and anuria and finally leading to uraemia which may be fatal. Renal damage and a number of other conditions also produce oliguria. In general anything that is known to produce uraemia is likely to produce oliguria and anuria.

Polyuria refers to the formation of large volumes of urine and not to an increase in the frequency of voiding urine. Excessive intake of water, diabetes insipidus and diabetes mellitus are some of the common diseases associated with polyuria. The most marked polyuria occurs in diabetes insipidus where values as high as 50 litres of urine per day have been reported. A number of drugs known as *diuretics* cause polyuria and are often used in the treatment of oedema.

(b) *Colour of urine:*   The colour of an average normal urine sample is described as amber. Dilute urine samples may appear pale yellow and the concentrated one dark brown in colour. The colour of urine is mainly due to a pigment, urochrome, a compound of urobilin and a peptide. In many diseases and normal conditions urinary colour changes to characteristic shades. Tables 131.2(a) and (b) show the colour of urine under various conditions.

(c) *Odour of urine:*   Urine possesses a characteristic and faintly aromatic odour. It is due to the presence of certain volatile organic acids and some of it is contributed by a foul-smelling neutral compound *urinod.* Many foods impart a rather characteristic smell to urine. On decomposition, the smell of ammonia, and in severe diabetes mellitus associated with ketosis, the smell of acetone may be recognised in urine.

TABLE 131.2(a)   Colour of urine and associated pathological factors

Colour	Causes
Red, reddish brown	Haematuria, haemoglobinuria, black-water fever, severe burns, various dyes, tomato juice, growth of B. prodigiosus.
Brown; on standing turning to brown-black	Alkaptonuria.
Dark brown	Urobilin in liver infection, malaria, pernicious anaemia.
Black	Visible haematuria, melanin (melanoma).
Greenish yellow	Bile pigments (jaundice).
Green	P. aeruginosa infection.
Dirty green or blue	Indigo pigments, methylene blue.
Turbid	Fat globules, pus corpuscles, bacteria, protein, phosphates, urates.

TABLE 131.2(b)   Colour of urine and associated non-pathological factors (Chemicals, drugs and dyes)

Colour of urine	Cause
Dark brown to black	Poisoning due to carbolic acid, salol, guaiacol, creosote or resorcin.
Deep red	Pyramidon, rose bengal, prontylin.
Green-blue	Thymol, methylene blue.
Pink (alkaline reaction)	Phenolsulphonaphthalein (used in the renal function test).
Yellow	Santonin, pyridium, tetracyclin, acriflavin.

(d) *Specific gravity:*   It ranges normally from 1.015 to 1.025 but is subjected to wide variations under various physiological and pathological conditions. It may be decreased to 1.003 (or even less) as a result of excessive water intake and increased to 1.040 (or even more) due to hemoconcentration or excessive perspiration. Specific gravity of urine is directly proportional to the solute content and usually varies inversely with the volume. In diabetes insipidus, the urine is very dilute and the specific gravity is close to that of water (1.000).

(e) *Reaction (pH) of urine:*   Fresh normal urine is weakly acidic but the pH varies with the diet. Non-vegetarian diets (meat, etc.) tend to make the urine acidic while most vegetables and fruits cause an alkaline reaction. Average pH of fresh normal urine is 4.8–7.8. On standing urine becomes alkaline due to bacterial decomposition with the formation of ammonia. In many diseases associated with *acidosis and alkalosis,* and bacteria causing diseases in the urinary tract, the pH of urine is changed. Bacteria themselves are influenced by a change in pH and for this reason acidic or alkaline diuretics are prescribed for treatment.

2. Chemical analysis: The constituents of normal and abnormal urine are dealt with below.

(a) *Normal constituents:*   These are the substances present in normal urine in physiological concentrations and have been described in Chapter 16/34.

(b) *Abnormal constituents:*   These refer to substances that are not normally found in urine and include proteins, carbohydrates, ketone bodies, blood and its products, and bile salts. These are termed as pathological constituents of urine as they appear in urine, in disease states. Abnormal urine or abnormal constituents of urine do not necessarily mean the presence of only these substances. The lack or absence of any normal constituent or excretion of any normal constituent in amounts more than the accepted normal range also refers to an abnormality in urine. The chemical analysis of the urine, therefore, refers to the analysis of (1) pathological constituents, (2) normal compounds excreted in either lower or higher concentration than normal levels, and (3) normal compounds that are absent in the urine.

CARBOHYDRATES IN URINE (GLYCOSURIA): *Glycosuria* refers to the condition when more than normal amounts of sugar are found in the urine. The quantity of glucose usually averages about 140 mg per 24 hour specimen. The presence of more than this quantity of glucose in urine is termed *glucosuria.* The most marked glucosuria is seen in severe diabetes mellitus. Urine may contain as high as 10 per cent (or even more) of glucose in this disease. Such glucosurias are either due to (i) elevated blood glucose level (hyperglycaemia) exceeding the renal threshold as in diabetes mellitus, (ii) *renal diabetes* in which the blood glucose level is normal but the renal threshold is below normal resulting in the excretion of some glucose in urine, (iii) emotional glucosuria causing an increased breakdown of liver glycogen due to adrenal action, producing thereby hyperglycaemia which exceeds the renal threshold, (iv) *alimentary glycosuria* due to the intake of large amounts of carbohydrates which cause sugar to pass into urine because of a higher rate of absorption, from intestine into blood, than body tissues can cope with, (v) *phlorhizin diabetes,* is a glycosuria which is only transitory. Phlorhizin (a glucoside from the bark of the cherry tree) injections cause a lowering of the renal threshold for glucose. This glycosuria does not occur in patients, of course, but phlorhizin is usually used in experimental work on animals, (vi) *liver damage,* i.e. failure of liver to remove glucose from blood at a normal rate due to moderate damage, results in glucosuria if the renal threshold is exceeded. This may occur in cirrhosis, phosphorus, and carbon tetrachloride poisoning.

Beside glucose, a number of other carbohydrates are also excreted in urine under a variety of normal and pathological conditions. They include fructose, lactose, galactose, and pentoses. It is of great clinical importance to distinguish glucose from other carbohydrates to diagnose the various conditions. The commonest test to detect sugar in urine is that *of Benedict's* which measures all the reducing substances and is not specific. The *Glucose oxidase* test must, therefore, be used to distinguish glucose from other sugars. Glucose oxidase is a specific bacterial enzyme that promotes the oxidation of glucose with the production of hydrogen peroxide. In the presence of another enzyme, peroxidase, oxygen from hydrogen peroxide is transferred to an acceptor dye (ortho-dianisidine or ortho-tolidine), producing a coloured product. The test has been simplified and reagent strips are commercially available, making testing possible at or near the site of patient care.

Other important sugars to clinically differentiate are glucose and lactose in pregnancy, and galactose and other sugars in galactosuria and galactosaemia. In the advanced stages of pregnancy, lactose is commonly present in urine with mild glucosuria which may result due to reduced renal threshold. Differential diagnosis is, therefore, important to distinguish between lactosuria and glucosuria. In galactosaemia, an inborn error of metabolism, high levels of this monosaccharide are found in blood and urine. Early diagnosis is very important so that a lactose-free diet may be started. Fructosaemia may, occur as a benign metabolic disorder, in association with Wilson's disease, or liver disease. Progressive muscular dystrophy may give rise to *ribosuria.*

Since many sugars appear in urine in a wide variety of diseases. Benedict's and glucose oxidase tests are not enough, and for maximum aid to diagnosis various carbohydrates should be distinguished from each other. This is routinely done by chromatography of urine on paper. Figure 131.1 shows the separation and characterization of various carbohydrates that are encountered in urine.

PROTEINURIA: Normally urine contains little or no protein (0–100 mg in 24 hour specimen). Some protein may be excreted in urine after a high protein diet, severe exercise and cold bath, or as a consequence of some temporary renal impairment when one stands erect (orthostatic or postural proteinuria). However, in a variety of disease states, large

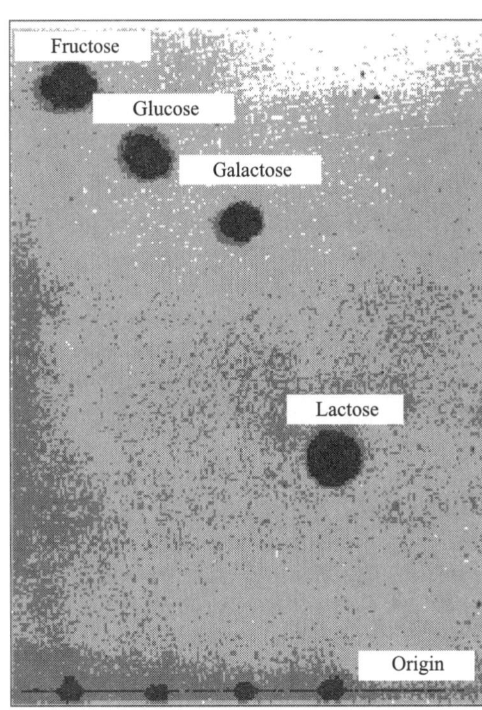

Figure 131.1 Paper chromatography of reducing sugars. The chromatogram is developed in butanolpyridine-water (3:2:1) solvent and the spots made visible with aniline-diphenylamine reagent. Glucose and galactose give gray-green, lactose blue-gray and fructose brown spots.

amounts of proteins are lost in urine. In nephrosis as high as 20 g protein may be found daily. The presence of protein detectable by usual methods in urine is referred to as proteinuria. A number of proteins including serum albumin and globulin, nucleoproteins, fibrin, myoglobin, haemoglobin, proteoses, peptones and Bence Jones protein may be present in abnormal urine. Sometimes, almost all the proteins of serum may be found in non-selective proteinuria.

ALBUMINURIA: The term implies the occurrence of abnormally high concentration of albumin in urine. Albumin may appear in urine in a variety of pathological conditions. It may result from (i) kidney changes secondary to abnormalities in organs other than urinary tract (prerenal albuminuria), (ii) primary kidney disorder (renal albuminuria), and (iii) disease of the lower urinary tract (post-renal albuminuria).

Pre-renal albuminuria: This term is used rather loosely. Most of the pre-renal causes of albuminuria are associated with changes in the kidneys. Nonetheless, the term appears advantageous in various ways for clinical assessment of the primary condition. Ascites, because of local intra-abdominal disorders unrelated to nephritis and heart failure, may result in the appearance of albumin in urine. It may be because of pressure upon the renal veins, causing congestion.

Heart disease, e.g., in decompensated myocardial disorder, with passive renal congestion, albumin appears in urine in quantities proportional to the degree of renal circulatory embarrassment.

Drugs causing irritation in the kidneys usually result in albuminuria. They include turpentine, quinine, naphthols, barbiturates, sulphonamides, arsenic, mercury, phosphorus, bismuth, lead, oxalates, etc.

Other conditions which are also associated with pre-renal albuminuria to varying degrees are intra-abdominal tumours, convulsive disorders, disease of blood and liver, hyperthyroidism, etc.

Renal albuminuria as in glomerulonephritis: Albumin is constantly present in the urine in acute glomerulonephritis. It is marked in the degenerative phase. Some evidence is available to suggest that one of the reasons associated with the pathogenesis of renal insufficiency in many cases of glomerulonephritis is simply due to mechanical obstruction of glomeruli and tubules by deposition of fibrin and hyalin protein coagula composed mainly of precipitated globulins.

Nephrosis (nephrotic syndrome) is a condition that signifies primarily degenerative lesions of the renal parenchyma. It is characterized by massive albuminuria, hypoalbuminaemia and hypercholesterolaemia (hyperlipaemia) oliguria, and almost invariably oedema. Frequently hypocalcaemia, hypo-gammaglobulinaemia and reduced basal metabolic rate, hypotransferrinaemia, and low protein bound iodine values are also seen in this condition.

Other renal disease in which albuminuria occurs are nephrosclerosis and eclampsia.

Post-renal albuminuria: In this category are studied the causes of so called *false albuminuria,* a term that is applied to the addition of albumin in urine at a point beyond the uriniferous tubule, including inflammatory and degenerative lesions of the kidney, pelvis, ureter, bladder, prostate, and urethra. The common tests for albumin will show a positive reaction if blood (the source of which usually resides in the lower urinary tract) or pus is present in the urine.

BENCE JONES PROTEINURIA: Bence Jones proteins are a heterogeneous group of proteins of low molecular weight (below 45,000) and behave in a peculiar manner when their solutions are heated. They precipitate when urine is heated to 50–60°C and redissolve almost completely at 100°C. These proteins are closely related to immunoglobulin light chains but are not the products of cleavage of immunoglobulin or other proteins of the blood and tissues.

Bence Jones proteins are specific products of abnormal plasma cells (myeloma cells) and occur in the urine most commonly in multiple myeloma and only rarely in leukaemia, Hodgkin's disease, and lymphosarcoma. They are of significant diagnostic value in multiple myeloma.

KETONE BODIES: Ketone bodies comprise acetoacetic acid, β-hydroxybutyric acid and acetone. Normally they are absent or found in low quantities (100 mg in 24 hour) in urine. The presence of ketone bodies in urine is termed *ketonuria.* The condition in which ketonuria occurs is known as *ketosis,* a term applied to the accumulation of ketone bodies in excessive quantities in the body.

The site of the synthesis of ketone bodies is mainly the liver. Muscle and other peripheral tissues are chiefly associated with their oxidation to carbon dioxide and water. When the capacity of the body tissues for utilizing acetone bodies is exceeded, the excess is excreted in urine. If the amount of acetone bodies is excessive than the kidneys can excrete in the urine, they accumulate in blood (hyperketonaemia). This occurs in severe ketosis which is usually accompanied by acidosis and not vice versa. Ketosis and ketonuria both occur in alkalosis due to persistent vomiting, unless glucose is given parenterally.

Ketosis and ketonuria occur in many pathological conditions which are associated with inadequate supply or utilization of carbohydrates, e.g. diabetes mellitus, starvation.

BLOOD AND ITS PRODUCTS: *Haematuria is* the term used for the presence of erythrocytes in urine in contradistinction to *haemoglobinuria* which signifies the presence of haemoglobin in urine.

HAEMOGLOBINURIA: When the level of haemoglobin rises above 90–150 mg per 100 ml of plasma, it appears in urine and may persist until the plasma level falls to 50 mg per 100 ml.

Any situation that causes the rapid destruction of large amounts of red blood cells results in the excretion of haemoglobin particularly when the haemolysis occurs intravascularly. It may occur due to the following factors:

1. *Bacteria:*   Haemolytic streptococcaemia, yellow fever, clostridia infections, scarlet fever and severe typhoid.
2. *Chemical agents:*   Poisoning with arseni-uretted hydrogen, arsphenamines, phenyl hydrazine, pyrogallic acid, saponin and sulphonamides.
3. *Vegetable poisons:*   Poisoning with ricin, crotin, and favism due to hypersensitivity to feva beans.
4. *Animal poisoning:*   Poisoning with certain snake venoms.

Haematuria: Normally less than 130,000 red blood cells are present in a 24 hour urine specimen. This number increases because of bleeding anywhere into the urinary tract. Of clinical interest is the inflammatory disease of the glomeruli in which marked haematuria occurs.

BILE PIGMENTS AND SALT: These are discussed below.

Bile salts: Sodium salts of taurocholic acid and glycocholic acid are called bile salts. When present they reduce the surface tension of urine. On this basis, Hay developed a test to detect their presence in urine by sprinkling powdered sulphur. Due to reduced surface tension, sulphur particles settle down. Bile salts in urine usually occur in obstructive jaundice.

Bile pigments: The principal bile pigments are bilirubin and biliverdin. Oxidations and reduction of these give rise to a great variety of other pigments. Some of these pigments occur in urine, in faeces, in bile, and sometimes in the skin and other parts of the body.

Bilirubinuria: This term refers to the occurrence of bilirubin in urine in amounts detectable by usual methods. It is normally not present in urine and occurs in many types of jaundice, but not in uncomplicated haemolytic jaundice. At equal serum concentrations, more bilirubin is excreted in urine in obstructive jaundice than in hepatocellular jaundice. When kidney functions are impaired, bilirubin excretion in urine is reduced. In advance kidney failure, the urine may not contain any bilirubin even in severely jaundiced cases. The testing of bilirubinuria is very useful for (1) differential diagnosis of uncomplicated haemolytic jaundice from other types, (2) for early detection of toxic or viral hepatitis, and (3) as an aid for recognizing kidney function impairment in jaundiced patients.

Urobilinogen in urine: About 1 to 4 mg of urobilinogen are excreted normally in urine in 24 hours. Absence of urobilinogen in the urine of untreated jaundiced patients indicates complete obstruction of the common duct or of complete suppression of hepatic excretion of bile pigments. Higher amounts of urobilinogen are found in the urine of patients suffering with (1) excessive haemolysis, (2) hepatocellular jaundice, and (3) hepatic diseases without jaundice.

Urobilinogenuria does not indicate any single disorder. Its main diagnostic significance apart from its presence in pernicious anaemia, haemolytic anaemia, and chronic haemolytic jaundice is that, (a) the liver functions are considerably impaired, (b) there is active infection in the bile ducts, and (c) it can distinguish between complete and incomplete obstruction of the flow of bile.

Aminoacidurias: A number of amino acids have been detected in the normal and abnormal urine. Aminoacidurias may occur in conditions in which plasma amino acid concentration is high (*overflow aminoacidurias*) or in which renal tubular reabsorption is impaired (*renal aminoacidurias*). The urinary amino acid pattern is sometimes so characteristic that it identifies a specific disease or syndrome.

Renal aminoacidurias: These are identified on the basis of (1) normal plasma amino acid content and (2) increased kidney clearance of the involved amino acid (i.e. lowered renal tubular reabsorption). It may occur as a result of congenital disorder of renal tubular function or in association with the acquired disease. A number of clinically distinct conditions of this type have been reported. Some of them are cystinuria, Fanconi syndrome, Hartnup syndrome, Wilson's disease, glycinuria, etc.

Overflow Aminoacidurias: These are characterized by (1) normal renal clearance and (2) high plasma concentration of one or more amino acids. This includes (a) specific metabolic disorders of an amino acid, (b) metabolic disorders due to liver damage, and (c) increased protein break down.

A summary of the various aminoacidurias and associated causes is provided in Table 131.3.

(c) ***Human chorionic gonadotropin (hCG) in embryo/placenta urine (Pregnancy detection)***: hCG, a product of the early placenta, occurs in urine primarily during pregnancy. Its occurrence in urine is specifically related to pregnancy and on this basis, biological and immunological methods have been developed to detect hCG or diagnose an early pregnancy within the first week after the last missed period. Biological methods (Aschheim-Zondek's and Friedman's tests) in general are lengthy and require animal care and are, therefore, not the methods of choice. Immunological methods, being rapid, sensitive, and specific, are now in routine practice to detect pregnancy by testing urine for the presence of hCG. Haemagglutination inhibition, particle agglutination inhibition, radioimmunoassay, and enzyme linked immunoassays are commonly used. The technology involved is oversimplified and many 'do yourself' or 'home tests' have been commercialized. Handy pregnancy testing kits have been developed by Talwar, and others in India for fast specific and sensitive detection of pregnancy.

(d) ***Human luteinizing hormone in urine (prediction of the time of ovulation)***: Human luteinizing hormone (hLH), a glycoprotein hormone, appears in large quantities in urine (at least 2–3 times of basal level) 30 hours (mean time) before ovulation. The analysis of hLH is an important tool in the prediction of ovulation and diagnosis and treatment of infertility in the female. The analysis of urinary hLH has been used successfully to time oocyte retrieval for *in vitro* fertilization and would similarly assist timing of artificial insemination. 'Do yourself immunologic tests' are available commercially to detect hLH in urine. Genetic, anatomical or physiological conditions and drug therapies may affect the normal synthesis of hLH and cyclic pattern of hLH secretion. Ovarian failure, pituitary and hypothalamic tumours, gonadotropin-producing neoplasms, hyperthyroidism and renal failure usually result

TABLE 131.3 Aminoacidurias

Aminoaciduria	*Causes*
Excessive protein breakdown.	Extensive tissue injury, protracted febrile illness, malignancy, thyrotoxicosis, muscular atrophy.
Liver disease.	Severe impairment of liver cell function, portal cirrhosis.
Alkaptonuria (characterized by the urinary excretion of homogentisic acid)	Congenital lack of homogentisic acid oxidase.
Tyrosinosis (characterized by the urinary secretion of p-hydroxyphenylpyruvic acid)	Congenital lack of an enzyme converting hydroxyphenylpyruvic acid to homogentisic acid.
Phenylketonuria (occurring in mentally defective subjects)	Congenital deficiency of phenylalanine hydroxylase.
Maple sugar urine disease (characterized by the urinary excretion of a substance having a smell resembling that of maple sugar)	Congenital, familial disease.

in chronically elevated hLH levels. Low hLH levels may be associated with pituitary failure, hypothalamic defects, use of oral contraceptives and estrogen or testosterone treatment. Such factors must be considered when evaluating the results.

(e) *17-Ketosteroids:*   The term "17-ketosteroids" refers to those steroids having a ketone group on the 17th carbon atom (see Chapter 16). These steroids are present in the urine as conjugates. They may be divided into three general classes: acidic, neutral, and phenolic. The neutral 17-ketosteroids are mainly the excretion of metabolic products from the endocrine secretion of the gonads and the adrenal cortex. This fraction is chiefly constituted by androsterone and its isomers, etiocholanol-3 α-17-one, isoandrosterone, and dehydroepiandrosterone. Neutral 17-ketosteroids are divided into two types, the α and the β. These isomers differ in the stereochemical configuration at carbon 3 of the steroid nucleus and are distinguished from each other with the help of digitonin which precipitates only the β-fraction. The α-fraction is not precipitated by digitonin and normally predominates (85–95%), consisting of androstrone and etiocholanol-3 α-17-one. β-17 ketosteroids are present in small amounts and consist of epiandrosterone and dehydroepiandrosterone.

The total 17-ketosteroids are estimated in urine usually by the Zimmermann reaction and more specifically by the calorimetric method developed by Pincus. In the Zimmermann reaction, 17-ketosteroids produce a purple colour with an absorption maxima at 520 mμ when treated with m-dinitrobenzene in strong alkali. In the Pincus procedure, a blue colour is produced with antimony trichloride. 17-ketosteroids are present in urine as esters; they are first hydrolysed with an acid and the free steroids are extracted in organic solvent.

The normal values for neutral 17-keto steroids in a 24 hour urine sample in men are 8 to 20 mg. The values are lower in adult women, and in children. There is a gradual increase in the values with age till the adult stage.

There are wide variations in the urinary 17-ketosteroid values. Higher values are seen during night and in late pregnancy. Subnormal values may result in primary or secondary hypofunctions of the adrenal cortex and testis. Lower values may also occur in many nonendocrine disorders in which the relation to adrenocortical or testicular function is not apparent. In conditions of hypopituitarism, male hypogonadism, female hypogonadism, adrenocortical deficiency (Addison's disease), and hypothyroidism, very low values are obtained and sometimes they are not excreted in the urine at all. Urinary 17-ketosteroids are usually lowered in acute or chronic liver diseases, chronic illness of all kinds including the common cold, malnutrition, anaemias, malignancy, and after marked physical fatigue.

Very high levels of urinary 17-ketosteroids occur in adrenal cortical tumours, adrenocortical hyperplasia and interstitial cell tumours of the testis. Of clinical interest is the determination and fractionation of 17-ketosteroids in the differential diagnosis of the testicular tumours, and adrenal hyperfunction. Adrenal cortical hyperfunction may occur in a variety of clinical conditions, including Cushing's syndrome, feminizing syndrome in males and adrenogenital syndrome. The highest values of urinary 17-ketosteroids have been reported in patients with adrenal cortical carcinoma (upto 350 mg per 24 hour specimen), but there is significant overlapping with adrenal cortical hyperplasia. These two conditions are usually distinguished by fractionation of urinary ketosteroids into the α- and β-fractions. The latter may be disproportionately higher in adrenal cortical carcinoma.

(f) *Antibodies:*   Antibody activity in urine determined against a number of infectious agents and other molecules including human immunodeficiency virus (HIV), hepatitis-A virus, P. falciparum, H. pylori, Vibrio cholera, polio virus, S, typhi, diphtheria, tetanus toxoid, rubella, etc. They are detected by enzyme immunoassays (EIA) (for details refer to Chapter 87). HIV-1 urinary EIA and Western blots have been commercialized for the detection of immunoglobulin-G and immunoglobulin-A types of antibodies as well as HIV-1 vau group-0 antibody. Urinary EIA results match well with those of blood EIA and eliminate several disadvantages associated with the handling of potentially infectious blood. Urine collection is user-friendly, easy, economic and safe (noninfectious and nonhazardous). There is no volume restriction and can be used under situations where blood collection is not possible.

(g) *Trans-renal DNA:*   Trans-renal DNA (Tr-DNA) is a class of extracellular urinary DNA that originates from cells dying throughout the body. The postapoptotic DNA appears in the circulating plasma and a portion of these fragments cross the kidney barrier and appear in urine in the form of 150-200 bp fragments. Polymerase chain reaction based urinary assays have been used for the detection of K-ras mutations in Tr-DNA of patients with pancreatic cancer and colorectal carcinoma, male specific Tr-DNA sequence in the urine of females transfused with male blood as well as in the urine of women pregnant with male fetuses, HIV, tuberculosis, etc.

Salivary HIV antibodies: Saliva has now attained the position of an alternate and safe body fluid to detect antibodies, hormones, cancers, infectious agents, nucleic acids and a number of other molecules. Saliva is suitable for the development of point-of-care tests as its collection and transport are simple, non-invasive and non-hazardous. A rapid strip immunoassay for the detection of antibodies to HIV-1 and HIV-2 named ORAQuick was commercialized in 2004. The test can also be performed on finger stick (finger-prick) sample of blood or plasma. The results are available in 20 minutes and fairly match with those obtained with blood or serum or plasma based assays.

3. Sediment analysis (Microscopic analysis):   Normal urine is usually clear but sometimes a turbidity is observed which may be due to the precipitation of phosphates in alkaline urine or on heating. Such a turbidity disappears on addition of small quantities of dilute acetic acid. In abnormal urine, turbidity signifies a pathological condition. Turbid urine produces sediment (pellet) on centrifugation. On microscopic examination, these sediments display a variety of organized and unorganized structures. The chief organized structures in the urinary sediment are casts, cells, spermatozoa, bacteria and animal parasites. Amongst the unorganized (crystalline material), calcium oxalate, uric acid, ammonium urates, phosphates, etc., are frequently observed. These constituents of the sediment are shown in Figure 131.2.

(a) *Casts:* These are cylindrical structures composed of some type of protein precipitated as matrix in which is included any substance present in the lumen of the renal tubules at the time of formation. When these structures are extruded from the urine, they retain their shape and may be regarded as models or casts of the tubules. For casts to form, protein must be precipitated in the lumen of the tubule. Proteinuria, acidic pH and high salt concentration in the tubular urine favour the formation of a cast. The appearance of casts in urine is termed as cylindruria. The appearance of various types of casts and the associated renal diseases are given in Table 131.4.

(b) *Crystals:* Urine contains a variety of crystalline substances in normal and pathological conditions. They are classified into two types as those appearing in acidic and alkaline urine. Figure 131.2 shows these crystalline substances in acidic and alkaline urine when examined microscopically. Some of these crystalline structures appear in large quantities in subjects having a tendency to form renal stones.

(c) *Renal stones (calculi):* The causes of the formation of renal stones are not fully understood. In most cases multiple factors operate. In a majority of cases, renal stones contain calcium (or magnesium) phosphates or oxalates and organic compounds (cystine, uric acid, etc.). If a stone does not cause obstruction, it is unlikely to cause inury or symptoms. At a later stage, however, if it causes obstruction and blocks a urinary passage, as usually happens, it leads to severe symptoms and kidney damage. It is, therefore, important to investigate the cause of the first stone in the prevention of later renal injury.

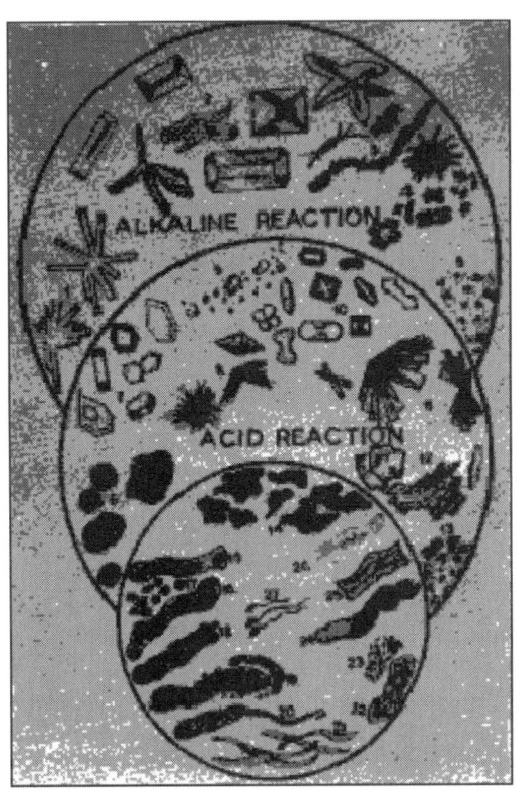

Figure 131.2 Some organized and crystalline structures found by microscopic examination of the urinary sediment. In most cases a large number of these structures indicate a disease: 1. Calcium phosphate, 2. Triple phosphate, 3. Ammonium urate, 4. Calcium carbonate, 5. Amorphous phosphates, 6. Leucine spheres, 7. Cystine, 8. Tyrosine needles, 9. Fat droplets, 10. Calcium oxalate, 11. Sodium acid urate, 12. Uric acid, 13. Amorphous urates, 14. Epithelial cells, 15. Epithelial cast, 16. Pus cast, 17. Leukocytes, 18. Granular cast, 19. Blood cast and erythrocytes, 20. Cylindroid, 21. Mucus threads, 22. Bacterial cast, 23. Bacteria, 24. Hyaline cast, 25. Waxy cast, 26. Fatty cast, and 27. Spermatozoa.

TABLE 131.4 Casts, their appearance and related pathological condition

Cast	Appearance/Organization	Pathological Conditions
Hyaline casts	Colourless, semi-transparent, homogeneous, cylindric structures having parallel sides and usually rounded ends.	Glomerulitis, polyarteritis malignant arteriolitis.
Waxy casts	More opaque, shorter and broader than hyaline variety, homogeneous, irregular broken ends, frequently contain a few granules or occasionally a cell. They are colourless or grayish having a dull waxy appearance.	Advanced nephritis, amyloid disease of the kidney.
Fatty casts	Contain numerous small droplets of fat.	Tubular degeneration malignant arteriolitis, polyarteritis.
Epithelial casts	Contain epithelial cells	Parenchymatous inflammation, acute nephritis (if the cells are well preserved).
Blood casts	Contain erythrocytes	Acute nephritis, acute exacerbation of chronic nephritis.
Pus casts	Composed almost wholly of pus cells.	Pyelonephritis.
Bacterial casts	True bacterial casts are rare.	Septic condition in the kidney.

(d) *Calcium oxalate:* About 50% of kidney stones are composed of calcium oxalate. Foods rich in oxalate (rhubarb, spinach, tomatoes, bathua, sorrel) are the major factors that are responsible for the formation of calcium oxalate. Calcium oxalate crystals occur in octahedral, envelope or dumb-bell shapes usually in acidic urine, but may also be formed in alkaline urine.

(e) *Cystine:* Cystinuria is a hereditary disease. Only a small percentage of these patients form stones.

(f) *Uric acid:* These stones are formed in acidic urine or there is rapid tissue breakdown such as occurs in the treatment of leukaemia, polycythaemia and carcinoma. It is important for these reasons to maintain an alkaline reaction during the treatment of these diseases. Many patients with gout form uric acid stones usually when undergoing treatment for arthritis.

III. NORMAL VALUES OF CONSTITUENT

Normal values refer to the amount of any substance or quantitative capacity of any property present in the body fluids or excretions of a healthy subject suffering from no disease. These values vary widely even in normal persons because of the marked influence of age, sex, race, genetic, constitution, weather, socio-economic status, and a variety of other physiological factors. Normal values are usually expressed as a range. Most (about 90 per cent) but not all healthy individuals in a population can be included in the range of accepted normal values, while some possess highly divergent values. These exceptional humans may be absolutely normal in all the possible investigated respects. Whether a given analytical value is normal or not, thus, depends on the experience and the entire data at the command of the interpreter of the results.

Table 131.5 shows the normal values for the different constituents in blood (plasma or serum) urine, and cerebrospinal fluid and the interpretation of the results. The values given in this table are taken from literature and from our own results. These values, in most cases, apply to the method of analysis or to an identical procedure.

The abbreviations and certain units used in these tables are defined as follows:

P = Plasma, B = Blood
meq/1 = milliequivalent per litre

$$= \frac{mg \text{ per } 100 \text{ ml} \times 10 \times valency}{atomic\ weight}$$

K.A.U. = King-Armstrong Unit = production of 1 mg of phenol in 15 minutes at 37°C for alkaline phosphatase, and in 60 minutes for acid phosphatase.

Somogyi units (for amylase) = Amount of the enzyme digesting 5 mg of starch at 37°C in 15 minutes.

I.U. = International Units of enzyme activity;

μ mole per minute per litre.

Pseudo-cholinesterase activity is expressed as the change in optical density at 412 mμ during the hydrolysis of butrylthiocholine iodide.

SUMMARY

Biological fluids are liquids originating from inside the human body and include both secretary and excretory products. Their composition reflects the changes that may occur in health and disease. Biological fluids are also widely recognized as vehicles for the transmission of infectious agents. Their analysis is, therefore, important in the diagnosis and follow-up of treatment and establishing their safety particularly blood for transfusion. Most frequently blood or plasma or serum, urine, cerebrospinal fluid and saliva are analyzed. Speacialized tests are also performed in tears, vaginal discharge, semen, amniotic fluid, bile, gastric juice, pleural fluid, peritoneal fluid etc. This chapter describes some of the important biological fluids, their preparations and physical, microscopic, chemical and immunologic analysis of diagnostic significance. Normal values of the constituents in blood or plasma or serum, urine and cerebrospinal fluid and their variations in diseases and other conditions are given in Table 131.5.

TABLE 131.5 Normal values of clinically important constituents of blood, urine and cerebrospinal fluid and interpretations of the results

Constituent	Units	Reference range*	Interpetation
In Serum/Blood			
Glucose Fasting	mg/dL	70–100	High in diabetes mellitus, glucagonomas, Cushings syndrome (increased cortisol), steroid therapy and low in hyperinsulinemia (exogenously on insulin management or insulin producing tumours)
Glucose PP (Postprandil)	mg/dL	90–140	
Glucose R (Random)	mg/dL	70–160	
HbAlC (Glycated Hemoglobin)	%	< 6	High in Diabetic mellitus , 6–6.4– Risk for Diabetes, ≥ 6.5– Diabetes, < 7 Good Diabetic control, > 7 Need for corrective action
BUN	mg/dL	5–23	High– increased protein catabolism, gastrointestinal hemorrhage, chronic nephritis, cardiac failure, prostatic obstruction, nephrolithiasis,
Creatinine	mg/dL	0.6–1.3	Marker of kidney function test, high in prostatic obstruction, nephrolithiasis, malignancy

(Contd.)

TABLE 131.5 Normal values of clinically important constituents of blood, urine and cerebrospinal fluid and interpretations of the results (*Contd.*)

Constituent	Units	Reference range*	Interpetation
Uric acid	mg/dL	M 2.4–7.2; F–2.0–5.7	High in renal disease, gout, leukemia, toxemia of pregnancy
Calcium total	mg/dL	8.2–10.4	High in hyperparathyroidism
Calcium free	mg/dL	3.5–5.5	Low in infantile tetany
Phosphorus	mg/dL	2.5–4.6	High in chronic nephritis, and hypoparathyroidism, Low in hyperparathyroidism
Sodium	mmol/L	132–148	Low in excessive loss of body fluid, Addison's disease
Potassium	mmol/L	3.5–5.5	High in pneumonia, acute infection, uremia
Chloride	mmol/L	98–108	High in nephritis, congestive heart failure and low in diabetes, pneumonia, fever
Magnesium	mg/dL	1.7–2.7	High in chronic nephritis, liver diseases, low in tetany, diarrhea, chronic alcoholism
Bilirubin total	mg/dL	0.2–1.0	High in hemolytic anemia, hepatitis, biliary obstruction and neonatal jaundice
Bilirubin direct	mg/dL	0.0 0.2	High in biliary obstruction
Magnesium	mg/dL	1.7–2.7	High in chronic nephritis, liver diseases, low in tetany, diarrhea, chronic alcoholism
Bilirubin total	mg/dL	0.2–1.0	High in hemolytic anemia, hepatitis, biliary obstruction and neonatal jaundice
Bilirubin direct	mg/dL	0.0 0.2	High in biliary obstruction
Total protein	g/dL	6.6–8.7	Albumin usually does not increase in clinical conditions. An increase may occur in dehydration and the rise is parallel to the rise in globulins. A/G remains constant. Decreased values occur in nephritis, nephrosis, myeloma, impaired synthesis, increased catabolism and inadequate supply.
Albumin	g/dL	3.5–5.0	
Globulin	g/dL	1.8–3.6	
A/G Ratio		2:1	
Fibrinogen	g/dL	0.2–0.4	High in nephrosis, multiple myeloma, low in hepatic necrosis and congenital afibrinogenemia
α1 Globulin	g/dL	0.1–0.4	High in acute and chronic infections, rheumatic diseases and low in hepatocell ular damage
α2 Globulin	g/dL	0.5–1.0	High in Wilsons disease
β–Globulin	g/dL	0.6–1.2	High in nephritic syndrome, obstructive jaundice, biliary and portal cirrhosis and acute hepatitis
γ– globulin	g/dL	600–1500	High in immunological disorders and multiple myeloma
Iodine	µg/dL	3.5–8.0	High in hyperthyroidism and low in hypothyroidism
Ascorbic acid	mg/dL	0.7–1.5	Low in scurvy
Copper	µg/dL	70–140	Paradoxically low in Wilsons disease, high in rheumatoid arthritis
AST	IU/L	0–42	High in myocardial infarction and hepatocellular damage
ALT	IU/L	0–60	High in infective hepatitis, Laennec's cirrhosis and biliary cirrhosis
ALP	IU/L	A–39–117; C–117–390	High in children, late pregnancy, bone disorders and hepatobiliary obstruction
GGT	IU/L	0–64	High in alcoholic cirrhosis
Amylase	IU/L	5–100	High in acute pancreatitis, inflammation of gall bladder, and tumour of pancreas, prostate and lung
Lipase	IU/L	22–51	High in acute pancreatitis, coelic disease, duodenal ulcer, pancreatic cancer, and sometimes in gastroenteritis
Acid phosphatase	ng/mL	< 3.5	High in metastasizing prostatic carcinoma

(Contd.)

TABLE 131.5   Normal values of clinically important constituents of blood, urine and cerebrospinal fluid and interpretations of the results (*Contd.*)

Constituent	Units	Reference range*	Interpetation
Aldolase	μmol/min/L	1.5–7.2	High in skeletal muscle dystrophy, megaloblastic anemia
Isocitrate dehydrogenase	μmol/min/L	0.8–4.4	High in kwashiorkor, and megaloblastic anemia
LDH	U/L	91–180	High in infective hepatitis, myocardial infarction, untreated pernicious anemia, progressive muscular dystrophy
5′–nucleotidase	IU/L	1.6–17	High in obstructive damage
Pseudocholinestrase	∇A/min/mL	25–40	Low in liver diseases, malnutrition, dermatomyositis
Ammonia	μmol/L	9–35	High in hepatic encephalopathy, liver and kidney damage
Creatine kinase	IU/L	22–195	It stores energy in the form of creatine phosphate in muscles. High in muscular damage, progressive muscular dystrophy, rhabdomyolysis. It has three isoenzyme forms CK–MM, CK–MB, CK–BB. CK–MB activity increases during myocardial infarction. CK–MB mass is more specific during myocardial infarction.
CK–MB activity	IU/L	< 22	
CK–MB Mass	ng/mL	< 5	
Troponin I	μg/L	< 0.01	High in myocardial infarction
hsCRP	mg/dL	< 1	High in myocardial infarction. CRP is C reactive protein. It is a nonspecific marker of inflammation. In any kind of inflammation CRP level in blood increases >10. High sensitivity CRP detects inflammation at lower concentration, i.e., > 0.4 mg/dL. This is highly specific for cardiovascular diseases. It is used as a risk marker for myocardial infarction.
NT–proBNP	pg/mL	< 125 in < 75 yrs < 450 in ≥ 75 yrs	High in myocardial infarction, heart failure, left ventricular dysfunction
Total cholesterol	mg/dL	120–190	High in nephritis, nephrosis, diabetes mellitus, hypothyroidism
HDL cholesterol	mg/dL	> 40	High HDL cholesterol in cardioprotective while high LDL cholesterol poses a risk for cardiovascular diseases. High LDL cholesterol is seen in hypothyroidism
LDL cholesterol	mg/dL	< 100	
VLDL cholesterol	mg/dL	< 45	
Triglycerides	mg/dL	<150	High in excessive intake of fats in diet and hereditary disorders of lipoprotein metabolism
Apolipoprotein A1	mg/dL	120–176	In heart disease , decreased ApoAl level and increase Apo B levels have been observed
Apolipoprotein B	mg/dL	63–114	
Lipoprotein (a)	mg/dL	< 30	High levels seen in congestive heart disease, cerebro vascular disease, uncontrolled diabetes mellitus and CRF
Homocysteine	μmol/L	5–15	Higher levels lead to risk of cardiac disorders
Total antioxidant status	mmol/L	1.1–2.0	Increases in acute exercise and decreases in diseases which increase oxidative stress in the body
Iron	μg/dL	50–160; F–40–150	Low intake of iron, hemolytic anemia
TIBC	μg/dL	250–400	High in iron deficiency anemia
Ferritin	ng/dL	20–300; F–11–307	High in repeated blood transfusion
Folic acid	ng/dL	3–17	Low in megaloblastic anemia
Vitamin B 12	pg/mL	220–960	Low in megaloblastic anemia, pernicious anemia
Free PSA	ng/mL	< 0.25	High in benign prostate hypertrophy
CA125	U/mL	< 35	High in ovarian tumours
CA15–3	U/mL	0–31	High in GIT tumours, hepatic tumours
CA19–9	U/mL	0–33	High in GIT tumours
Lactate	mg/dL	4.5–20	High in shock, congestive heart failure, renal failure, diabetes mellitus, neoplasia, drug like ethanol, methanol toxicity, Leukemia and inborn errors of metabolism like fatty acid oxidation defect, methylmalonic aciduria, vongierkes disease.

(Contd.)

TABLE 131.5 Normal values of clinically important constituents of blood, urine and cerebrospinal fluid and interpretations of the results (*Contd.*)

Constituent	Units	Reference range*	Interpetation
Cortisol	μg/dL	08.00 hr : 5–23 16 pm : 3–16	High in Cushing's syndrome
Cerebrospinal Fluid(Lumbar puncture)			
Calcium (Ionized)	mg/dL	4.5–5.5	High in pyogenic and tuberculous meningitis, brain tumours, low in uremia
Chloride (as NaCl)	mmol/L	700–760	Parallel to serum serum chloride in diseases
Cholesterol	mg/dL	0–0.22	High in pyogenic meningitis, cerebral haemorrhage, brain tumour
Glucose (Adults)	mg/dL	50–80	High in epidemic encephalitis, anterior poliomyelitis, diabetes mellitus
Glucose (Children)	mg/dL	70–90	and low in purulent and tuberculous meningitis
Lactic acid	mg/dL	8–27	High in meningitis
pH		7.35–7.40	Low in various types of pyogenic meningitis
Phosphate	mg/dL	1.5–2.0	High in uraemia, pyogenic and tuberculous meningitis, hydrocephalus
Protein (Lumbar)	mg/dL	15–40	High in pyogenic tuberculous meningitis
Cisternal		10–25	Haemorrhage and thrombosis.
Ventricular		5–15	
IgG	mg/dL	1.44 ± 0.53 (SD)	High in multiple sclerosis, meningitis and low in epilepsy
Urine (mg/24 hour specimen unless otherwise stated)			
Albumin (single specimen)		Negative	High in nephritis, nephrosis, tuberculosis, pyelonehritis, and pyelitis
Albumin (24 h specimen)		10–100	
Aldosterone	μg	2–23	High in primary aldosteronism
Ammonia	mmol/L	700	High in diabetic acidosis and liver damage
Amylase somogyi units		Up to 5000	High in first 24 hour of acute pancreatitis, obstruction of common duct, low in liver necrosis and severe burns
Bence Jones Protein		Negative	Appear in multiple myeloma
Calcium (Average Diet)	mg/dL	30–150	High in osteitis fibrosa cystica and low in tetany and myxoedema
Calcium (High Calcium Diet)	mg/day	100–300	
Catecholamines	μg	<230	High in pheochromocytoma
Chloride	mmol/L	110–250	High in Addison's disease and low in starvation, vomiting, pneumonia, ascites, pleuritic effusion, heart failure, nehritis
Coproporphyrin	μg	50–200	High in congenital porphyria
Coproporphyrin (Children)	μg	0–80	
Creatinine (Males)	g	1–1.9	High in typhoid fever, typhus, tetanus, debilitating disease and low in
Creatinine (Females)	g	0.8–1.7	renal insufficiency, anemia, paralysis, leukemia
Estrogen (Males)	μg	4–25	
Estrogen (Females)	μg	4–60	
Glucose	mg/100 mL	< 100	High in Diabetic mellitus, hyperthyroidism, hyperpitutarism
hLH	mlU/mL	≤ 20	≥40 at the time of hLH surge. Chronically high in ovarian failure, pituitary and hypothalamic tumour, gonadotrophic producing neoplasm, hyperthyroidism and renal failure. Low levels may be with the use of oral contraceptives.
17-hydroxy corticosteroid			< 3 mg in 24 hour indicate adrenal insufficiency and above 12
Males		3–10	mg strongly suggest hyperactivity of the adrenal cortex. High in
Females		2–8	adrenocortical carcinoma and adenoma.
5-hydroxy indoleacetic acid		2–9	High in carcinoid syndrome.
Indican		4–20	High in paralytic ileus, peritonitis, cholera, achlorhydria, pernicious anaemia, gangrene of the lungs, pelvic abscess.

(*Contd.*)

TABLE 131.5   Normal values of clinically important constituents of blood, urine and cerebrospinal fluid and interpretations of the results (*Contd.*)

Constituent	Units	Reference range*	Interpetation
17-Ketosteroids			Low in hypoadrenalism, hypopituitarism, hypogonadism, high in adrenal cortical carcinoma and hyperplasia
Age Yrs–10		1–4	
20–30		M–4–16; F–6–26	
50		M–3–9; F –5–18	
70		M–1–7; F–2–10	
Lead	µg/L	21–38	High on exposure; values up to 52 mg/L is safe, 97 v is toxic level, 175 v is dangerous 352 µg/L is a very hazardous exposure
pH		4.8–7.8	Low in diseases associated with acidosis and high in alkalosis.
Phosphorus	g	0.4–1.3	High in ostetis fibrosa cystic and alkalosis and low in nephritis with acidosis, rickets
Potassium	mEq/day	25–125	Effected in renal disorders
Proteins	mg/day	<150	High in nephritic syndrome
Sodium	mEq/day	40–220	Sodium level disturbs in renal Disorders
Urea		20–35	Low in hepatitismyxoedema, acidosis and high in increased catabolism, pneumonia
Uric acid	mg/dL	250–750	High in leukemia, gout, eclampsia, febrile diseases
Urobilinogen		1–4	High in hemolytic jaundice, pernicious anaemia, portal cirrhosis, decompensated heart, lobar pneumonia, malaria
Specific gravity		1.015–1.025	Low in diabetes insipidus
Microalbumin	µg/mg creatinine	< 30	High in diabetic nephropathy creatinine
VMA	mg/day	1–8	High in pheochromocytoma
Osmolality	mOsm	300–900	
Creatinine clearance	ml/min	90–120	—
Inulin clearance	ml	120–130	—
Urea clearance	Cm–ml/min	64–99	—
	Cs–ml/min	40–68	
Volume	ml	600–1800	High in diabetes insipidus, diabetes mellitus and low in uremic conditions
TMg tubular reabsorptive capacity for glucose	mg/min	375	Low in renal tutular disorders

*From - Sir Ganga Ram Hospital, New Delhi (2012) and Burtis CA, Ashwood ER, Bruns DE. *"Tietz Text Book of Clinical Chemistry and Molecular Diagnostics."* Fourth Edition. Elsevier, 2006.

SUGGESTIONS FOR FURTHER READING

Botezatu I., et al. (2000), Genetic Analysis of DNA Excreted in Urine: New Approach for Detecting Specific Genomic DNA Sequences from Cells Dying in an Organism, *Clin. Chem.*, *46*, 1078–1084.

Hongbao M. (2006), Western Blotting Method, *Jour. Amer. Sci.*, 2, 23–27.

Kaufman E. and Lambert I.B. (2002), Diagnostic Applications of Saliva: A Review, *Critical Rev. Oral Biol.* Med., 13, 197–212.

Lay L.A.R., Diaz D.L., Casanueva R.M. and Moreno A.G. (2002), Anti Hepatitis A Virus Immunoglobulin M Antibodies in Urine, *Clin. Vaccine Immunol.,* 3, 492–494.

Saiki R. et al. (1988), Primer Directed Enzymatic Amplification of DNA with a Thermostable DNA Polymerase, *Science*, 239, 487–491.

Urnowitz H.B. et al. (1999), Urine Antibody Tests: New Insights into the Dynamics of HIV-1 Infection, *Clin. Chem.*, 45, 1602–1613.

132

Enzymes and Isoenzymes in Clinical Diagnosis

B.K. Goel, Venkat Parameswaran, S.K. Sood and L.M. Srivastava

CONTENTS

I. INTRODUCTION

During the last two decades there has been a spurt in knowledge in the field of clinical enzymology. Enzyme determinations are now being performed routinely in hospital laboratories to assist in clinical diagnosis.

The structure and functions of enzymes, their mode of action and genetic control have been discussed in the previous chapters. Only those aspects which are related to the understanding of their importance in disease states are emphasized in this chapter.

Enzymes are biocatalysts which enhance the multiple reactions of cell metabolism and play a vital role in normal cellular function. They are protein in nature, unlike organic catalysts, are highly specific in their action, and are present in all cells. Due to regular wear and tear of cells, cellular enzymes are leaked into the extra cellular fluid and thence to the general circulation–possibly at a constant rate. Hence, levels at either side of the normal range signifies deviation from the norm. Certain enzymes leak into the circulation or other extra-cellular fluids only when the cells are damaged. The concentration of these enzymes in serum, urine or other fluids, to an extent, is related to the degree of cell damage and the number of cells involved.

Practically all enzymes occur in different molecular forms which differ in their physicochemical properties but catalyze the same reaction. The term 'isoenzymes' or 'isozymes' has been applied to each of these forms. They can be isolated and identified by differences in their mobility on electrophoresis, column chromatography or by differences in their kinetics, inhibition pattern in presence of certain chemicals, chemical constitution or immune specificity. The different enzymes and their isoenzymes occur in different concentrations in various tissues depending on their functional needs. As will be discussed in the following paragraphs this property of relative organ specificity of enzymes and isoenzymes is being exploited extensively for arriving at a laboratory diagnosis of the diseases of organs.

II.   POLYMORPHISM IN ISOENZYMES

Recent advances in instrumentation and molecular biology—especially protein chemistry—have enabled us to measure and isolate enzymes even when they are present in small quantities. This has led to an understanding of the mechanisms by which the enzymes occur in their multiple forms.

A.   Genetic Factors

Some enzymes are formed by a combination of polypeptide chains which may widely differ in their amino acid composition. Synthesis of different polypeptide chains is controlled by separate genes which occur at different loci. Varying combinations of the polypeptide chains give rise to different protein molecules which have similar enzymatic activity but different kinetics and different physico-chemical properties. One such classical enzyme is lactate dehydrogenase (LDH). The enzyme molecule consists of 4 sub-units each with a molecular weight of 32,000. The monomeric sub-units are of two different types, H and M, which on combination can give 5 possible tetramers, H_4, H_3M, H_2M_2, HM_3 and M_4. Thus LDH has five isoenzyme forms (1–5) which separate into distinct bands on starch gel electrophoresis. At a pH of 8.6 LDH_1 moves fastest and LDH_5 slowest with 2, 3 and 4 types in between in that order (Figure 132.1). Freezing of LDH_1 isoenzyme in 0.1 M sodium phosphate buffer at pH 7 in 1.0 M NaCl results in its dissociation into its sub-units (polypeptide chains). Reconstitution of the enzyme by thawing produces pure substance identical to the original enzyme in electrophoretic mobility and other properties. Splitting and recombination of LDH_5 fraction in a similar fashion results in production of a pure substance identical to the original enzymes. However, when a mixture of LDH_1 and LDH_5 isoenzymes is similarly frozen and thawed, it gives rise to formation of all the five isoenzymes as expected. There is a regular gradation in the properties of LDH isoenzymes from the most negatively charged LDH_1 to the least negatively charged LDH_5, in stability to heat and 2M urea, in vulnerability to inhibitors like oxamate and oxalate, and in affinity for 2-oxobutyrate as a substrate.

The polymorphism of the enzymes, creatine phosphokinase, aldolase, glucose 6-phosphate dehydrogenase, tryptophan synthetase and cerulo-plasmin has also been explained on sub-unit structures like those of LDH. Creatine phospho-kinase is made up of two sub-units having different polypeptide structures, and therefore three possible isoenzymes can be found. Two main pure types characteristic of muscle and brain, and the third, a hybrid type, have all been identified in man.

B.   Polymerization

Multiple molecular forms may also result from polymerization. Enzyme cholinesterase from human pooled sera separates into five bands on starch gel electrophoresis (CHE 1–5). Most of the enzyme activity resides in CHE–5. Concentration of the enzyme preparation results in an increase in CHE–5, whereas dilution of the preparation leads to progressive depolymerization of CHE–5 into four different smaller molecules (CHE 1–4). Different polymers will have different surface charges explaining the differences in their electrophoretic mobility.

C.   Conformational Isomerism

All enzymes are proteins with the exception of Ribozyme (they are RNA molecule not protein). Their catalytic properties depend upon the specific conformational structure in the folded peptide chain having least energy content. Alteration in the tertiary structures (folding of the chain) may occur without appreciable change in the free energy of the molecule in such a manner that it does not change the active sites and the enzymatic property of the protein but alters its electrophoretic mobility and other properties. Cytoplasmic and microsomal aspartate transaminases are conformational isomers.

D.   Presence of Charged Groups

Serum alkaline phosphatase (ALP) separates into three bands on agar gel electrophoresis corresponding to the α_2 pre-β and γ regions of serum proteins. The isoenzyme at α_2 position is of liver origin, the pre-β is from the bone and the γ is from the intestine. During pregnancy placental ALP also appears in the serum and on electrophoresis migrates slightly slower than α_2 ALP. On cellulose acetate electrophoresis an extra band in the α_1 region has been obtained from patients

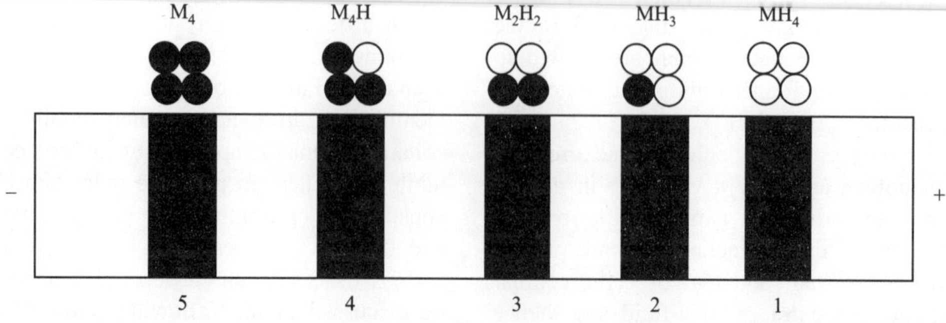

Figure 132.1   Diagrammatic representation of electrophoretogram of LDH on starch gel.

with hepatobiliary disorders (Figure 132.2). The placental isoenzyme is heat stable whereas the bone isoenzyme is heat labile. L-phenylalanine inhibits the placental and intestinal isoenzymes. Both bone and liver isoenzymes are inhibited with L-homoarginine.

The molecular heterogeneity of mammalian alkaline phosphatase has been demonstrated to be due to differences in the number of sialic acid residues (charged groups) in placental, liver and bone ALP. Intestinal isoenzyme does not contain any siliac acid residue.

III. ENZYMES OF DIAGNOSTIC SIGNIFICANCE

Some of the enzymes which have wide diagnostic application are discussed below:

A. Amylases

Amylase (α-1, 4-glucan-4-glucanohydrolase, EC 3.2.1.1) splits starch into glucose. The salivary glands and the pancreas secrete the enzyme into saliva and pancreatic fluid respectively. Damage to the glandular cells or the secretory pathway of these organs may cause amylase to enter the blood stream.

1. Serum amylase: In acute pancreatitis the enzyme concentration may reach (four to six-fold of the reference limit). The levels are highest at 5–12 hours and return to normal in 2–4 days. In patients with severe renal disease, levels may remain elevated for an abnormally long period. Chronic pancreatitis causes only a moderate rise in enzyme activity. Morphine and certain radio-opaque substances used in cholecystography induce spasm of the sphincter of Oddi and cause a moderate increase in serum amylase activity. A transient rise in enzyme levels may accompany perforated peptic ulcer, empyema of gall bladder, intestinal obstruction, ruptured ectopic pregnancy,

or peritonitis. Serum amylase levels as high as 2–3 fold may result from disease of the parotid gland.

Hyperlipemia suppress serum amylase. Estimation of serum lipase increases the specificity of S. amylase in the diagnosis of acute pancreatitis.

Normal range: 80–180 Somogyi units/100 ml.
25–200 U/L at 37°C

2. Urinary amylase: Following an attack of acute pancreatitis the urinary amylase reaches a very high level during the first 24 hours and remains elevated for 7–10 days or more.

Reference range: 1600–6000 Somogyi units/24 hours.
24–410 U/day.

B. Transaminases

Two types of transaminases are found in human tissues and both have glutamic acid as one of the substrates.

1. Glutamic oxalacetic transaminase (GOT, aspartate transaminase, L-aspartate, 2-oxogluta-rate amino-transferase, EC 2.6.1.1, AST)
2. Glutamic pyruvic transaminase (GPT, alanine trans-aminase, L-alanine: 2-oxoglutarate aminotransferase, EC 2.6.1.2, ALT)

Both the enzymes catalyze the interconversion of amino acids and α-ketoacids by transfer of amino groups.

$$\text{L-asparate} + \alpha\text{-oxoglutarate} \xrightarrow{\text{GOT}} \text{Oxalacetate} + \text{L-glutamate}$$

$$\text{L-alanine} + \alpha\text{-oxoglutarate} \xrightarrow{\text{GPT}} \text{Pyruvate} + \text{L-glutamate}$$

The GOT and GPT activities in different tissues in relation to serum, which are taken as unity, are as follows:

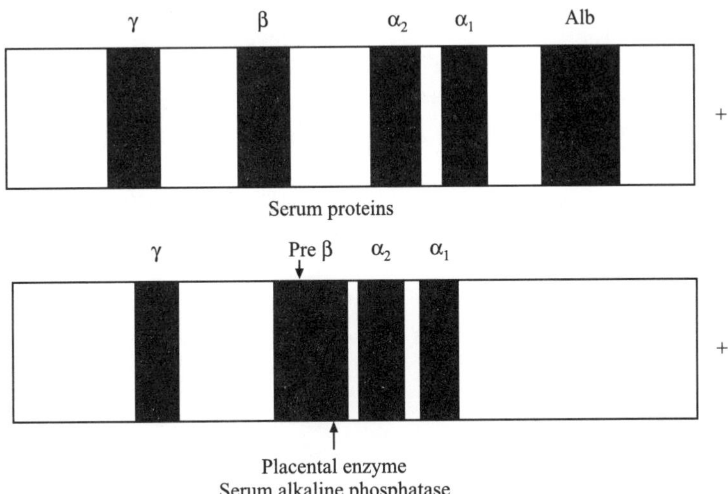

Figure 132.2 Diagrammatic representation of electrophoretogram serum proteins and ALP on cellulose acetate.

Source	GOT	GPT
Serum	1	1
Heart	7800	450
Liver	7100	2850
Skeletal muscle	5000	300
Pancreas	1400	130
Lungs	500	45
Red cells	15	7

As concentration of GOT is very high in the heart, in myocardial infarction very large amounts of enzyme are released into circulation. The serum GOT (SGOT) starts rising in 6–8 hours after the onset of pain and reaches a peak level in 24–48 hours returning to normal by the 4th or 5th day (Figure 132.3). A fresh infarct causes a further rise in the serum enzyme values. The peak activity is, in general, proportional to the degree of myocardial damage. The rise in level of 15–25 times the normal is often associated with fatal infarct. The GPT content of the heart is almost one twentieth of GOT content, therefore, the rise in serum GPT (SGPT) in myocardial infarction is less marked and less consistent.

Cardiac surgery, angiocardiography, cardiac catheterization and cardiac massage cause considerable elevation in enzyme level.

Both transaminases are present in high concentrations in the liver. In hepatitis and other diseases associated with liver necrosis, both SGOT and SGPT are elevated. The levels may rise several days before jaundice manifests. In viral hepatitis the rise in serum transaminase activity roughly parallels the degree of liver necrosis and with extensive necrosis may reach 100 times the upper reference limit. With moderate necrosis the levels may reach 20–50 times. In toxic hepatitis and viral hepatitis, there is extensive necrosis of liver cells. The ratio of SGOT and SGPT becomes 1.5–2.0 as against the normal value of normal > 1.0. Obstructive jaundice (without liver cell

necrosis) and malignant infiltration of liver, enzyme activity may be as high as 2–5 fold. In liver disease the rise in SGPT is generally higher than in SGOT but in severe cell necrosis SGOT may be higher than SGPT. In crushing muscle injury, muscular dystrophy and dermatomyositis (but not in muscular disease of neurogenic origin) the SGOT level may reach upto 10 fold. In acute pancreatitis and pulmonary embolism moderately elevated levels of SGOT are seen.

Reference range:

Serum GOT (SGOT) 5–30 U/L (less than 40 Karmen units/ml)

Serum GPT (SGPT) 5–35 U/L (less than 30 Karmen units/ml)

C. Creatine Phosphokinase

The enzyme creatine phosphokinase (CPK, creatine kinase, adenosine triphosphate: creatine phosphotransferase, EC 2.7.3.2) catalyzes the reaction:

$$\text{Creatine} + \text{ATP} = \text{Creatine phosphate} + \text{ADP}$$

Heart, striated muscle and brain have the highest CPK activity. Myocardial infarction raises the CPK levels within 3 hours, reaching a peak level of 6–20 times the upper normal limit in 24 hours and sharply declining to normal value by 72 hours (Figure 132.3). Liver and lungs have negligible CPK content and the serum activity does not rise in liver necrosis or pulmonary embolism and infarction. Rise of CPK in 4–6 hours signifies myocardial injury.

In progressive muscular dystrophy of Duchenne type, consistent and marked elevation in serum CPK upto 50–100 times the normal is seen. The rise occurs in infancy or early childhood even before the disease is clinically manifest. Later in the disease, with decrease in muscle mass, the rise is less pronounced. In about 80 per cent of symptom-less female carriers of the sex-linked muscular dystrophy (Duchenne type), serum CPK is moderately raised. In muscular atrophy of neurogenic origin serum CPK values are often within normal range or even lower. However, muscle trauma, fractures and even intramuscular injection can raise the enzyme to very high levels.

Serum CPK is also elevated in acute cerebro-vascular disease, hypothyroides, chronic alcoholism and after high doses of salicylates.

Three isoenzymes of creatine phosphokinase are known which are composed as dimeres of two subunits M (muscle) and B (brain).

1. In skeletal muscle 85%–90% is CK-MM 10%–15% CK-MB.
2. In heart muscle 30-40% is CK-MB. The rest is CK-MM.

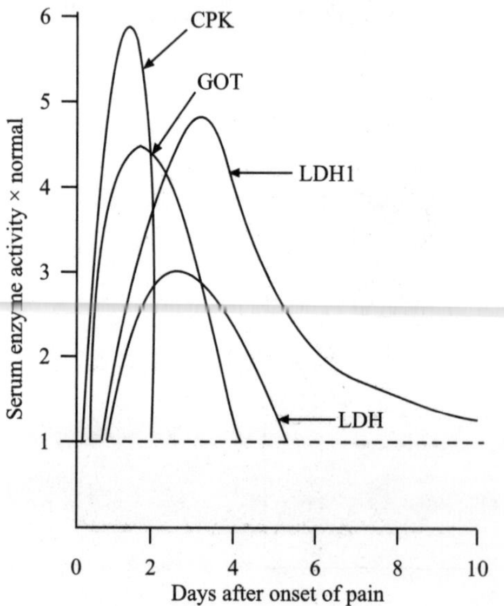

Figure 132.3   Various serum enzyme levels at different time periods in myocardial infarction.

while the brain contains CK-BB. CK-MB form 30 to 40% total CPK activity of the heart muscle but is almost specific for this organ. Rise in CK-MB in excess of 6% of the total serum CPK activity is of more diagnostic significance than the rise in total CPK activity. Peak activity of CKMB reaches 12 hours earlier than CPK in MI. Total CPK and CK-MB activities are also raised in myocarditis, severe arrhythmia and myocardial trauma. There is no elevation of serum CK-MB isoenzyme following hypothyroidism, physical exertion, intramuscular injection, muscle injury, hypoxic or other muscle damage, while total CPK activity does increases in these cases.

Reference range (in serum): 40–170 U/L (37°)

CKMB < 6% of total CPK

Recent methods for estimating CKMB actually estimate the mass rather than the enzymatic activity, thus avoiding the measurement of macrokinases (e.g. CK linked to IgG and dimmers of mitochondrial CK). This is found to correlate better with the diagnosis of acute myocardial infaraction.

In addition, estimation of the troponins (I and T) are more specific for any myocardial injury and hence more reliable in the diagnosis of myocardial infarction.

D. Cholinesterase

Two types of cholinesterases based on differences in substrate specificity and susceptibility to inhibitors, have been recognized in man.

1. Acetylcholinesterase (acetylcholine hydrolase, EC 3.1.1.7). The enzyme is also referred to as red cell, true, specific, or type I cholinesterase. This enzyme occurs in the red cells, the nerve endings, the lungs, the spleen, and gray matter the brain.
2. Pseudocholinesterase (acylcholine acyl hydrolase, EC 3.1.1.8). The enzyme is also referred to as serum, non-specific, or type II cholinesterase. It occurs in the serum, the liver, the pancreas, the heart, and white matter of the brain.

The red cell enzyme hydrolyzes acetylcholine and acetyl-β-methylcholine but not other types of choline esters. The serum enzyme hydrolyzes acetyl choline and other acyl esters but has no effect on acetyl-β-methylcholine.

Serum cholinesterase is synthesized in the liver and is a sensitive indicator of liver function. Enzyme levels are reduced in viral hepatitis, advanced cirrhosis and metastatic carcinoma of the liver, malnutrition and chronic debilitating diseases. Enzyme levels also fall but to a lesser degree in acute infections, pulmonary embolism, muscular dystrophy and following myocardial infarction and surgical procedures.

Many phosphorus containing organic insecticides inhibit both serum and red cell choline-sterases. The inhibition being complete and irreversible, the enzyme levels in red cells do not return to normal till the new cells replace the old ones. Though the serum levels fall more rapidly, it is the red cell enzyme which is essential for life. Measurement of the red cell cholinesterase at intervals is an important aid to determine the risk to individuals, who are exposed to these insecticides.

Serum cholinesterase exists in several genetically determined variants. Genetic heterogeneity results in biosynthesis of five types of cholinesterase which vary in their interaction with succinylcholine, dibucaine, fluoride ions, and other agents. Some individuals have subnormal levels of enzyme or have a genetically determined, atypical, weakly active, variant. These patients cannot hydrolyze succinyldicholine, a muscle relaxant used in surgery. Administration of this drug to these individuals may result in prolonged apnea and muscle relaxation.

Reference range: Serum cholinesterase 8–18 U/L (37°)

The most common form of a typical cholinesterase occurs due to a point mutation (aspartate to glycine) at nucleotide 209 of the gene on chromosome 3. Dibucaine is an amide local anaesthetic that greatly inhibits normal serum cholinesterase activity, but has limited inhibiton of the activity of a typical serum cholinesterase. This property of Dibucaine has been used to diagnose the presence of this a typical enzyme. It is expressed as a number termed 'Dibucaine number'. It is the percentage inhibition of the serium cholinesterase by dibucaine.

In the presence of normal cholinesterase, Dibucaine number = 80.

Individuals with heterozygosity for the variant, exhibit a Dibucaine number = 40-60 (Incidence is 1 in 480).

Individuals homozyogous for the variant, exhibit a Dibucaine number = 20 (Incidence is 1 in 3200).

E. Lactate Dehydrogenase

Lactate dehydrogenase (LDH, LD, L-lactate, NAD oxidoreductase, EC 1.1.1.27) catalyzes the oxidation of L-lactate to pyruvate. The reaction involves the transfer of hydrogen with NAD as the hydrogen acceptor.

L-lactate + NAD$^+$ $\rightleftharpoons$ Pyruvate + NADH + H$^+$ LDH is one of the glycolytic enzymes and as such it occurs in nearly all cells, but highest concentrations are found in the liver, the myocardium, the skeletal muscle, the kidney and the erythrocytes. Tissue activity is about a thousand fold higher than the serum activity, and leakage of enzyme from even small areas of necrosis can result in considerable increase in enzyme activity. The LDH isoenzyme pattern of each tissue is different from that of others. The distribution in some of the tissues is as shown in Table 132.1.

Several factors affect the serum LDH level. Values in children are generally higher than in adults. *Diurnal variations ranging from 30% to 200% have been reported.* Levels are higher by about 20% in summer months. Strenuous exercise produces a significant rise in enzyme activity.

In myocardial infarction the total LDH levels may rise 5–10 times the normal. The rise begins 6–12 hours after the onset of pain and persists for 6–8 days. LDH$_1$ and LDH$_2$ fractions especially the former show a much greater rise which can be seen as early as 6–12 hours and persists for 1–2 weeks. Renewed chest pain due to a second infarct results in a further rise in the cardiac fractions (LDH$_1$ and LDH$_2$). The magnitude of rise in LDH$_1$ values roughly parallels the size of infarct.

TABLE 132.1

	Percentage distribution of LDH isoenzymes				
	1	2	3	4	5
Heart	35–70	28–45	2–6	0–6	0–5
Liver	0–8	2–10	3–33	6–27	30–85
Brain	21–25	21–126	36–54	15–20	2–8
Skeletal muscle	1–10	4–18	8–38	9–36	40–97
Erythrocytes	39–46	36–56	11–15	4–5	2
Serum (electrophoresis)	18–32	28–40	18–30	6–15	2–13
Serum (heat stability)	20–40	—	33–55	—	10–30

If myocardial infarction is followed by congestive cardiac failure, resulting in liver damage, a rise in LDH_5 superimposes the existing pattern. In myocarditis and active rheumatic carditis, if the muscle is damaged, the LDH is elevated and the isoenzyme pattern is that of cardiac origin. The isoenzyme pattern is normal in angina pectoris, pericarditis, arrhythmias and pulmonary embolism.

In viral hepatitis, infectious mononucleosis and toxic jaundice—diseases associated with hepatocellular damage—a marked increase in LDH_5 and some increase in LDH_4 fractions is seen. In chronic liver diseases such as obstructive jaundice and cirrhosis the rise in LDH is not as pronounced.

Elevations in serum LDH are also seen in progressive muscular dystrophy, renal disease, carcinomatosis, megaloblastic and hemolytic anemias, leukemias and lymphomas.

Reference range (in serum): forward L → P 100–225 U/L (37°)

Reverse P → L   80–280 U/L (37°)

F.   5′-nucleotidase

The enzyme 5′-nucleotidase (5′-Nase, 5′-ribonucleotide phosphohydrolase, EC 3.1.3.5) specifically catalyzes the hydrolysis of 5′-nucleotides to the corresponding nucleosides and the phosphate ion.

The enzyme activity is increased in hepatitis and biliary obstruction. Unlike alkaline phosphatase the 5′-nucleotidase levels are unrelated to osteoblastic activity and are unaffected in osteomalacia, metastatic carcinoma of bone and Paget's disease.

Reference range (in serum): 2–17 U/L (37°)

G. Alkaline Phosphatase

Alkaline phosphatases (ALP, Orthophosphoric monoester phosphohydrolases, EC 3.1.3.1) are a group of non-specific enzymes which hydrolyse aliphatic, aromatic and hetrocyclic phosphoric acid esters with a pH optimum between 9 and 10.

Alkaline phosphatase is widely distributed in the human body notably in the intestinal mucosa, the kidney tubules, the osteoblasts of bone, the liver, the placenta and the lactating mammary glands. It is mainly localized at the cell membranes where it appears to be associated with transport mechanism. In the synthesis of bone it is associated with calcification process.

Diagnostic significance of serum alkaline phosphatase determinations is mainly in the heptobiliary disease and the bone diseases.

Serum alkaline phosphatase is very high, 10–12 times the upper limit of normal, in liver diseases associated with impairment of bile flow (Cholestasis) which could be intrahepatic or extra hepatic. Intra hepatic cholestasis may be due to viral hepatitis, damage by alcohol or drugs or mechanical obstruction by neoplacia. Extra hepatic cholestasis is due to the obstruction of common hepatic or common bile ducts due to gall stones, stricture or neoplasm. Increase in the activity of serum alkaline phosphatase in cholestasis is considered to be due to increased synthesis of alkaline phosphatase in epithelial cells of biliary canalculi and not due to the failure to excrete the enzyme through the bile as was believed earlier.

Serum alkaline phosphatase is generally mildly elevated in viral hepatitis, infectious mononucleosis and cirrhosis of the liver. The elevation may be marked (2–5 times the upper limit of normal) if there is occlusion of biliary canaliculi.

Among the bone disease serum ALP is highest in Paget's disease (10–25 times the upper limit of normal). It is moderately increased in osteomalacia and rickets, and hyperparathyroidism. Levels are very high in bone cancer.

Serum ALP is also moderately increased in Hodgkin's disease, congestive heart failure and ulcerative colitis.

Serum ALP is normally high in growing children and pregnant women (2–3 times). These could be beingn hyper-phosphatemia in growing children. This needs to be noted and reported to see if clinical significance exists.

The total activity of alkaline phosphatase fails to indicate satisfactorily the tissue of its origin which may, however, be pinpointed by its isoenzyme pattern. Raised α points to liver damage, pre β to increased osteoblastic activity and γ to an intestinal lesion.

Placental isoenzyme is seen only during pregnancy. Placental and bone isoenzyme bands are generally very diffused and migrate very close behind liver band (α region) and are therefore difficult to distinguish. These isoenzymes can however be distinguished by heat treatment and inhibitory action of L-phenylalanine. In hepatobiliary disorders, two distinct bands of ALP in regions of α_1 and α_2 globulin have been identified by electrophoresis on cellulose acetate paper. α_2 isoenzyme is derived directly from liver cells whereas α_1, band represents the alkaline phosphatase synthesized by the epithelial cells of biliary canalculi. The α_1 fraction is markedly elevated in metastatic carcinoma of liver, obstructive jaundice and cholangitis. Absence of α_1 band with an exaggerated α_2 band suggests hepatitis.

Reference range (in serum):
Adults 25–100 U/L (37°)
Growing children up to 500 U/L (37°)
(4, NPP as substrate and AMP as buffer)

H. Acid Phosphatase

The acid phosphatase, a class of non-specific phosphomono-esterases (orthophosphoric monoester phosphohydrolase, EC 3.1.3.2), hydrolyzes esters of phosphoric acid at acid pH (4–7).

Acid phosphatase occurs in red blood cells, platelets, leucocytes and prostate. Its presence in small amounts has also been shown in other tissues like the liver, bone, and the kidney.

Many different isoenzyme fractions of the serum acid phosphatase have been separated by electrophoresis but none except phosphatase of prostate, has found clinical diagnostic application. Prostatic serum acid phosphatase can be distinguished from erythrocyte and platelet phosphatases by its stability to formaldehyde and lability to L-tartrate.

Serum acid phosphatase is significantly raised in prostatic carcinoma with metastasis to bone and degenerating bone diseases. The enzyme levels return to normal after successful removal of the cancer or 3–4 weeks after estrogen therapy. In absence of bone metastasis enzyme levels are elevated only in 25% cases of prostatic carcinoma. Benign prostatic hyperplasia or prostatitis do not cause any elevation of the serum acid phosphatase level.

Reference range:
(4, NPP as substrate)

Total	
Males	2.5–11.7
Females	0.3–9.2
Tartrate-inhibited fraction	
Males	0.2–3.5
Females	0–0.8

I. Glucose-6-Phosphate Dehydrogenase

Glucose-6-phosphate dehydrogenase (G6-PD) (D-glucose-6-phosphate NADP oxidoreductase, EC 1.1.1.49) catalyzes the oxidation of glucose-6-phosphate to 6-phosphogluconolactone in the pentose phosphate shunt. The reaction involves reduction of NADP to NADPH. Though enzyme is present in all cells it is of special significance in the red cell. NADPH is necessary for conversion of oxidized glutathione (GSSG) to reduced form (GSH) Patients who are deficient in G6-PD, when exposed to oxidants, cannot generate adequate amounts of reduced glutathione to detoxify peroxides and to protect sulphydryl groups from oxidation. This results in denaturation of hemoglobin which forms aggregates in red cells. These aggregates can be seen when stained with brilliant cresyl blue as purplish inclusions called 'Heinz bodies'. If the oxidant

challenge is severe, irreversible damage to red cells is produced resulting in hemolytic anemia.

Over one hundred biochemical variants of G6-PD have been identified. These differ in their eletrophoretic mobility, kinetics, heat stability and affinity to different substrates. Some of the variants have abnormal activity and produce no disease while others are associated with drug induced hemolytic anemia. The gene for G6-PD is located on X-chromosome. Therefore, hemizygous males and homozygous females with genetic deficiency of the enzyme manifest the disease.

Hemolytic anemia due to deficiencies of many other red cell enzymes, both in the Embden-Meyerhof pathway and pentose phosphate shunt, have been reported. The deficiency of importance, in terms of its higher prevalence of occurrence, is pyruvate kinase deficiency.

G6PD can be determined in the red blood cell haemolysate after wasting with saline. The result is expressed as $mU/10^9$ cells. It is higher in young erythrocytes. Its main clinical significance is in detecting deficiency in suspected patients prescribing drugs like antimalarials, nalidixic acid, nitrofurantoin and phenytoin etc.

Reference range: 131 ± 13 $mU/10^9$ RBC

IV. GENETIC DEFICIENCIES

A large number of diseases with inherited deficiency of enzymes resulting in a generalized metabolic defect have been recognized. They are generally referred to as 'inborn errors of metabolism, a term introduced by Garrod. Some of the diseases which fall in this category are phenylketonuria, alkaptonuria, albinism, Lesch-Nyhan: syndrome, galactosemia, pentosuria, maple syrup urine disease, glycogen storage disease, deficiencies of glycolytic enzymes of importance to red cells and deficiency of hemoglobin reductase. It has now been recognized that in several instances the normal enzyme is replaced by its genetic variant which differs from the normal enzyme in its metabolic functions and many physico-chemical properties. Further advances in technology will reveal many more disorders which are due to genetic heterogeneity of enzymes.

V. SOME CONSIDERATIONS IN CLINICAL INTERPRETATION OF ENZYME DETERMINATIONS

1. Enzymes appear in the plasma by three different mechanisms

   (a) Certain enzymes such as pseudocholin-esterase, fibrinolytic and enzymes of blood coagulation are present in much higher concentration in the plasma than in the tissues. They perform important physiological functions in blood and are often referred to as 'functional plasma enzymes'. Most of these enzymes are synthesized in the liver and are actively secreted into the circulation. A rise in their levels is of little diagnostic importance. On the other hand, the serum levels of some of these enzymes such as pseudocholinesterase and thrombin may fall in liver disease due to their decreased synthesis. Determination of the serum levels of the two enzymes is often included in the profile of liver function tests. Deficiency of one or more of the enzymes involved in blood coagulation is known to occur both in acquired and inherited disorders associated with increased bleeding tendency. Measurement of these enzymes in plasma is of great value in diagnosis of bleeding disorders.

(b) Some enzymes such as transaminases, CPK, LDH, etc., are found in very high concentrations in the tissues with very low values in the plasma. These enzymes are referred to as 'non-functional plasma enzymes'. A rise in serum of these intracellular enzymes is an index of cell damage. These enzymes and their isoenzymes have a high organ specificity. This increases their diagnostic importance. Thus determination of serum activity of many enzymes and isoenzymes such as GOT, GPT, LDH, phosphatases and CPK is often of importance in clinical practice. More and more enzymes and their isoenzymes of diagnostic importance are being recognized.

(c) An obstruction in the secretary pathway may result in passive diffusion of certain enzymes into the circulation. The notable examples are pancreatic enzymes and alkaline phosphatases, levels of which are raised following obstruction of the pancreatic duct and obstructive jaundice respectively. An increase in alkaline phosphatase activity is in fact due to increased synthetic actvity in hepatocytes due to obstruction of bile flow.

2. Enzymes are cleared from the circulation at different rates. The biological half lives of CPK, GOT, GPT and LDH are 1.4, 2.0, 6.3 and 6.8 days, respectively. Thus in myocardial infarction, three days after the onset of pain, serum CPK returns to normal values, whereas GOT, GPT and LDH levels remain elevated. Therefore, for judicial inter-pretation of the enzyme values in serum, one should keep in mind the time of sampling in relation to the onset of symptoms, especially in diseases where the release of enzyme into circulation is periodic.

3. Several factors such as age, sex, time of sampling, exercise, and pregnancy influence enzyme levels. For example, CPK activity may rise following vigorous muscular effort or even intramuscular injection.

4. The enzymes are highly sensitive to changes in pH, temperature and concentrations of the substrate and the product. The conditions for enzyme assays need strict control as minor changes in these can produce erroneous results. Further, the presence of certain inhibitors or activators (drugs) in the serum of patients may give fallacious values. Kinetic analysis of an enzyme in a thermostatically controlled reaction cuvette with a recording spectrophotometer has greater reliability than the commonly used fixed time assay. Such techniques are now being used in some of the good labs.

SUMMARY

Enzymes and isoenzymes are present in all cells. Certain enzymes leak into the circulation or other extracellular fluids only when the cells are damaged. These enzymes and isoenzymes in serum, urine or other fluids are related to the degree of cell damage and the number of cells involved. The property of relative organ specificity of enzymes and isoenzymes as elaborated in the present chapter has been exploited extensively for arriving at a laboratory diagnosis of the diseases of specific organs.

SUGGESTIONS FOR FURTHER READING

Burtis C.A and Ashwood E.R. and Bruns E.E. (Eds.) (2006), Tietz Textbook of Clinical Chemistry, 4th ed., W.B. Saunders Company, Philadelphia, Pa, USA.

Henry Richard J., Cannon Donald C. and Winkleman James W. (Eds.) (1974), Clinical Chemistry Principles and Techniques, Harper and Row Publishers, New York, 818.

Huijgen H.B., Saunders G.T., Koster R.W., Vreeken J. and Bossuyt P.M. (1997), The Clinical Value of Lactate Dehydrogenase in Serum: A Quantitative Review, *Eru. J. Clin. Chem. Clin Biochem.*, 35:569–579.

Miller R.D., (2000), *Anesthesia*, 5th ed., Philadelphia, Churchill Livingstone, 419–421.

Ross G., Bever F.N., Uddin Z. and Hockman E.M. (2000), Troponin I Sensititvity and Specificity for the Diagnosis of Acute Myocardial Infarction, *J. Am Osteopath Assoc.*, 100:29–32.

Stoelting R.K. and Miller R.D. (2000), Basics of Anesthesia, 4th ed. Philadelphia, Churchill Livingstone 93–94.

Widmann Frances K. (Ed.) (1973), Goodale's Clinical Interpretation of Laboratory Tests, F.A. Davis Company, Philadelphia, 165.

Wilkinson J. Henry (1970), *Isoenzymes*, Chapman and Hall Ltd., London.

Zimmermann J., Fromm R., Meyer D., Boudreaux A., Wun C.C.C., Smalling R., Davis B., Habib G. and Roberts R. (1999), Diagnostic Marker Cooperative Study for the Diagnosis of Myocardial Infarction, *Circulation*, 99:1671–1677.

SECTION XX

DEVELOPING TECHNOLOGY TRENDS

Technology is advancing at an unprecedented rate based on the leads from basic sciences. Examples of such technological advancements include the birth of Nanotechnology as a very powerful tool for drug delivery, tissue specific expression, slow release, etc. Learning from nature is the best way to engineer, and the chapter on 'Biomaterials in Tissue Engineering' describes the technology behind creation of new smart materials for use in Medicine.

The chapter on Bioterrorism gives an overview of dual aspects of safe technology that is useful for defence and security but at the same time it can be misused as a weapon for mass destruction.

133

Nanotechnology in Medicine

Rinti Baneerjee

CONTENTS

I. INTRODUCTION

Nanotechnology refers to the development and application of structures < 100 nm in size in any one dimension. The size of these structures imparts certain unique features to the nanostructures by virtue of the large surface area to volume exposed. Nanotechnology has wide applications in medical sciences. Nanomedicine refers to the application of structures that are < 1 micron in size for diagnostic, therapeutic or regenerative applications. The size of the nanostructures allows certain advantages in terms of the cellular responses and biodistribution which are not seen with larger particles of the same composition. This chapter introduces the basic concepts of designing such nanostructures and outlines their advantages and applications in different areas of medical sciences.

II. BASIC CONCEPTS

Nanostructures need to be designed with precision such that the size, charge and physicochemical properties are within in a narrow distribution and are highly reproducible. Broadly, there are two ways of synthesizing nanostructures (a) the *top down approach* and (b) the *bottom up approach*. The top down approach involves breaking down larger particles into smaller and smaller ones by specific processes. The bottom up approach involves controlling the assembly of the materials to form small sized particles.

Since the size of the nanostructure plays an important role in determining the cellular responses, it is very important to have a narrow distribution of particle sizes. *Nanoparticle characterisation* includes determination of the size and structure by dynamic light scattering and transmission electron microscopy respectively. Further, it is important that the nanoparticles maintain their original size, even on storage and do not aggregate to form clumps of larger particles. This property of the nanoparticle is referred to as the *physical stability*. Imparting a sufficient charge to the surface of the nanoparticles causes them to repel each other leading to stable nanoparticles. This property is evaluated by a measure of the *zeta potential* of the nanoparticles. A positive or negative zeta potential of 20 mV or more is usually considered adequate to form stable nanoparticles.

III. CLASSIFICATION OF NANOPARTICLES

Nanoparticles may be classified based on their application, their structure, composition and degradability. Based on application, nanoparticles may be therapeutic, diagnostic, preventive, theranostic, or regenerative. Based on structure they may be classified as nanospheres, vesicles. nanocapsules, core-shell nanoparticles, nanotubes, nanowires, nanofibers, nanofilms or dendrimers. Based on their chemical composition they may be lipid, protein, polymeric, metallic, ceramic or hybrid nanostructures. Depending on whether they can be degraded in the body, nanoparticles may be classified as biodegradable or non-biodegradable. *Biodegradable nanoparticles* are preferred for drug delivery applications. Figure 133.1 depicts the commonly used nanostructures for biomedical applications.

IV. ADVANTAGES OF NANOTECHNOLOGY IN MEDICINE

There are several advantages of nanotechnology in biomedical applications. These differ depending on whether the nanostructures are designed for therapeutic, diagnostic or regenerative purposes.

From the point of view of therapeutics, nanoparticles of 100 nm which certain specially designed surfaces can *reduce their recognition by the macrophages of the reticuloendothelial system*. Commonly, hydrophilic long chain molecules are attached to the surface of nanoparticles for example polyethylene glycol. This causes steric hindrance, preventing the opsonisation of the nanoparticles and reducing the uptake by macrophages. This is beneficial as the nanoparticle can

then circulate in plasma for long periods. Such nanoparticles are referred to as stealth particles.

Another advantage of nanoparticles is that they can pass through different anatomical barriers to reach sites that are otherwise unreachable by conventional medications. For example, the blood brain barrier prevents the entry of drugs and toxins into the brain. Specially designed nanoparticles coated with surfactants can penetrate the blood brain barrier to allow efficient delivery of drugs into the brain.

For therapy of cancers, nanoparticles exploit the gap between the endothelial cells of leaky capillaries associated with most cancers, and can permeate preferentially to the tumor site. This along with the sluggish lymphatic drainage allows accumulation of the nanoparticles at the tumor site. This phenomenon is called the *enhanced permeation and retention (EPR) effect* and is widely exploited in cancer nanotechnology.

At the target sites, nanoparticles can be designed such that they are actively taken up by cells via *endocytosis*. The surface of the nanoparticles can be modulated to exploit different pathways of endocytosis. Further, nanoparticles can be designed to target different subcellular compartments within cells.

Certain nanoparticles have unique properties that can be exploited for therapeutic purposes. For example, magnetic nanoparticles < 10 nm can be *superparamagnetic* in nature,

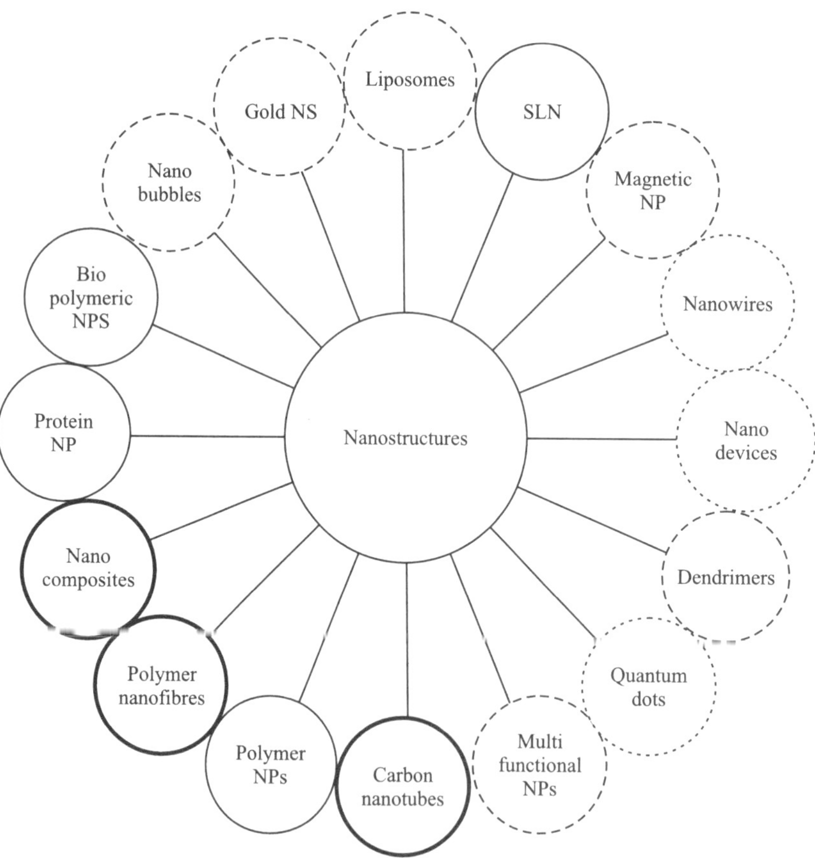

Figure 133.1 Nanostructures commonly used in biomedical applications. Dotted line indicates predominant diagnostic application, Continuous regular line indicates predominant therapeutic application, Bold thick line indicates predominant regenerative application, Dashed line indicates predominant theranostic application. NP: nanoparticle, NS: nanostructures SLN: solid lipid nanoparticles.

such that their magnetisation is high in the presence of a magnetic field and the nanoparticles are non-magnetic in the absence of the field. Such nanoparticles when subjected to an alternating magnetic field, convert the magnetic energy losses in the form of heat. Suitably designed magnetic nanoparticles can raise the temperature to 40-45 °C which is referred to as *hyperthermia*. Such localised hyperthermia is beneficial in killing cancer cells.

Nanoparticles can allow *access through several routes* that were otherwise not accessible. For example, nanoparticles can protect the encapsulated drugs from the surrounding environment and hence can be stable in the harsh gastric environment. Even labile molecules like peptides can be delivered by oral route when suitably packaged in nanoparticles. Similarly, nanoparticle aerosols can reach the deepest regions of the lungs allowing drugs to be delivered in a non-invasive manner.

The *biodistribution* of drugs can be altered favourably by encapsulating them within nanoparticles. The nanoparticles can be targeted to reach certain sites, increasing efficiency of drug delivery. This also implies that there would be less amounts of undesired deposition of drugs in other organs, reducing unwanted side effects. This is particularly useful for anticancer drugs.

The *pharmacokinetics* of drugs can also be altered as the nanoparticles act as depots of the drug which allow slow, sustained delivery of drugs over longer periods. This allows less frequent administration of the drugs. On the other hand, smart nanoparticles can be designed such that the nanoparticle degrades in the presence of certain stimuli or triggers, allowing a trigger responsive burst of drug delivery under specific conditions. Table 133.1 summarises the therapeutic advantage of nanoparticles.

TABLE 133.1   Therapeutic advantages of nanoparticles

Advantages: Therapeutic
Minimal recognition by reticuloendothelial system: long circulating
Penetrates anatomical barriers
Stabilises labile actives
Reduces frequency of doses
Access through patient compliant routes
Intrinsic therapeutic property (magnetic NP)
Enhanced permeation and retention effect
Active endocytosis by target cells
Altered biodistribution, targeted therapy

In diagnostic and imaging applications, nanotechnology has advantages of allowing the development of point of care diagnostics and enhancing contrasts in imaging modalities.

Microfluidic based technologies allow the miniaturisation of detection devices, such that they require small sample volumes for detection of analytes. Further, the entire device can be miniaturised such that it is portable and hence can be used at the site of the patient and in rural areas by community health workers. Nanowires, cantilevers, gold nanostructures and quantum dots are commonly used in miniaturised diagnostic devices.

Nanoparticles like magnetic nanoparticles, gold nanostructures and quantum dots have unique properties that allow *enhancing imaging* modalities. For example, iron oxide nanoparticles enhance contrast in T2 weighted images in magnetic resonance imaging. Gold nanostructures and quantum dots have a size dependant, intrinsic fluorescence which may be exploited for optical detection. The fluorescence of these nanostructures is stable over long periods of time and is not subjected to photobleaching unlike conventional fluorescent dyes. Table 133.2 summarises the advantages of nanoparticles and nanodevices for diagnostic applications.

TABLE 133.2   Diagnostic advantages of nanoparticles and nanodevices

Advantages: diagnostic
Low sample volumes for detection
Miniature, portable devices
Allows point of care testing
Rapid detection
Simultaneous analysis of multiple analytes
Enhanced contrast in imaging
Stable fluorescence over long periods, no photobleaching
Theranostics possible

In regenerative medicine, nanostructured materials are used to design scaffolds which can mimic the natural extracellular matrix and have superior cellular interactions than conventional scaffolds. *Nanocomposites* are commonly used to design bone scaffolds and have higher mechanical properties than conventional scaffolds of the same composition. Nanoparticles can also be used to stabilise and increase the half life of labile growth factors that are required to maintain threshold levels over a long period of time within the scaffold, during tissue regeneration. Table 133.3 summarises the advantages of nanomaterials for regenerative applications.

TABLE 133.3   Regenerative advantages of nanoparticles

Advantages: regenerative
Mimics extracellular matrix
Enhanced mechanical strength
Enhanced cell adhesion
Enhanced cell differentiation
Nanofibers, nanotubes guide directional cell growth
Stabilisation of growth factors
Sustained delivery of growth factors

V. POINT OF CARE DIAGNOSTICS

Point of care diagnostics refers to the development of devices that can be used for detection of analytes or diseases at the site of the patient using small, portable devices. Nanotechnology and microfluidics have enabled us to miniaturise the processes required for the detection of analytes from the samples. In fact, chips are designed where multiple analytes are detected using the same platform from a single sample of low volumes. The devices integrate all the processing required of the sample like separations, enrichments, the detection technology, processing of the signals obtained, and final readouts of results. The term "lab on a chip" is used to describe such devices. Such devices are very useful for global health applications and in epidemics, as they are portable and allow detection in a short span of time, at the site of the patient. Ideally a point of care device should fulfil the ASSURED criteria (Mabey et al. 2004) which are summarised in Table 133.4.

TABLE 133.4 Criteria for point of care diagnostics (based on Mabey et al., 2004)

Criteria
A: affordable by those at risk
S: sensitive
S: specific
U: user friendly (requiring minimal training)
R: rapid and robust
E: equipment free or requiring minimal equipments
D: delivered to those who need it

VI. NANOPARTICLE BASED IMAGING

Nanoparticles can be used for enhancing the contrasts of several imaging modalities. In magnetic resonance imaging (MRI), magnetic nanoparticles are commonly used. Iron oxide nanoparticles < 10 nm in diameter, reduce T2 and hence cause contrast enhancements in MRI (Artemov et al., 2003). In order to be useful as an image contrast agent, the iron oxide nanoparticle should be biocompatible and should be stable in an aqueous medium and in vivo, in plasma. These two features are usually obtained by coating the iron oxide nanoparticles with suitable lipid or biopolymeric shells. Lauric acid and dextran have been commonly used for these purposes. Several iron oxide nanoparticles have been clinically approved for imaging. For example Endorem® is a dextran coated iron oxide nanoparticle that is approved for liver imaging. Lumirem® is a silica coated iron oxide nanoparticle, given orally or rectally and is approved for abdominal imaging.

Gold nanostructures have unique size dependant optical properties. Nanostructures of gold lose their golden lustre and may appear red, blue, violet etc. depending on their size. Gold nanocages, nanoshells and nanorods are commonly used for imaging purposes. The advantage of gold nanostructures is their inertness which allows them to be biocompatible. Gold nanostructures show fluorescence due to a phenomenon called *surface plasmon resonance* (Eustis and El Sayed, 2006). For nanoparticles much smaller than the wavelength of the incident light, electromagnetic fields of certain frequencies cause a resonant oscillation of the surface free electrons of the metallic nanoparticle. This leads to enhanced optical properties in terms of absorption and scattering and can exploited as fluorescent nanoparticles for imaging. Such nanoparticles do not require any other fluorophore as a tag and the fluorescence due to these nanostructures are stable for long periods of time without photobleaching. Gold nanostructures are often tuned to obtain near infra red fluorescence hence allowing maximal penetration into deeper tissues for imaging. Gold nanoparticles also allow good CT contrast and may also be utilised for this purpose.

Quantum dots are semiconductor nanoparticles that have unique optical properties. They are in the form of core-shell structures where the core is composed of the semiconductor nanomaterial and has an insulating shell surrounding it. For example, one commonly used quantum dot has a cadmium selenide core and a zinc sulphide shell. Quantum dots also have a size dependant fluorescence. Quantum dots act as artificial atoms having discrete energy levels. Optical excitation from one level to another followed by relaxation to the lower energy state leads to fluorescent emission called *photoluminescence*. The energy gap in quantum dots is size dependant and hence quantum dots with different colours of emission can be prepared of the same material by finetuning their sizes. Quantum dots absorb light of a wide range of wavelengths and emit almost monochromatic light, the wavelength of which is determined by the size of the quantum dot. Many kind of quantum dots can be excited by the same light source and their emissions can be easily separated. The fluorescence of quantum dots is stable over a long period and is not subject to photobleaching (Gao et al., 2004). However, quantum dots show toxicity in cells and are thus preferred for in vitro imaging. In order to use quantum dots for in vivo imaging, suitable coatings on the nanoparticle are required to make them non-toxic and biocompatible.

Nanobubbles and ultrasound responsive nanocapsules can be used for enhancing ultrasound contrasts. These are usually bubbles formed of perfluorocarbon or other approved gases which are protected by a shell formed of biocompatible lipids or polymers. Nanobubbles and nanocapsules offer higher stability, longer half life and deeper penetration as compared to conventional microbubbles for ultrasound imaging.

All of the nanostructures described in for diagnostic imaging can also be used in combination with drugs for both diagnostic imaging and image guided drug delivery. Such a concept is referred to as theranostics and is often utilised in cancer therapy.

VII. NANOPARTICLES FOR DRUG DELIVERY

Nanoparticles that are used for drug delivery applications are usually designed of biodegradable materials. The type of nanoparticle that needs to be used differs depending on the nature of the drug to be encapsulated, the route through which the drug is to be administered and the desired duration of effect of the drug. Some of the commonly used types of nanoparticles are described in this section.

Liposomes are self assembled lipid bilayered nanostructures that closely resemble the cell membranes in their structure. Unilamellar liposomes having a single bilayer are also referred to as nanovesicles. Liposomes are formed by various self assembly processes and have advantages of lack of toxicity and versatility to encapsulate both hydrophilic and hydrophobic drugs. Conventional liposomes composed of phospholipids and cholesterol had a very short circulation time. Later, stealth liposomes have been developed which have polyethylene glycol corona, making them long circulating. The disadvantages of liposomes are usually of stability and high cost. Recent advances in liposomal technologies are overcoming some of these disadvantages. Smart trigger responsive nanovesicles which respond to changed in pH or enzymes or temperature are an exciting advancement to nanovesicle technology than can be exploited for site specific delivery of drugs (Andresen et al., 2005). Due to their similarity with cell membranes, lipid nanovesicles are efficiently internalised in target cells by fusion or endocytosis.

Several liposomes have been approved for clinical applications for example Doxil®, Caelyx®, Lipocyte® for delivery of various anticancer drugs.

Solid lipid nanoparticles and non-lamellar lipid phases have the potential to be exploited asstable platforms for the encapsulation of hydrophobic anticancer drugs as well as targeting to sentinel lymph nodes. They provide the advantages of internalisation seen with lipid nanovesicles along with the stability associated with solid polymeric nanostructures. They are very useful for the encapsulation of hydrophobic drugs.

Synthetic polymeric nanoparticles can be developed with wide varying properties andsurface characteristics making them suitable for different therapeutic applications (Couvrier and Vauthier, 2006). The degradation of the polymers determine the sustained release possible dueto the polymer nanoparticles. Biodegradable polymers like polylactide-co-glycolide (PLGA) are often used for their synthesis. Smart polymeric nanoparticles can be designed of It is possible to design these nanoparticles to be trigger responsive. Commonly, poly N isopropyl acrylamide (pNIPAAM) based polymers and poloxamers are used to impart thermosensitivity to polymeric nanoparticles.

Dendrimers are unique branched polymeric nanostructures that have a dendritic or tree branch like structure. The number of branches present, vary in dendrimers and are determined by the generation of the dendrimer with each generation having more branches.The unique branched hypernetworked structure of dendrimers make them very useful for the simultaneous entrapment or linkage of multiple agents, namely one or more drugs, animaging agent and a targeting moiety (Backer et al., 2005). However, the advantage of the increase in the number of surface groups and branches needs to weighed with the increase in size due to the branching generations. Further,

dendrimers need to be designed of biodegradable polymers for successful translation as theranostic agents.

Protein and biopolymeric nanostructures are designed to exploit the advantages of cellular interactions between the nanoparticles and the target cells. Protein nanoparticles have been clinically approved, for example, Abraxane® for the delivery of paclitaxel, an anticancer drug. The *albumin nanoparticles* have the advantage of being targeted nanocarriers which exploit the increased need of the protein in the cancer cells (Blum et al., 2007). Biopolymeric nanoparticles of chitosan and alginate are most commonly used and have the advantages of high mucosal residence due to their mucoadhesion. This feature is particularly useful for regional applications of drugs in mucosal surfaces.

Nanoparticles that are designed for drug delivery applications need to be optimised keeping the target organ and the route of administration in mind. For neurological conditions, nanoparticles are designed of lipophilic materials and are usually coated with surfactants which enable them to penetrate the *blood brain barrier*. Another strategy that is employed is that of bypassing certain efflux pumps like the p glycoprotein pump which is overexpressed in the blood brain barrier. Nanoparticles may be designed such that the drug is not recognised by the efflux pump, thereby preventing its efflux. Biomimetic nanoparticles are also designed for enhanced entry through the blood brain barrier, for example by mimicking certain nutrient molecules and exploiting the specific receptors present for allowing entry of the nutrient into the brain.

Ocular drug delivery is another application where nanotechnology can play an important role. Conventional eyedrops and ointments have very poor reach in the posterior segments of the eye. Further, nasolacrimal drainage causes washout of drugs administered topically. Specially designed *mucoadhesive nanoparticles* can enhance the ocular residence time of the drugs and enhance their effects. Membrane penetrating nanoparticles can be used to deliver drugs in the deeper tissues of the eye.

Nanoparticle aerosols are another important platform that can be exploited for efficient drug delivery to the respiratory system. While the deposition of aerosols within the respiratory tract is dependant on several parameters, the *mass median aerodynamic diameter (MMAD)* is a very important one. Aerodynamic diameter refers to the diameter of a unit-density sphere having the same terminal settling velocity as the particle in question. It depends on size, shape and density of the particle. MMAD refers to the geometric mean aerodynamic diameter of an aerosol distribution. For deeper lung deposition, the MMAD needs to be < 1–3 microns. Further, the distribution of drugs within the aerosol droplets need to be uniform. This is achieved by generating nanoparticles loaded with drug several of which can be incorporated within the aerosol droplet. Such nanoparticle aerosols can be designed to allow deeper and more uniform distribution of drugs in the lungs when administered as aerosols. In the lungs, the size and surface of the nanoparticle determines its uptake by the alveolar macrophages or epithelial cells. Nanoparticle aerosols serve as a non-invasive route of delivery in several respiratory diseases like pneumonia, tuberculosis and respiratory distress syndromes.

Nanotechnology plays an important role in the oral delivery of labile agents. Polymeric nanoparticles which can withstand low pH and proteolytic digestion can be used to encapsulate the drugs for delivery through the oral route. One area where this technology can provide large value addition is in diabetes mellitus. Insulin is currently delivered subcutaneously by injections several times a day. Packaging in suitable nanoparticles can open up ways to deliver insulin orally. A major challenge in this approach is to maintain the biological activity of the peptides during nanoparticle formation and storage.

Another interesting and emerging application of nanotechnology is the exploitation of the *transdermal route for drug delivery*. The stratum corneum or the outermost layer of the skin forms a protective barrier and prevents the entry of harmful substances and also of drugs. Nanoparticles of various lipids have been found to fluidise the packing of the ceramide rich stratum corneum layer, allowing the penetration of the drug loaded nanoparticle deeper into the dermis. This is useful for local delivery of drugs like antibacterials or antipsoriatic in diseases of the skin. Further, on reaching the dermis, it is possible for the nanoparticles to enter the adjacent rich capillary network to gain systemic access. Hence, nanoparticle based topical formulations can also be used for systemic delivery of drugs in a non-invasive and patient compliant manner.

Apart from drugs, other actives like genes and *siRNA delivery* can also be achieved efficiently using nanoparticles. Nanoparticles for gene delivery usually use cationic nanoparticles which allow efficient complexation of DNA. A challenge is to prevent agglomeration of the cationic nanoparticles in plasma and to maintain the stability and biological activity of DNA within the nanoparticles.

siRNA has been successfully delivered using various nanoparticle strategies. The nanoparticles need to be designed such that the nanoparticles are internalised within the cells. The nanoparticles are often designed to trigger the release of the siRNA intracellularly within the endosomes, or on escape from the endosome within the cytoplasm. pH sensitive nanoparticles that have increased release at pH 5, which is expected in the endosomes, and minimal release at pH 7.4 as expected in the systemic circulation, are designed (Stigliano et al., 2013). The nanoparticles protect the siRNA from degradation and are targeted intracellularly at the site of action.

VIII. NANOTECHNOLOGY IN CANCERS

By far the most widely researched application of nanotechnology in biomedical sciences is its application in the diagnosis and treatment of cancers. Nanotechnology can be used to enhance the diagnosis of cancers, for site specific targeted delivery of drugs, for preferential uptake of drugs in cancers, for trigger responsive delivery of drugs, for simultaneous imaging and drug delivery as well as by intrinsic cytotoxic effects of certain types of nanoparticles (Ferrari, 2005).

Targeted nanoparticles can be used for early detection of cancers. For example, suitably conjugated magnetic nanoparticles can be used for detection and separation of circulating tumor cells (CTCs) which may prove beneficial for early detection of specific cancers.

Targeted delivery refers to the concept of specific delivery of drugs to the cancers with minimal effect of the drugs on the surrounding normal tissues. The most common strategy of targeting seen in nanoparticles for cancer therapy is the enhanced permeation and retention effect (EPR) as explained earlier. The gaps between endothelial cells of cancerous capillaries are exploited by designing nanoparticles of 100-200 nm in diameter.

Active targeting strategies involve the conjugation or surface modification of the nanoparticles with targeting ligands which attach preferentially to cancer cells by exploiting antigen-antibody or specific receptor mediated endocytosis. Small molecules and peptides overexpressed on the surface of cancers are candidates for active targeting.

Trigger responsive, "smart" nanoparticles are an important advancement in designing site specific therapy in cancers. Trigger responsive nanoparticles become more leaky or degrade in the presence of certain stimuli. These stimuli or triggers may be internal or external. Internal triggers are those that are already present in the body and are associated only with the cancerous site, while extrinsic triggers are focussed at the desired site from an external source.

Commonly used *internal triggers* for cancers are low pH as the extracellular pH of most cancers has a pH of 6 to 6.5 due to the metabolic acids. pH sensitive nanoparticles are designed to degrade and release their cargo preferentially in the presence of low pH. Enzymes are another type of intrinsic triggers. Several enzymes like phospholipase A2, proteases, hydrolases and matrix metalloproteinasesare over expressed in various cancers. Nanoparticles can be designed to specifically degrade in the presence of these enzymes.

On the other hand, *triggers like temperature can be externally generated*. For example, the ability of magnetic nanoparticles to raise the local temperature when placed in an alternating magnetic field can be exploited as an external trigger. This is done by linking the magnetic nanoparticle to a temperature sensitive nanoparticle encapsulating drugs. For example, *magnetoliposomes* of dipalmitoylphosphatidylcholine have been developed which co-encapsulate iron oxide nanoparticles and an anticancer drug. In an alternating magnetic field, the iron oxide nanoparticles cause a rise of temperature to 41–42 °C (Pradhan et al., 2010). This is equal to or exceeds the phase transition temperature of the phospholipid causing it to change to a more fluid like state making the liposome leaky to drugs. Hence, release of the anticancer drug occurs leading to a temperature triggered drug delivery. External triggers are easier to control that internal triggers, but suffer from limited penetration of tissues in certain instances.

Similarly, light sensitive nanoparticles and ultrasound sensitive nanoparticles can be designed which release drugs in the presence of activation by a specific near infrared laser or a specific strength focussed ultrasound beam. Trigger responsive nanoparticles allow release of cargo preferentially at the site of the trigger and hence improve the specificity of anticancer therapy without the need to use expensive antibodies and aptamers for targeting.

A related aspect of nanotechnology based cancer therapy is the development of *theranostics*. The combination of magnetic hyperthermia and MRI imaging using drug loaded magnetoliposomes is one such example. Similarly, gold nanoshell-nanoparticle complexes can be used for photothermal therapy and near infrared optical imaging. A third theranostic modality is that of nanobubbles and nanocapsules loaded with drugs, which have an ultrasound sensitive core of gases. The presence of perfluorocarbon or other approved gases lead to an ultrasound contrast whereas sonoporation and focused ultrasound mediated increase in temperature and membrane permeability cause cell killing and release of drugs in response to ultrasound. Figure 133.2 depicts commonly used strategies for targeted and triggered delivery of drugs in cancers.

Multidrug resistant cancers are another area where nanotechnology can improve the efficacy of anticancer drugs. While multidrug resistance occurs due to multiple factors, it is often associated with overexpression of efflux pumps on the surface of the cancers like *p glycoprotein efflux pump (Pgp)*. Nanoparticles are designed such that they act as "Trojan horses" where they mask the drug and prevent its recognition by the Pgp efflux pump. The matrix of the nanoparticle is composed of biocompatible materials which allow the endocytosis of the nanoparticle containing the drug directly into the cancerous cell without its release extracellularly. This prevents the recognition by Pgp and prevents the efflux of the drug (Kang et al., 2010). Hence, the sensitivity of tumors to anticancer drugs can be increased in multidrug resistant cancers.

Another area of importance of nanotechnology in cancer therapy is the development of *preventive and therapeutic vaccines and immunotherapy* for cancers. Nanoparticles can be designed to be preferentially taken up by macrophages and

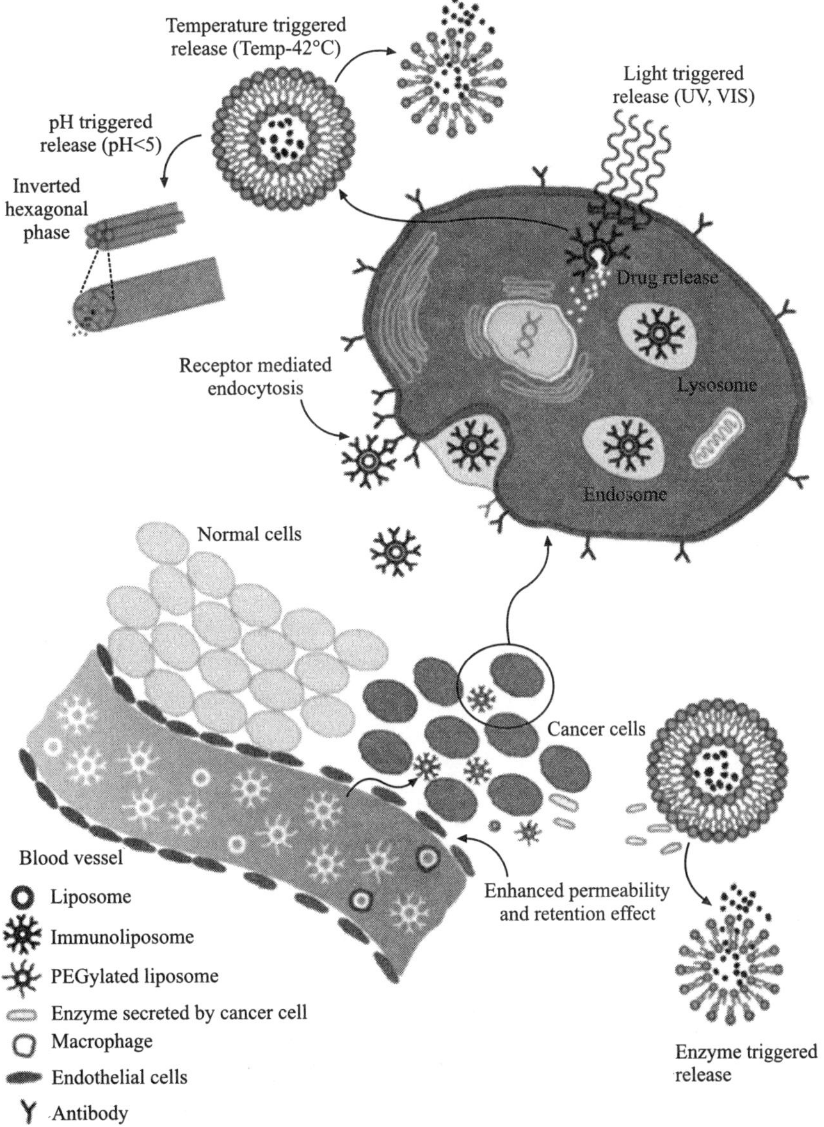

Figure 133.2 Strategies for targeted and triggered delivery in cancers using liposomes. (Reproduced from Joshi and Banerjee, 2012) (*see Plate 53 for colour figure*)

dendritic cells and stimulating a suitable immune response. In addition, mucosal immunity can be exploited by activating the mucosa associated lymphatic tissues through oral, intranasal and vaginal administration (Borges et al., 2006). Further, most antigens are labile (DNA, RNA or peptide). The stabilisation of these agents and increasing their biological half life using nanoparticles enhances the antigenicity and efficacy of these vaccines.

IX. NANOMATERIALS FOR REGENERATIVE MEDICINE

Nanostructured materials are preferred for designing scaffolds in many tissue regenerative technologies. Such scaffolds mimic the topography of fibrils within the natural extracellular matrix making them very suitable for growth and differentiation of cells (Shi et al., 2010).

Electrospinning is a process by which very thin fibers of polymers can be developed. These *nanofibers* serve as scaffolds to entrap cells and mimic the collagen fibrils of the extracellular matrix. The chemistry, size and orientation of the nanofibers can determine the cellular growth and differentiation.

Nanocomposites of ceramics and polymers for example hydroxyapatite and PLGA have been commonly used as scaffolds for bone tissue engineering. The mechanical strengths of nanocomposite scaffolds has been found to be enhanced as compared to the conventional scaffolds of the same composition. This is due to the increased surface area and close packing achieved by the nanocomposites.

Nanocomposities also mimic the natural orientation of hydroxyapatite nanocrystals along the collagen I fibrils. The response of osteoblasts to the nanocomposites also differs from that of conventional scaffolds. Multiple outgrowths of the osteoblasts are seen when grown in nanocomposite scaffolds which allow a more favourable and active state of the osteoblasts.

Nanotubes are often used to help orient the growth of neurons in nerve regeneration scaffolds. Successful nerve regeneration is dependant on biological and electrical cues and needs to be directional, towards the opposite end of the severed neuron. Random, misaligned sprouting of axons is undesirable and can lead to several neurological side effects. Nanotubes are used to align the growth of the neuron from one end to the other. Carbon nanotubes which are electrically conducting, have been used for enhancing neuronal regeneration within nerve regeneration tubes or scaffolds. Carbon nanotubes have also been explored for therapeutic (cytotoxic) effects and diagnostic (imaging) effects.

Nanofilms, nanoparticles and nanocapsules can be used to *stabilise and cause a sustained delivery of growth factors* within scaffolds at specific sites. This is important such that the half life of the growth factors is increased and their biological activity can be maintained over prolonged periods during tissue regeneration. Nanostructures that simultaneously or sequentially deliver multiple growth factors are also important as more than one growth factor is commonly required for regeneration of tissues. Further, in many instances the sequence of the various growth factors matters as some factors are overexpressed in the early differentiation stage while others are increased in the late differentiation stage. These can be controlled precisely using various core-shell nanostructures.

X. NANOTOXICOLOGY

There have been many toxicological concerns with the application of nanotechnology. The very properties that make nanoparticles advantageous to the body can lead to toxic concerns if they are not suitably controlled. For example, the ability of cross the blood brain barrier can be a disadvantage if it is undesired or if the nanoparticle leads to toxic effects in the brain.

Two important points that are to be kept in mind while detecting the effects of nanoparticles are (1) the toxicity or safety of each type of nanoparticle needs to be specifically determined and cannot be generalized and (2) the safety of a material in the bulk state does not imply that it is non-toxic when used in the form of nanoparticles.

Fortunately, a detailed understanding of the cellular effects and distribution of nanoparticles can lead to surface modifications to prevent unwanted toxicities.

Most of the toxic concerns are associated with non-degradable, synthetic nanoparticles. Biodegradable nanoparticles composed of materials widely present in the body like lipids and biopolymers have generally been found to be safe. Amongst non degradable nanomaterials, carbon nanotubes and metallic nanoparticles have been widely reported to show toxicities by specific routes of administration or in specific doses. *Carbon nanotubes* have been implicated in inflammation, fibrosis and the formation of epitheloid granulomas in the lungs on inhalational exposure (Lam et al., 2006).

Carbon nanotubes and *quantum dots* have been found to increase the reactive oxygen species leading to cellular toxicity.

In a ninety day inhalational study in rats, the exposure to multiwalled carbon nanotubes, were found to lead to toxic effects in the respiratory system (Pauluhn, 2010). Both the *composition and the shape of the nanoparticle* appears to be responsive for this toxic effect. For example, nanoparticles with high aspect ratios typically lead to fibrosis and cancers similar to asbestos fibres.

The clearance of nanoparticles by macrophages is size dependant and can determine their residence time. In alveoli, nanoparticles less than 100 nm are cleared less efficiently than particles of 1-3 micronssize (Takenaka et al., 2001). A study demonstrated that titanium dioxide nanoparticles of 20 nm had a longer retention period in the lungs than that of 250 nm sized particles and was more likely to be associated with toxic effects (Oberdorster et al., 1994).

The specific effects of each types of nanoparticle needs to be determined and the structures need to be optimised in terms of size, shape and charge to reduce toxic effects, prior to *in vivo* biomedical applications.

SUMMARY

Nanotechnology refers to the development and application of nanostructured materials and nanoparticles having at least one dimension < or equal to 100 nm. For biomedical applications, this definition can be extended to include nanomaterials or nanoparticles < 1 micron in size. Nanoparticles may be formed by top down or bottom up approaches and need to be characterised for a narrow size distribution and stability. Nanoparticles have several advantages in diagnosis and therapy, with many nanoparticles having intrinsic properties which can be exploited for these purposes. Nanoparticles can be designed of biodegradable materials for therapy and can be modulated such that they have a long circulation time and can reach otherwise inaccessible sites of the body by penetrating through several anatomical barriers. Magnetic nanoparticles, gold nanoshells and quantum dots are frequently used as contrast enhancing agents for magnetic resonance imaging and optical imaging respectively. Nanoparticles can show preferential accumulation at cancer sites by the enhanced permeation and retention effect. This reduces the toxicity associated with the anticancer drugs. Nanoparticles can bypass efflux pumps and be specifically endocytosed by cancerous cells. Trigger responsive nanoparticles can release their cargo in response to certain internal or external stimuli thereby causing specificity. Nanostructured materials mimic the extracellular matrix and enhance the cellular adhesion and differentiation in several tissue regeneration technologies. Nanoparticles having high aspect ratio and nondegradable in nature can be associated with toxic effects. The nanoparticles need to be designed and modulated to exploit the beneficial properties while avoiding the toxic effects. Biomimetic nanoparticles of degradable materials like lipids and biopolymers are one such strategy. Nanotechnology has several potential applications in medicine and more research is required to help translate this technology more widely in clinics.

SUGGESTIONS FOR FURTHER READING

Andresen T.L., Jensen S.S. and Jorgensen K. (2005), Advanced Strategies in Liposomal Cancer Therapy: Problems and Prospects of Active and Tumor Specific Drug Release, *Prog. Lipid Res.*, 44:68–97.

Artemov D., Mori N., Ravi R. and Bhujwalla Z.M. (2003), Magnetic Resonance Molecular Imaging of the HER-2/neu Receptor, *Cancer Research*, 63:2723–2727.

Backer M.V., Gaynutdinov T.I., Patel V., Bandyopadhyaya A.K., Thirumamagal B.T.S., Tjarks W., Barth R.F., Claffey K. and Backer J.M. (2005), Vascular Endothelial Growth Factor Selectively Targets Boronateddendrimers to Tumor Vasculature, *Molecular Cancer Therapeutics*, 4(9):1423–1429.

Blum J.L., Savin M.A., Edelman G., Pippen J.E., Robert N.J., Geister B.V., Kirby R.L., Clawson A. and O'Shaughnessy J.A. (2007), Phase II Study of Weekly Albumin-bound Paclitaxel for Patients with Metastatic Breast Cancer Heavily Pretreated with Taxanes, *Clin. Breast Cancer*, 7:850–856.

Borges O., Cordeiro-da-Silva A., Romeijn S.G., Amidi M., De Sousa A., Borchard G. and Junginger H.E (2006), Uptake Studies in Rat Peyer's Patches, Cytotoxicity and Release Studies of Alginate Coated Chitosan Nanoparticles for Mucosal Vaccination, *J. Control. Rel.*, 114:348–358.

Couvreur P., Vauthier C. (2006), Nanotechnology: Intelligent Design to Treat Complex Disease, *Pharm. Res.*, 23:1417–1450

Eustis S. and El-Sayed M.A. (2006), Why Gold Nanoparticles are More Precious than Pretty Gold: Noble Metal Surface Plasmon Resonance and Its Enhancement of the Radiative and Nonradiative Properties of Nanocrystals of Different Shapes, *Chem. Society Review*, 35(3):209–217.

Ferrari M. (2005), Cancer Nanotechnology: Opportunities and Challenges, *Nature Reviews Cancer*, 5:161–171.

Gao X., Cui Y., Levenson R.M., Chung L.W.K. and Nie S. (2004), *In vivo* Cancer Targeting and Imaging with Semiconductor Quantum Dots, *Nature Biotech.*, 22:969–979.

Joshi N. and Banerjee R. (2012), Liposomes for Anticancer Drug Delivery In: *Nanotechnology: Perspectives in Cancer* Banerjee R. (Ed.), Narosa International, New Delhi, 2012.

Kang K.W., Chun M.K., Kim O., Subedi R.K., Ahn S.G., Yoon J.H. and Choi H.K. (2010), Doxorubicin-loaded Solid Lipid Nanoparticles to Overcome Multidrug Resistance in Cancer Therapy, *Nanomedicine Nanotech. Biol. Med.*, 6(2):210–213.

Lam C.W., James J.T., McCluskey R., Arepalli S. and Hunter R.L. (2006), A Review of Carbon Nanotube Toxicity and Assessment of Potential Occupational and Environmental Health Risks, *Crit. Rev. Toxicol.*, 36 (3):189–217.

Mabey D., Peeling R.W., Ustianowski A. and Perkins M.D. (2004), Diagnostics for the Developing World, *Nature Rev.* 2:231–240.

Oberdorster G., Ferin J. and Lehnert B.E. (1994), Correlation Between Particle Size, *In vivo* Particle Persistence and Lung Injury, *Environ. Health Persp.*,102(5):173–179.

Pauluhn J. (2010), Subchronic 13 Week Inhalation Exposure of Rats to Multiwalled Carbon Nanotubes: Toxic Effects are Determined by Density of Agglomerate Structures, Not Fibrillar Structures, *Toxicol. Sci.*, 113:226–242

Pradhan P., Giri J., Rieken F., Koch C., Döblinger M., Banerjee R., Bahadur D. and Plank C. (2010), Targeted Temperature Sensitive Magnetic Liposomes for Thermo-chemotherapy, *Journal of Controlled Release*, 142(1):108–121.

Shi J., Votruba A.R., Farokhzad O.C. and Langer R. (2010), Nanotechnology in Drug Delivery and Tissue Engineering: From Discovery to Applications, *Nano Lett.*, 10 (9):3233–3230.

Takenaka S., Karg E., Roth C., Schulz H., Ziesenis A., Heinzmann U., Schramel P. and Heyder J. (2001), Pulmonary and Systemic Distribution of Inhaled Ultrafine Silver Particles in Rats, *Environ. Health Persp.*,109(4):547–551.

Stigliano C., Aryal S., deTullio M.D., Nicchia G.P., Pascazio G., Svelto M. and Decuzzi P. (2013), Si-RNA Chitosan Complexes in Poly (lactide Co-glycolic Acid) Nanoparticles for the Silencing of Aquaporin 1 in Cancer Cells, *Mol. Pharmaceutics*, Doi: 10.1021/mp400224u.

134

Biomaterials in Tissue Engineering

Ashok Kumar and Sumrita Bhat

CONTENTS

I. INTRODUCTION

Developments in the area of biomaterial sciences dates back to about fifty years. Biomaterial science being a very diverse field, different authors has put forward various definitions of it. Most primordial definition was put forward in the first Consensus Conference of the European Society for Biomaterials (ESB) in year 1976. According to ESB a biomaterial was defined as "a nonviable material used in a medical device, intended to interact with biological systems". However, as the time passed EBS has modified the definition of the biomaterials and is now defined as "material intended to interface with biological systems to evaluate, treat, augment or replace any tissue, organ or function of the body". This amendment in the definition of biomaterials is indicative of evolution in the field of biomaterials with time. With time biomaterials have now evolved from merely interacting with living tissue to applicability to the area of regenerative medicine or tissue engineering. There are other definitions of the biomaterials, those are also being frequently used. In general it can be "any material, natural or synthetic that comprises whole or a part of living structure or a biomedical device which performs, amplifies or replaces the function that has been lost by either accident or injury". According to Park and Lakes "biomaterial is any synthetic material which is used to replace part of a living system or to function in intimate contact with the living tissue". In accordance to Clemson University Advisory Board for biomaterials "it is a material which is synthetically and pharmacologically inert and is designed to be implanted within or incorporated with living system". As proposed by Dee et al. these "are the materials that constitutes part of medical implants, extracorporeal devices and disposables, those have been utilized in medicine, surgery, dentistry and veterinary medicine as well as in every aspect of patient health care". In accordance to National Institute of Health (NIH) "it is any substance (other than drug) or a combination of substances, synthetic or natural in origin which can be used for any period of time as a whole or as a part of a system which treats, augments, or replaces any tissue, organ or function of the body". According to Williams (1987) "a biomaterial is a material used in implants or medical device, intended to interact with biological systems". Usage of biomaterials in the form of three dimensional (3–D) scaffolds as implant devices and tissue repair tools has emerged due to their similarity to the native tissue. Normal native tissue is composed of cells those are embedded in extracellular matrix and combination of both forms a 3–D structure which resembles scaffold. Figure 134.1 illustrates the similarity of native tissue with the 3–D scaffold.

Biomaterials play a very critical role in the area of tissue engineering and regenerative medicine. "Tissue engineering

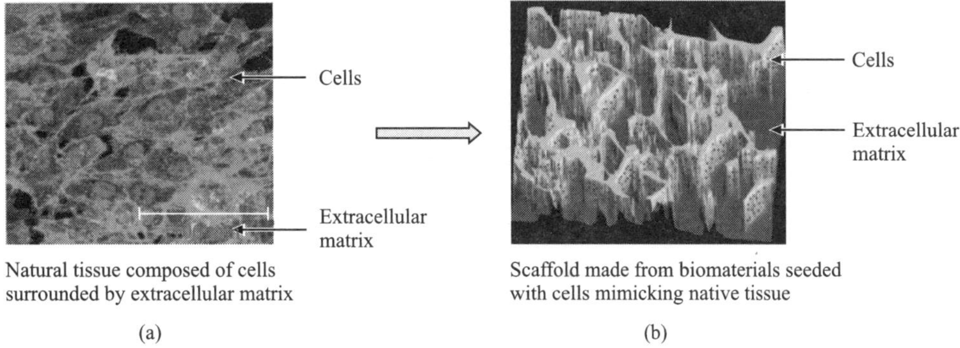

Figure 134.1　Similarity between native tissue and three-dimensional biomaterial scaffold. Biomaterial scaffold (b), mimic the native tissue (a), thus provides a favourable matrix for the cell growth and proliferation. (Reproduced with permission from (http://archive.nrc-cnrc.gc.ca/eng/projects/imi/functional-polymers.html) (*see Plate 54 for colour figure*)

combines the principles and methods of the life science with those of engineering to elucidate the fundamental understanding of structure-function relationships in normal and diseased tissue and to create entire tissue replacements". So phenomenon of tissue engineering ranges from regulating the cellular behaviour towards implants, enhancing the process of healing to maintain the integrity of the damaged tissue or organ, developing the understanding of many biological processes, etc. Biomaterials play a pivotal role in many of these processes like they serve as 3–D matrices to allow the cell migration and proliferation, block antibody penetration into transplants from other species.

Figure 134.2 illustrate the concept of tissue engineering where biopsy tissue is removed from the patient for the isolation of cells. Isolated cells are expanded under *in vitro* laboratory conditions to increase the cell number. Implantation is done in three ways, cells seeded with the scaffold (tissue construct) can be implanted in the patient or neo-tissue generated on the scaffold *in vitro* can be used for the implantation. Another alternative approach is the implantation of the scaffold only without cells and under *in vivo* conditions cells are recruited in the scaffold which enhances the healing process. This concept works for the repair of tissues like skin, liver,

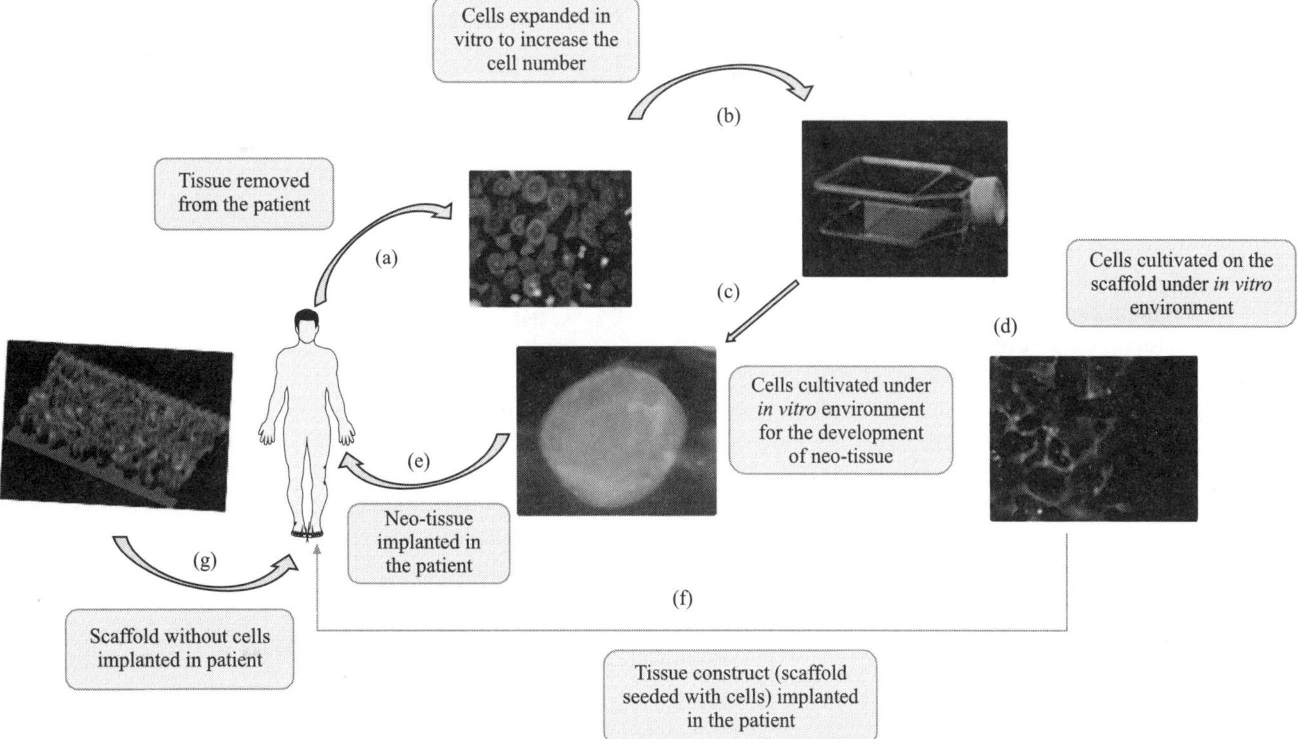

Figure 134.2　Concept of tissue engineering: Tissue sample from the patient is digested to recover the healthy cells (A). Isolated cells are expanded *in vitro* in the laboratory condition to increase the cell number (B). Cells are then cultivated on the three-dimensional scaffolds either for the generation of a neo-tissue (C) or tissue construct (scaffold seeded with cells) (D), for implantation in the patient (E and F). Other tissue engineering strategy is to implant a plain scaffold i.e. without cells in the patient (G). (*see Plate 54 for colour figure*)

etc. where native cells can move into the scaffold to enhance the healing process.

A. History and Evolution of Biomaterials

Biomaterials were initially introduced as 'nonbiological materials' those were used for the benefit of human health. Looking back into the history Mayan people fashioned nacre teeth using sea shells around 600 AD and apparently achieved what is now referred to as bone integration. Furthermore, Egyptians introduced linen sutures for the surgical purpose. Concept of construction of contact lenses using biomaterials dates back to year 1508 with the thoughts of Leo bnardo Do Vinci. Proposal for the construction of artificial heart and organ perfusion dates back to 4th century BC, however, due to limited technological advances during that time such apparatus could not be constructed. In year 1891, German surgeon Theodor Gluck performed first hip replacement using cemented ivory ball as an implant. Furthermore, improvements with the technology led to the development of chrome-based alloys and stainless steel implants with improved mechanical properties. Experimentation with the dental implants dates back to 1809, when Maggiolo implanted gold postanchor into the fresh extraction socket. In year 1901 John Jacob attempted to construct the dialysis unit for kidney ailments particularly for the removal of toxins from the patient's blood. However, major advances in kidney dialysis were made by Dr Scribner during 1921 to 2003 at the University of Washington. First silicon breast implant was constructed during early 1960's by Thomas Cornin and Frank Gerow at the University of Texas. After the initial developments with these implant devices, many variants have been tried over the years. First fully implantable pacemaker was developed by the fusion of engineering and medical concepts by Wilson Greatbatch an engineer by profession and WM Chardack a cardiologist in the year 1959. Use of metals as biomaterials is on decline now due to the availability of other alternatives with the advancement in the technology. Potential alternatives include the use of more natural materials (polymers) those resemble the native tissue matrix as implant medical devices and the transition from the concept of tissue replacement to tissue regeneration with the help of a novel concept of tissue engineering. Both natural and synthetic polymers are now being used as biomaterial for implant devices. Examples include polymethyl-methacrylate (PMMA) as contact lens material, polyvinylchloride as catheters, tubing and blood bags, etc. But search for better and more apposite polymeric biomaterial as medical implant or device is still a focal aspiration for tissue engineers.

B. Market of Biomaterials

Metals and ceramics are the major type of biomaterials those are being used widely for the clinical purpose. The global market for biomaterials is expected to reach $64.7 billion in 2015 from $25.6 billion in 2008 with a compound annual growth rate (CAGR) of 15% from 2010 to 2015. In 2009,

the orthopedic reconstructive biomaterial market contributed 31.8% to the global orthopedic biomaterial market. This market includes artificial knee, artificial hip, and other reconstructive products. The artificial knee domain dominated the market with a 50.4% share. Furthermore, biomaterial market is expected to grow 14% annually through 2017 according to the reports of "Biomaterials - A Global Market Overview" from MarketResearch.com. The global biomaterials market for 2011 is estimated at $37.6 billion and is projected to increase at a CAGR of 14% (2007–2017) to reach $83.9 billion by 2017. Figure 134.3 represents a global market for different tissue engineering products during year 2009–2018.

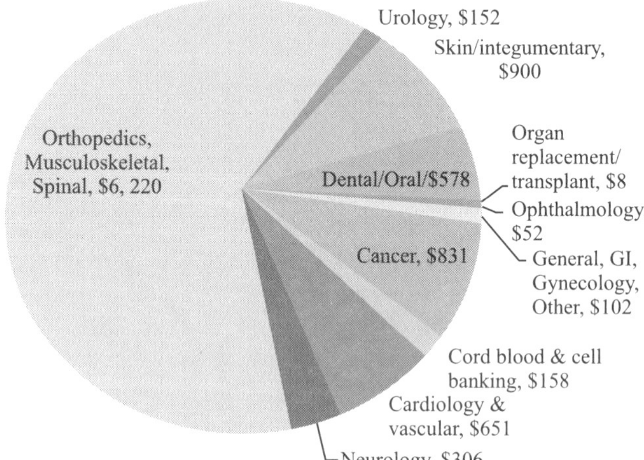

Figure 134.3 Status of biomaterial applications for various tissue/organ regeneration and its market value. Source: MedMarket Diligence, LLC; Report #S520, "Tissue Engineering, Cell Therapy and Transplantation: Products, Technologies & Market Opportunities, Worldwide, 2009–2018." (Reproduced with permission from http://archive. feedblitz.com/248640/~4076446)

C. Design of the Scaffold from Appropriate Biomaterials

As mentioned earlier concept of tissue engineering works on the use of three-dimensional (3–D) scaffolds for the cultivation of appropriate cells on it. The term 'scaffold' refers to the three-dimensional biomaterial before the cultivation of cells on it. These 3–D scaffolds can be fabricated from the appropriate biomaterial for example natural or synthetic polymers, synthetic or natural inorganic ceramics, etc. Suitable biomaterial can be fabricated into a scaffold using the following fabrication techniques:

Solvent-casting particulate-leaching

In this technique solution of polymers like, polylactic acid (PLA) is made in chloroform followed by the addition of salt particles of specific diameter resulting in the formation of a uniform suspension. Solvent is then allowed to evaporate leaving behind the polymeric matrix with salt particles embedded uniformly. Composite is then immersed in water which leaches out salt particles leaving behind a porous 3–D scaffold.

Gas foaming

This technique employs biodegradable polymer like poly (lactic-*co*-glycolic acid) (PLGA), which is saturated with carbon dioxide at a high pressure. Solubility of the gas in the polymer is then decreased rapidly by reducing the pressure of CO_2 to atmospheric level. This reduction in pressure results in nucleation and growth of gas bubbles resulting in the formation of a porous 3–D structure.

Phase separation

During this process a biodegradable synthetic polymer is dissolved in molten phenol or naphthalene followed by the addition of biologically active molecules like, alkaline phosphatase. Temperature of whole reaction mixture is then lowered to produce a liquid-liquid phase separation and quenched to form two phase solid. Solvent is then removed by the process of sublimation resulting in the formation of a porous 3–D scaffold.

Freeze drying

Process uses synthetic polymers like poly(lactic-*co*-glycolic acid) (PLGA) are dissolved in glacial acetic acid or benzene. The resulting mixture is then frozen and freeze dried, resulting in the formation of porous 3–D matrices. Such matrices can also be fabricated using natural polymers like collagen.

Cryogelation

Above mentioned conventional scaffold fabrication techniques have limitations, these are to some extent resolved by newer techniques like 'cryogelation'. Process of cryogelation synthesises three-dimensional scaffolds at a sub-zero temperature. The reaction mixture containing gel forming agents (e.g., polymeric solution) is frozen below few degrees of solvent crystallization point. The resultant frozen system, despite looking as a single solid block, is actually a heterogeneous

system, and contains so called unfrozen liquid microphase (UFLMP) along with the crystals of the frozen solvent. Gel-forming reagents are concentrated in UFLMP, and the process is termed as cryoconcentration. As UFLMP makes up a small proportion of the whole gel, the concentration of gel precursor increases significantly promoting the gel formation. The crystals of frozen solvent acts as porogens (pore forming agents). When these crystals are melted by thawing of the gel at room temperature, they leave voids or macropores filled with the solvent. During the freezing process the solvent crystals grow till they meet the facets of the other crystal, so that after the thawing process the whole system is made up of interconnected pores. The whole process leads to the formation of porous 3–D scaffold. Process of cryogelation is depicted in Figure 134.4.

D. Ideal Properties of the Scaffold for Tissue Engineering Applications

Different scaffolds synthesised using a range of biomaterials and a variety of fabrication techniques have been used in the field of tissue engineering for the regeneration of damaged tissues and organs. During the fabrication of the scaffold there are some key properties those should be considered:

Biocompatibility

The first decisive factor of any scaffold for tissue engineering applications is its biocompatibility both under *in vitro* and *in vivo* conditions. An ideal scaffold should allow the cells to grow and proliferate on it without eliciting any toxic effect on the cells. When used under *in vivo* conditions an ideal scaffold should elicit a negligible immune response and should not degrade into any toxic by-products.

Biodegradability

Any biomaterial or scaffolds intended for tissue engineering applications should have a regulated biodegradability which should match with the formation of a neo-tissue. The by-products released after the degradation should not elicit any toxic effects on the host tissues and organs. As most of the

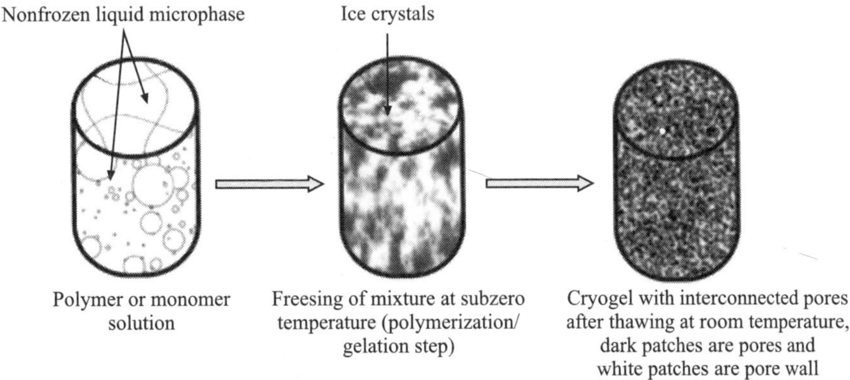

Figure 134.4   Cryogelation process: polymer or monomer is mixed in aqueous solvent and then the whole system is incubated at subzero temperature for cryopolymerization and/or gelation, along with ice crystal formation (ice crystals function as a porogen). (Reproduced with permission from Kumar and Srivastava, 2010)

tissue engineering strategies are entering into the clinical trials now, field of immunology plays an imperative role.

Mechanical properties

An ideal scaffold should have the mechanical properties matching with the native tissue. Looking from the practical perspective an ideal scaffold should have mechanical properties to allow surgical handling during the process of implantation. Fabrication of scaffolds for tissues like cartilage and bone is a major challenge for tissue engineers. For these applications scaffold should have an adequate mechanical stability which should match with the native tissue. From clinical viewpoint one major challenge is that the healing rates vary with the age, for example, in case of young individual's rate of repair and integration is faster which slows down in elderly patients. So an ideal scaffold should exhibit a balance between mechanical strength and porosity to allow cell infiltration and vascularisation.

Surface topology of the scaffold

Design of the scaffold used for tissue engineering applications is of critical importance. An ideal scaffold should have large and interconnected pores to allow adequate cell proliferation and waste exchange along with the deposition of extracellular matrix. Another key condition of an ideal scaffold is the presence favourable chemical groups (ligands) on the surface of the scaffold. Scaffolds fabricated using natural polymers like collagen possess Arg-Gly-Asp (RGD) sequences those favour the cell attachment. In case of scaffolds fabricated from synthetic polymers such ligands should be incorporated. So cell attachment and subsequent proliferation depends on the surface characteristics of the scaffold.

Choice of biomaterial used for fabrication

Final criteria of any scaffold to be used for tissue engineering applications is the choice of biomaterial used for its fabrication. Type of biomaterial used for the fabrication determines the cell attachment and proliferation on the scaffold which further decides its applicability to the area of tissue engineering.

II. CLASSES OF BIOMATERIALS

Biomaterials are classified into different classes depending on various criteria's for example (a) classification based on the host response, (b) conventional classes of biomaterials, and (c) novel classes of biomaterials.

A. Classification of Biomaterials Based on the Host Response

When any foreign material is implanted or introduced into the human body, host tissue reacts to it in a variety of ways depending on the type of material. In general there are three classes in which biomaterials are classified depending on the tissue response:

Bioinert biomaterials

Term bioinert refers to any material which once placed in the host body has minimal interaction with the surrounding tissue, for example, stainless steel, titanium, ultra high molecular weight polyethylene, etc. Generally a fibrous capsule might form around such implants so functionality relies on the tissue integration through the implant.

Bioactive biomaterials

Bioactive refers to the materials those when implanted into the human body interacts with the bone and soft tissue. Examples of such biomaterials include synthetic hydroxyapatite, glass ceramic, bioglass, etc.

Bioresorbable biomaterials

Such materials upon placement within the body start to dissolve and are slowly replaced by advancing tissue such as bone. Examples of bioresorbable biomaterials include tricalcium phosphate, polylactic-polyglycolic acid copolymers, etc. Figure 134.5 is the schematic representation of different types of biomaterials based on the host response.

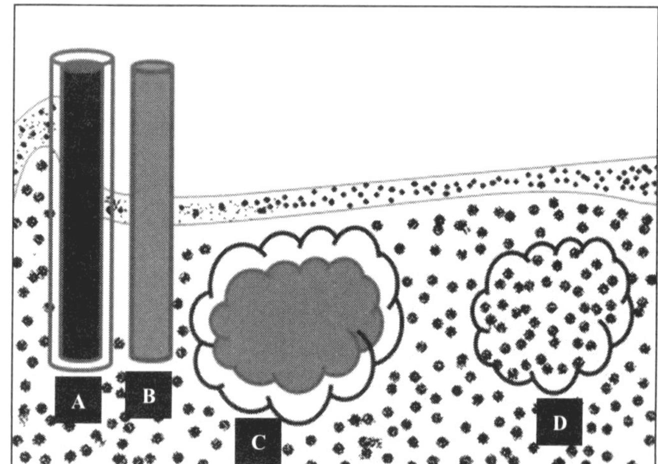

Figure 134.5 Classification of biomaterials according to their bioactivity: (a) bioinert; (b) bioactive hydroxyapatite; (c) surface active; and (d) bioresorbable. (Reproduced with permission from http://www.azom.com/article.aspx?ArticleID=2630). (*see Plate 54 for colour figure*)

B. Conventional Classes of Biomaterials

Conventionally biomaterials can be divided into three major classes; (a) polymers (natural and synthetic), (b) metals, and (c) ceramics:

Polymers

Polymeric biomaterials have many other applications in addition to their applicability in the area of tissue engineering. Examples of polymers those have been conventionally used in the area of regenerative medicine include poly(methyl methacrylate) as bone cements, poly(glycolic acid) as degradable surgical sutures, poly(glycolic-*co*-lactic acid) as bone screws or poly(vinyl siloxane) as dental implants, poly(hydroxyethyl methacrylate) (PHEMA) as soft contact lenses. Polymers having applications in the area of tissue engineering can be categorised as either natural or synthetic. Synthetic polymeric biomaterials being used in the area of regenerative medicine include; silicone rubber (SR), polyethylene (PE), polypropylene (PP), poly(ethylene terephthalate) (PET), polytetrafluoroethylene (PTFE), poly(methyl methacrylate) (PMMA), poly(vinyl chloride) (PVC), poly(hydroxy ethylmethacrylate) (PHEMA), poly(ethylene glycol) (PEG), etc. PMMA a hydrophobic polymer has shown applicability as bone cement for dental applications. Due to the excellent light transmittance of PMMA it has shown applicability as a good material for intraocular lenses (IOL's) and as hard contact lenses. For the fabrication of the soft contact lenses PHEMA is cross-linked with ethylene glycol dimethacrylate (EGDMA). During the fabrication process of contact lenses, polymers like, polymethacrylic acid is added in small quantities to improve the wetability. Polymers like polyacrylic acids are being used as dental cements and as mucoadhesive additives in mucosal drug delivery formulations. Polypropylene (PP) is an isotactic crystalline polymer with high rigidity and good tensile strength and has thus found applicability as surgical sutures. Polyvinyl alcohol (PVA) is being used as tubing (blood transfusion tubes, dialysis tubing, etc) and blood storage bags in health care industry. Biodegradable polymer, like PLGA is used as resorbable surgical sutures, drug delivery systems and in orthopaedic applications as fixation devices. Furthermore, the *in vivo* usage of PLGA is enhanced as the degradation products of PLGA are lactic acid and glycolic acid which are as such nontoxic for the host. Copolymers made from various monomeric units symbolize an important class of material design which has a significant applicability in the area of biomedical science and tissue engineering. A copolymer of tetrafluoroethylene with small amount of hexafluoro-proplylene (FEP Teflon) is used as a tubing connector and as catheters. In addition to the above mentioned synthetic polymers, natural polymers have also shown applications in the area of biomedical science and tissue engineering. There is a range of naturally occurring polymers those can be fabricated into 3–D tissue engineering scaffolds, among them commonly investigated materials include alginate, collagen, hyaluronic acid, fibrin gels, etc. Alginate consists of repeating monosaccharide units, i.e., L-guluronic acid and D-mannuronic acid. Exposure of alginate to calcium ions results in the formation of a 3–D gel. This gel can be used for the encapsulation of growth factors or specific cell type. Scaffolds fabricated from alginate have shown applicability in the area of cartilage regeneration, as chondrocytes are reported to maintain their phenotype in presence of alginate. Alginate gels have also shown enormous applicability for the storage of chondrocytes, as cells can survive for up to 8 months in these gels. Gels made from natural polymers like agarose has shown chondrocytes cultivated together with the production of significant amount of cartilage ECM

which enhances the mechanical strength of the construct. But major limitation with the use of agarose for cartilage tissue engineering is its poor biodegradabiltiy due to the lack of enzymes in mammalian tissues and may also elicit immune response in the host. Natural polymer like collagen has been used for various tissue engineering applications for example, collagen type I is being utilized as a carrier for chondrocytes and mesenchymal stem cells. Many preparations of collagen type I are soluble under acidic conditions and the subsequent neutralization of collagen solution results in the formation of a hydrated collagen gel. Major advantage with the collagen in cartilage tissue engineering applications is that it is a natural physical constituent whose fibrils have the cell adhesion capacity and carry the required biological cues which enhance the cell proliferation and growth. Further degradation products of collagen are nontoxic to host tissue. Collagen scaffolds have been extensively used for the *in vitro* characterization of chondrocytes and stem cell behaviour. Studies have reported that MSCs in collagen gels repaired the full thickness in rabbits with the formation of hyaline cartilage after 12 weeks time. Further collagen fibre scaffolds have been used to deliver chondrocytes treated with gene therapy to injury site. Collagen matrices containing glycosaminoglycans are also under investigation for gene therapy applications. One of the major limitation with the use of collagen as a biomaterial for cartilage tissue engineering is that the scaffolds made from collagen have very less mechanical strength. Other concern with the use of type I collagen as a scaffold material is that most cartilages do not contain type I collagen, so few authors have suggested type II collagen as a better alternative for cartilage tissue engineering. Bovine collagen combined with cellular components like human autologous keratinocytes and fibroblasts has shown good results when applied to severe burn victims. Collagen is also being used in the form of bilayered artificial skin for the purpose of skin regeneration. This template is acellular and is composed of collagen and glycosaminoglycans which is commercially available as integra artificial skin or integra. Other types of skin regeneration templates, e.g. matriderm uses bovine dermal collagen type I, III and elastin. Fibrin is a natural component of animal skin and plays a major role during the process of wound healing. So fibrin in the form of a biomaterial has a potential to be utilized as a scaffold for tissue regeneration. Major disadvantage with fibrin gels is their fast biodegradation which occurs due to the phenomenon of fibrinolysis. Other natural polymers those has shown potential to be used as scaffolding materials for regenerative medicine include cellulose, agarose, chitosan, gelatin, etc.

Metals

Metallic implants are being used traditionally in the area of tissue regeneration particularly for orthopaedic applications. Examples of metallic implants include steels 316, 316L, vitallium, silver, tantalum, cobalt, F-75 and alloys of Ti, Cr + Co, Cr + Co + Mo, etc. Metallic implants manifest few limitations like low biocompatibility, susceptibility to corrosion under physiological environment and large variations in the mechanical properties from the biological tissue. Metallic implants have an enormous applicability in the area of biomedical science as they are used as staples, plaques and wires. They are also being used as tooth implants, as mesh for facial reconstruction, etc.

Ceramics

Ceramics, glasses and glass-ceramics comes under the broad range of inorganic/nonmetallic composites. This class has shown applicability in different areas of regenerative medicine like as eye glasses, diagnostics instruments, chemical ware, thermometers, etc. Ceramics demonstrate characteristics like high biocompatibility, high resistance to corrosion, low electrical and thermal conductivity which makes them suitable as implants. Ceramics like hydroxyapatite (HAp) has shown potential for the formation of new bone tissue and other orthopaedic applications. Other examples of ceramics used as biomaterials include aluminum oxide, calcium aluminates, titanium oxides, calcium phosphate, carbon, bioglass, etc. Disadvantages associated with ceramics include low impact resistance, processing and fabrication difficulties, etc. Furthermore, ceramic composites, i.e., metals with ceramic coatings or materials coated with carbon have found applicability in the area of biomedical science and tissue engineering. Their high biocompatibility and resistance to corrosion make them appropriate implanting devices without eliciting any immune rejection. But their disadvantages associated with ceramics include the lack of consistency and a difficulty in the fabrication process which limits their applicability to the area of biomaterials. Their major biomedical applications include heart valve implants, knee implants, hip implants, etc.

C. Novel Classes of Biomaterials

In addition to the above mentioned classes there are some emerging classes of biomaterials:

Hydrogels as biomaterials

Hydrogels exhibiting a significant biocompatibility and similarity with the tissue components have demonstrated great potential as a novel group of biomaterials. In general hydrogels are 3–D materials those have ability to absorb large amount of water together with the maintenance of structural integrity. Among these major class of hydrogels is stimuli responsive hydrogels. Hydrogels falling under this category undergo relatively large and abrupt changes in their swelling behaviour, permeability, and mechanical properties in response to the small changes in the surrounding environment. Stimuli-responsive hydrogels are further classified as can either be physical or chemical responsive hydrogels. Examples of physical responsive hydrogels include temperature responsive, pressure responsive, magnetic field responsive etc. and examples of chemical responsive hydrogels are pH and glucose responsive, etc.

pH responsive hydrogels: These hydrogels are made up of polymeric backbones those possess ionic groups those can accept or donate electrons in response to the pH changes in the surrounding environment. Examples of pH responsive hydrogels include polyvinyl sulfonic acid (PVSA), polymethacrylic acid (PMAA), polydiethylaminoethyl methacrylate (PDEAEMA) and polydimethylaminoethyl methacrylate (PDMAEMA).

Temperature responsive hydrogels: These hydrogels are further classified as positive or negative responsive systems. Positive temperature responsive systems show phase transition at a critical temperature which is called as upper critical solution temperature (UCST). Hydrogels made from polymers with UCST shrink when they are cooled below their UCST. In case of negative responsive systems they have a lower critical solution temperature (LCST), these hydrogels shrink when heated above LCST. Examples of temperature responsive hyrogels include methylcellulose, chitosan, *N*-isopropylacrylamide (NIPAAm) etc. This class of hyrogels have shown vast applicability in the area of tissue engineering and regenerative medicine. They have shown application in the area of smart drug delivery systems, as injectable scaffolds, etc.

Glucose responsive hyrogels: Most extensively studied system is cationic hydrogels those carry a mixture of insulin and glucose oxidase. In presence of oxygen, glucose oxidase converts glucose to gluconic acid which reduces the local pH thus increases the swelling of cationic hydrogels which subsequently releases insulin. These responsive systems have shown potential for intelligent drug delivery systems for diabetic patients. Hydrogels manifest certain advantages those have made them ideal for tissue engineering applications. Hydrogels represent an aqueous environment which mimics the internal milieu of the cell. Most of the hyrogels are porous 3–D structures which allow the nutrient and waste exchange. Hydrogels fabricated using natural and synthetic polymers have shown applicability in the area of tissue engineering for the repair of cartilage, tendon, ligament, skin, etc.

Cryogels as biomaterials

Recently a new class of polymeric biomaterials called cryogels have shown potential in the area of tissue engineering and regenerative medicine. Cryogel is a system made up of large interconnected pores with spongy morphology. The pore system in such sponge like gels ensures unhindered convectional transport of solutes contrary to the diffusion mechanism of solutes in traditional homophasic gels. The pore size of macroporos within the cryogels varies 10 to 100 μm which depends on the cryogenic regime. Due to the interconnected system of large pores cryogels have the potential to be used as chromatographic matrices those can be used for the separation of biological nano- and microparticles. Cryogels being macroporous support for the infiltration of mammalian cells which suggests that these matrices can be used for the tissue engineering applications. Cryogel matrices are synthesised at a sub-zero temperature using either natural or synthetic polymers. As the process of cryogelation is simple, such matrices can be synthesised from range of natural and synthetic polymers. Cryogels have shown potential in the area of tissue engineering, biotechnology, regenerative medicine, etc. Organic/inorganic composite cryogel scaffolds of polyvinyl alcohol–tetraethylorthosilicate–alginate–calcium oxide (PTAC) have shown potential to support the growth and proliferation of human osteoblast cell lines which qualifies its applicability to bone tissue engineering. On similar lines cryogels matrices fabricated using natural polymers like gelatine, agarose, chitosan, etc. have shown the potential for cartilage tissue engineering. Furthermore ongoing research on cryogels is exploring their potential in other tissue engineering areas like skin, cardiac, liver, etc. Figure 134.6 (A, B and c) illustrate the different formats of cryogels, Figure 134.6 (D, E and F) demonstrate the internal architecture of cryogel matrices.

Functional biomaterials

Term 'functional biomaterials' refers to a broad class of biomaterials those carry biologically relevant functions. This class of biomaterials combine biological molecules such as proteins, peptides, nucleic acids, etc. those can interact with biological entities like cells in order to modulate the biological processes. Such modified materials are also termed as 'biomimetic biomaterials'. During earlier times long chains of extracellular matrix proteins such as fibronectin, vitronectin, and laminin were used for the surface modification. Biomaterials coated with these molecules promote cell adhesion and proliferation. Such modified biomimetic materials have shown potential as ideal tissue engineering scaffolds, they serve as artificial ECM which supports the cell proliferation and formation of a new tissue. Functionalization of a biomaterial surface with a peptide is extensively used in the area of tissue engineering. Commonly used peptides for surface modification include Arg-Gly-Asp (RGD), Tyr-Ile-Gly-Ser-Arg (YIGSR), Arg-Glu-Asp-Val (REDV), and Ile-Lys-Val-Ala-Val (IKVAV). These peptides are immobilized on the biomaterial surface using a bi-functional crosslinker that has a long spacer arm, which can further enable the free movement of the immobilized peptide in biological environment. Efficiency of the modification can further be enhanced by the bulk modification of biomaterials. With bulk modification signalling peptides are incorporated into the biomaterials, so recognition sites are present not only on the surface but also in the bulk material. Such bulk modified materials have applicability as injectable biomaterials where they are required to maintain a complex shape like native tissue at the defect site. Ideal biomaterial for bone tissue engineering should have osteoconductive and osteoinductive properties. To explore the role of RGD peptides in bone tissue engineering a RGD sequence derived from bone sialoprotein was immobilized on glass plates. Osteoblasts seeded on such functionalized surfaces exhibited a rapid cell attachment and proliferation. Growth factors like osteopontin-1 (OP-1) play a very crucial role in bone tissue engineering, they exhibited synergistic enhancement of mineralization of osteoblasts when cultured on RGD modified surfaces. In addition to the RGD peptides other adhesive peptides recognised by the polysaccharides on the cell membrane can also be used to modify the biomaterial surfaces for bone tissue

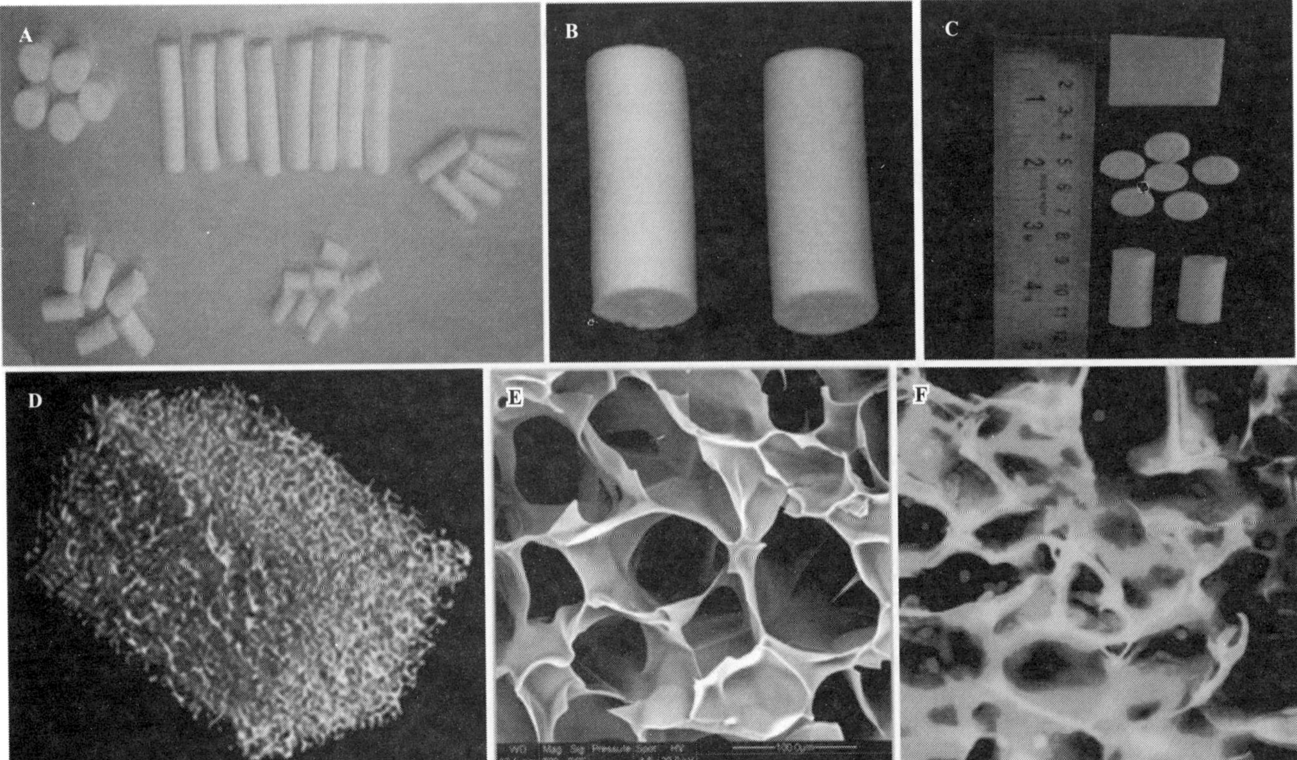

Figure 134.6 Cryogels in different formats (A, B and C); three-dimensional micro CT structure of cryogel scaffold (D); scanning electron micrograph of a typical cryogel scaffold (E); and fibroblasts on the surface of three-dimensional cryogel surface (F). (Figure 134.3F reproduced with permission from Bhat and Kumar 2012). (*see Plate 54 for colour figure*)

engineering. Recently heparin binding domain in fibronectin has been investigated for the attachment of osteoblasts and other bone derived cells. Commonly used osteogenic growth factors include members of transforming growth factor-β (TGF-β), bone morphogenetic proteins (BMPs) etc. Functinal biomaterials have shown utility in the area of neural tissue engineering. Extensive research work is being carried out for the delivery of neurotrophic growth factors or genetically modified cells at the site of nerve lesion to stimulate the neurite regeneration. Additionally biomimetic scaffolds those possess biological cues for neural tissues are being constructed for neural tissue engineering. Laminin is being widely employed to generate the biomimetic surfaces because it is active promoter of neural cell attachment, differentiation and neurite outgrowth. Other peptides like YIGSR when covalently incorporated into agarose gels have shown potential for the regeneration of nerves under *in vivo* environment. Biomimetic surfaces have shown immense potential in the area of cardiovascular tissue engineering. ECM proteins and other types of peptides have been used to generate biomimetic surfaces to allow the binding of endothelial cells (EC). For example expanded polytetrafluoroethylene (ePTFE) coated with fibronectin and RGD improve the EC attachment. Other types of peptides for example YIGSR immobilized on the surface of poly(ethyelene terephtalate) showed enhanced EC attachment. Other area which is being explored for cardiovascular tissue engineering is the modification of vascular grafts to modulate production of ECM proteins during tissue formation by providing biological signals which can regulate cellular interactions. Area of genetic engineering is also being explored for the fabrication of biomimetic surfaces those have an ability to mimic a complex biological functions in a regulated manner. Example of genetically modified surfaces includes the preparation of artificial ECM proteins by using specific genes to direct the synthesis of polymers with a controlled architecture addressing problems associated with synthetic vascular grafts.

Smart biomaterials

Stimuli-sensitive or smart biomaterials are the systems those may undergo significant property changes in response to minor changes in the external environment. These materials show large conformational changes in response to external stimuli like temperature, ionic strength, pH, etc. Response of the polymer may include precipitation, gelation or alteration between hydrophilic and hydrophobic states, etc. Smart polymers have shown applicability in the areas of bioseparation, biosensors, microfluidics, tissue engineering, etc. One approach to use smart biomaterials for tissue engineering application is to combine them with proteins. Specific protein can be conjugated either randomly or in a site specific manner. If a conjugated site is introduced near ligand binding domain of a protein it induces conformational changes in smart polymer which further regulates the activity of the protein. Smart polymeric biomaterials have shown applications in the area of bone and cartilage tissue engineering. In the area of tissue engineering, smart polymers are also used in the form of smart surfaces to coat cell culture dishes. During cell culture

experiments mammalian cells are cultivated on hydrophobic solid surfaces and they are detached from the plate by treatment with proteases which often damages the cellular integrity by hydrolysing various membrane associated proteins. This can be avoided by growing the cells on smart surfaces for example poly(*N*-isopropylacrylamide) (PNiPAAm) grafted surface. Such surfaces are hydrophobic at 37 °C at which cells grow and decrease in temperature results in transition of the surface to hydrophilic state which dislodges the cells without the use of harmful enzymes. As cell to cell contacts are maintained, cells are recovered in the form of a cell sheet which can be used in the repair of damaged tissues. Cardiomyocyte sheets cultured on such grafted surfaces have been used for the constriction of myocardial tissue for cardiac tissue engineering. Such sheet can also be used in the form of double layered co-culture sheets. Representative example is cell sheet of confluent human aortic endothelial cells recovered from PNiPAAm-grafted surfaces which was sandwiched with hepatocyte layer. It was observed that hepatocytes maintained functionality for a longer duration when cultivated in the form of a double layered co-culture system. Figure 134.7 represent different types of cell sheets prepared on smart polymeric surfaces.

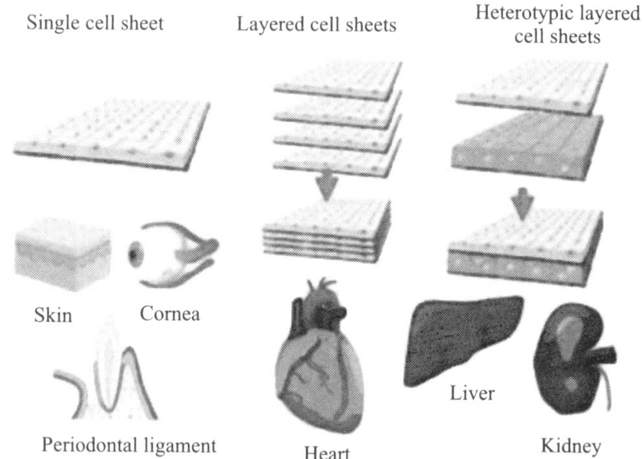

Figure 134.7 Tissue constructs using cell sheet engineering. These sheets are prepared on smart polymeric surfaces. Single cell sheets are prepared using a single monolayer from one type of tissue or organ. Layered cell sheets are made by stacking different monolayers of cells derived from one type of tissue or organ over each other. While heterotypic cell sheets are prepared by stacking the cells derived from different organs or tissues over each other. (Reproduced with permission from http://www.spsj.or.jp/c5/kobunshi/kobu2009/hottopics0901.html). (*see Plate 55 for colour figure*)

Nanobiomaterials

Field of biomaterials has grown and evolved from their usage as implant devices to their utility as vehicles to deliver nano and large bioactive molecules to a specific tissue site in order to restore its functionality. Accomplishments in the area of small scale technologies have seen clinical relevance for the diagnosis and treatment of diseases. Biomaterials when reduced in the size to a scale of nano are termed as nanobiomaterials. Advances in this area are possible due to the development of devices like BioMEMS (Biological Microelectrical Mechanical Systems) and BioNEMS (Biological Nanoelectrical Mechanical Systems). These devices are being used as intraocular pressure sensors, microneedles, lab-on-a-chip, biochips, etc. Nanobiomaterials have also been used as drug carriers and for tissue engineering applications. These nano sized carriers are used for the delivery of highly toxic, hydrophobic therapeutic agents.

Types of nanobiomaterials

Metal nanobiomaterials: Multifunctional metal nanoparticles show promising applicability for medical advancements like drug delivery, biomedical diagnosis and therapeutics, etc. due to their small size and opposite properties. Metal nanoparticels like gold nanoparticels (GNP) have shown application in the delivery of peptides, proteins and nucleic acids (DNA or RNA). Conjugation of DNA with GNP has shown an efficient delivery in mammalian cells like 293T. Other type of metal nanobiomaterials are quantum dots, it is a portion of matter e.g. semiconductors whose excitons are confined in all three spatial dimensions. Quantum dots are used for the detection of tumors in the area of biomedical diagnosis. They provide a nanoscle scaffold for the delivery of siRNA and for imaging. Quantum dots are also being used in the drug delivery and treatment. Aquasomes (carbohydrate-cerami nanoparticles) are the type of metal nanobiomaterials those contain particle core which is composed of nanocrystalline calcium phosphate or ceramic diamond whole structure is covered by a polyhyroxyl oligomeric film. They range in size from 60–300 nm and are used for drug and antigen delivery. They are also used for the delivery of other molecules like peptides, proteins, hormones, antigens, etc. Aquasomes can be fashioned using tin oxide, nanocrystalline carbon ceramics and calcium phosphate dihydrate. Haemoglobin loaded aquasomes (in hydroxyapatite core) has shown potential as an artificial oxygen carrying system.

Non-metal nanobiomaterials: Among the non-metallic nanobiomaterials silica materials with defined structure and surface properties exhibit biocompatibility which approves their usage for biomedical applications like artificial implants, etc. Other types of non-metal nanobiomaterials have application as nanobiosensors. Examples include carbon nanofibers (CNFs) with diameter ranging from 3–150 nm. They are arranged as stacked cones, cups or plates. CNFs have shown application in skin tissue engineering for the healing of wounds in diabetic ulcers, electrospun NFs cultivated with human epidermal growth factor (EGF) was used for this purpose. CNFs are also being used as neural and orthopaedic implants. Other class of non-metal biomaterials are carbon nanotubes (CNT). CNFs with graphene layers wrapped into perfect cylinders are termed as CNT. They have been used as vaccines, delivery systems or protein transporters. CNTs with suitable functionalization and

intrinsic optical properties can show a potential in the area of nanomedicines by combining drug delivery and cancer therapy.

Polymeric nanobiomaterials: Representative example in this class of nanobiomaterials is dendrimers. Multiple functionalization on the surface of dendrimers makes them ideal carriers for the targeted drug delivery. They range in size (diameter) from 10–100 nm. Other example of polymeric nanobiomaterial is polymeric micelles which have shown potential as drug carriers. They are usually less than 50 nm in diameter. Polymeric micelles are extensively used for the delivery of anticancer drugs at the targeted site.

Lipid nanobiomaterials: Examples includes liposomes, these are spherical vesicles those possess different targeting ligands attached to their surface. Liposomes being biocompatible are used as delivery systems. Commonly used materials for the fabrication of liposomes are phospholipids, glycolipids, sterols, cationic lipids, etc. Liposomes are also being studied as non-viral carrier for genes. Lipid nanobiomaterials like lipopolyplexes or polyplexes are used as carrier systems designed for site specific delivery and controlled release of macromolecules like proteins and DNA. They are generally used for the transfection of cells *in vitro* and *in vivo*.

Virus based nanobiomaterials: Viruses are being used for nanotechnological applications as biomaterials, vaccines, imaging tools, etc. Examples include cowpea mosaic virus (CPMV), cowpea chlorotic mottle virus (CCMV), MS2 bacteriophage, M13 bacteriophage, etc. CCMV particles are being studied for the development of image contrast agents. Furthermore deeper understanding of the structure, dynamics and *in vivo* behaviour of virus particles can open new avenues for biomedical nanotechnological applications. Figure 134.8

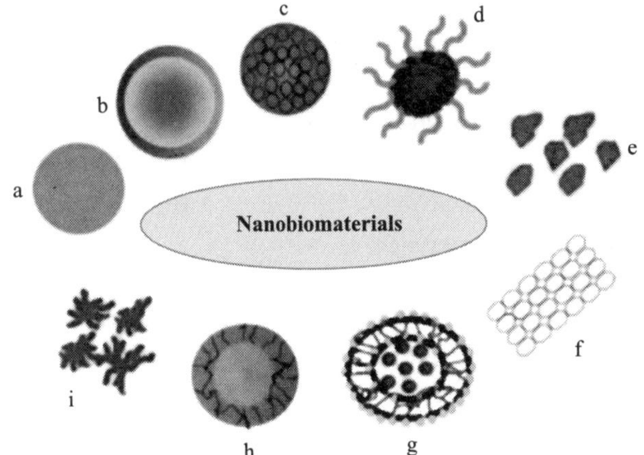

Figure 134.8   Different types of nanobiomaterials: (a) multifunctional nanoparticle; (b) quantum dot; (c) aquasomes; (d) polyplexes/lipopolyplexes; (e) superparamagnetic iron oxide crystals; (f) carbon nanomaterial; (g) liposomes; (h) polymeric micelles; and (i) dendrimers. (Reproduced with permission from Veerapandian and Yun 2009). (*see Plate 55 for colour figure*)

represents different types of nanobiomaterials used in the area of drug delivery, tissue engineering, regenerative medicine, etc. Table 134.1 lists down the nanobiomaterials those have application in the area of medicine and biology.

III. CELL AND BIOMATERIAL INTERACTIONS

Cultivation of cells out of their natural environment i.e., *in vitro* involves a deep understanding of the biochemical and biophysical aspects of cell growth and differentiation. Under *in vivo* conditions cell is linked to its external environment which governs the development of tissue and organs. Cells cultivated on the surface of three dimensional biomaterial scaffolds are more close to the native tissue architecture, therefore they tend to manifest their natural behaviour. Major challenge during the *in vitro* cultivation of cells is the supply of soluble molecules or growth factors within 3–D scaffold. In order to answer this challenge, plain surfaces are modified by the incorporation of growth factors those may be released in a controlled fashion to regulate the cell growth and differentiation. Advances in the area of biomaterials have led to the development of newer biomaterials those allow stem cells to infer the biomaterial signals and modify their fate accordingly. To direct the stem cells to a particular fate certain properties of biomaterials like surface topography, mechanical strength etc., play a very critical role.

A. Properties of Biomaterials

Surface properties

Different research groups have reported that surface topography of the biomaterial can drive cell adhesion, proliferation, migration, etc. Physical processes like radiofrequency oxygen plasma treatment has shown to change the surface properties of polylactic acid (PLLA) scaffolds. These modifications in the surface properties increase the roughness and alter the wettability which improves the stem cell attachment. Presence of ridges and groves on the scaffold can also influence the stem cell behaviour.

Mechanical properties

Material compliance or stiffness influences the behaviour of stem cells. During last two decades developments in the area of nanocomposites has played a major role in upgrading the structural and functional properties of synthetic polymers. Nanocomposite biomaterials are made by the combination of polymers and organic/inorganic fillers. Process is influenced by several microstructural parameters like matrix properties, distribution of fillers, synthesis process, etc. For example hydroxyapatite (HA) nanoparticles grafted with polylactic acid (PLA) exhibited improved mechanical properties.

TABLE 134.1 List of nanobiomaterials used for medical and biological applications (Modified from Veerapandian and Yun, 2009).

Nanobiomaterials	Medical/biological applications
Metal Nanobiomaterials	
1. Multifunctional metal nanoparticles	Delivery of biomolecules, (proteins, peptides, nucleic acid) therapeutics, biosensor, etc.
2. Quantum dots	Optical coding in gene expression studies, high throughput screening, in vivo imaging.
3. Aquasomes	Delivery of xenobiotics and antigens.
Non-metal nanobiomaterials	
1. Mesoporous silica nanoparticles (MSN)	Intracellular controlled release delivery.
2. Selenium nanocomposites	High photoconducting, xerography, nano-building block for nanomedicine.
Carbon nanobiomaterials	
1. Carbon nanofibers	Biopolymer composites, biocatalysts, chemical/biochemical sensing, neural and orthopedic implants.
2. Carbon nanotube	Delivery of drugs to specific cells.
Polymer nanobiomaterials	
1. Dendrimers	Delivery of therapeutics.
2. Polymeric micelles	Systemic delivery of water-insoluble xenobiotics.
3. Polysccharide nanobiomaterials	Biocompatible, biodegradable, highly stable, safe, non-toxic for variety of drug delivery (hemoglobin, anti-cancer agents, vitamins, lipids, fatty acids, amino acids, etc.)
Lipid nanobiomaterials	
1. Liposomes	Specific delivery of xenobiotics.
2. Lipopolyplexes/polyplexes	Transfection, gene therapy, gene transfer.
Virus nanobiomaterials	Tissue-specific targeting and imaging agents *in vivo*.

Electrostrictive properties

Electrostrictive or electroactive properties refer to changes in the size or shape of the polymer after electrical stimulation. These polymers are referred to as electroactive polymers (EAPs). Example of electroactive polymer which is used as a scaffold for tissue engineering application is a copolymer prepared by the condensation and polymerization of hydroxyl-capped poly(L-lactide) (PLA) and carboxyl-capped aniline pentamer (AP). These polymers influence the differentiation fate of the stem cells.

Scaffold morphology

Properties of the scaffold like interconnectivity pore-size and shape play a crucial role in stem cell biomaterial interaction. Scaffolds with large and interconnected pores are ideal for tissue engineering applications. These scaffolds allow the proper diffusion of nutrients and exchange of wastes.

Polymeric nanomaterials

Polymeric nanoparticles include smart nanosystems, nanoparticles, and nanoshells those are fabricated using biodegradable polymers. Smart polymeric nanosystems show large conformational changes in response to changes in the external environment. So smart polymeric nanosystems have immense potential in the area of tissue engineering and drug delivery. Figure 134.9 is the diagrammatic representation of properties of biomaterials and its effect on the stem cell behaviour.

B. Utility of Stem Cell-biomaterial Interactions in Tissue Engineering

Bone tissue engineering

Due to limitations with the conventional treatment for the repair of damaged bones tissue engineering has emerged as a probable solution. As cell source, stem cells are generally being used for such applications. Procedure consists of isolation and expansion of mesenchymal stem cells (MSCs) from patients and their seeding on a porous biodegradable 3–D scaffolds. During the *in vitro* cultivation of the cells on the scaffold they are supplied with growth factors (stimulators) those play a role in their differentiation to osteogenic lineage. Ideal biomaterials for bone tissue engineering are hydroxyapatite (HA), ceramics and tricalcium phosphate (TCP), etc.

Neural tissue engineering

As axons have a limited self repair capacity, there is need for novel therapies for nerve tissue repair. Current treatment strategies include the use of nerve autografts which has its limitations like donor site morbidity, limited availability, limited recovery, etc. So neural tissue engineering has emerged as a prospective approach for the repair of nerve tissue. Number of natural and synthetic biopolymers such as PLLA, PLGA, PCL, collagen and chitosan are being used for nerve repair.

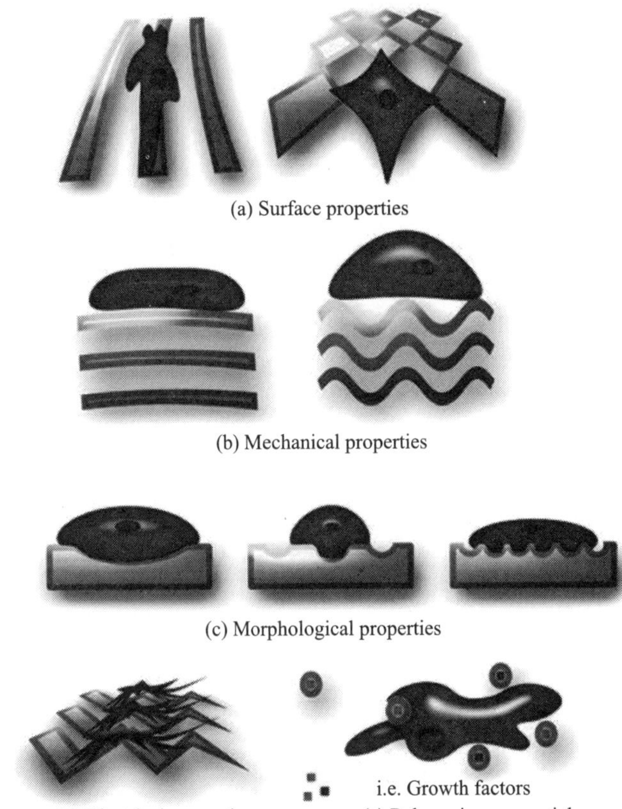

Figure 134.9 Scaffold properties; (a) surface properties; (b) mechanical properties; (c) morphological properties; (d) electrical properties influence the behaviour of stem cells; (e) polymeric nanoparticles (smart nanosystems). (Reproduced with permission from Martino et al., 2012). (*see Plate 55 for colour figure*)

Scaffolds with controlled surface topology are being used for the differentiation of stem cells to neural lineage. In this direction hydrogenated amorphous carbon (a-C:H) groove topographies with width/spacing ridges ranging from 80/40 µm, 40/30 µm and 30/20 µm and depths of 24 nm were used to generate neural cells from hBM-MSCs under *in vitro* environment. Recently a class of nanobiomaterials, i.e., carbon nanotubes (CNTs) are being used for the neural regeneration. Positively charged CNTs have shown to promote greater neurite outgrowth of hippocampal neurons under *in vitro* conditions. CNTs are biocompatible and can act a permissive scaffold for neural tissue growth, so have huge potential for neural tissue engineering.

Skeletal and cardiac tissue engineering

Accidental injuries can interfere with the process of muscle contraction by damaging skeletal muscles. Natural healing process can lead to the formation of a scar tissue which can impede muscle functions. So tissue engineering approach which uses 3–D scaffolds is being used for the repair of skeletal tissue. Electrospum chitosan microfibers are being used for the muscle repair together with other types of biomaterials like PLLA, etc. Cardiac tissue engineering aims to repair the damaged cardiac tissue using tissue engineering approach. Functionalized scaffolds are being used as scaffolds for cardiac tissue engineering; these scaffolds are fabricated in such a manner that they mimic the microarchitecture of the native tissue. Other types of modified scaffolds (possessing proteins or peptides on the surface) are also being used in the area of cardiac tissue engineering.

SUMMARY

Tissue engineering which is an emerging research area has shown a huge potential for the repair of damaged and diseased tissues. Tissue engineering uses 3–D scaffolds on which particular cell type grows and proliferates. These 3–D scaffolds are fabricated using different biomaterials. Biomaterial is any material intended to interface with biological systems to evaluate, treat, augment or replace any tissue, organ or function of the body. Biomaterials include ceramics, metals or polymers (both natural and synthetic). These biomaterials are fabricated into 3–D scaffolds with the aid of various fabrication technologies. But due to the limitations with the conventional fabrication technologies *cryogelation* has emerged as a novel and potential technique for the fabrication of 3–D scaffolds for tissue engineering applications. Cryogel matrices being highly porous and interconnected allow proper cell proliferation, nutrient diffusion and waste removal. Other novel classes of biomaterials those have shown potential in the area of tissue engineering are smart and functional polymers, those have shown applications for tissue engineering and drug delivery. Biomaterials when reduced in size to nano-scale are called nanobiomaterials have shown prospectus in the area of drug delivery, tissue engineering, etc. Furthermore, biomaterials interact with stem cells and control their fate to a particular lineage. Properties of the scaffold play a very crucial role in deciding the lineage, so scaffolding material is crucial in the area of stem cell usage in tissue engineering. Stem cells are being used along with the biomaterials for the repair of tissues like bone, nerves etc. Stem cells hold a huge potential to be used for the repair of other tissue which is a major goal for tissue engineers in future.

REFERENCES AND FURTHER READINGS

1. O'Brien F.J. (2011), Biomaterials and Scaffolds for Tissue Engineering, *Material Today*, 14:88–95.

2. http://www.u.arizona.edu/~trouard/courses/bme516/biomater_lec1.pdf

3. Williams D.F. (Ed). (1987), Definitions in Biomaterials: Proceedings of a Consensus Conference of the European Society for Biomaterials. *Chester, Progress in Biomedical Engineering*, Elsevier, Amsterdam.

4. Ratner B.D., Hoffman A.S., Schoen F.J. and Lemons J.E. (Eds.) (2004), *Biomaterials Science: An Introduction to Materials in Medicine*, 2nd ed., Academic Press.

5. Williams D.F. (2003), Biomaterials and Tissue Engineering in Reconstructive Surgery, *Sadhana*, 28, 563–574.

6. Mikos A.G., Thorsen A.J., Czerwonka L.A., Bao Y. and Langer R. (1994), Preparation and Characterisation of Poly(L-lactic Acid) Foams, *Polymer*, 35:1068–1077.

7. Mooney D.J., Baldwin D.F., Suh N.P., Vacanti J.P. and Langer R. Novel Approach to Fabricate Porous Sponges of Poly(D,L-lactic Co-glycolic Acid) Without the Use of Organic Solvents, *Biomaterials*, 17:1417–1422.

8. Lo H., Ponticiello M.S. and Leong K.W. (1995), Fabrication of Controlled Release Biodegradable Foams by Phase Separation, *Tissue Engineering*, 1:15–28.

9. Hsu Y.Y., Gresser J.D., Trantolo D.J., Lyons C.M., Gangadharam P.R.J. and Wise D.L. (1997), Effect of Polymer Foam Morphology and Density on Kinetics of *in vitro* Controlled Release of Isoniazid from Compressed Foam Matrices, *Journal of Biomedical Material Science*, 35:107–116.

10. Lozinsky V.I., Galaev I.Y., Plieva F.M., Savina I.N., Jungvid H. and Mattiason B. (2003), Polymeric Cryogels as Promising Materials of Biotechnological Interest, *Trends in Biotechnology*, 21:445–451.

11. Kuamr A., Mishra R., Reinwald Y. and Bhat S. (2009), Cryogels: Freezing Unveiled by Thawing, *Materials Today*, 13:42–44.

12. Rogero S.O., Malmonge S.M., Lugao A.B., Ikeda T.I., Miyamaru I. and Cruz A.S. (2003), Biocompatibility Study of Polymeric Biomaterials, *Artificial Organs*, 27:424–427.

13. Bushetti S., Singh V., Raju S. and Atharjaved Veermaram (2009), Stimuli Sensitive Hydrogels: A Review, *Indian Journal of Pharmaceutical Education and Research*, 43:241–250.

14. Patel A., Fine B., Sandig M. and Mequanint K. (2006), Elastin Biosynthesis: The Missing Link in Tissue-engineered Blood Vessels, *Cardiovascular Research*, 71:40–49.

15. Mishra R. and Kumar A. (2011), Inorganic/organic Biocomposite Cryogels for Regeneration of Bone Tissue, *Journal of Biomaterial Science*, Polymer Edition, 22:2107–2126.

16. Bhat S., Tripathi A. and Kumar A. (2010), Supermacroprous Chitosan-agarose-gelatin Cryogels: *In vitro* Characterization and *In vivo* Assessment for Cartilage Tissue Engineering, *Journal of Royal Society Interface*, 8:540–554.

17. Kumar A. and Srivastava A. (2010), Cell Separation Using Cryogel-based Affinity Chromatography, *Nature Protocol*, 5:1737–1747.

18. Humphries M.J., Akiyama S.K., Komoriya A., Olden K. and Yamada K.M. (1986), Identification of an Alternatively Spliced Site in Human Plasma Fibronectin that Mediates Cell Type Specific Adhesion, *Journal of Cell Biology*, 103:2637–2647.

19. Hubbell J.A., Massia S.P., Desai N.P. and Drumheller P.D., Endothelial Cell-selective Materials for Tissue Engineering in the Vascular Graft via A New Receptor, *Nature Biotechnology*, 9:568–572.

20. Sakiyama-Elbert S.E. and Hubbell J.A. (2000), Controlled Release of Nerve Growth Factor from a Heparin-containing Fibrin-based Cell Ingrowth Matrix, *Journal of Controlled Release*, 69:149–158.

21. Panitch A., Yamaoka T., Fournier M.J., Mason T.L. and Tirrell D.A. (1999), Design and Biosynthesis of Elastin-like Artificial Extracellular Matrix Proteins Containing Periodically Spaced Fibronectin CS5 Domains, *Macromolecules*, 32:1701–3.

22. Stayton P.S., Shimoboji T., Long C., Chilkoti A., Chen G., Harris J.M. and Hoffman A.S. (1995), Control of Protein–ligand Recognition Using A Stimuli-responsive Polymer, *Nature*, 378:472–474.

23. Yang J., Yamato M., Kohno C., Nishimoto A., Sekine H., Fukai F. and Okano T. (2005), Cell Sheet Engineering: Recreating Tissues Without Biodegradable Scaffolds, *Biomaterials,* 26:6415–6422.

24. Sandhu K.K., McIntosh C.M., Simard J.M., Smith S.W. and Rotello V.M. (2002), Gold Nanoparticle-mediated Transfection of Mammalian Cells, *Bioconjugate Chemistry*, 13:3–6.

25. Dahan M., Lévi S., Luccardini C., Rostaing P., Riveau B. and Triller A. (2003), Diffusion Dynamics of Glycine Receptors Revealed by Single-quantum Dot Tracking, *Science*, 302:442–445.

26. Manna L., Scher E.C., Li L.S. and Alivisatos A.P. (2002), Epitaxial Growth and Photochemical Annealing of Graded CdS/ZnS Shells on Colloidal CdSe Nanorods, *Journal of American Chemical Society*, 124:7136–7145.

27. Bianco A. and Prato M. (2003), Can Carbon Nanotubes be Considered Useful Tools for Biological Applications, Advanced Materials, 15:1765–1768.

28. Kam N.S.W. and Dai H. (2005), Carbon Nanotubes as Intracellular Protein Transporters: Generality and Biological Functionality, *Journal of the American Chemical Society*, 127:6021–6026.

29. Bao A., Phillips W.T., Goins B., Zheng X., Sabour S., Natarajan M., Ross Woolley F., Zavaleta C.R.A. and Otto R.A. (2006), Potential Use of Drug Carried-liposomes for Cancer Therapy via Direct Intratumoral Injection, *International Journal of Pharmacy*, 316:162–169.

30. Li S. and Huang L. (1997), *In vivo* Gene Transfer via Intravenous Adminis—Tration of Cationic Lipid-protamine-DNA (LPD) Complexes, *Gene Therapy*, 4:891–900.

31. Manchester M. and Singh P. (2006), Virus-based Nanoparticles (VNPs): Platform Technologies for Diagnostic Imaging, *Advanced Drug Delivery Reviews*, 58:1505–1522.

32. Sands R.W. and Mooney D.J. (2007), Polymers to Direct Cell Fate by Controlling the Microenvironment, *Current Opinion in Biotechnology*, 18:448–453.

33. Kumari A., Kumar Y.S. and Yadav S.C. (2010), Biodegradable Polymeric Nanoparticles-based Drug Delivery Systems, Review, *Colloids and Surface B: Biointerfaces*, 75:1–18.

34. D'Angelo F., Armentano I., Mattioli S., Crispoltoni L., Tiribuzi R., Cerulli G.G., Palmerini C.A., Kenny J.M., Martino S. and Orlacchio A. (2010), Micropatterned Hydrogenated Amorphous Carbon Guides Mesenchymal Stem Cells Towards Neuronal Differentiation, *European Cells and Materials*, 20:231–44.

35. Martino S., D'Angelo F., Armentano I., Kenny J.M. and Orlacchio A. (2012), Stem Cell-biomaterial Interactions for Regenerative Medicine, *Biotechnology Advances*, 30:338–351.

36. Langer R. and Vacanti J.P. (1993), Tissue Engineering, *Science*, 260:920–926.

37. Hubbell J.A. (1995), Biomaterials in Tissue Engineering, *Nature Biotechnology*, 13:565–575.

38. Bhat S., Kumar A. (2012), Cell Proliferation on Three-dimensional Chitosan-agarose-gelatin Cryogel Scaffolds for Tissue Engineering Applications, *Journal of Bioscience and Bioengineering*, 114:663–670.

39. Veerapandian M. and Yun K. (2009), The State of the Art in Biomaterials as Nanobiopharmaceuticals, *Digest Journal of Nanomaterials and Biostructures*, 4:243–262.

40. Bhat S. and Kumar A. (2012), Biomaterials in Regenerative Medicine, *Journal of postgraduate medicine education training research*, 46: 81–89.

41. Kumar A., Bhat S., Singh V. and Kumar R., *Bioengineering: Its Applications in Medicine and Agriculture.* (In Press)

42. Basu B., Katti D. and Kumar A. (Eds.). (2009), Advanced Biomaterials: Fundamentals, Processing and Applications, John Wiley and Sons Inc. NJ, USA.

43. Mattiasson B., Kumar A. and Galaev I. Yu. (Eds.). (2009), *Macroporous Polymeric Materials: Production, Properties and Biotechnological/ Biomedical Applications*, CRC Press, Taylor and Francis Group, Boca Raton, USA.

44. Kumar A. and Tripathi A. (2011), Biopolymeric Scaffolds for Tissue Engineering, In: Tiwari A. and Shrivastava R.B. (Eds.), *Biotechnology in Biopolymers,* i-Smithers Repra Publication Ltd., UK, 233–285.

135

Dual-use of Technology in Bioterrorism
Issues in Biosecurity and Biosafety

W. Selvamurthy, Rajesh Arora and PVL Rao

CONTENTS

I. TERRORISM: A MODERN DAY SCOURGE WITH DIFFERENT FACES

Terrorism is fast becoming a modern-day scourge. It has several faces in today's changing world. While terrorists continue to gain access to latest technologies they are in a position to use modern technologies and products as a weapon of mass destruction, as well as mass disruption. Terrorism in many ways can be considered a subversive activity that aims to accomplish certain political, financial ideological, religious, monetary or personal motives. The terrorists are nowadays more interested in causing widespread panic and are not just satisfied with inflicting mass casualties. The main aim of terrorists is to gain the attention of the world; a lot of people should be watching their actions and the repercussions of their actions so that they can get adequate attention worldwide for their violent acts, which are only the means to achieve their main objective of being heard loud and clear. The fear of widespread retaliation is making terrorists move on and focus on areas where their identity and involvement can be withheld to the extent they desire; detection by investigation agencies is difficult and place of origin of the plot is difficult to establish. Besides, the chances of getting caught are also fewer when biological agents are employed.

One of the new forces of terror is Chemical, Biological, Radiological and Nuclear (CBRN). However, since accessing conventional chemical and nuclear weapons of mass destruction (WMDs) is difficult, laborious, risky as well as expensive, bioterror agents are agents of choice in view of the ease with which they can be procured, grown and disseminated. Some of the most advanced molecular techniques have become routine today and several advances in life sciences have opened new arenas that were previously inconceivable. Rapid advances in Life Sciences have taken place around the world and consequently technology and expertise is fairly dispersed and easily available, often to terrorists as well. This makes bioterrorism a weapon of choice in the hands of terrorists, less responsible state and non-state actors.

II. THE EMERGENCE OF BIOTERRORISM: FIGHTING DIRTY

In the 2006 report of the then Secretary-General, United Nations, Mr. Kofi Annan stated that "the most important under-addressed threat relating to terrorism ... is that of terrorists using a biological weapon." The United States Commission on the Prevention of Weapons of Mass Destruction (WMD) Proliferation and Terrorism consider that a weapon of mass destruction (WMD) might be used in a terrorist attack before 2013, and that an attack with a biological weapon is more likely than one with a nuclear weapon. Several others believe that biological weapons would be preferred by terrorists as compared to chemical or nuclear weapons The European Commission too opines that biological weapons "may have particular attractions for terrorists".The WMD Commission emphasized the need for a "bio-specific strategy" in terms of preventing acts of terrorism.

Since antiquity, infectious diseases have taken a heavy toll on humans and an estimated 500 million people have succumbed to these diseases in the last century. Several tens of thousands of these deaths were due to the deliberate release of pathogens or toxins during the Second World War. Two International treaties outlawed biological weapons in 1925 and 1972, but they have largely failed to thwart countries from conducting offensive weapons research and large-scale production of biological weapons. As our knowledge of the biology of disease-causing agents – viruses, bacteria and toxins-increases, it is legitimate to fear that modified pathogens could constitute devastating agents for biological warfare. The emergence of highly virulent novel strains of pathogens (bacteria and viruses) in the recent past has highlighted the threat of such agents being used as biological weapons by terrorists. The new methods are opening up entirely new areas for investigation, including new physiological targets of biological warfare agents.

However, bioterrorism is not new; it has been used since centuries mainly by the military against enemies such as hostile armies or populations under siege e.g., ancient civilizations were frequently infected by plague, varicella, Leptospira, or Rickettsiae prowazekii. There is evidence that indicates that in ancient times, the Greeks, Persians and Romans used animal carcasses to pollute enemy water supplies leading to spread of diseases amongst populations, including soldiers. Similarly, dead bodies of animals were used in the United States during the Civil War. During the 18th century, British troops in North America gave blankets and handkerchiefs used by British smallpox victims to hostile Native Americans with the intention to cause mortality.

The 20th century saw a spurt in the development of biological weapons around the world. During World War I, Germany used cholera and plague against humans, and anthrax and glanders against livestock and grain with fungi. Such instances led to the formulation of the Geneva Protocol (Protocol for the Prohibition of the Use in War of Asphyxiating, Poisonous or Other Gases and of Bacteriological Methods of Warfare).

In the aftermath of the World War, a number of countries like the Soviet Union, Japan, the United Kingdom (UK) stepped up their research on biological weapons. Japan conducted at least a dozen field tests in Manchuria and China, which included the contamination of water and food supplies, aerial spraying, and the dropping of small bombs containing plague-infected fleas. Outbreaks of plague, cholera, and typhus were attributed to these activities. Unit 731 was a covert biological R&D unit of the Imperial Japanese Army that undertook lethal human experimentation during World War II and was notoriously involved in these activities resulting in the thousands of deaths. During World War II, the Germans developed vaccines and antimicrobial drugs and tested them on concentration camp inmates. The Allied forces are also known to have produced anthrax bombs that were tested near Scotland.

The Biological Weapons Facility in Sverdlovsk, Former Soviet Union, was established initially after World War II. The first smallpox factory was established in 1947 in the city of Zagoisk near to Moscow. An unusually virulent strain (India-1967 or India-I) was brought by the Soviet medical group from India, where they had gone to help eradicate the disease, which had resulted in millions of casualties. The strain was subsequently manufactured and stockpiled in huge quantities during the early 70s and 80s. Despite being a signatory to the 1972, Biological Weapons Convention, fearing non-compliance by the US, the USSR augmented their bio warfare program, which came to be known as Biopreparat. The large-scale biopreparat program pursued offensive research, development and production of biological agents under cover of legitimate civil biotechnology research. The Biological warfare program also resulted in some civilian casualties. In 1979, several citizens of Sverdlovsk became victims of the biological weapons manufacturing program in the former Soviet Union. Initially, the authorities claimed that consumption of contaminated meat was the reason; however, a decade later it was admitted that the problem was in fact caused due to the unintentional release of anthrax because of malfunction of filters in the biological weapons facility. The US government kept reiterating that underpaid/unpaid scientific staff, impoverished guards, and lack of security could increase the vulnerability of Russian arsenals and many of these could be passed on to less responsible state actors. It is suspected that at least 12 countries had tried to acquire bioweapons from the former Soviet Union, including North Korea and Iraq.

The biological weapons program of the United States of America was started in 1942. After World War II the US military established a program for the research and development of possible biological agents and also made some unannounced test on American civilians. In 1969, President Nixon declared biological weapons disarmament. This was followed by the 1972 Biological and Toxin Weapon Convention (BTWC), after which Western governments stalled the development of biological weapons considering them nasty. With the signing of the Biological Weapons Convention (BWC) by 165 countries, most have transparently destroyed their BW arsenal, however, the possession and use of rogue nations and state/non-state actors, particularly terrorists poses concern.

III. BIOLOGICAL AGENTS AS WEAPONS

Agents that can be used as biological weapons include human pathogens, food and water borne pathogens, plant and animal pathogens, and toxins of various types (Figure 135.1).

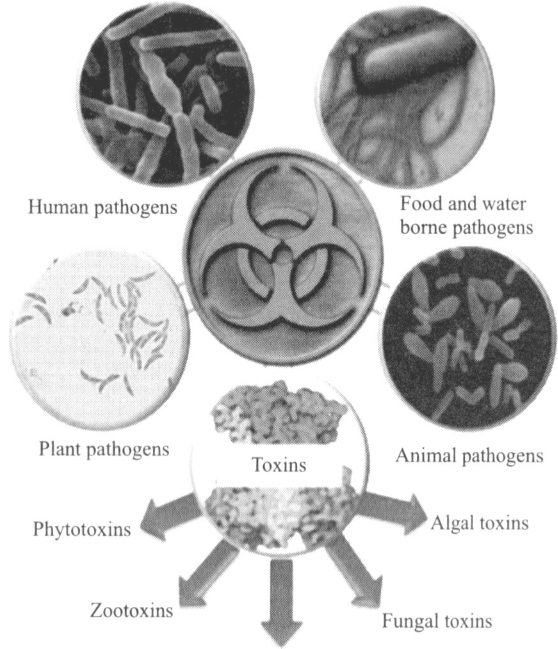

Figure 135.1   Several agents can be used as Biological weapons. (*see Plate 55 for colour figure*)

According to the Center for Disease Control (CDC), USA a bioterrorism attack is the deliberate release of viruses, bacteria, or other agents to cause illness or death in people, animals, or plants.

These agents are typically found in nature, but it is possible that they could be altered to increase their ability to cause disease, make them resistant to current drugs, or to increase their ability to be spread into the environment. Biological agents can be spread through the air, through water, or in food. Terrorists may use biological agents because they can be extremely difficult to detect and do not cause illness for several hours to several days and are, therefore, difficult to detect. Some bioterrorism agents, like the smallpox virus, can be spread from person to person, while others like anthrax can not.

Some examples of use of bioweapon agents in history are highlighted:

- **Anthrax:**   The best known biological weapon is anthrax, the *Bacillus anthracis* bacterium. Research to weaponize anthrax was performed by the UK during World War II. The UK produced million of cattle cakes spiked with anthrax to retaliate in case the Nazis would have used biological weapons. After field trials with anthrax bombs on Gruinard Island in Scotland, this isle was lethally contaminated and off limits for any human being for nearly 50 years.

- **Tularemia** is another first choice biological weapon agent, as it is like anthrax relatively stable in the environment and can be delivered through airborne particles (aerosols).
- **Plague**, caused by the bacterium *Yersinia pestis*, was one of the first infectious diseases that have been used for military purposes. In 1346, after three years of blockading the city of Kaffa, the Tartars catapulted their plague-victims into the city, causing a deadly epidemic within weeks.
- **Smallpox** has been globally eradicated in 1977. However, smallpox virus (called Variola) is still kept in research institutions. Bio warfare with smallpox goes back to the 18th century, when the British sold horse blankets contaminated with smallpox to Native Americans in the USA. The former USSR is alleged to have produced weaponized smallpox virus in a facility called Vector near Novosibirsk. As smallpox vaccinations were stopped some 20 years ago, especially the youth is extremely vulnerable to smallpox today.
- **Ebola**, one of the most deadly viruses, is a potential bioweapon. The Japanese Aum sect, known for the poisonous gas attack on the Tokyo Metro, allegedly attempted to get hold of Ebola samples by sending cult members to Zaire during an Ebola outbreak.
- **Foot and mouth disease** is an example for anti-animal weapons. Biological weapons are not restricted to human pathogens. Any living agent that is used for hostile purposes – regardless of its origin and target—is considered a biological weapon under the BTWC.
- Also **anti-crop bioweapons** have been developed. For example, the USA had an arsenal of 900 kg of a rice pathogen during the Vietnam war, before the US bio warfare program was dismantled in the late 1960s.
- **Toxins** are also covered by the Biological and Toxin Weapons Convention. Toxins are highly lethal substances that are usually produced by living organisms, such as fungi, algae or bacteria. One of most deadly toxins known to humankind is botulinum toxin, which has been weaponized in the past in offensive biological weapons programs.

Key characteristics of biothreat agents

In 1999, the Working Group on Civilian Biodefense expert panel reached consensus on the following list of key features of biological agents that pose particularly serious risks if used as weapons against civilian populations:

- High morbidity and mortality
- Potential for person-to-person transmission
- Low infective dose and highly infectious by aerosol dissemination, with a commensurate ability to cause large outbreaks
- Effective vaccine unavailable or available only in limited supply
- Potential to cause public and healthcare worker anxiety
- Availability of pathogen or toxin

- Feasibility of large-scale production
- Environmental stability
- Prior research and development as a biological weapon.

The U.S. Centers for Disease Control & Prevention (CDC) applied the following four criteria to assess and characterize potential biological weapon agents that cause infections in humans into Categories A, B, and C:

1. Potential public health impact, which is based on an agent's ability to cause illness and death
2. Dissemination potential, which is based on an agent's stability, mass production and/or delivery potential, and the contagiousness, or the degree to which it will spread from person to person
3. Public perception, which is an estimation of how much fear is associated with an agent and how much civil disruption might ensue
4. Special public health preparations required, which is related to the stockpile, surveillance, and diagnostic requirements associated with an agent.

Agroterrorism

Agroterrorism has been recognized as a potent threat that can severely disrupt the socioeconomy of a nation. On the one hand to achieve optimal farm output, attention has been directed towards utilization of conventional approaches, on the other modern biotechnological tools have been utilized to usher in a green revolution. People are advocating the ushering in of an evergreen revolution to feed the burgeoning population, which has now crossed the 7 billion mark. To feed the increasing population there is a need to harness the wide spectrum of biotechnology and utilize modern approaches such as breeding through marker-assisted selection, improve the quality of crops, tagging genes of interest using molecular markers genetic fingerprinting of cultivars. Detection of pests/pathogens making use of DNA-based diagnostics, productive utilization of land available. With the rising availability of genetically modified (GM) crops, crops having agronomic advantages such as herbicide tolerance or insect resistance can be cultivated. However, the use of these modern technological advancement in order to increase the quality and quantity of farm products and food products raises the possibilities of their use for dual use modifications which may lead to agro-terrorist activities.

A number of plant pathogens exist and infect limited areas of plants each year, making outbreaks and control efforts more routine. However, plant pathogens are technically more difficult to manipulate genetically. Some plant pathogens require specific environmental conditions like humidity, temperature, or wind to take hold or spread, while other plant diseases may take a longer time than an animal disease to become established or achieve destruction on the scale that a terrorist may desire. Nonetheless, they do pose a plausible threat.

The rate of increase in human population and global warming will lead to increased problems related to pests and disease, particularly the ones caused due to invasive species. A big challenge in coming times will be to control development of alien pests and diseases through use of dual use technology. The detrimental effects of deliberately releasing genetically modified organisms in the environment are going to be an area of concern in future. Through genetic engineering it might be possible to release potent biological agents with augmented infectivity and pathogenicity, persistency and environmental stability, therefore nullifying the available sero-diagnostic techniques as well as the vaccines.

Since 1986, a number of genetically modified (GM) crops and transgenic plants have been developed. Advances in the field of GN crops have led to their potential for dual-use for e.g., designing of anti-crop agents with improved properties. The deliberate introduction of GM seeds or crops within a country that has not approved such products could have serious socioeconomic consequences caused as a result of the efforts required in detection and cleanup operations.

Technological advancements in expression technology for the utilization of some plant viruses like cowpea mosaic virus, tobacco mosaic virus, potato virus X, and tobacco rattle virus as vectors for the expression of foreign proteins in plants. These vectors may also have applications in expression of vaccines and pharma compounds, particularly toxic compounds.

The use of plants as bioreactors (molecular farming) is an emerging area where significant advances have been made in recent years. Plant cells are ideal bioreactors for the production and oral delivery of vaccines and biopharmaceuticals. The advantages include low production cost, product safety and easy scale up. Besides, the use of plants obviates the need for purification, cold storage, transportation and sterile delivery. Large quantities of bio-regulatory or other toxic proteins having potential to be used as biological agents can be produced in a short time, eliminating the risk of detection. Such dual use technology is a cause of concern.

Molecular technologies have also been utilized to enhance the insecticidal toxicity of *Bacillus thuringiensis* (Bt) toxins by combining the attributes of the Bt toxin with other microorganisms, including baculovirus-based systems and other bacteria. It is possible to deliver toxins via microbial expression systems as a possible alternative to the production of transgenic plants. These developments also have the dual use potential in a BW programme. Expertise in the transfer of Bacillus genes among closely related species could be utilized for malicious purposes by people with nefarious intentions utilizing the available manufacturing expertise, facilities and field delivery systems.

Research in biological pest control has also increased the hunt for development of refined aerosol models for effective dispersal of biological agents. Genetic engineering technologies also make it possible to programme the survival of a released bacterial population and its effect in the environment. The knowledge of persistence and ecological effects would also be of value while releasing the genetically modified knowledge gained in the areas of dissemination technology of bio-pesticides and could be misused by an aggressor to release organisms or toxin harmful to crops, animals and to some extent humans. Conversely, bio-prospecting could be used as databank regarding the biological activity of dangerous pathogens and novel agents having severe effects and toxicity for human, could be used with the intent of introducing them into an immunologically naïve population.

Animal diseases

A number of initiatives have been taken worldwide for improvement of animal health and protection against diseases. Major initiatives have also been taken in areas of genomic analysis, molecular characterization including sequencing, genetic mapping, expressed sequence tags, comparative sequence analysis, etc. Outbreaks of infectious diseases in animals have been regularly linked to terrorist activities or offensive BW programmes. Examples include: bird influenza in 2004 and 2005, the 2003 epidemic of atypical pneumonia, and the 2001 outbreak of FMD. Consequently, there is a need to safeguard animal health and improve surveillance and management of exotic diseases should be based on simplistic differentiation of natural, accidental or deliberate release of biological agents.

Chief characteristics of a potential anti livestock agent include the following features: (i) highly infectious and contagious, pathogenic for livestock, (ii) easy to acquire or produce, (iii) good ability to survive in the environment, (iv) attributable to a natural outbreak, (v) not harmful to perpetrator, and (vi) easily disseminated. Diseases caused by agents that fulfill the above criteria can be found on lists A and B introduced by the Office International des Epizooties (OIE). The OIE is an intergovernmental organization created by international agreement on 25 January 1924 and signed by 28 countries. In March 2004, the OIE totaled 166 member countries, including Poland. The OIE serves world animal health organization.

The OIE develops procedures, guidelines, and shares information to prevent the spread and introduction of infectious diseases of livestock and poultry. The OIE introduced a list of animal diseases, where they are categorized into two levels named List A and List B. List A comprises 15 transmissible diseases that have the potential for very serious and rapid spread, irrespective of national borders. They are of serious socio-economic or public health consequence and are of major importance in the international trade of animals and animal products. The diseases from List A comprises: foot-and-mouth disease, vesicular stomatitis, swine vesicular disease, rinderpest, peste des petits ruminants, contagious bovine pleuropneumonia, lumpy skin disease, Rift Valley fever, bluetongue, sheep pox and goat pox, African horse sickness, African swine fever, classical swine fever, highly pathogenic avian influenza, and Newcastle disease.

List B includes over 90 transmissible diseases that are considered to be of socio-economic and/or public health importance within countries and that are significant in the international trade of animals and animal products. One could say that the List B diseases are "less important" than those from list A and therefore they do not pose as much threat on international trade. This is not always true; especially as far as bovine spongiform encephalopathy (BSE) is concerned. This disease can cause long-lasting barriers to trade meat and meat products from affected. Other examples of List B diseases are: anthrax, glanders, Q fever, rabies, brucellosis or Marek's disease.

The animal pathogen list has only a few agents common with the list of antihuman bioweapons agents. Actually, the exceptions are Venezuelan equine encephalitis virus, Rift Valley fever virus, Chlamydia psittaci, and B. anthracis. This is worth noting because one could assume that the same pathogens can serve as biological weapon against both human and livestock. The list of antilivestock pathogens is based on economic trade impact and ease of transmissibility, while that of human is based on high mortality or fear.

Many of the agents from List A and List B may be used as bio weapons, however the following viruses pose the greatest threat as potential antilivestock weapons: foot-and-mouth disease, classical swine, African swine, rinderpest, avian influenza, and Newcastle disease. Foot-and-mouth disease is one of the most contagious diseases of mammals. It affects cloven-hoofed animals. The primary mode of spread is through respiratory aerosols. The disease can be introduced by direct or indirect contact with other animals, contaminated objects (e.g. footwear, clothes fodder), or artificial insemination. It spreads very rapidly in pigs, but may not be diagnosed quickly when sheep are infected. Classical swine fever (hog cholera) is a highly contagious disease of swine, which are the natural reservoir of the virus. It is transmitted mostly by oral route and can be introduced into herd by an infected animal. African swine fever affects domestic and wild pigs. The disease occurs in several African countries and in Sardinia. It is tickborne; soft ticks Ornithodoros sp. are vectors of the virus. African swine fever can be introduced into a farm by fodder containing the virus. Rinderpest is a disease with high mortality. It affects cattle, sheep, and some breeds of pigs. Rinderpest spreads by contact with infected animals, but also can be transmitted by contaminated fodder or water. Avian influenza (fowl plague) affects poultry and a wide range of other bird species. Wild birds are a reservoir of the virus. It spreads by contact with other birds (mostly fecal-to-oral) and humans can be infected as well. Newcastle disease, especially velogenic form, is a highly contagious disease of poultry. Exotic birds might be a carrier of the virus. The disease spreads by contact with other birds and contaminated objects. Newcastle disease is endemic in many countries of Africa and Asia.

The recent epidemics of SARS, Avian influenza, Swine flu, and FMD outbreak in the UK has forced nations to rethink and develop their own veterinary surveillance strategies and policies and operational guidelines for control and management of outbreaks of exotic animal diseases, including standardization of laboratory technologies for detection of zoonotic diseases of public health importance. The veterinary surveillance strategy also records patterns of emerging diseases and alterations in endemic diseases. The knowledge gained in development of vaccines and other intervention strategies for safeguarding of animal health and effects of global animal and zoonotic disease outbreaks have raised public awareness further, as also the potential for use of such agents for hostile purposes. Genetic alteration or modified vaccines developed using dual use technology might make such outbreaks impossible to manage and pose a grave public health risk.

Humans: Categories of Biothreat Agents

Biothreat agents can be separated into three categories, depending on how easily they can be spread and the severity of illness or death they cause. Category A agents are considered the highest risk, while Category C agents are those that are considered emerging threats for disease.

Category A: These high-priority agents include organisms or toxins that pose the highest risk to the public and national security.

Category B: Agents are the second highest priority because of the following characteristics, while **Category C** are third highest priority agents and include emerging pathogens that could be engineered for mass spread in the future.

The biological agents can be water borne, air borne or food borne. A number of bacteria, particularly from a food perspective, pose concern as they pose risks in terms of food safety and are potent bio threat agents. Some such bacterial genera include *Escherichia*, *Salmonella*, *Shigella*, *Vibrio*, *Yersinia*, and *Francisella*. By extensive mining of the whole genome and protein databases of diverse, closely and distantly related bacterial species and strains, Woubit et al. (2012) have recently identified novel genome regions, which were utilized to develop a rapid detection platform for these human pathogens.

IV. SOURCES OF BIOTERROR THREATS AND THEIR SOCIO-ECONOMIC IMPACT

Bioterrorism threats occur in today's age since such pathogens can be easily isolated/obtained from naturally occurring diseases, accidental outbreaks or deliberately occurring diseases (Figure 135.2).

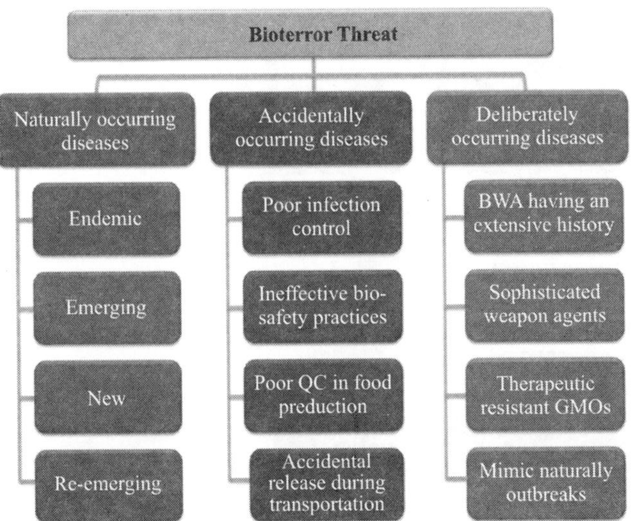

Figure 135.2　Some sources of bioterror threat.

The socio-economic impact of bioterrorism around the globe can be enormous and can be felt across borders (Figure 135.3).

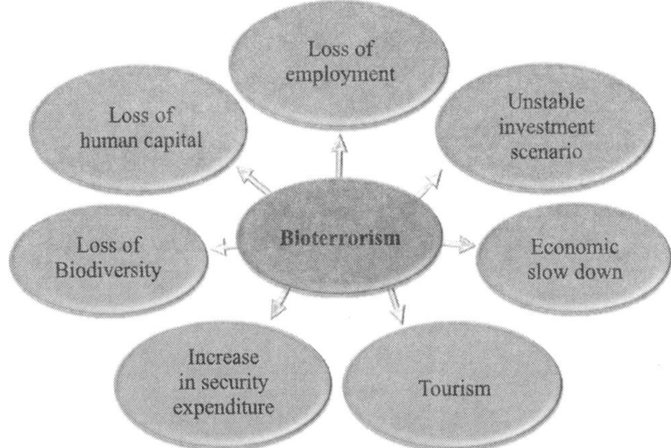

Figure 135.3　Bioterrorism: Socio-economic impact.

Biothreats are a major cause for concern since such agents can be easily accessed/obtained (Figure 135.4).

A number of disease outbreaks have taken place in the Indian subcontinent over the years (Figure 135.5), several of which could wreck havoc if they were to be weaponized.

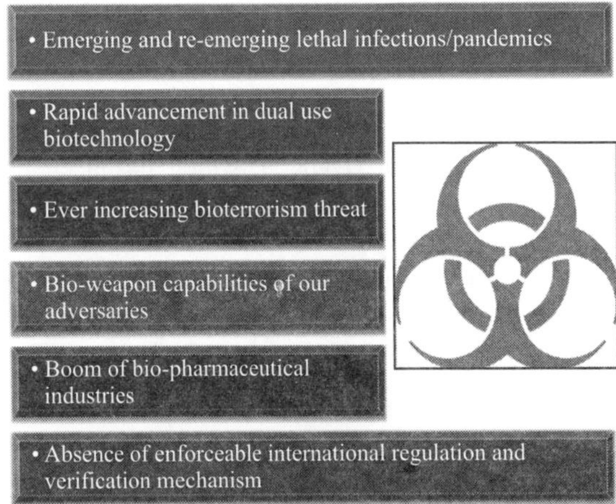

Figure 135.4　Why biothreats are a genuine concern?

V. WHY FOCUS ON BIOTERRORISM?

The biological warfare threat became more realistic in the aftermath of the 9/11 terrorist attacks on the World Trade Center in the United States. Not only did the terrorists demonstrate that they were prepared to murder large numbers of people indiscriminately, they also exposed the vulnerability of many societies. As alluded to earlier, several facts contribute to the growing danger of bioterrorism: Minute amounts of select biological agents can cause mass causalities; agents for biological weapons can be easily acquired; technology for production and weaponization is readily available; only limited financing and training are needed to establish a biological weapons programme; biological weapons have low visibility and are relatively easy to deliver; and the motivations of terrorists have changed.

Bioterrorism incidents have generated considerable political interest, often disproportionate to the size of events. For example, it was suggested that the geopolitical impact from the US anthrax letters in 2001, which resulted in 22 cases and five deaths, was of a similar scale to the 2002–03 SARS outbreak, which caused an estimated 8098 cases and 774 deaths. Public perception of the risk posed by bioterrorism is vital, as this feeds into the geopolitical response to such incidents. The main negative impact of a biological weapons attack or bioterrorism incident is panic, with consequent disruption of civilian services.

VI. SCIENTIFIC ADVANCES AND THE RISKS POSED DUE TO DUAL USE TECHNOLOGY

That modern technology can be used to abet bioterrorism particularly in view of rapid advances in the field is evident from several examples which include: chemical synthesis of polio virus and the fx174 bacteriophage by Craig Venter and colleagues, demonstration of the variola virus gene for its virulence, reconstitution of the 1918 influenza virus, research establishing the conversion of H5N1 avian influenza into a version that is contagious among ferrets and might transmit from person to person etc.

These scientific advances have brought to the fore, dual use research of concern (DURC), which has been defined by the US National Science Advisory Board for Biosecurity (NSABB) as "research that, based on current understanding, can be reasonably anticipated to provide knowledge, products or technologies that could be directly misapplied by others". DURC poses a concern since it can have widespread implications for bioterrorism.

VII. DUAL-USE TECHNOLOGY: MANIFOLD MEANINGS AND THE FINE DIVIDE

The term 'dual-use technology' initially referred to technology that could be employed for civilian as well as military purposes. Ever since the term has been commonly used to refer to research, knowledge, technology or materials that can be used for both beneficial and malefic purposes. Modern discussions regarding dual-use science and technology refer to science and technology that has legitimate uses for example in medicine, but which might also be used by for malevolent purposes for example in bioterrorism.

Three plausible definitions of 'dual-use technology' have been put forth:

1. Technology that has both civilian and military applications
2. Technology that can be used for both beneficial/good and malafide purposes, and
3. Technology which has both beneficial and harmful purposes—where the harmful purposes involve weapons, usually weapons of mass destruction (WMDs) in particular.

"Dual-use research" is defined as a biological research with legitimate scientific purpose that may be misused to pose a biologic threat to public health and/or national security, which has been monitored and evaluated by the NSABB of the USA (http://oba.od.nih.gov/biosecurity/ biosecurity.html). 'Dual use' refers to the tangible and intangible features of technologies that enable them to be applied to both hostile and peaceful ends with little or no modification, like a double-edged sword. The term is used to highlight how the same upstream activities, materials, information and equipment can potentially have both hostile and peaceful downstream applications. The dual use dilemma arises, because sometimes beneficial scientific research can also have harmful consequences. The dual use phenomenon poses a dilemma for scientists and policy makers alike.

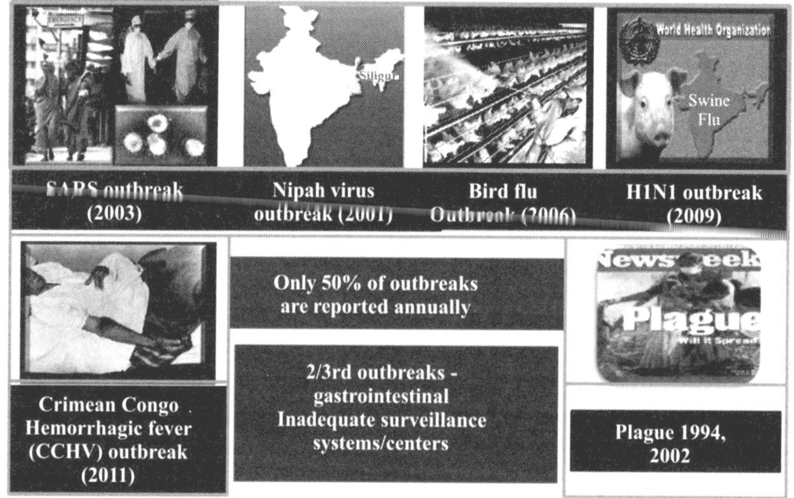

Figure 135.5 Some examples of disease outbreaks in the Indian subcontinent in recent years, which have posed a real threat.

When the term 'dual use' was first used, it had positive connotations and focused on the benefits accruing due to reusing civilian technologies to develop military technologies. However, the term 'dual use' has over the time gained a negative connotation and now refers to the potential of technologies with legitimate civilian uses to aid the proliferation of prohibited military technologies. The term commonly denotes and refers to technologies that have legitimate application in scientific research, drug/vaccine production, agriculture or industrial processing and which can be utilized to produce chemical or biological weapons. The publication of three papers – (i) on the synthesis of polio virus cDNA without a natural template (ii) how the variola virus (smallpox) can evade the immune system and (iii) how to overcome resistance to mouse-pox has raised widespread concerns.

Continuing advances in the life sciences over the last 50 years, supported by new enabling technologies, have brought great benefits for health, the economy and the environment. Many believe that the life sciences hold far greater promise for the future. Powerful tools are allowing biologists to probe complex systems in evergreater detail, from molecular events in individual cells to global biogeochemical cycles. Along with the achievements and hopes, however, have come a range of concerns about the implications and impacts of current and potential advances. Among these specific concerns is the potential security risk that states or terrorists groups or even individuals could misuse the knowledge, tools and techniques gained through life sciences research for biological weapons and terrorism.

The possibilities and attendant uncertainties regarding whether and how advances in the life sciences intended for legitimate and beneficial purposes might also be used for malevolent ends has come to be called the "dual-use dilemma". This highlights the problem of defining dual use in life sciences. One definition is provided by the US National Science Advisory Board for Biosecurity (NSABB) ; "Biological research which may provide knowledge, products, or technology that can be directly misapplied with sufficient scope so as to threaten public health or other aspects of national security, such as agriculture, plants, animals, the environment and material (NSABB 2006). The broad array of life sciences research that might be of proliferation concern covers many fields and types of research institutions; commercial research and applications are equally diverse, so that monitoring potentially relevant activities would be a formidable task. Following are some of the activities, which can be categorized into activities of low, moderate and extreme concern.

Activities of potential concern

- Work with listed agent-or exempt avirulent, attenuated or vaccine strain of select agent.
- Increasing virulence of non-listed agent
- Increasing transmissibility or environmental stability of non-listed agent
- Powder or aerosol production of non-listed agent
- Powder or aerosol dispersal of non-listed agent

- De novo synthesis of non-listed agent
- Genome transfer, genome replacement, or cellular reconstruction of non-listed agent

Activities of moderate concern

- Increasing virulence of listed agent or related agent
- Insertion of host genes into listed agent or related agent
- Increasing transmissibility or environmental stability of listed agent
- Powder or aerosol dispersal of listed agent
- De novo synthesis of listed agent
- Genome transfer, genome replacement, or cellular reconstruction of listed agent

Activities of extreme concern

- Any work on the variola virus (smallpox) or comparably virulent agent that has been eradicated in nature
- Any spontaneously infectious agent requiring BSL4/ABSL4 level of containment
- De novo synthesis of any agent matching the above characteristics
- Expanding the host range of an agent or changing the tissue range of an agent
- Constructing vaccine resistant or antibiotic resistant strains

A. Dual Use: Does it Purport Malafide Intentions?

It was stressed that 'dual use' relates to the threat of misapplying information or technologies rather than the carrying out of research itself. This highlights the extent of dual use dilemmas since many types of research may be dual use by implication. A 2004 NRC report, *"Biotechnology Research in an Age Terrorism"* tackled the critical question of how to define what is dangerous and how to design a system that contains that danger while allowing the legitimate biomedical research to proceed in a manner acceptable to society. The NRC Committee, chaired by Gerald Fink proposed a multistage system under which scientists would review experiments and their results to provide public reassurance that advances in biotechnology with potential applications for bioterrorism or bioweapons development were receiving responsible over-sight.

B. Examples of Research with Dual-Use Applications

The debates sparked by the publication of data related to the reconstruction of the 1918 influenza virus illustrated how scientific achievements may also generate security concerns. The additional research endeavors given below were all identified as having the potential for misuse.

(a) Dual-use of targeted delivery systems

Targeted delivery systems include both viral and non-viral systems. Systems designed to deliver a biological agent such as DNA or protein effectively to target cells and tissues have many important applications in vaccines, cancer treatment and other therapies. Multiple viruses have been explored for use as delivery of vectors, including adenoviruses, adeno-associated viruses, vaccinia virus and lentivirus, etc. Viral based delivery system has led to improvements in targeting to specific tissues, gene transfer and expression levels in those tissues and vector stabilization. Advances in aerosol delivery may lead to improved protection of the delivered agents or drugs from environmental stress and other detrimental factors.

Targeted delivery may also be employed to deliver bio regulatory molecules such as fentanyl, insulin oxytocin or orexin, which can alter the functions of physiological systems and have been previously delivered in aerosol form. Scientists developing targeted drug delivery systems also have a responsibility to consider dual-use implications of their technologies.

(b) Synthesis of polio virus

Researchers sought to resolve the unusual nature of polio virus, which behaves as both chemical and a living entity. Scientists succeeded in creating the virus chemically synthesizing a cDNA of its genome. Some critics opine that the publication of their methods provide a recipe for terrorists showing how one could create any virus from chemical reagents which can be purchased from open market.

(c) Viruses which could evade human immune system

Development of "stealth" viruses to serve as molecular means for introducing curative genes into patients with inherited diseases represent significant advancement. However, the same technology could potentially be induced to express dangerous proteins, such as toxins.

(d) Construction of fusion toxins

A method was developed for the construction of "fusion toxins" derived from two distinct nontoxic chemical predecessors. This technique was originally investigated for the purpose of killing cancer cells. But the same technique can be used to produce toxins that can kill normal cells of any tissue when introduced into a human host.

(e) Genetic engineering of the tobacco plant to produce subunits of cholera toxin

As tobacco is easy to manipulate by genetic engineering, it is a likely candidate for producing plant-based vaccines. The technique could be used to produce large quantities of cholera toxin cheaply and relatively easily, paving the way for fast and efficient vaccine production. Concerns have been raised that it might also have a potential for misuse.

(f) Generation of new highly virulent virus strains

Recently, Fouchier and colleagues at the Erasmus Medical Center in Rotterdam, The Netherlands, and Kawaoka's group at the School of Veterinary Medicine, University of Wisconsin-Madison, have generated several avian IAV H5N1 variants with mutations in HA protein and/or PB2. They found that

these H5N1 mutants became more transmissible among ferrets. According to Kawaoka's study, a reassortant H5HA/H1N1 virus – which comprised four mutations of H5 HA at N158D, N224K, Q226L, and T318I, and the remaining seven gene segments from a 2009 pandemic H1N1 virus – was identified to replicate efficiently among ferrets with the ability to spread through droplets in a ferret model. In Fouchier's case, they generated a modified A/H5N1 virus, which was able to acquire mutations during passage in ferrets, ultimately obtaining the ability to spread through the air among these mammals. They showed that four substitutions at H103Y, T156A, Q222L, and G224S of the host receptor-binding protein HA and one mutation at E627K of PB2 were able to consistently present in the airborne-transmitted viruses.

The transmissibility of these mutant H5N1 viruses between mammals constitutes a significant risk for influenza pandemic in humans. Thus, these studies related to mutant influenza viruses were considered as a typical dual-use research by the NSABB, a US government advisory panel. Accordingly, a variety of discussions and controversies occurred surrounding the dual-use research and the necessity for publication of the dual-use research-related findings. The submitted papers by Fouchier and Kawaoka's groups have been thoroughly reviewed and evaluated by the NSABB for the benefits and potential negative consequences before these manuscripts were approved for publication in *Nature* and *Science*. After a month-long debate on the controversy of the mammalian-transmissible mutant H5N1 influenza, these scientific findings representing the dual-use research were eventually published in the two aforementioned journals. There is no doubt that the findings from these studies will help in understanding the mechanism of viral transmission, improving international surveillance, and identifying pandemic flu threats because the flu surveillance is currently lacking. However, the laboratory-created viral mutants with increased airborne transmissibility may be misused to pose threats to national biosecurity, laboratory biosafety, and/or public health.

C. Synthetic Biology and Dual-Use Research

Synthetic biology is concerned with applying the engineering paradigm of systems design to biological systems in order to produce predictable and robust systems with novel functions that do not exist in nature. The field has been defined as "the engineering of biological components and systems that do not exist in nature and the re-engineering of existing biological elements; it determined on the intentional design of artificial biological systems, rather than understanding of natural biology." Synthetic biology is at the intersection of biology, chemistry and physics, overlapping and cross-fertilizing with a range other fields of research and technology development. It can be anticipated that the major change that the field of synthetic biology will bring is the synergistic integration of existing disciplines: not just biology and engineering, but also computer modeling, information technology, control theory, chemistry and nanotechnology.

Bringing the engineering paradigm to biology will allow us to apply existing biological knowledge to biotechnological problems in a much more rational and systematic way than has previously been possible. While several of the fundamental scientific issues and current applied objectives of the synthetic biology overlap with those in other, more mature fields, especially biotechnology and systems biology, synthetic biology should be properly seen as a completely new discipline, which bring a systematic, application driven engineering perspective to biology.

Synthetic biology will drive the industry, research, education and employment in the life sciences in a way that might rival the computer industry's development during the 1970 and 1990s. Due to fundamental change in the methodology that it entails for the modifications of living organisms, synthetic biology may be able to fulfill many of the promises that traditional biotechnology industry is still struggling to fulfill, in important areas such as:

- Biomedicine
- Synthesis of biopharmaceuticals
- Sustainable chemical industry
- Environment and energy
- Production of smart materials and biomaterials
- Security: counter bio-terrorism

The medical applications of synthetic biology include:

- Complex molecular devices for tissue repair—sense damage and repair
- Smart drugs – become active in cells affected by disease
- Biological delivery system
- Vectors for therapy – viruses to deliver healthy genes
- Personalized medicine
- Cells with new properties that improve human heath

The areas of materials synthesis and processing that might benefit from synthetic biology include:

- Biomedical materials
- Microelectronics and information technology
- Development of tough composites
- Sensors and actuators
- Materials for energy conversion

D. Ethical Aspects of Synthetic Biology

Synthetic biology is a promising new field of research with huge promises for society, but also large potential perils. It is at this stage in the development of synthetic biology that ethical and legal analysis should be attempted, based on interdisciplinary research which includes expertise in ethics, law, science and technology foresight.

(a) *Biosafety and Biosecurity Risks of Synthetic Biology*
Biosafety: "Laboratory biosafety describes the containment

principles, technologies and practices that are implemented to prevent the unintentional exposure to pathogens and toxins or their accidental release".

Biosecurity: "The objective of biosecurity is to prevent loss, theft or misuse of microorganisms, biological materials, and research related information".

Risk of negative environmental impact: The potential for biosecurity compromise has increased as knowledge and capability have deepened, as life scientists have dispersed from well-regulated training institutions to employment sites worldwide, and as environmentally provocative activities have reached new scales and complexities. Also complicating matters is the inevitability of unexpected and occasionally hard-to-control outcomes in normal science. Some experiments now involve organisms altered in ways making release, escape, or theft a biosecurity concern. Synthetically created organisms can have unintended side effects. Synthetic organisms could help to solve environmental problems like contaminated soil but entails release of synthetic organisms, which could have negative side-effects.

Risk of natural genome pool contamination: Synthetic organisms could transfer genes to natural organisms. One of the main aims of synthetic biology is the creation of novel genetically modified organisms (GMOs), which may have utility in the production of energy and bioremediation. However, such a prospect raises concerns about their accidental release into the environment, as by their very nature such biological machines could evolve, proliferate and produce unexpected interactions that might alter the ecosystem. Researchers have shown that it is possible to create or recreate deadly viruses such as polio virus and the Spanish flu.

(b) *Notions of life and other broader aspects*

Several philosophical questions are raised by synthetic biology, some of which highly relevant for ethical reflection and societal debate. One of the most potent promises of synthetic biology is the creation of 'artificial life'. This has provoked fears about scientists 'playing God' and raises philosophical and religious concerns about the nature of life and the process of creation. If a naturally occurring organism is recreated with synthetic elements what is its status: should it be regarded as natural or as artificial? What about a naturally existing organism that is recreated with minimal genome?

IX. ADDRESSING THE DUAL-USE RESEARCH CHALLENGE

Scientists, security professionals, researchers, policy makers and society as a whole support adopting active approaches to biosecurity challenges without hindering scientific innovation. Biology is advancing rapidly and powerful materials, tools and knowledge are widely accessible throughout the world. The free exchange of scientific information was an essential factor enabling the many scientific achievement of the 20th century.

Researchers, policy makers, regulators and the security experts are all grappling with biosecurity issues.

(a) Role of scientists

Scientists are the front line of defense against misuse of biological research. Working together scientists can develop effective guidelines and standards that deter misuse without inhibiting exploration of new and important lines of research.

As noted in the National Research Council report Biotechnology Research in an Age of Terrorism (NRC, 2004), biologists have a moral duty to avoid contributing to the advancement of bio warfare and bioterrorism. It is the scientist's responsibility to consider the dual use potential of his or her own work, and make informed decisions on whether or how to proceed, and conduct the work using the principles of responsible scientific practice.

(b) Role of research institutions

Research institutions providing biological research facilities like academic institutes, universities, Government funded research laboratories, private biotechnology R&D establishments, private industries have a responsibility to protect the safety and security of their workers, those living in the surrounding environment and general public at large. These Institutions should be aware of and comply with governing research activities and also National and international legally binding legislations and laws. The institutions should be prepared to actively educate their employees and students about relevant requirements and responsible practices related to biosafety and biosecurity.

(c) Role of professional societies

Professional societies should create educational programs about the potential risks of dual use research. The idea to incorporate dual use issues into professional codes of conduct has also gathered considerable attention in recent years. While codes of conduct cannot prevent those who are dedicated to carrying out malevolent acts from doing so, they can sensitize and educate active scientists to the risks of working with hazardous pathogens.

(d) Role of journal editors

Editors of peer reviewed journals serve as gatekeepers that determine what work should be publishes or what should not be published. To address biosecurity issues, government agencies, academic institutions, and professional societies have developed policies concerning the publication of "dual-use" biomedical research-that is, research that could be readily applied to cause significant harm to the public, the environment, or national security. Group of editors and scientists accepted the responsibility for screening manuscripts to reduce the risk of misuse of scientific information. In December of 2011 the U.S. National Science Advisory Board for Biosecurity (NSABB) took the unprecedented step of requesting that the top-tier journals *Science* and *Nature* remove certain methodological details and the identity of the key mutations from the results to be published of two H5N1 avian influenza virus studies.

Biosecurity needs change rapidly as science develops, and lot depends on both scientists who conduct research with dual use applications and the editors who publish it.

(e) Educational strategies

Many scientists and stake holders believe that dual use issues could be appropriately introduced to multiple potential audiences, including undergraduates, graduate students, laboratory technicians, post-doctoral fellows, professionals and faculty members. These materials would provide exposure to responsible conduct of science concepts such as dual use at different career stages.

(f) Role of government agencies

Because of the global distribution of pathogens that could be used for bioterrorism or bio warfare, it is important that all the countries adopt appropriate regulatory oversight measures to ensure that individuals with access to agents are trustworthy and that the agents are protected from potential misuse. Government agencies have important role to play in developing legislation related to biosecurity and biosafety issues, export and exchange of highly pathogenic and infectious organisms, potent toxins, chemical with dual applications and issues related to release of genetically modified organisms.

X. BIOLOGICAL AND TOXIN WEAPONS CONVENTION (BTWC) AND BIOTERRORISM

The first international treaty in modern law banning the use of biological weapons was the 1925 Protocol for the Prohibition of the Use of Asphyxiating, Poisonous or Other Gases, and of Bacteriological Methods of Warfare, known as the Geneva Protocol. Negotiated under the auspices of the League of Nations after the First World War, the Geneva Protocol had significant shortcomings: its prohibition did not cover production, development and stockpiling of biological (and chemical) weapons and many countries reserved the right to retaliate with biological weapons. So, even after ratifying the Protocol, many industrialized countries continued to build arsenals of chemical and biological weapons. After UK and USA renounced the use and development of biological weapons, a major stumbling block to the development of a more comprehensive multilateral ban was removed. The result was the 1972 Convention on the Prohibition of the Development, Production and Stockpiling of Bacteriological (Biological) and Toxin Weapons and on their Destruction (BTWC) (Figure 135.6). Unlike the Geneva Protocol's ban on use, the BTWC bans development, production and stockpiling, acquisition or retention of biological agents or toxins "whatever their method of production, of types and in quantities that have no justification for prophylactic, protective or other peaceful purposes". The ban also extends to "weapons, equipment or means of delivery designed to use such agents or toxins for hostile purposes or in armed conflict."

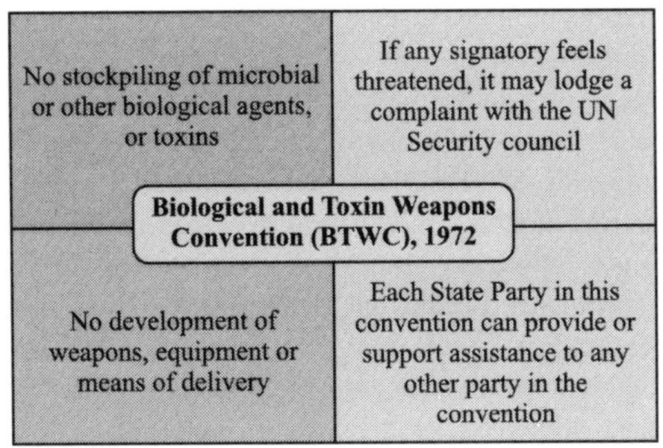

Figure 135.6 The Biological and Toxin Weapons Convention (BTWC), 1972.

A key problem in biological arms control is the dual use of the know-how and equipment involved in civilian research, biodefense projects and in offensive bio warfare programs. It is especially difficult to draw clear lines between offensive and defensive research. Emerging biotechnologies and their commercialization will facilitate research, development, and production of biological agents through means that involve smaller technical footprints and lower costs. This trajectory will make detecting violations of the BTWC more difficult and potentially enable proliferation of weaponization capabilities to terrorist organizations. Although the BTWC process has included bioterrorism as an issue, aspects of bioterrorism and the BTWC limit the role that the BTWC can play in preventing, protecting against, and responding to bioterrorism.

The BTWC plays a positive and crucial role as a component of the development and implementation of mutually supportive international strategies focused against bioterrorism. The Biological and Toxin Weapons Convention (BTWC) entails prohibition of the development, production and stockpiling and acquisition of biological and toxin weapons. Sustained efforts over a period of years led to negotiation of the protocol to include detailed investigation and verification to ensure that participating countries fulfilled their obligations of non-acquisition or retention of microbial or other biological agents or toxins harmful to plants, animals and humans, in types and in quantities that would have no justification for prophylactic, protective and other peaceful purposes.

BTWC, ever since its inception, has been quite skeptical of developments in dual-use technologies in several contiguous areas of modern life sciences like genetic engineering, biotechnology and microbiology, particularly for large-scale, high-volume products and processes that are capable of being used for purposes inconsistent with the original objectives and provisions of the BTWC. These include all microbial and biological agents or toxins, of natural or artificial nature, irrespective of their origin or method of production. The rapid pace of developments in life sciences research and improved ability to utilize the various pathogenic species of micro-organisms, ease with which extraction of toxins and other biological agents is possible today and the rapid developments in biotechnology further accentuate the possibilities of production and hostile use of biological agents. Dual-use technologies,

even though they may not in principle contravene offensive purposes, pose a potential risk from a biothreat perspective as it can be used for nefarious purposes anytime. Endeavors by most countries focus on technologies for creating new means of protection against biological threats, rather than their use for terror purposes; but how does one distinguish the fine line between the good, the bad and the ugly!

XI. R&D AREAS OF POTENTIAL CONCERN TO BTWC

Some of the research and development areas of potential concern to the BTWC are highlighted in the ensuing sections.

Bio-defense programmes of less-responsible state actors

BW agents are capable of inflicting mass casualties in civil as well as military domain. It is now practically feasible for several less responsible State as well as Non-state actors to clandestinely develop novel BW agents that can evade detection through deployed sensors, decontamination is impossible using conventional approaches, against which there are currently no vaccines or treatments available, and their efficiency is such that they can counter the currently available therapeutics effectively. In addition, placing embargoes on such nations and monitoring are issues of great concern that have to be addressed.

Microbiology and microbial biotechnological advances

R&D in the field of microbiology/biotechnology can lead to transformation of innocuous organisms into potent BW agents. Some ways by which this can be accomplished is highlighted in Figure 135.7.

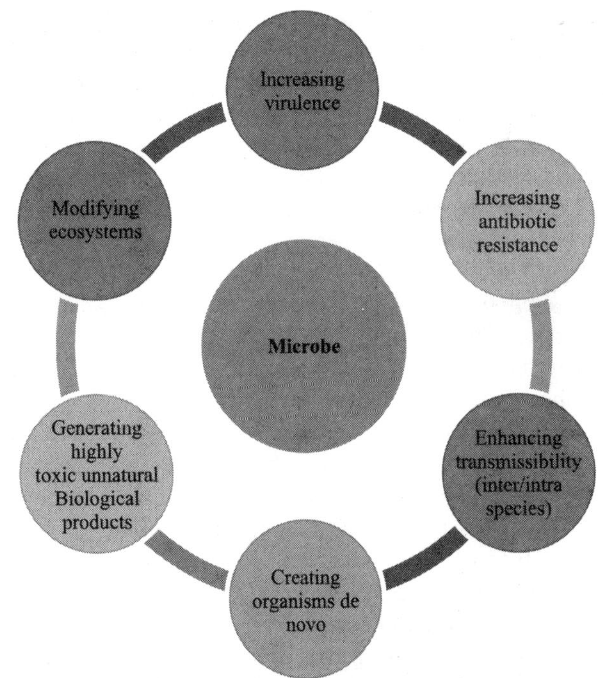

Figure 135.7 Ways and Means of weaponizing innocuous microbes that pose grave threat to humanity.

Advances in multiple disciplines, coupled with developments in areas like physics and chemistry and engineering, computational and material sciences, rDNA technology, toxicology, molecular biology, nanotechnology and other related sciences have also made it possible to create a new generation of biological weapons (BW).

Understanding the mechanism of action of microorganisms gives direct access to the mechanism of action of potential BW agents. Molecular biological approaches have led to identification of mechanisms of virulence and infections have been identified; the same techniques may also permit deliberate manipulation of these mechanisms. Transferring certain genetic traits into naturally infectious microorganisms can potentially create organisms of greater virulence, antibiotic resistance and environmental stability. Changing the microbes genetically could alter their immunogenicity, thereby invalidating vaccines and sero-diagnostic techniques. Otherwise innocuous microorganisms could be altered to produce toxemia or disease: the host would continue to recognize them as innocuous and therefore not defend against them.

New Directions from the Human Genome Project (HGP)

The human genome was sequenced in 2003 and led to the identification of far fewer genes than those envisaged earlier. The human genome is now thought to have 20,000-30,000 human genes and this makes them convenient and easily accessible for genetic modifications. The HGP also led to the elucidation of the entire sequence of the three billion DNA subunits bases in the human genome. The significant strides in research that were made during the HGP might take decades to fully comprehend, analyze and interpret. The immediate benefits of HGP include, but are not limited to, the identification and localization of genes responsible for inherited diseases and simplification of the drug development for treatment of a variety of illnesses, ease in prenatal diagnosis of diseases that have a genetic basis, reason for ethnic genetic differences between populations. It has been anticipated that the information generated by the human genome project would have multifarious applications in life sciences research leading to improved diagnosis of diseases, better and rapid drug design, pharmacogenomics, gene therapy etc.

HGP has also raised serious concerns. Elucidation of the structure of human DNA and the functioning and regulation of genes makes it possible to identify markers which determine the racial and ethnic differences between populations. Such data could be used for developing methods to selectively disturb cellular respiration and energy exchange, sexual reproduction and a number of other important functions including immune responses. The use of such crucial data for dual-use may be disastrous and is a reason to be worried about.

The International HapMap Project, a multi-nation endeavour was taken to identify and categorize genetic similarities and differences in human beings and a major effort toward identifying genes affecting health, disease, and individual responses to drugs and environmental factors. Information generated under the project was made available in the public domain to enable targeting of specific groups with harmful agents.

Analysis of the information obtained in the Human Genome project has revealed that the human genome has hundreds and even thousands of sequences that could serve as targets for selective BW agents, and differences exist in frequency of genetic markers between races and difference in the frequency of polymorphisms. Studies in Taiwan have shown that Severe Acute Respiratory Syndrome (SARS) can be associated with specific genetic profiles.

Use of advanced sequencing technologies has also led to mapping of the molecular signatures of the bioregulatory systems of the body and how these regulatory pathways respond to disease-induced disturbances. The information, which is critical as target for therapeutic and preventive intervention or manipulation, can also be used for a novel biological attack. Significant advances have also been made in genomic medicine, where patient patient-tailored treatment of diseases is prescribed based on analysis of his/her genetic makeup.

The study of immune-dominant proteins which trigger immune response to infection and that of the interaction of antibodies and antigens can possibly lead to an understanding of the humoral immune response and identification of antigenic determinants for inclusion in future vaccines. The introduction of protein arrays enables the rapid analysis of hundreds of proteins in parallel. These techniques are being used to detect and diagnose biomarkers as well as possible vaccine candidates against a number of diseases. At the same time, these technologies have opened avenues for dual use for nefarious purposes.

Rapid scientific advances in new and emerging areas, including disruptive technologies, have resulted in developments that are of concern to the BTWC. Some other dual-use technologies that are of concern include aerosol delivery of bacteria.

XII. NATIONAL INITIATIVES FOR MITIGATING DUAL USE OF TECHNOLOGY FOR BIOTERRORISM

The threat emanating from bioterrorism has necessitated that R&D be carried out to develop antidotes to novel toxins and agents that can be used as potential biological weapons. It is imperative that the Nation remains cognizant of the modalities and carefully monitors the developments in the field to avoid the use of dual use technology for nefarious purposes. Development of Biothreat Mitigation strategies is imperative since the threat of covert biological warfare and bioterrorism looms large. India is unique in having the world's second largest human population (1.23 billion) and the largest animal bio resource available anywhere in the world. Immense bio diversity exists in the form of 45000 species and hot spots distributed in the Northeast and Western Ghats. More than 18% of GDP contribution comes from the agriculture sector, which is susceptible to agroterrorism. The country has human resource expertise in academia, R&D as well as built pharmaceutical industry base in various parts of the country. The use of such centers for dual use technology poses real concern. There is a need to take care of such issues. Keeping this in mind, and the need to develop preventive and mitigative strategies against the threat of bioterrorism,the Government of India has decided to take up a holistic National Biothreat Mitigation programme to tackle any eventuality involving biological agents.

Networking of stakeholders

To handle the threats from dual use of technology in bioterrorism it is essential to have linkages with various ministries, departments and institutions within the country (National Crisis Management, National Disaster Management Authority, Ministry of Home Affairs, Ministry of Defense, Ministry of Environment & Forests, Ministry of Agriculture, Ministry of External Affairs and Department of Ports, Drinking Water Mission, Rural Development Ministry etc.) so that such activities can be forestalled.

CONCLUSION

The modern life sciences revolution, especially in the field of agricultural, medical and pharmaceutical sciences has made the lives of humans easier on the one hand; however, simultaneously it has led to development of full-fledged use of the same technology for dual-use applications. Dual-use technology poses great concern as it can be used to develop weapons of mass destruction (WMD). It is practically feasible to conceal a BW programme easily amidst programmes that are of dual-use. With the convergence of bio-nano-IT technologies, the ability to weaponize a biological programme has become much more easier. This raises the specter of new generation of biological weapons, as well as a sophisticated capability to manipulate the physiology of human beings for offensive purposes. Utilizing dual-use technology, it is also possible to produce non-lethal biological weapons that can incapacitate defence personnel or civil population without mortality. Such bioweapons could be used for potential military applications, causing slow-onset infections and creating novel anti-animal or anti-plant BW agents resulting in huge socioeconomic losses. Life scientists have a proactive moral duty to avoid contributing to the advancement of bio warfare or bioterrorism. Scientists, researchers and academicians have an important role to play in informing the public and policy makers concerning actions that need to be taken to prevent biological warfare and bioterrorism and also to attenuate the impact of use of biological weapons and dual-use technologies. Dual use technologies can be effectively used for creating a safer world that can protect itself from the onslaught of bioterrorism. The double-edged sword-dual use technology is in our hands. It is up to us how as to how we use it responsibly for the benefit of the human race!

Although all efforts have been made to discuss scientific information in a general way, the inadvertent use of dual-use knowledge for malicious use is a remote possibility. Hence, the authors will be in no way be responsible for the malicious use of the information discussed in any way, though it is hoped that the dual-use technologies will be used for beneficial purposes only

REFERENCES

Atlas Ronald M. (2002), Bioterrorism: From Threat to Reality, *Annual Review of Microbiology*.

Caitríona McLeish,Paul Nightingale (2007), Biosecurity, Bioterrorism and the Governance of Science: The Increasing Convergence of Science and Security Policy, *Research Policy*, Vol. 36, Issue 10, 1635–1654.

Gandhi B.M. (2011), An Overview of the Advances Made in Biotechnology and Related BTWC Concerns, *CBW Magazine*, Vol. 4, Issue 3

Lanying Du, Ye Li, Jimin Gao, Yusen Zhou and Shibo Jiang (2012), Potential Strategies and Biosafety Protocols Used for Dual-use Research on Highly Pathogenic Influenza Viruses, *Reviews in Medical Virology*, DOI: 10.1002/rmv.1729, Vol. 22 Issue 6.

Riede Stefan (2004), Biological Warfare and Bioterrorism: A Historical Review, *Proc (Bayl Univ Med Cent)*, 17(4): 400–406.

Sławomir Letkiewicz Andrzej Górski (2010), The Potential Dual Use of Online Pharmacies, Science and Engineering Ethics, Vol. 16, Issue 1, 59–75.

SUGGESTIONS FOR FURTHER READING

Arora R., Chawla R., Marwah R., Arora P., Sharma R.K., Kaushik V., Goel R., Kaur A., Silambarasan M., Tripathi R.P. and Bhardwaj J.R. (2011), Potential of Complementary and Alternative Medicine in Preventive Management of Novel H1N1 Flu (Swine Flu) Pandemic: Thwarting Potential Disasters in the Bud, Evid Based Complement Alternat Med. 2011;2011:586506. Doi: 10.1155/2011/586506. Epub 2010 Oct

Baililie L. Dyson H. and Simpson A. (2012), Dual Use of Biotechnology, Encyclopedia of Applied Ethics, 2nd ed., 876–883.

Cello J., Paul A.V. and Wimmer E. (2002), Chemical Synthesis of Polio Virus CDNA: Generation of Infectious Virus in the Absence of Natural Template, *Science*, 297;1016–1018.

Centers for Disease Control and Prevention and National Institutes of Health (2007), *Biosafety in Microbiological and Biomedical Laboratories*, 5th ed. (L.C. Chosewood and D.E.Wilson, Eds). Washington DC.

Cole L.A. (2003), *The Anthrax Letters: A Medical Detective Story*, Joseph Hendry Press: Washington DC.

Danzig R. and Berkowsky P.B. (1997), Why Should We be Concerned about Biological Warfare? JAM, 278:431–432.

NRC (2004), *Biotechnology Research in an Age of Terrorism*, Washington, DC. National Academies Press.

Responsible Research with Biological Agents and Toxins (2009), National Research Council, USA.

Schmidt M., Agomoni Gangul-Mitra, Torgersen H., Kelle A., Anna Deplazes and Nikola Biller-Abdorno (2009), A Priority Paper for the Societal and Ethical Aspects of Synthetic Biology, *Sys Synth Biol.*, 3:3–7.

Service R.F. (2006), Biosecurity: Synthetic Biologists Debate Policing Themselves, *Science*, 312:1116.

Synbiology (2005), Synbiology, an Analysis of Synthetic Biology Research in Europe and North America European Commission Framework Programme 6 Reference Contract 15357 (NEST), October 2005.

Trends in Science and Technology Relevant to The BTWC (2011), *Summary of International Workshop*, NRC, National Academic Press.

Weiss R.A. and McLean A.R. (2004), What Have We Learnt from SARS ? *Phil Trans R Soc Lond B* 359, 1137–1140.

WHO (2007), Biorisk Management: Laboratory Biosecurity Guidance, Geneva, WHO.

Woubit A., Yehualaeshet T., Habtemariam T. and Samuel T. (2012), Novel Genomic Tools for Specific and Real-time Detection of Biothreat and Frequently Encountered Food Borne Pathogens, *J. Food Prot.*, 660–70. Doi: 10.4315/0362–028X.JFP-11–480.

136

Bioinformatics, Genomics, Proteomics and Metabolomics in Biomedical Research

Seema Mishra and Syed Asad Rahman

CONTENTS

"DNA is like a computer program, but far, far more advanced than any software ever created."

— Bill Gates, *The Road Ahead*

I. BIOINFORMATICS

Imagine, you have in your hands, a really long stretch of some random nucleotide or amino acid sequence/s tangled together and interwoven like a woollen yarn waiting to be knitted into a clearly *understandable tapestry*. Like the powerful, simple tools of two knitting needles, scientists need the *science* and *powerful tools* of Bioinformatics to spin together their understanding of Nature through the story of these random sequences. Mastering the skill of Bioinformatics is an exciting scientific pursuit and as much arduous and meticulous as is mastering the skill of knitting.

The field of Bioinformatics (*Bio: Biology and Informatics: Information Handling*) is marriage between these two fields and arose as 'genes' and 'computers' started to become established in their respective roles around 1960s. It evolved further as a need to analyze *high throughput sequencing (HTS)* information churning out sequences in numbers too large to be deciphered easily by a human mind. As the technologies changed rapidly, *Bioinformatics and Computers* virtually forced each other to co-evolve for a mutually beneficial association and applications. As an example, *massively parallel* sequencing capabilities using Next Generation Sequencing (NGS) Technologies, require computers with Massively Parallel Processing (MPP) capabilities computers to make sense of the voluminous data. The term 'massively parallel' indicates very large number of sequences generated simultaneously and computer processors working in parallel to perform analytical tasks. Since this field is constantly evolving, there is a plethora of definitions for Bioinformatics. Scientifically speaking, majority of these definitions define it as the *interdisciplinary* field that uses *computational technologies/methods* to store, retrieve, analyze, solve, and provide a conceptual understanding of *high-throughput* (generated in large numbers) biological data. Intuitively, the *basic* fields that contribute to Bioinformatics' evolution and 'revolution', in addition to Biology, are Chemistry (e.g., Intermolecular forces in studying protein-ligand binding and molecular mechanisms), Physics (e.g., Newton's laws of motion in studying the motions of a protein and ligand towards each other), Mathematics and Statistics (e.g., differential equations to understand biochemical reactions or transcriptional regulatory networks) and Computer science (e.g., programming or cluster computing for managing large clusters of computers to develop methods and their applications for problem-solving). Although Bioinformatics and Computational biology have specific definitions, in many cases these are used interchangeably, and the latter holds true in this book chapter as well for ease of understanding.

Synergy of these different fields, while providing us with insights into deeply philosophical questions in biology such as 'what is life?', has also been instrumental in solving several important and crucial biological problems. In fact, a book entitled '*What is Life?*' written by a physicist Erwin Schrödinger, which delves on examining genetics through the lens of physics, was the inspiration behind the discovery of DNA as double-helical structure. The *Human Genome Project* for sequencing complete human genome, is one of the many successful applications of computational biology. A potent inhibitor of tyrosine kinases and one of the miracle drugs for the treatment of cancers, *Gleevec (Imatinib)*, was discovered through computational techniques. Two key *Nobel prize winning discoverie*s in recent times, have extensively used computational tools: one by Venkatraman Ramakrishnan, Thomas A. Steitz and Ada E. Yonath jointly for studies on structure and function of ribosomes, and for solving the *ribosome structure*, which is complex in terms of the numbers of proteins and RNA molecules located throughout; and another by Martin Karplus, Michael Levitt and Arieh Warshel for developing *multiscale models* of *complex chemical systems* (e.g., to study protein-drug interactions or photosynthesis). Multi-scale computational models of various types of cancers and other diseases, in terms of *network* of genes and proteins, have led to the elucidation of *novel drug targets* and their *interventions*. Computational models are being used to optimise metabolic pathways for the production of biofuels, optimising biomass production and enzyme engineering. Myriad examples exist in literature where computational biology has played a key, significant role.

II. INTERPLAYERS IN BIOINFORMATICS RESEARCH

Key macromolecules and small molecules inside a cell functioning together in a dynamic interplay constitute the main points of research in Bioinformatics. These key molecules include but are not limited to *Genes, Non-genes, RNAs, non-coding RNAs, Proteins, Metabolites and Small molecules*. Interactions between these key molecules are co-dependent in a *temporal* (time-dependent) and *spatial* (location-dependent) manner, and, therefore, can be understood in parts. However, a *whole* picture complete with all the parts and puzzles solved together, is possible only when taking into account the *myriad* molecular and cellular interactions taking place at each time and

space, and *integrating* these partial processes. Bioinformatics opens our avenues towards such a *condensed* picture leading us to understand and explore hitherto unimagined arenas of science. The integrative field arising as a result and which makes this level of integration possible is widely known as 'Systems Biology', relying heavily on computational models.

The ensembles of *data, information, characteristics* and *attributes* of these key molecules which are widely exploited in our scientific explorations in Bioinformatics are as follows:

1. Sequences
2. Structures
3. Pathways
4. Gene expression profiles
5. Protein expression profiles
6. Metabolite profiles

The various *areas* of active research utilizing these data, information and attributes are summarized below:

A. Sequences

- Gene finding to identify genes or hypothetical genes in an unknown sequence.
- Sequence alignments to understand the function and information content of a given sequence and protein structure prediction.
- Phylogenetic analyses to deduce the evolutionary history of an unknown sequence or organism.
- Genome analysis, to identify all of the genes, non-gene regions, gene orders, regulatory sequences present in a genome and comparing genomes of different organisms to analyze their respective features

B. Structures

- Protein structure prediction
- Nucleic Acids structure prediction
- Protein functional role
- Nucleic Acids functional role
- Structural alignments to understand the function of a given structure and protein structure prediction
- Protein folding
- Small molecules structure
- Drug designing
- Intermolecular interactions
- Drug-Receptor docking
- Protein-Protein docking

C. Pathways

- Protein Interaction Network
- Gene Regulatory Network
- Transcriptional Network
- Signalling pathways

D. Gene Expression Profiles

- mRNA expression analyses using microarrays
- Gene expression analyses using RNA sequencing

E. Protein Expression Profiles

- Protein expression analyses using microarrays
- Protein expression analyses and identification of unknown proteins using 2-dimensional polyacrylamide gel electrophoresis (2D-PAGE) and mass spectrometry

F. Metabolite Profiles

- Mass spectral analyses of metabolites

III. BIOINFORMATICS DATABASES, TOOLS AND WEB SERVERS

The present era is an era of high-throughput data mining, storage and exchange driven by the availability of high speed internet connectivity and robust computing technologies. Bioinformatics, in order to decode the biological phenomena, uses *data* and *information* (in the form of databases) to decipher and generate *knowledge*, this knowledge is also generated *cumulatively* with other forms of *a priori* information, skills and experiences (in the form of tools); and *learning and unlearning*, i.e., reflecting on several notions learnt along the way and focussing on the combined knowledge, and as a result, imparting *wisdom*.

As per the Data, Information, Knowledge and Wisdom (DIKW) hierarchical pyramid widely taught in Information Science and Knowledge Management courses, Bioinformatics thrives on this structure to fully serve as a science in its existential role. As an example (Figure 136.1), an *unknown* sequence **"A"** (data) deduced from a next generation sequencing technology, is identified through a high percentage of its identity or sequence similarity with another sequence **"B"** that is already known as encoding a particular protein

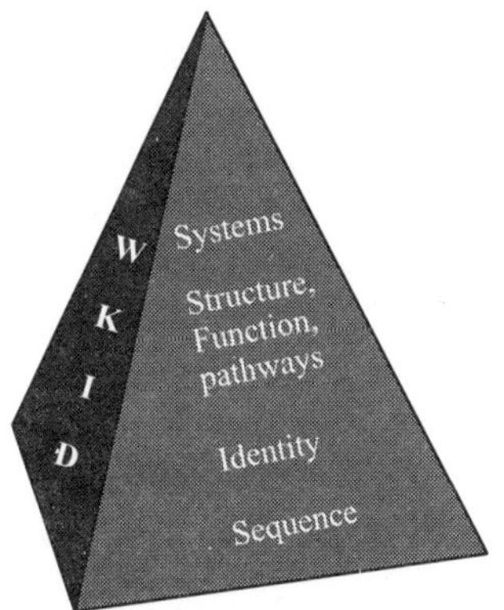

Figure 136.1 A representation of Bioinformatics concepts and understanding in terms of Data, Information, Knowledge and Wisdom (DIKW) hierarchical pyramid.

(information). However, the attempt does not stop here, for as Albert Einstein quoted, *"Information is not knowledge."* Further attempts to characterize this sequence "A" lead to a particular structure and function of its encoded protein. Through its interaction with other genes/proteins, this protein derived from sequence "A" may play a probable role/s in a pathway and hence, may be functionally useful. Combining this with *a priori* knowledge through information and experiences with working on a related set of proteins, an understanding of its contextual or real existence (knowledge) comes into fruition. Moreover, as Einstein quoted, *"The only source of knowledge is experience"*. For a deeper understanding, a whole picture of several such molecules interplaying with each other, in or out of context, as an organized whole (wisdom), constitutes understanding at systems level (systems biology). A lot of imagination and creativity also plays a role, for here is Einstein again: *"Imagination is more important than knowledge. For knowledge is limited to all we now know and understand, while imagination embraces the entire world, and all there ever will be to know and understand"*.

A. Bioinformatics Databases

A biological database is defined as an organized collection of biological data and all the relevant information associated with the data, for easier access, management, retrieval and analyses. The contents of the database need to be updated periodically to facilitate recent-most information access. To pursue the active research areas as detailed in the section on "Interplayers in Bioinformatics Research", there are several database resources pertaining to each molecule and its data and attributes; such as nucleotide sequence database, amino acid sequence database, protein structure database, pathways database, enzyme database, gene expression database and so on. Some of the most commonly used databases in Bioinformatics research and the websites on which these are hosted are listed in Table 136.1. Literature mentions that the first collection of protein sequences was compiled by Margaret O. Dayhoff, entitled "Atlas of protein sequence and structure" in 1965, and this database grew to be known later as Protein Information Resource (PIR).

The databases are usually maintained in a high performance computing (HPC) environment with huge storage spaces to account for terabytes (10^{12} bytes, a byte is a unit of information used by computers to represent a letter such as 'g' or a number) or petabytes (10^{15} bytes) of data storage capacity.

The National Center for Biotechnology Information (NCBI), USA, one of the front-runners in Bioinformatics, maintains a very large compute farm. The collection and deposition of data in these databases, e.g., nucleotide sequences generated using next generation sequencing technology, is usually done by researchers through online deposition the data collected along with appropriate annotation. Annotation refers to the process of collecting, collating and disseminating every possible information available for the respective data, and painstaking efforts are put in place for as much accuracy and reliability

TABLE 136.1   List of biological databases and websites

S. No.	Database	Name of the Database	Website
1.	Vertebrate Genome Sequence	Ensembl	http://www.ensembl.org/index.html
	Microbial Genome Sequence	UCSC Microbial Genome	http://microbes.ucsc.edu/
2.	Nucleotide Sequence	National Center for Biotechnology Information (NCBI)	http://www.ncbi.nlm.nih.gov
		European Molecular Biology Laboratory (EMBL)	http://www.embl.org
		DNA Databank of Japan (DDBJ)	http://www.ddbj.nig.ac.jp/
	Protein Sequence	UniProtKB	http://www.uniprot.org/
		National Center for Biotechnology Information (NCBI)	http://www.ncbi.nlm.nih.gov
3.	Structure	Research Collaboratory for Structural Bioinformatics Protein Data Bank (RCSB PDB)	http://www.rcsb.org/pdb/home/home.do
		Nucleic Acid Database (NDB)	http://ndbserver.rutgers.edu/
4.	Protein Domains and Folds	Prosite	http://prosite.expasy.org/
		ProDom	http://prodom.prabi.fr/prodom/current/html/home.php
		InterPro	https://www.ebi.ac.uk/interpro/
		CATH	http://www.cathdb.info
		SCOP	http://scop.berkeley.edu
5.	Pathways	Kyoto Encyclopedia of Genes and Genomes (KEGG)	http://www.genome.jp/kegg/kegg1.html
		Reactome	http://www.reactome.org/
6.	Literature	PubMed	http://www.ncbi.nlm.nih.gov/pubmed/
7.	Interaction	Molecular Interaction Database (MINT)	http://mint.bio.uniroma2.it/mint/Welcome.do
		Database of Interacting Proteins (DIP)	http://dip.doe-mbi.ucla.edu/dip/Main.cgi
		IntAct	http://www.ebi.ac.uk/intact/
8.	Gene expression profiles	Gene Expression Omnibus (GEO)	http://www.ncbi.nlm.nih.gov/geo/
	Protein expression profiles	Human Proteinpedia	http://www.humanproteinpedia.org/
	Metabolite profiles	Human Metabolome database	http://www.hmdb.ca/
		Biochemical, Genetic and Genomic Metabolic network reconstruction (BiGG) database	http://bigg.ucsd.edu/
		METLIN metabolite database	http://metlin.scripps.edu/
9.	Enzymes	BRENDA	http://www.brenda-enzymes.info
		IUBMB-EC	http://www.chem.qmul.ac.uk/iubmb/enzyme/

of annotation. NCBI collaborates with other laboratories such as European Molecular Biology Laboratory (EMBL) and DNA Databank of Japan (DDBJ) in an International Nucleotide Sequence Database Collaboration effort for data exchange. According to latest release in April 2014, GenBank, the comprehensive nucleotide sequence database at NCBI, contains over 159,813,411,760 base pairs of sequences for non-genomic sequences and 621,015,432,437 base pairs for genomic sequences, a huge number.

Usually exchange of information is optimised either as a human-readable format or is machine-friendly but in many cases it has to be kept in the balance. In Bioinformatics, most of the existing data models are user friendly (human-readable). An example of a user-friendly format in the form of a typical GenBank entry is shown in Figure 136.2. The contents are self explanatory with the headings in the first columns. Readers are suggested to visit the website http://www.ncbi.nlm.nih.gov/guide/all/ for further information on several databases and tools maintained at NCBI. The present data exchange models are mostly available in various formats depending on the computation needs like Extensible Markup

Language (XML), Systems Biology Markup Language (SBML) formats among others.

Three-dimensional (3D) structures of macromolecules and small molecules derived from X-ray crystallography, NMR spectroscopy and and electron microscopy are deposited and maintained at Research Collaboratory for Structural Bioinformatics Protein Data Bank (RCSB PDB) at Rutgers University and University of California, San Diego. As of April 15, 2014, PDB database has a total of 99472 structures. Compared to the number of nucleotide sequences above, the number of structures is low. The protein structures are varied and diverse [Figure 136.3(a) to (d)], each suited to a particular function, reflecting the diversity of structures deposited in PDB.

B.   Bioinformatics Tools and Web Servers

Bioinformatics softwares developed as tools provide us the means to measure, analyze and interpret the biological information. Bioinformatics tools are tailored on biological databases which act as the foundation for most of the software tools. As is obvious, sequence alignment tools are based on

Saccharomyces cerevisiae TCP1-beta gene, partial cds; and Axl2p (AXL2) and Rev7p (REV7) genes, complete cds
GenBank: U49845.1
FASTA Graphics
LOCUS   SCU49845   5028 bp   DNA   linear   PLN 23-MAR-2010
DEFINITION Saccharomyces cerevisiae TCP1-beta gene, partial cds; and Axl2p
            (AXL2) and Rev7p (REV7) genes, complete cds.
ACCESSION   U49845
VERSION   U49845.1 GI:1293613
KEYWORDS   .
SOURCE   Saccharomyces cerevisiae (baker's yeast)
ORGANISM   Saccharomyces cerevisiae
            Eukaryota; Fungi; Dikarya; Ascomycota; Saccharomycotina;
            Saccharomycetes; Saccharomycetales; Saccharomycetaceae;
            Saccharomyces.
REFERENCE   1 (bases 1 to 5028)
AUTHORS   Roemer,T., Madden,K., Chang,J. and Snyder,M.
TITLE   Selection of axial growth sites in yeast requires Axl2p, a novel plasma membrane glycoprotein
JOURNAL   Genes Dev. 10 (7), 777-793 (1996)
PUBMED   8846915
REFERENCE   2 (bases 1 to 5028)
AUTHORS   Roemer,T.
TITLE   Direct Submission
JOURNAL   Submitted (22-FEB-1996) Biology, Yale University, New Haven, CT
            06520, USA
FEATURES   Location/Qualifiers
     source   1..5028
                     /organism="Saccharomyces cerevisiae"
                     /mol_type="genomic DNA"
                     /db_xref="taxon:4932"
                     /chromosome="IX"
     mRNA   <1..>206
                     /product="TCP1-beta"
     CDS   <1..206
                     /codon_start=3
                     /product="TCP1-beta"
                     /protein_id="AAA98665.1"
                     /db_xref="GI:1293614"
                     /translation="SSIYNGISTSGLDLNNGTIADMRQLGIVESYKLKRAVVSSASEA
                     AEVLLRVDNIIRARPRTANRQHM"
     gene   <687..>3158
                     /gene="AXL2"
     mRNA   <687..>3158
                     /gene="AXL2"
                     /product="Axl2p"
     CDS   687..3158
                     /gene="AXL2"
                     /note="plasma membrane glycoprotein"
                     /codon_start=1
                     /product="Axl2p"
                     /protein_id="AAA98666.1"
                     /db_xref="GI:1293615"
                     /translation="MTQLQISLLLTATISLLHLVVATPYEAYPIGKQYPPVARVNESF
                     TFQISNDTYKSSVDKTAQITYNCFDLPSWLSFDSSSRTFSGEPSSDLLSDANTTLYFN
                     VILEGTDSADSTSLNNTYQFVVTNRPSISLSSDFNLLALLKNYGYTNGKNALKLDPNE
                     VFNVTFDRSMFTNEESIVSYYGRSQLYNAPLPNWLFFDSGELKFTGTAPVINSAIAPE
                     TSYSFVIIATDIEGFSAVEVEFELVIGAHQLTTSIQNSLIINVTDTGNVSYDLPLNYV
                     YLDDDPISSDKLGSINLLDAPDWVALDNATISGSVPDELLGKNSNPANFSVSIYDTYG
                     DVIYFNFEVVSTTDLFAISSLPNINATRGEWFSYYFLPSQFTDYVNTNVSLEFTNSSQ
                     DHDWVKFQSSNLTLAGEVPKNFDKLSLGLKANQGSQSQELYFNIIGMDSKITHSNHSA
                     NATSTRSSHHSTSTSSYTSSTYTAKISSTSAAATSSAPAALPAANKTSSHNKKAVAIA
                     CGVAIPLGVILVALICFLIFWRRRRENPDDENLPHAISGPDLNNPANKPNQENATPLN
                     NPFDDDASSYDDTSIARRLAALNTLKLDNHSATESDISSVDEKRDSLSGMNTYNDQFQ
                     SQSKEELLAKPPVQPPESPFFDPQNRSSSVYMDSEPAVNKSWRYTGNLSPVSDIVRDS
                     YGSQKTVDTEKLFDLEAPEKEKRTSRDVTMSSLDPWNSNISPSPVRKSVTPSPYNVTK
                     HRNRHLQNIQDSQSGKNGITPTTMSTSSSDDFVPVKDGENFCWVHSMEPDRRPSKKRL
                     VDFSNKSNVNVGQVKDIHGRIPEML"

Figure 136.2 (*Contd.*)

```
gene            complement(<3300..>4037)
                /gene="REV7"
mRNA            complement(<3300..>4037)
                /gene="REV7"
                /product="Rev7p"
CDS             complement(3300..4037)
                /gene="REV7"
                /codon_start=1
                /product="Rev7p"
                /protein_id="AAA98667.1"
                /db_xref="GI:1293616"
                /translation="MNRWVEKWLRVYLKCYINLILFYRNVYPPQSFDYTTYQSFNLPQ
                FVPINRHPALIDYIEELILDVLSKLTHVYRFSICIINKKNDLCIEKYVLDFSELQHVD
                KDDQIITETEVFDEFRSSLNSLIMHLEKLPKVNDDTITFEAVINAIELELGHKLDRNR
                RVDSLEEKAEIERDSNWVKCQEDENLPDNNGFQPPKIKLTSLVGSDVGPLIIHQFSEK
                LISGDDKILNGVYSQYEEGESIFGSLF"
ORIGIN
        1 gatcctccat atacaacggt atctccacct caggtttaga tctcaacaac ggaaccattg
       61 ccgacatgag acagttaggt atcgtcgaga gttacaagct aaaacgagca gtagtcagct
      121 ctgcatctga agccgctgaa gttctactaa gggtggataa catcatccgt gcaagaccaa
      181 gaaccgccaa tagacaacat atgtaacata tttaggatat acctcgaaaa taataaaccg
      241 ccacactgtc attattataa ttagaaacag aacgcaaaaa ttatccacta tataattcaa
      301 agacgcgaaa aaaaaagaac aacgcgtcat agaacttttg gcaattcgcg tcacaaataa
      361 attttggcaa cttatgtttc ctcttcgagc agtactcgag ccctgtctca agaatgtaat
      421 aatacccatc gtaggtatgg ttaaagatag catctccaca acctcaaagc tccttgccga
      481 gagtcgccct cctttgtcga gtaattttca ctttcatat gagaacttat tttcttattc
      541 tttactctca catcctgtag ......................................................
     4921 ttttcagtgt tagattgctc taattctttg agctgttctc tcagctcctc atatttttct
     4981 tgccatgact cagattctaa ttttaagcta ttcaatttct ctttgatc
//
```

Figure 136.2

Figure 136.2 An example of a GenBank entry for *Saccharomyces cerevisiae* TCP1-beta, Axl2p (AXL2) and Rev7p (REV7) genes, GenBank Accession number U49845.1

sequence databases, structure prediction tools on structure databases and so on. These tools are developed extensively, with a great deal of manual labor involved, through the use of computational programming languages such as C++, Java or PERL. Although their usage is simple enough, such that a layman is able to use them, their correct usage and interpretation invariably requires a skilled and knowledgeable person.

Bioinformatics tools are available either as a web server / web services (usage over the web) or as a standalone application (downloadable and used on one's desktop computer). A list of the commonly used tools is provided in Table 136.2 along with their websites to be consulted for further information. Some of the useful Bioinformatics tools are described below:

Gene finding

Gene finding tools are based on the observations of *unique characteristics genomic patterns and/or frequency of occurrence of DNA bases* that particular regions of a genome have into order to function as a gene. For example, in the case of prokaryotic genes, *an open reading frame of 90 bp or longer, matches to promoter sequences at –35: 5'-TTGACA-3' and at –10: 5'-TATAAT-3' positions upstream from the transcription start site, presence of Shine-Dalgarno sequences* are indicative of a gene region. Eukaryotic genes are more complex, and characterized by sequence features such as *presence of TATA box at –25 and CAAT box at –80 positions upstream from the transcription start site, presence of CpG islands, presence of splice sites at exon-intron junction* among others. Tools such as

Gene **L**ocator and **I**nterpolated **M**arkov **M**odell**ER** (GLIMMER) and **Pro**karyotic **Dy**namic Programming Genefinding **Al**gorithm (Prodigal) for prokaryotic gene prediction and Glimmer-Hidden Markov Model (GlimmerHMM) for eukaryotic gene prediction have been developed.

Sequence alignment

Sequence alignments are done to predict the identity of an unknown sequence, to deduce the evolutionary history of an unknown sequence or groups of sequences or to determine conserved regions. Throughout evolution, some parts of sequences or structures are conserved, primarily because these are essential to an organism's survival or these play a significant role in the proper functioning of a molecule. As an example, the His-Asp-Ser catalytic triad is conserved in serine proteases and subtilisin, two unrelated proteins that have different sequences (*analogous* proteins). Also, *proper relatedness* of organisms can be easily determined based on identity of their molecular features alone (*homologous sequences*) as compared to phenotypic features. As an example, using phenotypes such as eyes, that are present in divergent species such as humans and fishes, comes with the risk of improper classification of unrelated organisms into a related group. Molecular features are not subjected to environmental changes and natural selection asmuchas phenotypes are. Moreover, phenotypes such as colony morphology cannot distinguish between different species of bacteria, and this makes using molecular data to classify species and organisms useful.

Sequence alignments may be of the following types:

Pairwise alignment: Alignment is done between two sequences (a pair of sequences). One example of such a tool is Basic Local Alignment Search Tool (BLAST).

Multiple sequence alignment: Alignment is done between more than two sequences, e.g., ClustalW2 tool.

The accuracy of these alignments is also influenced by the evolutionary relationship between amino acids/DNA sequences supplied as an input matrix. This input matrix is characterized based on the *rate of substitution* from one amino acid/nucleotide to another in two or more sequences, as well as divergence times. Two major types of matrices in use are Blocks Substitution Matrix (BLOSUM) and Point Accepted Mutation (PAM) matrices, the former takes into account *only the conserved regions* while the latter uses *both highly conserved and highly mutable sequence characteristics* to align the sequences.

Phylogenetic analyses

Phylogenetic analyses are done to predict evolutionary history of a species occurring over time due to *changes in gene pool* through mutations, or through *changes in gene frequency*. An evolutionary "Tree of Life" constructed using molecular sequence data of 16S ribosomal RNA (rRNA) molecule revealed three major kingdoms: Eukarya, Archaea and Bacteria. Archaea was separated from Bacteria on the basis of molecular phylogeny, from the earlier notion where these were grouped together. This hypothesis has been refuted by some researchers in later molecular phylogenetic analyses using many conserved proteins. In these studies, grouping Archaea were grouped with

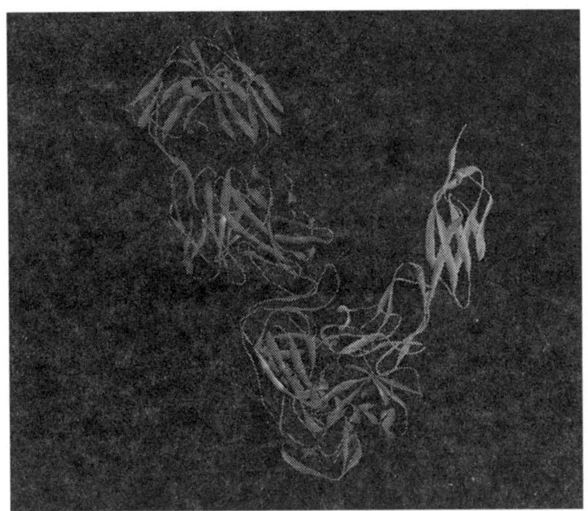

(a) Structure of a HIV envelope glycoprotein, GP120 (in pink) in complex with CD4 (in green) and neutralizing antibody (in purple), composed mostly of beta-sheets. Structure downloaded from PDB with PDB ID: 1GC1 (Reference: Kwong et al., 1998, *Nature* 393:648–659)

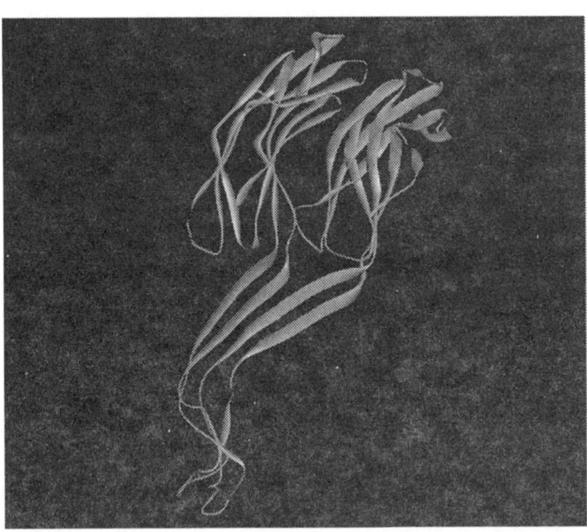

(b) Porin protein with beta-barrel shape, structure downloaded from PDB with PDB ID: 1UUN (Reference: Faller et al., 2004, *Science* **303**:1189)

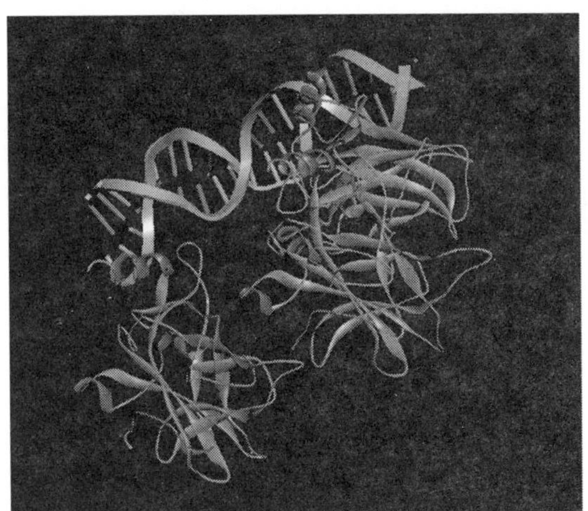

(c) p53 protein bound to DNA, structure downloaded from PDB with PDB ID: 1TUP (Reference: Cho et al., 1994, *Science* 265: 346–355

(d) PKD2 phosphopeptide (in yellow) bound to human MHC Class I molecule (in red), structure downloaded from PDB with PDB ID: 3BGM (Reference: Mohammed et al., 2008, *Nat. Immunol.* **9**: 1236-1243

Figure 136.3 Diverse types of 3-dimensional protein structures as visualized using visualization software. (*see Plate 56 for colour figure*)

TABLE 136.2   List of Bioinformatics tools and webservers

S. No.	Areas of Research	Tool	Website
1.	Gene Finding	Glimmer	http://www.ncbi.nlm.nih.gov/genomes/MICROBES/glimmer_3.cgi
		GlimmerHMM	http://ccb.jhu.edu/software/glimmerhmm/
		ORF Finder	http://www.ncbi.nlm.nih.gov/gorf/gorf.html
		GenScan	http://genes.mit.edu/GENSCAN.html
		GeneMark.hmm	http://opal.biology.gatech.edu/
		Prodigal	http://prodigal.ornl.gov
2.	Sequence alignment	Basic Local Alignment Search Tool (BLAST)	http://blast.ncbi.nlm.nih.gov/Blast.cgi
		ClustalW 2	https://www.ebi.ac.uk/Tools/msa/clustalw2/
		Multiple Sequence Comparison by Log-Expectation (MUSCLE)	https://www.ebi.ac.uk/Tools/msa/muscle/
3.	Phylogenetic analyses	Phylogeny Inference Package (Phylip)	http://evolution.genetics.washington.edu/phylip.html
		Phylogenetic Analysis Using Parsimony (PAUP)	http://paup.csit.fsu.edu/
4.	Structure prediction	SWISS-MODEL	http://swissmodel.expasy.org/
		RAPTOR-X	http://raptorx.uchicago.edu/
		I-TASSER	http://zhanglab.ccmb.med.umich.edu/I-TASSER/
		Robetta	http://robetta.bakerlab.org/
5.	Drug Design	Discovery Studio	http://accelrys.com/products/discovery-studio/
		MacroModel	http://www.schrodinger.com/smdd/
		GLIDE	http://www.schrodinger.com/smdd/
		AutoDock	http://autodock.scripps.edu/
		GOLD	http://www.ccdc.cam.ac.uk/Solutions/GoldSuite/Pages/GOLD.aspx
6.	Pathways	KEGG Pathway	http://www.genome.jp/kegg/
		Cytoscape	http://www.cytoscape.org/
		Pathway Tools	http://www.metacyc.org
		BioTapestry	http://www.biotapestry.org/
7.	Enzymes	EC-BLAST	http://www.ebi.ac.uk/thornton-srv/software/rbl/

Gram positive bacteria separate from Gram negative bacterial group. One of the commonly used phylogenetic analysis tools is Phylogeny Inference Package (Phylip). Availability of next generation sequencing data has led to network-based approaches rather than phylogenetic trees to accommodate the high volume of information from this technology. Further, studies on evolutionary history are bound to utilize genome-scale information and so, phylogenomics will come into play.

Structure prediction

Structure prediction also takes into fact the laws of evolution to predict a structure and a possible function. This possible function is deduced on the basis of Anfinsen's hypothesis proposed by Christian B. Anfinsen in 1961, that postulates that the structure of a protein, i.e., the way a proteins folds into a 3-dimensional structure, is determined by its amino acid sequence.

Secondary structure prediction: Proteins form secondary structures on the basis of amino acid propensities, i.e., some amino acids favor the alpha-helical conformation and others the beta-sheet or loop conformation. Based on these and other physicochemical parameters such as accessible surface area and the local environment of amino acid residues, and the probabilities inherent in the formation of secondary structures, several secondary structure prediction tools have been developed. Some examples are JPred3, PSIPRED and JUFO3D tools.

Tertiary structure prediction: During the determination of a probable structure of a protein, herein called the target protein, through computational tools like Modeller, a template protein is required. This template protein already has a known structure determined through X-ray crystallography or NMR, and should be highly similar to the target protein for which the structure is to be determined. Situations may arise when there are no template proteins to model primarily because of unavailability of similar proteins, and such situations need the target protein to be modeled on its own. Such targets are difficult to model computationally. Based on the percentage of similarity between sequences, three types of protein structure prediction tools are known:

Homology Modeling Tools: Used when sequence similarity between target and template proteins is greater than 30%. Most parts of a protein can be modeled.

Threading or Fold-prediction Tools: Used when sequence similarity between target and template proteins is less than 30%. Only some parts such as a fold of a protein can be modeled.

Ab-initio Modeling: '*Ab-initio*' means 'from the beginning'. Used when there are no similar template sequences at all on which to model the target protein.

Drug design

In silico drug designing tools enable us to design drugs for novel or older druggable targets and to accelerate drug discovery process for challenging diseases such as HIV, tuberculosis, malaria and cancer. Even existing drugs can be modified or optimized for more specific action or for increased bioavailability and good Absorption, Distribution, Metabolism, Excretion and Toxicity (ADMET) properties, in order to derive a drug molecule with enhanced efficacy, potency and safety properties. Molecular docking provides us with a picture of drug binding to its intended target and the extent to which it binds, provided by drug-ligand binding energy calculations. Databases such as PubChem hosting small molecules library for high-throughput virtual screening at NCBI and tools such as Accelrys' Discovery Studio and Schrodinger's Small Molecule Drug Discovery Suite are powerful in helping us improvise and innovate in the drug discovery space. Cases in point are FDA-approved anti-HIV drug nelfinavir (Viracept) [Figure 136.4(a) and (b)] and anti-influenza drug zanamivir (Relenza), which have been discovered computationally through rational drug design.

Pathways

Pathways or biological networks are a series of proteins or molecular actions connected together to transduce signal from one molecule to another through a signal transduction process, or to exert a regulatory effect through protein-protein or protein-DNA interactions and thereby affect gene expression or metabolism. The pathways focus on several molecules acting in concert to play a functional role inside a cell. One of the better known Pathways databases and tool, Kyoto Encyclopedia of Genes and Genomes (KEGG), is a comprehensive resource which focuses on metabolic and genetic pathways, cellular processes, organismal processes and environmental information processing. Further, tools such as Search & Color Pathway in KEGG help interpret biological phenomena through the mapping of large-scale molecular datasets in genomics, transcriptomics, proteomics, and metabolomics, to the KEGG pathway maps. There are some more tools for pathway reconstruction like Pathway Tools from MetaCyc, allowing us to draw pathways based on our findings and generate hypothesis.

IV. GENOMICS, PROTEOMICS AND METABOLOMICS

Focussed on *large scale* study of *key molecules* inside a cell, Bioinformatics encompasses several sub-fields such as genomics (studying genes and RNAs), proteomics (studying proteins) and metabolomics (studying metabolites). The complete set of DNA, proteins and metabolites in a cell, tissue, organ or an organism is termed a genome, a proteome and a metabolome, respectively. At the outset, it is *important* to note that each of these subfields relies heavily on Bioinformatics databases and tools to glean essential information and transform it into knowledge.

A. Genomics

Broadly speaking, the field of genomics utilizes *genes and/ or RNAs* to understand and decipher cellular processes. DNA and RNA *sequences*, *gene expression* data utilizing mRNA *amounts* present in cells (transcriptomics) are the main types of data generated. While DNA and RNA sequences provide an insight into *specific identification* and *evolutionary processes* operating at the *cellular* or *organismal* level, gene expression data can provide a clue to the *functional* aspects inside a cell in an organism as well as in disease states at a *systems* level.

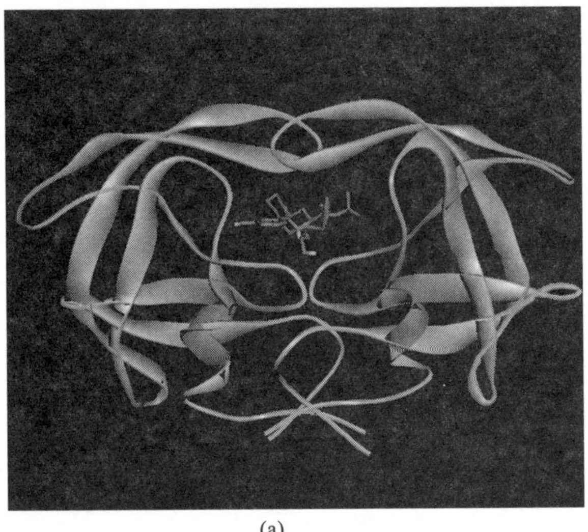

(a)

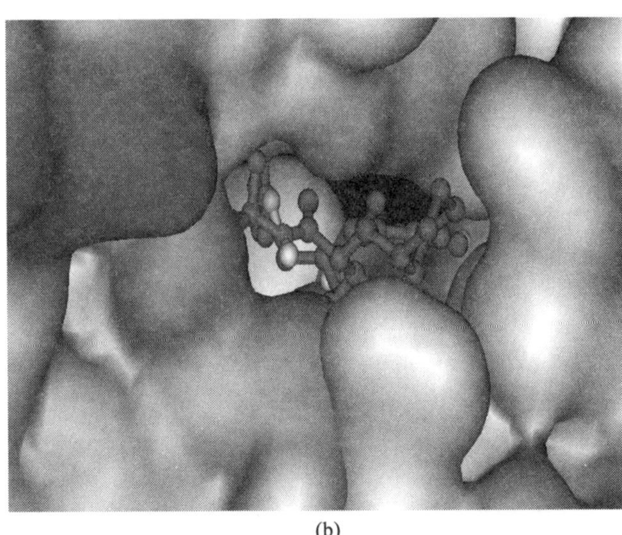

(b)

Figure 136.4 (a) Nelfinavir (Viracept) in complex with HIV-1 protease, structure downloaded from PDB with PDB ID: 1OHR (Reference: Kaldor et al., 1997. *J. Med. Chem.* 40: 3979-3985) (b) Nelfinavir (Viracept) in complex with HIV-1 protease seen bound inside the cavity, structure downloaded from PDB with PDB ID: 1OHR. (Reference: Kaldor et al., 1997. *J. Med. Chem.* 40: 3979-3985). (*see Plate 56 for colour figure*)

Major technologies in use include Next Generation Sequencing (NGS) for both DNA sequencing and RNA sequencing (RNA-Seq), microarrays, quantitative real time-PCR; while newer technologies such as NanoString nCounter gene expression CodeSets, Cap analysis of gene expression (CAGE) followed by single-molecule sequencing are being developed.

B. Proteomics

Proteomics utilizes the functional counterpart of genes, i.e., proteins, to understand biological processes. Since proteins are the molecules through which many genes function, and many proteins are modified post-translationally in order to play their biological roles, specificity and functional relevance is achieved mainly through proteins. Further, the longevity of mRNA and proteins alongwith their respective amounts inside a cell differs, and hence, a study of proteins at the global and cellular scale (proteomics) is necessary to contrast, complement or validate results arising from genomics. The information is gleaned on the abundance of proteins in a specific proteome (whole complement of proteins), as well as their structure, function, variation with time and stress, post-translational modifications and their effects, alongwith their interactions with other proteins and/or nucleic acids. The powerhorse technologies of proteomics are the 2D-PAGE followed by mass spectrometry with variations in the form of peptide mass fingerprinting (PMF), matrix-assisted laser desorption/ionization-time of flight (MALDI-TOF) mass spectrometry, surface-enhanced laser desorption/ionization-time of flight (SELDI-TOF) mass spectrometry and tandem mass spectrometry (*de novo* sequencing). Recently developed technologies include protein microarrays and nanotechnologies.

Protein evolution is an important phenomena in biology which gives rise to families and superfamilies of structurally related proteins, within which sequence identities can be extremely low. A whole protein molecule can be separated into domains or folds that can evolve independently of the rest of the protein chain. Tools like Pfam, InterPro, CATH, SCOP etc. help us to identify conserved regions in protein sequences which might be functionally related to other sequences even when sharing low sequence identities.

C. Metabolomics

Metabolomics is the study of metabolites or small molecules (with molecular weight *less than 1500 Da*), more specifically the identification, measurements, cataloguing and understanding of their respective functions and changes during cellular processes. Metabolites such as peptides, amino acids, sugars, lipids, steroids, ethanol etc. are the *end-products of chemical processes* inside a cell and can play a role as *sources of energy, antibiotics, pigments, co-factors, drugs, signalling molecules or pheromones*. Their *levels* inside a cell change in response to external stimuli such as drug action or as a result of physiological, pathophysiological processes or in response to genetic manipulation, e.g., insertion or deletion of a gene/s in a *dynamic* manner. The measurement of these levels provides a picture of a changing metabolome response. In a metabolome network, a large network of metabolites is involved in reactions where the end products of one reaction serve as inputs for another set of chemical reactions. The techniques widely used in metabolomics are the separation techniques such as high performance liquid chromatography (HPLC), gas chromatography (GC), capillary electrophoresis (CE) and detection techniques such as nuclear magnetic resonance (NMR) spectroscopy and mass spectrometry (MS). Combination of these such as GC-MS and LC-MS is also used. Databases such as human metabolome database, biochemical, genetic and genomic metabolic network reconstruction (BiGG) database, METLIN metabolite database store many of the relevant information on metabolites for querying and analysis purposes. Usually, after 1000's of metabolites have been identified and quantified, dimensionality reduction techniques such as Principal Components Analysis (PCA) to find major constituents is carried out followed by computational network/pathways modelling and subsequent biological interpretation. The main applications of metabolomics are in the area of nutrigenomics (the impact of nutrition on genome or gene expression), drug studies, clinical toxicology studies, functional genomics and the likes.

Linking natural products and drug-like molecules to the repertoire of metagenomics data will find many applications and use cases in pharma and biotech industry.

SUGGESTIONS FOR FURTHER READING

1. Searls D B (2010), The Roots of Bioinformatics, *PLoS Comput Biol.,* 6(6): E1000809.

2. Information from the Websites, with URLs Mentioned in Tables 1 and 2.

3. PLoS Computational Biology Collections Online http://www.ploscollections.org/static/pcbiCollections

4. NCBI Website http://www.ncbi.nlm.nih.gov

5. Dan E. Krane and Michael L. Raymer (2003), *Fundamental Concepts of Bioinformatics*, Pearson ISBN-13:978–8177587579.

6. Teresa K. Attwood and David Parry-Smith (2003), *Introduction to Bioinformatics*, Pearson Education. ISBN-13:978–8178085074.

7. David W. Mount (2013), *Bioinformatics: Sequence and Genome Analysis*, Cold Spring Harbor Laboratory Press. ISBN-13:978–0879697129.

8. Pennington S. and Dunn M. (2001), *Proteomics: From Protein Sequence to Function*, Springer-Verlag Telos. ISBN-13:978–0387915890.

9. John C. Lindon, Jeremy K. Nicholson, Elaine Holmes (2007), *The Handbook of Metabonomics and Metabolomics*, Elsevier Science, ISBN-13:978–0444528414.

Index

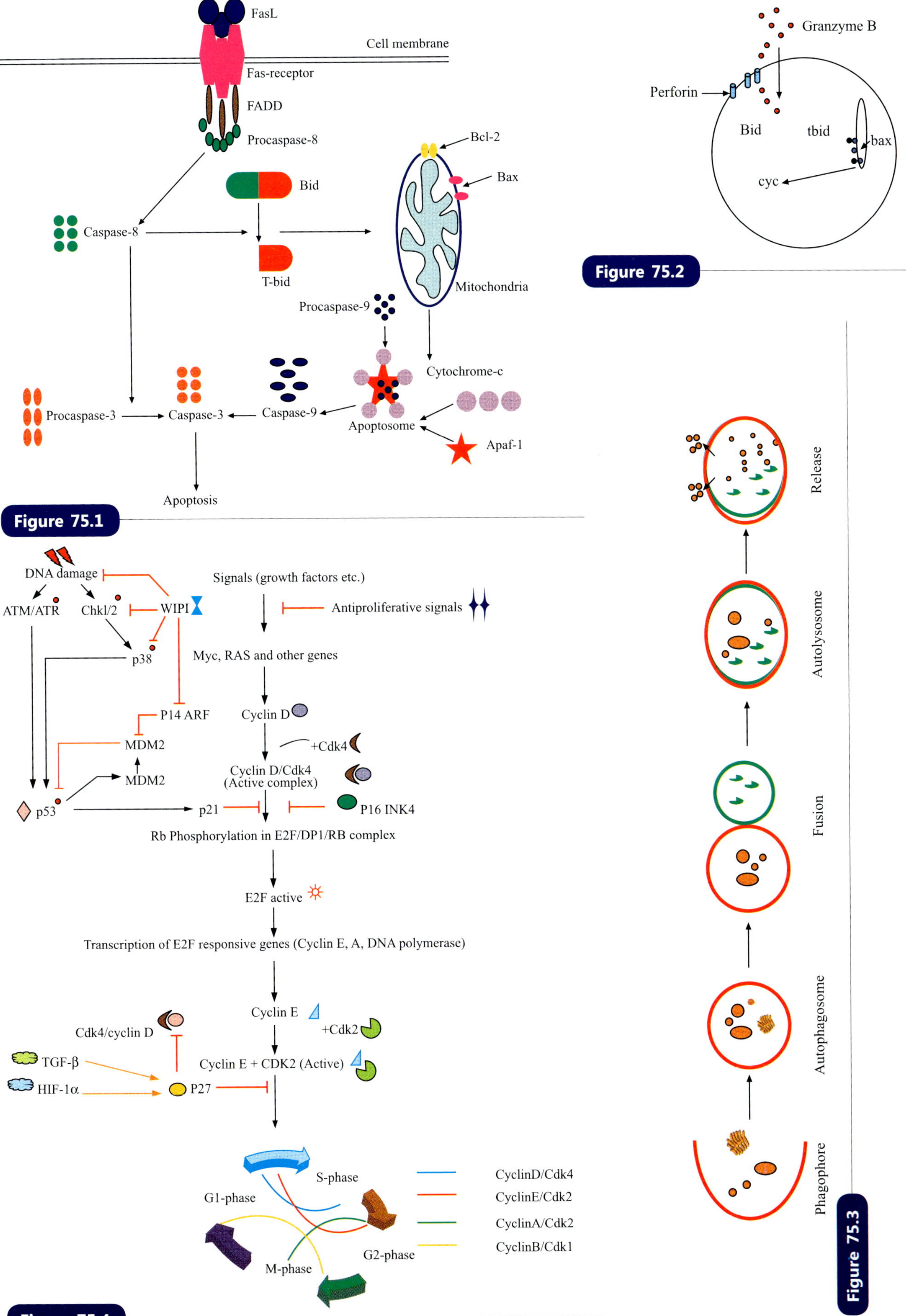

FasL
Cell membrane
Fas-receptor
FADD
Procaspase-8
Bcl-2
Bid
Bax
Caspase-8
T-bid
Mitochondria
Procaspase-9
Cytochrome-c
Procaspase-3
Caspase-3
Caspase-9
Apoptosome
Apaf-1
Apoptosis
Figure 75.1
Granzyme B
Perforin
Bid
tbid
bax
cyc
Figure 75.2
Release
Autolysosome
Fusion
Autophagosome
Phagophore
Figure 75.3
DNA damage
Signals (growth factors etc.)
ATM/ATR
Chkl/2
WIPI
Antiproliferative signals
p38
Myc, RAS and other genes
P14 ARF
Cyclin D
MDM2
+Cdk4
MDM2
Cyclin D/Cdk4
(Active complex)
p53
p21
P16 INK4
Rb Phosphorylation in E2F/DP1/RB complex
E2F active
Transcription of E2F responsive genes (Cyclin E, A, DNA polymerase)
Cdk4/cyclin D
Cyclin E
+Cdk2
TGF-β
Cyclin E + CDK2 (Active)
HIF-1α
P27
S-phase
G1-phase
CyclinD/Cdk4
CyclinE/Cdk2
CyclinA/Cdk2
M-phase
G2-phase
CyclinB/Cdk1
Figure 75.4

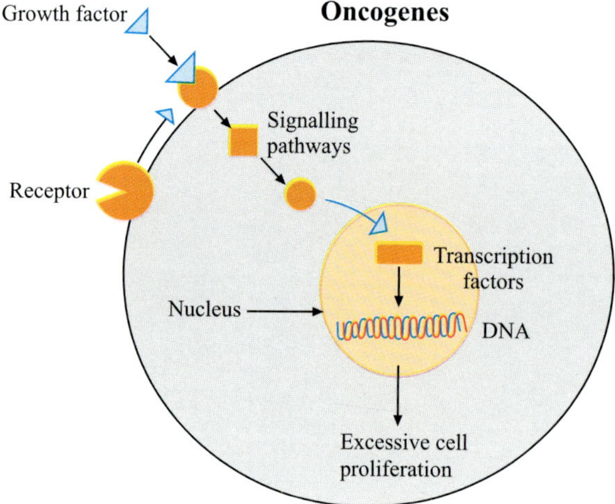

Figure 76.5

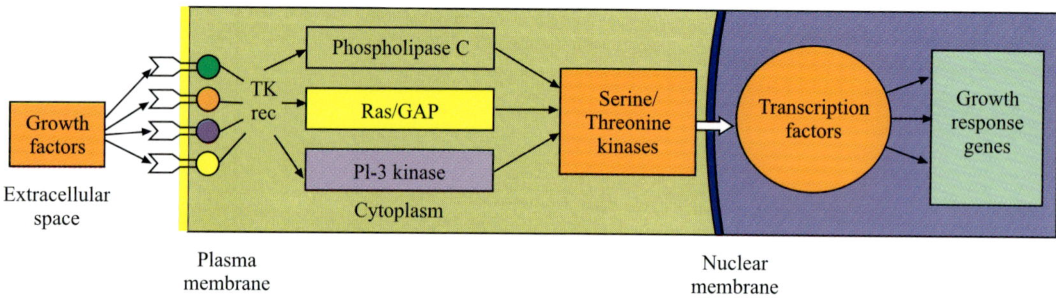

Figure 76.6

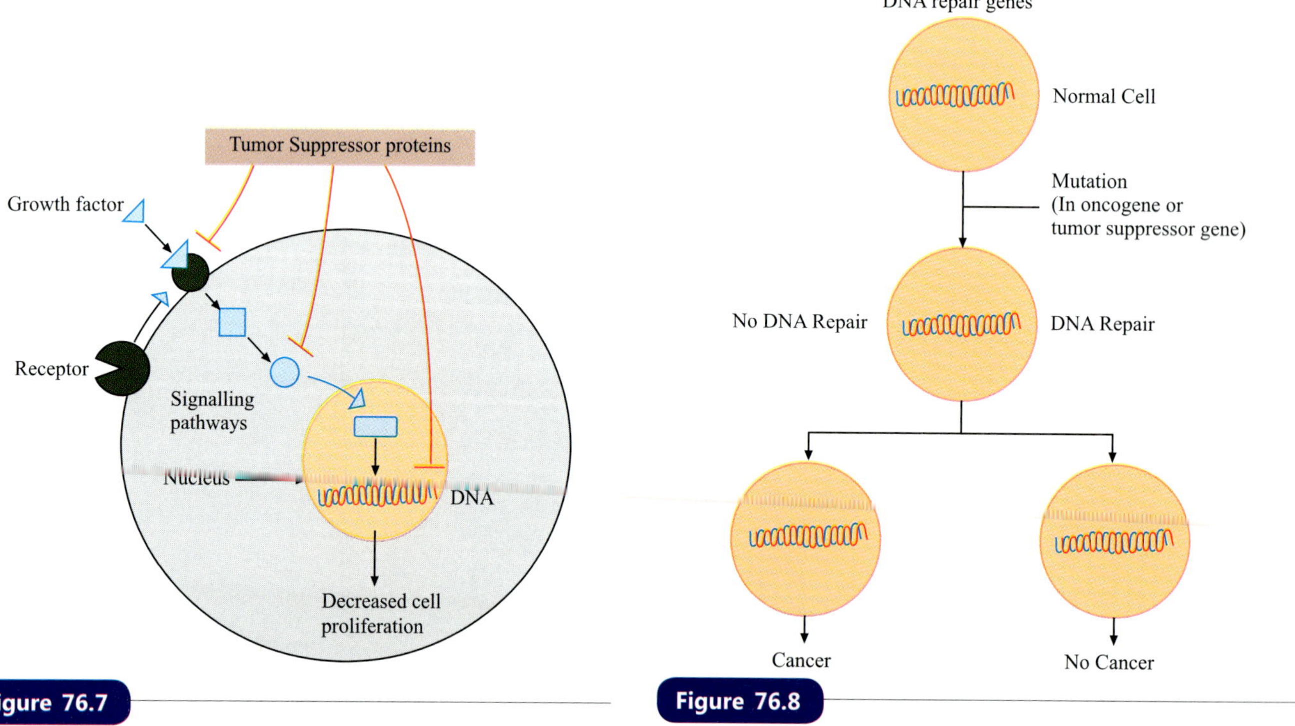

Figure 76.7   **Figure 76.8**

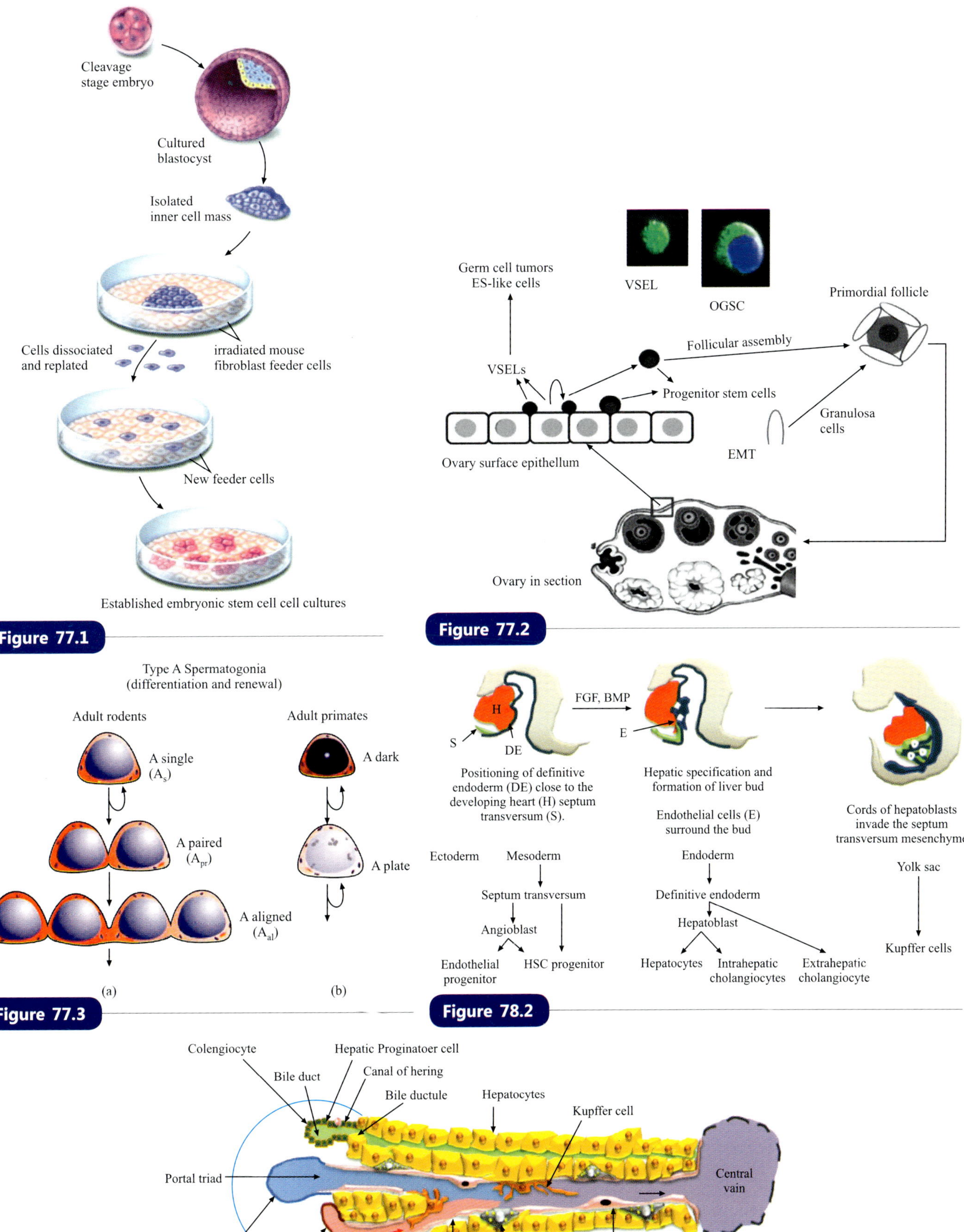

Figure 77.1

Figure 77.2

Figure 77.3

Figure 78.2

Figure 78.1

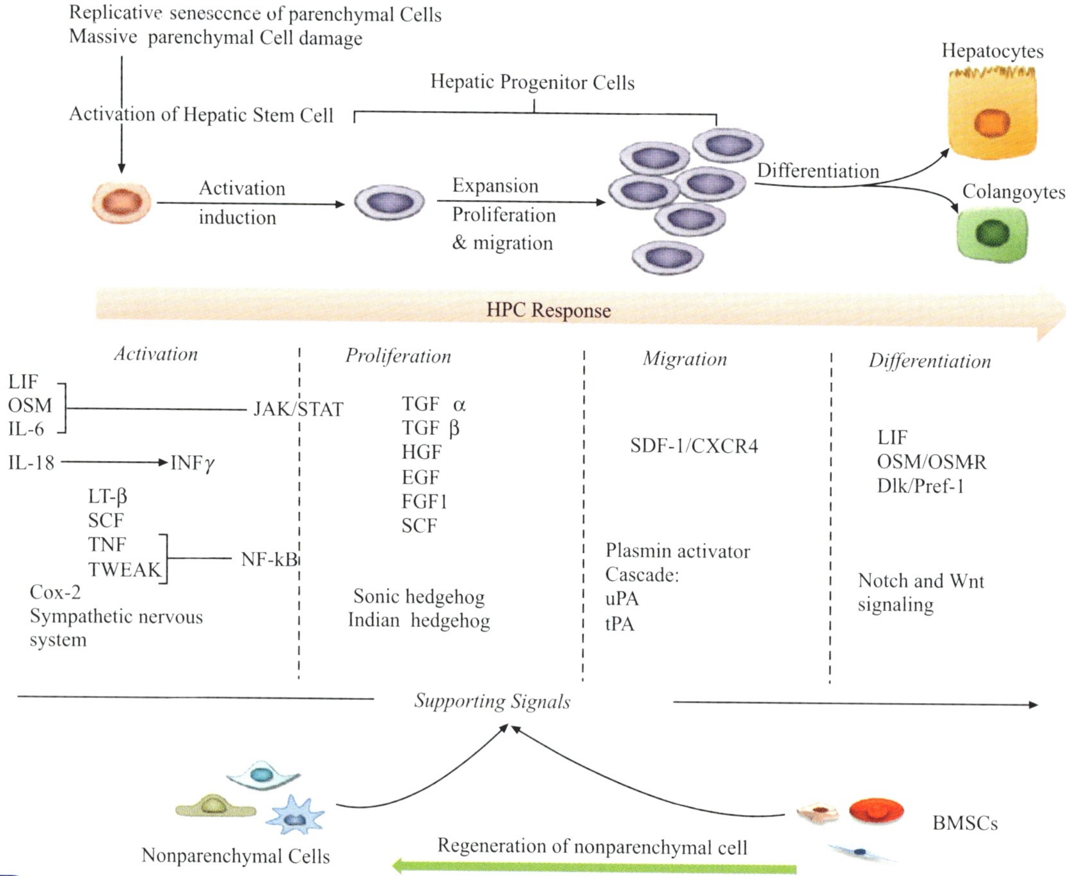

Figure 78.4

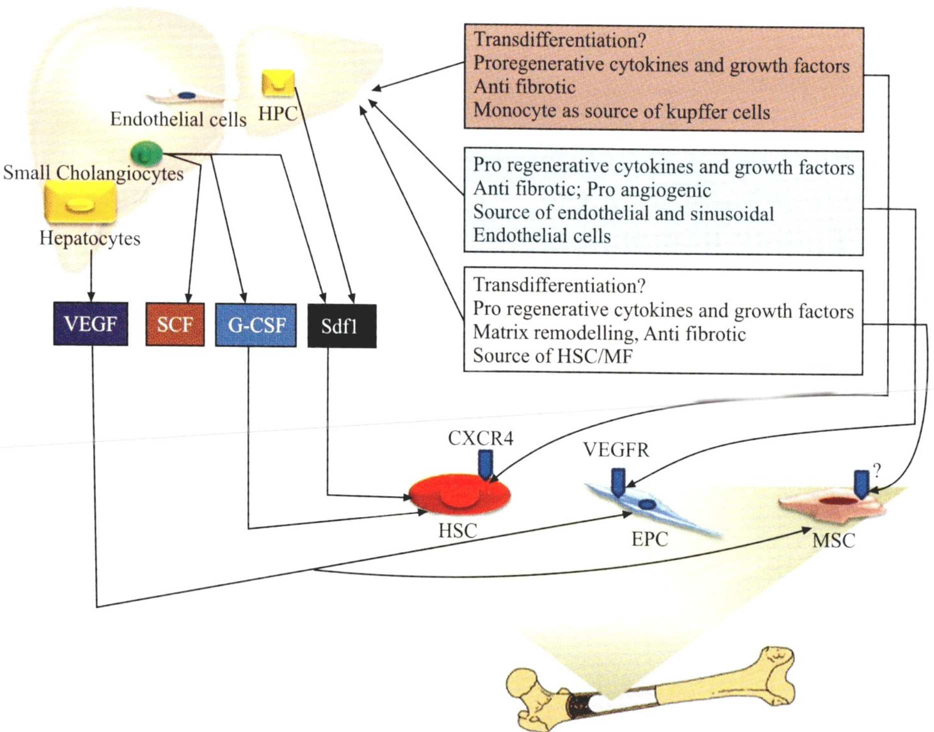

Figure 78.5

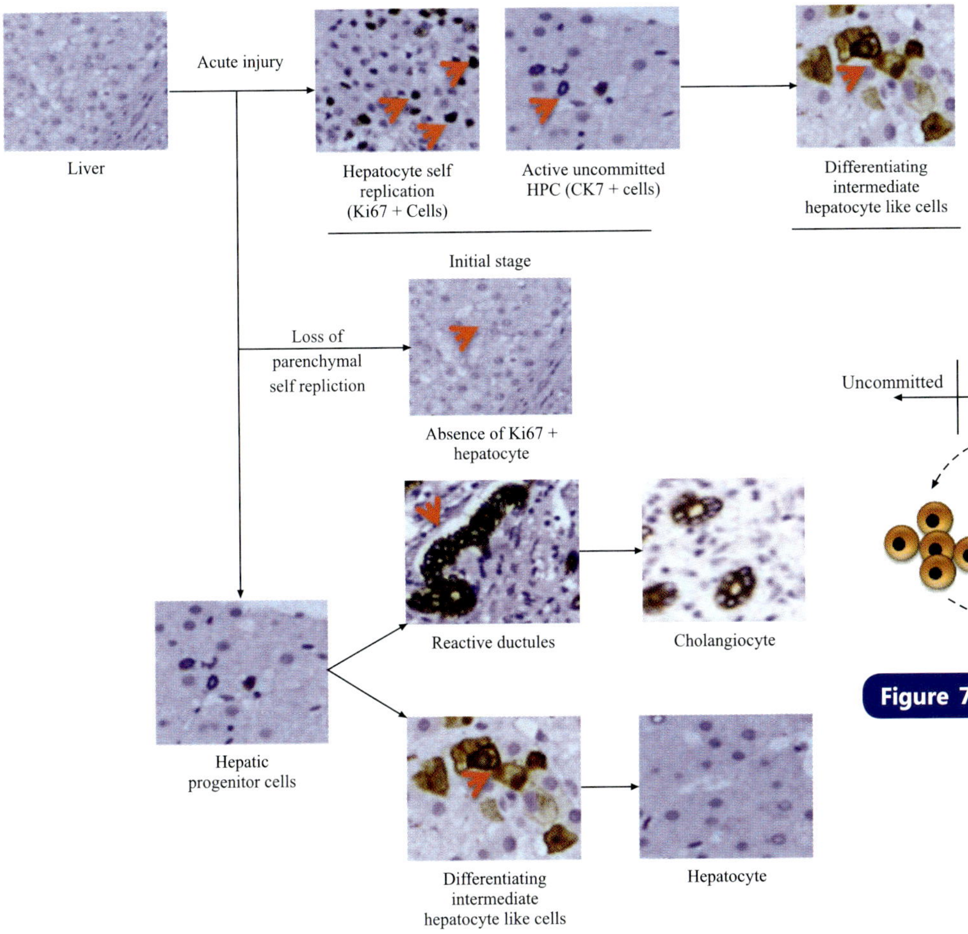

Liver
Acute injury
Hepatocyte self replication (Ki67 + Cells)
Active uncommitted HPC (CK7 + cells)
Differentiating intermediate hepatocyte like cells
Initial stage
Loss of parenchymal self repliction
Absence of Ki67 + hepatocyte
Reactive ductules
Cholangiocyte
Hepatic progenitor cells
Differentiating intermediate hepatocyte like cells
Hepatocyte

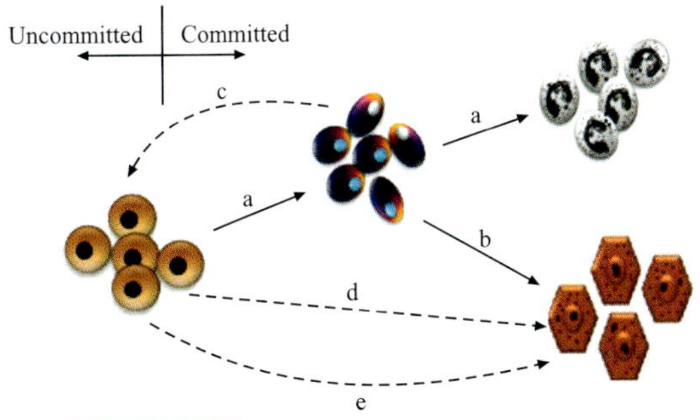

Uncommitted
Committed
c
a
a
b
d
e

Figure 79.3

Figure 78.6

H
1
Hormone binds to specific receptor (stimulatory external signal)
2
2
The conformational change in the receptor causes replacement of GDP by GTP
4
Adenylyl cyclase is activated
Plasma membrane
Ac
R
γ
β
GDP Gsα
GTP Gsα
3
GTP
GDP
Gsα moves to adenylyl cyclase
ATP
Cyclic AMP activates protein kinases
Cyclic AMP
5′ AMP
Phosphorylation of cellular proteins by protein kinases
Cellular response

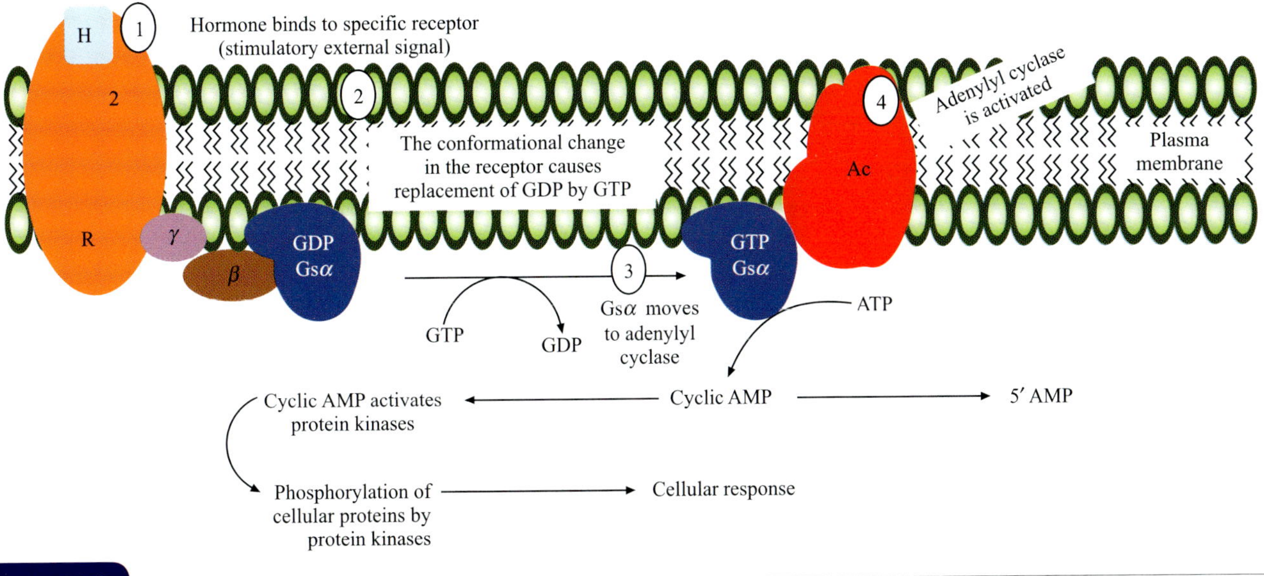

Figure 81.3

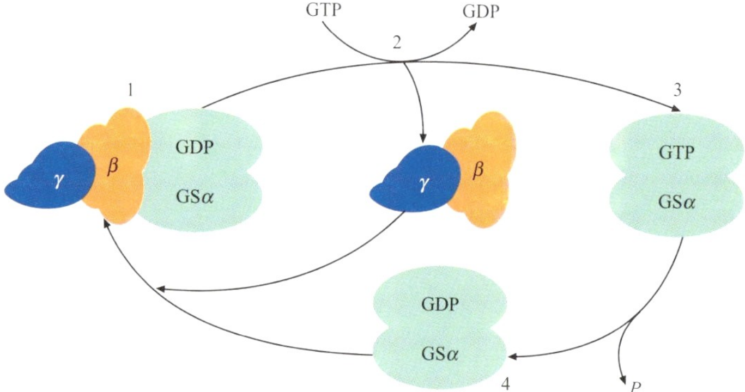

Figure 81.4

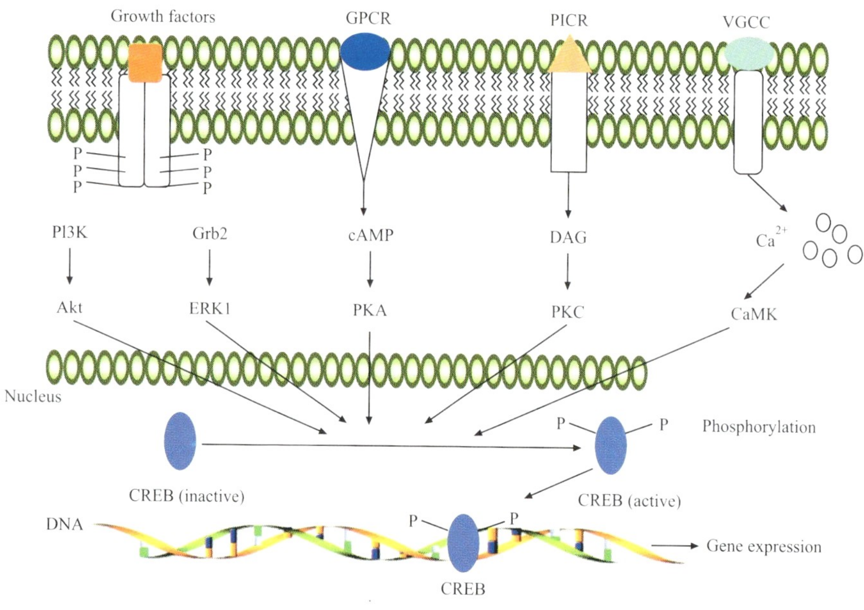

Figure 81.5

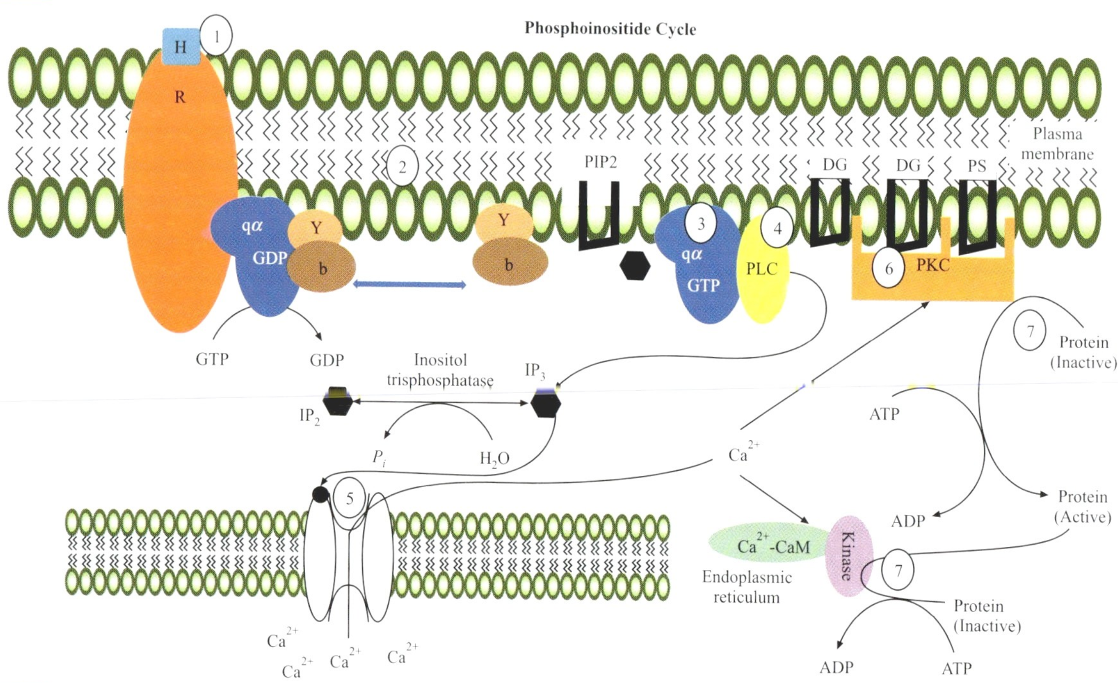

Figure 81.6

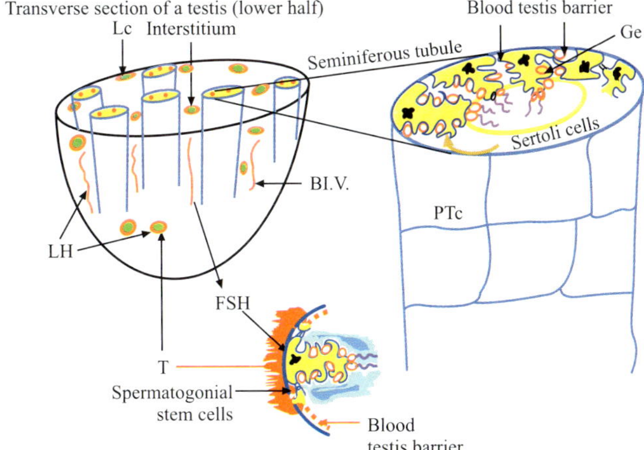

Figure 92.2A

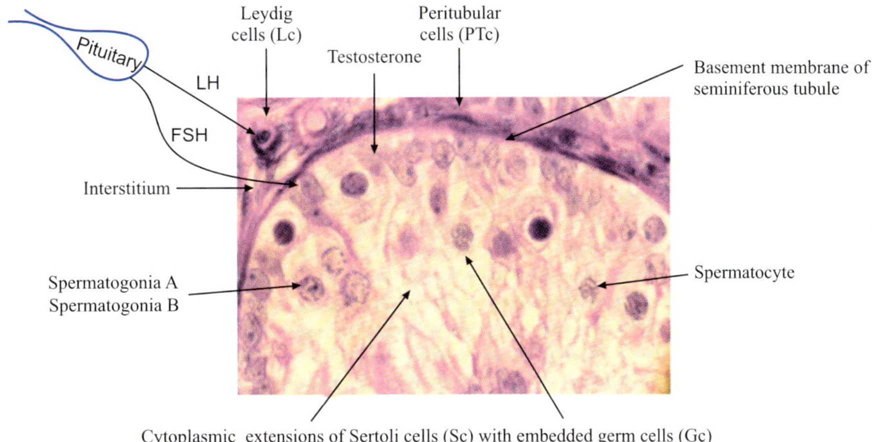

Figure 92.2B

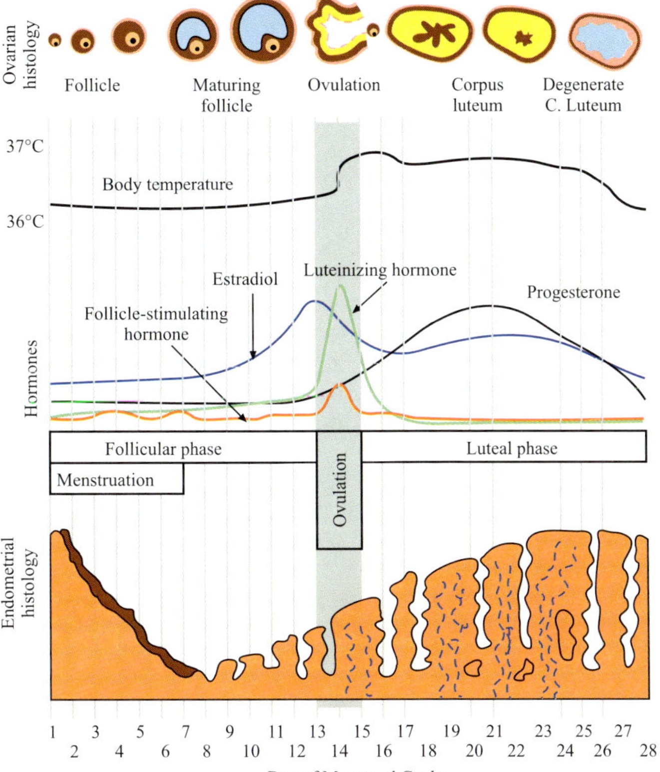

Figure 94.5

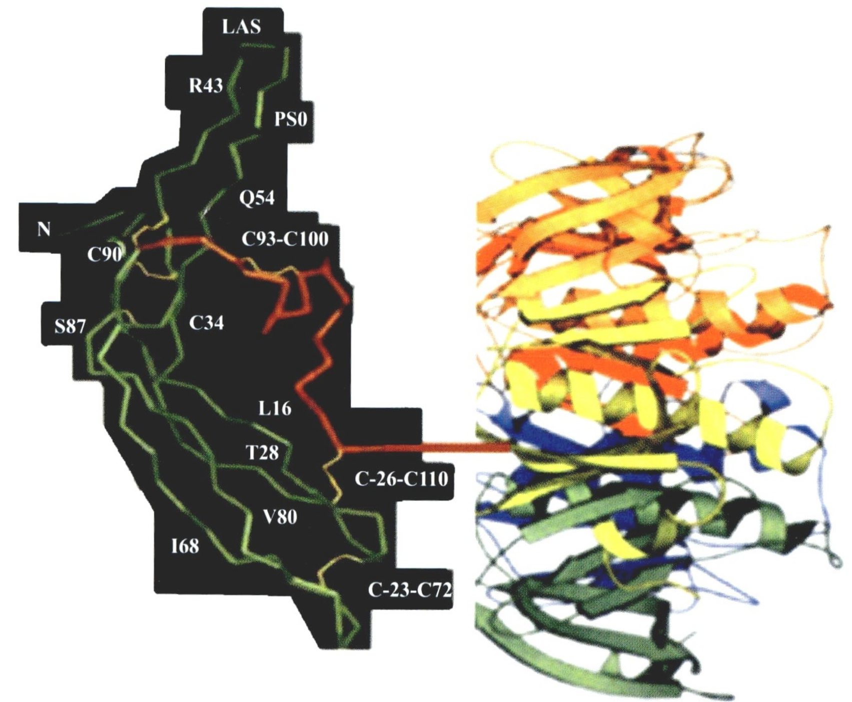

Figure 99.3

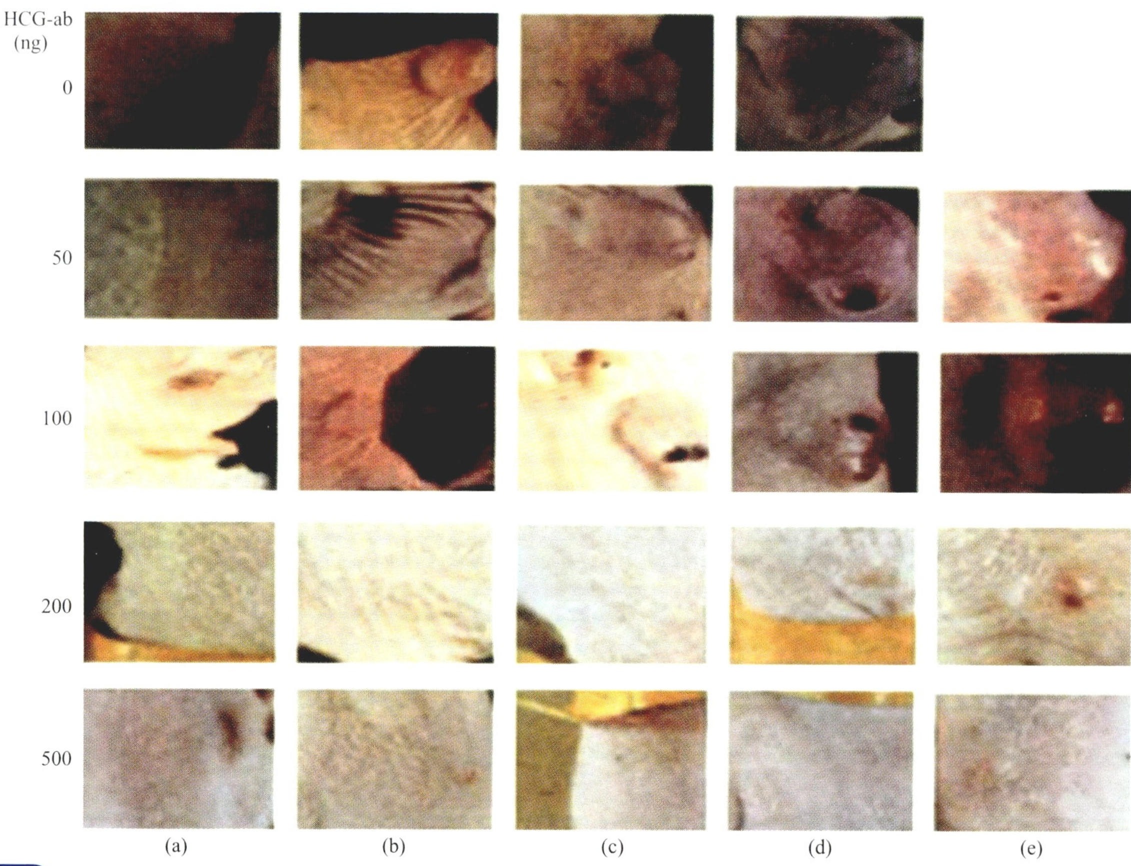

Figure 99.6

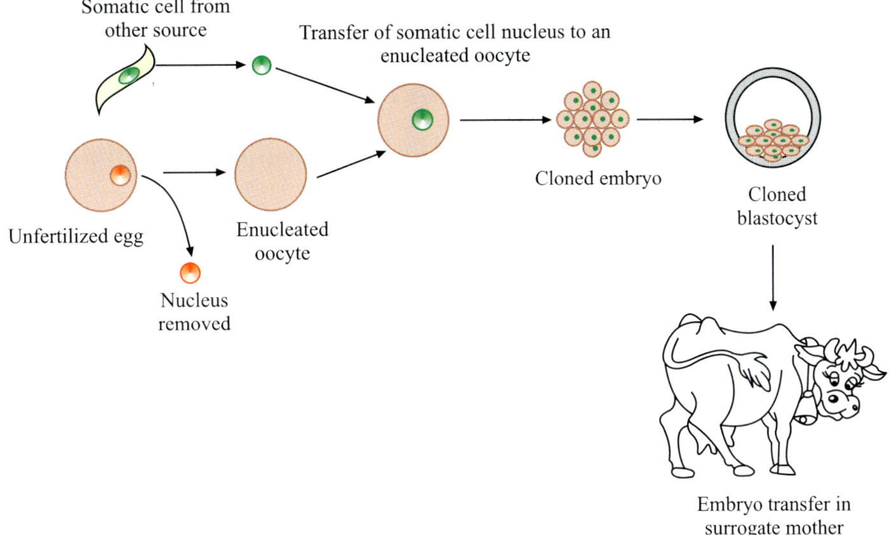

Figure 100.1

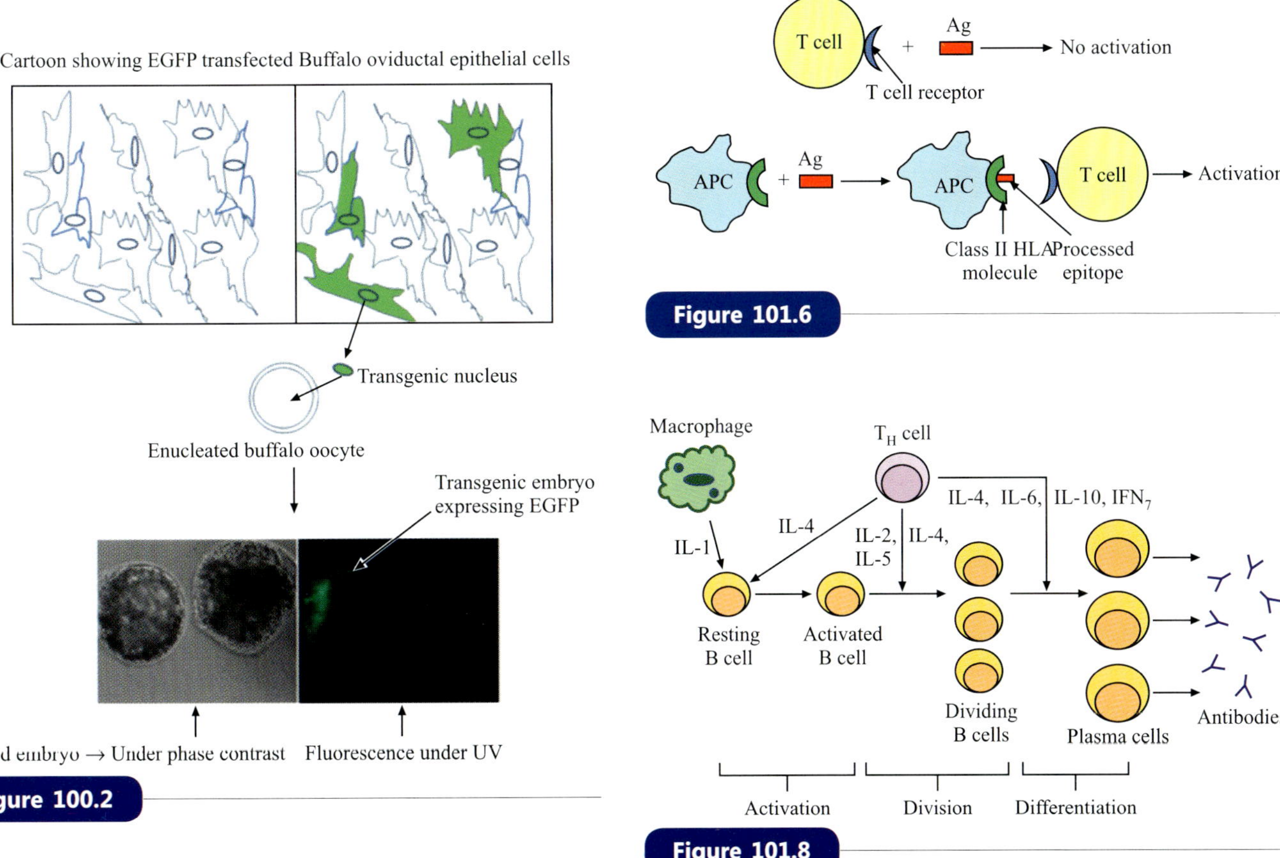

Figure 100.2

Figure 101.6

Figure 101.8

Complex	Major Histocompatibility Complex							
Class	II			III		I		
Gene	DP	DQ	DR	C	TNF	B	C	A
Molecule	HLA-DP	HLA-DQ	HLA-DR	C	TNF	HLA-B	HLA-C	HLA-A

Figure 101.5

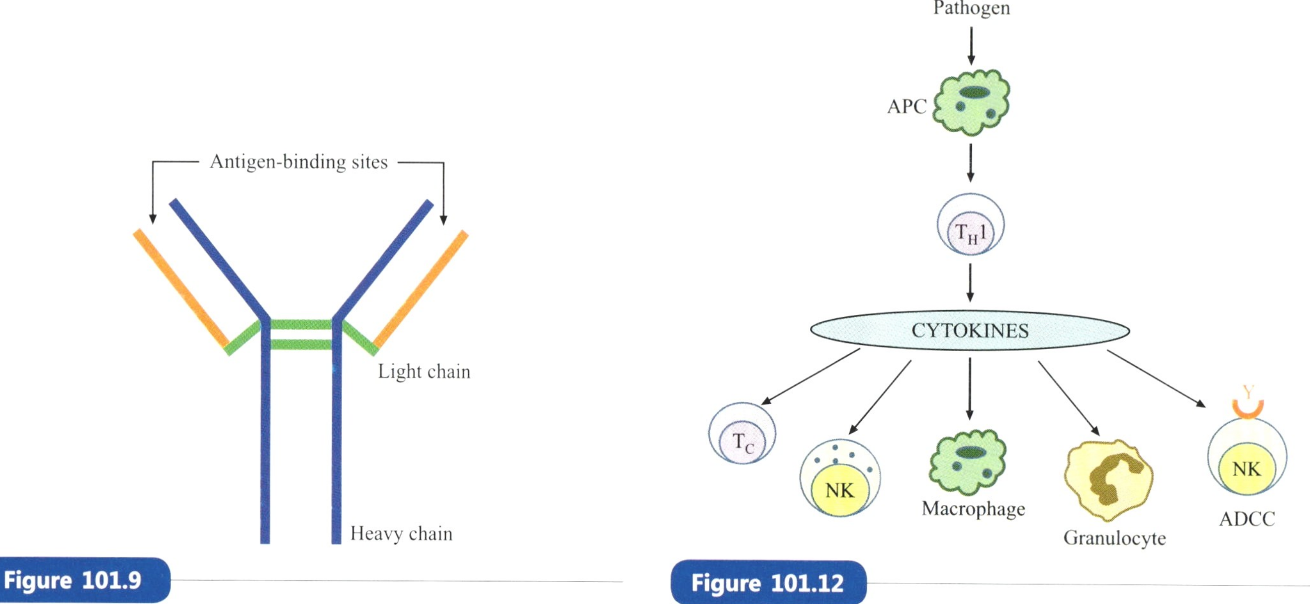
Antigen-binding sites
Light chain
Heavy chain
Figure 101.9
Pathogen
APC
$T_H 1$
CYTOKINES
T_C
NK
Macrophage
Granulocyte
NK
ADCC
Figure 101.12 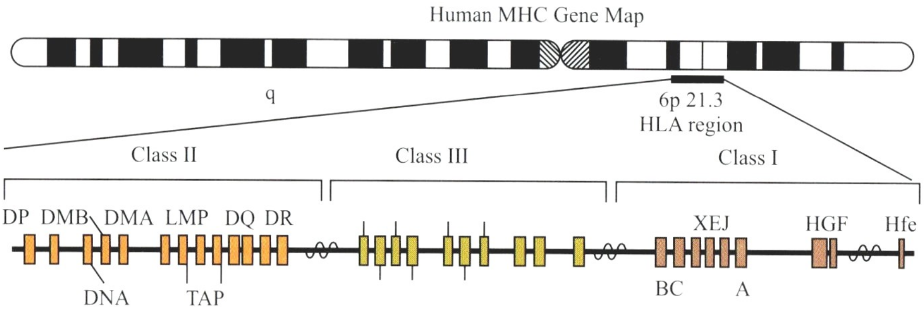
Human MHC Gene Map
q
6p 21.3
HLA region
Class II
Class III
Class I
DP DMB DMA LMP DQ DR
XEJ
HGF
Hfe
DNA
TAP
BC
A
Figure 104.1

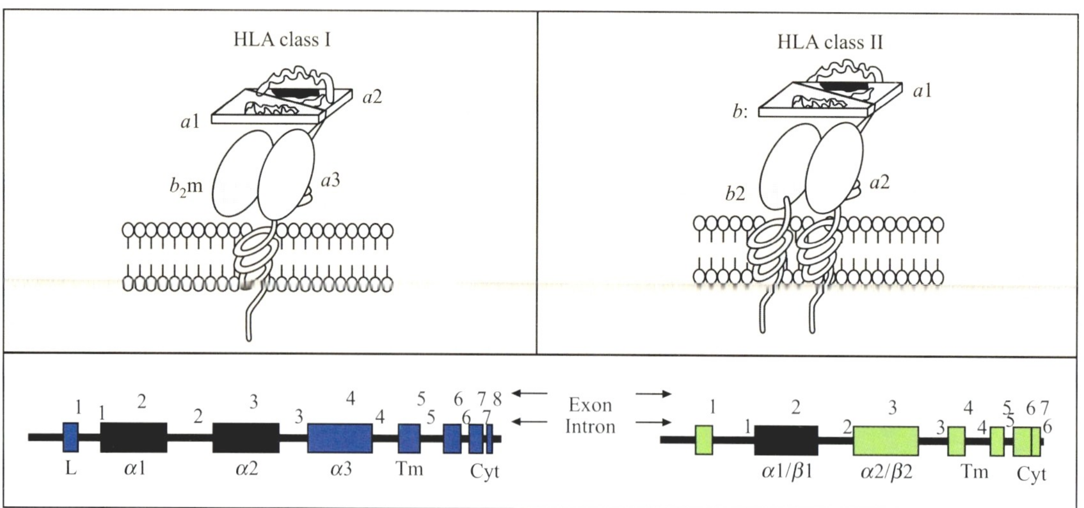
HLA class I
a2
a1
b_2m
a3
HLA class II
a1
b:
b2
a2
1 1 2 2 3 3 4 4 5 5 6 7 8 6 7
L $\alpha 1$ $\alpha 2$ $\alpha 3$ Tm Cyt
Exon
Intron
1 1 2 2 3 3 4 5 6 7 4 5 6
$\alpha 1/\beta 1$ $\alpha 2/\beta 2$ Tm Cyt
Figure 104.2

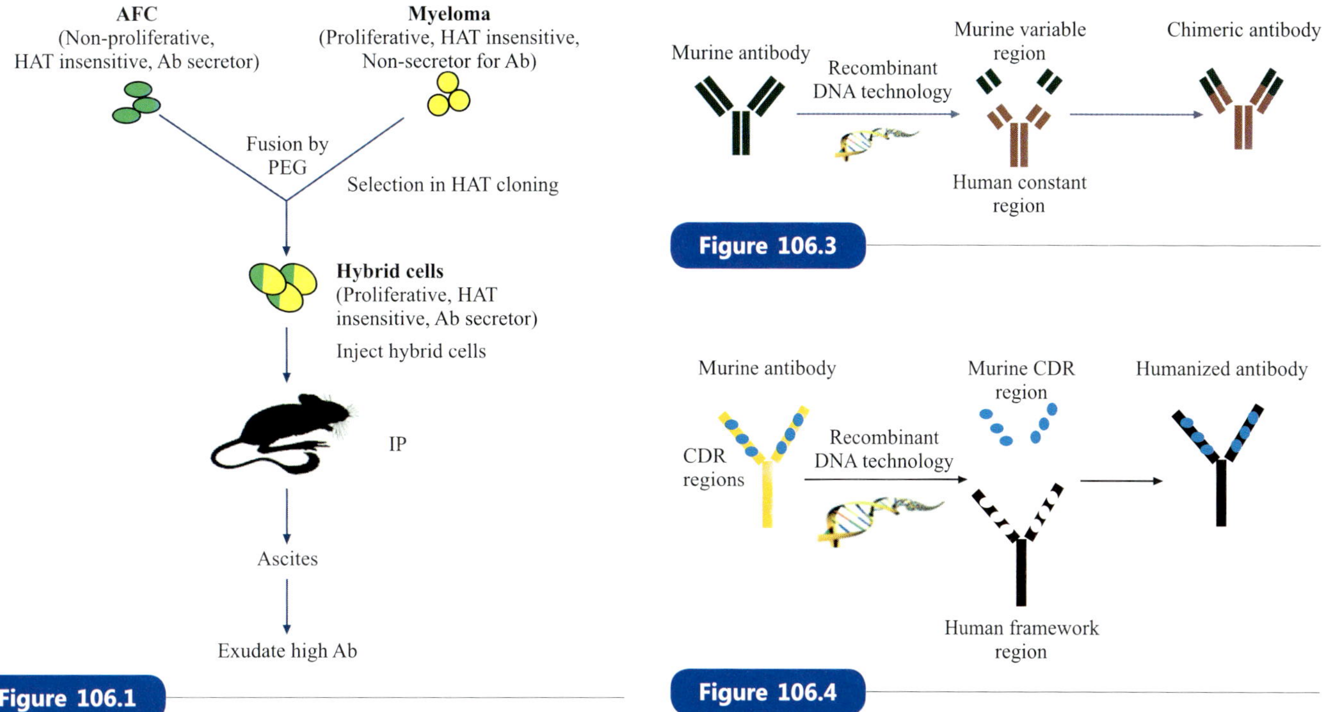

Figure 106.1

Figure 106.3

Figure 106.4

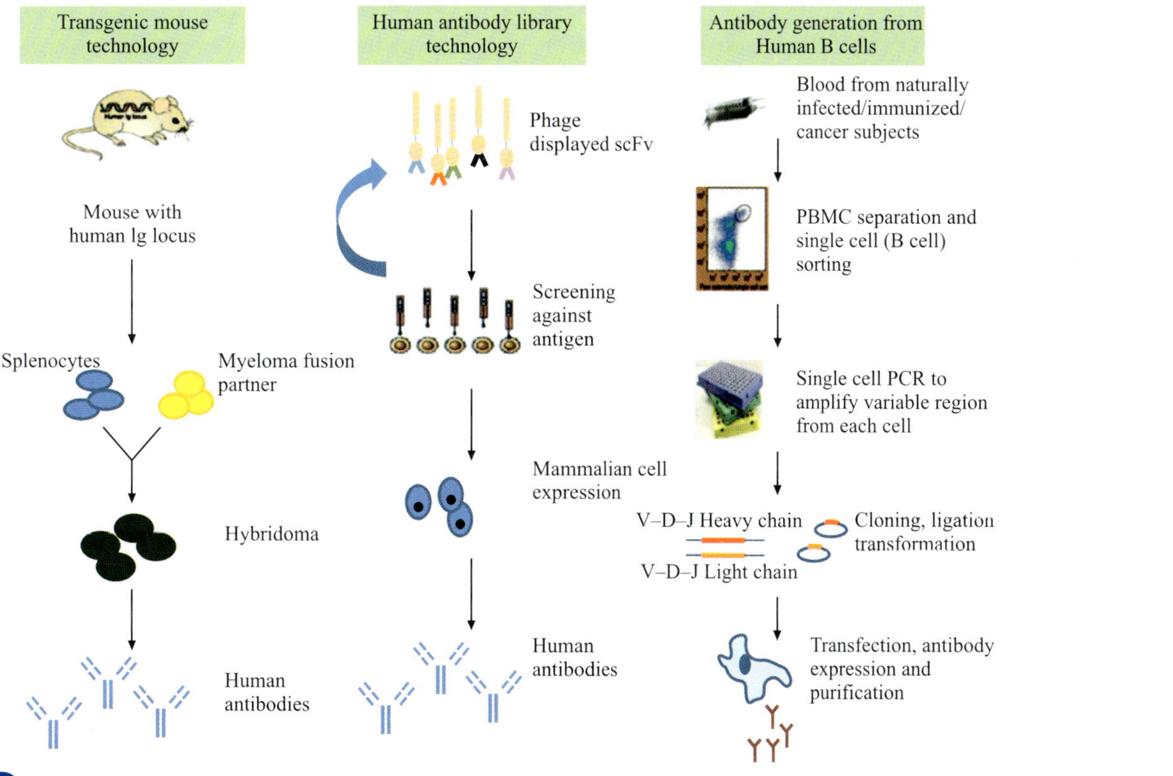

Figure 106.5

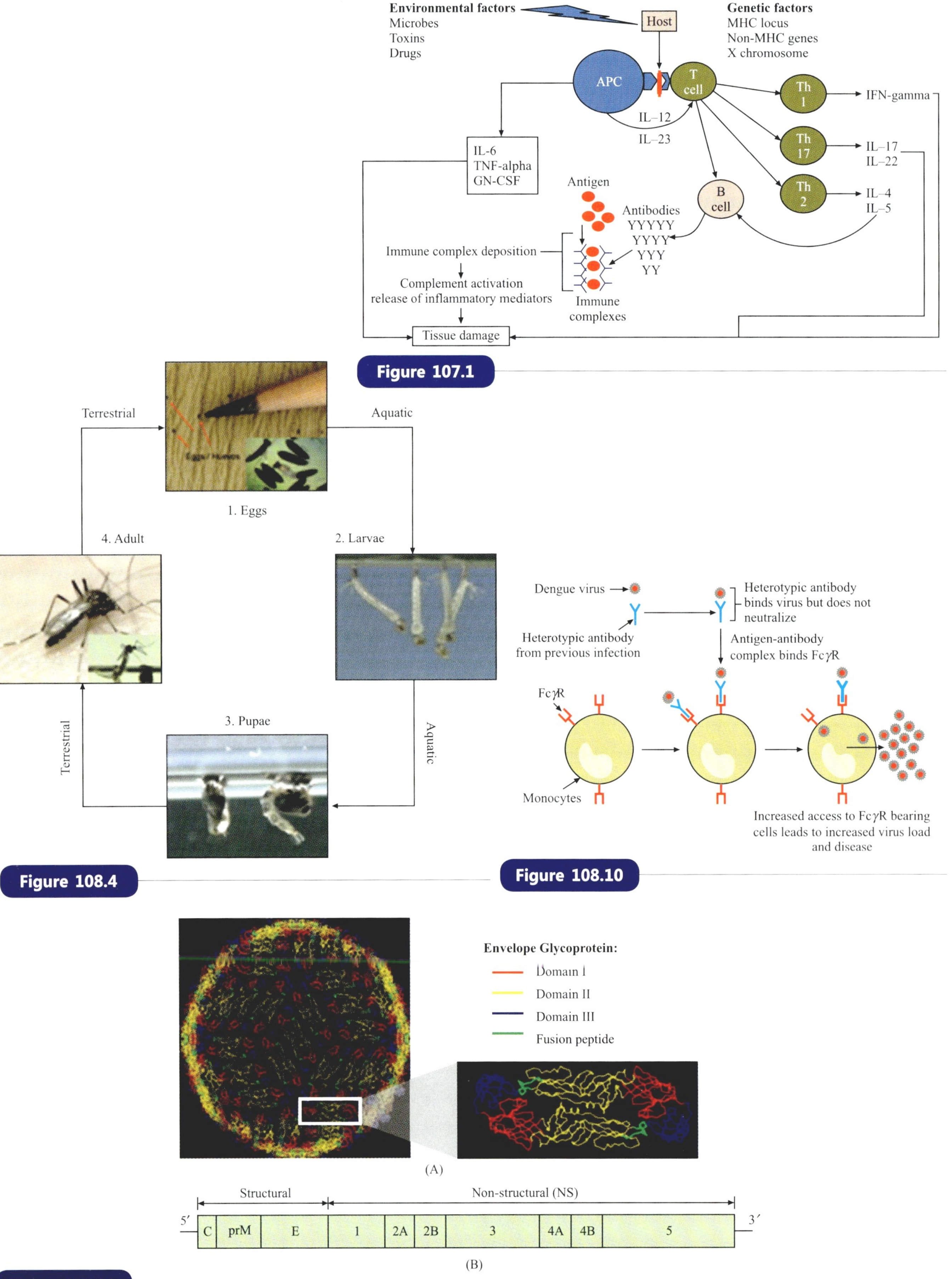

Figure 107.1

Figure 108.4

Figure 108.10

Figure 108.11

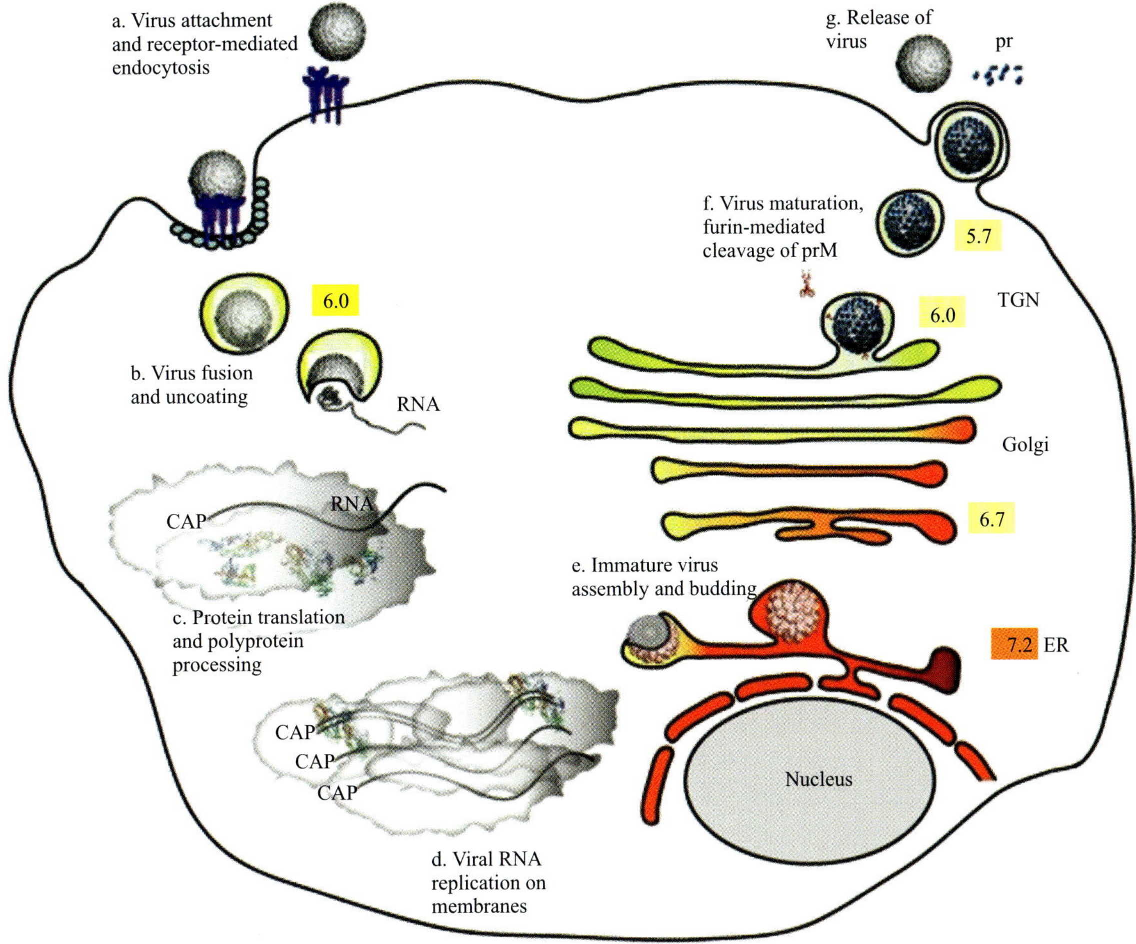

Figure 108.12

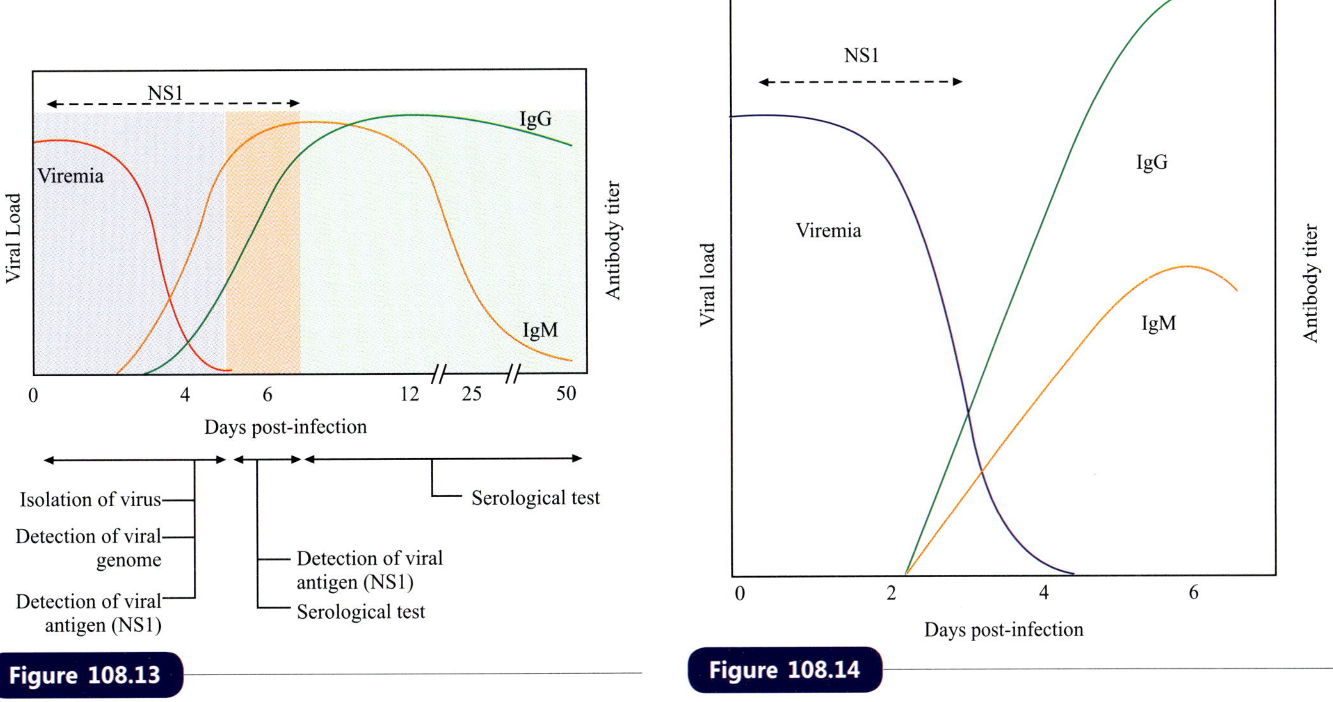

Figure 108.13

Figure 108.14

TABLE 108.7   Dengue vaccine candidates in clinical trials

Developer	Strategy	Clinical phase			
		I	II	III	IV
Sanofi Pasteur	YF17D backbone carrying DENV prM and E genes				
GSK/WRAIR	Empirical attenuation of DENVs				
Inviragen, CDC, Shantha Biotechnics	DENV-2 attenuated virus backbone carrying prM and E genes of remaining 3 DENV serotypes, separately				
NIH, NIAID, LID, Biological E, Panacea Biotech, Butantan, Vabiotech	DENVs with a 30 nucleotide attenuating deletion in the 3′NCR				
Merck/Hawaii Biotech	Recombinant E ectodomain expressed in insect cells				
GSK/WRAIR	Inactivated DENV				
Vical/WRAIR/NMRC	Plasmid DNA encoding DENV prM and E genes				

Blocks in green, blue and red denote phase completed, under study and on hold, respectively.
Abbreviations used: GSK: GlaxoSmithKline; WRAIR: Walter Reed Army Institute of Research, USA; CDC: Centers for Disease Control and Prevention, USA; NIH: National Institute of Health, USA; NIAID: National Institute of Allergy and Infectious Diseases, USA; LID: Laboratory of Infectious Diseases; NMRC: Naval Medical Research Center, USA.

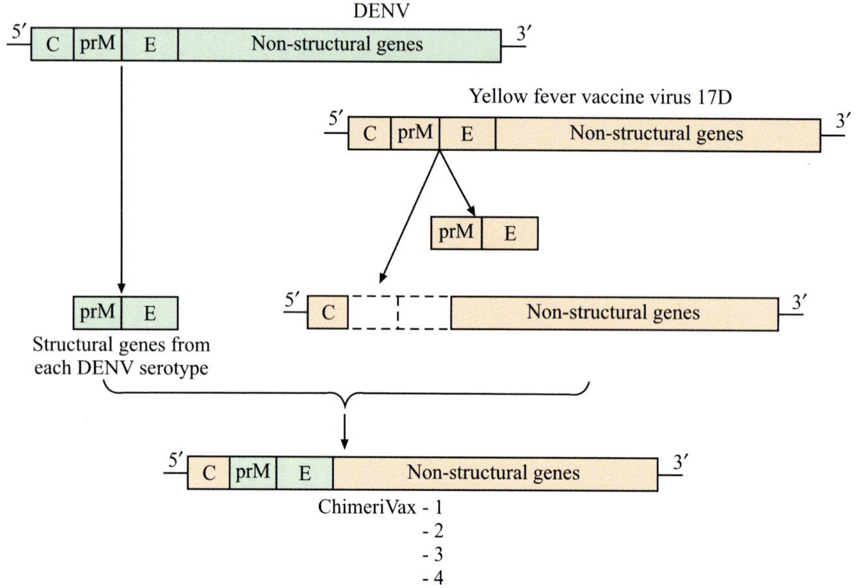

Figure 108.16

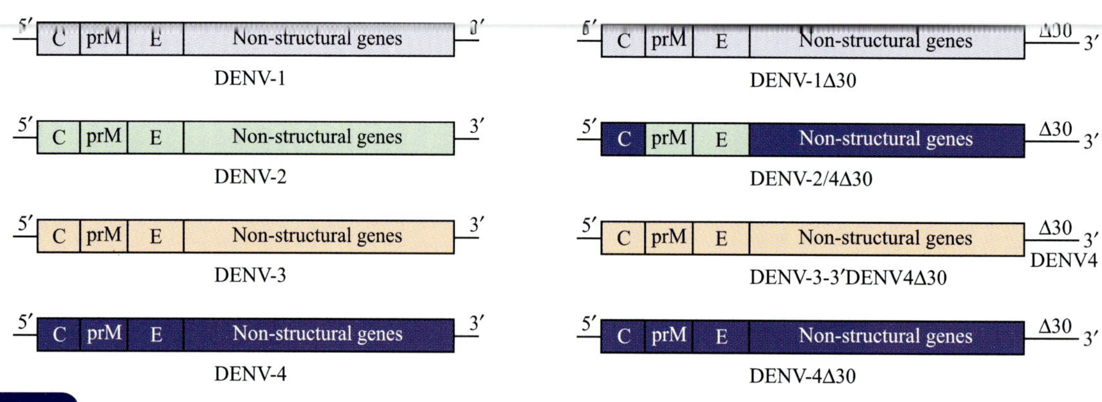

Figure 108.17

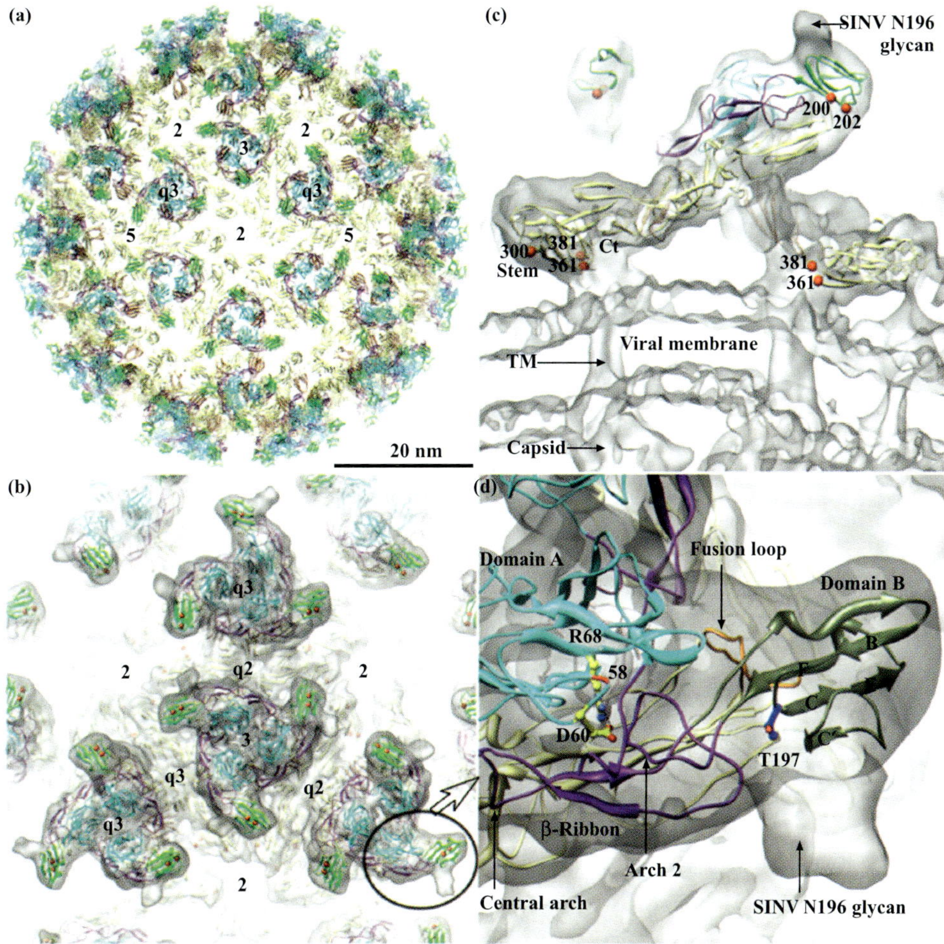
(a)
2
3
2
q3
q3
5
2
5
20 nm
(c)
SINV N196 glycan
200
202
300
381
Ct
Stem
361
381
361
Viral membrane
TM
Capsid
(b)
q3
2
2
q2
2
3
q3
q2
q3
2
(d)
Domain A
Fusion loop
Domain B
R68
B
58
D60
T197
C
β-Ribbon
Arch 2
Central arch
SINV N196 glycan

Figure 109.4

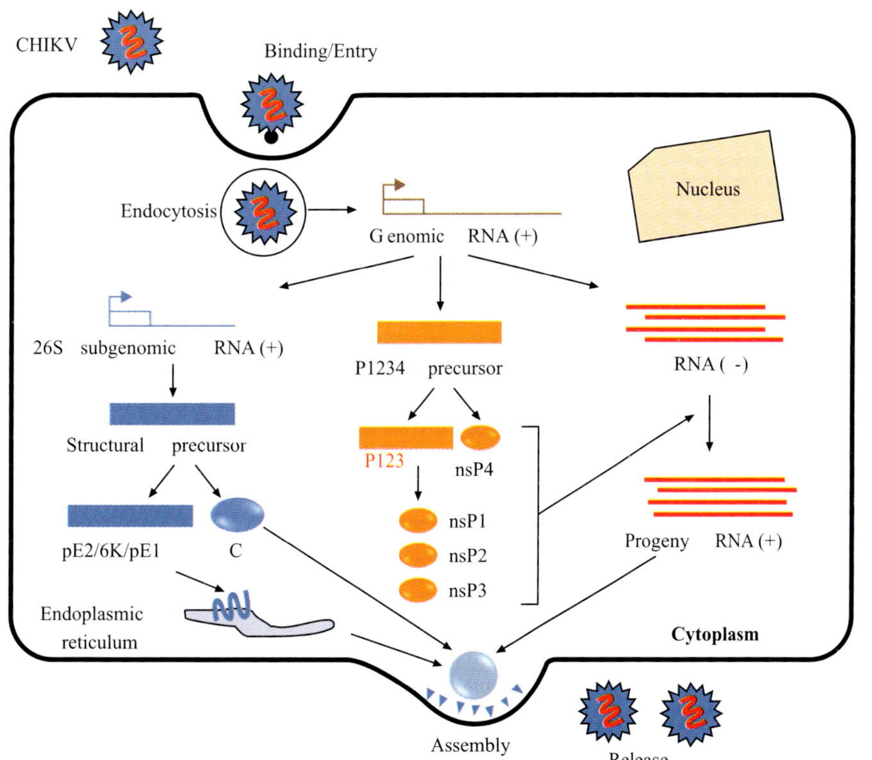
CHIKV
Binding/Entry
Endocytosis
Genomic RNA (+)
Nucleus
26S subgenomic RNA (+)
P1234 precursor
RNA (-)
Structural precursor
P123
nsP4
pE2/6K/pE1
C
nsP1
nsP2
nsP3
Progeny RNA (+)
Endoplasmic reticulum
Cytoplasm
Assembly
Release

Figure 109.6

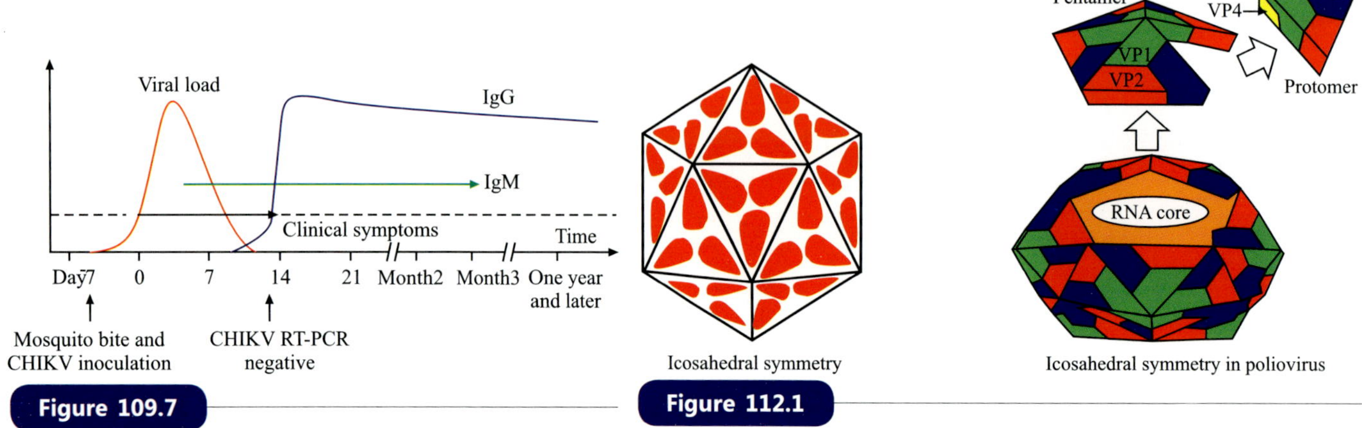

Figure 109.7

Figure 112.1

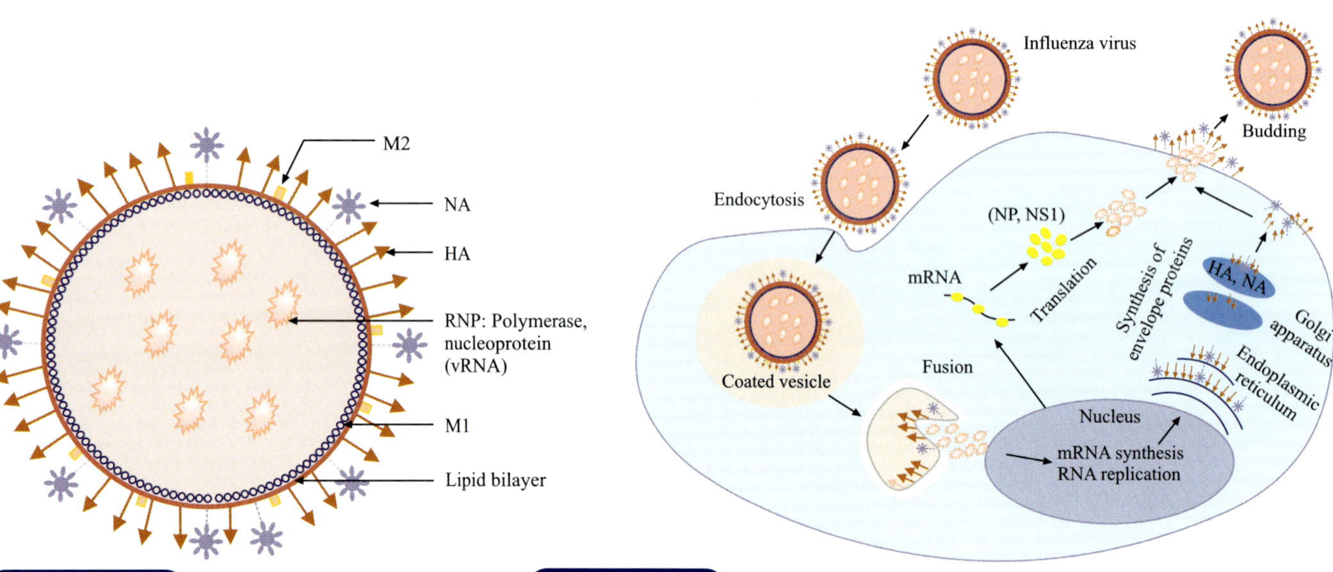

Figure 113.1

Figure 113.2

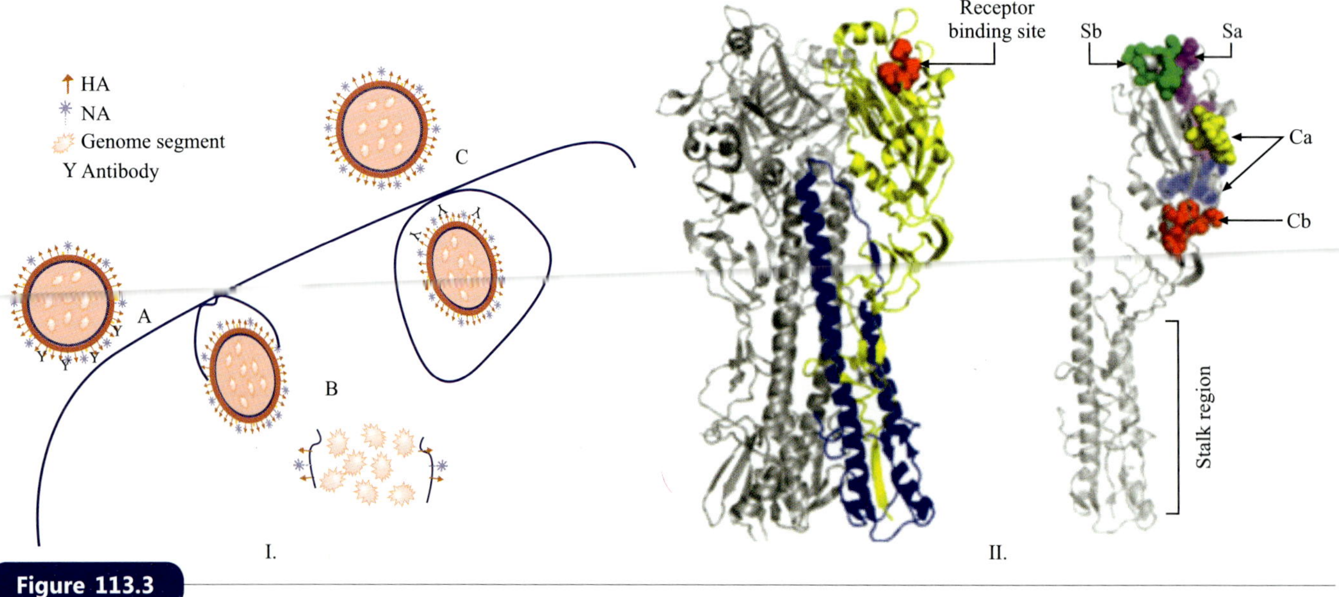

Figure 113.3

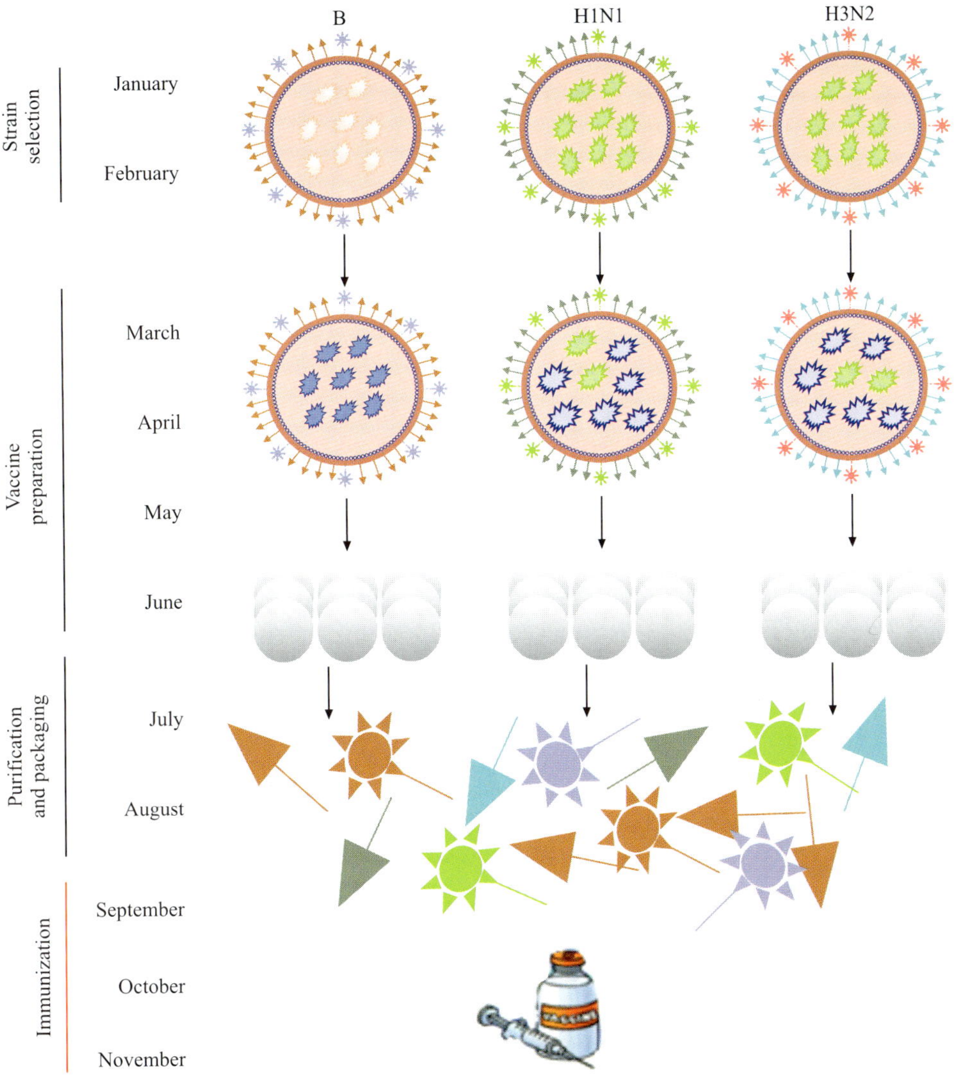

Figure 113.4

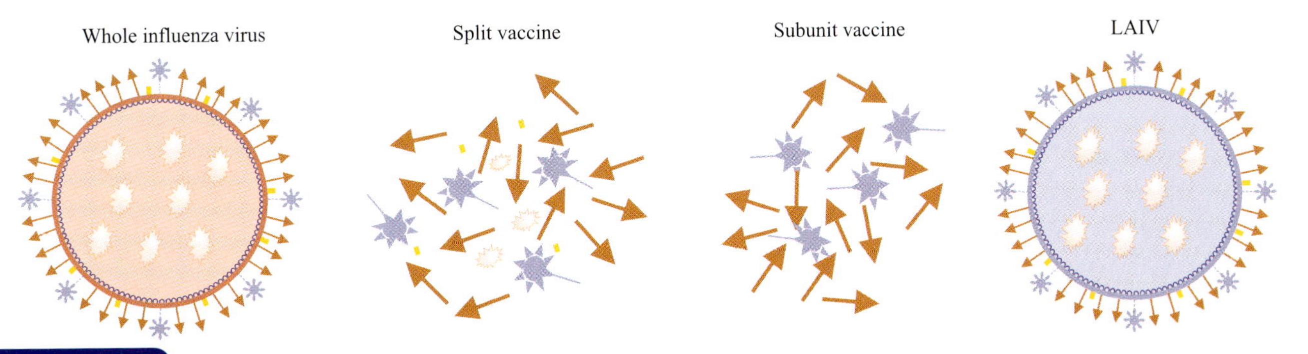

Figure 113.5

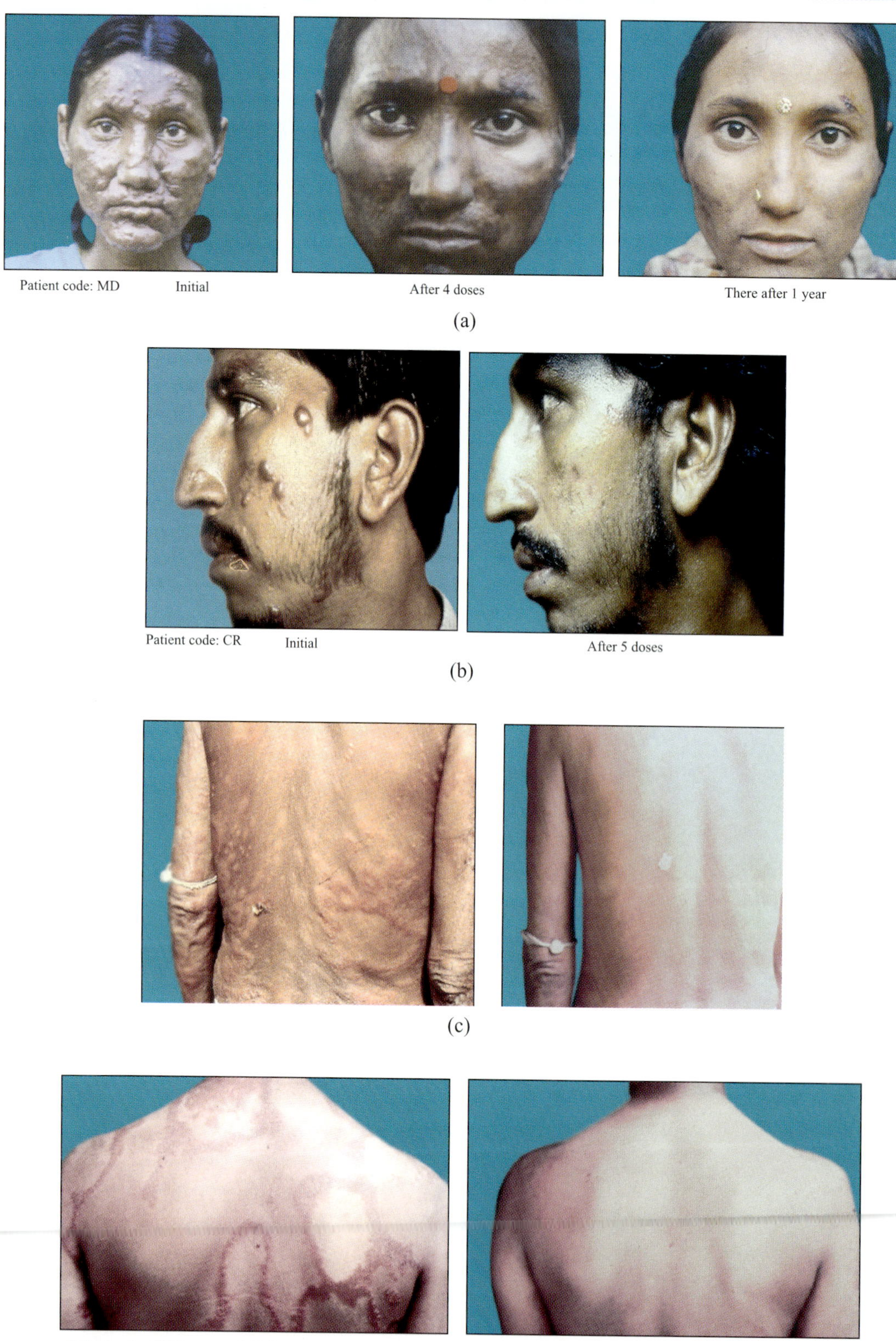

Figure 118.2 (Contd.)

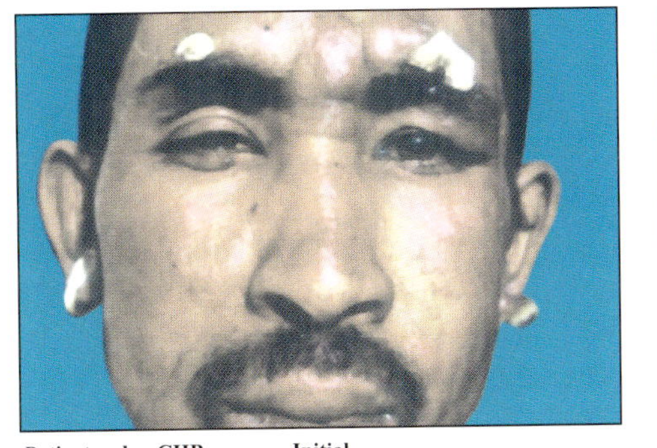
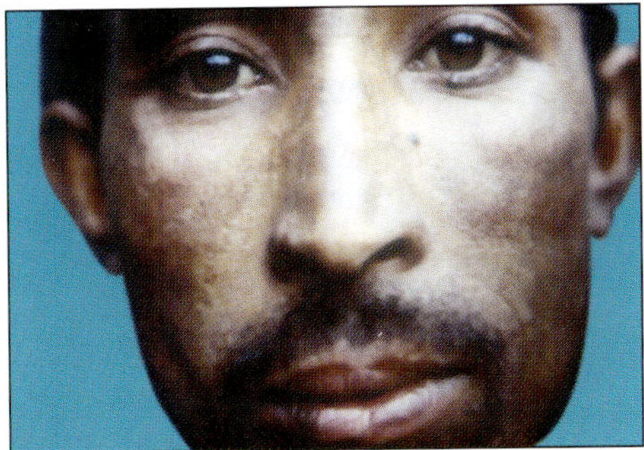

Patient code: CHB   Initial   After 1 Year

(e)

Patient code: RAB

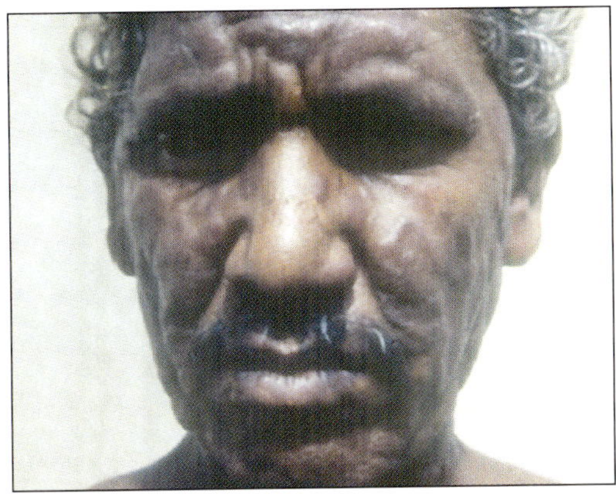
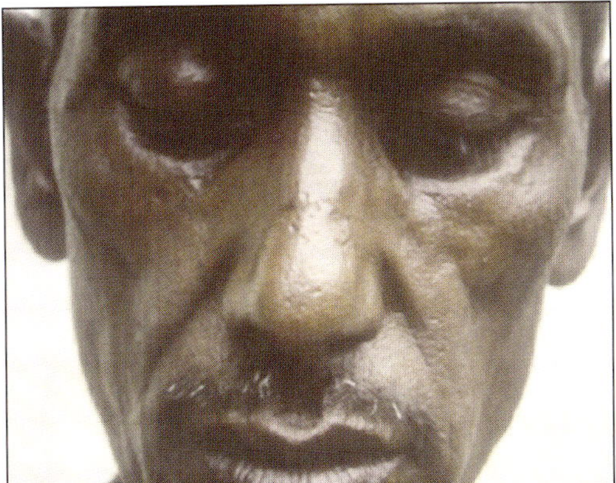

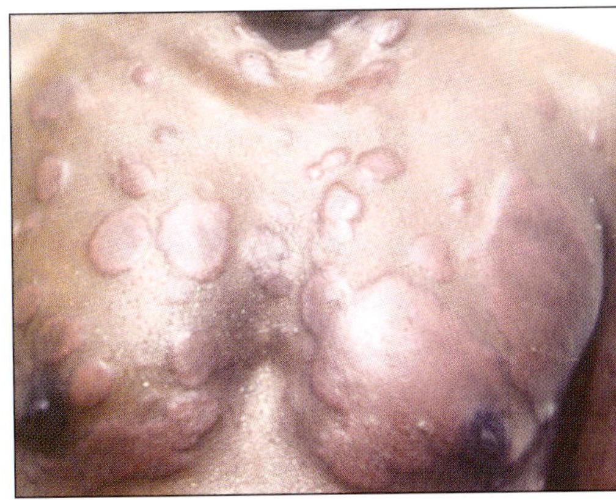
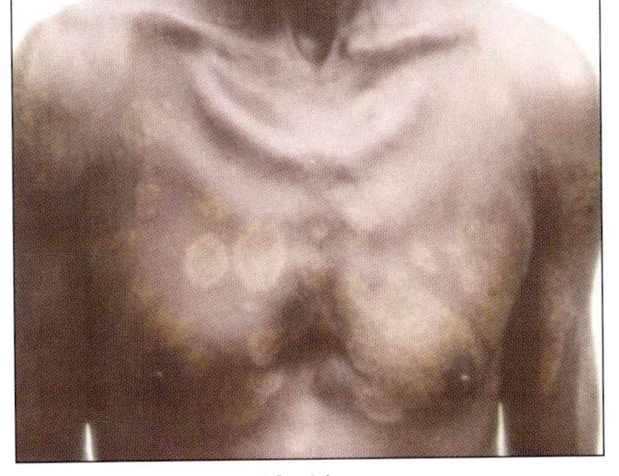

Initial   After 4 doses

(f)

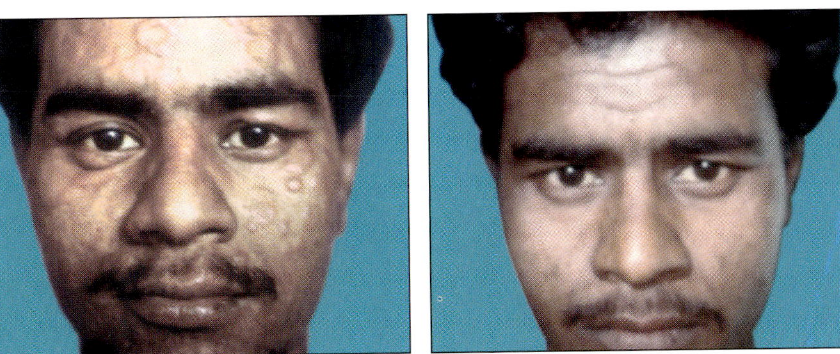

Patient code: SNC   Initial   After 2 doses (six months)

(g)

Figure 118.2

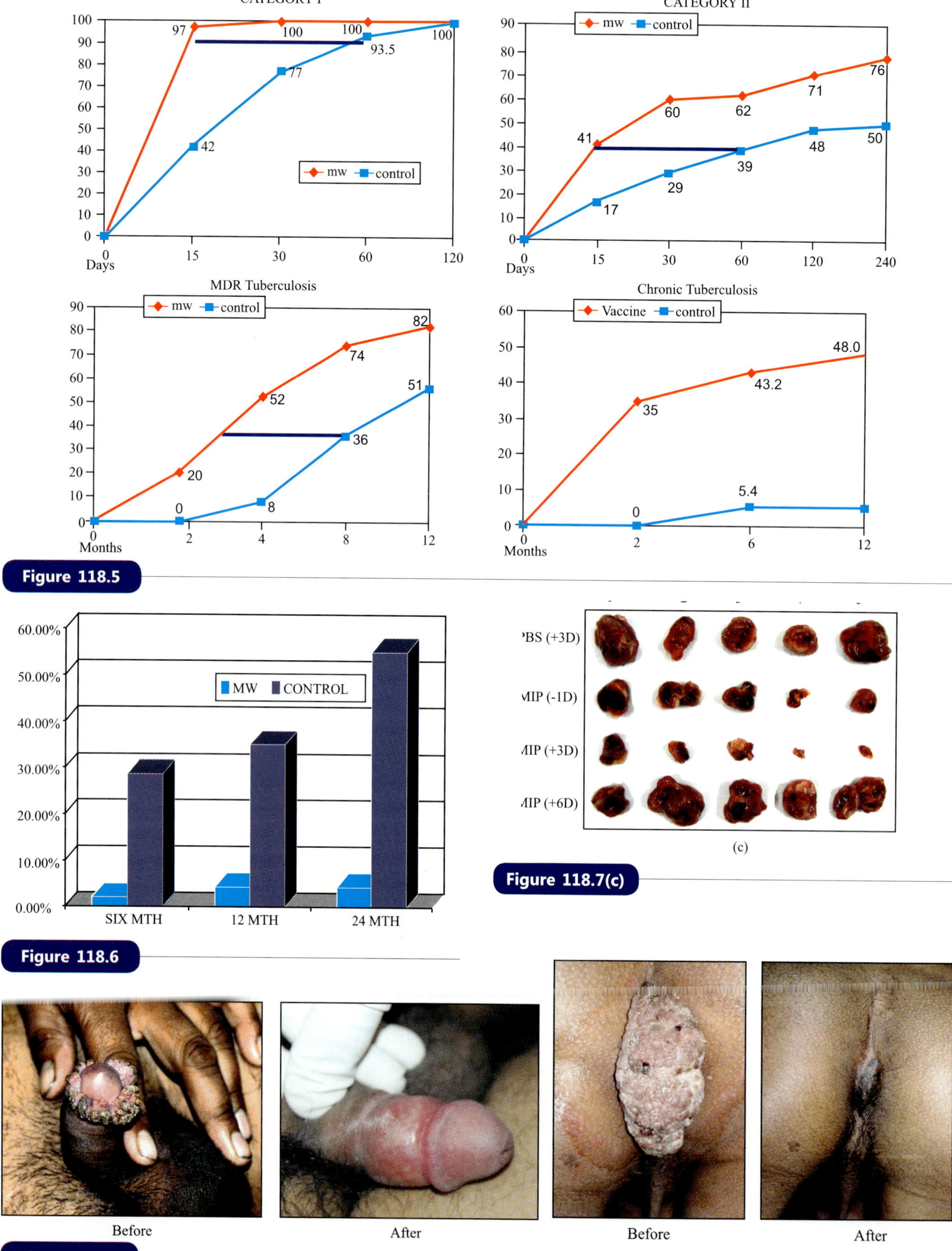

Figure 118.5

Figure 118.6

Figure 118.7(c)

Figure 118.8

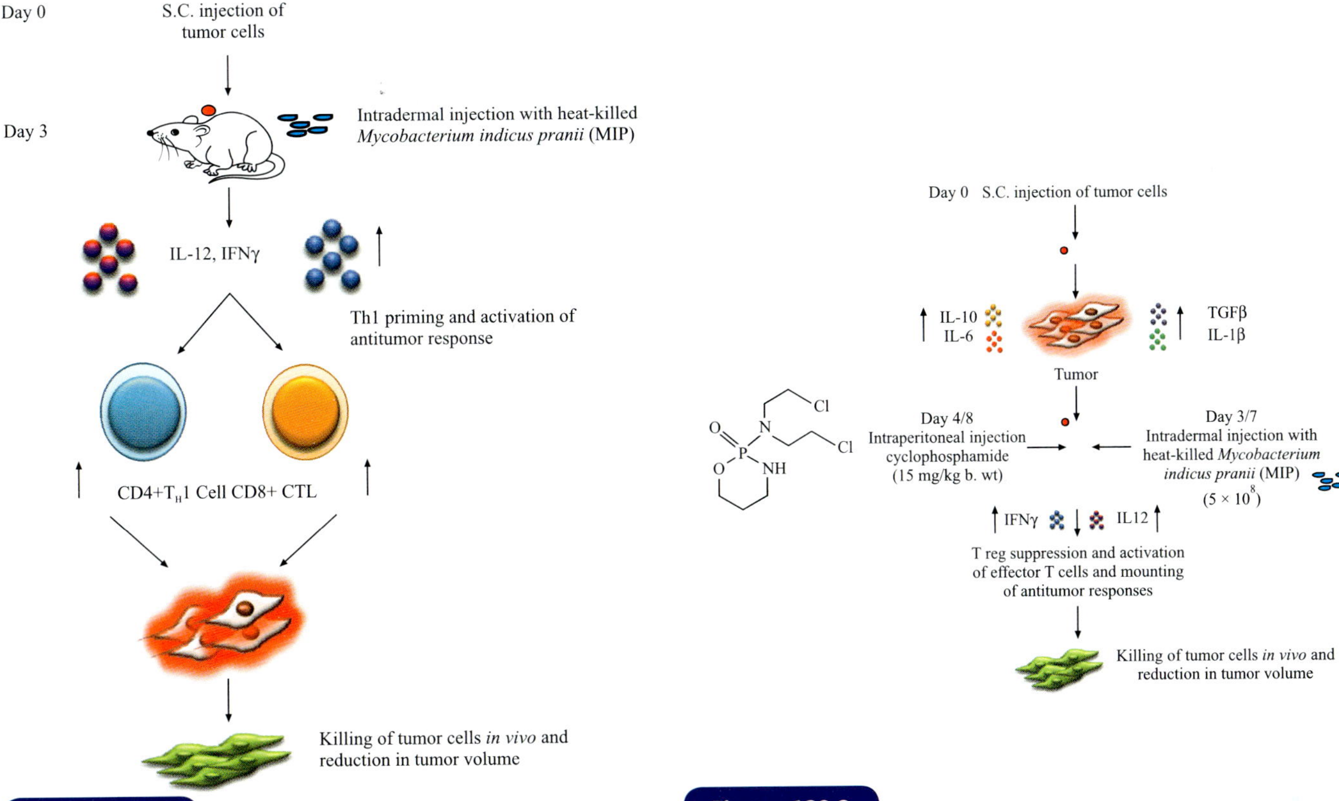

Figure 120.1

Figure 120.2

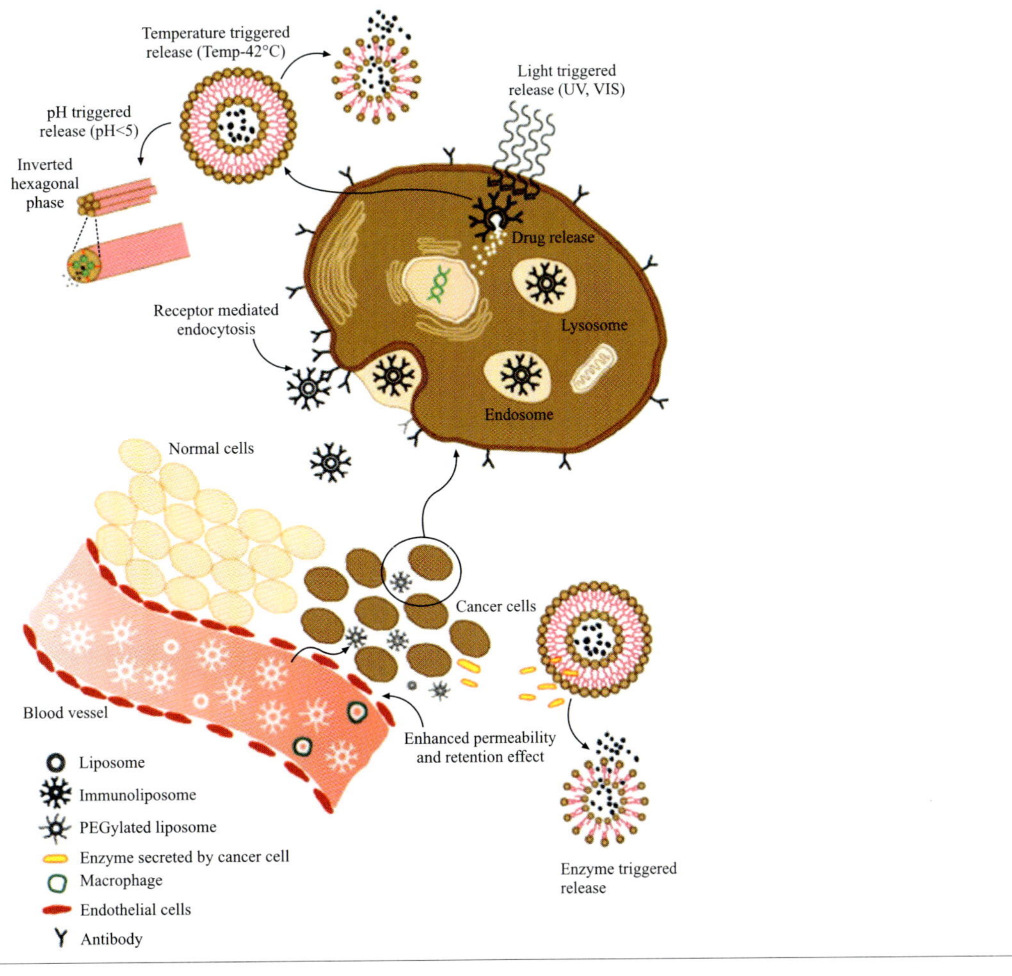

Figure 133.2

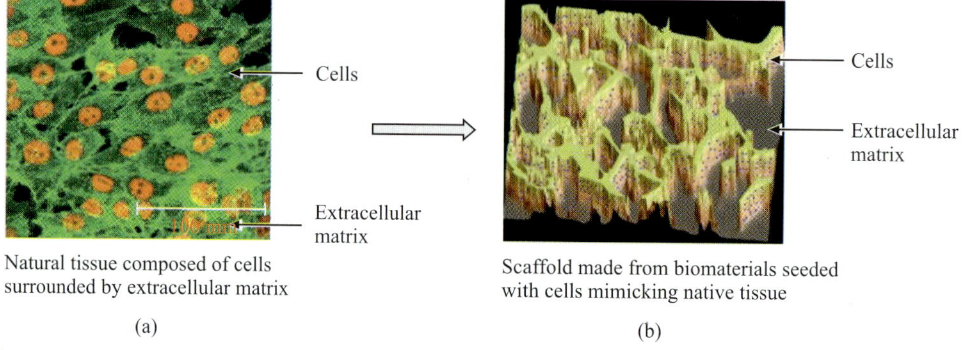

Natural tissue composed of cells
surrounded by extracellular matrix

(a)

Scaffold made from biomaterials seeded
with cells mimicking native tissue

(b)

Figure 134.1

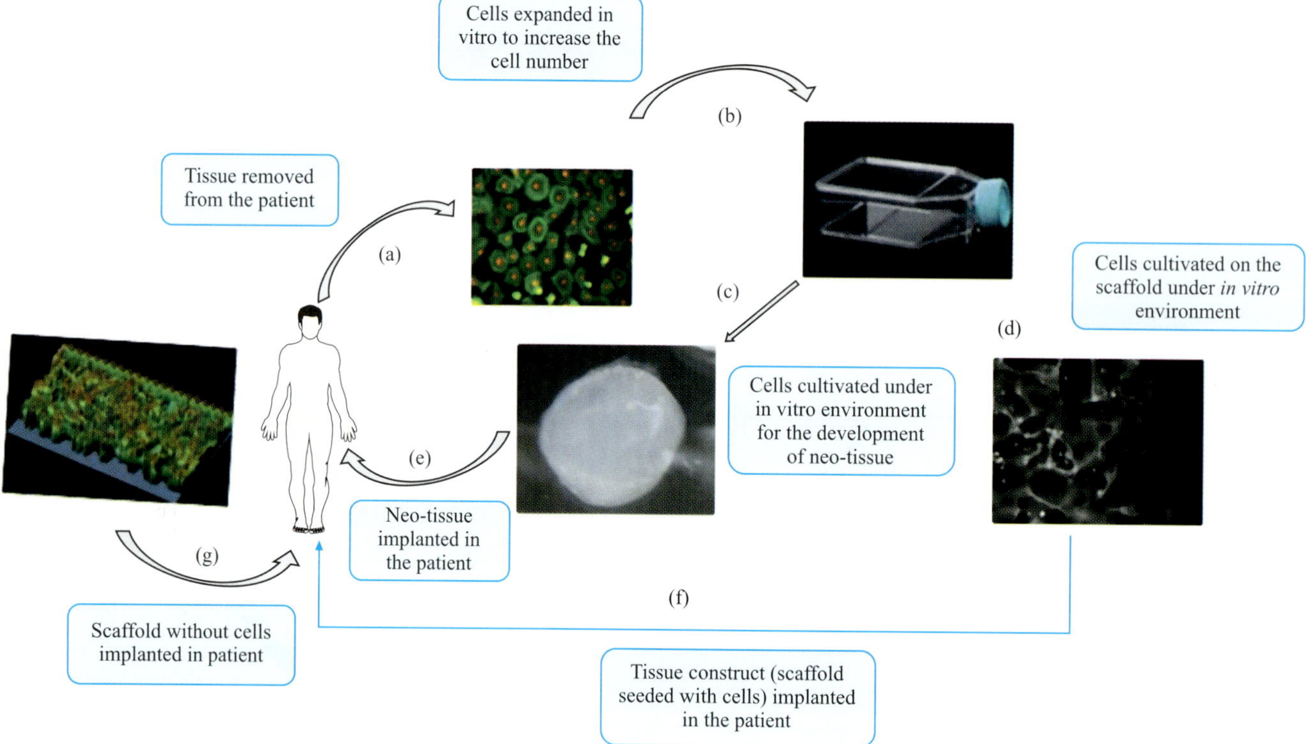

Figure 134.2

Figure 134.5

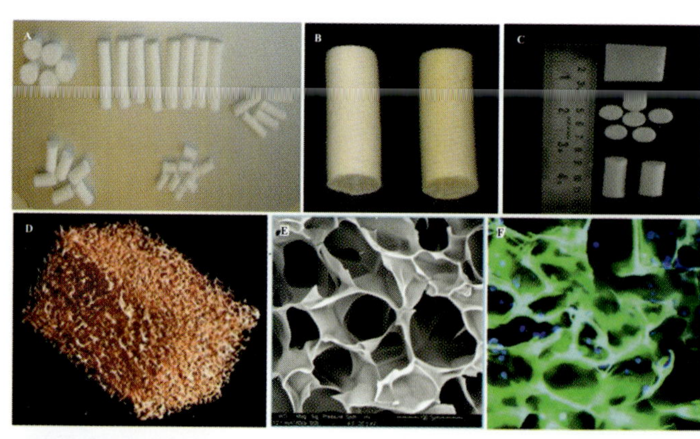

Figure 134.6

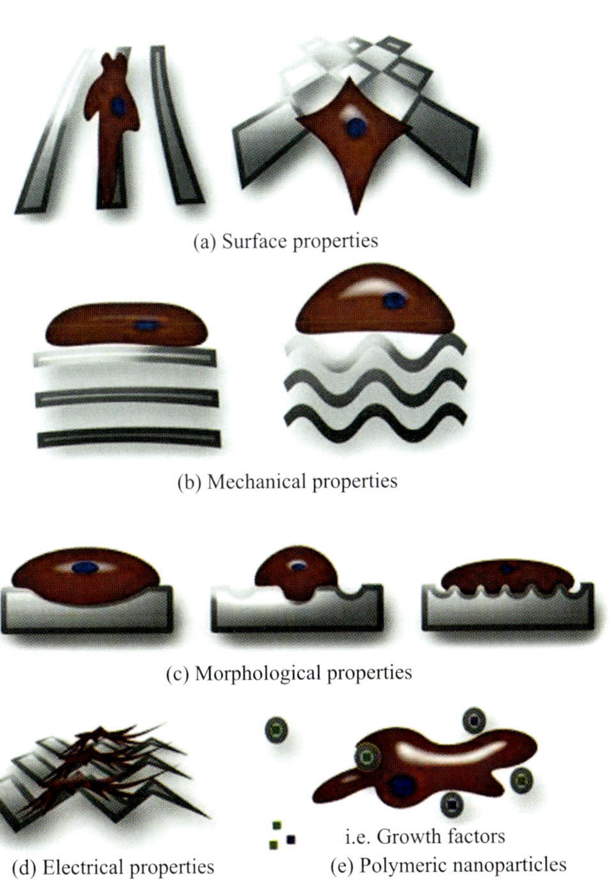

Figure 134.7

Figure 134.8

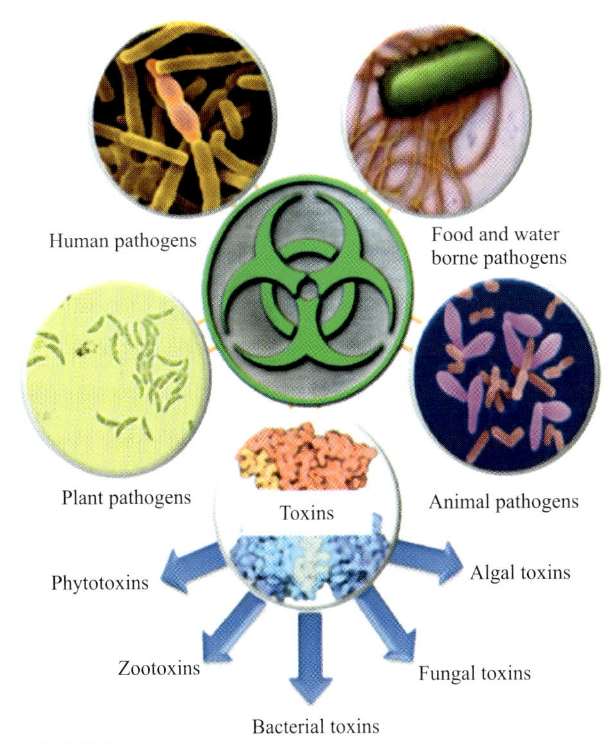

Figure 134.9

Figure 135.1

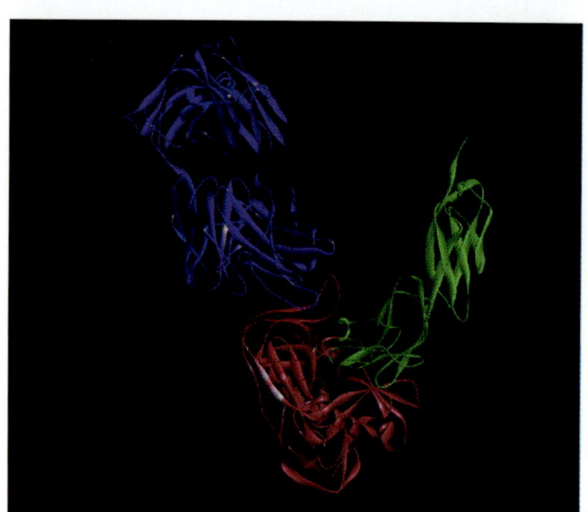

(a) Structure of a HIV envelope glycoprotein, GP120 (in pink) in complex with CD4 (in green) and neutralizing antibody (in purple), composed mostly of beta-sheets. Structure downloaded from PDB with PDB ID: 1GC1 (Reference: Kwong et al., 1998, *Nature* 393:648–659)

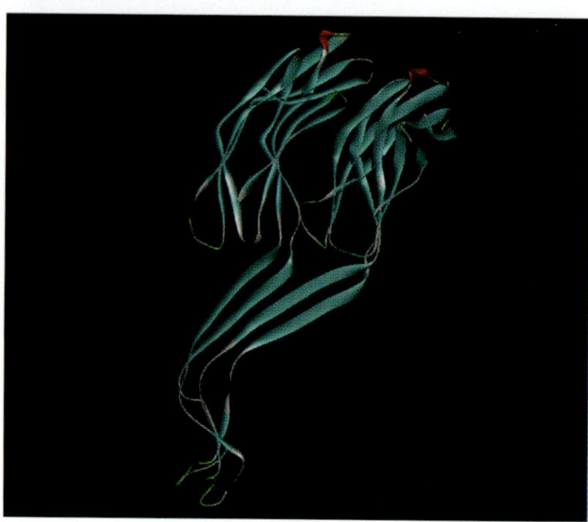

(b) Porin protein with beta-barrel shape, structure downloaded from PDB with PDB ID: 1UUN (Reference: Faller et al., 2004, *Science* **303**:1189)

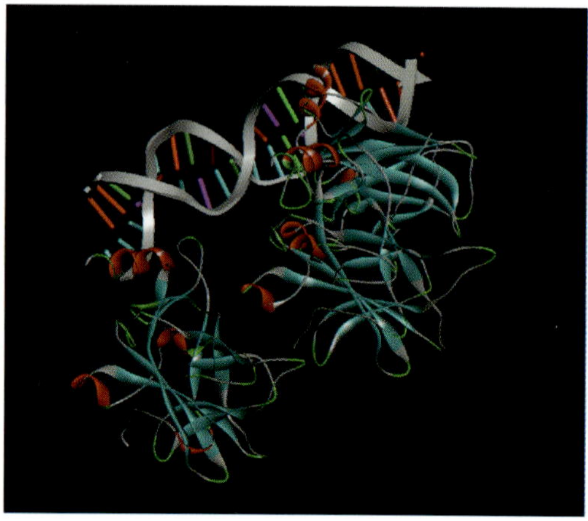

(c) p53 protein bound to DNA, structure downloaded from PDB with PDB ID: 1TUP (Reference: Cho et al., 1994, *Science* 265: 346–355

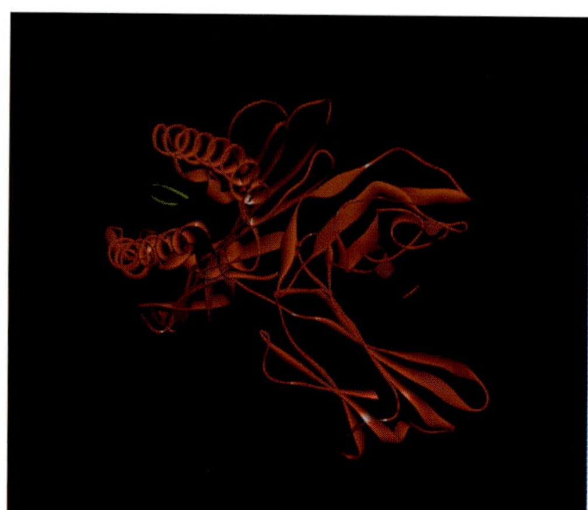

(d) PKD2 phosphopeptide (in yellow) bound to human MHC Class I molecule (in red), structure downloaded from PDB with PDB ID: 3BGM (Reference: Mohammed et al., 2008, *Nat. Immunol.* **9:** 1236-1243

Figure 136.3

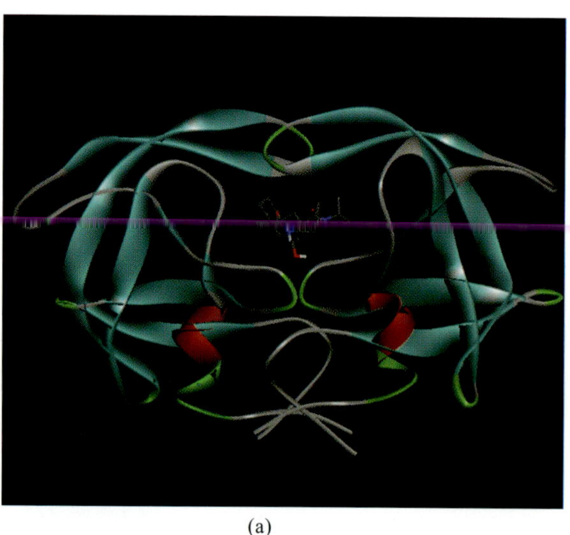

(a)

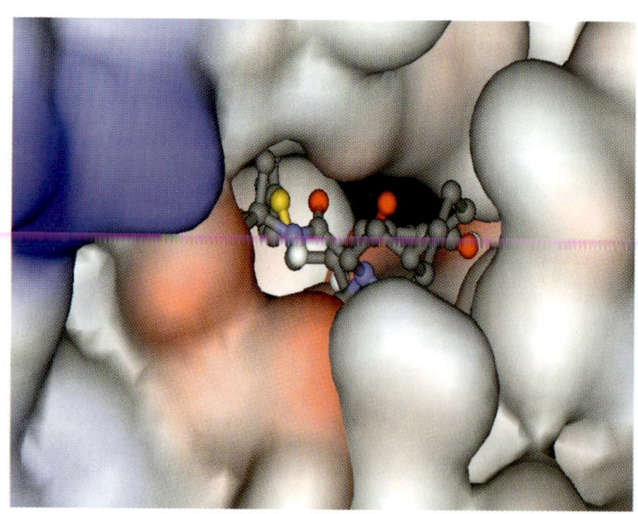

(b)

Figure 136.4